AA004532

Summer Biomechanics, Bioengineering and Biotransport Conference 2015

Snowbird, Utah, USA
17-20 June 2015

Volume 1 of 2

ISBN: 978-1-5108-2438-6

Printed from e-media with permission by:

Curran Associates, Inc.
57 Morehouse Lane
Red Hook, NY 12571

Some format issues inherent in the e-media version may also appear in this print version.

Printed by Curran Associates, Inc. (2016)

Summer Biomechanics, Bioengineering and Biotransport Organizing Committee
201 Waterfront St
National Harbor, MD 20745
USA

info@sb3c.org

Additional copies of this publication are available from:

Curran Associates, Inc.
57 Morehouse Lane
Red Hook, NY 12571 USA
Phone: 845-758-0400
Fax: 845-758-2633
Email: curran@proceedings.com
Web: www.proceedings.com

Proceedings of the 2015 Summer Biomechanics, Bioengineering and Biotransport Conference

June 17-20, 2015
Snowbird Utah

SB3C
biomechanics.
bioengineering.
biotransport.

5:00PM **Biohybrid Swimming at Low Reynolds Number** SB^{3}C2015-1183 pg.1311
Brian Williams[1], Sandeep Anand[1], Jagannathan Rajagopalan[2], M. Taher A. Saif[1], [1]*University of Illinois at Urbana-Champaign, Urbana, IL, United States,*[2]*Arizona State University, Tempe, AZ, United States*

WEDNESDAY, JUNE 17	**3:45pm - 5:15pm**

Multiscale Modeling in Biotransport (Special Session) — Golden Cliff / Eagle's Nest

Session Chair: Bumsoo Han, *Purdue University, West Lafayette, IN, United States*
Session Co-Chair: Ronghui Ma, *University of Maryland Baltimore County, Baltimore, MD, United States*

3:45PM **Energetics of Water Permeation across Stretched Phospholipid/Cholesterol Bilayer: Molecular Dynamics Simulation** SB^{3}C2015-200 pg.184
Taiki Shigematsu, Kenichiro Koshiyama, Ryotaro Kurumatani, Shigeo Wada, *Osaka University, Toyonaka, Osaka, Japan*

4:00PM **Numerical Modeling of Scalp Cooling Devices for the Prevention of Chemotherapy-Induced Alopecia** SB^{3}C2015-635
Bradley Pliskow, Mehmet Kaya, Kunal Mitra, *Florida Institute of Technology, Indialantic, FL, United States* pg.939

4:15PM **Cross Capillary Mesangial Transport** SB^{3}C2015-1066 pg.1100
Sarah E. Hunt, Yoav Segal, Kevin D. Dorfman, Victor H. Barocas, *University of Minnesota, Minneapolis, MN, United States*

4:30PM **Multiscale Simulation of Shear-Induced Platelet Activation: Correlating Numerical with Experimental Results** SB^{3}C2015-302 pg.351
Peng Zhang[1], Chao Gao[1], Na Zhang[1], Seetha Pothapragada[1], Marvin J. Slepian[1,2], Yuefan Deng[1], Danny Bluestein[1], [1]*Stony Brook University, Stony Brook, NY, United States,*[2]*University of Arizona, Tucson, AZ, United States*

4:45PM **Numerical Study Of A New Thermalplasty Treatment System For Atherosclerosis** SB^{3}C2015-1110 pg.1183
Shiqing Zhao, **Aili Zhang**, Lisa X. Xu, *Shanghai Jiao Tong University, Shanghai, China*

5:00PM **An Approach to Include Pre-Stress in a Finite Element Model of Convection-Enhanced Delivery With Backflow** SB^{3}C2015-383 pg.475
Gustavo A. Orozco[1], Fabian A. Urrea[1], **Fernando Casanova**[1], Joshua H. Smith[2], Jose J. Garcia[1], [1]*Universidad del Valle, Cali, Colombia,*[2]*Lafayette College, Easton, PA, United States*

WEDNESDAY, JUNE 17	**3:45pm - 5:15pm**

Muscle and Joint Loading — Primrose B

Session Chair: Jun Liao, *Mississippi State, Mississippi State, MS, United States*
Session Co-Chair: Andrew Anderson, *University of Utah, Salt Lake City, UT, United States*

3:45PM **Medial and Lateral Contact Pressure Distribution Following the Implantation of a Novel Medial Meniscus Implant** SB^{3}C2015-185 pg.156
Maoz Shemesh[1], Noa Cohen[1], Eyal Zylberberg[1], Adaya Shefy-Peleg[1], Ron Arbel[2], Vincenzo Condello[3], Nogah Shabshin[4], Eran Linder-Ganz[1], **Jonathan J. Elsner**[5], [1]*Active Implants, Netanya, Israel,*[2]*Ichilov Medical Center, Tel Aviv, Israel,*[3]*Sacro Cuore Hospital, Verona, Italy,*[4]*Carmel Medical Center, Haifa, Israel,*[5]*Active Implants, Cambridge, MA, United States*

4:00PM **Effect of Meniscal Properties and Patient Variables on Knee Joint Contact Mechanics** SB^{3}C2015-213 pg.206
Hongqiang Guo[1], Thomas J. Santner[2], Po-Hsu Chen[2], Amy L. Lerner[3], Suzanne A. Maher[1], [1]*Hospital for Special Surgery, New York, NY, United States,*[2]*The Ohio State University, Columbus, OH, United States,*[3]*University of Rochester, Rochester, NY, United States*

4:15PM **Non-Invasive Method for Investigating the Relationship between Muscle Strain and Post-Exercise Blood Perfusion Using MRI** SB^{3}C2015-1175 pg.1299
Amanda K. W. Buck, Christopher P. Elder, Bruce M. Damon, *Vanderbilt University, Nashville, TN, United States*

WEDNESDAY, JUNE 17	3:45pm - 5:15pm

Nano, Micro and Multiscale Mechanics — Magpie

Session Chair: Raffaella De Vita, *Virginia Tech, Blacksburg, VA, United States*
Session Co-Chair: Steven Abramowitch, *University of Pittsburgh, Pittsburgh, PA, United States*

3:45PM **Direct Measurement of Energy Landscape of Biological Interactions** SB^{3}C2015-1181 pg.1307
Ahmad Haider, Daniel Potter, Todd Sulchek, *Georgia Institute of Technology, Atlanta, GA, United States*

4:00PM **Constructing Rudimentary Limit Curves For Neuronal Phospholipid Bilayer Failure And Theoretical Calcium Penetration** SB^{3}C2015-1025 pg.1028
M. A. Murphy[1], Sungkwang Mun[1], M. F. Horstemeyer[1], Steven R. Gwaltney[1], Tonya W. Stone[1], Michelle C. LaPlaca[2], Jun Liao[1], Lakiesha N. Williams[1], **R. Prabhu**[1], [1]*Mississippi State, MS, United States,*[2]*Georgia Institute of Technology, GA, United States*

4:15PM **Deterioration Of Trabecular And Cortical Microarchitecture And Reduced Bone Stiffness At Distal Radius And Tibia In Postmenopausal Women With Vertebral Fractures** SB^{3}C2015-231 pg.236
Ji Wang[1], Emily Stein[2], Bin Zhou[1], Kyle Nishiyama[2], Elizabeth Shane[2], X.Edward Guo[1], [1]*Columbia University, New York, NY, United States,*[2]*College of Physicians and Surgeons, Columbia University, New York, NY, United States*

4:30PM **Correlating Molecular Structure With The Mechanical Response Of Aquatic Caddisworm Silk** SB^{3}C2015-325 pg.389
Nicholas N. Ashton, Russell J. Stewart, *The University of Utah, Salt Lake City, UT, United States*

4:45PM **Numerical Modeling of Human Fibrous Cap Delamination** SB^{3}C2015-181 pg.148
Xiaochang Leng, Xiaomin Deng, Michael A. Sutton, Susan M. Lessner, *University of South Carolina, Columbia, SC, United States*

5:00PM **Nonlinear Viscoelastic Response of the Spinal Cord In Vivo** SB^{3}C2015-150 pg.90
Snehal S. Shetye[1], Femke Streijger[2], Christopher Strickland[1], Jae H. T. Lee[2], Brian K. Kwon[2], Peter Cripton[2], Patrick D. Shipman[1], Christian M. Puttlitz[1], [1]*Colorado State University, Fort Collins, CO, United States,*[2]*University of British Columbia, Vancouver, BC, Canada*

WEDNESDAY, JUNE 17	3:45pm - 5:15pm

Organs, Morphogenesis, and Collective Cell Behavior — Maybird

Session Chair: Nandan L. Nerurkar, *Harvard Medical School, Boston, MA, United States*
Session Co-Chair: Celeste Nelson, *Princeton University, Princeton, NJ, United States*

3:45PM **Fgf8 Establishes A Contractile Gradient To Drive Directed Cell Movements In The Developing Avian Gut** SB^{3}C2015-143 pg.80
Nandan L. Nerurkar, Cliff Tabin, *Harvard Medical School, Boston, MA, United States*

4:00PM **Mechanics of Early Eye Development: Interactions Between the Retina and Lens** SB^{3}C2015-193 pg.170
Alina Oltean[1], Jie Huang[2], David C. Beebe[2], Larry A. Taber[1], [1]*Washington University in St. Louis, St. Louis, MO, United States,*[2]*Washington University School of Medicine, St. Louis, MO, United States*

4:15PM **Magnetic Micropillar Array To Study Epithelial Microtissue Morphogenesis And Local Mechanics** SB^{3}C2015-341 pg.413
Mohammadnabi Asmani, Yan Li, Christopher Kotei, David Olsen, Zhaowei Chen, Fanjie Meng, **Ruogang Zhao**, *State University of New York at Buffalo, Buffalo, NY, United States*

4:30PM **Viscoelastic Folding Instability Controls Airway Branching Morphogenesis** SB^{3}C2015-386 pg.481
Victor D. Varner, Jason P. Gleghorn, Celeste M. Nelson, *Princeton University, Princeton, NJ, United States*

4:45PM **In Silico Predictions Of Angiogenic Growth Coupled To Matrix Mechanics Within Heterogeneous Extracellular Environments** SB^{3}C2015-546 pg.776
Lowell T. Edgar[1], James B. Hoying[2], Jeffrey A. Weiss[1], [1]*University of Utah, Salt Lake City, UT, United States,*[2]*University of Louisville, Louisville, KY, United States*

4:30PM **Numerical Modeling of Skeletal Muscle Under High Strain and Stress Relaxation Compression Conditions** SB[3]C2015-247 pg.265
Benjamin B. Wheatley[1], Renee Pietsch[2], Tammy L. Haut Donahue[1], Lakiesha N. Williams[2], [1]*Colorado State University, Fort Collins, CO, United States,*[2]*Mississippi State University, Starkville, MS, United States*

4:45PM **Biomechanical Characterization of Porcine Skeletal Muscle Extracellular Matrix** SB[3]C2015-512 pg.710
Bryn Brazile, Sourav S. Patnaik, Sallie Lin, Xiaodan Shi, Shengfa Liao, Raj Prabhu, Hongjoo Rhee, Lakiesha N. Williams, Jun Liao, *Mississippi State University, Mississippi State, MS, United States*

5:00PM **Predicting the Stress and Intramuscular Pressure Response of Whole Skeletal Muscle Through Optimized Finite Element Analysis** SB[3]C2015-249 pg.269
Benjamin B. Wheatley[1], Duane A. Morrow[2], Gregory M. Odegard[3], Kenton R. Kaufman[2], Tammy L. Haut Donahue[1], [1]*Colorado State University, Fort Collins, CO, United States,*[2]*Mayo Clinic, Rochester, MN, United States,*[3]*Michigan Tech University, Houghton, MI, United States*

WEDNESDAY, JUNE 17	5:30pm - 7:00pm

Ocular Biomechanics II — Primrose A

Session Chair: Jun Liu, *Ohio State, Columbus, OH, United States*
Session Co-Chair: Ian A. Sigal, *University of Pittsburgh, Pittsburgh, PA, United States*

5:30PM **Equilibrium Shape of the Aqueous Humor-Vitreous Substitue Interface** SB[3]C2015-445 pg.598
Krystyna Isakova, Rodolfo Repetto, Jan Oscar Pralits, *University of Genova, Genova, Italy*

5:45PM **A Computational Model To Explore The Role Of Experimentally Determined Scleral Microstructure On Lamina Cribrosa Deformation** SB[3]C2015-1087 pg.1140
Avinash Ayyalasomayajula, Forest L. Danford, Jonathan P. Vande Geest, *University of Arizona, Tucson, AZ, United States*

6:00PM **Effects Of Hydration And Riboflavin/uva Collagen Crosslinking On Bovine Corneal Tensile Properties** SB[3]C2015-1074 pg.1114
Hamed Hatami-Marbini, *Oklahoma State University, Stillwater, OK, United States*

6:15PM **Intraocular Pressure Increases In Patients With Intraocular Gas Bubbles Following A Descent And Subsequent Ascent** SB[3]C2015-394 pg.497
Lucas A. Gsellman, Rouzbeh Amini, *The University of Akron, Akron, OH, United States*

6:30PM **The Effect Of Elevated Intraocular Pressure On Convective Flow In The Vitreous** SB[3]C2015-323 pg.385
Julie E. Whitcomb[1], Mohammad R. Kazemi[2], Michael R. Robinson[1], Mayssa Attar[1], Anita Penkova[3], Sati Sadhal[3], Susan S. Lee[1], [1]*Allergan, Irvine, CA, United States,*[2]*Consultant, San Jose, CA, United States,*[3]*University of Southern California, Los Angeles, CA, United States*

6:45PM **A Parametrized Model Of The Lamina Cribrosa For Studying Oxygen Transport** SB[3]C2015-312 pg.367
Fabian A. Bräu[1,2], Ian C. Campbell[1,3], Baptiste Coudrillier[1], **C. Ross Ethier**[1,3], [1]*Georgia Institute of Technology, Atlanta, GA, United States,*[2]*Technical University Munich, Munich, Germany,*[3]*Atlanta VA Medical Center, Decatur, GA, United States*

WEDNESDAY, JUNE 17	5:30pm - 7:00pm

Design and Devices II - Modeling and Simulation of Cardiovascular Therapies — Superior

Session Chair: Michael Moreno, *Texas A&M University, College Station, TX, United States*
Session Co-Chair: Lucas H. Timmins, *Georgia Institute of Technology, Atlanta, GA, United States*

5:30PM **Hemocompatibility Assessment Of Hyaluronan Enhanced Linear Low Density Polyethylene For Use In Aortic Heart Valve Leaflets** SB[3]C2015-1041 pg.1056
Rachael Simon-Walker, John Cavicchia, David Bark, Susan James, Lakshmi Prasad Dasi, Popat Ketul, *Colorado State University, Fort Collins, CO, United States*

5:45PM **Virtual Transplantation To Establish Donor Selection Criteria For Undersized And Complex Heart Recipients**
SB[3]C2015-335 pg.401
Justin Ryan[1], Randy Richardson[2], Erik Ellsworth[3], John J. Nigro[3], David Frakes[1], Stephen Pophal[3], [1]*Arizona State University, Tempe, AZ, United States,*[2]*St. Joseph's Hospital and Medical Center, Phoenix, AZ, United States,*[3]*Phoenix Children's Hospital, Phoenix, AZ, United States*

6:00PM **Physical Patient Specific Simulation For Ascending Aortic Aneurysms Surgery Pre-procedural Training**
SB[3]C2015-338 pg.407
Justine Garcia[1], ZhiLin Yang[1], Kevin Lachapelle[2], Rosaire Mongrain[1], Richard Leask[1], [1]*McGill University, Montreal, QC, Canada,*[2]*McGill University Health Network, Montreal, QC, Canada*

6:15PM **Computer Simulator of the Coil Insertion into Aneurysm: The Effect of Mechanical Property and Reference Configuration of the Coil** SB[3]C2015-343 pg.417
Tomohiro Otani[1], Satoshi Ii[1], Tomoyoshi Shigematsu[2], Toshiyuki Fujinaka[2], Masayuki Hirata[2], Tomohiko Ozaki[2], Shigeo Wada[1], [1]*Graduate School of Engineering Science, Osaka University, Toyonaka, Japan,*[2]*Graduate School of Medicine, Osaka University, Suita, Japan*

6:30PM **Computational Hemodynamic Assessment of a Novel Modular Anastomotic Valve Device for Hemodialysis Vascular Access** SB[3]C2015-1069 pg.1106
Andrew McNally[1], Philippe Sucosky[1], A. George Akingba[2], [1]*University of Notre Dame, South Bend, IN, United States,*[2]*Indiana University School of Medicine, Indianapolis, IN, United States*

6:45PM **Computational Modeling Of A Percutaneous Transvenous Mitral Annuloplasty Device Deployment Into The Coronary Sinus Vessel For Treatment Of Mitral Regurgitation: Analysis Of Anchor Sizes** SB[3]C2015-1068 pg.1104
Thuy Pham[1], Fanwei Kong[1], Milton Deherrera[2], Wei Sun[1], [1]*Georgia Institute of Technology, Atlanta, GA, United States,*[2]*Edwards Lifesciences, Irvine, CA, United States*

WEDNESDAY, JUNE 17	**5:30pm - 7:00pm**

Cardiovascular Diagnostics and Imaging — Wasatch

Session Chair: David Steinman, *University of Toronto, Toronto, ON, Canada*
Session Co-Chair: Matt Gounis, *University of Massachusetts, Worcester, MA, United States*

5:30PM **Invasive Measurement of Pulse Wave Velocity in Anesthetized Mice: a Computational-Experimental Study**
SB[3]C2015-600 pg.875
Federica Cuomo[1], Jacopo Ferruzzi[2], Jay D. Humphrey[2], Carlos A. Figueroa[1], [1]*University of Michigan, Ann Arbor, MI, United States,*[2]*Yale University, New Haven, CT, United States*

5:45PM **Modeling Hemodynamic Effects Of Cerebral Vasospasm** SB[3]C2015-1049 pg.1072
Jaiyoung Ryu[1], Xiao Hu[2], Shawn C. Shadden[1], [1]*University of California, Berkeley, Berkeley, CA, United States,*[2]*University of California, San Francisco, San Francisco, CA, United States*

6:00PM **A Novel Approach to Calculating Isovolumic Pressure for Single-Beat Estimation of Right Ventricular End-Systolic Elastance** SB[3]C2015-544 pg.772
Alessandro Bellofiore[1], Eric Dinges[2], Sanjiv J. Shah[3], Michael J. Cuttica[3], Ranya Sweis[3], Hamorabi Mkrdichian[3], Lauren Beussink-Nelson[3], Melissa Bailey[2], James R. Runo[2], Jon G. Keevil[2], Christopher J. Francois[2], Naomi C. Chesler[2], [1]*San Jose State University, San Jose, CA, United States,*[2]*University of Wisconsin-Madison, Madison, WI, United States,*[3]*Northwestern University, Chicago, IL, United States*

6:15PM **A Method for Rapid, Accurate Calculation of Wall Shear Stress from 2D Phase Contrast Magnetic Resonance Image Data** SB[3]C2015-562 pg.805
Elizabeth Iffrig[1,2], William R. Taylor[1,2], John N. Oshinski[1,2], [1]*Emory University, Atlanta, GA, United States,*[2]*Georgia Institute of Technology, Atlanta, GA, United States*

6:30PM **Tracking Mass Transport Of A Drug Surrogate In Porcine Coronary Tissue Using Photoacoustic Imaging And Spectroscopy** SB[3]C2015-1083 pg.1132
Kenneth J. Furdella, Jonathan P. Vande Geest, Russell S. Witte, *University of Arizona, Tucson, AZ, United States*

4:30PM **Quantification of Plaque Shift after Coronary Bifurcation Stenting** SB^{3}C2015-465 pg.632

Francesco Iannaccone[1,2], **Claudio Chiastra**[1,3], Francesco Migliavacca[3], Frank J. H. Gijsen[1], Patrick Segers[2], Evelyn Regar[4], Matthieu De Beule[2,5], Jolanda J. Wentzel[1], [1]*Thoraxcenter, Erasmus University Medical Center, Rotterdam, Netherlands,*[2]*Ghent University, Ghent, Belgium,*[3]*Politecnico di Milano, Milan, Italy,*[4]*Erasmus MC, Rotterdam, Netherlands,*[5]*FEops bvba, Ghent, Belgium*

4:45PM **Management of Distal Embolization during Interventional Treatment of Acute Ischemic Stroke in a Simulated Vascular Phantom** SB^{3}C2015-589 pg.857

Juyu Chueh, Olivia W. Brooks, Ajit S. Puri, Ajay K. Wakhloo, Matthew J. Gounis, *University of Massachusetts Medical School, Worcester, MA, United States*

5:00PM **Temperature Comparison Of Catheter Ablation Between Open Irrigation And Vibration** SB^{3}C2015-159 pg.108

Kaihong Yu[1], Tetsui Yamashita[2], Shigeaki Shingyochi[3], Kazuo Matsumoto[4], Makoto Ohta[5], [1]*Tohoku University, Sendai, Japan,*[2]*Cardiovascular Internal Medicine System Sec., JMS Co., Ltd., Tokyo, Japan,*[3]*Nidec Copal Electronics Corp., Tochigi, Japan,*[4]*Saitama Medical University, Hidaka, Japan,*[5]*Tohoku University, Sendai, Japan*

WEDNESDAY, JUNE 17	3:45pm - 5:15pm

Atherosclerosis — Wasatch

Session Chair: Michael Walsh, *University of Limerick, Ireland*
Session Co-Chair: Francis Loth, *University of Akron, OH, United States*

3:45PM **Human Coronary Plaque Morphological and Stress Vulnerability Indices Using IVUS-Based Fluid-Structure Interaction Models: A Multi-Patient Study** SB^{3}C2015-33 pg.18

Liang Wang[1], Jie Zheng[2], Akiko Maehara[3], Chun Yang[4], Richard Bach[2], David Muccigrosso[2], Gary Mintz[5], **Dalin Tang**[1,6], [1]*WPI, Worcester, MA, United States,*[2]*Washington University, St. Louis, St. Louis, MO, United States,*[3]*Columbia University, New York, NY, United States,*[4]*China United Network Communications Co., Ltd., Beijing, China,*[5]*Columbia University, The Cardiovascular Research Foundation, New York, NY, United States,*[6]*Southeast University, Nanjing, China*

4:00PM **The Effect of Head Rotation to the Geometry and Hemodynamics of Healthy Vertebral Arteries** SB^{3}C2015-495 pg.678

Nicolas Aristokleous[1], Ioannis Seimenis[2], Georgios C. Georgiou[3], Andreas S. Anayiotos[1], [1]*Cyprus University of Technology, Lemesos, Cyprus,*[2]*Democritus University of Thrace, Alexandroupoli, Greece,*[3]*University of Cyprus, Nicosia, Cyprus*

4:15PM **Development Of A Framework To Characterize The Role Of Wall Shear Stress In Atherosclerotic Plaque Transformation Through The Combined Use Of OCT And Vh-IVUS** SB^{3}C2015-1064 pg.1098

David Molony[1], Lucas Timmins[1], Emad Rasoul-Arzrumly[2], Olivia Hung[2], Bill Gogas[2], Habib Samady[2], Don Giddens[1], [1]*Georgia Institute of Technology, Atlanta, GA, United States,*[2]*Emory University, Atlanta, GA, United States*

4:30PM **Phenotypic Differences in Coronary Artery Disease Progression in the Clinical Setting and Dependence on a Focal Oscillatory Hemodynamic Environment.** SB^{3}C2015-426 pg.561

Lucas H. Timmins[1,2], David S. Molony[1,2], Parham Eshtehardi[3], Michael C. McDaniel[2], John N. Oshinski[1,2], Habib Samady[2], Don P. Giddens[1,2], [1]*Georgia Institute of Technology, Atlanta, GA, United States,*[2]*Emory University School of Medicine, Atlanta, GA, United States,*[3]*Albert Einstein College of Medicine, Bronx, NY, United States*

4:45PM **Relationship among Disturbed Shear Descriptors, the Extent of Flow Recirculation and Helicity in Carotid Bifurcation** SB^{3}C2015-547 pg.778

Diego Gallo[1], David A. Steinman[2], Umberto Morbiducci[1], [1]*Politecnico di Torino, Turin, Italy,*[2]*University of Toronto, Toronto, ON, Canada*

5:00PM **Application of Gold Particle Enhanced CT for Vulnerable Plaque Detection and Quantification in Atherosclerotic Mouse Models** SB^{3}C2015-366 pg.451

David De Wilde[1], Bram Trachet[1,2], Carole Van der Donckt[3], Bert Vandeghinste[1], Benedicte Descamps[1], Christian Vanhove[1], Guido R. Y. De Meyer[3], Patrick Segers[1], [1]*Ghent University, Gent, Belgium,*[2]*Institute for Bioengineering Ecole Polytechnique Fédérale de Lausanne, Lausanne, Switzerland,*[3]*University of Antwerp, Wilrijk, Belgium*

WEDNESDAY, JUNE 17	3:45pm - 5:15pm

Ocular Biomechanics I — Primrose A

Session Chair: Rouzbeh Amini, *The University of Akron, Akron, OH, United States*
Session Co-Chair: Jun Liu, *Ohio State, Columbus, OH, United States*

3:45PM **Development Of A Platform For Studying Astrocyte Mechanobiology: Compression Of Astrocytes In 3D Alginate Gels** SB^{3}C2015-655 pg. 977
John J. Mulvihill[1,2], Lisa A. Schildmeyer[1], Baptiste Coudrillier[1], Danny J. Kelly[2], C. Ross Ethier[1,3], *[1]Georgia Tech, Atlanta, GA, United States,[2]Trinity College Dublin, Dublin, Ireland,[3]Atlanta VA Medical Center, Atlanta, GA, United States*

4:00PM **Cellular Young's Modulus as a Novel Stemness Marker in the Corneal Limbus** SB^{3}C2015-216 pg. 212
Tom Bongiorno[1], Jena Chojnowski[2], James D. Lauderdale[2], Todd Sulchek[1,3], *[1]Georgia Institute of Technology, Atlanta, GA, United States,[2]University of Georgia, Athens, GA, United States,[3]The Parker H. Petit Institute for Bioengineering and Bioscience, Atlanta, GA, United States*

4:15PM **Experimental Measurement Of Collagen Fiber Uncrimping And Recruitment With Increases In Intraocular Pressure** SB^{3}C2015-417 pg. 543
Ian A. Sigal, Ning-Jiun Jan, Jonathan Grimm, Huong Tran, Hiroshi Ishikawa, Katherine A. Davoli, Larry Kagemann, Joel S. Schuman, Gadi Wollstein, Kira Lathrop, *University of Pittsburgh, Pittsburgh, PA, United States*

4:30PM **Phase-contrast Micro-tomography Measurements Of Intraocular Pressure-induced Deformation Of The Porcine Lamina Cribrosa** SB^{3}C2015-147 pg.86
Baptiste Coudrillier[1], Diogo M. Geraldes[2], Nghia Vo[3], Ian C. Campbell[1], Julie Albon[4], Richard L. Abel[2], C. Ross Ethier[1], *[1]Georgia Institute of Technology, Atlanta, GA, United States,[2]Imperial College, London, United Kingdom,[3]Diamond Light Source, Didcot, United Kingdom,[4]Cardiff University, Cardiff, United Kingdom*

4:45PM **Ocular Compliance in Mice** SB^{3}C2015-111 pg.34
Stephen A. S. Schwaner[1], Joseph M. Sherwood[2], Eric J. Snider[1], Eldon E. Geisert[3], Darryl R. Overby[4], C. Ross Ethier[1,3], *[1]Georgia Institute of Technology, Atlanta, GA, United States,[2]Imperial College London, London, United Kingdom,[3]Emory University, Atlanta, GA, United States,[4]Imperial College of London, London, United Kingdom*

5:00PM **Time And Pressure-Dependent Hydraulic Resistance Across Schlemm's Canal Endothelial Cell Layers** SB^{3}C2015-1045 pg.1064
Alice Spenlehauer, Joseph M. Sherwood, Darryl R. Overby, *Imperial College London, London, United Kingdom*

WEDNESDAY, JUNE 17	3:45pm - 5:15pm

Design and Devices I - Vascular Disease and Therapeutic Intervention — Superior

Session Chair: James Moore, *Imperial College London, London, United Kingdom*
Session Co-Chair: Joao S. Soares, *University of Texas at Austin, Austin, TX, United States*

3:45PM **Corrosion Behavior of a Novel Biodegradable Metallic Stent** SB^{3}C2015-240 pg.251
Jennifer Frattolin[1,2], Luca Gottellini[1,3], Olivier F. Bertrand[1,4], Rosaire Mongrain[1,2], *[1]McGill University, Montreal, QC, Canada,[2]Montreal Heart Institute, Montreal, QC, Canada,[3]Politecnico di Milano, Milan, Italy,[4]Quebec Heart and Lung Institute, Laval University, Quebec City, QC, Canada*

4:00PM **Geometric Analysis of Iliac Artery Tortuosity: Comparison to Current Clinical Practice** SB^{3}C2015-248 pg.267
Matthew G. Doyle[1,2], Elrasheed Osman[2], Naomi Eisenberg[2], Cristina H. Amon[1], Leonard W. Tse[2], *[1]University of Toronto, Toronto, ON, Canada,[2]Toronto General Hospital, Toronto, ON, Canada*

4:15PM **In Vitro Coronary Artery Model for the Study of Atherosclerosis** SB^{3}C2015-368 pg.453
Elizabeth E. Antoine, Abdul I. Barakat, *Ecole Polytechnique, Palaiseau, France*

6:45PM **Wave Intensity Analysis from Magnetic Resonance Imaging: Experimental Validation and New Insights into Single Ventricle Physiology** SB³C2015-459 pg. 622
Nikesh Arya[1], Silvia Schievano[1], Catriona Baker[1], Tain-Yen Hsia[2], Alessandro Giardini[2], Sachin Khambadkone[2], Andrew M. Taylor[1], **Giovanni Biglino**[1], *[1]University College London, London, United Kingdom,[2]Great Ormond Street Hospital for Children, London, United Kingdom*

WEDNESDAY, JUNE 17	5:30pm - 7:00pm

Multiscale Biomechanics - Coupling Muscuoskeletal, Joint, and Tissue Level Models (Special Session) — Magpie

Session Chair: Mariana Kersh, *University of Illinois, Urbana, IL, United States*
Session Co-Chair: Darryl Thelen, *University of Wisconsin-Madison, Madison, WI, United States*

5:30PM **Gait Simulation Using a Contact Force Feedback Controller** SB³C2015-416 pg.541
Jonathan P. Walter, Marcus G. Pandy, *University of Melbourne, Parkville, Australia*

5:45PM **A Multi-scale Finite Element Framework for Modeling Natural Knee Mechanics** SB³C2015-314 pg.371
Michael D. Harris, Ali Azhar, Alessandro Navacchia, Adam J. Cyr, Donald Hume, Clare K. Fitzpatrick, Paul J. Rullkoetter, Kevin B. Shelburne, *University of Denver, Denver, CO, United States*

6:00PM **Prediction Of Elbow Joint Contact Pressures In The Multibody Framework** SB³C2015-389 pg.487
Munsur Rahman[1], Akin Cil[1,2], Antonis P. Stylianou[1], *[1]University of Missouri-Kansas City, Kansas City, MO, United States,[2]Truman Medical Centers, Kansas City, MO, United States*

6:15PM **Multi-scale Models of Skeletal Muscle Reveal the Complex Effects of Muscular Dystrophy on Tissue Mechanics and Damage Susceptibility.** SB³C2015-478 pg.656
Kelley M. Virgilio, Kyle M. Martin, Shayn M. Peirce, Silvia S. Blemker, *University of Virginia, Charlottesville, VA, United States*

6:30PM **Multiscale Syntheses Of Articular Cartilage Stress In Post Aclr Model Under Physiological Loading** SB³C2015-605
Malek Adouni, Yasin Dhaher, *Rehabilitation Institute of Chicago/Biomedical Engineering, Northwestern University, Chicago, IL, United States pg.885*

6:45PM **Towards The Validation of a Multiscale Chemo-Electro-Mechanical Finite Element Model Using Electromyography** SB³C2015-475 pg.650
Thomas Heidlauf, Mylena Mordhorst, **Sook-Yee Chong**, Oliver Röhrle, *University of Stuttgart, Stuttgart, Germany*

WEDNESDAY, JUNE 17	5:30pm - 7:00pm

Musculoskeletal Tissue Engineering - Molecular, Soluble, and Mechanical Regulation of Tissue Development — Maybird

Session Chair: Danny Kelly, *Trinity College, Dublin, Ireland*
Session Co-Chair: Pen-hsiu Grace Chao, *National Taiwan University, Taipei, Taiwan*

5:30PM **Osteocyte Sclerostin Expression Is Directed By Substrate Composition And Dimensionality In Addition To Fluid Flow** SB³C2015-241 pg.253
Robert Thomas Brady[1], Andrew R. Cameron[1], David A. Hoey[2], Fergal J. O'Brien[1], *[1]Royal College of Surgeons in Ireland, Dublin, Ireland,[2]University of Limerick, Limerick, Ireland*

5:45PM **Investigating CRISPRi Cell-Engineering Methods For Treatment Of Intervertebral Disc Degeneration** SB³C2015-395
Niloofar Farhang[1], Jonathan M. Brunger[2], Joshua D. Stover[1], Pratishka I. Thakore[2], Brandon Lawrence[1], Farshid Guilak[2], Charles A. Gersbach[2], Lori A. Setton[2], Robby D. Bowles[1], *[1]University of Utah, Salt Lake City, UT, United States,[2]Duke University, Durham, NC, United States pg.499*

6:00PM **Label-Free Protein Profiling for Functional Characterization of Engineered Cartilage Following Inflammatory Cytokine Exposure** SB³C2015-450 pg.608
Andrea R. Tan[1], Shujuan Tao[2], David Chen[2], Lewis Brown[2], Clark T. Hung[1], [1]*Columbia University, New York, NY, United States,*[2]*Comparative Proteomics Center, Columbia University, New York, NY, United States*

6:15PM **Bisphosphonate Gives Trauma Damaged Chondrocytes a New Life via Inhibition of Mevalonate Pathway** SB³C2015-570 pg.819
Yilu Zhou, Miri Park, Liyun Wang, Enoch Cheung, X. Lucas Lu, *University of Delaware, Newark, DE, United States*

6:30PM **The Impact of Physical and Enzymatic Treatment on the Development of Tissue-Engineered Articular Cartilage Generated from Adult Human Chondrocytes** SB³C2015-1099 pg.1161
Brendan L. Roach[1], Terri-Ann N. Kelly[1], Michael K. Dermksian[1], Sonia Bansal[1], Paola A. Lopez[2], Aaron M. Stoker[3], James L. Cook[3], Gerard A. Ateshian[1], Clark T. Hung[1], [1]*Columbia University, New York, NY, United States,*[2]*University of Arizona, Tucson, AZ, United States,*[3]*University of Missouri at Columbia, Columbia, MO, United States*

6:45PM **Effects Of Structural Vs. Non-structural Scaffolds On Bmp-2- And Mechanical Load-mediated Bone Regeneration** SB³C2015-1180 pg.1305
Anna McDermott[1], Angela Lin[2], Robert Guldberg[2], **Joel D. Boerckel**[1], [1]*University of Notre Dame, Notre Dame, IN, United States,*[2]*Georgia Institute of Technology, Atlanta, GA, United States*

WEDNESDAY, JUNE 17	**5:30pm - 7:00pm**

Transport at the Nano- and Micro- scale — Golden Cliff / Eagle's Nest

Session Chair: Rubén Diaz-Rivera, *University of Puerto Rico at Mayagüez, PR, United States*
Session Co-Chair: Chris Rylander, *University of Texas at Austin, Austin, TX, United States*

5:30PM **Biophysical Fractionation Of Cells With Multiple Microfluidic Outlets** SB³C2015-294 pg.335
Todd Sulchek, Gonghao Wang, Cory Turbyfield, Kaci Crawford, Alexander Alexeev, *Georgia Tech, Atlanta, GA, United States*

5:45PM **Pressure Drop Approach to Estimate the Apparent Slip Length in a PDMS Micro-channel** SB³C2015-633 pg.935
Stephanie E. González Jiménez[1], Carlos R. Romero Peñaloza[1], Rubén E. Díaz-Rivera[1,2], [1]*University of Puerto Rico at Mayagüez, Mayagüez, PR, United States,*[2]*Purdue University, West Lafayette, IN, United States*

6:00PM **Mechano-activatable Microcapsules For Tunable Drug Delivery** SB³C2015-597 pg.869
Bhavana Mohanraj, Fuquan Tu, Daeyeon Lee, George R. Dodge, Robert L. Mauck, *University of Pennsylvania, Philadelphia, PA, United States*

6:15PM **Rate of Transport of Circulating Macromolecules Into The Arterial Wall May Determine Atherosclerotic Plaque Stability** SB³C2015-358 pg. 435
Peter Weinberg, **Zahra Mohri**, *Imperial College London, London, United Kingdom*

6:30PM **Gold Nanoparticle Mediated Enhanced Gene Delivery With High Selectivity For Breast Cancer Treatment** SB³C2015-659 pg.985
Binita Shrestha, Liang Tang, *University of Texas at San Antonio, San Antonio, TX, United States*

6:45PM **The Effect Of Particle Deformability On The Transport Of Soft Colloids Through Porous Media** SB³C2015-1148 pg.251
Eduard Benet, Louis Foucard, Franck J. Vernerey, John Pellegrino, *University of Colorado at Boulder, Boulder, CO, United States*

WEDNESDAY, JUNE 17	**5:30pm - 7:00pm**

Ligament and Tendon — Primrose B

Session Chair: Spencer Lake, *Washington University, St. Louis, MO, United States*
Session Co-Chair: Ozan Akkus, *Case Western Reserve University, Cleveland, OH, United States*

5:30PM **Two Bundles of the Anterior Cruciate Ligament Exhibit Different Microstructural Properties and Mechanics During Stress-Relaxation** SB³C2015-484 pg. 662
Ryan Castile, Nathan Skelley, Robert Brophy, Spencer Lake, *Washington University, St. Louis, MO, United States*

5:45PM **Composition-dependent Mechanical Properties and Multiscale Strain Transfer in Tendon Subjected to Shear Loading** SB³C2015-173 pg.132
Fei Fang, Spencer P. Lake, *Washington University in St. Louis, St. Louis, MO, United States*

6:00PM **Evaluation of Tendon Injuries Using Shear Wave Elastography: Preliminary In-vivo Results in Human Achilles and Semitendinosus Tendons** SB³C2015-160 pg.110
Daniel H. Cortes, Stephen M. Suydam, Jennifer A. Zellers, Karin Gravare-Silbernagel, Thomas S. Buchanan, Dawn M. Elliott, *University of Delaware, Newark, DE, United States*

6:15PM **Modeling Ligament Mechanics Using Microstructural Parameters Measured from Confocal Images of Collagen Networks** SB³C2015-223 pg.222
Christina J. Stender, Evan Rust, Erica E. Morrill, Roshani Lamichhane, Raquel J. Brown, Trevor J. Lujan, *Boise State University, Boise, ID, United States*

6:30PM **A Transversely Isotropic Constitutive Model Describes And Predicts The Contribution Of Elastin To The Multiaxial Mechanics Of Ligament** SB³C2015-441 pg.590
Heath B. Henninger, Sara A. Scott, Benjamin J. Ellis, Jeffrey A. Weiss, *University of Utah, Salt Lake City, UT, United States*

6:45PM **Detection, Quantification, and Localization of Subfailure Damage in Tendon Fascicles using Collagen Mimetic Peptides** SB³C2015-628 pg.925
Jared L. Zitnay, Yang Li, Shawn P. Reese, Boi-Hoa San, S. Michael Yu, Jeffrey A. Weiss, *University of Utah, Salt Lake City, UT, United States*

THURSDAY, JUNE 18	8:00am - 9:00am

PLENARY SESSION I – Margaret Gardel — Ballrooms 1-3

THURSDAY, JUNE 18	9:15am - 10:45am

Brain Injury Biomechanics — Primrose A

Session Chair: Brittany Coats, *University of Utah, Salt Lake City, UT, United States*
Session Co-Chair: Francis Gayzik, *Wake Forest University School of Medicine, Winston-Salem, NC, United States*

9:15AM **The Effect of Facemasks on the Impact Performance of Helmets** SB³C2015-1101 pg.1165
Steven Rowson, Evan Terrell, *Virginia Tech, Blacksburg, VA, United States*

9:30AM **Estimating Axonal Strain in White Matter Using Paranodal Proteins as Fiduciary Markers** SB³C2015-468 pg.638
Sagar Singh, Assimina A. Pelegri, David I. Shreiber, *Rutgers, The State University of New Jersey, Piscataway, NJ, United States*

9:45AM **In Vivo Comparison Of Wearable Head Impact Sensors** SB³C2015-1152 pg.1259
Lyndia C. Wu, Vaibhav Nangia, Kevin Bui, Bradley Hammoor, Calvin Kuo, Fidel Hernandez, David B. Camarillo, *Stanford University, Stanford, CA, United States*

10:00AM **Image-Based Dynamic Analysis of Brain Deformation on Porcine Brain Injury Model** SB³C2015-1163 pg.1279
Arnold D. Gomez, Gregory G. Scott, Boston C. Terry, Brittany Coats, *University of Utah, Salt Lake City, UT, United States*

10:15AM **Preliminary Development and Validation of an Atlas-Based Finite Element Brain Model** SB³C2015-1145 pg.1245
Logan Miller, Jillian Urban, Elizabeth Lillie, Joel Stitzel, *Virginia Tech-Wake Forest University School of Biomedical Engineering and Sciences, Winston-Salem, NC, United States*

10:30AM **Influence of Recent Head Impact History on Biomechanics of Concussion Sustained in College Football Athletes** SB³C2015-289 pg.327
Daniel J. Sjoquist, Brian D. Stemper, Alok S. Shah, James Murtha, John R. Humm, Ashley LaRoche, Adam Pfaller, Steven Broglio, Kevin Guskiewicz, Michael McCrea, *Medical College of Wisconsin, Milwaukee, WI, United States*

THURSDAY, JUNE 18	9:15am - 10:45am

Dynamics and Rehabilitation — Superior

Session Chair: Alan Eberhardt, *University of Alabama at Birmingham, Birmingham, AL, United States*
Session Co-Chair: Tammy Bush, *Michigan State University, East Lansing, MI, United States*

9:15AM **A Support Vector Machine Based On Vertical Ground Reaction Force To Supplement Observational Gait Evaluation** SB^{3}C2015-568 pg.817
Craig J. Simons[1], Cory L. Christiansen[2], Jennifer E. Stevens-Lapsley[2], Kevin B. Shelburne[1], Bradley S. Davidson[1], [1]*University of Denver, Denver, CO, United States,*[2]*University of Colorado Denver, Denver, CO, United States*

9:30AM **A Bootstrapping Method to Assess the Influence of Age, Obesity, and Gender on Probability of Tripping as a Function of Obstacle Height** SB^{3}C2015-506 pg.698
Christina M. R. Garman[1], Christopher T. Franck[1], Maury A. Nussbaum[1], Michael L. Madigan[2], [1]*Virginia Tech, Blacksburg, VA, United States,*[2]*Texas A & M University, College Station, TX, United States*

9:45AM **Metabolic Consumption Using Different Repetitive Lifting Strategies** SB^{3}C2015-1039 pg.1052
Timothy D. Craig[1], Alice E. Riley[1], Sandra A. Billinger[2], Neena K. Sharma[2], **Sara E. Wilson**[1], [1]*University of Kansas, Lawrence, KS, United States,*[2]*University of Kansas, Kansas City, KS, United States*

10:00AM **Design, Calibration, and Validation of a Novel In Vitro Tibial Force Sensor** SB^{3}C2015-29 pg.10
Joshua D. Roth, Maury L. Hull, Stephen M. Howell, *University of California, Davis, Davis, CA, United States*

10:15AM **Oh Deer! Morphological and Biomechanical Evaluation of Cervine Femora** SB^{3}C2015-415 539
Mark J. Hedgeland, Morgan A. Libruk, Nicole C. Corbiere, Mario J. Ciani, Laurel Kuxhaus, *Clarkson University, Potsdam, NY, United States*

10:30AM **The Effects of Low Intensity Vibration on Bone Mineral Density in the Intact Limb of Animals with a Percutaneously Attached Endoprosthesis.** SB^{3}C2015-502 pg.692
Kyle Bodnyk[1], Garrett Noble[1], Matthew Allen[2], Noel Fitzpatrick[3], Gabriel Pagnotti[4], Richard Hart[1], [1]*The Ohio State University, Columbus, OH, United States,*[2]*University of Cambridge, Cambridge, United Kingdom,*[3]*Fitzpatrick Referrals, Surrey, United Kingdom,*[4]*Stony Brook University, New York, NY, United States*

THURSDAY, JUNE 18	9:15am - 10:45am

Cerebral and Aortic Aneurysms — Wasatch

Session Chair: Ender A. Finol, *University of Texas at San Antonio, San Antonio, TX, United States*
Session Co-Chair: Naomi Chesler, *University of Wisconsin, Madison, WI, United States*

9:15AM **Effect of Branched and Fenestrated Stent-Grafts on Renal Blood Flow.** SB^{3}C2015-107 pg. 26
Harkamaljot S. Kandail[1], Mohamad S. Hamady[2], Xiao Y. Xu[1], [1]*Imperial College London, London, United Kingdom,*[2]*Imperial College Healthcare NHS Trust, London, United Kingdom*

9:30AM **Hemodynamic Changes in Treated Cerebral Aneurysms and Correlations with Long-Term Outcomes** SB^{3}C2015-557
Michael C. Barbour[1], Patrick M. McGah[1], Michael R. Levitt[2], Kurt Sansom[1], Ryan P. Morton[2], John D. Nevra[2], Pierre D. Mourad[2], Basavaraj V. Ghodke[3], Danial K. Hallam[3], Laligam N. Sekhar[2], Louis J. Kim[2], Alberto Aliseda[1], [1]*University of Washington, Seattle, WA, United States,*[2]*University of Washington, Neurological Surgery, Seattle, WA, United States,*[3]*Dept of Radiology, Harborview Medical Center, Seattle, WA, United States* pg.797

9:45AM **Anuerysm MRI Phantoms For Direct, Ex Vivo Fluid Dynamics** SB^{3}C2015-283 pg.319
Jeff R. Anderson[1], Orlando Diaz[2], Richard Klucznik[2], Yi J. Zhang[2], Gavin W. Britz[2], Robert G. Grossman[2], Christof Karmonik[1,2], [1]*Houston Methodist Research Institute, Houston, TX, United States,*[2]*Houston Methodist Hospital, Houston, TX, United States*

10:00AM **Endovascular Treatment of Intracranial Aneurysms: Finite Element Modeling of Various Intervention Strategies** SB^{3}C2015-518 pg.723
Robert Damiano, Ding Ma, Jianping Xiang, Adnan Siddiqui, Kenneth Snyder, Hui Meng, *University at Buffalo, State University of New York, Buffalo, NY, United States*

10:15AM **Experimental Evaluation of Hemodynamics in Patient-Specific Model of Type B Aortic Dissection** SB³C2015-608 pg.889
Joav Birjiniuk[1], Jean M. Ruddy[2], Mark Young[3], Lucas H. Timmins[1], Ravi K. Veeraswamy[4], David N. Ku[5], [1]*Georgia Institute of Technology and Emory University School of Medicine, Atlanta, GA, United States,*[2]*Medical University of South Carolina, Charleston, SC, United States,*[3]*Medtronic, Inc., Santa Rosa, CA, United States,*[4]*Emory University School of Medicine, Atlanta, GA, United States,*[5]*Georgia Institute of Technology, Atlanta, GA, United States*

10:30AM **Assessment And Quantification Of Transitional Flow In Intracranial Aneurysms - Highly Resolved Simulations Below The Kolmogorov Scales** SB³C2015-588 pg. 855
Kartik Jain[1,2], Kristian Valen-Sendstad[2], Sabine Roller[1], Kent-Andre Mardal[2,3], [1]*University of Siegen, Siegen, Germany,*[2]*Simula Research Laboratory, Oslo, Norway,*[3]*University of Oslo, Oslo, Norway*

THURSDAY, JUNE 18	9:15am - 10:45am

Microstructure of Aneurysms — Magpie

Session Chair: Alison Marsden, *University of California San Diego, San Diego, CA, United States*
Session Co-Chair: Hai-Chao Han, *University of Texas at San Antonio, San Antonio, TX, United States*

9:15AM **Stiffer Arterial Wall Enhances Aortic Aneurysm Formation In A Mouse Model Via Elastase Infusion** SB³C2015-537 pg.758
Zhijie Wang, Stephanie Morgan, Mark Golob, Zhenjie Liu, Bo Liu, Naomi C. Chesler, *University of Wisconsin - Madison, Madison, WI, United States*

9:30AM **Bayesian Calibration Of A Growth And Remodeling Computational Model Of Abdominal Aortic Aneurysms** SB³C2015-564 pg.809
Seungik Baek, Liangliang Zhang, Sajjad SeyedSalehi, Jongeun Choi, Chae Young Lim, Tapabrata Maiti, *Michigan State University, East Lansing, MI, United States*

9:45AM **Role Of Aneurysm On Failure Properties Of "Radially-oriented" Collagen Fibers In Human Ascending Thoracic Aortic Media** SB³C2015-572 pg.823
Siladitya Pal, Spandan Maiti, Alkiviadis Tsamis, Julie A. Phillippi, Thomas G. Gleason, David A. Vorp, *University of Pittsburgh, Pittsburgh, PA, United States*

10:00AM **Elastin Production Slows Aneurysm Enlargement in a Constrained Mixture Models of Aneurysm Growth and Remodelling** SB³C2015-404 pg.517
Kory J. Blose, Justin S. Weinbaum, Anne M. Robertson, David A. Vorp, *University of Pittsburgh, Pittsburgh, PA, United States*

10:15AM **Novel Image-based Analysis Of Patient-specific Abdominal Aortic Aneurysm Wall Stress Using A Membrane Mechanics Model** SB³C2015-1043 pg.1060
Mirunalini Thirugnanasambandam[1], Prahlad G. Menon[1,2], Stéphane Avril[3], Ender A. Finol[1], [1]*University of Texas at San Antonio, San Antonio, TX, United States,*[2]*Sun Yat-sen University - Carnegie Mellon University Joint Institute of Engineering, Pittsburgh, PA, United States,*[3]*Ecole Nationale Supérieure des Mines, Saint-Etienne, France*

10:30AM **Heterogeneous Elastic Properties Of Ascending Thoracic Aneurysms** SB³C2015-122 pg.46
Jia Lu[1], Yuanming Luo[1], Frances M. Davis[2], Stéphane Avril[2], [1]*The University of Iowa, Iowa City, IA, United States,*[2]*Ecole Nationale Suprieure des Mines de Saint-Etienne, St. Étienne, France*

THURSDAY, JUNE 18	9:15am - 10:45am

Musculoskeletal Tissue Engineering - Matrices and Interfaces — Maybird

Session Chair: Matt Fisher, *North Carolina State University, Raleigh, NC, United States*
Session Co-Chair: Mariana Kersh, *University of Illinois, Urbana, IL, United States*

9:15AM **Perfusion Decellularized Skeletal Muscle As A Scaffold For The Repair Of Volumetric Muscle Loss** SB³C2015-432 pg.574
Ben Kasukonis, John Kim, Tyrone Washington, **Jeffrey C. Wolchok**, *University of Arkansas, Fayetteville, AR, United States*

9:30AM **Fabrication Of Dense Porous Aligned Collagen Scaffolds Using 2D Plastic Compression And Porogens**
SB^{3}C2015-1034 pg.1042
Shawn Reese, Jared Zitnay, Jeffrey Weiss, *University of Utah, Salt Lake City, UT, United States*

9:45AM **Wrinkled, Wavelength-Tunable Graphene-Based Surface Topographies for Directing Cell Alignment and Morphology**
SB^{3}C2015-295 pg.337
Daniel F. Tonderys, Susan Leggett, Zhongying Wang, *Brown University, Providence, RI, United States*

10:00AM **Network Stiffening of Nanofiber Scaffolds by Mineral** SB^{3}C2015-402 pg.513
Justin H. Lipner[1], John J. Boyle[1], Victor Birman[2], Guy M. Genin[1], Stavros Thomopoulos[1], [1]*Washington University in St. Louis, Saint Louis, MO, United States,*[2]*Engineering Education Center, Missouri University of Science and Technology, Saint Louis, MO, United States*

10:15AM **Micrometer-scale Mechanical Properties Of The Tendon-to-bone Attachment** SB^{3}C2015-594 pg.865
Alix C. Deymier-Black[1], Yiran An[2], Andrea G. Schwartz[1], Guy M. Genin[1], Stavros Thomopoulos[1], Asa H. Barber[2], [1]*Washington University in St Louis, St Louis, MO, United States,*[2]*Queen Mary University of London, London, United Kingdom*

10:30AM **Local Microenvironment Response of Chondrocytes in Gradient Collagen Matrices** SB^{3}C2015-639 pg.947
Tyler Novak, Benjamin Seelbinder, Celina M. Twitchell, Sherry L. Voytik-Harbin, Corey P. Neu, *Purdue University, Lafayette, IN, United States*

THURSDAY, JUNE 18	9:15am - 10:45am

Transport in Tissue and Tumor Microenvironments — Golden Cliff / Eagle's Nest

Session Chair: John Pearce, *The University of Texas at Austin, Austin, TX, United States*
Session Co-Chair: Aili Zhang, *Shanghai Jiao Tong University, Shanghai, China*

9:15AM **Temperature Control of Hydrogel Delivered Through Endoscopic Needle as Potential Treatment for Pancreatic Cancer** SB^{3}C2015-477 pg.654
Tom M. Merrill[1], Jennifer Mitchell[2], Denise Merrill[2], Matthew Short[1], [1]*Rowan University, Glassboro, NJ, United States,*[2]*FocalCool, LLC, Mullica Hill, NJ, United States*

9:30AM **Shear Activated Nanoparticle Aggregate Delivery of Tissue Plasminogen Activator with Adjunctive Endovascular Bypass for Revascularization in Acute Ischemic Stroke** SB^{3}C2015-303 pg.353
Matthew J. Gounis[1], Netanel Korin[2], Miklos G. Marosfoi[1], Oktay Uzun[2], Erin T. Langan[1], Anne-Laure Papa[2], Olivia W. Brooks[1], Chris Johnson[2], Ajit S. Puri[1], Deen Bhatta[2], Ajay K. Wakhloo[1], Donald E. Ingber[2], [1]*University of Massachusetts Medical School, Worcester, MA, United States,*[2]*Wyss Institute for Biologically Inspired Engineering, Boston, MA, United States*

9:45AM **Characterization Of Endothelial Cell Permeability On A Biomimetic Blood Vessel Platform** SB^{3}C2015-217 pg.214
Yaling Liu, Antony Thomas, Christopher Uhl, *Lehigh University, Bethlehem, PA, United States*

10:00AM **Salmon Thrombin Inhibits Endothelial Barrier Disruption Through its Activation of Protein C** SB^{3}C2015-273 pg.301
Jenell Smith, Peter Galie, David Slochower, Paul Janmey, Beth Winkelstein, *University of Pennsylvania, Philadelphia, PA, United States*

10:15AM **An In Vitro Microfluidic Tumor Model to Mimic Ductal Carcinoma In Situ** SB^{3}C2015-673 pg.1011
Victoria Noe-Kim, Altug Ozcelikkale, Bumsoo Han, *Purdue University, West Lafayette, IN, United States*

10:30AM **Novel 3D Co-culture Assay Reveals Cancer Cell Migration Driven By Matrix-dependent Autonomous Formation Of CXCL12 Gradients** SB^{3}C2015-532 pg.750
Christopher Moraes[1], Taisuke Kojima[2], Stephen P. Cavnar[2], Gary D. Luker[2], Shuichi Takayama[2], [1]*McGill University, Montreal, QC, Canada,*[2]*University of Michigan, Ann Arbor, MI, United States*

THURSDAY, JUNE 18	9:15am - 10:45am

Cartilage and Intervertebral Disc — Primrose B

Session Chair: Clark Hung, *Columbia University, New York, NY, United States*
Session Co-Chair: Corey Neu, *Purdue University, West Lafayette, IN, United States*

9:15AM **Quantitative Analysis Of Raman Spectra For Assessment Of Crosslink Concentrations Toward Diagnosing Early Osteoarthritis** SB³C2015-452 pg.612
Chao Wang, **Krista M. Durney**, Jonathan L. Kuo, Jack R. Norton, Gerard A. Ateshian, Sinisa Vukelic, *Columbia University, New York, NY, United States*

9:30AM **An In-situ Cartilage Characterization Method With Reduced Time And Computation Demands: Comparison Against Linear And Nonlinear Biphasic Theories** SB³C2015-226 pg.228
David L. Burris, Axel C. Moore, Xingyu Chen, Brandon K. Zimmerman, X. Lucas Lu, *University of Delaware, Newark, DE, United States*

9:45AM **Maintenance of F-Spondin (Spon1) Knockout Phenotype in 3D Cultured Murine Chondrocytes** SB³C2015-1050 pg.1074
Eben G. Estell[1], Qing Yang[2], Andrea R. Tan[1], Mukundan Attur[2], Steven B. Abramson[2], Clark T. Hung[1], [1]*Columbia University, New York, NY, United States,* [2]*New York University School of Medicine and Langone Medical Center, New York, NY, United States*

10:00AM **Impact-induced Fissuring In Articular Cartilage: Bulk And Tissue-scale Mechanics** SB³C2015-663 pg. 993
Corinne R. Henak, Lena R. Bartell, Itai Cohen, Lawrence J. Bonassar, *Cornell University, Ithaca, NY, United States*

10:15AM **Modeling The Intervertebral Discs Repair With Cellular Therapies** SB³C2015-563 pg.807
Qiaoqiao Zhu, Xin Gao, Weiyong Gu, *University of Miami, Coral Gables, FL, United States*

10:30AM **The Contribution of Collagen Fibers in Human Cartilage Endplate Mechanics** SB³C2015-660 pg.987
John DeLucca[1], Mary Boggs[1], Edward Vresilovic[2], Randall Duncan[1], Dawn Elliott[1], [1]*University of Delaware, Newark, DE, United States,* [2]*Pennsylvania State University, Hershey, PA, United States*

THURSDAY, JUNE 18	11:00am - 12:30pm

Biomechanics of Microcirculation: From Cells to Networks and Changes During Disease States (joint with JSME) — Primrose A

Session Chair: Ken-ichi Tsubota, *Chiba University, Chiba, Japan*
Session Co-Chair: Walter Lee Murfee, *Tulane University, New Orleans, LA, United States*

11:00AM **Molecular Dynamics Simulations of Lipid Bilayer Failure under Mechanical Stresses: Toward Understanding Mechanical Hemolysis** SB³C2015-214 pg.208
Kenichiro Koshiyama, *Osaka University, Toyonaka, Japan*

11:15AM **A Computational Study Of Red Blood Cells In Stokes Flow: From Single Cellular Mechanics To Suspension Rheology** SB³C2015-256 pg.277
Toshihiro Omori, Yohsuke Imai, Takami Yamaguchi, Takuji Ishikawa, *Tohoku University, Sendai, Japan*

11:30AM **Microsystems Engineering for Bridging Angiogenesis with Mechanopathology** SB³C2015-28 pg.8
Jonathan W. Song, *The Ohio State University, Columbus, OH, United States*

11:45AM **Autodigestion And Its Manifestations In The Microcirculation.** SB³C2015-292 pg.331
Geert W. Schmid-Schoenbein, Frank De Iano, Angelina Altshuler, Erik Kistler, Michael Richter, *University of California, San Diego, La Jolla, CA, United States*

12:00PM **Quantifying Hemodynamics And Wall Mechanics In Patient-specific Coronary Artery Bypass Grafts** SB³C2015-1119
Abhay B. Ramachandra, Andrew Kahn, Alison L. Marsden, *University of California San Diego, La Jolla, CA, United States* pg.1201

12:15PM **The Effects of Aging on Microvascular Network Resistance and Flow Heterogeneity in the Rat Mesentery**
SB³C2015-644 pg.957
David C. Sloas, Scott A. Stewart, Richard S. Sweat, Walter L. Murfee, *Tulane University, New Orleans, LA, United States*

THURSDAY, JUNE 18	11:00am - 12:30pm

Human Dynamics — Superior

Session Chair: Rita M. Patterson, *University of North Texas Health Science Center, Fort Worth, TX, United States*
Session Co-Chair: Tom Gardner, *Columbia University, New York, NY, United States*

11:00AM **Lumbar-pelvic Coordination Differences Between Novice And Experienced Lifters In Repetitive Lifting**
SB³C2015-1029 pg.1034
Timothy D. Craig[1], Alice E. Riley[1], Sandra A. Billinger[2], Neena K. Sharma[2], **Sara E. Wilson**[1], [1]*University of Kansas, Lawrence, KS, United States,*[2]*University of Kansas, Kansas City, KS, United States*

11:15AM **Design and Simulation of an Ankle Foot Simulator** SB³C2015-1171 pg.1291
Jonathan J. Miller, *University of Utah, Salt Lake City, UT, United States*

11:30AM **Improved Detection Of Dynamic Balance Using Normalized COM-COP Inclination Angular Jerk During A Golf Swing**
SB³C2015-23 pg.3
Ahnryul Choi[1], Joung Hwan Mun[2], [1]*The University of Texas Health Science Center at Houston, Houston, TX, United States,*[2]*Sungkyunkwan University, Suwon, Korea*

11:45AM **Investigating the Effect of Ball Impact Location** SB³C2015-117 pg.42
Ainhoa Iglesias-Díaz, **Martin L. Tanaka**, Paul M. Yanik, Aaron K. Ball, *Western Carolina University, Cullowhee, NC, United States*

12:00PM **Designing Biomechanical Models of the Ankle: How Many Degrees of Freedom are Necessary to Reflect In Vivo Joint Kinematics?** SB³C2015-237 pg.246
Jennifer A. Nichols, Koren E. Roach, Charles L. Saltzman, Andrew E. Anderson, *University of Utah, Salt Lake City, UT, United States*

12:15PM **Obesity and Age Affect Trip Outcome and Severity Following a Laboratory-Induced Trip** SB³C2015-510 pg.706
Christina M. R. Garman[1], Maury A. Nussbaum[1], Michael L. Madigan[2], [1]*Virginia Tech, Blacksburg, VA, United States,*[2]*Texas A & M University, College Station, TX, United States*

THURSDAY, JUNE 18	11:00am - 12:30pm

Thrombus Prediction — Wasatch

Session Co-Chair: Shawn Shadden, *UC Berkeley, Berkeley, CA, United States*
Session Co-Chair: Dalin Tang, *WPI, Worcester, MA, United States*

11:00AM **Activated Expression And Shedding Of Platelet Glycoprotein IIb/IIIa Under Non-physiological Shear Stress: A Paradoxical Effect** SB³C2015-280 pg.313
Zengsheng Chen, Jun Ding, Zhongjun J Wu, *University of Louisville School of Medicine, Louisville, KY, United States*

11:15AM **Development of A Predictive Computational Model For Device-Induced Thrombosis** SB³C2015-243 pg.257
Joshua O. Taylor[1,2], Richard S. Meyer[2], Steven Deutsch[2], **Keefe Manning**[1,3], [1]*The Pennsylvania State University, University Park, PA, United States,*[2]*Applied Research Laboratory, State College, PA, United States,*[3]*Penn State Hershey Medical Center, Hershey, PA, United States*

11:30AM **Thrombotic Risk Assessment Using Transluminal Attenuation Gradient and Computational Modelling in Kawasaki Disease Patients with Coronary Artery Aneurysms** SB³C2015-1082 pg.1130
Noelia Grande Gutierrez[1], Andrew M. Kahn[2], Olga Shirinsky[3], Nina V. Gagarina[3], Galina A. Lyskina[3], Ryuji Fukazawa[4], Shunichi Ogawa[4], Jane C. Burns[2], Alison L. Marsden[1], [1]*University of California San Diego, La Jolla, CA, United States,*[2]*School of Medicine, University of California San Diego, La Jolla, CA, United States,*[3]*Sechenov First Moscow State University, Moscow, Russian Federation,*[4]*Nippon Medical School Hospital, Tokyo, Japan*

11:45AM **Hemodynamic Prediction of Thrombus Prone Regions in Abdominal Aortic Aneurysms** SB³C2015-455 pg.618
Konstantinos Tzirakis[1], **Eleni Metaxa**[1], Nikolaos Kontopodis[2], Christos V. Ioannou[2], Yannis Papaharilaou[1], *[1]Foundation of Research and Technology - Hellas, Heraklion, Greece,[2]University of Crete, Heraklion, Greece*

12:00PM **Thrombogenic Properties Of A Superhydrophobic Surface For Use As A Prosthetic Heart Valve Material** SB³C2015-652 pg.971
David L. Bark, Jr., Sanli Movafaghi, Brandon L. Moore, Arun K. Kota, Ketul C. Popat, Lakshmi P. Dasi, *Colorado State University, Fort Collins, CO, United States*

12:15PM **The Impact of VAD Surgical Implantation Configurations on its Thrombogenic Potential** SB³C2015-548 pg.780
Wei C. Chiu[1], Allison J. McLarty[1], Shmuel Einav[1], Marvin J. Slepian[2], Danny Bluestein[1], *[1]Stony Brook University, Stony Brook, NY, United States,[2]Sarver Heart Center, Tucson, AZ, United States*

THURSDAY, JUNE 18	11:00am - 12:30pm

Heart Valve Structure and Function — Magpie

Session Chair: Michael Sacks, *University of Texas, Austin, TX, United States*
Session Co-Chair: Shamik Bhattacharya, *St.Mary's University, San Antonio, TX, United States*

11:00AM **Age-dependent Changes In Stress And Strain In The Human Native Heart Valve And Their Relation With Collagen Remodeling** SB³C2015-523 pg.731
Pim J. A. Oomen[1], Sandra Loerakker[1], Daphne van Geemen[1], Jan Neggers[1], Marie-José T. H. Goumans[2], Antoon J. van den Bogaerdt[3], Ad J. J. C. Bogers[3], Carlijn V. C. Bouten[1], Frank P. T. Baaijens[1], *[1]Eindhoven University of Technology, Eindhoven, Netherlands,[2]Leiden University Medical Center, Leiden, Netherlands,[3]Erasmus Medical Center, Rotterdam, Netherlands*

11:15AM **Effects Of Leaflet Microstructure And Constitutive Model On The Closing Behavior Of The Mitral Valve** SB³C2015-339 pg.409
Chung-Hao Lee[1], Jean-Pierre Rabbah[2], Ajit P. Yoganathan[2], Robert C. Gorman[3], Joseph H. Gorman[3], Michael S. Sacks[1], *[1]The University of Texas at Austin, Austin, TX, United States,[2]Georgia Institute of Technology, Atlanta, GA, United States,[3]University of Pennsylvania, Philadelphia, PA, United States*

11:30AM **Capturing Detailed 3D Mitral Valve Geometry For Computational Valve Modeling** SB³C2015-462 pg.626
Charles H. Bloodworth, Eric L. Pierce, Thomas F. Easley, Milan Toma, Morten O. Jensen, Ajit P. Yoganathan, *Wallace H Coulter Department of Biomedical Engineering, Georgia Tech and Emory University, Atlanta, GA, United States*

11:45AM **True 3D Stresses in Heart Valve Leaflets** SB³C2015-1032 pg.1038
Bruno V. Rego, Michael S. Sacks, *The University of Texas at Austin, Austin, TX, United States*

12:00PM **A Framework For Parameter Estimation Of Heart Valves Using Inverse-Modeling Approach** SB³C2015-603 pg.881
Ankush Aggarwal, Michael S. Sacks, *University of Texas at Austin, Austin, TX, United States*

12:15PM **Immersogeometric Fluid-Structure Interaction Analysis of Bioprosthetic Heart Valves: Realistic Material Modeling and Experimental Validation** SB³C2015-1150 pg.1255
David Kamensky[1], Ming-Chen Hsu[2], Michael S. Sacks[1], Thomas J. R. Hughes[1], *[1]University of Texas at Austin, Austin, TX, United States,[2]Iowa State University, Ames, IA, United States*

THURSDAY, JUNE 18	11:00am - 12:30pm

Nano-, Micro-, and Multi-Scale Mechanics of Cells and Tissues — Maybird

Session Chair: Nadeen Chahine, *Feinstein Institute of Medical Research, Manhasset, NY, United States*
Session Co-Chair: Grace D. O'Connell, *University of California, Berkeley, Berkeley, CA, United States*

11:00AM **Micromechanical Heterogeneity And Anisotropy Of The Meniscus Extracellular Matrix** SB³C2015-128 pg.58
Qing Li[1], Feini Qu[2,3], Biao Han[1], Robert L. Mauck[2,3], **Lin Han**[1], *[1]Drexel University, Philadelphia, PA, United States,[2]University of Pennsylvania, Philadelphia, PA, United States,[3]Philadelphia Veterans Administration Medical Center, Philadelphia, PA, United States*

11:15AM **A Micromechanical Viscoelastic Computational Model Incorporating Progressive Detachment Of The Tau Proteins Predicting Microtubule Breaking Following Axonal Injury.** SB³C2015-153 pg. 96
Hossein Ahmadzadeh[1], Douglas H. Smith[2], Vivek B. Shenoy[1], [1]*Department of Materials Science and Engineering, University of Pennsylvania, Philadelphia, PA, United States,*[2]*Penn Center for Brain Injury and Repair and Department of Neurosurgery, University of Pennsylvania, Philadelphia, PA, United States*

11:30AM **Layer-specific Mitral Valve Interstitial Cell Deformations Under Physiological Loading** SB³C2015-340 pg. 411
Chung-Hao Lee[1], Christopher A. Carruthers[2], Ayoub Salma[1], Robert C. Gorman[3], Joseph H. Gorman[3], Michael S. Sacks[1], [1]*The University of Texas at Austin, Austin, TX, United States,*[2]*Medtronic, Minneapolis, MN, United States,*[3]*University of Pennsylvania, Philadelphia, PA, United States*

11:45AM **Relaxation of Simulated Viscoelastic Fiber Networks** SB³C2015-377 pg.467
Rohit Y. Dhume, Victor H. Barocas, *University of Minnesota, Minneapolis, MN, United States*

12:00PM **A Multi-Scale Approach in Analyzing Fluid/Solute Flow in Mechanically Loaded Bone** SB³C2015-499 pg.686
Lixia Fan[1], Shaopeng Pei[2], Xiaohan Lai[2], Xin Lu[2], **Liyun Wang**[2], [1]*Nanjing University of Science and Technology, Nanjing, China,*[2]*University of Delaware, Newark, DE, United States*

12:15PM **Blocking Inflammation Protects Against Alterations In Cell Biomecahnics, Morphology And Cytoskeleton** SB³C2015-1155 pg.1265
Timothy Jacobsen, Paula Hernandez, Victoria Wei, Nadeen O. Chahine, *Feinstein Institute for Medical Research, Manhasset, NY, United States*

THURSDAY, JUNE 18	**11:00am - 12:30pm**

Cryotherapy and Hyperthermia: 70+ Birthday Celebration for Prof. Avraham Shitzer

Golden Cliff / Eagle's Nest

Session Chair: John Bischof, *University of Minnesota at Minneapolis, Minneapolis, MN, United States*
Session Co-Chair: Liang Z. Zhu, *University of Maryland Baltimore County, Baltimore, MD, United States*

11:00AM **A Novel Method For Positioning And Operating Cryo-surgical Probes In 2-D Convex Target Areas Based On Super-positioning Of Unit Circles** SB³C2015-25 pg. 7
Avraham Shitzer, *Technion, Isreal Inst. Tech., Haifa, Israel*

11:15AM **Meeting The Need For Improved Monitoring Of Cryotherapy In Thin Cardiovascular Tissues With A New Micro-thermal Sensor** SB³C2015-1137 pg.1233
Harishankar Natesan[1], Jeunghwan Choi[1], Wyatt Hodges[2], Sean Lubner[2], Chris Dames[2], John Bischof[1], [1]*University of Minnesota, Minneapolis, MN, United States,*[2]*University of California, Berkeley, CA, United States*

11:45AM **Determination of the Convective Heat Transfer Coefficients on the Outer Surfaces of the Cryovivals Plunged in Liquid Nitrogen and 37°C Water Bath** SB³C2015-433 pg. 576
Tao Wang, Gang Zhao, *University of Science and Technology of China, Hefei, China*

11:45AM **Correcting Arrhenius Models for Application to Cell Death Processes Involving Intrinsic Protein Cascades** SB³C2015-142 pg.78
John Pearce, *The University of Texas at Austin, Austin, TX, United States*

12:00AM **Tumor Engineering to Elucidate the Effect of Mild Hyperthermia on the Transport of Single Walled Carbon Nanohorns in the Tumor Microenvironment** SB³C2015-586 pg.851
Matthew R. DeWitt[1], M. Nichole Rylander[2], [1]*Virginia Tech, Blacksburg, VA, United States,*[2]*The University of Texas at Austin, Austin, TX, United States*

12:15AM **Strategies Of Injecting Ferrofluid Into Tumors To Achieve Repeatable Nanoparticle Deposition And Thermal Dosage In Mangetic Nanoparticle Hyperthermia** SB³C2015-286 pg.323
Alexander LeBrun, Charles Bieberich, Ronghui Ma, Liang Zhu, *University of Maryland Baltimore County, Baltimore, MD, United States*

THURSDAY, JUNE 18	11:00am - 12:30pm

Cartilage Mechanics and Repair (joint with JSME) — Primrose B

Session Chair: **Hiromichi Fujie**, *Tokyo Metropolitan University, Tokyo, Japan*
Session Co-Chair: **Jennifer Wayne**, *Virginia Commonwealth University, Richmond, VA, United States*

11:00AM **Anisotropic Hydraulic Permeability In Articular Cartilage And Menisci: A Direct Measurement Using Biopsy Punches** SB[3]C2015-420 pg.549
Hiromichi Fujie, Ryosuke Nakamura, *Tokyo Metropolitan University, Tokyo, Japan*

11:15AM **The Influence of Cartilage Morphology and Elasticity on Tibiofemoral Contact Pressures During Walking** SB[3]C2015-667 pg.1001
Colin R. Smith, Rachel L. Lenhart, Mike F. Vignos, Jarred Kaiser, Darryl G. Thelen, *University of Wisconsin-Madison, Madison, WI, United States*

11:30AM **The Effect of Impact and Genipin Crosslinking on the Friction and Wear of Articular Cartilage** SB[3]C2015-24 pg.5
Craig M. Bonitsky[1], Michael Selep[1], Megan E. McGann[1], Timothy C. Ovaert[1], Stephen B. Trippel[2], Diane R. Wagner[1], [1]*University of Notre Dame, Notre Dame, IN, United States,*[2]*Indiana University, Indianapolis, IN, United States*

11:45AM **Development Of Artificial Cartilage Using Two Phase Collagen/Scaffold With Mesenchymal Stem Cells** SB[3]C2015-268 pg.297
Mitsugu Todo[1], Yusuke Nakamuta[1], Takaaki Arahira[2], [1]*Kyushu University, Kasuga, Japan,*[2]*Fukuoka Dental College, Fukuoka, Japan*

12:00PM **Evaluation Of Friction And Wear Properties Of Poly(vinyl Alcohol) Hydrogels As Artificial Articular Cartilage** SB[3]C2015-1078 pg.1122
Seido Yarimitsu[1], Ayumi Yoshida[1], Kazuhiro Nakashima[1], Teruo Murakami[1], Saori Sasaki[2], Atsushi Suzuki[2], [1]*Kyushu University, Fukuoka, Japan,*[2]*Yokohama National University, Yokohama, Japan*

12:15PM **Relationship Of Wear Particles Of Poly(vinyl Alcohol) Hydrogel And Immune Response Of Macrophage** SB[3]C2015-1079 pg.1124
Seiji Omata, Yoshinori Sawae, Teruo Murakami, *Kyushu University, Fukuoka, Japan*

THURSDAY, JUNE 18	12:30pm - 2:30pm

Poster Session I — BS Competition - Biofluid Mechanics — Event Center Tent

1 **Predicting Regions Of Low Wall Shear Stress In The Carotid Artery Bifurcation** SB[3]C2015-419 pg.547
Haley M. King, Emma J. Lichtenfels, Christian R. Fahrenbruck, Brandon Moore, Lakshmi P. Dasi, *Colorado State University, Fort Collins, CO, United States*

THURSDAY, JUNE 18	12:30pm - 2:30pm

Poster Session I — BS Competition - Biotransport — Event Center Tent

2 **Generating Tumor Models From MicroCT Scan Images For Simulating Temperature Elevation During Magnetic Nanoparticle Hyperthermia** SB[3]C2015-300 pg.347
Andrew S. Lee, Alexander LeBrun, Ronghui Ma, Liang Zhu, *University of Maryland Baltimore County, Baltimore, MD, United States*

3 **Evaluating Accuracy Of The Algorithm/formula Used In Store Brand Digital Thermometers To Predict Body Temperature Based On The First Few Seconds of The Temperature Measurements** SB[3]C2015-507 pg.700
Amirreza Saharkhiz[1], Oleg Vesnovsky[2], Jon P. Casamento[2], Laurence W. Grossman[2], L. D. Timmie Topoleski[1], Liang Zhu[1], [1]*University of Maryland Baltimore County, Baltimore, MD, United States,*[2]*U.S. Food and Drug Administration, Silver Spring, MD, United States*

4 **Development Of A Tissue Phantom To Mimic The Thermal Environment Of A Human Arm To Test Digital Thermometers** SB^3C2015-501 pg.690

Peter Dillon[1], Oleg Vesnovsky[2], L. D. Timmie Topoleski[1], Jon P. Casamento[2], Laurence W. Grossman[2], Liang Zhu[1], *[1]University of Maryland Baltimore County, Baltimore, MD, United States,[2]U.S. Food and Drug Administration, Silver Spring, MD, United States*

5 **Quantitative Visualization of Drug Response of Breast Cancer Cells Within a Three-Dimensional Extracellular Matrix** SB^3C2015-234 pg.240

Brett S. Klosterhoff[1], Kyeonggon Shin[1], J. Craig Dutton[2], Bumsoo Han[1], *[1]Purdue University, West Lafayette, IN, United States,[2]University of Illinois at Urbana-Champaign, Urbana, IL, United States*

THURSDAY, JUNE 18	12:30pm - 2:30pm

Poster Session I — BS Competition - Cellular and Tissue Engineering — Event Center Tent

6 **Fabrication and Evaluation of Biodegradable Tissue Scaffolds for Osteochondral Defect Repair** SB^3C2015-1154 pg.1263

Carly R. Garrow, Andrew J. Polk, Ferris M. Pfeiffer, *University of Missouri - Columbia, Columbia, MO, United States*

THURSDAY, JUNE 18	12:30pm - 2:30pm

Poster Session I — BS Competition - Design and Devices — Event Center Tent

7 **Wear Simulator for Canine Total Hip Replacements** SB^3C2015-1174 pg.1297

Kevin J. Warburton[1], Evan Rust[1], Jeff Brourman[2], Trevor J. Lujan[1], *[1]Boise State University, Boise, ID, United States,[2]Veterinary Surgeon WestVet Animal Emergency & Specialty Center, Garden City, ID, United States*

8 **Material Property Testing of Carboxymethylated Hyaluronic Acid Hydrogel Polymer** SB^3C2015-1186 pg.1315

McKenna Drysdale[1], Hee-Kyoung Lee[1,2], Barbara Wirostko[2], Brittany Coats[1], *[1]University of Utah, Salt Lake City, UT, United States,[2]Jade Therapeutics, Salt Lake City, UT, United States*

9 **Design of a 3-D Bioreactor for Simulation of Cerebrospinal Fluid Flow in the Third Ventricle and Aqueduct of Sylvius** SB^3C2015-1164 pg.1281

Michael J. Majcher, Matthew R. Dailey, David P. Lemmer, Joseph T. Havrilak, Nic Leipzig, Bryn A. Martin, *University of Akron, Akron, OH, United States*

11 **A New Method For Determining Cross-sectional Shape And Area Of Soft Tissues Using 3d Laser Scanning** SB^3C2015-176 pg.138

Yaniv Michaeli[1,2], Richard E. Debski[2], *[1]ORT Braude College, Nahariya, Israel,[2]University of Pittsburgh, Pittsburgh, PA, United States*

12 **Rapid Quantification Of Femoral Head Geometry From Magnetic Resonance Imaging Of Femoroacetabular Impingement** SB^3C2015-643 pg. 955

Haley Ehlers[1], Michael Roberts[1], Garry Gold[2], Saikat Pal[1], *[1]California Polytechnic State University, San Luis Obispo, CA, United States,[2]Stanford University, Stanford, CA, United States*

13 **Design and Implementation of an Instrumented Pedal for Cycling Biomechanics Research** SB^3C2015-1033 pg.1040

Luke I. Kraemer, Juan D. Gutierrez-Franco, Jake E. Deschamps, Karim C. Dudum, Eshan M. Dandekar, Scott J. Hazelwood, Hemanth V. Porumamilla, Stephen M. Klisch, *California Polytechnic State University, San Luis Obispo, CA, United States*

THURSDAY, JUNE 18	12:30pm - 2:30pm

Poster Session I | **BS Competition - Human Dynamics** | **Event Center Tent**

14 **Can Measured Muscle Synergies Reconstruct Unmeasured Muscle Excitations?** SB[3]C2015-573 pg.825
Nicholas A. Bianco, Carolynn Patten, Benjamin J. Fregly, *University of Florida, Gainesville, FL, United States*

15 **Using OpenSim to Predict Knee Joint Moments During Cycling** SB[3]C2015-1023 pg.1024
Karim C. Dudum, Jake E. Deschamps, Juan D. Gutierrez-Franco, Luke I. Kraemer, Alejandro M. Gonzalez-Smith, Eshan M. Dandekar, Scott J. Hazelwood, Stephen M. Klisch, *California Polytechnic State University, San Luis Obispo, CA, United States*

THURSDAY, JUNE 18	12:30pm - 2:30pm

Poster Session I | **BS Competition - Injury** | **Event Center Tent**

16 **The Temporal Change in Protein Biomarkers of Vitreous Humor following Blast Trauma of the Visual System** SB[3]C2015-1098 pg.1159
Justin A. Jones, Daniel F. Shedd, Brittany Coats, *University of Utah, Salt Lake City, UT, United States*

17 **Effect of Sclerostin Antibody on Tendon-to-Bone Healing in a Rotator Cuff Animal Model** SB[3]C2015-1051 pg.1076
Shivam A. Shah[1], Ioannis Kormpakis[1], Necat Havlioglu[2], Michael S. Ominsky[3], Leesa M. Galatz[1], Stavros Thomopoulos[1], [1]*Washington University of St. Louis, Saint Louis, MO, United States,*[2]*John Cochran VA Medical Center, Saint Louis, MO, United States,*[3]*Amgen Inc., Thousand Oaks, CA, United States*

THURSDAY, JUNE 18	12:30pm - 2:30pm

Poster Session I | **BS Competition - Tissue Mechanics** | **Event Center Tent**

18 **Comparison of Pulmonary Artery Material Properties between a Congenital Heart Disease Patient and a Normal Subject using Cardiac Magnetic Resonance: A Feasibility Study** SB[3]C2015-606 pg.887
Rajit Banerjee[1], Gavin A. D'Souza[2], Namheon Lee[3], Michael D. Taylor[3], [1]*University of Toledo, Toledo, OH, United States,*[2]*University of Cincinnati, Cincinnati, OH, United States,*[3]*Cincinnati Children's Hospital Medical Center, Cincinnati, OH, United States*

19 **Finite Element Analysis of the Biomechanical Consequences of Schmorl's Nodes** SB[3]C2015-1018 pg.1014
Boston B. Barham, Anton E. Bowden, *Brigham Young University, Provo, UT, United States*

20 **A Mechanical Model For Cortical Folding During Brain Development** SB[3]C2015-575 pg.829
Nithya Vijayakumar, Maria Holland, Ellen Kuhl, *Stanford University, Stanford, CA, United States*

THURSDAY, JUNE 18	12:30pm - 2:30pm

Poster Session I | **Multi-scale Modeling in Biotransport** | **Event Center Tent**

21 **Comparison Of MRI-measured Tracer Transport In The Rat Brain With Computational Model Prediction** SB[3]C2015-1149 pg.1253
Wei Dai, Garrett W. Astary, Aditya K. Kasinadhuni, Paul R. Carney, Thomas H. Mareci, Malisa Sarntinoranont, *University of Florida, Gainesville, FL, United States*

22 **A Reduced-Dimensional Transport Model for Thrombogenic Species in Large Arteries** SB[3]C2015-195 pg.174
Kirk B. Hansen, Shawn C. Shadden, *University of California, Berkeley, Berkeley, CA, United States*

23 **Low-Concentration Salmonella Detection Using Orbiting Magnetic Microbeads in a Continuous-Flow Microfluidic Device** SB³C2015-593 pg.863
Matthew S. Ballard, Drew L. Owen, Zachary G. Mills, Srinivas K. G. Hanasoge, Peter J. Hesketh, Alexander Alexeev, *Georgia Institute of Technology, Atlanta, GA, United States*

24 **Computational Multiphysics Model of Clot Lysis in a Completely Occluded Stenotic Artery** SB³C2015-139 pg.74
Andris Piebalgs, Xiao Y. Xu, *Imperial College London, London, United Kingdom*

25 **Effect of Intervertebral Disc Size on Nutrient Distributions: A Finite Element Analysis with Implications for In Vivo Models** SB³C2015-1077 pg.1120
Alicia R. Jackson, *University of Miami, Coral Gables, FL, United States*

26 **Reaction/Diffusion Modelling of AtzA Biocatalyst Encapsulated in a Silica Gel Matrix** SB³C2015-175 pg.136
Baris Ragip Mutlu, *University of Minnesota, Minneapolis, MN, United States*

27 **Finite Element Model of Mixed Porohyperelastic Transport in an Axisymmetric Porcine Coronary Artery** SB³C2015-184 pg.154
Michelle A. H. Armstrong, Bruce R. Simon, Jonathan P. Vande Geest, *University of Arizona, Tucson, AZ, United States*

THURSDAY, JUNE 18	12:30pm - 2:30pm

Poster Session I — Design, Dynamics and Rehab — Event Center Tent

28 **Mechanical Design of a Customizable Self-Expanding Endovascular Stent** SB³C2015-1189 pg.1321
Joel C. R. Scott, Clifton R. Johnston, Darrel A. Doman, *Dalhousie University, Halifax, NS, Canada*

29 **Which Of Four Commercially-Available Laser Scanner Systems Generates The Most Accurate Bone Model?** SB³C2015-152 pg.94
Valentina Campanelli[1,2], Stephen M. Howell[1], Maury Hull[1], *[1]UC Davis, Davis, CA, United States,[2]University of Verona, Verona, Italy*

30 **Deformation and Flow of Arterial Stenosis Model Regarding the Change in Curvature of Coronary Artery for Percutaneous Transluminal Coronary Angioplasty** SB³C2015-1142 pg.1239
Shunichi Kobayashi, *Shinshu University, Ueda, Japan*

31 **Computational Modeling Of Mechanical Stresses In Stents: Implications For The Optimization Of Drug-eluting Stents** SB³C2015-458 pg.620
Francois P. M. Cornat, Franz Bozsak, Abdul I. Barakat, *Ecole Polytechnique, Palaiseau cedex, France*

32 **Damage Analysis of Retrieved Oxidized Zirconium Femoral Components for TKA: Can Wear Breach the Oxide Layer?** SB³C2015-1135 pg.1229
Noah Bonnheim[1], Michael Ries[2], Sanjai Shukla[2], Lisa Pruitt[1], *[1]University of California, Berkeley, Berkeley, CA, United States,[2]Tahoe Fracture and Orthopaedic Clinic, Carson City, NV, United States*

33 **Design and Optimization of an In-Vitro Emboli Detector for Flow-Induced Thrombogenicity Evaluation** SB³C2015-344 pg.419
Ram Shtoltz, *Tel Aviv University, Tel Aviv, Israel*

34 **Development of Micro-scale Ultrasound Imaging for Tissue Characterization.** SB³C2015-1165 pg.1283
Jeremy Stromer, Leila Ladani, *University of Connecticut, Storrs, CT, United States*

35 **Bearing Surface Damage Analysis Of Total Shoulder Replacement Retrievals Across Fixation Designs And UHMWPE Composition** SB³C2015-531 pg.748
Louis G. Malito[1], Farzana Ansari[1], Lulu Li[1], Taylor Lee[1], Helen Park[1], Steve Gunther[2], Tom Norris[3], Mike Ries[4], Lisa Pruitt[1], *[1]University of California, Berkeley, Berkeley, CA, United States,[2]Martha Jefferson Hospital, Charlottesville, VA, United States,[3]San Francisco Shoulder, Elbow & Hand Clinic, San Francisco, CA, United States,[4]Tahoe Fracture and Orthopaedic Clinic, Carson City, NV, United States*

36 **Stress Angle Device: A Novel System For Reproducing The Mechanical Conditions Associated With Regions Most Susceptible To Vascular Disease** SB³C2015-406 pg.521
Steve Zambrano, Michael J. Draper, Brendan L. Swain, Caleb A. Davis, Michael R. Moreno, *Texas A&M University, Bryan, TX, United States*

37 **On The Characterization Of Mitral Valve Geometry And Development Of A Population-averaged Model** SB^{3}C2015-1022 pg.1022
Amir Khalighi[1], Andrew Drach[1], Fleur ter Huurne[2], Chung-Hao Lee[1], Charles Bloodworth[3], Eric Pierce[3], Morten Jensen[3], Ajit Yoganathan[3], Michael Sacks[1], [1]*University of Texas at Austin, Austin, TX, United States,*[2]*Eindhoven University, Eindhoven, Netherlands,*[3]*Georgia Institute of Technology, Atlanta, GA, United States*

38 **Modeling the Mechanical Behavior of Polymeric Bioresorbable Stents: a Finite Element Approach** SB^{3}C2015-658 pg.983
Nic Debusschere, Patrick Segers, Peter Dubruel, Benedict Verhegghe, Matthieu De Beule, *Ghent University, Ghent, Belgium*

THURSDAY, JUNE 18	12:30pm - 2:30pm

Poster Session I — Neuromuscular Control and Motion Analysis — Event Center Tent

39 **Artificial Neural Networks for the Optimization of Ligament Stiffnesses in a Computational Foot/Ankle Model** SB^{3}C2015-438 pg.584
Ruchi Chande, Norma Ortiz-Robinson, Jennifer Wayne, *Virginia Commonwealth University, Richmond, VA, United States*

40 **Movement Analysis Based on the Separation of Angular Momentum: Gait Progression and Segment Rotation in Patients with TKA** SB^{3}C2015-382 pg.473
Brecca M. Gaffney[1], Will M. Johnston[1], Cory L. Christiansen[2], Jennifer E. Stevens-Lapsley[2], Kevin B. Shelburne[1], Bradley S. Davidson[1], [1]*University of Denver, Denver, CO, United States,*[2]*University of Colorado Denver, Aurora, CO, United States*

41 **A Description of Segmental Angular Momentum Synergies using Independent Component Analysis During Gait** SB^{3}C2015-543 pg.770
Brecca M. Gaffney[1], Cory L. Christiansen[2], Kevin B. Shelburne[1], Bradley S. Davidson[1], [1]*University of Denver, Denver, CO, United States,*[2]*University of Colorado Denver, Aurora, CO, United States*

42 **Joint Angle and Muscle Activity during Hippotherapy: A Case Study Using Motion Capture Analysis and EMG** SB^{3}C2015-632 pg.933
Mary C. Baker, Stephen Wester, Timothy Monday, *Texas Tech University, Lubbock, TX, United States*

43 **Changes In The Range Of Motion Envelope Of The Lumbar Spine With Repetitive Lifting** SB^{3}C2015-618 pg.907
Muhammad I. Gul[1], Timothy D. Craig[1], Neena K. Sharma[2], **Sara E. Wilson**[1], [1]*University of Kansas, Lawrence, KS, United States,*[2]*University of Kansas, Kansas City, KS, United States*

THURSDAY, JUNE 18	12:30pm - 2:30pm

Poster Session I — Best Practices in Biomechanics, Bioengineering, and Biotransport Education — Event Center Tent

44 **Best Practices in Teaching Biomechanics: Integrating Reflective Learning Activities** SB^{3}C2015-138 pg.72
Laurel Kuxhaus, *Clarkson University, Potsdam, NY, United States*

45 **Assessment of a Bioengineering Innovation Program for Middle School Girls** SB^{3}C2015-577 pg.833
Kristen Billiar[1], Amanda Reidinger[1], Jeanne Hubelbank[2], Helen Vassallo[1], [1]*Worcester Polytechnic Institute, Worcester, MA, United States,*[2]*Independent Program Evaluation Consultant, Sudbury, MA, United States*

46 **Teaching Undergraduate Design: An Approach Based on Industrial Experience** SB^{3}C2015-118 pg.44
Martin L. Tanaka, *Western Carolina University, Cullowhee, NC, United States*

47 **A Progress Report From a Multi-disciplinary Capstone Experience Involving Engineering and Business Students** SB^{3}C2015-110 pg.32
Alan Eberhardt, Joel Dobbs, *University of Alabama at Birmingham, Birmingham, AL, United States*

THURSDAY, JUNE 18	12:30pm - 2:30pm

Poster Session I — Cardiovascular Devices and Imaging in Fluid Mechanics — Event Center Tent

48 **Retrospective Analysis of Echocardiography in Those Suspected of Pulmonary Hypertension with Comparison to Right Heart Catheterization** SB³C2015-439 pg.586
Travis B. Eason[1], Melinda V. Pyler[1], Lavanya Alapati[1], Sanjay Mehra[2], John M. Cahill[1], Stephanie M. George[1], [1]*East Carolina University, Greenville, NC, United States,*[2]*Vidant Medical Center, Greenville, NC, United States*

49 **The Influence Of Sub-Optimal Acquisition Delay In A C-arm Cone-Beam CT Perfusion Study** SB³C2015-1067 pg.1102
Antonius M. de Korte, Kajo van der Marel, Juyu Chueh, Olivia W. Brooks, Ajit S. Puri, Ajay K. Wakhloo, Matthew J. Gounis, *New England Center for Stroke Research, University of Massachusetts Medical School, Worcester, MA, United States*

50 **The Relationship Of Wall Shear Stress With Clinically Relevant Metrics In Pulmonary Arterial Hypertension** SB³C2015-648 pg.963
Alifer D. Bordones[1,2], Vitaly O. Kheyfets[3], Ender A. Finol[1], [1]*University of Texas at San Antonio, San Antonio, TX, United States,*[2]*University of Texas Health Science Center at San Antonio, San Antonio, TX, United States,*[3]*University of Colorado, Denver, CO, United States*

51 **Quantitative Coronary Angiography Based Reconstructions for Wall Shear Stress Calculations in Bifurcations** SB³C2015-545 pg. 774
Jelle T. C. Schrauwen[1], Antonios Karanasos[1], Nienke S. Ditzhuijzen[1], Jean-Paul Aben[2], Jolanda J. Wentzel[1], Antonius F. W. van der Steen[1,3], Frank J. H. Gijsen[1], [1]*Thoraxcenter, Erasmus Medical Center, Rotterdam, Netherlands,*[2]*Pie Medical Imaging, Maastricht, Netherlands,*[3]*Delft University of Technology, Netherlands*

52 **Swirling Flows in Arterial Hemodynamics** SB³C2015-1072 pg. 1112
Kartik Bulusu[1], Christopher Elkins[2], John Eaton[2], **Michael Plesniak**[1], [1]*George Washington University, Washington, DC, United States,*[2]*Stanford University, Stanford, CA, United States*

53 **Integrating an Open Source Meshing Alternative into SimVascular 2.0** SB³C2015-1062 pg.1094
Adam R. Updegrove[1], Nathan M. Wilson[2], Shawn C. Shadden[1], [1]*University of California, Berkeley, Berkeley, CA, United States,*[2]*Open Source Medical Software Corporation, Santa Monica, CA, United States*

54 **Microbubble Void Imaging - A Novel Technique For Flow Visualisation and Quantitative Assessment of Intravascular Mixing in Larger Vessels Using Ultrasound** SB³C2015-388 pg.485
Chee Hau Leow, **Francesco Iori**, Richard W. Corbett, Neill Duncan, Colin G. Caro, Peter E. Vincent, Mengxing Tang, *Imperial College London, London, United Kingdom*

55 **In-vivo Validation Of The In Silico Predicted Pressure Drop Across An Arteriovenous Fistula** SB³C2015-536 pg.756
Leonard Browne[1], Khalid Bashar[2], Philip Griffin[1], Eamon Kavanagh[2], Michael Walsh[1], [1]*University of Limerick, Limerick, Ireland,*[2]*Univeristy Hospital Limerick, Limerick, Ireland*

56 **Triangulated Surface Boolean Operations for Combining 2-D and 3-D Image Segmentation for Patient-Specific Blood Flow Analysis** SB³C2015-1052 pg.1078
Adam R. Updegrove[1], Nathan M. Wilson[2], Shawn C. Shadden[1], [1]*University of California, Berkeley, Berkeley, CA, United States,*[2]*Open Source Medical Software Corporation, Santa Monica, CA, United States*

57 **Role of Wall Thickness and Tethering in the Assessment of Arterial Stiffness** SB³C2015-156 pg.102
Simona Hodis[1], Mair Zamir[2], [1]*Texas A&M University - Kingsville, Kingsville, TX, United States,*[2]*University of Western Ontario, London, ON, Canada*

58 **SimVascular 2.0: an Open Source Pipeline for Cardiovascular Modeling and Simulation** SB³C2015-656 pg.979
Hongzhi Lan[1], Nathan M. Wilson[2], Daniele Schiavazzi[1], Jameson Merkow[1], Adam Updegrove[3], Shawn C. Shadden[3], Alison L. Marsden[1], [1]*University of California - San Diego, La Jolla, CA, United States,*[2]*Open Source Medical Software Corporation, Los Angeles, CA, United States,*[3]*University of California - Berkeley, Berkeley, CA, United States*

59 **Nt-proBNP Expression Originating From The RV Myocardium In Pediatric Pulmonary Hypertension Patients Is Correlated With Both The Reactive And Resistive Components Of Vascular Impedance** SB³C2015-1130 pg.1219
Vitaly O. Kheyfets[1], Jamie Dunning[1], Uyen Truong[2], Kendall Hunter[1], Dunbar Ivy[2], Robin Shandas[1], [1]*University of Colorado, Aurora, CO, United States,*[2]*Children's Hospital Colorado, Aurora, CO, United States*

THURSDAY, JUNE 18	12:30pm - 2:30pm

Poster Session I — Embryonic, Pediatric Cardiology, and Other Fluid Mechanics — Event Center Tent

60 **Valveless Flow in a Thick Elastic Tube** SB³C2015-582 pg. 843
Pavel Kozlovsky[1], Moshe Rosenfeld[1], Ariel Jaffa[2], **David Elad**[1], [1]*Tel Aviv Univesity, Tel Aviv, Israel,*[2]*Tel Aviv Medical Center, Tel Aviv, Israel*

61 **Prediction of Downstream Velocity Waveforms for In Vitro Aortic Flow Experiments** SB³C2015-229 pg.234
Rafeed A. Chaudhury, Justin R. Ryan, David H. Frakes, Ronald J. Adrian, *Arizona State University, Tempe, AZ, United States*

62 **In Vitro Optimization Of The Nozzle Used In Assisted Bidirectional Glenn Procedure For Single Ventricle Stage 1 Palliation** SB³C2015-626 pg.921
Jian Zhou[1], Mahdi Esmaily-Moghadam[2], Timothy Conover[1], Tain-Yen Hsia[3], Alison Marsden[4], **Richard Figliola**[1], [1]*Clemson University, Clemson, SC, United States,*[2]*Stanford University, Palo Alto, CA, United States,*[3]*Great Ormond Street Hospital, London, United Kingdom,*[4]*University of California, San Diego, La Jolla, CA, United States*

63 **Mean Flow Umbilical Doppler Indices Perform Better Than Pulsatility Indices In Determining Small For Gestational Age Pregnancies** SB³C2015-201 pg.186
Shier Nee Saw[1], Citra Nurfarah Zaini Mattar[2], Seow Heong Yeo[3], Shu-E Soh[4,5], Yap-Seng Chong[2,6], Peter David Gluckman[6,7], Keith Godfrey[8], Seang Mei Saw[5,9], Arijit Biswas[2], **Choon Hwai Yap**[1], [1]*National University of Singapore, Singapore,*[2]*Yong Loo Lin School of Medicine, NUS & NUHS, Singapore,*[3]*KK Women's and Children's Hospital, Singapore,*[4]*Yong Loo Lin School of Medicine, NUS, Singapore,*[5]*Saw Swee Hock School of Public Health, NUS, Singapore,*[6]*Singapore Institute for Clinical Sciences, A*STAR, Singapore,*[7]*Liggins Institute, University of Auckland, Auckland, New Zealand,*[8]*University of Southampton & University Hospital Southampton NHS Foundation Trust, Southampton, United Kingdom,*[9]*Singapore Eye Research Institute, Singapore*

64 **Accounting For Clinical Data Uncertainty In Multiscale Numerical Simulation Of Single Ventricle Palliation Surgery** SB³C2015-309 pg.363
Daniele E. Schiavazzi[1], Tain-Yen Hsia[2], Alison L. Marsden[1], [1]*University of California at San Diego, San Diego, CA, United States,*[2]*Great Ormond Street Hospital for Children and UCL Institute of Cardiovascular Science, London, United Kingdom*

65 **Automated Tuning for Parameter Identification in Multi-scale Coronary Simulations** SB³C2015-541 pg. 766
Justin S. Tran, Daniele Schiavazzi, Abhay Ramachandra, Andrew Kahn, Alison Marsden, *University of California, San Diego, La Jolla, CA, United States*

66 **Investigating Environmental Causes of Congenital Heart Diseases: A Subject-Specific Computational Fluid Dynamics Study** SB³C2015-204 pg.192
Venkat Keshav Chivukula[1], Sevan Goenezen[2], Sandra Rugonyi[1], [1]*Oregon Health and Sciences University, Portland, OR, United States,*[2]*Texas A&M University, College Station, TX, United States*

67 **Characterization of Transition to Turbulence for Blood in an S-Shaped Pipe Under Steady Flow Conditions** SB³C2015-1053 pg.1080
Dipankar Biswas[1], David M. Casey[1], Douglas C. Crowder[1], Kristian Valen-Sendstad[2], David A. Steinman[2], Yang H. Yun[1], Francis Loth[1], [1]*The University of Akron, Akron, OH, United States,*[2]*University of Toronto, Toronto, ON, Canada*

68 **Quantitative Analysis Of Heart Function In Embryonic Zebrafish: Retrograde Flow In The Atrioventricular Junction** SB³C2015-661 pg.989
Alexander T. Bulk, David L. Bark Jr., Brennan Johnson, Deborah Gogarty, **Lakshmi P. D. Dasi**, *Colorado State University, Fort Collins, CO, United States*

69 **The Impact of Parameter Variation, Experimental Data and Uncertainty Quantification for Complex Biomechanical Problems Examplified for AAA** SB³C2015-364 pg. 447
Wolfgang A. Wall, Jonas Biehler, Michael W. Gee, *Technische Universität München, Garching b. München, Germany*

70 **Novel Method For 3D Reconstruction Of The Chick Embryo Cardiovascular Anatomy From Non-invasive Ultrasound Scans For Longitudinal Studies** SB³C2015-109 pg.30
Germaine X. Y. Tan[1], Muhammad Jamil[1], Nicole G. Z. Tee[2], Liang Zhong[1,2], **Choon Hwai Yap**[1], [1]*National University of Singapore, Singapore,*[2]*National Heart Centre Singapore, Singapore*

71 **Investigating The Two-phase Nature Of Blood Flow** SB³C2015-1102 pg.1167
Joseph M. Sherwood[1], Xuejin Li[2], George Karniadakis[2], Stavroula Balabani[3], [1]*Imperial College London, London, United Kingdom,*[2]*Brown University, Providence, RI, United States,*[3]*University College London, London, United Kingdom*

72 **Tissue Hypoxia and Murray's Law of Minimum Work Control Neovascular Growth and Remodeling** SB³C2015-1176
Sean Moore, David Hoelzle, **Joel D. Boerckel**, *University of Notre Dame, Notre Dame, IN, United States* pg.1301

THURSDAY, JUNE 18	12:30pm - 2:30pm

Poster Session I — **MS Competition - Biofluid Mechanics** — **Event Center Tent**

73 **Elevated Wall Shear Stress Predicts Branch Graft Failure Following Chimney Endovascular Aortic Aneurysm Repair** SB³C2015-443 pg.594
Rosamaria Tricarico[1], He Yong[1,2], Adam Beck[1], Salvatore Scali[1,2], Roger Tran-Son-Tay[1], Scott Berceli[1,2], [1]*University of Florida, Gainesville, FL, United States,*[2]*North Florida/South Georgia Veterans Health System, Gainesville, FL, United States*

74 **Better Assessment Of Arteriovenous Fistula Patency Using Functional Diagnostic Endpoints** SB³C2015-442 pg.592
Krishna Subramony Anantha, Ehsan Rajabi-Jaghargh, Rupak Banerjee, *University of Cincinnati, Cincinnati, OH, United States*

75 **Establishing In Vivo Hemodynamic Baseline In A Normotensive Rat Model** SB³C2015-1059 pg.1090
Daniela Velez-Rendon, Erica R. Pursell, Daniela Valdez-Jasso, *University of Illinois at Chicago, Chicago, IL, United States*

THURSDAY, JUNE 18	12:30pm - 2:30pm

Poster Session I — **MS Competition - Biotransport** — **Event Center Tent**

76 **Imaging Processing Algorithms For Detecting Subtle Morphological Changes In Mole Images Over Time** SB³C2015-235 pg. 242
Alireza Chamani[1], Neera Nathan[2], Thomas Hornyak[3], Liang Zhu[1], [1]*University of Maryland Baltimore County, Baltimore, MD, United States,*[2]*National Cancer Institute / NIH, Bethesda, MD, United States,*[3]*University of Maryland at Baltimore, Baltimore, MD, United States*

77 **Evaluating the Influence of Tissue Properties on the Core Temperature Using a 3D Whole Body Model** SB³C2015-1037 pg.1048
Robins T. Kalathil[1], Swarup Zachariah[1], Amit Bhattacharya[2], Rupak Banerjee[1], [1]*University of Cincinnati, Cincinnati, OH, United States,*[2]*University of Cincinnati College of Medicine, Cincinnati, OH, United States*

THURSDAY, JUNE 18	12:30pm - 2:30pm

Poster Session I — **MS Competition - Cellular and Tissue Engineering** — **Event Center Tent**

78 **Individual Cell-Based Morphological Analysis to Determine Chirality of Epithelial Morphogenesis** SB³C2015-371 pg.459
Michael J. Raymond, Poulomi Ray, Leo Wan, *Rensselaer Polytechnic Institute, Troy, NY, United States*

THURSDAY, JUNE 18 | 12:30pm - 2:30pm

Poster Session I | MS Competition - Design and Devices | Event Center Tent

79 **Fetal Monitoring with Silicone Nanocomposite Strain Gauges** SB[3]C2015-334 pg.339
Daniel Baradoy, Anton Bowden, David Fullwood, *Brigham Young University, Provo, UT, United States*

80 **Development of 3D Printed Patient Specific Ascending Aortic Training Models for Cardiac Surgery** SB[3]C2015-446 pg. 600
ZhiLin Yang[1], Justine Garcia[1], Kevin Lachapelle[2], Rosaire Mongrain[1], Richard L. Leask[1], [1]*McGill University, Montreal, QC, Canada,* [2]*Royal Victoria Hospital, Montreal, QC, Canada*

81 **Image-based 3D Morphometric Analysis of the Clavicle Intramedullary Canal in Male Population** SB[3]C2015-494 pg.676
Jazmine Aira[1,2], Sergio Gutierrez[2], Brandon G. Santoni[1,2], Mark A. Frankle[1,3], Peter Simon[1,2], [1]*University of South Florida, Tampa, FL, United States,* [2]*Foundation for Orthopaedic Research and Education, Tampa, FL, United States,* [3]*Florida Orthopaedic Institute, Tampa, FL, United States*

THURSDAY, JUNE 18 | 12:30pm - 2:30pm

Poster Session I | MS Competition - Human Dynamics | Event Center Tent

82 **Pseudo-Rigid Body Method for Reducing Soft Tissue Artifact: Validation and Application to Gait** SB[3]C2015-1147 pg.1249
Jake Deschamps, Karim Dudum, Eshan Dandekar, Scott Hazelwood, Stephen Klisch, *California Polytechnic State University, San Luis Obispo, CA, United States*

THURSDAY, JUNE 18 | 12:30pm - 2:30pm

Poster Session I | MS Competition - Injury | Event Center Tent

83 **Assessing Head Impact Sensor Validity in the Laboratory** SB[3]C2015-631 pg.931
Abigail M. Zadnik, Steven Rowson, Stefan M. Duma, *Virginia Tech, Blacksburg, VA, United States*

84 **Acute Failure Modes in Small Diameter Spinal Segments** SB[3]C2015-282 pg.317
Aubrie L. Taylor, Cassandra Bell, Anton E. Bowden, *Brigham Young University, Provo, UT, United States*

85 **Human Knee Joint Finite Element Model Using a Two Bundle Anterior Cruciate Ligament: Validation and Gait Analysis** SB[3]C2015-610 pg.893
Nicholas A. Czapla[1], Meghan K. Sylvia[1], Zachary F. Lerner[2], David J. Tuttle[3], Otto S. Schueckler[4], Scott J. Hazelwood[1], Stephen M. Klisch[1], [1]*California Polytechnic State University at San Luis Obispo, San Luis Obispos, CA, United States,* [2]*Colorado State University at Fort Collins, Fort Collins, CO, United States,* [3]*Radiology Associates, Inc., San Luis Obispos, CA, United States,* [4]*Central Coast Orthopedic Medical Group, San Luis Obispos, CA, United States*

86 **Quantifying Head Impact Exposure in Collegiate Women's Soccer** SB[3]C2015-555 pg. 793
Jaclyn Press, Steven Rowson, *Virginia Tech, Blacksburg, VA, United States*

THURSDAY, JUNE 18	12:30pm - 2:30pm

Poster Session I | **MS Competition - Tissue Mechanics** | **Event Center Tent**

87 **Development of a Human Knee Joint Finite Element Model to Investigate Cartilage Stress During Walking in Obese and Normal Weight Adults** SB³C2015-611 pg. 895
Meghan Sylvia[1], Nicholas Czapla[1], Zachary Lerner[2], David Tuttle[3], Otto Schueckler[4], Scott Hazelwood[1], Stephen Klisch[1], [1]*California Polytechnic State University, San Luis Obispo, San Luis Obispo, CA, United States,*[2]*Colorado State University, Fort Collins, Fort Collins, CO, United States,*[3]*Radiology Associates, Inc., San Luis Obispo, CA, United States,*[4]*Central Coast Orthopedic Medical Group, San Luis Obispo, CA, United States*

88 **Accurate Prediction of Collagen Fiber Distribution using FFT: A Validation Study** SB³C2015-431 pg. 572
Erica E. Morrill, Christina Stender, Roshani Lamichhane, Raquel Brown, Trevor Lujan, *Boise State University, Boise, ID, United States*

THURSDAY, JUNE 18	12:30pm - 2:30pm

Poster Session I | **Joint and Spine Mechanics** | **Event Center Tent**

89 **Influence of Bone Density and Insertion Angle on Fixation Strength of Pedicle Screws** SB³C2015-186 pg.158
Giovanni Solitro, **Farid Amirouche**, Brenden Magnan, *University of Illinois at Chicago, Chicago, IL, United States*

90 **An Open-source Toolbox for Surrogate Modeling of Joint Contact Mechanics** SB³C2015-578 pg.835
Ilan Eskinazi, Benjamin J. Fregly, *University of Florida, Gainesville, FL, United States*

91 **Effect Of Axial Compression Preload On Intervertebral Disc Torsional Mechanics** SB³C2015-127 pg.56
Semih E. Bezci, **Grace D. O'Connell**, *University of California, Berkeley, Berkeley, CA, United States*

92 **Using Dynamic Community Detection to Map Collagen Fiber Network Reorganization During Tensile Loading of the Human Facet Capsular Ligament** SB³C2015-115 pg.40
Sijia Zhang, Danielle Bassett, Beth Winkelstein, *University of Pennsylvania, Philadelphia, PA, United States*

93 **Application of a Novel Robotically Simulated Pivot Shift for Anterior Cruciate Ligament Reconstruction: Comparison of the All-Epiphyseal and Over-The-Top Techniques** SB³C2015-500 pg.688
Robb Colbrunn, Tara Bonner, Joel Kolmodin, Paul Saluan, *Cleveland Clinic, Cleveland, OH, United States*

94 **Effect Of Sacroiliac Joint Fixation On Segmental Kinematics Of Lumbar Spine: A Finite Element Analysis** SB³C2015-1146 pg.1247
Ali Kiapour[1], Derek Lindsey[2], Scott Yerby[2], Vijay Goel[1], [1]*ECORE, Toledo, OH, United States,*[2]*Si-Bone Inc, San Jose, CA, United States*

95 **Effect of Variability in Anatomical Landmarks on Ankle Kinematic Descriptions** SB³C2015-513 pg. 712
Ednah G. Louie, Fallon G. Fitzwater, Matthew H. Dickinson, William M. Eboch, Bardiya Akhbari, Lorin P. Maletsky, *University of Kansas, Lawrence, KS, United States*

96 **Biomechanical Evaluation of Proximal Junctional Failure and Kyphosis in Lumbar Spinal Fusion Surgery** SB³C2015-155 pg.100
Won Man Park[1], Kyungsoo Kim[1], Yongjung J. Kim[2], **Yoon Hyuk Kim**[1], [1]*Kyung Hee University, Yongin, Korea, Republic of,*[2]*Columbia University College of Physicians and Surgeons, New York, NY, United States*

97 **Biomechanical Comparison Between Facet Sparing Laminectomy and Laminectomy with Facetectomy in Lumbar Spine: A Finite Element Analysis** SB³C2015-653 pg.973
Shady Elmasry[1], Shihab Asfour[1], Joseph Gjolaj[2], Loren Latta[2], Frank Eismont[2], **Francesco Travascio**[1], [1]*University of Miami, Coral Gables, FL, United States,*[2]*University of Miami, Miami, FL, United States*

98 **Open Knee(s): Comprehensive Tibiofemoral Joint Testing For Specimen-specific Next Generation Knee Models** SB³C2015-530 pg.746
Tara F. Bonner, Robb W. Colbrunn, Snehal Chokhandre, Craig Bennetts, Ahmet Erdemir, *Cleveland Clinic, Cleveland, OH, United States*

99 **Applying Mean Soft Tissue Properties To A Subpopulation Of Knee Models Reveals Inability To Capture Variations In Knee Stability** SB³C2015-613 pg.899
Kevin Schafer, Mohammad Kia, Daniel Green, Andrew Pearle, Thomas Wickiewicz, Timothy Wright, **Carl Imhauser**, *Hospital for Special Surgery, New York, NY, United States*

100 **2D/3D Registration To Find Host Bone Coverage Of Rsa Implants** SB³C2015-1156 pg.1267
Jonathan W. Keimel[1], Kristi L. Krebes[1], **Andres F. Cabezas**[1], Adam Lorenzetti[2], Brandon G. Santoni[3], Mark A. Frankle[2], Peter Simon[4], [1]*University of South Florida, Clearwater, FL, United States,*[2]*Florida Orthopaedic Institute, Tampa, FL, United States,*[3]*Foundation for Orthopaedic Research and Education, Clearwater, FL, United States,*[4]*Foundation for Orthopaedic Research and Education, Tampa, FL, United States*

101 **Do External Load Measures Predict Knee Contact Force Changes Due To Weight Loss?** SB³C2015-1158 pg.1269
Nathan R. Sauder[1], James C. Coburn[2], Melinda K. Harman[3], Heather K. Vincent[4], Darryl D. D'Lima[5], Benjamin J. Fregly[1], [1]*University of Florida, Gainesville, FL, United States,*[2]*Center for Devices and Radiological Health, Food and Drug Administration, Silver Spring, MD, United States,*[3]*Clemson University, Clemson, SC, United States,*[4]*University of Florida, Gainesville, FL, United States,*[5]*Shiley Center for Orthopaedic Research & Education, Scripps Clinic, La Jolla, CA, United States*

102 **Mechanical Characterization and a Computational Wear Model for Polycarbonate Urethane as a Bearing Material** SB³C2015-396 pg.501
Hannah Gramling, Amrita Srinivasan, Lisa Pruitt, *University of California, Berkeley, Berkeley, CA, United States*

103 **Does the Cylindrical or Spherical Axis More Accurately Locate the Flexion-Extension Axis of the Tibia of the Natural Knee?** SB³C2015-349 pg.425
Abheetinder S. Brar[1], Stephen M. Howell[1], Maury L. Hull[1], Mohamed R. Mahfouz[2], [1]*University of California, Davis, Davis, CA, United States,*[2]*University of Tennessee, Knoxville, TN, United States*

104 **Development Of A Finite Element Model Of The Pediatric Occipito-atlantoaxial Complex For Studying Osodontoideum And Atlanto-occipital Dislocation** SB³C2015-1070 pg.1108
Rinchen Phuntsok[1,2], Marcus D. Mazur[1,3], Vijay M. Ravindra[1,3], Douglas L. Brockmeyer[1,3], Benjamin J. Ellis[2,3], [1]*University of Utah, Salt Lake City, UT, United States,*[2]*Scientific Computing and Imaging, Salt Lake City, UT, United States,*[3]*Salt Lake City, UT, United States*

105 **A Hybrid Risk Model for Hip Fracture Prediction using Clinical and Stochastic Finite Element Data** SB³C2015-293
Peng Jiang, Samy Missoum, Zhao Chen, *University of Arizona, Tucson, AZ, United States pg.333*

106 **The Effect of Muscle Loading on Ankle Joint Complex Kinematics and Achilles Load: A Cadaveric Study** SB³C2015-525 pg.735
Bardiya Akhbari, Matthew H. Dickinson, Ednah G. Louie, Sami Shalhoub, Lorin P. Maletsky, *University of Kansas, Lawrence, KS, United States*

107 **The Hip Joint Estimates from Skin-Marker-Based Methods Do Not Correspond with Measurements using Dual Fluoroscopy** SB³C2015-1094 pg.1151
Niccolo M. Fiorentino, Penny R. Atkins, Michael J. Kutschke, Ashley L. Kapron, K. Bo Foreman, Andrew E. Anderson, *University of Utah, Salt Lake City, UT, United States*

109 **Effect Of Interspinous Device On Lumbar Spine: A Finite Element Study** SB³C2015-1035 pg.1044
Deniz U. Erbulut, Iman Zafarparandeh, Chaudhry R. Hassan, Ismail Lazolu, Ali F. Ozer, *Koc University, Istanbul, Turkey*

110 **The Effects Of Axial Compressive Loading On The Intersegmental Rotation Of A Virtual Cervical Spine** SB³C2015-308 pg.361
Ryan J. Moss, Kevin M. Bell, *Orthopaedic Robotics Laboratory, Pittsburgh, PA, United States*

111 **A Computational Method for Visualizing Femoral Range of Motion for Patients with Slipped Capital Femoral Epiphysis** SB³C2015-384 pg.477
Ferris Pfeiffer, David Tager, Sumit Gupta, *University of Missouri, Columbia, MO, United States*

THURSDAY, JUNE 18	12:30pm - 2:30pm

Poster Session I **Tissue Mechanics - General** **Event Center Tent**

112 **Extracting Mechanical Properties Of The Corneoscleral Shell From Whole Eye Perfusions** SB³C2015-629 pg.927
Joseph M. Sherwood[1], Ester Reina-Torres[1], Jacques Bertrand[1], C. Ross Ethier[2], Darryl Overby[1], [1]*Imperial College London, London, United Kingdom,*[2]*Georgia Institute of Technology/Emory, Atlanta, GA, United States*

113 **Strain Rate Dependency Of The Intracellular Calcium Ion Concentration During Neuronal Membrane Mechanoporation** SB³C2015-621 pg.911
Amirhamed Bakhtiarydavijani[1], Anna E. Florence[1], Michael A. Murphy[1], Sungkwang Mun[1], Jun Liao[1], Lakiesha N. Williams[1], M. F. Horstemeyer[1], Michelle C. LaPlaca[2], Raj Prabhu[1], [1]*Mississipi State University, Starkville, MS, United States,*[2]*Georgia Institute of Technology, Atlanta, GA, United States*

114 **Quasilinear and Non-Quasilinear Viscoelastic Behavior of Collagen Gels During Stress Relaxation** SB³C2015-551 pg.784
Christopher E. Korenczuk[1], Victor K. Lai[2], Victor H. Barocas[1], [1]*University of Minnesota, Minneapolis, MN, United States,*[2]*University of Minnesota Duluth, Duluth, MN, United States*

115 **Direct Estimation of Three-dimensional Deformation Gradient Tensors from Volumetric Ultrasound Data** SB³C2015-671 pg.1007
John J. Boyle[1], Roger Rowe[1], Frederick Damen[2], Arvin Soepriatna[2], Robert B. Pless[1], Craig Goergen[2], Stavros Thomopoulos[1], Guy M. Genin[1], [1]*Washington University in St Louis, Saint Louis, MO, United States,*[2]*Purdue University, West Lafayette, IN, United States*

116 **A Structural Constitutive Model for the Active and Passive Behavior of Biological Tissues** SB³C2015-330 pg.397
Ting Tan, **Raffaella De Vita**, *Virginia Tech, Blacksburg, VA, United States*

117 **Computational Modeling Of Synthetic Mesh Materials: Simulation And Experimental Assesment Of Model Predictions** SB³C2015-511 pg.708
William R. Barone[1], Katrina M. Knight[1], Pamela A. Moalli[2], Steven D. Abramowitch[1], [1]*University of Pittsburgh, Pittsburgh, PA, United States,*[2]*Magee-Womens Research Institute, University of Pittsburgh, Pittsburgh, PA, United States*

118 **Tough, Self-recovering Hydrogels Inspired By Caddisfly Silk** SB³C2015-337 pg.405
Dwight D. Lane, G. Mahika Weerasekare, Sarbjit Kaur, Russell J. Stewart, *University of Utah, Salt Lake City, UT, United States*

119 **Determining The Compressive Modulus Of Mouse Trabecular Meshwork With Atomic Force Microscopy** SB³C2015-1047 pg.1068
Ke Wang, Todd Sulchek, C. Ross Ethier, *Georgia Institute of Technology, Atlanta, GA, United States*

120 **Development of Shoe Sole Design Using Material Natural Frequencies** SB³C2015-251 pg.271
Brett D. Steineman[1], Ted Barber[2], Tammy L. Haut Donahue[1], [1]*Colorado State University, Fort Collins, CO, United States,*[2]*Pearl Izumi, Inc., Louisville, CO, United States*

121 **The Need for Validation in Soft Tissue Constitutive Models** SB³C2015-1058 pg.1088
Sandeep Madireddy, Kumar Vemaganti, *University of Cincinnati, Cincinnati, OH, United States*

122 **A Computational Study of A Simple, Transversely Isotropic Model of Soft Tissue, with Shear and Tensile Anisotropy, in Large Strain** SB³C2015-202 pg.188
Yuan Feng[1], Ruth J. Okamoto[2], Guy M. Genin[2], Larry A. Taber[2], Philip V. Bayly[2], [1]*Soochow University, Suzhou, China,*[2]*Washington University in St. Louis, St. Louis, MO, United States*

123 **Development Of A Clinical Ultrasound Technique for Analysis of Protein Content Within Hydrogel** SB³C2015-1109
Jessica Stukel[1], Monika Goss[2], Agata Exner[2], Rebecca Willits[1], [1]*The University of Akron, Akron, OH, United States,*[2]*Case Western Reserve University, Cleveland, OH, United States* *pg.1181*

124 **Validation Of High Rate Strip Biaxial Tension Deformations Of The Neuronal Phospholipid Bilayer Using Empirical Data** SB³C2015-1120 pg.1203
M. A. Murphy[1], M. F. Horstemeyer[1], Steven R. Gwaltney[1], Tonya W. Stone[1], Michelle C. LaPlaca[2], Jun Liao[1], Lakiesha N. Williams[1], **R. Prabhu**[1], [1]*Mississippi State, MS, United States,*[2]*Georgia Institute of Technology, Atlanta, GA, United States*

125 **Collective Chiral Rotation of Epithelial Microtissues Within a Three-Dimensional Matrigel System** SB³C2015-232
Amanda S. Chin, Kathryn E. Worley, Leo Q. Wan, *Rensselaer Polytechnic Institute, Troy, NY, United States* *pg.238*

126 **Finite Element Model of Cervical Pessary in Use: Evaluating Mechanical Interventions for Preterm Birth** SB³C2015-490 pg.668
Michael J. Fernandez[1], Michael D. House[2], Noelia M. Zork[3], Joy S. Vink[3], Ronald J. Wapner[3], Sachin R. Jambawalikar[3], Kristin M. Myers[1], [1]*Columbia University, New York, NY, United States,*[2]*Tufts Medical Center, Boston, MA, United States,*[3]*Columbia University Medical Center, New York, NY, United States*

127 **Impact of Urinary Bladder Matrix on Vaginal Smooth Muscle Function and Structure in the Nonhuman Primate Model** SB³C2015-164 pg.116
Katrina Knight[1], Zegbeh Jallah[1], Rui Liang[2], Stacy Palcsey[2], Pamela Moalli[1,2], Steven Abramowitch[1,2], [1]*Musculoskeletal Research Center, University of Pittsburgh, Pittsburgh, PA, United States,*[2]*Magee-Womens Research Institute, Magee-Womens Hospital, Pittsburgh, PA, United States*

128 **Biaxial Creep of Swine Cardinal and Uterosacral Ligaments** SB³C2015-299 pg. 345
Ting Tan, Nathan M. Cholewa, Scott W. Case, **Raffaella De Vita**, *Virginia Tech, Blacksburg, VA, United States*

129 **Characterization of Soft Tissue Microstructure via Transmural SALS** SB³C2015-1044 pg.1062
John G. Lesicko, Kristen R. Feaver, Michael S. Sacks, *University of Texas at Austin, Austin, TX, United States*

130 **Finite Element Modeling of the Posterior Eye in Microgravity** SB³C2015-348 pg.423
Andrew Feola[1], Julia Raykin[1], Lealem Mulugeta[2], Rudolph Gleason[1], Jerry G. Myers[3], Emily S. Nelson[3], Brian Samuels[4], Ross Ethier[1], [1]*Georgia Institute of Technology, Atlanta, GA, United States,*[2]*Universities Space Research Association, Houston, TX, United States,*[3]*NASA Glenn Research Center, Cleveland, OH, United States,*[4]*University of Alabama at Birmingham, Birmingham, AL, United States*

131 **Posterior Scleral Stiffening: How is Scleral Canal Expansion Affected by the Size of Stiffening Region?** SB³C2015-1048 pg.1070
Ian C. Campbell[1,2], Baptiste Coudrillier[1], C. Ross Ethier[1,2], [1]*Georgia Institute of Technology/Emory University, Atlanta, GA, United States,*[2]*Atlanta VA Medical Center, Decatur, GA, United States*

132 **Biomechanical Behavior Of Cornea When Subjected To Tension And Compression Loads** SB³C2015-1076 pg.1118
Hamed Hatami-Marbini, *Oklahoma State University, Stillwater, OK, United States*

133 **Correlating Urethral Rupture with Distension of the Urethra during the Inflation of a Misplaced Transurethral Catheter Balloon** SB³C2015-370 pg. 457
Connor V. Cunnane, *University of Limerick, Limerick, Ireland*

134 **Tuning Silk Fibroin Hydrogels: Genipin Crosslinking Pre-gelation Decreases Time-dependent Properties** SB³C2015-1125 pg.1211
Winston H. Elliott[1], Walter Bonani[2,3], Devid Maniglio[2,3], Antonella Motta[2,3], Wei Tan[1], Claudio Migliaresi[2,3], [1]*University of Colorado- Boulder, Boulder, CO, United States,*[2]*University of Trento, Trento, Italy,*[3]*European Institute of Excellence on Tissue Engineering and Regenerative Medicine, and INSTM Trento Research Unit, Trento, Italy*

135 **Spherical Shell Mechanical Model of the Pacinian Corpuscle** SB³C2015-411 pg.531
Julia C. Quindlen, Henryk K. Stolarski, Victor H. Barocas, *University of Minnesota, Minneapolis, MN, United States*

136 **Measuring Tortuosity Changes due to Central Retinal Vein Occlusion** SB³C2015-1182 pg.1309
Kendall McMillan[1], Shaun Evans[1], Gil Binenbaum[2], Brittany Coats[1], [1]*University of Utah, Salt Lake City, UT, United States,*[2]*The Children's Hospital of Philadelphia, Philadelphia, PA, United States*

138 **Mechanical Properties Of Human Placenta In Normal Pregnancies And During Intrauterine Growth Restriction** SB³C2015-206 pg.196
Jeanette Shifen Lau, Shier Nee Saw, Martin Lindsay Buist, **Choon Hwai Yap**, *National University of Singapore, Singapore*

139 **Determination of the Mechanical Properties of the Iris Using Inverse Finite Element Modeling** SB³C2015-424 pg.557
Anup D. Pant, Rouzbeh Amini, *The University of Akron, Akron, OH, United States*

140 **A Poroelastic High Fidelity Finite Element Model Of The Osteochondral Unit To Evaluate Changes In Permeability With Osteoarthritis** SB³C2015-144 pg.82
Michael E. Stender, Richard A. Regueiro, Virginia L. Ferguson, *University of Colorado, Boulder, Boulder, CO, United States*

141 **The Effects Of Stress State On The Mechanical Response And Failure Of The Neuronal Phospholipid Bilayer: A Molecular Dynamics Study** SB³C2015-1132 pg.1223
M. A. Murphy[1], M. F. Horstemeyer[1], Steven R. Gwaltney[1], Tonya W. Stone[1], Michelle C. LaPlaca[2], Jun Liao[1], Lakiesha N. Williams[1], **R. Prabhu**[1], *[1]Mississippi State, MS, United States,[2]Georgia Institute of Technology, Atlanta, GA, United States*

142 **Racial Differences In The Load-dependent Area Of The Lamina Cribrosa** SB³C2015-1172 pg.1293
Stephen J. Howerton, Forest L. Danford, Jonathan P. Vande Geest, Avinash Ayyalasomayajula, *The University of Arizona, Tucson, AZ, United States*

143 **Optic Nerve Sheath Mechanics in VIIP Syndrome** SB³C2015-488 pg.666
Julia Raykin[1], Andrew Feola[1], Rudy Gleason[1], Lealem Mulugeta[2], Jerry Myers[3], Emily Nelson[3], Brian Samuels[4], C. Ross Ethier[1], *[1]Georgia Institute of Technology, Atlanta, GA, United States,[2]Universities Space Research Association, Houston, TX, United States,[3]NASA Glenn Research Center, Cleveland, OH, United States,[4]University of Alabama at Birmingham, Birmingham, AL, United States*

144 **Modeling the Biaxial Mechanics of Brain White Matter** SB³C2015-177 pg.140
Kevin M. Labus[1], José J. García[2], Christian M. Puttlitz[1], *[1]Colorado State University, Fort Collins, CO, United States,[2]Universidad del Valle, Cali, Colombia*

145 **An In Silico Biomechanical Analysis of the Stent-Esophagus Interaction.** SB³C2015-365 pg. 449
Mathias Peirlinck[1], Benedict Verhegghe[1,2], Patrick Segers[1], Matthieu De Beule[1,2], *[1]Ghent University, Ghent, Belgium,[2]FEops bvba, Ghent, Belgium*

THURSDAY, JUNE 18	12:30pm - 2:30pm

Poster Session I — Bone Biomechanics — Event Center Tent

146 **Fabrication And Characterization Of Artificial Bone-cartilage Tissue Construction Using Mesenchymal Stem Cells** SB³C2015-207 pg.198
Takaaki Arahira, *Fukuoka Dental College, Fukuoka, Japan*

147 **Mechanical Analysis of Bone Tissue as Mineral and Organic Composite by Raman Spectroscopy** SB³C2015-447 pg.602
Masahiro Todoh, Shigeru Tadano, *Hokkaido University, Sapporo, Japan*

148 **The Effect of Pore Size on Bone Strain in the Proximal Femur** SB³C2015-1167 pg.1287
Mariana Kersh[1], Afrodite Zendelli[2], Yohann Bala[2], Ali Ghasem-Zadeh[2], Ego Seeman[2], Roger Zebaze[2], *[1]University of Illinois, Urbana, IL, United States,[2]Austin Health, Heidelberg, Australia*

149 **Multimodal Assessment of Bone Quality of the Human Rib** SB³C2015-38 pg.20
Lauren M. Mangano[1], Jean-Paul Roux[2], François Duboeuf[2], Delphine Farlay[2], David Mitton[3], Hélène Follet[2], *[1]Boston University, Brookline, MA, United States,[2]INSERM, Université Lyon 1, Lyon, France,[3]IFSTTAR, Université Lyon 1, Bron, France*

150 **The Characterization Of The Bone Marrow Mechanical Environment Using Poroelastic Finite Element Models** SB³C2015-1116 pg.1195
Joshua Gargac, Thomas Metzger, Tyler Kreipke, Hansel Weihs, Glen Niebur, *University of Notre Dame, South Bend, IN, United States*

151 **Spring Mediated Cranioplasty: a Patient Specific Numerical Model** SB³C2015-210 pg.202
Alessandro Borghi[1], Silvia Schievano[1], William Rodgers[1], Freida Angullia[2], Allan Ponniah[2], David Dunaway[2], Owase Jeelani[2], *[1]University College London, London, United Kingdom,[2]Great Ormond Street Hospital, London, United Kingdom*

152 **Preliminary Finite Element Analysis of Subchondral Bone Cysts in the Stifle of the Horse** SB³C2015-393 pg.495
Lance F. Frazer[1], Kenneth Fischer[1], Garrett Noble[2], Elizabeth Santschi[3], *[1]University of Kansas, Lawrence, KS, United States,[2]Ohio State University, Columbus, OH, United States,[3]Kansas State University, Manhattan, KS, United States*

THURSDAY, JUNE 18	12:30pm - 2:30pm

Poster Session I — Fabrication and Manipulation of the Cellular Microenvironment — Event Center Tent

153 **Mechanics of Interstitial Growth in Enzymatically Degradable Hydrogels: Characterization of Degradation Front** SB³C2015-1124 pg.1209
Umut Akalp, Stephanie J. Bryant, Stacey C. Skaalure, Franck J. Vernerey, *University of Colorado at Boulder, Boulder, CO, United States*

154 **Microgeometry and Microenvironment of Mitral Valve Interstitial Cells Under Physiological Loads** SB³C2015-287 pg.325
Salma Ayoub, Chung-Hao Lee, Michael S. Sacks, *The University of Texas at Austin, Austin, TX, United States*

155 **Understanding and Optimizing Electrical Stimulation of Neurons for Improving Regeneration: A Finite Element Simulation with Experimental Verification** SB³C2015-520 pg.727
Robert D. Adams[1], Rebecca K. Willits[2], Amy B. Harkins[1], [1]*Saint Louis University, Saint Louis, MO, United States,*[2]*University of Akron, Akron, OH, United States*

156 **Schwann Cell Proliferation in Scaffolds with Decoupled Mechanical and Biochemical Properties** SB³C2015-428 pg.565
Jessica Stukel, Wenda Zhou, Rebecca Willits, *The University of Akron, Akron, OH, United States*

157 **Wrinkled, Wavelength-Tunable Graphene-Based Surface Topographies for Directing Cell Alignment and Morphology** SB³C2015-297 pg.341
Daniel F. Tonderys, *Brown University, Providence, RI, United States*

158 **Biomimetic Ex Vivo Model for Tracking Stem Cells During Microvascular Network Growth** SB³C2015-1100 pg.1163
Mohammad S. Azimi, Amy L. Strong, Theresa B. Phamduy, Douglas B. Chrisey, Bruce A. Bunnell, Walter L. Murfee, *Tulane University, New Orleans, LA, United States*

THURSDAY, JUNE 18	12:30pm - 2:30pm

Poster Session I — Multi-Scale Mechanics in Cell and Tissue Engineering — Event Center Tent

159 **Optimization of Test Methods and Burst Property Characterization of Alginate Hydrogel Lung Sealants** SB³C2015-630 pg.929
Patrick N. Charron, Spencer L. Fenn, Rachael A. Oldinski, *University of Vermont, Burlington, VT, United States*

160 **Numerical Simulations of Fibrous Biomaterial with Randomly Distributed Fiber Network** SB³C2015-130 pg.62
Tao Jin, Ilinca Stanciulescu, *Rice University, Houston, TX, United States*

161 **Using Simulations with Realistic Fibrous Network Geometry to Find the Achievable Range of Mechanical Behaviors of Elastomeric Scaffolds** SB³C2015-535 pg.754
James B. Carleton, Gregory J. Rodin, Michael S. Sacks, *University of Texas at Austin, Austin, TX, United States*

162 **A Strain Based Approach To Quantify Non-affine Behavior Of Three Dimensional Random Network** SB³C2015-1080
Hamed Hatami-Marbini, *Oklahoma State University, Stillwater, OK, United States pg.1126*

163 **Empirically Determined Vascular Smooth Muscle Mechano-adaptation Laws** SB³C2015-275 pg.305
Kerianne E. Steucke, Patrick W. Alford, *University of Minnesota, Minneapolis, MN, United States*

164 **Microstructure And Dynamics Of Crosslinked Collagen Gel** SB³C2015-397 pg.503
Shengmao Lin[1], Linxia Gu[1,2], [1]*University of Nebraska-Lincoln, Lincoln, NE, United States,*[2]*Nebraska Center for Materials and Nanoscience, Lincoln, NE, United States*

165 **Probing Of A Complex Multi-layer Embryonic Tissue Through Novel 3D Bio-etching** SB³C2015-352 pg.431
Melis Hazar[1], YongTae Kim[2], Philip R. LeDuc[1], William C. Messner[3], Lance A. Davidson[4], [1]*Carnegie Mellon University, Pittsburgh, PA, United States,*[2]*Georgia Institute of Technology, Atlanta, GA, United States,*[3]*Tufts University, Medford, MA, United States,*[4]*University of Pittsburgh, Pittsburgh, PA, United States*

166 **Numerical Simulation Of Effects Of Membrane Surface Viscosity On Tank-treading Motion Of Red Blood Cell** SB³C2015-203 pg.190
Ken-ichi Tsubota, *Chiba University, Chiba, Japan*

167 **Wnt/beta-catenin Signaling Pathways Contributes To Dynamic Fluid Flow Loading Induced In Situ Osteocytic Calcium Oscillations In An Intact Mouse Femur** SB³C2015-1084 pg.1134
Minyi Hu, Guowei Tian, Yi-Xian Qin, *Stony Brook University, Stony Brook, NY, United States*

168 **Nonlinear Anisotropic Mechanical Properties of Vascular Smooth Muscle Cells** SB³C2015-650 pg.967
Zaw Win, Patrick W. Alford, *University of Minnesota, Minneapolis, MN, United States*

169 **Dynamic Fluid Flow Loading Induced In Situ Osteocytic Calcium Oscillations In An Intact Mouse Femur** SB³C2015-1081 pg.1128
Minyi Hu, Guowei Tian, Yi-Xian Qin, *Stony Brook University, Stony Brook, NY, United States*

FRIDAY, JUNE 19	8:00am - 9:30am

Workshop: Mow and Fung Lectures - Dawn M. Elliott, Adam J. Engler — Primrose A/B

Session Chair: Louis Soslowsky, *University of Pennsylvania, Philadelphia, PA, United States*
Session Co-Chair: Guy Genin, *Washington University, St. Louis, MO, United States*

FRIDAY, JUNE 19	8:00am - 9:30am

Workshop: Problem-based Learning in Biomechanics — Superior

Session Chair: Alisa Morss Clyne, *Drexel University, Philadelphia, PA, United States*
Session Co-Chair: Kristen Billiar, *Worcester Polytechnic Institute, Worcester, MA, United States*

FRIDAY, JUNE 19	8:00am - 9:30am

Workshop: CFD Challenge 2015 — Wasatch

Session Chair: Kenichi Kono, *Wakayama Rosai Hospital, Japan*
Session Co-Chair: Kristian Valen-Sendstad, *University of Toronto, Toronto, ON, Canada*

FRIDAY, JUNE 19	8:00am - 9:30am

Workshop: Mentee-mentor Matching Mixer and Best Practices in Mentoring — Magpie

Session Chair: Naomi Chesler, *University of Wisconsin, Madison, WI, United States*
Session Co-Chair: Lakiesha Williams, *Mississippi State University, Mississippi State, MS, United States*
Session Co-Chair: Victor Barocas, *University of Minnesota, Minneapolis, MN, United States*

FRIDAY, JUNE 19	8:00am - 9:30am

Workshop: Strategies for a Successful Postdoctoral Experience — Maybird

Organizers: ASME Bioengineering Division Student Leadership Committee (special thanks to Kathryn Drzewiecki, Samira Jamalian, Paola Jaramillo, and Samantha Schoell)

FRIDAY, JUNE 19	9:45am - 10:45am

PLENARY SESSION II – Andrew McCulloch **Ballrooms 1-3**

FRIDAY, JUNE 19	11:00am - 12:30pm

PhD Competition - Biofluid Mechanics — Primrose A

Session Chair: Stephanie George, *East Carolina University, Greenville, NC, United States*
Session Co-Chair: Brandon Dixon, *Georgia Institute of Technology, Atlanta, GA, United States*

11:00AM Increased Red Blood Cell Stiffness Increases Pulmonary Vascular Resistance And Pulmonary Arterial Pressure
SB[3]C2015-587 pg.853
David A. Schreier, Omid Forouzan, Timothy Hacker, John Sheehan, Naomi C. Chesler, *University of Wisconsin-Madison, Madison, WI, United States*

11:15AM Effect of Bicuspid Aortic Valve Morphotype on Ascending Aorta Hemodynamics: a Computational Study
SB[3]C2015-304 pg.355
Kai Cao, Philippe Sucosky, *University of Notre Dame, Notre Dame, IN, United States*

11:30AM Hemodynamic Characterization of Different Basilar Tip Aneurysm Templates Using Computational Fluid Dynamics
SB[3]C2015-1114 pg.1191
Priya Nair[1], Brian W. Chong[2], David Frakes[1], *[1]Arizona State University, Tempe, AZ, United States,[2]Mayo Clinic Hospital, Phoenix, AZ, United States*

11:45AM Modulation of Lymph Node Resistance during Inflammation: Experimental Measurement and Computational Modeling SB[3]C2015-566 pg.813
Mohammad Jafarnejad[1], Joshua P. Scallan[2], Joseph M. Sherwood[1], Darryl R. Overby[1], David C. Zawieja[3], Michael J. Davis[2], James E. Moore[1], *[1]Imperial College London, London, United Kingdom,[2]University of Missouri, Columbia, MO, United States,[3]Texas A&M Health Science Center, Temple, TX, United States*

12:00PM Inlet Flow Rate Variation and Onset of Flow Instabilities in the Carotid Siphon SB[3]C2015-276 pg.307
Resmi KrishnankuttyRema[1], Kristian Valen-Sendstad[1,2], David Steinman[1], *[1]University of Toronto, Toronto, ON, Canada,[2]Simula Research Laboratory, Lysaker, Norway*

12:15PM Suction Effect Produced by Active Contraction of Collecting Lymphatic Vessels Facilitates Lymphatic Filling
SB[3]C2015-539 pg.762
Samira Jamalian[1], Mohammad Jafarnejad[1], Christopher D. Bertram[2], David C. Zawieja[3], J. Davis[4], James E. Moore[1], *[1]Imperial College London, London, United Kingdom,[2]University of Sydney, New South Wales, Australia,[3]Texas A&M Health Science Center, Temple, TX, United States,[4]University of Missouri School of Medicine, Columbia, MO, United States*

FRIDAY, JUNE 19	11:00am - 12:30pm

PhD Competition - Biotransport and Simulation — Superior

Session Chair: Victor Varner, *Princeton University, Princeton, NJ, United States*
Session Co-Chair: Alisa Morss, *Drexel University, Philadelphia, PA, United States*

11:00AM Lymphatic Disease Phenotyping With Near-Infrared Imaging SB[3]C2015-1104 pg.1171
Tyler S. Nelson[1], Michael J. Weiler[1], Ira L. Savetsky[2], Xiaolei Liu[3], Babak J. Mehrara[2], Hong Chen[4], J. Brandon Dixon[1], *[1]Georgia Institute of Technology, Atlanta, GA, United States,[2]Memorial Sloan Kettering Cancer Center, New York, NY, United States,[3]St. Jude Children's Research Hospital, Memphis, TN, United States,[4]Oklahoma Medical Research Foundation, Oklahoma City, OK, United States*

11:15AM **Red Blood Cell Dynamic Deformability and Adhesion in Microscale Flow Determine Cellular Heterogeneity in Sickle Cell Disease** SB³C2015-504 pg.696
Yunus Alapan[1], Yumi Matsuyama[1], Jane Little[1,2], Umut A. Gurkan[1,3], [1]*Case Western Reserve University, Cleveland, OH, United States,*[2]*University Hospitals, Cleveland, OH, United States,*[3]*Louis Stokes Cleveland Veterans Affairs Medical Center, Cleveland, OH, United States*

11:30AM **Reduced Lymphatic Function Correlates With Disease Progression In a Novel Single Vessel Ligation Model of Lymphedema** SB³C2015-410 pg.529
Michael J. Weiler, Tyler S. Nelson, J. Brandon Dixon, *Georgia Tech, Atlanta, GA, United States*

11:45AM **Finite Element Modeling Of Active Transmembrane Cell Transport** SB³C2015-356 pg.433
Chieh Hou, Kelly Terlizzi, Gerard A. Ateshian, *Columbia University, New York, NY, United States*

12:00PM **Combined Experimental and Finite Element Analysis to Determine the Diffusion Coefficient Within and Between Human Skin Layers** SB³C2015-274 pg.303
Anne M. Römgens[1], Dan L. Bader[1,2], Frank P. T. Baaijens[1], Cees W. J. Oomens[1], [1]*Eindhoven University of Technology, Eindhoven, Netherlands,*[2]*University of Southampton, Southampton, United Kingdom*

12:15PM **CFD Simulation of Transition to Turbulence for Newtonian vs. Non-Newtonian Flow Through a Stenosis** SB³C2015-363 pg.445
M. Owais Khan[1], Kristian Valen-Sendstad[1,2], Dipankar Biswas[3], David M. Casey[3], Francis Loth[3], David Steinman[1], [1]*University of Toronto, Toronto, ON, Canada,*[2]*Simula Research Laboratory, Lysaker, Norway,*[3]*University of Akron, Akron, OH, United States*

FRIDAY, JUNE 19	11:00am - 12:30pm

PhD Competition - Cellular and Tissue Engineering — Wasatch

Session Chair: Victor Lai, *University of Minnesota, Duluth, MN, United States*
Session Co-Chair: Colleen Witzenburg, *University of Virginia, Charlottesville, VA, United States*

11:00AM **Interstitial Cell Migration in Dense Connective Tissues is Modulated by Matrix Microstructure and Micromechanics** SB³C2015-129 pg. 60
Feini Qu[1,2], Miltiadis H. Zgonis[1,2], Robert L. Mauck[1,2], [1]*University of Pennsylvania, Philadelphia, PA, United States,*[2]*Philadelphia VA Medical Center, Philadelphia, PA, United States*

11:15AM **Cells Alter Traction Force and Orientation in Response to Long-term Cyclic Stretch** SB³C2015-461 pg.624
Heather A. Cirka, Qi Wen, Kristen L. Billiar, *Worcester Polytechnic Institute, Worcester, MA, United States*

11:30AM **Dedifferentiaton of Chondrocytes Influences Strain Transfer Measured by Deformable Image Registration** SB³C2015-561 pg.803
Jonathan T. Henderson[1], Benjamin Seelbinder[1], Alexander Veress[2], Corey Neu[1], [1]*Purdue University, West Lafayette, IN, United States,*[2]*University of Washington, Seattle, WA, United States*

11:45AM **An Active Contraction Model Of Valvular Interstitial Cells** SB³C2015-614 pg.901
Yusuke Sakamoto, Michael Sacks, *The University of Texas at Austin, Austin, TX, United States*

12:00PM **A Predictive 3D High-Content/High Throughput Screening Platform to Elucidate and Enhance Multilineage Stem Cell Differentiation** SB³C2015-601 pg.877
Amit Paul, Bo Chen, Elise DeBruyn, Michael Cho, *University of Illinois at Chicago, Chicago, IL, United States*

12:15PM **In Vitro Growth Trajectory And In Vivo Implantation Of Cell-seeded Disc-like Angle Ply Structures For Total Disc Replacement** SB³C2015-131 pg.64
John T. Martin[1,2], Dong Hwa Kim[1,2], Kensuke Ikuta[1,2], Christian G. Pfeifer[1,2], Lachlan J. Smith[1,2], Dawn M. Elliott[3], Harvey E. Smith[1,2], Robert L. Mauck[1,2], [1]*University of Pennsylvania, Philadelphia, PA, United States,*[2]*Philadelphia VA Medical Center, Philadelphia, PA, United States,*[3]*University of Delaware, Newark, DE, United States*

FRIDAY, JUNE 19	11:00am - 12:30pm

PhD Competition - Mechanics and Rehabilitation — Magpie

Session Chair: Bradley Davidson, *University of Denver, Denver, CO, United States*
Session Co-Chair: Darryl Thelen, *University of Wisconsin-Madison, Madison, WI, United States*

11:00AM **Blocking Blood-Spinal Cord Barrier Breakdown Prevents the Development of Pain Following Nerve Root Compression Injury** SB³C2015-319 pg.379
Jenell Smith, Paul Janmey, Beth Winkelstein, *University of Pennsylvania, Philadelphia, PA, United States*

11:15AM **Defining Collagen Fiber Mechanics in Neuron-Collagen Constructs Under Stretch Using Integrated Experimental & Modeling Approaches** SB³C2015-281 pg.315
Sijia Zhang, Xuan Cao, Vivek Shenoy, Beth Winkelstein, *University of Pennsylvania, Philadelphia, PA, United States*

11:30AM **Redistribution of Knee Loads Using Auditory Feedback from Pressure Detecting Shoe Insoles** SB³C2015-245 pg.261
Christopher F. Ferrigno, Ina S. Stoller, Laura E. Thorp, Najia Shakoor, Markus M. Wimmer, *Rush University, Chicago, IL, United States*

11:45AM **Pre-Clinical Assesment of a Percutaneous Leaflet Resection Device for Treatment of Degenerative Mitral Valve Disease** SB³C2015-559 pg.799
Steven Boronyak, Brett Byram, Joseph Fredi, Michael Young, W. David Merryman, *Vanderbilt University, Nashville, TN, United States*

12:00PM **Subject-specific Calibration Of Geometric Neuromusculoskeltal Models** SB³C2015-585 pg.849
Andrew J. Meyer, Carolynn Patten, Benjamin J. Fregly, *University of Florida, Gainesville, FL, United States*

12:15PM **In-Vivo Dynamic Measurement of Tibiotalar and Subtalar Joint Kinematics Using Dual Fluoroscopy: A Framework for Studying OA.** SB³C2015-527 pg.739
Koren E. Roach, Bibo Wang, Ashley L. Kapron, Niccolo M. Fiorentino, Charles L. Saltzman, Madeline Singer, Andrew E. Anderson, *University of Utah, Salt Lake City, UT, United States*

FRIDAY, JUNE 19	11:00am - 12:30pm

PhD Competition - Characterization of Tissue Mechanics — Maybird

Session Chair: Sarah Kieweg, *University of Kansas, Lawrence, KS, United States*
Session Co-Chair: Ian A. Sigal, *University of Pittsburgh, Pittsburgh, PA, United States*

11:00AM **Regional Contraction Shapes the Three-Dimensional Morphogenesis of the Embryonic Forebrain** SB³C2015-1030 pg.1036
Kara E. Garcia, Philip V. Bayly, Larry A. Taber, *Washington University in St. Louis, St. Louis, MO, United States*

11:15AM **The Dynamic Mechanical Response is Severely Altered in Collagen V Deficient Mouse Supraspinatus Tendons** SB³C2015-104 pg.24
Brianne K. Connizzo[1], Mei Sun[2], David E. Birk[2], Louis J. Soslowsky[1], [1]*University of Pennsylvania, Philadelphia, PA, United States,*[2]*University of South Florida, Tampa, FL, United States*

11:30AM **Evidence that Interfibrillar Load Transfer in Tendon is Supported by a Network of Small Diameter Collagen Fibrils** SB³C2015-132 pg.66
Spencer E. Szczesny[1], Kristen L. Fetchko[2], Jeffrey L. Caplan[3], Pal Pedersen[4], Dawn M. Elliott[2], [1]*University of Pennsylvania, Philadelphia, PA, United States,*[2]*University of Delaware, Newark, DE, United States,*[3]*Delaware Biotechnology Institute, Newark, DE, United States,*[4]*Carl Zeiss Microscopy, Thornwood, NY, United States*

11:45AM **3D Strains in Posterior Sclera Using Ultrasound Speckle Tracking** SB³C2015-515 pg.716
Elias R. Pavlatos, Benjamin Cruz-Perez, Hugh J. Morris, Hong Chen, Richard T. Hart, Jun Liu, *The Ohio State University, Columbus, OH, United States*

12:00PM **The Collagen Directionality and Dispersion and Mechanical Indentation Response in Nonpregnant Human Cervical Tissue** SB³C2015-1090 pg.1144
Wang Yao[1], Yu Gan[1], Christine Hendon[1], Joy Vink[2], Ronald Wapner[2], Kristin Myers[1], [1]*Columbia University, New York, NY, United States,*[2]*Columbia University Medical Center, New York, NY, United States*

12:15PM **Tribological Rehydration: Maintaining and Rebuilding Interstitial Fluid Pressure in Cartilage** SB^{3}C2015-133 pg.68
Axel C. Moore, David L. Burris, *University of Delaware, Newark, DE, United States*

FRIDAY, JUNE 19	11:00am - 12:30pm

Undergraduate Design Competition — Golden Cliff / Eagle's Nest

Session Co-Chair: Martin L. Tanaka, *Western Carolina University, Cullowhee, NC, United States*
Session Co-Chair: Sara Roccabianca, *Michigan State University, East Lansing, MI, United States*

11:00AM **BioView: A Wearable Sensor Array For Rehabilitation Biofeedback** SB^{3}C2015-1420 pg.1329
Tyler Maydew, Brett Donnermeyer, Kathryn Thompson, Alwyn Johnson, Bradley Davidson, *University of Denver, Denver, CO, United States*

11:15AM **The Tension Assisted Device: An Orthotic Alternative to High Tone Plantar Flexion** SB^{3}C2015-298 pg. 343
Ron V. Perrone, Elizabeth Duncan, Cory Jeanes, Mike Martorano, Gary Bowlin, John Williams, Susan Anderson, *University of Memphis, Memphis, TN, United States*

11:30AM **DermaShift: Diagnostic Device For Pressure Ulcer Formation** SB^{3}C2015-1421 pg.1331
Francisca Acosta[1], Hope Atina[1], Kim Le[1], Andrea Pinto[1], William Wilson[1], Erice Richardson[1], Catherine Ambrose[2], Lex Frieden[2], [1]*Rice University, Houston, TX, United States,*[2]*Texas Health Science Center, Houston, TX, United States*

11:45AM **Smartboot: An Instrumented Clinical Walking Boot for Partial Weight Bearing Training** SB^{3}C2015-328 pg.395
Dustyn Roberts, Tim West, Michael Schenk, David Schnall, Margaret O'Brien, Melissa Groome, Brian Knarr, Jill Higginson, Anita Singh, *University of Delaware, Newark, DE, United States*

12:00PM **Soft Ankle-foot Orthotic** SB^{3}C2015-187 pg.160
Adam Podolec, Megan Erhart, Noah Schadt, Geni Giannotti, Jared Green, Tyler Leichenberger, *Rochester Institute of Technology, Rochester, NY, United States*

12:15PM **Design of a Low-cost Haptic Assistive Handwriting Device** SB^{3}C2015-225 pg.226
Eamon Campolettano, Alexander Croft, Kevin Fasano, Stephen Hodge, Kevin Myers, Brian Pinkard, Jessica Ross, Allison Scoular, Thomas S. Todd, Alexander A. Brown, *Lafayette College, Easton, PA, United States*

FRIDAY, JUNE 19	11:00am - 12:30pm

PhD Competition - Mechanics of Injury and Repair — Primrose B

Session Chair: Ramesh Ragupathy, *Drexel University, Philadelphia, PA, United States*
Session Co-Chair: Corey Neu, *Purdue University, West Lafayette, IN, United States*

11:00AM **Thermoresponsive, Redox-crosslinked Cellulosic Hydrogels Undergo In Situ Gelation And Restore Nucleus Pulposus Biomechanical Properties Post Nucleotomy** SB^{3}C2015-263 pg.287
Devika M. Varma[1], Huizi A. Lin[1], Rose G. Long[2], Carine Rognon[3], Andrew C. Hecht[2], James C. Iatridis[2], Steven B. Nicoll[1], [1]*The City College of New York, CUNY, New York, NY, United States,*[2]*Icahn School of Medicine at Mount Sinai, New York, NY, United States,*[3]*Swiss Federal Institute of Technology of Zurich (ETHZ), Zurich, Switzerland*

11:15AM **A Multigenerational Collagen Damage Model Explains Engineered Cartilage Growth and Remodeling Phenomena** SB^{3}C2015-642 pg.953
Robert J. Nims, Alexander D. Cigan, Brian K. Jones, Krista M. Durney, Clark T. Hung, Gerard A. Ateshian, *Columbia University, New York, NY, United States*

11:30AM **Influence of Intracortical Porosity on the Fracture Susceptibility of Human Cortical Bone** SB^{3}C2015-672 pg.1009
Andrew P. Baumann, Travis L. Turnbull, Glen L. Niebur, Ryan K. Roeder, *University of Notre Dame, Notre Dame, IN, United States*

11:45AM **Effective Remodeling in Cerebral Aneurysm: a Case Study** SB^{3}C2015-1159 pg.1271
Xinjie Duan[1], Bong Jae Chung[2], Juan R. Cebral[2], Khaled Aziz[3], Anne M. Robertson[1], [1]*University of Pittsburgh, Pittsburgh, PA, United States,*[2]*George Mason University, Fairfax, VA, United States,*[3]*Allegheny General Hospital, Pittsburgh, PA, United States*

12:00PM **Cartilage Wear Initiated by Fatigue Damage Under Physiologic Loading when Fluid Load Support and Boundary Lubrication are Compromised** SB³C2015-1160 pg.1273
Krista M. Durney, Sevan R. Oungoulian, Brian K. Jones, Jason T. Suh, Clark T. Hung, Gerard A. Ateshian, *Columbia University, New York, NY, United States*

12:15PM **Cornea Damage Progression following Blast Exposure** SB³C2015-1169 pg.1289
Dan F. Shedd[1], Justin A. Jones[1], Brian Zaugg[2], Brittany Coats[1], *[1]University of Utah, Salt Lake City, UT, United States,[2]John A. Moran Eye Center, Salt Lake City, UT, United States*

FRIDAY, JUNE 19	12:30pm - 3:00pm

Poster Session II — Thermal Effects and Nanoparticles — Event Center Tent

170 **Determination of the Biophysical Parameters of HUVECs and Their Application in Optimization of the Addition and Removal of Cryoprotective Agents** SB³C2015-434 pg.578
Yuntian Zhang, Dan Niu, Gang Zhao, *University of Science and Technology of China, Hefei, China*

171 **Nonlinear Derating of High-Intensity Focused Ultrasound using Hydrophone Measurements in Water** SB³C2015-542
Seyed Ahmad Reza Dibaji[1], Yunbo Liu[2], Joshua E. Soneson[2], Rupak K. Banerjee[1], Matthew R. Myers[2], *[1]University of Cincinnati, Cincinnati, OH, United States,[2]US Food and Drug Administration, Silver Spring, MD, United States* *pg.768*

172 **Effect of Hydroxyapatite Nanoparticles on Cryopreservation of HUVECs** SB³C2015-427 *pg.*563
Yuanyuan Zheng, Jianye Wang, **Gang Zhao**, Tao Wang, *University of Science and Technology of China, Hefei, China*

173 **Feasibility of Utilizing Thermal Images for Melanoma Screening** SB³C2015-126 *pg.*54
Alexander LeBrun, **Liang Zhu**, *University of Maryland Baltimore County, Baltimore, MD, United States*

174 **Application Of Mesoporous Silica Nanoparticle At Drug Delivery System** SB³C2015-1372 *pg.*1327
M. Titirinli, Sevil Yücel, B. Karakuzu, Y. Basarab, *Yildiz Technical University, Istanbul, Turkey*

FRIDAY, JUNE 19	12:30pm - 3:00pm

Poster Session II — Transport at the Cell and Tissue Level — Event Center Tent

176 **Cancer-Associated Fibroblasts Promote Vascularization in Collagen and Fibrin Matrices** SB³C2015-199 *pg.*182
M.K. Sewell-Loftin, Samantha van Hove, Gregory Longmore, Steven George, *Washington University in St. Louis, St. Louis, MO, United States*

177 **Insights Into the Hemodynamic Factors Affecting Embolus Transport for Stroke** SB³C2015-584 *pg.*847
Debanjan Mukherjee, Shawn C. Shadden, *U.C. Berkeley, Berkeley, CA, United States*

178 **A Peristaltic Mechanism For Clearance Of Solutes In Periarterial Basement Membranes** SB³C2015-391 *pg.* 491
M Keith Sharp[1], Alexandra Keith Diem[2], Roy O. Weller[2], Roxana O. Carare[3], *[1]University of Louisville, Louisville, KY, United States,[2]University of Southampton, Southampton, United Kingdom,[3]University of Southampton, Suthampton, United Kingdom*

179 **Trapping of Tumor Cells Using Rapid Electrokinetic Patterning (REP)** SB³C2015-163 *pg.*114
Katherine N. Clayton, Seungman Park, Steven Wereley, Bumsoo Han, *Purdue University, West Lafayette, IN, United States*

180 **Visco-Hyperelastic and Biphasic Properties of a Brain Phantom Agarose Gel** SB³C2015-114 *pg.*38
Gerson Cordoba[1], Gustavo Orozco[1], **Fernando Casanova**[1], Joshua H. Smith[2], Jose J. Garcia[1], *[1]Universidad del Valle, Cali, Colombia,[2]Lafayette College, Easton, PA, United States*

FRIDAY, JUNE 19	12:30pm - 3:00pm

Poster Session II — Design, Dynamics and Rehab — Event Center Tent

181 **The Effect Of Bone Defect Size And Position On Cementless Acetabular Cup Stability : A Finite Element Analysis** SB³C2015-1091 *pg.*1146
Mark H. Gonzalez, Farid Amirouche, Gianfranco Solitro, *University of Illinois at Chicago, Chicago, IL, United States*

182 **Simulation-Based Design of a Hip Actuator for Running a Mile Sprint** SB³C2015-641 *pg.* 951
John R. Rogers, Julie E. Dillon, Cameron I. McDonald, Gabriela C. Barrera-Gutierez, *United States Military Academy, West Point, NY, United States*

183 **Comparison of Human Walking Backward and Forward Using Optimization Method** SB³C2015-351 *pg.*429
Yujiang Xiang[1], Hyun-Jung Kwon[2], [1]*University of Alaska Fairbanks, Fairbanks, AK, United States,*[2]*The Ohio State University, Columbus, OH, United States*

184 **Beginning Braille Learning Device** SB³C2015-252 *pg.*273
Kelton Gubler, Jason Castillo, Adam Daly, Austin Eastman, Kay B. Freckleton, Andrew Silotti, *University of Utah, Salt Lake City, UT, United States*

185 **Ocular Bobbing Compensation System** SB³C2015-3803 *pg.*1335
Yucong Gu, Yuchen Yan, You Chen, Tai Kim, Ken Fischer, *University of Kansas, Lawrence, KS, United States*

186 **A Bioengineering Solution To Cure Spinal Cord Injury** SB³C2015-1161 *pg.*1275
Anita Singh[1], Jacklyn Witko[1], Brittany King[1], Alexander Herman[1], Andrea Vernengo[1], Babitha Tom[2], [1]*Rowan University, Glassboro, NJ, United States,*[2]*Widener University, Chester, PA, United States*

187 **High Performance Luxury 4WD (or All Terrain) Wheel Chair** SB³C2015-3731 *pg.*1333
Jason McCurry, Philip Stykes, Alex Wilfong, Martin Tanaka, *Western Carolina University, Cullowhee, NC, United States*

188 **Dynamic Balance Using The COM And COP Inclination Angle During A Golf Swing** SB³C2015-22 *pg.*1
Ahnryul Choi[1], Joung Hwan Mun[2], [1]*The University of Texas Health Science Center at Houston, Houston, TX, United States,*[2]*Sungkyunkwan University, Suwon, Korea, Republic of*

FRIDAY, JUNE 19	12:30pm - 3:00pm

Poster Session II — Fluid Mechanics of Atherosclerosis and Aneurysms — Event Center Tent

189 **FSI Simulations for the Hemodynamic Assessment of the Carotid Bifurcation in an Atherosclerotic Mouse Model** SB³C2015-528 *pg.*741
David De Wilde[1], Bram Trachet[1,2], Nic Debusschere[1], Francesco Iannaccone[1], Abigaïl Swillens[1], Joris Degroote[1], Jan Vierendeels[1], Guido R. Y. De Meyer[3], Patrick Segers[1], [1]*Ghent University, Gent, Belgium,*[2]*Ecole Polytechnique Fédérale de Lausanne, Lausanne, Switzerland,*[3]*University of Antwerp, Wilrijk, Belgium*

190 **Integrating Morphologic, Biomechanic, Biological And Clinical Risk Factors To Improve Decision Making In The Management Of Abdominal Aortic Aneurysm Disease** SB³C2015-454 *pg.*616
Eleni Metaxa[1], Nikolaos Kontopodis[2], Christos V. Ioannou[2], Yannis Papaharilaou[1], [1]*FORTH, Heraklion, Greece,*[2]*University of Crete, Heraklion, Greece*

191 **Biomechanical Response And Fiber Microstructure Coupled With Localized Protease Activity And Inhibition In The Angiotensin Ii Infused Apoe-/- Mouse Model Of Aneurysm** SB³C2015-188 *pg.*162
Darren G. Haskett[1], Tyler S. Smith[1], D. Catalina Ardilia[1], Tom C. Doetschman[1], Oliver J. McIntyre[2], Dominic V. McGrath[1], Urs Utzinger[1], Jonathan P. Vande Geest[1], [1]*University of Arizona, Tucson, AZ, United States,*[2]*Vanderbilt University, Nashville, TN, United States*

192 **Geometric and Hemodynamics Implications of Moyamoya Disease on Carotid Siphon** SB³C2015-264 *pg.*289
Muhammad Jamil[1], Mehnaz Haq[1], Heidi Kang[2], Zhi Rui Lee[2], Phua Hwee Tang[3], **Choon Hwai Yap**[1], [1]*National University of Singapore, Singapore,*[2]*Hwa Chong Institution, Singapore,*[3]*KKH Women's and Children's Hospital, Singapore*

193 **Coherent Wall Shear Stress Structures Determine the Near Wall Transport in Aneurysms.** SB^{3}C2015-423 *pg.*555
Amirhossein Arzani[1], Guoning Chen[2], Alberto M. Gambaruto[3], Shawn C. Shadden[1], [1]*University of California, Berkeley, Berkeley, CA, United States,*[2]*University of Houston, Houston, TX, United States,*[3]*Barcelona Supercomputing Center, Barcelona, Spain*

194 **RhoA Mediated Effects of Statin Therapy on Endothelial Cells: A Comparison to Fasudil and Latrunculin A** SB^{3}C2015-538 *pg.*760
Melissa L. Dick[1,2], Katherine N. MacDonald[1], Jean-Claude Tardif[2], Richard L. Leask[1,2], [1]*McGill University, Montreal, QC, Canada,*[2]*Montreal Heart Institute, Montreal, QC, Canada*

195 **In Vitro And Computational Fluid Dynamics Comparison Of The Flow Diversion Efficacy Of Five Commercial Stents** SB^{3}C2015-636 *pg.*941
Ronak J. Dholakia[1], Andrew Pagano[1], Fotis Drakopoulos[2], Ari Kappel[1], Chander Sadasivan[1], Xiangmin Jiao[1], David J. Fiorella[1], Nikos Chrisochoides[2], Henry H. Woo[1], Baruch B. Lieber[1], [1]*Stony Brook University, Stony Brook, NY, United States,*[2]*Old Dominion University, Norfolk, VA, United States*

196 **Effect of Red Blood Cells on Endothelial Cell Shear Stress Studied Using Discrete-Nature Blood Flow Simulations** SB^{3}C2015-1024 *pg.*1026
Brenna E. Hogan[1], Zaiyi Shen[2], Chaouqi Misbah[2], Abdul Barakat[1], [1]*Ecole Polytechnique, Palaiseau Cedex, France,*[2]*Universite Joseph Fourier, Grenoble, France*

197 **FRED versus Pipeline: An In-vitro Comparison of Cerebral Aneurysm Hemodynamics Using Particle Image Velocimetry** SB^{3}C2015-1108 *pg.*1179
Priya Nair[1], Brian W. Chong[2], Haithem Babiker[1], Justin Ryan[1], L. Fernando Gonzalez[3], David Frakes[1], [1]*Arizona State University, Tempe, AZ, United States,*[2]*Mayo Clinic Hospital, Phoenix, AZ, United States,*[3]*Duke University School of Medicine, Durham, NC, United States*

198 **In Vitro Validation of Endovascular Doppler-derived Flow Rates in Cerebral Vessels** SB^{3}C2015-526 *pg.*737
Patrick McGah, John Nerva, Ryan Morton, Michael Barbour, Pierre Mourad, Michael Levitt, Louis Kim, Alberto Aliseda, *University of Washington, Seattle, WA, United States*

199 **The Role of the Glycocalyx in Leukocyte Adhesion to the Endothelium** SB^{3}C2015-612 *pg.* 897
Karli K. McDonald, Scott E. Cooper, Richard L. Leask, *McGill University, Montreal, QC, Canada*

200 **Validation Of CFD Solver Of A Clinical Tool Using PIV On A Patient-Specific Intracranial Aneurysm.** SB^{3}C2015-623
Nikhil Paliwal, Christopher Martensen, Nicole Varble, Robert Damiano, Adnan Siddiqui, Elad Levy, Jianping Xiang, Hui Meng, *University at Buffalo, State University of New York, Buffalo, NY, United States* *pg.915*

201 **The Quantification of Blood Flow Patterns Induced by Endovascular Stent Grafts Using a Non-Newtonian Blood Analog** SB^{3}C2015-637 *pg.*943
Amanda Colella Centazzo, Clifton R. Johnston, *Dalhousie University, Halifax, NS, Canada*

FRIDAY, JUNE 19	**12:30pm - 3:00pm**

Poster Session II — Heart Valves, Mechanical Circulatory Support, Thrombosis, and Hemolysis — Event Center Tent

202 **Modulation of Platelet Microtubule Function Alters Platelet Stiffness and Mechanotransductive Responsiveness to Shear Stress** SB^{3}C2015-1036 *pg.*1046
Siu Ling Leung[1], Yi Lu[1], Danny Bluestein[2], Marvin J. Slepian[1,2], [1]*the University of Arizona, Tucson, AZ, United States,*[2]*Stony Brook University, Stony Brook, NY, United States*

203 **Convective Leakage Makes Heparin Locking of Central Venous Catheters Ineffective within Seconds: Experimental Measurements in an In Vitro Model of the Pediatric Superior Vena Cava** SB^{3}C2015-1042 *pg.*1058
Alberto Aliseda[1], Michael Barbour[1], Patrick M. McGah[1], Kurt R. Sansom[1], Kenneth Gow[2], [1]*University of Washington, Seattle, WA, United States,*[2]*Seattle Children's Hospital, Seattle, WA, United States*

204 **Aortic Flow Waveform Shape Regulates Valvular Hemodynamics And Energy Losses** SB^{3}C2015-1063 *pg.*1096
Brandon L. Moore, Lakshmi P. Dasi, *Colorado State University, Fort Collins, CO, United States*

205 **Analyzing The Effect Of Stent Geometry For Polymeric Aortic Valves** SB³C2015-1071 *pg.*1110
Atieh Yousefi Koupaei, Brandon L. Moore, David L. Bark Jr., David Prawel, Lakshmi P. Dasi, *Colorado State University, Fort Collins, CO, United States*

206 **Mathematical Modeling of Thrombus Formation in Idealized Models of Aortic Dissection** SB³C2015-170 *pg.*126
Claudia Menichini, Xiao Y. Xu, *Imperial College London, London, United Kingdom*

207 **Characterizing The Inflammatory And Coagulation Response To Trauma And Resuscitation** SB³C2015-579 *pg.*837
Elaheh Rahbar[1], Jessica C. Cardenas[2], Nena Matijevic[2], Deborah J. del Junco[2], Jeanette M. Podbielski[2], Mitchell J. Cohen[3], Bryan A. Cotton[2], John B. Holcomb[2], Charles E. Wade[2], [1]*Wake Forest University, Winston-Salem, NC, United States,*[2]*University of Texas Health Science Center, Houston, TX, United States,*[3]*University of California San Francisco, San Francisco, CA, United States*

208 **Valve Interstitial Cell Shape Regulates Cell Function And Phenotype.** SB³C2015-619 *pg.*909
Atefeh Razavi, Rachel Reynolds, Kartik Balachandran, *University of Arkansas, Fayetteville, AR, United States*

209 **Aortic Valve Calcification: Geometric And Biomechanical Analysis** SB³C2015-1055 *pg.*1084
Banafsheh Zebhi[1], Brandon L. Moore[1], Gary Luckasen[2], Lakshmi P. Dasi[1], [1]*Colorado State University, Fort Collins, CO, United States,*[2]*Medical Center of Rockies, Loveland, CO, United States*

211 **Effect Of Assymetric Deployment On The Function And Leaflet Mechanics Of Bioprosthetic TAVI Valves** SB³C2015-524 *pg.*733
Michael B. Gogarty[1], Pablo Maureira[2], Lakshmi P. Dasi[1], [1]*Colorado State University, Fort Collins, CO, United States,*[2]*Lorraine University Hospital of Nancy, Nancy, France*

212 **Vascular Compliance Effect On Endothelial-to-mesenchymal Transdifferentiation Under Highly Pulsatile Fluid Shear** SB³C2015-1111 *pg.*1185
Winston H. Elliott[1], Yan Tan[2], Nancy Tseng[2], Wei Tan[1,2], [1]*University of Colorado- Boulder, Boulder, CO, United States,*[2]*University of Colorado- Denver, Denver, CO, United States*

213 **Intimal Hyperplasia and its Implications in the Cephalic Arch: A Numerical Study of Non-Physiological Hemodynamics in Patients with Brachiocephalic Fistulae** SB³C2015-315 *pg.* 373
Seyed Mohammad Javid Mahmoudzadeh Akherat[1], Michael E. Boghosian[1], Kevin W. Cassel[1], Mary S. Hammes[2], [1]*Illinois Institute of Technology, Chicago, IL, United States,*[2]*University of Chicago, Chicago, IL, United States*

214 **In Vitro Pulsatile Flow Loop Using Human Blood To Mimic Physiological Flow Conditions** SB³C2015-664 *pg.*995
Ryan W. Oba, David Bark, Ketul Popat, **Lakshmi P. D. Dasi**, *Colorado State University, Fort Collins, CO, United States*

FRIDAY, JUNE 19	**12:30pm - 3:00pm**

Poster Session II — Respiratory and Cerebrospinal Fluid Motion — Event Center Tent

215 **Effects of Inhalation Transience on Flow Structures During Numerical Simulation of Airflow Through a CT-based Airway Geometry** SB³C2015-649 *pg.*965
Richard R. Gruetzemacher, *The University of Tennessee at Chattanooga, Chattanooga, TN, United States*

216 **Bariatric Surgery Improvements For OSA Patients** SB³C2015-151 *pg.*92
Ahmed M. Al-Jumaily, *Auckland University of Technology, Auckland, New Zealand*

217 **Acoustic Detection of Respiratory Sounds in Silicone Lung Airway Model Using Microphone Array System** SB³C2015-360 *pg.*439
Gabriel Pramudita Saputra[1], Kazunori Nozaki[2], Satoshi Ii[1], Chizu Habukawa[1,3], Shigeo Wada[1], [1]*Osaka University, Toyonaka, Osaka, Japan,*[2]*Osaka University Dental Hospital, Suita, Osaka, Japan,*[3]*Minami Wakayama Medical Center, Tanabe, Wakayama, Japan*

218 **Pressure Modulation Improves OSA Patient Therapy** SB³C2015-246 *pg.*263
Ahmed M. Al-Jumaily, *Auckland University of Technology, Auckland, New Zealand*

219 **Multicenter Comparison Of 4D Phase Contrast MRI Measurement Of Cerebrospinal Fluid Dynamics In The Cervical Spine** SB³C2015-583 *pg.*845
Suraj Thyagaraj[1], Daniel Giese[2], Francesco Santini[3], Eleonora Fornari[4], Alexander C. Bunck[2], Francis Loth[1], Bryn A. Martin[1], *[1]University of Akron, Akron, OH, United States,[2]University of Cologne, Cologne, Germany,[3]University of Basel, Basel, Switzerland,[4]Lausanne University Hospital, Lausanne, Switzerland*

220 **Integration Of A Spontaneous Respiratory Driver With Blood Gas Feedback Into Biogears, An Open-source, Whole-body Physiology Model** SB³C2015-327 *pg.*393
Yeshitila Gebremichael, Rachel Clipp, Jeffrey Webb, Aaron Bray, Cameron Thames, Zack Swarm, Jennifer Carter, Jeremiah Heneghan, *Applied Research Associates, Inc., Raleigh, NC, United States*

221 **A Patient-Specific Computational Model to Characterize The Impact of Neural Tissue Motion on Cerebrospinal Fluid Dynamics at the Cervical-Medullary Junction** SB³C2015-166 *pg.*120
Soroush Heidari Pahlavian[1,2], Francis Loth[1,2], Mark Luciano[3], Bryn Martin[1,2], *[1]The University of Akron, Akron, OH, United States,[2]Conquer Chiari Research Center, Akron, OH, United States,[3]Cleveland Clinic Foundation, Cleveland, OH, United States*

222 **Complexity Of The Cerebrospinal Fluid Flow In Patients Suffering From Chiari Malformation Type I - A Computational Study** SB³C2015-209 *pg.*200
Kartik Jain[1,2], Kent-Andre Mardal[2,3], *[1]University of Siegen, Siegen, Germany,[2]Simula Research Laboratory, Oslo, Norway,[3]University of Oslo, Oslo, Norway*

223 **Peclet Number Of Ciliary Transport On The Surface Of The Tracheal Lumen** SB³C2015-157 *pg.*104
Kouki Kiyota[1], Hironori Ueno[2], Keiko Numayama-Tsuruta[1], Yohsuke Imai[1], Takami Yamaguchi[1], **Takuji Ishikawa**[1], *[1]Tohoku University, Sendai, Japan,[2]Aichi University of Education, Aichi, Japan*

FRIDAY, JUNE 19	12:30pm - 3:00pm

Poster Session II — Musculoskeletal Soft Tissue Mechanics — Event Center Tent

224 **A Phenomenological Model to Describe The Viscoelastic Behavior in Multiple Loading** SB³C2015-440 *pg.*588
Behzad R. Babaei, *Washington University in St. Louis, St. Louis, MO, United States*

225 **Effect of Osteoarthritis on the Mechanical Properties of Human Articular Cartilage** SB³C2015-509 *pg.*704
Blair E. Larson, Kristine M. Fischenich, Kirk A. Kindsfater, Tammy L. Haut Donahue, *Colorado State University, Fort Collins, CO, United States*

226 **Conditioned Media from Degenerative Vertebral Discs Sensitizes Dorsal Root Ganglion Neurons to Heat Stimuli** SB³C2015-1040 *pg.*1054
Joshua D. Stover[1], Ibrahima Bah[2], Alexander Kotelsky[2], Mark R. Buckley[2], Brandon Lawrence[1], Robert Bowles[1], *[1]University of Utah, Salt Lake City, UT, United States,[2]University of Rochester, Rochester, NY, United States*

227 **Inter and Intra Variation in the Tensile Properties of the Porcine Temporomandibular Joint Disc** SB³C2015-498 *pg.*684
Jesse Lowe[1], Alejandro Almarza[1,2], *[1]University of Pittsburgh, Pittsburgh, PA, United States,[2]McGowan Institute of Regenerative Medicine, Pittsburgh, PA, United States,[3]*

228 **A Quantitative Evaluation of the Role of Cell Senescence in Intervertebral Disc Degeneration** SB³C2015-640 *pg.*949
Shady Elmasry, Shihab Asfour, Juan Pablo de Rivero Vaccari, **Francesco Travascio**, *University of Miami, Coral Gables, FL, United States*

229 **Characterizing the Change in Ankle Constraint Following Grade II and III Sprains** SB³C2015-392 *pg.*493
Matthew H. Dickinson, Ednah G. Louie, Bardiya Akhbari, William M. Eboch, Sami Shalhoub, Lorin P. Maletsky, *University of Kansas, Lawrence, KS, United States*

230 **Three-dimentional Strain Distribution In The Anterior Cruciate Ligament During Anterior Translation Of The Knee** SB³C2015-444 *pg.*596
Satoshi Yamakawa[1], Richard Debski[2], Hiromichi Fujie[1], *[1]Tokyo Metropolitan University, Hino, Japan,[2]University of Pittsburgh, Pittsburgh, PA, United States*

231 **A Finite Element Model to Evaluate the Role of the Medial Meniscotibial Attachment in Knee Biomechanics** SB^{3}C2015-1153 *pg.*1261
Andrew J. Polk, Ferris M. Pfeiffer, James L. Cook, James P. Stannard, Patrick A. Smith, *University of Missouri, Columbia, MO, United States*

232 **Characterization of Fatigue Failure in Bovine Meniscus.** SB^{3}C2015-1187 *pg.*1317
Jaremy J. Creechley[1], Trevor J. Lujan[2], [1]*Materials Science and Engineering, Boise State University, Boise, ID, United States,*[2]*Boise State University, Boise, ID, United States*

233 **Quantifying Skeletal Muscle Deformation in Three Dimensions** SB^{3}C2015-666 *pg.*999
Elisabeth Jensen, Kenton Kaufman, Duane Morrow, Joel Felmlee, *Mayo Clinic, Rochester, MN, United States*

234 **Treatment of Focal Cartilage Defects Using a Metal Implant: New Biomechanical Insights Using Finite Element Modeling** SB^{3}C2015-361 *pg.*441
Ashley Heuijerjans, Wouter Wilson, Keita Ito, Corrinus C. van Donkelaar, *Eindhoven University of Technology, Eindhoven, Netherlands*

235 **Progressive Changes In Cervical Spine Intervertebral Disc Properties During Cyclic Compressive Fatigue Loading** SB^{3}C2015-665 *pg.*997
Sagar Umale[1], Brian Stemper[1,2], Mingxin Zheng[3], Aidin Masoudi[3], Daniel Fama[1,2], Narayan Yoganandan[1,2], Brian Snyder[3,4,5], [1]*Medical College of Wisconsin, Milwaukee, WI, United States,*[2]*Clement J. Zablocki VA Medical Center, Milwaukee, WI, United States,*[3]*Beth Israel Deaconess Medical Centre, Boston, MA, United States,*[4]*Harvard Medical School, Boston, MA, United States,*[5]*Cerebral Palsy Clinic Children's Hospital, Boston, MA, United States*

236 **Structural Inhomogeneity Enhances Interstitial Fluid Pressurization in TMJ Condylar Cartilage** SB^{3}C2015-1107 *pg.*1177
Brandon Zimmerman[1], Leonardo Ruggiero[1,2], Miri Park[1], Lin Han[3], Liyun Wang[1], David L. Burris[1], Xin L. Lu[1], [1]*University of Delaware, Newark, DE, United States,*[2]*Vrije Universiteit Brussel, Brussels, Belgium,*[3]*Drexel University, Philadelphia, PA, United States*

237 **A Model To Study Articular Cartilage Mechanical And Biological Responses To Rolling And Sliding Loads** SB^{3}C2015-162 *pg.112*
Oliver R. Schätti[1,2,3], Luigi M. Gallo[2], Peter A. Torzilli[1], [1]*Laboratory for Soft Tissue Research, Hospital for Special Surgery, New York City, NY, United States,*[2]*Center for Dental Medicine, University of Zürich, Zürich, Switzerland,*[3]*Institute for Biomechanics, Swiss Federal Institute of Technology, ETH, Zürich, Switzerland*

238 **Ibuprofen Does Not Adversely Affect Supraspinatus Tendon Mechanical Adaptations in a Rat Model of Exercise** SB^{3}C2015-269 *pg.*299
Sarah I. Rooney, Rachel Baskin, Andrew F. Kuntz, Louis J. Soslowsky, *University of Pennsylvania, Philadelphia, PA, United States*

239 **Toward Understanding the Mechanisms by Which Microparticles Induce Synovial Inflammation in Osteoarthritis** SB^{3}C2015-374 *pg.*463
Amy M. Silverstein[1], Robert M. Stefani[1], Sevan R. Oungoulian[1], Eric L. Tong[1], Mukundan G. Attur[2], Steven B. Abramson[2], Christopher S. Ahmad[1], James L. Cook[3], Gerard A. Ateshian[1], J. Chloe Bulinski[1], Clark T. Hung[1], [1]*Columbia University, New York, NY, United States,*[2]*New York University, New York, NY, United States,*[3]*University of Missouri, Columbia, MO, United States*

240 **Regional Mechanical Properties of the Long Head of the Biceps Tendon** SB^{3}C2015-148 *pg.*88
Christopher W. Kolz[1], Thomas Suter[1,2], Heath B. Henninger[1], [1]*University of Utah, Salt Lake City, UT, United States,*[2]*Clinic of Orthopaedic Surgery, Kantonsspital Baselland, Liestal, Switzerland*

241 **Biomechanical Effects of Menisco-Tibial Repair** SB^{3}C2015-399 *pg.*507
Ferris Pfeiffer[1], James Stannard[1], James Cook[1], Matthew Bollier[2], Patrick Smith[3], [1]*University of Missouri, Columbia, MO, United States,*[2]*University of Iowa, Iowa City, IA, United States,*[3]*Columbia Orthopaedic Group, Columbia, MO, United States*

242 **Mechanical And Adhesive Properties Of Hydrogels In Tension And Shear** SB^{3}C2015-1097 *pg.*1157
Jennifer Kadlowec, Daniel Collins, Patrick Myers, Thomas Christiani, Jennifer Vernengo, *Rowan University, Glassboro, NJ, United States*

243 **Dynamic Viscoelastic Properties of Porcine Patellar Tendon Tissue: A Study of Regional Variation and Frequency Dependent Behaviour** SB³C2015-449 *pg.*606
Sourav S. Patnaik[1,2], Taylor Szasz[1,2], **Raj Prabhu**[1,2], Hongjoo Rhee[2], Mark F. Horstemeyer[2], Jun Liao[1,2], Lakiesha Williams[1,2], [1]*Mississippi State University, Mississippi State, MS, United States,*[2]*Center for Advanced Vehicular Systems, Mississippi State, MS, United States*

244 **Targeting Collagen Strands by Triple Helix Hybridization** SB³C2015-194 *pg.*172
Michael Yu, **Yang Li**, *University of Utah, Salt Lake City, UT, United States*

245 **Functional Tensile Properties of a Split Quadriceps Graft for Double-Bundle ACL Reconstruction** SB³C2015-180 *pg.*146
Robert Matthew Miller, Amir Ata Rahnemai-Azar, Todd Jasinski, Fabio V. Arilla, Levent Surer, Freddie H. Fu, Richard E. Debski, Volker Musahl, *University of Pittsburgh, Pittsburgh, PA, United States*

246 **A New Method for Measuring Stress Concentrations in Finite Element Analysis of Fibrocartilage Predicts Greater Fracture Risk for Angled Center Cracks** SB³C2015-326 *pg.*391
John M. Peloquin[1], Dawn M. Elliott[2], [1]*University of Pennsylvania, Philadelphia, PA, United States,*[2]*University of Delaware, Newark, DE, United States*

247 **The Effect of Anatomical Variability on Temporomandibular Joint Mechanics.** SB³C2015-403 *pg.*515
Jessica Coogan[1], Travis Eliason[1], Mark Wong[2], **Daniel Nicolella**[1], [1]*Southwest Research Institute, San Antonio, TX, United States,*[2]*The University of Texas School of Dentistry at Houston, Houston, TX, United States*

FRIDAY, JUNE 19	12:30pm - 3:00pm

Poster Session II — Injury Mechanics — Event Center Tent

248 **Sensitivity Study of Head Impact Parameters on Intracranial Dynamics** SB³C2015-413 *pg.*535
Yi Hua[1], Praveen Akula[1], Matthew Kelso[2], Linxia Gu[1,3], [1]*University of Nebraska-Lincoln, Lincoln, NE, United States,*[2]*University of Nebraska Medical Center, Omaha, NE, United States,*[3]*Nebraska Center for Materials and Nanoscience, Lincoln, NE, United States*

249 **Improving Brain-Skull Interface Through Application of Mesh Smoothing Algorithm** SB³C2015-1093 *pg.*1149
Mireille Kelley[1,2], Logan Miller[1,2], Jillian Urban[1,2], Joel Stitzel[1,2], [1]*Wake Forest University School of Medicine, Winston-Salem, NC, United States,*[2]*Virginia Tech - Wake Forest University, Winston-Salem, NC, United States*

250 **Mouse Model Of Blast Traumatic Brain Injury: An Imaging, Behavior And Pathological Assessment Study** SB³C2015-380 *pg.*471
Sujith Sajja[1], Jiangyang Zhang[1], Jeff Bulte[1], Joseph Long[2], Robert Stevens[1], Piotr Walczak[1], Miroslaw Janowski[1,3], [1]*Johns Hopkins School of Medicine, Baltimore, MD, United States,*[2]*Walter Reed Army Institute of Reseatch, Silver Spring, MD, United States,*[3]*NeuroRepair Department, MMRC, PAS, Warsaw, Poland*

251 **The Effects of the Impact of a Soccer Ball on a Human Head** SB³C2015-517 *pg.*720
Kimberly A. Brown, Aalaap Desai, Yuxiong Mao, Mark Horstemeyer, Jun Liao, Lakiesha Williams, Hongjoo Rhee, Raj Prabhu, *Mississippi State University, Mississippi State, MS, United States*

252 **Biomechanics of Human Tibia and Fibula Fracture Caused by a Mixed Martial Arts Kick** SB³C2015-436 *pg.* 582
Andrew Lamont[1,2], Robbin Bertucci[1,3], Youssef Hammi[2], Mark Horstemeyer[2], Jun Liao[1,2], Hongjoo Rhee[3], Lakiesha Williams[1,2], Rajkumar Prabhu[1,2], [1]*Agricultural and Biological Engineering, Mississippi State University, Starkville, MS, United States,*[2]*Center for Advanced Vehicular Systems, Starkville, MS, United States,*[3]*Center for Advanced Vehicular Systems, Mississippi State, MS, United States*

253 **Influence Of Sulci On Mechanical Response Of The Brain And Injury Prediction Under High-rate Impact** SB³C2015-473 *pg.*646
Alan Leung[1], Nithyanand Kota[2], Amit Bagchi[3], **Siddiq Qidwai**[3], [1]*Advanced Technology & Research Corporation, Columbia, MD, United States,*[2]*Leidos Corporation, Arlington, VA, United States,*[3]*US Naval Research Laboratory, Washington, DC, United States*

254 **Finite Element Analysis of Lower Extremity Military Boot Protection at Blast Conditions** SB³C2015-1143 *pg.*1241
Robbin Bertucci, **R. Prabhu**, M. F. Horstemeyer, Jun Liao, Lakiesha N. Williams, *Mississippi State University, Starkville, MS, United States*

255 **Development of a Computationally Efficient Full Human Body Finite Element Model** SB^{3}C2015-638 *pg.*945
Doron Schwartz[1,2], Berkan Guleyupoglu[1,2], Bharath Koya[1,2], Joel D. Stitzel[1,2], F. Scott Gayzik[1,2], [1]*Wake Forest School of Medicine, Winston Salem, NC, United States,*[2]*Virginia Tech – Wake Forest University Center for Injury Biomechanics, Winston Salem, NC, United States*

256 **An Efficient and Reliable Biomechanical Testing Device to Perform Torsion Testing in Long Bones with Locking Compression Plates** SB^{3}C2015-574 *pg.*827
Joseph P. Loftus[1], **Anita Singh**[1], Lindsay Stoy[1], Douglas J. Patterson[2], [1]*Widener University, Chester, PA, United States,*[2]*Christiana Care, Newark, DE, United States*

257 **A Computational Model Of Blast Loading To The Eye: A Comparison With Field Tests** SB^{3}C2015-481 *pg.* 660
Thao D. Nguyen[1], Rajneesh Bhardwaj[2], Shantanu Bailoor[2], [1]*Johns Hopkins University, Baltimore, MD, United States,*[2]*Indian Institute of Technology Bombay, Mumbai, India*

258 **Supine to Prone Thoraco-abdominal Deformation and Organ Migration in a Set of Healthy Young Adults** SB^{3}C2015-1027 *pg.*1030
Berkan Guleyupoglu, Josh C. Tan, Craig A. Hamilton, F. Scott Gayzik, *Wake Forest University School of Medicine, Winston Salem, NC, United States*

259 **Design of a Novel Shock Tube System for Blast Induced Traumatic Brain Injury** SB^{3}C2015-598 *pg.*871
Andrew B. Robbins[1], Raoul Van Loon[2], Ashok K. Shetty[3], Michael R. Moreno[1], [1]*Texas A&M University, College Station, TX, United States,*[2]*Swansea University, Swansea, United Kingdom,*[3]*Texas A&M Health Science Center, Temple, TX, United States*

261 **A Computational Head Model Validated Against Pressure Responses Only May Not Be Used To Estimate Brain Strain Responses** SB^{3}C2015-662 *pg.*991
Wei Zhao, Songbai Ji, *Dartmouth College, Hanover, NH, United States*

262 **Head Impact Response Resulting from Forceful Impact with Toy Swords by Pediatric Males** SB^{3}C2015-604 *pg.*883
Stephanie M. Beeman, Steven Rowson, Stefan M. Duma, *Virginia Tech, Blacksburg, VA, United States*

263 **Behavioral And Inflammatory Consequences Of Cerebrovascual Dysfunction In Primary Blast Injury** SB^{3}C2015 *pg.*1319
Stewart Yeoh, Kenneth L. Monson, *University of Utah, Salt Lake City, UT, United States*

264 **Pelvic Response of a Total Human Body Finite Element (FE) Model During Simulated Under Body Blast (UBB) Impacts** SB^{3}C2015-1028 *pg.*1032
Caitlin M. Weaver[1,2], Randolph S. Coates[2], Andrew S. Merkle[3], Joel D. Stitzel[1], [1]*Wake Forest University, Winston-Salem, NC, United States,*[2]*US Army Research Laboratory, Aberdeen Proving Ground, MD, United States,*[3]*Johns Hopkins Applied Physics Lab, Laurel, MD, United States*

265 **The Effect of Pre-Crash Velocity Reduction on Occupant Response Using a Finite Element Model** SB^{3}C2015-301 *pg.*349
Nicholas A. Vavalle[1,2], Berkan Guleyupoglu[1,2], Jeremy M. Schap[1,2], Kristofer D. Kusano[3,4], F. Scott Gayzik[1,2], [1]*Virginia Tech – Wake Forest Center for Injury Biomechanics, Winston-Salem, NC, United States,*[2]*Wake Forest School of Medicine, Winston-Salem, NC, United States,*[3]*Virginia Tech – Wake Forest Center for Injury Biomechanics, Blacksburg, VA, United States,*[4]*Virginia Tech, Blacksburg, VA, United States*

266 **Alteration And Failure Of Cerebral Artery Internal Elastic Lamina Following Axial Overstretch** SB^{3}C2015-1140 *pg.*1235
Matthew I. Converse, Tessa Sommer, Kenneth L. Monson, *University of Utah, Salt Lake City, UT, United States*

268 **Prediction Of Extra-axial Injury Location From Real World Motor Vehicle Crash And Occupant Data** SB^{3}C2015-1086
Jillian E. Urban[1], Sarah Lynch[1], Ervin Lowther[1,2], Christopher Whitlow[1,2,3], Joel Stitzel[1], [1]*Wake Forest University, Winston-Salem, NC, United States,*[2]*Wake Forest School of Medicine, Winston-Salem, NC, United States,*[3]*Translational Science Institute, Winston-Salem, NC, United States* *pg.1138*

269 **Effect of Neck Cable Tension During Helmeted Head Impact** SB^{3}C2015-1113 *pg.*1189
Bethany Rowson, Steven Rowson, Stefan M. Duma, *Virginia Tech, Blacksburg, VA, United States*

270 **Traumatic Brain Injury Resulted in Increased Aquaporin-4 Expression - Relevance to Post Injury Edema** SB^{3}C2015-647 *pg.* 961
Nasya Sturdivant, Jeffrey Wolchok, Kartik Balachandran, *University of Arkansas, Fayetteville, AR, United States*

FRIDAY, JUNE 19	12:30pm - 3:00pm

Poster Session II **Mechanics of Growth, Remodeling and Repair** **Event Center Tent**

271 **Damage Accumulation Modeling and Rate Dependency of Spinal Dura Mater** SB^3C2015-178 *pg.* 142
Nicole Ramo, Snehal S. Shetye, Christian M. Puttlitz, *Colorado State University, Fort Collins, CO, United States*

272 **Analytical Approximation for Predicting Stress Fiber Remodeling Due to Dynamic Mechanical Stimuli** SB^3C2015-215
Tommaso Ristori[1,2], [1]*Eindhoven University of Technology, Eindhoven, Netherlands,*[2]*Institute for Complex Molecular Systems, Eindhoven, Netherlands* *pg.*210

273 **Multiscale Model of Strain-Dependent Glomerular Basement Membrane Remodeling** SB^3C2015-624 *pg.*917
Lazarina Gyoneva[1], Yoav Segal[1,2], Kevin D. Dorfman[1], Victor H. Barocas[1], [1]*University of Minnesota, Minneapolis, MN, United States,*[2]*VA Medical Center, Minneapolis, MN, United States*

274 **Quantification Of Transient Temperature And Thermal Damage In An Established Burn Model** SB^3C2015-1166 *pg.*1285
Stephanie Lindow, **F. Scott Gayzik**, *Wake Forest University School of Medicine, Winston-Salem, NC, United States*

275 **Simulated Collagen Network Remodeling in Response to Stress.** SB^3C2015-591 *pg.*859
Carley B. Hovell, *University of Minnesota, Twin Cities, Minneapolis, MN, United States*

276 **Isolation of Subarachnoid Hemhorrage Factors on a Single Chip for Understanding Cerebral Vasospasm Progression** SB^3C2015-514 *pg.*714
Eric S. Hald, Kerianne E. Steucke, Connor Timm, Patrick W. Alford, *University of Minnesota, Minneapolis, MN, United States*

277 **A Finite Element Investigation of Fracture Healing Under Simulated Microgravity Loading Conditions** SB^3C2015-422
Benjamin C. Gadomski, Zachary F. Lerner, Raymond C. Browning, Christian M. Puttlitz, *Colorado State University, Fort Collins, CO, United States* *pg.*553

FRIDAY, JUNE 19	12:30pm - 3:00pm

Poster Session II **Cardiovascular Tissue Mechanics** **Event Center Tent**

278 **Novel Technique for Assessment of Mechanical Properties of Carotid Arteries** SB^3C2015-476 *pg.*652
Stefan Sanders, Frans van de Vosse, Marcel Rutten, *Eindhoven University of Technology, Eindhoven, Netherlands*

279 **Morphological Characterization Of Collagen Fibers At The Crack Initiation Sites In Biaxially Stretched Porcine Thoracic Aortas Toward Clarification Of Aneurysm Rupture Mechanism** SB^3C2015-425 *pg.*559
Shukei Sugita, Takeo Matsumoto, *Nagoya Institute of Technology, Nagoya, Japan*

280 **Biaxial Mechanical Characterization of Non-Uniform Thermal Shrinkage Deformations to Guide Ablative Therapy** SB^3C2015-565 *pg.*811
Steven Boronyak, W. David Merryman, *Vanderbilt University, Nashville, TN, United States*

281 **Characterization of the Fatigue Life, Dynamic Creep and Modes of Damage Accumulation within Mitral Valve Chordae Tendineae** SB^3C2015-533 *pg.*752
Gillian M. Gunning, Bruce P. Murphy, *Trinity College Dublin, Dublin, Ireland*

282 **Finite Element Modeling Of Cardiac Muscle Contraction** SB^3C2015-408 *pg.* 525
Xiaoyan Zhang, Kenneth S. Campbell, **Jonathan F. Wenk**, *University of Kentucky, Lexington, KY, United States*

283 **Novel Micro-Computed Tomography Technique for Soft Tissue Deformation Tracking - Application to the Mitral Valve** SB^3C2015-182 *pg.*150
Eric L. Pierce[1], **Charles H. Bloodworth**[2], Ajay Naran[2], Thomas F. Easley[2], Morten O. Jensen[1], Ajit P. Yoganathan[1], [1]*Georgia Institute of Technology and Emory University, Atlanta, GA, United States,*[2]*Georgia Institute of Technology, Atlanta, GA, United States*

284 **Identification and Quantification of Extracellular Matrix Proteins at the Plaque - Internal Elastic Lamina Interface in a Mouse Model of Atherosclerosis** SB^{3}C2015-291 *pg.*329
Lindsey Davis, Susan Lessner, *University of South Carolina, Columbia, SC, United States*

285 **The Number of Lesions Does Not Govern the Functionality of Coronary Bifurcation Lesions: A Study of the Effect of Relative Stenosis Severity** SB^{3}C2015-224 *pg.*224
Catherine Pagiatakis[1,2], Jean-Claude Tardif[2,3], Philippe L. L'Allier[2,3], **Jennifer Frattolin**[1], Rosaire Mongrain[1,2], [1]*McGill University, Montreal, QC, Canada,*[2]*Montreal Heart Institute, Montreal, QC, Canada,*[3]*University of Montreal, Montreal, QC, Canada*

286 **Force Required to Cinch the Tricuspid Annulus: An Ex Vivo Study** SB^{3}C2015-137 *pg.*70
Amy N. Adkins[1], Jesus Aleman[1], Edward Sako[2], Lori Boies[1], Shamik Bhattacharya[1], [1]*St.Mary's University, San Antonio, TX, United States,*[2]*University of Texas Health Science Center San Antonio, San Antonio, TX, United States*

287 **Infarcted Left Ventricles Have Stiffer Material Properties and Lower Stiffness Variation: 3D Echo-Based Modeling to Quantify In Vivo Ventricle Material Properties** SB^{3}C2015-32 *pg.*16
Longling Fan[1], Jing Yao[2], Chun Yang[3], Di Xu[2], **Dalin Tang**[1,4], [1]*Southeast University, Nanjing, China,*[2]*Nanjing Medical University, Nanjing, China,*[3]*China United Network Communications Co., Ltd., Beijing, China,*[4]*Worcester Polytechnic Institute, Worcester, MA, United States*

288 **Development Of An Estimation Method Of Blood Vessel Configuration At No Load State For FE Patient-specific Simulation** SB^{3}C2015-1133 *pg.*1225
Ming Yu[1], Sota Yamamoto[1], Mrie Oshima[2], [1]*Shibaura Institute of technology, Tokyo, Japan,*[2]*The University of Tokyo, Tokyo, Japan*

289 **Post Endovascular Stent Repair Of Descending Aorta: Side Effects And Development Of Aneurysm In The Ascending Aorta** SB^{3}C2015-1095 *pg.*1153
Giampaolo Martufi, **Manal Altamimi**, Raied Aburashed, Cyrus Fiori, Jehangir J. Appoo, Elena S. Di Martino, *University of Calgary, Calgary, AB, Canada*

290 **Bicuspid Aortic Valve Hemodynamics Induce Acute Asymmetric Remodeling Of Porcine Ascending Aortas: An Ex Vivo Study** SB^{3}C2015-1112 *pg.*1187
Samantha K. Atkins, Philippe Sucosky, *University of Notre Dame, Notre Dame, IN, United States*

291 **The Effect of Vascular Curvature on Blood Flow and Oxygen Transport in Arterio-Venous Fistulae** SB^{3}C2015-807
Francesco Iori, Lorenza Grechy, Richard W. Corbett, Wladyslaw Gedroyc, Neill Duncan, Colin G. Caro, **Peter E. Vincent**, *Imperial College London, London, United Kingdom pg. 1013*

292 **Measuring Vessel Wall Displacement and Circumferential Strain Using Displacement Encoded with Stimulated Echo (DENSE) MRI Sequence** SB^{3}C2015-592 *pg.*861
Elizabeth Iffrig[1,2], Xiaodong Zhong[1,3], William R. Taylor[1,2], John N. Oshinski[1,2], [1]*Emory University, Atlanta, GA, United States,*[2]*Georgia Institute of Technology, Atlanta, GA, United States,*[3]*Siemens Medical Solutions, Malvern, PA, United States*

293 **Computational Analysis Of The Effect Of Sequential Coiling On The Wall Stress Of Cerebral Aneurysms** SB^{3}C2015-1190 *pg.*1323
Joseph E. Pichamuthu, Brian T. Jankowitz, David A. Vorp, *University of Pittsburgh, Philadelphia, PA, United States*

294 **Smooth Muscle Cell Elastin Generation Stimulated by Adipose-Derived Mesenchymal Stem Cells** SB^{3}C2015-496
Aneesh Ramaswamy, Kory Blose, Justin Weinbaum, David Vorp, *University of Pittsburgh, Pittsburgh, PA, United States pg.680*

295 **Computational Modeling Of Passive Myocardium:A Comparison Between Two Constitutive Models** SB^{3}C2015-347
Amir Nikou[1], Shauna M. Dorsey[2], Jeremy R. McGarvey[3], Joseph H. Gorman III[2], Jason A. Burdick[2], James J. Pilla[2], Robert C. Gorman[2], Jonathan F. Wenk[1], [1]*University of Kentucky, Lexington, KY, United States,*[2]*University of Pennsylvania, Philadelphia, PA, United States,* [3]*Gorman Cardiovascular Research Group, University of Pennsylvania, Philadelphia, PA, United States pg.421*

296 **Toward an Experimentally Validated Immersed Boundary Model of Left Ventricular Fluid Dynamics Using In Vitro Experiments** SB^{3}C2015-1123 *pg.*1207
Boyce E. Griffith[1], Jae Ho Lee[1], Pritam Mekala[2], Arvind Santhanakrishnan[2], [1]*University of North Carolina at Chapel Hill, Chapel Hill, NC, United States,*[2]*Oklahoma State University, Stillwater, OK, United States*

297 **Role Of Cyclic Strain On Calcific Nodule Formation Among Aortic Heart Valve Cusps** SB^{3}C2015-1085 *pg.*1136
Ying Lei, Zannatul Ferdous, *The University of Tennessee, Knoxville, TN, United States*

298 **Impact of Partial Intraluminal Thrombus Attachment on Peak Stresses on Abdominal Aortic Aneurysm Wall** SB³C2015-277 *pg.*309
Juan S. Stockle, David A. Romero, Cristina H. Amon, *University of Toronto, Toronto, ON, Canada*

299 **Stability Analysis of the Continuum Constrained Mixture Model for Vascular Growth and Remodeling** SB³C2015-1096 *pg.*1155
Jiacheng Wu, Shawn C. Shadden, *University of California, Berkeley, Berkeley, CA, United States*

300 **Ultrasound Monitoring of Abdominal Aortic Aneurysm Progression in a Murine Model** SB³C2015-1046 *pg.*1066
Arvin H. Soepriatna[1], Gurneet S. Sangha[1], Amelia R. Adelsperger[1], Evan H. Phillips[1], Clifford M. Babbey[2], Michael P. Murphy[2], Pavlos P. Vlachos[1], Craig J. Goergen[1], [1]*Purdue University, West Lafayette, IN, United States,*[2]*Indiana University - Purdue University Indianapolis, Indianapolis, IN, United States*

301 **Aortic Peak Stress Induced By Antihypertensive Medications In Aortic Dissection Patients** SB³C2015-552 *pg.*786
Vittoria Flamini, *New York University, Brooklyn, NY, United States*

302 **Mechanical Parameters Characterization of Embryonic Mouse Hearts for Studying Human Congenital Heart Disease** SB³C2015-492 *pg.*672
Andres Rubiano, Kyle G. Rowe, W. Gregory Sawyer, Chelsey S. Simmons, *University of Florida, Gainesville, FL, United States*

303 **Machine Learning Based Structured Edge Detection for Cardiovascular Modeling** SB³C2015-1021 *pg.*1020
Jameson T. Merkow[1], Zhouwen Tu[1], David Kriegman[1], Nathan Wilson[2], Alison L. Marsden[1], [1]*University of California San Diego, San Diego, CA, United States,*[2]*Open Source Medical Software Corporation, San Diego, CA, United States*

304 **A Study On The Interplay Of Sex And Cyclic Stretch In Aortic Valve Calcification** SB³C2015-466 *pg.*634
Shirin Masjedi, **Ying Lei**, Zannatul Ferdous, *University of Tennessee, Knoxville, TN, United States*

FRIDAY, JUNE 19	12:30pm - 3:00pm

Poster Session II — Tissue Engineering and Regenerative Medicine: Materials and Interfaces — Event Center Tent

306 **Genipin Cross-linking Silk Fibroin Post-gelation Increases Gel Mechanical Stiffness** SB³C2015-1134 *pg.*1227
Winston H. Elliott[1], Walter Bonani[2,3], Devid Maniglio[2,3], Antonella Motta[2,3], Wei Tan[1], Claudio Migliaresi[2,3], [1]*University of Colorado- Boulder, Boulder, CO, United States,*[2]*University of Trento, Trento, Italy,*[3]*European Institute of Excellence on Tissue Engineering and Regenerative Medicine, and INSTM Trento Research Unit, Trento, Italy*

307 **Silica-Collagen Hydrogel for Corneal Replacement** SB³C2015-259 *pg.*281
Michael DiVito, *University of Minnesota, Minneapolis, MN, United States*

308 **In Situ Polymerization Of Thiol-acrylate Nanocomposite Foam For Bone Defects** SB³C2015-253 *pg.*275
Anoosha Forghani, *Louisiana State University, Baton Rouge, LA, United States*

309 **Fabrication Of Scaffolds From Different Silica Based Bioactive Glass And Investigation Of Bioactivity And Biodegredation Behaviors** SB³C2015-1371 *pg.*1325
Sevil Yücel[1], A. C. Ozarslan[1], B. S. Oduncu[1], P. Terzioglu[2], [1]*Yildiz Technical University, Istanbul, Turkey,*[2]*Muğla Sıtkı Koçman University, Muğla, Turkey*

310 **Evaluation Of Two Formulations Of Polycaprolactone For Use In Tissue Scaffold Applications** SB³C2015-390 *pg.*489
Jacob D. Harris, A. Sharif El-Gizawy, Ferris M. Pfeiffer, *University of Missouri, Columbia, MO, United States*

311 **Aqueous Biphasic Micro-printing Of Tumor Spheroids** SB³C2015-654 *pg.*975
Stephanie Lemmo Ham, Ehsan Atefi, Hossein Tavana, *University of Akron, Akron, OH, United States*

312 **Using Finite Element Analysis to Study the Mechanical Advantages of a Turtle's Shell** SB³C2015-1105 *pg.*1173
John W. Wood, Rajkumar Prabhu, *Mississippi State University, Starkville, MS, United States*

313 **Characterization of a Thermoreversible Collagen for Free-Form Fabrication of Scaffolds** SB^{3}C2015-474 *pg.*648
Kathryn E. Drzewiecki, David I. Shreiber, *Rutgers University, Piscataway, NJ, United States*

314 **3D Printing Of B. Mori Silk Proteins For Implantable Devices** SB^{3}C2015-320 *pg.*381
Tom M. Merrill, Maria Torculas, Jethro Medina, Brenton Boszczuk, Kyle Meehan, Ian Miller, Xiao Hu, Wei Xue, *Rowan University, Glassboro, NJ, United States*

315 **Influence of Nano- and Micro-Scale Structure of Aligned Electrospun Scaffolds on Mechanical Properties and Cell Response** SB^{3}C2015-141 *pg.*76
Hannah M. Pauly[1], Ketul C. Popat[1], Daniel J. Kelly[2], Tammy L. Haut Donahue[1], [1]*Colorado State University, Fort Collins, CO, United States,*[2]*Trinity College Dublin, Dublin, Ireland*

316 **An Assessment of the Native Species in Articular Cartilage and Synovial Fluid as Potential Absorptive Barriers to UV-Initiated Scaffold Polymerization** SB^{3}C2015-429 *pg.*567
Anthony Finch, Patrick Donnelly, Peter Torzilli, *Hospital for Special Surgery, New York, NY, United States*

FRIDAY, JUNE 19	12:30pm - 3:00pm

Poster Session II — Tissue Engineering and Regenerative Medicine: Cells, Constructs, Culture Systems, and Regeneration — Event Center Tent

317 **Effects of Mechanical Constraints on Cell-Generated Stress and Collagen Remodeling in Statically Cultured Microtissues** SB^{3}C2015-369 *pg.*455
Mathieu A. J. van Kelle, Sandra Loerakker, Inge A. E. W. van Loosdregt, Carlijn V. C. Bouten, Frank P. T. Baaijens, *Eindhoven University of Technology, Eindhoven, Netherlands*

318 **Optimization of Parameters For Long-Term Storage of Tissue Engineered Articular Cartilage** SB^{3}C2015-385 *pg.*479
Adam B. Nover[1], Robert M. Stefani[1], Stephanie L. Lee[1], Rebecca A. Peyser[1], Daniel R. Howard[2], Gerard A. Ateshian[1], Aaron M. Stoker[3], James L. Cook[3], Clark T. Hung[1], [1]*Columbia University, New York, NY, United States,*[2]*Mount Sinai St. Luke's, New York, NY, United States,*[3]*University of Missouri, Columbia, MO, United States*

319 **Maturation Of Human Stem Cell-derived Cardiomyocytes In 3D Tissues Through Increasing Collagen Concentrations** SB^{3}C2015-1131 *pg.*1221
Aric Q. Pahnke, *University of Toronto, Toronto, ON, Canada*

320 **Use Of Kartogenin To Augment The Tendon-bone Tunnel Healing** SB^{3}C2015-310 *pg.*365
Yiqin Zhou, Jianying Zhang, Guangyi Zhao, **James H-C. Wang**, *University of Pittsburgh, Pittsburgh, PA, United States*

321 **Modular Tissue Engineered Cartilage Surfaces** SB^{3}C2015-324 *pg.*387
Audrey C. Ford, Kayla Wolf, Aditya Nandy, Anne Y. Zeng, Grace D. O'Connell, *University of California Berkeley, Berkeley, CA, United States*

322 **Tensile Properties of Stem Cell-Based Self-Assembled Tissue (scSAT) Biosynthesized on Nanoperiodic Structured Substrate** SB^{3}C2015-448 *pg.*604
Kei Oya[1], Yuki Tani[2], Kota Koizumi[3], Norihiko Sugita[3], Kenji Suzuki[4], Norimasa Nakamura[3], Hiromichi Fujie[2], [1]*Tokai University, Hiratsuka, Kanagawa, Japan,*[2]*Tokyo Metropolitan University, Hino, Tokyo, Japan,*[3]*Osaka University, Suita, Osaka, Japan,*[4]*Kogakuin University, Hachioji, Tokyo, Japan*

323 **Ultrasound Assisted Human Mesenchymal Stem Cell Chondrogenesis: Engineering Large-scale Cartilage Grafts** SB^{3}C2015-669 *pg.*1003
Anu Subramanian, Sanjukta Guha Thakurta, Neety Sahu, Abdul Qadir Chama, Hendrik J. Viljoen, *University of Nebraska, Lincoln, NE, United States*

324 **Osteogenic Induction Of Human Adipose Derived Stem Cells Cultured On Poly (l-lactic Acid) Scaffolds Prepared By Thermally Induced Phase Separation Method** SB^{3}C2015-172 *pg.*130
Harish Chinnasami, Ram Devireddy, Dan Hayes, *Louisiana State University, Baton Rouge, LA, United States*

325 **Investigating the Role of Osteoactivin in Muscle Regeneration** SB^{3}C2015-625 *pg.* 919
Jinjin Ma[1], Bing Yu[2], Andrew Baker[1], Min-Ho Kim[2], Anthony Calabro[1], Fayez Safadi[3], Christopher Malcuit[2], Kathleen Derwin[1], [1]*Cleveland Clinic, Cleveland, OH, United States,*[2]*Kent State University, Kent, OH, United States,*[3]*Northeast Ohio Medical University, Rootstown, OH, United States*

326 **Effect Of Heat Shock On Cryopreservation Of Adipose Tissue Derived Stem Cells** SB³C2015-165 *pg.*118
Mulla S. Shaik[1], Jeffrey M. Gimble[2], Ram Devireddy[1], [1]*Louisiana State University, Baton Rouge, LA, United States,*[2]*LaCell Incorporation, New Orleans, LA, United States*

327 **Evaluating the Consistency of Cardiomyocyte Self-assembly** SB³C2015-316 *pg.*375
Nancy K. Drew, Danny B. Baldo, Jason Q. Core, Anna Grosberg, *University of California, Irvine, Irvine, CA, United States*

FRIDAY, JUNE 19	12:30pm - 3:00pm

Poster Session II Mechanotransduction and Sub-Cellular Biophysics Event Center Tent

328 **Comparison Between Nonlinear Material Models And Fiber Network Models Reveals Importance Of Fiber Interactions In Cell Mechanosensing On Fibrous Substrates** SB³C2015-609 *pg.* 891
Maziar Aghvami[1], Kristen L. Billiar[2], Edward A. Sander[1], [1]*University of Iowa, Iowa City, IA, United States,*[2]*Worcester Polytechnic Institute, Worcester, MA, United States*

329 **Role of SRC in Electric Field-Induced Directed Cell Migration** SB³C2015-469 *pg.*640
Shun-Hao Tsao, Pen-Hsiu Grace Chao, *National Taiwan University, Taipei, Taiwan*

330 **Exploring the Response of Astrocytes to Traumatic Brain Injuries Using a Novel Bench Top Crash Tester** SB³C2015-435 *pg.* 580
Joe Wyatt, Addison Walker, Kartik Balachandran, **Jeff Wolchok**, *University of Arkansas, Fayetteville, AR, United States*

331 **Endoplasmic Reticulum Calcium Dynamics in Osteocyte Mechanobiology** SB³C2015-179 *pg.*144
Genevieve N. Brown, Prajesh Desai, X. Edward Guo, *Columbia University, New York, NY, United States*

333 **Movement In Engineered Valvular Tissues In Relation To Regional Flow Physics And Nutrient Transport** SB³C2015-279 *pg.*311
Manuel Salinas, *Florida International University, Davie, FL, United States*

334 **Cellular Cholesterol Content Modulates Monocyte Interaction With E-selectin** SB³C2015-412 *pg.*533
Amit K. Saha, Anand K. Ramasubramanian, *University of Texas at San Antonio, San antonio, TX, United States*

335 **Time Evolution of Contractility in Fibroblasts as a Measure of Cell Photodamage** SB³C2015-1177 *pg.*1303
Samantha G. Knoll[1], Wylie W. Ahmed[2], Taher A. Saif[1], [1]*University of Illinois at Urbana-Champaign, Urbana, IL, United States,*[2]*Institut Curie, Paris, France*

336 **Topology of Prestin Expressed in the CHO Cell Membrane -Atomic Force Microscopy Study-** SB³C2015-221 *pg.*218
Michio Murakoshi[1,2], Hiroshi Wada[3], [1]*Kagoshima University, Kagoshima, Japan,*[2]*JST, Kawaguchi, Japan,*[3]*Tohoku Bunka Gakuen University, Sendai, Japan*

337 **Role of Cell-Cell Interaction in Tensional Homeostasis** SB³C2015-1118 *pg.*1199
Alicia Zollinger[1], Elizabeth Canovic[2], Michael Smith[1], Dimitrije Stamenovic[1], [1]*Boston University, Boston, MA, United States,*[2]*Massachusetts Institute of Technology, Cambridge, MA, United States*

338 **Computational And Experimental Analysis Of Intracellular Motion, Forces, And Structure** SB³C2015-1184 *pg.*1313
Michael Mak[1], Taeyoon Kim[2], Muhammad H. Zaman[3], Roger D. Kamm[1], [1]*Massachusetts Institute of Technology, Cambridge, MA, United States,*[2]*Purdue University, West Lafayette, IN, United States,*[3]*Boston University, Boston, MA, United States*

339 **Epithelial to Mesenchymal Transition Alters Cellular Chiral Behavior** SB³C2015-238 *pg.*249
Kathryn E. Worley, Andrew K. Watrobski, David Shieh, Leo Q. Wan, *Rensselaer Polytechnic Institute, Troy, NY, United States*

340 **Modeling Cellular Contraction on Biohybrid Devices Using Thermal Contraction Capabilities of Finite Element Analysis Tools** SB³C2015-400 *pg.*509
Victoria A. Webster, Ozan Akkus, Hillel J. Chiel, Roger D. Quinn, *Case Western Reserve University, Cleveland, OH, United States*

FRIDAY, JUNE 19	3:00pm - 4:30pm

Bone Structure, Mechanics, and Function — Primrose A

Session Chair: Amy Wagoner Johnson, *University of Illinois at Urbana-Champaign, Urbana, IL, United States*
Session Co-Chair: Shigeo Tanaka, *Kanazawa University, Kanazawa, Japan*

3:00PM **Collagen Bound Water Is A Strong Correlate Of Bone's Toughness** SB[3]C2015-262 *pg.*285
Mustafa Unal, Ozan Akkus, *Case Western Reserve University, Cleveland, OH, United States*

3:15PM **Sequential Bisphosphonate and Parathyroid Hormone Treatment Decouples Bone Remodeling in Favor of Bone Formation** SB[3]C2015-521 *pg.*729
Allison R. Altman, Carina Lott, Chantal M. de Bakker, Wei-Ju Tseng, Ling Qin, **X. Sherry Liu**, *University of Pennsylvania, Philadelphia, PA, United States*

3:30PM **Tissue Mineral Density Dependent Mechanical Properties of Individual Trabecula Plates and Rods Do Not Differ in Anatomic Directions but Individual Trabecular Directions** SB[3]C2015-242 *pg.*255
Y. Eric Yu, Ji Wang, Bin Zhou, X. Edward Guo, *Columbia University, new york, NY, United States*

3:45PM **Post-yield Damage Denatures Bone's Collagen As Determined By A Novel Molecular Spectroscopic Biomarker** SB[3]C2015-261 *pg.* 283
Mustafa Unal, Hyungjin Jung, Ozan Akkus, *Case Western Reserve University, Cleveland, OH, United States*

4:00PM **Measurement of Bone Mineral Density in Motor Vehicle Crash Occupants and Correlation with Age and Fracture Incidence** SB[3]C2015-236 *pg.*244
Ashley A. Weaver, Richarlette C. Hightower, Sarah K. Lynch, Kristen M. Beavers, Anna N. Miller, Joel D. Stitzel, *Wake Forest University, Winston-Salem, NC, United States*

4:15PM **Proximal Femoral Cortical Bone Thickness in Patients with Femoroacetabular Impingement and Normal Hips Analyzed using Statistical Shape Modeling** SB[3]C2015-1115 *pg.*1193
Penny R. Atkins, Prateep Mukherjee, Shireen Y. Elhabian, Sumedha Singla, Michael D. Harris, Jeffrey A. Weiss, Ross T. Whitaker, Andrew E. Anderson, *University of Utah, Salt Lake City, UT, United States*

FRIDAY, JUNE 19	3:00pm - 4:30pm

Joint Motion and Rehabilitation — Superior

Session Chair: Laurel Kuxhaus, *Clarkson University, Potsdam, NY, United States*
Session Co-Chair: Bradley Davidson, *University of Denver, Denver, CO, United States*

3:00PM **A Long-Term Simulated Degradation Study of a Synthetic Meniscus Implant: Material and Functional Properties** SB[3]C2015-191 *pg.*166
Maoz Shemesh[1], Adaya Shefy-Peleg[1], Eyal Zylberberg[1], Eran Linder-Ganz[1], **Jonathan J. Elsner**[2], *[1]Active Implants, Netanya, Israel,[2]Active Implants, Cambridge, MA, United States*

3:15PM **A Robotic Knee Orthosis for Locomotive Assistance** SB[3]C2015-409 *pg.*527
Saroj Thapa[1], Hao Zheng[1], Geza Kogler[2], Xiangrong Shen[1], *[1]University of Alabama, Tuscaloosa, AL, United States,[2]Georgia Institute of Georgia, Atlanta, GA, United States*

3:30PM **Towards Vertebral Compression Fracture Prevention: Simulating Physiologic Fracture During Small Movement ADLs.** SB[3]C2015-414 *pg.*537
Nicole C. Corbiere, Stacey L. Zeigler, Kathleen A. Issen, Arthur J. Michalek, Laurel Kuxhaus, *Clarkson University, Potsdam, NY, United States*

3:45PM **Effect of Arm Posture on Voluntary Activation and Moments Generated by Individuals with Tendon Transfer and Quadriplegia** SB[3]C2015-1151 *pg.*1257
Carrie L. Peterson[1,2,3], Michael S. Bednar[4], Anne M. Bryden[5,6], Michael W. Keith[5,6], Eric J. Perreault[1,2], Wendy M. Murray[1,2,3], *[1]Rehabilitation Institute of Chicago, Chicago, IL, United States,[2]Northwestern University, Chicago, IL, United States,[3]Edward Hines Jr., VA Hospital, Hines, IL, United States,[4]Loyola University, Maywood, IL, United States,[5]MetroHealth Medical Center, Cleveland, OH, United States,[6]Case Western Reserve University, Cleveland, OH, United States*

4:00PM **Use of a Torque Range of Motion Device to Teach Evaluation of Joint Dysfunction** SB³C2015-296 *pg.*339
Rita Patterson[1], Jeongsik Shin[2], Aditya Das[2], Vanneise Collins[1], Carol Kominski[1], Katelyn Rockenbach[1], Robert Longnecker[1], David Mason[1], *[1]University of North Texas Health Science Center, Fort Worth, TX, United States,[2]University of Texas at Arlington Research Institute, Fort Worth, TX, United States*

4:15PM **Development of Virtual Environment Navigation Options for Individuals Post-stroke or With Cerebral Palsy Using Kinect** SB³C2015-124 *pg.*50
Alan Eberhardt[1], Sean Pool[1], Scott Bickel[1], Gerald McGwin[1], James Rimmer[1], Laurie Malone[2], *[1]University of Alabama Birmingham, Birmingham, AL, United States,[2]Lakeshore Foundation, Birmingham, AL, United States*

FRIDAY, JUNE 19	3:00pm - 4:30pm

Cardiovascular Imaging — Wasatch

Session Chair: Frank Gijsen, *University of Rotterdam, Netherlands*
Session Co-Chair: Christof Karmonik, *Houston Methodist Research Institute, Houston, TX, United States*

3:00PM **Differential Hemodynamic Changes And Lumen Remodeling In The Artery And Vein Of Porcine Arteriovenous Graft And Fistula** SB³C2015-123 *pg.*48
Daniel B. Pike[1], Yong He[2], Christi M. Terry[1], Alfred K. Cheung[1,3], Yan-Ting Shiu[1], *[1]University of Utah, Salt Lake City, UT, United States,[2]University of Florida, Gainesville, FL, United States,[3]VA SLC Health Care System, Salt Lake City, UT, United States*

3:15PM **Does Aortic Open Distal and Hemi-arch Procedure Remove All Tissue Suspected for Progression of Bicuspid Valve Aortopathy?** SB³C2015-322 *pg.* 383
Alex J. Barker[1], Pim van Ooij[1], David Guzzardi[2], Emilie Bollache[1], S. Chris Malaisrie[1], Patrick M. McCarthy[1], James Carr[1], Jeremy Collins[1], Michael Markl[1], Paul W. M. Fedak[1,2], *[1]Northwestern University, Chicago, IL, United States,[2]University of Calgary, Calgary, AB, Canada*

3:30PM **Quantification of Helical Flow Patterns in Left Ventricles of Healthy Subjects and Patients with Dilated Cardiomyopathy** SB³C2015-471 *pg.*642
Jonas Lantz, Carljohan Carlhäll, Tino Ebbers, *Linköping University, Linköping, Sweden*

3:45PM **Preparation of a Hydrogel Phantom of Human Athrosclerotic Plaque for Medical Simulation and Imaging** SB³C2015-622 *pg.*913
Juyu Chueh[1], Tanya N. Turan[2], Truman R. Brown[2], Todd LeMatty[2], Hui Mao[3], Olivia W. Brooks[1], Matthew J. Gounis[1], *[1]University of Massachusetts Medical School, Worcester, MA, United States,[2]Medical University of South Carolina, Charleston, SC, United States,[3]Emory University School of Medicine, Atlanta, GA, United States*

4:00PM **Microcirculation-Induced MRI Signal Anisotropy in Organized Tissues: Finite Element Modeling and Validation on Perfused Hearts** SB³C2015-1162 *pg.*1277
Osama Abdullah[1], Arnold David Gomez[1], Adam Schmidt[1], Edward Hsu[2], *[1]University of Utah, Salt Lake City, UT, United States,[2]University of Utah, Salt Lake City, UT, United States*

4:15PM **Functional and Anatomical Measures for Outflow Boundary Conditions in Atherosclerotic Coronary Bifurcations** SB³C2015-549 *pg.*782
Jelle Schrauwen[1], Adriaan Coenen[1], Akira Kurata[1], Jolanda J. Wentzel[1], Antonius F. W. van der Steen[1,2], Koen Nieman[1], Frank J. H. Gijsen[1], *[1]Thoraxcenter, Erasmus Medical Center, Rotterdam, Netherlands,[2]Delft University of Technology, Delft, Netherlands*

FRIDAY, JUNE 19	3:00pm - 4:30pm

Vascular Remodeling and Stented Flow — Magpie

Session Chair: Lucas H. Timmins, *Georgia Institute of Technology, Atlanta, GA, United States*
Session Co-Chair: Zhijie Wang, *University of Wisconsin - Madison, Madison, WI, United States*

3:00PM **Effects Of Estrogen On Pulmonary Wave Reflection And Energy Transmission In Pulmonary Arterial Hypertension** SB³C2015-627 *pg.*923
Aiping Liu, Naomi Chesler, *University of Wisconsin-Madison, Madison, WI, United States*

3:15PM **Biomechanical Comparison of Glutaraldehyde-crosslinked Gelatin/Fibrinogen Electrospun Cylindrical Scaffolds to Porcine Native Vascular Tissue** SB³C2015-1128 *pg.*1217
Ehab Tamimi, Catalina D. Ardila, Darren G. Haskett, Thomas Doetschman, Jonathan P. Vande Geest, *University of Arizona, Tucson, AZ, United States*

3:30PM **Hemodynamics of Healthy Vs. Pathological Venous Valve - Fluid-Structure Interaction Computational Model** SB³C2015-212 *pg.*204
Elina Soifer[1], Dar Weiss[1], Oren Rotman[1], Uri Zaretsky[1], Shmuel Einav[1,2], [1]*Tel Aviv University, Tel Aviv, Israel,*[2]*Stony Brook University, Stony Brook, NY, United States*

3:45PM **Patient-specific Treatment Of Intracranial Aneurysms: An Automatic CFD-based Flow-diverter Optimization Principle** SB³C2015-265 *pg.*291
Philipp Berg[1], László Daróczy[1], Oliver Beuing[2], Gábor Janiga[1], [1]*University of Magdeburg, Magdeburg, Germany,*[2]*University Hospital Magdeburg, Magdeburg, Germany*

4:00PM **Vaso-CT for Quantitative Mapping of Vessel Wall Apposition in a Flow-Diverter Implant** SB³C2015-493 *pg.*674
Kajo van der Marel, Ajay K. Wakhloo, Matthew J. Gounis, Ajit S. Puri, *University of Massachusetts Medical School, Worcester, MA, United States*

4:15PM **Biomechanical Testing to Improve Constitutive Models of the Two-Layered Carotid Artery Media** SB³C2015-453 *pg.*614
Caleb Davis[1,2], Ashish Pandya[2], Stephen E. Greenwald[2], [1]*Texas A&M University, College Station, TX, United States,*[2]*Queen Mary University of London, London, United Kingdom*

FRIDAY, JUNE 19	3:00pm - 4:30pm

Cardiovascular Tissue Engineering — Maybird

Session Chair: Alisa Morss, *Drexel University, Philadelphia, PA, United States*
Session Co-Chair: Pat Alford, *University of Minnesota, Minneapolis, MN, United States*

3:00PM **Spatiotemporal Cell-matrix Interactions During En Masse Migration of Fibroblasts on Collagen Matrices** SB³C2015-491 *pg.*670
Altug Ozcelikkale[1], Frederick Grinnell[2], Bumsoo Han[1], [1]*Purdue University, West Lafayette, IN, United States,*[2]*University of Texas Southwestern Medical Center, Dallas, TX, United States*

3:15PM **Modeling the Enhancement of Extracellular Matrix Quantity and Quality in Large-Deformation Mechanically-Conditioned Heart Valve Tissue Engineering** SB³C2015-169 *pg.*124
Joao S. Soares[1], John A. Stella[2], Antonio D'Amore[2], Will Zhang[1], William R. Wagner[2], John E. Mayer[3], Michael S. Sacks[1], [1]*University of Texas at Austin, Austin, TX, United States,*[2]*University of Pittsburgh, Pittsburgh, PA, United States,*[3]*Harvard Medical School, Boston, MA, United States*

3:30PM **Tubular Pediatric Pulmonary Valves By Suturing Decellularized Engineered Tissue Tubes** SB³C2015-183 *pg.*152
Jay Reimer, Zeeshan Syedain, Bee Haynie, **Robert Tranquillo**, *University of Minnesota, Minneapolis, MN, United States*

3:45PM **Adipose-Derived Stem Cells From Diabetic Donors Cause Thrombotic Failure of Autologous Tissue Engineered Blood Vessels** SB³C2015-407 *pg.*523
Jeffrey T. Krawiec[1], Han T. Liao[1,2], Justin S. Weinbaum[1], Dominic J. Pezzone[1], Antonio D'Amore[1], J. P. Rubin[1], William R. Wagner[1], David A. Vorp[1], [1]*University of Pittsburgh, Pittsburgh, PA, United States,*[2]*Chang Gung University, Guishan District, Taiwan*

4:00PM **Biomechanical Characterizations of Scar ECM During the Acute to Chronic Stages of Myocardial Infarction** SB³C2015-503 *pg.*694
Bryn Brazile[1], Ryan Butler[1], Sourav S. Patnaik[1], Yanyi Xu[2], Andrew Claude[1], Raj Prabhu[1], Lakiesha N. Williams[1], Jianjun Guan[2], Jun Liao[1], [1]*Mississippi State University, Mississippi State, MS, United States,*[2]*Ohio State University, Columbus, OH, United States*

4:15PM **Development of a Induced Pluripotent Stem Cell Derived Cardiomyocyte Seeded Fibrin Suture for Cardiac Regeneration** SB³C2015-599 *pg.*873
Katrina J. Hansen[1], Michael A. Laflamme[2], Glenn R. Gaudette[1], [1]*Worcester Polytechnic Institute, Worcester, MA, United States,*[2]*University of Washington, Seattle, WA, United States*

FRIDAY, JUNE 19	3:00pm - 4:30pm

The Cellular Microenvironment (joint with JSME) — Golden Cliff / Eagle's Nest

Session Chair: Wei Tan, *University of Colorado, Boulder, CO, United States*
Session Co-Chair: Brendon Baker, *Boston University, Boston, MA, United States*

3:00PM **MicroRNAs and Related Tissue Remodeling Genes in Rotator Cuff With Delayed Repair in A Rat Model** SB^{3}C2015-1089 *pg.*1142
Christopher T. Chen, Fuxin Wei, Erik Contreras, Lucas M. Chen, Zachary Shirley, William Shelton, Michael Khazzam, *UT Southwestern Medical Center, Dallas, TX, United States*

3:15PM **Soft 3-Dimensional Neotissue Microarrays as High-throughput Platforms for Interrogating Stem Cell Instructional Microenvironments** SB^{3}C2015-1173 *pg.*1295
Michael Floren, Wei Tan, *University of Colorado, Boulder, CO, United States*

3:30PM **Long-range Communication Between Cells in Fibrous Matrices Enabled by Tension-driven Alignment of Fibers** SB^{3}C2015-113 *pg.*36
Hailong Wang[1], Abhilash Nair[1], Brendon M. Baker[2], Britta Trappmann[2], Christopher S. Chen[2], Rebecca G. Wells[1], Vivek B. Shenoy[1], [1]*University of Pennsylvania, Philadelphia, PA, United States,*[2]*Boston University, Boston, MA, United States*

3:45PM **Engineered Cardiac Model System Reveals Fibroblast Threshold for Synchronized Beating** SB^{3}C2015-158 *pg.*106
Ariane C. C. van Spreeuwel[1], Noortje A. M. Bax[1], Christopher S. Chen[2], Carlijn V. C. Bouten[1], [1]*Eindhoven University of Technology, Eindhoven, Netherlands,*[2]*Boston University, Boston, MA, United States*

4:00PM **Cadherin-Specific Extracellular Interactions Alter Stem Cell Sensation and Interpretation of Soft Tissue Microenvironments** SB^{3}C2015-190 *pg.*164
Brian D. Cosgrove[1,2], Kush D. Mehta[1], Tristan P. Driscoll[1,2], Jason A. Burdick[1,2], Robert L. Mauck[1,2], [1]*University of Pennsylvania, Philadelphia, PA, United States,*[2]*Philadelphia VA Medical Center, Philadelphia, PA, United States*

4:15PM **Microscopic Heterogeneity in the Aortic Wall: Correlation between Mechanical Environment and Protein Expression** SB^{3}C2015-567 *pg.*815
Takeo Matsumoto[1], Yohei Uno[1], Shintaro Iijima[1], Yoshitaka Moriyama[1], Shukei Sugita[1], Kazuaki Nagayama[1,2], Akio Matsumoto[3], [1]*Nagoya Institute of Technology, Nagoya, Japan,*[2]*Ibaraki University (present), Hitachi, Japan,*[3]*Chiba University, Chiba, Japan*

FRIDAY, JUNE 19	3:00pm - 4:30pm

Soft Tissue Mechanics — Primrose B

Session Chair: Heath Henninger, *University of Utah, Salt Lake City, UT, United States*
Session Co-Chair: Trevor Lujan, *Boise State University, Boise, ID, United States*

3:00PM **Computational Framework For Application Of Residual Stress When The Stress-free Configuration Is Unknown.** SB^{3}C2015-1075 *pg.*1116
Steve Maas[1], Ahmet Erdemir[2], Jason P. Halloran[3], Jeffrey A. Weiss[1], [1]*University of Utah, Salt Lake City, UT, United States,*[2]*Cleveland Clinic, Cleveland, OH, United States,*[3]*Cleveland State University, Cleveland, OH, United States*

3:15PM **Using Slow and Fast Shear Waves to Estimate Shear and Tensile Moduli: Results from Simulations of Anisotropic Tissue.** SB^{3}C2015-317 *pg.*377
Dennis J. Tweten, John L. Schmidt, Ruth J. Okamoto, Philip V. Bayly, *Washington University, St. Louis, MO, United States*

3:30PM **Experimental Measurement of Shear and Tensile Moduli In Anisotropic Tissue Using Magnetic Resonance Elastography** SB^{3}C2015-244 *pg.*259
John L. Schmidt[1], Dennis J. Tweten[1], Maisie M. Mahoney[1], Tally Portnoi[2], Ruth J. Okamoto[1], Joel R. Garbow[1], Philip V. Bayly[1], [1]*Washington University, St. Louis, MO, United States,*[2]*Massachusetts Institute of Technology, Cambridge, MA, United States*

3:45PM **A Nonlinear Anisotropic Inverse Mechanics Method for Computational Dissection of Inhomogeneous Planar Soft Tissues** SB³C2015-171 *pg.128*
Colleen M. Witzenburg[1], Victor H. Barocas[2], [1]*University of Virginia, Charlottesville, VA, United States,*[2]*University of Minnesota, Minneapolis, MN, United States*

4:00PM **Structural Properties of the Anterolateral Structures of the Knee** SB³C2015-284 *pg.321*
Amir Ata Rahnemai-Azar, **R Matthew Miller**, Daniel Guenther, Freddie H. Fu, Bryson P. Lesniak, Volker Musahl, Richard E. Debski, *University of Pittsburgh, Pittsburgh, PA, United States*

4:15PM **Stiffness Characterization in Biological Materials Based on Deformation Imaging and Topology Optimization** SB³C2015-556 *pg.795*
Luyao Cai[1], Claus Pedersen[2], Ross Mclendon[2], Manuel Biedermann[2], Gergana Dimitrova[2], Jiang Yao[2], Corey P. Neu[1], [1]*Purdue University, West Lafayette, IN, United States,*[2]*Dassault Systèmes, Johnston, RI, United States*

SATURDAY, JUNE 20	11:30am - 1:00pm

Workshop: Critical Steps in Composing a Successful Mentorship Plan — Primrose A

Session Chair: Sara E. Wilson, *University of Kansas, Lawrence, KS, United States*
Session Co-Chair: Rouzbeh Amini, *The University of Akron, Akron, OH, United States*

SATURDAY, JUNE 20	11:30am - 1:00pm

Workshop: Teaching Undergraduate Design — Superior

Session Chair: Martin Tanaka, *Western Carolina University, Cullowee, NC, United States*
Session Co-Chair: Ken Fischer, *University of Kansas, Lawrence, KS, United States*

SATURDAY, JUNE 20	11:30am - 1:00pm

Workshop: SimVascular Workshop and New User Training — Wasatch

Session Chair: Alison Marsden, *University of California, San Diego, CA, United States*
Session Co-Chair: Shawn Shadden, *UC Berkeley, Berkeley, CA, United States*
Session Co-Chair: Nathan Wilson, *Open Source Medical Software Corporation, CA, United States*

SATURDAY, JUNE 20	11:30am - 1:00pm

Workshop: FEBio Workshop and Discussion — Magpie

Session Chair: Jeffrey A. Weiss, *University of Utah, Salt Lake City, UT, United States*
Session Co-Chair: Gerard A. Ateshian, *Columbia University, New York, NY, United States*
Session Co-Chair: Steve Maas, *University of Utah, Salt Lake City, UT, United States*

SATURDAY, JUNE 20	11:30am - 1:00pm

Workshop: Experimental and Computational Frameworks for Biotransport in Tumors — Maybird

Session Chair: M. Nichole Rylander, *University of Texas at Austin, Austin, TX, United States*
Session Co-Chair: Malisa Sarntinoranont, *University of Florida, Gainesville, FL, United States*

SATURDAY, JUNE 20	11:30am - 1:00pm

Workshop: Taking the Guesswork out of the Interview Process — Golden Cliff / Eagle's Nest

Organizers: ASME Bioengineering Division Student Leadership Committee (special thanks to Justine Garcia, Elizabeth Iffrig, and Corinne Riggin)

SATURDAY, JUNE 20	11:30am - 1:00pm

Workshop: Robotic Testing Systems to Study Joint and Tissue Function — Primrose B

Session Chair: Hiromichi Fujie, *Tokyo Metropolitan University, Tokyo, Japan*
Session Chair: Richard Debski, *University of Pittsburgh, Pittsburgh, PA, United States*

SATURDAY, JUNE 20	1:30pm - 3:00pm

Heart Valves and Cardiovascular Devices — Primrose A

Session Chair: Sarah Vigmostad, *University of Iowa, Iowa City, IA, United States*
Session Co-Chair: Keefe Manning, *Pennsylvania State University, State College, PA, United States*

1:30PM **Fluid-Structure Interaction Analysis of Mitral Valve Forces using a Comprehensive Model with 3D Chordal Structure: Synergy of Modeling and Experiments** SB^{3}C2015-387 *pg.*483
Milan Toma[1], **Morten O. Jensen**[1], Daniel R. Einstein[2], Ajit P. Yoganathan[1], Richard P. Cochran[3], Karyn S. Kunzelman[3], *[1]Georgia Institute of Technology, Atlanta, GA, United States,[2]Pacific Northwest National Laboratory, Richland, WA, United States,[3]University of Maine, Orono, ME, United States*

1:45PM **Transcatheter Aortic Valve Replacement Model: Crimping And Deploying In Patient-pathology Specific Roots** SB^{3}C2015-418 *pg.*545
Matteo Bianchi[1], Ram P. Ghosh[1], Debapria Das[1], Gil Marom[1], Thomas Claiborne[1], Marvin Slepian[2], Danny Bluestein[1], *[1]Stony Brook University, Stony Brook, NY, United States,[2]Sarver Heart Center,University of Arizona, Tucson, AZ, United States*

2:00PM **Computational Assessment of Hemodynamics in Tricuspid and Bicuspid Aortic Valves** SB^{3}C2015-305 *pg.*357
Kai Cao, Philippe Sucosky, *University of Notre Dame, Notre Dame, IN, United States*

2:15PM **Effects Of Coronary Flow On Sinus Hemodynamics In The Presence Of A Transcatheter Aortic Valve Implantation** SB^{3}C2015-1060 *pg.*1092
Brandon Moore[1], Pablo Maureira[2], **Lakshmi P. Dasi**[1], *[1]Colorado State University, Fort Collins, CO, United States,[2]Lorraine University Hospital of Nancy, Nancy, France*

2:30PM **Left Ventricle Assist Device Anastomosis Hemodynamic Analysis Using Direct Numerical Simulations (DNS) And Large Eddy Simulations (LES)** SB^{3}C2015-463 *pg.*628
Ricardo J. Bonilla-Alicea, *Georgia Institute of Technology, Atlanta, GA, United States*

2:45PM **Microfluidic Facsimile Of Ventricular Assist Device Shear Stress Patterns: Towards Point-of-care Devices To Monitor Patient Thrombotic Risk** SB^{3}C2015-1126 *pg.*1213
Annalisa Dimasi[1], Filippo Consolo[1], Lorenzo Valerio[1], Marco Rasponi[1], Danny Bluestein[2], Gianfranco B. Fiore[1], Alberto Redaelli[1], Marvin Slepian[3], *[1]Politecnico di Milano, Milano, Italy,[2]Stony Brook University, Stony Brook, NY, United States,[3]University of Arizona, Tucson, AZ, United States*

SATURDAY, JUNE 20	1:30pm - 3:00pm

Growth, Remodeling and Repair — Superior

Session Chair: Kristin Miller, *Tulane University, New Orleans, LA, United States*
Session Co-Chair: C. Alberto Figueroa, *University of Michigan, Ann Arbor, MI, United States*

1:30PM **Cell-mediated Compaction and Collagen Remodeling in Tissue-engineered Heart Valves due to Dynamic Loading Conditions** SB³C2015-219 *pg.*216
Sandra Loerakker, Tommaso Ristori, Frank P. T. Baaijens, *Eindhoven University of Technology, Eindhoven, Netherlands*

1:45PM **Non-Linear Optical Characterization of the Extracellular Matrix Changes Following Myocardial Infarction Predicts Alterations in Mechanical Properties** SB³C2015-1127 *pg.*1215
Kyle P. Quinn[1], Kelly E. Sullivan[1], Carlo A. Alonzo[1], Zachary Ballard[1], Irene Georgakoudi[1], **Lauren D. Black**[1,2], [1]*Tufts University, Medford, MA, United States,*[2]*Tufts University School of Medicine, Boston, MA, United States*

2:00PM **Tensile Equilibrium Material Response of Pregnant Mouse Cervical Tissue During Normal Remodeling** SB³C2015-372 *pg.*461
Kyoko Yoshida[1], Mala Mahendroo[2], Joy Vink[3], Ronald Wapner[3], Kristin Myers[1], [1]*Columbia University, New York, NY, United States,*[2]*UT Southwestern Medical Center, Dallas, TX, United States,*[3]*Columbia University Medical Center, New York, NY, United States*

2:15PM **Strain-dependent Degradation as a Mechanism for the Paradoxical Effects of Mechanical Loading on Collagen Fiber Alignment in Healing Tendon** SB³C2015-467 *pg.*636
William J. Richardson[1], Stavros Thomopoulos[2], Jeffrey W. Holmes[1], [1]*University of Virginia, Charlottesville, VA, United States,*[2]*Washington University, St. Louis, MO, United States*

2:30PM **A Computational Growth and Remodelling Approach to Study Adaptation and Tortuosity in a Buckled Artery** SB³C2015-580 *pg.*839
Mehdi Farsad[1], Qin Liu[2], Hai-Chao Han[2], Seungik Baek[1], [1]*Michigan State University, East Lansing, MI, United States,*[2]*University of Texas at San Antonio, San Antonio, TX, United States*

2:45PM **Quantification Of Mechanical Properties Of Rat Thoracic Ducts For Long-term Prediction Of Mechanically-mediated Growth And Remodeling** SB³C2015-401 *pg.* 511
Alexander W. Caulk, Zhanna Nepiyushchikh, Ryan Shaw, J. Brandon Dixon, Rudolph L. Gleason, *Georgia Institute of Technology, Atlanta, GA, United States*

SATURDAY, JUNE 20	1:30pm - 3:00pm

Pediatric/Embryonic Hemodynamics — Wasatch

Session Chair: Morbiducci Umberto, *Politecnico di Torino, Turin, Italy*
Session Co-Chair: Anayiotos Andreas, *Cyprus University of Technology, Cyprus*

1:30PM **In Vitro Multi-scale Patient Specific Study of the Effects of Coarctation in the Norwood Circulation** SB³C2015-430
Tianqi Hang[1], Alessandro Giardini[2], Giovanni Biglino[3], **Richard Figliola**[1], [1]*Clemson University, Clemson, SC, United States,*[2]*Great Ormond Street Hospital for Children, London, United Kingdom,*[3]*Great Ormond Street Hospital, London, United Kingdom* *pg.569*

1:45PM **Improved Post-operative Flow Modeling For Complex Peripheral Pulmonary Artery Stenosis** SB³C2015-1103
Weiguang Yang[1], Jeffrey A. Feinstein[1], Alison L. Marsden[2], Frank L. Hanley[1], Frandics P. Chan[1], Lisa W. Faberowski[1], Irene E. Vignon-Clementel[3], [1]*Stanford University, Palo Alto, CA, United States,*[2]*University of California, San Diego, La Jolla, CA, United States,*[3]*INRIA Paris-Rocquencourt, Paris, France* *pg.1169*

2:00PM **The Effect of Resolution on Viscous Dissipation Measured with 4D Flow MRI in Patients with Fontan Circulation: Evaluation Using Computational Fluid Dynamics** SB³C2015-421 *pg.*551
Merih Cibis[1], Kelly Jarvis[2], Michael Markl[2], Michael Rose[2], Cynthia Rigsby[2], Alex Barker[2], Jolanda Wentzel[1], [1]*Erasmus MC, Rotterdam, Netherlands,*[2]*Northwestern University, Chicago, IL, United States*

2:15PM **Modeling Blood Flow in Embryo-Specific Geometry of the Zebrafish Heart** SB³C2015-266 *pg.*293
Pavel Kozlovsky[1], Robert Bryson-Richardson[2], Moshe Rosenfeld[1], Ariel Jaffa[3], **David Elad**[1], [1]*Tel Aviv Univesity, Tel Aviv, Israel,*[2]*Monash University, Melbourne, Australia,*[3]*Tel Aviv Medical Center, Tel Aviv, Israel*

2:30PM **Biomechanical Role For Cardiac Jelly In Pumping Mechanics Of Developing Heart During Looping** SB³C2015-657
David L. Bark, Jr., Brennan M. Johnson, Bryce W. Schroder, Deborah M. Garrity, Diego Krapf, Lakshmi P. D. Dasi, *Colorado State University, Fort Collins, CO, United States* *pg.981*

2:45PM **Developmental Hemodynamics in the Embryonic Heart Outflow Tract** SB³C2015-205 *pg.*194
Venkat Keshav Chivukula, Madeline Midgett, Sandra Rugonyi, *Oregon Health and Sciences University, Portland, OR, United States*

SATURDAY, JUNE 20	1:30pm - 3:00pm

Biomechanics in Treatment of Heart Disease — Magpie

Session Chair: Jonathan Wenk, *University of Kentucky, Lexington, KY, United States*
Session Co-Chair: Lik Chuan Lee, *Michigan State University, East Lansing, MI, United States*

1:30PM **Cardiac Reversible Growth & Remodeling Model: Predicting and Understanding the Chronic Effects of Bioinjection Therapy** SB³C2015-31 *pg.*14
Lik Chuan Lee[1], Martin Genet[2], Jonathan Wenk[3], Joakim Sundnes[4], Samuel Wall[4], [1]*Michigan State University, East Lansing, MI, United States,*[2]*ETH Zurich, Zurich, Switzerland,*[3]*University of Kentucky, Lexington, KY, United States,*[4]*Simula Research Laboratory, Oslo, Norway*

1:45PM **The Effect Of Trabeculae Carnae On The Passive Compliance Of Left Ventricle** SB³C2015-516 *pg.*718
Arnav Sanyal[1], David Halaney[2], Marc D. Feldman[2], Hai-Chao Han[1], [1]*University of Texas at San Antonio, San Antonio, TX, United States,*[2]*University of Texas Health Science Center at San Antonio, San Antonio, TX, United States*

2:00PM **Effect of Scar Compaction on Therapeutic Efficacy of Anisotropic Reinforcement Following Myocardial Infarction** SB³C2015-615 *pg.*903
Samantha A. Clarke, Gorav Ailawadi, Jeffrey W. Holmes, *University of Virginia, Charlottesville, VA, United States*

2:15PM **Impaired Collagen Degradation Prevents Right Ventricular Hypertrophy And Dysfunction With Development Of Pulmonary Arterial Hypertension** SB³C2015-645 *pg.*959
Mark Golob, Zhijie Wang, Anthony Prostrollo, Timothy Hacker, Gaoussou Diarra, Naomi Chesler, *University of Wisconsin-Madison, Madison, WI, United States*

2:30PM **The Degree Of Outflow Tract Banding Predicts Cardiac Remodeling In Chicken Embryos** SB³C2015-108 *pg.*28
Madeline Midgett, Sandra Rugonyi, *Oregon Health & Science University, Portland, OR, United States*

2:45PM **Mechanical Decoupling of Nuclei from the Cytoskeleton Indicate Mechanosensitivity in a Myocardiocyte Pathology Model** SB³C2015-540 *pg.*764
Benjamin Seelbinder, Sarah Calve, Corey P. Neu, *Purdue University, West Lafayette, IN, United States*

SATURDAY, JUNE 20	1:30pm - 3:00pm

Mechanotransduction I - Cellular and Sub-Cellular Biophysics (joint with JSME) — Maybird

Session Chair: Leo Wan, *Rensselaer Polytechnic Institute, Troy, NY, United States*
Session Co-Chair: James H. Wang, *University of Pittsburgh, Pittsburgh, PA, United States*

1:30PM **TGF-beta and BMP Signaling Pathways Regulate Chromatin Condensation in Mesenchymal Stem Cells in Response to Dynamic Loading** SB³C2015-398 *pg.*505
Su-Jin Heo[1], Woojin M. Han[1,2], Tristan P. Driscoll[1], Dawn M. Elliott[2], Randall L. Duncan[2], Robert L. Mauck[1], [1]*University of Pennsylvania, Philadelphia, PA, United States,*[2]*University of Delaware, Newark, DE, United States*

1:45PM **Temperature Rise Causes Upregulation Of Tenocyte Catabolism And Enhances Gap Junctional Intercellular Communications** SB^{3}C2015-342 *pg.*415
Masataka Tashiro, Eijiro Maeda, **Toshiro Ohashi**, *Hokkaido University, Sapporo, Japan*

2:00PM **Changes In Mechanosensitivity Of Children Cells: An Approach To Investigate Bone Repair And Bone Developmental Disease** SB^{3}C2015-553 *pg.*789
Sara Barreto[1,2,3], Andrew R. Cameron[1,2,3], Dylan Murray[4], Fergal J. O'Brien[1,2,3], [1]*Tissue Engineering Research Group, Royal College of Surgeons in Ireland, Dublin, Ireland,*[2]*Trinity Centre for Bioengineering, Trinity College Dublin, Dublin, Ireland,*[3]*Advanced Materials and Bio-Engineering Research (AMBER) Centre, Ireland, Dublin, Ireland,*[4]*National Paediatric Craniofacial Center, Children's University Hospital, Temple Street, Dublin, Ireland*

2:15PM **Notch1 Mutation Leads to Valvular Calcification Through Enhanced Cadherin-11 Mechanotransduction** SB^{3}C2015-651 *pg.*969
Joseph Chen[1], Larisa Ryzhova[1], M.K. Sewell-Loftin[1], Christopher Brown[1], Stacey Huppert[2], H. Scott Baldwin[1], W. David Merryman[1], [1]*Vanderbilt University, Nashville, TN, United States,*[2]*Cincinnati Children's Hospital, Cincinnati, OH, United States*

2:30PM **Vinculin-network Mediated Cytoskeletal Remodeling Regulates Contractile Function In The Aging Heart** SB^{3}C2015-1057 *pg.*1086
Gaurav Kaushik[1], Anthony Cammarato[2], **Adam J. Engler**[1], [1]*UC San Diego, La Jolla, CA, United States,*[2]*Johns Hopkins University, Baltimore, MD, United States*

2:45PM **Acto myosin Catch Bonds And Mechano-sensitivity In Non-muscle Cells** SB^{3}C2015-1144 *pg.*1243
Franck J. Vernerey, Umut Akalp, *University of Colorado at Boulder, Boulder, CO, United States*

SATURDAY, JUNE 20	1:30pm - 3:00pm

Injury Biomechanics I: Spine, Military, Modeling — Golden Cliff / Eagle's Nest

Session Chair: Liming Voo, *Johns Hopkins University Applied Physics Laboratory, Laurel, MD, United States*
Session Co-Chair: Reuben H. Kraft, *The Pennsylvania State University, State College, PA, United States*

1:30PM **A Point-Wise Normalization Method for Development of Biofidelity Response Corridors** SB^{3}C2015-497 *pg.* 682
Ian Marcus[1], Scott Gayzik[1], Kerry Danelson[1], Jonathan Rupp[2], Cameron Bass[3], Narayan Yoganandan[4], JiangYue Zhang[5], [1]*Wake Forest Univeristy, Winston Salem, NC, United States,*[2]*University of Michigan Transportation Research Institute, Ann Arbor, MI, United States,*[3]*Duke University, Durham, NC, United States,*[4]*Medical College of Wisconsin, Milwaukee, WI, United States,*[5]*Applied Physics Laboratory Johns Hopkins Univeristy, Laurel, MD, United States*

1:45PM **A Comparison Of Brain Injury Predictors Based On Four Benchmark Impact Studies** SB^{3}C2015-479 *pg.*658
Siddiq Qidwai, Nithyanand Kota, Amit Bagchi, *US Naval Research Laboratory, Washington, DC, United States*

2:00PM **Effect Of Geometric And Material Property Changes In The Thoracic Skeleton For An Older Occupant Finite Element Model** SB^{3}C2015-125 *pg.*52
Samantha Schoell[1,2], Ashley Weaver[1,2], Nicholas Vavalle[1,2], Joel Stitzel[1,2], [1]*Virginia Tech- Wake Forest University, Winston-Salem, NC, United States,*[2]*Wake Forest School of Medicine, Winston-Salem, NC, United States*

2:15PM **Collagen Mimetic Peptide as a Marker of Mechanical Damage in Lamb Middle Cerebral Arteries** SB^{3}C2015-1020 *pg.*1018
Raymond G. Walther, Matthew I. Converse, Kenneth L. Monson, *University of Utah, Salt Lake City, UT, United States*

2:30PM **Injuries And Failure Biomechanical Responses Of Artificial Discs In The Cervical Spine: Potential Applications To Military Environments** SB^{3}C2015-519 *pg.*725
Narayan Yoganandan[1], Frank A. Pintar[1], Jamie L. Baisden[1], Joseph B. McEntire[2], Valeta Carol Chancey[2], [1]*Medical College of Wisconsin, Milwaukee, WI, United States,*[2]*U.S. Army Aeromedical Research Laboratory, Fort Rucker, AL, United States*

2:45PM **Investigation Of Possible Correlation Between Brain Tissue Response And Head Kinematics For Blast-induced Brain Injury** SB^{3}C2015-1136 *pg.*1231
Hesam Sarvghad-Moghaddam, Mariusz Ziejewski, Ghodrat Karami, *North Dakota State University, Fargo, ND, United States*

SATURDAY, JUNE 20	1:30pm - 3:00pm

Shoulder Mechanics — Primrose B

Session Chair: Richard Debski, *University of Pittsburgh, Pittsburgh, PA, United States*
Session Co-Chair: Antonis Stylianou, *University of Missouri-Kansas City, Kansas City, MO, United States*

1:30PM **The Effect Of Size And Location Of Tears In The Supraspinatus Tendon On Potential Tear Propagation** SB³C2015-486 *pg 664*
James R. Thunes, Siladitya Pal, R. M. Miller, Richard E. Debski, Spandan Maiti, *University of Pittsburgh, Pittsburgh, PA, United States*

1:45PM **3D Quantification of Osteophyte Distribution on the Humeral Head** SB³C2015-576 *pg.831*
Shea K. Taylor[1,2], Brandon G. Santoni[1,2], Mark A. Frankle[1,3], Peter Simon[1,2], [1]*University of South Florida, Tampa, FL, United States,*[2]*Foundation for Orthopaedic Research and Education, Tampa, FL, United States,*[3]*Florida Orthopaedic Institute, Tampa, FL, United States*

2:00PM **Surgical Accuracy of Traditional Humeral Head Osteotomy in Shoulder Arthroplasty** SB³C2015-146 *pg.84*
Thomas Suter[1,2], **Christopher W. Kolz**[1], Sean T. Tagge[1], Robert Z. Tashjian[1], Ariane Gerber Popp[2], Heath B. Henninger[1], [1]*University of Utah, Salt Lake City, UT, United States,*[2]*Clinic of Orthopaedic Surgery, Kantonsspital Baselland, Liestal, Switzerland*

2:15PM **Validation of a Subject-Specific Computer Model of Glenohumeral Instability and Capsular Plication** SB³C2015-616
Charlie Yongpravat, David Kovacevic, T Sean Lynch, Charles M. Jobin, William N. Levine, Gerard A. Ateshian, Thomas R. Gardner, Christopher S. Ahmad, *Columbia University, New York, NY, United States* *pg.905*

2:30PM **Design Considerations for Glenoid Components: A Computational Stress Analysis of Translational Motion in Shoulder Replacements** SB³C2015-258 *pg.279*
Christopher Berthelet[1], Farzana Ansari[1], Lisa Pruitt[1], Tom Norris[2], Steve Gunther[3], Michael Ries[4], [1]*University of California, Berkeley, Berkeley, CA, United States,*[2]*San Francisco Shoulder, Elbow & Hand Clinic, San Francisco, CA, United States,*[3]*Martha Jefferson Hospital, Charlottesville, VA, United States,*[4]*Tahoe Fracture and Orthopedic Clinic, Carson City, NV, United States*

2:45PM **Damage Analysis Of Metallic And Polymeric Bearings Used In Reverse Total Shoulder Arthroplasty** SB³C2015-192
Suzanne Chou[1], Isabel Yang[1], **Noah Bonnheim**[1], Farzana Ansari[1], Steve Gunther[2], Tom Norris[3], Michael Ries[4], Lisa Pruitt[1], [1]*University of California, Berkeley, Berkeley, CA, United States,*[2]*Martha Jefferson Hospital, Charlottesville, VA, United States,*[3]*San Francisco Shoulder, Elbow & Hand Clinic, San Francisco, CA, United States,*[4]*Tahoe Fracture and Orthopaedic Clinic, Carson City, NV, United States* *pg.168*

SATURDAY, JUNE 20	3:15pm - 4:45pm

Bone Tissue Engineering (joint with JSME) — Primrose A

Session Chair: Shigeo Tanaka, *Kanazawa University, Kanazawa, Japan*
Session Co-Chair: Ryan K. Roeder, *University of Notre Dame, Notre Dame, IN, United States*

3:15PM **Electromagnetic Field Stimulation Enhances Mechanical Properties Of Tissue-engineered Bone Constructed With Calcined Bovine Trabecular Bone Scaffold** SB³C2015-336 *pg. 403*
Shigeo Tanaka, Yuki Yamashita, *Kanazawa University, Kanazawa, Japan*

3:30PM **Development And Characterization Of Novel Bone Substitutes Using Composite Scaffolds With Mesenchymal Stem Cells** SB³C2015-267 *pg. 295*
Mitsugu Todo[1], Phanny Yos[2], [1]*Kyushu University, Kasuga, Japan,*[2]*Institute of Technology of Cambodia, Phnom Penh, Cambodia*

3:45PM **Hydroxyapatite Reinforced Collagen Scaffolds Designed for Improved Architecture, Mechanical Properties, and Tunable Growth Factor Delivery** SB³C2015-174 *pg.134*
Ryan K. Roeder, Matthew J. Meagher, Holly E. Weiss-Bilka, Diane R. Wagner, Robert J. Kane, *University of Notre Dame, Notre Dame, IN, United States*

4:00PM **Microporosity Dominates Growth for Large and Small Macropores in BCP Scaffolds** SB³C2015-670 *pg.*1005
Amy Wagoner Johnson[1], Laurie Rustom[1], David Hoelzle[2], Mark Markel[3], Brett Nemke[3], Yan Lu[3], [1]*University of Illinois at Urbana-Champaign, Urbana, IL, United States,*[2]*University of Notre Dame, Notre Dame, IN, United States,*[3]*University of Wisconsin, Madison, WI, United States*

4:15PM **Fabrication Of Collagen Based 3-d Complex Constructs With Electrochemical Compaction Method** SB³C2015-154
Mousa Younesi[1], Vipuil Kishore[2], **Ozan Akkus**[1], [1]*Case Western Reserve University, Cleveland, OH, United States,*[2]*Florida institute of technology, Melborne, FL, United States* *pg.*98

SATURDAY, JUNE 20	3:15pm - 4:45pm

Spine Mechanics — Superior

Session Chair: Brian D. Stemper, *Medical College of Wisconsin, Milwaukee, WI, United States*
Session Co-Chair: Farid Amiriouch, *University of Illinois, Chicago, IL, United States*

3:15PM **SEM-based Multi-scale Mechanical Modeling of Lumbar Spine Facet Capsular Ligament During Biaxial Extension** SB³C2015-1054 *pg.* 1082
Vahhab Zarei, Amy A. Claeson, Victor H. Barocas, *University of Minnesota, Minneapolis, MN, United States*

3:30PM **Simulation-Directed Design of Planar Biaxial Tests on the Lumbar Facet Capsular Ligament** SB³C2015-168 *pg.*122
Amy A. Claeson, Victor H. Barocas, *University of Minnesota, Minneapolis, MN, United States*

3:45PM **Effect Of Implant Length And Orientation On Biomechanics Of Sacroiliac Joint Stabilization: A Finite Element Analysis** SB³C2015-1141 *pg.*1237
Ali Kiapour[1], Derek Lindsey[2], Scott Yerby[2], Vijay Goel[1], [1]*ECORE, Toledo, OH, United States,*[2]*Si-Bone Inc, San Jose, CA, United States*

4:00PM **Internal Disc Strain Template Highlights Regions of High Local Strain During Compression Loading and Validates a Finite Element Model** SB³C2015-1019 *pg.*1016
Brent L. Showalter[1], John F. DeLucca[2], John M. Peloquin[1], Daniel H. Cortes[2], Alexander C. Wright[1], James C. Gee[1], Edward J. Vresilovic[3], Dawn M. Elliott[2], [1]*University of Pennsylvania, Philadelphia, PA, United States,*[2]*University of Delaware, Newark, DE, United States,*[3]*Pennsylvania State University, Hershey, PA, United States*

4:15PM **Accuracy of QCT-Based Finite Element Predictions of Vertebral Fracture When Boundary Conditions are Based on Intradiscal Pressure Profiles** SB³C2015-313 *pg.*369
Timothy Jackman, Alexander DelMonaco, Elise Morgan, *Boston University, Boston, MA, United States*

4:30PM **Differences In The Intradiscal Pressure In The L4-L5 And L5-S1 Lumbar Segments** SB³C2015-196 *pg.* 176
Hector E. Jaramillo[1], Christian M. Puttlitz[2], Jose J. Garcia[3], Kirk McGilvray[2], [1]*Universidad Autonoma de Occidente, Cali, Colombia,*[2]*Colorado State University, Fort Collins, CO, United States,*[3]*Universidad del Valle, Cali, Colombia*

SATURDAY, JUNE 20	3:15pm - 4:45pm

Biological Flows in the Interstitium and Lymphatics — Wasatch

Session Chair: Brandon Dixon, *Georgia Institute of Technology, Atlanta, GA, United States*
Session Co-Chair: Walter Lee Murfee, *Tulane University, New Orleans, LA,United States*

3:15PM **A Poroelastic Fluid/Structure-Interaction Model of the Spinal Cord and Surrounding Structures with a Cord Syrinx and Associated Stenosis of the Subarachnoid Space** SB³C2015-197 *pg.*178
Christopher D. Bertram[1], Matthias Heil[2], [1]*University of Sydney, New South Wales, Australia,*[2]*University of Manchester, Manchester, United Kingdom*

3:30PM **Neural Tissue Deformation And Cerebrospinal Fluid Flow Impedance Are Positevely Correlated At The Craniocervical Junction** SB³C2015-634 *pg.*937
Bryn A. Martin[1], Nicholas Shaffer[1], John N. Oshinski[2], Mark Luciano[3], Francis Loth[1], [1]*The University of Akron, Akron, OH, United States,*[2]*Emory University, Atlanta, GA, United States,*[3]*Cleveland Clinic Foundation, Cleveland, OH, United States*

3:45PM **In Silico and In Vitro Modelling of Flow Behaviour in Lymphatic Vessels.** SB³C2015-222 *pg.220*
Sinéad T. Morley[1,2,3], David T. Newport[1,4], Michael T. Walsh[1,2,3], *[1]University of Limerick, Limerick, Ireland,[2]Centre for Applied Biomedical Engineering Research, Limerick, Ireland,[3]Materials and Surface Science Institute, Limerick, Ireland,[4]Stokes Institute, Limerick, Ireland*

4:00PM **Quantification of Lymphatic Valve Resistance to Forward Flow Using Computational Fluid and Solid Modeling** SB³C2015-560 *pg.801*
John T. Wilson[1], Raoul van Loon[2], James E. Moore[1], *[1]Imperial College London, London, United Kingdom,[2]Swansea University, Swansea, United Kingdom*

4:15PM **A New Paradigm for the Contribution of Active Tension to the Constitutive Relation in Small Lymphatic Vessels** SB³C2015-198 *pg.180*
Christopher D. Bertram[1], Charlie Macaskill[1], Michael J. Davis[2], James E. Moore[3], *[1]University of Sydney, New South Wales, Australia,[2]University of Missouri School of Medicine, Columbia, MO, United States,[3]Imperial College, London, United Kingdom*

4:30PM **Spatiotemporal Image Correlation Spectroscopy Techniques for Quantifying Fluid Flow in Microfluidic Channels and Porous Tissues** SB³C2015-472 *pg.644*
Brian T. Graham, Christopher Price, *University of Delaware, Newark, DE, United States*

SATURDAY, JUNE 20	3:15pm - 4:45pm

Micromechanics of Atherosclerosis — Magpie

Session Chair: Dalin Tang, *Worcester Polytechnic Institute, Worcester, MA, United States*
Session Co-Chair: Susan Lessner, *University of South Carolina, Columbia, SC,United States*

3:15PM **On the Effect of Calcific Content on the Mechanical Behaviour of Carotid Plaque Tissue** SB³C2015-359 *pg.437*
Hilary E. Barrett, *University of Limerick, Limerick, Ireland*

3:30PM **Intima Heterogeneity In Atherosclerotic Plaque Stress Calculations** SB³C2015-451 *pg. 610*
Lambert Speelman[1], Bas Van Velzen[2], Anton F. W. Van der Steen[1,2], Jolanda J. Wentzel[1], Frank J. H. Gijsen[1], *[1]Erasmus MC, Rotterdam, Netherlands,[2]Delft University of Technology, Delft, Netherlands*

3:45PM **Diet-induced Vascular Remodeling Produces a Shift in Collagen Fiber Angle Distribution in a Mouse Model of Atherosclerosis** SB³C2015-508 *pg.702*
Shana R. Watson, Piaomu Liu, Edsel A. Pena, Michael A. Sutton, John F. Eberth, Susan M. Lessner, *University of South Carolina, Columbia, SC, United States*

4:00PM **Towards the Development of an Atherosclerotic Plaque Stratification Parameter to Predict Arterial Restenotic Response following Endovascular Treatment** SB³C2015-379 *pg.469*
Eoghan M. Cunnane[1], Hilary E. Barrett[1], Eamon G. Kavanagh[2], Michael T. Walsh[1], *[1]University of Limerick, Limerick, Ireland,[2]University Hospital Limerick, Limerick, Ireland*

4:15PM **Reconstruction of Incomplete Lipid Pool Geometry for Stress Calculations in Atherosclerotic Arteries** SB³C2015-362
Annette M. Kok, Lambert Speelman, Frank J. H. Gijsen, Jolanda J. Wentzel, *Erasmus MC, Rotterdam, Netherlands* *pg.443*

4:30PM **MMP-2 Expression and the Tissue Mechanics of Human Ascending Thoracic Aortic Aneurysms** SB³C2015-405
Alexander A. Emmott[1,2], Nastaran Shahmansouri[1,2], Mohammed Alreshidan[3], Stefanie Pohlod[1], Rosaire Mongrain[1], Raymond Cartier[2], Kevin Lachapelle[3], Richard Leask[1,2], *[1]McGill University, Montreal, QC, Canada,[2]Montreal Heart Institute, Montreal, QC, Canada,[3]Royal Victoria Hospital, Montreal, QC, Canada* *pg.519*

SATURDAY, JUNE 20	3:15pm - 4:45pm

Mechanotransduction II - Interactions between Cells and Their Environment (joint with JSME) — Maybird

Session Chair: Toshiro Ohashi, *Hokkaido University, Sapporo, Japan*
Session Co-Chair: Chelsey Simmons, *University of Florida, Gainesville, FL, United States*

3:15PM **A Microtissue Array Device To Screen The Lung Fibrogenic Potential Of Carbon Nanotubes** SB^{3}C2015-350 *pg.* 427
Zhaowei Chen, Qixin Wang, Mohammadnabi Asmani, Yan Li, Yun Wu, **Ruogang Zhao**, *State University of New York at Buffalo, Buffalo, NY, United States*

3:30PM **The Impact of Pre-stretch Induced Surface Anisotropy on Axonal Regeneration** SB^{3}C2015-464 *pg.*630
Chun Liu, Seungik Baek, **Christina Chan**, *Michigan State University, East Lansing, MI, United States*

3:45PM **Excessive Mechanical Loading Causes Aberrant Differentiation Of Tendon Stem Cells (TSCs) That Leads To The Development Of Degenerative Tendinopathy** SB^{3}C2015-227 *pg.* 230
Jianying Zhang, **James H-C. Wang**, *University of Pittsburgh, Pittsburgh, PA, United States*

4:00PM **Both Nuclear And Extracellular Matrix Rigidity Determine The Stable Size Of Focal Adhesion Plaques** SB^{3}C2015-228
Xuan Cao, Tristan P. Driscoll, Robert L. Mauck, Vivek B. Shenoy, *University of Pennsylvania, Philadelphia, PA, United States* *pg.232*

4:15PM **Role Of Cell Tension On Anisotropic Mechanosensing** SB^{3}C2015-1106 *pg.*1175
Shin Min Wen, Pen Hsiu Grace Chao, *National Taiwan university, TAIPEI, Taiwan*

4:30PM **Osteocyte Mechanotransduction During Low Magnitude Mechanical Stimulation** SB^{3}C2015-1121 *pg.*1205
Thomas R. Coughlin, Tyler C. Kreipke, Glen L. Niebur, *University of Notre Dame, Notre Dame, IN, United States*

SATURDAY, JUNE 20	3:15pm - 4:45pm

Injury Biomechanics II - Head to Foot, Modeling, Risk — Golden Cliff / Eagle's Nest

Session Chair: Steven Rowson, *Virginia Tech, Blacksburg, VA, United States*
Session Co-Chair: Ken Monson, *University of Utah, Salt Lake City, UT, United States*

3:15PM **The Effect of Grade II and Grade III Ankle Injury on the Ankle Joint Complex Kinematics and Achilles Load: A Cadaveric Study** SB^{3}C2015-529 *pg.*743
Bardiya Akhbari, Matthew H. Dickinson, Ednah G. Louie, Sami Shalhoub, Lorin P. Maletsky, *University of Kansas, Lawrence, KS, United States*

3:30PM **Driver Injury Risk Sensitivity in Finite Element Model Reconstructions of Real World Motor Vehicle Crashes** SB^{3}C2015-602 *pg.* 879
James P. Gaewsky, Ashley A. Weaver, Bharath Koya, Joel D. Stitzel, *Wake Forest University, Winston-Salem, NC, United States*

3:45PM **Head-neck Relative Posture Affects Fracture Outcome of the Basilar Skull and Upper Cervical Spine under Helmeted Head Crown Impact** SB^{3}C2015-571 *pg.*821
Liming Voo, Kyle Ott, Christopher Dooley, Andrew Merkle, *Johns Hopkins University Applied Physics Laboratory, Laurel, MD, United States*

4:00PM **Investigation of Head Rotational Impulse Characteristics on Brain Strain Responses** SB^{3}C2015-39 *pg.*22
Wei Zhao, Songbai Ji, *Dartmouth College, Hanover, NH, United States*

4:15PM **Development of a 5th Percentile Female Finite Element Model Using a Multi-Modality Image Dataset** SB^{3}C2015-1038
Matthew L. Davis, Bharath Koya, Jeremy M. Schap, F. Scott Gayzik, *Virginia Tech-Wake Forest University, Winston Salem, NC, United States* *pg. 1050*

4:30PM **Embedded Finite Elements For Modeling Traumatic Axonal Injury** SB^{3}C2015-1117 *pg.*1197
Reuben H. Kraft, Harsha T. Garimella, *The Pennsylvania State University, University Park, PA, United States*

SATURDAY, JUNE 20	3:15pm - 4:45pm

Lower Extremity Mechanics — Primrose B

Session Chair: Lorin Maletsky, *University of Kansas, Lawrence, KS, United States*
Session Co-Chair: Ferris Pfeiffer, *University of Missouri, Columbia, MO, United States*

3:15PM **Understanding the Mechanics of Focal Chondral Defects in the Hip: a Framework to Advance Treatment Options** SB³C2015-596 *pg.*867
Brenden J. Klennert, Benjamin J. Ellis, Travis G. Maak, Ashley Kapron, Tyler O. Kaiser, Jeffrey A. Weiss, *University of Utah, Salt Lake City, UT, United States*

3:30PM **Specimen-Specific Evaluation of a Multi-Body Model Predicts Instability and Increased Meniscal Load in the Anterior Cruciate Ligament-Deficient Knee** SB³C2015-375 *pg.*465
Mohammad Kia, Kevin Schafer, Daniel Green, Andrew Pearle, Thomas Wickiewicz, Timothy Wright, Carl Imhauser, *Hospital for Special Surgery, New York, NY, United States*

3:45PM **In-Vivo Kinematics of the Asymptomatic Hip during Dynamic Pivoting: Foundations for the Evaluation of Femoroacetabular Impingement** SB³C2015-306 *pg.*359
Penny R. Atkins, Niccolo M. Fiorentino, Michael J. Kutschke, Sara J. Fauver, Ashley L. Kapron, Christopher L. Peters, Stephen K. Aoki, Andrew E. Anderson, *University of Utah, Salt Lake City, UT, United States*

4:00PM **Open Knee(s): Magnetic Resonance Imaging for Specimen-specific Next Generation Knee Models** SB³C2015-581
Craig Bennetts[1], Snehal Chokhandre[1], Shannon Donnola[2], Chris Flask[2], Tara Bonner[1], Robb Colbrunn[1], Ahmet Erdemir[1], [1]*Cleveland Clinic, Cleveland, OH, United States,*[2]*Case Western Reserve University, Cleveland, OH, United States* *pg.841*

4:15PM **A New Joint Coordinate System for Robotic Testing of Cadaveric Knee Specimens** SB³C2015-554 *pg.*791
Daniel Boguszewski, Nirav Joshi, Edward Cheung, Paul Yang, Keith Markolf, David McAllister, *UCLA, Los Angeles, CA, United States*

4:30PM **Do The Laxities Of The Normal Knee At 0° And 90° Of Flexion Support The Goal Of Gap-balancing A Total Knee Arthroplasty?** SB³C2015-30 *pg.*12
Joshua D. Roth, Stephen M. Howell, Maury L. Hull, *University of California, Davis, Davis, CA, United States*

SATURDAY, JUNE 20	5:00pm - 6:00pm

LISSNER LECTURE – James A. Ashton-Miller — Ballrooms 1-3

DYNAMIC BALANCE USING THE COM AND COP INCLINATION ANGLE DURING A GOLF SWING

Ahnryul Choi (1), Joung Hwan Mun (2)

(1) Department of Internal Medicine
The University of Texas
Health Science Center at Houston,
Houston, TX, USA

(2) Department of Bio-mechatronic Engineering
Sungkyunkwan University
Suwon, Republic of Korea

INTRODUCTION

Golf requires complex movement with fine motor control, and requires proper dynamic balance in order to accurately control the club head through a harmonious coordination of each segment and joint in the body. Traditionally, the training for a balance ability is part of the rehabilitation program for injuries in the joints of lower extremity [1]. Since the injury mechanisms differ between skilled and unskilled golfers, a quantitative evaluation of the ability for balance with respect to the skill level should provide further insight into golf swing-specific injuries as well as into rehabilitation strategies [2].

Dynamic balance is quantified and evaluated by utilizing the center of mass (COM), the center of pressure (COP), and the separation distance between COM and COP [3]. However, these data should be normalized according to leg length in order to make a comparison between groups since the magnitudes of these variables are affected by the subject's stature. Recently, Lee et al. proposed a COM-COP inclination angle (IA) method that removes the influence that the difference in height has in the gait of different subjects [4]. Also, the rate of change for COM-COP IA provides further insight to evaluate the ability for dynamic balance since it is difficult to recover the COM within the base of support without maintaining a proper COM-COP inclination angular velocity (IAV) [5].

Although many studies have been conducted to evaluate the relationship between sports performance and the ability to maintain balance, it has not yet been clearly understood whether the ability for dynamic balance has a direct effect on the performance of a golf swing. Therefore, the purpose of this study is to evaluate the ability for dynamic balance during a golf swing according to skill level by using the COM-COP IA and COM-COP IAV. It was hypothesized that skilled golfers would exhibit better ability for dynamic balance with proper postural control compared to amateur golfers.

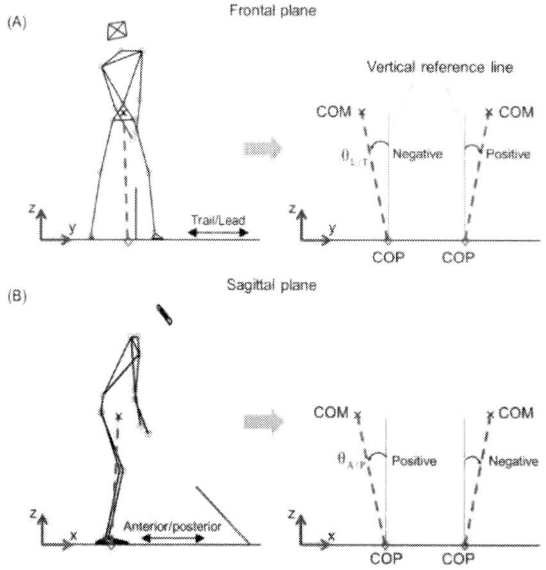

Fig. 1. Configuration of frontal (A) and sagittal (B) COM-COP IA

METHODS

Participants and experimental setup

Twelve professional golfers were recruited from the Korea Professional Golf Association, and 23 amateur golfers were recruited and then further divided into two groups, advanced and novice, according to their Korea Golf Association handicap score. The advanced group included golfers with handicaps ranging from 1-14, and the novice group was composed of golfers with handicap scores greater than or equal to 15.

Six infrared cameras captured the swing motion of each participant at a sampling rate of 120Hz. The ground reaction force data were collected at 1080Hz by two force platforms, and the timing of the swing motion trajectories obtained from the cameras and the ground reaction force data were synchronized by suing a VICON 460 system. The SB-Clinic software (SWINGBANK Ltd, Korea) was used to extract the raw data.

Data reduction

Thirty-five reflective optical markers were placed on the bony landmarks of the golfers (modified Helen Hayes markerset). Five trials were conducted for each participant, and the average of the two best shots was used in the analysis. Each marker trajectory was filtered utilizing a 4th order, zero phase-lag, low-pass Butterworth filter, and the cut-off frequency was set to 10 Hz. The swing was divided into four events (address, backswing top, impact and finish) and three phases (backswing: BS, downswing: DS and follow-through: FT) according to the trajectories of the marker attached to the clubhead.

The net COM for the entire body was calculated by using the kinematic method. This method requires anthropometric model and full kinematic descriptions. The net COP of the entire body was calculated as the weighted sum of the trajectories for each COP under the foot, which accounted for the proportion of vertical ground reaction force for each limb. The COM-COP IA in the lead/trail (L/T) and in the forward/backward (F/B) direction were determined to be the instantaneous orientation of the line connecting the COM and COP with respect to the vertical line through the COP (Fig. 1). The COM-COP IAV, as the rates of changes for COM-COP IA, is calculated by using the central difference method.

Statistical analysis

We applied the Kruskal-Wallis H test, a non-parametric method, to make a comparison between the groups. A pairwise Tukey's-HSD post hoc test was performed on each variable, and the significance level was set to $p<0.05$.

RESULTS

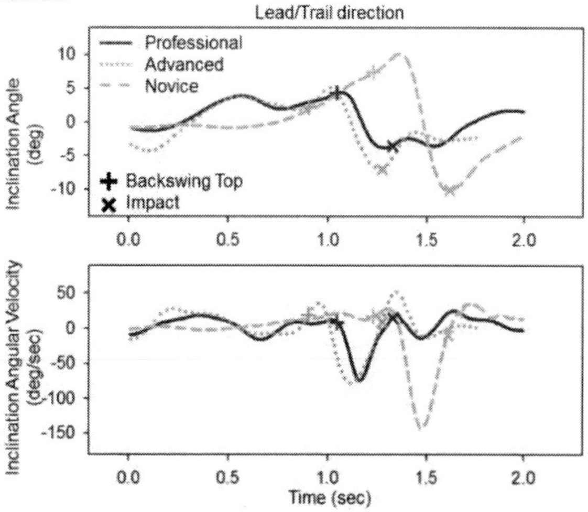

Fig. 2. Patterns of COM-COP IAs and IAVs during a complete golf swing cycle in representative participants of each group

In the L/T direction, the COM-COP IA increased to a positive value for all three participants during the BS phase, and a remarkable change into a negative value was observed during the DS phase. The maximum and minimum COM-COP IA values for all three participants appeared in the DS phase, and the maximum and minimum values occur right after the backswing top and near the impact event, respectively. Fluctuations in the pattern appeared during the BS and FT phases for professional and for advanced golfers, but none such patterns were observed in the novice golfers. The minimum value for COM-COP IAV was observed in the DS phase for the three participants, and a significantly lower velocity was observed for the novice golfer relative to the professional and the advanced golfers.

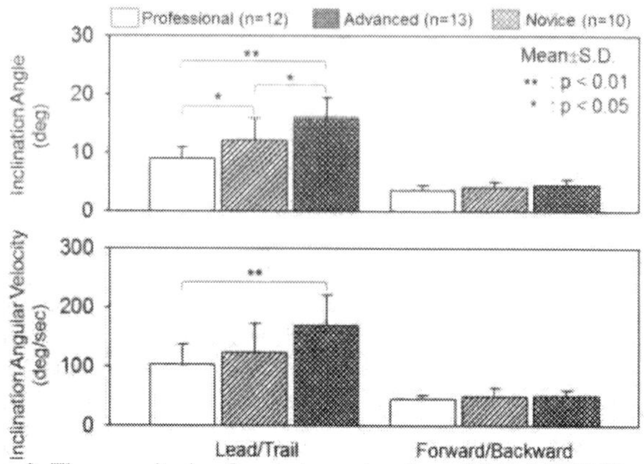

Fig. 3. The magnitude of peak-to-peak value of COM-COP IA and IAV in a complete golf swing cycle

For COM-COP IA, the professional group showed the smallest angular displacement within 10 degrees in the L/T direction, and all the three groups showed a statistically significant difference (Fig. 3). The peak-to-peak change in the COM-COP IAV in the L/T direction appeared in an increasing order of professional, advanced, and novice groups, and a significant difference was observed between the professional and novice groups ($p<0.01$). However, there was no statistical difference in both the COM-COP IA and COM-COP IAV in the F/B direction.

DISCUSSION AND CONCLUSIONS

Skilled golfers tend to exhibit improved dynamic balance, and this can be attributed to the neuromusculoskeletal system that maintains balance with proper postural control through repetitive swing training. In addition, COM-COP IA and COM-COP IAV are valuable parameters that can be used to evaluate dynamic balance. This study has the potential to allow for an evaluation of the dynamic balance mechanism and will provide useful basic information for swing training and prevention of golf injuries.

ACKNOWLEDGEMENTS

This research was supported by the Basic Science Research Program through the National Research Foundation of Korea (NRF) funded by the Ministry of Science, ICT and Future Planning (NRF-2013R1A2A2A03068269).

REFERENCES

[1] Hrysomallis, C., *Sports Med*, 37:547-556, 2007.
[2] Zheng, N. et al., *Int J Sports Med*, 29:487-493, 2008.
[3] Mackinnon, C.D. et al., *J Biomech*, 26:633-644, 1993.
[4] Lee, H.J. et al., *Arch Phys Med Rehabil*, 87:569-575, 2006.
[5] Chien H.L. et al., *Gait Posture*, 38:391-396, 2013.

IMPROVED DETECTION OF DYNAMIC BALANCE USING NORMALIZED COM-COP INCLINATION ANGULAR JERK DURING A GOLF SWING

Ahnryul Choi (1), Joung Hwan Mun (2)

(1) Department of Internal Medicine
The University of Texas
Health Science Center at Houston,
Houston, TX, USA

(2) Department of Bio-mechatronic Engineering
Sungkyunkwan University
Suwon, Republic of Korea

INTRODUCTION

Balance is the ability to obtain a state of equilibrium by adjusting the displacement of the center of mass (COM) within the center of pressure (COP) or base of support of the body [1]. A balanced posture is retained through continuous feedback between visual, vestibular, and somatosensory inputs; accordingly, harmonious neuromuscular behaviors follow [2]. Specifically, dynamic balance is regarded as the ability to maintain posture on an unstable surface with minimal movement or to perform a certain task while retaining a stable posture. Dynamic balance ability is quantified by the COM-COP inclination angle (IA) method that removes the effect of participants' stature [3].

Golf is a complex activity requiring dynamic balance ability with proper motor control. During a golf swing, the weight on both feet predominantly shifts to one leg while the upper extremity rotates, and this movement requires a high level of balance recovery ability. Balance control ability has been revealed to have a close relationship with sports performance and also with injury of the lower extremity joints such as the ankle and knee [4]. Since the injury mechanisms and swing characteristics differ between skilled and unskilled golfers, a quantitative evaluation of balance ability in accordance with skill level should provide further insight into golf swing-specific injury and rehabilitation as well as appropriate coaching programs to improve performance [5].

The smoothness in movement for COM within COP might provide useful information for the evaluation of dynamic balance ability. The enhanced co-contraction of the agonist-antagonist muscle group around the human joints increases joint stiffness and allows joints to be stabilized against perturbations, leading to an improved ability for balance [6]. A temporally coordinated execution due to proper co-contraction of the muscles is considered to be a direct cause for smooth movement [7]. Therefore, a close relationship between movement smoothness and dynamic balance ability can be considered to exist in terms of efficient muscular coordination. The smoothness of movement can be quantified as a jerk cost, defined as the rate of change of the acceleration over time [8].

The purpose of this study is to evaluate the ability for dynamic balance during a golf swing according to skill level by using COM-COP inclination angular jerk cost variables.

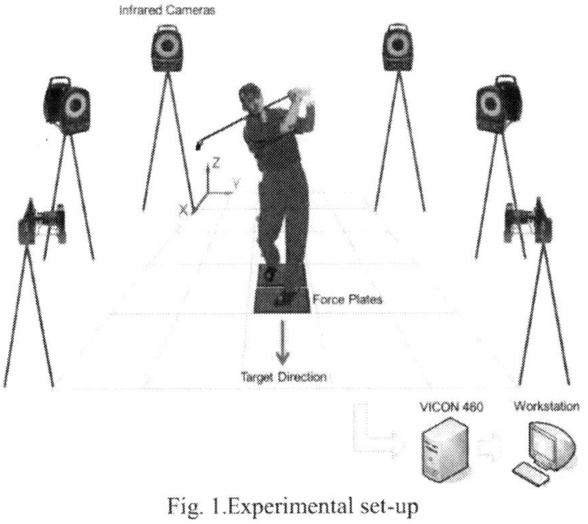

Fig. 1.Experimental set-up

METHODS

Participants and apparatus

Twelve professional golfers (handicap score < 0) were served from the Korea Professional Golf Association, and 23 amateur golfers were recruited and then further divided into two groups, advanced and novice, according to their Korea Golf Association handicap score. The advanced group included golfers with handicaps ranging from 1-14, and the novice group was composed of golfers with handicap scores greater than or equal to 15.

Six infrared cameras captured the swing motion of each participant at a sampling rate of 120Hz. The ground reaction force data were collected at 1080Hz by two OR6-6-2000 force platforms, and the timing of the swing motion trajectories obtained from the cameras and the ground reaction force data were synchronized by suing a VICON 460 system. The SB-Clinic software (SWINGBANK Ltd, Korea) was used to extract the raw data.

Data processing

Thirty-five reflective optical markers were placed on the bony landmarks of the golfers, and the location of the markers was based on the modified Helen Hayes markerset protocol. Five trials were conducted for each participant, and the average of the two best shots, as determined by the response of the participant and the quality of the reconstructed three-dimensional data, was used in the analysis. Each marker trajectory was filtered utilizing a 4th order, zero phase-lag, low-pass Butterworth filter, and the cut-off frequency was set to 10 Hz.

The net COM for the entire body was calculated by using the kinematic method. The net COP of the entire body was calculated as the weighted sum of the trajectories for each COP under the foot, which accounted for the proportion of vertical ground reaction force for each limb. The COM-COP IA in the lead/trail (L/T) and in the forward/backward (F/B) direction was determined to be the instantaneous orientation of the line connecting the COM and COP with respect to the vertical line through the COP. The COM-COP inclination angular jerk (IAJ) was calculated by differentiating the COM-COP IA 3 times in the time domain. In order to remove the effects of the duration and angle of movement, the COM-COP IAJ was normalized by using the following equation.

$$\text{Normalized COM-COP IAJ} = = \sqrt{\frac{1}{2}\int j(t)^2 \times \frac{duration^5}{length^2}\, dt}$$

The normalized COM-COP IAJ in the swing major phases (Backswing: BS, downswing: DS, and follow-through: FT) were computed.

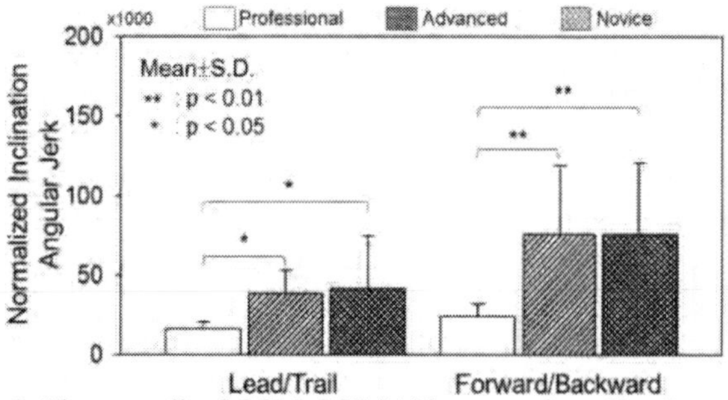

Fig. 2. The normalized COM-COP IAJ in a complet golf swing cycle

RESULTS

Fig. 2 demonstrated the normalized COM-COP IAJ in each direction during a complete golf swing cycle. The professional group showed a lower value than the other two groups in the L/T and F/B directions, with a very small standard deviation. The advanced and novice group did not exhibit a statistically significant difference and also had a larger standard deviation than the professional group.

Table 1 lists the detailed results for the normalized COM-COP IAJ for the major swing phases. The professional group exhibited a statistically lower value than the other two groups in all directions during the DS phase (p<0.05).

Table 1. Normalized COM-COP IAJ in the major swing phases (Mean±SD).

(×1000)		Professional	Advanced	Novice
BS	L/T	2.50±0.9	7.44±6.4	7.21±6.6
	F/B	4.42±2.6	24.29±19.3*	27.19±22.0§
DS	L/T	0.33±0.1	0.62±0.3*	0.51±0.2§
	F/B	0.29±0.1	1.85±1.8*	1.35±0.9§
FT	L/T	2.66±1.4	6.84±4.5*	5.25±4.9
	F/B	4.07±2.1	7.11±6.1	7.38±6.6

L/T: Lead/trail, F/B: Forward/backward direction
*p<0.05 professional vs. Advanced group
§p<0.05 Professional vs. Novice group

DISCUSSION

The professional group exhibited a higher level of coordination in movement through the appropriate muscular co-contraction during a golf swing. A temporally coordinated execution can result in rapid, accurate, consistent, and smooth movement. Specifically, there was significant difference between the groups in all directions (Table 1), and professional golfers were observed to have a higher level of temporal coordination between segments, between joints, and between muscle contractions. The temporal coordination for the body segments and the joints appeared clearly in the rotational [9] and translational [10] movement of skilled golfers during the DS phase, which provides a clue for the close relationship between smooth movement and an improved balance due to better neuromuscular coordination.

Superior dynamic balance ability plays an important role in the execution of a golf swing. The repetition that results from practicing golf swings leads to swing-specific neuromuscular adaptation and stabilized the joints and their neighboring tissues. In addition, the normalized COM-COP IAJ is valuable parameters that can be used to evaluate dynamic balance, and there is a close relationship between the smoothness of movement and the dynamic balance. More skilled golfers were observed to have a more smooth swing strategy, leading to a more stable swing with better dynamic balance.

ACKNOWLEDGEMENTS

This research was supported by the Basic Science Research Program through the National Research Foundation of Korea (NRF) funded by the Ministry of Science, ICT and Future Planning (NRF-2013R1A2A2A03068269).

REFERENCES

[1] Winter, D.A., *Gait Posture*, 3:193-214, 1995.
[2] Hrysomallis, C., *Sports Med*, 37:547-556, 2007.
[3] Lee, H.J., *Arch Phys Med Rehabil*, 87:569-575, 2006.
[4] Tropp, H. et al., *Med Sci Sports Exer*, 16:64-66, 1984.
[5] Zheng, N. et al., *Int J Sports Med*, 29:487-493, 2008.
[6] Lloyd, D.G., *J Orthop Sports Phys Ther*, 31:645-654, 2001.
[7] Yan, J.H. et al, *J Appl Biomech*, 16:196-203, 2000.
[8] Choi, A. et al., *Biomed Eng Online*, 13:20, 2014.
[9] Horan, S.A. et al., *Sports Biomech*, 11:165-174, 2012.
[10] Beak, S.H. et al., *Biomed Eng Online*, 12:13, 2013.

THE EFFECT OF IMPACT AND GENIPIN CROSSLINKING ON THE FRICTION AND WEAR OF ARTICULAR CARTILAGE

C.M. Bonitsky (1), M. Selep (1), M.E. McGann (1), T.C. Ovaert (1), S.B. Trippel (2), D.R. Wagner (1)

(1) Aerospace and Mechanical Engineering
University of Notre Dame
Notre Dame, Indiana, USA

(2) Department of Orthopaedic Surgery
Indiana University
Indianapolis, Indiana, USA

INTRODUCTION

High energy trauma to a synovial joint causes an array of mechanical, cellular, and biochemical responses that can ultimately lead to posttraumatic osteoarthritis (PTOA). Mechanical damage to the cartilage, such as surface fissures [1] and microstructural damage [2], have been observed at impact sites immediately following trauma. However, the degree to which this damage renders the tissue more susceptible to wear and contributes to the progression of PTOA is unknown. Additionally, no treatments are currently available to strengthen cartilage after joint trauma and protect the tissue from subsequent wear. Our previous work has investigated collagen crosslinking of cartilage using genipin as a potential therapeutic treatment [3]. We showed that genipin crosslinking improves the wear resistance of healthy, intact cartilage *in vitro* and significantly protects the tissue from collagenase digestion at doses that are non-toxic to chondrocytes [3]. The purposes of this study were to investigate the role of mechanical damage in cartilage wear and friction after an impact, and to determine whether genipin crosslinking is an effective treatment to improve the wear of impacted cartilage.

METHODS

Cartilage/bone specimens from thawed bovine stifles were impacted via a drop tower instrumented with a load cell and accelerometer, using a 3.2 cm spherical impact head from a height of 25 cm with a total mass of 499 g. Alcian blue staining was performed to investigate glycosaminoglycan (GAG) levels post-impact. Immunohistochemistry was performed to assess denatured collagen (COL2-3/4M) and collagenase cleavage (COL2-3/4C_{short}). Crosslinked and control specimens were incubated post-impact for 15 min in 0, 2, or 10 mM genipin (Challenge Bioproducts, Taiwan) solutions in PBS at 37 °C, then were transferred to genipin-free PBS at 37 °C to bring the total time to 24 h. Stress-relaxation tests were performed via indentation initially, then again after impact, and finally after crosslinking with a Hysitron TI950 TriboIndenter (Minneapolis, MN) equipped with a 3-D OmniProbe® transducer and a 750 µm diameter flat punch probe. The indentation protocol, which had previously been shown to produce repeatable stiffness measurements of cartilage that were sensitive to changes due to crosslinking [4], consisted of a 20-sec loading period, a 50-sec hold at a peak total displacement of 70 µm, and a 1-sec unloading period; three separate locations on each of four specimens were indented. The unloading stiffness was calculated for each indentation, and a standard linear solid model was fit to the holding phase of the stress-relaxation tests (n = 12) [5]. Impacted/crosslinked specimens and control specimens underwent reciprocating pin-on-disk friction tests for 30 minutes using a Universal Micro-Tribometer (Bruker, Campbell, CA); the initial coefficient of friction was reported (n = 4). Separate specimens underwent identical wear tests for 48 hours in a square pattern (n = 6) against stainless steel discs using an Orthopod (AMTI, Watertown, MA) as previously described [6]. To visualize damage from the wear test, surfaces were stained with india ink. To quantify the amount of collagen released to the hydrating fluid during the wear test of each specimen, a hydroxyproline assay was performed. Differences were determined using a one-way ANOVA with Tukey's post-hoc test and significance set at $p<0.05$.

RESULTS

The impact energy from the drop tower was 0.89 ± 0.30 J and resulted in shallow fissures at the articular surface (Fig. 1A). Alcian blue staining of impacted cartilage revealed that a large amount of GAG's were retained by the cartilage post-impact (Fig. 1A). Immunohistochemistry indicated that collagen became denatured under the impact (Fig. 1B), which was not due to collagenase cleavage (data not shown). The cartilage unloading stiffness decreased by 40.2% after impact ($p<0.001$) and was not significantly different after crosslinking in either genipin concentration. Similarly, equilibrium modulus from the standard linear solid model decreased 42.4% upon impact, and did not significantly change with crosslinking. The initial modulus from the standard linear solid model also decreased after impact, was not significantly changed with 2 mM crosslinking, but increased with the 10 mM treatment (data not shown). The initial coefficient of friction rose by 88% after impact and was not significantly changed by crosslinking. India ink staining suggested that impacted specimens with 0 mM genipin treatment sustained the most wear while the fissures due to impact did not noticeably worsen in the specimens that had been crosslinked in 2 and 10 mM genipin (Fig. 2A). The wear quantification confirmed these results, indicating that the impacted specimens released significantly more collagen to the hydrating fluid during the wear test than those that had not been impacted, and that the 2 and 10 mM genipin crosslinking treatments reduced the wear of the impacted specimens to levels that were comparable to that of the non-impacted controls (Fig. 2B).

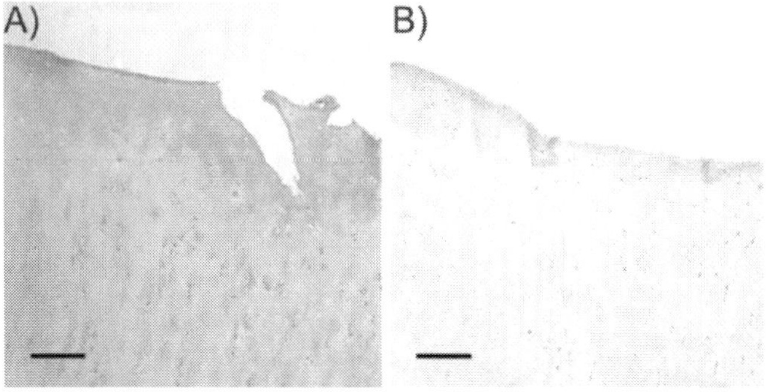

FIGURE 1: A) ALCIAN BLUE STAINED ARTICULAR CARTILAGE AFTER IMPACTING THEN INCUBATING IN PBS AT 37°C FOR 24 HRS. B) IMMUNOHISTOCHEMISTRY OF SURFACE DAMAGE DUE TO A SINGLE IMPACT SHOWING DENATURED COLLAGEN (COL2-3/4M). BAR REPRESENTS 200 µm.

DISCUSSION

This study investigated a novel method to improve articular cartilage wear resistance after a traumatic injury. We used genipin, a natural plant extract, to crosslink the collagen network in a clinically relevant 15 minute time span. The impact protocol caused fissures and microstructural damage at the articular surface and decreased the cartilage stiffness via indentation testing. Although we have previously shown that our crosslinking protocol in 2 and 10 mM genipin increases the unloading stiffness of intact cartilage by 16 and 64%, respectively [3], no significant increase in unloading stiffness was observed when the impacted specimens were crosslinked with these same treatments. The results suggest that the impact damage dominates the indentation stiffness measurement, perhaps due to the rupture of collagen fibers in the superficial zone. Immunohistochemistry staining demonstrated collagen network denaturation and microarchitectural damage in the superficial zone, while Alcian blue straining revealed that a large amount of GAGs were retained by the cartilage post-impact. Microarchitectural damage coupled with GAG retention may cause the swelling and increased water content that has been reported in impacted cartilage [7] due to the decreased ability of the damaged collagen network to restrict proteoglycan expansion through the absorption of water. The microarchitectural damage and subsequent swelling may also be observed in the decreased moduli of the cartilage post-impact and may lead to decreased fluid pressurization and increased cartilage friction [8], as was observed in the current study. The wear resistance of impacted articular cartilage that had not been crosslinked was less than that of undamaged cartilage in the *in vitro* wear test, suggesting that mechanical damage that is immediately induced by impact loading may directly contribute to the progression of PTOA. The wear resistance was improved by the crosslinking treatments, and was about equal with the 2 and 10 mM genipin protocols. As previous studies have demonstrated that the 2 mM genipin crosslinking treatment is non-toxic to chondrocytes [3], it may be appropriate to adapt this protocol for clinical use to prevent accelerated wear after a patient experiences injurious joint trauma.

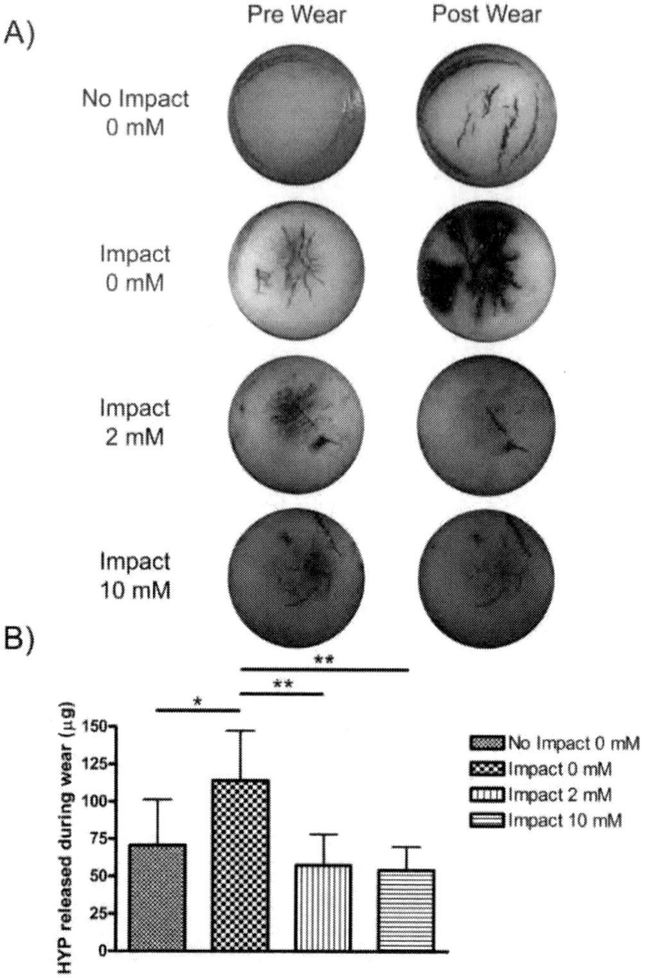

FIGURE 2: A) INDIA INK STAINING BEFORE AND AFTER WEAR TESTING. B) EFFECT OF IMPACT AND GENIPIN CROSSLINKING ON HYDROXYPROLINE (HYP) RELEASED TO THE HYDRATING FLUID DURING WEAR TESTING (*:p<0.05, **:p<0.01).

ACKNOWLEDGEMENTS

Supported by the US Army Medical Research & Materiel Command W81XWH-07-0662. We thank Malcolm Cabello for his help in fabricating test specimens.

REFERENCES

[1] Repo, R, et al., *J. Bone Joint Surg. Am.*, 59(8):1068-1076, 1977.
[2] Wilson, W et al., *J. Orth. Res.*, 24(2):220-228, 2006.
[3] McGann, M et al., in review.
[4] McGann, M et al., *J. Mech. Beh. Biomed. Mat.*, 34:264-272, 2014.
[5] Cheng, L et al., *J. Polymer Physics*, 38(1):10-22, 2000
[6] McGann, M et al., *Proc Inst Mech Eng Part H J Eng Med*, 226(8):612-622, 2012.
[7] Torzilli, P et al., *J. Biomech. Eng.*, 121(5):433-441, 1999.
[8] Ateshian, G, *J. of Biomech.*, 42:1163-1176, 2009.

A novel method for positioning and operating cryo-surgical probes in 2-D convex target areas based on super-positioning of unit circles

Avraham Shitzer
Department of Mechanical Engineering
Technion, Israel Institute of Technology
Haifa, Israel 32000

Abstract

A novel method for positioning and operating needle-like cryo-surgical probes in 2-D convex target areas is presented. The method is based on the performance of a single probe that is embedded in a semi-infinitely large, tissue-like medium. All cryo-probes are inserted into the same depth and are operated uniformly. Initially, the probes are arranged inside the target area such that the "lethal temperature" circles (LTCs) produced by them are placed tangent to its contour and to both adjacent LTCs. Subsequently, all probes are withdrawn inwardly, each at a specific distance that depends on the local curvature of the target area. Actual insertion and uniform operation of the withdrawn cryo-probes produce temperatures below the "lethal temperature" in the entire inner region of the target area. Total "defect areas," of tissue-specific temperatures either above or below the "lethal" temperature, amounted to between 7.6% and 2.5% of the target area, the lower values representing greater numbers of inserted probes coupled with shorter operating times. Possible freezing damages outside the target area enclosed between the "lethal" and the freezing temperatures, are also reduced by up to about 30% for these cases. Mixed convex-concave target areas are treated with externally inserted heating probes. Computations of the initial arrangement, withdrawal and determination of the final probes' insertion map are straight forward and require less than a minute on a PC.

MICROSYSTEMS ENGINEERING FOR BRIDGING ANGIOGENESIS WITH MECHANOPATHOLOGY

Jonathan W. Song (1,2)

(1) Mechanical and Aerospace Engineering, The Ohio State University, Columbus, OH 43210

(2) Solid Tumor Biology Program, The Ohio State University Comprehensive Cancer Center, Columbus, OH 43210

INTRODUCTION

Sustained angiogenesis is a hallmark of cancer [1] and disrupting angiogenesis is a clinically validated target for oncology [2, 3]. The seminal contributions by Jain et al. established that the determinants of tumor blood flow contribute significantly to tumor pathophysiology [4]. Blood vessels within tumors are in general more leaky compared to normal ones [5] due in large part to the excess production of vascular endothelial growth factor (VEGF) by tumor cells [6]. In addition, blood flow patterns in tumors are heterogeneous where it moves briskly in some vessels and is static in others [7]. Moreover, leaky tumor blood vessels enhances transvascular fluid exchange into the interstitial space which produces elevated interstitial fluid pressure (IFP) throughout a tumor that drops precipitously at the peri-tumor margin to drive elevated interstitial flow out from the tumor periphery and into the surrounding normal tissue [8].

Established model systems for studying angiogenesis have provided us a rich understanding of the molecular and genetic basis [9], but they have provided little insight on the role of the mechanical forces associated with the unique flow patterns of tumor blood vessels that is systematic and quantitative. Conventional *in vitro* culture systems conducted under static conditions and/or 2-D planar surfaces cannot couple controlled fluid mechanical stimuli with 3-D extracellular matrix (ECM) remodeling that is central to angiogenesis, whereas *in vivo* systems present numerous challenges for experimentally manipulating flow dynamics while observing angiogenesis in complex tissue.

METHODS

To address these challenges, we used a microfluidic tissue analogue that features human umbilical vein endothelial cell (HUVEC) lined channels in contact with, and separated by, a 3-D collagenous matrix, which function as pre-formed vessel analogs from which ECs can sprout and undergo morphogenesis [10]. The experimental system allows precise and independent specification of vascular endothelial growth factor (VEGF) gradients, transendothelial interstitial flow, and intraluminal shear rates (Fig. 1).

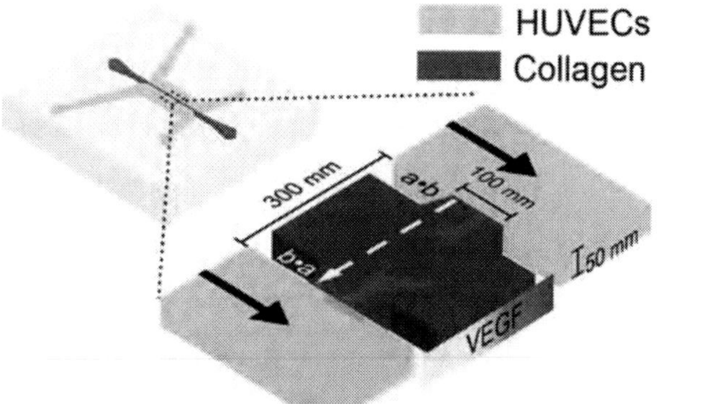

Fig. 1. Microsystem with multiple apertures allowing endothelial invasion into a central collagen gel. HUVECs are exposed to combinations of shear stress (solid arrows), transendothelial interstitial flow (dashed arrow), and VEGF gradients (blue gradient bar). Interstitial flow can be applied to the endothelial cells lining the collagen with orientation either apical to- basal (a-b; flow leaving the vessel) or basal-to-apical (b-a; flow entering the vessel). Reproduced from Ref [11].

RESULTS

We found that physiological fluid shear stress (3 dyn/cm^2) inhibits EC morphogenesis (sprouting and invasion into the 3-D matrix) and that interstitial flow (2.5-35 μm/s) increases the rate of morphogenesis [10] (Fig. 2). We also discovered that like VEGF gradients, the orientation of interstitial flow relative to the endothelium is crucial for triggering polarized vessel sprouting in a RhoA dependent manner [11]. Moreover, our system provides the conditions necessary for anastomosis: extension of endothelial sprouts away from an existing vessel into a 3D matrix, and connection of this sprout with another endothelial-lined structure to form a new circuit for flow [12]. This process is both enhanced by interstitial flow and sensitive to anti-VEGF treatment [12].

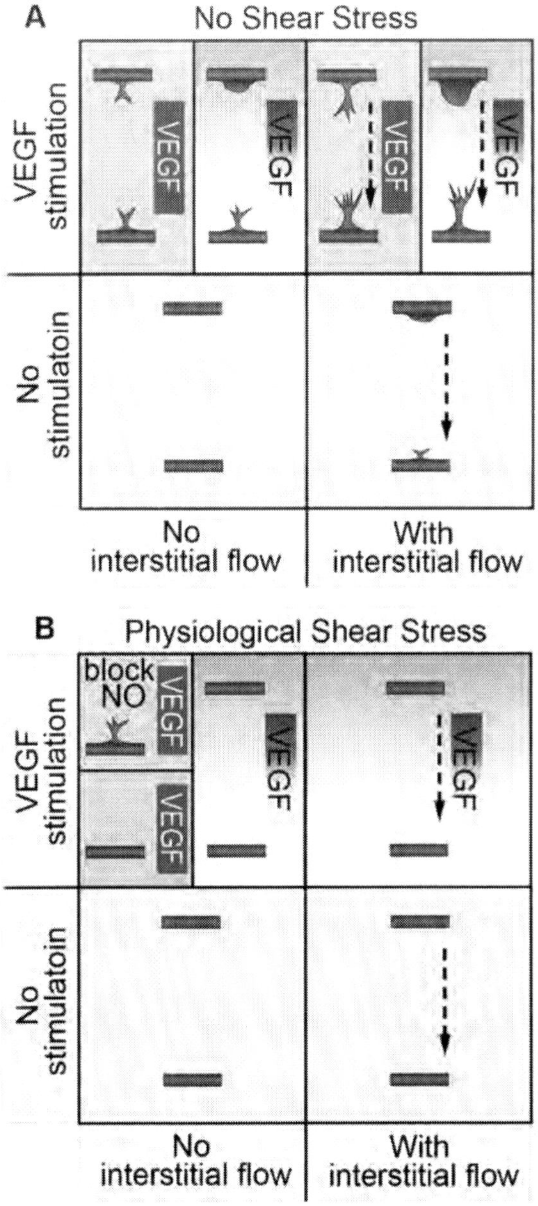

Fig. 2. Fluid forces guide VEGF-induced sprouting. (A and B) Schematic summary of the various combinations of VEGF concentrations, gradients, and interstitial flow regimes under (A) no shear stress and (B) physiological shear stress conditions. Reproduced from Ref [10].

DISCUSSION

The findings from this approach have helped establish a mechanical basis for the characteristics of tumor blood vessels. We showed that physiological shear stress (3 dyn/cm^2) attenuates VEGF-induced angiogenesis compared to static conditions [10]. This result may explain how heterogeneous blood flow patterns with local regions of flow stasis contribute to the tortuous nature of tumor blood vessels [5]. In addition, we showed that interstitial flow helps potentiate VEGF-induced angiogenesis and anastomosis [10-12]. .

The implications of fluid mechanical stimuli co-directing angiogenesis with molecular determinants such as VEGF are immense for tumor biology and oncology. Clinically approved modalities for disrupting tumor angiogenesis only target VEGF [2]. Unfortunately, the therapeutic benefits have been modest with many patients becoming resistant to therapy [13]. These results from the clinic indicate that other options need to be explored to complement and maximize the anticancer effects of targeting tumor angiogenesis. Mechanically sensitive targets [11, 14] offer a largely unexplored avenue for discovery, and microsystems engineering presents opportunities to dissect the mechanistic basis of fluid mechanical forces on angiogenesis that can be exploited for therapeutic benefit for oncology.

In summary, our microfluidic-based approach accurately reproduces the dynamics of angiogenic sprouting, morphogenesis, and anastomosis. These capabilities enabled us to systematically study the role of fluid flow in vascular formation. Insights from these studies have the potential to provide a greater understanding of how physical forces can contribute to pathological angiogenesis in solid tumors.

ACKNOWLEDGEMENTS

I want to acknowledge the Edwin L. Steele Laboratory at Massachusetts General Hospital where this research was conducted. In particular, I want to thank my mentor in the Steele Lab, Dr. Lance Munn as well as the director of the lab, Dr. Rakesh Jain.

REFERENCES

[1] D. Hanahan and R.A. Weinberg, *Cell*, 2011.
[2] N. Ferrara, K.J. Hillan, H.P. Gerber, *et al*, *Nat Rev Drug Discov*, 2004.
[3] J. Folkman, *N Eng J Med*, 1995.
[4] R,K. Jain, *Cancer Research*, 1988.
[5] E.B. Brown, R.B. Campbell, Y. Tsuzuki, *Nat Med*, 2001.
[6] H.F. Dvorak, *J Clin Oncol*, 2002.
[7] P.A. Netti, S. Roberge, Y. Boucher, *et al*, *Microvas Res*, 1996.
[8] Y. Boucher, L.T. Baxter, and R.K. Jain, *Cancer Research*, 1990.
[9] J. Folkman, P.A. D'Amore, *Cell*, 1996.
[10] J.W. Song and L.L. Munn, *PNAS*, 2011.
[11] J.W. Song, J. Daubriac, J.M. Tse, *et al*, *Lab Chip*, 2012.
[12] J.W. Song, D. Bazou, and L.L. Munn, *Integ. Biol*, 2012.
[13] G. Bergers, D. Hanahan, *Nat Rev Canc*, 2008
[14] C. Kim, H. Yang, Y. Fukushima, *et al*, *Cancer Cell*, 2014.

DESIGN, CALIBRATION, AND VALIDATION OF A NOVEL *IN VITRO* TIBIAL FORCE SENSOR

Joshua D. Roth (1), Maury L. Hull (1,2,3), Stephen M. Howell (1,3)

(1) Biomedical Engineering Graduate Group
University of California, Davis
Davis, CA, USA

(2) Department of Biomedical Engineering
University of California, Davis
Davis, CA, USA

(3) Department of Mechanical Engineering
University of California, Davis
Davis, CA, USA

INTRODUCTION

Soft tissue balancing is a critical step during total knee arthroplasty (TKA) that involves adjusting the position and orientation of the bone cuts, altering the lengths of the soft tissue restraints through releases, or both until the desired tension is restored to these restraints. Soft tissue imbalance may have two postoperative consequences: (1) imbalance of the compartmental contact forces, and (2) abnormal contact kinematics; both of these consequences have been associated with decreased implant longevity and decreased patient satisfaction [1-4]. Imbalance of the compartmental contact forces may lead to flexion instability and condylar lift-off [5, 6]. Abnormal contact kinematics including reverse axial rotation and anterior sliding have been linked to abnormal function and accelerated wear [7, 8].

Although several groups have developed sensors to assess soft tissue balance by measuring the loads between the femur and tibia *in vitro*, each sensor had limitations [9-14]. These limitations include (1) inability to decouple the medial and lateral contributions to the overall contact force [9-11], (2) inability to measure the contact force and locate the center of pressure anywhere on the articular surface [12, 13], (3) different geometry than the standard tibial component [12-14].

Accordingly, the first objective was to design and manufacture a novel tibial force sensor which overcomes the limitations of previous designs. The second objective was to calibrate and then validate the novel tibial force sensor for *in vitro* measurements of compartmental contact force and contact kinematics after TKA.

METHODS

Design description: This tibial force sensor is made up of five layers (Figure 1). The tibial base plate is a modified tibial component (Zimmer Persona, Zimmer Inc.). Fixed to the tibial base plate are two triangular arrays each consisting of three transducers. Each transducer is instrumented with four uniaxial strain gages wired into a full Wheatstone bridge (Figure 2). Also mounted to the tibial base plate is a printed circuit board that interconnects each of the six full Wheatstone bridges (three bridges in each of the medial and lateral compartments). Fixed to the top of each array of transducers is an interface tray that serves as the connection between the transducers and the custom tibial liner. The liner has the same geometry on the proximal surface as a standard tibial insert, but has custom distal geometry allowing it to be fixed to the interface tray.

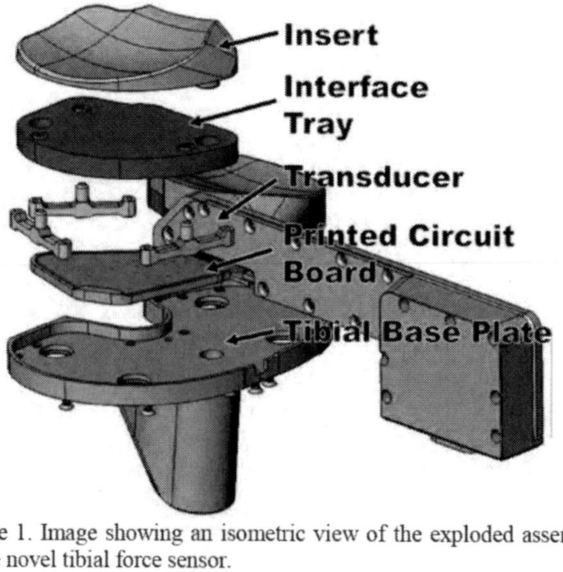

Figure 1. Image showing an isometric view of the exploded assembly of the novel tibial force sensor.

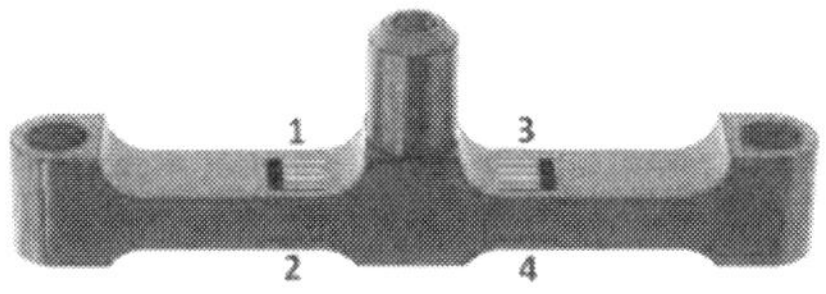

Figure 2. Rendering showing the mounting locations of the four uniaxial strain gages that are wired into a full Wheatstone bridge.

Calibration: Forces of 22 N and from 45 N to 445 N in 45 N increments were applied in each of the medial and lateral compartments at evenly spaced locations defined by metal spheres seated in hemi-spherical detents in the custom tibial liner covering the entire articular surface (Figure 3). The loading locations were randomized. In each compartment, the three voltage signals along with the applied force were recorded at each force level. Hence, twenty-one signals were recorded for each loading location. Nine inverse sensitivity coefficients were determined using a non-linear least squares optimization to minimize the error between the computed force and computed center of pressure location (Equations 1-3) and the applied force and applied center of pressure location (*lsqnonlin*, MATLAB R2014b, MathWorks, Natik, MA).

$$F_{computed,i} = c_{11}V_{1,i} + c_{12}V_{2,i} + c_{13}V_{3,i} \quad (1)$$

$$ML_{computed,i} = \frac{c_{11}V_{1,i}c_{21}x_1 + c_{12}V_{2,i}c_{22}x_2 + c_{13}V_{3,i}c_{23}x_3}{c_{11}V_{1,i} + c_{12}V_{2,i} + c_{13}V_{3,i}} \quad (2)$$

$$AP_{computed,i} = \frac{c_{11}V_{1,i}c_{31}y_1 + c_{12}V_{2,i}c_{32}y_2 + c_{13}V_{3,i}c_{33}y_3}{c_{11}V_{1,i} + c_{12}V_{2,i} + c_{13}V_{3,i}} \quad (3)$$

Validation: Maximum forces were randomly selected from the range of loads used for calibration. Each randomly selected maximum force was applied to one of the metal spheres that was not used for calibration. The accuracy of each measurement was quantified as the bias, precision, and root-mean-square error (RMSE) of the force, medial-lateral (M-L) location of the center of pressure (COP), and the anterior-posterior (A-P) location of the center of pressure.

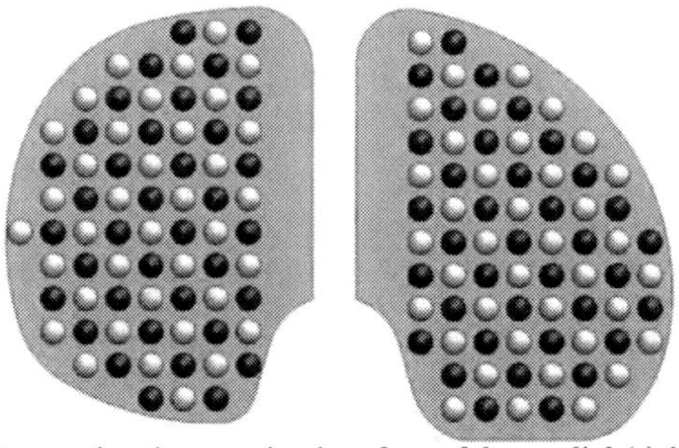

Figure 3. Image showing proximal surface of the medial (right) and lateral (left) calibration/validation inserts. The detents in black served as the loading locations for calibration, and those in white served as the loading locations for validation.

RESULTS

Validation: The RMSE of the force in each compartment was about 10 N (Table 1). The RMSE of the ML location of the center of pressure was about 2 mm. The RMSE of the AP location of the center of pressure was about 4 mm in the medial compartment and about 2 mm in the lateral compartment.

Table 1. Validation results for the medial and lateral compartments of the tibial force sensor.

		Bias	Precision	RMSE
Medial Compartment	Force	-7.7 N	5.8 N	9.6 N
	M-L COP	0.1 mm	2.0 mm	2.0 mm
	A-P COP	1.5 mm	3.5 mm	3.8 mm
Lateral Compartment	Force	2.6 N	10.2 N	10.6 N
	M-L COP	1.1 mm	1.8 mm	2.1 mm
	A-P COP	-1.4 mm	1.5 mm	2.1 mm

DISCUSSION

The novel tibial force sensor designed, manufactured, calibrated, and validated in the present study overcomes the main limitations of previous sensors. First, it mechanically decouples the measurements in the medial and lateral compartments. Second, it measures the force and locates the center of pressure anywhere on the custom insert. Third, it has the same distal and proximal geometry as the standard tibial component and insert.

The accuracy of the tibial force sensor is sufficient to measure clinically-important force imbalances. A previous study found that a contact force imbalance of 67 N between the compartments led to decreased patient satisfaction [3]. The tibial force sensor can measure this clinically important difference because the RMSE of the force measurement is seven fold less than the clinically-important imbalance.

One limitation to this tibial force sensor is the relatively large errors in the location of the center of pressure. Despite these errors, this tibial force sensor should be able to detect abnormal patterns such as reverse axial rotation because the typical A-P excursion of the contact location after TKA is 13 mm in the lateral compartment and 3 mm in the medial compartment [15].

This novel tibial force sensor will have at least two impacts on the study of TKA. First, it will allow for the *in vitro* evaluation of new alignment goals, new surgical techniques, and new component designs. Second, its measurements will serve as a standard for validating computational models of the knee.

ACKNOWLEDGEMENTS

We acknowledge the support of the National Science Foundation, Award No. CBET-1067527 and Zimmer, Inc.

REFERENCES

[1] Fehring, T.K., et al., *Clin Orthop Relat Res*, 392:315-8, 2001.
[2] Sharkey, P.F., et al., *Clin Orthop Relat Res*, 404:7-13, 2002.
[3] Gustke, K.A., et al., *Bone Joint J*, 96-B:1333-8, 2014.
[4] McGloughlin, T.M., et al., *Proc Inst Mech Eng H*, 214:349-59, 2000.
[5] Wasielewski, R.C., et al., *J Biomech*, 38:333-9, 2005.
[6] Romero, J., et al., *J Arthroplasty*, 22:235-40, 2007.
[7] Scuderi, G.R., et al., *Clin Orthop Relat Res*, 410:148-54, 2003.
[8] Kretzer, J.P., et al., *J Biomech*, 43:1092-6, 2010.
[9] Kaufman, K.R., et al., *J Biomech*, 29:667-71, 1996.
[10] Gustke, K., *J Bone Joint Surg Br*, 94B:147-150, 2012.
[11] Kirking, B., et al., *J Biomech*, 39:1744-51, 2006.
[12] Crottet, D., et al., *IEEE Trans Biomed Eng*, 52:1609-11, 2005.
[13] Nicholls, R.L., et al., *J Orthop Res*, 25:1506-12, 2007.
[14] Wasielewski, R.C., et al., *Clin Orthop Relat R*, 427:171-8, 2004.
[15] Suggs, J.F., et al., *Knee Surg Sports Traumatol Arthrosc*, 17:150-6, 2009.

DO THE LAXITIES OF THE NORMAL KNEE AT 0° AND 90° OF FLEXION SUPPORT THE GOAL OF GAP-BALANCING A TKA?

Joshua D. Roth (1), Stephen M. Howell (1,3), Maury L. Hull (1,2,3)

(1) Biomedical Engineering Graduate Group
University of California, Davis
Davis, CA, USA

(2) Department of Biomedical Engineering
University of California, Davis
Davis, CA, USA

(3) Department of Mechanical Engineering
University of California, Davis
Davis, CA, USA

INTRODUCTION

There is general agreement that optimum function of a total knee arthroplasty (TKA) requires correct tension of the soft tissues, Overly tight soft tissues are associated with pain [1], stiffness [2], and loss of flexion [2]. Overly lax soft tissues are associated with instability [3] and accelerated polyethylene wear [4]. Overly tight and loose soft tissues account for up to 54% of revision procedures [5].

One common procedure for TKA is gap-balancing where the goal is to create rectangular gaps and equal laxities at 0° and 90° of flexion [1, 6]. Although the desired tension and laxity for the soft tissues around the knee are unknown, gap-balancing assumes that equal varus, valgus, and distraction laxities are desired at 0° and 90° of flexion. However, there is very little direct evidence to support this assumption [7].

If the goal of TKA is restoration of normal knee laxity, then the laxities of the normal knee should be used to determine whether the soft tissues of a TKA are overly tight or overly loose. Hence, the laxities of a gap-balanced TKA at 0° and 90° of flexion should equal those of the normal knee.

Although, several studies have measured the laxities of the normal knee [8-18], none have determined whether the laxities are equal at 0° and 90° of flexion. Because there is a lack of direct evidence to support the goal of gap-balancing a TKA, the present study determined whether the varus-valgus (V-V) and internal-external (I-E) rotation and distraction (D) and anterior-posterior (A-P) translation laxities of the normal knee are equal at 0° and 90° of flexion.

METHODS

Seven laxities were measured at 0° and 90° of flexion using a six degree-of-freedom load application system (Figure 1) in ten fresh-frozen normal cadaveric knees. Laxities were determined by applying moments in V-V (± 5 N-m) and in I-E (± 3 N-m) rotation and forces in D (100 N) and in A-P (± 45 N) translation

The mean and standard deviation described each laxity at 0° and 90° of flexion, and a paired Student's t-test determined whether each laxity were equal at 0° and 90° of flexion (JMP Version 11.2.0; SAS Institute Inc., Cary, NC; www.jmp.com).

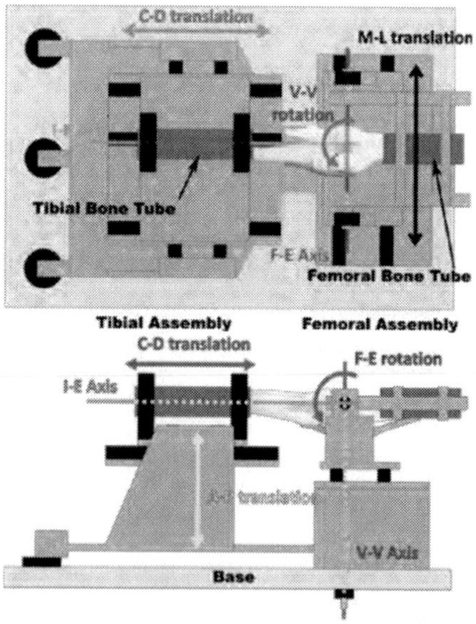

Figure 1. Schematic showing the six degree-of-freedom load application system, which measures the laxities of the normal knee [10].

RESULTS

At 90° of flexion, the mean laxity was fivefold greater in varus ($p<0.0001$), fourfold greater in distraction ($p<0.0001$), threefold greater in valgus ($p=0.0003$), internal ($p=0.0002$), and external rotation ($p<0.0001$), and twofold greater in anterior translation ($p=0.003$) than at 0° of flexion (Figure 2).

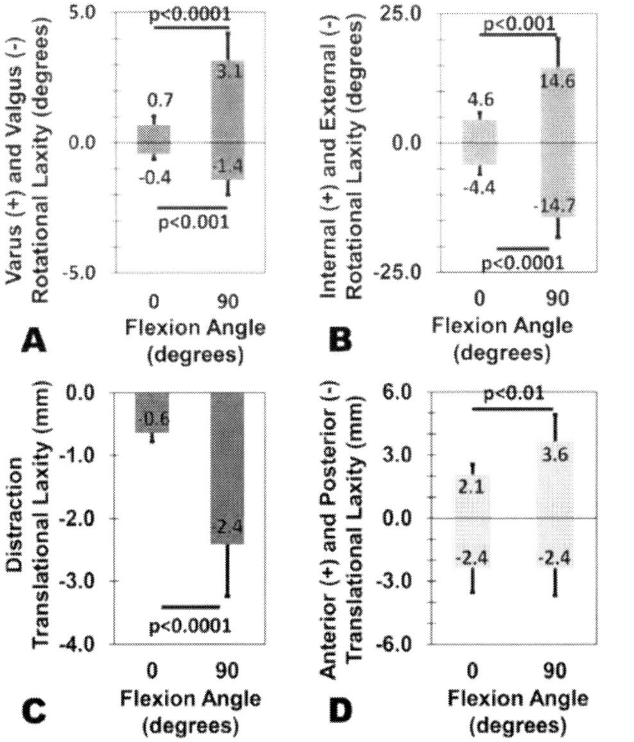

Figure 2. Bar chart **(A)** shows varus (+) and valgus (-) laxity, **(B)** shows internal (+) and external (-) laxity, **(C)** shows distraction laxity, and **(D)** shows anterior (+) and posterior (-) laxity at 0° and 90° of flexion of the normal knee. Error bars show ±1 standard deviation. A $p < 0.05$ indicates the laxity at 90° of flexion is different than at 0° of flexion.

DISCUSSION

The result that six of seven laxities were five to two times greater at 90° than at 0° of flexion in the normal knee calls into question the goal behind gap-balancing TKA. The penalty for creating gaps with equal laxities at 0° and 90° of flexion is that the soft-tissues of the TKA are overly tight relative to normal at 90° of flexion, which patients may perceive as pain, stiffness, and limited flexion.

The finding that six of the seven laxities were greater at 90° of flexion than those at 0° of flexion affects the understanding of how soft tissue tension influences internal and external rotation of a TKA during flexion. In the normal knee, the medial contact point on the tibia remains nearly fixed as the knee flexes, but the lateral contact point on the tibia translates posteriorly [19, 20] (often referred to as medial pivot), which causes internal rotation of the tibia with flexion and posterior roll-back of the lateral femoral condyle on the tibia. In a gap-balanced TKA, the components are positioned and the soft tissues are adjusted until the laxities at 90° of flexion equal those at 0° of flexion, which makes the soft tissues overly tight at 90° of flexion relative to those of the normal knee. The penalty for creating overly tight soft tissues at 90° of flexion in a TKA, especially in the lateral compartment, is a paradoxical external tibial rotation with flexion (often referred to as lateral pivot) and anterior sliding of the lateral femoral condyle on the tibia in flexion reported to occur in some TKAs [21, 22].

Another penalty of gap-balancing a TKA is that the release of one of the collateral ligaments is often needed at 0° of flexion to create a rectangular gap at 0° of flexion [23]. A release should be made with caution because a release made to balance one laxity may lead to other instabilities that are uncorrectable [23-26].

One strategy for avoiding overly tight soft tissues and unnecessary releases is to strive to restore the laxities of the knee to normal as is done in kinematically aligned TKA [27]. One recent study found better outcomes and improved function after kinematically aligned TKA than after mechanically aligned TKA, in which surgeons often strive for balanced gaps and equal laxities [28]. Another recent study found that kinematically aligned TKA minimized paradoxical kinematics during knee flexion [29]. These studies suggest that striving to restore the laxities of the knee to normal may contribute to better outcomes and improved function after TKA.

In summary, the result that six laxities are five to two times greater at 90° than at 0° flexion in the normal knee calls into question the goal behind gap-balancing TKA.

ACKNOWLEDGEMENTS

We acknowledge the support of the National Science Foundation, Award No. CBET-1067527 and Zimmer, Inc. Additionally, the authors wish to thank individuals who donate their bodies and tissues for the advancement of education and research.

REFERENCES

[1] Babazadeh, S., et al., *Orthop Rev*, 1:e26, 2009.
[2] Heesterbeek, P.J., et al., *Knee*, 15:45-9, 2008.
[3] Athwal, K.K., et al., *Clin Biomech*, 29:119-28, 2014.
[4] Kretzer, J.P., et al., *J Biomech*, 43:1092-6, 2010.
[5] Mulhall, K.J., et al., *Clin Orthop Relat Res*, 446:45-50, 2006.
[6] Griffin, F.M., et al., *J Arthroplasty*, 15:970-3, 2000.
[7] Sikorski, J.M., *J Bone Joint Surg Br*, 90:1121-7, 2008.
[8] Markolf, K.L., et al., *J Bone Joint Surg Am*, 58-A:583-594, 1976.
[9] Blankevoort, L., et al., *J Biomech*, 21:705-20, 1988.
[10] Eagar, P., et al., *J Biomech*, 34:1655-1660, 2001.
[11] Okazaki, K., et al., *J Orthop Sci*, 11:264-6, 2006.
[12] Bull, A.M., et al., *Clin Orthop Relat Res*, 466:2491-9, 2008.
[13] Delport, H., et al., *Clin Biomech*, 28:777-82, 2013.
[14] Moewis, P., et al., *Med Eng Phys*, 36:889-95, 2014.
[15] Wang, X., et al., *J Biomech Eng*, 136:081005, 2014.
[16] Haimes, J.L., et al., *Am J Sports Med*, 22:402-9, 1994.
[17] Nowakowski, A.M., et al., *J Orthop Res*, 30:522-7, 2012.
[18] Ghosh, K.M., et al., *Knee Surg Sports Traumatol Arthrosc*, 22:1736-43, 2014.
[19] Freeman, M.A., et al., *Clin Orthop Relat Res*, 410:35-43, 2003.
[20] Dennis, D.A., et al., *J Biomech*, 38:241-53, 2005.
[21] Dennis, D.A., et al., *Clin Orthop Relat Res*, 428:180-9, 2004.
[22] Dennis, D.A., et al., *Clin Orthop Relat Res*, 356:47-57, 1998.
[23] Gu, Y., et al., *J Bone Joint Surg Am*, 96:e101, 2014.
[24] Kanamiya, T., et al., *Clin Orthop Relat Res*, 404:24-31, 2002.
[25] Saeki, K., et al., *Clin Orthop Relat Res*, 392:184-9, 2001.
[26] Whiteside, L.A., *J Arthroplasty*, 17:23-7, 2002.
[27] Howell, S.M., et al., *Insall & Scott Surgery of the Knee*, 5 Ed:1255-69 2011.
[28] Dossett, H.G., et al., *Bone Joint J*, 96-B:907-13, 2014.
[29] Howell, S.M., et al., *Knee Surg Sports Traumatol Arthrosc*, 21:2281-7, 2013.

SB^{3}C2015
Summer Biomechanics, Bioengineering and Biotransport Conference
June 17-20, 2015,Snowbird Resort, Utah, USA

SB^{3}C2015-031

CARDIAC REVERSIBLE GROWTH & REMODELING MODEL: PREDICTING AND UNDERSTANDING THE CHRONIC EFFECTS OF BIOINJECTION THERAPY

Lik Chuan, Lee
Michigan State University
Lansing, MI, USA
lclee@egr.msu.edu

Martin Genet
ETH Zurich
Zurich, Switzerland

Jonathan Wenk
University of Kentucky
Lexington, KY, USA

Joakim Sundnes
Simula Research Laboratory
Oslo, Norway

Samuel Wall
Simula Research Laboratory
Oslo, Norway

INTRODUCTION

Computational modeling has increasingly been used to elucidate the effects of heart failure (HF) treatments. Most such models are, however, confined to simulating the acute treatment effects, and do not directly simulate the effects of chronic LV remodeling (and reverse remodeling). Since left ventricular (LV) remodeling and reverse remodeling are hallmarks of many cardiac diseases and favorable response to HF treatments, the ability to simulate and predict these chronic effects would have significant impact on the use of computational modeling for patient care. Here, we apply a recently proposed reversible cardiac volumetric growth model [1] to simulate and predict the chronic effects of an emerging bioinjection therapy that is intended to stiffen the infarct [2].

METHODS

Volumetric growth was described by multiplicatively decomposing the deformation gradient $\mathbf{F}$ into a growth component $\mathbf{F}_g$ and an elastic component $\mathbf{F}_e$, i.e.,

$$\mathbf{F} = \mathbf{F}_e \cdot \mathbf{F}_g \,. \tag{1}$$

Growth in the fiber $\mathbf{f}$ direction was described by parameterizing $\mathbf{F}_g$ with a scalar growth multiplier θ (i.e., $\mathbf{F}_g = (\theta - 1)\mathbf{f}\otimes\mathbf{f} + \mathbf{I}$) that was described using the following constitutive relation [1]. Denoting $\bar{\lambda}_h$ as the average homeostatic elastic myofiber stretch for each cycle, the local evolution of the growth multiplier θ was described by

$$\dot{\theta} = k(\theta)(\bar{\lambda}_e - \bar{\lambda}_h)\,, \tag{2}$$

where $k(\theta) \geq 0$ is a rate limiting function whose main purpose is to ensure that θ lies within some physiological values (i.e, $\theta_{\min} \leq \theta \leq \theta_{\max}$). It has the feature that $k(\theta) \to 0$ as $\theta \to \theta_{\min}$ or $\theta \to \theta_{\max}$. In Eq. (2), growth ($\dot{\theta} > 0$) or shrinkage ($\dot{\theta} < 0$) occurs when the average myofiber stretch $\bar{\lambda}_e$ exceeds or falls below the homeostatic stretch $\bar{\lambda}_h$, respectively.

The growth constitutive model was integrated into a cardiac electromechanics model described by Sundnes et al. [3]. Briefly, cardiac electrophysiology described by the bidomain equations and a cellular electrophysiology model was coupled to a cardiac mechanics model comprising of a Fung-type passive constitutive model and an active stress constitutive model. The elastic deformation gradient tensor $\mathbf{F}_e$ in Eq. (1) enters the Cauchy stress tensor which is given as follows:

$$\boldsymbol{\sigma} = \frac{1}{\det\mathbf{F}_e}\mathbf{F}_e \cdot \frac{dW(\mathbf{E}_e)}{\mathbf{E}_e} \cdot \mathbf{F}_e^{\,T} + \frac{1}{\det\mathbf{F}_g}\boldsymbol{\sigma}^a(\mathbf{s}, \lambda_e, \dot{\lambda}_e) - p\mathbf{I}\,. \tag{3}$$

In Eq. (3), W is the Fung-type strain energy function, $\mathbf{E}_e = \frac{1}{2}(\mathbf{F}_e^{\,T}\mathbf{F}_e - \mathbf{I})$ is the elastic Green strain tensor, $\boldsymbol{\sigma}^a$ is the active stress that is a function of (i): state variables $\mathbf{s}$ describing membrane channels and intracellular ionic concentration, and (ii): elastic myofiber stretch λ_e and rate of stretch $\dot{\lambda}_e$. The active stress $\boldsymbol{\sigma}^a$ is scaled by $1/\det\mathbf{F}_g$ to ensure that the active force per unit cell remained constant and does not increase or decrease with cellular growth and shrinkage, respectively.

The coupled growth-electromechanics model was applied to a human LV geometry using standard finite element method with an operator splitting scheme [3]. The epicardial edge of the LV was fixed and pressure was applied to the LV endocardial surfaces. Fiber helix angle with a transmural variation from -60° (endocardium) to 60° (epicardium) was prescribed to the LV [4]. The cardiac cycle was simulated by stimulating the LV at the apex. End-diastolic and aortic valve opening pressures of 13.5 mmHg and 52.5 mmHg were prescribed to the LV model, respectively, and the ejection phase was described using a 3-parameter Windkessel model. Timescale between growth and acute hemo-

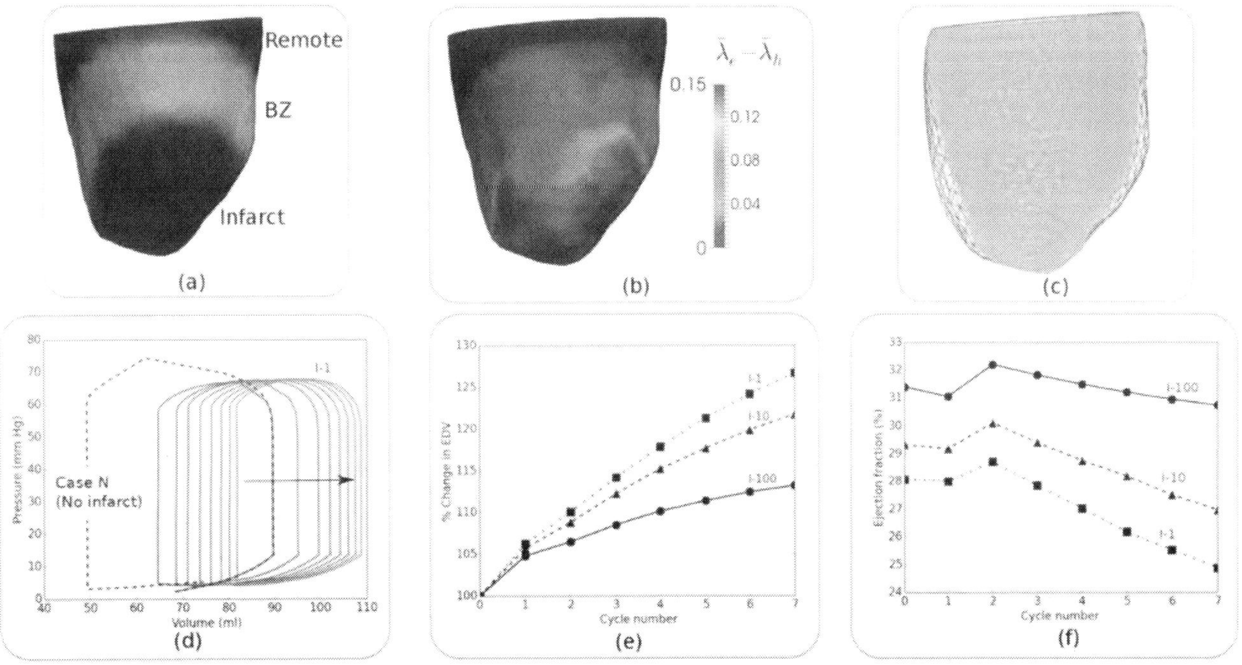

FIGURE 1. (a): Material regions in the LV. (b): Difference between average elastic myofiber stretch and its homeostatic value - driver of the remodeling process (I-1). Note: only BZ and remote regions are allowed to grow. (c): Remodeled LV (red) superimposed onto the original LV (blue) at ED (I-1). (d) Rightward shifting of PV loop during remodeling (I-1). Effects of increasing infarct stiffness on (e): rate of change of EDV and (f): EF.

dynamic loading was separated by locally updating the growth multiplier θ at the end of each cardiac cycle.

Four cases were simulated, namely, LV with homogeneous normal contractility (N), an apical non-contracting and non-growing transmural infarct that was (I-1) unstiffened, (I-10) 10 times stiffer and (I-100) 100 times stiffer. A borderzone (BZ) with half the normal contractile strength was prescribed in the last 3 cases (Fig. 1a). The homeostatic stretch field $\bar{\lambda}_h$ was prescribed using the average elastic myofiber stretch from case N.

RESULTS

In the presence of infarct (I-1, I-10 and I-100), the average cycle myofiber elastic stretch became greater than its homeostatic value at the BZ and infarct (Fig. 1b). As a result, the LV remodeled and became more spherical (Fig. 1c). This result is manifested by a rightward shift of the pressure-volume (PV) loop with both an increase in the end-systolic volume (ESV) and end-diastolic volume (EDV) (Fig. 1d).

With increasing infarct stiffness, however, the rate of increase in ESV and EDV was reduced substantially. A 100-times increase in infarct stiffness led to about a 10 percentage point reduction of the increase in EDV (Fig. 1e) and ESV (not shown) by the 7th growth cycle. Correspondingly, LV ejection fraction (LVEF) was less reduced with stiffer infarct (Fig. 1f). In I-100, LVEF remained fairly constant at about 30%.

DISCUSSION

We have integrated a reversible growth model to a cardiac coupled electromechanics model and have applied it to study the effects of material-induced stiffening therapy by varying the degree of infarct stiffness. Our results suggest that therapeutic benefits in the form of attenuating the remodeling process can be achieved via stiffening the infarct. Specifically, the rate of increase in both EDV and ESV during remodeling were found to be reduced with increasing infarct stiffness. Similarly, our results also indicate that infarct stiffening can attenuate the rate of decrease in EF during remodeling.

These results are consistent with experimental studies using ovine MI model [2]. In that study, transmural infarct was stiffened and thickened by injecting it with a reactive tissue filler agent within hours of coronary occlusion. At 8-week follow up, LV remodeling was found to be attenuated in animals that underwent the therapy when compared to those that did not. The increase in LV volumes were found to be lower in treated animals (ESV : 39% and EDV : 28%) compared to untreated animals (ESV : 83% and EDV : 65%), which is qualitatively consistent with our findings (Fig. 1e). And LVEF was found to decrease at a slower rate in treated animals (from 36% to 31%) compared to untreated animals (from 35% to 28%), which is also qualitatively consistent with our findings (Fig. 1f).

In summary, we have shown that the coupled growth-electromechanics model's prediction of the effects of infarct stiffening is generally consistent with existing experimental results on material-induced infarct stiffening therapy. Future works will include calibration of the model using animal-specific data.

REFERENCES

[1] Lee, LC et al. *Biomech. Model. Mechan.*, 2014 (In press)

[2] Morita, M et al. *Ann. Thorac. Surg.*, 92(2):617–24, 2011.

[3] Sundnes, J et al. *Comput. Methods Biomech. Biomed. Engin.*, 17(6):604–15, 2014.

[4] Streeter, DD et al., *Circ. Res.*, 24(3):339–47, 1969.

SB³C2015
Summer Biomechanics, Bioengineering and Biotransport Conference
June 17-20, 2015,Snowbird Resort, Utah, USA
SB³C2015-032

INFARCTED LEFT VENTRICLES HAVE STIFFER MATERIAL PROPERTIES AND LOWER STIFFNESS VARIATION: 3D ECHO-BASED MODELING TO QUANTIFY IN VIVO VENTRICLE MATERIAL PROPERTIES

Longling Fan (1), Jing Yao (2), Chun Yang (3), Di Xu (2), Dalin Tang (1,4)

(1) Department of Mathematics,
Southeast University
Nanjing, 210096, China

(2) Department of Cardiology
First Affiliated Hospital of Nanjing Medical University
Nanjing 210029, China

(3) Network Technology Research Institute
China United Network Communications Co., Ltd.
Beijing, 100048, China

(4) Mathematical Sciences Department
Worcester Polytechnic Institute
Worcester MA 01609, USA

INTRODUCTION

Medical images and computational modeling have been used more and more in cardiovascular research [1]. Early 3D models for blood flow in the heart included Peskin's model which introduced fiber-based LV model and the celebrated immersed-boundary method to study blood flow features in an idealized geometry with fluid-structure interactions [2]. Patient-specific CMR-based computational RV/LV models with fluid-structure interactions were introduced to assess outcomes of various RV reconstruction techniques with different scar tissue trimming and patch sizes [3,4]. In this paper, echo-based 3D LV models were introduced to quantify ventricle material properties, and to investigate morphological and mechanical stress/strain differences between ventricle with and without infarct. This will serve as a starting point to use computational models for infarct differentiation and surgical planning.

3D ECHO DATA ACQUISITION, MODELS AND METHODS

Echo image data were acquired from ten patients with informed consent obtained (8M, 2F, mean age 58.3 years). Group 1 (G1) included five patients with recent infarction (P1-P5). Group 2 (G2) included five healthy volunteers (P6-P10). Details of the data acquisition procedures were previously described and are omitted here [4]. Figures 1 show the echo images and re-constructed 3D LV geometries from the group 1 patients. The ventricle material was assumed to be hyperelastic, anisotropic, nearly-incompressible and homogeneous. Infarct tissue was assumed to be hyperelastic, isotropic, nearly-incompressible and homogeneous. The nonlinear Mooney-Rivlin model was used to describe the nonlinear anisotropic (for normal tissue) and isotropic (for infarct tissue) material properties. Parameter values for infarct tissue were chosen from [3]. Parameter values for regular LV tissue were chosen to fit LV maximum and minimum volume data obtained from Echo.

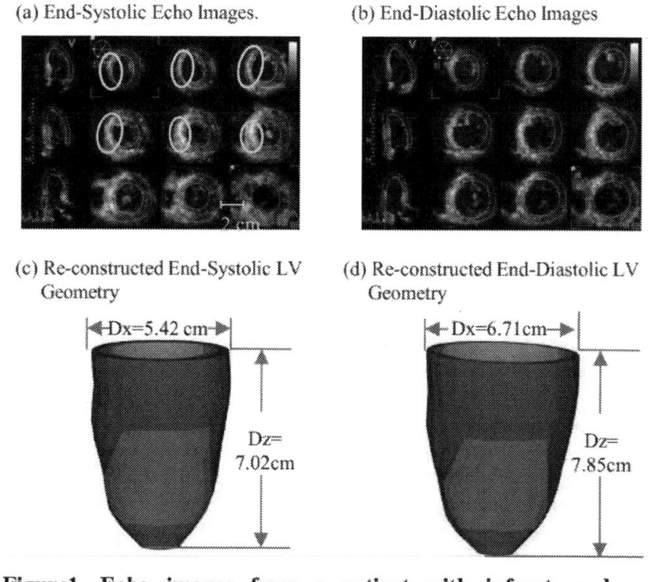

Figure1. Echo images from a patient with infarct, and re-constructed geometry. Infarct locations were marked by yellow circles on the echo images.

As patient-specific fiber orientation data was not available from these patients, we chose to construct a two-layer LV model and set fiber orientation angles using fiber angles given in [3]. Active contraction and expansion of myocardium were modeled by material stiffening and softening in our model. In our model construction process, a pre-shrink process was applied to in vivo end-systolic ventricular geometries to generate the starting shape (zero ventricle pressure) for the computational simulation. The effect of the pre-shrink process was demonstrated in [4]. Modeling active LV contraction and relaxation is very difficult and some simplifications are needed to obtain proper models to serve our purposes: a) quantify material properties under maximum and minimum pressure conditions; b) compare morphological and mechanical stress/strain parameter values under maximum and minimum pressure conditions. Material stiffening and softening were achieved by adjusting parameter values at each Echo-time step (28 Echo frames per cycle) to simulate active contraction and expansion and match LV volume data. The least-squares method was used to find the equivalent Young's moduli (YM) for the material curves for easy comparison. LV wall thickness, circumferential and longitudinal curvatures (C- and L-curvature), material parameter values, and stress/strain values were recorded for analysis. The anisotropic LV computational models were constructed for the two groups and the models were solved by ADINA (ADINA R&D, Watertown, MA, USA) using unstructured finite elements and the Newton-Raphson iteration method.

RESULTS

Figures 2 gives the stress-stretch plots for two patients, one from each group for illustration. Using the mean YM value of Group 2 as the base value, at end-diastole, Group 1 mean YM value for the fiber direction stress-strain curve was 54% stiffer than that of Group 2 (136.24 kPa vs. 88.68 kPa). At end-systole, the mean YM values from the two groups were similar (175.84kPa vs. 200.2kPa). More interestingly, Group 2 end-systole mean YM was 126% higher that its end-diastole value, while Group 1 end-systole mean YM was only 29% higher that its end-diastole value. This indicated that Group 2 had much greater systole-diastole material stiffness variations. At beginning-of-ejection (LV pressure and volume at maximum), LV ejection fraction (LVEF) showed positive correlation with C-curvature, stress and strain and negative correlation with LV volume, respectively. At beginning-of-filling (LV pressure and volume at minimum), LVEF showed positive correlation with C-curvature, strain, but negative correlation with stress and LV volume, respectively. Using averaged values of two groups at beginning-of-ejection, Group 1 stress, strain, and wall thickness were 32%, 29%, and 18% lower (thinner), respectively, compared to those of Group 2. L-curvature from Group 1 was 61% higher than that from Group 2. Difference in C-curvature between the two groups was not statistically significant.

DISCUSSION

Identification of infarct area is of great important in clinical applications. Now that we demonstrated that ventricles with and without infarct have considerably large differences in contractibility and material stiffness variations, proper inverse methods could be introduced to determine if a ventricle had infarct based on its contractibility and material parameter values predicted by our models. This could serve as the basis for people to develop accurate and automatic methods to identify infarct area based on image data, which is of great clinical relevance. An echo-based computational modeling approach was proposed to investigate left ventricle material properties and stress/strain conditions. With ten patients studied, our results indicated that our modeling approach has the potential to be used determine in vivo ventricle material properties, which in turn could lead to inverse methods to infer presence of infarct from LV contractibility and material stiffness variations. Quantitative differences in LV volume, curvatures, stress, strain and wall thickness between Group 1 and Group 2 were provided.

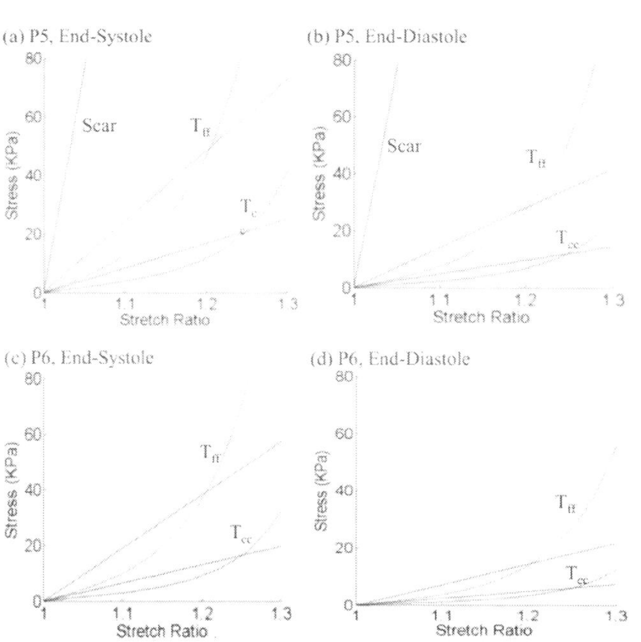

Figure 2. LV material stress-stretch curves from P5 (infarct) and P6 (no infarct) in fiber coordinates showing P6 tissue stiffness has larger variations between systole and diastole. T_{ff}: stress in fiber direction; T_{cc}: stress in circumferential direction of the fiber.

ACKNOWLEDGEMENTS

This research was supported in part by NIH-1R01-HL 089269. Chun Yang's research is supported in part by National Sciences Foundation of China 11171030.

REFERENCES

[1] Desmond-Hellmann, S., Sawyers, C. L., Cox, D. R., Fraser-Liggett, C.,Galli, S. J., Goldstein, D. B., Hunter, D., Kohane, I. S., Lo, B., Misteli, T., Morrison, S. J., Nichols, D. G., Olson, M. V., Royal, C. D., Yamamoto, K. R. Toward Precision Medicine: Building a Knowledge Network for Biomedical Research and a New Taxonomy of Disease, Committee on a Framework for Development a New Taxonomy of Disease, National Research Council, The National Academies Press, http://www.nap.edu/catalog.php?record_id=13284. 2011.

[2] Peskin, C. S., Mathematical Aspects of Heart Physiology, Lecture Notes of Courant Institute of Mathematical Sciences, New York. 1975.

[3] Tang, D., Yang, C., Geva, T., Gaudette, G., del Nido, P. J., Multi-physics MRI-based two-layer fluid-structure interaction anisotropic models of human right and left ventricles with different patch materials: cardiac function assessment and mechanical stress analysis, Computers & Structures, 89:1059-1068, 2011.

[4] Fan, R., Tang, D., Yao, J., Yang, C., Xu, D., 3D Echo-Based Patient-Specific Computational Left Ventricle Models to Quantify Material Properties and Stress/Strain Differences between Ventricles with and without Infarct, Computer Modeling in Engineering & Sciences, 99(6):491-508, 2014.

HUMAN CORONARY PLAQUE MORPHOLOGICAL AND STRESS VULNERABILITY INDICES USING IVUS-BASED FLUID-STRUCTURE INTERACTION MODELS: A MULTI-PATIENT STUDY

Liang Wang (1), Jie Zheng (2), Akiko Maehara (3), Chun Yang (1,4), Richard Bach (5), David Muccigrosso (2), Gary S. Mintz (3), Dalin Tang (1,6)

(1) Math Sciences Department
Worcester Polytechnic Institute
Worcester, MA 01609

(2) Mallinckrodt Inst of Radiology,
Washington University
St. Louis, MO USA 63110

(3) The Cardiovascular Research
Foundation, Columbia University,
New York, NY 10022

(4) Network Technology Research Institute
China United Network Comm. Co., Ltd.
Beijing, China 100048

(5) Cardiovascular Division,
Washington University
Saint Louis, MO, USA 63110

(6) School of Biological Science
and Medical Engineering
Southeast University
Nanjing, China, 210096

INTRODUCTION

Atherosclerotic plaque progression is believed to be associated with mechanical stress conditions [1-2]. Increasing evidences showed that the mechanical forces play an important role in plaque progression and rupture and should be considered in an integrated way with plaque morphology and composition for possible improvement of plaque vulnerability assessment. Magnetic resonance image (MRI)-based multi-component models with fluid-structure interactions (FSI) were introduced by Tang et al. to obtain both flow shear stress and structural plaque stress/strain for analysis [3]. Effect of key plaque morphological features such as cap thickness and lipid core size were investigated in [4]. Our previous result based on 34 2D MRI slices from 14 human carotid plaque samples showed the computational plaque vulnerability index (CPVI) assigned according to the local maximum Stress-P_1 at critical sites had an 85% agreement rate with Histopathological Plaque Vulnerability Index (HPVI) [5]. In this paper, patient-specific intravascular ultrasound (IVUS)-based FSI models of 14 human coronary plaques were constructed to obtain flow shear stress and plaque wall stress data for plaque assessment. Critical flow shear stress (CFSS), plaque stress (CPWS) and strain (CPWSn) and key morphological features (minimal cap thickness and lipid percentage) for all 617 slices from the 14 patients were calculated to identify possible associations between morphological and mechanical stress features. Stress index and morphological plaque vulnerability index (MPVI) for the 617 slices were calculated and compared.

Methods

IVUS Data. 3D IVUS data were acquired during cardiac catheterization from 14 patients (11M, 3F, mean age: 59) with informed consent obtained. Figure 1 gives plots of a sample IVUS slice image, segmented contours, contour after smoothing, X-ray angiography and the reconstructed 3D geometry of a coronary plaque.

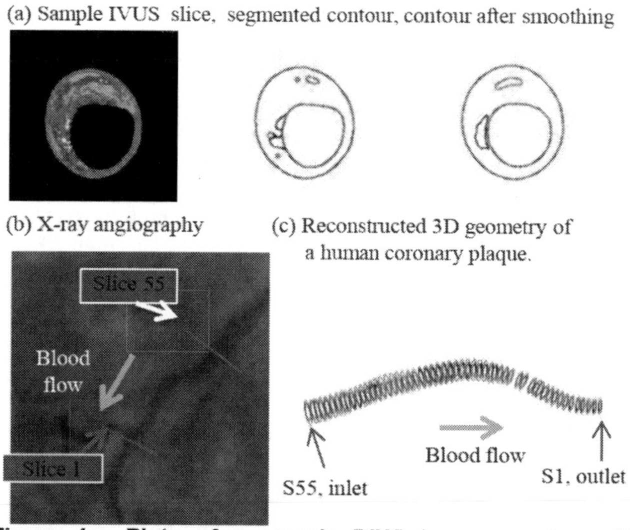

Figure 1: Plots of a sample IVUS image, contour, 3D reconstructed plaque, X-ray angiographic image, vessel curvature, and on-site blood pressure and flow velocity.

FSI Model. 3D computational FSI models with geometries re-constructed from patient-specific IVUS data and X-ray angiograph data were constructed for 14 plaques with cyclic bending due to cardiac contraction and subjected to the pressure loading taken into consideration to better simulate the mechanical conditions in human coronary [6]. A modified Mooney-Rivlin (M-R) model was used to describe the anisotropic material properties of blood vessel:

$W=c_1(I_1-3)+c_2(I_2-3)+D_1\{\exp(D_2(I_1-3))-1\}+K_1/2K_2\{\exp(K_2(I_4-1)^2-1)\}$.

The full 3D FSI models were solved by ADINA to calculate flow and stress/strain distributions for analysis [6].

Data Extraction and Index Assignments. For all 617 IVUS slices from 14 patients, CPWS, CPWSn, CFSS, cap thickness and lipid percentage were calculated and recorded for analysis following established procedures [5]. Analogue to AHA classification standard of morphological plaque vulnerability, three indices with value form 0 to 4 were introduced in this paper based on the morophological charactertics for each slice. A lipid index was assigned depend on lipid percentage value of the slice and a cap index was also assigned according to its min cap thickness value. Furthermore, a morphological plaque vulnerability index (MPVI) was defined as the maximum of lipid index and cap index. Each slice was assigned a stress index (0, 1, 2, 3, or 4) according to its CPWS value by using five stress intervals determined to have the best match with MPVI. Correlation between CFSS, CPWS/CPWSn and lipid percentage, min cap thickness and those three morphological related indices were analyzed and the agreement between stress index and those three indices was also performed.

RESULTS

Tables 1 summarized correlation results between the three mechanical factors (CPWS, CPWSn, CFSS) and 5 morphological factors/indices (cap thickness, cap index, lipid percentage, lipid index and MPVI) using 617 slices. Results for each individual plaque changed from patient to patient. CPWS had negative correlation with cap thickness, and positive correlations with cap index, lipid percentage, lipid index and MPVI. Since higher cap index was assigned to thinner cap, it is reasonable that the correlation coefficients between CPWS and cap index had an opposite signs. The correlation results between CPWSn and cap thickness, cap index and MPVI were basically similar to those of CPWS (except lipid and lipid index), but much weaker. The correlation results between CFSS and five morphological factors were much weaker, compared to those CPWS results. CPWS had the best correlation with MPVI among the three mechanical factors.

Table 1. CPWS had best correlation with MPVI among the three mechanical risk factors.

CPWS v.s.	Cap Thickness	Cap Index	Lipid Percent	Lipid Index	MPVI
r	-0.6570	0.8016	0.2209	0.2304	0.7725
P	<0.001	<0.001	<0.001	<0.001	<0.001
CPWSn v.s.	Cap thickness	Cap index	Lipid percent	Lipid index	MPVI
r	-0.1823	0.2477	-0.1798	-0.2087	0.1690
P	<0.001	<0.001	<0.001	<0.001	<0.001
CFSS v.s.	Cap thickness	Cap index	Lipid percent	Lipid index	MPVI
r	0.1336	-0.1910	-0.1569	-0.1215	-0.1833
P	<0.001	<0.001	0.0001	0.0025	<0.001

It is of interest to see if plaque assessment using stress would be different or consistent with morphological assessment. The comparison between stress index and cap index, lipid index, MPVI were preformed and the agreement rates are given in Table 2. When the stress index and the morphological related indices are the same for a slice, we say these two indices are matched. The agreement rates between stress index and cap index, lipid index and MPVI are 57.37%, 25.93% and 66.77% respectively. The result demonstrated that stress index has a strong agreement with morphological feature closely related to plaque vulnerability, and could be used to aid the assessment of plaque vulnerability and plaque rupture.

Table 2. Stress index has good agreement rates (total slices=617) with cap index, lipid index and MPVI.

Stress Index v.s.	Cap Index	Lipid Index	MPVI
Matched #	354	160	412
Unmatched #	263	457	205
Matching Rate	0.5737	0.2593	0.6677

DISCUSSION AND CONCLUSION

Our results indicate that CPWS had the best correlation results with the morphological characteristics comparing to CPWSn and CFSS, and the stress index had a 66.77% agreement with the MPVI. The variation between stress index and MPVI provided room for improvements when both indices were utilized for optimal plaque assessment. Limitations of this study include: a) lack of gold standard to validate stress-based plaque vulnerability assessment; b) better image resolution will definitely improve data accuracy and accuracy of modeling predictions; c) patient-specific plaque material properties were not available for our study; d) angiographic data with two different (preferably orthogonal) view angles are needed to obtain accurate 3D curvature variations; e) obtaining zero-stress conditions remains a challenge. More patient follow-up data and large-scale studies are needed to obtain more support for new stress-based plaque assessment schemes.

Acknowledgement: This research was supported by NIH grant NIH/NIBIB R01 EB004759. Chun Yang's research is supported in part by National Sciences Foundation of China 11171030.

REFERENCES

[1] Samady H, Eshtehardi P, McDaniel MC, Suo J, Dhawan SS, Maynard C, Timmins LH, Quyyumi AA, Giddens DP. Coronary artery wall shear stress is associated with progression and transformation of atherosclerotic plaque and arterial remodeling in patients with coronary artery disease. Circulation.124:779-788, 2011.

[2] Tang D, Kamm RD, Yang C, Zheng J, Canton G, Bach R, Huang X, Hatsukami TS, Zhu J, Ma G, Maehara A, Mintz GS, Yuan C. Image-based modeling for better understanding and assessment of atherosclerotic plaque progression and vulnerability: data, modeling, validation, uncertainty and predictions. J Biomech. 47(4):834-46. 2014.

[3] Tang D, Yang C, Zheng J, Woodard PK, Sicard GA, Saffitz JE, Yuan C. 3D MRI-based multicomponent FSI models for atherosclerotic plaques. Ann Biomed Eng. 32(7):947-60, 2004.

[4] Tang D, Yang C, Zheng J, Woodard PK, Saffitz JE, Sicard GA, Pilgram TK, Yuan C. Quantifying effects of plaque structure and material properties on stress distributions in human atherosclerotic plaques using 3D FSI models. J Biomech Eng. 127(7):1185-94, 2005.

[5] Tang D, Yang C, Zheng J, Woodard PK, Saffitz JE, Petruccelli JD, Sicard GA, Yuan C. Local Maximal Stress Hypothesis and Computational Plaque Vulnerability Index for Atherosclerotic Plaque Assessment, Ann Biomed Eng. 33(12):1789-1801, 2005.

[6] Yang C, Bach RG, Zheng J, Naqa IE, Woodard PK, Teng Z, Billiar K, Tang D. In vivo IVUS-based 3D fluid structure interaction models with cyclic bending and anisotropic vessel properties for human atherosclerotic coronary plaque mechanical analysis, *IEEE Trans. Biomed. Engineering,* 56(10):2420-2428, 2009.

SB^{3}C2015
Summer Biomechanics, Bioengineering and Biotransport Conference
June 17-20, 2015,Snowbird Resort, Utah, USA
SB^{3}C2015-038

MULTIMODAL ASSESSMENT OF BONE QUALITY OF THE HUMAN RIB

Lauren M. Mangano (1), Jean-Paul Roux (2), François Duboeuf (2), Delphine Farlay (2), David Mitton (3), Hélène Follet (2)

(1) Biomedical Engineering
Boston University
Boston, MA, USA

(2) UMR 1033
INSERM, Université Lyon 1
Lyon, France

(3) LBMC UMR_T 9406
IFSTTAR, Université Lyon 1
Bron, France

INTRODUCTION

Osteoporosis is a bone disease characterized by an imbalance in bone remodeling, leading to a loss of bone and increased susceptibility to fracture. Studies of osteoporotic bone loss have focused on the spine, femoral neck, and radius—load-bearing sites susceptible to osteoporotic fracture. In contrast, there has been little investigation of the ribs, which experience only small bending loads during respiration. However, bone turnover is higher in the ribs than at other sites, and women with a history of rib fracture have increased risk of fracture at the rib and other sites. [1,2,3] The rib may be an important side for considering fracture risk in osteoporosis.

To improve prediction of an individual's fracture risk, it is necessary to understand which factors of bone quality contribute to mechanical properties. Bone loss in the ribs occurs due to increased endosteal resorption and increased intracortical porosity, both of which may affect mechanical properties. [4,5] Clinical modalities such as quantitative computed tomography (QCT) and ultrasound predict mechanical properties of the ribs. [6,7] However, bone quality of the ribs at the tissue level has not yet been linked to mechanical properties.

The goal of this study is to determine which bone quality factors contribute to mechanical properties of the cortical bone of the rib. This was accomplished using three-point bending, micro-computed tomography, microradiography, fourier-transform infrared microspectroscopy, and hardness testing. Understanding which factors contribute to mechanical properties will enable more effective diagnosis and treatment of osteoporosis in the rib.

METHODS

Sample Preparation: The 8th and 6th ribs were taken from 17 human donors (76.2±11.6 years old, 10 male, 7 female). Twelve-cm-long sections of the ribs were used for three-point bending testing. Adjacent one-cm-long sections were embedded in methyl methacrylate for subsequent characterization.

Three-Point Bending: Sections of the ribs were tested in three-point bending until failure (INSTRON 8802, High Wycombe, England) with a sampling rate of 1000 Hz, a 1 kN load cell, and a distance of 10 cm between the bottom supports. The modulus of elasticity and maximum stress (strength) were calculated analytically by modeling the rib as a beam with constant cross-section.

µCT (µCT): Scans of embedded rib samples were performed with a 19.1 µm isotropic voxel size (SkyScan 1174, Kontich, Belgium). For a manually-selected region of interest in the cortical bone, cortical porosity (Ct.Po) was determined using an adaptive threshold, while cortical bone mineral density (BMD) and tissue mineral density (TMD) were determined using a global threshold of 0.699 g/cm^3.

Microradiography: Images of 100±10 µm-thick sections were acquired using a quantitative digital micrograph at 40 kV and 50 µA, and equipped with a digital detector (Photonic Science, St Etienne de Geoirs, France). Images were calibrated using a step-wedge aluminum reference. Degree of mineralization (DMB) of the cortical bone was determined using custom software (Matlab, Mathworks,Natick, MA).

Fourier-transform infrared microspectroscopy (FTIRM): FTIRM was performed on 2 µm-thick sections in transmission mode (Perkin-Elmer, Courtaboeuf, France). Spectra were collected at 4 cm^{-1} resolution over the range 4000-450 cm^{-1} and were the average of 50 measurements. Spectra were collected at 20 locations in the cortical bone and with an aperture size of 100x100 µm. The spectra were analyzed using GRAMS/AI software (Thermo Galactic, Salem, NH). The following variables were calculated: 1) crystallinity index (CI, inversely proportional to FWHM of v_4PO_4); 2) mineral maturity ($v_1v_3PO_4$); 3) carbonation (v_2CO_3/ v_1 v_3PO_4); 4) mineral to matrix ratio (MM, $v_1v_3PO_4$/amide I).

Microhardness: Testing was performed using a Vickers indenter (Micromet 5104, Buehler, Lake Bluff, Illinois) with load of 25 g applied for 10 s at 20 locations in the cortical bone. Indents were imaged using optical software (OmniMet HMS, Buehler, Lake Bluff, Illinois). Vicker's hardness was determined based on the length of the diagonals of the indents.

Statistics: Normality of the distributions was checked and non-parametric tests were used. Differences between the 6th and 8th ribs and between male and female samples were identified using a Wilcoxon ranked sum test and a Mann-Whiteney test, respectively. Correlations among parameters were identified using the Spearman coefficient. Outliers (±2SD) were eliminated only in cases of unusual experimental conditions.

RESULTS

No differences were found between the 6th and 8th ribs for any parameters. No difference were found between males and females, except for heterogeneity index of DMB in the 8th rib, which was higher in males than in females ($p=0.03$).

In both ribs modulus and strength were highly correlated (6th: $r'=0.85$, $p<0.001$; 8th: $r'=0.82$, $p=0.001$). Neither modulus nor strength was correlated to the bone quality factors measured in this study. In the 8th rib, hardness was positively correlated with both mineral to matrix ratio and crystallinity index and was negatively correlated with carbonation. **(Figure 1)** In the 6th rib, there was a trend towards a correlation between hardness and mineral to matrix ratio ($r'=0.45$, $p=0.069$), but no correlation between hardness and crystallinity index or carbonation.

In both ribs, mineral to matrix ratio was positively correlated with crystallinity index (6th: $r'=0.53$, $p=0.027$, 8th: $r'=0.73$, $p=0.001$) and negatively correlated with carbonation (6th: $r'=-0.52$, $p=0.035$, 8th: $r'=-0.90$, $p<0.001$), while crystallinity index and carbonation were negatively correlated with each other (6th: $r'=-0.59$, $p=0.013$, 8th: $r'=-0.51$, $p=0.043$). In the 6th rib, DMB and mineral to matrix ratio were positively correlated ($r'=0.54$, $p=0.026$), while in the 8th rib, there was a borderline correlation between these parameters ($r'=0.44$, $p=0.087$). In the 6th rib, DMB was also correlated to BMD ($r'=0.53$, $p=0.031$). In both ribs, TMD and BMD are highly correlated ($r'=0.85$, $p<0.001$, $r'=0.78$, $p<0.001$).

DISCUSSION

In this study, we identified bone quality factors that contributed to microhardness, but not the modulus or hardness, of the rib. Although the mechanical properties of the 6th and 8th rib are not different, different factors are correlated to microhardness in each rib.

Bone quality factors measured by FTIRM influenced mechanical properties at the length scale of a hardness test. The hardness of the 8th rib is influenced not only by mineral to matrix ratio, which quantifies the amount of mineral, but also by crystallinity and carbonation, which describe the structure and composition of mineral. In contrast, the hardness of the 6th showed only a tentative dependence on mineral to matrix ratio, indicating a less nuanced relationship between mineral composition and hardness. These hardness results are consistent with studies of bone at other sites. [8, 9] Generally increased amount of mineral, represented by mineral to matrix ratio, and increased crystal size and perfection, represented by crystallinity index, are associated with increased hardness. The negative correlation between carbonation and hardness is likely due to decreased crystallinity index associated with increased carbonation.

In contrast, neither modulus nor strength was correlated to the intrinsic bone quality factors examined in this study, despite the fact that bone density is typically associated with these mechanical properties. In fact, a previous study on a subset of this data demonstrated that BMD measured by QCT is correlated to strength and modulus, calculated using beam theory (unpublished data) and numerical methods. [6] This discrepancy may be due to the different resolutions of the CT measurements or to the fact that measurements of bone quality in this study were performed on sections of ribs adjacent to the section used for three-point bending.

This study indicates that the mechanical properties of the 6th and 8th rib are dependent upon distinct bone quality factors. These differences may be caused by the different *in vivo* stresses experienced by the 6th rib and the 8th rib. The results from this study further indicate bone quality factors may influence mechanical properties at specific length scales.

ACKNOWLEDGEMENTS

This material is based upon work supported by a National Science Foundation Graduate Research Fellowship under Grant No. DGE-1247312 and National Science Foundation Graduate Research Opportunities Worldwide. The authors acknowledge Sébastien Rizzo, Camille Ponçon, Brigitte Burt-Pichat, and Nathalie Portero-Muzy for their technical expertise and Doris Tran, Charlene Delimoge, and Yuan Fang for their biomechanical results.

REFERENCES

[1] Frost, H, *Calcified Tissue Research,* 3:211-237, 1969.
[2] Ismail, A et al., *Osteoporosis Int,* 17:41-45, 2006.
[3] Sajjan, S et al., *Osteoporosis Int,* 23:821-828, 2012.
[4] Sedin; ED et al., *J Gerontol,* 18:9-13, 1963.
[5] Agnew, AM et al., *Am J Phys Anthropol,* 148;462-466, 2013.
[6] Zhu, YM et al., *Comput Method Biomech Biomed Engin,* 16:1-2, 2013.
[7] Mitton, D et al., *J Biomech,* 47:1548-1553.
[8] Bala, Y et al., *J Mech Behav Biomed Mater,* 4:1473-1482, 2011.
[9] Burket, J et al., *J Biomech,* 44:277-284, 2011.

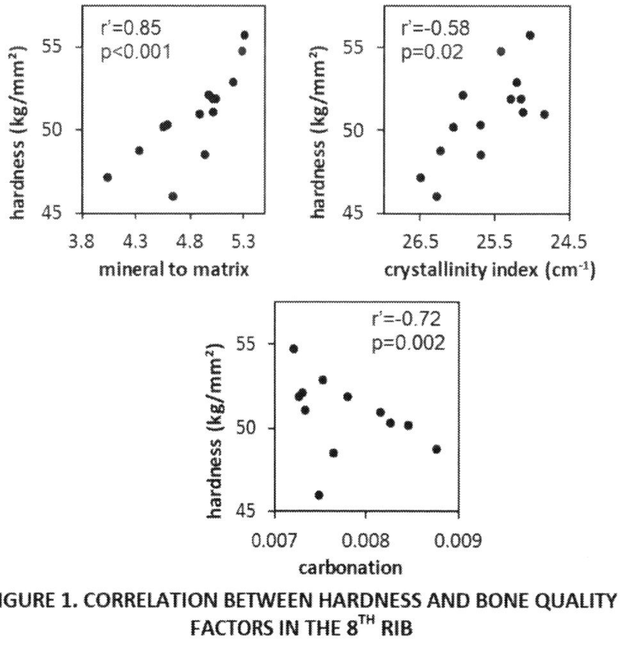

FIGURE 1. CORRELATION BETWEEN HARDNESS AND BONE QUALITY FACTORS IN THE 8TH RIB

INVESTIGATION OF HEAD ROTATIONAL IMPULSE CHARACTERISTICS ON BRAIN STRAIN RESPONSES

W. Zhao (1), S. Ji (1,2)

(1) Thayer School of Engineering
Dartmouth College
Hanover, NH, USA

(2) Department of Surgery and of
Orthopaedic Surgery
Dartmouth College
Hanover, NH, USA

INTRODUCTION

Sports-related concussion remains a major socioeconomic and public health problem. An urgent need exists to investigate its biomechanical mechanisms and to understand its injury etiology. Currently, head injury criterion (HIC) based on the temporal profile of resultant linear acceleration ($\boldsymbol{a}_{lin}$) measured at the head center of gravity is widely used in government agencies and industries as the head injury standard. While it has been effective in mitigating injury risks in vehicular crash impacts where severe brain injury often dominates, its use in the mild spectrum of brain injury has been criticized. This is because tissue-level strain responses that are believed to initiate brain injury in the first place are significantly related to rotational acceleration ($\boldsymbol{a}_{rot}$) but not to $\boldsymbol{a}_{lin}$ [1]. Consequently, recent attempts have been made to devise injury risk metrics by combining the peak magnitudes of both $\boldsymbol{a}_{rot}$ (a_{rot}^p) and $\boldsymbol{a}_{lin}$ (a_{lin}^p) or even based on a_{rot}^p alone, in order to assess the concussion risk in contact sports [2–4].

However, brain strain responses are significantly correlated to rotational velocity ($\boldsymbol{v}_{rot}$) as opposed to $\boldsymbol{a}_{rot}^p$ alone [1], which was further confirmed recently when simulating different $\boldsymbol{a}_{rot}$ temporal profiles with an identical velocity based on the Dartmouth Head Injury Model (DHIM [5]). Because $\boldsymbol{v}_{rot}$ is more informative for brain strain responses, a brain injury criterion (BrIC) was proposed that was based on $\boldsymbol{v}_{rot}$ peak magnitude (v_{rot}^p) in order to assess the risk of strain-induced injury [6]. These studies suggest that a potential limitation of injury risk metrics defined based on the peak magnitude of rotational impulses only is the lack of the consideration of its duration, Δt_{rot}. Therefore, in order to enhance our understanding of the rotational brain injury mechanism, we parametrically investigated the significance of rotational impulse characteristics on brain strain responses in this study.

METHODS

All brain strain responses were simulated using DHIM, which has been successfully validated against relative brain-skull displacement and intracranial pressure from cadaveric experiments as well as strain from a live human volunteer [1,7,8].

Rotational impulses with a sine, a triangular, and a haversine temporal shape were used as model inputs rotating about the medial-lateral axis, effectively simulating head frontal impacts (rotation within the sagittal plane). To design the test matrix, we first kept both a_{rot}^p and Δt_{rot} constant (5000 rad/s^2 and 10 ms; **Fig 1** a) with these impulse shapes. Then we kept both v_{rot}^p and Δt_{rot} constant (25 rad/s and 10 ms; **Fig 1** b). Finally, using the triangular impulse shape, a_{rot}^p and Δt_{rot} were adjusted to maintain an identical v_{rot}^p (25 rad/s; **Fig 1** c). These $\boldsymbol{a}_{rot}$ temporal profiles were used as DHIM inputs.

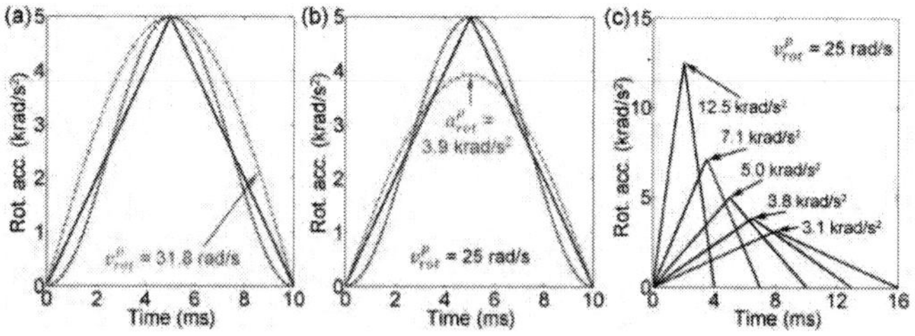

Fig. 1 Temporal $\boldsymbol{a}_{rot}$ profiles using different impulse shapes with equal a_{rot}^p (a) or equal v_{rot}^p (b). For the triangular impulse, a_{rot}^p and Δt_{rot} were adjusted to maintain an identical v_{rot}^p (c)

Element-wise peak strain values during the entire simulation regardless of the time of occurrence (ε^p) were extracted from each simulation. Regional volume-weighted averages of ε^p (ε^p_{VWA}) in generic brain regions of interest (ROIs) were also calculated.

RESULTS

By maintaining the same a^p_{rot} and Δt_{rot}, the sine impulse resulted in the highest strain (**Fig. 2**a), while the triangular and the haversine impulses led to nearly identical strains (**Fig. 2** b and c). Maintaining an identical v^p_{rot} and Δt_{rot}, the impulse shape had little effect on strain responses (**Fig. 3**), even when the a^p_{rot} differed by 22% (**Fig. 1**b). Using the triangular impulse shapes, increasing Δt_{rot} lowered ε^p (**Fig. 4** a and b) for an identical v^p_{rot}. The regional ε^p_{VWA} in all ROIs (**Fig. 4**c) differed by 13.2% in the cerebrum to 29.7% in the cerebellum (when Δt_{rot} varied from 4 to 16 ms; **Fig. 4**c).

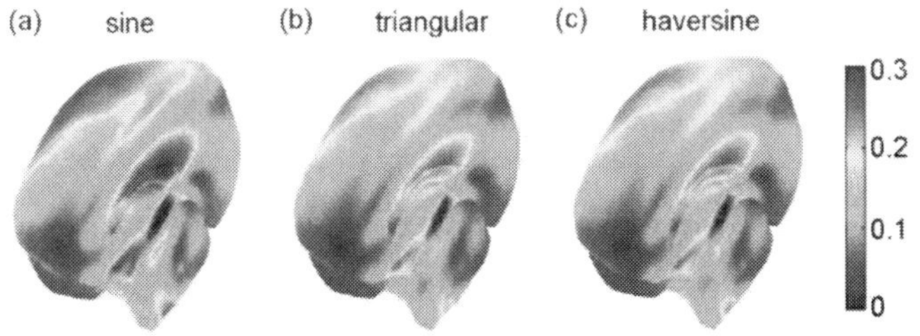

Fig. 2 Comparison of ε^p resulted from different impulse shapes with an identical a^p_{rot} and Δt_{rot}

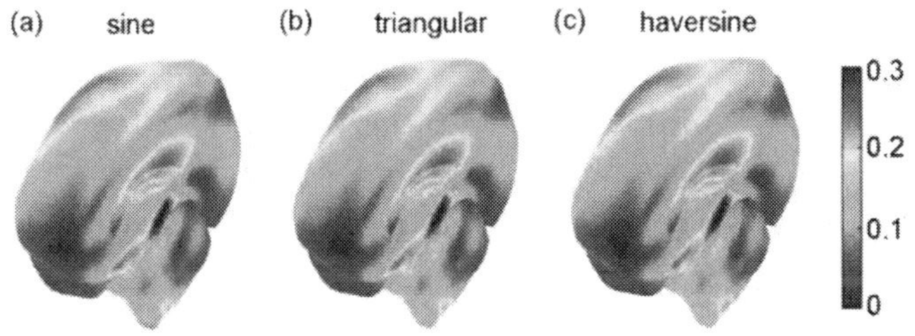

Fig. 3 Comparison of ε^p resulted from different impulse shapes with an identical v^p_{rot} and Δt_{rot}

DISCUSSION

Once again, our current study confirms that v^p_{rot} is more informative for brain strain responses than a^p_{rot} alone, highlighting the significance of the impulse duration, Δt_{rot}. With the same v^p_{rot} and Δt_{rot}, brain strain appears insensitive to the $\boldsymbol{a}_{rot}$ temporal shapes. Maintaining the same Δt_{rot}, nearly identical brain strain responses were obtained as long as the same v^p_{rot} was maintained, even though the difference between a^p_{rot} values could be significant (i.e., by more than 22%; **Fig. 1**). This finding is in agreement with a previous parametric study [9], and underscores the importance of relating $\boldsymbol{v}_{rot}$ to brain regional strain response levels. An important implication from these results is that the accuracy of v^p_{rot} measurement, as opposed to that of a^p_{rot}, should be adopted to characterize the effectiveness of a head impact measurement system [10,11].

As expected, Δt_{rot} also significantly affects brain strain responses. Even with an identical v^p_{rot}, the brain strain levels were lowered with the increase in Δt_{rot} for a given impulse shape, as previously observed [9]. Such a behavior is likely a result of the brain's viscoelasticity because longer Δt_{rot} increases the dissipation of brain strain energy [5]. This finding suggests that for a given v^p_{rot}, extending the impact duration would likely reduce the risk of brain injury, which may have important practical implications when designing head protection gears (e.g., helmet, airbag, etc.).

To summarize, this study parametrically investigated the significance of head rotational impulse characteristics on brain strain responses. We found that both v^p_{rot} and Δt_{rot} play an important role in determining brain strain responses. Results confirm and further extend other parametrical investigations of rotational impulses based on mono-axis, mono-phasic impulses [5,9]. Future studies involving multi-phasic acceleration impulses may be warranted.

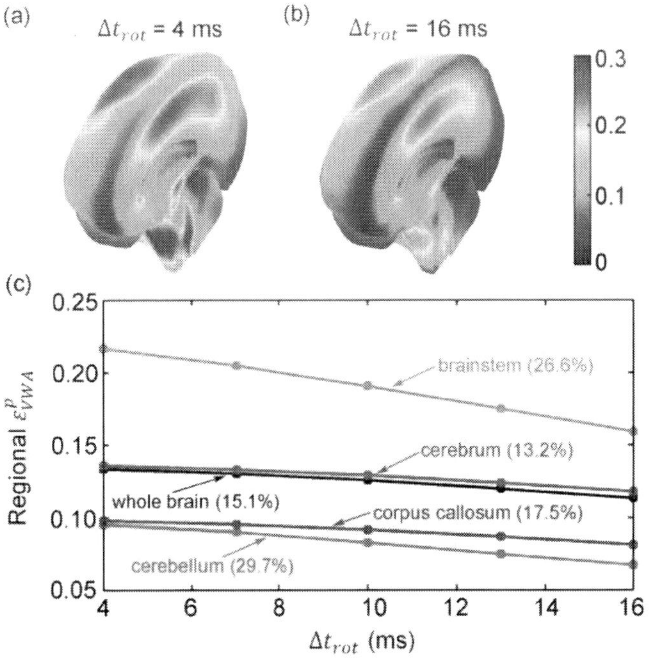

Fig. 4 Comparison of ε^p for the triangular impulses with an identical v^p_{rot} when Δt_{rot} was 4 ms (a) and 16 ms (b). Changes in regional ε^p_{VWA} in different ROIs due to the change in Δt_{rot} are shown in (c). The maximum absolute regional differences in ε^p_{VWA} relative to that obtained when Δt_{rot} was 4 ms are shown in parenthesis

ACKNOWLEDGEMENTS

This work was sponsored, in part, by the NIH Grant R21 NS078607 and the Dartmouth Hitchcock Foundation.

REFERENCES

[1] Ji et al., *Biomech. Model. Mechanobiol.*, 13(5): 1121–1136, 2014.
[2] Rowson et al., *Ann. Biomed. Eng.*, 40(1): 1–13, 2012
[3] Rowson et al., *Ann. Biomed. Eng.*, 41(5): 873–882, 2013.
[4] Zhang et al., *J. Biomech. Eng.*, 126(2): 226–236, 2004.
[5] Ji et al., *Ann. Biomed. Eng.*, 2014. DOI: 10.1007/s10439-014-1193-3.
[6] Takhounts et al., *Stapp Car Crash J.*, 57: 243–266, 2013.
[7] Ji et al., *J. Neurotrauma.*, 2014. DOI:10.1089/neu.2013.3268.
[8] Zhao et al., *Biomech. Model. Mechanobiol.*, 2014. DOI: 10.1007/s10237-014-0634-0.
[9] Kleiven, *Int. J. Crashworthiness*, 11(1): 65–79, 2006.
[10] Camarillo et al., *Ann. Biomed. Eng.*, 41(9): 1939–1949, 2013.
[11] Bartsch et al., *Stapp Car Crash J.*, 58: 1–28, 2014.

THE DYNAMIC MECHANICAL RESPONSE IS SEVERELY ALTERED IN COLLAGEN V DEFICIENT MOUSE SUPRASPINATUS TENDONS

Brianne K. Connizzo (1), Mei Sun (2), David E. Birk (2), Louis J. Soslowsky (1)

(1) McKay Orthopaedic Research Laboratory
University of Pennsylvania
Philadelphia, PA, United States

(2) Department of Molecular Pharmacology & Physiology
University of South Florida
Tampa, FL, United States

INTRODUCTION

Classic (type I) Ehlers-Danlos Syndrome (EDS) is a rare connective tissue disorder characterized by mutations in collagen V, typically a single *Col5a1* null allele. Patients with classic EDS exhibit connective tissue hyperelasticity, joint hypermobility and recurrent dislocations [1]. Although collagen V is a quantitatively minor component of many collagen I-rich connective tissues, its removal results in drastic changes in tissue structure and function due to its critical role in the regulation of collagen fibrillogenesis [2,3].

In tendon, collagen V haploinsufficiency or removal results in significantly fewer fibrils with larger diameters as well as altered mechanical properties in the mouse flexor digitorum longus [4]. Recent studies also have investigated the effect of alterations in collagen V expression on the mouse anterior cruciate ligament, Achilles tendon, and supraspinatus tendon, reporting that the magnitude of the effect of is tendon-specific, with larger effects on tendons that directly contribute to joint stability [5,6].

However, these studies only measured the quasi-static mechanical response. The purpose of this study was to investigate the dynamic mechanical response in tendons with reduced or absent collagen V expression. This was accomplished using the classic EDS mouse model and a tendon/ligament specific collagen V null mouse. We hypothesized that both heterozygous and null animals would have reduced area, maximum load and stress, quasi-static and dynamic moduli, as well as increased viscous properties (tanδ).

METHODS

Sixty mice at 120 days of age from three genotypes, *Col5a1*$^{+/+}$ (WT, n=20), *Col5a1*$^{+/-}$ (Het, n=20), and Scx-*Cre*$^{+}$/*Col5a1*$^{flox/flox}$ (Null, n=20) [6], were used in this IACUC-approved study. Following sacrifice, body weight was measured and supraspinatus tendons were dissected and prepared for mechanical testing [7].

All soft tissue adjacent to the tendon was removed, leaving the supraspinatus tendon attached to the humerus. Verhoeff's stain lines were placed on the tendon denoting the tendon insertion site and midsubstance for regions of strain analysis. Tendon cross-sectional area was measured using a custom laser-based device. The humerus was then embedded in a 1-inch diameter acrylic tube piece with the use of PMMA. The proximal end of the tendon was glued between two pieces of sandpaper with an initial gauge length of 2.5mm and both the tendon and the acrylic pot were placed in custom grips for tensile testing.

Specimens were tested using a standard dynamic testing protocol, consisting of 1) preload, 2) 10 cycles of preconditioning, 3) stress relaxation to 4% strain, 4) sinusoidal frequency sweep, 5) stress relaxation to 6% strain, 6) sinusoidal frequency sweep, 7) stress relaxation to 8% strain, 8) sinusoidal frequency sweep, (9) return to zero and 10) ramp to failure at 0.05%/sec. Each frequency sweep consisted of 10 cycles of sinusoidal strain at frequencies of 0.1, 1.0, and 10 Hz. Images also were obtained during the ramp to failure for optical strain analysis.

The dynamic modulus $|E^*|$ (the stress amplitude divided by the strain amplitude) and the tangent of the phase angle δ between stress and strain were computed using a custom program (Matlab, Natick, MA) [8]. $|E^*|$ represents the material strength under dynamic loading and tanδ is a measure of viscous dissipation in the tissue. Stain lines were tracked optically to measure regional mechanical strain. A bilinear curve fit was applied to the data to obtain measures of toe- and linear-region moduli as well as the transition stress and strain as previously described [7]. Our specific hypotheses were best tested in a manner utilizing a series of one-way ANOVAs with comparisons only across genotype while keeping all other factors constant. Post-hoc Bonferroni corrected t-tests were performed with significance set at $p<0.01667$.

RESULTS

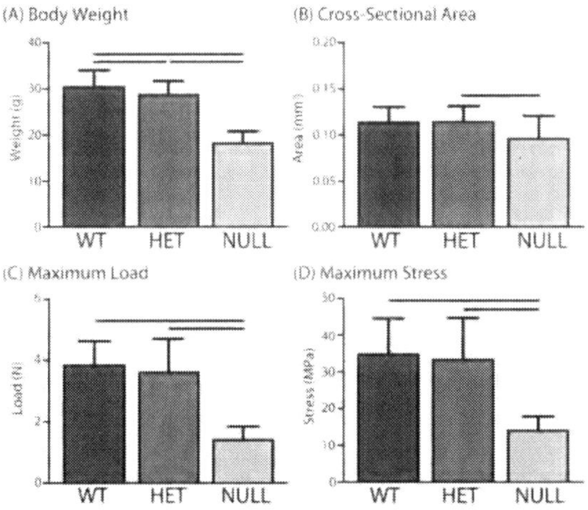

Figure 1. (A) Body weight is decreased in both Het and Null animals, while (B) tendon area is only decreased in the Null group when compared to the Het group. (C) Maximum load and (D) Maximum stress were both severely decreased in the Null group.

The Het and Null groups both had decreased body mass when compared to the Wild Type (WT) group (Fig. 1A). The Null group also exhibited decreased tendon area (Fig. 1B). There were no differences in transition strain across groups (Fig. 2A), but transition stress was decreased for the Null group at the insertion site (Fig. 2B). Maximum load and maximum stress (Figure 1C,D) as well as toe and linear modulus at both locations (Fig. 2C-D) were significantly and severely decreased in the Null group. The Het group had decreased midsubstance toe modulus and insertion site linear modulus. For dynamic properties, the Null group exhibited increased tanδ (Fig. 3A) at all but one frequency/strain combination compared to the WT group and at all frequencies at 6% and 8% strain when compared to the Het group. The Null group had a lower dynamic modulus compared to the WT group at 6% strain and to both groups at 8% strain (Fig. 3B).

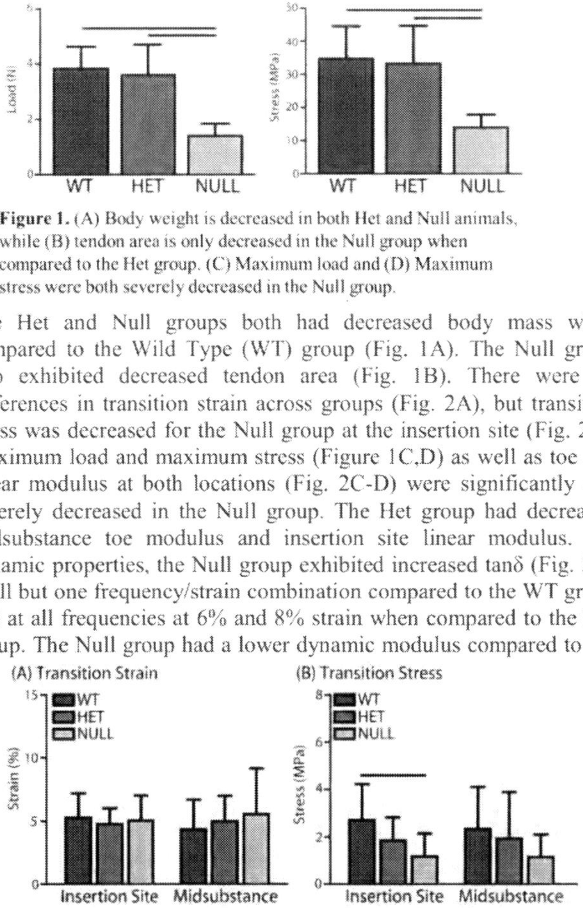

Figure 2. (A) No differences in transition strain were found between groups, but (B) transition stress was decreased in the Null group at the insertion site. (C) Toe and (D) linear moduli were both severely decreased in the Null group when compared to WT at both locations. The Het group also had a reduced toe modulus at the midsubstance and reduced linear modulus at the insertion.

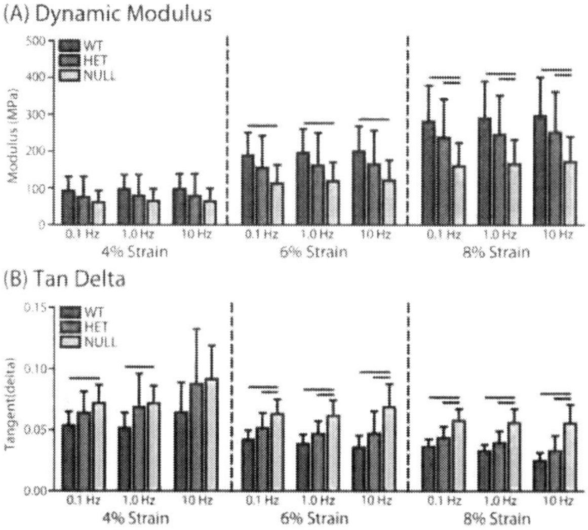

Figure 3. (A) Dynamic modulus was significantly decreased in the Null group at 6% and 8% strain for all frequencies. (B) Tan delta was significantly increased in the Null group at all frequencies and strain levels except at 4% strain and 10 Hz.

DISCUSSION

Collagen V null tendons exhibit inferior mechanical properties when compared to wild type tendons in almost every mechanical parameter. On the other hand, the classic EDS tendons (Het group) exhibited smaller mechanical changes than the null group. Although many of the parameters appear to have a dose-dependent response, the changes were small between the WT and Het groups. However, the Het group did have a decreased toe modulus in the midsubstance and a decreased linear modulus at the insertion site, confirming the functional deficit.

The cause of these mechanical insufficiencies is still unknown. While several studies have attempted to correlate mechanical properties with tissue composition and/or structure, many have been unable to fully explain the mechanical behavior. Particularly interesting is the dynamic response at 4% strain, where there were no differences between any of the groups in dynamic modulus at any frequency. This could suggest that dynamic responses during the transition and linear region of the mechanical curve (after 4% strain) are responsible for the mechanical insufficiency. Future studies will assess the mechanisms behind this severe deficiency by fully defining the structure, composition and dynamic response to load in tendons with altered collagen V contents.

ACKNOWLEDGEMENTS

This study was supported by NIH/NIAMS (AR065995, AR044745) and the Penn Center for Musculoskeletal Disorders (NIH, P30 AR050950). We thank Qingmei Yao for technical expertise.

REFERENCES

[1] Malfait, F et al. *Genet Med,* 12:597-605, 2010. [2] Birk, DE. *Micron,* 32:223-37, 2001. [3] Wenstrup, RJ et al. *J Biol Chem,* 279:53331-7, 2004. [4] Wenstrup, RJ et al. *J Biol Chem,* 286:20455-65, 2011. [5] Connizzo, BK et al. Manuscript under review, 2014. [6] Sun, M et al. *Am J Pathol,* In press, 2015. [7] Connizzo, BK et al. *J Biomech Eng,* 135:021019, 2013. [8] Dunkman, AA et al. *Matrix Biol,* 32:3-13, 2013.

SB³C2015
Summer Biomechanics, Bioengineering and Biotransport Conference
June 17-20, 2015, Snowbird Resort, Utah, USA
SB³C2015-107

EFFECT OF BRANCHED AND FENESTRATED STENT-GRAFTS ON RENAL BLOOD FLOW

Harkamaljot S. Kandail (1), Mohamad S. Hamady (2), Xiao Y. Xu (1)

(1) Department of Chemical Engineering
Imperial College London
London, UK

(2) Department of Interventional Radiology
Imperial College London
London, UK

INTRODUCTION

Fenestrated stent-grafts (FSG) were introduced to make endovascular aneurysm repair (EVAR) feasible for patients not suitable for standard EVAR. FSG allow for total exclusion of aneurysm sac from any blood perfusion along with providing additional anchoring force for safe deployment. A slight misalignment of renal side branches in FSG can lead to serious complications such as renal occlusion. Antegrade and retrograde stent-grafts, collectively known as branched stent-grafts (BSG), can be seen as successors to FSG.

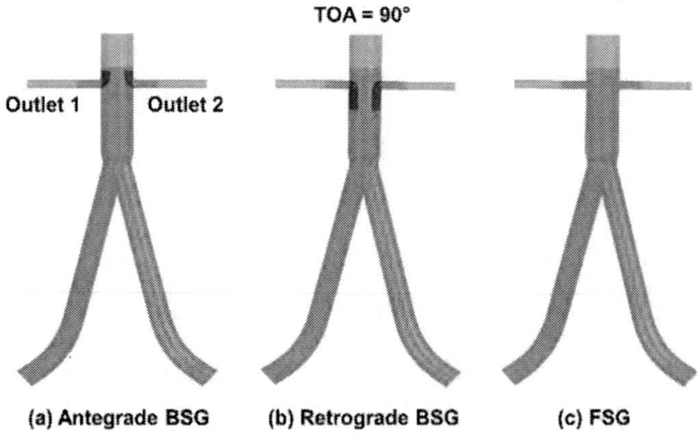

FIGURE 1: SCHEMATICS OF IDEAL BRANCHED AND FENESTRATED STENT-GRAFTS.

As shown in Figure 1, BSG have inner docking lumen in the main endograft body, and the docking capability of BSG not only make it easy to cannulate the renal arteries, but also has the potential to reduce the chances of renal occlusion. Although BSG are conceptually more appealing than FSG with respect to renal stability and ease of deployment, there are certain critical questions which must be answered beforehand in order to ensure durable device design. Some of these key questions are: (a) the impact of antegrade and retrograde BSG on renal blood flow, (b) the consequence of misaligned renal branches on renal blood flow and (c) displacement forces acting on all stent-grafts. Answering these key questions will help us predict the likelihood of future complications such as occlusion. The objective of this study is to address the aforementioned questions using computational fluid dynamics (CFD).

METHODS

Ideal models of BSG and FSG were constructed with different take-off angles (TOA) and planar lateral aortic neck angle of 60°. TOA was defined as the angle between the centerlines of the main endograft body and side branches (outlets 1 and 2), with 90° representing normal alignment, while 30° and 120° representing severely and mildly misaligned side branches, respectively.

CFD simulations were performed by solving the mass and momentum conservation equations governing the laminar blood flow [1]. As blood is a non-Newtonian fluid, it was modelled using experimentally proven Quemada viscosity model. In order to obtain clinically relevant results, physiologically realistic flow waveform was prescribed at the inlet [2] along with Womersley velocity profiles. Outlet pressure waveforms were obtained by coupling all the outlets with a 3-Element Windkessel model (3-EWM). The parameters of 3-EWM were obtained using Nelder-Mead Simplex algorithm. No slip conditions were specified at the stent-graft walls, which were assumed to be rigid.

RESULTS

For stent-grafts with a straight aortic neck, at every TOA, maximum renal flow was observed in FSG, followed by antegrade BSG while minimum renal flow was observed in retrograde BSG. Cycle-averaged percentage split of the aortic inflow into each renal artery is summarized in Figure 2. As these stent-grafts were planar and symmetric, flow split between the left and right renal arteries (outlet 1 and 2) was 50:50.

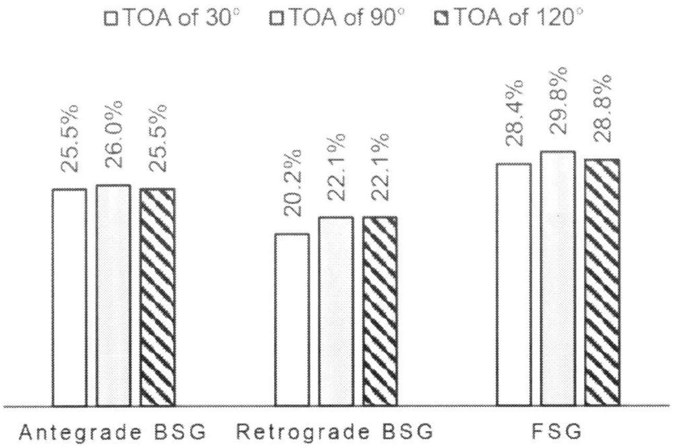

FIGURE 2: PERCENTAGE FLOW SPLIT OF THE AORTIC INFLOW INTO EACH RENAL ARTERY FOR STENT-GRAFTS WITH STRAIGHT AORTIC NECK AND VARYING TOA.

For stent-grafts with a lateral aortic neck angulation of 60° and TOA of 90°, under identical inflow conditions, cycle-averaged outflow into each renal artery is summarized in Figure 3. Due to aortic neck angulation, these models become asymmetric, thus the flow split between outlet 1 and 2 was no longer 50:50, with the exception of antegrade BSG.

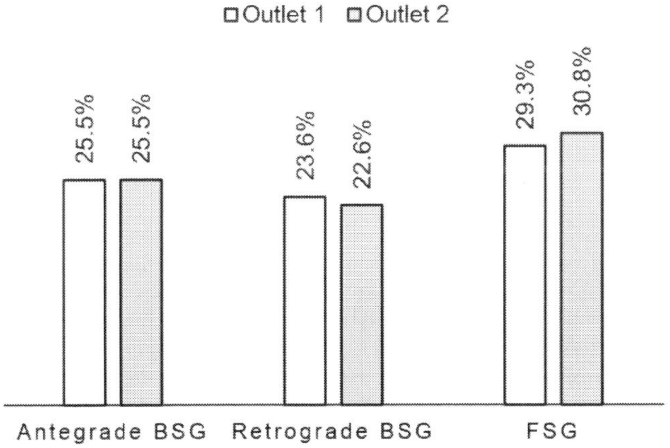

FIGURE 3: PERCENTAGE FLOW SPLIT OF THE AORTIC INFLOW INTO EACH RENAL ARTERY FOR STENT-GRAFTS WITH AORTC NECK ANGLE OF 60° AND TOA OF 90°.

In terms of displacement forces, at TOA of 30°, 90° and 120°, all stent-grafts with a straight aortic neck experienced equal cycle-averaged displacement forces of 1.25 N, 1.69 N and 1.95 N, respectively. All stent-grafts with a lateral aortic neck angle of 60° and TOA of 90° experienced equal cycle-averaged displacement force of 3.60 N.

DISCUSSION

For BSG and FSG to be effective for EVAR, along with excluding aneurysm sac from any blood flow, they must also maintain sufficient perfusion to kidneys throughout the entire cardiac cycle. From *in vivo* ultrasound measurements, it has been found that under normal physiological conditions, cycle-averaged blood flow to each renal artery is between 15-25% of the aortic inflow [3]. Based on these measurements, it was assumed that a minimum of 15% aortic flow would be required to maintain sufficient blood perfusion to kidneys. As shown in Figure 2 and Figure 3, cycle-averaged blood flow in each renal artery for antegrade BSG, retrograde BSG and FSG was higher than the threshold of 15%, therefore they can all be used for EVAR without compromising renal perfusion.

It can also be observed from Figure 2 that percentage renal flow depends not only on the configuration of the stent-graft but also on renal alignment, as lower renal flow was obtained in severely misaligned renal branch (TOA of 30°), especially in retrograde BSG. However even in this case, which corresponded to minimum renal flow of 20.2%, there was sufficient perfusion to the kidneys. As summarized in Figure 3, sufficient renal perfusion was also observed in stent-grafts with a lateral aortic neck angle of 60° and TOA of 90°. Different from the cases with a straight aortic neck, aortic neck angulation increased the flow to renal branches by 4.5% and 0.8% in retrograde BSG and FSG, respectively.

It is also important to quantify the displacement forces experienced by all the stent-grafts because displacement force is strongly correlated with device migration [1], which can lead to Type Ia and Type Ib endoleaks as well as in case of BSG and FSG, renal occlusion. Our results show that the magnitude of cycle-averaged displacement force is dependent more on TOA rather than the configuration of the stent-graft. Linear regression analysis between TOA and displacement forces yielded an R^2 value of 0.99 for BSG and FSG, suggesting that displacement force increased almost linearly with TOA. Nevertheless, the increase in the magnitude of displacement force with TOA was relatively small compared to the change brought on by aortic neck angulation. When TOA increased from 30° to 90°, the cycle-averaged displacement force increased from 1.25 N to 1.69 N, but when neck angle changed from straight to 60° for a TOA of 90°, the mean displacement force was more than doubled from 1.69 N to 3.60 N. Therefore, the results suggest that the displacement force experienced by stent-grafts is more sensitive to lateral aortic neck angle than TOA.

This is an on-going research aimed at evaluating the hemodynamic and biomechanical effects of stent-grafts for endovascular repair of abdominal aortic aneurysms.

ACKNOWLEDGEMENTS

This research project is funded by Engineering and Physical Sciences Research Council, UK.

REFERENCES

1. Kandail H, Hamady M and Xu XY. Patient-specific analysis of displacement forces acting on fenestrated stent-grafts for endovascular aneurysm repair. *J Biomech*, 47:3546-3554, 2014.

2. Fraser KH, Meagher S, Blake JR, Easson WJ and Hoskins PR. Characterization of an abdominal aortic velocity waveform in patients with abdominal aortic aneurysm. *Ultrasound Med Biol*, 34:73-80, 2008.

3. Hennerici, MG and Neuerburg-Heusler D. *Vascular diagnosis with ultrasound: Clinical references with case studies*. 1st edition. Stuttgart: Thieme Medical Publishers; 1998.

THE DEGREE OF OUTFLOW TRACT BANDING PREDICTS CARDIAC REMODELING IN CHICKEN EMBRYOS

Madeline Midgett (1), Sandra Rugonyi (1)

(1) Biomedical Engineering
Oregon Health & Science University
Portland, Oregon, USA

INTRODUCTION

Congenital heart defects affect nearly 1% of newborns and are the leading cause of infant death in the United States [1, 2]. Embryonic heart formation is a finely orchestrated interplay between genetic and environmental factors. Blood flow during early embryonic stages plays a critical role in heart development, as constant interactions between blood flow and cardiac tissues generate biomechanical forces that modulate cardiac growth and remodeling. Normal hemodynamic conditions are essential for proper cardiac development, while altered blood flow is known to result in cardiac malformation and defects seen in congenital heart disease. Although it is clear that biomechanical forces are fundamental components of heart morphogenesis, the mechanisms that relate altered blood flow to congenital heart defects are unknown.

Development of congenital heart defects can be mimicked by surgically modifying blood flow in animal models. Chicken embryos, in particular, are often used as a biological model of human heart development because of ease of accessibility in the egg for surgical manipulation and imaging. Cardiac development among vertebrate species is highly conserved, and consists of a straight tube that transforms through looping and septation processes into a four-chambered heart [3]. A well-established intervention called outflow tract (OFT) banding is used to constrict the diameter of the early embryonic OFT of the heart, which increases the hemodynamic load (pressure and shear stresses) on cardiac tissues. These altered forces are known to ultimately lead to a wide spectrum of congenital heart defects. Using a chick embryo model, we have recently determined that the amount of increased pressure and blood flow velocity due to OFT banding can be controlled by the degree of band constriction [4-5]. Varied OFT constriction produces a range of hemodynamic perturbation that can be used to evaluate how the developing cardiovascular system adjusts and remodels in response to diverse hemodynamics loads. The current study extends this work in the context of characterizing subsequent cellular and tissue remodeling. Specifically this research determines the effects of altered blood flow on cellular structure of the OFT wall, and characterizes how varied hemodynamics during development affect mature cardiac formation.

METHODS

Chick embryo preparation and hemodynamic intervention. Fertilized White Leghorn chicken eggs were incubated blunt end up at 38°C and 80% humidity until stage HH18 (~3 days). Three groups of embryos were studied: 1) normal group, no interventions were performed; 2) control group, a 10-0 nylon suture was passed under the OFT but not tightened; 3) OFT banded group, a 10-0 nylon suture was passed under the OFT and tied in a knot around the mid-section of the OFT to constrict the cross-sectional area. Band tightness ranged from 0-60% in the banded group. Following interventions, eggs were sealed with saran wrap and incubated until evaluation. The OFT was longitudinally imaged with optical coherence tomography before and after banding to calculate the band tightness of each banded embryo using the change in OFT diameter.

Electron microscopy imaging and analysis of the OFT wall. The ultrastructure of OFT tissue was imaged with electron microscopy at HH24 after 24 hours of manipulation to determine the effect of altered hemodynamics on cellular remodeling. Tissue was collected and processed for transmission electron microscopy (TEM) and focused ion beam scanning electron microscopy (FIB-SEM). TEM was used to identify areas of interest, and FIB-SEM was used to acquire serial (3D), high-resolution images through the endocardium. Image stacks were then aligned to create approximately ~0.8 mm^3 isotropic image volumes. Cellular endocardial structure was manually segmented from the images using Amira software in order to reconstruct and quantify the 3D cellular orientation. Results were characterized based on the degree of band constriction.

Ultrasound biomicroscopy and micro-CT imaging and analysis of mature cardiac formation. Hearts were imaged with ultrasound biomicroscopy (UBM) and X-ray micro-computed tomography (micro-CT) in late embryonic development at HH36 after blood flow was manipulated for 24 hours during early development stages to determine the effects of varied hemodynamics on mature cardiac function and structure. Blood flow was detected *in vivo* using Color Doppler UBM images to identify flow throughout the heart. After UBM imaging, the hearts were collected and stained for micro-CT to identify major structural cardiac defects. Results were characterized based on the degree of band constriction

RESULTS

After 24 hours of banding, transmission electron microscopy (TEM) images demonstrate a disrupted OFT endocardium in banded embryos compared to controls. Control embryos have a tightly arranged endocardium with plump and round cells (Figure 1A), while the altered arrangement in banded embryos displays hallmarks of a more mature configuration in the process of endothelial mesenchymal transition (EMT). Banded embryo endocardial cells have projections that invade the cardiac jelly layer with large gaps between cells (Figure 1C). Cellular reconstructions display dramatic cell projections and gaps between cells in the banded embryo compared to the control (Figure 1 B, D). The degree of cell elongation was quantified using Amira reconstruction software where the elongation factor of 3D cellular surfaces (based on a ratio of eigenvector values) of banded endocardium with 35% band tightness (n=1) was 60% larger than control endocardium (n=1).

Ventricular septal defect (VSD) was identified in 33% of banded embryos (n=9) at HH36, once the heart was four-chambered with mature valves. This defect, which is a hole in the ventricular septum, only occurred in banded embryos with greater than 36% constriction. Flow was detected through the ventricular septum with Color Doppler UBM and the hole in the septum was confirmed with micro-CT (Figure 2).

DISCUSSION

Blood flow manipulation in early development triggers abnormal tissue remodeling that eventually leads to cardiac defects in the mature heart. Quantification of the alterations in cellular shape and organization that precede cardiac defects characterizes how increased hemodynamic load interferes with development. In early development, a portion of OFT endocardial cells transform into mesenchymal cells that invade the cardiac jelly through the process of EMT. As EMT progresses, the original plump and tightly organized endocardium disassembles and rearranges to form an intact flattened layer [6]. This study displays altered endocardium organization in banded tissue compared to control at HH24, indicating that increased hemodynamic load after banding modifies the progression of EMT in the OFT.

This study also evaluated whether the diverse biomechanical conditions that result from a range of band constrictions generate distinct cardiac defects in the mature heart. Previous studies have reported a spectrum of structural cardiac defects following OFT banding, including VSD, double outlet right ventricle, aortic arch interruption, and valve malformation, however the degree of band constriction was not reported. This study found that VSD develops with band tightness greater than 36%, indicating that the altered hemodynamics created by tight bands trigger detrimental development processes that result in VSD.

Future research plans include characterizing mitochondrial and myofibril organelle shape and orientation in acquired FIB-SEM data, as well as cardiac dysfunction *in vivo* with UBM.

ACKNOWLEDGEMENTS

This work has been supported by grants NIH R01 HL094570 and internal OHSU Center for Spatial Systems Biomedicine.

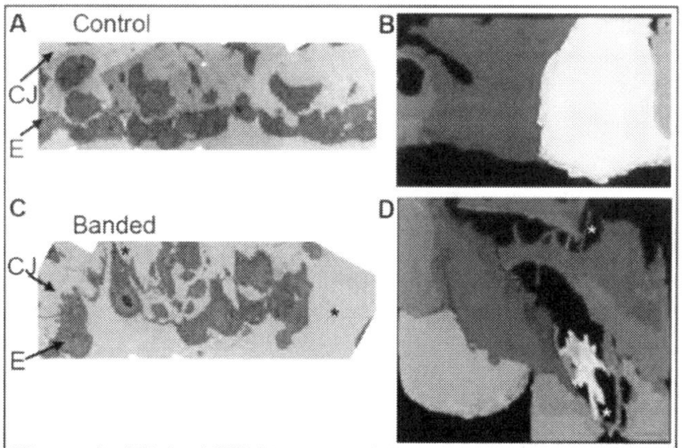

Figure 1. Stitched TEM images of the endocardium and adjacent cardiac jelly layer at 1700x (A&C), and corresponding cell segmentations from 3D image stacks displaying each cell in a different color (B&D) from a control embryo (A&B) and a banded embryo with 35% band tightness (C&D). CJ, cardiac jelly; E, endocardium; cell projections and gaps, *.

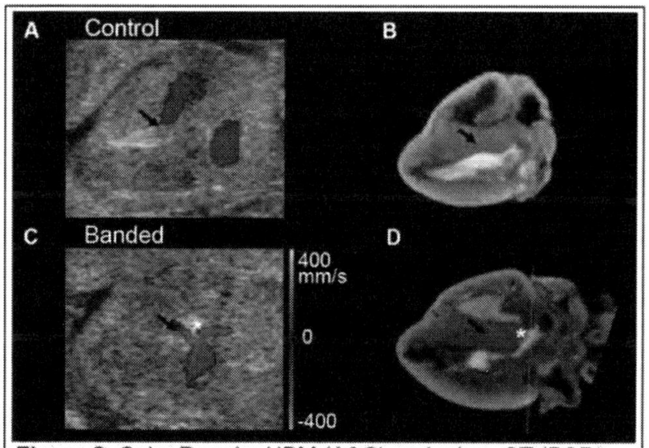

Figure 2. Color Doppler UBM (A&C) and micro-CT (B&D) images from a control embryo (A&B) and a banded embryo (C&D). VSD is detected in the banded embryo with UBM as color across the ventricles designating blood flow (C) and micro-CT as a hole in the interventricular septum (D). The ventricular septum is marked by an arrow, and the VSD is marked by (*).

REFERENCES

1. Hoffman, JI et al., *JACC*, 39:1890-900, 2002.
2. Yang, Q et al., *Birth Defects Res A*. 76:706-13, 2006.
3. Warkman, AS et al. *Sem Cell Dev Biol*. 18:46-53, 2007.
4. Shi, L et al. *Am J* Physiol-*Heart C*. 305:H386-96, 2013.
5. Midgett, M et al. *FPHYS*. 5, 2014.
6. Person, AD et al. *Int Rev Cytol*. 243:287-335, 2005.

NOVEL METHOD FOR 3D RECONSTRUCTION OF THE CHICK EMBRYO CARDIOVASCULAR ANATOMY FROM NON-INVASIVE ULTRASOUND SCANS FOR LONGITUDINAL STUDIES

Germaine X.Y. Tan (1), Muhammad Jamil (1), Nicole G.Z. Tee (2), Liang Zhong (2, 3), Choon Hwai Yap (1)

(1) Department of Biomedical Engineering, National University of Singapore, Singapore, Singapore

(2) National Heart Research Institute of Singapore, National Heart Centre Singapore, Singapore, Singapore

(3) Duke-NUS Graduate Medical School, National University of Singapore, Singapore, Singapore

INTRODUCTION

Congenital cardiovascular defects are the most common type of birth defects, affecting approximately 0.6-1.9% of newborns each year [1]. Recent small animal studies provided evidence that blood flow fluid mechanics may play a role in the pathogenesis of these malformations [2-5]. The imaging of the small animal embryonic cardiovascular anatomy is an important endeavor to support furthering these researches. To date, however, scan techniques are destructive or invasive, rendering it challenging to reliably perform longitudinal studies on the same embryo [6-8]. In the current study, the aim is to establish such a method to enable serial imaging and 3D reconstruction of cardiovascular anatomy of the chick embryo, based on non-invasive high-frequency ultrasound. We show that our technique has sufficient resolution to produce accurate geometries for Computational Fluid Dynamics simulations to quantify embryonic cardiovascular fluid mechanics.

METHODS

Fertilized eggs were opened at the blunt end to expose the chick embryo, and imaged in 3D over time with 45MHz high-frequency ultrasound (Vevo2100, VisualSonics Inc., Canada), non-invasively. Chick embryos at Hamburger Hamilton stages 21, 25, and 27 were imaged. Motion cancellation algorithms were applied to negate embryonic bodily motion for embryos past the 5th gestation day. Quadratic temporal averaging was performed in order to differentiate blood space from tissue space. The vessels of interest, namely the internal carotid arteries, dorsal aorta and pharyngeal aortic arches, were obtained by 3D reconstruction through image-based modelling using the open-source software, VMTK, and the accuracy of the reconstructed vessel sizes was validated with direct stereoscopic measurements. Finally, Computational Fluid Dynamics (CFD) simulation using ANSYS Workbench® (Ansys Inc., PA, USA) was used to model fluid flow in the generated construct of a HH27 embryo.

RESULTS

From the raw Cine B-mode images, it was difficult to distinguish between blood and tissue (Figure 1A), but after taking quadratic mean over time, distinction between the two was clear (Figure 1B). The motion cancellation algorithm, when applied to the images of embryos past the 5th gestation day, played a significant role in reducing motion artifacts in the image processing (Figure 1C-D).

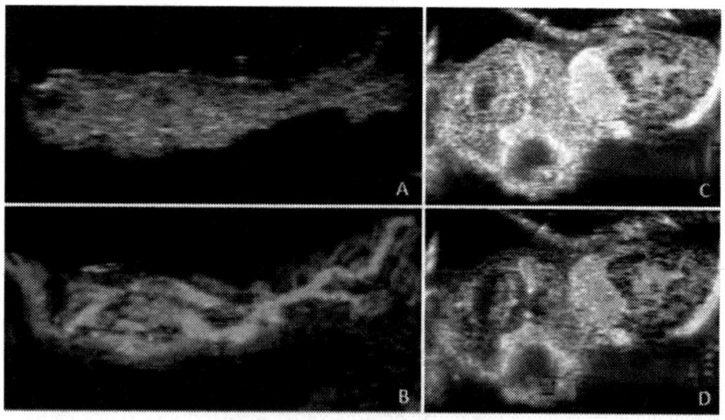

FIGURE 1: (A) ORIGINAL UBM STILL-IMAGE OF A HH21 CHICK EMBRYO. (B) SAME IMAGE AFTER TEMPORAL AVERAGING. (C) IMAGE OF A HH28 EMBRYO WITHOUT MOTION CANCELLATION. (D) SAME IMAGE WITH MOTION CANCELLATION.

The processed images were successfully stacked and underwent 3D reconstructions of the blood space geometries. Figure 2 shows the simple reconstructed geometries of a HH21, HH25 and HH27 embryo, consisting of the dorsal aortae and internal carotid arteries. To demonstrate that our technique has sufficient resolution for detailed CFD simulations to quantify embryonic cardiovascular mechanics, the dorsal aorta, internal carotid arteries and the three pairs of aortic arches in a HH27 embryo were isolated and geometrical smoothing was performed to obtain the finalized construct for the CFD simulations. From the simulations, a region of divergent streamlines and low shear stress is observed at the carotid duct (Figure 3).

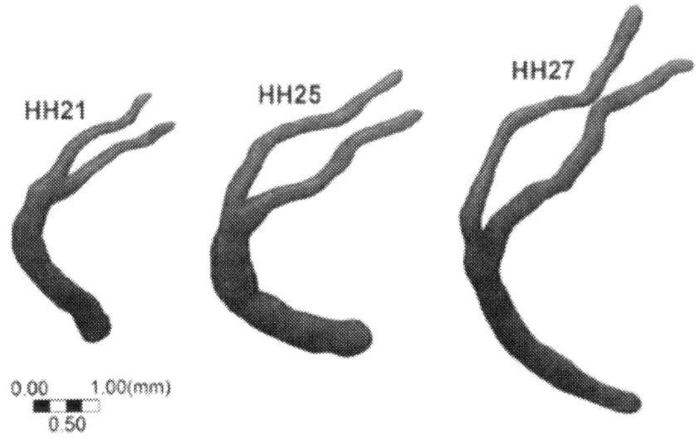

FIGURE 2: RECONSTRUCTS OF THE DORSAL AORTA AND INTERNAL CAROTID ARTERIES OF A HH21, HH25 AND HH27 EMBRYO.

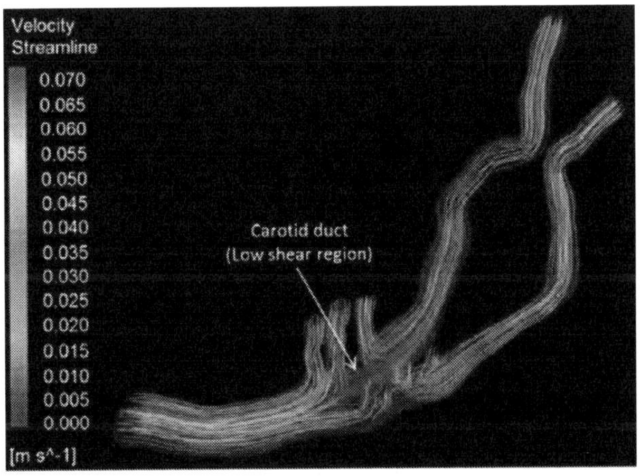

FIGURE 3: VELOCITY STREAMLINES IN THE VESSELS OF INTEREST OF A HH27 EMBRYO.

DISCUSSION

Ultrasound was the chosen imaging modality for this research due to its capabilities in acquiring dynamic data while being completely non-invasive [9]. It avoids preparation steps such as the injection of contrast agents which may disturb native physiology. Further, it has a greater scanning depth than Optical Coherence Tomography (OCT) and does not rely on optical access [9]. Ultrasound, however, produces much more noise and reduced resolution from micro-CT and OCT, bringing to mind the question of whether using ultrasound for small animal embryonic cardiovascular fluid mechanics is feasible. In the current study, we demonstrate the feasibility of this approach, but also point out limitations to this approach.

Our results showed that it is possible to reconstruct detailed geometries of chick embryonic cardiovascular vessels with ultrasound and image processing. The sizes of the dorsal aortae measured in our reconstructed models matched direct stereoscopic measurements satisfactorily, indicating reasonable accuracy of our technique.

We further showed that performing CFD with the reconstructed geometry was possible and the CFD led to an interesting finding. At HH 27, streamlines from the third aortic arch lead into the internal carotid arteries, while streamlines from the fourth and sixth aortic arches leads to the dorsal aorta, causing a diversion of flow, and resulting in the carotid duct experiencing little flow and low shear stresses. We hypothesize that this low shear environment plays a role in the natural regression and disappearance of the carotid duct at the later stages (HH28 onwards).

However, compared to micro-CT or OCT, our technique offers lower resolution, and some vasculature remained difficult to reconstruct. For example, the proximal parts of the arches connected to the outflow tract could not be visualized as our image processing techniques relies on time averaged echo signal but the proximal arches were overly dynamic, being close to the heart. During the reconstruction of the HH27 embryo, the left fourth and sixth aortic arches were slightly merged, and had to be separated manually by cutting through the groove between them. Hence, the main limitations of our technique are that (1) vessels that are close together may appear to be joined, and (2) vessels which overly dynamic cannot be segmented. We believe, however, that these limitations can be resolved with improved motion cancellation algorithm or improved phase averaging algorithm.

In addition, even though high frequency ultrasound has much greater scanning depths greater than OCT, there are still limits. Higher frequency transducers allow greater resolution, but reduce the penetration and scanning depth. Depth limits sets an upper limit to the age of embryos that can be scanned (less than 7 days gestation), and the resolution sets a lower limit to the age of embryos that can be scanned (more than 3 days gestation).

In conclusion, this study has proposed a non-invasive technique using high frequency ultrasound to reconstruct the 3D anatomic geometry of the cardiovascular system accurately, which can enable downstream CFD analysis. This method allows repeated live scanning of a single chick embryo to be conducted, permitting embryo-specific studies of the influence of hemodynamic forces on cardiovascular development for greater insights into congenital cardiovascular malformation progression. However, there remain resolution and scanning depth limitations to the approach.

ACKNOWLEDGEMENTS

This study was supported by the MOE AcRF grant # R397-000-163-133 (National University of Singapore startup fund, PI: Yap). We thank Seep, Keith Chong Zheng Tao, Foo Toon Jin and Dewi Susanti for their assistance in the scanning experiments and image processing.

REFERENCES

[1] Hoffman, J.I., et al., *J Am Coll Cardiol*, **39**(12):1890-900, 2002.
[2] Hogers, B., et al., *Circ Res*, **76**(5):871-7, 1995.
[3] Tobita, K., et al., *Am J Physiol Heart Circ Physiol*, **279**(3):H959-69, 2000.
[4] Hove, J.R., et al., *Nature*, **421**(6919):172-7, 2003.
[5] Broekhuizen, M.L., et al., *Ultrasound Obstet Gynecol*, **13**(6):437-45, 1999.
[6] Hiruma, T., et al., *Anat Embryol*, **191**(5):415-423, 1995.
[7] Butcher, J.T., et al., *Dev Dyn*, **236**(3):802-9, 2007.
[8] Yelbuz, T.M., et al., *Circulation*, **106**(22):2771-4, 2002.
[9] Gregg, C.L., et al., Differentiation, **84**(1):149-62, 2012.
[10] Schellpfeffer, M.A., et al., *Birth Defects Res A Clin Mol Teratol*, **73**(1):39-49, 2005.

A PROGRESS REPORT FROM A MULTI-DISCIPLINARY CAPSTONE EXPERIENCE INVOLVING ENGINEERING AND BUSINESS STUDENTS

Alan W. Eberhardt (1) and Joel H. Dobbs (2)

(1) Department of Biomedical Engineering
School of Engineering
University of Alabama at Birmingham
Birmingham, Alabama

(2) Collat School of Business
University of Alabama at Birmingham
Birmingham, Alabama
USA

INTRODUCTION

For the past 17 years, the lead author has received support through the NSF RAPD program (now GARDE) that has financed senior design projects to aid people with disabilities. Working with physical and occupational therapists as well as disabled individuals, students teams designed, constructed and delivered prototype devices that benefited children and adults. Therapists and physicians from United Cerebral Palsy of Greater Birmingham, the Bell Center of Homewood (a special care facility for children), the Children's Hospital of Alabama, and UAB Hospital, among others, helped guide the student teams working under engineering faculty supervision.

In the fall of 2012, with funding from an NIH-R25 award, the lead author began a collaborative effort with the School of Business (SoB) around our capstone Biomedical Engineering (BME) course. Faculty with business expertise were brought in to teach business topics, and business students joined the engineering student teams to create a truly multi-disciplinary learning environment. We expanded the project scope beyond the development of functional prototypes and began to explore the potential for new intellectual property and product commercialization.

The course was re-structured somewhat to better follow the Stanford Biodesign [1] approach and includes clinical rotations in the first stages of the design sequence. Participating medical faculty direct student exposure to biomedical device ideation, which enhances the team-based experiential learning activities. The students perform extensive market analyses and patent/prior art searches, study the regulatory processes, develop business models and work on commercialization strategies for their devices. The potential for new intellectual property (IP) is explored and disclosed as deemed appropriate. The current abstract summarizes the approach to the course, and provides details regarding how we have used university resources to help generate IP aimed at increasing the potential for commercialization of prototyped devices.

METHODS

We coordinate class times to allow students in the BME capstone sequence to be joined by select business students enrolled in a senior entrepreneurship elective. At the beginning of the fall term, the students shadow clinicians, therapists, and research scientists to determine device needs and ideate potential solutions. Each Business student is joined with 3-4 BME students to create interdisciplinary teams around vetted need statements. Device design constraints are established around both prototype and marketable designs. Engineering lectures are provided with a "Just in Time" (JIT) approach, including brainstorming and evaluation of design alternatives, a review of computer aided drawing (CAD), and an introduction material selection using CES software.

Concurrently, the students begin their investigation into development issues for their medical devices through readings from the Biodesign text. A series of guest speakers from the SoB and industry cover market analysis, IP, and entrepreneurship. Regulatory topics included device classes and FDA approval processes (510(k), PMA, etc.). The students are introduced to different business models through the "Business Model Canvas" [2]. A series of writing assignments culminate in a formal Design Proposal, for which client signature approval is required, along with that of a registered professional engineer who is a member of our Engineering faculty. The fall semester ends with an oral presentation of their proposed designs to faculty from the new UAB Institute for Innovation and Entrepreneurship (IIE) for a first glance at potential material for IP disclosure.

In the spring, students take part in a manufacturing rotation where they gain hands-on experience associated with state-of-the-art manufacturing methods for metals, plastics, and composites, quality assessment and control, and good manufacturing practices (GMP). Options include student-accessible machine shops, where students may gain experience with end mills, drill presses, band saws, lathes, and hand tools. CNC machining and 3D printing, plastics molding, extrusion, pultrusion and compression molding are also available. Electronics and control systems may be developed with assistance in a new dedicated lab. Thus, we have ensured that students have adequate access to equipment and expertise to help fabricate their prototype devices.

The students revisit the topic of IP strategy, and write up an IP disclosure for their device. This IPD is submitted to the IIE well in advance of any public disclosures (e.g., the students are required to participate in our Undergraduate Research Expo in March/April each year). Business topics are continued including project management, Gantt charts, economic analysis and development of a commercialization plan. Additional lectures on finite element analysis and software use are provided for students to use in their design and materials selection processes. The students may request other engineering lectures, such as manufacturing processes, bearing selection or frame design. The students begin purchasing, and ultimately construct and deliver a working prototype. As the projects are completed, the students perform verification and validation testing (as is feasible) and demonstrate some form of customer evaluation of the final working product. The Final Design Presentation includes and investor pitch and scientific poster, and is held at the last BME seminar for the term, which is attended by the BME faculty, students, alumni, clients and members of the IIE.

In the past two years, we have expanded our instructional efforts around how to work in teams. In the not-so-distant past, teams were provided little guidance regarding how to function on a team, how to confront difficult team members, etc. Lectures now include the classic "five dysfunctions" [3], project management, and how to have "crucial conversations" [4] to help students resolve issues that arise during the project development.

Assessment of student learning is performed each semester of the two-semester sequence, using a set of indirect assessment surveys (Senior Design Audience Reports, Instructor Evaluation of Class, IDEA Student Course Evaluation). Direct assessment of student work products include multi-faculty review of the oral and written forms of the Design Proposal and Final Design Report using rubrics established by the BME Undergraduate Committee. Design teams are self-assessed in the middle of the fall semester using a Team Assessment Questionnaire, which relates the five dysfunctions of teams. Peer assessment of team functionality and individual contribution is assessed multiple times over each semester, and counts as 10% of their course grade.

RESULTS/DISCUSSION

Direct assessment of student reports and presentations has revealed that students are successfully meeting numerous ABET student outcomes (e.g., design, teams, communication, modern tools, etc.). The guest lectures by UAB Business faculty and local entrepreneurs have resulted in enhanced student exposure to real-world product development topics including regulatory issues and business models.

IDEA surveys indicate that a strong majority of students (>85%) score a 4 or 5 (out of 5) agreeing that the course was useful in developing specific skills, competencies, and points of view needed by professionals in the field most closely related to this course. We have observed increasing percentages (>90%) of scores between 4 and 5 indicating that the course was useful in acquiring skills in working with others as a member of a team. The results of the manufacturing rotations have been seen in better-made devices and more content students, who no longer complain of lack of access to the tools needed for successful construction. On average, peer assessments indicate that a majority of teams are content with their overall team involvement. Student comments have indicated the new lectures have been beneficial in identifying problem behaviors and methods by which teams may correct those behaviors, without significant blood loss.

As a result of the 2012-2013 and 2013-2014 course offerings, we have successfully generated IP disclosures for seven projects (out of 12). Among these projects, four have been deemed worthy by the IIE to file provisional patents, and one has led to a PCT filing. Currently, we have 11 projects under development, and expect roughly half of these will generate new IP. These devices will be vetted by members of the IIE, along with engineering and business faculty, for continued development in our new Industrial Scholars Program, through which students will earn a Master of Engineering degree along with a certificate in entrepreneurship from the SoB, as they work to bring a device to a commercial success.

ACKNOWLEDGMENTS

The authors gratefully acknowledge support from the National Science Foundation GARDE Program, NSF CBET-1263941, and the National Institutes of Health, NIH 5R25HD078327-02.

REFERENCES

[1] Zenios, S., Makower, J., Yock, P. (2010) Biodesign: The Process of Innovating Medical Technologies. Cambridge.

[2] Osterwalder, A and Pigneur, Y. (2010) Business Model Generation: A Handbook for Visionaries, Game Changers, and Challengers. Wiley

[3] Lencioni, P. (2002). The five dysfunctions of a team. San Francisco: Jossey-Bass.

[4] Patterson, K., Grenny, J., Maxfield, D., McMillan, R., and Switzler, A., Crucial Accountability, 2nd edition, McGraw Hill, 2013.

OCULAR COMPLIANCE IN MICE

Stephen A. Schwaner (1), Joseph M. Sherwood (2), Eric J. Snider (3), Eldon E. Geisert (4), Darryl R. Overby (2), C. Ross Ethier (1,3,4)

(1) George W. Woodruff School of Mechanical Engineering, Georgia Institute of Technology, Atlanta, GA

(2) Department of Bioengineering, Imperial College London, London, United Kingdom

(3) Wallace H. Coulter Department of Biomedical Engineering, Georgia Institute of Technology and Emory University, Atlanta, GA

(4) Department of Ophthalmology, Emory University, Atlanta, GA

INTRODUCTION

Glaucoma is the second leading cause of blindness worldwide and is characterized by the death of retinal ganglion cells (RGC) which carry vision information to the brain [1]. A key risk factor in glaucoma is elevation of intraocular pressure (IOP), which induces increased stress and strain in the tissues of the optic nerve head, the region in the posterior eye where RGC axons exit the eye [2]. This, and other data, has suggested that biomechanics is important in glaucoma. Modelling has also shown that the properties of the sclera have a strong influence on the biomechanical response of the optic nerve head to elevated IOP [3]. Furthermore, scleral stiffness varies significantly between individuals [4]. However, it is not known which genes are responsible for this variation. Knowledge of the genes involved in coding for scleral stiffness may allow for the identification of patients at risk for glaucoma, as well as to help develop genetic therapies/tests targeted at modifying/identifying scleral properties.

The BXD mouse substrain set is a powerful tool for investigating the genetic origin of a given trait [5]. Collecting data on quantifiable traits for different BXD substrains allows one to define quantitative trait loci (QTLs), which can identify genes that modulate a given trait. We measured ocular compliance, the rise in ocular volume per unit rise in IOP, as a way to compare scleral stiffness of different BXD mouse substrains (Method 1). In addition, we determined the relationship between compliance and IOP (Method 2). This information provides a starting point for future work identifying the genetic factors that determine ocular compliance/scleral properties.

METHODS

Method 1: Ocular compliance was measured in C57BL/6J mice (B6), DBA/2J mice (D2), and 10 different BXD substrains (68 mice; 11-14 weeks old; Male & Female). Freshly enucleated eyes were cannulated through the cornea with a 30 gauge needle (Fig. 1). Each eye received multiple 500 nL PBS boluses (200 nL/s x 2.5 s) administered by a syringe pump. Each bolus occurred at a starting IOP of 8 mmHg, and an in-line pressure transducer recorded the resulting increase in IOP. Ocular compliance was calculated in Method 1 as

$$\varphi_{e1} = \Delta Volume/\Delta IOP. \qquad (1)$$

The effects of needle resistance and system compliance were accounted for before obtaining values for ΔIOP and $\Delta Volume$. To control for possible effects of eye size, we recorded eye diameter by analyzing overview micrographs using ImageJ software. QTLs were then determined to identify candidate loci modulating ocular compliance.

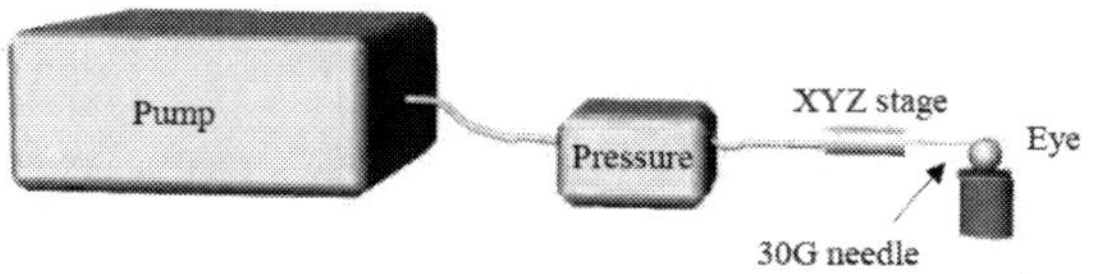

Figure 1: Experimental setup for Method 1.

Method 2: We determined the relationship between ocular compliance and IOP in B6 mice (42 mice; 10-14 weeks old; Male). Freshly enucleated eyes were cannulated through the cornea with a 33 gauge needle and submersed in PBS (Fig. 2). IOP was controlled by changing the height of an actuated open pressure reservoir of PBS. Flow rate of PBS into the eye and IOP were measured with a thermal flow sensor and a pressure transducer. Ocular compliance was calculated based on the step response of the hydraulic-electrical analogue circuit

$$\varphi_{e2} = \tau(R_e + R_q)/R_q R_e - \varphi_s \qquad (2)$$

where R_e and R_q are the resistances of the eye and the flow sensor, φ_s is the compliance of the system, and τ is the time constant of the system response. In method 2, the φ_{e2}-IOP relationship was fit to

$$\varphi_{e2}(P) = \varphi_0(P_0/P)^{\alpha} \qquad (3)$$

where φ_0 is the ocular compliance at a reference IOP, P_o = 8 mmHg.

RESULTS

Method 1: φ_{e1} differed between mouse substrains (Fig 3; Kruskal-Wallis, $p < 10^{-6}$) and was independent of eye diameter (ANCOVA, $p = 0.707$). Preliminary analysis identified suggestive QTLs associated with φ_{e1} on chromosomes 1, 7, and 11.

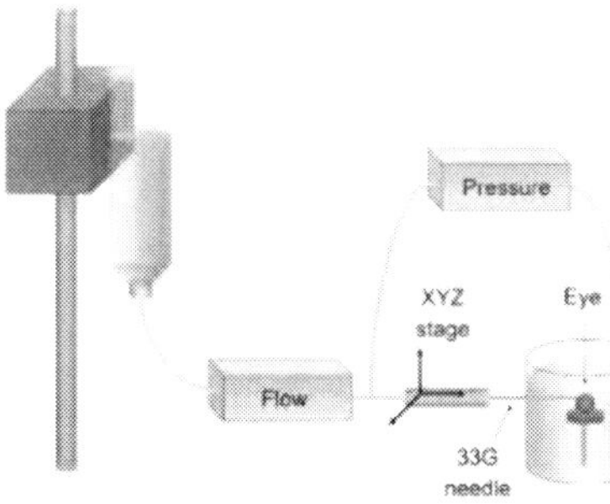

Figure 2: Experimental setup for Method 2.

Method 2: The average φ_0 and α for B6 mice were $106^{\times}/1.42$ nL/mmHg and 1.37 ± 0.48 (Fig. 4). Lower apparent ocular compliance values by Method 1 vs Method 2 were due to the IOP elevation inherent to Method 1.

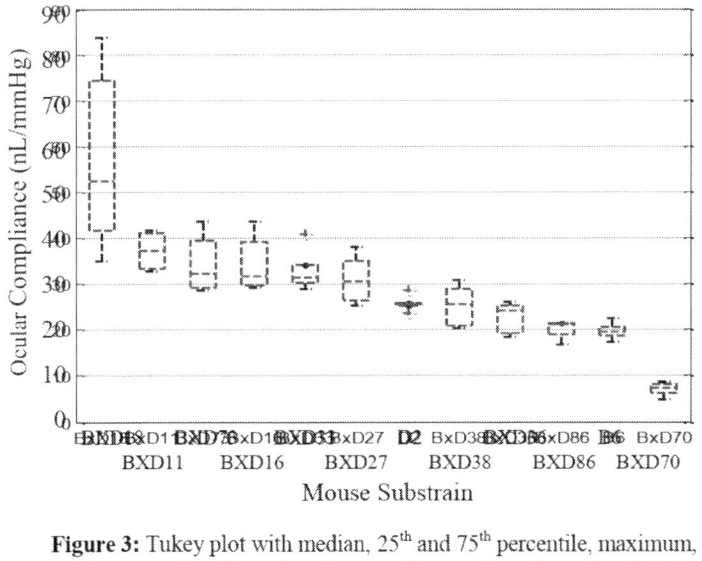

Figure 3: Tukey plot with median, 25th and 75th percentile, maximum, minimum, and outliers of ocular compliance values for each substrain ($n \geq 4$ mice per substrain).

DISCUSSION

As expected, there was a significant difference in ocular compliance due to BXD substrain. Because this difference did not depend on eye size, we conclude that it was due to a variation in the ocular structural stiffness between mice, which depends primarily on both corneoscleral envelope mechanical properties and thickness. One limitation of our study is that we measure overall ocular compliance, but cannot determine the relative contribution of the sclera and cornea to this overall measurement. For example, it is theoretically possible that the cornea of each BXD mouse eye is much more compliant than the sclera, and the differences shown in our tests are dominated by differences in corneal compliance/dimensions. Future digital image correlation studies of the scleral response during inflation will address this possibility.

Three QTLs were identified that showed association with ocular compliance; although these associations fell just short of reaching statistical significance, this is an encouraging preliminary result. Additional tests on more substrains are needed to obtain more statistical power and confirm these preliminary results.

Method 2 characterized the relationship between ocular compliance and IOP, which was well fit by a power law relationship. The results of Method 1 should thus be interpreted as a single compliance value acquired over a range of IOPs corresponding to the range of IOP elevation induced by the volume impulse, which explains the larger value in B6 mice for φ_0 using Method 2 compared to the average compliance value obtained in Method 1. However, results from Method 1 are still valid when used strictly as a comparison between substrains, e.g. for use with QTL analysis.

The ability to characterize specific aspects of the φ_{e2}-IOP relationship (via φ_0 and α) opens up opportunities to gain a deeper understanding of the genetic factors involved in coding for ocular compliance during future tests. For example, we assume that scleral thickness and collagen content both have an effect on ocular compliance. It might be the case that scleral thickness is the main factor that determines φ_0, but collagen content is the main factor that determines α. Future testing on BXD substrains using Method 2 and QTL analysis may allow us to pinpoint these relationships if they exist.

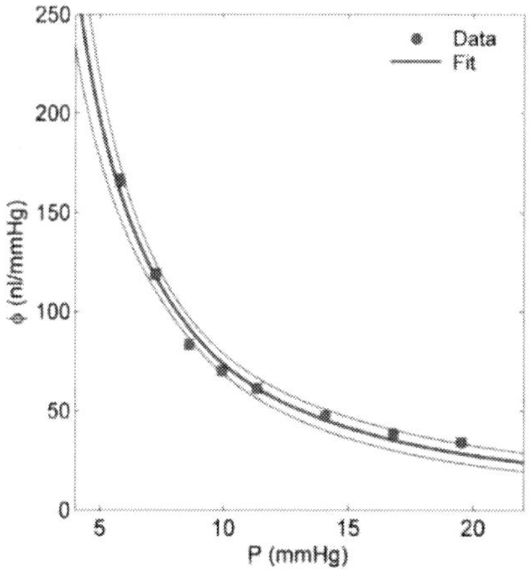

Figure 4: Ocular compliance vs. IOP for a B6 mouse eye with power law fit and 95% CI.

ACKNOWLEDGEMENTS

Georgia Research Alliance (CRE), Engineering and Physical Sciences Research Council EP/J010499/1 (DRO), NIH R01EY017841 (EEG), NIH P30EY06360 (Core)

REFERENCES

[1] Varma R, et al. *Am J Opthalmol.* 2011; 152(4): 515-22.
[2] Burgoyne CF. *Exp Eye Res.* 2011; 93(2): 120-132.
[3] Sigal IA, et al. *Invest Opthalmol Vis Sci.* 2005; 46(11): 4189-99.
[4] Eilaghi A, et al. *Ann Biomed Eng.* 2010; 38(4): 1586-92.
[5] Geisert EE, et al. *Mol Vis.* 2009; 15: 1730-63.

SB³C2015
Summer Biomechanics, Bioengineering and Biotransport Conference
June 17-20, 2015,Snowbird Resort, Utah, USA
SB³C2015-113

LONG-RANGE COMMUNICATION BETWEEN CELLS IN FIBROUS MATRICES ENABLED BY TENSION-DRIVEN ALIGNMENT OF FIBERS

Hailong Wang (1), A. S. Abhilash (1), Brendon M. Baker (2), Britta Trappmann (2), Christopher S. Chen (2), Rebecca G. Wells (3), Vivek Shenoy (1)

(1) Department of Materials Science and Engineering
University of Pennsylvania
Philadelphia, PA, US

(2) Department of Biomedical Engineering
Boston University
Boston, MA, US

(3) Departments of Medicine (GI) and Pathology and Laboratory Medicine
University of Pennsylvania
Philadelphia, PA, USA

INTRODUCTION

Cells in fibrous matrices sense and respond to mechanical forces over distances many times their diameter. Although cells cultured on polyacrylamide gels fail to sense substrate stiffness or the presence of other cells beyond a distance of~20–25 mm, long-range force sensing (250–1000 mm) between cells in fibrous gels has been appreciated for decades. Long-range force transmission has significant relevance in normal physiology and pathophysiology. Mechanically based communication between cells over long distances regulates patterning, including both tube formation and the detachment of cells from multicellular aggregates.

Although some investigators suggest that the strain-hardening properties of fibrous materials could explain long-range mechanical communication, more recent evidence suggests that the fibrous nature of the extracellular matrix, specifically the presence of cross-linked fibers, is critical for transmission of force over scales that are 10–20 times the diameters of the cells (1). It has also been observed that the shapes of cells play a crucial role in the transmission of forces, for example, tumor cells are elongated compared to their noninvasive counterparts and they observed, through displacements of beads in the matrices (2).

In this work, we first develop a discrete fiber network model to naturally accounts for the process of matrix reorganization resulting from cell-mediated fiber alignment and stretching, and use this model to predict the deformation zone around a contractile cell (3). Furthermore, a new nonlinear and anisotropic constitutive description of fibrous materials is developed to accounts for fibrous materials stiffening along the directions of tensile principal stretches (4). In both discrete and continue models, we show that collagen fiber alignment is critical and that anisotropy in cell shape and contraction result in significantly greater collagen alignment and force transmission.

METHODS

We developed a finite element based 2D discrete fiber model that captures all aspects of network mechanics including non-affine stiffening, fiber alignment and bending-stretching transitions following our earlier work on crosslinked biopolymer networks. The 2D random fiber networks representing collagen gels are created with linear elastic fibers and rigid crosslinks. Fibers are uniformly distributed in the computational domain and a crosslink is formed when two fibers intersect. Collagen fibers have diameter in the range of few 100 nanometers to few microns and moduli of few 100 kPa. As the persistence length of collagen fibers is in the range of few microns, these fibers are typically modeled as linear elastic. Fibers are modeled using shear flexible Timoshenko beam elements in the finite element package, ABAQUS.

We carried out discrete fiber simulations (Fig 1a) and found that there are two families of fibers as a fibrous matrix. The set of fibers that align with the direction of the maximum principal stretch as the material is loaded (Fig. 1a, red) and the set of fibers that do not align with the applied load and thus display an isotropic mechanical response. To capture the presence of these two distinct families of aligned and isotropic fibers, we assume that the overall energy density of the collagen network consists of two contributions, The isotropic response using the neo Hookean hyperelastic model, and the last term is the contribution from the aligned fibers. The function form is chosen such that the system stiffens only in the direction of tensile principal stretches (beyond a critical value of tensile stretch, as observed in experiments and discrete fiber simulations (5)) (Fig 1b). All simulations were carried out using the finite-element package Abaqus by implementing the material model of the new fibrous constitutive law in a user material subroutine.

RESULTS

(a) (b)

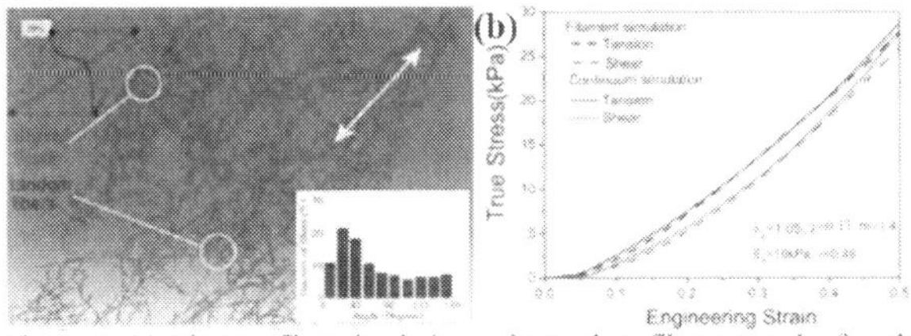

Figure 1: (a) Discrete fiber simulations of a random fiber network after shear deformation (50% shear strain). Insets show that the initial random distributions of fibers develop a peak close to the 45 degree orientation. (b) Stress-strain curves under uniaxial tension (black) and shear (red) deformations from discrete fiber simulations and from our constitutive law.

Cellular forces of a uniformly contracting elliptical cell lead to large deformations along the long axis where fibers are under tension compared with the short axis where fibers are predominantly in compression (Fig 2a). The response at large strains is predominantly dictated by the constitutive response of the aligned fibers undergoing axial stretching. The bending resistance of fibers is low compared with stretching and under applied tractions, fibers easily bend and reorient along the principal loading direction. In our simulations, we have observed that the fibers that align with the long axial get stretched excessively (Fig. 2b). The image also shows the increase in the fraction of aligned fibers along long axis as the cell shape anisotropy increases.

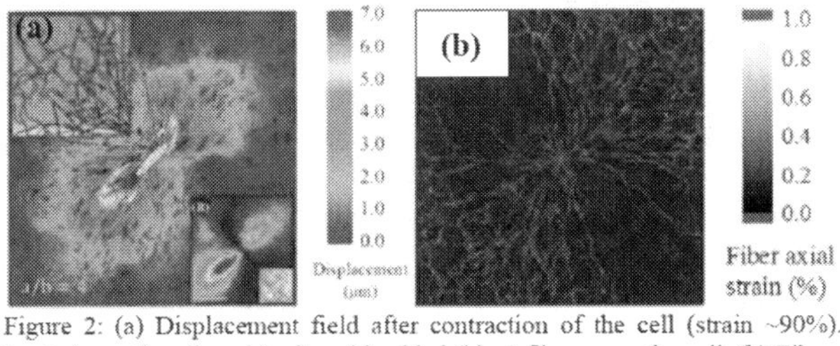

Figure 2: (a) Displacement field after contraction of the cell (strain ~90%). Inset shows the aligned (red) and buckled (blue) fibers near the cell. (b) Fibers along the longer cell axis experience large axial stretching and preferential alignment. Axial strains of red fibers exceed 1%, blue fibers are in compression and black to white denote fibers with strains in the range 0% to 1%.

A heat map of the zone of influence based on the energy criterion around a single cell for different cell aspect ratios and contractility is developed (Fig. 3a). This map predicts the size of the deformation zone for cells of shapes ranging from circular to highly spindle-shaped and for contractile strains as high as 55%. A similar heat map of the range of force transmission as a function of these parameters using continue model is given in Fig. 3b. We find that the extent of the fibrous region can be as high as 20 times the characteristic size of the fully polarized cells with large aspect ratio, as has been observed in several studies (1).

(a) (b)

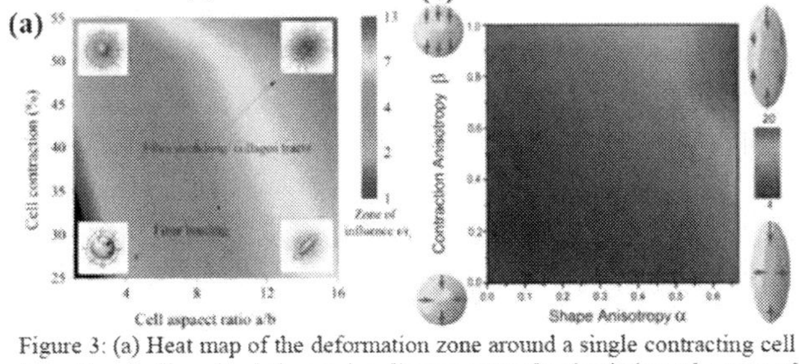

Figure 3: (a) Heat map of the deformation zone around a single contracting cell from discrete fiber simulations using discrete networks simulation. The zone of influence increases with both contractility and the shape anisotropy. (b) Contour plots of force transmission distance as a function of shape anisotropy and contraction anisotropy using our continue model.

We simulated the interactions between two cells in 3D fibrous and non-fibrous matrices, as well as on substrates and verified that cell-cell interactions are significant when their separations are of the order of twice the sensing distance of a single cell. The clear role of fibrous matrices in mediating cell-cell interactions is shown in Fig. 4, where significant overlap and alignment of strain fields are observed for pairs of cells located in fibrous matrices at a distance of 10 times their size.

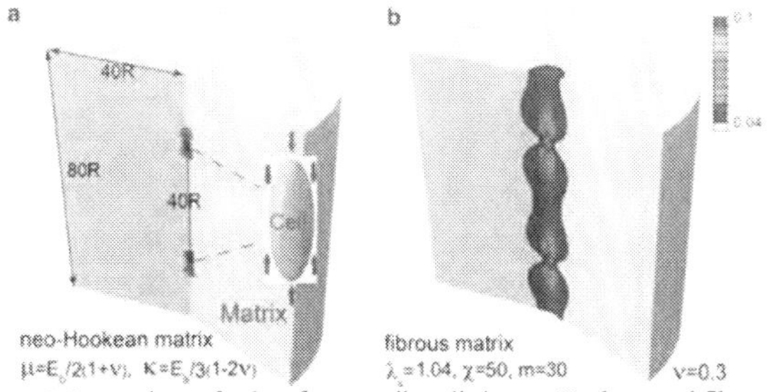

Figure 4: Interactions of pairs of contractile cells in neo-Hookean and fibrous matrices: Contour plots of maximum principle strain in 3D matrices.

DISCUSSION

Our results clearly show that the long-range displacement fields within matrices can be captured by tension-driven local fiber alignment along principal directions of stretch, and that the heterogeneities result from the anisotropic shape of the cell domain. The circular cell has a uniform deformation zone around it whereas a large tensile deformation zone is identified along the poles of the elongated spindle-shaped cells. Fibers along the short axes of elongated cells are found to be in compression as observed in experiments. We have shown that the long-range force transmission observed in collagen gel is attributable to the fibrous nature of the matrices and conventional strain hardening material models are incapable of predicting this.

Both shape and contraction anisotropy are important for long-range force transmission. Tension-driven fiber alignment plays a crucial role in mechanosensing: small critical stretch for fiber alignment, large fiber stiffness, and fiber strain-hardening behavior enable long-range interactions. Cells in 3D fibrous matrices and cells on 2D fibrous substrates sense rigid boundaries and other cells over relatively long distances compared to cells in and on linear and strain-hardening isotropic materials. Cells in 3D fibrous matrices sense rigid boundaries over 10 times their diameters and other cells over 20 times their diameters. Cells on 2D fibrous substrates sense radial rigid boundaries up to 8 times their radii and thicknesses up to 3.5 times their radii.

ACKNOWLEDGEMENTS

Research was supported by the National Institute of Biomedical Imaging and Bioengineering of the National Institutes of Health under award number R01EB017753 and the U.S. National Science Foundation Grant CMMI-1312392.

REFERENCES

1. Ma, X. et al. 2013, Biophys. J. 104: 1410–1418.
2. Koch, T.M.et al. 2013, PLoS One. 7: e33476.
3. Abhilash, A.S., B.M. Baker, B. Trappmann, C.S. Chen, and V.B. Shenoy.. 2014. Biophys. J. 107: 1829–1840
4. Wang, H., A. Nair, C.S. Chen, R.G. Wells, and V.B. Shenoy. 2014, Biophys. J. 107: 1829–1840.
5. Roeder, B.A et al. 2002, J. Biomech. Eng. 124: 214–222.

VISCO-HYPERELASTIC AND BIPHASIC PROPERTIES OF A BRAIN PHANTOM AGAROSE GEL

Gerson Córdoba (1), Gustavo A. Orozco (1), Fernando Casanova (1), Joshua H. Smith (2), José J. García (3)

(1) Escuela de Ingeniería Mecánica
Universidad del Valle
Cali, Colombia

(2) Department of Mechanical Engineering
Lafayette College
Easton, PA, USA

(3) Escuela de Ingeniería Civil y Geomática
Universidad del Valle
Cali, Colombia

INTRODUCTION

Agarose gel have been used as a brain tissue phantom material to study convection-enhanced delivery (CED), which is a technique to deliver controlled amounts of drugs directly into the brain for the treatment of tumors and diseases [1]. Meaningful translation of experimental results from agarose gel to clinical applications requires understanding the differences between the brain tissue phantom and brain tissue. Further, experimental studies of CED are complemented with theoretical models aimed to understand the physics of the problem, and proper analysis with the theoretical models requires knowing the properties of the material.

Few studies have been carried out to determine the material properties of agarose gels. Raghavan et al. [2] present —agarose gel material properties using a linear elastic and constant hydraulic conductivity model, while Sasson et al. [3] present properties of chitosan hydrogels represented as hyperelastic materials. However, to the best of our knowledge, there are no reported properties of agarose gels including the nonlinear elastic behavior of the material, the intrinsic viscous effects of the solid phase, and the variation of the hydraulic conductivity with strain. Hence, this study was aimed to determine the biphasic visco-hyperelastic properties of agarose gels commonly used as brain tissue phantoms, taking into account the aforementioned effects.

METHODS

Experiments were conducted with a solution of 0.4% (w/v) Sigma-Aldrich A6013 agarose gel. The material properties of the agarose gel were determined experimentally through unconfined and confined relaxation compression tests. Twenty cylindrical samples of 10 mm height and 63.5 mm diameter were used, 10 for each of the confined and unconfined tests. The samples were subjected to two sequential ramp-and-hold nominal strains of 0.05 using a dynamic testing machine, where the ramps were applied over 180 s. For the confined and unconfined tests, respectively, the first strain increment was held for 1200 s and 3600 s, while the second increment was held for 1500 s and 3900 s. A load cell of 50 ± 0.074 N (Omega Engineering, Inc., USA) was used to measure the force and a LVDT LD 610-15 (Omega Engineering, Inc., USA) was used to measure the displacement. In the confined tests, a stainless steel ring laterally confined the sample while two porous filters allowed the flow in the axial direction. In the unconfined tests, the sample was allowed to expand laterally and two porous filters were in contact with the plane surfaces of the samples.

As there is no closed-form solution that takes into account the nonlinear effects included in these problems, axi-symmetric finite element models were developed to fit the experimental stress-time curves. After a mesh convergence study, the confined and unconfined models used 100 and 400 CAX4P elements, respectively. Model boundary conditions were consistent with those of the experiments.

The agarose gel was represented as a biphasic material composed of a fluid and a visco-hyperelastic solid phase, described by the Ogden-type energy function

$$W(\bar{\lambda}_1,\bar{\lambda}_2,\bar{\lambda}_3)=\sum_{i=1}^{N}\frac{2\mu_i}{\alpha_i^2}\left(\bar{\lambda}_1^{\alpha_i}+\bar{\lambda}_2^{\alpha_i}+\bar{\lambda}_3^{\alpha_i}-3\right)+\sum_{i=1}^{N}\frac{1}{D_1}(J-1)^{2i} \quad (1)$$

where $\bar{\lambda}_1,\bar{\lambda}_2,\bar{\lambda}_3$ are the principal deviatoric stretch ratios, α_i, μ_i, and D_1 are material parameters, and J is the determinant of the deformation gradient tensor. As the experimental force-time curves displayed a high ratio of peak force-to-equilibrium force, which cannot be explained by an isotropic elastic biphasic model, we included the intrinsic viscous effects of the solid using the Prony series

$$K(\tau)=K_\infty+\textstyle\sum_{i=1}^{2}k_i e^{-\tau/\tau_i},\quad G(\tau)=G_\infty+\sum_{i=1}^{2}g_i e^{-\tau/\tau_i} \quad (2)$$

where $K(\tau)$ and $G(\tau)$ are relaxation functions that represents the volumetric and deviatoric components of the material, respectively, τ_i

are the relaxation times, and K_∞ and G_∞ are the bulk and shear moduli at equilibrium, respectively. The sum of each of the series terms k_i and g_i compared to 1 represents the amount of viscous contribution, e.g., if k_1 plus k_2 is 0.9, the equilibrium bulk modulus is only 10% of the instantaneous value. The hydraulic conductivity κ was allowed to vary with strain as

$$\kappa = \kappa_0 \left(\frac{J-\varphi_0}{1-\varphi_0}\right)^2 exp\left(\frac{M(J^2-1)}{2}\right) \quad (3)$$

where κ_0 is the initial hydraulic conductivity, M is a dimensionless coefficient, and φ_0 is the solid volume fraction.

The fitting of the experimental curves was accomplished by manually varying model parameters. To facilitate this task, a sensitivity analysis was performed first to determine the parameters that affected the equilibrium, peak, and shape of the curves from the two tests. The following objective function *FIT* was used to determine the best fit:

$$FIT = 100\sum_{i=1}^{N} \frac{\sqrt{(F_m^{(i)}-F_{ex}^{(i)})^2}}{\sqrt{(F_m^{(i)}-F_{pr})^2}} \quad (4)$$

where $F_m^{(i)}$ are the model values, $F_{ex}^{(i)}$ the experimental values, and F_{pr} the mean of experimental values. A value of *FIT* equal to 0 represents a perfect fit whereas a value near 100 describes a large error.

In order to validate model results, an independent test was performed by indenting four agarose gel samples with a steel spherical indenter of 63.5 mm diameter at a load velocity of 0.0028 mm/s and a nominal strain of 0.15. Predictions from an axi-symmetric model using 900 CAX4P elements were compared with the indentation results.

RESULTS

The experimental curves were characterized by large ratios between the peak and equilibrium forces of about 20 and 13 for the unconfined and confined tests, respectively (Fig. 1). The model provided a relatively good fitting of the experimental curves described

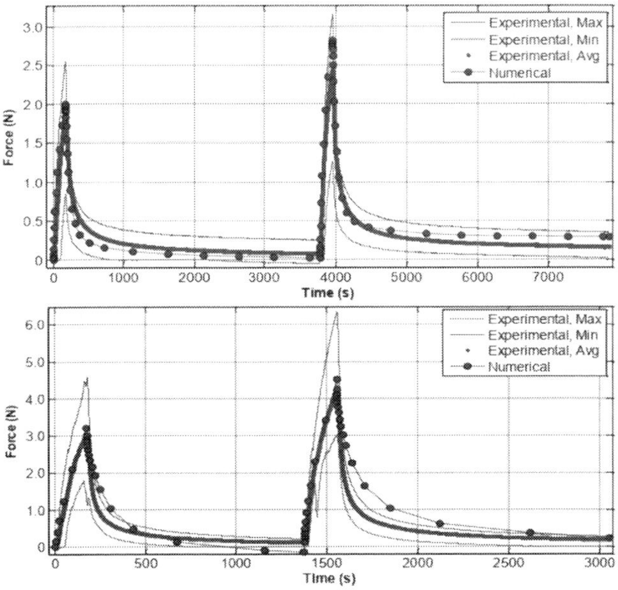

Figure 1. Experimental force-time curves and model predictions for the unconfined (above) and confined (below) tests.

with a *FIT* coefficient of 25.6% and 29.6% for the unconfined and confined tests, respectively, and coefficients of determination (R^2) higher than 0.97. Validation of model parameters with the independent

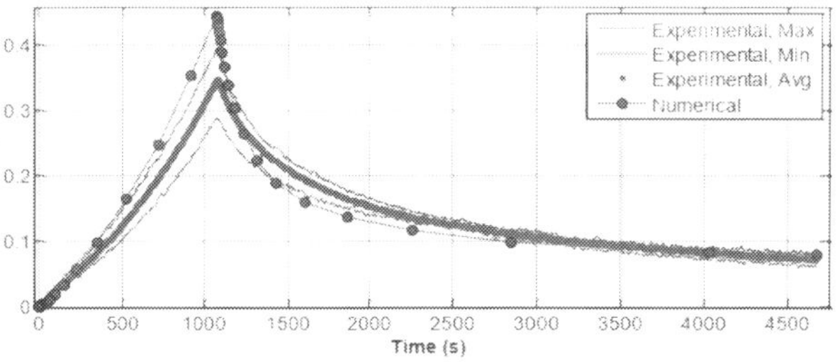

Figure 2. Experimental force-time curves and model predictions for the indentation validation tests.

Table1. Material properties obtained in this study

Parameter	Value
Energy function, μ_1, μ_2	1100 , −327 Pa
Equilibrium shear modulus, G_∞	773 Pa
Energy function exponents, α_1, α_2	−0.1 , −1.3
Poisson's ratio, ν	0.15
Initial hydraulic conductivity, κ_0	15 $mm^4 N^{-1} s^{-1}$
Nonlinear conductivity, M	7
Solid volume fraction, φ_0	0.0062
Prony series terms, g_1, g_2	0.775, 0.170
Prony series terms, k_1, k_2	0.830, 0.100
Relaxation times, τ_1, τ_2	120, 1000 s

indentation tests yielded a *FIT* coefficient of 17.2% and a coefficient of determination of 0.99 (Fig. 2). The equilibrium shear modulus (Table 1) was of the same order of magnitude as that of brain tissue, whereas the initial hydraulic conductivity was relatively high with respect to 4 $mm^4 N^{-1} s^{-1}$ reported for white matter tissue [4].

DISCUSSION

A methodology was presented to obtain the nonlinear material parameters of agarose gel intended to mimic the biphasic behavior of brain tissue. High viscous effects of agarose gel, characterized by sums of k_i and g_i of 0.945 and 0.93, respectively, are similar to those observed under unconfined tests of white matter in which the sum of the g_i terms was 0.86 [4]. In this respect, agarose gel is a feasible alternative to conduct experiments aimed to understand the mechanics of brain tissue. The high contribution of the viscoelastic effects of the solid phase indicates that there may be important differences in the results of experiments undertaken at different velocities. This may explain the high dispersion observed in the measurement of the shear modulus of agarose gel in a previous study [2]. These results, and others for brain tissue [4], also indicate the need to include the intrinsic viscous effects in finite element simulations intended to study CED.

ACKNOWLEDGEMENTS

The authors appreciate the support of the Universidad del Valle and Lafayette College to undertake this study. Thanks are also due to Colciencias program 516-2012 (contract #110656933826).

REFERENCES

[1] Bobo, R. H., et al., *Proc. Natl. Acad. Sci.*, 91:2076–2080, 1994.
[2] Raghavan, R., et al., *Phys. Med. Biol.*, 55:281–304, 2010.
[3] Sasson, A., et al., *J. Mech. Behav. Biomed. Mater.*, 8:143–153, 2012.
[4] Cheng, S., and Bilston, L.E., *J. Biomech.*, 40:117–124, 2007.

USING DYNAMIC COMMUNITY DETECTION TO MAP COLLAGEN FIBER NETWORK REORGANIZATION DURING TENSILE LOADING OF THE HUMAN FACET CAPSULAR LIGAMENT

Sijia Zhang (1), Danielle S. Bassett (1,2), Beth A. Winkelstein (1,3)

(1) Department of Bioengineering
University of Pennsylvania
Philadelphia, PA, United States

(2) Department of Electrical and Systems Engineering
University of Pennsylvania
Philadelphia, PA, United States

(3) Department of Neurosurgery
University of Pennsylvania
Philadelphia, PA, United States

INTRODUCTION

The cervical facet capsular ligament (FCL) is vulnerable to injury during trauma and other spine pathologies. The FCL's collagen matrix accommodates the macroscopic deformations during loading and exerts local forces via fiber motion and network reorganization [1-3]. Anomalous collagen fiber realignment occurs at FCL strains sufficient to induce pain [4], prior to tissue rupture, and at sites of eventual tissue failure [1]. Previous analytic methods detecting anomalous fiber realignment [1] neglect the potential for larger-scale realignment of fiber groups as they adapt to FCL loading. Thus, it is as yet unknown if, and to what extent, local collagen networks display meso-scale organizational changes whose complexity alters ligament mechanics.

Networks provide a useful representational framework to study a system's topological characteristics and dynamics [5-8]. System components are represented as network nodes and component interactions are represented as network edges. In many biological systems, this formalism reveals important structural and organizational properties [5]. Community detection techniques, for example, uncover meso-scale structure, where components cluster into modules [6-8]. Dynamic community detection techniques reveal changes in meso-scale structure over time, often summarized by diagnostics including modularity, number of modules, stationarity and flexibility [8].

We hypothesize that network analysis can resolve the complexity of the coordinated multivariate pattern of collagen fiber realignment. We used network analysis to investigate how stretch-induced fiber kinematics modulates organization of the local collagen matrix in the FCL under tension. We defined collagen networks based on similarity in fiber alignment directions, and then applied community detection techniques to probe organizational changes due to anomalous realignment. Dynamic properties of the network architecture were then defined and their associations with FCL mechanics were investigated.

METHODS

Collagen fiber alignment data from specimens (n=7; age 63±15 yrs) sustaining anomalous realignment were obtained previously using quantitative polarized light imaging (QPLI) during uniaxial tensile loading of human FCLs [1]. For each specimen, two regions of interest (ROIs) were created with: (1) anomalous fiber realignment and (2) normal realignment (Fig. 1). Using each ROI, fiber alignment maps were generated from the QPLI data at 1s increments during FCL distraction from the unloaded pre-stressed resting reference position through the stretched position at the onset of anomalous fiber realignment (Fig. 1). The nodes of the collagen network were defined as non-overlapping 3x3 pixel windows in each ROI, and the edges of the network linking the nodes were defined based on the similarity in fiber alignment directions. The average of the fiber orientation angles within one standard deviation of all angles in the 3x3 pixel window defined the mean alignment direction in each window; the difference in alignment angles between nodes established the edges. Two nodes were considered "connected" when their angle difference was less than 90°. Edge weights were assigned by scoring the 90° difference evenly from 1 to 10, with smaller differences weighing more (Fig. 1).

Static and dynamic community detections were used to study changes in modular structure of the collagen networks. Static community detection tested organizational differences in both the anomalously and normally realigned regions in relation to their respective unloaded networks. Modules were defined as those sets of nodes that are highly inter-connected [6]. Optimized modularity and the number of modules were computed using the Louvain algorithm [6] for each sample at the resting and the stretched positions. Those configurations were compared, as well as each to randomized null models, using separate paired t-tests. In order to define organizational features of the dynamic fiber realignment process during FCL loading, dynamic community detection was implemented on multilayer networks created by linking the same nodes between consecutive 1s increments up to the onset of anomalous realignment [7]. Modularity, the number of modules, flexibility and stationarity were computed for each multilayer network [8]. Flexibility characterizes the altered nodal allegiance to modules and stationarity measures the maintenance of nodal composition of modules over time [8]. To verify the reliability of the measured modular organization, differences in network diagnostics between multilayer networks and their nodal null models, where nodal connections between layers were randomized, were compared to zero using a t-test [8]. Lastly, separate linear regressions tested the correlations between stationarity and each of the tissue forces and stresses at anomalous realignment. The association between stationarity and the distraction magnitude at tissue rupture was similarly tested, with significance evaluated by a Chi-squared test.

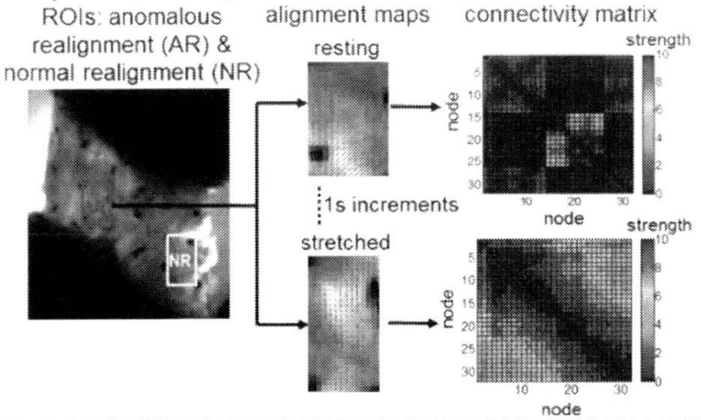

FIG. 1. NETWORK CONSTRUCTION USING ANOMALOUS REALIGNMENT (AR) & NORMAL REALIGNMENT (NR) ROIS. WEIGHTED NETWORKS WERE CREATED USING FIBER ALIGNMENT MAPS BASED ON SIMILARITY IN FIBER DIRECTION.

RESULTS

Only regions with anomalous fiber realignment exhibit heightened modular structure at rest than when stretched. Specifically, anomalously realigned regions display decreases in both the modularity ($p=0.03$) and the number of modules ($p=0.02$) after loading compared to rest (Fig. 2); no difference is observed in regions with normal realignment. Only anomalously realigned regions under stretch show similar modularity and number of modules as randomized null models, while the other three networks show significantly greater ($p<0.04$) modularity than randomized null models. The decreased modularity and number of modules comparable to random levels in the anomalously realigned region after stretch is likely driven by increasing inter-modular connections as the alignment directions become more similar between modules after loading (Fig. 2).

Dynamic community detection reveals that the collagen network reorganizes smoothly and adaptively over time. Collagen fiber networks exhibit significantly higher modularity ($p<0.01$) than the null model, indicating complex structure. Significantly increased stationarity ($p<0.01$) and decreased flexibility ($p=0.01$) compared to the nodal null models suggest a smooth temporal transition between the resting state and the onset of anomalous fiber realignment.

Tight associations exist between dynamic organizational features of local collagen networks and parameters of the mechanical response of the FCL. One outlier was removed using the robust regression. Stationarity is significantly correlated with both the force ($R=-0.90$; $p=0.02$) and stress developed in the ligament ($R=-0.94$; $p<0.01$) at the onset of anomalous fiber realignment. In addition, a significant negative correlation ($R=-0.82$; $p=0.04$) is observed between the stationarity and the displacement at initial FCL rupture.

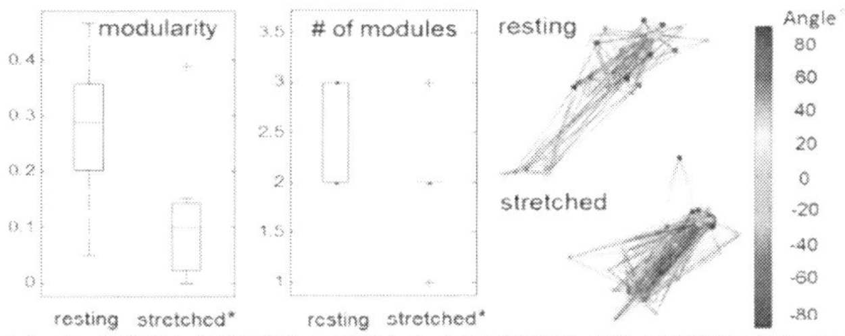

FIG. 2. BOTH MODULARITY & NUMBER OF MODULES OF ANOMALOUSLY REALIGNED NETWORKS ARE SIGNIFICANTLY DIFFERENT (*P<0.03) AT THE ONSET OF FIBER REALIGNMENT (STRETCHED) FROM THE RESTING STATE, AS IS EVIDENT IN THE VISUALIZATION OF MODULAR STRUCTURES (NODE SHAPE SHOWS MODULE ASSIGNMENT; COLOR IS MEAN ALIGNMENT DIRECTION).

DISCUSSION

Our finding of heightened modular structure at rest compared to the onset of anomalous fiber realignment (Fig. 2) agrees with previous reports of non-random FCL morphology [9] and collagen fiber realignment towards the loading direction [3,10]. This study also shows that regions with anomalous fiber realignment correspond to more evident stretch-induced changes than regions with normal reorganization, as has been shown using vector correlation [5], validating network analysis as a means to study collagen fiber reorganization. Higher modular stationarity, lower nodal flexibility and heightened modular structure of dynamic collagen networks compared to nodal null models indicates the existence of long-lasting modules, which reflect coordinated movement of fibers to smoothly adapt to FCL loading. The strong associations between stationarity and FCL mechanics suggest that dynamic changes of small-scale collagen networks may contribute to tissue-level FCL mechanical responses. The dynamic architecture of the anomalously realigned local collagen network may be able to predict tissue failure, as indicated by correlation between stationarity and displacement at tissue rupture.

Defining the time-dependent fiber kinematics of collagen-based tissues continues to become more feasible due to advancing imaging techniques and computational models [1-3]. Yet, quantification of the temporally coordinated reorganization across spatially-distributed fibers remains a challenge due to a lack of analytical tools. This study demonstrates the relevance and significance of using dynamic community detection to define and understand loading-induced collagen reorganization. Although network construction techniques may be further tested and improved in future studies, the findings that the dynamic collagen network reorganizes adaptively with coordinated fiber motions in relation to macroscopic FCL mechanical behaviors provide new insights for understanding how micro- and macro-scale tissue mechanics evolve across time and length scales.

ACKNOWLEDGEMENTS

This work was supported by grants from the NIH (U01EB016633) and the Alfred P. Sloan Foundation.

REFERENCES

[1] Quinn K & Winkelstein B. *J Biomed Opt* 14:054010, 2009. [2] Quinn K et al. *J Appl Physiol* 105:1881-8, 2008. [3] Sander E et al. *PNAS* 106:17675-80, 2009. [4] Dong L et al. *Brain Res* 1461:51-63, 2012. [5] Zhu X et al. *Genes Dev* 21:1010-24, 2007. [6] Blondel V et al. *J Stat Mech* 2008:P10008, 2008. [7] Mucha P et al. *Science* 328:876-8, 2010. [8] Bassett D et al. *PNAS* 108:7641-6, 2011. [9] Yamashita T et al. *Spine* 21:538-43, 1996. [10] Vader D et al. *PloS One* 4:e5902, 2009.

SB^{3}C2015
Summer Biomechanics, Bioengineering and Biotransport Conference
June 17-20, 2015,Snowbird Resort, Utah, USA
SB^{3}C2015-117

INVESTIGATING THE EFFECT OF TENNIS BALL IMPACT LOCATION

Ainhoa Iglesias-Díaz, Martin L. Tanaka, Paul M. Yanik, Aaron K. Ball

Department of Engineering and Technology
Western Carolina University
Cullowhee, NC, USA

INTRODUCTION

Tennis is a competitive sport played by millions of people worldwide. The characteristics of the game of tennis produce stress on the musculoskeletal system, especially in the upper extremity. Upper extremity injuries often occur when the arm is highly accelerated, as happens in tennis. These high accelerations require large forces to be applied to the wrist, elbow and shoulder. Upon ball impact, a large amount of force is transferred to the ball from the tennis racket. However, depending on the impact location, large reaction forces can also be produced in the body. These forces must pass through the kinematic chain from the hand to the wrist, elbow and shoulder joints and into the torso. As a result, wrist, elbow and shoulder joint injuries are common. Motion capture systems using retroflective markers are the most common devices used to study the overhead motion in tennis [1]. Movement is measured before and after impact determine what happens at the moment the ball and racket collide, but collision forces are not directly measured.

In this research an instrumented tennis racket and data acquisition system was developed to directly measure the forces and moments caused by ball impact during game play.

METHODS

A custom instrumented tennis racket was designed to measure the forces and torques transferred to the wrist at ball impact. A commercial racket was purchased and a unidirectional surface mounted strain gauge [2] was attached. However, stiffness of the racket was found to be too high for accurate strain measurements, so a modified tennis racket was developed (Figure 1). The regular handle was replaced with a 2.54 cm diameter acetyl rod that allowed the handle to flex at ball impact. The new handle was notched using a Haas VF-1 4 axis mill and was attached to the frame of the racket by two 20 gauge mild steel plates (one on each side) previously cut using a 500W CO_2 Haas laser cutter. The pieces of steel were secured by bolts to rigidly attach the end of the handle to the racket frame. The material used to build the new instrumented tennis racket added additional weight, increasing the mass of the racket from 320 g to 680 g. Unidirectional and triaxial strain gauges were attached to the handle of the racket to measure handle flexion and twist at ball impact.

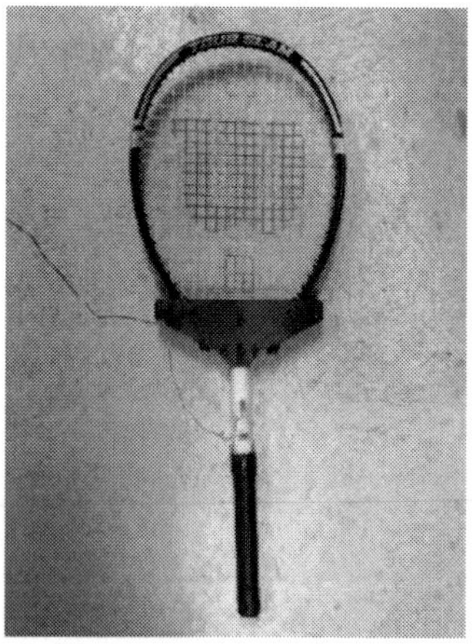

Figure 1: Instrumented tennis racket designed for the experiment.

A Wheatstone Bridge circuit was designed to measure the strain in the gauge attached to the racket handle (Figure 2). The strain signal from the bridge circuit was amplified using the LMC6484 operational amplifier. An additional non-inverting amplifier was added to increase the voltage level to a range that was easily readable by the oscilloscope.

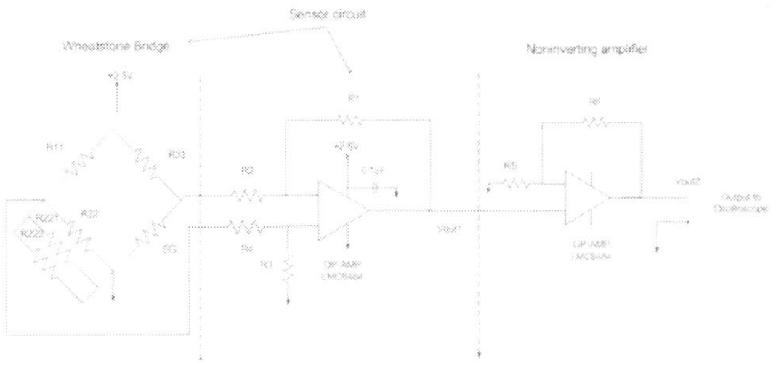

Figure 2: Schematic of the analog circuit used to measure the strain obtained by the strain gauge attached to the racket.

Calibration tests were performed by hanging weights (3 kg, 1.773 kg, 1.182 kg) from the racket at various distances to determine the relationship between externally applied loads and the strain within the racket shaft caused by bending (Figure 3). Trendlines show the linear relationship between the distance and voltage.

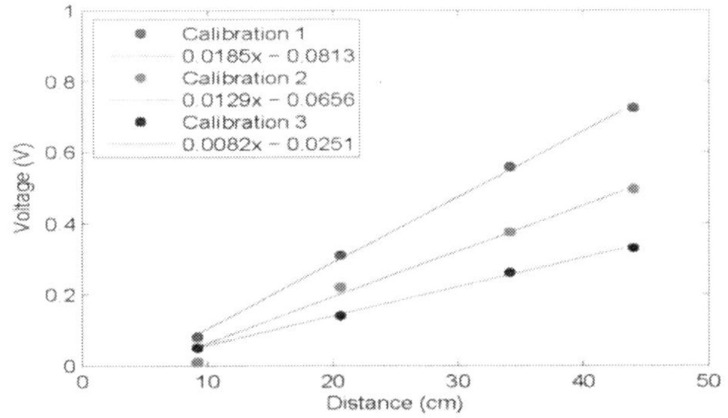

Figure 3: Scaled voltages as a function of distance

Two participants were recruited for the study, one male recreational player and one female athlete tennis player. Participants were provided with the instrumented racket and were asked to stand behind the service line (Figure 4). The test consisted of hitting a total of 20 regular serves. Participants were photographed during the serve using a GoPro Hero3 high speed camera at 120 frames per second.

Figure 4: Participant performing the testing

Each hit was analyzed to identify the frame where the ball made contact with the racket. A screenshot of each impact was generated and ball impact location was calculated (Figure 5). Strain data collected using the oscilloscope was imported into Matlab to produce the graphs corresponding to the strain signal of each impact (Figure 6). Forces and bending moments were calculated for each serve.

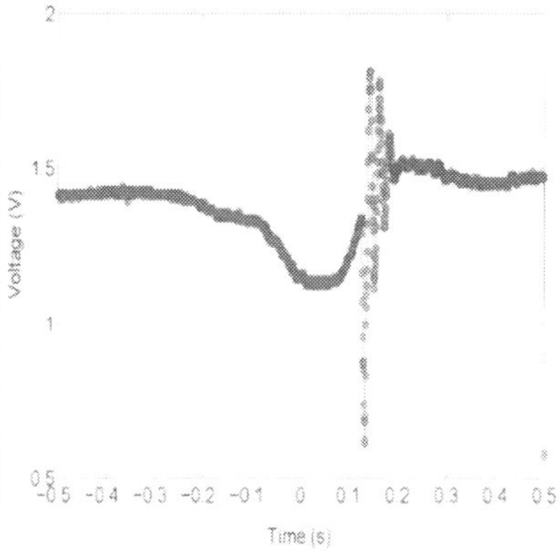

Figure 5: Screenshot of ball impact Figure 6: Strain signal at impact

RESULTS

The instrumented tennis racket system met the design goals. Strain within the racket was captured before, during, and after ball impact. A dip in voltage between -0.3 and 0.1 seconds in figure 8 show the strain caused when the racket is accelerated at the beginning of the serve. Ball impact occurs at 1.2 seconds causing a sharp spike downward followed by rapid oscillations that are quickly damped within 0.1 seconds. The wrist reaction force increase with distance from the handle (Figure 7). Higher forces were observed for the athlete than the recreational player.

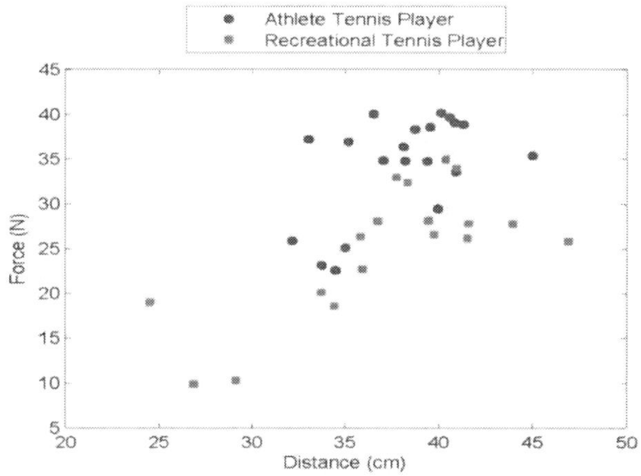

Figure 7: Wrist Reaction Forces at ball impact

DISCUSSION

The system was found to be effective at measuring forces and moments before, during, and after ball impact. Although this research only evaluated wrist reaction forces it could easily be combined with motion capture to develop a richer understanding on impact mechanics. This device or an improved version may also be useful to develop a better understanding of the forces and moments created from different types of movements during tennis play. Ultimately this research may lead to a reduction in injuries or improved performance.

REFERENCES

1. A. Cappozzo et. al, "Position and orientation in space of bones during move-ment: Experimental artifacts," Clinical Biomechanics, vol. 11, no. 2, pp. 90-100, December 1996.
2. N. K. Savage, "Vibration absorption in the tennis grip and the e ects on raquet dynamics", Ph.D. thesis, School of Aerospace, Mechanical and Manufacturing Engineering, RMIT University.

TEACHING UNDERGRADUATE DESIGN: AN APPROACH BASED ON INDUSTRIAL EXPERIENCE

Martin L. Tanaka

Department of Engineering and Technology
Western Carolina University
Cullowhee, NC, USA

INTRODUCTION

The design of products is an important area in bioengineering and the field of engineering in general. In fact, it is this desire to create something new that attracts many people to pursue careers as engineers. Using the foundational skills learned in the classroom, adding specialized knowledge gained during employment, and some creativity, amazing new products have been developed that have changed the world. The purpose of this abstract is to share the basic principles of engineering design and show how these can be effectively taught to undergraduate students.

The author has over 11 year of experience as a design engineer working for several different companies including Texas Instrument, Siemens VDO, and ABB. Products designed include pressure switches for control of air conditioning and refrigeration systems, an automotive cruise control unit for Saturn, an instrument Cluster for Mercedes Benz, a clock and outside temperature module for Volkswagen, and a mass spectrometer for ABB. Experience was gained working as a junior engineer, senior engineer, project manager, mechanical engineering supervisor, and engineering manager.

METHODS

The design of an engineered product is both similar to and different from non-engineering design. Both approaches require an understanding of the problem, creativity, and processing skills to transform raw materials into a completed product. However, engineering design also utilizes engineering calculations to predict the performance of a new design prior to building and testing the product. An examples of these types of calculations are to determine the size of the motor, pump, and pipes needed to transport water from one reservoir to another for an artificial waterfall. The pressure head, volumetric flow rate, and flow resistance can be taken into account prior to purchasing the motor, pump, and piping. The initial conceptual design can be modified (scaled) appropriately so that the finished product has a good chance of operating correctly the first time. An alternative to the engineering approach is to arbitrarily pick the motor, pump, and pipes without the benefit of calculations. After the system is built it can be tested. If the "guess" was correct, the system will function properly. If not, new components must be purchased, the design rebuilt, and the system tested again. This "trial and error" approach is undesirable for several reasons. Equipment may need to be purchased multiple times increasing the material cost of the project. In addition, each trail requires labor to rebuild and retest the system. To overcome these problems designer will tend to choose larger and more expensive components, but this leads to overdesign and wastes money. In addition to the increased cost, ordering new parts, rebuilding, and retesting all take time. This increases the time needed to complete the project. In this simple example it is clear that taking some time to perform calculations can dramatically improve project success.

In other cases, there are multiple variables that affect product performance and achieving a small window of parameter values is required to produce a functional product. An example of this is the design of an aircraft. It is not feasible to simply increase the thickness of components to overcome stress failures because this also adds weight. Creative solutions, calculations, advanced materials and computer simulations are critical to reducing product development time and cost.

Although these principles seem rather simple and logical, engineering students (and even working engineers) often use the "trial and error" approach. Inexperienced engineers generally have the skills necessary to perform the calculations, but lack practical experience having only performed these calculations in the classroom environment. This is readily apparent in senior design courses where student generally want to purchase components of a system that were selected arbitrarily. By helping the team to perform an often simple calculation, the proper components can be specified saving both time and money.

RESULTS

Below are other product design principles, tips, and best practices that may be helpful when conducting and/or teaching engineering product design.

Overview of the Entire Product Design Process

- Understand the product that you are trying to design including customer specifications, competitive products already in the market, government or regulatory constraints, etc.
- Develop a conceptual design (or several). Be creative and take the best aspects of all the designs and combine them. Evaluate the pros and cons. Consider the risks in each approach. Select to a single design concept to take to the next phase.
- Create a detailed design. A detailed design includes all the information that can be put "on paper" without actually building a prototype. It includes calculations and specification of all components. In theory, the design engineer should be able to hand the document to someone else and they could build the prototype without personal involvement.
- Build a prototype and test it to verify the design. Optimize the design, rebuild, and retest.
- Build production tooling and create the manufacturing line.
- Build parts on the line and test again to validate the design before going into full scale production.

Mentoring Students and Managing your Design Team

- Select a team leader to coordinate activities of the team & write meeting minutes (not the faculty member).
- For senior design projects, your product is a design (CAD drawings, electrical schematics, board layouts, test results, etc.). The physical prototype is only part of it.
- Teach critical thinking. In our fast paced world it is easy to just glance over information and only respond to something that "jumps out" at us. This is not a good approach in engineering design. It is better to think about your design, all possible outcomes and worry a little.
- Teach students that the reason products often fail is due to an unanticipated mode not taken into account in the design.
- Spend time with your students. Seeing how you approach a problem and how you interact with a customer will help them greatly. Be a good role model.
- Initially you should show the students what to do and have them follow your lead. As they gain experience, begin shift the responsibility to the students. Let them lead. You should follow, support, and provide guidance.

Manage the Scope of the Project

- Be realistic about what can be accomplished in each phase of development.
    - Sr. Designs: part time work by inexperienced people over ½ - 1 year → proof of concept prototype
    - Graduate Student: dedicated effort by an inexperienced person → working prototype
    - Faculty: Part time work → results will vary
- Be careful not to allow your student to do "Lego" designs for senior projects where they simply assemble existing products into a system. There needs to be a custom design component in order to teach them how to do this type of work.
- Break projects into phases, each with a defined and achievable scope (applies to all projects, not just student projects).
- Breaking the design into phases can be helpful to maintain focus on measureable and verifiable results.
- Select a project that is challenging, but not so much that it cannot be accomplished.
- Watch for scope creep by both the customer and the students.
- In contract engineering something extra can often be squeezed if not difficult for "good will". However, if the proposed change becomes involved, a contract amendment will be needed that can effect engineering design cost and piece price.

Business, Business Culture & Professionalism

- Design of engineered products is closely tied to business.
- Teach students how to communicate with a customer in a professional manner.
- Teach students to set achievable goals for yourself and achieve them. It is better to negotiate the scope of work or the deadline than to accept more than can be accomplished, miss the deadline, and let down the team.
- Use meeting minutes to track each assignment and due date.
- Teach students the basics of intellectual property (patents) – Inventor, Assignee, Costs of a Patent, Quality is important (loop holes), Types Utility vs. Design.
- Cost is always an important factor to consider in the design of commercial products.
- Basic knowledge of economics is important including return on investment, amortization of fixed costs (engineering, tools and equipment), design cost estimations, product cost estimation (costed Bill of Materials), development timeline estimation.
- It is exciting to watch people grow from naïve students into professional colleagues able to handle themselves in the business world. This applies to senior design students, graduate students, and junior engineers.

DISCUSSION

The above list is only a small portion of the entire body on knowledge available on product design and the teaching of this subject. It highlights some important basic ideas and acts as a guide for using engineering in product design. It includes lessons learned, warns of possible pitfalls, and presents best practices to train student for entry into the work world. It is a pleasure to teach practical product design principles to students and colleagues. I encourage people who are interested in product design to contact me or others with industrial experience before beginning your next engineering design project.

ACKNOWLEDGEMENTS

I would like to express my appreciation to the faculty an NCSU and Virginia Tech who taught me the principles of engineering. Furthermore, I would like to thank my colleagues at Texas Instruments and VDO for helping me to learn the necessary skills of engineering practice. Finally, I would like to express my sincere gratitude to Werner Eckardt for being an excellent role model, mentor, and friend to me during the early years of my career.

SB³C2015
Summer Biomechanics, Bioengineering and Biotransport Conference
June 17-20, 2015, Snowbird Resort, Utah, USA
SB³C2015-122

HETEROGENEOUS ELASTIC PROPERTIES OF ASCENDING THORACIC ANEURYSMS

Yuanming Luo (1), Frances M. Davis (2), Stéphane Avril (2), Jia Lu (1)

(1) Department of Mechanical & Industrial Engineering
The University of Iowa
Iowa City, IA, USA

(2) Ecole Nationale Suprieure des Mines de Saint-Étienne, CIS-EMSE,
CNRS: UMR5307, LGF
St. Étienne, France

INTRODUCTION

In the past decades there has been a growing interest in the mechanical properties of human ascending thoracic aortic aneurysms (ATAA). The most commonly used method to estimate the tissue properties is uniaxial tensile testing [1,2,3]. Yet it is well known that the *in vivo* loading experienced by arterial tissues is much better approximated by biaxial tests [4]. These methods involve the homogeneous parameters assumption and few reports discussed spatial variations of heterogeneous properties. However, spatial variations in the mechanical properties occur naturally due to local remodeling of the vessel wall [5]. To understand the interplay between mechanical stress and aneurysmal remodeling, it is imperative to consider the regional variation in elastic properties.

In this study, we utilize a Digital Image Correlation (DIC) based point-wise method [6] to delineate the heterogeneous properties in 12 ATAA samples. The ATAAs are subjected to a bulge inflation test whereby the surface displacements were recorded using a DIC system. The displacement and pressure data are used as inputs to an inverse membrane analysis to predict the stress distributions in the wall. Based on the acquired full field stress-strain data, the constitutive properties of a hyperelastic material model selected to describe the tissue are identified at every Gauss point. To validate the identified parameters, a forward finite element analysis is performed using the distributive properties to simulate the inflation experiment. The resulting displacements, strains and stresses are compared to the experimental data and a close match is observed in all 12 cases.

METHODS

The ATAA samples of ~4 cm in diameter were clamped to an inflation device and were pressurized incrementally to rupture. Images of the deforming samples were collected every 3 kPa until the samples ruptured. The collected images were analyzed using the commercial correlation software ARAMIS (GOM, v. 6.2.0) to determine the three dimensional displacement of the tissue surfaces. Details of the DIC specifications were reported in [6]. The central region of DIC point clouds of each sample was fitted to a NURBS mesh of initial diameter ~2 cm. The (normal) distance deviation between the DIC points and the NURBS meshes were found to be 0.001 to 0.036 mm, indicating a high accuracy of the geometric reconstruction. The surface strains were computed from the NURBS mesh.

Membrane tensions were predicted using an isogeometric inverse membrane method [6,7]. The deformed configurations and boundary conditions are given as inputs to the finite element model, and the wall stress was determined by way of finding a load-free reference configuration. The finite element formulation is based on the standard weak form but utilized to solve the reference configuration.

Once the stress and strain distribution in all pressure steps were obtained, the elastic properties of the wall tissue were identified at Gauss points. A hyperelastic model akin to that of Gasser, Ogden, and Holzapfel (GOH)[8] was used to describe the ATAA tissue. The energy function reads:

$$w = \frac{\mu_1}{2}(I_1 - \ln(I_2) - 2) + \frac{\mu_2}{4\gamma}(e^{\gamma(I_k - 1)^2} - 1) \qquad (1)$$

Here $I_1 = tr\mathbf{C}$, $I_2 = \det\mathbf{C}$ are the principal invariants of the Cauchy-Green deformation tensor, and $I_k = C:(\kappa\mathbf{I} + (1-2\kappa)\mathbf{M}\otimes\mathbf{M})$. The symmetry axes $\mathbf{M}$ was defined locally, $\mathbf{M} = \cos\theta\mathbf{G}_1 + \sin\theta\mathbf{G}_2$, where ($\mathbf{G}_1$, $\mathbf{G}_2$) are local surface basis derived from the NURBS mesh.

Values of the model parameters μ_1, μ_2, γ, κ, and θ were determined by minimizing the objective function:

$$f = \sum_{i=1}^{n} \left\| \mathbf{S}_i^{inv} - \mathbf{S}_i^{mod} \right\|^2 \qquad (2)$$

Where $\mathbf{S}_i^{inv}$ is the 2nd Piola-Kirchhoff stress computed from the inverse analysis and $\mathbf{S}_i^{mod}$ is the modeled stress.

RESULTS

Domain average values and the standard deviations of the material parameters are summarized in Table 1. As a representative example the parameters distribution of μ_2, γ, κ, and the fiber angle in a sample were presented in Figure 1.

Table 1: Identified material parameters

Case	$\mu_1(N/mm)$	$\mu_2(N/mm)$	γ	κ	θ
40	0.62E-02	0.60	7.01	0.41	1.67
42	0.56E-02	0.91	4.90	0.39	1.64
48	0.34E-02	1.11	13.86	0.42	1.76
50	0.26E-02	0.70	12.25	0.40	0.62
54	0.20E-02	0.87	12.19	0.42	1.39
55	0.22E-02	0.28	7.75	0.41	1.39
56	0.69E-01	1.75	23.94	0.37	1.54
57	0.72E-01	1.06	7.53	0.41	1.28
58	0.28E-02	1.37	12.36	0.43	1.56
60	0.51E-01	1.10	15.19	0.37	1.74
65	0.12E-01	0.27	1.61	0.40	1.56
66	0.84E-01	0.28	2.21	0.41	1.60
Mean	5.30E-02	0.94	11.40	0.40	1.50

The heterogeneous properties were validated using a forward analysis. The pointwise properties were applied to the mesh and the inflation deformation was simulated. The percentage errors of displacement, stress, and strain were 0.12-0.95%, 1.73-5.86%, and 0.81-7.61%, respectively in all sample over all pressure states. As an example of stress comparison, the inversely computed stress and the forward prediction in a sample under the highest pressure are shown in Figure 2.

DISCUSSION

In this study, we have delineated the heterogeneous properties in 12 human ATAA samples using full-field stress and strain data acquired from the DIC measurements and inverse stress analysis. The purpose of the study is to investigate the heterogeneity of ATAA tissues, albeit still over samples of moderate size. We have resolved the material properties to a spatial resolution of ~0.3-0.4 mm.

From the results, it is evident that ATAAs are highly heterogeneous. There is a significant level of property variation even within regions of 2 cm in diameter. This is a length scale over which aortic tissue specimens are typically treated as homogeneous in traditional tests. In terms of the average property values, the present results are consistent with the findings of other studies [9, 10]. The average value of κ is 0.4, suggesting that the ATAAs are moderately anisotropic. The ratio of D_{11}/D_{22} where D_{11} and D_{22} are the 11- and 22-component of the material tangent tensor at the ground state, were found to range from 0.94 to 1.29 (average value), confirming a lack of prominent strong direction. This finding is in contrary to some studies, which concluded that the circumferential direction is stronger [1,8]. The fibers appear to align in four dominating directions: circumferential, longitudinal, and two helical directions at 45° and 135°.

In closing, we characterized the heterogeneous nonlinear elastic properties of 12 ATAA samples. The value of this work lies in that it resolved the heterogeneous properties to a spatial resolution comparable to that of microscopic imaging. This opens the possibility of quantitatively correlating the local properties to the underlying microstructures, a subject that the authors will pursue in future.

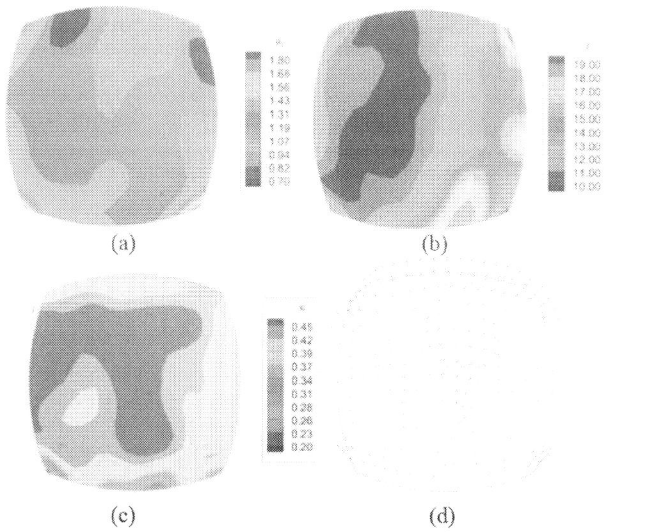

Figure1: Parameters distributions. (a) μ_2, (b) γ, (c) κ, (d) fiber direction

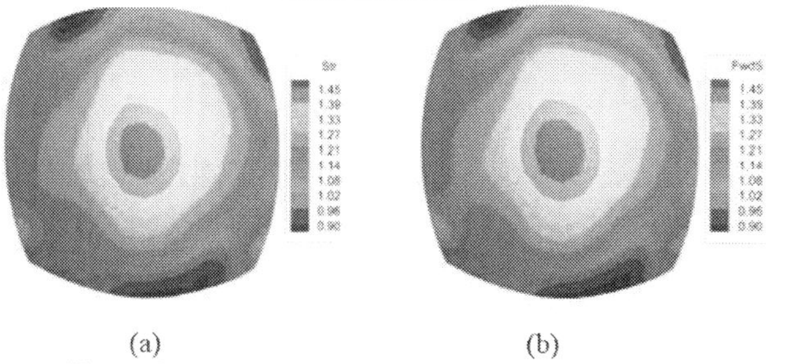

Figure2: (a) Inverse stress; (b) Forward stress

ACKNOWLEDGEMENTS

The work was supported, in part, by The University of Iowa Mathematical and Physical Sciences Funding Program. Dr. Davis was supported by the Whitaker International Scholars Program.

REFERENCES

[1] Iliopoulos, D.C. et al., Med. Eng. Phys. 31 (2009) 1-9.
[2] Vorp, D.A. et al., Thorac. Surg. 75 (2003) 1210-1214.
[3] Sokolis, D.P. et al., Med. Biol. Eng. Comput. 50 (2012) 1227-1237.
[4] Wilson, J.S. et al., J. R. Soc. Interface 9 (2012) 2047-2058.
[5] Müller-Schweinitzer, E. et al., Organogenesis 5 (3) (2009) 97-104.
[6] Davis, F.M. et al., Biomech Model Mechanobiolog, in press(2014).
[7] Lu, J. et al., J. Appl. Mech. –Transactions ASME 76 (6) (2009) 061013/1061013/10.
[8] Pham, T. et al., Acta Biomater. 9 (2013) 7927-7936.
[9] Pierce, D. M. et al., Journal of the Mechanical Behavior of Biomedical Materials.
[10] Romo, A. et al., Journal of biomechanics 47.3 (2014): 607-616.

DIFFERENTIAL HEMODYNAMIC CHANGES AND LUMEN REMODELING IN THE ARTERY AND VEIN OF PORCINE ARTERIOVENOUS GRAFT AND FISTULA

Daniel B. Pike (1), Yong He (2), Christi M. Terry (3), Alfred K. Cheung (3,4), Yan-Ting Shiu (3)
(1) Bioengineering, University of Utah, Salt Lake City, UT, USA
(2) Surgery, University of Florida, Gainesville, FL, USA
(3) Nephrology, University of Utah, Salt Lake City, UT, USA
(4) Nephrology, VA SLC Health Care System, Salt Lake City, UT, USA

INTRODUCTION

Chronic hemodialysis requires a long-term form of vascular access, typically either an arteriovenous graft (AVG) or arteriovenous fistula (AVF) [1]. Both access methods suffer from stenosis (i.e., decrease in the lumen area) over time, especially the AVG. In the AVG, stenosis is primarily a result of neointimal hyperplasia (NH) at the vein-graft anastomosis [2]. In the AVF, failure primarily results from stenosis in the fistula vein and/or the inability of the venous lumen to expand large enough to support the higher, arterial level of flow and hemodynamic wall shear stress (WSS) as compared to the low flow and WSS normally seen in veins [3]. Evidence from the literature suggests that WSS affects the changes in lumen area, both narrowing and expanding, in both vascular accesses [4-6]. However, detailed WSS and lumen changes have not been reported. Our group developed a magnetic resonance imaging-computational fluid dynamics (MRI-CFD) method to monitor lumen area remodeling and characterize WSS over time [7].

In the AVG and AVF, both vein and artery undergo lumen area changes but the vein enlarges to a much greater degree and yet experiences greater incidence of stenosis than artery. It is unclear if the relationships between WSS and subsequent lumen area changes are different between the artery and vein or are different between the AVG and AVF. To better understand the role of WSS and lumen changes in both AVG and AVF, we compared the WSS and subsequent changes in lumen area (Δarea) between the vein and artery in pig models of AVG and AVF. Finally, previous CFD simulations of blood flow in AV accesses have generally assumed a rigid vessel wall, because of computational time and power limitations as well as difficulties in measuring the thickness and mechanical properties of veins. To explore the influence of wall deformation on CFD modeling, we included wall deformation in CFD simulations, using 2-way coupled fluid structure interaction (FSI) modeling between the fluid (blood) and solid (vessel wall) domains in idealized AVG and AVF models.

METHODS

For AVG, synthetic vascular grafts were implanted between the common carotid artery and external jugular vein in swine. For AVF, the external jugular vein was anastomosed to the side of the carotid artery. At 1, 2, 3, 4, or 6 wk (n=5 at each wk) after graft implantation or 1, 2, 6, or 12 wk after fistula creation (n=4 at each wk), contrast-free MR images were acquired to assess the lumen geometry and measure blood inflow and outflow rates, which were then used as boundary conditions for CFD simulations wherein rigid vessel walls were assumed [7]. WSS magnitudes and lumen area in the AVG were averaged over 2 cm lengths of arterial and venous segments both upstream and downstream from their respective anastomoses. WSS and lumen area in the AVF were averaged over 2 cm of the fistula vein near the anastomosis, and 2 cm of both the feeding and draining arteries from the anastomosis. In the deformable wall simulations, we used an idealized AVG or AVF geometry and average lumen diameters measured from our pig models at later time points. We assumed that both artery and vein were 300-μm thick (measured from artery [8]) and had an elastic modulus of 800 kPa (measured from vein [9]). As a first approximation, the rigid graft was also assumed to be 300-μm thick. The inflow and outflow blood flow rates were actual values obtained also by MRI in our porcine models at later time points. FSI simulation was run in ANSYS Workbench.

RESULTS

In our pig AVG model, artery lumen area changed little over 6 wk after graft implantation, while in the vein the area expanded from wk1 to wk4, and then decreased at wk6 (Figure 1A). In examining the later lumen narrowing, no significant linear or nonlinear association was found in the artery between WSS at wk3 and Δarea from wk3 to wk6, while a nonlinear, positive association was found in the vein between WSS at wk3 and Δarea from wk3 to wk6 (Figure 1C). That is, lower wk3 WSS was associated with greater decreases in wk6 lumen area and higher WSS was associated with lesser decreases in wk6 lumen area.

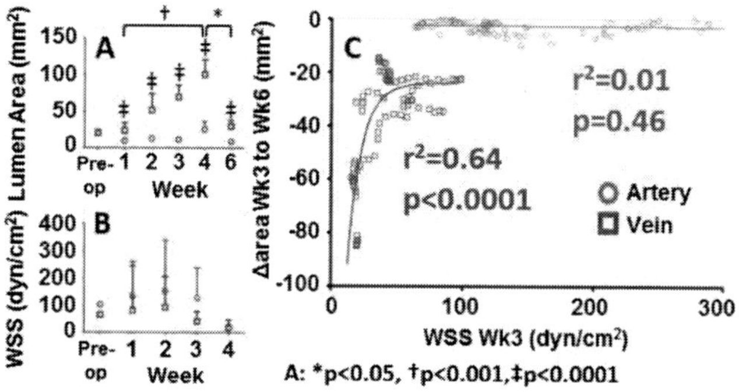

FIG 1: (A) THE LUMEN AREA, SINGLE SYMBOLS ARE ARTERY TO VEIN COMPARISONS, BRACKETS ARE VEIN TO VEIN, (B) WALL SHEAR STRESS (WSS), AND (C) CORRELATION BETWEEN WK3 WSS AND LUMEN AREA CHANGE BETWEEN WK3 & WK6 IN A PIG AVG MODEL.

In our pig AVF model, average lumen area increased from wk1 to 6 in both the artery and the vein but remained steady thereafter (Figure 2A). The absolute and percentage area increases were larger in vein than artery. Average WSS at wk1, 2, and 6 was also significantly larger in the vein as compared to the artery (Figure 2B). To examine the lumen expansion, we regressed wk1 WSS with Δarea from wk1 to wk6. In the artery, WSS at wk1 and Δarea from wk1 to wk6 had a positive exponential relation with Δarea increasing with increasing WSS at wk1 (Figure 2C). In contrast, WSS at wk1 and Δarea from wk1 to wk6 had a negative exponential relation in the vein, with Δarea generally decreasing with increasing WSS at wk1 (Figure 2C).

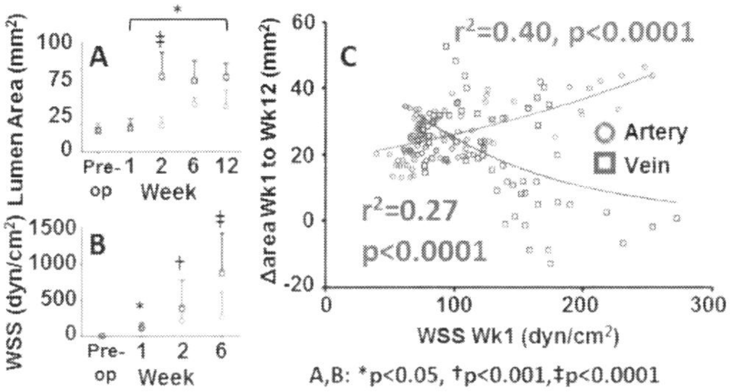

FIGURE 2: (A) LUMEN AREA, (B) WALL SHEAR STRESS (WSS), AND (C) CORRELATION BETWEEN WK1 WSS AND LUMEN AREA CHANGE BETWEEN WK1 & WK6 IN A PIG AVF MODEL. IN (A, B), SINGLE SYMBOLS ARE ARTERY TO VEIN COMPARISONS. BRACKETS ARE VEIN TO VEIN.

When wall deformation was simulated, the differences in modeled blood velocity and WSS were more pronounced in AVF than in AVG (Figure 3). The deformable and rigid AVG models have similar velocity (A,B) and WSS (C,D), but the deformable AVF model has higher velocity and WSS (F,H) than the rigid model (E,G). However, flow patterns in both deformable models (B,F) exhibit slightly more complex recirculation when compared to their respective rigid wall models (A,E).

DISCUSSION

Our demonstration of differential WSS-lumen area relationships in the artery and vein of both AVG and AVF, and our comparison between the effects of wall deformability in AVG and AVF is novel.

In the *pig AVG model*, wk3 WSS smaller than 50 dyn/cm^2 was associated with large decreases in lumen area from wk3 to wk6 in the vein and wk3 WSS of approximately 80-110 dyn/cm^2 was associated with a smaller area decrease in the vein; in contrast, area was unchanged in the artery in these WSS ranges. Thus, in our AVG model, the same level of WSS does not correlate with the same lumen area response between the artery and vein, both in terms of absolute area change, and in the correlation between WSS and subsequent lumen area change.

In the *pig AVF model*, the relationship between WSS and lumen expansion is significant in both artery and vein, but is negative in the vein and positive in the artery. In the artery, larger wk1 WSS correlated with larger lumen area increases from wk1 to wk12, which is consistent with what has been reported in the literature. In contrast, in the vein, similarly large wk1 WSS correlated with much smaller lumen area increases and even with area decreases from wk1 to wk12.

In our FSI simulations, the different effects of deformable vessels on velocity and WSS between AVG and AVF may be because motion in the AVG is constrained by the presence of the rigid graft. Arterial- and venous-specific wall thickness and elastic properties will next be incorporated to further improve the physiological relevance of our CFD models.

Our AVG and AVF simulation results reveal a different response to the same levels of WSS by venous and arterial vascular wall cells. A lot of research has been focused on the potentially beneficial changes of flow-induced dilation and/or 'arterialization' seen in successful AVF and arterial bypass veins. One of the prevailing paradigms in vascular access has been that 'arterializing' the vein (i.e., thickening of the venous wall) may lead to better outcomes, and is brought on by the arterial levels of flow and WSS introduced into the vein through the creation of an AVG or AVF [10]. Our results suggest

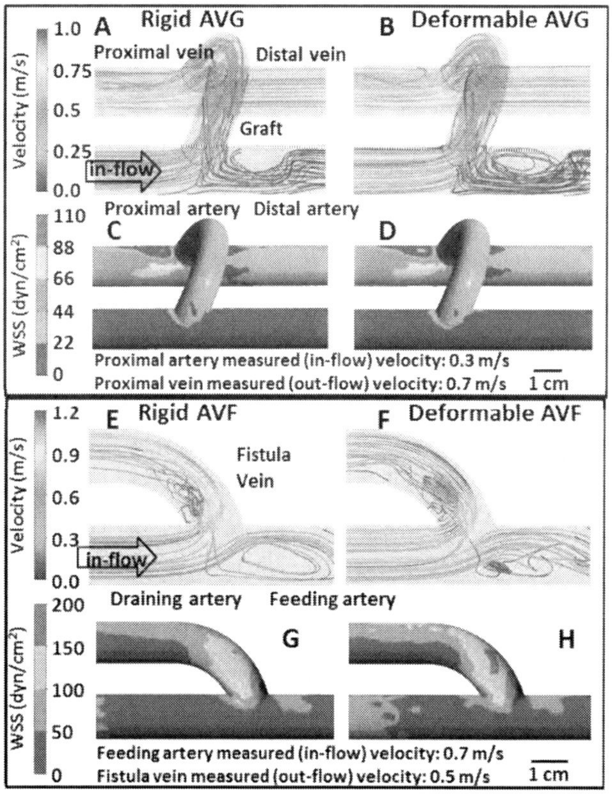

FIGURE 3: CFD RESULTS FROM RIGID AND DEFORMABLE MODELS OF AVG (A-D) AND AVF (E-F).

that this current dogma is overly simplified. In the AVG vein, higher WSS led to smaller area decreases from wk3 to wk6, indicating that less stenosis has occurred, and possibly leading to better outcomes. However, the range of wk3 WSS values was smaller in the vein than the artery, so it is unclear how the WSS-Δarea correlation behaves at larger vein WSS values. In the AVF, in contrast, smaller WSS led to greater Δarea in the vein, but smaller Δarea in the artery. The results from our AVF model suggest that higher levels of WSS in the vein may actually be counterproductive.

Our study provides a much more detailed characterization of hemodynamics in AVG and AVF than what has previously been reported. In summary, similar magnitudes of WSS are associated with different lumen area changes in the artery and vein of AVG and AVF. In addition, the vessel responses to similar WSS can be markedly different in AVG and AVF. These findings provide an impetus for further study into the differences between arterial and venous lumen remodeling in the context of vascular access.

ACKNOWLEDGMENTS

The work was supported in part by NIH R01DK088777 and R01DK100505.

REFERENCES

[1] Kidney Int, 59:358-362, 2001.; [2] J Am Soc Nephrol, 17:1112-1127, 2006.; [3] Kidney Int, 70:1413-1422, 2006.; [4] Kidney Int, 59:2325-2334, 2001.; [5] Exp Mol Pathol, 72:150-160, 2002.; [6] Nephrol Dial Transplant, 27:358-368, 2012.; [7] J Biomech, 46:165-169, 2013.; [9] Biomech Model Mechanobiol, 13:679-695, 2014.; [8] Nephrol Dial Transplant, 17:1057-1062, 2002.; [10] Am J Kidney Dis, 42:151-157, 2003.

DEVELOPMENT OF VIRTUAL ENVIRONMENT NAVIGATION OPTIONS FOR INDIVIDUALS POST-STROKE OR WITH CEREBRAL PALSY USING KINECT

Sean Pool (1), Laurie A. Malone (2), C. Scott Bickel (2,3), Gerald McGwin (4), James H. Rimmer (2,3), Alan W. Eberhardt (1,3)

(1) Department of Biomedical Engineering,
University of Alabama at Birmingham,
Birmingham, AL, USA

(2) UAB/Lakeshore Research Collaborative,
University of Alabama at Birmingham,
Birmingham, AL, USA

(3) School of Health Professions,
University of Alabama at Birmingham,
Birmingham, AL, USA

(4) Department of Epidemiology,
University of Alabama at Birmingham,
Birmingham, AL, USA

INTRODUCTION

Coincident with mobility impairment resulting from stroke or cerebral palsy (CP) is a tendency towards decreased physical activity. Secondary conditions commonly associated with low levels of physical activity are pain, fatigue, depression, and anxiety, as well as obesity, diabetes and cardiovascular disease. Exercise has been shown to decrease the prevalence and magnitude of secondary conditions, increase self-esteem, and improve personal relations.

Among people with disabilities, only 25% attain a moderate amount of physical activity each week, compared to 43% of adults without disability [1]. Interventions seeking to increase physical activity in individuals with mobility impairments must address many commonly reported barriers including physical, economic, psychological, and knowledge-related [2,3,4]. Our research team is currently working toward development of a device that permits both arm and leg cycling integrated with a virtual exercise environment. The system aims to overcome barriers that inhibit exercise among persons with disabilities, while capitalizing on attentional capture in order to distract from the tedium and fatigue of exercising.

Microsoft Kinect (Microsoft Corporation, Redmond, Washington) has been used to navigate Google Earth and Google Street View for healthy individuals [5], while others have used Kinect to train wheelchair users how to drive a manual or motorized wheelchair [6]. The purpose of the present study was to develop directional navigation options for movement within a virtual environment using Kinect by individuals with CP and post-stroke. Various body movements that allowed the user to choose between different paths were evaluated for degree of success.

METHODS

The Microsoft Kinect is composed of an RGB camera and an infrared emitter and receiver that capture color and depth image data. Kinect for Windows includes a "Near Range" tracking mode (range 0.4-0.8 m) and adds a "Seated" tracking mode that tracks the upper body only, including head, shoulders, arms, wrists, and hands, and is comparable to mainstream solutions commonly used for gait training and pose estimation [7]. A comparison of Kinect to a laser scanner established its accuracy as <1 cm at the 0.4-0.8 m depth range [8].

A virtual environment was created that moves along the path until it arrives at an intersection. Input from the user determines left or right turns, or continuing straight. When a gesture is recognized indicating directional choice, the corresponding video continues (Fig 1). The simulation lasts thirty seconds which allows for multiple gestures to be quickly tested. The virtual environment utilizes real-world video captured in 720p resolution using a Canon T1i. These videos were motion corrected using the Deshaker (http://www.guthspot.se/video/deshaker.htm) plugin for VirtualDub (http://www.virtualdub.org) to eliminate camera shake caused by walking while recording.

The program that connects the videos with the Kinect and user input was developed in Microsoft Visual C# 2010 Express. Upon start-up, the Kinect and real-world environment videos are loaded and the Kinect RGB and Skeleton streams are enabled. RGB video is captured in 1280x960 resolution at 12 frames per second (FPS). When the virtual environment is initialized, a fade transition occurs from the Kinect RGB video to displaying the virtual environment. The first video is played and once the intersection is reached, arrows flash on screen prompting the user to make a directional decision. When the user performs the indicated gesture, the next video plays corresponding to the indicated direction. If the user has not performed the appropriate gesture after 10 seconds, the program restarts. The pre-choice video plays again and the program loops through the directional logic until all gestures are exhausted. Eleven gestures are recognized: a left or right vertical hand raise over the head, left, right, or both arms extended horizontally to the side, a torso twist to the left or the right, a vertical shrug of the left or right shoulder, or a head nod to the left or the right (Fig 2).

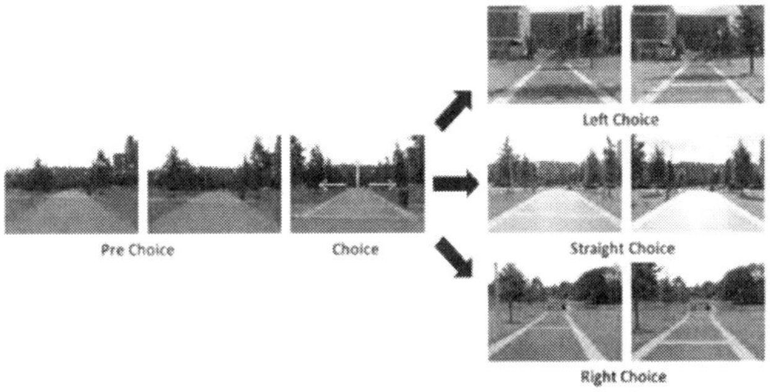

Figure 1. Video progression of virtual environment where arrows prompt the user when a choice is required leading them down left, right, or straight paths.

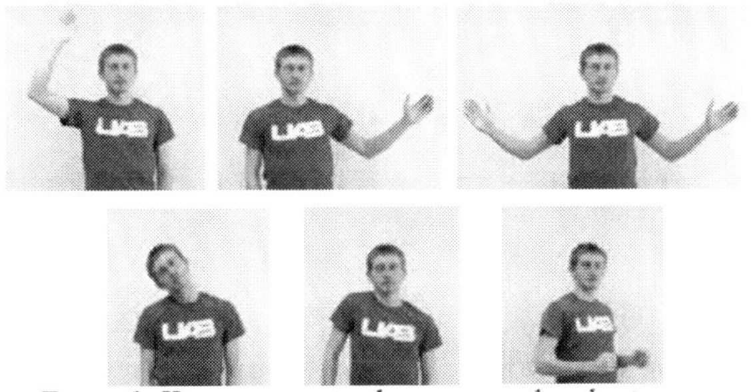

Figure 2. Kinect-recognized movements hand raise, arm extend, both arm extend, head nod, shoulder shrug, torso twist.

Thirty participants (14 male) aged 19 years or older, including 15 with CP and 15 post-stroke, were recruited at Lakeshore Foundation. Approval was obtained from the UAB Institutional Review Board. A Box and Block Test [9] was used to measure upper extremity function and to establish dominant hand. Each participant performed the Kinect computer interface movement test, repeating each of the fifteen unique gestures three times. 3-D joint position data were collected at 12 frames/sec across 10 data points. A seven-point moving average filter smoothed joint coordinate data and removed motion-capture noise. These data were output along with movement time in milliseconds, which gesture the user was instructed to perform, and at what time the Kinect registered the gesture. The success score for each movement was defined as the number of movements successfully performed divided by the number of movements attempted, multiplied by 100%. For successful movements, preparation and performance times were calculated. Generalized estimating equations were used to compare success scores between the CP and stroke groups. For successful gestures only, mixed statistical models were used to compare preparation and performance time between the groups.

RESULTS

Three participants post-stroke were excluded from the analyses due to incomplete motion data; thus, analyses included 12 post-stroke and 15 CP. Individuals with CP showed a significant effect of movement type on success ($p = 0.02$, Fig 3). Success scores were significantly lower for shrug and twist as compared to arm extend, head nod, and hand raise. Individuals post-stroke did not exhibit any significant effect of movement type on success. Individuals with CP demonstrated lower success performing the twist gesture ($p = 0.0185$) and a higher success performing the hand raise gesture ($p = 0.044$) as compared to those post-stroke. No differences in preparation time were observed between the groups. Individuals with CP showed a significant effect of movement type on preparation time ($p < 0.001$). Individuals post-stroke did not exhibit any significant effect of movement type on preparation time. Individuals with CP prepared for the nod gesture faster than those post-stroke ($p = 0.01$). Among individuals with CP, there was an interaction between movement type and hand dominance on performance time ($p = 0.01$). The dominant direction was faster when performing arm extend and head nod movements and slower when performing the twist and shrug movements. Individuals post-stroke did not show an effect of movement type on performance time.

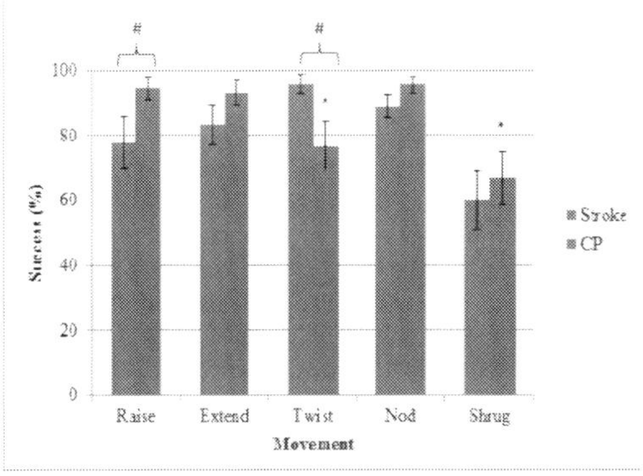

Figure. 3. Success scores (%) for CP and post stroke, indicating differences within (*) and between (#) groups ($p < 0.05$).

DISCUSSION

The present study demonstrated that all gestures may be viable for the post-stroke population; however, for individuals paralyzed in one arm, only one of the two directions will be feasible for certain movements. Future development should allow a single arm to make both left and right movements for these gestures. For those with CP, the twist and shrug movements were less feasible than the head and hand gestures. A more comprehensive functional ability test may allow for libraries to be designated according to functional ability level rather than specific disability group.

On-the-fly smoothing of joint position data may improve movement recognition and the perception of the virtual environment, but at the potential cost of introducing processing time delays. The benefits of improving the functioning of movement recognition, however, likely outweigh the decrease in usability. Kinect technology is many years old; use of more precise tracking methods may benefit future development of virtual exercise environment.

ACKNOWLEDGEMENTS

NIDRR Grant #H133E120005

REFERENCES

[1] Boslaugh, SE, & Andresen, EM, *Prev Chron Dis*, 3(3): A78, 2006.
[2] Malone, LA, et al,. *Disability Health Journal*, 5(4):254–60, 2012.
[3] Rimmer, JH, et al., *Amer J Prevent Med*, 26(5): 419–25, 2004.
[4] Rimmer, JH, et al., *J Rehab Res Dev*, 45(2): 315–22, 2008.
[5] Boulos, MNK, et al., *Inter J Health Geo*, 10(1): 45, 2011.
[6] Zondervan, DK, & Reinkensmeyer, DJ http://lifesciences.ieee.org/publications, 2012)
[7] Obdrzálek, S, et al., *IEEE Eng Med Bio Soc*, 1188–93, 2012.
[8] Khoshelham, K, & Elberink, SO, *Sensors* 12(2): 1437–54, 2012.
[9] Mathiowetz, V, et al., *Amer J Occ Ther*, 39(6): 386–91, 1985.

EFFECT OF GEOMETRIC AND MATERIAL PROPERTY CHANGES IN THE THORACIC SKELETON FOR AN OLDER OCCUPANT FINITE ELEMENT MODEL

Samantha L. Schoell (1,2), Ashley A. Weaver (1,2), Nicholas A. Vavalle (1,2), Joel D. Stitzel (1,2)

(1) Wake Forest University School of Medicine, Winston-Salem, NC, USA

(2) Virginia Tech - Wake Forest Center for Injury Biomechanics, Winston-Salem, NC, USA

INTRODUCTION

In motor vehicle crashes (MVCs), thoracic injury ranks second only to head injury in terms of the number of fatalities and serious injuries, the body region most often injured, and the overall economic cost [1,2]. The shape, size, bone density, and cortical thickness of the thoracic skeleton vary significantly with age, which can affect the injury tolerance especially in at-risk populations such as the elderly. Computational modeling has emerged as a powerful and versatile tool to assess injury risk. However, current computational models only represent certain ages in the population. One current finite element (FE) model is the Global Human Body Models Consortium (GHBMC) v4.2 average male occupant model. The GHBMC is representative of a 50th percentile male (M50) and was based on medical images of a 26 year old (YO) individual (height- 174.9 cm, weight- 78.6 ± 0.77 kg, and body mass index (BMI)- 25.7 ± 0.25) [3]. The objective of this study is to develop an older occupant GHBMC model representative of a 65 YO male with a focus on the thoracic skeleton. The biomechanical response of the 65 YO model will be studied through simulations of a frontal hub impact to assess the effects of geometric and material property changes with age.

METHODS

Development of the 65 YO model involved morphing the baseline GHBMC model using thin-plate spline interpolation to accurately capture the geometric changes in the thoracic skeleton. In order to execute the thin-plate spline interpolation, homologous landmarks on the reference, target, and FE model are required. An image segmentation and registration algorithm was previously used to collect homologous rib and sternum landmark data from males and females aged 0-100 years [4, 5]. Although homologous landmark data was only collected for the ribs and sternum, associated structures such as the costal cartilage, intercostal muscles, spine, and thoracic organs as well as external body surface geometry are morphed along with the skeletal structures due to the connection of the elements.

To evaluate the role of material property changes with age, techniques and data from the literature were implemented. Golman et al. utilized a statistically significant correlation between age and ultimate strain based on two previous studies to adjust the ultimate plastic strain of the ribs [6]. El-Jawahri et al. developed a 75 YO FE model and implemented rib material properties based on aging functions for femur data [7]. Ito et al. also developed a 75 YO FE model based on both aging functions for rib and femur data [8]. Kent et al. developed an elderly FE model based on aging functions for another set of femur data [9]. The material properties for the cortical bone of the ribs based on each study are summarized in Table 1.

Table 1: Material properties for the rib cortical bone.

Literature Source	E (GPa)	σ_{yield} (MPa)	ϵ_{yield} (%)	σ_{UT} (MPa)	ϵ_{UT} (%)
Golman et al. 2014	11.5	88.0	0.77	117.0	2.30
El-Jawahri et al. 2010	13.0	90.0	0.69	123.0	2.76
Ito et al. 2009	7.73	54.0	0.70	68.3	1.01
Kent et al. 2005	10.3	61.6	0.60	66.6	0.42

E = Elastic Modulus; σ_{yield} = Yield Stress; ϵ_{yield} = Yield Strain; σ_{UT} = Ultimate Stress; ϵ_{UT} = Ultimate Strain

To evaluate the thoracic response, a frontal hub impact was simulated. The frontal hub impact was conducted at 6.7 m/s with a 23.4 kg cylindrical impactor striking the sternum at the fourth rib interspace. A total of 9 simulations were performed to study the influence of geometric and material property changes alone as well as the combination of both geometric and material property changes. The single factors studied were the geometric changes and the four variations of the material property changes as previously described. To compare the simulation results, the maximum thoracic force (Fmax) and maximum percent compression of the chest were evaluated. In addition, the number of predicted rib fractures was also compared across the simulations.

RESULTS

Figure 1 shows the morphing results comparing the baseline GHBMC model and the 65 YO male. The 65 YO model possesses increased thoracic kyphosis, increased roundedness of the rib cage, and an upward rotation of the ribs which influences the biomechanical response of the thorax.

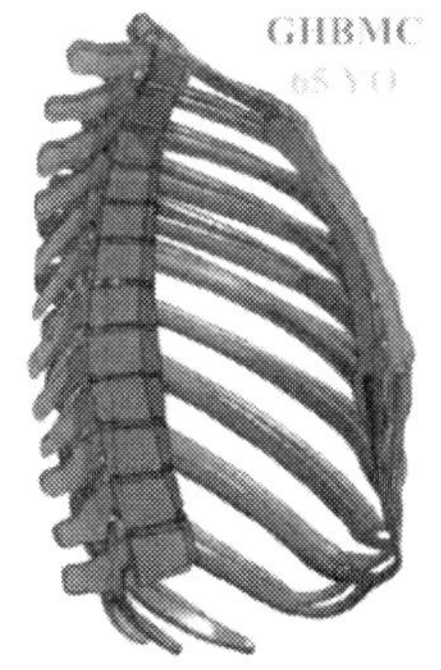

Figure 1: Comparison of GHBMC (grey) and 65 YO (green) FE models.

The percent difference of the 65 YO model relative to the baseline GHBMC model for the Fmax and maximum percent compression is summarized in Figure 2 for geometric changes only, material property changes only, and combined geometric and material property changes. The geometric changes alone resulted in a 3.79% decrease in Fmax and a 2.68% increase in the maximum percent compression relative to the baseline GHBMC model. Material property changes alone had little effect on Fmax. Material property changes alone had up to a 3.69% increase in the maximum percent compression. The combined geometric and material property models showed the most variation in the response with up to a 5.11% decrease in Fmax and up to a 4.95% increase in maximum percent compression.

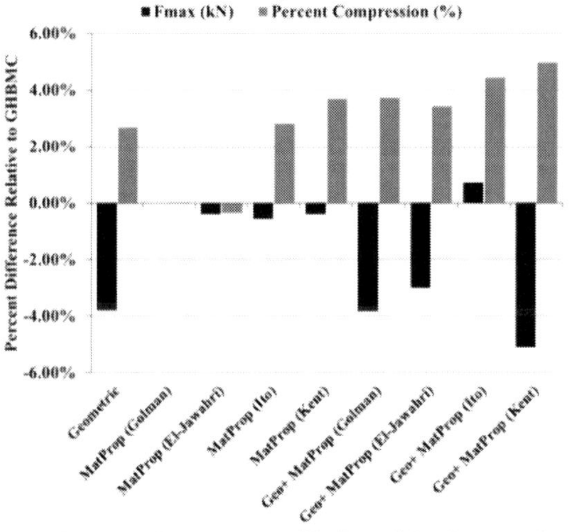

Figure 2: Percent difference of 65 YO model relative to baseline GHBMC. Percent difference was calculated as follows: (65 YO-GHBMC)/GHBMC.

A summary of the predicted rib fractures is provided in Figure 3. Abbreviated Injury Scale (AIS) 98 injury severity for each simulation was determined using the number of predicted rib fractures. In the experimental data for the frontal hub impact tests, eight subjects with a mean and standard deviation age of 55.5 ± 23.4 sustained 9.4 ± 7.2 rib fractures. The computational models underpredicted the number of rib fractures in comparison to the experimental data. The combined geometric and material properties derived from the Kent et al. model predicted the most rib fractures with 5 left and 5 right fractures resulting in an AIS 4 injury. In all cases, adding geometric changes in addition to material property changes increased the number of rib fractures predicted on average by 219%.

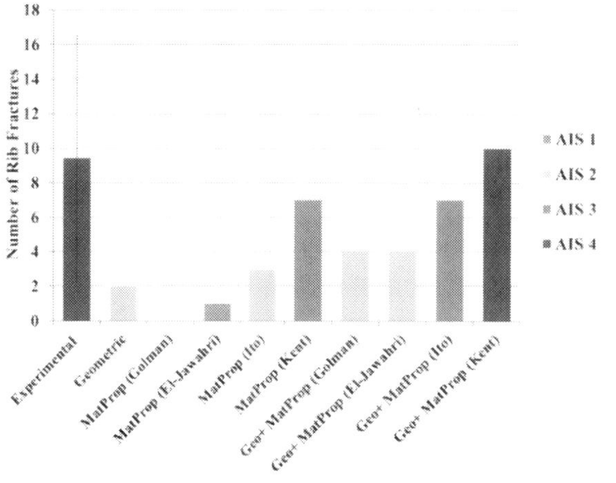

Figure 3: Summary of predicted rib fractures.

DISCUSSION

In this study, the thoracic response of a 65 YO male was evaluated using a frontal hub impact to assess the importance of geometric and material property changes. The geometric changes with age contribute to the differences in thoracic stiffness. With age, the ribs rotate upward which results in less joint rotation and a stiffer thoracic response. As a result, loading in the anterior-posterior direction results in increased deflection and an increased risk of rib fracture. Material property changes alone had little effect on Fmax and some effect on the maximum percent compression. The combined geometric and material property change models resulted in the most variation in Fmax and maximum percent compression. This variation is also noticeable on a more local and tissue level in regards to the strain values and the number of rib fractures that are predicted. Due to the decrease in ultimate plastic strain with age, the 65 YO geometric model with material properties derived from Kent et al. resulted in the most predicted rib fractures. The most appropriate material property assumption in this estimation is the Golman et al. data which utilizes rib data. Both geometric and material property changes are necessary to accurately predict rib fractures. Future implementation of cortical thickness changes with age will further improve the prediction accuracy. The development of these age-specific FE models will lead to an improved understanding of the complex relationship between thoracic geometry, age, and injury risk. The improved understanding gained from these models will aid in changes in restraint design and use to not only reduce injuries but also save lives.

ACKNOWLEDGEMENTS

Funding was provided by the National Highway Traffic Safety Administration under Cooperative Agreement Number DTN22-09-H-00242. Views expressed are those of the authors and do not represent the views of NHTSA.

REFERENCES

[1] Cavanaugh, JM. *Accidental Injury: Biomech and Prevention*, 2002.
[2] Ruan, J et al., *Stapp Car Crash*, 47:299-321, 2003.
[3] Vavalle, NA et al., *Ann Biomed Eng*, 41(3): 497-512, 2013.
[4] Weaver, AA et al., *J Morphol*, 275(11): 1284-1299, 2014.
[5] Weaver, AA et al., *J Anat*, 225(2): 246-261, 2014.
[6] Golman, AJ et al., *Accident Analysis & Prevention*, 64:1-8, 2014.
[7] El-Jawahri, RE et al., *Stapp Car Crash*, 54:407-430, 2010.
[8] Ito, O et al., *SAE Technical Paper*, 2009-01-0381, 2009.
[9] Kent, R et al., *Stapp Car Crash*, 49:231-249, 2005.

SB^{3}C2015
Summer Biomechanics, Bioengineering and Biotransport Conference
June 17-20, 2015,Snowbird Resort, Utah, USA
SB^{3}C2015-126

FEASIBILITY OF UTILIZING THERMAL IMAGES FOR MELANOMA SCREENING

Alexander LeBrun, and Liang Zhu

Department of Mechanical Engineering
University of Maryland Baltimore County
Baltimore, MD, USA

INTRODUCTION

Previous efforts have also been focused on developing non-invasive methods to measure or monitor local blood perfusion via imposing heating or cooling on skin surfaces, then record how the skin temperature recovers to its baseline value. The blood perfusion rate in the Pennes bioheat equation is then adjusted until the theoretically predicted temperature transients agree well with the recorded temperature recovery after the initial heating or cooling stimulation. The blood perfusion device by Mudaliar et al. [1] injects cold air to the skin surface and heat flux data are used to extract local blood perfusion rate in the tissue. The device developed by Liu et al. [2] also requires direct physical contact to the skin surface, where oscillatory resistor heating is generated to heat the skin surface and skin temperature is monitored by thermocouples. Following similar approaches, the study by Centingul and Herman [3] introduces infrared thermography imaging to record temperature rises recovering from air cooling on skin surface. They found that during the recovery duration the maximal temperature differences between healthy tissue and normal mole are all below 0.3°C, while the value between melanoma and healthy tissue in three melanoma cases vary between 0.5 and 2.2°C. All of the previous researches have demonstrated that dynamic responses of skin temperature after heating or cooling may be sensitive to local blood perfusion rate in targeted tissue region.

This study is focused on combining the approaches of using oscillatory heat flux at skin surface [2] and infrared thermography [3] to examine key physical factors most sensitive to the local blood perfusion rate in targeted tissue, and to identify sensitivity of the proposed approach for screening melanoma lesion associated with high blood perfusion rates. A 2-D heat transfer model consisting of a melanoma lesion embedded in a cylindrical tissue is developed to derive the surface temperature oscillations when the skin surface is subject to a convection and incident oscillatory heat flux. Key components in the skin temperature are identified to show their sensitivity to the blood perfusion rate in the melanoma lesion, therefore, to be used for melanoma screening.

METHODS

Figure 1 shows the schematic diagram of the thermal model consisting of a melanoma embedded in tissue. The diameter of the mole is selected as 10 mm in diameter and its thickness varies from 1 mm to 5 mm. A large mole is one of the risk factors leading to melanoma. The tissue cylinder with a diameter of 100 mm and a height of 60 mm is selected sufficiently large that the boundary conditions at the bottom and side surfaces do not affect the temperature field of the melanoma significantly. The top surface is subject to a combination of a convection boundary condition (convection coefficient h and air temperature T_{air}) and a periodic heat flux condition of $q_o''cos(2\pi ft)$, where q_o'' is the average heat flux, t is time, and f is the frequency of the periodic heat flux. The side surface of the tissue cylinder is adiabatic and the bottom surface is a prescribed temperature of 37°C. The initial condition is the steady state temperature field without imposing the periodic heat flux.

The Pennes bioheat equation is used to simulate the temperature field inside both the melanoma and the normal tissue, and is written as

$$\rho c\frac{\partial T}{\partial t} = k\nabla^2 T + q_m + \omega\rho c(37 - T) \qquad (1)$$

where k is thermal conductivity, ω is the local blood perfusion rate, ρ is density, c is specific heat, and q_m is metabolic heat generation rate. Simulation of the transient temperature field is carried out by COMSOL 4.3 software package. Sensitivity to meshing and time steps has been examined to ensure that less than 1% changes on the maximal temperature when the size of the elements and/or time step are decreased by 50%.

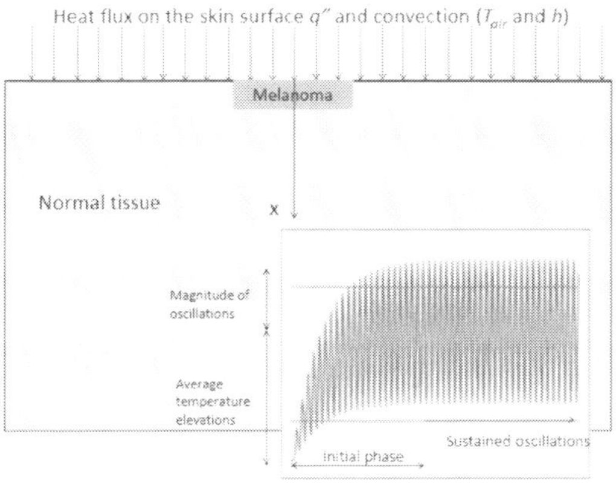

Figure 1 Schematic diagram of the 2-D axisymmetric geometry of a melanoma embedded in a tissue cylinder. The inset represents possible temperature transients at the skin surface, where the later part of the temperature transients denotes to the sustained oscillations.

RESULTS

All the properties used in the model are given in Table 1. Note that the large difference of the blood perfusion rate between the melanoma and the normal tissue. As shown in Figure 1, in general, the temperature transient at any location x after a while will be sustained temperature oscillations, which are usually referred to unsteady temperature changes repeating themselves steadily. It has an average temperature elevation and a magnitude of oscillation.

Table 1 Thermal and physiological properties of normal skin tissue, melanoma, and blood.

	ρ (kg/m^3)	c (J/kg°C)	k (W/m°C)	q_m (W/m^3)	ω (1/s)
Normal tissue	1000	3800	0.5	342	0.0003
Melanoma	1000	3800	0.5	9120	0.008
Blood	1000	3800	0.5	n/a	n/a

The temperature contours of the 2-D models are illustrated by the three images in Figure 2 with the average heat flux q_o'' as 140 W/m^2. Due to the large blood perfusion rate in the melanoma, temperature elevations in the melanoma lesion are limited compared with that in the tissue region. In other words, local blood perfusion rate serves as a cooling mechanism to limit the temperature elevation. When the melanoma is 5 mm thick, the average temperature difference between the melanoma and the surrounding tissue at the skin surface is 1.46°C. However, as the size of the melanoma is getting smaller, one notes that the cooling becomes less powerful, resulting in a more uniform temperature field on the skin surface. The temperature difference is only 0.6°C when the melanoma is 1 mm in thickness.

Simulation results also suggest that the oscillation frequency of the skin surface temperatures follow the frequency of the heat flux imposed on the skin surface, however, with a time delay of approximately $\pi/4$. It also shows that the time delay is slightly different between the normal tissue and the melanoma due to their different blood perfusion rates, and the difference is more evident when the frequency of the heat flux is smaller than 0.01 Hz. If an infrared camera can be used to record the surface temperature transients during the sustained oscillation phase, it is possible that time delay difference can be interpreted as an indication of a much large blood perfusion rate in the melanoma. Figure 3 provides quantitative results of the time delay difference. When the frequency is 0.01 Hz, the time delay difference between the normal tissue and the melanoma is 0.96 s for the 5 mm thick melanoma case, while it increases to 3.2 s when the frequency is 0.005 Hz. The size of the melanoma also influences the sensitivity of this approach. Shown in Figure 3, the time delay difference decreases with the melanoma thickness, however, it still is a detectable 0.5 s for the 1 mm thick melanoma case when the frequency is 0.01 Hz.

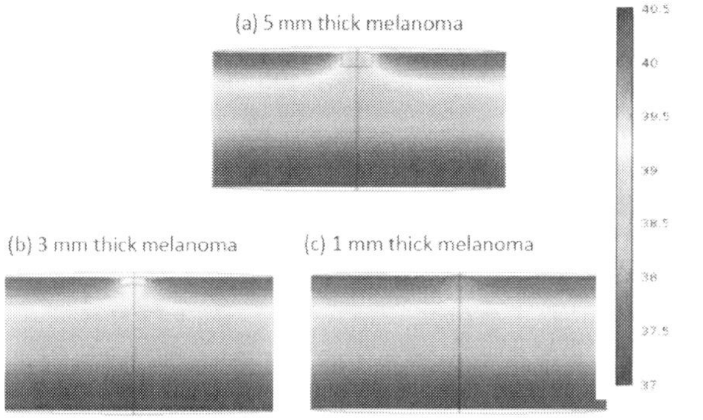

Figure 2 Temperature contours simulation in the tissue region with a melanoma embedded. The contours represent the average temperature values during the sustained oscillations.

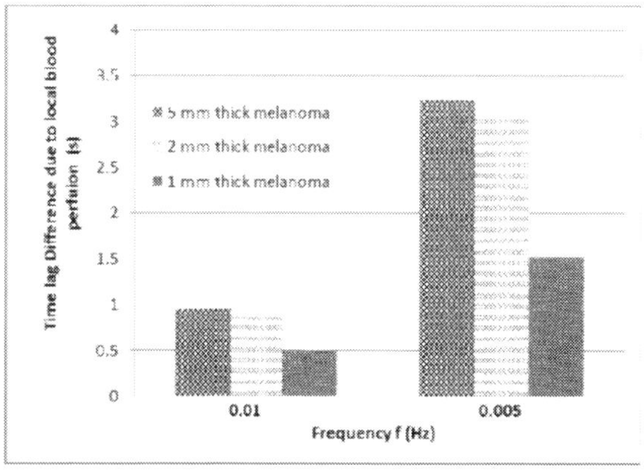

Figure 3 Effects of the frequency of the heat flux and the thickness of the melanoma on the time delay difference between the normal tissue and the melanoma at the skin surface.

In summary, we evaluate temperature fields in tissue with an embedded melanoma subject to a periodic heat flux on its skin surface using 2-D heat transfer analyses. It has been shown that the limited average temperature rise may be used as an indication of a large local blood perfusion rate in the melanoma. We also identify another parameter, the difference of the time delays in the temperature oscillations between a melanoma and its surrounding tissue to extract local blood perfusion rate, if the time lag difference can be accurately captured by commercial infrared cameras. This study provides theoretical basis and rationales to implement thermal imaging technique in screening melanoma or other diseases having high blood perfusion on the skin surface.

ACKNOWLEDGEMENTS

This study was supported by UMBC.

REFERENCES

[1] Mudaliar, A. V., Ellis, B., Ricketts, P. L., Lanz, O. I., Lee, C. Y., Diller, T. E., and Scott, E. P., 2008, ASME *Journal of Biomechanical Engineering*, 130(6), pp. 061013(1-8).

[2] Liu, J., Zhou, Y. X., and Deng, Z. S., 2002, IEEE *Trans. Biomed. Engineering*, 49, pp. 867-877.

[3] Centingul, M. P., and Herman, C., 2011, ASME *Journal of Thermal Science and Engineering Applications*, 3, pp. 031006(1-9).

EFFECT OF AXIAL COMPRESSION PRELOAD ON INTERVERTEBRAL DISC TORSIONAL MECHANICS

Semih E. Bezci and Grace D. O'Connell

Department of Mechanical Engineering
University of California, Berkeley
Berkeley, CA, USA

INTRODUCTION

The primary function of the intervertebral disc is to absorb and transfer large complex loads placed on the spine, including combinations of compression, tension, bending, and torsion [1]. Collagen fibers in the annulus fibrosus are oriented at ±30° to the horizontal plane, suggesting an important role in load distribution during torsional loading. Moreover, the disc's mechanical function under torsion is important for understanding the cause and progression of annular tears, such as rim lesions. Rim lesions are tears between the disc boundary and surrounding vertebral bodies, and have been observed in over 50% of discs [2]. The incidence rate of rim lesions increases significantly with age, such that 80% of adults over 50 years have some type of rim lesion [2]. The cause for rim lesions is unknown; however, axial compression in combination with torsional loading, where strains are greatest at the disc edge, may cause the disc to be more susceptible to annular damage and rim lesions.

Extensive research has been performed on disc mechanical function under axial compression [3-5]; however, little is known about the disc behavior under axial torsion [6, 7]. Furthermore, there remains a lack of information concerning the coupled effect between axial compression, a primary loading configuration, and axial torsion. Therefore, the objective of this study was to evaluate the compression-torsional mechanical function of healthy intact bovine discs. The findings from this study describe disc torsion mechanics as a function of disc geometry and axial compression.

METHODS

Bovine caudal spines were acquired from the local abattoir (n = 6 spines, ~18 months). The musculature and facet joints were removed and motion segments were prepared by cutting through the vertebral bodies (n = 15 samples). Motion segments were potted in polymethylmethacrylate (PMMA) bone cement and stored in a freezer at -20°C. All samples were hydrated overnight in 0.15 M saline solution and allowed to equilibrate to ambient room temperature prior to mechanical testing. The motion segments were mounted in the biomechanical testing apparatus with screws evenly spaced 60° apart in a saline bath (Figure 1). Samples were preloaded under axial compression (20 – 500 N) for 10 minutes, which was followed by 10 cycles of torsion (± 5°, 0.05 Hz). Samples were tested under four different axial loads applied in a random order with full recovery between experiments. The y-direction force, y-direction displacement, rotation, and torque were recorded during all tests.

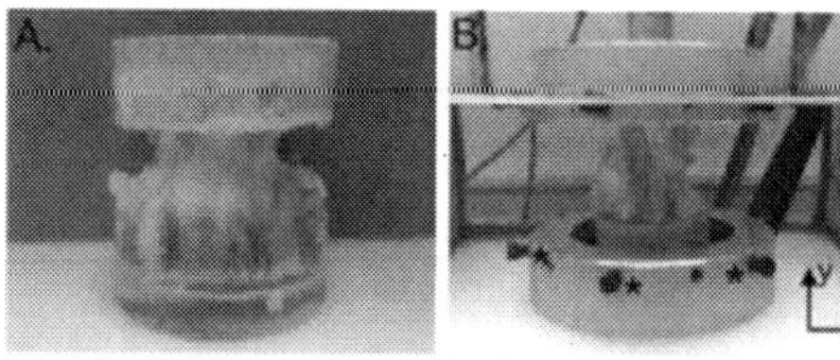

Figure 1. (A) Sample potted to ensure flat parallel surfaces. (B) Experimental set-up for compression-torsion testing. Offset screws were used to fix samples for torsion testing (asterisks).

Following mechanical testing, discs were removed from the vertebral bodies using a scalpel and, then, disc geometry (i.e. disc height and cross sectional area) was measured using a custom-written algorithm in Matlab (Mathworks, Inc.) [8]. The height (h), area (A), volume (V), and the polar moment of inertia (eqn. 1 [6]) was calculated for each disc and used to normalize mechanical properties measured from axial-torque tests. The applied axial compressive stress was calculated as the axial load divided by the total disc area.

$$J = \frac{\pi(W_{AP}W_L^3 + W_{AP}^3 W_L - N_{AP}N_L^3 + N_{AP}^3 N_L)}{64} \qquad (1)$$

Torque-rotation data was analyzed using the last cycle of torsional loading. Torsional stiffness, k_T, was defined as the slope of the torque-rotation curve at rotation range of 0-2°. Energy loss was defined as the area between loading and unloading torque-rotation curve. The normalized torsional stiffness (MPa/deg), T, was calculated using equation (2). Shear stress, τ, at a rotation of 2 degrees was calculated from the torque (T), averaged disc radius (r_{disc}), and polar moment of inertia (eqn. 3).

$$T = k_T * \frac{h}{J} \quad (2)$$

$$\tau = T * r_{disc}/J \quad (3)$$

The Pearson's correlation test was used to determine the effects of axial compressive stress on torsional disc mechanics. Significance was considered for parameters that demonstrated a high r- and a low p-value ($|r| \geq 0.7$ and $p \leq 0.05$).

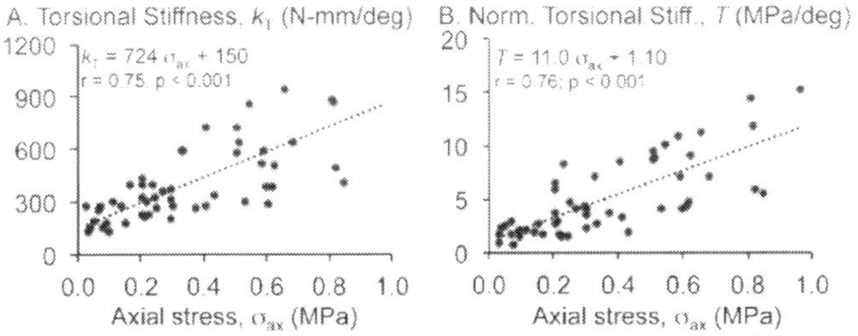

Figure 2. (A) Direct and (B) normalized torsional stiffness as a function of axial compressive stress.

RESULTS

The disc is a viscoelastic material with solid-like and fluid-like behavior. The solid-like behaviors measured, including torsional stiffness and shear stress, were strongly correlated with the axial compressive stress (Figure 2). The shear stress calculated based on torsion and disc geometry followed a linear response with the axial compressive stress ($\tau = 0.54 * \sigma_{ax} + 0.10$ MPa, $r = 0.69$, $p < 0.01$).

There was no significant correlation between the energy loss (i.e. hysteresis) and axial compressive stress (Figure 3A). When normalized by disc geometric based parameters (i.e. moment of inertia or volume), the energy loss was strongly correlated with axial compressive stress (Figure 3B; $r = 0.5$ & $p < 0.001$ for E_h normalized by J). To determine whether disc cross-sectional area or height had a greater effect, the initial energy loss, which is the energy loss under no axial compression (Figure 3A - red box), and the rate of energy loss with axial stress (Figure 3A - slope of red line) were correlated with geometric parameters. The initial energy loss was strongly correlated with disc height, while the rate of energy loss was strongly correlated with disc area (Figure 3C & D; $p < 0.001$, $r \geq 0.08$).

DISCUSSION

The intervertebral disc is a highly viscoelastic material that experiences complex loading during daily activities. This study demonstrates and quantifies the coupled effect of axial compression and disc geometry on disc mechanical function under torsional loading. Within the physiological range of axial compressive stress (< 1.0 MPa), torsional disc mechanics exhibited up to 5-fold change in behavior (e.g. torsional stiffness, shear stress, and hysteresis).

Elastic torsion properties were strongly correlated with the axial compressive stress (Figure 2). Previous work by Showalter et al demonstrated that the normalized torsional stiffness of bovine discs is comparable to healthy human discs [6]. Therefore, it is possible that the coupled torsion-compression mechanics of healthy human intervertebral discs may follow a similar linear relationship, but will need to be confirmed with further studies (Figure 2B).

The energy loss during torsion (i.e. hysteresis), which is a measure of the fluid-like properties of the disc, was not strongly correlated with axial compressive stress. However, there was strongly positive correlation between the energy loss between loading and unloading and disc geometry. This finding demonstrates a functional difference between the elastic solid-like properties (i.e. torsional stiffness), which was correlated with axial compression, and the fluid-like properties of the disc under torsion. The data in this study will be valuable in accurately describing the viscoelastic function of the disc.

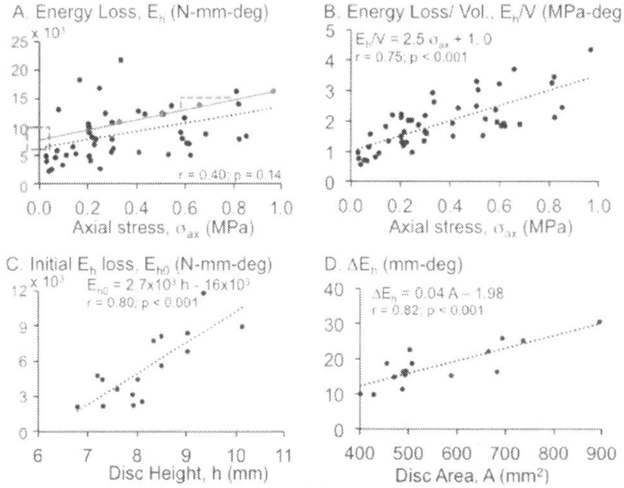

Figure 3. (A) Energy loss with respect to axial stress. (B) Energy loss normalized by disc volume with axial stress. (C) Initial energy loss (i.e. no axial load) with disc height and (D) rate of energy loss with respect to disc area.

Directly comparing values between studies is often a challenge for biomechanical properties due to studies measuring properties under a single test parameter (i.e. loading rate or applied axial compressive stress) or differences in boundary conditions. Therefore, we aimed to describe the mechanical response as a functional relationship based on compressive stress or disc geometry. The properties reported here agree with previously reported values under an axial compressive stress of 0.48 MPa [6]. However, the shear stress reported here is an order of magnitude greater than previously reported values of annulus tissue *ex situ* [9]. These differences are likely due to difference in boundary conditions at the vertebral bodies.

Annular failure from torsional loading is thought to contribute to low back pain and disc degeneration [10]. We observed significant differences in the disc's fluid-like behavior under combined compression-torsion loading due to disc geometry, which changes significantly with degeneration [1]. In contrast, the disc's solid-like response was not dependent on disc geometry, but was correlated to axial compressive stress, which is dependent on body weight and loads applied by surrounding musculature. In conclusion, structural and geometric properties play a crucial role in torsional-compression properties, which is important to accurately model the disc function or develop a repair strategy that mimics the native function of a healthy disc. Future work will evaluate torsion-compression behavior of healthy, degenerated, and injured (e.g. rim lesions) human discs to validate the coupled response reported here for human discs.

ACKNOWLEDGEMENTS

This work was supported in part by a University of California, Berkeley Junior Faculty Research Award (2013-2014).

REFERENCES

[1] Adams, M.A. *Mech. Prop of Soft Tissues (Ch2), 2015*; [2] Vernon-Roberts, B.+ *Spine: 32(25), 2007*; [3] Koeller, W.+ *Spine: 9(7), 1984*; [4] Beckstein J.C.+ *Spine: 33(6), 2008*; [5] O'Connell, G.D.+ *JOR 29, 2011*; [6] Showalter, B.L.+ *Spine: 37(15), 2012*; [7] Bisschop, A+ *Eur Spine J: 22, 2013*; [8] O'Connell, G.D.+ *Spine: 32(3), 2007*; [9] Iatridis J.C.+ *JOR: 17, 1999*; [10] Farfan, H.F.+ *JBJS: 52A(3), 1970*.

MICROMECHANICAL HETEROGENEITY AND ANISOTROPY OF THE MENISCUS EXTRACELLULAR MATRIX

Q. Li (1), F. Qu (2,3), B. Han (1), R. L. Mauck (2,3), L. Han (1)

(1) School of Biomedical Engineering, Science and Health Systems
Drexel University
Philadelphia, PA, United States.

(2) McKay Orthopaedic Research Laboratory
University of Pennsylvania
Philadelphia, United States.

(3) Translational Musculoskeletal Research Center
Philadelphia Veterans Administration Medical Center
Philadelphia, PA, United States.

INTRODUCTION

The meniscus is an essential component of the knee joint, responsible for stability, load transmission, and shock absorption [1]. To enable efficient transmission of the complex tensile, compressive, and shear stresses experienced by this tissue, the type I collagen-rich meniscus extracellular matrix (ECM) possesses a hierarchical structural framework that is both heterogeneous and anisotropic [2,3]. Meniscal damage or maceration is a common antecedent of osteoarthritis and occurs with joint injury [1]. As restoring proper meniscal function is key to successful joint repair, it is imperative to understand the structure-mechanical function design principles of the native meniscus [4]. In this study, we used atomic force microscopy (AFM)-based nanoindentation in combination with nano-to-micro scale structural analyses to reveal the region, depth, and direction-dependent micromechanical properties of the bovine meniscus.

METHODS

Sample preparation. Menisci were dissected from juvenile (≤ 3 months old [5]) cows. The outer, middle, and inner zones of the meniscus body were isolated and embedded in Optimal Cutting Temperature (OCT) media. Each zone was cryosectioned transversely and radially onto glass slides to produce 20 μm thick horizontal and vertical cross-sections, respectively (Fig. 1). For horizontal sections, samples from the intact surface, superficial (~100 μm from the femoral surface) and deep zones (>500 μm into the tissue) were taken. Samples were stored at –20°C until micromechanical testing.

AFM-based nanoindentation was performed on each sample using a colloidal spherical tip ($R \approx 5$ μm, nominal $k \approx 0.2$ N/m) and a Dimension Icon AFM (BrukerNano), following our established procedures [6]. For each microstructural region of each section, indentation was repeated on at least 10 locations to a maximum indentation depth of ~1 μm. Effective indentation modulus, E_{ind}, was calculated using finite thickness-corrected Hertz model [7] (Fig. 2). After indentation, each section was treated with Karonorsky's fixative for 3 hours, dehydrated in a series of graded ethanol washes, and dried in hexamethyldisilazane to preserve the 3D architecture of the collagen network [8].

Structural Analysis. Scanning electron micrographs (***SEM***, Zeiss Supra 50VP) were acquired to visualize collagen fibril structure (Fig. 3). For ***histology***, separate deep zone sections were stained with Picrosirius Red (PSR) and imaged under polarized light to visualize collagen fiber bundle organization (Fig. 3).

RESULTS

For both the horizontal and vertical cross-sections, we found significant micromechanical heterogeneity in the outer/middle regions, but not in the inner region (Fig. 2). In the outer/middle regions, circumferential fiber bundles showed significantly higher E_{ind} than the superficial zone and deep zone radial fibers. On the contrary, in the inner region, all three structural units had similar E_{ind} (Fig. 2). For the deep zone circumferential fiber bundles, E_{ind} was significantly higher when indented along the direction of fiber alignment (vertical cross-section) compared to normal to the direction of fiber alignment (horizontal cross-section) (Mann-Whitney U test, $p < 0.01$ for all outer, middle and inner regions). The data in Fig. 2 are from one animal, though the trends were consistent across multiple donors. For both the circumferential and radial fibers in the deep zone, collagen appeared to be highly aligned. Collagen crimping was also observed at both the individual fibril and the fiber bundle levels (Fig. 3).

DISCUSSION

This study highlights the heterogeneous and anisotropic mechanical design of meniscus ECM. The observed mechanical heterogeneity is consistent with its known structural heterogeneity [1]. The meniscus superficial zone consists of a thin, transversely aligned collagen fibril mesh layer (~5 – 10 μm thick) covering radially aligned fibers (~ 150 μm thick) [2,9]. The deeper zone includes mostly circumferential fiber bundles held together by interdigitating radial fibers [2] (Figs. 1,3). During joint motion, fiber bundles in the outer/middle regions provide circumferential tensile resistance, while radial fibers in both the superficial and deep zones are responsible for structural stability and integrity [10]. Variations in E_{ind} (Fig. 2) thus reflect different mechanical functions associated with these structural units. In contrast, the inner region of meniscus primarily resists compression in an isotropic manner [1]. Correspondingly, this region displayed less structural organization and did not exhibit significant mechanical heterogeneity.

Indentation normal to the circumferential fibrils (horizontal cross-section, CF) should create tensile stress along fibrils, while indentation along the fibrils (vertical cross-section, CFc) should lead to fibril compression. The observation that the CFc unit was stiffer than the CF unit thus seems contradictory to the fact that meniscus is stiffer in tension than compression [10]. However, when collagen crimping is present (arrowheads, Fig. 3), local stress induced by nanoindentation normal to fibrils could mainly result in fibril uncrimping and sliding [11,12], rather than tensile stretching, until a threshold in tensile strain is reached. In fact, values of E_{ind} measured here are ~10^3× lower than macroscopic circumferential tensile moduli of meniscus (~100 MPa) [13], but are consistent with previous meniscus nanoindentation studies [3,14]. This length scale-associated softening may correspond to the toe-region of meniscus tensile stress-strain curves [10]. It is likely a critical material design principle that benefits joint stability and shock absorption. Further ongoing investigation of meniscal mechanical heterogeneity and anisotropy associated with regions, length scales, deformation rates, as well variations with age and disease, will provide crucial insights into the developmental and pathological changes seen by this complex tissue.

This study highlighted the regional heterogeneity and anisotropy in the micromechanical properties of meniscus. Study of the nanostructural/ nanomechanical features of the tissue is essential for understanding meniscus tissue properties across length scales and for optimizing functional repair strategies.

ACKNOWLEDGEMENTS

This work is supported by the Drexel Faculty Startup Grant (LH) and NIH R01 EB02425 (RLM).

REFERENCES

[1] Makris, EA et al., *Biomaterials* 32:7411-7431, 2011.
[2] Petersen, W et al., *Anat. Embryol.* 197:317-324, 1998.
[3] Sanchez-Adams, J et al., *J. Orthop. Res.* 31:1218-1225, 2013.
[4] Han, WM et al., *Biophys. J.* 105:807-817, 2013.
[5] Ionescu, LC et al., *Tissue Eng. A* 17:193-204, 2011.
[6] Batista, MA et al., *Matrix Biol.* 38:84-90, 2014.
[7] Dimitriadis, EK et al., *Biophys. J.* 82:2798-2810, 2002.
[8] Bray, DF et al., *Microsc. Res. Tech.* 26:489-495, 1993.
[9] Ling, CH et al., *Trans. Orthop. Res. Soc.* 58:767, 2012.
[10] Proctor, CS et al., *J. Orthop. Res.* 7:771-782, 1989.
[11] Miller, KS et al., *J. Biomech.* 45:2061-2065, 2012.
[12] Rigozzi, S et al., *J. Biomech.* 46:813-818, 2013.
[13] Fithian, DC et al., *Clin. Orthop. Relat. Res.* 252:19-31, 1990.
[14] Kwok, J et al., *Nanomed. Nanotech. Biol. Med.* 10:1777-1785, 2014.

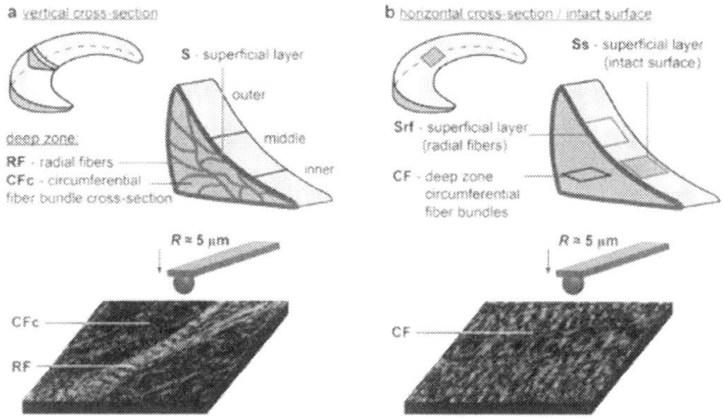

Fig. 1 Schematics of juvenile bovine meniscus tissue cross-sections and regions of interest for nanoindentation. a) For vertical cross-sections, indentation was performed on the superficial zone (mostly radial fibers, S), radial fibers in the deep zone (RF) and circumferential fiber bundle cross-sections (CFc). b) For horizontal cross-sections/intact surface, indentation was carried out on the intact surface (Ss), horizontal cross-section at 100 μm below the surface containing mostly radial fibers in the superficial zone (Srf) and circumferential fibers in the deep zone (CF). All nanoindentation tests were performed on the outer, middle and inner regions.

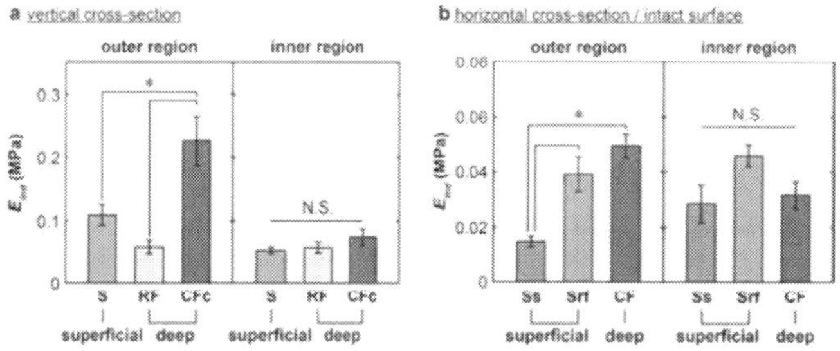

Fig. 2 Effective AFM-nanoindentation modulus, E_{ind}, measured within a) a vertical cross-section and b) a horizontal cross-section/intact surface of juvenile bovine meniscus ($R \approx 5$ μm, 10 μm/s indentation rate, ≥ 10 locations for each region, mean ± SEM, *: $p < 0.05$ from Kruskal-Wallis test followed by Tukey-Kramer multiple comparison). Abbreviations for different structural units are described in Fig. 1. The moduli in the middle and outer regions were similar.

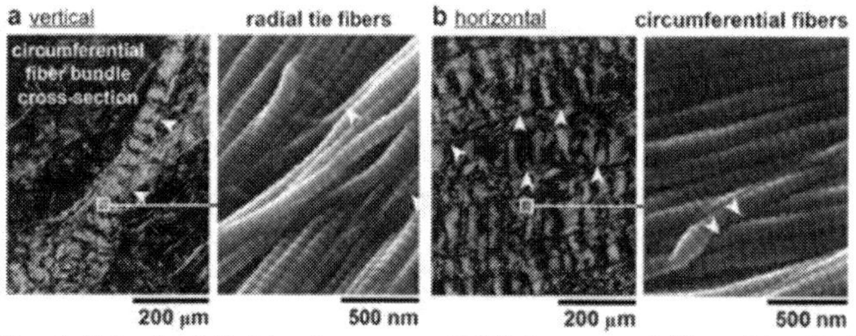

Fig. 3 Polarized light microscopy (LM) images of Picrosirius Red (PSR)-stained a) vertical and b) horizontal cross-sections, as well as corresponding SEM images of collagen fibril nanostructure in the radial and circumferential fibers of the deeper zone (arrowheads: fibril and fiber crimp). To increase clarity, boxes in LM images do not correspond to the scale of SEM images.

INTERSTITIAL CELL MIGRATION IN DENSE CONNECTIVE TISSUES IS MODULATED BY MATRIX MICROSTRUCTURE AND MICROMECHANICS

Feini Qu (1,2,3), Miltiadis H. Zgonis (1,3), Robert L. Mauck (1,2,3)

(1) McKay Orthopaedic Research Laboratory
Department of Orthopaedic Surgery
University of Pennsylvania
Philadelphia, PA, USA

(2) Department of Bioengineering
University of Pennsylvania
Philadelphia, PA, USA

(3) Translational Musculoskeletal Research Center
Philadelphia VA Medical Center
Philadelphia, PA, USA

INTRODUCTION

Few regenerative approaches exist for the treatment of injuries to adult dense connective tissues. Compared to fetal tissues, these adult connective tissues are hypocellular and show limited endogenous cell-mediated healing after injury [1,2]. However, it is unknown whether this deficiency in repair is due to the limited cell number itself, by steric constraints to migration imposed by the extracellular matrix (ECM), or by a combination of the two. We hypothesized that robust repair occurs in fetal tissues due to an immature ECM conducive to cell migration and fails in adults due to a dense and stiff ECM that acts as a biophysical impediment [3,4]. Using the knee meniscus as a test platform, this study evaluated the microstructure and micromechanics of fetal and adult tissues, with a particular focus on the inter-fibrillar regions through which cells might migrate. Further, we monitored the morphology and interstitial migration of adult meniscal cells through fetal and adult tissue microenvironments. Finally, building from our recent studies showing that local pre-treatment with collagenase improves cellular colonization of the wound interface in the adult meniscus [5], we measured adult cell migration through a pre-digested adult ECM. Our results show that the ECM microstructure and micromechanics of the adult meniscus act as physical barriers to cell mobility, and that partial digestion of this otherwise impenetrable collagenous network expedites cell migration to improve repair.

METHODS

Microstructural and micromechanical analysis: Tissue explants (8 mm diameter) were excised from fetal and adult bovine meniscal bodies in the axial plane and cryo-sectioned onto glass slides (~30 μm thick). To visualize fibrillar collagen, second harmonic generation (SHG) imaging was performed [5]. Maximum z-stack projections were converted into binary images using Fiji to identify discrete areas of positive and negative signal, representing organized fiber bundles parallel to the substrate and inter-fibrillar material that constitute the remaining tissue ECM (n=7–9 stacks/group). The average diameter of each inter-fibrillar region was quantified via the Local Thickness plugin (n=533–788 regions/group). To assess micromechanics, force spectroscopy (n=20 indentations/group) was performed on hydrated sections using an atomic force microscope (AFM). Elastic moduli were computed using a Hertzian model [5].

Analysis of interstitial cell migration: Devitalized sections were UV sterilized and hydrated. Three substrate conditions were tested: fetal (fECM), adult (aECM), and aECM pre-treated with 0.1 mg/mL of collagenase for 1 hour (aECM-C). To visualize cell morphology and invasion, adult meniscal explants (4 mm diameter, 3 mm height) were incubated in CellTracker Green (5 μg/mL) for 1 hour and then placed atop sections to allow for cell egress. After 48 hours in basal media, explants were removed and the egressed cells were stained with 4', 6-diamidino-2-phenylindole (DAPI) to visualize nuclei. Confocal z-stacks were obtained in the GFP and DAPI channels to visualize cells, nuclei, and devitalized matrix (which was autofluorescent in the DAPI channel). Cell area, aspect ratio (a measure of elongation), circularity, and solidity (a measure of boundary irregularity) were determined from maximum projections of z-stacks (n=71–97 cells/group). Cell infiltration depth was defined as the distance between the apical surface of the tissue and basal surface of the cell (n=15 cells/group).

Statistical analysis: Significance was assessed by one-way ANOVA with Tukey's HSD post hoc for matrix stiffness; inter-fibrillar area fraction and average diameter; cell area and shape descriptors; and infiltration depth ($p \leq 0.05$). A cumulative distribution plot, coupled with the Kolmogorov-Smirnov test, was used to determine whether the distribution of infiltration was different between groups ($p \leq 0.05$). Data are presented as mean ± standard error.

RESULTS

SHG imaging and AFM analysis of fetal and adult meniscal sections revealed distinct microenvironments (Fig. 1B). Qualitatively, adult ECM (aECM) had thicker and more organized collagen bundles than fetal ECM (fECM). The area fraction of inter-fibrillar ECM, indicated by the absence of SHG signal, was lower for aECM compared to fECM (Fig. 1C, $p \leq 0.05$). However, discrete inter-fibrillar areas were on average larger for aECM compared to fECM (Fig. 1C, $p \leq 0.05$). The average local modulus of aECM was approximately 3 times greater than that of fECM (Fig. 1D, $p \leq 0.05$).

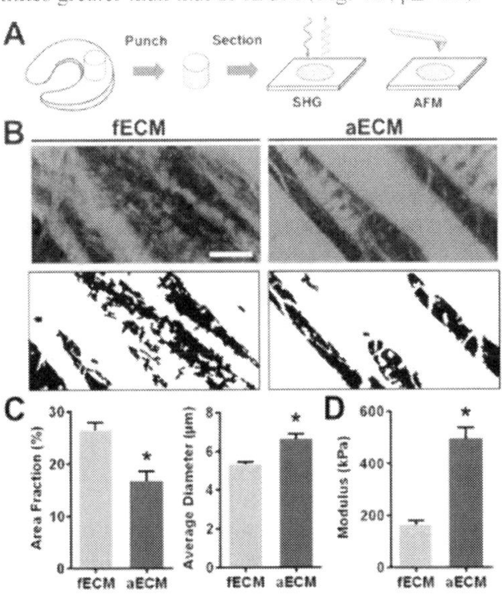

FIG. 1. ADULT MENISCUS IS DENSER AND STIFFER THAN FETAL MENISCUS. (**A**) Experimental schematic. (**B**) Top: SHG signal (green) of fECM and aECM. Scale = 50 µm. Bottom: Binarized images depicting inter-fibrillar areas in black. (**C**) Area fraction (left) and average diameter (right) of inter-fibrillar regions. (**D**) Tissue micromechanics via AFM. * = $p \leq 0.05$ vs. fECM.

Adult meniscal cells from explants adhered, spread, and infiltrated into devitalized tissue substrates within 48 hours (Fig. 2B). Cells on aECM were 3x larger than on fECM (Fig. 2C, $p \leq 0.05$). Cells on both substrates aligned in the fiber direction of the underlying tissue, although cells on aECM had a higher aspect ratio and exhibited lower circularity and solidity compared to cells on fECM (Fig. 2C, $p \leq 0.05$). Cells that egressed onto aECM pre-treated with collagenase (aECM-C) were smaller and rounder compared to aECM, better approximating those on fECM (Fig. 2C, $p \leq 0.05$).

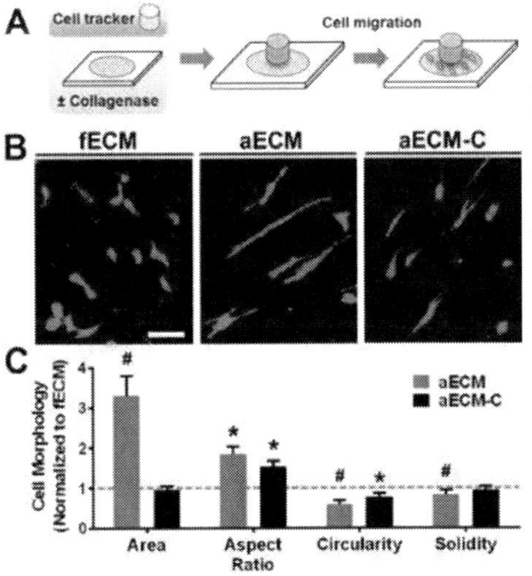

FIG. 2. CELL MORPHOLOGY IS DEPENDENT ON TISSUE AGE AND TREATMENT. (**A**) Experimental schematic. (**B**) 2D projections of cells on fECM, aECM, and aECM-C. Scale = 50 µm. (**C**) Morphology of adult meniscal cells on aECM and aECM-C, normalized to fECM values (dashed red line). * = $p \leq 0.05$ vs. fECM. # = $p \leq 0.05$ vs. all other groups.

Cells migrating through aECM were visibly more deformed compared to those within fECM and aECM-C, and formed narrow protrusions into the surrounding matrix (Fig. 3A). Cell infiltration depth was greater for cells on fECM compared to cells on aECM (Figs. 3B & 3C, $p \leq 0.05$). When the adult ECM was digested, cells on aECM-C migrated to a similar extent as on fECM, with 20% reaching depths of ≥25 µm from the surface (Figs. 3B & 3C, $p \leq 0.05$).

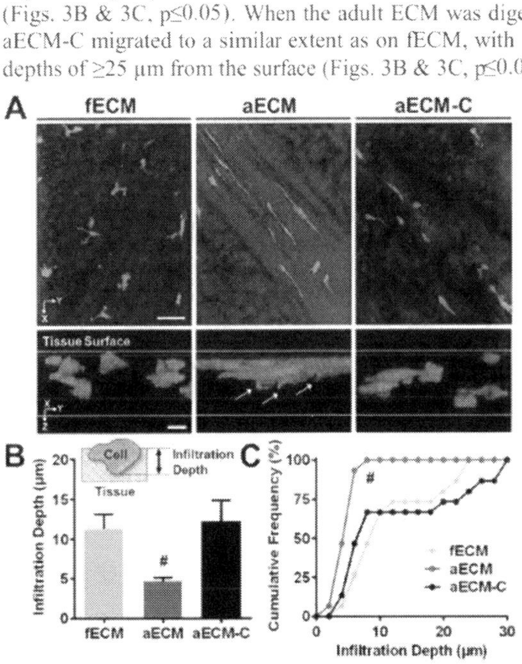

FIG. 3. CELL MIGRATION IS MODULATED BY ECM DENSITY. (**A**) 2D and 3D images of cells within tissues. Arrows indicate protrusions. Scale = 50 µm (top) and 10 µm (bottom). (**B**) Cell infiltration depth with schematic inset. (**C**) Cumulative frequency of infiltration depth. # = $p \leq 0.05$ vs. all other groups.

DISCUSSION

Our findings suggest that cell mobility through dense connective tissues decreases with tissue maturation. Compared to the fetal ECM, adult ECM is denser, stiffer, and more organized, resulting in a physically restrictive microenvironment that limits interstitial cell migration. Meniscal cells on devitalized fetal tissue sections are smaller and more invasive than the same cells on adult tissue, which exhibit enhanced cell spreading on the substrate surface and elongation along the collagen fiber direction. Importantly, infiltration occurred primarily at inter-fibrillar regions, which are more numerous and compliant in immature fetal tissue. In adult tissue, migrating cells must deform through narrow crevices between rigid, aligned collagen bundles, similar to the steric constraints imposed by decreasing pore size in Transwell assays [3]. While adult tissue has larger areas of inter-fibrillar matrix, this is due to the presence of well-developed, interpenetrating fibers orthogonal to the substrate plane, which have low SHG signal but are autofluorescent. These age-related, biophysical changes to the microenvironment likely contribute to the dearth of reparative cells at the wound site and subsequent poor repair in the adult meniscus [1,2]. Recently, we showed that reprogramming the meniscal wound interface with collagenase reduces ECM density and stiffness, increases interfacial cellularity, and facilitates repair [5]. Here, we provide further evidence that expedited migration to the injury site is responsible for this superior healing response. By clarifying the role of local matrix properties in wound repair, these studies may establish new clinical strategies to promote endogenous regeneration of the meniscus and other dense connective tissues.

ACKNOWLEDGEMENTS: This work was supported by the NIH (R01 AR056624, T32 AR007132), the Department of Veterans Affairs (I01 RX000174), and the Montague Award.

REFERENCES: [1] Ionescu LC et al., *Tissue Eng Part A*, 17(1-2):193-204, 2011. [2] Mesiha M et al., *Am J Sports Med*, 35(1):103-12, 2007. [3] Wolf K et al., *J Cell Bio*, 201(7):1069-84, 2013. [4] Zaman MH et al., *Proc Natl Acad Sci U S A*, 103(29):10889-94, 2006. [5] Qu F et al., *Biomaterials*, 39:85-94, 2015.

NUMERICAL SIMULATIONS OF FIBROUS BIOMATERIAL WITH RANDOMLY DISTRIBUTED FIBER NETWORK

T. Jin, I. Stanciulescu

Civil and Environmental Engineering
Rice University
Houston, TX 77005, USA

INTRODUCTION

Understanding the connection between the structure and the function of fibrous biomaterials is an important topic in tissue engineering and can advance the biomedical field. Many fibrous biomaterials can be categorized as fiber-reinforced material, and have randomly distributed fiber network structures in the microscopic scale. These fiber networks can undergo large deformation and exhibit significantly nonlinear behavior. The distribution and the randomness of fiber network in the microscopic level can largely influence the mechanical behavior of the material in the macroscopic level, for example, making the material isotropic, transversely isotropic or anisotropic.

One way to model the mechanical behavior of fibrous biomaterial is to follow the phenomenological approach and use the concept of strain energy function, for example, see Fung (1967, 1983, 1986), Humphrey (2003, 2012), Holzapfel (2000, 2004, 2006) and Balzani (2006). This methodology belongs to the category of homogenization, is relatively easy to implement and has nice mathematical properties, such as polyconvexity, if the form of strain energy function is properly chosen. However, most of the model parameters lack physical meaning. Furthermore, this methodology is deterministic and cannot reflect the influence of the randomness in the microscopic level on the material behavior. Another way to model the mechanical behavior of fibrous biomaterial is to include detailed information of the fiber network in the modeling process. This methodology, adopted in this work, can not only establish the connection between the topological structure of the fiber network with the macro-mechanical material behavior, but can also ensure that model parameters have physical meaning. More importantly, this modeling methodology can be easily combined with stochastic methods, such as Monte Carlo (MC) simulation, to study the statistical behavior of fibrous biomaterials.

The objective of this research is to study how the randomly distributed fiber network influences the macroscopic behavior of fibrous biomaterial. This study is helpful to understand the connection between the microscopic structure and the function of fibrous biomaterials. The numerical simulation methods used in this research can be easily modified for other types of fiber-reinforced material.

METHODS

To fulfill the objective of this research, in the first step, a 2D randomly distributed fiber network is generated with a random walk algorithm, with fiber volume ratio, length of individual fiber segment as the control parameters. Then, this fiber network is embedded into a finite element mesh, and a special data structure is used to record the topological information of the above random fiber network. Through a variation approach, the finite element residual force vector and the stiffness matrix can be derived. The effects of large deformation and material nonlinearity can be both considered.

Since the aforementioned fiber network is randomly generated, to get the average material behavior, Monte Carlo method combined with finite element simulations are used to evaluate the average responses of samples with embedded random fiber network. Several numerical examples are run to show the ability of this methodology to capture the nonlinearity and anisotropy of fibrous biomaterials.

RESULTS

Fig.1 shows an example of randomly distributed fiber network generated by the random walk algorithm with a fiber volume ratio of 0.35%. The fiber network is embedded inside a 2D 4-node finite element mesh, and a special data structure is used to record the parent coordinates of fiber segments contained in each element.

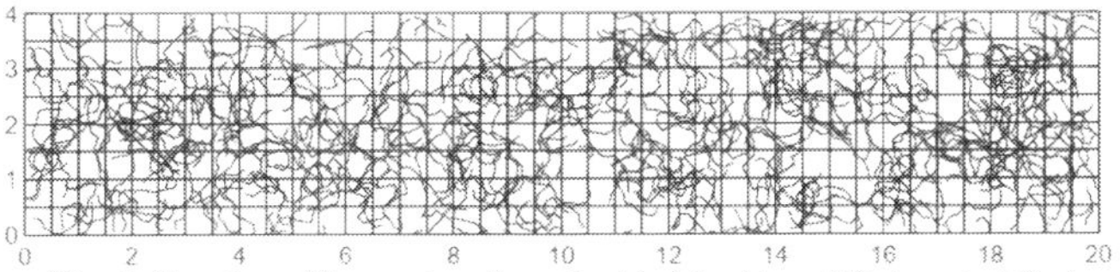

Fig. 1. Random fiber network embedded inside a 2D 4-node 20x4 finite element mesh

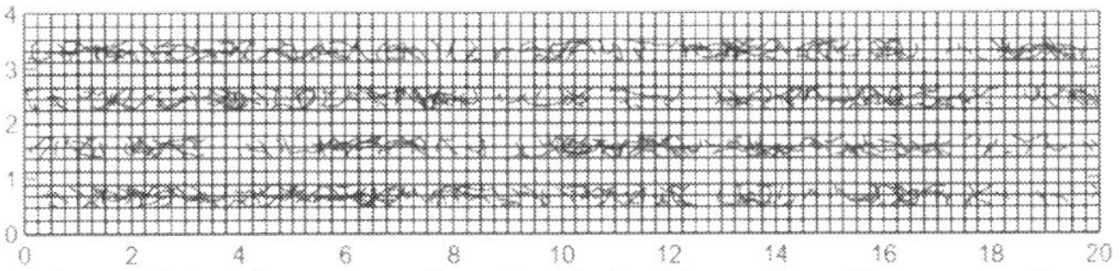

Fig.2. Finite element mesh embedded with random fiber network to model the layered structure of patterned PEGDA hydrogel sample with 40% pattern ratio

Patterned PEGDA hydrogel is a bio-inspired material and can be potentially used to fabricate artificial heart valve. This type of hydrogel has a layered structure with cross-linked fiber network inside each layer. Here, patterned PEGDA hydrogel is used as an example to show the capability of the above modeling methodology to capture the material anisotropy and nonlinearity. Fig.2 shows the finite element mesh embedded with random fiber network to model the layered structure of the patterned hydrogel. The layers without any fiber are pure matrix material, while the layers with fiber network are matrix material embedded with fibers. Fig. 3 shows the stress distribution of the matrix material when the stretch ratio equals to 2.0 (large deformation). Since the fiber network in the layers are random, MC simulations are used to get the average behavior of 40% pattern ratio of hydrogel under stretch. From Fig.4 we can see that the average reaction forces converge after around 300 MC steps. The model captures material anisotropy: the material is stiffer when the loading direction is parallel to the pattern orientation than when the loading direction is perpendicular the pattern orientation.

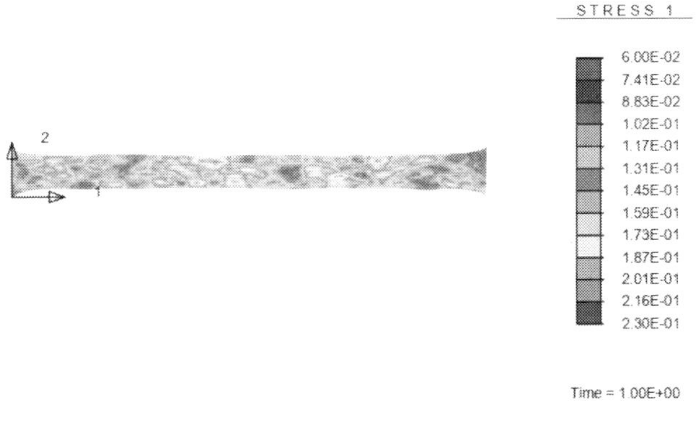

Fig.3. The stress distribution of the matrix material when the stretch ratio $\lambda = 2.0$

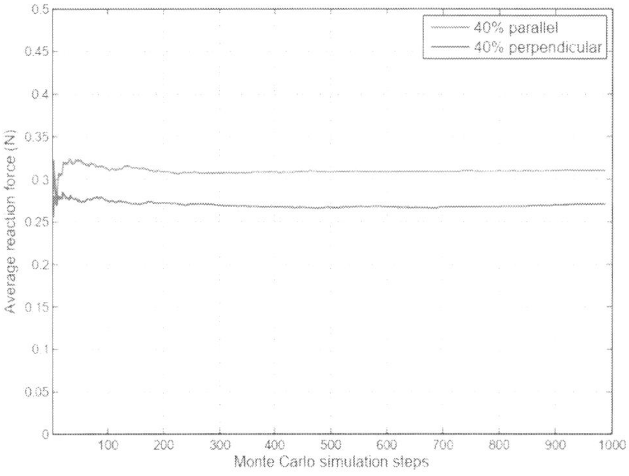

Fig.4. Monte Carlo simulations of 40% patterned hydrogel when the loading direction is parallel or perpendicular to the layer, the stretch ratio $\lambda = 1.5$

DISCUSSION

From the above results we can see that the modeling strategy proposed in this research is able to establish connections between the microscopic structure and the macroscopic behavior of fibrous biomaterials. The material anisotropy and the nonlinearity can both be properly captured. In fact, the macro-mechanical behavior strongly depends on the topological structure of the fiber network. In this research, the random fiber network is generated by a random work algorithm, which may not truly reflect the real topology of fiber network. Another limitation of this modeling method is that the interaction between fiber chains is ignored. In order to address these limitations, an algorithm that characterizes the realistic topology of fiber network, including fiber orientation, intersection spatial density and so on should be considered in future work.

ACKNOWLEDGEMENTS

This work was made possible in part through support from the Ken Kennedy Institute for Information technology at Rice University under the Collaborative Advances in Biomedical Computing seed funding program supported by the John and Ann Doerr Fund for Computational Biomedicine.

REFERENCES

Selected Reference:

[1] Fung, Y.C., *American Journal of Physiology*, Vol. 23, No. 6, 1967.

[2] Holzapfel, G.A. and Gasser,T.C., *J Elasticity*, 61:1-48, 2000.

[3] Balzani, D., *Int J Solid and Structure*, 43:6052-6070, 2006.

[4] Nemir, S. et al., *Biotech Bioeng*, Vol. 105, No. 3, 2009.

[5] Carleton, J.B. et al., *Acta Biomaterialia*, in press, 2014.

IN VITRO GROWTH TRAJECTORY AND IN VIVO IMPLANTATION OF CELL-SEEDED DISC-LIKE ANGLE PLY STRUCTURES FOR TOTAL DISC REPLACEMENT

John T. Martin (1,2), Dong Hwa Kim (1,2), Kensuke Ikuta (1,2), Christian G. Pfeifer (1,2), Lachlan J. Smith (1,2,3), Dawn M. Elliott (4), Harvey E. Smith (1,2), Robert L. Mauck (1,2)

(1) Department of Orthopaedic Surgery
University of Pennsylvania
Philadelphia, PA, USA

(2) Translational Musculoskeletal Research Center
Philadelphia VA Medical Center
Philadelphia, PA, USA

(3) Department of Neurosurgery
University of Pennsylvania
Philadelphia, PA, USA

(4) Department of Biomedical Engineering
University of Delaware
Newark, DE, USA

INTRODUCTION

Low back pain presents a significant socioeconomic burden in the US and lumbar intervertebral disc degeneration is often implicated as a causative factor [1]. Current surgical strategies for treating disc degeneration are aimed only at alleviating back pain, however, not at restoring disc function. To treat severely degenerated discs, our group has developed engineered implants that replicate the hierarchical structure of the native disc. These implants are comprised of both an organized multi-lamellar electrospun nanofibrous scaffold that replicates the outer annulus fibrosus (AF) and a hydrogel that replicates the central nucleus pulposus (NP). In combination, these subunits form disc-like angle ply structures (DAPS) for total disc replacement. When seeded with mesenchymal stem cells (MSCs) or disc cells, DAPS mature with *in vitro* culture [2]. In addition, we have developed and validated a rat tail model of total disc replacement in which to evaluate the *in vivo* integration and function of DAPS [3]. In this study, we hypothesized that the *in vivo* integration potential is related to the *in vitro* growth trajectory and rate of functional maturation of the engineered tissue. To test this hypothesis, we established the *in vitro* growth trajectory of DAPS seeded with either AF and NP cells or MSCs over a 15 week period. Then, after short- or long-term preculture durations, we tested the ability of DAPS to integrate and function in the rat tail *in vivo*.

METHODS

AF Fabrication: Poly(ε-caprolactone) (PCL) nanofibers were electrospun onto a rotating mandrel as aligned fibrous sheets. Strips were excised at 30° to the fiber direction and two strips with alternating ±30° alignment were wrapped concentrically to form the AF. These were fabricated with one layer (th=125µm) of water-soluble poly(ethylene oxide) for every two layers of PCL (th=125µm) to improve initial cell colonization [3]. Formed AF regions were sterilized in ethanol, coated with fibronectin and seeded with either bovine AF cells or bovine MSCs (2M cells/construct). NP Fabrication: Methacrylated hyaluronic acid (MeHA) was produced by reacting HA with methacrylic anhydride. A photocrosslinkable 1% w/v solution was produced by dissolving MeHA and 0.05% Irgacure 2959 in phosphate-buffered saline. Bovine NP cells or bovine MSCs were suspended in the MeHA solution (600K cells/construct), which was poured into a mold and photo-polymerized with UV light. *In Vitro* Growth Trajectory (n=4/group): AF and NP regions were cultured separately in chemically defined media containing TGFβ3, and then combined into DAPS at 2 weeks. At regular intervals up to 15 weeks, DAPS were harvested, and mechanical properties (cyclic compression (20 cycles, 0 to -3N, 0.5Hz), creep (-3N, 10 minutes)) and composition (glycosaminoglycan (GAG), collagen via biochemical assays and histological staining) were evaluated. *In Vivo* Implantation (n=6/group): AF/NP cell-seeded DAPS were implanted into the rat caudal disc space after preculture for either 5 weeks (immature) or 10 weeks (mature). Athymic retired breeder rats were prepared for implantation by installing an external fixator designed to unload and stabilize the caudal disc space [3]. Two surgical wires were passed through both the C8 and C9 vertebrae and the external fixator was mounted. Following a dorsal skin incision, the native C8/C9 disc was removed and DAPS were implanted into the disc space. Rats were returned to normal cage activity and euthanized 5 weeks after implantation. Vertebra-DAPS-vertebra segments were scanned by µCT, and then sectioned and stained with alcian blue/picrosirius red.

RESULTS

In Vitro Growth Trajectory: Both MSC- and disc cell-seeded DAPS matured compositionally and functionally with increasing time in culture (Fig 1). Collagen- and GAG-positive staining initiated at the periphery of the AF at 5 weeks and gradually infiltrated to deeper regions by 10 weeks. At 15 weeks, a peripheral concentration of dense extracellular matrix indicated encapsulation. Both collagen and GAG-positive staining in the NP increased with time, with collagen staining concentrated at the center of the NP, and GAG staining evenly distributed throughout the NP. Measurements of GAG and collagen (Fig 2a) showed a progressive increase in matrix deposition in the AF and NP regions for both cell types, with deposition rate by MSC-seeded DAPS outpacing that by disc-cell seeded DAPS. There were no significant changes in elastic mechanical properties (toe and linear region compressive moduli) over 15 weeks, and no mechanical differences between MSC and disc cell-seeded DAPS (Fig 2b). There was a significant increase in creep strain over the first 10 weeks that plateaued between 10 and 15 weeks, with no differences between cell types (Fig 2b). <u>*In Vivo* Implantation</u>: After implantation, there were no differences between immature and mature DAPS. μCT reconstructions revealed normal vertebral morphology, with mild bone formation around the surgical wire holes (Fig 3). Alcian blue/picrosirius red stained vertebra-DAPS-vertebra sections showed that DAPS maintained their structure after 5 weeks of implantation and integrated well into the adjacent soft tissue. However, integration into the adjacent vertebra was less robust (Fig 3). Furthermore, there was a marked loss in GAG-positive stain in the NP region in both groups.

DISCUSSION

This study details the compositional and functional growth trajectory of an engineered disc composed of an electrospun AF and a hydrogel NP. We found that a skeletally mature immunocompromised rat served as a suitable model with which to evaluate engineered discs, and that DAPS maintained their structure after 5 weeks *in vivo*.

With extended *in vitro* growth, the compositional and functional properties of DAPS reached levels comparable to native human and rat discs [4,5]. GAG content of MSC/MSC and AF/NP seeded DAPS approached human levels in the AF and NP regions, while collagen content was 40% of the human AF and exceeded or matched that of the human NP. DAPS demonstrated nonlinear stress/strain behavior and moduli within 70% of the native rat disc; however, creep strain exceeded native levels. Thus, while DAPS resembled native tissue in elastic mechanical behavior, there were differences in time dependent mechanisms, perhaps driven by differences in the ability to retain water under load, or by plastic deformations in the AF region. Finally, the most sensitive metrics of implant maturity were collagen content and creep strain, both of which peaked at 10 weeks.

Upon implantation, DAPS retained their gross structure and were not associated with intervertebral fusion; however there was a marked depletion of GAGs in the NP region, suggesting a shift in cell behavior. Phenotypic transformations following transition from culture media to the *in vivo* environment were documented previously for both fully differentiated and stem cell populations [6]. Such changes may be related to the abrupt removal of media components such as glucose or TGFβ3 upon implantation. A strategy for transitioning engineered tissues from *in vitro* to *in vivo* may be required for tissue replacement and sustained function post-implantation.

ACKNOWLEDGEMENTS

This study was supported by the Departments of Defense and Veterans' Affairs and the Penn Center for Musculoskeletal Disorders.

REFERENCES

[1] Katz et al., *JBJS*, 88-A:21-24, 2006. **[2]** Nerukar et al., *Spine*, 35:867-73, 2010. **[3]** Martin et al., *Acta Biomat*, 6:2473-81, 2014. **[4]** Bowles et al., *PNAS*, 32:13106-11, 2011. **[5]** Beckstein et al., *Spine*, 33:E166-73, 2008. **[6]** Vinardell et al., *Tiss Eng A*, 18:1161-70, 2012

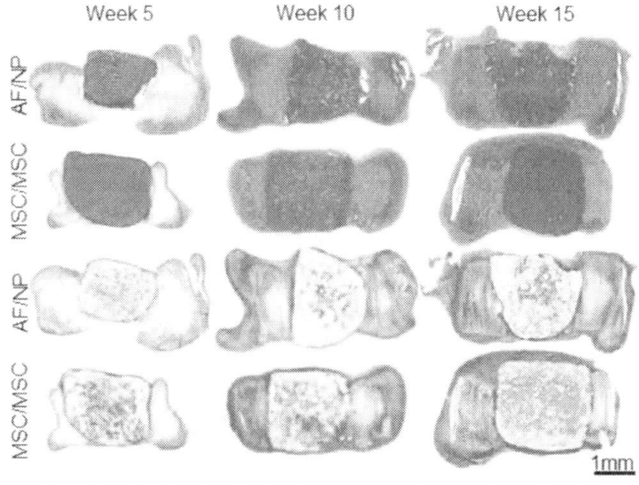

FIG.1 – AF/NP AND MSC/MSC DAPS STAINED FOR GAG (TOP ROWS) AND COLLAGEN (BOTTOM ROWS) AT 5, 10, AND 15 WEEKS *IN VITRO*.

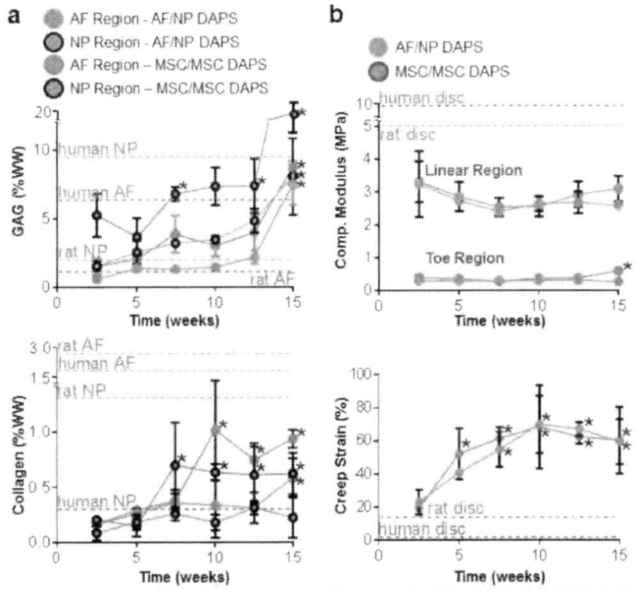

FIG.2 – **(a)** GAG AND COLLAGEN, AND **(b)** COMPRESSIVE MODULI AND CREEP STRAIN OF DAPS OVER 15 WEEKS *IN VITRO*. *, P<0.05 V. WK 2.5. NATIVE VALUES FROM [4,5].

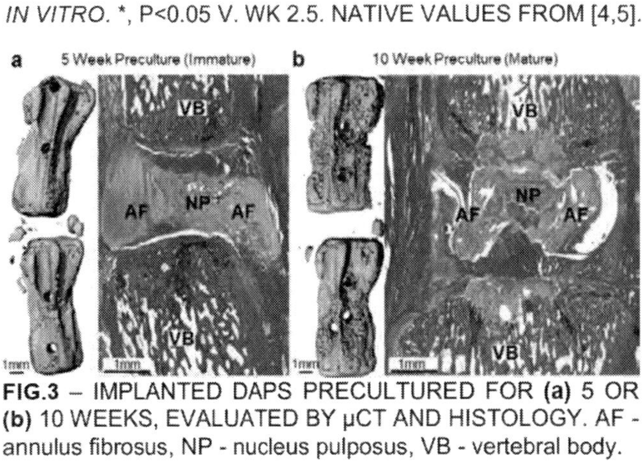

FIG.3 – IMPLANTED DAPS PRECULTURED FOR **(a)** 5 OR **(b)** 10 WEEKS, EVALUATED BY μCT AND HISTOLOGY. AF - annulus fibrosus, NP - nucleus pulposus, VB - vertebral body.

EVIDENCE THAT INTERFIBRILLAR LOAD TRANSFER IN TENDON IS SUPPORTED BY A NETWORK OF SMALL DIAMETER COLLAGEN FIBRILS

Spencer E. Szczesny (1), Kristen L. Fetchko (2), Jeffrey L. Caplan (3), Pal Pedersen (4), Dawn M. Elliott (2)

(1) Department of Bioengineering
University of Pennsylvania
Philadelphia, PA, USA

(2) Biomedical Engineering
University of Delaware
Newark, DE, USA

(3) Delaware Biotechnology Institute
University of Delaware
Newark, DE, USA

(4) Carl Zeiss Microscopy
Thornwood, NY, USA

INTRODUCTION

The hierarchical collagenous structure within tendon is the primary determinant of tendon mechanics. However, the specific mechanisms that govern this structure-function relationship are unclear. Specifically, it is still debated whether collagen fibrils in tendon bear load independently or if the applied load is transferred across the fibrils through interfibrillar shear forces. Multiscale studies of tendon under uniaxial tension indicate that collagen fibrils slide relative to each other with only a portion of the total tissue strain transmitted to the fibrils themselves [1,2]. Furthermore, the observed multiscale mechanics can be explained by a shear lag model, which assumes that the fibrils are discontinuous and transmit load via shearing of the interfibrillar matrix [3]. However, the physical structures that transmit load between fibrils are unknown. Glycosaminoglycans (GAGs) have been traditionally suggested as interfibrillar mechanical links, but selective removal of GAGs via enzymatic digestion failed to show any effect on tendon mechanics [4,5]. The objective of this study is to investigate whether any non-collagenous proteins are involved in interfibrillar load transfer or if such load transfer occurs through the collagenous structures themselves. Serial block-face scanning electron microscopy (SBF-SEM) was performed to observe potential interactions between collagen fibrils in rat tail tendon fascicles. Additionally, trypsin digestions were used to remove non-collagenous proteins and evaluate their contribution to the multiscale tendon mechanics.

METHODS

Tracking of Fibril Trajectories

A tail tendon fascicle from a 7-month Sprague-Dawley rat was fixed, stained, and embedded for SBF-SEM [6]. Eighty-eight serial cross-sectional images (2000 x 2000 pixels, 5 nm/pixel) were taken every 100 nm along the fascicle length with an SEM fitted with an *in situ* ultramicrotome. The raw images were processed using Fiji and 1,213 fibrils were tracked with a custom Matlab algorithm. The angle of each fibril with respect to the fascicle axis was calculated and linearly correlated with fibril diameter. Representative fibrils were segmented and reconstructed in three-dimensions using Amira.

Trypsin Digestions

For all analyses, samples were either incubated with 1 mg/ml acetylated trypsin (Sigma) in 0.15 M NaCl, 0.05 M phosphate buffer (pH 7.2) for 24 h at room temperature or in buffer alone without tryspin (controls). To determine if trypsin solubilized any collagen by either damaging the collagen structure or disrupting intermolecular crosslinks, rat tail tendon fascicles (n=3) were digested, removed from the digestion media, reduced with sodium borohydride, placed in 0.05 M acetic acid for 20 h at 4 °C, and centrifuged. The percent solubility of collagen was determined by dividing the hydroxyproline content in the digestion media and acid supernatant by the total hydroxyproline.

Digestion of non-collagenous proteins was assessed via immunohistochemistry. Samples were fixed in 4% paraformaldehyde, frozen and cryosectioned. Non-specific binding was avoided with a 0.1% sodium borohydride treatment and 5% goat serum block. Sections were incubated with either decorin (1:50), collagen XII (1:100), or COMP (1:50) primary antibodies in goat serum overnight at 4 °C and then labeled with Alexa Fluor 488 secondary antibodies. Negative controls followed the same procedure while omitting primary antibodies. The percent drop in mean signal intensity between digested samples and controls was calculated with Image J. Additionally, removal of GAGs from tissue was assessed via DMMB assays (n=3).

Multiscale mechanical testing was used to determine the effect of trypsin digestion on fascicle mechanics [3]. Trypsin digested (n=4) and control fascicles (n=4) were stained with DTAF at 10 μg/ml and loaded into a custom uniaxial tensile device mounted on an inverted confocal microscope. A 1 mN preload was applied to the tissue to define the reference length (~30 mm) and a set of four lines separated by 125 μm were photobleached onto the tissue surface. The testing protocol consisted of incremental ramps to 2, 4, 6, and 8% applied grip strains at 1%/s followed by a 15 min relaxation. Fibril strains were calculated as the change in the distance between pairs of photobleached lines and macroscale tissue strains were measured optically. The amount of fibril loading within the tissue was determined by calculating the fibril:tissue strain ratio.

RESULTS

While the collagen fibrils were closely aligned with the fascicle axis, with an average angle of 1.7 ± 1.0 deg (mean ± s.d.), the smaller diameter fibrils (< 150 nm) tended to lie at greater angles to the fascicle axis (Fig. 1A) ($r = -0.17$, $p<10^{-8}$). This agreed with our visual observations of the smaller diameter fibrils winding around and weaving between their neighboring larger diameter fibrils (Fig. 1B). Furthermore, we found abundant evidence of the smaller diameter fibrils fusing with and integrating directly into the larger diameter fibrils (Fig. 2). These data suggest that smaller diameter collagen fibrils may be the source of interfibrillar shear stress through entanglement and fusion with the larger diameter load-bearing fibrils.

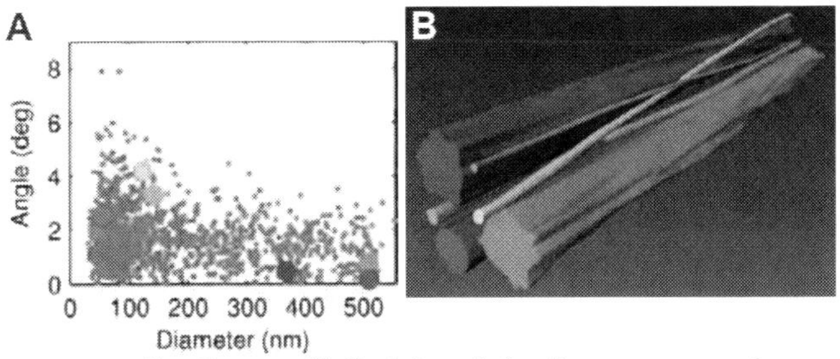

Fig. 1: Smaller diameter fibrils (A) tended to lie at greater angles to the fascicle axis as (B) they wound around the larger diameter fibrils.

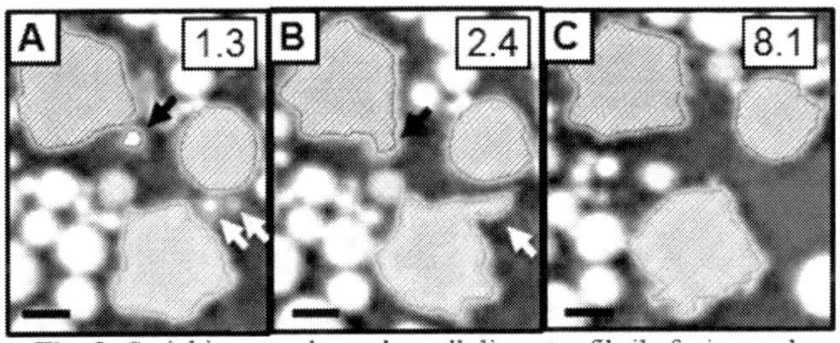

Fig. 2: Serial images showed small diameter fibrils fusing and ultimately fully incorporating within the larger diameter fibrils. Annotated distances from initial image (μm). Scale bar = 200 nm.

Digestion of rat tail tendon fascicles with trypsin produced no additional hydroxyproline in either the digestion buffer or in acid compared to non-digested controls (solubility: 0.5 ± 0.9% (trypsin) vs 0.6 ± 0.9% (control)). This demonstrates that acetylated trypsin does not digest the intact collagen molecules or disrupt the intermolecular crosslinks in intact rat tail fascicles, which is consistent with similar digestions of bovine tail tendons [7]. However, trypsin did successfully remove a substantial amount of non-collagenous protein. Immunostaining intensity for decorin was reduced evenly throughout the tissue cross-section by 76% with trypsin digestion (Fig. 3), which matched an 83% reduction in GAG content. No collagen XII or COMP was observed in undigested fascicles. Digestion with trypsin had no effect on the macroscale fascicle mechanics or on the amount of strain transmitted to the fibrils (Fig. 4).

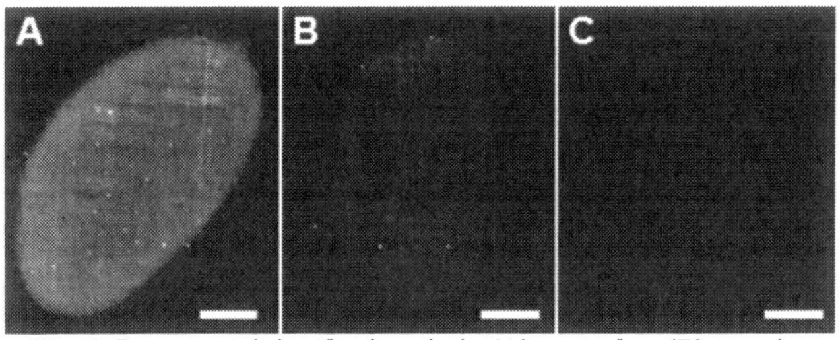

Fig. 3: Immunostaining for decorin in (A) control vs (B) trypsin treated samples demonstrating a loss of signal intensity with digestion. (C) Neg. control. Scale bar = 100 μm.

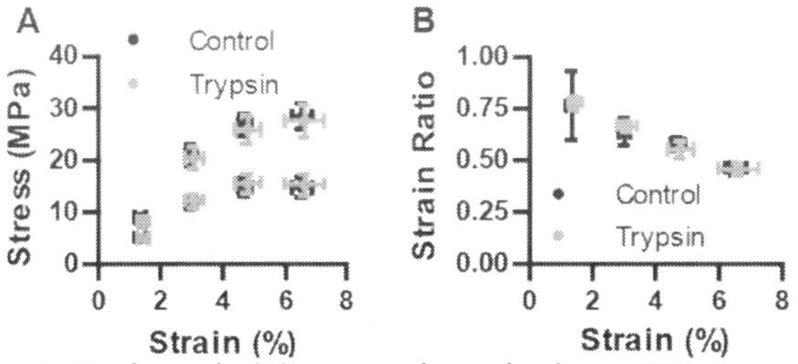

Fig. 4: No change in (A) macroscale mechanics or (B) amount of strain transmitted to the fibrils was observed with trypsin digestion.

DISCUSSION

This work investigated whether non-collagenous proteins are involved in interfibrillar load transfer or if such load transfer occurs through the collagenous structures themselves. Data from SBF-SEM suggests that interfibrillar load transfer is produced by a network of smaller diameter collagen fibrils, which weave around and fuse with the larger diameter load-bearing fibrils. This is consistent with the fibril fusion and branching described previously in embryonic tendon [8]. Furthermore, digestion of non-collagenous proteins using acetylated trypsin failed to produce any changes in fascicle macroscale mechanics or in the underlying fibrillar deformation mechanisms. This further supports the suggestion that smaller diameter fibrils in tendon, rather than non-collagenous tissue components, act to transmit load between the larger load-bearing fibrils. While future work is necessary to confirm this hypothesis, it is consistent with reports that GAGs may help to promote interfibrillar sliding by preventing excessive fusion between fibrils [9]. One limitation to this study is the relatively short length used to track fibrils (8.7 μm); however, this represents an aspect ratio greater than 20 even for the largest diameter fibrils. Additionally, given the limited amount of non-collagenous protein in rat tail tendons, it is difficult to quantify the success of trypsin in removing proteins other than decorin. Nevertheless, the fact that trypsin was able to digest the decorin core protein evenly throughout the tissue cross-section suggests that other non-collagenous proteins were digested as well. The suggestion that one set of collagen fibrils may transmit load between another population of fibrils represents a novel hypothesis in tendon hierarchical structure-function relationships that may resolve the controversy regarding fibrillar loading mechanisms.

ACKNOWLEDGEMENTS

We thank Jean Ross for her help with the immunohistochemistry.

REFERENCES

[1] Puxkandl *Phil Trans R Soc B* 2002 [2] Cheng *J Mater Sci* 2007 [3] Szczesny *Acta Biomat* 2014 [4] Lujan *JOR* 2007 [5] Fessel *Matrix Biol* 2009 [6] Starborg *Nat Protocols* 2013 [7] Willett *Ann Biomed Eng* 2007 [8] Birk *Dev Dyn* 1995 [9] Rigozzi *J Biomech* 2013

TRIBOLOGICAL REHYDRATION: MAINTAINING AND REBUILDING INTERSTITIAL FLUID PRESSURE IN CARTILAGE

Axel C. Moore (1) and David L. Burris (1,2)

(1) Department of Biomedical Engineering
University of Delaware
Newark, DE, USA

(2) Department of Mechanical Engineering
University of Delaware
Newark, DE, USA

INTRODUCTION

In 1959, it was first discovered that the unusual frictional properties of cartilage are fundamentally linked to its multi-phasic (solid+fluid+charge) structure; Charnley called cartilage 'the synovial sponge' and McCutchen called the mechanism 'weeping lubrication'. During loading, fluid is squeezed through the porous solid, which, as described by Darcy's Law, results in a pressure gradient with maximal pressure at the contact location and zero pressure at the free-boundaries. The pressurized fluid hydrostatically supports the contact force, thus reducing friction just as air pressure supports the load of a hovercraft. However, this pressure gradient drives fluid out of the tissue resulting in the time-dependent loss of hydration, fluid pressure, and lubrication.

Today, it remains unclear how interstitial pressure and weeping lubrication are sustainable in the long-term. However, joint-space and joint contact stress measurements have proven that hydration levels and interstitial pressures are maintained in the long term despite consistent laboratory evidence to the contrary. Ateshian's group was the first to demonstrate that fluid pressure can be sustained over long durations if sliding speeds exceed diffusive speeds. However, this idea requires pre-existing hydration and fails to explain how our joints restore hydration after inevitable periods of static contact during standing and sitting. Understanding how hydration is restored is, in our opinion, the key to long-term lubrication.

Interestingly, several clinical measurements in the literature have definitively shown that exercise thickens cartilage [1]. In this paper, we demonstrate that loaded sliding contacts can actively push fluid back into cartilage via a special form of elastohydrodynamics for porous surfaces. To our knowledge, this is the first paper demonstrating active rehydration of cartilage. The phenomenon, which we call tribological rehydration, suggests that activity (which is tribological in nature) is essential for maintaining joint lubrication and health.

METHODS

Cartilage samples, Ø12mm osteochondral cores, were taken from mature bovine stifles. Immediately after removal, samples were rinsed and stored in phosphate buffered saline (PBS). Samples that did not undergo immediate testing were placed in a laboratory refrigerator for no more than 3 days, a time frame for which we find no statistically detectable changes in mechanical or tribological properties [2]. The cartilage was placed in stationary contact (SCA) [3] against an optically smooth/flat glass microscope slide [4, 5] in a PBS flooded contact. PBS was used to eliminate the confounding boundary lubricating and shear thinning effects of synovial fluid [4].

Under normal laboratory circumstances, SCA drives fluid out of the tissue over time. An *in-situ* elasto-hydrodynamic tribometer was used to assess exudation. The relevant elements of the tribological system are shown in Figure 1. A 6-axis load cell measured all forces and moments on the cartilage during sliding and an LVDT measured changes in distance between the tidemark and counterface, which was interpreted in terms of fluid exudation or imbibition. All tribological experiments used a track length of 20mm, speeds which ranged from 0 – 50mm/s and normal loads from 1 to 10N.

RESULTS

As described by biphasic theory, loading cartilage against a flat impermeable platen results in a time dependent loss of fluid pressure, hydration, and thickness. Since fluid pressure is the major mechanism of load support and friction reduction in cartilage this leads to a progressive increase in friction with time [5]; this response is well known in the literature and can be observed on the left of Figure 1.

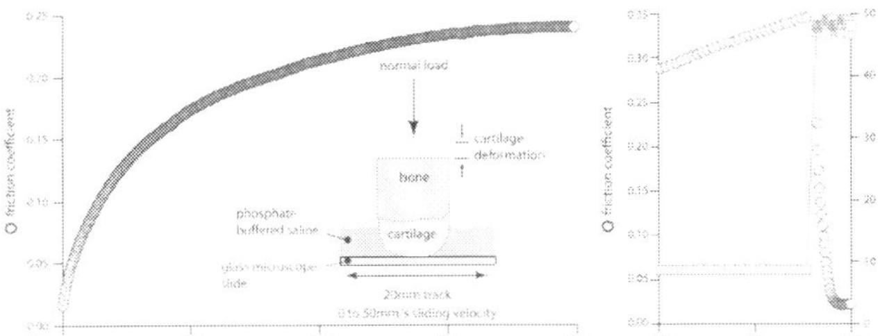

Figure 1. Cartilage frictional responses for the relevant tribological setup (center). Left: a typical time dependent frictional response for articular cartilage; sliding velocity was 20mm/s. Right: the frictional response of cartilage as a function of speed. Note that the cartilage sample had undergone 1000s of continuous sliding prior to the start of the data shown.

Figure 1 (right) demonstrates the phenomenon of interest in this paper. At 10mm/s, friction increases with time in the usual manner. However, when sliding speeds increased to 50 mm/s, friction decreased abruptly and continued to decrease until equilibrating at μ=0.02 after 50 seconds of sliding.

Cartilage friction coefficients of 0.02 have only been reported for: 1) cartilage on cartilage, 2) a rigid sphere sliding against cartilage at large Peclet numbers (migratory contact area), and 3) the beginning of a stationary contact experiment at time scales much less than characteristic exudation time scales. The surprising fact in this case is that low friction is quickly restored at walking speeds in a stationary contact ***following*** significant exudation of interstitial fluid.

The deformation, friction coefficient and speed are plotted versus time in Figure 2 to illustrate the transition from stationary contact to stationary sliding at walking speed. The cartilage compressed by ~70μm due to the exudation of interstitial fluid. Reciprocated sliding at 50mm/s was initiated after ~360 seconds, at which point friction and deformation began to decrease with time. Such a dramatic drop in friction with increased speed is often attributed to the development of hydrodynamic films as first proposed by MacConnaill for cartilage [6]. Qualitatively, the decreased deformation is consistent with the formation of a fluid film. Quantitatively, the results are inconsistent with the fluid film theory. First, a 12μm thick film of water being sheared at 50 mm/s should produce a shear stress on the order of 4Pa or a friction coefficient of 0.0001. Second, no current theory, empirical or otherwise, would predict the development of such a thick film under these conditions. Third, Hlavacek showed that porous surfaces provide a flow pathway into the surface that defeat the development of fluid films [7].

DISCUSSION

Theoretical and experimental results suggest that friction in this system is not reduced by the formation of a fluid film. We propose a hybrid mechanism, which we call tribological rehydration. The phases (1-4) of tribological rehydration can be observed in Figure 2 (top). The presence of a convergent wedge at the leading edge provides the environment necessary for hydrodynamic pressure to build (phase 1). Rather than separating surfaces as would be the case for a traditional surface, hydrodynamic pressure drives fluid back into the porous and depressurized surface. It is interesting to note that friction consistently stabilizes faster than deformation. Friction is modified the instant fluid enters the surface (phase 2), but a significant volume must be imbibed (phase 3) before tissue thickening becomes detectable. McCutchen reported similar effects with dramatically reduced friction but minimal thickening after various unloading intervals.

The most significant point here is the demonstration that sliding actively restores hydration in cartilage, even in stationary contacts. To our knowledge this paper is the first to report on such a mechanism and represents a first attempt to explain this phenomenon using both theory and in-situ measures of friction and tissue growth.

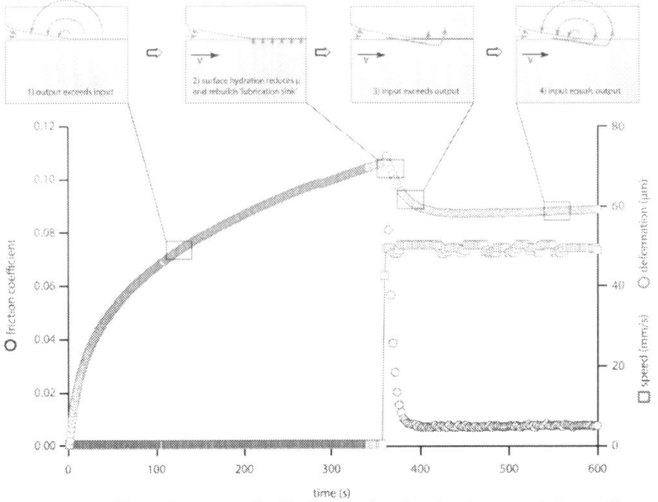

Figure 2. The phases of tribological rehydration and the effect of speed on the frictional and deformation response of articular cartilage. Note that the cartilage sample was loaded to ~5N.

ACKNOWLEDGEMENTS

The authors acknowledge financial support from the NIH (grant P20-RR016458).

REFERENCES

[1] Ekholm, R., and Norbäck, B., 1951, "On the Relationship Between Articular Changes and Function," ort Acta Orthopaedica, 21(2-4), pp. 81-98.

[2] Moore, A. C., and Burris, D. L., "Tribological and material properties for cartilage of and throughout the bovine stifle: support for the altered joint kinematics hypothesis of osteoarthritis," Osteoarthritis Cartilage.

[3] Caligaris, M., and Ateshian, G. A., 2008, "Effects of sustained interstitial fluid pressurization under migrating contact area, and boundary lubrication by synovial fluid, on cartilage friction," Osteoarthritis and Cartilage, 16(10), pp. 1220-1227.

[4] Forster, H., and Fisher, J., 1996, "The influence of loading time and lubricant on the friction of articular cartilage," Proceedings of the Institution of Mechanical Engineers Part H-Journal of Engineering in Medicine, 210, pp. 109-119.

[5] McCutchen, C. W., 1962, "The frictional properties of animal joints," Wear, 5, pp. 1-17.

[6] Macconaill, M. A., 1932, "The function of intra-articular fibrocartilages, with special reference to the knee and inferior radio-ulnar joints," Journal of Anatomy, 66, pp. 210-227.

[7] Hlavacek, M., 1993, "The Role of Synovial-Fluid Filtration by Cartilage in Lubrication of Synovial Joints .1. Mixture Model of Synovial-Fluid," Journal of Biomechanics, 26(10), pp. 1145-1150.

FORCE REQUIRED TO CINCH THE TRICUSPID ANNULUS: AN EX VIVO STUDY

Amy Adkins (1)*, Jesus Aleman (1)*, Lori Boies (1), Edward Sako (2), Shamik Bhattacharya (1)

(1) School of Science, Engineering and Technology
St. Mary's University
San Antonio, Texas, United States

(2) Cardiothoracic Surgery
University of Texas Health Science Center at San Antonio
San Antonio, Texas, United States

INTRODUCTION

The tricuspid valve is a three leaflet valve located between the right atrium and right ventricle of the heart and prevents the back flow (regurgitation) of blood into the right atrium. Functional tricuspid regurgitation (TR) is increasingly recognized as a source of morbidity in patients particularly those with chronic mitral valve regurgitation[1]. Annulus dilation is a major feature in functional TR. However, the mechanism of annulus dilation is not understood clearly. Mitral valve annulus mechanics affects annulus dilation in addition to left ventricle remodeling [2]. This mechanism is likely to be true for the right heart. Tricuspid annuloplasty is the most preferred surgical therapy to repair TR. In this technique, tricuspid annulus (TA) is sutured to a rigid or semi-rigid annuloplasty ring which results in a "cinching up" or shortening of a portion of the circumference of the TA thereby decreasing the severity of TR. However, recurrent regurgitation and ring dehiscence has been reported afterwards. The long-term efficacy of tricuspid annuloplasty demands a deeper understanding of the interactive tissue mechanics between tricuspid annulus and ring annuloplasty. The objective of the current study is to quantify the force required to cinch the normal and dilated tricuspid annulus. The cinching force is important to understand the interaction between native tricuspid annulus and annuloplasty and can contribute to the long term durability of tricuspid annuloplasty.

METHODS

This method was modeled after a previous study of mitral valve annulus cinching [3]. Sheep hearts were collected from a local slaughter house and transported in a cooler. The hearts were cryopreserved until the day of the experiment upon which hearts were thawed in a water chamber at 37.5°C. The right atrium was removed and a suture was inserted into the adjacent tissue just above the right coronary artery (Figure 1), next to the tricuspid annulus. One end of the suture was tied and anchored at septal-anterior commissure and the other end exited through the passage in between the mitral valve and aorta (Figure 1).

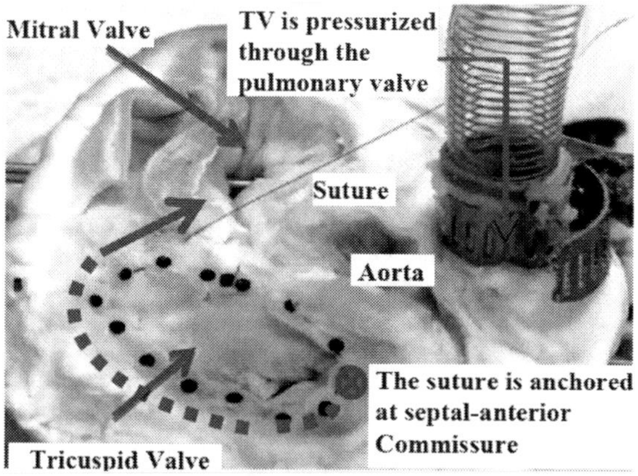

FIGURE 1- SUTURE POSITIONING

The free-end (Figure1) of the suture was kept parallel to the septa-lateral direction without touching the tissue and was attached to a lever and force transducer mounted on a slider system (Figure 2A). The slider moved the force transducer in a given direction at regular intervals as shown (Figure 2A). The tricuspid valve was closed by pressurizing the right ventricle through the pulmonary valve, using a custom built static pressure head system (Figure 1 & 2A).The whole system was immersed in a Plexiglas chamber filled with physiological saline (Figure 2A). The suture was pulled at 2 mm increments to cinch the tricuspid annulus (TA) after the valve closure at 32.5 ± 2.5 mmHg. A digital camera was placed perpendicular to the TA plane at a close distance. The TA area was measured from the image (Figure 2B) at each increment and the corresponding force and leakage data were collected for 3 conditions: 1) no pressure-(NP) 2) pressurized-(P) and 3) dilated-pressurized (DP). Annulus dilation was created by injecting 95% phenol around the annulus tissue [4].

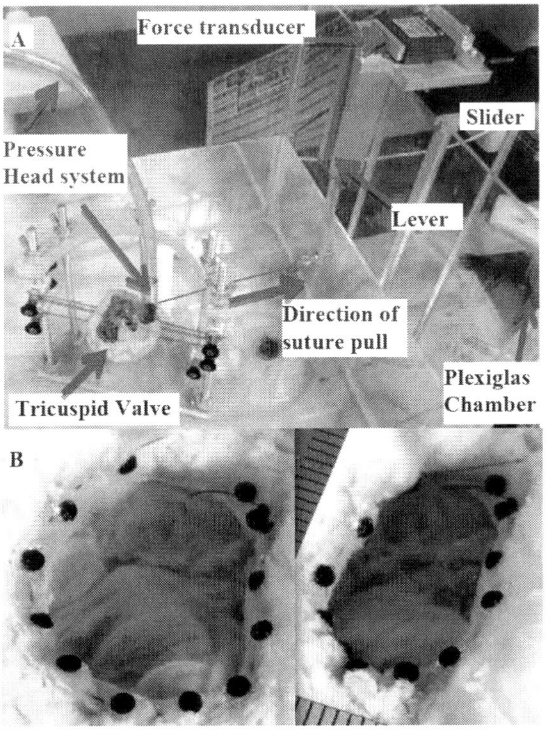

FIGURE 2- A) EXPERIMENTAL SET UP B) IMAGE OF TRICUSPID ANNULUS SHOWING CHANGE IN AREA FROM 0 MM TO 24 MM WHEN PRESSURIZED.

RESULTS

The results are based on three sheep hearts and the mean value of the data is presented. As pulling length was increased for all three protocol conditions, the cinching force increased (Figure 3).

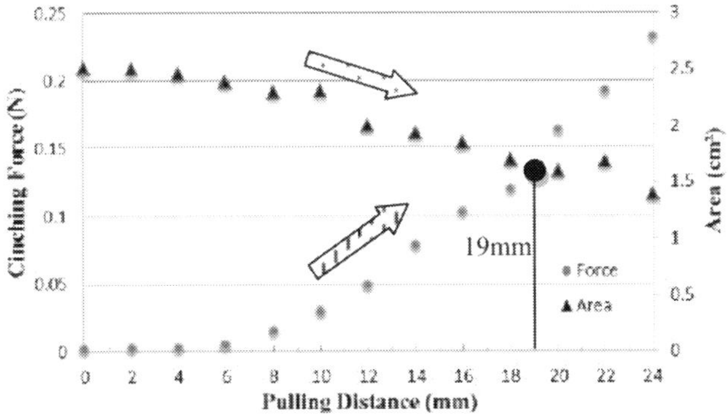

FIGURE 3- THE CHANGE OF FORCE AND AREA IN THE TRICUSPID ANNULUS WITH INCREASE IN PULLING DISTANCE -PRESSURIZED CONDITION.

The DP condition resulted in the maximum cinching force followed by the P condition. In Figure 3, the change in TA area and cinching force with pulling distance for the P condition is shown. The force and the area were found to be linearly related. At 19 mm, 0.14N of force is required to reduce the area by 40.5%. At 0 mm or without any pull, the TA area at P condition increased by 9.1% after phenol was injected to create the DP condition. The force and the TA area at two different pulling distances are shown in Table 1. After 12 mm pull, the force at P condition was 59.2% greater than NP condition and the force on the DP condition was 39.9% greater than the force on the P condition. After 24mm pull, the TA area at DP condition was 8.14% larger than P condition. The leakage was greater in the DP condition compared to the P condition. The leakage reduced in both the P and DP conditions after cinching. The reduction in leakage due to cinching was greater in DP condition than in P condition.

TABLE 1-CINCHING FORCE AND VALVE AREA

	Cinching Force (N)		Area (cm²)	
	12 mm	24 mm	12 mm	24 mm
NP	0.02	0.08	2.64	1.56
P	0.05	0.23	1.97	1.53
DP	0.08	0.31	2.10	1.67

DISCUSSION

The force required to cinch the tricuspid annulus is the sum of the right ventricular tissue resistance force and force due to transvalvular pressure. The cinching force at NP condition represented the right ventricular tissue resistance force alone whereas the cinching force in P and DP conditions is the sum of both the forces. The 9.1 % increase in area and the increase in leakage between the pressurized tricuspid valve and the dilated valve validate the method of annulus dilation. The cinching of the tricuspid annulus at the DP condition requires a greater force compared to the P and NP condition but the reduction in area is less compared to P and NP condition. It seems that the force due to transvalvular pressure is the dominating force resisting the cinching. The contraction of right ventricular myocardium was not considered in this study as the tissue was passive. Though the leakage was clinically insignificant, the cinching of the TA did reduce the leakage amount for the pressurized conditions. Therefore, cinching force needs to be considered when restoring a dilated TA to the normal shape and size. More experiments need to be done to demonstrate the effect of cinching force on annulus dilation.

ACKNOWLEDGEMENTS

*Amy Adkins and Jesus Aleman equally contributed to this work. The work is supported by St. Mary's Research grant, IIMS-CTSA grant (UTHSCA), and MARC fellow training aid. We thank Ray Jaszcz, Rey Balli, Evelia Salinas, and Miguel Pelayo for their help in this work and appreciate all the support given by Dr. Tim Raabe.

REFERENCES

1. Izumi, C., K. Iga, and T. Konishi, *Progression of isolated tricuspid regurgitation late after mitral valve surgery for rheumatic mitral valve disease.* J Heart Valve Dis, 2002. **11**(3): p. 353-6.
2. Bhattacharya, S. and Z. He, *Role of annulus tension in annular dilatation.* J Heart Valve Dis, 2009. **18**(5): p. 481-7.
3. Bhattacharya, S., et al., *Tension to passively cinch the mitral annulus through coronary sinus access.* J Biomech, 2014. **47**(6): p. 1382 - 1388
4. Green, G.R., et al., *Mitral annular dilatation and papillary muscle dislocation without mitral regurgitation in sheep.* Circulation, 1999. **100**(19 Suppl): p. II95-102.

BEST PRACTICES IN TEACHING BIOMECHANICS: INTEGRATING REFLECTIVE LEARNING ACTIVITIES

Laurel Kuxhaus

Department of Mechanical & Aeronautical Engineering
Clarkson University
Potsdam, NY USA

INTRODUCTION

Reflective learning asks students to reflect on their educational experiences under the basic premise that the process of reflection will inform students' future actions and choices both in and outside of the classroom. For example, students may struggle with a particular problem set (homework assignment) and, upon reflection, conclude that they either need to study more and/or are not well-suited to an engineering career. When asked to reflect on the challenge of this experience compared to their motivation for studying engineering, the same student reflecting on the same experience may conclude that engineering is just the challenging career they seek. By encouraging guided reflection in the classroom, students are coached to think broadly and view their learning and learning experiences from different perspectives, which influences their future actions, learning, and careers.

Reflection exercises in the engineering classroom are becoming more prominent [1], particularly with the establishment of the Consortium to Promote Reflection in Engineering Education (CPREE, pronounced as "spree".) This consortium seeks to develop "...diverse, adaptive, broad-thinking and nimble engineering experts who can respond to the ever-increasing challenges that society faces" [2] via reflection in engineering education. Over the next several years, the CPREE program will document and promote reflective learning at 12 universities in the United States, including Clarkson University.

Reflection in the classroom can be open-ended (as in the example above), or may ask students to reflect on their experiences through a particular lens. For example, students may be asked to reflect upon that same difficult homework assignment through a 'lens' of their current understanding of the topic, their current understanding of what it means to learn, or their assumptions of the relevance of the homework assignment to their future career. [2]

The undergraduate biomechanics course can attract a wide variety of students. For example, the author routinely teaches courses consisting of approximately equal proportions of Chemical and Mechanical Engineering students, with about 10% non-engineering students. As such, students have diverse mechanics backgrounds and, despite a variety of learning assessments offered, the author observes that students with strong mechanics backgrounds often wish the relevant mechanics lessons were presented more quickly and more deeply, while those students with less-strong mechanics backgrounds often feel overwhelmed by this same content. Reflective exercises can help all students to articulate the value of both the technical content and skills learned through the course, and help form broad-thinking habits which will be carried throughout these young engineers' careers.

This abstract presents some ready-to-implement strategies which promote reflective learning in an undergraduate Biomechanics class. These activities are intended to be customized to individual classrooms as needed.

METHODS

The activities described below have been implemented in six years of teaching biomechanics. It should be noted that, while these activities can be used as graded assessments of student learning, they are also effective as ungraded exercises.

Technique: Feedback cards.
Estimated class time: Less than 1 minute.

Feedback cards can easily be used for reflective activities. Index cards (or quarter-sheets of paper) are easy to use. These may be pre-printed with a prompt, or the instructor may dictate a prompt to the students. Example prompts can range from general ("Today in class, I...." or

"The muddiest point from today's lecture was....") to specific ("One question I have about bone mechanics is..."). Ideal prompts follow a "complete-the-sentence" structure. The author uses feedback cards occasionally in her biomechanics class, and daily (doubling as an attendance record) in other basic mechanics courses.

Technique: Minute papers. [3]
Estimated class time: 1 minute.

Minute papers are a standard active learning technique in which students are given one minute to write a short summary of a topic. Applied to the biomechanics classroom with the goal of a reflective experience for the students, example prompts may be, "Explain how you might use what you know about the structure-function relationship of cartilage to design a new material for bearing surfaces" or "Reflect on a time that you (or someone you know) experienced a soft tissue injury, such as an ankle sprain. Using what you've learned in this class, describe how the doctor's recommended course of action agrees (or disagrees) with your knowledge of ligaments." Questions also might ask students to reflect on their learning process such as "How did today's assigned article prepare you for the lecture, and what will you look for when you re-read it after class?".

To implement the Minute Paper, ask students to take out a blank sheet of paper and spend one minute writing to address the prompt. (The author sometimes extends this to 90 seconds or 2 minutes.)

Technique: In-class reflective writing.
Estimated class time: 5-10 minutes (can also be done as a take-home activity).

Inspired by "Emergency Assessments" [4], this technique guides students to reflect on their course-related experiences through the 'lens' of their future selves. Students spend a few minutes writing in response to a prompt, such as:

You've applied for a job as a technical consultant for the next Batman movie. They liked your application and have called you in for an interview. The person doing the hiring says "I see that you have taken a course in Biomechanics. How will a background in that make you a better member of our crew?" What will you reply? (modified from [4])

Or

You've been asked to help write part of a webpage advertisement for a science camp for middle-school girls, discussing the session in which the biomechanics of sports and injury will be presented. Write a one-paragraph introduction to this audience that will motivate them to sign up for this camp. (modified from [4])

Technique: Think-pair-share jigsaw activity.
Estimated class time: 5 minutes or less.

A jigsaw activity expands on the "think-pair-share" activity [5] with an additional layer of peer interaction. The author commonly uses this in conjunction with a review session for an exam. In this exercise, students each start with their own piece of paper. An example is illustrated below:

- Divide the students into 3 (or more) groups.
  - Example: those with birthdays January-April, May-August, and September-December.
- Each group is assigned a topic to write at the top of their page.
  - Example: ligaments, bone, muscle
- In 30 seconds, students are asked to write one item related to their topic.
  - Example: how structure and function are related.
- Students then trade papers with a neighbor, and have 30 seconds to add an item to the list they just received.
- Students trade papers with a *different* neighbor, and add an item to that list.
- A brief class discussion (in which the instructor calls on students randomly to share items from their lists) summarizes main points.

Technique: Dedicated reflective assignment.
Estimated class time: Completed outside of class.

Dedicated assignments may guide students to reflect on their learning experiences. For example, the author uses a "Popular Press" assignment to encourage students to reflect on the way scientific results are portrayed in the media. Students must identify a news story published during the current semester which reports the results of a scientific study related to biomechanics. They must then find the original journal article before writing a short (1-2 page) discussion of how the results were accurately (or not) portrayed in the media.

Technique: Reflective questions on quizzes, exams, or other existing assignments.
Estimated class time: No additional time (included in quizzes or exams)

These questions may be short-answer (as part of a quiz or exam), longer (as a take-home component of a quiz or exam), or simply a required section on a homework assignment or report write-up. Example questions might include: "How will your knowledge of musculotendon mechanics affect your warm-up routine before you next go for a run?" (specific); "What skills in this course will you take with you?" (general); "What is the most surprising thing you learned in this course/unit/topic?" (more general); or "In 5 years, what will you remember most about this course?"

For reports, papers, or projects, reflection can be a required section with instructions such as, "In your report, you must have a section entitled "Reflection" in which you reflect on the learning process in this project, and the skills you will take away from the experience."

DISCUSSION

Reflective learning encourages broad thinking skills and deepens the learning experience for students. Simple activities, such as those described above, can encourage reflection both in and outside of the classroom. Instructors can use the feedback from students in the reflective exercises to inform their own teaching and adjust the course accordingly.

ACKNOWLEDGEMENTS

The author thanks her students for their willingness to try these activities in class.

REFERENCES

[1] Turns, J. et al. (2014) "Integrating Reflection on Experience into Engineering Education" ASEE Conference.

[2] Consortium to promote reflection in engineering education. http://depts.washington.edu/celtweb/cpree/ accessed 1/12/15.

[3] Johnson, D.W. et al. (2006). "Active Learning: Cooperation in the College Classroom". Interaction Book Co., 316pp.

[4] Huston, T. (2012). "Teaching what you don't know." Harvard University Press. 320pp.

[5] Mazur, K. (1997). "Peer instruction: a user's manual." Prentice Hall, 253pp.

COMPUTATIONAL MULTIPHYSICS MODEL OF CLOT LYSIS IN A COMPLETELY OCCLUDED STENOTIC ARTERY

Andris Piebalgs (1) and Xiao Y. Xu (1)

(1) Department of Chemical Engineering
Imperial College London
London, United Kingdom

INTRODUCTION

Cardiovascular disease in the form of myocardial infarction and ischaemic stroke accounted for over 20% of global mortality rates in 2011. Thrombolytic therapy involves the infusion of a drug known as tissue plasminogen activator (tPA) that can lyse the occluding blood clot and restore blood flow in the artery. However, since the drug cannot discriminate between a healthy haemostatic clot and thrombus, internal bleeding such as intracerebral hemorrhage can occur. Clot lysis is a very complex process that involves the interplay between many different proteins in the blood stream [1]. It is hoped that a complete mathematical description and modelling of this process could lead to a better understanding of clot lysis and improved treatment.

Diamond and Anand were among the first researchers who attempted to model the process of clot lysis through the solution of a system of partial differential equations describing both the transport and reaction of the drug and other important proteins on the clot surface [2]. The clot was treated as a fibrous porous medium and the one-dimensional (1D) fluid flow field was solved with Darcy's equation. The material properties of the porous clot such as the voidage and permeability were determined as functions of the concentration of the drug. Modifications were made to this model in the subsequent years to take into account more complex reaction kinetics [3] and simple 2D geometries [4].

Wooton extended the model developed by Diamond and Anand to examine the effect of permeation on the lysis of mural blood clots and the effect of increasing binding sites due to nicked fibrin [5]. They found that clot lysis was accelerated by permeation outside the clot, and suggested that at very high arterial pressures other important processes might play a fundamental role in accelerating clot lysis.

Bannish developed a 1D model for clot lysis by describing the transverse cutting of fibrin by plasmin instead of a shrinking fibre radius as assumed by Diamond and Anand [6]. Adsorption was modelled in terms of binding site doublets where only tPA and plasminogen in a similar vicinity would allow for plasmin production. Although the model gave similar results to the reaction-diffusion model of Diamond, it couldn't explain why the lysis of coarser clots is faster than that of thinner and finer clots.

To the authors' knowledge, 2D analysis of flow in a fully-occluded clot has not yet been reported in the literature. It is the aim of this study to extend the original Diamond and Anand model to account for fluid convection and monitor how a variety of pressure drops across a fully occluded clot may affect the pattern of lysis. In this abstract, we present a multiphysics computational model capable of modelling the lysis of fully occluded blood clots upon application of tPA.

METHODS

Equations describing the motion of blood throughout the domain both inside and outside the clot can be found by applying mass and momentum conservations over an arbitrary control volume. Blood was assumed to behave as an incompressible Newtonian fluid [7].

$$\frac{\partial \epsilon}{\partial t} = -\nabla \cdot u \tag{1}$$

$$\rho\left(\frac{\partial u}{\partial t} + u\nabla \cdot u\right) = -\nabla p + \mu \nabla^2 u - \frac{\mu}{k}u \tag{2}$$

where ε is the voidage, u is the velocity, μ is the viscosity, p is the pressure, ρ is the density and k is the permeability of the clot. The last term in the momentum equation represents the resistance to flow imposed by the clot. The permeability of the clot is a function of the fibre radius (R_f) and the voidage [8].

$$k \propto \frac{R_f^2}{fn((1-\epsilon))} \tag{3}$$

The drug tPA adsorbs onto a fibrin surface and reacts with adsorbed plasminogen to produce the enzyme plasmin that goes on to cleave the fibrin fibres. The generic transport equation for a free phase and bound species can be written as [7]

$$\frac{\partial \epsilon C_i}{\partial t} + u\nabla \cdot \epsilon C_i = D\nabla^2 \epsilon C_i - (1-\epsilon)\frac{\partial A_i}{\partial t} + (S_i)\left(\frac{\partial \epsilon}{\partial t}\right) \tag{4}$$

$$\frac{\partial S_i}{\partial t} = \frac{\partial A_i}{\partial t} + P_i \tag{5}$$

where C_i is the concentration of the free phase species i, S_i is the concentration of the bound phase species. D is the diffusion coefficient, A_i is the amount that has been adsorbed onto the fibrin surface and P_i is the amount that has been produced in the bound phase. The amount that has been adsorbed can be expressed as

$$\frac{\partial A_i}{\partial t} = k_{ads} C_i(\theta_i - S_i) - k_{rev} S_i \tag{6}$$

where θ_i is the concentration of clot binding sites and k_{ads} and k_{rev} are the rates of reaction for adsorption and desorption, respectively. Plasmin generation and plasminogen consumption can be expressed via Michaelis-Menten kinetics as [9]

$$P_{pls} = -P_{plg} = \frac{k_2 S_{plg} S_{tPa}}{K_M + S_{plg}} \tag{7}$$

where K_M and k_2 are Michaelis-Menten reaction coefficients. The concentration of fibrin lysed can be described by a differential equation as

$$\frac{\partial L}{\partial t} = k_{cat} \gamma S_{pls} \tag{8}$$

where L is the concentration of fibrin lysed, k_{cat} is the reaction rate coefficient of fibrin lysis and γ is the solubilisation parameter. The fibrin fibre radius and the voidage can be expressed as a function of the concentration of fibrin lysed [2].

$$R_f = R_f(L, \epsilon), \qquad \epsilon = \epsilon(R_f) \tag{9}$$

The above equations were implemented in Ansys Fluent v 14.5 through user defined functions (UDFs). The model geometry is a 2D channel with a width of 10 mm and length of 200 mm that is fully occluded in the center by a 5 mm rectangular clot. Mesh independence was attained for 100,000 elements and time step independence for 0.1 s. The convergence criterion was set as 1E-6.

RESULTS

A constant pressure drop was applied across the clot with 2 μmol/L of plasminogen and 0.3 μmol/L of tPA being introduced at the inlet. Figure 1 shows lysis contours displaying the fraction of clot that has been dissolved for a pressure drop of 1 Pa. The red and blue colours correspond to 100% and 0% lysis, respectively. Velocity vectors of the blood flow are also included. The time points have been chosen to coincide with the more interesting aspects of clot lysis. Corresponding results for a pressure drop of 10 Pa are shown in Figure 2.

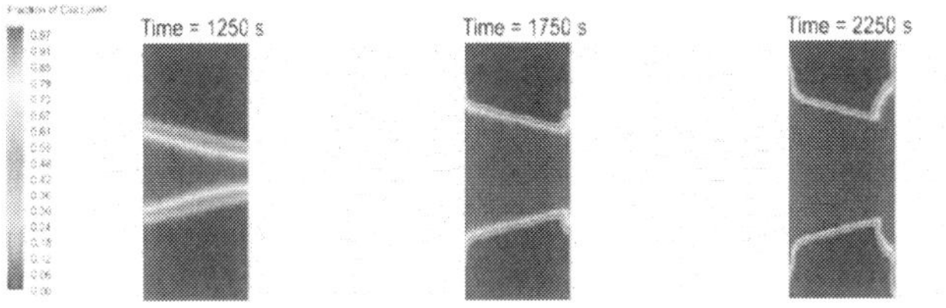

FIGURE 1: LYSIS CONTOURS FOR 1 PA

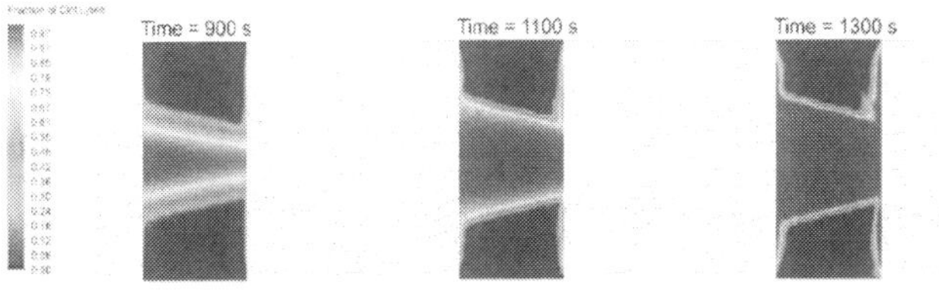

FIGURE 2: LYSIS CONTOURS FOR 10 PA

Numerical results show that a higher pressure drop allows for a higher velocity and hence more rapid transportation of drug through the clot. As expected, clot lysis is considerably faster for a larger pressure drop but the patterns of clot lysis are markedly different.

DISCUSSION

By applying a steady pressure boundary condition at the inlet, a parabolic poiseuille flow field was established throughout the domain outside the clot. As such the velocity is the highest at the centre of the channel and zero at the wall due to the no-slip conditions. The tPA drug thus reaches the centre of the clot first and lyses straight through it forming an open funnel-shaped space surrounded by two symmetric mural clots. Residual mural clots from partial lysis of a fully-occluded artery are commonly seen in fibrinolytic therapy.

The narrow channel opened up at the centre of the clot also introduces a recirculation region behind each residual mural clot where tPA and plasmin are entrapped. This promotes lysis downstream of the clot, which is more noticeable as the pressure drop increases. This can be seen by comparing figures 1 and 2 where the residual clot for the higher pressure drop seems taller and thinner when compared to its lower pressure drop counterpart.

The key strength of this model is that it can easily be applied to a variety of arbitrary 2D and 3D geometries to investigate patterns of clot lysis. The major limitation lies in describing the initial state of the clot. The porosity and permeability of the clot is likely to be heterogeneous and is not normally well known. It has been shown that large changes in these properties can severely affect the observed patterns of clot lysis [4].

REFERENCES

1. Tiefenbrunn, A.J et al., *Circ.*, 73(6): 1291-1299, 1986.
2. Diamond, S.L., S. Anand, *Biophys J*, 65(6): 2622-2643, 1993.
3. Anand, S., S.L. Diamond, *Circ.*, 94(4): 763-774, 1996.
4. Anand, S et al., *Ann Biomed Eng*, 25(6): 964-974, 1997.
5. Wootton, D.M et al., *Biotechnol Bioeng* 77(4): 405-419, 2002.
6. Bannish, B.E et al., *Math Med Biol*, 31(1): 45-64, 2014.
7. Bear, J.B., Bachmat, Y, *Introduction to Modelling of Transport Phenomena in Porous Media*, Dordrecht, The Netherlands: Kluwer Academic Publishers, 1991
8. Wufsus, A.R et al., *Biophy J.*, 104(8): 1812-1823, 2013.
9. Hoylaerts, M et al., *J. Biol. Chem*, 257(6): 2912-2919, 1982.

SB^{3}C2015-141

INFLUENCE OF NANO- AND MICRO-SCALE STRUCTURE OF ALIGNED ELECTROSPUN SCAFFOLDS ON MECHANICAL PROPERTIES AND CELL RESPONSE

Hannah M. Pauly (1), Ketul C. Popat (1,2), Daniel J. Kelly (3) Tammy L. Haut Donahue (1,2)

(1) School of Biomedical Engineering
Colorado State University
Fort Collins, CO, USA

(2) Department of Mechanical Engineering
Colorado State University
Fort Collins, CO, USA

(3) Trinity Centre for Bioengineering
Trinity College Dublin
Dublin, Ireland

INTRODUCTION

Electrospinning is a polymer nanofabrication technique that involves the ejection of charged polymer particles from a syringe to create a non-woven mat of nanofibers [1]. The nanoscale fibers produced by the electrospinning process mimic the native extracellular matrix, encouraging cell adhesion and proliferation as well as matrix deposition [2]. The electrospinning process can be modified to achieve variable nanofiber diameter as well as nanofiber alignment. Aligning electrospun nanofibers along a primary axis allows the scaffolds to better mimic the anisotropy of collagen fibers in tissues such as tendons and ligaments [3,4]. The presence of aligned nanofibers also significantly alters the mechanical properties of scaffolds, primarily by increasing the tensile modulus [3,4].

Despite the promise of aligned electrospun scaffolds for tissue engineering of tendons and ligaments, a major drawback is the inherent fragility of traditional flat electrospun sheets. Although flat sheets of electrospinning are useful for assessing the response of cells, they will likely be challenging to implement in a clinical setting. Additionally, although the mechanical properties of aligned electrospun sheets are superior to random electrospun sheets, they are still far inferior to the mechanical properties of native tendons and ligaments [2]. To overcome these challenges several research groups are investigating methods for creating more robust electrospun scaffolds including using lamination techniques and combining electrospinning with hydrogels [5]. With more complex scaffolds, encouraging cell adhesion and proliferation while simultaneously achieving robust mechanical properties can be challenging.

The goal of this work was to assess the influence of micro-scale structure and nanoscale fiber structure on the mechanical properties of aligned electrospun scaffolds as well as the ability of the scaffolds to support cell adhesion and proliferation. The impact of micro-scale scaffold architecture was determined by creating a novel cylindrical bundle of nanofibers.

METHODS

A 10% w/v solution of polycaproloactone (PCL, M_w = 80,000) was prepared in a 3:1 v/v mixture of chloroform and methanol. Oleic acid sodium salt (OLA) was added to the PCL solution at a ratio of 97:3 PCL: OLA to ensure uniform fiber diameter. The polymer solution was ejected at a rate of 2 ml/hr from a horizontally oriented glass syringe fitted with an 18-guage blunt tip needle. The collecting surface was a rotating aluminum drum placed 10 cm from the needle tip. To create a sheet of aligned electrospun fibers the aluminum drum was rotated at ~3000 rpm, corresponding to a linear velocity of ~12 m/s. Nanofibers were collected for 15 minutes, resulting in a mat approximately 20 μm thick. Aligned electrospun sheets were used to create aligned nanofiber bundles (NFB) by excising a rectangular section of the electrospun sheet and rolling it into a solid 3-dimensional cylindrical bundle. Scaffolds were imaged with scanning electron microscopy (SEM) to quantify nanofiber diameter.

Uniaxial tensile testing was performed on flat sheets and nanofiber bundles using a servo-hydraulic mechanical test system equipped with an 8.9N load cell. Samples were punched from flat sheets using a dogbone cutter with a central width of 3.175 mm and a length of 10 mm. After applying a pre-load of 0.1 N samples were strained until midsubstance failure. Four strain rates were tested: 0.1% strain/s, 1% strain/s, 5% strain/s, and 25% strain/s. All scaffolds were strained parallel to the direction of primary fiber alignment.

After sterilization with ethanol and ultraviolet light, adult human adipose derived stem cells (ADSC) at passage 4 were seeded at a density of 10,000 cells/cm^2 on the scaffolds. Cell seeded scaffolds were stained with 5-Chloromethylfluourescein Diacetate (CMFDA – live cells), rhodamine-phalloidin (actin – cytoskeleton), and 4'6-diamidino-2-phenylindole (DAPI – nucleus) after 1, 4, and 7 days of culture to assess cell adhesion and proliferation. Cell morphology and spatial organization was assessed at the same time points using scanning electron microscopy (SEM) and cell viability was measured using a commercially available CellTiter Blue assay.

RESULTS

The mean fiber diameter of the scaffolds was 292 nm ± 19 nm. The average diameter of the nanofiber bundles was 500 μm ± 95 μm. When strained to failure, scaffolds exhibited a toe region followed by a small linear elastic region and finally a large region of plastic deformation. Aligned nanofiber bundles had a significantly lower elastic modulus at a strain rate of 1%/s compared to aligned sheets, whereas at all other strain rates of interest, no difference in elastic modulus was measured (Figure 1A). At all strain rates, aligned nanofiber bundles had significantly a higher yield stress and yield strain compared to the aligned sheets (Figure 1B, 1C). Nanofiber bundles were also capable of sustaining significantly higher ultimate loads and ultimate displacements compared to flat sheets (data not shown).

For all scaffolds the CellTiter-Blue assay indicated that there were no cytotoxic effects of the scaffolds and that cell viability increased with time in culture. Live cell counts of DAPI stained nuclei from immunofluorescence images indicated significant cell proliferation (Figure 2A, 2B). SEM images showed that cell morphology was significantly impacted by the aligned nanofibers. Cells tended to elongate and preferentially align along the primary direction of nanofibers (Figure 2C, 2D). Additionally, SEM images revealed numerous cellular projections interacting with the nanofibers as well as neighboring cells.

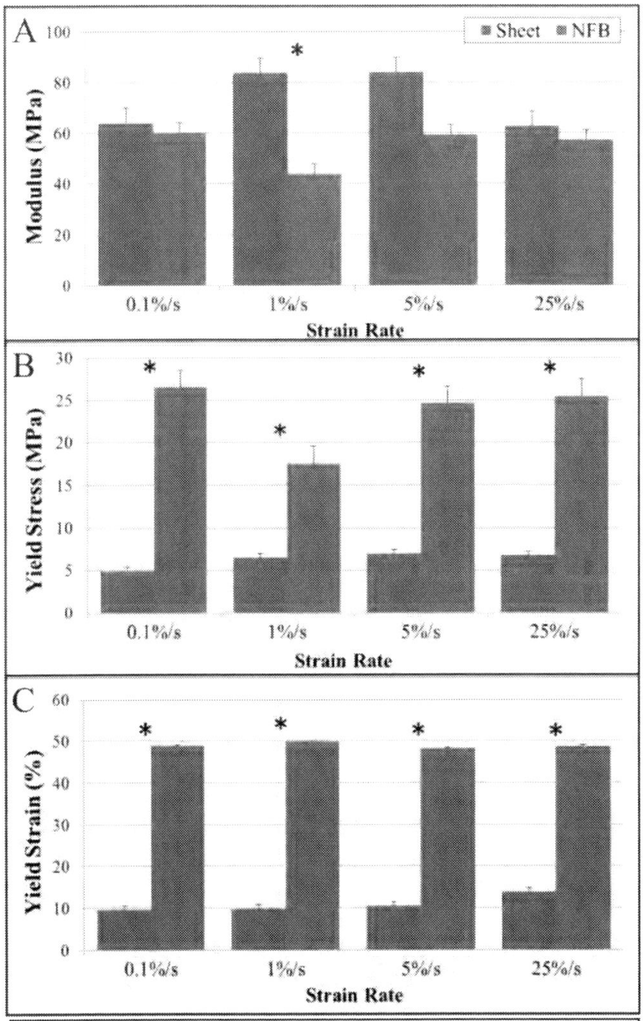

Figure 1: (A) Modulus, (B) yield stress, and (C) yield strain of electrospun scaffolds. * Indicates significantly difference between flat sheets and nanofiber bundles.

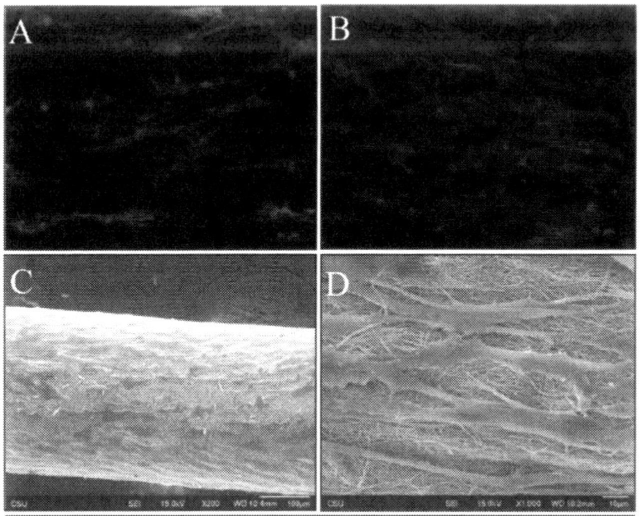

Figure 2: Fluorescence staining at day 4 (A) and day 7 (B) showing cell proliferation. SEM images of cells on nanofiber bundles on day 7 at 200x magnification (C) and 1000x magnification (D).

DISCUSSION

Using the electrospinning technique described above an aligned mat of nanofibers was created with relatively uniform fiber diameter. The flat electrospun sheets were also successfully manipulated into a micro-scale cylindrical scaffold comprised entirely of nanofibers.

The micro-scale structure of the scaffolds had a significant influence on mechanical behavior. Although they were fabricated from the same material, the cylindrical nanofiber bundles had significantly different mechanical behavior than the flat sheets (Figure 1). The altered mechanical properties may be due to shear forces within the nanofiber bundles as well as an increased number of nanofiber interactions. It is promising that the ultimate load and yield stress and strain were significantly higher for nanofiber bundles compared to flat sheets, suggesting that modifying the micro-scale structure of electrospun sheets may make them more applicable for load bearing tissue engineering applications.

The results of cell viability, adhesion, and proliferation indicated that flat sheets and nanofiber bundles both provided a favorable scaffold for the adhesion and proliferation of ADSCs. SEM images showed that cells tended to align in the primary direction of fiber alignment (Figure 2), indicating that nanofiber structure has a significant effect on cell morphology. Interestingly, previous work by our group has shown that regardless of nanofiber alignment, cells on nanofiber bundles align in the direction of the long axis of nanofiber bundles [6]. This indicates that both nanoscale and micro-scale structure of electrospun scaffolds plays an important role in dictating cellular response.

Overall these results suggest that the nanoscale and micro-scale structure of electrospun scaffolds has a significant influence on the mechanical properties of the scaffold as well as the cellular response. Future work will investigate the mechanical properties of scaffolds after extended cell culture to determine the influence of cells on mechanical properties. Additionally, combining multiple nanofiber bundles into a larger scaffold may further improve mechanical properties and increase the applicability of the scaffold for ligament or tendon repair.

ACKNOWLEDGEMENTS

This material is based upon work supported by the National Science Foundation under Grant No. DMR 1306741.

REFERENCES

[1] Pharm, Q et al., *Tissue Eng*, 12:1197-211, 2009.
[2] Lee, C et al. *Biomaterials*, 26:1261-70, 2005.
[3] Nerukar, N et al. *J Orthop Res*, 25:1018-28, 2007.
[4] Baker, B et al. *Biomaterials*, 28:1967-77, 2007.
[5] Xu, W et al. *Acta Biomater*, 6:1992-2002, 2010.
[6] Pauly, H et al. *ORS Annual Meeting*, #PS2-115, 2015.

SB^{3}C2015
Summer Biomechanics, Bioengineering and Biotransport Conference
June 17-20, 2015,Snowbird Resort, Utah, USA
SB^{3}C2015-142

CORRECTING ARRHENIUS MODELS FOR APPLICATION TO CELL DEATH PROCESSES INVOLVING INTRINSIC PROTEIN CASCADES

John A. Pearce (1)

(1) Department of Electrical & Computer Engineering
The University of Texas at Austin
Austin, TX, USA

INTRODUCTION

Irreversible Arrhenius kinetic models have frequently been applied to model cell survival curves for many decades. The form of the model can be either a prediction of surviving fraction — Eqn (1) — or a calculation of the Cummulative Equivalent Minutes at 43 C (CEM43), as is used in tumor hyperthermia studies. However, as described by Wright,[1] the simple exponential decay of the irreversible Arrhenius approach grossly over-predicts cell death in the early stages of exposures at typical hyperthermia temperatures, 43 to 50 C. Over-prediction leads to an unacceptably optimistic assessment of thermal treatment effectiveness.

$$\ln\left\{\frac{C(0)}{C(\tau)}\right\} = \int_0^\tau A \exp\left\{\frac{E_a}{RT}\right\} dt \qquad (1)$$

where: C = the initial and final fraction of live cells, A = the "frequency factor" (s^{-1}), E_a = the activation energy (J/mole), R = the gas constant (8.3143 J/mole-K), and T = temperature (K).

The prediction error derives from the inability of the classical Arrhenius model to represent the initial slowly developing "shoulder" region of the curve. This is because the Arrhenius model is derived from the constant-rate region of the survival curve, which often appears only after a significant exposure time has elapsed. The reason for the initial delay is that intrinsic cell death processes, such as apoptosis and necroptosis, involve cascades of functional protein activations and binding processes.[2] Typically, the protein sequences include a signaling block followed by a decision-making block and a subsequent execution block. The decision block comprises a variable delay (depending on the strength of the input signal) that prevents cell death when the cell might be repaired. The delay in the decision block arises from second- and third-order reactions in the dynamic state equation model of the overall reaction kinetics.[2, 3]

While the detailed kinetic models are accurate, including them in standard numerical model calculations is cumbersome and may be unnecessary. Feng *et al.* [4] introduced a two-compartment model that effectively represented their experimental measurements in PC3 cells. Their work delineated the failure of the Arrhenius model to represent their experiments at lower temperatures, but noted that it was more accurate at higher temperatures, above 54 C. We seek simpler models that can provide acceptable accuracy to represent cell survival curves. Since the primary difficulty in the Arrhenius model is the initial delay period, a model that adds a temperature-dependent time delay might prove useful and practical. This work applies a temperature-dependent time delay to model the experiments in the PC3 cells.

METHODS

The original experimental PC3 cell survival curve data were interpolated from the publication.[4] At each of the experiment temperatures that exhibited a measurable time delay — 44, 46, 48, 50, and 52C — the constant rate region was identified and the applicable reaction velocity determined (Figure 1). The exponential fit also yielded the time delay. The collected reaction velocities, k (s^{-1}) were plotted on Arrhenius axes — $\ln(k)$ vs $1/T$ — and values for E_a and A determined from the slope and intercept of the linear regression fit, respectively.

RESULTS

The required time delay, t_d, closely fits a straight line (Eqn 2) that crosses zero at T = 54.5 C. This result is in keeping with the common wisdom (and with the Feng *et al.* observation) that the Arrhenius model fits quite well at higher temperatures. Presumably, the decision-making protein block has been overwhelmed by the strength of the input signal, elevated temperature in this case.

$$t_d(s) = 2703 - 49.6T(C) \qquad r^2 = 0.953 \tag{2}$$

For the constant rate region of the PC3 experiments $E_a = 2.222 \times 10^5$ (J/mole) and $A = 6.751 \times 10^{31}$ (s^{-1}). These differ significantly from the values obtained by matching entire curves with an Arrhenius model: $E_a = 2.318 \times 10^5$ (J/mole) and $A = 1.19 \times 10^{35}$ (s^{-1}).[4] The time delay Arrhenius calculations fit the PC3 experiments exceptionally well (Figure 2).

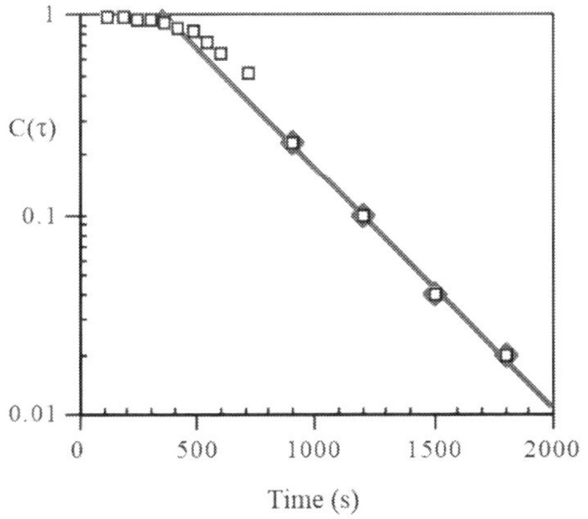

Figure 1. PC3 cell curve at 46 C (open squares) and fit to the constant rate region. The fit line (solid diamonds) has reaction velocity, $k_{46} = 2.747 \times 10^{-3}$ (s^{-1}), delay time = 358 s and $r^2 = 0.997$.

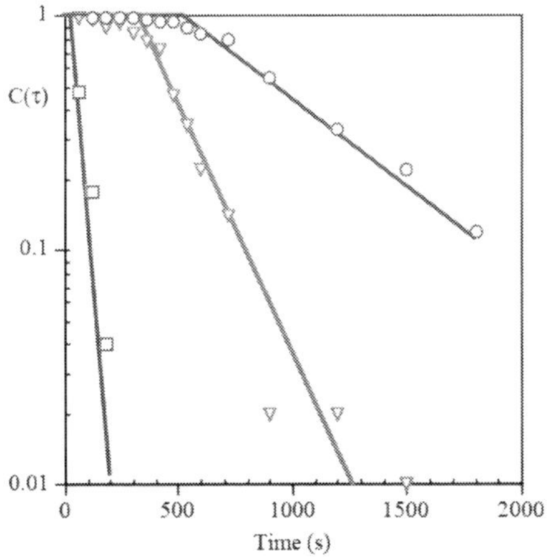

Figure 2. Fit with time delay to PC3 data at 44, 48 and 54 C.

DISCUSSION

In many, but certainly not all, cases the classical irreversible Arrhenius model for cell death is inadequate to describe the survival curve at hyperthermic temperatures. In other studies the irreversible Arrhenius model fits the measured results extremely well.[5, 6] This may be partially explained by differences in the particular assay used to determine cell death. In the case of the PC3[4] and SN12[6] experiments the measure of cell death was uptake of Propidium Iodide, which stains cells with compromised membranes. Therefore, the difference in survival curve functional form between them must be due to differences in cell type rather than death assay. In the case of the AT-1 cell experiments[5] a clonogenicity assay was used: EthD-1, Ethidium Homodimer-1. That these cells are well fit by the simple Arrhenius model is interesting because, like the classical CHO cell model, they are thermally sensitive, while the SN12 cells are quite thermally robust by comparison. These two cell types thus provide realistic limiting cases and require only the simple Arrhenius calculation.

For cases in which the survival curve exhibits a significant initial shoulder region a temperature-dependent time delay followed by standard Arrhenius calculation may provide an adequate representation with minimal complexity. To be sure, any information in the shoulder region is discarded by this approach; however, in the wide sense the shoulder region is, in fact, of little interest or consequence. We want to know when a thermal treatment begins to be effective, and do not want to over-estimate its effectiveness. The time-delay approach can be used to provide a more realistic prediction for treatment planning and assessment purposes when required by cellular dynamic response.

ACKNOWLEDGEMENTS

The author wishes to acknowledge the partial support for these studies provided by the T.L.L. Temple Foundation, and by the O'Donnell Foundation.

REFERENCES

[1] N.T. Wright, "Comparison of Models of Post-Hyperthermia Cell Survival," *Journal of Biomechanical Engineering*, vol. 135, (no. 5), 2013.

[2] T. Eissing, H. Conzelmann, E.D. Gilles, F. Allgower, E. Bullinger, and P. Scheurich, "Bistability analyses of a caspase activation model for receptor-induced apoptosis. ," *Journal of Biological Chemistry*, vol. 279, (no. 35), pp. 36892-36897, 2004.

[3] J.A. Pearce, "Comparative analysis of mathematical models of cell death and thermal damage processes," *International Journal of Hyperthermia*, vol. 29, pp. 262-280, 2013.

[4] Y. Feng, J.T. Oden, and M.N. Rylander, "A two-state cell damage model under hyperthermic conditions: Theory and in vitro experiments," *Journal of Biomechanical Engineering*, vol. 130, (no. 4), pp. 041016, 2008.

[5] P. Bhowmick, J.E. Coad, S. Bhowmick, J.L. Pryor, T. Larson, d.l.R. J., and J.C. Bischof, "In vitro assessment of the efficacy of thermal therapy in human benign prostatic hyperplasia," *International J. of Hyperthermia*, vol. 20, (no. 4), pp. 412-439, 2004.

[6] X. He and J.C. Bischof, "The Kinetics of Thermal Injury in Human Renal Carcinoma Cells," *Annals of Biomedical Engineering*, vol. 33, (no. 4), pp. 502-510, 2005.

FGF8 ESTABLISHES A CONTRACTILE GRADIENT TO DRIVE DIRECTED CELL MOVEMENTS IN THE DEVELOPING AVIAN GUT

Nandan L. Nerurkar, Cliff J. Tabin

Department of Genetics
Harvard Medical School
Boston, MA, USA

INTRODUCTION

The construction of tissues and organs during embryogenesis has been widely studied in the context of gene expression and cell signaling. What is lacking, however, is an understanding of how these cues generate and regulate the physical forces necessary to shape tissues. Here we use the developing chick endoderm as a model to study how molecular cues control tissue level forces, combining live imaging and gene misexpression with biophysical approaches and mathematical modeling. In particular, we identified that a spatial gradient of FGF signaling in the endoderm results in a gradient in cytoskeletal contraction. This contractile gradient is responsible for generating the directed cell movements that underlie formation of the hindgut.

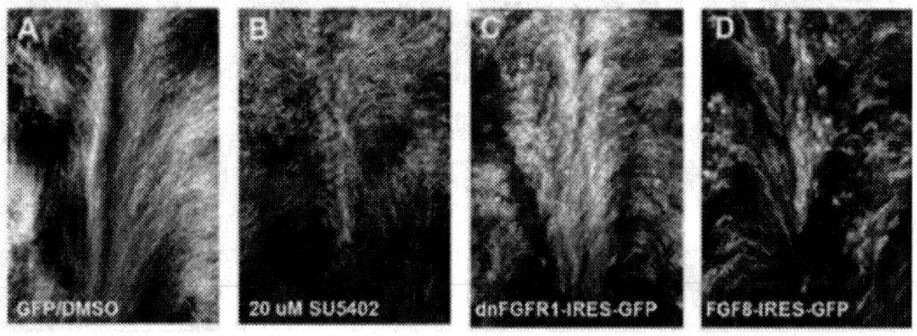

Fig. 1 Time lapse cell tracks for control (A) and inhibition (B,C) or activation (D) of FGF signaling.

METHODS

Fertilized eggs were incubated 48 hours to Hamburger Hamilton stage 12 [1]. Nuclear GFP was expressed in the endoderm by electroporation, and cell movements were observed by in vivo live imaging for 24 hours. During this time the posterior endoderm transforms from an epithelial sheet to a tube, the hindgut. Cell tracking was used to compute strain during cell movements. Relative tissue forces were measured by introducing small cuts into the endoderm and measuring their opening. FGF signaling was disrupted by culturing embryos in 20 uM SU5402, or by expression of a dominant negative (dn) receptor FGFR1. FGF8 was misexpressed by electroporation as well. Both constructs were expressed from the pCAAGIG vector.

To study contractile gradients as a mode for collective cell movements, the endoderm was modeled as a 1D viscoelastic Maxwell solid, whose movements are opposed by a basement membrane of stiffness k_{ecm}. The following constitutive law was assumed:

$$\sigma = \sigma_e + \sigma_v + \sigma_a = E\frac{\partial u}{\partial x} + \eta\frac{\partial^2 u}{\partial t \partial x} + \alpha[FGF8] \qquad (1)$$

where σ is stress, and the subscripts *e*, *v*, and *a* denote elastic, viscous, and active/contractile components of the stress, E = modulus, η = viscosity, and α is the constant relating FGF8 concentration with stress generation. When (1) is substituted into the equilibrium equation the result is a mixed third order partial differential equation (PDE) relating displacement *u* to time and position [2]:

$$E\frac{\partial^2 u}{\partial x2} + \eta\frac{\partial^3 u}{\partial x^2 \partial t} + \alpha\frac{\partial}{\partial x}[FGF8] - k_{ecm}u = 0 \qquad (2)$$

with boundary conditions $u(0,t) = u(L,t) = 0$ and initial condition $u(x,0) = 0$. This PDE was solved using Fast Fourier Transforms [3], yielding the exact solution:

$$\bar{u}(\bar{x},\bar{t}) = \sum_{n=1}^{\infty} \frac{n\pi\Lambda\iota(e^{-\frac{1}{\iota}}(-1)n-1)}{((n\pi)2+\kappa)(1+(n\pi\iota)2)}\left[1 - e^{-\frac{(n\pi)2+\kappa}{(n\pi)2}\bar{t}}\right]\sin n\pi\bar{x} \quad (3)$$

where the overbar indicates nondimensionalized variables, and the three resulting dimensionless constants are given by

$$\Lambda = \frac{\alpha C_o}{E};\ \iota = \frac{\sqrt{D/R_d}}{L};\ \kappa = \frac{k_{ecm}L^2}{E} \qquad (4a,b,c)$$

where C_o is the maximum [FGF8], and D and R_d are FGF8 diffusivity and degradation rate.

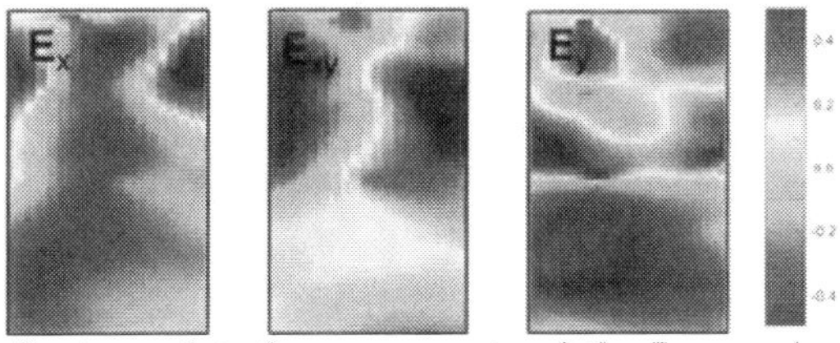

Fig. 2 Contour plots for components of the Lagrangian strain, where x and y are the medio-lateral and antero-posterior directions, respectively.

RESULTS

Live imaging revealed a polarized flow of cells from anterior to posterior along the embryonic midline (Fig. 1A). Along the direction of cell movement (y-axis), posterior endoderm (leading cells) demonstrated compressive strain, compared to tensile strain in the anterior (trailing) cell population (Fig. 2). A tension gradient was observed from posterior to anterior (Fig. 3A), and was opposed by a cell size gradient (Fig. 3B). This gradient was lost upon inhibition of cytoskeletal contraction. The coincidence of tensile stresses with compressive strains suggests a mechanism of cell movement by which posterior cells contract to pull anterior cells toward them.

FGF8 is expressed in a posterior to anterior gradient [4].Inhibition (Fig. 1B,C) and activation (Fig.1D) of the FGF pathway similarly resulted in a loss of cell movements and failure to form hindgut. Activation of FGF signaling resulted in an increase in anterior tension and reduced cell size, while inhibition resulted in decreased tension (Fig. 3C,D). These findings suggest that the FGF8 signaling gradient establishes a contractile gradient to drive cell movements.

The continuum model for contraction in response to a diffusible activating signal was consistent with experimental observations, including directed and efficient cell movements, neighboring zones of compressive and tensile strains (Fig. 4A-C), and a tensional gradient (Fig. 4D). These qualitative behaviors were robust, as parameters could be altered by several orders of magnitude without loss of directed movements. The contractile gradient works most efficiently when the characteristic length of the transport problem ($\sqrt{D/R_d}$) is small relative to the length separating the fixed ends (L). As diffusivity of FGF8 increases, or degradation decreases, the contraction becomes progressively homogeneous, and directional cell movements are lost.

DISCUSSION

The present study revealed a novel mode of collective cell movements in the developing chick embryo, whereby isotropic, yet heterogeneous cytoskeletal contractions generate anisotropic cell movements. Although FGF8 has been widely implicated in morphogenesis of several organ systems[5], the precise mechanism by which FGF signaling elicits a physical transformation at the tissue level is unknown. Based on our current findings, one appealing hypothesis is that FGF8 modulates cytoskeletal contractility in many contexts, and the resulting multicellular response is simply a function of the boundary constraints and cell-cell contacts.

The model successfully captured key experimental observations, further supporting the contractile gradient hypothesis. The ability to enact directional movement across a wide range of parameter values was due in large part to the positive feedback loop that arises, as cells that are passively displaced up the FGF8 gradient become active, increasing tissue tension and cell movements. As such even small contractile forces are sufficient to create large cellular flows.

The present work represents one of the few examples of how molecular cues organize mechanical forces to drive morphogenesis. The study also sheds new light on how the gut tube, an essential embryonic structure that forms the respiratory and gastrointestinal tracts, is formed.

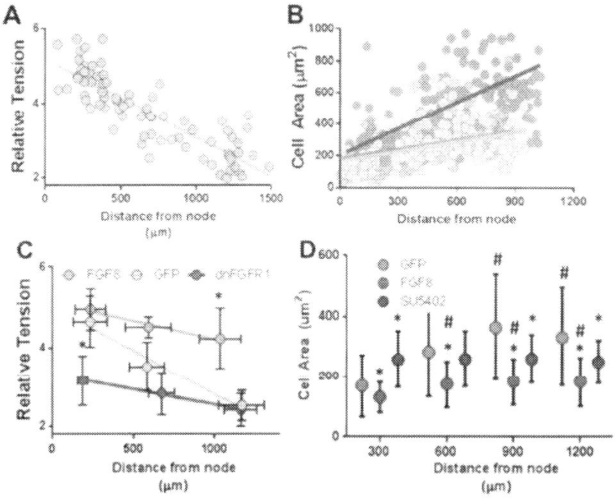

Fig. 3 A) Relative tension in the endoderm decreases from posterior to anterior. B) Cell size shown for three embryos reveals a cell size gradient that opposes the tensional gradient. C) Activation by FGF8 or inhibition by dnFGFR1 reduce the tensional gradient by elevating anterior tension and decreasing posterior tension, respectively. D) Cell size decreases upon FGF8 overexpression, while inhibition of FGFs causes a loss of cell size gradient. *=p <0.05 vs. GFP; # = p<0.05 vs. SU5402.

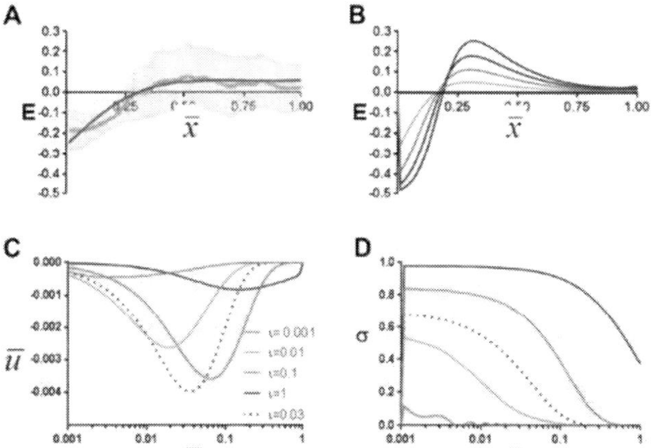

Fig. 4 A) At 8h, the model (red) closely approximates experimentally measured strain distributions (green). B) Strain as a function of position at 2, 4, 6, and 8h using average parameter values from literature. Parametric variation of ι strongly alters displacement (C) and stress (D) distributions. Note loss of convergence for very low ι.

ACKNOWLEDGEMENTS

This work was funded by the NICHD (F32HD069074)

REFERENCES

[1] Hamberger+ *J Morph* 1951; [2] Taber *Acta Biotheor.* 2000; [3] Deen *Analys. Trans. Phen* 1998; [4] Dubrulle+ *Nature* 2003; [5] Dorey+ *Dev* 2013;

A POROELASTIC HIGH FIDELITY FINITE ELEMENT MODEL OF THE OSTEOCHONDRAL UNIT TO EVALUATE CHANGES IN PERMEABILITY WITH OSTEOARTHRITIS

Michael E. Stender (1), Richard A. Regueiro (2), Virginia L. Ferguson (1)

(1) Department of Mechanical Engineering
University of Colorado Boulder
Boulder, CO, USA

(2) Department of Civil, Environmental, and Architectural Engineering
University of Colorado Boulder
Boulder, CO, USA

INTRODUCTION

The osteochondral unit (OU) is defined as the functional combination of articular cartilage (AC), calcified cartilage (CC), subchondral cortical bone (SCB), and subchondral trabecular bone (STB). *In vivo*, all tissues of the OU are saturated with fluid and thus, exhibit a poroelastic response to mechanical loading that plays a critical role in the loading response of the OU [1,2]. Additionally, the poroelastic properties of OU tissues have been shown to vary greatly with the onset and progression of osteoarthritis (OA). With OA, hydraulic permeabilities increase for all tissues in the OU [1,2]. To improve upon the current understanding of the function of the OU in health and with OA, this study seeks to combine appropriate constitutive material modeling, finite element modeling, and published experimental data with the goal of elucidating the consequences of changes in permeability in tissues of the OU as observed experimentally with OA.

METHODS

A 3-D poroelastic finite element model of a quarter symmetry OU plug in unconfined compression (UCC) was developed in Abaqus (v6.13, Simulia). A depth-dependent anisotropic model of AC [3] was used including discrete superficial zone (SZ), middle zone (MZ), and deep zones (DZ) of AC was implemented with anisotropic collagen fibril distributions from [4]. AC mechanical constants in the SZ and MZ were used from [3] with DZ properties calculated as a linear extrapolation of SZ and MZ values. All tissues of the OU were assumed to be deformable, permeable, and initially saturated with fluid. All discrete tissue regions were joined using a tied contact constraint and a contact permeability set high enough such that permeability between regions had no effect on results. Due to the relatively large deformations experienced by AC, a strain-dependent permeability was implemented for the SZ, MZ, and DZ of AC. The following relationship as used by [5] was implemented to define the deformation-dependent permeability in AC

$$\hat{k} = \hat{k}_0 \left[\frac{e}{e_0}\right]^2 exp\left(\frac{M}{2}\left(\left(\frac{1+e}{1+e_0}\right)^2 - 1\right)\right) \quad (1)$$

where $\hat{k}$ is the permeability in the current configuration $\left(\frac{mm^2}{Pa{\cdot}s}\right)$, $\hat{k}_0$ is the permeability at zero strain $\left(\frac{mm^2}{Pa{\cdot}s}\right)$, e and e_0 are the void ratios of the tissue in the current configuration and at zero strain, respectively, and M is a dimensionless material parameter controlling the nonlinearity of the permeability dependence [6]. Permeability values were taken from two published, experimental studies across various tissue sources for healthy (Normal) and osteoarthritic (All OA) tissues as listed in Table 1 [1,2].

Table 1. Properties as simulated of the tissues of the healthy (Normal) and osteoarthritic (All OA) osteochondral unit [1,2].

Tissue	$\hat{k}_0$ Normal $\left(\frac{mm^2}{Pa{\cdot}s}\right)$	$\hat{k}_0$ All OA $\left(\frac{mm^2}{Pa \cdot s}\right)$	e_0	M	Thickness (cm)
AC SZ	4.55e-3	6.81e-3	4.5	5.48	0.15
AC MZ	1.46e-3	2.16e-3	4.0	5.49	0.50
AC DZ	0.50e-3	0.74e-3	3.2	7.38	0.35
CC	90.0	140	1.1	-	0.05
SCB	90.0	140	1.1	-	1.0
STB	90.0	140	3.7	-	1.0

CC, SCB, and STB were assumed to be linear isotropic elastic with Young's moduli of 320, 3900, and 1091 MPa respectively. CC, SCB, and STB Poisson's ratios were all assumed to be 0.3. Boundary conditions were applied as appropriate to simulate UCC of an OU plug. UCC loading was applied *via* an impermeable platen which was ramped to a 0.1 mm compressive displacement over the first 400 seconds, and subsequently held at a constant displacement for an additional 800 seconds.

RESULTS

UCC simulations demonstrated differences in peak load and relaxation between Normal and All OA simulations (Figure 1). The nonlinear loading response in UCC was a result of material nonlinearity, anisotropy, nonlinear permeability, and pore fluid diffusion.

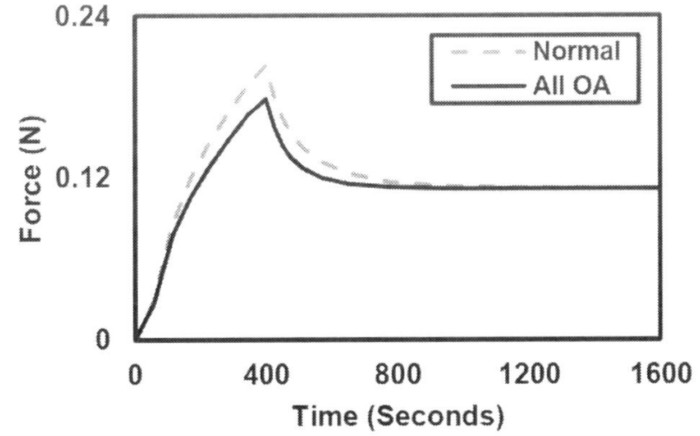

Figure 1. Plot of force (N) vs. time (seconds) for unconfined compression of osteochondral plugs for healthy (Normal) and, osteoarthritic permeabilities (All OA).

Pore pressure magnitudes for the UCC model were higher in Normal simulations, compared with All OA simulations (Figure 2). Fluid velocities following peak loading were similar in both Normal and All OA UCC simulations (Figure 3).

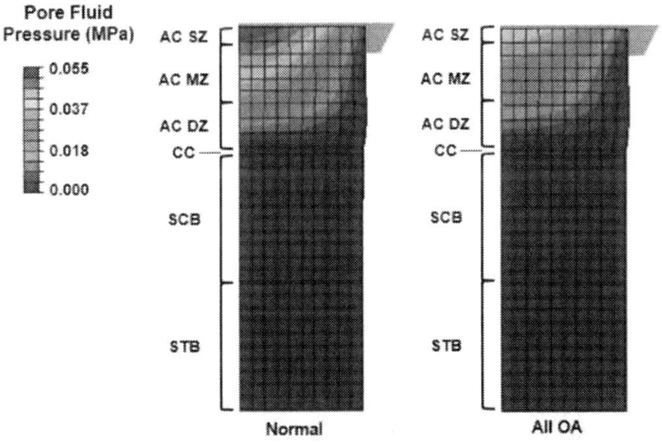

Figure 2. Contour plots showing finite element simulation results for pore fluid pressure (MPa) at peak compressive loading prior to relaxation for osteochondral plugs with healthy (Normal) and osteoarthritic (All OA) permeability values. Brackets denote differential material regions including superficial, middle, and deep zone articular cartilage (AC SZ, AC MZ, and AC DZ), calcified cartilage (CC), subchondral trabecular bone (SCB), and subchondral trabecular bone (STB).

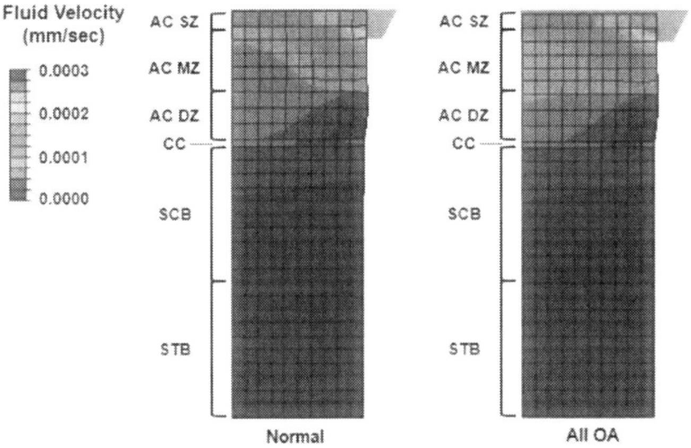

Figure 3. Contour plots showing finite element simulation results for fluid velocity (mm/sec) at peak compressive loading prior to relaxation for osteochondral plugs with healthy (Normal) and osteoarthritic (All OA) permeability values. Brackets denote tissue regions as in Figure 2.

DISCUSSION

This work presents an approximation of the behavior of the OU as a complete poromechanical unit, and the corresponding consequences of alterations in permeability in tissues of the OU as observed experimentally with OA. To precisely determine the specific effects of evolutions of tissue permeability in the OU, changes to the solid mechanical properties of OU tissues with OA such as cartilage damage/wear and bone remodeling were not considered. As a consequence of permeability increase, the OU demonstrated a diminished load supporting capacity (Figure 1) as a result of a decreased pore fluid pressure (Figure 2). Fluid pressurization was primarily present in the SZ, MZ, and DZ of AC. The decreased pore fluid pressure in All OA simulations compared with Normal simulations may lead to a progressive worsening of OA symptoms as additional solid matrix deformations are required to support normal short term loading.

This model is intended to represent a general example of a poromechanical OU rather than an OU from a specific source as parameters and assumptions were obtained from a variety of previously published sources [1,2,3,4]. Future studies that measure, more precisely, the depth-dependent changes in the permeability of AC with OA may improve the accuracy of these results. Presently, these results help to elucidate the role that permeability plays in the OU as well as to guide future experimental and computational studies aimed at preventing and/or alleviating the symptoms of OA.

ACKNOWLEDGEMENTS

This work was supported by NSF *CAREER* award #1055989, and the Innovative Grant Program at the University of Colorado Boulder.

REFERENCES

[1] Hwang, J. *et al.* (2008). *Arthritis & Rheumatism*, *58*(12), 3831–3842

[2] Setton, L. A. *et al.* (1995). *Journal of Orthopaedic Research*, *13*(4), 473-482.

[3] Stender, M. E. *et al.* (2013). *Biomechanics and modeling in mechanobiology*, *12*(6), 1073-1088.

[4] Aspden, R. M., & Hukins, D. W. L. (1981). *Proceedings of the Royal Society of London. Series B, Biological Sciences*, 299-304.

[5] Ficklin, T. P. *et al.* (2009). *Journal of biomechanical engineering*, *131*(4), 041008.

[6] Holmes, M. H., & Mow, V. C. (1990). *Journal of Biomechanics*, *23*(11), 1145–1156.

SB^{3}C2015
Summer Biomechanics, Bioengineering and Biotransport Conference
June 17-20, 2015,Snowbird Resort, Utah, USA
SB^{3}C2015-146

SURGICAL ACCURACY OF TRADITIONAL HUMERAL HEAD OSTEOTOMY IN SHOULDER ARTHROPLASTY

Thomas Suter (1,3), Christopher W. Kolz (1,2), Sean T. Tagge (1), Robert Z. Tashjian (1), Ariane Gerber Popp (3), Heath B. Henninger (1,2)

(1) Orthopaedic Research Laboratory
University of Utah
Salt Lake City, UT, USA

(2) Department of Bioengineering
University of Utah
Salt Lake City, UT, USA

(3) Clinic of Orthopaedic Surgery
Kantonsspital Baselland
Liestal, Switzerland

INTRODUCTION

Shoulder arthroplasty (SA) is an effective treatment for reducing pain and restoring function in patients with end-stage shoulder arthritis [1,2]. Short-term pain relief, restoration of function, and implant survival are highly satisfactory, but long-term outcomes can be greatly improved. Reports suggest that component malpositioning might play an important role in glenoid failure in the long-term timespan [3,4] and that surgical experience and intra-operative assessment are required to accurately orient and implant the prosthesis [5,6].

A critical step in SA is to locate anatomic landmarks that define the osteotomy plane on the humeral head. Modern prosthetic systems are based on an osteotomy along the anterosuperior anatomic neck, and assume that the resected segment corresponds to a spherical segment oriented similarly to the native anatomic neck [7-10]. Recent studies use computational models to determine the best-fit arthroplasty configuration to replicate the humeral anatomy [11,12]. These models neglect the importance of surgical skill and intra-operative assessment. The surgeon's role in implementing these optimized resections is unclear. To date, no laboratory assessments quantify the accuracy of surgically implementing the humeral osteotomy in SA as compared to an optimized resection plane from computational models.

Therefore, the purpose of this study was to quantify the surgical accuracy of performing an anterosuperior oriented osteotomy in cadaver shoulders. A virtually performed osteotomy (VPOT) from a 3D computational model was compared to a surgically performed osteotomy (SPOT) on the same cadaver shoulder. Comparisons of orientation of the osteotomy plane and geometry of the resected head were made. We hypothesized that the orientation of the osteotomy plane and geometry of the resected head would not differ between VPOT and SPOT, or the native anatomy.

METHODS

Eight fresh-frozen upper extremities (4 M, 4 F; Age: 60 ± 5 years) with no history of shoulder pathology were used for this study. Board-certified orthopaedic surgeons performed a deltopectoral approach to expose the humeral head in order to mimic a clinical anterosuperior osteotomy. Marker clusters for defining local coordinate systems were affixed to the anterolateral aspect of the proximal humerus using two nylon screws. The anatomic neck was divided into 12 discrete points representing a clock face with 12 o'clock corresponding to the most superior point of the anatomic neck.

Axial CT images were obtained before and after the osteotomy and segmented from the CT slices using Amira (v5.4, Visage Imaging, San Diego, CA), and 3D reconstructions were generated. All digitized data were referenced back to the CT reconstruction via a Matlab (MathWorks, Natick, MA) algorithm that transformed data between laboratory and CT coordinate systems using the marker clusters. The VPOT was defined as the anterosuperior anatomic neck from 12 to 6 o'clock [13,14]. To provide a fixed reference for a Microscribe G2x digitizer (±0.23 mm accuracy, Solution Technologies, Oella, MD), specimens were rigidly mounted on a custom jig fixing both the humerus and scapula. The humeral head cartilage, the circumference of the anatomic neck, and each bead on the marker clusters were digitized. To access the anatomic landmarks, the deltoid was removed from its origin on the scapula. The SPOT was then made freehand by an orthopaedic surgeon along the anterosuperior neck aiming to hit the digitized points that defined the VPOT.

Inclination and retroversion angles were determined based on the humeral coordinate systems (Fig. 1). Additionally, the resected heads were digitized, imaged, and compared to a best-fit sphere for both the VPOT and SPOT. Humeral head length, width, height, width-length ratio, and RMS error between the partial sphere and digitized points were calculated.

By comparing the outcome parameters of the VPOT and SPOT, the surgical accuracy in performing the desired osteotomy was quantified. All statistical comparisons were conducted with paired *t* tests for each parameter where $P \leq 0.05$ was considered significant. All the analyses were performed using SPSS 17.0 software (IBM Corp., Armonk, NY).

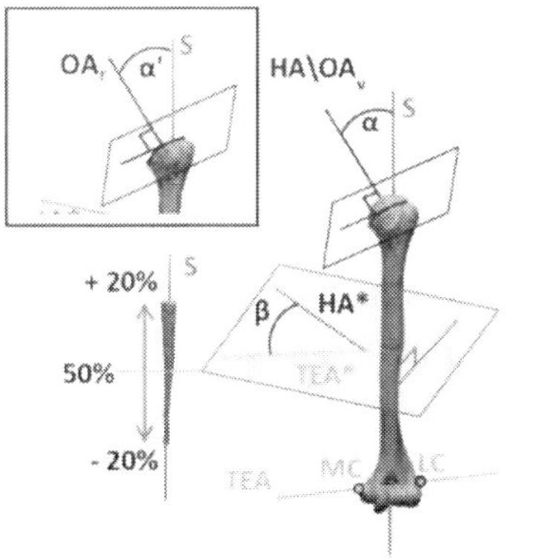

Figure 1. 3D reconstruction of the left humerus (anterolateral view). Inclination (α) and retroversion (β) angles were determined by the following parameters as calculated using Matlab: Humeral shaft axis (S), transepicondylar axis (TEA), humeral head axis (HA), and the real/virtual osteotomy axis ($OA_{r/v}$). Inclination was calculated as the 3D angle between axes S and HA for the native anatomy and between axes S and OA_v for the osteotomy. Retroversion was the 2D projection of the angle between TEA and HA/OA_v on a plane perpendicular to S (middle). To assess the orientation of the SPOT, the OA_r was calculated as the line perpendicular to the best fit plane of the digitized points along the performed osteotomy. The inclination angle (α') and retroversion angle of the osteotomy plane (β') were calculating using OA_r in place of OA_v (inset).

RESULTS

No significant differences were found in inclination and retroversion between the VPOT and SPOT (Tables 1, 2). The VPOT significantly increased the mean head height and RMS error when compared to the SPOT (Tables 1, 3). In contrast, no differences between the VPOT and SPOT were found for the mean head width, length, and length-width ratio. The VPOT and SPOT differed in inclination and retroversion compared to the orientation of the native humeral head (Table 2). Additionally, the VPOT resulted in significant changes to the head height and the RMS error when compared with the native humeral head, but no changes were found for any other metrics (Table 3). No differences were detected for any parameters between the SPOT and the native head geometry.

No specimens showed any damage to the supraspinatus, infraspinatus, or teres minor tendons after the osteotomies.

Table 1. Absolute differences between VPOT and SPOT

Characteristics	Absolute Difference		Significance†
	Mean	SD	
Inclination (°)	3	3	0.561
Retroversion (°)	6	5	0.112
Head height (mm)	3	2	***0.002***
Head width (mm)	1	1	0.528
Head length (mm)	1	1	0.180
Width-length ratio	0.0	0.0	0.175
RMS error‡ (mm)	0.7	0.6	***0.009***

† Paired t-test between VPOT and SPOT; ‡ RMS error: between cartilage and virtual sphere surfaces

Table 2. VPOT and SPOT orientation compared to the native anatomy

	Native		**VPOT**			**SPOT**		
Units (°)	Mean	SD	Mean	SD	Sig.*	Mean	SD	Sig.*
Inclination	137	3	131	3	***<0.001***	132	5	***0.002***
Retroversion	35	14	50	15	***<0.001***	46	16	***0.001***

* Paired t-test compared to native anatomy.

Table 3. VPOT and SPOT geometry compared to the native anatomy

	Native		**VPOT**			**SPOT**		
	Mean	SD	Mean	SD	Sig.*	Mean	SD	Sig.*
Head height (mm)	16	2	20	2	***<0.001***	16	2	0.732
Head width (mm)	42	4	42	3	0.351	42	4	0.685
Head length (mm)	49	4	49	4	0.351	48	4	0.080
W-L ratio^	0.9	0.0	0.9	0.0	0.951	0.9	0.0	0.168
RMS error‡ (mm)	1.1	0.4	1.7	0.3	***0.034***	1.1	0.2	0.926

* Paired t-test compared to native anatomy; ^ W-L: Length-width ratio
‡ RMS error: between cartilage and virtual sphere surfaces

DISCUSSION

This study is the first to show surgical accuracy in performing a common osteotomy technique in SA. No differences were found for inclination or retroversion when VPOT was compared to SPOT, and no specimen showed damage to the posterosuperior rotator cuff insertions after the osteotomy. This demonstrates that the osteotomy of the anatomic neck during SA represents a safe and accurate procedure when compared to the orientation of an idealized plane of resection. In contrast, significant differences were found for head height and RMS error when the VPOT and SPOT were compared. This indicates that computational models must be implemented with caution in daily surgical practice, and that a slightly altered orientation of the osteotomy influences geometry metrics of the resected humeral head, impacting the selection of the correct humeral head implant size. Finally, the present study found significant differences between the native humeral head orientation and both the VPOT and SPOT. This is one possible reason for altered biomechanics and component failure in SA, where the native anatomy is not directly replicated by the commonly performed osteotomy. This motivates future studies to investigate the biomechanical consequences due to these aberrations of the prosthetic head orientation, and investigate alternative resection planes that may more accurately reproduce native humeral anatomy.

ACKNOWLEDGEMENTS

Funding from the LS Peery Foundation Discovery Program in Musculoskeletal Restoration (U of U Orthopaedics) and from the Swiss Orthopaedic Society is gratefully acknowledged.

REFERENCES

[1] Gartsman GM, *JBJS Am*, 82:26-34, 2000. [2] Norros TR, *JSES*, 11:130-5, 2002. [3] Papadonikolakis A, *JBJS Am*, 95:2205-12, 2013. [4] Spencer EE, *JBJS Am*, 87:808-14, 2005. [5] Boileau P, *JBJS Br*, 79:857-65, 1997. [6] Fenlin JM, *Semin Arthroplasty*, 1:102-11, 1990. [7] Surgical Technique Arthrex UITSS. 2012. [8] Surgical Technique Biomet B-MCSS. 2008. [9] Surgical Technique DePuy GA. 2010. [10] Surgical Technique Zimmer ASS. 2011. [11] Pearl ML, *JOR*, 20:192-7, 2002. [12] Youderian AR, *JSES*, 23:955-63, 2014. [13] Gerber A, *ORS Meeting*, [poster], 2003. [14] Harrold F, *JSES*, 22:115-21, 2013.

PHASE-CONTRAST MICRO-TOMOGRAPHY MEASUREMENTS OF INTRAOCULAR PRESSURE-INDUCED DEFORMATION OF THE PORCINE LAMINA CRIBROSA

Baptiste Coudrillier (1), Diogo M. Geraldes (2), Nghia Vo (3), Ian C. Campbell (1, 6), Julie Albon (4), Richard L. Abel (5), C. Ross Ethier (1, 6)

(1) Wallace H. Coulter Department of Biomedical Engineering, Georgia Institute of Technology and Emory University, Atlanta, GA

(2) Biomechanics Group and (5) Surgery, Imperial College, London, United Kingdom

(3) Diamond Light Source, Didcot, United Kingdom

(4) Cardiff Centre for Vision Science, School of Optometry and Vision Sciences, Cardiff University, Cardiff, Wales, United Kingdom.

(6) Atlanta VA Medical Center, Decatur, GA.

INTRODUCTION

Glaucoma is the second leading cause of blindness worldwide [1]. Loss of vision in the disease is caused by the progressive death of retinal ganglion cell (RGC) axons, occurring primarily within the lamina cribrosa (LC). Biomechanics plays an important role in glaucoma: elevated intraocular pressure (IOP) is the main risk factor for developing glaucoma, and drugs or surgical procedures to reduce IOP are the only treatments currently effective at slowing the progression of the disease. Because the LC provides structural support to RGC axons, LC biomechanics is believed to be centrally involved in RGC death in glaucoma [2]. However, the anatomical complexity, small size, and relative inaccessibility of the LC made high resolution measurements of LC tissue deformations impossible until recent imaging advances, e.g. optical coherence tomography [3] and multiphoton microscopy [4]. Unfortunately, the limited light penetration into the LC prevents optical methods from being able to generate 3D volume images of the entire tissue. Compared with light-based methods, X-ray imaging methods such as micro-computed tomography (μCT) offer superior tissue penetration and similar spatial resolution. In conventional μCT, contrast arises from variations in X-ray attenuation by the different structures within the tissue. The connective and neural tissues of the LC have similarly weak X-ray attenuation properties, which makes the tissue almost impossible to differentiate using X-ray absorption. To overcome this limitation, phase-contrast techniques can indirectly measure the phase-shift that X-rays undergo when traversing a specimen. In soft tissues, X-ray phase-shift is typically orders of magnitude larger than X-ray absorption and therefore, phase-contrast X-ray computed tomography (PC μCT) represents a promising alternative to noninvasively image the full thickness of the LC at high resolution. In this study, we investigate the use of an inflation test and synchrotron radiation PC μCT imaging to measure the IOP-induced deformation of the porcine LC.

METHODS

Three porcine eyes were harvested from animals aged approximately 9 months (W.T. Maddock, Wales, UK) and stored at -4°C. Before testing, the eyes were slowly thawed, cleaned of extra-orbital fat and muscle, and the cornea and internal ocular tissues were excised. The ocular shell was mounted on a 3D-printed specimen holder by gluing the limbus to a mounting apparatus, so that the optic nerve was vertically centered (Fig. 1A). The specimen and holder were placed in a custom-made acrylic pressure chamber filled with PBS (Fig. 1B). Pressure was hydrostatically controlled by adjusting the height of a PBS reservoir connected to the base of the chamber. The pressure chamber was immersed in a PBS bath and the inflation apparatus was placed on a high-precision rotating stage for PC μCT interrogation.

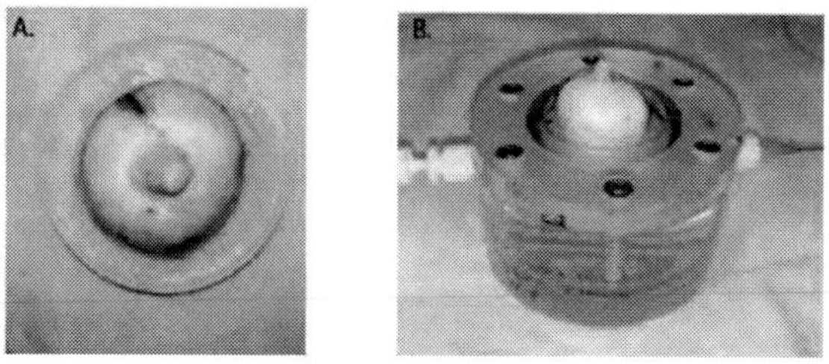

Figure 1: Specimen glued to the holder (A) and placed on the pressure chamber (B).

PC μCT image series were collected at Beamline I12-JEEP, Diamond Light Source (53 keV, 1200 projections, 1.2 s exposure time, 2.56 μm isotropic resolution, 1 m sample-detector distance) at pressure elevations of 6, 15, 22, and 30 mmHg. The phase retrieval method was based on a free propagation technique and implemented using the algorithm of Paganin et al. [5].

Frangi's vesselness filter image processing was applied to the reconstructed images to segment the beams of the LC in 3D [6]. IOP-induced changes in LC shape and thickness, anterior LC stretch, and scleral thickness were estimated by manually delineating the LC and the sclera in the central sagittal section of the scans. Full-field deformation of the LC was calculated using digital volume correlation (DVC) on the image series (DVC, LaVision).

RESULTS

Fig. 2 (left) shows a transverse PC μCT slice through a porcine LC. The location of the transverse section is indicated with a black line in the sagittal view at right. It was possible to visually identify the anterior and posterior surface of the LC. Fig. 3 maps the probability of a pixel being within a beam as calculated by Frangi's vesselness filter, and the resulting 3D rendering of the LC.

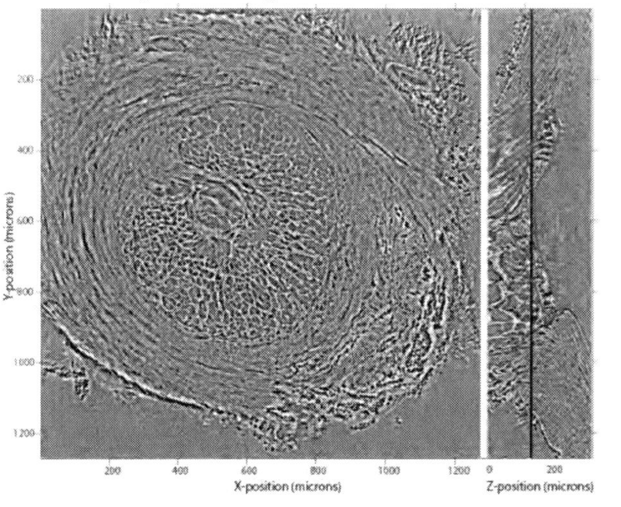

Figure 2: Left: Transverse section of a PC μCT scan of a porcine LC at the baseline pressure of 6 mmHg. Right: Sagittal section through the center of the LC. The black line indicates the anterior-posterior location of the transverse slice. The LC is delineated in red. This view was used to measure the average LC and scleral deformation shown in Fig. 4. V represents the location of the ventral pole (inferior).

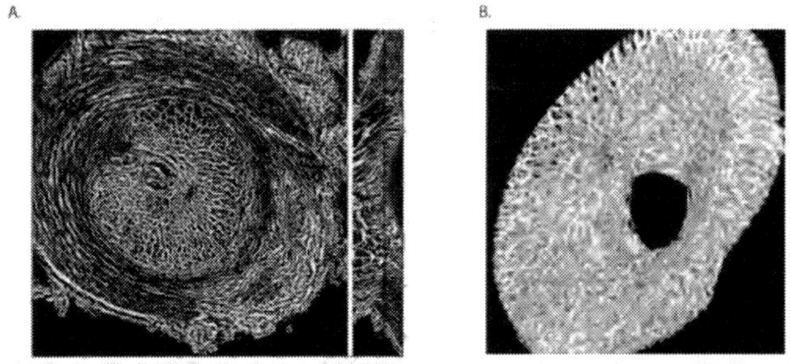

Figure 3: Frangi's vesselness filter was applied to the 3D scan from Fig. 2. (A) shows the transverse and sagittal cross-sections from Figure 2. Each voxel's greyscale value represents the probability that the voxel is a member of a beam. (B) The beams were segmented to generate a 3D rendering of the LC.

Matching sagittal sections were manually delineated at the four IOP levels (Fig. 4A) to estimate anterior LC surface stretch (Fig. 4B), LC thickness (Fig. 4C), and scleral thickness (Fig. 4D). At 30 mmHg, average porcine LC tensile and compressive strains were 6.5% and 21% respectively. Average scleral compressive strain was 7.95%.

Fig. 5 shows a map of the first principal strain within a transverse section of the LC calculated using DVC. Using scans with synthetic deformations applied, the uncertainty in strain calculation was estimated to be 0.05 %. Mean first principal strain in the porcine LC due to IOP increasing from 6 to 15 mmHg was 3.3% (range: 1.2-6.9%).

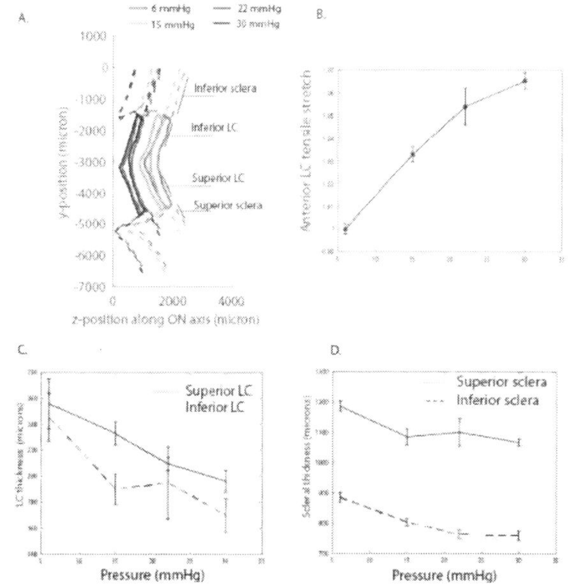

Figure 4: (A) Delineation of the sclera and LC at 4 IOP levels. (B) Anterior LC stretch, defined as the stretch of the line delineating the anterior LC, versus IOP. (C) LC thickness versus IOP. (D) Scleral thickness versus IOP. Plotted is the mean and standard deviation of the thickness over the regions shown in (A). Error bars represent the standard deviation from 3 delineations of a single LC.

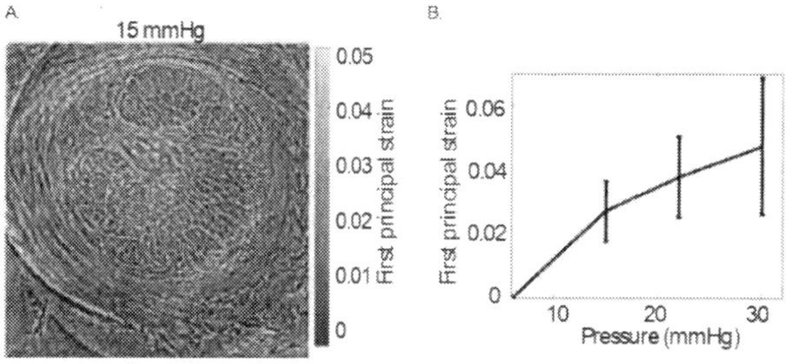

Figure 5: (A) Map of the first principal strain at 15 mmHg. (B) Average first principal strain versus IOP.

DISCUSSION

PC μCT is a powerful tool to measure 3D IOP-induced changes in shape and microarchitecture of the porcine LC. Segmentation methods and algorithms to compute IOP-induced deformations from the 3D image datasets were tested on three porcine LCs and are currently being applied to human LCs.

ACKNOWLEDGEMENTS

This research was funded by Georgia Research Alliance and the UK Diamond Light Source synchrotron

REFERENCES

[1] Quigley HA et al., *Br J Ophthalmol,* 90:262-7, 2006. **[2]** Burgoyne CF, et al., *Prog Retin Eye Res,* 24:39-73, 2005. **[3]** Girard MJ, et al., *J R Soc Interface,*10-20130459, 2013. **[4]** Sigal IA, et al., *Invest Ophthalmol Vis Sci,* 55:1-15, 2014. **[5]** Paganin et al, *Journal of Microscopy,* 206:33-40, 2002. **[6]** Campbell, IC et al., *J. R. Soc. Interface,* 2015.

REGIONAL MECHANICAL PROPERTIES OF THE LONG HEAD OF THE BICEPS TENDON

Christopher W. Kolz (1,2), Thomas Suter (1,3), Heath B. Henninger (1,2)

(1) Department of Orthopaedics
University of Utah
Salt Lake City, UT, USA

(2) Department of Bioengineering
University of Utah
Salt Lake City, UT, USA

(3) Clinic of Orthopaedic Surgery
Kantonsspital Baselland
Liestal, Switzerland

INTRODUCTION

The long head of the biceps (LHB) tendon is a secondary shoulder stabilizer that contributes to elbow flexion and supination strength and limits superior humeral head migration [1]. In the event of tendinopathies such as tears, subluxation, or synovitis, LHB tenodesis reliably relieves pain, restores strength, stability, and normal appearance of the upper extremity [2].

Interference screws, suture anchors, and bone tunnels are commonly used long head tenodesis constructs that secure the tendon to the humerus, eliminating pathologic tendon proximal to the construct [3,4]. In addition to the repair technique, the location of the repair placement is still debated [5]. Subpectoral tenodesis is effective at alleviating pain and is associated with stronger bone for fixation in the humerus, but requires detaching and reattaching the pectoralis major to the humerus and can be challenging in arthroscopic repairs [6]. Suprapectoral tenodesis eliminates the need to detach the pectoralis major by placing the repair proximal to the pectoralis insertion, but presents challenges with "groove pain" and thinner cortical bone [5].

One aspect that may affect the success of the tenodesis repair is the mechanical integrity of the tendon. Only one study has investigated the material properties of the LHB tendon [7], but no studies have examined if regional differences in material or structural properties of the LHB tendon exist. Regional differences in tendon strength or stiffness may provide surgeons with valuable guidance in the selection or placement of the tenodesis repair construct.

Therefore, the purpose of the present study was to compare the mechanical properties of the LHB tendon for three different functional regions of the tendon. We hypothesized that there would be differences for the material and structural properties between the intra-articular (proximal), suprapectoral (middle), and subpectoral (distal) regions of the LHB tendon.

METHODS

The LHB tendon was harvested from 12 pairs of fresh-frozen human cadaver shoulders with no history of shoulder pathology (10 M, 2F; Age: 58 ± 6 years). Each tendon was divided into three functional regions designated as the intra-articular, suprapectoral, and subpectoral regions (Fig. 1). Segments from one tendon of each pair were punched into dogbone shaped test specimens to characterize the material properties. The contralateral tendon from the pair remained as intact segments to examine the structural properties of the LHB tendon, where the clamp boundary conditions simulated the common "gold standard" tenodesis, the interference screw. The cross sectional area was recorded for all tendon segments.

The experimental protocol was adapted from previous tissue characterization studies [8,9]. Tendon segments were clamped in a servo-hydraulic materials testing machine (Instron 1331 Load Frame, Model 8800 controller; Instron Corp, Norwood, MA) equipped with a 1-kN tension compression load cell (Dynacell Model 2527-130, Instron Corp) in custom soft tissue clamps. Three 300-μm fiducial markers were adhered along the gauge length of the tissue, allowing tissue deformation to be monitored with high resolution video tracking software (DMAS v6.5; Spica Technology Corporation, Maui, HI) using a Prosilica GC1350 Gigabit Ethernet camera (Allied Vision Technologies, Vancouver, BC, Canada).

The initial conditions of the tissue were normalized by applying a preload equivalent to 0.05 MPa stress. The tissue was loaded in a triangle waveform to 8% clamp-to-clamp strain at 1%/s for 10 cycles. Immediately following the cyclic phase, the tendon was pulled to failure at 1mm/s.

Material properties of peak stress, elastic modulus, and hysteresis were calculated for dogbone punched specimens using data from the 10th cycle. From the failure test, the ultimate stress and strain were calculated. DMAS was used to quantify the marker displacement and tissue strain. Structural properties of peak load and stiffness were quantified for intact specimens using data from the 10th cycle of the loading curve. Tensile strength was quantified from the failure test.

Material and structural properties were compared between the three functional regions using paired *t* tests and were adjusted for multiple comparisons using the Holm-Bonferroni method. Statistical significance was considered to be P≤0.05 for all comparisons.

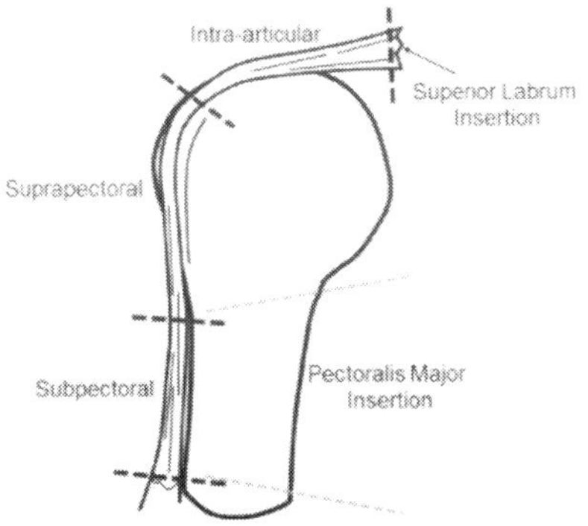

Figure 1. Demarcation of the three functional regions of the LHB tendon. The intra-articular, suprapectoral, and subpectoral regions were defined by: the insertion into the superior labrum, proximal margin of the bicipital groove, proximal edge of the pectoralis major insertion, musculotendinous junction. This equated to 30%, 40%, and 30% of the total tendon length, respectively.

RESULTS

Material properties for the three functional regions were obtained from dogbone specimens (Table 1). Common to tendon and ligament behavior, a nonlinear toe region was followed by a linear response in all stress-strain curves from the cyclic data (Fig. 2). All dogbone specimens failed in the tissue mid-substance.

Structural properties for the three functional regions were obtained based on intact specimens (Table 2). The failure mechanism for intact specimens was tearing at the clamp, similar to interference screws where tissue failure occurs due to tearing at the bone/tendon/screw interface [4].

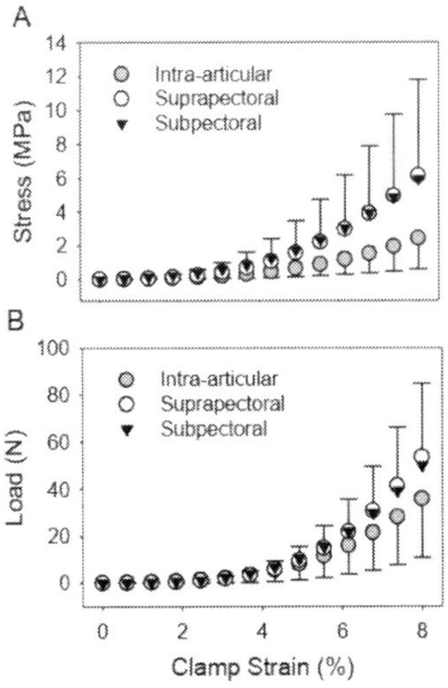

Figure 2. Average stress-strain curves from material tests (A) and average loading curves from structural tests (B). Mean±SD.

Table 1. Material Properties of the LHB Tendon by Region

(mean ± SD)	Intra-articular	Suprapectoral	Subpectoral
Cross-Sectional Area (mm^2)	5.7 ± 1.5*	4.9 ± 0.4†	4.4 ± 0.6
Peak Cyclic Stress (MPa)	2.5 ± 1.9	6.0 ± 5.6	6.2 ± 4.8
Elastic Modulus (MPa)	161 ± 88	283 ± 144^	281 ± 123^
Hysteresis (%)	25.3 ± 3.2	17.6 ± 4.3	17.3 ± 4.4
Ultimate Stress (MPa)	22.3 ± 9.3	31.7 ± 15.4	25.5 ± 10.5
Ultimate Strain (%)	12.5 ± 5.1	11.6 ± 6.6	11.6 ± 6.3

*Significant versus all other regions (P≤0.011)
†Significant versus the subpectoral region (P≤0.043).
^Significant versus the intra-articular region (P≤0.048)

Table 2. Structural Properties of the LHB Tendon by Region

(mean ± SD)	Intra-articular	Suprapectoral	Subpectoral
Cross-Sectional Area (mm^2)	22.6 ± 4.9*	15.1 ± 2.4	15.3 ± 2.9
Peak Cyclic Load (N)	36.0 ± 24.9	53.6 ± 31.2	50.1 ± 33.4
Stiffness (N/mm)	76.5 ± 46.5	105.6 ± 49.0	112.3 ± 63.0
Hysteresis (%)	16.5 ± 5.2	18.9 ± 2.8	17.9 ± 2.7
Tensile Strength (N)	337.3 ± 57.0†	345.3 ± 99.8†	282.8 ± 53.8

*Significant versus all other functional regions (P ≤ 0.001).
†Marginally significant versus the subpectoral region (P = 0.051).

DISCUSSION

The purpose of this study was to compare the material and structural properties of the LHB tendon between the intra-articular, suprapectoral, and subpectoral regions. The material properties, including ultimate stress, ultimate strain, and elastic modulus in the present study were in good agreement with a prior LHB tendon study that examined the properties of the intact tendon [7]. The higher tensile strength of the suprapectoral region, as compared to the subpectoral region, may be an important consideration since interference screw constructs often fail at the tendon-bone interface [4]. The structural tests performed in the present study mimic the compression of the intact tendon between the screw teeth and cortical bone. A prior laboratory study comparing tenodesis sites found that suprapectoral fixation tended to have higher failure loads compared to subpectoral fixation with interference screws, although no statistical significance was reached [10]. Our findings reinforce the suprapectoral region exhibits superior tissue strength, which may be a more desirable location for tenodesis placement. In addition to superior strength, placing the tenodesis site towards the proximal end of the tendon allows ample tendon length for revision surgery in the event of construct failure. While tissue properties are an important consideration, it is still critical to account for the composite construct strength and bone quality of the patient.

REFERENCES

[1] Hwang E, *JOR*, 32:1424-29, 2014. [2] Wittstein JR, *AJSM*, 39:857-65, 2011. [3] Mazzocca AD, *Arthroplasty*, 21:1296-1306, 2005. [4] Tashjian RZ, *JSES*, 22:1408-12, 2013. [5] Johannsen AM, *AJSM*, 41:2919-24, 2013. [6] Werner BC, *Arthroscopy*, 30:1075-84, 2014. [7] McGough RL, *KSSTA*, 3:226-29, 1996.
[8] Henninger HB, *JOR*, 31:1226-33, 2013. [9] Jackson JB, *F&AI*, 35:922-28, 2014. [10] Patzer T, *JSES*, 21:116-25, 2012.

THE IN VIVO VISCOELASTIC RESPONSE OF THE PORCINE SPINAL CORD

Snehal S. Shetye (1), Femke Streijger (2), Christopher Strickland (3), Jae H. T. Lee (2), Brian K. Kwon (2), Peter Cripton (2), Patrick D. Shipman (3), Christian M. Puttlitz (1)

(1) Mechanical Engineering
Colorado State University
Fort Collins, CO, USA

(2) Orthopaedic Surgery
University of British Columbia
Vancouver, BC, Canada

(3) Mathematics
Colorado State University
Fort Collins, CO, USA

INTRODUCTION

Computational models can allow for detailed examinations of the mechanical response of biological constructs during regular activity or traumatic loading. The accuracy of these models is highly dependent on the available material property descriptions of the tissues being investigated.

Simulation of traumatic loading events requires the use of viscoelastic material properties. For example, traumatic loading during motor vehicle accidents can cause significant injury to the spinal cord complex and finite element modeling of these injury scenarios require a temporal description of the spinal cord. Thus, several studies have investigated the *in vitro* quasi-static and quasilinear viscoelastic mechanical properties of the spinal cord. The resultant mechanical descriptions have been used extensively to describe spinal cord behavior in finite element models. However, relatively few studies have assessed the *in vivo* viscoelastic behavior of neural tissue. Specifically, one study has found rat brain tissue to be similar *in vivo* and *in situ* [1]. Conversely, another study found the dynamic mechanical response of brain tissue *in vivo, in situ,* and *in vitro* to be divergent [2]. Surprisingly, the spinal cord, which can experience significant transient loading during traumatic events, has not been extensively investigated for its *in vitro* and *in vivo* viscoelastic behavior. We recently reported on the *in vitro* nonlinear viscoelastic response of the spinal cord [3]. The current study now reports on the *in vivo* viscoelastic response of the porcine spinal cord.

METHODS

In vivo stress relaxation and dynamic cyclic tests were performed on thoracic spinal cord tissues of six-month old Yorkshire female pigs (N = 6, BW = 26.5 to 34.5kg). The entire *in vivo* experimental protocol was approved by the Animal Care Committee of the University of British Columbia. All animals were induced with Telazol, xylazine and atropine given intramuscularly and intubated endotracheally. Following a midline skin incision and retraction of the muscles, the spinous processes, laminae, and pedicles were exposed. Laminectomy was performed on low thoracic and lumbar levels T12 - L7 to expose the spinal cord. After removing the dura mater, the nerve root insertions and dentate ligaments on the exposed levels were carefully excised. Using a digital caliper, a 70 mm segment of the spinal cord (lumbar level L3-L6) was measured and the rostral and caudal ends of the segment were fixed with two small circular pieces of PVC tubing (20 mm). Next, two surgical Babcock clamps, slightly modified to fit around the tubing were locked in place and connected rigidly to the actuating fixture. The caudal clamp was attached to the immobile platform of testing machine, while the rostral clamp was attached to the mobile actuator of the machine. A high resolution DVRT was then carefully inserted at the mid-substance of the exposed spinal cord [Figure 1]. A custom LabVIEW code controlled the actuator and simultaneously collected the DVRT and load cell data (100Hz). The cord was then immediately bathed in surgical saline to prevent desiccation. Each cord was preconditioned for 50 cycles (1Hz) at a strain magnitude of 3%. The test protocol involved four stress relaxation experiments at 1%, 2%, 2.5% and 3% strain magnitudes. The ramp speed was kept constant at 6mm/s. This resulted in varying rise times for each specimen. A dynamic cyclic test was also performed at 2% strain magnitude for 20 cycles (1Hz). To allow for maximum tissue recovery, the wait time between each test was set to 1000 seconds. Further, the aforementioned tests were randomized for each specimen to minimize order effects. Based on our previous work, the stress relaxation data were fit with a fully nonlinear viscoelastic formulation of the form:

$$\sigma[\varepsilon(t),t] = \int_0^t E[\varepsilon(\tau),t-\tau]\frac{d\varepsilon(\tau)}{d\tau}d\tau + \sigma_0, \quad (1)$$

where, τ represents the continuous time variable of integration, σ_0 represents the initial stress and $E[\varepsilon(\tau), t-\tau]$ represents the nonlinear (strain and time-dependent) relaxation modulus. The relaxation modulus was approximated by the following Prony series:

$$E(\varepsilon,t) = E_\infty(\varepsilon) + \sum_{i=1}^{4} E_i(\varepsilon)e^{-t/\tau_i} \quad (2)$$

where $E_\infty(\varepsilon)$ represents the strain-dependent steady-state modulus and $E_i(\varepsilon)$ represent the strain -dependent moduli corresponding to the τ_i time constants. The time constants were set to decadal values ranging from 0.1 to 100 s and were selected to encompass the important features of soft tissue viscoelastic behavior within the bounds of the experiment (100s). To interrogate the validity of the fitted moduli for describing dynamic viscoelastic behavior of the spinal cord, the parameters were used to predict the 2% dynamic cyclic experiments using the following modified form of equation (1):

$$\sigma[\varepsilon(t),t] = -\int_0^t \frac{dE[\varepsilon(t),t-\tau]}{d\tau}\varepsilon(\tau)d\tau + E[\varepsilon(t),0]\varepsilon(t) + \sigma_0 \quad (3)$$

To quantify the ability of the fitted parameters to predict the cyclic behavior, a non-weighted root mean squared error (RMSE) equation and the percent error were calculated for each value of strain amplitude. To compare the model predictions to the experimental variability, one standard deviation from the experimental mean was calculated as a percentage of the mean and its corresponding RMS and percent errors were determined.

RESULTS

The nonlinear comprehensive viscoelastic characterization method (CVC) was able to fit all stress relaxation curves very well (Figure 1). In comparison, a traditional Heaviside ramp assumption over-predicted the relaxation response.

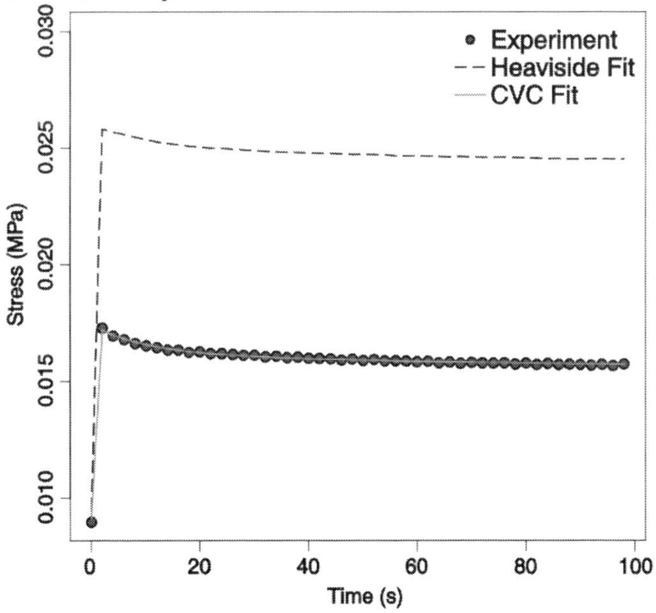

FIGURE 1 - A 2% STRAIN AMPLITUDE STRESS RELAXATION EXPERIMENT FIT VIA THE CVC METHOD.

The obtained nonlinear viscoelastic parameters were validated by predicting the stress response of the spinal cord tissue during a 2% strain magnitude cyclic experiment. The predicted stress response fell well within the ±1 standard deviation variability of the experimental average cyclic response (Figure 2). The predicted stress had an RMSE of 16.8 kPa, which was substantially lower than the experimental RMSE of 27.6 kPa.

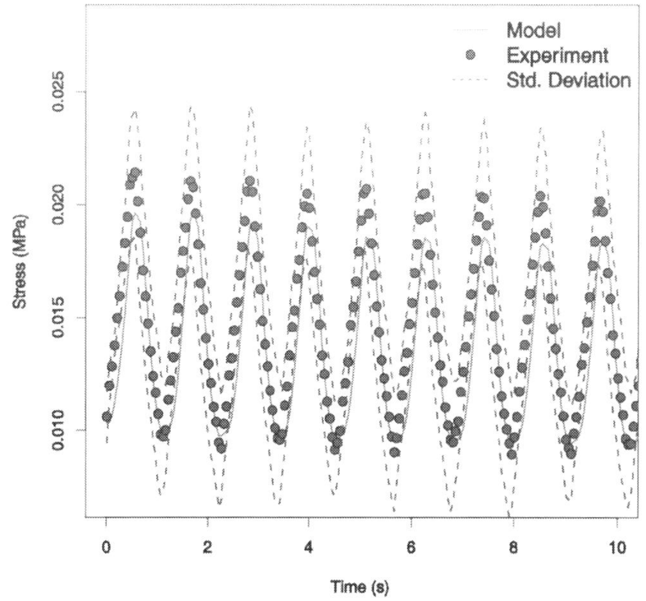

FIGURE 2 - PREDICTED CYCLIC STRESS RESPONSE (GREEN LINE) OF THE AVERAGE 2% STRAIN MAGNITUDE DYNAMIC EXPERIMENTS. THE DASHED LINES REPRESENT ONE STANDARD DEVIATION OF THE EXPERIMENTAL VARIABILITY

DISCUSSION

This study comprehensively described the nonlinear viscoelastic response of the porcine spinal cord *in vivo*. The robustness of the fitted parameters was validated by predicting the dynamic cyclic stress response of the tissue. These data will be crucial for investigators examining new strategies for minimizing trauma during dynamic loading events such as motor vehicle accidents, falls, sports and violence. Further, these results can help improve the predictive fidelity of currently available finite element models of the spinal cord.

In summary, this study was the first to report the *in vivo* viscoelastic response of the porcine spinal cord at moderate strain rates. Future work will involve comparison of *in vivo* and *in vitro* properties of the spinal cord. Additionally, the highly dynamic etiology of spinal cord trauma necessitates fast strain rate viscoelastic experiments.

ACKNOWLEDGEMENTS

Funding provided by the National Institutes of Health (R01EB012048) and the ICORD Distinguished Visiting Scholar Program 2012.

REFERENCES

[1] Gefen, A et al., *J Biomech*, 37:1339-1352, 2004.
[2] Prevost, T et al., *Acta Biomaterialia*, 7:4090-4101, 2011
[3] Shetye, S et al., *Acta Biomaterialia*, 10:792-797, 2014

SB^{3}C2015
Summer Biomechanics, Bioengineering and Biotransport Conference
June 17-20, 2015,Snowbird Resort, Utah, USA
SB^{3}C2015-151

BARIATRIC SURGERY IMPROVEMENTS FOR OSA PATIENTS

Ahmed M. Al-Jumaily (1), Sherif Ashaat (1), Bryn A. Martin(2), Rachael Pohle-Krauza (3), Matthew L. Krauza (3)

(1) Institute of Biomedical Technology, Auckland Technical University, New Zealand

(2) Dept. of Mechanical Engineering, Conquer Chiari Research Center, University of Akron, Ohio, USA

(3) Respiratory, Critical Care and Sleep Associates, Akron City Hospital, Akron, Ohio, USA.

INTRODUCTION

This paper assesses the improvements obtained by bariatric surgery (BS) to obese patients with obstructive sleep apnea (OSA) by comparing the physiological upper airway (UA) parameters before and after surgery. Ten morbidly obese women with a body mass index (BMI=weight/height) above 35 underwent BS (mean BMI 48.5 kg/m^2). Polysomnographic (PSG) and Computerized Tomography (CT) data were conducted and assessed before surgery and 6 and 12-month intervals after surgery. OSA symptoms were completely resolved in some patients and worsened in others. Although the results are not statistically conclusive, they do suggest bariatric surgery does have improvements for some of the patients.

OSA occurs due to obstruction (narrowing) in the UA region [1] and characterized by repeated collapse and/or obstruction of the pharyngeal airway during sleep. Its severity is determined according to the value of the Apnea-Hypopnea Index (AHI), which is defined as the total number of apneas (complete obstruction) and hypopneas (partial obstruction) occurring per hour of sleep (using overnight polysomnogram) [2]. OSA is classified as mild, moderate and severe for AHI of 5-15, 15-30 and above 30, respectively [3]. In addition to the fact that obesity may have negative effects on health, leading to reduced life expectancy and/or increased health problems, it is considered one of the main risk factors affecting OSA patients [2]. Those patients have narrower airways and reduced chest wall compliance compared with normal weight subjects. BS is a commonly used treatment to reduce weight and consequently help alleviate OSA symptoms [4]. It improved OSA symptoms from 33% preoperatively to 2% postoperatively [5] and either improved or completely resolved OSA symptoms [6]. Little is understood about quantifying the effect of this surgery on OSA patients' physiological parameters.

This paper investigates the effectiveness and biomechanical impact of bariatric surgery on the UA and how various patients' physiological parameters are affected after the surgery.

METHODS

Baseline PSG and CT data for the UA from nasopharynx to hypopharynx were conducted on 10 morbidly obese women, aged between 28-62 years with mild to severe OSA promoted for BS. Three dimensional UA models were constructed from baseline, 6-month and 12-month data for each subject using 3D-DOCTOR (Version 5, Able Software Corp, MA, USA), Fig. 1.

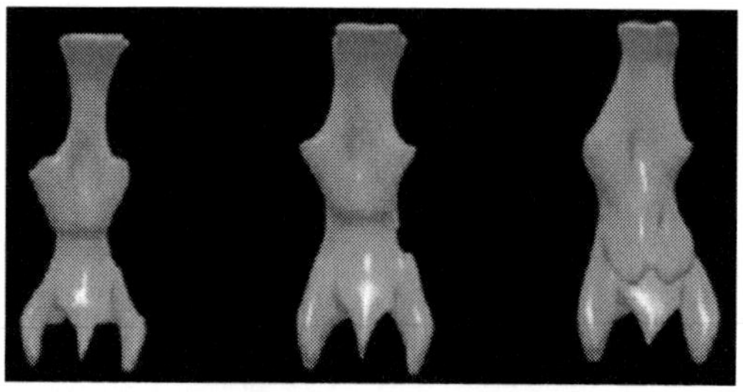

Fig. 1 UA volume at (a) Baseline, (b) 6-Months and (c) 12-Months after surgical intervention obtained from 3D-DOCTOR

Measurements of the UA expanding from the beginning of nasopharynx to the end of hypopharynx were made, including cross sectional area (A), wetted perimeter (P) and upper airway volume (V). The hydraulic diameter (HD) was calculated at each transverse location (slice) Fig. 2; HD=$4A/P$.

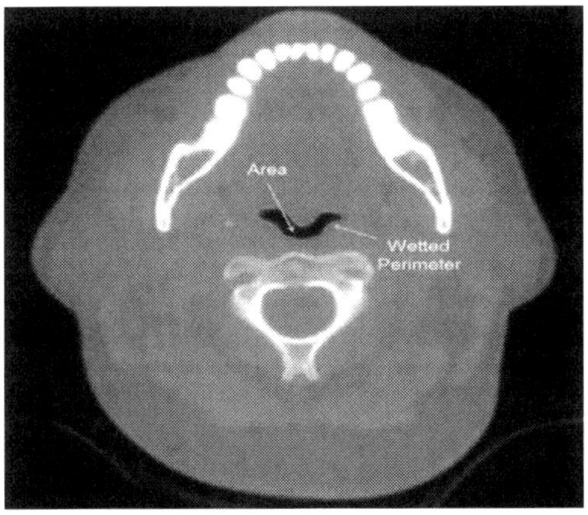

Fig. 2 Cross sectional area and wetted perimeter of air gap

Univariate statistics statistically assess the obtained data. A standard paired t-test was performed to check the significance of the results and assess the differences between means assuming equal variance of the study groups (baseline, 6 months post and 12 months post). All statistical analyses were considered significant when $p \leq 0.05$. Changes in the minimum HD, BMI, AHI, and V were assessed by the analysis of variance (ANOVA).

RESULTS

BMI, AHI, UA volume and maximum and minimum HD at baseline, as well as 6 and 12 months after BS, were determined. For brevity average data rather than full data are presented in Table 1. This table shows a mean BMI of 48.5 kg/m^2 at baseline, reduced to 33.7 kg/m^2 after 6 months and then further reduced to 30.3 kg/m2 after 12 months of surgical intervention. Also this table shows significant changes in the mean AHI which was 38.1 events/h (range 2-95) at baseline, reduced by 59% to 15.7 events/h (range 0-44) 6 months following surgery, and further reduced to 5.6 events/h (range 0-29) one year following surgical intervention (85% reduction compared with baseline).

Values for the UA volume show an increase after 6 months and a further increase after one year. For each patient the maximum and minimum HD were determined and the average values are shown in Table 1. Although these values show no specific trend of variation, at the average they increase after surgery.

Table 1 Average parameters for 10 subjects before and after BS

Parameter	Baseline	6-Months	12-Months
BMI (kg/m^2)	48.5	33.7	30.3
AHI (events/h)	38.1	15.7	5.6
Volume (mm^3)	17032	17749	18262
Min. HD (mm)	6.4	7.4	7.2
Max. HD (mm)	19.9	25.4	21.1

DISCUSSION

Significant decrease in BMI for all patients occurred during the first year after bariatric surgery ($n=10$, $p<0.05$). Morbid obesity was resolved in all subjects, five patients became overweight and the remaining five were classified as obese.

The volumetric changes increased in all subjects (compared to baseline values) one year after surgery except for one patient where BS had no significant effect on the minimum HD in morbidly obese middle-aged OSA patients (< 50 years) although OSA was completely resolved in this age group. For patients with age range 50-60 years, minimum HD increased at 6 months after surgical intervention and reduced after one year associated with remarkable improvements in OSA. Minimum HD increased in patients ≥ 60 years of age while OSA was worsened and became more severe. Also, minimum HD was observed to reduce in one patient who was non apneic at baseline.

AHI reduced in all patients except those with age ≥ 60 years where it increased. OSA was completely resolved in patients under 50 years (diagnosed with moderate-to-severe OSA at baseline) which might be related to the good response of UA muscles in the young age, improved in patients with age range 50-60 years (less response of UA muscles), worsened in patients ≥ 60 years and developed in the originally non-apneic patient (was considered to have a healthy airway at baseline). Consequently, sleep architecture was improved in patients where OSA was either resolved and/or improved.

It is believed that the good response of UA muscles for younger subjects in this study is the main contributor to resolve OSA symptoms regardless of the changes in the hydraulic diameter, which explains the complete resolution of OSA in the age group < 50 years. Also, as the response of the UA muscles decreased with age, the surgery would only improve OSA symptoms but not completely resolve them; thus helping explain the results for the middle age range in this study of 50-60 years. In patients age 60 and older, the results support that it is difficult to stimulate the UA muscles even with improvement in all other parameters analyzed (BMI, HD, V, and VBMI).

The outcomes of this preliminary study with 10 subjects show that BS improved BMI, AHI, V and HD and is recommended for middle-aged (under the age of 50) morbidly obese patients diagnosed with moderate to severe OSA. For this group BS reduced weight and resolved OSA.

ACKNOWLEDGEMENTS

All data were collected after an ethical approval obtained at the Akron City Hospital.

REFERENCES

[1] Mihaescu, M., et al., Computational modeling of upper airway before and after adenotonsillectomy for obstructive sleep apnea. Laryngoscope, 2008. 118(2): p. 360-362.
[2] Moos, D.D. and J.D. Cuddeford, Implications of Obstructive Sleep Apnea Syndrome for the Perianesthesia Nurse. Journal of Perianesthesia Nursing, 2006. 21(2): p. 103-118.
[3] Emsellem, H.A. and K.E. Murtagh, Sleep apnea and sports performance. Clinics in Sports Medicine, 2005. 24(2): p. 329-341.[4]
[4] Rasheid, S., et al., Gastric Bypass is an Effective Treatment for Obstructive Sleep Apnea in Patients with Clinically Significant Obesity. Obesity Surgery, 2003. 13(1): p. 58-61.
[5] Dixon, J.B., L.M. Schachter, and P.E. O'Brien, Sleep disturbance and obesity: Changes following surgically induced weight loss. Archives of Internal Medicine, 2001. 161(1): p. 102-106.
[6] Frigg, A., et al., Reduction in Co-morbidities 4 Years after Laparoscopic Adjustable Gastric Banding. Obesity Surgery, 2004. 14(2): p. 216-223.

SB^{3}C2015-152

WHICH OF FOUR COMMERCIALLY-AVAILABLE LASER SCANNER SYSTEMS GENERATES THE MOST ACCURATE BONE MODEL?

Valentina Campanelli (1,3), Stephen M. Howell (1), Maury L. Hull (1,2),

(1) Department of Mechanical Engineering
University of California, Davis
Davis, CA, USA

(2) Department of Biomedical Engineering
University of California, Davis
Davis, CA, USA

(3) Department of Neurological and
Movement Sciences
University of Verona
Verona, Italy

INTRODUCTION

Laser scanning is one of the various imaging modalities used to create 3D bone models. A 3D laser scanner system is composed by the laser sensor, which determines the distance between the laser source and the surface of the object being scanned, and the motion tracking device that determines the position and orientation of the laser sensor in the 3D space. The most recent types of motion tracking devices for laser scanning include the manual articulating measuring arm, which tracks the position and orientation of the laser sensor device using encoders placed at the various joints of the arm, and the stereo photogrammetric system, which tracks the position and orientation of markers placed on the laser sensor using two or three cameras.

Knowing the accuracy and limitations of these laser scanner systems is imperative to pick the best system for scanning bones. However the only available information about the accuracy of these systems is the accuracy value stated by the manufacturer. This value is obtained using standardized tests that typically require the scanning of satin titanium gage blocks or bars with spheres and it often refers to the laser sensor or the tracking device separately. In those cases where overall system accuracy is available, the use of non-uniform testing methods among the manufacturers makes it difficult to compare the accuracies stated for different laser scanning systems. Most importantly, because the accuracy of these systems highly depends on the shape, texture, and material reflectivity of the object being scanned, the accuracy of the laser scanners as stated by the manufacturer may not apply to bone scanning. For these reasons, other studies investigated the accuracy of different laser scanner systems for applications such as face scanning [1], orthodontics [2] , and ice shapes scanning [3]. However, no studies investigated the accuracy of laser scanner systems for bone scanning.

Hence, the aim of this study was to determine the scanning repeatability of four commercially-available portable laser scanner systems in the scanning of the distal femur.

METHODS

The same bovine distal femur specimen was sent to each of the manufacturers of four commercially-available portable laser scanner systems. The four selected portable laser scanner systems were: 1) Nikon hand-held laser scanner (MMD50) and manual articulating measuring arm (MCA30); 2) LDI hand-held laser scanner (SLP) and manual articulating measuring arm (Space Arm 3.2); 3) Creaform hand-held laser scanner (Metrascan 3D) and photogrammetric system (C-Track 780); and 4) NDI hand-held laser scanner (Scantrack) and photogrammetric system (Pro CMM 1000). These laser scanner systems were selected among the others because the accuracies stated by the manufacturers were the highest for portable laser scanner systems that allow the scanning of objects of the size of a bone.

Prior to sending the bovine specimen, the frozen specimen was thawed for 12 hours and then dissected to remove soft tissues. Four spherical fiducial markers were attached to the bone to enable a precise registration of the bone models. Next, the bone was submerged in 6% sodium hypochlorite solution for 10 hours to remove soft tissues remains and the articular cartilage. The femur was then dried by oven-cooking for 2 hours at 180°C, sterilized, and potted into a hollow metal cylinder (Figure 1A).

Each manufacturer was asked to perform ten scans of the specimen using the laser scanner settings that he believes result in the highest level of accuracy of the bone models. The distal femur was chosen because of its complicated morphology, so that the scanning repeatability found for the femur can be easily translated to the other bones in the body. The ten unprocessed scans of the specimen were then sent to the authors in stereolithography format (STL, Figure 1B) and point cloud format (i.e. text file).

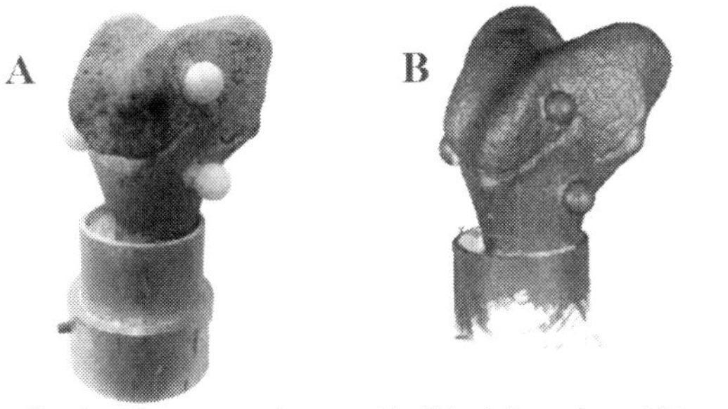

Figure 1. Bovine femur specimen with fiducial markers (A) and corresponding 3D bone model generated with Creaform laser scanner system (B).

The scanning repeatability for each laser scanner system was computed by performing a pairwise comparison of the ten scans (i.e. 45 comparisons for each system) using Geomagic (3D System, USA). The paired scans were registered using the centers of the spheres obtained through a best-fit of the fiducial markers. Next, each model was cleaned by removing the polygons corresponding to the fiducial markers and to the metal cylinder to limit the repeatability analysis to the bone surface only. Finally, the deviation of two bone models was computed and reported in term of average positive/negative deviations and root mean square error (RMSE, Figure 2).

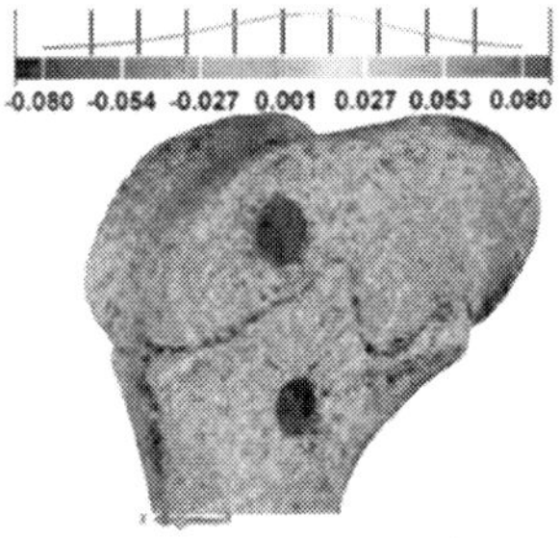

Figure 2. Bone model deviation distribution (mm).

The scanning repeatability of each laser scanner system was quantified by computing the followings:

$$\text{Mean Absolute Deviation} = \frac{\dfrac{\left|\sum_{i=1}^{45} pd_i\right|}{45} + \dfrac{\left|\sum_{i=1}^{45} nd_i\right|}{45}}{2} \quad (1)$$

$$\text{Mean Standard Deviation} = \frac{\sqrt{\dfrac{\sum_{i=1}^{45}\left(pd_i-\overline{pd}\right)^2}{45} + \dfrac{\sum_{i=1}^{45}\left(nd_i-\overline{nd}\right)^2}{45}}}{2} \quad (2)$$

$$\text{Mean RMS} = \sqrt{\frac{\sum_{i=1}^{45}\left(\frac{1}{n}\sum_{j=1}^{n}\sqrt{(x_j-\hat{x}_j)^2+(y_j-\hat{y}_j)^2+(z_j-\hat{z}_j)^2}\right)_i^2}{45}} \quad (3)$$

where pd and nd are the average positive and negative deviations for each comparison, $\overline{pd}$ and $\overline{nd}$ are the mean of the 45 average positive and negative deviations, $x-y-z$ and $\hat{x}-\hat{y}-\hat{z}$ are the coordinates of paired points on the two bone models (i.e. test and reference bone models respectively), and n is the total number of points on the test bone model.

RESULTS

As expected all the laser scanner systems performed well at scanning repeatability with the mean RMSEs and mean absolute deviations being well below 1 mm for all systems (Table 1). The Creaform laser scanner system was the best performing in terms of scanning repeatability, having the lowest RMSE and mean absolute deviation, followed by the Nikon system. All bone models presented a closed surface excepting that from the LDI system whose 3D models presented many holes.

	Mean RMSE (μm)	Mean Absolute Deviation (μm)	Mean Standard Deviation (μm)
Creaform	66.8	36.3	2.9
Nikon	69.7	46.9	8.7
NDI	79.6	48.8	10
LDI	97.2	64.3	6.7

Table 1. Scanning repeatability for the four laser scanner systems.

DISCUSSION

In the present study, we evaluated the repeatability of four laser scanner systems in scanning the same bovine femur. The Creaform system performance was ranked as the best of all scanner systems evaluated because it generated the most repeatable bone models. The Creaform laser scanner system had the highest number of polygons in the bone models thanks to the use of an optimized custom method for meshing the point cloud (Table 2). Presumably, the high number of polygons accounts at least in part for the best accuracy.

The main limitation of this study is that the systematic error could not be computed for these systems because of the unavailability of a gold standard bone model with an order of magnitude better accuracy than the ones stated for these systems. However, we believe that the systematic error would be negligible compared to the random error (i.e. repeatability) of these laser scanning systems.

The results of the present study indicate that the best portable laser scanning systems evaluated can be used to scan bony anatomy with repeatability below 70 μm.

	Average Number of Polygons	Reported Scanning Time
Creaform	1,810,779	20 min
Nikon	1,294,367	3 min
NDI	302,689	10 min
LDI	249,899	30 min

Table 2. Additional comparison parameters for the laser scanner systems.

ACKNOWLEDGEMENTS

We acknowledge the support of the National Science Foundation, Award No. CBET-1067527.

REFERENCES

[1] Boehnen, C., et al., *Fifth International Conference on 3-D Digital Imaging and Modeling, Proceedings*, 310-317, 2005.

[2] Kusnoto, B., et al., *American Journal of Orthodontics and Dentofacial Orthopedics*, 122:342-348, 2002.

[3] Sam Lee, R.S., Ellen Pifer, *Development of 3D Ice Accretion Measurement Method* in *AIAA Atmospheric and Space Environment Conference*. 2012.

SB³C2015
Summer Biomechanics, Bioengineering and Biotransport Conference
June 17-20, 2015,Snowbird Resort, Utah, USA
SB³C2015-153

A micromechanical viscoelastic computational model incorporating progressive detachment of the tau proteins predicting microtubule breaking following axonal injury.

Hossein Ahmadzadeh (1), Douglas H. Smith (2), Vivek B. Shenoy (1)

(1) Department of Materials Science and Engineering
University of Pennsylvania
Philadelphia, Pennsylvania, USA

(2) Penn Center for Brain Injury and Repair and Department of Neurosurgery
University of Pennsylvania
Philadelphia, Pennsylvania, USA

INTRODUCTION

The unique viscoelastic nature of axons is thought to underlie selective vulnerability to damage during axonal injury. In particular, dynamic loading of axons has been shown to mechanically break microtubules and create axonal swellings at the time of injury. Alternatively, under slow pulling rates, axon has been shown to resist large tensile strain without any evidence of microtubule rupture or cytoskeletal damage. In our previous work[1] we presented a computational model for uniaxial tension of axon where a bundle of microtubules were interconnected by tau proteins and the applied load was transferred through the sliding and stretching of the microtubules, regulated by the tau proteins. There we showed that the viscoelastic behavior of the tau proteins could increase the chance of the microtubule rupture at the time of the injury[1]. Although the mentioned model was successful to predict the rate dependent behavior of the axons observed in experimental studies, here in the next step we present a more comprehensive model where by including the rupture of the connections between the linking tau proteins, the pathway of the load transfer within the axon can be obstructed and correspondingly the axon response can be altered. In this model, under different pulling rates, we monitor the rupture of the connections between the attached tau proteins and the tensile force exerted on the microtubules and at the end, present conditions leading to the rupture of the microtubules or the complete disassembly of the microtubule bundle. By including the rupture of the interconnecting elements within the axon, our current model can provide a picture of the damaged axon following traumatic brain injury where distinct regions including ruptured microtubules and disassembled microtubule bundles can be distinguished explicitly.

METHODS

Our axon model consists of staggered array of MTs with length L, outer and inner radii 12.5 nm, 7 nm enclosed by the axon lipid membrane. Tau proteins are bonded to the outer surface of the MTs at their MT binding domains while the distance between the adjacent tau proteins on a single MT is 30nm. Also the zipper bond dimer configuration is adopted for the connection between the tau proteins belonging to neighboring MTs. To include the viscoelastic response of the tau proteins, we modeled the tau proteins with two parallel springs, representing the backbone stiffness and the rate-dependent stiffness that arises from the intramolecular bonds within the tau protein.

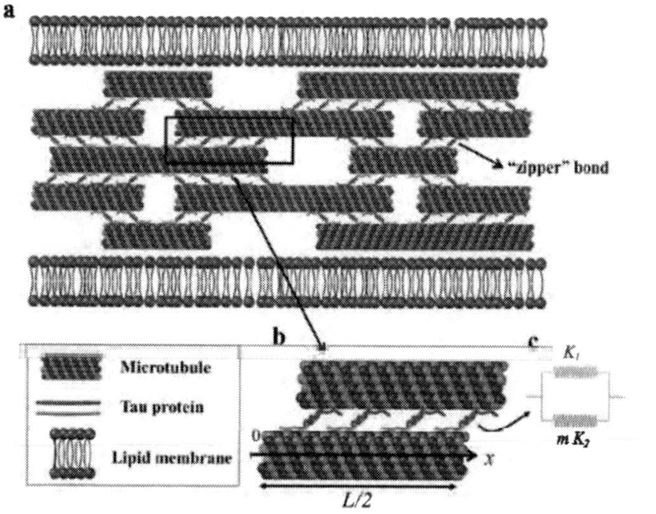

Fig. 1. (a,b) Axon model consists of MTs and tau proteins enclosed by axon lipid membrane. (c) A mechanical element for the tau proteins.

While the connection between the two tau proteins can rupture at a critical force, floating end of a free tau protein can still reattach to the tail of the next available tau protein and reengage in the load transfer mechanism if the applied loading rate is sufficiently slow (quasi-static loading). As illustrated in Fig. 2, this recruitment process of the free tau proteins can be repeated until the ends of the adjacent MTs pass each other.

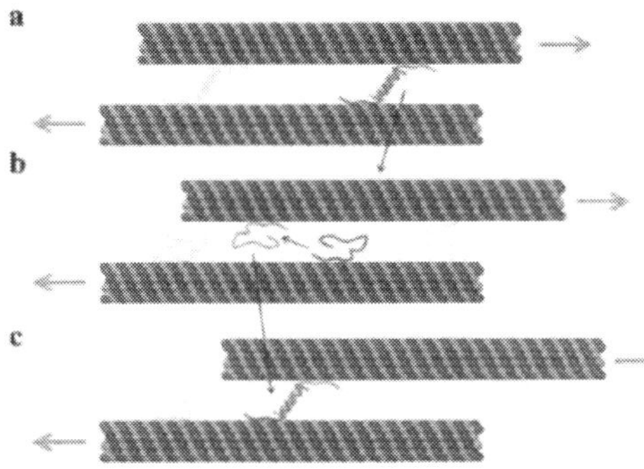

Fig. 2. (a-c)Recruitment of the free tau proteins after rupture incorporated in our model when the loading rate is sufficiently slow (quasi-static loading).

RESULTS

MT rupture is sensitive to strain rate compared to strain

As shown in Fig. 3, under dynamic loading rates (1–50 sec-1), MTs are under a significantly larger force (almost 3–4 times) than the quasi-static rate. In this case, the large force can lead to the rupture of the MTs and the subsequent interruption of the cargo transport similar to the observations for the axons injured by TBI. For the unruptured MTs, due to the large extension of the tau proteins, the zipper bond connecting tau proteins start to break. Eventually by the rupture of the majority of the interconnections, the unruptured MT bundle disassembles and the force-strain curve displays a rapid drop as the axon loses its load bearing capacity completely.

For the quasi-static loading the axon behavior is considerably different. While the force exerted on the MTs is significantly small compared to the dynamic loading, with the help of the reformation process (explained in Fig. 2), the free tau proteins can still reengage in the load transfer mechanism and avert the drop of the force. As a result, we observe a slow decline in the force as well as the concentration of the connected tau proteins with strain.

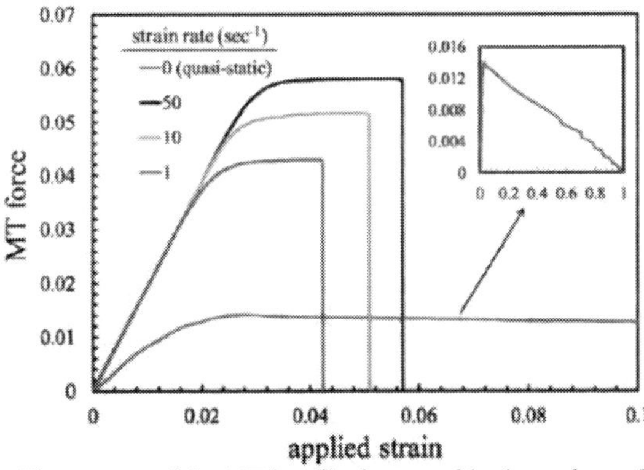

Fig. 3. (left) The force exerted on the MTs at different strain rates. The MT length is 200μm in both figures.

The pattern of the MT bundle disassembly depends on the strain rate

By increasing the applied tensile strain, the connections between the attached tau proteins start to rupture. As shown in Fig 4 the model predicts that under the fast loading rate, the rupture of the tau-tau bonds starts from the ends of the MTs and progresses sequentially toward the interior points. In the contrary under slow loading rate, these bonds rupture simultaneously and cause a rapid drop in the stiffness of the axon at a threshold strain.

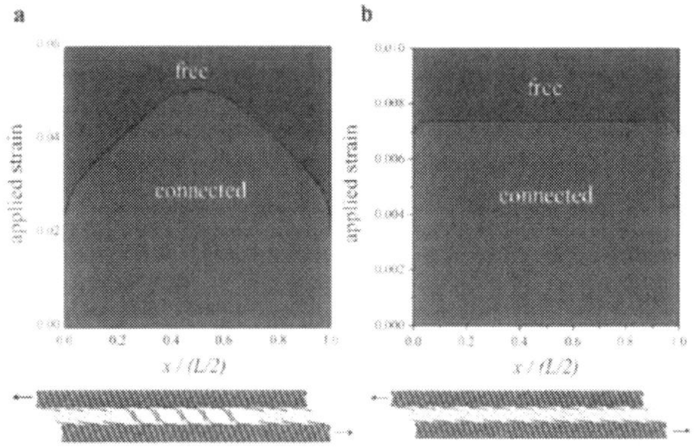

Fig.4. Breaking of the tau-tau connection by (a): fast (50 sec-1) and (b): slow (quasi-static) loading rates. MT length is 50μm.

Following TBI, the long MTs inside the axon break while the short MTs detach from the MT bundle.

For varying lengths of for the microtubules under a fast pulling rate, we obtained the force exerted on the microtubules with strain. Our model predicts that the long MTs inside the axon experience large tensile forces while the short MTs undergo relative sliding with respect to the neighboring MTs without any significant force. On the other hand, the tau proteins attached to long MTs undergo different degrees of elongation depending on their position relative to the MT ends while, the tau proteins connecting the short MTs are extended equally to accommodate the relative sliding. As a result, while the large force on the long MTs would eventually break the MT, the large sliding of the short MTs would result in the MT disassembly from the rest of the MTs by an instantaneous rupture of the interconnection bonds between the tau proteins. Based on this prediction, our model presents a picture of the damaged axonal cytoskeleton by injury. In this picture (Fig 4) the long microtubules are broken while the short microtubules are detached from the microtubule bundle and remain unaffected.

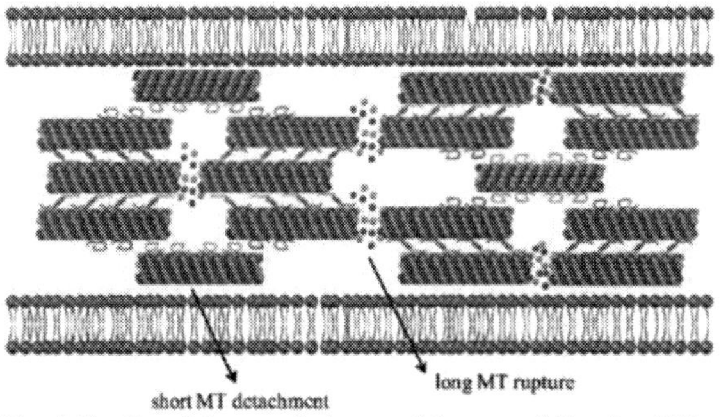

Fig. 5. Predicted damaged picture of the axon following TBI.

Discussion

In this study we presented a computational model to explain the axonal pathology observed in traumatic brain injury. While the fast loading rates rupture the long microtubules inside the axon, the slow loading rates cause disassembly of the unruptured microtubules.

REFERENCES

[1] Ahmadzadeh, et al., Bio Phys J., 106:1123–1133,2014.

SB³C2015
Summer Biomechanics, Bioengineering and Biotransport Conference
June 17-20, 2015, Snowbird Resort, Utah, USA
SB³C2015-154

FABRICATION OF COLLAGEN BASED 3-D COMPLEX CONSTRUCTS WITH ELECTROCHEMICAL COMPACTION METHOD

Mousa Younesi (1), Vipuil Kishore (2), Ozan Akkus (1,3)

(1) Dept. of Mechaical and Aerospace Engineering
Case Western Reserve University
Cleveland, Ohio, US.

(2) Dept. of Chemical Engineering
Florida Institute of Technology
Melbourne, Florida, US.

(3) Dept. of Biomedical Engineering
Case Western Reserve University
Cleveland, Ohio, US.

INTRODUCTION

Tissue engineering of anatomically complex forms, such as joint surfaces or craniofacial features require scaffolds with curved topographies. To date, there is no reliable method that can pattern pure collagen solutions as curved topographies that can retain their shape. Rapid 3D deposition of biological proteins in curved topographies with robust mechanical properties for shape retention would be significant. Recent studies have shown that collagen molecules in solution form can be compacted by applying electric current [1-4]. Thus far, simple linear threads have been obtained using electrochemical compaction (ECOMP). The current study introduces the electrochemical compaction method for fabrication of collagen based scaffolds with different dimensions and topographies. Effects of electrocompaction process on the mechanical properties of collagen products and response of resident cells with respect to the standard uncompacted collagen gel are investigated. The capacity of electrocompaction as a materials processing modality is demonstrated by fabrication of multilayered hemispherical construct to emulate a striated structure of the curved joint surface; and fabrication of a geometrically complex nasal shaped scaffold.

METHODS

Electrochemical Fabrication: A 12.5 mm diameter hemispherical indent was milled in a carbon electrode which was connected as the cathode. The indent was filled with type-I collagen solution (Advanced Biomatrix, CA). An aluminum ball of 10 mm diameter, positioned concentrically within the hemispherical indent, served as the anode. Electrical current compacts the molecules close to the cathode under the effect of mechanisms published before [1-4]. On the top of the electrocompacted pure collagen layer, a mixture of hydroxyapatite (HA) and type-I collagen solution (60% w/w HA) was compacted as a second layer. The more complex structure of nose was fabricated by molding liquid metal as electrodes.

Effects of ECOMP on Mechanical Properties: Type-I and II collagen solutions were loaded between two parallel planar carbon sheet electrodes to obtain disc shaped sheets. Sheets were crosslinked in 0.625% genipin in 90% ethanol for 3 days, tested under tension and compression (1%/sec) to obtain the strength and modulus (RSAII, Rheometrics Inc., Piscataway, NJ). The size and strength of cells focal adhesions were analyzed with talin transfected cardiomyocytes after one day of culture on electrocompacted collagen sheet and collagen gel.

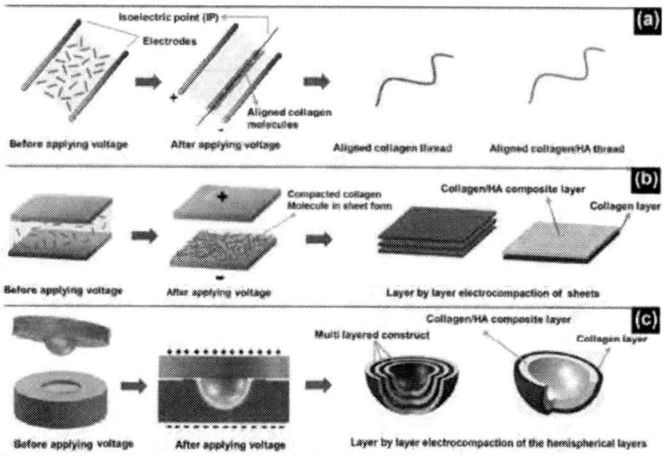

Fig 1. Schematic process of ECOM: (a) linear electrodes for making collagen threads, (b) planar electrodes for making 2-D multiphasic construct, and (c) hemispherical or curviplanar electrode for making 3-D multiphasic constructs.

RESULTS

As it presented schematically in fig.1a-c electrochemical compaction method enables fabrication of 1, 2, and 3 dimensional constructs with type I and II collagen which can have inclusions like mineral particles or other biomolecules like Chondroitin sulfate (CS) and growth factors. As the ECOM compact the collagenmolecules and increase the efficiency of crosslinking the mechanical properties of compacted sheet improve dramatically compare to collagen gel. Fig. 2a and b show the compressive stress and tensile strength of compacted type I and II collagen sheets and also collagen gel.

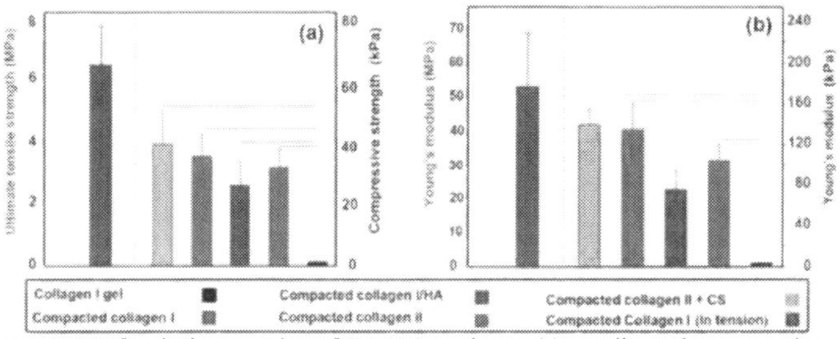

Fig 2. Mechanical properties of ECOM products: (a) tensile and compressive stress of compacted type I & II collagen sheets, (b) Young's modulus of compacted collagen I and II sheets.

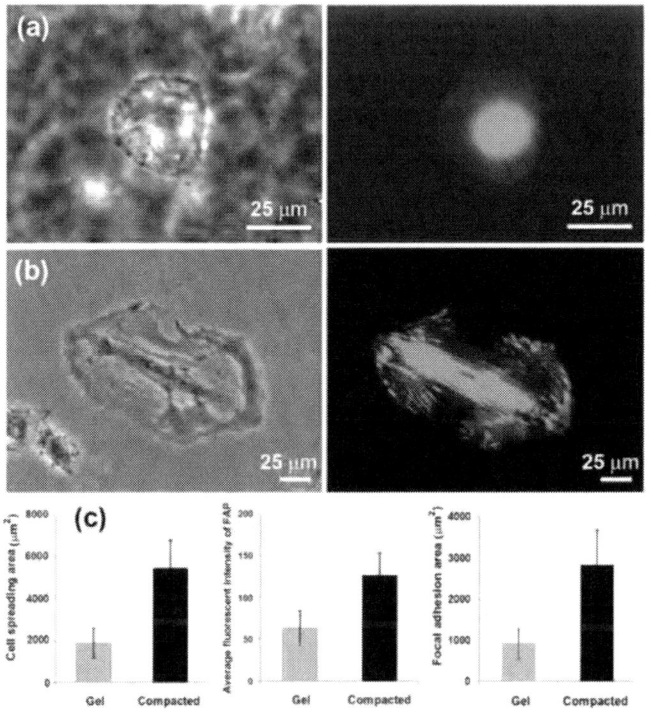

Fig. 3. Focal adhesions of cells on compacted collagen sheet are more strongly present than uncompacted collagen gel. Focal adhesions are visualized by talin-green fluorescent protein fusion in situ. Focal adhesions of cells on random collagen gel (a) and compacted collagen sample (b). Cells are more polarized on electrocompacted sheet (a vs. b). Focal adhesions are more mature, stronger, and present in greater amounts on electrocompacted sheets than they are on collagen gel as reflected by greater amount of fluorescent intensity values (c). (All scale bars: 25µm).

The cells seeded on the uncompacted collagen sheets were round and poorly adherent with low levels of focal adhesions while the cells on the compact sheet were readily with greater number and bigger focal adhesions and the strength of focal adhesions established by cells were higher (Figure 4a-c). To illustrate the ability of ECOM in fabricating different construct with different composition, fig.4 a-c present some of the product like compacted collagen II sheet incorporated with CS, a semispherical construct with an inner layer of compact collagen I/hydroxyapatite and out layer of pure compact collagen I, a nose construct made with type I collagen.

DISCUSSIONS

In compare to other method of making collagen scaffold the main advantage of electrochemical compaction is its ability to produce products with complex geometry rapidly and directly from aqueous collagen solution which can contain other inclusion such as mineral particles, other biomolecules and bioinductive growth factors without harming the collagen microstructure. Fast compaction of the collagen molecules up to the level of native tissue with similar topographical form is the main feature of electrochemical compacted collagen product. Higher density of collagen molecules in electrocompacted collagen samples enables cells to establish stronger, bigger and higher number of focal adhesion points on substrate. This higher density and stiffness of electrocompacted collagen samples improve the cell proliferation compare to standard collagen gel.

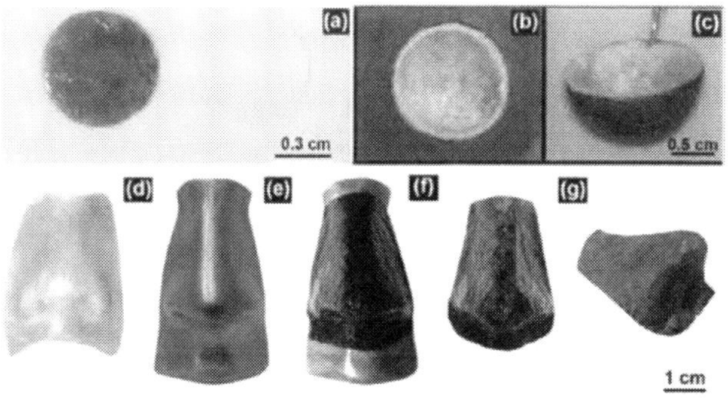

Fig 4. ECOMP product: (a) compacted collagen II (left) and collagen II incorporated Cs (right), (b) semispherical multiphasic collagen construct compacted onto cathode, (b) electrochemically fabricated, (c) semispherical scaffolds with HA-collagen (white) and pure collagen (black) layers standing alone. (d-f) the female, male electrode counterparts and nasal shell layer.

CONCLUSIONS

Electrochemical compaction method is a solid freeform fabrication (SFF) method for making mechanically strong 3-D complex geometry constructs with a variety in composition and geometry relevant to targeted tissues and organs. Therefore, electrochemical compaction is a cutting-edge technology with significant potential to be used for the joints and craniofacial defects, and dental complications.

ACKNOWLEDGEMENTS

The study was partially funded by the research grant 1306665 co-funded by the CBET and BMAT programs of the National Science Foundation and National Institutes of Health (Grant Number R01 AR063701).

REFERENCES

[1] Cheng et al., Biomaterials, 2008, 29(22), 3278-88.
[2] Kishore et al., Biomaterials, 2012;33(7):2137-44.
[3] Uquillas et al. Ann Biomed Eng, 2012;40 (8):1641-53.
[4] Younesi et al. Adv. Func. Mater, 2014; 24(36): 5762-70.

BIOMECHANICAL EVALUATION OF PROXIMAL JUNCTIONAL FAILURE AND KYPHOSIS IN LUMBAR SPINAL FUSION SURGERY

W. M. Park (1), K. Kim (2), Y. J. Kim (3), Y. H. Kim (1)

(1) Department of Mechanical Engineering
Kyung Hee University
Yongin, Gyeonggi-do, S. Korea

(2) Department of Applied Mathematics
Kyung Hee University
Yongin, Gyeonggi-do, S. Korea

(3) Department of Orthopaedic Surgery
Columbia University
New York, New York, USA

INTRODUCTION

Spinal fusion surgery with or without corrective osteotomy is a widely used surgical procedure for deformity correction of the spine. However, adjacent segment disease (ASD) such disc failure, vertebral fracture, and implant-bone interface failure at the adjacent motion segments have been reported in many clinical studies [1-3]. Even though it has been reported that increase in stiffness at the fusion level is a possible reason of the ASD, the effects of fusion level and location on the ASD are not clear. In this study, we evaluated the risk of the proximal junctional failure and kyphosis at the adjacent motion segments using finite element (FE) models of the lumbar spine with various spinal fusion cases.

METHODS

A three dimensional FE model of the lumbar spine from T12 to S1 was developed from 1-mm thickness of computed tomography (CT) images. The CT images were taken from a healthy male volunteer whose height and age were 175 cm and 21 years old. The FE model which consisted of seven vertebrae, six intervertebral discs, and seven major ligaments was developed to be symmetric across the mid-sagittal plane. Clinical data and results of previous studies were taken for determining attachment points of ligaments and material properties of the model. Pedicle screw system was inserted to the L2-L3, L3-L4, L4-L5, L5-S, L2-L4, L3-L5, L4-S, L2-L5, and L3-S spinal motion segment, respectively, based on their surgical protocols using developed FE model of T12-S (Fig. 1). Facet joints, capsular and flaval ligaments were resected in the fusion models. Effects of fusion level and location on the biomechanical behaviors such as intersegmental rotation and stress on the upper adjacent vertebrae were analyzed in flexion-extension movement. Inferior plane of the sacrum was fixed in all directions and flexion-extension moments of 7.5 Nm were applied on the superior plane of T12 vertebra in intact case with a compressive force of 400 N along the follower load direction. For the fusion models, we applied hybrid loading condition which generates the same overall rotation with the intact model with a compressive force of 400 N. In this study, a commercial FE analysis software ABAQUS Standard™ ver. 6.10 (Simulia Corp., USA) was used.

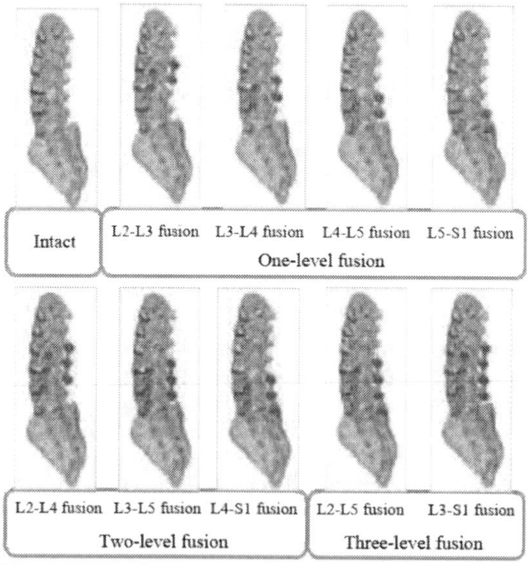

FIG. 1 INTACT LUMBAR SPINE AND NINE SPINAL FUSION MODELS WITH PEDICLE SCREW SYSTEM

RESULTS

Range of motions (ROMs) in flexion-extension were 7.1 °, 7.1 °, 7.5 °, 7.3 °, 9.4 °, and 12.6 ° at the T12-L1, L1-L2, L2-L3, L3-L4, L4-L5, and L5-S1 motion segments in the intact model. The ROMs of adjacent segment were increased by 111.1%-133.8%, 126.1%-166.3%, and 159.5%-185.4% in the one-level, two-level, and three-level fusion models, respectively in comparison with the intact (Fig. 2). Similarly, The maximum von-Mises stresses of vertebra were increased by 111.9%-130.8%, 135.8%-228.0%, and 233.0%-309.7% in the one-level, two-level, and three-level fusion models, respectively in comparison with the intact in flexion, and 119.8%-149.1%, 217.9%-295.5%, and 290.3%-827.0% in the one-level, two-level, and three-level fusion models, respectively in comparison with the intact in extension (Fig. 3). Increase in the number of fusion levels and change in fusion location from proximal to distal direction led to increase in both the ROMs of flexion-extension and stresses on the upper adjacent segment.

DISCUSSION

Several previous studies have demonstrated that the increased ROMs at the upper adjacent level accelerate degeneration of the adjacent segments [4, 5]. The stress on the vertebra is associated with a fracture risk of the spinal bone. The results of this study showed that a larger fusion range and more distal fusion location led increase in

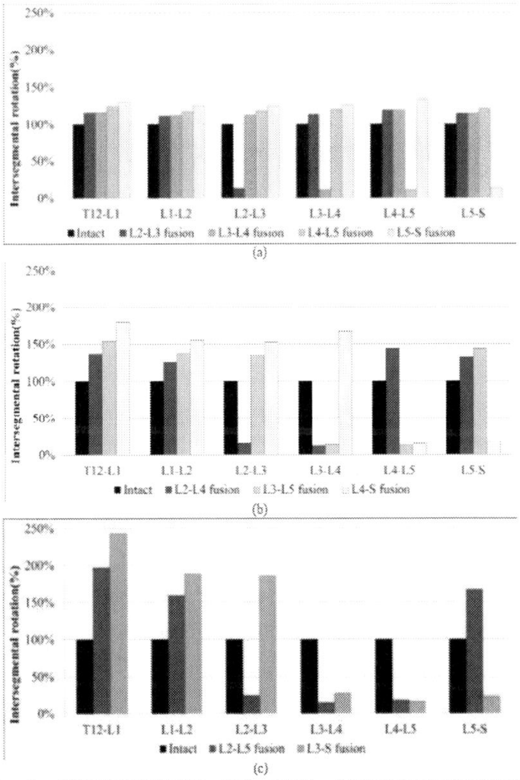

FIG. 2 PERCENTAGE OF THE INTERSEGMENTAL ROTATION ANGLES OF THE LUMBAR SPINE ON FLEXION-EXTENSION DURING 7.5 NM MOMENT WITH 400 N COMPRESSION: (A) ONE-LEVEL FUSIONS, (B) TWO-LEVEL FUSIONS, (C) THREE-LEVEL FUSIONS

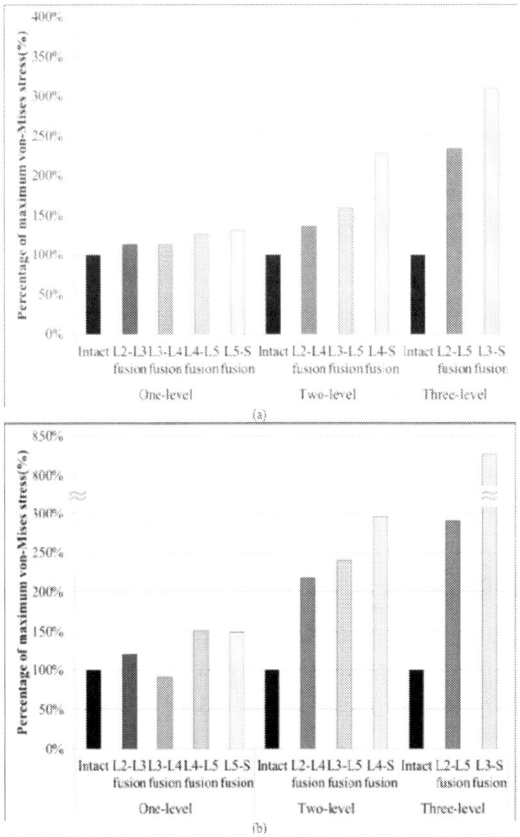

FIG. 3 PERCENTAGE OF THE MAXIMUM VON-MISES STRESS IN THE UPPER ADJACENT VERTEBRA : (A) FLEXION AND (B) EXTENSION

ROM at the adjacent segments, which would accelerate degeneration of the adjacent segments. Furthermore, the risk of compression fracture could increase when spinal fusion is applied at a more distal level. We could conclude that larger fusion range and more distal fusion level increase the risk of proximal junctional failure and kyphosis, based on the results of this study. Therefore, fusion range and location should be carefully selected for preventing proximal junctional failure and kyphosis in the spinal fusion to achieve satisfactory sagittal realignment.

ACKNOWLEDGEMENTS

This work was supported by National Agenda Project (NAP) in Convergence R&D Project funded by National Research Council of Fundamental Science & Technology (NAP-09-2-KISTI).

REFERENCES

[1] Watanabe, K. et al., *Spine*, 35:138-144, 2010.
[2] Yagi, M. et al., *Spine*, 36:E60-E68, 2011.
[3] Kim, Y.J. et al., *Spine*, 31:2359-2366, 2006.
[4] Hilibrand, A.S. et al., *Spine J*, 4:190S-194S, 2004.
[5] Lo, C.C. et al., *Comput Methods Biomech Biomed Engin*, 14:969-978, 2011.

SB³C2015
Summer Biomechanics, Bioengineering and Biotransport Conference
June 17-20, 2015, Snowbird Resort, Utah, USA
SB³C2015-156

ROLE OF WALL THICKNESS AND TETHERING IN THE ASSESSMENT OF ARTERIAL STIFFNESS

Simona Hodis
Department of Mathematics
Texas A&M University - Kingsville
Kingsville, Texas
USA

Mair Zamir
Department of Applied Mathematics
University of Western Ontario
London, ON
Canada

INTRODUCTION

Arterial stiffening as a clinical condition cannot be measured directly but is usually assessed indirectly by measuring the pulse wave velocity (PWV). Using PWV as a marker for the diagnosis of vascular stiffening has its origin in the theory of wave propagation where the fluid travels faster in a stiffer tube. The main hypotheses on which this theory is based are the assumptions that the tube wall material is purely elastic, the wall thickness is negligible and tethering to surrounding tissue can be neglected. The extent to which these assumptions are satisfied in the cardiovascular system is not known. Using the PWV as a diagnostic test without including all these variables, could lead to error when interpreting the PWV value as an index of vessel wall compliance. We propose a model of the arterial wall where the wall thickness is considered non-negligible and tethering to surrounding tissue is included. The results indicate that the relationship between PWV and arterial stiffness can be severely altered when wall thickness and/or tethering is involved, therefore leading to possible error when PWV is used as a diagnostic tool.

METHODS

Analytically, the wave speed of a pressure or flow wave in an axisymmetric straight tube is calculated by coupling the fluid flow with the wall motion at the fluid-wall interface. The fluid-wall compatibility conditions are expressed in terms of velocity and stress continuity, which are written as [1]:

$$\begin{cases} u(a,x,t) = \dfrac{\partial \xi}{\partial t}(a,x,t) \\ v(a,x,t) = \dfrac{\partial \eta}{\partial t}(a,x,t) \\ \mu\left[\dfrac{\partial v}{\partial x} + \dfrac{\partial u}{\partial r}\right] = E\left[\dfrac{\partial \eta}{\partial x} + \dfrac{\partial \xi}{\partial r}\right] \\ -p + 2\mu\dfrac{\partial v}{\partial r} = -K + 2E\dfrac{\partial \eta}{\partial r} \end{cases} \tag{1}$$

where a is the tube radius, x, t are the axial and temporal variables, u, v are the axial and radial fluid velocities, ξ, η are the axial and radial wall displacements, μ is the fluid viscosity, E is the Young modulus for the elastic wall material, p is the fluid pressure and K is the bulk modulus of elasticity of he wall.
The classical thin-wall solution of the fluid flow problem in terms of pressure and axial and radial velocities was given analytically by Womersley [1]. The solution for the wall motion in a more general form in which wall thickness is included can be found in [2]. This solution requires the displacement or tethering condition at the outer boundary of the wall.

Fully tethered wall

In a fully tethered wall with non-negligible thickness h, it is assumed that the displacements at the outer layer of the wall are zero:

$$\xi(a+h,x,t) = 0 \text{ and } \eta(a+h,x,t) = 0 \tag{2}$$

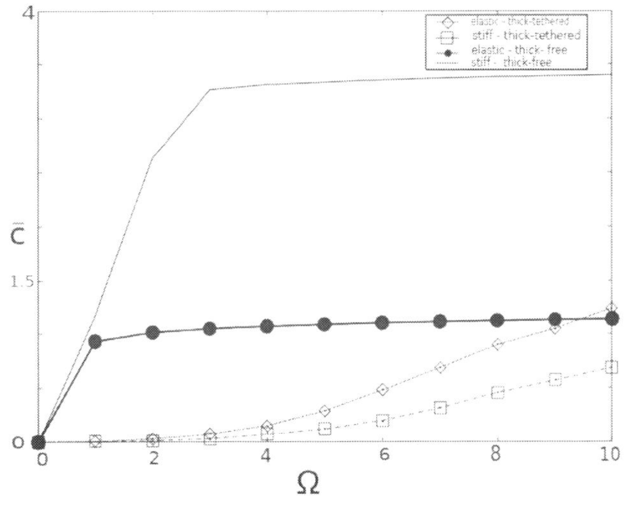

Figure 1: Non-dimensional wave speeds for thick-free (Eqs.1 and 3) and thick-tethered (Eqs.1 and 2) wall cases corresponding to elastic and stiff wall materials.

Equations 1 and 2 can be used to calculate the wave speed corresponding to a fully tethered wall with thickness h.

Free wall

To solve the boundary value problem for a thick wall with no tethering, we assume zero displacements at infinity [3]:

$$\lim_{r\to\infty} \xi(r,x,t) = 0 \text{ and } \lim_{r\to\infty} \eta(r,x,t) = 0 \qquad (3)$$

Equations 1 and 3 can be used to calculate the wave speed corresponding to a non-tethered wall with thickness h.

RESULTS

We compared the classical wave speed obtained by Womersley [1] for a thin wall with the wave speeds obtained with our mathematical model where wall thickness and external tethering were included. We shall use the terms *thick-tethered* wave speed for the solution of Eqs.1 and 2 and *thick-free* wave speed for the solution of Eqs.1 and 3. The classical Womersley wave speed will be referred to as *Womersley*. Figure 1 presents the wave speeds for the thick-free and thick-tethered wall cases for two wall materials: elastic and stiff. Figure 2 shows the Womersley wave speed together with the wave speed corresponding to the thick-free wall case. In both figures the wave speeds are normalized with respect to Moens-Korteweg wave speed and the frequency parameter Ω is a nondimensional parameter given as: $\Omega = a\sqrt{\rho\omega/\mu}$.

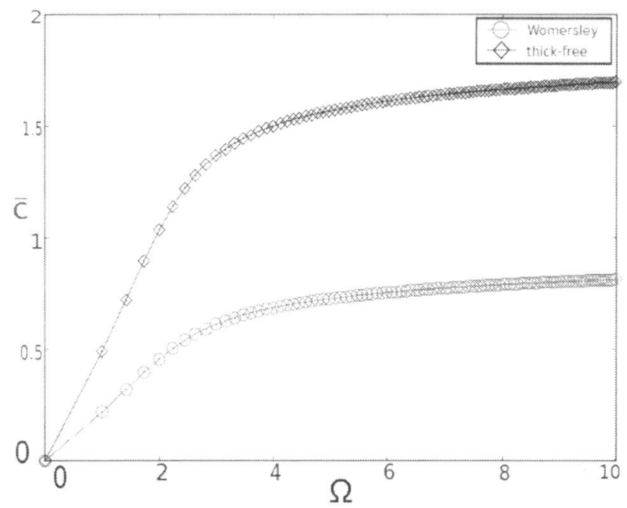

Figure 2: Wave speed corresponding to thick-free wall (Eqs.1 and 3) (diamonds), which includes the wall thickness and Womersley wave speed [1] (circles), which neglects the wall thickness with respect to frequency parameter Ω. Both wave speeds are normalized with respect to Moens-Korteweg wave speed.

DISCUSSION

Comparing our results with the classic method to calculate the wave speed, where the wall thickness is neglected, we found differences as large as 2.5 times of the Womersley wave speed values. Including external tethering in the mathematical model, we found the wave speed to be lower compared with the free wall case. The physical principle of higher pulse wave velocity in a stiffer tube is therefore strictly valid only when the wall is free from tethering, which is not always the case in the vascular system. The assumption of a 'thin' wall when calculating the pulse wave velocity, underestimates the actual PWV, which could generate a false positive vascular stiffening. On the other hand, a normal value of PWV may hide an arterial stiffening if there is strong external tethering present along the vessel wall. In conclusion, a more accurate and comprehensive technique for the assessment of arterial stiffness is needed.

REFERENCES

[1] J.R. Womersley,1957. *An elastic tube theory of pulse transmission and oscillatory flow in mammalian arteries*, WADC TR 56-614, USA.

[2] S. Hodis and M. Zamir, 2011. *Coupled radial and longitudinal displacements and stresses within the arterial wall in pulsatile flow under tethered and free-wall conditions*, Physical Review E 83.

[3] S. Hodis and M. Zamir, 2009. *Arterial wall tethering as a distant boundary condition*, Physical Review E 80, 051913.

PECLET NUMBER OF CILIARY TRANSPORT ON THE SURFACE OF THE TRACHEAL LUMEN

Kouki Kiyota (1), Hironori Ueno (2), Keiko Numayama-Tsuruta (1)
Yohsuke Imai (3), Takami Yamaguchi (1), Takuji Ishikawa (1,3)

(1) Dept. Biomedical Engineering
Tohoku University
Sendai, Japan

(2) Faculty of Education
Aichi University of Education
Aich, Japan

(3) Dept. Bioengineering and Robotics
Tohoku University
Sendai, Japan

INTRODUCTION

When we inhale air to obtain oxygen, we sometimes also inhale small particles, such as viruses and dust. These particles are trapped on the surface of the tracheal lumen and transported towards the larynx by cilia-generated mucous flow, known as the clearance function. If the clearance function does not work properly, harmful substances may remain in the body and may cause various diseases and infections.

The clearance function has been investigated by observing cilia-generated flow on the surface of the tracheal lumen. Some former studies reported the time- and space-average flow field; however, the time- and space-fluctuations of the flow have received little attention to date. Flow fluctuation is one of the key quantities in particle transport, given that transport phenomena are governed by advection, due to the bulk flow, and diffusion, due to flow fluctuations and Brownian motion.

In this paper, we report our recent results on the time- and space-fluctuations of cilia-generated flow in mice [1]. We measured the in-plane flow field at different heights of observation using a confocal micro-PTV system. Then, we quantified the effect of flow fluctuation on bulk flow by evaluating the Peclet number of the system.

METHODS

Experiments were conducted in accordance with the animal ethics review board of Tohoku University. Tracheas were obtained exclusively from wild-type mice, 4–16-weeks-old. The mice were sacrificed by cervical dislocation. The trachea was then dissected from the larynx to the bronchial main branches and doused with L-15 medium + 10% fetal bovine serum. The trachea was opened longitudinally, and 3-mm square specimens were excised. After incubating in L-15 medium with DTT, the specimens were washed with L-15 + 10% FBS. Then, we added a drop of working solution including fluorescent particles. To measure the flow field, tracers of 1-μm diameter were used.

When we measured the flow field, a confocal micro-PTV system was used, similar to that in our former study [2]. The system consists of a fluorescence upright microscope, a Nipkow lens-type confocal unit, a high-speed camera, and a ×100 objective lens. The temperature of the sample was regulated at 37°C using a thermo plate.

The trajectories of tracer particles were recorded using the confocal micro-PTV system at a frame rate of 60 fps. The observation plane was controlled by the piezo actuator. The observation height H was measured from the plane of the tips of the cilia, and particle motions in the planes with $H = 0$, 5, 10, 15, and 20 μm were measured. We note that the focal plane thickness is about 0.7μm in the present set-up. In each movie, motions of five particles during an at least 1-s period were analyzed using the ImageJ software.

RESULTS

We first tracked tracer particle motion at $H = 0$ μm in the fluorescent field, and then identified the positions of epithelial cells and cilia in the bright field at the same location. The results are shown in Figure 1. We see that particles moved from the lungs side (bottom of the figure) to the larynx side (top of the figure), on average. The trajectories were not a straight line, but erratic and unsteady. We think that the fluctuations in the trajectories were caused by the reciprocal motion of the ciliary beat and the spatial inhomogeneity of ciliated cells.

The average velocity of all particles at each height of observation plane is shown in Figure 2. The velocity was largest at the tip of cilia (*i.e.*, $H = 0$ μm) at 13.8±8.6 μm/s ($n = 20$). The velocity decays rapidly as the height increases.

Next, we quantified the spreading of particles relative to the bulk motion. We define the dispersion coefficient D', i.e. the ensemble average of square displacements of beads relative to the bulk motion, as:

$$D' = \frac{\langle R^2 \rangle}{4k\Delta t} , \qquad (1)$$

where $k\Delta t$ is the time interval. R_i is the position vector of bead i relative to the bulk motion, defined as:

$$R_i(k\Delta t) = \int_0^{k\Delta t} \left[v_{i,m}(t) - V_m(t) \right] dt , \qquad (2)$$

where $v_i(t)$ is the velocity of particle i at time t, and V_m is the average velocity of particles in movie m. $\langle R^2 \rangle$ is the ensemble average of the square displacements relative to the bulk motion. By making the time interval sufficiently large, D' may converge to the well-known diffusion coefficient if the spreading process can be described as a diffusion process.

The results of D' are plotted in Figure 3 at each height. The results of pure Brownian motion (*i.e.*, in the absence of cilia-generated flow) were also plotted in the figure for comparison. We see that the values of D' converged well when the time interval was larger than about 0.5 s. In this time scale, the beads could experience several directional changes, because the beat frequency of cilia was about 12 Hz. The beads lost directional memory and the spreading became diffusive. We can conclude that the spreading of particles, relative to bulk motion, becomes diffusive if the time scale is sufficiently larger than the beat period.

Once the dispersion coefficient is obtained, we can evaluate the Peclet number (Pe) of the system, which is the ratio of advection to diffusion, defined by:

$$\mathrm{Pe} = \frac{VL}{D'} , \qquad (3)$$

where V is the characteristic velocity, and L is the characteristic length. Pe indicates the ratio of advection to diffusion. When Pe >> 1, mass transport is dominated by advection, and particles move with the bulk fluid motion. When Pe << 1, on the other hand, mass transport is dominated by diffusion. Here, we take the characteristic velocity V as the average velocity at H = 0 μm (i.e., V = 13.8 μm/s) and the characteristic dispersion constant at H = 0μm of D' = 1.85×10^{-11} m²/s. The characteristic length, L, is taken as the typical thickness of the mucous layer on the surface of tracheal lumen, and L = 10 μm is assumed. This is because the film thickness is usually taken as the characteristic length scale for film flow in the field of fluid mechanics. Using these values, the Pe of the system can be derived as about 7.5. Because Pe is larger than unity, particles are transported more by advection than diffusion. This explains why tracer particles in the present study could be transported as a bulk even in the presence of such strong flow fluctuations.

CONCLUSIONS

In this study, we investigated the fluctuation of cilia-generated flow on the surface of the tracheal lumen. We quantified the effect of flow fluctuation on bulk flow by evaluating the Peclet number of the system. The results of Pe = 7.5, indicating that the directional transport is one order of magnitude larger than the isotropic diffusion. These results are important for understanding the transport phenomena of airways on the cellular scale.

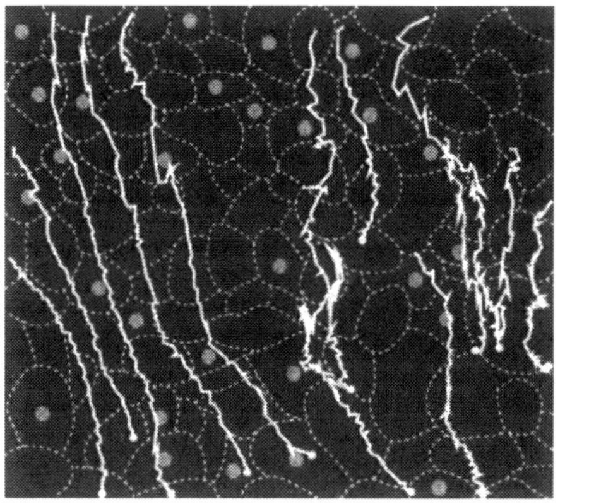

Fig.1 Trajectories of tracer particles at H = 0 μm. Borders of epithelial cells are shown by white dashed lines, and ciliated cells are indicated by gray circles. Particle trajectories are shown by white lines [1].

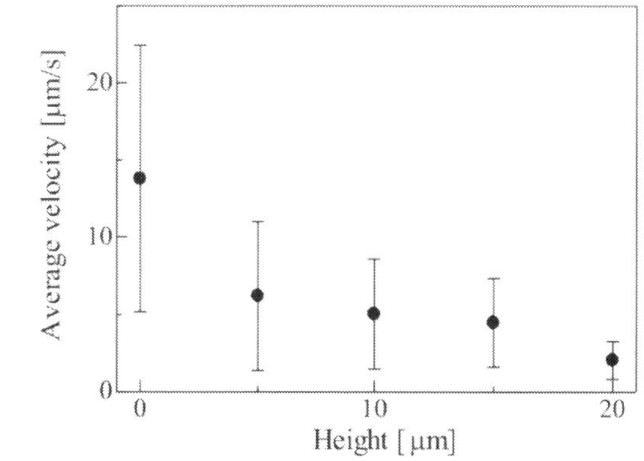

Fig. 2 Relationship between the average velocity and the height of the observation plane [1].

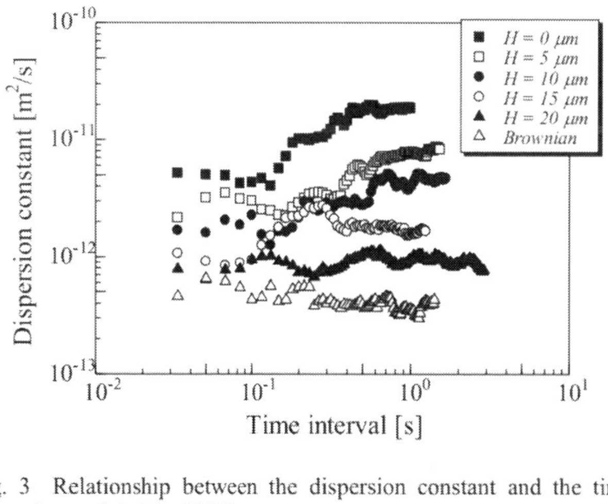

Fig. 3 Relationship between the dispersion constant and the time interval at each height H [1].

REFERENCES

[1] Kiyota, K. et al., *Am J Physiol Lung Mol Physiol*, 306:L144-L151, 2014.
[2] Saadatmand, M. et al., *J Biomech*, 44:170-175, 2011.

SB³C2015
Summer Biomechanics, Bioengineering and Biotransport Conference
June 17-20, 2015,Snowbird Resort, Utah, USA
SB³C2015-158

ENGINEERED CARDIAC MICROTISSUE MODEL REVEALS FIBROBLAST THRESHOLD FOR SYNCHRONIZED BEATING

Ariane C.C. van Spreeuwel (1,2), Noortje A.M. Bax (1,2), Christopher S. Chen (3,4,5), Carlijn V.C. Bouten (1,2)

(1) Eindhoven University of Technology
Eindhoven, the Netherlands

(2) Institute for Complex Molecular Systems
Eindhoven, the Netherlands

(3) University of Pennsylvania
Philadelphia, USA

(4) Boston University
Boston, USA

(5) Wyss Institute for Biologically
Inspired Engineering
Boston, USA

INTRODUCTION

The cardiac microenvironment is mainly composed of cardiomyocytes (CMs) and cardiac fibroblasts (cFBs), which are embedded in an extracellular matrix (ECM). Cardiac pathologies are accompanied by changes in the microenvironment of CMs, usually caused by proliferation of cFBs and increased ECM production (fibrosis). These microenvironmental changes disturb the normal contractile function of the CMs, and thereby hamper the coordinated contraction of the heart. Fibrosis changes tissue stiffness, which affects both CM [1] and cFB [2] behavior. Furthermore, an increase in cFBs causes arrhythmic contractions [3]. However, it remains unclear how the microenvironment of the CMs is affected by cFB proliferation and fibrosis and how this influences cardiac tissue contractility on the cellular level. An *in vitro* cardiac model system [4] in which both the number of cFBs and the collagen content of these microtissues were manipulated, was used to unravel the effect of cFBs and matrix accumulation on cardiac beating.

METHODS

A mixture of CMs and cFBs was isolated from neonatal mouse hearts as described before [5]. To engineer cardiac microtissues, the cardiac cells were resuspended in collagen/Matrigel and seeded in polydimethylsioxane (PDMS) microfabricated tissue gauges (μTUGs) with uniaxial constraints [4]. To create microtissues with increasing amounts of cFBs, to mimic cFB proliferation, extra cFBs (15%, 30%, and 45%) from a previous isolation were added to the cell suspension before seeding in the μTUGs. To study the effect of matrix accumulation, collagen concentration of the gel mixture was varied between 0.5, 1.5, and 2.5 mg/ml.

Displacement of the flexible microposts, caused by rhythmic contraction of the microtissues at day two of culture, was monitored using a Zeiss Observer microscope. Dynamic contraction force and beating frequency were calculated from the micropost displacements as described previously [4]. Microtissues were fixed afterwards for fluorescent staining and confocal microscopy to analyze cellular distribution and organization. A-actinin (CMs) and vimentin (cFBs) positive cells were counted in ImageJ to determine the CM/cFB ratio of the microtissues.

Data are represented as mean ± standard deviation. Statistical differences ($P<0.05$) were determined using two-way analysis of variance (ANOVA) with Bonferroni post-hoc test using GraphPad Prism Software.

RESULTS

Microtissues compacted around the microposts (Figure 1A) and started beating rhythmically at day two of culture (figure 1B).

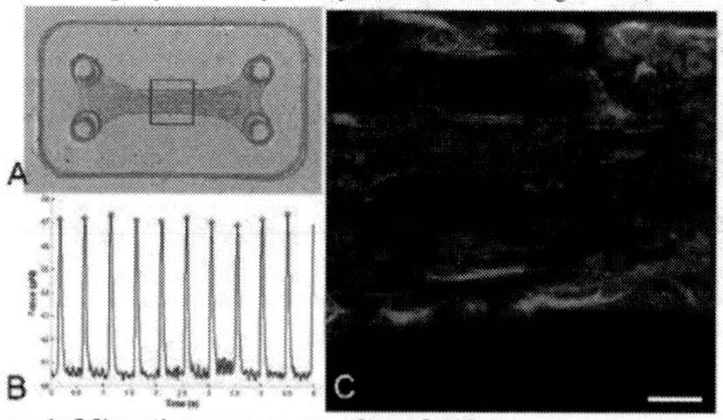

Figure 1: Microtissues were monitored (A) to measure contraction force and beating frequency (B). CMs (α-actinin, red) and cFBs (vimentin, green) aligned in the direction of the constraints (C, nuclei in blue). Scalebar represents 20μm.

Due to the uniaxial constraints, both CMs and cFBs aligned in one direction (figure 1C). Cardiac microtissues without additional cFBs (0% group) consisted of 68±5% CMs and 32±5% cFBs at day two of culture. Addition of extra fibroblasts resulted in groups with 42±6% cFBs (15% group), 62±7% cFBs (30% group), and 66±9% cFBs (45% group).

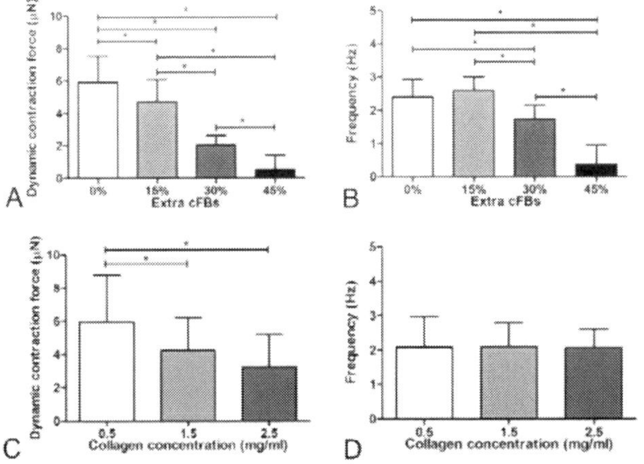

Figure 2: Addition of extra cFBs to the microtissues caused a decrease in both force (A) and frequency (B), while increasing the collagen concentration only affected force (C) but not frequency (D). N≥10 samples per group.

Microtissues of the 0% group had a mean dynamic contraction force of 5.9μN (Figure 2A) and a mean beating frequency of 2.4Hz (figure 2B). Increasing the amount of cFBs in the microtissues, caused a significant decrease in both contraction force (figure 2A) and frequency (figure 2B) reducing to almost no contraction in the 45% group. Analyzing the percentage of cFBs and frequency per microtissue revealed that the beating frequency remained ~2Hz or higher until the percentage of cFBs reached 55% and then dropped rapidly to 0Hz for most tissues (Figure 3).

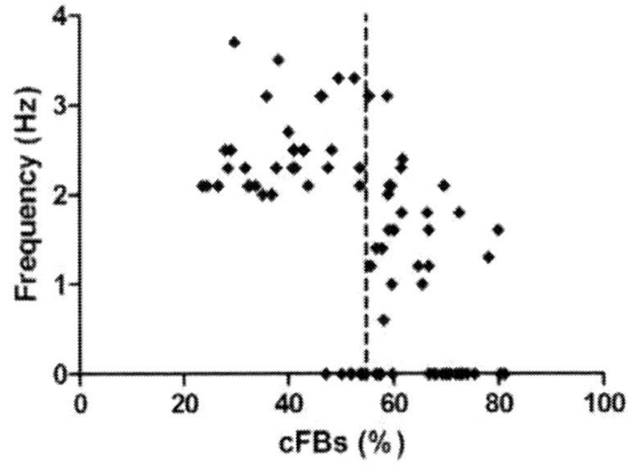

Figure 3: Beating frequency (Hz) of the cardiac microtissues dropped rapidly when the percentage of cFBs reached more than 55% (dashed line).

Next to manipulating the amount of cFBs in the microtissues, collagen concentration was also varied, mimicking various stages of fibrosis. Microtissues with the lowest collagen concentration (0.5mg/ml) had an average dynamic contraction force of 6μN and with increasing collagen content this significantly decreased to 4.3μN (1.5mg/ml) and 3.3μN (2.5mg/ml) (figure 2C). Beating frequency however, was not affected by collagen concentration of the tissue (Figure 2D).

DISCUSSION

An *in vitro* cardiac model system was used to study the effect cFB proliferation and fibrosis on cardiac tissue contractility. Our results showed that high numbers of cFBs decrease the beating frequency of cardiac microtissues. This corresponds with 2D studies that correlated a higher amount of cFBs with a decrease in conduction velocity [3]. Interestingly, our results revealed a threshold of 55% cFBs, after which synchronized beating of the microtissues is severely decreased. Like in 2D [3] it is expected that in the microtissues conduction was slowed down or completely blocked by the cFBs, which prevented contraction. Furthermore, we showed that increasing the number of cFBs not only affects beating frequency, but also decreased the dynamic contraction force that the microtissues can exert. Mimicking matrix accumulation as a characteristic of fibrosis, by increasing the collagen content of the microtissues, revealed a decrease in dynamic contraction force, but not in beating frequency. By increasing the collagen content even more, future experiments will elucidate if there is also a threshold for the collagen concentration that will affect synchronized beating. Future research is also needed to clarify if the decrease in contraction force due to accumulation of collagen is caused by increased tissue stiffness. Together, our results suggest that in treatment of heart disease, both cellular and ECM composition should be restored to regain coordinated and rhythmic contraction of the heart. This study presented a model system that is suitable for investigating pathophysiological events associated with cardiac fibrosis and may thereby facilitate in optimization of new therapies.

ACKNOWLEDGEMENTS

The authors thank Prof. Marie-Jose Goumans and Kirsten Lodder from the Leiden University Medical Center for assistance with the mouse neonatal cardiac cell isolations. This work was supported by a grant from the Dutch government to the Netherlands Institute for Regenerative Medicine (NIRM, grant No. FES0908).

REFERENCES

[1] Rodriguez A.G., *Biophys J,* 101: 2455-2464, 2011.
[2] Zhao H. et al., *Biofabrication,* 6, 2014.
[3] Askar S.F.A et al., *Cardiovasc Res,* 90:295-304, 2011.
[4] van Spreeuwel A.C.C. et al. *Integr Biol,* 6:422-429, 2014.
[5] Weeke-Klimp A. et al., *J Mol Cell Cardiol,* 49(4):606-616, 2010.

SB^{3}C2015
Summer Biomechanics, Bioengineering and Biotransport Conference
June 17-20, 2015,Snowbird Resort, Utah, USA
SB^{3}C2015-159

TEMPERATURE COMPARISON OF CATHETER ABLATION BETWEEN OPEN IRRIGATION AND VIBRATION

Kaihong Yu (1), Tetsui Yamashita (2), Shigeaki Shingyochi (3), Kazuo Matsumoto (4), Makoto Ohta (5)

(1) Graduate School of Biomedical Engineering, Tohoku University Sendai, Miyagi, Japan

(2) Cardiovascular Internal Medicine System Sec., JMS Co., Ltd. Tokyo, Japan

(3) R&D CENTER, NIDEC COPAL ELECTRONICS CORP. Sano, Tochigi, Japan

(4) Cardiology Department, Saitama Medical University Hidaka, Saitama, Japan

(5) Institute of Fluid Science, Tohoku University Sendai, Miyagi, Japan

INTRODUCTION

Radiofrequency (RF) catheter ablation is a highly effective treatment for many cardiac tachycardias. During catheter ablation, a catheter with an electrode on its tip is inserted into the heart through blood vessels and placed at the target place which causing the abnormal rhythm. Then RF current is delivered to the target tissue through the electrode on the catheter. The contacted cardiac tissue is heated by Joule heat and the deeper tissue is heated by thermal conduction. A tissue temperature ≥ 50°C causes cells lose excitability irreversibly and forms a lesion [1].

For postinfarction ventricular tachycardia, a deeper lesion is necessary because the target place may be located deeply to the endocardium [2]. Using high power is the only way to increase lesion depth. However, high power delivery causes the electrode-tissue interface overheating (interface temperature ≥ 80°C) which may lead to serious complications such as thrombus. Therefore, allowing high power delivery to deeper tissue without interface overheating is a very important issue.

Three strategies: temperature-control, larger electrode and active electrode cooling by open irrigation, have been generally accepted to prevent the interface overheating during high power delivery. However, based on the mechanism, both temperature-control and larger electrode is not effective in low blood flow because the power needs to decrease when the blood flow decrease. In open irrigation strategy, the saline in room temperature is flushing from the holes arranged in the electrode. The saline cools electrode, surrounding blood and tissue surface [3]. This active cooling by irrigation allows high power delivery with significantly low incidence of thrombus formation even in low blood flow [4]. However, about 1500 ml saline infusion during open irrigation may aggravate heart failure or pulmonary edema [5].

When open irrigation cannot be used under the conditions of low blood flow, a new method for preventing interface overheating may be required. Based on the thinking that active electrode cooling can be increased without using cooling fluid, we have proposed a new active electrode cooling by making the catheter vibrating to increase the convective cooling from blood flow. In our previous work [6], the effect of vibration frequency on electrode temperature was investigated by temperature measurement using an in-vitro system with PVA-H as ablated tissue and saline flow in an open channel. RF ablation with vibration shows a decrease on electrode temperature which indicating that RF ablation with a vibrating electrode improves electrode cooling. And the higher frequency leads to the lower temperature, indicating the higher frequency increase electrode cooling more. We also observed the flow around the catheter by Particle Image Velocimetry (PIV) method to investigate the mechanism of the cooling effect by vibration. The PIV analysis indicated that the vibrations disturbed the flow around the catheter, and the flow velocity around the catheter is increased with the higher frequency. The vibration increases the electrode cooling may result form that the vibration could increase the flow velocity around the catheter.

Both irrigation and vibration can make a cooling effect around electrode. To deeply comprehend the cooling effect of vibration, in this study, we compared these two cooling methods by measuring the temperature of electrode and internal tissue during ablation with a no flow in-vitro ablation system using PVA-H as ablated tissue.

METHODS

Figure 1 shows the schematic diagram of the experimental system. The PVA-H (70×70×20 mm) and dispersive plate were mounted on a platform and placed in a chamber filled with saline (about 3.5 L) maintained at 37°C±1°C. The advantages of using PVA-H as ablated tissue can be found in our other previous work [7]. Three thermocouples (K-type, 0.5 mm diameter, Hayashi Denko Co., Ltd., Japan) were inserted at depths of 2, 5, 8 mm. A 7-Fr open irrigation catheter (Triguy, APT MEDICAL INC., China) with a 4-mm ablation electrode was set perpendicular to the upper surface of the PVA-H. Catheter contact force was set at 2.2 gf which measured by the electronic balance (resolution 0.01 g, maximum weight 4000 g, AS PRO ASP4001F, ASONE Co., Japan) set under the chamber.

The catheter was fixed to a customized vibration unit. Vibration direction was perpendicular to the direction of thermocouple. The vibration amplitude of the catheter was set within ± 0.5 mm and the maximum frequency was 63 Hz.

Ablations were performed in three modes: normal ablation, ablation with irrigation (normal saline at room temperature, irrigation rates of 5 and 17 ml/min were used) and ablation with vibration (frequencies of 31 and 63Hz were used). Irrigations and vibrations commenced at 5 s before the onset RF ablation and were maintained until 5 s after completion of RF application. All RF applications were delivered at a constant power of 2 W for 60 s. RF energy (480 kHz) was generated by an ablator (VATION-100C, APT MEDICAL INC., China) which also allowed the recording of power, impedance, and electrode temperature at 0.1 s intervals. The averages of maximum electrode and internal temperatures during ablation were compared among the normal ablation, ablation with 17 ml/min irrigation, ablation with 5 ml/min irrigation, ablation with 31 Hz and ablation with 63 Hz. The measurement was carried out three times for each condition.

RESULTS

Figure 2 shows the comparison of the maximum electrode temperatures and internal temperatures during ablation among normal ablation, ablation with irrigation and ablation with vibration. For the temperature of electrode and internal 2 mm depth, ablation with irrigation is lower than ablation with vibration, and ablation with vibration is lower than normal ablation. The decrease on electrode and internal 2 mm depth temperature of ablation with irrigation is greater than ablation with vibration. For the temperature of internal 5 mm depth, ablation with vibration shows no change but ablation with irrigation decrease a little. All three ablation modes show no increase on the temperature of internal 8 mm depth during RF application.

The electrode temperature decreases with increasing vibration frequency or irrigation rate. The temperature of internal 2 mm depth decreases with increasing vibration frequency but shows no significant change between 5 and 17 ml/min. The temperature of internal 5 mm depth shows no difference with increasing vibration frequency or irrigation rate.

DISCUSSION

In the present study, the cooling effect on electrode and internal temperature of ablation with vibration and ablation with irrigation were compared. Although compared to irrigation the decrease of vibration on electrode and internal 2 mm depth is lower, these decrease should also be enough to cool the electrode-tissue interface. The cooling effect of vibration results from the vibration could increase the flow velocity around the catheter, the higher flow velocity increase the convective cooling surrounding electrode [6]. Irrigation directly cools the electrode, surrounding blood and tissue surface by the lower temperature of saline (room temperature) [3]. As a result, the maximum electrode temperature during ablation with irrigation is even lower than the environment temperature (37°C). And this excessive cooling by irrigation also decreases the temperature of internal 5 mm depth while vibration shows the same temperature as normal ablation. This suggests that cooling by vibration may create deeper lesion than cooling by irrigation, because the stronger cooling by irrigation may decrease the heat conducted to the deeper tissue.

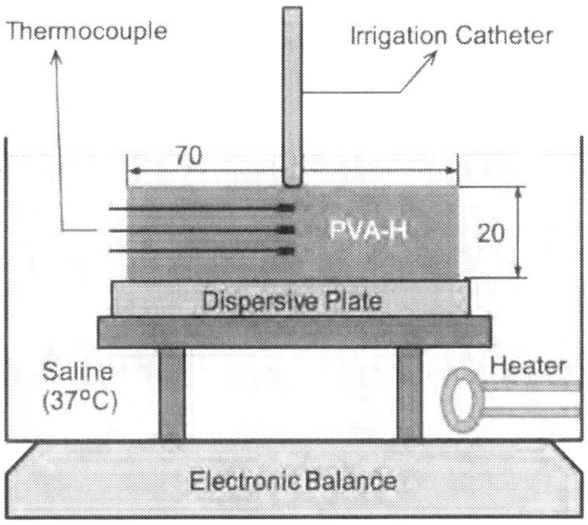

Fig. 1 Schematic Diagram of Experimental System.

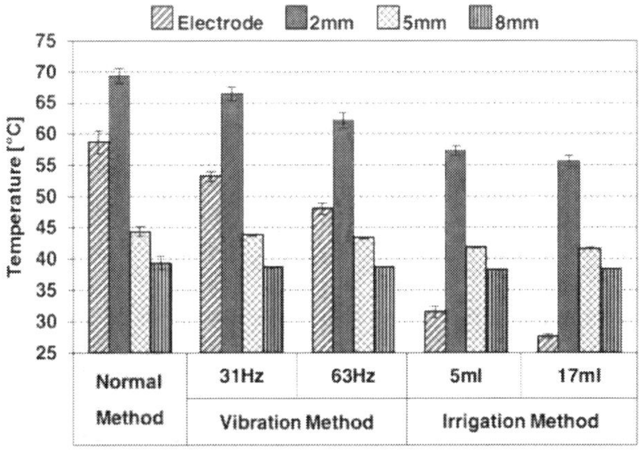

Fig. 2 Comparison of Max. Temperatures during Ablation.

Compared with vibration, open irrigation is more effective on temperature decrease. A flow rate with 5 ml/ min can provide a great temperature decrease on electrode and even a little decrease on internal 5 mm depth. However, the cooling by irrigation may be excessive then affect the tissue heating. The effect of cooling is evaluated not only by temperature decrease, but also by lesion size and incidence of thrombus formation. However, PVA-H and saline do not allow evaluation of the effects of lesion size or the subsequent incidence of thrombus formation. These parameters require further studies.

ACKNOWLEDGEMENTS

Mr. Kenshi Sakaguchi and Mr. Chikara Shindo, R&D Center, Nidec Copal Electronics Corp., all supported the development of the present cooling system.

REFERENCES

[1] Nath, S et al., Circulation, 88:1826–1831, 1993.
[2] Svenson, RH et al., J. Am. Coll. Cardiol., 15:163–170, 1990.
[3] Wittkampf, FHM et al., PACE, 29:1285–1297, 2006.
[4] Yokoyama, K et al., Circulation, 113:11–192, 2006.
[5] Stevenson, WG et al., Circulation, 118:2773–2782, 2008.
[6] Yu K., 11th International Conference on Flow Dynamics, IFS, Tohoku University, Sendai, Japan, pp. 306–307, 2014.
[7] Yu, K et al., Volume 3A: Biomedical and Biotechnology Engineering, ASME, San Diego, USA, pp. V03AT03A025, 2013.

SB³C2015
Summer Biomechanics, Bioengineering and Biotransport Conference
June 17-20, 2015,Snowbird Resort, Utah, USA
SB³C2015-160

EVALUATION OF TENDON INJURIES USING SHEAR WAVE ELASTOGRAPHY: PRELIMINARY IN-VIVO RESULTS IN HUMAN ACHILLES AND SEMITENDINOSUS TENDONS

Daniel H. Cortes (1), Stephen M. Suydam (2), Jennifer A. Zellers (3), Karin Gravare-Silbernagel (3), Thomas S. Buchanan (2), Dawn M. Elliott (1)

(1) Biomedical Engineering Program
University of Delaware
Newark, DE, USA

(2) Department of Mechanical Engineering
University of Delaware
Newark, DE, USA

(3) Department of Physical Therapy
University of Delaware
Newark, DE, USA

INTRODUCTION

Tendon injuries, such as tendinopathy and ruptures, often result in altered cross-sectional area, abnormal collagen-fiber structure, and increased hydration. These compositional and structural changes result in changes of viscoelastic properties (e.g., elasticity and viscosity) [1]. Therefore, viscoelastic properties can be used as biomarkers to quantify changes in tendon structure and function after injury and during healing. Elastic properties of the Achilles tendon have been measured in-vivo using ultrasound elastography [2, 3, 4]. However, the high stiffness of tendons, compared to other soft tissues in the body, often exceeds the range of commercial scanners [3]. Additionally, commercial systems do not provide measurements of viscosity or other viscoelastic properties.

We recently developed a technique to measure viscoelastic properties in tendons [5]. The proposed technique, named continuous Shear Wave Elastography (cSWE), calculates viscoelastic properties by applying continuous shear waves using an external actuator and measuring the wave speed at different excitation frequencies. The objective of this study is to show the feasibility of using viscoelastic properties, measured via cSWE, as biomarkers for tendon structure and function. First, measurements of viscoelastic properties in healthy and tendinopathic Achilles tendons are presented. Second, the recovery of viscoelastic properties of re-growing semitendinosus tendons (after being used as autograft in ACL reconstruction) is quantified as a function of time after surgery.

METHODS

Description and validation: cSWE is a technique designed to measure local variations, or maps, of viscoelastic properties. To do this, shear waves at several frequencies are excited in the tissue using a mechanical actuator. Displacements in the tissue are measured using ultrasound plane wave imaging. From the measured displacements, maps of shear wave speed are calculated for each of the applied frequencies using a method called Local Frequency Estimation [6]. Pixel-wise viscoelastic properties are obtained by curve-fitting a viscoelastic model to the wave speed as function of frequency. This technique was validated by comparing the values of wave speed to those measured using Magnetic Resonance Elastography.

Healthy and Tendinopathic Achilles Tendons: Viscoelastic properties of human Achilles and were measured in 29 healthy volunteers (age: 29 yrs ± 9.5, height: 173 cm ± 8.8, body weight: 70 kg ± 13.2). In addition to healthy Achilles tendons, measurements were applied to a patient (61 year old, male) with chronic tendinopathy in the right Achilles tendon. The region imaged corresponded to the location of the symptomatic area. Imaging was applied in a longitudinal view along the length of the tendon. Shear waves were excited at 323, 340, 358, 379, 403 and 430 Hz. Subjects read and signed an informed consent approved by the University of Delaware institutional review board prior to study participation.

Semitendinosus tendons: Nine subjects who sustained an ACL rupture were recruited to participate in this study. Each patient, following an ACL tear, received a semitendinosus tendon autograft and was 6-24 months post-op. For each subject, shear modulus of the semitendinosus tendons was measured in the donor tendon and intact contralateral tendon using cSWE. The elasticity recovery was defined as the ratio of shear modulus between the donor and contralateral tendon.

RESULTS

Validation: An excellent agreement between cSWE and MRE was obtained for each of the gels. The ratio of the shear modulus between cSWE and MRE was in average 0.94.

Achilles tendon measurements: cSWE applied to healthy volunteers produced nominal shear moduli from 37.8 kPa to 67.4 kPa, with a mean of 51.8 kPa and viscosity from 64.7 Pa-s to 116.5 Pa-s with a mean of 88.1 Pa-s. These values provide a range of normal viscoelastic properties that can be used as reference to compare injured tendons. For the chronic tendinopathy patient, cSWE measurements (shear elasticity and viscosity) are overlaid on a panoramic B-mode image. It can be observed that the size of the injury is larger than the field of view. The speed map and the maps for shear elasticity and viscosity show spatial variation the in the imaged region (Fig. 1).

Semitendinosus tendons: The shear modulus and viscosity of the donor tendons were often lower than those of the contralateral tendon (Figures 2 and 3). A positive correlation between time post-op and elasticity recovery was observed ($r = 0.77$) (Fig 3). This demonstrates not only tendon regrowth, but also recovery of its functional mechanical properties.

DISCUSSION

This study presents a novel ultrasound method, continuous Shear Wave Elastography, designed to measure local viscoelastic properties of tendons. This technique was validated by comparing the wave speed to that measured with magnetic resonance elastography in agarose gels of different concentrations. Measurements in the re-growing semitendinosus tendons show that it is possible to use viscoelastic properties of the tendons as biomarkers. One advantage of cSWE compared to other elastographic techniques used for tendons is that parameters such as excitation frequency can be tuned to evaluate tissues with high stiffness such as the Achilles tendon. Previous studies have shown that healthy Achilles tendon often exceeds the upper limit of commercial systems. cSWE overcomes this limitation by adjusting the frequency of the external excitation to higher values. Another advantage of cSWE is that maps of viscoelastic properties can be measured to show local variation of viscoelastic properties due to injury. Therefore, it is possible to evaluate healthy tendons as well as localized lesions such as tendinopathy, tears or ruptures using cSWE.

This study showed the feasibility of applying this technique to the evaluation of viscoelastic changes within injured tendons. cSWE was applied to a patient with chronic Achilles tendinopathy and patients with a re-growing hamstring tendon, resulting from a resection as part of an ACL autograft. In both cases, the size of the affected region was larger than the lateral field of view (38 mm). However, 0ther pathologies such as ruptures or tears, with more localized changes in structure and composition can also be monitored with cSWE.

The recovery of the shear modulus is a measure showing the ability for the semitendinosus muscle to regain its functionality and contribute to internal rotation and deep knee flexion strength. The restoration of this function has the potential to restore lost strength to the knee and prevent kinematic and loading alterations of the joint. By following the regrowth of the tendon and the restoration of the mechanical properties, a rehabilitation regimen could be developed to strengthen the once again functional semitendinosus muscle.

In conclusion, cSWE has the potential of evaluating localized changes in tendon viscoelastic properties due to injury, provide insights on the relations between mechanical properties of tendon and clinical, functional and patient-reported outcomes, and evaluate the effect of treatments on the mechanics and function of tendons.

ACKNOWLEDGEMENTS

NIH grants: R01AR050052, R21AR067390, P30-GM103333.

REFERENCES

[1] Soslowsky et al., *J Shoulder Elb Surg Am Shoulder Elb Surg Al* 9:79–84, 2000. [2] Aubry et al. *Skeletal Radiol* ;42:1143–1150, 2013. [3] DeWall et al. *J Biomech*; 47:2685–2692, 2014. [4] Brum et al., *Phys Med Biol*; 59:505–523, 2014. [5] Cortes et al., *Ultra. Med. Biol.* Submitted. [6] Knutsson et al. *Proc ICIP-94 IEEE Int Conf 1994*. pp. 36–40 vol.1.

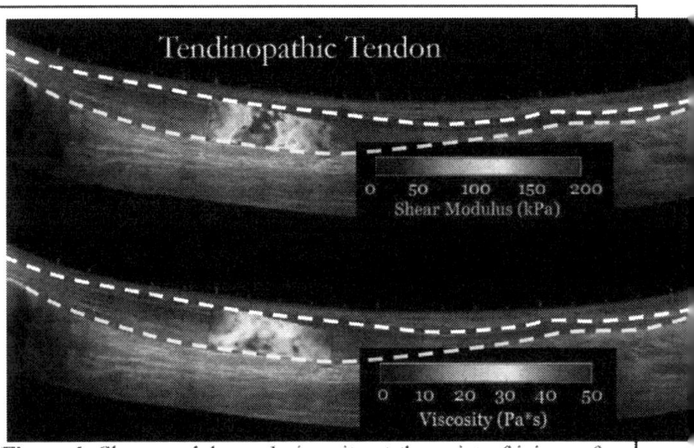

Figure 1. Shear modulus and viscosity at the point of injury of a patient (Male, 61 years old) with chronic Achilles tendinopathy.

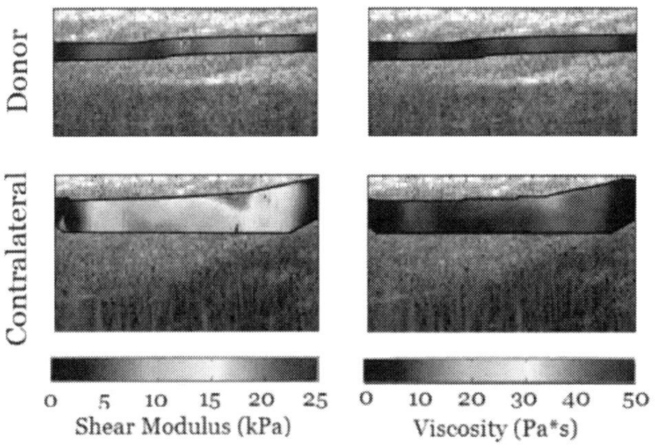

Figure 2. cSWE was applied to a re-growing (Donor) semitendinosus tendon that was used as a graft to reconstruct a ACL rupture patient (female, 19 years old). The contralateral tendon is also shown for comparison.

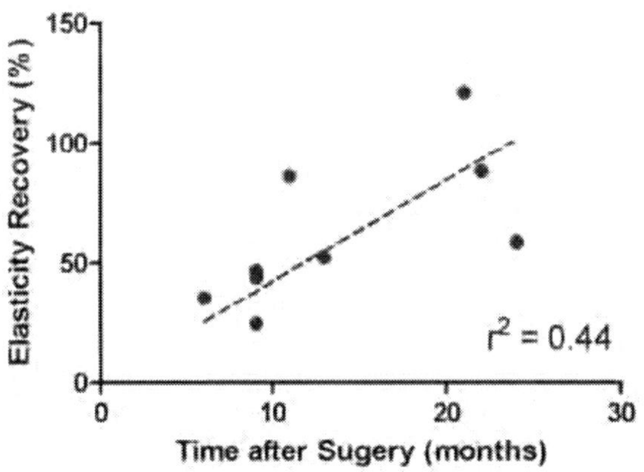

Figure 3. The elasticity recovery (ratio of shear modulus between donor and contralateral tendons) of the semitendinosus tendon increases to values close to 100% approximately 24 months after ACL reconstruction surgery.

A MODEL TO STUDY ARTICULAR CARTILAGE MECHANICAL AND BIOLOGICAL RESPONSES TO ROLLING AND SLIDING LOADS

Oliver R. Schätti (1,2,3), Luigi M. Gallo (2), Peter A. Torzilli (1)

(1) Laboratory for Soft Tissue Research
Hospital for Special Surgery
New York City, NY, USA

(2) Laboratory of Physiology and
Biomechanics of the Masticatory System
University of Zürich
Zürich, Switzerland

(3) Institute for Biomechanics
Swiss Federal Institute of Technology, ETH
Zürich, Switzerland

INTRODUCTION

The biological and mechanical responses of articular cartilage to stationary forces have been extensively studied [1,2]. In physiological conditions joint locomotion involves continuously changing contact areas and stresses over the tissue surface [3]. Such migrating contacts play an important role for the durability and wear-resistance of the tissue [4]. It is also well known that in pathological joints, such as with anterior cruciate ligament (ACL) tears, the motion paths and contact areas change [3], resulting in exposure of unconditioned cartilage to contact stresses [5]. Nevertheless, limited information exists on the impact of such physiological and pathophysiological dynamic contact stresses on cartilage biology.

The aim of this study was to develop and validate a mechanical device capable of simultaneously applying compression and rolling/sliding forces onto cartilage explants to simulate migrating joint contact. Preliminary tests were performed on mature bovine knee condyles and the mechanical and biological responses to different loads investigated. We hypothesized that higher forces and contact stresses would cause chondrocytes to increase the gene expression (mRNA) associated with tissue remodeling.

METHODS

Design The "Dynamic Articular Cartilage Test System" (DACTS) uses a spherical ball to apply a sliding or rolling axial force onto the surface of the cartilage layer on a femoral condyle. DACTS is composed of dead-weight load-frame attached to two parallel, vertically aligned rods and movable linear bearings, the frame suspended by a 111.2 N load cell and air cylinder to control the applied dead–weight force. The delrin sphere (diameter = 2.54 cm) is attached to a horizontal rod that can freely rotate on bearings or be fixed to not rotate, thus allowing for rolling or sliding contact with the cartilage (Figure 1). The condyle is fixed with a vise clamp mounted in a tank attached to a linear stage driven by a stepper motor. The stepper motor displacement (X-direction, 1 um resolution), velocity and acceleration is controlled via a computer-interface. The X-displacement was monitored using a ± 50.8 mm LVDT and the vertical Y-displacement using two ± 5.08 mm LVDTs.

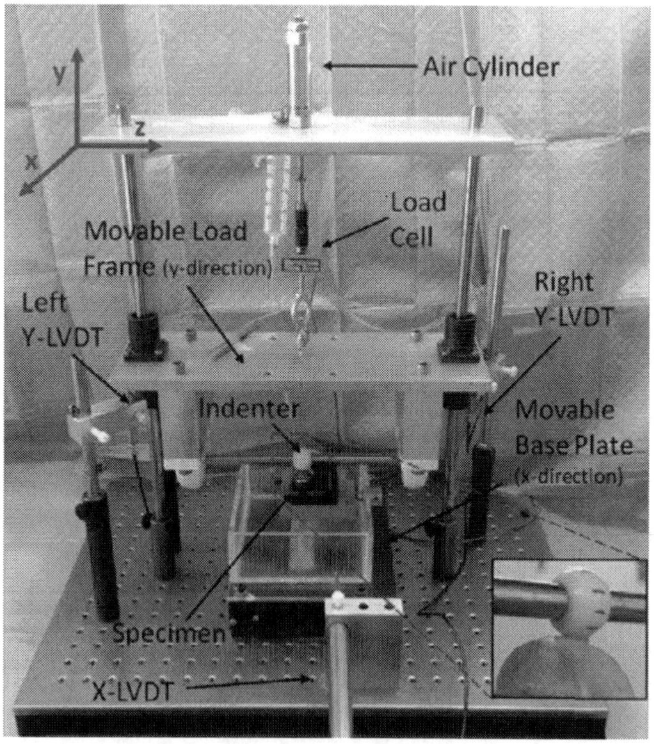

FIGURE 1. Dynamic Articular Cartilage Test System "DACTS".

Validation To validate the systems performance (accuracy and precision), test cases were run with different loads (~ 1, 2, 3 and 4 kg) and varying speeds (1, 10 and 40 mm/s) applied to a metal femoral condyle, with the x-displacement = ± 12 mm..

Mechanobiology Mature, viable bovine femoral condyles were removed and mounted into the DACTS. The cartilage was kept wet by a continuous flow of culture medium during the entire experiment. The cartilage surface was mapped by repeatedly lowering the sphere onto the cartilage with a 12 N load, yielding force-deformation and stress-strain curves which were use to determine the surface geometry (at contact) and elastic modulus (Hertzian), respectively. Following a 20-minute recovery period, the sphere was cyclically rolled on the cartilage with ± 18 mm amplitude, 10 mm/s and 12, 24, 36 or 48 N for 400 cycles. After each test, cartilage samples were harvested for mRNA analysis and then the cartilage layer removed to map the bone surface, as described above. The cartilage thickness was calculated from the surface and bone mappings. All mechanical parameters were corrected for the curved geometry of the condyle, allowing continuous calculation of the mechanical parameters (force, deformation, stress, strain, dynamic modulus) at each location on the articular surface.

For each specimen harvested, the gene response was analyzed by RT-qPCR. Total strain, stress and dynamic modulus after 400 cycles were calculated according to Hertzian theory of elastic deformation. A linear regression model was used to correlate strain, stress and dynamic modulus with mRNA for the following genes: MMP-3, fibronectin, collagen type IIA, aggrecan and sox-9. Fold expression was calculated with the ΔΔCt-method and normalized to RPLA13a and GAPDH. Adjacent unloaded cartilage specimens were used as controls. Significant interactions between the mechanical and biological parameters were determined from the Pearson's correlation coefficient (r-value) based on α-level = 0.05.

RESULTS

Validation The displacement profile used was a saw tooth with 400 mm/s^2 acceleration. Input and measured speed differed by < 2%. The air cylinder was able to apply and control the dead-weight loads for all speeds. For the velocity of 10 mm/s, the 1, 2, 3 and 4 kg loads were = 0.972 ± 0.118 kg, 1.915± 0.094 kg, 3.016± 0.067 kg, 4.026± 0.036 kg, or a COV = 12.1%, 4.9%, 2.2% and 0.9%, respectively.

Mechanobiology Independent of the load, all condyles undergo creep deformation, reaching "dynamic equilibrium" after 15-25 cycles of loading (Figure 2). The application of four different axial forces led to a variety of strains, contact stresses and dynamic moduli. The regression analysis between mechanical parameters and gene expression (N = 16) found significant positive correlations (increases in mRNA) for collagen IIA with increasing contact stress (p = 0.021) and dynamic modulus (p = 0.047), and aggrecan with increasing dynamic modulus (p = 0.027).

DISCUSSION

DACTS is capable of applying a constant rolling or sliding load to the surface of articular cartilage of osteochondral explants of different geometric shape. It can also apply a constant cyclic sliding speed over the cartilage surface while measuring the creep deformation of the extracellular matrix (ECM).

As shown by other researchers [6], this study found that a short-term application of cyclic rolling compression applied to the articular surface of cartilage resulted in substantial location-dependent changes in strain, contact stress and dynamic modulus at different locations on the condyle. Amongst other factors, this location-dependence is probably due to varying intrinsic geometric and mechanical properties of the cartilage ECM along the cartilage surface, which also caused different magnitudes of the applied force due to different sphere-surface contact angles. Chondrocytes were therefore subjected to different strains and stresses depending on their location and the mechanical properties of their surrounding ECM, resulting in altered location-dependent cellular responses in terms of collagen type IIA and aggrecan gene regulation. Both genes showed significant positive correlations with mechanical parameters. This suggests an increase in matrix turnover with higher mechanical forces. Similar trends with mechanical load have already been found [7]. Additional experiments with different axial loads and velocities will be necessary to further quantify the biological response. Modeling dynamic contact parameters to mimic gait is complex, and our DACTS is a simplified model of the effect of rolling and sliding cyclic contact mechanics applied to articular cartilage. Depending on the location, reciprocating motion introduces changes in loading/unloading frequencies that can influence the tissue's ability to reach equilibrium. This will influence the tissue's mechanical response and also its biological response, both of which need to be taken into account. In addition, we did not measure the amount of deformation recovery of the tissue's ECM between loading cycles (impacts), and therefore we had to use the total ECM deformation to calculate strains and stresses. Local spatial and temporal contact stresses and strains will need to be determined and correlated with cell spatial and temporal biological responses to fully characterize and understand the mechanobiology of chondrocytes during physiological joint motion. To summarize, this study describes a new approach to investigate the effect of rolling and sliding joint contact on cartilage mechanics and biology. It provides valuable insight into how cartilage responds to reciprocating motion, as occurs during gait. This information will be helpful in identifying mechanical and/or biological mechanisms that lead to the initiation and development of cartilage degeneration and ultimately OA, as after traumatic events such as ACL tears.

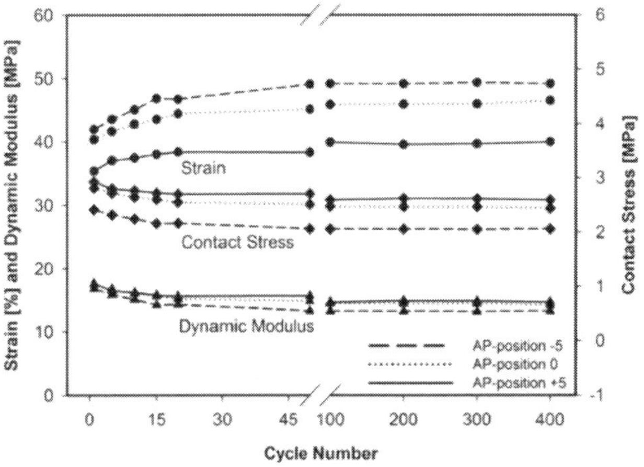

FIGURE 2. Typical strain, contact stress and dynamic modulus over 400 cycles of dynamic sliding with 10 mm/s and 36 N axial load. AP-position represents the location along the femoral condyle with respect to the apex (AP-position = 0) in centimeter. The tissue reaches a state of dynamic equilibrium within approximately 15-25 cycles.

ACKNOWLEDGEMENTS

The study was supported by the Swiss National Science Foundation Grant 325230-130715 (LMG) and National Institutes of Health Grants (NIAMS) R21-AR059203 (PAT) and (NCRR) C06-RR12538-01.

REFERENCES

[1] Buschmann, M. D. et al., *Arch. Biochem. Biophys*, 366(1):1-7, 1999.

[2] Jin, M. et al., *Arch. Biochem. Biophys*, 395(1):41-8, 2001.

[3] Kizuki, S et al., *The Knee*, 2(3):151–155, 1995.

[4] Caligaris, M and Ateshian G.A., *Osteoarthr. Cartil.*, 16(10):1220–1227, 2008.

[5] Andriacchi, T. et al., *Ann. Biomed. Eng.*, 32(3):447–457, 2004

[6] Moore, A.C. and Burris D.L., *Osteoarthr. Cartil*, 23(1):161–9, 2015

[7] Correro-Shahgaldian M. R., et al., *J Biomech*, 44(13):2472–6, 2011.

TRAPPING OF TUMOR CELLS USING RAPID ELECTROKINETIC PATTERNING (REP)

K. Clayton (1), S. Park (1), S. Wereley (1,2), and B. Han (1,2,3)

(1) School of Mechanical Engineering,
Purdue University, West Lafayette, IN, U.S

(2) Birck Nanotechnology Center
Purdue University, West Lafayette, IN, U.S

(3) Weldon School of Biomedical Engineering
Purdue University, West Lafayette, IN, U.S

INTRODUCTION

Currently, tumor spheroids are often made using techniques such as hanging drop tissue method [1]. Albeit a simple and scaffold-free approach in tissue engineering, this procedure still has its limitations. Firstly, hanging drop tissue method does not allow for controlled cell orientation. Secondly, without this control, the user cannot distinctly structure their organoid. Here, we propose an alternative method for scaffold-free three-dimensional microtissues.

The purpose of this study was to use an opto-electrokinetic technique that traps immortal cells into a cluster formation. Using electric fields for cell-based work is not new. Previously Park *et al.* used Electrowetting on Dielectric for cryopreservation methods [2]. Additionally, techniques such as dielectrophoreis (DEP) and optoelectronic tweezers (OET) have been used in cell work [3,4]. However, these electrokinetic platforms require intricate fabrication processes. Alternatively, the hybrid opto-electrokinetic technique, Rapid Electrokinetic Patterning (REP) requires minimal fabrication. In this study, REP was used in order to form cell clusters. Immortal cells are swept into a trapping area on a microfluidic chip using a 1064 nm wavelength laser beam combined with an AC electric field. This technique produces an electrothermal vortex that traps particles where the laser is focused. Not only is the trapping rapid but also easily manipulable. By trapping these cells they were able to self-assemble, forming a permanent microtissue.

MATERIALS AND METHODS

Cell Preparation

A human breast carcinoma cell line (MCF7) was maintained as an attached monolayer in commercially defined culture medium (DMEM/F12, Invitrogen, Grand Island, NY) with 5% fetal bovine serum, 2 mM L-glutamine, 100 μg/ml penicillin/streptomycin, and 10 μg/ml insulin [5]. The cells were incubated in 10-20 ml of supplemented culture medium on 75 cm^2 T-flasks at 37 °C and 5% CO_2 and were split by using 0.05% trypsin and 0.53 mM ethylenediaminetetraacetic acid at 70-80% confluency. Finally, the cultured cells were counted manually using a hemocytometer, then diluted with 1 mL of low-conductive solution (8.5% sucrose and 0.3% dextrose) with a final cell concentration of 3×10^6 cells/mL.

REP Set-Up

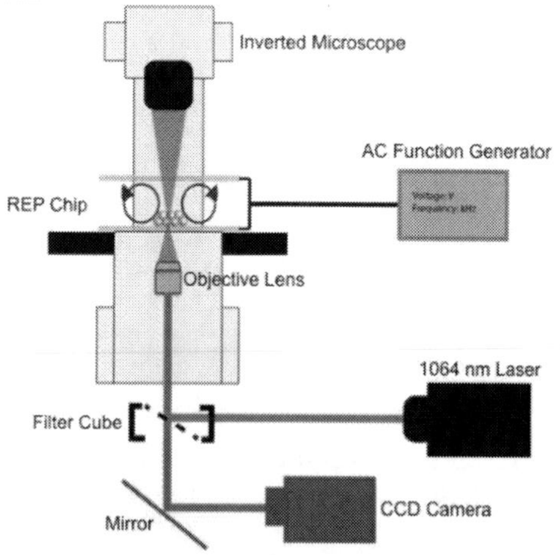

Figure 1. Experimental setup.

A microfluidic chip was prepared using Indium Tin Oxide (ITO) coated glass slides. The substrate was pre-treated with Pluronic F68 for several hours at room temperature, covered, in order to prevent cell attachment on the glass slides. Excess solution was aspirated off of the glass pieces and then dried using compressed air. After treatment, the two substrates were placed with the electrodes facing one another. A non-reactive spacer material was adhered between the electrodes (the side coated with ITO). REP was performed using an inverted microscope (Nikon TE2000U) assembled with an Nd:YAG laser illumination system (Bioryx® 200, Arryx Inc., Chicago, USA) with a low magnification objective.

REP Experiments

Prepared MCF-7 solution was fed into the REP microfluidic chip and placed on the microscope stage. The laser beam was focused on the bottom electrode of the chip at a low laser power. Additionally, the AC electric field was applied to the apparatus. After turning on the trap, images were taken in ten-minute intervals for one hour. This enabled us to see the number of individual MCF-7's that were captured and maintained within the cell cluster. After an hour, the trap was turned off. Images were recorded at ten-minute intervals, thirty minutes post-trapping. This data was analyzed to determine the maintenance of the organoid structure. Finally, to ensure cells were adhered to one another rather than the ITO substrate, the cluster was washed out of the apparatus. Visual cue would determine whether or not its overall structure was maintained.

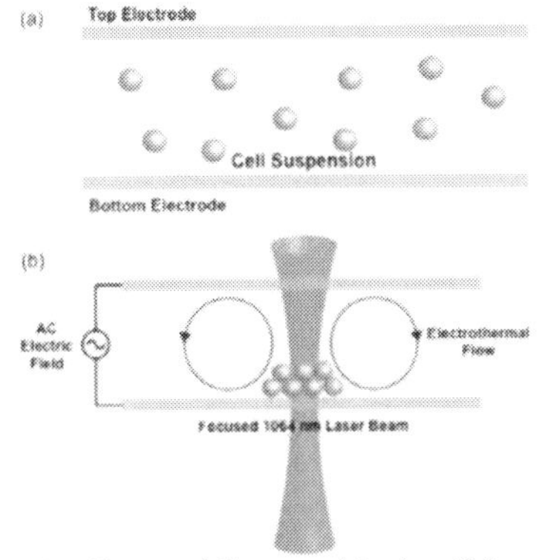

Figure 2. Electrothermal flow guided cell trapping using REP.

RESULTS AND DISCUSSIONS

Experimental Results

Within the first few minutes, eight MCF-7 cells were trapped using REP. Over time, more cells slowly traveled to the trapping site. After 20 minutes, the majority of cells were captured. For the remainder of the trapping time, the cell structure rotated counter-clockwise, yet maintained its overall shape.

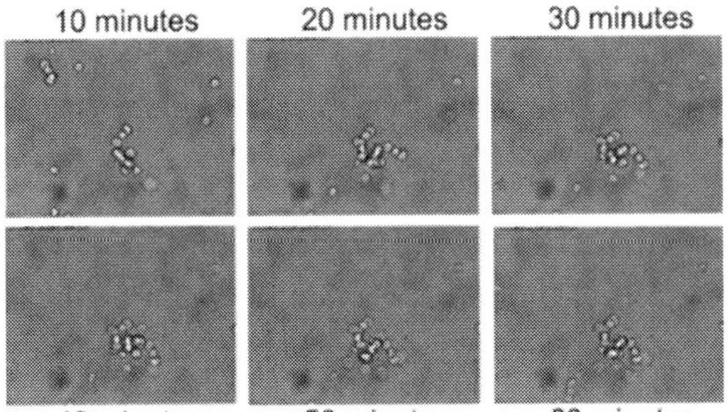

Figure 3. Time-lapse images of human breast cancer cells.

After the hour-long duration, the REP trap was released, and the cell cluster was imaged every 10 minutes. Overlaid images show that there was minimal disassembly of the cell cluster post-trap. In Figure 4, green indicates cell placement at 0 minutes and red for MCF-7 displacement at 30 minutes post- trapping. This indicates that the MCF-7 cells began to network with one another. Therefore, a microtissue structure was built and maintained during and after the hour long trapping period.

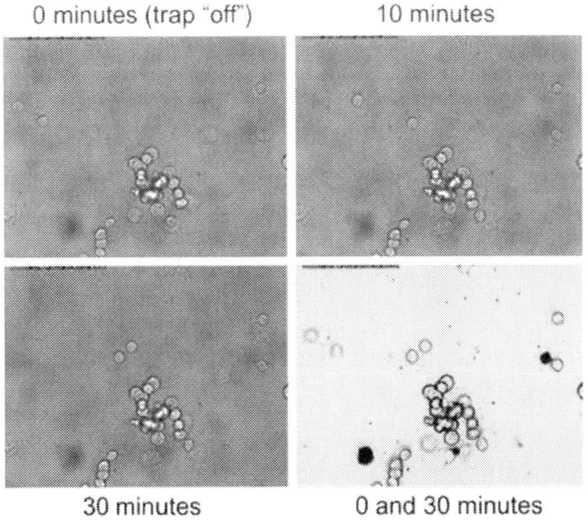

Figure 4. Post-trapping time-lapse images of MCF-7 cell line and overlaid image showing cell migration.

Cells were washed away from the chip with their respective isotonic solution. Recordings show that the MCF-7 cell cluster moved in its entirety off of the ITO substrate. This maintenance of the organoid structure during the washing step implied that a MCF-7 microtissue was formed rather than nonspecifically adhering to the electrode surface.

ACKNOWLEDGEMENTS

This work was partially supported by Research Incentive Grant and Discovery Park Fellowship from Purdue University.

REFERENCES

[1] Kelm, J. M., Timmins, N. E., Brown, C. J., Fussenegger, M., and Nielsen, L. K., 2003, "Method for generation of homogeneous multicellular tumor spheroids applicable to a wide variety of cell types.," Biotechnol. Bioeng., **83**(2), pp. 173–80.

[2] Park, S., Wijethunga, P. A. L., Moon, H., and Han, B., 2011, "On-chip characterization of cryoprotective agent mixtures using an EWOD-based digital microfluidic device.," Lab Chip, **11**(13), pp. 2212–21.

[3] Chiou, P. Y., Ohta, A. T., and Wu, M. C., 2005, "Massively parallel manipulation of single cells and microparticles using optical images.," Nature, **436**(7049), pp. 370–2.

[4] Jen, C.-P., Huang, C.-T., and Weng, C.-H., 2010, "Focusing of biological cells utilizing negative dielectrophoretic force generated by insulating structures," Microelectron. Eng., **87**(5-8), pp. 773–777.

[5] Seawright, A., Ozcelikkale, A., Dutton, C., and Han, B., 2013, "Role of cells in freezing-induced cell-fluid-matrix interactions within engineered tissues.," J. Biomech. Eng., **135**(9), p. 91001.

IMPACT OF URINARY BLADDER MATRIX (MATRISTEM™) ON VAGINAL SMOOTH MUSCLE STRUCTURE AND FUNCTION IN THE NONHUMAN PRIMATE MODEL

Katrina Knight (1), Zegbeh Jallah, PhD (1), Rui Liang, MD (2), Stacy Palcsey (2), Pamela Moalli MD, PhD (1,2), Steven D. Abramowitch, PhD (1,2)

(1) Musculoskeletal Research Center, Department of Bioengineering University of Pittsburgh Pittsburgh, PA, USA

(2) Magee Womens Research Institute, Department of Obstetrics & Gynecology Magee Womens Hospital Pittsburgh, PA, USA

INTRODUCTION

Pelvic organ prolapse (POP) is a gynecologic condition that negatively impacts a woman's quality of life and is observed in nearly half of all women over the age of 50 [1]. It is characterized by the descent of the pelvic organs into the vaginal canal, and in severe cases these organs may invert the vagina causing it to protrude outside of the body. Synthetic mesh is commonly used in the repair of POP due to the high recurrence rates observed with native tissue repairs. In prolapse repair, the purpose of synthetic mesh is to restore normal function and support to the vagina and the surrounding pelvic organs while concomitantly providing a repair that is more or equally durable to autologous tissues [2]. Although successful, mesh repairs have been associated with complications prompting the FDA to issue two public health warnings (2008 & 2011). Complications include infection, mesh shrinkage/contraction, mesh erosion or exposure, and pain, with the latter two being the most commonly reported.

The etiology of mesh complications is currently unknown. In an effort to better understand the pathogenesis of these complications, there has been an emphasis placed on understanding how the textile and structural properties of synthetic mesh impact surgical outcomes. Previously in our lab, we implanted synthetic meshes with different values of structural stiffness into non-human primates. The results of this study showed that synthetic mesh negatively impacts the structural and mechanical properties of the underlying and newly incorporated vagina, particularly the smooth muscle. The contractility, organization, and volume of vaginal smooth muscle were compromised and this effect was related to the stiffness of the mesh [4-5]. The greatest negative response occurred following implantation with Gynemesh PS, the "gold-standard" mesh for POP repair and the stiffest mesh that was evaluated in the study; nevertheless, all meshes had a negative effect on the smooth muscle. This result is significant given that studies have shown that women with POP have a decreased fractional area of smooth muscle, increased smooth muscle apoptosis, and disorganization of the smooth muscle layer [5-6]. Thus, implanting current synthetic meshes into an already comprised vagina may increase the chances of inducing a maladaptive remodeling response, which could ultimately lead to complications, such as mesh erosion.

Urinary bladder matrix (UBM- MatriStem™, ACell, Inc.) has been widely used in various tissue engineering applications for its ability to remodel into the host tissue types to which it is in contact and for its ability to positively impact the immune response. With the idea that UBM may be able to preserve and/or regenerate smooth muscle, the objective of this study was to assess the effect of UBM, when placed between the mesh and vagina, on vaginal smooth muscle contractility, organization, volume, and apoptosis at 3 months following synthetic mesh implantation. We hypothesize that combining UBM with synthetic meshes will induce a more favorable tissue remodeling and regenerative response, as measured by an increase in smooth muscle contractility and volume, smooth muscle orientation similar to sham controls, and decreased apoptosis relative to mesh implantation without UBM. To test this hypothesis, active smooth muscle function was analyzed via a smooth muscle contractility assay and histology was used to observe changes in smooth muscle structure and apoptosis.

METHODS

Surgery: The University of Pittsburgh Institutional Animal Care and Use Committee approved the protocol in this study. Animals were housed in a temperature and light-controlled room with access to food and water *ad libitum*. A total of 16 parous non-human primates (Rhesus Macaques, age 7 to 16) were used in this study. On the day of surgery, animals were randomly assigned to the following groups:

Sham-operated controls (n=8), Gynemesh PS (n=6), Gynemesh PS+UBM, 2-ply (n=8), and UBM, 6-ply (n=8). Sham-operated controls and Gynemesh PS are groups from our previously published work [3-4]. Meshes were implanted via abdominal sacral colpopexy after hysterectomy. After three months, animals were euthanized and the mesh-tissue complexes were explanted. Specimens were taken from the proximal vagina where the mesh was implanted. The viability of the explanted specimens was maintained by submerging the specimens in oxygenated (95% O_2-5% CO_2) Krebs'-bicarbonate buffer. For functional analysis of the smooth muscle, specimens were tested within 30 minutes of tissue harvesting. For histological analysis, samples from the explanted mesh-tissue complex were embedded in O.C.T. compound (Sakura Finetek USA, Inc., Torrance, CA, USA) and cryosectioned at 7 μm thick.

Functional Analysis: Two to three circumferentially oriented strips, approximately 5 mm wide x 8 mm long, were cut from each explanted specimen and clamped using custom tissue clamps. Clamped strips were then attached to a force transducer (Transbridge 4M, World Precision Instruments) and suspended in a 37°C tissue organ bath containing oxygenated Kreb's-bicarbonate buffer. The forces generated throughout testing were recorded using Chart 7 software with a PowerLab system (AD Instruments) at a sampling rate of 40 Hz. To begin functional analysis, each strip was preloaded to 1.0 g of tension along its length and allowed to equilibrate for 60 minutes. The buffer solution was changed every 20 minutes throughout this equilibration period. Immediately following equilibration, a single dose of 120 mM KCl, a muscle myofiber stimulant, was applied and the resulting isometric force generated was measured and used to assess smooth muscle function. This force was normalized by the tissue volume (mN/mm^3). Strips were then washed and exposed to electrical field stimulation (EFS:1-64 Hz) as a means to measure nerve mediated contractility. For EFS, the contractile force was normalized by the 120 mM KCl response and expressed as a percentage. Only the maximum force generated for each strip was reported.

Histological Analysis: Tissue sections were fixed, blocked with 2% bovine serum albumin, and a primary antibody was added. A mouse anti-smooth muscle actin secondary antibody (α-SMA; Sigma, St Louis, MO, USA) was used to label the smooth muscle. The TUNEL assay (Roche, Branchburg, NJ, USA) was then conducted according to the manufacturer's protocol to assess apoptosis. Images were taken at 10X using a 90i Nikon Light-Fluorescence microscope.

Statistical Analysis: Statistical analysis was performed using SPSS 22.0 statistical software (IBM, Armonk, NY). Differences in the KCl response between groups were compared using the Welch alternative for the F-ratio followed by a Games-Howell post-hoc test. The maximum EFS response was compared between groups using one-way ANOVA with a Gabriel's post-hoc test. The significance level was set to $p<0.05$.

RESULTS

In comparison to sham, the force generated in response to 120 mM KCl was significantly decrease by 70% following implantation with Gynemesh PS, p=.009. There was no significant difference in force generated in response to 120 mM KCl for Gynemesh PS+UBM (p=.107) and UBM (p=.592) groups in comparison to sham-operated controls. Additionally, the force generated following EFS was approximately 50% lower when Gynemesh PS was implanted. There was no significant difference observed between Gynemesh PS+UBM and UBM relative to sham in response to EFS. The smooth muscle layer for the Gynemesh PS+UBM and UBM groups consisted of dense and organized smooth muscle bundles similar to that of the sham-operated controls (Figure 1). This is in contrast to the disorganized and thin smooth muscle layer following implantation with Gynemesh PS, as seen previously [4]. In this study, cell apoptosis was also significantly increased around the fibers of Gynemesh PS (p=.024). This response was not observed when Gynemesh PS was implanted with UBM.

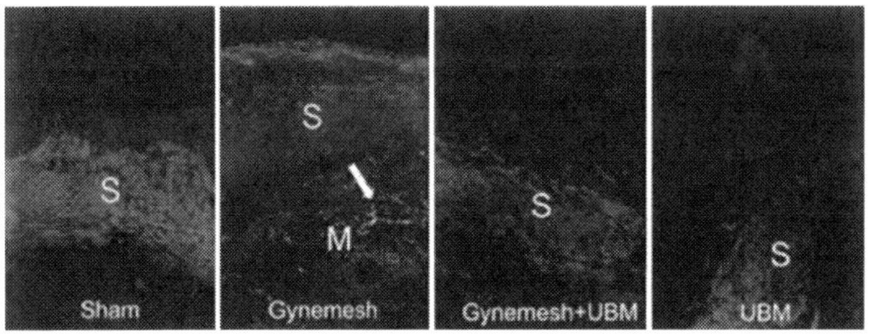

FIGURE 1: Images depict fluorescently labeled smooth muscle layer (S, red signal), cell nuclei (blue signal), and cell apoptosis (arrow, green signal). M indicates mesh fiber area.

DISCUSSION

Overall, the results of this study suggest that UBM, when used in combination with synthetic mesh, has the ability to maintain the functionality, organization, and structure of vaginal smooth muscle over the short term. Previously, implantation with Gynemesh PS was shown to significantly lower the contractile ability of smooth muscle by approximately 70%. In this study, it was demonstrated that when Gynemesh PS is used in combination with UBM, the contractility of the smooth muscle can be restored to levels similar to that of sham-operated controls. Additionally, UBM was shown to not interfere with vaginal smooth muscle organization and structure.

The mechanism that is responsible for the positive effects of UBM in this study is unclear; however, it is believed that UBM initially provides a buffer between the mesh and the vagina thus shielding the vagina from direct contact with the mesh. As the UBM slowly degrades, it is able to regenerate tissue and change the local host response to a more reconstructive remodeling response that is favorable for tissue incorporation as described by Badylak [7].

The major strength of this study is the use of the non-human primate model, an accepted model for studying pelvic floor disorders. Despite this strength, there are some limitations including the small sample size and the short implantation time. Nevertheless, given the current controversy over the use of synthetic meshes due to mesh complications, this study is very timely and relevant. Future studies will aim to quantify changes in the innervation density and to assess the immune response in response to UBM.

ACKNOWLEDGEMENTS

We are grateful to the NSF Graduate Research Fellowship (NSF GRF DGE-1247842), to the NIH for funding support R01 HD061811, and to ACell, Incorporated for providing the UBM and funding.

REFERENCES

[1] Subak, LL et al., *Obstet Gynecol*, 98:646-651, 2001. [2] Chen, CC et al., *Clin Obstet Gynecol*, 50:383-411, 2007. [3] Feola, A et al., *BJOG*, 120:224-232, 2013. [4] Liang R et al., *BJOG*, 120:233-243, 2012. [5] Takacs, P et al., *Int Urogynecol J Pelvic Floor Dysfunct*, 19:1559-1564, 2008. [6] Boreham, MK et al., *Am J Obstet Gynecol*, 187:56-63, 2002. [7] Badylak, SF. *Ann Biomed*, 42:1517-1527, 2014.

EFFECT OF HEAT SHOCK ON CRYOPRESERVATION OF ADIPOSE TISSUE DERIVED STEM CELLS

M. Shaik[1], J. Gimble[2], R. Devireddy[1]

[1]Department of Mechanical Engineering
Louisiana State University
Baton Rouge, Louisiana, USA

[2]LaCell Incorporation
New Orleans, Louisiana, USA

INTRODUCTION

Human adipose-derived stromal/stem cells (huASCs) can differentiate along multiple lineage pathways, including adipocytes, chondrocytes, myocytes, neurons and osteoblasts [1]. As new stem-cell-based therapies become the standard of care for a wide variety of diseases, appropriate cryopreservation protocols will be necessary to increase patient access, reduce cost, and enhance the safety and effectiveness of huASCs. Recent studies from our laboratories have described the in vitro cryopreservation of huASCs frozen in a media consisting of minimal amounts of dimethyl sulfoxide (DMSO) and serum (animal or human) [2]. However, the infusion of cells cryopreserved in DMSO has been associated in nearly all patients with nausea, chills, hypotension, dyspnea and cardiac arrhythmia [3]. Therefore, the practical goals of this study are the development and optimization of a serum-free and DMSO-free cryoprotective medium.

Instigating the cell's natural cytoprotective mechanisms like heat shock response, Keap-Nrf2 pathway, ubiquitin protease system could help in alleviating the damages associated with cryopreservation on cells thus reducing the amount of cryoprotectants used. Numerous reports in literature suggest that the up regulation of Heat Shock Proteins (HSPs) can provide cytoprotection in various diseases [4, 5]. Therefore, this study is aimed to evaluate the effect of heat shock response on cryopreservation of huASCs by determining the post thaw cell viability of heat shocked huASCs cryopreserved in the presence and absence of cryoprotectants (CPAs).

METHODS

Cell culture: The huASCs were obtained from Lacell incorporation, New Orleans and cultured till passage 2 using the medium containing DMEM, 10% FBS, and 1% antibiotic in a cell culture incubator at 37 °C.

Heat shock: The cells were given heat shock by exposing them to 43°C for 1 hour in a cell culture incubator. After the heat shock the media was replaced for both heat shocked cells and their controls followed by culturing at 37°C for the next 16 hours before further analysis.

Western blot: RIPA buffer is used to lyse the cells and the total protein content is estimated using BCA assay kit. For the western blot, 25 ug of protein is loaded per lane in polyacrylamide gel and SDS PAGE is carried out followed by the transferring of proteins on to nitrocellulose membrane. The antibodies (Santa Cruz Biotechnology) for HSP90, HSP70, HSP27, and β-actin in the dilutions of 1: 200, 1: 200, 1: 5000, and 1: 200 were used for blotting.

Cryopreservation/Thawing: The huASCs are cryopreserved at a cooling rate of -1°C /min in different groups with 10% CPA, 1% CPA, and no CPA with CPAs either DMSO or polyvinylpyrrolidone (PVP) using control rate freezer till -35°C and plunged in liquid nitrogen. Thawing was performed using 37°C water bath incubator by shaking gently for 1-2 minutes.

Cell viability assay: After thawing the cells were recovered by culturing for 24 hours. The live cells attached to surface of culture flask were obtained by trypsinisation and the dead floating cells in the medium were collected. The post thaw cell viability is assessed using the apoptosis kit (BD Biosciences) that uses annexin V/ propidium iodide.

RESULTS

Western blotting: The expression of HSPs is found to be increased following the heat shock (43°C for 1 hour).

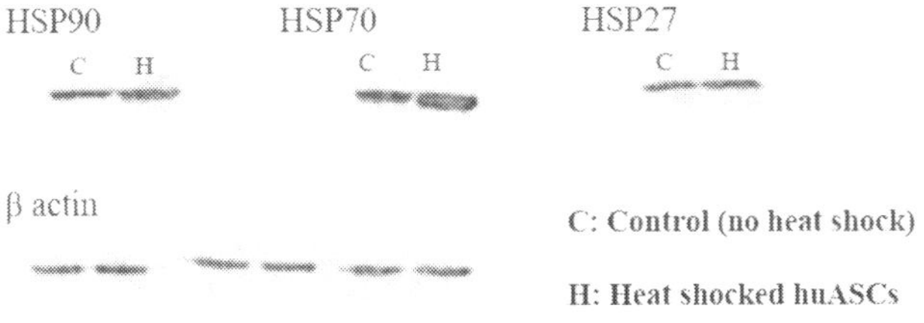

The HSP expression remains unchanged in the post-thawed heat shocked huASCs

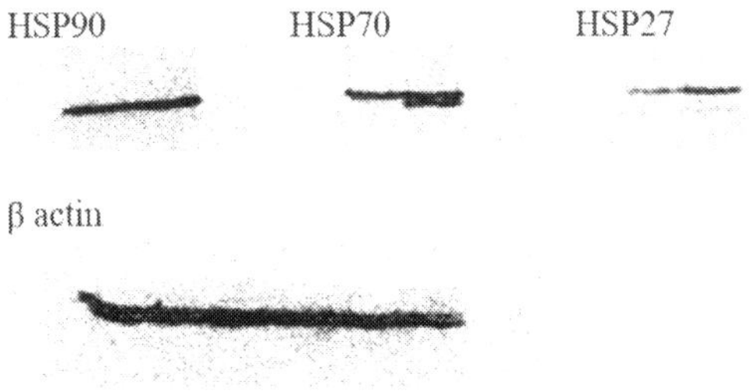

Post thaw cell viability: The bar graph below indicates the post thaw cell viability of heat shocked huASCs cryopreserved either with DMSO or PVP or without any CPA.

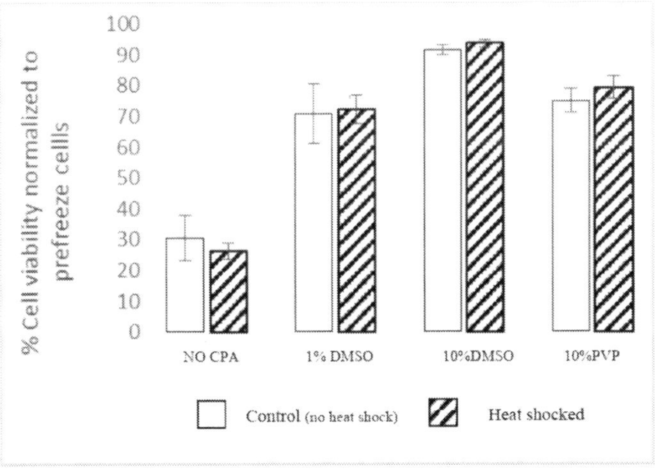

DISCUSSION

It is well known that HSPs can be overexpressed by mild thermal stress and act as molecular chaperones which assist in proper folding of misfolded proteins. Several reports suggest that preconditioning the cells to induce the HSPs expression offers the protection against apoptosis, ischemic/reperfusion injuries of heart and brain tissues, and oxidative stress [4].

The western blotting results indicate the increase in expression of HSPs 90 and 70 after heat shock. Comparing the post thaw cell viability of heat shocked huASCs cryopreserved without CPAs with the controls indicate that overexpression of HSPs alone cannot provide cytoprotection from the freezing damage caused due to cryopreservation. Therefore, the CPAs were included during the cryopreservation of heat shocked huASCs to determine whether the overexpression of HSPs improve the post thaw survival rate in the presence CPAs. It was found that induction of HSPs has slightly improved the cell viability rate than the controls but not significantly. Both intracellular CPA, DMSO, and extracellular CPA, PVP are used to see if the type of CPA effects the cell viability while HSPs are overexpressed. We found that the type of CPA either intracellular or extracellular does not seem to effect cell viability rate in a significant way in heat shocked huASCs in comparison to controls. Therefore, the next step is to enhance cell's other cytoprotective pathways like Keap-Nrf2 pathway or ubiquitin protease system or optimal enhancement of all the different pathways together or in different combinations so that a better cryopreservation protocol to replace CPAs or to minimize their usage could be developed.

ACKNOWLEDGEMENTS

Research reported in this was supported by institute of Diabetes and Digestive and Kidney Diseases of the National Institutes of Health under award number R21DK91852. The content is solely the responsibility of the authors and does not necessarily represent the official views of the National Institutes of Health.

REFERENCES

[1] Zuk, PA et al., *Tissue Eng,*7 (2):211-28, 2001
[2] Thirumula, S et al., *J Tissue Eng Regen Med.* 4(3):224-32, 2010
[3] David, B et al., *Am J Hematol.* 82(6): 463–472, 2007.
[4] Kalmar, B et al., *Adv Drug Deliv Rev.* 28;61(4):310-8, 2009
[5] Sharp, FR et al., *Transl Stroke Res.* 4(6):685-92, 2013

SB³C2015
Summer Biomechanics, Bioengineering and Biotransport Conference
June 17-20, 2015,Snowbird Resort, Utah, USA
SB³C2015-166

A PATIENT-SPECIFIC COMPUTATIONAL MODEL TO CHARECTRIZE THE IMPACT OF NEURAL TISSUE MOTION ON CEREBROSPINAL FLUID DYNAMICS AT THE CERVICAL-MEDULLARY JUNCTION

S. H. Pahlavian (1, 2), F. Loth (1, 2), M. Luciano (3), B. A. Martin (1, 2)

(1) Conquer Chiari Research Center, The University of Akron, Akron, OH, U.S.A.

(2) Department of Mechanical Engineering, The University of Akron, Akron, OH, U.S.A.

(3) Department of Pediatric Neurosurgery, Cleveland Clinic Foundation, Cleveland, OH, U.S.A.

INTRODUCTION

Abnormal morphology of cervical subarachnoid space (SAS) and the resulting alterations in cerebrospinal fluid (CSF) hydrodynamics have been thought to be important in pathophysiology of central nervous system (CNS) disorders such as of type 1 Chiai malformation (CMI) and syringomyelia [1,2]. Near the cervical-medullary junction (CMJ), up to ~2 mm bulk motion of the spinal cord and cerebral tonsils has been observed *in vivo* [3] and this motion has been found to be greater in in CMI patients. The impact of this motion on CSF dynamics is not clear. Subject-specific computational fluid dynamics (CFD) simulations have been used to non-invasively understand CSF flow dynamics by application of variational analysis techniques that incorporated various anatomical features and physiological functions [4,5]. The goal of this study was to investigate the importance of the CNS tissue motion on CSF dynamics using a subject-specific moving-boundary CFD model of the CMJ of a CMI patient.

METHODS

T2-weighted MRI images of an adult CMI patient (with tonsillar descent of 7.9 mm) were manually segmented to reconstruct the 3D anatomical geometry of the SAS near the CMJ extending from foramen magnum (FM) to C2 axial level. The segmented geometry was post-

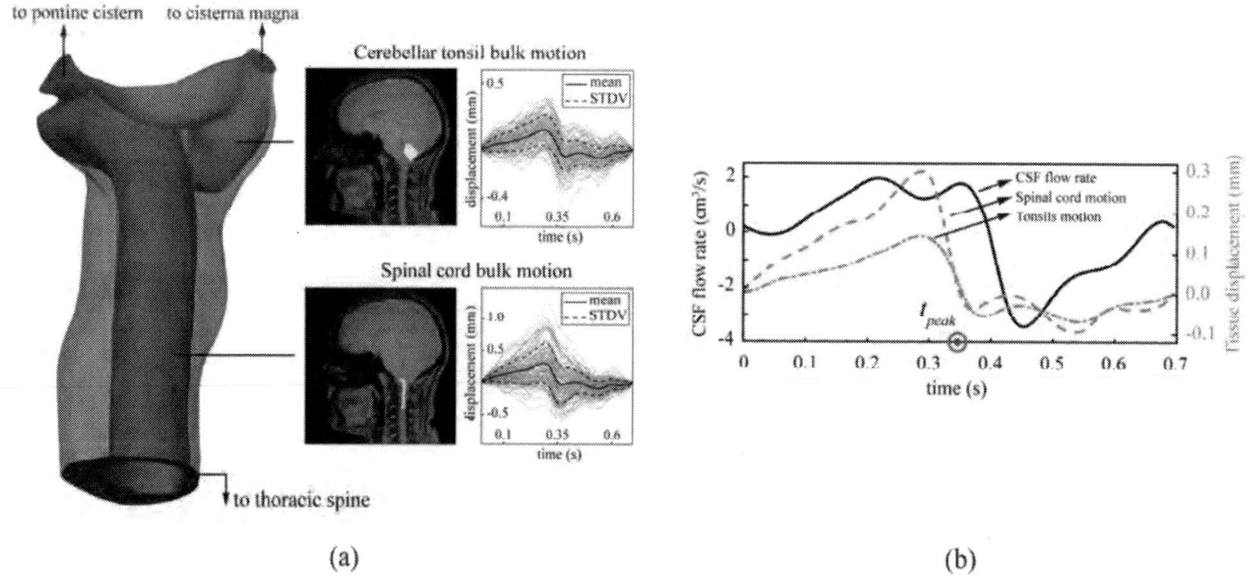

Figure 1. a) Rendering of the reconstructed geometry of the upper cervical spinal subarachnoid space. Insets show the PCMRI mask selections and the tissue voxel displacement waveforms (mean and STDV) prescribed at the tonsil and spinal cord. b) Mean motion and volumetric flow rate waveforms used for the CFD simulation. Time point t_{peak} corresponds to the peak velocity of neural tissues.

processed to remove the pixilation artifacts and its cranial and caudal ends were extended to minimize the impact of inlet and outlet boundary conditions on the domain of interest. Maya software (Autodesk Inc. CA, USA) was used to extract and define the surface of the cerebellar tonsils and spinal cord from the overall geometry so that the motion profiles could be imposed on each structure separately (figure 1 (a)). CFD simulations were performed using the commercially available finite volume solver ANSYS FLUENT (ANSYS Inc., Canonsburg, USA). The velocity inlet boundary condition was imposed at the caudal end of the geometry and was obtained from through-plane phase contrast MRI (PCMRI) measurements acquired on the same axial plane. The caudocranial motion of CNS tissues were obtained from a high-temporal resolution PCMRI sequence and the resulting waveforms were imposed on their corresponding tissues. Various hydrodynamic parameters were compared between fully rigid (baseline) and dynamic models on six axial planes between FM and C2. An additional model was simulated to investigate the impact of maximum static alteration in the SAS geometry corresponding to the peak diastolic ascent of cerebellar tonsils and spinal cord in the cranial direction (diastolic model).

RESULTS

The impact of CNS tissue motion on CSF hydrodynamic parameters was more pronounced during peak tissue velocity in the caudal direction (t_{peak} in Figure 1 (b)). CNS tissue motion increased the pressure dissociation between the FM and C2 axial planes up to 120% and also caused a transient increase of pressure on the surfaces of tonsils and spinal cord (Figure 2 (a)). This increase was most noticeable on the surface of the tonsils. Inclusion of CNS tissue motion in the dynamic model increased the peak CSF velocities near the craniovertebral junction up to 60% (Figure 2 (b)). Differences in peak velocities were less pronounced at planes near the caudal end of the geometry. Inclusion of the CNS tissue motion increased the duration and magnitude of the bidirectional CSF flow at the six axial planes analyzed by 77±50% and 79±39% (mean ± standard deviation), respectively. The static geometric alteration in diastolic model had relatively small impact on the variation of pressure dissociation (<5%) and peak velocities (<15%).

DISCUSSION

The findings of this study show that a small cardiac related CNS tissue motion (150 - 300 μm) has an important impact on CSF dynamics in the upper cervical SAS, in terms of pressure and velocity-based parameters. The transient increase in hydrodynamic parameters was more evident during the movement of CNS tissues with their peak velocities. This was due to the great alteration of CSF flow in the dynamic model during the same time range. The increased pressure acting on the surface of CNS tissues in the dynamic model may change the mechanical properties of CNS tissues by mechanotrasduction. This could contribute to neurological symptoms in various craniospinal disorders such as CMI [6].

The dynamic pulsatile action of the CNS tissues was found to have a major impact on CSF velocities. The driving factor leading to alterations in CSF velocities was the CNS tissue velocity (e.g. abruptness of tissue motion) and not tissue static displacement. Maximum impact of CNS tissue motion on CSF velocities occurred at the time point of peak CNS tissue velocity (t_{peak}, Figure 1 (b)) and *not* peak CNS tissue displacement. This was the case, despite having a smaller SAS cross-sectional area for the CSF to flow at the latter time point.

The usage of CFD modeling was helpful to understand alterations in CSF hydrodynamics around the brain and spinal cord due to CNS tissue motion. Future studies are needed to investigate the possible long-term pathophysiological effects of altered CSF hydrodynamics that act on the delicate CNS tissue and meninges.

ACKNOWLEDGEMENTS

The authors acknowledge Conquer Chiari and National Institutes of Health (NIH) Grant No. 1R15NS071455-01 for support of this work.

REFERENCES

[1] Shaffer N, et al. *Neurological research*, 33(3):247-60, 2011.
[2] Martin BA, et al., *PLoS ONE*, 8(10):e75335, 2013.
[3] Cousins J, et al., *Am J of Neuro Biomech*, 30(8):1587-8, 2009.
[4] Heidari Pahlavian S, et al., *PLoS ONE*, 9(4):e91888, 2014.
[5] Cheng S, et al., *J Biomech*, 45(7):1186-91, 2012.
[6] Heiss JD,et al., *J Neurosurg*, 91(4):553-62, 1999.

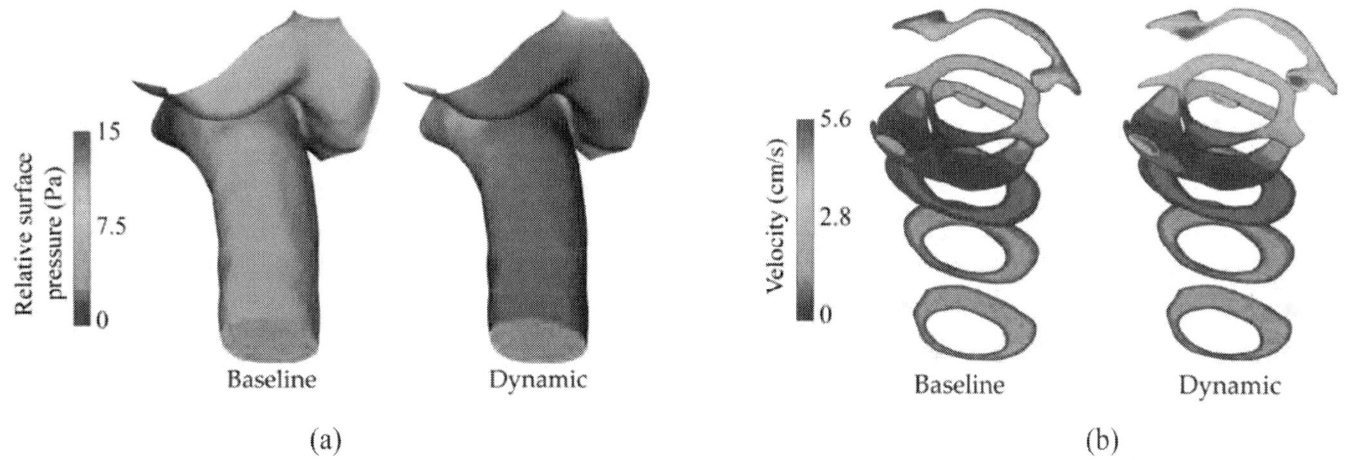

Figure 2. a) Relative pressure distribution acting on the surface of spinal cord and tonsils, plotted at the time corresponding to the peak difference between the two models (t_{peak}). b) Comparison of CSF velocity distributions around the spinal cord at different axial planes in the baseline and dynamic models; plotted at the time corresponding to peak neural tissue velocity (t_{peak}).

SB^{3}C2015
Summer Biomechanics, Bioengineering and Biotransport Conference
June 17-20, 2015, Snowbird Resort, Utah, USA
SB^{3}C2015-168

SIMULATION-DIRECTED DESIGN OF PLANAR BIAXIAL TESTS ON THE LUMBAR FACET CAPSULAR LIGAMENT

Amy A. Claeson and Victor H. Barocas

Department of Biomedical Engineering
University of Minnesota
Minneapolis, MN, 55455

INTRODUCTION

The lumbar facet joint (FJ) is composed of the facet capsular ligament (FCL), a flexible, strong and highly aligned collagenous tissue, which guides the relative motion of the articular facets (facets) from adjacent vertebrae. The posteriorly located bilateral FJs flank the spinous processes at every spinal level and lie on the same plane as the IVDs. Embedded within the highly aligned collagen fiber bundles are mechanoreceptors that may be involved in proprioceptive and nociceptive signaling pathways[1]. Differential stimuli, including localized tissue strain, may activate the mechanoreceptors to initiate pain- or position-based signals.

Uniaxial and planar biaxial mechanical tests are routinely performed to determine material properties of anisotropic tissues. Because of the lumbar FCL's highly aligned and medially-laterally oriented collagen fiber bundles[2], it is intuitive to implement a biaxial test of the FCL with the fiber axis aligned with a testing axis. This geometry does not, however, apply the shear deformation that occurs during lumbar flexion. Little et al. reported[3] the mechanical properties of cadaveric lumbar FCLs determined by two sets of planar uniaxial tests, parallel and perpendicular to the aligned fiber bundles. To our knowledge, mechanical data from planar biaxial tests of cadaveric lumbar FCLs, whether on-axis or shear, have not been published outside of our group, which performed on-axis experiments[4].

The aim of this study is to determine the ideal orientation for a lumbar FCL sample in planar biaxial mechanical testing to match the deformation that occurs during lumbar FCL motion during flexion.

METHODS

FJ displacements and rotations from healthy subjects and mechanical properties from planar biaxial tests of cadaveric lumbar FCLs were implemented into a lumbar FCL FE model (from micro-CT of a healthy cadaveric FJ) to elucidate the magnitude and angle of the resulting principal strains. Figure 1 outlines the general procedure, and the following paragraphs expand upon each boxed section.

Vertebral Motion: Sagittal plane fluoroscopic data of the lumbar spine were obtained from fifteen young, healthy subjects bending forward in flexion and returning to a neutral position. Coordinates of the L3-S1 vertebral bodies were recorded in ImageJ to extract the deformation of the L3-L4 posterior IVD margin at the maximum lumbar flexion angle. Vertebral displacements and rotations were calculated[5] and translated from the local coordinate axis to the facet coordinate axis through rigid body assumptions, because individual facet features were indistinguishable in the sagittal plane.

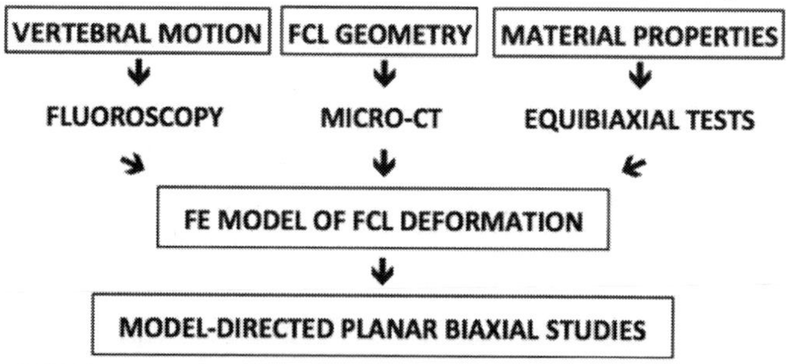

FIGURE 1. OBTAINING MODEL-DIRECTED PARAMETERS

Facet Capsule Geometry: A cadaveric L3-L4 facet capsule was isolated from a motion segment for micro-CT imaging. Image segmentation in Mimics (Figure 2A) facilitated surface area, thickness (Figure 2B) and curvature measurements. A simplified FE model with these dimensions (Figure 2C) was created and meshed in Abaqus for importation into FEBio[6]. The mesh was partitioned to reflect the anterior attachment of the FCL to the rigid facets. The facets were treated as rigid, and the FCL was defined as a transversely-isotropic Neo-Hookean (NH) solid with horizontally aligned fibers. The L4 facet was held fixed while superior and anterior displacements, and sagittal plane rotation were prescribed to the L3 facet.

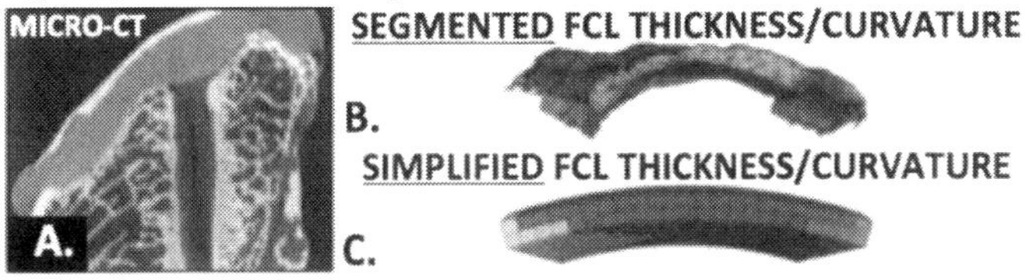

FIGURE 2. (A) SAGITTAL PLANE MICRO-CT OF A L3-L4 FACET CAPSULE GIVES THE (B) SEGMENTED GEOMETRY USED TO CREATE THE (C) SIMPLIFIED FCL FE MODEL

FCL Material Property Estimation: Cleared L3-L4 FCLs (n=3) were resected from the motion segments with facet attachments remaining intact. The samples were mounted with the fibers aligned with one test axis, and the surface was speckled for displacement tracking. A preload of 1.5N was applied before implementing ten cycles of preconditioning followed by a range of one arm, two arm, and equibiaxial extensions. Grip force-stretch data and motion-capture were recorded. Surface displacements were calculated[7], and the displacement fields and grip forces for an equibiaxial extension were fit to a four-parameter structural model[4]. The planar equibiaxial geometry for each test was simulated in PreView and FEBio was used to tune the parameters of the transversely-isotropic NH model to the structural model. Parameter values c1 = 80.5 kPa, c3 = 15.9 MPa, c4 = 0.800, and k = 0.805 MPa, were selected as the mean of the lowest residual fits.

FE Model of FCL Deformation During Joint Motion: The subject-specific FJ motion data (n=10) and the average material fit from the cadaveric biaxial tests were implemented in the FE model. The *en face* axis of the first and second principal strains were recorded from the simulation results.

Model-Directed Planar Equibiaxial Studies: Additional L3-L4 cadaveric FCLs (n=3) were dissected from all bony attachments and trimmed to a cruciform geometry such that the fiber axis was aligned with the average first principal strain (E1) axis. The specimens were equibiaxially extended to the average middle-body E1 magnitude and a test from each sample was fit to the four-parameter structural model. These cruciform tests were also simulated with the above parameters to determine if the model-directed equibiaxial tests were necessary.

RESULTS

In the FE FCL simulations, flexion was guided by the L3 and L4 facets from adjacent vertebrae (Figure 3A) and produced average middle-body first principal strains (E1) of 14.9 ± 5.30% across the posterior surface (Figure 3B). The medial and lateral edges of the FCL model exhibited a lower magnitude E1 because these elements were directly above the rigid facets. The surface projection of the three-dimensional (3D) E1 vector was 50.8 ± 1.80° (Figure 3B).

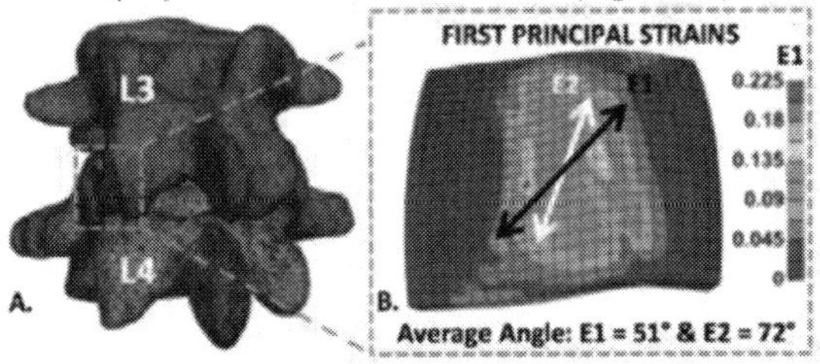

FIGURE 3. (A) FCL ATTACHED TO ADJACENT FACETS OF THE L3-L4 MOTION SEGMENT. (B) REPRESENTATIVE FCL SIMULATION E1 FIELD IN FLEXION

The 3D E2 vector was primarily oriented perpendicular to the FCL surface. The average FCL surface projection was 71.5 ± 5.60° (Figure 3B), although average magnitude was small 0.43 ± 0.37%.

Cadaveric L3-L4 FCLs, with rotated fiber orientation, equibiaxially extended at 15% grip strain (Figure 4A) produced tightly bound low magnitude forces (Figure 4B). Once the FCL was detached from the facets, its small size posed difficulty in creating cruciforms with the desired fiber rotation. Thus, the resulting samples had a fiber orientation of 30° rather than the 50° of the mean E1 vector. Nonetheless, the off-axis fiber rotation and detachment from the facets reduced the stiffness of the L3-L4 FCLs compared to the on-axis fibers with the physiologic bony attachments[4]. In addition, anisotropic effects were also minimized, in that both axes exhibited relatively similar force-stretch responses. The fits to the four-parameter model agreed with the experimental data, outputting low values for stiffness, non-linearity and degree of fiber alignment.

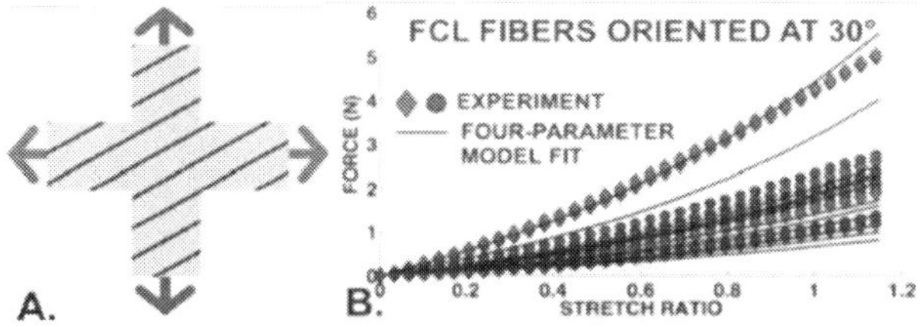

FIGURE 4. (A) SCHEMATIC OF A LUMBAR FCL WITH A 30° FIBER ORIENTATION. EACH AXIS WAS DISPLACED AT 15% GRIP STRAIN. (B) FORCE-STRETCH EXPERIMENTAL AND FITTED DATA FOR ALL ROTATED FIBER SAMPLES

Simulations of the rotated fiber equibiaxial tests with the fitted transverse-isotropic NH parameters were inconclusive; hence, these mechanical tests are necessary to mitigate shear deformation effects.

DISCUSSION

FE simulations of lumbar FCL deformation via flexion show the first principal strain (E1) axes are not aligned with the horizontally aligned collagen fiber bundles. Thus, planar mechanical tests aiming to extract the mechanical properties of the lumbar FCL in flexion should rotate the fiber axis approximately 50°. In contrast, reorientation of the fiber axis is unnecessary if aiming to replicate lateral bending or axial rotation because deformation occurs inline with the horizontal fibers. Additionally, calculated FJ displacements and rotations prescribed in the FE model were within range of previously reported measurements[8] of FJ kinematics during lumbar spine motions. A limitation of this study is the lack of lumbar spine CT data for subjects participating in the analyzed fluoroscopic study. The lumbar FJ varies notably in size and orientation between individuals, therefore, geometry-matched simulations may alter the E1 fields reported here. Nonetheless, this study affirms the importance of kinematic analysis and FE simulations in the design of planar biaxial experiments.

ACKNOWLEDGEMENTS

This work is supported by NIH/NIAMS T32 AR050938 Musculoskeletal Training Grant and NIH grant U01-EB016638. We also thank the University of Minnesota Anatomy Bequest Program, the Minnesota Supercomputing Institute, and Dr. Alex Fok.

REFERENCES

[1] Johnson, GM et al., J Man Ther, 9:71-76, 2004. [2] Yamashita, T et al., Spine, 21:538-543, 1996. [3] Little, JS et al., J Biomech Eng, 127:15-24, 2005. [4] Nagel, TM et al., J Biomech Eng, 136:114501, 2014. [5] Nagel, TM et al., Eur Spine, J 23:754-761, 2014. [6] Maas, SA et al., J Biomech Eng, 134:011005, 2012. [7] Ragupathy, R et al., J Biomech Eng, 133:091011, 2011. [8] Kozanek, M et al., Spine, 19:E689-E696, 2009.

MODELING THE ENHANCEMENT OF EXTRACELLULAR MATRIX QUANTITY AND QUALITY IN LARGE-DEFORMATION MECHANICALLY-CONDITIONED HEART VALVE TISSUE ENGINEERING

Joao S. Soares (1), John A. Stella (2), Antonio D'Amore (2), Will Zhang (1)
Willam R. Wagner (2), John E. Mayer, Jr. (3), Michael S. Sacks (1)

(1) Center for Cardiovascular Simulation
Institute for Computational Engineering and Sciences
University of Texas at Austin
Austin, Texas, USA

(2) Department of Bioengineering
McGowan Institute for Regenerative Medicine
University of Pittsburgh
Pittsburgh, Pennsylvania, USA

(3) Department of Cardiac Surgery
Boston Children's Hospital
Harvard Medical School
Boston, Massachusetts, USA

INTRODUCTION

To date, our understanding of how various forms of biophysical stimulation promote newly formed extracellular matrix (ECM) mass accretion and bulk mechanical behavior remains limited. More specifically, mechanical conditioning is widely accepted as one of the most relevant physical cues for influencing ECM mass accretion, mechanical response, and topology. Several experimental studies have been conducted to investigate how cyclic flexion enhances matrix stiffness and the effect of strain magnitude and modality on de novo collagen architecture [1,2]. Despite these efforts, a major goal remains to develop a thorough understanding of the underlying structural and mechanical factors responsible for the initial formation and functional development of optimal ECM structures in engineered tissues.

METHODS

Construct Fabrication and Cell Source. Poly(ester urethane)urea (PEEU) scaffolds micro-integrated with vascular smooth muscle cells (VSMCs) were manufactured by electrospraying and electrospinning [3]. Electrospinning produces continuous fibrous scaffolds with mechanical behavior very similar to native soft collagenous tissues and suitable environment for cell proliferation and growth. Control of fiber alignment is obtained by variations of the rotation velocity of the mandrel, resulting in anisotropic scaffolds with preferred and cross-preferred fiber directions. VSMCs were isolated from Lewis rat aortas, expanded, and micro-integrated through electrospraying. The tubular construct was placed in static culture media and cut in the longitudinal direction into individual ring specimens.

Stretch Bioreactor Study and Conditioning Regimes. The tension bioreactor employed in this study is similar to the bioreactor of Merryman et al. [4], with modifications enabling better deformation control. The bioreactor applies uni-directional tension to 8 ring specimens simultaneously in separate wells to limit the risk of cross-contamination (Figure 1). The biosynthetic response of micro-integrated VSMC ring constructs was investigated at 1 Hz cycles for three culture durations of 7, 14 and 21 days and 3 strain levels: low (15%), intermediate (30%) and high (50%). Acellular, 0 day, and statically cultured groups were carried as controls.

Biochemical Assays and Histology. DNA was quantified with a technique adapted from Kim et al. [5]. Collagen and glycosaminoglycans (GAG) were assayed with techniques adapted from Brown et al. [6]. Histological evaluations were carried out with hematoxylin and eosin (H&E) stain for morphology and with picrosirius red, a collagen-specific stain. An image analysis based protocol was implemented to determine ECM area fraction from digitized histological sections.

Mechanical Evaluation with Biaxial Testing. Square specimens were cut from micro-integrated constructs mechanically conditioned for 21 days at 30% strain and prepared for planar biaxial mechanical testing under stress control (methodology described in detail in [7]). The method in Sacks [8] was employed to analyze the experimental data collected (fiducial marker displacement and forces). Specimens

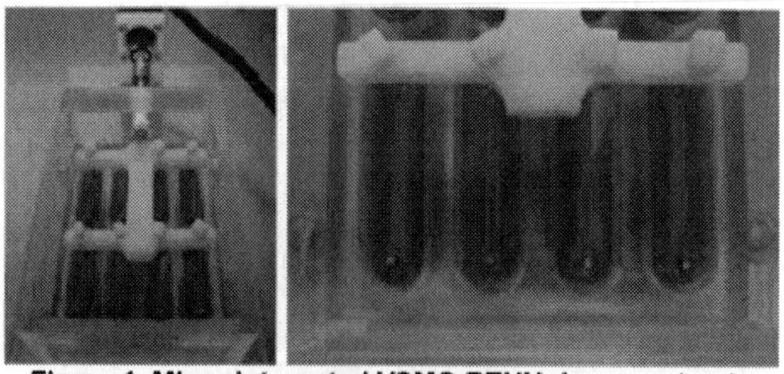

Figure 1. Micro-integrated VSMC-PEUU ring constructs inside the uniaxial tension bioreactor

were preconditioned for 10 cycles under equibiaxial stress up to the maximum testing stress. The post-preconditioning 0.5g tare configuration was chosen as the reference state and the testing protocol consisted of axial stress ratio controlled paths (t_{11}:t_{22} = 30:300, 150:300, 225:300, 300:300, 300:225, 300:150, 300:30).

Extracellular Matrix Degradation. The polymer phase of cultured constructs was isolated by placing the specimens in 0.25% Trypsin EDTA for 24 hours at 37 °C. The ECM degradation protocol was validated by visual comparison of H&E stained sections before and after degradation. The same biaxial testing protocol was again used after degradation to quantify the PEUU scaffold mechanical behavior. This approach enables the direct comparison of the biaxial mechanical response of the engineered construct with its polymeric scaffold.

Modeling the De Novo ECM Mechanical Behavior. Assuming zero energy is stored at initial configuration, the stored energy density is

$$\psi = \int S_{11} dE_{11} + \int S_{22} dE_{22} + \int S_{12} dE_{12}$$

where S_{ij} and E_{ij} are the components of the 2nd Piola-Kirchhoff stress and Green strain. Due to the nature of the biaxial experiment, the shear component can be neglected. The stress-controlled loading paths define consistent convex surfaces in (E_{11}, E_{22}, ψ)-space, which we fit using a 2x2 2D-biquintic finite element interpolating patches [9]. A Sobolev norm-based penalty function was added to ensure stability and regularize the fitted surface by restricting its length and curvature. Assuming that no stored energy is associated with ECM-scaffold interactions, direct subtraction of fitted-surfaces corresponding to before and after ECM degradation, i.e. ψ_{S+M} and ψ_S respectively, results in the stored energy density surface of the de novo ECM

$$\psi_M = \psi_{S+M} - \psi_S$$

defined on the common (E_{11}, E_{22}) range. To obtain further insight of the mechanical response of the de novo ECM, we fit the ψ_M surface to a structural constitutive model of the form [10]

$$\psi = \phi_{ECM} \frac{\eta}{2} \int_{-\pi/2}^{\pi/2} \Gamma(\theta) \left[\int_0^{E_{ens}} D(x) \frac{E_{ens} - x}{(1+2x)^2} dx \right] d\theta$$

where ϕ_{ECM} is the mass fraction of ECM, η is the stiffness of the collagen fibers, $\Gamma(\theta)$ is the fiber orientation distribution defined on $\theta \in [-\pi/2, \pi/2]$, and $D(x)$ is the recruitment distribution function defined on $x \in [0, E_{ens}]$ with the upper bound of integration being the fiber ensemble strain $E_{ens} = E_{11} + E_{22}$. The fiber orientation distribution Γ is a linear combination of a Gaussian distribution and uniform distribution and the recruitment function D is a Beta distribution. A set of structural model parameters was obtained for each individual sample. Subsequently, all responses were coalesced into an average response of the de novo ECM (n=4), and compared with experimental results of untreated bovine pericardium (BP) [11].

RESULTS & DISCUSSION

Mechanical training induced dimensional changes: (i) construct thickening due to ECM deposition was prevalent in the 30% strain regime with average thickening of 18.8 ± 3.1 % at 21 days while acellular specimens exhibited a decrease of 25.0 ± 7.6 %; (ii) measurable elongation occurred, with the acellular scaffolds creeping significantly more than micro-integrated constructs.

Augmented collagen synthesis at 30% cyclic strain was observed, whereas constructs subjected to 15 or 50% strain showed minimal increases over static and day 0 specimens (Figure 2). A steady GAG production was observed in all conditioning regimes, which seems to be driven primarily by the application of cyclic deformation regardless of strain level. Biochemical assay results were corroborated by histological assessment. After 14 days cell and ECM rich regions were present, formed lamellae, and reduced construct delamination. Histological evaluation confirmed the biological component removal with the degradation protocol. Acellular specimens subjected to the degradation protocol showed no significant change in the biaxial response, ensuring that it does not adversely affect scaffold structure and mechanical behavior.

Systematic mechanical evaluation of constructs before and after degradation on a per specimen basis showed consistent compliance increases and allowed the determination of the mechanical contribution of the de novo ECM (Figure 3). Subsequent modeling indicates that the ECM phase exhibits a highly nonlinear response, however it is mildly anisotropic. Fits to the structural model resulted in collagen fiber stiffness of 15.36 ± 5.23 MPa, about half of 30.99 MPa of BP but much higher than ever obtained with infinitesimal deformations [1]. Comparable fiber orientations were observed; however, fiber recruitment occurred at minimal or at zero strains (about 8% strain in BP). Lack of undulation and mild anisotropy indicates inferior microstructural organization and the insufficient collagen maturation of highly specific dense connective tissues.

DISCUSSION

Improved means of understanding tissue formation and remodeling in engineered tissues is needed to achieve functionally equivalent tissues. Empirically based work has yet to elucidate and evaluate tissue formation and function. Here we demonstrate the role of large-deformation mechanical conditioning in de novo ECM quantity and quality with a two-pronged approach connecting critical experimentation and experimentally-driven modeling.

ACKNOWLEDGEMENTS

NIH/NHLBI Grants R01 HL- 068816 and R01 HL-089750.

REFERENCES

[1] Engelmayr GC, et al., *Biomech Model Mechanobiol*, 7:309-21, 2008; [2] de Jonge N, et al., *Ann Biomed Eng*, 41:763-74, 2013; [3] Stankus JJ, et al., *Biomaterials*, 27:735-44, 2006; [4] Merryman WD, et al., *Cardiovasc Pathol*, 16:268-76, 2007; [5] Kim YJ, et al., *Anal Biochem*, 174:168-76, 1988; [6] Brown AN, et al., *Tissue Eng*, 6:297-305, 2000; [7] Billiar KL, et al., *J Biomech Eng*, 122:23-30, 2000; [8] Sacks M, *J Elasticity*, 61:199-246, 2000; [9] Smith DB, et al., *Ann Biomed Eng*, 28:598-611, 2000; [10] Fan R, et al., *J Biomech* 47:2043-54, 2014; [11] Sacks MS, et al., *Ann Biomed Eng*, 26:892-902, 1998.

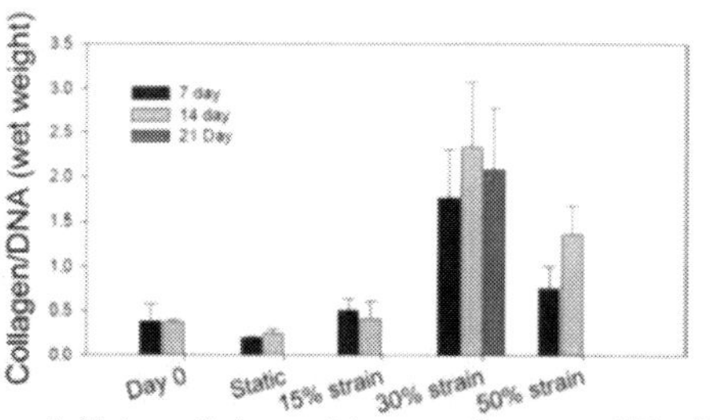

Figure 2. Extracellular matrix accretion quantification

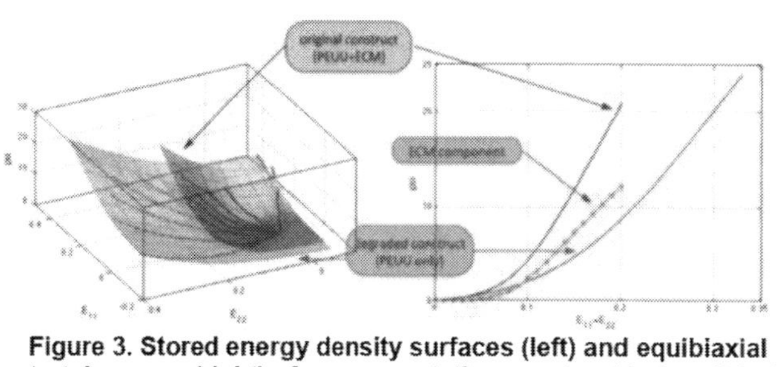

Figure 3. Stored energy density surfaces (left) and equibiaxial stretch curves (right) of a representative construct before (blue) and after (red) trypsin degradation. Stored energy density of the ECM component (green) is obtained as their subtraction.

SB³C2015
Summer Biomechanics, Bioengineering and Biotransport Conference
June 17-20, 2015,Snowbird Resort, Utah, USA
SB³C2015-170

MATHEMATICAL MODELING OF THROMBUS FORMATION IN IDEALIZED MODELS OF AORTIC DISSECTION

Claudia Menichini (1), Xiao Y. Xu (1)

(1) Department of Chemical Engineering
Imperial College London
London, UK

INTRODUCTION

Aortic dissection is one of the most catastrophic diseases that can affect the aorta. This disease is triggered by the formation of a tear on the inner surface of the aorta which allows blood to flow into the aortic wall. The surge of diverted blood causes the separation of the inner and outer layers of the aortic wall and leads to the formation of a second blood-filled channel defined as the "false lumen". Dissection of the aortic wall compromises the normal function of the circulatory system and can cause severe complications such as aortic insufficiency, malperfusion syndromes, or rupture, resulting in high mortality rates [1]. Depending on morphological features of the dissection and the risk for complications, patients will be given either surgical, medical, or endovascular treatments.

The development and progression of aortic dissection causes pronounced morphological changes which impact blood flow and hemodynamic parameters such as shear stress and pressure, and creates local conditions which are likely to promote thrombus formation. Previous research has highlighted that patient prognosis is influenced by the level of thrombosis in the false lumen for both endovascular and medical treatments [2]. While partial thrombosis of the false lumen has been associated with higher mortality risks, complete thrombosis has been shown to improve the effectiveness of medical and endovascular treatments and survival chances. However, the dynamics related to thrombus formation are still poorly understood, and it is not possible to predict which patients will develop false lumen thrombosis and to what extent.

A preliminary computational model is proposed in this study to predict thrombus formation in aortic dissection and its effects on blood flow. The aim of this study is to elucidate the relationship between hemodynamics and thrombosis of aortic dissection, with an ultimate goal of using the model to predict possible outcomes and optimize treatment strategies.

METHODS

The computational model proposed describes both blood flow by means of computational fluid dynamics (CFD) and mass transfer to simulate thrombus growth. The model is based on the assumption that the formation of thrombus is predominantly influenced by local flow patterns and shear distribution [3,4], with the initial activation of platelets being triggered by high shear stress levels, while adhesion and growth taking place in low shear regions.

The key variables included in the model are activated and resting platelets, residence time, and a coagulant which takes into account the processes involved in the coagulation cascade. All the transported variables are modeled through convection-diffusion-reaction (CDR) transport equations:

$$\frac{\partial c_i}{\partial t} + \boldsymbol{v} \cdot \nabla c_i = D_i \nabla^2 c_i + S_i \tag{1}$$

where t is time, $\boldsymbol{v}$ is the velocity of blood, c_i and D_i represent the concentration and the diffusivity of the i-th transported species and S_i is the reactive source term.

The growth of thrombus is tracked through the concentration of an additional variable defined as bound platelets (BPs). This variable is formed in areas where the local conditions consent the formation of thrombus, which are identified through high concentrations of activated platelets, coagulant, and residence time, and low-shear flow. The variable BP is related to a momentum source which impedes fluid flow in regions of thrombus growth and enhances local viscosity by up to 100,000 fold. This method allows us to track the formation of solid thrombus over time.

The model has been implemented in CFX 15 (Ansys Inc.) and applied to an idealized model of aortic dissection (Fig. 1 and 2) with six different geometric variations by varying the number and location of tears in order to determine the influence of geometric features on the formation of thrombus. Realistic boundary conditions are applied to simulate blood flow, with physiological concentrations of transported species, a patient specific velocity waveform at the inlet [5], and a pressure waveform at the outlet calculated through a three-element Windekessel model. Walls are assumed to be rigid. The results have been analysed with CEI Ensight 10 (CEI Inc.).

RESULTS

The time-averaged distribution of activated platelets over a full cardiac cycle in one of the geometries studied is presented in Fig. 1A. As expected, higher concentrations of activated platelets are observed in the false lumen with local peaks in regions of stasis and flow recirculation. A similar pattern is observed for time-averaged residence time (Fig. 1B), with particularly high values in regions of recirculation and stasis.

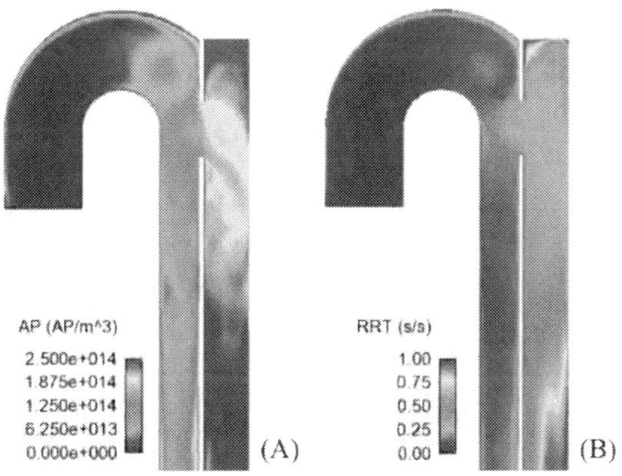

FIGURE 1: (A) ACTIVATED PLATELET DISTRIBUTION AVERAGED OVER A FULL CARDIAC CYCLE. (B) AVERAGE RESIDENCE TIME DISTRIBUTION, NORMALIZED WITH RESPECT TO TIME.

The growth of thrombus is tracked over time through the concentration of bound platelets, which is shown in Fig. 2. The high level of bound platelets activates a momentum source which gradually stops the flow in thrombosed regions and causes local increase in viscosity to simulate the solid nature of thrombus.

DISCUSSION

The study of thrombosis in aortic dissection is of particular relevance as patient outcomes are affected by the formation of blood clots in the false lumen. Although several efforts have been made in the literature to predict thrombus formation in simplified geometries, the high computational costs of existing models limit their applicability to complex 3D geometries [6].

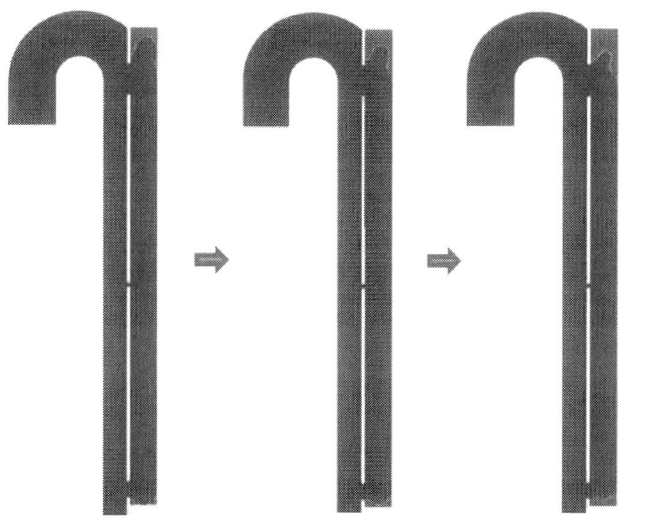

FIGURE 2: THROMBUS GROWTH OVER TIME IN ONE OF THE IDEALIZED PHANTOM MODELS. THE REGION IN RED REPRESENTS THE THROMBOSED AREA, EXPANDING OVER TIME.

The model proposed in this study is less numerically intensive and involves the identification of zones where thrombus formation is likely to happen through the analysis of flow patterns and key variables such as shear stress and residence time.

The described numerical model allows the simulation of thrombus growth over time and accounts for its effects on blood flow, as well as the effects of fluid flow on thrombus growth. Preliminary results are in qualitative agreement with in vivo observations [7], and show thrombus formation in regions of slow recirculation and stasis where a high concentration of activated platelets and long residence time are observed. Results from models of different geometric variations show that the extent of thrombosis is affected by the variation of flow patterns and shear distribution related to changes in the number and size of tears. Further investigations will be carried out to validate the model with patient-specific case studies and assess the assumptions made, including neglecting the full complexity of the biochemical reactions involved in the coagulation cascade and adapting time-scales.

ACKNOWLEDGEMENTS

This research project is funded by a PhD scholarship awarded by the Department of Chemical Engineering at Imperial College London.

REFERENCES

[1] Elefteriades, JA et al., *Ann Thorac Surg,* 53.1:11-21, 1992.
[2] Tsai, TT et al., *New Eng J Med,* 357.4: 349-359, 2007.
[3] Bluestein, D et al., *Ann Biomed Eng,* 25.2: 344-356, 1997.
[4] Hellums, JD, *Ann Biomed Eng,* 22.5:445-455, 1994.
[5] Cheng, Z et al., *Med Eng Phys,* 36.9:1176-1184, 2014.
[6] Cito, S et al., *Thromb Res,* 131.2: 116-124, 2013.
[7] Cheng, Z et al., *J Vasc Surg,* 57.2: 35S-43S, 2013.

A NONLINEAR ANISOTROPIC INVERSE METHOD FOR COMPUTATIONAL DISSECTION OF INHOMOGENEOUS PLANAR SOFT TISSUES

Colleen M. Witzenburg (1) and Victor H. Barocas (2)

(1) Biomedical Engineering
University of Virginia
Charlottesville, VA, USA

(2) Biomedical Engineering
University of Minnesota
Minneapolis, MN, USA

INTRODUCTION

The intrinsic complexity of soft tissues presents a tremendous challenge in constitutive modeling. Soft tissues exhibit mechanical behavior that is anisotropic and nonlinear. In addition they are often naturally inhomogeneous, particularly at insertion or connection points and in cases of disease or injury. We have previously used a linear Generalized Anisotropic Inverse Mechanics (GAIM) approach [1,2] to estimate material properties of anisotropic, heterogeneous soft tissues. GAIM uses the displacement field of the sample domain as well as the traction forces on the sample boundary to directly solve the finite element representation of the Cauchy stress balance for the unknown components of the general linear elasticity tensor. Heterogeneity is captured by segmenting the sample area into many partitions, assumed to have uniform properties. The linear constitutive equation was modified [3] to accommodate the large-deformation kinematics exhibited by many soft tissues, but the need to describe nonlinear material properties was not addressed.

One possible approach to tissue nonlinearity would be to replace the neo-Hookean constitutive law employed with a higher order constitutive equation (e.g. [4]). The use of high-order constitutive equations however, typically requires additional assumptions about the material (isotropy and homogeneity in [4]) and increases computation time. Therefore, while still accounting for the nonlinear kinematics of the deformation as a whole, we approach the problem by considering the mechanical behavior of the tissue in a step-wise fashion.

METHODS

Overall Strategy: An extensive biaxial loading [2,3] protocol was used. The more and varied the loading protocol the wider the span of strain-space investigated for each subregion. Experimentally, full-field deformation was captured using digital image correlation (DIC) and

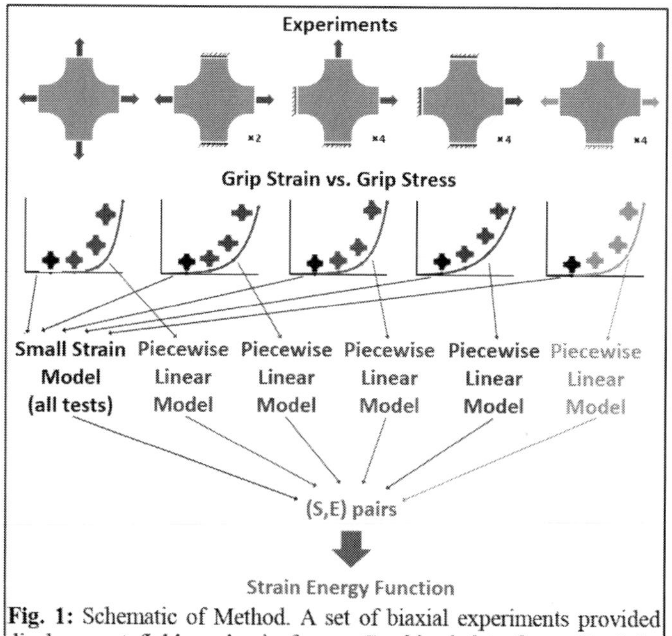

Fig. 1: Schematic of Method. A set of biaxial experiments provided displacement fields and grip forces. Combined data from all of the experiments at small-strain (black segments) were used to specify a linear small-strain model. Data from each experiment at larger strains (colored segments) were used to generate a piecewise linear model for the experiment's strain path. Using the models the stress at each strain value for each experiment was determined and integration along each path generated the strain energy density function.

both normal and shear forces at the sample grips were measured using 6-degree-of-freedom load cells (JR3 Inc.).

Small-Strain Analysis: For the first step (black segments in **Fig. 1**), homogeneous subregions were identified [5]) and data from all loading protocols were used [3], to get an anisotropic, inhomogeneous ***linear*** model of the material behavior at small strain. The force and displacement data for the first step from all extensions were considered simultaneously, and a neo-Hookean elasticity tensor was determined for each subdomain by solving the Cauchy stress balance.

Large-Strain Analysis: For larger strains, each finite element was considered a separate material and data from each individual protocol were used to generate a piecewise linear model of the material following that protocol's experimental strain path. GAIM could not be applied concurrently to steps beyond the first because of the changing reference configuration (cf. [6]).

The PK2 stress at the end of step N was given by

$$S_{ij}{}^{N} = \sum_{n=1}^{N} \mathbb{S}_{ij}{}^{n} = \sum_{n=1}^{N} \mathbb{K}_{ijkl}{}^{n}\mathbb{E}_{ij}{}^{n} \qquad (1)$$

where the differential neo-Hookean elasticity tensor, $\mathbb{K}$, and Green strain, $\mathbb{E}$, are relative to the end of the previous step, $S_{ij}{}^{N}$ is the PK2 stress after step N in reference to the undeformed sample, and the PK2 stress for the first step is from the small-strain analysis. The change in $\mathbb{K}$ from step to step was minimized subject to the constraint that the residual force vector remain zero.

Generating an Elemental Constitutive Data Set: The solution of the global matrix equation at each step of each protocol yielded a unique neo-Hookean elasticity tensor for each element. By combining the PK2 stresses calculated over all segments of all protocols for an element, along with the Green strain from DIC, a set of ordered pairs, was generated from which the strain energy was obtained using the trapezoid rule. In this way, a unique data-driven strain energy density function was generated for each element within the sample domain.

Simulated Experiments: The nonlinear GAIM method, NGAIM, was applied to simulated tissues using the closed-form nonlinear fiber-based structural model (NFSM) for soft tissues [7]. For the first simulation, a cruciform sample was generated with a central triangular inclusion that differed from the bulk only in its fiber orientation. For the second simulation sample orientation was set with anisotropy vectors radially aligned about the sample center.

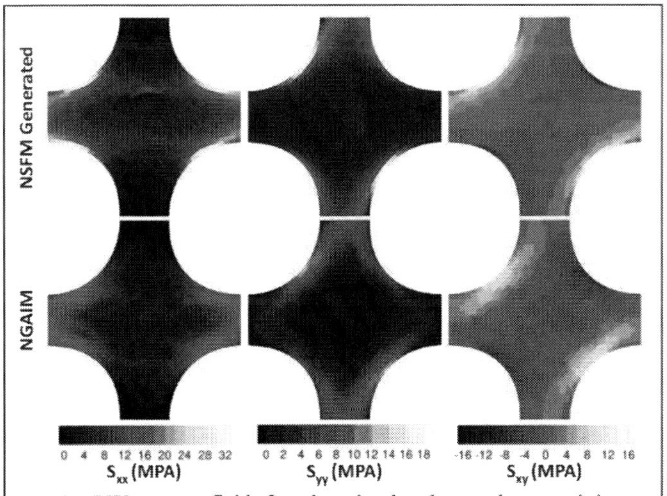

Fig. 2: PK2 stress field for the simulated sample containing an inclusion with rotated fiber alignment at the end of the final time step of the equibiaxial extension.

RESULTS

The full stress-fields for all steps of all protocols for both simulations (Fig. 2 shows selected example) show qualitative agreement between the NSFM simulation and NGAIM. Two elements within each sample were selected for further analysis; for the sample with the inclusion an element located in the middle of the inclusion and one outside it and for the radially symmetric sample an element located near the middle and one towards the edge. Fig. 3 shows a direct comparison between the strain energy generated from the NFSM and that determined by NGAIM. There is excellent agreement between the NGAIM prediction and NFSM generated strain energies.

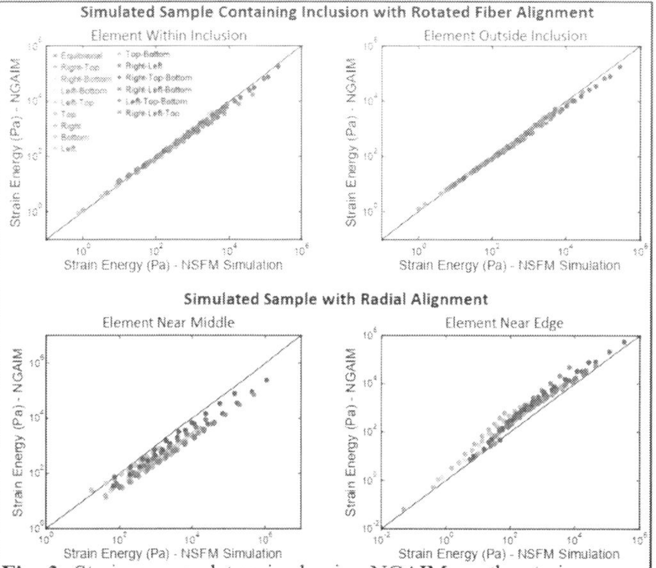

Fig. 3: Strain energy determined using NGAIM vs. the strain energy generated by the NSFM for all extensions (each step is a dot in a different color protocol). The black line indicates a perfect prediction.

DISCUSSION

The NGAIM approach is, to our knowledge, the first to be able to estimate a full stress field within a sample based solely on the sample strain and the edge forces, and it is the first to be able to generate a useful, data-driven strain energy function at every point in the material. Thus, it has the potential to be used in a wide range of tissue characterization applications, particularly in the case of tissues that are two small to be dissected or do not lend themselves to non-invasive identification of the appropriate dissection strategy.

There are, of course, challenges. Signal-to-noise ratio will be a major issue in real tissue experiments, especially if the sample has both compliant and stiff regions because the latter will experience little deformation. The different experimental protocols only interact at small strain, so large-strain errors could go unchecked and pollute the strain energy function calculations. These concerns notwithstanding, the NGAIM approach is a new way to study the mechanics of complex tissues, allowing us to explore inhomogeneity, anisotropy, and nonlinearity on an elementwise basis.

ACKNOWLEDGEMENTS AND REFERENCES

This work was supported by NIH Grant R01-EB005813. CMW was supported by a UMN Doctoral Dissertation Fellowship. Thanks to the Minnesota Supercomputing Institute for computing resources.

[1] Raghupathy R., 2010., J. Biomech. Eng. [2] Raghupathy R., 2011, J. Biomech. Eng. [3] Witzenburg C. M., 2012, J. Biomech., [4] Tönük E., 2003, IEEE Trans. Neural Syst. Rehabil. Eng. [5] Witzenburg, C.M., Submitted, IEEE Trans. Med. Imag. [6] Flynn D. M., 1998, J. Biomech. Eng. [7] Raghupathy R., 2009, J. Biomech.

OSTEOGENIC INDUCTION OF HUMAN ADIPOSE DERIVED STEM CELLS CULTURED ON POLY (L-LACTIC ACID) SCAFFOLDS PREPARED BY THERMALLY INDUCED PHASE SEPARATION METHOD

Harish Chinnasami[1], Daniel Hayes[2] and Ram Devireddy[1]

[1]Department of Mechanical Engineering,
Louisiana State University,
Baton rouge, LA 70803, USA

[2]Department of Biological and Agricultural Engineering,
Louisiana State University,
Baton Rouge, LA 70803, USA

INTRODUCTION

A tissue-engineered scaffold structurally mimics the intended tissue to be developed with adequate pore sizes for the stem cells to proliferate and differentiate. The initial scaffolds developed were primarily composed of ceramic compounds like hydroxyapatite, akermanite and tricalcium phosphate that are structurally rigid and resemble the mineral composition in bones. In recent years, numerous researchers stressed on the advantages of using bio-compatible polymers such as poly (l-lactic acid) (PLLA), Polycaprolactone (PCL), poly (l-glycolic acid) (PLGA) as a scaffold material. Some of the prominent methods used to make polymer scaffolds are solvent casting, fiber bonding, membrane lamination and particulate leaching [1,2]. In this research we use thermally induced phase separation (TIPS) process to synthesize a porous PLLA scaffold. The crystalline PLLA was dissolved in 1,4 Dioxane at 3, 7 and 10 (wt/vol)% to form a homogenous clear solution. These solutions were each frozen to a low temperature (-60°C), at 1, 10 and 40°C/min cooling rates (C.R.). The frozen solutions were lyophilized (freeze-dried) to remove the solvent resulting in the final scaffold, which was 3D porous PLLA structure. The porous morphology of the scaffold was partly determined by the crystal formation of dioxane in the solution which was influenced by the applied freezing profile. The objective of this study was to (i) select a suitable PLLA scaffold for cell culture, (ii) to observe the viability of human adipose stem cells (hASCs) when cultured on these scaffold and (iii) to determine the osteo-inductive potential of the PLLA scaffold selected.

MATERIALS AND METHODS

Porosity (%) of a scaffold was calculated at two positions (0-5 mm and 5-10 mm from bottom) by weighing a known volume of the scaffold and the density of the polymer as:

$$Porosity\ (\%) = \left(1 - \frac{Scaffold\ Density}{Polymer\ Density}\right) \times 100 \quad \text{--- (1)}$$

Table 1: Porosities of the scaffolds made using the specified PLLA:Dioxane (wt/v) composition and the C.R.

PLLA: Dioxane	-1°C/min		-10°C/min		-40°C/min	
	(0-5) mm	(5-10) mm	(0-5) mm	(5-10) mm	(0-5) mm	(5-10) mm
3% (wt/v)	96.96%	96.89%	96.34%	96.75%	95.56%	95.50%
7% (wt/v)	94.16%	93.98%	93.92%	93.96%	91.81%	93.28%
10% (wt/v)	90.47%	89.46%	89.25%	89.69%	88.82%	88.39%

Scaffold sections of 12mm diameter and 5mm height were coated with Platinum (15nm thick) using EMXS550X Sputter Coater. JEOL JSM-6610LV Scanning Electron Microscope was used to scan the samples and capture SEM images to qualitatively observe the pore structure and measure the pore sizes (see Fig. 1).

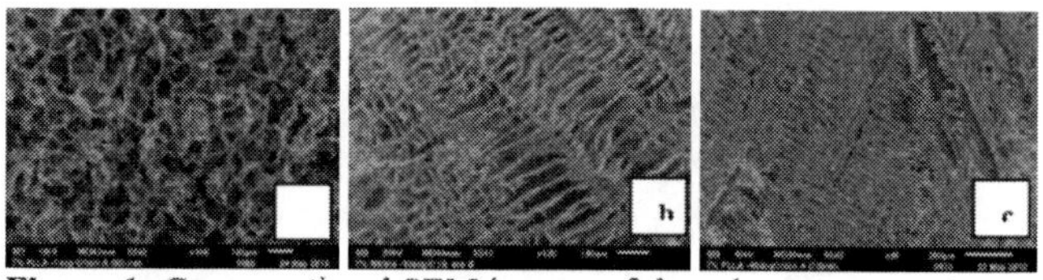

Figure 1: Cross-sectional SEM images of the micro-pore structures of the 7% (wt/vol) scaffolds cooled at (a) 1°C/min, (b) 10°C/min and (c) 40°C/min

INSTRON 5900 Series – Advanced Mechanical Testing Systems Compression tester (Instron Industrial Products, Liberty Street, Grove City, PA, USA) was used to compress scaffolds of 12mm in diameter and 10mm in height to 20% of its initial height, while simultaneously recording the applied load and strain. Compressive modulus was calculated as follows:

$$Compressive\ Modulus = \frac{\left(\frac{Applied\ Load}{Cross\ sectional\ area\ of\ scaffold}\right)}{Compressive\ strain} \quad \text{---(2)}$$

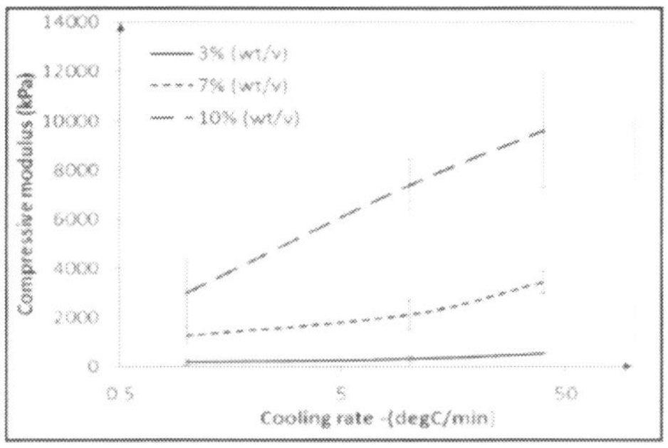

Figure 2: Compressive Moduli of PLLA scaffolds made with PLLA in Dioxane in 3, 7 and 10% (wt/vol) initially and also their variation with increasing cooling rate.

From the compression test results it is evident that the 3% (wt/vol) scaffolds have very low modulus resulting in a weak scaffold that is difficult to manipulate during cell culture. Whereas the 10% (wt/vol) scaffolds were practically difficult to trim using razor blades (VWR). Therefore 7% (wt/vol) was decided to be optimal. Figure. 1 suggests that 7% (wt/vol) frozen at -10°C/min gave more uniform pore structure patterns than other C.Rs and therefore was selected for cell culture.

Cell Culture: Cryo-preserved hASCs were thawed and cultured in traditional Stromal Medium (SM) containing 90%(v/v) DMEM/F-12 (Gibco), 10%(v/v) Fetal Bovine Serum (Gibco), Penicillin - 90 IU/mL, Streptomycin – 90 µg/mL and Amphotericin B – 225 ng/mL. Known number of hASCs (~10^4) were loaded on PLLA Scaffolds ⌀12mm and 5mm in height. The PLLA Scaffolds were attached to the bottom of the 6-well plates using 1% Agarose (Sigma) as an adhesive. Equal number of hASCs was added to wells without scaffold to be cultured as a control.

RESULTS

DNA Quantification: The PLLA Scaffolds, homogenized with Dounce tissue grinder pestle (7 mL) (Sigma) as well as the control samples were incubated at 37°C with Lysis buffer (0.2 mg/mL Sodium Dodecyl Sulfate and 0.2 mg/mL Proteinase K) for 20 mins. The DNA present in the Lysis buffer was quantified using Quant-iT™ PicoGreen® dsDNA Assay Kit (Invitrogen) according to the manufacturer's protocol.

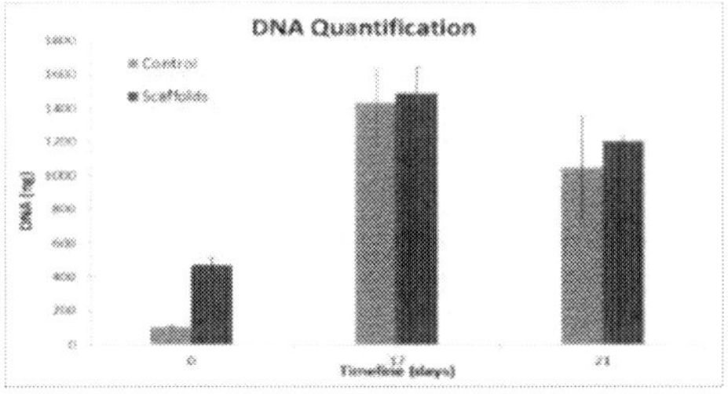

Figure 3: Quantity of DNA (ng) for PLLA scaffolds and control samples at three different time points with standard deviations.

Live/Dead Viability Assay: Samples were incubated in staining solution of 2 µM Calcein AM and 4 µM Ethidium homodimer-1 (EthD-1) was prepared using LIVE/DEAD® Viability/Cytotoxicity Kit (Molecular Probes) stock solutions in DPBS for 20 mins at 37°C and were taken for fluorescence imaging using SteREO Lumar.V12 (Zeiss).

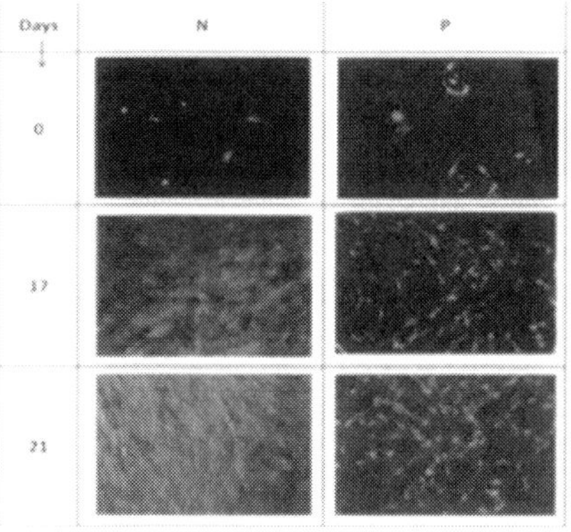

Figure 4: Live Cells (green) and dead cells (red) in Control samples (N) and PLLA Scaffolds (P) at 0, 17 and 21 days.

Alizarin Red S (ARS) Quantification: Cells in control and scaffolds were fixed with 4% paraformaldehyde and stained with 40mM of ARS for 30 mins at 37°C. Excess dye was washed with deionized water. 10% (wt/v) cetylpyridinium chloride (CPC) was used to elute the stain. Optical density (OD) of eluted stain from each sample along with known concentrations was measured at 531nm using Wallac Victor2 Plate reader.

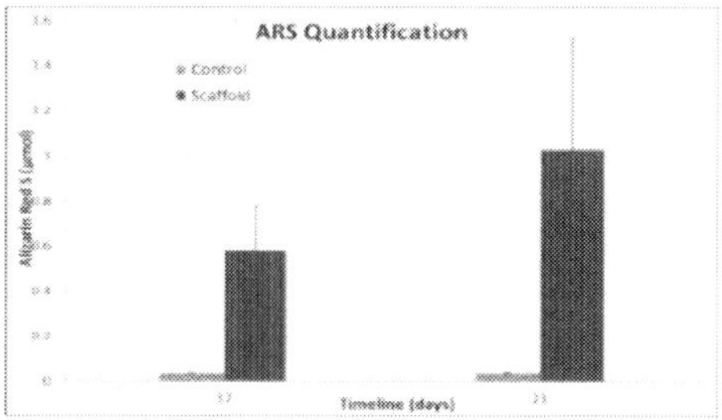

Figure 5: Amount of ARS (µmol) bound to control and scaffold after 17 and 21 days shown with standard deviations.

DISCUSSION

The DNA results show that cell population doubles in 17 days on scaffolds and decreases by ~27% from 17 days to 21 days on control. This may be because of the confluence that can be observed in Figure 4 at day 17. Though scaffolds had more space to grow we find that the cell number have decreased by ~20%. Live Dead stain shows most of the cells were alive on the scaffold. This suggests that PLLA is very bio-friendly and there is no cytotoxicity to the cells. Alizarin Red S quantification from Figure 5 shows an increase in calcium deposition from 17 days to 21 days. Hence PLLA could be capable of inducing osteogenesis even when cultured in SM. More data is required to completely validate these findings and to explain the anomalous DNA quantification data.

ACKNOWLEDGEMENTS: LSU Economic Development Assistantship (EDA) awarded to HC.

REFERENCES

1. Qingpu Hou et al., *J.Biomed Mater Res. Part B: Applied biomaterials*, 67(2):732-740, 2005.
2. Scott J. Hollister, *Nature Materials*, 4:518-524, 2005.

COMPOSITION-DEPENDENT MECHANICAL PROPERTIES AND MULTISCALE STRAIN TRANSFER IN TENDON SUBJECTED TO SHEAR LOADING

Fei Fang (1), Spencer P. Lake (1-3)

(1) Department of Mechanical Engineering
(2) Department of Biomedical Engineering
(3) Department of Orthopaedic Surgery
Washington University in St. Louis
St. Louis, MO

INTRODUCTION

Tendons in different anatomical locations can experience complicated multi-axial *in vivo* loading, such as tension, shear, compression, and torsion. In response, compositional and structural properties of the extracellular matrix are modulated by cells to enable proper function. For example, distal regions of bovine deep digital flexor tendon (DDFT) have more disorganized collagen fibers and more proteoglycan than proximal regions, because the anterior side of distal DDFT is compressed against navicular bone [1]. Similarly, human supraspinatus tendon shows location-dependent fiber alignment and tensile modulus likely resulting from complex *in vivo* loading [2, 3]. To understand the influence of loading on tendon properties, previous studies of tendon under uniaxial tension demonstrate strain attenuation from tissue to local extracellular matrix to cell nuclei by tracking photobleached grids or nuclei centroids [4, 5]. The deflection of photobleached grids and significantly smaller local strains along fibers compared to applied strains indicated that sliding occurs between collagen fibers when tendon is loaded in tension [6]. However, to the best of our knowledge, no studies have evaluated the mechanical response of tendon to shear and the physical mechanisms governing tendon behavior under multi-axial loading. Therefore, this study performed incremental stress-relaxation shear tests on bovine DDFTs and evaluated strain transfer among different hierarchical levels to evaluate mechanisms responsible for tendon behavior under shear.

METHODS

Shear testing. Distal and proximal samples (n=5/group) were harvested from regions 4 cm and 2 cm away from the bifurcation of DDFTs, respectively [1]. A freezing-stage sliding microtome was used to smooth lateral surfaces, and sample dimensions were measured using a non-contact laser scanning system. For collecting images at the tissue level, four small stainless steel balls (diameter = 1/32'') were glued to one lateral surface as fiducial markers, and samples were loaded in a custom micromechanical test device equipped with a digital camera [1]. For acquiring images at the micro level, an additional set of distal and proximal samples (n=5/group) were prepared by the same method above and incubated in a solution of a 10 µg/ml 5-(4,6-dichlorotriazinyl) aminofluorescein (5-DTAF) solution of 0.1M sodium bicarbonate buffer for 25 min to stain collagen [7]. Following PBS rinses, cell nuclei were stained by 1µg/ml 4′,6-diamidino-2-phenylindole (DAPI) solution for 20 min. All samples were mounted in the same test device, then secured to the stage of a Zeiss multichannel multiphoton-confocal microscope system. A 0.05N shear pre-load was applied to define the clamp position at zero axial strain. Samples were loaded in shear to three incremental steps of 8% strain (i.e., applied displacement divided by sample thickness) at a rate of 20mm/s, with 12 min relaxation time between each step [1].

Image acquisition. Digital photos were acquired by camera before the loading and after 8 minutes relaxation for each step. For tracking 2D strain at the local matrix level, four adjacent squares (100µm x 100µm) were photobleached at three locations (anterior, medial, posterior) at a depth of ~50µm [5]. Multichannel two-photon z-stack images of photobleached grids and nuclei at three locations were collected through the stained depth by a 20X objective (field of view: 224.8 µm × 224.8 µm; DAPI: 750 nm/BP 400-490 nm; 5-DTAF: 750 nm/LP 500 nm) at the same time points as digital photos (Fig. 1A).

Dimethylmethylene blue assay. Glycosaminoglycan (GAG) content in distal and proximal samples (n=5/group) was evaluated using the dimethylmethylene blue (DMMB) assay. Three segments (anterior, medial, posterior) cut from each sample were weighed, minced, and digested in 250 µL papain suspension at 64 °C for 18 hours, followed by centrifugation to obtain supernatant [8]. DMMB dye solution was added to 40 µL quantities of samples and chondroitin sulfate standards and absorbance was read at 525 nm. GAG content were normalized by sample wet weight [9].

Data analysis. Peak and equilibrium stresses of distal and proximal samples were computed at each strain step (0.08, 0.16, 0.24) using measured cross-sectional areas. Two-photon z-stacks were collapsed to form single, summed 2D multichannel images using ImageJ (NIH). A custom Matlab program was used to compute 2D Lagrangian tissue and local matrix strains using digital photos of surface markers and two-photon images of photobleached girds, respectively. After the threshold of 2D DAPI images, nuclei centroids and lengths of long and short axes were tracked by ImageJ to compute rotations and nuclei aspect ratios: (1) "rotation" within rows of fibers (i.e., relative motion perpendicular to fiber direction, indicative of uncrimping and/or collagen reorganization), (2) "rotation" between rows of fibers (i.e., a measure of fiber sliding), and (3) the ratio of the long axis to the short axis for nuclei (i.e., nuclear aspect ratio or NAR). At each strain level, significant differences of parameters (i.e., stresses, strains, rotations, NAR) were determined by student t-tests between distal and proximal samples. Within distal or proximal locations, unpaired t-tests were used to evaluate differences between strains and rotations. Significance was set at $p < 0.05$.

RESULTS

At each strain step, distal samples showed peak and equilibrium stresses approximately 3X greater than proximal samples (Fig.1B). For all samples, stresses increased with increasing clamp strains. Equilibrium stresses were much smaller than peak stresses, demonstrating significant relaxation at each strain level for both distal and proximal samples. Measured tissue strains were smaller than applied clamp-to-clamp strains, but were not significantly different than local matrix strains (Fig. 1C). Rotations computed by tracking nuclei centroids were significantly greater between-rows of collagen fibers than those measured within-rows for all samples, and between-row rotations were significantly greater in proximal samples than in distal samples (Fig. 1D). NAR values of distal samples were significantly smaller than those of proximal samples before loading (not shown), while ΔNAR values were not different between samples and did not increase with increasing applied shear strain (Fig. 1E). GAG content in distal samples was 3.6 X greater than proximal, with anterior and medial locations of distal samples significantly greater than similar locations in proximal samples. Within the distal region, the posterior location had significantly smaller GAG content than anterior and medial locations, but similar to all the three locations in proximal samples (Fig. 1F).

DISCUSSION

Strains were moderately attenuated from tissue to local matrix to nuclei-based measures, concurring with previous finding [4, 10]. Non-zero within-row rotations for both distal and proximal samples illustrate the contribution of uncrimping/reorganization to the mechanical response under shear loading. However, compared with within-row rotations, between-row rotations in both regions were significantly larger, demonstrating that sliding between collagen fibers is more prominent than uncrimping or reorganization behavior. Changes of NAR show macroscopically applied strains transferred down the hierarchical length scales to the cell nuclei. Decreased NAR values under shear loading suggest that sliding between fibers possibly causes cell rotation leading to further rounding of nuclei rather than elongation of nuclei reported in tensile loading [10]. Comparing different tendon regions, distal and proximal regions exhibited different shear stresses, between-row rotation, and NAR values, but similar strain responses at tissue and local matrix levels and within-row rotation in cell level. Thus, different compositions (i.e., GAG content) contribute to different shear stresses in these two regions from the tissue level. The varying local matrix environment in distal and proximal regions also facilitates cells responding differently to shear, causing different NAR and between-row rotations. The two-photon images in our experiments revealed breaking of photobleached grids at the interface of subfascicles (not shown), potentially attributed to less stiff interfascicular material compared to fascicles themselves. Additional studies should be performed to understand how non-fibrillar matrix distributes amongst subfascicles and how it assists force transfer between subfascicles under multiaxial loading. This study provides additional insight into the mechanobiologic response of tendon to physiologically-relevant non-tensile loading and further defines structure-function relationships of tendon.

REFERENCES

[1] Fang, F et al., J Biomech, 47(12): 2869-77, 2014. [2] Lake, SP et al., J OR, 27(12): 2009. [3] Lake, SP et al., J Biomech, 43(4): 727-32, 2010. [4] Cheng, VW et al., J Mater Sci, 42(21): 8957-65, 2007. [5] Thorpe, CT et al., Int J Exp Pathol, 94(4): 248-59, 2013. [6] Thorpe, CT et al., Acta Biomater, 9(8): 7948-56, 2013. [7] Szczesny, SE et al., PLoS one, 9(6): e99588, 2014. [8] Scott, JE et al., J Biochem, 195(3): 573-81, 1981. [9] Adekanmbi, I et al., Med Eng Phys, 35(2): 205-10, 2013. [10] Han, WM et al., Biophys J, 105(3): 807-17.

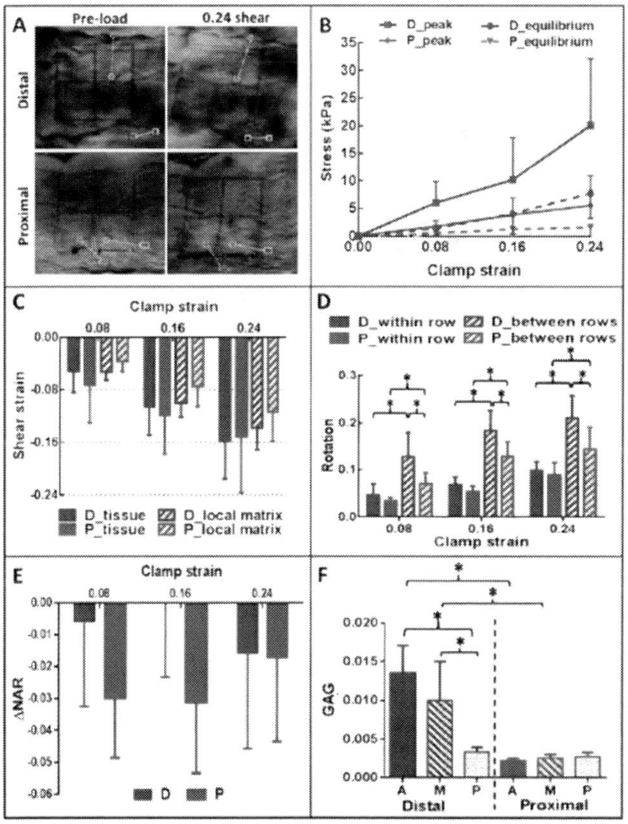

Figure 1. (A) Two-photon images before loading and after 0.24 shear strain, nuclei outlined by rectangles and by circles illustrate rotation between fibers and rotation within fiber, respectively; (B) Peak and equilibrium stresses under shear stress-relaxation testing; (C) Multiscale strain values vs. clamp strain; (D) Rotation within row and between rows vs. clamp strain; (E) Change of nuclear aspect ratio (NAR) normalized by undeformed NAR vs. clamp strain; (F) GAG content normalized by sample wet weight. (D and P mean distal and proximal, respectively; data are shown as mean ± 95% confidence intervals; * is for $p<0.05$).

HYDROXYAPATITE REINFORCED COLLAGEN SCAFFOLDS DESIGNED FOR IMPROVED ARCHITECTURE, MECHANICAL PROPERTIES, AND TUNABLE GROWTH FACTOR DELIVERY

Ryan K. Roeder, Matthew J. Meagher, Holly E. Weiss-Bilka, Diane R. Wagner, and Robert J. Kane

Department of Aerospace and Mechanical Engineering
Bioengineering Graduate Program
University of Notre Dame
Notre Dame, IN, USA

INTRODUCTION

Collagen and collagen-hydroxyapatite (col-HA) scaffolds are used clinically for bone regeneration as a synthetic bone graft substitute and/or delivery vehicle for BMP-2 [1]. Collagen and col-HA scaffolds are typically prepared by freeze-drying or lyophilization to achieve high levels of porosity [1,2] and have exhibited osteogenicity *in vivo* [3,4], but are limited by an inherent scaffold architecture that results in relatively small pores, thin scaffolds struts, and weak mechanical properties (Fig. 1a) [2-5]. Thus, freeze-dried col-HA scaffolds are less than ideal for surgical handling, fixation (e.g., with a pin), and bearing osteogenic loads during healing.

Growth factors are delivered from freeze-dried collagen and col-HA scaffolds via physical entrainment and/or non-covalent (electrostatic) adsorption which inevitably results in burst release [6,7]. After burst release, growth factors rapidly dissociate from the scaffold, diffuse away from the injury site, and are metabolized. Thus, the majority of the dose is ineffective necessitating the use of superphysiological doses. In the case of BMP-2, this problem has resulted in poor outcomes, adverse side effects, and high treatment costs [1,6,8]. Thus, current methods for growth factor delivery do not permit control over the dosing and release kinetics of a single growth factor from these scaffolds, let alone multiple growth factors.

Therefore, col-HA scaffolds were designed for improved architecture, mechanical properties, and growth factor delivery compared with freeze-dried scaffolds (Fig. 1). Effects of the scaffold architecture and composition on the scaffold permeability, mechanical properties, *in vitro* bioactivity, and *in vivo* osteogenicity were evaluated. Spatiotemporal control over the delivery of VEGF and BMP-2 along physiologically relevant timelines was demonstrated using independent loading and delivery mechanisms.

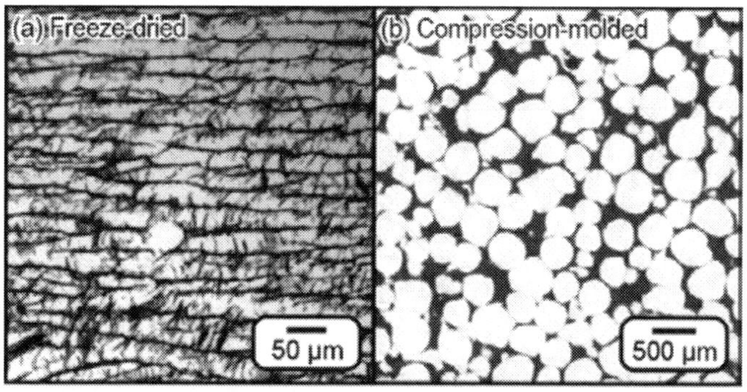

collagen + 0-60 vol% HA	collagen + 0-80 vol% HA
>90% lamellar porosity	85-90% spherical porosity
<100 µm pore width	~300-400 µm pore diameter
<5 µm strut thickness	~1-200 µm strut thickness
1-250 kPa compressive *E*	60-1000 kPa compressive *E*

Figure 1. Comparison of col-HA scaffolds prepared by freeze-drying [5] versus those prepared by compression molding [9] in this study highlighting differences in scaffold architecture and mechanical properties.

METHODS

Col-HA scaffolds with 85 or 90% porosity were prepared by concentrating a suspension of collagen fibrils to ~180 mg/ml, compression molding a mixture of HA (0-80 vol% powder or whiskers), collagen, and paraffin microspheres (300-425 µm) at 1 MPa, cross-linking the collagen matrix using EDC/NHS chemistry, and leaching the paraffin porogen [9]. The scaffold architecture and microstructure was characterized using micro-CT and SEM.

Mechanical properties were measured in unconfined uniaxial compression in PBS and cyclic compression at 1 Hz to 50% strain for up to 100,000 cycles. Fluid permeability was measured using a forced flow apparatus and Darcy's law. *In vitro* bioactivity was investigated by seeding murine adipose-derived stromal cells (mASCs), culturing in growth media, and measuring total DNA and alkaline phosphatase (ALP) activity for 14 d. *In vitro* cell infiltration was investigated by seeding human ASCs, culturing in osteogenic media for 14 d, and examining H&E stained histologic sections. *In vivo* angiogenesis and osteogenesis were investigated by implantation in ectopic subcutaneous pockets in the dorsal cervical region of 4 w/o, female athymic nude mice for 8 weeks. Osteogenesis and angiogenesis were characterized by micro-CT, H&E stained histologic sections, and immunolabeling for osteocalcin (OCN) and CD31, respectively.

VEGF and BMP-2 were loaded in col-HA scaffolds by hydrogel encapsulation and covalent conjugation, respectively. For BMP-2 loading, scaffolds were prepared from biotinylated collagen fibrils and subsequently soaked in a solution of streptavidin, followed by a solution of biotinylated BMP-2 for binding to collagen fibrils. For VEGF loading, scaffold pore spaces were infiltrated with a collagen hydrogel containing VEGF and gelled. Release kinetics were measured *in vitro* using ELISA.

RESULTS AND DISCUSSION

Col-HA scaffolds prepared by compression molding and porogen leaching exhibited an architecture with high porosity, but with larger interconnected pores and struts than freeze-dried scaffolds (Fig. 1), and a matrix with a relatively high collagen fibril density (~180 mg/cm^3) containing 0-80 vol% HA whisker or powder reinforcements. Col-HA scaffolds exhibited compressive moduli (60-1000 kPa) which were 1-2 orders of magnitude greater than comparable freeze-dried scaffolds (Fig. 1), and fully recoverable elastic deformation upon loading to 50% compressive strain for at least 100,000 cycles. Thus, the scaffold mechanical properties were well-suited for surgical handling, fixation (e.g., with a pin), and bearing osteogenic loads during bone regeneration.

The scaffold architecture, permeability, and composition were shown to be conducive to the infiltration and differentiation of ASCs *in vitro*. ALP activity was greater in col-HA scaffolds compared with collagen alone at days 7 and 14. Cells were able to completely infiltrate 3 mm scaffolds after 14 d and col-HA scaffolds were able to resist cell-mediated contraction. Acellular scaffolds were demonstrated to induce angiogenesis and osteogenesis *in vivo* after subcutaneous ectopic implantation by recruiting endogenous cell populations even without exogenous growth factors (Fig. 2). *De novo* bone formation was evidenced by a 61% increased in the thresholded bone volume measure by micro-CT and positive immunolabeling for OCN. These results suggest the col-HA scaffolds are bioactive and osteoinductive.

Encapsulated VEGF was gradually released over 3 weeks *in vitro*, while the entire dose of biotinylated BMP-2 was completely immobilized within scaffolds for 31 d and was only released after collagenase digestion of the scaffold (Fig. 3), suggesting that biotinylated BMP-2 will also be immobilized *in vivo* until the collagen matrix undergoes cell-mediated enzymatic degradation. In contrast, delivery of BMP-2 from freeze-dried collagen sponges occurs in <14 d either *in vitro* or *in vivo* [1,7].

In summary, the col-HA scaffolds in this study appear to be suitable for clinical use as a synthetic bone graft substitute and tissue engineering scaffold and offer improvements over existing freeze-dried HA-collagen scaffolds and absorbable collagen sponges used clinically. A new growth factor delivery platform was also developed to enable tunable, independent control over the dosing, spatial presentation, and release kinetics of multiple growth factors.

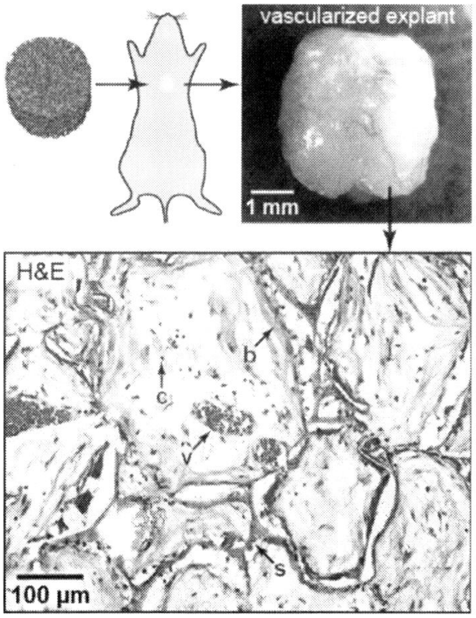

Figure 2. Col-HA scaffolds with 85% porosity and 40 vol% HA whiskers were implanted in ectopic subcutaneous pockets of 4 w/o, female athymic nude mice for 8 w. Explants showed visually apparent capillaries. Histologic sections (H&E) of explants showed widespread vascularization (v), cellular invasion (c), and immature bone (b) or osteoid adjacent to the scaffold (s), which was confirmed by positive immunolabeling for CD31 and OCN.

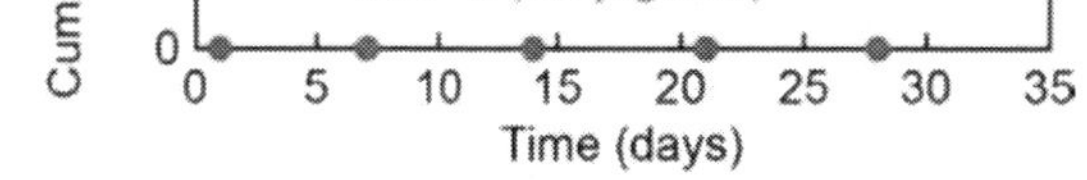

Figure 3. Cumulative *in vitro* release (% dose) of a 2.5 µg dose of VEGF encapsulated within a collagen hydrogel filling col-HA scaffold pore spaces and a 1.0 µg dose of biotinylated BMP-2 conjugated to collagen fibrils in the col-HA scaffold matrix. BMP-2 was immobilized for 31 d and released only after collagenase digestion of the scaffold.

ACKNOWLEDGMENTS

This research was partially supported by the U.S. Army Medical Research and Materiel Command (W81XWH-09-1-0741).

REFERENCES

[1] Geiger M, *et al., Adv Drug Delivery Rev* 55:1613-29, 2003. [2] O'Brien FJ, *et al., Biomaterials* 26:433-41, 2005. [3] Gleeson JP, *et al., Eur Cells Mater* 20:218-30, 2010. [4] Lyons FG, *et al., Biomaterials* 31:9232-43, 2010. [5] Kane RJ, *et al., J Mech Behav Biomed Mater* 7:41-9, 2012. [6] Santo VE, *et al., Tissue Eng* 19B:308-52, 2013. [7] King WJ, *et al., Adv Drug Delivery Rev* 64:1239-56, 2012. [8] Cahill KS, *et al., JAMA* 302:58-66, 2009. [9] Kane RJ, *et al., Acta Biomaterialia*, in press, 2015.

SB³C2015
Summer Biomechanics, Bioengineering and Biotransport Conference
June 17-20, 2015,Snowbird Resort, Utah, USA
SB³C2015-175

REACTION/DIFFUSION MODELLING OF AtzA BIOCATALYST ENCAPSULATED IN A SILICA GEL MATRIX

Baris R. Mutlu (1), Sujin Yeom (2), Lawrence P. Wackett (2,3), Alptekin Aksan (1,3)

(1) Department of Mechanical Engineering, University of Minnesota, Minneapolis, MN 55455, USA

(2) Department of Biochemistry, Molecular Biology and Biophysics, University of Minnesota, Minneapolis, MN 55455, USA

(3) BioTechnology Institute, University of Minnesota, St Paul, MN 55108, USA

INTRODUCTION

Silica gel encapsulation of biocatalysts (i.e. bioencapsulation) have been extensively studied in the last decade, including applications for water bioremediation. Efficient utilization of a biocatalytic material (i.e. synthesized by bioencapsulation) in a water bioremediation application requires optimization of engineering parameters such as biocatalyst loading density (ρ) and characteristic length (i.e. size) of the material (L_c). Varying these two design parameters (ρ and L_c) can significantly increase the cost of the system without substantial improvement in performance, primarily due to transport limitations. Thus, a reaction/transport model is required for optimization of these parameters. As an example, the mass transport resistances in a packed bed reactor utilizing encapsulated whole-cell biocatalysts are shown in Figure 1.

In this study, we developed a model which can be used to optimize these design parameters: ρ and L_c. Catalytic activity of the free cells was characterized as a function of cell membrane permeability (membrane permeabilized by ethanol, a by-product of silica gel encapsulation). Permeability of the gel was investigated as a function of ρ, since it was expected that the micron scale size of the free cell biocatalyst can significantly affect the microstructure of the mesoporous silica network. Using the characterization results, a steady-state reaction/diffusion model was developed for the encapsulated cells which was also verified experimentally.

The biocatalyst used in this study was a recombinant *Escherichia coli (E.coli)* strain overexpressing the atrazine dechlorinating enzyme AtzA that transforms atrazine to hydroxyatrazine. Atrazine (2-chloro-4-ethylamine-6-isopropylamino-s-triazine) is a widely used herbicide in the U.S, up to 36,000 tons annually. Its concentration in drinking water is regulated by the U.S. Environmental Protection Agency (EPA), and thus, bioremediation is of great practical importance.

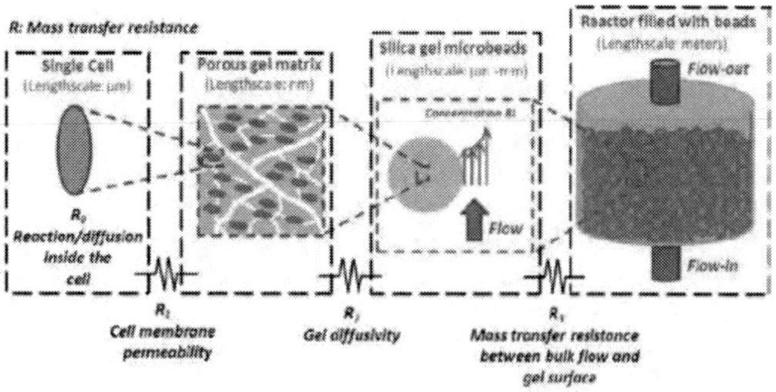

Figure 1 – Mass transport resistances in a packed bed reactor utilizing encapsulated whole-cell biocatalysts

METHODS

Silica gel synthesis and encapsulation of cells

A previously reported silica gel matrix (NS125-7:1-TEOS) and encapsulation method was used in this study [1].

Characterization of free cell AtzA catalyst reaction kinetics

An aliquot of cells suspended in PBS was added to atrazine solution prepared with 0.1 M potassium phosphate buffer (at pH 7.0) in a scintillation vial. Vials were incubated for 20 minutes on a shaker. After incubation, samples were collected and immediately immersed in a 90°C water bath to stop the catalytic activity by denaturing the enzyme. The concentrations of atrazine and its metabolite, hydroxyatrazine, were measured by using high performance liquid chromatography (HPLC). In order to evaluate the effects of ethanol concentration, buffer solutions were prepared with varying ethanol concentrations. Sonicated cell suspension was obtained by immersing a sonicator tip into the free cell suspension. Sonicated cell suspension was centrifuged to separate the supernatant with cytoplasmic proteins from the cell membrane fragments. Then, non-sonicated free cells and supernatant solution from sonicated cells were added to the buffer solutions with ethanol and were incubated for an hour. After incubation, atrazine solution was added and vials were incubated for 20 minutes. The remainder of the assay procedure was identical.

Determination of atrazine transport in silica gel matrix

The silica gel matrix was synthesized as a thin cylindrical disk (diameter = 7.5 mm, thickness = 3 mm). One side of the disk was exposed to a reservoir filled with 150 μM atrazine solution in 0.1 M potassium phosphate buffer, and the other side was exposed to water. After 24 hours of exposure, samples were collected and analyzed using HPLC as previously described.

RESULTS

Characterization of whole-cell AtzA biocatalyst

It was shown that up to the solubility limit of atrazine in water (150 μM), the enzymatic reaction transforming atrazine into hydroxyatrazine is of first order. Therefore, the free cell model takes the following form:

$$\mathbf{r = k_{free}(\rho) \times C_{Atrazine} = k'_{free} \times \rho \times C_{Atrazine}} \quad \mathbf{(1)}$$

where $C_{Atrazine}$ is the atrazine concentration in [μM], $k_{free}(\rho)$ is the degradation rate constant for free cells at a given biocatalyst loading density [s^{-1}], k'_{free} is the degradation rate constant for free cells per biocatalyst loading density [s^{-1}/(g cells/mL)] and r is the reaction rate in [μmole/L-s]. Using the experimental parameters ($\rho = 6 \times 10^{-4}$) and a least-squares-regression analysis, we determined that $k_{free} = 4.43 \times 10^{-5}$ s^{-1} and incorporating ρ, we get $k'_{free} = 7.38 \times 10^{-2}$ s^{-1}/(g cells/mL).

It was observed that both k'_{free} (free cells) and $k'_{sonicated}$ (sonicated cells) changed significantly with changing ethanol concentration (C_{EtOH}). At C_{EtOH} = 0%, $k'_{sonicated}$ (182 x 10^{-2} s^{-1}/(g cells/mL)) was 20-fold higher than k'_{free} (7.38 x 10^{-2} s^{-1}/(g cells/mL)), which suggests a significant transport limitation caused by the cellular membrane and intracellular transport. It was also observed that the activity loss with increasing C_{EtOH} was more severe with sonicated cells compared to free cells, indicating the importance of the protection offered to the enzyme by the intracellular environment. A peak was observed around C_{EtOH} = 30% for free cells (17.9 x 10^{-2} s^{-1}/(g cells/mL)), which was attributed to increased permeability of the cell membrane due to ethanol damage. No activity was observed at that ethanol concentration with sonicated cells.

Characterization of gel permeability

It was shown that atrazine has a diffusivity of D_{eff} = 3.51 x 10^{-4} mm^2/s within the silica gel matrix without encapsulated cells. It was observed that increasing biocatalyst loading density (ρ) increased the permeability of the gel to atrazine. To incorporate this into the model, the following correlation, derived from the experimental data was used to estimate the diffusivity of the matrix for a given ρ as follows:

$$D_{eff} = (7.23 \times 10^{-4})\,\rho + (3.46 \times 10^{-4})\ mm^2/s \quad (2)$$

Experimental verification of the model

Two dimensionless numbers were used to estimate the steady-state performance of encapsulated biocatalysts: Thiele modulus (Φ) and the effectiveness factor (η). Mathematically, these two dimensionless numbers are expressed as follows:

$$\Phi \equiv L_c \sqrt{(k'_{free}*\rho)/D_{eff}} \ (3.a) \text{ and } \eta \equiv r_{obs}/r_{max} \ (3.b)$$

Note that k'_{free} and D_{eff} were determined in characterization studies and r_{max} can be evaluated from Eqn. (1). η can also be analytically derived as a function of Φ, for different encapsulation geometries. For slab and spherical geometries, η can be written in terms of Φ as:

$$\eta = (\tanh(\Phi))/\Phi \quad \text{(for slab)} \ (4.a)$$
$$\eta = 3/\Phi\,[1/(\tanh(\Phi)) - 1/\Phi] \quad \text{(for sphere)} \ (4.b)$$

Using Eqn. 4.a or 4.b, η can be determined and inserted in Eqn. 3.b to get an estimation for the observed degradation rate.

For verification of this model, we compared this estimated rate by the model to the experimentally observed degradation rates. It was shown that the experimentally observed activity was lower than the expected activity (81.3% ± 10.2%) of the model when ethanol-free rate constant (k'_{free}) was used. This result suggests that around 19.7% of cell degradation activity was lost during encapsulation. When 5% ethanol medium rate constant (k'_{free} at 5%EtOH) is used, the observed activity was higher than the expected activity (116.7% ± 14.7%). This result suggests that the alcohol content of the gel is overestimated or some alcohol evaporated during gelation. After experimental verification of the model, Thiele modulus (Φ) and the effectiveness factor (η) were plotted against the engineering design parameters: biocatalyst loading density (ρ) and characteristic length of the material (L_c) (Fig 2).

DISCUSSION

It was previously reported that free AtzA enzyme obeys Michaelis-Menten (MM) kinetics and its K_m and V_{max} values were determined [2]. To apply these values for kinetic modeling of an intact cell containing AtzA, one would need to have good estimates of parameters such as: cell membrane permeability, enzyme concentration in the cell and diffusivity of atrazine in the cell. In this study, we have shown that a free cell model that does not require explicit knowledge of these parameters can be used by characterizing the free cell kinetics and the change in cell membrane permeability induced by encapsulation.

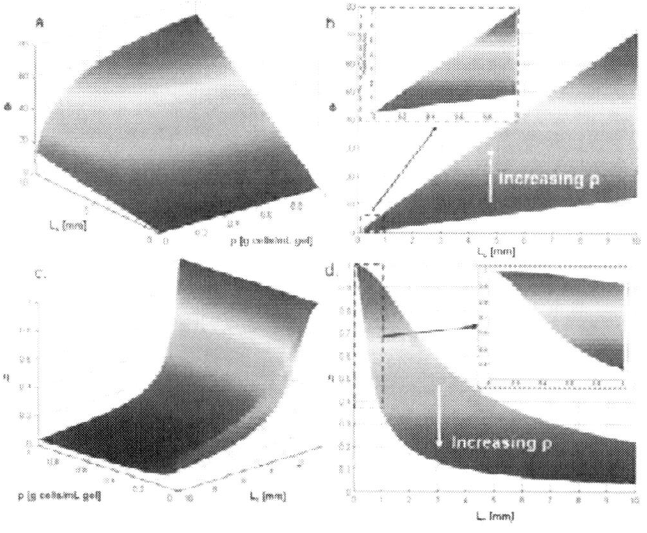

Figure 2 – Variation of dimensionless numbers Φ (a,b) and η (c,d), based on design parameters: ρ and L_c

An unexpected result of the study was the significant contribution of the biocatalyst loading to the gel membrane permeability, shown in Eqn. (2). We suspect that this phenomenon could be due to increased transport of atrazine through: a) Cell-silica interface within the matrix, b) The cells themselves. Due to this effect, ρ not only affects the biocatalytic activity rate constant, but also the transport of atrazine in the matrix (*i.e.* D_{eff} is a function of ρ in Eqn 3.a).

Using the characterization results, a reaction/transport model was developed and this model was used to plot the dimensionless numbers against design parameters (Fig 2). These graphs can be used to draw some constraints on the design parameters, such as it is not possible to operate at $\eta > 0.6$, when L_c is greater than 3 mm. This implies that more than 40% of the material is not utilized, which is undesirable since the extra cost of the material does not improve the degradation performance of the system. It should also be noted that a model is not sufficient by itself for optimization of these design parameters, since other constraints such as mechanical properties of the material and pressure drop across the reactor also need to be considered [3].

ACKNOWLEDGEMENTS

We acknowledge the support of an NSF-IIP/PFI grant (#1237754), an NSF-CAREER award (#0644784), a research grant from Syngenta Crop Protection and MnDrive fellowship to BRM from the BioTechnology Institute in University of Minnesota.

REFERENCES

[1] Mutlu BR *et al.* Silicon alkoxide cross-linked silica nanoparticle gels for encapsulation of bacterial biocatalysts. Journal of Materials Chemistry A 2013;1(36):11051-60.

[2] deSouza ML *et al.* Atrazine chlorohydrolase from Pseudomonas sp strain ADP: Gene sequence, enzyme purification, and protein characterization. Journal of Bacteriology 1996;178(16).

[3] Mutlu BR *et al.* Modelling and optimization of a bioremediation system utilizing silica gel encapsulated whole-cell biocatalyst. Chemical Engineering Journal 2015;259:574-80.

A NEW METHOD FOR DETERMINING CROSS-SECTIONAL SHAPE AND AREA OF SOFT TISSUES USING 3D LASER SCANNING

Y. Michaeli (1,3), R. E. Debski (1,2)

(1) Department of Bioengineering
University of Pittsburgh
Pittsburgh, Pennsylvania
United States

(2) Department of Orthopaedic Surgery
University of Pittsburgh
Pittsburgh, Pennsylvania
United States

(3) Department of Mechanical Engineering
ORT Braude College of Engineering
Karmiel, Israel

INTRODUCTION

An accurate measurement of the cross-sectional area is required for evaluation of the mechanical properties of any soft tissue. Tissue contractions can cause complex geometries and deformations within soft tissues, and pose difficulties when measuring cross-sectional areas [1, 2]. Techniques developed to measure the cross-sectional area are separated into either contact or noncontact methods. Contact methods create known errors either by forcing the tissue into a geometric shape using pressure, or by assuming geometric cross-sections [4]. The objective of this study was to design and validate a non-contact device to measure the cross-sectional shape and area of soft tissues. The design constraints were 3% accuracy and 3% repeatability while using a low cost system. In addition the noncontact method needed to be simple to perform, require minimum time, and be able to determine the cross-sectional area of multiple soft tissue types and sizes.

METHODS

A non-contact method for determining soft tissue cross-sectional area was developed and validated. It was then used to determine the cross-sectional areas of in vitro porcine ACL and human MCL. The noncontact method relies on the NextEngine® 3D laser scanner (NextEngine, Inc., Santa Monica, CA). The technology uses MultiStripe Laser Triangulation (MLT) with its array of solid state lasers to create a 3D model. The NextEngine® system is composed of a laser scanner and the AutoDrive, a rotary servo positioner, auto-incremented under scanner control. (Figure 1) The NextEngine® 3D has been previously validated for automated and rapid calculation of cross-sectional area of long bone diaphyses [3]. To examine a specimen, it is first clamped to the AutoDrive, then the laser scan is initiated. After the 3D image is created, it is imported as an STL to Geomagic Verify Viewer software that is used to calculate the cross-sectional area of the specimen.

Figure 1: Experimental setup showing medial collateral ligament clamped to the AutoDrive. After clamping a specimen, laser scan is initiated.

The process typically takes approximately 6.5 minutes, and displays graphical results. The NextEngine® system is controlled using two parameters, (1) number of rotation increments to complete a 360° scan (4-16 increments), and (2) resolution of the scan which is determined by the laser scanning speed (speed levels 1-9). Accuracy and repeatability of the NextEngine® system was determined by using five objects of known cross-sectional areas: 2 cylinders with known concavities (121.4, 471.1mm^2), a cuboid (163.2mm^2), a triangular prism (212mm^2) and a cylinder with a square keyway (115.3mm^2) (Delrin, Dupont, Inc., Wilmington, DE). To minimize scanning time while maximizing accuracy, an optimized combination of rotation increments and scanning speed was determined. The first object was scanned forty eight (48) times and the Geomagic Verify Viewer calculated the area at a constant cross-section on the object. Control parameters were tested using combinations of 6, 7 and 8 rotation increments and speed levels 1-7. To further optimize scan time, the subsequent four objects were measured using 42 scans. Not all combinations were tested in these ranges and each combination was repeated three times (first object was tested using 16 combinations and 14 combinations for the other four objects). After the system was optimized and validated using standardized shapes, it was then used to measure human MCL and porcine ACL tissues. The ligaments were clamped to the AutoDrive and minimal tension was applied. An additional test was performed on the ACL to verify that the laser beams did not penetrate the tissue surface. The ACL was stained with India ink and the cross-sectional area was remeasured. Six scans were taken before and after staining (12 scans overall) the ACL. Six scans were taken on the human MCL and two different locations were measured on each scan. The samples were sprayed with saline to prevent dehydration.

RESULTS

System optimization trials yielded maximum accuracy and repeatability with minimum scanning time at speed level 1 and 7 rotation increments. The NextEngine® system was able to determine the cross-sectional area and detect concavities of the standardized geometric objects. (Figure 2) The mean accuracy of four objects was 2.0%. The fifth object (cylinder with a square keyway) was excluded due to cross-sectional shape that does not demonstrate a typical concavity of soft tissues. The repeatability of the laser scans was 0.5mm^2 and was not affected by the range of the object size (16.5-91.7mm^2). For human MCL at two locations, cross sectional area was 16.5±0.3mm^2 and 37.1±0.5mm2. For porcine ACL, cross-sectional area was 91.1±0.6mm^2 and 91.7±2.2mm^2 after staining India ink on the tissue. (Figure 3) The accuracy of the system is 2.0% for standardized shapes and repeatability of 1.8% for soft tissues.

DISCUSSION

This study introduces a novel method for the determination of cross-sectional shape and area of soft tissues using laser scan data. This method produces an accurate and rapid production of a 3D image. Advantages of this method include automation, control over section orientation and the ability to examine multiple section locations across a soft tissue sample. Moreover, 3D images can be used to create 3D printed samples for future studies. For further examination of the system, testing range of sizes and types of tissues is recommended. NextEngine 3D laser scanner had met most of the design constraints as an accurate and affordable lab device (less than $5k and overall $6k including PC and clamping jig shown in Figure 1). Previous studies [1, 2] using non-contact laser systems determined cross-sectional areas within 1-5 minutes. Our goal was to similarly measure cross-sectional area within this same time interval. However, previous systems only measured the cross-section at a single location; the laser telemetric system and the laser reflectance system return the cross-sectional area as a function of its position along the axis of the object. Our system determined the cross-sectional area of the entire 3D model in 6.5 minutes. This study establishes a new method and the use of a laser scan protocol for the automated and rapid calculation of cross-sectional shape and area of soft tissues. It provides accurate results that can be compared to other methods. The results of this study show that the NextEngine® overcomes the complex issue of semitransparent surfaces by not penetrating the surface and thus not underestimating the cross-sectional area. This is a low cost system that has a quick technique and simple to carry out for determining the cross-sectional shape and area of soft tissues.

Cross-sectional Shape	Cross-sectional area measured [mm^2]	Accuracy %	Photo	Geomagic
1" Diameter cylinder with cutouts	472.9	0.4		
0.5" Diameter cylinder with cutouts	125.3	3.2		
0.5" Cuboid	167.1	2.4		
Triangular prism	215	1.9		
0.5" Diameter cylinder with square keyway	123.9	7.4		

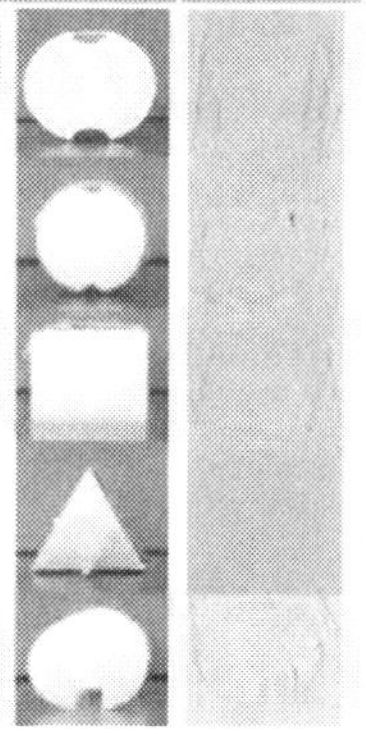

Figure 2: Cross-sections of standardized shapes

Tissue	Cross-sectional area measured [mm^2]	Photo	Geomagic
Porcine ACL	91.1± 0.6		
Porcine ACL with India Ink	91.7± 2.2		
Human MCL 1	37.1± 0.5		
Human MCL 2	16.5± 0.3		

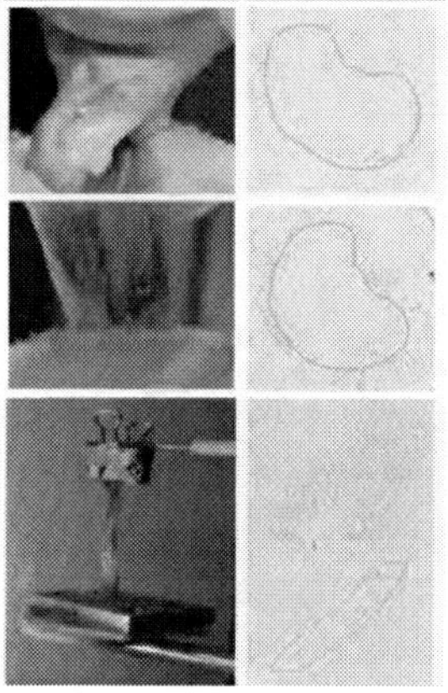

Figure 3: Cross-sections of soft tissues

ACKNOWLEDGEMENTS

Support from the Department of Orthopaedic Surgery and the Department of Bioengineering is gratefully acknowledged.

REFERENCES

[1] Lee, T.Q., Woo, S.L.-Y. A new method for determining cross-sectional shape and area of soft tissues. *J. Biomech.*, 110(2), 110-114, 1998.

[2] Moon, D.K., Abramowitch, S. D., and Woo, S. L. The development and validation of a charge-coupled device laser reflectance system to measure the complex cross-sectional shape and area of soft tissues. *J. Biomech.*, 39(16), 3071-3075, 2006.

[3] Davies, T.G., Shaw, C.N., Stock J.T. A test of a new method and software for the rapid estimation of cross-sectional geometric properties of long bone diaphysis from 3D laser surface scans. *Archaeol Antropol Sci*, 4: 277-290, 2012.

[4] Woo, Lee, Abramowitch, Gilbert. Structure and Function of ligaments and tendons. *Lippincott Williams & Wilkins Publishing*, Philadelphia, 7: 304-306, 2005.

MODELING THE BIAXIAL MECHANICS OF BRAIN WHITE MATTER

Kevin M. Labus (1), José Jaime García (3), Christian M. Puttlitz (1,2)

(1) School of Biomedical Engineering
Colorado State University
Fort Collins, CO, USA

(2) Department of Mechanical Engineering
Colorado State University
Fort Collins, CO, USA

(3) Escuela de Ingeniería Civil y Geomática
Universidad del Valle
Cali, Colombia

INTRODUCTION

Computational models of the brain rely on the accuracy of the constitutive equations defining the materials. It is important to develop grey and white matter models derived from experiments using a variety of loading states. White matter is structurally anisotropic due to the alignment of its axon tracts. White matter has also been shown to be mechanically anisotropic via experiments in uniaxial tension, shear, and indentation [1,2]. Each of these experiments have demonstrated anisotropy by testing the tissue in directions parallel to and perpendicular to the axons in one direction at a time. However, there is a lack of data describing the effects of multi-axial loading, even though *in vivo* brain tissue is in a constant multi-axial stress state due to physiologically applied loads such as intracranial pressure. Data from uniaxial experiments do not sufficiently describe simultaneously applied multi-axial stresses. Therefore, the purpose of this study was to perform a series of biaxial tests on brain white matter and to model the mechanical behavior using two anisotropic hyperelastic continuum models, elucidating any interaction effects arising from a biaxial state of stress.

METHODS

Ovine brain tissue specimens were dissected immediately after animals were euthanized for unrelated studies. Cruciform-shaped specimens were dissected from sagittal slices of the brain in the region of the corona radiata, such that the axons were in line with one of the loading directions (Figure 1). The experimental setup utilized specimen grips floating on a saline bath, attached to two linear actuators and two load cells for stress calculations (Figure 1). This setup was designed to minimize shear and obtain a homogeneous strain field in the center of the specimen.

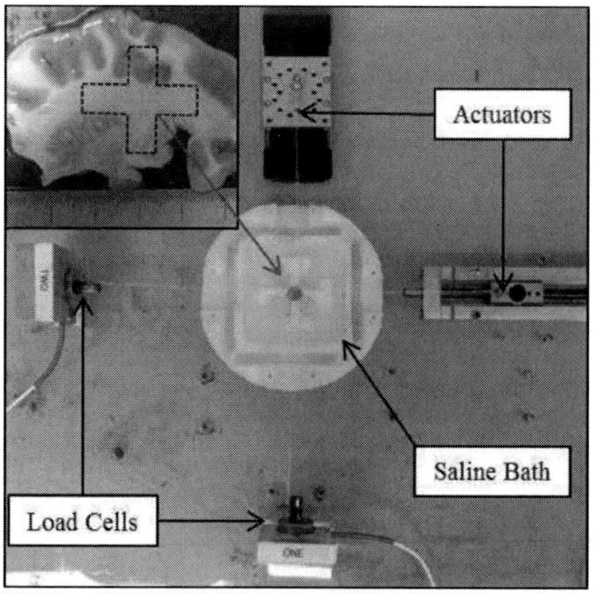

FIGURE 1: CRUCIFORM SPECIMENS WERE DISSECTED FROM THE CORONA RADIATA (INSET) AND ATTACHED TO FLOATING GRIPS. TWO LINEAR ACTUATORS AND TWO LOAD CELLS WERE ATTACHED TO THE GRIPS VIA STRINGS.

Specimens were subjected to a preload of 1.0 mN (mean stress of 71 Pa), followed by an initial preconditioning test of five cycles at a 1:1 (axonal: transverse) displacement ratio. The experiment consisted of three biaxial tests at 1:1, 1:0.5, and 0.5:1 displacement ratios, and two uniaxial tests. Each test was performed for five cycles at a maximum rate of 0.05 mm/s (strain rate of 0.0008 s^{-1}). In order to track the deformations, graphite powder was dusted on the top surface of the specimens. A MATBLAB-based digital image correlation program was used to calculate average stretch values in the region of interest, defined as a 2.4 mm square in the center of the specimen.

The experimental stress-stretch data were modeled using two hyperelastic strain energy density functions. The first model (*No Int.*) took the form of a two-term Yeoh model with an added anisotropic term. The second model (*Int.*) took the same form, but with an added interaction term [3]. The two models were given as:

$$W_{NoInt} = W_{iso} + W_{aniso} \quad (1)$$

$$W_{Interaction} = W_{iso} + W_{aniso} + W_{int} \quad (2)$$

such that:

$$W_{iso} = C_1(I_1 - 3) + C_2(I_1 - 3)^2 \quad (3)$$

$$W_{aniso} = C_3(I_4 - 1) \quad (4)$$

$$W_{int} = C_4(I_1 - 3)(I_4 - 1) \quad (5)$$

The coefficients C_1, C_2, C_3, and C_4 represent half the infinitesimal shear modulus, the isotropic nonlinearity, the anisotropy, and the interaction, respectively. I_1 and I_4 are the isotropic and anisotropic strain invariants, respectively. Each of the models was fitted to all 5 tests (3 biaxial and 2 uniaxial) of each given specimen simultaneously using a least squares fit.

RESULTS

The digital image correlation results demonstrated a high degree of homogeneity of the strain field in the central region of the specimens. Combining all tests, the mean longitudinal stretch was 1.082 and the average of the standard deviations of stretch in the region of interest was 0.016. The testing setup also effectively resulted in minimal shear, with a mean peak Lagrangian shear strain of 0.011.

The stress-stretch data from the uniaxial tests exhibited anisotropy, with the axonal direction stiffer than the transverse direction. However, the white matter appeared to be nearly isotropic when tested biaxially. The ability of the two models to predict this behavior was assessed by finding the mean fitted coefficients for each model. The mean coefficients for the *No Int.* model were C_1 = 1240 Pa, C_2 = 2430 Pa, and C_3 = 885 Pa. The mean coefficients for *Int.* model were C_1 = 1270 Pa, C_2 = 7460 Pa, C_3 = 560, and C_4 = 7300 Pa. The mean coefficients were used to predict the stress responses for uniaxial stretches in the axonal and transverse directions, and for equibiaxial stretches (Figure 2). The *Int.* model was able to better fit the experimental mean than the *No Int.* model. The mean RMSE of the fits were 151 Pa and 170 Pa for the *Int.* and *No Int.* models, respectively. The standard deviations of the RMSE were 49 Pa and 107 Pa, respectively, demonstrating that the *Int.* model more consistently fit the experiments well across the five tests performed on each specimen. Similar to the experimental mean data, the *Int.* model clearly shows a greater degree of anisotropy in uniaxial stretch compared to biaxial stretch, which is not captured in the *No Int.* model. The degree of anisotropy was defined as the axonal stress divided by the transverse stress at a stretch of 1.05. For the mean experimental data, the degree of anisotropy in the uniaxial and equi-biaxial tests were 3.7 and 1.2, respectively. The *Int* model matched this relatively closely, with a degree of anisotropy in the uniaxial and equi-biaxial tests of 3.3 and 1.3, respectively. However, for the *No Int* model, the uniaxial and equi-biaxial degree of anisotropy were both 1.7.

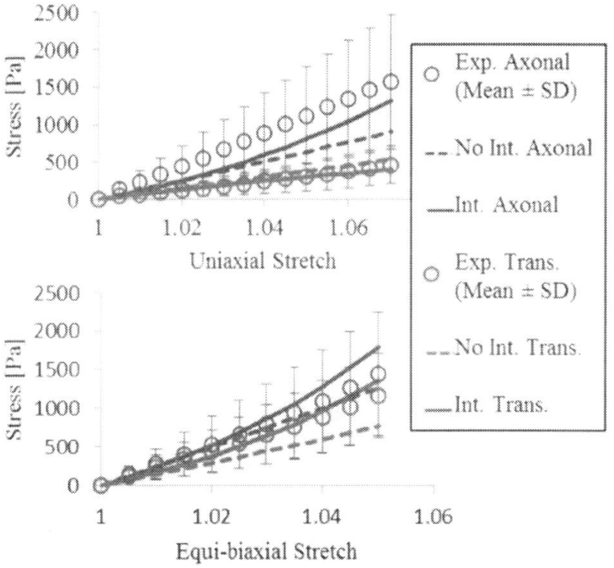

FIGURE 2: THE MEAN EXPERIMENTAL STRESS-STRETCH DATA COMPARED TO PREDICTIONS OF THE TWO MODELS FOR THE AXONAL AND TRANSVERSE UNIAXIAL TESTS (PLOTTED TOGETHER, TOP), AND THE EQUI-BIAXIAL TEST (BOTTOM).

DISCUSSION

The white matter exhibited an apparent near-isotropic behavior when tested biaxially. This was unexpected due to the axonal alignment present in white matter, as well as experiments on corona radiata in uniaxial tension which have found the axonal direction to be stiffer [1]. When uniaxial loading conditions were applied to the same specimens in the current study, an anisotropic behavior was demonstrated. The degree of anisotropy appears to be dependent on the relative stretch in each direction. This behavior may be due to oligodendrocyte-axon connectivity, which contributes to the tensile stiffness of white matter [4]. It is possible that there is a mechanical interaction between the axons and the interconnected glial cells that results in stiffening of the tissue in the transverse direction when the axonal direction is also loaded.

The first model (*No Int.*) follows a common approach in modeling white matter by including an isotropic term representing the glial matrix and an anisotropic term representing the axons. However, this model was unable to capture the variable degree of anisotropy. Due to the interaction term included in the second model (*Int.*), the degree of anisotropy is dependent on the relative stretches in each direction. The *Int.* model was able to better fit the experimental data in biaxial and uniaxial tests.

The current data indicate that it is important to perform both uniaxial and biaxial tests to more fully characterize the tensile mechanical behavior of white matter, and the inclusion of an interaction term in the constitutive model can capture the variable anisotropy in both loading states.

ACKNOWLEDGEMENTS

Funding was provided internally by Colorado State University.

REFERENCES

[1] Velardi, F et al., *Biomech Model Mechanbiol.* 5:53-61, 2006.
[2] Feng, Y et al., *J Mech Behavior Biomed Mat.* 23:117-132, 2013.
[3] Novak, VP et al., *J Biomech.* 27(4):403-412, 1994.
[4] Shreiber, DI et al., *Biomech Model Mechanobiol.* 8:311-321, 2009.

SB³C2015
Summer Biomechanics, Bioengineering and Biotransport Conference
June 17-20, 2015, Snowbird Resort, Utah, USA
SB³C2015-178

DAMAGE ACCUMULATION MODELING AND RATE DEPENDENCY OF SPINAL DURA MATER

Nicole L. Ramo (1), Snehal S. Shetye (2), Christian M. Puttlitz (1,2,3)

(1) School of Biomedical Engineering
Colorado State University
Fort Collins, CO, USA

(2) Department of Mechanical Engineering
Colorado State University
Fort Collins, CO, USA

(3) Department of Clinical Sciences
Colorado State University
Fort Collins, CO, USA

INTRODUCTION

As the outermost and strongest of the three spinal meninges, the dura mater plays an important stability role in the overall behavior of the spinal cord-meningeal complex (SCM), especially during traumatic loading scenarios such as vertebral burst fracture events [1]. Current investigations of spinal cord injury etiology which utilize computational techniques typically do not include information on the post-yield behavior of the SCM tissues. To comprehensively investigate spinal cord injuries, a thorough description of the material behavior of these tissues, including damage characterization during the sub-failure and failure regimes, is requisite.

Continuum viscoelastic damage theories have been developed and applied to a variety of biological tissues [2-4], but this work is the first to use a damage theory to describe the failure behavior of any tissue of the SCM. Due to its functional importance in injury prevention, the objective of this study is to characterize the damage accumulation behavior of the spinal dura mater under uniaxial loading.

METHODS

Longitudinal ovine dura mater samples (n = 9) measuring 35mm by 5mm were tensioned-to-failure at a 1mm/s (representing a strain rate of 0.051 ± 0.007 sec^{-1}) using a custom-built uniaxial test stand (Figure 1). The resulting stretch-stress curves were fit to a hyperelastic continuum non-linear directional damage model for fibrous biological soft tissues first proposed by Calvo et al. in 2007 [5] and further developed in 2008 [6] and 2012 [4]. This is an uncoupled directional damage model which defines different characteristics for the isotropic ground material/matrix ($\widehat{\boldsymbol{\psi}}^m_{iso}$) and anisotropic collagen fibers ($\widehat{\boldsymbol{\psi}}^f_{ani}$):

$$\boldsymbol{\psi} = (1 - D_m)\widehat{\boldsymbol{\psi}}^m_{iso}(\boldsymbol{C}) + (1 - D_f)\widehat{\boldsymbol{\psi}}^f_{ani}(\boldsymbol{C}, \boldsymbol{M}), \quad (1)$$

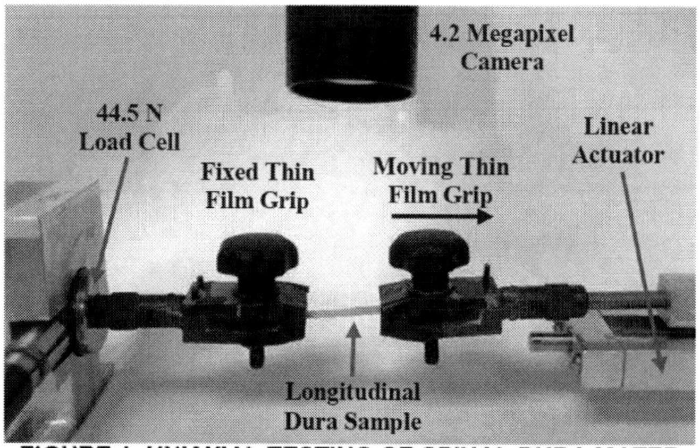

FIGURE 1. UNIAXIAL TESTING OF SPINAL DURA MATER

where **C** is the right Cauchy-Green tensor and **M** is the tensor product of the fiber direction vector. The terms $(1 - D_m)$ and $(1 - D_f)$ represent the reduction factors for damage as a function of the deformation (stretch); the damage parameters, D_m and D_f, are normalized [0,1] values related to how the matrix and fibers, respectively, accumulate damage. The evolution of damage is given by the following with k representing the matrix (m) or fibers (f):

$$D_k(\Xi^k_t) = \begin{cases} 0, & if\ \Xi^k_t < \Xi^{0_k}_{min} \\ \xi^2[1 - \beta_k(\xi^2 - 1)], & if\ \Xi^{0_k}_{min} \le \Xi^k_t \le \Xi^{0_k}_{max} \\ 1, & if\ \Xi^k_t > \Xi^{0_k}_{max} \end{cases} \quad (2)$$

In Eq. 2, $\xi = \frac{\Xi_t^k - \Xi_{min}^{0k}}{\Xi_{max}^{0k} - \Xi_{min}^{0k}}$ and is a dimensionless variable, and $\Xi_{min}^{0_k}$ and $\Xi_{max}^{0_k}$ are variables associated with the strain energy at which initial damage and total damage, respectively, occur during the loading procedure.

The isotropic response of the dural matrix was modeled with an exponential strain energy function [7], while the anisotropic response was modeled with a piecewise function [8].

$$\hat{\psi}_{iso}^m = \frac{c_1}{c_2}\{\exp[\frac{c_2}{2}(I_1 - 3)] - 1\} \quad (3)$$

$$\hat{\psi}_{ani}^f = \begin{cases} 0, & if\ I_4 < I_{4_0} \\ \frac{c_3}{c_4}\{\exp[c_4(I_4 - I_{4_0})] - c_4(I_4 - I_{4_0}) - 1\}, & if\ I_{4_0} < I_4 < I_{4_{ref}} \\ c_5\sqrt{I_4 - I_{4_0}} + c_6 \ln(I_4 - I_{4_0}) + c_7, & if\ I_4 > I_{4_{ref}} \end{cases} \quad (4)$$

It is assumed that the anisotropic fiber term only contributes to the total strain energy when the fibers are stretched ($I_4 > I_{4_0}$). The terms I_{4_0} and $I_{4_{ref}}$ characterize the location and length of the toe region of the response. In the above equations, c_1, $c_3 > 0$ represent parameters that are analogous to the stiffnesses of the matrix and fibers, respectively, while c_2, $c_4 > 0$ are dimensionless parameters that characterize the matrix and fiber non-linearity, respectively.

Fitting of the experimental data was performed via constrained optimization of c_1, c_2, c_3, c_4, I_{4_0}, $I_{4_{ref}}$, Ξ_{min}^m, Ξ_{max}^m, Ξ_{min}^f, Ξ_{max}^f, β_m and β_f with the fit quality assessed by the root mean standard error (RMSE) between experimental Cauchy stress and stress given by fitted damage model strain energy function. Paired t-tests were used to determine differences between related parameters (i.e. c_1 and c_3); a p-value of 0.05 defined statistical significance.

RESULTS

The data fit the continuum damage model well, with a maximum RMSE of 0.46 MPa (<3% of average maximum stress). The average value and standard deviation of the twelve fitted parameters are given in Table 1 grouped by related terms. Significant differences are indicated with the associated p-value given in the third column.

TABLE 1. FITTED PARAMETERS

Fitted Parameter	Average ± Std. Deviation	p-Value
c_1 (MPa)	2.92 ± 2.14	$c_1 > c_3$ $p < 0.001$
c_3 (MPa)	4.66 ± 2.47	
c_2	9.59 ± 2.52	$c_2 > c_4$ $p < 0.001$
c_4	1.03 ± 0.24	
I_{4_0}	1.21 ± 0.10	N/A (constrained)
$I_{4_{ref}}$	1.80 ± 0.27	
Ξ_{min}^m (MPa$^{1/2}$)	0.51 ± 0.08	$\Xi_{min}^m < \Xi_{min}^f$ $p < 0.001$
Ξ_{min}^f (MPa$^{1/2}$)	1.52 ± 0.19	
Ξ_{max}^m (MPa$^{1/2}$)	1.29 ± 0.24	$\Xi_{max}^m < \Xi_{max}^f$ $p = 0.002$
Ξ_{max}^f (MPa$^{1/2}$)	1.54 ± 0.20	
β_m	0.16 ± 0.20	N/A
β_f	0.44 ± 0.54	

As shown in Table 1, significant differences were found between the following pairs: stiffness values c_1 and c_3, non-linearity values c_2 and c_4, initiation of damage parameters Ξ_{min}^m and Ξ_{min}^f, and total damage parameters Ξ_{max}^m and Ξ_{max}^f. No significance differences were found between the other two sets of parameters.

DISCUSSION

As expected, the stiffness of the fibers (c_3) was significantly greater than that of the matrix (c_1). The significant difference between c_2 (matrix non-linearity) and c_4 (fiber non-linearity) was also expected since the dura represents a fiber-reinforced composite; while collagen fiber straightening (i.e. uncrimping) does contribute to composite non-linearity, the majority of composite non-linearity can usually be attributed to the intrinsic properties of the matrix [9].

The values of I_{4_0} and $I_{4_{ref}}$ are similar to those reported for rectus sheath [4] and vaginal tissue [8]. Also in agreement with these previous soft tissue damage studies, the results imply that the matrix material of the dura mater incurs damage before the collagen fibers and fails before complete fiber failure. Figure 3 shows these results in graphical form by plotting the damage parameters D_m and D_f (defined in Eq. 2 as functions of Ξ_{min}^m, Ξ_{max}^m, β_m and Ξ_{min}^f, Ξ_{max}^f, β_f, respectively). The initiation of damage relates to the deviation of the damage parameter from zero, while complete damage relates to the maximum value of the damage parameter (typically one). For the majority of tests D_m departs from zero and reaches its max value at lower stretch values than D_f. It also can be seen from Figure 3 that the matrix seems to fail more gradually, while the fibers experience more brittle-like failure.

Despite multiple studies in the literature on the tension-to-failure properties of spinal dura mater from both cadaveric and animal samples, this is the first study to apply a damage constitutive model to the dura. These data and damage formulation can be implemented into finite element computational models of the SCM to improve the accuracy of simulations of spinal cord injury scenarios and dynamics.

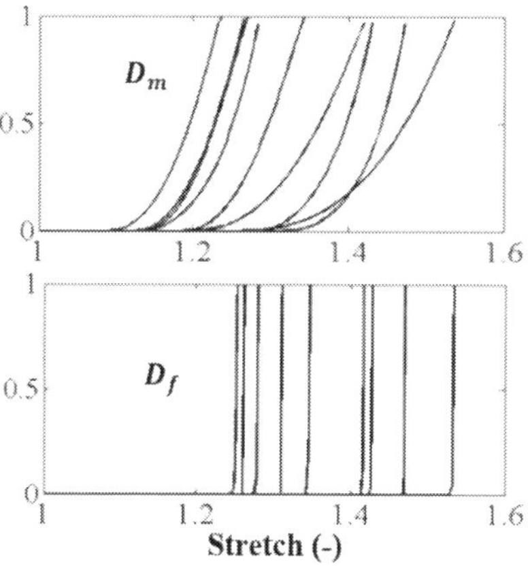

FIGURE 3. DAMAGE ACCUMULATION IN DURAL MATRIX AND FIBERS

ACKNOWLEDGEMENTS

Funding for this work was provided by a grant from the National Institutes of Health (EB012048).

REFERENCES

[1] Hall, R et al., *J Neurosurg: Spine,* 5:243-250, 2006.
[2] Gasser, TC et al., *Acta Biomater*, 7:2457-2466, 2011.
[3] Liao, H & Belkoff, SM, *J Biomech*, 32:183-188, 1999.
[4] Martins, P et al., *J Mech Behav Biomed Mater*, 8:111-122, 2012.
[5] Calvo, B & Pena, E, *Int J Numer Meth Engng*, 69:2036-2057, 2007.
[6] Pena, E et al., *Int J Numer Meth Engng*, 74:1198-1218, 2008.
[7] Demiray, H et al., *J Biomech*, 21:369-374, 1988.
[8] Calvo, B et al., *J Biomech*, 42:642-651, 2009.
[9] Hirsekorn, M & Grail, G, *IOM3 Int Conf Compos Mater*, 1-6, 2011.

ENDOPLASMIC RETICULUM CALCIUM DYNAMICS IN OSTEOCYTE MECHANOBIOLOGY

Genevieve N. Brown (1), Prajesh H. Desai (1), X. Edward Guo (1)

(1) Department of Biomedical Engineering
Columbia University
New York, NY, USA

INTRODUCTION

Due to their abundance and unique arrangement within bone tissue, osteocytes are considered the primary mechanosensing cells in bone. One of the earliest responses of bone cells to mechanical stimulation is a rise in intracellular calcium ($[Ca^{2+}]_i$), and previous studies have demonstrated that osteocytes exhibit unique $[Ca^{2+}]_i$ patterns. Both *in vitro* [1] and *ex vivo* [2], osteocytes exhibit spike-like oscillations in $[Ca^{2+}]_i$ in response to mechanical stimuli, whereas osteoblasts (OB) demonstrate fewer, weaker responses. Inhibitor studies have implicated the release of Ca^{2+} from endoplasmic reticulum (ER) stores as being critical to subsequent responses. Thus, it is possible the mechanisms of Ca^{2+} release and reuptake by the ER are important to osteocyte mechanobiology.

The differentiation of osteoblasts to osteocytes is characterized by many phenotypic changes, including changes in expression of membrane calcium channels. Osteoblasts express both T- and L-type voltage-sensitive calcium channels (VSCC), whereas L-type VSCC are barely detectable in osteocytes. We hypothesized that the predominant expression of T-type channels in osteocytes may contribute to their unique $[Ca^{2+}]_i$ patterns and further speculated that T-type VSCC in osteocytes may interact with ER stores. We sought to visualize the dynamics of Ca^{2+} signaling within bone cells under fluid flow by simultaneously monitoring Ca^{2+} separately in the cytosolic ($[Ca^{2+}]_i$) and endoplasmic ($[Ca^{2+}]_{ER}$) spaces in osteocytes and osteoblasts. The purpose of this study is to determine the mechanisms by which osteocytes generate $[Ca^{2+}]_i$ oscillations and their dependence on VSCC.

METHODS

The osteocyte-like MLO-Y4 and osteoblastic MC3T3-E1 cell lines were maintained at 37°C and 5% CO_2 in serum-supplemented medium. For flow experiments, cells were grown to ~80% confluency onto fibronectin-coated glass slides to establish cell-cell contact. To visualize intracellular calcium changes, cells were incubated with Fluo-8 AM and assembled into a parallel plate flow chamber. Cells were pre-treated with the T-type VSCC inhibitor NNC 55-0396 or the L-type inhibitor nifedipine prior to flow. Time-lapse images were collected at 20x magnification for 10 minute periods of fluid flow stimulation at 35 dynes/cm^2. To simultaneously visualize intracellular and ER calcium, cells were transiently transfected with the D1ER plasmid [3] using Fugene 6. Thirty-six to forty-eight hours post transfection, cells were incubated with 20μm Fura Red-AM to indicate intracellular calcium. Time-lapse images were collected at 40x magnification. Single excitation (430/24) was used to excite the dye and FRET biosensor. Fluorescence emissions of YFP (530/30), CFP (470/28), and Fura Red (641/75) were captured separately and simultaneously using a quadview beamsplitter and custom quad-band polychroic [4]. To monitor ER depletion, transfected cells were imaged for 20 minutes following the addition of VSCC inhibitors. The FRET ratio was calculated on a pixel-by-pixel basis using image registration of the FRET and donor emissions. All images underwent background subtraction and median filtering.

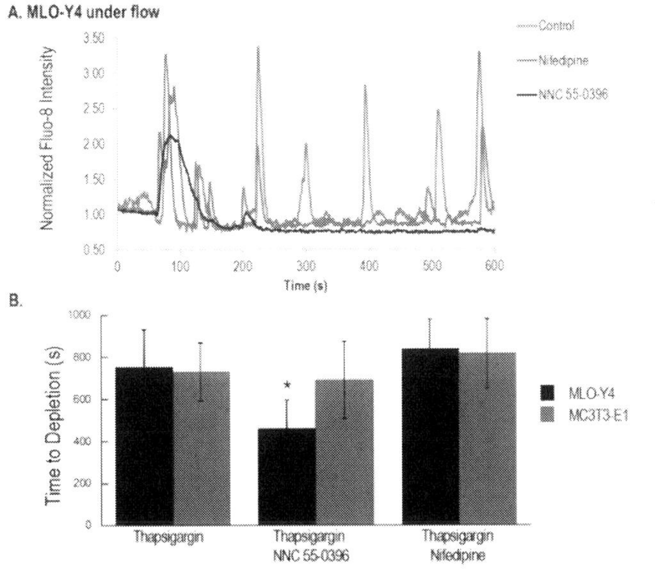

FIGURE 1. Effects of T- and L-type VSCC inhibitors on intracellular and ER calcium. *p<0.01

RESULTS

Consistent with previous reports, osteocytes exhibited numerous $[Ca^{2+}]_i$ oscillations under fluid flow, and these multiple responses were abolished by treatment with the T-type inhibitor NNC 55-0396 (Fig. 1A). The L-type inhibitor had no significant effect on the number of $[Ca^{2+}]_i$ responses in osteocytes, consistent with the limited expression of L-type VSCC in these cells (Fig. 1A) [5]. Osteoblasts demonstrated fewer responses under fluid flow, and these responses were not affected by T- or L-type inhibition (Fig. 1B). Treatment of D1ER transfected cells with thapsigargin depleted the ER stores, and depletion times were not significantly different in MLO-Y4 or MC3T3-E1 (Fig. 1C). Inhibition of VSCC had no effect on ER depletion times in osteoblasts, and L-type inhibition had no effect on osteocytes. However, inhibition of T-type VSCC resulted in faster depletion of ER stores in osteocytes, suggesting an interaction between T-type channels and the ER in osteocytes. In D1ER-transfected osteocytes stained with Fura Red-AM, elevations of $[Ca^{2+}]_i$ coincided with depression of $[Ca^{2+}]_{ER}$, with subsequent peaks occurring after recovery of $[Ca^{2+}]_{ER}$ levels (Fig. 2A). In osteoblasts, only a single $[Ca^{2+}]_i$ response was observed, and while the ER contributed to this response, it did not refill in the time course of the experiment (Fig. 2B).

DISCUSSION

These studies demonstrate the ER is more intimately involved in osteocyte calcium signaling than previously considered and support the hypothesis that OCY are capable of generating characteristic multiple responses by an ability to refill the ER stores. Future studies will further explore the role of VSCC and other channels on release and reuptake dynamics of ER Ca^{2+} trafficking in osteocytes under flow, which may contribute to their mechanobiological function.

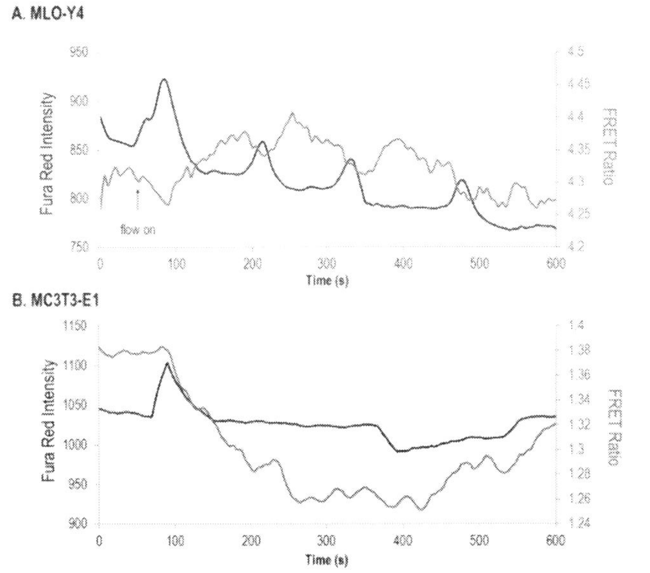

FIGURE 2. Simultaneous monitoring of intracellular and ER calcium levels under fluid flow.

ACKNOWLEDGEMENTS

This work was supported by NIH grant R01 AR052461. GB was supported by a NSF graduate research fellowship.

REFERENCES

[1] Lu, XL et al., *JBMR*, 2012.
[2] Jing, D et al., *FASEB J*, 2014.
[3] Palmer, AE et al., *PNAS*, 2004.
[4] Baik, AD et al., *Biophys J*, 2010.
[5] Thompson, W.R. et al., *JBMR*, 2011.

SB^{3}C2015
Summer Biomechanics, Bioengineering and Biotransport Conference
June 17-20, 2015,Snowbird Resort, Utah, USA
SB^{3}C2015-180

FUNCTIONAL TENSILE PROPERTIES OF A SPLIT QUADRICEPS GRAFT FOR DOUBLE-BUNDLE ACL RECONSTRUCTION

R Matthew Miller (1,3), Amir Ata Rahnemai-Azar (2,1,3), Todd Jasinski (1,3), Fabio V Arilla (2,3), Levent Sürer (2,3), Freddie H Fu (2,3), Richard E Debski (1,2,3), Volker Musahl (2,1,3)

(1) Department of Bioengineering
University of Pittsburgh
Pittsburgh, PA, United States

(2) Department of Orthopaedic Surgery
University of Pittsburgh
Pittsburgh, PA, United States

(3) Orthopaedic Robotics Laboratory
Departments of Orthopaedic Surgery and Bioengineering
University of Pittsburgh
Pittsburgh, PA, United States

INTRODUCTION

ACL reconstruction is one of the most common orthopedic procedures due to an increased incidence of knee trauma and injury of the anterior cruciate ligament (ACL) in sports [1]. Anatomic double-bundle ACL reconstruction techniques allow for improved anteroposterior stability, rotational stability, and native anatomy [2]. Quadriceps tendon with bone block has become an increasingly popular alternative graft choice for ACL reconstruction [3]. When splitting the tendinous portion of the autograft into a "Y-shape", quadriceps tendon has been shown to be an excellent option for anatomic double-bundle ACL reconstruction that exhibits the advantages of osseous healing, less perioperative donor site morbidity, and minimal re-rupture rate [4].

Quadriceps grafts for double-bundle ACL reconstruction can be split in either sagittal or coronal planes, which results in different portions of the three layers of quadriceps tendon [5] being included in the graft halves, depending on the splitting plane. Furthermore, double-bundle ACL reconstruction can sometimes result in differential failure of either the AM or PL bundle due to graft rupture or elongation [6], indicating the importance of understanding the functional tensile properties of the graft halves after splitting. Therefore, the objective of this study is to evaluate the functional tensile properties of quadriceps tendon grafts split in coronal and sagittal planes. It was hypothesized that tendons split in the coronal plane will have superior functional properties due to preserving the natural fiber split of the tendon.

METHODS

Twenty quadriceps tendons (specimen age 60 ± 7.3 years, 14 male, 6 female) with superior patellar bone were harvested from fresh-frozen cadaveric knees. Full-thickness quadriceps tendon grafts of approximate dimensions of 1 cm by 10 cm were prepared with attached bone block to mimic grafts for double-bundle ACL reconstruction surgery. Ten grafts were split in the sagittal ($n = 10$) plane, and ten grafts were split in the coronal ($n = 10$) plane to create Y-shaped grafts with tendon halves of approximately equal width. The bone block of each graft was then potted in polymethyl methacrylate and mounted in specialized clamps (Fig 1A). Both halves of each graft were placed in a 3D laser scanning system (NextEngine 3D Scanner HD, Santa Monica, CA) with 2% accuracy to measure cross-sectional area (CSA). Grafts were then mounted to a materials testing machine (Instron Model 5965, Norwood, MA) for tensile testing (Fig 1B).

After application of a 10 N preload, graft halves were preconditioned from 10 – 50 N for 20 cycles at 10 mm/min, followed by cyclic creep testing for 100 cycles between 10 and 100 N at 10 mm/min. Order of testing (medial versus lateral graft halves and anterior versus posterior graft halves) was randomized. After completion of creep testing, specimens were removed from clamps and allowed 30 minutes of recovery time. After reapplication of preload and preconditioning, five of each half from each group (i.e. medial, lateral, anterior, and posterior) were loaded to failure.

Percent creep, ultimate load, ultimate elongation, and stiffness of the linear region of the load-elongation curve were determined from load-elongation data for each graft half. Creep was calculated as the difference in maximum elongation between the peaks of the 1st and 100th cycles during cyclic creep testing. Related-samples Wilcoxon signed rank tests were used to compare all variables for anterior versus posterior and medial versus lateral. Independent-samples Mann-Whitney U tests were used to compare samples outside of pairs (i.e. medial versus anterior). Significance was set at $p < 0.05$.

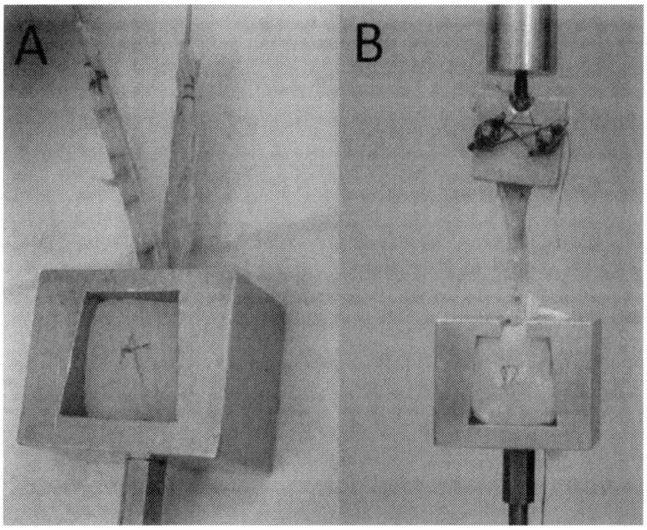

FIGURE 1: A) Quadriceps tendon graft split in sagittal plane into Y-shape for tensile testing. The patellar bone block is potted and clamed with the tendon ends left free. B) Testing set up for one half of Y-shaped graft.

RESULTS

During testing, two tendon halves (one lateral, one posterior) partially slipped out of the soft tissue clamps during cyclic creep testing, resulting in abnormally large amounts of creep. These two samples were therefore removed from calculations of creep and percent creep. Lateral and Medial groups showed similar properties for all variables. The Anterior group had slightly larger CSA, creep, and elongation than the Posterior group. However, no statistically significant differences in functional properties were found between the groups ($p>0.05$) (Table 1).

Failure modes were also similar among groups. In the anterior group, 2 failed near the clamp, 1 failed mid-substance, and 2 failed in the substance near the insertion to the patella. In the posterior group, 2 failed near the clamp, and 3 failed mid-substance. The medial group showed 2 grafts that failed near the clamp, 1 that failed mid-substance, and 2 that failed in the substance near the patellar insertion. In the lateral group, 3 failed near the clamp, 1 failed mid-substance, and 1 failed near the insertion. Large variation in ultimate load and ultimate elongation was observed within groups (Figure 2).

Table 1: Graft Functional Properties for Coronal and Sagittal Split Groups (Average ± Standard Deviation)

	Lat	Med	Ant	Post
CSA (mm^2)	48.7 ± 11.5	48.1 ± 9.6	56.2 ± 11.7	42.5 ± 15.1
% Creep	42.5 ± 12.4	41.8 ± 16.0	53.4 ± 13.6	43.0 ± 13.3
Ultimate Load (N)	445 ± 210	496 ± 128	555 ± 144	426 ± 127
Ult. Elong. (mm)	7.7 ± 2.0	7.3 ± 1.9	9.1 ± 2.2	6.5 ± 1.8
Stiffness (N/mm)	75.7 ± 19.9	90.8 ± 15.9	80.2 ± 24.8	93.1 ± 16.1

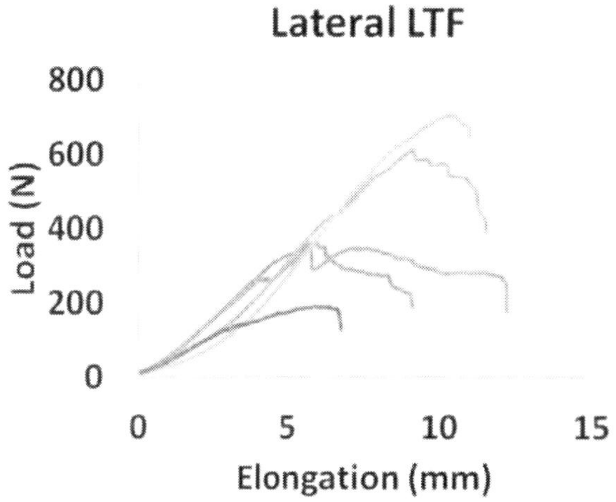

FIGURE 2: Example load-elongation curves for the lateral group, load to failure tests.

DISCUSSION

Overall, splitting quadriceps tendon grafts with bone block for anatomic double-bundle ACL reconstruction results in similar functional tensile properties regardless of splitting plane. Due to the similar properties between groups, splitting in either the coronal or sagittal plane may result in a suitable graft for double-bundle ACL reconstruction. Notably, the combined stiffness and ultimate load values for both halves of each graft are considerably lower than the properties of a whole quadriceps tendon graft at similar loading magnitude (570 N/mm and 2353 N, respectively) [3]. Ultimate elongation of a single graft is also lower than that of the halves of the graft (5.6 mm) [3], as is percent creep (2.2%) [7]. The act of physically splitting the graft into halves may therefore lead to significant reductions in graft strength and enable the graft to elongate more than if it were left intact, possibly resulting in improper function of the graft for double-bundle ACL reconstructions. However, the specimens in this study were significantly older than those used in the study by Stäubli et al (average age 25 years), which may account for the observed differences. In the future, robotic testing of ACL reconstructed knees using split and intact grafts to investigate the effects of graft splitting on knee kinematics will be performed. This study provides data that supports the use of both coronal and sagittal splits for anatomic double-bundle ACL reconstruction due to similar functional tensile properties between groups.

ACKNOWLEDGEMENTS

Support from the Department of Orthopaedic Surgery is gratefully acknowledged. The authors would also like to thank Dr. James Irrgang for assistance with statistical analysis.

REFERENCES

[1] Gordon and Steiner. Orthopaedic knowledge update: sports medicine. AAOS. 2004, pp169-81.
[2] Fu et al. Am J Sports Med 2008;36:1263-74.
[3] Stäubli et al. Knee Surg Sports Traumatol Arthroscopy 1996; 4:100-110.
[4] Kato et al. J Orthop Sci 2010; 15:269-276.
[5] Waligora et al. Clin Orthop Relat Res 2009, 467(12):3297-306.
[6] van Eck et al. Am J Sports Med 2012; 40(4):800-7.
[7] Stäubli et al. Am J Sports Med 1999; 27(1):27-34.

SB³C2015
Summer Biomechanics, Bioengineering and Biotransport Conference
June 17-20, 2015,Snowbird Resort, Utah, USA
SB³C2015-181

NUMERICAL MODELING OF HUMAN FIBROUS CAP DELAMINATION

X. Leng (1), X. Deng (1), M.A. Sutton (1), S.M. Lessner (2)

(1) Department of Mechanical Engineering
University of South Carolina
Columbia, SC 29208, USA

(2) Department of Cell Biology & Anatomy
University of South Carolina
Columbia, SC 29208, USA

INTRODUCTION

Atherosclerotic plaque rupture is a serious complication of advanced atherosclerosis, often leading to life-threatening clinical consequences such as myocardial infarction or stroke [1]. Plaque rupture is, in essence, a mechanical failure phenomenon, often in the form of fibrous cap delamination from the underlying tissue. Fibrous cap failure typically occurs at one of two locations: one is at the center of the plaque, where the fibrous cap may be very thin and is prone to rupture under large deformation and stress concentration; the other is at the shoulder region, where the interface between the plaque and the underlying vascular wall is located, which has the potential to delaminate when the arterial wall suffers from large expansion [2]. As such, it is important to understand these failure modes in order to develop effective approaches for treatment and intervention.

In the current study, a finite element modeling and simulation approach for fibrous cap delamination experiments has been developed. Two types of material models have been implemented. The first type is the viscoelastic fiber-reinforced material model for the bulk mechanical behavior of the arterial wall, and the second type is the exponential cohesive zone model used for the delamination behavior at the fibrous cap-plaque interface.

METHODS

The current study is focused on the numerical modeling of fibrous cap delamination experiments performed on human carotid artery plaques obtained from endarterectomies. At the start of the experiment, a small initial delamination was carefully introduced by a scalpel at the proximal end of the fibrous cap, so that the microclamp can clamp the fibrous cap, as shown in Fig.1. The carotid artery was glued with DermaBond to a glass plate connected to the load cell of a Bose ELF 3200 for load data recording. The delaminated tissue tab on the proximal end of the fibrous cap was gripped by a pair of microclamps connected to the Bose ELF 3200 actuator, which applies sequential loading-delamination-unloading cycles, as shown in Fig. 1. For each experiment, a load-displacement curve with multiple loading-delamination–unloading cycles was obtained, and three sequential loading-unloading cycles are shown in Fig. 2.

A cohesive zone model (CZM) approach is applied to simulate delamination of the fibrous cap from the underlying plaque tissue. We extended the Holzapfel-Gasser-Ogden (HGO) model [3] by incorporating viscoelasticity [4], and adopted this model for the bulk material behavior. Meanwhile, we implemented a cohesive zone model (CZM) for the delamination behavior along the fibrous cap-plaque interface. In the viscoelastic HGO model, each artery layer is treated as a fiber-reinforced material with the fibers symmetrically disposed with respect to the axial direction of the aorta. In the CZM, delamination is governed by an exponential traction-separation law which utilizes critical energy release rates obtained directly from the experiments. A proper set of viscoelastic HGO parameter values and CZM parameter values is determined based on values suggested in the literature and through matching simulation predictions of the load vs. load-point displacement curve with experimental measurements for one fibrous cap delamination cycle [5] (cycle #1, as shown in Fig. 2).

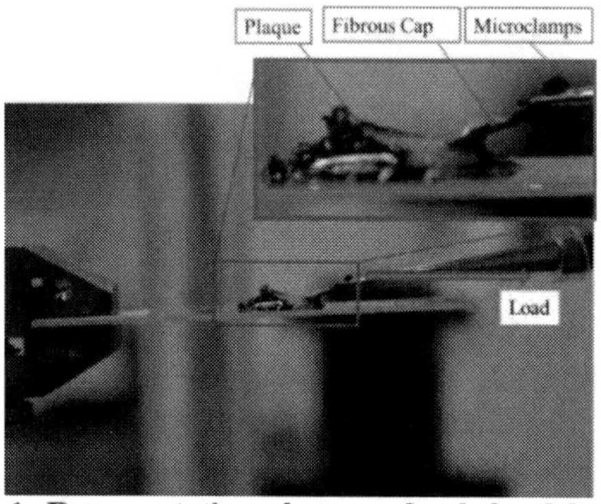

Fig. 1. Representative photograph of the delamination stage of fibrous cap delamination test. Scale bar=5 mm.

Bulk material for the human carotid artery

The Holzapfel-Gasser-Ogden (HGO) model assumes that collagen fibers are oriented parallel to the arterial wall [3], thus neglecting the out-of-plane component. The strain energy function in the HGO model is given by [3]:

$$\Psi = \Psi_{vol} + \bar{\Psi}_{\text{iso}} \tag{1}$$

$$\bar{\Psi}_{\text{iso}} = \frac{\mu}{2}(\bar{I}_1 - 3) + \sum_{i=1}^{2} \frac{k_1}{2k_2}\left[e^{k_2[\kappa \bar{I}_1 + (1-3\kappa)\bar{I}_{4i} - 1]^2} - 1\right] \tag{2}$$

$$\Psi_{vol} = \frac{1}{D}\left(\frac{J^2 - 1}{2} - \ln J\right) \tag{3}$$

where $\bar{I}_1 = \text{tr}(\bar{\mathbf{C}})$ denotes the first invariant of $\bar{\mathbf{C}}$ and $\bar{\mathbf{C}}$ denotes a modified counterpart of the right Cauchy-Green strain tensor $\mathbf{C}$; $\bar{\mathbf{C}} = \overline{\mathbf{F}^{\mathrm{T}}}\bar{\mathbf{F}}$ and $\bar{\mathbf{F}} = J^{-1/3}\mathbf{F}$, where $\mathbf{F}$ is the deformation gradient and $J = \det(\mathbf{F})$. The tensor invariants $\bar{I}_{41} = \mathbf{a}_{01} \cdot \bar{\mathbf{C}}\mathbf{a}_{01}$ and $\bar{I}_{42} = \mathbf{a}_{02} \cdot \bar{\mathbf{C}}\mathbf{a}_{02}$ are equal to the square of the stretch in the direction of $\mathbf{a}_{01}$ and $\mathbf{a}_{02}$, respectively. The constitutive parameter k_1 is related to the relative stiffness of fibers in the small strain region, which is determined from mechanical tests of tissue; k_2 is a dimensionless stiffness related to the large strain behavior of the two families of fibers; κ ($0 \le \kappa \le 1/3$) is the dispersion parameter; μ is the neo-Hookean parameter and $\frac{1}{D}$ is analogous to the bulk modulus.

The second Piola-Kirchhoff stress can be derived from the Clausius-Planck inequality and is given by:

$$\boldsymbol{S}^{\infty} = 2\frac{\partial \Psi}{\partial C}, \boldsymbol{S}^{\infty}_{vol} = 2\frac{\partial \Psi_{vol}}{\partial C}, \boldsymbol{S}^{\infty}_{iso} = 2\frac{\partial \bar{\Psi}_{iso}}{\partial C} \tag{4}$$

The second Piola-Kirchhoff stress tensor of the viscoelastic HGO model $\boldsymbol{S}_{n+1}$ at time t_{n+1} [4] is given by:

$$S_{n+1} = \left(S^{\infty}_{vol} + S^{\infty}_{iso} + \sum_{\alpha=1}^{n} Q_{\alpha}\right)_{n+1} \quad (5)$$

The non-equilibrium stress tensor $Q_{\alpha\, n+1}$ at time t_{n+1} is given by:

$$Q_{\alpha\, n+1} = \mathcal{H}_{\alpha\, n} + \beta_{\alpha}\, exp\left(-\frac{\Delta t}{2T_{\alpha}}\right)\left(S^{\infty}_{iso}\right)_{n+1} \quad (6)$$

where the history term is defined as:

$$\mathcal{H}_{\alpha\, n} = \exp\left(-\frac{\Delta t}{2T_{\alpha}}\right)\left[\exp\left(-\frac{\Delta t}{2T_{\alpha}}\right) Q_{\alpha\, n} - \beta_{\alpha}\left(S^{\infty}_{iso}\right)_{n}\right] \quad (7)$$

In the above equations, Δt is the time increment at time t_{n+1}; T_{α} is the relaxation time and β_{α} is a non-dimensional parameter.

Exponential Cohesive Zone Model

An exponential form of the free energy potential (φ) per unit (undeformed) area can be expressed as [6, 7]:

$$\varphi = e\sigma_c\delta_c\left[1 - \left(1 + \frac{\delta}{\delta_c}\right)\exp\left(-\frac{\delta}{\delta_c}\right)\right] \quad (8)$$

where $e = \exp(1) \approx 2.71828$ and σ_c denotes the strength of the material; δ_c is the maximum effective displacement at $t = \sigma_c$ and δ is the displacement jump across the cohesive surfaces. Considering the loading conditions, the cohesive traction is given by:

$$t = \frac{\partial\varphi}{\partial\delta} = e\sigma_c\frac{\delta}{\delta_c}\exp\left(-\frac{\delta}{\delta_c}\right) \quad (9)$$

The energy release rate per unit cohesive surface follows the form:

$$G_c = e\sigma_c\delta_c \quad (10)$$

The viscoelastic HGO model and exponential CZM are implemented in the current study through a user UMAT and UEL subroutine in ABAQUS [8], respectively.

RESULTS

Simulations for cycle #2 and #3 (as shown in Fig. 2) are carried out based on the material parameter values identified from cycle #1. It can be seen from Fig.2 that the simulation predictions match well with the experimental data, especially the hysteresis behavior for the unloading phase of cycle#1 and the loading phase of cycle #2. Meanwhile, there are some differences between the predictions and measurements, such as the shape of the curve during the delamination phase.

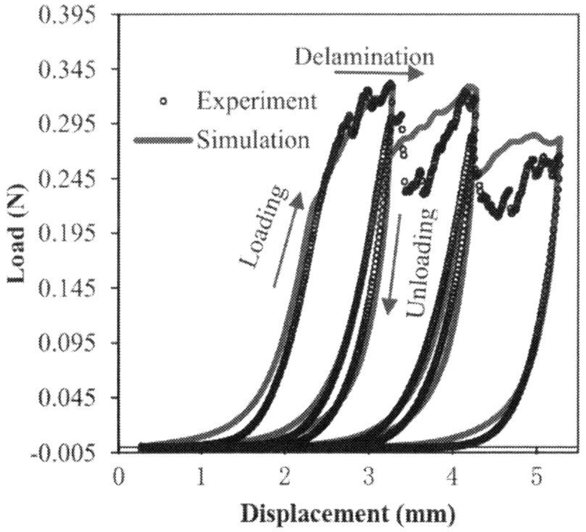

Fig. 2. The simulation-predicted load-displacement curves of three loading-unloading cycles are compared with the measured curves. Each cycle includes a loading phase in which the fibrous cap is pulled from the atherosclerotic plaque, a delamination phase in which the fibrous cap undergoes further delamination, and an unloading phase in which the fibrous cap is returned to its initial position.

The von Mises stress contours for three typical points of the loading-delamination-unloading cycle of fibrous cap delamination simulation for cycle #3 are shown in Fig. 3.

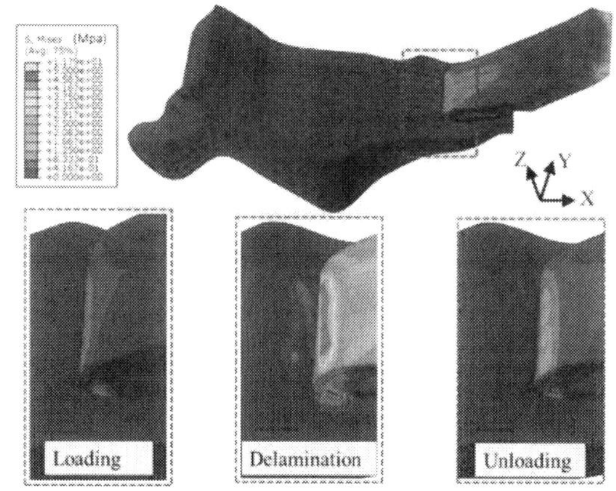

Fig. 3. von Mises stress (MPa) contours for three typical points of the loading-unloading cycle of fibrous cap delamination simulation.

DISCUSSION

This study shows that a 3D finite element model of human fibrous cap delamination, with the implementation of a viscoelastic HGO model for the bulk material and an exponential CZM for the fibrous cap-plaque interface, can be used to predict the delamination phenomenon in carotid artery plaques. Similar studies have not been found in the literature.

Differences between the simulation predictions and experimental results may be due to several simplifications, approximations and uncertainties: First, the approximation for the geometries of the delamination path and the specimen surface; Second, the simplification that the local components of the atherosclerotic plaque (e.g. lipid core and tissue calcification) have the same material properties; Third, the simplification that the critical energy release rates for cycles #2 and #3 are equal to that for cycle #1 (the critical energy release rate is observed experimentally to vary along the delamination path); Fourth, the uncertainty of the material parameter values used for the viscoelastic HGO model, which may not be the best values to describe the complex bulk mechanical response.

ACKNOWLEDGEMENTS

Research reported in this publication was supported by the National Science Foundation (NSF) under Award Number CMMI-1200358. The authors thank Aaron McCann and Ms Lindsey Davis for their assistance with the use of the experimental data.

REFERENCES

[1] Assemat, P et al., *ChronoPhysiol. Ther.*, 3:23–40, 2013.

[2] Loree, M et al., *Circ Res.*, 71:850-8, 1992.

[3] Gasser, T. C et al., *Journal of the Royal Society Interface*, 3:15–35, 2006.

[4] Holzapfel, G.A. et al., *Comput. Methods Appl. Mech. Engr.* 190: 4379-4403, 2001.

[5] Leng, X et al., *Proceedings of the Society for Experimental Mechanics Series*, 68:81-88, 2015.

[6] Ortiz, M et al., *International Journal for Numerical Methods in Engineering*, 44:1267–1282, 1999.

[7] Arun Roy Y et al., *International Journal of Fracture*, 110:21–45, 2001.

[8] ABAQUS. Analysis user's manual version 6.12, Dassault Systemes Corp. 2013

SB³C2015
Summer Biomechanics, Bioengineering and Biotransport Conference
June 17-20, 2015, Snowbird Resort, Utah, USA
SB³C2015-182

NOVEL MICRO-COMPUTED TOMOGRAPHY TECHNIQUE FOR SOFT TISSUE DEFORMATION TRACKING – APPLICATION TO THE MITRAL VALVE

Eric L. Pierce (1, 2), Charles H. Bloodworth IV (1), Ajay Naran (1), Thomas F. Easley (1), Morten O. Jensen (1, 2), Ajit P. Yoganathan (1, 2)

(1) Department of Biomedical Engineering
Georgia Institute of Technology
Atlanta, Georgia, USA

(2) Department of Biomedical Engineering
Emory University
Atlanta, Georgia, USA

INTRODUCTION

Understanding a biological tissue in terms of its material properties and the deformations it experiences is critical to studying its associated pathologies. Abnormal tissue deformation is a well-established driver of pathogenesis via mechanotransduction. Alterations to a tissue's material properties invariably lead to abnormal function (e.g., arterial stiffening promotes numerous hypertension and other cardiovascular co-morbidities [1]). Moreover, when modeling biological systems computationally, whether by Finite Element, Fluid-Structure Interaction, or other methods, mechanical moduli and spatial boundary conditions govern dynamic simulations.

For numerous soft tissues, improved material and kinematic descriptions are under active investigation, with ample room for improved physiologic understanding and/or simulation accuracy. Supporting these pursuits, the increased availability of computed tomography (CT), and especially micro-computed tomography (μCT), offers unprecedented structural detail. However, standalone structures are necessary but not sufficient in this pursuit. Tracking movement of structures across time can provide essential insights into the associated material properties and local deformation fields.

Indeed, recent studies have used μCT-based digital volume correlation approaches to track features and compute material stiffness and/or strain fields in tissues such as bone or wood [2-3]. However, these techniques are less applicable to soft tissues that undergo large strains and displacements and that generally lack identifiable internal structures. Our group has successfully tracked heart valve leaflet motion using stereophotogrammetry (SP) [4-5], whereby a series of ink dots on the tissue is observed by two high-speed cameras, and the three-dimensional (3D) coordinates are computed. This work has provided highly valued [6] insights into leaflet strains throughout the cardiac cycle. Yet this method suffers key limitations, including the lack of a proper unstrained state, limited optical access to certain anatomic features, and user error during post-processing.

Therefore, this study aimed to develop a novel technique to adhere μCT-compatible fiducial markers to soft tissue and to track them across large deformations. We hypothesized that such a technique could be applied to gain novel insights about mitral valve (MV) kinematics. To validate such a technique, a process was sought in which (a) small markers could be fixed to the valve in a desired configuration with high precision and accuracy, (b) the instrumented MV could be imaged in both open and closed states, and (c) marker materials could be distinguished from the soft tissue following 3D reconstruction.

METHODS

Experimental Setup and Data Processing

MVs were excised from ovine hearts acquired from a local market. Markers of various sizes and materials were attached to the MVs using tissue adhesive (Loctite 4014, Henkel Corporation, Westlake, OH). For marker material testing, valve samples were glued to acrylic sheet before imaging; markers were grouped by type along the scan axis to prevent interference. All other MVs were mounted in the extensively validated Georgia Tech Left Heart Simulator (GTLHS) [4, 7]. Healthy MV geometry was established in the GTLHS as previously described [7]. μCT imaging used an Inveon scanner (Siemens Medical Solutions USA, Inc., Malvern, PA) set to image soft tissue (80 kV energy, 500 μA intensity, 650 ms integration time) at 43.29 μm isotropic voxel size.

Following μCT, segmentation and 3D reconstruction were performed using InVesalius (InVesalius, São Paulo, Brazil), and marker centroid coordinates were found using Geomagic Studio (3D Systems, Cary, NC). For SP, 3D coordinates were extracted from each image pair using a custom MATLAB package (MathWorks, Natick, MA). Strains were computed using a custom MATLAB package and visualized using Tecplot 360 (Tecplot, Inc., Bellevue, WA).

μCT-based Marker Tracking Validation

To validate the technique, seven markers were attached to an optically clear, low-radiopacity phantom. The phantom was imaged by SP using two A504k cameras (Basler AG, Ahrensburg, Germany) and XCAP acquisition software (Epix Inc., Buffalo Grove, IL). The phantom was also imaged by μCT. Both datasets were re-oriented to a common coordinate system with one point constrained to the origin. Positional error between methods was computed from the six unconstrained points. Additionally, nine markers were attached to the central anterior leaflet of an MV mounted in the GTLHS. μCT was performed in open and closed valve configurations, and areal Almansi strain was computed as proof-of-principle.

RESULTS

A technique was sought to (a) fix small markers to MVs with high precision and accuracy, (b) scan the instrumented MV in both unpressurized and pressurized states, and (c) distinguish the markers from the soft tissue following 3D reconstruction.

Marker Application Technique

All markers types could be attached by gripping with fine forceps, dredging through a well of adhesive, and pressing to the MV leaflet. However, this method could only accommodate markers with diameter >1mm. Moreover, stability while transporting the gripped markers was tenuous, and additional adhesive was carried on the forceps tips. This resulted in imprecise placement, and excess glue that caused tissue stiffening within the marker attachment zone.

A custom marker applicator was designed to address these issues. Using this device, markers as small as ~500 μm in average diameter could be lifted, held indefinitely, and released with ease. Moreover, markers could be passed through adhesive with no excess adhesive pickup onto the applicator. Attachment could therefore be performed quickly, with high repeatability, with accuracy equivalent to pressing a pen to paper, and with negligible adverse effects on tissue stiffness.

Scan Conditions

Our group has recently reported on the capability to conduct μCT imaging on the MV when open and resting in air [7]. However, for the present study it was found that the previous protocol caused leaflet and chordal bunching that did not adequately represent the MV in a natural fluid environment. A simple tissue treatment protocol was thus developed to effectively reduce this bunching and yield more accurate open MV geometries. In addition, novel modifications to the GTLHS chamber were made to facilitate imaging of a closed MV using steady air pressure.

Marker Viability in Post-processing

Eight candidate marker types were applied to MV leaflets, including various metals, glass, sand, clay, or organic matter. μCT was conducted. During post-processing, raw DICOMs were thresholded to isolate the soft tissue intensity range and/or higher ranges. Due to their high radiopacity, all metallic markers were observed to cause X-ray artifacts at lower intensities, such that soft tissue reconstruction quality was severely compromised. These were rejected. Smoother glass markers demonstrated weak attachment strength and were also rejected. In total, four viable materials were identified, all of which were available in a range of sizes as small as ~500 μm diameter (meaningfully smaller than can be reliably captured by SP). Some materials were more consistently spherical than others, and one could be resolved in a distinct intensity range from the others.

Technique Validation

The six unconstrained markers on the rigid phantom had mean diameter 1.9±0.1 mm; the maximum distance between markers was 51.1 mm. Reconstructed positions of marker centroids agreed well between SP and μCT (Euclidean distance error: 1.0±0.2 mm). Nine markers were also successfully fixed to the mounted MV's central anterior leaflet, and were clearly resolved from the tissue by μCT. Average areal Almansi strain in this region was 38±13% (**Figure 1**).

DISCUSSION

A novel technique was developed whereby markers as small as 500 μm in diameter can be attached to soft tissue and used to track its displacement and strain through μCT. Accuracy of ≈1 mm was demonstrated against the gold standard stereophotogrammetry technique, with greatly improved ability to identify small/optically obstructed markers, and with reduced opportunity for user error. Four viable materials were identified, which greatly simplifies marker tracking during volume reconstruction in certain applications. Validation using MV closure *ex vivo* demonstrated the ability to measure strain fields whose magnitudes agree well with previous reports [6]. The technique holds significant potential to describe strains in smaller regions of the MV such as the commissural leaflets, which are poorly visualized by SP. It may be equally applicable in other settings, particularly other soft tissues or engineered biomaterials. Further exploration is warranted to identify marker materials for other imaging modalities. Application of this technique to magnetic resonance or ultrasound imaging, for example, would enable description of interactions between solid and fluid phases.

ACKNOWLEDGEMENTS

This work was supported by the National Science Foundation Graduate Research Fellowship (ELP) under Grant DGE-1148903, and by the National Institute of Health under Grant R01 HL119297.

REFERENCES

[1] Yambe, M et al., *Hypertens Res*, 27:625-631, 2004.
[2] Liu, L et al., *J Biomech*, 40:3516-3520, 2007.
[3] Forsberg, F et al., *Strain*, 46: 47-60, 2010.
[4] Siefert, AW et al., *Ann Thorac Surg*, 95:825-830, 2013.
[5] Yap, CH et al., *Am J Phys Heart Circ Phys*, 298:H395-405, 2010.
[6] Rabbah, JPM et al., *J Biomech Eng*, 135:021007, 2013.
[7] Rabbah, JPM et al., *Ann Biomed Eng*, 41:305-315, 2012.

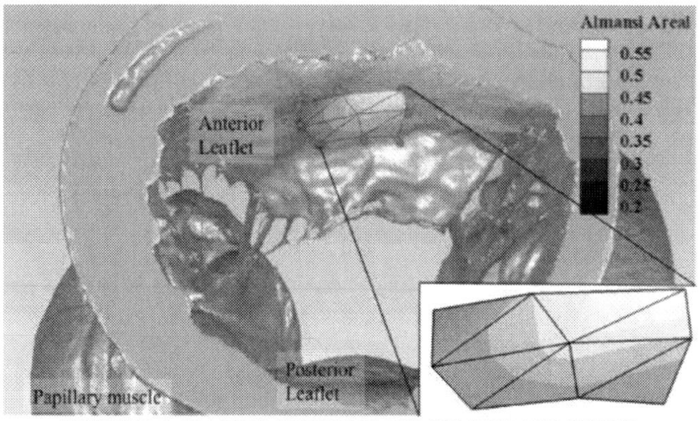

FIGURE 1. VALVE WITH LEAFLET STRAIN FIELD DERIVED FROM MICRO-COMPUTED TOMOGRAPHY

SB³C2015
Summer Biomechanics, Bioengineering and Biotransport Conference
June 17-20, 2015,Snowbird Resort, Utah, USA
SB³C2015-183

TUBULAR PEDIATRIC PULMONARY VALVES BY SUTURING DECELLULARIZED ENGINEERED TISSUE TUBES

J. Reimer (1), Z. Syedain (1), B. Haynie, R. Tranquillo (1, 2)

(1) Biomedical Engineering
University of Minnesota
Minneapolis, Minnesota, United States

(2) Chemical Engineering & Materials Science
University of Minnesota
Minneapolis, Minnesota, United States

INTRODUCTION

No FDA-approved heart valve prosthesis possesses somatic growth potential. Therefore, all current heart valve replacements are poorly suited for pediatric patients, where malformation of the pulmonary valve is the most common congenital defect. We recently reported a novel tissue-engineered tubular heart valve based on a collapsing tube of allogeneic, decellularized engineered tissue mounted on a three-pronged frame [1]. In this conception, it is unsuitable as a pediatric valve because of an inert frame material. Herein is presented a tissue-engineered pediatric heart valve design again based on the tubular heart valve design, but the constraints needed for a collapsing tube are achieved using a resorbable suture line, which attaches one tissue tube within another. *In vitro* and *in vivo* evaluation of its valvular function using a custom pulse duplicator system and a growing animal model, respectively, is described.

METHODS

Engineered grafts were fabricated by injecting a solution of ovine dermal fibroblasts, bovine fibrinogen, and thrombin into a cylindrical glass mold and allowing it to gel. Following a static culture period, the tubular constructs were transferred to a custom bioreactor in order to induce tissue remodeling and collagen production. The grafts were then decellularized using sequential detergent treatments. Tissue samples were analyzed mechanically, histologically, and for collagen and DNA content. The above methods are detailed in [1].

Tissue-engineered tubular pulmonary valves (TPVs) were fabricated by sewing two concentric, engineered tissue tubes together using two resorbable suture lines (Figure 1a). The TPVs, and a commercial pulmonary valve, were mounted within a fluid-filled chamber and tested under pulmonary conditions in a modified pulse duplicator, which was previously in [1]. The fluid-filled chamber was connected to an outflow line that allowed for valve distension and transmural flow, as would occur in vivo. Flowrate and pressure profiles recorded using LabVIEW leaflet and root motion were captured using a digital camera. Valve function, recellularization, and growth potential were assessed in the pulmonary valve position of a growing lamb for two TPVs. Valve function and hemodynamic parameters were assessed using echocardiography.

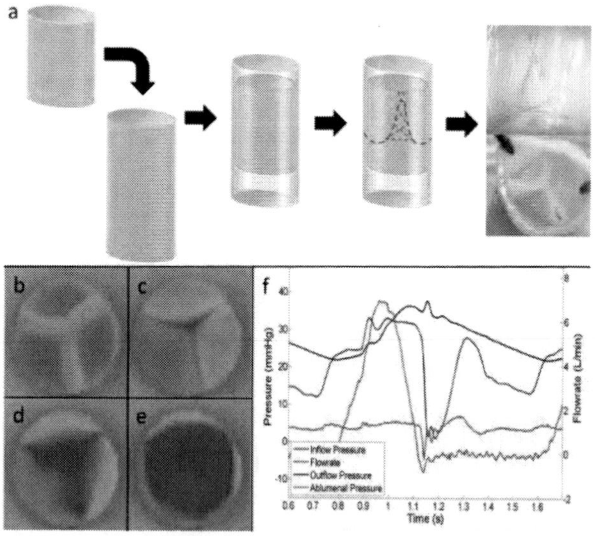

Figure 1. (a) Schematic representation of the TPV fabrication process. Pulse duplicator testing during (b) peak diastole. The "leaflets" (c, d) rapidly and (e) fully open during systole. (f) A representative pressure-flow rate trace.

RESULTS

A TPV was successfully fabricated by sewing two engineered tissue tubes together (Figure 1a). The first suture line attached the two tubes together around their entire circumference and generated three "commissure" and "leaflet" (Figure 1a, purple line). A second suture line was added to reinforce the three commissure regions (Figure 1a, green line). The three "leaflets", regions between the commissures, collapsed inward when exposed to back pressure. Pulse duplicator testing indicated good valve function under pulmonary conditions with full leaflet opening and closing each cycle. (Figure 1b-e). One cycle of a representative pressure (inflow, outflow, and ablumenal) and flowrate trace is shown in Figure 1f. The TPVs exhibited lower regurgitation and systolic pressure drops compared to the commercial bioprosthetic valve when tested at identical conditions. TPV regurgitation and systolic pressure drop were ~5% and ~2.5 mmHg, respectively. TPV function was maintained after fatiguing for more than 2 million cycles. Distention of the outer tube, or root, was captured using a digital camera over numerous physiologically relevant trans-wall pressure gradients by manipulating the ablumenal pressure on the TEHV.

Two TPVs were successfully implanted into ~5 week old lambs (Figure 1a). Ultrasound immediately and 1 week after implant showed full valve opening and closing, with minimal (< 3 mmHg) systolic pressure drop and trace regurgitation. A subsequent ultrasound after 4 weeks showed no change in leaflet motion, thickness, systolic pressure drop, or regurgitation. Full leaflet closing (Figure 2b) and opening (Figure 2c) were observed and flow occupied the entire conduit, as evidenced by color Doppler (Figure 2d). No clotting or calcification was evident in either of the animals. These studies are ongoing.

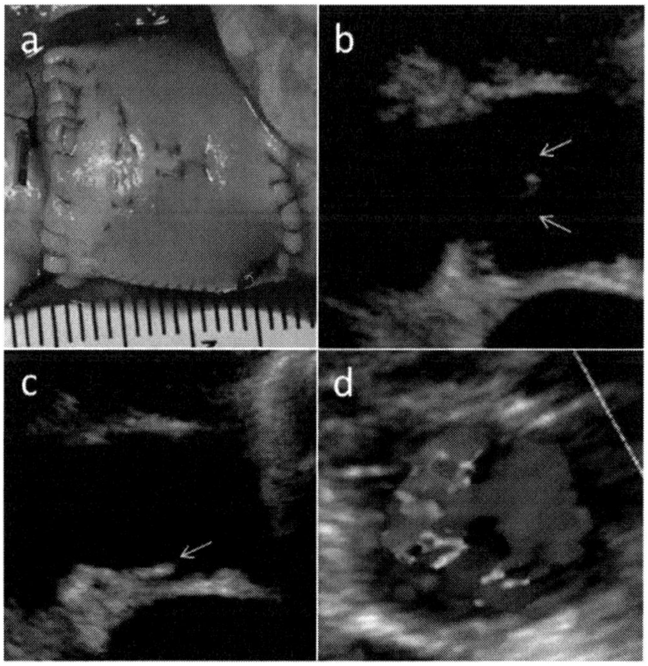

Figure 2. (a) TPV implanted in a lamb. Side view of the "leaflets" (arrows) shown in their (b) closed and (c) open positions. (d) Color Doppler of the open TPV in the ventricular view.

DISCUSSION

A completely biological TPV was fabricated by suturing two concentric, engineered tissue tubes together. This iteration of our tissue-engineered tubular valve lacks a frame, which was previously reported as an adult valve replacement [1]. Instead, the outer tube and the resorbable suture lines provide the mechanical support for the inner tube to collapse inward and perform a valve-like action. Testing in a pulse duplicator system under pulmonary conditions showed full leaflet opening and closing, a low regurgitant fraction, and a low systolic pressure drop. For comparison, a commercially-available bioprosthetic valve was tested as well and the TPVs met or exceeded its performance. Fatigue testing was successfully performed to demonstrate the durability of the tissue and the suture lines holding the two engineered tissue tubes together for over 2 million cycles. Notably, pulse duplicator testing allowed for valve root distention in a fluid-filled chamber under physiologically relevant trans-wall pressure gradients.

Animal studies were initiated to assess valve performance, graft recellularization, and growth in a growing lamb model. Both implanted TPVs maintained normal valve function at 4 weeks (trials currently ongoing) despite the animals nearly doubling in weight and their cardiac output increasing by 20% over that time period. Host cell invasion and fusion between the two tubes will be critical to the sustained performance of the implanted TPVs. Valve function will be assessed as the animal continues to grow and the suture lines further resorb.

While many tissue-engineered pulmonary valves have been implanted [2], these studies were not performed in a growing animal model. This is the first study to assess both tissue remodeling and growth potential of a novel TPV.

ACKNOWLEDGEMENTS

Authors acknowledge technical assistance from Naomi Ferguson and Sandy Johnson and funding from NIH R01 HL107572 to R.T.T.

REFERENCES

[1] Syedain, G et al., Tubular heart valves from decellularized engineered tissue. *Ann Biomed Eng*, 41: 2645-2654, 2013.

[2] Schmidt, J. and Tranquillo, R. (2013). Tissue-engineered heart valves. In P. Iaizzo, A. Hill, R. Bianco, & J. St. Louis (Eds.), *Heart valves: from design to clinical implantation* (pp. 261-280).

SB^{3}C2015
Summer Biomechanics, Bioengineering and Biotransport Conference
June 17-20, 2015, Snowbird Resort, Utah, USA
SB^{3}C2015-184

FINITE ELEMENT MODEL OF MIXED POROHYPERELASTIC TRANSPORT IN AN AXISYMMETRIC PORCINE CORONARY ARTERY

Michelle Hine Armstrong (1), Bruce R. Simon (2), Jonathan P. Vande Geest (1,2,3,4)

(1) Graduate Interdisciplinary Program in Applied Mathematics,
University of Arizona,
Tucson, AZ, USA

(2) Department of Aerospace and Mechanical Engineering,
University of Arizona,
Tucson, AZ, USA

(3) Department of Biomedical Engineering,
University of Arizona,
Tucson, AZ, USA

(4) BIO5 Institute for Biocollaborative Research
University of Arizona,
Tucson, AZ, USA

INTRODUCTION

It is known that the hydrophobicity and hydrophicility of a drug play an important part in delivery from a drug-eluting stent (DES) [1]. Recent research has modeled the complex response of soft tissue via a porous media formulation [2]. The effective stress experienced by a porous media is significantly less than the stress experienced by a purely solid material. The addition of a neutral mobile species also allows the model to track transport of a chemical concentration, which could represent drug concentration post-delivery from a DES.

The current work builds upon previous work done by our lab creating an in-house finite element code with large-strain Mixed Porohyperelastic Transport and Swelling (MPHETS), as well as investigating the diffusive and convective mass-transport properties of a coronary artery [3,4,5]. The purpose of this current work is to develop an axisymmetric finite element code to model a porcine left anterior descending coronary (LADC) artery and investigate the effect of changing the convection coefficient, which represents the changing hydrophobicity or hydrophilicity of a drug.

METHODS

An in-house finite element code was developed in MATLAB to simulate the time evolution and deformation of soft tissue represented with MPHETS. For the purposes of this study, osmosis is neglected.

The model is a continuum with an incompressible solid skeleton, an incompressible fluid, and a neutral mobile species. Together these form a compressible material. The system is solved over the primary variables $u_i, \tilde{\mu}^{f*}, \tilde{\mu}^{c*}$, which represent displacement, fluid potential, and chemical potential, respectively.

The elements are constant strain triangles, with linear interpolation of the primary variables. To avoid element locking and to handle jumps in fluid pressure due to osmosis, the code uses secondary variables defined at the Gauss points. These scalar secondary variables are coupled to the primary variables via the classical mechano-chemical-potentials as written in equations (1), (2)

$$\tilde{\mu}^{f*} = p^o + p^f \,, \quad \tilde{\mu}^{c*} = \mu_0^c + R\theta \log(\gamma c) \qquad (1),(2)$$

Equation (1) states that the fluid potential is the sum of the osmotic pressure p^o and the pore fluid pressure p^f (both in Pa). For this paper, osmosis is set to zero, and then fluid potential is equal to pore fluid pressure. Equation (2) relates the chemical potential to the chemical concentration c (in mol/m^3), where μ_0^c represents a datum chemical potential that prevents the chemical concentration from reaching zero, R is the universal gas constant, θ is the temperature (assumed to be constant), and γ is the activity coefficient of the material.

Conservation Equations: The finite element equations are solved in Lagrangian coordinates, denoted with a tilde above the variables. The material must respect conservation of total momentum (neglecting body forces) and conservation of mass of the fluid and species, as in equations (3)-(5).

$$\frac{\partial T_{ij}}{\partial X_j} = 0 \qquad (3)$$

$$\frac{\tilde{j}_i^{fr}}{\partial X_i} + JH_{ij}\dot{E}_{ij} = 0 \,, \quad \frac{\tilde{j}_i^{cr}}{\partial X_i} + cJH_{ij}\dot{E}_{ij} + Jn\dot{c} = 0 \qquad (4),(5)$$

A dot above a variable represents the total time derivative. The variables $\tilde{j}_i^{fr}, \tilde{j}_i^{cr}$ represent the Lagrangian apparent relative fluid and species velocities, respectively; $H_{ij} = F_{ik}^{-1}F_{jk}^{-1}$ is Finger's strain tensor; and n is the porosity of the material.

Constitutive Equations: The effective stress principle allows separation of total stress into effective and fluid, as in equation (6). This relationship then couples the total stress to the time-dependent fluid and chemical equations.

$$\sigma_{ij}^{tot} = \sigma_{ij}^{eff} - p^f \delta_{ij} \tag{6}$$

The Eulerian relative fluid and species velocities j_i^{fr} , j_i^{cr} are governed by equations (7),(8)

$$j_i^{fr} = -k^{ff}\left(\frac{\partial p^f}{\partial x_i} + R\theta\left(b^{cf} - \phi^c\right)\frac{\partial c}{\partial x_i}\right) \tag{7}$$

$$j_i^{cr} = -d^{cc}\frac{\partial c}{\partial x_i} + b^{cf} c k^{ff}\frac{\partial p^f}{\partial x_i} \tag{8}$$

where k^{ff} is the permeability, p^f is the pore pressure, R is the universal gas constant, θ is the temperature, b^{cf} is a convection coupling coefficient, ϕ^c is the osmotic coefficient (set to zero by the assumption of no osmosis), c is the concentration, and d^{cc} is the diffusivity. Note that convection coupling coefficient appears in both equations (7),(8).

We assume a Neo-Hookean stress-strain relation, as in Abaqus 6.12, with strain energy density function as in equation (9).

$$\psi = C_{10}(\bar{I}_1 - 3) + \frac{1}{D_1}(J-1)^2, \quad \bar{I}_1 = J^{-2/3} I_1 \tag{9}$$

The geometry and material parameters for this paper were developed from our lab's previous study of a porcine LADC [4,5]. Material parameters may be found in **Table 1**.

Table 1: Material Parameters.

k^{ff} [m^4/N·s]	d^{cc} [m^2/s]	b^{cf} [unitless]	C_{10} [Pa]	D_1 [Pa^{-1}]
4.46e-11	4.55e-14	6e-6, 6e-7, 6e-8	1e6	7.6e-9

At each time step, Newton-Raphson iterations solve the nonlinear system of coupled equations using Backward Euler for stability. This yields the time evolution of the system to quasi-steady state.

Problem Description: The LADC is modeled as an axisymmetric tube, with inner radius of 2mm and thickness of 250μm. Motion in the axial direction is constrained. The inside of the tube is pressurized with fluid to 100 mmHg while the outside is maintained at zero pressure. A gradient in chemical potential is also imposed, with a high concentration at the inside of the tube (0.0128 mol/m^3), and a low concentration outside of the tube (0.00128 mol/m^3). The material initially has zero displacement, zero pore fluid pressure, and a very low constant chemical concentration, equal to the concentration imposed on the outside of the tube.

RESULTS

The material is modeled with 240 axisymmetric linear triangles, with a mesh as pictured in **Figure 1A**. Numerical results are plotted at the Gauss points for the secondary variables of pore fluid pressure and concentration, in **Figures 1B-1D**.

Displacements (not pictured) are small because of the neo-Hookean parameters derived from the model in [4,5]. Due to the set of material parameters obtained from the tissue, equation (7) is dominated by the first term and acts similarly to Darcy's law; the pore fluid pressure goes almost immediately to steady state as seen in **Figure 1B**. This behavior appears very similar for all values of b^{cf} modeled. Species diffusion occurs slowly, as depicted in **Figure 1C** for a value of b^{cf}=6e-8. For larger values of b^{cf}, concentration reaches steady-state more quickly (not pictured).

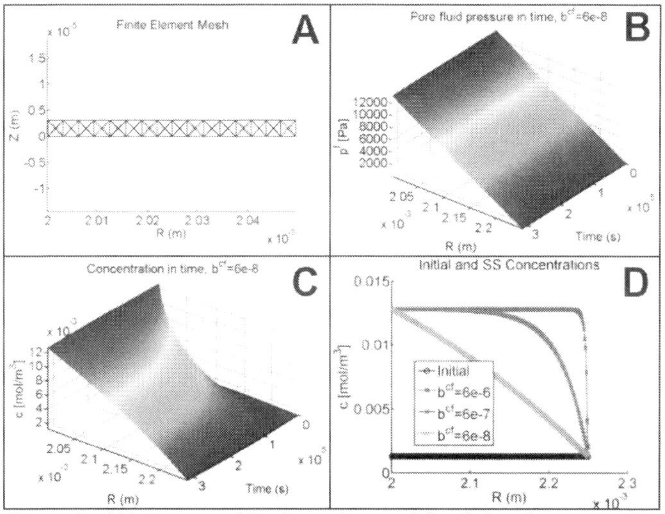

Figure 1: (A) A zoomed-in view of part of the finite element mesh; (B) evolution of pore fluid pressure in time for b^{cf}=6e-8; (C) evolution of concentration profile in time for b^{cf}=6e-8; (D) comparison of quasi-steady-state concentration profiles for varying values of b^{cf}.

Changing the coupling coefficient changes the Peclet-like balance of k^{ff}, d^{cc}, and b^{cf}. Increasing b^{cf} increases the effect of convection, so that the final concentration profile is more nonlinear. As evidenced by **Figure 1D**, a higher b^{cf} causes the quasi-steady-state concentration to be distributed more evenly across the tissue.

DISCUSSION

This work is an initial step to creating a computational model to simulate the drug distribution post-treatment with a DES. As expected, the value of coupling coefficient has a large effect on the steady-state concentration profile of the drug. Furthermore, while our lab has simulated chemical transport in the LADC in the past, previous simulations used PHEXPT, a combination of Abaqus and MATLAB that did not have the capability of back-coupling the species to the pore fluid pressure equation [6]. For certain material parameter sets, simplification to the PHEXPT model is not valid. This new model allows back-coupling, enabling study of a wider variety of problems. Future work could add swelling from osmotic effects or model growth due to mechanical stimulation.

ACKNOWLEDGEMENTS

Funding for this work was provided by the ARCS Foundation to MHA and NIH Grant R01EY020890 to JPVG. An allocation of computer time from the UA Research Computing High Performance Computing (HPC) and High Throughput Computing (HTC) at the University of Arizona is gratefully acknowledged.

REFERENCES

[1] Yang, C et al. *Adv Drug Deliv Rev* 58.3:402-11, 2006.
[2] Simon, BR et al., *Int J Solids Struct*, 35.3:5021-31, 1998.
[3] Harper, JL et al., *J Mech Behav Biomed Mater*, 29:663-75, 2014.
[4] Keyes, J et al., *J Mech Behav Biomed Mater*, 17:296-306, 2013.
[5] Keyes, J et al., *Pharmaceut Res*, 30.4:1147-60, 2013.
[6] Vande Geest, J et al. *J Biomech Eng* 133.4:044502, 2011.

MEDIAL AND LATERAL CONTACT PRESSURE DISTRIBUTION FOLLOWING THE IMPLANTATION OF A NOVEL MEDIAL MENISCUS IMPLANT

Maoz Shemesh (1), Noa Cohen (1), Eyal Zylberberg (1), Adaya Shefy-Peleg (1), Ron Arbel (2), Vincenzo Condello (4), Nogah Shabshin (3), Eran Linder-Ganz (1), Jonathan J. Elsner (1)

(1) Research and Development, Active Implants, Netanya, Israel

(2) Orthopaedics, Ichilov Tel-Aviv Medical Cener, Tel-Aviv, Israel

(3) Radiology, Carmel Hospital, Haifa, Israel

(4) Orthopaedics, Sacro Cuore Hospital, Verona, Italy

INTRODUCTION

Meniscal tears are common knee injuries that have been shown to alter the stress distribution in the knee, which can subsequently lead to degenerative arthritis [1]. Often, osteoarthritis (OA) of the medial compartment of the knee is associated with varus malalignment. Shifting the mechanical axis from relative varus to valgus may help to reduce the load in the medial compartment when significant malalignment exists. However, the invasive nature of this procedure entails a long recovery period and resection of bone material osteotomy can make later knee replacement surgery more challenging at a later stage.

Recently, a medial meniscus implant was developed as a bridge treatment for middle-aged patients, suffering from joint pain associated with loss of meniscal function. The concept of the implant follows that of early metallic hemiarthroplasty as described by the McKeever and MacIntosh inter-positional spacers [1]. In contrast to its hard-bearing metal predecessors that differed substantially from the native meniscus tissue, the proposed polycarbonate-urethane (PCU) implant was designed to be suited to the mechanical properties of the natural meniscus [2]. Preliminary work has shown that the implantation of the artificial device can reduce loads on the medial articular surface by distributing the load over a larger area [3]. The improved distribution of loads alone could be the reason why patients report pain relief following implantation [4], yet we hypothesize that there is a more complex effect dictated by the presence of the implant in the medial compartment which increases the joint space and also changes the balance of loads between the medial and lateral compartments. Thus, the purpose of this study was to measure load distribution patterns on the medial and lateral tibial plateau under normal native conditions, following a subtotal meniscectomy, and subsequently to implantation of the meniscus implant, and by doing so to shed light on a possible mechanism of action.

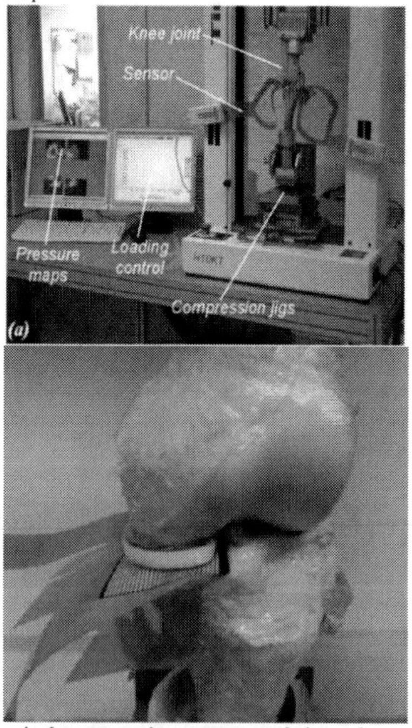

Figure 1. Mechanical compression setup consisting of a specially designed loading stage for correct anatomical positioning, pressure sensors and data acquisition and analysis system.

METHODS

Contact pressures under the intact medial and lateral menisci were measured in normal human cadaveric knees (n=5) under compression. Knees were positioned on specially designed stage for mechanical compression testing (Fig. 1), then all degrees of freedom were fixed, and MCL/ LCL bone plugs were released to allow for the insertion of the contact pressure sensors (Tekscan Inc., Boston MA). With the medial meniscus intact, knees were subjected to a 1200N load at a flexion angle of 0° and the medial pressure map was recorded. Next, a subtotal medial meniscectomy was then performed and the protocol was repeated. The PCU meniscus implant was inserted into the medial compartment and the protocol was repeated. An example of the attained pressure maps is shown in Fig.2. The maps were analyzed using the Tekscan dedicated software to examine the overall tibio-femoral contact area and vertical force excreted on the tibia, and the average and peak contact pressures on the tibial plateau. The load distribution between the medial and lateral compartments was compared between cases. To complement the measurements in cadavers, MRI's of 5 patients who underwent implantation of the device were analyzed retrospectively to evaluate joint space changes. The medial and lateral joint space were calculated from a baseline MRI taken prior to implantation, and from an MRI taken 1.5 months after implantation.

RESULTS

The contact areas measured under the medial and lateral menisci in the native joint state were 682 ± 241 mm^2 and 402 ± 160 mm^2, respectively. Thus, the natural medial meniscus occupies an area 70% larger than the lateral meniscus when loaded. After performing a subtotal meniscectomy of the medial meniscus, the contact areas were 402 ± 160 mm^2 and 512 ± 163 mm^2 for the medial and lateral compartments, respectively. While the contact area in the medial compartment was reduced by 41%, that of the lateral compartment remained unchanged. Following implantation of the artificial meniscus, the medial contact area increased to 701 ± 193 mm^2, and the lateral contact area was reduced to 478 ± 142 mm^2. This is a 3% increase in the medial contact area compared to the natural medial meniscus and a 74% increase in contact area compared to the meniscectomized medial compartment. The laterally loaded area with the implant in place was 6% lower than in the other cases. Force distribution in the knees was found to distribute 60:40% between the medial and lateral tibial plateaus. After a medial meniscecotmy this ratio was altered to 52:48%, and after implantation of the device this ratio was 57:43%.

The average contact pressure measured on the tibia under the natural menisci was 1.6 ± 0.2 MPa and 1.8 ± 0.6 MPa for the medial and lateral menisci, respectively. Upon subtotal medial meniscectomy, the average contact pressure was 2.6 ± 0.8 MPa (63% higher) and 1.9 ± 0.7 MPa (6% higher), respectively. Both compartments were loaded more than in the intact state. After implantation of the medial meniscus implant, the average contact pressures was 1.4 ± 0.1 MPa in the medial compartment (13% lower than the natural and 46% lower than the meniscectomized conditions) and 1.5 ± 0.3 MP in the lateral compartment (17% lower than the natural and 21% lower than the meniscectomized conditions).

The peak contact pressures showed similar trends. At the baseline, the medial and lateral peak pressures were 5.9 ± 2.1 MPa and 5.8 ± 1.3 MPa, respectively. Following a medial meniscectomy they increased to 7.4 ± 1.3 MPa and 6.1 ± 2.0 MPa, respectively. Following implantation of the device they were 5.1 ± 1.4 MPa and 5.7 ± 0.7 MPa, respectively.

The average joint space measured for five randomly selected patients, who were implanted with the medial meniscus implant as part of a clinical study, was found to be 4.4 ± 0.9 and 7.4 ± 1.1 mm prior to implantation and 7.4 ± 1.1 and 7.0 ± 0.7 after implantation. This is a 3 mm increase in joint space on the medial side and 0.4 mm decrease on the lateral side.

DISCUSSION

The implantation of the medial meniscus implant was found to improve the load distribution between the medial and lateral compartments, following a meniscectomy. On the medial side, it restored the contact area and reduced cartilage loads to near normal, while on the lateral side there was a slight decrease in the contact area, but also a decrease in the average and peak stresses, both compared to the normal state and meniscectomized condition. This may be explained by the shift in load from the lateral to the medial side compared to the baseline and more importantly, compared to the meniscectomized knee. The mechanism by which these changes occur is likely linked to the changes noted in the joint space between the medial and lateral compartments. While the medial joint space is restored by the implant from approximately 4 to 7 mm, there is very little change in the lateral space, which is normal to start with. We conclude that the implantation of the PCU implant may benefit patients not only by restoring contact mechanics in the medial compartment, but also by enabling better load distribution between medial and lateral compartments.

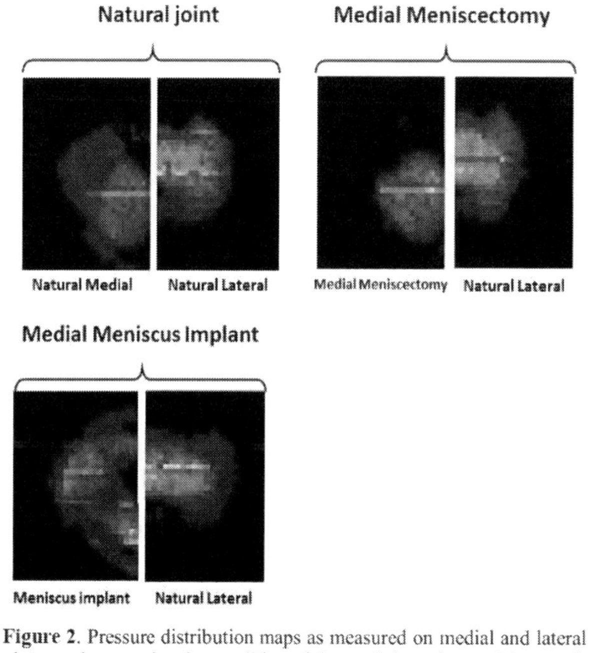

Figure 2. Pressure distribution maps as measured on medial and lateral tibial plateau when varying the condition of the medial meniscus: (a) natural native medial meniscus, (b) a subtotal meniscectomy of the medial meniscus and (c) following implantation of an artificial meniscus implant

ACKNOWLEDGEMENTS

This Study was funded by Active Implants.

REFERENCES

1. McKeever, DC Clin Orthop Relat Res 18:86–95, 1960.
2. Shemesh, M et al., J Mech Behav Biomed Mater 29:42-55, 2014.
3. Linder-Ganz, E, et al. J Biomech Eng 132(2):024501, 2010.
4. Verdonk, P et al., ePoster #823 - 2013 ISAKOS Biennial Congress.

Influence of Bone Density and Insertion Angle on Fixation Strength of Pedicle Screws

Giovanni Solitro (1), Farid Amirouche (1, 2), Brenden P. Magnan (3)

(1) Department of Orthopaedics
University of Illinois at Chicago
Chicago, Illinois, USA

(2) Department of Mechanical Engineering
University of Illinois at Chicago
Chicago, Illinois, USA

(3) College of Medicine
University of Illinois at Chicago
Chicago, Illinois, USA

INTRODUCTION

In the treatment of acute and chronic instabilities or deformities of the spine, instrumentation is required for stabilization and immobilization of the spine during recovery. However, instrument failure related to pedicle screws has also been clinically reported [1][2] and proves that the screw fixation has its flaws. The stability of pedicle screws depends mainly on the interface between screw and bone, and when this interface is negatively altered, the stability of the screw can become compromised [3]. The influence of bone mineral density on the fixation strength of pedicle screws in the thoracolumbar spine has been studied extensively over the years. However, differences in the testing conditions, methods and materials have made the data difficult to compare. Differences range from the design of screws used, insertion technique, cadaver vertebrae or polyurethane foam, range of bone density, age of vertebrae, segment of spine, testing setup, pullout rate and units of measurement. A correlation between BMD and pullout has been found in human cadaveric vertebrae and polyurethane foam [7]. Another critical factor known to influence the fixation strength of pedicle screws is the angle of screw insertion into the pedicle (Figure 1). Kilincer, using lumbar spinal vertebrae pullout strength was influenced by the angle between two pedicle screws [6]. The objective of this study is to analyze the relationship between pedicle screw fixation strength and bone mineral density for different screw insertion angle to determine if adjustment in the insertion angle significantly increases fixation strength.

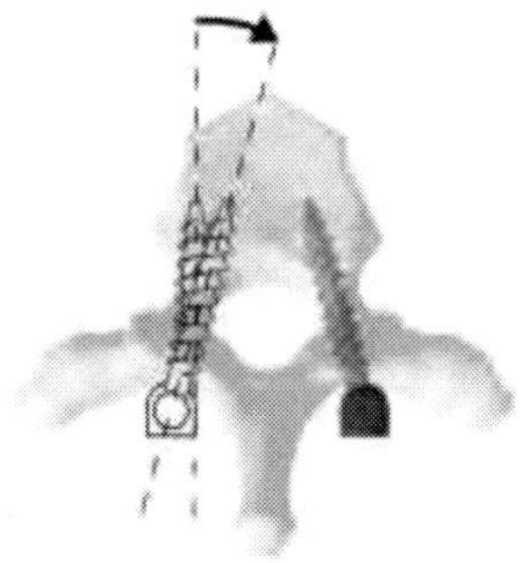

FIGURE 1: PLACEMENT OF SCREW INTO THE VERTEBRAL PEDICLE AND DIRECTION OF THE IMPOSED INSERTION ANGLES IN THE TRANSVERSE PLANE.

METHODS

We have tested a posterior fixation construct characterizing two pedicle screws connected at a distance of 40mm as suggested by the ASTM F1717. Using the ASTM F543 (Test Methods for Metallic Medical Bone Screws) as reference for the performed tests we used a grip span equivalent to seven (ASTM suggested 5) times the diameter of the screw tested. We performed tests on four densities of polyurethane foam in accordance with the ASTM F1839-08 standard to simulate bone densities ranging from osteoporotic (5 pcf) to higher than normal (20 pcf) as already used in previous study. The insertion torque was measure for each screw using a Mark-10 MTT03-100 digital torque gauge with resolution of 1 Ncm. Insertion torque required to make each screw flush with the brick varied by density and was limited at 20 Ncm for grade 5, 60 Ncm for grade 10, 120 Ncm for grade 15 and 300 Ncm for grade 20 foam. After screw insertion, an electromechanical tensile-testing Instron Machine 5569 (Instron®, Norwood, MA) was used to pull the custom designed jig (Figure 2) at a rate of 5 mm/minute as indicated by the ASTM F543 and recording reaction force and displacement at a frequency of 50 Hz or 1 N increments.

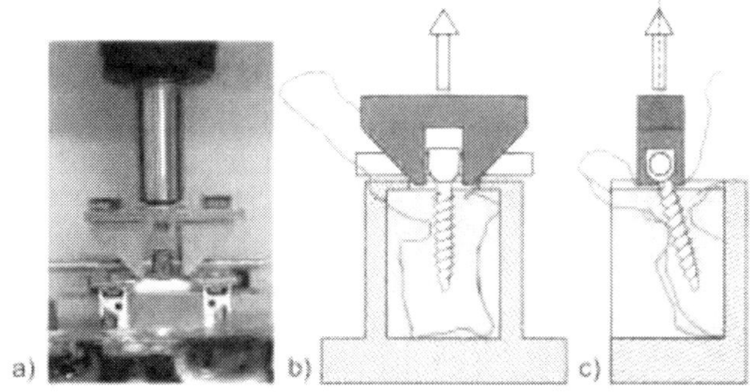

FIGURE 3: (A) THE CUSTOM DESIGNED JIG USED FOR THE AXIAL PULLOUT TESTS. (B) SCHEMATIC SHOWING DIRECTION OF APPLIED FORCE TO SCREW WITH SUPERIMPOSED VERTEBRA IN BOTH SAGITTAL AND (C) TRANSVERSE PLANES.

RESULTS

The mechanism of screw failure was dependent on both foam density and insertion angle of the screw. In the grade 20 foam with angles greater than 0°, the observed mechanism of failure involved fracture of the foam block into one or more fragments. In all other densities of foam and angles of 0° in the Grade 20 foam, the observed mechanism of failure was stripping of the internal screw threads. The average pullout force in the Grade 5 foam was 126.41 N (± 16.54 N). Similarly for Grades 10 and 15 the pullout forces were respectively of 360.97 N (± 41.81 N) and 753.10 N (± 86.24 N). The highest average pullout force value of 1538.26 N (±207.77 N) was found for the Grade 20 foam. At each insertion angle, there was a significant increase in pullout strength with increasing polyurethane foam density (Fig.4). In the Grade 5 foam, there was no significant difference in pullout strength with increasing angle of insertion. In foams of grades 10, 15 and 20, there was a significant difference in pullout strength with increasing angle of insertion (Figure 3). A multivariable linear regression was performed on the data to determine the influence of both foam density and insertion angle on pullout strength (Fig.5). A linear regression equation of $Y = -354.8812 + 91.8102 * X_1 - 6.8747 * X_2$ (X_1 = density [pcf], X_2 = angle [degrees]) was found with a correlation coefficient of 0.95.

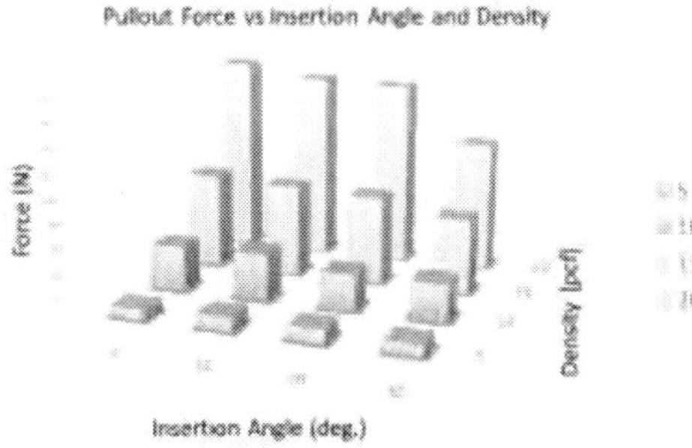

FIGURE 3: 3D COLUMN CHART PLOTTING PULLOUT FORCE AGAINST BOTH FOAM DENSITY AND INSERTION ANGLE.

DISCUSSION

The pullout strength of the tested screw is dependent on both host material density and insertion angle. Pullout force is greatest when an insertion angle of 0° is used and decreases at each successively higher angle until its lowest value at an insertion angle of 30°. Density of host material influences pullout strength to a much greater extent than does insertion angle. Insertion angle does have a mechanical advantage or an influence on pedicle screw pullout strength. The differences seen in our experiment are however small for the healthy bone. This minimal influence is most evident in osteoporotic bone models, where changes in insertion angle are found to have little to no effect on pullout strength. For osteoporotic patients the results support the fact that insertion angle as depicted in this study has no influence on the pullout strength and might be considered as a convenience when placing the pedicle screws.

ACKNOWLEDGEMENTS

This work was partially supported by Aurelio M. Caccomo Family Foundation.

REFERENCES

[1] K. E. Ponnusamy, S. Iyer, G. Gupta, and A. J. Khanna, "Instrumentation of the osteoporotic spine: Biomechanical and clinical considerations," *Spine Journal*, vol. 11, no. 1, pp. 54–63, 2011.

[2] C. J. DeWald and T. Stanley, "Instrumentation-related complications of multilevel fusions for adult spinal deformity patients over age 65: surgical considerations and treatment options in patients with poor bone quality.," *Spine (Phila. Pa. 1976).*, vol. 31, no. 19 Suppl, pp. S144–51, Sep. 2006.

[3] K. Okuyama, K. Sato, E. Abe, H. Inaba, Y. Shimada, and H. Murai, "Stability of transpedicle screwing for the osteoporotic spine. An in vitro study of the mechanical stability.," *Spine (Phila. Pa. 1976).*, vol. 18, no. 15, pp. 2240–2245, 1993.

[4] J. D. Coe, K. E. Warden, M. A. Herzig, and P. C. McAfee, "Influence of Bone Mineral Density on the Fixation of Thoracolumbar Implants: A Comparative Study of Transpedicular Screws, Laminar Hooks, and Spinous Process Wires," 1989.

[5] C. Hsu, C. Chao, and J. Wang, "Increase of pullout strength of spinal pedicle screws with conical core : biomechanical tests and finite element analyses," *J. ...*, vol. 23, pp. 788–794, 2005.

[6] C. Kilinçer, S. Inceoglu, M. J. Sohn, L. a Ferrara, and E. C. Benzel, "Effects of angle and laminectomy on triangulated pedicle screws.," *J. Clin. Neurosci.*, vol. 14, no. 12, pp. 1186–91, Dec. 2007.

SB^3C2015
Summer Biomechanics, Bioengineering and Biotransport Conference
June 17-20, 2015, Snowbird Resort, Utah, USA
SB³C2015-187

SOFT ANKLE-FOOT ORTHORIC

Adam Podolec
Rochester Institute of Technology
Rochester, NY, United States

Megan Ehrhart
Rochester Institute of Technology
Rochester, NY, United States

Noah Schadt
Rochester Institute of Technology
Rochester, NY, United States

Geni Giannotti
Rochester Institute of Technology
Rochester, NY, United States

Tyler Leichtenberger
Rochester Institute of Technology
Rochester, NY, United States

Jared Green
Rochester Institute of Technology
Rochester, NY, United States

Faculty Advisor(s)

Dr. Elizabeth DeBartolo
Rochester Institute of Technology
Rochester, NY, United States

Mr. Edward Hanzlik
Rochester Institute of Technology
Rochester, NY, United States

INTRODUCTION

Foot drop is a neurological disorder which impairs the ability of an individual to dorsiflex the foot (i.e., point the toe upward). This condition is a common side effect of a stroke, ALS, Multiple Sclerosis, or a peroneal nerve injury. Patients who experience foot drop often utilize an assistive device known as an ankle-foot orthotic (AFO) which provides a stable and comfortable support for their foot and ankle and thus mitigates the condition effects.

Current AFOs are bulky, rigid, and disrupt the user's natural gait by providing assistance at all times, regardless of need. An active AFO will provide users with assistance only during appropriate times in the gait cycle. This will be particularly useful when the active AFO is used on non-level terrain which is difficult to navigate using the current AFOs. Additionally, this active AFO should comfortably fit into a user's existing footwear. The inability to do so is an often voiced user complaint of current AFOs.

The specific goal of our team is to incorporate previous work done, including using a McKibben muscle and a terrain sensing system, into an un-tethered AFO. It should also have an aesthetically pleasing flexible exoskeleton made from allergy conscious materials. The exoskeleton needs to be integrated with the actuation device, sensing system, and microcontroller. The AFO must also be capable of lifting the user's foot and should be designed to endure an entire day use un-tethered. The sensors and microcontroller system should incorporate the existing terrain sensing system as well as implementing more suitable heel strike sensing.

PRODUCT DESIGN PROCESS

A multidisciplinary design team used a thorough, meticulous, documented, and iterated engineering process to produce an AFO prototype design. First, the problem being addressed was clearly defined based on a project summary and guidelines provided by the customer as well as further customer interviews. Next, a detailed table of requirements was created from given and implied customer needs and verified with the customer. Based on the customer requirement table, a table of engineering requirements was derived and mapped to specific, feasible testing to verify functionality.

Research and benchmarking was then conducted on existing AFOs. This included research into the past AFO senior design projects at RIT and professional AFOs in order to address advantages and disadvantages of the AFO technology. This drove an initial design decision to place the Mckibben muscle on the front of the leg, where the tibialis anterior is located; the muscle in the leg that provides dorsiflexion. A functional decomposition of the AFO was conducted in order to identify the functions of the system that required design decisions to be made. Multiple possible design solutions were made for each identified function. They were then placed into a morphological chart, a design tool used to identify the best solutions. Selection criteria were then defined, driven by both the customer and engineering requirements. The selection criteria were used to select four system concepts to be compared in an iterated Pugh analysis, resulting in a final system concept selection agreed upon by the design team and the customer.

Feasibility testing was planned and conducted, further defining and refining the system, resulting in a final system design. This testing included but was not limited to creating a full electrical prototype as well as extensive testing on the muscle design and attachment using a soft brace as a mechanical prototype. A bill of materials (BOM), build/assembly plan, debug plan, and prototype testing plan were all made in order to build and test a prototype in the Spring of 2015. The prototype test plan was mapped to each engineering requirement to ensure each one was met and therefore achieving each customer requirements.

PRODUCT DESIGN

The design of the AFO is split into two pieces, an upper component and a lower component. The upper component consists of a small backpack worn by the user. Housed within the backpack is a compressed air tank with a regulator, a solenoid valve, an Arduino PCB board and batteries. A pressurized air hose, power cable, and signal wiring are tethered together in a mesh sleeve and routed from the backpack to the lower component. The lower component is comprised of the soft orthotic, the McKibben muscle with lower and upper attachments, the lower component housing with the sensor PCB board and IR sensor, the heel sensor and the toe sensor. A drawing of the lower component and the full system design are shown in figure 1 and 2, respectively.

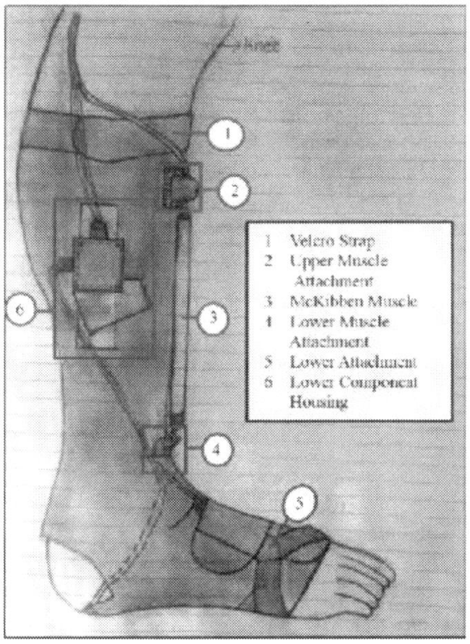

FIGURE 1: LOWER COMPONENT

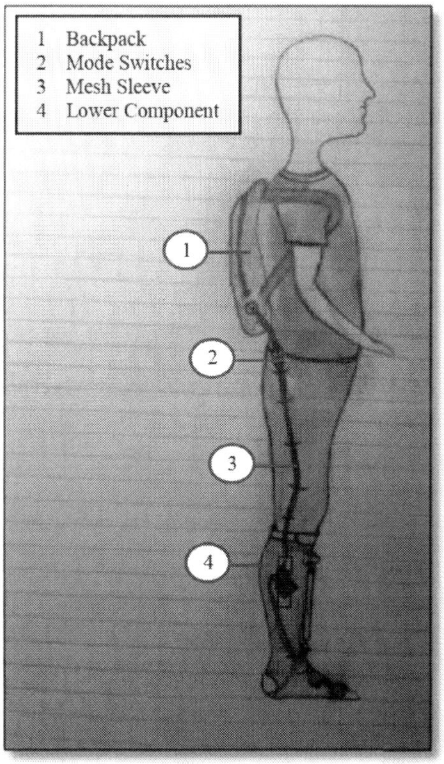

FIGURE 2: INTEGRATED SYSTEM, COMBINED LOWER AND UPPER COMPONENTS

To use the device, an individual would apply the lower component to their foot, making all adjustments as necessary. Once the air hose and power lines are connected, the user will turn on the main power switch in the backpack. To activate the AFO, the user will press a button near their hip on the power supply line. When the device is active it will begin sensing and recording the user's gait data to a micro SD card. Using data from the heel strike sensor, the Arduino will articulate the double acting solenoid, which controls the pressurized air flowing into the muscle from the air tank. This in turn, will contract and release the McKibben muscle, lifting the toe during the gait cycle only when needed. The system will recognize when the user is on a staircase and the IR sensor will determine whether the user is ascending or descending the stairs. Using this information the Arduino will actuate the muscle when needed. When the user is sitting, driving, or anytime they do not need dorsiflexion support, they can turn the active mode off via the switch near the hip. The device will constrict the McKibben muscle and become a passive orthotic, constantly providing dorsiflexion or the user can press a separate button releasing the muscle so the device is not providing any support.

BUDGET & MARKET ANALYSIS

The allotted budget for the project is five hundred dollars. To ensure that our team does not exceed this amount, a BOM, detailing the cost of each part of the prototype, was created and maintained. In addition to the prototype cost, the cost of feasibility testing was included in this budget as well. Currently, the projected cost of the project is \$456.96. A feasibility analysis was performed to determine the appropriate size of an air tank that would supply the AFO with enough air to last for an entire day's use. A 90cu.in 4,500psi air tank was found to meet the requirements; however, this tank would cost an additional \$160.95, which would exceed the budget. After a brief customer meeting, about possible trade off options, it was decided that purchasing PCB equipment for the prototype was a higher priority than buying a larger tank. Therefore, for the purpose of the prototype, we will be using a smaller tank; however, we created a model that proves that the larger tank would last for an entire day's use.

There is a large market potential for a soft active AFO due to a relatively large foot drop population. Specifically, foot drop is a common side-effect of a stroke, affecting approximately 20% of survivors (~1.3 million people each year). (1) Foot drop can also occur as a side effect of ALS (Lou Gehrig's Disease), Multiple Sclerosis, radiculopathy, or injury to the peroneal nerve, increasing the number of people affected. Furthermore, there is a drive for a soft orthotic, as many users are unhappy with current rigid AFO's because they are uncomfortable and do not allow for a natural gait cycle. The market potential was identified, but a full manufacturing and marketing analysis was not within the scope of the project: the focus was a proof of concept driven by a market potential.

ACKNOWLEDGEMENTS

- Dr. Kathleen Lamkin-Kennard, McKibben Muscle Expert and Facility Provider
- Rochester Institute of Technology, Project Funding.

REFERENCES

[1] Kesar, Trisha M., Ramu Perumal, Angela Jancosko, Darcy S. Reisman, Katherine S. Rudolph, Jill S. Higginson, and Stuart A. Binder-Macleod. "Novel Patterns of Functional Electrical Stimulation Have an Immediate Effect on Dorsiflexor Muscle Function During Gait for People Poststroke." *Journal of the American Physical Therapy Association* (2010): 55-66. *US National Library of Medicine National Institutes of Health*. Web. 14 Dec. 2014.

BIOMECHANICAL RESPONSE AND FIBER MICROSTRUCTURE COUPLED WITH LOCALIZED PROTEASE ACTIVITY AND INHIBITION IN THE ANGIOTENSIN II INFUSED APOE$^{-/-}$ MOUSE MODEL OF ANEURYSM

Darren G. Haskett (1), Tyler S. Smith (2), D. Catalina Ardilia (1), Tom C. Doetschman (3,4), Oliver J. McIntyre (5), Dominic V. McGrath (2,4), Urs Utzinger (4,6), Jonathan P. Vande Geest (4,6)

(1) Graduate Interdisciplinary Program of Biomedical Engineering
University of Arizona
Tucson, AZ USA

(2) Chemistry and Biochemistry
University of Arizona
Tucson, AZ USA

(3) Cellular and Molecular Medicine
University of Arizona
Tucson, AZ USA

(4) Bio 5 Institute
University of Arizona
Tucson, AZ USA

(5) Radiology and Radiological Sciences
Vanderbilt University
Nashville, TN USA

(6) Biomedical Engineering
University of Arizona
Tucson, AZ USA

INTRODUCTION

Abdominal aortic aneurysm (AAA) has become an increasingly prevalent disease found in up to 9% of adults over 65 years of age and carries with it an 85% mortality rate when ruptured. Each year approximately 33,000 patients undergo elective repair, and deaths due to ruptured AAAs and complications after surgical treatment account for an estimated 15,000 to 30,000 deaths each year. For this reason significant research has been put into improving treatment options for repair and rupture prediction, and recently a pharmacologic method of treatment using doxycycline to reduce aneurysmal expansion has started clinical trials. However, basic issues regarding the etiology and progression of aneurysmal disease have received relatively less attention due to the difficulties of studying such a complex and multifactorial disease, leaving important questions about AAA pathophysiology unanswered.

In order to overcome the obstacles of studying AAA, the Angiotensin II (AngII) infused ApoE$^{-/-}$ mouse model has been used as a test bed, for it shares many histological and environmental characteristics of native human AAA. The model has allowed recent work to focus on understanding the underlying mechanisms of disease development and progression, specifically the role of increased matrix metalloproteinase (MMP) production in aneurismal tissue that leads to extensive extracellular matrix (ECM) remodeling. However, changes in localized MMP expression and activity have not yet been coupled to the resulting functional tissue damage or changes in fiber structure and mechanical integrity. Also unknown is any beneficial mechanism and underlying impact of targeted treatment on vascular function and returned or retained mechanical function. The aim of this study is to investigate the correlation of MMP activity with subsequent alterations in vessel mechanical properties leading to aneurysm formation and develop a mechanism for the reducing pathologic MMP activity.

METHODS

Prior to animal testing, cultures of murine vascular smooth muscle cells (VSMCs) were done for both ApoE$^{-/-}$ and control mice. AngII was added in culture at both high and low doses. A FRET based proteolytic beacon as described by McIntyre *et al* [1] was first evaluated based on sensor and reference signal using a two-photon spectral sweep. Once an appropriate wavelength was selected, the beacon was added to the media aspirate from culture and changes in fluorescence were measured using two-photon microscopy. Changes in the ratio of the fluorescent signals indicated the presence and quantity of active MMPs. The results were then compared to gel zymography.

Culture of VSMCs was also used to determine reduction of MMP activity and expression using a molecule providing a positive feedback mechanism for protease inhibition preventing aneurysm progression in the AngII infused ApoE mouse model of aneurysm. The molecule consists of a doxycycline derivative attached to a peptide chain cleavable by active MMPs. Based on a design developed by Albright *et al* [2], the compound was synthesized and confirmed by mass spectrometry and purified using high-performance liquid chromatography. To determine competency of the doxycycline derivative prodrug as a substrate to MMPs, the prodrug was placed in assays with activated MMP-2, -9, and -14. After the activation assays were completed, VSMC assays were conducted to determine the prodrug's ability to inhibit MMP activation and expression in culture. Proliferation assays were then conducted to determine the toxicity of the new prodrug and doxycycline derivative released based on the LC50.

RESULTS

Two-photon spectral scans showed sensor peaks at 780 and 920 nm. When the beacons were added to an MMP solution with different active concentrations, the results showed an increase in sensor to reference ratio (S/R) that also corresponded with increased MMP concentration (S/R of 1.8x for 1 μg/ml and 4.4x for 10 μg/ml). For the cell culture aspirate, the average increase in signal to noise ratio was 16.6 times that of the non-activated beacons (Figure 1: Top). Gel zymography was then used to confirm the presence of MMPs in the culture aspirate.

Gel zymography also showed a decrease in MMP activity when doxycycline, the 9-amino doxycycline derivative, and the prodrug were added to culture (Figure 1 Bottom). The results as expected showed the doxycycline, the 9-amino derivative, and the prodrug decrease the amount of MMP activity, and this decrease was significant when using the prodrug. Proliferation assays using doxycycline and the 9-amino doxycycline derivative then showed that the LC50 was over 10 times the effective dose (Table 1).

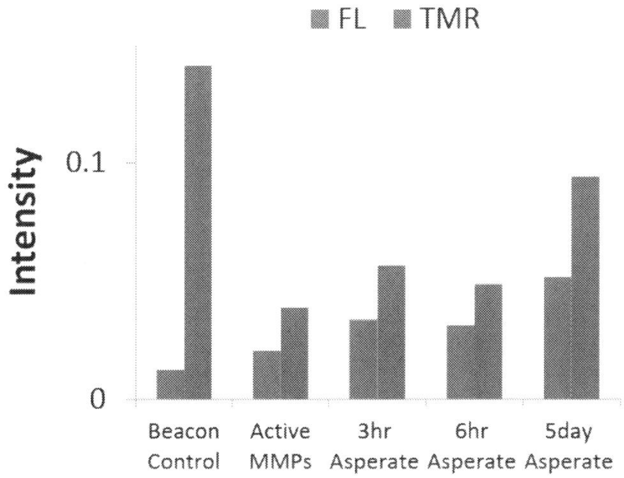

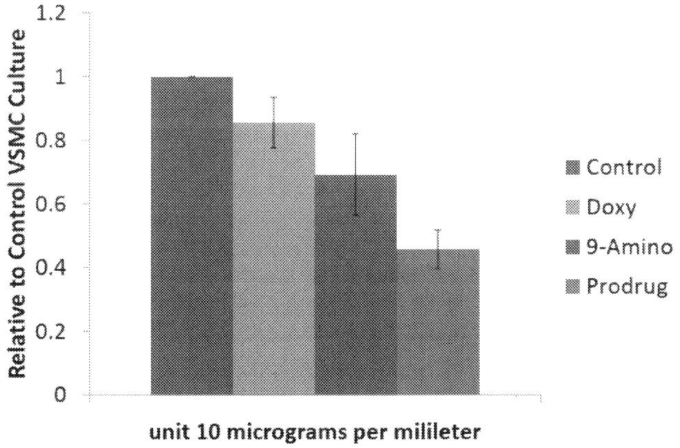

FIGURE 1: TOP) FLUORESCENT INTENSITY IN BOTH THE SENSOR (FL) AND REFERENCE (TMR) CHANNELS SHOWING A DISTINCT INCREASE IN THE SENSOR SIGNAL AND DECREASE IN REFERENCE SIGNAL WHEN ADDED TO MMP OR ASPIRATE SOLUTION. BOTTOM) MMP ACTIVITY IN VSMC ASPIRATE WAS SHOWN TO HAVE DECREASED WITH THE ADDITION OF DOXYCYCLINE, THE DERIVATIVE, AND THE PRODRUG.

LC50	Doxycycline	9-Amino Doxycycline	Ethanol
μg/ml	156.8	148.85	55

TABLE 1: THE LC50 FOR STANDARD DOXYCYCLINE, THE DOXYCYCLINE DERIVATIVE USED IN THE PRODRUG, AND ETHANOL CONTROL.

Preparations have been made and initial *ex vivo* and *in vivo* testing of MMP activity conducted in the ApoE$^{-/-}$ mouse model of aneurysm (Figure 2).

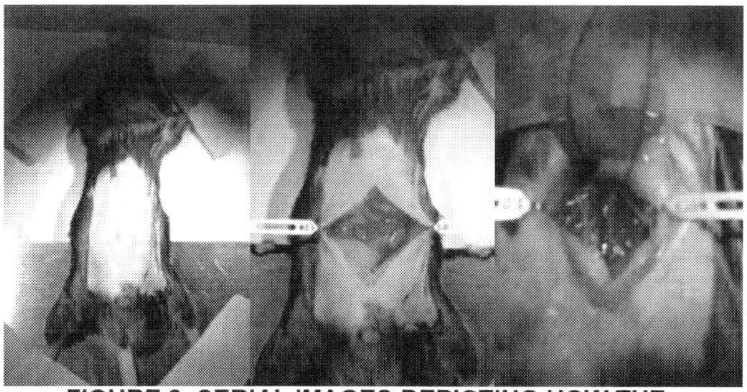

FIGURE 2: SERIAL IMAGES DEPICTING HOW THE ABDOMINAL AORTA IS ACCESSED AND IMAGED.

DISCUSSION

Results from the *in vitro* experiments demonstrate the ability to observe differences in MMP activity with the proteolytic beacons. Now that the use of the proteolytic beacons to determine MMP activity has been established, the next step will be to observe the activity *in vivo* and combine the activity with previously observed biomechanics [3] in order to connect changes in microstructure to the underlying biochemistry and effects on function.

Additionally, work with the prodrug offers a novel way to prevent disease formation or limit progression in a manner that is more beneficial than the current treatment.

ACKNOWLEDGEMENTS

The authors would like to acknowledge sources of support: Cardiovascular Training Grant to Dr. Jennifer Barton (NIH 232HL007955), American Heart Association (GIA #10GRNT4580045 to JPVG), NIH research grant (1RO1EY020890-02A1 to JPVG). Imaging was done on an NIH sponsored shared device (NIH/NCRR S10RR023737).

REFERENCES

1. McIntyre, J.O., R.L. Scherer, and L.M. Matrisian, *Near-infrared optical proteolytic beacons for in vivo imaging of matrix metalloproteinase activity*. Methods Mol Biol, 2010. **622**: p. 279-304.
2. Albright, C.F., et al., *Matrix metalloproteinase-activated doxorubicin prodrugs inhibit HT1080 xenograft growth better than doxorubicin with less toxicity*. Mol Cancer Ther, 2005. **4**(5): p. 751-60.
3. Haskett, D., et al., *Progressive alterations in microstructural organization and biomechanical response in the ApoE mouse model of aneurysm*. Biomatter, 2013. **3**(3). e24648.

CADHERIN-SPECIFIC EXTRACELLULAR INTERACTIONS ALTER STEM CELL SENSATION AND INTERPRETATION OF SOFT TISSUE MICROENVIRONMENTS

Brian D. Cosgrove (1,2,3), Kush D. Mehta (1), Tristan P. Driscoll (1,2,3), Jason A. Burdick (1,3), Robert L. Mauck (1,2,3)

(1) Department of Bioengineering
University of Pennsylvania
Philadelphia, PA, 19104

(2) McKay Orthopaedic Research Laboratory
Department of Orthopaedic Surgery
University of Pennsylvania
Philadelphia, PA, 19104

(3) Translational Musculoskeletal Research Center
Philadelphia VA Medical Center
Philadelphia, PA, USA 19104

INTRODUCTION. Numerous studies over the past decade have highlighted the importance of the extracellular microenvironment in regulating stem cell differentiation[1,2]. During development of connective tissues, the microenvironment changes from one that is rich in cell-cell interactions to one that is typified by cell-extracellular matrix (ECM) interactions[3]. These interactions are similar in that both start with the engagement of a surface receptor (cadherin or integrin, respectively) and converge on the contractile machinery of the cell. One central pathway with signficant interplay between cell-cell and cell-ECM signaling is the YAP (Yes-associated protein) / TAZ (transcriptional co-activator) complex. Here, two main branches are operative: the Hippo pathway initiated by cadherin-cadherin adhesion that results in YAP sequestration in the cytosol, and an independent mechanosensitive pathway wherein cytoskeletal tension promotes the nuclear translocation of YAP and activation of downstream transcription pathways that govern proliferation, migration, and differentiation[4,5]. As such, YAP/TAZ has been thought of as a mechanical rheostat, a signaling pathway that changes as the cell senses microenvironments of different stiffness. To query how cell-cell and cell-ECM interactions govern mechanosensation, we utilized a methacrylated hyaluronic acid (MeHA) hydrogel system to decouple and present both N-Cadherin (scrambled or un-scrambled HAVDI peptides[6]) and fibronectin (RGD peptide) adhesion moieties randomly interspersed on a hydrogel surface of changing elasticity. Our goal was to determine the influence of N-Cadherin presentation on mesenchymal stem cell (MSC) mechanosensing.

METHODS. <u>Hyaluronic Acid Hydrogels:</u> Methacrylate-modified hyaluronic acid was synthesized as in [3]. Fibronectin adhesive domain ("RGD", GCGYG**RGD**SSPG-), N-Cadherin adhesive domain ("NCad", **HAVDI**GGGC-), and non-functional scrambled N-Cadherin ("scram", AGVGDHIGC-) peptide sequences were incorporated into a 3% MeHA solution (**Fig. 1A**). Gels were polymerized using I-2959 and 4.5mW/cm^2 UV light, and the time of UV exposure vs. Young's modulus analyzed using AFM (**Fig. 1B**). **<u>Traction Force Microscopy (TFM):</u>** TFM was performed on 10kPa MeHA gels modified with 1mM NCad or scram and 1mM RGD. Gels also contained 1% v/v 0.2μm fluorescent microspheres. Phase contrast images of cells and beads were taken both before and after cell lysis with SDS. TFM analysis was performed using a plugin suite for ImageJ by Q. Tseng [7]. **<u>Cell Culture and Immunofluorescence Imaging:</u>** Juvenile bovine MSCs were isolated from bone marrow and cultured for one passage in growth media before being plated overnight on gels. Cells were allowed to culture for 24 hours before fixation and subsequent staining. Antibodies used included YAP/TAZ (#sc-101199) and an N-Cadherin neutralizing Ab (Sigma# GC4). YAP/TAZ ratios were determined by measuring the average fluorescent intensity within the nuclear and cytoplasm domains using ImageJ. Proliferation over 48hrs was assayed using an EdU+ quantification kit (Life Technologies). In some studies, ROCK was inhibited with 10μM Y-27632, while active Rac1 was inhibited with 50μM NSC-23766, both for 1hr. **<u>Statistics:</u>** Analysis was performed in GraphPad, as described in figure legends.

RESULTS. MSCs cultured on MeHA gels presenting RGD and NCad (at a 1:1 ratio) produced significantly lower traction forces and stresses than cells on control gels presenting RGD and non-functional scrambled NCad (**Fig. 1C**). Quantification of the YAP/TAZ nuclear to cytoplasmic ratio across a physiological range of substrate stiffnesses revealed that N-Cadherin presentation significantly lowered YAP/TAZ ratios at intermediate stiffnesses, while at higher and lower stiffnesses, this ratio was not different (**Fig. 2A-B**). Blocking external N-Cadherin on MSCs prior to seeding (using a neutralizing antibody) resulted in complete abrogation of the differences in YAP/TAZ (**Fig. 2C**). Consistent with the changes in YAP localization on NCad gels, proliferation was greatly reduced (**Fig. 2D**).

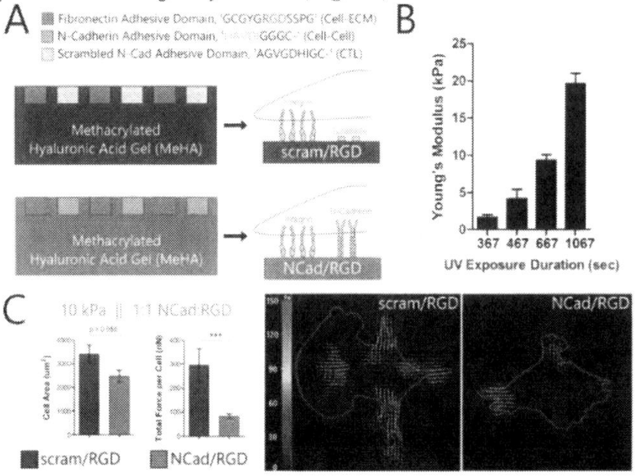

Figure 1- (**A**) Schematic of MeHA gel system with a 1:1 scram or NCad to RGD ratio (2mM total ligand). (**B**) UV exposure times @ 4.5mW/cm^2 and resulting moduli measured by AFM (+SD). (**C**) TFM of MSCs (±SE, *** indicates $p<0.001$ by Student's t-test).

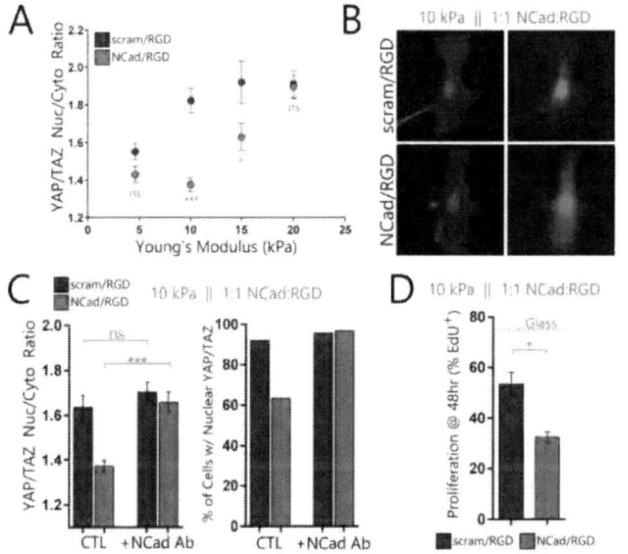

Figure 2- (**A**) YAP/TAZ ratios over a range of stiffnesses on 1:1 scram/RGD and NCad/RGD gels (±SE, *=$p<0.05$, ***=$p<0.001$ by 2-way ANOVA). (**B**) Representative images for cells in (A), red=F-actin, blue=DAPI, green=YAP/TAZ. (**C**) YAP/TAZ ratios for MSCs with or with/out treatment with N-Cadherin blocking antibody prior to seeding (±SE, ***=$p<0.001$ via 1-way ANOVA w/Dunn's post hoc). (**D**) MSC proliferation (+/-SD, * indicates $p<0.05$ by Student's t-test).

Next, we examined how changing the ratio of peptides influenced the response. As NCad availability decreased and RGD availability increased, there was a significant increase in YAP/TAZ nuclear signal. This increase was not observed when there was no NCad present (**Fig. 3B**). Treatment of MSCs with 50μM NSC-23766 (a pharmacological inhibitor of GTP-Rac1) reduced differences in YAP/TAZ signal between scram/RGD and NCad/RGD conditions (**Fig. 3C-D**). Treatment with 10μM Y-27632, a ROCK inhibitor that decreases contractility, resulted in less nuclear YAP/TAZ for both groups, without reducing the effect of NCad (**Fig. 3D**). Combination of both inhibitors further decreased nuclear YAP/TAZ, and abrogated the differences seen with NCad presentation.

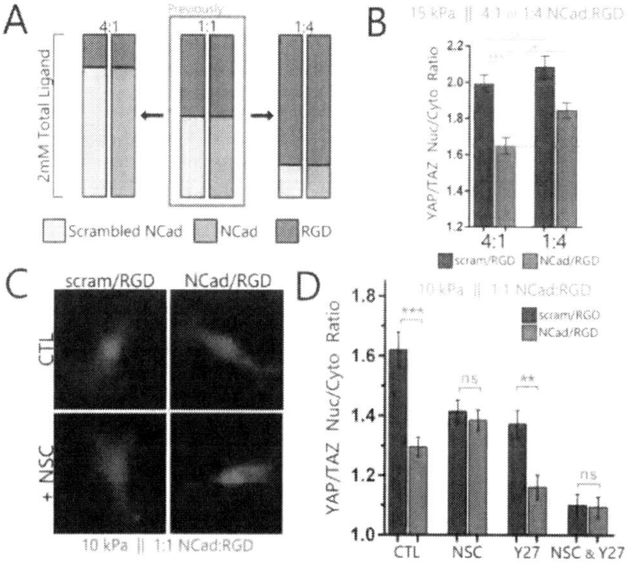

Figure 3- (**A**) Schematic for peptide ratios. (**B**) YAP/TAZ on 4:1 or 1:4 gels (± SE, ***=$p<0.001$, *=$p<0.05$ via 1-way ANOVA w/Dunn's post hoc). (**C**) YAP/TAZ staining on 10kPa 1:1 gels, with or without 50μM NSC-23766 treatment. (**D**) YAP/TAZ for cells treated with 50μM NSC-23766, 10μM Y-27632 (ROCK inhibitor), or both (±SE, ***=$p<0.001$, **=$p<0.01$ by 1-way ANOVA w/Bonferonni post hoc).

DISCUSSION. This study suggests that N-Cadherin alters stem cell sensation and interpretation of the physical microenvironment, even in the presence of traditional ECM moieties. In the context of a constant level of RGD, presentation of N-Cadherin adhesion peptide at a 1:1 ratio lowered the YAP nuclear/cytoplasmic signal on substrates of intermediate stiffness. Along with lower traction forces, this decreased nuclear YAP/TAZ suggests a shift in the interpretation of stiffness: MSCs on 15kPa gels with both NCad and RGD behaved as though they were on 6kPa gels with only RGD. Pharmacologic inhibition of Rac completely abrogated these differences in YAP/TAZ with N-Cadherin. This suggests that rather than solely working through Hippo, N-Cadherin presentation may suppress Rac activation. This in turn reduces actin organization and traction force, lowering the YAP/TAZ nuclear to cytosolic ratio. Blocking cellular N-Cadherin prior to seeding (with blocking antibodies) confirmed that this phenomenon was mediated by N-Cadherin. Altering the ratio of NCad to RGD illustrated the dose-dependent nature of this interaction. Interestingly, previous studies on stiff substrates showed no effect of Rac signaling on YAP/TAZ nuclear localization[5]. Consistent with this, our data shows that Rac activity is less important in mechanosensing at a high stiffness (>20kPa), but our data additionally showed an emergent role for Rac in YAP-based mechanosensing at intermediate stiffnesses (2-20kPa). Taken together, these findings provide new insight into the interplay between cell-cell and cell-ECM signaling that dictate how stem cells interpret dynamic and soft tissue microenvironments that are characteristic of developing and/or regenerating tissues.

ACKNOWLEDGMENTS. This work was supported by the National Institutes of Health (R01 EB008722) and the Penn Center for Musculoskeletal Disorders. **REFERENCES.** [1] McBeath+, *Dev Cell*, 2004. [2] Engler+, *Cell*, 2006. [3] Bian+, *PNAS*, 2013. [4] Low+, *FEBS Letters*, 2014. [5] Dupont+, *Nature*, 2011. [6] Williams+, *J Biol Chem*, 2000 [7] Tseng+, *PNAS*, 2012.

A LONG-TERM SIMULATED DEGRADATION STUDY OF A SYNTHETIC MENISCUS IMPLANT: MATERIAL AND FUNCTIONAL PROPERTIES

Maoz Shemesh (1), Adaya Shefy-Peleg (1), Eyal Zylberberg (1), Eran Linder-Ganz (1), Jonathan J. Elsner (1)

(1) Research and Development, Active Implants, Netanya, Israel

INTRODUCTION

Long-term wear experiments are an important tool to evaluate the potential of success of joint replacement devices during the stages of research and development. A synthetic and readily functional non-anchored polycarbonate-urethane (PCU) meniscus implant, reinforced with UHMWPE fibers, was developed for medial meniscus replacement (NUsurface®, Active Implants, Memphis TN, USA). The implant is distinctively different from most orthopaedic implants in two distinct features. It was designed as a soft bearing (composed of a compliant material), and it is not fixed to any of the soft or hard tissues. Its femur-conforming design enables implantation of the device through a minimal arthrotomy, without need for bone resection or attachment [1]. Due to these unique features of the implant, it is important to show that the material and design of the device can tolerate the environmental and loading conditions in the knee joint. Previous studies of this device were focused on its ability to distribute joint loads under static loading conditions [2], and its viscoelastic characteristics after short-term use [3]. This study aimed to evaluate its long-term performance focusing on potential degradation of the device by subjecting it to a full mixed-mode loading regime for a total of 5 million load cycles (Mc). A low wear rate as well as the implant's ability to maintain its structure, chemical stability, and function following repeated long-term loading in joint-like environment were considered key indicators for success.

METHODS

Six implants were removed from their original sterile packaging, placed in a dry oven (40°C) for 3 days to remove fluid exudates and their initial dry weight was recorded. Following, the implants were pre-soaked for 7 days in a lubricant (bovine serum), cleaned and patted dry, and their weight was recorded for reference. Five implants were then assigned to undergo full mixed-mode simulation. The sixth implant served as a load-soak control and was only loaded with the vertical load component (based on ISO-14243). The simulation was conducted for a total of 5 million cycles (Mc), with stops at 0.5, 1, 1.5, 2, 3, 4 Mc. Each specimen was placed in a closed cell containing lubricant (bovine serum) at 37°C and loaded with the previously described load pattern at 1 Hz. The lubricant was completely replaced at every stop for the continuing simulation, but samples were maintained for particle isolation and characterization. The implants were then cleaned, patted dry, and weighed. After reaching the 5 Mc time point, the implants were weighed once again, and then cleaned and placed in a dry oven (40°C) for 3 days. Their final dry weight was used to evaluate the overall loss of mass.

The load distribution capacity of the meniscus is thought to be its most important role. This function of the artificial implant was monitored throughout the simulation by measuring the pressure distribution maps under the implant when subjected to a 1500 N compressive load, at a 10 mm/min deformation rate, at 0° flexion, while located between polyethylene replicas of femur and tibia cartilages. This loading condition corresponds to the maximal loading conditions, which are applied to the medial compartment of the knee during gait, based on ISO-14243-3 (Wear of total knee-joint prostheses).

Upon completion of the simulation, the implants were further analyzed by semi-destructive tests. The superior and inferior articulating surfaces were observed with an environmental SEM (ESEM, Quanta 200 FEG, FEI, Eindhoven, The Netherlands). Atomic Force Microscopy (AFM) was used to measure the surface roughness of the articulating surfaces after 5 Mc. Oxidation of the implant material was measured by recording the Fourier Transform Infra-Red (FTIR) Spectra on the surface of the implant, which is exposed to the lubricant, using a Perkin Elmer Spectrum 65 FTIR (Waltham, MA), equipped with an Attenuated Total Reflectance (ATR) accessory. Finally, samples of the material were taken for molecular characterization, which was conducted using viscosimetry and Gel Permeation Chromatography (GPC).

RESULTS

All of the implants maintained their position between the tibia and femur throughout the simulation. None of the implants showed signs of significant damage or wear when examined visually at the completion of the test. Gravimetric measurements demonstrated an overall wear of 75.9 ± 8.0 mg after 5 Mc with an average wear rate of ~15 mg/Mc.

Microscopic (SEM) images of the implants demonstrated minor damage on the articulating surfaces. The superior articulating surfaces had some regions with mild abrasive wear, which consists of a uniform scatter of superficial depressions and flakes in the direction of articulation. Quantitative analysis by AFM showed that following use, there was an increase in the average surface roughness and an increase in the peak-to-valley roughness (10 to 20 nm and 82 to 200 nm, respectively). There were no signs of cracks or pits on the superior and inferior surfaces which could indicate the onset of fatigue.

The results of FTIR analysis of these implants show very little change in the chemical footprint on the surface of the implants. Average increase in oxidation compared to baseline measurements was found to be less than 6%, and not any different from control specimens placed in open air. There was a slight increase in the number average (Mn) and weight average (Mw) molecular weights (92 to 96 kg/mol and 173 to 178 kg/mol, respectively) after simulation, when compared to an unused control implant (Fig. 1). There was no difference in the polydispersity index (Mw /Mn) between the tested specimens and unused control implant (1.78).

Functionality tests, which were focused on the measurement of pressure distribution on the tibial plateau when loaded with a 1500N load, demonstrated good preservation of function following use (Fig. 2). Interestingly, the pressure distribution capacity of the implant improved over time, as a result of use (Fig. 1). Baseline measurements demonstrated a total average contact area of 722±34 mm^2, which corresponds to 61.6% of the total inferior surface of the device. The mean contact pressure was 3.3±0.03 MPa, and the peak contact pressure that occurred on the anterior side of the tibial plateau and was found to be 10.1±0.5 MPa. After 5 Mc, the implant occupied a 1.4 larger contact area (1011±30.2 mm^2), compared to the baseline measurement (86.2% of the total inferior surface). Also, the mean contact pressure and peak contact pressure were reduced by 20.6% (2.62 ± 0.03 MPa) and 10.9% (to 9.03 ± 0.17 MPa), respectively.

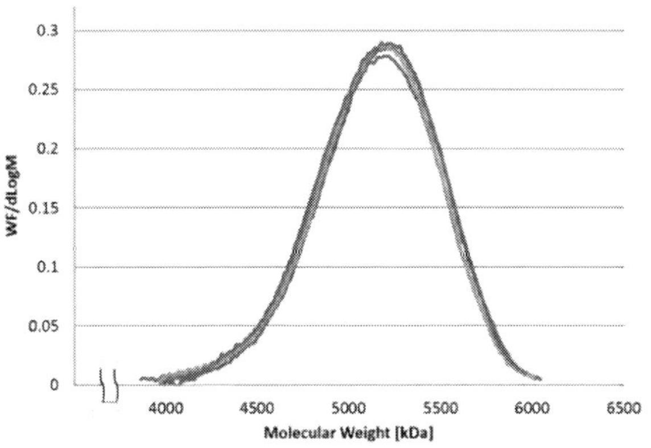

Figure 1. Gel permation chromatography measurements of the molecular weight of three implants after 5 Mc (red, green and purple) compared to a control (blue)

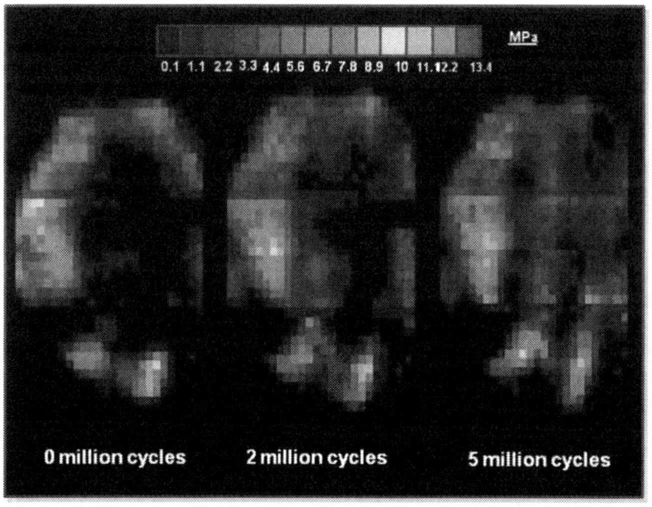

Figure 2. Pressure distribution maps developed under peak gait load (1500N) at the baseline time point, and following 2 and 5 million loading cycles.

DISCUSSION

This is the first long-term degradation study of a meniscus replacement device composted of polycarbonate-urethane. We aimed at providing reasonable assurance that the device can be used safely and fulfill its biomechanical role as a load distributor in the knee after prolonged use. Mechanical failure is the principal mode of failure that have been reported in previous meniscus implant solutions [4, 5], yet in this study, the PCU implant was found to be resilient enough to maintain its form, and its material properties throughout the long-term simulation. We found no signs of degradation of the material or in the implants overall functional properties. Surprisingly, the implants ability to distribute load improved over time, compared to its initial condition as it underwent moderate creep, and was able to increase the contact area with the articulating surfaces.

The average wear rate measured for the device is low, and comparable to that of hard bearing polymers such as polyethylene. Thus, we believe that the implant has good potential for long-term clinical use.

ACKNOWLEDGEMENTS

This Study was funded by Active Implants.

REFERENCES

[1] Elsner J, et al., J Biomech Eng. 132(9):095001, 2010.
[2] Linder-Ganz E, et al., J Biomech Eng. 132(2):02450, .2010.
[3] Shemesh M, et al., J Mech Behav Biomed Mat. 29:42-55, 2014.
[4] Kobayashi, M et al., Biomaterials, 24:639–647, 2003.
[5] Hannink, G et al., Knee Surg Sports Traumatol Arthrosc. 19(3):441-51, 2011.

DAMAGE ANALYSIS OF METALLIC AND POLYMERIC BEARINGS USED IN REVERSE TOTAL SHOULDER ARTHROPLASTY

S. Chou (1), I. Yang (1), N. Bonnheim (1), F. Ansari, MS (1), S. Gunther, MD (2), T. Norris, MD (3), M. Ries, MD (4), L. Pruitt, PhD (1)

(1) Department of Mechanical Engineering
University of California, Berkeley
Berkeley, CA, USA

(2) Department of Orthopedic Surgery
Martha Jefferson Hospital
Charlottesville, VA, USA

(3) San Francisco Shoulder, Elbow & Hand Clinic
San Francisco, CA, USA

(4) Tahoe Fracture and Orthopaedic Clinic
Carson City, NV, USA

INTRODUCTION

Total shoulder replacement (TSR) is an established treatment to alleviate pain and restore compromised function associated with osteoarthritis and traumatic shoulder fractures. A metal-on-polymer bearing system is implanted to either mimic the natural shoulder anatomy or reverse the orientation of the ball and socket components (reverse shoulder replacement, RSR). Roughly 53,000 TSRs were performed in the United States in 2011, making the shoulder the third most common total joint replacement behind knee and hip arthroplasty [1]. A TSR complication rate of roughly 15% at five to ten years - relatively high compared to the knee and hip - led to a recent surge in RSR popularity [2]. The RSR system is designed to increase implant stability and surface area contact, thereby minimizing eccentric loading and contact stresses. However, complication rates of RSRs have been reported to be as high as 26% [3]. Retrieval analysis is essential to illuminate the *in vivo* performance RSRs. This work presents an assessment of damage on both cobalt chrome (CoCr) and ultrahigh molecular weight polyethylene (UHMWPE) bearings for RSRs to better understand material behavior and damage mechanics.

METHODS

The bearing and rim surfaces of 20 retrieved UHMWPE humeral cups manufactured by Tornier (n=19) or Depuy (n=1) were analyzed in this study. Average *in vivo* duration was 41 months (n=13, range: 5 days-113 months), and average patient age at implantation was 62.2 years (n=15, range: 37-79 years). The bearing surfaces of the humeral cups were divided into 4 regions of equal area, similar to the method reported by Nam et al. [4]. Six observers classified the type of UHMWPE damage present based on published classification methods of polyethylene implants [5], as shown in **Figure 1**.

The existence of each damage mode was noted through visual inspection of each quadrant. Careful attention was paid to discriminating between tool and *in vivo* damage. Two observers classified the severity of damage on the rim surface for each humeral cup based on the depth of material loss. A score of 0 corresponded to no observable material loss while a score of 3 corresponded to complete material loss through the thickness of the humeral cup. In addition to severity, other characteristics of the damage were noted, including percentage of the quadrant damaged and surface roughness.

The bearing surfaces of 14 retrieved CoCr glenospheres manufactured by Tornier (n=13) or Depuy (n=1) were analyzed. Average *in vivo* duration was 39 months (n=10, range: 1-92 months), and average patient age at implantation was 61.2 years (range: 37-79 years). 6 observers classified the damage present on the glenospheres based on published classification methods [5]. The assessed CoCr damage modes are shown in **Figure 2**. Two observers then verified and classified the damage modes present on the entire bearing surface. Each damage mode was scored based on extent of surface coverage from not present (score of 0) to 75-100% coverage (score of 4).

RESULTS

The results of the UHMWPE damage mode scoring analysis are shown in **Figure 3,** where the numbers correlate to the percentage of implants that have damage in each quadrant. Of the assessed damage modes, only subsurface cracking was not observed. The most common damage modes were scratching and striations, which occurred on over 50% of the samples in each quadrant. The results of the rim surface analysis are shown in **Table 1**.

Twelve samples showed abrasion on the rim of the humeral cup, especially on the inferior end, where 55% of samples showed damage. For 25% of the samples (5 of 20), roughly half of the rim was missing on one side of the component. Though most of the damaged rims had rough and abrasive surfaces, one of the samples had a burnished and smooth surface. One sample contained significant metal debris embedded in the surface of the polyethylene humeral cup.

All CoCr damage modes evaluated were present on at least 1 of the 14 samples. All samples showed signs of hairline scratching, curvilinear abrasion, and striated scratching. Curvilinear abrasion was the most severe and prominent damage mode, with an average severity score of 2.71 (nearly half of the glenosphere). Linear abrasion was also common, with 85% showing signs of this damage mode (covering over 25% of the surface, on average).

DISCUSSION

To our knowledge, this work is the first to assess surface damage modes on both metal and polymer bearings from RSR retrievals, and thus more thoroughly elucidates the occurrence of *in vivo* RSR damage. Our findings indicate that both scratching and striations are prominent in all quadrants of the bearing surface of the polyethylene humeral cup. Damage is most frequent and severe in the inferior quadrant of the rim surface, consistent with findings by Nam et al [4]. It is hypothesized that embedded metal debris came from the metaglene screw, likely due to impingement of the humeral cup on the scapula, resulting in severe scapular notching as seen in Day et al [6]. Our findings also suggest that UHMWPE damage seen in highly constrained and conforming RSR components is more consistent with surface wear mechanisms (abrasion and embedded debris) than fatigue wear mechanisms (delamination, pitting, and fracture), which are commonly seen in conventional less constrained and less conforming TSR implants [7]. Our analysis indicates that both curvilinear abrasion and linear abrasion are prominent on the CoCr glenosphere, as is also the case for TSRs [5]. Continued retrieval assessment to evaluate oxidation levels, surface and rim damage and debris composition using SEM and EDS techniques will be valuable to further understand *in vivo* dynamics and damage modes.

ACKNOWLEDGEMENTS

We would like to acknowledge Louis Malito, Taylor Lee, Ally Belinski, and Jessica Sov for their help with these studies.

REFERENCES

[1] American Academy of Orthopaedic Surgeons. "Shoulder Joint Replacement", www.orthoinfo.aaos.org, 2011.
[2] Bohsali, K, et al. *J Bone Joint Surg*, 88-A(10), 2279-2292, 2006.
[3] Roberts, C, et al. *Radiographics*, 27(1), 223-235, 2007.
[4] Nam, D, et al. *J Shoulder Elb Surg*, 19(7), 1003-1012, 2010.
[5] Malito, L, et al. "Bearing Surface Damage Analysis of Coupled Total Shoulder Replacement Retrievals" Presented at UHMWPE Lexicon, 2013
[6] Day, J, et al. *J Shoulder Elb Surg*, 21(5), 667-674, 2012.
[7] Gunther, S, et al. *J Arthroplasty*, 17(1), 95-100, 2002.

TABLE 1. GENERAL TRENDS IN UHMWPE DAMAGE.

	% Implants with Damage	Severity of Damage in Damaged Implants (0-3)	% Coverage of Damage in Damaged Implants
Superior (n=20)	20	1.25	71.25
Posterior (n=17)	47.06	1.5	42.5
Inferior (n=20)	55	1.91	77.27
Anterior (n=17)	35.29	1.17	65

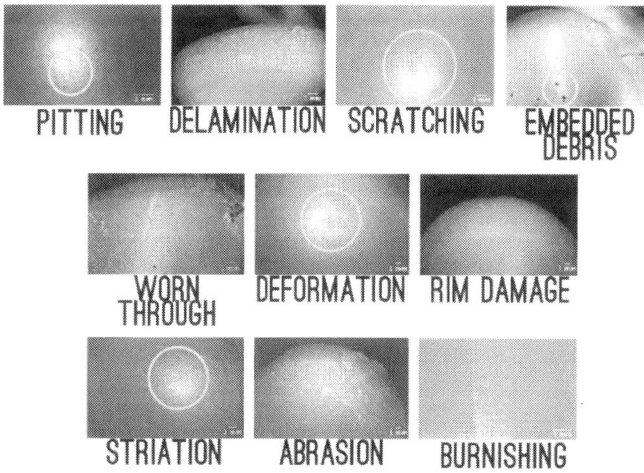

FIGURE 1. DAMAGE MODES OBSERVED ON UHMWPE HUMERAL CUP RETRIEVALS.

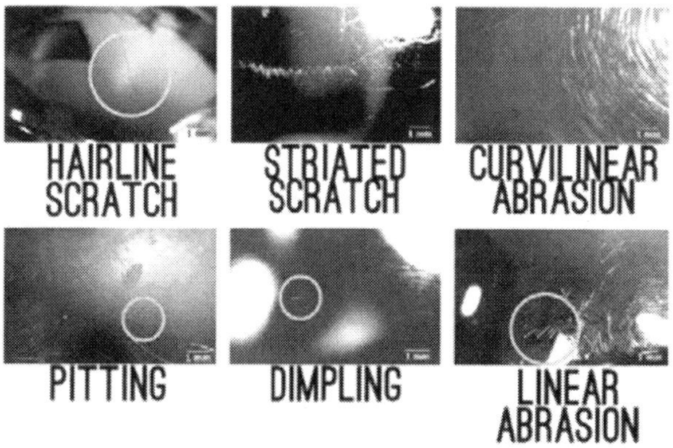

FIGURE 2. DAMAGE MODES OBSERVED ON COCR GLENOSPHERE RETRIEVALS.

FIGURE 3. TRENDS IN QUANTITY AND LOCATION OF UHMWPE DAMAGE.

MECHANICS OF EARLY EYE DEVELOPMENT: INTERACTIONS BETWEEN THE RETINA AND LENS

Alina Oltean (1), Jie Huang (2), David C. Beebe (2), Larry A. Taber (1)

(1) Dept. of Biomedical Engineering
Washington University
St. Louis, MO USA

(2) Dept. of Ophthalmology and Visual Sciences
Washington University School of Medicine
St. Louis, MO USA

INTRODUCTION

Invagination is an essential mechanism in the early development of many sensory structures such as the eye, inner ear, and olfactory epithelium [1]. The mechanics of both retina and lens invagination are not yet understood [1, 2]. The eyes begin as two spherical optic vesicles that grow outwards from the forebrain and contact the surface ectoderm. Both the surface ectoderm (SE) and optic vesicle (OV) become tightly adhered and thicken, forming the lens placode and retinal placode, respectively. The two placodes then bend inward or invaginate forming the lens vesicle (LV) and optic cup (OC) [2, 3].

Research in the mouse embryo suggests an important role for the extracellular matrix (ECM) that is deposited between the surface ectoderm and optic vesicle during placode formation. The ECM may prevent expansion of the cell layers as they grow, creating cell crowding, and forcing lens placode formation [2]. We extend this theory to formation of the retinal placode and invagination of the OV as it grows. For lens development after placode formation, apical constriction has been shown to play a role, but the mechanisms of lens invagination and vesicle closure are not known [1]. The tight adhesion followed by detachment of the retina and lens was studied in chick embryos to determine how the two organs coordinate invagination. The mechanics of various mechanisms were tested with finite element models, mechanical perturbation, and imaging techniques.

METHODS

Experimental Setup

Fertilized white leghorn chicken eggs were incubated at 38°C approximately 46-56 hours until the appropriate Hamburger and Hamilton stage (HH12 to HH16) was reached [4]. Whole embryos were transferred from each egg into culture dishes using a filter paper carrier method [5]. Embryos were incubated with culture media (89% DMEM, 10% chick serum, 1% penicillin/streptomycin/neomycin) at 38°C in 95% O2 and 5% CO2.

The embryos were transferred and imaged with the dorsal surface up. The vitelline membrane was cut near the optic vesicle using a pulled glass needle to access the eye. In perturbation experiments, the surface ectoderm was removed by applying 1-3 μl Nile Blue Sulfate (1.5% in water) with a pulled glass capillary tube as described by Hyer, et al. (2003) [3]. The Nile Blue Sulfate loosens the surface ectoderm to allow removal with a glass needle.

Optical Coherence Tomography & Imaging

Optical coherence tomography images were captured using a Thorlabs (Newton, NJ) Swept Source OCT Imaging System with a Nikon FN1 microscope. Images can be collected from live embryos up to a depth of approximately 2 mm [6]. Image stacks were then viewed, filtered to reduce noise, and reoriented using ImageJ software (National Institutes of Health, NIH). Changes in morphology such as placode thickness were measured in ImageJ. Embryos were also imaged live with brightfield microscopy. Fixed embryos were stained for cytoskeletal proteins and ECM components (filamentous actin and fibronectin) then imaged with a Zeiss LSM 510 or 710 confocal microscope (Carl Zeiss Microimaging, Inc., Thornwood, NY).

Finite Element Models

Mechanisms for placode formation and invagination were tested with finite element models of the retina or lens. For optic cup invagination, an axisymmetric finite element model was created using COMSOL Multiphysics (Version 4.2a, COMSOL AB, Providence, RI, USA). A hyperelastic material was chosen with a specialized Blatz-Ko strain energy density function. The OV was represented as a spherical shell with a thickness to outer radius ratio of 0.2, a normalized shear modulus of 1, and Poisson's ratio of 0.45. The ECM layer on the future placode region was 1/10 the thickness of the OV and stiffer by a factor of 25. Growth was specified in the placode region and outer OV by modifying the governing equations within COMSOL. The deformation gradient tensor, F, is defined as

$$F = F_e * G \qquad (1)$$

where F_e is the elastic deformation gradient tensor and G is the growth tensor as described previously [7]. Circumferential and meridional growth began with a value of 1 (no growth) and was increased linearly in the placode region and the outer OV region, while remaining constant in the ECM region.

Lens formation was simulated with separate models for placode formation and invagination. First the pre-placode lens was modeled as 80 degrees of a spherical shell with the same thickness, modulus, and Poisson's ratio as the OV. In this axisymmetric model, a thin stiff ECM with a modulus of 25 was located on the basal side of the lens. The lens grew as described for the OV while the ECM did not grow. To model invagination, instead of a basal ECM, a layer of actin with 1/8 the thickness of the lens was added apically on top. The actin began 15 times stiffer than the lens and increased in stiffness to a final factor of 55 as the actin contracted down from a growth value of 1 to 0.6. The lens tissue grew linearly over time with more growth along the basal side to create cell wedging. The lens invagination model was tested in 2D and 2D axisymmetric models.

RESULTS

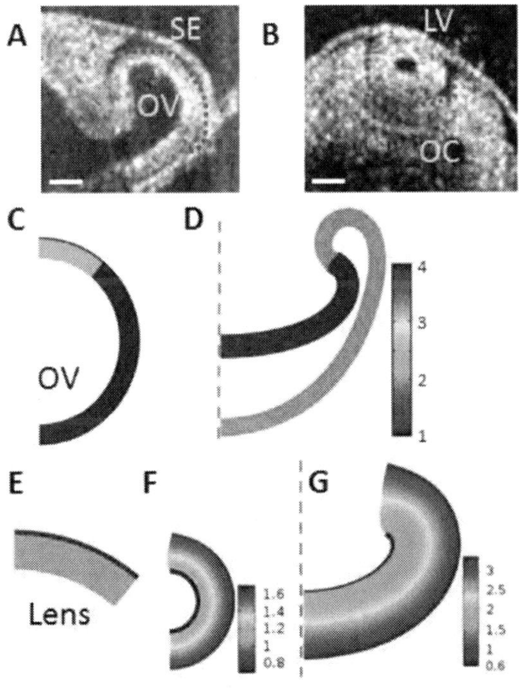

Figure 1. (A) OCT image of the OV and SE before placode formation. (B) OCT image of bilayered optic cup (OC) and lens vesicle (LV). Red dashed lines mark the boundary between the two cell layers. (C) Undeformed OV finite element model with the future placode region in light blue and stiff ECM in red. (D) Invaginated OV model after the OV grew. Legend shows growth value where 1 equals no growth. (E) Initial lens geometry with SE in blue and a stiff, contracting actin layer in red. (F) The 2D lens model invaginates and closes with the actin and wedge-shaped cell growth. (G) The 2D axisymmetric model does not close even with higher growth. Scale bars: 100 μm.

Retina and lens development was imaged live using OCT as shown in Figure 1A and 1B to observe placode formation and invagination. The dissection experiments of Hyer, et al. (2003) were repeated by removing the surface ectoderm before or after lens placode thickening and observing if optic cup formation occurred. In embryos where the surface ectoderm was removed before placode formation, the optic vesicle did not invaginate. However, if the lens placode had formed before dissection, the optic vesicle was typically able to invaginate as described previously [3]. OCT imaging showed the geometry of the optic cup developing without the lens placode. In early OV invagination, the lens has a strong effect on OV geometry. During later invagination, the lens had little to no effect on OV concavity.

Staining for filamentous actin showed the distribution of actin along the apical side of the lens. Fibronectin staining revealed a thick ECM between the invaginating OV and lens. To model how the optic vesicle invaginates after lens placode removal, we created a finite element model of the OV with a thin, stiff ECM layer as seen in Figure 1C-D. As the OV grew, the ECM restricted the top surface of the OV enough to cause the OV to bend inward or invaginate. In a simulation of lens placode formation with a stiff basal ECM, the growing lens thickened due to the restricting ECM (image not shown). In 2D and 2D axisymmetric models of lens invagination where the lens grows in a cell-wedge shape and an actin layer stiffens and contracts, the lens was able to invaginate (Figure 1E-G). However, in the 2D axisymmetric model, the constriction and growth was not enough to cause the lens to close (Figure 1G).

DISCUSSION

We were able to recreate the dissection experiments of Hyer, et al. (2003), and with OCT we saw how the geometry of the OV changes over time without the neighboring lens tissue. We hypothesize that the early lens is needed to allow the ECM to be fully deposited on the basal OV for invagination. The OV and lens coordinate placode formation and early invagination through their tight adhesion with the ECM. Our finite element models showed that the ECM is sufficient to cause the growing OV to invaginate and for the growing lens to form a placode. For invagination of the lens, wedge-shaped growth and apical constriction were able to bend the lens inward significantly. However, other mechanisms are probably required to help the lens close into a vesicle. Future models will include additional mechanisms between the OV and lens such as simulating the growth of both layers at once.

ACKNOWLEDGEMENTS

We thank the NIH and the Imaging Sciences Pathway Fellowship from Washington University in St. Louis for funding.

REFERENCES

[1] Breau, MA, Schneider-Maunoury S, *Dev. Biol.* pii: S0012-1606(14)00638-1, 2014.
[2] Huang, J et al., *Dev Biol.* 355(1), pp. 32-42, 2011.
[3] Hyer, J et al., *Dev Biol.*, 259(2), pp. 351-63, 2003.
[4] Hamburger, V, Hamilton, HL, *Dev Dyn.* 195(4), pp. 231-72, 1992.
[5] Voronov, DA, Taber, LA, *Dev. Dyn.* 224(4), pp. 413–421, 2002.
[6] Fujimoto, JG, *Nat Biotechnol.* 21(11), pp. 1361-7, 2003.
[7] Taber, LA, *Biomech. Model. Mechanobiol.* 7(6), pp. 427-441, 2008.

TARGETING COLLAGEN STRANDS BY TRIPLE HELIX HYBRIDIZATION

S. Michael Yu (1), Yang Li (1)

(1) Department of Bioengineering
University of Utah
Salt Lake City, UT, USA

INTRODUCTION

As the most abundant protein in mammals, collagens play a crucial role in tissue development and regeneration, and their structural or metabolic abnormalities are associated with debilitating genetic diseases and various pathologic conditions. For example, excess collagen remodeling activity is commonly seen in tumors, arthritis, and many other chronic wounds, as well as in heritable connective tissue diseases such as Marfan's syndrome, Ehlers-Danlos syndrome. During collagen remodeling, large portions of collagens are degraded by proteolytic enzymes and their triple helical structures become unfolded at body temperature, which can be explored for diagnostic and therapeutic purposes. Since unstructured proteins are not ideal targets for rational drug design, it is not easy to develop molecules that specifically bind to cryptic sites in denatured collagen strands by traditional approaches. We envisioned that the triple helix, the hallmark structural feature of collagen, could provide a unique targeting mechanism for the denatured collagens. Since the triple helix is nearly exclusively seen in collagens [1], we thought that a small peptide sequence with strong triple-helix propensity could specifically hybridize to denatured collagen strands by forming triple helix in a way similar to DNA fragments binding to complementary DNA strands.

Previously we discovered that small collagen mimetic peptides (CMPs, molecular weight: 2–3 kDa) capable of reversibly forming collagen triple helices are able to hybridize with type I collagen strands [1]. Binding was observed only when CMPs [$(GPO)_x$, $x = 6\text{-}10$, O: hydroxyproline] in a thermally melted, single-strand form were allowed to fold by cooling in the presence of type I collagen fibers [1] (Fig. a, left panel). No binding to collagen strands was detected for folded trimeric CMPs since they have little driving force for collagen hybridization. Because addition of hot CMP solution can damage tissues for ex vivo and in vivo applications, non-thermal means to control the triple helix folding and collagen binding have been developed: we developed a caged CMP with a photo-cleavable nitrobenzyl (NB) group attached to the central glycine [sequence: $(GPO)_4{}^{NB}GPO(GPO)_4$]. The NB cage group sterically prevented the CMP from folding into triple helix; yet removal of the cage group by UV irradiation immediately triggered the triple helical folding of the peptide and its hybridization with unfolded collagen strands [2] (Fig. a, right panel).

Here, we present an overview of the hybridization interactions of CMP with collagen strands produced by heat or by matrix metalloproteinase (MMP) digestion. We demonstrate that fluorescently labeled CMP allows direct detection of collagens in SDS-PAGE, as well as in processed tissue sections, even where collagen antibody fails to detect collagens. We further show that CMPs can be used in vivo to target and image denatured collagen in remodeling tissues associated with many pathologic conditions, such as tumor progression and musculoskeletal diseases (e.g., Marfan syndrome).

METHODS

Peptide Synthesis. CMPs and caged CMPs were synthesized using standard solid-phase chemistry, with the exception of the amino acid following ^{NB}Gly, which was conjugated by PyBroP. The peptides were labeled with carboxyfluorescein (CF) or near infrared fluorophores (NIRF), purified by HPLC, and analyzed by MALDI-ToF and circular dichroism spectroscopy.

SDS-PAGE staining. Heat denatured type I collagen protein (0.7 μg/lane) was resolved by SDS-PAGE under denaturating conditions. After electrophoresis, the gels were washed by deionized water briefly to remove the remaining SDS. PBS solutions containing 6 μM of fluorescently labeled CMP [CF(GPO)9] were heated to 85 °C for 10 min and immediately pipetted onto the gels. The gels were soaked in this CMP solutions for 3 h at room temperature, followed by washing with water. The stained gels were scanned using a Typhoon™ 9410 imager. Then the gels were further stained with coomassie brilliant blue (CB) followed by imaging.

Fluorescence Tissue Staining. Fixed mouse tissue slides were blocked by 10% v/v goat serum in 1×PBS for 30 min at room temperature. Fluorescent caged CMP CF^{NB}(GPO)9 was dissolved in 1×PBS solution containing anti-collagen I antibody (1:100 dilution). After blocking, CMP solutions (15 μM) were applied to the tissue sections and incubate for 10 min. The CMP folding and collagen binding were triggered by exposing the samples to UV light (~15 mW/cm^2) for 10 min. After irradiation, the tissue sections were incubated at 4 °C for over 1.5 h. The tissue slides were then washed by 1×PBS solutions and probed by an Alexafluor594 labeled secondary antibody (1:300) for 1 h. The slides were washed by PBS again before imaging by a Nikon Eclipse TE2000-E microscope.

In Vivo Targeting. Saline solution (100 μL) containing 3.7 nmol of NIRF labeled caged CMP [IRDye800-Ahx-^{NB}(GPO)9], activated by UV exposure (365 nm, > 25 mW/cm^2, 5 min) was immediately injected via lateral tail vein to a NOD/SCID mouse bearing prostate cancer tumors (PC-3) 5-7 mm in diameter behind forearms. Marfan mice were dosed with 4 nmol of UV-activated [IRDye680-Ahx-^{NB}(GPO)9] via tail vein as described. Images were acquired at designated times using a LI-COR Pearl Impulse Imager at 800 nm.

RESULTS

The image of the SDS-PAGE gel stained by heated CF(GPO)9 showed distinct fluorescent emission from all bands of type I collagen chains, whereas no protein band of the HUVEC cell lysate was visualized (Fig. b). This result demonstrates CMP's remarkable affinity and specificity for collagen hybridization in vitro [3].

Because heated CMP solutions may cause undesired tissue damage in ex vivo and in vivo studies, caged CMPs were developed which can be triggered by UV light to bind to collagen strands (Fig. a). We tested the ability of CMP to visualize collagen in processed tissue sections, which contain denatured collagen strands due to the chemical and thermal fixation processes. As shown in Fig. c, photo-triggered caged CF(GPO)9 effectively reveals the parallel organization of collagen fibrils in the stroma of a fixed mouse cornea section. The fluorescence signals from CF(GPO)9 overlap completely with the signals from the anti-collagen I antibody, confirming the high specificity of the CMP peptide. We also found that processing of the tissue seems to enhance the CMP binding. In particular, the mouse tibia bone sections that have undergone an acidic demineralization process as well as fixation and paraffin embedding exhibited strong CF(GPO)9 signals, whereas almost no collagen antibody signals was detected (Fig. c, lower panels). It is highly likely that structure of the epitopes for the antibody recognition has been compromised by the tissue preservation processes (e.g., paraffin embedding). This result shows that the CMP is superior to antibodies in detecting collagens in highly processed tissues since it recognizes the unfolded collagen chains that are prevalent in such samples.

We performed in vivo targeting experiments to examine the ability of CMP to target tissues with high collagen remodeling activities. NIRFs were attached to CMPs to allow live fluorescence imaging. The NIRF-labeled caged CMPs were rapidly decaged immediately before being injected into the mouse models. In vivo imaging of a mouse bearing PC-3 prostate tumors at forward flanks (circled) administered with the NIRF-CMP indicated stable and tumor specific CMP uptake for over four days post injection (Fig. d), while a labeled scrambled sequence lacking the triple-helix folding capacity showed little tumor accumulation (data not shown). Similarly, in a mouse model of Marfan syndrome, a genetic disorder of the ECM exhibiting marked skeletal pathology [4], strong CMP signal was detected at the spine and ribs of the mutant mouse with at least four times the intensity of the wild-type control mouse 96 h after NIRF-CMP injection (Fig. e).

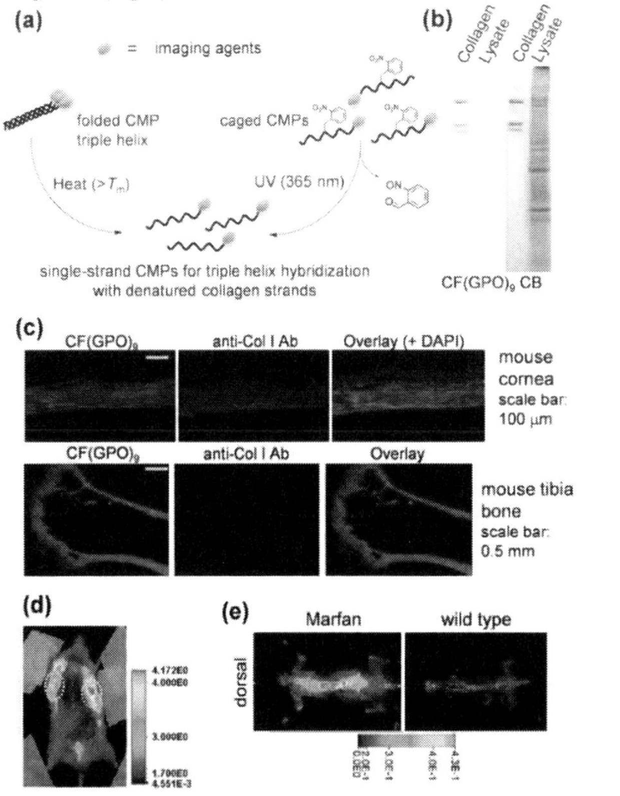

Figure. **(a)** Schematic illustration of two approaches (heat and UV activation) to generate single strand CMPs that hybridize with denatured collagen strands. **(b-e)** Representative images showing applications of CMP-collagen strand hybridization in vitro, ex vivo and in vivo.

DISCUSSION

To our knowledge, this is the first time that the structural remodeling of ECM has been directly interrogated by peptide hybridization. CMP is a simple peptide that can be readily conjugated to imaging and therapeutic moieties. Thus, CMP-mediated collagen strand targeting opens new avenues, beyond tumor and skeleton imaging, for applications in detecting and treatment of a wide range of pathologic conditions associated with collagen damage and remodeling, such as wound healing, fibrosis, as well as musculoskeletal and connective tissue injuries.

ACKNOWLEDGEMENTS

This work was supported by grants from NIAMS/NIH (R01-AR060484 and R21-AR065124) and DOD (W81XWH-12-1-0555).

REFERENCES

[1] Li, Y et al., *Curr Opin Chem Biol*, 17:968-975, 2013.
[2] Li, Y et al., *PNAS*, 109: 14767-14772, 2012
[3] Li, Y et al., *Bioconj Chem*, 24:9-16, 2013
[4] Judge, DP et al., *J Clin Invest* 114:172-181, 2004

SB^{3}C2015
Summer Biomechanics, Bioengineering and Biotransport Conference
June 17–20, 2015, Snowbird Resort, Utah, USA
SB^{3}C2015-195

A REDUCED-DIMENSIONAL TRANSPORT MODEL FOR THROMBOGENIC SPECIES IN LARGE ARTERIES

Kirk B. Hansen and Shawn C. Shadden

Department of Mechanical Engineering
University of California, Berkeley
Berkeley, CA, US

INTRODUCTION

In the absence of biomechanical platelet activation, the coagulation process is typically driven by the exposure of reactive species at the vessel wall. Although coagulation is a complex process involving dozens of reactive species, the *initiation* phase of thrombin formation can be modeled as a rate-limited linear first-order kinetic reaction [1]. Thrombogenic potential can thus be approximated as the surface concentration of a single species released from the vessel wall at a constant rate.

In large arteries, this transport problem is characterized by high Prandtl numbers (Pr $= \nu/D \approx 10^3$) and therefore thin concentration boundary layers (BL). As opposed to the flow problem, which is fully three-dimensional (3D), the transport problem is primarily two-dimensional (2D) and limited to a small near-wall region. This disparity suggests that a multiscale approach may be appropriate. Herein we develop a model that reduces the 3D transport problem to a 2D partial differential equation that can be solved on the manifold defining the vessel lumen surface of large arteries.

METHODS

Three-dimensional mass transport of a dilute species with uniform diffusivity is governed by

$$\frac{\partial c}{\partial t} = -\nabla \cdot (c\mathbf{u}) + D\Delta c, \tag{1}$$

where $c(\mathbf{x}, t)$ is concentration, D is diffusivity, and $\mathbf{u}(\mathbf{x}, t)$ is velocity. We now choose an intrinsic, orthonormal reference frame such that $\hat{t}$, $\hat{n}$, and $\hat{b}$ are the tangent, inward normal, and binormal directions relative to the surface, respectively, and $\hat{n} = 0$ corresponds to the location of the wall. In order to reduce the 3D transport problem to the 2D surface, both sides of (1) are integrated with respect to $\hat{n}$:

$$\int_0^\infty \frac{\partial c}{\partial t}\, d\hat{n} = -\int_0^\infty \nabla \cdot (c\mathbf{u})\, d\hat{n} + \int_0^\infty D\Delta c\, d\hat{n}. \tag{2}$$

Assuming fixed walls and that the vessel radius is much larger than the concentration BL, and defining

$$\phi(\hat{t}, \hat{b}, t) = \int_0^\infty c(\hat{t}, \hat{b}, \hat{n}, t)\, d\hat{n}, \tag{3}$$

it can be shown that (2) becomes

$$\frac{\partial \phi}{\partial t} = -\nabla_\Gamma \cdot \int_0^\infty c\mathbf{u}_\Gamma\, d\hat{n} + D\Delta_\Gamma \phi + q, \tag{4}$$

where q is the flux at the wall. This is the transport equation for the new 2D surface variable ϕ.

In order to calculate ϕ based on surface measures alone, $\mathbf{u}$ and c must be decoupled from their $\hat{n}$ dependencies. Assuming the concentration BL is much smaller than the momentum BL (valid for Pr $\gg 1$), we can make the approximation that

$$\mathbf{u}(\hat{t}, \hat{b}, \hat{n}, t) = \frac{\boldsymbol{\tau}_w(\hat{t}, \hat{b}, t)\hat{n}}{\mu}, \tag{5}$$

where $\boldsymbol{\tau}_w(\hat{t}, \hat{b}, t)$ is wall shear stress (WSS) and μ is dynamic viscosity. The transport equation then becomes

$$\frac{\partial \phi}{\partial t} = -\nabla_\Gamma \cdot \left(\frac{\boldsymbol{\tau}_w}{\mu} \int_0^\infty c\hat{n}\, d\hat{n}\right) + D\Delta_\Gamma \phi + q. \tag{6}$$

To decouple the integral of $c\hat{n}$ we must assume some functional profile for $c(\hat{n})$. As one approximation, c can take the form of the solution to the two-dimensional, steady, simple shear problem. In this case it can be shown that

$$\int_0^\infty c\hat{n}\, d\hat{n} = \beta\phi^{3/2}, \tag{7}$$

where $\beta = 0.5868(D/q)^{1/2}/\mu$. Substituting (7) into (6) gives the 2D surface transport equation,

$$\frac{\partial \phi}{\partial t} = -\beta\nabla_\Gamma \cdot \left(\boldsymbol{\tau}_w \phi^{3/2}\right) + D\Delta_\Gamma \phi + q. \tag{8}$$

This model assumes that all transport is parallel to the wall and that mass never leaves the BL region. This approximation breaks down in flow separation regions, where there is a significant wall-normal transport component. To account for this, the contribution to the advective term from negative WSS divergence was removed.

To model thrombogenic potential in diseased vasculature, a stabilized finite element method was used to solve (8) in patient-specific models of abdominal aortic aneurysms (AAA). The WSS field was obtained using image-based CFD as described in [2]. A value of $D = 5 \times 10^{-6}\,\mathrm{cm}^2/\mathrm{s}$ was chosen to represent a typical value for the biochemical species involved in the coagulation process. The source term was set to $q = 1$ in the aneurysmal region, and $q = 0$ elsewhere. These results were compared to those obtained by solving (1) in the entire 3D domain, as well as to the WSS-based near-wall particle residence time metric [3].

RESULTS

Representative time-averaged results in a patient-specific AAA model are shown in Fig. 1. For this patient, the 2D model was solved on a 38k element triangular mesh, while the 3D model was solved on a 9.5M element tetrahedral mesh. The results obtained using the 2D and 3D models demonstrate similar topological features. While some of these features are also seen in the near-wall particle residence time results, there are also some regions that appear quite different, especially on the anterior side.

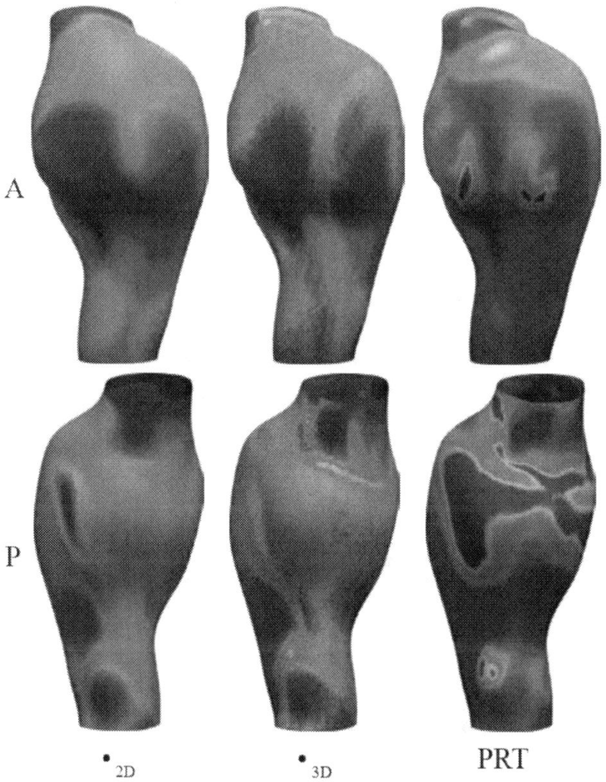

FIGURE 1: NORMALIZED COMPARISON OF TIME-AVERAGED THROMBOGENIC POTENTIAL CALCULATED USING 2D MANIFOLD MODEL (ϕ_{2D}), FULL 3D SIMULATION (ϕ_{3D}), AND NEAR-WALL PARTICLE RESIDENCE TIME (PRT) ON THE ANTERIOR (A) AND POSTERIOR (P) SURFACES OF A PATIENT-SPECIFIC ABDOMINAL AORTIC ANEURYSM MODEL.

DISCUSSION

In this work, a multiscale approximation was used to reduce the 3D transport equations in large arteries to a 2D manifold model. Because this model provides a direct link between WSS and near-wall transport, it can be used to help elucidate the nature of this dependence. The results presented herein show that previous attempts to describe the relationship between WSS and transport, e.g. the development of metrics such as near-wall particle residence time, may not adequately describe the conditions that promote stasis-driven coagulation.

This model has the potential to significantly reduce computational cost in solving these types of transport problems. For the presented example, the computation time for the 3D model to reach periodic steady state exceeded 800 processor hours, while the time for the 2D model was less than 0.5 processor hours. This is despite the fact that mesh convergence studies showed that the 3D model likely still required additional mesh resolution near the vessel walls to properly resolve the thin boundary layers present. The vast majority of patient-specific blood flow simulations are performed below the resolution required to understand near-wall transport, and previous work on this problem has typically been limited to simplified, 2D geometries [4]. The use of a multiscale method such as presented herein allows for the investigation of transport effects in realistic, 3D, patient-specific geometries.

In future work, the presented formulation for near-wall transport will be incorporated into a complete continuum model of thrombus formation. This model will include features such as WSS-dependent platelet adhesion rates and reactions between the specific biochemicals involved in the coagulation cascade. Such a surface model is incapable of discerning the influence of a growing thrombus on the flow field, which is an important consideration in studying thrombus formation in the smaller arteries. It should prove useful, however, in the study of thrombus formation in patient-specific large arteries, where the multiple spatial and temporal scales involved render previous mathematical models computationally prohibitive.

ACKNOWLEDGMENTS

This work was supported by the NSF, award 1150947, and by the NIH NHLBI, award R21HL108272.

REFERENCES

[1] Papadopoulos, K et al., *Med Eng Phys*, 36(2):196–204, 2014.
[2] Les, AS et al., *Ann Biomed Eng*, 38(4):1288–1313, 2010.
[3] Himburg, HA et al., *Heart Circ Phys*, 286(5):H1916–H1922, 2004.
[4] Biasetti, J et al., *Frontier Phys*, 3:266, 2012.

SB^{3}C2015
Summer Biomechanics, Bioengineering and Biotransport Conference
June 17-20, 2015,Snowbird Resort, Utah, USA
SB^{3}C2015-196

DIFFERENCES IN THE INTRADISCAL PRESSURE IN THE L4-L5 AND L5-S1 LUMBAR SEGMENTS

H.E. Jaramillo (1,4), C.M. Puttlitz (2), K.C. McGilvray (3), J.J. García (1)

(1) Escuela de Ingeniería Civil y Geomática
Facultad de Ingeniería
Universidad del Valle
Cali, Valle, Colombia

(2) Engineering and School of Biomedical Engineering
Colorado State University
Fort Collins, CO, USA

(3) Orthopaedic Bioengineering Research Laboratory
Colorado State University
Fort Collins, CO, USA

(4) Departamento de Energética y Mecánica
Facultad de Ingeniería
Universidad Autónoma de Occidente
Cali, Valle, Colombia

INTRODUCTION

Intradiscal pressure (IDP) is been considered to be one of the most important parameters used to evaluate disc function and to estimate the stress in the annulus fibrosus [1]. Various studies have been carried out to obtain the IDP using *in vivo* experiments [2-4] and *in vitro* tests of intact L4-L5 functional spine segments [2]. The IDP is frequently used to validate numerical models [3,4]. In order to take into account the contribution of each component of the segment in the IDP behavior, Heuer et al. [1,5–7] developed the stepwise reduction method (SRM) and applied it to the L4-L5 segment. However, this method has not been applied to the lumbosacral junction (L5-S1 segment). Hence, the objective of this *in vitro* study was to determine the IDP in the L5-S1 segment using the SRM, and compare the IDP between the L4-L5 and L5-S1 segments.

METHODS

Seven (n=7) lumbosacral segments (L4-L5-S1) were harvested from frozen human donor spines (median: 68 years, range: 53-84 years, SD=12.5 years). The specimens were stored at -20° C, thawed at 8° C for 8 hours before the tests and kept moist with periodic 0.9% saline sprays during the tests. For each sample, the entire experimental protocol was conducted during two nonconsecutive days. In the first day, the whole L4-L5-S1 unit was tested, where the L4-L5 segment was sequentially reduced using SRM while the L5-S1 segment remained intact. In the second day, the SRM was applied to the L5-S1 segment. To support the specimens and transmit the load, the cranial face of vertebra L4 and the vertebral body of S1 were embedded in poly-methylmethacrylate (PMMA). At the end of the first day, the L4 vertebrae and L4-L5 intervertebral disc were removed and the cranial face of the L5 was embedded in PMMA.

The testing apparatus consisted of a servomotor actuator (Danaher Controls, Gurnee, IL) that applies pure moments in a specified direction [8]. Four sinusoidal moment cycles were applied continuously up to of 8.5 N-m in flexion, extension, lateral flexion, and axial rotation. A standard torque cell (0-22.6 N-m range, TRT-200 model, Transducer Techniques Co, Temecula, CA, USA) was used to measure the moment and the IDP was measured using a miniature pressure transducer (Precision Measurement Co., MI, USA) of 1.5 mm diameter and 0.3 mm thickness, which was inserted into the center of the nucleus and its position was verified by radiographic observation.

The structures of the specimens were successively removed starting from the posterior aspect of the segments using the SRM and following the protocol proposed by Heuer et al [1,5,6]. First the sample was tested in the intact condition (INTACT), next the supraspinous ligament (Wout_SSL) was cut and the segment was tested. This procedure was repeated after cutting the interspinous ligament (Wout_ISL), the flavum ligament (Wout_FL), the facet capsules (Wout_FC), the vertebral arches (Wout_VA), the posterior longitudinal ligament (Wout_PLL), the anterior longitudinal ligament (Wout_ALL), and finally, after removing the nucleus pulposus (Wout_NUC).

Data were statistically evaluated for each anatomical reduction stage using Minitab software (State College, PA, USA). As the data were non-normally distributed, a non-parametric test was used. If the p-value in the Friedman test was less than 0.05, then a Wilcoxon's U-test (signed rank test) was used to delineate statistically significant paired differences between two consecutives reduction stages and between the L4-L5 and L5-S1 segments. The correlation factor (R) was calculated in order to know if the IDP increases with the moment.

RESULTS

With respect to the IDP between successive stages, significant differences were detected for some reduction stages. For example, at the L4-L5 segment, there were significant differences in the IDP for two reduction stages under the 2 N-m in flexion, while for L5-S1 segment, significant differences were detected for two reduction stages under the 2 N-m and 6 N-m moments. This trend was similar for the other bending moments, for brevity we present a median of the IDP from all reduced stages (Fig. 1, 2).

The median IDP for the L5-S1 segment was 1.7 (p=0.005) and 1.4 times (p=0.034) higher than that obtained for the L4-L5 segment in extension (Fig. 1) and axial rotation (Fig. 2), respectively. No significant differences between segments were found for the other two movements.

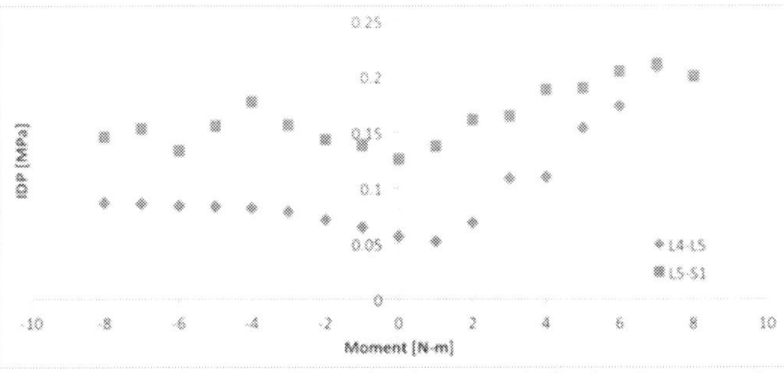

Fig. 1. IDP vs. Moment. Extension (moment -), flexion (moment +).

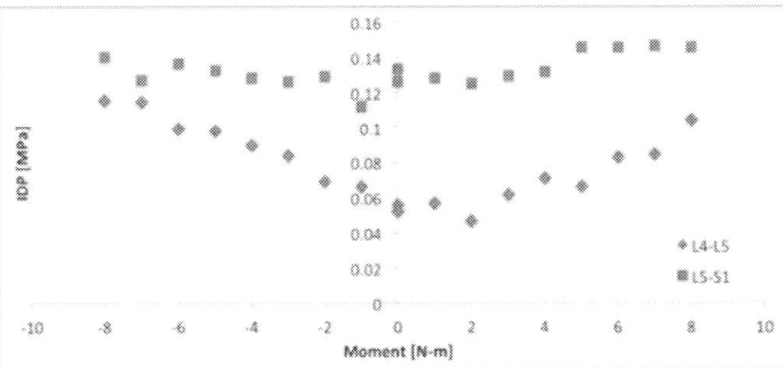

Fig. 2. IDP vs. Moment. Lateral flexion (moment -), axial rotation (moment +).

For the L4-L5 segment under flexion in the INTACT, Wout_SSl, Wout_ISL, Wout_FL, and Wout_FC stages, the IDP (Fig. 3) aproximated previously reported values and was 30% lower for the Wout_VA, Wout_PLL and Wout_ALL stages [1]. In extension, the IDP (Fig. 3) for all reduction stages was about 1.6 to 4.6 lower than that reported by Heuer et al. [1]. No comparison was possible for other movements since Heuer et al. [1] did not report IDP as a function of reduction stage for lateral bending and axial rotation.

For the median of all reduction stages and the L4-L5 segment, an increase of moment from 2 N-m to 8 N-m caused increases in the IDP from 0.07 MPa to 0.20 MPa in flexion (R=0.98), from 0.07 MPa to 0.09MPa in extension (R=0.93), from 0.06 MPa to 0.12 MPa in lateral bending (R=0.99) and, from 0.05 MPa to 0.10 MPa in axial rotation (R=0.92). For the median of all reduction stages at the L5-S1 segment, an increase of moment from 2 N-m to 8 N-m produced increases in the IDP from 0.16 MPa to 0.20 MPa in flexion (R=0.95), and from 0.16 MPa to 0.20 MPa in axial rotation (R=0.91), whereas it remained approximately constant in extension (R=0.31) and lateral bending (R=0.53).

DISCUSSION

The IDP differences found between the L4-L5 and L5-S1 segments in extension and axial rotation, which may be attributed to the anatomical differences between the segments, demonstrate that lumbar spine mechanics data cannot be reliably extrapolated to the lumbosacral segment and that the distinct studies of the lumbosacral region are required in order to improve the characterization of the L5-S1 segment.

We did not find a strong correlation between the increase of the IDP and the anatomical reduction stage, which is consistent with lumbar data previously reported by Heuer et al. [1]. Some differences in the IDP with respect to previous studies may be due to the experimental protocol, i.e. while in the study [1] the pure moment was applied and maintained step by step, our protocol applied the moment continuously. We hypothesize that when the moment is applied continuously, the intradiscal pressure does not have the opportunity to fully relax to an equilibrium state before the measurement is obtained. Our IDP values for both the L4-L5 and L5-S1 segments under axial rotation approximate the previously reported predictions obtained with a finite element model [9].

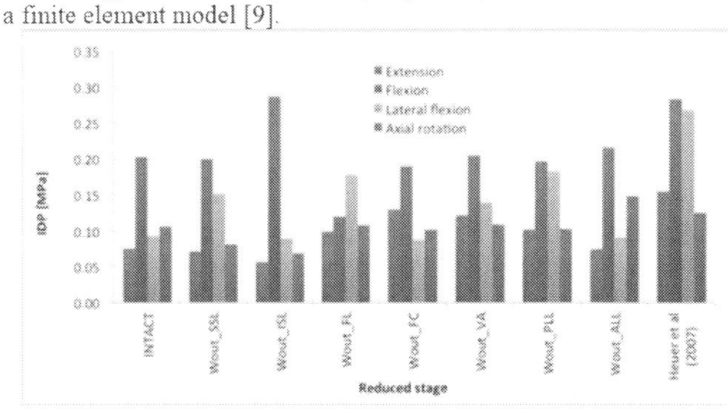

Fig. 3. IDP for all reduced stages of the L4-L5 segment at 8 N-m compared with median results obtained by Heuer et al [1].

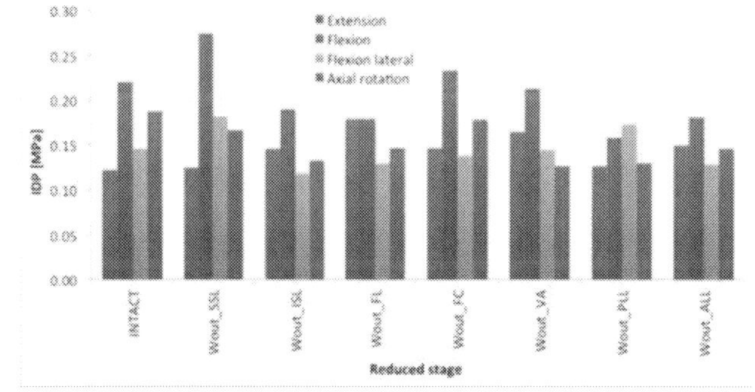

Fig. 4. IDP for all reduced stages of the L5-S1 segment and 8 N-m.

ACKNOWLEDGEMENTS

The authors appreciate the support of the Universidad del Valle, Universidad Autónoma de Occidente, and Colorado State University to undertake this study.

REFERENCES

[1] Heuer F. et al., *J Biomech*, 40:795–803, 2007.
[2] Brinckmann P. et al., *Spine*,16:641–6, 1991.
[3] Goto K. et al., *J Orthop Sci*, 7:243–6, 2002.
[4] Shirazi-Adl SA. et al., *Spine*, 9:120–34, 1984.
[5] Heuer F. et al., *J Biomech*, 41:1953–60, 2008.
[6] Heuer F. et al., *J Biomech*, 40:271–80, 2007.
[7] Noailly J. et al., J Biomech, 40:2414–25, 2007.
[8] Leahy D. et al., *J Biomech*, 45:2668–72, 2012.
[9] Shirazi-Adl A. *J Biomech*, 27:289–99, 1994.

A POROELASTIC FLUID/STRUCTURE-INTERACTION MODEL OF THE SPINAL CORD AND SURROUNDING STRUCTURES WITH A CORD SYRINX AND ASSOCIATED STENOSIS OF THE SUBARACHNOID SPACE

Christopher D. Bertram (1), Matthias Heil (2)

(1) School of Mathematics and Statistics
University of Sydney
Sydney, New South Wales, Australia

(2) School of Mathematics
University of Manchester
Manchester, UK

INTRODUCTION

The causes of the neurological condition syringomyelia remain obscure, despite much research in recent years. Syringomyelia is defined by the existence of one or more macroscopic fluid-filled cavities in the spinal cord. The vast majority of cases are associated with either congenital anatomical abnormality or past spinal trauma. The fluid is thought to originate in the subarachnoid space (SAS) as cerebrospinal fluid (CSF), but may possibly be derived from interstitial fluid (ISF). Variable symptoms include pain and loss of sensory or neuromotor transmission. Surgical intervention to drain the syrinx or to relieve attendant CSF flow constrictions has a relatively poor long-term prognosis.

Bioengineers have contributed to the effort to understand and improve treatment of this illness through measurements of mechanical properties of tissues, and simulation of the tissue/fluid system consisting of the cord and its surrounding structures, usually *in silico* but occasionally in hardware. Over time, the neurosurgical community has evolved many hypotheses to try to explain the condition, and a principal contribution of engineers has been to subject these to rigorous scrutiny for their physical feasibility or otherwise [1].

As part of that effort, an idealised axisymmetric fluid/structure-interaction (FSI) model of the cord, SAS and dura has been developed. The model has been used to define the available modes of wave travel and CSF bulk motion [2], to explore a new syringogenesis hypothesis involving cord tethering [3], and to define the syrinx fluid motions and resulting cord stresses associated with coughing [4], especially when a syrinx occurs with SAS stenosis [5]. However, until now, the model has used impermeable (visco)elastic solids; fluid seepage was therefore excluded. We have now developed a poroelastic version which allows examination of possible syrinx volume change and other porous-medium phenomena. We set out to test the hypothesis (developed from observation of results from the impermeable model) that pressure gradients across the annular cord and pia over the syrinx would favour syrinx filling if the cord material was porous.

METHODS

The model includes geometrically idealized representations of the spinal cord and its enclosing *pia mater* membrane, the *filum terminale* ligament anchoring the cord sacrally, and the tough *dura mater* membrane which encloses the CSF around the cord. A substantial syrinx at the cervicothoracic boundary is included, and most of the SAS is obstructed locally by an adjacent partial blockage. The system is excited by idealized pulse waves at the cranial end of the cervical SAS. The Navier-Stokes equations are integrated over time for the two fluid spaces (SAS and syrinx), and the equations of mass and momentum conservation, and a constitutive equation, are solved for both the poroelastic medium and the pore fluid in each of the solids. FSI boundary conditions match the stresses and velocities at the interfaces between the poroelastic media and the bulk fluids. The model is implemented in open-source finite-element software (www.oomph-lib.org) [6].

Besides the parameters defined in the impermeable version of the model [5], it is necessary to specify the cord porosity (0.3), the drained Poisson ratio for the porous matrix (0.35), the Biot-Willis parameter for a fully saturated medium ($\alpha = 1$), and the permeability k. We examined values of $k = 10^{-12}$, 10^{-13}, and 10^{-14} m^2, both when permeability was restricted to the annular cord and pia over the syrinx and when the whole cord and pia were porous We also compared a model allowing Darcy flow but no other permeate interactions with the solids ($\alpha = 0$) and an impermeable version ($k = 0$). The latter two models reverted to the Poisson ratio 0.49 used previously [5]. Unlike previous models, the solids here were perfectly elastic except insofar as poroelastic behaviour mimics the effect of viscoelasticity. However viscoelastic effects are important here largely only in damping short-wavelength transients, which were not employed.

RESULTS

The gross behaviour of the model was unchanged by the addition of poroelasticity. Both deformations and pressures were similar to those seen in the impermeable model when excited cranially by sine waves of frequency 2.5 Hz. Thus the previously observed elevation of SAS mean pressure caudal to the (flexible) stenosis was preserved, and syrinx mean pressure continued to be elevated above mean cranial SAS pressure about one-third as much as caudal SAS pressure. Although the stenosis occluded 90% of the SAS area locally, the axial CSF flow through the stenotic gap was much larger than the parallel shunt flow via the syrinx (passing through the annular cord and pia and back again).

Since the model does not allow for remodeling in response to altered stresses, any porous-seepage tendency to favour further syrinx filling beyond the initial volume, or to drain it, leads from the starting condition (via a transient which may extend over many cycles of excitation) to an equilibrium in which altered syrinx mean pressure is again balanced by local solid stresses, as in the impermeable model. But the new porous-media equilibrium includes a possibly changed syrinx mean volume, and continuous (cycle-average) seepage from place to place within the model. We found that the changes in syrinx mean volume were always small. When permeate interaction with the porous media was turned off ($\alpha = 0$), mean syrinx volume increased by on the order of 0.01%. With full poroelasticity, mean syrinx volume change was yet smaller. There was a well-defined slow steady streaming of fluid round a circuit involving the stenotic gap, radially across the pia and cord over the syrinx, and back via the syrinx.

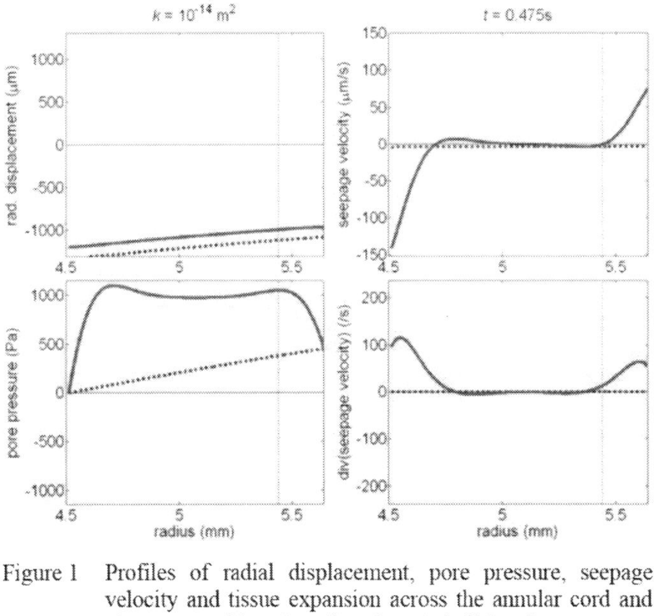

Figure 1 Profiles of radial displacement, pore pressure, seepage velocity and tissue expansion across the annular cord and pia, at $\alpha = 0$ (blue, dotted) and $\alpha = 1$ (solid, red).

The pulsatile seepage of fluid radially out of the SAS into the pia over the syrinx equaled that out of the annular segment of cord into the syrinx when $\alpha = 0$ (Darcy only), but with $\alpha = 1$ (full poroelasticity), most of the (now amplified) SAS outflow was absorbed in radial swelling of the annular cord, such that the syrinx received less pulsatile seepage flow than at $\alpha = 0$. The cyclic swelling of the tissue was confined to an oscillatory porous-medium boundary layer near each FSI boundary, as exemplified in Fig. 1. This shows one instant from a simulation focusing on the annular cord and pia over the syrinx, in which SAS pressure was varied sinusoidally and syrinx pressure was held at zero. At $\alpha = 0$ the pore pressure varies linearly between these limits, but at $\alpha = 1$, cyclic tissue swelling causes marked deviation. At this value of k and frequency of excitation, the porous-medium boundary layer is about 0.5 mm thick; its width increases with k. The lower right panel shows the divergence of the radial seepage velocity, a measure of the rate at which the tissue is imbibing or giving up fluid locally. In Fig. 1 the pia ($5.44 \leq r \leq 5.64$ mm) has the same stiffness as the cord. When the pia is given its true stiffness (250 times that of the cord), swelling is inhibited therein, and the SAS-side boundary layer is displaced inside the pia. The stiff pia also greatly reduces the magnitude of radial displacement and cyclic swelling everywhere.

DISCUSSION

The main purpose of this investigation was to test the hypothesis that the cycle-average pressure gradients set up by the flexible SAS stenosis would favour pore fluid transport from the SAS to the syrinx, causing a slight increase in (cycle-average) syrinx volume relative to the starting condition. Our results show that the hypothesis, essentially based on consideration of Darcy flow only ($\alpha = 0$), underestimated the complexity of the processes occurring in a poroelastic medium in unsteady flow, and in particular neglected the importance of cyclic pore fluid retention within the porous matrix. The several interacting processes simultaneously in play in this model combine to create a situation which is too complex for the outcomes to be predicted by thought alone; only the detailed computations are reliable. Specifically, our results indicate that the proposed mechanism of syrinx filling is unlikely, at least with the stenosis in its current position and in the absence of other factors such as spatially varying permeability which is not modeled here. However, the cyclic tissue swelling has major implications for mixing of CSF (and syrinx fluid) with ISF in poroelastic tissue boundary layers. This mechanism, based on bulk tissue properties only (i.e. in the absence of the defined conduits of the Virchow-Robin perivascular spaces) may explain the apparently conflicting empirical observations from different groups of both cord tissue ingress of CSF tracers and egress of ISF tracers to the SAS. In turn this suggests that the usually rather rigid distinction between ISF and CSF maintained in discussion about central-nervous-system fluids may not be realistic, for either the cord or the brain.

ACKNOWLEDGEMENTS

The porous model development was made possible by a grant from the Chiari and Syringomyelia Foundation. Matthew J. Russell (student supervised by MH) contributed to the initial implementation of the model.

REFERENCES

[1] Elliott, NSJ et al., *J Fluids & Structures*, 40:1-24, 2013.

[2] Bertram, CD et al., *J Biomech Eng*, 127:1099-1109, 2005.

[3] Bertram, CD et al., *Med Biol Eng & Comput*, 46:701-707, 2008.

[4] Bertram, CD, *J Fluids & Structures*, 25:1189-1205, 2009.

[5] Bertram, CD, *J Biomech Eng*, 132:061009, 2010.

[6] Heil, M, Hazel, AL, *Fluid-Structure Interaction*, eds. Schafer, M, Bungartz, H-J, pp. 19-49, Springer, 2006.

SB³C2015
Summer Biomechanics, Bioengineering and Biotransport Conference
June 17-20, 2015,Snowbird Resort, Utah, USA
SB³C2015-198

A NEW PARADIGM FOR THE CONTRIBUTION OF ACTIVE TENSION TO THE CONSTITUTIVE RELATION IN SMALL LYMPHATIC VESSELS

Christopher D. Bertram (1), Charlie Macaskill (1), Michael J. Davis (2), James E. Moore Jr (3)

(1) School of Mathematics & Statistics
University of Sydney
Sydney, New South Wales, Australia

(2) Department of Medical Pharmacology & Physiology
University of Missouri School of Medicine
Columbia, MO, USA

(3) Department of Bioengineering
Imperial College of Science, Technology & Medicine
London, UK

INTRODUCTION

Models of biomechanical systems generally involve the simultaneous solution of equations of mass and momentum conservation along with a constitutive relation. The presence of actively contracting muscle in heart, the musculoskeletal system, and the walls of body conduit systems, brings unique problems in the formulation of a constitutive relation. A long-established paradigm is to treat the active contribution as additive to the passive mechanical properties. Experimentally this leads to the evaluation of the active contribution as the difference between two measured relations, one with full muscle activation and one with muscle deactivated. Usually for a conduit such as a blood vessel, an airway, or a lymphatic, the active tension is thus derived as a function of diameter.

This model has been successfully adapted from skeletal muscle and myocardium to the description of length/active-tension relations in bovine mesenteric lymphatic vessels [1]. However, our experimental data from small lymphatic vessels in the rat mesentery suggest that it may not be adequate to deal with their contractile properties, especially at shorter sarcomere lengths (smaller diameters).

We have followed an alternative path as used in blood vessels by Cox [2]: the derivation of the active contribution to diameter reduction as a function of distending pressure. However, other than in prescribing the time-course of active tension, we work with the total length/active-tension relation $D = D(p_{tm}, t)$ that combines active and passive properties (D = diameter, p_{tm} = transmural pressure, t = time). Our measurements of spontaneous twitch contractions show that this time-varying relation is a monotonic function, whereas the corresponding constitutive relation expressing p_{tm} as a function of D is in general not monotonic. Thus there is no one-to-one relation between D and p_{tm}; sometimes there are multiple values of p_{tm} corresponding to a single D.

To avoid this ambiguity, after each integration step forward in D we can evaluate p_{tm} iteratively from $D(p_{tm})$, subject to the constraint that the new (p_{tm}, D) root be the closest to that at the last time-step. We also tried integrating dp_{tm}/dt and then finding $D(p_{tm})$, instead of integrating dD/dt and hence finding $p_{tm}(D)$. However, at turning points of the $D(p_{tm})$ relation, dp_{tm}/dt becomes infinite, which can lead to erroneous solutions by either procedure when integrating forward in time. We have minimised this, by integrating with respect to arc length instead of time.

METHODS

The twitch contractions were recorded in experiments on isolated segments of rat mesenteric lymphatic vessel, following long-established protocols [3]. At each of a series of pressures from 0 to 60 cmH_2O, maintained by servo control of pressure in reservoirs at each end of the cannulated segment, multiple spontaneous twitch contractions were recorded by video. The video record was subsequently analysed for segment diameter as a function of time. Twitch contraction amplitude in terms of diameter reduction was well repeated across many twitches. The diameter reductions vs. pressure were fitted by a semi-analytical curve, and the diameters vs. pressure at the start of contraction were likewise fitted. It was necessary to extrapolate the data to negative pressures where diameter could not be measured. To maximize the fidelity to the original data, the initial analytical fits were adjusted and the results stored as a pair of look-up tables which could be interpolated for data at any desired pressure.

The time-course of contraction onset and decay was imposed as a proportional variation of the twitch amplitudes, yielding an instantaneous curve of total $D(p)$ incorporating an active contribution which varied smoothly from zero to its maximum and back again to zero. This constitutive relation was solved along with equations of conservation of mass and momentum in our otherwise pre-existing model of a single lymphatic segment or lymphangion including two one-way valves [4].

RESULTS

Twitch contractions start from a diameter which is less than the maximally dilated diameter for that pressure (Fig. 1). They reduce diameter to a pressure-dependent value which is greater than the minimum which can be achieved by pharmacological intervention.

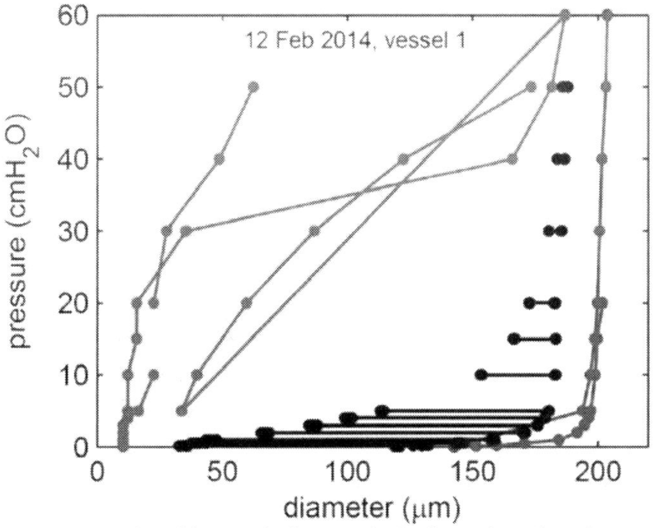

Figure 1 Passive (blue) and pharmacologically activated (red), pre-twitch and peak-twitch p/D data-points (black), at a series of steady pressures. Red/blue lines show sequence, but were not used. Twitch amplitude: horizontal black lines.

The pre-twitch and peak-twitch p/D data were ultimately reduced to a pair of curves bounding the locus of actual contraction (Fig. 2).

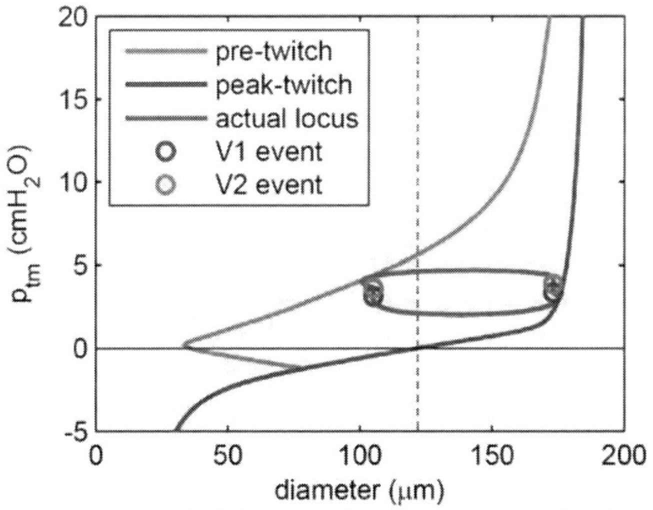

Figure 2 The anti-clockwise loop of cyclic contraction (red) and the pre-twitch (blue) and peak-twitch (green) p/D relations constructed from the experimental data. Circles: the instants of opening and closure of the two valves.

Integration with respect to time was successful for the majority of operating points. However, for certain combinations of inlet, outlet and external pressure, particularly those which caused the locus to dwell near the 'nose' of the instantaneous p_{tm}/D relation, a smooth solution could not be maintained; instead the solution jumped discontinuously down to the passive curve, eventually reacquiring the correct solution path. For these, the alternative mathematical technique of integration with respect to (modified) arc length on the p_{tm}/D locus was successfully adopted, taking advantage of the relatively simple form of the instantaneous p_{tm}/D relation, $D = g_1(p_{tm}) - f(t).g_2(p_{tm})$, where $f(t)$ is a known analytical function of time (g_1 and $g_1 - g_2$ are the blue and green curves respectively in Fig. 2). An increment of arc length ds is defined as

$$ds = dt\sqrt{\left(\frac{dD}{dt}\right)^2 + \left(\frac{dh_{tm}}{dt}\right)^2},$$

where $h_{tm} = p_{tm}/\alpha$ is p_{tm} scaled so as to yield approximately equal increments on the x- and y-axes. An equation in dh_{tm}/ds is integrated with respect to s, instead of one for dD/dt with respect to t. To reconstruct t, dt/ds is also integrated in parallel. The technique has the disadvantage that the end-time cannot be stipulated in advance, since dt varies, becoming very small at certain sites around the p_{tm}/D locus.

DISCUSSION

Published data on the relation between muscle length and active tension for lymphatic muscle are scarce. Up to now our model of a lymphatic vessel has utilized a length/active-tension relation that was an analytical approximation based on generalized muscle properties and semi-qualitative deductions from lymphatic properties and comparison of model behaviour. For the first time our model can now incorporate experimental data from an individual lymphatic segment in respect of both passive and active mechanical properties. The measured active properties for rat mesenteric lymphatic segments indicate that the p/D relation covers a different D-range when muscle is active, and thus that the active contribution to tension cannot be derived by subtracting activated and passive curves at corresponding values of D. We have developed an alternative formulation that allows for this behaviour. In a cardiac context, the new methodology allows contraction to reduce volume below the volume intercept V_0 about which the time-varying elastance pivots (the dotted vertical line in Fig. 2 indicates the equivalent D here). A minority of operating points proved difficult to handle, but these have been dealt with by moving to integration with respect to arc length instead of time. The methodology is applicable to areas of physiology beyond lymphatic vessels if the measured properties of the contractile system so warrant.

ACKNOWLEDGEMENTS

MJD's laboratory was supported by NIH grant R01-HL-120867, and JEM's by grant R01-HL-094269. All authors acknowledge newly awarded NIH grant U01-HL-123420. JEM gratefully acknowledges support from the Royal Academy of Engineering and a Royal Society-Wolfson Research Merit Award.

REFERENCES

[1] Ohhashi, T et al., *Am J Physiol*, 239:H88-H95, 1980.
[2] Cox, RH, *J Biomech*, 5:135-152, 1972.
[3] Davis, MJ et al., *J Physiol*, 587:165-182, 2009.
[4] Bertram, CD et al., *Biom Model Mechanobiol*, 13:401-416, 2014.

SB^{3}C2015
Summer Biomechanics, Bioengineering and Biotransport Conference
June 17-20, 2015,Snowbird Resort, Utah, USA
SB³C2015-199

CANCER-ASSOCIATED FIBROBLASTS PROMOTE VASCULARIZATION IN COLLAGEN AND FIBRIN MATRICES

M.K. Sewell-Loftin (1), Samantha van Hove (2), Gregory Longmore (2,3), Steven C. George (1)

(1) Department of Biomedical Engineering
Washington University in St. Louis
St. Louis, MO, USA

(2) Department of Cell Biology and Physiology
Washington University in St. Louis
St. Louis, MO, USA

(3) Department of Medicine, Oncology Division
Washington University in St. Louis
St. Louis, MO, USA

INTRODUCTION

Improving our understanding of the micromechanics of the peritumoral environment could lead to novel therapeutic strategies. Currently, anti-angiogenic therapies that target the developing blood vessels near tumors demonstrate limited effectiveness against certain types of cancers. This lack of efficiency could be due in part to our limited understanding of biomechanical forces that regulate angiogenesis near tumors. Carcinoma- or cancer-associated fibroblasts (CAFs) have been shown to play a crucial role in the regulation of tumor development, spread, and response to therapies through secretion of a variety of signaling factors including VEGF and SDF-1 (1). Recent work has demonstrated that CAFs are mechanosensitive, probing the biomechanical environment of the ECM through the YAP pathway to modulate cell phenotype (2). While it is well-known that CAFs display a more myofibroblast-like phenotype, including increased expression of alpha smooth muscle actin (αSMA), compared to noncancerous fibroblasts, studies have demonstrated no appreciable genetic differences in these cells within a single patient (3). This observation implicates environmental regulation in the control of the CAF phenotype, and therefore also increased angiogenesis and ultimately metastasis. The origin of CAFs remains poorly understood, and it is likely that several different mechanisms can lead to their formation. Of significant interest is epithelial-to-mesenchymal transformation (EMT) which can be initiated by active mechanical forces (4, 5). Mechanical forces are also sufficient to induce increased blood vessel growth in developing avian models (6). Interestingly, as the process that gives rise to tip cells during angiogenesis mimics an EMT-like transformation; thus, forces generated by surrounding cells may drive blood vessel formation in the peritumoral environment. Our global hypothesis is that mechanical forces generated by CAFs in the peritumoral environment actively initiate and regulate tumor-associated angiogenesis. Initial studies demonstrated the enhanced ability of CAFs to support vascularization of a 3D tissue model when compared to normal fibroblasts. Future experiments are planned to determine αSMA expression and YAP regulation.

METHODS

We employed an angiogenesis assay consisting of a fibrin gel embedded with endothelial cells (ECs) and fibroblasts. We have previously used this assay to demonstrate the formation of robust interconnected vascular networks in 3D when normal lung fibroblasts (NHLFs) are used as the necessary stromal cell (7, 8). For this study, fibroblasts used were either human normal breast fibroblasts (NBFs) or human breast cancer-associated fibroblasts (CAFs) with NHLFs acting as a positive control (9). Briefly, a 2:1 mixture of fibroblasts and ECs were mixed into either a 5 mg/mL fibrin or a combination collagen (2.5 mg/mL) and fibrin (5 mg/mL) gel before addition of thrombin. Media was replaced every 2 days. After 7 days, gels were fixed and stained for CD31 to identify vasculature. Gels were then analyzed for blood vessel network formation.

RESULTS

As shown in Figure 1, CAFs support the formation of longer and more numerous blood vessels in both fibrin and collagen/fibrin combination gels compared with NBF samples. NLHF positive control samples demonstrated vessel formation in fibrin only gels and limited vessel growth within combination gels. Normalizing the total length of blood vessels to the area of the explant demonstrates that CAFs support increased vessel formation in the presence of collagen (Figure 2). Vascularization supported by NHLFs and CAFs is similar in combination gels. Interestingly, a significant decrease in vessel length is observed in sample groups with NHLFs when comparing a fibrin/collagen gel to a fibrin only gel.

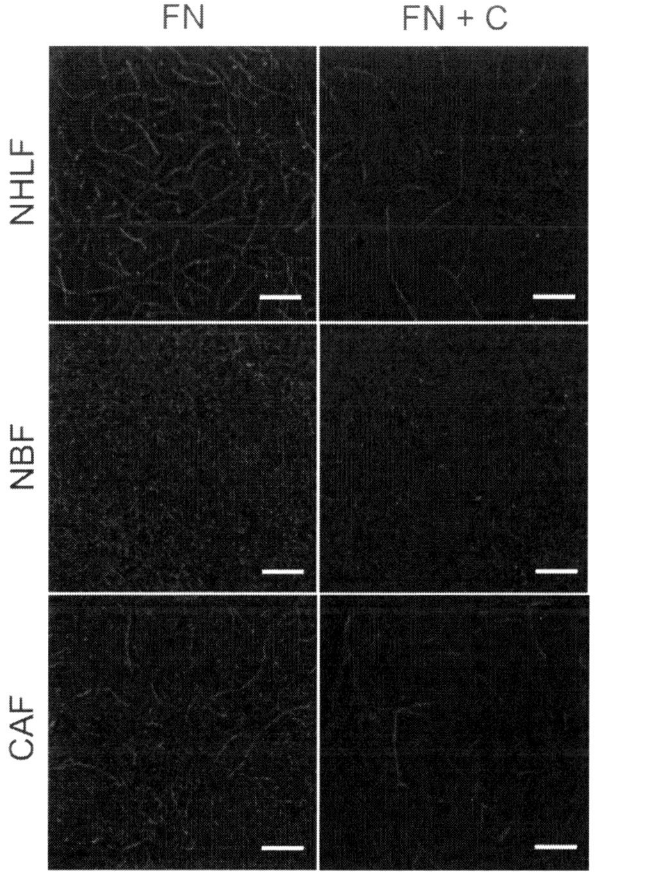

Figure 1. Representative images of blood vessels formed in 5 mg/mL fibrin (FN) or 5 mg/mL fibrin + 2.5 mg/mL collagen (FN + C) gels. Scale bar = 100 μm.

DISCUSSION

Taken together, these data indicate that CAFs can promote angiogenesis in a 3D tissue model relative to normal fibroblasts. In addition, we also observe enhanced vascularization in combination collagen/fibrin gels by the CAFs relative to both NBFs and NHLFs. Since the microenvironment of the tumor ECM often contains increased collagen I depositions and associated increase in stiffness, this suggests that CAFs may play a particularly important role in tumor angiogenesis. Since CAFs support the formation of longer vessels in both gel types compared to NBFs, this demonstrates they are promoting enhanced angiogenesis. In fact, no vessel development was observed in either fibrin or collagen/fibrin gels that contained NBFs, which highlights the angiogenic potential of CAFs when compared to normal stromal cells. In the presence of collagen, CAFs were able to slightly increase vessel formation, perhaps indicating that CAFs can enhance angiogenesis based solely on the peritumoral microenvironment even in the absence of tumor cells. This is further supported by the limited vascularization observed in collagen/fibrin gels supported by NHLFs compared to fibrin only samples.

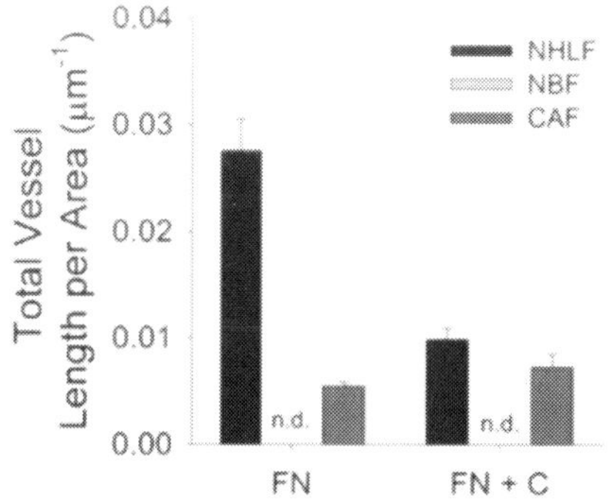

Figure 2. Total blood vessel length per unit area of vascular networks supported by NHLFs, NBFs, or CAFs embedded in different ECMs. No vessels were observed in NBF sample groups (n.d. = none detected). FN = 5 mg/mL fibrin; FN+C = 5 mg/mL fibrin + 2.5 mg/mL collagen.

Future work will focus on quantifying NBF and CAF generated forces in a 3D matrix as well as determining the mechanisms of regulation of this process. While abundant studies demonstrate the role of soluble factors, especially VEGF, in tumor-associated angiogenesis we also postulate that active biomechanical forces induced by CAFs promote enhanced vascularization of peritumoral tissues. By understanding the nature of these forces, we hope to develop strategies to target and limit blood vessel growth *in vivo*.

ACKNOWLEDGEMENTS

The authors wish to acknowledge the following funding from the NIH: UH3-TR00048 and R01-CA170879.

REFERENCES

(1) Orimo, A et al. *Cell*, 121(3):335-348, 2005.
(2) Calon, A et al. *Semin Cancer Biol*, 25:15-22, 2014.
(3) Corver, W et al. *Cell Oncol*, 34(6):553-563, 2011.
(4) Gjorevski, N et al. *Cancer microenvironment*, 5(1): 29-38, 2012.
(5) Sewell-Loftin, M et al. *Biomaterials*, 35(9):2809-2815, 2014.
(6) Kilarski, W et al. *Nature Medicine*, 15(6):657-U145, 2009.
(7) Moya, M et al. *Methods Mol Biol*, 1202:21-27, 2014.
(8) Moya, M et al. *Tiss Eng Part C Methods*, 19(9):730-737, 2013.
(9) Alspach, E et al. *Cancer discovery*, 4(6):716-729, 2014.

SB³C2015
Summer Biomechanics, Bioengineering and Biotransport Conference
June 17-20, 2015,Snowbird Resort, Utah, USA
SB³C2015-200

ENERGETICS OF WATER PERMEATION ACROSS STRETCHED PHOSPHOLIPID/CHOLESTEROL BILAYER: MOLECULAR DYNAMICS SIMULATION

Taiki Shigematsu, Kenichiro Koshiyama, Ryotaro Kurumatani, Shigeo Wada

Department of Mechanical Science & Engineering
Graduate School of Engineering Science, Osaka University
Toyonaka, Osaka, Japan

INTRODUCTION

Mechanical hemolysis is the rupture of red blood cell (RBC) membrane under mechanical stresses and the subsequent leakage of intracellular substances [1]. In order to understand the mechanism of the hemolysis, many experimental, theoretical, and computational studies for the rupture of the RBC membrane have been performed [2-4]. The experimental and theoretical studies [2,3] have shown that the membrane rupture is triggered by the formation of a transient pore penetrating the phospholipid bilayer, which is a primary structure of biological cell membranes. Computational studies, for example, molecular dynamics (MD) simulations of stretched phospholipid bilayers, have shown that the pore formation itself is initiated by the penetration of water molecules into the hydrophobic inside of the stretched bilayer [4]. Considering these findings, the water permeability of the stretched bilayer is expected to be closely related with the stretch-induced pore formation and the bilayer toughness to the subsequent rupture. However, the water permeability of the stretched bilayer is not fully understood.

Furthermore, we have performed a series of MD simulations of stretched phospholipid bilayer containing cholesterol molecules [5], which occur richly in RBC membranes. We found that the stretched phospholipid/cholesterol bilayer forms an interdigitated gel-phase-like structure, in which phospholipid and cholesterol molecules in the monolayer of the bilayer penetrate into opposite monolayer. The stretch-induced structural changes of the cholesterol-containing bilayer should involve changes of the bilayer properties, e.g. water permeability.

Inhomogeneous solubility-diffusion model [6] is one of the widely-used model to evaluate the permeability of water molecules through phospholipid bilayers. In this model, permeation process is divided into two steps: water molecules dissolve in and diffuse across the phospholipid bilayer. The contribution of the "dissolving step" to the water permeability is quantified by calculation of potential of mean force (PMF) of the water molecule along the bilayer normal. In this article, in order to understand the stretching effects on the water permeability of the bilayer containing cholesterol molecules, we performed MD simulations of stretched phospholipid bilayer containing cholesterol molecules and calculated the PMF of the water molecule along the bilayer normal.

METHODS

SYSTEM We employed a DPPC bilayer system containing cholesterol molecules at about 40 mol%, which is almost same to the cholesterol concentration of the RBC membrane. The system was composed of 32 DPPC, 20 cholesterol, and 2512 water molecules in rectangular simulation box with periodic boundary conditions. The system was well-equilibrated by 120-ns MD simulation at a constant temperature (323 K) and pressure (0.1 MPa). At equilibrium state, the bilayer area in the simulation box A_0 was 10.0 nm².

STRETCHING SIMULATION The equilibrated system was stretched by equibiaxial stretching MD simulation [7], in which the stretching is expressed by the proportional scaling of both the atom positions and the simulation box lengths. From the trajectory of the stretching simulation, we took the stretched bilayer, whose simulation box area is $A_{||}$, and which satisfy the areal strain $\varepsilon_A = 0.2, 0.4, 0.6, 0.8$, or 1.0, where ε_A is defined as $A_{||}/A_0 - 1$. Using the five stretched bilayers as the starting configurations, MD simulations for at least 100 ns at a constant area $A_{||}$ are performed.

PMF CALCULATION The PMF of the water molecules across the bilayer can be calculated by the density distribution of water molecule along the bilayer normal. However, it is difficult to accurately estimate the distribution inside the bilayer, where water molecules rarely sneak in, in conventional MD simulations due to the limitation of the available computer time. To avoid the difficulties, we employed the umbrella sampling method [8]. In this method, MD simulation with a biasing potential is performed. The biased MD simulation simply produces the biased local density distribution of the constrained water molecule around the interested position. Changing the interested position along the bilayer normal, same biased MD simulations are performed and we obtain the biased local density distributions from individual simulations. Combining and unbiasing these biased local distributions by the weighted histogram analysis method (WHAM)[9], we can obtain the density distribution of the water molecule along the bilayer normal and the PMF of the water molecules along the bilayer normal can be calculated. In this article, we performed a series of the biased MD simulations of the equilibrium bilayer and five stretched bilayers listed above. Each biased MD simulation was performed for 10 ns and the trajectories during latter 9 ns were used for analyses.

RESULTS

The representative snapshots of the stretched bilayers are shown in Fig. 1. With increasing ε_A, the bilayer thickness decreases monotonically. On the other hand, the orientations of DPPC and cholesterol molecules become disordered (Fig. 1A-D), but those become ordered when ε_A exceeds 0.6. (Fig. 1D-F). In the bilayers under $\varepsilon_A = 0.8$ and 1.0 (Fig. 1E and F), DPPC and cholesterol molecules in the monolayer of the bilayer penetrate into the opposite monolayer.

Fig. 2 shows the PMF of the water molecule across the stretched bilayers. The PMF in bulk water is set to 0 kJ/mol. The PMFs in the bilayer are significantly larger than that in bulk water. With increasing ε_A, the peak position of the PMF shifts to the bilayer center. The peak height decreases with the increase of ε_A and reaches a minimum at $\varepsilon_A = 0.4$ (inset in Fig. 2). After that, the peak height increases, peaks at $\varepsilon_A = 0.8$, and slightly decreases.

DISCUSSION

From the visual inspections of the stretched bilayers, DPPC and cholesterol molecules in the bilayer under $\varepsilon_A = 0.0$ are well-ordered and form a bilayer structure (Fig. 1A). These structural characteristics indicate that the DPPC and cholesterol molecules are in the liquid-ordered phase [10]. Although the DPPC and cholesterol molecules in the bilayer under $\varepsilon_A = 0.8$ and 1.0 are also well-ordered, the two monolayer of the bilayer interpenetrate (Fig. 1E and F). Such the structure of the stretched bilayer indicates that the DPPC and cholesterol molecules in the bilayer under $\varepsilon_A = 0.8$ and 1.0 are in the interdigitated gel-like phase [11]. The phase difference clearly shows that the increase of the areal strain up to 0.8 induces the phase transition from the liquid-ordered to the interdigitated gel-like phase.

The peak height of the PMF means the height of the energetic barrier for the water penetration across the bilayer. The energetic barrier is non-monotonically altered in accordance with the applied areal strain (inset in Fig. 2). Interestingly, the height of the barrier correlates with the phase state of the stretched bilayer. The heights of the barrier in the interdigitated gel-like phase are higher than those in the liquid-ordered phase and the transient phase between the liquid-ordered and the interdigitated gel-like phase. Moreover, that in the liquid-ordered phase is higher than those in the transient phase.

These results imply that the energetic cost for the water penetration across the stretched DPPC/cholesterol bilayer might be maintained at a higher level by the phase transition to the interdigitated gel-like phase and impede the initiation of the pore formation. Our recent MD simulation study [5] have shown that the phase transition to the interdigitated gel-like phase occurs most prominently in the bilayer containing cholesterol molecules at 40 mol%. The concentration is almost same to that of the RBC membrane. Considering these facts, the high deformability of the RBC membrane without rupturing might be because of the phase transition to the interdigitated gel-like phase.

ACKNOWLEDGEMENTS

This work was supported in part by JSPS Grants-in-Aid for Scientific Research (No. 23686031 and 24240073) and the MEXT project, "Creating Hybrid Organs of the Future" at Osaka University.

REFERENCES

[1] Rother, R. et al., *J. Am. Med. Assoc.*, 293:1653-1662, 2005.
[2] Rand, R., *Biophys. J.*, 4:303-316, 1964.
[3] Melikov, K.C. et al, *Biophys. J.*, 80:1829-1836, 2001.
[4] Koshiyama, K. et al., *Phys. Rev. Lett.*, 105:018105, 2010.
[5] Shigematsu, T. et al., *Chem. Phys. Lipids*, 183:43-49, 2014.
[6] Marrink, S. and H. Berendsen, *J. Phys. Chem.*, 98:4155-4168, 1994.
[7] Koshiyama, K. and S. Wada, *J. Biomech.*, 44:2053-2058, 2011.
[8] Torrie, G. and J. Valleau, *Chem. Phys. Lett.*, 28:578-581, 1974.
[9] Kumar, S. et al., *J. Comp. Chem.* 13:1011-1021, 1992.
[10] Sankaram, M. and T. Thompson, *Proc. Natl. Acad. Sci. USA*, 88:8686-8690, 1991.
[11] Slater, J. and C. Huang, *Prog. Lipid Res.*, 27:325-359, 1988.

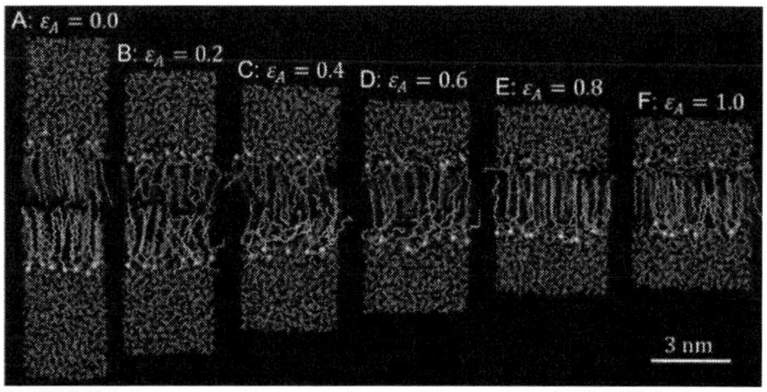

FIGURE 1 Representative snapshots of the stretched DPPC/cholesterol bilayers. The applied areal strains are 0.0 (A), 0.2 (B), 0.4 (C), 0.6 (D), 0.8 (E), and 1.0 (F). The DPPC molecules in the upper and lower monolayers of the bilayer are shown in pink and yellow, respectively. The cholesterol molecules in the upper and lower monolayers are shown in red and green, respectively. The water molecules are shown in blue.

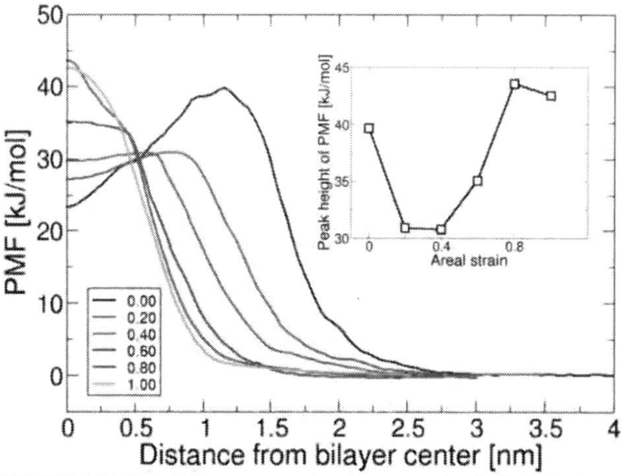

FIGURE 2 PMFs of the water molecule across the DPPC/cholesterol bilayers under $\varepsilon_A = 0.0$ (black), 0.2 (red), 0.4 (green), 0.6 (blue), 0.8 (violet), and 1.0 (cyan). The inset shows the relationship between the peak heights of the PMFs and the applied areal strains.

SB³C2015
Summer Biomechanics, Bioengineering and Biotransport Conference
June 17-20, 2015,Snowbird Resort, Utah, USA
SB³C2015-201

MEAN FLOW UMBILICAL DOPPLER INDICES PERFORM BETTER THAN PULSATILITY INDICES IN DETERMINING SMALL FOR GESTATIONAL AGE PREGNANCIES

Shier Nee Saw (1), Citra Nurfarah Zaini Mattar (2), Seow Heong Yeo (3), Shu-E Soh (5,7), Yap-Seng Chong (2,4), Peter D. Gluckman (4,6), Keith M. Godfrey (8), Seang Mei Saw (7,9), Arijit Biswas (2), Choon Hwai Yap (1)

(1) Department of Biomedical Engineering, NUS, Singapore, Singapore

(2) Department of Obstetrics and Gynaecology, Yong Loo Lin School of Medicine, NUS & NUHS, Singapore, Singapore

(3) Department of Maternal Fetal Medicine, KKH, Singapore, Singapore

(4) Singapore Institute for Clinical Sciences, A*STAR, Singapore, Singapore

(5) Department of Paediatrics, Yong Loo Lin School of Medicine, NUS, Singapore, Singapore

(6) Liggins Institute, University of Auckland, Auckland, New Zealand

(7) Saw Swee Hock School of Public Health, NUS, Singapore, Singapore

(8) NIHR Southampton Biomedical Research Centre, University of Southampton & University Hospital Southampton NHS Foundation Trust, Southampton, UK

(9) Singapore Eye Research Institute, Singapore, Singapore

INTRODUCTION

Small for gestational age (SGA) fetuses refer to fetuses which are smaller in size compared to normal ones due to insufficient blood flow in utero-placental circulation [1]. Pulsatility indices computed from ultrasound Doppler scans of the umbilical cord blood vessels are commonly used for detecting SGA fetuses. However, recent experiments have shown that pulsatility indices are influenced by both vascular resistance and compliance and not resistance alone [2]. Therefore, the effectiveness of various umbilical Doppler indices for the determination of SGA pregnancies is investigated here, using a computational model and clinical data.

METHODS

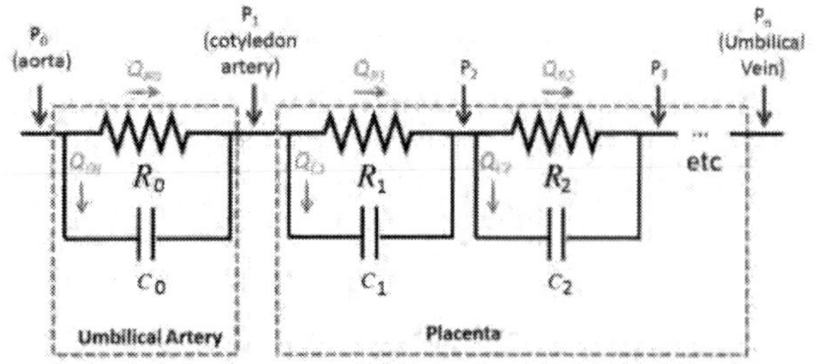

FIGURE 1: SCHEMATIC OF THE LUMPED PARAMETER MODEL.

Lumped-parameter numerical modeling was performed to study flow in the umbilical-placenta circulation [3]. Various umbilical Doppler indices were evaluated for their reliability in detecting high placenta vascular resistance. Clinical Doppler ultrasound scans of 537 normal and 13 SGA human fetuses from the GUSTO study were also investigated [4]. The Resistance Index (RI), Pulsatility Index (PI), umbilical artery time-averaged velocity (UATAV), umbilical vein area (UVA), time-averaged velocity (UVV) and volumetric flow (UVVF) were calculated, and investigated for their capabilities for distinguishing SGA from normal fetuses.

RESULTS

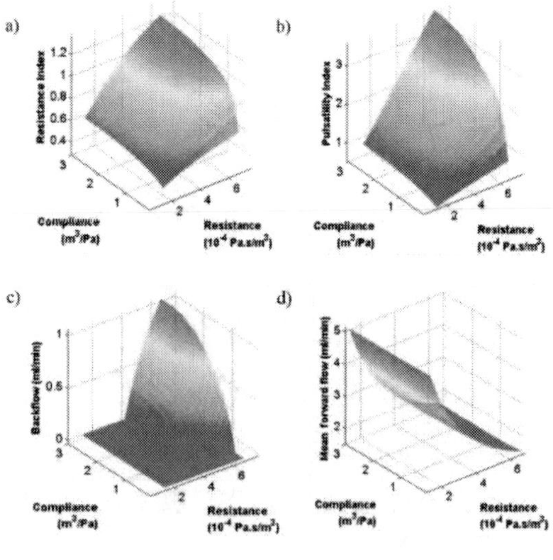

FIGURE 2: PLOTS OF (a) RI, (b) PI, (c) PEAK BACKFLOW VELOCITY, AND (d) MEAN FORWARD FLOW RATE AS A FUCTIONS OF PLACENTA VASCULAR RESISTANCE AND VASCULAR COMPLIANCE.

Numerical results showed that the pulsatility indices (RI and PI) and umbilical artery backflow showed significant dependence on both placenta vascular resistance and compliance, and could not be unique representations of vascular resistance and of placenta insufficiency (Figure 1). Mean forward flow through the umbilical arteries, however, is independent of compliance, and dependent only on resistance, indicating improved ability to represent placenta insufficiency. Clinical data showed that mean flow parameters such as UATAV ($p=0.012$) and UVVF ($p=0.049$) were more significantly different between small and normally grown babies, and pulsatility indices were less significantly different between the two groups (RI: $p=0.059$, PI: $p=0.025$). The area under receiver operating characteristic curve (ROC), which is a measure of positive predictive value, was the largest for UATAV compared to other indices (Table 1).

TABLE 1: AREA UNDER ROC CURVE (POSITIVE PREDICTIVE VALUE) OF VARIOUS DOPPLER INDICES IN DISTINGUISHING NORMAL FETUSES FROM SGA FETUSES; AND P-VALUE RESULTS OF STATISTICAL TESTS OF WHETHER EACH INDEX SIGNIFICANTLY CHANGED BETWEEN NORMAL AND SGA FETUSES.

Doppler Indices	Area Under ROC Curve	P-value
Umbilical Artery Time-Averaged Velocity (UATAV)	0.617	0.012
Pulsatility Index (PI)	0.616	0.025
Resistance Index (RI)	0.612	0.059
Umbilical Vein Area (UVA)	0.600	0.024
Umbilical Vein Velocity (UVV)	0.582	0.175
Umbilical Vein Volumetric Flow (UVVF)	0.582	0.049

DISCUSSION

Our modeling results showed that the RI and PI are dependent on both the vascular compliance and vascular resistance of the placenta, and are not uniquely dependent on resistance, and may not be good surrogate representations of the high vascular resistance that is the hallmark of IUGR. This result agrees with previous experimental work by Bude et al. [2], where it was shown that compliance is an important determinant for RI values, and was required for the occurrence of pulsatile flow (as opposed to steady flow) and flow reversal. This demonstrates the important of understanding placenta vascular compliance, which is, to date, under-explored.

Our study indicates that mean flow indices, may be a better indices in IUGR detection, since it is independent of compliance, and can be a good surrogate representation of high vascular resistance in IUGR. With the advent of 3D ultrasound, accurate acquisition of these measurements that allows proper quantification of Doppler angles should be accurately achievable. Clinically, there is also evidence for the utility of mean placenta blood flow rate as a determinant for placenta insufficiency: Tchirikov et al. [5] showed that reduced umbilical vein flow is significantly correlated to compromised fetal development resulting in an SGA fetus.

UATAV, a mean flow parameter, was observed to have a lower p-value and higher AUC value than pulsatility parameters (RI and PI), indicating that UATAV is a better parameter for distinguishing between SGA and normally grown fetuses than RI and PI. This showed that UATAV would be more accurate in detecting SGA fetuses compared to pulsatility indices. UVVF on the other hand, had lower p-value than RI, but not PI. On the whole, we observed that flow-related parameters perform slightly better than pulsatility indices, corroborating data from our numerical model.

Our computational work suggests that mean forward flow is a better parameter in gauging SGA severity. Clinical results corroborate this notion: UATAV showed slightly greater differences between normal and SGA fetuses than RI and PI.

ACKNOWLEDGEMENTS

This research is supported in part by the Singapore Ministry of Health's National Medical Research Council (NMRC), grant number BNIG13nov010; and in part by the Singapore National Research Foundation under its Translational and Clinical Research (TCR) Flagship Programme and administered by the Singapore Ministry of Health's National Medical Research Council (NMRC), Singapore - NMRC/TCR/004-NUS/2008; NMRC/TCR/012-NUHS/2014. The GUSTO study group includes Pratibha Agarwal, Arijit Biswas, Choon Looi Bong, Birit FP Broekman, Shirong Cai, Jerry Kok Yen Chan, Yiong Huak Chan, Cornelia Yin Ing Chee, Helen Y. H Chen, Yin Bun Cheung, Audrey Chia, Amutha Chinnadurai, Chai Kiat Chng, Mary Foong-Fong Chong, Yap-Seng Chong, Shang Chee Chong, Mei Chien Chua, Chun Ming Ding, Eric Andrew Finkelstein, Doris Fok, Marielle Fortier, Peter D. Gluckman, Keith M. Godfrey, Anne Eng Neo Goh, Yam Thiam Daniel Goh, Joshua J. Gooley, Wee Meng Han, Mark Hanson, Christiani Jeyakumar Henry, Joanna D. Holbrook, Chin-Ying Hsu, Hazel Inskip, Jeevesh Kapur, Kenneth Kwek, Ivy Yee-Man Lau, Bee Wah Lee, Yung Seng Lee, Ngee Lek, Sok Bee Lim, Yen-Ling Low, Iliana Magiati, Lourdes Mary Daniel, Michael Meaney, Cheryl Ngo, Krishnamoorthy Naiduvaje, Wei Wei Pang, Anqi Qiu, Boon Long Quah, Victor Samuel Rajadurai, Mary Rauff, Salome A. Rebello, Jenny L. Richmond, Anne Rifkin-Graboi, Seang-Mei Saw, Lynette Pei-Chi Shek, Allan Sheppard, Borys Shuter, Leher Singh, Shu-E Soh, Walter Stunkel, Lin Lin Su, Kok Hian Tan, Oon Hoe Teoh, Mya Thway Tint, Hugo P S van Bever, Rob M. van Dam, Inez Bik Yun Wong, P. C. Wong, Fabian Yap, Ge orge Seow Heong Yeo.

REFERENCES

[1] Lausman, A., et al., *Screening, diagnosis, and management of intrauterine growth restriction.* J Obstet Gynaecol Can, 2012. **34**(1): p. 17-28.

[2] Bude, R.O. and J.M. Rubin, *Relationship between the resistive index and vascular compliance and resistance.* Radiology, 1999. **211**(2): p. 411-7.

[3] Adamson, S.L., K.J. Whiteley, and B.L. Langille, *Pulsatile pressure-flow relations and pulse-wave propagation in the umbilical circulation of fetal sheep.* Circ Res, 1992. **70**(4): p. 761-72.

[4] Shu-E Soh, 2 Mya Thway Tint,3 Peter D Gluckman,4,5 Keith M Godfrey,6,7 Anne Rifkin-Graboi,4, et al., *Cohort Profile: Growing Up in Singapore Towards healthy Outcomes (GUSTO) birth cohort study.* International Journal of Epidemiology, 2013. **1**(9).

[5] Tchirikov, M., et al., *Umbilical vein blood volume flow rate and umbilical artery pulsatility as 'venous-arterial index' in the prediction of neonatal compromise.* Ultrasound Obstet Gynecol, 2002. **20**(6): p. 580-5.

A COMPUTATIONAL STUDY OF A SIMPLE, TRANSVERSELY ISOTROPIC MODEL OF SOFT TISSUE, WITH SHEAR AND TENSILE ANISOTROPY, IN LARGE STRAIN

Y. Feng (1), R. J. Okamoto (2), G. M. Genin (2, 3), L. A. Taber (2, 4), P. V. Bayly (2, 3, 4)

(1) School of Mechanical and Electronic Engineering
Soochow University
Suzhou, Jiangsu, China

(2) Department of Mechanical Engineering and Materials Science
Washington University
St. Louis, MO, USA

(3) Department of Neurological Surgery
Washington University
St. Louis, MO, USA

(4) Department of Biomedical Engineering
Washington University
St. Louis, MO, USA

INTRODUCTION

Structural anisotropy exists in many soft biological tissues. Transverse isotropy, which is the essential feature of fiber reinforced materials with a single dominant fiber direction, is one of the most common anisotropic forms. Typical soft tissues with fiber reinforced structures include skeletal muscle, myocardium, brain stem, white matter, ligament, and tendons. Hyperelastic models are widely used to study transversely isotropic materials. Most hyperelastic transversely isotropic models use only one anisotropic invariant, typically I_4. However, recent studies have shown that at two anisotropic invariants (e.g. I_4 and I_5) are needed to fully characterize both shear and tensile anisotropic behavior [1,2,3] of a transversely isotropic material.

Although many strain energy functions (SEF) have been studied analytically, few of them have been implemented in finite element (FE) simulation. In a previous study, we have shown that a simple hyperelastic model with both anisotropic invariants I_4 and I_5 can explain the anisotropic behavior of white matter in both shear and indentation tests involving small strains [3]. In the current study, we implement the transversely isotropic material model with both I_4 and I_5 invariants into FE simulation. The elasticity tensor was derived and anisotropic behavior of the model was demonstrated for canonical deformations in the large strain regime.

METHODS

If deformation gradient $\mathbf{F}$ is defined as $\mathbf{F} = \partial \mathbf{x} / \partial \mathbf{X}$, where $\mathbf{x}$ is the position vector in the deformed coordinate and $\mathbf{X}$ is the position vector in the reference coordinate, the right Cauchy-Green tensor is $\mathbf{C} = \mathbf{F}^T \mathbf{F}$, and the volume ratio is $J = \det \mathbf{F}$. A unit vector $\mathbf{A}$ defines the fiber direction. For the computational implementation of this study, the transversely isotropic SEF was:

$$\psi = \frac{\mu}{2}\left[(I_1 - 3) + \zeta(I_4 - 1)^2 + \phi I_5^*\right] \tag{1}$$

where $I_1 = tr\mathbf{C}$, $I_4 = \mathbf{A}\cdot\mathbf{CA}$, $I_5 = \mathbf{A}\cdot\mathbf{C}^2\mathbf{A}$, and $I_5^* = I_5 - I_4^2$. For nearly incompressible materials, a decoupled SEF was used:

$$\psi = \frac{\mu}{2}\left[(\bar{I}_1 - 3) + \zeta(\bar{I}_4 - 1)^2 + \phi \bar{I}_5^*\right] + U(J), \tag{2}$$

where $\bar{I}_1 = J^{-2/3} I_1$, $\bar{I}_4 = J^{-2/3} I_4$, $\bar{I}_5 = J^{-4/3} I_5$, $\bar{I}_5^* = \bar{I}_5 - \bar{I}_4^2$, and $U(J)$ is the volumetric constraint function. The corresponding Cauchy stress tensor was given in [3].

The material elasticity tensor C is needed for the FE simulation. Following the formulation by Holzapfel [4]

$$C = C_{vol} + C_{iso}, \tag{3}$$

where the volumetric part is

$$C_{vol} = \left(JdU/dJ + J^2 d^2U/dJ^2\right)\mathbf{C}^{-1} \otimes \mathbf{C}^{-1} - 2JdU/dJ \cdot \partial \mathbf{C}^{-1}/\partial \mathbf{C} \tag{4}$$

and the isochoric part is given in Eq. (5).

Commercial FE software ABAQUS 6.12 (Simulia, Providence, RI) was used. The material elasticity tensor was transformed into the spatial frame and finally to the corrotational elasticity tensor. The 6x6 DDSDDE array was generated by an automatic subroutine generation function [5]. To study anisotropic behavior of the model, two tests were carried out. (1) A single C3D8H type element was sheared either parallel (Fig. 1a) or perpendicular (Fig. 1b) to the fiber direction. The shear displacement applied to the cube surface was set to k=1.5. (2) A rectangular plate was constructed with a circular hole at the center. The dimensions of the plate were 32 by 16 with a thickness of 1. The radius of the hole was 3. The geometry was meshed with 902 C3D8H type elements. The material parameters for the above two simulations were μ=2.5,ζ=5,ϕ=0.5, near those estimated for brain white matter [3].

An x-symmetry boundary condition was applied to the left side of the plate where the right side was stretched with a displacement of 5. The displacement at the top and bottom surface of the plate was confined in the y direction so that the plate was stretched only in the x direction.

$$
\begin{aligned}
\mathbf{C}_{iso} = \\
& -\frac{2}{3}J^{-2/3}\Big[\mu \boldsymbol{I}\otimes\boldsymbol{C}^{-1}+\left(2\varsigma\left(\bar{I}_4-1\right)-2\phi\bar{I}_4\right)\boldsymbol{A}\otimes\boldsymbol{A}\otimes\boldsymbol{C}^{-1} \\
& \quad +\phi\left(\boldsymbol{A}\otimes\bar{\boldsymbol{C}}\boldsymbol{A}\otimes\boldsymbol{C}^{-1}+\boldsymbol{A}\bar{\boldsymbol{C}}\otimes\boldsymbol{A}\otimes\boldsymbol{C}^{-1}\right) \\
& \quad -\frac{1}{3}\left(\mu I_1+2J^{2/3}\bar{I}_4\left(\varsigma\left(\bar{I}_4-1\right)-\phi\right)-2\phi J^{2/3}\bar{I}_5\right)\boldsymbol{C}^{-1}\otimes\boldsymbol{C}^{-1}\Big] \\
& +2J^{-4/3}\Big[2(\varsigma-\phi)\boldsymbol{A}\otimes\boldsymbol{A}\otimes\boldsymbol{A}\otimes\boldsymbol{A}+\phi\frac{\partial^2\bar{I}_5}{\partial\bar{\boldsymbol{C}}\partial\bar{\boldsymbol{C}}}-\frac{2}{3}(\varsigma-\phi)I_4\boldsymbol{C}^{-1}\otimes\boldsymbol{A}\otimes\boldsymbol{A} \\
& \quad -\frac{1}{3}\phi\boldsymbol{C}^{-1}\otimes\left(\boldsymbol{C}:\frac{\partial^2\bar{I}_5}{\partial\bar{\boldsymbol{C}}\partial\bar{\boldsymbol{C}}}\right) \\
& \quad -\frac{1}{3}J^{2/3}\mu\boldsymbol{C}^{-1}\otimes\boldsymbol{I}-\frac{2}{3}J^{2/3}\left(\varsigma\left(\bar{I}_4-1\right)-\phi\bar{I}_4\right)\boldsymbol{C}^{-1}\otimes\boldsymbol{A}\otimes\boldsymbol{A} \\
& \quad -\frac{1}{3}\phi J^{2/3}\left(\boldsymbol{C}^{-1}\otimes\boldsymbol{A}\otimes\bar{\boldsymbol{C}}\boldsymbol{A}+\boldsymbol{C}^{-1}\otimes\bar{\boldsymbol{C}}\boldsymbol{A}\otimes\boldsymbol{A}\right) \\
& \quad -\frac{1}{3}J^{2/3}\left(\mu\,\mathrm{tr}(\boldsymbol{C})+2J^{2/3}\bar{I}_4\left(\varsigma\left(\bar{I}_4-1\right)-\phi\right)+2J^{2/3}\phi\bar{I}_5\right)\frac{\partial\boldsymbol{C}^{-1}}{\partial\boldsymbol{C}}\Big]:\mathbf{P}^T
\end{aligned} \tag{5}
$$

where the projector tenor $\mathbf{P}=\mathbf{I}-1/3\mathbf{C}\otimes\mathbf{C}^{-1}$.

RESULTS

In test (1), when simple shear displacement was applied parallel to the fiber direction, the magnitude of the shear stress was lower than when shear was applied perpendicular to the fiber direction (Fig. 1c). At the largest shear displacement point when k=1.5, the shear stress when sheared perpendicular to the fiber direction was 16 times greater. Shear stresses in both cases exceed the stress for shear in the plane of isotropy. This compliments the difference of shear modulus in and out of the isotropy plane by ϕ in the small strain regime [3].

In test (2), when the fiber was along the x-axis, the circular hole was deformed into an ellipse with its major axis along the x-axis (Fig. 2a). The highest stress occurred at the tip ends of the minor axis of the ellipse with a magnitude of 24μ. The ratio of the major-to-minor axis was 1.63. When the fiber was along the y-axis, the circular hole deformed similarity (Fig. 2b). However, the ratio of the major-to-minor axis of the ellipse was 1.36. The maximum stress also occurred at the tip ends of the short axis with a magnitude about 1.5μ.

When the fiber was 45 degrees with respect to the x-axis, the circle was deformed into an ellipse with the major axis tilted about -23 degrees to x-axis (Fig. 2c). When the fiber was -45 degrees with respect to the x-axis, the deformation responses showed to be symmetric with respect to the centerline along the x-axis (Fig. 2d), as expected. The expected stress concentrations arose at the points of material discontinuity in the corners.

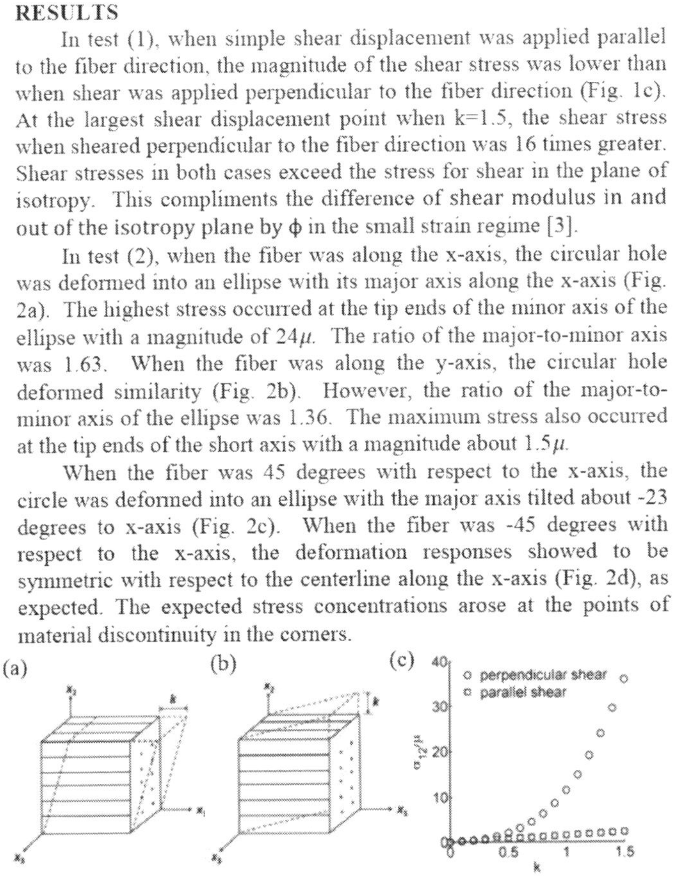

Figure 1. Simple shear applied (a) parallel or (b) perpendicular to the fiber direction in a unit cube. (c) A comparison of the shear stress from (a) and (b). The stress was normalized with respect to μ.

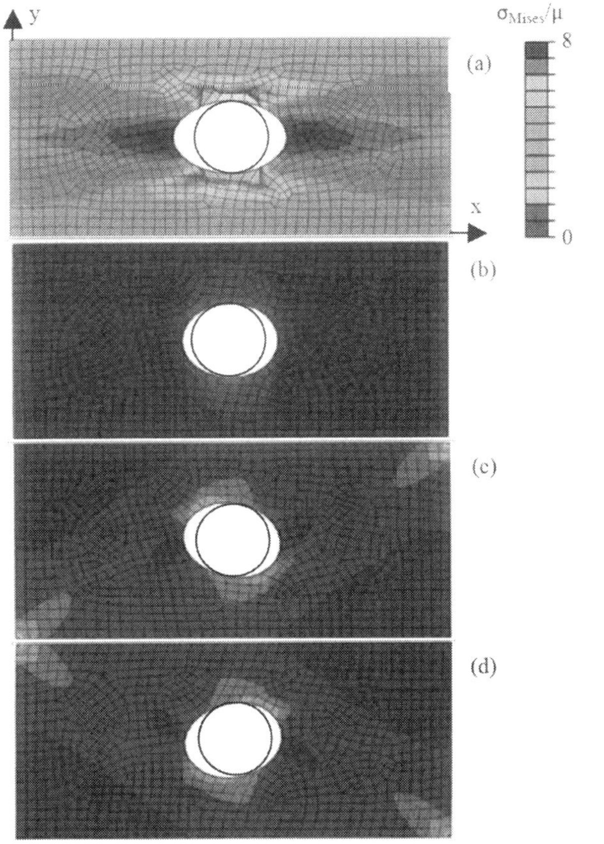

Figure 2. Uniaxial stretch of a plate-hole structure with the transversely isotropic material fiber along (a) x-axis, (b) y-axis, (c) 45 degrees with respect to x-axis, and (d) -45 degrees with respect to x-axis. The circle in dark color was the original shape of the hole. The von Mises stress (σ_{Mises}) was normalized with respect to μ.

DISCUSSION

Anisotropic behavior in shear deformation indicates that the model is capable of modeling tissues with shear anisotropy, such as white matter in the brain [3]. Under uniaxial stretch of the plate-hole structure, fibers at angles to the stretch direction modulate the deformation of the circular hole (Fig. 2b). As would be expected for a strongly anisotropic solid, the fiber direction is dominant in determining the deformation of the hole and the associated stress concentrations.

Future studies will focus on the use of this simple material model to explain experiment data from soft, transversely isotropic materials.

ACKNOWLEDGEMENTS

The authors thank Dr. Yunfei Shi, Dr. Chunghao Lee, and Dr. Barna Szabo for their helpful discussions. Grant BK20140356 (YF) from Jiangsu Province, NIH Grant NS055951 (PVB), and NSF Grant CMMI-1332433 (PVB) are acknowledged.

REFERENCES

[1] Destrade, M. et al., *CompMech*, 52(4):959-969, 2013.
[2] Murphy, J.G., *Euro J Mech - A*: 42:90-96, 2013.
[3] Feng, Y. et. al., *J MBBM*: 23:117-132, 2013.
[4] Holzapfel, G.A., Nonlinear Solid Mechanics, 2000.
[5] Young, J.M., et. al., *J BME*, 132(10):104505, 2010

SB³C2015
Summer Biomechanics, Bioengineering and Biotransport Conference
June 17-20, 2015,Snowbird Resort, Utah, USA
SB³C2015-203

NUMERICAL SIMULATION OF EFFECTS OF MEMBRANE SURFACE VISCOSITY ON TANK-TREADING MOTION OF RED BLOOD CELL

Ken-ichi Tsubota (1)

(1) Department of Mechanical Engineering,
Chiba University
Chiba, Japan

INTRODUCTION

A red blood cell (RBC) comprises an outer viscoelastic cell membrane and an inner viscous fluid; therefore, the mechanics of an RBC exposed to fluid flow involves the mechanics associated with the deformation of the cell membrane and those of the flow of the inner and suspending fluids. These complex mechanics have been so far assessed on the basis of experimental and analytical methods; however, a numerical simulation greatly helps understanding and quantifying different parameters involved in RBC mechanics. Recent studies have clarified some details related to the elastic deformation of the cell membrane. In contrast, studies aiming to simulate the viscosity of the cell membrane are scarce, and comprehensible simulations are lacking with only few exceptions [1,2]. In this study, a direct numerical simulation is performed to investigate how membrane surface viscosity influences the mechanics of tank-treading (TT) motion of RBC under simple shear flow, where TT is defined as a steady membrane rotation around the interior of the RBC, accompanied with an elongation along the direction of shear.

METHODS

An RBC was represented by an incompressible viscous fluid encapsulated by a membrane with an in-plane viscoelasticity and out-of-plane bending elasticity [3]. For the membrane mechanics, the in-plane elasticity was modeled on the basis of Skalak law, and shear and area-dilatation deformations were expressed as follows:

$$T_1 = G\frac{\lambda_1}{\lambda_2}\left[\left(\lambda_1^2-1\right)+C_a\lambda_2^2\left(\lambda_1^2\lambda_2^2-1\right)\right], \qquad (1)$$

where T_1 is the principal tension, λ_1 and λ_2 ($\lambda_1 \geq \lambda_2$) are the principal stretches, G is the shear modulus, and C_a is the elastic constant that determines the modulus of local area dilatation. The second principal tension T_2 is calculated by interchanging the indices 1 and 2. The in-plane viscosity was modeled on the basis of a two-dimensional Newtonian fluid [4] as follows:

$$\boldsymbol{T} = 2\eta \boldsymbol{V}, \qquad (2)$$

where $\boldsymbol{T}$ is the tension, $\boldsymbol{V}$ is the rate of the first-order deformation tensor (i.e., the rate of the small strain tensor), and η is the surface viscosity. The out-of-plane elasticity was modeled with the Helfrich isotropic bending energy as follows;

$$W_B = \frac{1}{2}B\int_A\left(2H - C_0\right)^2 dA, \qquad (3)$$

where $H = C_1 + C_2$ is the spontaneous mean curvature determined by the two principal curvatures C_1 and C_2, C_0 is the reference curvature, and B is the bending rigidity. For the total membrane area (A), the elastic energy associated with a change in the area A from its reference A_0 was introduced with the area dilatation modulus K_A as follows:

$$\Gamma_A = \frac{k_A}{2}\left(\frac{A-A_0}{A_0}\right)^2 A_0. \qquad (4)$$

If the modulus K_A is significantly larger than the elastic constants G, C_aG, and B, Γ_A can operate as a penalty function for the areal constraint conditions. In addition to membrane mechanics, a constraint in the volume of RBC was imposed [3].

The cell membrane was discretized with triangular elements (Fig.1). A biconcave discoid shape of a normal RBC at equilibrium

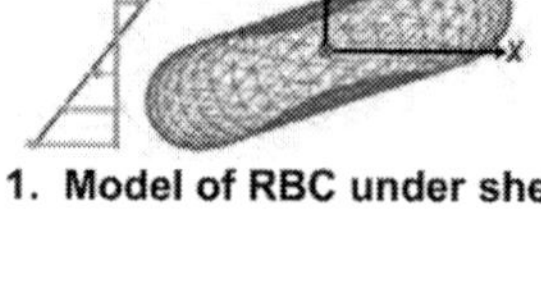

Figure 1. Model of RBC under shear flow

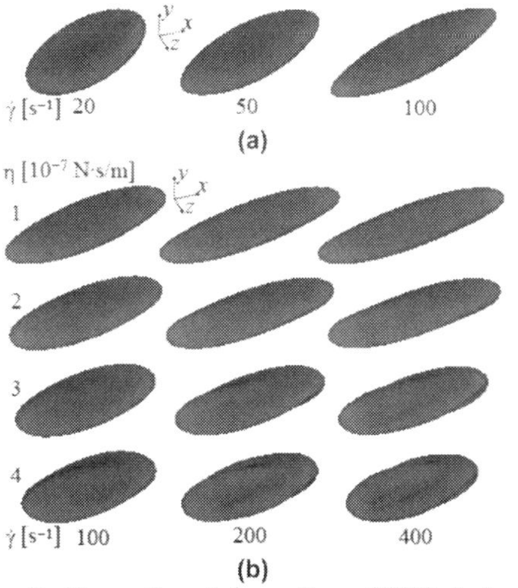

Figure 2. Elongation deformation of RBC during TT motion (a) without and (b) with surface viscosity

was created by solving a minimum energy problem [3,5], on the basis of a set of model parameters obtained from previously published studies, including constant global area A_0 = 134.1 μm^2, volume V_0 = 94.1 μm^3, shear modulus $G = 2.5 \times 10^{-6}$ N/m, constant for local area expansion C_a = 2, which allows changes in the local membrane area, and bending rigidity $B = 5 \times 10^{-19}$ Nm. The cell membrane surface viscosity was varied within the range of $0 \leq \eta \leq 4 \times 10^{-7}$ [Ns/m] [4,6]. The natural state of an RBC membrane for in-plane deformation (Eq.(1)) adopts an intermediate shape between spherical and biconcave disk shapes [3], while the natural state, represented by C_0 in Eq. 3, for the out-of-plane bending deformation is set constant over the membrane as $C_0 = 8/a$. Here, $a = (3V_0/4\pi)^{1/3}$ = 2.82 μm is the characteristic radius of the RBC. The xy plane as the shear plane, the undisturbed velocity was set as ($\dot{\gamma}y$, 0, 0) for infinite domain, and $\dot{\gamma}$ defined as the shear rate; in these settings, the RBC was placed under simple shear flow (Fig. 1). The normal direction of the RBC discoid plane was set along the shear flow direction at the initial state of the simulation. Parametric studies were performed for suspension viscosity μ and shear rate $\dot{\gamma}$ on the basis of published experiments [7]. Boundary integral equations were numerically solved with the triangulated membrane surface for the viscoelastic deformation of the cell membrane coupled with the flow of viscous fluids [8]. The force on the membrane was determined by a finite-element approximation for the tension Eqs. (1) and (2). In contrast, it was based on the partial derivative with respect to the nodal position for the energy Eqs. (3), (4), and (5) [3,5]. The number of triangular elements was 1246 and 4604 for small and large deformation, respectively.

RESULTS

TT motion of RBC was successfully simulated at fluid shear stress $\mu\dot{\gamma}$ greater than 0.1-0.2 Pa. In the absence of surface viscosity (η = 0 Ns/m), the RBC became more elongated along the direction of the shear flow, as the shear rate $\dot{\gamma}$ increased (Fig. 2a). When the membrane surface viscosity reached $\eta = 1 \times 10^{-7}$ Ns/m, these results did not change; however, deformation decreased below that for η = 0 Ns/m. When η increased even further, buckling deformation was observed in the membrane (Fig. 2b), which continues to increase with the increase in $\dot{\gamma}$ and the decrease in suspension viscosity μ.

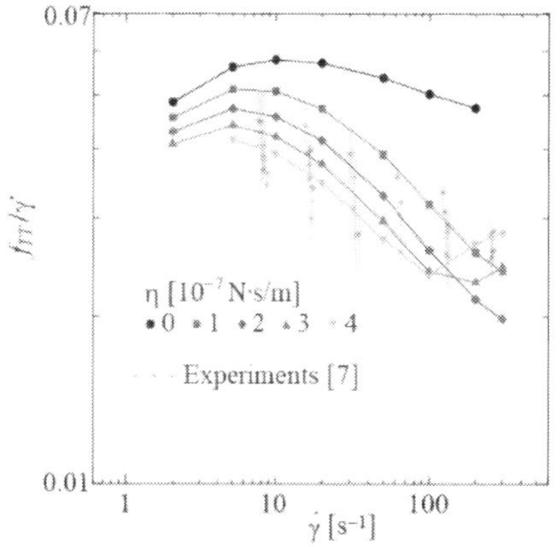

Figure 3. The ratio of TT frequency to shear rate as function of shear rate

The TT frequency f_{TT} increased with shear rate $\dot{\gamma}$, consistent with previously published experimental data [7] and numerical simulations [9]. For details, the ratio between f_{TT} and $\dot{\gamma}$ at μ = 28.9 mPas is represented as a function of $\dot{\gamma}$ in Fig. 3. In the experiment [7], $f_{TT}/\dot{\gamma}$ decreased with increasing $\dot{\gamma}$. This was reproduced in numerical simulations for $\eta \leq 2 \times 10^{-7}$ Ns/m, while $f_{TT}/\dot{\gamma}$ increased with $\dot{\gamma} \geq 100$ s^{-1} for $\eta \geq 3 \times 10^{-7}$ Ns/m at which membrane buckling (Fig. 2b) occurred. Experimentally measured $f_{TT}/\dot{\gamma}$ at high (small) $\dot{\gamma}$ was better reproduced by numerical simulations with small (high) η. This result indicates a shear-thinning behavior of cell membrane, corresponding to a published experimental study [6].

DISCUSSION

The effect of membrane surface viscosity on the RBC tank-treading mechanics was numerically simulated. The results demonstrated that, for large shear rates, the cell membrane of an RBC with large surface viscosity showed buckling deformation during tank-treading motion. The membrane buckling prevented the reproduction of the ratio between TT frequency and shear rate of published experiments. The numerical results are consistent with a shear-thinning behavior of the cell membrane which has been suggested in published experimental studies. Toward quantitative identification of the cell membrane viscoelasticity, a numerical simulation of TT using a modified continuum model of the cell membrane is underway.

ACKNOWLEDGEMENTS

The author thanks Dr. Fischer for providing the numeric data obtained in his experiment [7]. This study was partly funded by JSPS (25630046). The author also thanks RIKEN for supporting the computing facilities essential for the completion of this study.

REFERENCES

[1] Peng, Z et al., *J Fluid Mech*, 686:299-337, 2011.
[2] Yazdani, A, and Bagchi, P, *Phys Rev E*, 84:026314, 2013.
[3] Tsubota, K et al., *Biomech Model Mechanobiol*, 13:735-746, 2014.
[4] Evans, EA, and Hochmuth, RM, *Biophys J*, 16:1-11, 1976.
[5] Tsubota, K, *J Compt Phys*, 277:320-328, 2014.
[6] Tran-Son-Tay, R et al., *Biophys J*, 46:65-72, 1984.
[7] Fischer, TM, *Biophys J*, 93:2553-2561, 2007.
[8] Pozrikidis, C, *J Non-Newtonian Fluid Mech*, 51:161-178, 1994.
[9] Dodson, WR 3rd, and Dimitrakopoulos, P, *Biophys J*, 99:2906-2916, 2010.

SB^{3}C2015
Summer Biomechanics, Bioengineering and Biotransport Conference
June 17-20, 2015,Snowbird Resort, Utah, USA
SB^{3}C2015-204

INVESTIGATING ENVIRONMENTAL CAUSES OF CONGENITAL HEART DISEASES: A SUBJECT-SPECIFIC COMPUTATIONAL FLUID DYNAMICS STUDY

Venkat Keshav Chivukula (1), Sevan Goenezen (2), Sandra Rugonyi (1)

(1) Department of Biomedical Engineering
Oregon Health and Science University
Portland, Oregon, USA

(2) Department of Mechanical Engineering
Texas A&M University
College Station, Texas, USA

INTRODUCTION

In the US alone, Congenital Heart Disease (CHD) affects 1% (~32,000) of newborns amounting to a hospitalization cost of $2.6 billion, with 25% of these infants requiring surgery within their first year [1]. While both genetic and environmental factors can cause CHD, the latter is considered to be more prevalent (>95%) [2]; thereby suggesting a crucial role for environmental factors such as hemodynamics in CHD. Initiation, formation and subdivision of the heart is genetically controlled; however, blood flow plays a critical role in regulating cardiac growth by providing hemodynamic stimuli such as blood pressure and fluid shear stress for cardiac tissues.

Altered hemodynamics can lead to CHD by exposing cardiac walls to atypical fluid stresses, which affect cell proliferation, migration and extracellular matrix deposition via mechanotransduction mechanisms. Abnormal blood flow induced by surgical interventions in developing animal models have been shown to cause a variety of cardiac developmental defects [3]. However, the exact mechanisms by which altered hemodynamics in the embryonic heart lead to CHD are still unclear.

The current study focuses on the hemodynamics in the embryonic cardiac outflow tract (OFT) in the chick heart, during the tubular developmental stage. The OFT connects the embryonic ventricle to the arterial circulation and cardiac structures such as the interventricular septum and semilunar valves arise from the OFT region. The OFT is also known to be very sensitive to hemodynamic changes. Chicken embryos are frequently used to study cardiac development as morphological changes are similar to those of human heart development. Additionally, the chicken heart provides easy access for imaging and intervention. Hemodynamics in a normal and a hemodynamically altered embryo are studied and various hemodynamic parameters are compared. Understanding the changes in hemodynamics after intervention can provide insights into hemodynamic regulation of cardiogenesis.

METHODS

Our group has successfully performed surgical interventions in a developing heart by performing outflow tract banding where a suture is tied around the embryonic OFT. OFT banding restricts motion and constricts the lumen area of the OFT leading to increased blood flow and corresponding stresses through the banded region. Cardiac outflow tract banding was performed at the Hamburger and Hamilton (HH) developmental stage HH18 [4]. Structural images and flow velocities from the OFT were acquired within ~2 hours of banding using a custom-built Optical Coherence Tomography (OCT) system. Embryo-specific 4D (3D + time) computational fluid dynamics (CFD) models were generated and an inverse-method based optimization algorithm was used to simulate the blood flow over an entire cardiac cycle in a normal and a banded embryonic heart OFT. Various hemodynamic parameters were compared between the normal and banded OFT to elucidate the developmental hemodynamics under normal and altered circumstances.

Embryo-specific 4D model creation: The OCT system collects simultaneous structure and Doppler velocity images over 3-4 cardiac cycles at multiple locations along the length of the OFT (typically ~100 locations). The structural images obtained undergo an in-house developed synchronization procedure to obtain a 4D image set which is synchronized for a single cardiac cycle at every location. The synchronized images were analyzed by an in-house segmentation algorithm to extract 4D surface meshes of the OFT lumen. The surface meshes were converted into volume meshes for simulating blood flow within the lumen. Configurations of the lumen were obtained at multiple time points over the cardiac cycle.

CFD simulations: An inverse-method based optimization algorithm was used to replicate subject-specific hemodynamics for each embryo using ADINA. The procedure to obtain appropriate boundary conditions optimized velocities at a specific point inside the lumen within 1% of the experimentally measured velocities, which provided validation of our CFD models to replicate embryo-specific hemodynamics. A convergence study was also performed to ensure accuracy and confidence in the CFD simulations. Velocity distributions, pressure drop across the OFT and wall shear stress (WSS) were obtained and compared.

RESULTS

4D geometries for a representative normal and banded embryo were successfully segmented from structural images (Figure 2, TOP) and CFD simulations were performed using ADINA for the active portion of the cardiac cycle (when flow occurs). The banded embryo represented constriction of the OFT lumen area by ~30%, which resulted in doubling of maximum velocities through the lumen at peak systole. This closely matched the experimental observations of Midgett et al.[5]. (Figure 1, LEFT). The pressure drop for achieving the experimentally observed flow through the lumen was higher by 50% for the banded embryo compared to the normal embryo. The maximal WSS doubled for the banded embryo over the normal embryo at peak systole (Figure 1, RIGHT), with the WSS distribution over the OFT being highly heterogeneous for both embryos. (See Figure 2, BOTTOM)

DISCUSSION

In this work, we have characterized the immediate hemodynamic response to outflow tract banding using subject-specific CFD simulations. CFD showed a marked increase in pressure drop, blood flow velocities and WSS for the banded embryo. Additionally, the distribution of WSS was heteregenous for both embryos, with a large portion of the heterogeneity in the OFT cushion region. These results suggest that an abnormal hemodynamic environment in the developing heart exposes the cardiac cells to altered levels of hemodynamic stimuli which could potentially lead to CHD. The computational models developed by our research group are being used to characterize 4D blood flow and to examine the effect of altered hemodynamics on cardiogenesis.

ACKNOWLEDGEMENTS

This study has been supported by NIH grant RO1 HL094570.

REFERENCES

[1] Go, AS et al., *Circulation*, 127(1), 6-245, 2013
[2] Hove, JR et al., *Nature*, 421(6919):172-177, 2003
[3] Midgett, M and Rugonyi, S, *Frontiers in Physiology*, 5:287, 2014
[4] Hamburger, V and Hamilton, H, *J Morphol*, 88(1): 49-92, 1951
[5] Midgett, M et al., *J R Soc Interface*, 11(100), 2014

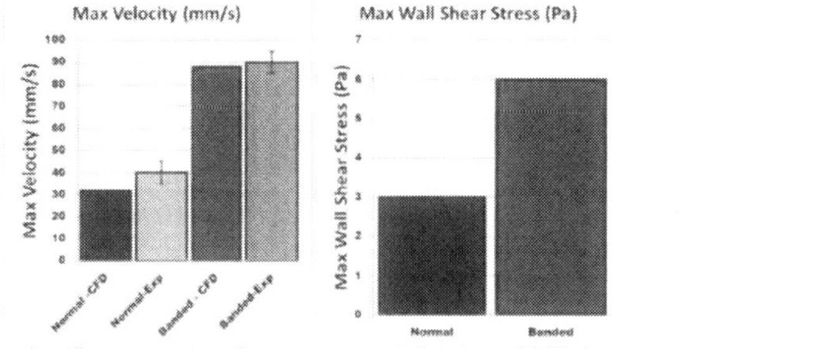

FIGURE 1: VELOCITIES AND WALL SHEAR STRESS COMPARISON FOR NORMAL AND BANDED EMBRYOS _AT PEAK SYSTOLE_. LEFT: COMPARISON OF EXPERIMENTALLY MEASURED [5] PEAK VELOCITIES (EXP) AND PEAK VELOCITIES OBTAINED FROM CFD SIMULATIONS (CFD) FOR NORMAL AND BANDED EMBRYOS. RIGHT: COMPARISON OF WSS OBTAINED BY CFD SIMULATIONS IN THE OFT CUSHION ZONE (SEE FIGURE 1) FOR NORMAL AND BANDED EMBRYO. VELOCITIES AND WSS IN THE BANDED EMBRYO ARE MORE THAN TWICE THAT OF THE NORMAL EMBRYO AT PEAK SYSTOLE.

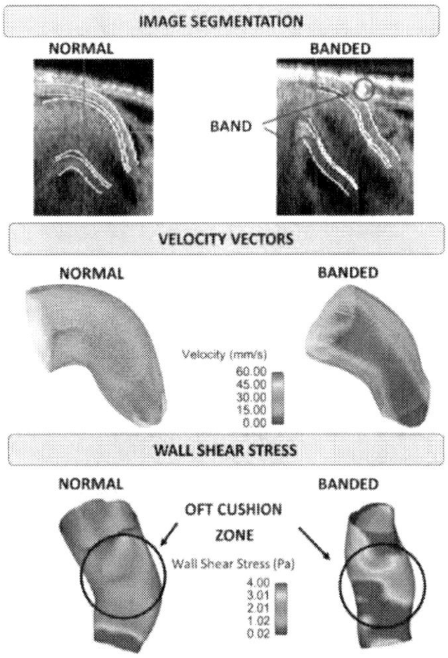

FIGURE 2: EMBRYO-SPECIFIC CFD SIMULATION RESULTS _AT PEAK SYSTOLE_. TOP: IMAGE SEGMENTATION OF A NORMAL (LEFT) AND BANDED (RIGHT) EMBRYO, WITH THE BAND LOCATIONS SHOWN IN RED CIRCLES. MIDDLE: VELOCITY VECTORS FOR NORMAL AND BANDED EMBRYO FROM CFD SIMULATIONS, SHOWING DOUBLING OF THE PEAK VELOCITIES FOR THE BANDED EMBRYO. BOTTOM: WALL SHEAR STRESS (WSS) COMPUTATIONS FOR A NORMAL AND BANDED EMBRYO. THE WSS IN THE OFT CUSHION ZONE DOUBLED FOR THE BANDED EMBRYO AND IS HETEROGENEOUSLY DISTRIBUTED, INDICATING A SIGNIFICANT HEMODYNAMIC EFFECT OF OUTFLOW TRACT BANDING.

SB^{3}C2015
Summer Biomechanics, Bioengineering and Biotransport Conference
June 17-20, 2015, Snowbird Resort, Utah, USA
SB^{3}C2015-205

DEVELOPMENTAL HEMODYNAMICS IN THE EMBRYONIC HEART OUTFLOW TRACT

Venkat Keshav Chivukula (1), Madeline Midgett (1), Sandra Rugonyi (1)

(1) Department of Biomedical Engineering
Oregon Health and Science University
Portland, Oregon, USA

INTRODUCTION

Cardiac development is a complex interplay of both genetic and environmental factors. Congenital heart disease (CHD) or defects in the heart at birth afflict nearly 1% of all newborns in the US [1]. Environmental factors such as hemodynamics are known to be a major contributor to the occurrence of CHD (>95%) [2]. Hemodynamics regulate cardiac morphogenesis by providing mechanical stimuli to which cardiac cells respond. The flow of blood in the developing heart controls, and is controlled by, the development of cardiac structures which culminate in the four-chambered heart. While structural changes of the heart over development have been thoroughly studied, the hemodynamic processes driving growth and remodeling are still poorly understood.

The current study focuses on the developmental hemodynamics in the embryonic cardiac outflow tract (OFT) in the chick heart. At the early stages of heart development studied here, the heart has a tubular structure. The OFT is situated between the embryonic ventricle and the aortic sac which connects to the arterial circulation and cardiac structures such as the interventricular septum and semilunar valves arise from the OFT region. Moreover, the OFT is known to be very sensitive to hemodynamic changes. Chicken embryos are frequently used to study cardiac development as the developmental process is highly conserved. Additionally, the chicken heart provides easy access for imaging and surgical intervention.

The hemodynamic environment in the developing heart at five developmental stages is characterized in detail using a combination of imaging and subject-specific computational fluid dynamic models. Understanding the hemodynamics in a developing heart can provide insights into hemodynamic regulation of cardiogenesis.

METHODS

Our group has successfully imaged *in vivo* the chicken heart OFT at the Hamburger and Hamilton (HH) developmental stages HH13 through HH18 (which correspond to 48-69 hours of incubation) [3]. Structural images and flow velocities from the OFT were acquired using a custom-built Optical Coherence Tomography (OCT) system. Image analysis was performed to calculate peak flow rate, stroke volume, cardiac cycle, and the percentage of time in the cardiac cycle with flow at each developmental stage. Embryo-specific 4D (3D + time) computational fluid dynamics (CFD) models were generated to simulate blood flow during systole and beginning diastole in the OFT of the various hearts. Various hemodynamic parameters were obtained and compared across the developmental stages to elucidate developmental hemodynamics.

OCT Image Analysis: The OCT system acquires simultaneous structure and Doppler velocity images, at high frame rates necessary to capture the dynamic and fast beating of the embryonic chick heart (140 frames per second). Images from a 2D longitudinal plane down the center of each outflow tract were first used to calculate hemodynamic parameters using Newtonian Poiseuille assumptions. 10 embryos each at HH13, 14, 15, 16, 17, and 18 stages were imaged with 2D OCT. Custom Matlab programs were used to process images and hemodynamic calculations. M-mode images were generated from the velocity measurement location to display cardiac wall motion over time in order to calculate OFT diameter for blood volume calculations. Peak flow rate, stroke volume, cardiac cycle and the time of flow were calculated from the centerline of the OFT.

CFD simulations: The OCT system also collected images over 3-4 cardiac cycles at multiple locations along the OFT geometry (typically ~100 locations spaced 7.5 microns apart). The structural images obtained underwent an in-house developed synchronization procedure to obtain a 4D image set, followed by an in-house segmentation procedure to segment the lumen for CFD. An inverse-method based optimization algorithm was used to replicate subject-specific hemodynamics for each embryo using ADINA. The velocity data obtained from the Doppler-OCT system was used as an optimization parameter in the CFD simulation A convergence study was also performed to ensure accuracy and confidence in the CFD simulations. Velocity distributions, pressure drop across the OFT and wall shear stress (WSS) were characterized.

RESULTS

Experimental Results: We have successfully measured blood flow characteristics over an influential period of embryonic heart growth (stages HH13 to HH18, n=10 at each stage) using structural and Doppler/OCT images. Flow rate increased 2.4 fold over the imaged developmental stages from 1.8 ± 0.3 mm^3/s at HH13 to 4.4 ± 0.9 mm^3/s at HH18 (Figure 1). Stroke volume also increased, 2.2 fold from 0.22 ± 0.1 mm^3/beat at HH13 to 0.47 ± 0.1 mm^3/beat at HH18. Cardiac period decreased from 599 ± 59 ms at HH13 to 419 ± 23 ms at HH18, while time of flow increased from 41.9 ± 4.4% at HH13 to 53.3 ± 8.4% at HH18.

CFD Results: CFD simulations provided further insights into the developmental hemodynamics of chicken embryo OFT. The flow velocities and pressure drops across the OFT increased by ~20% from HH13 to HH18, with the velocity profiles being strongly 3-D in nature. The overall WSS magnitudes remained similar; however the distribution of WSS was highly heterogeneous along the OFT. There were regions of increased WSS at specific locations near the OFT cushions and in regions of changing curvature (Figure 2(E)). The value of the increased WSS in the "hot spots" were more than three times the average, indicating a strong effect of hemodynamics on the mechanical stimuli observed by the surrounding cells.

DISCUSSION

These results show the normal changes in hemodynamics through the outflow tract as the cardiovascular system develops to meet the demands of the growing embryo. Hemodynamic assessment is the first step in understanding the complex processes that control cardiac development. Since the OFT cushions play an important role in the development of valves and other cardiac structures, it is necessary to determine the hemodynamic environment during normal development to understand the effect of hemodynamics on cardiogenesis. Hemodynamics in the developing heart is extremely dynamic in both spatial and temporal domains. Our results indicate that stoke volume, time of flow, flow rates and pressure drops increase while cardiac period reduces during early development. WSS is heterogeneously distributed along the OFT, with higher vales observed near the OFT cushions. Together, these results provide valuable insights into normal cardiac development. Variances from these characterized hemodynamic environments are known to result in CHD. This study focused on thoroughly defining the changes in normal hemodynamics needed to maintain proper cardiac development.

ACKNOWLEDGEMENTS

This study has been supported by NIH grant RO1 HL094570.

REFERENCES

[1] Go, AS et al., *Circulation*, 127(1), 6-245,2013

[2] Hove, JR et al., *Nature*, 421(6919):172-177,2003

[3] Hamburger, V and Hamilton, H, *J Morphol*, 88(1): 49-92, 1951

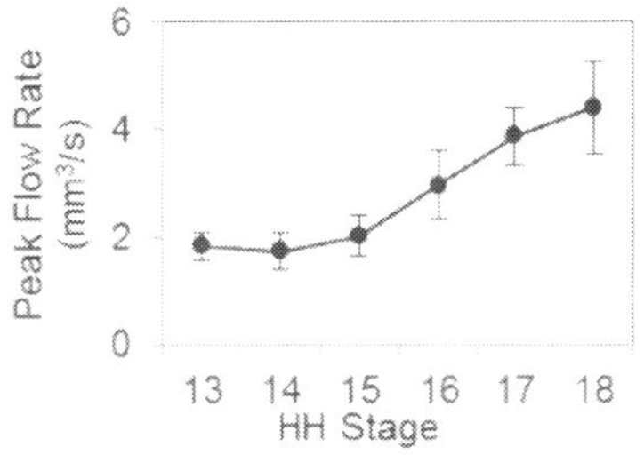

FIGURE 1: BEHAVIOR OF FLOW RATE BASED ON DEVELOPMENTAL STAGE. THE FLOW RATE INCREASED 2.4 FOLD FROM HH13 TO HH18.

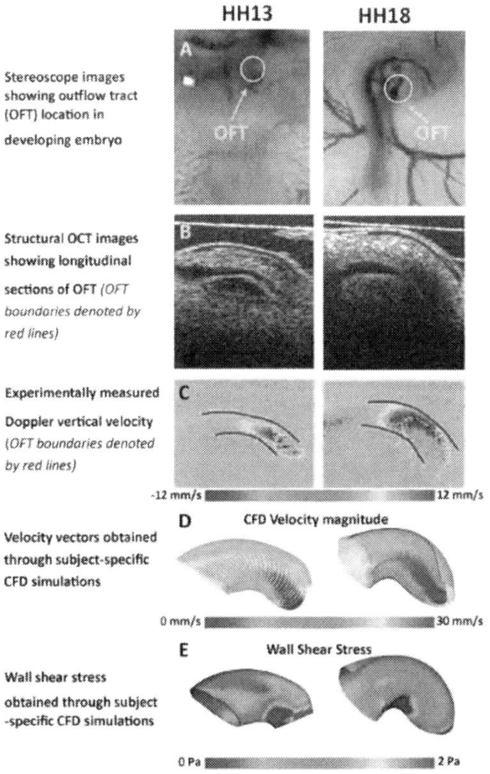

FIGURE 2: SUBJECT-SPECIFIC CHARACTERIZATION OF HEMODYNAMICS IN THE OUTFLOW TRACT. A: STEREOSCOPIC IMAGES, B: STRUCTURAL IMAGES FROM OCT SYSTEM, C: DOPPLER VELOCITY OBTAINED FROM DOPPLER-OCT (NOTE THAT THE REGIONS OF "NEGATIVE" VELOCITY WITHIN THE DOWNWARD PORTION OF THE OFT IS DUE TO "VELOCITY WRAPPING" WHERE VELOCITIES HIGHER THAN A SPECIFIC VALUE GET WRAPPED BACK TO THE MINIMUM VALUE (I.E. -12MM/S)), D: VELOCITY VECTORS AND E: WALL SHEAR STRESS OBTAINED FROM SUBJECT-SPECIFIC CFD SIMULATIONS.

MECHANICAL PROPERTIES OF HUMAN PLACENTA IN NORMAL PREGNANCIES AND DURING INTRAUTERINE GROWTH RESTRICTION

Jeanette S. Lau (1), Shier Nee Saw (2), Martin L. Buist (2), Choon Hwai Yap (2)

(1) Department of Mechanical Engineering
National University of Singapore
Singapore, Singapore

(2) Department of Biomedical Engineering
National University of Singapore
Singapore, Singapore

INTRODUCTION

Intrauterine Growth Restriction (IUGR) is a complex vascular disease of the placenta that results in insufficient nutrient transfer to support potential fetal growth. It is the most important condition affecting fetus viability [1], is responsible for life-long morbidity such as neuro mal-development and diabetes [2], and has a high incidence rate of 3% in developed countries and 10-15% in developing countries [3]. Current IUGR detection methods have been found to be poor [3], and investigations are ongoing to determine if ultrasound elastography, which tests for tissue stiffness non-invasively, could be used as an improved detection tool [4]. Thus, this paper aims to investigate the differences in mechanical properties of IUGR and normal placenta tissues to ascertain if elastography could be used to detect IUGR.

METHODS

All human sample testing protocols were approved by the Singapore National Heathcare Group Domain Specific Review Boards, and informed consent were obtained from all subjects.

Cyclic uniaxial compression mechanical testing was conducted on post-delivery human placenta samples from the National University Hospital, Singapore. 9 samples from 3 normal placentae and 8 samples from 2 IUGR placentae were tested, and viscoelastic mathematical models describing both normal and IUGR placentae were derived by iterative non-linear fitting experimental data to the models.

Mechanical Testing

Samples were placed in PBS and tested within 48 hours of delivery. 3-4 slices of about 15mm thick were excised from each whole placenta, turned laterally and mechanically tested in a PBS bath. The tester used was model 3345 (Instron Corp., Norwood, MA, USA), with a 10N force transducer (Instron-2519-101 load cell, Instron Corp., Norwood, MA, USA). The indenter shape was circular, with a radius of 12mm. Each specimen was subjected to cyclic loading on 3 positions, namely, the fetal, middle and maternal side (Figure 1a). Tests were conducted at rates of 40mm/min and 20mm/min. Figure 1b shows a schematic of the experimental set-up.

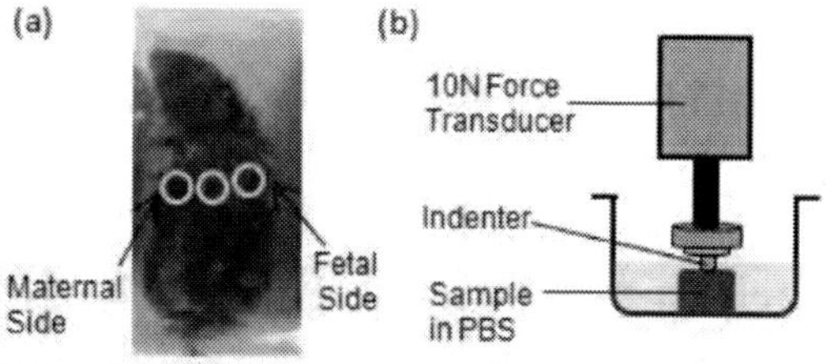

FIGURE 1. (a) DISSECTED SAMPLE TO BE TESTED ON FETAL, MIDDLE AND MATERNAL SIDE. (b) SCHEMATIC OF EXPERIMENTAL SET-UP

Each specimen was tested with strains of between 2-20%. 7 cycles of preconditioning was applied before 3 cycles of actual mechanical testing. Stress, time and strain were recorded at a data acquisition rate of 20Hz.

Mathematical Modeling

To characterize the stress strain interactions of the placenta tissues, several viscoelastic models were explored in order to obtain the model that best fits the experimental stress-strain curve. Model parameters were determined by minimizing the distance between the experimental data and the model results using "Unfit", a non-linear optimisation tool [5].

3 viscoelastic models, the Maxwell model, Voigt model and the

Standard Linear Solid (SLS) model, with linear component springs and dampers, were first explored. The same models, with exponential component spring and dampers [6] were tested next. Finally, the exponential SLS model was modified to consist of 6 permutations of linear and exponential components, where 3 bi-exponential models, along with another 3 uni-exponential SLS models were explored.

5 random sets of experimental stress-strain data were selected and iterative curve fitting experimental data to the above 12 models was done. Their respective R^2 values were computed and the experimental and modeled curves were plotted on the same graph in order to gauge the success of the fit. Model parameters of the best-fit model were then obtained for every specimen at the compression rate of 40mm/min and the results were compared by statistical tests.

RESULTS

Mechanical Testing

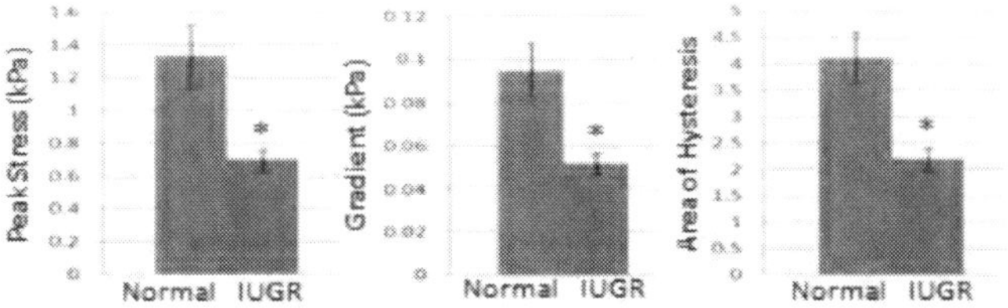

FIGURE 2. GRAPHS OF (a) PEAK STRESS, (b) AREA OF HYSTERESIS, (c) GRADIENT AT PEAK, WITH STANDARD ERROR BAR FOR MIDDLE PORTION AT 40MM/MIN

Results showed that the peak stress, area of hysteresis and gradient at peak stress were lower in the IUGR case for all 3 portions, fetal, middle, maternal side, and for both compression rates. A representative set of results can be found in Figure 2 where (*) represents p-value <0.05 for a difference between normal and IUGR placentae. Statistical tests were done and it was found that at the fetal side, p-values for all the above variables showed no significant difference between the normal and IUGR placenta tissues at a 0.05 significance level. As for the middle and maternal side, p-values for peak stress and gradient at peak stress showed a difference for both compression rates. However, p-values for area of hysteresis only showed significant difference between the two tissues for the 40mm/min compression rate, but not for 20mm/min.

Mathematical Modeling

The five parameter bi-exponential model (Figure 3) with exponential serial elastic component (SEC), serial viscous component (SVC) and linear parallel elastic component (PEC) was ultimately chosen due to its lowest R^2 value, ease in obtaining a solution and the fact that the solution was unique. Equation (I) describes this model.

$$\sigma = \int \alpha\dot{\varepsilon} + \gamma(\delta + \sigma - \alpha\varepsilon)\left[\dot{\varepsilon} - \frac{1}{\lambda}\ln\left(1 + \frac{\sigma - \alpha\varepsilon}{\mu}\right)\right] dt \qquad \text{(I)}$$

where α, γ, δ, λ, μ are constants, σ is stress, ε is strain, $\dot{\varepsilon}$ is strain rate and *t* is time.

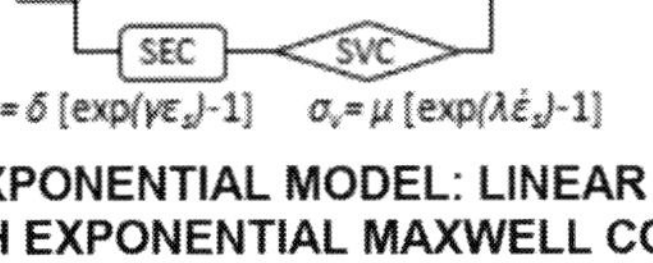

FIGURE 3. BI-EXPONENTIAL MODEL: LINEAR SPRING IN PARALLEL WITH EXPONENTIAL MAXWELL COMPONENT

The model parameters for all specimens were then obtained for the 40mm/min compression rate. Findings generally showed that the values of coefficients α, δ and μ for the IUGR cases were quite significantly less than the normal case, while exponents, γ and λ were found to be roughly the same, with p-values indicating no significant differences. A representative set of model results can be found in Figure 4 where (*) represents p-value <0.05 for a difference between normal and IUGR placentae. It was found that for the fetal side, the p-value of α indicated a significant difference between the IUGR and normal cases. The p-values for δ showed a significant difference for the middle and maternal side while p-values for μ showed a significant difference for the middle side and a weak difference for the maternal side at a significance level of 0.1.

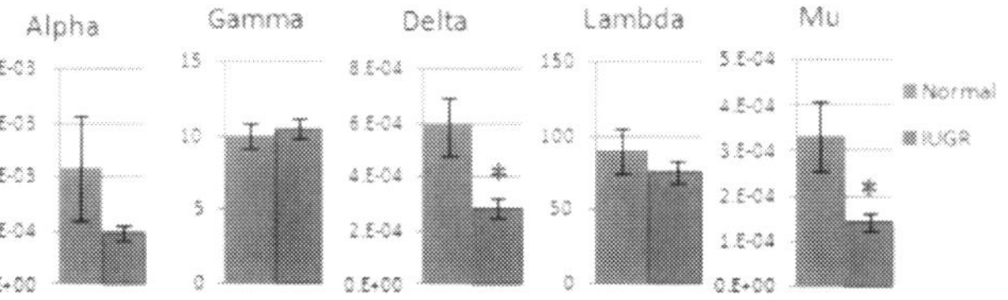

FIGURE 4. GRAPHS OF MODEL PARAMETERS FOR MIDDLE SIDE (40MM/MIN)

DISCUSSION

Results showed that there is a difference in properties between IUGR and normal placenta tissues. The peak stress and gradient at peak stress for the IUGR cases was found to be lower than the normal cases on all three positions (fetal, middle, maternal side) and at both compression rates, suggesting a decreased in stiffness and viscoelastic nature of the IUGR placenta. Area of hysteresis was also found to be lower for the IUGR placenta tissues, indicating reduced viscoelasticity, and viscous energy losses during mechanical deformations.

Results also showed that generally, model parameters describing the Maxwell spring and damper decreases in IUGR samples as compared to normal samples. This is especially so for coefficients α, δ and μ, belonging to the Maxwell spring and damper respectively, which seems to be about two times stronger in the normal model as compared to the IUGR model, agreeing with experimental results that normal placenta tissues are indeed stiffer than that of IUGR.

This study confirmed that the placenta mechanical properties changes significantly during IUGR and suggests that ultrasound elastography is a possible IUGR detection tool.

ACKNOWLEDGEMENTS

This research is supported in part by the Singapore Ministry of Health's National Medical Research Council (NMRC), grant number NMRC/BNIG/2020/2014. In addition, we would like to thank Cecille Arquillo Laureano and Cynthia Pamela Zapata Tagarino of the Obstetrics & Gynecology Department at National University Hospital for their assistance with patient eligibility and tissue procurement. We would also like to extend our appreciation to Alberto Corrias, Chung Cheuk Wang, Nicholas Cheng Zi Yi, Yeoh Jing Wui and Lai Chang Quan for their help rendered.

REFERENCES

[1] Figueras, F., et al., *Am J Obstet Gynecol*, 204(4): 288-300, 2011.

[2] Cosmi, E., et al., *J Pregnancy*, 2011: 364381, 2011.

[3] Barut, F., et al.,. *Diagn Pathol*, 5: 24, 2010.

[4] Quibel, T., et al., *Ultrasound in Obstetrics & Gynecology*, 40(S1): 169-169, 2012.

[5] Buist, M.L. and Corrias, A., *Unfit: data fitting and optimization software*, 2012.

[6] Chung, C.W. and Buist, M.L., *Nonlinear Analysis: Real World Applications*, 13(3): p. 1480-1488, 2012

FABRICATION AND CHARACTERIZATION OF ARTIFICIAL BONE-CARTILAGE TISSUE CONSTRUCTION USING MESENCHYMAL STEM CELLS

T. Arahira (1), M. Maruta (1), S. Matsuya (1), M. Todo (2)

(1) Department of Dental Engineering
Fukuoka Dental College
Fukuoka, Fukuoka, Japan

(2) Research Institute for Applied Mechanics
Kyushu University
Kasuga, Fukuoka, Japan

INTRODUCTION

Porous structures of natural polymer such as collagen have widely been used as scaffolds in tissue engineering. Collagen guarantees excellent biological conditions, for example, it stimulates generation and differentiation of cells as extracellular matrix [1]. Collagen scaffolds usually have high porosity [2,3] and interconnected porous structures for cell proliferation [4]. On the other hand, bioactive ceramics such as β-tricalcium phosphate (β-TCP) and hydroxyapatite has also widely been used in bone tissue engineering because of excellent oseteoconductivity, cellular adhesion, accelerated differentiation and mechanical property. It is also noted that β-TCP has faster degradation rate than crystalline hydroxyapatite [5].

Recently, in the field of bone and cartilage tissue engineering, regenerated bone graft has been one of the primary concerns instead of autografts and allografts. A regenerated graft may be developed by culturing and differentiating mesenchymal stem cells in porous scaffold [6]. In this case, the biochemical and biomechanical culture conditions and the structures and compositions of the scaffold are important factors controlling the quality of the regenerated bone and cartilage graft. It is therefore ideal that such grafts have mechanical compatibility with surrounding tissues such that mechanical stimulus are transferred into the bone and cartilage grafts to complete regeneration native tissue. However, detailed mechanisms of proliferation, differentiation and ECM formation of MSCs in a composite scaffold have not been well understood yet.

In the present study, a two-phase porous composite scaffold consisting of β-tricalcium phosphate (β-TCP) and type I collagen was developed for bone tissue engineering. The β-TCP porous structure was first constructed by the template method, and then the collagen porous structure was introduced into the continuous pores of the β-TCP structure. Type II collagen sponge was developed for cartilage tissue engineering. Rat mesenchymal stem cells (rMSC) were cultured in both scaffolds for 28 days.

METHODS

β-TCP scaffold was fabricated by template method using polyurethane sponge [7]. The polyurethane sponge (HR-20, Bridgestone, Japan) was used in this study. 5wt% poly vinyl alcohol (PVA) solution was used as the binder. β-TCP powder (β-TCP-100, Taihei Chemical Industrial Co., Japan) and PVA solution (10g/10ml) were mixed by centrifuge. The polyurethane sponge (10×10×5mm) were immersed into this slurry, and then, removed excess slurry. The sponge coated with slurry was dried at room temperature. After drying, the β-TCP coated polyurethane sponge was heated from room temperature to 400°C for 6 hours and 40 minutes, and then, from 400°C to 1,100°C for 5 hours. After sintering, the β-TCP scaffolds were allowed to cool down slowly inside the furnace until they reached room temperature. β-TCP/collagen scaffold for bone layer and collagen scaffold for cartilage layer were fabricated by freeze drying method. β-TCP/collagen and collagen solution were frozen at -80°C in a freezer and freeze-dried using a vacuum pump. The freeze-dried scaffolds were cross linked by glutaraldehyde vapor at 37°C for 4 hours. After cross-linking, the scaffolds were treated with 0.1M glycine water solution to block unreacted aldehyde, afterwards, they were washed by deionized water and lyophilized [8].

rMSC (DS Pharma Biomedical Co.) were cultured in cell growth medium consisting of alpha-minimal essential medium (α-MEM) supplemented with 10% fetal bovine serum and 1% penicillin-streptomycin. 100,000 cells suspended in 10μl of α-MEM were seeded in each of the scaffolds and then they were incubated for 1 hour to make cells adhered. After 1 hour incubation, these scaffolds were transferred to a 24-well plate containing 1 ml of differentiation medium per well. The plate was then incubated at 37 °C in a humidified atmosphere of 5% CO_2. The medium was changed twice per week. After 14 days culture, collagen scaffold with chondrocytes was put on β-TCP/collagen scaffold with osteoblasts. Artificial bone-cartilage constructs were continued to culture for another 28 days.

Compression tests of the scaffolds with proliferated cells were conducted periodically by using a conventional testing machine at a loading-rate of 1 mm/min to evaluate compressive mechanical properties such as the stress-strain relation and the elastic modulus. Gene expression of osteogenesis and chondrogenesis was analyzed using the real-time PCR reactions with primers of β-actin, collagen type I and aggrecan. RNA of cells proliferated in each scaffold was isolated from the powder from each scaffold using NucleoSpin® RNA II (MACHEREY-NAGEL GmbH & Co.). For each scaffold, three samples were prepared for the RT-PCR measurement. β-actin was used as the house keeping control and results were quantified for collagen I and aggrecan using the ΔΔCt relative quantification method[9]. All data were presented as means ± SD and derived from five independent samples.

RESULTS

Optical micrographs of β-TCP/collagen and collagen scaffold are shown in Fig.1. β-TCP frame of β-TCP/collagen scaffold was similar in the structure of cancerous bone. Collagen scaffold was porous structure and very easy to operate.

Compressive moduli are shown in Fig.2. Compressive modulus of bone layer tended to increase with rMSC culture up to 14 days and decrease up to 28 days. On the other hand, that of cartilage layer increase up to 7 days, and then, decrease up to 28 days.

Gene expressions are shown in Fig.3. In the case of bone layer, the expression of type-I collagen is high up to 14 days, after that, low up to 28 days compared with cartilage layer. In that of cartilage layer, that of aggrecan is high except for 21 days compared with bone layer.

DISCUSSION

In this study, artificial bone-cartilage tissue was fabricated by combining β-TCP/collagen with osteoblast and collagen with chondrocyte. In the previous study, it was reported that the variational behavior of compressive modulus of the β-TCP/collagen composite scaffold during cell culture were affected by both the material degradation and the proliferation of cells and the ECM formation[10]. TypeI collagen and aggrecan were used as osteoblastic maker, chondrogenesis maker, respectively[11]. Compressive mechanical properties and bone or cartilage-related markers such as type I collagen and aggrecan were evaluated every 7 days up to 28 days. It was found from the experimental results that the compressive modulus of both scaffolds tended to increase with rMSC culture up to 14 days and decrease up to 28 days. It was concluded that the two layered scaffolds can be a good structural stability as regenerated bone-cartilage layered tissue.

ACKNOWLEDGEMENTS

This work was supported by MEXT-Supported Program for the Strategic Research Foundation at Private Universities, 2010-2015)

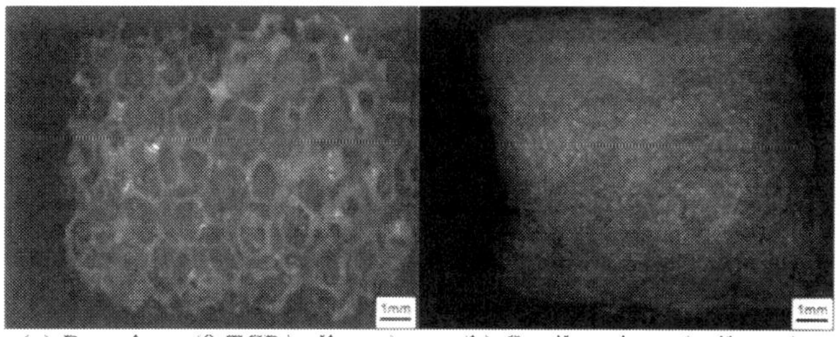

(a) Bone layer(β-TCP/collagen). (b) Cartilage layer(collagen).

Fig.1 optical micrographs of β-TCP/collagen and collagen scaffold.

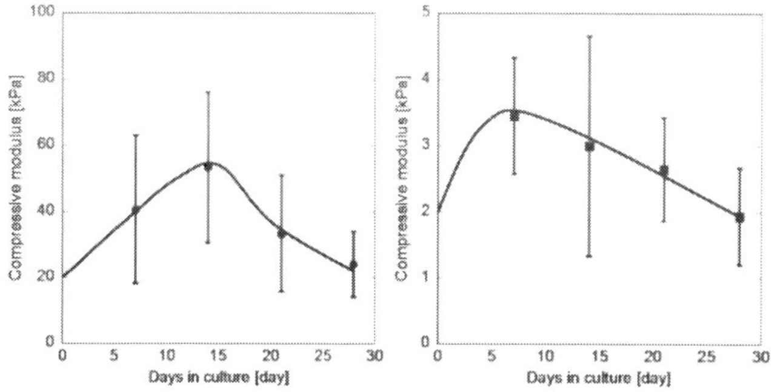

(a) Bone layer. (b) Cartilage layer.

Fig.2 Variation of compressive modulus during cell culture.

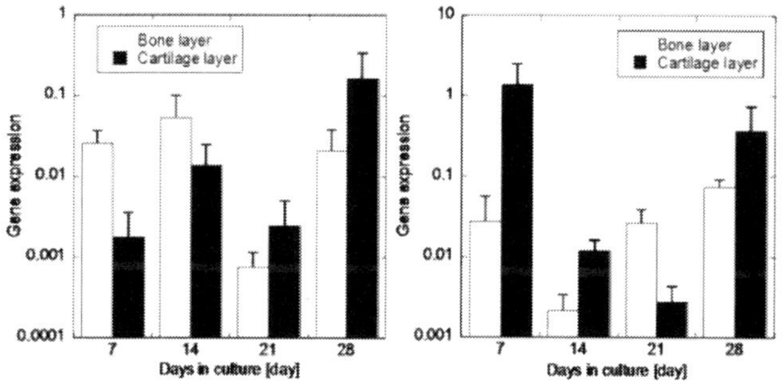

(a) Type-I collagen (b) Aggrecan

Fig.3 Variation of real time PCR analysis during cell culture.

REFERENCES

[1] Arahira, T et al., *IFMBE Proc*, 31:1234-1237, 2010.
[2] Keogh, M.B et al., *Acta Biomaterialia*, 6:4305-4313, 2010.
[3] Liang, W.-H et al., *J Biomed Mater Res Part A*, 94:2743-2753, 2010.
[4] Ma, L et al., *Biomateirals*, 24:4833-4841, 2003.
[5] Cao, H et al., *Bone*, 46:386-395, 2010.
[6] Oliveira, J.M et al., *Bone*, 46:1424-1435, 2010.
[7] Udoh, K et al., *Dent Mater J*, 29:154-159, 2010.
[8] Chen, G et al., *J Biomed Mater Res*, 51:273-279, 2000.
[9] Bonanomi, A et al., *J Immunol Methods*, 283:27– 43, 2003.
[10] Arahira, T et al., *J Mech Behav Biomed Mater*, 39:218-230, 2014.
[11] Dawson, J.I et al., *Biomaterials*, 29:3105-3116, 2008.

COMPLEXITY OF THE CEREBROSPINAL FLUID FLOW IN PATIENTS SUFFERING FROM CHIARI MALFORMATION TYPE I – A COMPUTATIONAL STUDY

Kartik Jain (1,2) and Kent-Andre Mardal (2,3)

(1) Simulation Techniques and Scientific Computing,
University of Siegen,
57076 - Siegen, Germany

(2) Center for Biomedical Computing,
Simula Research Laboratory,
P.O. Box 134, N-1325, Lysaker, Norway

(3) Department of Mathematics,
University of Oslo,
0316, Oslo, Norway

INTRODUCTION

Chiari malformation type I is a congenital disease characterized by the herniation of cerebellar tonsils through the foramen magnum that results in the obstruction of the Cerebrospinal Fluid (CSF) outflow. The dynamics of the CSF in the Subarachnoid Space (SAS) have been thought to play an important role in Chiari malformation. Computational Fluid Dynamics (CFD) allow for an efficient tool to assess CSF dynamics in the SAS. Most of the CFD simulations are however based on the assumption of laminar CSF flow – an assumption that advocates the choice of resolutions and accuracy in the simulation.

The CSF flow varies significantly between healthy subjects and Chiari patients. A recent study suggested that the CSF flow might exhibit transition to turbulence by simulating CSF in a patient compliant model of SAS [1]. The aim of the present study was to investigate and quantify the flow fluctuations in CSF by performing highly resolved numerical simulations below the Kolmogorov Micro-scales in order to elucidate whether the tonsillar herniation caused by Chiari I malformation results in the transition of CSF towards turbulence.

The CSF flow was simulated in patient-specific models of the SAS of two Chiari I patients and one healthy subject using direct numerical simulations (DNS) based on the Lattice Boltzmann Method (LBM).

The results revealed the presence of regions where flow fluctuates with high frequency similar to what is observed in transitional flows. These regions were mostly confined near the Chiari I malformation. The frequency of fluctuations was correspondent with the severity of obstruction while the CSF flow remained laminar for the healthy subject.

METHODS

MR images of one control subject and two Chiari patients were available from the Oslo University Hospital, which were segmented to obtain 3D models [2]. These are referred to as control, P1 and P2 in the following. The PCMR data from the SAS was used as inflow/outflow boundary conditions. The models were discretized with cubical cells of 32μm that resulted in ~ 800 million cells for each model. The time step of the simulations was set to 22μs. Simulations were performed using the open-source parallel simulation tool chain APES [3], and were executed on 12000 cores of the *SuperMUC* petascale system installed at the LRZ Munich, Germany.

The CSF was modeled with physical properties of water at 37°C i.e. ρ = 1000 Kg/m^3 and ν = 0.7 x 10^{-6} m^2/s. The LBM simulations were conducted for 15 cardiac cycles after damping out of the effects of initial conditions. Phase averages were gathered for 15 cardiac cycles to enable the quantification of CSF flow characteristics like the flow velocity, Turbulent Kinetic Energy (TKE) and the Kolmogorov micro scales.

RESULTS

The flow remained laminar in all parts of the SAS of control subject. Minor fluctuations were observed near the Chiari malformation in P1 that were highly intense in the P2 due to the difference in the anatomical complexity of each model. The fluctuations were most intense during the peak systole, which dampened as the flow began to decelerate. The flow field across horizontal planes at the peak systole is shown in figure 1 where the velocity vectors are warped to depict the areas with fluctuations. The figure shows phase-averaged velocity over 15 cardiac cycles to average out the *cycle-to-cycle variations*, which reduced with each subsequent cycle, but were persistent up to 15 cardiac cycles, mainly near the Chiari malformations. The peak velocities in control, P1 and P2 were about 60 mm/s, 120 mm/s and 380 mm/s respectively.

The phase averaged TKE in the three models is shown in figure 2. The magnitude of TKE is maximal in P2 with the presence of peaks and huge instantaneous variations as compared to P1 and control. Fluctuations of up to 10 KHz were observed in P2 and up to ~ 4 KHz in P1.

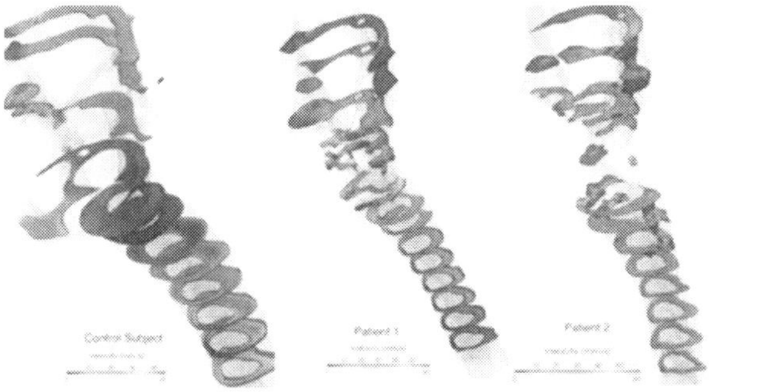

FIGURE 1: FLOW FIELD IN THE SAS OF CONTROL SUBJECT AND TWO CHIARI PATIENTS AT PEAK SYSTOLE

The l^+ and t^+ remained below 1.0 in all parts of the computational domain of each model. This confirms that the current simulations were conducted using sufficient spatial and temporal resolution to resolve the smallest structures expected in the flow field. A mesh sensitivity analysis was performed by conducting simulations at δx=48μm which exhibited errors of ~ 4-5% in P2 and ~ 2% in control and P1; the errors were maximum near the Chiari malformation. Kolmogorov micro scales were not fully resolved by the coarser resolution in the major areas of interest.

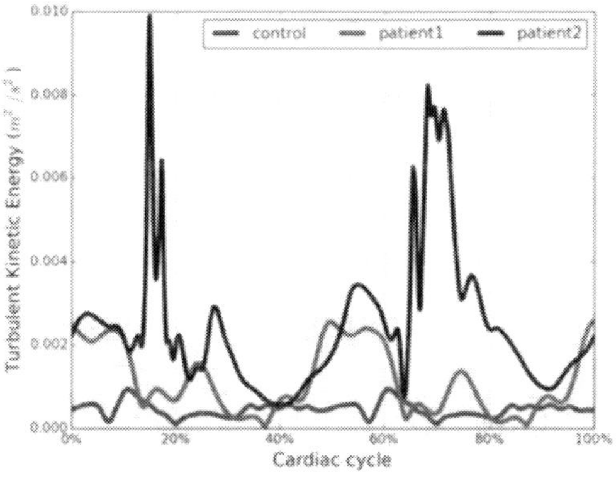

FIGURE 2: THE TURBULENT KINETIC ENERGY IN THE THREE MODELS PHASE AVERAGED OVER 15 CARDIAC CYCLES

DISCUSSION

The present investigations suggest that the tonsillar herniation caused by the Chiari I malformation induces fluctuations in the dynamics of CSF in the SAS as a result of obstruction to the CSF outflow. These fluctuations are not fully developed turbulence but resemble transitional flow, and are focused mainly near the craniovertebral junction.

The TKE was higher in the Chiari patient with most sever obstruction as the *stenosed* area near the craniovertebral junction in P2 intensified the fluctuations. Whereas the magnitudes of the peak systolic velocities distinguished Chiari patients from the control subject, it seems that the TKE might provide as an efficient means for the assessment of the extent of herniation.

The complexity of the oscillating CSF flow exhibited fully three-dimensional velocity field with manifold fluctuations in the two Chiari patients whereas in the control subject these fluctuations were not observed, albeit the CSF appeared to mix during the reversal of flow. The bi-directional nature of the CSF flow contributed significantly to the occurrence of opposing flow structures. The fluctuations in the Chiari patients were promoted by the absence of cerebellomedullary cistern, which added to the complexity of the geometry, and resulted in an enhanced flow separation. The elevation of velocities in Chiari patients is also pertinent from a fluid mechanical point of view as the absence of the Cerebellomedullary cistern reduces the area of the SAS whereas the same volume of CSF passes through that area.

Most of the computational studies of CSF are based on the assumption of laminar flow because of the relatively low Reynolds number (typically < 300) in non-obstructed areas. The low Re and relatively small computational domain allowed for Direct Numerical Simulations with LBM and quantification of the Kolmogorov micro scales. It is hard to extrapolate this resolution requirement to other methods but it is clear that to accurately capture the highly complex nature of bi-directional CSF and to quantify its turbulent characteristics, simulations need to be conducted with sufficient resolution and phase averages over a large number of cardiac cycles need to be gathered to account for cycle-to-cycle variations.

The mesh convergence analysis revealed that a coarse resolution was able to capture the transition, though with compromised accuracy. The complexity of the SAS models under study did not allow LBM simulations at a δx larger than 48μm due to the Mach number restriction in the method [4] and the local elevations in the velocities of the CSF.

The present study recommends that the common assumption of laminar CSF flow might need re-assessment in simulations and suggests that transition occurs in Chiari patients, which is commensurate with the extremeness of the pathology of SAS. The inclusion of a small part of the SAS in the present study and negligence of the fine structures of the nerve roots, denticulate ligaments and arachnoid trabeculae would not let extrapolation of these findings *in vivo,* and further analysis is required into this. Moreover, the study was limited to two patients and one control subject and analysis on a larger cohort is required to increase the confidence on the occurrence of transition in Chiari patients.

ACKNOWLEDGEMENTS

We are thankful to *Prof. Dr. Per Kristian Eide, MD, PhD, Dr. Geir Ringstad, MD* and *Dr. Karen-Helene Støverud, PhD* for the clinical data and fruitful discussions.
Computer resources for this study were provided by the Leibniz Supercomputing Center, Munich.

REFERENCES

[1] Helgeland, A., Mardal, K. A., Haughton, V., & Reif, B. A. P. (2014). Numerical simulations of the pulsating flow of cerebrospinal fluid flow in the cervical spinal canal of a Chiari patient. *J. of Biomechanics*, 47(5), 1082-1090.

[2] Støverud, Karen-Helene (2014). Relation between the Chiari I malformation and syringomyelia from a mechanical perspective. PhD Thesis, *University of Oslo*.

[3] Klimach, H., Jain, K., & Roller, S. (2013). End-to-end Parallel Simulations with APES. In *PARCO* (pp. 703-711).

[4] Junk, M., Klar, A., Luo, L.S., 2005. Asymptotic analysis of the lattice Boltzmann equation. *J. of Computational Physics* 210, 676-704.

SPRING MEDIATED CRANIOPLASTY: A PATIENT SPECIFIC NUMERICAL MODEL

Alessandro Borghi[1,2], Silvia Schievano[1,2], Will Rodgers [1,2], Freida Angullia [2], Allan Ponniah [2], David Dunaway [2], Owase Jeelani [2].

(1) UCL Institute of Child Health,
30 Guilford Street, London WC1N 1EH.

(2) Great Ormond Street Hospital for Children,
Great Ormond Street, London WC1N 3JH

INTRODUCTION

Craniosynostosis consists of premature fusion (ossification) of one or more cranial sutures during infancy; it affects 1 in 1,700 live births [1]. The most common presentation is sagittal craniosynostosis (SC), which occurs when the sagittal suture fuses. SC causes craniofacial deformity and may cause functional problems as well as raised intracranial pressure. Conventional treatment for this condition involves surgical removal of the fused sagittal suture, placement of several cuts along the skull bones to allow for appropriate brain growth, and, following surgery, in some instances customized moulding helmet therapy for 4-6 months. The surgery usually takes 4-5 hours and complications are mainly related to bleeding and need for a transfusion [2].

At Great Ormond Street Hospital for Children (GOSH) we have pioneered a less invasive procedure using metal spring-like distractors to achieve the desired change in skull shape. This operation entails one or possibly two smaller cuts over the top of the head, two bone cuts to free the fused suture and then placement of the springs. Over the following weeks, the springs expand and the skull gradually remolds to the desired shape (Figure 1). Once this has been achieved, the springs are removed under a short general anesthetic (usually a day-case). This technique has shortened considerably not only the time of surgery, but also the patient postoperative recovery. The procedure has proven functionally and aesthetically effective in correcting skull deformities caused by SC, but final outcomes remain unpredictable because of an incomplete understanding of normal craniofacial surface mechanics. In this work, a computational model of spring cranioplasty is presented and validated: the results show that spring expansion is predictable by means of computational modelling.

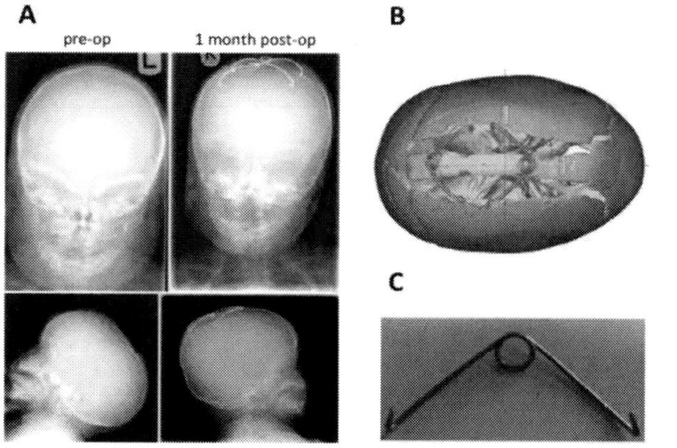

Figure 1: A) Preoperative scans (on the left) and postoperative scans (on the right) of a patient with craniosynostosis, treated with springs. B) CT Reconstruction of pediatric skull after spring insertion: parasagittal (vertical) cuts are performed and the springs are inserted in notches and left to expand over time. C) Sample of cranioplasty spring.

METHODS

A 5 month old boy diagnosed with scaphocephaly (elongated head shape) due to early fusion of the sagittal suture was surgically treated with springs in March 2014, subsequently removed 6 months later.

Computed Tomography (CT) was performed before the implantation in order to assess the severity of the pathology and plan the surgery. In theatre, a sagittal osteotomy running from the coronal to the lambdoid suture was performed; two sets of notches were created to accommodate the springs. Two springs were inserted, one in the anterior and one in the posterior position. During surgery, local skull thickness, position and dimension of the osteotomies, type and model of the springs were recorded.

CT data were processed in order to create a 3D model of the patient: the scans were imported into Mimics (Materialise) and thresholding of the grey level was performed in order to isolate hard tissue from soft tissue A 3D volume was calculated and the region of interest related to upper portion of the skull was isolated (Figure 2A). The mesh of this volume was imported into 3Matic (Materialise) for geometry as well as mesh improvement (Figure 2B): skull holes were filled in order to achieve a continuous volume and smoothing was performed. An IGS file was created, encompassing the whole volume. Such file was imported into Solidworks (Dassault Systems) where surgical osteotomies as well as bony notches were created using measurements collected during surgery (Figure 2B). The model was then imported into Ansys Mechanical (Ansys Inc.) for simulation (Figure 2C). A tetrahedral mesh was created, composed of 101,144 elements and 175,062 nodes. Boundary conditions were applied: the bottom surface of the model was restrained in the vertical direction and a node was fully constrained to prevent rigid body movement.

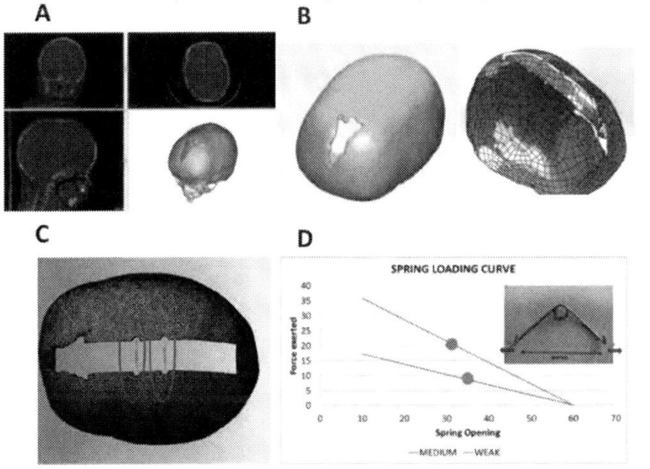

Figure 2: A) Segmentation of CT images using Mimics B) Creation of CAD model (left) and modelling of craniotomy and surgical notches. C) FE model with spring connections D) characteristic curves of the cranioplastly springs.

To simulate the effect of spring expansion, two predefined linear spring connections were applied in the locations equivalent to the slots performed during the surgery. The cranioplasty spring loading curves were retrieved using an analytical method previously developed and reported elsewhere [3] (Figure 2D): each spring exerts a force which is proportional to the amount of crimping performed. Both springs had an opening at rest of 60mm and calculated stiffness was 0.715 N/mm for the MEDIUM spring (used in anterior position) and 0.345 N/mm for the WEAK spring (used in posterior position).

The spring expansion was simulated and the equilibrium configuration was calculated. The values of in-silico spring opening were compared with the values retrieved during surgery.

RESULTS

The expansion of the sagittal springs was modeled, using a patient specific 3D geometry where surgical osteotomies were replicated, using measurements taken in theatre. Stress and displacement pattern were retrieved and analyzed: peak displacement occurs around the area of the spring placement (Figure 3A), although the whole skull undergoes the effect of spring expansion.

The change in anterior-posterior dimension (AP, measured from forehead to back of the skull, figure 3B) was calculated and compared with the measurement taken on the operating table: good agreement was found with a predicted value of 5.8mm versus an on-table measurement of 6mm. The opening of the anterior and posterior spring were calculated and compared with the measurements: for the anterior spring, the model predicted an opening of 35mm vs a 30mm on table measurement (16% difference); for the posterior spring, the model predicted an opening of 36mm vs 35mm on table opening (3% difference) (Figure 3C).

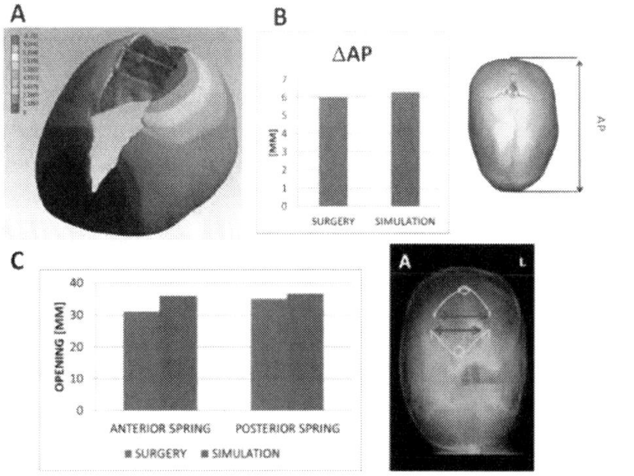

Figure 3: A) Displacement pattern of the skull after cranioplasty spring implantation. B) Variation of Antero-posterior dimension, comparison of computational values with on-table measurement. C) Spring opening, comparison of computational values and on-table measurement for anterior and posterior spring.

DISCUSSION

A patient specific model of spring cranioplasty was created, using 3D data from medical imaging and surgical information gathered in theatre. The model accurately predicted the amount of expansion experienced in theatre and overall variation of skull dimensions.

Further refinement of this model, including the application of viscoelastic material properties, will allow modeling the time-dependent changes in skull dimensions, help with surgical planning and provide an important tool for designing new distractors.

ACKNOWLEDGEMENTS

The work has been funded by the Great Ormond Street Hospital for Children Charity. (Grant n. 12SG15).

REFERENCES

[1] Fearon, J et al., *Plastic and Reconstructive Surgery*, 133:1261-75, 2014.

[2] David, LR et al., *J Craniofacial Surgery*, 21:3-9, 2010.

[3] Borghi, A et al., *Proceedings of 7th World Congress of Biomechanics (WCB)*, 2014.

SB³C2015
Summer Biomechanics, Bioengineering and Biotransport Conference
June 17-20, 2015,Snowbird Resort, Utah, USA
SB³C2015-212

HEMODYNAMICS OF HEALTHY VS. PATHOLOGICAL VENOUS VALVE – FLUID-STRUCTURE INTERACTION COMPUTATIONAL MODEL

Elina Soifer (1), Dar Weiss (1), Oren Rotman (1), Uri Zaretsky (1), Shmuel Einav (1,2)

(1) The Department of Biomedical Engineering,
Tel-Aviv University, Tel-Aviv, Israel
(2) Stony Brook University, Stony Brook, NY, USA

INTRODUCTION

Venous valve leaflets are mainly composed of valvular interstitial cells which have characteristics of smooth muscle cells, and thus respond to mechanical stimuli. Stretching of these cells and fibers may induce mechanotransduction, resulting in tissue remodeling. Therefore, pathological perturbations in the biomechanical environment i.e. wall shear stress (WSS) and blood pressure, may lead to various valvular disease. These diseases are usually characterized with fibrosis induced by phlebitis, in which the leaflets' structure stiffens.

The objective of the current study was to computationally examine the effect of the leaflets' stiffening process on the hemodynamics of the venous valve. The hypothesis suggests that the biomechanical condition of the valve, characterized by leaflet wall stress and fluid WSS, becomes hemodynamically less favorable as the leaflets become stiffer.

METHODS

A two-dimensional computational model of typical venous-valve geometry was studied, and a fully coupled, two-way fluid-structure interaction (FSI) analysis was implemented, using the commercial finite-element package of ADINA (version 9.0, Watertown, NY, USA).

A parabolic spatial flow function with a maximal velocity of 7 cm/s was assigned at the model inlet [1], while a uniform fluid pressure of 0 Pa was set at the outlet. The wall was fixed near the inlet and the outlet boundaries (**Fig 1**). Fluid acceleration due to gravity was considered in the Z-direction. FSI interfaces were defined at the surfaces of the wall, sinus and leaflets.

Blood was assumed to be slightly compressible with density of 1060 kg/m^3 and viscosity of 0.0035 Pa·s. The vein was modeled with diameter of 4.3 mm and wall thickness of 0.5 mm, which are characteristic dimensions of the most proximal end of the Greater Saphenous Vein [2, 3]. Both vein wall and valve leaflets were assumed to be isotropic hyperelastic materials with Poisson's ratio of 0.45, and Young's modulus (E) of 1 MPa and 0.2 MPa for vein wall [4] and "healthy" valve leaflets, respectively. In order to model abnormal cases of fibrotic leaflets, increased values of the leaflets' elastic modulus, range between 200%-1000% higher than the healthy reference state, were examined.

For evaluation of the valve condition, leaflet's wall stress and fluid WSS were analyzed for the base region of the back side of the leaflet, as this is a critical area prone to valve pathology [5], and is expected to develop the maximal wall stress values.

Furthermore, cross-sectional area between the leaflets at the maximal opened state of the valve was compared with the cross-sectional area distal to the valve. Cross-sectional area calculations assumed that the effective cross-sectional area of the vein was circular above and below the valve and elliptic in a maximally-open valve state, with the longer axis equal to the diameter of the sinus and the shorter axis equal to the distance between the leaflets [1].

All of the simulations were constructed under identical flow conditions to assess the fluid and solid forces on the valve leaflet.

RESULTS

The geometry of the model, the flow field and the fluid shear stress for the healthy valve are shown in **Fig 1**. A stasis region can be identified behind the leaflets, in the sinus pockets. The stasis region implies a lower fluid WSS in the sinus pockets. In all four cases of valve elasticity, a jet was formed between the upper parts of the leaflets, as that is the narrowest region for the flow. **Fig 2** shows the effective cross-sectional area between the leaflets in a maximally-open valve state as a percent of the cross-sectional area distal to the valve for each of the tested values of elasticity. As the leaflets become more rigid, their range of motion becomes smaller. As a result, the cross-sectional area between the leaflets decreases and that causes a stronger jet.

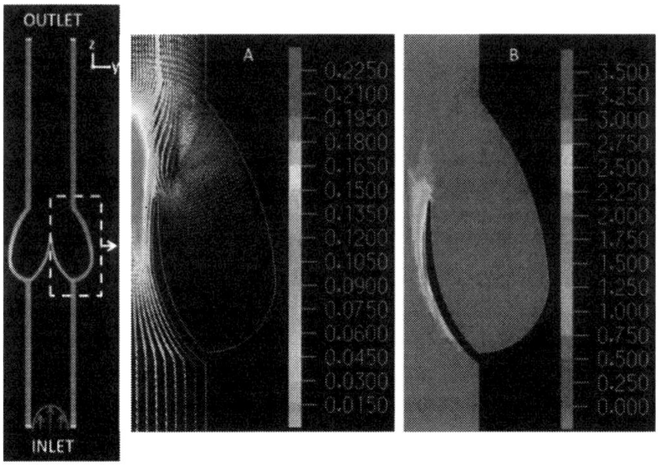

FIG 1. THE GEOMETRY OF THE MODEL IS DESCRIBED ON THE LEFT. YELLOW LINES INDICATE THE FIXED WALL. (A) THE VECTOR FLOW FIELD OF THE HEALTHY VALVE [M/SEC]. (B) FLUID WSS FOR THE HEALTHY VALVE [PA].

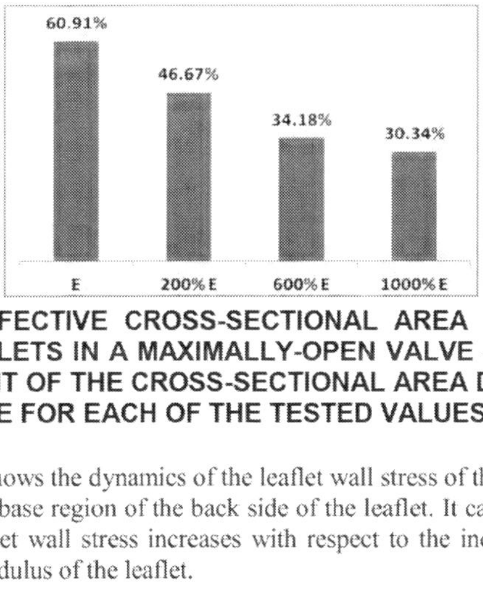

FIG 2. EFFECTIVE CROSS-SECTIONAL AREA BETWEEN THE LEAFLETS IN A MAXIMALLY-OPEN VALVE STATE AS A PERCENT OF THE CROSS-SECTIONAL AREA DISTAL TO THE VALVE FOR EACH OF THE TESTED VALUES.

Fig 3 shows the dynamics of the leaflet wall stress of the four valve cases, at the base region of the back side of the leaflet. It can be noticed that the leaflet wall stress increases with respect to the increase of the elasticity modulus of the leaflet.

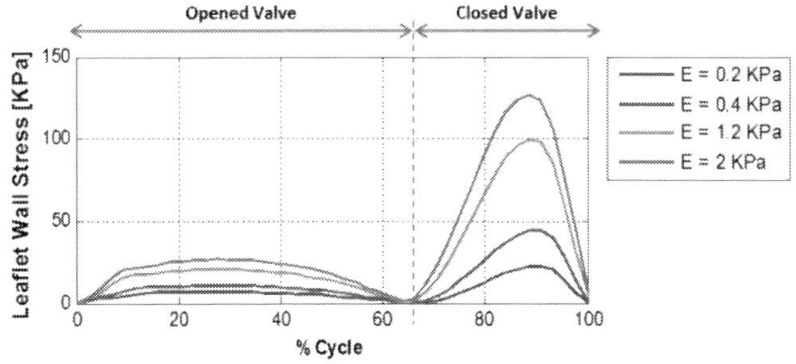

FIG 3. THE DYNAMICS OF LEAFLET WALL STRESS OF THE FOUR VALVE CASES, AT THE BASE REGION OF THE BACK SIDE OF THE LEAFLET.

The wall stress for the healthy valve is shown in **Fig 4**. It can be noticed that the maximal wall stress is developed at the base region of the back side of the leaflet, as expected. The sinus pocket region experiences the lowest fluid shear, and the base of the leaflet experiences the highest wall stress value. Thus, this region is vulnerable for thrombosis and hyperplasia.

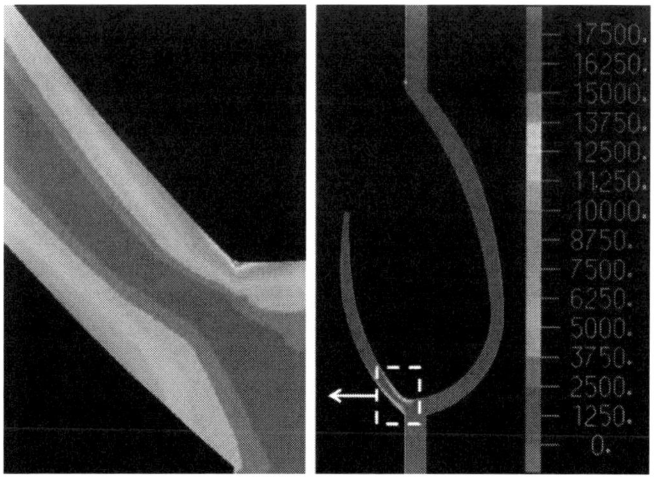

FIG 4. WALL STRESS FOR THE HEALTHY VALVE [PA]. THE MAXIMAL WALL STRESS IS DEVELOPED AT THE BASE REGION OF THE BACK SIDE OF THE LEAFLET, AS EXPECTED.

DISCUSSION

The current study showed that the sinus pocket region experiences the lowest fluid shear, and the leaflet wall stress increases as the leaflets become stiffer. Valve failure in the superficial venous system may be secondary to superficial phlebitis [6]. As a consequence, the valve becomes fibrotic and its elasticity is damaged. Higher fluid shear stress and lower wall stress on the leaflets are hemodynamically more favorable. Areas with relative stagnation are prone to phlebitis because of increased adhesion of thrombotic and inflammatory cells [5]. Phlebitis leads to stiffening of the valve leaflets; the stiffening leads to a worse biomechanical condition of the valve; the worse biomechanical condition (lower fluid shear and higher wall stress) indicates that the valve is vulnerable for phlebitis, and this process activates itself until the valve is totally damaged. Understanding the described mechanism is helpful for elucidating the venous valve stiffness-function relationship in nature and for designing better prosthetic valves.

REFERENCES

[1] Lurie, F et al., *Journal of Vascular Surgery*, Vol. 38, No. 5, Pages 955–961, 2003.
[2] Spivack, DE et al., *Ultrasound in Med & Biol,* Vol. 38, No. 2, Pages 190–194, 2012.
[3] Buxton, GA, Clarke, N, *J Biol Phys*, 32:507–521, 2006.
[4] Buescher, CD et al., *Biotechnol Prog*, 21:938-945, 2005
[5] Chen, HY et al., *Journal of Vascular Surgery: Venous and Lymphatic Disorders*, Vol. 2, No. 2, Pages 188–193, 2014
[6] Eberhardt, RT, Raffetto, JD, *Circulation*, 111:2398-2409, 2005.

SB[3]C2015-213

EFFECTS OF MENISCAL PROPERTIES AND PATIENT VARIABLES ON KNEE JOINT CONTACT MECHANICS

Hongqiang Guo (1), Thomas J. Santner (2), Po-Hsu Chen (2), Amy L. Lerner (3), Suzanne A. Maher (1)

(1) Department of Biomechanics
Hospital for Special Surgery
New York, NY, United States

(2) Department of Statistics
The Ohio State University
Columbus, OH, United States

(3) Department of Biomedical Engineering
University of Rochester
Rochester, NY, United States

INTRODUCTION

The menisci of the knee are C-shaped fibrocartilage disks with a wedge-shape, whose main function is to distribute load across the articulating surface. Although surgical treatment of the damaged meniscus is the most commonly performed orthopaedic procedure, few treatment options exist [1]. Meniscal implants to replace removed tissue have yielded mixed clinical results in Europe [2]. This is in part because the structural characteristics of candidate meniscal implants necessary to ensure function across a wide patient population have not been established. Our goal was to assess the effect of meniscal variables (controllable at manufacture or implantation), and patient variables (which can be set in computational models but cannot be controlled clinically), on knee joint contact mechanics during the activity of walking. To achieve this goal, a statistically-augmented computational platform was used.

METHODS

A knee-specific finite element model (FEM) created from an MRI scan of a cadaveric knee [3] was used. The FEM was driven by knee-specific kinematics as measured from a cadaveric test in which gait was simulated. Specifically, inputs to the FEM were: anterior-posterior translation, medial-lateral translation, varus-valgus angle, internal-external angle, flexion-extension angle, and axial force, which were applied to the mid-point of the femoral epicondylar axis. Articular cartilage and menisci were modeled as elastic materials, and menisci were assumed to be transversely isotropic. Varus-valgus angle (VVA) was used as a tuning parameter so that the balance of force between the medial and lateral compartments in the FEM matched that of the physical model, as previously reported [3]. The effects of seven meniscal variables and two patient variables (Table 1) were studied at three points in the gait cycle where the axial force magnitude peaked – at 4, 14 and 45% (peaks 1-3, respectively) of the gait cycle (Fig. 1). To assess the potential impact of the uncertainty of the experimental VVA measurement, VVA was varied ± 0.2° around tuned values [3].

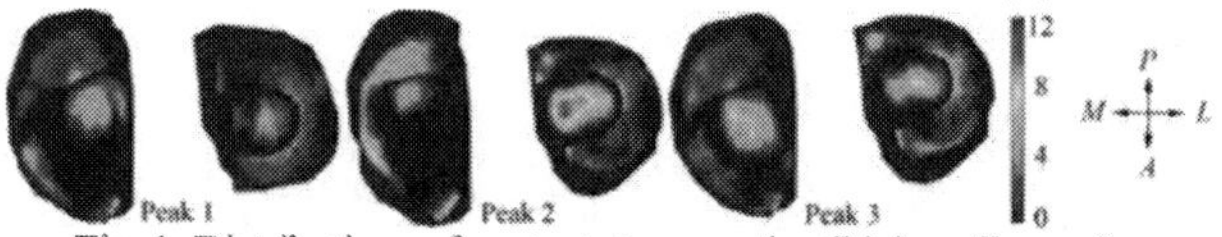

Fig. 1. Distributions of contact stress on the tibial cartilage of a representative model at three peaks of the axial force in the gait cycle.

Table 1. Patient (first two inputs) and meniscal (last seven inputs) variables that were varied in the FEM

Input Variable	*Lower Bound*	*Upper Bound*
Elastic modulus of the articular cartilage, $\mathbf{E_c}$ (MPa) [4]	11.5	35
Body Mass Index (**BMI**) [5]	22	32
Circumferential modulus of the meniscus, $\mathbf{E_{cm}}$ (MPa) [6]	80	220
Radial/axial modulus of the meniscus, $\mathbf{E_{rm}}$ (MPa) [6]	11.5	35
Stiffness of the meniscal attachments, $\mathbf{S_{ma}}$ (N/mm) [7]	140	220
Anterior-Posterior position of the bony insertion point of the anterior meniscal attachments, $\mathbf{IP_a}$ (mm) [8]	-2.4	2.4
Anterior-Posterior position of the bony insertion point of the posterior meniscal attachments, $\mathbf{IP_p}$ (mm) [8]	-2.4	2.4
Length of the anterior meniscal attachments, $\mathbf{L_a}$ (mm) [7]	10	17
Length of the posterior meniscal attachments, $\mathbf{L_p}$ (mm) [7]	6	11

A set of 21 FEM runs was made in which the values of the input variables were determined using a space-filling algorithm [9]. An additional 180 runs were added using a grid pattern approach to ensure adequate coverage of the design space. Main Effect sensitivity indices (SIs) were computed for VVA and the Table 1 variables on the load sharing between the medial and lateral (M/L) compartments; on the load sharing between the cartilage-cartilage (C-C) and cartilage-meniscus (C-M) zones within each compartment; and on the forces in each individual zone [10].

RESULTS

Load sharing between the medial and lateral (M/L) compartments: At peak 1 of the gait cycle, 89% of the variation was due to VVA, while, at peaks 2 and 3, 42-48% of the variation was due to VVA; at peak 2, 16% was due to elastic modulus of the articular cartilage (E_c) and at peak 3, 44% was due to the AP position of the bony insertion points of the posterior meniscal attachments (IP_p).

Load sharing between the C-C and C-M zones in the medial compartment: Of the meniscal variables, the circumferential modulus of the meniscus (E_{cm}) and the AP position of the bony insertion point of the meniscal attachments (IP_a and IP_p) had the greatest effect on variation of load sharing between the C-C and C-M zones (Fig 2. and Table 2); 18-35% of the variation was due to E_{cm}, while 5-19% was due to IP_a and IP_p. Of the patient variables, BMI had the greatest effect. 18% of the variation at peak 2 was due to BMI, while at peaks 1 and 3 9-11% was due to BMI. 18% of the variation at peaks 1 and 2 and 7% of the variation at peak 3 was due to VVA.

Load sharing between the C-C and C-M zones in the lateral compartment: Of the meniscal variables, IP_p had the greatest effect (18-58%) on the variation of load sharing between the C-C and C-M zone (Fig. 2 and Table 2). Of the patient variables, BMI contributed 11% at peak 1, while their effects were less than 5% for peaks 2 and 3. VVA contributed 27% at peak 1. Other variables had minimal effects.

Table 2. Main effect SIs of input variables for load sharing between C-C and C-M zones.

		Patient Variables		Meniscal Variables						
	VVA	E_c	BMI	E_{cm}	E_{rm}	S_{ma}	IP_a	IP_p	L_a	L_p
Medial compartment										
Peak 1	**0.18**	0.03	0.09	**0.21**	0.11	0.01	0.13	0.13	0.03	0.01
Peak 2	**0.18**	0.05	**0.18**	**0.18**	0.09	0.01	0.05	**0.17**	0.01	0.02
Peak 3	0.07	0.02	0.11	**0.35**	0.07	0.01	**0.19**	0.07	0.07	0.00
Lateral compartment										
Peak 1	**0.27**	0.03	0.11	0.12	0.07	0.01	0.14	**0.18**	0.00	0.00
Peak 2	0.00	0.03	0.00	0.01	0.01	0.00	0.04	**0.78**	0.00	0.01
Peak 3	0.05	0.04	0.02	0.04	0.03	0.00	0.01	**0.58**	0.05	0.04

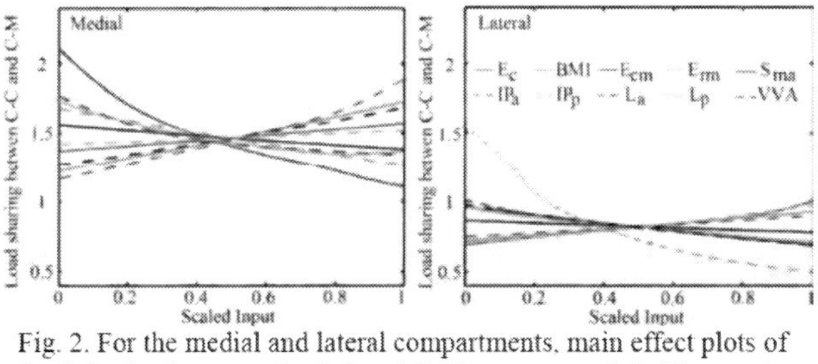

Fig. 2. For the medial and lateral compartments, main effect plots of load sharing between the C-C and C-M zones at peak 3. The x-axis shows scaled inputs, where 0 is the min of each input and 1 the max.

Force in C-C zones: Of the meniscal variables, 16-19% of the variation of force in the medial C-C zone at peaks 2 and 3 was due to IP_p and 15% was due to IP_p in the lateral C-C zone at peak 2. Effects of the other variables were minimal. Of the patient variables, 22-37% of the variation was due to BMI. E_c contributed 19% of the variation in lateral C-C zone force at peak 2. 13-55% variation was due to VVA.

Force in C-M zones: Of the meniscal variables, E_{cm} and IP_a had the greatest effect on the variation of force in the medial C-M zone; 35-43% of the variation was due to E_{cm} and 14-29% due to IP_a. E_{rm} contributed 15% of the variation in the force in medical C-M zone at peak 2. For the variation of force in lateral C-M zone, IP_p had the greatest effect (36-74%). IP_a contributed 18% of the variation of force in the lateral C-M zone at peak 1. Effect of patient variables was minimal (<12%). VVA contributed 16% of the variation of force in the lateral C-M zone at peak 1.

DISCUSSION

Using a statistically-augmented computational platform, the effects of seven meniscal variables, two patient variables, and VVA on knee joint mechanics during simulated gait were studied. While some results were intuitive, for example, VVA was the dominant parameter (89% effect) affecting load sharing between the medial and lateral compartment at peak 1 of the gait cycle, results throughout the remainder of gait were more complex. For example, the effect of VVA decreased to 42% at peak 3 of the gait cycle, while 44% of the variation was due to IP_p. It should be noted that, while all other variables were chosen to span the range of the normal population, VVA was restricted to ± 0.2° about the tuned VVA at each phase of gait that was studied. Such a restriction was necessary because larger changes in VVA required changes in kinematics that could not be realistically modeled in our knees. Thus, the results from the current analysis cannot necessarily be expanded to the goal of understanding the effect of VVA across the wide population on joint mechanics.

Of the meniscal variables that were studied, E_{cm}, IP_a, and IP_p had greatest effect on the knee joint contact mechanics, but the effects of IP_a and IP_p were not consistent throughout gait. E_{cm} had greater effects for the load sharing between the C-C and C-M zones in the medial compartment than in the lateral compartment, which indicates that matching the native anisotropy of the meniscus may be more critical in a medial implant than in the lateral implant. IP_a and IP_p had greater effects on the joint contact mechanics than the length and stiffness of the meniscal attachments. This finding is consistent with a cadaveric study [11] which found that fixation method of the meniscal allograft has a significant effect on the knee joint mechanics. Clinically this suggests that the location of the insertion points of the meniscal attachments and surgical implantation is critical to recover the normal contact mechanics of the patient's knee.

Of the patient variables, BMI had the greatest effect on the knee joint contact mechanics. Increasing BMI had a greater effect in the force variation in the C-C zones than in the C-M zones (Table 3). These findings suggest that by decreasing their weight (hence BMI) a patient can help to lessen the knee contact forces, especially, in the C-C zones. Our statistical model also offers perspective on the importance of BMI relative to other meniscal and patient variables.

In summary, using a statistically-augmented computational platform we assessed the effect of meniscal and patient variables on knee joint contact mechanics throughout the activity of simulated gait. Our approach can be used to understand the mechanical consequences of meniscal variables intended for meniscal repair and replacement.

ACKNOWLEDGEMENTS: This study was supported by NIH R01 AR057343 and Russell Warren Chair in Tissue Engineering.

REFERENCES: [1] Hutchinson, D et al., Am J Sports Med, 42:987-998, 2014. [2] Vrancken, A et al., Int Orth, 37:291-299, 2013. [3] Guo, H et al., J Biomech, Under Review. [4] Li, L et al., J Biomech Eng, 125: 161-168, 2003. [5] Flegal, K et al., J Am Med Assoc, 307:491-497, 2012. [6] Tissakht, M et al., J Biomech, 28:411-422, 1995. [7] Hauch, K et al., J Biomech, 43:463-468, 2010. [8] Johannsen, A et al., Am J Sports Med, 40:2342-2347, 2012. [9] Draguljić, D et al., Technometrics, 54:169-178, 2012. [10] Svenson, J, The Ohio State University, 2011. [11] Wang, H et al., Am J Sports Med, 42:1682-1689, 2014.

SB^{3}C2015
Summer Biomechanics, Bioengineering and Biotransport Conference
June 17-20, 2015,Snowbird Resort, Utah, USA
SB^{3}C2015-214

MOLECULAR DYNAMICS SIMULATIONS OF LIPID BILAYER FAILURE UNDER MECHANICAL STRESSES: TOWARD UNDERSTANDING MECHANICAL HEMOLYSIS

Kenichiro Koshiyama, Taiki Shigematsu, Shigeo Wada

Department of Mechanical Science & Bioengineering,
Graduate School of Engineering Science, Osaka University
Toyonaka, Osaka, Japan

INTRODUCTION

The lipid bilayer is the fundamental structure of all biological cell membranes and acts as a barrier between the inside and outside of a cell. In critical conditions induced by using medical devices (e.g., high shear flows in ventricular assisted devices, extensional flows with shock wave lithotripter, with ultrasound drug delivery, etc.), the tension exerted on the bilayer will cause pore formation and increase the permeability. On the positive side, the transient permealization can be used for effective drug delivery. On the negative side, the permeabilization causes the failure of the bilayer resulted in cell death. Hence, understanding and controlling the failure are an important issue for the effective and safe use of medical devices.

So far, various experiments have been conducted to understand the failure of the bilayer. The studies have found that the critical (i.e. breaking) tension on the bilayer, where the failure occurs, increase with the stretching speeds [1]. Also, it has been found that, in the bilayer including cholesterol, which is abundant especially in red blood cell's (RBC) membranes, the critical areal strain changes nonlinearly depending on the cholesterol concentrations [2]. The failure of the bilayer has been considered to be triggered by the formation of small pore in the bilayer. However, the detailed mechanisms how the pore forms under stretching and how the pore grows into catastrophic size are still unknown because such pore formation is extremely fast and is elusive in experiments.

Hemolysis is the leakage of hemoglobin molecules from RBCs. Stretch-induced hemolysis (i.e., mechanical hemolysis) is an important issue in the development of the blood flow pump. The RBC membrane consists of a lipid bilayer and its underlying cytoskeletal layer of spectrin. The combination of these two layers results in the flexible nature of the RBC membrane. The spectrin network confers high resistance to shear deformation. In contrast, the lipid bilayer offers resistance to membrane area change because lipid molecules are packed densely in the membrane. In the normal deformation, the mechanical behavior of RBC's membranes is governed by the mechanical properties of the spectrin network and various RBC models have been presented based on the mechanics of the membranes. On the other hand, in hemolysis, as the mesh of the spectrin network is larger than that of hemoglobin molecules, the failure of the outmost shell of the RBC's membranes, i.e. breakage of lipid bilayers, is essential. In the past four years, adding to the continuous-level RBC model [3], we have investigated the molecular-level pore formation in stretched lipid bilayers by using molecular dynamics (MD) simulations [4,5]. Especially, we have examined the effects of stretching speeds [4] and the cholesterol concentration [5] on the pore formation. In this paper, we introduce our recent findings about the pore formation in the stretched bilayers.

METHODS

Phospholipid/cholesterol bilayer systems with various cholesterol concentrations were used. The lipid molecules were represented by the united-atom models and the water molecules by the three-cite and rigid-body model. The equation of motions was numerically integrated by Leap-frog algorithm. The equilibrated bilayer systems in liquid-crystal phase were prepared by performing constant number of particles in the system, pressure, and temperature MD simulations with velocity-rescaling thermostat and Parrinello-Rahman barostat. To investigate the molecular details of the pore formation in stretched bilayers, we performed MD simulations of the phospholipid bilayers under two different equibiaxial stretching conditions: first, unsteady stretching with various stretching speeds, and second, quasistatic stretching. Unsteady stretching of the bilayer was expressed by a proportional scaling of molecular coordinates and simulation box lengths [4]. Quasistatic stretching of the bilayer was expressed by a series of constant area, temperature, and the pressure perpendicular to the bilayer plane MD simulations with various simulation box areas. The details of the simulation algorithm and conditions can be found in Ref. [4] and [5].

RESULTS AND DISCUSSION

Figure 1 shows the equilibrated bilayer structure for different concentrations of cholesterol obtained by MD calculations. For the bilayer including cholesterol, the lipid molecules tend to be straight, which is well known phenomenon induced by including cholesterol, so called ordering effects [5].

The pore structure penetrating the bilayer is formed under unsteady stretching, when the applied areal strains exceed critical values (Panels A and B in Fig. 2). For the high-speed stretching, the critical areal strain becomes larger and the small pores (multipore) temporarily form in a small area, unlike a single pore formation for the low-speed stretching. The dependency of the critical areal strain on the stretching speeds implies the viscoelastic behavior of the bilayer [1].

The critical areal strain of phospholipid/cholesterol bilayer is larger than that of pure lipid bilayers (Fig. 3). This qualitatively agrees with results obtained by pipet aspiration experiments for phospholipid/cholesterol vesicles [2]. Under stretching, the two monolayers of the phospholipid/cholesterol bilayer interpenetrate and form an interdigitated gel-phase-like structure just before the pore formation (Fig. 4). For the pore formation, water molecules must penetrate into the hydrophobic region of the bilayer [6]. Hence, permeability of water in the bilayer governs the pore formation. As the bilayer in gel-like phase is known as less permeable, we deduce that the transient phase transition in stretched phospholipid/cholesterol bilayer may be a factor for the large critical areal strain. Interestingly, the transient phase transition is pronounced for 40% cholesterol concentration, which corresponds to the cholesterol concentration of RBC's membrane. Currently, we have quantified the permeability changes of the stretched phospholipid/cholesterol bilayers by using inhomogeneous solubility-diffusion theory [7] and this will be reported in the near future.

SUMMARY

We presented our recent findings about the pore formation in the stretched bilayers obtained by stretching MD simulations. Especially, we showed the effects of stretching speeds and the cholesterol concentration on the pore formation. The findings presented here may help explain the toughness of RBC's membranes and mechanical hemolysis from molecular scales.

ACKNOWLEDGEMENTS

We would like to thank Prof. M. Nakamura for his useful comments and suggestions. This work was partly supported by the Japanese Ministry of Education, Culture, Sports, Science and Technology (22300155, 23686031, and 24240073) and the MEXT project "Creating Hybrid Organs of the future" at Osaka University.

REFERENCES

[1] Evans et al., *Biophys. J.*, 85: 2342-2350, 2003.
[2] Needham and Nunn, *Biophys. J.*, 58: 997–1009, 1990.
[3] Nakamura et al., *Int. J. Numer. Meth. Biomed. Engng.*, 30:42–54, 2014.
[4] Koshiyama and Wada, *J. Biomech.*, 44: 2053–2058, 2011.
[5] Shigematsu et al., *Chem. Phys. Lipid*, 183:43–49, 2014.
[6] Koshiyama et al., *Phys. Rev. Lett.*, 105:018105, 2010.
[7] Marrink et al., *J. Phys. Chem.*, 98:4155-4168, 1994.

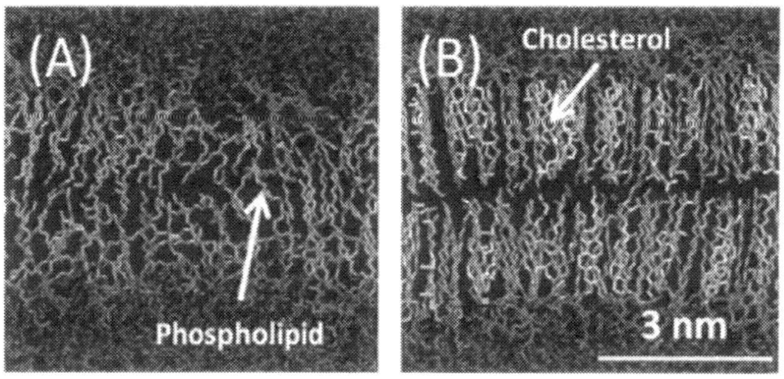

FIGURE 1 Side views of lipid bilayers for pure (A) and 40% cholesterol concentration (B).

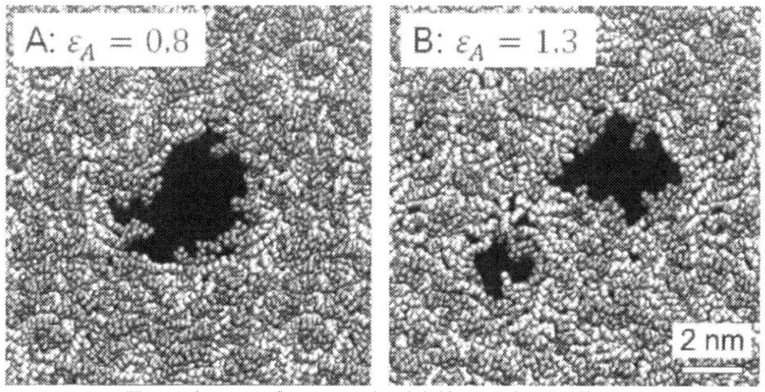

FIGURE 2 Top views of a stretched bilayer including pore for low-speed (A) and high-speed stretching (B). ϵ_A is a areal strain.

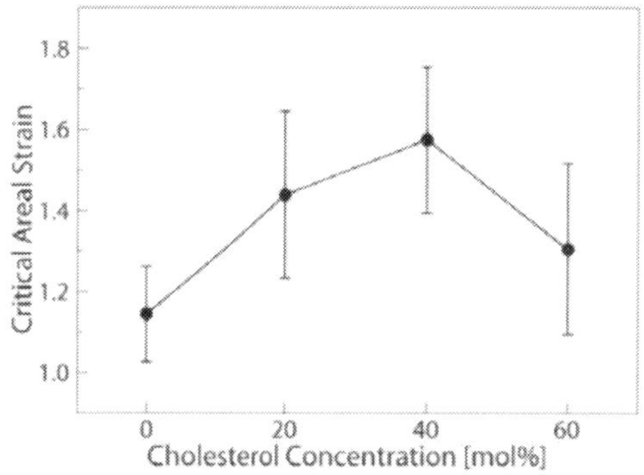

FIGURE 3 Variation of critical areal strains for different cholesterol concentrations obtained by unsteady stretching simulations. The error bars indicate the standard deviation for 20 samples.

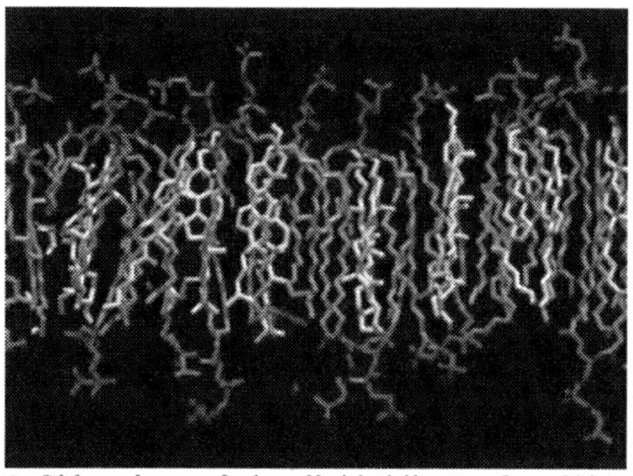

FIGURE 4 Side view of the lipid bilayer for 40% cholesterol concentration just before the pore formation.

ANALYTICAL APPROXIMATION FOR PREDICTING STRESS FIBER REMODELING DUE TO DYNAMIC MECHANICAL STIMULI

Tommaso Ristori (1,2), Sandra Loerakker (1,2), Cees W.J. Oomens (1), Frank P.T. Baaijens (1,2)

(1) Department of Biomedical Engineering
Eindhoven University of Technology
Eindhoven, The Netherlands

(2) Institute for Complex Molecular Systems
Eindhoven University of Technology
Eindhoven, The Netherlands

INTRODUCTION

The mechanical properties of soft tissues depend to a large extent on the organization of the collagen network, which is the main load-bearing component in most tissues. Next to external loads, the collagen network is remodeled by contractile forces exerted by the cells, which are determined by the organization of actin stress fibers [1][2][3]. Understanding the processes of collagen and stress fiber remodeling in response to mechanical stimuli is essential for the development of functional engineered tissues, where adverse remodeling effects such as leaflet retraction in tissue-engineered heart valves (TEHVs) [4] are prevented. Due to the complexity of the phenomena involved, computational models are necessary to obtain a mechanistic understanding of tissue remodeling.

Recently, we have developed two computational models for the simulation of collagen and stress fiber reorganization [5][6]. We have applied these models to predict remodeling of soft tissues under static conditions. However, predicting tissue remodeling due to dynamic mechanical stimuli using direct integration of the differential equations is not feasible due to the excessive computational costs associated with the different time scales of the mechanical stimuli and the remodeling processes.

In this study, we derived an analytical approximation of the stress fiber remodeling algorithm, which allows for predicting stress fiber remodeling in response to dynamic conditions while only a small part of the loading history has to be applied. Using this approach, we could drastically reduce the computational costs while the prediction of the stress fiber orientations was similar to the results using direct numerical integration. Consequently, this method provides a valuable technique for future investigations on in vivo tissue remodeling due to dynamic mechanical stimuli.

METHODS

The stress fiber response to mechanical stimuli was modeled using a system of ordinary differential equations [5][6], with the amount of stress fibers in each direction as unknown variables. The coefficients of this system are constant in case of static conditions, and periodic with dynamic mechanical stimuli. While an analytical solution can easily be found for the first case, this is not possible for the second. Hence, we estimated the amount of stress fibers during dynamic conditions using an approximation deduced from the analysis of the static solution. In each simulation, dynamic mechanical loads were only applied during a small part of the analysis. In the remainder of the simulation, the dynamic loads were substituted with a constant mechanical load, while the dynamic features of the load were accounted for in the analytical approximation. For verification, we performed the same simulations presented in the study of Obbink-Huizer et al. [6] using both direct numerical integration and the analytical approximation. Then, the model predictions and the computational time were compared between both approaches.

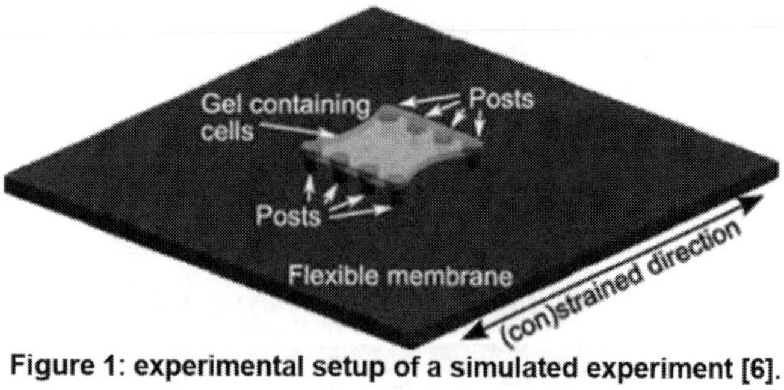

Figure 1: experimental setup of a simulated experiment [6].

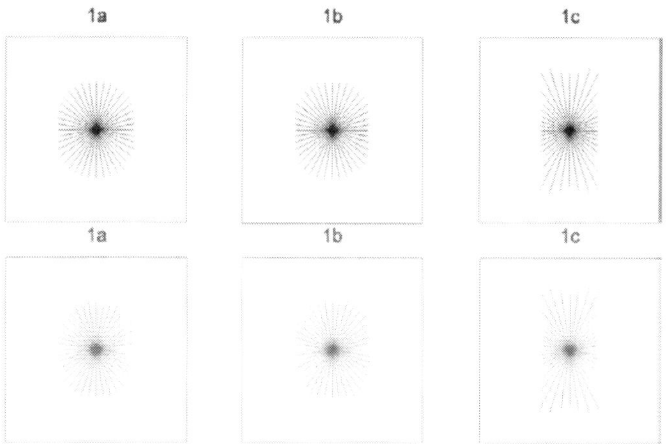

Figure 2: Prediction of the stress fiber orientations using direct numerical integration (black [6]) and the analytical approximation (blue) for cells seeded on a high stiffness substrate cyclically strained with increasing frequencies from left to right. The load was applied in the horizontal direction.

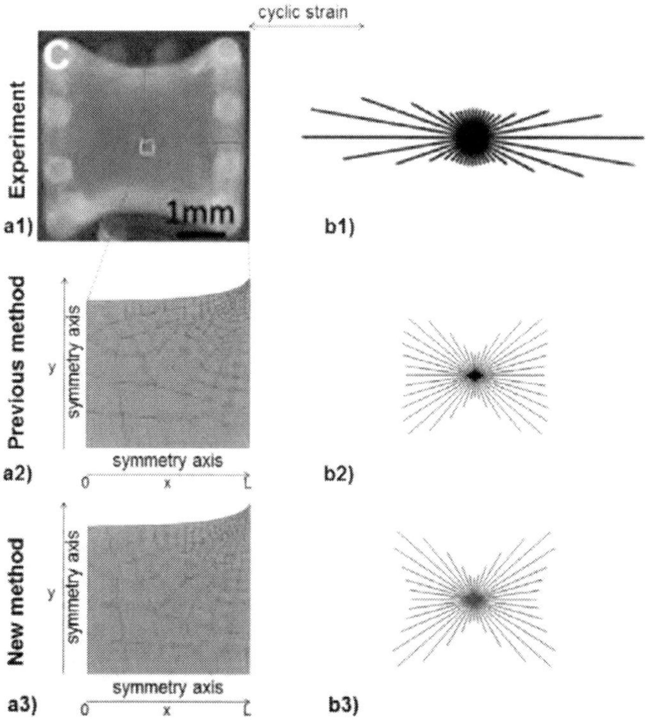

Figure 3: Experimental results (a1, b1 [6]) and prediction of the compacted tissue (a) and stress fiber orientation (b) using the old (a2, b2 [6]) and herein proposed method (a3, b3) for cells seeded on a low stiffness substrate cyclically strained in the horizontal direction.

RESULTS

In general, the results obtained with both approaches were very similar, although the accuracy of the analytical approximation appeared to depend on the deformation profile (Fig. 2, 3). In particular, when the cells were subjected to small displacements, the approximation was almost perfect (Fig. 2). When the displacements were higher, the difference in results between the two strategies increased. Nevertheless, the main stress fiber orientation observed experimentally was always well predicted (Fig. 3). Furthermore, the differences between the experimental results and the simulation results with direct numerical integration were more significant compared to the differences between the two computational approaches (Fig. 3).

Most importantly, there was a tremendous decrease in the computational time required to perform the simulations when the analytical approximation was used instead of direct integration (Fig. 4). This decrease was particularly evident in simulations with higher or more complex deformation profiles.

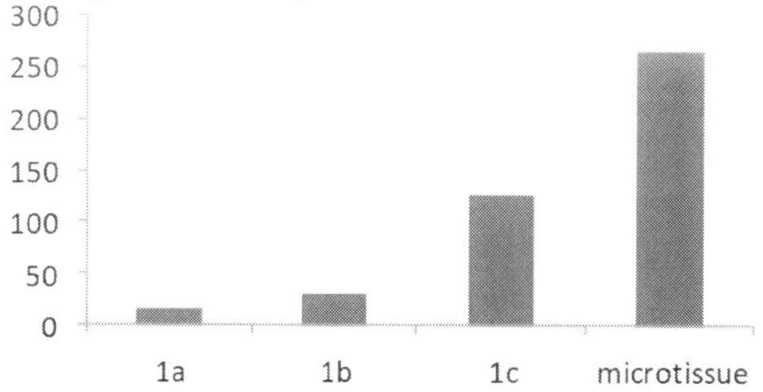

Figure 4: Decrease in computational time, represented by the ratio of the computational time using direct integration to the computational time using the analytical approximation.

DISCUSSION

The aim of this research was to develop an approach where stress fiber remodeling due to dynamic mechanical stimuli can be predicted, without the requirement of applying the complete load history. We derived an approximation for predicting stress fiber remodeling, and compared the results of this approach with the results obtained using direct numerical integration [6]. The results obtained with both approaches were similar, while the computational time was much lower when the analytical approximation was used (Fig. 2,3,4). Given the fact that the difference in results between the computational and experimental results was much larger than the difference in results between the two computational approaches, we consider the loss in accuracy due to the approximation procedure as rather small.

On the other hand, the tremendous decrease in computational time using this new approach has important implications. Without the requirement of applying the complete load history, this approach will allow for bridging the different time scales of dynamic mechanical stimuli and remodeling processes, such that predicting tissue remodeling in response to cyclic mechanical stimuli will be feasible.

ACKNOWLEDGEMENTS

This study has been funded by the European Union's Seventh Framework Programme, as part of the TECAS-ITN project.

REFERENCES

[1] Ghibaudo, M et al., *Soft Matter*, 4:1836-1843, 2008.
[2] Wang, JH et al., *J Biomech*, 36(1):97-102, 2003.
[3] Faust, U et al., *Plos One*, 6(12):e28963, 2011.
[4] Driessen-Mol, A et al., *J Am Coll Cardiol*, 63(13):1320-9, 2014.
[5] Loerakker, S et al., Biomech Model Mechanobiol, 13(5):985-1001, 2014.
[6] Obbink-Huizer, C et al., *Biomech Model Mechanobiol*, 13(1):227-36, 2014.

SB^{3}C2015
Summer Biomechanics, Bioengineering and Biotransport Conference
June 17-20, 2015,Snowbird Resort, Utah, USA
SB^{3}C2015-216

CELLULAR YOUNG'S MODULUS AS A NOVEL STEMNESS MARKER IN THE CORNEAL LIMBUS

Tom Bongiorno (1), Jena Chojnowski (2), James D. Lauderdale (2), Todd Sulchek (1,3)

(1) The G. W. Woodruff School of Mechanical Engineering
Georgia Institute of Technology
Atlanta, GA, USA

(2) Department of Cellular Biology
University of Georgia
Athens, GA, USA

(3) The Parker H. Petit Institute for Bioengineering and Bioscience
Georgia Institute of Technology
Atlanta, GA, USA

INTRODUCTION

Each year, over 1 million Americans suffer corneal damage, which is typically treated via corneal or limbal tissue transplantation. Recently, limbal stem cells (LSCs), which are known to continuously renew the corneal epithelium [1], have shown promise in cellular regeneration and functional improvement following corneal damage. Cell implants containing greater than 3% putative LSCs were 7-fold more clinically successful than implants containing lower LSC fractions [2]. However, improved therapy via LSC-enriched tissue implantation is precluded by a poor understanding of the precise molecular signature of LSC phenotype. Therefore, the objective of this study was to investigate mechanical properties as novel stemness indicators for limbal cells. The results of this study can be used to enrich for LSCs within implanted tissues, thereby enhancing clinical success for corneal damage patients.

METHODS

Purified human LSCs ("limbal") were obtained from Life Technologies and aseptically cultured in stem cell maintenance media. Human eyes were obtained from the Georgia Eye Bank, and the corneas were dissected and dissociated to directly obtain all cell types from the cornea, including stem cells and differentiated cells ("corneal"). Differentiated corneal cells were obtained by removing the stem cell maintenance media from the LSCs for 3 weeks ("differentiated LSC").

Cells were plated on poly-L-lysine-coated glass dishes and adhered for 16-24 h. Atomic force microscopy (AFM, Fig. 1)

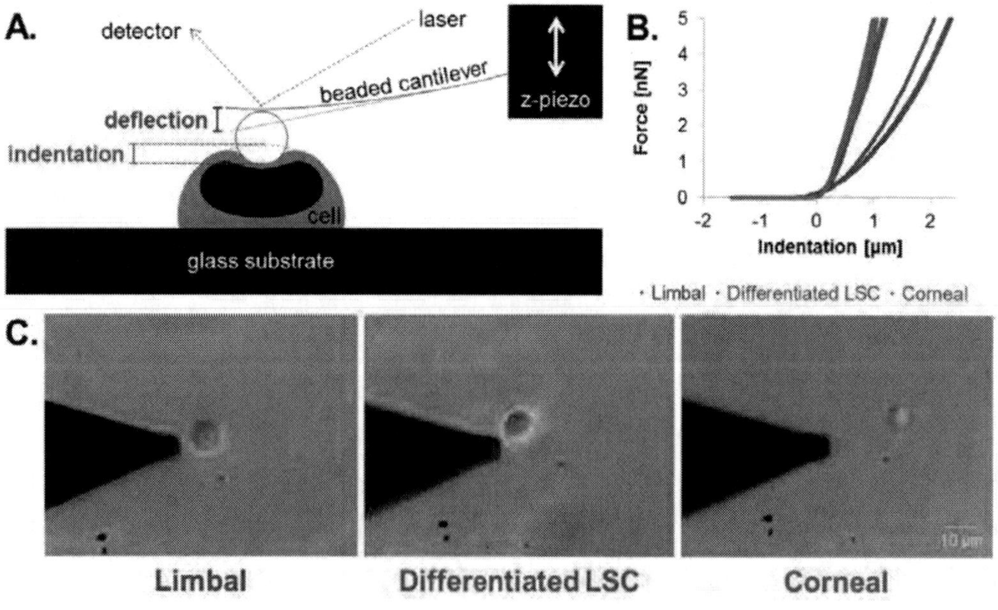

Figure 1. Mechanical Phenotyping via Atomic Force Microscopy. A) A beaded cantilever was used to probe the mechanical properties of individual cells. By comparing the cantilever deflection (blue) to the indentation (green), force-indentation plots were obtained for each cell and analyzed to determine the Young's modulus. **B)** Representative force-indentation curves indicated that the limbal cells (blue) were softer than the differentiated LSCs (green) and corneal cells (red). **C)** Phase contrast microscopy was used to both position the cantilever over each individual cell and measure the diameter and aspect ratio of each cell. Scale bar, 10 μm.

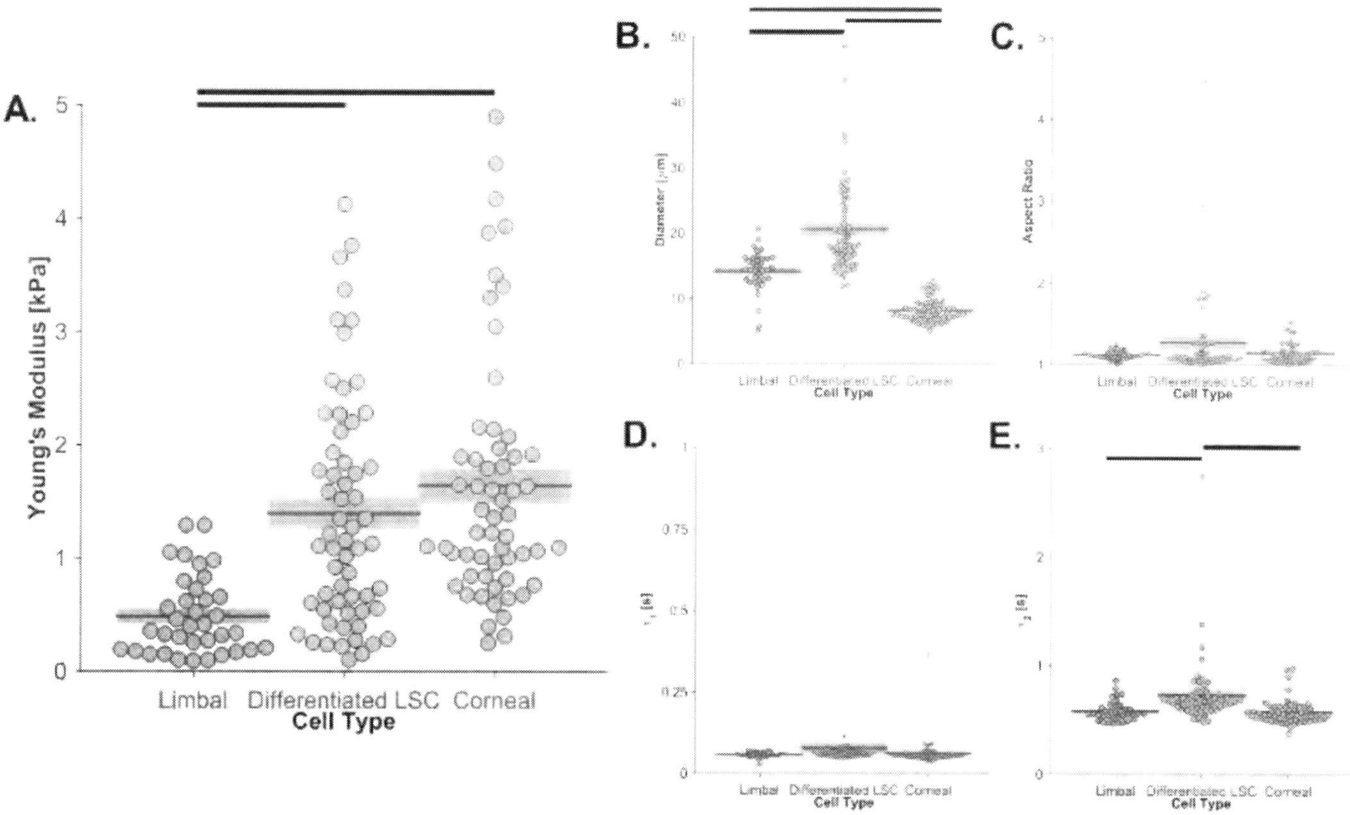

Figure 2. Cell Mechanics. A) The limbal cells had a significantly lower Young's modulus than both differentiated LSCs and corneal cells, suggesting that Young's modulus may be used as a phenotype indicator for limbal cells. **B)** The diameters of limbal cells, differentiated LSCs, and corneal cells were significantly different from each other. **C-D)** There were no significant differences in aspect ratio or the first viscoelastic time constant, $\tau1$. **E)** The differentiated LSCs had a significantly higher second viscoelastic time constant, $\tau2$, than both limbal and corneal cells. Colored bars indicate mean ± standard error. Populations connected by black bars were significantly different (Holm's adjusted p-values, α=0.10).

was employed to investigate cellular mechanical properties. A thermally calibrated cantilever with a beaded tip (k≈15 pN/nm, d=4.7 μm) was used to measure the Young's modulus (via the Hertzian model) and viscoelastic time constants of each cell. Cellular diameter and aspect ratio were obtained from ImageJ analysis of phase contrast images captured during AFM. Due to the non-normal distribution of each parameter (Shapiro-Wilk W Test, α=0.05), bootstrapping ANOVA was used to discern statistically significant differences.

RESULTS

The limbal cells had a significantly lower Young's modulus than both differentiated LSCs and harvested corneal cells (Fig. 2), suggesting that Young's modulus may be used as a phenotype indicator for cells derived from the cornea. Neither cell diameter, aspect ratio, nor the viscoelastic time constants serve as specific markers of LSC phenotype.

DISCUSSION

For the first time, the mechanical properties of corneal cells were characterized. Importantly, cellular Young's modulus is reported as a phenotype indicator of LSCs relative to other cells found in the corneal limbus. The limbal cells were smaller than the differentiated LSCs, which is consistent with a previous finding [3] that naive LSCs are smaller than the more differentiated corneal cells. However, since the corneal cells obtained from tissue do not reflect the published finding, the size data presented here may not accurately represent cell size immediately after dissociation from a freshly harvested tissue.

The findings of this study suggest that the microfluidic device developed in our lab [4] can be used to quickly enrich LSCs from a harvested cornea via cellular Young's modulus-driven sorting. The enriched LSC population can then be used to generate corneal tissue implants, which are expected to yield superior clinical outcomes compared to tissue implants that are directly harvested from the cornea.

ACKNOWLEDGMENTS

The authors gratefully acknowledge Wenwei Xu for assistance with AFM. Funding was provided by the University of Georgia-Georgia Tech Partners Seed Grant.

REFERENCES

[1]Ahmad, S., *Stem Cells Transl Med,* 1:110-115, 2012.
[2]Rama, P. et al., *N Engl J Med,* 363:147-155, 2010.
[3]Romano, A. et al., *Ophthalmol Vis Sci,* 44:5125-5129, 2003.
[4]Wang, G. et al., *PLoS One,* 8:e75901, 2013.

SB³C2015
Summer Biomechanics, Bioengineering and Blotransport Conference
June 17-20, 2015,Snowbird Resort, Utah, USA
SB³C2015-217

Characterization of endothelial cell permeability on a biomimetic blood vessel platform

Antony Thomas (1), Chris Uhl (2), Yaling Liu (1,2)

(1) Bioengineering Program
Lehigh University, Bethlehem, PA, 18015, USA

(2) Department of Mechanical Engineering and Mechanics
Lehigh University, Bethlehem, PA, 18015, USA

INTRODUCTION

Blood vessels are permeable, which allow for transportation of small molecules such as water, ions, nutrients, even a whole cell across vessels. This capability is called vessel permeability [1]. Endothelial cells (ECs) line the blood vessel lumen and form a semi-permeable barrier through cell-cell conjunction. This basal level vascular permeability is essential in the selective transport between blood and the interstitial space of all organs. Such permeability barrier is maintained through tight cell-cell junctions and is controlled by growth factors, cytokines and other stress related molecules [2]. Disruptions in the EC layer barrier can result in increased permeability.The permeability of blood vessels increaseon detection of a potentially harmful substance, foreign organism, or injured tissue cells, to supply plasma proteins to the extravascular compartment in need of repair. This is mediated by several inflammatory mediators (e.g., Vascular endothelial growth factor- (VEGF)-A, Tumor necrosis factor- (TNF)-α, Interleukin- (IL)-6, and IL-1β) and delivers immunoglobulins, antiproteases, constituents of the complement and coagulation systems and other acute phase proteins to the site to act in local host defense and initiate tissue repair. Characterizing permeability of the ECsthat form the inner layer of blood vessels provides fundamental physiological information and is essential for evaluating drug and other biomolecules uptake. By understanding thiswe can engineer nanodrug carriers with optimized ligands and coating densities to facilitate trans-endothelial drug delivery with high efficacy at sites of inflammation[17].

The dynamic nature of vascular permeability and the environmental factors involved is still elusive. The complex cellular mechanisms contributing to increase in permeability of blood vessels is partially understood. Scientists have employed in vivo and in vitro platforms to understand the underlying mechanism. In vitro works are also mostly performed under static conditions without consideration to the fluid shear stress (FSS) conditions the endothelium are exposed to in vivo. In this work we used a versatile biomimetic blood vessel platform that facilitate the study of independent and combined interplay of various mechanisms contributing to increased permeability in blood vasculature. A fluorescent nanosphere based assay system is used to quantify the vascular permeability.

METHODS

Design of microfluidic device and On-chip cell culture

The microfluidic device channels are made of Polydimethylsiloxane through a standard photolithography process [58]. The upper channel is 20 mm long and has a width of 350 μm. The lower channel is 5 mm long and 1 mm wide and is arranged perpendicular to the top channel (Figure 1 A). The top channel is bonded to the bottom channel with the semipermeable membrane in between through O_2 plasma bonding and liquid PDMS along the channel edges. The porous PC membrane is cell culture friendly and provides the base for cell growth and proliferation. Such membrane based mimicking of biological platforms has been widely demonstrated [59-62]. Transport of materials between the upper and lower channels is limited to diffusion through the membrane pores. This allows for spatially controlled and independent transport between the channels without leakage.

ECs are constantly exposed to blood flow as they form the 'tunica intima' or the inner coat of a blood vessel. The exposure of ECs to this shear stress activates a number of cellular mechano-sensors, which along with adaptor molecules regulates the expression of genes and proteins relevant to EC functions in healthy and disease conditions [63-64]. Maintaining the quiescent phenotype of ECs is critical to the success of a biomimetic blood vessel model and subjecting the ECs to native *in vivo* FSS is important [45]. Pro-inflammatory and proliferative pathways of vascular ECs become down-regulated when the EC layer is exposed to directed and sustained FSS [46]. To validate the bio-mimetic nature of our device, confluent layer of BAOECs were subjected to physiological levels of FSS (12 dyne/cm2). BAEOCs were exposed to steady and directed laminar FSS for a time period of 5 days to check the longevity of maintaining a confluent cell layer. Confluent and flow aligned BAOEC layers after 5 days looked similar to the F-actin stained illustrative image of BAOECs at 12 dyne/cm² FSS after 24 hrs in Figure 1B (b), confirming the maintenance of a quiescent and sustaining phenotype by the BAOECs.

1 A) was developed by studying the absorption of these particles under different concentrations ($5*10^{12}$-10^{14}/ml) using a UV-Vis spectrophotometer (Nanodrop 2000). These particles have a characteristic absorption curve ~450 nm as it is close to the particle's fluorescence excitation wavelength (Fig. 2 B).

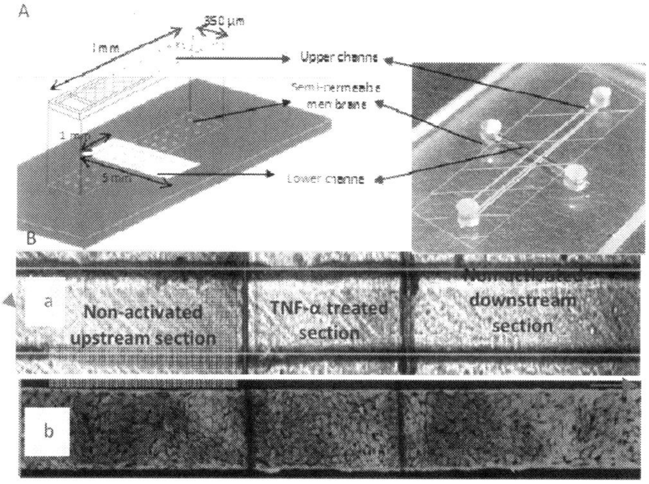

Figure 1: A, Graphical schematic and photograph of the bi-layer device showing the upper (marked with yellow lines) and lower (marked with red lines) channel separated by the semi-permeable membrane. B, (a) Bright field image of the top (marked with yellow lines) and bottom (marked with red lines) channel defining the upstream, TNF-α activated and downstream sections, (b) Fluorescence labeled actin cytoskeleton (FITC phalloidin) images of confluent BAOEC layer aligned to flow (12 dyne/cm^2 FSS for 24 hrs). The cells grow on the semi-permeable membrane in the top channel of the device. Arrow shows flow direction

Before using the nanobead based assay to quantify the vessel permeability the integrity and confluence of the fixed monolayer of endothelial cells were examined under a fluorescence microscope. Then 30 nm fluorescent spheres were flowed in the upper channel for 5 minutes under a physiologically relevant FSS of 12 dyne/cm^2. NPs that permeate to the lower channel through the intercellular gaps and the membrane are measured/collected from the lower channel. The quantity of buffer available in the lower channel is fixed. On performing UV-Vis spectroscopy on this NP sample and comparing to the particle concentration curve we quantified the amount of particles that passed from the upper channel to the lower channel through the inter cellular gaps (Fig. 2 B).

By evaluating the amount of particles that permeated to the lower channel for different TNF-α treatment time we characterized the dynamic nature of increase in vascular permeability and cell-cell barrier dysfunction which is important in regulation of fluid and solute balance between the vascular space and the surrounding tissue (Fig. 2 C). Vascular permeability is lowest for a confluent layer of endothelial cells not subjected to flow. On subjecting the cells to physiological levels of flow for 24 hrs the permeability increases as documented [7]. On TNF-α treatment It is observed that the endothelial cell layer permeability increases after 3 hrs of cytokine activation at 10 U/ml concentration, and the trend continues till 6 hrs. A comprehensive and quantified analysis of this vascular permeability dynamics will be performed using particles of a range in size (10-100 nm), different TNF-α concentrations (0-100 U/ml) and treatment times (0-12 hrs).

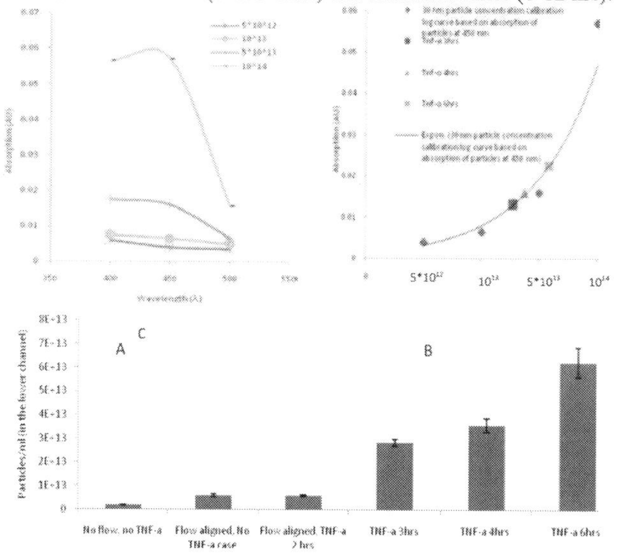

Figure 2: Quantifying the amount of 30 nm fluorescent particles that permeated through the endothelial cell layer for different TNF-α and control cases. A, Particle concentration control curve based on the absorption of 30 nm particles from 400-500 nm wavelength for different particle concentrations ($5*10^{12}$-10^{14}/ml). B, Quantifying the amount of particles that passed from the upper channel to the lower channel by comparing data sample to particle concentration control curve at 450 nm. C, Particle concentration of the sample collected from the lower channel for control cases and after TNF-α treatment for 2, 3, 4 and 6 hrs.

TNF-α treatment to induce inflammation thus increase permeability

TNF-α and other cytokines are known to induce a time and dose dependent increase in EC monolayer permeability [1-2]. TNF-α is found to produce a concentration and time dependent increase in permeability from 1 to 100 U/ml which peaked around 12 hr [3]. BAOECs were treated with TNF-α to locally induce cytokine mediated increase in permeability on the EC layer. DMEM media containing 10 Units (U)/ml of TNF-α was introduced from the lower channel of the device while the cells were subjected to flow at 12 dyne/cm^2 in the upper channel. TNF-α diffuses through the pores of the membrane separating the lower from the upper channel and spatially controls the direct activation of BAOECs only present above the lower channel. TNF-α induced increase in vessel permeability was studied for 2, 3,4 and 6 hr of treatment time. Once the test condition was fulfilled the cells were stained (CellMask™, Life technologies) to fluorescence tag the BAOECs and fixed with 3.7% paraformaldehyde.

Quantifying vascular permeability

Vascular permeability occurs through intercellular gap formation. Our goal here is to apply an assay that can quantify cell permeability based on the dimension and frequency of intercellular gap formation for different dosage and time based cytokine treatments.

RESULTS

We demonstrate a proof of concept of our nanobead based assay to quantify vessel permeability using 30 nm carboxylate coated fluorescent nanosphere (Sigma). The size range of intercellular gap diameter formed on inflammatory triggering is between 30-50 nm [4-6], making our 30 nm particle an ideal candidate. The carboxylate coating resists particle attraction towards cells, and the particles being fluorescent (λ_{ex} ~470 nm; λ_{em} ~505 nm) enabled quantified measurement of the permeability. A particle concentration curve (Fig.

CELL-MEDIATED COMPACTION AND COLLAGEN REMODELING IN TISSUE-ENGINEERED HEART VALVES DUE TO DYNAMIC LOADING CONDITIONS

Sandra Loerakker (1,2), Tommaso Ristori (1,2), Frank P.T. Baaijens (1,2)

(1) Department of Biomedical Engineering
Eindhoven University of Technology
Eindhoven, The Netherlands

(2) Institute for Complex Molecular Systems
Eindhoven University of Technology
Eindhoven, The Netherlands

INTRODUCTION

The currently available heart valve prostheses do not consist of living tissue and, hence, they cannot remodel in response to changing demands. Tissue-engineered heart valves (TEHVs) may be able to overcome this limitation due to their inherent remodeling capacity. Unfortunately, the remodeling process remains poorly understood, which has become evident in many pre-clinical studies where TEHVs lost their functionality over time due to leaflet retraction (Fig. 1) [1-5].

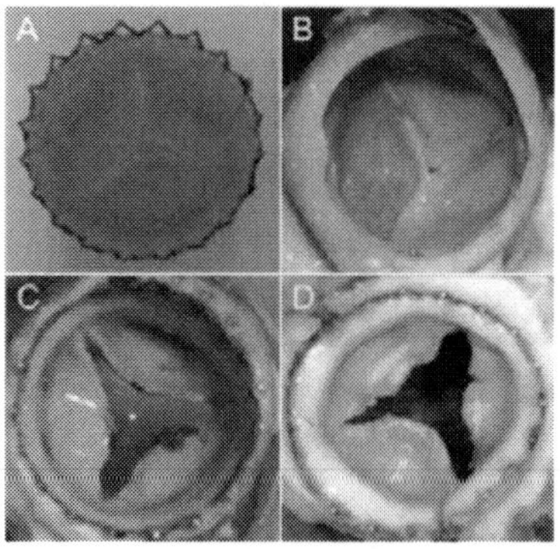

Figure 1: TEHV before (A) and 8 (B), 16 (C), and 24 (D) weeks after implantation, demonstrating the gradual development of leaflet retraction (adapted from [5]).

Leaflet retraction develops due to contractile forces exerted by the cells in combination with cell-mediated collagen contraction and remodeling. The goal of the present study was to analyze these processes in TEHVs using computational modeling, and investigate the development of valvular insufficiency as a function of the dynamic pressure difference over the valve. In addition, different degrees of cell contractility were studied to determine the potential influence of different cell phenotypes.

METHODS

The engineered tissue was modeled as a mixture of contractile stress fibers, collagen fibers, and isotropic tissue components [6]. The following remodeling processes were included:

- Stress fiber remodeling in response to the local strains and strain rates in the valve [7].
- Cell-mediated collagen contraction depending on the stress exerted by the cells [6].
- Collagen remodeling due to strain-dependent degradation and oriented production in the directions of the cells [6].

The remodeling algorithm was applied to a heart valve geometry, where the pressure was applied to the arterial side of the leaflets. Two different dynamic loading conditions were investigated: pulmonary pressure conditions (0-2 kPa), and aortic pressure conditions (0-10 kPa). In addition, the maximum cell contractility was varied between 10 and 100 % of the value in the original model [7], where 100 % corresponds with the myofibroblast phenotype.

To reduce the computational costs associated with dynamic loading conditions during remodeling, we used an analytical approximation to predict stress fiber remodeling. In this way, the number of load cycles in the simulations could be drastically reduced, and only the amplitude of the pressure had to be applied during a large part of every simulation. Collagen contraction and remodeling both depended on the average deformations in the valve during each load cycle.

RESULTS

Most importantly, the loaded configurations of the valves after remodeling (Fig. 2) indicate that valve closure is not complete for pulmonary pressure conditions when myofibroblasts are present in the valve (Fig. 2, top left), which corresponds with the results of many pre-clinical studies. Valve closure under these low-pressure loading conditions was only present when the maximum cell contractility was decreased to 10 % of the original value (Fig. 2, bottom left). For aortic pressure conditions, valve closure was predicted to be present for all degrees of cell contractility (Fig. 2, right column). Unsurprisingly, the strains in the valve were predicted to increase with the amount of pressure applied to the valve, as well as when cell contractility is decreased.

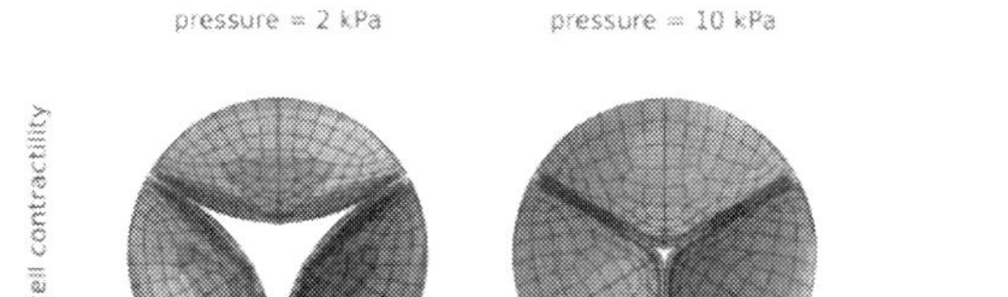

Figure 2: Radial strains in TEHVs at pulmonary and aortic diastolic pressure conditions after remodeling due to different degrees of cell contractility.

Quantification of the valve opening as well as the coaptation area in the loaded configuration (Fig. 3), confirms that valve closure with sufficient coaptation is only present at pulmonary loading conditions when cell contractility is very low (10 %). For aortic pressure conditions, valve closure and sufficient coaptation was predicted for the complete range of cell contractility.

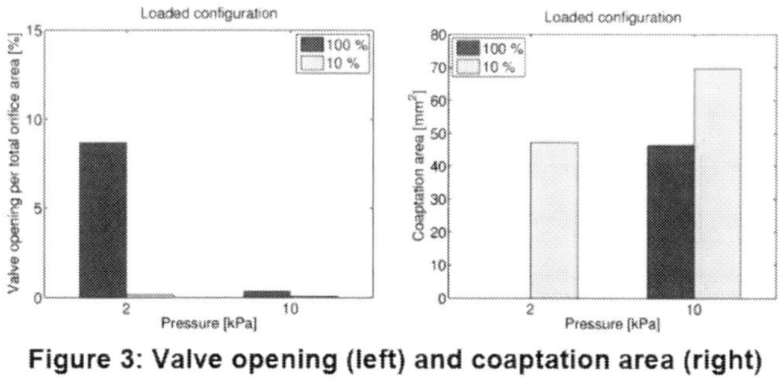

Figure 3: Valve opening (left) and coaptation area (right) during diastolic pressure conditions.

The different loading conditions and degrees of cell contractility also had an effect on the collagen architecture in the valve after remodeling (Fig. 4). For pulmonary pressure conditions, a circumferentially aligned collagen network was predicted when cell contractility was high (Fig. 4, top left). However, a large decrease in cell contractility resulted in a radial alignment of the collagen network (Fig. 4, bottom left). For aortic pressure conditions, a circumferential alignment of the collagen network was predicted for all degrees of cell contractility (Fig. 4, right column).

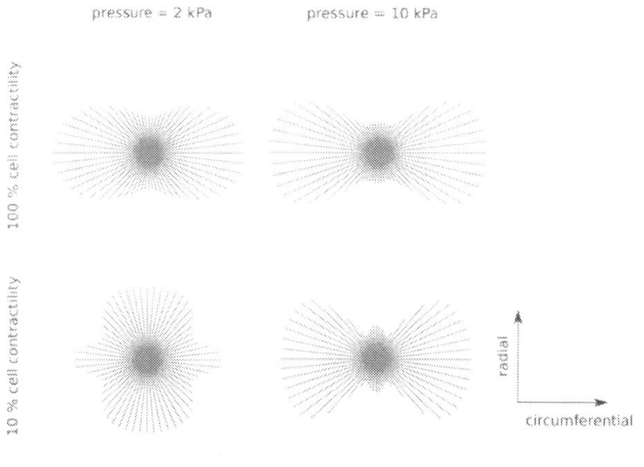

Figure 4: Collagen network in the belly of the valve after tissue remodeling at different pressure levels and degrees of cell contractility.

DISCUSSION

The goal of the present study was to analyze tissue remodeling in TEHVs due to the in vivo dynamic pressure conditions, and assess when there would be a risk for the development of valvular insufficiency after implantation. For pulmonary pressure conditions, the model predicted that valve closure could only be obtained when cell traction was very low compared to the myofibroblast phenotype. However, this low degree of cell contractility was associated with the development of a radially aligned collagen network, which is not mimicking the native heart valve. For aortic pressure conditions, on the other hand, valve closure was present in all simulations, together with a circumferential alignment of the collagen network.

In summary, to preserve valve functionality in the pulmonary position, the model predicts that it is crucial to attract cells with low intrinsic cell contractility. When TEHVs are implanted in the aortic position, the model predicts that valvular insufficiency is less likely to occur, even at relatively high cell contractility.

ACKNOWLEDGEMENTS

We gratefully acknowledge the support from the European Union's Seventh Framework Program ([FP7/2007-2013], grant agreement no. 242008), and the Netherlands CardioVascular Research Initiative (CVON2012-01): the Dutch Heart Foundation, Dutch Federation of University Medical Centers, the Netherlands Organization for Health Research and Development, and the Royal Netherlands Academy of Sciences.

REFERENCES

[1] Hoerstrup, SP et al., *Circulation,* 102:III44-III49, 2000.
[2] Flanagan, TC et al., *Tissue Eng Part A,* 15:2965-2976, 2009.
[3] Gottlieb, D et al, *J Thorac Cardiovasc Surg,* 139:723-731, 2010.
[4] Schmidt, D et al., *J Am Coll Cardiol,* 56:510-520, 2010.
[5] Driessen-Mol, A et al., *J Am Coll Cardiol,* 63:1320-1329, 2014.
[6] Loerakker, S et al., *Biomech Model Mechanobiol,* 13:985-1001, 2014.
[7] Obbink-Huizer, C et al., *Biomech Model Mechanobiol,* 13:227-236, 2014.

SB³C2015
Summer Biomechanics, Bioengineering and Biotransport Conference
June 17-20, 2015, Snowbird Resort, Utah, USA
SB³C2015-221

TOPOLOGY OF PRESTIN EXPRESSED IN THE CHO CELL MEMBRANE –ATOMIC FORCE MICROSCOPY STUDY–

Michio Murakoshi (1, 2), Hiroshi Wada (3)

(1) Department of Mechanical Engineering
Kagoshima University
Kagoshima, Japan

(2) PRESTO, JST
Kawaguchi, Japan

(3) Department of Intelligent Information System
Tohoku Bunka Gakuen University
Sendai, Japan

INTRODUCTION

The outer hair cells (OHCs) are the sensory cells in the mammalian cochlea of the inner ear. They exhibit elongation and contraction in response to acoustical stimulation. Somatic motility of OHCs is thought to be driven by a membrane protein prestin [1]. Due to this motility, the basilar membrane is subjected to force, resulting in cochlear amplification and thus leading to the high sensitivity of mammalian hearing (Fig. 1).

Prestin consists of 744 amino acids with a molecular weight of about 81.4 kDa. In a series of attempts to clarify its membrane topology, hydrophobicity analysis in conjunction with the prediction of the conserved phosphorylation site has suggested that prestin is a 10 [2] or 12 transmembrane protein [3, 4] with cytoplasmic N- and C-termini. Regarding the size of prestin, it was reported to be about 10 nm in diameter [5-12], presumably forming tetramer [13, 14]. Although extensive researches and experiments have been conducted, the topology of prestin in the cell membrane has not yet been clarified.

In the present study, a direct single-molecule technique producing detailed information about the membrane topology of prestin molecules using an atomic force microscope (AFM), so-called single molecule force spectroscopy, was developed. Analysis of the topology of prestin expressed in Chinese hamster ovary (CHO) cell membrane was then carried out.

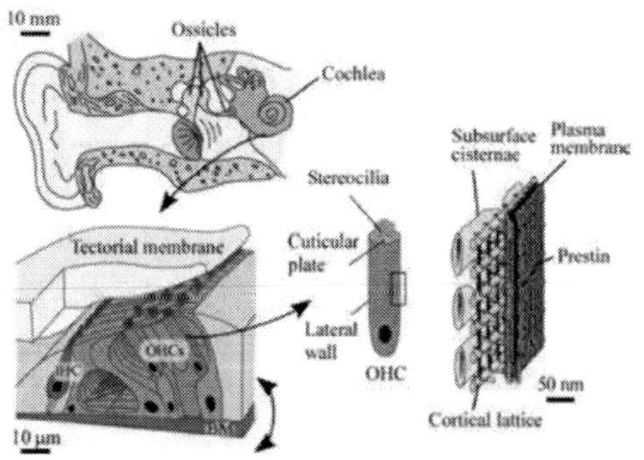

Fig. 1 Human auditory system. Sensory cells, such as outer hair cells (OHCs), inner hair cells (IHCs) and various kinds of the other cells, sit on the basilar membrane (BM). OHCs subject the membrane to force, leading to cochlear amplification, resulting in the high sensitivity of mammalian hearing. Prestin is thought to be the origin of the motility of OHCs.

METHODS

CHO cell lines which stably express prestin tagged with Avi-tag, a receptor peptide which is enzymatically biotinylated by a biotin ligase, were constructed.

Gold-coated AFM cantilevers were incubated with a drop of mixture of two alkane thiols, i.e., protein resistant oligoethylene glycol (OEG) thiols and biotinylated alkane thiols. After incubation with the mixture, the surfaces were incubated with streptavidin. The cells were then grown on a dish and the inside-out plasma membranes were isolated. Such membranes were incubated with the biotin ligase and prestin was biotinylated so that prestin connects with the streptavidin-coated AFM cantilever via biotin-streptavidin binding (Fig. 2). Finally, Avi-tagged prestin was pulled out from the plasma membrane of CHO cell and force curves were obtained.

In the single molecule force spectroscopy, viscous elasticity of protein is required to be allowed into the calculation. In this study, therefore, each force peak detected in the obtained force curves was

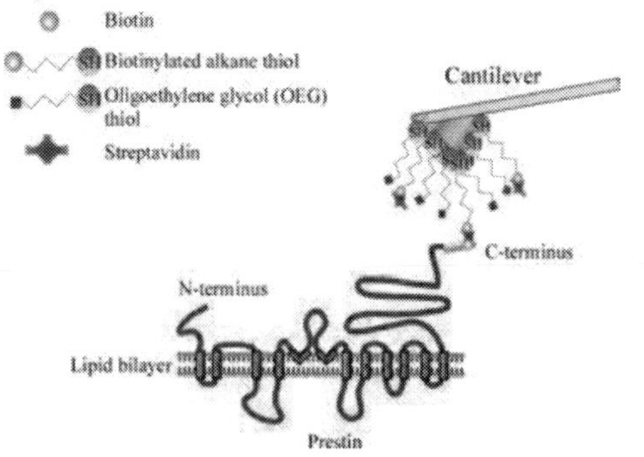

Fig. 2 Experimental scheme of single molecule force spectroscopy of Avi-tagged prestin in the CHO cells. Streptavidin-coated cantilever is brought close to the biotin-conjugated C-terminus of Avi-tagged prestin in the plasma membrane of the CHO cell, resulting in connection between the cantilever and an individual prestin via streptavidin.

fitted by the worm-like chain (WLC) model [15, 16], which is known to well fit to the force curves which represents the extension of unfolded amino acid chain well above 50 pN. The WLC model is represented by the following formula:

$$F(z) = \frac{kT}{b}\left[\frac{1}{4(1 - z/L)^2} - \frac{1}{4} + \frac{z}{L}\right], \qquad (1)$$

where F is the tensile force, z is the extension length, k is Boltzmann's constant, T is the temperature (in Kelvin) and b is the persistence length, which reflects the polymer stiffness. In this study, the persistence length of 0.4 nm was employed [16]. By using this formula, the contour length L of each force peak, which describes the length of stretched portion of the molecule, was calculated.

RESULTS

Figure 3 shows the fluorescence images of the prestin and biotin labeling of the isolated plasma membranes of the Avi-tagged prestin expressing CHO cell. Both prestin labeling (red, left panel) and biotin labeling (green, right panel) were confirmed. In addition, prestin labeling was colocalized with the biotin labeling, indicating that Avi-tagged prestin was biotinylated in the plasma membrane.

Figure 4 shows representative force curves obtained from the isolated plasma membrane of Avi-tagged prestin expressing CHO cells (n = 800). These force curves have one to five peaks. The force curves were fitted by the WLC model. Figure 5 indicates a histogram of the contour length L (n = 243). Multi-peak Gaussian fitting (blue line) revealed that this histogram had 15 peaks (circles).

DISCUSSION

The peak positions predicted by the 12 transmembrane model (arrows, Fig. 5) [3, 4] were similar to those detected in the histogram, suggesting that prestin may possibly have 12 transmembrane domains.

The average forces at around peak position of the histogram were shown in Fig. 6. These forces are considered to represent unbinding force between each transmembrane domain and the plasma membrane. The average unbinding force of transmembrane domain 2 was statistically weaker than such forces of transmembrane domains 1, 5, 8, 11 and 12 (P < 0.05, Student's t-test). Since the transmembrane domain 2 of prestin is known to contain a highly conserved sulfate transport motif, which is essential for normal prestin activity [4], this weak unbinding force is thought to indicate the flexibility of this domain, possibly being related to the conformational change of prestin.

ACKNOWLEDGEMENTS

This research was supported by Grant-in-Aid for Scientific Research on Priority Areas 15086202 from the Ministry of Education, Culture, Sports, Science and Technology of Japan to H.W., and by Grant-in-Aid for Young Scientists (B) 22791564 from the Ministry of Education, Culture, Sports, Science and Technology of Japan to M.M.

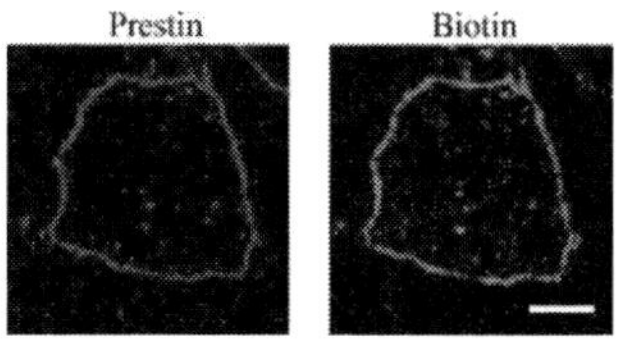

Fig. 3 Fluorescence images of prestin and biotin labelings of the isolated plasma membrane of the Avi-tagged prestin expressing CHO cell. Scale bar is 5.0 μm.

Fig. 4 Representative 10 force curves obtained from Avi-tagged prestin expressing CHO cells (n = 800).

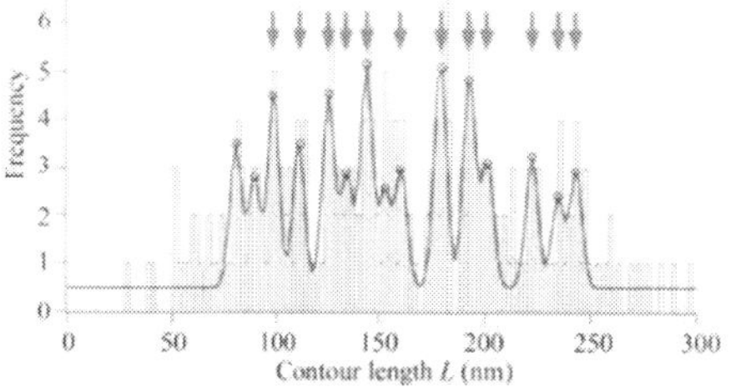

Fig. 5 Histogram of contour lengths L (n = 243). Line shows multi-peak Gaussian fit. The histogram has 15 peaks (circles). Arrows represent the peak positions predicted from the 12 transmembrane model.

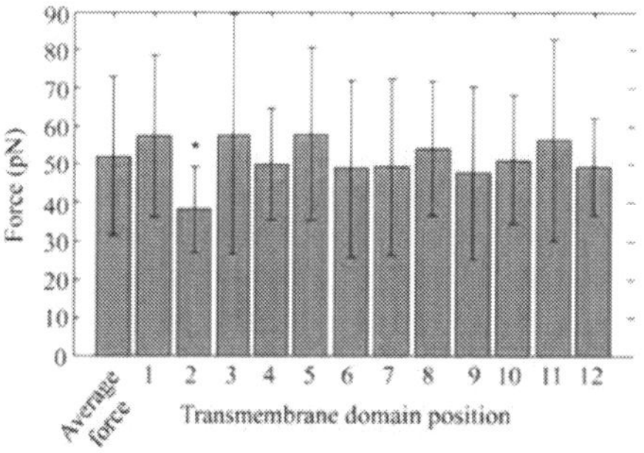

Fig. 6 The average force of each force peak shown in Fig. 5. Asterisk indicates the statistical difference between the average unbinding force of transmembrane domain 2 and such forces of transmembrane domains 1, 5, 8, 11 and 12 (P < 0.05, Student's t-test).

REFERENCES

[1] Zheng J et al. *Nature* 405:149-155, 2000.
[2] Navaratnam D et al. *Biophys J* 89:3345-3352, 2005.
[3] Zheng J et al. *Neuroreport* 12:1929-1935, 2001.
[4] Deak L et al. *J Physiol* 563:483-496, 2005.
[5] Arima T et al. *Cell Tissue Res* 263:91-97, 1991.
[6] Forge A. *Cell Tissue Res* 265:473-483, 1991.
[7] Kalinec F et al. *Proc Natl Acad Sci USA* 89:8671-8675, 1992.
[8] Souter M et al. *Hear Res* 91:43-62, 1995.
[9] Le Grimellec C et al. *J Comp Neurol* 451:62-69, 2002.
[10] Murakoshi M et al. *J Assoc Res Otolaryngol* 7:267-278, 2006.
[11] Murakoshi M et al. *Pflugers Arch* 457:885-898, 2009.
[12] Kumano S et al. *FEBS Lett* 584:2872-2876, 2010.
[13] Zheng J et al. *J Biol Chem* 281:19916-19924, 2006.
[14] Wang X et al. *Brain Res* 1333:28-35, 2010.
[15] Bustamante C et al., *Science* 265:1599-1600, 1994.
[16] Rief M et al. *Science* 276:1109-1112, 1997.

IN SILICO AND IN VITRO MODELLING OF FLOW BEHAVIOUR IN LYMPHATIC VESSELS

Sinéad T. Morley (1,2), David T. Newport (3), Michael T. Walsh (1,2)

(1) Centre for Applied Biomedical Engineering Research (CABER), Dept. of Mechanical Aeronautical & Biomedical Engineering, University of Limerick, Limerick, Ireland.

(2) Material and Surface Science Institute (MSSI), University of Limerick, Limerick, Ireland.

(3) Stokes Institute, University of Limerick, Limerick, Ireland.

INTRODUCTION

Breast cancer is the most common cancer form in women, with an estimated 1.2 million new cases worldwide each year and it is the second most common cause of death in women in the United States [1]. Tumour metastasis to regional lymph nodes is a crucial step in the progression of breast cancer and approximately 90% of cancer related deaths are due to metastasis [2]. It is not yet known why some cancerous cells metastasize from the lymph nodes and others do not, therefore, a fundamental understanding of the response of cells to the forces in the lymph vessels needs to be established. Lymph contains cells that are approximately 6μm and the diameter of breast cancer cells varies from 20-40μm [3]. The dynamics by which breast cancer cells enter the lymphatic system, travel to distant sites, and eventually establish metastatic tumours, remains poorly understood.

There is a considerable wealth of research that examines transport properties of biological cells/particles in microfluidic channels [3-5]. The majority of these studies examine very small particles (0.2-1μm) under typical inertial focusing flow rates (30< Re <300) and very limited data is available on the response of large particles (20-50% the channel width) to low Reynolds number Poiseuille flow such as that seen in the lymphatic vessels [6]. Inertial focusing of biological cells/particles in channels is dependent on properties such as particle size, shape and deformability. However, when the channel height approaches the same order of magnitude as the cells, the cells are influenced by the channel walls and the assumption that the cells travel with the local fluid velocity is no longer valid. It has been shown that larger particles tend to lag the fluid velocity but the exact forces that these particles are subjected to are not yet fully understood [7].

Using both computational and experimental methods, the effect of breast cancer cells within the lymph vessels and how they respond to the forces generated by the flow field is investigated in this study. Cancer cells can change morphologically in response to local flow conditions. The objective is to determine whether such cells migrate due to morphological changes imposed by the local flow environment.

This paper models the fluidic transport behaviour of breast cancer cells through lymph channels by investigating both cell flow (MCF-7's and MDA-MB-231 cells) and particle flow (Bangs Laboratories and Cospheric) in a microchannel at low flow rates (by HTS / LoC standards) which are representative of *in-vivo* conditions in the collecting lymphatics. The longer term goal of this work is to expose metastatic breast cancer cells to an appropriate fluidic context, mimicking their migration through the collecting lymphatics to the sentinel nodal sites downstream of a cancerous tumour.

METHODS

Experimental Methods: Experimental methods, using micro-Particle Image Velocimetry (μPIV) techniques, have been employed to model the flow behaviour of microparticles and breast cancer cells in a polymer 100x100μm channel. Microparticles have been used to simulate the flow of various immune cells contained within the lymph. A range of separate particle sizes and cell sizes are exposed to a low Reynolds number Poiseuille flow and their response to the flow in terms of speed and spatial distribution are analysed and compared. The Reynolds number is 0.08 which is considerably lower than that used in inertial sorting devices, and the particle/cell concentration is representative of that experienced in the collecting lymph *in-vivo*.

The experimental set-up consists of a high speed digital camera (PowerView TSI, UK) connected to an inverted microscope (IX50; Olympus, Tokyo) with illumination provided by dual Nd: YAG lasers (New Wave Research, UK) using an excitation wavelength of 532nm. The channel was placed on the stage of the microscope and the flow rate was kept constant at 0.233 μL/min using a syringe pump (Harvard

Apparatus, Holliston, MA). Microparticles of 5µm, 10µm, 20µm, (Bangs Laboratories, IN) and 27-32µm (Cospheric, CA) were mixed at a 0.08% wt fraction in distilled water, with 1% volume Tween-20 added to the solution. These particle sizes and their density were chosen to correspond to the cell sizes and density seen in the lymph vessels *in vivo* [8]. The breast cancer cells were cultured in Dulbecco's Modified Eagle's medium and incubated at 37°c in 5% CO_2 until near confluence before being used for testing.

Images were captured with a resolution of 4008 x 2672 pixels and 12-bit greyscale, at a rate of 4.8 frames per second. The µPIV images of the particles and cells were processed and the flow velocity was determined using a combination of Insight 4G software and MATLAB. The images were evaluated using a cross-correlation method in which the time between two images was set to a known value for all working fluids. Using a multi-pass interrogation algorithm with a 50% overlap the corresponding velocity fields were obtained. To evaluate the performance of the imaging system in measuring the velocity fields of fluid through the square microchannel, the experimental results for distilled water seeded with 1µm particles were compared to an established analytical solution for steady flow through a long straight square microchannel [9].

Computational Methods: Numerical simulations were performed using the commercially available software Star CCM+ Version 7.06.012 (CD-adapco, NY). A three-dimensional model of the microchannel used in the experiments was created and computational fluid dynamic analysis was carried out. The flow field was modelled as steady and laminar and the inlet/outlet boundary conditions were set as velocity inlet/pressure outlet respectively. The Lagrangian Multiphase model was employed to solve the equations of motion of the particles/cells computationally as they pass through the system.

RESULTS

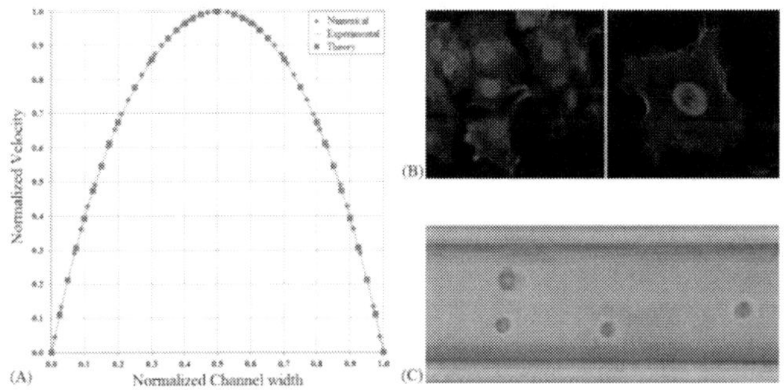

FIGURE 1: (A) FLOW PROFILES AT 50µM DEPTH (B) IMMUNOFLUORESCENT MCF-7 CELLS (C) MCF-7S IN THE 100µm CHANNEL

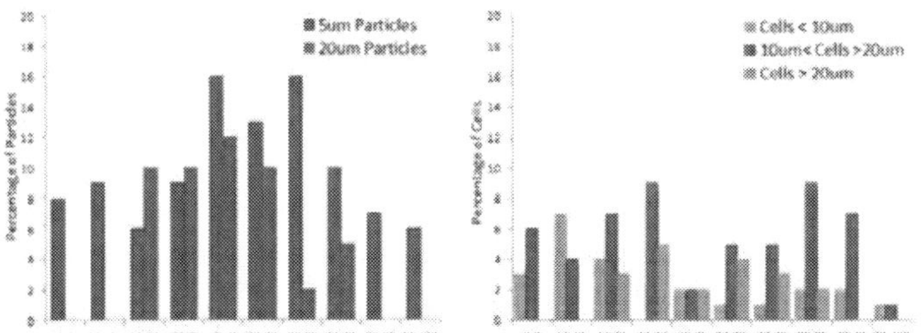

FIGURE 2: FREQUENCY DISTRIBUTION OF (A) PARTICLES AND (B) MCF-7 CELLS IN A 100um CHANNEL

DISCUSSION

The work carried out in this study investigates the migration behaviour of MCF-7 (Fig. 1 (B)) and MDA-MB-231 breast cancer cells to Poiseuille flow within a 100µm square channel. The objective is to determine whether such cells migrate due to morphological changes imposed by the local flow environment, and are compared to solid particles of different sizes as a reference case. The experimental velocity profile for the 1µm particles was compared to both the theoretical and computational solutions (Fig. 1 (A)). Predictions using the theoretical model for a square microchannel corresponded well with the experimental and numerical results.

Particles and cells flowing in a microchannel experience drag and lift forces that result in complex behaviour: particle velocities may lag or exceed the undisturbed flow velocity, and migrate to different equilibrium positions across streamlines. The corresponding underlying mechanism for this phenomenon is not yet clear. Experimental data illustrates a parabolic form of the horizontal velocity profiles for the 5µm and 10µm particles. However, visualisation of the particle motion through the microchannel demonstrated fluctuations present in the flow caused by both rotational and interaction forces between neighbouring particles. Elements of fluid at different distances from the particles are subjected to different forces due to the particle interaction with the fluid and the effect of particle size on the flow-cell interaction is elucidated.

The frequency distribution of the 5µm and 20µm particles in the channel is shown in Fig 2 (A). The 5µm particles are very dispersed throughout the width of the channel, with the majority of the cells aligning in the centre of the channel. However, the 20µm particles travelled only in the centre of the channel with no particles present in the near wall region which indicates that particle position within the channel is dependent not only on flow but also on size. The flow behaviour of the MCF-7s, (Fig. 1(C)), is largely dispersed and observations indicate that the size of the cell plays an important role in cell motility. The cells ranged in diameter from 8µm to 35µm. All cells larger than 20µm travelled in the centre region, Fig. 2 (B), of the channel, similar to the behaviour of the 20µm particles.

Good correlation between measured and predicted velocity profiles was seen with the accuracy of the prediction decreasing with radius. Particle sizes (1-32µm) corresponding to different cell types did affect the results suggesting that cell size, relative to lymph vessel diameter, is an important factor in cell behavior in the flow.

A fundamental understanding of the flow behaviour of particles in lymphatic vessels has been established that has provided a research platform for future experimental and simulation analysis. Future work will involve obtaining images of breast lymphatic vessels in order to estimate and examine the lymph vessel physiology and pathway.

ACKNOWLEDGEMENTS

EMBARK Initiative from the Irish Research Council and the Mid-Western Cancer Foundation.

REFERENCES

[1] Lee, J. H. and Nan A., *Journal of Drug Delivery*, 2012.
[2] Chandrasekaran, S., *et al.*, *Lab on a Chip*, 14(1):118-127, 2014.
[3] Faivre, M., *et al.*, *Biorheology*, 43(2):147-159, 2006.
[4] Peng, Z., *Mechanical Engineering*, 126, 2011.
[5] Hur, S. C., *et al.*, Biomicrofluidics, 5(2):022206, 2011.
[6] Rahbar, E. and Moore, J. E., *Journal of Biomechanics*, 44: 1001-1007, 2011.
[7] Staben, M. *et al.*, *Physics of Fluids*, 15(6):1711-1733, 2003.
[8] Dixon, J. B., *et al.*, *Biomedical Microdevices*, 13(7):597-610, 2006.
[9] Bruus, H., *Theoretical Microfluidics*, 346, 2008.

SB^{3}C2015
Summer Biomechanics, Bioengineering and Biotransport Conference
June 17-20, 2015, Snowbird Resort, Utah, USA
SB^{3}C2015-223

MODELING LIGAMENT MECHANICS USING MICROSTRUCTURAL PARAMETERS MEASURED FROM CONFOCAL IMAGES OF COLLAGEN NETWORKS

Christina J. Stender, Evan Rust, Erica E. Morrill, Roshani Lamichhane, Raquel J. Brown, Trevor J. Lujan

Department of Mechanical and Biomedical Engineering
Boise State University
Boise, Idaho, USA

INTRODUCTION

Ligament mechanical behavior is governed by the composition and organization of its fibrous constituents. Ligament's primary fibrous constituent is type I collagen, which assembles hierarchically into fibrils, fibers and fascicles. The fibrous organization typically exhibits unidirectional alignment along the loading axis but can also exhibit disorganized or multidirectional fiber architecture [1,2]. These multidirectional networks are pervasive in granulation and scar tissue, and may be associated with the inferior mechanics of repaired ligament [1]. Studying the effect of collagen organization on ligament mechanical behavior could advance a mechanistic understanding of fibrous networks and clarify the functional impact of microstructural adaptations.

The mechanical contribution of fibrous networks can be tested using constitutive models that incorporate structural parameters into the mathematical stress-strain relationship. These structural models can directly link predicted mechanical behavior to material microstructure. Although structural models have previously been developed that incorporate a fiber angle distribution to model multidirectional networks in ocular tissue, heart valves, and arteries [3-5], these models have not been applied to ligament. Moreover, there is a scarcity of experimental studies that use imaging data to validate the use of a fiber angle distribution for modeling soft fibrous tissue [6].

The objective of this study is to determine if collagen fibril organization is predictive of healthy ligament mechanical behavior. It is hypothesized that the mechanical behavior of ligament with disorganized fibril architecture can be better predicted by incorporating structural parameters from confocal imaging into an anisotropic constitutive model.

METHODS

Overview. An experimental and computational approach was used to test the research hypothesis. Forty ligament specimens from ten adult bovine hooves were mechanically tested in uniaxial tension. The collagen fibril networks in the tested specimens were imaged using confocal microscopy. Images were digitally processed using the fast Fourier transform (FFT) to quantify the fibril angle distribution parameters. These microstructural parameters were input into constitutive models as specimen-specific material coefficients. Deformations were applied to these models to simulate the uniaxial tensile tests. Finally, statistical analysis assessed if incorporating the microstructural parameters from the imaging data improved model predictions.

Tensile Testing. The lateral collateral sesamoid ligament (LCSL) and palmar annular ligament (PAL) were harvested from ten left bovine hooves. The visible fascicle networks are unidirectional in the LCSL and disorganized in the PAL. Dog-bone shaped specimens were punched longitudinal and transverse to the primary loading axis for uniaxial tensile tests with an Instron E10000 (Fig. 1). After preconditioning, a 0.05 N preload was applied followed by ten 1 Hz triangular cycles at 10% and 6% clamp-to-clamp strain for longitudinal and transverse samples, respectively. Data from the 10th cycle was used for mechanical analysis.

Confocal Imaging and Image Processing. Mechanically tested samples were sliced to a thickness of 50 μm using a cryostat at -20°C. The autofluorescence of collagen fibrils was captured at 63× using the 405 nm laser of a confocal microscope and ZEN imaging software [7]. Three z-stack projections were obtained for all PAL specimens and all LCSL longitudinal specimens. In total, 90 z-stacks were acquired.

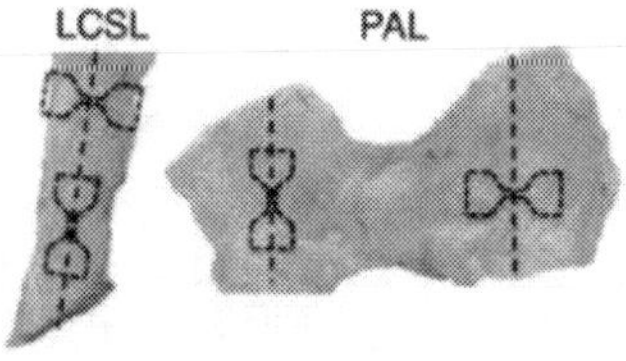

FIGURE 1. SPECIMENS HARVESTED TRANSVERSE OR LONGITUDINAL TO FASCICLE DIRECTION (DASHED LINES).

Filtered binary confocal images were input into a custom Matlab program to perform FFT.

A minimum threshold was applied to the power intensity to remove noise and was normalized to an area of 1 [8]. To calculate the preferred fibril orientation θ_p, an ellipse was fit to the data in polar coordinates, where the semi-major axis signifies θ_p. To calculate the fibril distribution κ, the semicircular von Mises distribution was fit in Cartesian coordinates with θ_p centered along the x-axis.

Modeling. The microstructural parameters θ_p and κ were embedded into a hyperelastic strain energy formulation W that accounted for multidirectional fiber orientation [5]:

$$W = W_{gs} + \int_{\theta_p-\pi/2}^{\theta_p+\pi/2} P(\theta \mid \theta_p, \kappa)\, W_{fibers} d\theta \tag{1}$$

where W_{gs} is the strain energy density of the ground substance using a Mooney-Rivlin material model [10,11], and W_{fibers} is the strain energy density of the fibers oriented along θ and is described by a piecewise function [9]. The semicircular von Mises distribution P is defined as:

$$P(\theta \mid \theta_p, \kappa) = \frac{1}{\pi I_o(\kappa)} e^{\kappa \cos(2(\theta-\theta_p))} \tag{2}$$

For comparison with this multidirectional fiber model, a transversely isotropic hyperelastic formulation was used to model unidirectional fiber orientation [9]. To test the impact of each microstructural parameter on model predictions, the ligament fiber networks were modeled with three approaches: (1) unidirectional fiber model with assumed θ_p based on visible fascicle direction; (2) unidirectional fiber model using θ_p from confocal images; (3) multidirectional fiber model using θ_p and κ from confocal images. All models were analyzed with a custom Matlab program.

Curve Fitting. The non-structural material coefficients within the W_{gs} and W_{fiber} functions were selected to best fit the experimental LCSL stress for each tested specimen. These non-structural coefficients were then held constant to determine if changes in the fibrous microstructure could predict the PAL mechanical behavior. The fitting of the non-structural coefficients within the W_{gs} and W_{fiber} functions for the multidirectional fiber model required a fitting constraint estimating the fiber stress contribution of the transverse LCSL tests and is a function of θ_p and κ.

RESULTS

Imaging. The fibril distribution κ was significantly greater in the LCSL than in the PAL, which indicates greater fiber alignment in the LCSL ($p < 0.001$) (Table 1). A significant difference existed in θ_p between transverse tests of the LCSL and PAL ($p = 0.021$).

TABLE 1. MICROSTRUCTURAL PARAMETERS MEASURED FROM CONFOCAL IMAGES

Specimen Type	Fibril Distribution Mean $\kappa \pm$ SD	Preferred Fibril Orientation Mean $\theta_p \pm$ SD (°)
LCSL Longitudinal	4.0 ± 1.8	2.5 ± 11.8
LCSL Transverse		92.5 ± 11.8
PAL Longitudinal	1.9 ± 0.5	9.7 ± 24.7
PAL Transverse	1.0 ± 0.3	57.1 ± 40.8

Note: 0° is along the loading axis.

Tensile Tests. The LCSL had a significantly smaller tangent modulus than the PAL under transverse loading (2.3 ± 1.4 MPa and 9.9 ± 4.5 MPa, respectively; $p < 0.001$). The LCSL and PAL had no significant difference in tangent modulus under longitudinal loading (80 ± 32 MPa and 65 ± 38 MPa, respectively).

Modeling. All models predicted excellent stress-strain curve fits to LCSL longitudinal and transverse experiments ($R^2 \geq 0.96$), and to PAL longitudinal experiments ($R^2 \geq 0.92$). The multidirectional fiber model, which used microstructural parameters from confocal imaging data, gave the best fit to PAL transverse experiments (Fig. 2).

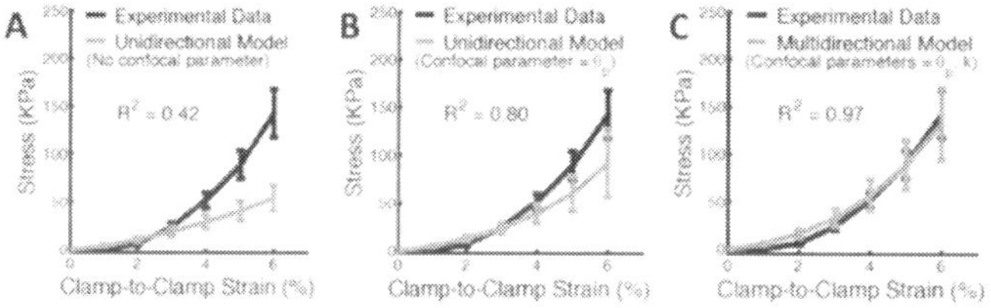

FIGURE 2. STRESS-STRAIN CURVES BETWEEN PAL TRANSVERSE EXPERIMENTS AND MODEL PREDICTIONS WHEN MICROSTRUCTURAL PARAMETERS FROM CONFOCAL IMAGES WERE (A) NOT USED, (B) USED FOR θ_p OR (C) USED FOR θ_p AND κ.

DISCUSSION

This study investigated whether differences in collagen fibril organization are predictive of macroscopic ligament mechanical behavior. By using confocal imaging and FFT, fiber angle distribution parameters were quantified and input into constitutive models. This study found that incorporating these microstructural parameters improved the model's predictive accuracy when ligament has multidirectional fiber networks. These results support our hypothesis and emphasize the importance of fibril angle distribution in ligament mechanics.

It is important to note the predictive improvement of using experimentally measured θ_p within the unidirectional model. Furthermore, results from this study indicate that a multidirectional fiber model could accurately predict the mechanics of ligament and tendon with disorganized fiber architecture, such as the supraspinatus tendon or reparative ligament.

This study introduced two novel methodologies. First, a new model validation method was used to test if differences in fibril microstructure between the LCSL and PAL were predictive of differences in the mechanical behavior between the LCSL and PAL. Second, this study imaged collagen fibrils, while previous studies that quantify soft tissue structure imaged collagen fibers. An advantage of imaging fibrils is that they are distinct entities ideal for image processing.

This study had limitations. Human ligament, over bovine ligament, testing was preferable though the structures are comparable and results from this study are similar to published bovine mechanical properties [12,13]. The imaging methodology captured the autofluorescence of collagen and elastin. Although elastin constitutes only 5% of ligament dry-weight, it still likely influences the calculation of θ_p and κ [14]. Additionally, the effect of the microstructural parameters on viscoelastic behavior was not considered.

In summary, the proposed microstructural model represents a valid mathematical framework to model ligament with multidirectional collagen networks. This multidirectional fiber model can be utilized to advance a mechanistic understanding of ligament physiology.

ACKNOWLEDGEMENTS

This research was supported by NIH NIGMS P20 GM109095 and the Biomolecular Research Center. Kind thanks to Phil Boysen.

REFERENCES

[1] Frank, C et al., *J Musculoskel Neuron Interact*, 7:130-140, 1999. **[2]** Provenzano, P and Vanderby, R, *Matrix Biol*, 25:71-84, 2006. **[3]** Holzapfel, G et al., *J Elast*, 61:1-48, 2000. **[4]** Freed, A et al., *Biomech Model Mechano*, 4:100-17, 2005. **[5]** Girard, M et al., *J Biomech*, 131, 2009. **[6]** Wan, W et al., *Biophys J*, 102:2916-25, 2012. **[7]** Monici, M, *Biotech Annu Rev*, 11:227-56, 2005. **[8]** Ayres, C et al., *Biomaterials*, 27:5524-34, 2006. **[9]** Wiess, J et al., *Comput Methods Appl Mech Engrg*, 135:107-28, 1996. **[10]** Mooney, M, *J Appl Phys*, 11, 1940. **[11]** Rivlin, R, *Math Physc Sci*, 241:379-397, 1948. **[12]** Pini, M et al., *J Periodontal Res*, 37:237-44, 2002. **[13]** Eleswarapu, S et al., *PLoS ONE*, 6, 2011. **[14]** Gentleman, et al., *Biomaterials*, 24:3805-3813, 2003.

THE NUMBER OF LESIONS DOES NOT GOVERN THE FUNCTIONALITY OF CORONARY BIFURCATION LESIONS: A STUDY OF THE EFFECT OF RELATIVE STENOSIS SEVERITY

Catherine Pagiatakis (1,2), Jean-Claude Tardif (2,3), Philippe L.-L'Allier (2,3), Rosaire Mongrain (1,2)

(1) Department of Mechanical Engineering
McGill University
Montreal, Quebec, Canada

(2) Montreal Heart Institute
Montreal, Quebec, Canada

(3) Faculty of Medicine
University of Montreal
Montreal, Quebec, Canada

INTRODUCTION

Procedures associated with the treatment of coronary bifurcation lesions (CBLs) are high risk. To date, the factors that influence their functional impact are still not understood. Results from a preliminary study investigating different multi-lesion configurations with uniform stenosis severities suggested that certain arrangements could experience branch steal, therefore rendering them more haemodynamically critical than other configurations [1]. As such, the objective of the study was to investigate the impact of relative stenosis severity on the functionality of different CBL configurations.

METHODS

Seven planar, synthetic CBL configurations based on the Medina classification [2] were investigated (FIGURE 1).

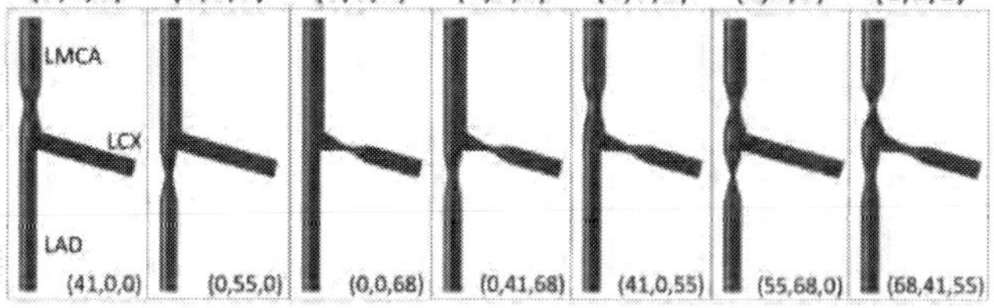

FIGURE 1: CBL CONFIGURATIONS WITH EXAMPLES OF THE APPLIED STENOSIS SEVERITIES

Three stenosis severities, corresponding to mild (41%), moderate (55%) and severe (68%) diameter reductions were considered; a total of 63 cases were simulated. A geometric multiscale (MS) model (Figure 2) of the bifurcation of the left main coronary artery (LMCA) into the left anterior descending artery (LAD) and left circumflex artery (LCX), was employed to quantify and compare the functionality between the different configurations [1].

The commercial software package ANSYS Fluent (Canonsburg, PA, USA) was utilized to solve the transient, incompressible and Newtonian Navier Stokes equations (NSE) modelling the 3D fluid domain. The dynamics of the lumped parameter model (LPM) of the heart and systemic, pulmonary and coronary circulations were described by a nonlinear system of ordinary differential equations. It was solved using a 4th-order Runge-Kutta method and coupled, at each timestep, to the NSE through average uniform boundary pressures at the inlet and outlets, using an in-house subroutine written in C++ that was implemented in ANSYS by a fully-automated external user-defined function written in C. The heterogeneous coupling between the 3D and 0D models was achieved using an iterative, partitioned approach (Quarteroni *et al.* [3]).

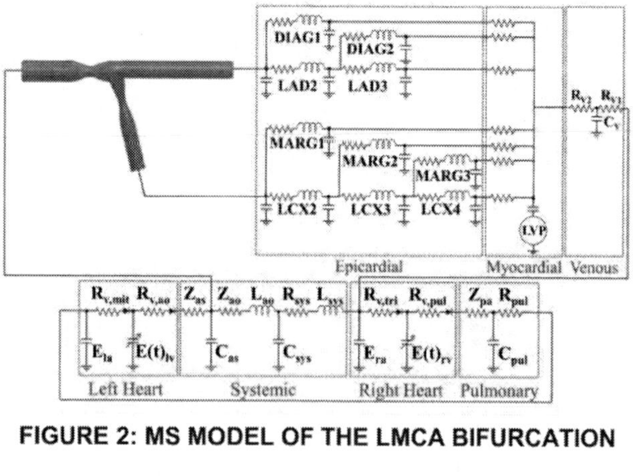

FIGURE 2: MS MODEL OF THE LMCA BIFURCATION

For each case, the flow-based fractional flow reserve (FFR):

$$FFR = \frac{Q_{d,i}}{Q_{n,i}} \tag{1}$$

where $Q_{d,i}$ and $Q_{n,i}$ are the diseased and healthy flow in vessel i respectively, was calculated. The FFR is the gold-standard clinical diagnostic index characterizing the functionality of coronary lesions. FFR values below 0.8 are considered to be haemodynamically critical.

RESULTS

The FFR in the LCX is presented as a function of LCX stenosis severity for all LMCA and LAD stenosis severities (FIGURE 3).

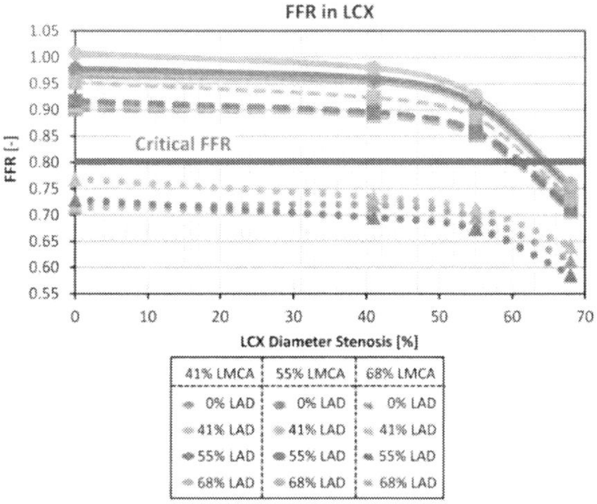

FIGURE 3: LCX FFR WITH RESPECT TO LCX SEVERITY

In the LCX, when the LMCA stenosis is severe (68%), the FFR is below the critical value for all LAD and LCX stenosis severities. When the LMCA stenosis is either mild (41%) or intermediate (55%), a relatively severe stenosis (~60%) in the LCX is required to generate a critical FFR, for all LAD stenosis severities. For example, the (68,0,0), (68,0,55) and (68,41,0) cases are found to be haemodynamically critical, whereas the (41,68,55) and (55, 68, 55) cases remain above the critical threshold. Interestingly, for severe LMCA stenoses, the (1,1,0) configuration, can result in a critical FFR in the LCX, regardless of the fact that the LCX is free of disease.

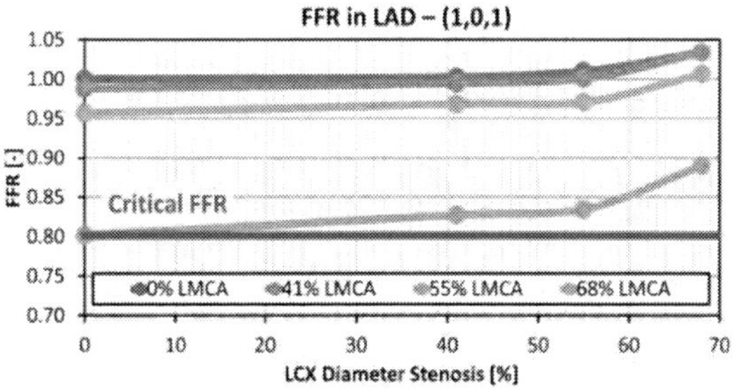

FIGURE 4: FFR IN LAD FOR THE (1,0,1) CONFIGURATION

FIGURE 4 displays the FFR in the LAD for the (1,0,1) configuration, with increasing LMCA and LCX stenosis severities. Here, branch steal [4], wherein a low-resistance daughter vessel diverts flow away from a high-resistance daughter vessel, is evident; the healthy LAD diverts flow away from the diseased LCX and therefore the corresponding FFR increases with increasing LCX severity. It is clear that as the severity of the LMCA stenosis increases, the LAD increasingly diverts flow away from the LCX. Interestingly, branch steal is also observed for the (1,1,1) configuration (FIGURE 3): for all LMCA stenosis severities, more severe LAD stenoses result in a higher FFR in the LCX compared to those with lower LAD severities; as such, a healthy vessel is not required in order to induce branch steal.

DISCUSSION

To the best of the authors' knowledge, this is the first study to systematically investigate the effect of relative stenosis severity on the functionality of CBLs; the findings have potentially significant clinical implications. Over a wide range of LAD and LCX stenoses, for a severe stenosis in the mother vessel, one-lesion and two-lesion configurations were found to produce a haemodynamically critical environment whereas three-lesion configurations with mild to moderate LMCA stenoses did not. This finding is significant because it deviates from logical expectations; more specifically, it signifies that clinically, single- and double-lesion configurations with severe stenoses in the supplying artery should be considered just as, if not more significant than three-lesion configurations with milder supplying artery stenoses. The idea that the number of stenoses does not govern the severity of the configuration is further displayed in cases that experience branch steal. Although previously reported to occur between a healthy and a diseased daughter vessel [4], this study showed that branch steal can also occur between two diseased daughter vessels whose relative stenosis severity is non-zero, with notable impact on the functionality. From a clinical perspective, this is important because it suggests that a large relative severity between two daughter vessels haemodynamically de-stabilizes the configuration, regardless of the absolute severity of the stenoses. Consequently, such configurations could be at higher risk for disease progression, as was suggested in [5]. Furthermore, by increasing the relative daughter vessel resistance (for example, by treating one of the stenoses), a more haemodynamically critical environment could be generated.

This study was subject to certain limitations. Linearized, synthetic geometries were employed whereas *in vivo*, coronary geometry is highly tortuous and curved. However, by utilizing idealized geometries, it was possible to isolate the contribution of relative stenosis severity on the functionality of the different configurations. Furthermore, myocardial dynamics play a significant role in the overall functionality of coronary lesions, yet a single myocardial model was considered in the study. Nevertheless, the model that was employed produced physiological flow and pressure curves in both the healthy and diseased states and therefore was deemed acceptable. From a biophysical perspective, collateral flow and myocardial remodeling due to prolonged exposure to disease were not considered. These factors vary significantly between patients and are not well understood and thus, their exclusion simplified numerics.

ACKNOWLEDGEMENTS

The authors would like to acknowledge financial support provided by the Natural Sciences and Engineering Research Council of Canada Collaborative Health Research Projects grant (NSERC-CHRP grant no. 385833-10), the Canadian Institutes of Health Research grant (CIHR grant no. CPG-104292) and the McGill Engineering Doctoral Award.

REFERENCES

[1] Pagiatakis, C et al., *WCCM XI-ECCM V-ECFD VI,* ID:1852, 2014.
[2] Medina, A et al., *Rev Esp Cardiol,* 59(2):183-184, 2006.
[3] Quarteroni, A et al., *Comput Visual Sci,* 4:111-124, 2001.
[4] Yong, A et al., *Circ Cardiovasc Interv,* 6:161-165, 2013.
[5] Zarandi, M et al., *Model Simul Eng,* 2012: ID: 419087, 2012.

SB^{3}C2015
Summer Biomechanics, Bioengineering and Biotransport Conference
June 17-20, 2015, Snowbird Resort, Utah, USA
SB^{3}C2015-225

DESIGN OF A LOW-COST HAPTIC ASSISTIVE HANDWRITING DEVICE

Eamon T. Campolettano
Lafayette College
Easton, PA, USA

Alexander A. Croft
Lafayette College
Easton, PA, USA

Kevin J. Fasano
Lafayette College
Easton, PA, USA

Stephen H. Hodge
Lafayette College
Easton, PA, USA

Kevin J. Myers
Lafayette College
Easton, PA, USA

Brian R. Pinkard
Lafayette College
Easton, PA, USA

Jessica L. Ross
Lafayette College
Easton, PA, USA

Allison A. Scoular
Lafayette College
Easton, PA, USA

Thomas S. Todd
Lafayette College
Easton, PA, USA

Faculty Advisor
Alexander A. Brown
Lafayette College
Easton, PA, USA

INTRODUCTION

It is estimated that over 3.5 million Americans are currently living with Autism Spectrum Disorder (ASD), a pervasive developmental disorder that costs US citizens $236-262 billion annually. However, research has found that the cost of lifelong care of ASD can be reduced by roughly 66% with early diagnosis and treatment of its associated disabilities [1]. One of the many varied disabilities associated with ASD is dysgraphia, "a writing disability that manifests itself in poor-quality script" [2, 3]. Effective handwriting has many social implications; it has been shown to be critical for academic progress, the development of communication skills, and the building of self-esteem in elementary-aged children. ASD negatively affects interpersonal skills [4], so it is vital that children with ASD develop alternative ways to communicate; handwriting is one such method of communication.

Based on the prevalence of handwriting difficulties for individuals with ASD, along with stroke victims and those suffering from Parkinson's disease, it is clear that a significant market exists for an assistive handwriting device that is both effective and reasonably priced. Various attempts have been made to design a haptic handwriting device, but most have left room for improvement in terms of practicality or availability. In [5], the "Telemaque", a visuo-haptic device consisting of a tablet and accompanying haptic robot, only required participants to trace over desired letters and not create original script. This may limit development of generalized, unassisted writing. The I-TOUCH haptic guidance software, explored in [6], developed a four stage progressive teaching methodology to limit student dependence on the teacher during the learning process. However, in this paper, there was no explicit comparative exploration of the effectiveness of this teaching style. To address a large potential market of children with ASD diagnoses and their care providers, the authors are developing a low-cost haptic handwriting aid that will be financially accessible to care providers and families alike. This novel design, leveraging data-driven teaching methods and handwriting profiles, will be effective in teaching and correcting handwriting to treat dysgraphia. Specifically, the authors' design will leverage position, velocity, and force feedback while employing research-supported teaching methods and a multi-input, multi-output control architecture.

PRODUCT DESIGN

The Design Goal is to produce a device capable of guiding a subject's hand along the path of every letter in the alphabet, and of providing force feedback to keep the hand on the path. To offer a distinct advantage over state-of-the-art devices, it will use low-cost hardware and open-source software tools where possible. Similarly to methods used in the I-TOUCH system [6], the team intends to implement a progressive teaching methodology into the device. By using research-supported teaching methods, possibly based on Applied Behavior Analysis (ABA), the device will allow for the user to be completely or partially guided along the path of the letter, on a continuum spanning full guidance to simple visual feedback.

Device design metrics were quantified by analyzing the biomechanical forces, velocities, and accelerations required for a hand to produce script. Given that Roman characters are only comprised of two-dimensional lines and arcs, representing two degrees of freedom, an initial prototype was developed, comprised of two linkages controlled by two independent motors for position and velocity control of an instrumented pen. Based on a coarse dynamic analysis and data from relevant literature, motors were sized to control the hand with 15 in-lbs of torque at 25 in/sec [7]. Two Pittman GM9236S025 brushed

DC motors with integral encoders were selected, capable of 31 in-lbs of continuous torque.

To quantify the exact kinematic and kinetic requirements of normal handwriting, an Arduino, three force-sensitive resistors, a digitizing tablet, and a custom data collection program were used to collect position and force data for each team member writing with his or her dominant hand. To date, data have been collected for each team member drawing the letter 'g' for ten trials and each individual's statistically mean 'g' position and velocity profile has been calculated, along with additional joint position data from a state-of-the-art motion capture system. The lateral offset of each trial's position data from a tangential vector at each point along the average 'g' was calculated to provide bounds of expected variance for use in the controller of the device. Additionally, position and velocity standard deviation bounds were calculated as shown in Figure 1, which will be used as design specifications for the physical device and the control/correction algorithm. Data for each letter in the alphabet will be collected and analyzed in the same fashion.

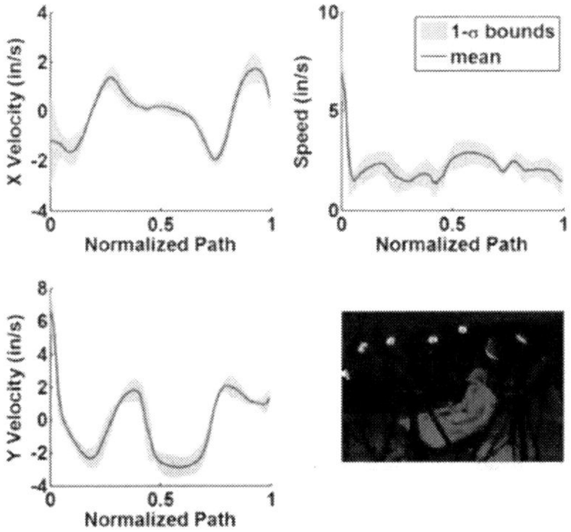

Figure 1. One team member's mean and 1-sigma velocity profiles for ten trials of writing a print "g," with tablet, instrumented pen, and motion capture setup used to collect data

Using data-driven reference letter profiles such as the one shown in Figure 1, the feedfoward/PID control law on the haptic device will combine encoder-based position and velocity feedback with force data from each plane of a rubber triangular pen grip to determine corrective action. As trials progress for a participant, she/he will experience varying levels of corrective force based on past performance. Should an individual deviate beyond the expected variance from the standardized path for the desired letter, the controller will provide a force that will move the pen and hand back onto the desired letter trajectory. The algorithm determining the amount of correction will vary from trial to trial based on subject performance. The control system is implemented using an Arduino Mega 2560 controlling a Seeed Studio H-bridge for motor control. The controls software was developed using Simulink and MATLAB with a loop frequency of 100 Hz. With the kinematic data collected from characterization experiments as in Figure 1, a reference force and velocity profile can be calculated for the per-trial controller. The controller will compare the desired path with the actual trajectory of the pen to determine what, if any, is the proper corrective force for the participant. The per-trial outline of the controller's structure is shown in block diagram form in Figure 2.

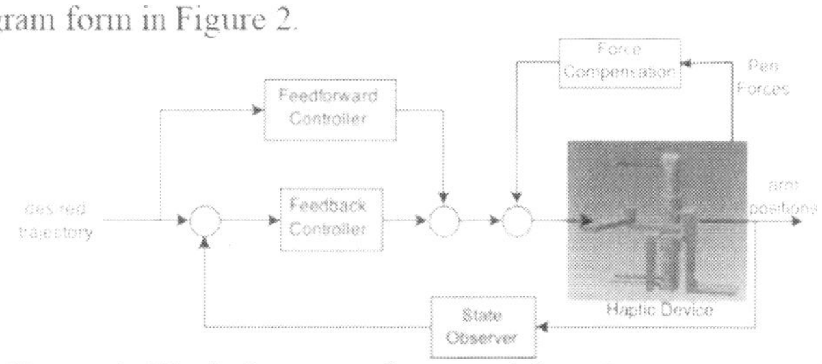

Figure 2. Block diagram of per-trial force/position control algorithm (algorithm modifications from trial-trial not shown)

BUDGET & MARKET ANALYSIS

The current exploratory prototype device utilizes two Pittman GM9236S025 brush motors that cost $260.00 each to accommodate the two degrees of freedom. Adding the costs of the Arduino board at $25.00, the Wacom Intuos Pro Tablet at $500.00, and the motor mount at $200.00, the total material costs for the haptic system, including components for initial data collection and handwriting characterization, amount to $985.00. Compared to other assistive handwriting devices currently on the market, this system already represents a more cost-effective option. Comparatively, the Haptic Handwriting Aid presented in [8] costs about $8000. Final design cost calculations are in progress, but are expected to land at around $250.

ACKNOWLEDGEMENTS

The team would like to acknowledge the significant financial support provided by Lafayette College, which has allowed for invaluable personal exploration and collective design experience for the entire team.

REFERENCES

[1] "Facts and Statistics | Autism Society." *Autism Society Facts and Statistics Comments*. N.p., n.d. Web. 10 Dec. 2014.

[2] Hamstra-Bletz, L, & Blöte, A, *Journal of Learning Disabilities*, 26(10): 689-699, 1993.

[3] Fuentes, C et al., *Neurology* 73.19: 1532-1537, 2009.

[4] Mundy, P et al., *Journal of child psychology and psychiatry*, 27(5): 657-669, 1986.

[5] Palluel-Germain, R et al., *EuroHaptics Conference, 2007 and Symposium on Haptic Interfaces for Virtual Environment and Teleoperator Systems. World Haptics 2007. Second Joint.* IEEE, 2007.

[6] Bayart, B et al., *Haptic Audio Visual Environments and their Applications. IEEE International Workshop on*, IEEE, 2005.

[7] Knoblich, G et al., *The Quarterly Journal of Experimental Psychology: Section A* 55.3: 1027-1046, 2002.

[8] Mullins, J et al, *Haptic handwriting aid for training and rehabilitation. Systems, Man and Cybernetics,* IEEE, 2005.

AN IN-SITU CARTILAGE CHARACTERIZATION METHOD WITH REDUCED TIME AND COMPUTATION DEMANDS: COMPARISON AGAINST LINEAR AND NONLINEAR BIPHASIC THEORIES

D. Burris (1,2), A. Moore (2), X. Chen (1), B. Zimmerman (1), X. Lu (1,2)

(1) Mechanical Engineering
University of Delaware
Newark, DE, USA

(2) Biomedical Engineering
University of Delaware
Newark, DE, USA

INTRODUCTION

Hertzian contacts (of general ellipsoids) are found in virtually all bearing-applications, including diarthrodial joints. Unfortunately, the classical solution for linearly elastic solids does not apply to biphasic materials like cartilage. Agbezuge and Deresiewicz first solved the Hertzian contact problem for linearly biphasic materials [1] and Oyen curve-fit numerical results to their solution to improve its utility for materials characterization [2].

Models based on linear biphasic theory (LBT) are known to grossly under-predict the interstitial fluid pressure responsible for bearing capacity and lubricity. Armstrong *et al.* [3] first recognized the issue of under-prediction when they measured far greater stresses than predicted in unconfined compression, a loading condition in which fluid pressure is balanced only by tensile stress. Volume is initially conserved, so tensile strains must equal one-half the compressive strains. Therefore, tensile stresses and fluid pressures are one-half the compressive elastic stress in a linearly elastic material. The fluid pressures they measured were clearly unbalanced by relatively smaller tensile stresses, so they reasoned that adhesion at the platen surface must have provided additional confinement [3]. However, in addition to external confinement by a platen, a chamber, or the underlying bone, a stiff collagen matrix can produce internal confinement. Since volume conservation fixes the ratio of tensile strain to compressive strain, tensile stresses and fluid pressures build relative to compressive elastic stresses as the tensile modulus builds relative to the compressive modulus; this was first shown by Soltz and Ateshian [4]. Given that cartilage is much stiffer in tension than compression and the low adhesive friction coefficient of cartilage, the discrepancy between experiment and linear theory in unconfined compression is almost certainly driven by the inherent tension-compression nonlinearity of cartilage. This suggests that the tensile modulus, which is not typically reported by *in-situ* cartilage characterization protocols, is as important to function as permeability or aggregate modulus.

Moore and Burris [5] recently developed an analytical solution to the Hertzian contact problem for tension-compression nonlinear biphasic materials to address the under-prediction limitation of solutions based on LBT; here on we call the solution HBT and the most important form for the purposes of this paper is given by Eq. 1, where E_c is the effective contact modulus (with fluid pressurization), E_{c0} is the equilibrium contact modulus (and approximately equal to the aggregate modulus for cartilage), k is permeability, $\dot{\delta}$ is the indentation rate, and R is the radius of the indenter. This equation analytically relates the functional response to the mechanical conditions and material properties. This paper aims to: 1) demonstrate the use of this equation for characterizing the tensile modulus, contact modulus, and permeability of cartilage and other biphasic materials; 2) validate the results against accepted linear and nonlinear biphasic approaches.

$$E_c = \frac{E_{co}}{1-\left(\frac{E_t}{E_t+E_{co}}\cdot\frac{\dot{\delta}\cdot R}{\dot{\delta}\cdot R+E_{co}\cdot k}\right)} \quad (1)$$

METHODS

A single 12mm diameter osteochondral core having a 1.3mm thick layer of cartilage was used to demonstrate and validate the method. The sample was removed from the center of the medial femoral condyle of a mature bovine stifle. The custom microtribometer reported previously was used to perform the indentation measurements with smooth (R_a<80nm) stainless steel spheres (ϕ=6.4mm). The sample was submerged in phosphate buffered saline (PBS) during the tests.

The sample was indented to 75 μm at 100, 50, 10, 5, 1, 0.5, and 0.1μm/s, in that order. Following the last indent, the probe remained in contact to determine E_{c0} at equilibrium; in this study, equilibrium occurs when the indentation depth changed by less than 1μm over a five-minute window. Before fitting, each contact modulus was modified to eliminate the substrate effect using the correction factor from Stevanovic *et al.* [6]. The substrate-corrected contact modulus was fit as a function of indentation rate to Eq. 1 to determine E_t and k. To assist other researchers in implementing the method, we have developed a fitting template and user guide, both of which can be downloaded from our website at: http://research.me.udel.edu/~dlburris/publicationsOther.html.

To test the results against the gold standards for cartilage characterization, a creep test was performed on the same sample at the same location with a ϕ=1.6mm porous plane-ended indenter. Following surface alignment and equilibration at a 50mN preload, the load was increased and held at 200mN load until equilibrium (less than 1μm of creep in five minutes). The creep data were then curve-fit using a custom LBT curve-fitting program to simultaneously determine Poisson's ratio, aggregate modulus, and permeability based on LBT. The same data were fit using a custom tension-compression nonlinearly elastic (TCN) finite element model using FEBio [7] to determine aggregate modulus, tensile modulus and permeability based on TCN.

RESULTS

The effective contact modulus for Hertzian contact is plotted as a function of indentation rate in Figure 1 (top). The effective contact modulus is a sigmoidal function of indentation rate. According to Eq. 1, the effective contact modulus approaches the equilibrium contact modulus at zero speed and the sum of the contact modulus and the tensile modulus at infinite speed (Pe>>1). When the product of indentation rate and radius equals the product of aggregate modulus and permeability (Pe=1), the effective modulus is halfway between these limits. Direct measures give E_{co}=0.47MPa and the fit of the data to Eq. 1 gives E_t=8.7MPa and k=0.0026mm^4/Ns with R^2=0.994.

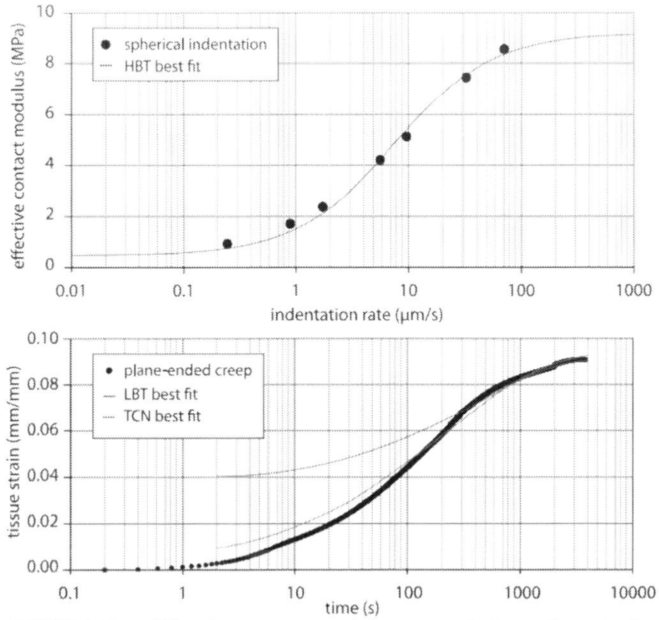

FIGURE 1. Top: Effective contact modulus versus indentation rate for Hertzian indentation with the HBT fit to the dataset. Bottom: Strain versus time for plane-ended indentation measurements with LBT and TCN fits to the dataset. Both measurements correspond to the same location on the same sample.

Creep test results for plane-ended indentation of the same sample in the same location are shown in Figure 1 (bottom). The fit to LBT for plane-ended indentation gives a Poisson's ratio ν=0.1, aggregate modulus H_A=0.47 MPa and permeability k=0.0014 mm^4/Ns ($E=E_t=E_{c0}$=0.46 and H_A=0.47MPa for a linearly elastic material with ν= 0.1). The fit illuminates the under-prediction of fluid pressure and stiffness in the early time response (which is equivalent to the order of 10 hours in the knee) discussed in the introduction. We have restricted the fit to the upper 30% of strain measurements, as is best practice for LBT.

TABLE I. Material properties obtained for a single location of the cartilage sample from HBT, LBT, and TCN characterization methods. The asterisk indicates a fit involving the upper 30% of strain data.

	H_A (MPa)	k (mm^4/Ns)	E_t (MPa)	R^2
HBT	0.47	0.0026	8.7	0.994
LBT	0.47	0.0014	0.5	*0.979
TCN	0.40	0.0016	10.3	0.984

The TCN model better reflects the actual mechanical response of cartilage under indentation loading with R^2>0.98 for a fit to the entire dataset. The properties were H_A=0.40 MPa, k=0.0016 mm^4/Ns, and E_t=10.3 MPa. The mean results from each of the three models are given in Table I for direct comparison.

DISCUSSION

The most important goal of this paper is to establish confidence in the HBT method by direct comparison to the gold standards. The most obvious discrepancy between the models is the assumption by LBT that the tensile modulus equals the compressive modulus. Although HBT and TCN do predict tensile modulus independently, they differ by ~15%. This is, in our opinion, strong agreement and both are consistent with published values for healthy cartilage, which range from 3.5-14 MPa [4, 8-10]. The slightly higher value from TCN likely reflects the increased stresses and stiffness (further from the stress-strain toe) near the stress concentrator. The aggregate modulus and permeability exhibited stronger agreement across methods. LBT and HBT, being based on direct measurements and the same thin-layer mechanics, agree with each other. TCN gave a slightly lower modulus to compensate for the stiffening effect of internal fibers in tension, an effect that is embedded into the LBT and HBT results. LBT gave the smallest permeability (k=0.0014 mm^4/Ns), TCN gives a slightly larger permeability (k=0.0016 mm^4/Ns), and HBT gives the largest permeability (k=0.0026 mm^4/Ns). Each of these falls in a relatively narrow band within the wide range of values found in the literature (0.0004-0.008 mm^4/Ns) [4, 9, 11-16]. Our analysis showed that these variations are mostly attributable to the wide variations in pressure present for each method. For example, fitting only the HBT data at low pressures comparable to those from LBT (0.1-1μm/s) decreases permeability to k=0.0014 mm^4/Ns.

The HBT method can be performed *in-situ*, characterizes tensile modulus, avoids stress concentrators, can be conducted with relatively little time, and without significant computational demands. The results support the HBT method and encourage its use, especially in light of its practical advantages.

ACKNOWLEDGEMENTS

The authors acknowledge financial support from the NIH (grant P20-RR016458) for the development of the experimental and theoretical methods described in the paper.

REFERENCES

[1] Agbezuge, L. and Deresiewicz H, *Israel J Technol*, 12:322-338, 1974.
[2] Oyen, M., *J Mater Res*, 23:1307-1314, 2008.
[3] Armstrong, C et al., *J Biomech Eng*, 106:165-173, 1984.
[4] Soltz, M and Ateshian A, *J Biomech Eng*, 122:576-586, 2000.
[5] Moore, A and Burris D, *J Biomech*, 47:148-153, 2014.
[6] Stevanovic, M et al., *IEEE Trans. Compon. Packag. Technol*, 24: 207-212, 2001.
[7] Maas, S, et al., *J Biomech Eng*, 134:011005, 2012.
[8] Akizuki, S., et al., *J Orthop Res*, 4:379-392, 1986.
[9] Basalo, I, et al., *J Biomech Eng*, 126:779-786, 2004.
[10] Elliott, D et al., *J Orthop Res*, 1999. **17**(4): p. 503-508.
[11] Athanasiou, K et al., *J Orthop Res*, 9:330-340, 1991.
[12] Chen, A et al., *J Biomech*, 34: 1-12, 2001.
[13] Macconaill, M, *J Anat*, 66:210-277, 1932.
[14] Mow, V et al., J *Biomech*, 22:853-861, 1989.
[15] Mow, V et al., *J Biomech Eng*, 102:73-84, 1980.
[16] Williamson, A et al., *J Orthop Res*, 19:1113-1121, 2001.
[17] Brand, R, Iowa Orthop. J., 25:82-94, 2005.

EXCESSIVE MECHANICAL LOADING CAUSES ABERRANT DIFFERENTIATION OF TENDON STEM CELLS (TSCS) THAT LEADS TO THE DEVELOPMENT OF DEGENERATIVE TENDINOPATHY

J. Zhang, JH-C. Wang

MechanoBiology Laboratory
Department of Orthopaedic Surgery, University of Pittsburgh
School of Medicine, Pittsburgh, PA, USA

INTRODUCTION

Advanced stage tendinopathy often manifests as degenerative changes in tendons, and affects millions of Americans each year. Since weight bearing is the primary function of tendons, it is believed that excessive mechanical loading could be a major etiological factor of tendinopathy. However, the precise pathogenic mechanisms leading to tendinopathy remain unclear. In a previous *in vitro* study, we found that small mechanical loading induces differentiation of tendon stem/progenitor cells (TSCs) into tenocytes, but large mechanical loading also causes TSCs to differentiate into non-tenocytes including adipocytes, chondrocytes, and osteocytes [1], which are observed in tendinopathic tendons. *In vivo,* moderate treadmill running (MTR) of mice upregulated tenocyte-related genes but intensive treadmill running (ITR) induced both non-tenocyte and tenocyte differentiation of TSCs [2]. However, it is not certain whether the tenocytes, which are the resident tendon cells, also undergo such aberrant differentiation thus contributing to tendinopathy. In this study, we tested the hypothesis that TSCs *in vivo* undergo non-tenocyte differentiation when subjected to intensive mechanical loading.

METHODS

Irradiation and Injection of GFP TSCs – To precisely determine whether TSCs in vivo differentiates aberrantly, we first eliminated resident tendon cells from the patellar tendons of 12 adult C57BL/6J mice by irradiating with 6 Gy from a Gamma cell 40 (cerium). A live/dead cell viability assay (Life Technologies, Cat. # L-7013) revealed that such irradiation killed > 99% of tendon cells. Then, GFP-TSCs (P1, 1×10^4/μl) obtained from C57BL/6 transgenic mice (GFP-mice) [C57BL/6-TgN (ACTbEGFP)1Osb mice, Jackson Labs] were injected into the irradiated tendons using a 30G syringe.

Mouse treadmill running experiment – Six irradiated and GFP-TSC injected mice were subjected to an intensive treadmill running (ITR) regimen (1 week training, then running at 13 meters/min, 1-5 hrs/day, 5 days/week for 3 weeks). Then, tendon sections were immunostained to evaluate the presence of GFP-TSCs and non-tenocyte proteins in the mouse patellar tendons. Tendon cells in the injected area were also isolated for qRT-PCR analyses to determine tenocyte and non-tenocyte gene expression.

RESULTS

Irradiation of the mouse patellar tendons successfully eliminated > 99% of the native tendon cells. Fluorescence microscopic observations and immunohistochemical analysis revealed that the injected GFP-TSCs migrated from the injected region into the irradiated region 4 weeks post-injection (**Fig. 1, Top panel**). Immunohistochemistry further showed that in these mice, the ITR regimen increased the expression of non-tenocyte specific proteins (PPARγ, Sox-9 and Runx-2) (**Fig. 1, Bottom panel**) when compared to mice without ITR (**Fig. 1, Top panel**). Further, qRT-PCR analysis demonstrated that ITR upregulated the expression of non-tenocyte related genes, LPL, Sox-9 and Runx-2, and tenocyte related genes, collagen I and Scx (**Fig. 2A,B**) in contrast to MTR, which increased only tenocyte-related gene expression [2].

DISCUSSION

Using immunohistochemistry and qRT-PCR, we have shown that subjecting mice to an ITR regimen enhances non-tenocyte differentiation of TSCs *in vivo* similar to our previous *in vitro* findings [1]. More importantly, ITR not MTR induced the aberrant differentiation of TSCs in mice. Our irradiation and injection approach also proved feasible to test our hypothesis that TSCs indeed are the tendon cells that specifically induce non-tenocyte differentiation as evidenced by the presence of abundant GFP-TSCs in the irradiated tendon region. Together, these findings indicate that TSCs are likely the primary contributors to the development of degenerative tendinopathy.

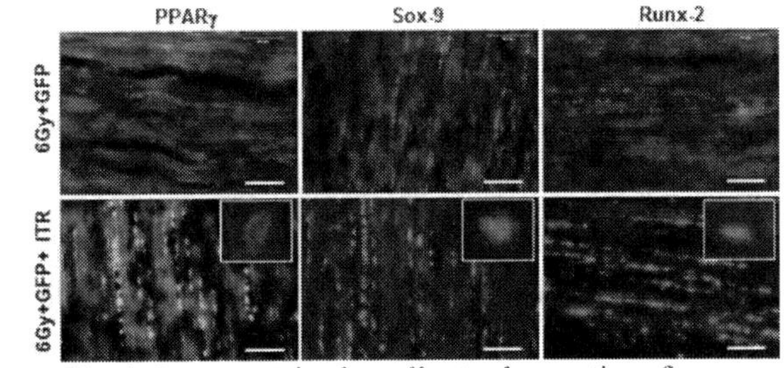

Fig. 1 Immunostained patellar tendon sections from irradiated and GFP-TSCs injected mice subjected to Intensive treadmill running (ITR). Four weeks after injecting GFP-TSCs, the tendon region was filled with GFP positive cells (**Top panel - 6Gy+GFP**). Expression of non-tenocyte related proteins, PPARγ, Sox-9 and Runx-2, is higher in mice on an ITR regimen (**Bottom panel - 6Gy+GFP+ITR**). Inset shows magnification at 20x.

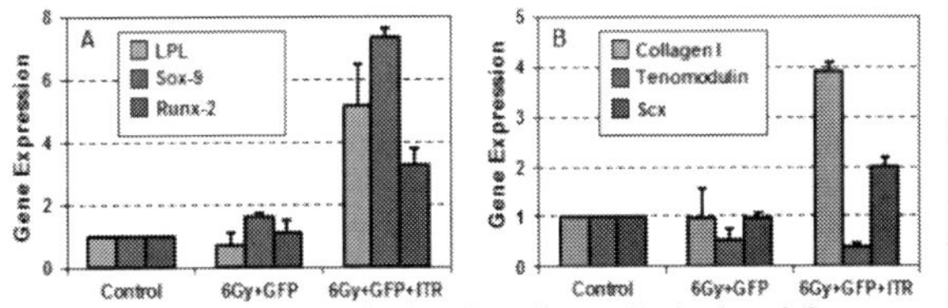

Fig. 2 qRT-PCR analysis of tendon cells isolated from irradiated and GFP-TSCs injected mice subjected to Intensive treadmill running (ITR). ITR upregulated the expression of non-tenocyte related genes (LPL, Sox-9, and Runx-2) (**A**). Tenocyte related genes, collagen I and scleraxis (Scx) also had enhanced expression (**B**). Control: wild-type mice; 6Gy+GFP: mice irradiated and injected with GFP-TSCs; and 6Gy+GFP+ITR: Irradiated and GFP-TSCs injected mice subjected to ITR. All data were normalized to control mice.

ACKNOWLEDGEMENTS

This work was supported in part by the NIH awards AR061395 and AR060920 (JHW).

REFERENCES

[1] Zhang, J. and J.H. Wang, Mechanobiological response of tendon stem cells: implications of tendon homeostasis and pathogenesis of tendinopathy. *J Orthop Res* 28(5): p. 639-43, 2010.

[2] Zhang, J. and J.H. Wang, The effects of mechanical loading on tendons - an in vivo and in vitro model study. *PLoS ONE* 8(8): p. e71740, 2013.

SB³C2015
Summer Biomechanics, Bioengineering and Biotransport Conference
June 17-20, 2015, Snowbird Resort, Utah, USA
SB³C2015-228

BOTH NUCLEAR AND EXTRACELLULAR MATRIX RIGIDITY DETERMINE THE STABLE SIZE OF FOCAL ADHESION PLAQUES

Xuan Cao (1), Tristan P. Driscoll (2, 3), Robert L. Mauck (2,3), Vivek B. Shenoy (1,3)

(1) Department of Materials Science and Engineering,
University of Pennsylvania
Philadelphia, PA 19104, USA

(2) Department of Orthopaedic Sugery,
University of Pennsylvania
Philadelphia, PA 19104, USA

(3) Department of Bioengineering,
University of Pennsylvania
Philadelphia, PA 19104, USA

INTRODUCTION

To perform functions such as proliferation (1), differentiation (2) and locomotion (3), living cells establish stable attachments to the extracellular matrix (ECM) via the formation of specialized receptor mediated contact foci, referred to as focal adhesions (FAs). Accumulating evidence has demonstrated that, besides biochemical factors, the formation and function of FAs is tightly regulated by mechanical cues (4, 5). Theoretically, various FA models have been proposed (6–8) to test theories regarding the physics of FA formation. Despite these aforementioned efforts, several important issues remain unsettled: in most existing models, the size of the adhesion plaque has often been taken as a constant (6) or did not enter the formulation at all (7, 8); additionally, the impact of extracellular matrix structure and stiffness on FA assembly has also been reported (9); finally, recent observations also suggest that the physical properties of cell nucleus (10) can influence the size of FAs, however, to the best of our knowledge, no theoretical explanation has been provided. Aiming to address these problems, we developed a chemo-mechanical model to describe the growth dynamics of adhesion plaques as a function of substrate and nuclear stiffness. In particular, we will demonstrate the existence of a critical nucleation size for nascent patches to evolve into mature ones, and the extent to which stable plaque size is influenced by both the substrate and the nuclear rigidity.

METHODS

As described in the literature, a focal adhesion (FA) consists of a cluster of proteins (including individual integrin molecules that bind to ECM), that is then connected to the cell nucleus via the actomyosin stress fibers (Fig. 1a). A schematic of our proposed structural model describing the full mechanical feedback loop regulating FAs, the cytoskeleton and the nucleus is shown in Fig. 1b. The FA and ECM complex are treated as a spring (green box in Fig. 1b) whose stiffness is dependent on the FA structure and ECM mechanical properties. The actomyosin responsible for contractile force generation is illustrated in the blue box, and the nucleus is modeled as an elastic component to capture nuclear deformability (yellow box, Fig. 1b). The structure of the FA-ECM complex allows for sensing of ECM mechanical properties through force dependent assembly or disassembly of adhesions. Since cells have been shown to sense the ECM as fibers or non-fibers depending on ECM structure, we considered both fibrous and non-fibrous ECM types. If the ECM protein tethering spacing were larger than the FA size, a cell would sense the material as fibrous (Fig. 1c). However, if the ECM protein tethering spacing were smaller than the FA size, a cell would sense the ECM as a continuous elastic material (Fig. 1d). The model was combined with energy criteria for FA plaque recruitment to study the FA size dependence on the model parameters.

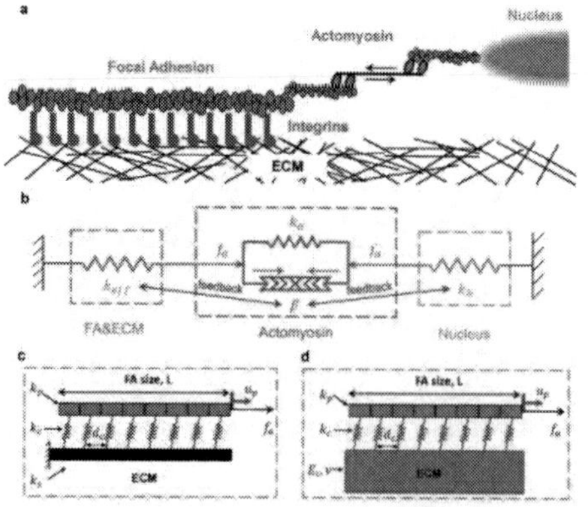

Figure 1: Illustration of cell-ECM adhesion (a). Schematic of model (b) and structures of FA-ECM complex on different kinds of ECM, fibrous (c) and continuous elastic (d).

RESULTS

The stiffness a cell senses increases first and then decreases as FA becomes larger

As discussed above, the FA effective stiffness (k_{eff}) can be determined by solving the mechanical governing equations. This physical parameter indicates the actual mechanical stiffness of the extracellular environment that the cell senses. This k_{eff} is plotted as a function of both the FA size and ECM rigidity in Fig. 2a and 2b for ECMs treated as fibrous or continuous elastic materials respectively. Interestingly, as FA size (L) increases, k_{eff} first climbs to a maximum before undergoing a monotonic decrease for both ECM types. The difference between the two models is that for gels there is a saturated effective stiffness, which is related to gel modulus: the stiffer the gel, the higher the effective stiffness. In conclusion, the effective stiffness of the extracellular environment that cells sense, first increases and then decreases as FAs becomes larger for both fibrous and non-fibrous ECMs.

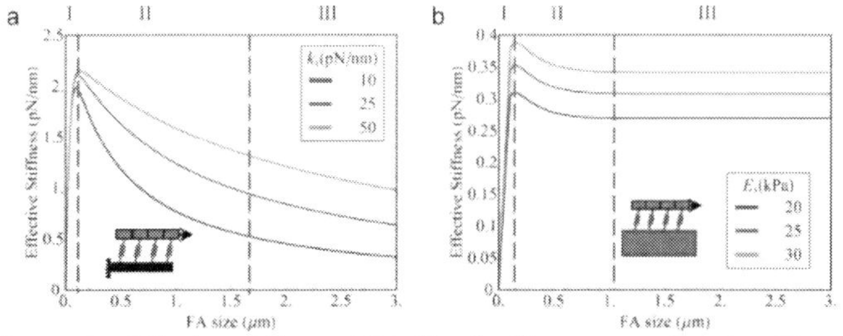

Figure 2: Effective stiffness of FA (k_{eff}) with respect to FA size (L) on different fibers with stiffness k_s (a) or on non-fibrous elastic materials with modulus E_s (b).

FA plaque recruitment is divided into three regimes by two important sizes: nucleation size and stable size.

Both models show the generic shape of the plaque recruitment flux J, as a function of FA size as shown in Fig. 3a. An immediate observation is that the value of J is positive only when $L_{nu} < L < L_{st}$ where L_{st} stands for the stable size of the plaque while L_{nu} can be considered the critical size a nascent plaque must overcome in order to grow and become fully developed. As such, the plaque dynamics can be divided into three regimes as depicted in Fig. 3a. Newly nucleated FAs with sizes smaller than L_{nu} will dissemble and eventually disappear. In comparison, a nascent FA will develop into a stable one once its size stochastically reaches this critical value. Larger FAs ($L > L_{st}$) are expected to shrink until reaching the predicted stable size. Fig. 3b shows the extent to which ECM rigidity influences the growth rate of FAs. Our results suggest that the stable size increases monotonically as the substrate becomes more rigid.

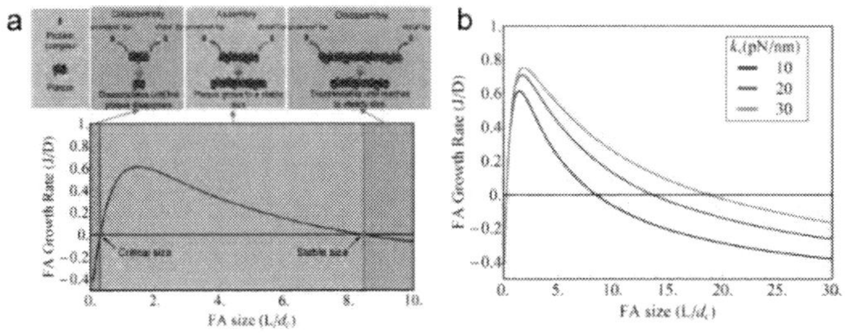

Figure 3: Two essential sizes for a FA plaque: critical size and stable size are predicted by the model (a). FA growth rate for different substrate stiffnesses (b) indicates stable FA size is larger on stiffer substrate.

Cells with stiffer nuclei have a lower barrier to form FAs and would assemble larger FAs, while higher contractility leads to a small drop in the FA nucleation barrier and a significant increase in stable FA size.

Our model predicts the dependence of FA size on nuclear stiffness. As shown in Fig. 4a, cells with stiffer nucleus would have a smaller critical FA recruitment size, which means the FA recruitment barrier is lower for these cells. Meanwhile, the model predicts that stiffer nuclei would also lead to a larger FA stable size, allowing for more and larger FAs to form.

Cell contractility is essential for FA formation and regulates FA size. Our model predicts the extent to which contractility regulates FA growth. For lower levels of cell contractility (20% lower actin pulling force) our model predicts that the FA nucleation barrier is larger and the stable size is smaller (Fig. 4b).

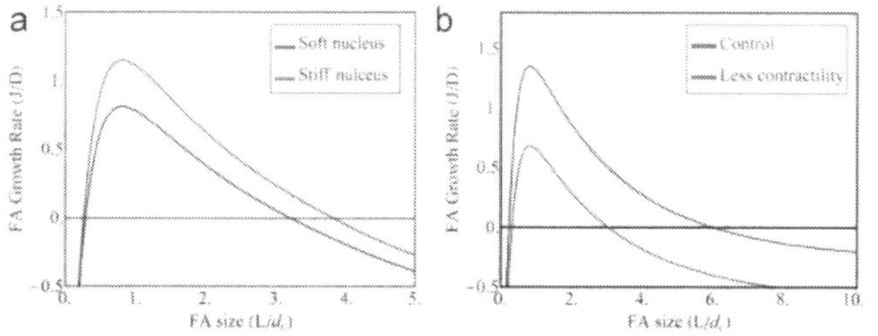

Figure 4: Effect of nuclear stiffness on FA recruitment (a). Stronger feedback effects with increased contractility result in larger stable FA and smaller FA nucleation barrier (b).

DISCUSSION

In this study, a chemo-mechanical model was developed to predict the growth of focal adhesion plaques, a process heavily influenced by the assembly of adhesion proteins as well as the stress build-up in the plaque itself (induced by actomyosin contractility). Main findings obtained here are summarized as follows:

- Two quantities of key interest, i.e. the stable size of adhesion plaques and the critical size for newly nucleated plaques to grow into mature ones, have been identified.
- A stiffer substrate/nucleus will lead to larger plaques. The prediction is consistent with experimental findings that large focal adhesions are formed on rigid surface (5).
- Interestingly, it was also found that the mechanosensitivity of the plaque is significantly influenced by contractility. Specifically, high levels of contractility will lead to bigger plaques.

We believe that the present model can also be used in studying phenomena such as cell locomotion and stretch-induced cell reorientation where contractile forces are generated inside the cell and the turnover of FAs play pivotal roles, indeed efforts along these lines are currently underway.

ACKNOWLEDGEMENTS

This work was supported by the University Research Foundation of the University of Pennsylvania and the NIH (R01 AR056624).

REFERENCES

[1] Nelson, C.M, et al., *PNAS*. 102: 11594–9, 2005.
[2] Engler, A.J, et al., *Cell*. 126: 677–89, 2006
[3] Janmey, P. A, et al., *Cell Motil. Cytoskeleton*. 66: 597–605, 2009.
[4] Oakes, P.W., et al., *J. Cell Biol*. 196: 363–74, 2012
[5]Plotnikov, S. V, et al., *Cell*. 151: 1513-1527, 2012
[6] Nicolas, A., et al., *PNAS*. 102: 12520–12525, 2004
[7] Shemesh, T., et al., *PNAS*. 102: 12383–8, 2005
[8] Chan, C., et al., *Science*. 322: 1687–1691, 2008
[9] Trappmann, B., et al., *Nat. Mater*. 11: 642–9, 2012
[10] Chancellor, T.J., et al., *Biophys. J*. 99: 115–23, 2010

SB^{3}C2015
Summer Biomechanics, Bioengineering and Biotransport Conference
June 17-20, 2015,Snowbird Resort, Utah, USA
SB^{3}C2015-229

PREDICTION OF DOWNSTREAM VELOCITY WAVEFORMS FOR IN VITRO AORTIC FLOW EXPERIMENTS

Rafeed A. Chaudhury (1), Justin R. Ryan (1), David H. Frakes (1,2), Ronald J. Adrian (3)

(1) School of Biological and Health Systems Engineering
Arizona State University
Tempe, AZ, USA

(2) School of Electrical, Computer, and Energy Engineering
Arizona State University
Tempe, AZ, USA

(3) School for Engineering of Matter, Transport and Energy
Arizona State University
Tempe, AZ, USA

INTRODUCTION

The American Heart Association recently reported that an estimated 85.6 million Americans (>1 in 3 adults) suffer from cardiovascular disease (CVD) [1]. Further, CVD accounts for 17.3 million deaths per year globally, a number that is expected to grow to >23.6 million by 2030 [2]. Aortic diseases account for significant cardiovascular morbidity and mortality worldwide. Specifically, atherosclerosis and aneurysms play a major role in diseases of the aortic arch, descending thoracic, and abdominal aorta [3]. Due to the highly complex nature of in vivo cardiovascular flow, the development of transitional and turbulent flows under normal physiologic conditions is still not fully understood. However, these flows have been linked to the development of CVD [4-6]. Pulsatile flow piston pumps are advantageous for fluid dynamic investigations of anatomical phantom models because they can replicate flow waveforms like those generated by the human heart [7-11]. Image-based computational fluid dynamics (CFD) and direct numerical simulations (DNS) are also widely used to obtain patient-specific hemodynamic fields in atherosclerotic and aneurysmal cases. However, these approaches make simplifying assumptions on the system dynamics and/or parameters due to limited computational resources and time. Additionally, the downstream waveform seldom matches the intended waveform at the phantom model. Prediction of the inlet velocity waveform at the test section during steady and unsteady flow is not only critical for obtaining meaningful experimental results but is also useful for CFD simulations because of the potential to reduce the computational domain when using patient-specific geometries.

In this study, we present the use of linear and time-invariant (LTI) system theory to predict the response to a piston-driven pump in order to establish inlet velocity waveforms at a phantom test section. Specifically, mass control flux [12] and the step response of a piston-driven pump are used to find the transfer function of the entire system with respect to flow conditions at locations downstream of the pump. The flows are examined experimentally, using stereographic particle image velocimetry (PIV), and are compared to the pump input and predicted velocity waveform.

METHODS

Experiments were carried out using a flow loop driven by a custom computer-controlled piston pump capable of imposing both steady and unsteady flow rates described in an earlier study [12]. The flow loop consisted of the piston pump, a urethane block test section, and a reservoir tank to receive the flow. An aqueous sodium iodide solution was used as the working fluid.

Starting flow data at a Reynolds (Re) number of 1,000 and pulsatile flow data were obtained using previously published techniques [12-14]. Velocity fields were captured using PIV in accordance with Adrian et al. [15]. The urethane model was positioned at 55 diameters (D) downstream of the pump at the model inlet.

The unit impulse response of the system, $h(t)$, is the derivative of the unit step response, $s(t)$:

$$h(t) = \frac{ds(t)}{dt} \quad (1).$$

Thus, the system output, $y(t)$, can be predicted by convolving the input velocity waveform, $x(t)$, and the unit impulse response of the system, or transfer function, $h(t)$, described by:

$$y(t) = x(t) * h(t) \quad (2).$$

Here, the system is defined as the pump and the flow loop since flow response depends upon the inertia of the fluid in the flow loop.

Piston face position is defined by $x(t)$, and $y(t)$ is the centerline velocity of the fluid 55D downstream of the pump. Image processing motion estimation techniques were used to find the step response of the pump system by tracking piston motion using high-speed cameras. The process for predicting and validating the input waveform is summarized in Figure 1. The subscripts s, p, and e, denote step, predicted, and experimental, respectively.

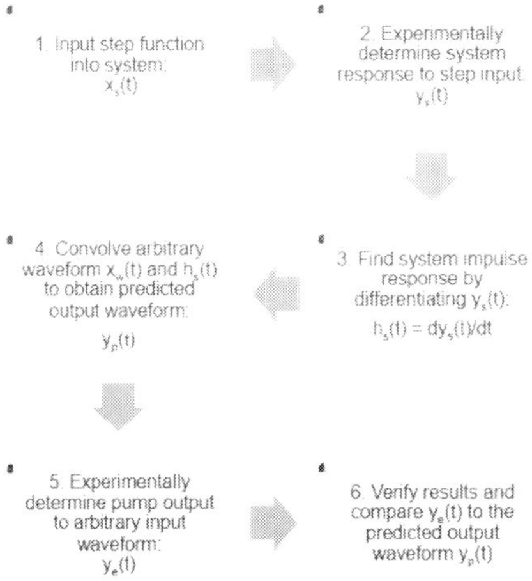

Figure 1. Process diagram for predicting the output waveform to an arbitrary waveform

RESULTS

Figure 2 simultaneously plots the step response (left) and impulse response (right) of the pump system during starting flows at $Re_D = 1{,}000$. The impulse response is obtained from taking the derivative of the experimentally measured step response.

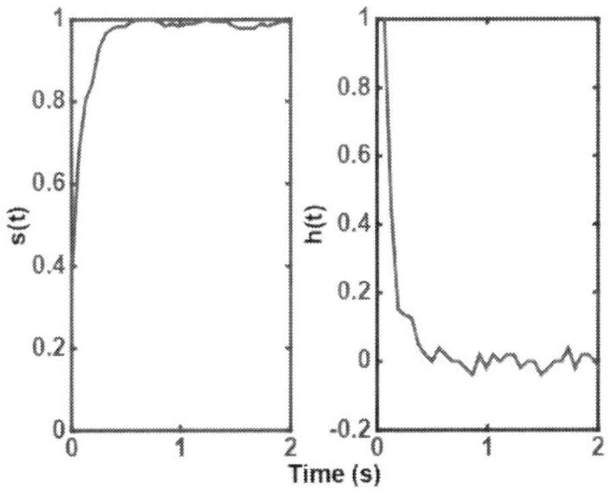

Figure 2. Pump system step response (left) and impulse response (right) versus time (s)

A comparison of the input aortic flow waveform, predicted pump response waveform based on LTI system theory, and phase-locked PIV measurements of the pump output relative to the start of the cardiac cycle is presented in Figure 3. The peak velocity normalizes all data presented. The pump input is an aortic flow waveform with a 0.7 s period. The predicted pump response is obtained by convolving the pump input waveform with the calculated system impulse response obtained in Figure 2. The predicted pump output curve closely follows the input and measured pump output curves during systole to within 2.5%. The predicted piston lag time is confirmed experimentally using PIV. The maximum error between the predicted pump output and the measured pump output is 3% at the end of the cardiac cycle.

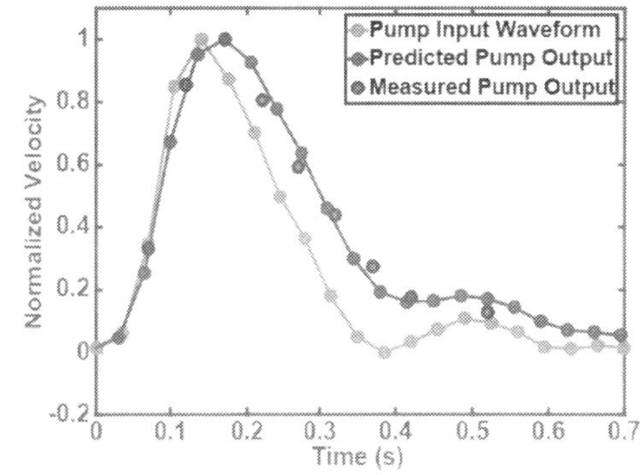

Figure 3. Comparison of the pump input waveform, predicted output waveform, and measured pump output relative to the start of the cardiac cycle

DISCUSSION

Downstream velocity waveforms are generally very difficult to predict in a phantom model for both experimental and computational work. Most flow pump systems are able to follow and generate physiological waveforms downstream of the pump, at the area of interest. Although the mean flow rates may be equivalent to the desired mass and volume flux, the velocity waveforms themselves are dissimilar. The input waveform needed to achieve the intended waveform at the test section is largely unknown without repeated input tweaking and downstream measurements and experimentation. This study demonstrates a straightforward method to predict downstream velocity waveforms at the test section based on an arbitrary pump input waveform. Not only is this useful for experimentalists, but computational groups may leverage these results to minimize the computational domain, especially when using patient-specific geometries. Starting simulations at the inlet of the test section near the area of interest rather than waiting for fully developed conditions will also reduce computation time. Future work will focus on expanding these results to other aortic waveforms and determining the input waveform needed to achieve the intended waveform at the test section.

ACKNOWLEDGEMENTS

The authors gratefully acknowledge financial support from the Arizona Biomedical Research Commission (ABRC Grant #11028752), National Science Foundation (NSF CAREER Award #1151232), the Ira A. Fulton Endowment, and ARCS Foundation.

REFERENCES

[1] Mozaffarian, D et al., *Circulation*, 2014.
[2] Alwan, A, World Health Organization, 2011.
[3] Novaro, GM, Cleveland Clinic CCE, 2014.
[4] Stein, PD et al., *Circulation Research*, **39**:58–65, 1976.
[5] Lasheras, JC, *J of Fluid Mechanics*, **664**:1–4, 2010.
[6] Stalder, AF et al., *J MRI*, **33**:839–846 2011.
[7] Lieber, BB et al., *Ann Biomed Eng*, **30**:768–777, 2002.
[8] Ku, JP et al., *Ann Biomed Eng* **33**:257–269, 2005.
[9] Hoskins, PR, *Ultrasound Med Biol*, **34**, 693–717, 2008.
[10] Taylor, CE et al., *J. Med. Devices*, **6**, 045002,2012.
[11] Babiker, MH et al., *J Biomech*, **45**, 440–447, 2012.
[12] Chaudhury, RA et al., *Exp Fluids*, **56**:22, 2015.
[13] Roszelle, BN et al. *J Biomech Eng*, 132:2, 2014.
[14] MICCAI, CFD Challenge, 2012.
[15] Adrian, RJ et al., Cambridge University Press, New York, 2010.

DETERIORATION OF TRABECULAR AND CORTICAL MICROARCHITECTURE AND REDUCED BONE STIFFNESS AT DISTAL RADIUS AND TIBIA IN POSTMENOPAUSAL WOMEN WITH VERTEBRAL FRACTURES

Ji Wang (1), Emily M. Stein (2), Bin Zhou (1), Kyle K. Nishiyama (2), Elizabeth Shane (2), X. Edward Guo (1)

(1) Department of Biomedical Engineering, Columbia University
New York, New York, USA

(2) Department of Medicine, College of Physicians and Surgeons, Columbia University,
New York, New York, USA

INTRODUCTION

Vertebral fractures, which directly result from collapse of weakened vertebral bodies, are also reflective of extensive deterioration of trabecular and cortical microarchitecture at peripheral skeletal sites [1-3]. At distal radius and distal tibia, postmenopausal women with vertebral fractures have fewer and more widely spaced trabeculae, and thinner cortices compared with women without fractures. Studies segmenting plate- and rod-like trabeculae in the trabecular network suggest that increased bone fragility in postmenopausal women is associated with disrupted trabecular plate microarchitecture, the predominant component of trabecular bone strength [4-6]. However, it is unknown whether women with vertebral fractures have such deficiencies in trabecular plate and rod microarchitecture. This study aims to characterize explicitly the alterations of trabecular and cortical microarchitecture and bone stiffness at distal radius and distal tibia in postmenopausal women with vertebral fractures.

METHODS

Postmenopausal women with a history of low trauma vertebral fractures (n=45) and controls with no history of fracture and normal spine x-rays (n=45) had areal bone mineral density (aBMD) of lumbar spine, total hip, femoral neck, 1/3 radius, ultradistal radius measured by duel-energy x-ray absorptiometry. Microarchitecture of distal radius and tibia were imaged by high-resolution peripheral quantitative computed tomography (HR-pQCT). We applied an automatic segmentation algorithm to divide the trabecular and cortical compartments [7], which were subjected to individual trabecula segmentation (ITS)-based trabecular plate and rod morphological analysis [6, 8] and a customized cortical evaluation script [9, 10], respectively. Finite element analysis estimated whole bone and trabecular bone stiffness.

RESULTS

The fracture group and the control group were similar in age (70±8 and 70±7 yrs), race (82% and 84% Caucasian, respectively), and body mass index (28±10 and 27±6). Mean aBMD T-scores of the fracture group were 0.5-0.75 SD lower ($p<0.05$) than controls at total hip, femoral neck, and ultradistal radius, but did not differ at lumbar spine and 1/3 radius (Figure 1). The prevalence of osteoporosis and osteopenia in this cohort were 45% and 47%, respectively. Marked differences in trabecular and cortical microarchitecture were observed between the two groups (Figure 2). At distal tibia, trabecular microarchitecture of the fracture subjects was characterized by preferentially lower trabecular plate bone volume, number, and connectivity over rod-related parameters, in addition to decreased axially aligned trabeculae and lower plate-rod ratio. At distal radius, both trabecular plate and rod microarchitecture showed significantly lower volume, number, and connectivity. Additionally, differences in trabecular plate and rod sizes were observed. Differences in trabecular plate volume, number, connectivity, and axial trabecular bone volume remained significant after adjustment for aBMD, but trabecular rod parameters did not. Compared to controls, women with vertebral fractures had lower cortical thickness, smaller cortical area, and larger trabecular area. Cortical porosity and pore size did not differ. Group differences in trabecular microarchitecture appeared greater at the radius than at the tibia, whereas differences in cortical microarchitecture were more pronounced at the tibia. Whole bone stiffness and trabecular bone stiffness were reduced in women with vertebral fractures by 18% and 22%, respectively at the radius, and by 19% and 16% at the tibia.

DISCUSSION

Postmenopausal women with vertebral fractures showed deterioration of both trabecular and cortical microarchitecture at distal radius and tibia. ITS analysis in the vertebral fracture subjects detected major depletion of trabecular plates, loss of axially aligned trabeculae, and trend to convert into more rod-like trabecular network. Trabecular microarchitecture at distal radius tended to be more severely affected than tibia, which is consistent with the finding that the association of microarchitecture of distal radius with stiffness at the spine was stronger than associations between the tibia parameters and the spine [11]. Moreover, the thinner cortex and expanded trabecular and total bone area suggested more endocortical bone resorption in women with vertebral fractures. These results suggest that this pattern of bone quality deterioration contributes to bone fragility and increased susceptibility to vertebral fractures in postmenopausal women.

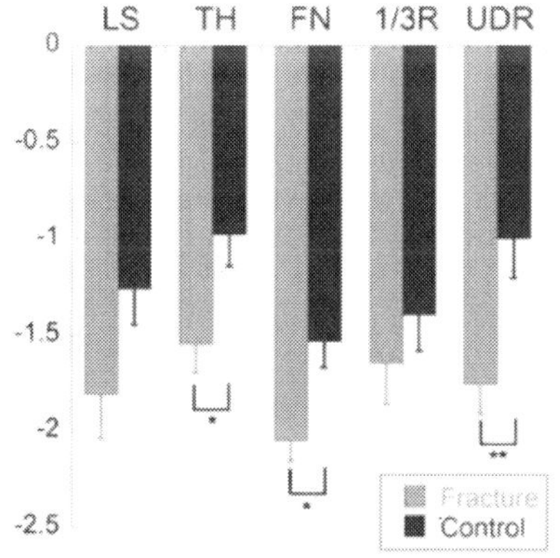

FIGURE 1. COMPARISON OF ABMD T-SCORES BETWEEN WOMEN WITH VERTEBRAL FRACTURES AND CONTROLS. LS, LUMBAR SPINE; TH, TOTAL HIP; FN, FEMORAL NECK; 1/3R, 1/3 RADIUS; UDR, ULTRADISTAL RADIUS. (* P<0.05, ** P<0.01)

ACKNOWLEDGEMENTS

This work is supported by the NIH AR058004 and AR051376. The author, Ji Wang, is supported by HHMI International Predoctoral Student Fellowship

REFERENCES

1. Melton, L.J., 3rd, et al., *J Bone Miner Res*, 25: 1922-30, 2010.
2. Sornay-Rendu, E., et al., *J Bone Miner Res*, 24: 737-43, 2009.
3. Stein, E.M., et al., *J Clin Endocrinol Metab*, 97: 1918-26, 2012.
4. Liu, X.S., et al., *J Bone Miner Res*, 27: 263-72, 2007.
5. Stein, E.M., et al., *J Bone Miner Res*,29: 1101-9, 2014.
6. Liu, X.S., et al., *J Biomech*, 42: 249-56, 2009.
7. Buie, H.R., et al., *Bone*, 41: 505-15, 2007.
8. Liu, X.S., et al., *J Bone Miner Res*, 23: 223-35, 2008.
9. Burghardt, A.J., et al., *J Bone Miner Res*, 25: 983-93, 2010.
10. Nishiyama, K.K., et al., *J Bone Miner Res*, 25: 882-90, 2010.
11. Liu, X.S., et al., J Bone Miner Res, 25: 2229-38, 2010.

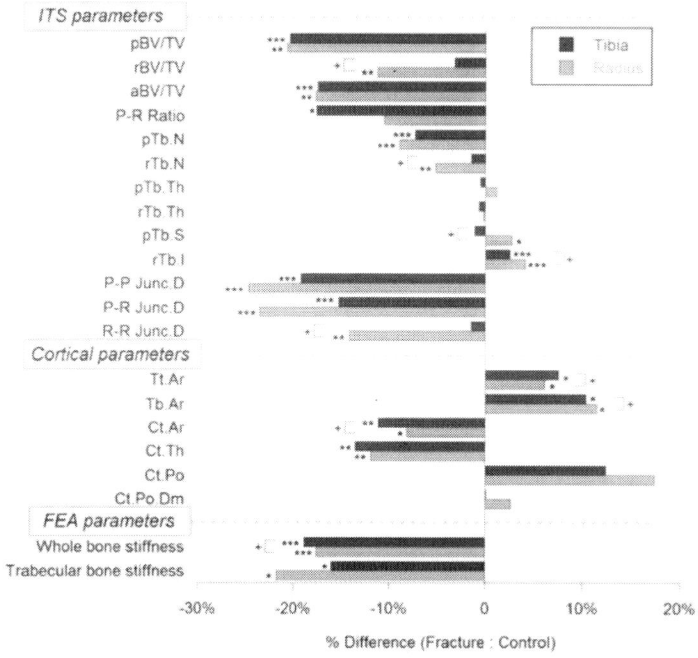

FIGURE 2. DIFFERENCE IN ITS, CORTICAL, AND FEA PARAMETERS BETWEEN WOMEN WITH VERTEBRAL FRACTURES AND CONTROLS. PBV/TV, PLATE BONE VOLUME FRACTION; RBV/TV, ROD BONE VOLUME FRACTION; ABV/TV, AXIAL BONE VOLUME FRACTION; P-R RATIO, PLATE-ROD VOLUME RATIO; PTB.N, TRABECULAR PLATE NUMBER; RTB.N, TRABECULAR ROD NUMBER; PTB.TH, TRABECULAR PLATE THICKNESS; RTB.TH, TRABECULAR ROD THICKNESS; PTB.S, TRABECULAR PLATE SURFACE AREA; RTB.L, TRABECULAR ROD LENGTH; P-P JUNC. D, PLATE-PLATE JUNCTION DENSITY; P-R JUNC.D, PLATE-ROD JUNCTION DENSITY; R-R JUNC.D, ROD-ROD JUNCTION DENSITY; TT.AR, TOTAL AREA; TB.AR, TRABECULAR AREA; CT.AR, CORTICAL AREA; CT.TH, CORTICAL THICKNESS; CT.PO, CORTICAL POROSITY; CT.PO.DM, CORTICAL PORE DIAMETER. (* P<0.05, ** P<0.01, *** P<0.01, FRACTURE VS. CONTROL DIFFERENCE; + P<0.05, DISTAL RADIUS DIFFERS FROM DISTAL TIBIA IN FRACTURE VS. CONTROL DIFFERENCE)

SB³C2015
Summer Biomechanics, Bioengineering and Biotransport Conference
June 17-20, 2015,Snowbird Resort, Utah, USA
SB³C2015-232

COLLECTIVE CHIRAL ROTATION OF EPITHELIAL MICROTISSUES WITHIN A THREE-DIMENSIONAL MATRIGEL SYSTEM

Amanda S. Chin (1), Kathryn E. Worley (1), Leo Q. Wan (1)

(1) Department of Biomedical Engineering
Rensselaer Polytechnic Institute
Troy, NY, United States

INTRODUCTION

Chirality, also referred to as left-right (LR) asymmetry, has emerged as an intrinsic property of mammalian cells that is vital for normal tissue morphogenesis. LR asymmetry is observed throughout nature from whole organisms to organ development. Nevertheless, the expression of chirality at the cellular level has yet to be fully understood. Through the use of ring-shaped 2D micropatterned surfaces and image analysis to study cell alignment and directional migration, it has been determined that chirality is phenotype specific [1]. However, 2D cultures may not fully represent the dynamics of tissue polarization and cell-cell interactions during tissue formation *in vivo*. To more accurately simulate the complexity of cellular behavior and morphology during natural tissue development, we cultivated epithelial cells in 3D layered hydrogel, which revealed emergent rotational behavior.

We found that individual cells exhibited clear LR biased rotation and that as the cells proliferated and self-organized into microtissues with a polarized lumen structure, the directional rotation persisted. The role of actin was confirmed in this 3D system, similar to what was found on 2D micropatterned cultures [1]. We propose that the bias of collective rotation of microtissues could be attributed to cellular chirality and may play a vital role in tissue morphogenesis.

METHODS

Cells and Cell Culture: Madin-Darby canine kidney (MDCK; ATCC) epithelial cells were cultured in Dulbecco's modified Eagle's medium (DMEM) containing 10% fetal bovine serum, 1% penicillin streptomycin (Sigma-Aldrich) and 1% sodium pyruvate (Invitrogen).

3D Multilayer Matrigel System: Using an 8-chambered glass slide (Ibidi) a base layer was formed by coating the bottom of the wells with 50 µl of 100% Matrigel (Corning) and incubating at 37°C for 25 min to solidify the Matrigel (Figure 1). MDCK cells were seeded onto the base Matrigel layer at a seeding density of 6,000 cells/well. After incubation for 1 hr the cells attached and the media was replaced with fresh media containing 2% Matrigel to generate a top layer, thus embedding the cells between a bilayer of Matrigel. The bilayer system facilitates a flat plane for imaging and also creates a gel gradient due to the two different concentrations of Matrigel which establishes a z-axis orientation on the embedded cells promoting directional rotation in the x-y plane (Figure 2). The media was replaced every 2 days and the culture was maintained for 5 to 10 days.

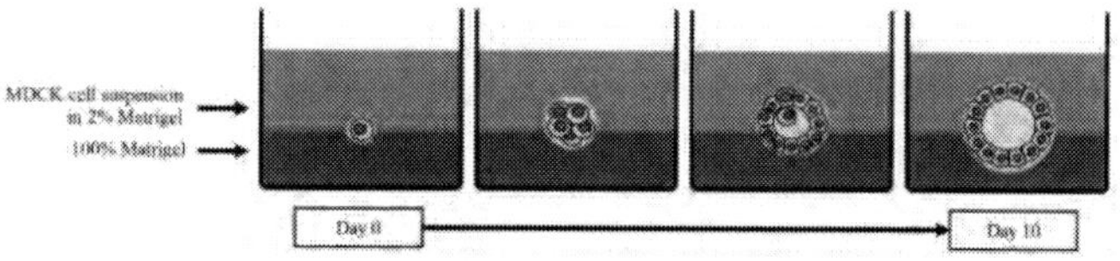

FIGURE 1. SCHEMATIC OF BILAYER 3D CULTURE CONFIGURATION.

Morphology and Rotation Analysis: Both phase-contrast and fluorescence time-lapse images were gathered for at least 2 hrs with an interval of 1 min. Time-lapse videos were processed and analyzed using ImageJ where the direction of cellular rotation was categorized into clockwise (CW), counterclockwise (CCW), no rotation, and complex rotation in which the cell rotates about several planes or undergoes directional switching.

Immunofluorescence: Self-organization of cells during microtissue development was observed using confocal microscopy. Cells were fixed with paraformaldehyde, and stained with DAPI for nuclei, Phalloidin for actin filaments and ZO-1 antibody for tight junctions.

2D Chirality Characterization: MDCK cultures on ring-shaped micropatterns were prepared as described previously [1]. An array of ring-shaped molds were prepared with microfabrication and cast to form PDMS stamps. The stamps were used to micro-contact print rings onto gold-coated glass upon which fibronectin was applied to promote cell attachment. Once near confluency, patterned cells were imaged using phase contrast and analyzed with a MATLAB program to determine chiral alignment.

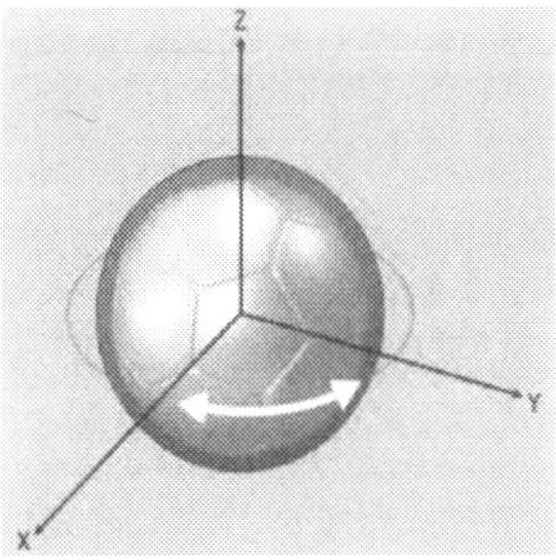

FIGURE 2. ROTATIONAL PATH IN 3D. WHITE ARROW INDICATES ROTATIONAL BIAS OF EITHER CLOCKWISE OR COUNTERCLOCKWISE.

RESULTS

When the MDCK cells were cultured in 3D Matrigel, the cells proliferated, migrated, and self-assembled into hollow spheroids. Confocal microscopy imaging of the microtissues revealed the architecture of the microspheres (Figure 3). The luminal spheroid structure exhibited higher expression of actin along the apical surface, indicative of a polarized acinar formation. The polarization of the assembly was further confirmed by the expression of laminin V only at the exterior surface of the spheroid [2].

MDCK cells were cultured on ring-shaped micropatterns to analyze left-right asymmetry based on the directional migration and alignment of the epithelial cells in 2D. In order to investigate the chirality of MDCK cell in 3D, the cells were also embedded within the multilayer Matrigel system to characterize collective migration. The initial single cells embedded between the layers had biased rotation and as the cells proliferated, the rotation persisted in the collective microtissues that self-assembled into hollow spheroids.

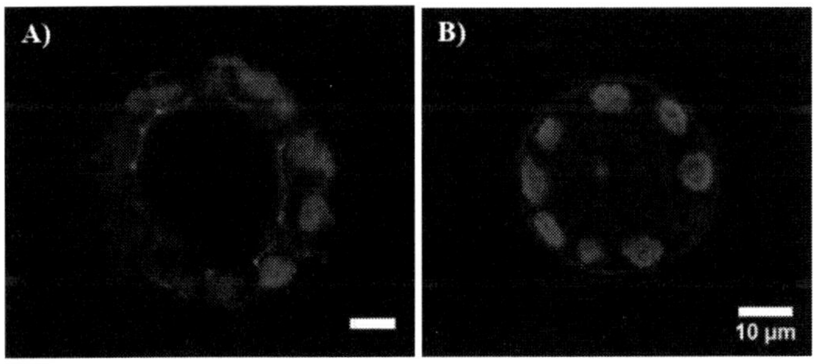

FIGURE 3. CROSS-SECTIONS OF THE MULTICELLULAR MICROTISSUES REVEALED A DEFINED LUMEN FORMATION ON DAY 10. (A.) RED: PHALLOIDIN, GREEN: ZO-1, AND BLUE: DAPI. (B.) RED: LAMININ V AND BLUE: DAPI.

The microspheres of MDCKs underwent spontaneous collective rotation, as previously reported [3, 4]. More interestingly, the rotation is chiral, with slight counterclockwise dominance (Figure 4). Such a bias is consistent with what was observed on 2D micropatterned substrates. Intriguingly, such chirality arises from single cells, since initial cells embedded between the layers had spontaneous rotation with the same bias (Data not shown).

Given that little is understood about the underlying factors that drive cellular chirality, we utilized Latrunculin A (Lat A) to study the role of actin function in left-right asymmetry. Latrunculin A is a drug which inhibits actin polymerization. On micropatterned rings, MDCK cells exhibit a slight bias towards counterclockwise alignment (Figure 4B). Treatments with Latrunculin A caused a reversal of chirality in which the MDCK cells were greatly biased towards clockwise direction. Similarly, the 3D microtissues were initially biased towards counterclockwise but upon disrupting the actin, their rotation switched to be increasingly clockwise with higher concentrations of Latrunculin A (Figure 4A).

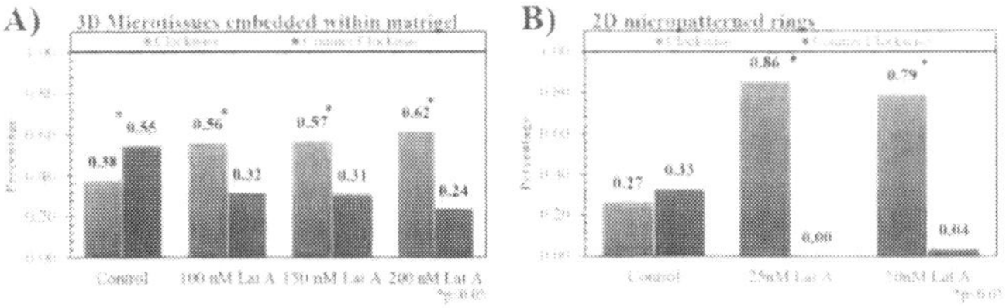

Rotation	Control	100nM Latrunculin A	150nM Latrunculin A	200nM Latrunculin A	LR Bias	Control	25nM Latrunculin A	50nM Latrunculin A
Clockwise	46	64	66	68	Clockwise	50	18	65
Counter-clockwise	66	36	36	26	Counter-clockwise	60	0	3
Total	121	114	116	110	Total	186	21	82

FIGURE 4. CHIRALITY OF MDCK CELLS WITH LATRUNCULIN A TREATMENT IN 3D (A) VS 2D (B).

DISCUSSION

In this study, we determined that there is an asymmetrical bias in the direction of rotation of epithelial microspheroids in layered Matrigel. This bias was consistent with 2D chiral determination results, and was also observed for single cells in 3D. Therefore, we conclude that chirality is an inherent cellular property that is an acute factor in coordinated rotation. We infer that such biased rotation could play a critical role in directional looping of developing organs such as the heart tube and mid- and hind- guts.

The success of observing rotational bias of epithelial cells is largely attributed to the layered design of hydrogel. Without the gradient of Matrigel, the spheres would not be able to perform the in-plane (x-y) rotation, not to mention the consistent bias (i.e., count clockwise). The strong cell-cell adhesion and formation of the apical-basal polarity further enhanced the collective cell rotation, so that the bias could be easily appreciated.

Inhibition of actin polymerization by treatment with Latrunculin A revealed a dose-dependent switching of rotational directionality suggesting the imperative role of actin functionality in cellular chirality. To further study the mechanisms underlying 3D morphogenesis, we intend to quantify rotation rate and examine cell-cell interactions. The dynamics of collective rotation by microtissue spheroids may be attributed to inherent cellular chirality and could be a driving force behind the coordinated migration of cell populations during tissue development.

ACKNOWLEDGEMENTS

The authors would like to thank NIH Director's New Innovator Award, National Science Foundation, American Heart Association, and March of Dimes for funding Support. Leo Q. Wan is a Pew Scholar in Biomedical Sciences, supported by the Pew Charitable Trusts.

REFERENCES

[1] Wan, LQ et al., *PNAS*, 108:12295-12300, 2011.
[2] Debnath, J et al., *Methods*, 30:256-268, 2003.
[3] Wang H et al., *PNAS*, 110:163-168, 2013.
[4] Tanner K et al., *PNAS*, 109:1973-1978, 2011.

SB³C2015
Summer Biomechanics, Bioengineering and Biotransport Conference
June 17-20, 2015,Snowbird Resort, Utah, USA
SB³C2015-234

QUANTITATIVE VISUALIZATION OF DRUG RESPONSE OF BREAST CANCER CELLS WITHIN A THREE-DIMENSIONAL EXTRACELLULAR MATRIX

Brett S. Klosterhoff (1), Kyeonggon Shin (1), J. Craig Dutton (2), Bumsoo Han (1,3)

(1) School of Mechanical Engineering
Purdue University
West Lafayette, Indiana, United States

(2) Department of Aerospace Engineering
University of Illinois at Urbana-Champaign
Urbana, Illinois, United States

(3) Weldon School of Biomedical Engineering
Purdue University
West Lafayette, Indiana, United States

INTRODUCTION

The tumor *in vivo* is a dynamic, heterogeneous tissue comprised of multiple subpopulations of cancerous cells, often with divergent genomes driven by clonal evolution[1]. Additionally, tumors are surrounded by fibrous stroma and abnormal vasculature which result in increased interstitial pressure, poor tissue perfusion, and hindered intratumoral penetration by drug macromolecules[2]. In clinical settings, the emergent properties of the complex three-dimensional tumor structure are characterized by spatially heterogeneous and transient responses to therapeutic agents[2], posing significant challenges to effective treatment. Thus, an improved understanding of the dynamic response of tumor cells to physiologically appropriate drug treatment conditions will significantly improve treatment planning.

To improve this understanding, novel methods capable of characterizing the temporal cell response during therapeutic treatment are needed to establish a mechanistic understanding of heterogeneous chemoresistance, and to quantify the extent of drug efficacy. In this study, we quantitatively visualized the response of human breast cancer cells to the chemotherapeutic agent, Doxorubicin. To recapitulate the complexity of the tumor *in vivo*, cells were cultured in a previously developed, three-dimensional microfluidic platform, termed the tumor-microenvironment-on-chip (T-MOC), which assimilates the complex transport interactions of fluids, ECM, cells and drug macromolecules *in vitro*[3].

In order to characterize the drug response, we quantified cell motility by analyzing time-lapse microscopy images with a digital image cross-correlation algorithm[4]. The objectives of this study were to evaluate the feasibility of a digital image cross-correlation analysis method as a novel, non-invasive technique to quantify cellular motility, and to assess cell motility as a temporal measure of human breast cancer cell response to drug treatment.

METHODS

Fabrication of T-MOC: A description of the T-MOC has been reported in detail elsewhere[3]. Briefly, the T-MOC was fabricated via standard photolithography. The microfluidic geometry consists of an interstitial chamber where tumor cells were cultured within a collagen matrix, with two side channels to perfuse medium across the matrix.

Cells and Reagents: Briefly, human breast cancer epithelial cells (MDA-MB-231) were maintained in a culture medium (DMEM/F12, Invitrogen) supplemented with 10% FBS, 2 mM L-glutamine, 100 units/ml penicillin, and 100 μg/ml streptomycin. Collagen was prepared as previously reported[5]. Cells were seeded in collagen, then loaded into the T-MOC and polymerized in an incubator (0.5 h at 37 °C). The concentrations used for the cells and collagen were 10^7 cells/ml and 6 mg/ml, respectively. Then, the T-MOC was perfused with culture medium and incubated (24 h at 37 °C and 5% CO_2).

Time-Lapse Microscopy: After 24 hours of incubation, the culture medium was refreshed and the T-MOC was placed in an incubation stage under an inverted microscope (IX71, Olympus). Time-lapse bright-field images were acquired using a 20x objective. Imaging was conducted every 20 min for 3 consecutive 24-hour spans. At the end of the first 24-hour pre-culture image sequence, drug treatment commenced with media plus Doxorubicin at a concentration of 4 μM being perfused through the T-MOC. After 24 hours imaging the drug treatment sequence, standard culture medium was perfused and imaging was conducted for a final 24-hour post-culture sequence. Microscope down-time between sequences was minimized and only halted to refresh the culture medium. At the conclusion of time-lapse imaging, cell death was confirmed by a DAPI/TRITC image assay.

Digital Image Analysis: Time-lapse images were cross-correlated to estimate intracellular motility utilizing a commercial software package (DaVis 7.2, LaVision). A user-defined mask was applied to each image sequence by drawing circular regions of interest to isolate 16 individual cells whose motility was monitored temporally throughout the culture experiment. The circular mask was drawn larger than the cell to account for translational motion and to provide full coverage of the cell by the square interrogation windows. A multi-pass particle image velocimetry (PIV) time-series analysis was utilized. The cross-correlation algorithm computes the displacement of each 32x32 pixel interrogation window between sequential images to generate spatiotemporal vector fields mapping the chronological motility of individual tumor cells[4].

RESULTS

Bright-field images were obtained throughout time-lapse imaging (Figure 1(a)). A sub-region of the bright-field image (red rectangle in Figure 1(a)) after the motility analysis is provided to illustrate example displacement field frames before (Figure 1(b)) and after drug perfusion (Figure 1(c)). Samples of three consecutive displacement fields are also illustrated for a single cell (yellow rectangle in Figure 1(a)) before (Figure 2(a)) and after drug perfusion (Figure 2(b)). The magnitude of the vectors are magnified in the images (pixel displacement x 10) to improve clarity. Even with magnification, due to the practically negligible motility of cells after drug perfusion, the computed vectors are visualized as small blue dots on the cells (Figures 1(c) and 2(b)).

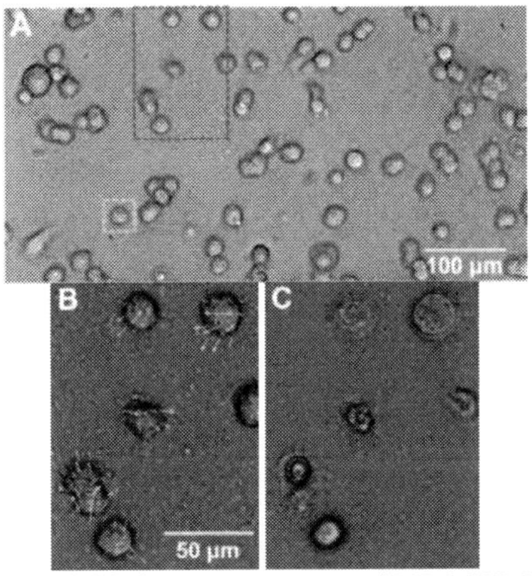

Figure 1: Bright-Field Time-Lapse Image of MDA-MB-231 Cells (A), Example Displacement Field Correlation for Image Sub-Region Prior to Doxorubicin Perfusion (B), and After Doxorubicin Perfusion (C)

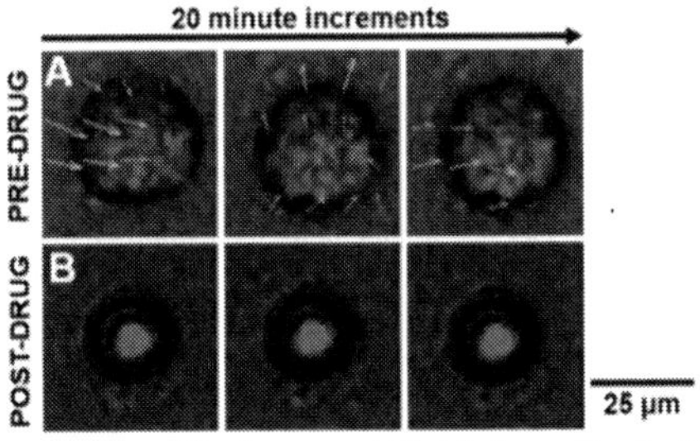

Figure 2: Consecutive Displacement Fields for a Single Cell Prior to Doxorubicin Perfusion (A), and After Doxorubicin Perfusion (B)

A plainly observable decline in displacement vector magnitude was demonstrated between pre-culture and post-culture conditions, indicating that cell death had occurred. This was confirmed by a DAPI/TRITC viability assay. By performing the analysis procedure throughout the entire 72-hour culture sequence, a temporal measure of tumor cell motility was computed.

DISCUSSION

The results substantiate the feasibility of the cross-correlation analysis as a novel cell motility evaluation method. Additionally, the computed decline in cell motility indicates that the technique is able to capture salient trends of the cellular response during drug perfusion and apoptosis such as translation, rotation, dilation, shrinkage and formation of apoptotic bodies (membrane blebbing). Culture sequence movies created by concatenating the analysis output from each of the consecutive culture sequences provide a temporal characterization of changes in cell motility before, during and after chemotherapeutic treatment. Additionally, a custom analysis script spatially averaging the computed vector fields on a per-cell basis quantifies the individual cell motility versus time (results not shown).

Tumor cell motility exhibits a positive correlation with the *in vivo* metastatic potential of the cell[6]. Thus, the ability to quantify and characterize temporal changes in tumor cell motility during drug treatment and subsequent cell death without perturbing the physiological cell response is beneficial to the evaluation of drug efficacy. Fluorescent labels are the current standard to assess cellular viability for culture experiments. However, labels are often cytotoxic due to chemical staining. Therefore, labels can only provide a discrete, post-treatment assay of viability before killing all of the stained cells, lending no information to analyze the temporal, heterogeneous response to the experimental treatment conditions of interest.

A notable strength of the current method is that the use of an optical analysis technique does not perturb the tumor microenvironment. Furthermore, cell motility can be evaluated on a single-cell basis, enabling temporal quantification of heterogeneous chemoresistance. The method developed in this study complements the growing body of literature leveraging computational image analysis techniques (PIV[7], digital image correlation[5], particle tracking[8]) to quantify cellular micromechanics *in vitro*.

A limitation of this study is that the T-MOC was seeded at a relatively low cell concentration (10^7 cells/ml), and cells were not aggregated. Further studies applying the method to different cell concentrations, cell lines, drug concentrations, and drug types are warranted for a rigorous validation of the technique.

We report a preliminary study confirming proof-of-concept for a correlation-based image analysis as a temporal, non-invasive method to quantitatively visualize tumor cell response to drug perfusion in a physiologically appropriate *in vitro* platform. We speculate that future development of the presented method could help to establish mechanistic understanding of heterogeneous tumor chemoresistance under numerous experimental treatment conditions.

ACKNOWLEDGEMENTS

This work was supported by a Collaboration in Translational Research Award from the Indiana Clinical and Translational Sciences Institute, funded in part by UL1 TR000006 from NIH, and the B.S.F. Schaefer Award, the Discovery Park Fellowship, and an Incentive Grant Program from Purdue University.

REFERENCES

[1] Anderson, K et al., *Nature*, 469:356–361, 2011.
[2] Trédan, O et al., *J Natl Cancer Inst*, 99:1441–1454, 2007.
[3] Kwak, B et al., *J of Controlled Release*, 194:157–167, 2014.
[4] Raffel, M et al., *Particle Image Velocimetry: A Practical Guide*, 1998.
[5] Teo, K et al., *J Biomech Eng*, 132:031003-1-8, 2010.
[6] Partin, A et al., *Proc Natl Acad Sci*, 86:1254–1258, 1989.
[7] Puliafito, A et al., *Proc Natl Acad Sci*, 109:739–744, 2012.
[8] Tseng, Y et al., *Biophys J*, 83:3162–3176, 2002.

SB^{3}C2015
Summer Biomechanics, Bioengineering and Biotransport Conference
June 17-20, 2015,Snowbird Resort, Utah, USA
SB^{3}C2015-235

IMAGING PROCESSING ALGORITHMS FOR DETECTING SUBTLE MORPHOLOGICAL CHANGES IN MOLE IMAGES OVER TIME

Alireza Chamani (1), Neera Nathan (2), Thomas J. Hornyak (2, 3), Liang Zhu (1)

(1) Department of Mechanical Engineering
University of Maryland Baltimore County
Baltimore, MD, USA

(2) Department of Dermatology
National Cancer Institute/NIH
Bethesda, MD USA

(3) Department of Dermatology
Department of Biochemistry and Molecular Biology
University of Maryland, School of Medicine
Baltimore, MD, USA

INTRODUCTION

Treatment of skin cancer depends upon the timely detection of abnormal skin lesions [1]. Unfortunately, in the early stages of skin cancer, patients may not notice changes in skin lesions that would prompt them to visit their primary physician or dermatologist. Once patients detect the symptoms of skin cancer, particularly melanoma, it may have metastasized, by which time treatment options are limited and costs are high.

Currently, early skin cancer detection depends upon findings from visual examination. Although diagnosis of melanoma is confirmed by biopsy, initial visual assessment is essential. Unfortunately, visual assessment may not be sufficiently informed or objective, and may be dependent on the experience and specialization of the diagnosing clinician. The objective of this study is to apply our previously developed computational algorithms to evaluate quantitative changes of morphological features of mole images with time. The algorithms tested on moles are expected to assist dermatologists in quantifying early subtle changes that may not be easily detected by naked eyes.

METHODS

Mole images were acquired using a digital dermatoscope (Fotofinder Systems) from patients enrolled in an IRB-approved clinical research protocol in the Dermatology Branch, National Institutes of Health, from 2005-2011. Standardized techniques were used to obtain sequential images with similar magnification, illumination, and pixel density. Lesions selected for analysis were those determined clinically to exhibit significant change, defined as either an asymmetric enlargement or change in pigmentary characteristics, in one or more sequential images. Initial images were obtained either at the first visit or at a subsequent visit during the study, with successive images generally obtained every 3 to 12 months. Figure 1 shows the initial set of images of 10 of 13 lesions that exhibited significant change. All lesions demonstrating significant change were biopsied at or after the time point that the change was initially measured. For the 10 lesions shown in Figure 1, the mean period of lesional observation prior to biopsy was 450 days (range 98-889 days). Lesion #1 was the only one determined histopathologically to represent a melanoma.

MATLAB software was used for imaging segmentation to divide the image into constituent regions of the mole and its background. Specifically, the border of the mole boundary was identified via selecting a threshold grayscale number to minimize the "within cluster variance" [2]. Three dimensionless ratios were defined to quantify the border irregularity, the asymmetry of the shape, the color variation inside the mole, and the size change of the mole over time. Derivations of the three ratios can be found in details in our previous study [3]. Higher ratios correspond to greater mole irregularity. The following are the definitions of the three ratios:

$$R_{\text{irregularity}} = \frac{\text{the total number of the boundary pixels of the mole}}{\text{the total number of the boundary pixels of a circle having the same area}}$$

$$R_{\text{asymmetry}} = \frac{\text{total number of pixels having no mirrored counterparts}}{\text{total number of pixels inside the mole}} \qquad (1)$$

$$R_{\text{color variation}} = \frac{\text{standard deviation of the pixel gray scale values}}{\text{mean value of the pixel gray scale values}}$$

The size change of a mole can be assessed using the change of the diameter of an equivalent circular shape having the same area. Once a ruler is placed with the mole when the image is taken, the pixel size can be determined to then identify the size of the mole.

RESULTS

We first calculated $R_{irregularity}$ of the initial and final. Results are listed in Table 1. The mean image irregularity ratios (initial=3.23 and final= 3.25) are larger than the threshold irregularity ratio of 1.96

identified in our previous study [3]. 4/10 mole boundaries became more irregular over time, indicated by the positive values of the variation percentage in Table 1, while the other six moles have a lower irregularity ratio of their final images compared to the initial images. The mean change of the irregularity ratio is 0.20%, with the absolute percentage change of the irregularity ratio 8.79±6.27%.

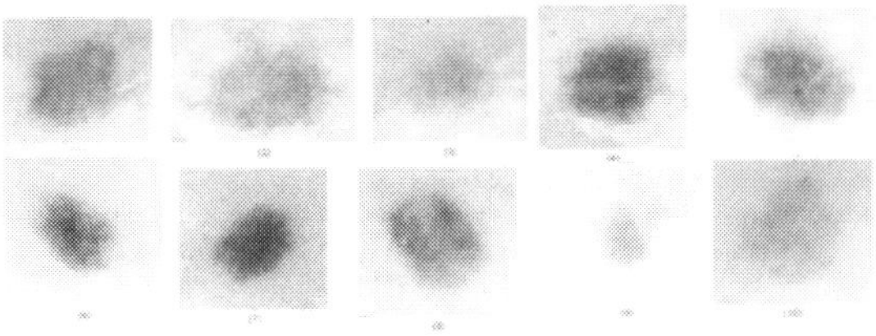

Figure 1 Initial images of 10 moles analyzed

Table 1 Comparison of the irregularity ratios of the ten moles from their initial images to their images after the observation period.

Sample	Initial	Final	Percentage change	Absolute Change
1	3.7811	3.3927	-10.27%	10.27%
2	3.8468	3.9669	3.12%	3.12%
3	4.1696	4.6966	12.64%	12.64%
4	2.5545	2.762	8.12%	8.12%
5	3.2645	3.0784	-5.70%	5.70%
6	2.3871	2.3026	-3.54%	3.54%
7	2.4975	2.0513	-17.87%	17.87%
8	2.5644	2.5065	-2.26%	2.26%
9	2.9344	3.5533	21.09%	21.09%
10	4.3025	4.1596	-3.32%	3.32%
Mean	**3.23**	**3.25**	**0.20%**	**8.79%**
SD	**0.70**	**0.82**	**10.80%**	**6.27%**

Similar trends can be observed when changes of asymmetry ratios are assessed. Almost all the asymmetry ratios are higher than the cutoff asymmetry ratio identified in the previous study [3]. The absolute percentage change of $R_{asymmetry}$ is calculated as 9.13%±6.54%, with the largest change over 18%.

It is surprising that almost all the color variation ratios are below our cutoff color variation ratio of 0.334 [3]. The average values of the color variation ratio in both groups are 0.141 and 0.162, respectively. One possible reason is that our algorithm filters the color components [3]. Nevertheless, our computational results indicate large changes in the color variation ratios over time. 8 of the 10 moles appear to have more grayscale variations over time, i.e., up to 81%. For the only two moles with a "negative" variation percentage, the change is less than 14%. The calculated absolute variation percentage and its standard deviation are 23.58%±21.45%.

Size changes of the moles can be quantified as described above. 8/10 moles increased in size. One mole (#9) nearly quadrupled its size. For the two moles which size is smaller after the observation period, the size decrease is less than 6%.

For some moles, not every ratio shows substantial changes over the period of the observation. In these cases, subtle changes in pigmentation or size were likely identified as the criterion for significant lesional change. Table 2 identifies the substantial changes of each mole among the four ratios. Here, we define the threshold of "substantial changes" as ±14%. It is interesting that the majority (12/17) of the significant changes occur in the categories of the R_{size} and the $R_{color\ variation}$ correlating most closely with the clinical parameters used to define significant changes, namely an asymmetric increase in size or a change in pigmentary features.

Lesion #8, whose initial and final images are shown in Figure 2, is an outlier since none of the ratios are higher than 8%. This is the only sample that our algorithm cannot validate the dermatologist's perception of change in the mole image. The proprietary algorithm used on the digital dermatoscope detected a slight increase in size with this lesion, suggesting that algorithms used to detect nevus size may vary in their sensitivity. In the other nine moles, there is at least one ratio with a percentage variation change larger than 14%. For example, mole #9 has all four ratios larger than 14%. There is a very large increase in size of this mole over time, shown in Figure 2. When the size changes, on can see that mole #9 also shows substantial changes in its shape, leading to observed variations in its boundary irregularity and asymmetry. It becomes patchier over time, ending with an increase in the color variation ratio over time.

Table 2 Substantial changes in the morphological ratios of all the samples over time.

Sample	$R_{irregularity}$	$R_{asymmetry}$	$R_{color\ variation}$	R_{size}
1	-	-	14.61%	-
2	-	-	17.35%	-
3	-	-	81.27%	-
4	-	16.35%	14.03%	23.05%
5	-	-	15.65%	29.09%
6	-	-	-	29.06%
7	-17.87%	-	-	40.05%
8	-	-	-	-
9	21.09%	15.45%	44.63%	268.57%
10	-	-18.79%	-	-

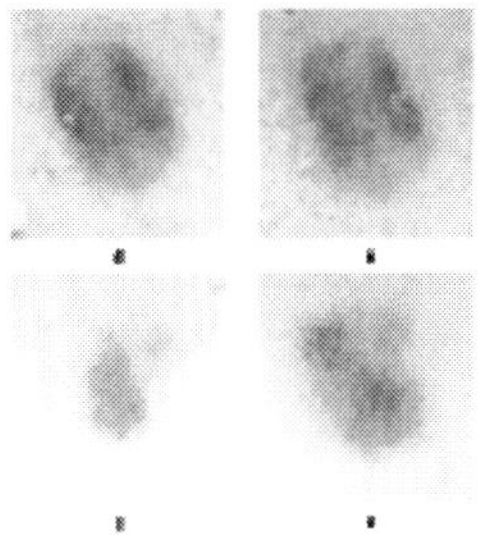

Figure 2. (a) the initial image of mole #8; (b) mole #8 later; (c) The initial image of mole #9; and (d) mole #9 later.

In summary, this study is focused on advancing the image processing algorithms used for screening pigmented skin lesions based on photos taken from regular cameras and digital dermatoscopes. The algorithms have been implemented to assess "changes" of mole images over time observed by a dermatologist. Using a ±14% as the definition of "substantial changes", the algorithm identifies 9 of the 10 mole images as changed over time. Among the irregularity, asymmetry, color variation, and size ratios, 5 out of the 9 moles have shown changes in one ratio, 2 out the 9 moles have experience changes in two ratios, 1 mole has shown changes in three ratios, and only 1 mole shows changes in all four ratios. The computational results are consistent with the general observations that human eyes are sensitive to size changes and color variation changes, and may not be as sensitive to changes in border irregularity and asymmetry.

ACKNOWLEDGEMENTS

This study was supported by the University of Maryland Baltimore County.

REFERENCES

[1] Saraiya, M, Glanz, K., Briss, P.A., Nichols, P., White, C., et al. *Am. J. Prev. Med.* 27(5): 422-466, 2004.

[2] Garnavi, R. Aldeen, M., Celebi, M. Varigos, E. G., and Finch S., *Computerized Medical Imaging and Graphics,* 35:105-115, 2011.

[3] Chamani, A. MS thesis, University of Maryland Baltimore County, 2014.

SB^{3}C2015
Summer Biomechanics, Bioengineering and Biotransport Conference
June 17-20, 2015,Snowbird Resort, Utah, USA
SB^{3}C2015-236

MEASUREMENT OF BONE MINERAL DENSITY IN MOTOR VEHICLE CRASH OCCUPANTS AND CORRELATION WITH AGE AND FRACTURE INCIDENCE

Ashley A. Weaver (1), R. Caresse Hightower (1), Sarah K. Lynch (1), Kristen M. Beavers (2), Anna N. Miller (3), Joel D. Stitzel (1)

(1) Biomedical Engineering
Wake Forest University School of Medicine
Winston-Salem, NC, United States

(2) Health and Exercise Science
Wake Forest University
Winston-Salem, NC, United States

(3) Orthopaedics
Wake Forest University School of Medicine
Winston-Salem, NC, United States

INTRODUCTION

Approximately 54 million Americans have osteoporosis and low bone mass and the incidence of osteoporosis and fragility fractures is expected to increase as the population continues to age [1]. Osteoporosis is characterized by porous bone of decreased density and quality. Bone quality, an amalgamation of all the factors that determine how well the skeleton can resist fracturing, can be measured by strength, fragility, bone thickness, biomarkers, and bone mineral density (BMD) [2]. Areal BMD (aBMD) is traditionally measured with dual-energy X-ray absorptiometry (DXA) and used to diagnose osteopenia and osteoporosis from T-scores derived from comparative young, normal BMD means [3]. Volumetric BMD (vBMD) is traditionally measured with quantitative computed tomography (qCT) using a scanned phantom to calibrate BMD measures to known mg/cc calcium hydroxyapatite (CaHA) values.

Low bone quality is a known contributing factor to injury in motor vehicle crashes (MVCs) and quantification of an occupant's BMD is important from an injury causation standpoint. Since most trauma CT scans do not contain a phantom, the first aim of this study was to validate a technique for measuring lumbar vBMD from phantom-less clinical CT scans. The second aim was to apply the validated technique to quantify lumbar vBMD for 281 seriously injured Crash Injury Research and Engineering Network (CIREN) occupants from phantom-less clinical CT scans and to correlate vBMD to age, fracture incidence, and osteopenia/osteoporosis diagnoses.

METHODS

DXA and qCT scans were collected prospectively for 50 subjects and used to validate a technique to measure vBMD from 281 phantom-less CT scans of CIREN occupants (ages 16+; 109 males, 172 females). Mean Hounsfield Unit (HU) values were collected from regions of interest on the L1-L5 vertebrae, right psoas muscle, and anterior subcutaneous fat for all subjects and from three phantom ports with known mg/cc CaHA values for the group of 50 subjects used for validation (Figure 1). HU measurements were collected using the public domain open source software program Image J (U.S. National Institutes of Health, Bethesda, MD).

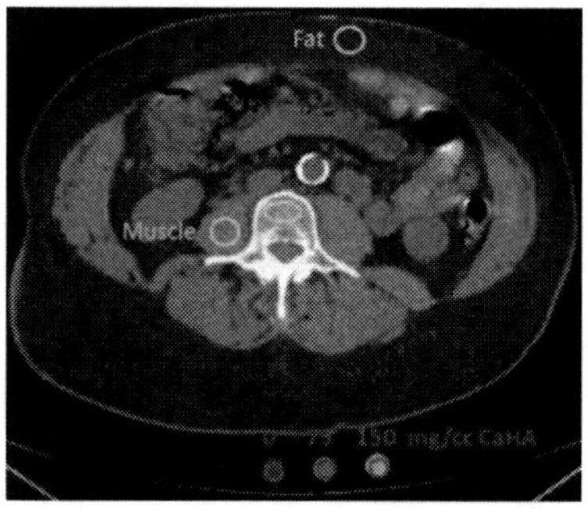

Figure 1. Mean HU measured within the illustrated regions of interest on the L1-L5 vertebrae, right psoas muscle, and anterior subcutaneous fat for the validation and CIREN groups and from 0, 75, and 150 mg/cc phantom ports for the validation group.

qCT calibration was accomplished using regressions between the phantom HU and mg/cc values to convert L1-L5 HU values to mg/cc for the 50 subject validation group. A phantom-less calibration technique was developed where the fat and muscle HU values were linearly regressed against fat (-69 mg/cc) and muscle (77 mg/cc) to establish a conversion for L1-L5 HU measurements to mg/cc. Mean lumbar vBMD calculated using the two techniques (qCT versus phantom-less) was compared for the 50 subjects to assess agreement. The mg/cc threshold that best discriminated for osteopenia was established by performing a Receiver Operating Characteristic (ROC) analysis with DXA-derived T-scores. CIREN mean lumbar HU measurements were converted to mg/cc using the validated phantom-less calibration technique and the mg/cc threshold for osteopenia was used to compare vBMD measured from CT to osteopenia comorbidity classifications in CIREN. Correlations between vBMD measures and age and fracture incidence were explored for the CIREN occupants.

RESULTS

Excellent agreement between the mean lumbar vBMD mg/cc values was achieved when comparing the qCT and phantom-less calibrations for the 50 subjects used for validation (Figure 2). A 145 mg/cc threshold for osteopenia was established from the ROC analysis of DXA-derived T-scores (sensitivity=1, specificity=0.57).

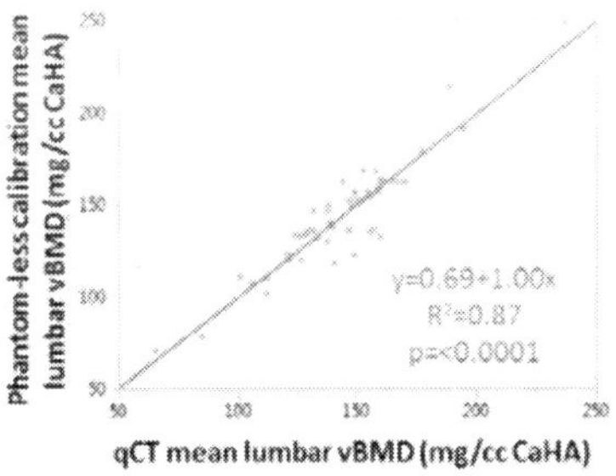

Figure 2. Statistically significant positive correlation of the lumbar vBMD mg/cc values obtained for 50 subjects using qCT versus the fat and muscle phantom-less calibration technique.

Among the CIREN occupants, 44 (16%) had vBMD measures below the 145 mg/cc threshold (Figure 3). Some occupants with low vBMD were documented as osteopenic in CIREN, such as a 40 year old male paraplegic. Other occupants, including 44 and 46 year old females that sustained a multitude of fractures, had low vBMD with no documented osteopenia comorbidity in CIREN. Age was negatively correlated with vBMD as expected (p=<0.0001).

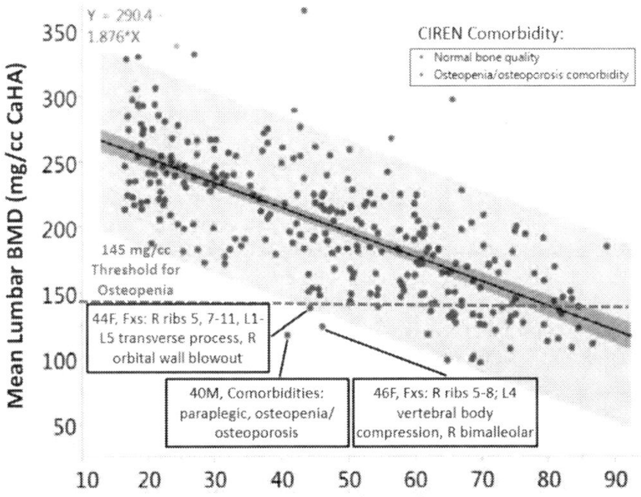

Figure 3. Mean lumbar vBMD measurements for 281 CIREN occupants were negatively correlated with age. Comorbidities documented in the CIREN database are indicated by the color legend. vBMD measurements for 44 CIREN occupants were below the 145 mg/cc threshold for osteopenia. Injuries are highlighted for three occupants aged 40-46 years with particularly low vBMD.

Of the 44 occupants with vBMD below 145 mg/cc, 28 (63%) were not classified as osteopenic in CIREN, but vBMD measures suggested undiagnosed osteopenia (Figure 4). Of the 29 occupants classified as osteopenic in CIREN, 13 (45%) had normal vBMD measures (>=145 mg/cc), suggesting possible misclassification in CIREN. CIREN occupants with <145 mg/cc vBMD sustained an average 1.7 additional rib/sternum fractures (3.5 versus 5.2 rib/sternum fractures, p=0.036). Of the occupants with no documented bone quality comorbidity, those with <145 mg/cc vBMD sustained a higher number of rib/sternum fractures (4.1 versus 3.4 fractures), which further supports the notion of undiagnosed osteopenia in the occupants with low vBMD. Of the occupants with >=145 mg/cc vBMD, those classified as osteopenic in CIREN had a higher number of rib/sternum fractures (5.5 versus 3.4 fractures). This finding may suggest that these occupants were classified as osteopenic in CIREN because they sustained a high number of fractures and not because of low bone quality indicated on CT.

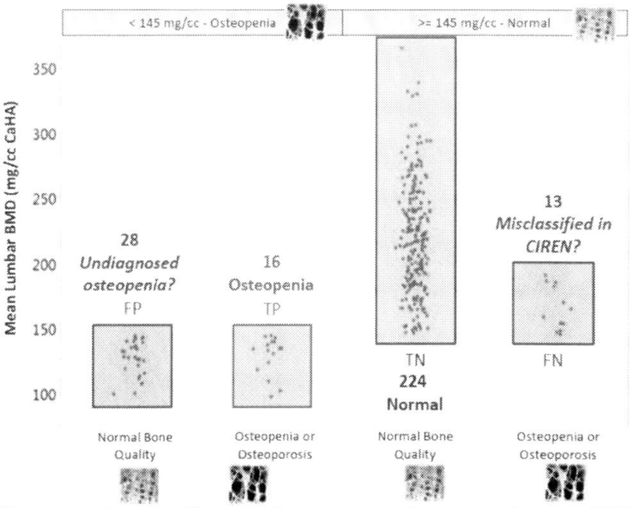

Figure 4. Truth table comparing mg/cc vBMD measurements to documented osteopenia/osteoporosis for the 281 CIREN occupants.

DISCUSSION

The technique developed and validated to measure lumbar vBMD from phantom-less clinical CT scans can be used to classify osteopenia in CIREN occupants. It is a valuable quantitative tool for CIREN center case reviews and for other injury causation analyses to assess the contribution of bone quality to injury causation. As lumbar vBMD was estimated from phantom-less CT scans with accuracy similar to qCT, the phantom-less technique can be broadly applied to both prospectively and retrospectively assess patient bone quality for research and clinical studies related to MVCs, falls, aging, and other applications.

ACKNOWLEDGEMENTS

Funding was provided by the National Highway Traffic Safety Administration under Cooperative Agreement Number DTN22-10-H-00294. Views expressed are those of the authors and do not represent the views of NHTSA.

REFERENCES

[1] National Osteoporosis Foundataion, 2014.
[2] Licata, A et al., *Cleve Clin J Med*, 76(6):331-6, 2009.
[3] World Health Organization, *WHO Technical Report Series*, 921, 2003.

DESIGNING BIOMECHANICAL MODELS OF THE ANKLE: HOW MANY DEGREES OF FREEDOM ARE NECESSARY TO REFLECT IN VIVO JOINT KINEMATICS?

Jennifer A. Nichols (1), Koren E. Roach (1, 2), Charles L. Saltzman (1), Andrew E. Anderson (1, 2, 3, 4)

(1) Department of Orthopaedics
University of Utah
Salt Lake City, UT, USA

(2) Department of Bioengineering
University of Utah
Salt Lake City, UT, USA

(3) Department of Physical Therapy
University of Utah
Salt Lake City, UT, USA

(4) Scientific Computing and Imaging Institute
University of Utah
Salt Lake City, UT, USA

INTRODUCTION

Defining a joint model is a critical step to analyze human motion. The analysis of experimental data collected using motion capture techniques often requires the definition of *a priori* joint models, which are fit to the measured data [1]. Additionally, many computational models prescribe motion by defining joint axes of rotation (e.g., [2]). Thus, the validity of the assumed joint model can influence the findings of both experimental and computational studies.

The suitability of user-defined simplifications to a joint model ultimately depends on the scope of the biomechanical problem being investigated. However, simplifications are often perpetuated in the biomechanical literature for joints that have been shown to exhibit complex motion, such as the ankle. The ankle joint complex, consisting of the subtalar and tibiotalar joints, has been shown in both cadaveric and *in-vivo* studies to provide near six degree-of-freedom (DOF) motion [3]. Yet, it is often simplified to one DOF per joint. Though the use of a single DOF ankle model decreases complexity and eases data interpretation, it may not capture subtle, but clinically relevant differences between nonimpaired and pathological ankles.

In this study, we evaluated the importance of developing a more complex ankle joint model that describes *in vivo* motion at the tibiotalar and subtalar joints. Specifically, we (i) examined how incorporating one versus three DOFs at each ankle joint influenced the results of inverse kinematic simulations and (ii) assessed to what extent the simulation results reflected tibiotalar and subtalar joint motion, which was independently measured by dual fluoroscopy (DF).

METHODS

Inverse kinematic simulation results were compared using two models of the ankle joint: 1 DOF and 3 DOF. Both models were based on a validated model of the lower limb [2]. The 1 DOF model defined the ankle as a series of one DOF joints, as described in [2], with dorsiflexion-plantarflexion and inversion-eversion at the tibiotalar and subtalar joints, respectively. The 3 DOF model defined the ankle as a series of three DOF joints with dorsiflexion-plantarflexion, inversion-eversion, and internal-external rotation included at both joints.

For each model, inverse kinematic simulations were performed in OpenSim v.3.2 [4] using motion data for 5 subjects (age 32.2 ± 4.0 years; 2 females and 3 males) and 3 activities: slow walk (0.5 m/s), fast walk (1.0 m/s), and one-leg heel rise. Subjects provided informed consent for this IRB approved study. Data were collected using a 10-camera Vicon system and instrumented split-belt treadmill. A plug-in gait marker set with additional markers and clusters was used to track the torso, pelvis, thigh, shank, and foot segments [5]. Following data collection, marker data were filtered and transformed into a model-based coordinate system. Simulations were performed using the 1 DOF and 3 DOF models scaled to each subject's height and weight.

Differences in joint angles between the 1 and 3 DOF models were computed both within and across the ankle. Within the ankle, we quantified joint angles at the tibiotalar and subtalar joints. Across the ankle, the orientation of the foot relative to the tibia was quantified. Joint angles were analyzed for 5 randomly selected normalized gait cycles (right heelstrike to right heelstrike) for the slow and fast walk activities, and for 1 normalized heel rise (heel leaves and returns to the ground). Joint angles estimated by the inverse kinematic simulations were also compared with those measured *in vivo* for the same subjects using our validated DF system [6]. Briefly, the DF system collected high-speed images of the tibia, talus, and calcaneus as subjects performed each activity. The 3D bone positions were quantified using model-based tracking [7] and then converted into joint angles. Importantly, the DF data is independent of the marker data; thus the DF data provide an independent validation of the simulation results.

RESULTS

For all activities, the orientation of the foot relative to the tibia was similar regardless of whether the inverse kinematic simulations were performed using the 1 DOF or 3 DOF models (see Fig. 1 for representative activity). However, tibiotalar and subtalar joint angles were distinct in all simulations using the 1 DOF versus the 3 DOF models. Notably, for all activities, the 3 DOF model required less dorsiflexion-plantarflexion at the tibiotalar joint and less inversion-eversion at the subtalar joint than the 1 DOF model (Table 1). At each joint, this decrease in range of motion at one DOF was accompanied by an increase in range of motion at the other two DOFs. For example, when using the 3 DOF model, the inverse kinematic simulations predicted more than 5° inversion-eversion at the tibiotalar joint and more than 9° dorsiflexion-plantarflexion at the subtalar joint (Table 1).

The tibiotalar and subtalar motion predicted by simulations using the 3 DOF model reflected the motion measured *in vivo* using DF more closely than simulations using the 1 DOF model (Table 1, bold). Interestingly, for the 3 DOF model, the tibiotalar joint matched the experimental data more closely than the subtalar joint.

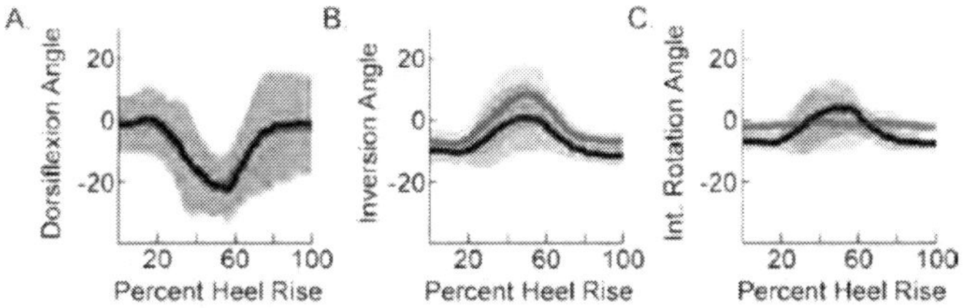

Figure 1. Orientation of the foot relative to the tibia during one-leg heel rise for the 1 DOF (black) and 3 DOF (red) models. Orientation depicted using roll-pitch-yaw angles (in degrees) defining rotation about the (A) medial-lateral, (B) anterior-posterior, and (C) proximal-distal axes of the tibia. These angles represent dorsiflexion, inversion, and internal rotation, respectively. Shaded regions indicate standard deviation across subjects.

DISCUSSION

Our simulation results demonstrate that both the 1 DOF and 3 DOF models are appropriate for studying overall ankle motion (foot relative to tibia). However, when explicitly studying motion of the ankle joint complex (tibiotalar and subtalar motion), data from the 3 DOF model more closely replicated joint kinematics measured *in vivo* by DF. The ability of the 3 DOF model to predict *in vivo* joint motion is intriguing because it suggests that an inverse kinematics pipeline that utilizes only the 3 DOF model and the positions of skin-based markers may be adequate for predicting tibiotalar and subtalar motion (i.e., joint motions that skin-based markers cannot directly measure as there is no external location to place markers on the talus).

Choosing the most appropriate ankle model to obtain robust inverse kinematics results is essential to accurately predict joint torques and muscle activations with inverse dynamics. Preliminary work extending our comparison of the 1 DOF and 3 DOF models to inverse dynamic simulations suggests that the chosen kinematic model substantially alters the estimation of the torques required to stabilize the ankle during a one-leg heel rise (Fig. 2). We acknowledge that further work is necessary to understand how small a change in ankle angle or joint torque is important. However, the results reported here suggest that developing a validated, multi-DOF ankle model that fully describes tibiotalar and subtalar motion could yield more accurate predictions of gait and muscle metrics.

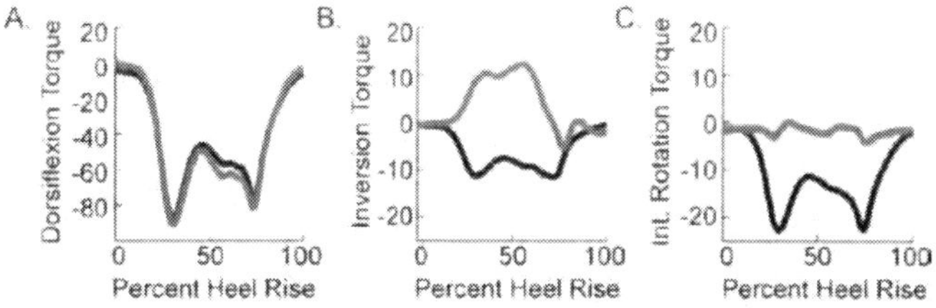

Figure 2. Tibiotalar joint torque estimated by inverse dynamics for one subject (male, weight 68 kg, height 1.630 m) during one-leg heel rise for the 1 DOF (black) and 3 DOF (red) models. The torque (in Nm) is decomposed into (A) medial-lateral, (B) anterior-posterior, and (C) proximal-distal components of tibia reference frame, representing dorsiflexion, inversion, and internal rotation torques respectively.

ACKNOWLEDGEMENTS

Funding from the LS-Peery Discovery Program in Musculoskeletal Restoration and the American Orthopaedic Foot & Ankle Society is acknowledged. We thank Niccolo Fiorentino, Ashley Kapron, Bo Foreman, Bibo Wang, Madeline Singer, Justine Goebel, and Austin West for their assistance with data analysis, and Glen Litchwark and Tim Dorn for their open source OpenSim toolboxes.

REFERENCES

[1] Mundermann, L. et al., *J Neuroeng Rehab*, 3:6, 2006.
[2] Arnold, EM. et al., *Ann Biomed Eng*, 38(2):269-279, 2010.
[3] Lundgren P. et al., *Gait & Posture*, 28(1): 93-100, 2008.
[4] Delp, SL. et al., *IEEE Trans Biomed Eng*, 54(11):1940-1950, 2007.
[5] Davis, R. et al., *IEEE Eng Med Eng Mag*, 7(3):35-40, 1988
[6] Bey, MJ et al., *J Biomech Eng*, 128:604-609, 2006.
[7] Kapron, AL. et al. *J Appl. Biomech*. 30(3):461-470, 2014.

Table 1. Activity-Based Range of Motion*

Activity	Model	Tibiotalar Joint Range of Motion (degrees)			Subtalar Joint Range of Motion (degrees)		
		Dorsiflexion	Inversion	Internal Rotation	Dorsiflexion	Inversion	Internal Rotation
Slow Walk (0.5 m/s)	1 DOF Model	18.3 (3.9)				17.1 (5.7)	
	3 DOF Model	9.5 (2.6)	6.4 (1.7)	9.3 (2.4)	9.2 (1.2)	6.2 (1.8)	7.4 (3.9)
Fast Walk (1.0 m/s)	1 DOF Model	23.7 (5.1)				19.2 (5.4)	
	3 DOF Model	12.5 (3.5)	6.9 (1.6)	9.8 (1.9)	10.5 (1.7)	6.9 (1.7)	8.3 (4.3)
One-Leg Heel Rise	1 DOF Model	25.9 (14.6)				16.1 (9.4)	
	3 DOF Model	8.9 (5.6)	5.9 (3.5)	2.1 (0.3)	14.0 (8.2)	6.0 (3.4)	3.7 (0.8)
Simulation Averages	**1 DOF**	**22.7 (8.9)**				**17.4 (6.5)**	
	3 DOF	**10.3 (3.9)**	**6.4 (2.2)**	**7.0 (3.9)**	**11.2 (5.0)**	**6.4 (2.1)**	**6.4 (3.8)**
Experiment	***in vivo* DF**	**14.0 (9.1)**	**3.5 (1.9)**	**5.5 (2.8)**	**4.7 (2.9)**	**9.5 (4.9)**	**4.6 (2.1)**

*Standard deviations across subjects are reported in parentheses. Gray hashing denotes degree of freedom not included in the 1 DOF model. Bold box at bottom of table indicates average range of motion across all subject and activities for both the inverse kinematic simulations and the dual fluoroscopy experiment.

SB³C2015
Summer Biomechanics, Bioengineering and Biotransport Conference
June 17-20, 2015, Snowbird Resort, Utah, USA
SB³C2015-238

EPITHELIAL TO MESENCHYMAL TRANSITION ALTERS CELLULAR CHIRAL BEHAVIOR

Kathryn E. Worley (1), Andrew K. Watrobski (1), David Shieh (2), Leo Q. Wan (1)

(1) Department of Biomedical Engineering
Rensselaer Polytechnic Institute
Troy, NY, USA

(2) Department of Biology
Rensselaer Polytechnic Institute
Troy, NY, USA

INTRODUCTION

Development of an organism is a complex process culminating in the formation of distinct and properly placed tissues. Left-right asymmetry, also known as chirality, is an aspect of embryonic development that is of particular interest. Chirality is found in both plant and animal systems. A prominent example in human physiology is the looping of the aorta connecting to the heart. However, chirality is also a notable feature in other areas of development as well. In embryonic development, chirality and epithelial migration are required for proper function, positioning, and orientation of various organs and organ systems [1]. In order to alter migration rate and establish proper cell type(s) during development a conversion of cell phenotype from epithelial to mesenchymal (EMT) and/or from mesenchymal to epithelial (MET) is often necessary [2]. EMT is known to alter cell properties, such as cell-cell adhesion, migration rate, and transcription factor expression.

This research seeks to examine the intrinsic biomechanical properties of epithelial cells and determine how a transition from an epithelial state to a mesenchymal state influences these properties. We hypothesize that cells going through an EMT will, when grown on microcontact printed ring patterns, exhibit alternations in the observed chiral bias.

METHODS

Cell Culture: Madin-Darby canine kidney epithelial cells (MDCK) (ATCC® CCL-34™) were maintained in Dulbecco's Modified Eagle Medium (DMEM; Life Technologies) supplemented with 10% fetal bovine serum (Sigma) in tissue culture flasks. Epithelial mesenchymal transition was achieved through treatment with transforming growth factor-β1 (TGF-β1) and epidermal growth factor (EGF) [3]. Cells were cultured in DMEM until 80% confluence was reached and then cultured in serum free medium for 24 hours. After 24 hours the medium was replaced with either DMEM, or a low serum medium containing 1% FBS [4]. The treatment groups received either 10 ng/mL of TGF-β1 alone or a combination of TGF-β1 and 10 ng/mL of EGF. The cells were cultured in this medium for 48 hours before being trypsinized and seeded onto microcontact printed rings.

Immunofluorescence: The extent of transition was determined through immunofluorescence staining using anti-E-cadherin-FITC antibody (5 μg/mL, BD Bioscience) for epithelial phenotype and anti-vimentin for mesenchymal phenotype (abcam).

Micro-Contact Printing: Polydimethylsiloxane (PDMS; Corning) stamps were created from a silicon wafer master mold and coated with 50 μg/mL of fibronectin for 30 minutes. The stamps were dried and the pattern transferred onto plasma treated suspension culture dishes. The background was blocked using 100 μg/mL of poly-L-lysine-g-polyethylene glycol (PLL-g-PEG) for 30 minutes [5]. Cells were then seeded and allowed to reach approximately 50% confluence.

Chiral Alignment Analysis: Once the rings became confluent they were fixed in 4% paraformaldehyde and imaged. Phase contrast images were analyzed using a custom-written MATLAB program [5]. The program is based on the automated detection of intensity gradient of phase contrast images of cells [6] allowing for the determination of the angle of alignment with the ring boundary. A culmination of alignment angles results in a determination of clockwise (CW), counterclockwise (CCW) or non-chiral (NC) for each ring.

RESULTS

The extent of the epithelial to mesenchymal transition in MDCK cells was determined through immunofluorescence staining for E-cadherin and vimentin. E-cadherin is known to be expressed in MDCK cells while vimentin is not. Transitioning to a mesenchymal state through the use of TGF-β1 and/or EGF is shown to alter these markers as well as the general morphology (FIGURE 1).

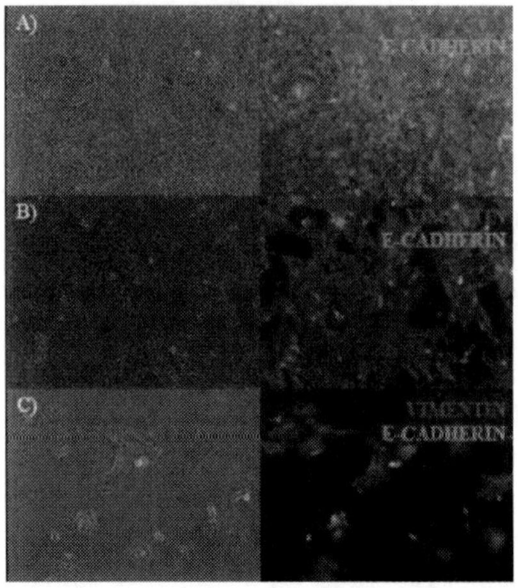

FIGURE 1. EPITHELIAL TO MESENCHYMAL TRANSITION ALTERS THE EXPRSSION OF E-CADHERIN AND VIMENTIN IN MDCK CELLS. A) 10% FBS (MDCK-T) B) 1% FBS C) TGF-β1 + EGF

Alterations in the general morphology include an increase in elongation as well as a decrease in the formation of cell clusters. Expression of vimentin increased and localization of E-cadherin on the boundaries between cells decreased with treatment. MDCK cells were placed under several conditions either no treatment (MDCK_NT), or serum starvation and then 48 hours of culture in DMEM; with 10% FBS (MDCK_T), 1% FBS, 1% FBS and TGF-β1 (TGF- β1 group) or 1% FBS, TGF-β1 and EGF (TGF- β1+EGF). (FIGURE 2)

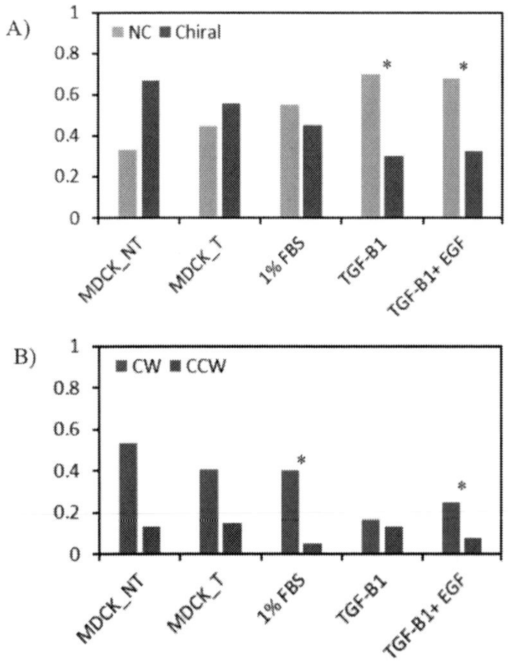

FIGURE 2. TRANSITION OF EPITHELIAL CELLS TO A MESENCHYMAL STATE RESULTS IN AN OVERALL DECREASE IN CHIRALITY. COMPARISON OF A) NON-CHIRAL VS. CHIRAL AND B) CLOCKWISE VS. COUNTERCLOCKWISE; * p < 0.5

MDCK cells transitioned to a mesenchymal state and then seeded onto microcontact printed rings exhibit a decrease in the number of chiral rings (FIGURE 2A), but the bias towards the clockwise direction appears to be still dominant over the counter clockwise direction. (FIGURE 2B).

DISCUSSION

In this study, we used serum starvation and the treatment of growth factors to induce EMT, and found significant alterations in cellular adhesion, morphology, and chiral behavior on micropatterned rings. These results suggest that EMT might be used for regulating the left-right symmetry breaking events during embryonic development.

Transitioning MDCK cells from their epithelial state to a more mesenchymal state resulted in a decrease in E-cadherin and an increase in vimentin. Decreasing E-cadherin expression disrupts adherens junctions linking the cells together resulting in a decrease in collective migration [7]. Vimentin expression is seen primarily in mesenchymal cells and is therefore a reasonable marker for evaluating EMT [8].

When patterned onto microcontact printed rings transitioned MDCK cells exhibit alterations in the observed chirality. While the transition does not alter general trends in clockwise verses counterclockwise there is an overall decrease in the number of chiral rings. This is likely due to the alterations in cell-cell adhesion resulting from changes in collective migration due to the decrease in E-cadherin expression. Further work involving gene expression profiling, traction force microscopy and cell modeling will allow us to quantify the changes in cell properties resulting in alterations to chirality.

Examination of migration, focal adhesions and force generation in conjunction with an examination of the cytoskeleton can further our understanding of how the differentiation process alters the cytoskeleton leading to more in depth understanding of chirality. Quantification of these aspects of the cells can lead to enhanced computer models wherein migration and cell-cell interactions can be more accurately evaluated in both 2D and 3D simulations.

ACKNOWLEDGEMENTS

The authors would like to thank the NIH Director's New Innovator Award, National Science Foundation, American Heart Association, and March of Dimes for funding Support. Leo Q. Wan is a Pew Scholar in Biomedical Sciences, supported by the Pew Charitable Trusts.

REFERENCES

[1] Thiery, JP. *Cur Op Cell Bio.* 15:740–746, 2003.
[2] Kalluri, R & Weinberg, RA., *J Clin Invest.* 119(6):1420-1428, 2009.
[3] Grande, M et al., *J Cell Sci.* 115:3227-4236, 2002.
[4] Park, SH et al., *J Am Soc Nephrol.* 18:1497-1507, 2007.
[5] Wan, LQ et al., *PNAS.* 108(30): 12295-12300, 2011.
[6] Karlon, WJ et al., *Ann Biomed Eng.* 27:712–720, 1999.
[7] Li, L et al., *Cell Mol Life Sci.* 69:2779-2789, 2012.
[8] Sudo, T et al., *Ann Surg Oncol.* 20(3):S324-35, 2013.

CORROSION BEHAVIOR OF A NOVEL BIODEGRADABLE METALLIC STENT

Jennifer Frattolin (1,2), Luca Gottellini (3,1), Olivier F. Bertrand (4,1), Rosaire Mongrain (1,2)

(1) Department of Mechanical Engineering, McGill University, Montreal, QC, Canada

(2) Montreal Heart Institute Montreal, QC, Canada

(3) Department of Chemistry, Materials and Chemical Engineering Politecnico di Milano Milan, Italy

(4) Department of Interventional Cardiology Quebec Heart and Lung Institute Laval University Quebec City, QC, Canada

INTRODUCTION

Coronary artery disease (CAD) is the leading cause of death in the developed world. The current trend of CAD treatment has seen a shift from permanent metallic stents to the implementation of novel biodegradable materials. In order for biodegradable stents to be implemented in a clinical setting, it is critical that a controlled corrosion rate be achieved. In general, little degradation should occur within the first six months following insertion, with the stent fully degraded in approximately 12-24 months [1].

A novel nanostructured biomaterial for stent application is proposed, fabricated using cold gas-dynamic spray (CGDS) technology with iron and stainless steel 316L powders. The application of cold spray to coronary stent design is particularly useful, due to a significant reduction in grain size during fabrication. This leads to an improvement of the material's strength, fatigue and wear properties. This is critical as device failure due to stent fracture is attributed to high cyclic fatigue, and cold spray processing may alleviate current concerns of fatigue-related failure. The intent of this work is to investigate the corrosion behavior of this novel material through the use of static and dynamic corrosion tests, utilizing stent specimens.

METHODS

Cylindrical specimens were fabricated using cold spray [2], as shown in Figure 1(a). Commercially available iron and stainless steel 316L powders, with weight percentages of 80% and 20%, respectively, were mixed. Coatings were sprayed on 5 mm diameter rods (Figure 1(b)), and then annealed in a tube furnace at 1150°C for one hour. The coated substrates were grinded to a thickness of 250 μm (Figure 1(c)), and electrical discharge machining was performed to remove the substrate from the coating (Figure 1(d)). The stent was then laser cut with the desired stent cell design (Figure 1(e)), utilizing a femto laser. The fabricated stent is 0.8 cm in length with a strut thickness of 145 μm.

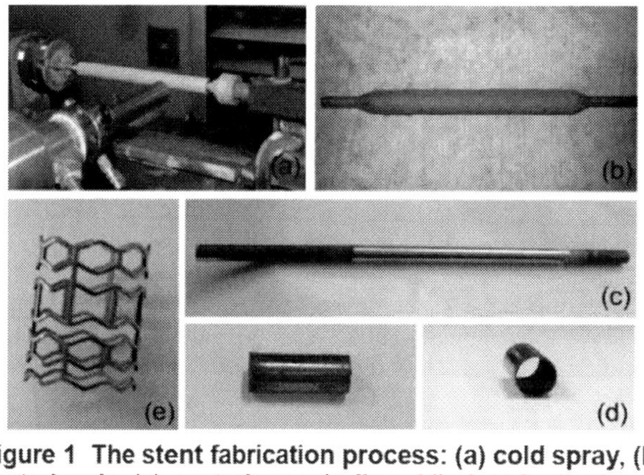

Figure 1 The stent fabrication process: (a) cold spray, (b) coated rods, (c) centreless grinding, (d) electric discharge machining (EDM), and (e) laser cutting of stent.

Static immersion tests were conducted according to ASTM G31-72. Beakers were prepared with Hanks' Balanced Salt Solution (HBSS), which has an ionic composition similar to blood plasma. Two stents were utilized for the static degradation test. The stent specimens were removed from the test bench initially after the first 7 days [3], and every 48 hours thereafter, for a total test period of 31 days. Mass loss was measured using a high precision scale. The solution was maintained at 37°C ± 1°C, and the pH was recorded every 24 hours, with the HBSS changed every 48 hours (after the initial 7 days). The corrosion rate, in mm year^{-1}, was calculated from Faraday's Law by the following equation:

$$Corrosion\ Rate = \frac{KW}{ADt} \quad (1)$$

Where K is a unit constant equal to 8.76×10^4, W is the mass loss in grams, A is the surface area in cm^2, D is the density in g cm^{-3}, and t is the total exposure time in hours. The density was calculated to be 7.886 g cm^{-3}, based on the respective weight percentages of 316L and Fe within the stent material, disregarding porosity.

Dynamic corrosion tests will be conducted in a similar manner, utilizing a novel test bench (Figure 2). A Masterflex© peristaltic pump (Cole-Parmer Instrument Co., Illinois, USA) is used to provide flow consistent with that found in the left anterior descending (LAD) artery. Stent specimens are crimped, and subsequently expanded to a diameter of 4 mm within the specimen holder. Physiological conditions are replicated, with a modified HBSS used to preserve physiological pH. The solution is maintained at 37°C ± 1°C, and the pH is monitored every 24 hours. Similar to the immersion tests, the stent specimens are removed from the test bench initially after the first 7 days, and every 48 hours thereafter, for a total of test period of one month. The corrosion rate of the specimens will then be determined and compared.

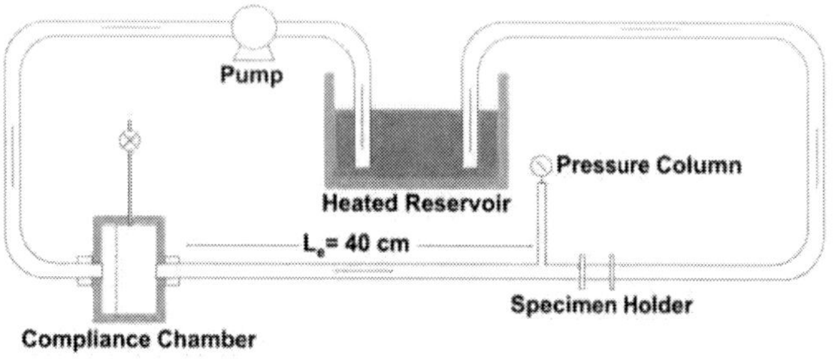

Figure 2 Dynamic corrosion test bench.

Scanning electron microscopy (SEM) will be utilized to analyze the corrosion progression of the samples.

RESULTS

The corrosion behavior of the stent was initially examined through static immersion tests. Preliminary results of these tests, with standard deviation, are presented in Figure 3. Typically, the peak corrosion rate is observed within the initial phase of the test, with the corrosion rate subsequently stabilizing over time. The increase seen in the corrosion rate after 408 hours (day 17) was due to stent fracture in multiple struts, and, as a result, an overall increase in the exposed surface area. It is theorized that high stent porosity contributed to failure. This hypothesis will be confirmed through SEM. After the corrosion rate stabilized, an average corrosion rate of approximately 0.31 mm $year^{-1}$ was calculated.

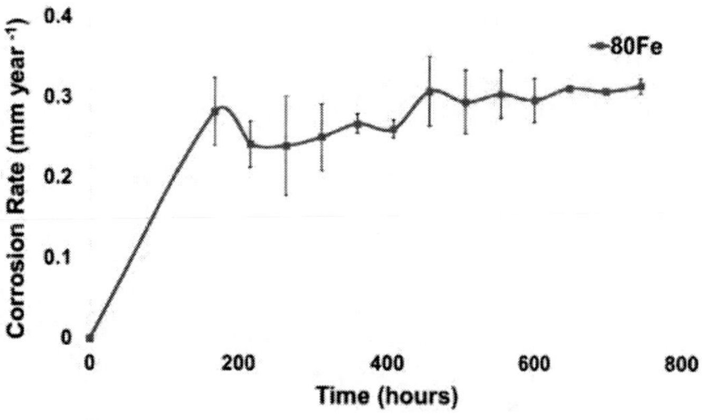

Figure 3 Results of the static immersion test.

A comparison of the corrosion rate from the static immersion tests to other biodegradable stent materials is presented in Table 1. In comparison to other iron-based stent materials, a slightly elevated degradation rate resulted, likely due to the observed stent fracture.

Table 1 Corrosion rates of metallic biodegradable stents.

Material	Corrosion Rate (mm $year^{-1}$)	Reference
80Fe Stent	0.31 ± 0.03	-
Pure Iron	0.20 ± 0.03	[4]
Fe-35Mn	0.23±0.03	[4]
Electroformed Iron	0.51 ± 0.006*	[5]
Mg WE 43	1.057	[6]

*Corrosion rate determined from potentiodynamic polarization test.

Dynamic corrosion tests are ongoing, but it is expected that the results will provide more comparable corrosion rates with other developing iron-based stents. According to Zhen *et al.* [3], static tests tend to overestimate the actual *in vivo* corrosion rate. The pseudo-physiological conditions of the dynamic corrosion test is expected to provide a corrosion rate more representative of that observed *in vivo*.

DISCUSSION

The intent of these investigations was to assess the corrosion behavior of a novel biodegradable stent composed of iron and stainless steel 316L. The primary benefit of utilizing two dissimilar metals is the ability to control the corrosion rate by means of the resulting galvanic couple, leading to a galvanic corrosion reaction at the micro and nano-scale. As a result, multiple corrosion mechanisms contribute to the overall degradation behavior of this material.

The conducted tests investigated the general corrosion behavior of the 80Fe material. Static and dynamic corrosion tests were conducted utilizing stent specimens under pseudo-physiological flow conditions. The corrosion rate determined from the degradation tests of 0.31 mm $year^{-1}$ indicated that the overall corrosion behavior of this novel stent is comparable with other currently developing biodegradable stents, and is within the feasible range for clinical implementation.

There are some limitations associated with the presented investigations. The implementation of a pseudo-physiological environment is critical to accurately mimic the degradation process that occurs *in vivo*. The physical reactions that occur on the surface of the stent after implantation, including protein adsorption and cell attachment, were not considered. These processes will slow the stent's overall degradation with time. Additional corrosion investigations, such as polarization and galvanic corrosion tests, should be conducted to obtain a more thorough basis of the material's corrosion properties.

The results of these investigations provide insight into the corrosion behavior of a novel biodegradable stent. The preliminary results support the use of cold spray technology as an innovative means for biodegradable stent fabrication.

ACKNOWLEDGEMENTS

This research was supported by the Natural Sciences and Engineering Research Council of Canada (NSERC).

REFERENCES

[1] Serruys, P W et al. *N. Engl. J. Med.*, 354:483-495, 2006.
[2] Al-Mangour, B et al. *Surf. Coat. Tech.*, 216:297-307, 2013.
[3] Zhen, Z et al. *Trans. Nonferr. Met. Soc. China*, 23:2283-2293, 2013.
[4] Hermawan, H. Biodegradable Metals: From Concept to Application. Berlin; New York: Springer, 2012.
[5] Moravej, M et al. *Acta Biomater.*, 6:1843-1851, 2010.
[6] Li, N et al. *Corros. Eng. Sci. Technol.*, 47:346-351, 2012.

OSTEOCYTE SCLEROSTIN EXPRESSION IS DIRECTED BY SUBSTRATE COMPOSITION AND DIMENSIONALITY IN ADDITION TO FLUID FLOW

Robert T Brady (1,2,3,4), Andrew R Cameron (1,2,3), David A Hoey (2,4,5,6), Fergal J O'Brien (1,2,3)

[1] Tissue Engineering Research Group, Department of Anatomy, Royal College of Surgeons in Ireland
[2] Trinity Centre for Bioengineering, School of Engineering, Trinity College Dublin, Ireland
[3] Advanced Materials and BioEngineering Research Centre (AMBER), Trinity College Dublin, Ireland
[4] Dept. Mechanical, Aeronautical and Biomedical Engineering, CABER, MSSI, University of Limerick, Ireland
[5] The Centre for Applied Biomedical Engineering Research, University of Limerick, Ireland
[6] Materials & Surface Science Institute, University of Limerick, Ireland

INTRODUCTION

Bone formation is a complex process regulated in part by the osteocyte-specific protein, sclerostin, which demonstrates anti-anabolic effects through antagonism of Wnt signaling pathway [1]. Whilst this protein has recently become a pivotal focus of research in the development of novel anabolic therapeutics for metabolic bone disease [1], MLO-Y4 cells, which are the most commonly used cell-line for *in vitro* osteocyte studies, lack the capacity to express sclerostin at physiological levels under standard culture conditions [2]. Although MLO-Y4 cells have indeed contributed significantly to our understanding of bone biology, the development of an *in vitro* model that expresses physiological levels of sclerostin would provide substantial opportunities for furthering this understanding.

The hypothesis of this study was that these cells would more closely match the behaviour of native osteocytes when presented with a microenvironment that better mimics *in vivo* conditions. Thus, the aims of the study were to; a) investigate if substrate composition could direct expression of osteocyte-specific sclerostin (encoded by the Sost gene) in MLO-Y4 cells, b) compare any changes in expression between 2D and 3D culture conditions using collagen-based, biomimetic (i.e. comparable to native ECM[3]) biomaterials, c) compare sclerostin expression in primary bone cells with MLO-Y4 cells cultured under physiologically relevant conditions, and d) demonstrate the ability of the developed MLO-Y4 model to respond appropriately to physical stimuli (i.e. reduced sclerostin expression under fluid flow).

METHODS

MLO-Y4 osteocytes were cultured for 5 days on,2D thin-films comprising type I collagen and other components native to the bone extracellular matrix (ECM), including; (i) collagen, (ii) collagen-glycosaminoglycan (GAG), (iii) collagen-hydroxyapatite (HA) and, (iv) collagen-GAG-HA. MLO-Y4 osteocytes were also cultured on porous 3D scaffolds of identical compositions for 5 days. Real time (RT)-PCR was used to determine composition-induced differences in Sost gene expression, normalised to the collagen only group, for 2D and 3D cultures respectively.

For comparisons with primary osteocytes, sclerostin expressing positive control cells were isolated from the femora of skeletally mature male mice. For fluid flow studies, 3D collagen-HA scaffolds were seeded with MLO-Y4 cells for 72 hours before being subjected to fluid shear using a perfusion bioreactor for a period of 48 hours[4, 5]. RT-PCR was used to compare Sost gene expression in these mechanically stimulated cells, with MLO-Y4 cells seeded on collagen-HA scaffolds and statically cultured in parallel as a control.

RESULTS

The addition of GAG and hydroxyapatite to the collagen-based films resulted in significant increases in MLO-Y4 Sost expression compared to collagen alone films. This increase was most robust in the collagen-HA films (Figure 1). Regarding dimensionality, results showed significant increases in Sost expression in cells cultured on porous, 3D scaffolds compared to those cultured on thin-film, 2D scaffolds for all compositions (Figure 2).

Furthermore, Sost expression in MLO-Y4 cells on 3D collagen-HA scaffolds was not only found to be comparable to that of primary bone cells (Figure 3), but it also responded to stimulation in a predicted manner, with a significant decrease in expression when cells were exposed to flow in a 3D perfusion bioreactor (Figure 4).

DISCUSSION

Given its key role in the regulation of bone formation, the ability to study the expression of sclerostin in osteocytes under *in vitro* culture conditions is of great importance to the field of bone biology. As such, a significant outcome of this study was the discovery that substrate composition can significantly affect sclerostin expression, with MLO-Y4 cells cultured on collagen-HA and collagen-GAG thin-film scaffolds producing robust increases in Sost expression relative to those on collagen only films. Interestingly, expression was further increased with culture on porous, 3D scaffolds, irrespective of composition. Perhaps most notable were the findings that Sost expression in MLO-Y4 cells was comparable to primary osteocytes when cultured on 3D, collagen-HA scaffolds, and these cells were able to respond to flow stimulation in a predicted manner.

Ultimately, this study further validates the use of the MLO-Y4 cell line as an *in vitro* model for osteocytes, by demonstrating their capacity to express sclerostin under physiologically relevant culture conditions. This is of great importance to furthering our understanding of load-induced bone formation and has substantial implications for the development of novel bone therapeutics.

ACKNOWLEDGEMENTS

1. European Research Council grant agreement n° 239685 awarded to Fergal O'Brien, and n°336882 awarded to David Hoey.
2. Dept of Anatomy, RCSI.
3. Tissue Engineering Research Group, RCSI.

Twitter @TissueEngDublin

REFERENCES

1. Nakamura, S. and S. Tanaka, *[Biological therapy for osteoporosis]*. Clin Calcium., 2014. **24**(6): p. 919-25. doi CliCa1406919925.
2. Woo, S., et al., *Cell line IDG-SW3 replicates osteoblast-to-late-osteocyte differentiation in vitro and accelerates bone formation in vivo*. J Bone Miner Res., 2011. **26**(11): p. 2634-46. doi 10.1002/jbmr.465.
3. O'Brien, F.J., *Biomaterials & scaffolds for tissue engineering*. Materials Today, 2011. **14**(3): p. 88-95.
4. McCoy, R. and F. O'Brien, *Visualizing feasible operating ranges within tissue engineering systems using a "windows of operation" approach: a perfusion-scaffold bioreactor case study*. Biotechnol Bioeng., 2012. **109**(12): p. 3161-71. doi 10.1002/bit.24566. Epub 2012 Jun 11.
5. McCoy, R., et al., *Orchestrating osteogenic differentiation of mesenchymal stem cells - identification of placental growth factor as a mechanosensitive gene with a pro-osteogenic role*. Stem Cells, 2013. **29**(10).

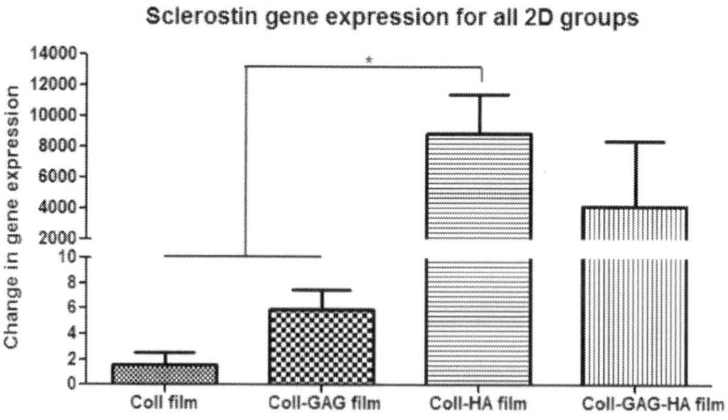

FIGURE 1. Sost EXPRESSION WAS FOUND TO BE HIGHEST IN THE COLLAGEN-HA FILM WHICH WAS STATISTICALLY GREATER THAN THE COLLAGEN-GAG AND COLLAGEN ONLY FILMS (*, p<0.05).

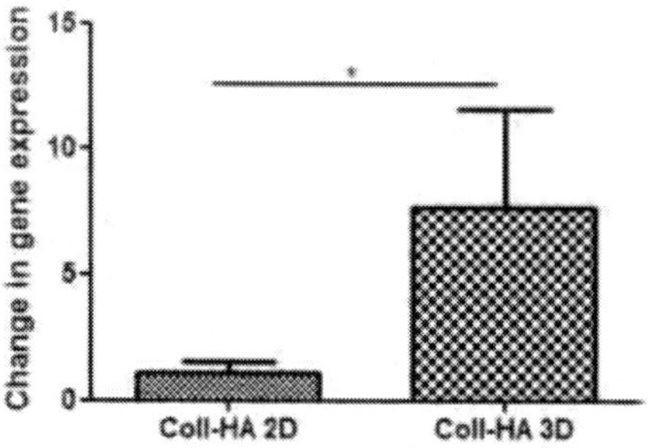

FIGURE 2. Sost EXPRESSION WAS HIGHER IN THE 3D SEEDED CONSTRUCT RELATIVE TO THE SEEDED 2D FILM (*, p<0.05).

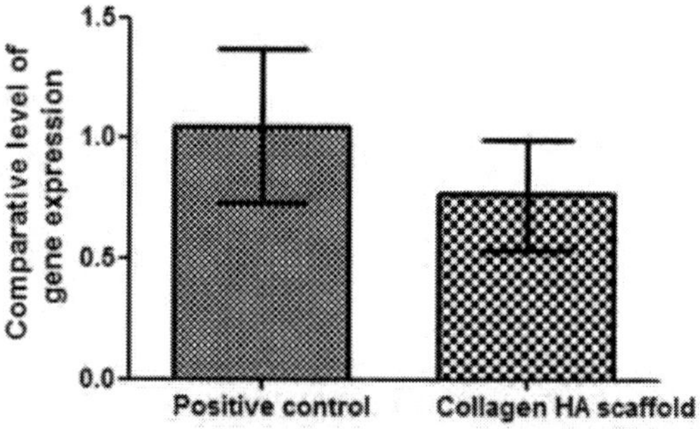

FIGURE 3. Sost EXPRESSION WAS COMPARABLE BETWEEN THE TWO GROUPS.

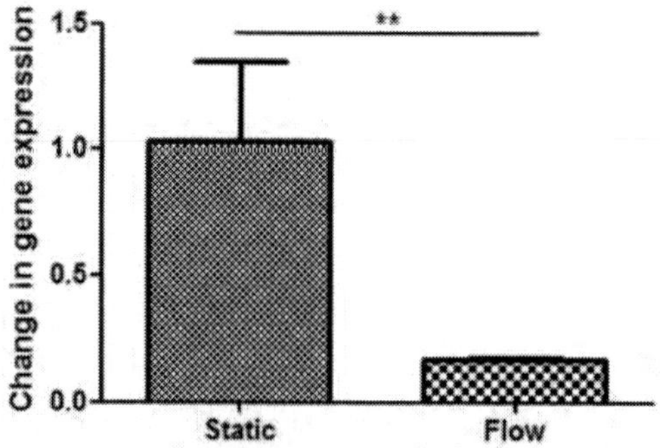

FIGURE 4. Sost EXPRESSION DECREASED SIGNIFICANTLY UPON FLOW STIMULATION (**, p<0.01)

SB^{3}C2015
Summer Biomechanics, Bioengineering and Biotransport Conference
June 17-20, 2015,Snowbird Resort, Utah, USA
SB^{3}C2015-242

TISSUE MINERAL DENSITY DEPENDENT MECHANICAL PROPERTIES OF INDIVIDUAL TRABECULA PLATES AND RODS DO NOT DIFFER IN ANATOMIC BUT INDIVIDUAL TRABECULAR DIRECTIONS

Y. Yu (1), J. Wang (1), B. Zhou (1), X. Guo (1)

(1) Biomedical Engineering, Columbia
University, New York, New York, USA

INTRODUCTION

Human trabecular bone, a crucial bone compartment susceptible to osteoporosis, consists of individual trabecular plates and rods. These trabecular plates and rods are distributed distinctly along the longitudinal, transverse, or oblique anatomic directions of the skeleton (Fig 1A). In addition, individual trabecular plates and rods are formed in preferred lamellar layers, parallel to their trabecular directions (Fig 1). Therefore, it is expected that mechanical properties of bone tissue are different in alignment with (axial) or against (lateral) the direction of individual trabeculae, *i.e.*, having anisotropic mechanical properties. The tissue properties of individual trabeculae and their dependence on tissue mineral density (TMD) are important bone quality measures. However, anisotropic mechanical properties of individual trabeculae along various anatomic directions are currently not available. The objectives of this study are as follows: to measure anisotropic tissue modulus and TMD of individual trabeculae; to examine their dependence on trabecular type and anatomic direction; and to determine the relationship between anisotropic tissue modulus and TMD of individual trabeculae.

METHODS

Twelve cylindrical trabecular bone cores were harvested from human proximal femurs. Trabecular cores were embedded in a weak exothermal plastic (Vishay MicroMeasurements). After curing, the samples were rehydrated and imaged at 25μm resolution by μCT (Scanco Medical) and then analyzed by individual trabecula segmentation (ITS) technique to provide the individual trabecular types and their anatomic directional information [1] (Fig1D). Subsequently, the samples were cut and polished with progressive grades of silicon carbide paper (800, 1200, 2400, 4000 grit) and finished with a 0.3μm alumina suspension under wet condition. The polished samples were scanned with five density calibration phantoms (0, 100, 500, 750, 1000 mg HA/cm3) at 12.5μm resolution and 500ms integration time to reduce the noise. Individual trabeculae on the polished surface were subjected to microindentation. The slope at the initiation of unloading on the force-displacement curve was used to calculate the tissue modulus [2]. Three indentation tests were performed under wet condition on both axial and lateral cross-sections (Fig 1.C) of each selected plates and rods in longitudinal (L), transverse (T), and oblique (O) directions, respectively (Fig 1.B, E). The microscope images of selected trabeculae were registered with the corresponding 12.5μm resolution μCT images using a landmark initialized mutual information-based registration toolkit (national library of medicine insight segmentation and registration toolkit, Fig1.F). The point-by-point registered grayscale values of the μCT image at the indentation sites (average of 4 voxels) were converted to TMD using calibration phantoms. In total, 432 indentations were made and linear regressions were used to correlate anisotropic tissue modulus with TMD, according to trabecular types or anatomic directions.

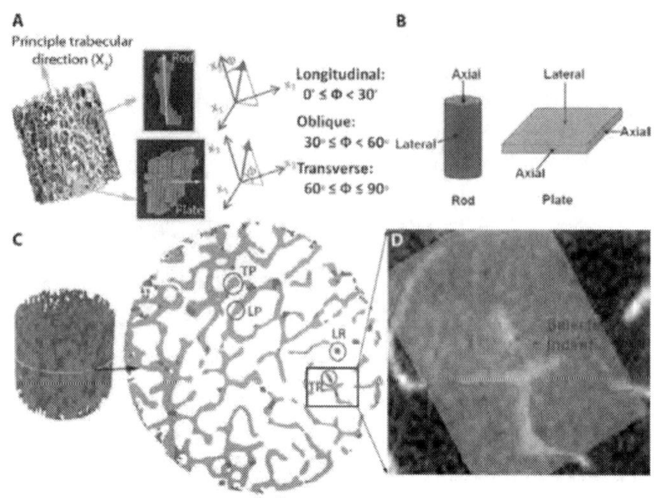

FIG.1 ILLUSTRATION OF THE ANATOMIC DIRECTIONS, THE AXIAL AND LATERAL TRABECULAR DIRECTIONS OF INDIVIDUAL TRABECULAR PLACES AND RODS IN PROXIMAL FMUR AND tmd MEASUREMENTS

A. CLASSIFICATION OF ANATOMIC DIRECTIONS OF PALTES AND RODS
B. AXIAL AND LATERAL DIRECTIONS OF TRABECUALR PLATES AND RODS
C. SELECTION OF INDIVIDUAL TRABECULAE OF DIFFERENT TYPES AND ANATOMIC DIRECTIONS
D. IMAGE REGISTRATION OF MICROSCOPE PHOTO (FRONT) AND μCT IMAGE (BACK) TO LOCATE THE INDENTATION SITES IN μCT IMAGE

RESULTS

The tissue modulus and the co-localized TMD of trabecular plates were significantly higher than those of trabecular rods ($p<0.001$, Fig. 2A, B). The axial tissue modulus of individual trabeculae was significantly higher than the lateral tissue modulus for both trabecular plates and rods in all the three anatomic directions ($p\leq0.05$, Fig. 2C). In general, the tissue modulus correlated strongly with TMD ($R^2=0.52$). These correlations did not differ significantly between plates and rods or different anatomic directions. However, the axial modulus-TMD correlation was significantly different from that in lateral direction (Fig. 2D).

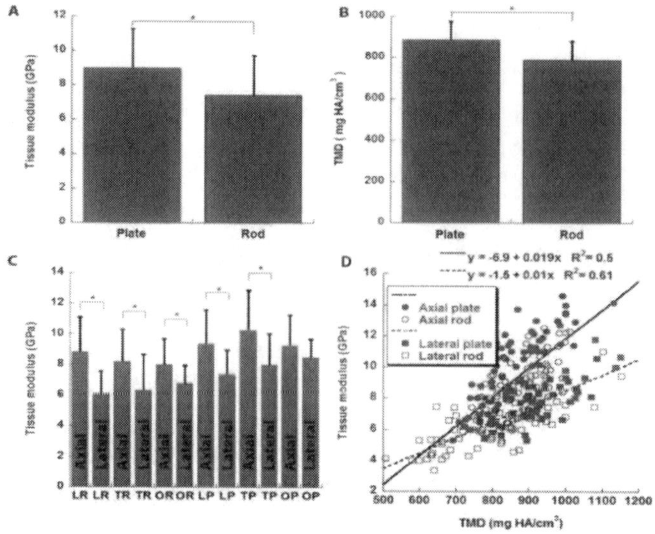

FIG.2 TISSUE MODULUS AND TMD VARY IN DIFFERENT TRABECULAR TYPES, ANATOMIC DIRECTIONS AND INDIVIDUAL TRABECULAR DIRECTIONS. (PLATE: N=216; ROD: N=216)

A. PLATES HAVE HIGHER TISSUE MODULUS THAN RODS. *p<0.05
B. PALTES HAVE HIGHER TMD THAN RODS. *p<0.05
C. TISSUE MODULUS IN AXIAL DIRECTION IS HIGHER THAN IN LATERAL DIRECTIONS. * p<0.05. P AND R DENOTE PLATE AND ROD; L,T,O DENOTE LONGITUDINAL, TRANSVERSE AND OBLIQUE (p=0.053 FOR OP_AXIAL vs. OP_LATERAL
D. TISSUE MODULUS CLOSELY CORRELATES WITH TMD. HOWEVER, THE CORRELATIONS DIFFER BETWEEN AXIAL AND LATERAL INDENTATION DIRECITONS. *p<0.05

DISCUSSION

The study reported novel findings in mechanical properties of individual trabecular plates and rods. Despite trabecular plates possessing distinct morphology from rods; the differences in their microscopic mechanical properties are mainly a result of the mineralization. For both plates and rods, we confirmed anisotropic elastic moduli dependent on different anatomic directions. Surprisingly, the heterogeneous tissue modulus correlated with TMD regardless of trabecular type and anatomic direction, suggesting that distinctly oriented trabecular plates and rods may have similar mineral/collagen ultrastructure. However, the correlation significantly differs in the axial and lateral trabecular directions, which further indicates that the underlying ultrastructure may play a critical role in the tissue mechanical properties of the individual trabecula. It remains to be seen how these heterogeneous anisotropic tissue properties play a role in whole bone mechanical competence.

ACKNOWLEDGEMENTS

The work is supported by NIH grant R01 AR051376.

REFERENCES

[1] XS Liu et. al. *JBMR*, 23(2):223-35, 2008
[2] WC Oliver et. al. *J Mat Res*, 7(6): 1564, 1992

DEVELOPMENT OF A PREDICTIVE COMPUTATIONAL MODEL FOR DEVICE-INDUCED THROMBOSIS

Joshua O. Taylor (1,2), Richard S. Meyer (2), Steven Deutsch (2), Keefe B. Manning (1,3)

(1) Department of Biomedical Engineering
The Pennsylvania State University
University Park, PA, USA

(2) Applied Research Laboratory
The Pennsylvania State University
State College, PA, USA

(3) Department of Surgery
Penn State Hershey Medical Center
Hershey, PA, USA

INTRODUCTION

Blood-contacting devices are regularly used by clinicians to treat a range of cardiovascular conditions. While complications relating to many of these devices have seen a decrease over the years, thrombosis still remains an obstacle. To help better understand and predict thrombus deposition and growth, many computational thrombosis models have been developed for use in computational fluid dynamics (CFD) simulations. However, due to the complexity of the thrombotic process, the most thorough models need to track tens of species and their interactions, making them too computationally expensive to use in the complex, three-dimensional flows associated with cardiac devices. In addition, they often rely on *a priori* knowledge to dictate locations of initial thrombus deposition and/or are not rigorously validated due to a lack of *in vitro* or *in vivo* thrombus growth data.

One of the primary causes of device-induced thrombosis is disrupted blood flow through the device [1]. Regions of high shear stress can activate platelets, while regions of low wall shear rate (WSR) can provide locations for platelet deposition. A thrombosis model for use in blood-contacting devices must be able to simulate the entire thrombotic process; platelet activation, platelet adhesion, and thrombus growth; at both high and low shear conditions, relying purely on the fluid dynamics of the system to dictate locations of thrombus deposition.

METHODS

A single-scale thrombosis model was taken from literature and modified for use in cardiac devices [2]. Terms were added to simulate platelet activation in regions of high shear stress and platelet deposition in regions of low WSR, and the model was implemented using the open-source CFD toolbox OpenFOAM. It is unnecessary to track individual cells and particles, as they are treated as continuous concentration distributions rather than discrete objects. The considered species in the model are non-activated platelets (Φ_n, Eqn 1), activated platelets (Φ_a, Eqn 2), and a chemical activator, adenosine diphosphate (ADP, Eqn 3). In addition, the density of platelet-platelet links (l_{pp}) is tracked to quantify aggregation (Eqn 4). The incompressible continuity (Eqn 5) and Navier-Stokes (Eqn 6) equations are used to calculate the velocity and pressure fields, with a momentum sink, $F(l_{pp})$, incorporated into Equation (6) to couple the velocity field with a growing thrombus. The momentum sink is a modified version of the Brinkman term [3], used to model flow through porous media.

$$\frac{\partial\phi_n}{\partial t} + (\boldsymbol{u}\cdot\nabla)\phi_n = D_n\nabla^2\phi_n - \left[R(ADP) + C\tau^\alpha\Delta t^\beta\right]\phi_n \quad (1)$$

$$\frac{\partial\phi_a}{\partial t} + (\boldsymbol{u}\cdot\nabla)\phi_a = D_a\nabla^2\phi_a + \left[R(ADP) + C\tau^\alpha\Delta t^\beta\right]\phi_n \quad (2)$$

$$\frac{\partial ADP}{\partial t} + (\boldsymbol{u}\cdot\nabla)ADP = D_{ADP}\nabla^2 ADP + A\phi_a - K\cdot ADP \quad (3)$$

$$\frac{\partial l_{pp}}{\partial t} + (\boldsymbol{u}\cdot\nabla)l_{pp} = \alpha_0 P(\dot{\gamma}_w)\phi_a^2 - \beta l_{pp} \quad (4)$$

$$\nabla\cdot\boldsymbol{u} = 0 \quad (5)$$

$$\rho\frac{\partial \boldsymbol{u}}{\partial t} + \rho(\boldsymbol{u}\cdot\nabla)\boldsymbol{u} = -\nabla p + \mu\nabla^2\boldsymbol{u} - F(l_{pp})\boldsymbol{u} \quad (6)$$

The term used to predict platelet adhesion and aggregation in low shear regions, $P(\gamma_w)$ in Equation (4), is adapted from a term developed to assess potential thrombosis in pulsatile blood pumps [4]. Here,

$P(\gamma_w)$ gives an instantaneous measure of thrombus deposition based on material-dependent high and low WSR thresholds, along with a critical adhesion time for platelets (Eqn 7). Thrombus growth is simulated using the same term (platelet adhesion to an already formed thrombus). A power law model ($C\tau^{\alpha}\Delta t^{\beta}$) is used to quantify platelet activation, using the scalar shear stress, τ (Eqn 8), the computational time step, Δt, and empirical coefficients estimated from literature [5]. In (Eqn 8), σ is the viscous stress tensor.

$$P(\dot{\gamma}_w) = \frac{\Delta t}{t_{adhesion}} \left(1 - \frac{|\dot{\gamma}_w|}{\dot{\gamma}_{high}} \cdot \frac{e^{\frac{|\dot{\gamma}_w| - \dot{\gamma}_{low}}{\dot{\gamma}_{high} - \dot{\gamma}_{low}}} - 1}{e^1 - 1} \right) \tag{7}$$

$$\tau = \left[\frac{1}{6} \sum_{i,j=1}^{3} (\sigma_{ii} - \sigma_{jj})^2 + \sigma_{ij}^{\,2} \right]^{\frac{1}{2}} \tag{8}$$

The model was tested in a two-dimensional backward-facing step (BFS) geometry with dimensions matching a previous *in vitro* thrombus growth study by Taylor *et al.* [6]. The step height was 2.5 mm, and the downstream diameter was 10 mm. The fluid was assumed to be Newtonian with a kinematic viscosity of 3.5 cSt, and the upstream Reynolds number was 490, ensuring the flow was laminar.

RESULTS

The BFS geometry produces a recirculation region downstream of the BFS (Fig. 1) and mimics the flow conditions that have proved troublesome in cardiac devices.

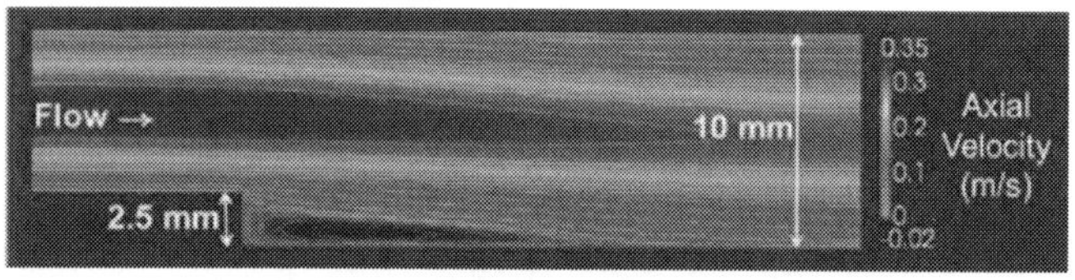

Figure 1: Contour map of the initial velocity field near the BFS. Velocity contour lines have been added to help illustrate the region of separated flow downstream of the asymmetric sudden expansion of the BFS.

A major improvement of the presented thrombosis model on existing models is its ability to simulate the entire thrombotic process; platelet activation, platelet adhesion, and thrombus growth; based purely on the fluid dynamics of the system. As expected, the model predicts the highest rates of mechanical platelet activation in the regions of high shear (Fig. 2), and it captures the established phenomenon of activated platelets becoming trapped in regions of flow separation (Fig. 3) [7].

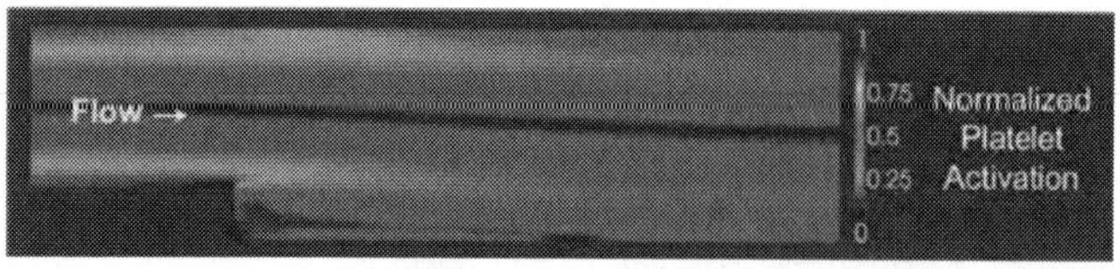

Figure 2: Plot of initial mechanical platelet activation, normalized by the maximum value. Regions of high shear stress produce the highest rates of platelet activation.

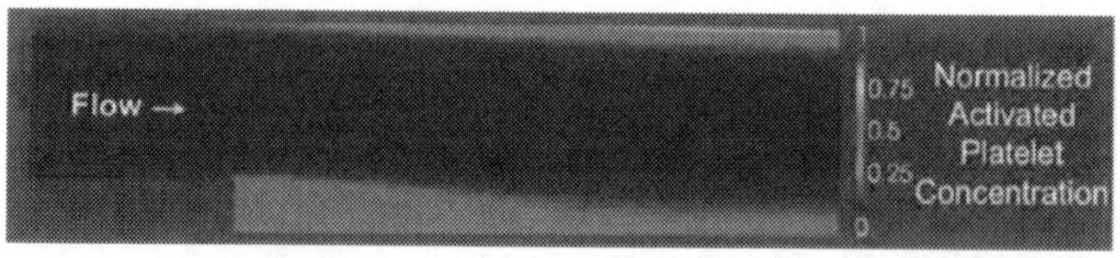

Figure 3: Plot of early activated platelet concentration (Φ_a), normalized by the maximum value. Agreeing with previous research [7], high concentrations of activated platelets are trapped in regions of flow separation, providing ideal conditions for thrombosis.

With increasing simulation time, the thrombosis model predicts platelet deposition and thrombus growth in the region of flow separation downstream of the BFS. The simulated thrombus is thickest near the BFS and decreases in cross sectional area in the downstream direction (Fig. 4).

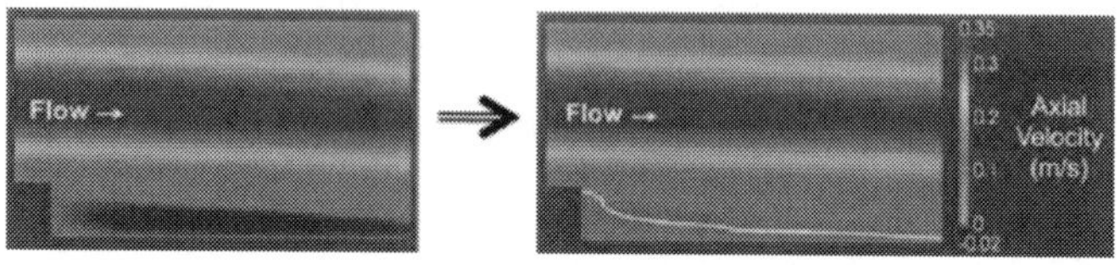

Figure 4: Visualization of the initial velocity field (left) and the velocity field after a thrombus was formed (right). The thrombus is visualized using a contour line based on the density of platelet-platelet links. The formed thrombus fills much of the initial recirculation region, and flow is greatly restricted through the thrombus due to the momentum sink in (Eqn 6).

DISCUSSION

Using the presented thrombosis model, the simulated thrombus (Fig. 4) resembles the *in vitro* thrombi from a previous study [6] in both size and shape. As thrombus deposition and growth is dictated by the shear rates in the computational domain, no boundary conditions are required to initiate thrombosis. This is essential for future predictions of device-induced thrombosis. Also, using a modified version of the Brinkman term as a momentum sink in (Eqn 6) allowed for the entire domain (both blood and thrombus) to be simulated as a fluid. Not only does this simplify the governing equations, it also allows the same mesh to be used for the duration of the simulation. As many medical devices cannot be simulated using simple, two-dimensional geometries, these simplifying aspects of the model are important for the eventual extension to complex, three-dimensional geometries.

ACKNOWLEDGEMENTS

This research was supported by a Walker Graduate Assistantship from the Applied Research Laboratory at the Pennsylvania State University and a Penn State Grace Woodward Foundation grant.

REFERENCES

[1] Deutsch S, *et al.*, *Annu Rev Fluid Mech*, 38:65-86, 2006.
[2] Fogelson AL, *SIAM J Appl Math*, 52(4):1089-1110, 1992.
[3] Leiderman K and Fogelson AL, *Math Med Biol*, 28(1):47-84, 2011.
[4] Medvitz RM, Ph.D. Thesis, The Pennsylvania State University, University Park, PA, 2008.
[5] Nobili, M, *et al.*, *ASAIO J*, 54(1):64, 2008.
[6] Taylor JO, *et al.*, *J Biomech Engr*, 136:071012, 2014.
[7] Bluestein D, *et al.*, *J Biomech Engr*, 118(3):280-286, 1996.

SB³C2015
Summer Biomechanics, Bioengineering and Biotransport Conference
June 17-20, 2015,Snowbird Resort, Utah, USA
SB³C2015-244

EXPERIMENTAL MEASUREMENT OF SHEAR AND TENSILE MODULI IN ANISOTROPIC TISSUE USING MAGNETIC RESONANCE ELASTOGRAPHY

JL Schmidt (1), DJ Tweten (1), MM Mahoney (2), T Portnoi (3), RJ Okamoto (1), JR Garbow (4), PV Bayly (1,2)

(1) Mechanical Engineering and Materials Science
Washington University
St. Louis, MO, United States

(2) Biomedical Engineering
Washington University
St. Louis, MO, United States

(3) Electrical Engineering
Massachusetts Institute of Technology
Cambridge, MA, United States

(4) Biomedical Magnetic Resonance Laboratory, Department of Radiology
Washington University
St. Louis, MO, United States

INTRODUCTION

Mechanical anisotropy is a key property of fibrous tissues; the anisotropic mechanical properties of brain white matter may play a key role in the mechanics of traumatic brain injury (TBI). The simplest anisotropic material model for small deformations of soft tissue is a nearly incompressible, transversely isotropic (ITI) material with three parameters: minimum shear modulus (μ_2), shear anisotropy ($\phi=\mu_1/\mu_2 - 1$) and tensile anisotropy ($\zeta=E_1/E_2 - 1$)[1]. These parameters, and the angle between the propagation and fiber directions, can be determined using magnetic resonance elastography (MRE). Using MRE, mechanical properties are estimated from phase-contrast MR images of motion during propagation of shear waves. Most MRE studies assume isotropic material models with single shear (μ) or tensile (E) modulus. Recent studies have provided parameters of anisotropic (transversely isotropic[2,3] or orthotropic[4]) material models. Here, two types of shear waves are analyzed, "fast" and "slow", for a given propagation direction, to estimate anisotropic shear and tensile moduli.

METHODS

Images of shear wave propagation in turkey breast (muscle) tissue *ex vivo* were acquired using MRE. Samples were obtained from a retail butcher and tested 3-5 days post-mortem. Two different sample geometries were used (Fig. 1): **Cylinder – axial excitation** (Fig. 1a): A cylindrical sample ~45 mm diameter was embedded in gelatin in a ~48 mm ID container. A 3-mm diameter plastic rod provided axial (z) excitation at 400 and 800 Hz, leading to shear waves with radial propagation direction ($\boldsymbol{n} \approx \boldsymbol{e_R}$). **Cube – transverse excitation** (Fig. 1b-c): A ~25mm cube of tissue was embedded in gelatin in a ~30×30×30 mm³ plastic container. Horizontal (z) 800 Hz excitation led to shear waves propagating downward ($\boldsymbol{n} \approx \boldsymbol{e_Y}$). The two excitation scenarios (Fig. 1b-c) differ in the direction of the excitation relative to the fiber orientation. To verify the average fiber orientation of the sample, diffusion tensor imaging (DTI) was performed (30 directions, b=2050 s/mm²).

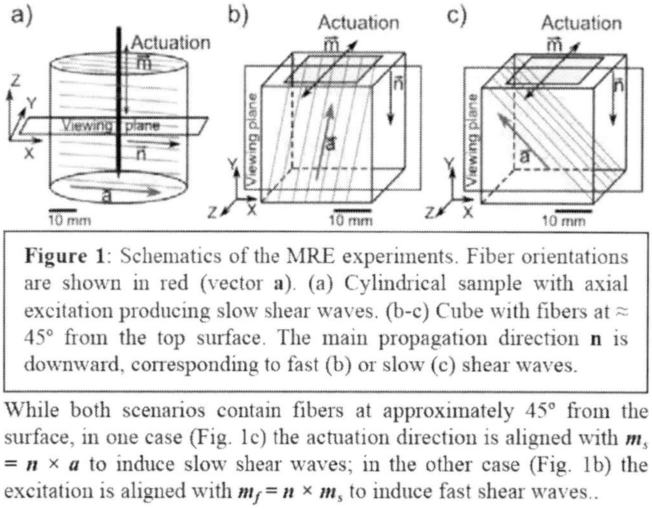

Figure 1: Schematics of the MRE experiments. Fiber orientations are shown in red (vector **a**). (a) Cylindrical sample with axial excitation producing slow shear waves. (b-c) Cube with fibers at ≈ 45° from the top surface. The main propagation direction **n** is downward, corresponding to fast (b) or slow (c) shear waves.

While both scenarios contain fibers at approximately 45° from the surface, in one case (Fig. 1c) the actuation direction is aligned with $\boldsymbol{m}_s = \boldsymbol{n} \times \boldsymbol{a}$ to induce slow shear waves; in the other case (Fig. 1b) the excitation is aligned with $\boldsymbol{m}_f = \boldsymbol{n} \times \boldsymbol{m}_s$ to induce fast shear waves..

THEORY

Assuming a harmonic plane-wave displacement field:

$$\mathbf{u}(\mathbf{x},t) = \mathbf{u}_0 \exp\left[i(k\mathbf{n}\cdot\mathbf{x} - \omega t)\right] = u_0 \mathbf{m} \exp\left[ik(\mathbf{n}\cdot\mathbf{x} - ct)\right], \quad (1)$$

The following wavespeed equations can be derived for slow (c_s) and fast (c_f) shear waves:

$$c_s^2 = \frac{\mu_2}{\rho}\left[1 + \phi\cos^2(\theta)\right], \quad (2)$$

$$c_f^2 = \frac{\mu_2}{\rho}\left[1 + \phi\cos^2(2\theta) + \zeta\sin^2(2\theta)\right]. \quad (3)$$

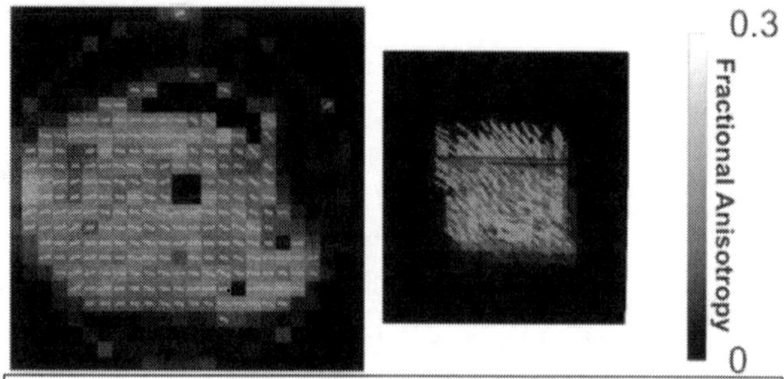

Figure 2: Fiber orientation in two experimental setups, acquired by DTI. (left) Cylindrical sample fiber orientations. (right) Cube sample fiber orientations. Fiber orientation vectors (blue) are superimposed on fractional anisotropy (FA, grey) for each voxel.

RESULTS

Figure 3 shows the displacement fields in the cylinder and cube samples for 400 and 800 Hz excitation.

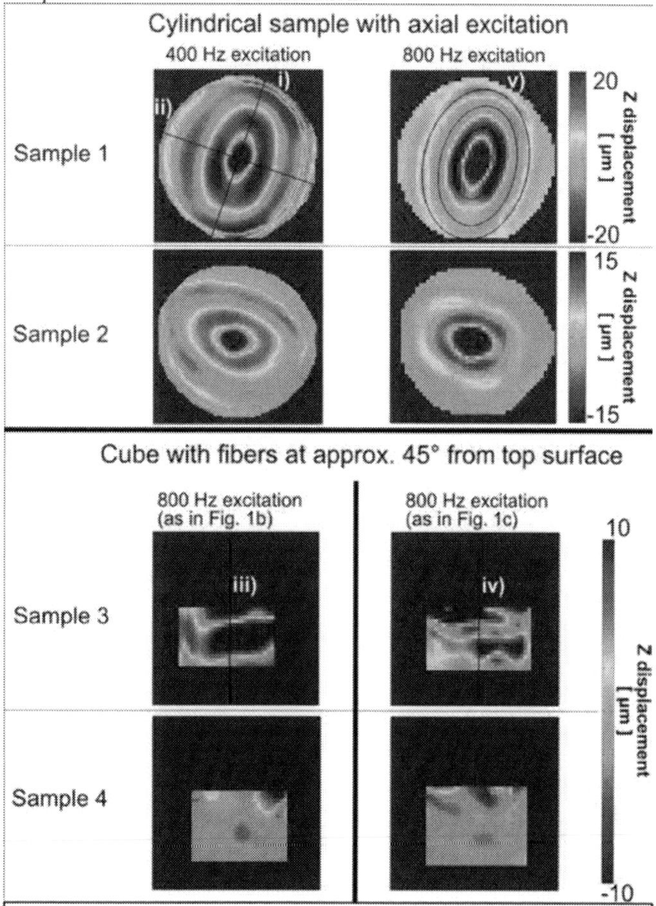

Figure 3: Out-of-plane displacement fields from 400 and 800 Hz MRE of samples in Fig. 1. Lines (i-iv) show propagation (i) $\theta = 0$ from fiber axis (excitation as in Fig. 1a); (ii) $\theta = 90°$ from fiber axis (excitation as in Fig. 1a); (iii) $\theta = 49°$ from fiber axis, fast shear mode (excitation as in Fig. 1b); (iv) $\theta = 49°$, slow shear mode (excitation as in Fig. 1c). (v) Shows elliptical slow shear waves with longer wavelength parallel to the fiber axis ($\theta = 0$).

In the cylindrical sample under axial excitation (Fig. 3, Sample 1,2), slow shear waves dominate. Elliptical waves (Fig. 3v) are evident with a longer wavelength (Fig. 3i) along the fiber direction (θ=0) and shorter wavelength (Fig. 3ii) perpendicular to the fiber axis ($\theta = 90°$). In the cube with $\theta \approx 45°$, when the excitation direction is parallel to $\boldsymbol{m}_s = \boldsymbol{n} \times \boldsymbol{a}$, slow shear waves (Fig. 3iv) also dominate, but when the excitation direction is aligned with $\boldsymbol{m}_f = \boldsymbol{n} \times \boldsymbol{m}_s$, fast shear waves (Fig. 3iii) dominate. The difference in slow shear-wave speeds in the cylinder at $\theta = 0$ and $\theta = 90°$ (Fig. 4) indicates shear anisotropy $\phi \approx 1.6$, from the relationship for c_s. The difference between fast and slow shear-wave speeds at $\theta = \sim 45°$ indicates tensile anisotropy $\zeta \approx 2.6$, from the equation for c_f. The baseline shear modulus $\mu_2 \approx 54$ kPa.

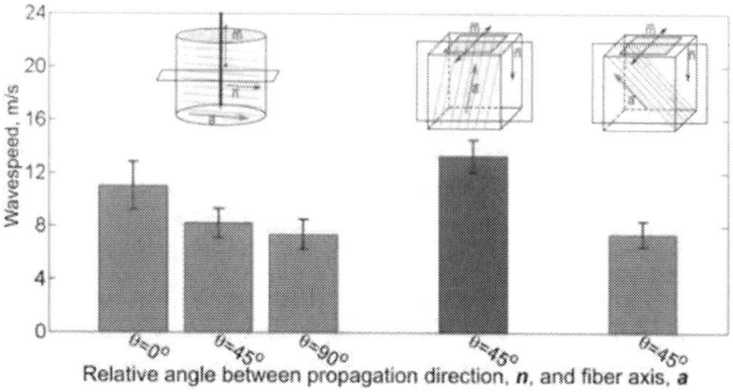

Figure 4: Mean wave-propagation speed (±std. deviation) at 800 Hz excitation in cylindrical and cubic specimens for different values of θ, the angle between the fiber axis and propagation direction. Color: slow (green) and fast (red) wave speeds. N=3 samples each.

DISCUSSION

Shear-wave speed in soft tissue is affected by anisotropy in both shear and tensile moduli, and by the angle between propagation and fiber directions. Both fast and slow shear waves are observed in MRE studies of muscle. Estimates of a minimal set of three parameters: shear modulus (μ_2), shear anisotropy (ϕ), and tensile anisotropy (ζ) can be extracted from fast and slow shear-wave speeds. This extends the methods in reference[2,3] for estimation of shear anisotropy but is simpler than other approaches[4] and may be easier to apply and interpret. The existence of both types of shear waves may be responsible for variations in parameter estimates in MRE of anisotropic tissue if isotropic material models are assumed, or if only slow shear waves are assumed. The current approach may enable greater understanding and more accurate measurement of normal and pathological mechanical properties of white matter in the brain.

ACKNOWLEDGEMENTS

Support from NIH grant NS055951 (PVB) and NSF grant CMMI-1332433 (PVB) is gratefully acknowledged.

REFERENCES

[1] Feng Y, et al., *J Mech Behav Biomed Mater*, 23:117-132, 2013
[2] Sinkus R, et al., *Magn Reson Med*, 53(2): 372-387, 2005.
[3] Qin EC, et al., *J Magn Reson Imaging*, 37(1): 217-226, 2013.
[4] Romano A, et al., *Magn Reson Med*, 68(5):1410-1422, 2012.

Redistribution of Knee Loads Using Auditory Feedback from Pressure Detecting Shoe Insoles

Chris Ferrigno (1), Ina S. Stoller (2), Laura E. Thorp (1,3), Najia Shakoor (3), Markus A. Wimmer (1,2)

(1) Department of Anatomy and Cell Biology
Rush University
Chicago, IL USA

(2) Department of Orthopedic Surgery
Rush University
Chicago, IL USA

(3) Divison of Rheumatology
Rush University
Chicago, IL USA

INTRODUCTION

Biomechanical factors contribute significantly to knee osteoarthritis (KOA), a debilitating and progressive joint disease. Through an increased understanding of aberrant joint loading and its association with joint disease and progression, research has focused on methods to restore normal load distribution across the knee joint. The knee adduction moment (KAM), a surrogate marker of medio-lateral load distribution, has been linked to KOA incidence, severity, pain, and progression. Gait alterations, such as incorporating a medial thrust, have therefore focused on KAM reduction to change the local biomechanics at the knee joint [1].

Despite its initial effectiveness, learning a gait modification such as medial thrust gait and incorporating the gait method into a daily routine can be challenging, uncomfortable, time consuming, and requires oversight by a physical therapist or physician. Since individual gait patterns are inherently pre-programmed, they are difficult to permanently change. While effective at reducing the KAM, medial thrust gait may compromise the overall effect of a diminished medial compartment load by increasing the sagittal plane moment as shown in instrumented knee prosthesis studies [2-3]. Altered gait patterns such as medial thrust also tend to affect spatiotemporal gait parameters [2]. New intervention strategies are necessary to target the KAM while preserving an individual's normal gait speed and minimizing unwanted mechanical side effects.

In the current study, we investigate a ground-up approach to reduce the KAM using auditory feedback provided from a plantar pressure-based shoe insole to gait train healthy subjects. The insoles are programmed to provide negative feedback with lateral foot pressure during gait. This feedback should result in a shift towards increased medial foot pressure and is designed to subtly alter an individual's gait to reduce the KAM. In the present work, we compare subjects' peak KAMs, peak knee flexion moments (KFM), and spatiotemporal variables recorded during medial thrust gait and while walking with pressure-based auditory feedback.

The overall hypothesis of this study is that walking with medialized plantar pressure using feedback will result in a significant reduction in the KAM and that the magnitude of this reduction will be similar to medial thrust gait without an increase in the KFM. Our secondary hypothesis is that average speed, stride length, and cadence will not be affected when walking with pressure-based feedback. Based on speculation that the mechanisms of load reduction are inherently different between medial thrust and feedback gait, we further hypothesized that both gait interventions should not affect each other, even if taught in the same session.

METHODS

Following IRB approval and informed consent, 28 healthy subjects underwent a single session of 3-D gait testing. The study design is outlined in Figure 1. Five normal walking trials were acquired followed by 5 trials each of the two gait alterations: (1) walking with medial thrust gait and (2) walking while receiving plantar pressure based audible feedback that cued the subjects to indirectly medialize plantar pressure by unloading the lateral foot by 25%. Each subject completed 15 total walking trials. The order which subjects performed the two gait alterations was randomized. For all trials, subjects wore standardized, flexible walking shoes [4] and walked at a self-selected normal speed. Only right sided data were analyzed.

For medial thrust gait, all subjects were trained by a licensed physical therapist for approximately 5 minutes until the subject was comfortable with the new gait strategy and the therapist observed a fluid gait pattern. During the pressure-feedback training, no verbal cues were offered to subjects other than instructions to avoid the audible cues while walking.

Kinetic and kinematic data were acquired using a modified Helen Hayes marker set, 12 optoelectric cameras (Qualysis, Gothenberg, Sweden), and 1 ground-embedded force plate (Bertec, Columbus, OH, USA). A Pedar Insole System (Novel, Munich) was used to acquire plantar pressure and provide pressure feedback programmed to medialize foot pressure, a strategy based on previous studies demonstrating an association between a decrease in KAM and a medializaton of foot pressure [5-6]. All systems were time-synced and data were acquired at 100Hz.

Repeated measures ANOVA was conducted to evaluate the null hypothesis that there is no difference in participants' spatiotemporal parameters including speed, stride length, and cadence, and relevant peak moments, including KAM and KFM during the three walking conditions. A between-subject factor of intervention order was applied to assess whether the order in which the interventions were performed played a role. Post-hoc Bonferonni comparisons were performed after ensuring that order was not a factor.

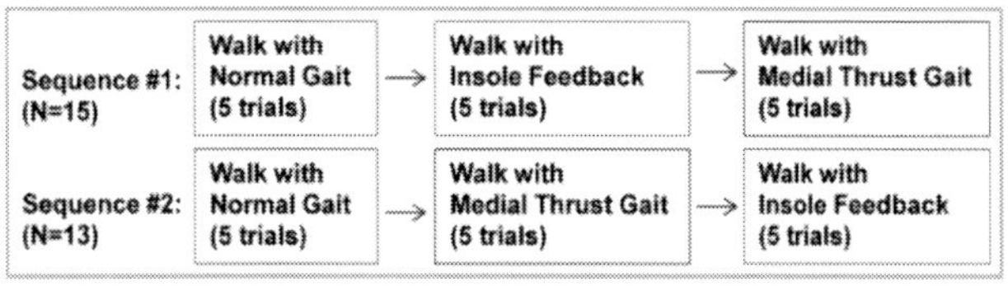

Figure 1: Study Design.

RESULTS

All 28 subjects (26.6 yrs ± 3.7 years, 14F, 14M, N=15 Sequence 1, N=13 Sequence 2) have been analyzed. Means and SDs of the test results are presented in Table 1. The between-subject factor of intervention in the repeated measures ANOVA indicated that the order in which subjects performed the gait alterations had no effect on the speed (p=0.438), stride length (p=0.535), cadence (p=0.664), KFM (p=0.317) and KAM (p=0.565). Post-hoc Bonferroni comparisons indicated that compared with normal walking, the KAM was significantly lowered with both medial thrust gait (p<0.001) and while receiving pressure-based feedback (p<0.001). Compared to normal gait, walking with medial thrust gait resulted in a KAM reduction of 0.386 %Bw*Ht (p<0.001) and walking with pressure-based feedback resulted in a KAM reduction of 0.400 %Bw*Ht (p<0.001); these reductions corresponded with a 13.2% and 13.7% KAM reduction, respectively. 21 of 28 subjects reduced their KAM executing medial thrust gait, while pressure-based feedback resulted in 24 of 28 subjects reducing their peak KAM.

In contrast, compared to normal gait, walking with a medial thrust gait resulted in an increase in the KFM of 0.843 %Bw*Ht, while walking with pressure-based feedback resulted in a KFM decrease of 0.280% Bw*Ht; these changes corresponded to a 29% increase with medial thrust gait (p=0.04) while and a 9.1% decrease for the feedback group (p=0.370). Subjects walked significantly slower with medial thrust gait (p<0.001) but speed remained constant with feedback gait (p=0.188). Stride length shortened significantly with medial thrust gait (p=0.011) and a trend towards a shorter stride was observed with feedback gait (p=0.082). Cadence decreased significantly with medial thrust gait (p<0.001) but not significantly with feedback, though a trend was also observed (p=0.089).

Table 1. Knee moments and spatiotemporal parameters for subjects walking in three conditions.

	Descriptive Statistics					
	Normal Walking		Medial Thrust Walking		Walking with Feedback	
	Mean	SD	Mean	SD	Mean	SD
Speed (m/s)	1.45	0.17	1.28*	0.15	1.40†	0.19
Stride (m)	1.36	0.10	1.31*	0.11	1.33	0.10
Cadence (steps/min)	55.44	3.59	51.31*	4.10	54.14†	4.21
KFM (%BW*Ht)	2.98	1.44	3.85*	2.04	2.71†	1.18
KAM (%BW*Ht)	2.92	0.82	2.53*	0.88	2.52*	0.84

* P<0.05 compared with normal gait

† P<0.05 compared with medial thrust gait

KFM=peak knee flexion moment KAM=peak knee adduction moment

DISCUSSION

This study supports the primary hypothesis that the use of a pressure-detecting insole providing feedback to reduce lateral foot pressure can significantly reduce the KAM. The magnitude of reduction is similar to the magnitude observed during medial thrust gait. This concept of manipulating lateral plantar pressure builds on previous research in our laboratory that found a significant association between a medialization of plantar pressure and a reduction in KAM [5]. In contrast to medial thrust gait, this reduction is not accompanied by an increase the KFM. Although not directly measured in this study, this suggests that the overall load on the medial knee compartment is likely reduced more during gait with pressure feedback compared with medial thrust gait. These findings agree with previous work suggesting that while medial thrust gait lowers the peak KAM it causes a potentially detrimental increase in the KFM [2-3].

The results of this study also support the secondary hypothesis that speed, stride length, and cadence would not be compromised with pressure-based feedback. Although subjects tended to walk slower compared to their normal gait, this difference was not statistically significant. The goal of a gait alteration as a treatment for knee OA is solely to target a reduction of the KAM. Since loss of speed is strongly associated with loss of functional mobility, gait efficiency must be conserved in order for the gait alteration to be efficacious in the long term. This is especially true in individuals with knee OA who may already have pronounced disability and orthopedic comorbidities which could become exacerbated if mobility is compromised.

In conclusion, the use of pressure based feedback lowers the peak KAM without affecting the KFM, speed, stride length, or cadence. This suggests that pressure-based feedback is an effective treatment modality to reduce medial compartment loading and the short-term and long-term efficacy should be explored further in subjects with medial compartment knee OA.

ACKNOWLEDGEMENTS

We acknowledge the data acquisition contributions of Renee Kawecki, BS and the time and efforts of our volunteer study subjects.

REFERENCES

[1]Simic, M et al, Arthritis Care & Research, 63 (3): 405–426, 2011.
[2]Gerbrands, TA et al, Clin Biomech, 29:828-834, 2014.
[3]Walter, JP et al, J Orth Res, 28(10):1348-1354, 2010.
[4]Shakoor, N et al, Arthritis Rheum, 59(9); 1214-20, 2008.
[5]Ferrigno, C et al, Trans of Orth Res Soc, 0777, 2014.
[6]Dowling, AV et al, J of Biomech Eng, 132, 0710071-5, 2010.

SB^{3}C2015
Summer Biomechanics, Bioengineering and Biotransport Conference
June 17-20, 2015,Snowbird Resort, Utah, USA
SB^{3}C2015-246

Pressure Modulation Improves OSA Patient Therapy

Ahmed M. Al-Jumaily (1), Raphaël Heinzerar (2), José Haba-Rubio (2)

(1) Institute of Biomedical Technology, Auckland Technical University, New Zealand

(2) Center for Investigation and Research in Sleep, Lausanne University Hospital (CHUV), Lausanne, Switzerland

INTRODUCTION

Over the last three decades the continuous positive airway pressure (CPAP) method has been used as an effective tool for the treatment of obstructive sleep apnea (OSA) [1, 2]. It provides air at an elevated pressure, which is normally called the titration pressure, through a nasal or facial mask, creating a pneumatic splint that keeps the pharyngeal airway open during inspiration and expiration. In spite of the CPAP therapy effectiveness, a number of drawbacks and negative impacts have been reported with its use. Over 45% of CPAP patients report negative side effects including discomfort, nasal irritation problems and some possible negative physiological impacts caused by the effect of CPAP on cerebral blood flow [3]. However, modulating the titration pressure has been reported to introduce some improvement to remedy some of these problems.

It has been reported that the pressure waves resulting during human snoring have high frequencies ranging between 30 to 50 Hz [1]. Plowman et al [4] conducted an experimental study on three sleeping tracheotomised dogs. They used pressure oscillation waves similar to those occurring during human snoring (30 Hz) combined with an amplitude of ± 3 cm of water and this has resulted in an increase in the activity of the genioglossus muscle. Zhang et al, on the other hand conducted an experimental study on ten anesthetized supine dogs using pressure oscillations of 10, 20, and 30 Hz, combined with amplitude of ± 2.5 cm of water and concluded that the pressure oscillations increased the activity of the pharyngeal receptors [5]. Further, Henke et al [2] experimentally studied the effects of pressure oscillations on the UA muscles during the inspiratory breathing phase, on 6 normal subjects and 10 subjects with OSA. It was reported that the activity of the UA muscles was increased as a result of the effect of pressure oscillations, in both normal and OSA subjects. Several other studies on human subjects have demonstrated that pressure oscillation does affect the UA stiffness and dynamic characteristics of the UA [6]. However, those studies were limited to certain breathing conditions and/or sleeping positions. This paper presents a general principle which could be adopted to improve the OSA patients' response to CPAP treatment.

METHODS

The main principle of this work is to superimpose a small sinusoidal pressure oscillation over a mean titration pressure to modulate the breath cycle and stimulate temporary deformation of the UA which is anticipated to allow better breathing capability, Fig. 1. It is believed that the titration pressure would stretch the UA tissue and improve its dynamic characteristics (stiffness and frequency) while the pressure oscillation at certain frequencies stimulates some tissue deformation which results in having better airflow conditions (less obstruction).

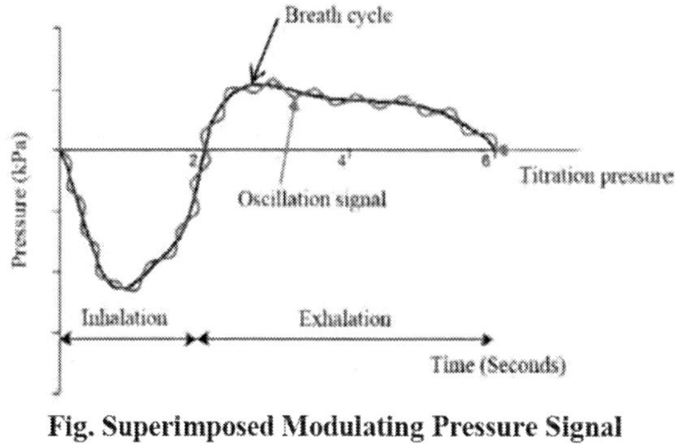

Fig. Superimposed Modulating Pressure Signal

The type of signal above is generated by an oscillator with controllable frequencies and amplitude. The constant pressure is produced by the CPAP device. Although some of the modulation frequencies required to stimulate the UA can be predicted [7], they are patient dependent and have to be adjusted. To do that a sensor is located at the throat to measure the vibration induced by the oscillator. Normally a physician or a nurse will need to alter those frequencies until maximum vibration is felt at the throat, see Figure 2 for details.

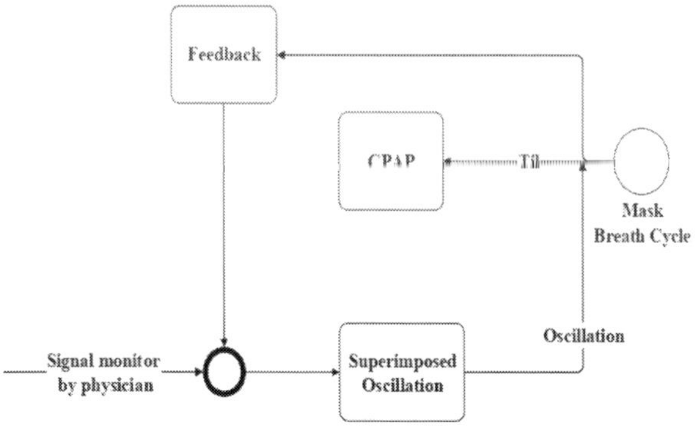

Fig. 2 Superimposed signal generation-Experimental Setup

RESULTS

Subjects aged 37-73 years with initial Apnea Hypopnea Index (AHI) above 15/hour were recruited among CPAP-treated OSA outpatients at the Centre for Investigation and Research in Sleep. Those patients had good tolerance to CPAP, mean use of CPAP of more than 4 hours per night in the last 3 months.

Polysomnography was used for two nights. In the first night a CPAP therapy was performed in sleep at 70% of the optimal titration pressure. The second night the modulating signal was added to the CPAP treatment at 70% titration pressure. The modulating signal was monitored by a technician. The amplitude was around 1 cm and the starting frequency was 15 Hz. The frequency was then altered depending on the response measured at the throat during testing. The polysomnographies were performed using an Embla N7000TM and analyzed with the SomnologicaTM software (Embla Systems, Inc., Broomfield, CO, USA). Sleep stages were scored by one investigator blinded to treatment modality, according to the American Academy of Sleep Medicine (AASM) 2007 recommendations [8]. Respiratory events were scored using the criteria from the AASM [9].

Most of the tests have resulted in significant reduction in AHI by using modulation. In some cases this reduction could be up to one third its value during pure CPAP treatment.

Discussion

Vibration principles have demonstrated that any element with some elasticity can be deformed if oscillated at specific resonant frequencies. The upper airways collapse during OSA makes the tissues in a loose stage which jeopardize breathing conditions by blocking the airways. Physically the AHI indicates how severe this blockage is. We believe modulating the pressure helps to temporarily deform these tissue and allow blockage reduction.

This preliminary study has indicated significant reduction in AHI in addition to reducing the pressure to 70% during testing. The applied modulated signal was well tolerated by patients. Considering these positive results further research needs to be conducted to demonstrate the actual tissue deformation and its significance in improving the CPAP treatment.

As this is a first stage investigation we cannot exclude the absence of a significant result which was due to a lack of statistical power. Many factors need to be considered in future studies in order to improve the outcomes of this research. These include but are not limited to (i) The UA tissue and genioglussus muscle stiffness may change during sleep and the oscillating pressure may need to be adjusted during the course of the investigation. (ii) It is possible that some OSA patients cannot respond sufficiently to the applied vibration.

In conclusion, pressure modulation seems to have positive improvement in the use of CPAP and is well tolerated by patients. However, further investigations are necessary to prove the principles as well as to improve the methodology.

ACKNOWLEDGEMENTS

We acknowledge the support of "Marie Curie Incoming Fellowship" for supporting the first author during the clinical part of this investigation. We acknowledge the contribution of Françoise Cornettea and Nadia Tobbacka from the Center for Investigation and Research in Sleep in collecting the data during this research.

REFERENCES

[1] Robin, I.G., Snoring. Proceedings of the Royal Society of Medicine, 1968. 61(6): p. 575-582.

[2] Henke, K.G. and C.E. Sullivan, Effects of high-frequency oscillating pressures on upper airway muscles in humans. Journal of Applied Physiology, 1993. 75(2): p. 856-862.

[3] Mohsenin V (2001) Sleep-related breathing disorders and risk of stroke. Stroke 32 (6):1271-1278 [4]

[4] Plowman, L., et al., Waking and genioglossus muscle responses to upper airway pressure oscillation in sleeping dogs. Journal of Applied Physiology, 1990. 68(6): p. 2564-2573.

[5] Zhang, S. and O.P. Mathew, Response of laryngeal mechanoreceptors to high-frequency pressure oscillation. Journal of Applied Physiology, 1992. 73(1): p. 219-223.

[6] Vanderveken, O.M., et al., Quantification of pharyngeal patency in patients with sleep-disordered breathing. ORL; journal for oto-rhino-laryngology and its related specialties, 2005. 67(3): p. 168-179.

[7] Al-Jumaily,A.M.,et al "uvula dynamic characteristics",Paper Number SBC2013-14019, Proceedings of the ASME 2013 Summer Bioengineering, SBC2013, June 26-29, Sunriver, Oregon, USA.

[8] Iber C A-IS, Chesson A, Quan SF. for the American Academy of Sleep Medicine, ed. The AASM Manual for the Scoring of Sleep and Associated Events: Rules, Terminology and Technical Specifications. Westchester: American Academy of Sleep Medicine, 2007.

[9] Berry RB, et al. Rules for scoring respiratory events in sleep: update of the 2007 AASM Manual for the Scoring of Sleep and Associated Events. Journal of clinical sleep medicine : JCSM : official publication of the American Academy of Sleep Medicine 2012;8:597-619.

NUMERICAL MODELING OF SKELETAL MUSCLE UNDER HIGH STRAIN AND STRESS RELAXATION COMPRESSION CONDITIONS

Benjamin B. Wheatley (1), Renee Pietsch (2), Tammy L. Haut Donahue (1, 3), Lakiesha N. Williams (2)

(1) Department of Mechanical Engineering
Colorado State University
Fort Collins, CO, USA

(2) Department of Agricultural and Biological Engineering
Mississippi State University
Starkville, MS, USA

(3) School of Biomedical Engineering
Colorado State University
Fort Collins, CO, USA

INTRODUCTION

Skeletal muscle, in addition to providing locomotion, acts to disperse compressive loads during high impact and resting. Current *in vitro* methods to investigate how these loads affect the human body are dependent on utilizing materials with a similar behavior to muscle or developing computational models that can accurately re-create muscular mechanics. Previous investigations have highlighted muscle properties [1] and developed numerical models [2, 3], but often employ difficult approaches to implement into computational methods. Additionally, there still lacks a complete understanding of how muscle behaves under compression, specifically how the time dependent properties of the tissue may change as a function of strain level.

The goal of this work was to utilize an explicit, simple strain energy density (SED) function, which can be easily incorporated into other models utilizing continuum mechanics, and to determine any strain dependence of the viscoelastic behavior of skeletal muscle.

METHODS

Porcine hind limbs were locally obtained from an abattoir (Sansing Meat Services, Maben, MS) from adult pigs, and testing was completed within five hours post mortem to avoid the effects of rigor mortis [2]. The semimembranosus muscle was dissected into cylindrical samples 10 mm in height and 20 mm in diameter and stored moist in 0.01 M phosphate buffered saline. All samples were cut either perpendicular to the muscle fiber direction (transverse, N=13), or parallel to the fiber direction (longitudinal, N=13).

Compressive stress relaxation was performed on each sample with a MACH-1™ uniaxial testing system (Biomomentum Inc., Canada). Each sample was compressed in five steps of 5% engineering strain at a rate of 5%/s until a rate of relaxation of 1g/min was reached, for a total of 25% strain. The peak responses, equilibrium responses, and time to relaxation were analyzed between longitudinal and transverse samples with a two sample t-test ($p<0.05$).

A transversely isotropic hyperleastic strain energy density (SED) function (Equation 1, where a_i and b_i are material parameters, $\boldsymbol{I_1}$ is the first invariant of the right Cauchy-Green deformation tensor, and $\boldsymbol{I_4}$ is a pseudo-invariant accounting for transverse isotropy) was simultaneously fit to one longitudinal sample (where $\boldsymbol{I_4}$ is set to one and this term becomes zero) and one transverse sample (where both terms exist). This was completed using a non-linear least squares optimization procedure in Matlab (The Mathworks Inc, Natick, MA) with the "lsqnonlin" function. The function varied the six hyperleastic parameters in Equation 1 to match experimental data at each of the five strain levels. As the peak and equilibrium responses were clearly different, they were fit using separate optimizations. The initial parameter guesses for each specimen were determined by varying each parameter across five values to locate the closest initial agreement with experimental data.

$$\Psi = \sum_{i=1}^{3} a_i\,(\boldsymbol{I_1}-3)^{2*i} + b_i\,(\boldsymbol{I_4}-1)^{2*i} \tag{1}$$

Each load-relaxation step was isolated and fit to a three-term linear viscoelastic Prony series approximation (Equation 2, where E_0 is the instantaneous modulus, e_n are relaxation moduli, and τ_n are time constants) with a similar least squares optimization, resulting in five sets of viscoelastic parameters for longitudinal and five for transverse samples. A second-order polynomial [4] was then applied to the Equation 2 parameters as a function of strain level to investigate any viscoelastic non-linearity. All viscoelastic parameters were compared between longitudinal and transverse with a two sample t-test ($p<0.05$).

$$E(t) = E_0\left(1 - \sum_{i=1}^{3} e_n\left[1 - exp\left(-\frac{t}{\tau_n}\right)\right]\right) \tag{2}$$

RESULTS

The transverse peak stresses were found to be larger in steps 2-5 ($p=0.173$, 0.004, 0.001, 0.001, <0.001, for steps 1-5), and the first three steps for equilibrium stress, ($p<0.001$, <0.001, 0.005, 0.075, 0.081 for steps 1-5) (Figure 1). Relaxation time was found to be higher in transverse samples for all steps ($p=0.050$, <0.001, <0.001, <0.001, 0.007 for steps 1-5, Figure 1).

Non-linear fitting of both the peak and equilibrium responses showed good agreement for longitudinal and transverse data (Figure 2, Table 1 for parameter values). Optimized Prony series terms for each load-relaxation step were averaged together to form a mean set of parameters (Table 2). Fitting these viscoelastic Prony terms for each step to a polynomial showed an inconsistent dependence on strain level. The first and second time constants were found to be longer in transverse specimens ($p=0.013$, 0.015, respectively), while the instantaneous modulus was also higher in transverse ($p<0.001$).

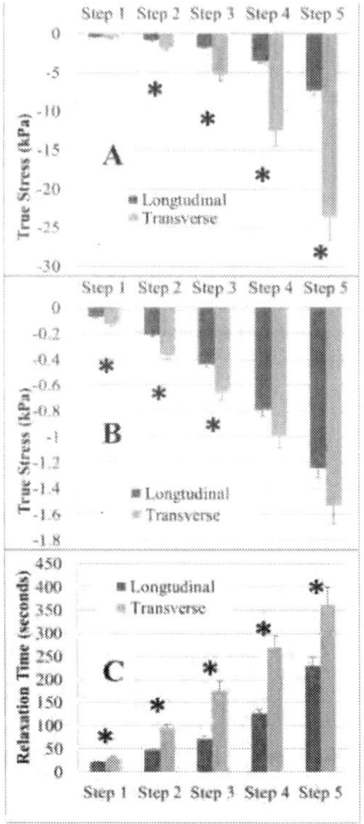

Figure 1: Transverse and longitudinal peak stress (A), equilibrium stress (B), and relaxation time (C) (+ SE) for each stress relaxation step (* denotes $p<0.05$).

Table 1: Hyperleastic optimized parameters [kPa].

Parameter	a_2	a_4	a_6	b_2	b_4	b_6
Peak Mean (± SE)	1.53 (0.17)	25.1 (6.1)	901 (67)	0.0845 (0.046)	22.5 (6.2)	81.0 (17)
Equil. Mean (± SE)	0.550 (0.058)	7.19 (1.6)	71.3 (16)	0.174 (0.027)	0.219 (0.16)	0.0521 (0.029)

Table 2: Viscoelastic Prony parameter fitted values (* denotes $p<0.05$) [Unit-less].

Parameter	e_1	e_2	e_3	τ_1*	τ_2*	τ_3	E_0*
Long Mean (± SE)	0.455 (0.0053)	0.289 (0.0049)	0.192 (0.0046)	0.546 (0.039)	4.95 (0.65)	50.6 (5.73)	79.9 (9.16)
Trans Mean (± SE)	0.453 (0.0062)	0.296 (0.0059)	0.201 (0.0030)	0.718 (0.054)	7.12 (0.57)	61.5 (3.86)	231 (32.6)

DISCUSSION

The observed highly non-linear behavior of skeletal muscle in compression (Figures 1 and 2) agrees with previous investigations [1, 2]. The increasing value of hyperelastic parameters with order also supports this previously identified non-linearity. The utilized SED function from this study (Equation 1) showed an excellent ability to match both the longitudinal and transverse experimental response (Figure 2) while being straightforward enough to be easily implemented into a finite element analysis or coupled with other models such as viscoelasticity. To the author's knowledge, this was the first time skeletal muscle was modeled with an explicit SED function in stress relaxation. Furthermore, the structure of this function allows for increasing non-linearity, shear behavior, and compressibility by simply adding similar invariant based terms.

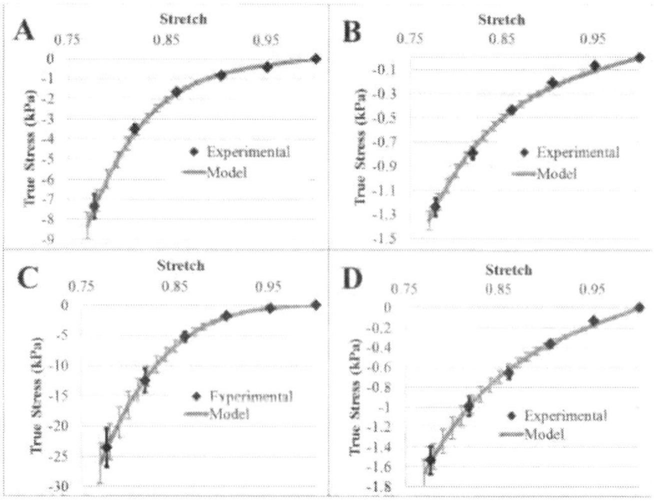

Figure 2: Hyperelastic fitting of longitudinal (A for peak, B for equilibrium) and transverse (C for peak, D for equilibrium) experimental data of muscle (mean ± SE bars).

The observed increase in stiffness of transverse samples has been previously documented [1, 2], however this work was able to investigate further relaxation behavior. Longer relaxation times in transverse versus longitudinal samples suggest that relaxation rate has a directional dependence. This is also supported by the similarity between equilibrium stiffness at higher strains (fourth and fifth steps, Figure 1B). An analysis of either current literature [3] or new data could elucidate how muscle behaves at very long relaxation times, as any differences between transverse and longitudinal stress may disappear when the tissue has reached an infinite relaxation time.

While there were some increasing trends in viscoelastic terms as a function of strain (R^2 values between 0.60 and 0.82 for τ_2, τ_3, and E_0 for both transverse and longitudinal samples fit to a second order polynomial), the data set as a whole does not show enough consistency to accurately develop a non-linear set of Prony terms. However, future investigations to utilize a single non-linear function may prove successful. Our investigation did suggest that the rate of relaxation and stiffness played a larger role in the differences between longitudinal and transverse responses than relaxation moduli. Our optimized Prony parameters do follow a previously determined trend of decreasing relaxation moduli with longer time constants [3].

While these data provide valuable insight into the behavior of skeletal muscle in compression, the true value of this investigation can only be met with accurately utilizing a three dimensional model to predict tissue behavior.

ACKNOWLEDGEMENTS

This work was supported in part by the U.S. Army TACOM Life Cycle Command under Contract No. W56HZV-08-C-0236.

REFERENCES

[1] Pietsch, R et al., *J Boimech Eng,* 136:111003-9, 2014
[2] Van Loocke et al, *J Boimech,*39:2999-3009, 2006
[3] Van Loocke et al., *J Boimech,* 41:1555-66, 2008
[4] Shetye et al., *Acta Biomaterialia,* 10:792-7, 2014

SB^{3}C2015
Summer Biomechanics, Bioengineering and Biotransport Conference
June 17-20, 2015,Snowbird Resort, Utah, USA
SB^{3}C2015-248

GEOMETRIC ANALYSIS OF ILIAC ARTERY TORTUOSITY: COMPARISON TO CURRENT CLINICAL PRACTICE

Matthew G. Doyle (1,2), Elrasheed Osman (2), Naomi Eisenberg (2), Cristina H. Amon (1,3), Leonard W. Tse (2)

(1) Department of Mechanical and Industrial Engineering
University of Toronto
Toronto, ON, Canada

(2) Division of Vascular Surgery
Peter Munk Cardiac Centre, UHN,
Toronto General Hospital
Toronto, ON, Canada

(3) Institute of Biomaterials and Biomedical Engineering
University of Toronto
Toronto, ON, Canada

INTRODUCTION

Endovascular medical procedures are often carried out by inserting a catheter or other medical device, such as a stent graft, into the femoral artery, through the iliac artery, and into position in the aorta or further along the vascular tree. Navigating a medical device through the iliac artery can be challenging as this vessel can be tortuous (twisting), calcified (stiff), and/or stenosed (narrow). Currently, assessment of the suitability of the iliac artery for a particular device is done in a qualitative or semi-qualitative manner by the clinician prior to the procedure [1-2].

Recently, three geometric parameters have been proposed [3-4] for the quantitative assessment of tortuosity of a blood vessel centerline, namely tortuosity itself, which is the ratio of the length of a line to the Euclidean distance between its endpoints, curvature, which is a local measure of the deviation of a curve from a straight line, and torsion, which is a local measure of the twisting of a curve. The degree of calcification can be quantified by a calcium score calculated using proprietary software on board commercial CT scanners [5-6]. The degree of vessel stenosis can be quantified using the maximum inscribed sphere radius (MISR), which is a measure of the local vessel radius calculated from an inscribed sphere [4].

The overall objective of our study is to quantitatively assess the tortuosity, degree of calcification, and degree of stenosis in iliac arteries to provide a more objective method to assess the suitability of these arteries for medical device insertion and to better understand device behavior as they traverse the iliac arteries. More specifically, we are interested in understanding the build-up of rotational energy in stent grafts during delivery through iliac arteries, because these devices are known to rotate unexpectedly during deployment [7]. This abstract describes the results of the first part of this study, which focuses on vessel tortuosity.

METHODS

Following approval from the Research Ethics Board at Toronto General Hospital, we retrospectively examined preoperative CT scans of 43 patients (31 male and 12 female with an average age of 75 years) who had undergone implantation of a fenestrated aortic stent graft at our hospital between January 15, 2007 and April 30, 2014. These CT scans were anonymized to remove patient information, and then imported into the software package The Vascular Modeling Toolkit (VMTK) for analysis.

Within VMTK, the left and right iliac arteries were separately segmented from the aortic bifurcation to the femoral artery. Segmentation was carried out using a level set method, initialized by a colliding fronts approach. Once segmented, a vessel centerline was calculated, resampled at 0.3 mm intervals, and then smoothed. Local values of curvature and torsion were then calculated along the length of each centerline using the approach described in Ref. [4].

Briefly, the centerline is parameterized by a variable s, which is a measure of the Euclidean distance of each point on the centerline $c(s)$ from the origin [3-4]. Local curvature $\kappa(s)$ and local torsion $\tau(s)$ are then calculated as [3-4]

$$\kappa(s) = \frac{|c'(s) \times c''(s)|}{|c'(s)|^3}, \tag{1}$$

$$\tau(s) = \frac{[c'(s) \times c''(s)] \cdot c'''(s)}{|c'(s) \times c''(s)|^2}. \tag{2}$$

Tortuosity χ was calculated globally as

$$\chi = \frac{L}{D} - 1, \tag{3}$$

where L is the centerline length and D is the Euclidean distance between the endpoints [4].

RESULTS

Left and right iliac artery centerline tortuosity, local curvature, and local torsion were calculated for all cases. An example of centerline curvature and torsion in the left iliac artery is shown in Figure 1. For each case, total curvatures and torsions were calculated as the sums of the absolute values at each data point and average curvatures and torsions were calculated as the totals divided by the number of data points [3].

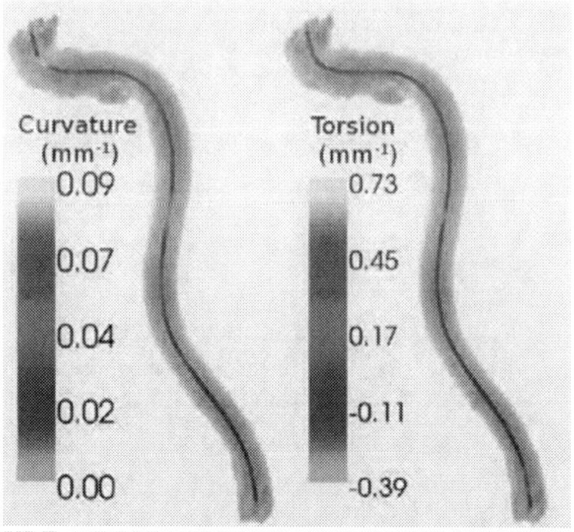

FIGURE 1. REPRESENTATIVE LEFT ILIAC ARTERY CENTERLINE CURVATURE AND TORSION.

Table 1 shows the results for left and right iliac arteries, presented as the average plus or minus the standard deviation. Unpaired two-sample t-tests showed no significant differences ($p \geq 0.4197$) between the left and right iliac arteries for any of these parameters.

TABLE 1. AVERAGE LEFT AND RIGHT ILIAC ARTERY TORTUOSITIES, CURVATURES, AND TORSIONS.

	Left	Right
Tortuosity	0.278 ± 0.143	0.302 ± 0.129
Total curvature (mm^{-1})	21.2 ± 5.37	21.7 ± 5.24
Average curvature (mm^{-1})	0.0312 ± 0.0066	0.0317 ± 0.0074
Total torsion (mm^{-1})	65.8 ± 15.7	68.6 ± 20.4
Average torsion (mm^{-1})	0.0972 ± 0.0217	0.0999 ± 0.0265

In our hospital, clinicians qualitatively assess iliac artery tortuosity, insert the stent graft into the more tortuous side, and use the less tortuous side for branch manipulation. If both sides have roughly the same tortuosity, the right side is selected for stent graft insertion due to operating room ergonomics. To study whether our analysis aligned with this qualitative procedure, we reordered the data into two groups, the side through which the stent graft was inserted ("Chosen") and the side that was not chosen ("Not"). Table 2 summarizes the average values for the calculated parameters for the "Chosen" and "Not" sides. As with the left and right iliac arteries, unpaired two-sample t-tests showed no significant differences ($p \geq 0.4598$) between the "Chosen" and "Not" sides for any of the calculated parameters.

TABLE 2. AVERAGE "CHOSEN" AND "NOT" ILIAC ARTERY TORTUOSITIES, CURVATURES, AND TORSIONS.

	"Chosen"	"Not"
Tortuosity	0.291 ± 0.122	0.289 ± 0.150
Total curvature (mm^{-1})	21.0 ± 5.06	21.8 ± 5.52
Average curvature (mm^{-1})	0.0311 ± 0.0067	0.0319 ± 0.0073
Total torsion (mm^{-1})	68.3 ± 20.2	66.0 ± 16.0
Average torsion (mm^{-1})	0.101 ± 0.026	0.0966 ± 0.0225

To gain further insight into the choice of side for stent graft insertion, we analyzed each case to determine whether each parameter was greater on the left or the right. To account for small differences between the left and right, we set a buffer of 25% difference, such that values less than 25% different were taken to be equal. For each parameter, we assigned a value of -1 if the left side was greater than the right side, 0 if they were equal, and +1 if the right side was greater than the left side. By summing these values, we determined whether the clinician should have selected the left side (totals from -5 to -1) or the right side (totals from 0 to +5).

Based on this analysis, we found that the clinicians chose the "correct" side 26 out of 43 times. Assuming values from -1 to +1 are too close to distinguish between the left and right sides, we removed these values from the calculations and found clinicians chose the "correct" side 8 out of 17 times. Both of these results clearly indicate that we can gain additional insight into iliac artery tortuosity through quantitative assessment compared to qualitative assessment.

DISCUSSION

No statistically significant difference was found between any of the parameters for the left and right iliac arteries or for the "Chosen" or "Not" sides. When each case was analyzed to determine which side should have been chosen based on the standard that the most tortuous side is chosen, it was found that the correct side was chosen only about 60% of the time, and when only large differences between sides were considered, the correct side was chosen only about 47% of the time. This suggests that quantitative assessment of iliac artery tortuosity could aid clinicians in identifying which side to choose during their procedures.

This study represents the first step in the development of quantitative tools to assess the tortuosity of iliac arteries to determine their suitability for medical device insertion. Quantitative assessment of tortuosity will have implications for determining access, i.e. will a particular device fit through a particular iliac artery, as well as control, i.e. how will the device behave in the artery. The latter is of interest to us in the context of fenestrated aortic stent grafts, which are known [7] to occasionally rotate upon deployment due to the build-up of rotational energy as they travel through tortuous iliac arteries.

To enhance the utility of these quantitative tools, ongoing work will develop tools to assess the degree of calcification and degree of stenosis in iliac arteries. Combining vessel tortuosity, stiffness, and narrowing into a single set of tools will allow for a more complete picture of the vessel to aid in determining whether a medical device can fit through the vessel as well as address any possible issues related to unwanted device behavior, such as rotation.

ACKNOWLEDGEMENTS

We are grateful to Raymond Wong for preparing the anonymized patient CT scans for use in our study and David Romero for his help with the statistical analysis.

REFERENCES

[1] Kristmundsson, T et al., *J Endovasc Ther*, 19:157-164, 2012.
[2] Ghatwary, T et al., *J Endovasc Ther*, 19:400-412, 2012.
[3] O'Flynn, PM et al., *Ann Biomed Eng*, 35:1368-1381, 2007.
[4] Piccinelli, M et al., *IEEE Trans Med Imag*, 28:1141-1155, 2009.
[5] Komen, N et al., *Acta Radiol*, 52:943-950, 2011.
[6] Willemink et al., *Eur Radiol*, 24:2201-2212, 2014.
[7] Bicknell, CD et al., *Eur J Vasc Endovasc Surg*, 37:175-181, 2007.

PREDICTING THE STRESS AND INTRAMUSCULAR PRESSURE RESPONSE OF WHOLE SKELETAL MUSCLE THROUGH OPTIMIZED FINITE ELEMENT ANALYSIS

Benjamin B. Wheatley (1), Duane A. Morrow (2), Gregory M. Odegard (3), Kenton R. Kaufman (2), Tammy L. Haut Donahue (1, 4)

(1) Department of Mechanical Engineering
Colorado State University
Fort Collins, CO, USA

(2) Department of Orthopedic Surgery
Mayo Clinic
Rochester, MN, USA

(3) Department of Mechanical Engineering – Engineering Mechanics
Michigan Technological University
Houghton, MI, USA

(4) School of Biomedical Engineering
Colorado State University
Fort Collins, CO, USA

INTRODUCTION

Skeletal muscle provides movement by generating force through activation. However, passive mechanics play a key role in both force transmission and maximum contractile force. Specifically, resting muscle length, which can be disturbed as a result of tendon transfer surgery [1], is vital to proper muscle function. The ability to properly determine ideal muscle resting length is reliant upon evaluating passive muscle tension. There are no current clinical approaches to measure individual muscle force, however intramuscular pressure (IMP) correlates with both passive and active muscle force [2].

A method to accurately predict both passive muscle force and IMP could aid in clinical practices to avoid muscle weakness as a result of improper resting length following surgical intervention. Thus, the goal of this work is to develop a finite element model of skeletal muscle capable of characterizing both muscle force and IMP.

METHODS

Two New Zealand White Rabbit (NZWR) whole tibialis anterior muscles were dissected out and frozen until 24 hours prior to testing. Each specimen underwent load relaxation testing of 5% strain at a rate of 20%/s or 0.25%/s and 300s of relaxation in a material testing system (MTS Corporation, Eden Prairie, MN) with a 1 kg load cell (Futek, Irvine, CA) at a 100 Hz sampling rate. A pressure microsensor (Fiso Technologies Inc, Canada) was inserted into the muscle midbelly to evaluate intramuscular pressure at a sample rate of 125 Hz.

Since skeletal muscle tissue contains roughly 75% fluid, the material was modeled as poroelastic, (Equations 1, where the total model stress $\boldsymbol{\sigma}^{total}$ is the sum of the stresses in the solid and fluid pressure p multiplied by the identity matrix $\boldsymbol{I}$) with an incompressible fluid phase and hyperelastic solid phase (Equation 2, where a_2, b_2, c_2, and d_2 are material properties, $\boldsymbol{I}_1$ and $\boldsymbol{I}_2$ are the first and second invariants of the right Cauchy-Green deformation tensor, J is the volume ratio, and $\boldsymbol{I}_4$ is a pseudo-invariant for transverse isotropy). Additionally, as high collagen content tissues exhibit viscoelastic behavior [3] and the collagen rich extracellular matrix of muscle plays a key role in passive force transmission [4], the solid phase of muscle included a three term Prony series (Equation 3, where g_R is the shear relaxation modulus, g_n are fractions of that modulus, and τ_n are time constants). Equation 3 was implemented for shear (g_n) and bulk (where k_n replaces the g_n terms) behavior, which control the stress and pressure relaxation responses, respectively. Thus a total of six relaxation moduli (three shear, three bulk) and six time constants (three shear, three bulk) were used.

$$\boldsymbol{\sigma}^{total} = \boldsymbol{\sigma}^{solid} + p\boldsymbol{I} \tag{1}$$

$$\Psi = a_2(\bar{\boldsymbol{I}}_1 - 3)^2 + b_2(\bar{\boldsymbol{I}}_2 - 3)^2 + c_2(J-1)^2 + d_2(\boldsymbol{I}_4 - 1)^2 \tag{2}$$

$$g_R(t) = 1 - \sum_{n=1}^{3} g_n \left[1 - exp\left(-\frac{t}{\tau_n}\right)\right] \tag{3}$$

A finite element model of the NZWR tibialis anterior was generated using a commercial package (Abaqus, Dassault Systèmes, France). Mesh generation software (TrueGrid, XYZ Scientific Applications, Inc, Livermore, CA) was utilized to manually develop a three dimensional, hexahedral mesh. The model was distally displaced to match experimental strain and strain rate, while keeping the proximal end pinned. Equation 2 was incorporated using a custom user subroutine. Aponeuroses were modeled as isotropic shell elements with a tensile and shear moduli of 200 and 50 MPa, respectively, and negligible bending stiffness. The distal tendon was modeled as a linear spring with a stiffness of 1.3 MN/m. Mean model cross section was calculated similar to experimental specimens and used with reaction force at the distal tendon to calculate engineering stress.

Permeability was fixed at 1E-9 m4/N-s with the assumption that muscle permeability is significantly higher that other orthopedic tissues, such as cartilage, which rely on saturating fluid to resist deformation. No fluid was allowed to pass across the muscle boundary and the tissue was assumed to be fully saturated. IMP was interpreted as averaged hydrostatic pressure of four midbelly elements [5].

Two separate non-linear least squares optimization procedures were employed to fit experimental data using the "lsqnonlin" function in Matlab (The Mathworks Inc, Natick, MA). First, a linear Prony series (Equation 3) approximation was applied to stress and IMP relaxation data separately (12 parameters, Table 1). These terms were then fixed relative to one another for a coupled finite element-least squares optimization. The overall stiffness, compressibility, and viscoelastic properties were then varied to match stress and IMP data simultaneously (Figure 1 and Table 1). A total of six parameters were optimized for this portion of the analysis. The "goodnessOfFit" function in Matlab was utilized to evaluate the model fit (-∞ to 1) to experimental data with normalized mean square error (NMSE).

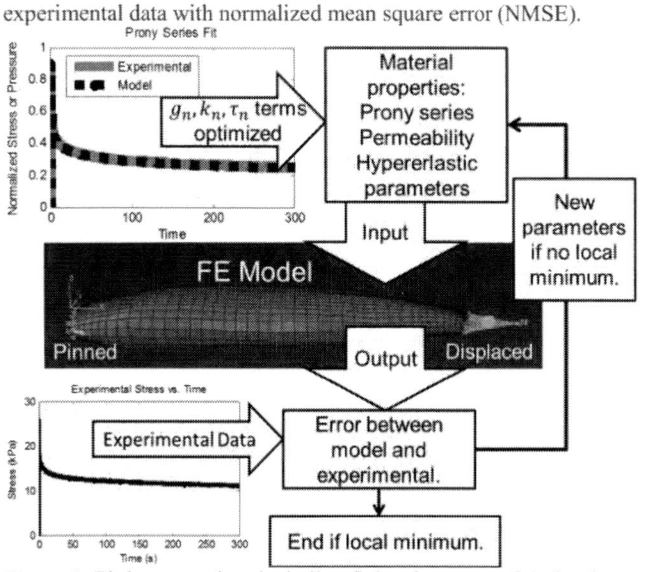

Figure 1: Fitting procedure including finite element model visual.

Table 1: Model parameters. Hyperelastic parameters a_2, b_2, and d_2 were coupled per previous experimental work [6].

Model Component	Parameter	First Optimization	Second Optimization	Units
SED Function	a_2*	N/A	Stiffness (a_2, b_2, d_2 parameters) and compressibility (c_2)	kPa
	b_2*			
	c_2			
	d_2*			
Prony Series	g_n	N=3 optimized	Four parameters: all g_n, all k_n, shear τ_n, bulk τ_n	Unit-less
	k_n	N=3 optimized		
	τ_n	N=6 optimized		
Permeability	k	Fixed at 1E-9		m^4/N-s

RESULTS

The model showed excellent agreement to experimental data in stress and pressure responses (Figure 2). The mean peak error between model and experiment was 0.018% and the mean fit was 0.77.

DISCUSSION

This model tackles the challenging problem of capturing both the stress and IMP response of passive skeletal muscle under a range of strain rates. To the author's knowledge, this is the first time a model of skeletal muscle has accurately captured the time dependent behavior of IMP. Furthermore, while previous modeling efforts have investigated IMP in a 2D geometry [5], this novel implementation modeled the whole tissue. This allows for the expansion into a more accurate *in vivo* environment with multiple muscles, bone, and skin to investigate how these surroundings affect IMP. Furthermore, this whole muscle model is able to investigate the specific relationship between muscle force, which is a global output, and local IMP, while sectioned models rely on local stress outputs.

The use of two separate optimizations to fit experimental data is advantageous by reducing the total number of parameters to be optimized when coupling a finite element model, improving whole optimization run time. As the relaxation behavior is largely driven by the viscoelastic Prony parameters, the general shape of the relaxation curve can be individually fitted before implementing a finite element analysis. This proved to be not only a faster approach, but yielded an excellent fit to experimental data (Figure 2).

While this work proves promising, the expansion to various strains and loading conditions would greatly improve implementation. However, the clear robustness, as evident by the fit across strain rates differing by nearly two orders of magnitude, suggests this model may be effective in doing so.

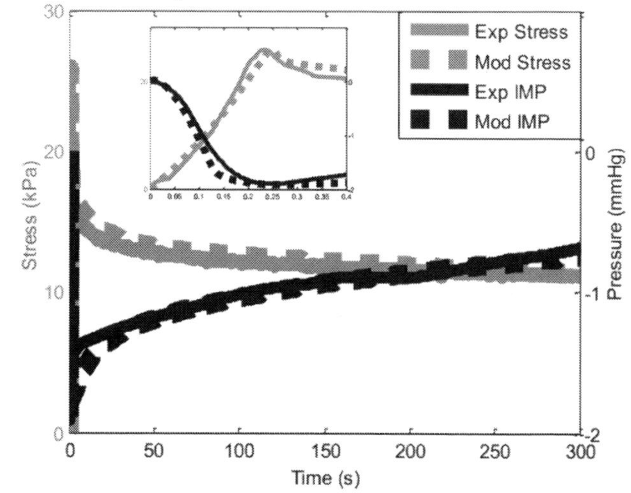

Figure 2: Model agreement to experimental stress and pressure relaxation data at a 20%/s strain rate with loading response inset.

ACKNOWLEDGEMENTS

This work was funded by the National Institutes of Health: National Institute of Child Health & Human Development [R01HD31476-12].

REFERENCES

[1] Fridén, J et al., *J Hand Surg*, 23:105-10, 1998
[2] Davis, J et al., *J Boimech*, 36:505-12, 2003
[3] Li, L et al., *Med Eng Phys*, 27:51-7, 2005
[4] Meyer, G and Lieber, R, *J Biomech*, 44:771-3, 2011
[5] Jenkyn, T et al., *PhysMed Biol*, 47:4043-61, 2002
[6] Morrow, D et al., *J Mech Behav Biomed*, 3: 124-9, 2010
[7] Wheatley, B et al., World Congress of Biomechanics, 2014, Poster

SB[3]C2015
Summer Biomechanics, Bioengineering and Biotransport Conference
June 17-20, 2015,Snowbird Resort, Utah, USA
SB³C2015-251

DEVELOPMENT OF SHOE MIDSOLE DESIGN USING MATERIAL NATURAL FREQUENCIES

Brett D. Steineman (1), Ted Barber (2), Tammy L. Haut-Donahue (1, 3)

(1) School of Biomedical Engineering
Colorado State University
Fort Collins, CO 80523
United States

(2) Pearl Izumi, Inc.
Louisville, CO 80027
United States

(3) Department of Mechanical Engineering
Colorado State University
Fort Collins, CO 80523
United States

NTRODUCTION

Shoe manufacturers have attempted to develop midsoles that return energy during the propulsion phase to enhance athletic performance. High performance runners talk about some shoes feeling responsive which could be the effect of efficient energy return. A finite element model was developed to measure the net energy return of midsole materials across the entire step however there are no tests or studies that evaluate the material during strictly the propulsion phase [1].

When mechanical systems, such as viscoelastic materials commonly used in shoe midsoles, are subjected to vibrations matching the natural frequency, a mechanical resonance is reached. In an under-damped system, the force transmitted through the material is amplified with respect to the input force as the material reaches mechanical resonance. If the natural frequency of a shoe midsole could be better matched with the ground reaction force frequency of walking or running, less energy would be lost through damping and the transmitted force through the material could be amplified to improve athletic performance.

In this study we utilized the frequency-domain concept of force transmissibility to determine the natural frequencies of midsole materials currently used in athletic shoes and compare them to the propulsion-phase frequencies found in human walking and running. Our hypothesis was that the frequency response could be used by shoe manufacturers to develop midsole materials that optimize the efficiency of walking and running propulsion phases.

METHODS

Pearl Izumi, Inc. provided a total of 42 sheets of midsole materials currently used in their shoes with seven different materials and six sheets of each. Half of the sheets for each material were 10 mm thick and the other half were 5 mm thick. All sheets had a surface area of approximately 200 mm by 200 mm.

The sheets were subjected to uniaxial sinusoidal compression tests (Bionic Model 370.02 MTS Corp., Eden Prairie, MN). The acrylic sphere (50 mm dia.) was lowered until contact was made with the sheet (0.1 N), and then the sheets were compressed to 10% of the original thickness before subjecting it to the sinusoidal displacement cycling between 10% thickness and 60% thickness for 100 cycles. The displacement of 60% thickness was determined after testing different values and choosing the maximum force amplitudes with minimal plastic deformation of the material. Each sheet was tested at a sinusoidal frequency of 1 Hz and tested again (in a new location on the sheet) in 0.25 Hz increments up to 5 Hz to represent the walking to sprinting propulsion-phase frequencies [2, 3]. The input and transmitted force responses with respect to time were recorded from the load cells with a sampling frequency of 101.4 Hz.

Before processing the raw data, the initial 50 cycles were filtered out to avoid the transient response from the viscoelastic material. A MATLAB code was generated to perform fast Fourier transforms of the input and transmitted force responses, identify the force amplitudes at the driven frequency, and use the force amplitudes to calculate the force transmissibility

$$T = \left|\frac{F_t}{F_i}\right| \qquad (1)$$

where F_t is the amplitude at the driven frequency of the force transmitted through the material and F_i is the amplitude at the driven frequency of the input force [4].

A nonlinear least-squares optimization technique was used to fit the mean force transmissibility points of each material to determine an approximate natural frequency value using the force transmissibility.

The equation of motion for the viscoelastic materials was used to derive an equation for force transmissibility with respect to the natural frequency

$$T = \sqrt{\frac{1 + \left(2\zeta\frac{f_d}{f_n}\right)^2}{\left[1-\left(\frac{f_d}{f_n}\right)^2\right]^2 + \left(2\zeta\frac{f_d}{f_n}\right)^2}} \quad (2)$$

where ζ is the damping ratio, f_d is the driving frequency, and f_n is the natural frequency [4]. The coefficient of determination was used after optimizing the experimental data to determine the quality of fit of the model.

ANOVA was used to assess differences in force transmissibility between the seven different materials for each driving frequency tested at a significance level of $p < 0.05$.

RESULTS

The mean force transmissibility values for each material in the range of human propulsion-phase frequencies were compared (Figure 1). A small amplification region was found for each material within the tested range.

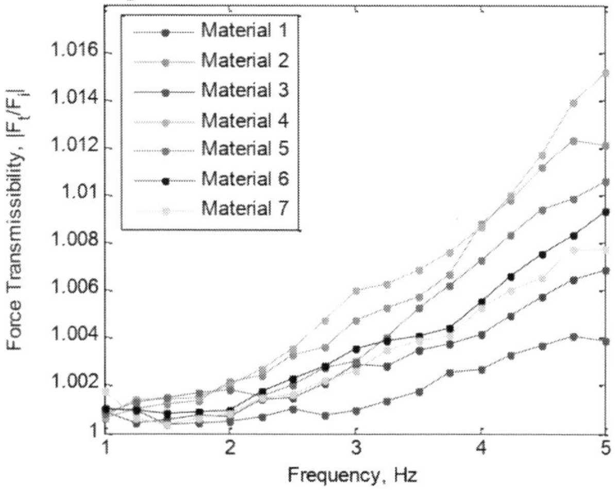

Figure 1 - Mean force transmissibility for all provided materials with respect to the human propulsion-phase frequencies tested.

The nonlinear least-squares optimization technique used with the force transmissibility equation converged on an approximate value for the natural frequency associated with each material (Table 1). Adjusting the initial values before optimization resulted in consistent natural frequencies but convergence for the damping ratio varied. After optimization, it appeared there was a small increase of each force transmissibility value that was independent of the system. This independent error was optimized for each material. The coefficient of determinations suggest a high quality of fit between theory and the experimental data. Also, the coefficients of determination remained unchanged or only varied slightly with the inconsistent convergence of damping ratio.

Table 1 – Nonlinear least-squares optimization results of the experimental force transmissibility values for each tested material.

	Natural Frequency (Hz)	Constant, k	R^2
Material 1	60.3318	0.0161	0.9870
Material 2	44.1907	0.0175	0.9887
Material 3	77.6192	0.0144	0.9108
Material 4	41.5716	0.0144	0.9903
Material 5	48.1084	0.0128	0.9789
Material 6	52.8698	0.0154	0.9856
Material 7	55.9351	0.0140	0.9664

Despite minor amplifications of the transmitted forces through all material, ANOVA showed significant differences between some of the materials tested at the higher propulsion-phase frequencies (Figure 2). ANOVA showed that none of the materials were significantly different between the frequencies of 1 Hz and 2.5 Hz.

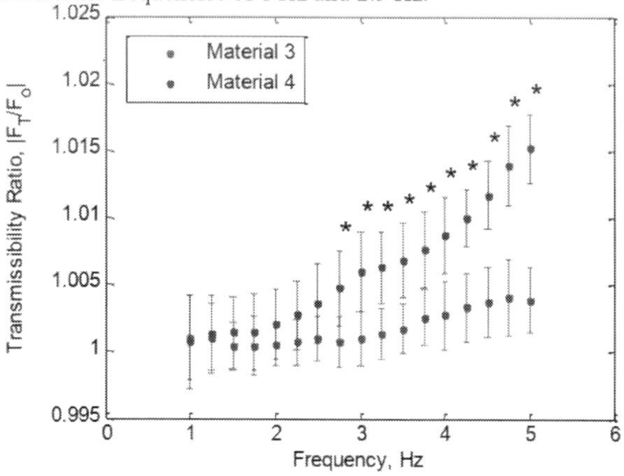

Figure 2 - Material 3 and Material 4 were significantly different at the most frequencies; * denotes significant difference between materials

DISCUSSION

Optimization of the experimental force transmissibility data resulted in successful convergence of natural frequency values for each material. The frequency response of interest can be used to compare materials. This test could be used in the future to develop new materials based on the frequency response of interest and to easily compare them.

The results of this study are similar to another study using finite element modeling with current running shoe cushioning systems that suggests differences in well-designed shoes are small [1]. The materials tested in this study show only a small difference in the transmitted force with an increase in driven frequency.

Despite the small differences in amplification, some materials were found to be significantly different than others at the higher propulsion-phase frequencies. The results of this study could be used with market research of the tested materials to determine if there are correlations between the significant difference in materials for running propulsion-phase frequencies and the preference of runners.

ACKNOWLEDGEMENTS

This work was supported by the funding of Pearl Izumi, Inc.

REFERENCES

[1] Shorten MR, *J Biomechanics*, 26:41-51, 1993.
[2] Nilsson J et al., *Acta Physiol Scand*, 136:217-227, 1989.
[3] Weyand PG et al., *J Appl Physiol*, 89:1991-1999, 2000.
[4] Ogata K, *System Dynamics*, 1978.

BEGINNING BRAILLE LEARNING DEVICE

Jason Castillo
University of Utah
Salt Lake City, Utah, United States

Adam Daly
University of Utah
Salt Lake City, Utah, United States

Austin Eastman
University of Utah
Salt Lake City, Utah, United States

Kay Beau Freckleton
University of Utah
Salt Lake City, Utah, United States

Kelton Gubler
University of Utah
Salt Lake City, Utah, United States

Andrew Silotti
University of Utah
Salt Lake City, Utah, United States

Faculty Advisor(s)

Andrew S. Merryweather, PhD
Project Advisor
Salt Lake City, Utah, United States

INTRODUCTION

There is a lack in available technologies to assist people trying to learn braille. The number of people with vision loss and impairment who could benefit from a system to teach beginning braille is significant. According to NBC news, 56% of people with impaired vision who can read braille are employed while only 23% of people with impaired vision that cannot read braille are employed[1]. Currently there are few devices on the market that do not require supervision and help from a person with sight to facilitate learning. There are currently not enough teachers and resources to sufficiently supply the needs for visually impaired students. Most available products are thousands of dollars and do not have the technological ability to self-teach. There are some devices that contain audio input, audio output, and a

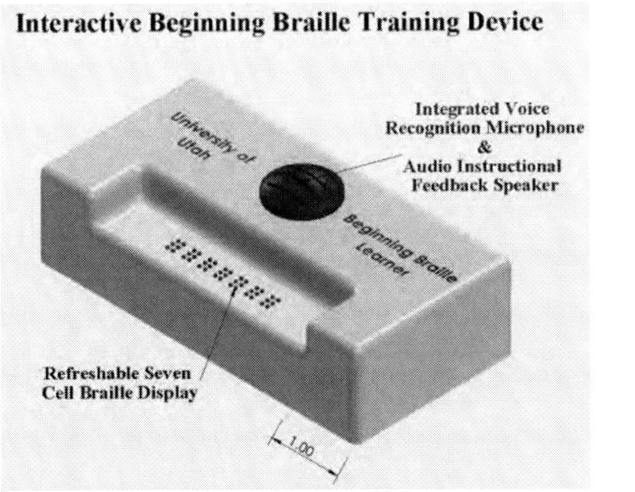

Figure 1 Descriptive Title

refreshable braille screen, but there is no device on the market that incorporates all of these technologies into one. These deficiencies are directly related to increased costs and limited use for these devices. The proposed project is a beginning braille learning device to enable a person with impaired vision to learn how to read braille on their own. The device (illustrated in **FIGURE 1**) will provide audio feedback, sensory feedback through a refreshable screen that displays braille characters, and basic learning modules to teach an individual how to feel and interpret braille characters. The braille learning device will not require a teacher or additional supervision to facilitate learning. The device will also be designed to be handheld and portable.

PRODUCT DESIGN

The most important design aspect of the device is the ability to refresh the haptic display based on user inputs during near real-time user interaction. The device's refreshable display must be able to correctly display the braille characters based on user inputs or else the user will not be able to interact successfully with the device. Along with the refreshable screen, the voice recognition software must be able to identify the specified word. These two metrics are directly related because the audio input will determine the braille characters display, based on the voice recognition and text to braille conversion software. The design of this device contains three main subsystems, the audio input, audio output, and the refreshable braille display. The final product will incorporate these three subsystems working together to interact with the user.

The main customer for this product is someone who is visually impaired and is trying to learn how to read braille. However, this project is not limited to those who are visually impaired, rather anyone trying to learn to read braille. This device will also be useful for teachers and schools for the visually impaired. Each of these customers contributes a different need. These needs have been determined through online research and have been verified by communicating with students visually impaired and their teachers at the University's Center for Disabilities.

BUDGET & MARKET ANALYSIS

The total projected budget for the Beginning Braille Training Device is $1,232.41 as seen in Table 1.This represents the total cost to fabricate the prototype.

Table 1: Expense Report

Expense Report	
Hardware	$799.82
Software	$261.71
Miscellaneous	$170.88
Total	**$1,232.41**

There are currently 60,393 blind students registered in the United States[2]. Learning devices exist that have an audio feedback system, or a refreshable display but there are no affordable devices that incorporate both systems. One of the major goals of this projects was to develop a system that would be affordable to more users. An advantage of this device over others is that it is designed to require virtually no supervision. The controls are designed for tactile recognition, and the entire system can be voice controlled. The need for constant teacher supervision is greatly reduced. This reduction makes our device a perfect candidate for classroom teaching, as well as independent learning. Additional users include parents, grandparents, and employers of people who are blind. People who are losing their vision due to age or medical reasons are also target users. Because of this, the estimated market size in the United States alone is in excess of 120,000. At this market size the projected cost of manufacturing would be greatly reduced from our prototyping budget. A single device will cost $450, after assuming a 40 percent reduction in manufacturing cost and improved methods to design for manufacturability. Each device will be sold at a price of $600. Because a large component of the device is already produced by an outside manufacturer (braille cell unit) the need for tooling, and engineering investment will be greatly reduced. The single device return will be $150, but for an overall return we expect to break even after 5000 units so all units after are pure profit.

****Note**** additional information and data from the project will be included in the final abstract submission.

ACKNOWLEDGEMENTS

This work was supported in part by the Selective Innovative Technology and Engineering (SITE): Senior Design Projects Grant funded by NSF under Grant 1159885 and the University of Utah Senior Design Program.

REFERENCES

[1]"Fewer Blind Americans Learning to Use Braille." Msnbc.com. NBC News, 26 Mar. 2009. Web. 08 Sept. 2014. <http://www.nbcnews.com/id/29882719/ns/us_news-life/t/fewer-blind-americans-learning-use-braille/#.VA3s8_ldWLk>

[2]"Blindness Statistics." National Federation of the Blind. 6 Aug. 2014. Web. 14 Jan. 2015. <https://nfb.org/blindness-statistics>.

SB³C2015
Summer Biomechanics, Bioengineering and Biotransport Conference
June 17-20, 2015,Snowbird Resort, Utah, USA
SB³C2015-253

IN SITU POLYMERIZATION OF THIOL-ACRYLATE NANOCOMPOSITE FOAM FOR BONE DEFECTS

Anoosha Forghani[1], Leah Garber[2], Cong Chen[3], John Pojman[3], Dan Hayes[3], and Ram Devireddy[1]

[1]Department of Mechanical Engineering, Louisiana State University, Baton Rouge, Louisiana, United States

[2]Department of Chemistry, Louisiana State University, Baton Rouge, Louisiana, United States

[3]Department of Biological and Agricultural Engineering, Louisiana State University, Baton Rouge, Louisiana, United States

INTRODUCTION

Tissue engineering has attracted attentions due to its advantages over traditional methods such as allografting and autographing. Using patients own cells eliminate the risk of disease transfer and infections [1]. Recently, in situ generated scaffolds have received significant attention in biomedical applications due to their advantages over ex vivo-fabricated scaffolds. For example, injectable scaffolds do not need invasive surgery and they also bond to the surrounding tissue during the hardening process which leads to a better fixation. They also have the ability to fill defects with irregular size and shape [2]. On the other hand, unlike cell seeding process, introducing the stem cells into the scaffold precursor leads to a homogenous cell distribution and tissue formation which can be reached by gas foaming method [3]. Foaming fabrication method makes "polymerizing while encapsulating cells simultaneously" possible and introduces gaseous bubbles during injection which forms interconnected pores into the scaffolds.

Thiol-ene chemistry contributes to the properties of an ideal scaffold that includes suitable mechanical properties, an adequate sized porous structure and interconnectivity [4]. Pentaerythritol triacrylate-co-trimethylolpropane (PETA) scaffolds, which can be fabricated through thiol-ene chemistry and gas foaming, displayed its potential for bone augments and grafts [4]. However, delivering cells along with PETA polymerization remains a challenge due to the toxic monomers and excess heat generated during the reaction. Cell friendly processes for fabricating scaffolds via photopolymerization of PEG-based hydrogel has already been reported for bone tissue engineering applications [5]. The alternative method presents a base catalyzed Michael addition reaction which proceeds by a step-growth polymerization of the thiol and PEG acrylate groups eliminating the danger of unreacted monomers and radicals compared to photopolymerization.

The purpose of this study is to develop a novel, cell-friendly fabrication process of bone scaffolds through in situ polymerization of PEGDA and TMPTMP (10% HA) composite polymer. Initial characterization studies were performed to determine the temperature profile during the exothermic reaction indicating a maximum temperature of 50°C. To prevent hyperthermic cell damage and death during the exothermic polymerization procedure, the ASCs were encapsulated in Alginate particles that are widely studied in tissue engineering applications as a cytocompatible carrier material [6].

METHODS

Alginate 1.2% by weight was dissolved in standard stromal media solution and mixed with, 3×10^6 ASCs at passage 2. In order to make Alginate beads, the mixture was extruded into a Calcium chloride-Stromal media solution using an electrical syringe set at 2 ml/min .The 50% hASCs-Alginate beads were combined with 50% of the extruded polymeric foam to complete the polymerization process. The foam was prepared by adding Trimethylolpropane tris(3-mercaptopropianate) (TMPTMP) to Poly(ethylene glycol) diacrylate (PEGDA) in the presence of Diethylamine (DEA) as the catalyst . HA (10% wt/wt) was added to the precursor monomers and poured into a 250 mL pressurized spray canister using 7g-compressed nitrous oxide as a gaseous porogen. Physical Characterization of the composite foam was performed using FIB-SEM and Micro-CT in order to measure pore size and the interconnectivity of the pores. Mechanical strength including compressive strength was measured using instron testing at 90% strain. Live/dead staining was performed at days 7,14,21 to assess hASCs attachment and viability inside the scaffolds.

RESULTS

Physical Characterizations: The morphology of PEGDA:HA composite was analyzed using FIB-SEM (Fig 1)

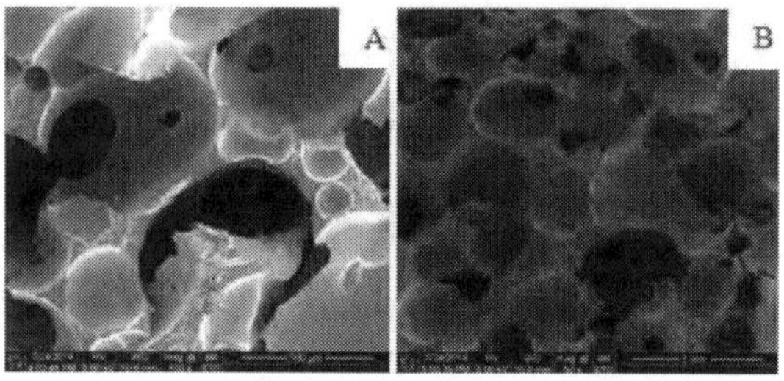

Fig1. FIB-SEM images of PEGDA TMPTMP 10%HA composite in two different magnifications

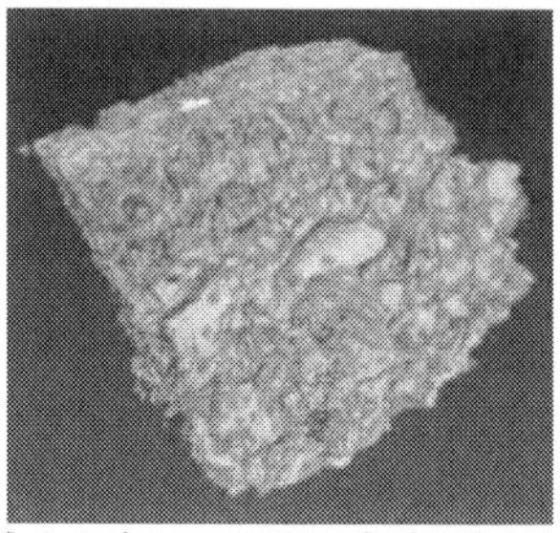

Fig2. Micro- CT Rendering image

Live/Dead Viability Assay: A fluorescence microscope (Confocal Microscopy) was used to observe the cells. Figure (3A-5A) show viability of the cells entrapped inside the Alginate beads as positive control, while the viability of the cells inside the PEGDA polymer have been shown in Figure (3B-5B).

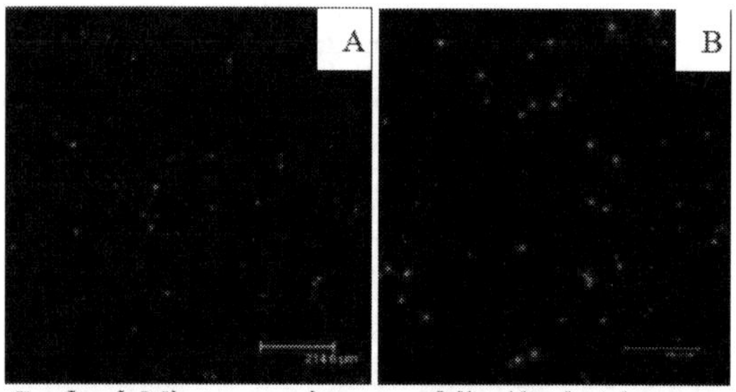

Fig 3. Confocal Microscopy images of live/dead stain assay. A&B Positive control and PEGDA polymer respectively after 7 days

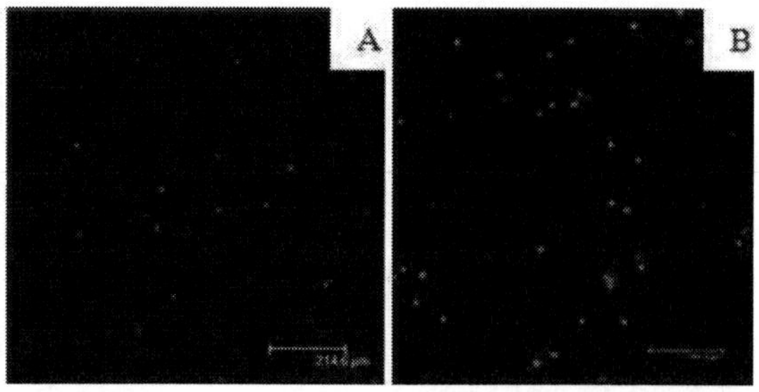

Fig 4. Confocal Microscopy images of live/dead stain assay. A&B Positive control and PEGDA polymer respectively after 14 days

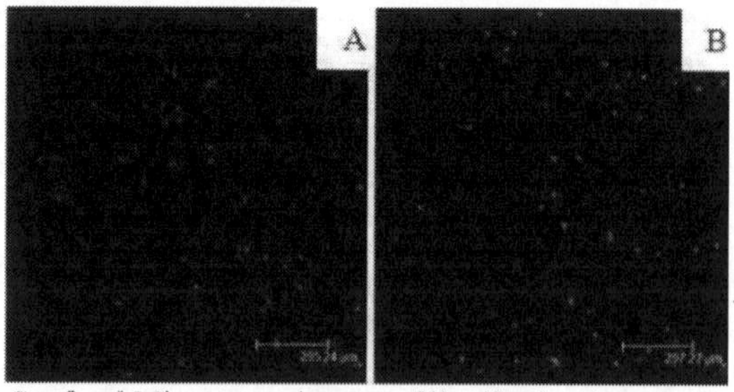

Fig 5. Confocal Microscopy images of live/dead stain assay. A&B Positive control and PEGDA polymer respectively after 21 days

DISCUSSION

SEM images of PEGDA composite (Fig1) show the pore size of PEGDA in the range of 500-600μm which is within an ideal bone scaffold pore size ranging from 100 to 500 μm [7]. Large surface tension of PEGDA-co-TMPTMP caused larger pore sizes in PEGDA:HA scaffolds during the foaming process. Micro CT analysis of the samples demonstrates interconnectivity of the polymer which is necessary for cells, nutrients and blood to migrate through the scaffold (Fig 2). Mechanical properties play an important role in biocompatible materials for tissue engineering. Despite having good biocompatibility, injectable hydrogels synthesized using Aliginate exhibit poor mechanical strength [8]. Alginate scaffolds have compressive strength and of 1.5 MPa which is less than that of the PEGDA-co-TMPTMP (9.5±0.4 MPa). Fig 3-5 show that PEGDA-co-TMPTMP composite has an increased number of live cells indicated by green specks, and less dead cells shown with red dots. Wang et al reported the viability of encapsulated cells with oligo(poly(ethylene glycol) fumarate) ,OPF, hydrogel for up to 16 days while encapsulated cells in the PEGDA polymer survived 21 days which shows a better biocompatibility of PEGDA co TMPTMTP polymer [3]. Having adequate mechanical properties and good biocompatibility, PEGDA polymer is a good candidate as an injectable scaffold for bone tissue engineering applications.

ACKNOWLEDGEMENTS

The current work is funded by the NSF EPSCoR LA-SiGMA project under award #EPS-1003897.

REFERENCES

[1] Shrivates, A et al., *Tissue therapeutics and regenerative medicine,*19: 781–786, 2014.

[2] Gariepy, E et al., *European Journal of Pharmaceutics and Biopharmaceutics*, 58: 409–426, 2004.

[3] Wang, L et al., *Tissue Engineering part C,* (doi: 10.1089/ten.TEC.2014.0224.

[4] Garber, L et al., *Journal of Biomedical Materials Research Part* 101:3531-41. 2013

[5] Han, LH et al., *Advanced Materials*, 26:1757-62, 2014.

[6] Awad, HA et al., *Biomaterials*, 25:3211-22. 2004.

[7] Yoshito, I et al.,. *Biological Abstracts. Present. Web. 28 Aug. 2014.*

[8] Bharatham, B et al, *Biomed Research International,* Web. 27 Aug. 2014.

SB³C2015
Summer Biomechanics, Bioengineering and Biotransport Conference
June 17-20, 2015, Snowbird Resort, Utah, USA
SB³C2015-256

A COMPUTATIONAL STUDY OF RED BLOOD CELLS IN STOKES FLOW: FROM SINGLE CELLULAR MECHANICS TO SUSPENSION RHEOLOGY

Toshihiro Omori (1), Yohsuke Imai (1), Takami Yamaguchi (2), Takuji Ishikawa (1,2)

(1) Dept. Bioengineering and Robotics
Tohoku University
Sendai, Miyagi, Japan

(2) Dept. Biomedical Engineering
Tohoku University
Sendai, Miyagi, Japan

INTRODUCTION

Blood is a dense suspension of red blood cells (RBCs), and its main function is mass transport to the body. Transport phenomena of blood flow are thus important in discussing the physiology of the cardiovascular system. To understand the mass transport of blood flow, we must clarify detailed hemodynamics, as well as flow-induced chaotic motions of RBCs within blood vessels, because flow field is strongly dependent on the microscopic RBC behavior.

In former studies, hemodynamics of large arteries and microcirculation have been investigated by many groups. However, there are still few studies that connect cellular- and continuum-scale hemodynamics. The vascular system is ranged from micro-meter scale capillaries to centi-meter scale arteries, and we need to develop a mechanical blood model, which can be applied to arbitrary scale physics of blood flow.

Objective of this study is development of a new continuum blood model, which is based on cellular-scale RBC behaviors. Especially, in this paper, we will clarify relationship between kinematic cellular motions and macroscopic rheological properties of an RBC suspension.

METHODS

We first explain the governing equations and numerical methods of fluid mechanics and membrane mechanics.

Assume that RBCs are filled with a Newtonian liquid and are freely suspended in another Newtonian liquid. Due to the small size of RBCs, the inertia effect of external and internal fluid motions can be neglected. The fluid motion is then governed by the Stokes equation, and the velocity field $\boldsymbol{v}$ is given by the boundary integral equation over the deformed membrane:

$$\boldsymbol{v} = \boldsymbol{v}^{\infty} - \sum_{i=1}^{N}\left(\frac{1}{8\pi\mu_{out}}\int \boldsymbol{J}\cdot\boldsymbol{q}\,dA_i + \frac{1-\lambda}{8\pi}\int \boldsymbol{v}\cdot\boldsymbol{T}\cdot\boldsymbol{n}\,dA_i\right), \quad (1)$$

where $\boldsymbol{q}$ is the surface traction, $\boldsymbol{v}^{\infty}$ is the background flow, N is the number of cells, $\mu_{\rm out}$ is the viscosity of the external liquid, λ is the viscosity ratio between internal cytoplasm and external liquids, and $\boldsymbol{J}$, and $\boldsymbol{T}$ are the single- and double-layer potentials of the Green function, respectively. Equation (1) is solved by a boundary element method [1]. We also used the Ewald summation method to set periodic boundary conditions.

In order to connect microscopic particle behaviors and macroscopic rheological properties of a suspension, Batchelor [2] introduced an additional stress term in the viscous stress of a suspension:

$$\sigma = -p\boldsymbol{I} + 2\mu_{out}\boldsymbol{e} + \boldsymbol{\Sigma}^{(p)}, \quad (2)$$

where $\boldsymbol{\sigma}$ is the viscous stress of the suspension, p is the pressure, $\boldsymbol{e}$ is the rate of strain, and $\boldsymbol{\Sigma}^{(p)}$ is the particle stress tensor, which represents the additional bulk stress of the suspension due to the presence of cells. When force-free particles are immersed in a suspension, the particle stress tensor is given by

$$\boldsymbol{\Sigma}^{(p)} = \frac{1}{V}\sum_{i=1}^{N}\boldsymbol{S}_i, \quad (3)$$

where V is the bulk volume, $\boldsymbol{S}_i$ is the stresslet on the i-th RBC.

Next, we derive the membrane mechanics. Since the membrane thickness is much smaller than the cell size, the membrane can be assumed as a two-dimensional material. In this study, the membrane is modeled as a 2D isotropic hyper-elastic material with finite bending resistance. The in-plane elastic tensions $\boldsymbol{\tau}$ is given by

$$\boldsymbol{\tau} = \frac{1}{J_s}\boldsymbol{F}_s \cdot \frac{\partial w}{\partial \boldsymbol{E}} \cdot \boldsymbol{F}_s^t, \tag{4}$$

where J_s is the Jacobian, $\boldsymbol{F}_s$ is the surface deformation gradient tensor, $\boldsymbol{E}$ is the Green-Lagrange strain, and w is an elastic strain energy function. To model local area incompressibility of an RBC membrane, we used the Skalak's constitutive law [3]. Neglecting inertia effects on the membrane deformation, the equilibrium equation of the membrane in the weak form is given by

$$\int \hat{\boldsymbol{u}} \cdot \boldsymbol{q}_p \, dA = \int \hat{\boldsymbol{\varepsilon}} : \boldsymbol{\tau} \, dA, \tag{5}$$

where $\hat{\boldsymbol{u}}$ and $\hat{\boldsymbol{\varepsilon}}$ are virtual displacement and strain, respectively. Eq.(5) is solved by a finite element method with respect to $\boldsymbol{q}_p$. For the bending rigidity, we employed the Helfrich bending energy function [4] and the bending force is calculated from the following equation:

$$\boldsymbol{q}_b = E_b\left[(2H + c_0)(2H^2 - 2K - c_0 H) + 2\Delta H\right]\boldsymbol{n}, \tag{6}$$

where E_b is the bending resistance, H is the mean curvature, K is the Gaussian curvature, c_0 is spontaneous curvature, and Δ is the Laplace-Beltrami operator. The surface traction is then calculated by

$$\boldsymbol{q} = \boldsymbol{q}_p + \boldsymbol{q}_b. \tag{7}$$

Once the velocity field is computed, the membrane material point $\boldsymbol{x}$ is updated by using kinematic condition $d\boldsymbol{x}/dt = \boldsymbol{v}$, which is solved by a second order explicit Runge-Kutta method.

RESULTS AND DISCUSSION

Typical results of an RBC in shear flow are shown in Figures 1(a) and (b). In this study, shear rate is normalized by the capillary number, which is defined as

$$Ca = \frac{\mu_{out}\dot{\gamma}a}{G_s}, \tag{8}$$

where a is radius of an RBC, G_s is membrane shear elastic modulus, and $\dot{\gamma}$ is shear rate. In moderate viscosity ratio regime ($\lambda \approx 1$), the RBC shows a tank-treading motion, as shown in Fig.1(a). While, in high viscosity ratio (cf. Fig.1b), the RBC shows a tumbling-like rotation with repeating stretch and compression during the rotation. In this regime, the RBC does not show a tank-treading motion even in higher Ca. This kind of motion is similar to a vacillating-breathing (VB) motion, which has been reported in former studies of vesicles. Due to the trembling VB motion in high λ, the stresslet acting on the membrane becomes large and the bulk stress tends to be higher than that of small or moderate viscosity ratio cases.

We next investigate the effect of cell-cell interactions on the macroscopic suspension rheology. Figure 2 shows a typical result of cell-cell interaction in shear flow. 50 RBCs (volume fraction $c = 0.05$) are immersed in the unit domain and capillary number and the viscosity ratio are set to Ca = 0.6 and $\lambda = 1$, respectively. In the dilute to semi-dilute regime, RBCs rarely interact to each other, and the cell-cell interactions are dominated by 2-body interactions. In Figure 3, apparent viscosity of the suspension with various volume fractions is shown. Broken line appeared in the figure represents result of the dilute-limit theory [2], which is given by

$$\Sigma^{(p)} = \frac{3c}{4\pi a^3}\boldsymbol{S}^{sol} + O\left(c^2\right), \tag{9}$$

where $\boldsymbol{S}^{sol}$ is the stresslet of solitary immersed RBC. From Fig.3, we see that the apparent viscosity increases linearly when the volume fraction is enough small and corresponds to the analytical prediction. When the volume fraction reaches $c = 0.1$, the cell-cell interaction effect becomes large, and the apparent viscosity seems to nonlinearly increase with respect to the volume fraction.

In more dense suspension, the cell-cell interaction effect is expected to be significant, we are now calculating the cellular flow with more large scale and high viscosity ratio conditions.

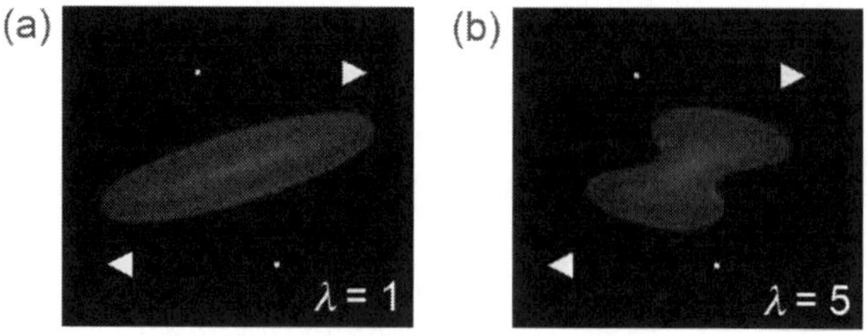

Fig.1 A red blood cell in shear flow with different viscosity ratio; (a) λ = 1, and (b) λ = 5. In both cases, capillary number is set as Ca = 0.6.

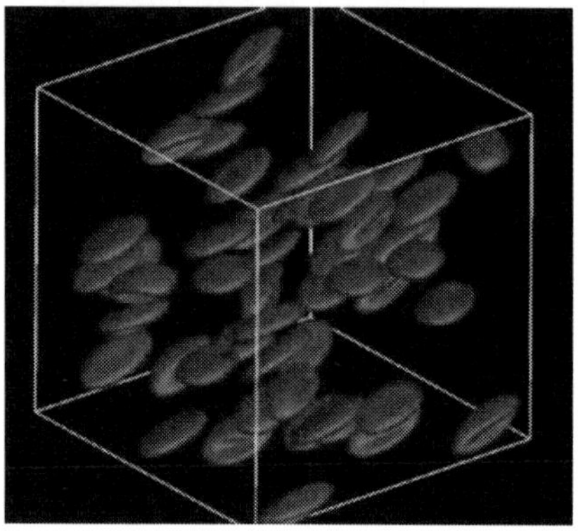

Fig.2 50 RBCs in shear flow (volume fraction c = 0.05, capillary number Ca = 0.6, the viscosity ratio λ = 1).

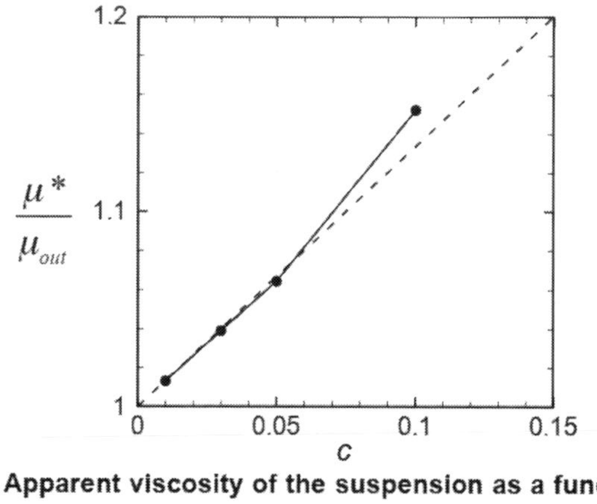

Fig.3 Apparent viscosity of the suspension as a function of the volume fraction c (Ca = 0.6, λ = 1). The broken line represents result of the dilute-limit theory.

REFERENCES

[1] Omori, T et al., *J Fluid Mech*, 724:154-174, 2014.
[2] Batchelor, G K, *J Fluid Mech*, 41: 545-570, 1970
[3] Skalak, R et al., *Biophys J*, 13: 245-264, 1973
[4] Zhong-Can & Helfrich, *Phys Rev A*, 39:5280, 1989

SB³C2015
Summer Biomechanics, Bioengineering and Biotransport Conference
June 17-20, 2015,Snowbird Resort, Utah, USA
SB³C2015-258

DESIGN CONSIDERATIONS FOR GLENOID COMPONENTS: A COMPUTATIONAL STRESS ANALYSIS OF TRANSLATIONAL MOTION IN SHOULDER REPLACEMENTS

C. Berthelet (1), F. Ansari, MS (1), T. Norris, MD (2), S. Gunther, MD (3), M. Ries, MD (4), L. Pruitt, PhD (1)

(1) Department of Mechanical Engineering
University of California, Berkeley
Berkeley, CA, USA

(2) San Francisco Shoulder, Elbow & Hand Clinic
San Francisco, CA, USA

(3) Department of Orthopedic Surgery
Martha Jefferson Hospital
Charlottesville, VA, USA

(4) Tahoe Fracture and Orthopedic Clinic
Carson City, NV, USA

INTRODUCTION

Total shoulder replacement (TSR) is the third leading reconstructive orthopedic procedure behind hip and knee replacements, and the fastest growing arthroplasty device in the market today [1]. TSRs have largely derived their design and material composition from well-characterized devices used in the hip and knee: the ball-and-socket design typically consists of a cobalt chrome (CoCr) humeral head that articulates against an ultrahigh molecular weight polyethylene (UHMWPE) glenoid. Despite similarities with other joint replacements, TSRs encounter complex rotational and translational joint kinematics that have led to highly variable reasons for failure, including implant loosening, joint instability, infection, implant wear and device fracture [2-4]. Polymer fracture along the glenoid rim is a particularly unique failure mode [5]; while shoulders adopt a nonconforming bearing system similar to that seen in the knee, excessive translational motion results in eccentric loading that can cause subsurface cracking and eventual fracture along the circumference (**Fig. 1**) [5]. Decreased device conformity and limited UHMWPE component thickness contribute to elevated contact stresses that lead to wear and fracture in TSRs [6]. Yet, the interplay of these design factors with the properties of modern UHMWPE formulations, or their combined influence on contact stresses, especially under translational motion often associated with glenoid rim fracture remains limited [7]. This study presents a computational assessment of contact stress in the UHMWPE glenoid as a function of device geometry, material properties, and translational motion.

METHODS

The glenoid contact stresses that develop due to translation of the humeral head were computationally investigated using MATLAB (v.R2013a Natick, Massachusetts). A ball-in-socket model of a shoulder prosthesis was examined, consisting of a cobalt chrome (CoCr) humeral head component indenting into an all-polyethylene glenoid component. A standard elasticity solution (SES) developed by Bartel et al. [6] was utilized in order to calculate contact stresses. This method assumes a quasi-static, frictionless system in which deformation of the elastic layer (UHMWPE) only occurs in the radial direction. Following Sweiszkowski et al. [8], the translation of the humeral head from the central axis was modeled as a rotation of the head about the origin of the glenoid. This displacement was considered as an angular rotation of the humeral head through a range of -36 to 36 degrees [8]. Maximum glenoid contact stress data was generated by the MATLAB code at each degree of angular translation.

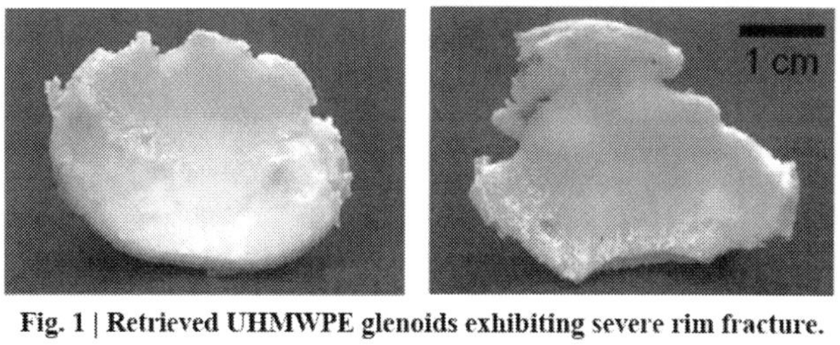

Fig. 1 | Retrieved UHMWPE glenoids exhibiting severe rim fracture.

In addition to testing the effects of humeral head translation on contact stresses in the glenoid component, the effects of material composition (elastic modulus), glenoid thickness, component conformity, and specific humeral-glenoid radii combinations were investigated. All variables were assessed under an axial load of 700 N. Baseline controls were set to 3 mm for humeral-glenoid conformity; 4 mm for glenoid thickness, and virgin UHMWPE (GUR 1020, $E = 386$ MPa, $\nu = 0.46$) for material composition. Effects of material composition were evaluated using a controlled thickness and conformity for virgin UHMWPE, moderately crosslinked and remelted UHMWPE (E = 413.9 MPa), highly crosslinked and remelted UHMWPE (E = 424.6 MPa), highly crosslinked and sequentially annealed UHMWPE (E = 536.8 MPa), and highly crosslinked and annealed UHMWPE (E = 548.2 MPa) [9]. Conformity effects were evaluated at 0.6, 3, 4.5, 5.5, 6.5, and 10 mm of radial mismatch. In addition, the influence of device size and glenoid thickness on contact stress for a controlled thickness and conformity was investigated; **Table 1** lists the values investigated.

RESULTS

For all variables evaluated, higher contact stresses were seen farther away from the central axis (i.e. greater translation). Contact stresses near the rim of the glenoid were on average 11.3% greater than those at the center of the component for all conformities, material compositions and thicknesses evaluated. As seen in **Fig. 2**, decreasing conformity (increasing radial mismatch) caused an overall increase in contact stress and enhanced the effect of humeral head translation. Contact stresses were greatest for highly crosslinked and sub melt annealed formulations (**Fig. 3**). Contact stress also increased as glenoid thickness decreased or as device size decreased (**Table 1**).

DISCUSSION

TSRs have been shown to suffer a unique array of damage modalities *in vivo* that span that seen in hips and knees [5]. Rim fracture (**Fig. 1**) demonstrates a particular severe damage type that results from excessive translational motion that can occur in the body. Our study presents a thorough evaluation of contact stresses as a result of translational motion using an established protocol for glenoid components [8]. Our finding that translation can result in an 11% increase in contact stress on average has clinical implications considering that neutral positioning results in stresses ranging from 12 to 40 MPa. Such elevated stresses can contribute to crack initiation that under cyclic loading can propagate to fracture, especially as crosslinked formulations have reduced fatigue resistance [9].

We show that such contact stresses are elevated with increasing translation, decreasing thickness, decreasing conformity and increasing modulus. Our findings regarding conformity and thickness are consistent with previous studies [6,8]. However, it should be noted that many glenoids are often well below the suggested 6 mm thickness limit suggested by Bartel [6]. Typical glenoid thicknesses of 3-4 mm results in contact stresses on par with UHMWPE yield limit (20 MPa). While low conformity dramatically increases contact stresses for a radial mismatch as low as 3 mm (**Fig. 2**), it should be recognized that increased conformity has the tradeoff of limiting shoulder motion and potentially enhancing bending stresses that lead to rocking horse motion [7]. Highly crosslinked and remelted formulations appeared to maintain stresses below yield (**Fig. 3**), which is consistent with their lower stiffness. However, higher crosslinking does come with a reduced resistance to fatigue crack propagation [9], which can increase the incidence of fracture under cyclic loading. Finally, radial mismatch cannot necessarily be isolated from the overall implant size (**Table 1**). Future work will incorporate a full-scale finite element model to investigate edge loading, as well as to validate this current analysis. Overall, our findings demonstrate that TSRs experience a complex stress state that subjects the UHMWPE to both yield and fracture, and that the assumption that UHMWPE glenoid components experience stresses similar to that seen in knees and hips, may not be accurate.

ACKNOWLEDGEMENTS

We would like to acknowledge Robin Parrish for her assistance with code development and analysis for this project.

REFERENCES

[1] Day, JS, et al., *J Shoulder Elbow Surg*, 19(8):1115–20, 2010.
[2] Bohsali, K., et al. *J Bone Joint Surg*, 88-*A*(10): 2279-2292, 2006.
[3] Wirth, MA., et al, *J Bone Joint Surgery Am.*, 78(4):603-16, 1996.
[4] Chin, PYK et al., *J Shoulder Elbow Surg.*, 15(1):19–22, 2006.
[5] Ansari, F, et al. *World Congress on Biomechanics*, June 2014
[6] Bartel, DL, et al. *J Bone Joint Surg Am.*, 68(7):1041–51, 1986.
[7] Armstrong, AD et al., *JBJS Reviews*, 1.2, 2013.
[8] Swieszkowski, W et al., *J Engr in Medicine*, 217:49-57, 2003.
[9] Atwood, SA et al., *J Mech Beh Biomaterials*, 4.7:1033-1045, 2011.

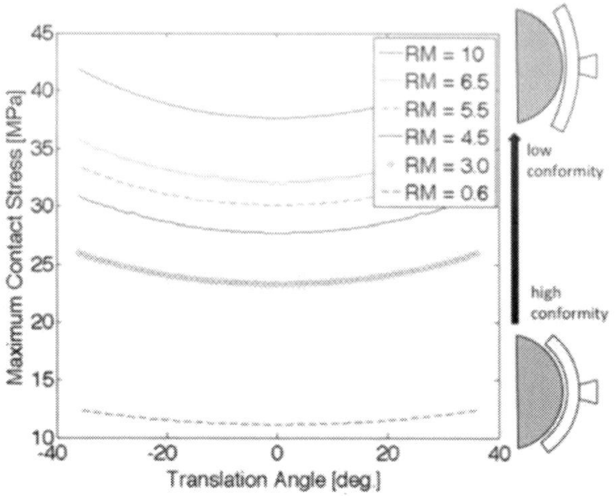

Fig. 2 | Contact stress as a function of translation and radial mismatch (RM). High RM is equivalent to low conformity.

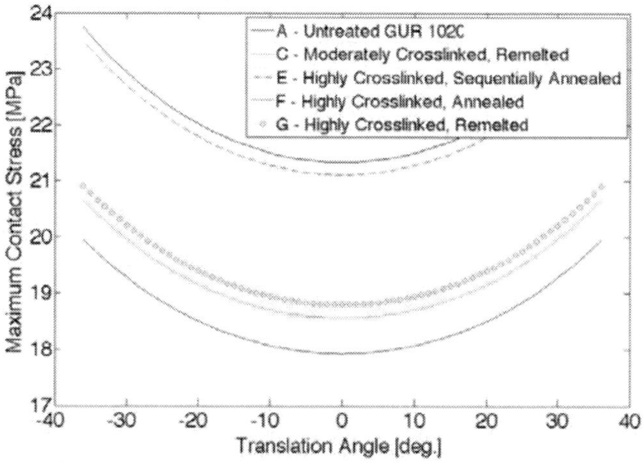

Fig. 3 | Contact stress as a function of translation and material composition. Conformity and thickness are controlled.

Table 1. Contact stress for varying glenoid thickness & device sizes (gray shading)

Humeral Head Radius [mm]	Glenoid Radius [mm]	Radial Mismatch [mm]	Thickness [mm]	Max Contact Stress at 0° [MPa]	Max Contact Stress at 36° [MPa]
28	31	3	3	21.4	23.2
28	31	3	4	18.4	20.0
28	31	3	5	16.3	17.7
28	31	3	6	14.8	16.0
28	31	3	7	13.6	14.7
28	31	3	4	18.4	20.0
26	29	3	4	19.7	21.4
24	27	3	4	21.2	23.0
22	25	3	4	23.0	24.9

SB³C2015
Summer Biomechanics, Bioengineering and Biotransport Conference
June 17-20, 2015,Snowbird Resort, Utah, USA
SB³C2015-259

SILICA-COLLAGEN HYDROGEL FOR CORNEAL REPLACEMENT

Michael D. DiVito (1), Stephen G. Rudisill (2), Andreas Stein (2), Sanjay V. Patel (3), Jay W. McLaren (3), & Allison Hubel (1)

(1) Dept. of Mechanical Engineering
University of Minnesota
Minneapolis, MN, USA

(2) Dept. of Chemistry
University of Minnesota
Minneapolis, MN, USA

(3) Dept. of Ophthalmology
Mayo Clinic
Rochester, MN, USA

INTRODUCTION

Corneal diseases are the second leading cause of blindness. In the United States, the cornea is the most commonly transplanted tissue, with over 40,000 transplants performed annually.[1] Due to factors such as the lack of proper eye banking facilities and the prevalence of refractive eye surgery which render tissue ineligible for donation, the supply of donor corneas has never met the demand.[2] Because of this reality, the long-term goal of this project is to develop an artificial cornea that would reduce the dependency on donor eye banking facilities.

While many groups have contributed to the artificial cornea field, there are no implants clinically available which truly integrate with the surrounding native tissue.[3,4] In addition to biocompatibility, the implant must match the cornea's biomechanical and optical properties. Silica-collagen hybrid materials have shown great potential to mimic typical corneal function. Therefore, this work utilizes a bottom-up approach where microstructural features of silica-collagen hydrogels are modulated to yield a material suitable for corneal replacement.

METHODS

Two main approaches were taken to synthesize silica-collagen constructs: copolymerization method and Stober method (Figure 1). The copolymerization method was used initially for preliminary studies. Briefly, an acidified silica precursor is mixed with a soluble collagen solution. The silica and collagen polymerize simultaneously to form a hydrogel. The main benefit of this method is the short amount of time it takes to make a sample since the collagen and silica are polymerized simultaneously. However, microstructural features such as collagen fibril diameter and silica-coating thicknesses are difficult to control independent of each other because they are both dependent on chemical conditions such as pH, temperature, and ionic strength.

In order to fine tune microstructural properties selectively, a two-step Stober method was adopted. First, collagen solutions were concentrated via dialysis, and fibril formation was induced by exposing the solution to ammonia vapor. A layer-by-layer approach that has been used to deposit silica on various nanoparticles such as gold spheres was then used on the collagen hydrogel. In order to make gels that are composed of collagen fibrils uniformly coated in silica, the collagen gels are pretreated with a poly-L-lysine (PLL) solution making the collagen fibril surfaces more cationic. This makes interactions between the fibril surface and negatively charged silica precursor molecules favorable and reduces the formation of large homogeneous silica aggregates that scatter light in the visible spectrum.

Scanning and transmission electron microscopy(SEM,TEM) were used to characterize the microstructural features of the gels: fibril diameter, spacing, bundling, silica coating thickness, and aggregation. These features were then related to the bulk properties relevant to corneal function: ultraviolet/visible light transmittance, modulus of elasticity, ultimate tensile strength, and enzymatic degradation rate. Spectral transmittance of gels was carried out on a spectrophotometer, and the biomechanical properties were quantified using a low force Instron uniaxial tensile tester. A collagenase assay was used in conjunction with Raman spectroscopy to characterize the weight loss of samples when exposed to collagenase and the chemical properties of the product of degradation. Fourier transform infrared spectroscopy and Raman spectroscopy was used for chemical characterization of samples before and after silica deposition. To test the biocompatibility of samples, *in vitro* and *ex vivo* experiments were carried out.

RESULTS

Lens implants were made using the copolymerization and characterized. While the biomechanical properties were significantly lower than the cornea, the spectral transmittance mimicked that of a rabbit cornea. Additionally, the samples were significantly stiffer and stronger than the pure collagen control, and were durable enough to withstand implantation of an explanted rabbit cornea. The figures below display the favorable biocompatibility of the silica-collagen with a 3:1 silica to collagen weight ratio. The ex-vivo tissue section showed that an epithelial layer was able to form on the implant surface, which is vital for proper corneal function. (Figure 2).[5]

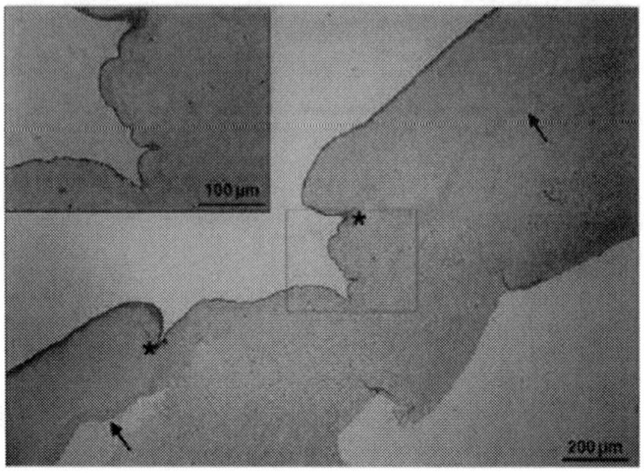

FIGURE 2: EPITHELIALIZATION OF IMPLANT LABELLED BY ARROWS AND ASTERICKS.

In order to make the gels stronger without reducing transparency, the initial collagen concentration was raised by an order of magnitude (5 to 50 mg/mL) via dialysis. When working with highly concentrated collagen solutions, it is difficult to mix in other components without introducing air bubbles. Therefore, a Stober soaking step was performed on fibrillar collagen to deposit silica on the fibril surface. The spectral transmittance of a 50 mg/mL collagen sample that underwent silica deposition for 3 days is similar to that of a rabbit cornea (Figure 3). With SEM, an increase in fibril diameter by nearly an order of magnitude is observed (Figure 4). The chemical characteristics and degradation behavior are currently being quantify. However, qualitative collagenase experiments have shown the enzymatic degradation rate of silica coated samples are significantly slower than the collagen control.

DISCUSSION

The materials used currently for corneal replacement do not truly integrate with the host tissue. Collagen gels have good biocompatibility, but require chemical modification to increase the degradation resistance and biomechanical properties. The simplest way to approach the biomechanical properties of the cornea is to raise collagen concentration. However, this is usually at the cost reducing light transmittance. The research presented has shown that a concentration of 50 mg/mL can be reached and a thick coating of silica can be deposited onto fibrils without reducing the optical quality. While degradation techniques are still being developed, silica coated samples have shown no reduction in size after 3 hours, and the collagen controls have shown up to a 40% reduction in the same time period. This is what is expected when considering the literature that shows highly condensed silica degrades over months when exposed to physiological conditions.[6] The main difference between the work presented and what is already published is the ability to coat the fibrils uniformly and in the absence of large aggregates. Treating the collagen with PLL prior to soaking in Stober solution allows for favorable silica formation around the fibrils.

While this work has proven a novel method for depositing silica onto collagen fibrils, more work must be carried out to optimize deposition conditions such as silica precursor concentration, deposition time, and ethanol to water ratio. Additionally, the degradation rates of silica-collagen samples need to be characterized relative to the cornea and other collagen based replacements made *in vitro*.

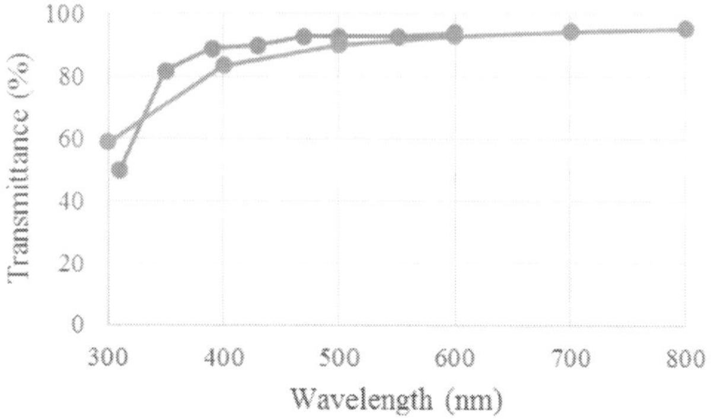

FIGURE 3: SPECTRAL TRANSMITTANCE OF (BLUE) RABBIT CORNEA AND (RED) SILICA-COLLAGEN HYDROGEL.

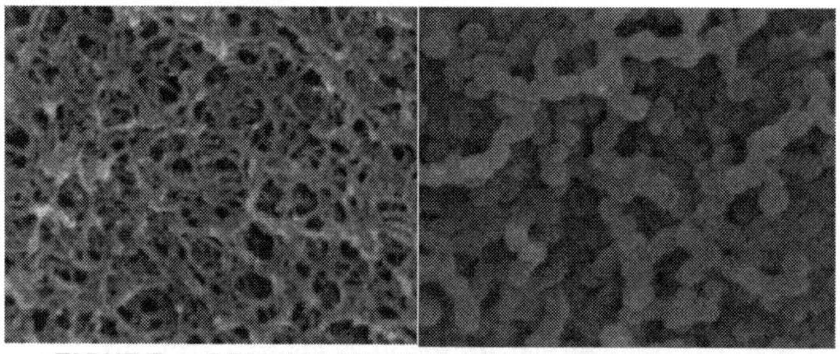

FIGURE 4: SEM IMAGE OF (LEFT) PURE COLLAGEN AND (RIGHT) STOBER SOAKED HYDROGEL. BOTH IMAGES ARE 1 MICRON WIDE.

ACKNOWLEDGEMENTS

Supported by the Minnesota Partnership for Genomics and Biotechnology; Research to Prevent Blindness, New York, (unrestricted grant to the Department of Ophthalmology at Mayo Clinic, and SVP as Olga Keith Wiess Special Scholar); and Mayo Foundation, Rochester, Minnesota.

REFERENCES

[1] EBAA. Eye Bank Association of America. 2010–2011, A Year in Review. Washington, DC: Eye Bank Association of America; 2011.

[2] EBAA. Eye Bank Association of America. Annual Statistical Report. Washington, DC: Eye Bank Association of America; 2010.

[3] Hicks CR, Crawford GJ, Lou X, et al. Corneal replacement using a synthetic hydrogel cornea, AlphaCor: device, preliminary outcomes and complications. Eye. 2003;17:385–392.

[4] Ma JJK, Graney JM, Dohlman CH. Repeat penetrating keratoplasty versus the Boston keratoprosthesis in graft failure. Int Opthalmol Clin. 2005;45:49–59.

[5] DiVito, M.D. et al., *IOVS*, 53(13): 8192-8199,2012.

[6] Viitala, R et al., *IJOP*, 336: 382-390, 2007.

SB³C2015
Summer Biomechanics, Bioengineering and Biotransport Conference
June 17-20, 2015, Snowbird Resort, Utah, USA
SB³C2015-261

POST-YIELD DAMAGE DENATURES BONE'S COLLAGEN AS DETERMINED BY A NOVEL MOLECULAR SPECTROSCOPIC BIOMARKER

Mustafa Unal (1), Hyungjin Jung (1), Ozan Akkus (1,2,3)

(1) Department of Mechanical and Aerospace Engineering
Case Western Reserve University
Cleveland, OH, USA

(2) Department of Biomedical Engineering
Case Western Reserve University
Cleveland, OH, USA

(3) Department of Orthopaedics
Case Western Reserve University
Cleveland, OH, USA

INTRODUCTION

Vibrational spectroscopy (FTIR and Raman) has become a powerful tool in the assessment of bone quality since they provide information on chemical composition and bonding environments of both mineral and organic moieties of bone tissue simultaneously. Although FTIR and Raman techniques provide much the same information, only Raman spectroscopy has potential to be used for noninvasive and nondestructive measurements *in vivo*. General outcomes of these techniques include: mineral/matrix ratios, mineral maturity/crystallinity, carbonate content and enzymatic collagen cross-links which are shown to be associated with bone mechanical integrity to varying degrees. The use of Raman spectroscopy to assess collagen quality is less established than mineral quality. The only specific spectroscopic biomarker for collagen quality in bone is the ratio of underlying sub-bands of amide I region (1660/1690 cm^{-1}) corresponding to enzymatic collagen cross-links [1]; however, the validation of this ratio is still controversial [2]. Toughness and post-yield toughness (PYT) of bone are mostly determined by collagen rather than the mineral phase. Therefore, it is essential to identify new spectroscopic biomarkers which help infer the status of collagen quality in bone and other collagen-rich tissues. Amide I and amide III bands are uniquely useful for the analysis of protein/collagen secondary structures such as α helix, β sheet, random coil and triple helix structures. The **first aim** of this work was to identify the regions of amide bands which are sensitive to denaturation. Pure collagen sheets are thermally denatured to attain this aim. The **second aim** was to assess whether mechanical damage denatures bone, as reflected by the molecular spectroscopic markers identified in the first aim. The **third aim** was to assess the correlation between the new spectroscopic biomarkers and mechanical properties (toughness and PYT) of cortical bone.

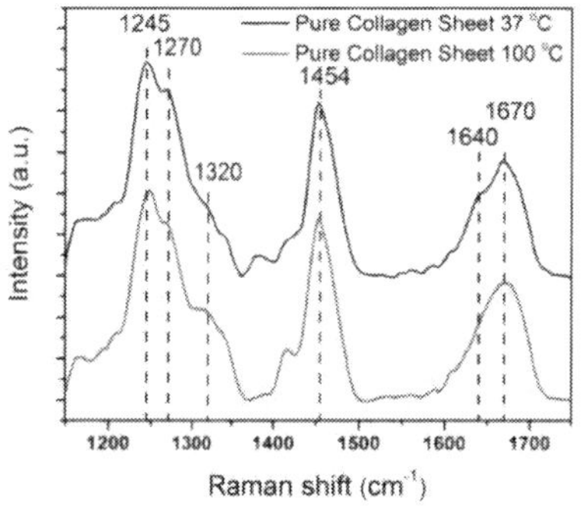

FIGURE 1: CHANGES IN RAMAN PEAKS OF PURE COLLAGEN SHEETS DUE TO THE THERMAL DENATURATION.

METHODS

Pure bovine collagen type I solution was used as reference for the identification of spectral signatures which change under thermal denaturation. Collagen sheets were prepared by bovine collagen solution at 37°C. Collagen sheets were placed in an oven at 100 °C for 2 h to induce thermal denaturation. The bovine bone samples were prepared under two schemes: 1) cortical bone wafers (40 mm × 4 mm, 200 μm thickness) from bovine femur with V-notch (1.5 mm in depth) to be used in thermal and mechanical denaturation experiments (N=3 in each experimental group), and 2) cortical beams (30 x 3.0 x 1.1 mm) for three-point bending test (N=25) to assess the correlation between mechanical properties and new spectral biomarkers. Thermal denaturation of bones was attained by boiling the samples in water for ~ 90 min. Mechanical denaturation was attained by loading V-notch bone wafers using a miniaturized tensile loading device to induce an extensive damage zone at the notch tip. The notch tip was observed during loading for the emergence of damage zone which is manifested by an opaque zone as we demonstrated earlier [3]. The loading is stopped upon damage initiation.

Raman spectra of pure collagen sheets were collected from 10 points randomly using commercial 785 nm Raman system (Xplora, Horiba). Raman mapping of each cortical bone wafer was conducted before (control) and after the denaturation (thermal or mechanical) on the same area of the notched region to investigate changes in target spectroscopic ratios due to denaturation. On three-point bending specimens, an average of 12 Raman spectra was collected in the vicinity of the region where tensile fracture occurred. Three V-notch bone wafers were utilized as negative controls from which spectra were collected in a similar time frame with mechanically loaded specimens, but without any mechanical loading. Spectral ratios of experimental groups were compared using one-way ANOVA followed by Tukey's post-hoc test. Linear regression analyses were performed between Raman spectroscopic biomarkers and the mechanical properties. Statistical significance was set at $p< 0.05$

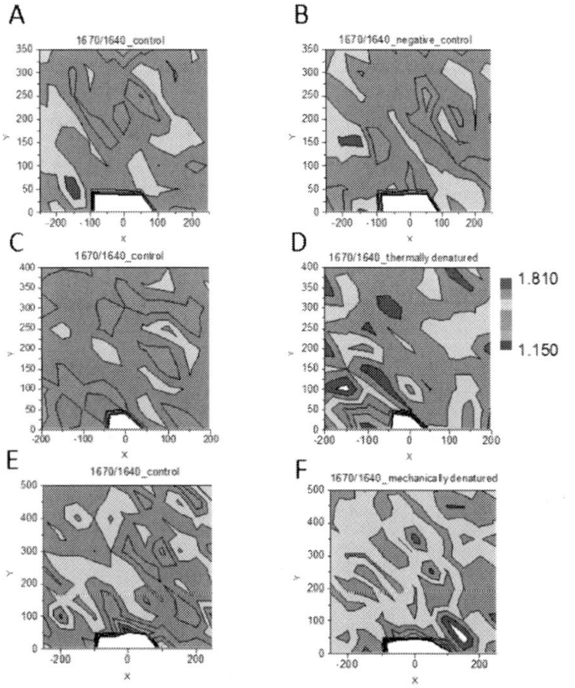

FIGURE 2: RAMAN 2D MAPPING OF BONE WAFERS CORRESPONDING TO 1670/1640 RATIO OF (A-C-E) BEFORE (CONTROL), AND (B) NEGATIVE CONTROL, (D) THERMALLY AND (F) MECHANICALLY DENATURED

RESULTS

The Raman spectra of pure collagen sheets revealed five peaks whose intensities were sensitive to thermal denaturation: 1245, 1270 and 1320 cm^{-1} in amide III region and, 1640 and 1670 cm^{-1} in amide I region(Fig. 1). Four intensity ratios were calculated using these peaks: 1670/1640, 1320/1454, 1245/1270 and 1245/1454. These ratios were significantly affected by thermal denaturation ($p < 0.001$ compared to pure collagen at 37 °C). Raman analysis of negative controls (same regions of bone samples scanned at two discrete times without any intermediate treatment) revealed that there were no significant differences in Raman spectral markers ($p>0.05$) (Fig. 2A-B). The results of thermal and mechanical denaturation experiments of cortical bone wafers showed that all these four ratios to be sensitive to both thermal and mechanical denaturation in cortical bone samples (Fig. 2C-F). The ratio of 1670/1640 had a strong correlation with toughness and PYT of bone samples whereas 1245/1454 and 1320/1454 had relatively weaker correlation with PYT of bone samples. 1245/1270 did not correlate with toughness and PYT (Table 1).

DISCUSSION

To the best of our knowledge, the current work is among the first thorough analysis of the Raman amide bands in terms of providing spectroscopic manifestations of collagen integrity and bone toughness. The results showed the 1670/1640 peak intensity ratio as a novel biomarker to assess collagen quality. The changes in this ratio with denaturation may reflect alterations in the collagen secondary structures, specifically a transition from ordered to unordered structure since the Raman study of collagen type I indicated that the peak at ~1640 cm^{-1} is associated with ordered (triple helix or α-helix) structure whereas ~1670 cm^{-1} is associated with random/unordered structure [4]. So far, the literature utilized the ratio of ~ 1245/1454 ratio based on FTIR studies. However, current results indicate that 1245/1454 ratio have a weak correlation with PYT and no correlation with toughness. 1320/1454 ratio, which is associated with ordered structure in total amount of collagen, had a weak correlation with PYT and no significant correlation with toughness indicating that 1320/1454 may be required to have a critical level of denatured collagen to properly reflect them as in our mechanically denatured bone wafers. Same discussion may be also thought for 1245/1270. In conclusion, these new spectral information can be used to understand the involvement of collagen quality in the fragility of aging and in diseased bone.

TABLE 1: CORRELATIONS BETWEEN MECHANICAL PROPERTIES AND RAMAN RATIOS

Linear Regression Analyses (R^2)	Raman Analysis			
	1670/1640	1245/1454	1245/1270	1320/1454
Toughness	0.38 $p<0.001$	N.S	N.S	N.S
PYT	0.54 $p<0.001$	0.22 $p<0.05$	N.S	0.22 $p<0.05$

N.S = not significant

ACKNOWLEDGEMENTS

This research is partially supported by think[box] Student Research Fund at CWRU and NSF CMMI-1233413. We thank to Mousa Younesi for his help to prepare collagen sheets.

REFERENCES

[1] Paschalis E, Verdelis K, Doty S, Boskey A, Mendelsohn R, Yamauchi M. Spectroscopic Characterization of Collagen Cross-Links in Bone. Journal of Bone and Mineral Research 2001;16: 1821-1828.

[2] Farlay D, Duclos M-E, Gineyts E, Bertholon C, Viguet-Carrin S, Nallala J, Sockalingum GD, Bertrand D, Roger T, Hartmann DJ. The ratio 1660/1690 cm− 1 measured by infrared microspectroscopy is not specific of enzymatic collagen cross-links in bone tissue. PLoS One 2011;6: e28736.

[3] Sun X, Hoon Jeon J, Blendell J, Akkus O. Visualization of a phantom post-yield deformation process in cortical bone. J Biomech 2010;43: 1989-1996.

[4] Wisniewski M, Sionkowska A, Kaczmarek H, Lazare S, Tokarev V, Belin C. Spectroscopic study of a KrF excimer laser treated surface of the thin collagen films. Journal of Photochemistry and Photobiology A: Chemistry 2007;188: 192-199.

SB³C2015
Summer Biomechanics, Bioengineering and Biotransport Conference
June 17-20, 2015,Snowbird Resort, Utah, USA
SB³C2015-262

COLLAGEN BOUND WATER IS A STRONG CORRELATE OF BONE'S TOUGHNESS

Mustafa Unal (1), Ozan Akkus (1,2,3)

(1) Department of Mechanical and Aerospace Engineering
Case Western Reserve University
Cleveland, OH, USA

(2) Department of Biomedical Engineering
Case Western Reserve University
Cleveland, OH, USA

(3) Department of Orthopaedics
Case Western Reserve University
Cleveland, OH, USA

INTRODUCTION

Water in bone exists in unbound state in the pore spaces or in he bound state to the matrix. Assessment of bound water is emerging as a novel measure of bone's mechanical integrity [1]. Despite the implications of water in bone fragility, there is a dearth of nondestructive methods available for studying bone hydration status such as NMR; however, assignment of NMR data to specific water compartments of bone is indirect. Therefore, a broader range of methods is needed to infer the status of water in bone. Raman spectroscopy is one of the few nondestructive modalities to assess the hydration status in bone; however, it has not been used to study the OH-band in bone. Recently, we developed a custom designed short-wave infrared (SWIR) Raman spectroscopy to identify the bound and unbound water compartments of bone [2] as a measure of skeletal fragility. Standard Raman systems have low sensitivity in the OH-region and the custom system provided greater sensitivity while minimizing fluorescence. The spectrometer was optimized to probe the water-related OH region of the vibrational spectrum. Using this technique, we previously identified four peaks which were sensitive to dehydration. The first peak is at 3220 cm^{-1} and belongs to only water, the second peak at 3325 cm^{-1} belongs to NH groups in collagen besides water, the third one at 3453 cm^{-1} belongs to OH of hydroxyproline (Hyp) besides water, and the last one at 3584 cm^{-1} belongs to OH of mineral and water [2] (Figure 1). The current study assessed the relationship between these water related Raman markers (I_{OH}/I_{CH}) and the mechanical properties of bone.

METHODS

Mechanical properties: Stiffness, maximum flexural strength, toughness and post-yield toughness (PYT) of cortical bovine bone samples (N=30) were measured via three-point bending under displacement control (1 mm/sec).

Sequential dehydration: Contributions of different water compartments were probed by a sequential dehydration process: 1) low energy dehydration at 40 °C for 48 h to remove unbound water [3], 2) partial replacement of remaining bound water by ethanol for 36 h following vacuum process for 40 h to remove the remnant ethanol. Water loss was also recorded gravimetrically during dehydration.

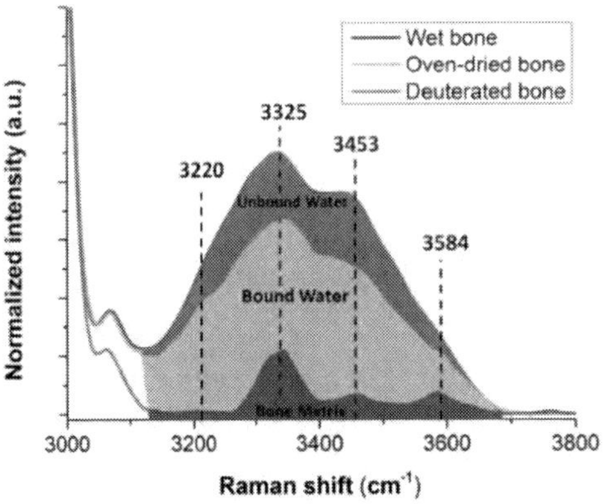

FIGURE 1: THE BREAKDOWN OF BAND INTENSITIES BETWEEN BOUND WATER, UNBOUND WATER AND BONE MATRIX (SEE REFERENCE FOR DETAILS [2])

Porosity measurement: The microstructural porosity of each sample was measured near the fracture surface. Two digital images covering most of the surface were captured and converted to binary form for determining pores (i.e., vascular and lacunar porosity). Pore area was normalized by the total area of the image.

SWIR Raman spectroscopy: The custom SWIR system involved excitation at 847 nm and signal collection using a shortwave InGaAs IR spectrometer optimized for 1000-1300 wavenumber suited for OH-stretch range. Raman spectra were collected immediately after each step of dehydration episode to record the reductions in OH-band intensities. All spectra were normalized to the protein related CH stretch band. Intensity ratios (I_{OH}/I_{CH}) of different OH peaks in the range of 3100-3800 cm^{-1} were qualified in terms of correlation with mechanical properties of bone.

Statistical analysis: Linear regression analyses were conducted to study associations between mechanical properties and Raman OH peaks and water loss measured by gravimetric analysis. Statistical significance was set at $p < 0.05$ for the resulting R^2 value.

RESULTS

The results of gravimetric analysis showed that unbound water portion (8.65±0.62 % by weight) of bone correlated negatively with modulus only (R^2 =0.42) (Figure 2B). Bound water portion (1.53±0.09 % by weight) of bone had stronger positive correlations (R^2 = 0.85 for toughness and 0.92 for PYT) (Figure 2A-C). Previously, we showed that the Raman intensity reduction in OH region was well correlated with water loss measured by gravimetric analysis [2]. The overall results showed that the peak at 3220, which is solely associated with collagen related water (Figure 1), is the best predictor (explaining up to 81% of total variance independently from Bone Mineral Density (BMD) and porosity for toughness, PYT as well as maximum strength of bone. Furthermore, 3584, which is associated with mineral and mineral water, is the best predictor for stiffness (Table 1).

DISCUSSION

We demonstrated a new method to assess bone quality in the perspective of bone hydration. The overall results showed that greater amount of bound water indicated increased toughness and strength. Therefore, bound water is more intimately related to mechanics than unbound water compartment (Figure 2A-C). The overall results also indicated that there are different sorts of bound water compartments in bone such that the bound water layer that was partially removed by ethanol treatment is the critical portion for the mechanics since R^2 pairwise correlations between mechanical properties and Raman OH peaks peaked after oven dried and decreased after ethanol treatment (Table1). Furthermore, we found that intensities of Raman OH peaks have different degrees of correlations with different mechanical properties of bone (Table 1) indicating different water compartments may be associated with different mechanical properties since each peak was previously assigned as different types of water compartment in bone such as collagen-bound water (loosely or tightly) and mineral-bound water [2].

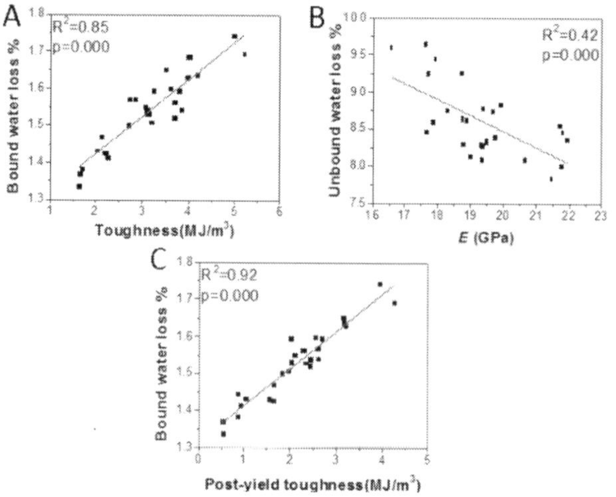

FIGURE 2: R^2 PAIRWISE CORRELATIONS BETWEEN MECHANICAL PROPERTIES AND WATER LOSS MEASURED BY GRAVIMETRIC ANALYSIS

Another important finding of this study is that even in wet situation (bone's natural condition), mechanical integrity of bone can be explained up to 37% of total variance independently from BMD and porosity (Table 1). In respect to the clinical importance of bone material composition, for the first time, it was revealed that Raman-based water measurement has a strong correlation with mechanical integrity; therefore, it is a novel approach to assess bone quality which may help to improve the current diagnosis method for those who may be at risk of bone fracture due to aging and diseases.

ACKNOWLEDGEMENTS

This research is partially supported by think[box] Student Research Fund at CWRU. We also thank to Shan Yang and Bolan Li for their valuable help to set up our customized SWIR Raman system and thank to Cedric Hansen for his help to prepare bone samples.

REFERENCES

[1] Nyman JS, Roy A, Shen X, Acuna RL, Tyler JH, Wang X. The influence of water removal on the strength and toughness of cortical bone. J Biomech 2006;39: 931-8.

[2] Unal M, Yang S, Akkus O. Molecular spectroscopic identification of the water compartments in bone. Bone 2014;67: 228-236.

[3] Mkukuma L, Imrie C, Skakle J, Hukins D, Aspden R. Thermal stability and structure of cancellous bone mineral from the femoral head of patients with osteoarthritis or osteoporosis. Ann Rheum Dis 2005;64: 222-225.

TABLE 1: R^2 PAIRWISE CORRELATIONS BETWEEN MECHANICAL PROPERTIES AND RAMAN OH PEAKS

Linear Regression Analyses(R^2)	Raman Analysis											
	Wet				Oven-dried				Ethanol treatment			
	3220	**3325**	**3453**	**3584**	**3220**	**3325**	**3453**	**3584**	**3220**	**3325**	**3453**	**3584**
Toughness	0.37#	0.23*	0.22*	N.S	0.81#	0.79#	0.28*	N.S	0.48#	0.47#	0.39#	0.43#
Post-Yield Toughness	0.32#	0.17*	0.19*	N.S	0.65#	0.73#	0.18*	N.S.	0.44#	0.48#	0.40#	0.47#
Stiffness	N.S	N.S	N.S	0.39#	N.S	N.S	N.S	-0.78#	N.S	N.S	N.S	N.S
Max. Flexural Strength	0.23*	0.18*	0.20*	0.45#	0.40#	0.29*	0.48#	0.46#	N.S	N.S	N.S	0.15*

*p<0.05, #p<0.001, N.S= not significant

THERMORESPONSIVE, REDOX-CROSSLINKED CELLULOSIC HYDROGELS UNDERGO *IN SITU* GELATION AND RESTORE NUCLEUS PULPOSUS BIOMECHANICAL PROPERTIES POST NUCLEOTOMY

Devika M. Varma (1), Huizi A. Lin (1), Rose G. Long (2), Carine Rognon (3), Andrew C. Hecht (2), James C. Iatridis (2), Steven B. Nicoll (1)

(1) Department of Biomedical Engineering
The City College of New York, City University of New York
New York, New York, 10031
USA

(2) Leni and Peter W. May Department of Orthopaedics
Icahn School of Medicine at Mount Sinai
New York, NY
USA

(3) Department of Information Technology and Electrical Engineering
Swiss Federal Institute of Technology of Zurich (ETHZ)
Zürich, Switzerland

INTRODUCTION

Intervertebral disc disease is strongly associated with back pain, affecting 15-30% of the US population at an estimated annual cost of $194 billion [1,2]. Discectomy, which involves removal of nucleus pulposus (NP) tissue, is employed when non-operative treatments fail to alleviate pain. Although discectomy often provides immediate pain relief, it does not restore disc biomechanical function due to loss in disc height and intradiscal pressure [3]. Hydrogels are widely studied as NP replacement materials [4,5] as a result of their ability to attract water and resist compressive loads. *In vitro* characterization of photocrosslinked methacrylated carboxymethylcellulose (CMC) hydrogels demonstrated that these materials possess functional properties comparable to native NP and support stem cell differentiation towards an NP phenotype [6]. A recent advance employed redox crosslinking, thereby providing an important step in the development of an injectable CMC hydrogel formulation [7]. However, injection of the CMC/redox initiator mixture into large bovine motion segments resulted in extravasation of the polymer solution and limited polymerization within the disc. Thus, the hydrogel formulation required modification to enable stable polymerization. Methacrylated methylcellulose (MC) may be incorporated to create CMC-MC copolymers, capitalizing on the inherent thermogelation property of MC to achieve higher viscosities at 37°C. The increased viscosity would allow local retention of the injected material and subsequent curing *in situ*. Hence, the objectives of this study were to develop an injectable, thermogelling, redox-crosslinked CMC-MC copolymer with material properties akin to the native NP, and determine if the hydrogel implant could restore native biomechanics of the bovine disc post nucleotomy.

METHODS

Polymer Synthesis and Material Characterization: *Macromer synthesis:* Methacrylation of 90 kDa CMC, and a low and medium viscosity blend of MC polymer was performed as previously described [8]. *Hydrogel preparation:* Solutions of CMC and MC in DPBS, both at 3% (w/v) macromer concentration, were prepared in dual-barrel syringes and mixed with ammonium persulfate (20 mM) and tetramethyl-ethylenediamine (20 mM) in separate syringe barrels.

***In Situ* Gelation:** *Motion segment preparation:* Bone-disc-bone motion segments were harvested from skeletally mature bovine tails and frozen at -20°C until testing. *Nucleotomy:* Approximately 0.15-0.25 g of NP tissue was removed with a pituitary rongeur via a cruciate incision [4]. *In situ gelation protocol:* CMC-MC solutions (containing trypan blue for visualization) were injected into the NP void with a 20G syringe needle. Post 30 min. incubation, the motion segments were dissected along the endplate and the hydrogel was carefully extracted.

Mechanical testing: Disks (3-mm diameter, ~2-mm thickness) were cored from extracted hydrogels and tested in unconfined compression [8] to determine equilibrium Young's modulus (E_y) and % relaxation (n=7). *Swelling ratio:* Swelling ratio (Q_w) of hydrogels was measured at day 1 (n=7) after overnight incubation in DPBS at 37°C [8].

Biomechanical Restoration Post Nucleotomy: *Study design and specimen preparation:* Seven bovine tails were used for the study with two motion segments per tail assigned to an experimental and a sham group. Each motion segment underwent a mechanical loading regimen under Intact, Injured and Repaired/Sham conditions with repeated measures design to eliminate variations between animals and disc levels [4]. Motion segments were potted in poly(methylmethacrylate) [9]. The height and average diameters were determined via X-Ray imaging and caliper measurements, respectively. For repair, CMC-MC solutions were injected into the void space of the injured disc with a radiopaque dye to visualize implant location. *Mechanical testing:* Mechanical testing was performed on an MTS Bionix Servohydraulic Test system in a saline DPBS bath. Samples underwent an initial 30 N preload followed by 25 cycles of tension-compression at 0.1 Hz between -0.5 MPa [10] and 0.25 MPa with a final slow ramp from 0 to 170 N at 1 N/s [4,11]. *Analyses:* The 25th cycle of cyclic loading was used for data analysis to ensure dynamic equilibrium was attained [12]. Range of Motion (ROM) was calculated as peak-to-peak displacement. Neutral zone (NZ) stiffness was calculated as described [13]. The compressive and tensile stiffness were calculated from a linear regression of the load-displacement curve between 60-100% and 80-100% of the maximum load, respectively [4]. A linear regression of the slow ramp load-displacement curve was used to obtain the slow ramp compressive stiffness [11,12]. *Statistical analysis:* Data are presented as the mean ± s.d. Student's paired t-tests were performed to compare biomechanical properties between all paired combinations (i.e., Injured vs. Repaired) ($p<0.05$ considered significant).

RESULTS

The CMC-MC hydrogels had a swelling ratio of 25.11± 5.36, an E_y of 16.62 ± 3.57 kPa and a percent relaxation of 33.83 ± 3.11. The materials formed stable gels *in situ* (Fig. 1 A-B) and restored disc height (Fig. 1 C-E). Moreover, CMC-MC hydrogels restored several biomechanical parameters in bovine motion segments that were altered by nucleotomy (Table 1). There was a significant increase in the ROM, compressive and slow-ramp stiffness, and a decrease in tensile and neutral zone stiffness in motion segments post nucleotomy compared to Intact samples (Fig. 2, Table 1). Upon repair, the ROM, and compressive and slow-ramp stiffness significantly decreased with respect to Injured values, while the tensile and NZ stiffness were significantly higher than Injured discs. All of these parameters for the Repaired condition were not significantly different from Intact values except for the NZ stiffness, which was higher than Intact specimens. Conversely, these properties were significantly different in the Sham repair with respect to Intact samples. The hydrogel implant did not extrude at any point during testing or incubation.

DISCUSSION

This study is the first to engineer an injectable, copolymer system of CMC and MC for NP replacement with the ability to restore disc biomechanics post nucleotomy. The thermogelling MC component was an important addition allowing the hydrogel to remain localized within the disc tissue for curing *in situ*. The specific formulation of CMC-MC was based on prior optimization of injectability, gelation and material properties. The E_y and Q_w of the gels were comparable to values for native NP tissue (~5 kPa and ~19, respectively) [14,15].

The CMC-MC hydrogels restored key biomechanical properties of the injured disc to intact values. The ROM and neutral zone have previously been found to be most dependent on NP integrity, while both NP and the annulus fibrosus (AF) impact the compressive, tensile and slow-ramp stiffness [13]. The significant increase in the ROM and decrease in the NZ stiffness corroborated these findings. The compressive and slow-ramp stiffness were found to increase with removal of the NP, which suggests the transfer of load to the stiffer AF component of the disc [9]. Along with the loss of NP tissue, nucleotomy resulted in destabilization of the AF and a significant decrease in tensile stiffness [9]. Importantly, the CMC-MC implant was able to restore essential parameters associated with NP integrity without AF repair, and did not herniate or extrude. Interestingly, the NZ stiffness increased significantly post repair, which may be attributed to the higher E_y of the CMC-MC gels (~16 kPa) compared to native NP (~5 kPa) and possible overfilling of the implant in the disc. Also, the disc height, suggestive of the level of hydration and hydrostatic pressure, was restored with repair. Future fatigue loading studies will be conducted to assess resistance to herniation. Overall, this study demonstrates the potential application of this CMC-MC hydrogel system for functional NP replacement.

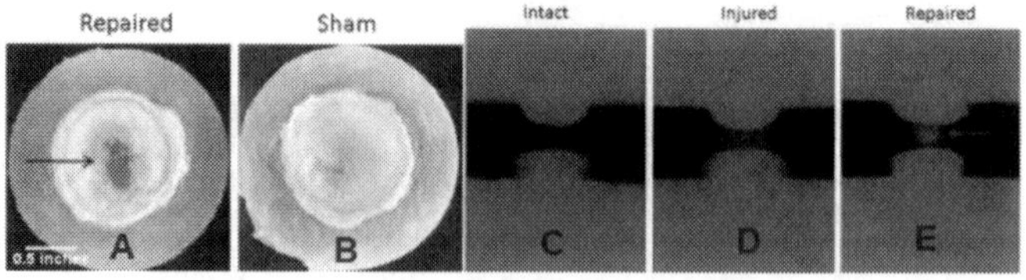

FIG. 1: (A-B) Transverse images of the motion segments after mechanical testing post repair. Repaired sample contains the blue CMC-MC hydrogel (black arrow) while the sham exhibits a void NP (Scale: 0.5 inches). (C-E) Sagittal radiographs of Intact, Injured and Repaired motion segments. CMC-MC hydrogel visible in the Repaired condition (red arrow).

TABLE 1: Comparison of biomechanical parameters between Intact, Injured and Repaired samples in experimental and sham groups. * Sig. Diff. W.R.T Intact, $ Sig. Diff. W.R.T Injured, # Sig. Diff. W.R.T No Implant. P<0.05

	Sham			Experimental		
	Intact	Injured	No implant	Intact	Injured	Implant
Compressive Stiffness (kN/mm)	0.544 ± 0.156	0.582 ± 0.140 *	0.604 ± 0.160 *	0.543 ± 0.150	0.594 ± 0.132 *	0.500 ± 0.126 $
Tensile Stiffness (kN/mm)	0.186 ± 0.038	0.170 ± 0.032 *	0.172 ± 0.025	0.189 ± 0.041	0.174 ± 0.048 *	0.192 ±0.044 $
Slow Ramp Stiffness (kN/mm)	0.236 ± 0.057	0.262 ± 0.057 *	0.275 ± 0.066 *, $	0.220 ± 0.062	0.247 ± 0.056 *	0.209 ± 0.051 #, $

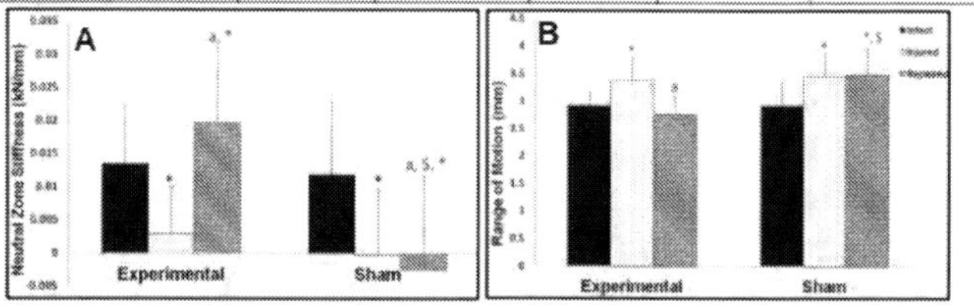

FIG. 2: Comparison of (A) NZ stiffness and (B) ROM between Intact, Injured and Repaired conditions in experimental and sham groups. *- Sig. Diff. W.R.T Intact within Group, a- Sig. Diff. W.R.T Injured within Group, $- Sig. Diff. W.R.T Experimental Repaired. (P<0 05)

ACKNOWLEDGEMENTS

We would like to thank Dr. Svenja Illien-Junger for training and Gittel T. Gold for technical contributions. Funding provided by NSF DMR 1207480 (S.B.N.) and NIH 1R01 AR057397 (J.C.I.)

REFERENCES

[1] J.J. Jacobs, et al., *Amer Acad Orthop Surgeons*, 247: 2008. [2] A. Becker, et al., *Spine*, 35: 2010. [3] K. Kim, et al. *Spine*, 34: 2674-8, 2009. [4] N.R. Malhotra, et al., *Spine*, 37: E1099-105, 2012. [5] M.S. Gupta, S.B. Nicoll, *Cell Tissue Res*, 358: 527-39, 2014. [6] L.J. Smith, et al., *Tissue Eng Part A*, 20:1841-9, 2014. [7] D.M. Varma et al., *Acta Biomater* 10: 4996-5004, 2014. [8] A.T. Reza, S.B. Nicoll, *Acta Biomater*, 6: 179-86, 2010. [9] M. Likhitpanichkul et al., *Eur Cells Mater*, 28: 25-38, 2014. [10] H.J. Wilke et al., *Spine*, 24: 755-62, 1999. [11] W. Johannessen et al., *Ann Biomed Eng*, 32: 70-6, 2004. [12] W. Johannessen et al., *Ann Biomed Eng*, 34: 687-696, 2006. [13] B.L. Showalter et al., *J Biomech*, 47:2633-40, 2014. [14] J.M. Cloyd et al., *Eur Spine J*, 16:1892-8, 2007. [15] A.T. Reza, S.B. Nicoll, *Biotechnol and Bioeng*, 105:384-95, 2010.

SB3C2015
Summer Biomechanics, Bioengineering and Biotransport Conference
June 17-20, 2015, Snowbird Resort, Utah, USA
SB³C2015-264

GEOMETRIC AND HEMODYNAMICS IMPLICATIONS OF MOYAMOYA DISEASE ON CAROTID SIPHON

Muhammad Jamil (1), Mehnaz Haq (1), Heidi Kang (2), Lee Zhi Rui (2), Tang Phua Hwee (3), Yap Choon Hwai (1)

(1) Department of Biomedical Engineering
National University of Singapore
Singapore

(2) Hwa Chong Institution
Singapore

(3) KK Women's and Children's Hospital
Singapore

INTRODUCTION

Moyamoya is a disease which is characterized by the progressive occlusion of the distal end of the carotid artery and leads to formation of multiple small arteries to compensate for the blood supply to the brain. When viewed in CT scans, the collateral network of arteries resembles a "puff of smoke" and the direct translation in Japanese is "Moyamoya", as it was first reported in Japan in 1957. Moyamoya's prevalence is 3 cases per 100,000 persons in Japan, affecting females almost twice as often than males [1], and about 0.57 per 100,000 individuals [2] in US. To date, the Moyamoya disease is considered idiopathic as the conclusive pathogenesis is still unknown [1]. It has been hypothesized that mechanical stress owing to changed hemodynamics plays a role in the pathogenesis of Moyamoya [3]. Further, the changes in vascular anatomy and flow associated with Moyamoya may also influence future vascular health. It is thus important to study the Moyamoya vascular geometry and fluid mechanics to understand pathogenesis, inform diagnosis efforts and monitor progression of disease.

METHODS

Analysis of the Magnetic Resonance Angiography (MRA) image data from 5 Moyamoya and 10 control patients was performed retroactively. Moyamoya Data was available for both pre- and post-revascularization operations. The revascularization was done by encephaloduroarteriosynangiosis (EDAS) procedure. Image data was segmented to obtain the 3D geometry of the carotid artery and geometric parameters such as curvature, tortuosity, cross sectional area and torsion were analyzed. Special interest was given to carotid siphon which has a sinuating shape and is characterized by proximal (B2) and distal (B1) bends in the carotid artery. Hemodynamic changes were studied using Murray's law and Computational Fluid Dynamics (CFD) was used to quantitatively observe the changes in hemodynamic parameters such as pressure losses and secondary flows.

RESULTS

We observed 4 major changes in the carotid siphon under Moyamoya disease. Firstly, from geometric analysis it was revealed that moyamoya cases had significantly smaller carotid siphons and internal carotid arteries ($p<0.05$). Even after the resvascularization surgery, the vessel sizes were smaller as compared to control cases. Secondly, we observed that in the normal carotid siphon, normalized curvature was independent of vascular size but in Moyamoya carotid siphons, the normalized curvature decreases with vascular size (Figure 1). This indicated that normal carotid siphons maintained its non-dimensional overall "S"-shape as it grew, but moyamoya carotid siphons did not, as it regresses. Thirdly, we observed that the geodesic torsion of the B2 segment significantly increased Moyamoya patients, and the tortuosity of the same segment increased for post-op Moyamoya patients. Finally, hemodynamic analysis showed that, compared to normal cases, there was higher pressure drop in the B1 segment and lower secondary flows in both B1 and B2 segments in Moyamoya carotid siphons (Figure 2).

DISCUSSION

The statistical analysis showed significant reduction in the artery sizes for Moyamoya cases. It has been reported in previous studies that arterial growth is dictated so as to maintain a homeostatic wall shear stress [4]. We hypothesize that occlusion at the distal end of the carotid artery resulted in reduced flow which caused the regression of the vessels hence significantly smaller cross sectional areas.

A change in growth-curvature relationship between control and Moyamoya cases was also observed. In control cases the curvature decreases with vessel size but the normalized curvature (curvature normalized by vascular cross-sectional radius) was independent of vessel size. This means that the entire s-shape of the carotid siphon grows at the same pace so as to keep the normalized curvature and non-dimensional shape constant. Whereas in the Moyamoya, curvature was independent of vessel size, but normalized curvature decreased with size, indicating that as the carotid artery regressed in vascular diameter, the shape of the artery didn't change correspondingly to keep the normalized curvature and non-dimensional shape constant.

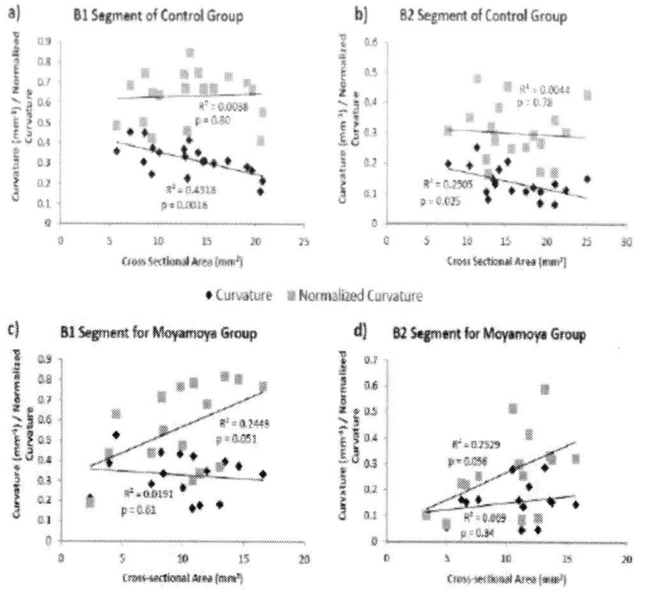

Figure 1: Scatter plots showing the curvature and normalized curvature versus average cross sectional area of sections B1 & B2 respectively for Control (a-b) and Moyamoya (c-d) cases

An increased torsion was observed in both pre and post-operative cases. In terms of hemodynamics an increase in torsion affects the velocity profile and the secondary flows which can lead to redistribution of the shear stress. Whether a change in torsion is caused by disease or a remodeling response to disease is unclear at this point.

Our simulations based on Murray's law show significantly lowered secondary flows in Moyamoya carotid siphons, which is most likely a result of restricted flow owing to smaller size. Moreover, the increased pressure drop at the segment B1 in Moyamoya is associated with increased frictional losses and area changes.

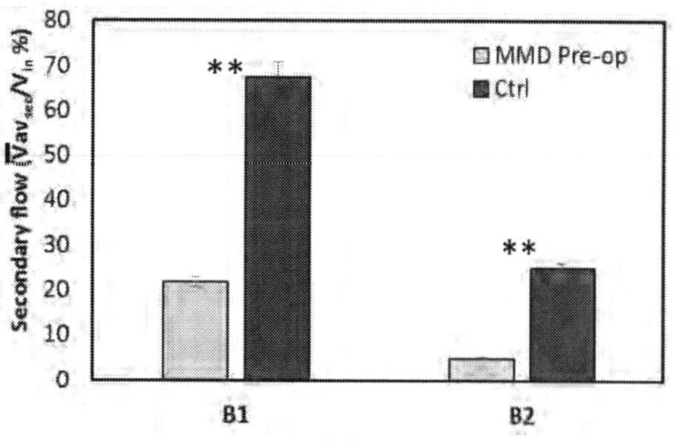

(a)

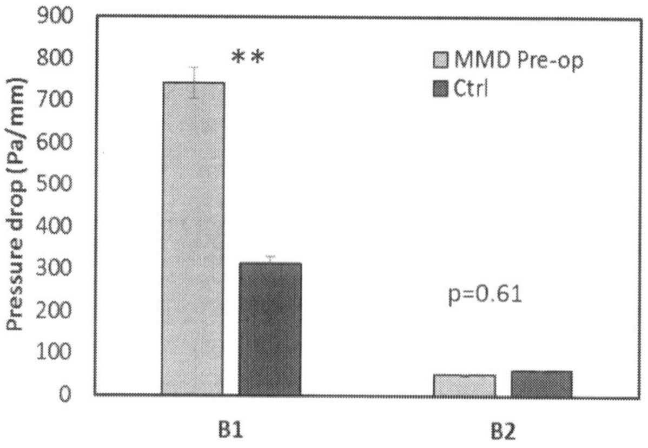

(b)

Figure 2: Bar plots comparing the secondary flows (a) and pressure drop (b) respectively in section B1 and B2 for Ctrl and MMD Pre-op and corresponding p values; * – $p < 0.1$; ** – $p<0.05$.

ACKNOWLEDGEMENTS

This study is approved by the SingHealth Centralised Institutional Review Board (CIRB), under the reference number CIRB/2014/047/C.

REFERENCES

[1] Ganesan V., Moyamoya, Journal of Pediatric Neurology, 8 (2010) 93-95.

[2] Starke R.M., Crowley R.W., Maltenfort M., Jabbour P.M., Gonzalez L.F., Tjoumakaris S.I., Randazzo C.G., Rosenwasser R.H., Dumont A.S., Moyamoya disorder in the United States, Neurosurgery, 71 (2012) 93.

[3] Shin D., Seol H., Kim S.-K., Wang K.-C., Cho B.-K., Shim E., Computational analysis of the hemodynamics in cerebral arteries related to Moyamoya disease, in: Magjarevic R., Nagel J.H. (Eds.) World Congress on Medical Physics and Biomedical Engineering 2006, Springer Berlin Heidelberg, 2007, pp. 155-158.

[4] Painter P.R., Eden P., Bengtsson H.U., Pulsatile blood flow, shear force, energy dissipation and Murray's Law, Theor. Biol. Med. Model., 3 (2006) 10.

SB^{3}C2015
Summer Biomechanics, Bioengineering and Biotransport Conference
June 17-20, 2015,Snowbird Resort, Utah, USA
SB^{3}C2015-265

PATIENT-SPECIFIC TREATMENT OF INTRACRANIAL ANEURYSMS: AN AUTOMATIC CFD-BASED FLOW-DIVERTER OPTIMIZATION PRINCIPLE

Philipp Berg (1), László Daróczy (1), Oliver Beuing (2), Gábor Janiga (1)

(1) Department of Fluid Dynamics and Technical Flows
University of Magdeburg
Magdeburg 39106, SA, Germany

(2) Department of Neuroradiology
University Hospital Magdeburg
Magdeburg 39120, SA, Magdeburg

INTRODUCTION

Intracranial aneurysms can be treated with a standalone flow-diverting device, which ensures a drastic reduction of entering blood flow and initiates a self-healing thrombosis [1].

Unfortunately, unsuccessful treatments raised concerns whether those stents lead to the desired effects [2, 3]. Since for existing stent designs only the diameter, length and porosity differ, the variety of patient-specific aneurysm scenarios is not covered. Therefore, several numerical studies investigated the influence of flow-diverters to improve the patient's outcome [4, 5]. However, most approaches consider just a single or a small number of configurations per patient and the proposed treatment might not be the global optimum.

This study introduces a new automatic approach based on CFD (Computational Fluid Dynamics) to optimize the flow-diverter shape and identify the ideal patient-specific treatment configuration. Nine different stent configurations are investigated and compared to the untreated scenario.

METHODS

A 53-year-old female with a giant aneurysm of the right internal carotid artery was treated with a SILK flow diverter. MRI six months after the implantation showed complete resolution of the aneurysm.

The 3D-DSA data were segmented and reconstructed using MeVisLab 2.3. Based on the reconstructed patient-specific aneurysm model the virtual stent deployment was carried out with an in-house software [6]. As presented in Fig. 1, nine different stent configurations were considered, whereas one contains no compression (UC), seven are locally compressed but differ in their position below the ostium (C1-7) and one represents a fully compressed stent (HC).

The finite volume meshes, which consider every single stent strut, have been generated using an intelligent JAVA macro for the commercial tool CD-Adapco Star-CCM+ 9 (Melville, NY, USA). The number of polyhedral cells ranged between 9 and 11 million cells for the treated cases and 1.9 million for the untreated case.

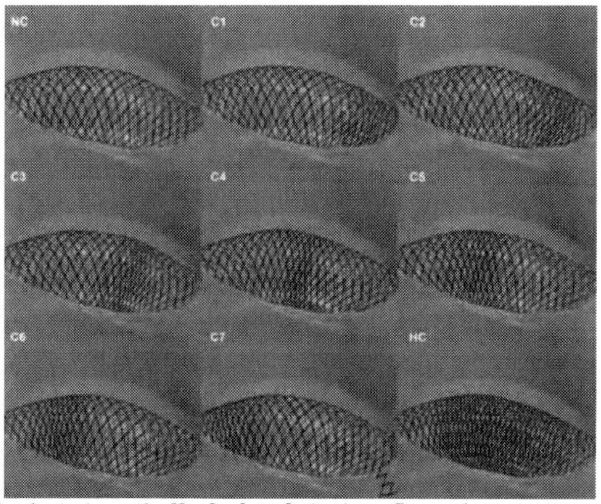

Fig. 1: Automatically deployed stent configurations: untreated case (UN) is not shown. NC - no compression, C1-7 - equal compression at different locations, HC - full compression

Steady-state blood flow simulations (u_{inlet} = 0.5 m/s) were carried out using Star-CCM+ 9 as well and blood was treated as a laminar, incompressible (ρ=1055 kg/m^3) and Newtonian fluid (η=0.004 Pa s). The placement of a partial compression on the stent was optimized for minimal mass inflow (single objective). The whole optimization was carried out in a single day, which demonstrates that this approach would be compatible with a clinical use.

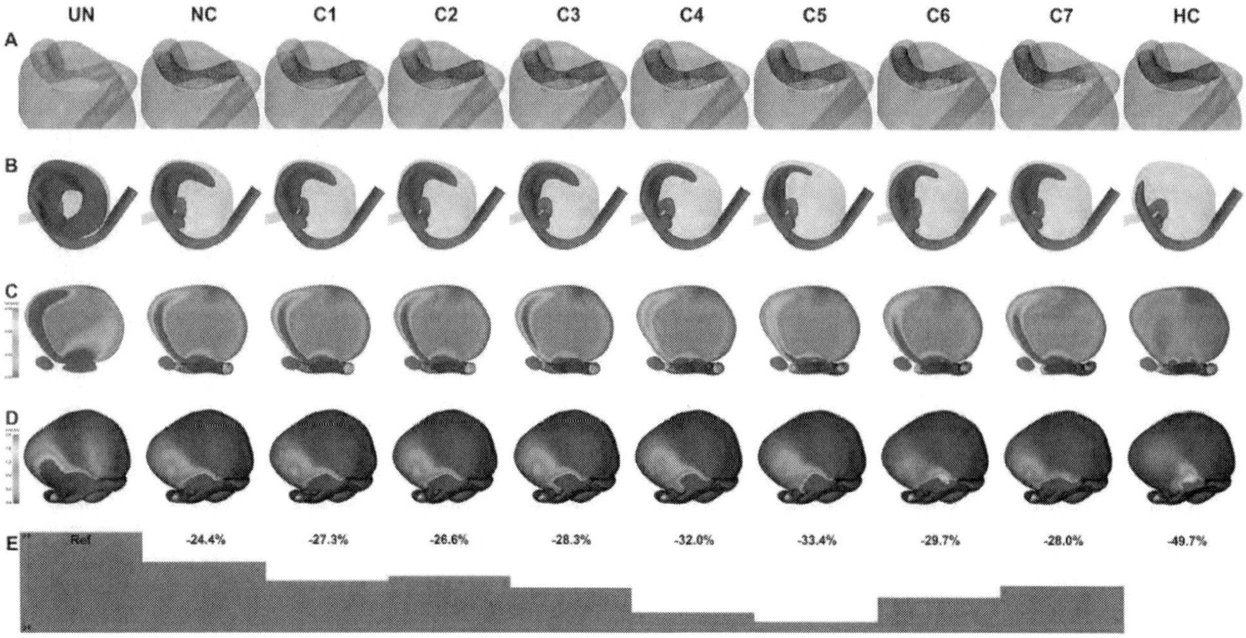

Fig. 2: A: Visualization of the configurations below the ostium (yellow), B: Iso-surface contour of the velocity magnitude for 0.04 m/s, C: Cut-plane representation containing the velocity magnitude, D: Wall shear stress distribution at the impingement zone of the blood flow jet, E: Quantification of the entering flow rate through the aneurysm ostium.

RESULTS

Fig. 2 demonstrates the results for the iso-surface velocities, cut-plane representations, wall shear stresses (WSS), as well as the quantitative analysis of the inflow reduction.

The deployment of flow-diverting devices clearly reduces the entering blood velocity and the iso-surface expansion decreases within the dilation. Additionally, the inflow jet gets much narrower as well as less intense. Using the two-dimensional illustration, cases C4 and C5 can be identified as the most effective regarding flow reduction (without consideration of the fully compressed case HC).

Increased WSS values are visible especially close to the aneurysm neck as well as along the impingement zone (see Fig. 2D). This elevated WSS of the reference case can clearly be reduced by all treatment configurations. However, the highest decrease is associated to case C5 and C6, since the impingement zone appears to be much smaller compared to the other locally compressed scenarios. Additionally, the neck regions of these cases, which can be locations of high rupture risk aside from the dome, experience a smaller area of high WSS as well.

The virtual deployment of the uncompressed flow-diverter (NC) reduced the inflow into the aneurysm by 24.4% and could further be reduced ranging from 27.3% (C1) to 33.3% (C5). This confirms the previous observations and case C5 represents the best treatment scenario for the locally compressed FDs. With increased stent compression a decrease of even 49.7% could be achieved (HC). Overall, a broad range of possible treatment outcomes was identified, illustrating the variability of this patient-specific flow-diverter deployment.

DISCUSSION

This study is the first of its kind that combines three-dimensional blood flow simulations, a realistic virtual stent deployment as well as the optimization of stent location and compression in patient-specific intracranial aneurysms. Especially, the integration of all these steps into an automated simulation loop allows the consideration of arbitrary treatment scenarios by only providing a reconstructed aneurysm shape and the desired stent model. For this particular case, it could be shown how minor differences in the stent deployment influence the resulting blood flow and therefore the initiation of a thrombotic cascade.

Relevant limitations include a rigid vessel and stent wall assumption, steady-state simulations and the consideration of only one test case.

Overall, the introduced optimization principle is applicable in a clinical context and may assist physician during patient-specific treatment planning.

Further advancements of the proposed method include an extension of the investigated configuration number. Additionally, competing objective functions will also be considered in the optimization process to identify the optimal treatment for bifurcation aneurysms, e.g., at the basilar tip or the middle cerebral artery.

ACKNOWLEDGEMENTS

We thank the International Max Planck Research School for Analysis, Design and Optimization in Chemical and Biomedical Process Engineering Magdeburg and the Federal Ministry of Education and Research in Germany within the Research Campus STIMULATE under grant number 03FO16102A for financial support.

REFERENCES

[1] Bergmann, G et al., *J Biomech*, 34:859-871, 2001.
[2] Cebral, J R et al., *Am J Neuroradiol*, 32:27-33, 2011.
[3] Schneiders, J J et al., *Am J Neuroradiol*, 34:E1-4, 2013.
[4] De Bock, S et al., *J Biomech*, 45:1353-1359, 2012.
[5] Ma, D et al., *Ann Biomed Eng*, 41:2143-2156, 2013.
[6] Janiga, G et al., *J Biomech*, 46:7-12, 2013.

SB^{3}C2015
Summer Biomechanics, Bioengineering and Biotransport Conference
June 17-20, 2015,Snowbird Resort, Utah, USA
SB^{3}C2015-266

MODELING BLOOD FLOW IN EMBRYO-SPECIFIC GEOMETRY OF THE ZEBRAFISH HEART

Pavel Kozlovsky (1), Robert Bryson-Richardson (2), Moshe Rosenfeld (3), Ariel Jaffa (4), David Elad (1)

(1) Department of Biomedical Engineering
Tel Aviv University
Tel Aviv, Israel

(2) Department of Biological Sciences
Monash University
Melbourne, Victoria, Australia

(3) School of Mechanical Engineering
Tel Aviv University
Tel Aviv, Israel

(4) Ultrasound in Obstetrics and Gynecology
Lis Maternity Hospital, Tel Aviv Medical Center
Tel Aviv, Israel

INTRODUCTION

It is widely acknowledged that flow-induced forces are critical for proper development and maintenance of many physiological organs [1]. The flow-derived pressures and wall shear stresses (WSS) affect the distribution of genes and growth materials, and thus, they play crucial roles during embryonic development of the cardiovascular system. There is a consensus that WSS plays an important role not only during cardiogenesis, but also imposes important factors in the pathogenesis of cardiovascular diseases [2]. Quantification of the biophysical environment (i.e., flow field and WSS) during the critical early stages of cardiogenesis is extremely complex due to the three-dimensional (3D) fluid-structure interaction (FSI) nature of this problem.

The major objective of this study was to develop a computational method for investigating the flow field and biophysical environment of anatomical-specific 3D replicas of the zebrafish embryonic heart during the critical first stages of cardiogenesis.

METHODS

The zebrafish embryos at an early stage of 48 hour post fertilization (hpf) were prepared according to the developed protocol [3]. The inner endocardial cells' layer and outer myocardial cells' layer of the heart tube wall were marked with cell-specific biological markers and high-resolution images of consecutive cross-sections of the embryonic heart were acquired using a confocal microscope (Fig. 1a). These images were automatically segmented using the Active Contour Toolbox of Matlab [4] in order to extract the geometry of the heart wall (Fig. 1b). The 3D geometry of the heart was reconstructed using the ScanIP module of Simpleware commercial software package (Bradninch Hall, UK) (Fig. 1c, d).

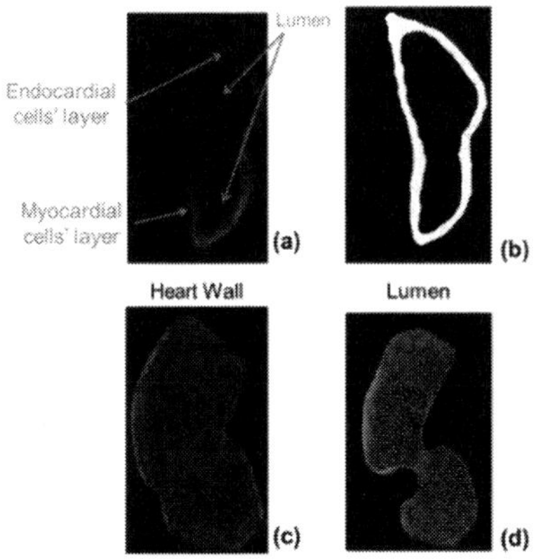

Figure 1. (a) Cross-section of the embryonic zebrafish heart 48 hpf acquired using confocal microscope; **(b)** The detected heart wall; The 3D reconstructed geometries of the **(c)** Heart wall and **(d)** Lumen.

The heart wall was assumed to be composed of a single-layer uniform hyperelastic material and the contraction was achieved by applying periodic external pressure on the atrium and the ventricle.

The governing equation for the motion of the heart wall was calculated using the balance of momentum,

$$\mathrm{M}\ddot{\mathbf{U}} + \mathrm{C}\dot{\mathbf{U}} + \mathrm{K}\mathbf{U} = \mathbf{R} \tag{1}$$

where M, C, K are the mass, damping and stiffness matrices; $\mathbf{R}$ is the vector of externally applied loads; $\mathbf{U}$ is the displacement vector and $\dot{\mathbf{U}}$ and $\ddot{\mathbf{U}}$ are the velocity and acceleration vectors of the heart wall, respectively.

Heart wall undergoes large displacements and strains during the cardiac cycle, and thus, it was assumed to behave as Neo-Hookean hyperelastic material for which the strain energy density function was given by,

$$W = \frac{G}{2}\left(\lambda_1^2 + \lambda_2^2 + \lambda_3^2 - 3\right) + \frac{\kappa}{2}\left(\lambda_1\lambda_2\lambda_3 - 1\right)^2 \tag{2}$$

where λ is the stretch ratio, $G=E/2(1+\nu)$ is the shear modulus, $\kappa=E/3(1-2\nu)$ is the bulk modulus, $E=100$ dyne/cm^2 is the Young's modulus and $\nu=0.45$ is the Poisson's ratio of the heart wall [5].

The embryonic blood was assumed to be incompressible and Newtonian fluid [6] due to low values of hematocrit. The blood flow was governed by the following Navier-Stokes and continuity equations:

$$\rho\left(\frac{\partial \mathbf{V}}{\partial t} + \mathbf{V}\cdot\nabla\mathbf{V}\right) = -\nabla P + \mu\nabla^2\mathbf{V} \tag{3}$$

$$\nabla\cdot\mathbf{V} = 0 \tag{4}$$

where $\mathbf{V}$ is velocity vector, P is a pressure, $\rho=1.035$ gr/cm^3 is the density and $\mu=0.04$ Poise is the dynamic viscosity of the blood [6, 7].

The constructed geometries were meshed and the governing Eqs. 1 - 4 were solved using commercial finite-element package of ADINA (Watertown, NY, USA), which can handle FSI problems.

RESULTS

The deformation of the heart wall and blood velocity vectors during ventricle contraction is depicted in Fig. 2a. The calculated results yielded atrium and ventricle pressure patters similar to those experimentally measured in zebrafish embryos [8, 9]. The magnitudes of calculated maximal atrium and ventricle pressure were ~0.018 (Fig. 2c) and ~0.024 mmHg (Fig. 2d), respectively, and the maximal calculated blood flow was ~4.5·10^{-7} cm^3/s (Fig. 2b). The calculated values were very close to experimentally collected data in zebrafish embryos [8, 9, 10].

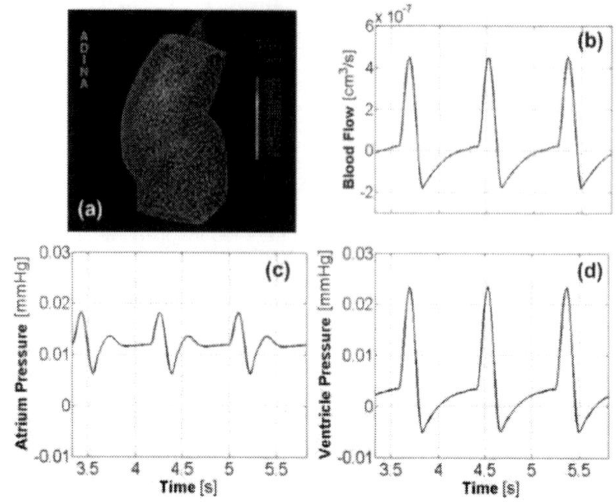

Figure 2. Computational results of blood flow in 48 hpf heart. **(a)** The deformation of the heart wall and blood velocity vectors during ventricle contraction; **(b)** Blood flow at heart outflow; **(c)** Atrium pressure; **(d)** Ventricular pressure.

DISCUSSION

The present computational method provided for the first time full and realistic hemodynamic characteristics in the tubular embryonic heart. This computational approach may serve as a platform for studying flow patterns and the resulting pressures and WSS distributions within the embryonic heart, and for a reversed engineering analysis of the physical loads on the embryonic heart due to perturbations that may mediate cardiac teratogenesis. Moreover, applying this method on both healthy and defected hearts may allow the correlation between blood flow patterns and cardiac malformations.

REFERENCES

[1] Hove, JH, *Pediatr. Res.*, 60: 6-13, 2006.

[2] Santhanakrishnan, A, Miller, LA, *Cell Biochem. Biophys*, 61, 1–22, 2011.

[3] Jamison, RA, *J Biomed Opt.*, 17, 2012.

[4] Li, B, Acton, ST, *IEEE Trans. Image Process.* 16, 2096–2106, 2007.

[5] Taber, LA, *Int. J. Dev. Biol.* 50, 323–332, 2006.

[6] Rideout, VC., *Prentice Hall, Englewood Cliffs, N.J*, 1991.

[7] Meier, G, *Embryologische Hefte* 1, 1–19, 1987.

[8] Hu, N et al., *Anat. Rec.* 260, 148–157, 2000.

[9] Hu, N et al., *Anat. Rec.* 264, 1–12, 2001.

[10] Pelster, B, Burggren, WW, *Circ. Res.* 79, 358–362, 1996.

SB³C2015
Summer Biomechanics, Bioengineering and Biotransport Conference
June 17-20, 2015,Snowbird Resort, Utah, USA
SB³C2015-267

DEVELOPMENT AND CHARACTERIZATION OF NOVEL BONE SUBSTITUTES USING COMPOSITE SCAFFOLDS WITH MESENCHYMAL STEM CELLS

M. Todo (1), Ph. Yos (2)

(1) Research Institute for Applied Mechanics
Kyushu University
Kasuga, Fukuoka, Japan

(2) Geo-resources and Geotechnical
Engineering Department,
Institute of Technology of Cambodia
Phnom Penh, Cambodia

INTRODUCTION

In bone tissue engineering, scaffolds play an important role for regenerating artificial bone tissues in vitro as the matrix for tissue formation and is required to have three-dimensional porous structure with high porosity, pore interconnectivity, uniform pore distribution, surface properties permitting cell adhesion, differentiation, non-cytotoxicity and osteoconductivity [1]. In order to develop porous structural scaffolds, different kinds of fabrication method such as use of organic porosifiers, sponge template [2,3], foaming, and freeze-drying [4] have been applied. Among which, the template method using polyurethane foam as template is known to produce a three dimensional fully interconnecting porous structure similar to spongy bone. In our previous study, we successfully developed two different structures of scaffold for bone tissue engineering by using a bioactive ceramics such as hydroxyapatite (HA) and biodegradable polymers such as PLLA and PCL [2]. However, their biocompatibilities have not been well understood yet.

The objective of this study is therefore to characterize the effects of cell culture within the composite scaffolds on the mechanical and biological properties of the scaffold/cell systems. Human mesenchymal stem cells (hMSC) were utilized as the cell source, and cultured in a standard medium for 28 days. Variational behavior of compressive mechanical properties, cell proliferation, and ALP activity were evaluated quantitatively. Effectiveness of two different kinds of scaffold, namely, HA/PCL 2-phase and coated scaffolds was discussed and compared on the basis of those properties.

EXPERIMENTAL

HA porous materials were fabricated by using the template method. HA slurry was prepared from commercial micro-HA powder mixed with PVA: PU sponge templates of 1x1x1 cm dimensions were immersed into the slurry, and then fully compressed to remove excess slurry. The slurry-immersed templates were dried at 60 °C for 24 hours and then sintered at 400 °C for 6 hours as the calcination process followed by the sintering process at 1300°C for 4 hours. PCL solution was prepared using 1,4-dioxane with a concentration of 5wt%. The sintered HA porous materials were dipped into the solution and vacuumed for 1 hour to remove the air. Two different conditions were then employed in the drying process. In the first drying process, the soaked HA porous materials were taken out of the PCL solution and then dried in an air oven at a room temperature for 24 hours. These samples were called 'HA/PCL coated scaffolds'. In the second drying process, the freeze-drying procedure was utilized to obtain 2-phase porous structure. These specimens were named as 'HA/PCL 2-phase scaffolds'.

hMSCs from passage 4 were cultured with the composite scaffolds. 1×10^5 cells in 100μl of medium were seeded on each scaffolds and vacuumed for 1minute then incubated for another hour to allow cells attachment. A standard medium consisting of MEMα, FBS and Pen-Strep was added to the samples, and incubated under the standard environment of cell culture for up to 4weeks. The medium was changed twice per week. Proliferation behavior of cells was determined using the cell counting kit (Dojindo). The optical densities of the prepared samples were evaluated using a plate reader at 37°C at wavelength of 450nm. Alkaline phosphatase activity (ALP) assay was also conducted using the LabAssay™ ALP kit (Wako). Reaction solutions containing p-nitrophenol were evaluated by the plate reader at 37°C at a wavelength of 450nm.

Morphology of the scaffolds was observed with FE-SEM. The samples were freeze-dried using a freeze drying machine, and the surfaces were coated with a thin layer of Pt/Pd using an ion sputter. Compression tests of the scaffolds with cells were conducted in order to understand the effects of cell culture on the mechanical properties. Compressive mechanical properties were evaluated from the stress-strain curves obtained from the tests.

MRESULTS AND DISCUSSION

Micro-porous structures of the composite scaffolds are shown in Fig.1. In the coated scaffold, HA surface was totally covered by PCL phase and membrane structures are also observed as shown in Fig.1(a). On the contrary, in the 2-phase scaffold, PLA porous phase was created inside of the HA porous structures.

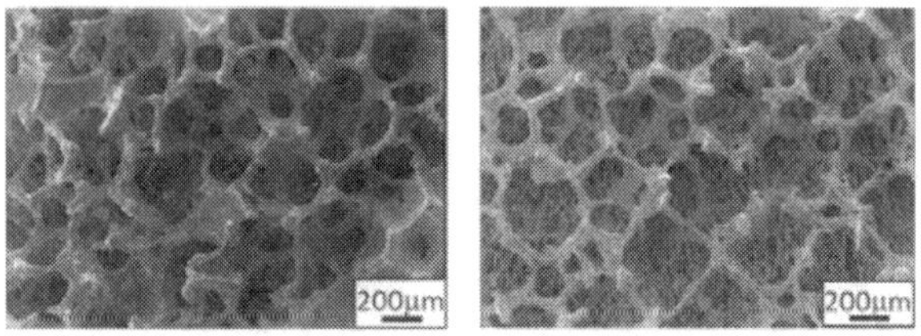

(a) Coated (b) 2-phase

Fig.1: Micro-porous structures of scaffolds.

FE-SEM micrographs of morphology of cell proliferation behavior after 1 and 4 weeks culture are shown in Fig.2. After 1 week, pores of the scaffolds were partially covered by cells with extra-cellular matrices (ECM). After 4 weeks, the most of the surfaces were totally covered by the biological structures created by cells and ECMs.

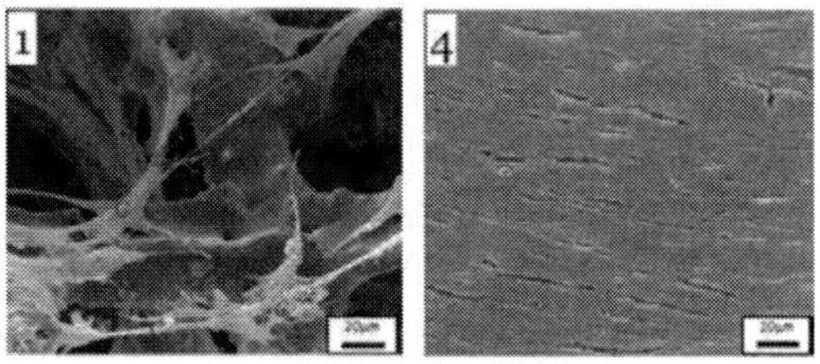

(a) Coated scaffold

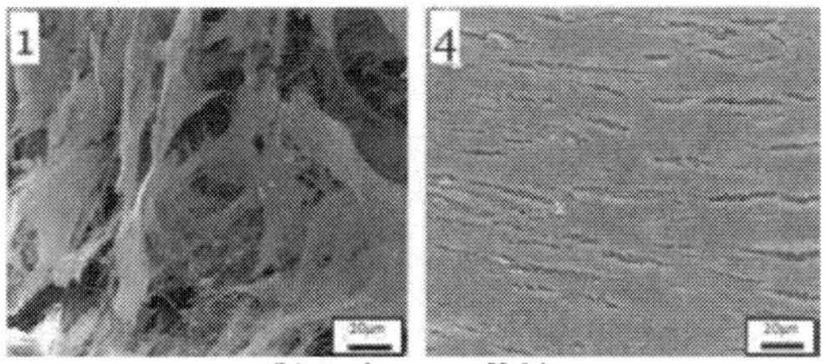

(b) 2-phase scaffold

Fig.2: FE-SEM micrographs of cell growth.

Cell viability and ALP activity of each scaffold are shown in Fig.3. It is clearly seen that the 2-phase scaffold exhibited much better viability than the coated scaffold. It is thought that in the 2-phase scaffold, cells can be proliferated on the surfaces both HA and PLC porous structures, while in the coated, only on the surfaces of HA covered with PCL layer. For both cases, cells monotonically increased up to 3 weeks of culture and then slightly decreased up to 4 weeks. ALP activities of both scaffolds were almost identical as shown in the figure. ALP activity increased up to 2 weeks and reached the peak value, then decreased up to 4 weeks.

Compressive mechanical properties such as elastic modulus and compressive strength are shown in Figs.4 and 5, respectively. These properties of scaffolds only (no cells were seeded) are also shown in the figures. It is obvious that the modulus of the coated scaffold is higher than that of the 2-phase, however, after 4 weeks, the modulus of the coated was almost identical with that of the 2-phase. For both scaffolds, cell culture improved the modulus effectively. In the coated with cells, the modulus firstly decreased up to 2 weeks, and then kept increased until 4 weeks. On the other hand, in the 2-phase with cells, the modulus monotonically increased up to 4 weeks. The compressive strength of the coated tended to be higher than that of the 2-phase. The strengths of both the scaffolds without cells tended to decrease with increase of culture days. It is obviously seen that the strengths were improved by cell culture. This improvement of the compressive mechanical properties is mainly due to cell proliferation with ECM formation that creates membrane structures enhancing structural stability and preventing fracture initiation.

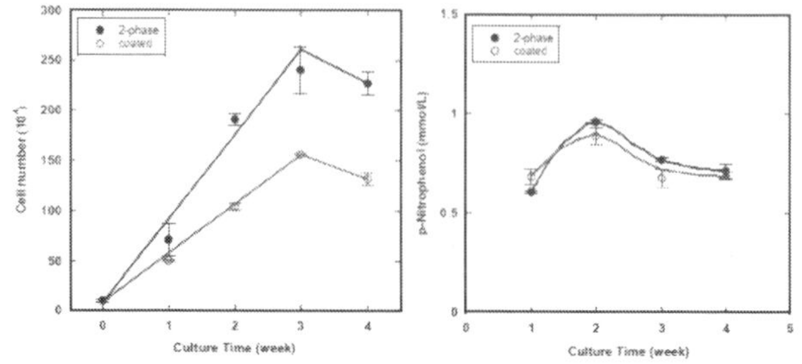

Fig.3: Variation of cell viability and ALP activity.

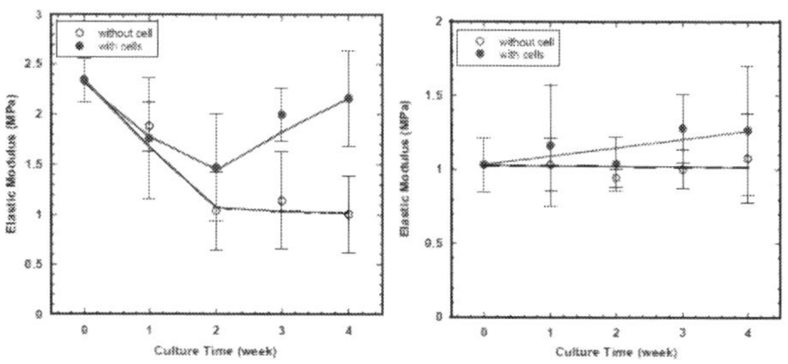

(a) Coated (b) 2-phase

Fig.4: Variation of compressive modulus.

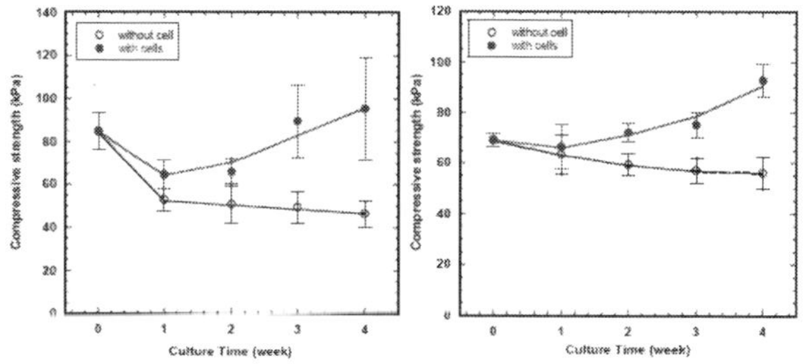

(a) Coated (b) 2-phase

Fig.5: Variation of compressive strength.

CONCLUSION

Two kinds of HA/PCL composite scaffolds were developed and hMSCs were cultured within the scaffolds in order to examine the biocompatibility. The 2-phase scaffold showed better cell proliferation behavior than the coated scaffold, while the compressive mechanical properties of the coated were higher than those of the 2-phase. It is worth noting that cell culture improved the mechanical properties through proliferation and ECM formation that enhance the structures of the scaffolds.

REFERENCES

[1] Lee, S.H., and Shin, H., Adv. Drug. Deliv. Rev., 59: 339, 2007.
[2] Yos, Ph., and Todo, M., Key. Eng. Mater., 529-530: 447, 2013.
[3] Yos, Ph., and Todo, M., Evergreen, 01:1, 2014.
[4] Whang, K., et al., Tissue Eng., 5:35, 1999.

DEVELOPMENT OF ARTIFICIAL CARTILAGE USING TWO PHASE COLLAGEN SCAFFOLD WITH MESENCHYMAL STEM CELLS

M. Todo (1), U. Nakamuta (2), T. Arahira (3)

(1) Research Institute for Applied Mechanics
Kyushu University
Kasuga, Fukuoka, Japan

(2) Interdisciplinary Graduate School of
Engineering Science, Kyushu University
Kasuga, Fukuoka, Japan

(3) Fukuoka Dental College
Fukuoka, Japan

INTRODUCTION

Tissue engineering approach for cartilage regeneration has clinically been conducted by using chondrocytes with collagen gel scaffold [1-3]. However, such engineered cartilage tissue cannot be implanted into the load-bearing part of joint surface because of luck of mechanical properties such as elastic modulus, strength and viscoelastic properties. Therefore, engineering scaffolds with better mechanical properties has actively been developed in the field of cartilage tissue engineering. In previous research, we have prepared collagen gel/sponge two-phase scaffold that was a composite of type I collagen gel and type II collagen sponge. The study showed that a gel/sponge composite scaffold has excellent mechanical and biological properties than pure gel and sponge scaffolds. However, chondrogenic medium cannot increase cell number. Compressive modulus tends to decrease due to low cell viability. In the present study, combined culture with normal and chodrogenic medium was introduced in order to enhance cell proliferation. Human mesenchymal stem cells (hMSCs) were cultured using composite scaffold up to 28days.

EXPERIMENTAL

Type I collagen gel as the basic scaffold was produced using Cellmatrix TypeI-A. It was gelled after mixing collagen solution, 10 times concentrated MEM-α, and buffer solution. The mixing ratio was selected as 8:1:1(Fig.1 (a)). Collagen sponge was fabricated by the freeze-drying method using chicken cartilage collagen solution (type II collagen). Gel/sponge composite scaffold was produced by immersing the collagen sponge into the collagen gel (Fig.1 (b)). hMSCs were cultured using gel/sponge composite scaffold up to 28days. Firstly, we seeded 1.0×105 cells per scaffold. Next, we added the normal medium(MEM-α , 10%FBS , 1% penicillin-streptomycin), and performed 14 days culture. We then added the differentiation medium (RPMI1640, FBS, ascorbic acid, amino acids, antibiotics), and performed 14 days culture. Compressive mechanical properties, cell viability and chondrogenic marker were measured. Also the micro-structures were observed using FE-SEM.

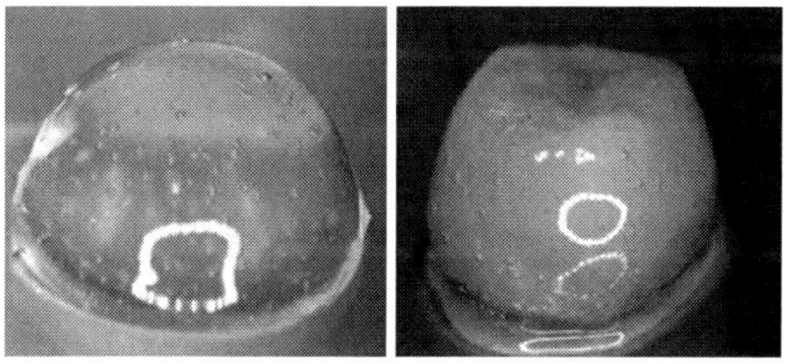

(a) Gel (b) Gel/sponge

Fig.1: Over views of collagen based scaffolds.

MRESULTS AND DISCUSSION

Variation of cell viability is shown in Fig.2. For the condition of the chondrogenic differentiation culture of 14 days, the cell number kept constant. However, for the combined culture condition, cell number increased by the normal culture for 14 days and stayed at a constant value during the chondrogenic culture.

Gene expression of type II collagen and aggrecan which are typical proteins in cartilage tissue are shown in Fig.3 and Fig.4. Type II collagen and aggrecan do not increase in the normal medium. However, they tended to increase with days in culture in the differentiation medium. Increasing chondrogenic marker implies that hMSCs were differentiated into chondrocytes.

The compressive mechanical properties of the scaffolds with cells are shown in Fig.5. Compression modulus increased by the normal culture for 14 days, and the modulus still increased during the chondrogenic differentiation culture. It is thought that the structural stability was improved by cell proliferation and extracellular matrix formation, resulting in the recovery of the compressive modulus by the normal culture for 14 days. It is thought that if the cell number was increased by the normal culture for 14 days at the beginning, scaffold can retain a high elastic modulus after 14 days chondrogenic differentiation culture.

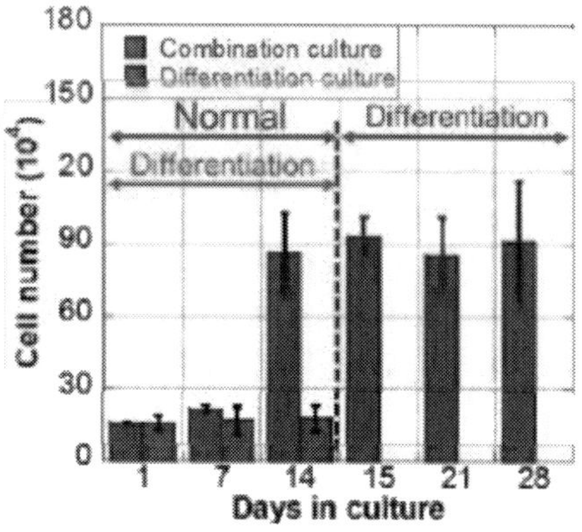

Fig.2: Variation of cell viability.

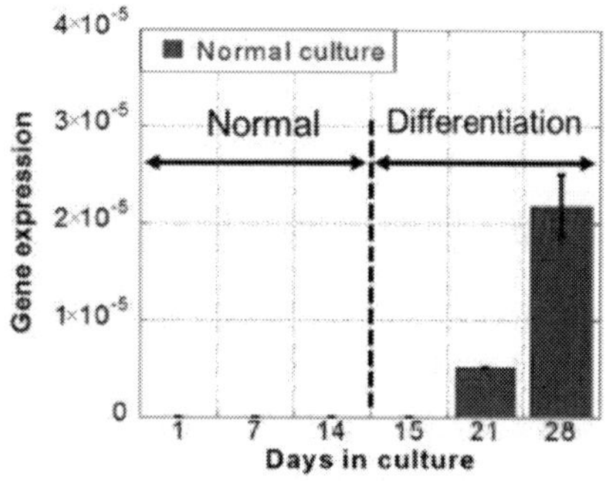

Fig.3: Variation of type II collagen.

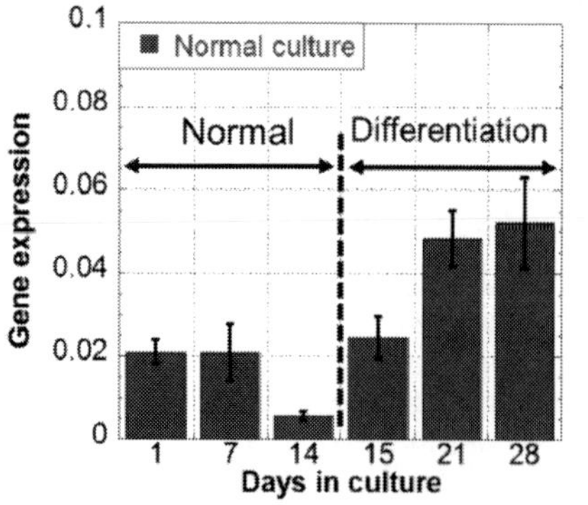

Fig.4: Variation of aggrecan.

FE-SEM micrographs of cells on the scaffold surface and cross-section after 28 days culture are shown in Fig.6. The spherical shape of the cells implies that MSCs were differentiated into chondrocytes. The fibril structures observed in the vicinity of cells are thought to be type II collagen fibrils created by the cells.

CONCLUSION

Combined culture with normal and chodrogenic medium was introduced in order to enhance cell proliferation. Cell number effectively increased during the normal culture up to 90 x 10^4 and still slightly increased during the chondrogenic culture with increasing type II collagen and aggrecan. The compressive modulus increased up to 4.5 kPa after the combined culture, while the modulus decreased to 2 kPa with only chondrogenic culture. It is thus concluded that the gel/sponge composite material developed in this study can be one of the promising candidates as a scaffold with hMSCs for cartilage tissue engineering.

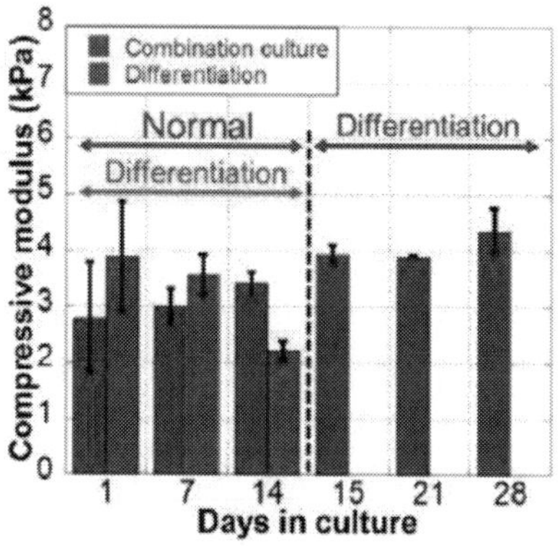

Fig.5: Compressive mechanical properties.

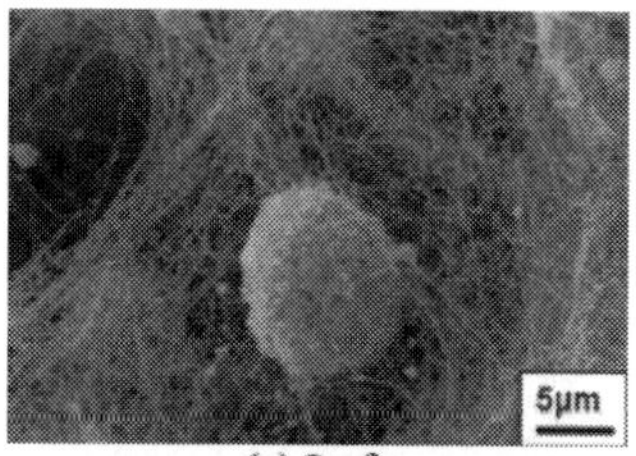

(a) Surface

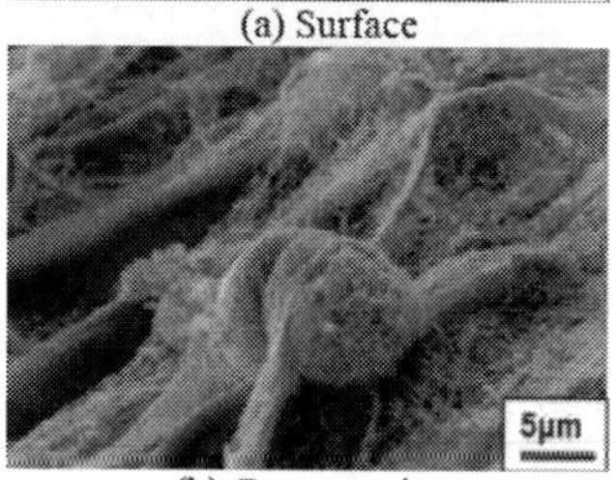

(b) Cross-section

Fig.6: FE-SEM micrographs of cells after 28 days culture.

REFERENCES

[1] Lee, H.H., e al., Osteoarthritis and Cartilage,21:385-93, 2013.

[2] Schaefer, D., et al., Biomaterials, 21: 2599-606, 2000.

[3] Chen, J., et al., Biomaterials, 32:4793-805, 2011.

IBUPROFEN DOES NOT ADVERSELY AFFECT SUPRASPINATUS TENDON MECHANICAL ADAPTATIONS IN A RAT MODEL OF EXERCISE

Sarah I. Rooney, Rachel Baskin, Andrew F. Kuntz, Louis J. Soslowsky

McKay Orthopaedic Research Laboratory
University of Pennsylvania
Philadelphia, PA, USA

INTRODUCTION

In the United States, acetaminophen, ibuprofen, and aspirin are the top 3 consumed drugs by adults [1]. Specifically, many athletes self-diagnose and "treat" themselves with these over-the-counter medications or use them prophylactically: for example, 95.7% of college football players have used nonsteroidal anti-inflammatory drugs (NSAIDs) such as ibuprofen [2]. Previous studies have shown that following acute tendon injury, administration of ibuprofen, which blocks COX-1 and COX-2 of the arachidonic acid inflammatory cascade, is detrimental [e.g., 3], suggesting that inflammation is a required component of successful tissue healing. A mild inflammatory response may also be important in the adaptation of musculoskeletal tissues to chronic, non-injurious exercise; however current literature on the effects of ibuprofen on tissue adaptations is equivocal. For example, in a plantaris overload model in rats, one study found 50% decreased muscle hypertrophy with administration of ibuprofen [4], and ibuprofen administered to mice undergoing 4 weeks of wheel running eliminated the correlation between run distance and skeletal muscle adaptations [5]. In contrast, following chronic resistance training in older adults, ibuprofen administration led to increased muscle hypertrophy and strength but did not affect tendon cross-sectional area or strain [6]. Overall, little is known about the interaction between ibuprofen, a commonly consumed drug, and the response of tendons to exercise. Therefore, the objective of this study was to determine the effects of ibuprofen on the mechanical properties of supraspinatus tendon in a rat model of exercise. We hypothesized that administration of ibuprofen would abolish the beneficial tendon adaptations found with exercise but have no effect on sedentary tendon mechanical properties.

METHODS

Following IACUC approval, 104 male, Sprague-Dawley rats (400-450g) were divided into cage activity (CA) or exercise (EX) groups. These groups were further divided into 2 or 8 week time points (CA2, EX2, CA8, EX8) with half of these animals receiving ibuprofen (IBU) for this duration (n=12-17 per group). Animals in the EX groups underwent 2 weeks of treadmill acclimation followed by 2 or 8 weeks of flat treadmill walking at 10m/min, 1 hour/day, 5 days/week as previously described [7]. CA animals maintained normal cage activity for the duration of the study. Animals in the IBU groups were administered liquid ibuprofen at 20mg/kg orally via syringe every 12 hours, 7 days per week for 2 or 8 weeks as previously described [3]. Upon completion of the protocol, animals were sacrificed, weighed, and stored at -20°C.

For mechanical testing, animals were thawed, and the left shoulder of each animal was dissected, keeping the supraspinatus tendon attached to its bony insertion on the humerus. Verhoeff's stain lines for optical strain measurements were placed at the bony insertion site and 2, 4, and 8 mm proximally. Tendon cross-sectional area was measured with a custom laser device [8]. Cyanoacrylate was used to fix the tendon between two pieces of sandpaper at the 8 mm stain line and gripped with a custom screw clamp. Humeri were potted in PMMA, and a custom fixture secured the potted humerus for testing with an Instron ElectroPuls E3000 affixed with a 250 N load cell and a LVDT for increased displacement resolution. Tendons were submerged in a 37°C PBS bath and underwent the following testing protocol: 1) preconditioning (10 triangle cycles from 1-1.5% strain at 0.25 Hz, followed by hold at gauge length for 300s), 2) stress-relaxation to 4% strain held for 300s, 3) frequency sweep (0.1, 1, 2, 10 Hz) of 10 sine cycles with amplitude of 0.125% strain, 4) return to gauge length and hold for 60s, 5) stress-relaxation to 8% strain held for 300s, 6) recovery to 4% strain held for 300s, 7) return to 8% strain and held for 300s, 8) frequency sweep (0.1, 1, 2, 10 Hz) of 10 sine cycles with amplitude of 0.125% strain, 9) return to gauge length and hold for 180s, 10) ramp to failure at 0.3%/s. Only the 8% stress-relaxation (step 5) and the ramp to failure (step 10) were used for the present analysis. Local 2D Lagrangian strain of the tendon relative to the bone during the ramp to failure was measured by tracking the stain lines with a custom MATLAB program. Elastic modulus was calculated as the slope of the linear region of the stress-strain curve.

To determine the effects of exercise and ibuprofen, separate 2-way ANOVAs were performed for the 2 and 8 week time points. Each ANOVA had 2 factors (activity and drug) with 2 levels each (CA/EX and none/IBU) and an interaction term. Significance of the ANOVA effects was set to $p \leq 0.05$ and trends at $p \leq 0.1$. If an interaction effect was significant or a trend, pairwise comparisons were performed with t-tests. A Bonferroni correction for the 4 planned comparisons set significance to $p \leq 0.01$ and trends at $p \leq 0.03$ for these contrasts. Figures are shown with mean and standard deviation (* indicates significance and + indicates trends).

RESULTS

IBU had a significant main effect to decrease body weight gain (not shown) and tendon cross-sectional area (CSA, **Fig 1**) but increase tendon modulus (**Fig 2**) at 2 and 8 weeks. Modulus trended toward increased with EX at 2 weeks (**Fig 2**). Stiffness trended toward increased with EX at 8 weeks (not shown). Pairwise comparisons revealed trends toward increased stiffness with IBU at 2 weeks in the CA group and increased stiffness with EX in the non-drug treated group (not shown). At 8 weeks, max load trended toward decreased with IBU and increased with EX (not shown). A significant interaction effect for max stress at 2 weeks followed by pairwise comparisons revealed that IBU increased max stress in the CA group, and EX increased max stress in the non-drug treated group (**Fig 3**). Percent relaxation showed no differences at 8 weeks but decreased with EX in the non-drug treated group at 2 weeks (**Fig 4**).

DISCUSSION

Results suggest that chronic intake of ibuprofen at pharmacologic doses does not detrimentally affect supraspinatus tendon mechanical adaptation to exercise. Few interaction effects were detected, suggesting that ibuprofen does not differentially impact sedentary and exercised animals. This finding is in contrast to our hypothesis and suggests that the arachidonic acid cascade may not play a major role in the adaptation of tendon to load in an exercise model, in contrast to that seen in an acute injury model. Other factors, such as matrix metalloproteinases that regulate matrix turnover, may play a more important role. Surprisingly, animals receiving ibuprofen gained less weight over the course of the study, which could be due to decreased animal growth, increased metabolic rates, or decreased food intake.

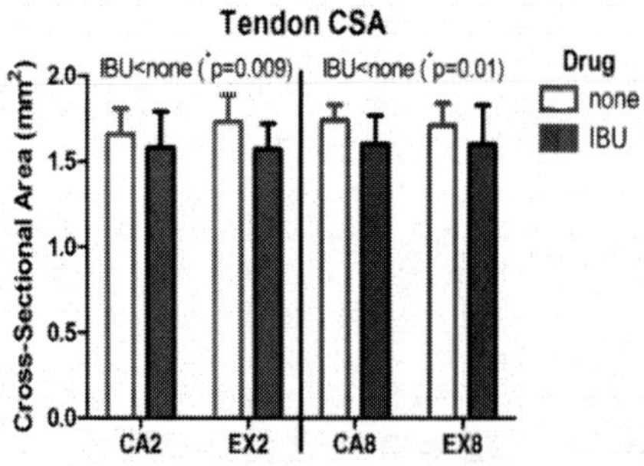

FIG 1. IBU DECREASED TENDON CSA AT BOTH TIME POINTS.

These parameters were not measured in this study but would be of interest in future studies. Future studies will also investigate how ibuprofen affects inflammatory (prostaglandins, cytokines) and matrix (collagen, proteoglycans, MMP/TIMP) proteins in tendons acutely and chronically. In conclusion, ibuprofen does not adversely affect the adaptation of supraspinatus tendon in a rat model of exercise.

ACKNOWLEDGEMENTS

The authors thank R. Vafa. This study was funded by the Penn Center for Musculoskeletal Disorders (NIH/NIAMS P30 AR050950).

REFERENCES

[1] Kaufman DW et al., *JAMA*, 287:337-44, 2002. **[2]** Holmes N et al., *Clin J Sport Med*, 23:283-6, 2013. **[3]** Connizzo BK et al., *Clin Orthop Relat Res*, 472:2433-9, 2014. **[4]** Soltow QA et al., *Med Sci Sports Exerc*, 38:840-6, 2006. **[5]** Machida M et al., *J Physiol Pharmacol*, 61:559-63, 2010. **[6]** Carroll C et al., *J Appl Physiol*, 111:508-15, 2011. **[7]** Rooney SI et al., *Muscles Ligaments Tendons J*, 4:413-9, 2014. **[8]** Peltz CD et al., *J Orthop Res*, 27:416-20, 2009.

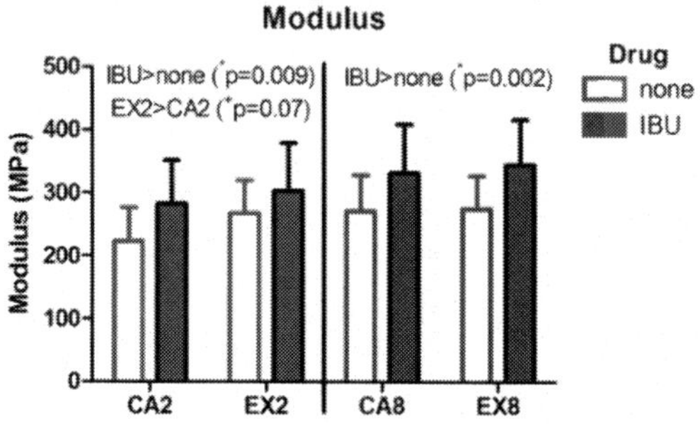

FIG 2. IBU INCREASED TENDON MODULUS AT BOTH TIME POINTS, AND EXERCISE TRENDED TOWARD INCREASING MODULUS AT 2 WEEKS.

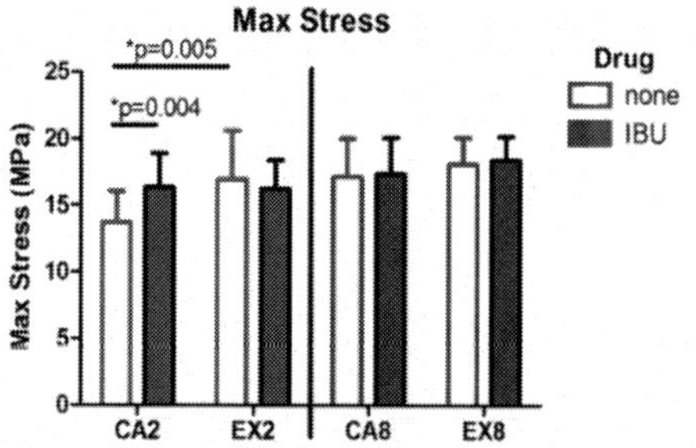

FIG 3. IBU INCREASED FAILURE STRESS WITH CAGE ACTIVITY, AND EXERCISE INCREASED FAILURE STRESS IN THE NON-DRUG TREATED GROUP, BOTH AT 2 WEEKS.

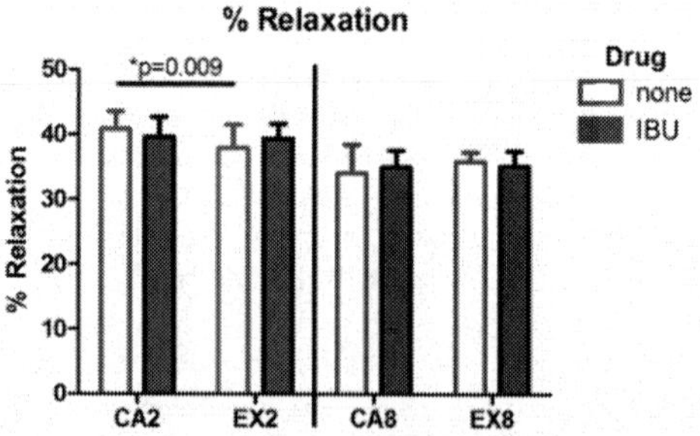

FIG 4. PERCENT RELAXATION DECREASED WITH EXERCISE IN THE NON-DRUG TREATED GROUP AT 2 WEEKS.

SB³C2015
Summer Biomechanics, Bioengineering and Biotransport Conference
June 17-20, 2015,Snowbird Resort, Utah, USA
SB³C2015-273

SALMON THROMBIN INHIBITS ENDOTHELIAL BARRIER DISRUPTION THROUGH ITS ACTIVATION OF PROTEIN C

Jenell R. Smith (1), Peter A. Galie (1,2), David R. Slochower (2), Paul A. Janmey (1,2), Beth A. Winkelstein (1,3)

(1) Department of Bioengineering
University of Pennsylvania
Philadelphia, PA, USA

(2) Department of Physiology
University of Pennsylvania
Philadelphia, PA, USA

(3) Department of Neurosurgery
University of Pennsylvania
Philadelphia, PA, USA

INTRODUCTION

A variety of diseases are characterized by modified vascular permeability. In cancerous tumors, leaky blood vessels provide nutrients to tumor cells [1]. During central nervous system edema, blood-brain barrier breakdown enables infiltration of immune cells, which fuel neuroinflammation [2]. Thrombin is the prototypical mediator of vascular permeability with the capability to both permeabilize and strengthen endothelial junctions, depending on the substrate it activates [3,4]. Mammalian thrombin at low concentrations binds to cofactors at the endothelial cell surface facilitating activation of protein C (PC), a serum component that also binds to surface receptors [5-7]. Activated protein C (aPC) then further activates cell-bound receptors that initiate signaling cascades to stabilize endothelial integrity [5-7]. At high concentrations, thrombin shifts its preference to potently activate protease-activated receptor-1 (PAR-1), which induces mechanisms that increase vascular permeability [4,8]. It is not known whether other thrombin properties, such as its species of origin, also influence substrate preference in addition to concentration.

Thrombin derived from salmon blood has been reported to have different enzymatic effects than its mammalian counterpart, believed to be due to the co-evolution of the coagulation and inflammatory cascades in fish and mammals [9-12]. We recently found that salmon thrombin activates PAR-1 substantially less potently than does human thrombin [11], suggesting that the two species of thrombin might also have different affinities for other target substrates, such as PC. Salmon thrombin also initiates different inflammatory responses than human thrombin [11,12]; astrocytic cultures treated with human thrombin transcribe substantially more pro-inflammatory cytokines, TNF-α and IL-1β, and secrete more IL-6, than salmon thrombin-treated cultures [11]. Although salmon and human thrombins exhibit differences in receptor activation and inflammatory cascade initiation, it is unknown if they also regulate endothelial permeability differently and if it is controlled by protein C activation.

This work investigates whether salmon and human thrombins differentially regulate endothelial permeability through their level of PC activation, using two coordinated studies. In the first study, the effects of each species of thrombin on endothelial barrier permeability are measured using an in vitro system of endothelial-lined microchannels to quantify the extent of diffusion out of channels that have been exposed to thrombin. Because thrombin protects endothelial barriers indirectly through the serum component PC, experiments were performed under serum and serum-free conditions to determine if any detected differences might be PC-dependent. In the second study, the activation of PC by salmon thrombin and human thrombin separately was quantified using a fluorogenic substrate assay. This work aims to further measure the enzymatic differences between salmon and human thrombin and their effects on cellular signaling.

METHODS

To test the relative effects of salmon and human thrombin on endothelial permeability, HUVECs were seeded on the inner surface of microchannels that were embedded within collagen (Fig. 1) [13]. Initially impermeable cell-lined channels were subjected to flow (5μl/min) containing human TNF-α (100ng/ml, Sigma) to induce permeability [7]. Fluorescent dextran (70kDa, Sigma) enabled tracking of flow leakage into the surrounding collagen (Fig. 1). Under serum-containing conditions, cells were treated with salmon thrombin (STh, 1U/ml, Sea Run Holdings) or

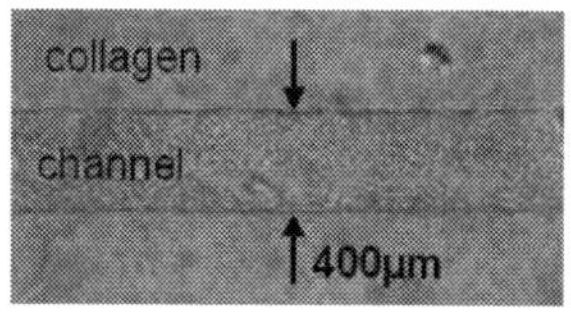

FIG. 1. HUVEC-LINED MICROCHANNELS USED TO QUANTIFY TNF-α-INDUCED PERMEABILITY.

human thrombin (HTh, 1U/ml, Sigma) in addition to the TNF-α. Fluorescent images of the channel were acquired over 10 minutes to measure the flux of dextran into the collagen. To test whether serum components, such as PC, are required for the action of STh, the experiment was repeated after serum-starving the cells for 4 hours. All fluxes were normalized to that for TNF-α with serum; groups were compared (n=3/group) using a two-way ANOVA with Tukey's test.

To determine whether the thrombin species enzymatically convert PC into aPC at different rates, either STh or HTh and PC were added to solutions containing the substrate for aPC. Each thrombin species (1U/ml) was separately added to human PC (19μg/ml, Haematologic Technologies) and a fluorogenic substrate (50μg/ml, Sigma) that maps the peptide cleaved by aPC (Leu-Ser-Thr-Arg). Fluorescence was measured for 5 minutes; assays were run in triplicate. The slopes of the fluorescence-time curves were corrected by subtracting the slope from the respective thrombin species and the substrate alone to account for any substrate cleavage by thrombin. Corrected slopes were compared between STh and HTh by an ANOVA with Tukey's test.

RESULTS

TNF-α increases the flux of dextran out of the HUVEC-lined channels regardless of the serum content (Fig. 2). Dextran flux is significantly blunted ($p<0.01$) for STh in the presence of serum compared to all other groups (Fig. 2). STh does not reduce the flux in the serum free environment, and the flux induced by TNF-α and STh is significantly higher ($p<0.01$) relative to flow with serum (Fig. 2). Conversely, TNF-α and HTh significantly increases ($p=0.03$) dextran flux compared to TNF-α alone (Fig. 2). STh is significantly more effective ($p=0.008$) at activating PC (18.29±0.9 intensity/sec) compared to HTh (0.42±1.95 intensity/sec).

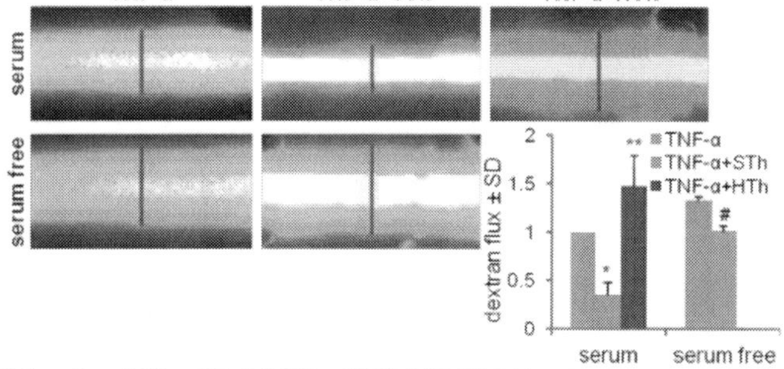

FIG. 2. STh BLOCKS ENDOTHELIAL PERMEABILITY INDUCED BY TNF-α COMPARED TO ALL GROUPS (*P<0.01). HTh INCREASES DEXTRAN FLUX (P=0.03) OVER TNF-α. TNF-α+STh WITHOUT SERUM IS HIGHER (#P<0.01) THAN WITH SERUM.**

DISCUSSION

These studies are the first to demonstrate that salmon thrombin induces the opposite effects on vascular permeability as human thrombin. When serum components are present, salmon thrombin actively blunts the breakdown of barrier disruption induced by TNF-α (Fig. 2). However, those effects of barrier protection provided by salmon thrombin are diminished under serum free conditions, which may suggest that salmon thrombin indirectly strengthens endothelial integrity via a component in serum. Salmon thrombin also activates PC faster than does human thrombin. Since the serum component, PC, bound to the endothelial protein C receptor can be activated by thrombin to initiate signaling cascades that strengthen transendothelial junctions [5-7], these data support the hypothesis that salmon thrombin protects endothelial barrier integrity through PC.

Previous mutagenesis studies manipulating the structure of mammalian thrombin defined key residues and regions in its native amino acid structure that direct the enzymatic activity of thrombin towards substrates [14,15]. The deletion of residues 146-154 (chromotrypsin numbering) drastically shifts its enzymatic activity from PAR-1 to PC [14,16]. This sequence of residues is highly divergent in fish and human thrombin [14]. Accordingly, to determine if the disparity in primary structures translates into a difference in predicted tertiary structure and if that region interacts with PC, we modeled that loop (UniProt: P00734) into a previously validated crystal structure for human thrombin (PDB: 4DT7 chain B). The fully sequenced trout thrombin (UniProt: Q5NKF9), which is most similar in primary structure to salmon thrombin and identical in the 146-154 loop [14], was threaded into the human thrombin model. Tertiary structure was conserved between species for most regions, as predicted by only conservative differences in their primary structures. However, the 146-154 loop varies between the two thrombin species, both in the constituent residues and the structure, and closely interacts with PC (Fig. 3); which may provide structural evidence for the differences in substrate specificity observed between them. Together with previous reports [10-12], these findings provide evidence for salmon thrombin having unique effects on cellular function compared to mammalian thrombin through its activation of PC and support its use to treat disorders that are associated with leaky vasculature.

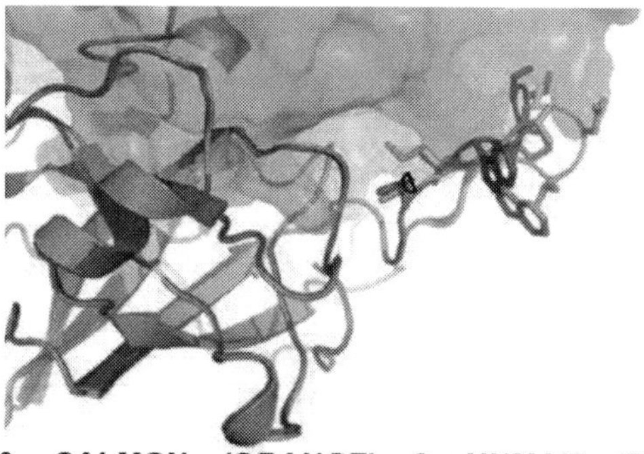

FIG. 3. SALMON (ORANGE) & HUMAN (PURPLE) THROMBIN DIFFER IN PRIMARY & TERTIARY STRUCTURE IN THE 146-154 LOOP (CIRCLED) THAT INTERACTS WITH PC (GREEN).

ACKNOWLEDGEMENTS

Funded by DOD (W81XWH-10-1-1002), CSRS, Sharpe Foundation, Ashton Foundation, and NIH (T32-AR007132).

REFERENCES

[1] Zhou, W et al., *Cancer Cell*, 25:501-515, 2014.
[2] Ghosh, A et al., *PLoS One*, 9:e98143, 2014.
[3] Griffin, J, *Nature*, 378:337-338, 1995.
[4] Komarova, Y et al., *Sci STKE*, 2007:re8, 2007.
[5] Bouwens, E et al., *J Thromb Haemost*, 11:242-253, 2013.
[6] Feistritzer, C et al., *J Thromb Haemost*, 3:2798-2805, 2005.
[7] Riewald, M et al., *J Biol Chem*, 280:19808-19814, 2005.
[8] Coughlin, S, *Nature*, 407:258-264, 2000.
[9] Doolittle, R, *J Innate Immun*, 3:9-16, 2011.
[10] Michaud, S et al., *Thromb Res*, 107:245-254, 2002.
[11] Smith, J et al., *PLoS One*, 8:e80006, 2013.
[12] Weisshaar, C et al., *Biomaterials*, 32:9738-9746, 2011.
[13] Galie, P et al., *Proc Natl Acad Sci*, 111:7968-7973, 2014.
[14] Dang, Q et al., *J Biol Chem*, 272:19649-19651, 1997.
[15] Marino, F et al., *J Biol Chem*, 285:19145-19152, 2010.
[16] Fuentes-Prior, P et al., *Nature*, 404:518-525, 2000.

COMBINED EXPERIMENTAL AND FINITE ELEMENT ANALYSIS TO DETERMINE THE DIFFUSION COEFFICIENT WITHIN AND BETWEEN HUMAN SKIN LAYERS

Anne M. Römgens (1), Dan L. Bader (1,2), Frank P.T. Baaijens (1), Cees W. J. Oomens (1)

(1) Department of Biomedical Engineering
Eindhoven University of Technology
Eindhoven, The Netherlands

(2) Faculty of Health Sciences
University of Southampton
Southampton, UK

INTRODUCTION

Administration of various drugs in or through the skin can have advantages above more conventional administration routes, like intramuscular injection or delivery via the gastrointestinal tract. The intradermal route may be preferred particularly for vaccine delivery, because the skin is rich in immune-responsive cells and, therefore, a lower dose may be required compared to intramuscular injection to achieve an equivalent immune response. Techniques like microneedle arrays and microjet systems are being developed to breach the upper layer of the skin, the stratum corneum, and subsequently administrate the drug. However, to controllably release drugs and target specific cells in various skin layers, it is important to understand the transport of these molecules in skin after intradermal delivery. This is essential to improve the design of delivery systems or delivery strategy.

The diffusion coefficient is a key parameter to describe passive transport of a molecule in a material. A previous study determined the diffusion coefficient of macromolecules within various skin layers of mice [1]. However, their method is not feasible for thicker human skin. Diffusion in human skin was previously determined for small molecules or for a single layer. However, the diffusion coefficient may not only vary between skin layers, but also within skin layers [1]. These data are still lacking for human skin.

Accordingly, the aim of this study was to determine the diffusion coefficient of human skin throughout its different layers by using fluorescent recovery after photobleaching (FRAP) experiments [2] in combination with finite element (FE) analysis.

METHODS

Human skin samples of 0.8 x 0.8 cm were prepared and cut perpendicular to the skin surface. Subsequently, the samples were submerged in 0.8-1.2 mg/ml 500 kDa fluorescein dextran and stored at 4°C for at least 20 hours to permit complete diffusion of the fluorescent macromolecules into the skin.

Each sample was placed on a microscope slide with the perpendicular cutting plane down. Then, FRAP experiments were performed at 7 locations at various depths in the epidermis, papillary and reticular dermis. At each location a global image visualizing the various skin layers was made (Fig 1a). Subsequently, 10 images were recorded at 10x higher magnification, hereafter called pre-bleach images. Immediately after that, a circular area with a diameter of 30 μm was bleached at 100% laser power. The recovery of the fluorescence was monitored 40-50 μm inside the sample for 100 sec at 4 Hz, resulting in a spatiotemporal intensity distribution $I(x,y,t)$.

To determine the diffusion coefficient from the experiments, each post-bleach image was subtracted from the average intensity distribution of the 10 pre-bleach images, resulting in relative changes in intensity (Fig 1b). Because the intensity was linearly related to the concentration within the used range, the set of images provided the spatiotemporal change in the concentration distribution $C_{exp}(x,y,t)$. These FRAP experiments were subsequently simulated using a two-dimensional FE model in Abaqus/Standard, which solved the time-dependent diffusion equation for dextran. The finite element mesh had the same dimensions as the global image (Fig 1a). The concentrations of the first post-bleach image $C_{exp}(x,y,0)$ were used as initial values for the center of the FE model. The mean concentration of the outer pixels of this post-bleach image (> 67.5 μm from the center) was assigned as the initial concentration of the rest of the mesh. The outer boundaries of the model were kept at this concentration during the entire simulation. Different skin layers, which can be seen in Fig 1a, may have different diffusion coefficients. This was taken into account by manually dividing the global image (Fig 1a), based on stains, into the epidermal, papillary dermal, and reticular dermal part and a part which contained the stratum

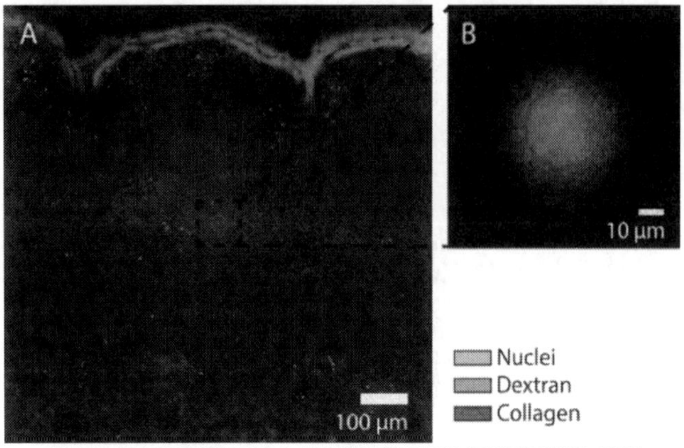

FIG 1. A) EXAMPLE OF A GLOBAL IMAGE OF THE SKIN. THE BLEACHING LOCATION IS IN THE CENTER OF THIS IMAGE. B) A FIRST POST-BLEACH IMAGE SUBSTRACTED FROM THE AVERAGE PRE-BLEACH IMAGE. A HIGH INTENSITY REPRESENTS A LARGE CHANGE IN INTENSITY COMPARED TO THE PRE-BLEACHED SITUATION. THE RED CIRCLES REPRESENT THE BLEACHED AREA

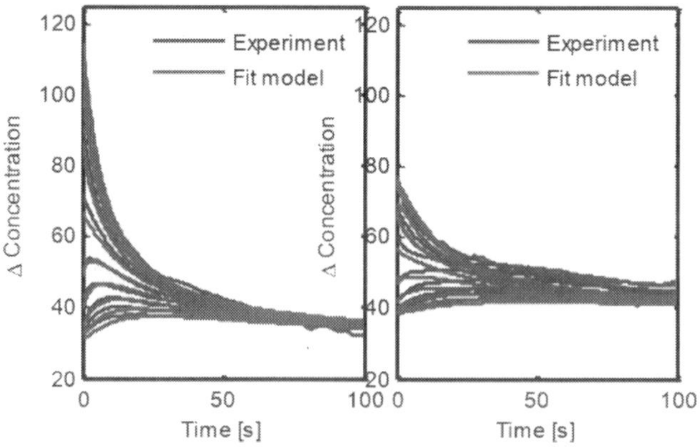

FIG 2. EXAMPLE OF THE CHANGE IN CONCENTRATION OF A SINGLE EXPERIMENT IN THE PAPILLARY (LEFT) AND RETICULAR DERMIS (RIGHT), AND THE FINITE ELEMENT ANALYSIS FITTED TO THE EXPERIMENT. EACH LINE REPRESENTS THE MEAN INTENSITY OF A CIRCLE WITH A WIDTH OF 2.1 µM AROUND THE CENTER OF THE IMAGE. NOTE THAT A HIGH DIFFERENCE IN CONCENTRATION REPRESENTS THE BLEACHED AREA, BECAUSE THE INTENSITIES WERE SUBSTRACTED FROM THE PRE-BLEACH IMAGE

corneum and air around the sample. Each element in the model was assigned to the region in which the center of the element was located.

After running the simulation with an initial value for the diffusion coefficient, the spatiotemporal change in the concentration distribution of the analysis $C_{sim}(x,y,t)$ was compared with the experimental concentration $C_{exp}(x,y,t)$. By iteratively changing the diffusion coefficient in the model, the difference between C_{sim} and C_{exp} was minimized using the following optimization function f (1):

$$f = \Sigma\left(\left(C_{exp}(x,y,t) - C_{sim}(x,y,t)\right)^2\right). \quad (1)$$

During optimization, the diffusion coefficient of the current layer was iteratively adapted, while keeping the diffusion coefficients of other layers constant. The diffusion coefficient at different locations in the reticular dermis was determined first. This was followed by the locations in the papillary dermis, using the previously determined coefficients of the reticular dermis. Last, the diffusion coefficient of the epidermis was fitted, again using previous coefficients for other layers. This process was repeated until all diffusion coefficients did not change by more than 1 percent compared to the previous iteration. Based on the quality of fit some experiments were not used for further analysis.

RESULTS

The spatiotemporal change in concentration simulated with the finite element model corresponded well with the change in concentration of the experiments (Fig. 2).

The estimated diffusion coefficient from all the experiments of 500 kDa fluorescein dextran at various depths in the epidermis, papillary and reticular dermis ranged between 5.3 and 13.3 $\mu m^2/s$ with a mean value of 8.4 $\mu m^2/s$. Although there was a slight variation in diffusion coefficients, no statistical difference in diffusion coefficient was found between different locations in the skin ($p > 0.05$, Kruskal-Wallis test).

DISCUSSION

To facilitate the design of drug delivery systems for administrating drugs in and through the skin, the transport of macromolecules in human skin was assessed by means of the diffusion coefficient. The diffusion coefficient of 500 kDa dextran minimally varied within and between various skin layers, with a mean value of 8.4 $\mu m^2/s$.

Previous studies showed that the diffusion coefficient gradually increased with depth in murine and porcine dermis [1,3]. This difference was not found for human dermis. The higher values estimated for human skin compared to murine skin suggests that the diffusion coefficient is different for different species, and hence, has to be considered when studying transport in skin. In addition, it must be noted that the current study was performed on unstretched ex vivo skin, which may have influenced the structure of the skin and its resulting properties.

The method presented can be of interest for other inhomogeneous materials or macromolecules. Analyzing the experiment with a finite element model has advantages over analytical models, the latter of which require a stepwise change of the intensity in the bleached image. However, during bleaching diffusion already occurs, so a stepwise intensity change cannot be obtained. The FE model employs the first image after bleaching as the initial condition, and hence the important first phase of the experiment is described more accurately than with an analytical model. Although previous studies have used a combination of experiments and FE analysis [4,5], the current method benefits from its use of both the total spatiotemporal information and inhomogeneous materials.

This study investigated the transport of a macromolecule through the skin. These data are essential to optimize trans- and intradermal drug delivery systems to target specific layers or cells in the human skin.

ACKNOWLEDGEMENTS

This research is supported by the Dutch Technology Foundation STW, which is part of the Netherlands Organisation for Scientific Research (NWO), and which is partly funded by the Ministry of Economic Affairs.

REFERENCES

[1] Raphael, AP et al., *J Control Release*, 166:87-94, 2013.
[2] Deschout, H et al., *Pharm Res*, 31:255-270, 2014.
[3] Liu, P et al., *J Biomed Opt*, 18:020507-1:3, 2013.
[4] Sniekers, YH, van Donkelaar, CC, *Biophys J*, 89:813-823, 2005.
[5] Irrechukwu, ON et al., *Cell Mol Bioeng*, 2:104-117, 2009.

SB³C2015
Summer Biomechanics, Bioengineering and Biotransport Conference
June 17-20, 2015, Snowbird Resort, Utah, USA
SB³C2015-275

EMPIRICALLY DETERMINED VASCULAR SMOOTH MUSCLE MECHANO-ADAPTATION LAWS

Kerianne E. Steucke (1) and Patrick W. Alford (1)

(1) Biomedical Engineering
University of Minnesota
Minneapolis, MN, USA

INTRODUCTION
Vascular smooth muscle cells (VSMCs) are responsible for sustaining physiological vessel tone through active contraction and relaxation. Pathologies such as hypertension and atherosclerosis can perturb the VSMCs' mechanical environment, affecting VSMCs' contractile function. VSMCs respond to mechanical perturbations through growth and remodeling, or mechano-adaptation, which can involve morphological, migration, phenotype, and functional changes [1]. It has been proposed that VSMCs adapt their own contractility and remodel their surrounding extracellular matrix to maintain a 'target stress' [2, 3]. *In vivo* data suggest that early mechano-adaptation is primarily due to changes in contractility [4]. To date, mechano-adaptation laws for stress-induced changes in VSMC function have not been directly measured.

In our previous work we measured VSMC functional contractility using the vascular muscular thin film (vMTF) assay [5]. Here, we build upon the vMTF technology to engineer chronic strain muscular thin films (CSMTF) that apply a step change in strain to the tissue layer, and allow temporal measurement of stress in the tissue layer, to directly measure VSMC mechano-adaptation laws.

METHODS
Cell Culture
Human umbilical artery VSMCs were purchased from Lonza at passage three and cultured at 37°C and 5% CO_2. Medium 199 (GenDEPOT, Barker, TX) containing 10% FBS, 10 mM HEPES (GIBCO, Invitrogen, Grand Island, NY), 50 U/ml penicillin (GIBCO), 50 U/ml streptomycin (GIBCO), 3.5 g/L glucose (Sigma Aldrich, St. Louis, MO), 2 mM L-glutamine (Sigma Aldrich), 1X MEM Non-essential amino acids (GIBCO), and 2 mg/L vitamin B-12 (Sigma Aldrich) was changed every other day. Passages 5-7 were used for all experiments and all tissues were serum starved 24 h prior to data collection.

Chronic Strain vMTF Fabrication
Polydimethylsiloxane (PDMS) membranes were cleaned in 70% ethanol, stretched to 20% strain, and secured with clamps. A plastic spin board was bonded to the backside of the membranes. Everything except for a single strip along the length of the membrane surface was taped off and the entire construct was exposed to UVO for 8 min. Poly(N-isopropylacrylamide) (PIPAAm) was incubated on the surface of the exposed strip for 1 h. Excess PIPAAm and all tape were removed from the surface of the membranes. Then a layer of 1 MPa PDMS, doped with fluorescent microspheres, was spin coated on the membrane's surface and cured. Membranes were removed from the spin boards resulting in compression of the PDMS layer equal to the strain of the membrane as the membrane returned to its original configuration. Glass coverslips were cleaned in 70% ethanol, dried, and spin coated with PDMS. Before curing, the PDMS coated coverslip was bonded to the backside of the membrane and then cured (Fig. 1 A). The membrane is in its zero stress configuration, while the PDMS is in compression equivalent to the initial strain of the membrane. Microcontact printing was used to create 10 um lines of human fibronectin with 10 um gaps orthogonal to the PIPAAm strip on the surface of the construct. Human umbilical artery VSMCs were seeded onto the construct at a density of 62,500 cells/cm^2 and created an aligned, confluent arterial lamella tissue mimic following the guidance cues provided by the fibronectin pattern.

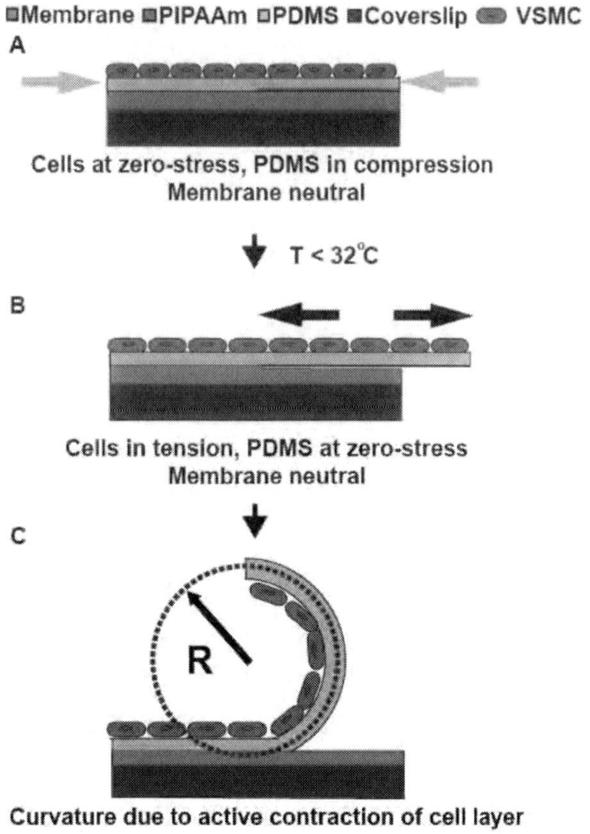

Figure 1. Chronic Strain vMTF. **A.** Pre-film release the cell layer is at zero-strain and the PDMS layer is in compression. **B.** Post-film release, the PDMS layer relieves its compression, thereby increasing in length and applying strain to the cell layer. **C.** Active contraction of the cell layer results in film curvature, which is measured using the radius of curvature (R).

Temporal Contractility Experiments

Constructs were transferred to 1X Tyrode's Buffer and 8 cuts parallel to tissue alignment and 2 cuts orthogonal to tissue alignment were made with a razor blade resulting in 2 rows of 4 films each. Cooling the construct to just below 32°C allowed the PIPAAm layer to dissolve releasing the 8 films and instantly relieving the compression in the PDMS layer resulting in an increase in film length and a strain to the tissue layer along the longitudinal axis of the cells (Fig. 1A, B). Active contraction of the tissue layer results in curvature of the film, which can be measured and used to calculate stress in the cell layer. The construct was transferred to a fresh solution of 1X Tyrode's Buffer maintained by a heating unit at 37°C. The films were allowed to equilibrate for 15min and then a single brightfield and fluorescent image were captured for each time point with a Lumar V12 stereomicroscope (Zeiss, Oberkochen, Germany). Data was collected at the following time points (relative to 24 h post serum starvation); every hour for the first 12 h, and then every 24 h for 6 days. Immediately after the final time point was collected, 100 uM HA-1077 (Sigma Aldrich), a rho-kinase inhibitor, was added to the construct to knock down all contraction and both brightfield and fluorescent images were taken every 30 s for 30 min.

Stress Analysis

Top down images of the film's projection length are used to determine each film's radius of curvature. Stress in the tissue layer was calculated as previously described [5]. Active contraction of the tissue layer results in curvature of the film. The film is modeled as a 2-layer beam in which the tissue layer undergoes pseudo-contraction and tissue stress is calculated from the observed curvature. The PDMS layer thickness was measured with a Tencor profilometer and tissue thickness was evaluated with actin z-stacks as previously described [7]. Temporal stress evolution following a step change in strain was measured for each construct for up to 6 d.

RESULTS AND DISCUSSION

Using the CSMTF system, we applied up to 20% strain to the engineered arterial tissue lamella and tracked the VSMC stress over the course of 6 d (Fig. 2). Preliminary results found that during the first 2 h there is a decrease in stress, possibly due to tissue injury. The tissue then seeks to return to its target stress, but overshoots. Finally, after day 3 the tissue stress begins to approach the original target stress. The change in stress over time demonstrates that mechano-adaptation is occurring within the vascular tissue due to mechanical perturbation. Observing the functional response of vascular tissue to mechanical perturbations allows us to directly compare theory to *in vitro* experiments; ultimately leading to a more robust model with improved physiological parameters for a better understanding of vascular disease initiation and progression.

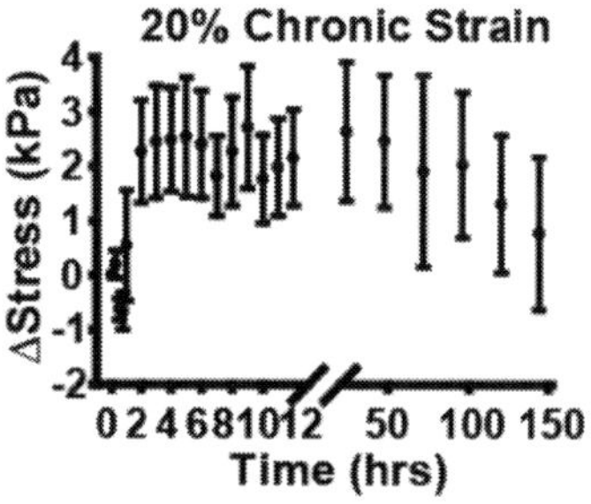

Figure 2. Temporal response of basal stress relative to the 0 h time point after a 20% chronic strain was applied (n = 13-16). Circles are data means and error bars are SEM.

ACKNOWLEDGEMENTS

Device fabrication was performed at the Minnesota NanoCenter at the University of Minnesota (a member of the National Nanotechnology Infrastructure Network), which is supported by the National Science Foundation. Parts of this work were carried out in the Characterization Facility, University of Minnesota, a member of the NSF-funded Materials Research Facilities Network via the MRSEC program.

REFERENCES

[1] Alexander, RW, *Hypertension*, 25:155-161, 1995.

[2] Fung, YC, Biomechanics: Mechanical Properties of Living Tissues Second Edition. *Springer Science and Business Media*. 2004.

[3] Taber, LA, *J Biomech Eng*, 120:348-354, 1998.

[4] Fridez, P et al, *Ann Biomed Eng*, 30:905-916, 2002.

[5] Alford PW, Feinberg AW, Sheehy SP, & Parker KK, *Biomaterials*, 31(13):3613-3621, 2010.

[6] Grosberg A, Alford PW, McCain ML, & Parker KK, *Lab chip*, 11(24):4165-73, 2011.

[7] Alford PW, Nesmith, AP, Seward NS, Grosberg A, Parker KK, *Integrative Biology*, 3:1063-1070, 2011.

SB[3]C2015
Summer Biomechanics, Bioengineering and Biotransport Conference
June 17-20, 2015, Snowbird Resort, Utah, USA
SB³C2015-276

INLET FLOW RATE VARIATION AND ONSET OF FLOW INSTABILITIES IN THE CAROTID SIPHON

Resmi KrishnankuttyRema (1), Kristian Valen-Sendstad (2,3), David A. Steinman (1,2)

(1) Inst. Biomaterials & Biomedical Engineering
University of Toronto
Toronto, ON, Canada

(2) Mechanical & Industrial Engineering
University of Toronto
Toronto, ON, Canada

(3) Centre for Biomedical Computing
Simula Research Laboratory
Lysaker, Norway

INTRODUCTION

Recent high-resolution CFD studies from our group have shown that flow instabilities may be occurring in the internal carotid artery (ICA) siphon [1], and these flow instabilities might be associated with aneurysm initiation and rupture. However, to proceed with this hypothesis, we should have an in-depth knowledge about the various parameters associated with the onset of these flow instabilities. There are thus two motivations behind the current study. One is to find the effect of varying the inlet flow rate on onset and propagation of flow instabilities in carotid siphons. Second is to set the stage for a flow experimental study using real blood and blood mimicking fluids in a carotid siphon model to understand the characteristics of transition to turbulence compared to Newtonian fluids.

METHODS

The Aneurisk ICA model C0002, which has been studied by CFD [1], was chosen because it showed highly unstable flow in the pulsatile simulation. As shown in Figure 1a, the aneurysm had previously been removed, using an automated approach [2]. Since a simpler and shorter version of the model is ultimately desired for the experimental studies, for the present study all branches were removed and the model inlet was truncated, as shown in Figure 1b. Preliminary tests revealed that the gross flow patterns were kept intact despite these geometrical changes.

High-resolution CFD simulations using fully-developed steady inflow boundary condition (but solving the transient Navier–Stokes equations) were performed with ~4M quadratic tetrahedral elements, equivalent to ~32M linear tetrahedra. The time step size was based on maintaining a CFL number of 0.5, and 12,000 steps were required for each model to complete 10 'flow-throughs' in order to wash out initial conditions. The simulations were conducted for Reynolds number (Re) varying from 100 to 1000 in increments of 100. Based on the inlet diameter of 4.41 mm and an assumed viscosity of 0.035 cm^2/s, the corresponding steady flow rates varied from 1.21 to 12.0 mL/s, and the resulting time-step sizes varied from 0.314 to 0.0314 ms.

The temporal evolution of flow instabilities, normalized by the local mean velocity, was recorded at selected locations along the model, as shown in Figure 1b. We also computed the Q-criterion [3] to illustrate a snapshot of the complex vortex cores associated with the different Reynolds numbers.

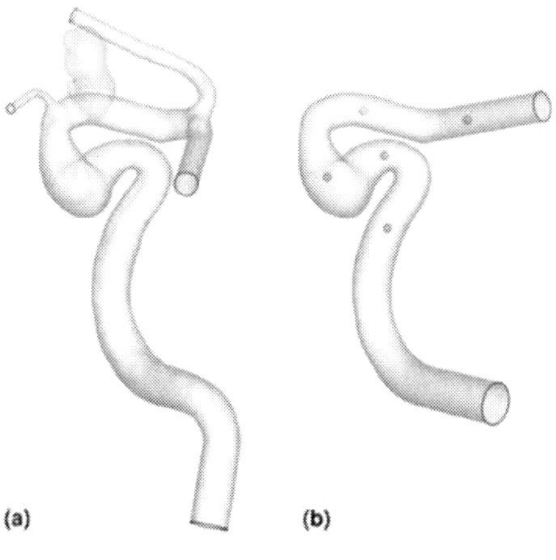

Figure 1: (a) Original carotid siphon CFD model used by [1]. Note (removed) aneurysm sac. (b) Modified model used in the present study, after smoothing, removal of side and outlet branches, and truncation of inlet. Note colour-coded velocity probe points.

RESULTS

Velocity traces for 'flow-throughs' 5-10 shown in Figure 2 reveal that flow instabilities are initiated at Re=600, and at the first bend proximal to the siphon, and propagate all the way to the outlet. Fluctuations are different at each location, and are maximum at the point distal to the flare (Location 4).

The intensity of flow instabilities with increasing Reynolds numbers is also evident from the iso-surface plots of Q criterion shown in Figure 3. Strong, persistent vortical structures are present from Re=500-600, above which intricate patterns of vortex shedding are evident, in broad agreement with the velocity traces shown in Figure 2. One can also notice that the bend proximal to the siphon, corresponding to Location 2, is where these flow instabilities are being initiated.

Time averaged wall shear stress (WSS) normalized to the respective inlet WSS for these cases are shown in Figure 4. Intensity of high WSS increases progressively with Re (i.e., flow rate) at low Re, but seems to be more constant at the higher flow rates, suggesting that patterns of low and high WSS are insensitive to flow rate variations.

DISCUSSION

To the authors' knowledge, Stehbens et al. [4] were the first to investigate and report flow instabilities in the carotid siphon. In glass model studies of idealized siphons exhibiting various tortuosity, flow disturbances were reported to occur at Re as low as ~500. Takeuchi et al. [5] later used patient specific casts and also reported the flow to be "highly disturbed". As recently pointed out by our group [1], such flow disturbances at the carotid siphon are largely unreported in the

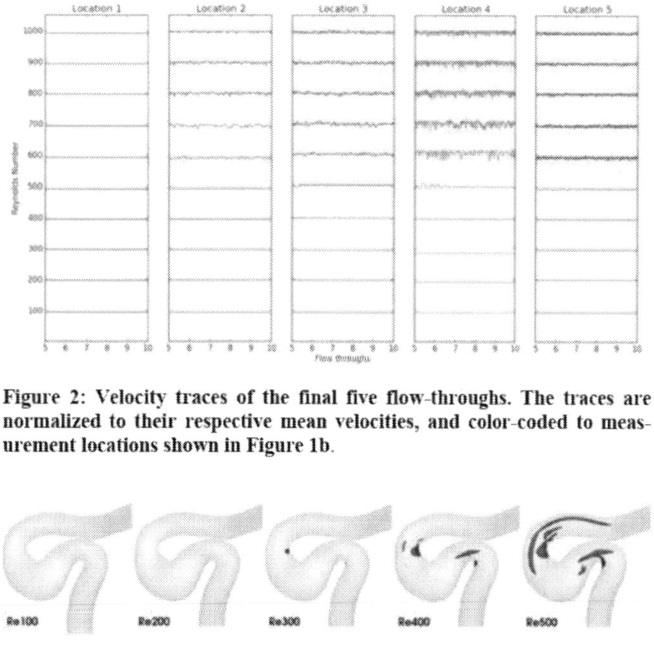

Figure 2: Velocity traces of the final five flow-throughs. The traces are normalized to their respective mean velocities, and color-coded to measurement locations shown in Figure 1b.

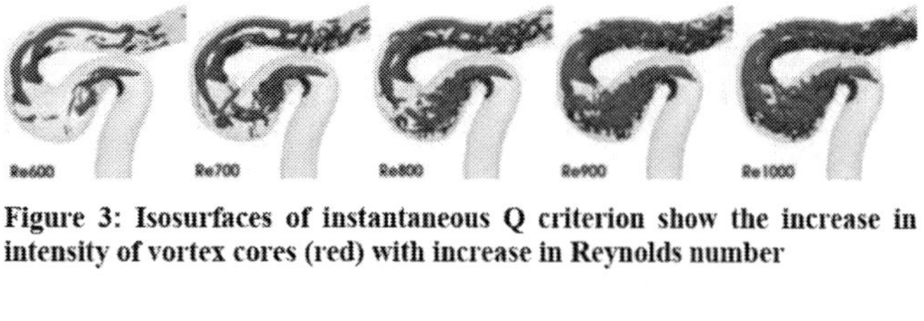

Figure 3: Isosurfaces of instantaneous Q criterion show the increase in intensity of vortex cores (red) with increase in Reynolds number

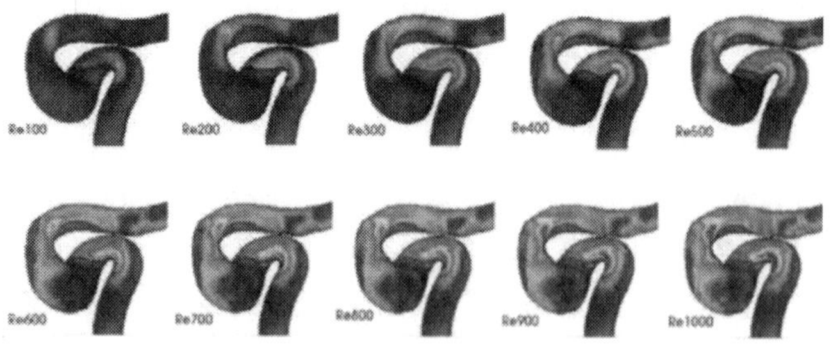

Figure 4: WSS maps, normalized to the respective inlet WSS, show relative consistency of patterns of low and high WSS.

CFD literature, which may be explained by the historical reliance on stabilized CFD methods and/or under-resolution of meshes and time-stepping. In [1], peak systolic Re was on the order of 500, and pronounced flow instabilities were evident during flow deceleration. The steady inflow inlet condition used in the present study could be overestimating the Re number at which transition may occur, owing to the fact that, during the deceleration phase of the cardiac cycle, flow is expected to be further destabilized. This may be why the velocity fluctuations shown in Figure 2 appear less intense than those reported by [1].

Our study also would seem to suggest that, at least in the unstable flow regime, non-negligible uncertainties in flow rate might not markedly impact the estimation of normalized WSS patterns. This is important, since patient-specific flow rates are rarely available, and so must be estimated, with some uncertainty, from population or model-based values.

The study has some potential limitations that make it difficult to draw general conclusions, chief among them being that we focused on a single case. As discussed above, pulsatility effects were not incorporated in the study. We also made the common assumptions of rigid walls and Newtonian fluid. Flow instabilities similar to those seen in our rigid-wall CFD models are broadly consistent with findings from experimental studies of a compliant carotid siphon [6]. Non-Newtonian effects may, however, not be negligible in light of recent work hypothesizing an impact of the discrete nature of blood for turbulent or turbulent-like flows [7].

In conclusion, this study shows that general patterns of flow instabilities and WSS extrema are relatively insensitive to Reynolds number and hence to uncertainty in imposed flow rates. These findings also set the stage for performing an experimental study for understanding the transition to turbulence in siphon models using blood and/or blood mimicking fluids.

ACKNOWLEDGMENTS

This work was supported by grant and salary awards from the Heart & Stroke Foundation of Canada, and a Resource Allocation from Compute Canada's SciNet HPC Consortium.

REFERENCES

[1] Valen-Sendstad, K et al., *J. Biomech.*, 47: 3210–6, 2014.
[2] Ford, M. et al., *Br. J. Radiol.*, 82: S55–61, 2009.
[3] Hunt, J.C.R et al., *Cent. Turbul. Res. Rep*, 12:193–208, 1988.
[4] Stehbens, W. E. et al., *Q. J. Exp. Physiol.*, 72: 201–213, 1987.
[5] Takeuchi, S. et al., *World neurosurgery* 73.3:174-185, 2010.
[6] Yagi, T. et al., *J. R. Soc. Interface*, 10: 20121031, 2013.
[7] Antiga, L et al., *Biorheology*, 46: 77-81, 2009.

IMPACT OF PARTIAL INTRALUMINAL THROMBUS ATTACHMENT ON PEAK STRESSES ON ABDOMINAL AORTIC ANEURYSM WALL

Juan S. Stockle, David A. Romero, and Cristina H. Amon
Department of Mechanical and Industrial Engineering
University of Toronto
Toronto, Ontario, Canada

INTRODUCTION

Abdominal Aortic Aneurysms (AAAs) are dilations of the aorta, in which the most common location is between the renal and iliac arteries. This degenerative disease is the 13th major cause of death in the USA [1]. If is not treated, aneurysms can progress to rupture and death. In order to understand the causes that lead to the rupture of the aneurysm wall and accurately predict the stress acting on this tissue, researchers have endeavored to relate the aneurysm failure with its mechanical environment. Interest has been given to the role played by the Intraluminal Thrombus (ILT), a natural buildup of layers of clotted blood, found in over 75% of AAAs [2]. Some authors have pointed out that ILT reduce peak stress and smooth the stress distribution [3]. In contrast, other researches have showed that departures from the idealized conditions used in previous studies, e.g., the presence of internal cracks on the ILT, could lead to larger stresses on the AAA wall [4-5]. Also, the presence of liquid-phases at the AAA-ILT interface has been observed in some patients, suggesting that ILT are not always completely attached to the aneurysm walls. This feature, typically found in thick ILTs, could play a critical role in the enlargement and rupture of AAAs [6-7]. As consequence of these observations, further studies that include these characteristics are warranted to help elucidate the mechanical role of ILTs.

METHODS

To meet the research objectives, FEM analysis was carried out with the ANSYS commercial software. Hypothetical AAA and ILT geometries were created to model the structures with full control of the geometrical variables. To evaluate the impact of ILT on AAA wall stress distribution, different ILT attachment modes were tested. According to the literature surveyed [6-7], there is evidence that ILT structures are not always completely attached to aneurysm walls. Based on this, an investigation considering different types of attachment, attachment areas and friction factors (varied from 0,1 to 0.3) was performed to understand their impact on AAA wall peak stress and stress distribution, and to elucidate under which attachment conditions ILTs protect aneurysm walls. The mechanical properties of the AAA and ILT were included in the computational model using hyperelastic constitutive equations from previous works [8-9]. To calculate the stress distribution and peak stress, a constant pressure of 120 mmHg acting along the luminal part of the system was applied. For comparison, a control case with no ILT structure was used as baseline.

RESULTS

Figure 1 shows an idealized AAA and ILT along with the boundary conditions considered in this study. Figure 1(b) and Fig. 1(c) represent complete and incomplete ILT attachment cases, respectively. The region in the AAA dome, (lighter-shaded region in Fig. 1 (c)), represents the portion of ILT that is not completely attached to the AAA wall, on which a friction coefficient is applied. To test the impact of these attachment types on the protection of AAA wall, a control case that does not include the ILT was also simulated, Fig. 1(a).

The results of these computational simulations are plotted in Fig. 2. More specifically, the values of the peak wall stress for different types of attachment area (incomplete 0%, 11.3%, 17.3%, 42% and 100%) and for different friction factor values ($f = 0.1$, $f = 0.2$, $f = 0.3$) are presented.

Figure 2 shows that the peak stress increases as the percentage of attachment area increases. For the cases studied, the friction factor does not seem to produce a significant difference in peak stress, this difference is only 0.87% at the lowest percentage of ILT attachment area (0%). If we compare the extreme cases (0% and 100%), there is a difference of approximately 3.5% between the peak stresses. It can be concluded that ILT with incomplete attachment areas in combination with low friction factor values exert lower stress on AAA walls.

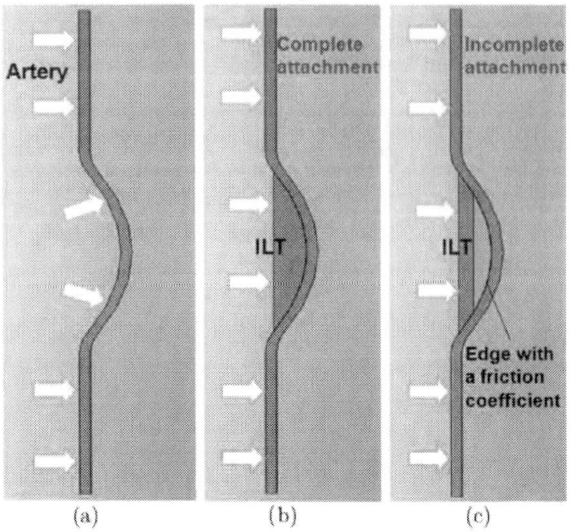

FIGURE 1: THREE CASES OF AAA. FROM LEFT TO RIGHT, (A) AAA ALONE, (B) FULLY ATTACHED ILT TO THE AAA WALL, AND (C) PARTIALLY ATTACHED ILT.

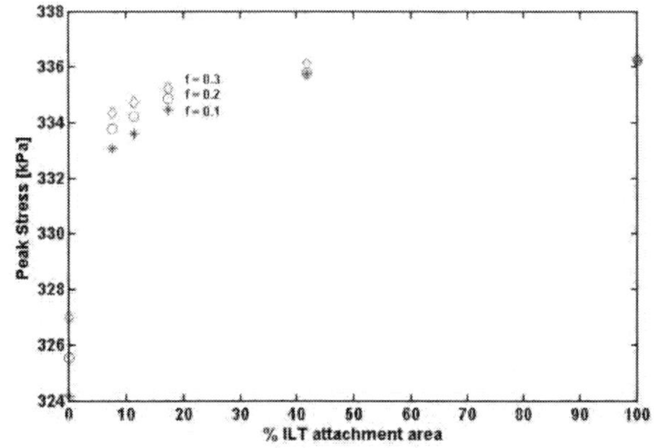

FIGURE 2: PEAK STRESS FOR DIFFERENT PERCENTAGES OF ILT ATTACHMENT AREAS AND FRICTION FACTORS.

The peak stress values assuming a friction factor of 0.1 are plotted again in Fig. 3, which also includes a case without ILT (aneurysm alone) for comparison purposes. A considerable reduction in the peak stress is observed when ILT structures are present on AAAs. There is 14.8% difference in the peak stress for an aneurysm without ILT and for the full incomplete attachment configuration. This suggests that the presence of ILTs could reduce the risk of rupture of AAA walls. These results confirm that ILT protects the AAA wall as found by Di Martino [3]. In addition, the effect of incomplete attachment areas is not harmful to the AAA wall because it does not promote a significant increase in the peak stress.

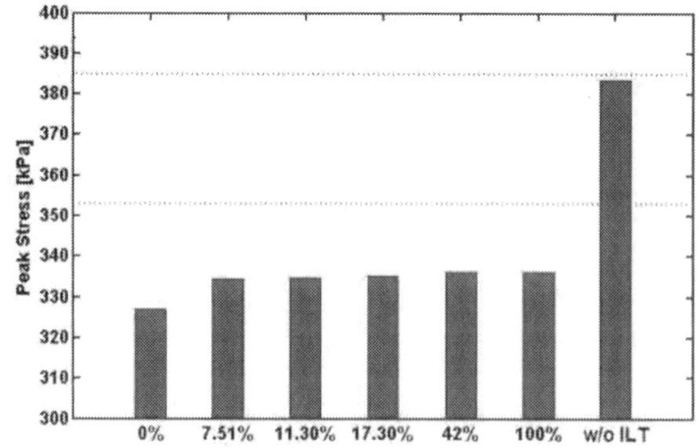

FIGURE 3: PEAK STRESS FOR FIVE ILT ATTACHMENT SCENARIOS AND CONTROL (W/O ILT). THE TWO DASHED LINES INDICATE THE RANGE OF PEAK STRESS VALUES FOUND (IN-VIVO) IN AAAS WHOSE MAXIMUM DIAMETER MATCHES THE THRESHOLD USED FOR ELECTIVE REPAIR [9].

DISCUSSION

The results obtained suggest that ILT structures play a protective role on AAAs. It is important to point out that the lower peak stresses were found in cases of fully detached ILTs that exert contact forces on the wall through a friction factor. However, since this study considered only pressure-induced forces, tangential forces from the fluid flow were neglected, and these could potentially dislodge fully detached, friction-held ILTs. Hence, to obtain more conclusive results, more comprehensive studies are needed including realistic flow environments, e. g. both under steady and pulsatile blood flows conditions. Results from the proposed work may have an impact in practice as tools for design/re-design of endovascular devices that promote ILT formation, as knowledge base for evaluating their effectiveness in AAA repair, and to formulate model-based guidelines both for elective repair and for site-specific deployment of such devices.

ACKNOWLEDGEMENTS

The authors would like to acknowledge the support of BECAS CHILE scholarship program.

REFERENCES

[1] Patel, M. et al., *Journal of the American College of Surgeons*, 181(4):371-82, 1995.
[2] Harter L. et al., *Journal of Ultrasound in Medicine*, 1:315-8, 1982.
[3] Di Martino E et al., *Annals of Biomedical Engineering*, 31(7):804-09, 2003.
[4] Roy J. et al. *Journal of Vascular Surgery*, 48(5):1108-13, 2008.
[5] Polzer S. et al., *European Journal of Vascular and Endovascular Surgery*, 41(4).467-473, 2011.
[6] Fontaine V. et al., *American Journal of Pathology*, 161(5):1701-10, 2002.
[7] Swedenborg J. et al., *Annals of the New York Academy of Sciences*, 1085:133-38, November 2006.
[8] Raghavan M et al., *Journal of Biomechanics*. 33(4):475-482, 2000.
[9] Vande Geest J. et al., *Journal of Biomechanics*, 39(13):2347-54, 2006.
[10] Fillinger, M. et al., *Journal of Vascular Surgery*, 36(3):589-597, September 2002.

MOVEMENT IN ENGINEERED VALVULAR TISSUES IN RELATION TO REGIONAL FLOW PHYSICS AND NUTRIENT TRANSPORT

Manuel Salinas (1), Sasmita Rath (1), Ana Villegas (1), Vinu Unnikrishnan(2), Sharan Ramaswamy (1)

(1) Department of Biomedical Engineering, Florida International University, Miami, FL, USA

(2) Department of Aerospace Engineering and Mechanics, The University of Alabama, Tuscaloosa, AL

INTRODUCTION

Previous efforts in heart valve tissue engineering demonstrated that the combined effect of cyclic flexure and steady flow on bone marrow derived stem cell-seeded scaffolds resulted in significant increases in engineered collagen formation [1]. In addition, another related investigation demonstrated the impact of fluid flow on DNA content and quantified the fluid-induced shear stresses on the engineered heart valve tissue specimens [2]. In this study, we built on our previous [2] computational fluid dynamics (CFD) analysis that focused on oscillatory wall shear stresses imparted on specimens when mechanically conditioned by a combination of cyclic flexure and steady flow. Here,, we specifically coupled a convective diffusive mass transport model to the flow field, in order to determine efficacy of nutrient delivery of the glucose and oxygen species to the growing engineered tissue specimens.

METHODS

Computational Fluid dynamic (CFD) simulations of a flow-stretch-flexure (FSF) bioreactor available in our laboratory were performed. The physical description of the FSF bioreactor has been provided in detail previously [3]. we computationally quantified the i) magnitude of fluid-induced shear stresses; ii) the extent of temporal fluid oscillations in the flow field using an oscillatory shear index (OSI) parameter [4], and iii) glucose and oxygen mass transport profiles. We incorporated moving boundary computational fluid dynamic (CFD) simulations of samples housed within a bioreactor (Fig.1) to consider the effects of: 1) No flow, no flexure (control group), 2) Steady flow-alone, 3) Cyclic flexure-alone, and 4) Combined steady flow and cyclic flexure environments.

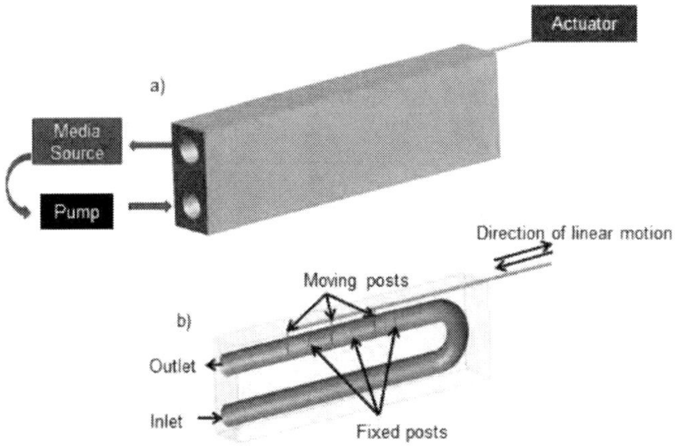

Figure 1. a) Shows a single bioreactor chamber connected to media source. The media is pumped into the bioreactor by a peristaltic pump. An external actuator attached to one end of the samples provides movement. b) Shows an inside look of the bioreactor. The chamber can hold three specimens. Each sample has one end fixed while the other is allowed to move linearly by means of the actuator [2].

The convective-diffusive model is characterized by the following equations [5]:

$$\Phi_f \frac{\delta C^\beta}{\delta t} + \left(v_f \gamma C^\beta - D^{\beta *} \nabla C^\beta\right) = q^\beta$$

$$D^{\beta*} = \frac{v_f}{(2-v_f)^2} D^{\beta}$$

Mackie and Meares Relation $\quad K_m^{\beta} \ll C^{\beta}$

$$q^{\beta} = \frac{V^{\beta}\rho^{cell}}{K_m^{\beta}+C^{\beta}} C^{\beta}$$

Michaelis-Menten Relation $\quad q^{\beta} = -V^{\beta}\rho^{cell}$

Where C β is the concentration of the solute in the medium; Φ_f is the fluid volume fraction; Υ is the sieving coefficient in a porous material; qβ is the source term for generation, consumption or degradation of solute mass; v_f is the velocity of the fluid phase; D^{β} is the effective diffusive coefficient in the porous medium; ρ cell is the density of cell seeded in the scaffold; V β is the maximum uptake rate, K_m^{β} is the concentration at half velocity. The diffusivity in the fluid was prescribed to be $5.40x10^{-6}$ cm^2/s for glucose and $5.40x10^{-6}$ cm^2/s for oxygen [5]. Meanwhile, the diffusivity in the scaffold was set to be $1.08x10^{-6}$ cm^2/s and $0.60x10^{-5}$ cm^2/s for glucose and oxygen respectively [6]. Finally, the glucose consumption rate was set to $3.83x10^{-16}$ mole/cell.s, while oxygen was assumed to be consumed at a rate of $3.75x10^{-17}$ mole/cell.s [5].

RESULTS

We previously found [2] that the co-existence of both OSS and appreciable shear stress magnitudes explained the high levels of engineered collagen observed from combining cyclic flexure and steady flow states [1]. We found here that oxygen and glucose mass distributions were significantly similar ($p < 0.05$) at 75% scaffold porosity (Fig. 2), but were significantly enhanced ($p<0.05$) by specimen movement when coupled to scaffolds of significantly low porosity (Fig. 3). A similar trend was observed for the oxygen species.

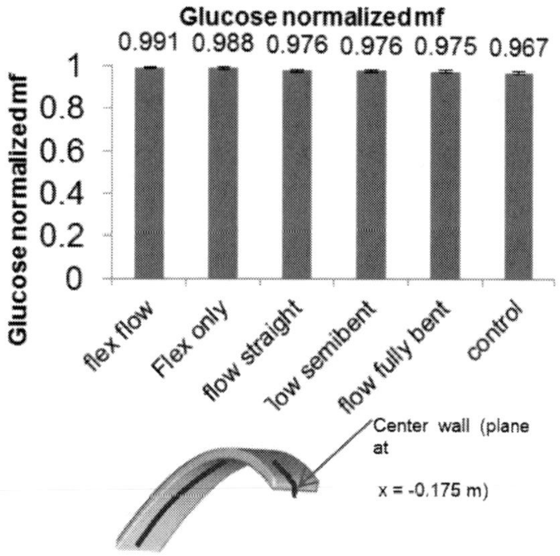

Figure 2. At 75% porosity: Area and time averaged normalized mass fraction at sample longitudinal wall at x = -0.0175m for glucose.

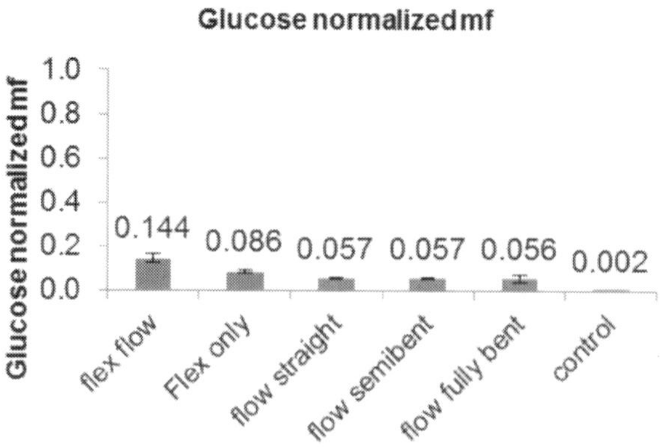

Figure 3. At 5% porosity: Area and time averaged normalized mass fraction at sample longitudinal wall at x = -0.0175m for glucose.

DISCUSSION

In this study the effects of combine steady flow and cyclic flexure on nutrient mass transport to developing engineered tissues was examined. The motivation for this lies in the importance of ensuring that cells within the growing constructs remain viable and can proliferate and differentiate robustly during the *in vitro* culturing process. There was low variability in the mass fraction of glucose and oxygen. This finding suggests that at 75% sample porosity there is adequate mass transport of nutrients even when no flow or flexure are present (control group), suggesting that convective effects resulting from flow or specimen movement are not essential. We note however that lower porosities hindered mass transport resulting in less nutrient delivery to local cells which could compromise engineered tissue production (Fig. 3); such porosities are not pertinent in typical heart valve tissue engineering studies but are relevant in native heart valve mechanobiology. In summary, we believe that for engineered heart valve development, measurable shear stresses when applied during the concomitant presence of localized oscillatory flow effects on the specimen surface are primarily responsible for mediating *in vitro* tissue growth and cell phenotype, rather than nutrient delivery to the cells in dynamic culture states.

ACKNOWLEDGEMENTS

M.S. supported by NIH/NIGMS *R25 GM61347*

REFERENCES

References

1. **Engelmayr et al. , Biomat. 2006.**
2. **Salinas et al. , J. Biomech. 2014.**
3. **Ramaswamy S, et al., J Biomech Eng. 2014.**
4. **He X, Ku D. J Biomech Eng. 1996.**
5. **Das et al. , Chem Eng Sci. 2007.**

SB³C2015
Summer Biomechanics, Bioengineering and Biotransport Conference
June 17-20, 2015,Snowbird Resort, Utah, USA
SB³C2015-280

ACTIVATED EXPRESSION AND SHEDDING OF PLATELET GLYCOPROTEIN IIB/IIIA UNDER NON-PHYSIOLOGICAL SHEAR STRESS: A PARADOXICAL EFFECT.

Zengsheng. Chen (1), Jun. Ding (1, 2), Nandan K. Mondal (1), Bartley P Griffiht (3), Zhongjun J. Wu (1)

(1) Department of Cardiovascular and Thoracic Surgery, University of Louisville School of Medicine, Louisville, KY 40202

(2) Department of Mechanical Engineering, University of Maryland Baltimore County, Baltimore, MD 21250, USA

(3) Department of Surgery, University of Maryland School of Medicine, Baltimore, MD 21201, USA

INTRODUCTION

The application of blood contacting medical devices (BCMDs) had provided clearly medical benefits, while unintended consequences often arise. In some cases the use of these devices in patients causes dangerous pathological complications, in particular thrombosis and bleeding. These complications have been investigated for many years but still remain the unsolved problems [1]. It is known that non-physiological high shear stresses exist in some of Blood contacting medical devices (BCMDs), such as in the blade region of rotary ventricular assist devices (VADs), and the hinge area of mechanical heart valves (MHVs), and can cause damage to blood. Blood damage manifests as morphologic changes, shortened life span, biochemical functional alterations, and complete rupture of blood cells and proteins [2]. All of these lead to alteration of normal blood function. It is well documented in the literature that the non-physiological high shear stress can induce platelet activation.

Platelets are anucleate blood cells that play a critical role in hemostasis. There are many receptors in the membrane of platelets, which are important to maintain platelet function. Many studies have been conducted to identify the level of non-physiological high mechanical shear stress and exposure time as the dominant factors causing platelet activation [3]. However, few studies on impact of the non-physiological shear stress on platelets have been conducted to examine other aspects of shear-induced platelet dysfunction, which may be associated with bleeding complication. Glycoprotein (GP) IIb/IIIa (or Integrin αIIb/β3) is the major membrane integrin complex on the surface of platelets and plays a key role in platelet stable adhesion, aggregation and physiological hemostasis [4]. Because of the importance of platelet GPIIb/IIIa receptors in hemostasis, it is critical to investigate the effects of non-physiological high shear stress on platelet activation and structural alteration of GPIIb/IIIa receptors to gain a profound understanding of non-physiological shear-induced hemostatic dysfunction. Non-physiological shear stress is known to induce the conformational change in platelet GPIIb/IIIa receptors, but less known about shear-induced shedding of this receptor [5]. The purpose of this study was to investigate the influence of non-physiological high shear stress on both the aspects of platelet GP IIb/IIIa receptors.

METHODS

The blood-shearing device used in this experiment was a centrifugal flow-through Couette device, adapted from the CentriMag magnetically levitated blood pump (Thoratec, Pleasanton, CA, USA). The rotor is magnetically suspended with the bearingless motor technology and can rotate between 500 and 5000 rpm. There is a narrow gap with a uniform width of 150 μm and a length of 2.5 mm, created between the housing and the rotor. The detail design of this device could be found in the previous work of our group [6].

The healthy donor blood was exposed to three levels of high shear stresses (25, 75, 125Pa) from the physiological to non-physiological status in combination with three short exposure times (0.05, 0.5, 1.5s), created by our blood shearing system. During the experiment, a baseline blood sample was collected from the inlet sampling port of this shearing device and used as the control for comparison with the sheared blood samples. For each combination of shear stress and exposure time, the sheared blood samples were collected from the outlet sampling port of the blood shearing device. For each condition, the blood shearing experiments were repeated for 5 times.

The expression of activated form and shedding of the platelet GPIIb/IIIa were analyzed using flow cytometry. In addition, platelet P-selectin expression of sheared blood, which is a marker for activated platelets, was also analyzed.

RESULTS

Figure 1 shows the mean channel fluorescence (FITC) of the total platelet population. The expression of the activated GPIIb/IIIa increases with increasing the shear stress level and exposure time. The increase of the mean fluorescence of activated GPIIb/IIIa surface expression becomes statistically significant ($P<0.05$) for the exposure time of 0.5 s and 1.5s when the shear stresses were 75Pa and 125Pa compared with those of the baseline blood. In addition, the number of activated platelets, as indicated by the expression of P-selectin, increases with increasing the shear stress level and exposure time.

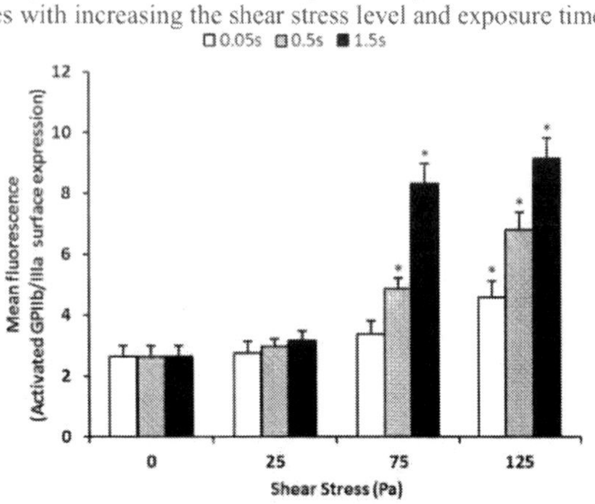

Figure 1: The quantification of mean fluorescence about activated GP IIb/IIIa surface expression indicated by FITC labeled PAC-1 in unsheared baseline sample and sheared samples under three levels of high shear stresses (25, 75, 125Pa) for three short durations (0.05, 0.5 and 1.5 sec.)(n=5).

Figure 2 shows the quantification of the platelet GPIIb/IIIa expression (PE mean fluorescence) in the baseline and sheared blood samples under the three levels of shear stresses (25, 75, 125 Pa) for the three exposure times (0.05, 0.5 and 1.5 sec.). The mean fluorescence, indicating the platelet GPIIb/IIIa expression, decreased in all the sheared blood samples. This decrease indicated that there was a reduction in the number of copies of the platelet GPIIb/IIIa receptors after the blood experienced the non-physiological shear stress. The reduction of the platelet GPIIb/IIIa receptors increased with increasing the shear stress level and exposure time. However, the reduction of the platelet GPIIb/IIIa receptors in the sheared blood samples became statistically significant ($P<0.05$) only when the shear stress level was above 75 Pa for the exposure time longer than 0.5 s compared to the baseline blood samples.

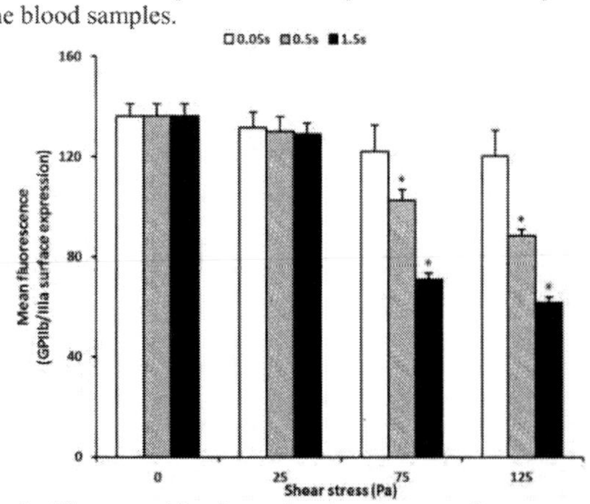

Figure 2: The quantification of reduced mean fluorescence about platelet GP IIb/IIIa surface expression (indicated by PE conjugated anti-CD41/61) induced by three different levels of high shear stress (25, 75, 125 Pa) for the three exposure times (0.05, 0.5 and 1.5 sec) (n=5).

In addition, the ADP stimulation test also proves that, compared with the baseline sample, the activated GPIIb/IIIa expression either induced by the non-physiological shear stress or late by ADP decreases in the platelets of the sheared blood samples. The reduction in the activated GPIIb/IIIa expression increases with increasing the shear stress and exposure time. This reduction in the activated GPIIb/IIIa expression indicates that the available copies of the GPIIb/IIIa receptors for ADP stimulation were reduced after the blood experienced the non-physiological shear stress. These results further confirmed the reduction of the GPIIb/IIIa on the platelet surface under the non-physiological high shear stress conditions.

DISCUSSION

In the present study, we investigated the influence of the non-physiological high shear stress on the platelet GPIIb/IIIa. The blood from healthy donor was exposed to three level high shear stress (25, 75, 125Pa) from physiological to non-physiological conditions with short exposure times (0.05, 0.5 and 1.5 sec.). The activation and shedding of the GPIIb/IIIa receptors and the expression of CD62P in the platelet surface were examined. The results clearly showed that the non-physiological shear stress caused the activation of the GPIIb/IIIa and CD62P expression in the platelets. The degree of activation of the GPIIb/IIIa and CD62P expression increased with increasing the shear stress level and exposure time. Interestingly, the GPIIb/IIIa expression in the platelets of the sheared blood samples decreased, indicating receptor reduction (shedding).The reduction in the copies of the GPIIb/IIIa receptors was further confirmed by the reduced activation of the GPII/IIIa by ADP stimulation.

Although we have not directly verified, the loss of the GPIIb/IIIa receptors on the platelet surface could lower the adhesion capacity of platelets for normal hemostasis. This is a paradoxical phenomenon in contrast to the common thinking that the non-physiological shear stress induces platelet activation, thus enhances platelet aggregation and adhesion. In our study, we did observe that with increasing the shear stress level and exposure time, the number of platelets with the activated GPIIb/IIIa increased. However the number of available copies of the GPIIb/IIIa on platelet surface reduced after being sheared by the non-physiological shear stress. Thus under the non-pathological high shear stress conditions, shear-induced activation and shedding of the GPIIb/IIIa receptors simultaneously occurred. From the aforementioned results, we believe that the shear-induced platelet activation and shedding of the GPIIb/IIIa receptors could have two opposite effects occurring simultaneously in patients accompanied with both thrombosis and bleeding induced by BCMDs. Above all, the study demonstrated that the non-physiological high shear stress could induce both activation and shedding of the GPIIb/IIIa receptors on the platelet surface. The activation and shedding increased with increasing the shear stress level and exposure time.

ACKNOWLEDGEMENTS

This work was partially supported by the National Institutes of Health (Grant number: R01 HL 088100).

REFERENCES

[1] Eckman, PM et al., Circulation, 125:3038-47, 2012.
[2] Bluestein, D et al., Expert Rev Med Devices, 1(1):65-80, 2004.
[3] Fischer, H et al., *Ann Thorac Surg,* 87(1):131-7,2009.
[4] Phillips, DR et al., *Blood,* 71(4):831-43, 1988.
[5] Reddy, KB et al., *J Cell Sci,* 15:1211641-8, 2008.
[6] Zhang, et al., *Artificial organs* 35:1180-6, 2011.

SB^{3}C2015
Summer Biomechanics, Bioengineering and Biotransport Conference
June 17-20, 2015,Snowbird Resort, Utah, USA
SB^{3}C2015-281

DEFINING COLLAGEN FIBER MECHANICS IN NEURON-COLLAGEN CONSTRUCTS UNDER STRETCH USING INTEGRATED EXPERIMENTAL & MODELING APPROACHES

Sijia Zhang (1), Xuan Cao (2), Vivek B. Shenoy (1,2,3), Beth A. Winkelstein (1,4)

(1) Department of Bioengineering
University of Pennsylvania
Philadelphia, PA, United States

(2) Department of Materials Science and Engineering
University of Pennsylvania
Philadelphia, PA, United States

(3) Department of Mechanical Engineering & Applied Mechanics
University of Pennsylvania
Philadelphia, PA, United States

(4) Department of Neurosurgery
University of Pennsylvania
Philadelphia, PA, United States

INTRODUCTION

The cervical facet capsular ligament (FCL) is a common source of pain from neck trauma. Since the FCL is richly innervated by pain fibers [1,2], its excessive stretch can activate the nociceptors and lead to pain [2-4]. Under macroscopic tissue loading, the collagen matrix of the FCL reorganizes and generates fiber forces [5,6]. Even subfailure tensile loading of the FCL can induce collagen disorganization and microstructural injury [5,7]. Although injurious FCL loading produces collagen fiber realignment, activates and/or injures nociceptors, and induces pain [3-5,7], the local biomechanical mechanisms by which collagen fibers reorganize under macro-scale loading and relate to fiber strain in the context of neural injury are unknown.

Computational models provide methods for predicting the microscopic fiber strains and stresses of the collagen matrix that otherwise cannot be measured experimentally. A finite element-based two dimensional (2D) discrete fiber network (DFN) model has been previously developed to systematically examine the large parameter space of properties of a fibrous matrix and cells embedded in it [8,9]. This 2D model is computationally simpler than models in three dimensions (3D), but captures important mechanics of the collagen matrix, including fiber realignment and non-affine deformations [8].

We have developed an in vitro neuron-collagen gel construct system mimicking the anatomy of FCL that enables studying the relationships between bulk and regional mechanics and fiber responses in real time during tensile loading [10]. Although integration of quantitative polarized light imaging (QPLI) in the system provides methods to evaluate the relationship between collagen fiber realignment and the macroscopic strain/stress response [5,7], it is not possible to assess fiber stress and strains experimentally. Since the neurons embedded in the FCL are directly influenced by strains of the surrounding collagen matrix [10], defining the fiber kinematics and local fiber stress and strain responses is necessary in order to fully understand the biomechanical mechanisms by which neurons undergo loading and respond to non-physiologic deformation. The goal of this study is to define collagen fiber mechanics in relation to macroscopic deformation of a neuron-collagen gel using a combined experimental and modeling approach. We first evaluated the relationship between collagen fiber realignment and matrix strain using QPLI during tensile loading of the neuron-collagen construct simulating painful FCL stretch. We used a DFN model to test if fiber reorientation under tension predicted in a 2D network agrees with 3D experiments; then using the DFN model, we predicted fiber strains under macroscopic loading that simulates FCL strain that produces pain in vivo [4].

METHODS

A solution of rat tail collagen I (2mg/ml) was cast in a dog-bone shaped Teflon mold to form gels with random fiber orientation [11]. Embryonic (day 18) Sprague Dawley rat cortical neurons were plated on the collagen gels at a density of 10^7cells/ml (~4500cells/mm2). After 2 days, additional collagen solution was added to encapsulate the neurons in the gel construct. On day 5, neuron-collagen gel constructs (n=5) underwent tensile loading at 0.5mm/s to 8mm (16% strain), corresponding to displacements simulating painful FCL loading [4,5,7] using an Instron 5865 integrated with a QPLI system. Each gel was divided into 10 elements by a grid of markers (Fig. 1) that was tracked during loading to measure the mean maximum principal strain (MPS) using LS-DYNA. Collagen fiber alignment maps were created from the QPLI data and the circular variance (CV), a measure of the spread of dihedral angles with lower values indicating tighter clustering around the mean, of fiber angles from the alignment maps was computed to quantify the degree of fiber realignment at 1sec increments during loading [12]. The mean CV and mean MPS were averaged for all gels and the relationship between them was tested using a two-phase linear regression model [13], which also estimated the breakpoint between the two phases. The significance of the regression was evaluated using Chi-squared tests.

In order to define the collagen fiber density and lengths for model creation, additional unstretched neuron-collagen gels (n=4) were immunolabeled for collagen I and imaged by confocal microscopy (Fig. 1). The fiber density, length and the distribution of fiber orientation measured experimentally were used as the network parameters to create a comparable fibrous network as described previously [8], for the neuron-collagen constructs (Fig. 1). Briefly, collagen fibers were treated as randomly organized linear elastic materials with rigid cross-links and modeled using shear flexible Timoshenko beam elements in ABAQUS. Volume conserved uniaxial tensile test was performed on the model (Fig.1) and network MPS, fiber orientation, and axial strain were computed.

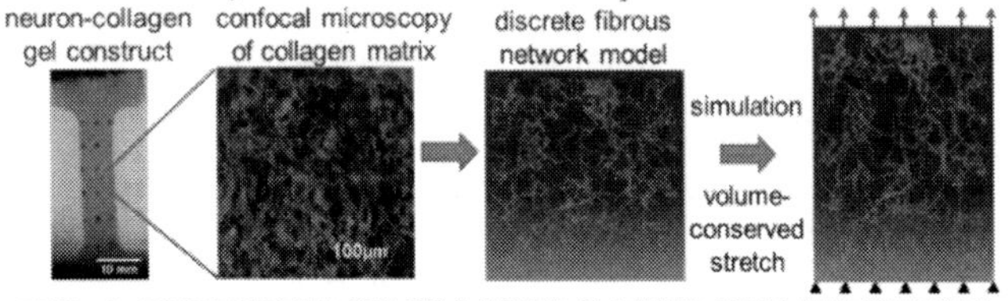

FIG. 1 OVERVIEW OF IMAGING-BASED DFN MODELING OF NEURON-COLLAGEN GEL CONSTRUCTS.

RESULTS

The circular variance of fiber orientation angles exhibits a biphasic response to increasing applied gel strain, with an initial negligible change followed by rapid increase (Fig. 2A). The random orientation of fibers in unstretched gels or those under small stain results in a tight clustering of fiber angles around zero, corresponding to a small-to-negligible CV (Fig. 2A). As strain increases, fibers realign towards the loading direction, resulting in two peaks at -90° and 90° and an increase in CV with applied strain (Fig. 2A). Using a two-phase linear regression model, a breakpoint is estimated at 11.3% strain, above which the CV increases sharply (Fig. 2A).

Fiber realignment predicted by the 2D DFN model is consistent with the 3D experimental observations. For small strains (<7%), the fiber alignment is only minimally different from the initial condition, with a uniform distribution (Fig. 2B). As strain increases, realignment towards the loading direction (±90°) is evident; this preferable re-orientation is more significant at larger network strains (Fig. 2B). These fiber orientations suggest that fibers in 2D networks realign first in a slowly-changing phase followed by rapid reorientation towards the loading direction, with a transition between 7 and 16% strain.

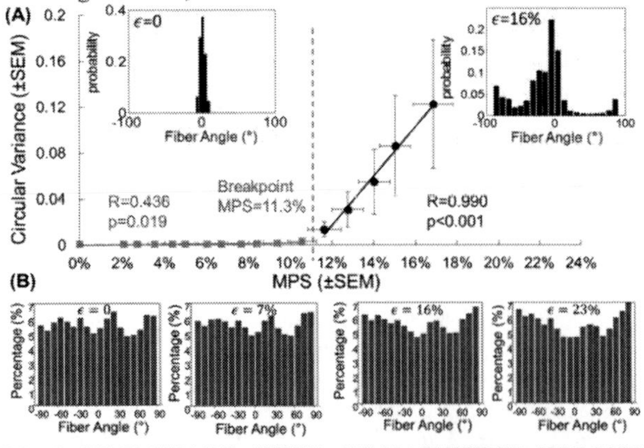

FIG. 2 TWO-PHASE FIBER REALIGNMENT MEASURED USING (A) EXPERIMENTS AND THE (B) 2D DFN MODEL.

The DFN model predicts heterogeneous fiber axial strains that are mostly less than the network loading strain. For example, at 16%, both tensile and compressive fiber strains are produced to accommodate the macroscopic deformation (Fig. 3). Most fibers undergo -5% to 10% strain, whereas load-bearing fibers aligned in the loading direction exhibit axial strains greater than the macroscopic strain (Fig. 3).

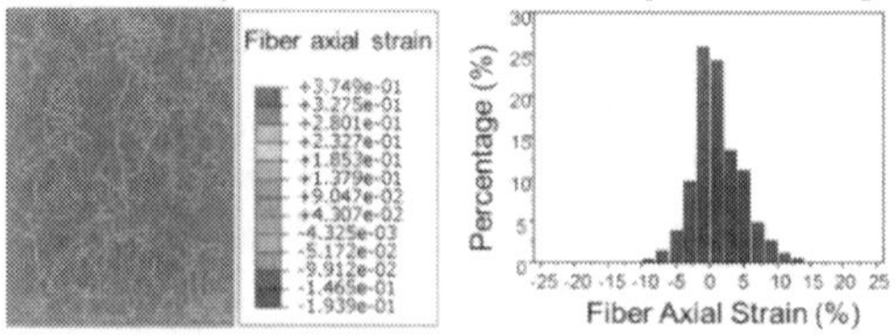

FIG. 3 HETEROGENEOUS FIBER AXIAL STRAIN PREDICTED AT 16% GEL STRAIN USING THE DFN MODEL.

DISCUSSION

The biphasic realignment of collagen fibers during uniaxial tension measured here (Fig. 2) suggest strain stiffening and may have implications for neuronal activation and/or painful injury. The initial phase where collagen fibers show minimal realignment towards the loading direction (Fig. 2) is consistent with the "toe region" of the stress-strain curve of uniaxially loaded collagen gels, which also exhibited strain stiffening at strains of 10% [9]. That strain stiffening threshold and the transition point (11.3%) between slow and rapid fiber realignment identified here (Fig. 2) are comparable with the strain threshold for pain production due to FCL tension [4], suggesting that changes in the local biomechanical environment of neurons at supraphysiological strains affect neuronal mechanosensation.

Heterogeneous fiber strains predicted by the DFN model agree with non-homogenous network restructuring predicted by a multi-scale model [6], and provide insight into the microenvironment of neurons embedded in this collagen matrix. Interestingly, most fiber strains predicted are much lower than the gel strain (Fig. 3), pointing to the importance of defining microscopic mechanics to fully understand the neuronal responses during macroscopic loading. Since neurons have passive mechanical contributions due to their lack of contractility and high compliance compared to collagen matrices and fibers [14-16], a simple fibrous network model without neuron inclusion is adequate to study the matrix and fiber mechanics of neuron-collagen gels. Although neuron morphology may be included in future models to simulate the FCL's complex architecture, our combined experimental and modeling approach provides new insights about the local biomechanical environment of neurons during painful FCL loading.

ACKNOWLEDGEMENTS

This work was supported the grants from the NIH (U01EB016633, R01EB017753) and the NSF (CMMI-1312392).

REFERENCES

[1] Kallakuri S et al. *Wold J Orthop*, 3:10-4, 2012. [2] Chen C et al. *J Bone Joint Surg Am*, 88:1807-16, 2006 [3] Kallakuri S et al. *Eur Spine J*, 17:556-63, 2008. [4] Dong L et al. *Brain Res*, 1461:51-63, 2012. [5] Quinn K & Winkelstein B. *J Biomed Opt*, 14:054010, 2009. [6] Sander E et al. *PNAS*, 106:17675-80, 2009. [7] Quinn K et al. *J Appl Physiol*, 105:1881-8, 2008. [8] Abhilash A et al. *Biophys J*, 107:1829-40, 2014. [9] Wang H et al. *Biophys J*, 107:2592-603, 2014. [10] Zhang S et al. *7th WCB Mtg*, 2014 [11] Lake S & Barocas V. *Ann Biomed Eng*, 39:1891-903, 2011. [12] Miller K et al. *Ann Biomed Eng*, 40:1102-10, 2012. [13] Stoimenova E et al. *Pliska Stud Math Bulgar*, 16:245-57, 2004. [14] Spedden E et al. *Biophys J*, 103:868-77, 2012. [15] Wenger M et al. *Biophys J*, 93:1255-63, 2007. [16] Lopez-Garcia M et al. *Biomed Mater Eng* 20:361-9, 2010.

ACUTE FAILURE MODES IN SMALL DIAMETER SPINAL SEGMENTS

Aubrie L. Taylor (1), Cassandra Bell (1), Anton E. Bowden (1)

(1) Mechanical Engineering
Brigham Young University
Provo, Utah, United States

INTRODUCTION

Acute failure modes of intervertebral discs (IVDs) have long been a topic of disagreement among the spine research community. There have been several studies evaluating acute failure modes in the human lumbar spine, including notable work by Adams and Hutton in the early 1980's that demonstrated that disc herniation was unlikely to be induced during traumatic overload events [1]. However, there continues to be substantial disagreement as to whether the same can be said of smaller diameter (e.g., cervical) intervertebral discs. To date, little work has been done in regards to evaluating acute failure modes for smaller IVDs. Biomechanically, there are reasonable arguments to be made that IVDs will fail differently depending on their size or location. The purpose of the present work was to investigate failure modes of small diameter IVDs under various acute loading conditions.

METHODS

Young, healthy, bovine tail segments were obtained from a local slaughterhouse. All extraneous soft tissue was removed from the segments, leaving the IVD and bony surroundings intact. The segments were then separated into single-level functional spinal unit (FSU) specimens. Specimens were selected for testing if the disc diameter was between 15-30 mm. Each specimen was potted using a two-part polyester resin. The segments then underwent various loading methods to determine failure modes. Segments were randomly assigned to one of 4 acute loading protocols and loaded to failure: Quasi-static hyperflexion, quasi-static flexion + lateral bending, quasi-static flexion + axial rotation, or sudden dynamic compressive overload (i.e., drop testing) (Table 1). Following testing, the specimens were carefully examined individually by the authors to identify damage initiation sites and failure modes.

Table 1. Test Setup

Loading Protocol	*Hyper-flexion*	*Flexion + Lateral Bending*	*Flexion + Axial Rotation*	*Sudden Dynamic Compressive Overload*
Number of Specimens	2	1	2	2
Flexion Magnitude (deg)		45	45	N/A
Compressive Load (N)	220	220	220	N/A
Applied Torque (Nm)	12, 24	12	12	N/A
Drop Weight Energy (J)	N/A	N/A	N/A	17
Observed Failure Mode	Vertebral failure below bony endplate	No visible damage	Vertebral failure below bony endplate + facet failure	Vertebral failure below bony endplate

RESULTS

Results demonstrated that regardless of loading protocol, the predominant failure mode and location were consistent across all tests. All failures occurred in the vertebrae, adjacent to the bony endplate. Photographs of the resulting failures can be seen in Figure 1. The

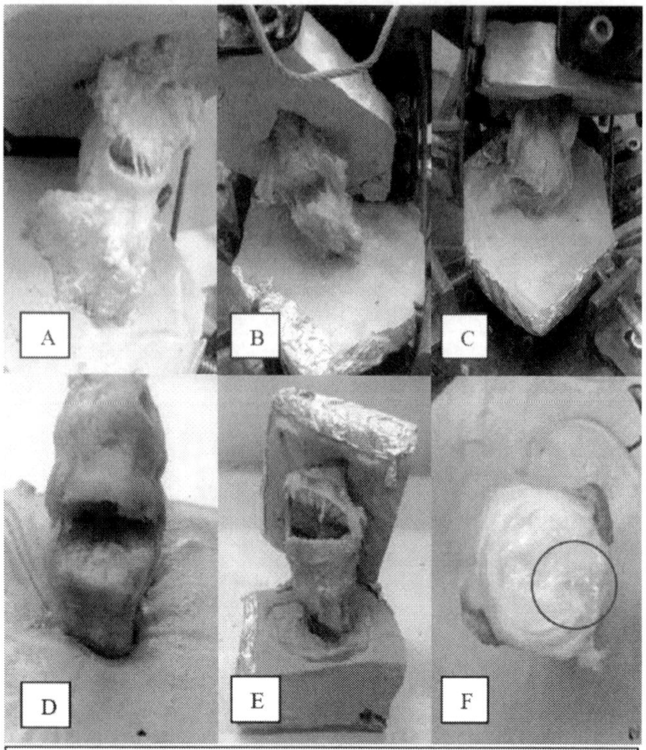

Figure 1: Failure mode results
A-F show the failure modes for the testing segments following acute overload. A) Hyperflexion failure. B-C) Hyperflexion + axial rotation resulted in endplate and facet failure. D-E) Sudden dynamic compressive overload failure. F) Disc herniation following hyperflexion + induced scalpel damage to annulus fibrosus.

resulting fracture surface was dimpled and superior to the fracture line. We also observed that both the cartilaginous and bony endplates remained strongly adhered to the intervertebral disc surface. None of the acute loading conditions resulted in spontaneous disc herniation. However, in the interest of investigating whether the health of the annulus fibrosus (AF) may be a contributing factor, we scored the anterior surface of the AF with a scalpel while it was under extreme hyperflexion (24 Nm of hyperflexion + 220 N compressive follower load). For these conditions, we observed an immediate traumatic herniation at the site of the AF damage (Figure 1F). Interestingly, we also observed a distinct reduction in the viscosity and a visible change in the composition of the nucleus pulposus tissue following acute overload. The pulposus was no longer a thick gelatinous substance safely contained in the AF but instead became a much less viscous substance that extruded outwards into the surrounding AF tissue.

DISCUSSION

Our testing confirmed that the healthy IVD is not the weakest mechanical portion of small diameter spinal segments. Acute overload failure uniformly occurred in the vertebrae, just below the bony endplate. Only when damage was applied to the AF were we able to induce IVD herniation. We were very intrigued by the change in composition of the nucleus pulposus tissue following traumatic overload as this may be a result of mechanical forces overcoming the hydrogen bonding between the proteoglycans and the water in the tissue. This dissociation could be a necessary precursor to disc herniation and may indeed allow acute disc herniation to occur, albeit only in the presence of pre-existing damage to the AF (e.g., as a result of fatigue damage to the AF tissue).

We did not expect to find such uniform failure modes across all four acute loading protocols. Previous studies on the lumbar spine have found that quasi-static compression overload of an intervertebral joint results in vertebral failure, while torsion damages the articular facets, and hyperflexion damages the ligaments of the neural arch but not the disc [2-6]. Adams and Hutton were able to induce disc herniation, but only through a non-physiologic combination of extreme hyperflexion + over 8000 N quasi-static compressive loading [1]. These loading conditions are even more unlikely to occur in the cervical spine as compared to the lumbar spine. Our results with small diameter bovine tail segments are consistent with these previous studies, but were more directly focused on the vertebra and disc without the confounding factors associated with the ligamentous complex and facet joints.

Previous studies have led some to believe that disc herniation does not occur during acute spinal injuries. Our data shows that this is likely an oversimplification of a complex phenomenon. Our results show that in smaller diameter IVDs acute (e.g., single-event) disc herniation of healthy IVDs is unlikely to occur, as the IVD is not the weakest mechanical point of the vertebral segment. However, pre-existing damage to the AF may allow acute disc herniation during acute overload events, and may be accompanied by an associated compositional change in the nucleus pulposus.

ACKNOWLEDGEMENTS

Funding for this project was provided by Armstrong Forensics Engineers and the National Science Foundation (CMMI-0952758). Any opinions, findings, and conclusions or recommendations expressed in this material are those of the authors and do not necessarily reflect the views of the National Science Foundation

REFERENCES

1. Adams, M.A. and W.C. Hutton, *Prolapsed intervertebral disc. A hyperflexion injury 1981 Volvo Award in Basic Science.* Spine (Phila Pa 1976), 1982. **7**(3): p. 184-91.
2. Eie, N., *Load capacity of the low back.* J Oslo City Hosp, 1966. **16**(4): p. 73-98.
3. Hardy, W.G., et al., *Repeated loading tests of the lumbar spine; a preliminary report.* Surg Forum, 1958. **9**: p. 690-5.
4. Perey, O., *Fracture of the vertebral end-plate in the lumbar spine; an experimental biochemical investigation.* Acta Orthop Scand Suppl, 1957. **25**: p. 1-101.
5. Adams, M.A., W.C. Hutton, and J.R. Stott, *The resistance to flexion of the lumbar intervertebral joint.* Spine (Phila Pa 1976), 1980. **5**(3): p. 245-53.
6. Farfan, H.F., et al., *The effects of torsion on the lumbar intervertebral joints: the role of torsion in the production of disc degeneration.* J Bone Joint Surg Am, 1970. **52**(3): p. 468-97.

SB³C2015
Summer Biomechanics, Bioengineering and Biotransport Conference
June 17-20, 2015,Snowbird Resort, Utah, USA
SB³C2015-283

ANUERYSM MRI PHANTOMS FOR DIRECT, EX VIVO FLUID DYNAMICS

J.R. Anderson (1), O. Diaz (2), R. Klucznik (2), Y.J. Zhang (2), G.W. Britz(2), R.G. Grossman (3), C. Karmonik (1,2)

(1) MRI Core Laboratory
Houston Methodist Research Institute
Houston, Texas, USA

(2) Cerebrovascular Center
Department of Neurosurgery
Houston Methodist Hospital
Houston, Texas, USA

(3) Department of Neurosurgery
Houston Methodist Hospital
Houston, Texas, USA

INTRODUCTION

Intense research interest has recently developed with respect to fluid dynamics measurements/calculations yielding predictive biomarkers of aneurysmgenesis, aneurysm prognosis, and treatment outcomes [1]. Computational fluid dynamics (CFD) approaches are notable for their ability to describe flow as well as other hemodynamic parameters such as wall shear stresses [2].

Medical imaging methods also exist for the measurement of hemodynamic properties. Recently, MRI pulse sequences and post-processing software have been developed by major vendors to visualize fluid dynamics in the heart, aorta and the cerebral vasculature (4D Flow – Siemens, Qflow – Philips, Inhance – GE). The possibility exists to apply similar techniques, after appropriate methodological variations, to intracranial aneurysms. Initial comparison studies reveal some level of agreement between *in vivo* MRI results and CFD [3].

Further validation of *in vivo* MRI phase contrast flow imaging of intracranial aneurysms is challenged by the oftentimes real threat of subarachnoid hemorrhage. The transition from diagnosis to surgical intervention can be swift. Thus, an alternative, *ex vivo* platform for primary validation has been developed in the form of 3D printed aneurysm models derived from diagnostic digital subtraction angiography (DSA) images [4]. Results from the first cohort of aneurysm phantoms (10 aneurysms) are herein presented and comprise of flow imaging results and validation of the accuracy of the aneurysm phantoms.

METHODS

Aneurysm MRI phantoms were created for 10 subjects. The aneurysm and surrounding vessel were segmented from digital subtraction angiography (DSA) images. Isosurfaces were created using the segmented images and printed in polylactic acid (PLA) using a MakerBot Replicator 2 3D printer (MakerBot Industries, LLC, Brooklyn, NY, USA). The phantoms were printed with a hollow interior. The accuracy of the aneurysm phantoms was assessed. Three observers measured aneurysm dimensions idenpendently on source DSA images and on the printed phantoms. The measurement results were directly compared with each other.

The aneurysm MRI phantoms were attached to a flow loop via tubing after coding their surface to close any cracks and imaged at 3T (Philips Ingenia, Philips Healthcare, Best, The Netherlands) using a flow loop with a constant flow pump. An aqueous solution containing 0.2 mM gadopentetic acid (Magnevist, Bayer, Leverkusen, Germany) was circulated in the loop. T1-weighted images and high-resolution quantitative flow images were acquired with isotropic 1 mm voxel resolution over a $64\times64\times64$ cm^3 volume. Flow images were acquired for each aneurysm with a velocity encoding gradient of 110 cm/s along each of the three primary axes.

Post-processing of quantitative flow images were carried out on a dedicated workstation (Macbook Pro, Apple Inc., Cupertino, CA, USA). Paraview (Kitware Inc., Clifton Park, NY, USA) was used to generate quantitative flow vector glyphs by combining the data from the various flow images.

RESULTS

All 10 aneurysms were successfully segmented and 3D printed as aneurysm MRI phantoms. The inner dimensions of the phantoms faithfully reproduced the true anatomy observed in the DSA images. The measurements of the two groups were found to have no statistically significant differences (paired t-test; $p = 0.4$). Furthermore, no bias was observed.

T1-weighted images highlighted aneurysm shape well and were used to plan the imaging planes for dissecting the aneurysm neck and perpendicular to the aneurysm neck dissecting the aneurysm dome (Figure 1). Through-plane flow images were created for each of the planes of reference in four representative aneurysms (Figure 1). Flow vector images were effective in visualizing, quantitatively, flow within aneurysms (Figure 2).

DISCUSSION

Realistic aneurysm MRI phantoms can be used to facilitate the transition of existing cardiovascular imaging methods to *in vivo* intracranial aneurysm imaging. The 3D printed phantoms presented herein mimic well aneurysm anatomy. Complex flow patterns can be captured via quantitative flow imaging (Figure 2).

The MRI phantoms presented have many advantages including the ability to carry out studies under many different mock physiological conditions. Another advantage of such phantoms, over for instance purchasable silicon aneurysm forms, is that the diversity of individual patient anatomy is addressed. Indeed, among the 10 subjects assessed for this study, variation in both aneurysm size and location were observed. Images can also be acquired with extremely high resolution and SNR while using phantoms as time constraints compared to in-vivo imaging does not exist. This is useful for assessing the accuracy of imaging methods and the effects of decreased scan times and resolution on resultant data. Currently, efforts are underway to compare the flow patterns observed in the phantoms with complimentary CFD results. Finally, the models described here can be created in a comparably short time of one hour or less in a very cost-effective manner.

REFERENCES

[1] Turjman, AS et al., *Circulation*, 129:373-382, 2014

[2] Mantha, A et al., *AJNR*, 27:1113-1118, 2006

[3] Karmonik, C et al., *Technol Health Care*, 16:119-128, 2008

[4] Anderson, JR et al., *Conf Proc IEEE Eng Med Biol Soc*, 6699-6701, 2014

Hemo1

Hemo2

Hemo3

Hemo4

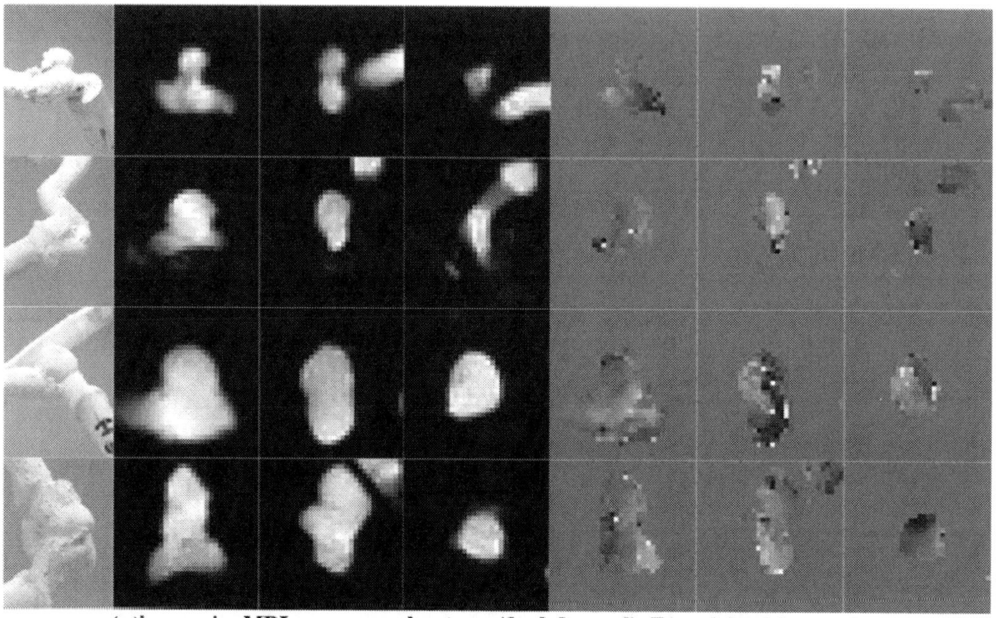

Figure 1. Four representative *ex vivo* MRI aneurysm phantoms (far left panel). T1-weighted images highlighting the three primary axis views centered on each aneurysm (left) and corresponding through-plane flow images (right).

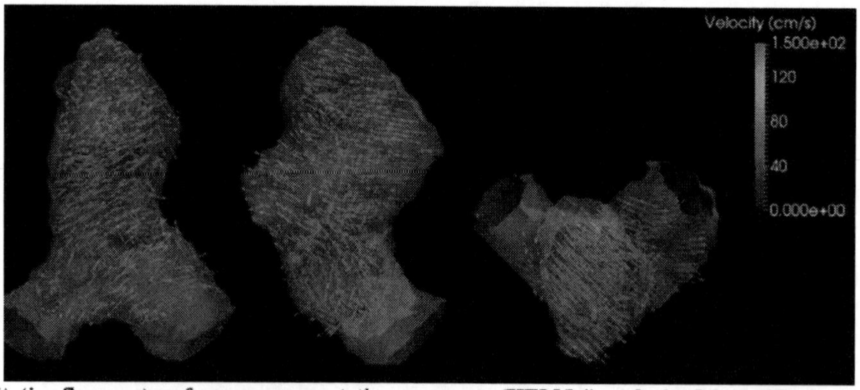

Figure 2. Quantitative flow vectors from a representative aneurysm (HEMO4) as derived from 3D phase contrast images.

SB³C2015
Summer Biomechanics, Bioengineering and Biotransport Conference
June 17-20, 2015,Snowbird Resort, Utah, USA
SB³C2015-284

STRUCTURAL PROPERTIES OF THE ANTEROLATERAL STRUCTURES OF THE KNEE

Amir Ata Rahnemai-Azar (1,2,3), R Matthew Miller (2,3), Daniel Guenther (1,3), Freddie H Fu (1,3), Bryson P Lesniak (1,3), Volker Musahl (1,2,3), Richard E Debski (2,1,3)

(1) Department of Orthopaedic Surgery
University of Pittsburgh
Pittsburgh, PA, United States

(2) Department of Bioengineering
University of Pittsburgh
Pittsburgh, PA, United States

(3) Orthopaedic Robotics Laboratory
Departments of Orthopaedic Surgery and Bioengineering
University of Pittsburgh
Pittsburgh, PA, United States

INTRODUCTION

The role of the anterolateral structures of the knee as secondary restraints for rotatory laxity in anterior cruciate ligament (ACL) deficient knees has been previously demonstrated [1,2]. Recently, the existence of a new structure, referred to as the anterolateral ligament (ALL), has been proposed to exist as an additional component in the anterolateral capsule of the knee that may contribute to rotational stability of the knee [3]. However, there is a lack of knowledge regarding the properties of the anterolateral capsule and it is unclear how these structures may individually contribute to knee stability.

Extra-articular reconstruction techniques using strips of iliotibial band (ITB) have been advocated for improving knee stability after concomitant injury to the ACL and anterolateral capsule of the knee [4]. To ensure that surgical repair using tissue allografts successfully restores normal joint function, it is desired for the choice of graft to have similar structural properties to the native tissue. In order to understand the roles of the anterolateral capsule and ITB in knee stability, it is important to quantify the structural properties of these tissues. Therefore, the objective of this study was to determine and compare the structural properties of the anterolateral capsule and ITB in human cadaveric knees. It was hypothesized that the tensile structural properties of the anterolateral capsule will be comparable to the ITB, due to the ITB being used as a graft for repair of anterolateral capsule during extra-articular reconstruction.

METHODS

Nine fresh frozen cadaveric knees (8 male, 1 female, average age 57±10 years) were dissected down to the anterolateral structures. Each knee was disarticulated from the medial side by dissecting away all medial structures. The ITB was dissected away from the underlying capsule, leaving the tibial insertion attached. The borders of the LCL were defined and the anterolateral capsule was carefully palpated and visually inspected for the presence of distinct fibers running from the lateral epicondyle to the anterior tibia that indicates the presence of a capsular thickening (Figure 1).

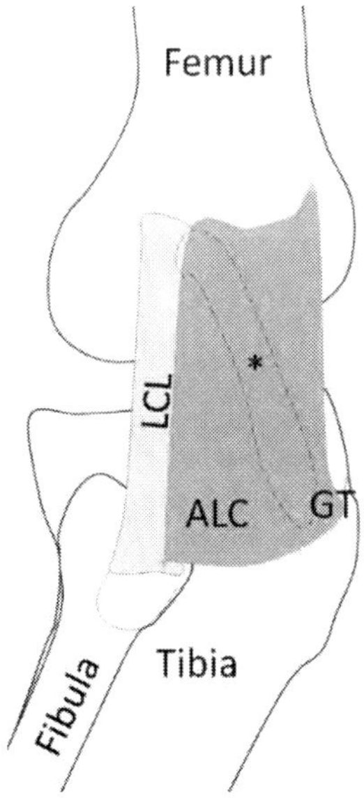

FIGURE 1: Schematic of anterolateral capsular structures. LCL: Lateral Collateral Ligament. ALC: Anterolateral capsule. GT: Gerdy's Tubercle. * Indicates capsular thickening that may represent ALL.

The capsule was sectioned 2 cm anterior to the LCL, creating a strip of anterolateral capsule 2 cm wide and remaining attached to the femur and tibia for mechanical testing. The lateral meniscus and other structures were dissected off the tibia but left attached to the capsule to avoid damaging the capsule during dissection. A strip of ITB was cut measuring approximately 15 cm long and 1 cm wide and left attached to Gerdy's tubercle for testing, with dimensions based on those similar to grafts used for extra-articular reconstruction techniques [4].

The femur and tibia of each specimen was potted in epoxy putty and affixed in a set of custom clamps to a materials testing machine (Instron Model 5965, Instron Corp., Norwood, MA, USA) for tensile testing (Figure 2A). The knee flexion angle for testing was set at 45° and was determined during preliminary testing as the angle that most evenly loaded the fibers of the bone-capsular tissue-bone complex with the axis of loading. Each specimen was preloaded to 5N and preconditioned for 10 cycles from 5 to 20N at 10 mm/min. Cyclic creep testing was then performed for 50 cycles from 5 to 50N at 10 mm/min. Cyclic creep was calculated as the difference in peak elongation between the first and last cycles to provide a measure of capsule extensibility after repeated loading. Specimens were allowed to recover for 30 minutes prior to reapplication of preload and preconditioning, and load to failure was performed at 10 mm/min to assess ultimate load, ultimate elongation, and stiffness of the linear region of the load-elongation curve.

After load to failure of the capsule, the ITB was clamped to the crosshead with the full tibia remaining clamped to the materials testing machine (Figure 2B). A preload of 10N was applied with preconditioning for 10 cycles between 10 and 50N at 10 mm/min, and cyclic creep was performed for 50 cycles at 10 mm/min between 10 and 100N. Otherwise, the testing procedure was the same as for the anterolateral capsular tissue. Statistical comparisons were made between capsule and ITB using paired t-tests. Statistical significance was set at $p < 0.05$.

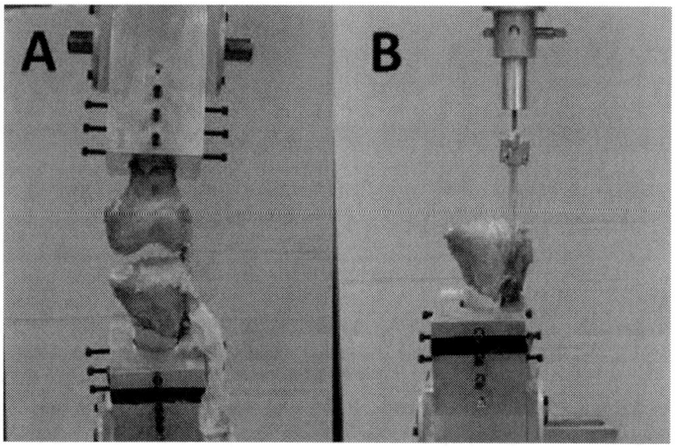

FIGURE 2: A) Anterior view of mechanical testing setup for the anterolateral capsule. B) Anterior view of mechanical testing setup for the ITB.

RESULTS

Only one of the specimens tested had an observable capsular thickening. One specimen slipped from the clamps during load to failure testing and was therefore excluded from analysis (n = 8 specimens). In general, the ITB showed superior tensile structural properties to anterolateral capsule, though significant variation in properties was observed in both groups (Table 1). The ultimate load of the ITB was approximately 55% greater than that of the capsule, and the creep and ultimate elongation of the capsule were nearly twice as large as those of the ITB. The stiffness of the ITB was nearly 2.75 times greater than the stiffness of the anterolateral capsule (Figure 2). Capsule specimens failed in multiple modes: 4 failed at the tibial insertion, 3 failed mid-substance, and 2 failed at the femoral side. ITB specimens failed either mid-substance (5) or near the clamp (4). All parameters were statistically significant except for % creep.

Table 1: Structural Properties of anterolateral capsule and iliotibial band (Average ± SD). *p < 0.05

	AL Capsule	ITB
% Creep	17 ± 5	12 ± 8
Ultimate Load (N)*	319.7 ± 227.3	494.5 ± 176.7
Ultimate Elongation (mm)*	15.5 ± 7.8	8.5 ± 1.5
Stiffness (N/mm)*	27.5 ± 12.3	75.4 ± 26.5

DISCUSSION

This study defined the structural properties of the anterolateral capsule and strips of iliotibial band typically used as grafts for extra-articular reconstruction of anterolateral capsule. It was determined that the lateral capsule is significantly less strong and less stiff than the ITB graft, and is capable of greater deformation compared to ITB, disproving the hypothesis under study. This indicates that the use of ITB as a graft for extra-articular reconstruction could potentially over-constrain the knee, leading to loss of range of motion. Also, the tensile properties obtained for the anterolateral capsule in this study are similar to those previously obtained for the posteriomedial capsule (ultimate load 425 ± 151 N, Stiffness 56 ± 20 N/mm), indicating that the anterolateral capsule does not appear to have localized increases in strength and stiffness compared to other parts of the capsule.

A limitation of this study is that due to the structure of the pieces of anterolateral capsule testing, it was not possible to evenly load all fibers of the tissue with the axis of loading. This resulted in specimens failing in multiple ways and precluding determination of material properties for the capsule. Additionally, a capsular thickening was observed in only one specimen, and so comparing properties of capsule with and without this thickening could not be accomplished. Regardless, the specimen with an observed thickening failed at similar loads to specimens without, indicating that the thickening may not substantially contribute to strength of the capsule. The results of this study suggest that when treating rotatory knee instability, it should be considered that ITB reconstructions exceed the structural properties of anterolateral capsule in patients with ACL injury. In the future, robotic analysis of knee kinematics will be performed to further assess the contributions of anterolateral capsule to knee instability.

ACKNOWLEDGEMENTS

Support from the Department of Orthopaedic Surgery is gratefully acknowledged.

REFERENCES

[1] Monaco E et al. *Am J Sports Med.* 2014;42(7):1669-74.
[2] Wroble RR et al. *Am J Sports Med* 1993;21(2):257-62.
[3] Claes S et al. *J Anat.* 2013;223(4):321-8.
[4] Ireland J and Trickey EL. *J Bone Joint Surg Br.* 1980; 62(3):340-5.
[5] Robinson et al. *J Biomech.* 2005;38(5):1067-74.

STRATEGIES OF INJECTING FERROFLUID INTO TUMORS TO ACHIEVE REPEATABLE NANOPARTICLE DEPOSITION AND THERMAL DOSAGE IN MANGETIC NANOPARTICLE HYPERTHERMIA

Alexander LeBrun (1), Charles Bieberich (2), Ronghui Ma (1), Liang Zhu (1)

(1) Department of Mechanical Engineering
University of Maryland Baltimore County
Baltimore, MD, USA

(2) Department of Biology
University of Maryland Baltimore County
Baltimore, MD, USA

INTRODUCTION

Previous studies have shown that magnetic nanoparticle hyperthermia via intratumoral infusions can be successfully used to treat various tumors [1-3]. Since the amount of thermal energy required to induce tumor cell necrosis is critical for predicting treatment outcomes, it is imperative to understand resulted temperature elevation history in the targeted tumors and its surrounding tissue. It is well known that the volumetric heat generation rate, often referred to as the specific absorption rate or SAR, is directly proportional to the local nanoparticle concentration in tumors in magnetic nanoparticle hyperthermia. The challenge is that nanoparticle concentration distribution in tumors is often not uniform and unpredictablé. Therefore, a one-size-fits-all approach in heating protocol designs may not result in consistent treatment outcomes when dealing with individual tumors.

The objective of this study is to perform microCT scan on resected prostate tumors after ferrofluid injection. The ferrofluid is injected intra-tumorally at its center with controlled influsion rates. The tumors are then scanned and reconstructed to determine the nanoparticle distribution volumes. The infusion rate with the least irregular nanoparticle distribution volumes is identified as the one leading to relatively controllable nanoparticle distribution in the tumors. Temperature elevations in the tumor are simulated based on the obtained nanoparticle distribution and a generated tumor model to design a heating protocol leading to 100% damage to the tumor.

METHODS

MicroCT scan of PC3 tumors

The tumors used in this study are PC3 prostatic tumors implanted on the flanks of BALB/c Nu/Nu male mice. Water based ferrofluids (EMG700 series, Ferrotec (USA) Corporation, Nashua, NH) with a concentration of 5.8% by volume and a particle size of 10 nm are loaded on a syringe pump (S230, Kd Scientific, Inc. Holliston, MA) to control the infusion rate (3, or 4, or 5 μl/min) and the amount injected (0.1 cc). A 26-gauge Hamilton needle (Fischer Scientific, Springfield, NJ) is used for injecting ferrofluid into the tumors. After the experiments, the mouse is euthanized and the tumor is surgically resected and scanned in a microCT imaging system (Skyscan1173, MicroPhotonics, PA), resulting in a total of approximately 1000 individual images and each image has a pixel size of 1052×1944. The nanoparticle distribution volume is defined as the summation of the volumes of pixels containing nanoparticles.

Determination of SAR

We preform experimental studies using gel specimen containing controlled magnetic nanoparticle concentrations to determine the relationship between the nanoparticle concentration and the microCT Hounsfield value. Later, heating experiments are conducted to correlate the nanoparticle concentration to the measured SAR values by placing the specimen inside an alternating magnetic field that is induced by a two-turn coil with a current of 400 A at a frequency of 190 kHz [4].

Generation of tumor geometry and SAR file

The details of the generation of the tumor geometry and the SAR file can be found in our previous study [4]. MATLAB codes, ImageJ, and ProE software are used for the image processing process.

Simulation of temperature elevations and determination of thermal dosage

The Pennes bioheat equation is used to simulate the temperature field inside the mouse model, and is written as

$$\rho c\frac{\partial T}{\partial t} = k\nabla^2 T + q_m + \omega\rho c(37-T) + SAR \qquad (1)$$

where k is thermal conductivity, ρ is density, c is specific heat, and q_m is metabolic heat generation rate. Note that the SAR file and the tumor geometry are imported to the COMSOL software. The boundary condition of the tumor is a convection boundary condition (T_∞=25°C and h = 5 W/m^2 °C). Thermal damage to the tumor tissue is assessed via the Arrhenius equation and damage is defined as the Arrhenius integral Ω≥4.

RESULTS

Figure 1 shows the nanoparticle concentration distribution illustrated by the light colors in the images. It appears that the nanoparticle can be visualized by the microCT system. When the infusion rate is lower, the nanoparticle distribution volume is smaller and the nanoparticles are more concentrated than that obtained with a higher infusion rate. The average volume of the tumors is approximately 920 mm^3. Shown in Figure 2 is the nanoparticle distribution volume around 77 mm^3 when the injection rate is 3 μl/min. It increases by 60% to 126 mm^3 when the infusion rate is 5 μl/min. Also noted in Figure 2 is a smaller standard deviation in the 3 μl/min infusion group (77±7 mm^3) than that in the other two groups. It suggests that using the 3 μl/min infusion rate would result in a relatively repeatable particle distribution volume. The calculated total heat generation rate in each group is also given in Figure 1. It is reasonable that the total energy rate deposited in the tumors are very similar, varying from 0.37 to 0.39 (W), since the total infusion amount is fixed in the study as 0.1 cc.

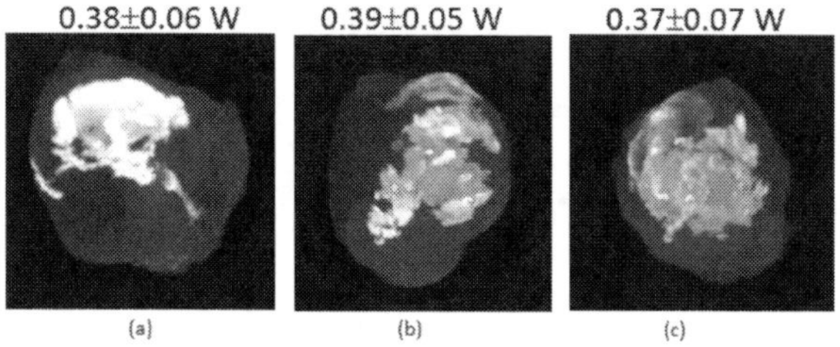

Figure 1 Nanoparticle distribution patterns as affected by the infusion rate (a) 3 μl/min, (b) 4 μl/min, and (c) 5 μl/min.

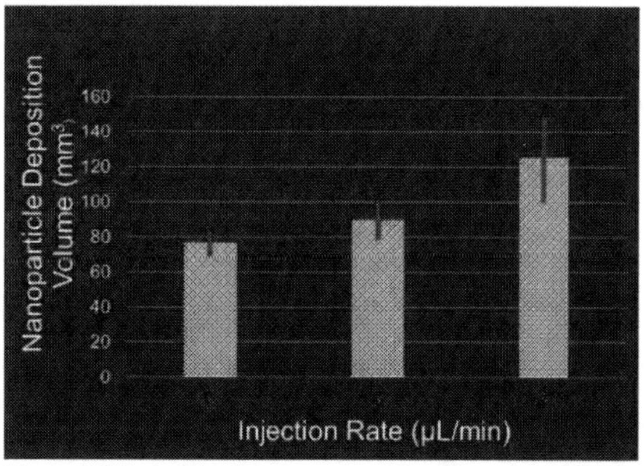

Figure 2 Effect of the infusion rate on the nanoparticle distribution volume. The sample size is 5 per group. The standard deviation is illustrated by the red bar.

The SAR distribution inside the generated tumor can be visualized using the COMSOL software. Figure 3 gives one slice of the microCT image and the SAR distribution inside the COMSOL geometry when the infusion rate is 3 μl/min. The SAR distribution looks very similar to the white cloud in the microCT image. The SAR values vary from 0 to 5.9x10^6 W/m^3. Once both the tumor geometry and the SAR file are imported to the COMSOL software, heat transfer simulation is carried out to determine the temperature field inside the tumor when the tumor is subject to the same alternating magnetic field as our previous heating experiment. Figure 3 also illustrates the temperature field with the nanoparticle injection. With the heat generated by the nanoparticles subject the magnetic field, the tumor temperatures are elevated up to 75°C from their baseline values.

The evolving thermal damage contours inside the tumor are illustrated in Figure 4. As heating time is getting longer, more and more tumor regions have their Ω values larger than 4. Our simulation has shown that the heating time should be longer than 21 minutes when 100% damage to the entire tumor (Ω≥4) is achieved.

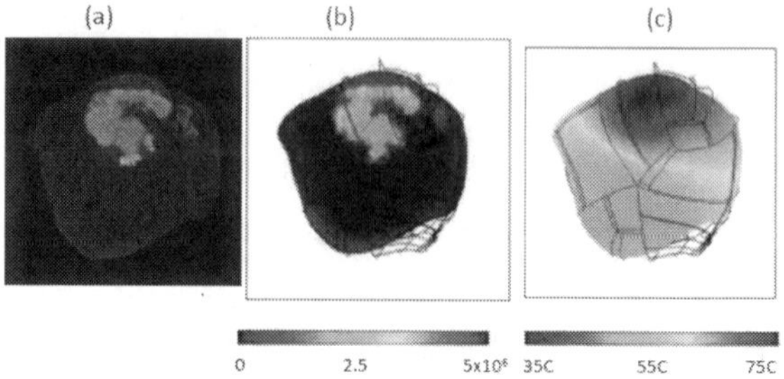

Figure 3 (a) A slice of microCT image; (b) the generate SAR color contours; and (c) simulated temperature contours on the same slice.

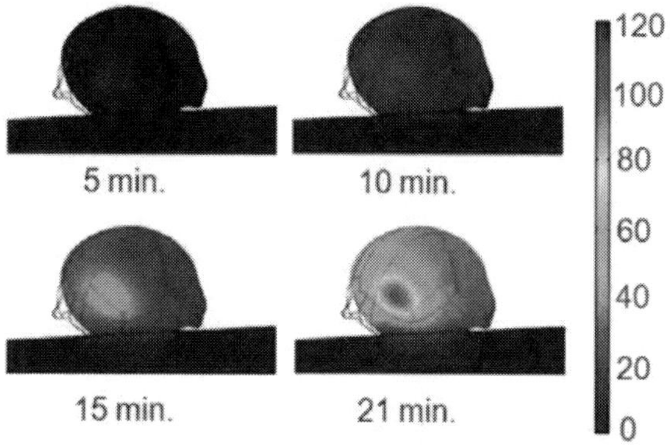

Figure 4 Simulated thermal damage distribution in the tumor (Ω) affected by the heating time.

In summary, we have performed microCT scanning of PC3 tumors injected with ferrofluid to identify an infusion strategy leading to repeatable and controllable nanoparticle particles distribution specific for the tumor type and ferrofluid. The feasibility of using our computer algorithms to generate tumor prototypes and SAR distribution has been demonstrated in the study. Further temperature elevation simulation and thermal damage assessment lead to design of a heating protocol to achieve 100% thermal damage to PC3 tumors.

ACKNOWLEDGEMENTS

This study was supported in part by an NSF research grant CBET-1335958 and by the GAANN Scholarship Program (DoE P200A100124) at UMBC. The research was performed in partial fulfilment of the requirements for the PhD degree from University of Maryland Baltimore County by Alexander LeBrun.

REFERENCES

[1] Jordan, A., Scholz,R., Wust, P., Fahling, H., and Felix, R., 1999, *J. Magnetism and Magnetic Materials*, 201:413-419.
[2] Salloum, M., R. Ma, D. Weeks, and Zhu, L., 2008, *International Journal of Hyperthermia*, 24(4):337-345.
[3] Attaluri, A., Ma, R., Qiu, Y., Li, W., and Zhu, L., 2011, *International Journal of Hyperthermia*, 27(5): 491–502.
[4] LeBrun, A., Manuchehrabadi, N., Attaluri, A., Wang, F., Ma, R., and Zhu, L., 2013,*International Journal of Hyperthermia*, 29:730-738.

SB³C2015
Summer Biomechanics, Bioengineering and Biotransport Conference
June 17-20, 2015, Snowbird Resort, Utah, USA
SB³C2015-287

MICROGEOMETRY AND MICROENVIRONMENT OF MITRAL VALVE INTERSTITIAL CELLS UNDER PHYSIOLOGICAL LOADS

Salma Ayoub, Chung-Hao Lee, Michael S. Sacks

Center for Cardiovascular Simulation
Institute for Computational Engineering and Science
Department of Biomedical Engineering
The University of Texas at Austin
Austin, Texas, USA

INTRODUCTION

It is estimated that five million people domestically are diagnosed with heart valve disease each year [1]. Of these, nearly 40,000 patients undergo mitral valve (MV) repair due to mitral regurgitation. Improvements of heart valve disease and development of novel therapies must rest on a strong fundamental knowledge of native valve function. Pathophysiological alterations in mechanical loading can lead to stress-induced changes in cellular function and tissue adaptation. In the MV, numerous pathological factors have been shown to affect tissue structure and composition.

Valve interstitial cells (VICs) are important in valve tissue homeostasis and pathophysiology: they maintain the structural integrity of the leaflet via protein synthesis and enzymatic degradation of the extracellular matrix (ECM) [2]. While cell phenotype and ECM regulation under physiological stress have been previously studied, little attention has been paid to the layer-specific structure and microenvironment of mitral VICs (MVICs).

The MV consists of four layers: the fibrosa is the primary load-bearing layer and consists mainly of Type I collagen fibrils, which are oriented in the circumferential direction; the atrialis and ventricularis are thin, predominantly elastic layers on the atrial and ventricular surfaces and the spongiosa, which lies between the atrialis and ventricularis, contains proteoglycans and glycosaminoglycans. Extracellular matrix (ECM) components are not exclusive to one layer and can be found across all four layers along with the VICs [3].

Our aim is to fully characterize the ultrastructure of MVICs under physiological loads and to measure and quantify the three-dimensional VIC microenvironment and geometric deformation under physiological loads for computational model development.

METHODS

Tissue Preparation: Porcine hearts were acquired from a local abattoir. The mitral valve was isolated and the anterior leaflet was fixed in Type I EM grade glutaraldehyde for 4 hours. Fixed tissue was cut into small 1 cm x 0.5 cm blocks, stained with osmium tetroxide for a total of 4 hours, dehydrated with alcohol and acetone (microwave-enhanced method). Resulting tissue blocks were infiltrated in epoxy resin overnight and cured for 48 h in a 60°C oven.

Block Trimming: Prior to sectioning, the resin block is trimmed using a razor blade into a trapezoid block face. A test thick section (1 μm) is imaged using a light microscope to ensure that there is tissue in the area of interest and to guide the serial sectioning.

Serial Sectioning: The block face is trimmed and smoothed using a Diatome Trimtool 45° set at 500 nm. Serial sectioning is carried out using a Diatome Ultra Diamond Knife 35° (70 nm). Formvar-coated slot grids are used to pick up the sections. The sections on the slot grid are imaged using transmission electron microscopy (TEM). The resulting images are registered and segmented using *Amira* (Fig. 1).

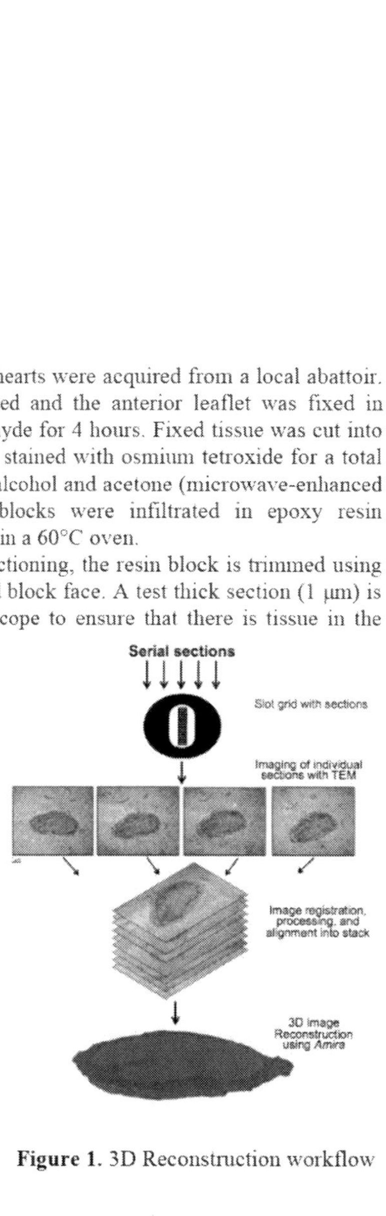

Figure 1. 3D Reconstruction workflow

Stimulating Physiological Stress: A biaxial tissue fixation system (Fig. 2) is used to stimulate physiological stress in the circumferential and radial directions. Excised MV anterior leaflet tissue specimens are fixed under a range of equibiaxial loads (0, 5, 10, 20, 50, 100 and 150 N/m). The 3D cellular and microenvironment deformations are quantified using the methods outlined above.

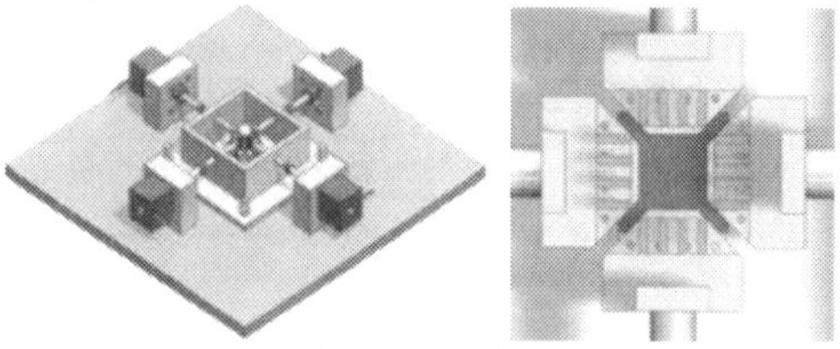

Figure 2. Tissue Fixation System

RESULTS

3D Reconstruction: A total of 135 serial sections were imaged using TEM (Fig. 3). The resulting images were filtered, registered, and aligned using *Amira*. 3D image reconstruction was carried out using *Amira*, resulting in the 3D microgeometry of the VIC's nucleus and cytoplasm after segmentation (Fig. 4). This ultrastructural study highlights the previously unrecognized complexity between ECM fiber networks and MVICs under physiological stress and provides important geometrical information on the deformation of the valve interstitial cell under physiological loads.

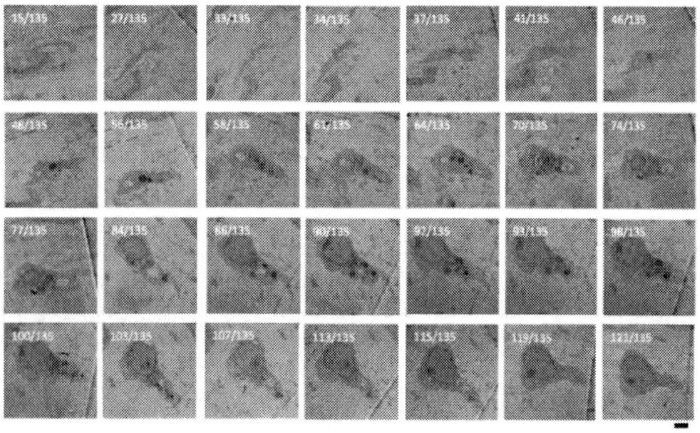

Figure 3. Representative MVIC images from a stack of 135 serial sections. Raw images are processed, registered, and aligned using *Amira* prior to segmentation and 3D reconstruction. Scale bar = 2 μm.

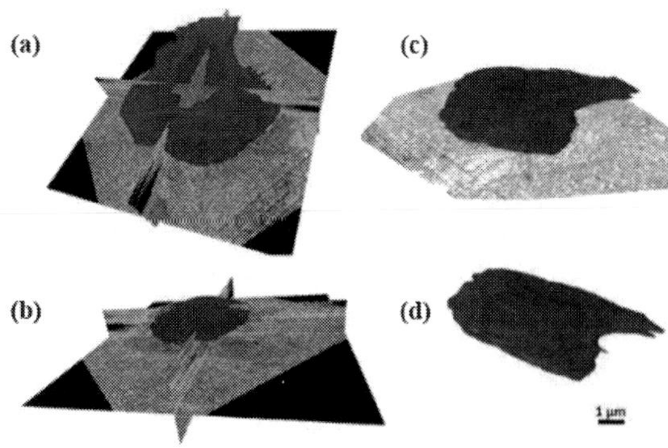

Figure 4. Surface generation post-segmentation using *Amira* of the (a) cytoplasm and (b) nucleus in each section of the image stack (c) and (d) 3D representation of the MVIC nucleus in red and cytoplasm in blue. Scale bar = 1 μm.

Resulting data from 3D reconstructions was then used to quantify cellular morphology and subsequent deformation under physiological loads. Improving the osmium staining protocol will result in enhanced staining of collagen fibrils (Fig. 5), which will allow the quantification of the cellular microenvironment—particularly, the circumferentially oriented collagen fibrils surrounding the VICs: (1) transmural-radial representation of fibrils from which the diameter of each can be determined as well as the distance from the cell membrane, (2) transmural-circumferential representations of fibrils are used to quantify the collagen fibril length and degree of crimp in both the loaded and unloaded configurations.

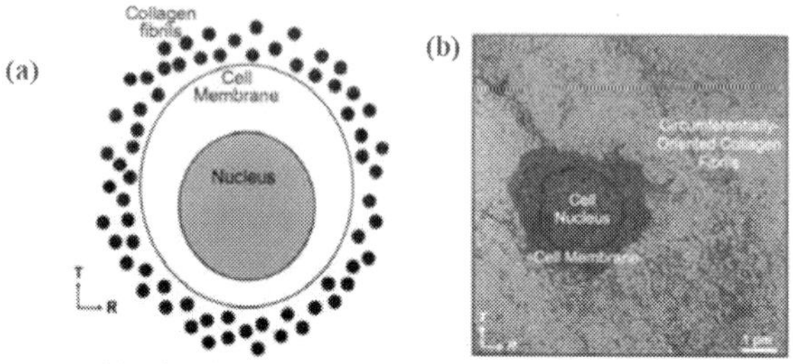

Figure 5. (a) Simplified representation of VIC and collagen fibrils (b) TEM micrograph of an MVIC from the fibrosa layer of the MV highlighting collagen fibril circumferential orientation and close interaction with the cell membrane, T: transmural, R: radial.

DISCUSSION

The technique outlined above allows us to capture the previously unknown VIC microgeometry and surrounding microenvironment with a high level of detail. An advantage of serial TEM over single section TEM and other microscopy techniques is that it allows us to track individual collagen fibrils, which have an average diameter of 300 nm. This detailed data will enable us to develop more structurally accurate models that incorporate the heterogeneous complexity of the mechanobiologically-stimulated microenvironment.

Currently in progress is the quantification of layer-specific 3D MVIC microgeometry and microenvironment deformations under physiological loads. Since each layer of the MV has different ECM components with specific orientations, using the technique outlined above will provide valuable data on layer-specific MVIC microenvironment. Integrins, which are transmembrane signal receptors that physically link the cell surface to the surrounding ECM, sort and process external signals and play a key role in transducing them to the cell interior, leading to a network of intracellular signaling pathways that result in the cell response. In the MV for example, the integrin $\alpha_2\beta_1$, which shows higher affinity for type I collagen, is the main collagen binding integrin. Using immunogold localization, we can stain for these integrins and create a 3D "map" that will give us new insights on the VIC mechanotransduction pathway. Using a macro-micro finite element computational framework, we will be able to link tissue-level mechanical behavior to layer-specific MVIC deformations and alterations in the ECM microenvironment. This will result in a highly accurate model that can be used to simulate MVIC behavior in various pathological and surgical repair scenarios.

ACKNOWLEDGEMENTS

Amy Hooper and John Lesicko (TFS), Dr. Dwight Romanovic (TEM), National Institute of Health (NIH) grant R01HL119297.

REFERENCES

[1] Go et al. *Circulation*, 129:28-292, 2014.
[2] Taylor et al. *Int J Bioc & Cell Bio*, 113-8, 2003.
[3] Grande-Allen et al. *Cur Cardio Reports*, 13:113-120, 2011.

SB[3]C2015
Summer Biomechanics, Bioengineering and Biotransport Conference
June 17-20, 2015, Snowbird Resort, Utah, USA
SB[3]C2015-289

INFLUENCE OF RECENT HEAD IMPACT HISTORY ON BIOMECHANICS OF CONCUSSION SUSTAINED IN COLLEGE FOOTBALL ATHLETES

Daniel J. Sjoquist (1), Brian D. Stemper (1,2), Alok S. Shah (1,2), James Murtha (1), John R. Humm (1,2), Ashley LaRoche (1), Adam Pfaller (1) Steven Broglio (3), Kevin Guskiewicz (4), Michael McCrea (1,2)

(1) Department of Neurosurgery
Medical College of Wisconsin
Milwaukee, WI, USA

(2) Neuroscience Research
Zablocki VA Medical Center
Milwaukee, WI, USA

(3) NeuroSport Research Laboratory
University of Michigan
Ann Arbor, MI, USA

(4) Department of Exercise and Sport Science
University of North Carolina at Chapel Hill
Chapel Hill, NC, USA

INTRODUCTION

Repetitive subconcussive head impacts are associated with cumulative changes in the brain that may contribute to decreased injury tolerance and more deleterious outcomes for the concussed athlete. A subconcussive impact is classified as a head impact that does not result in the signs and symptoms associated with the clinical diagnosis of a concussion. Incorporation of helmet- or nonhelmet-based head impact sensors has led to investigations of head impact biomechanics with clinical outcomes and imaging evidence of structural damage in the brain. Football players may sustain greater than 400 recordable subconcussive impacts over the course of a single season and have been a population commonly used in head impact sensor studies [1]. Accordingly, prior head impact exposure was linked to medical imaging evidence of structural changes and changes in post-season cognitive performance in non-concussed football and ice hockey athletes [2, 3].

Peak linear and rotational acceleration, as well as rotational velocity of the head have been associated with the onset and severity of concussion in human and animal studies [4]. Using scaled primate data, Gennarelli and colleagues identified a correlation between increasing head rotational acceleration and severity of diffuse brain injury [5], implying greater levels of brain tissue damage associated with greater accelerations. Repetitive subconcussive head impact exposures may also contribute to cumulative brain tissue damage. Accordingly, a higher number and higher severity of subconcussive impacts could contribute to greater levels of cumulative tissue damage and an associated decreased injury tolerance. This phenomenon has been demonstrated in other biological tissues with repetitive loading eventually contributing to vertebral failure according to fatigue principles at loading magnitudes below the injury tolerance for single cycle loads [6]. However, to date, the relationship between the number and severity of repetitive subconcussive head impacts and decreasing injury tolerance has not been clearly defined.

METHODS

The study protocol was approved by the IRB at the Medical College of Wisconsin. Informed consent was obtained from 438 college football players across four schools during the 2013 and 2014 football season, with 202 athletes participating in the study for both seasons. Each player was instrumented with an X-patch head impact sensor (X2 Biosystems, Inc., Seattle WA). The X-patch is a non helmet-based system that records head linear accelerations and rotational velocities at 1,000 Hz and transfers those metrics to an estimated head center of gravity, when a 10-g linear acceleration is exceeded by any of the accelerometer sensors. Concussions sustained by athletes during practice or game activities were diagnosed by a healthcare provider and were based on alteration of mental status, with symptoms including headache, nausea, vomiting, dizziness/balance problems, fatigue, insomnia/sleep disturbances, drowsiness, sensitivity to light/noise, blurred vision, difficulty remembering, and/or difficulty concentrating.

Peak linear acceleration and peak rotational velocity were obtained from time-based sensor outputs. Because numerous head impacts were recorded during each activity, concussions were identified using athlete and healthcare provider accounts of the event as well as audio and video footage. Injuries in which the concussive event could not be identified were excluded from data analysis. The magnitude of linear acceleration and rotational velocity for all concussions were correlated to the total number of recordable head impacts as well as the number of significant subconcussive head impacts during the session in which the concussion was sustained (game or practice), during the prior six days, and during the season leading up to the concussive event.. Significant subconcussive impacts were defined as 25%, 50%, or 75% of the median peak linear acceleration or rotational velocity across all sustained concussions. Statistically significant correlations ($p<0.05$) were graded according to Pearson's correlation coefficient.

RESULTS

Thirty-three concussions were sustained during the course of the study. 19 concussive events were able to be confirmed using event descriptives and audio/video analysis. Concussions occurred during impacts with peak linear acceleration between 39 and 167 g's (median: 68 g's) and rotational velocities between 542 and 3243 deg/s (median: 2663 deg/s). Injury biomechanics were not significantly correlated ($p<0.05$) to any measure with respect to the number of impacts on the day of concussion. Likewise, injury biomechanics were also not significantly correlated to the total number of recordable impacts or impacts with severity greater than 25% of the median concussion biomechanics threshold with respect to the prior six days or season. The remaining correlations are presented in Table 1.

Table 1: Correlations between recent head impact history and injury biomechanics for concussions sustained by college football athletes.

Injury Metric	Correlate	Percent Median	Period	p-value	Correlation Coefficient
$\ddot{x}$	$\ddot{x}$	50%	Season	NS	-0.27
$\ddot{x}$	$\ddot{x}$	75%	Season	NS	-0.28
$\dot{\theta}$	$\dot{\theta}$	50%	Season	NS	-0.30
$\dot{\theta}$	$\dot{\theta}$	75%	Season	$p<0.05$	-0.62
$\dot{\theta}$	$\ddot{x}$	50%	Season	$p<0.05$	-0.71
$\dot{\theta}$	$\ddot{x}$	75%	Season	$p<0.05$	-0.66
$\ddot{x}$	$\ddot{x}$	50%	six days	$p<0.05$	-0.47
$\ddot{x}$	$\ddot{x}$	75%	six days	$p<0.05$	-0.51
$\dot{\theta}$	$\dot{\theta}$	50%	six days	NS	-0.14
$\dot{\theta}$	$\dot{\theta}$	75%	six days	NS	-0.35
$\dot{\theta}$	$\ddot{x}$	50%	six days	$p<0.05$	-0.60
$\dot{\theta}$	$\ddot{x}$	75%	six days	$p<0.05$	-0.54

$\ddot{x}$: linear acceleration, $\dot{\theta}$: rotational velocity, NS: not significant

The strongest correlations were identified between rotational velocity of the injury impact and the number of significant subconcussive head impacts over the course of the season, assessed as 50% or 75% of the median linear acceleration. Those correlations demonstrated decreasing rotational velocity of the injury impact with a greater number of prior significant subconcussive head impacts. Specifically, rotational velocities decreased by 16% for players that had 75 or more prior impacts above 75% of the median linear acceleration compared to those with 10 or more prior impacts above that level during the season. In general, the number of significant subconcussive impacts over the course of a season was a better predictor of injury biomechanics compared to the previous six days, with decreasing injury biomechanics associated with a greater number of significant subconcussive impacts. Likewise, strongest correlations were for rotational velocity compared to linear acceleration.

DISCUSSION

This preliminary study has demonstrated the importance of subconcussive head impact history on concussion injury biomechanics for college football athletes. Results demonstrated decreasing injury biomechanics for players that sustained a greater number of subconcussive head impacts. Correlations also suggest the existence of a subconcussive impact threshold for lowering injury biomechanics, with the total number of recordable head impacts and number of head impacts above 25% of the median linear acceleration or rotational velocity not correlated to decreasing injury biomechanics. However, the number of impacts above both 50% and 75% of the median injury biomechanics were correlated with peak rotational velocity of the head impact leading to injury.

The finding of decreasing rotational velocities with a greater number of prior subconcussive head impacts is significant from a mechanistic standpoint. This study demonstrated that recent head impact history had the strongest correlations with rotational velocity. A majority of biomechanics-related information obtained from head impact sensors during sporting events has focused on measures of head linear acceleration. However, the importance of rotational metrics in the development of diffuse brain injury (i.e., concussion) has long been understood as outlined in Stemper and Pintar [4]. This is evident from the formation of rotational acceleration and velocity based metrics for concussion [7, 8]. Therefore, continued investigation into rotational-based metrics from head impact sensors may provide a better estimate of injury risk and tolerance, as well as the effect of repetitive subconcussive head impacts.

ACKNOWLEDGEMENTS

This work was supported by the U.S. Army Medical Research and Materiel Command under Award No. W81XWH-12-1-0004 and the Clinical and Translational Science Institute grant 1UL1-RR031973 (-01). Opinions, interpretations, conclusions and recommendations are those of the author and are not necessarily endorsed by the U.S. Army. This publication was also supported by the National Center for Advancing Translational Sciences, National Institutes of Health, through Grant Number 8UL1TR000055. Its contents are solely the responsibility of the authors and do not necessarily represent the official views of the NIH. Support of Veterans Affairs Medical Research is also acknowledged.

REFERENCES

1. Martini, D., et al., *Subconcussive head impact biomechanics: comparing differing offensive schemes.* Med Sci Sports Exerc, 2013. **45**(4): p. 755-61.
2. McAllister, T.W., et al., *Effect of head impacts on diffusivity measures in a cohort of collegiate contact sport athletes.* Neurology, 2014. **82**(1): p. 63-9.
3. Talavage, T.M., et al., *Functionally-detected cognitive impairment in high school football players without clinically-diagnosed concussion.* J Neurotrauma, 2014. **31**(4): p. 327-38.
4. Stemper, B.D. and F.A. Pintar, *Biomechanics of concussion.* Prog Neurol Surg, 2014. **28**: p. 14-27.
5. Gennarelli, T.A., et al. *Biomechanical tolerances for diffuse brain injury and a hypothesis for genotypic variability in response to trauma.* in *47th Proceedings of the Association for the Advancement of Automotive Medicine.* 2003.
6. Schmidt, A.L., et al., *Risk of lumbar spine injury from cyclic compressive loading.* Spine, 2012. **37**(26): p. E1614-21.
7. Margulies, S.S. and L.E. Thibault, *A proposed tolerance criterion for diffuse axonal injury in man.* J Biomech, 1992. **25**(8): p. 917-23.
8. Takhounts, E.G., et al., *Development of brain injury criteria (BrIC).* Stapp Car Crash J, 2013. **57**: p. 243-66.

IDENTIFICATION AND QUANTIFICATION OF EXTRACELLULAR MATRIX PROTEINS AT THE PLAQUE – INTERNAL ELASTIC LAMINA INTERFACE IN A MOUSE MODEL OF ATHEROSCLEROSIS

Lindsey A. Davis (1, 2), Susan M. Lessner (1, 2)

(1) Biomedical Engineering
University of South Carolina
Columbia, SC, USA

(2) Cell Biology and Anatomy
University of South Carolina School of Medicine
Columbia, SC, USA

INTRODUCTION

Atherosclerotic plaque failure is a leading cause of myocardial infarction and ischemic stroke. There are at least two potential failure mechanisms for atherosclerotic plaque. The first involves tensile rupture of the fibrous cap and the second, which is prominent during angioplasty, involves plaque delamination from the internal elastic lamina (IEL) at the plaque shoulder. During balloon angioplasty, a catheter is inserted into the blood vessel and a small balloon is inflated at the lesion. The goal is to flatten the plaque and increase the lumen area; however, in some cases this results in intimal dissection at the plaque shoulder. This is commonly treated with the implantation of a stent over the dissected tissue to juxtapose the intimal and medial surfaces to recreate a smooth lumen[123].

Identifying the proteins responsible for adhesion between the plaque and the IEL may provide insight into the delamination mechanism for atherosclerotic plaque failure at the shoulder. Our previous studies used SEM and cryo-SEM to examine intact or partially delaminated mouse atherosclerotic plaques, and we have observed perpendicular matrix fibers at the interface between the plaque and the IEL. The composition of these fibers is unknown, though they likely form from proteins found in the surrounding extracellular matrix (ECM). Collagen and elastin are major components of vascular ECM and may be key constituents of these fibers. Contributions to adhesion at the plaque-IEL interface may also come from cell adhesive glycoproteins. We aim to identify and quantify several key extracellular matrix proteins at the interface between atherosclerotic plaque and the underlying internal elastic lamina (IEL). Immunohistochemistry (IHC) along with immunoblotting of atherosclerotic lesions in transgenic mice will identify which proteins are present at the plaque-IEL interface. Relative concentrations of these proteins can be quantified using the plaque immunoblotting technique.

METHODS

Immunohistochemistry (IHC) was used to determine the presence and distribution of collagen I, collagen IV, and fibrillin-1, a key component of elastic fibers, in the plaque and the aorta wall. ApoE-KO mice were put on a Western (high-fat) diet for at least eight months to develop advanced atherosclerotic plaques in the descending aorta. Both formalin-fixed, paraffin embedded aortic plaque sections and frozen sections of mouse carotid arteries 28-days after partial carotid ligation surgery were used as mouse models of atherosclerosis. Collagen I, collagen IV, and fibrillin-1 were detected with the following antibodies, respectively: rabbit anti-collagen I polyclonal antibody (Calbiochem, San Diego, CA), rabbit anti-collagen IV polyclonal antibody (Abcam, Cambridge, MA), and rabbit anti-fibrillin-1 polyclonal antibody (Novus Biologicals, Littleton, CO). Donkey anti-rabbit rhodamine red X (RRX)-labeled secondary antibodies (Jackson Immuno Research Laboratories, West Grove, PA) were used in each case and a Zeiss LSM 510 confocal microscope was used to capture the images.

A novel plaque blotting technique was used to identify proteins present locally at the plaque-IEL interface. ApoE-KO mice were sacrificed used CO_2 asphyxiation and perfused with heparinized saline for five minutes. The descending aorta was opened longitudinally to visualize the atherosclerotic lesions. Plaques were peeled from the aorta wall and blotted onto a PVDF membrane that had been rinsed with methanol and transfer buffer. The underlying IEL was also blotted separately onto the PVDF membrane with each side of the interface contacting the membrane for approximately three seconds. A set of purified protein standards (rat tail collagen I, Millipore,

Billerica, MA; mouse collagen IV, BD Biosciences, Bedford, MA; partial recombinant fibrillin-1, Novus Biologicals, Littleton, CO) at known concentrations was dot-blotted onto the membrane to allow for approximate quantitation of the protein transferred from the plaque-IEL interface. The membranes were then placed in blocking solution, incubated with the appropriate primary and secondary antibodies and developed using either diaminobenzidine substrate (DAB) (Thermo Scientific, Waltham, MA) or West Pico chemiluminescent substrate (Thermo Scientific, Waltham, MA).

After incubation with chemiluminescent substrate, the membrane was placed in a cassette contacting autoradiography film for approximately five minutes. The film is then transferred to a processor to visualize the results. In order to determine the approximate concentration of the protein of interest, the first blot of each tissue was approximately matched to one of the standards with a known concentration.

For membranes incubated with DAB substrate, a digital image of the membrane was collected at the end of the experiment. ImageJ was used to measure the average pixel intensity of each region of interest. To compensate for uneven lighting, the pixel intensity of the unstained membrane directly next to the region of interest was also measured. The difference between these two pixel intensities, the net pixel intensity, was used as a metric for the level of staining. The net pixel intensities of the standards were used to obtain a logarithmic line of best fit. This line was used to determine the approximate concentration of the protein of interest in the tissue, using the measured average pixel intensity in the region of interest.

RESULTS

The IHC results (figure 1) demonstrate that collagen I was found in the adventitia of the arterial wall (triangles), in the body of the plaque (circles), and at the interface between the plaque and the IEL (arrows). Collagen IV was present in the body of the plaque, within the media of the artery (diamonds), and, to a lesser extent, at the plaque-IEL interface. Fibrillin-1 was present primarily in the media of the artery, in the body of the plaque, and at the plaque-IEL interface. Taken together, these results indicate that all three of these proteins are present at the plaque-IEL interface, though collagen IV is not continuously present along the entire plaque-IEL interface.

Representative results from the plaque blotting experiments (figure 2) confirm the presence of each protein at the plaque-IEL interface. The quantitative results (table 1) show that collagen I and fibrillin-1 are present in concentrations of about one order of magnitude higher than that of collagen IV. These results correspond well with our IHC results.

DISCUSSION

We have confirmed the presence of collagen I, collagen IV, and fibrillin-1 with both IHC and plaque immunoblotting. The results suggest that while collagen IV does not play a significant role in plaque adhesion to the arterial wall, collagen I and fibrillin-1 may contribute, at least in part, to the bridging fibers between the plaque and the IEL. This observation is consistent with our previous results demonstrating a correlation between plaque adhesion strength and local collagen content and elastin fragmentation[4]. These methods may also be used to determine the presence of adhesive glycoproteins, such as fibronectin, that may also contribute to the adhesion between the atherosclerotic plaque and the IEL.

ACKNOWLEDGEMENTS

This work is funded by NSF CMMI-1200358.

REFERENCES

1. Fischman, D. L. et al. *Am. Coll. Cardiol.* 18, 1445–51 (1991).
2. Schomig, A. et al. *Am. Coll. Cardiol.* 23, 1053–60 (1994).
3. Gonzalo, N. et al. *Int. J. Cardiol.* 150, 151–5 (2011).
4. Wang, Y., Johnson, J. A., Fulp, A., Sutton, M. A. & Lessner, S. M. *J. Biomech.* 46, 716–22 (2013).

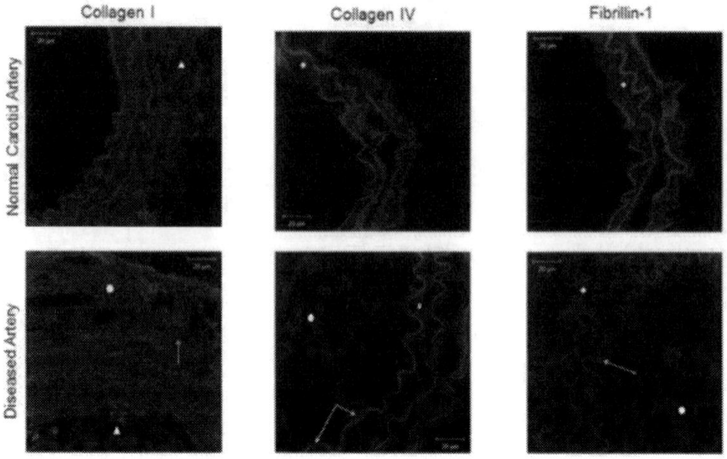

FIGURE 1: Confocal images of normal carotid arteries compared with diseased aorta (collagen I) or carotid artery (collagen IV and fibrillin-1). Red indicates the protein of interest and green is autofluorescence. All images are at 63X magnification.

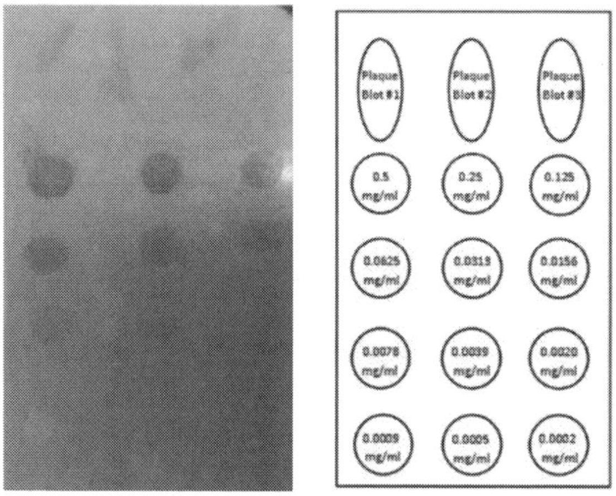

FIGURE 2: Protein presence at the plaque-IEL interface. The presence of collagen I is shown with a chromogenic substrate (DAB) (left) along with the blot diagram (right).

TABLE 1: Average concentration of protein at the plaque-IEL interface.

Protein	Average Approximate Concentration	
	Underside of Plaque	Underlying Aorta
Collagen I	0.11 mg/mL	0.013 mg/mL
Collagen IV	0.017 mg/mL	0.0058 mg/mL
Fibrillin-1	≥0.09 mg/mL	≥0.09 mg/mL

SB^{3}C2015
Summer Biomechanics, Bioengineering and Biotransport Conference
June 17-20, 2015,Snowbird Resort, Utah, USA
SB^{3}C2015-292

AUTODIGESTION AND ITS MANIFESTATIONS IN THE MICROCIRCULATION

Frank A. DeLano, Angelina E. Altshuler, Erik B. Kistler, Michael D. Richter, Geert W. Schmid-Schönbein

(1) Department of Bioengineering
University of California San Diego
La Jolla, CA, USA

INTRODUCTION

It has been recognized for decades that one of the earliest hallmarks of a disease is loss of microvascular perfusion and loss of cell functions. But the underlying molecular mechanisms remain speculative. An example is physiological shock in which a severe reduction of microcirculatory perfusion as well as many cell dysfunctions is observed. Shock is also associated with high levels of organ dysfunctions and mortality. An extensive body of evidence shows that in shock the microcirculation fails to the degree that entire regions of the microvascular network are no longer perfused, red cells and other blood cells escape into the interstitial tissues, various cell functions fail, including mechanotransduction mechanisms, metabolic mechanisms and many others.

While several tissue injury mechanisms (e.g. acute mechanical trauma, thermal or chemical exposures, infections) can lead to shock there are also mechanisms in which these particular injuries could not be documented. There is a need to explore alternative mechanisms for tissue injury.

The intestine plays a central role in shock. We recently proposed that cell and organ dysfunctions in shock are due to digestive enzymes. These degrading enzymes are synthesized in the pancreas, discharged into the lumen of the small intestine where they are and normally compartmentalized and responsible for digestion of food. We hypothesize that after injury to the intestine (e.g. by reduction of its blood flow, administration of bacterial endotoxins) the digestive enzymes escape from the lumen of the intestine into the systemic circulation and generate cell dysfunctions by cleavage of membrane receptors and microvascular perfusion deficiencies.

METHODS

After general anesthesia, rats were exposed to hemorrhagic shock by reduction of blood volume via the femoral vein. The mean arterial blood pressure was reduced to 40 mmHg for a period of two hours followed by return of all shed blood volume and observation over a period of 2 hours at which time tissues were collected for analysis. In a separate cohort of animals endotoxic shock was produced by intravenous administration of gram-negative lipopolysaccharide from *escherichia coli*, and tissue were collected for analysis at 4 hours after the administration.

In order to investigate the role of pancreatic digestive proteases, a serine protease inhibitor (ANGD, tranexamic acids) was administered directly into the small intestine before and also 1 hour after the start of the shock period [1].

At the end of the experiments organ structure and microvessel morphology was investigated in the small intestine, lung and heart. Zymography was used to determine protease activity in the plasma intestine and in the lung.

Mortality was determined in a separate group of rats without and with serine protease inhibition in the lumen of the small intestine that were observed over a period of 3 months after hemorrhagic and endotoxic shock [1]. As an independent marker for cell dysfunction we determined the level of insulin resistance in circulating cells by a bolus injection of glucose and of insulin while monitoring blood glucose levels.

RESULTS

Zymography shows that hemorrhagic shock leads to escape of digestive enzymes (e.g. trypsin, chymotrypsin, elastase) from the lumen of the intestine across the mucosal epithelial barrier into the intestinal villi and the wall of the intestine [2]. The protease activity is also detected in peritoneal fluid outside the intestine, in lymphatic fluid derived from mesentery lymph vessels and in plasma, indicating escape of digestive enzymes into the systemic circulation. In hemorrhagic shock there is also activation of matrix metalloproteinase activity, e.g. in the lung and heart [3].

The escape of the digestive enzymes from the lumen of the small intestine is accompanied by severe morphological damage of the villi and destruction of their microvascular and lymphatic structures. A similar destruction of the intestinal villi and their microvasculature is present after endotoxic shock [1, 4].

The heart and lung exhibit extensive evidence of escape of red blood cells from the microvasculature into the interstitial space and formation of tissue lesions in both hemorrhagic and endotoxic shock [1, 3].

Blockade of the digestive enzymes in the lumen of the intestine serves to significantly attenuate the protease activity in the wall of the small intestine, in the peritoneum, and in plasma and leads to a dramatic reduction of morphological damage of the villi and their microvasculature. Furthermore blockade of the digestive enzymes significantly reduces mortality in both hemorrhagic and endotoxic shock [1].

Furthermore, the acute insulin resistance after hemorrhagic shock, as seen by an attenuated reduction of the blood glucose levels, was abolished after blockade of digestive enzymes in the lumen of the intestine. The insulin resistance was accompanied by a reduced labeling density with an antibody against the extracellular domain of the insulin receptor after hemorrhagic shock and was restored by digestive enzyme inhibition in the small intestine [5].

DISCUSSION

These results suggest that after hemorrhagic or endotoxic shock the pancreatic digestive enzymes, which serve over a lifetime digestion in the small intestine, escape across the mucosal epithelium into the wall of the intestine where they destroy interstitial and microvascular structures, i.e. they cause autodigestion. The digestive enzymes escape into the systemic circulation where they cause destruction of the microvasculature and produce overt tissue lesion due to escape of red cells from the microvasculature into the interstitial space. They cause membrane receptor cleavage, such as in the case of the insulin receptor, and consequently generate a state of insulin resistance. Blockade of the digestive enzymes inside the lumen of the small intestine reduces the level of autodigestion, microvascular destruction, attenuates insulin resistance and reduces mortality.

These results point towards the digestive enzymes as key mediators for organ dysfunction and microvascular failure in shock. The digestive enzymes are synthesized in the pancreas and discharged in relatively high concentrations and as fully activated enzymes into the lumen of the small intestine to degrade any source of food. They are usually compartmentalized by the mucosal epithelial barrier inside the lumen of the intestine. But in shock after failure of the mucosal barrier they escape into wall of the intestine and the systemic circulation and initiate autodigestion. The evidence in the hemorrhagic and endotoxic shock models shows that they escape from the intestine into the portal venous circulation, the intestinal lymphatics and also directly into the peritoneum. They can cleave membrane receptors, such as the insulin receptor, and in the process generate a state of reduced response to insulin (i.e. insulin resistance). There are likely other receptors to be explored that are cleaved by the digestive proteases after their escape from the small intestine. This mechanisms may provide a potential explanation for the multitude of cell and microvascular functions encountered in shock.

Blockade of digestive enzymes in the lumen of the intestine may serve as a potential intervention to minimize autodigestion in patients and remains to be tested in a clinical trial.

ACKNOWLEDGEMENTS

Supported by an unrestricted gift from Leading BioSciences Inc.and by NIH grants HL 67825 and GM 85072.

REFERENCES

[1]. DeLano, FA et al., Science Translational Medicine, 5(169):169ra11, 2013.

[2] Chang, M. et al., PLoS ONE, 7(6):e40087, 2012.

[3] Altshuler, AE et al., PLoS ONE, 7: e32672, 2012.

[4] Chang, M, et al., Shock, 37(3):297-305, 2012

[5] DeLano, FA et al., Shock, 41: 56-62, 2014.

SB^{3}C2015
Summer Biomechanics, Bioengineering and Biotransport Conference
June 17-20, 2015,Snowbird Resort, Utah, USA
SB^{3}C2015-293

A HYBRID RISK MODEL FOR HIP FRACTURE PREDICTION USING CLINICAL AND STOCHASTIC FINITE ELEMENT DATA

Peng Jiang (1), Samy Missoum (1), Zhao Chen (2)

(1) Aerospace and Mechanical
Engineering Department
University of Arizona
Tucson, AZ, USA

(2) Mel and Enid Zuckerman
College of Public Health
University of Arizona
Tucson, AZ, USA

INTRODUCTION

Hip fracture is very common in older populations and is an important cause of disability and mortality. It is estimated that 50% of women aged 50 or older may suffer from a fragility fracture in their remaining life [1]. Therefore, it is of great importance to develop a hip fracture prediction model to better understand the causation of hip fracture and in the end help prevent or at least reduce the occurrence of hip fractures clinically.

Hip fracture is a complex process involving a large numbers of factors. Many studies have investigated influential factors as well as the construction of risk models based on clinical data [2, 3]. Of particular interest are clinical studies involving hip geometry factors [4]. Statistical techniques such as logistic regression [5], Cox-proportional model [6], and support vector machine classifiers (SVM) [7] have been used.

In order to improve the predictive ability of statistical risk models, the authors propose to combine clinical and finite element (FE) data. For this purpose a fully parameterized finite element model of a femur, which can represent any type of hip geometry, has been developed. In addition, clinical data were obtained from the Women's Health Initiative (WHI) [8].

The clinical and finite element data are combined by adding a mechanical quantity from computational simulations (e.g., maximum 1st principal strain) to the clinical set. This paper will compare the predictive ability of the risk models, constructed using SVM, before and after adding the computational data.

In addition, this paper will also present results in the case where the added quantity to the clinical database is stochastic. This is important, since bone material properties, as well as the loading scenarios of the femur are not known deterministically. Thus, accounting for uncertainties can provide a more realistic risk model.

METHODS

COMBINING CLINICAL AND FINITE ELEMENT DATA FOR HIP FRACTURE PREDICTION

One possible way of combining the data is to construct the prediction model in the "augmented space" where the outputs of the computational simulations are used as additional input parameters for hip fracture prediction (Fig. 1). The prediction model is thus constructed in a higher dimensional space than the clinical data.

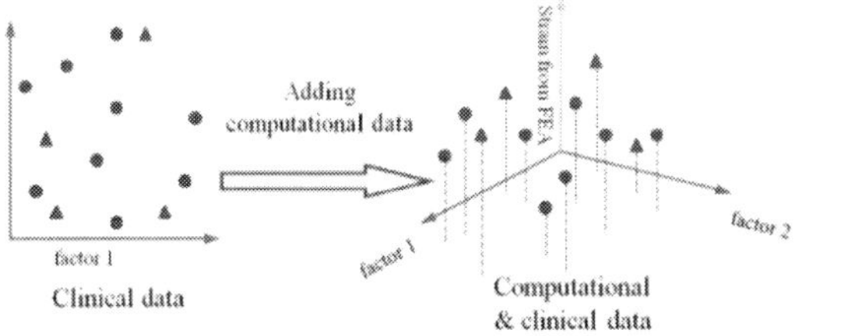

Fig. 1 COMBINING CLINICAL AND COMPUTATIONAL DATA IN THE "AUGMENTED SPACE".

PROPAGATING UNCERTAINTY THROUGH FEA FOR HIP FRACTURE PREDICTION In order to account for uncertainties in bone material properties as well as the loadings, the authors propose to assign random distributions to these variables instead of using deterministic values. Thus, a distribution of the output parameters (strains) can be obtained. In this paper, the uncertainty is propagated by approximating the maximum principal strain using a surrogate (Kriging [9]) constructed from a design of experiments.

Monte-Carlo simulations (MCS) are then used to generate the distribution (Fig. 2).

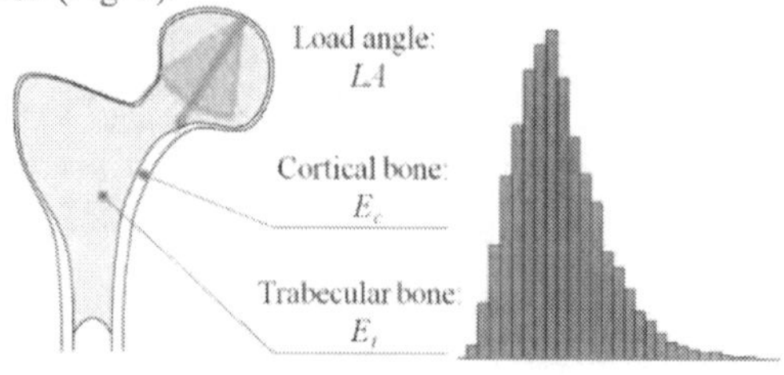

Fig. 2 UNCERTAINTY PROPAGATION THROUGH THE FEMUR FE MODEL. AN OUTPUT DISTRIBUTION IS OBTAINED INSTEAD OF A DETERMINISTIC VALUE.

From the strain distribution stochastic quantities can be calculated. For instance, the probability of the maximum principal strain ($\varepsilon_{1\,max}$) being larger than a threshold can be used to quantify the bone "strength". The technique used to identify the threshold is introduced in the result section.

FULLY PARAMETERIZED FINITE ELEMENT MODELS

FE models are used to study the bio-mechanical behaviors of the femur and provide measures of hip strength. Two fully parameterized FE models of different fidelities are constructed using ANSYS Parametric Design Language (APDL). Parameterization of the FE model enables simulation of a wide range of hip geometry. If the femur geometry of any individual is available, the patient-specific FE model can be created. Fig. 3 shows the responses of the femur FE models.

Fig. 3 CONTOUR OF 1ST PRINCIPAL STRAIN OF THE HIGH FIDELITY MODEL (LEFT). DEFORMED SHAPE OF A BEAM ELEMENT LOW FIDELITY MODEL (RIGHT).

RESULTS

VALIDATION OF THE FE MODELS Checked against the WHI clinical dataset (11497 participants with 168 fractures observed), both FE models show similar Area under the ROC curve (AUC) [10] scores (74% for the high fidelity FE model and 75% for the low fidelity one), which indicate a good predictive ability.

COMBINING CLINICAL AND COMPUTATIONAL DATA FOR HIP FRACTURE PREDICTION

DETERMINISTIC DATA Tab. 1 compares the predictive ability of the risk models constructed with SVMs before and after adding FE data. Comparison between the 1st and 2nd row indicates that the combination increases the AUC by 3%. The results confirm that computational data do add more information and thus improve the accuracy of the prediction.

STOCHASTIC DATA Normal distributions are assigned to load angle (*LA*) and Young's moduli of the cortical (E_c) and the trabecular bone (E_t). These uncertainties are propagated through MCS.

Based on the results of MCS, femur fracture risks (P_f) can be estimated for each participant in the WHI dataset. Fig. 4 shows the evolution of AUC based on P_f calculated with different thresholds of the maximum 1st principal strain. The maximum AUC is 77.8% and the corresponding strain is 0.84%.

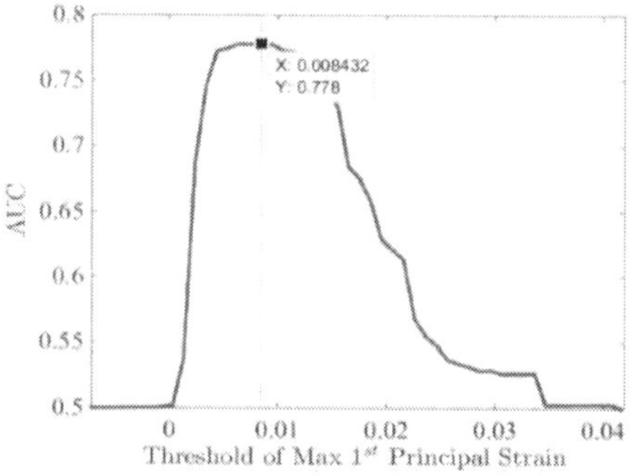

Fig. 4 EVOLUTION OF PREDICTIVE ABILITY OF THE FE MODEL WITH UNCERTAINTY AS THE STRAIN THRESHOLD INCREASES.

For this threshold, P_f (instead of the strain) is combined with the clinical data for hip fracture prediction. The corresponding SVM achieves 82% AUC (Tab. 1). In other words, accounting for uncertainties improves the hip fracture prediction model.

Tab. 1 COMPARISON OF HIP FRACTURE PREDICTION BY COMBINING FE DATA AND UNCERTAINTY PROPAGATION.

Parameters for SVM construction	Validation dataset	
	AUC	95% CI
Hip parameters	0.76	[0.70, 0.82]
Hip parameters + strains from FEA	0.79	[0.74, 0.85]
Hip parameters + P_f with uncertainties	0.82	[0.77, 0.86]

DISCUSSION

In this work, clinical and computational data are combined for hip fracture risk prediction. The computational data are generated from fully parameterized high and low fidelity FE models of a femur. These FE models exhibit good predictive ability checked against the clinical data. Both sources of data are combined in the "augmented space" and can help improve the prediction model. The improvements from high and low fidelity FE models are similar. Uncertainties from loads and materials are also accounted for, which makes the prediction more realistic and further increases the predictive capability of the prediction model. Different loading scenarios will be incorporated for future work.

ACKNOWLEDGEMENTS

The support of the National Institute of Health (grant NIAMS 1R21AR060811) is gratefully acknowledged.

REFERENCES

[1] Johnell et al. *Osteoporos Int.*, 16: S3-7, 2005
[2] Kanis, J. A. et al., *Bone.*, 44: 734–743, 2009
[3] Geusens, P et al. *Ther Adv Musculoskelet Dis.* 2(2): 63–77, 2010
[4] Beck, T. J. *Curr Osteoporos Rep.* 5: 49-55, 2007
[5] Gortmaker, S. L. et al. *Contemp Sociol.* 23: 159, 1994.
[6] Clinton J. et al. *J. Bone Joint Surg. Am.*91:503–11, 2009.
[7] Jiang, P. et al. *Struct Multidiscip Optim.* 50(4):523–35.2014
[8] Rossouw, J. et al. *J Am Med Womens Assoc* 50(2): 50–55, 1995.
[9] Sacks, J. et al. *Stat Sci.* 4 (4): 409–435. 1989
[10] Metz, C., *Semin Nucl Med*, Vol. 8:283–298. 1978.

SB³C2015
Summer Biomechanics, Bioengineering and Biotransport Conference
June 17-20, 2015, Snowbird Resort, Utah, USA
SB³C2015-294

BIOPHYSICAL FRACTIONATION OF CELLS WITH MULTIPLE MICROFLUIDIC OUTLETS

Gonghao Wang (1), Cory Turbyfield (2), Kaci Crawford (2), Alexander Alexeev (1), Todd Sulchek (1,2)

(1) The G. W. Woodruff School of Mechanical Engineering
Georgia Institute of Technology
Atlanta, GA, USA

(2) Department of Biomedical Engineering
Georgia Institute of Technology
Atlanta, GA, USA

INTRODUCTION

Biophysical properties, including size (1) and stiffness (2), are biomarkers of diseases such as cancer and malaria. Utilizing differences in cell biophysical properties for disease detection would be greatly aided by means to sort cells biophysically. Biophysical properties are intrinsic to the cell, and as such provide a label-free approach to enrichment. Recently, several micro-fluidic cell sorting approaches based on variations in cell biophysical properties have been demonstrated (3-5). However, the ultimate purity of the separated cells is limited by the intrinsic variability and overlap of the biophysical properties of different cell types. As a result, biophysical approaches to cell enrichment are in general inferior to the results achieved by antibody-based magnetic or fluorescent cell sorting (6).

We report a new microfluidic cell enrichment technology that continuously fractionates cells through differences in biophysical properties, resulting in highly pure cellular subpopulations. Cell fractionation is achieved in a microfluidic channel with multiple ridges that are designed to segregate biophysically distinct cells to different locations in the channel. Multiple outlets are then utilized to collect cells with finer gradation of differences in cell biophysical properties. The result is that cell fractionation dramatically improves cell separation purity compared to binary outputs. As a proof-of-concept, we mix two different leukemia cell lines and utilize cell fractionation to achieve over 45-fold enhancement of cell populations, with high purity cellular enrichment (90% to 99%) of each cell type. These results will enable label-free separation of cellular mixtures based upon physical differences.

METHODS

The microfluidic cell sorter utilizes a micro-channel decorated with diagonal ridges on the top wall. The ridges are designed to compress and translate cells normal to the flow direction. The ridges and the channel bottom surface form a gap which is smaller than the cell diameters (Fig. 1a.). Cells enter the microfluidic channel through a focused cell inlet. As cells flow through the microfluidic channel, the ridges compress the cells in rapid succession and alter their trajectory dependent on their biophysical properties. The three-outlet fractionation circumvents the major challenge in biophysical cell sorting caused by the inherent heterogeneity and significant overlap of cellular populations, greatly improving the purity of sorted cells.

The microfluidic cell sorter is made using replica molding of polydimethylsiloxane (PDMS) on a SU8 mold. The PDMS chip is then bonded to a glass substrate to form the microfluidic channel. The substantial improvement in the purity of sorted cells is benefited from the three outlets that are hydro-dynamically balanced (Fig. 1a). To improve the cell separation quality, we also incorporate two additional design elements: an expansion region and outlets with serpentine architecture. Computational fluid dynamics is used to design a balanced channel flow egress across all outlets (Fig. 1b). The central outlet is designed to collect the cells that have overlapping biophysical properties (Fig. 1c).

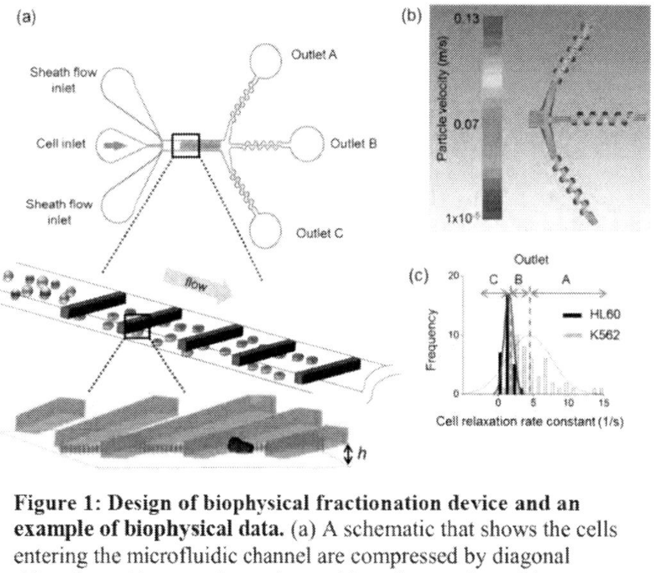

Figure 1: Design of biophysical fractionation device and an example of biophysical data. (a) A schematic that shows the cells entering the microfluidic channel are compressed by diagonal ridges in rapid succession and sorted into three outlets. (b) Computational fluid dynamics simulations are used to engineer hydrodynamics near the outlets to ensure even splitting of the channel flow. The color gradient represents flow velocity vectors of particles. (c) A histogram of biophysical data for HL60 and K562 cells and a schematic indicating the range of collection for each outlet.

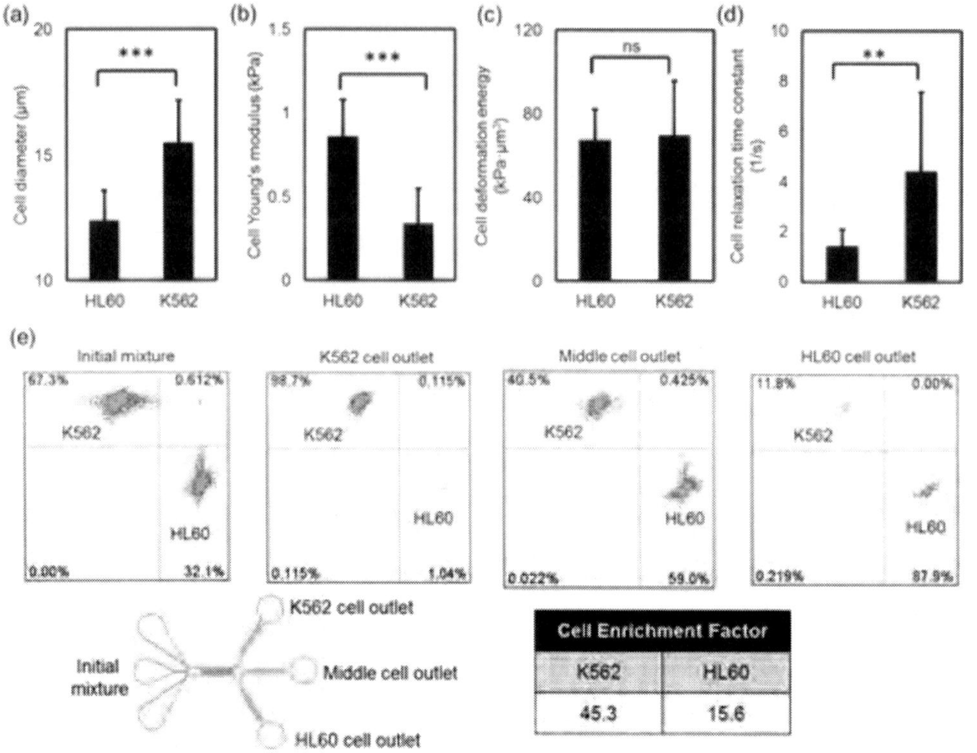

Figure 2: Biophysical data for two leukemia cell line models and separation results. The average a) cell diameter, b) modulus, c) deformation energy, and d) viscous relaxation time constant for K562 and HL60 cells. K562 and HL60 have similar deformation energy. AFM measurements of cell relaxation show HL60 (1.45 ± 0.65 s^{-1}, n = 30) is slower in size recovery than K562 (4.42 ± 3.15 s^{-1}, n = 52). (e) Flow cytometric analyses of the inlet mixture of cells and cell enrichment using a three-outlet channel. At the outlets, the cell enrichment factor for K562 and HL60 cells are 45.3 and 15.6 respectively. The purity of enriched K562 cells and HL60 cells are 98.7% and 87.9% respectively.

RESULTS

To demonstrate the cell sorting efficiency of the three-outlet channel, we tested cell sorting of two types of leukemia cell lines: K562 and HL60. The cell size, Young's modulus, size-adjusted elasticity (deformation energy) and the relaxation rate constants are shown in Fig. 2. The cell sorting result with the three-outlet device is evaluated by flow cytometry and shown in Fig. 2e. We report the ability to enrich the percentage of K562 and HL60 cells is 45-fold and 15-fold respectively, which is over an order of magnitude improvement compared to the results utilizing binary outlets. Therefore, the three-outlet channel enables a dramatic improvement in cell sorting purity.

In addition to enrichment of cell mixtures into individual cell types, the three-outlet channel can also be applied to fractionate a single cell type into distinct biophysical phenotypes. Since biological cells are inherently heterogeneous in nature, their biophysical properties may include large variations due to differences in cytoskeleton or nucleus. For example, K562 cells have average Young's modulus 0.34 kPa and a standard deviation 0.21 kPa (Fig. 1c). Utilizing the fractionation approach, we are able to obtain subpopulations of K562 cells with differing biophysical properties.

DISCUSSION

Diseased cells are shown to have differences in biophysical properties, yet the overlap of these biophysical properties can limit their use in disease cell enrichment and detection. We show that cell enrichment through multiple-outlet biophysical fractionation significantly improves the purity of sorted cells. The fractionation cell sorting circumvents the major challenge in biophysical cell sorting caused by the inherent heterogeneity and significant overlap of cellular populations. Biophysical fractionation may find use in preprocessing samples with low numbers of target cells for downstream analysis and detection, or to better understand the relationship of biophysical phenotypes and molecular content and function. Biophysical fractionation may find use in purifying samples with low numbers of target cells for downstream analysis and detection.

ACKNOWLEDGMENTS

This research was supported by NSF project number CBET-0932510, TI:GER program at Scheller College of Business at GT, the Regenerative Engineering and Medicine Seed Grant, and the GT's President's Undergraduate Research Award (PURA) program.

REFERENCES

1. Vona G, "Isolation by size of epithelial tumor cells - A new method for the immunomorphological and molecular characterization of circulating tumor cells." Am J Pathol. 2000;156(1):57-63.
2. Suresh S. "Biomechanics and biophysics of cancer cells." Acta Biomater. 2007;3(4):413-38.
3. Hou HW, "Deformability based cell margination-A simple microfluidic design for malaria-infected erythrocyte separation." Lab Chip. 2010;10(19):2605-13.
4. Wang G, "Stiffness dependent separation of cells in a microfluidic device." PloS one. 2013;8(10):e75901.
5. Wang GH, "Microfluidic cellular enrichment and separation through differences in viscoelastic deformation." Lab Chip. 2015;15(2):532-40.
6. Wolff A, "Integrating advanced functionality in a microfabricated high-throughput fluorescent-activated cell sorter." Lab Chip. 2003;3(1):22-7.

WRINKLED, WAVELENGTH-TUNABLE GRAPHENE-BASED SURFACE TOPOGRAPHIES FOR DIRECTING CELL ALIGNMENT AND MORPHOLOGY

Daniel Tonderys,[1,3] Zhongying Wang,[1,2] Susan E. Leggett,[1,3,4] Evelyn Kendall Williams,[1,3] Mehrdad T. Kiani,[1,3] Ruben Spitz-Steinberg[1], Yang Qiu[1], Ian Y. Wong,[1,3,4,5] Robert H. Hurt.[1,3,5]

[1]School of Engineering, [2]Department of Chemistry, [3]Center for Biomedical Engineering, [4]Pathobiology Graduate Program, [5]Institute for Molecular and Nanoscale Innovation.

INTRODUCTION

Textured surfaces with periodic topographical features and long-range order are highly attractive for tissue engineering. They mimic physiological environments more accurately than planar surfaces and can fundamentally alter cell alignment, shape, gene expression, and cellular assembly into superstructures or microtissues. Here we demonstrate for the first time that wrinkled graphene-based surfaces are suitable as textured cell attachment substrates, and that engineered wrinkling can dramatically alter cell alignment and morphology. The wrinkled surfaces are fabricated by graphene oxide wet deposition onto pre-stretched elastomers followed by relaxation and mild thermal treatment to stabilize the films in cell culture medium. Multilayer graphene oxide films form periodic, delaminated buckle textures whose wavelengths and amplitudes can be systematically tuned by variation in the wet deposition process. Human and murine fibroblasts attach to these textured films and remain viable, while developing pronounced alignment and elongation relative to those on planar graphene controls. Compared to lithographic patterning of nanogratings, this method has advantages in the simplicity and scalability of fabrication, as well as the opportunity to couple the use of topographic cues with the unique conductive, adsorptive, or barrier properties of graphene materials.

METHODS

Fabrication and Characterization of Textured Surfaces

Graphene oxide suspensions were prepared by a modified Hummer's as described previously [1]. Elastomer films were stretched uniaxially to a desired pre-strain (1.5%-50%) and secured with a fastener. A specific volume of graphene oxide aqueous suspension was pipetted onto the substrate, dried overnight, and relaxed to form the wrinkle patterns. For cell studies the films were rendered stable to re-dissolution in cell culture medium by mild thermal reduction through annealing overnight at 120° C in air. Morphologies of the textured surfaces were characterized by optical microscopy, atomic force microscopy (Asylum MFP-3D Origin AFM) in pulse mode, and field emission SEM (LEO 1530 VP) in the variable operating pressure mode.

Cell Culture and Staining

NIH-3T3 mouse fibroblasts and normal human fibroblasts (NHF cells) derived from neonatal foreskin were used in this study. Both cell lines were carried out under standard culture conditions. Once the rGO materials were prepared and sterilized a 100uL droplet with a concentration of 40,000 cells/mL was deposited on the materials, allowed to settle for ~1 hour, and then the well was filled with 1 mL of culture media. After 96 h of culture on wrinkled and flat rGO substrates, cells were fixed and then permeablized. For F-actin microfilament detection, cells were immunostained with a conjugated antibody, Alexa Fluor 647 Phalloidin. Counterstaining with Hoechst (bis-Benzimide) was then carried out to fluorescently label cell nuclei.

RESULTS

Film Morphology and Wavelength Tuning

The main objective of this research was to explore wrinkled graphene surfaces as anisotropic cell attachment substrates for control of cell alignment and morphology in tissue engineering applications. Relaxation of the dried GO produces a series of nearly parallel, delaminated ridges with high-curvature crests separated by broader valleys. These valleys adhere the films to the substrate, and the local delamination produces cavities with triangular cross-sections between the substrate and film. The wrinkle patterns can be removed by re-stretching to the initial state, and the process is reversible over multiple cycles. For manipulation of cell alignment and morphology, it is desirable to control the amplitude and wavelength of the wrinkle patterns in our multilayer graphene films. Approaches to this control are readily suggested by the significant literature on the theoretical and experimental mechanics of stiff thin films on pre-stretched compliant substrates [2-5], mainly they are film thickness, substrate stiffness and prestrain. For a given substrate and prestrain (50%), wavelengths can be changed by an order of magnitude (e.g. 3 to 30 um for the softer substrate) by varying GO film thickness from 20-200 nm through increasing GO suspension concentration from 0.02-0.2 mg/ml as shown in Figure 1.

Cell Alignment and Morphology

As a model system, we chose to study murine and human fibroblasts on matched samples of 25 um wavelength rGO surfaces and a planar rGO substrate. Fibroblasts displayed good viability on both substrates, comparable with conventional tissue culture plastic. Both NIH-3T3 and NHF fibroblasts appear highly aligned on the wrinkled substrates (Fig. 2A&C) and randomly oriented on the flat substrates (Fig. 2B&D). The quantitative results are summarized numerically in Figure 2E.

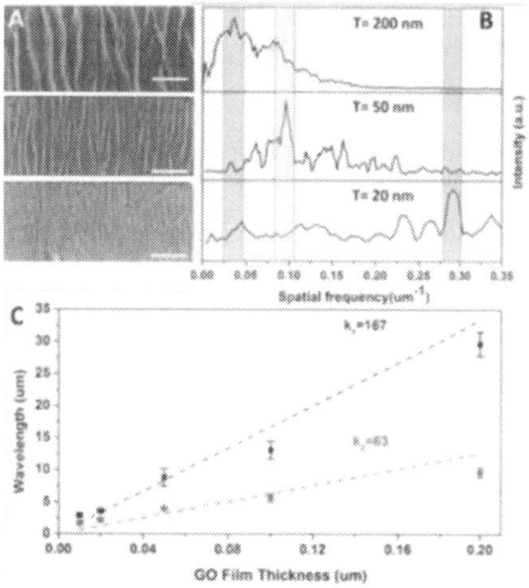

Figure 1. Wavelength tunability through control of multilayer GO film and through substrate selection. (A) SEM images of wrinkled GO films of thickness from 10 nm to 200 nm on the softer elastomer (modulus $5 \cdot 10^4$ Pa) and (B) their Fourier transform spectra with the dominant spatial frequency shaded. (C) Thickness-dependent wavelengths on softer (black squares) and stiffer elastomer (modulus $6 \cdot 10^5$ Pa; red circles) with linear regression slopes shown. The scale bars in (A) represent 50 nm.

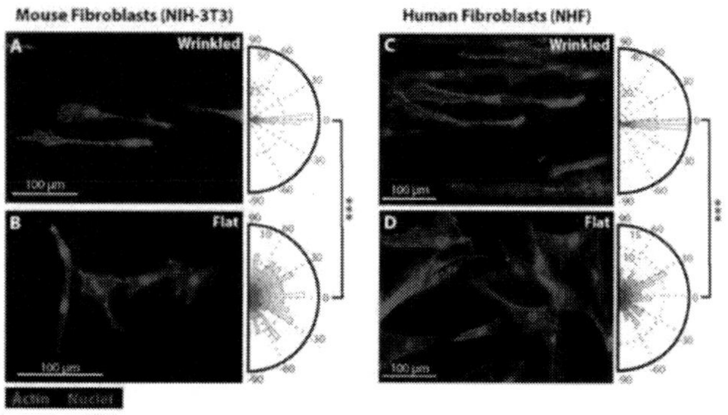

E

Variable	NIH-3T3 Cells		NHF Cells	
	Flat (n=137)	Wrinkled (n=164)	Flat (n=156)	Wrinkled (n=163)
Circular Standard Deviation	57°	15°	53°	18°
Rao's Spacing Test (p)	0.50 < p < 0.90	< 0.01	0.10 < p < 0.50	< 0.01
Mardia-Watson-Wheeler (p)	< 1E-12		< 1E-12	

Figure 2. Fibroblast culture on wrinkled graphene materials results in highly aligned cells. (A-D) Fluorescence images of Actin-phalloidin (red) and nuclei (blue) for NIH-3T3 and NHF cells cultured on 200 nm wrinkled rGO substrates (A, C) and on paired flat graphene controls (B,D). Circular histograms of fibroblast orientation, ranging from -90° to 90°, for cells on wrinkled and flat materials are displayed as orientation rose plots to the right of each corresponding fluorescence image (A-D). (E) Table of statistics summarizing circular standard deviation, and statistical significance for uniformity (one-sample test) and distribution (two-sample test) using Rao's spacing test and Mardia-Watson-Wheeler test, respectively.

The morphology of NIH-3T3 and NHF cells was analyzed by extracting cell shape based on F-actin immunostaining. Overall, both fibroblast lines displayed narrower morphologies with smaller standard deviations on wrinkled substrates compared to flat substrates, with highly significant differences ($p < 10^{-3}$). These narrower cell widths are comparable to the wrinkle spacing of ~25 um, suggesting that the elongation reflects some degree of cell micro-confinement in the inter-ridge valley spaces. Furthermore, fibroblasts are intrinsically quite elongated on flat substrates, but are even more elongated on our wrinkled graphene surfaces, where they have eccentricity values close to 1, and the effect of the wrinkle texture is statistically significant. In addition, NIH-3T3 cells on wrinkled substrates display increased solidity, while NHF cells display increased compactness. These results indicate that the topography of these wrinkled graphene substrates can strongly influence cell morphology

DISCUSSION

This study demonstrates the feasibility of wrinkled graphene as a surface texturing agent to direct cell alignment and morphology in tissue engineering. Wet deposition of graphene oxide multilayer tiled films on pre-stretched elastomers followed by relaxation and mild thermal reduction produces cell-culture-compatible textured substrates with long-range periodic topography in the form of parallel ridges and valleys. Both the spatial period and the wrinkle amplitude can by systematically tuned from 1 – 25 um by selection of the substrate stiffness, by variation of pre-strain, and by alteration of the GO precursor concentration in the feed suspension. Fibroblasts are observed to attach to these textured reduced graphene oxide surfaces and remain viable, while the wrinkles induce statistically significant cell alignment and elongation relative to flat rGO substrates in a manner similar to that seen on lithographically fabricated nano-gratings [6]. Together the results suggest that rGO wrinkle engineering is a promising new approach for creation of functional surfaces and scaffolds in tissue engineering. In comparison with competing methods of patterning, wrinkled rGO offers potential advantages in the ease, cost, and scale-up of fabrication. We envision a modular graphene-based platform for the application of orthogonal topographic stimuli, chemical stimuli (through functionalization or adsorption), electrical stimuli (enabled by conductive graphene), as well as dynamic mechanical actuation of topographic confinement. Furthermore, we believe these biologically inspired architectures will be widely applicable to other cell types, including neurons, skeletal, smooth muscle, cardiomyocytes and stem cells.

ACKNOWLEDGEMENTS

This work was supported by the U.S. National Science Foundation (Grants CBET-1132446 and INSPIRE Track 1 CBET-1344097), the Environmental Pathology Training Grant from the National Institute of Health (5T32ES007272-23, S.E.L.), the Vincent and Ruby DiMase Undergraduate Summer Fellowship (M.T.K.), the Karen T. Romer Undergraduate Teaching and Research Award (E.K.W.), seed funding from Jason and Donna McGraw Weiss '89 (I.Y.W.) and startup funds from Brown University (I.Y.W.). We thank Profs. A.B. Kane and J.R. Morgan for their gifts of NIH3T3 and normal human fibroblasts, respectively.

REFERENCES

[1] Qiu, Y et al., *Nanoscale* 2014, 6, 11744-11755.
[2] Kunz, D. A. et al., *Adv. Mater. (Weinheim, Ger.)* 2013, 25, 1337-1341.
[3] Chen, X. et al., *Journal of Applied Mechanics* 2004, 71, 597-603.
[4] Huang, Z. Y. et al., *Journal of the Mechanics and Physics of Solids* 2005, 53, 2101-2118.
[5] Song, J. et al., *International Journal of Solids and Structures* 2008, 45, 3107-3121.
[6] Charras, G. et al., *Nat. Rev. Mol. Cell Biol.* 2014, 1-12.

Use of a Torque Range of Motion Device to Teach Evaluation of Joint Dysfunction

Rita M. Patterson (1), Jeongsik Shin (2), Aditya Das (2), Robert Longnecker (1), Katelyn Rockenbach (1), Carol Kominski (3), Vanneise Collins (3), and David Mason (1)

(1) Osteopathic Manipulative Medicine
University of North Texas Health Science Center
Fort Worth, TX, USA

(2) UTARI
University of Texas at Arlington
Fort Worth, TX, USA

(3) Center for Innovative Learning
University of North Texas Health Science Center
Fort Worth, TX, USA

INTRODUCTION

Up to 1 in 4 people will experience shoulder problems in their lifetime [1]. Cell phone use (texting) and computer/desk work place the upper shoulder extremity in front of the body often resulting in poor posture and decreased strength in the upper back muscles. This results in a vicious cycle of muscle imbalance, pain and eventually *disability* due to abnormal kinematics and biomechanics of the joint. The effects of muscle atrophy that occur with aging will only add to the incidence of shoulder related problems in the general population.

The shoulder is particularly complicated because it must be mobile enough to produce the wide range of functional activities necessary for activities of daily living yet stable enough to support the forces required for lifting, pushing and pulling. The compromise between mobility and stability puts the shoulder at risk for injury.

Clinical practitioners are trained to assess the presence and severity of musculoskeletal dysfunction through direct palpation, passive and active range of motion testing to estimate joint stiffness dysfunction deficits. Clinically, joint stiffness is evaluated through subjective methods such as end feel. Cyriax [2] defined this as that sensation (force) imparted to the examiner as the end range of passive motion is reached. The more the examiner senses "give" at the end-range, the more compliant the structure is thought to be. Conversely, the harder or "less give", the stiffer the joint is thought to be.

Palpatory findings may differ depending on practitioners' skill level and experience [3]. Over time, with adequate practice, their skills improve and they become more comfortable assessing and are more reliably able to repeat this assessment. However, there are few commonly accepted objective measures to complement palpatory findings.

Knowledge gained from this study will allow for a better description of stiffness (function) so that future innovative advancements in rehabilitation, ergonomics for mechanical advantage, adaptive methods, and assistive aids can be provided to patients with a variety of disabilities. **Therefore the purpose of this project is to examine the effectiveness of a mechanical device that can objectively measure the force at the physiologic barrier (end feel) of the shoulder) and potentially accelerate the learning of this skill in first year medical students.**

METHODS

Students (127 Male, 71 female, age range 21-31) and experts (7 Male, 2 female with 5-20 years of active clinical experience) used the device during the fall semester, first year Osteopathic Manipulative Medicine lab course (MEDE 7410). Force was recorded while subjects performed a passive shoulder assessment aimed at moving the shoulder to the end of its range of motion.

Students were put into pairs and each student (examiner and simulated patient) used the device to evaluate their partner's shoulder stiffness. Examiners were instructed to passively move the shoulder throughout its full range of motion stopping at end feel. They repeated the entire protocol and assessment 3 times for each motion (shoulder extension, shoulder internal rotation, and shoulder external rotation) and side (left and right). The force sensor device, composed of a three-axis force sensor (Xiamen Elane Electronics, XiaMenShi China) attached to a plastic handle, was used to record the force.

After the initial evaluation session, expert evaluations of end feel force values were analyzed and used as the gold standard for training. Half of the medical student class had the device available for three training sessions. During the training sessions the students observed a visual display of force during the evaluation and were instructed to move the shoulder to its physiologic barrier until the "expert gold standard" end feel force value was reached. They were instructed to remember the "feel" of this resistance for future stiffness evaluations. After training all students were tested again using the device. Students also answered questions through an online questionnaire using Qualtrics software. These questions measured students' perception of the quality of their learning and their confidence in assessment of end feel.

Outcome Measures

Outcome measures included 1) students' assessments of end feel force, 2) repeatability in students' assessments of end feel force, 3) student satisfaction with learning how to assess end feel, and 4) students' confidence in assessing end feel.

RESULTS

The average force and repeatability (standard deviation) recorded by the experts and students in each category is listed in table 1. The average force recorded before any specific training was lower than the faculty standard. Both groups' assessments were improved (increased) during the semester. Repeatability between trials in the expert group was no different than in the student groups.

Table 1

	Pre Assessment		Post Assessment	
			Student only	
	Experts	Students	Control	Training
Force (kg)	2.77	2.10	2.58	3.16
STD (kg)	0.22	0.24	0.24	0.30

Students' satisfaction with their learning increased (ratio is positive) over the semester with the training group gaining more satisfaction than the control group, Figure 1.

Student's confidence in evaluating joint stiffness increased over the semester with the training group gaining more confidence than the control group, Figure 2.

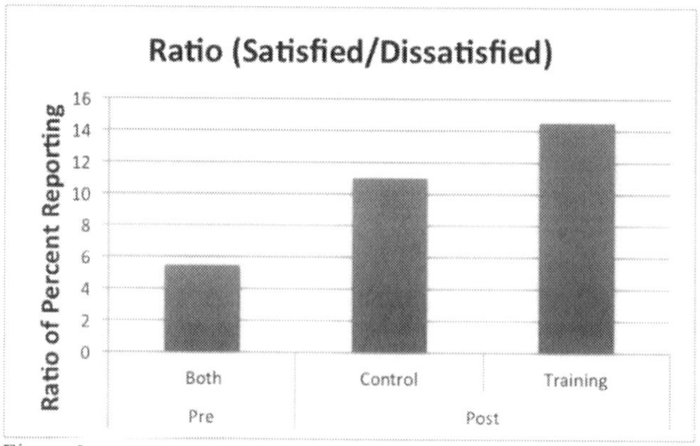

Figure 1

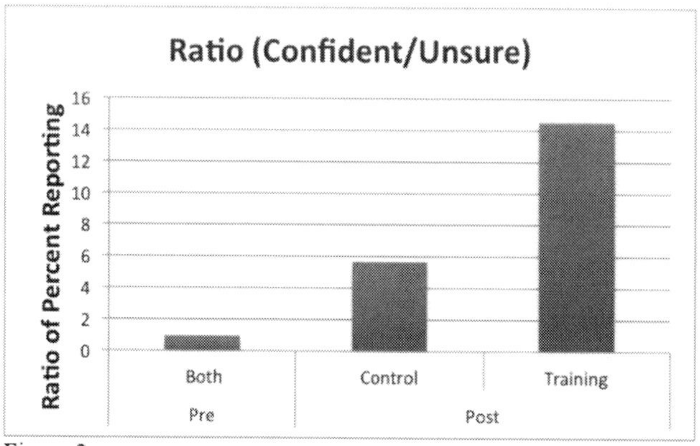

Figure 2

CONCLUSION

Use of a force-sensing device was successful in facilitating learning of the assessment of end feel. Students in the training group were able to more closely match expert assessment forces after only 3 training sessions over a 2-month period.

ACKNOWLEDGMENTS

Special thanks to Lindsay Appleby for her help with data collection. This project was supported by a local UNTHSC educational seed grant.

REFERENCES

1. Lentz, T.A., et al. The Relationship of Pain Intensity, Physical Impairment, and Pain-Related Fear to Function in Patients with Shoulder Pathology. J. Orthopaedic & Sports Physical Therapy. 39(4):270-277; 2009.

2. Cyriax J. Orthopaedic Medicine: Diagnosis of Soft Tissue Lesions, Baltimore, Williams and Wilkins, 76, 1975.

3. Bush, T.R., and Vorro, J., Kinematic Measures to Objectify Head and Neck Motions in Palpatory Diagnosis: A Pilot Study, JAOA, Vol 108, No 2, February 2008, 55-62.

SB[3]C2015
Summer Biomechanics, Bioengineering and Biotransport Conference
June 17-20, 2015,Snowbird Resort, Utah, USA
SB[3]C2015-297

WRINKLED, WAVELENGTH-TUNABLE GRAPHENE-BASED SURFACE TOPOGRAPHIES

Daniel Tonderys,[1,3] Zhongying Wang,[1,2] Susan E. Leggett,[1,3,4] Evelyn Kendall Williams,[1,3] Mehrdad T. Kiani,[1,3] Ruben Spitz-Steinberg[1], Yang Qiu[1], Ian Y. Wong,[1,3,4,5] Robert H. Hurt.[1,3,5]

[1] School of Engineering, [2]Department of Chemistry, [3]Center for Biomedical Engineering, [4]Pathobiology Graduate Program, [5]Institute for Molecular and Nanoscale Innovation.
Brown University, Providence, RI 02912.

INTRODUCTION

Textured surfaces with periodic topographical features and long-range order are highly attractive for tissue engineering. They mimic physiological environments more accurately than planar surfaces and can fundamentally alter cell alignment, shape, gene expression, and cellular assembly into superstructures or microtissues. Here we demonstrate for the first time that wrinkled graphene-based surfaces are suitable as textured cell attachment substrates, and that engineered wrinkling can dramatically alter cell alignment and morphology. The wrinkled surfaces are fabricated by graphene oxide wet deposition onto pre-stretched elastomers followed by relaxation and mild thermal treatment to stabilize the films in cell culture medium. Multilayer graphene oxide films form periodic, delaminated buckle textures whose wavelengths and amplitudes can be systematically tuned by variation in the wet deposition process. Human and murine fibroblasts attach to these textured films and remain viable, while developing pronounced alignment and elongation relative to those on planar graphene controls. Compared to lithographic patterning of nanogratings, this method has advantages in the simplicity and scalability of fabrication, as well as the opportunity to couple the use of topographic cues with the unique conductive, adsorptive, or barrier properties of graphene materials.

METHODS

Fabrication and Characterization of Textured Surfaces

Graphene oxide suspensions were prepared by a modified Hummer's as described previously [1]. Elastomer films were stretched uniaxially to a desired pre-strain (1.5%-50%) and secured with a fastener. A specific volume of graphene oxide aqueous suspension was pipetted onto the substrate, dried overnight, and relaxed to form the wrinkle patterns. For cell studies the films were rendered stable to re-dissolution in cell culture medium by mild thermal reduction through annealing overnight at 120° C in air. Morphologies of the textured surfaces were characterized by optical microscopy, atomic force microscopy (Asylum MFP-3D Origin AFM) in pulse mode, and field emission SEM (LEO 1530 VP) in the variable operating pressure mode.

Cell Culture and Staining

NIH-3T3 mouse fibroblasts and normal human fibroblasts (NHF cells) derived from neonatal foreskin were used in this study. Both cell lines were carried out under standard culture conditions. Once the rGO materials were prepared and sterilized a 100uL droplet with a concentration of 40,000 cells/mL was deposited on the materials, allowed to settle for ~1 hour, and then the well was filled with 1 mL of culture media. After 96 h of culture on wrinkled and flat rGO substrates, cells were fixed and then permeablized. For F-actin microfilament detection, cells were immunostained with a conjugated antibody, Alexa Fluor 647 Phalloidin. Counterstaining with Hoechst (bis-Benzimide) was then carried out to fluorescently label cell nuclei.

RESULTS

Film Morphology and Wavelength Tuning

The main objective of this research was to explore wrinkled graphene surfaces as anisotropic cell attachment substrates for control of cell alignment and morphology in tissue engineering applications. Relaxation of the dried GO produces a series of nearly parallel, delaminated ridges with high-curvature crests separated by broader valleys. These valleys adhere the films to the substrate, and the local delamination produces cavities with triangular cross-sections between the substrate and film. The wrinkle patterns can be removed by re-stretching to the initial state, and the process is reversible over multiple cycles. For manipulation of cell alignment and morphology, it is desirable to control the amplitude and wavelength of the wrinkle patterns in our multilayer graphene films. Approaches to this control are readily suggested by the significant literature on the theoretical and experimental mechanics of stiff thin films on pre-stretched compliant substrates [2-5], mainly they are film thickness, substrate stiffness and prestrain. For a given substrate and prestrain (50%), wavelengths can be changed by an order of magnitude (e.g. 3 to 30 um for the softer substrate) by varying GO film thickness from 20-200 nm through increasing GO suspension concentration from 0.02-0.2 mg/ml as shown in Figure 1.

Cell Alignment and Morphology

As a model system, we chose to study murine and human fibroblasts on matched samples of 25 um wavelength rGO surfaces and a planar rGO substrate. Fibroblasts displayed good viability on both substrates, comparable with conventional tissue culture plastic. Both NIH-3T3 and NHF fibroblasts appear highly aligned on the wrinkled substrates (Fig. 2A&C) and randomly oriented on the flat substrates (Fig. 2B&D). The quantitative results are summarized numerically in Figure 2E.

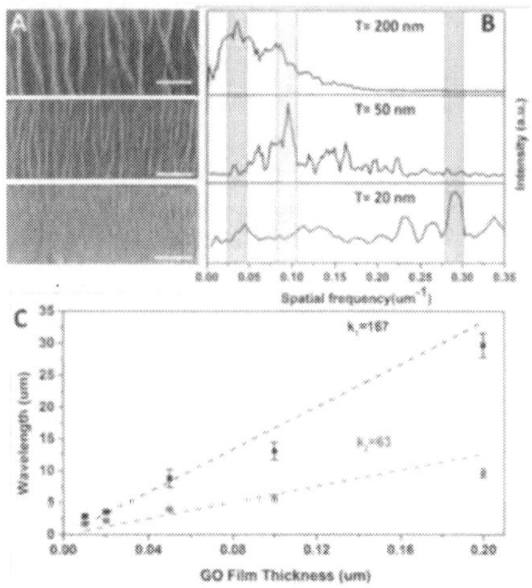

Figure 1. Wavelength tunability through control of multilayer GO film and through substrate selection. (A) SEM images of wrinkled GO films of thickness from 10 nm to 200 nm on the softer elastomer (modulus $5 \cdot 10^4$ Pa) and (B) their Fourier transform spectra with the dominant spatial frequency shaded. (C) Thickness-dependent wavelengths on softer (black squares) and stiffer elastomer (modulus $6 \cdot 10^5$ Pa; red circles) with linear regression slopes shown. The scale bars in (A) represent 50 nm.

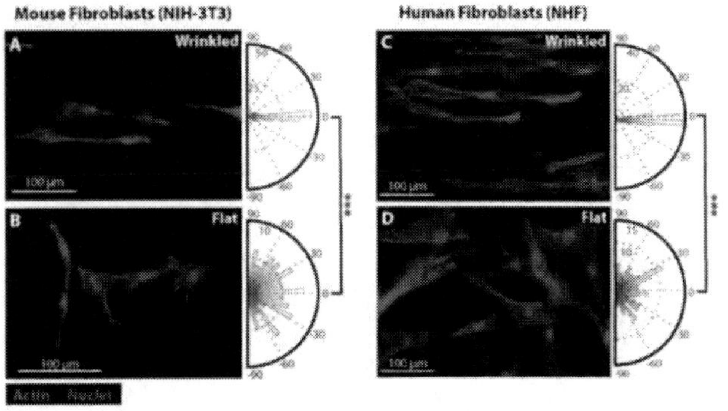

E

	NIH-3T3 Cells		NHF Cells	
Variable	Flat (n=137)	Wrinkled (n=164)	Flat (n=156)	Wrinkled (n=163)
Circular Standard Deviation	57°	15°	53°	18°
Rao's Spacing Test (p)	0.50 < p < 0.90	< 0.01	0.10 < p < 0.50	< 0.01
Mardia-Watson-Wheeler (p)	< 1E-12		< 1E-12	

Figure 2. Fibroblast culture on wrinkled graphene materials results in highly aligned cells. (A-D) Fluorescence images of Actin-phalloidin (red) and nuclei (blue) for NIH-3T3 and NHF cells cultured on 200 nm wrinkled rGO substrates (A, C) and on paired flat graphene controls (B,D). Circular histograms of fibroblast orientation, ranging from -90° to 90°, for cells on wrinkled and flat materials are displayed as orientation rose plots to the right of each corresponding fluorescence image (A-D). (E) Table of statistics summarizing circular standard deviation, and statistical significance for uniformity (one-sample test) and distribution (two-sample test) using Rao's spacing test and Mardia-Watson-Wheeler test, respectively.

The morphology of NIH-3T3 and NHF cells was analyzed by extracting cell shape based on F-actin immunostaining. Overall, both fibroblast lines displayed narrower morphologies with smaller standard deviations on wrinkled substrates compared to flat substrates, with highly significant differences ($p < 10^{-3}$). These narrower cell widths are comparable to the wrinkle spacing of ~25 um, suggesting that the elongation reflects some degree of cell micro-confinement in the inter-ridge valley spaces. Furthermore, fibroblasts are intrinsically quite elongated on flat substrates, but are even more elongated on our wrinkled graphene surfaces, where they have eccentricity values close to 1, and the effect of the wrinkle texture is statistically significant. In addition, NIH-3T3 cells on wrinkled substrates display increased solidity, while NHF cells display increased compactness. These results indicate that the topography of these wrinkled graphene substrates can strongly influence cell morphology

DISCUSSION

This study demonstrates the feasibility of wrinkled graphene as a surface texturing agent to direct cell alignment and morphology in tissue engineering. Wet deposition of graphene oxide multilayer tiled films on pre-stretched elastomers followed by relaxation and mild thermal reduction produces cell-culture-compatible textured substrates with long-range periodic topography in the form of parallel ridges and valleys. Both the spatial period and the wrinkle amplitude can by systematically tuned from 1 – 25 um by selection of the substrate stiffness, by variation of pre-strain, and by alteration of the GO precursor concentration in the feed suspension. Fibroblasts are observed to attach to these textured reduced graphene oxide surfaces and remain viable, while the wrinkles induce statistically significant cell alignment and elongation relative to flat rGO substrates in a manner similar to that seen on lithographically fabricated nano-gratings [6]. Together the results suggest that rGO wrinkle engineering is a promising new approach for creation of functional surfaces and scaffolds in tissue engineering. In comparison with competing methods of patterning, wrinkled rGO offers potential advantages in the ease, cost, and scale-up of fabrication. We envision a modular graphene-based platform for the application of orthogonal topographic stimuli, chemical stimuli (through functionalization or adsorption), electrical stimuli (enabled by conductive graphene), as well as dynamic mechanical actuation of topographic confinement. Furthermore, we believe these biologically inspired architectures will be widely applicable to other cell types, including neurons, skeletal, smooth muscle, cardiomyocytes and stem cells.

ACKNOWLEDGEMENTS

This work was supported by the U.S. National Science Foundation (Grants CBET-1132446 and INSPIRE Track 1 CBET-1344097), the Environmental Pathology Training Grant from the National Institute of Health (5T32ES007272-23, S.E.L.), the Vincent and Ruby DiMase Undergraduate Summer Fellowship (M.T.K.), the Karen T. Romer Undergraduate Teaching and Research Award (E.K.W.), seed funding from Jason and Donna McGraw Weiss '89 (I.Y.W.) and startup funds from Brown University (I.Y.W.). We thank Profs. A.B. Kane and J.R. Morgan for their gifts of NIH3T3 and normal human fibroblasts, respectively.

REFERENCES

[1] Qiu, Y et al., *Nanoscale* 2014, 6, 11744-11755.
[2] Kunz, D. A. et al., *Adv. Mater. (Weinheim, Ger.)* 2013, 25, 1337-1341.
[3] Chen, X. et al., *Journal of Applied Mechanics* 2004, 71, 597-603.
[4] Huang, Z. Y. et al., *Journal of the Mechanics and Physics of Solids* 2005, 53, 2101-2118.
[5] Song, J. et al., *International Journal of Solids and Structures* 2008, 45, 3107-3121.
[6] Charras, G. et al., *Nat. Rev. Mol. Cell Biol.* 2014, 1-12.

SB^3C2015
Summer Biomechanics, Bioengineering and Biotransport Conference
June 17-20, 2015, Snowbird Resort, Utah, USA
SB^3C2015-298

THE TENSION ASSISTED DEVICE: AN ORTHOTIC ALTERNATIVE TO HIGH TONE PLANTAR FLEXION

Elizabeth Duncan
Biomedical Engineering
University of Memphis
Memphis, TN, United States

Cory Jeanes
Biomedical Engineering
University of Memphis
Memphis, TN, United States

Mike Martorano
Biomedical Engineering
University of Memphis
Memphis, TN, United States

Ron Perrone
Biomedical Engineering
University of Memphis
Memphis, TN, United States

Faculty Adivsor
Gary L. Bowlin
Biomedical Engineering
University of Memphis
Memphis, TN, United States

Faculty Adivsor
John L. Williams
Biomedical Engineering
University of Memphis
Memphis, TN, United States

Clinical Advisor
Susan Anderson
LeBonheur Children's Hospital
Memphis, TN, United States

INTRODUCTION

Cerebral palsy (CP) is the most common pediatric motor disability and is estimated to affect roughly one in 323 children in the United States according to the Centers for Disease Control and Prevention [1-3]. Depending on the epidemiological study, prevalence reports vary, but one large meta-analysis study pooled results from 49 smaller studies to conclude that the overall prevalence of CP worldwide is 2.11 per 1,000 live births, which translates to one in 474 children [4]. Seventeen Million people worldwide live with a diagnosis of CP [3].

The term "cerebral palsy" is considered to be an umbrella term encompassing multiple possible symptoms, but overall refers to how a child's "ability to move and maintain balance and posture" is limited [1]. The symptoms of CP substantially affect function and quality of life. There is no cure for CP and it is considered a non-progressive, non-fatal lifetime disability. Symptoms exist due to permanent, unchanging brain, brainstem, or spinal cord lesions acquired either in utero, during delivery, or in the first two years of life. Symptoms of CP and other neurological conditions may include the following [5]:

- Spasticity: stiff, tight muscles with exaggerated movement
- Toe-walking, crouched gait and/or scissored gait
- Change in muscle tone: hypertonia or hypotonia
- Developmental milestone delays in motor development: i.e. sitting, crawling, standing and walking

The symptom we have been given to focus on is high tone plantar flexion, or hypertonia in the gastrocnemius and soleus muscles of the calf. These muscles may be shorter and/or feel tight in the back of the leg, which leads to toe-walking.

Clinical Problem: Children with neurological conditions exhibiting high tone plantar flexion, experience impaired control of the ankle, resulting in difficulty crawling, walking, and performing the associated transitions between walking and crawling.

Need Statement: There is a need to control the ankle position in children with high tone plantar flexion, in order to mitigate the difficulty of crawling, walking, and performing associated transitions.

Current braces either focus on blocking plantar flexion or dorsiflexion but fall short in how they accommodate transitions between the two. Children under the age of three need to be able to transition between walking, crawling, and standing without being restricted by "block" style braces. Our clinical contact (Susan Anderson) is interested in a device that can control ankle position without caregiver assistance. The targeted age range is approximately eight months to five years old.

PRODUCT DESIGN

Our preliminary design concept was a tension-assisted device (TAD), which hypothesized that the quadriceps femoris muscles working in conjunction with a tethered system could be used to counteract the spasticity of the gastrocnemius and soleus muscles. The system linked the toe region to the waistline with a fixed length tether. The intent was to counteract the high tone induced plantar flexion when in the standing position, and allow ankle rotational control during gait pattern phases. Blocking plantar flexion at the end of the swing phase of the gait pattern would promote leading with the heel, while allowing controlled plantar flexion permits the individual to take advantage of the torque force generated during the toe off phase of the gait pattern. Ankle rotation control can also create a more stable condition for the individual in a crawling position, allowing more of the tibialis anterior region to make contact with the floor by allowing plantar flexion. The device is designed to give the individual full control of the rotation of the ankle based on the position of their leg. When the leg is straight, maximum tension is produced in the tether, pulling the toe out of plantar flexion, into dorsiflexion. When the leg is bent, slack is provided in the tether, allowing the spastic muscle to regain control of the ankle angle, returning the foot to plantar flexion. A wooden cable assisted model (CAM) of the system was built for a proof of concept, which was converted into an adult sized single leg version of a wearable device (Figure 1a).

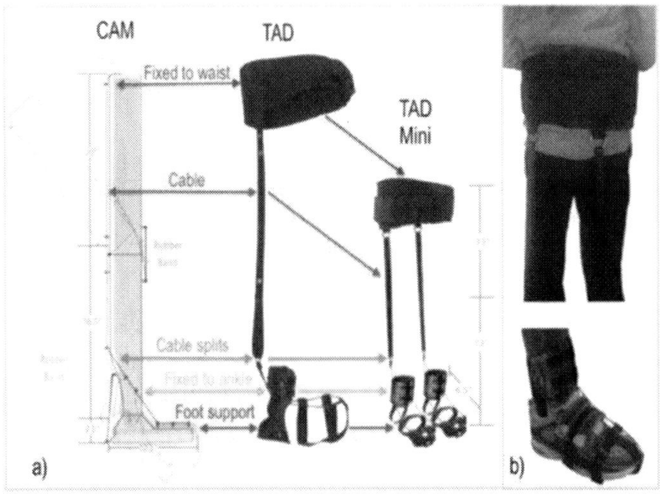

Figure 1a: (Left) Proof of concept model. (Middle) Adult sized prototype. (Right) TAD Mini® Figure 1b: The TAD Mini® worn by a 5-year-old child

The adult sized, single leg version of the device that was fabricated showed promise during preliminary "bench-side" testing. When the device was worn, and the leg in the straightened position, a noticeable force was generated that pulled the toe into dorsiflexion. When the leg was bent, slack in the tether allowed the foot to freely rotate from dorsiflexion to plantar flexion.

The device features an adjustable weightlifting style belt for the waste-line anchor point, a 2.54 cm nylon strap for the tether and foot support, which is linked with a 1.6 mm vinyl coated cable. The ankle is protected from the cable that wraps around the ankle to link up with the nylon strap, with a polymer ankle support (Figures 1a and 1b). The foot support features a Topy® pad on the bottom for traction and wear resistance.

Overall, we believe this design is an improvement on the available brace style orthotics currently in use and meets the stated needs of our clinical contact and the observed needs of children diagnosed with high tone plantar flexion. We believe it is a superior concept because it does not require caregiver assistance while in use, once it has been fitted. The design also allows the individual to produce the force required to counteract the hypertonia. Therefore, we believe there is a reduced potential for injury to due to overstraining the muscle by forcing it into an orthotic that completely blocks plantar flexion. This will also produce a more comfortable orthotic experience for the individual, as they are in control of the applied force, giving them the option to increase or decrease the force based on their specific condition and comfort level. As the device has yet to undergo any clinical trials, there may be unforeseen issues to be addressed once testing has begun. The team's prototype is designed to assist with sagittal movement only. It is not designed to control supination, pronation, varus or valgus ankle, or any rotation of the foot, such as eversion or inversion. The child's bone anatomy, foot position, and other ankle positioning may be co-symptoms that need to be considered. It's possible a child would need to wear a non-plantar flexion blocking ankle foot orthosis (AFO), supra-malleolar orthosis (SMO), or a foot positioning orthotic in side a shoe to help address his/her other positioning needs.

The team is currently working on a duel leg prototype using parameters of the target age range, to produce the TAD Mini® (Figure 1a). The new device will feature far more adjustability for a "fast fit", as the intended application is a therapeutic assisted device. The team is also currently working with The Center For Orthotics and Prosthetics, Inc. (COPI) in Memphis, TN to bring the components of the device up to industry standards for orthotics. The next generation prototype will feature components that are commonly used in on-the-spot custom orthotics, which consist of prefabricated off-the-shelf parts that meet industry standards regarding insurance reimbursable *current procedure terminology* (CPT) codes.

BUDGET & MARKET ANALYSIS

Approximately 10,000 children are born with CP each year in the United States. 70% of children with CP are ambulatory at some level. This presents a market of 7,000 new children per year in the United States who will require assistive devices. With a projected growth rate world wide of 1.043% for 2015, there will be approximately 158,250 children born with CP, which creates a total available market of approximately 110,000 children per year requiring an assistive device [6].

The intended market for our device is short duration therapy sessions versus all day assistance. Therefore, adjustability for a more universal fit takes precedence over a custom device, which also contributes to lower manufacturing costs. The adult prototype was developed at a cost of approximately $110, which excluded bulk pricing and labor costs.

Manufacturing the device entailed tailor style fabrication, which included prefabricated off-the-shelf polymer ankle protection cuffs. The device required minimal assembly that also utilized off-the-shelf hardware, such as the Hosmer® cable retainer system. Utilizing bulk pricing and in-house fabrication, we believe that we can keep the manufacturing cost to approximately $60 per unit with a markup of 65%, putting the sale price at approximately $100. Assuming a conservative estimate of acquiring 15% market share during the first year should yield a demand of 16,500 units. Running a 5-person operation with a starting capital and operating cost of approximately $1,256,000 should net a profit of approximately $394,000. Assuming a 10% increase in market share each year, at the end of the 3rd year a 35% market share should net $1,008,000. This estimate includes the increase in capital and operating costs required with the hiring of enough employees to realistically keep up with a demand of 38,500 units. We feel this will still allow for acceptable profit margins resulting in approximately a 150% return on the initial investment at the close of year 3, without contributing to an already inflated market place for orthotics.

The lifetime cost of caring for a person with CP is almost $1 million US dollars. The combined lifetime cost for all people with CP born in 2,000 has been estimated to be $11.5 billion US dollars [1]. Necessary physical therapy, medications, doctor visits, assistive devices, and caretaking all contribute to expenses. Our goal is to provide an affordable assistive device and not contribute unnecessarily to rising costs.

A sweep of the current orthotics market showed a gap for a product that can transition between plantar and dorsiflexion without caregiver assistance. Cascade DAFO Inc. is a leading manufacturer of pediatric AFOs. They produce a wide range of orthotics for different indications and diagnoses, which range from $150 for an off-the-shelf type, to $2,500 for a full custom AFO. Their dynamic AFOs do offer limited ankle flexion, however their line appears to lack a hands free device that allows full range of motion [7]. Considering the gap in the market for a device like the TAD or TAD Mini®, it is reasonable to assume at least an eventual market share of 35% or more. At some point it would be expected that competition would taper the market share increases until a plateau has been reached. Nevertheless, even if the market share never went above 25%, that would still yield a net of $736,000 per year. This annual figure is tantalizing enough to follow through with the endeavor, without losing site of the purpose of the device, which is to vastly increase the quality of life for as many individuals as possible.

ACKNOWLEDGEMENTS

We thank the following individuals for their time, assistance, and contributions to our project: Jack Steele, LPO, CO, FAAOP Jim Huddleston, CO, LO Cathie Baker, LPO, CPO. Funded in part by NSF GARDE award 1160107

REFERENCES

1. *Cerebral Palsy*. 2014 July 14, 2014 Available from: http://www.cdc.gov/ncbddd/cp/index.html
2. L Andrew Koman, *The Lancet*, 2004. **363**: p. 1619-31.
3. *About Cerebral Palsy*. 2014 Available from: https://en.worldcpday.org/about
4. Oskoui, M., et al., *An update on the prevalence of cerebral palsy: a systematic review and meta-analysis*. Dev Med Child Neurol, 2013. **55**(6): p. 509-19.
5. *NINDS Cerebral Palsy Information Page*. 2014 July 21, 2014 Available from:http://www.ninds.nih.gov/disorders/cerebral_palsy/cerebral_palsy.htm
6. *How common is cerebral palsy?* 2014; Available from: http://cerebralpalsy.org/about- cerebral-palsy/prevalence-and-incidence/
7. *DAFO Optimization Index*. September 30, 2014 Available from: http://www.dafo.com/library/dafo-optimization-index/

SB^{3}C2015
Summer Biomechanics, Bioengineering and Biotransport Conference
June 17-20, 2015, Snowbird Resort, Utah, USA
SB^{3}C2015-299

BIAXIAL CREEP OF SWINE CARDINAL AND UTEROSACRAL LIGAMENTS

Ting Tan, Nathan M. Cholewa, Scott W. Case, and Raffaella De Vita

Department of Biomedical Engineering and Mechanics
Virginia Tech
Blacksburg, VA, USA

INTRODUCTION

Pelvic floor disorders (PFDs) such as urinary incontinence, fecal incontinence, and pelvic organ prolapse affect millions of women every year. PFDs occur mainly due to structural and mechanical alterations of pelvic organs, muscles, and connective tissues. The two major suspensory ligaments of the uterus, cervix, and vagina complex are the uterosacral ligament (USL) and the cardinal ligament (CL). The importance of investigating the biomechanical properties of the USL and CL for the treatment of PFDs has been recognized only in recent years. Elastic and viscoelastic properties of these ligaments have been investigated by performing uniaxial and biaxial experiments [1-4]. However, no creep data for the CL and USL have been reported. In this study, a digital image correlation (DIC) system for non-contact strain measurement and a planar biaxial testing system were utilized to characterize the biaxial creep properties of these ligaments in the swine. Scanning electron microscopy (SEM) was also used to examine the microstructure of the ligaments, and relate it to the collected biaxial creep data.

METHODS

This study was conducted with approval of the Institutional Animal Care and Use Committee (IACUC) at Virginia Polytechnic Institute and State University. Four adult domestic swine were obtained from a slaughterhouse. The swine were 3–4 years old and had a weight of ~450 lbs. The CL and USL were carefully dissected, frozen at −20 °C, and thawed at room temperature before testing.

Specimens for scanning electron microscope (SEM) examination were fixed overnight in a 2% glutaraldehyde–0.01 M sodium cacodylate buffer. They were washed with PBS solution, post-fixed in osmium tetraoxide, and dried in a critical point dryer (Model 28000, LADD Research Industries). The specimens were immersed in liquid nitrogen and fractured with a sharp razor blade in order to reveal their cross-section. After being sputter coated with gold, the cross sections were examined using an environmental scanning electron microscope (SEM) (Quanta 600 FEG, FEI).

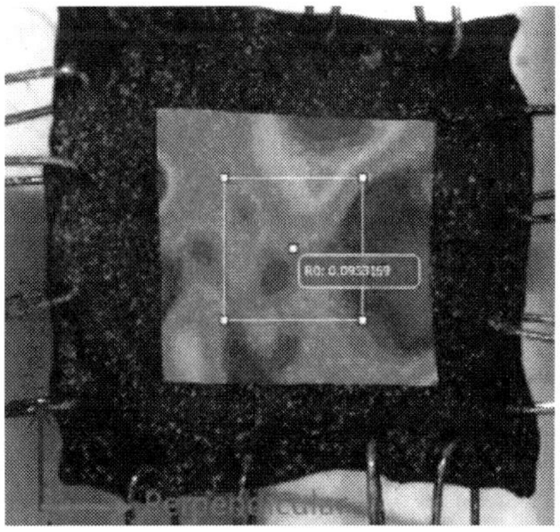

FIGURE 1: DIC STRAIN MAP IN THE PERPENDICULAR LOADING DIRECTION OF ONE CL SPECIMEN. RO DENOTES THE AVERAGE STRAIN (~9%) OF THE SELECTED REGION.

Specimens for biaxial creep test were cut into approximately 3×3 cm^2 squares. The thickness of each specimen was measured using a digital caliper (accuracy ±0.05 mm, Mitutoyo Series 573, Japan) under a 50-g compressive load. Specimens were first immersed in methylene blue 1% (Fisher Scientific) to obtain a dark blue background and then gripped by four safety pins on each edge. Finally, a speckle pattern was created with white paint. Specimens were then mounted into a planar biaxial system (Instron, UK) with four 20 N load cells by fishing lines connected to safety pins and pulleys in the test device. The two loading directions were selected to be parallel and perpendicular to the longitudinal axis of the vagina. The perpendicular direction coincided with the physiological loading direction of the ligaments. Each specimen was then submerged in a bath of phosphate-buffered saline (PBS, 0.5 M, pH 7.4) and preconditioned from 0.1 N to 0.6 N at 0.1 mm/s for 10 cycles. After preconditioning, the specimen was unloaded and allowed to recover for 10 minutes. Each specimen was then stretched at 0.1 mm/s and subjected to constant equi-biaxial loads of 2 N for 2 hours (n=15). Lagrangian strain was measured using the digital image correlation method (Correlated Solutions, VIC-3D v7). The average strain in a 1×1 cm^2 area during creep test was calculated for each specimen (Figure 1).

Mean and standard deviation values were computed for the constant load, corresponding constant nominal stress, initial strain during creep, and normalized strain, with respect to the initial strain, during the 2 hour-long creep test. One-way analysis of variance was conducted to compare the mean of the normalized strain in the two loading directions of the specimens. The student's *t* test was used and the threshold chosen for statistical significance was 0.05. Data were analyzed using the JMP statistical software (JMP, Version 10, SAS Institute Inc.).

RESULTS

One SEM micrograph of a swine CL planar-section is shown in Figure 2. Collagen fibrils appeared to be organized into bundles. These bundles were arranged primarily in the direction perpendicular to the longitudinal axis of the vagina.

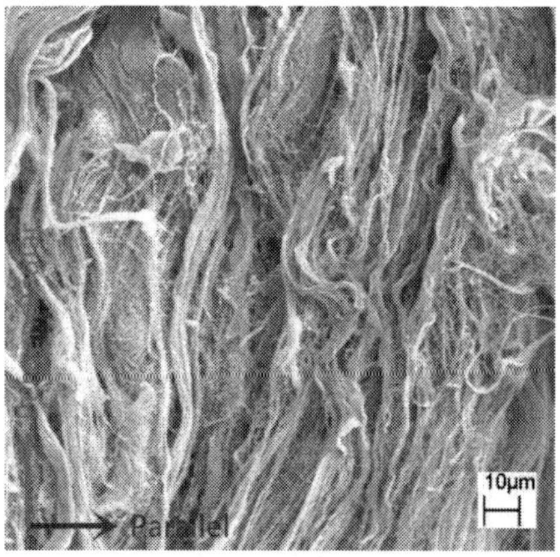

FIGURE 2: SEM IMAGE OF THE CL SPECIMEN.

During creep tests, the load, corresponding nominal stress, and initial strain of the specimens in the direction that was parallel to the longitudinal axis of the vagina were 1.997±0.069 N, 0.142±0.061 MPa, and 0.152±0.076, respectively. The load, corresponding nominal stress, and initial strain of the specimens in the direction that was perpendicular to such axis were 2.091±0.189 N, 0.153±0.066 MPa, and 0.116±0.065, respectively. The mean and standard deviation values of the normalized biaxial creep strain data over two hours under the afore-mentioned loads are presented in Figure 3. The creep behavior of the ligaments appeared to be different in the two loading directions. However, the statistical analysis showed that the normalized creep strain data were not significantly different between the two directions ($p>0.05$). The creep rate of the specimens in both loading directions showed a similar trend during 2 hour-long duration of the test. The normalized strain increased sharply over the first 20 minutes and very slowly over the last 40 minutes.

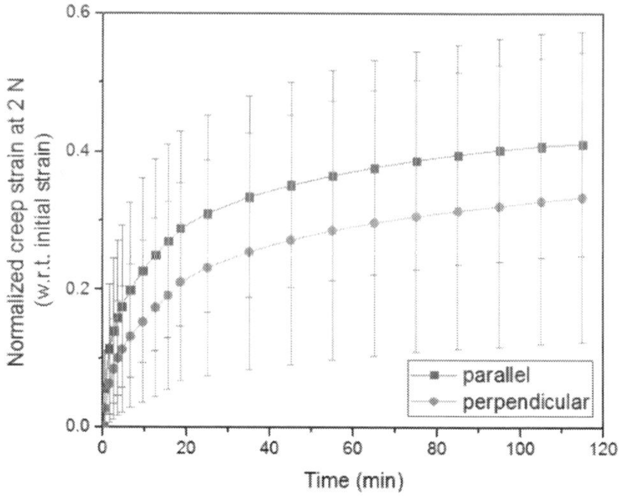

FIGURE 3: NORMALIZED CREEP STRAIN DATA OF CL AND USL WITH RESPECT TO INITIAL STRAIN OVER TIME.

DISCUSSION

The biaxial creep properties of the swine CL and USL were characterized for the first time. In this study, a planar biaxial system was used to perform creep tests and the DIC method was used to measure the strain field. A newly developed procedure to create effective speckle pattern on biological soft tissues with methylene blue and white paint was adopted [5]. The SEM analysis showed that the collagen fibers are wavy and arranged primarily along the physiological loading direction of these ligaments [6]. The creep behavior in the CL and USL were found to be slightly anisotropic: the ligaments creeped less along their physiological loading direction (i.e., perpendicular to the vaginal longitudinal axis). This is, perhaps, due to the larger number of collagen fibers oriented along such direction. Additional studies are currently being conducted to determine the nonlinearity in the creep response of the CL and USL.

ACKNOWLEDGEMENTS

Funding was provided by NSF CAREER Grant no. 1150397.

REFERENCES

[1] Vardy, M.D., Gardner, T.R., Cosman, F., Scotti, R.J., Mikhail, M.S., Preiss-Bloom, A.O., Williams, J.K., Cline, J.M., Lindsay, R., *Am J Obstet Gynecol*. 192: 1741–1751, 2005.
[2] Martins, P., Silva-Filho, A.L., Fonseca, A.M.R.M., Santos, A., Santos, L., Mascarenhas, T., Jorge, R.M.N., Ferreira, A.M., *Arch Gynecol Obstet*. 287: 313–318, 2013.
[3] Luo, J., Smith, T.M., Ashton-Miller, J.A., DeLancey, J.O.L., *J Biomech Eng*, 136, 021016, 2014.
[4] Becker, W.R., De Vita, R., *Biomech Model Mechanobiol*, DOI 10.1007/s10237-014-0621-5.
[5] Gionello, G., Sirieix, C., Baleani, M., *J Mech Behav Biomed Mater*, 39: 1-8, 2014.
[6] Tan, T., Davis, F.M., Gruber, D.D., Massengill, J.C., Robertson, J.L., De Vita, R, *J Mech Behav Biomed Mater*, 42: 129-137, 2015.

SB³C2015
Summer Biomechanics, Bioengineering and Biotransport Conference
June 17-20, 2015,Snowbird Resort, Utah, USA
SB³C2015-300

GENERATING TUMOR MODELS FROM MICROCT SCAN IMAGES FOR SIMULATING TEMPERATURE ELEVATION DURING MAGNETIC NANOPARTICLE HYPERTHERMIA

Andrew Lee, Alexander LeBrun, Ronghui Ma, Liang Zhu

Department of Mechanical Engineering
University of Maryland Baltimore County
Baltimore, MD, USA

INTRODUCTION

Previous investigations in magnetic nanoparticle hyperthermia for cancer treatments have demonstrated that particle size, particle coating, and magnetic field strength and frequency determine its heating generation capacity. However, once the nanoparticles are manufactured, the spatial distribution of the nanoparticles dispersed in tissue dominates the spatial temperature elevation during heating [1-3]. Therefore, understanding the distribution of magnetic nanoparticles in tumors is critical for developing theoretical models in order to predict temperature distribution in tumors for hyperthermia treatment design. Another important factor that may affect temperature elevations in tumors is the tumor geometry since not all tumors can be represented by a pre-described simple geometry such as a sphere. An accurate description of the nanoparticle distribution and the tumor geometry will greatly enhance the simulation accuracy of the heat transfer process in tumors, which is crucial for generating an optimal temperature distribution that can prevent the occurrence of under-dosage heating in the tumor and overheating in the healthy tissue. Recently studies by our group have demonstrated that nanoparticle concentration distribution in tumors can be visualized via microCT images due to density elevation in the presence of magnetic nanoparticles. The problem is that intensive memory is required to directly import the microCT images to numerical simulation software packages such as COMSOL. Although commercial software packages exist to handle detailed entities inside tumors, they are very expensive to purchase. In addition, having very small entities at the micrometer level inside the tumor geometry may provide challenge to numerical simulation software to accept the generated geometry.

The objective of this study is to extend our previous imaging processing algorithms to develop realistic 3-D tumor models from microCT images for heat transfer simulation during hyperthermia treatment. The tumor geometry and the SAR file will then be tested whether they are acceptable by the COMSOL software package for heat transfer simulation.

METHODS

MicroCT scans of PC3 tumors

The tumors used in this study are PC3 prostatic tumors implanted on the flanks of BALB/c Nu/Nu male mice. Water based ferrofluids (EMG700 series, Ferrotec (USA) Corporation, Nashua, NH) with a concentration of 5.8% by volume and a particle size of 10 nm are loaded on a syringe pump (S230, Kd Scientific, Inc. Holliston, MA to control the flow rate (3 μl/min) and the amount of the injection (0.1 cc). A 26-gauge Hamilton needle (Fischer Scientific, Springfield, NJ) is used for injecting ferrofluid into the two tumors. After the experiments, the mouse is euthanized and the tumors are resected and scanned in a microCT imaging system (Skyscan1172, MicroPhotonics, PA), resulting in a total of 523 individual images with a pixel resolution of 1052x1944.

Generation of the tumor geometry and the SAR file

A MATLAB code is written in order to identify the tumor boundary of each image, smooth boundary lines, and convert gray scale values inside the tumor boundary as uniform. When background noises are too significant to be filtered through MATLAB, they are removed manually using a standard paint editor. ImageJ 3-D viewer is used to stack all the modified slices together with a resampling factor of 10 in order to generate a "prototype" of the 3-D tumor with uniform properties. Assembling tumors through stacking images creates sharp boundaries between slices. Rhinoceros 5.0® was used to convert the ImageJ model into a series of 3-D spline curves in order achieve a smooth surface boundary. The new surface is not only a better representation of the actual tumor, but also provides a better model for conducting finite element analyses. The tumor is then attached to a mouse body generated in similar fashion for heat transfer analyses using COMSOL software. The Pennes bioheat equation with convection boundary conditions (T_∞=25°C and h = 5 W/m^2 °C) on the mouse surface is used to simulate the temperature elevations induced by the heat generation in the tumor (specific absorption rate or SAR) during hyperthermia treatment.

In this study, the pixel index numbers of individual microCT images are converted first to the SAR values at individual pixel locations, using the linear relationship described in our previous study [4]. In order to reduce the size of the SAR file, we average the SAR values over a 3-D cluster of pixels (7x7x7). This greatly decrease the size of the SAR file, while also ensuring that the amount of total energy deposited in the tumor is not affected by the averaging procedures. The cluster size can be adjusted to yield a reasonably sized SAR file without sacrificing the accuracy of the numerical simulation. The vast SAR data are processed to generate a single file that gave the SAR values in the form of a 3-D matrix using the software SAS (SAS Institute, Cary, NC), which is widely employed for data analysis and statistics.

RESULTS

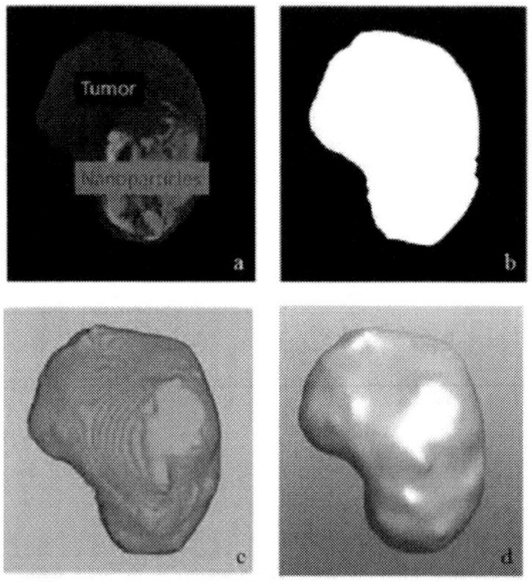

Figure 1. Processes of generating a tumor model from a microCT image (a), to a MATLAB generated tumor boundary (b), to an ImageJ tumor geometry (c), to a smoothed tumor model by Rhinoceros (d).

Figure 1a shows a slice of the 3-D microCT scan of a tumor. The tumor boundary is evident in the figure. The white cloud in the image represents the pixel locations, in which density is significantly higher than the tumor tissue density. Since the density elevation is due to the presence of the magnetic nanoparticles, the cloud can be viewed as pixel locations with nanoparticle depositions. The same microCT image after it has been processed by the MATLAB code is shown in Figure 1b. The image shows a clear tumor boundary and all gray scale values inside the boundary are made uniformly white while noise is filtered from the background and is made uniformly black. After processing all images in MATLAB, they are stacked together using ImageJ 3-D viewer with a resampling factor of 10. Figure 1c clearly illustrates visible layers and rough boundaries in the 3-D model. Rhinoceros 5.0® is used to smooth the surface of the ImageJ model and eliminate the appearance of layers and rough edges. The STL model created in ImageJ was converted into a series of 3-D splines resulting in the smooth tumor surface shown in Figure 1d.

We are successful in merging the tumor model with a mouse body using ProE software. Figure 2 shows the final geometry imported to COMSOL 4.3. The combined structure can be meshed in the COMSOL software in order to conduct a heat transfer simulation. The temperature contours inside the tumor and the mouse body are also illustrated in Figure 3, under the same heating conditions (~3 kA/m magnetic field) as that described in our previous study [4]. Most temperature elevations occur in the tumor, while the temperature field in the mouse body is relatively uniform (~37°C). The maximal temperature in the tumor is approximately 69°C at the center region of the tumor. The interface between the tumor and the mouse body experiences the lowest temperature elevation, and is only heated to approximately 52°C during steady state.

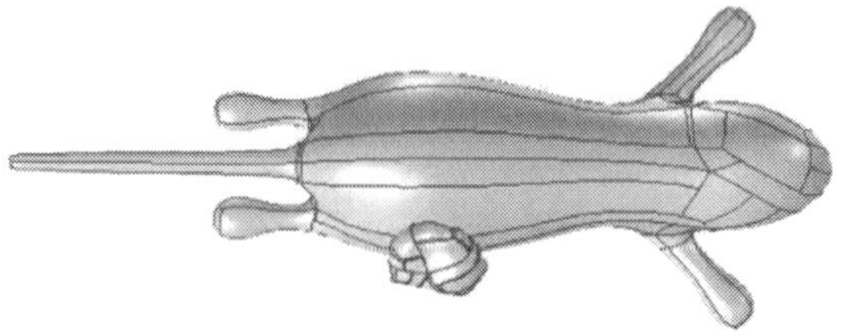

Figure 2 Computer generated mouse body with a tumor attached.

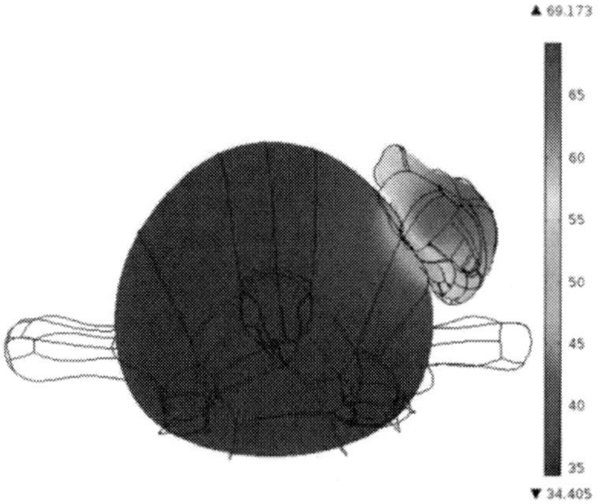

Figure 3 Temperature contours of the tumor and mouse body during steady state.

In summary, we have developed computational algorithms to simplify the large amount of microCT data of tumors. The feasibility of using this algorithm to generate a tumor model resembling its original shape and size has been demonstrated in the study. We have also developed an algorithm to simplify the huge SAR information contained in the microCT images to ensure that tumor model can be accepted by the COMSOL numerical software. The simulated temperature field using the generated tumor geometry and simplified SAR distribution agrees well with our experimental results [4].

ACKNOWLEDGEMENTS

This study was supported by an NSF research grant (CBET CBET-1335958.

REFERENCES

[1] Jordan, A., Scholz,R., Wust, P., Fahling, H., and Felix, R., 1999, *J. Magnetism and Magnetic Materials*, 201:413-419.

[2] Salloum, M., R. Ma, D. Weeks, and Zhu, L., 2008, *International Journal of Hyperthermia*, 24(4):337-345.

[3] Attaluri, A., Ma, R., Qiu, Y., Li, W., and Zhu, L., 2011, *International Journal of Hyperthermia*, 27(5): 491–502.

[4] LeBrun, A., Manuchehrabadi, N., Attaluri, A., Wang, F., Ma, R., and Zhu, L., 2013, *International Journal of Hyperthermia*, 29: 730-738.

THE EFFECT OF PRE-CRASH VELOCITY REDUCTION ON OCCUPANT RESPONSE USING A HUMAN BODY FINITE ELEMENT MODEL

Nicholas A. Vavalle (1,2), Berkan Guleyupoglu (1,2), Jeremy M. Schap (1,2), Kristofer D. Kusano (1,3), F. Scott Gayzik (1,2)

(1) Virginia Tech – Wake Forest
Center for Injury Biomechanics
Winston-Salem, NC, USA

(2) Department of Biomedical Engineering
Wake Forest School of Medicine
Winston-Salem, NC, USA

(3) Department of Biomedical Engineering
and Mechanics
Virginia Tech
Blacksburg, VA, USA

INTRODUCTION

Vehicles with active safety features, including pre-collision braking, are becoming more prevalent in the vehicle fleet. While decreasing collision speed reduces the energy in the system, the complete effects of pre-braking on the occupant kinematics and loads have not been studied. The objective of this study is to use a validated finite element model of the human body [1] to evaluate the effect of simulated pre-crash braking on driver kinematics, restraint loads, body loads and computed injury criteria in four commonly-injured body regions.

METHODS

The Global Human Body Models Consortium (GHBMC) 50th percentile male occupant (v. 4.3, 1.3M nodes, 2.2M elements, 77.0 kg) was used for all simulations. The model was gravity settled in the driver position of a generic interior designed for frontal crash studies. The environment was equipped with an advanced three point belt (including pretensioner and retractor) and driver airbag. Restraint properties were based on literature values [2]. Eight simulations were conducted, including four scenarios at two severity levels, median and severe. The four scenarios were: no pre-collision system (baseline), forward collision warning (FCW), FCW with pre-braking assist (FCW+PBA) and FCW and PBA with autonomous pre-crash braking (FCW+PBA+PB). The baseline delta-V were 17 kph and 34 kph for median and severe scenarios, respectively, based on crash reconstructions from NASS/CDS.

Delta-V values were then reduced according to simulations assuming that vehicles had been equipped with pre-crash systems. Additional driver and vehicle braking could reduce delta-V by up to 34% for the median and 20% for the severe case. The simulations with pre-collision braking had constant decelerations of 0.4 g, 0.6 g, and 0.8 g prior to the scaled crash pulse based on previous work modeling collision severity reduction of these systems [3]. The generic pulse was taken from the literature [4] and scaled to match target delta-V values. Simulations were run on a Linux cluster using 48 CPUs and MPP LS-DYNA. The test matrix used in the study can be seen in Table 1.

Table 1. The eight cases simulated of varying speed and pre-crash conditions.

Case	Speed	Pre-Crash	Delta-V (kph)
1	Median	None	17.0
2	Median	FCW	14.7
3	Median	FCW+PBA	13.0
4	Median	FCW+PBA+PB	11.3
5	Severe	None	34.0
6	Severe	FCW	30.2
7	Severe	FCW+PBA	29.0
8	Severe	FCW+PBA+PB	27.2

RESULTS

A representative time-lapse of the simulations can be seen in Figure 1. All injury criteria, belt forces, and occupant kinematics can be found in Table 2 and Figure 2. Shoulder and lap belt loads were maximal in the baseline case and reduced for each pre-collision scenario for both the median and severe sets. Maximum shoulder belt loads for the severe cases were 3.7 kN at baseline, and reduced 1%, 21%, and 22% respectively for the three pre-collision adjusted cases. Average femur load was relatively constant for all scenarios in the median case set (range: 0.86 to 1.1 kN), but showed a roughly linear decrease in the severe set (range: 3.6 kN, to 2.1 kN, 43% decrease for FCW+PBA+PB). While Nij values did not approach critical levels, they showed an increasing magnitude from the FCW+PBA case to the FCW+PBA+PB case (e.g. from 0.20 to 0.25 in the severe set). Sternal deflection also increased, from 58 mm to 70 mm in the median set and 62.3 mm to 80.0 mm in the severe set. HIC values showed a substantial increase in pre-collision braking cases for the median impact (HIC: baseline 13.7, FCW 11.6, FCW+PBA 140 and FCW+PBA+PB 147).

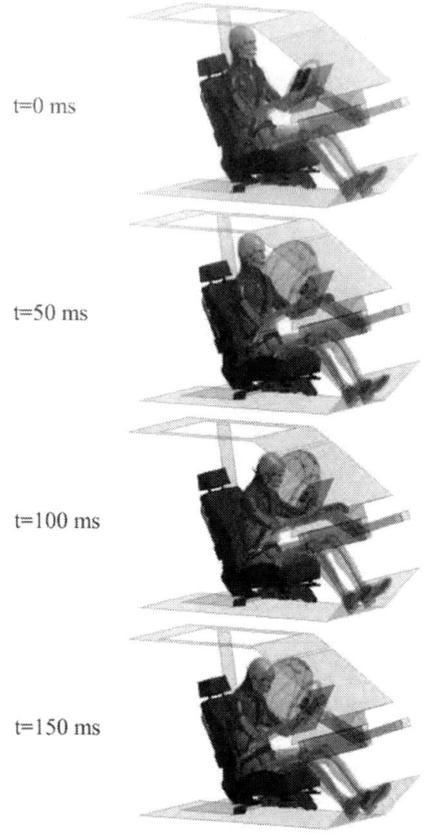

Figure 1. Time-lapse images of the Case 1 simulation.

DISCUSSION

Forward occupant excursion during pre-crash braking was thought to be responsible for the difference in HIC. Forward head CG excursion at time of airbag deployment was less than 1 mm for the baseline, but was 223 mm for the FCW+PBA+PB case. In the severe case, HIC was reduced in the pre-collision equipped cases (HIC: baseline 261, FCW 199, FCW+PBA 146, FCW+PBA+PB 182).

Reduction in delta-V caused reductions in some injury criteria, while others (Nij and HIC) increased due to autonomous pre-crash braking that modified the occupant position before the crash event. These results show the need for integrated safety systems that have restraints, for example a reversible pretensioner, which optimizes the occupant's position during braking and prior to impact. A limitation of this analysis is that pre-crash occupant bracing was not modeled, which could influence occupant position at the time of airbag deployment.

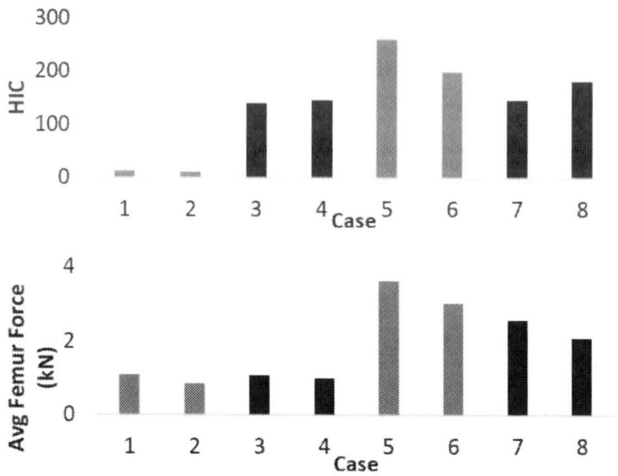

Figure 2. Baseline and FCW shown in gold, pre-crash braking cases in black. HIC increased with autonomous pre-braking while femur force was level, or decreased.

ACKNOWLEDGEMENTS

The authors would like to acknowledge the Global Human Body Models Consortium, LLC for funding and support. Wake Forest University is the Integration Center of Expertise of the GHBMC. The authors gratefully acknowledge the meshing and validation efforts by our academic partners in the GHBMC. All simulations were run on the DEAC cluster at Wake Forest University, with support provided by Drs. Timothy Miller and Damian Valles.

Dr. Gayzik is a member of Elemance, LLC., which is the sole licensor and distributor of academic and commercial licenses for use of the GHBMC, LLC. – owned M50 occupant model.

REFERENCES

[1] Vavalle, NA et al., *Ann Biomed Eng*, 41:497-512, 2013.

[2] Forman, JL et al., *Stapp Car Crash J*, 53:49-74, 2009.

[3] Kusano, KD and Gabler, HC, *IEEE Trans Intell Transp Syst*, 13:1546-1555, 2012.

[4] Barbat, SD et al., *SAE Int J Trans Safety*, 1:328-333, 2013.

Table 2. Model outcomes for the 8 simulations, including injury metrics, belt forces, and airbag deployment information.

Case	HIC	Nij	Sternal Deflection (mm)	Average Peak Femur Force (kN)	Peak Shoulder Belt Force (kN)	Peak Lap Belt Force (kN)	Head Excursion at Airbag Deployment (mm)
1	13.7	0.199	44.3	1.11	2.79	2.82	0.2
2	11.7	0.150	44.4	0.86	2.48	2.32	0.2
3	140.8	0.110	58.1	1.08	1.00	1.80	109.7
4	147.1	0.152	70.2	0.99	0.97	1.40	166.5
5	260.9	0.382	62.3	3.64	3.67	5.39	0.5
6	199.2	0.376	58.7	3.04	3.65	4.90	0.4
7	146.8	0.202	62.2	2.57	2.91	3.79	161.1
8	181.8	0.256	80.1	2.09	2.85	3.72	223.4

SB³C2015
Summer Biomechanics, Bioengineering and Biotransport Conference
June 17-20, 2015,Snowbird Resort, Utah, USA
SB³C2015-302

MULTISCALE SIMULATION OF SHEAR-INDUCED PLATELET ACTIVATION: CORRELATING NUMERICAL WITH EXPERIMENTAL RESULTS

Peng Zhang (1), Chao Gao (1), Na Zhang (2), Seetha Pothapragada (2), Marvin J. Slepian (1,3), Yuefan Deng (2), Danny Bluestein * (1)

(1) Biomedical Engineering Department
Stony Brook University
Stony Brook, NY, USA

(2) Applied Mathematics Department
Stony Brook University
Stony Brook, NY, USA

(3) Departments of Medicine and BioMedical Engineering
Sarver Heart Center, University of Arizona
Tucson, AZ, USA

INTRODUCTION

The coagulation cascade of blood may be initiated by shear-induced platelet activation, which prompts clot formation in prosthetic cardiovascular devices and in arterial disease processes. Upon activation platelets undergo complex biochemical and morphological changes. Activated platelets polymerize fibrinogen into a fibrin network that enmeshes red blood cells. Continuum methods fail to capture the smaller scale molecular mechanisms such as the filopodia formation during platelet activation. Utilizing molecular dynamics that can capture and model this is computationally prohibitive. To address these challenges, a multiscale approach may offer an effective means to bridge the gap between macroscopic flow and the cellular scales. We have developed a multiscale model which interfaces nanoscale microstructures of human platelets and mesoscale transport of blood flows, for providing a more accurate flow-induced dynamic stress mapping on platelet membranes and predict their activation and adhesion. Coarse-grained molecular dynamics (CGMD) and dissipative particle dynamics (DPD) have been employed to simulate complex processes at molecular scales, as well as low-to-high Reynolds numbers viscous flow dynamics at mesoscopic scales. Coupling CGMD and DPD systems needs to interface the disparate spatial and temporal scales. Spatial interface is established by a hybrid force field between surface membranes and surround flows, in which microstructural changes of platelets respond to extracellular viscous shear stresses transferred to them [1]. The temporal interface is formulated by a multiple time stepping algorithm, in which three orders of magnitude disparity at microseconds and nanoseconds are handled by four-level integrator [2]. The phenomena of filopodia formation could be mimicked [3]. Additionally, numerical results were directly correlated with differential interference contrast (DIC) microscopy observations of platelets flowing in microchannels [3].

METHODS

We consider two spatiotemporal scale methods: (i) top/microscale using dissipative particle dynamics (DPD) to describe viscous blood fluid flows; and, (ii) bottom/nanoscale using coarse-grained molecular dynamics (CGMD) to describe intra-platelet cytoskeleton, cytoplasm and membrane. CGMD is composed of nonbonded potentials (such as Lennard-Jones (LJ) and Morse) and bonded potentials (such as bonds, angles and dihedrals). Figure 1 shows the multitude of spatial and temporal scales integrated in the multiscale model.

In this model, the top-bottom spatial-scale interface is established by imposing a hybrid force field at fluid-membrane surface interface [1]. This hybrid force field uses stochastic thermostats (top-scale) to maintain the flow's local thermodynamic and exchange momentum between the platelets and the surrounding flow. Simultaneously, it uses hard-core LJ force to provide bounce-back reflection of flow particles on surface membrane, with no-slip achieved by slowing down fluid particles as the fluid particles are getting closer to membrane surface (using the same repulsive term). These repulsive-drag forces fully achieve a no-slip boundary condition between flipping platelets and viscous fluid flows and thus used to evaluate hemodynamic stresses on the flowing platelets membrane, using a force virial stress algorithm.

In addition, the top-bottom temporal-scale interface is improved by using a multiple time-stepping (MTS) algorithm [2]. This algorithm is developed for handling the disparity in integration stepping sizes between DPD and CGMD that requires time stepping at microseconds and nanoseconds, respectively. The new scheme utilized four different step sizes: the fluid system is advanced at the largest step size, the fluid-platelet interface at a middle step size, and non-bonded and bonded interactions of intra-platelet microstructural components at the two smallest step sizes. The 4-level integrator is adjustable to balance the relationship of biomedical accuracy vs. computational complexity.

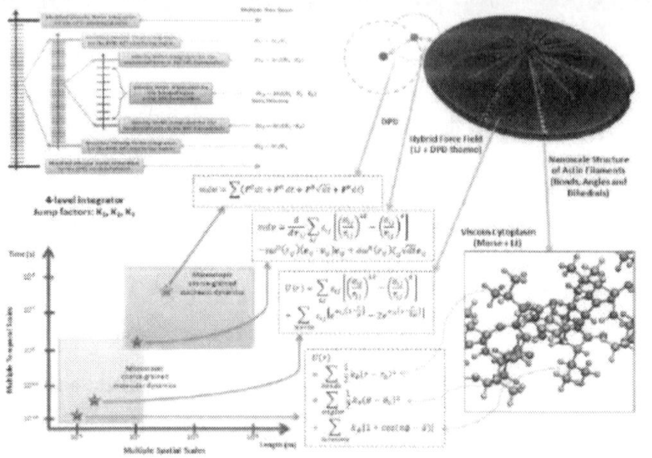

Figure 1: Multiscale Framework for Simulating Platelets in Blood Flows

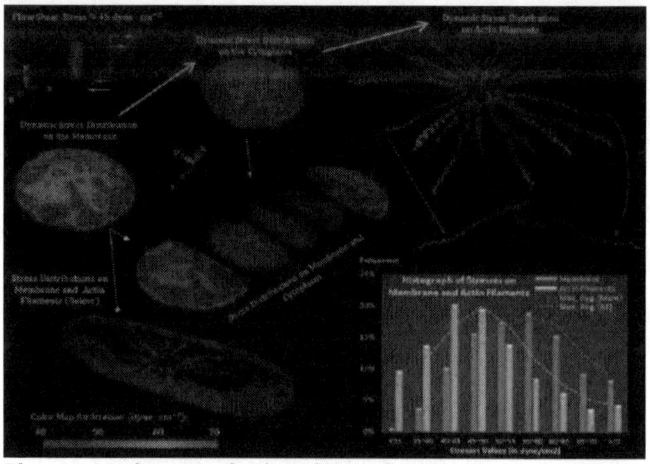

Figure 2: Mechanotransduction of Hemodynamic Stresses in Platelets

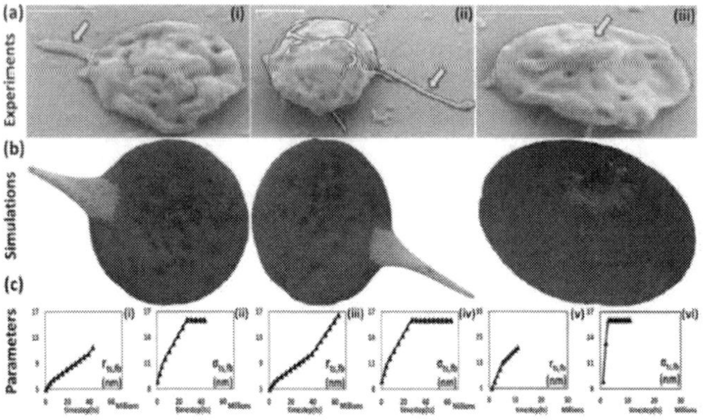

Figure 3: Filopodia Formation Correlating with Experiments

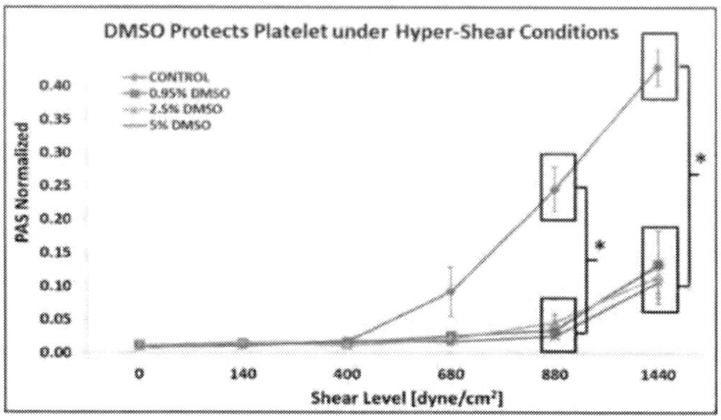

Figure 4: Certain levels of DMSO (2.5-5% v/v) can protect platelets from hyper sheared conditions, using a syringe capillary shearing device (SCSD)

RESULTS

This multiscale model describes nanoscopic mechanotransduction and biophysical characteristics of deformable platelets in complex viscous flows. The membrane was modeled with Young's modulus of 31.2 μN/m, cytoplasmic viscosity of 4.1 mPa·s, single actin filaments stiffness of 56.3±1.0 pN/nm [1, 4]. Fluid-platelet dynamic interaction preserved the dynamic properties of flowing platelet [1] which allows the study of the effects of the mechanotransduction processes of hemodynamic stresses on the flowing platelets (Figure 2). Comparative studies between rigid and deformable platelet models have indicated that rigid models significantly overestimate the stresses on the membrane, which in turn may lead to erroneous predictions of the platelet activation potential under shear conditions. The model also mimicked the formation of filopodia observed during early stage platelets activation [3] (Figure 3). Simulation results compared favorably with in vitro microchannel experiments.

DISCUSSION

Spatial-temporal interfaces have been achieved in the multiscale model, demonstrating the hallmarks of platelets under flow induced shear conditions: flipping trajectories, platelets deformability, viscous blood plasma flows, cytoplasm and elastoviscous membrane dynamics, and formation of filopodia upon platelet activation[1-5]. Flow induced stress information was transmitted to intracellular components. The morphological changes of platelets interactively influenced the surrounding flow field. Dynamical coupling of the micro-nano scales thus allowed the platelets to continuously change their shape during activation in response to the flow induced shear.

We are currently employing several microfluidic techniques and SEM imaging of platelets activated under shear flows to observe their interactions and corroborate our multiscale model predictions. We are also assessing the efficacy of antiplatelet agents in inhibiting shear-induced platelet activation. Work to date has demonstrated most conventional anti-platelet agents have limited efficacy under "hypershear conditions," whereas Dimethyl sulfoxide (DMSO) a polar aprotic solvent capable of modulating platelet membrane fluidity is effective in limiting shear-induced activation and aggregation [6] (Figure 4). We have applied our multiscale approach to model the mechanotransduction process driving the inhibition effect of the membrane fluidity modulation achieved by DMSO [6]. To the best of our knowledge this the first study of its kind.

CONCLUSIONS

Our multiscale approach offers a computationally feasible and highly resolved methodology for modeling platelet activation under shear conditions. Biophysical properties of the deformable platelets are accurately described down to the nanoscales. Hemodynamic stress is mapped on intra-platelet components. Phenomena of filopodia formation are mimicked and correlate with *in vitro* microchannel experiments. This model can be further employed to simulate other processes involved in platelet activation and thrombosis, offering a practical multiscale methodology for solving complex clinical problems at the juncture of biology and engineering.

ACKNOWLEDGEMENTS

This project was funded by NIH (NHLBI R21 HL096930-01, NIBIB Quantum U01EB012487, DB) and used computer resources at the National Supercomputing Center at Jinan (YD), and XSEDE computer resource award on TACC Stampede (TG-DMS140019, PZ).

REFERENCES

[1] Zhang, P, et al., *Cell Mol Bioeng*, 7:552-574, 2014.
[2] Zhang, P, et al., *J Comput Phys*, 2015, doi: 10.1016/j.jcp.2015.01.004.
[3] Pothapragada, S, et al., *Int J Numer Meth Biomed Engng*, 2015, doi: 10.1002/cnm.2702.
[4] Zhang, N, et al., *J Comput Phys*, 257, Part A: 726-736, 2014.
[5] Bluestein, D, et al., *Proceed. ASME 2nd Global Congress on Nanoengineering for Medicine and Biology, Nemb* 2013, 73-74, 2013.
[6] Valerio L, et al., *J Heart Lung Transpl*, 33:S143, 2014.

SHEAR ACTIVATED NANOPARTICLE AGGREGATE DELIVERY OF TISSUE PLASMINOGEN ACTIVATOR WITH ADJUNCTIVE ENDOVASCULAR BYPASS FOR REVASCULARIZATION IN ACUTE ISCHEMIC STROKE

Matthew J. Gounis (1), Netanel Korin (2), Miklos G. Marosfoi (1), Oktay Uzun (2), Erin T. Langan (1), Anne-Laure Papa (2), Olivia W. Brooks (1), Chris Johnson (2), Ajit S. Puri (1), Deen Bhatta (2), Mathumai Kanapathipillai (2), Ben Bronstein (2), Ajay K. Wakhoo (1), Donald E. Ingber (2,3,4)

(1) New England Center for Stroke Research
Department of Radiology
University of Massachusetts
Worcester, MA, USA

(2) Wyss Institute for Biologically Inspired Engineering
Harvard University
Boston, MA, USA

(3) Vascular Biology Program
Boston Children's Hospital and Harvard University
Boston, MA, USA

(4) School of Engineering and Applied Sciences
Harvard University
Boston, MA, USA

INTRODUCTION

Acute ischemic stroke (AIS) caused by an occlusion of a large intracranial vessel is associated with dismal clinical outcomes, making stroke the leading cause of disability in the United States. The recently published randomized clinical trial, MR CLEAN, has shown that stent-retriever mechanical thrombectomy (MT) alone or in combination with intravenous tissue plasminogen activator (tPA) leads to improved clinical outcomes as compared with standard medical therapy alone [1]. Although this technology represents an enormous advancement in the treatment of AIS, the majority of patients remained functionally dependent after 90-days, 88% of patients were not able to return to work, and 97% of patients retained a disability from the stroke. Poor outcomes following MT may be caused by futile recanalization in patients with large infarcts and unfavorable post-treatment events, such as early re-occlusion or reperfusion hemorrhage. In the latter two cases, vessel wall injury and inflammation with possible distal (brain parenchyma) involvement occur after MT procedures [2]. This observation is supported by results of histopathological exams from animal studies, where extensive endothelial damage was found after stent retriever usage [3]. Denudation of the endothelium results in a highly thrombogenic surface that facilitates local thrombus formation potentially serving as a nidus for the generation of distal emboli or vessel re-occlusion; in addition, it elicits an inflammatory response that theoretically worsens infarct progression. The combination of a less traumatic endovascular approach using temporary endovascular stent-bypass [4] and targeted thrombolytic drug delivery may help to avoid these complications and ultimately increase the potential for favorable outcomes.

Herein, we deploy a targeting strategy using a novel shear-activated nanoparticle aggregate (NPA) to deliver tPA[5]. tPA is adhered to the nanoparticles with targeted release dictated by the local hemodynamics. The high shear stress environment, necessary to release tPA, is achieved by temporarily opening a stent through the clot that creates a narrow flow channel.

To test this hypothesis, we used a previously published rabbit vascular occlusion model developed for studying new endovascular devices [6]. We assessed the combination of endovascular temporary stent-bypass technique with the targeted (shear stress-activated) intra-arterial NPA-tPA delivery system in a large vessel occlusion model. We report the rate of successful recanalization as the primary efficacy metric and vessel histology as the primary safety endpoint.

METHODS

After receiving institutional approval, 21 white New Zealand rabbits received bilateral common carotid artery (n=42 vessels) occlusions as previously described [6]. Briefly, bilateral common carotid arteries (CCA) were accessed and using external sutures, stenoses were applied in anesthetized animals. Endovascular access was achieved with insertion of a 6 Fr catheter in the common femoral artery. Thrombin-induced clots (10 mm length) were injected into each CCA through a 6 Fr catheter and became lodged at the site of the stenosis. After 45 minutes, the strictures were released and vessel occlusion was confirmed by digital subtraction angiography (Figure 1). Thereafter, vessel occlusion was randomly assigned to one of 6 treatments (n=7 vessels each): 1. Local infusion of 2 mg tPA, 2. Stent-bypass alone, 3. Stent-bypass with local infusion of 2 mg tPA, 4. Stent-bypass with local infusion of 20 mg tPA, 5. Stent-bypass with local infusion of NPA carrying an equivalent dose of 2 mg tPA, and 6. Stent-bypass with local infusion of NPA carrying an equivalent dose of 20 mg tPA. Infusions were administered for approximately 30 minutes. In all cases where a stent was deployed, the stent was resheathed at the conclusion of the experiment.

The primary efficacy endpoint was vessel recanalization assessed by digital subtraction angiography using the Thrombolysis in Cerebral Infarction (TICI) scale [7] applied by a blinded rater. Given evidence that TICI ≥2b is associated with improved clinical outcomes, we dichotomized scores at 2b. Vascular trauma was evaluated by cross-sectional histology.

RESULTS

The shear-activated NPA treatment delivering an equivalent dose of 20 mg tPA through a stent bypass was the only treatment to offer reproducible recanalization (TICI≥2b) in all cases, and generated complete recanalization in (TICI=3) in all but one of the experiments (Figure 2). But NPA delivering 2mg tPA through the endovascular bypass offered very high rates of recanalization (86% TICI≥2b) as well. The stent bypass with adjunctive delivery of free tPA (both 2 and 20 mg) provided successful recanalization in 71% of cases, but complete recanalization in less than half of the experiments. Stent bypass alone or local tPA infusion alone were not efficacious treatments.

Histological analysis revealed that damage to the vessel was restricted to the segment in which the stent was deployed, and confined to denudation of the endothelium (Figure 3). In our control group of intra-arterial tPA infusion only, endothelial denudation was limited and likely only along the path of the microcatheter. Gross pathological evaluation of the brain showed no gross or petechial hemorrhage.

DISCUSSION

Nanoparticle aggregates that are mechanically activated in high shear environments result in targeted release tPA to the surface of the embolus locally [5]. In this study, a stent-bypass was used to create a small central channel through an occlusive clot. In this model of large vessel occlusion representative in terms of arterial diameter and clot burden to a human middle cerebral artery stroke, the combination of the endovascular bypass with NPAs delivering 20mg of tPA produced high rates of complete revascularization (TICI=3) without evidence of intracranial hemorrhage. Moreover, NPAs delivering a lower dose of tPA (2 mg) resulted in successful recanalization (TICI≥2b) in nearly all cases (6 of 7 vessels) as well. Control groups deploying either selective intra-arterial free tPA or the endovascular bypass failed to reliably restore flow through the artery and distal vasculature. Free tPA infused through the endovascular bypass did provide TICI 2b or greater flow restoration in most cases, but only produced complete recanalization in less than half of the animals. Importantly, when dose matched, the stent-bypass with adjunctive NPA-tPA targeted delivery system was superior to free tPA in establishing successful recanalization.

In all cases using the temporary stent-bypass, focal injury to the endothelium was observed on histology that is confined to the vessel segment in which the device was deployed. Since the device is resheathed, rather than dragged through the vessel as is the case during thrombectomy, vascular injury is limited providing a potential reduction in the risk of re-occlusion, distal embolic events, and inflammation that may be a detriment to favorable clinical outcomes. Injury past the endothelium to the internal elastic lamina was rarely observed, and there was no vessel trauma beyond the tunica intima to deeper layers of the artery.

This novel nanotherapeutic drug targeting technology combined with modern endovascular techniques is readily translatable to the clinic where stent-based thrombectomy is a proven treatment for strokes caused by a large vessel occlusion. Reliable complete recanalization coupled with reduced vessel injury may translate to improved clinical outcomes.

ACKNOWLEDGEMENTS

This study was funded by the Wyss Institute for Biologically Inspired Engineering at Harvard University.

REFERENCES

[1] Berkhemer et al. *New Engl J Med*; 372:11-20, 2015.
[2] Power et al. *Stroke*, 45:2330-2334, 2014.
[3] Gounis et al. *Stroke*, 44:S7-S10, 2013.
[4] Kelly et al. *Stroke*, 39:1770-1773, 2008.
[5] Korin et al. *Science*, 337:738-742, 2012.
[6] Gounis et al. *J Neurointv Surg*, Suppl 1:85-89, 2013
[7] Higashida et al, *Stroke*, 34:e109-137, 2003.

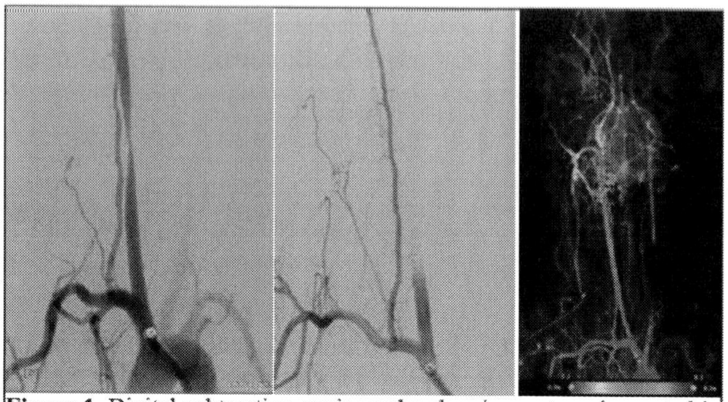

Figure 1. Digital subtraction angiography showing a stenosis created in the right common carotid artery (left) with subsequent occlusion upon clot introduction (middle). Following stent-bypass with NPA (2mg tPA) there is TICI 2b recanalization of the right carotid artery, whereas stent-bypass with 2 mg free tPA fails to revascularize the left carotid artery

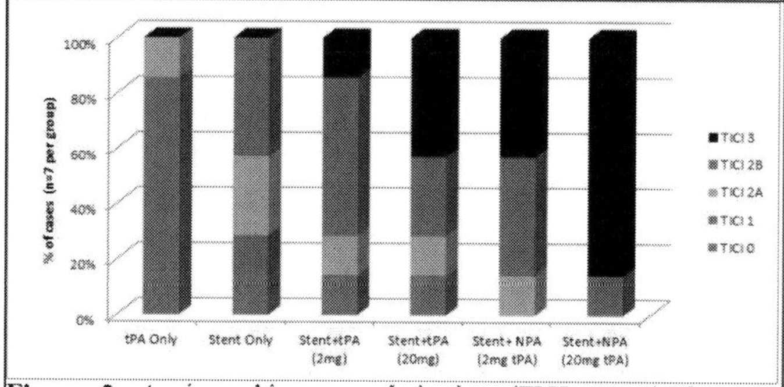

Figure 2. Angiographic revascularization (TICI scores) for each treatment.

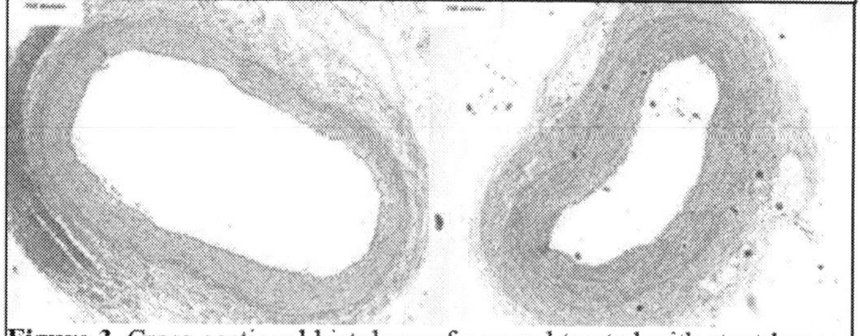

Figure 3. Cross-sectional histology of a vessel treated with stent-bypass and NPA infusion (20mg tPA, left) with disruption of the endothelium and intact internal elastic lamina. In the selective tPA infused vessels without stent-bypass, the endothelium remains intact (right, proximal to occlusion). (H&E stain, scale bar = 250μm)

SB^{3}C2015
Summer Biomechanics, Bioengineering and Biotransport Conference
June 17-20, 2015, Snowbird Resort, Utah, USA
SB^{3}C2015-304

EFFECT OF BICUSPID AORTIC VALVE MORPHOTYPE ON ASCENDING AORTIC HEMODYNAMICS: A COMPUTATIONAL STUDY

K. Cao, P. Sucosky

Department of Aerospace and Mechanical
Engineering
University of Notre Dame
Notre Dame, IN, 46556
USA

INTRODUCTION

The bicuspid aortic valve (BAV) is a valvular defect present in 1-2% of the general population. As compared to the normal tricuspid aortic valve (TAV) which consists of three leaflets, the most prevalent type-I BAV is characterized with one regularly sized leaflet and one larger leaflet. While 71% of type-I BAVs result from the fusion between the left- and right-coronary leaflets (LR subtype), 15% feature right- and non-coronary cusp fusion (RN subtype) and 3% present with non- and left-coronary cusp fusion (NL subtype). As compared to TAV patients, BAV patients have a much higher risk of developing secondary aortopathies such as aortic dilation.

While BAV aortopathy has been historically linked to a genetic predisposition, the abnormal BAV hemodynamics has emerged as a potential coincident etiology. Recent studies have revealed the existence of abnormal helical flow patterns in AA regions vulnerable to dilation [1] and the dependence of the local expression of dilation and thinning of the ascending aorta (AA) on the particular BAV leaflet fusion type. While those results collectively provide evidence for a possible role of hemodynamics in the pathogenesis of BAV aortopathy, the validation of this hemodynamic pathway requires the characterization of the local flow patterns and fluid wall shear stress (WSS) generated by each BAV subtype on the AA wall, which is still lacking. Therefore, the objective of this study was to quantify computationally the impact of different type-I BAV cusp fusions (i.e., LR, RN and NL) on AA hemodynamics and WSS using three-dimensional (3D) fluid-structure (FSI) modeling.

METHODS

<u>3D aorta model reconstruction:</u> The 3D aorta geometry was based on a human anatomy reconstructed from histological images obtained from the Visible Human Project (National Library of Medicine) and included the ascending aorta, the aortic arch (along with the proximal segments of the innominate, left common carotid and left subclavian arteries), and the thoracic descending aorta (Fig. 1).

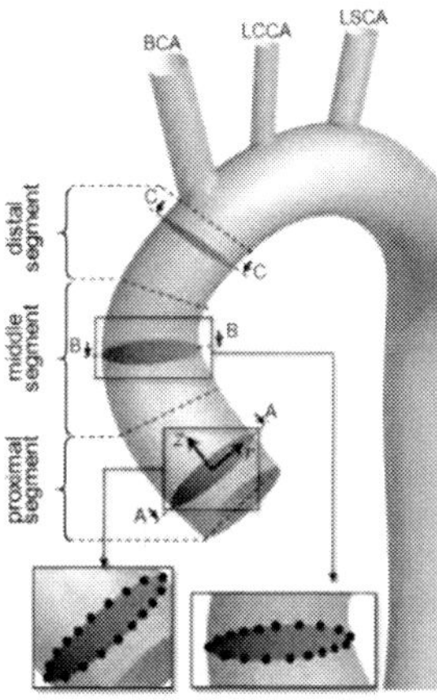

Figure 1: Aorta geometry

<u>Boundary conditions:</u> Four simulations were performed by imposing four different 3D pulsatile and spatially dependent flow velocity profiles (i.e., TAV, LR-BAV, RN-BAV and NL-BAV) at the inlet section of the AA. As a first approximation, the 3D pulsatile TAV and BAV flow velocity profiles prescribed at the inlet of the fluid domain were generated by extrapolating the 2D pulsatile velocity profiles predicted at the sinotubular junction by our previously published TAV and LR-BAV models [2]. Briefly, based on the near angular and near planar symmetry of the TAV and LR-BAV anatomy, respectively, the 3D TAV velocity profile was created by revolving the 2D profile about the inlet axis, while the 3D BAV velocity profile was obtained by extrapolating the 2D BAV velocity profile in the transverse direction as a parabolic profile satisfying the no-slip condition at the wall (Fig. 2a). The RN-BAV and NL-BAV velocity profiles were obtained by rotating the LR-BAV velocity profile by 120° and 240°, respectively, in order to reflect the different cusp fusion patterns (Fig. 2b). The boundary conditions at the four outlets of the aorta model (BCA, LCCA, LSCA and descending aorta) consisted of physiologic pressure waveforms.

Constitutive models: The aortic wall was treated as an incompressible, isotropic and linear elastic material with a Young's modulus of 2 MPa and a Poisson's ratio of 0.45. Blood was modeled as an incompressible Newtonian fluid, an acceptable assumption for large blood vessels in which shear rates typically exceed 100 s^{-1}.

FSI modeling approach: The 3D models were developed using the commercial software ANSYS (ANSYS Inc). The governing equations were solved in their arbitrary Lagrangian Eulerian (ALE) forms and consisted of the flow continuity and momentum equations, the aortic wall momentum equation and three coupling conditions enforcing continuity of displacements, continuity of velocities and compatibility of traction at the fluid-structure interface.

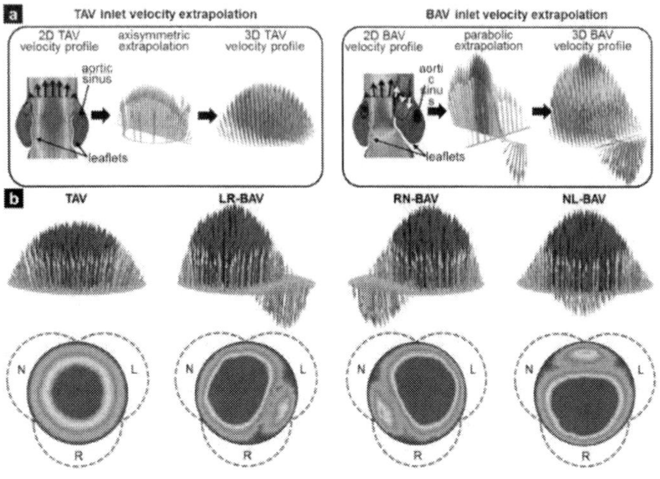

Figure 2. Aorta inlet flow conditions

Hemodynamic endpoints: Aorta hemodynamics was assessed in terms of streamline, velocity and WSS fields. The spatial variations of the velocity profile along the streamwise direction were investigated by characterizing the local velocity contour field over three cross-sections located 10 mm (proximal section), 35 mm (middle section) and 55 mm (distal section) downstream of the AA inlet (see Fig. 1). In addition, the angular distribution of the longitudinal WSS component (i.e., component acting in the direction of the main flow) was investigated in the proximal and middle AA sections at 18 equi-angularly spaced points (20° increment) along the wall circumference (see insets in Fig. 1).

RESULTS

Streamlines and total WSS distribution: In the TAV AA, the streamlines are essentially parallel to the aortic wall with only a few peripheral streamlines wrapping around the main flow (Fig. 3a). In contrast, the streamlines in the BAV AAs are marked by an extensive wrapping of the low-velocity peripheral streamlines around the main flow and their backward deflection toward the valve. In addition, the TAV AA velocity field remains nearly axisymmetric while the BAV AA velocity distribution exhibits more skewness whose direction depends on cusp fusion type. In all three BAV AAs, the intense flow helicity results in WSS overload (TAV AA: 34.9 dyn/cm^2; LR-BAV AA: 38.1 dyn/cm^2; RN-BAV AA: 43.1 dyn/cm^2; LR-BAV AA: 43.2 dyn/cm^2) (Fig. 3b). While those high WSS levels are localized to the distal segment of the LR-BAV AA, they diffuse over the entire wall of the RN-BAV and NL-BAV AAs.

Peak-systolic longitudinal WSS: In all models, the peak-systolic longitudinal WSS captured in the proximal AA reflects the characteristics of the inlet velocity profile (Fig. 3c). The WSS distribution is essentially uniform over the entire circumference of the TAV AA and more asymmetric in all BAV AAs. This asymmetry subjects the BAV AAs to a positive stress overload in the region impinged by the valvular jet and a negative stress in the diametrically opposite region. The flow development from the proximal to the middle AA section is accompanied by an overall increase in longitudinal WSS magnitude (Fig. 3d). At the same time, the curvature of the aortic wall tends to eliminate the effects of the initial discrepancies in jet skewness imposed at the AA inlet and to normalize the longitudinal WSS distributions among all models.

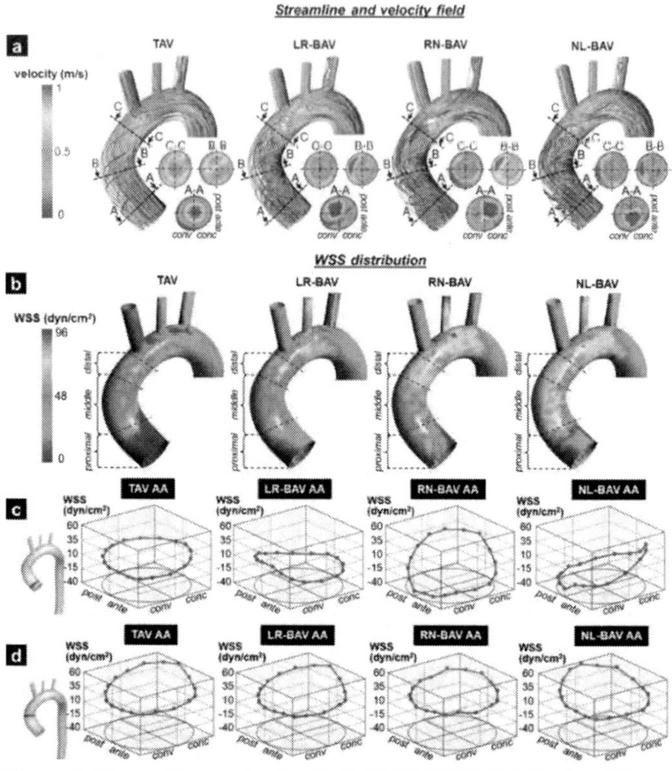

Figure 3: Streamlines and velocity field (a); WSS distribution (b); Local longitudinal WSS distributions at peak systole in the proximal section (c) and middle section (d).

DISCUSSION

In this study, we quantified the flow and WSS environments in a normal, non-dilated AA subjected to the flows of a TAV and three BAV morphotypes. The results demonstrate: 1) the existence of abnormalities in WSS magnitude and circumferential distribution on BAV AAs; and 2) the dependence of the locations of WSS abnormalities on the type of BAV cusp fusion. This flow characterization sheds new light on the potential role played by hemodynamic stresses in BAV disease and provides the mechanical data required for the rigorous assessment of the hemodynamic theory of BAV aortopathy. The FSI models could benefit from several improvements such as the implementations of more realistic anatomies and flow boundary conditions. The design of a unified 3D FSI model integrating a realistic aortic valve geometry featuring compliant leaflets in an aorta geometry is currently under way in our laboratory. This approach will eliminate the need for prescribing inlet flow conditions at the sinotubular junction and will effectively capture the dynamic interactions between the aortic root and the aortic wall.

REFERENCES

[1] Atkins,S et al., *Biomech Model Mechan,* 13(6):1209-1225.2014

[2] Chandra,S et al., *Biomech Model Mechan*, 11(7):1085-1096, 2012.

SB3C2015
Summer Biomechanics, Bioengineering and Biotransport Conference
June 17-20, 2015, Snowbird Resort, Utah, USA
SB3C2015-305

COMPUTATIONAL ASSESSMENT OF HEMODYNAMICS IN TRICUSPID AND BICUSPID AORTIC VALVES

K. Cao, P. Sucosky

Department of Aerospace and Mechanical Engineering
University of Notre Dame
Notre Dame, IN, 46556
USA

INTRODUCTION

The bicuspid aortic valve (BAV) is the most common congenital cardiac anomaly and has emerged as the first indication for surgical valve replacement due to its frequent association with calcific aortic valve disease (CAVD). As compared to the normal tricuspid aortic valve (TAV) which consists of three leaflets, the type-0 BAV consists of two equally-sized leaflets and two sinuses and the type-I BAV is characterized by asymmetry between a normal leaflet and a larger fused leaflet. Although CAVD also develops in the TAV, the calcification of the BAV is more rapid and severe. Historically, CAVD in BAV patients has been linked to the same genetic defect responsible for the BAV morphogenesis. However, recent studies have revealed the existence of abnormal hemodynamics in the BAV [1,2], which may contribute to the early and rapid development of CAVD. Despite recent evidence for the involvement of hemodynamic stress abnormalities in BAV disease [3], the investigation of causality between abnormal BAV stresses and pathological states requires the quantification of the local degree of hemodynamic abnormality experienced by BAV leaflets relative to a reference physiologic hemodynamic state, which remains largely unknown. Therefore, the objective of this study was to quantify and compare the regional WSS on TAV and BAV leaflets using fluid-structure interaction (FSI) modeling.

METHODS

Geometry and computational grid: The 3D aortic valve geometries (Fig.1) were designed in Solidworks. The TAV geometry consisted of three identical, equi-angularly spaced leaflets attached to the root of the aortic sinus. The type-0 BAV geometry consisted of two completely developed and symmetric leaflets and sinuses, the circumferential amplitude of a single leaflet and its sinus was 180°. The type-I BAV was created with one normal non-coronary leaflet and one larger leaflet reflecting the fusion between the two coronary leaflets. Consistent with previous morphologic studies, the circumferential amplitude of the non-coronary leaflet and the corresponding sinus equaled to 150°. The left- and right-coronary leaflets were assumed to be identical and the raphe was located along the free edge of these two leaflets. The dimensions of the aortic root, sinus and leaflets were obtained from published human valve data. The geometry was imported and meshed using tetrahedral elements in the commercial software ANSYS 15.0.

Figure 1. TAV, type-0 BAV and type-I BAV geometries

Boundary conditions: Valvular flow was simulated by imposing a physiologic transvalvular pressure waveform ranging between -94 and +6 mmHg as a traction condition at the inlet of the fluid domain. Leaflet and aortic wall surfaces in contact with blood flow were treated as fluid-structure interfaces satisfying the no-slip condition. Fixity conditions were prescribed at the inlet and outlet of the structural domain to prevent longitudinal motion. A frictionless contact condition was applied to the ventricular side of adjacent leaflets.

Constitutive models: The leaflet tissue was treated as a hyperelastic material with a three-parameter Mooney-Rivlin model.

The aortic wall was treated as an incompressible, isotropic and linear elastic material with a Young's modulus of 2 MPa and a Poisson's ratio of 0.45. Blood was modeled as an incompressible Newtonian fluid, an acceptable assumption for large blood vessels in which shear rates typically exceed 100 s^{-1}.

FSI modeling approach: An arbitrary Eulerian-Lagrangian approach with remeshing technique was implemented in ANSYS to design the valve FSI models. Briefly, the fluid (blood) domain was discretized using a moving grid fitted to the moving structure boundary (i.e., leaflets and aortic wall) and the motion of the fluid grid was accounted for by formulating the flow governing equations relative to the moving grid. At each time step, the fluid mesh was updated to conform to the moving solid boundary and the flow and structure equations were solved implicitly and iteratively until convergence was reached.

RESULTS

Global flow and leaflet dynamics: The TAV, type-0 BAV and type-I BAV models predicted leaflet and blood flow dynamics at a physiologic cardiac output of 4.2 L/min, 3.6 L/min and 3.5 L/min, respectively. Snapshots of the leaflet profile and flow velocity vector field computed at mid-systole (t = 100 ms), peak systole (t = 180 ms), early diastole (t = 300 ms) and mid-diastole (t = 450 ms) are shown in Figure 2. During systole (t=100 ms and 180 ms), the positive pressure drop generates central forward flow in the TAV and type-0 BAV but a skewed flow toward the non-coronary leaflet in the BAV. Additionally, under a similar pressure gradient, the predicted TAV orifice is wider than the BAV orifice, regardless of the BAV morphologies. During diastole (t=300 ms and 450 ms), inversion of the transvalvular pressure drop produces strong flow recirculation and vortical structures in the aortic sinus, which promotes leaflet coaptation and valve closure. While the TAV and type-0 BAV generates symmetrical vortices, the closure of the type-I BAV leaflets results in perturbed flow dynamics marked by increased vorticity on the aortic side of the non-coronary leaflet.

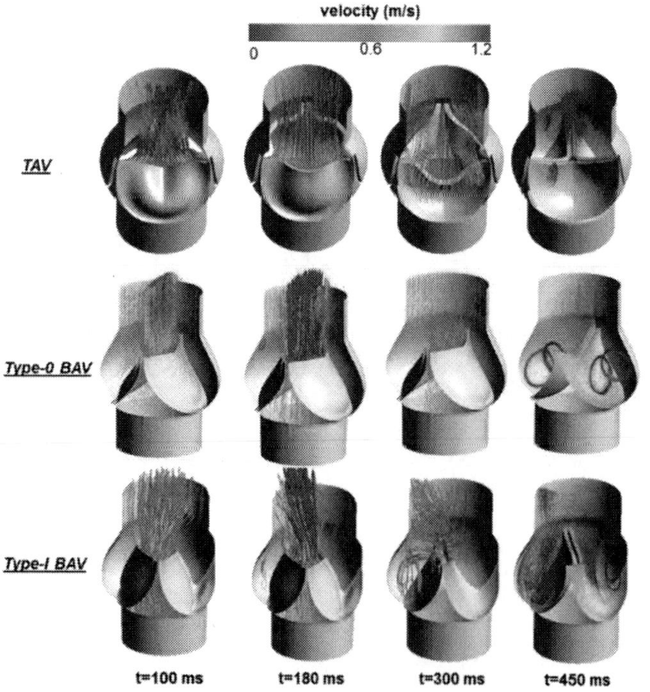

Figure 2. TAV and BAV velocity vectors

WSS distribution: Predictions of the total WSS distribution on the aortic and ventricular sides of the leaflet are shown in Figure 3. Regardless of the valve anatomy, TAV and BAV leaflets experience similar WSS environment marked by high magnitude during systole and low magnitude during diastole. In contrast, the WSS predicted on the aortic side of the leaflets exhibits significant variability between the different valve anatomies. At the beginning of diastole (t=300 ms), the flow vorticity and recirculation in the sinus region result in an overall increase in fibrosa WSS in all valve models. While high WSSs diffuse over the entire TAV leaflet, they remain concentrated in the belly region of the type-0 BAV leaflets and type-I BAV non-coronary leaflet. Additionally, the fibrosa WSS predicted on the type-I BAV fused leaflet is lower than that predicted on the TAV, type-0 BAV and type-I BAV non-coronary leaflets.

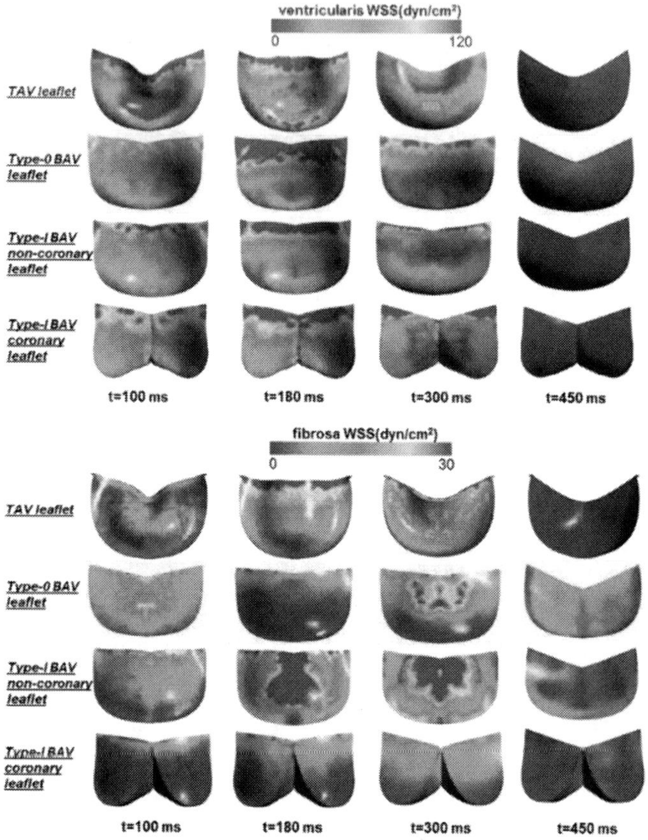

Figure 3. TAV and BAV leaflets WSS distribution

DISCUSSION

This study provided the first detailed WSS on the BAV leaflets and demonstrated the existence of abnormal WSS in terms of magnitude on the aortic side of BAV leaflets. These results will be critical for investigating the potential cause-and-effect relationship between the local hemodynamics and the local biology of the BAV.

REFERENCES

[1] Chandra, S., Rajamannan, N.M., Sucosky, P., *Biomech Model Mechan*, 11(7):1085-1096, 2012.

[2] Atkins, S., Cao, K., Rajamannan, N.M., Sucosky, P., *Biomech Model Mechan*, 13(6):1209-1225, 2014

[3] Sun, L., Rajamannan NM., Sucosky,P., *PLoS ONE*, 8(12):e84433, 2013

SB³C2015
Summer Biomechanics, Bioengineering and Biotransport Conference
June 17-20, 2015, Snowbird Resort, Utah, USA
SB³C2015-306

IN-VIVO KINEMATICS OF THE ASYMPTOMATIC HIP DURING DYNAMIC PIVOTING: FOUNDATIONS FOR THE EVALUATION OF FEMOROACETABULAR IMPINGEMENT

Penny R. Atkins (1,2), Niccolo M. Fiorentino (1), Michael J. Kutschke (2), Sara J. Fauver (2), Ashley L. Kapron (1), Christopher L. Peters (1), Stephen K. Aoki (1), Andrew E. Anderson (1-4)

(1) Department of Orthopaedics
University of Utah
Salt Lake City, UT, USA

(2) Department of Bioengineering
University of Utah
Salt Lake City, UT, USA

(3) Department of Physical Therapy
University of Utah
Salt Lake City, UT, USA

(4) Scientific Computing and Imaging Institute
University of Utah
Salt Lake City, UT, USA

INTRODUCTION

Femoroacetabular impingement (FAI) has been implicated as a primary cause of hip osteoarthritis [1]. FAI patients have pain and perceived limitations in hip range of motion (ROM), especially during flexion and internal rotation. Though hypothesized, we have yet to directly observe the mechanisms responsible for soft-tissue damage and reduced ROM in hips with FAI as it is a challenge to quantify dynamic in-vivo hip motion. With standard skin-marker-based motion analysis, soft-tissue artefact, inaccuracies in the estimation of the hip joint center, and poor consistency in the placement of skin markers lead to measurement errors. Dual fluoroscopy (DF) allows for quantification of in vivo hip joint motion without markers by tracking the position of the pelvis and femur bones. We have established the accuracy of DF of the hip to 0.5 mm and 1 degree, and have examined in-vivo hip kinematics during passively applied clinical exams using DF [2-4]. Ultimately, we hope to discover those mechanisms responsible for limited ROM and tissue damage in hips with FAI under dynamic loading. Toward this objective, the purpose of this study was to quantify hip joint kinematics in asymptomatic controls during a pivot maneuver that induced internal rotation.

METHODS

With Institutional Review Board approval, 17 volunteers without a history of hip pain or injury were recruited. Measures of the hip, including alpha angle and lateral center edge angle from an anteroposterior film screened for radiographic evidence of FAI or dysplasia. 7 subjects failed radiographic screening, leaving 10 subjects (5 male) to image with DF. The selected cohort had an average age, height, and weight of 23±2 years, 172.3±10.5 cm, and 62.0±9.7 kg, similar to the demographics of FAI patients. Volunteers were asked to complete a number of activities of daily living while being imaged under the DF system [3]. All trials were collected at 100 Hz; fluoroscopic energy settings were optimized for each subject. Two trials of each activity were recorded. Only the netural standing position and internal rotation pivot were analyzed for the work herein. The side to be imaged and order of activities performed were randomized.

Volumetric computed tomography (CT) images (Siemens SOMATOM 128 Definition) that included both hips and knees were acquired of each volunteer with 120 kVp tube voltage, 512 x 512 acquisition matrix, 1.0 mm slick thickness, 0.9-1.0 pitch, 250 mAs + Care-Dose tube current, and 300-400 mm field of view. The pelvis and proximal and distal ends of the studied femur were semi-automatically segmented from the CT image data using Amira (V5.6, Visage Imaging, San Diego, CA, USA).

Digitally reconstructed radiographs (DRRs) were generated by ray trace projection through the segmented CT image stack of each subject's pelvis and femur. The positions of the pelvis and femur were tracked in DF using semiautomatic registration of DRRs in each frame of the resulting DF video frame (Fig. 1) [4].

Landmarks on the pelvis and ipsilateral femur were defined to transform raw bone positions quantified by DF to joint angles. Using Pre- and PostView (FEBio, University of Utah, Salt Lake City, UT, USA), the articulating region of the femoral head and acetabulum was isolated automatically by principal curvature. Isolated regions were fit to a sphere to calculate the center of the acetabulum and femur.

Bone-to-bone distance was computed between the femur and pelvis for the standing position and the frame in which maximum rotation occurred during the pivot using PostView.

Due to the known anatomical differences of the pelvis and the understood variation in presentation of FAI in males and females [5], we analyzed kinematics and bone-to-bone distance by gender.

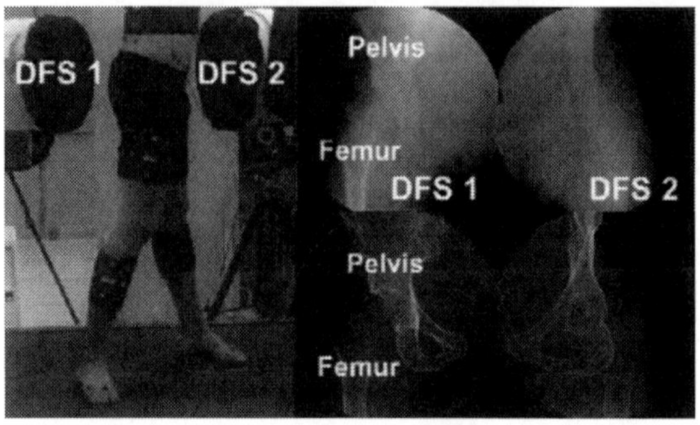

Figure 1: Imaging of a volunteer during a pivot (left), example frames from dual fluoroscopy (top right), and registration of DRRs using model based tracking (bottom right).

Maximum internal rotation was defined as the maximum positive rotation angle between the femur and the pelvis. Positional movement between the two centers quantified hip joint translation. Rotation angles and translations were referenced relative to static natural standing position to define ROM for the activity. Location of minimum distance between the two bones is described qualitatively, while the minimum distance between the two bones for each subject were used for analysis. Analysis of the differences in measurements between male and female populations were completed using a t-test.

RESULTS

Minimum bone-to-bone distance for the entire population was located inferiorly at 1.6±0.6 mm for the standing trial and more posterior-inferior at 1.4±0.6 mm at the frame of maximum rotation (Fig. 2). Though not statistically significant, females had a larger decrease in minimum distance between the standing and maximum rotation.

The mean ± standard deviation for all subjects at maximum rotation was 23.2 ± 6.7, -12.3 ± 5.7, and 38.6 ± 8.8 degrees for flexion, abduction, and internal rotation and -1.6 ± 1.0, -0.8 ± 1.5, -5.8 ± 2.1 mm for lateral, anterior, and superior translation (see Fig. 3 for representative findings). Overall, a total magnitude of 2.8 mm of translation was observed between the natural standing and maximum rotation positions. Maximum rotation was significantly larger in females than males (44.3±3.1 vs. 33.0±9.1 degrees, $p=0.030$). No other angular or translational data were significantly different between females and males.

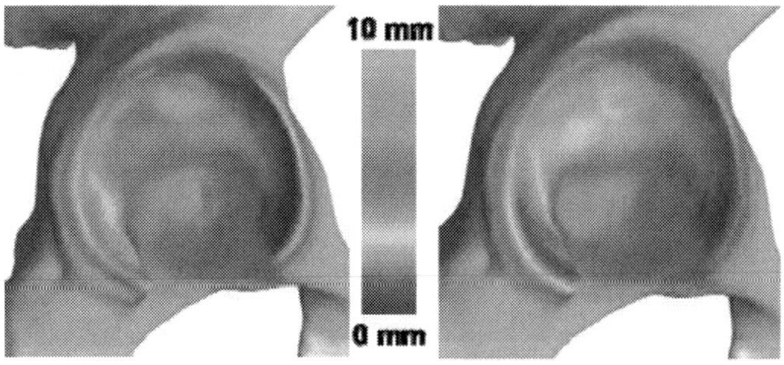

Figure 2: Distance between the femur and pelvis for a representative female volunteer for a standing (left) and maximum rotation (right) trial. Note that the femur was closest to the anterior acetabulum in the standing trial and closest to the posterior acetabulum at maximum rotation, which corresponded to an average posterior translation of 0.8 ± 1.5 from standing to maximum rotation in all subjects.

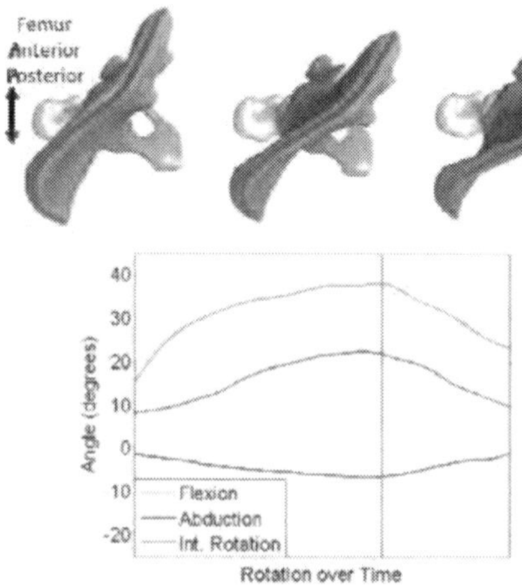

Figure 3: Three dimensional bone models with iliac crest highlighted of the pelvis rotating above the femur for initial rotation (top left), mid-rotation (top center), and maximum rotation (top right). Kinematic angles for the rotation over time, maximum rotation identified by vertical line (bottom).

DISCUSSION

To our knowledge, this is the first study to quantify dynamic hip motion during pivoting in living subjects. Our data suggest that motion of the hip is complex. Specifically, a pivot is expected to yield "pure" internal rotation, but we also observed substantial changes in hip flexion and abduction. Gender proved to be a significant factor in range of motion, which can likely be attributed to differences in the anatomical shape of the hip. With this logic, it is conceivable that patients with FAI, who have abnormal anatomy, will demonstrate significant changes in ROM. By visualizing how the 3D anatomy move temporally, the chosen protocol holds great potential to elucidate the pathomechanics of conditions such as FAI and dysplasia.

Minimum bone-to-bone distance plots agreed with calculated joint translations, showing a general trend of the minimum distance to be confined to the posteroinferior region at maximal rotation. As translations are calculated assuming a perfect sphere fit, aspherical femoral and/or acetabular anatomy may accentuate translational movement of the hip.

In conclusion, our findings for the internal pivot in an asymptomatic control group provide a valuable dataset of arthrokinematics necessary to study the pathomechanics of FAI. During the pivot maneuver, the hip joint experiences complex movements, incorporating six degree-of-freedom motion. We are currently processing data from other dynamic activities such as level and inclined walking on an instrumented treadmill. Having these data could prove fruitful to other orthopaedic studies (e.g. *in vitro* and *in silico*), where kinematic oversimplifications are often made.

ACKNOWLEDGEMENTS

Funding from NIH R01 EB016701 02 and the LS Peery Discovery Program in Musculoskeletal Restoration is acknowledged.

REFERENCES

[1] R Ganz, et al. *Clinical orthopaedics and related research,* 466: 264-72, 2008. [2] AL Kapron, et al. *Clinical Orthopaedics and Related Research,* 472: 3912-22, 2014. [3] AL Kapron, et al. *Journal of Applied Biomechanics,* 30: 461-70, 2014. [4] MJ Bey, et al. *Journal of biomechanical engineering,* 128: 604-9, 2006. [5] JJ Nepple, et al. *The Journal of Bone & Joint Surgery,* 96: 1683-9, 2014.

THE EFFECTS OF AXIAL COMPRESSIVE LOADING ON THE INTERSEGMENTAL ROTATION OF A VIRTUAL CERVICAL SPINE

Ryan J. Moss, James D. Kang, MD, Kevin M. Bell, PhD
Orthopaedic Robotics Laboratory, Department of Orthopaedic Surgery and Department of Bioengineering
University of Pittsburgh
Pittsburgh, Pa, United State

INTRODUCTION

Cervical spine degeneration is caused by the intervertebral discs wearing down. This can cause pain, stiffness, and eventually, cervical stenosis, causing compression of the spinal cord and the nerves.[1]

Currently, spinal fusion is the one of the main surgical technique to treat cervical spine degeneration. Spinal fusion accounts for 3.1% of all surgical operations.[2] The treatment helps to alleviate the pain and discomfort caused by the degenerated cervical vertebrae by immobilizing the damaged vertebrae. The process involves placing a bone graft between two or more vertebrae and securing it with either screws or wire.

In vitro biomechanical testing is important in the design and validation of surgical procedures. However, traditional testing methodologies are designed for comparative testing and have little input from in vivo data. Two separate studies have attempted to make in vitro biomechanical testing more accurately represent physiologic data by applying different loading conditions other than an unconstrained pure moment, which is the current gold standard.[3][4] The study by Miura focused on the analysis of a pure moment, while the study by Diangelo focused on applying a compressive loading to the cervical spine. A study in 2013 by Bell expanded the standard pure moment loading by combining it with axial loading and follower loading.[5]

However, currently the In Vitro compressive loading conditions required to mimic In Vivo loading are still unknown.

Objective

The objective of this study is to determine how the application of an axial compressive load changes the intersegmental kinematics of a virtual cervical spine model.

Hypothesis

The hypothesis is that a pure moment applied across the model will result in approximately equal (<+/- 1 STD) intersegmental rotations at each level. In addition, applying a moment in the presence of an axial compressive load will violate the pure moment assumption and will result in unequal intersegmental rotations (>+/- 1 STD) at each level. Finally, varying the line of action of the axial compressive load will enable the intersegmental kinematics to be differentially altered.

MATERIALS AND METHODS

CT scans of fresh-frozen human cervical cadaveric specimens were created and segmented in Mimics (Materialise, Leuven) and masks were created of each of the vertebrae. These masks were then smoothed and reconstructed into 3D models. The models were exported in STL format and imported into Simulink (Mathworks Inc. MA).

The properties of the ligaments were implemented in the model using a study by Hartman et al. in 2009.[6] Hartman et al.'s study collected data on the range of motion (ROM) when 2.5Nm of torque was applied while different ligaments were cut. The different ligament stiffnesses and joint stiffness were calculated by using the change in ROM resulting ROM from the sectioned structures. The resulting ROM was then compared to the ROM of the joint system with single ligaments and calculated the stiffness of the ligaments until the whole joint system was recreated.

A spherical joint was used to model the intervertebral disc with a different stiffness for each degree of freedom, allowing for pitch, roll, and yaw.

A flexion moment of 2.5Nm was applied to the model at the center and the resulting ROM (in degrees) was recorded. As seen in Figure 1, the attachment points for each of the ligaments is visualized with a tiny sphere. Secondly a 100 N axially compressive load was applied to the top vertebrae down through the center of rotation and the simulation was ran a second time. Lastly, the compressive loading was rotated 10, 20, and 30 degrees respectively about the flexion/extension angle, centered at the top segment joint.

RESULTS

The ROM of the model for each joint system was between 10.5 degrees and 12.5 degrees when only the pure moment was applied to the model. The standard deviation across all segments was 0.66.

Pure Moment	C3-C4 Joint	C4-C5 Joint	C5-C6 Joint	C6-C7 Joint
Angle (Degrees)	12.2811	10.6884	12.0152	11.2047

TABLE 1: RESULTING DATA FROM THE PURE MOMENT SIMULATION

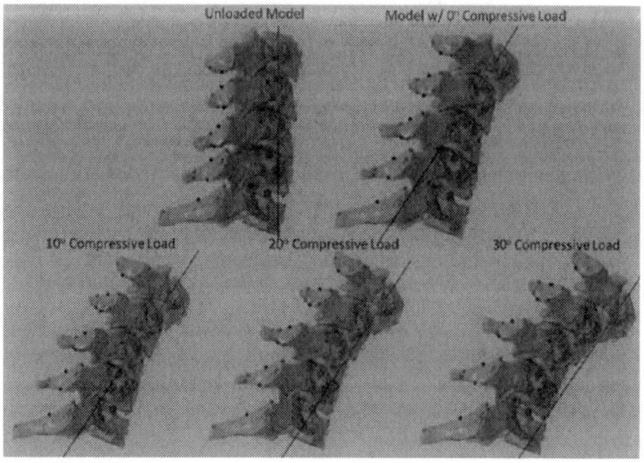

FIGURE 1: UNLOADED LATERAL PHOTO OF THE MODEL WITH MARKERS FOR EACH OF THE LIGAMENTS (TOPLEFT) MODEL WITH THE COMPRESSIVE LOADING AT 0 (TOPRIGHT) 10, 20, AND 30 DEGREES (BOTTOMLEFT, BOTTOMCENTER, BOTTOMRIGHT)

Applying the vertical compressive load disrupted the pure moment and the segments no longer had equal intersegmental rotations. The standard deviations of the angles were 2.34, 1.69, 1.42, and 1.70 for the model with the compressive loading at 0°, 10°, 20°, and 30° respectively. The difference of intersegmental rotation angles from the pure moment simulation to the compressed simulations was calculated and compared to the average In Vivo data presented in the Miura study.

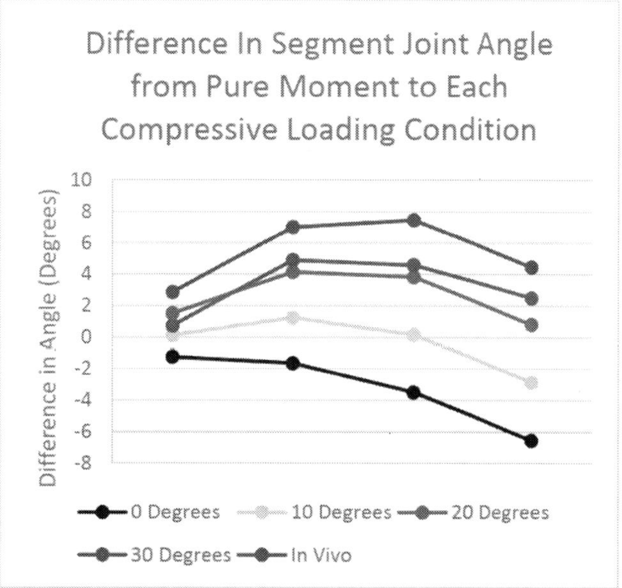

FIGURE 2: GRAPH OF ANGLE DIFFERENCE DATA FROM THE SIMULATION OF THE MODEL WITH 2.5NM OF TORQUE TO THE MODEL WITH APPLIED MOMENT AND 100 N OF COMPRESSIVE LOADING AT 0, 10, 20, 30 DEGREES

DISCUSSION

After the application of the pure moment, the resulting angles for each joint segment are within 1 standard deviation of the average In Vivo data. This is consistent with the hypothesis that applying a pure moment across the model with have approximately equal intersegmental rotations as compared to in vivo studies.

The compressive loading either increased or decreased the range of motion depending on the location and angle of the force, which matches the hypothesis. The compressive loading increased the standard deviation by at least 0.76 for the smallest case. The rotation of the compressive loading allowed it to be applied through both the C3-C4 joint and the C6-C7 joint. This resulted in intersegmental rotations that are distributed similarly to In Vivo intersegmental kinematics. As seen in figure 2, the average In Vivo results very closely fit the angle differences for the compressive loading rotated 20 degrees. However, this study only used 1 specimen and results may be atypical. Future work will need to be completed on additional specimens in order to confirm these findings.

CONCLUSION

With further development of the model, it can be used to study the effect of different compressive loading methods on intervertebral kinematics. The ultimate goal is to have the model be able to determine the compressive loading methods that best replicate in vivo loading. Possible future direction is to make the model comparable to in vivo kinematics, allowing more in-depth studies of the different medical procedures that help to reduce cervical spine degeneration such as cervical spine fusion.

Overall, the model will help to understand the forces acting on the human cervical spine, which will lead to better medical devices and procedures.

ACKNOWLEDGMENTS

I would also like to thank Dr.Richard Debski, the rest of the Orthopaedic Robotics Laboratory, the Department of Orthopaedic Surgery, and the Swanson School of Engineering. I would like to thank the Albert B. Ferguson fund of The Pittsburgh Foundation for providing financial support for this study.

REFERENCES

1. *Vokshoor A. (February 14, 2010). "Spinal Stenosis". eMedicine*
2. *Weiss AJ, Elixhauser A, Andrews RM. (February 2014). "Characteristics of Operating Room Procedures in U.S. Hospitals, 2011." HCUP Statistical Brief #170. Rockville, MD: Agency for Healthcare Research and Quality.*
3. *Miura, T., M.M. Panjabi, and P.A. Cripton, A method to simulate in vivo cervical spine kinematics using in vitro compressive preload. Spine (Phila Pa 1976), 2002. 27(1): p. 43-8.*
4. *DiAngelo, D.J. and K.T. Foley, An improved biomechanical testing protocol for evaluating spinal arthroplasty and motion preservation devices in a multilevel human cadaveric cervical model. Neurosurgery Focus, 2004. 17(3): p. E4.*
5. *Bell et al, Optimization of In-Vitro Compressive Loading Parameters in the Cervical Spine to Mimic In-Vivo Kinematics, 2013*
6. *Hartman, Detailed Analyses of the Components of the Posterior Column in a Distractive-Flexion Injury Model , 2009*

SB3C2015
Summer Biomechanics, Bioengineering and Biotransport Conference
June 17-20, 2015, Snowbird Resort, Utah, USA
SB³C2015-309

ACCOUNTING FOR CLINICAL DATA UNCERTAINTY IN MULTISCALE NUMERICAL SIMULATION OF SINGLE VENTRICLE PALLIATION SURGERY

Daniele E. Schiavazzi (1), Tain-Yen Hsia (2) and Alison L. Marsden (1), for the Modeling of Congenital Hearts Alliance (MOCHA) Investigators

(1) Cardiovascular Biomechanics Computation Lab
MAE Department
University of California, San Diego
La Jolla, California 92093

(2) Great Ormond Street Hospital for Children
UCL Institute of Cardiovascular Science
London WC1N 3JH, UK

INTRODUCTION

Cardiovascular simulation has shown potential value in the clinical decision-making process, providing a physically consistent framework to assess changes in blood flow and pressure produced by physiological and surgical alterations. For patients suffering from severe congenital pathologies, such as single ventricle heart disease, these models provide improved understanding of how competing surgical choices or patient specificity may affect clinical outcomes [1].

State-of-the-art predictions are provided by multiscale numerical models combining three dimensional anatomic models (e.g., 3D discrete models requiring a finite element NS solver) and a lumped parameter boundary circulation network. Making an analogy with electrical circuits in which flow and pressure drop replace current and voltage, a system of ordinary differential equations (ODEs) consisting of resistance, compliance and inductance elements is derived. Despite the significant recent advances in the numerical treatment of these problems [2], cardiovascular predictions are almost uniformly presented in terms of deterministic numerical simulation, i.e., solution of computer models where inputs are fully characterized without accounting for the variability in clinical data, material parameters or physiological response to surgery.

Our objective is therefore to develop methodologies to account for uncertainty in cardiovascular simulations, enriching deterministic results with measures of confidence that will ultimately increase their clinical impact [4]. We focus, in this work, on how clinical data uncertainty affects numerical predictions and consider the problem of tuning physiological boundary conditions (e.g., vessel resistance) and full stochastic circulation models using patient-specific data (i.e., average, maximum or minimum values) for blood flow and pressure.

FIGURE 1. Distribution of standard deviation of blood velocity for pre-operative and post-operative models of stage II Fontan palliation surgery. Significant variance in the velocity downstream the Blalock-Taussig (BT) shunt can be seen in the pre-operative model (left).

METHODS

We first consider the prediction of post-operative physiology in the cavo-pulmonary connection in stage II single ventricle palliation. Distributions of outlet resistance for left and right pulmonary arteries are inferred to match clinical variability (from patient-specific clinical data and literature review) in pre-operative flow split ratio and average pulmonary pressure. Principal boundary resistance modes are estimated using adaptive Markov chain Monte Carlo (MCMC) [3] on a Kriging approximant of the resistance-flow and resistance-pressure relationships produced by the three-dimensional finite element model, thus minimizing the associated computational cost. We use sparse grid stochastic collocation to propagate the random resistance forward to the result of multiscale models.

The problem of tuning a full stochastic boundary circulation model based on uncertain clinical measurements is considered next. Due to typical overparameterization of these models, conditions for identifiability are first assessed, by analysis of the Fisher information matrix rank and using systematic Maximum a Posteriori (MAP) estimation with random initial parameters. We then employ adaptive, multiple chain MCMC to sample from the joint distribution of the circulation model parameters. To improve estimation in presence of non identifiable parameter combinations we gather prior knowledge on the heart model parameters from an isolated heart model subject to prescribed aortic and venous flows.

The anatomical model is reduced to a pressure-flow relationship between inlets/outlets and included in the circulation model so that the tuning operations can be performed at a significantly reduced cost.

RESULTS

Using the procedures discussed in the previous section, we are able to identify the variability in local blood velocity over a range of boundary resistances consistent with the clinical measurement uncertainty. A significant variability is observed downstream of the BT shunt (Figure 1); this identifies regions where the precision of MRI velocity acquisitions maybe affected.

The effectiveness of parameter learning through adaptive MCMC is illustrated in Figure 2 for a stochastic circulation model of stage I Fontan palliation with 44 parameters, one systemic and one pulmonary loop. In particular, it can be seen how the aortic compliance is accurately determined for all the patients with a significant reduction in variance respect to its prior range. In addition to a formal study of identifiability, this shows how the estimated marginal distributions and correlations reveal which combination of parameters are effectively learned from the data.

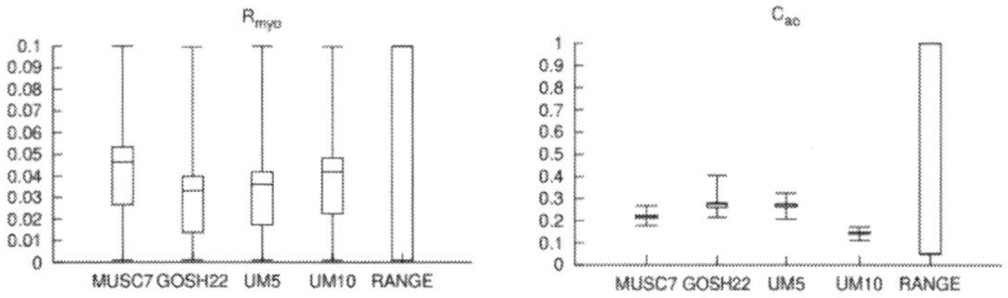

FIGURE 2. Box wiskers representation of circulation model parameter distributions R_{myo} and C_{ao} (myocardial resistance and aortic compliance, respectively) obtained using MCMC from uninformative prior knowledge ("RANGE") in four single ventricle patients.

Figure 3 illustrates pressure and flow time histories resulting from optimal MAP parameter estimates with 95% confidence intervals.

DISCUSSION

This work demonstrates how inclusion of clinical data uncertainty in cardiovascular simulation results is possible while maintaining reasonable computational cost. It also shows the effec-

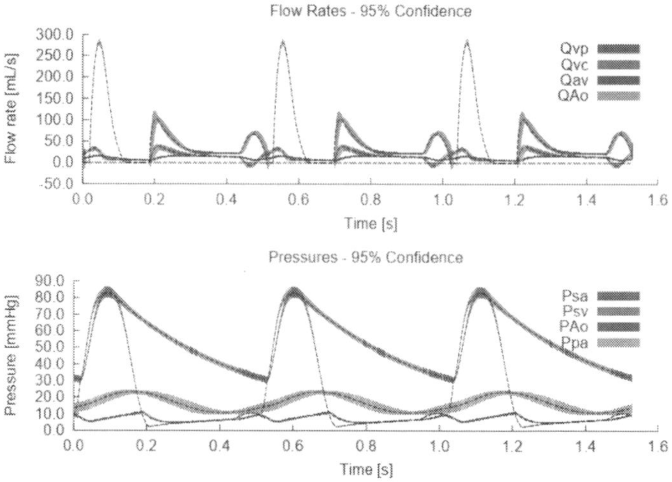

FIGURE 3. Pressure-flow curves obtained for optimal MAP parameter estimates and superimposed 95% confidence intervals.

tiveness of a systematic identifiability study followed by multilevel Bayesian estimation with adaptive MCMC in reducing non identifiable combinations through elimination of unimportant parameters and specification of prior knowledge.

Full multiscale model tuning is achieved by condensation of anatomical models through inlet/outlet flow-pressure relationships in a form that can be easily integrated within existing ODE solvers.

Other than the effects of clinical data uncertainty, as considered in this study, significant variability is also present in the vessel wall mechanical response, anatomy and physiological response to surgery. These aspects also affect numerical predictions and will be addressed in future studies.

ACKNOWLEDGEMENTS

This work was supported by American Heart Association Grant #15POST23010012 (Daniele Schiavazzi) and by the Leducq Foundation as part of a Transatlantic Network of Excellence for Cardiovascular Research, a Burroughs Wellcome Fund Career award at the Scientific Interface, NSF CAREER OCI-1150184 (Alison Marsden) and used computational resources from the Extreme Science and Engineering Discovery Environment (XSEDE), supported by National Science Foundation grant number ACI-1053575. We also acknowledge the open source SimVascular project at www.simvascular.org.

REFERENCES

[1] Corsini, C. et al., Comput Methods Biomech Biomed Engin, 17(14):1572-1589, 2014.

[2] Takizawa, K. et al., Arch Comput Method E, 19(2):171-225, 2012.

[3] Vrugt, J.A. et al., Int J Nonlinear Sci, 10(3):273-290, 2009.

[4] Schiavazzi, D.E. et al., submitted for review, 2014.

USE OF KARTOGENIN TO AUGMENT THE TENDON-BONE TUNNEL HEALING

Y. Zhou (1, 2), J. Zhang (1), G. Zhao (1), JH-C. Wang (1)

(1) MechanoBiology Laboratory, Department of Orthopaedic Surgery
University of Pittsburgh School of Medicine
Pittsburgh, PA, USA

(2) Joint Surgery and Sports Medicine Department
Shanghai Changzheng Hospital, Second Military Medical University
Shanghai, China.

INTRODUCTION

The normal tendon-bone junction is a strong structure protected by the *fibrocartilage transition zone*. This crucial tendon feature allows a gradual transition of mechanical forces between tendon and bone, thus decreasing stress effects, improving the strength of tissue bonding, and decreasing the chances of graft failure. Unfortunately, healing of the tendon-bone interface is slow, and the healed junction often lacks the transition zone. For example, a previous study on human patients used MRI and biopsy to show that even years after ACL reconstruction, patients had no fibrocartilage zone regeneration [1]. This and other studies show that surgical repair alone does not restore the unique protective mechanism provided by the fibrocartilage transition zone and, as a result, their clinical outcomes are compromised.

In recent years, great strides have been made in a pursuit to augment tissue healing. A small heterocyclic compound called kartogenin (KGN) was found by Johnson *et al.* [2] to stimulate endogenous differentiation of stem/progenitor cells into chondrocytes. In a recent study, we showed that when KGN was injected into experimentally injured rat Achilles tendon-bone junctions, wound healing in the TBJs was enhanced, as evidenced by the formation of extensive cartilage-like tissues [3]. Therefore, in this study, we aimed to determine the effect of KGN treatment on tendon-bone tunnel healing.

METHODS

Preparation of PRP and KGN – Blood was drawn from Sprague–Dawley rats to prepare PRP using a two-step centrifugation method. In the first step, the blood was centrifuged at 2,300g with the centrifugation time varying based on the blood volume (10 s per mL). The middle buffy-layer containing PRP was collected and centrifuged again at 2,200g and the supernatant was collected as the PRP solution. Platelet concentration in PRP was measured and adjusted to 3 times over the baseline platelet concentration in whole blood. KGN was prepared at the concentration of 100 μM in PRP solution to obtain the KGN + PRP solution. Thrombin, at the concentration of 1 kU/mL, was used as the PRP activator.

Surgical procedure – 18 female Sprague–Dawley rats (250g) were used for the *in vivo* study. The protocol for using rats was approved by the University of Pittsburgh IACUC. First, a surgical procedure was performed on the rats to create tendon-bone tunnel healing models. Briefly, a 10mm longitudinal skin incision was made on the medial side of the Achilles tendon and the fascia was cut longitudinally. One bundle of the Achilles tendon was dissected from the calf muscle with the proximal part of the bundle and calcaneal insertions left intact. A 1.5 mm hole was drilled at the distal end of tibia from posterior to anterior. The tendon was pulled through the tunnel and sutured to the anterior soft tissue with resorbable suture (Polysorb 4-0). Ankle joint was fixed at 90 degrees. The skin was closed with resorbable sutures.

Injection of PRP and KGN – After surgery, rats were randomized into three groups, which immediately received the following injections in their tendon-bone-tunnels. Group A received a mixture of 50μl KGN + PRP solution + 5μl thrombin solution into the tendon-bone tunnel interface; Group B was given only 50μl PRP solution + 5μl thrombin solution; and Group C was left untreated. Then, rats were sacrificed 4 and 8 weeks later for histological analysis. During the experiments cage activity was allowed for all rats.

Histological analysis – At each time point, three rats from each group were dissected and the tendon samples subjected to histological analysis. The whole tibia with the inserted tendon were harvested, fixed in 10% formalin for 24 hrs at room temperature, and decalcified in 10% EDTA. Decalcification was verified by needle testing of the specimens until they were free of calcium. Then, the tissues were embedded in paraffin and sectioned to 6 μm thickness in the sagittal plane. Finally, the sections were stained with hematoxylin and eosin (H&E), Safranin O and Fast green, and observed microscopically.

RESULTS

All animals were in good condition after surgery and complications were not encountered.

In the PRP+KGN group, the tendon-bone junction stained extensively with Safranin O at 8 weeks post treatment indicating the successful formation of the transitional zone at the tendon-bone interface **(Fig. 1A)**. Safranin O staining also indicates that the new transitional zone mainly consisted of cartilage-like tissues. In contrast, the transitional zone at 4 weeks stained less with Safranin O, suggesting that less matured cartilage-like tissues were formed (data not shown). Additionally, in the PRP only group, even at 8 weeks post treatment, no apparent tendon-bone interface was observed **(Fig. 1B)**. Finally, in the untreated group, large gaps between the tendon and bone tunnels remained at 4 and 8 weeks post surgery (data not shown).

DISCUSSION

Enhancing the tendon-bone tunnel healing possess a significant challenge to physicians much like the ACL reconstruction procedure. Using an animal model of tendon-bone tunnel healing, we showed in this study that KGN promotes the formation of "cartilage-like" interface between the tendon graft and bone tunnel. This result suggests that the delivery of KGN into the tendon-bone interface could be a promising approach to augment the tendon-bone interface healing. KGN may have a unique advantage when compared to many tissue engineering approaches used to heal the tendon-bone interface; that is, the application of KGN can be a convenient cell-free therapy that may be used with ease in clinics. Future research is required to determine the optimal KGN dosage regimens, the optimal delivery method, and to evaluate the effects of KGN-scaffold composites on tendon-bone interface healing.

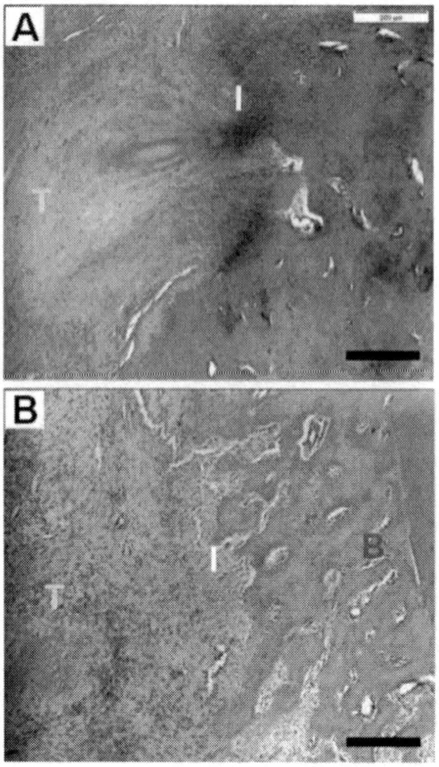

Fig. 1 PRP+KGN treatment promotes the formation of the transitional zone between the tendon graft and bone tunnel. **A**. 8 weeks after PRP+KGN treatment (T- tendon; B - bone; and I - interface between tendon and bone). Note that the heavily stained Safranin O regions indicate the extensive formation of the transitional zone (I). **B.** No apparent interface was seen in the group treated with PRP alone at 8 weeks. Bar: 100 μm.

ACKNOWLEDGEMENTS

This work was supported in part by the NIH awards AR061395 and AR060920 (JHW).

REFERENCES

[1] Liu SH, Hang DW, Gentili A, Finerman GA: MRI and morphology of the insertion of the patellar tendon after graft harvesting. *J Bone Joint Surg Br* 78(5):823-826, 1996.

[2] Johnson K, Zhu S, Tremblay MS, Payette JN, Wang J, Bouchez LC, Meeusen S, Althage A, Cho CY, Wu X *et al*: A stem cell-based approach to cartilage repair. *Science* 336(6082):717-721, 2012.

[3] Zhang J, Wang JH: Kartogenin induces cartilage-like tissue formation in tendon-bone junction. *Bone research* 2, 2013.

A PARAMETRIZED MODEL OF THE LAMINA CRIBROSA FOR STUDYING OXYGEN TRANSPORT

Fabian A. Bräu (1, 2), Ian C. Campbell (1, 3), Baptiste Coudrillier (1), C. Ross Ethier (1, 3)

(1) Wallace H. Coulter Department of Biomedical Engineering, Georgia Institute of Technology and Emory University, Atlanta, GA

(2) Institute for Computational Mechanics, Technical University Munich, Munich, Munich, Germany

(3) Atlanta VA Medical Center, Decatur, GA

INTRODUCTION

Glaucoma is the second most common cause of blindness [1] and is often associated with elevated intraocular pressure. The disease is characterized by slow and irreversible damage to the retinal ganglion cell (RGC) axons, thought to be initiated in the lamina cribrosa (LC). The LC is part of the optic nerve head (ONH) and fills the scleral canal with a fine porous structure, through which the RGC axons exit the eye. Capillaries, located in the collagenous beams of the LC, supply the RGC axons with oxygen. It has been hypothesized that a cascade of vascular, mechanical and biochemical events cause the damage and eventual loss of RGC axons in the LC. However, the exact mechanism(s) by which elevated intraocular pressure lead to vision loss are not fully understood. Although there are a number of finite element models which investigate the biomechanics in the LC, there is a lack of computational models evaluating oxygen transport in the LC. Here we create a parametrized model of the LC to study factors affecting oxygen transport from the capillaries to RGC axons.

METHODS

An idealized cross-section of the LC [2] was first used to create a parametrized 3D model of a subregion of the LC (Fig. 1). A hexagonal shape was selected because it allowed the creation of a model which matched morphological measurements (e.g. the area occupied by capillaries [3]) and because it permitted tessellation of single geometries to create a larger model. Only the components of the LC expected to influence oxygen transport were included in the model. The exact size of the different tissue constituents [4] and the mass transfer characteristics were controlled by a number of input parameters (Table 1). COMSOL, in combination with MATLAB, was used to generate models and simulate oxygen mass transport. The entire process was automated to allow rapid parametric studies, which were conducted to determine the influence of varying RGC axon bundle diameter on the oxygen tension in the LC.

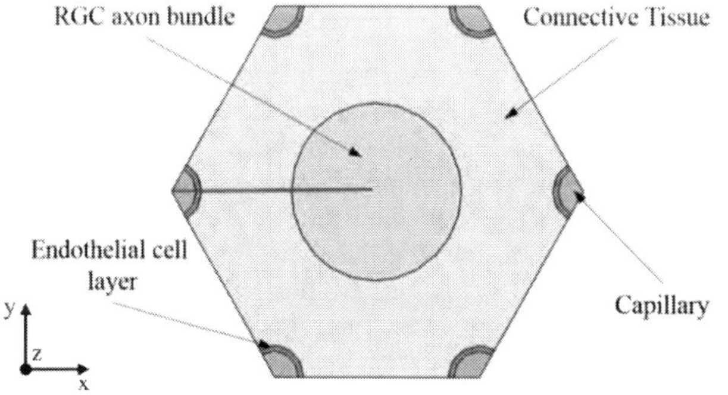

Figure 1: Model showing the anterior face of a small region of the LC, consisting of capillaries (red), endothelial cells (grey), connective tissue (yellow), and RGC axon bundle (green). The length of the geometry in the z-direction is the anterior-posterior thickness of the LC.

We then used micro-computed tomography (μCT) images of a porcine LC to calculate the volume fraction of neural and connective tissue (ConTis) in different subregions of the LC [4]. A parametrized model (Fig. 1) was created for each subregion, with local RGC axon bundle diameter adapted to the volume fractions of neural and ConTis as determined from the μCT scan. By tessellating these geometries, a specimen-specific model of an entire porcine LC was built (Fig. 2).

Number of capillaries	6
Capillary diameter [μm]	10
Endothelial cell (EC) layer thickness [μm]	1
Connective tissue (ConTis) thickness [μm]	36 – 54
RGC axon bundle diameter [μm]	30 – 48
Diffusion coefficient in blood [cm^2/s]	16.2×10^{-6}
Diffusion coefficient in EC layer [cm^2/s]	1.96×10^{-6}
Diffusion coefficient in ConTis [cm^2/s]	8.1×10^{-6}
Diffusion coefficient in RGC axon bundle [cm^2/s]	20×10^{-6}
RGC axon O_2 consumption rate of [$mol/(m^3s)$]	0.0322
Blood oxygen tension entering the LC [mol/m^3]	0.0915
Blood velocity [mm/s]	2
Capillary blood flow velocity profile (z direction)	Poiseuille
LC thickness [μm]	500

Table 1: Input parameters used for the parametric study with varying RGC axon bundle diameter. Diffusion coefficients are for oxygen.

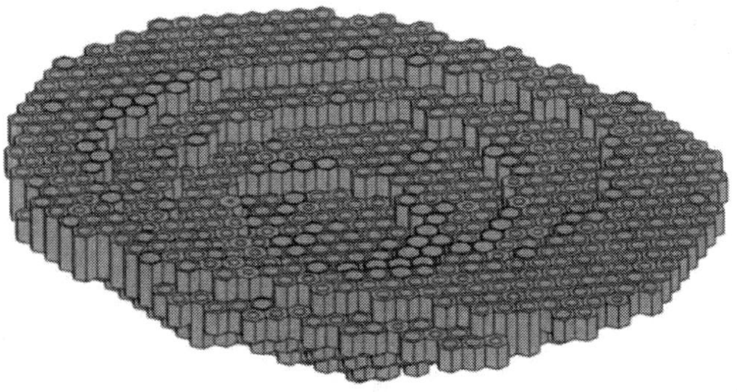

Figure 2: Model of a whole porcine LC; Created with the use of segmented μCT images.

RESULTS

Fig. 3 shows oxygen concentration profiles along the red line shown in Fig. 1, for varying RGC axon bundle diameters (30μm to 48μm). The oxygen concentration decreases continuously from the capillary to the RGC axons. For each study, the largest concentration drop occurs in the ConTis, which can be explained by the large thickness and the relatively small oxygen diffusion coefficient in this tissue component. Comparing the concentration profiles for bundle diameters of 30 μm and 48 μm shows that there is a large difference in the resulting oxygen concentration in the center of the RGC axons. This is because the RGC axons are the only tissue in the idealized model consuming oxygen, and because they consume oxygen at a relatively high rate.

We were able to successfully use a μCT scan of a porcine LC to create a detailed specimen-specific model (Fig. 2). Because of the large number of elements which are necessary to mesh the whole structure, parametric studies are ongoing to examine mass transfer processes in this large-scale model.

DISCUSSION

By using parametric studies of an idealized 3D model of a subset of the LC, we conclude that oxygen tension depends sensitively on RGC axon bundle diameter, RGC oxygen consumption rate and connective tissue thickness. This suggests that connective tissue thickening, known to occur in glaucoma, could lead to relative hypoxia in RGC bundles. As impaired vascular perfusion has already been suggested as a contributing factor in glaucoma, these two processes could work together to accelerate RGC damage in the end stages of glaucoma. It is clear that lamina cribrosa ConTis architecture strongly influences both structural and mass transfer effects, and future studies, including those with more specimen-specific models, will examine the interplay of these effects.

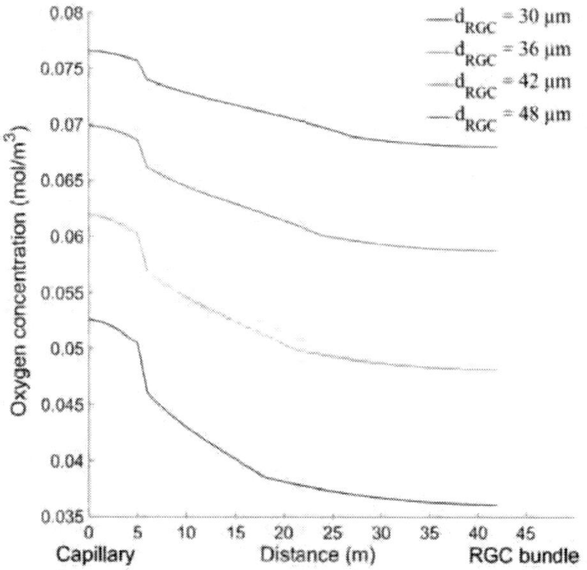

Figure 3: Oxygen concentration profiles (at z = 200 μm) from the center of a capillary (left) to the center of the RGC axon bundle (right) for different bundle diameters (d_{RGC}). Refer to red line in Figure 1 for path along which oxygen tensions are plotted.

ACKNOWLEDGEMENTS

We gratefully acknowledge financial support from the Georgia Research Alliance. FAB thanks Professor Wolfgang A. Wall, who established the contact to Professor C. Ross Ethier and therefore allowed this study to occur.

REFERENCES

[1] Quigley H. A. et al., *Br J Ophthalmol*, 90:262-7, 2006.
[2] Quigley, H. A. et al., *Investigative Ophthalmology & Visual Science*, 19:505–517, 1980.
[3] Balaratnasingam, C. et al., *Exp Eye Res*, 121:11-22, 2014.
[4] Nadler, Z. et al., *Biomed Opt Express*, 5:1114, 2014.
[5] Campbell, I. C. et al., *J. R. Soc. Interface: 201512 20141009*, DOI: 10.1098/rsif.2014.1009, 2015.

SB³C2015
Summer Biomechanics, Bioengineering and Biotransport Conference
June 17-20, 2015, Snowbird Resort, Utah, USA
SB³C2015-313

ACCURACY OF QCT-BASED FINITE ELEMENT PREDICTIONS OF VERTEBRAL FRACTURE WHEN BOUNDARY CONDITIONS ARE BASED ON INTRADISCAL PRESSURE PROFILES

Timothy M. Jackman (1), Alexander M. DelMonaco (2), Elise F. Morgan (1,2)

(1) Department of Biomedical Engineering
Boston University
Boston, MA, USA

(2) Department of Mechanical Engineering
Boston University
Boston, MA, USA

INTRODUCTION

Patient-specific finite element (FE) models built from quantitative computed tomography (QCT) scans allow for predictions of bone strength and fracture patterns [1]. Yet even for simple axial compression of the spine, loading across the vertebral endplate is highly non-uniform [2], suggesting that QCT-based FE simulations must carefully consider the choice of boundary conditions. Using measured deformation fields obtained from digital volume correlation (DVC) analyses of time-lapsed micro-computed tomography (μCT) images as a gold standard, a recent study found substantial inaccuracies in FE analyses that used highly idealized boundary conditions [3]. The overall goal of this study was to determine whether boundary conditions based on non-invasive assessments of intervertebral disc (IVD) health can improve FE predictions of vertebral failure. The objectives were: 1) to construct boundary conditions based on previously reported measurements of intradiscal pressure and non-invasive assessments of IVD health; and 2) to examine the effect of applying these boundary conditions on the accuracy of the FE simulations of vertebral failure.

METHODS

Specimen Preparation. Fourteen T7-T9 human spine segments were dissected from fresh-frozen spines (ages: 35-91 years; mean ± standard deviation: 69.4 ± 15.9 years; 9 male, 5 female). *Mechanical Testing and Imaging.* Specimens were first imaged with QCT (GE Lightspeed VCT; 0.31x0.31x0.625 mm/voxel). After preconditioning, the specimens were loaded in compression with anterior flexion in a stepwise manner (0.25mm/step and 0.5°/step) to failure, with a μCT scan (Scanco Medical; 37 μm/voxel) performed prior to each loading step. Loading was applied directly to the T7 and T9 vertebral bodies but not the T7 and T9 posterior elements. Images from all load increments were aligned using image registration. *DVC.* From the registered images, the displacements occurring throughout the T8 vertebral body at each load increment were measured experimentally at a spatial resolution of ~1.9 mm using a custom DVC technique [4]. *IVD Grading.* Using QCT images of a mid-transverse cross-section of each IVD, a trained observer assigned each disc an "apparent loss of disc integrity" (ALDI) score [5]. The ALDI score ranges from 0 (no or mild degeneration) to 2 (severe degeneration). *Nonlinear FE Analyses.* Voxel-based FE models of T8 were generated from QCT images using hexahedral elements (0.625mm side length). Isotropic, linear elastic material properties and yield strength were assigned based on local bone mineral density [6,7]. Post-yield material behavior was modeled as a crushable foam (Abaqus FEA 6.12, Dassault Systèmes) with a yield surface defined by dimensionless stress ratios of 1.181 and 0.540 in compression and tension, respectively [7]. *Boundary Conditions.* Four sets of boundary conditions were used: 1) "Experimentally Matched" boundary conditions were DVC-measured displacements at the superior and inferior endplates; 2) "Idealized" boundary conditions were a uniform compressive displacement and a uniform angular displacement applied to the superior endplate; 3) "IVD-Generic" boundary conditions were based on intradiscal pressure profiles measured in an additional set of specimens [8]; and 4) "IVD-Specific" boundary conditions were based on the measured intradiscal pressure profiles for only the IVDs with the same ALDI scoring as the IVDs adjacent to the vertebra being modeled. Briefly, pressure measurements were made along mid-sagittal lines in 13 T10-T12 human spine segments (26 IVDs) loaded to 500 N in compression with 5° of anterior flexion [8]. Profiles were resampled at fixed percentages of the width of the IVD, and then were normalized by a scaling factor such that the area underneath the profile was the same for every IVD. These normalized profiles were averaged for ALDI scores of 0 to 1, for scores of 2, and across all IVDs to obtain representative pressure distributions (Figure 1) for moderately healthy, degenerated, and generic (*i.e.*, all) IVDs. Mid-coronal profiles were modeled as an isosceles trapezoid based on pressure profiles observed experimentally and in FE simulations [8,9]. The mid-coronal and mid-sagittal profiles were interpolated in the transverse plane of the IVD to produce an applied distributed load on the superior endplate. The total applied force matched the force developed when Experimentally Matched boundary conditions were applied. Only small displacements (<0.2 mm) were observed at the inferior endplate, thus the inferior endplate was fixed in the Idealized, IVD-Generic, and IVD-Specific boundary conditions. *Statistics.* The experimentally measured displacements throughout the interior of T8 were regressed against the corresponding FE-computed displacements (JMP 11, SAS Institute). Points with measured displacements lower than the DVC detection limit (300% of the standard deviation of the DVC displacement error (0.0556 mm)) were excluded [10]. Median displacement errors per specimen were compared using a repeated-measures ANOVA.

RESULTS

All FE models captured some of the general, qualitative features of the deformation; however, the Idealized, IVD-Specific, and IVD-Generic models did not predict the localized deformation occurring superiorly (Figure 2). R^2 values for the regressions of FE-computed vs. measured displacements increased by 0.05 from Idealized to IVD-Generic, by 0.01 from IVD-Generic to IVD-Specific, and by 0.14 from IVD-Specific to Experimentally Matched boundary conditions (Figure 3). The median percent error in displacement was higher for Idealized vs. Experimentally Matched boundary conditions ($p<0.001$); no other differences were found ($p>0.05$).

DISCUSSION

QCT-based FE simulations using Idealized boundary conditions produced greater errors in predictions of vertebral deformation as compared to those that used Experimentally Matched boundary conditions. Use of boundary conditions informed by measurements of IVD pressure mitigated, but did not eliminate, this inaccuracy. Although the loading applied to the T7-T9 spine segments was comparable to the Idealized boundary conditions on T8, the current results suggest that the T8 vertebral bodies within these segments experienced more complex loading conditions that could be better modeled by considering intradiscal pressure measurements. FE analyses incorporating boundary conditions based on pressure profiles from IVDs of comparable disc health did not improve predictions of deformation over boundary conditions derived from generic pressure profiles. These results indicate that further work is needed to accurately model the physiological loading conditions of the vertebra. In addition, poor agreement was observed for some specimens between measured and FE-computed displacements even with Experimentally Matched boundary conditions, suggesting that further improvements in material modeling of the bone are needed to produce accurate predictions of vertebral failure by QCT-based FE models.

ACKNOWLEDGEMENTS

NIH R01 AR054620 (EFM). The authors also thank Glenn Barest, Paul Barbone, Ginu Unnikrishnan, Kamil Makhnejia, Alexander Adams, Paul Fein, Greg Janese, Anderson Camp, and Rachel Danzig.

REFERENCES

[1] Cheng XG et al. *J Bone Miner Res*, 1997; 12:1721-8. [2] Jackman et al. *J Orthop Res*, 2013; 32:880-6. [3] Jackman et al. *ORS* #1531, 2015. [4] Hussein AI et al. *Procedia IUTAM*, 2012; 4:116-25. [5] Hussein et al. *Osteoporosis Int*, 2013; 24:3021-30. [6] Kopperdahl et al. *J Orthop Res*, 2002; 20:801-5. [7] Kinzl et al. *J Mech Behav Biomed Mater*, 2013; 26:136-47. [8] DelMonaco et al. *ORS* #0694, 2015. [9] Jacobs et al. J Biomech, 2014; 47:2540-6. [10] Ghosh et al. *J Biomech Eng*, 2012; 134:081003.

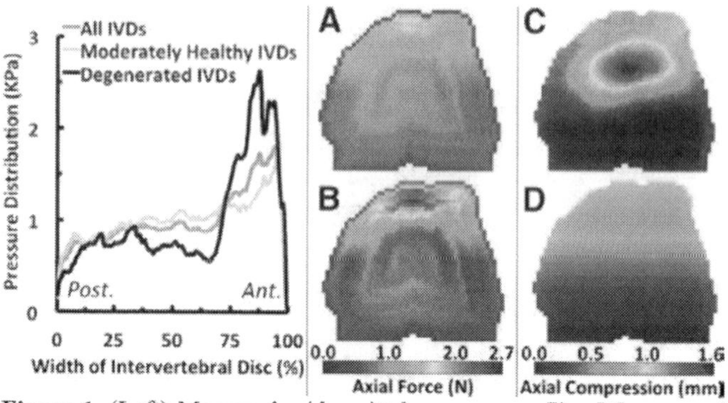

Figure 1. (Left) Measured mid-sagittal pressure profiles [7] averaged for moderately healthy (ALDI=0-1), degenerated (ALDI=2), and all (ALDI=0-2) IVDs. (Right) Color maps of the distributed load across the superior endplate for (A) IVD-Generic, (B) IVD-Specific, (C) Experimentally Matched, and (D) Idealized boundary conditions.

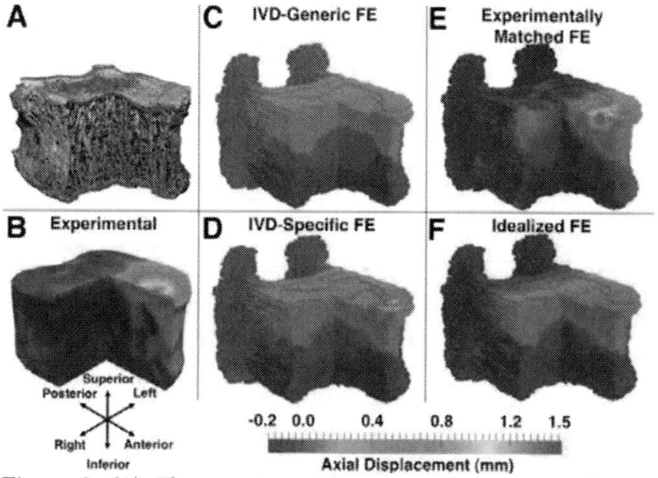

Figure 2. (A) Three-quarter section view of a representative T8 vertebral body before loading (gray) and at the load increment following peak of loading (blue); (B-F) Three-quarter section views of experimental and FE axial displacements (positive values are downward displacements) at the load increment following peak of loading.

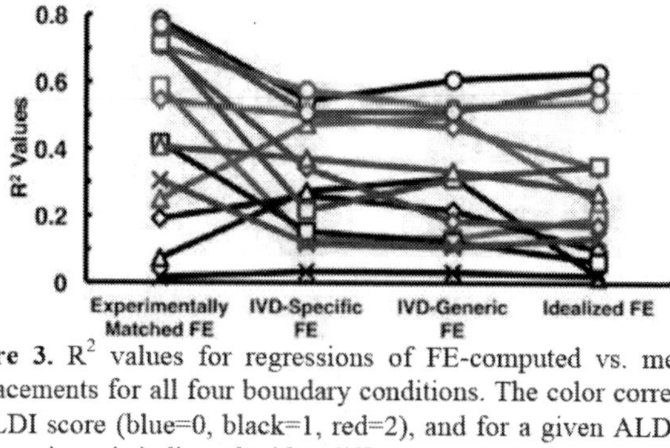

Figure 3. R^2 values for regressions of FE-computed vs. measured displacements for all four boundary conditions. The color corresponds to ALDI score (blue=0, black=1, red=2), and for a given ALDI score each specimen is indicated with a different symbol.

SB³C2015
Summer Biomechanics, Bioengineering and Biotransport Conference
June 17-20, 2015,Snowbird Resort, Utah, USA
SB³C2015-314

A MULTI-SCALE FINITE ELEMENT FRAMEWORK FOR MODELING NATURAL KNEE MECHANICS

Michael D. Harris, Ali Azhar, Alessandro Navacchia, Adam J. Cyr, Donald Hume, Clare K. Fitzpatrick, Paul J. Rullkoetter, Kevin B. Shelburne

Department of Mechanical and Materials Engineering
University of Denver
Denver, CO, USA

INTRODUCTION

Computational modeling provides non-invasive estimation of muscle, ligament, and joint loading and enables investigation of joint biomechanics that cannot be achieved through experimental means alone. A key challenge of the computational approach is maintaining sufficiently complex joint-level tissue representations while avoiding computational burden that prohibits muscle force calculations at the whole-body scale. Our overall objective is to create multi-scale finite element (FE) models of the human lower extremity that combine sophisticated joint-level models and whole-body dynamic muscle modeling in a single FE framework and then apply this framework to study the mechanics of natural and implanted knees. In the current study, two steps in the development and validation of a multi-level specimen-specific natural knee model are reported. Incorporation of the knee model into a musculoskeletal model of the lower-extremity is also demonstrated.

METHODS

In Vitro Experiments: Two experiments were used to provide knee loading and kinematic profiles for model validation. First, passive laxity tests were performed on a cadaveric knee (male, 21.0 BMI, 44 yrs) at 0, 15, 30, 45 and 60° knee flexion by applying ±8 Nm internal-external (IE) torque, ±10 Nm varus-valgus (VV) torque, and ±40 N anterior-posterior (AP) force in each direction (**Fig. 1**). Applied loads were measured with a 6-DOF triaxial load cell attached to the distal tibia [1]. Second, quadriceps and hamstring muscles were clamped and attached to lines that passed through a series of pulleys to maintain physiologic lines of action. A motor attached to the quadriceps (vastus intermedius and rectus femoris) drove flexion-extension and a load cell measured quadriceps force; 40 lb static weights were evenly distributed between two major hamstring groups (semimembranosus and biceps femoris). The contribution of individual intra-joint structures was assessed by sequentially resecting the anterior cruciate ligament (ACL), meniscus, and posterior cruciate ligament (PCL). The in-vitro tests were repeated after each resection. Tibiofemoral (TF) and patellofemoral (PF) kinematics were recorded using an Optotrak 3020 system (Northern Digital Inc.).

FE Model of the Knee Laxity: Cartilage, meniscus, and cortical bone geometry of the cadaveric knee were segmented from MR images (1x0.35x0.35mm) and 3D reconstructions were generated using ScanIP software (Simpleware). The bones were meshed with triangular shell elements and represented as rigid bodies. The cartilage surfaces of the tibia, femur, and patella were meshed with hexahedral elements, enabling deformable or rigid-body analysis of joint contact. Relative alignment of geometry was defined using the digitized points from the experimental marker tracking system.

From digitized ligament attachment sites, non-linear spring elements were created to represent the anteromedial and posterolateral ACL, posteromedial and anterolateral PCL, deep and superficial medial collateral ligament, posterior oblique ligament, and lateral collateral ligament. Additional springs were placed to represent the anterolateral ligament structure, popliteofemoral ligament, and the posterior capsule. The in-vitro experiments were simulated in Abaqus/Explicit (SIMULIA). An optimization of ligament stiffness, slack length, and attachment location was performed using Isight (SIMULIA) to match experimental kinematics. The allowed ligament parameter perturbations were derived from literature-based material properties and digitized footprints of the ligaments on the specimen [2]. Ligament parameters were optimized at all flexion angles for the fully resected condition using IE and VV experimental kinematics. Resulting parameter values were then applied to the intact knee model and the ACL and PCL were tuned using corresponding AP, IE, and

VV experimental kinematics. Root mean square errors (RMSE) between model and experimental kinematics were calculated.

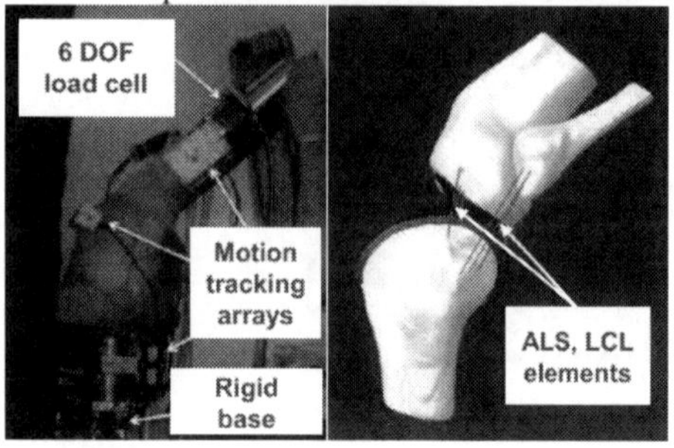

Figure 1. In-vitro setup for laxity testing and FE model showing the anterolateral ligament structure (ALS) and lateral collateral ligament (LCL).

Representation of the Patella and Quadriceps Mechanism: Next, the patella, point-to-point quadriceps and hamstring muscle connectors, and PF soft tissue (rectus-femoris, patellar ligament, medial and lateral PF ligaments) represented by fiber-reinforced membranes were added to the TF model. Muscle lines of action were defined using digitized muscle attachment points from the experiment. TF kinematics were prescribed based on the experiment, except for TF flexion, which was controlled by quadriceps load. To calibrate the model with experimental kinematics and quadriceps force, soft tissue attachment sites and muscle lines of action were perturbed based on experimental uncertainty. PF kinematics, quadriceps load and PF contact mechanics were predicted and compared to the experiments.

RESULTS

Torque- and force-displacement curves predicted by the models during laxity tests agreed well with experimental results (**Fig. 2**), with RMSE ≤2.3° for VV (ROM ±8-10°), ≤6° IE (ROM ±20-30°), and ≤2.5mm AP (ROM ±10-15mm). The timing of ligament tension agreed with literature reports [3,4].

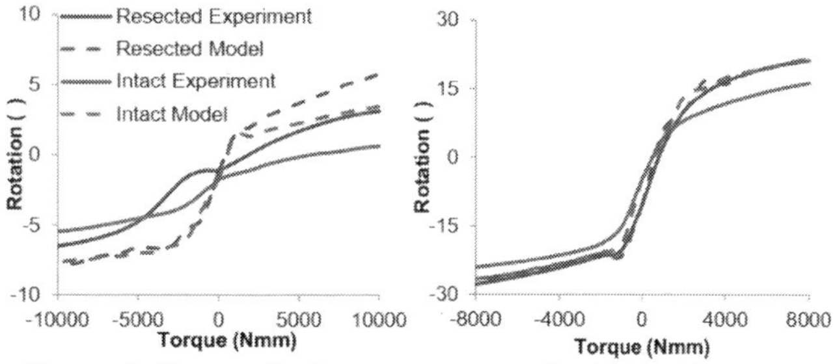

Figure 2. Torque-displacement curves for VV (left) and IE laxity tests at 30° flexion for the cruciates/meniscus intact (green) and resected (red) conditions.

Peak quadriceps forces through flexion-extension in the intact and resected trials had average errors of 4.3-7.2%. PF kinematic predictions also matched experimental data with RMSE differences of FLEX-EXT <5°, VV <3.5° and IE <5°, and AP <4mm (**Fig. 3**). Patellar tendon angle, effective moment arm, and patellar force ratio agreed with values reported for normal knees [5,6].

DISCUSSION

This study demonstrates development and calibration of a natural knee model spanning scales from ligament to whole-joint for inclusion in a musculoskeletal model of the lower extremity. Simulations of ligament-restrained knee laxity, quadriceps driven PF kinematics, and muscle geometry agreed well with experimental results and with

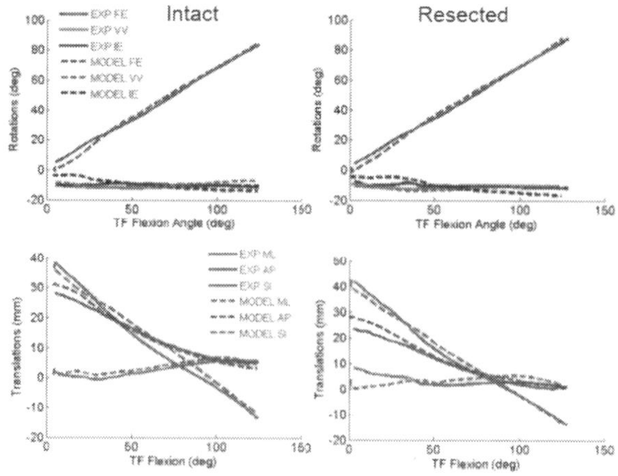

Figure 3. PF rotations and translations – experiment vs model.

literature reports. In addition, the model's adaptability in FE means certain elements can be replaced with more or less complex meshes of the collaterals, cruciates, or meniscus for detailed analyses.

The subject-specific knee model has also been incorporated in a FE musculoskeletal model of the complete lower extremity containing 13 muscles spanning the hip, knee, and ankle. Generic pelvis, femur, tibia, and foot geometry were added to represent a six segment, five joint (hip, tibiofemoral, patellofemoral, ankle, toe) mechanical system. (**Fig. 4**).

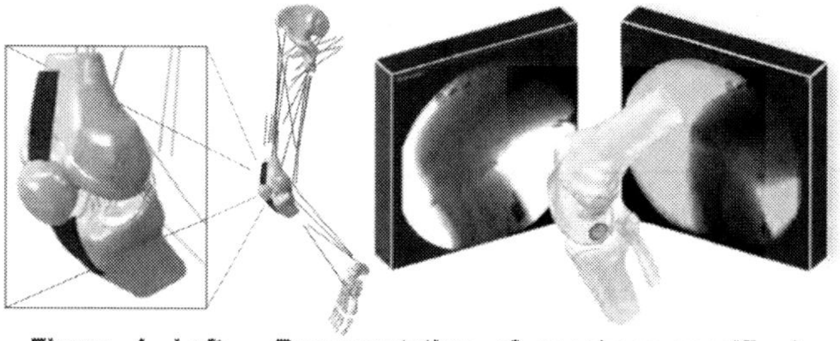

Figure 4. Left – Representation of specimen-specific knee incorporated in full FE musculoskeletal model of lower extremity. Right – Stereo radiography image and subject-specific geometry for a live subject.

The musculoskeletal model of a cadaveric knee enables predictive analyses or can be driven with kinematic profiles from constrained experiments or live subjects. To drive the model, we have collected motion data for 6 live subjects using a high-speed stereo radiography system, combined with marker-based whole-body motion capture (**Fig. 4**). Stereo radiography enables precise measurement of bone motion with sub-millimeter accuracy. The complete multi-scale musculoskeletal model can be used to understand the etiology of knee pathologies, evaluate implant designs, and assess the influence of physical therapy on joint mechanics.

ACKNOWLEDGEMENTS

Supported by the NIH NIBIB R01EB015497, NSF CISE MRI 12-29148 and DePuy Synthes.

REFERENCES

[1] Cyr, A.J. and Maletsky, L.P., *J Eng Med*, 228:494-500, 2014.
[2] Baldwin, M et al., *J Biomech*, 45:474-483, 2012.
[3] LaPrade, R.F. et al., *Am J Sports Med*, 31:854-860. 2003.
[4] Blankenvoort, L. Et al., *J Biomech Eng*, 113:94-103, 1991.
[5] Ahmed, A.M. et al *J Orthop Res*, 5:69-85, 1987.
[6] Buff, H. et al., *J Biomech* 21:17-23, 1988.

SB³C2015
Summer Biomechanics, Bioengineering and Biotransport Conference
June 17-20, 2015, Snowbird Resort, Utah, USA
SB³C2015-315

INTIMAL HYPERPLASIA AND ITS IMPLICATIONS IN THE CEPHALIC ARCH: A NUMERICAL STUDY OF NON-PHYSIOLOGICAL HEMODYNAMICS IN PATIENTS WITH BRACHIOCEPHALIC FISTULAE

S. M. Javid Mahmoudzadeh Akherat (1), M. E. Boghosian (1), K. W. Cassel (1), M. S. Hammes (2)

(1) Mechanical, Materials and
Aerospace Engineering Department
Illinois Institute of Technology
Chicago, IL, United States

(2) Section of Nephrology
Department of Medicine
University of Chicago
Chicago, IL, United States

INTRODUCTION

The study of progressive cardiovascular diseases has become very important recently due to a high rate of fatal cardiovascular incidents in the industrialized nations. One of the most common causes of vascular disease and abnormalities is neointimal hyperplasia, which occurs in the initial phases of atherosclerosis [1]. It is one of the major contributors to the development of stenosis in vasculature especially in patients receiving vascular access, namely Brachiocephalic Fistula (BCF), who undergo chronic dialysis treatments. The number of these patients is expected to double in the next 25 years and the related annual healthcare expenses will triple. BCF has an average patency of only 3.6 years. A leading cause of BCF failure is stenosis in the arch of the cephalic vein near its junction with the axillary vein, which is called cephalic arch stenosis (CAS). Coupling a clinical and computational investigation, we seek to improve our comprehension of the causes of CAS, and to develop a means of predicting intimal hyperplasia leading to CAS risk in patients with a planned BCF access. Therefore, a thorough non-invasive examination of the onset and progression of CAS will present a tremendous help to the community in order to provide alternative treatments as well as surgical and demographical intervention options. Computational fluid dynamics has been a viable tool for this purpose for decades and will be implemented therein. The ultimate goal is to present possible modifications to the existing vascular accesses to increase the patency and reduce the mortality rate caused by thrombi.

METHODS

A cohort of 133 End Stage Renal Disease (ESRD) patients has been under observation for 3.5 years. After the patient is enrolled, pre-operative venogram, Doppler velocity measurement of the cephalic arch and whole blood viscosity are obtained. This data is again obtained at 3, 12, 24 and 36 months. Image processing is preformed using a MATLAB code to reconstruct the patient specific geometries (figure1).

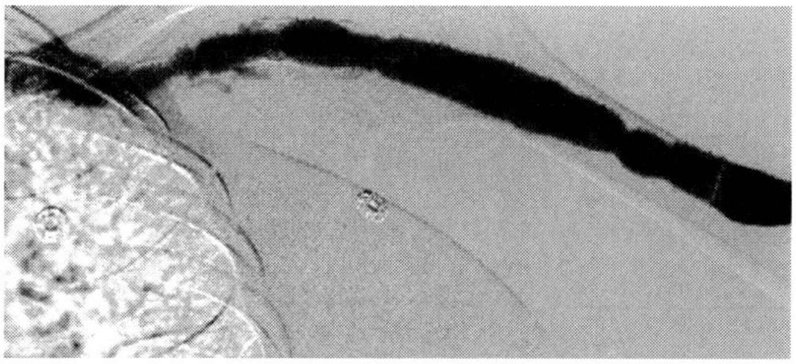

Figure 1 Image processing of a single cephalic arch stenosis venogram, subject 27 at 12 month. CAS is evident at the end of the arch, distal to the junction of axillary and the cephalic. The blood flow is from right to left.

This research details the methodology used to determine the hemodynamic consequences of the post-fistula environment and illustrates results for a representative sample of patient-specific anatomies, including a single, bifurcated, and trifurcated arch. A spectral elements code (NEK5000) and a finite elements code (COMSOL Multiphysics) are implemented to perform the 2D and 3D CFD analysis. The authors extensively investigated the non-Newtonian effects in hemodynamic simulations, especially in the case of ESRD patients [2]. Moreover, a patient specific pulsatile inlet excitation obtained from Doppler measurements will be applied in the numerical inlet domain, effects of which coupled with the non-Newtonian constitutive relations was studied on the variations in wall shear stresses and the frequency and strength of vortex shedding in the Womersley flow. Following the framework presented by Figueroa et. al. [3], a fluid-solid interaction (FSI) analysis will be performed to address the effect of wall parameters on the cephalic vein's remodeling, the results of which will set the stepping stone of the analysis on growth and remolding caused by the occurrence of intimal hyperplasia in the vasculature of ESRD patients.

RESULTS

The ultimate goal in this research is to eventually formulate and comprehend the mechanics behind the onset and progression of vascular incidents that lead to the failure of accesses. CFD simulations reveal that the dramatic increase in volumetric flow rates (10 times normal physiological values) and the natural bend of the arch itself create a fluid dynamically violent environment, resulting in recirculation regions and regions of wall shear stress (WSS) that are lower than the normal values. It has been hypothesized that the neointimal hyperplasia is the vessel's effort to bring back the non-physiological values of WSS within an acceptable range. This finding is also in agreement with other research where recent evidence suggests that the low wall shear stress hypothesis may play an important role in anastomoses failures [4] (figure 2). It has also been concluded that the volumetric concentration ratio of red blood cells (hematocrit) plays an essential role in the determination of whole blood viscosity in ESRD patients. Hence it was concluded that the non-Newtonian viscosity relation has to have both hematocrit and shear rate as the independent variables and the models given by Quemada and Walburn-Schneck [5-7 and 8] best represents the patient population under observation. Tangible differences in flow patterns were detected between the two models. See figure 3.

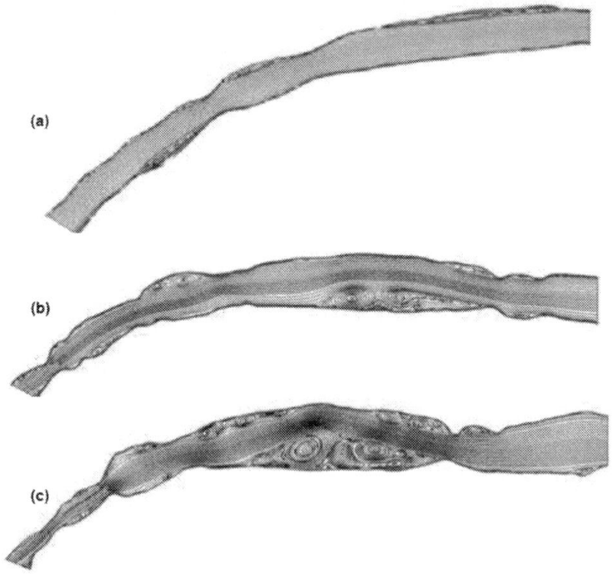

Figure 2 Streamline plot superimposed on WSS contours for subject 27 at 3 (a), 12 (b) and 24 (c) month. The progression of stenotic sites through time is evident. Red solid lines represent sites where WSS drops below the critical value (0.076 Pa) [9].

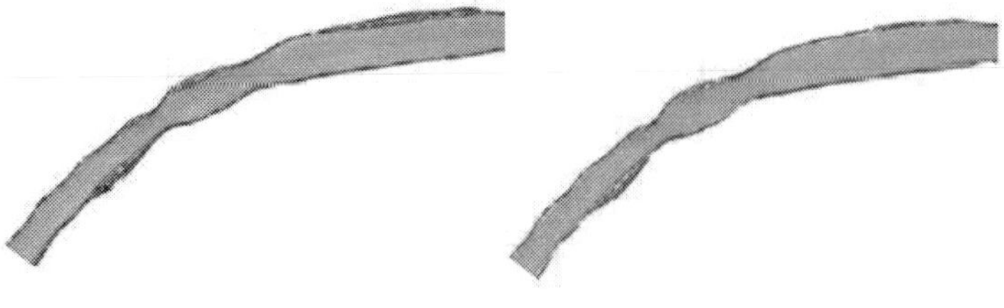

Figure 3 Newtonian (left) and non-Newtonian Quemada (right) flow fields. Recirculation zones are either not present or smaller and weaker in the non-Newtonian flow. Less solid red lines indicate less susceptibility to intimal hyperplasia [2,9].

DISCUSSION

Qualitative and quantitative differences have been observed between the Newtonian and non-Newtonian flows, indicating an over estimation of low WSS sites by the Newtonian assumption. Therefore, sites with potential susceptibility to intimal hyperplasia may be over predicted. The consideration of a Womersley flow further necessitated the implementation of non-Newtonian models for hemodynamic simulation of ESRD patients with highly unsteady flows. In order to predict how hemodynamic leads to intimal hyperplasia and possible stenosis growth and remodeling, models are being explored to determine how changes in the shape of the vessel influences the fluid domain, which thereby implies a two-way fluid-structural coupling using the ALE (Arbitrary Lagrangian-Eulerian) method. We are striving to elucidate the rate of constituent turnover, e.g. collagen matrix and smooth muscle cells, to be able to incorporate the constitutive relations representing intimal hyperplasia in to the FSI analysis.

ACKNOWLEDGEMENTS

Research reported in this dissertation was supported by the National Institute Of Diabetes And Digestive And Kidney Diseases of the National Institutes of Health under Award Number R01DK090769. The content is solely the responsibility of the authors and does not necessarily represent the official views of the National Institutes of Health.

REFERENCES

[1] C. A. Taylor, Open problems in computational vascular biomechanics: Hemodynamics and arterial wall mechanics, *Computer Methods in Applied Mechanics and Engineering*, 198 (2009) 3514-3523.

[2] S. M. Javid Mahmoudzadeh Akherat, Non-Newtonian Effects In Hemodynamic Simulations Of The Cephalic Vein In End Stage Renal Disease Patients, M.Sc. Thesis, Illinois Institute of Technology, 2013.

[3] C. Alberto Figueroa, et. al. A computational framework for fluid-solid-growth modeling in cardiovascular simulations, *Computer Methods in Applied Mechanics and Engineering*, 198 (2009) 3583-3602.

[4] McGah P, Leotta D, Beach K, Zierler R, Aliseda A. Incomplete restoration of homeostatic shear stress within arteriovenous fistulae, *Journal of Biomechanical Engineering*, 2013;135, http://dx.doi.org/10.1115/1.4023133.

[5] Quemada, D., Rheology of concentrated disperse systems and minimum energy dissipation principle, *Rheologica Acta*, 16:82-94, 1977.

[6] Quemada, D., Rheology of concentrated disperse systems II. A model for non- Newtonian shear viscosity in steady flows, *Rheologica Acta*, 17.632-642, 1978.

[7] Quemada, D., Rheology of concentrated disperse systems III. General features of the proposed non-Newtonian model. Comparison with experimental data, *Rheologica Acta*, 17:643-653, 1978.

[8] Walburn, F.J., Schneck, D., A constitutive equation for whole human blood, *Biorheology*, 13:201-210, 1976.

[9] M. E. Boghosian, K. W. Cassel, et. al., Hemodynamics in the cephalic arch of a brachiocephalic fistula, *Medical Engineering and Physics*, 36(2014) 822-830.

EVALUATING THE CONSISTENCY OF CARDIOMYCOTYE SELF-ASSEMBLY

Nancy K. Drew (1), Danny B. Baldo Jr. (1), Jason Q. Core (1), Anna Grosberg (1,2)

(1) Biomedical Engineering
University of California, Irvine
Irvine, CA, USA

(2) Chemical Engineering
University of California, Irvine
Irvine, CA, USA

INTRODUCTION

In biology, organization at multiple length scales drives functionality. Current advances in staining and imaging of biological cells provide a wealth of data, but there are few metrics to quantitatively describe these findings. Qualitatively some cardiomyocytes patterned into shapes produce consistent architecture. However there is a need for a quantitative characterization of consistency of orientation of different biological constructs within a cell. In order to investigate consistency, we need to establish a metric to quantify consistency of cytoskeleton architecture in cardiomyocytes over a range of length scales. The new metric, the co-orientational order parameter (COOP) [1], designed based on the mathematical framework of a classical parameter orientational order parameter [2], can be used to quantify cytoskeleton architecture consistency at multiple scales.

METHODS

Experimental Methods

Microcontact printing of extracellular matrix was used to engineer various shaped neonatal rat ventricular myocytes (NRVM). These myocytes were fixed and immunostained for DAPI, sarcomeric α-actinin, and actin. To quantify the orientation of sarcomeric z-lines and actin fibers we adapted a previous MATLAB code [3]. Then we used a custom MATLAB code to calculate the COOP between pairs of cells for both sarcomeric z-lines and actin.

Theoretical Methods

The COOP can be used to quantify consistency orientation of constructs in two images. To create this new metric we define a new vector f (eqns. 1, 2), which corresponds to the angle in between each vector pair (θ):

$$f_{i,x} = \vec{p}_i \cdot \vec{q}_i = p_{i,x}q_{i,x} + p_{i,y}q_{i,y} = \cos(\theta), \tag{1}$$

$$f_{i,y} = \left|\vec{p}_i \times \vec{q}_i\right| = p_{i,x}q_{i,y} - p_{i,y}q_{i,x} = \sin(\theta). \tag{2}$$

The COOP is calculated similarly to the OOP:

$$\mathbf{T}_{PQ} = \left\langle 2\begin{bmatrix} f_{i,x}f_{i,x} & f_{i,x}f_{i,y} \\ f_{i,x}f_{i,y} & f_{i,y}f_{i,y} \end{bmatrix} \right\rangle, \tag{3}$$

$$COOP_{PQ} = \max\left[eigenvalue(\mathbf{T}_{PQ})\right]. \tag{4}$$

The COOP ranges from zero, for completely non co-oriented images, to one, for perfectly co-oriented images.

RESULTS

Theoretical Results

We were able to show that when comparing the same construct to itself provides a COOP=1 (Fig. 1).

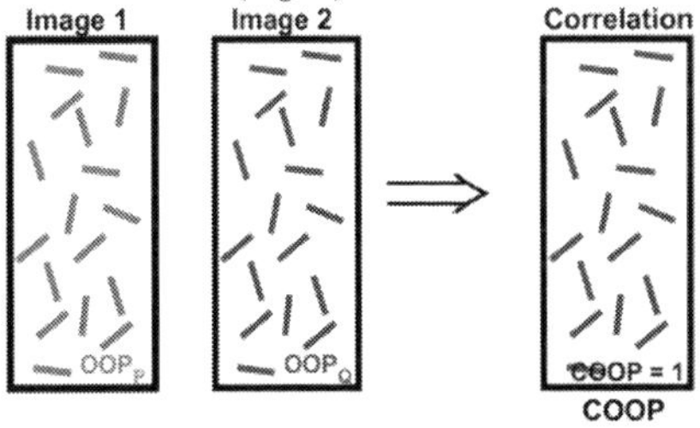

Figure 1: COOP=1 for two identical sets compared to each other.

To prove this, we imagine there are two identical pseudo vectors:

$\vec{p}_i = [\cos(\alpha), \sin(\alpha)]$, and $\vec{q}_i = [\cos(\alpha), \sin(\alpha)]$. (5)

Then we determine the mean tensor is

$$\mathbf{T}_{PQ} = \left\langle 2\begin{bmatrix} f_{i,x}f_{i,x} & f_{i,x}f_{i,y} \\ f_{i,x}f_{i,y} & f_{i,y}f_{i,y} \end{bmatrix} \right\rangle - \begin{bmatrix} 1 & 0 \\ 0 & -1 \end{bmatrix}, \quad (6)$$

and the COOP=1.

Experimental Results

To test the consistency of cytoskeleton architecture within patterned cardiomyocytes, we compared actin and sarcomeric z-lines distributions across cells. We determined each constructs' mean orientation in each grid or partition area. Qualitatively the cardiomyocytes constrained to a triangular extracellular matrix island have consistent architecture with the myofibril bundles spanning the edges of the cell (Fig. 2).

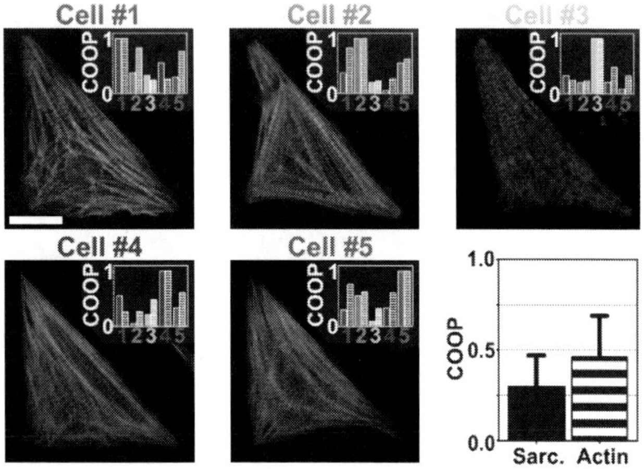

Figure 2: NRVMs cultured on identical fibronectin islands. Each image contains a histogram of COOP between that cell and all others (solid-sarcomere, stripe-actin). Stains are: green-actin, red-α-actinin, and blue-DAPI. Scale bar is 20μm. The histogram shows the average COOP for consistency of sarcomere (solid) and actin (stripe) organization.

However, when using a grid our quantitative results indicate that the architecture is not fully consistent on the length scale of a few sarcomeres (Fig. 2). We also used natural partitions that follow the original shape to calculate mean orientations of construct (Fig. 3).

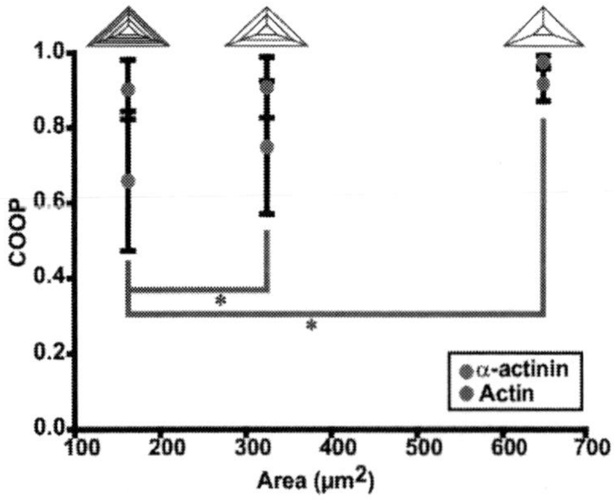

Figure 3: Consistency of actin and sarcomeric z-lines in triangular shaped NRVMS using partitions of various sizes.

Self-assembly of sarcomeric z-lines is consistent with large partition areas (Fig. 3). However, there is a loss of consistency with smaller partition areas for sarcomeric z-lines (Fig. 3). Similar to triangular NRVMs, qualitatively square shaped NRVMs produce consistent architecture. Quantitatively, square shaped NRVMs actin is consistent with large grid areas (Fig. 4).

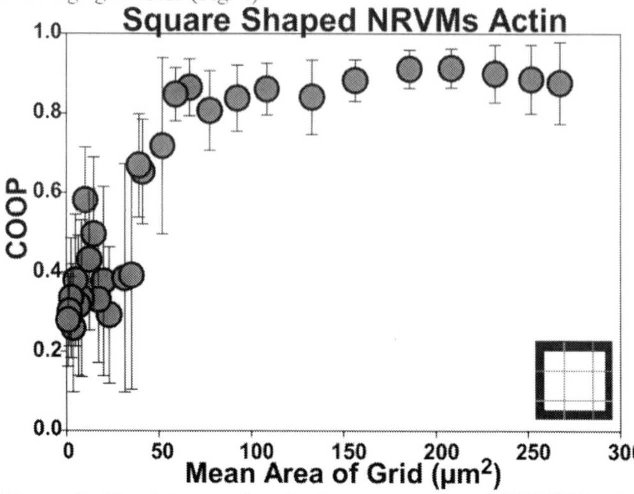

Figure 4: Consistency of actin in square shaped NRVMS using grids of various areas.

Overall, actin self-assembly consistency increases asymptotically with grid areas above 50 μm^2 in square shaped NRVMs (Fig. 4). For square shaped cells we analyzed construct consistency using grids as grids follow the original shape for square shaped cells. For triangular shaped cells natural partitions of shape can be used to quantify construct consistency.

DISCUSSION

We have used the COOP to show that cardiomyocytes self-assembly is only consistent at some length scales but not others. Triangular and square shaped NRVMS showed construct orientation consistency from cell to cell with large grids/partition areas. We plan to study this relationship for more shapes i.e. circles, ovals, and different partition areas. We aim to determine if length scales consistency is different for various shaped cells and the mechanisms behind the consistency. In the future, the parameter could be used to identify length scales at which cells of different origin are consistent in the organization of their cytoskeleton.

ACKNOWLEDGEMENTS

This research was partially funded by the NIBIB, National Research Service Award EB009418 from the UCI, CCBS and NSF Eager award (Award Number 1338609).

REFERENCES

[1]. Drew, NK et al., Plos Comput Biol Manuscript submitted for publication, 2014.

[2]. Grosberg A et al., Lab Chip 11: 4165–4173, 2011

[3]. Bray, MAP et al., Biomaterials 31: 5143–5150, 2010.

SB^{3}C2015
Summer Biomechanics, Bioengineering and Biotransport Conference
June 17-20, 2015,Snowbird Resort, Utah, USA
SB^{3}C2015-317

USING SLOW AND FAST SHEAR WAVES TO ESTIMATE SHEAR AND TENSILE MODULI: RESULTS FROM SIMULATIONS OF ANISOTROPIC TISSUE

D.J. Tweten (1), J.L. Schmidt (1), R.J. Okamoto (1), P.V. Bayly (1,2)

(1) Mechanical Engineering and Materials Science
Washington University
St. Louis, MO, United States

(2) Biomedical Engineering
Washington University
St. Louis, MO, United States

INTRODUCTION

The estimation of anisotropic material parameters is an important step in understanding traumatic brain injury (TBI) [1]. Of particular interest are techniques to estimate parameters of aligned fibrous materials such as white matter in the brain. Recently, researchers have presented anisotropic material models with either two [2,3] or more than five [4] elastic parameters. The simplest model for small elastic deformations of soft tissue is an incompressible, transversely isotropic (ITI) material with three parameters: minimum shear modulus (μ_2), shear anisotropy ($\phi = \mu_1/\mu_2 - 1$) and tensile anisotropy ($\zeta = E_1/E_2 - 1$) [1]. Slow and fast shear waves, phenomena specific to anisotropic elastic or viscoelastic media, can be exploited to estimate these material parameters. Validating estimation techniques using simulated data is an important step to applying these methods experimentally.

METHODS

For the case of harmonic, elastic, plane waves travelling in an ITI material, the assumed solution $\vec{u}(\vec{x},t) = u_0\vec{m}\exp[i(k\vec{n}\cdot\vec{x} - \omega t)]$ satisfies the equation of motion (EOM) $\mathrm{div}\,\boldsymbol{\sigma} = \rho\,\partial^2\vec{u}/\partial t^2$, where σ is Cauchy stress, u is displacement, $\vec{m}$ is polarization of the displacement, $\vec{n}$ is propagation direction, k is wavenumber, and ω is frequency. Substituting the assumed solution into the EOM results in the eigenvalue problem: $\boldsymbol{Q}(\vec{n})\cdot\vec{m} = \rho c^2\vec{m}$, where Q is the acoustic tensor, ρ is density, and c is wave speed. Solving the eigenvalue problem defines three wave speeds (eigenvalues are ρc^2) and eigenvectors ($\vec{m}$). Two of the eigenvectors define the slow ($\vec{m}_s = \vec{n}\times\vec{a}$) and fast ($\vec{m}_f = \vec{n}\times\vec{m}_s$) shear waves, where $\vec{a}$ is the fiber orientation. The associated speeds of the slow and fast shear waves are given by $c_s^2 = \mu_2/\rho(1+\phi\cos^2\theta)$ and $c_f^2 = \mu_2/\rho(1+\phi\cos^2 2\theta + \zeta\sin^2 2\theta)$, respectively, where θ is the direction angle. The third eigenvector defines a longitudinal wave which exhibits infinite wave speed in incompressible media. Figure 1 shows an example of the orientation of the shear wave polarizations.

A finite element (FE) simulation was created using the three parameter material model (μ_2=1000 Pa, φ=1, ζ=2, and ρ=1000 kg/m^3) with the results shown in Figure 2. The fiber direction is parallel to the xz-plane at an angle of 45° from the xy-plane. Note that the larger wavelength (wave speed) is in the direction of the fibers. The boundary conditions (BCs) include a 5 μm excitation at 200 Hz on the inner radius (1.6 mm); fixed displacement on the outer radius (23 mm); and free displacement on the top and bottom faces. The field size was 48x48x24 mm^3.

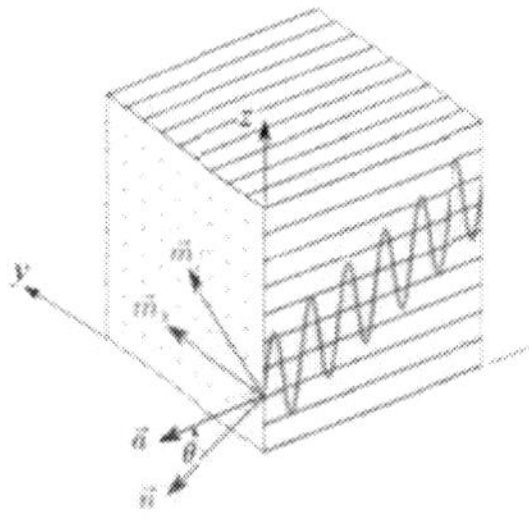

Figure 1
The fast component of a shear wave (red curve) propagating in fibrous material has direction ($\vec{n}$) with direction angle (θ) from fiber direction ($\vec{a}$) with both directions in the xz-plane. The polarizations of the slow and fast shear waves are in directions $\vec{m}_s$ (y-dir.) and $\vec{m}_f$ (xz-plane), respectively.

To estimate material parameters, a directional filter was applied to the data to isolate displacements with a specific propagation direction and polarized in the slow or fast orientation. Next, wave speed was estimated at each voxel (1 mm^3) for each direction using local frequency estimation (LFE). LFE also provides a "certainty" value for each estimate [5]. Figure 3 shows differences in slow and fast wave speeds using this approach. For all image data, xy-planes are shown with the lowest slice at 1a. Finally, weighted least squares (WLS) was used to calculate the material parameters at each voxel using the estimated wave speeds, fiber orientation, and propagation direction. The WLS algorithm used 27 voxels for each estimate.

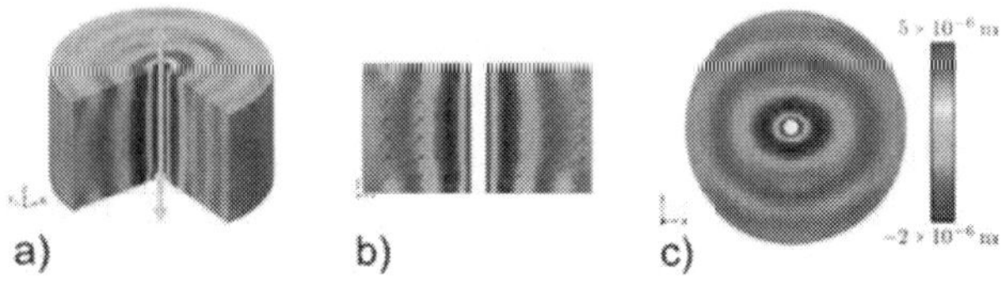

Figure 2
Finite element (Comsol^{TM}) simulation of ITI material with a) location and direction of excitation shown by arrow, b) arrows indicating fiber orientation (xz-plane), and c) resulting wave propagation (xy-plane).

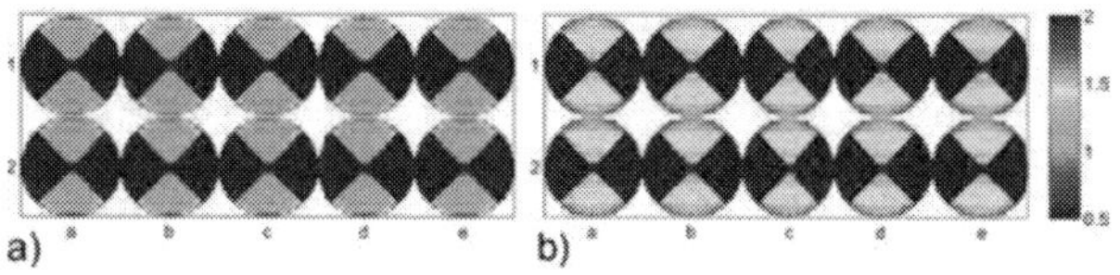

Figure 3
Estimated speeds of a) slow and b) fast shear waves propagating in the y-dir (θ=90°). Wave speed images are masked to take into account propagation direction. Note: middle ten slices are displayed.

RESULTS

Figure 4 shows analytical wave speeds and those estimated using directional filters and LFE. The estimated wave speeds are calculated from 32 propagation directions within the xy-plane. The angle between propagation and fiber directions varies between 45° and 135°.

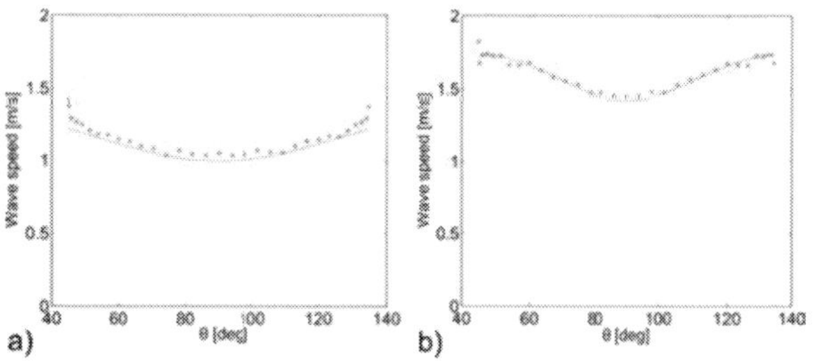

Figure 4
Analytical (solid line) and estimated (stars) speeds of a) slow and b) fast shear waves. Estimated speeds are peak values of histograms calculated from images similar to Figure 3 for each direction.

A total of 128 propagation directions were used to obtain estimates of shear modulus, shear anisotropy, and tensile anisotropy in Figure 5. For each voxel, wave speeds were included in the WLS algorithm if the LFE certainty was at least 0.95 and then weighted based on relative amplitude. For clarity, estimates within 20% of the parameter are shown, with the remaining voxels set to black.

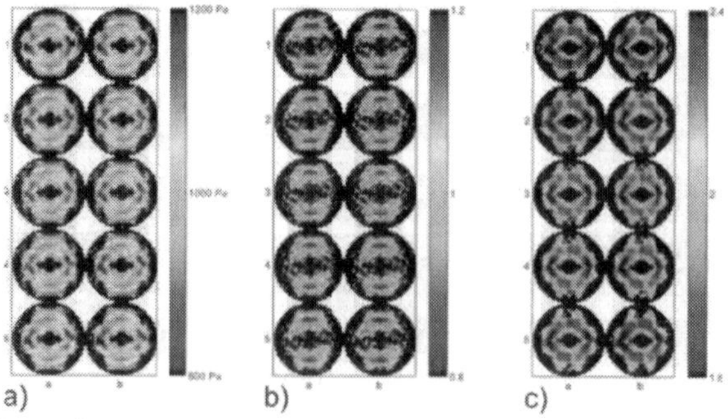

Figure 5
Estimated a) shear modulus, b) shear anisotropy and c) tensile anisotropy. Only estimated values within 20% of μ_2=1000 Pa, φ=1, and ζ=2 are shown. Note: middle ten slices are displayed.

DISCUSSION

Figure 4 highlights the difference in speed between slow and fast shear waves in different propagation directions. Excellent agreement is found between analytical predictions and estimated wave speeds. Figure 4 also demonstrates that the slow and fast shear waves can be successfully separated using the proposed directional filtering algorithm. This separation is critical if the three material parameters are to be accurately estimated. It should be noted that the orientation of the fibers and direction of excitation was selected in this simulation so that the amplitudes of both shear waves would be of similar order. It is possible to excite the material in such a way as to excite only slow shear waves (e.g. if the fibers were parallel to the xy-plane). It is important to note that if only slow shear waves are excited, it is impossible to estimate tensile anisotropy.

The agreement between estimated parameters μ_2, φ, and ζ and the actual values is good, as shown in Figure 5. Estimated values for all three parameters are within 20% for nearly all voxels within the cylinder. Estimation of the parameter ζ is more difficult than μ_2 or φ; this is likely due to the fact that ζ only affects fast shear waves. Discrepancies occur mainly at the boundaries and are likely due to reflections at the boundaries and limitations of LFE. Distortions of wave speed estimates by LFE are caused by sharp changes in the displacement fields at the boundaries. Using a windowing function on the field might improve the distortions but would also greatly reduce the number of voxels with valid estimates. Artificially extending the data might also improve parameter estimates at the boundaries.

The present simulation study has shown that knowledge of three parameters is sufficient to model shear wave propagation in nearly incompressible, transversely isotropic, fibrous materials. This study has demonstrated and validated a directional filter based approach for estimating material parameters using both slow and fast shear waves. Future work will focus on application of this technique to experimental data from magnetic resonance elastography (MRE).

ACKNOWLEDGEMENTS

The authors gratefully acknowledge funding from the NIH from grant no. NS055951 and the NSF from grant no. CMMI-1332433.

REFERENCES

[1] Feng, Y et al., J Mech Behav Biomed Mater, 23:117-132, 2013.
[2] Qin, E et al., J Magn Reson Imaging, 37:217-226, 2013.
[3] Sinkus, R et al, Magn Reson Med, 53:372-387, 2005.
[4] Romano, A et al., Magn Reson Med, 68:1410-1422, 2012.
[5] Okamoto, R et al., Proc of SPIE, 9038:90381E, 2014.

BLOCKING BLOOD-SPINAL CORD BARRIER BREAKDOWN PREVENTS THE DEVELOPMENT OF PAIN FOLLOWING NERVE ROOT COMPRESSION INJURY

Jenell R. Smith (1), Paul A. Janmey (1,2), Beth A. Winkelstein (1,3)

(1) Department of Bioengineering
University of Pennsylvania
Philadelphia, PA, USA

(2) Department of Physiology
University of Pennsylvania
Philadelphia, PA, USA

(3) Department of Neurosurgery
University of Pennsylvania
Philadelphia, PA, USA

INTRODUCTION

Chronic pain is a debilitating and costly disorder affecting up to one-third of the adult population [1]. Chronic pain is maintained by sustained neuroinflammation in the central nervous system (CNS) which is controlled by activated immune and glial cells and the factors they release [2]. In early stages of spinal neuroinflammation, the blood-spinal cord barrier (BSCB) becomes more permeable and facilitates the transmigration of blood-derived immune cells and the diffusion of serum molecules into the CNS [3,4]. The BSCB increases permeability even after peripheral neuropathic injuries remote from the CNS. For example, sciatic nerve compression increases the permeability of the lumbar BSCB for at least 3 days, which is accompanied by blood-derived IL-1β influx and precedes infiltration of bone marrow-derived monocytes into the spinal cord [3,5]. Despite being correlated with neuroinflammation after painful nerve injury, it is unknown if BSCB breakdown contributes to the onset of pain and if inhibiting it can prevent the onset of pain from neural trauma.

Blood-brain barrier (BBB) breakdown occurs in many conditions, such as hemorrhagic stroke and sepsis [6,7]. Activated protein C (aPC) was once a promising agent to treat these disorders because it fortifies the BBB and reduces the associated CNS inflammation that can lead to pain [8,9]; but, the potent anticoagulant properties of aPC decrease clotting, which presents an increased risk [10]. Since traumatic injury is often associated with bleeding from damaged vessels, aPC is not ideal for reversing BBB breakdown after neural injury. Salmon thrombin is a natural coagulant which has analgesic, neuroprotective and anti-inflammatory properties after nerve root injury [11-13], supporting its potential as an alternative to aPC. The effects of salmon thrombin on BBB/BSCB integrity have not been determined.

This study assesses whether BSCB breakdown leads to radicular pain and if blocking its disruption after neural trauma prevents pain development. First, the extent of BSCB breakdown after different nerve root compression injuries that are painful or non-painful [14] is defined in the rat. Based on those results, aPC is used to inhibit BSCB breakdown after painful root injury, but before BSCB disruption, to evaluate effects on the resulting pain. Lastly, salmon thrombin is investigated for reducing BSCB breakdown after painful root injury when administered immediately after injury and both pain and BSCB permeability are measured. Together, these studies begin to define whether BSCB breakdown occurs after painful nerve root injury and evaluate a potential treatment for pain and associated CNS pathology.

METHODS

All animal procedures were IACUC-approved. For characterization of BSCB breakdown, male Holtzman rats underwent a unilateral C7 dorsal root compression using a 10gf clip applied for 15 minutes (painful), 3 minutes (non-painful) or 0 minutes (sham) [14]. Behavioral sensitivity (i.e. pain) to a mechanical stimulus was measured in the ipsilateral forepaw before injury (day 0) and for up to 7 days after injury using von Frey filaments. Differences in sensitivity between injury groups were measured using a two-way repeated measures ANOVA (group X day) with Tukey's post-hoc test. To measure BSCB permeability, the C7 spinal cord was harvested from rats on either day 1 (painful n=5, non-painful n=7, sham n=5) or 7 (painful n=5, non-painful n=5, sham n=4), fixed with 4% paraformaldehyde, and labeled for rat IgG, a serum component not typically found in the spinal parenchyma [15]. The extent of IgG in the dorsal horn was quantified on the side ipsilateral to injury using a custom MATLAB program. The amount of IgG was compared between groups using a two-way ANOVA with Tukey's post-hoc test.

To test if blocking BSCB breakdown inhibits pain development, aPC was administered after painful root compression. Rats received a

15 minute root compression (painful n=5) or sham procedure (sham n=4), as described above. In a separate group of rats (n=6) 0.2mg/kg of aPC (Haematologic Technologies Inc.) was given at 5ml/kg in saline intravenously at 1 hour after compression. Behavioral sensitivity was measured before (day 0) and on day 1 after injury; the withdrawal threshold to a mechanical stimulus was normalized to pre-injury values and averaged for each group. Bilateral C7 spinal cord tissue was labeled for IgG and labeling was quantified as above.

To test whether salmon thrombin (STh) blocks BSCB breakdown, STh (1U/ml, Sea Run Holdings) in neurobasal (NB) media was applied to the nerve root immediately after its compression (STh n=5). Additional groups of rats served as controls, receiving either vehicle treatment of NB media immediately after root compression (n=5) or a sham compression with no treatment. Behavioral sensitivity was assessed on days 0 and 1; bilateral C7 spinal cord was harvested on day 1, labeled for IgG and quantified and averaged for each group.

Differences in pain responses between injury and treatment groups were separately assessed for the aPC and the STh studies using two-way repeated measures ANOVAs (group X day) with Tukey's post-hoc tests. Differences in IgG between the injury and treatment groups were also separately compared in those studies using two-way ANOVAs (group X side) with Tukey's post-hoc tests.

RESULTS

Painful root compression induces significantly more mechanical hyperalgesia over 7 days compared to either a non-painful compression ($p<0.001$) or sham ($p<0.001$), which themselves are not different from each other. IgG in the ipsilateral spinal cord is significantly elevated ($p<0.001$) on day 1 after a painful compression compared to both non-painful compression and sham (Fig. 1). By day 7 after a painful compression, IgG is significantly reduced ($p<0.001$) from its levels on day 1, and is not different from levels observed in the non-painful compression or sham at either time point (Fig. 1).

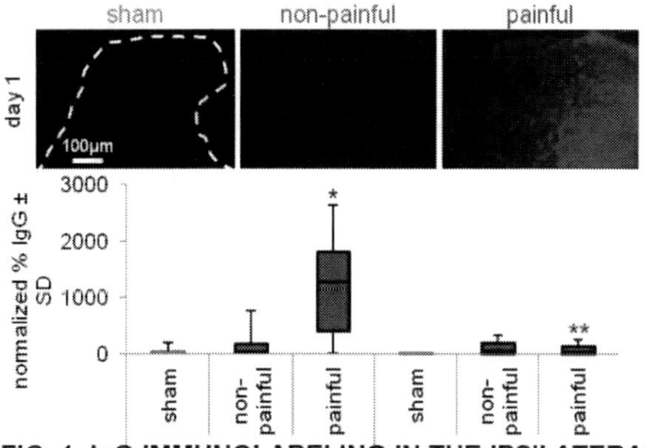

FIG. 1. IgG IMMUNOLABELING IN THE IPSILATERAL DORSAL HORN INCREASES (*P<0.001) ON DAY 1 AFTER PAINFUL COMPRESSION BUT SUBSIDES (P<0.001) BY DAY 7.**

On day 1 after a painful root injury, IgG increases only in the ipsilateral spinal cord and is significantly elevated ($p=0.028$) above levels in the contralateral side. Treatment with aPC significantly decreases ($p<0.001$) ipsilateral IgG labeling compared to untreated painful injury. In addition, aPC also significantly increases (*$p=0.034$) the withdrawal threshold to a mechanical stimulus on day 1, which is not different from sham (Fig. 2), supporting its preventing pain onset.

STh significantly prevents ($p=0.02$) the development of behavioral sensitivity on day 1, with pain responses being not different from sham. STh also significantly reduces ($p<0.001$) ipsilateral IgG compared to painful injury with the NB media vehicle. In addition, the IgG response on the side ipsilateral to injury after STh treatment is also not different from its corresponding uninjured contralateral side.

DISCUSSION

This work demonstrates that BSCB disruption occurs only after *painful* radicular injury and that inhibiting that BSCB disruption with aPC or salmon thrombin blocks the development of pain (Figs. 1 & 2). The induction of BSCB breakdown by day 1 after painful root compression is consistent with previous studies of increased BSCB permeability induced 6 to 24 hours after painful peripheral nerve injury, along the same time frame as pain [3,5]. BSCB permeability resolves by day 7 when pain is still present (Fig. 1) [14], suggesting that BSCB breakdown facilitates *only* the initiation of pain, rather than its maintenance. Since spinal astrocytes and microglia become activated by day 7 after painful root compression [14] in the same spinal region where BSCB breakdown occurs, it is possible that the early BSCB breakdown that is observed here may also regulate the sustained glial activation that influences the maintenance of pain.

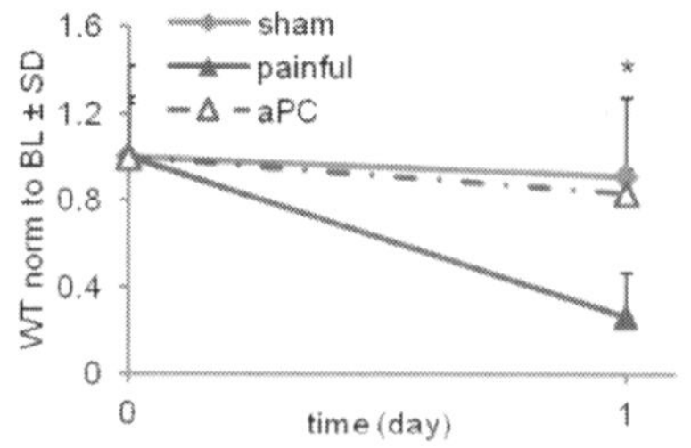

FIG. 2. BEHAVIORAL SENSITIVITY AT DAY 1 IS PREVENTED (*P=0.034) WITH aPC TREATMENT.

Treating painful radicular injury from tissue compression with aPC before the BSCB breaks down can block its disruption and inhibit the development of pain (Fig. 2). This effect is also observed with the analgesic salmon thrombin which inhibits both the early BSCB breakdown after painful injury and the development of persistent pain. Since both of these treatments elicit the same behavioral and BSCB responses, salmon thrombin may be an ideal alternative to block BBB breakdown after traumatic injury, but without contributing the same risk of increased bleeding that aPC presents [10]. The ability of salmon thrombin to block BBB breakdown without adversely affecting coagulation has utility as a treatment for a wide variety of pathological disease states, ranging from sepsis to leaky vasculature in the tumor microenvironment that fuels cancer cell proliferation [16].

ACKNOWLEDGEMENTS

Grant support from a DOD grant (W81XWH-10-1-1002), an NIH training grant (T32-AR007132) and an Ashton Fellowship.

REFERENCES

[1] Breivik, H et al., *Eur J Pain*, 10:287-333, 2006.
[2] DeLeo, J et al., *Pain*, 90:1-6, 2001.
[3] Echeverry, S et al., *J Neurosci*, 31:10819-10828, 2011.
[4] Webb, A et al., *J Vet Intern Med*, 14:399-411, 2000.
[5] Beggs, S et al., *Molecular Pain*, 6:74.
[6] Ballabh, P et al., *Neurobiol Dis*, 16:1-13, 2004.
[7] Yang, G et al., *J Neurosurg*, 81:93-102, 1994.
[8] Cheng, T et al., *Nature Medicine*, 12:1278-1285, 2006.
[9] Wang, Y et al., *Stroke*, 44:3529-3536, 2013.
[10] Bernard, G et al., *N Engl J Med*, 344:699-709, 2001.
[11] Michaud, S et al., *Thromb Res*, 107:245-254, 2002.
[12] Smith, J et al., *PLoS One*, 8:e80006, 2013.
[13] Weisshaar, C et al., *Biomaterials*, 32:9738-9746, 2011.
[14] Rothman, S et al., *J Neurotrauma*, 27:1-12, 2010.
[15] Winkler, E et al., *Proc Natl Acad Sci*, 111:E1035-E1042, 2014.
[16] Zhou, W et al., *Cancer Cell*, 25:501-515, 2014.

SB³C2015
Summer Biomechanics, Bioengineering and Biotransport Conference
June 17-20, 2015,Snowbird Resort, Utah, USA
SB³C2015-320

3D PRINTING OF *B. MORI* SILK PROTEINS FOR IMPLANTABLE DEVICES

M. Torculas (3), J. Medina (2), B. Boszczuck (1), K. Meehan (1), I. Miller (1), P. Vu (2)
X. Hu (4), W. Xue (1), T. Merrill (1)
(1,2,3,4)

(1) Mechanical Engineering
Rowan University
Glassboro, NJ, United States

(2) Chemical Engineering
Rowan University
Glassboro, NJ, United States

(3) Electrical and Computer Engineering
Rowan University
Glassboro, NJ, United States

(4) Physics
Rowan University
Glassboro, NJ, United States

INTRODUCTION

As 3D printing becomes an essential tool for manufacturing specialized biomedical devices, viable materials for 3D printing must be explored. Silk from the silkworm, *Bombyx mori*, has been used for centuries in biomedical applications such as suturing, and there have been recent studies showing how the core silk fibroin fibers show biocompatibility comparable to other common biomaterials, such as polylactic acid and collagen [1]. However, the silk fibroins hold many advantages over similar materials.. The composition of the silk can be adjusted depending on the need, and some silks have novel mechanical properties that are superior to many other natural fibers [1]. Unfortunately, extrusion of a solid silk crystal filament would require immense heat to liquefy making standard 3D printing methods nearly infeasible [2]. This means that a novel approach to 3D printing is needed.

METHODS

Silk fibers were separated from *B. mori* cocoons using a boiling water and sodium carbonate solution (5 g silk and 8.5 g sodium carbonate). After 30 min in boiling water, the fibers were rinsed three times in distilled water and left to dry overnight.

To get the silk into a usable form, a solution of silk and formic acid (FA) was used. With formic acid, "silk nanofibrils seemed also to be dispersed instead of being destroyed" [3]. The silk-FA solutions were 27 % (w/v) solutions (1 mL formic acid and 270 mg silk). An additional 40 mg of calcium carbonate was added to aid in silk dissolution. When the solution was heated, formic acid would evaporate, and the dissolved silk fibers would recrystallize. The amount of formic acid evaporated was determined (see Table 1 under *Results*). High FA loss in less time supports the solution as a feasible 3D printing material (shows quick layer formation).

To test 3D printing feasibility, silk-FA solutions were manually extruded from a standard syringe to Teflon strips. The strips were heated via hot plate and temperatures were measured using K-type thermocouple wire and a National Instruments cDAQ-9172. Extruded solutions were observed for resolidification of silk solutions and evaporation of formic acid while heating.

RESULTS

A silk layer about 50 μm thick was successfully extruded manually from a syringe. The evaporation of formic acid from the solutions resulted in a brittle product with each test. Figure 1 below shows the dried silk-FA solution after heating for 5 minutes. Sample thickness was measured with calipers.

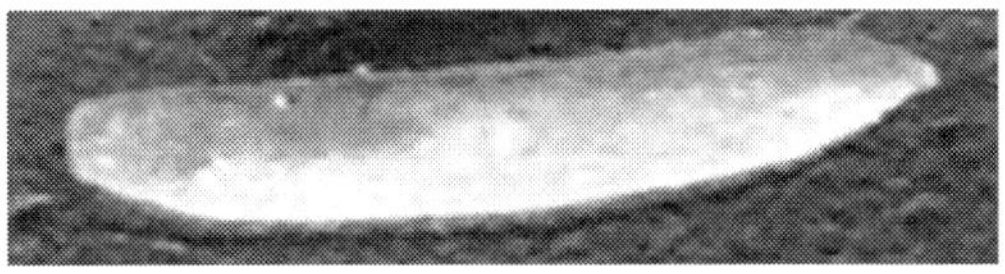

FIGURE 1: Dried silk-FA film about 50 μm thick post formic acid evaporation after 5 minutes heated at 97 °C.

FTIR (Fourier Transform Infrared Spectroscopy) was used to check for beta-sheet formation. Beta-sheets are a type of secondary protein structure caused by the hydrogen bonds in the backbone atoms of the protein. For silk-FA solutions, increased beta-sheet presence induces recrystallization in solutions as the formic acid evaporates, meaning products with increased beta-sheet presence are more solid when the formic acid has been completely evaporated [4]. The 27 % silk-FA solutions were subjected to varying conditions, such as heating the solution before extrusion or washing the product after FA evaporation, to determine ideal treatment for beta-sheet presence. Results are graphed below in Figure 2, and further descriptions of the processes tested can be found after the graph.

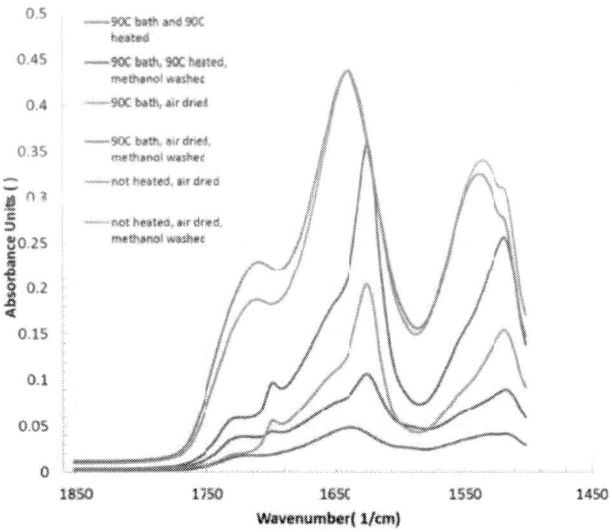

FIGURE 2: FTIR results for 27 % silk-FA solutions under different treatments. In four cases the solutions were warmed in a 90C bath before extrusion. Two cases used heated drying and three used a methanol wash.

The peaks of interest (high absorbance) are 1625 cm^{-1} and 1641 cm^{-1} (characteristic peaks of strong *intermolecular* and *intramolecular* beta-sheets respectively). The line colors in Figure 2 show the results based on six different treatments to the silk-FA solution.

Table 1 shows the silk-FA evaporation results. This data pertains to heating the solution for 5 minutes at 97 °C.

TABLE 1: - Formic Acid (FA) loss over 5 minutes at 97 °C

Sample	Mass FA, Initial [g]	Mass FA, After [g]	% loss of FA
1	0.0145	0.0000[a]	100.00
2	0.0163	0.0007	95.78
3	0.0279	0.0035	87.31
4	0.0276	0.0030	89.05
Average % loss FA			93.04

[a] 100% FA loss to four significant figures.

DISCUSSION

One of the most challenging hurdles to overcome was determining what form of the material to use for printing. Literature on using *B. mori* silk fibroin in 3D printing is currently non-existent, so the team needed to determine the best way to approach the problem. Originally, custom silk filaments were the chosen direction. This would make designing the final 3D printer simpler, since it could be modified from a standard model in the market. Rough experiments showed that this would be difficult to do with the crystallized silk samples, requiring a heat rate of 2000 K/s to melt. Instead, a method similar to that typically used in bio-printing was determined to be more convenient. Essentially, a reservoir of the silk-FA solution will be held in a chamber and printed in a manner similar to regular inkjet printing.

The results of the research prove that the *B. mori* silk-FA solution is an adequate printing material. A solidified layer about 50 μm thick was extruded manually from a syringe. Further experiments shortened the time to achieve this solidified layer to 5 minutes. Future tests aim to prove the possibility of consistent multi-layer extrusion controlled with microcontrollers. In preparation for these tests, a linear motion track was designed, simulated (using SOLIDWORKS), and constructed for automated and consistent solution extrusion for future tests.

Going forward we seek to design and fabricate a 3D printer that can extrude using the silk-FA solution as printer material and test various silkworm silks for 3D printer feasibility. Other silk types underwent formic acid evaporation tests similar to *B. mori* to obtain low FA concentration, solid films. The additional silks tested were: Muga, Eri, Thai, and Tussar. Conditions that need to be analyzed include the degradation of the printed product over time and its dissolution in water and in an environment that mimics that of the body. To design the overall printer, the best heating methods for multi-layer printing and the storage method for the silk-FA solution must also be explored. No matter the future steps taken, the overall conclusion is that *B. mori* can be used in 3D printing for biomedical applications.

ACKNOWLEDGEMENTS

Project research was primarily funded through the NIH Bioengineering Team Design Fund, Grant No.: 1R25EB014782-01A1. Acknowledgments must also be given to Dr. Xiao Hu for his significant contributions to the project. As an expert on the *B. mori* silk fibroin, the team was able to use his equipment and supplies for experiments to pinpoint the ideal composition of the printable silk solution.

REFERENCES

[1] Gregory H. Altman *et al.* "Silk-based biomaterials." Biomaterials 24 (2003) 401-416. Web.

[2] Cebe Peggy *et al.* "Beating the Heat - Fast Scanning Melts Silk Beta Sheet Crystals." Scientific Reports. 3:1130 (2013) 1-7. Web.

[3] In Chul Um *et al.* "The role of formic acid in solution stability and crystallization of silk protein polymer." International Journal of Biological Macromolecules 33 (2003) 203–213. Web.

[4] Hu, Xiao, David Kaplan, and Peggy Cebe. "Determining Beta-Sheet Crystallinity in Fibrous Proteins by Thermal Analysis and Infrared Spectroscopy." *Macromolecules* 39.18 (2006): 6161-170. Web. 20 Oct. 2014.

SB³C2015
Summer Biomechanics, Bioengineering and Biotransport Conference
June 17-20, 2015,Snowbird Resort, Utah, USA
SB³C2015-322

DOES AORTIC OPEN DISTAL AND HEMI-ARCH PROCEDURE REMOVE ALL TISSUE SUSPECTED FOR PROGRESSION OF BICUSPID VALVE AORTOPATHY?

Alex J. Barker (1), Pim van Ooij (1), David Guzzardi (2), Emilie Bollache (1), S. Chris Malaisrie (3), Patrick M. McCarthy (3), James Carr (1), Jeremy Collins (1), Michael Markl (1,4), Paul W.M. Fedak (2,3)

(1) Department of Radiology
Northwestern University
Chicago, IL USA

(2) Department of Cardiac Sciences
University of Calgary
Calgary, AB Canada

(3) Division of Surgery-Cardiac Surgery
Northwestern University
Chicago, IL USA

(4) Department of Biomedical Engineering
Northwestern University
Chicago, IL USA

INTRODUCTION

Bicuspid aortic valve (BAV) is associated with a high risk for aortopathy, including aortic dilation, aneurysm formation, and dissection (20-85% incidence)[1]. In cases with significant aortic disease, aggressive resection strategies of the ascending aorta are sometimes warranted. The American College of Cardiology and American Heart Association recommend surgical intervention in BAV subjects based on a consensus threshold for ascending aortic diameter (>5.5 cm, or >4.5-5.0 cm with additional risk factors)[2]. However, there are no consensus recommendations as to the extent or type of surgery and no data to support the extent of aortic resection. Thus, the question remains: should the surgeon risk (1) hypothermic circulatory arrest with open beveled hemi-arch anastomosis, thereby removing most of the ascending aorta native tissue, or (2) standard replacement of the proximal ascending aorta, leaving a remnant of the distal ascending aorta at the level of the cross-clamp? Option 1 presents a higher risk for near-term morbidity and mortality, but may completely remove tissue at risk for further complications. Option 2 is less risky in the near term, but may leave compromised tissue at risk for further aneurysm formation.

Recent data indicate that 4D flow MRI can identify regions of the aorta at risk of altered tissue architecture and signaling pathways through the use of wall shear stress (WSS) heatmaps[3, 4]. While additional validation is needed, it may be that regions of elevated WSS should be resected during aneurysm repair. Based on this critical assumption, this study surveys the efficacy of standard aortic resection practices at our institution to include areas hypothesized to be at risk for disease progression, as determined by WSS measured from preoperative 4D flow MRI (i.e. time-resolved 3D phase contrast MRI with 3-directional velocity encoding).

METHODS

33 subjects were enrolled for this study with IRB approval and informed consent. Of those, 23 BAV patients (49±13yrs) undergoing ascending aortic repair received preoperative standard of care contrast-enhanced MR angiography (CEMRA) and a research 4D flow MRI exam. 4D flow MRI was implemented with prospective ECG gating and a free-breathing navigator to image the thoracic aorta at 1.5 and 3T (Espree, Avanto, Skyra, Aera, Siemens, Erlangen, Germany). Spatial resolution was 1.7-3.6x1.7-2.7x2.2-3.0mm^3; temporal resolution was 36-43ms (13-23 timeframes per cardiac cycle); echo

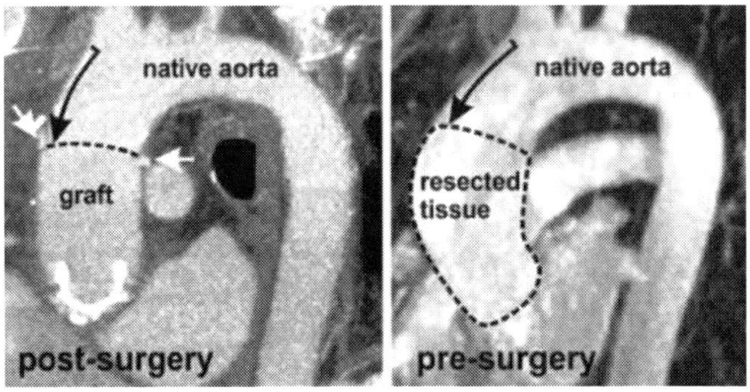

Fig. 1. The region of resected tissue was determined by pre- & post-surgical CEMRA / CTA. The anastomosis (left, white arrows) and surgical notes indicated the type of procedure and the location of removed tissue. The distance of the anastomosis from the brachiocephalic (left, black arrow) was used to determine the resection region (right).

time/repetition time/flip angle were 2.2-2.8ms/4.5-5.3ms/7-15° and the encoding velocity (VENC) ranged from 150 to 400cm/s depending on degree of valve disease. The surgeon was blinded to the results and resected tissue according to the procedure and standard of care. Tissue was labeled by aorta location and reserved for histology, biomechanical testing, and multiplex analysis (results previously reported[4]). Finally, postoperative CEMRA or CT angiography (CTA) identified the exact region of resection (Fig. 1). CTA imaging was performed on a dual-source 64-slice scanner (Somatom Definition, Siemens, Erlangen, Germany) with prospective ECG synchronization, and administration of 110 mL of iodinated contrast.

10 age-matched control subjects (50±14yrs) with healthy tricuspid aortic valves were examined using the aforementioned 4D flow MRI protocol (VENC=150 cm/s). For all BAV and control subjects, a 3D segmentation of the thoracic aorta was performed and WSS calculated. The control 3D WSS maps were registered, averaged and used to compute a 95% confidence interval (CI) representing physiologically normal WSS values[3]. The control-averaged values were registered to each BAV patient, thereby enabling the creation of a heatmap. These BAV heatmaps identified WSS values in the aorta that were outside of the 95% WSS CI (abnormally high values were highlighted in red while abnormally low values were highlighted in blue Fig. 2). WSS above the healthy 95% CI classified tissue at risk (as represented by dark gray shading, Fig. 2). Risk was only classified inside the region treatable by open distal/hemi-arch repair. A Wilcoxon rank-sum test was used to test if the difference of remaining at risk tissue between the two surgical procedures was significant.

RESULTS & DISCUSSION

Preoperative mean aortic diameter in the BAV sinus of valsalva (SOV) and mid-ascending aorta were 4.6±0.5 and 4.8±0.6 cm, respectively; control diameters were 3.0±0.5 and 2.9±0.5 cm, respectively (p<0.001). 50% of patients had severe aortic stenosis or insufficiency. All patients had aortic WSS above the physiologic norm (Fig. 2). All 12 patients who underwent open distal and hemi-arch repair had complete removal of "at-risk" tissue as defined by elevated WSS. In all but one case, resection with clamp on would have resulted in residual at-risk regions (1 case had 20% more tissue resected than at risk). Of the 11 patients with the occluding clamp left on: 4 had regions at risk that matched/were smaller than the resected regions (-9±6% of resection area) while 7 had remaining tissue at risk (33±18% of resection area). The average at-risk region remaining for the 'clamp on' cohort was 17±25% of the resection area. Between the two procedures, a significant difference existed regarding the absolute amount of at-risk tissue remaining (p=0.02, Fig. 3).

Preliminary results have demonstrated evidence for the contribution of WSS to tissue dysfunction[4]. Based on the hypothesis that WSS biomarkers could be predictive for aortopathy progression, we set out to survey the efficacy of resection strategies to remove "at-risk" tissue at our institution. These results demonstrate that in selected patients, aggressive resection using open distal/hemi-arch repair may be necessary. Indeed, we observed that less aggressive resections without an open distal anastomosis did not always completely remove at-risk regions. With further validation, 4D flow MRI could be useful for guiding patient-specific resection strategies.

ACKNOWLEDGEMENTS: AHA14POST20460151, NIH K25 HL119608, NIH R01 HL115828, & Bluhm Cardiovascular Institute.

REFERENCES [1] Verma, S, *et al.* JTCS, 2013. **[2]** Nishimura, RA, *et al.* JACC, 2014. **[3]** van Ooij, P, *et al.* Ann Biomed Eng, 2014. **[4]** Guzzardi, DG, *et al.*, in *AHA Scientific Sessions: Chicago* 2014.

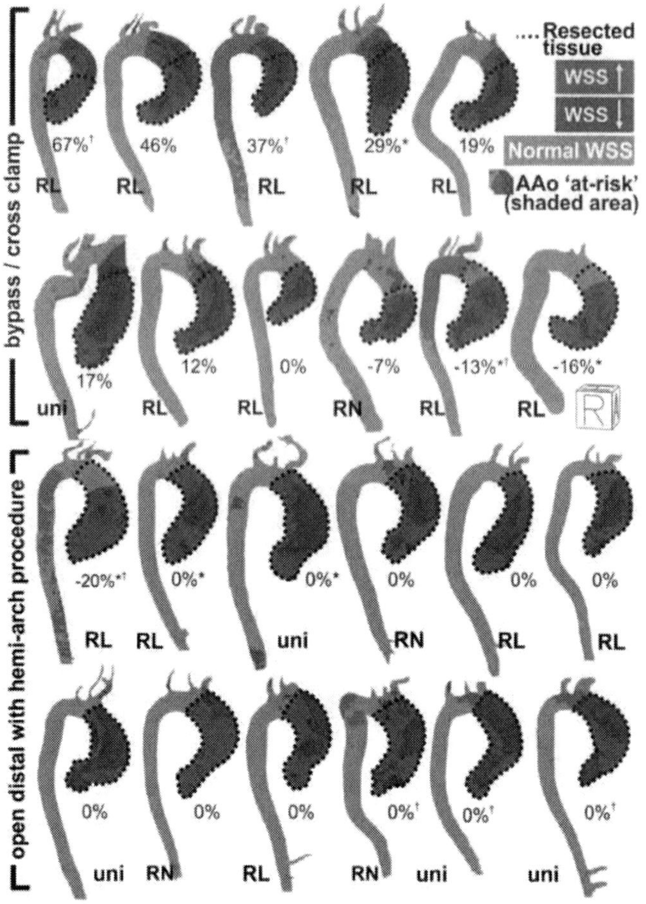

Fig. 2. Aortic 4D flow WSS heatmaps indicate ascending aorta regions at risk for aortopathy (dark shading) for each BAV patient. At-risk tissue regions are compared to surgically resected regions (dashed lines, see methods & Fig. 1). Differences between the area at risk and the resected area are provided in percentages. [RL: right-left coronary leaflet fusion; RN: right-non coronary leaflet fusion; uni: unicuspid; *at-risk region not fully visible in the illustrated view; † notable at-risk regions beyond hemi-arch repair.]

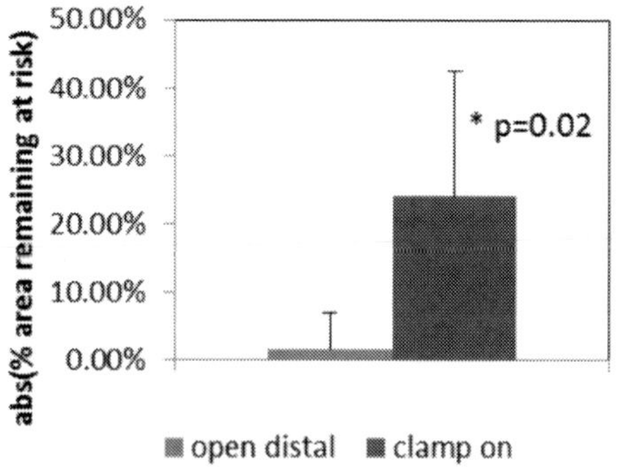

Fig. 3. Absolute percentage of remaining area of aortic tissue at risk in BAV patients, according to the performed surgical procedure.

SB³C2015
Summer Biomechanics, Bioengineering and Biotransport Conference
June 17-20, 2015, Snowbird Resort, Utah, USA
SB³C2015-323

THE EFFECT OF ELEVATED INTRAOCULAR PRESSURE ON CONVECTIVE FLOW IN THE VITREOUS

Julie E. Whitcomb (1), Mohammad R. Kazemi (2), Michael R. Robinson (1), Anita Penkova (3), Sati Sadhal (3); Mayssa Attar (1), Susan S. Lee (1)

(1) Non-Clinical Development
Allergan, Inc.
Irvine, CA, USA

(2) Mechanical Engineering
Independent CFD Consultant
San Jose, CA, USA

(3) Mechanical Engineering
University of Southern California
Los Angeles, CA, USA

INTRODUCTION

Intravitreal injections of corticosteroids and anti-vascular endothelial growth factor (VEGF) molecules are currently the standard of care for sight threatening retinal conditions such as diabetic macular edema and neovascular age-related macular degeneration.

Our major question was to answer how convection currents in the vitreous humor (VH) have an impact on the transport and clearance of large and small molecular weight compounds (i.e. anti-VEGF's and corticosteriods). We initially varied the intraocular pressure (IOP) experimentally and determined that a significant change in VH velocity was observed in different IOP conditions. This information was then utilized in the computational fluid dynamic (CFD) simulation for compounds with varying molecular weight that model corticosteroids and anti-VEGF.

METHODS

A perfusion model was developed based off of a previously described ocular perfusion system [1] in freshly enucleated bovine eyes to determine the physiologic velocity of AH movement in the vitreous. A contrast agent was perfused (MW 900 Daltons) in the ophthalmic artery and a serial set of MRI images were taken over time to calculate the changes in the contrast agent concentrations to determine AH velocity in the vitreous. Both normal and elevated IOP conditions were simulated and used to determine accurate input parameters for a CFD model.

A 3D axisymmetric CFD model of the human eye was developed and solved for diffusion and convection using ANSYS Fluent.

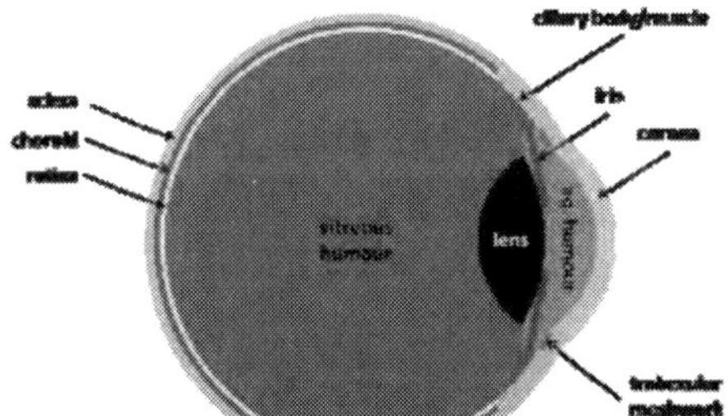

Figure 1: Finite Element Model of Anatomical Realistic Human Eye

Drug movement in the vitreous is dependent on both convection and diffusion currents driven by the intraocular pressure driving the velocity of the AH. The Darcy flow equation, which describes the flow through a static, incompressible porous medium, is applied to the AH fluid movement

(1) $$v = -\frac{K}{\mu}\nabla P$$

where K is the permeability of the vitreous and μ is the viscosity of the permeating AH. The term K/ μ is collectively referred to as the hydraulic conductivity and is modified appropriately in the other porous tissues in the eye. To model the drug distribution in the eye the mass balance equation was used for the drug:

$$\frac{\partial \phi}{\partial t} + \mathbf{v} \cdot \nabla \phi = D \nabla^2 \phi + S \quad (2)$$

where φ represents the concentration of the drug, v is velocity of AH humor permeation from equation 1, D is diffusion coefficient of the drug in the specified tissues, and S is the sink term. Pressure boundary conditions were defined on the sclera and cornea to simulate physiologicepiscleral venous pressure and atmospheric pressure, respectively. The viscous resistance on the trabecular meshwork was iteratively determined to maintain a IOP of 15 mmHg.

Simulations of a pars plana intravitreal injection of a small molecular weight molecule similar to a corticosteroid were conducted with IOPs of 10, 15, and 30 mmHg. Similarly, simulations of a pars plana injection of a large molecular weight biologic molecule similar to an anti-VEGF were also conducted at the same IOPs. Drug transport was distinguishably different in the mathematical model between small and large molecules by varying the diffusion coefficients in the porous tissues in the model.

RESULTS

The perfusion model successfully determined experimentally accurate convection parameters in the perfused eye using the MRI to visually distinguish velocity gradients (Figure 2a). Various velocity profiles were examined along the ocular axis (Figure 2b) and the average IOP for a normal eye (15 mmHg) was 0.47 mm/hr; whereas, an elevated IOP eye (30 mmHg) was 0.95 mm/hr.

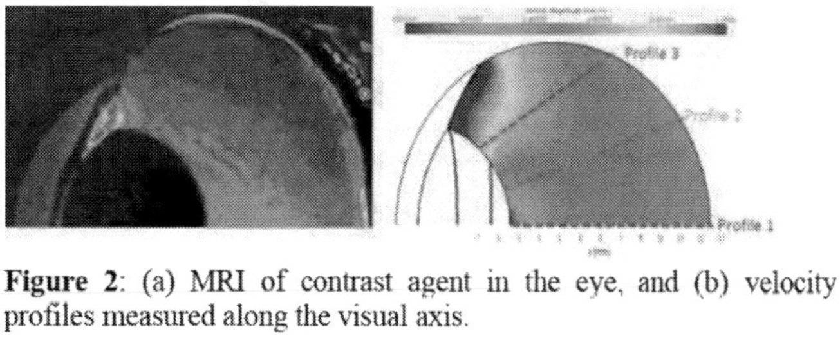

Figure 2: (a) MRI of contrast agent in the eye, and (b) velocity profiles measured along the visual axis.

In the CFD model the experimentally determined velocity measurements were utilized to determine the volume-averaged drug concentrations over a month duration for the varying IOP's. The volume-averaged velocity magnitude in the AH was (8.53, 8.24, and 7.42 mm/hr) and the trabecular meshwork (4.14, 4.00, 3.52 mm/hr) decreased with increasing IOP (10, 15, 30 mmHg), see Figure 3. Velocity in the cornea, sclera, vitreous, retina, and choroid increased with increasing IOP, and the latter anatomic regions demonstrated a significant increase in velocity when the IOP increased from 10 mmHg.

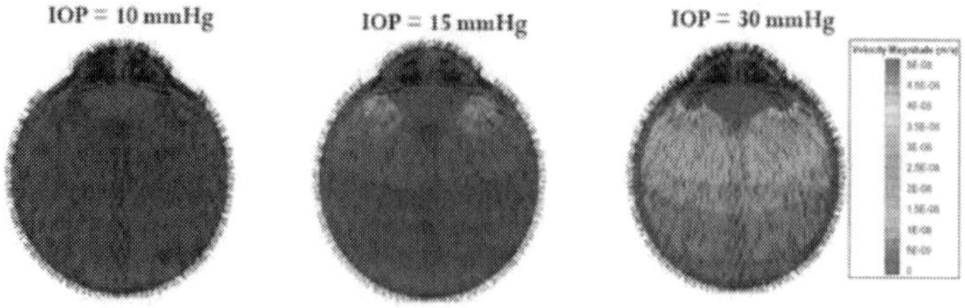

Figure 3: Velocity contour plots (m/s) with velocity vectors overlaid on the axisymmetric plane for 10, 15, and 30 mmHg

The drug residence time was prolonged in the vitreous (Figure 4) for lower IOP values suggesting increased clearance with increasing IOP.

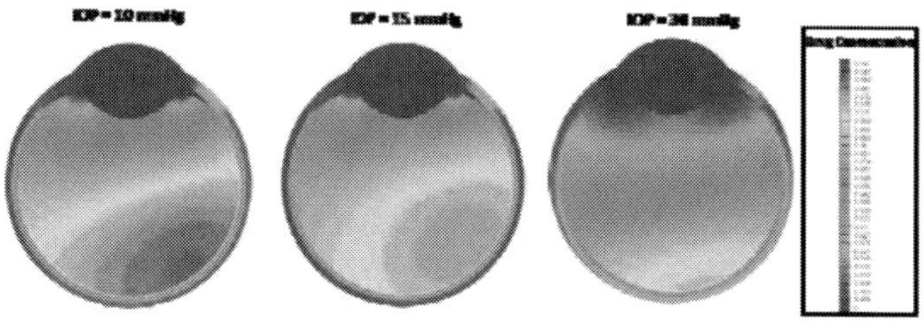

Figure 4: Drug concentration contour plots on the axisymmetric plane for 10, 15, and 30 mmHg. Drug concentration is represented as a scalar value.

The vitreous simulated half-life of the small and large simulated molecules were calculated and were 1.25 days for all 3 IOPs tested for the small molecule, whereas the biologic molecule vitreous half-life was 12 days at 10 and 15 mmHg and decreased to 8 days at 30 mmHg.

DISCUSSION

The in vitro perfusion and CFD model suggest that an elevated IOP increases velocity into the vitreous and may result in an increase of drug clearance specifically for large molecular weight compounds. CFD modeling of intravitreal injections of surrogate corticosteroids and anti-VEGFs give a better understanding of the effect of IOP and molecular weight on convective flow in the vitreous.

ACKNOWLEDGEMENTS

In vitro perfusion work was done in collaboration with University of Southern California and the imaging lab at the Children's Hospital of Los Angeles with Dr. R. Moats, Dr. H. Pollack, and Dr. Y. Tang.

REFERENCES

[1] Koeberle, G et al., *Pharmaceutical Research*, Vol 23. No. 12 2006
[2] Balachandran, R., Barocas,. *Pharmaceutical Research*, 25 (11): 2685-96

SB³C2015
Summer Biomechanics, Bioengineering and Biotransport Conference
June 17-20, 2015, Snowbird Resort, Utah, USA
SB³C2015-324

MODULAR TISSUE ENGINEERED CARTILAGE SURFACES

Audrey C. Ford (1), Kayla Wolf (2), Aditya Nandy (3), Anne Y. Zeng (2), and Grace D. O'Connell (1)

(1) Department of Mechanical Engineering
University of California Berkeley
Berkeley, California, USA

(2) Department of Bioengineering
University of California Berkeley
Berkeley, California, USA

(3) Department of Chemical Engineering
University of California Berkeley
Berkeley, California, USA

INTRODUCTION

Osteoarthritis is a debilitating degenerative disease of the articular cartilage, affecting more than 21% of adults [1]. Treatments, including total joint replacement, have a limited life span in the body, while allografts and autografts have had limited success due to tissue availability [2,3]. Cartilage tissue engineering techniques have shown promise as a potential implantable biological repair strategy. Our approach employs a well-established agarose hydrogel model with cells encapsulated within the biocompatible scaffold to develop functional cartilage *in vitro* [3,4]. This technique has been successful in developing small-engineered cartilage plugs (4 mm diameter) with glycosaminoglycan (GAG) composition and compressive mechanical properties similar to native cartilage [4,5].

Cultivating large-scale engineered tissues has been difficult due to limited nutrient diffusion in larger scaffolds, resulting in inhomogeneous matrix deposition and mechanical properties [5, 6]. Modeled from the native vasculature of cartilage development, micro- and macro-channels have been added to 3D scaffolds to improve nutrient diffusion [5]. Macro-channels, cut into the existing construct, have been shown to decrease spatial variation in larger constructs and maintain the mechanical properties similar to smaller constructs. However, the channels become occluded with deposited extracellular matrix within weeks, when nutrient diffusion is critical [5].

The objective of this study was to employ a novel modular tissue engineering technique to maximize nutrient diffusion during early formative stages, and, then, connect individual constructs to form larger engineered tissue surfaces (Figure 1). This technique takes advantage of the propensity of developing constructs to stick together in culture to synthesize bonds [7]. The negative space between joined constructs serve as macro-channels for increased nutrient diffusion. We hypothesize that the modular engineered tissue surfaces (METS) will maintain the same mechanical and biochemical properties as the individual control constructs.

METHODS

FIGURE 1. Modular engineered tissue surface (METS).

Articular cartilage was harvested from juvenile bovine knees (3 - 6wks, n = 3) and digested with Type IV collagenase serum medium (DMEM, Invitrogen Co., Carlsbad, CA, USA). Chondrocytes were expanded with serum media supplemented with a growth factor cocktail (10% FBS, 1% PSAM, 1ng/mL PDGF, 0.5 ng/mL FGF, and 0.5 ng/mL TGF) [4]. Cells were passaged twice to achieve sufficient cell numbers for tissue engineering studies and to improve matrix production in 3D cultures [8]. Once confluence was reached, passaged cells were encapsulated within agarose for a final concentration of 2% w/v agarose (Type VII, Sigma-Aldrich, St. Louis, MO, USA) at a cell density of $30x10^6$ cells/mL. Constructs were prepared using a 4 mm diameter biopsy punch, and cultured in chondrogenic media supplemented with 10 ng/mL TGF-β3 for the first 14 days in culture [4]. At day 7, constructs were placed into a 3D printed porous mold (9 mm x 9 mm) to cultivate constructs in groups of 4 (2 x 2; Figure 1). Individual constructs served as the control, and both groups were cultured for 28 days.

Compressive mechanical properties (i.e. equilibrium and dynamic moduli) were determined weekly for individual constructs and METS samples (n = 4/5 per group per time point). The equilibrium modulus was measured in unconfined compression at 10% strain, and the dynamic modulus was determined from sinusoidal ±1% strain at 0.5 Hz. Following bulk compressive testing, each group of four was cut in half to isolate a single bond. The tensile strength of the bond was assessed using a tensile strain rate of 1 mm/min until failure. The stiffness was calculated as the slope of the force-displacement curve, and the peak force at failure was recorded. After mechanical testing, the wet weight was recorded and samples were prepared for biochemical assays.

Biochemical characterizations measured the levels of DNA, GAG and collagen in the samples. DNA was measured using a PicoGreen (Invitrogen, Co.) assay kit. GAG was measured using a DMMB assay and collagen was measured using a hydroxyproline assay [4]. Cell viability and distribution was evaluated using a LIVE/DEAD assay (Invitrogen Co.).

A one-way ANOVA was used to compare changes in mechanical and biochemical properties over time. A Student's t-test was used to compare mechanical and biochemical properties of METS and individual constructs at day 28. Significance was set at $p \leq 0.05$. Data are presented as mean ± standard deviation.

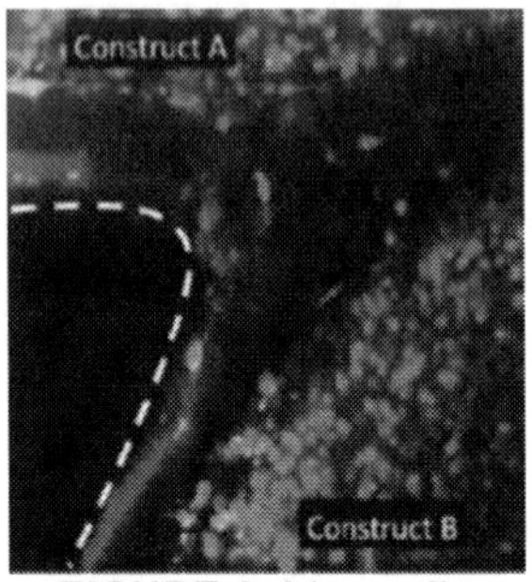

FIGURE 2. Live cells (green) in junction. Oversaturation of stain (red) used to mark contact tissue (line).

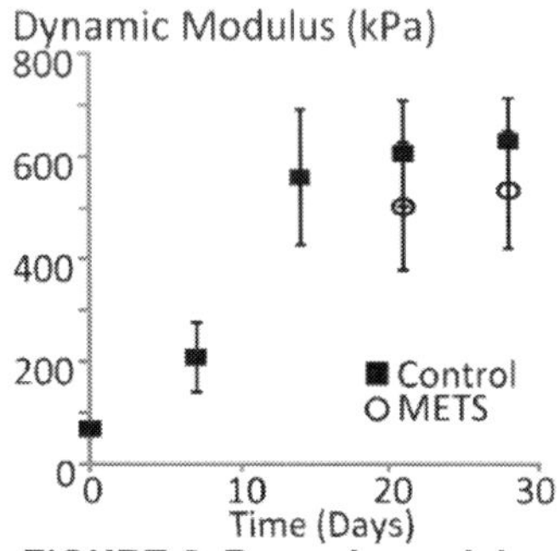

FIGURE 3. Dynamic modulus over time (p > 0.5).

RESULTS

By day 21, stable connections had formed between constructs in METS samples, allowing for bulk mechanical testing (final dimensions: 8 mm x 8 mm; Figure 1). Tissue deposition was observed optically with small fibers observed connecting constructs (data not shown). Images for cell viability demonstrated cell infiltration into the connections (Figure 2). At day 28, the bond width was 2.2 ± 0.3 mm.

Compressive mechanical properties of METS and individual constructs at day 28 were 3-9X greater than initial properties ($p < 0.01$, Figure 3). At day 28, there were no significant differences in dynamic and equilibrium moduli between individual constructs and METS samples ($p > 0.19$; Figure 4A & B). Tensile mechanical properties of the bonds increased over time (day 21 versus day 28, $p < 0.05$; Figure 4C & D).

Biochemical assays demonstrated no significant differences in the

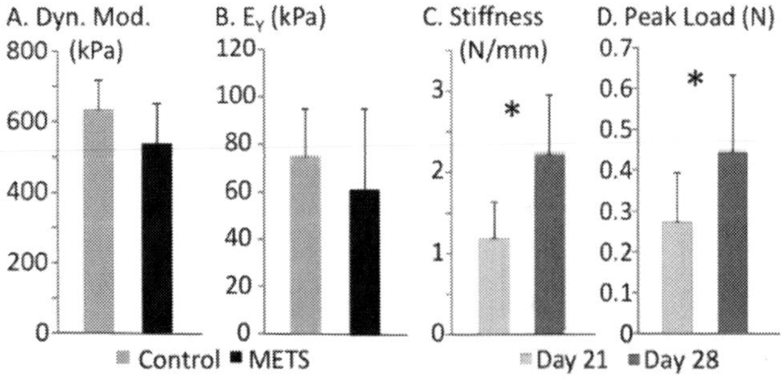

FIGURE 4. (A) Dynamic and (B) equilibrium (E_Y) moduli were not significantly different at day 28. (C) Stiffness and (D) peak force under tension. *p<0.05.

DNA and collagen contents of METS samples compared to the individual control constructs ($p > 0.05$; Figure 5A & B. Interestingly, the GAG content of METS samples was 29% greater than the control ($p < 0.01$; Figure 5C). Similar results were observed for GAG and collagen contents normalized by DNA content and dry weight.

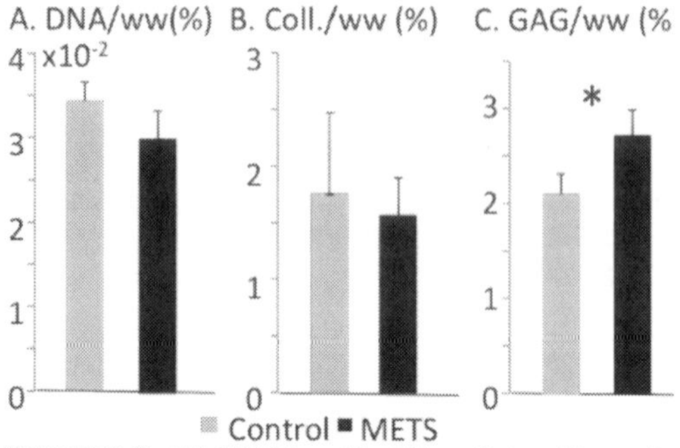

FIGURE 5. (A) DNA/ww%, (B) coll./ww%, and (C) GAG/ww% at day 28. * p< 0.01.

DISCUSSION

This study demonstrated feasibility of creating modular engineered cartilage using the inherent propensity of developing *de novo* cartilage to adhere to neighboring constructs. Mechanical and biochemical properties of METS samples were not significantly different from individual constructs, except for GAG content, which was greater in METS. These findings agree with previous work using macro-channels added to the scaffold during fabrication (i.e. day 0) [5, 6]. In this study, macro-channels remained free of extracellular matrix throughout the culture period (Figure 1). Moreover, the tensile stiffness of the bonds was within the range of values reported for bovine articular cartilage (0.3-2 N/mm) [9]. These results suggest that forming macro-channels later in culture is feasible and may be beneficial for maintaining long-term nutrient diffusion in the channel.

Cartilage tissue engineering has focused on optimizing properties of small tissue constructs, and researchers have been largely successful in achieving native properties for compressive modulus and GAG content [3-6]. However, these constructs are too small to be a clinically viable solution. Recent work in the literature has demonstrated methods for developing large-scale engineered cartilage tissues either through large scaffold fabrication on day 0 [5, 6] or through fractal fabrication techniques [10], where smaller engineered tissue samples are used to create larger surfaces. The findings of this study demonstrate that mature engineered cartilage plugs (4 mm diameter) can be used to develop larger surfaces with mechanical and biochemical properties of individual constructs. Further work will evaluate depth-dependent and location-dependent variations in mechanical and biochemical properties of larger METS samples. In conclusion, the METS approach presented in this study describes a method for scaling functional cartilage tissue engineering techniques towards clinically relevant dimensions.

ACKNOWLEDGEMENTS

This study was funded through the Rose Hills Innovator Program (UC Berkeley).

REFERENCES

[1] Helmick, C+, *Arthr. and Rheum.*, 2008; [2] Hunt, H+, *Amer. J. of Sp. Med.*, 2014; [3] O'Connell, G+, *JKS*, 2012; [4] O'Connell, G +, *eCells and Mat.*,, 2014; [5] Bian, L+, *Osteoarth. and Cart.*, 2009; [6] Nims, R+, *T. Eng. Part C*, 2015 – In Press. [7] Ng, S+, *A. Biomat.*, 2012. [8] Tan, AR+, *Osteoarth. and Cart.*, 2015 – In Press. [9] Danso, E+, *J. of Biomat.*, 2014; [10] Bhumiratana+, *PNAS*, 2014.

SB³C2015
Summer Biomechanics, Bioengineering and Biotransport Conference
June 17-20, 2015,Snowbird Resort, Utah, USA
SB³C2015-325

CORRELATING MOLECULAR STRUCTURE WITH THE MECHANICAL RESPONSE OF AQUATIC CADDISWORM SILK

Nicholas N. Ashton & Russell J. Stewart

Department of Bioengineering
University of Utah
Salt Lake, Utah, USA

INTRODUCTION

The insects known commonly as caddisflies, live most of their lives as larvae occupying numerous freshwater habitats worldwide. The species used in this study (*H. occidentalis)* builds a protective tubular case around its soft larval body by tethering stones with silken ties. The silk fibers comprise a thin sticky peripheral layer encasing a viscoelastic core. The silk is tough, absorbing ~17 MJ/m^3, a property consistent with its structural role in the protective case.[1] At low strain, the fibers are stiff with an initial modulus ranging from 80-140 MPa. At about 3% strain, the fibers abruptly strain soften (pseudo-yield) after which the stress plateaus briefly (Fig1). The fibers strain stiffen gradually before failing at ~130% strain and ~33 MPa stress.[1] Silk deformation beyond the pseudo-yield point is reversible; the fibers fully recover their initial length and stiffness as ruptured domains gradually refold.

The major protein of the viscoelastic core, H-fibroin, consists of repeating $(pSX)_n$ motifs, where pS is phosphoserine and n is 2-6.[2,1] In addition to the high charge density from the pS groups, the N-terminus of H-fibroin contains 17.7 mol% amino acids with carboxylate sidechains.[2] A second major structural component of the fiber core is a proline-rich protein in which 21.3 mol% of the amino acids have carboxylate sidechains.[3] The high negative charge density of the proteins is balanced by the presence of divalent metal ions, predominantly Ca^{2+}. The $(pSX)_n$ motifs can hypothetically assemble into repeating β-sheet nanodomains stabilized by Ca^{2+} (Fig 1B).[1] The carboxylate sidegroups can also form a crosslinked network through complexation with Ca^{2+} ions. Exchange of Ca^{2+} with Na^+-EDTA reduced the initial modulus, toughness, and pseudo-yield stress by over 98% demonstrating the critical mechanical role of the Ca^{2+}-phosphate and Ca^{2+}-carboxylate crosslinked networks.[1]

Our goal is to understand how the unusual mechanical response of caddisfly silk to strain is determined by its molecular structure. This understanding may lead to the development of tough, self-recovering underwater materials using the double Ca^{2+}-dependent network structure of caddisfly silk as a unique source of design principles.

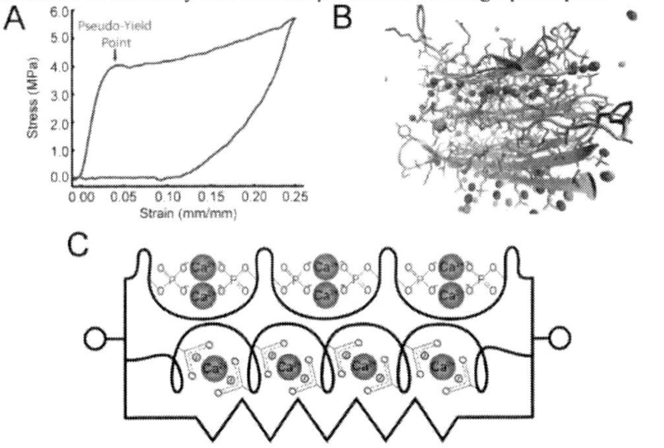

Figure 1: (A) A Representative stress strain curve of caddisfly silk cycled to 0.25 mm mm^{-1}. (B) Molecular structure of Ca^{2+} (Blue spheres)–crossbridged $(pSX)_4$ domains in caddisfly silk. (C) A model of the multiple network structure of self-recovering caddisfly silk, which includes two reversible Ca^{2+}-dependent and a covalent elastic network acting in parallel..

METHODS

pH dependent mechanics.[4] Three independent silk fibers were cyclically strained to 0.2 mm mm^{-1} at 0.03 s^{-1} as the pH was progressively decreased between strain cycles. The test solutions contained 5 mM CaCl2 and 1 mm of the following buffers: Tris–HCl (pH 8.0), ammonium bicarbonate (pH 7.0 and 6.0), acetic acid (pH 5.5

and 5.0), formic acid (pH 4.5 and 4.0) and glycine (pH 3.0). After each pH cycle the fiber was allowed to recover for one hour. Mechanical values including residual strain, and yield stress, were computed for each cycle of the three fibers tested and normalized as a percentage of the first pH 8 cycle values. These data were then fitted with a logistic function.

Determination of side chain pK$_a$ by IR spectroscopy.[4] Vibrational modes for dianionic phosphoserine between 953 and 1300 cm^{-1}, and an anionic carboxylate mode between 1305 and 1477 cm^{-1} were used to titrate the respective functional groups using ATR-FTIR. A batch of silk fibers was sequentially adjusted and scanned while lowering the pH as indicated: pH 8.0, 7.0, 6.0, 5.5, 5.0, 4.5, 4.0, 3.5, 3.0, 2.0, 1.5, and 1.0. Test solutions contained 5 mM $CaCl_2$ in 1 mM Tris–HCl (pH 8.0), ammonium bicarbonate (pH 7.0–6.0), acetic acid (pH 5.5–4.5), formic acid (pH 4.0–3.5) or glycine (pH 3.0–1.0). The spectra were corrected for water, baseline subtracted, and intensity adjusted using a pH insensitive vibrational mode.[4] The area under the vibration modes of interest was integrated, plotted against pH, and fitted with a logistic curve.

Recovery rate.[4] Two sets of fibers, three at each time point, were strained to 0.20 mm mm^{-1} at 0.03 s^{-1} in either 1 mM $CaCl_2$, 5 mM Tris, pH 8.0 or in 1 mM $CaCl_2$, 5 mM formate, pH 3.5. At pH 8.0, recovery times were 0, 0.033, 0.166, 0.5, 2, 5, 15, 30, 60.0, and 120 min. At pH 3.5, recovery times were 0, 0.5, 1.0, 2.0, 5.0, 10.0, and 20.0 min. The magnitude of the yield stress, and residual strain were computed for each cycle and normalized to the initial cycle values. [4]

RESULTS

Mechanical testing of silk fibers over the pH range 8.0-2.0 showed a marked drop in yield stress and fiber shortening that began at pH 5.0 (Fig 2A). Another drop in yield stress was observed to occur from pH 3.0-2.0 (Fig 2A). Logistic curve fitting to the magnitude of the pseudo-yield point and the residual strain vs pH showed inflection points at pH 4.6 and 4.4, respectively. To determine the functional groups responsible for the pH-dependence of fiber mechanics, caddis silk was analyzed by ATR-IR spectroscopy at pHs ranging from pH 8.0-1.0. Several pH sensitive absorption bands were identified. The integrated intensity for the phosphate dianionic degenerate and symmetric stretching modes (Fig 2A Green Trace) displayed inflections points at pH 4.7 and 4.5, respectively, corresponding to the pK$_a$ of the phosphate groups. Likewise, the pK$_a$ of the anionic carboxylate symmetric stretching mode was pH 3.0 (Fig 2A Blue Trace).

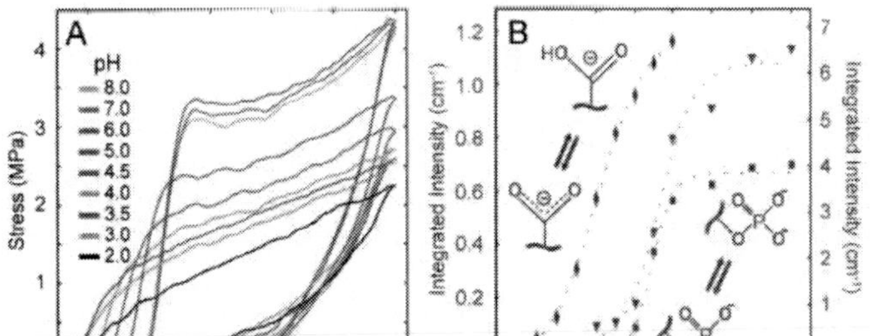

Figure 2: Effects of pH on silk mechanics and chemistry: (A) The mechanics of a caddis silk fiber at pHs from 8.0-2.0. (B) FTIR was used to probe pH dependent integrated intensity for the phosphoserine dianonic degenerate (triangles) and symmetric stretching (squares) and carboxylate symmetric stretching absorbance bands (diamonds).

At neutral pH, silk fibers cycled to 0.2 mm/mm nearly recovered their initial length and yield stress within 120 min. (Fig 3 A-B). Recovery was biphasic; a quick (0-10 min) recovery process was superimposed on a slower (10-120 min) recovery process (Fig 3B). The distinct recovery processes were chemically separated by adjusting the solution pH. At pH 3.5, below the pK$_a$ of phosphates and above the pK$_a$ of carboxylates, the recovery kinetics corresponded to the fast process with near complete recovery within 30 min (Fig 3C).

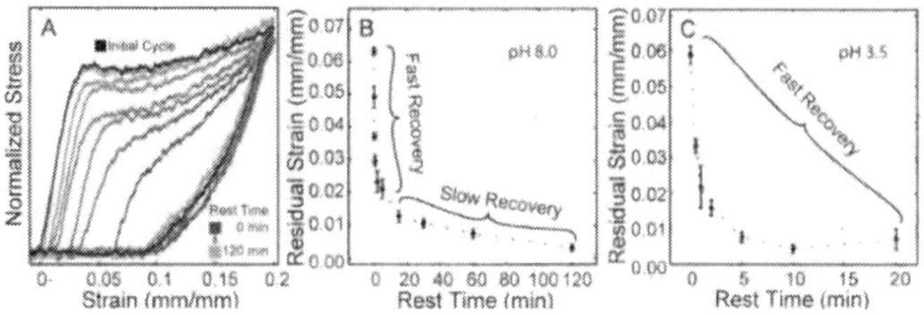

Figure 3: Fiber recovery kinetics. (A) Representative stress strain curves of fibers tested at pH 7.0 with varying amounts of rest time between cycles. Displayed stress-strain curves were normalized using initial cycle values. (B) Recovery of fiber length. For each rest time 3 fibers were cycled to 0.2 mm mm^{-1} and varying the inter-cycle rest time as indicated.

DISCUSSION

The spectroscopically determined pK$_a$ of silk fiber phosphate (~4.6) and carboxylate (~3.0) sidechains corresponded closely to the sigmoidal decreases in pseudo-yield stress with midpoints at pH ~4.6 and <3.0 determined mechanically. These results indicate a direct role for phosphate and carboxylate side groups in the pseudo-yield domain structures that produce the reversible strain softening of silk fibers under cyclical loads. The near complete recovery of mechanical properties fibers within 120 min after unloading, (Fig 3B) implicates the rupture of non-covalent reversible bonds rather than covalent bonds. Two independent recovery processes suggest the phosphate and carboxylates form independent Ca^{2+}crosslinked networks that act in parallel. At pH 3.5, only the fast recovering Ca^{2+}-carboxylate network is observed because the Ca^{2+}-phosphate network is disrupted by phosphate protonation.

The results and hypotheses were consolidated into a model (Fig. 1C) with two parallel Ca^{2+}-dependent networks, phosphate and carboxylate, in parallel with a soft elastic network. The elastic network guides the reformation of ruptured pseudo-yield domains when stress is relieved. Inter-protein covalent crosslinking, including disulfide[5] and dityrosine crosslinks,[3] have been identified which might form the elastic rubber-like covalent network.

ACKNOWLEDGEMENTS

Funding from the Army Research Office is gratefully acknowledged

REFERENCES

[1] N. N. Ashton, D. R. Roe, R. B. Weiss, T. E. Cheatham, and R. J. Stewart, *Biomacromolecules*, 2013, **14**, 3668–81.

[2] R. J. Stewart and C. S. Wang, *Biomacromolecules*, 2010, **11**, 969–974

[3] C.-S. Wang, N. N. Ashton, R. B. Weiss, and R. J. Stewart, *Insect Biochem. Mol. Biol.*, 2014, **54C**, 69–79.

[4] N. N. Ashton and R. J. Stewart, *Soft Matter*, 2014. DOI: 10.1039/c4sm02435d

[5] N. Yonemura, K. Mita, T. Tamura, and F. Sehnal, *J. Mol. Evol.*, 2009, **68**, 641–53.

A NEW METHOD FOR MEASURING STRESS CONCENTRATIONS IN FINITE ELEMENT ANALYSIS OF FIBROCARTILAGE PREDICTS GREATER FRACTURE RISK FOR ANGLED CENTER CRACKS

J.M. Peloquin (1), D.M. Elliott (2)

(1) Department of Bioengineering
University of Pennsylvania
Philadelphia, PA, USA

(2) Department of Biomedical Engineering
University of Delaware
Newark, DE, USA

INTRODUCTION

Cracks are a key driver of ubiquitous musculoskeletal pathologies such as intervertebral disc degeneration, meniscus tears, and osteoarthritis. A crack (or tear, fissure, etc.) is not just a potentially symptomatic failure, but also a stress concentrating defect that can cause further failure. The stress concentration at a crack tip can cause local failure, permitting failure by crack propagation (fracture) at loads that would otherwise be less than the material's strength.

Despite the importance of tissue cracks, they have received very little attention. Fewer than a dozen publications report fracture toughness for fibrous soft tissue, and in most cases the test configuration did not actually produce fracture [1]. These tests were done using edge-cracked specimens in uniaxial tension. Since cracks do grow in vivo, it should be possible to reproduce fracture if different conditions are used. In vivo cracks are often centrally located, biaxially loaded, or orientated at various angles. In this study, we evaluated the effects of these factors using finite element analysis (FEA), with the aim of identifying a test configuration more appropriate for fracture testing. Achieving this aim required the development of a method to quantify the crack tip stress from FEA.

METHODS

FEA was chosen as the investigational tool to permit rapid iteration through the parameters of interest. Cracked specimens were simulated in three tensile loading configurations: center cracked biaxial (CCB), center cracked uniaxial (CCU), and single edge notch (SENT) (Fig. 1). The crack angle was varied relative to the fiber axis from 0° (∥ to fibers) to 90° (⊥ to fibers) in 15° increments. FEBio was used for the simulations.

The specimen material was represented by a mixture model of Holmes-Mow matrix and unidirectional fibers representative of fibrocartilage [2]. The half-length of the notch was 1 mm, the half-width of the simulated specimen was 10 mm, and the thickness was 1 mm. The mesh was made of 8-node trilinear elements and was biased to place smaller elements near the crack (Fig. 1). Since the near-crack stress field is mesh sensitive, the element size and geometry near the crack tip were parameterized and varied in a mesh sensitivity analysis. The smallest element size used was 50 μm; this size was used for all quantification herein. A stretch ratio of 1.14 was applied via the perimeter nodes. In the uniaxial cases, the stretch was applied in the fiber direction. Fiber-∥ stress, fiber-⊥ stress, and fiber- matrix shear stress were computed and used for analysis.

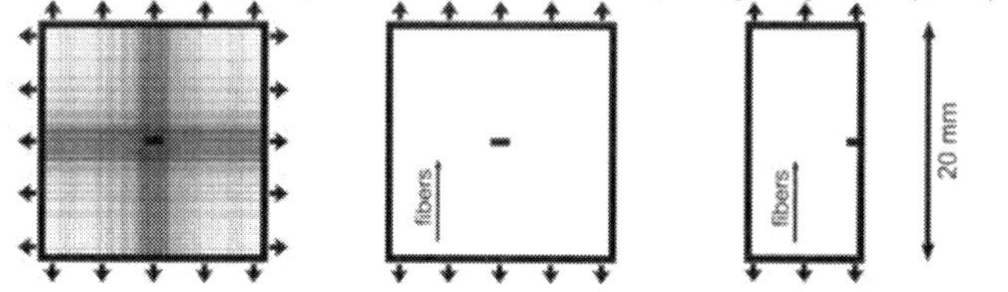

Figure 1. Three loading configurations were compared, illustrated here with 90° cracks. The mesh is shown for the CCB case only; the other cases are meshed in the same manner.

Fracture risk was quantified using the crack tip stress. Stress was sampled at evenly spaced points on a line extending from the crack tip in the same direction as the crack (Fig. 2a). Sub-elements defined on the Gauss points were used for interpolation. The near-tip part of this stress profile was fit with an exponential function to obtain the stress at the crack tip (Fig. 2b). This method was devised to improve upon the common use of maximum element-wise stress as a metric.

The J integral (energy release rate for a unit crack extension) was also calculated as an alternate measure of fracture risk [3].

Tensile tests of physical meniscus specimens were performed for the single edge notch configuration with 90° cracks. Specimens were cut from the mid-circumferential, mid-radial, mid-thickness region of adult (age > 36 months) bovine knee meniscus. Target specimen dimensions were 25 mm × 10 mm × 1 mm, with the circumferential fibers parallel to the long axis. Specimens were loaded in uniaxial tension at 0.5 mm/s, with the tensile axis parallel to the fiber axis. Ten cycles of preconditioning to 4% strain were applied followed by loading to failure. Comparison between cracked ($n = 3$) and control (crack-free; $n = 40$) specimens was done using the log-transformed peak stress and one-tailed Student's t-test (σ_{peak} cracked < σ_{peak} control). The type I error rate was set at 0.05.

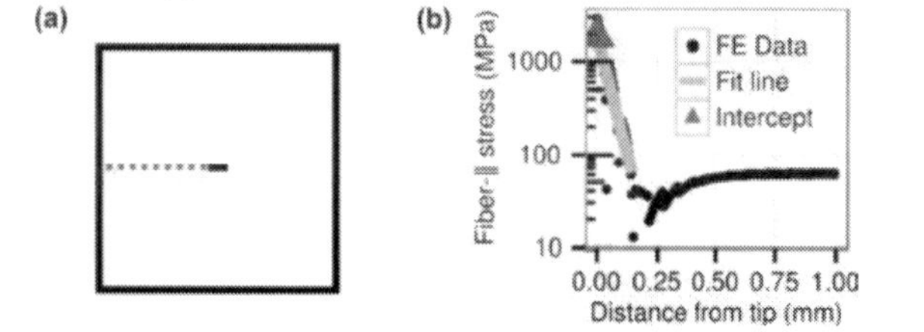

Figure 2. Stress profile quantification. (a) Stress was sampled along a line extending from the crack tip. (b) The stress profile was fit with an exponential function to find the stress at the crack tip.

RESULTS

Fiber-‖ stress at the crack tip varied with both crack angle (relative to the fiber axis) and loading configuration. Cracks ‖ to the fibers (0°) did not increase the stress above the crack-free configuration (Fig. 3). Cracks ⊥ to the fibers (90°) moderately increased the stress concentration. Acutely oriented cracks further increased the stress concentration for the SENT case, but again only by a moderate amount. In contrast, acutely oriented CCB and CCU cracks greatly increased the stress concentration, producing stresses an order of magnitude higher than any SENT specimen. CCB and CCU specimens had essentially equivalent fiber-‖ stress.

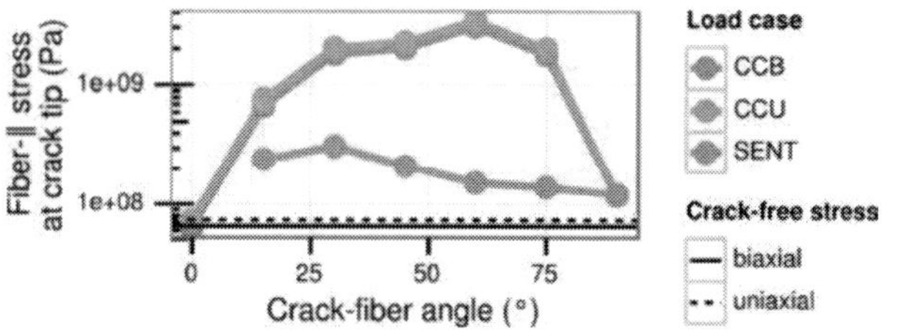

Figure 3. Center-cracked specimens with acutely angled cracks have greater fiber-‖ stress compared to edge-cracked specimens.

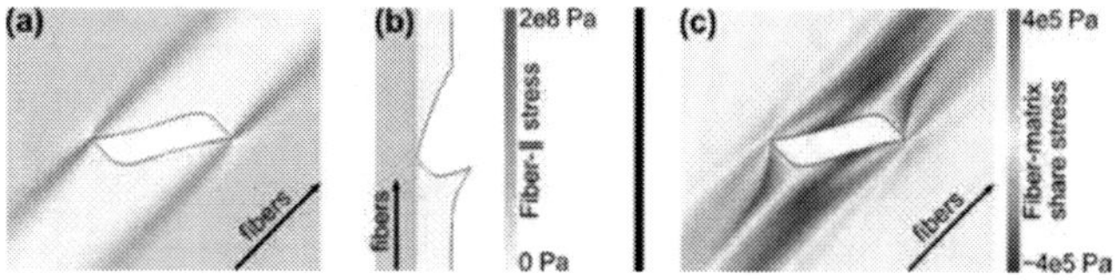

Figure 4. Representative stress fields, with 45° cracks used as examples. (a) CCB specimen, with sharp crack tips. (b) SENT specimen, with crack blunting. (c) CCB specimen, with fiber shear stress concentrations away from the crack tips. Panels a & c would be nearly identical for CCU specimens.

Each stress component behaved differently. Fiber-‖ stress concentrations were centered on the crack tip and followed the fiber axis (Fig. 4a,b). Fiber-matrix shear stress concentrations were located away from the crack tip in CCU and CCB specimens (Fig. 4c), but was centered on the crack tip and followed the fiber axis in SENT specimens (not shown). Fiber-⊥ stress concentrations tended to follow arcs roughly perpendicular to the crack (not shown).

The mesh sensitivity analysis showed an increase in crack tip stress of about half an order of magnitude for each ×0.5 reduction in element size, with no apparent convergence (data not shown). Consistent near-tip mesh geometry was therefore used to permit comparisons between cases. The J integral was less sensitive to element size and converged in some cases, but was numerically unstable in others (data not shown).

Physical testing of 90° SENT specimens showed no crack propagation, nor any significant decrease in peak stress compared with controls. The failure stress was 14 MPa (95% CI = 7–29 MPa) for control and 19 MPa (95% CI = 6–65 MPa) for cracked specimens.

DISCUSSION

The FEA predicted that center cracked specimens with acutely oriented cracks have greater crack tip stress, and therefore greater crack propagation risk, than edge cracked specimens. We attribute this difference to local twisting of the CCB and CCU cracks, which keeps the crack tips sharp (Fig. 4a). In contrast, edge cracks open widely and blunt (Fig. 4b), reducing the stress concentration. The low stress concentrations predicted for edge cracked specimens may be why all but two tissue fracture studies (of stratum corneum and superficial articular cartilage) show no evidence of fracture [1,4]. The 90° SENT meniscus specimens tested in this study were also crack insensitive. Tests with center cracked specimens, however, may produce fracture. Physical tests of this prediction are ongoing.

The CCB and CCU fiber-‖ stress and fiber-matrix shear stress fields were very similar (Fig. 3), so CCU specimens may be useful as a proxy for biaxial loading.

The development of robust methods to quantify stress concentrations in fibrous soft tissue is an ongoing process. Although methods exist for engineering materials, the standard formulas assume small strain or linearly elastic behavior, which limit their application to soft tissue. It is common therefore in biomechanics studies to use the maximum element-wise stress. The use of stress profiles in this study (Fig. 2) is preferable because it samples the stress field at specific points of interest. This greatly reduces random error when severe stress gradients exist, like those created by a crack.

Although the use of stress profiles is an important step for quantifying biomechanics, it is still limited. Profiles may not be fittable, and the crack tip stress is still mesh dependent. The use of consistent meshes (as in this study) makes relative comparisons possible, but the method is not suitable for defining failure criteria. The J integral may circumvent these issues. It was not successful in this study, but we believe this to be due to the use of trilinear elements. Work on a robust quantification method is ongoing.

Since only the fiber-‖ stress could be quantified by the profile method in this study, the predicted trends in fracture risk pertain only to failure by fiber rupture. Fiber-matrix shear and tensile rupture may have different trends. Based on the location and shape of the stress concentrations, each failure mechanism should cause failure at a different location and along a different trajectory.

Quantification of crack-induced stress concentrations is a challenging problem, but important for fibrous soft tissue pathology. This work demonstrates quantification of stress concentrations using a stress profile method, predicting that center-cracked specimens have greater risk of fracture by fiber rupture than edge-cracked specimens. This prediction will guide future physical experiments.

ACKNOWLEDGEMENTS

This work was supported by NIH grant R01AR050052.

REFERENCES

[1] Taylor, D et al., *J Mech Behav Biomed Mater*, 6:139–147, 2012.
[2] Jacobs, NT et al., *J Biomech Eng*, 135:021004, 2013.
[3] Shih, CF et al., *Int J Fract*, 30:79–102, 1986.
[4] Von Forell, GA et al., *J Mech Behav Biomed Mater*, 35C: 77–84, 2014.

SB³C2015
Summer Biomechanics, Bioengineering and Biotransport Conference
June 17-20, 2015, Snowbird Resort, Utah, USA
SB³C2015-327

INTEGRATION OF A SPONTANEOUS RESPIRATORY DRIVER WITH BLOOD GAS FEEDBACK INTO BIOGEARS, AN OPEN-SOURCE, WHOLE-BODY PHYSIOLOGY MODEL.

Y. Gebremichael, R. Clipp, J. Webb, A. Bray, M. Cameron Thames, Z. Swarm, J. Carter, and J. Heneghan

Applied Research Associates
Raleigh, NC, USA

INTRODUCTION

The BioGears® engine (www.biogearsengine.com) is an open-source, general-purpose, whole body human physiology model that allows consistent simulation across the medical training and research communities. The current version of BioGears consists of well-developed respiratory and cardiovascular (CV) models with substance transport modeled both throughout and between the systems. A model for an anesthesia machine is also available to allow simulation of positive-pressure ventilation.

The BioGears engine is designed to drive toward homeostasis similar to the human physiology. The BioGears respiratory system must interact with other physiological systems to receive feedback and adjust spontaneous breathing to maintain homeostasis. This is particularly important because the engine models disease and treatment, such as airway obstruction, tension pneumothorax, and needle decompression. To accurately model the respiratory response under these conditions, a robust respiratory model that responds to severe mechanical stresses and chemical stimuli is required.

The existing BioGears respiratory system directly calculated intrapleural pressure to drive spontaneous respiration. However, this provided limited feedback from the connected systems. This lack of feedback prevented the respiratory system from contributing to the maintenance of homeostasis through regulation of respiration rate and tidal volume. In particular, a chemical feedback mechanism that mimics the respiratory response to blood gas levels as sensed by central and peripheral chemoreceptors was lacking.

Fresnel, et al. provides a dynamic model of the respiratory muscle pressure source that allows variability in the muscle contraction and relaxation phases [1]. While based on clinical data, this model uses pre-selected ventilation frequencies to model various pathophysiological conditions. To address the limitations of the current BioGears model, the new driver extends the Fresnel, et al. model to include chemical stimuli feedback to create a robust ventilator driver that contributes to overall blood gas regulation.

METHODS

The BioGears respiratory model has its roots in the lumped parameter mathematical model of Fukui and Smith [2]. The more advanced model currently implemented includes the trachea, the right and left bronchi, the right and left lung compartments, as well as the right and left pleural compartments to handle pathological conditions and interventions. Each of these compartments uses resistors and/or capacitors to represent the resistance to airflow and compliance of the lungs and chest wall, respectively.

The respiratory muscle pressure source is based on the mathematical model proposed by Fresnel et al. [1]. The time series of the respiratory muscle pressure, P_{mus}, is given by Eq. (1).

$$P_{mus}(t) = \begin{cases} P_{\max}\left(1 - e^{-\frac{f_v + 4P_{0.1}}{10}t}\right), 0 < t \leq T_I \\ P_{\max}\left(e^{-\frac{f_v + \frac{P_{0.1}}{2}}{10}t}\right), T_I < t \leq T_{tot} \end{cases} \qquad (1)$$

Where f_v ($\equiv \frac{1}{T_{tot}}$) is the ventilation frequency, $P_{0.1}$ is the mouth occlusion pressure, and P_{max} is the maximum muscle pressure that specifies the amplitude of the pressure source signal. P_{max} is calculated by setting the muscle pressure at 0.1 s equal to $P_{0.1}$

The Fresnel model regulates respiration using only the ventilation frequency; however, the BioGears engine must also account for changes in tidal volume that naturally occur to attempt to balance the oxygen and carbon dioxide needs of the body. Therefore, a scaling parameter, $\alpha(t)$, was introduced, as shown in Eq. (2).

$$P_{max} = P_{0.1}\left(1 - \exp\left(-\frac{f_v + 4P_{0.1}}{10} \times 0.1\right)\right)^{-1} - \alpha(t) \qquad (2)$$

T_I and T_{tot} are the inspiratory and total respiration times, respectively. These times are related to f_v through Eq. (3) [1].

$$\frac{T_I}{T_{tot}} = 0.0125 f_v + 0.125 \qquad (3)$$

Previous work used relationships between minute ventilation, $\dot{V}_E$, or alveolar ventilation, $\dot{V}_A$, and the blood gas partial pressures to model the respiratory response to chemical stimuli at the peripheral and central chemoreceptors [3, 4]. The resulting mathematical relationship (Eq. (4)) was implemented in BioGears.

$$\dot{V}_A = G_p e^{-0.05P_{aO_2}} \max\left(0, P_{aCO_2} - I_p\right) + G_c \max\left(0, P_{aCO_2} - I_c\right) \quad (4)$$

P_{aO_2} and P_{aCO_2} are the arterial oxygen and carbon dioxide partial pressures, respectively. I_p and I_c are the cutoff threshold constants, and G_p and G_c are the peripheral and central controller gain constants, respectively.

The f_v used in the pressure source corresponds to the target respiration rate that the engine must attain to ensure accurate blood gas levels. It is related to alveolar ventilation, as shown in Eq. (5).

$$\dot{V}_A = f_v \cdot V_A \quad (5)$$

Where V_A is the target alveolar volume calculated using Eq. (6).

$$V_A = V_T - V_D \quad (6)$$

V_T is the target tidal volume of the patient (7 ml/kg [5]), and V_D is the dead space volume (volume that does not participate in gas exchange). The calculated f_v is then used in Eq. (1) to calculate the muscle pressure. The model was implemented in the BioGears engine with reference values, and the model parameters were tuned to meet validation data. The reference and tuned values for the model parameters are shown in Table 1.

Table 1: Respiratory driver model parameters and values.

Parameter (units)	Reference Value	Model value
Gp	30.24 [4]	30.24
Gc	1.44 [4]	1.44
Ip, Ic (mmHg)	35.5 [4]	37
$P_{0.1}$ (cmH$_2$0)	0.75[7] and 0.93[6]	0.43

RESULTS

Figure 1 demonstrates the integration of the respiratory driver in the BioGears engine. The blood gas partial pressure for the lungs and aorta are shown over four respiratory cycles for the standard patient (77 kg) during resting physiologic conditions.

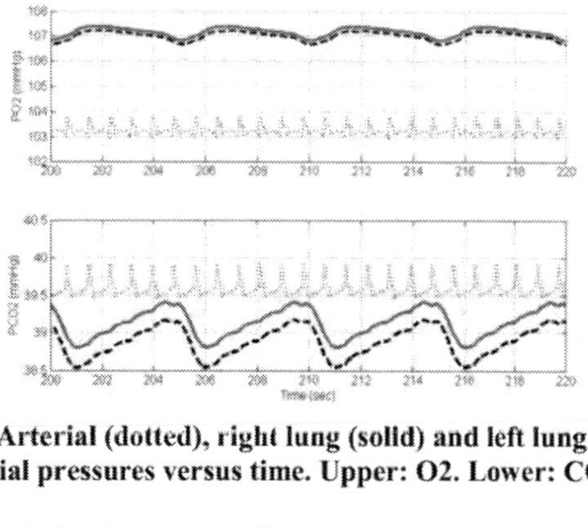

Figure 1: Arterial (dotted), right lung (solid) and left lung (dashed) partial pressures versus time. Upper: O2. Lower: CO2.

The calculated O_2 and CO_2 partial pressures are within the expected physiological values. The small variation between the lungs is a result of the volume difference between the right and left lungs. Table 2 shows the validation data outputs of the integrated respiratory driver.

The mechanical responses arising from the interplay of the driving muscle pressure and the compliances of the lung and chest wall are visible in the varying lung volume (Figure 2).

Table 2: Validation table for measured values – output refers to mean value from the engine. P_AO_2 and P_ACO_2 are alveolar O_2 and CO_2 partial pressures.

Parameters (unit)	Output	Reference	% Error
PaO_2 (mmHg)	103.3	95 [8]	8.7
P_AO_2 (mmHg)	107.1	104 [5]	2.98
$PaCO_2$(mmHg)	39.6	40 [5]	1.01
P_ACO_2 (mmHg)	39	40 [5]	2.5
RR (breaths/min)	12.1	12	0.5
V_T (ml)	522	539 (from 7ml/kg)	3.2

The patient respiration rate (RR) and tidal volume V_T are extracted from the breathing cycle, as shown in the figure. The calculated mean values are also shown in Table 2.

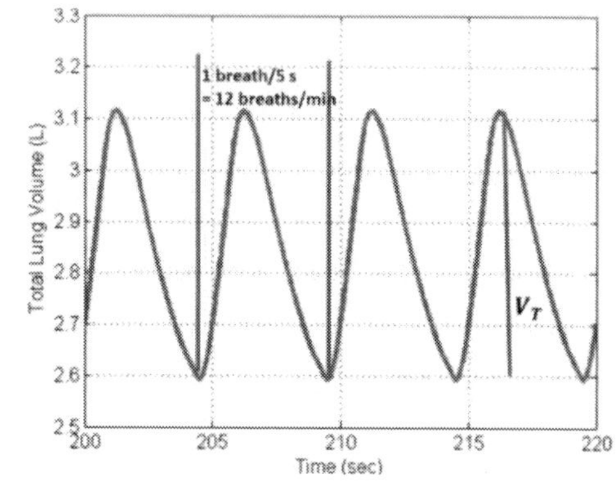

Figure 2: The total lung volume with respiration rate and tidal volume calculation shown.

DISCUSSION

A critical challenge in whole body physiology modeling is the development of mechanistic models that captures the interplay of connected systems and can dynamically respond to physiologic changes. The BioGears respiratory system meets this challenge by incorporating a respiratory driver with chemical feedback mechanisms that capture the dynamic respiratory changes, both frequency and tidal volume, that occur in response to blood gas tension. The data shown illustrates this advancement for a standard healthy patient at rest. Future work will include analysis of a variety of patients and pathological conditions.

ACKNOWLEDGEMENTS

This work is supported by the U.S. Army Medical Research and Materiel Command (USAMRMC) in Fort Detrick, MD under Contract Number: W81XWH-13-2-0068.

REFERENCES

[1] Fresnel, E et al., *EPJ Nonlinear Biomedical Physics,* 2:7, 2014.
[2] Fukui, Y. and N. Ty Smith, *Anesthesiology*, 54:107-118, 1981.
[3] Khoo, M et al., *J. Appl. Physiol. Respirat. Environ. Exercise Physiol.*, 53: 644-659, 1982.
[4] Batzel, J et al., *J. Math. Biol.*, 50:293-335, 2005.
[5] Levitzky, M., *Pulmonary Physiology*, 8th ed, 2013
[6] Scott, G. et al. *Chest*, 98(4):900–906, 1990
[7] Budweiser, S et al. *Respir Med*, 101:2343–2351, 2007
[8] Feher, J. *Quantitative Human Physiology: An Introduction*, 2012 pp588.

SMARTBOOT: AN INSTRUMENTED CLINICAL WALKING BOOT FOR PARTIAL WEIGHT BEARING TRAINING

Timothy K. West
University of Delaware,
Biomedical Engineering
Newark, DE, USA

Michael E. Schenk
University of Delaware,
Mechanical Engineering
Newark, DE, USA

David A. Schnall
University of Delaware,
Biomedical Engineering
Newark, DE, USA

Margaret E. O'Brien
University of Delaware,
Mechanical Engineering
Newark, DE, USA

Melissa A. Groome
University of Delaware,
Electrical Engineering
Newark, DE, USA

Faculty Advisor(s)

Dr. Brian Knarr
Sponsor
University of Delaware
Newark, DE, USA

Dr. Jill Higginson
Sponsor
University of Delaware
Newark, DE, USA

Dr. Anita Singh
Advisor
University of Delaware
Newark, DE, USA

INTRODUCTION

Following surgery or an injury to the ankle or foot, patients are often instructed by a clinician to wear a padded walking boot and to walk with only a specific portion of their body weight on the injured limb. The individual must use a crutch, cane, or other brace to offload the body weight from the affected limb[1]. The percentage of weight applied to the injured limb is prescribed by the clinician based on the severity and nature of the injury. It is important for the patient to apply only the prescribed weight to the leg; too little or too much weight on the limb can result in increased recovery time or further injury[1],[2]. One study questions the effectiveness of partial weight bearing after hip reconstruction surgery because it showed a significant difference between patient perception of partial weight bearing and the measured value[3].

There are three major methods currently applied in clinics to teach patients how to apply the target partial load to the affected limb using crutches: tactile feedback, weight scales, and biofeedback devices[4]. Tactile feedback requires physical therapists to qualitatively estimate the amount of weight the patient is applying by holding their hand or foot underneath the patient's injured limb. The use of weight scales and force plates are often used to more accurately and quantitatively instruct the patient whether more or less weight should be applied. Lastly, biofeedback devices are used to give the patient immediate knowledge of how much weight they are placing on the limb, and how close they are to applying the recommended amount. The use of biofeedback devices has shown to be the most successful method of teaching patients to accurately apply a specific portion of their weight because data can be observed while the patient is walking and real-time adjustments can be made[5]. However, biofeedback is often provided in the clinic with a visual aid which cannot be brought outside the clinic, and therefore partial weight bearing learning is still not retained accurately[1],[2].

Partial weight bearing is difficult for patients to quantify consistently and accurately, especially outside of the clinic over the period of several days. Providing patients and medical personnel with real-time and long-term force data will allow clinicians to monitor patient compliance and could also reduce recovery time. In order to accomplish this, this project modified a clinical padded walking boot to include a system that measures, records, and wirelessly transmits force data from the sole of the boot to a computer and a biofeedback system. The product will be marketed to physical therapy clinics for use with patients.

PRODUCT DESIGN

The design was divided into two overarching categories and various subcomponents for concept development. The categories are force sensing and electronics, which encompass the mechanical and electrical components, respectively. Subcomponents include: force sensor type, force sensor layout, logic board, biofeedback, wireless transmission, long-term storage, and power supply (Figure 1).

For the electrical components, an Arduino Uno microprocessor (SmartProjects, Italy) is implemented for prototyping due to ease-of-use and extensibility. A strip of light-emitting diodes provides feedback because it is more accurate, programmable, and tolerable than tactile or audible feedback. Wireless transmission is achieved through Bluetooth due to its compatibility with the microprocessor as well as integrated communication with most modern computers and portable devices. A secure digital (SD) card stores data while preserving compactness and compatibility. A standard 9-volt battery powers all of the electronic components.

For the force components, piezoresistive Flexiforce® A401 sensors (Flexiforce®, Netherlands) were chosen due to their good force range, acceptable accuracy, and extremely thin profile. The force sensor layout includes two pairs of aluminum plates with three force sensors inserted between each pair of plates (Figure 2).

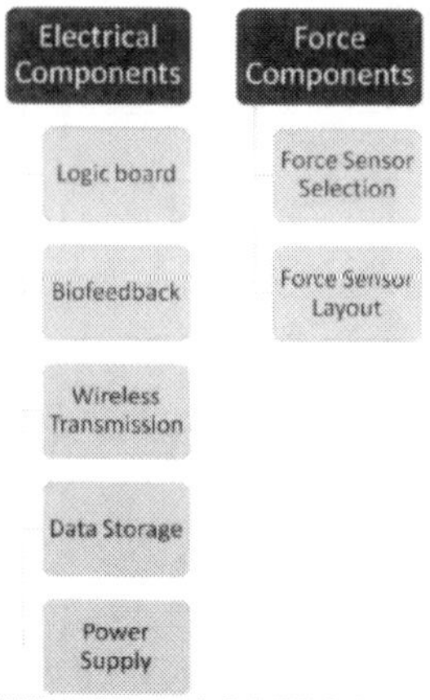

FIGURE 1: THE DESIGN CATEGORIES AND SUB COMPONENTS USED FOR CONCEPT GENERATION

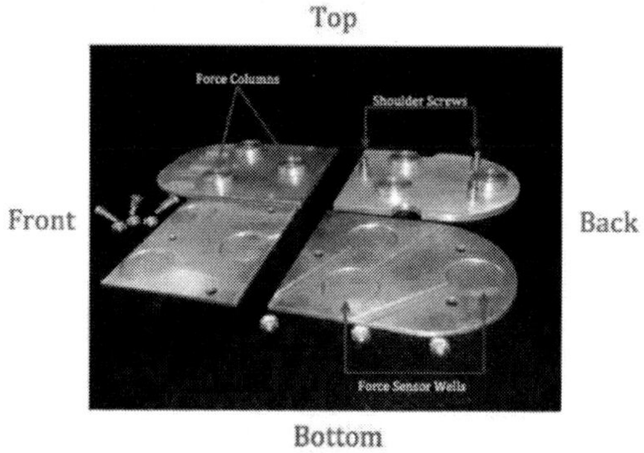

FIGURE 2. FORCE SENSOR PLATES

The electronics are placed in a custom housing created with acrylonitrile butadiene styrene (ABS) and attaches to the straps of the clinical boot. The LED strip is attached to the toe of the boot. The force plates and force sensors are placed under the soft cushion inside the boot (Figure 3).

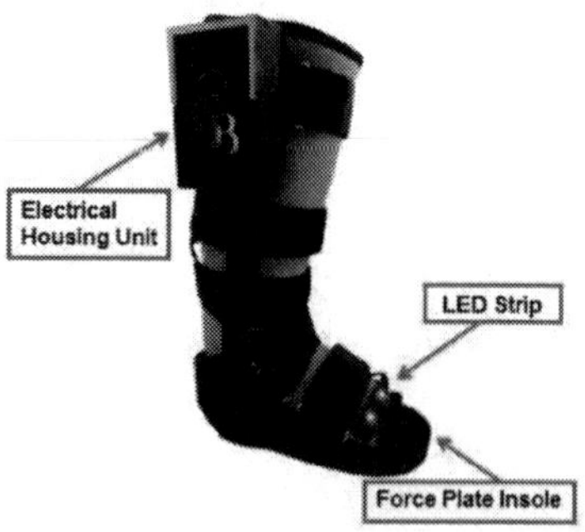

FIGURE 3. FINAL ASSEMBLED PROTOTYPE

BUDGET & MARKET ANALYSIS

The total expense for the project was $798.07. However, the cost of the materials used for the final prototype was $317.79. Those costs are detailed below in Table 1.

TABLE 1. FINAL PROTOTYPE EXPENSES

Item	Cost per Unit	Quanity	Cost
Arduino Uno	$24.95	1	$24.95
SD Card Shield	$19.95	1	$19.95
Bluetooth Shield	$19.95	1	$19.95
LED Strip	$29.95	2 LEDs	$2.00
Flexiforce Sensors	$17.50/sensor	6	$105
SD Card	$11.69	1	$11.69
9V Battery Case	$3.95	1	$3.95
9V Battery	$1.15	1	$1.15
Bluetooth USB	$10.95	1	$10.95
6061 Aluminum 3/32"	$0.09/in^2	34 in^2	$3.06
6061 Aluminum 3/16"	$0.15/in^2	41.5 in^2	$6.23
6061 Aluminum Rod	$0.46/in	27 in	$12.42
Stainless Steel Locknut	$0.09	6	$0.53
Steel Dowel Pins	$4.47	6	$26.82
10 KOhm Resistors	$0.10	7	$0.67
Pushbutton	$0.35	1	$0.35
Plastic Case	$66.14	1	$66.14
Slide Switch	$2.00	1	$2.00
		Total	**$317.79**

The short-term goal for the prototype will be an IRB-approved study to validate the significance of the design. Improvements to battery life and compactness will be easily achievable through iterative prototyping. This relatively inexpensive product could be a common tool for physical therapists who want to explore the benefits of partial weight bearing. Components such as the microprocessor will likely be custom manufactured in large quantities by an outside source for mass production.

ACKNOWLEDGEMENTS

Dr. David Logerstedt for his clinical feedback during development; Dr. Dustyn Roberts for her technical support; Kevin McGinnis for his assistance with data collection and processing; University of Delaware College of Health Sciences for their grant through the First Step program.

REFERENCES

[1] Hershko, E et al., *Am J Orthop*, 37:E92–E96, 2008.
[2] Winstein, C et al., *Phys Ther*, 76:985–993, 1996.
[3] Tveit, M et al., *J Rehabil Med*, 33:42–46, 2001.
[4] Malviya, A et al., *Injury*, 36:556–559, 2005.
[5] Hustedt, J et al., *Yale J Biol Med*, 85:119–125, 2012.

SB³C2015
Summer Biomechanics, Bioengineering and Biotransport Conference
June 17-20, 2015, Snowbird Resort, Utah, USA
SB³C2015-330

A STRUCTURAL CONSTITUTIVE MODEL FOR THE ACTIVE AND PASSIVE BEHAVIOR OF BIOLOGICAL TISSUES

Ting Tan and Raffaella De Vita
Mechanics of Soft Biological Systems Lab
Department of Biomedical Engineering and Mechanics
Virginia Tech
Blacksburg, Virginia 24061

INTRODUCTION

Smooth muscle cells (SMCs) in the walls of many hollow biological structures such as arteries, esophagus, uterus and vagina, are responsible for their *in-vivo* active deformations. In this study, we propose a novel constitutive model that takes into account both the active and passive mechanical response of the biological tissues that compose these structures. The model is validated using published experimental data on the media layers of pig carotid arteries [1]. This theoretical study will provide valuable information on the mechanical behavior of biological tissues containing SMCs that could enhance the management of medical conditions such as hypertension, asthma, and pelvic floor disorders.

MODEL DEVELOPMENT

Constitutive Model for Active Response

In the proposed model, SMCs and the connected collagen fibers are assumed to be the active structural components of the tissue. Each SMC is assumed to be viscoelastic with a residual motor force that initiates muscle contraction while each connected collagen fiber is assumed to be linear elastic.

The active stress of the tissue is equal to the stress of collagen fibers that are connected to the SMCs. The first Piola-Kirchhoff stress tensor that defines the active stress of the tissue, $\mathbf{P}_\mathrm{a}$, can be expressed as:

$$\mathbf{P}_\mathrm{a} = -p_\mathrm{s}\mathbf{F}_\mathrm{s}{}^{-\mathrm{T}} + 2\mathbf{F}_\mathrm{s} \cdot \frac{\partial}{\partial \mathbf{C}_\mathrm{s}} \int_\Sigma R(\mathbf{m}) w_\mathrm{s}(\lambda_\mathrm{s}(\mathbf{C}_\mathrm{s},\mathbf{m}))\mathrm{d}\Sigma, \quad (1)$$

where p_s is the Lagrange multiplier, $\mathbf{F}_\mathrm{s}$ is the deformation gradient, $\mathbf{C}_\mathrm{s}$ is the right Cauchy-Green deformation tensor, $\lambda_\mathrm{s}(\mathbf{C}_\mathrm{s},\mathbf{m})$ is the axial fiber stretch, $R(\mathbf{m})$ is the probability density of collagen fibers oriented along the unit vector $\mathbf{m}$ in the reference configuration. w_s and σ_s are the strain energy and the stress of the collagen fibers that are connected to the SMCs, respectively, and are assumed to be arranged in series with them along $\mathbf{m}$. They can be related and expressed as:

$$\frac{\mathrm{d}w_\mathrm{s}}{\mathrm{d}\lambda_\mathrm{s}} = \sigma_\mathrm{s} = \int_{\lambda_\mathrm{s}(0)}^{\lambda_\mathrm{s}(t_\mathrm{eq})} \gamma \mathrm{e}^{-\gamma(\lambda_\mathrm{st}-1)} K_\mathrm{s}\left(\frac{\lambda_\mathrm{s}}{\lambda_\mathrm{st}} - 1\right)\mathrm{d}\lambda_\mathrm{st}, \quad (2)$$

where the exponential probability function $\gamma \mathrm{e}^{-\gamma(\lambda_\mathrm{st}-1)}$, with γ constant, is introduced to describe the recruitment of taut collagen fibers. Each taut collagen fiber is assumed to be linear elastic so that its stress is $K_\mathrm{s}(\frac{\lambda_\mathrm{s}}{\lambda_\mathrm{st}} - 1)$ with elastic modulus K_s. $\lambda_\mathrm{s}(0)$ and $\lambda_\mathrm{s}(t_\mathrm{eq})$ are the axial fiber stretches when muscle contraction starts at $t = 0$ and ends at $t = t_\mathrm{eq}$, respectively.

Evolution Law

In order to determine the stretch of the collagen fibers that surround the SMCs over time, $\lambda_\mathrm{s}(t)$, in the $\mathbf{m}$ direction, we need to derive an evolution law. We assume that the force acting on a group of taut collagen fibers that are connected to the SMCs during muscle contraction is equal to the force acting on the SMCs. Thus, these fibers and the SMCs are assumed to be arranged in series so that

$$k_\mathrm{s}u_\mathrm{s}(t) = \eta \dot{u}_\mathrm{c}(t) + k_\mathrm{c}u_\mathrm{c}(t) + F_\mathrm{m}, \quad (3)$$

where k_c and $u_\mathrm{c}(t)$ are the spring stiffness and displacement of the collagen fibers, respectively, η, k_c, $u_\mathrm{c}(t)$, and F_m are the viscous coefficient, spring stiffness, displacement, and motor force of the SMC contractile units, respectively, and t is the contraction time. $u_s(t)$ and $u_c(t)$ are related to the stretch of the tissue, $\lambda(t)$, by:

$$\lambda(t) = \frac{L_0 + u_s(t) + u_c(t)}{L_0}, \tag{4}$$

where L_0 is the original length of the tissue, $u_s(t)$ is the deformation of the taut collagen fibers that are in series with the SMCs, $u_c(t)$ is the deformation of the SMCs.

Let $0 < a < 1$ be the ratio of the original length of the active components to the original length of the tissue. Substituting Eq. (4) into Eq. (3), the evolution law for $\lambda_s(t)$ under constant λ (in isometric experiments, $u_s(t)$ and $u_c(t)$ vary with the time while $\lambda(t)$ is kept constant) becomes:

$$\begin{aligned}\lambda_s(t) &= 1 + \frac{u_s(t)}{(1-a)L_0} \\ &= 1 + \frac{(\lambda-1)(1-ae^{-\alpha t}) + \beta[F-(\lambda-1)](1-e^{-\alpha t})}{1-a},\end{aligned} \tag{5}$$

where $\alpha = \frac{k_c+k_s}{\eta}$, $\beta = \frac{k_s}{k_c+k_s}$, and $F = \frac{F_m}{k_s L_0}$.

Constitutive Model for Passive Response

The passive stress of the tissue is assumed to be determined by the collagen fibers that are not connected to the SMCs. The first Piola-Kirchhoff stress tensor that defines the passive stress, $\mathbf{P}_p$, can be expressed as:

$$\mathbf{P}_p = -p_p \mathbf{F}_p{}^{-T} + 2\mathbf{F}_p \cdot \frac{\partial}{\partial \mathbf{C}_p} \int_\Sigma R(\mathbf{n}) w_p(\lambda(\mathbf{C}_p, \mathbf{n})) d\Sigma, \tag{6}$$

where p_p is the Lagrange multiplier, $\mathbf{F}_p$ is the deformation gradient, $\mathbf{C}_p$ is the right Cauchy-Green deformation tensor, $\lambda(\mathbf{C}_p, \mathbf{n})$ is the axial fiber stretch. $R(\mathbf{n})$ is the probability density of collagen fibers oriented along the unit vector $\mathbf{n}$ in the reference configuration. w_p and σ_p are the strain energy and the stress of the collagen fibers that are not connected to the SMCs, respectively, oriented along $\mathbf{n}$ and defined as done previously by

$$\frac{dw_p}{d\lambda} = \sigma_p = \int_1^\lambda v e^{-v(\lambda_{st}-1)} K_p \left(\frac{\lambda}{\lambda_{st}} - 1\right) d\lambda_{st}, \tag{7}$$

where v is a constant and K_p is the elastic modulus of taut collagen fibers not connected to SMCs.

RESULTS

Published active and passive stress stretch data collected from the media layers of pig carotid arteries [1] were used to test the constitutive model for an isochoric axisymmetric deformation. In the experiments, the tissue was loaded in one direction. The straightened collagen fibers and SMCs were assumed to be all oriented in the loading direction, say $\mathbf{e}_3$, so that $\mathbf{m} = \mathbf{n} = \mathbf{e}_3$. The active stress-stretch curve was fitted to isometric data collected at the equilibrium time ($t_{eq} = 300$s) for various constant

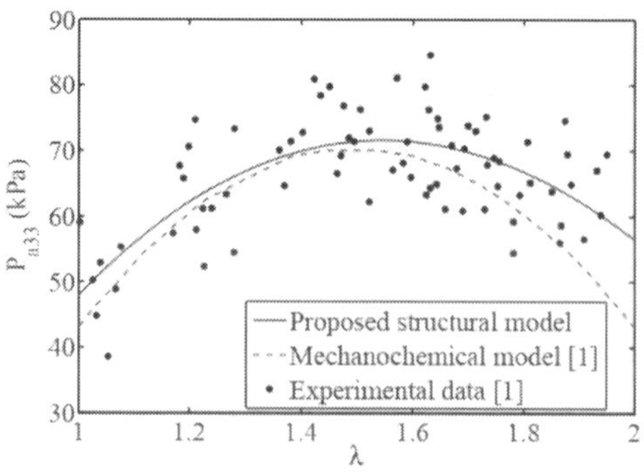

FIGURE 1. ACTIVE STRESS-STRETCH BEHAVIOR ($\alpha = 0.282$ S^{-1}, $\beta = 0.121$, $A = 0.486$, $F = 1.146$, $\gamma = 1.307$, $K_S = 4.289$ MPA)

stretches ($1 \leq \lambda \leq 2$) (Fig. 1). This model fit the data well with fewer parameters than the mechanochemical model presented in [1].

The constitutive law for the passive response of the tissue was used to fit the passive experimental data and compared to the Neo-Hookean and exponential model adopted by Murtada et al. [1] (Fig. 2). Both models fit the experimental data well.

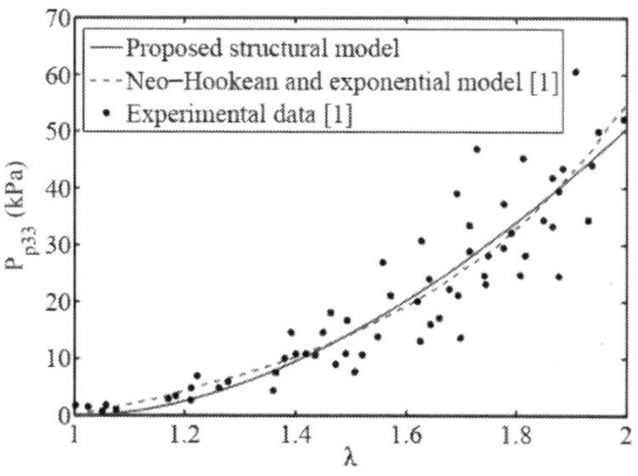

FIGURE 2. PASSIVE STRESS-STRETCH BEHAVIOR ($v = 0.139$, $K_P = 1.124$ MPA)

DISCUSSION

In this study, a novel constitutive model considering both the active and passive mechanical behavior of biological tissues containing SMCs is proposed and compared to experimental data on the pig carotid arteries [1]. Future experimental and theoretical work will be conducted to further validate the proposed model.

ACKNOWLEDGMENT

Supported by the NSF under CAREER Grant No. 1150397.

REFERENCES

[1] S. C. Murtada, A. Arner, G. A. Holzapfel, Experiments and Mechanochemical Modeling of Smooth Muscle Contraction: Significance of Filament Overlap, Journal of Theoretical Biology, 2012 (297): 176-186.

SB[3]C2015
Summer Biomechanics, Bioengineering and Biotransport Conference
June 17-20, 2015,Snowbird Resort, Utah, USA
SB³C2015-334

FETAL MONITORING WITH SILICONE NANOCOMPOSITE STRAIN GAUGES

Daniel A. Baradoy, Anton E. Bowden, David T. Fullwood

Mechanical Engineering
Brigham Young University
Provo, UT

INTRODUCTION

In the age of "smart things," low profile "on the go" data acquisition has become paramount. Sensors in mobile devices, automobiles and in other applications are being miniaturized and made easier to add into existing technology. Even shoes, wristbands and other wearables are joining the smart revolution with integrated accelerometers, gyroscopes, heart rate trackers and more. In recent work, our lab has developed a novel silicone-based strain sensor capable of accurately and repeatedly detecting up to 100% strain [1,2]. We have developed these sensors into highly flexible patches that can be used in human biometrics tracking.

The sensing material is a silicone with nickel nanostrand (NiNs) and nickel coated carbon fiber (NCCF) additives. Strain in the material is proportional to a change in the electrical resistance when a current is passed though it. The resistive response has been observed to be vary with the material composition of the sensor and the backing on which it is adhered. This customizability permits gauges to be designed for a variety of bandwidths and thresholds. The work to date utilized a structured block design of experiments approach to determine the characteristics of the sensor material composition with most effective properties for tracking the deformation of a pregnant woman's abdomen in response to fetal movement (e.g., "kicks") and simulated labor contractions [3]. In particular this study reports on the development of a fetal monitoring belt capable of real-time fetal monitoring and reporting.

METHODS

To determine the effect of different compositions of the silicone sensors, a block design experiment was designed to vary the proportions of NiNs and NCCF that were added as well as the lengths of NCCF. A total of 24 unique samples were manufactured. Samples were then mounted in a tensile testing machine (Figure 1) and stretched to 50, 60 or 70 percent strain depending on the amount of additives. Position was given by the tensile tester and electrical resistance was recorded with a 6.5 digit multimeter for analysis.

Results were examined to find the sensors that exhibited the most dramatic change in resistance over the largest amount of strain. This would provide the best resolution for situations in which continuous determination of position is desired.

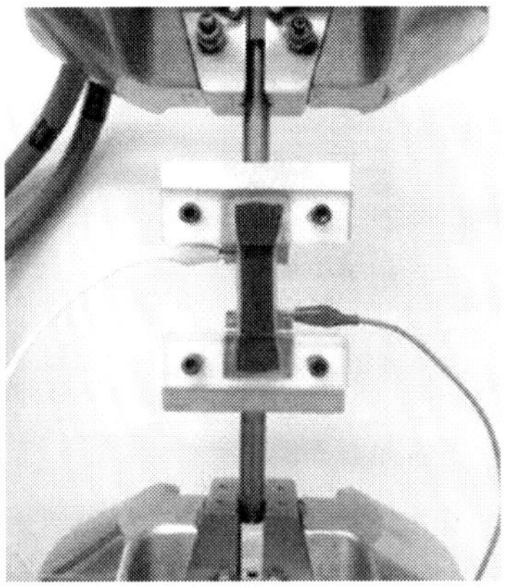

Figure 1 - Sensor sample mounted in tensile tester

Once this sensor was chosen, another series of tests were done with a variety of backing materials. Identical sensors were adhered onto fabric elastic, spandex, KT tape and molded silicone using a silicone RTV glue. A duplicate of each sample was made and preconditioned to determine if the effect of pre-straining the sensor would prevent the plastic deformation that the material is prone to a short way into its elastic region. Tensile testing with resistance measurements was then repeated with all 8 samples.

In order to choose a sensor suitable for the fetal monitoring belt it was necessary to match the bandwidth of the strain gauge as closely to the predicted deformation of the abdomen during contractions as possible. During labor it is expected that the abdomen may change in circumference 1-2 cm, which is within the effective range of the sensor. This range of deflection is also consistent with deflections observed in a pregnant woman's abdomen during fetal movement.

The prototype fetal monitor utilized a high-stiffness canvas strap with a gap in the anterior portion where the high-deflection strain gauge was positioned. Thus, all abdominal deflections were expected to express themselves as extension of the strain gauge. An Arduino micro-controller read the resistance and sent the data via USB or

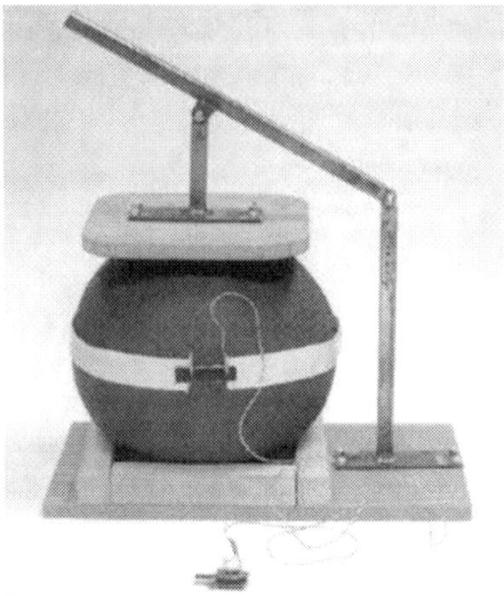

Figure 2 – Contraction simulator. The ball is depressed and released to simulate the change in circumference of the abdomen.

Bluetooth to another processing platform such as a smartphone or computer.

Proof of concept testing of the prototype was accomplished using a utility ball and a compression platform to mimic the changing shape of a mother's abdomen during contractions (Figure 2). The compression platform was actuated at intervals to give a 1 cm change in ball's circumference. Strain gauge results were recorded by the micro-controller and saved on a desktop computer.

RESULTS

The block design experiment produced a wide range of resistance curves (Figure 3). Most samples made using longer strands of NCCF stayed in a low resistance range and had relatively smooth curves. Samples containing shorter NCCF showed more drastic changes in resistance going from almost completely insolating to conductive in a very small amount of strain.

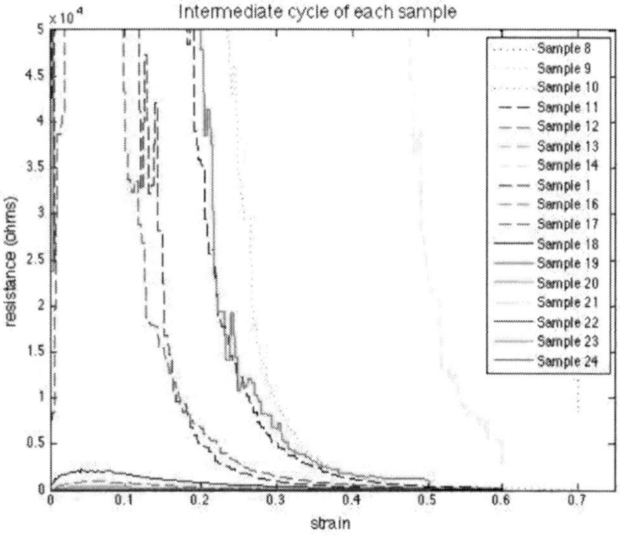

Figure 3 – Strain/resistance curves for samples the produced significant results.

Of the backings explored, none exhibited any problems with electrical signal "leakage", and each was found to be desirable for different situations. The primary difference between the tested backing options was simply mechanical stiffness. For applications adhering directly onto skin, the molded silicone seemed to be ideal; it exhibited a stiffness even less stiff than the underlying skin, and had a desirable non-permanent self-adhesion. KT tape is a good choice for one time uses of application directly onto skin. Fabric elastic was

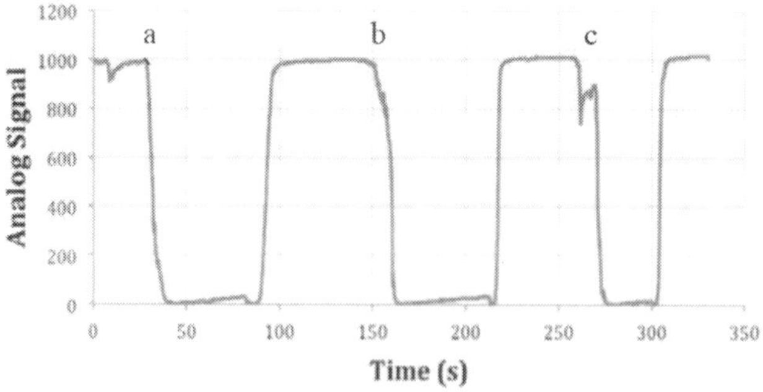

Figure 4 – Analog signal as contractions are simulated by fluctuating the circumference of the utility ball. a) 1 minute contraction initiated. b) 1 minute contraction. c) 30 second contraction.

found to be the stiffest of the backings tested and the stiffness could even be manipulated by the width of the elastic used.

Based on the block DOE, the fabric elastic backing was combined with a gauge that had a NiNs volume fraction of 7%, a NCCF volume fraction of 2% and an NCCF length of 1 mm. Using this sensor, signals differentiating contractions were obvious and accurate to the timing of the simulation. Minor noise fluctuations from the sensors were observed (e.g, Figure 4 at 20 and 260 seconds).

DISCUSSION

The wide range of strain-resistance curves exhibited by the different sensor compositions provided ample options for application use. The fetal monitoring belt performed well with the fabric elastic as the backing for the sensor and was stiff enough to keep the belt firmly in place during normal maternal movements such as sitting, standing and walking.

The contraction simulation data indicates that the fetal monitoring belt is a viable solution for detecting both fetal movements, as well as contractions during labor. Distinct jumps in the signal provided by the micro-controller obviously show the changes in circumference are quantifiable by the sensor. Future testing on human subjects during labor is planned to further verify the effectiveness of the belt.

ACKNOWLEDGEMENTS

Funding for this project was provided by the National Science Foundation (CMMI-1235365). Any opinions, findings, and conclusions or recommendations expressed in this material are those of the author and do not necessarily reflect the views of the National Science Foundation.

REFERENCES

[1] Hyatt, T., *Piezoresistive Nano-composites: Characterization and Applications*, in *Mechanical Engineering*. 2010, Brigham Young University: Provo.

[2] Koecher, M., et al., *Characterization of Nickel Nanostrand Nanocomposites through Dielectric Spectroscopy and Nanoindentation*. Polymer Eng & Sci, In Press.

[3] Johnson, O., et al., *Optimization of Nickel Nanocomposite for Large Strain Sensing Applications*. Sensors and Actuators A, 2010.

[4] Euliano, T.Y., et al., *Monitoring uterine activity during labor: a comparison of 3 methods*. Am J Obstet Gynecol, 2013. 208(1): p. 66 e1-6.

SB³C2015
Summer Biomechanics, Bioengineering and Biotransport Conference
June 17-20, 2015, Snowbird Resort, Utah, USA
SB³C2015-335

VIRTUAL TRANSPLANTATION TO ESTABLISH DONOR SELECTION CRITERIA FOR UNDERSIZED AND COMPLEX HEART RECIPIENTS

J. Ryan (1), R. Richardson (2), E. Ellsworth (3), J. Nigro (3), D. Frakes (1), S. Pophal (3)

(1) School of Biological & Health Systems Engineering
Arizona State University
Tempe, AZ, USA

(2) Radiology
St. Joseph's Hospital and Medical Center
Phoenix, AZ, USA

(3) Heart Center
Phoenix Children's Hospital
Phoenix, AZ, USA

INTRODUCTION

Each year in the United States, approximately 250,000 people are diagnosed as having advanced heart failure [1]. The American Heart Association defines heart failure as a chronic, progressive condition in which the heart is unable to pump enough to meet the body's needs. Advanced heart failure is the final stage in which neither palliative treatment nor symptom management is sufficient for patient survival [2]. Nearly 1% of that population is pediatric [3]. Despite these high incidence numbers, only 2,200 donor hearts are available each year, with 400 of these hearts available for pediatric transplantation [1,3]. Children listed for heart transplant are subject to the highest waiting list mortality in solid-organ transplant medicine in the United States [4]. According to Almond et al., 17% of pediatric patients died while awaiting heart transplantation from 1999 – 2006 [4]. Given the lack of available donor hearts, it is critical that pediatric patients with advanced heart failure be promptly identified for proper donor heart size. Proper sizing is expected to result in reduced morbidity and mortality from fit-related complications.

The current method for selecting appropriate donor heart size lacks patient-specificity and largely relies on the surgeon's prior experience. The surgeon must balance the emergent need of the recipient by expanding donor selection criteria as much as possible without accepting a heart too large for a patient. In the event a heart-size mismatch, the patient may fail to thrive due to fit complications.

CASE STUDY: A 2-year-old male with heterotaxy syndrome and complex congenital heart disease, including dextrocardia, transposition of the great arteries, developed advanced heart failure and was subsequently listed for heart transplantation. The unusual cardiac anatomy and small patient size made donor selection criteria difficult to establish. Image processing, virtual reconstruction, and virtual surgery aided in the selection of a donor heart with appropriate specifications.

METHODS

A contrast-enhanced magnetic resonance image dataset was acquired of the donor's thoracic anatomy. The medical images were imported into Mimics Innovation Suite (Materialise, Lueven, Belgium), a medical image processing software suite. The software facilitated medical image reconstruction of the patient's cardiac anatomy.

Concurrently, contrast-enhanced computer tomography (CT) datasets were obtained from patients presenting with normal cardiac anatomy with body weights of 22kg and 36kg. These datasets were also reconstructed. The patient reconstructions were imported into an engineering software suite, Geomagic (3DSystems, Rock Hill, SC, USA), for post-processing.

The reconstruction of the recipient's cardiac anatomy underwent virtual resections similar to the actual surgical procedure. The great vessels were terminated, and the atria and ventricles were removed. The remaining cardiac anatomy provided a template for the placement of donor heart anatomy. The reconstruction of the donors' cardiac anatomies underwent virtual resections similar to donor explantation. Great vessels were terminated near the atria and ventricles.

The recipient and donor reconstructions were imported back into Mimics for virtual implantation. Cardiothoracic surgeons guided the virtual placement of the donor anatomies in the recipient's virtual template. Extra consideration was given by the surgeons to the donor heart's placement near the sternum and to the recipient's remaining pulmonary arteries and veins. An image of the virtual implantation is seen in Figure 1.

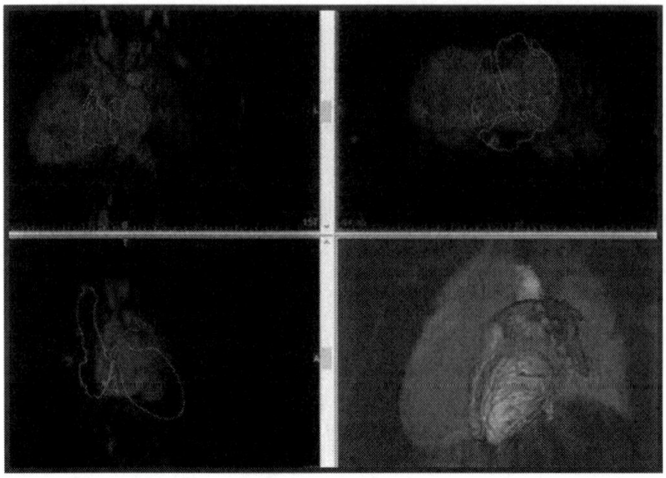

Figure 1: Virtual transplantation of a 22kg donor heart in the virtual recipient. Recipient anatomy is shown in red and partial transparency.

In addition to the virtual surgical procedure, the recipient template and truncated donor reconstructions were 3D printed in a gypsum-based powder media using a zPrinter 650 RP machine (3D Systems, Rock Hill, SC, USA). This 3D printing technology uses a full-color ink-cyanoacrylate infiltration system. The resulting models are life-size replications of the donor and recipients anatomy. The printer facilitated color differentiation of the cardiac structures. Images of the 3D printed models are seen in Figure 2.

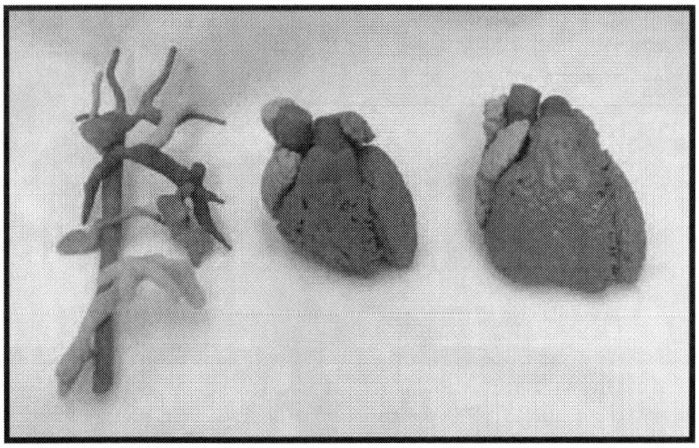

Figure 2: Truncated recipient anatomy (left) juxtaposed to a 22kg donor heart (center) and 36kg donor heart (right).

RESULTS

Surgeons utilized the virtual implantation process to assist in the establishment of selection criteria for the donor hearts. In addition, the virtual surgery assisted the surgeons in planning for the surgery and potential complications.

While the 3D printed models were printed after the virtual surgery was completed, anecdotal evidence from the surgeons illustrates promise for 3D printing as a clinical tool. The surgeon noted the ease of physically inserting a donor heart model into the recipient anatomy. In this case study of complex congenital heart disease, dextrocardia, and unusual systemic and pulmonary venous connections, visualizing a donor heart within the recipient patient was perceived to be helpful.

As of submission, the patient is listed for a donor heart.

DISCUSSION

Current means of establishing donor selection criteria rely strongly on prior experience and surgeon perception of recipient anatomy instead of on patient-derived metrics. The proposed virtual implantation process brings donor and recipient metrics to the forefront without excluding surgeon experience, as the surgeon still guides the virtual surgery process.

Anecdotally, the surgeons perceived that the virtual process can help to predict fit complications by observing overlap or proximity of virtual recipient structures with/to virtual donor morphology. In the case study presented here, the patient's abnormal morphology made this virtual planning technique critical to the clinical care process. A larger clinical trial is necessary to establish actual efficacy of the virtual implantation process.

This case study also reveals larger implications for the role of virtual surgery especially with regards to donor selection criteria. For this emergent scenario, only two normal anatomies were reconstructed and virtually implanted. These normal anatomies were assumed to be representative of the population with respects to weight/size. Should this virtual surgery process be further implemented into clinical care, a library of normal cardiac anatomy, representative of a standardized population, must be established.

The implications of a normal cardiac library would be that cardiothoracic surgeons could utilize patient-specific metrics to establish criteria for listing a patient for a donor heart. Pediatric patients may further benefit by potentially allowing a wider net to be cast for donor listing, while oversized hearts would be more likely to be excluded.

ACKNOWLEDGEMENTS

The authors would like to acknowledge the Lund Foundation for their continued support of the Cardiac 3D Print Lab at Phoenix Children's Hospital.

REFERENCES

[1] Givertz, M. M., 2011, "Ventricular Assist Devices Important Information for Patients and Families," Circulation, **124**(12), pp. e305 e311.

[2] Allen, L. A., Stevenson, L. W., Grady, K. L., Goldstein, N. E., Matlock, D. D., Arnold, R. M., Cook, N. R., Felker, G. M., Francis, G. S., Hauptman, P. J., Havranek, E. P., Krumholz, H. M., Mancini, D., Riegel, B., and Spertus, J. A., 2012, "Decision Making in Advanced Heart Failure," Circulation, p. CIR.0b013e31824f2173.

[3] Alboliras, E., 2014, Atlas of Neonatal Cardiology, John Wiley & Sons, Hoboken, NJ.

[4] Almond, C. S. D., Thiagarajan, R. R. M., Piercey, G. E., Gauvreau, K., Blume, E. D., Bastardi, H. J., Fynn-Thompson, F., and Singh, T. P., 2009, "Waiting List Mortality Among Children Listed for Heart Transplantation in the United States," Circulation, **119**(5), pp. 717–727.

ELECTROMAGNETIC FIELD STIMULATION ENHANCES MECHANICAL PROPERTIES OF TISSUE-ENGINEERED BONE CONSTRUCTED WITH CALCINED BOVINE TRABECULAR BONE SCAFFOLD

SM. Tanaka (1), Y. Yamashita (2)

(1) Institute of Nature and Environmental Technology,
Kanazawa University
Kanazawa, Ishikawa, Japan

(2) Graduate School of Natural Science and Technology
Kanazawa University
Kanazawa, Ishikawa, Japan

INTRODUCTION

Calcined bovine trabecular bone is attractive material as a tissue-engineered bone scaffold, because of natural porouse structure, absent of proteins causing immuno-responses, and low-cost in manufacturing comparing to current bone substitute materials such as hydroxyapatite and β-TCP. However, the mechanical properties of tissue-engineered bone with this scaffold are greatly inferior comparing to those of original trabecular bone [1], requiring improvements of the properties for clinical applications. Electromagnetic field stimulation has been known to activate osteoblasts *in vitro* [2]. We hypothesized that an electromagnetic field stimulates osteoblastic calcification in tissue-engineered bone constructed with a bovine trabecular bone scaffold *in vitro*, by which the mechanical properties of the tissue-engineered bone could be improved. The objective of this study is to verify the hypothesis by mechanical tests for the tissue-engineered bones after electromagnetic filed stimulation *in vitro*.

METHODS

Tissue-engineered bone

Bovine trabecular bone blocks (L5 × W5 × H6 mm) were cut out of the distal epiphysis of a frozen meat bovine femur frozen at -20°C using a diamond blade cutter. The longest axis direction of the block was determined corresponding to the bone axis. The blocks were boiled in deionized water for 30 min to eliminate bone marrow and naturally dried (Fig.1a). The blocks were calcined at 600°C for 22 hours in an electric furnace, and used as a scaffold (Fig.1b). Osteoblasts differentiated from bone marrow stem cells of 8-week-old female Sprague-Dawley rat were used. The osteoblasts were dispersed in 1 mg/ml type-I collagen solution at cell density of 10,000 cells/mm^3. The cell suspension solution was poured into a calcined trabecular scaffold and gelatinized overnight at 37°C in an incubator.

Electromagnetic field stimulation

Electromagnetic field stimulation was applied to tissue-engineered bones in culture for 20 min per day for 21 consecutive days using a solenoid coil (Fig.2). The pulse train of electromagnetic field consists of a series of single sinusoidal wave with a period of 1 ms and amplitude of 1 mT. Two types of pulse train were tested in this study; periodic EMF pulse train with a repetition frequency of 500 Hz (PrPT); and random EMF pulse train following a geometric probability density distribution up to 500 Hz (RdPT).

Quasi-static compression test

To evaluate the mechanical properties of the tissue-engineered bones, quasi-static compression tests were performed for the longest direction of sample at displacement speed of 1mm/min. From load-deformation curves, stress-strain curves were obtained using the apparent cross-sectional area and the height of sample. Compressive strength (maximum stress) and elastic modulus were determined on a stress-strain curve.

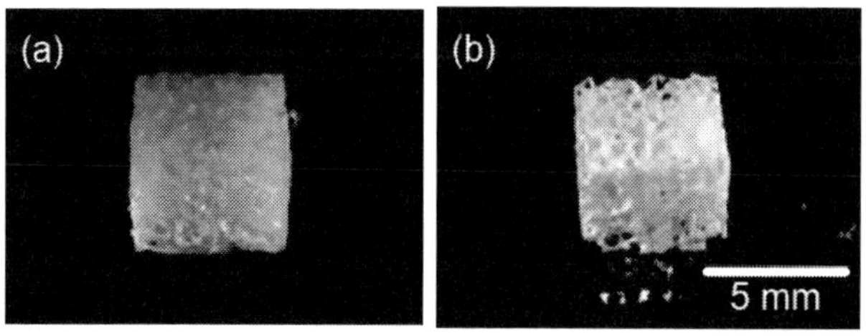

Figure 1 Close-up photographs of bovine trabecular bone blocks, (a) boiled and (b) calcined at 600 °C.

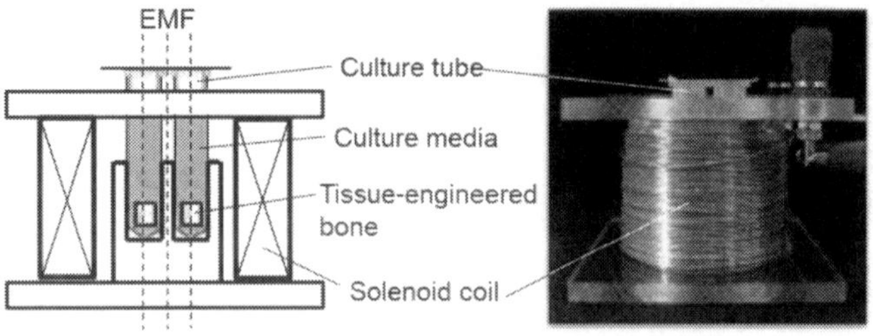

Figure 2 Equipment used for electromagnetic field (EMF) stimulation. Four tissue-engineered bones are cultured and simultaneously stimulated with electromagnetic field generated by a solenoid coil.

RESULTS

Figs 3 and 4 show typical stress-strain curves of each group and a comparison of mechanical properties among all the groups, respectively. The mechanical properties of the tissue-engineered bone immediately decreased after one-hour culture (BL-Control) from those of scaffold only before culture (Scaffold (Dry)). Non-stimulated sample (Control) did not exceed the values of scaffold (Dry) even after 21 days in culture. On the other hand, electromagnetic field stimulations increased mechanical properties either in maximum stress or in elastic modulus. There were no significant differences in mechanical properties between the two types of EMF pulse train (PrPT and RdPT).

DISCUSSION

The increases in mechanical properties in the stimulation groups seem to be brought on by the promotion of osteoblastic calcification by electromagnetic field simulations. We have confirmed that random pulse train is more stimulating waveform than periodic pulse train in electromagnetic field stimulation [3] in 2D cell culture studies. However, in this study, there were no obvious differences between random pulse train and periodic pulse train stimulations in the 3D culture environment. This suggests that the randomness of EMF pulse train might be effective more for the proliferation than the differentiation and calcification of osteoblast, because the latter stages could be more activated and dominant in 3D environment with collagen gel.

Mechanical strength and elastic modulus of human trabecular bone have been reported to be 4.4 MPa and 309.8 MPa, respectively [4], which are higher than those of the tissue-engineered bone stimulated by the electromagnetic fields, requesting further improvement of stimulation regimen to elicit more osteoblastic calcification. Fiber reinforcement of calcined bovine trabecular bone with biodegradable and biocompatible organic fibers [5] could be another option to give better mechano-compatibility to the tissue-engineered bones.

CONCLUSION

In conclusion, electromagnetic field stimulation is an effective way for enhancing the mechanical properties of tissue-engineered bone constructed with calcined bovine trabecular bone scaffold.

ACKNOWLEDGEMENTS

This study was partly supported by the Ministry of Education, Science, Sports and Culture, Grant-in-Aid for Challenging Exploratory Research, 26560206.

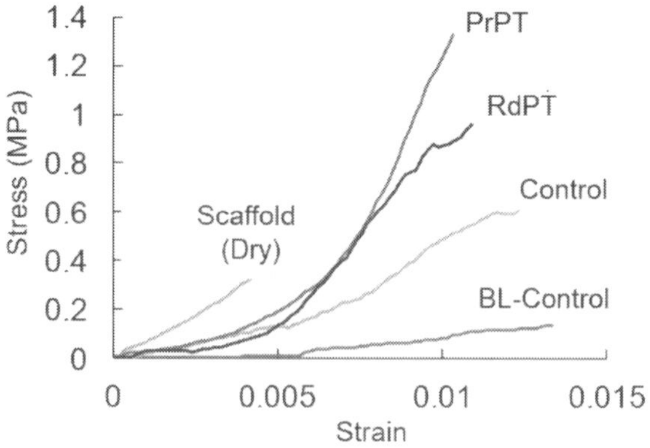

Figure 3 Typical stress-strain curves of tissue-engineered bones obtained from quasi-static compression test. The curves were displayed until maximum stresses.

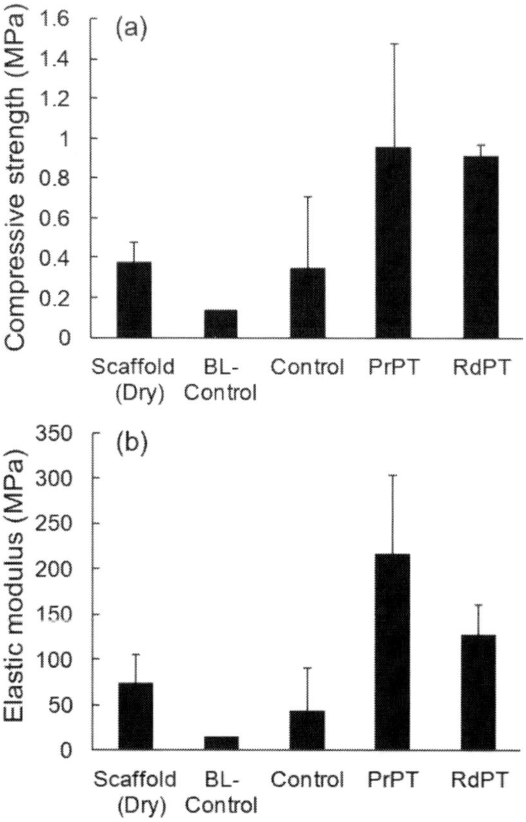

Figure 4 Comparison of mechanical properties among all the groups. (a) Compressive strength. (b) Elastic modulus. Scaffold (Dry) represents a sample group of scaffolds before seeding and culturing cells.

REFERENCES

[1] Yamashita, Y et al., *Proc. of the 26th Bioengineering Conference, 2014 Annual Meeting of BED/JSME*, 13-69:503-504, 2014.
[2] Chang, W.H.-S et al., *Bioelectromagnetics*, 25: 457-465, 2004.
[3] Matsumura, H et al., *Jpn J Clin Biomech*, 32:33-38, 2011.
[4] Dunham, C.E et al., *Clin Biomech*,, 20:834-838, 2005.
[5] Laird D.F et al., *J Biomed Mater Res A*, 100:2581-2592, 2012.

SB³C2015
Summer Biomechanics, Bioengineering and Biotransport Conference
June 17-20, 2015, Snowbird Resort, Utah, USA
SB³C2015-337

TOUGH, SELF-RECOVERING HYDROGELS INSPIRED BY CADDISFLY SILK

Dwight D. Lane, Sarbjit Kaur, G. Mahika Weerasekare, Russell J. Stewart

Department of Bioengineering
University of Utah
Salt Lake City, Utah, USA

INTRODUCTION

Aquatic caddisfly larvae use silk like an underwater adhesive tape to construct protective cases with adventitiously gathered rocks or sticks. Silk fibers comprise an adhesive outer coating on a viscoelastic core. The core is composed of multiple networks, including an elastic network and two independent Ca^{2+}-dependent phosphate and carboxylate networks. Phosphate or carboxyl groups bind Ca^{2+} to form reversible complexes that rupture at a critical stress, but reform when unloaded. Consequently, caddisfly silk is highly extensible, tough, and dissipates energy during cyclical loading.[1,2]

Caddisfly silk is a unique model for development of biomimetic energy-dissipating, tough, underwater materials. The goal of this project is to create synthetic hydrogels based on the double network structure of caddisfly silk that replicate the mechanical properties of the natural fibers. Hydrogels consist of cross-linked polymer networks with a high water content. Though appropriate for underwater applications, most hydrogels are weak and brittle. Recently described double network hydrogels have interpenetrating networks crosslinked through two types of bonds, reversible non-covalent bonds and covalent bonds. [3,4] The non-covalent network provides toughness, while in some cases the covalent network provides elasticity and self-recovery. We describe toughened, self-recovering hydrogels created with caddis silk-mimetic Ca^{2+}phosphate crosslinks in forming one network and covalently crosslinked polyacrylamide forming a second elastic network. Applications for these hydrogels could include biomaterials for wet physiological environments.

METHODS

MOEP monomer synthesis. Hydrogels were synthesized with a polyphosphate pre-polymer composed of 2-(Methaacryloyloxy)Ethyl Phosphate (MOEP). The MOEP monomer was synthesized by reaction of phosphorus oxychloride and hydroxy ethyl methacrylate (HEMA) in dry toluene and triethylamine (TEA).

Polymerization of MOEP to form polyMOEP. PolyMOEP was synthesized by free radical polymerization of 70 mol% MOEP, 15 mol% hydroxyethyl methacrylate (HEMA) and 15 mol% methacrylic acid (MA) in methanol initiated with AIBN. MA groups were grafted onto the MA sidechains with glycidyl methacrylate (GMA). The resulting phosphate prepolymer contained 15 mol% crosslinkable

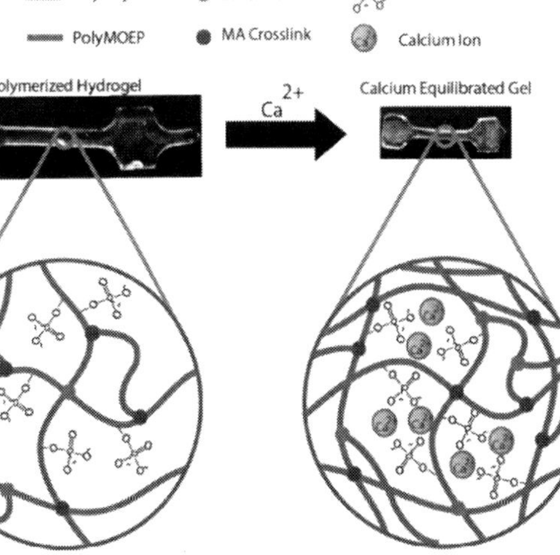

Figure 1: Hydrogel double network structure schematic: Complexation of Ca^{2+} ions phosphate sidechains creates a Ca^{2+}-phosphate crosslinked network, with a concomitant 40-50% decrease in volume.

methacrylate sidechains (MA prepolymer).

Hydrogel polymerization and phosphate-metal network equilibration. Hydrogels were formed by free radical polymerization of acrylamide and bis-acrylamide to crosslink the MA prepolymer (Fig. 1). The total weight percentage of MA prepolymer/monomer was held constant at 7.5% while the weight % of MA prepolymer was varied from 0.5% to 7.0%. Polymerization was initiated with APS and TEMED in dumbbell shaped molds. After 90 min, the gels were immersed in 150 mM NaCl, 50 mM metal ion, and 10 mM tris (pH 8) for 24 hrs. Next, the gels were soaked in 50 mM, 10 mM tris buffered metal solution for an additional 24 hrs. with frequent solution changes to remove salt from the gels. Metal ion selection included: Na^+, Ca^{2+}, Mg^{2+} and Zn^{2+}. Gels were considered fully equilibrated when the volume reached steady state.

Mechanical testing of hydrogels. Hydrogels were strained under water on an Instron 3342 tensile testing machine with a custom tank. Gels were strained cyclically and to failure in 5 mM metal ion and 5 mM Tris, pH 8. Gels were pulled at a strain rate of 0.15 s^{-1} and for cyclic tests to a strain of 0.5.

Processing of experimental data. Data was processed in matlab (MathWorks). Linear fits to the initial part of the stress strain curve were used to estimate Young's modulus. The yield point was determined using a 5% strain offset from the initial linear portion of the curve.

RESULTS

Increasing the ratio of the polyphosphate MA prepolymer to polyacrylamide resulted in higher initial modulus and yield stress. Hydrogels with polyphosphate < 4.0 wt % did not display sudden and dramatic strain softening (yield). The rate of metal induced volume change increased with ion species with $Mg^{2+} > Ca^{2+} > Zn^{2+}$. However, fully equilibrated volumes were equal between ion species.

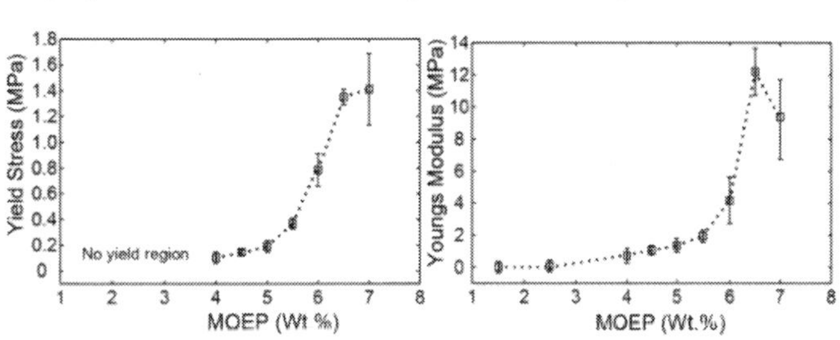

Figure 2: Left: Correlation between polyphosphate wt% and the presence and magnitude of yield point. Right: Relationship between polyphosphate wt% and Young's Modulus.

Young's modulus, yield stress, and stress at failure increased with ion species in the following order: Na^+ (0.04 MPa) < Mg^{2+} (0.14 MPa) < Ca^{2+} (10.3 MPa) < Zn^{2+} (34.2 MPa) (Fig 3). Hydrogels prepared with Na^+ and Mg^{2+} did not display yield. The yield stress for Ca^{2+} and Zn^{2+} was 1.80 and 3.46 MPa, respectively. Stress at failure was 0.05, 0.31, 1.94, and 3.82 MPa for Na^+, Mg^{2+}, Ca^{2+}, and Zn^{2+}, respectively. Increased stiffness came at the expense of extensibility; Zn^{2+} gels failed almost immediately following yield.

Cyclic testing showed hysteresis during loading and unloading and slow recovery of initial length and stiffness similar to natural caddisfly silk fibers (Fig 3B). Hydrogels prepared with Ca^{2+} were cyclically strained to 0.5, then allowed to recover for various times between strain cycles (Figure 3). Gels recovered over 90 % of the initial modulus, yield stress, hysteresis, and residual strain in 90 mins (Fig 3). Recovery kinetics fit an exponential profile.

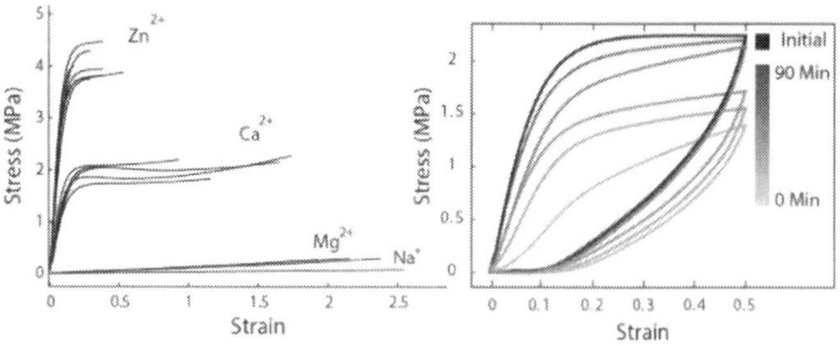

Figure 3: Left: Strain to failure of hydrogels prepared with Na^+, Mg^{2+}, Ca^{2+}, and Zn^{2+} ions. Right: Cyclic tensile tests with increasing recovery periods between cycles.

DISCUSSION

Polyphosphate/polyacrylamide double network hydrogels exhibited mechanical properties similar to caddisfly silk, including pseudo-yield, strain cycle hysteresis, and self-recovery. Gel toughness and the yield stress was dependent on the strength and density of the phosphate-metal complexes in the network. The higher density of phosphate-metal crosslinks with increased wt% of MA prepolymer, resulted in higher initial modulus and stress at yield. Different affinities between ion species and phosphates affects the bond strengths with Zn^{2+} strongest followed by Ca^{2+} then Mg^{2+}.[5,6] Equilibrated gel volumes were the same regardless of ion species, yet Mg^{2+} gels do not yield and are two orders of magnitude softer than Ca^{2+} or Zn^{2+} gels, indicating that the difference in mechanics is a result of phosphate-metal complexation and not dehydration of the gels. Recoil in the elastic network promotes reformation of the phosphate calcium network resulting in hydrogel recovery.

Several aspects of the natural silk fibers were recreated in the synthetic hydrogels, although some important differences remain. For one, structural organization of the caddisfly silk fibers leads to strain hardening after yielding. In contrast, the randomized structure of the gels results in a weak underlying elastic network with no hardening post yield. Additionally, the exponential recovery profile in the gels is a result of a single yielding network, caddisfly silk shows a biphasic recovery, a result of two recovering networks, Ca^{2+}phosphates and Ca^{2+}carboxylates.[2] Caddisfly silk has a 2MPa higher yield stress and a full 100% recovery following strain which may result from cooperativity in the fiber structure.[2] Future experiments will include combinations of ions or different structural organizations to further tune the gel mechanics as needed to meet material requirements.

ACKNOWLEDGEMENTS.

The authors thankfully acknowledge funding from the Army Research Office.

REFERENCES

1. Ashton, N. N., Roe, D. R., Weiss, R. B., Cheatham, T. E. & Stewart, R. J.. *Biomacromolecules* **14,** 3668–81 (2013).
2. Ashton, N. N. & Stewart, R. J. *Soft Matter* (2014). doi:10.1039/c4sm02435d
3. Gong, J. P., Katsuyama, Y., Kurokawa, T. & Osada, Y. *Adv. Mater.* **15,** 1155–1158 (2003).
4. Gong, J. P. Materials science. *Science* **344,** 161–2 (2014).
5. Brule, G. *Le Lait.* **68,** (1988).
6. Holm, R., Kennepohl, P. & Solomon, E. *Chem. Rev.* **2123,** (1996).

SB³C2015
Summer Biomechanics, Bioengineering and Biotransport Conference
June 17-20, 2015,Snowbird Resort, Utah, USA
SB³C2015-338

PHYSICAL PATIENT SPECIFIC SIMULATION FOR ASCENDING AORTIC ANEURYSMS SURGERY PRE-PROCEDURAL TRAINING

J. Garcia (1), Z. Yang (1), K. Lachapelle (2), R. Mongrain (1), R. L. Leask (3)

(1) Mechanical Engineering,
McGill University,
Montreal, Quebec, Canada

(2) Cardiovascular Surgery,
McGill University Health Network,
Montreal, Quebec, Canada

(3) Chemical Engineering,
McGill University,
Montreal, Quebec, Canada

INTRODUCTION

Cardiovascular disease is a leading cause of death in North America and thoracic aortic aneurysms (TAA) are the most common pathology requiring cardiothoracic surgery. Cardiac surgery requires high-levels of competency and precision. Simulation for open surgical procedures usually involves a phantom with normal anatomy composed of commercial tissue substitutes or expensive cadaveric models. These simulators lack patient pathologies and the living human tissue characteristics essential for realistic pre-procedural practice. There is therefore a need for realistic physical training simulators to help train the next generation of surgeons.

The objective of this study was to develop a series of anatomically correct ascending aorta pathological training phantoms composed of materials bio-mechanically similar to TAA and normal aortic tissue.

METHODS

General process

The *in vivo* anatomy of patient's from the McGill University Health Centre were acquired from cardiac gated 128-slice CT scans in standard Digital Imaging and COMmunication (DICOM) files. Files were segmented with commercial software (3D Doctor, Able Software). A three dimensions (3-D) model and a stereolithography file format (.stl) were created. The .stl file was edited with a Computer Aided Design software (Solidworks, Dassault System) prior to creating objects with a 3-D printer (Object Connex 3D, Stratasys, Eden Prairie, MN, USA) using photopolymer materials. All the steps are illustrated in Figure 1.

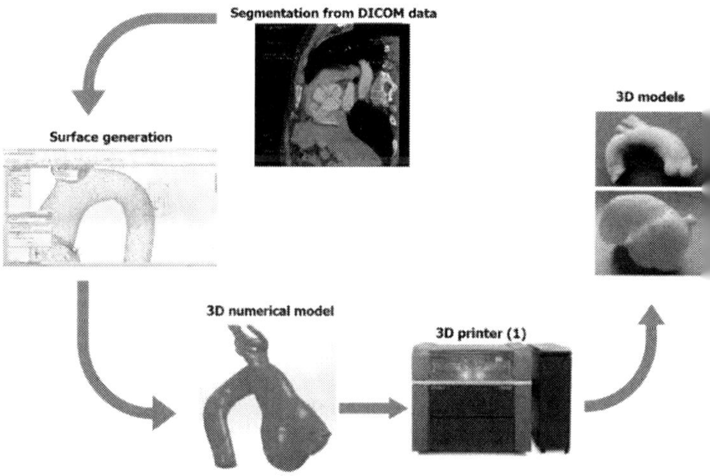

Figure 1 Steps of the general process

Mechanical testing

To identify suitable materials for the physical models, we evaluated candidate materials by mechanical testing with a biaxial tensile tissue tester (Electroforce ELF 3200, Bose, Figure 2). The results were compared to normal human ascending aortic tissue and TAA from patients undergoing ascending aortic replacement. Commercial elastomers commonly used in the simulation centers for mimicking human tissues were analyzed and called 1, 2 and 3. Furthermore, we designed and tested composites made of rubber-like flexible and rigid photopolymers (Tangoplus Fullcure 930, and VeroWhitePlus RGD835 respectively, Stratasys, Eden Prairie, MN, USA). The two photopolymers were printed simultaneously with the ridged polymer providing a fibrous structure similar to collagen to allow for tuned stiffness, anisotropy and suture retention strength. There was no pre-determined direction in the commercial artificial tissue composites.

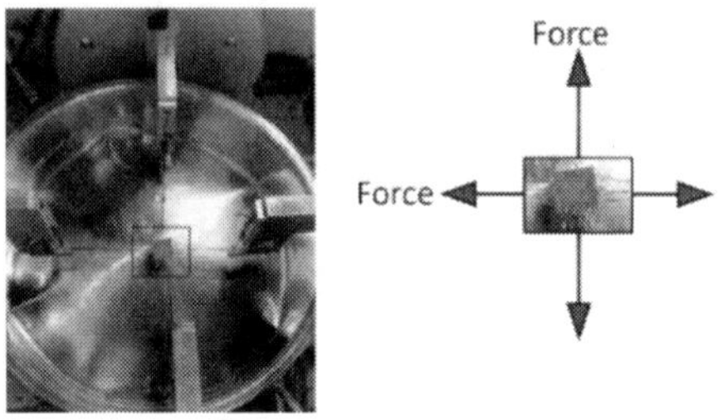

Figure 2 Biaxial tensile tests

Data analysis

Mathematical model

Unique stress-strain curves were created during loading and unloading for each material. We determined the stress-stretch behavior in order to analyze the coefficients (material parameters) of the Mooney-Rivlin model of the fitted curves:

$$\Psi = \sum^{n}_{\substack{i+j=1 \\ i,j=0,\ldots,n}} c_{ij}\,(I_1 - 3)^i (I_2 - 3)^j \quad (1)$$

c_{ij} are materials parameters, n is the order and I_1, I_2 the invariants of the right Cauchy-Green deformation tensor, such as:

$$\begin{cases} I_1 = \lambda_1^{\,2} + \lambda_2^{\,2} + \lambda_3^{\,2} \\ I_1 = \lambda_1^{\,2}\lambda_2^{\,2} + \lambda_2^{\,2}\lambda_3^{\,2} + \lambda_3^{\,2}\lambda_1^{\,2} \end{cases} \quad (2)$$

$\lambda_{i=1,2,3}$ are the principal stretches and because of the incompressibility of the material:

$$\lambda_1\lambda_2\lambda_3 = 1. \quad (3)$$

Apparent modulus of elasticity

To compare the samples we used the apparent modulus of elasticity (E) computed by calculating the slope values of the hyperelastic model fit of the loading stress-strain curves at 40% strain (Figure 3).

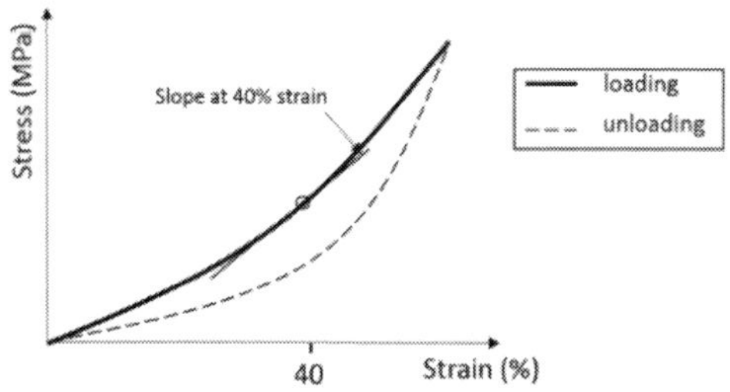

Figure 3 Typical stress-strain curve shape

Energy loss

The energy loss that described the relative amount of energy absorbed by the aorta during the loading and unloading cycles was determined with the original raw data. It was calculated by the difference between the integral of the stress-strain curve during loading and unloading.

RESULTS

All the results are shown in Table 1.

Apparent modulus of elasticity

Chung, et al. found an apparent elastic modulus of 0.10 ± 0.03 MPa for normal aortic human tissues and 0.15 ± 0.03 MPa for TAA at 40% strain (2). All the artificial tissues modulus lie within the range of the normal aortic tissue, but only a few composites do (1, 4, 5 and 9). Only the artificial tissues 1 and 2 had similar values to TAA tissue presented by Chung, et al (2). Composites 1, 4, 5, 7, 8, 9 and 10 also had an apparent modulus similar to TAA.

Energy loss

Chung, et al. calculated 28 ± 2% energy loss for normal aortic tissue and 34 ± 4% for TAA (2). All the artificial tissues and the composites 2 and 4 are in between the range of the normal aortic tissue and TAA energy loss. All of the 3-D printed composites had on average less energy loss.

Table 1 Apparent modulus of elasticity and energy loss

		Apparent modulus of elasticity E (MPa)				Energy loss (%)			
		Axis 1	Axis 2	Average	Standard deviation	Axis 1	Axis 2	Average	Standard deviation
Artificial tissues	1	0.12	0.10	0.11	0.01	39%	26%	32%	9%
	2	0.12	0.17	0.14	0.03	25%	29%	27%	3%
	3	0.11	0.09	0.10	0.01	23%	29%	26%	4%
Composites	1	0.10	0.14	0.12	0.03	19%	22%	20%	2%
	2	0.30	0.24	0.27	0.04	26%	24%	25%	1%
	3	0.26	0.25	0.26	0.01	17%	18%	17%	1%
	4	0.23	0.11	0.17	0.08	18%	26%	22%	6%
	5	0.22	0.14	0.18	0.05	17%	19%	18%	2%
	6	0.26	0.22	0.24	0.03	17%	18%	18%	1%
	7	0.32	0.19	0.26	0.09	24%	24%	24%	0%
	8	0.16	0.18	0.17	0.01	17%	29%	23%	9%
	9	0.16	0.08	0.12	0.05	16%	15%	15%	1%
	10	0.19	0.17	0.18	0.02	22%	25%	23%	2%

DISCUSSION

The commercial artificial tissues 1 and 2 are biomechanically similar to human AA tissue. Anisotropy and tuned stiffness was achieved in the 3-D tissue composites. Using these results, we are now adjusting the crimp and amplitude of the stiff polymer fibers to further tune the composite properties to match patient specific biomechanics. Printing of 3-D composites has the potential to produce patient specific physical models of human ascending aortic pathologies.

ACKNOWLEDGEMENTS

McGill Engineering Doctoral Awards (MEDA)

Natural Sciences and Engineering Research Council of Canada (NSERC)

REFERENCES

(1) Retrieved January 12, 2015, from http://www.stratasys.com/3d-printers/design-series/connex-systems

(2) Chung, J., Lachapelle, K., Wener, E., Cartier, R., De Varennes, B., Fraser, R., & Leask, R. L. (2014). Energy loss, a novel biomechanical parameter, correlates with aortic aneurysm size and histopathologic findings. The Journal of Thoracic and Cardiovascular Surgery, 148

SB[3]C2015
Summer Biomechanics, Bioengineering and Biotransport Conference
June 17-20, 2015, Snowbird Resort, Utah, USA
SB[3]C2015-339

EFFECTS OF LEAFLET MICROSTRUCTURE AND CONSTITUTIVE MODEL ON THE CLOSING BEHAVIOR OF THE MITRAL VALVE

Chung-Hao Lee (1), Jean-Pierre Rabbah (2), Ajit P. Yoganathan (2), Robert C. Gorman (3), Joseph H. Gorman, III (3), Michael S. Sacks (1)

(1) Institute for Computational Engineering and Sciences
Department of Biomedical Engineering
The University of Texas at Austin
Austin, TX 78712, USA

(2) Cardiovascular Fluid Mechanics Laboratory
Department of Biomedical Engineering
Georgia Institute of Technology
Atlanta, GA 30318, USA

(3) Gorman Cardiovascular Research Group
University of Pennsylvania
Philadelphia, PA 19104, USA

INTRODUCTION

Recent long-term studies [1] showed an unsatisfactory recurrence rate of severe mitral regurgitation after surgical repair, suggesting that excessive tissue stresses and the resulting strain-induced tissue failure are potential etiological factors controlling the success of surgical repair for treating mitral valve (MV) diseases. We hypothesized that restoring normal MV tissue stresses in MV repair techniques would ultimately lead to improved repair durability through the restoration of MV normal homeostatic state. The accuracy of the MV computational model is thus critical for understanding how MV tissue adaption maintains its normal homeostasis in response to stress overload, pathologies and surgical intervention. In this study, we developed an anatomically accurate MV finite element (FE) model (Fig. 1) for investigating the MV behavior during systolic closure.

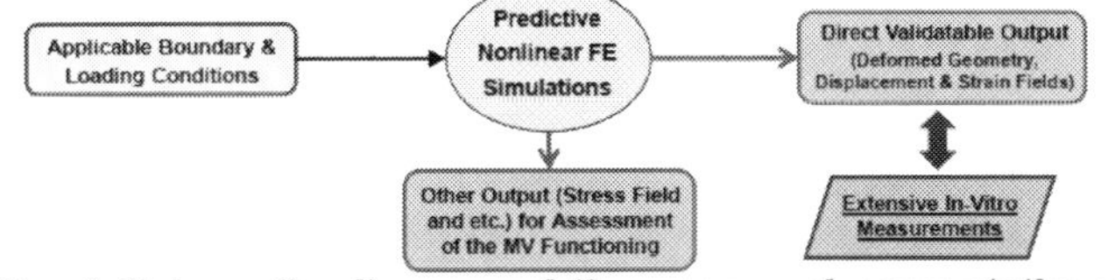

Fig. 1 Schematic diagram of the proposed computational framework for FE simulations of the MV closing behavior.

METHODS

Specimen preparation & micro-CT imaging: A fresh ovine heart from a 35 kg sheep was acquired from a local USDA approved abattoir. The MV apparatus, including the anterior leaflet (MVAL), posterior leaflet (MVPL), chordae tendineae and papillary muscles (PMs), were excised from the explanted ovine heart, and the surrounding myocardium was carefully trimmed. A modified left heart simulator developed previously [2] was utilized for acquiring the in-vitro MV geometry under a vivaCT40 system at the fully-open state chosen as the reference configuration in the FE simulations, and at 70 mmHg pressure loaded state for validations of the proposed computational models. Note that ebony glass beads (0.2-0.4 mm in diameter) were glued to the two leaflets which were served as fiducial markers in the strain and displacement error calculations.

Image segmentation & FE model development: The stack of micro-CT images were segmented using ScanIP (Simpleware Ltd., United Kingdom) to obtain anatomically accurate geometry of the MV apparatus. The atrial and ventricular leaflet surfaces were identified from the segmented MV geometry, and a median MV leaflet surface was reconstructed. FE mesh at the fully-open (reference) configuration associated with this MV leaflet median surface was generated in HyperMesh (Altair Engineering, Inc., MI, USA). For idealization of the MV chordae tendineae using truss elements, 3D locations of the landmark points were quantified from the micro-CT images.

Mapping of collagen fiber architecture & constitutive modeling: The MV leaflets were separated along the anterior and posterior commissures, and prepared for the measurement of the gross (effective) fiber microscopic architecture using the small angle light scattering (SALS) technique developed previously [3]. To incorporate the measured fiber microstructural architecture with the constructed MV FE model, we developed the mapping algorithm based on affine fiber kinematics [4] to obtain the fiber splay dispersion and local material axes at the reference configuration. This information was integrated into a transversely isotropic simplified structural model [5] to simulate the in-vitro mechanical behaviors of the MV leaflet tissues.

FE simulations of the MV closure: The interaction between blood flow and the MV apparatus is idealized by transvalvular surface pressure acting on the MV leaflets, and the in-vitro experimental conditions are mimicked by clamping the MV annulus and PMs. Alterations of the MV stress and strain fields in response to external loading were of the interest, and simulations of MV closure were performed in ABAQUS (Simulia, RI, USA). Displacement errors at fiducial marker positions and in-surface principal strains of the central leaflet regions were validated with the in-vitro measurements. Moreover, we investigated the effect of the leaflet structural property as well as local fiber mechanical responses on the predictions using perturbations with increasing levels of modeling fidelity and variations in the local ensemble fiber stress-strain relations (Fig. 2a-b).

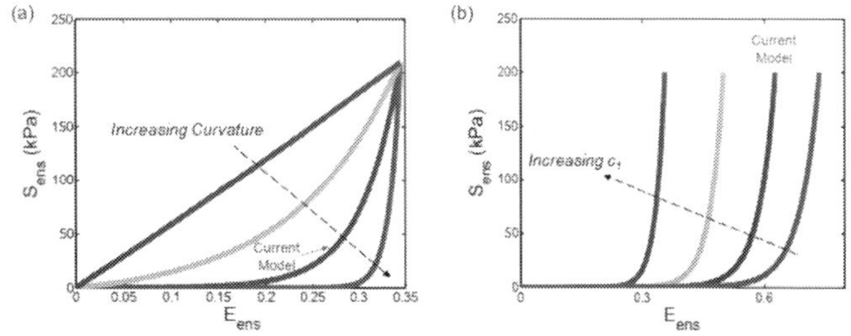

Fig. 2 Parametric studies considering: (a) variation in the curvature of the ensemble fiber stress-strain relation, and (b) variation in local fiber mechanical property.

RESULTS

In FE simulations of the MV closing behavior by the proposed computational framework, the displacement errors evaluated at the fiducial markers between the model predictions and in-vitro data were satisfactory: 0.09±0.05 mm and 0.19±0.06 mm at 30 mmHg transvalvular pressure and 0.13±0.09 mm and 0.26±0.09 mm at 70 mmHg transvalvular pressure for the MVAL and MVPL, respectively. FE predicted in-surface principal strains also agreed qualitatively and quantitatively well with the in-vitro measurements (Fig. 3a-b).

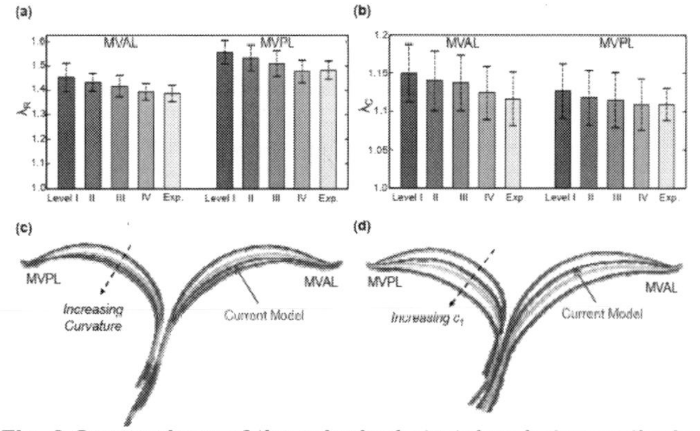

Fig. 3 Comparison of the principal stretches between the in-vitro measurements and the FE predictions with various levels of model fidelity: (a-b) radial and circumferential stretches. Comparisons of the deformed leaflet surfaces in the parameter studies considering: (c) variation in the curvature of the ensemble fiber stress-strain relation, and (d) variation in local fiber mechanical property.

In addition to the above in-vitro validations, we investigated how the fiber architecture of the MV leaflets changes in response to external pressure loading, in terms of the normalized orientation index (NOI) values in the deformed configurations (Fig. 4). These results provides us better insight into the important micro-mechanisms within MV leaflets for maintaining the functioning of the MV apparatus. We further evaluated the effect of leaflet microstructural information and local fiber mechanical properties on the predicted MV closure behavior (Fig. 3c-d). These results underscored the necessary model fidelity and information to be included in the simulations for achieving desirable accuracy as well as appropriate parameter ranges and micromechanical fiber responses for proper coaptation and natural functioning behaviors of the MV under physiological loading.

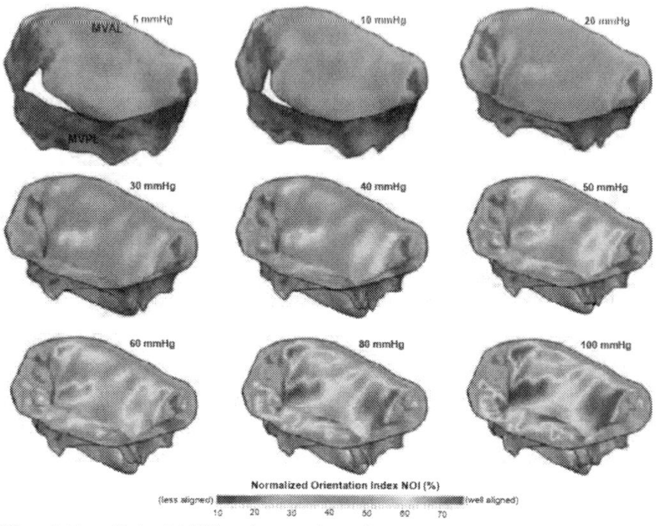

Fig. 4 Predicted NOI values at various pressure levels.

DISCUSSION

In summary, we have proposed an image-based computational framework for modeling of the functioning MV and successfully implemented a computational model into the nonlinear FE simulation procedure. This computational model, to our knowledge, is the first of this kind that has been thoroughly validated with extensive in-vitro data for its model accuracy and robustness. The proposed model would serve as an important first step toward our long-term modeling goals, which are to employ computer simulations for assessing the performance of existing repair techniques and to provide insightful guidance in the design of optimal repair strategies for treating diseased MVs with better restored functionality and improved durability.

ACKNOWLEDGEMENTS

Support from the National Institutes of Health (NIH) grants R01 HL119297, HL63954, HL103723, and HL73021 is greatly acknowledged. Dr. Lee was supported in part by the American Heart Association (AHA) Postdoctoral Fellowship (14POST18160013).

REFERENCES

[1] Braunberger *et al.*, *Circulation*, 104: 8-11, 2001.
[2] Rabbah *et al.*, *Ann Biomed Eng*, 41: 305-315, 2013.
[3] Sacks *et al.*, *Ann Biomed Eng*, 30: 1281-1290, 1997.
[4] Lee *et al.*, *Multiscale Modeling in Biomechanics and Mechanobiology*, Springer-Verlag, 225-255, 2015.
[5] Fan and Sacks, *J Biomech*, 47: 2043-2054, 2014.

LAYER-SPECIFIC MITRAL VALVE INTERSTITIAL CELL DEFORMATIONS UNDER PHYSIOLOGICAL LOADING

Chung-Hao Lee (1), Christopher A. Carruthers (2),Salma Ayoub (1), Robert C. Gorman (3), Joseph H. Gorman, III (3), Michael S. Sacks (1)

(1) Institute for Computational Engineering and Sciences
Department of Biomedical Engineering
The University of Texas at Austin
Austin, TX 78712, USA

(2) Cardiac Rhythm Disease Management
(CRDM) Clinical Specialist
Medtronic
Minneapolis, MN 55432, USA

(3) Gorman Cardiovascular Research Group
University of Pennsylvania
Philadelphia, PA 19104, USA

INTRODUCTION

Within each of the four layers of mitral valve (MV) leaflet tissues there resides a heterogeneous population of interstitial cells that maintain the structural integrity of the MV tissue via protein biosynthesis and enzymatic degradation. There is increasing evidence that tissue stress-induced MV interstitial cell (MVIC) deformations can have deleterious effects on their biosynthetic states that are potentially related to the reduction of tissue-level maintenance and to subsequent organ-level failure [1]. To better understand the interrelationships between tissue-level loading and cellular responses, we developed an integrated experimental-computational approach (i) to quantify the in-situ layer-specific MVIC deformations for each of the four layers using a controlled biaxial tension loading device coupled to multi-photon microscopy (MPM), and (ii) to explore the inter-relationship between the MVIC stiffness and deformation to layer-specific tissue mechanical and structural properties via the proposed macro-micro finite element (FE) computational model.

METHODS

Tissue preparation & biaxial testing: Fresh ovine hearts from 40 kg sheep were acquired from a local USDA approved abattoir. MV leaflets were dissected and the surrounding myocardium was carefully trimmed from the leaflets. Tissue samples of 10 mm x 10 mm (Fig. 1a) were then taken from the center region of the MV anterior leaflet (MVAL). The samples were loaded into a miniature biaxial testing system (Fig. 1b), in which the specimen's circumferential and radial directions were aligned with the device axes. Four graphite fiducial markers were placed in the center of each tissue sample and their positions were tracked optically in real time. The real-time tissue deformation gradient tensor **F** was computed by using standard surface strain calculation techniques [2].

Fixation & MPM imaging: Three tissue specimens (n=3) from the MVAL center region were measured at the unloaded state (0 N/m tension) and at 6 tension levels (5, 10, 20, 50, 100, 150 N/m). Each sample was fixed in 2% aqueous paraformaldehyde (PFA) solution for 2 hours in the loaded configuration, and then split into two sections for the analyses (Fig. 1b). The first 7 μm section was stained with Movat's Pentachrome to distinguish the histologic four layers of the MV leaflet: *atrialis*, *spongiosa*, *fibrosa*, and *ventricularis* layers. The second 30 μm section was paraffin-embedded and cryosectioned along the circumferential-transmural surface. The sample slides were taken for imaging under the multi-photon microscopy (MPM) system [3] at an excitation wavelength of 830 nm with an incremental z-step of 1 μm. Here, the MPM was used to image the MVIC Cytox green nuclei via two-photon excited fluorescence (TPEF) at 525±50 nm.

Determination of in-situ MVIC nuclear deformation: MVIC nuclear orientation and deformation for each tissue layer and tension level were measured and analyzed as follows. The MPM image stacks were first imported into Imaris (Bitplane AG, Zurich, Switzerland) for 3D nuclear deformation visualization and quantitative analysis (Fig. 2). Individual surface was then fitted to each nuclear point cloud using an automated maximum minimum voxel thresholds. The eigenvectors associated with each 3D nuclear surface were determined based on an assumed MVIC *ellipsoidal* geometry, and the major and minor axes of each ellipsoid were converted into the nuclear aspect ratio (NAR=L_a/L_c) as a dimensionless indicator of MVIC deformation.

Simulation of MVIC microenvironment: We developed a FE computational framework for modeling the MVIC microenvironment to investigate the interrelationship between layer-specific mechanical behaviors and the measured MVIC deformations. Here, the MVIC microenvironment for each of the four layers was idealized using a representative area element (100μm x 100μm) embedded with 37 uniformly distributed (Fig. 3a) with layer-specific extracellular matrix (ECM) microstructural compositions (Fig. 3b) and collagen fiber architecture (Fig. 3c). The MVIC regions were modeled using a Saint-Venant Kirchhoff material with a single parameter E_ξ for each layer, and the surrounding ECM regions were modeled by the following structural constitutive model:

$$\mathbf{S} = \phi_m^\xi \mu_m (\mathbf{I} - C_{33}\mathbf{C}^{-1}) + \int_{-\pi/2}^{\pi/2} \Gamma^\xi(\theta)\phi_f^\xi c_0 \{\exp[c_1 E_{ens}(\theta)] - 1\} \mathbf{N}(\theta) \otimes \mathbf{N}(\theta) d\theta \quad (1).$$

where $\mathbf{S}$ is the 2nd Piola-Kirchhoff stress tensor, ξ is the layer identifier, ϕ_m and ϕ_f are the volume fractions of the non-fibrous matrix and collagen fiber, respectively, μ_m, c_0, c_1 are material constants determined from the fit to the biaxial loading test, $\mathbf{N}$ is the preferred fiber direction, $\mathbf{I}$ is the identity matrix, $\mathbf{C}=\mathbf{F}^T\mathbf{F}$ is the right Cauchy deformation tensor, $E_{ens}=\mathbf{N}^T(\mathbf{F}^T\mathbf{F})\mathbf{N}$ is the ensemble fiber strain, and $\Gamma(\theta)$ is the fiber orientation distribution function represented by a Gaussian distribution. In the FE simulations, we considered that the interfaces between the MVICs and MVAL ECM were assumed to be perfectly bonded, and tissue-level stretches from the equi-biaxial tension test were applied as incremental nodal displacements to mimic the tissue forces acting on the MVIC microenvironment.

RESULTS

The MVICs in the fibrosa and ventricularis layers had a noticeably larger deformation than those in the atrialis and spongiosa layers, reaching a NAR of 3.3 at the maximum physiological loading (Fig. 4-a). The deviations for each layer were large, indicating that there was a large amount of variation in deformation that the MVICs were experiencing in each layer. Moreover, the MVIC moduli for the four layers were characterized by matching the NAR-tension curves to the experimental measurements, reported as follows: $E_{Atrialis}$=5.01 kPa, $E_{Spongiosa}$=4.71 kPa, $E_{Fibrosa}$=5.35 kPa (Fig. 4-d), and $E_{Ventricularis}$=5.17 kPa. These novel results suggest that while the MVICs may be phenotypically and biomechanically similar throughout the leaflet, they experience layer-specific mechanical stimulatory inputs due to their distinct ECM architecture and mechanical behaviors.

DISCUSSION

In this study, we for the first time employed an integrated computational-experimental approach to quantify and simulate layer-specific cellular deformations across the VMAL tissue under controlled physiological loading. The results elucidate the essential role played by layer-specific microstructural compositions and fiber architecture in the underlying MVIC deformations. This work can be utilized to provide insightful information about MV function and behavior across multiple length scales in response to mechanical stimuli, such as the stress overload induced by MV disease and surgery intervention, and eventually to investigate the effect of cellular deformations on collagen biosynthesis as well as tissue adaption. This important piece of information can in turn, be used to establish the interrelationship between cellular deformations and tissue-level mechanical behaviors that maintains the homeostasis of the functioning MV. The proposed macro-micro computational framework will be beneficial in the future when integrated with organ-level, structurally accurate FE models to investigate MV pathophysiology and disease progression as a function of MVIC mechanobiology and ECM architecture, and ultimately to improve MV surgical repair procedures by enhancing the long-term durability.

ACKNOWLEDGEMENTS

The assistance from Dr. Simon C. Watkins and Greg Gibson for the MPM imaging and Bryan Good for development of the biaxial device is greatly appreciated. This work is funded by the National Institutes of Health grants R01 HL119297, HL63954, HL103723 and HL73021. Dr. Chung-Hao Lee is supported in part by an American Heart Association Postdoctoral Fellowship 14POST18160013.

REFERENCES

[1] Grande-Allen *et al.*, *J Am Coll Cardio*, 42: 271-277, 2003.
[2] Sacks, *J Elasticity*, 61: 199-246, 2000.
[3] Watkins, *Immunohistochemistry*, John Wiley & Sons, Inc., 2001.

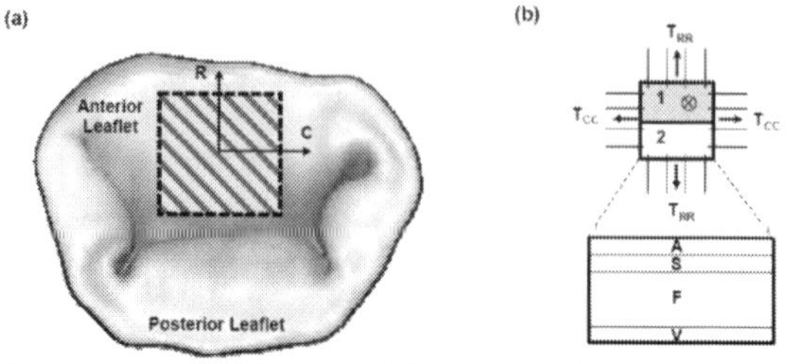

Fig. 1 (a) 10mmx10mm MVAL specimen (C and R: circumferential and radial directions), and (b) experimental setup for the mini-biaxial testing, followed by histologic analyses and measurements of the MVIC NARs.

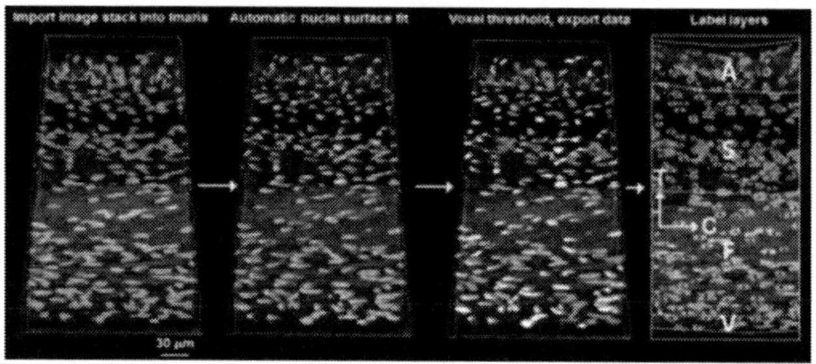

Fig. 2 Analysis of MPM images for determining MVIC deformations associated with the four distinct layers.

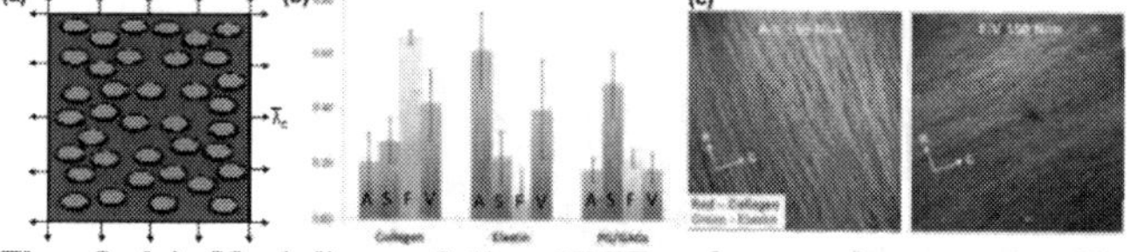

Fig. 3 (a) Modeling of the MVIC microenvironment with prescribed tissue-level deformations and 37 uniformly distributed ellipsoidal cell inclusions, (b) layer-specific microstructural compositions from histologic analysis, and (c) layer-specific preferred fiber direction N obtained from second harmonic generation (SHG) images.

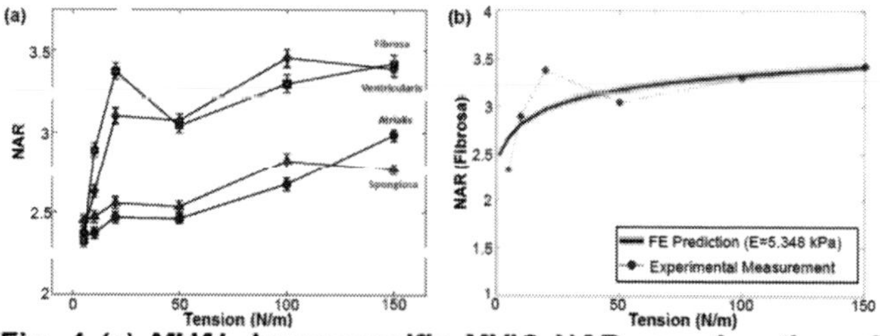

Fig. 4 (a) MVAL layer-specific MVIC NAR as a function of applied biaxial tension, derived from MPM analysis, (b) Comparisons of the measured and numerically predicted NARs as a function of tissue-level tension.

SB³C2015
Summer Biomechanics, Bioengineering and Biotransport Conference
June 17-20, 2015,Snowbird Resort, Utah, USA
SB³C2015-341

MAGNETIC MICROPILLAR ARRAY TO STUDY EPITHELIAL MICROTISSUE MORPHOGENESIS AND LOCAL MECHANICS

Mohammadnabi Asmani[1], Yan Li [1], Christopher Kotei[1], David Olsen[1], Zhaowei Chen[1], Fanjie Meng [2], Ruogang Zhao[1]

(1) Department of Biomedical Engineering
The State University of New York at Buffalo
Buffalo, New York, USA

(2) Department of Physiology and Biophysics
The State University of New York at Buffalo
Buffalo, New York, USA

INTRODUCTION

Cell generated contraction forces and force transmission through adhesive complexes are two main factors driving tissue morphogenesis during development [1]. The adaptive process by which cells spatially resolve and respond to non-local forces is critical to an integrated response of the tissue. Existing models for developmental biomechanics rely mainly on insect or amphibian embryos where both the manipulation of local forces and the measurement of force transmission are difficult. Furthermore, such models significantly lag behind the fast growing field of human and mammalian stem cells where the importance of mechanics has been clearly demonstrated in stem cell-driven tissue and organ formation [2]. Therefore, a new model system, which can allow the manipulation and measurement of tissue mechanics at both local and global levels with the potential to incorporate mammalian stem cells, is greatly needed. Here we describe a magnetic micropillar array system that can apply magnetic force to the local region of a self-assembled multilayered epithelial microtissue and simultaneously measure the global response of such a microtissue through pillar array detectors and genetically encoded FRET stress sensors. Such a system will open a novel window for the study of developmental biomechanics.

METHODS

We fabricated microtissue pillar array device (tPAD) by PDMS replica molding from masters made using a multilayer microlithography technique [3]. As shown in Fig. 1A to C, an array of microwells (8 columns by 8 rows) was casted in a PDMS substrate contained in a standard P35 petri-dish. Each microwell (1000 x 1000 μm^2) contains an array of PDMS micropillars (8, 9 or 16). When a mixture of cells and collagen is placed in the microwell, the contractile action of the cells causes the compaction of the matrix hydrogel. We used MDCK canine kidney epithelial cells in the seeding and

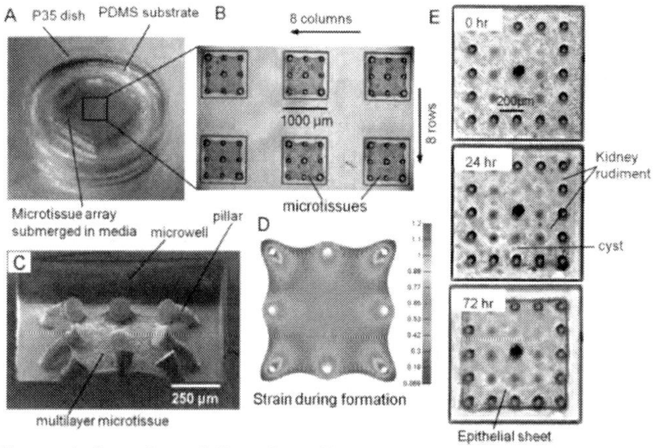

Figure 1. Overview of the micropillar array device. (a) A P35 petri-dish contains a microtissue array fabricated in a PDMS substrate. (b) A portion of the microtissue array (3x2). (c) SEM image of one of the microtissues anchored to the neck section of PDMS micropillars. (d) Finite element modeling of the microtissue formation process. (e) Microtissue formed through the self-assembly of epithelial cells and matrix during a 3-day period.

monitored the tissue formation process daily. To analyze the tissue formation process, finite element modeling was performed using a cell growth constitutive model in the FEBio package [4].

We then developed an electromagnetic actuation system that can apply mechanical stretching to local region of single microtissue through magnetic forces generated by a magnetic tweezer (Fig. 2A). The magnetic tweezer consisted of a computer-controlled electromagnet with a sharpened pole tip, which could be brought in close proximity to the Nickel (Ni) sphere pre-attached to the top of the micropillar (Fig. 2A), as previously described [3]. We applied ramped local stretching in the center of the microtissue to observe its short term response and hold the loading for 15 mins to observe tissue's creep response. The deflection of the edge micropillars was measured to analyze the transduced force to the boundary. We also used an optical flow algorithm to track the displacement field of the microtissue. Finite element modeling of the loading process was performed using a Neo-Hookean Standard Linear Solid material model in ABAQUS 6.8. The elastic modulus of the microltissue was assumed to be 30kPa, based on previous measurement of similar tissue [3].

To obtain a refined mapping of the microtissue response to local forces, we adopted a recently developed FRET stress sensor that is linked with α-actinin, an actin filament binding protein [5]. This new stress sensor, called cpstFRET, is designed to be modulated by the angles between the donor (CFP) and acceptor (YFP). By measuring the CFP/YFP fluorescence ratio in the sample, we are able to detect the intracellular force changes. A wound healing assay was performed to test the utility of the FRET sensor in microtissues populated with sensor encoded MDCK cells. A ~ 200 μm long wound was created manually using a scalpel in the microtissue. Wound area immediately after injury (Fig. 3A) and 6 hours after injury (Fig. 3B) was imaged. CFP and YFP images of the microtissue were processed to calculate the fluorescence ratio.

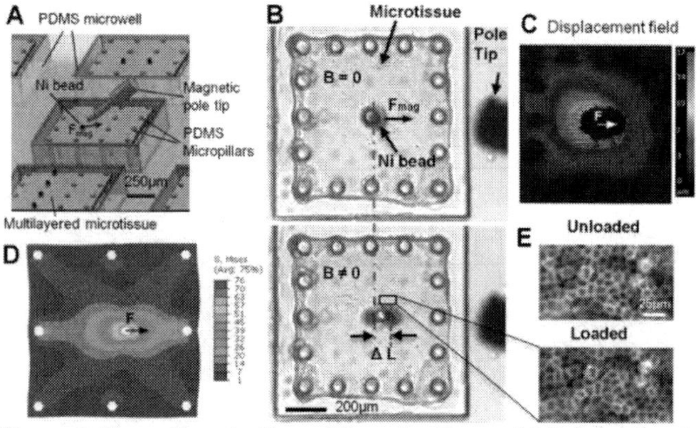

Figure 2. Magnetic actuation of the microtissue. (a) Schematic of the magnetic microtissue pillar array device. (b) A microtissue without (B=0) and with (B≠0) magnetic actuation in the center region. ΔL shows pillar head displacement. (c) Displacement field of the microtissue under local magnetic stretching. (d) FE model predicted stress distribution of the microtissue under local magnetic stretching. Unit is kPa. (e) Epithelial sheet was compressed in the local region close to loading micropillar. Red arrows show significant shape change of representative cells.

RESULTS

After seeding, the collective contractile forces of embedded cells caused the collagen gel to compact. During this process, MDCK cells migrate and proliferate to form cyst or larger aggregates of kidney rudiments (0-24 hrs, Fig. 1E), which is consistent with the observation of epithelial culture in conventional collagen matrix. However, in the current model the mechanical constraints of the micropillars counterbalanced cellular contraction forces. Such a mechanical balance provided a cue for the formation of epithelial microtissue. By 72 hrs, we observed the formation of a thin flat microtissue wrapping around the neck of micropillars (Fig. 1C, E). The finite element modeling of collagen gel deformation under hydrostatic compressive force and micropillar constraints recapitulated the microtissue formation process (Fig. 1D), proving the balance between contraction forces and boundary conditions drove the tissue formation.

We calibrated the loading capacity of the magnetic tweezer system using empty Ni bead-adhered micropillars without tissue attachment. Magnetic forces up to ~105 μN can be generated at the micropillar head. With microtissue attached, such forces can cause large deformation in the local region of the microtissue (Fig. 2B). The deformation in the vicinity of the loading pillar propagated to the surrounding tissue area and dissipated at the tissue edge, as a result of the counterbalance by the edge micropillars (Fig. 2C). The reaction forces of the edge micropillars directly resulted in the formation of force transduction pathways from the loading center to the edge pillars, as shown by the stress distribution estimated by the finite element model (Fig. 2D). Magnetic actuation caused significant cell shape change in the epithelial sheet adjacent to the loading micropillar (Fig. 2E).

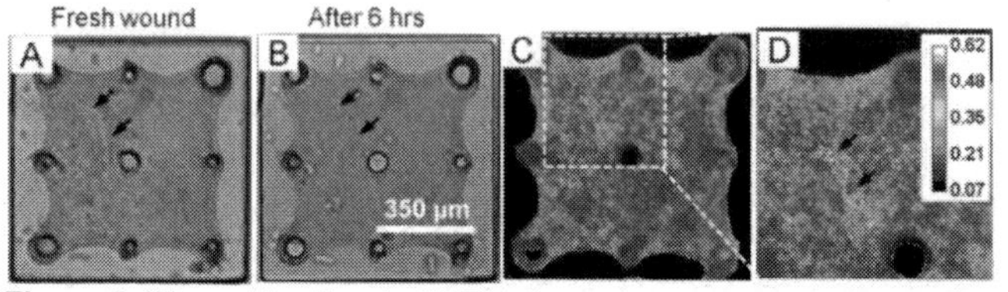

Figure 3. Wound healing in FRET stress sensor encoded epithelial microtissue. (a) Fresh wound created in microtissue. Arrows show wound location. (b) Wound healing 6 hours later. (c) CFP/YFP ratio map of the whole microtissue. (d) Enlarged view in the wound area. The high CFP/YFP ratio at the wound edge (red) indicates high mechanical stress.

It can be seen that high CFP/YFP fluorescence ratio existed in the wound healing area (red color, arrows, Fig. 3C, D), indicating low FRET state and high mechanical stress. The distribution of the high mechanical stress matched with the shape of the healing wound (arrows, Fig. 3C, D). This finding is consistent with the previous studies on wound healing showing that cells at the wound edge generating high contraction force to help the closure of the wound.

DISCUSSION

In this study, a suspending 3D epithelial microtissue self-assembled from a mixture of cell and matrix under the mechanical boundary conditions provided by the micropillar array. This opens a new pathway towards engineering 3D tissue sheets using mammalian stem cells, which had been traditionally performed on 2D culture models that are known to be a deviation from the physiological 3D tissue environment. Local magnetic loading caused both tension and compression zones within a single microtissue, which is a technological improvement over conventional tissue stretching experiments. The changes in the cell shape in the vicinity of the loading zone recapitulated similar morphological events in the developing epithelium under local force. However, measurements at longer time points are desired. Together, this system provides a promising research tool for the study of developmental mechanics.

ACKNOWLEDGEMENTS

University at Buffalo Clinical and Translational Pilot Studies Fund

REFERENCES

[1] C. Guillot, et al. *Science*, 340, no. 6137, pp. 1185-1189, (2013).
[2] M. Eiraku. et al. *Nature*, 472, 51-56, 2011
[3] R. Zhao, et al. *Advanced Materials*, 25(12): 1699-1705 (2013).
[4] S.A. Maas, et al. *J Biomech Eng*, 134(1) 2012.
[5] J. Guo, et al. *Proc Natl Acad Sci U S A*, 2014 9;111(49):E5252-61

TEMPERATURE RISE CAUSES UPREGULATION OF TENOCYTE CATABOLISM AND ENHANCES GAP JUNCTIONAL INTERCELLULAR COMMUNICATIONS

Masataka Tashiro (1), Eijiro Maeda (2), Toshiro Ohashi (2)

(1) Graduate School of Engineering
Hokkaido University
Sapporo, Hokkaido, Japan

(2) Faculty of Engineering
Hokkaido University
Sapporo, Hokkaido, Japan

INTRODUCTION

It is known that the temperature of the central core of equine digital flexor tendon could rise to 43-45°C during a sustained gallop [1]. The similar phenomenon is also estimated in human Achilles tendon [2]. Such temperature rise may indicate that both overloading and hyperthermia are involved in the onset of tendinopathy. Although a large number of studies have been performed on effects of overloading on tenocyte functions, only a few studies have been performed on effects of hyperthermia on tenocyte activities. In these studies, lethal effects of hyperthermia on tenocytes were mainly reported. A 60 minutes exposure of equine tenocytes to 45°C resulted in the cell viability of approximately 55% [3]. However, little has been known about effects of hyperthermia on tenocyte functions, in particular regulation of catabolism.

Tenocytes establish intercellular communication through gap junctions (GJIC), and the inhibition of GJIC resulted in the suppression of collagen synthesis from tenocytes in response to an application of cyclic tensile loading [4]. In addition, GJIC itself is regulated by mechanical loading [5]. However, how GJIC in tenocytes is changed at a high temperature remains unknown.

As tenocyte GJIC is involved in the mechano-regulation of its functions, both tenocyte metabolism and GJIC could be altered when inflammatory responses from tenocytes are induced under non-physiological conditions. Therefore, it was hypothesized that hyperthermia upregulates tenocyte catabolism and downregulates GJIC. The present study was performed to evaluate changes in tenocyte metabolism and GJIC in vitro in a quantitative manner.

METHODS

Tenocytes were isolated from rabbit Achilles tendons and maintained in DMEM+10% FBS. The cells were then cultured in a microgroove device [6] for 24 hours, at a cell density of 2,000 cells/cm^2, prior to subsequent experiments described below. The device was improved to equip cylindrical flow paths to circulate warm water to maintain/change the temperature of cell culture environment (Fig. 1).

We firstly evaluated the cell viability following a hyperthermia treatment. Culture medium within the microgroove device was heated up to 41°C or 43°C from 37°C for 30 minutes, followed by cooling down to 37°C, using a water circulation system. The device was then placed in a 5% CO_2 incubator at 37°C for 24 hours as a post incubation. At the end of incubation period, tenocytes within the device were stained using calcein-AM and ethidium homodimer for live and dead staining. Cell viability was determined using a fluorescence microscope. Tenocytes within the device maintained at 37°C throughout the above-mentioned procedures were used as

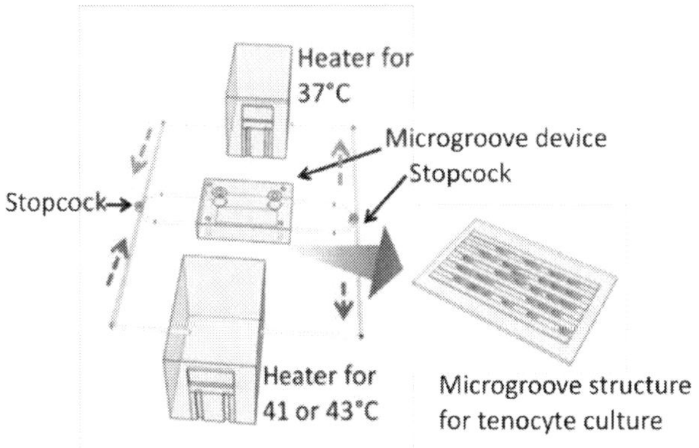

Fig.1 Experimental set-up for heat treatment.

controls. The temperature of culture medium within the device was monitored using thermocouples during the heat treatment.

To assess the effects of the heat treatment on tenocyte metabolism, mRNA expressions for type I collagen (COL1), MMP-1, IL-1β and caspase-3 (Casp3) were analysed. Heat stimulation was applied to tenocytes within the device in the same manner to the cell viability assay. Total RNA was extracted at the end of the post incubation and reverse-transcribed to cDNA, which was then subjected to real-time qPCR.

For a quantitative evaluation of alteration of GJIC between tenocytes, we adopted FLIP protocol [5]. We modified the mathematical model proposed in [5] to evaluate both intra- and inter-cellular diffusion coefficients from the same experimental data. Following the heat treatment (37 and 43°C only) as described above, tenocytes were labelled with calcein-AM for 30 minutes. FLIP protocol, modified appropriately from [5], was performed; it consisted of a series of 100 consecutive laser photobleaching of one of tenocytes connected in series along the microgroove (target cell) during a 300 second period. Fluorescence intensity of tenocytes adjacent to the target cell was measured, and the mathematical model, developed using one-dimensional diffusion theory, was fitted to the intensity profiles to determine diffusion coefficients.

RESULTS

There was a decreasing trend in the viability of tenocytes with an increase in heating temperature. Average cell viability at 37, 41, 43°C were 95.5, 76.0 and 68.9% respectively, and the changes were not statistically significant.

It was clearly demonstrated that the heat treatment affected expression patterns of tenocyte genes related to matrix metabolism (Fig. 2). Gene expression level of MMP-1 was elevated at 43°C compared to the level at 37°C, while the expression level of COL1 was decreased at 41 and 43°C from the level at 37°C. In addition, gene expression of IL-1β, an inflammation marker, was also upregulated at 43°C when compared to 37°C level. For a marker of cell apoptosis, caspase-3 gene expression was decreased at 43°C. Please note that these changes in gene expressions were not statistically significant.

Average intercellular diffusion coefficient at 37 and 43°C was 1.46 and 3.26 $\mu m^2/s$, respectively; the latter was significantly higher than the former (Fig. 3). Average intracellular diffusion coefficient at 37 and 43°C was 16.4 and 25.6 $\mu m^2/s$, respectively. The difference was not statistically significant.

DISCUSSION

The observed tend of the decreased cell viability by exposure to high temperatures is in an agreement with results in a previous report [3]. Decrease in expression of Casp3 mRNA by hyperthermia indicates that cell death induced by hyperthermia was not due to apoptosis but necrosis.

The increases in MMP-1 and IL-1β gene expression and the decrease in COL1 gene expression indicate that temperature rises in cellular environment upregulates tenocyte catabolism and downregulates anabolism. In addition, as MMP-1 and IL-1β are markers for inflammation, the upregulation of these genes may suggest that temperature rise in the tendon core could be a trigger of inflammatory responses from tenocytes.

It was against our hypothesis that intercellular diffusion coefficient (i.e. Gap junction diffusion coefficient) increased by exposure to 43°C. Although the exact mechanism of the increased GJIC is unknown, it may partly be due to involvement of heat-sensitive ion channels, such as transient receptor potential channel species (TRP channels). TRP channels have been known as heat-sensing machinery in neurons, and have increasingly been recongnised their involvements in other types of tissues. Therefore, influx of cation (e.g. Ca^{2+}) following the heat treatment may affect membrane potential, resulting in opening of existing gap junctions.

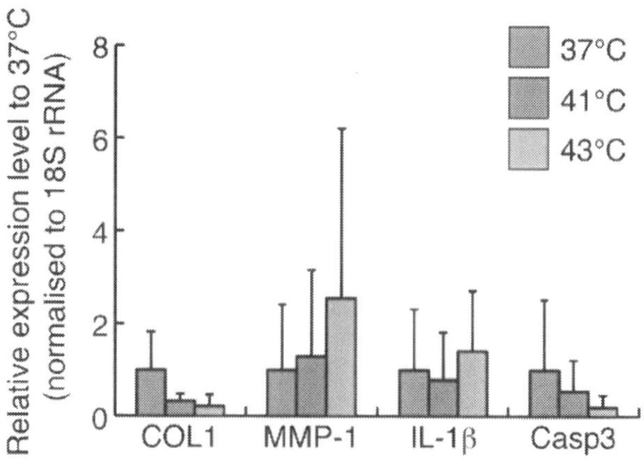

Fig. 2 mRNA expression for COL1, MMP-1, IL-1b and Casp3 from tenocytes subjected to heat treatment. Mean ± SD.

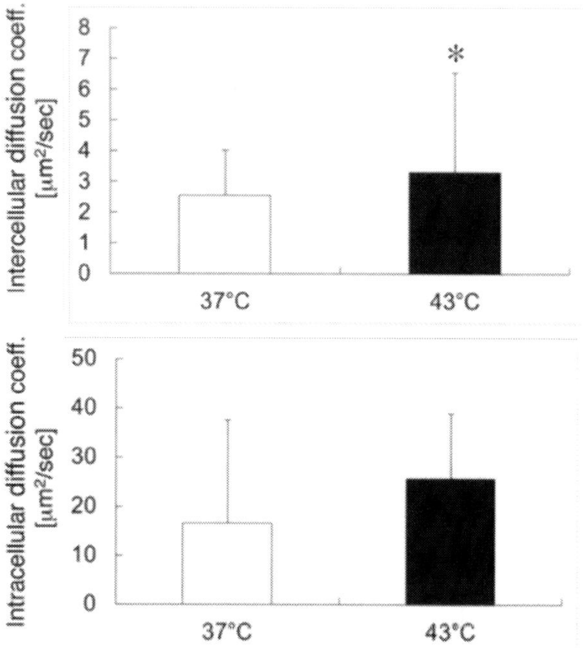

Fig. 3 Intrercellular and intracellular diffusion coefficient determined using FLIP experiment. Mean ± SD. * denotes $p < 0.05$ vs. 37°C.

ACKNOWLEDGEMENTS

JSPS KAKENHI (Grant-in-aid for young scientist) No. 25702022

REFERENCES

[1] Wilson and Goodship, *J Biomech*, 27:889-905, 1994.
[2] Farris, DJ et al., *J Biomech*, 44:822-826, 2011.
[3] Birch, H L et al, *J Exp Biol*, 200:1703-1708, 1997.
[4] Banes, A et al., *Clin Orthop Relat Res*, 367S:S356-S370, 1999.
[5] Maeda, E et al, *Biomech Modeling Mechanobiol*, 11:439-447, 2012.
[6] Maeda, E et al, *Biomed Microdev*, 15:1067-1075, 2013.

SB³C2015
Summer Biomechanics, Bioengineering and Biotransport Conference
June 17-20, 2015,Snowbird Resort, Utah, USA
SB³C2015-343

COMPUTER SIMULATOR OF THE COIL INSERTION INTO ANEURYSM : THE EFFECT OF MECHANICAL PROPERTY AND REFERENCE CONFIGURATION OF THE COIL

Tomohiro Otani (1), Satoshi Ii (1), Tomoyoshi Shigematsu (2), Toshiyuki Fujinaka (2), Masayuki Hirata (2), Tomohiko Ozaki (2), Shigeo Wada (1)

(1) Graduate School of Engineering Science
Osaka University
1-3 Machikaneyamacho, Toyonaka, Osaka, Japan

(2) Graduate School of Medicine
Osaka University
2-2 Yamadaoka, Suita, Osaka, Japan

INTRODUCTION

A coil embolization is one of the popular treatment methods to prevent the growth and rupture of the cerebral aneurysm. In this procedure, helical coils consisted of a platinum wire are manually inserted to the aneurysm by a micro-catheter. The inserted coils induce blood flow stagnation that enhances the blood clot formation, and finally lead to the occlusion of the aneurysm.

In the clinical practice, the low packing density or inhomogeneous distribution of the coils is considered to cause the recanalization of the aneurysm [1, 2]. To aim for well-packed and homogeneous coil embolization, various kinds of coils have been developed by changing the reference configuration (shape memory) and mechanical property (e.g., coil stiffness) [3]. The optimal coil is empirically selected by surgeons in each clinical case and manually inserted to the aneurysm through a trial and error process.

Since the coil packed in the aneurysm forms complex structures, actual coil configuration in the aneurysm is still unclear. To address this issue, several computational approaches have been proposed to express the realistic coil configuration in the aneurysm [4-6]. In our previous study, we have developed a mechanical coil model for the coil insertion to the aneurysm with consideration for the stretching and bending resistances of the coil [6]. The proposed model enables to obtain the realistic coil configuration in the aneurysm, whereas it has limitations to express the physical behavior of the coil due to ignoring the torsional resistance and the reference configuration of the coil. It is expected that more physically-accurate coil insertion simulation leads to support the surgeons for predicting the optimal coil properties and insertion conditions depending on the situation.

In this study, we present a computational model of the coil insertion simulation including the mechanical properties (i.e., stretching bending and torsional resistances) and reference configuration of the coil. The coil insertion simulation is conducted by solving a variational problem based on the minimum energy principle with satisfying geometric constraints for the coil-aneurysm and coil-coil contacts.

METHODS

Coil kinematics The coil is expressed by a set of degenerated beam elements based on the Timoshenko beam theory (e.g., [7]). Schematic description of the beam element is shown in Fig. 1. We adopt the second order beam element constituted by three nodes. The shape of the beam element is assumed to be cylindrical configuration with the outer diameter of D and thickness of d. Local coordinate system (r_1, r_2, r_3) is defined in each element. The orthonormal basis (V_1, V_2, V_3) is assigned to the node corresponding to the local coordinate axes at the reference state. Hence, the position vector a the beam element, $\mathbf{X}$ is described by

$$\mathbf{X} = N^p\left(\mathbf{X}^p + \frac{D}{2}r_2\mathbf{V}_2^p + \frac{D}{2}r_3\mathbf{V}_3^p\right), \quad (1)$$

where p is the dummy index and N is the shape function expressed by the function of r_1. $\mathbf{X}$ has six degrees expressed by the translation of $\mathbf{X}^p$ and rotation of $\mathbf{V}^p$.

Geometric constraints for the contacts The geometric constraint for the contact is expressed by the Signorini (KKT) condition which is given by

$$g\lambda = 0, \quad (2)$$

where g is the minimum distance between contacted surfaces and λ is the magnitude of the contact force in the normal direction being understood as the Lagrange multiplier [8]. Frictional force is treated by the penalty method obeying the Coulomb friction law [9] with consideration for the slip and static condition of the friction.

To treat the contact between the coil and aneurysm, the aneurysm wall is assumed to be fixed even if any interaction force is exerted on the surface from coil. The wall is divided to a set of triangle element, and the minimum distance between the contacted element and node of the coil is computed. To treat the coil-coil contact, the computation of the minimum distance g between beam elements is conducted by the idea of [9].

Governing equations The functional of the system, Π, is given by

$$\Pi = W + g\lambda, \tag{3}$$

where W is the potential energy of the coil expressed as the function of the displacement vector $\mathbf{u}$. The stationary conditions of eq. (3) are given by

$$\frac{\partial \Pi}{\partial \mathbf{u}} = \frac{\partial W}{\partial \mathbf{u}} + \frac{\partial g}{\partial \mathbf{u}}\lambda, \tag{4}$$

$$\frac{\partial \Pi}{\partial \lambda} = g. \tag{5}$$

The principle of the virtual work and total Lagrange formulation are used for solving the derivative of the W with respect to the $\mathbf{u}$ in eq.(3) is given by

$$\frac{\partial W}{\partial \mathbf{u}} = \rho \int_V \ddot{\mathbf{u}} \delta \mathbf{u} dV + \int_V \mathbf{S} : \delta \mathbf{E} dV. \tag{6}$$

Introducing the Galerkin finite element discretization, the semi-discrete forms of the stationary conditions of eq. (4) and (5) including the damping and friction terms are expressed as

$$\mathbf{M}\ddot{\mathbf{U}} + \mathbf{C}\dot{\mathbf{U}} + \mathbf{Q} + \lambda\mathbf{G} + \mathbf{T} = \mathbf{0}, \tag{7}$$

$$\mathbf{Gx} = \mathbf{r}, \tag{8}$$

where **M** is the mass matrix, **C** is the damping matrix, **Q** is the inner force vector, **U** is the displacement vector, **G** is the contact boundary condition matrix [8] and **T** is the friction force vector. **x** is the position vector of nodes associating the contact and **r** is the minimum distance between the contacted surfaces. Time integration is treated by the idea of [11] based on the multi-step central differential scheme.

Numerical setup Density of the coil is set corresponding to the physical properties of Pt/W (8%) [12]. Artificial viscosity coefficient and friction coefficients are empirically defined to produce the stability of the simulation. Elastic resistance of the coil is independently determined with respect to the stretching, shearing and torsion of the coil based on the theoretical value of the spring coefficients of the coil spring [13].

For the validation of the proposed coil model, coil insertion simulation is performed using an idealized geometry of the aneurysm with a radius of 2.5 mm (Fig. 2). Geometry of the catheter is set to be straight. The coil with the length and diameter of 12 cm and 254 μm is inserted at a constant speed of 1.0 mm/s [14]. To confirm the effects coil stiffness on the coil behavior in the aneurysm, two kinds of coil stiffness (hard and soft) are employed corresponding to the hardest and softest coils with the wire diameters of 76.2 μm and 44.5 μm, respectively, used in clinical practice reported in [5]. The reference configuration of the coil is assumed to be a 2d helical coil with a curvature radius of 3 mm.

RESULTS AND DISCUSSIONS

Representative snapshots of the insertion process of the hard coil to the aneurysm are illustrated in Fig. 3. The coil was curved and tangled around the aneurysmal wall during insertion and partly moved out of the aneurysm. There were no penetration of the aneurysmal wall by the coil and overlap between coils, and thus it exhibited that the proposed model enables to treat the multiple contacts precisely. Figure 4 shows the snapshots of the insertion process of the soft coil. The coil remained in the aneurysm with the complex tangling.

The present results show that the insertion of the hard coil resulted in the coil protrusion from the aneurysm. It is explained that the reference curvature radius of the coil set in the present study is bigger than the aneurysm radius. According to the clinical requirements, the first inserted coil is required not to deform by the interaction with further inserted coils, and thus its stiffness is expected to be relatively hard. It is required to investigate the optimal conditions to insert the hard coil focusing on the relationship between the reference curvature of the coil and aneurysm radius.

SUMMARY

The present study proposed the computational simulation of the coil insertion into the aneurysm taking account of the stretching, bending, torsion and reference configuration of the coils. The significant effects of coil stiffness on the behavior in the aneurysm are presented. For the estimation of the optimal coil property and insertion conditions, it is beneficial that trial experiments of the coil insertion simulation with various calculation conditions.

ACKNOWLEDGEMENTS

This work was supported in part by JSPS Grants-in-Aid for Scientific Research (No.23650261, A2616220).

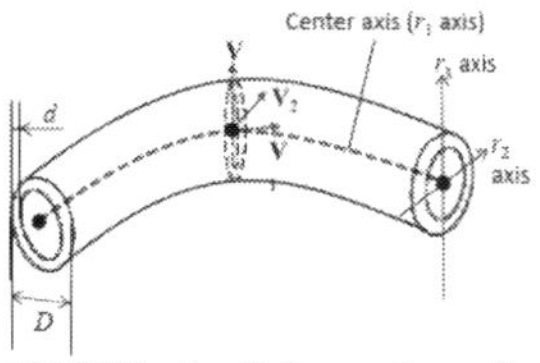

FIGURE 1 Schematic of the beam element.

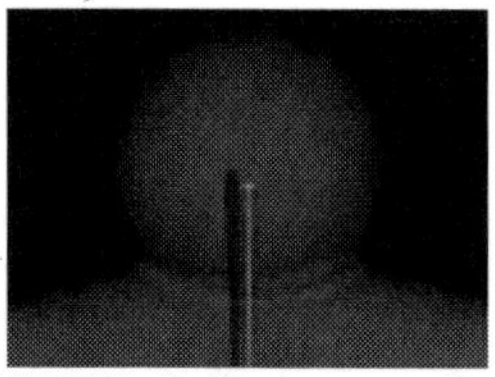

FIGURE 2 Geometry of the aneurysm and the catheter.

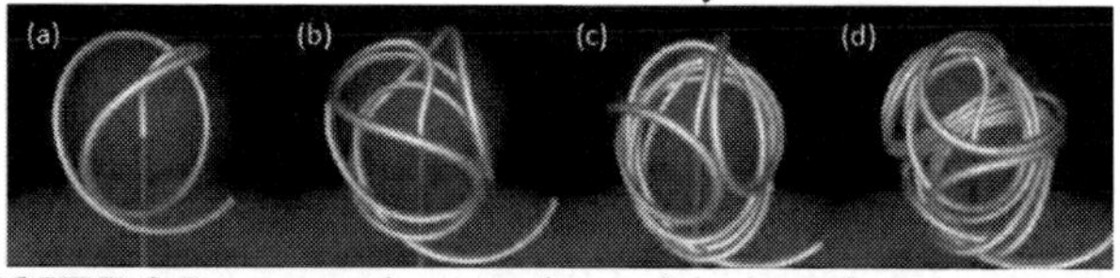

FIGURE 3 Representative snapshots of the insertion process of the hard coil at t = 30 s (a), 60 s (b), 90 s (c) and 120 s (d).

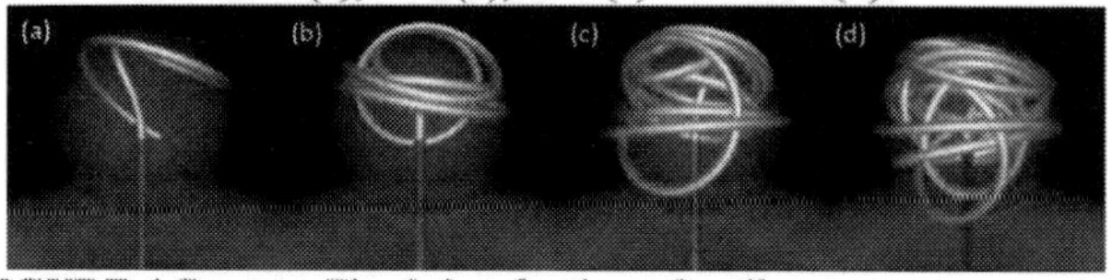

FIGURE 4 Same as Fig. 3, but for the soft coil.

REFERENCES

[1] Sluzewski, M et al., *Radiology*, 231:653–658, 2004.
[2] Roy, D et al., *Stroke*, 32: 1998–2004, 2001.
[3] White, B et al., *AJNR. Am. J. Neuroradiol*, 29: 1242–1246, 2008.
[4] Morales, HG et al., *IEEE Trans. Med. Imaging*, 32:119-129, 2012.
[5] Babiker, MH et al., *J. Biomech.*, 46:2809-2816, 2013.
[6] Otani, T et al., *ASME IMECE2013-64435*, 2013.
[7] Bathe K, *Prentice Hall*, 1996.
[8] Wriggers P, *Springer Berlin*, 2006.
[9] Wriggers P and Zavarise G, *Commun. Numer. Meth. Eng.*, 13:429-438, 1997.
[10] Kikuuwe, R et al., *IEEE Trans. Robotics*, 22:1176-1188, 2006.
[11] Carpenter, NJ et al., *Int. J. Num. Methods in Eng.*, 32:103-128, 1991.
[12] The PGM Database: http://www.pgmdatabase.com
[13] Wahl AM, *New York McGraw-Hill*, 1963.
[14] Matsubara, N et al., *Neuroradiology*, 53:169-175, 2011.

DESIGN AND OPTIMIZATION OF AN IN-VITRO EMBOLI DETECTOR FOR FLOW-INDUCED THROMBOGENICITY EVALUATION

Ram Shtoltz (1), Oren Rotman (1), Dar Weiss (1), Uri Zaretsky (1), Shmuel Einav (1, 2)

(1) Dept. of Bio-Medical Engineering, Tel-Aviv University, Tel-Aviv, Israel

(2) Stony Brook University, Stony Brook, NY, USA

INTRODUCTION

Coronary heart disease (CHD) is the most common type of heart disease, responsible for the death of nearly 380,000 people annually in the US alone [1]. The victims of CHD have invariably been found to have platelet and fibrin thrombi in their myocardium [2]. Knowing these numbers, it comes as no surprise that the process of approving an implanted cardiovascular device with the FDA requires extensive thrombogenicity assessment tests.

It is a well-established fact for a couple of decades now, that abnormal shear stress initiates thrombosis [3]. Flow-induced thrombosis is one of the main areas of interest when evaluating an implantable cardiovascular device, as it is crucial to know whether the device will affect the blood flow regime in a way that would initiate thrombosis.

Currently, most of the FDA requirements for thrombogenicity tests are in- and ex-vivo on large animal models, mainly canine. These kinds of tests are very costly, time consuming, and involve harming a large number of animals. Therefore, a relatively simple, yet accurate, in-vitro thrombogenicity evaluation experimental setup could prove to be quite efficient for the R&D process of any cardiovascular implantable devices.

One of the major difficulties when planning such a system is the positive feedback mechanisms that exist in the coagulation cascade. An activated platelet will promote the recruitment and activation of many additional platelets, essentially meaning that a thrombus or an embolus will be assembled even though not all of the participating platelets have been affected by abnormal shear stresses. The result of these feedback mechanisms is an immediate growth of an embolus within the blood flow.

The positive feedbacks of platelet activation are mediated, among other substances, by thrombin, adenosine triphosphate (ADP) and thromboxane A2 (TXA2). K. Hosokawa et al. [4] recently showed that by reducing the effect of these three substances, the process of thrombus formation can become more gradual.

Thrombin is a serine protease that has an important role in the formation and buildup of a thrombus or an embolus. Platelets have three specific receptors for thrombin: protease-activated receptors number 1, 3 and 4 (PAR1, PAR3 and PAR4), of which PAR1 is accounted for most of the activation [5-6].

To be able to assess accurately the thrombogenicity profile of a cardiovascular device, and be sensitive enough to differentiate between different thrombogenicity profiles (e.g., emboli growth), platelets activation mechanisms should be restricted to flow disturbances (e.g., abnormal shear stresses) only. By adding a PAR1 receptor antagonist, along with Aspirin (blocks the production of TXA2) and a $P2Y_{12}$ receptor antagonist (inhibits ADP-induced platelet aggregation), the feedback ability of activated platelets can be significantly reduced. Therefore, platelets activation will be mostly limited to abnormal blood shear stresses.

In the present study an in-vitro blood flow loop setup was built for the assessments of flow-induced thrombogenicity of implantable cardiovascular devices. Platelets feedback mechanisms in this system are impaired pharmacologically, and therefore the thrombus formation is more gradual, and the flow-induced thrombogenicity footprint of a device can then be estimated with increased sensitivity.

METHODS

The flow induced thrombogenicity simulator was comprised of a closed flow loop of 3 mm diameter infusion set tubing (Teva Medical, Israel), and driven by a peristaltic pump (MasterFlex® L/S® Easy-Load® II, Cole-Parmer, Vernon Hills, IL, USA). The working fluid was human platelet-rich plasma (PRP) at 37C (from subjects who had not taken any medication that might have affected platelet function or coagulation in the two weeks preceding the blood draw, and had no history of coagulation disorders). A short segment of the flow loop was comprised of a crystal clear silicone (RTV615, Momentive Performance Materials Inc., Waterford, NY, USA), from which images of embolism formation and buildup were continuously captured using a digital camera (uEye®, IDS Imaging Development Systems GmbH, Obersulm, Germany) and recorded using a custom made software (LabView®, NI Corporation, Austin, Texas, USA). For rapid and powerful illumination of the embolism, a pulsatile NdYag 532 nm laser was utilized (New Wave Research, Fremont, California, USA). Post-processing included detection and sizing of emboli and statistical analysis was performed on emboli count and size over time (MATLAB, MathWorks®, Natick, MA, USA). The sizing algorithm was calibrated and validated using polymer microspheres (ChromoSpheres®, Thermo Scientific, Waltham, MA, USA).

Preparation of the PRP samples included dilution of the sample with saline to reach a platelet count of 180,000 platelets/μl. Prior to each test, 2.5mM of $CaCl_2$ (Sigma-Aldrich, St. Louis, MO, USA) was added to the PRP sample. In order to impair the positive feedback mechanisms of platelet activation, a cocktail of PAR1 receptor antagonist (SCH79797, Tocris Bioscience, Bristol, UK), $P2Y_{12}$ receptor antagonist (AR-C66096, Tocris Bioscience, Bristol, UK), and Aspirin (Tocris Bioscience, Bristol, UK) was also added to the PRP solution prior testing.

In Vitro tests were performed under a constant flow rate of 19 ml/min, which accounts for mean coronary flow rate at rest (Re~118) [7]. In these tests various compositions of the anti-platelet cocktail were investigated, for optimization of reduction of the platelets' positive feedback.

RESULTS

Figure 1 depicts an example of the work done by the size computing algorithm, on actual emboli, at different stages of the emboli growth.

In Vitro tests evaluated the effect of various concentrations of SCH79797 on the thrombogenicity profile, calculated as the normalized average area of the emboli (Figure 2). With a concentration of 80μM SCH79797, the thrombogenic process was noticeably more gradual than with smaller concentrations and with the control samples.

Other experiments evaluated the thrombogenicity profile achieved with a combination of Aspirin, AR-C66096 and SCH79797.

DISCUSSION

Cardiovascular device flow-induced thrombogenicity evaluation is of high importance. In the present study a new in-vitro test method is introduced with the aim of rapid and accurate assessment of flow-induced thrombogenic footprint. Over the past years several other systems have been proposed [8-9]. These systems are highly effective for design optimization of the cardiovascular device, yet operating such systems is costly and time consuming. In contrast, the proposed method is aimed to act as a preliminary quick and inexpensive test for rapid evaluation of a device thrombogenicity, before considering stepping into a costly optimization process.

Preliminary experience with the simulator showed very promising results in the detection of emboli and following their build-up, starting from the very early stages of the coagulation onset. Subsequent work is focused on evaluation of the thrombogenicity footprint of various known arterial pathologies, as well as coronary stents, for proof of concept and the efficacy of the proposed method.

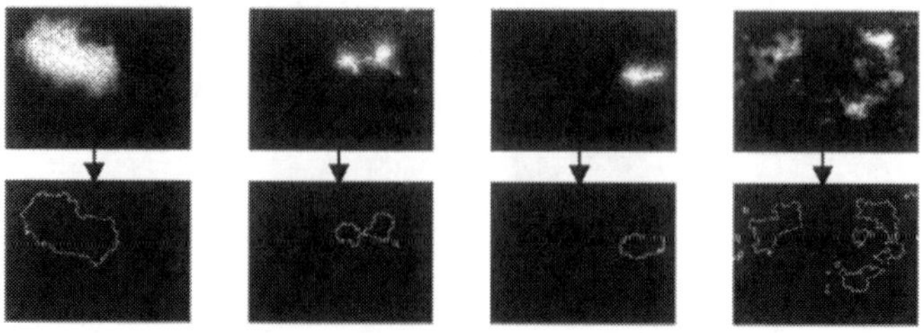

FIGURE 1: EMBOLI AT DIFFERENT GROWTH STAGES AS CAPTURED IN THE IN-VITRO FLOW LOOP, AND THE AUTOMATIC DETECTION ALGORITHM.

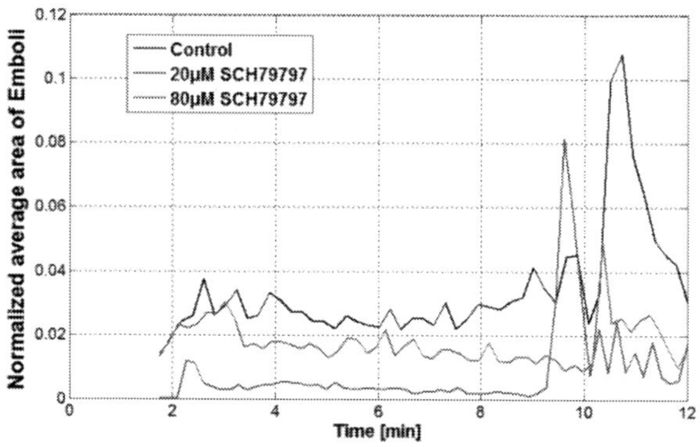

FIGURE 2: THE EFFECT OF DIFFERENT SCH79797 CONCENTRATIONS ON THE THROMBOGENICITY PROFILE IN THE IN-VITRO FLOW LOOP.

REFERENCES

[1] Murphy, S.L et al. *Natl Vital Stat Rep. 61(4), 2013.*
[2] Frink, R.J et al. *Br Heart J*, 59:196-200, 1988.
[3] Miyazaki, Y et al. *Blood*, 88: 3456-3464, 1996.
[4] Hosokawa, K et al. *Thrombosis Research*, 133:66–72, 2014.
[5] Ahn, H.S et al. *Biochemical Pharmacology*, 60: 1425–1434, 2000.
[6] Kahn, M.L et al. *J Clin Invest*, 103(6):879–887, 1999.
[7] Konala, B.C et al. *Molecular and Cellular Biology*, 8(1):1-20, 2011.
[8] Jesty, J et al. *Analytical Biochemistry*, 272: 64–70, 1999.
[9] Jesty, J et al. *Platelets*, 14:143–149, 2003.

SB³C2015
Summer Biomechanics, Bioengineering and Biotransport Conference
June 17-20, 2015,Snowbird Resort, Utah, USA
SB³C2015-347

COMPUTATIONAL MODELING OF PASSIVE MYOCARDIUM: A COMPARISON BETWEEN TWO CONSTITUTIVE MODELS

Amir Nikou (1), Shauna M. Dorsey (2), Jeremy R. McGarvey (3), Joseph H. Gorman III (3), Jason A. Burdick (2), James J. Pilla (3,4), Robert C, Gorman (3), Jonathan F. Wenk (1,5)

(1) Department of Mechanical Engineering
University of Kentucky
Lexington, KY, USA

(2) Department of Bioengineering
University of Pennsylvania
Philadelphia, PA, USA

(3) Gorman Cardiovascular Research Group
University of Pennsylvania
Philadelphia, PA, USA

(4) Department of Radiology
University of Pennsylvania
Philadelphia, PA, USA

(5) Department of Surgery
University of Kentucky
Lexington, KY, USA

INTRODUCTION

Engineers increasingly use the finite element method (FEM) in order to better understand the mechanics of the heart and its disorders. However, FEM is only reliable when the incorporated constitutive model carries a realistic mathematical description of the tissue. In this study, a recently developed orthotropic constitutive law by Holzapfel and Ogden (H-O) [1] was implemented to model left ventricular myocardium in diastole. In-vivo strains were measured from magnetic resonance images (MRI) of four healthy pig hearts, along with simultaneous catheterization pressure data, and used for parameter identification of the constitutive model. Optimization was performed by minimizing the difference between MRI measured and FE predicted strains and cavity volumes. A similar approach was followed for the parameter identification of a transversely isotropic constitutive law by Guccione et al. [2] which is frequently used in previous studies. The goal of the study was to determine which constitutive law yields the most realistic description of healthy myocardium with the available data.

METHODS

The LV strain, volume, and contour data were all generated from the 3D SPAMM MRI and matched with the pressure measurements. FE models were generated by fitting the endocardial and epicardial contours derived from MRI with 3D surfaces. Hexahedral trilinear elements were used to generate the finite element mesh and a custom MATLAB script was used to assign a linear distribution of myofiber orientation and myocyte sheet angles to each element [3].

The strain energy function per unit reference volume proposed for the H-O model is given by

$$\psi = \frac{a}{2b}\{exp[b(I_1-3)]-1\} + \sum_{i=f,s}\frac{a_i}{2b_i}\{exp[b_i(I_{4i}-1)^2]-1\} + \frac{a_{fs}}{2b_{fs}}\{exp(b_{fs}I_{8fs}^2)-1\} \quad (1)$$

where $a, b, a_f, b_f, a_s, b_s, a_{fs}$ and b_{fs} are the eight positive material constants. The four invariants used in the strain energy function are defined as follows:

$$I_1 = tr(\mathbf{C}), I_{4f} = \mathbf{f_0}.(\mathbf{Cf_0}), I_{4s} = \mathbf{s_0}.(\mathbf{Cs_0}), I_{8fs} = \mathbf{f_0}.(\mathbf{Cs_0})$$

where f_0 and s_0 are unit vectors that define the myofiber direction and myocyte sheet direction, respectively, and $\mathbf{C}$ is the right Cauchy-Green deformation tensor [2]. For the transversely isotropic model of Guccione et al. the strain energy function is

$$\psi = \frac{1}{2}C(e^Q-1) \quad (2)$$

where

$$Q = b_f E_{11}{}^2 + b_t(E_{22}{}^2 + E_{33}{}^2 + E_{23}{}^2 + E_{32}{}^2) + b_{fs}(E_{12}{}^2 + E_{21}{}^2 + E_{13}{}^2 + E_{31}{}^2)$$

where the constants C, b_f, b_t and b_{fs} are material parameters, E_{11} is the Green-Lagrange strain in the fiber direction, E_{22} is the sheet normal strain, E_{33} is the strain in the sheet direction and the rest are shear strains [2].

The concept of multiplicative decomposition of the deformation gradient tensor **F** is used to derive stresses from strain energy functions and a user defined material subroutine was developed to be implemented in the nonlinear FE solver LS-DYNA (Livermore Software Technology Corporation, Livermore, CA). An explicit (central difference) time integration scheme with adaptive time stepping for stability was used to conduct the FE simulations.

The multi-step optimization scheme proposed by Gao et al. [4] was used to estimate the eight material parameters of the H-O model. Optimization was performed in three steps by minimizing the two following objective functions in a specific sequence

$$f_{O1} = \sum_{n=1}^{N} \sum_{i,j=1,2,3} \left(E_{ij,n} - \bar{E}_{ij,n}\right)^2 + (V - \bar{V})^2 \quad (3)$$

$$f_{O2} = \sum_{n=1}^{N} \sum_{i,j=1,2,3} \left(E_{ij,n} - \bar{E}_{ij,n}\right)^2 + \left(\frac{V - \bar{V}}{\bar{V}}\right)^2 \quad (4)$$

where n is the strain point within the myocardium, N is the total number of strain points, $E_{ij,n}$ and V are the FE predicted end-diastolic strain and end-diastolic LV cavity volume, respectively, and the corresponding over bar variables represent in-vivo measured values. After running the optimization, the final optimized material parameters from both constitutive models were used to generate end-diastolic pressure-volume relationship (EDPVR) curves for each case. To do this, the FE model of the LV was loaded with different end-diastolic pressures above and below the measured end-diastolic pressure to obtain the corresponding end-diastolic volumes. The method proposed by Klotz et al.[5] to estimate the EDPVR curve from a measured end-diastolic pressure and volume point, was used to generate a physiological benchmark for comparison with the FE predicted EDPVR curves.

RESULTS

Tables 1 and 2 show the final results of the parameter optimizations for two of the four cases under investigation (case 1 and case 2) using H-O model and model of Guccione et al. respectively. Figure 1 shows the EDPVR curves of one of case 1 obtained using the H-O model, model of Guccione et al., and the physiological benchmark curve using the Klotz et al. method. It is very clear that the H-O model gives better agreement with the Klotz curve.

Table 1 Parameters of H-O model for two cases (a parameters are in kPa)

a	b	a_f	b_f	a_s	b_s	a_{fs}	b_{fs}
1.05	20.85	4.65	0.23	0.22	0.11	1.83	21.80
2.40	54.05	0.27	151	0.12	47.33	4.17	56.50

Table 2 Parameters of Guccione et al. model (C parameters are in kPa)

C	B_f	B_t	B_{fs}
0.493	38.06	4.11	45.07
0.589	98.44	12.36	26.45

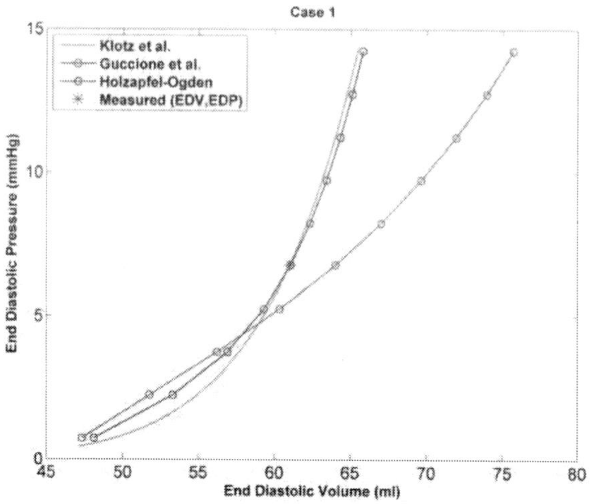

Figure 1

DISCUSSION

In this study, material parameters for recently developed structurally based constitutive model, as well as a widely used phenomenological constitutive model, were quantified. The final results indicate that given the rule-based and homogeneous fiber and sheet architecture used for the cases under investigation, the phenomenological model is able to predict a slightly better strain field within the ventricle at end-diastole, but the structurally based model results in a more realistic estimate of the ventricular mechanics in terms of generating a more physiological EDPVR curve. In order to better understand the influence of the sheet angles on the fit of the H-O models, all of the LV simulations were run with sheet angles that either aligned with the radial direction or varied by ±45 degrees. As expected, the fit with the Guccione material law were unchanged, but the fit with the H-O material law was significantly improved when the sheet angles varied. This indicates that more realistic sheet angles lead to a better fit between the H-O model and experimental data. In future studies, animal-specific, real fiber and sheet angles from DTMRI will be used in order to fully utilize this structurally based constitutive model for healthy and diseased myocardium.

ACKNOWLEDGEMENTS

This study was supported by National Institutes of Health grants R01 HL063954 (R. Gorman), R01 HL111090 (J. Burdick), R01 HL73021 (J. Gorman), and by a grant from the American Heart Association 14BGIA18850020 (J. Wenk).

REFERENCES

[1] Holzapfel, Ogden, R. W. Philos Trans A Math Phys Eng Sci, 367:3445-75, 2009.
[2] Guccione, J. M. et al. J of Biomechanical Eng. 113:42-55, 1991.
[3] Mojsejenko, D. W. et al. Biomech Model Mechanobiol, 2014
[4] Gao, H. Et al. J of Eng Mathematics, (In Press) 2014.
[5] Klotz, S et al. AJP-Heart and Circ, 291:403-412, 2006.

SB³C2015
Summer Biomechanics, Bioengineering and Biotransport Conference
June 17-20, 2015,Snowbird Resort, Utah, USA
SB³C2015-348

FINITE ELEMENT MODELING OF THE POSTERIOR EYE IN MICROGRAVITY

A. Feola (1), J. Raykin (1), L. Mulugeta (2), R. Gleason (1), J.G. Myers (3), E.S. Nelson (3), B. Samuels (4), C.R. Ethier (1)

(1) Department of Biomedical Engineering, Georgia Institute of Technology/Emory University, Atlanta GA

(2) Universities Space Research Association, Division of Space Life Sciences Houston, TX

(3) NASA Glenn Research Center, Cleveland, OH

(4) Department of Ophthalmology, U. Alabama at Birmingham, Birmingham, AL

INTRODUCTION

Microgravity experienced during spaceflight affects astronauts in various ways, including loss of muscle mass and bone density. Recently, visual impairment and intracranial pressure (VIIP) syndrome has become a major human health concern for space missions lasting longer than 30 days. Astronauts affected by VIIP syndrome have changes in ocular anatomy and suffer from visual impairment, both of which persist after returning to earth (1). It is hypothesized that the cephalad fluid shift that occurs in microgravity increases the intracranial pressure (ICP), which leads to an altered biomechanical environment in the posterior globe and optic nerve sheath (ONS), and subsequent vision impairment.

However, it is not currently well understood how elevated ICP may lead to vision impairment and connective tissue changes in VIIP syndrome. Our goal was to develop a finite element model to simulate the acute effects of elevated ICP on the posterior eye and optic nerve sheath. We used finite element (FE) analysis to understand the response of the lamina cribrosa and optic nerve to the elevations in ICP thought to occur in microgravity, and to identify which tissue features have the greatest impact on strain experienced by optic nerve head tissues.

METHODS

To simulate the effects of different pressures on strains in the tissues of the posterior eye, we extended on an existing geometric model of the posterior eye (2). Specifically, we elongated the optic nerve and optic nerve sheath, the latter of which included both the pia mater and dura mater. We also included a central retinal vessel that represented the combined effects of the central retinal artery and vein. The open-source program Gmsh was used for geometric model creation and mesh generation (3). For all simulations we used the FE solver FEBio (4). The geometry of the eye for FE analysis was considered as axisymmetric; however, due to constraints of the FE solver, computations were carried out using a three-dimensional FE model in which the model cross-section was rotated through 3° about an axis of symmetry passing though the central retinal vessel (Figure 1).

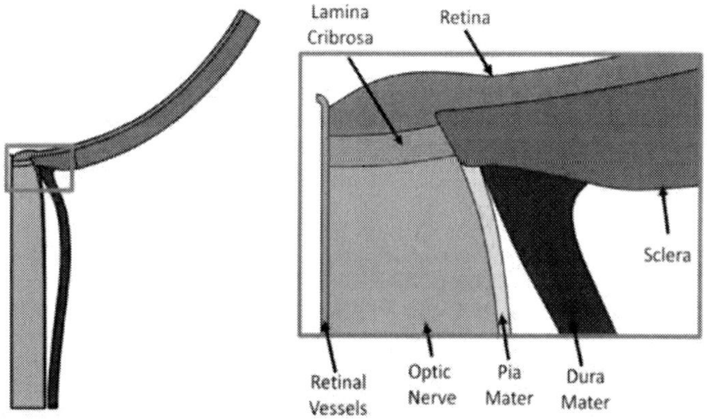

Figure 1: Left: overview of the geometric model for FE analysis. Right: A zoomed image of the boxed region, identifying key components of the finite element model in the optic nerve head region.

All tissues were modeled as isotropic, linear-elastic and nearly incompressible (ν=0.49). The baseline Young's modulus for each tissue component was based on previously reported values (2). The intraocular pressure (IOP) was fixed at 15 mmHg, while the central retinal vessel pressure was set to 55 mmHg, representing the terrestrially hydrostatically-corrected average pressure in the central retinal artery. The ICP was allowed to vary from 0 mmHg (suitable for standing on earth) to 30 mmHg (thought to represent severe intracranial hypertension in microgravity). Results were confirmed to be mesh-independent in a pilot mesh refinement study. The primary outcome measures were strains in the lamina cribrosa and post-laminar optic nerve (defined as the optic nerve region 1 mm posterior to the posterior lamina cribrosa boundary), since deformations in these areas are hypothesized to play a role in visual dysfunction in VIIP syndrome. We report the 95th percentile value of first principal strain and 5th percentile value of third principal strains, i.e. the extreme values of strain occurring within the tissues of interest. For linear-elastic and nearly incompressible materials, the first and third principal strains represent the maximum tensile and compressive strains in a tissue, respectively.

Since the Young's modulus of each tissue affects computed stresses and strains, we also performed a parametric study in which the mechanical properties of the tissues were changed. Specifically, we individually varied the Young's modulus of each tissue independently over two orders of magnitude, i.e. a ten-fold increase and a ten-fold decrease compared to baseline values. To identify which tissue components were most influential in affecting lamina cribrosa strain, we calculated the 95th percentile of the first principal strain in the lamina cribrosa at each modulus value.

RESULTS

Utilizing the baseline Young's modulus values, we found that strain distributions in the lamina cribrosa and the post-laminar optic nerve were drastically altered as ICP was elevated from 0 to 30 mmHg (Figure 2). Specifically, increasing ICP induced 2.7- and 2.5-fold increases in the extreme tensile (95th percentile) and compressive (5th percentile) strains in the lamina cribrosa, respectively. Similarly, elevating ICP increased extreme tensile and compressive strains by 3.8-fold and 3.3-fold, respectively, within the post-laminar optic nerve.

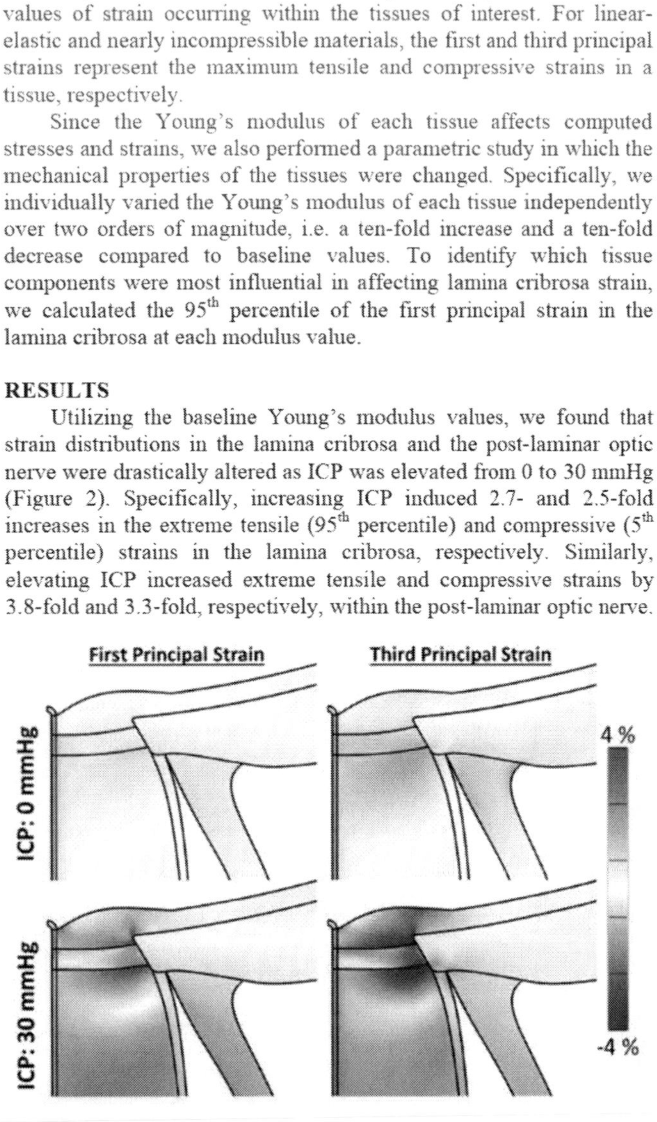

Figure 2: Computed strain distributions in the optic nerve head. Columns represent the first (left) and third (right) principal strains. The rows show simulations at different levels of ICP (Upper: ICP=0 mmHg; Lower: ICP=30 mmHg. The intraocular pressure (15 mmHg) and retinal vessel pressure (55 mmHg) were held constant during all simulations.

Examining the 95th percentile first principal strain within the lamina cribrosa, we found that the Young's modulus of the lamina cribrosa had the largest effect on lamina cribrosa strain (Figure 3). Specifically, altering the Young's modulus of the lamina cribrosa over two orders of magnitude led to a 6-fold change in the 95th percentile of the first principal strain in the lamina cribrosa. Interestingly, altering the Young's modulus of the sclera only altered the strain in the lamina cribrosa by 1.2-fold over the range of moduli we examined.

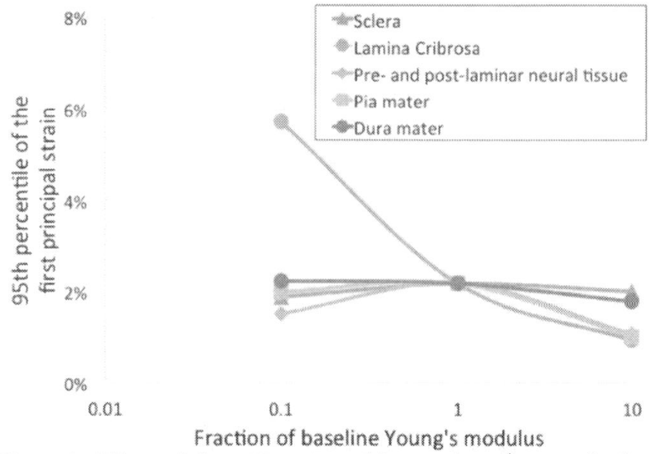

Figure 3: Effects of tissue Young's modulus on the 95th percentile first principal strain in the lamina cribrosa at an ICP of 30 mmHg. The intraocular pressure and retinal vessel pressure were fixed at 15 mmHg and 55 mmHg for the parametric study.

DISCUSSION

Our data shows that ICP elevations (e.g. believed to occur in spaceflight) lead to large acute changes in the biomechanical environment of the lamina cribrosa and post-laminar nerve. Our parametric study demonstrated that the modulus of the lamina cribrosa had the largest effect on strains in the lamina cribrosa, while the Young's modulus of the sclera had only a minor effect on the computed strains within the lamina cribrosa as ICP increased. In contrast, Sigal and colleagues showed that when elevating intraocular pressure, the modulus of the sclera had the most impact on lamina cribrosa strains (2). The difference between these two results indicates that elevating ICP does not significantly deform the sclera, at least when compared to the effects of increasing intraocular pressure, as expected intuitively.

This study had several limitations. First, we assumed an axisymmetric geometry for the posterior eye, which is a simplification and may affect the strains in the optic nerve head tissues. A second limitation was the use of linear-elastic and isotropic tissue models; future work will aim to incorporate more advanced material models.

In summary, as ICP increased there were differences in the strain distributions within the optic nerve head tissues, which we hypothesize may activate local cells and induce tissue remodeling, thus contributing to VIIP syndrome. We also identified that the material properties of the lamina cribrosa could greatly alter the lamina cribrosa strains as ICP increased.

ACKNOWLEDGEMENTS

Financial support from the Georgia Research Alliance and NASA (grant NNX13AP91G) is gratefully acknowledged. We also thank DeVon Griffin for administrative support.

REFERENCES

[1] Alexander, DJ, et al., Evidence Report: Risk of spaceflight-induced intracranial hypertension and vision alterations. NASA, 2012.

[2] Sigal, I et al., IOVS, 45(12):4378-87, 2004.

[3] Geuzaine, CR, et al., Int J Num Method Eng, 79(11):1309-31, 2009.

[4] Maas, SA, et al., J Biomech Eng. 134(1):011005, 2012.

SB^{3}C2015
Summer Biomechanics, Bioengineering and Biotransport Conference
June 17-20, 2015,Snowbird Resort, Utah, USA
SB^{3}C2015-349

DOES THE CYLINDRICAL OR SPHERICAL AXIS MORE ACCURATELY LOCATE THE FLEXION-EXTENSION AXIS OF THE TIBIA OF THE NATURAL KNEE?

Abheetinder S. Brar (1), Stephen M. Howell (2), Maury L. Hull (1,2), Mohamed R. Mahfouz (3)

(1) Department of Biomedical Engineering
University of California, Davis
Davis, California, USA

(2) Department of Mechanical Engineering
University of California, Davis
Davis, California, USA

(3) Department of Biomedical Engineering
University of Tennessee
Knoxville, Tennessee, USA

INTRODUCTION

Accurately locating the flexion-extension (F-E) axis of the tibia on the femur is important when studying the kinematics of the natural knee, anterior cruciate ligament reconstructed knee, and total knee arthroplasty [1-3]. The F-E axis of the tibia penetrates the two centers of the circular portion of the posterior femoral condyles from about 20° to 115° like an axle passing through two wheels and is parallel to the natural joint line of the posterior femoral condyles [3-7].

Three methods are used to locate the F-E axis of the tibia [1, 2, 8]; however, the method that is most accurate and that has the highest reproducibility and repeatability has not been determined. The visual-fit cylindrical axis method visually best-fits a cylinder to the tibial articular surfaces of the medial and lateral posterior femoral condyles from 15° to 115° of flexion [1]. The visual-fit spherical axis and the computational-fit spherical axis are two methods that connect the centers of two spheres either visually or computationally best-fit to the medial and lateral posterior femoral condyles (Figure 1) [2, 8, 9]. A three-dimensional analysis called the dot product, though not familiar to most surgeons, computes the angle between the axis of a method and the natural joint line, which do not intersect [1]. The computation of the dot product at 0, 30, 60, 90, and 120° of flexion describes the angle between the axis of a method and the F-E axis of the tibia.

One indicator that a method accurately locates the F-E axis of the tibia is a mean of the dot product at 0, 30, 60, 90, and 120° of flexion that is small enough to not cause problems in clinical interpretation. The second indicator of accuracy is a standard deviation (SD) of the dot product at 0, 30, 60, 90, and 120° of flexion that is small relative to the mean. Reproducibility and repeatability can be computed with use of the dot product at 0, 30, 60, 90, and 120° of flexion measured once by two observers and measured twice by one observer, respectively, on randomly selected knees.

The primary purpose of the present study was to determine whether the visual-fit cylindrical axis, the visual-fit spherical axis, or the computational-fit spherical axis method has the most accuracy, highest reproducibility, and highest repeatability in locating the F-E axis of the tibia.

METHODS

Forty (fifteen males to date) three-dimensional bone models of natural knees constructed from thin slice (0.7 mm) 3.0 Tesla magnetic resonance images (MRI) were randomly selected from the Osteoarthritis Initiative database (www.oai.ucsf.edu). Each MRI was segmented and combined with free, open-source software (Paraview) to create a three-dimensional model.

A series of steps determined the orientation of the sagittal, axial, and coronal kinematic planes on each bone model. Superimposition of the distal and posterior femoral condyles from the sagittal view oriented the femur parallel to the sagittal kinematic plane. The axial kinematic plane was oriented perpendicular to the sagittal kinematic plane and parallel to the medial articular surface of the tibial plateau. The coronal kinematic plane was oriented perpendicular to both the sagittal kinematic plane and the tibial resection plane.

The natural joint line at 0, 30, 60, 90, and 120° of flexion was identified as the line connecting the most distal points of the medial and lateral femoral condyles when the femur was flexed to the corresponding flexion angle (Figure 1A).

The visual-fit cylindrical axis was found by visually best-fitting a cylinder to the tibial articular surfaces of the medial and lateral femoral condyles from 15° to 115° of flexion following the method described by Eckhoff et al. [1]. The axis passing through the center of the cylinder was the visual-fit cylindrical axis.

The two methods for identifying the spherical axis of the knee were developed based on different methods for best-fitting spheres to the medial and lateral posterior femoral condyles. The visual-fit spherical axis connects the centers of the spheres that were visually fit to the tibial articular surfaces of the medial and lateral femoral condyles. In a projection of the distal femur in the axial kinematic plane, a sphere was sized and positioned to fit the tibial articular surface of the medial femoral condyle. In a projection of the posterior femur in the coronal kinematic plane, the sphere was resized and repositioned to improve its fit to the tibial articular surface of the medial femoral condyle. In a projection of the medial femur in the sagittal kinematic plane, the sphere was resized and repositioned to improve its fit to the tibial articular surface of the medial femoral condyle. This process of iteratively using distal, posterior, and sagittal views to resize and reposition the sphere was repeated until its fit to the tibial articular surface of the medial femoral condyle could not be improved. The process was then replicated to visually fit a sphere to the tibial articular surface of the lateral femoral condyle.

The computational-fit spherical axis requires the use of a three dimensional measurement software (Geomagic Control) to best-fit spheres to the medial and lateral posterior femoral condyles. In a projection of the medial femur in the sagittal kinematic plane, a resection plane parallel to the axial kinematic plane was used to remove the femoral shaft and a resection plane parallel to the coronal kinematic plane was used to remove any bone anterior to the most distal point on the medial femoral condyle. In a projection of the posterior femur in the coronal kinematic plane, a resection plane parallel to the sagittal kinematic plane was used to remove the intercondylar notch and the lateral femoral condyle. A sphere was then best-fitted to the segmented medial femoral condyle. The process was then replicated to best-fit a sphere to the lateral posterior femoral condyle.

On each specimen, the dot product, which describes the angle between the axis and the natural joint line of the posterior femoral condyles at 0, 30, 60, 90, and 120° of flexion, was computed for the visual-fit cylindrical axis, the visual-fit spherical axis, and the computational-fit spherical axis. The accuracy for each method was determined by computing the mean and SD of the dot product at 0, 30, 60, 90, and 120° of flexion across all specimens. A one-factor analysis of variance was used to determine whether the method of finding the flexion-extension axis of the tibia affected its dot product.

Reproducibility was described by the mean ± standard deviation of the difference between the dot products measured by two observers at each flexion angle on fifteen randomly selected knees. Repeatability was described by the mean ± standard deviation of the difference between the dot products measured twice by one observer at each flexion angle on fifteen randomly selected knees. Two one-factor analyses of variance were used to determine whether the method of finding the flexion-extension axis of the tibia affected its reproducibility and repeatability. Tukey's post hoc test was used for the statistical analysis, with $p < 0.05$ considered significant.

RESULTS

Accuracy was described by the mean ± standard deviation of the dot products of all flexion angles. The measurement of the visual-fit cylindrical axis was 0.6° ± 0.4°, which was more accurate (smaller) than the visual-fit spherical axis of 0.9° ± 0.7°, and the computational-fit spherical axis of 1.5° ± 0.8°, respectively ($p = 0.0181$ and $p < 0.0001$, respectively, Tukey's post hoc test) (Figure 2).

The reproducibility of the visual-fit cylindrical axis was 0.0° ± 0.0°, which was more reproducible (smaller) than the visual-fit spherical axis of 0.7° ± 0.6°, and computational-fit spherical axis of 0.3° ± 0.2°, respectively ($p < 0.0001$ and $p = 0.0423$, respectively, Tukey's post hoc test).

The repeatability of the visual-fit cylindrical axis was 0.0° ± 0.0°, which was more repeatable (smaller) than the visual-fit spherical axis of 0.7° ± 0.6°, and computational-fit spherical axis of 0.3° ± 0.2°, respectively ($p < 0.0001$ and $p = 0.0423$, respectively, Tukey's post hoc test).

DISCUSSION

The visual-fit cylindrical axis was the method that was most accurate and that had the highest reproducibility and repeatability in locating the F-E axis of the tibia. Accordingly, the visual-fit cylindrical axis should be used when studying the kinematics of the natural knee, anterior cruciate ligament reconstructed knee, and total knee arthroplasty.

ACKNOWLEDGEMENTS

We thank National Science Foundation grant number CBET-1067527.

REFERENCES

[1] Eckhoff, D., et al., *Clin Orthop Relat Res*, 461:238-44, 2007.
[2] Incavo, S.J., et al., *Clin Biomech (Bristol, Avon)*, 2013.
[3] Hollister, A.M., et al., *Clin Orthop Relat Res*, 259-68, 1993.
[4] Pinskerova, V., et al., *Orthopade*, 29 Suppl 1:S3-5, 2000.
[5] Iwaki, H., et al., *J Bone Joint Surg Br*, 82:1189-95, 2000.
[6] Weber, W.E., et al., *Mechanik der menschlichen Gehwerkzeuge*. 1836, Göttingen, Germany: Verlag der Dietrichschen Buchhandlung.
[7] Eckhoff, D.G., et al., *J Bone Joint Surg Am*, 87 Suppl 2:71-80, 2005.
[8] Victor, J., et al., *J Bone Joint Surg Br*, 91:683-90, 2009.
[9] Bowers, A.L., et al., *Arthroscopy-the Journal of Arthroscopic and Related Surgery*, 27:1511-1522, 2011.

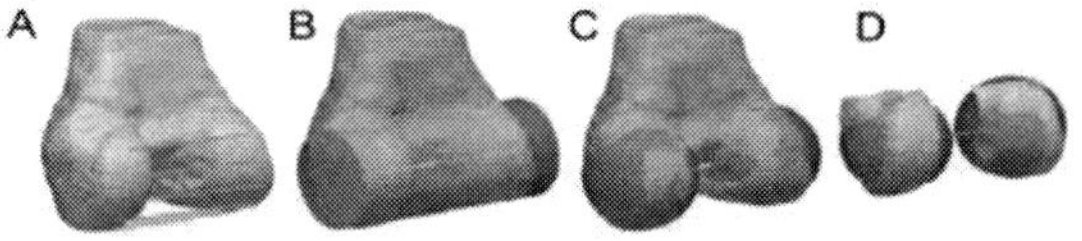

FIGURE 1. The composite shows a posterior view of a right femur. The joint line (green line) is shown at 0, 30, 60, 90, and 120° of flexion (A). The cylindrical axis (white line) is centered in the visual-fit cylinder (blue) (B). The visual-fit spherical axis (white line) is connecting the center of two spheres visually best-fit to the medial and lateral posterior femoral condyles (green) (C). The computational-fit spherical axis (white line) is connecting the center of two spheres visually best-fit to the medial and lateral posterior femoral condyles (green) (D).

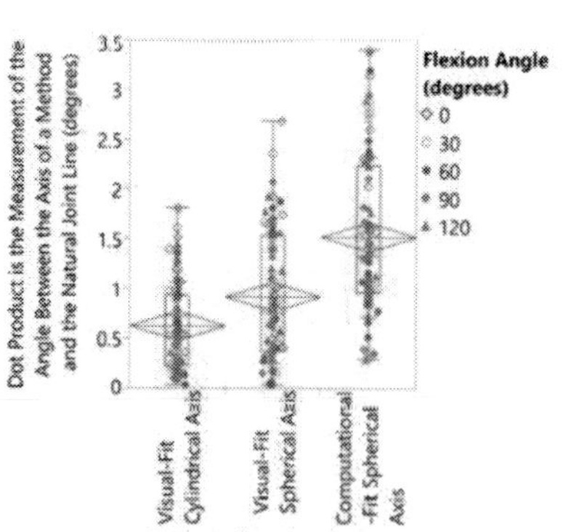

FIGURE 2. The graph shows the diamond (green), which illustrates the mean and 95% confidence interval, and an outlier quartile box plot (red) which shows the variability in the dot product for each method of locating the F-E axis of the tibia.

SB³C2015
Summer Biomechanics, Bioengineering and Biotransport Conference
June 17-20, 2015,Snowbird Resort, Utah, USA
SB³C2015-350

A MICROTISSUE ARRAY DEVICE TO SCREEN THE LUNG FIBROGENIC POTENTIAL OF CARBON NANOTUBES

Zhaowei Chen, Qixin Wang, Mohammadnabi Asmani, Yan Li, Yun Wu, Ruogang Zhao

Department of Biomedical Engineering
The State University of New York at Buffalo
Buffalo, New York, USA

INTRODUCTION

Due to their excellent physical and chemical characteristics, multi-wall carbon nanotubes (MWCNT) are widely used in many biomedical applications such as drug delivery and medical imaging. However, the application of MWCNT also raises health concerns. The small size and light weight of MWCNT allow them to enter the working environment as particulate matter (PM) of respirable sizes [1]. The suspended MWCNT are potentially an occupational inhalation exposure risk to those working with them. In vivo animal studies have shown that certain types of MWCNT cause cellular apoptosis and inflammation in the lung, which lead to lung fibrosis that is fatal and incurable [2]. Therefore, identifying the fibrogenic potential of specific types of MWCNT is important to the health safety in applications involving carbon nanotube. However, due to the high cost and low throughput of the animal models, the progress in this field is very slow.

The recent emergence of bio-microelectromechanical systems (bio-MEMS) technology allows realistic modeling of 3D human tissue and high throughput testing of many samples in a single device, thus provides a potential viable solution to screen the pathogenic factors. Here we report the development of a 3D human lung microtissue array device for the screening of the fibrogenic potential of two different types of MWCNT at different concentrations. Our device consists of arrays of PDMS microwells with each microwell containing a self-assembled microtissue hanging between a pair of micropillars (Fig. 1A and 1B). Using this device, we measured the changes in the mechanical properties and biomarkers related to the fibrogenesis in BEAS-2B (B2B) normal lung epithelial cell-populated microtissues. The results showed that short-MWCNT (0.5-2μm in length) at 50ng/ml induced the most fibrotic differentiation compared to MWCNT (10-50μm in length) at 50ng/ml and short-MWCNT at 5μg/ml in the lung tissue. These results demonstrated the utility of the microtissue array device in high throughput screening of pathogenic factors in realistic 3D environment.

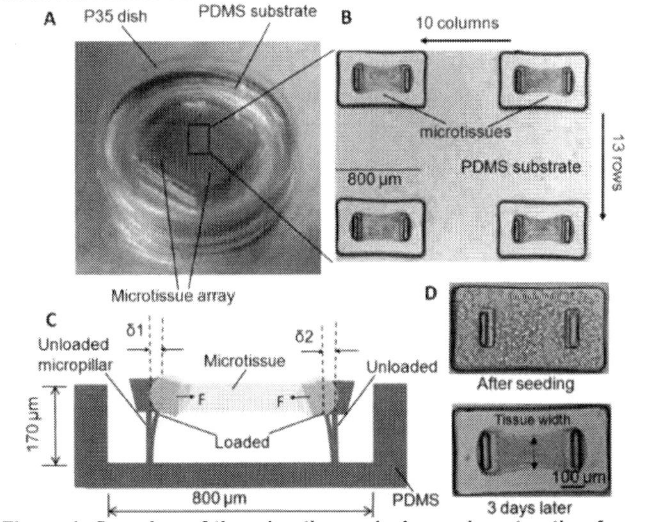

Figure 1. Overview of the microtissue device and contraction force measurement. (A) A P35 petri-dish contains a microtissue array fabricated in a PDMS substrate. (B) A portion of the microtissue array (2x2). (C) Schematic of microtissue contraction force measurement. (D) Microtissue formation during a 3 days period.

METHODS AND MATERIALS

Mechanical property measurement of microtissues under different treatment conditions

We fabricated microtissue array device by PDMS replica molding from masters made using a multilayer microlithography technique [3].

Each device contains an array of 10 x 13 microwells (Fig. 1A, B). The micropillar pair in the each of the microwell can both provide mechanical support to the microtissue and serve as an *in-situ* force gauge to measure the tissue contraction forces. It is well established that fibrotic tissues *in vivo*, such as those at the edges of a closing wound, generates elevated contraction force. Therefore, the contraction force can be used as a phonotypical indicator for fibrotic tissue. The contraction force F generated by the microtissues is calculated according to the cantilever bending theory as $F = k\delta$, where δ is the averaged deflection $\delta = (\delta 1+\delta 2)/2$ of the two pillars and $k = 0.9\mu N/\mu m$ is the spring constant of the micropillar (Fig. 1C). Micropillar deflection is determined by comparing the deflected position of the pillar top with its base. During cell seeding, suspensions of human B2B cells in unpolymerized type-I collagen were centrifuged into microwells and the collagen was polymerized under 37°C. Within 24 hours, the contractile action of the cells caused the compaction of the matrix hydrogel and formation of a dog-bone shaped microtissue between the two pillars (Fig. 1D).

To study the fibrogenic effect of MWCNT on B2B cell-populated microtissue, three different conditions (MWCNT 50ng/ml, short-MWCNT 50ng/ml and short-MWCNT 5ug/ml) were used to treat the microtissue for three continuous days. The changes in tissue contraction force and tissue width under different conditions were measured daily (Fig. 2).

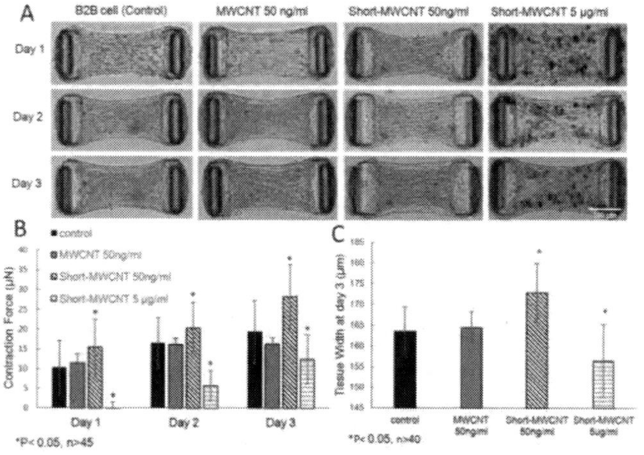

Figure 2. Measurement of mechanical properties of the microtissue under different MWCNT treatments. (A) The evolving morphology of microtissues during a 3 days culture period. (B) Contraction force measurement. (C) Microtissue width measurement at day 3.

Fibrogenic biomarkers identification by qRT-PCR

We also used qRT-PCR to measure fibrosis-related biomarkers in the microtissue after MWCNT treatements. MicroRNAs (miR) are small non-coding RNA molecules that are promising biomarkers in diagnosis of cancer and many other diseases [4]. MicroRNAs have been found to relate to the regulation of lung fibrosis and lung cancer. For example, up-regulation of miR-21 has been found to correlate with the increased expression of TGF-β, which is a potent stimulant for fibrotic differentiation of the lung cells [5]. In this study, we measured the expressions of miR-21 at day 3 after carbon nanotube treatment. To perform the qRT-PCR measurement, microtissues were first collected manually from the wells through repeated washing steps, and RNA was then extracted from the microtissues for miR-21 expression quantification.

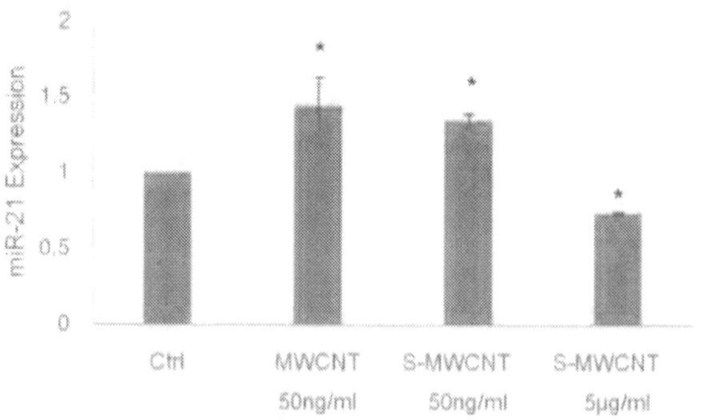

Figure 3. Expression levels of miR-21 in control and treated microtissue groups.

RESULTS AND DISCUSSION

At day 1 after initial seeding, the control and treated microtissues formed nice dog-bone shape and produced substantial contraction force, except for the short-MWCNT 5µg/ml group, which is likely due to the cytotoxicity caused by overdose treatment of the short-MWCNT. The contraction forces continued to increase for all the groups over a 3-day culture period with the short-MWCNT 50ng/ml group producing the largest contraction force (Fig. 2A, B). The elevated contraction force correlated with increased tissue width (Fig. 2C). This is because the squeezing of microtissue in the longitudinal direction under high contraction force caused increase in the cross-section area. Both of the contraction force and tissue width of MWCNT 50ng/ml group are close to the control group (Fig. 2A-C), indicating this carbon nanotube type and concentration did not affect the mechanical property of the microtissue.

Results of qRT-PCR showed significant up-regulation of miR-21 in short-MWCNT 50ng/ml group (Fig. 3). Since the expression level of miR21 was shown to correlate with the expression of TGF- β, this result indicated that short-MWCNT 50ng/ml group is likely stimulating the fibrotic differentiation of the microtissue. This is consistent with the high contraction force measurement of this treatment group in Fig. 2. Together, the above results suggested that short-MWCNT 50ng/ml is a potential pathogenic risk of lung fibrosis. Fig. 3 also showed significant down-regulation of miR-21 in Short-MWCNT 5µg/ml group, this is probably due to the loss of large number of cells induced by cytotoxicity of this particular type and concentration of carbon nanotube.

In this study, the utilization of a novel microtissue array device allowed high throughput screening of fibrogenic potential of three different types of carbon nanotubes and concentrations. The integrated 3D biomechanical capacity of the device also allowed in situ collection of mechanical properties of the model tissue under different treatment conditions. Such information was then used as a phonotypical marker to compare with the results obtained through qRT-PCR test. Together, the cross examination of the results obtained from this novel system enabled robust identification of the pathogenic factors for lung fibrosis.

REFERENCES

1. C. Lam, et al. *Crit Rev Toxicol* 2006 Jan;36(3):189–217.
2. S. Friedman, et al. *Sci Transl Med* 2013 5(167): 167sr161.
3. R. Zhao, et al. *Advanced Materials* 2013 25(12): 1699-1705.
4. H. Schwarzenbach, et al. *Nat Rev Clin Oncol* 2014 11(3):145-56
5. G. Liu, et al. *J Exp Med* 2010 vol. 207 no. 8 1589-1597.

COMPARISON OF HUMAN WALKING BACKWARD AND FORWARD USING OPTIMIZATION METHOD

Yujiang Xiang (1) , Hyun-Jung Kwon (2)

(1) Department of Mechanical Engineering
University of Alaska Fairbanks
Fairbanks, AK, USA

(2) Injury Biomechanics Research Center
The Ohio State University
Columbus, OH, USA

INTRODUCTION

In recent years, virtual human modeling has received considerable interest in many fields to simulate real-world tasks. Various motions of the model are required, especially the walking motion which has been studied in both robotics and biomechanics fields. In addition, there have been a growing number of studies in the use of backward walking and running for training and rehabilitation purposes. Previous research on backward walking primarily focused on investigating the motor patterns during motion using the experimental methods [1]. There has been a little work on the simulation side to study backward walking in the literature.

The goal of this study is to formulate, simulate and analyze backward walking of a full human-body model using the inverse dynamics based optimization method. For this case, the simulation problem is formulated as a nonlinear optimization problem. The results of the backward walking motion are compared to those of the forward walking motion in order to study the differences between the two walking patterns.

METHODS

A three dimensional human skeletal model with 55 degrees of freedom (DOF) is considered in this study as shown in Figure 1. The model consists of six physical branches and one virtual branch. The physical branches include the right leg, the left leg, the spine, the right arm, the left arm, and the head. The virtual branch is used to allow the global motion of the pelvis with respect to the origin (o-xyz) of the Cartesian coordinate system [2]. The anthropometric data of a 50th percentile male, generated using GEBOD software, are used in this study.

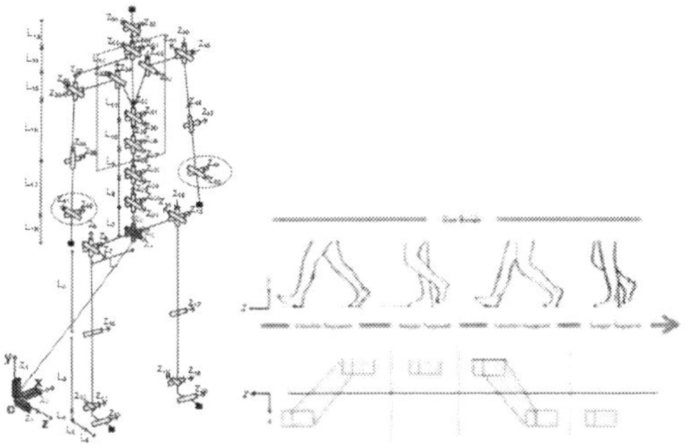

Figure 1. The 3D skeletal human model and backward walking

A complete gait cycle consisting of two continuous steps, i.e., a stride, is considered in the simulation. In the current work, backward walking is assumed to be a cyclic motion with continuity conditions between the strides as shown in Figure 1.

In this study, recursive formulation is used to calculate the kinematics of the skeletal model. The joint torques are computed using the recursive Lagrangian dynamics formulation because of its computational efficiency [3]. In addition, the closed-form analytical gradients are provided for the kinematics and dynamics variables for use in the optimization process. Detailed formulas to calculate the recursive kinematics, dynamics, and their sensitivities are provided by [4]. In addition, a two-step algorithm is used to calculate the zero moment point (ZMP) and ground reaction forces (GRF) to obtain the joint torques for the multibody human system [2].

For the optimization problem, the time domain is discretized by using cubic B-spline functions. With this representation, the control points become the optimization variables. In this study, the knot vector is specified and fixed in the optimization process. 7 control points are used for each DOF. Thus there are a total 385 optimization variables in the formulation. The joint torques are treated as dependent variables that are evaluated directly from the equations of motion. The dynamic effort, the time integral of the squares of all joint torques, is used as the performance criterion for the walking motion. Two types of constraints are considered for the walking backward optimization problem: time-dependent constraints that are imposed throughout the stride walking motion. The time-dependent constraints include (1) joint angle limits; (2) joint torque limits; (3) ground penetration; (4) foot contact position; (5) ZMP stability; (6) arm leg coupling; and (7) self-avoidance. The time-independent constraints are the second type of constraints that are imposed at specific time points during the stride; e.g., at the initial and final times to impose the continuity conditions for one stride to the next. Details formulations refer to [5].

The appropriate backward walking parameters are obtained from the motion capture experiments. The results of the backward walking are compared to those of the forward walking. A computer with an Intel(R) dual core 3.16 GHz CPU and 8 GB RAM is used to solve the optimization problem. The optimal solution is obtained in about 30 seconds of CPU time starting from a feasible solution [6].

RESULTS

In the backward walking case, two different initial data (step length, speed) are used based on the experimental data. Using the optimization formulation, the backward walking motion is obtained with different speeds and step lengths. The solution is simulated in Figure 2.

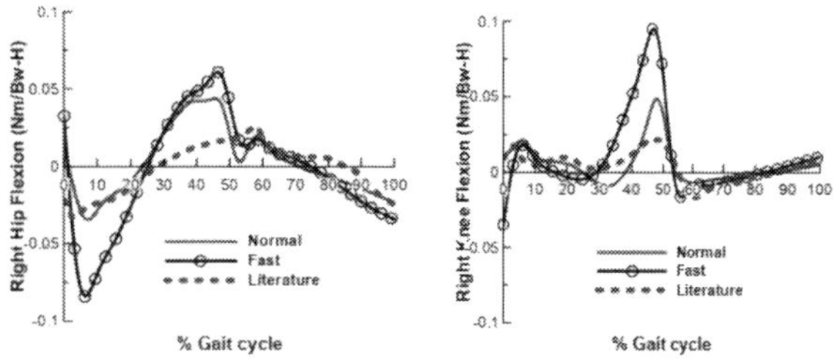

(a) (b)

Figure 2. Joint torque profiles: (a) hip; (b) knee

Comparison between different walking motions (forward and backward) has been studied with various experiments in the literature. Generally one cannot walk backward as fast as forward. This is due in part to the anatomical structure of the body, as well as aspects of skilled/learned performance. The joint torque profiles and vertical ground reaction force (normalized with respect to Bw, the body weight) for the forward walk (FW) and backward walk (BW) are compared in Figure 3.

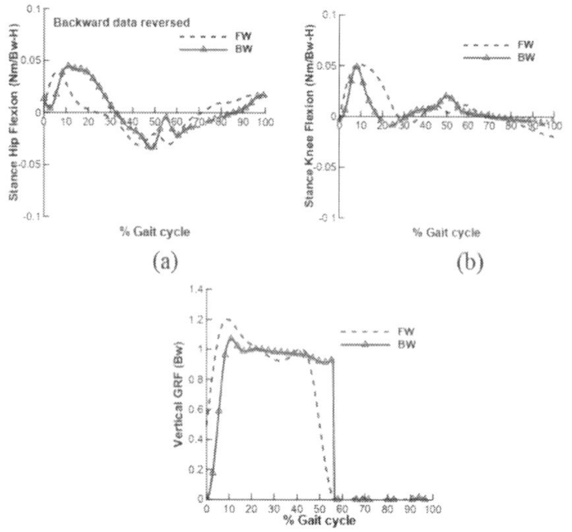

(c)

Figure 3. Comparison of joint torque profiles and vertical GRF for forward and backward walks: (a) hip; (b) knee; (c) vertical GRF

DISCUSSION

Optimization-based walking prediction revealed some insights for the real human gait.

1. For backward walking, the initial data (step length and speed) affected the final results considerably. For example, the peak torques at the knee and hip were much larger for the faster cadence compared to the slower walk.
2. The joint torques had similar patterns but different time activations in the backward walk compared to forward walk. Since muscle forces are computed from joint torque and joint profiles, this could lead to activation of different muscle groups during forward and backward walks which have implications in the rehabilitation fields [7]. In addition this could affect the whole body fatigue and recovery which could lead to different human performance.
3. The ground reaction forces had patterns that were similar for the forward and backward walks. However, the peak for the vertical ground reaction force was smaller for the backward walk compared to the forward walk. This could also affect the whole body fatigue.

ACKNOWLEDGEMENTS

The first author acknowledges the start-up fund from University of Alaska Fairbanks.

REFERENCES

[1] Winter, D.A. et al., *J Mot Behav*, 21(3):291-305, 1989
[2] Xiang, Y. et al., *Int J Numer Methods Eng*, 79(6):667-695, 2009
[3] Hollerbach, J.M., *IEEE Trans Syst Man Cybern*, 11(10):730-736, 1980
[4] Xiang, Y. et al., *Struct Multidiscip Optim*, 37(6):595-608, 2009
[5] Kwon, H.J. et al., *Struct Multidiscip Optim*, 50:169–179, 2014
[6] Xiang Y. et al., *J Biomech*, 44(4):683-693, 2011
[7] Zajac, F.E., *Crit Rev Biomed Eng*, 17(4):359-411, 1989

SB^{3}C2015
Summer Biomechanics, Bioengineering and Biotransport Conference
June 17-20, 2015, Snowbird Resort, Utah, USA
SB^{3}C2015-352

PROBING OF A COMPLEX MULTI-LAYER EMBRYONIC TISSUE THROUGH NOVEL 3D BIO-ETCHING

Melis Hazar (1), YongTae Kim (2), Philip R. LeDuc (1), William C. Messner (3), Lance A. Davidson (4)

(1) Mechanical Engineering
Carnegie Mellon University
Pittsburgh, PA, USA

(2) Mechanical Engineering
Georgia Institute of Technology
Atlanta, GA, USA

(3) Mechanical Engineering
Tufts University
Medford, MA, USA

(4) Bioengineering
University of Pittsburgh
Pittsburgh, PA, USA

INTRODUCTION

Morphogenesis consists of a complex series of cell signaling, migration and differentiation that are coordinated during both embryo development and cases of tissue self-assembly. Collective cell sheet migration is an important morphogenetic process that sculpts the shape of an organism and its internal tissues during early development. Studies of these collective cell movements have focused on the behavior of epithelial monolayers of cultured cells such as human umbilical vein endothelial cells (HUVEC) or Madin-Darby canine kidney (MDCK) cells. By contrast, most embryonic development and tissue self-assembly requires the integration of cell movements within multiple cell layers composed of different cell types. Although the role of cell mechanics in tissue self-assembly has been demonstrated, little is known about the mechanical response of the multi-layer tissues to chemical cues. This knowledge gap is due to a lack of technologies to control microenvironmental cues and to analyze the individual responses of epithelial and mesenchymal cell sheets in a multi cell layer tissue to those cues. Another reason for the knowledge gap is due to the difficulty in engineering multilayered cellular tissues. To investigate the collective movements within multilayered tissues our group has focused on developing a novel microfluidic technique capable of producing complex patterns of laminar multicellular structures. We call this technique "3D tissue-etching" because it is analogous to techniques used in the microelectromechanics (MEMS) field where complex 3D structures are built by successive steps to remove material from a monolithic solid. We use tissue etching to shape a complex multi-layered embryonic tissue and explore the dynamic collective responses of epithelial and mesenchymal cells in a single tissue.

METHODS

We use a custom-designed microfluidic control system to deliver a range of tissue etching reagents (detergents, chelators, proteases, etc.) over tissues microsurgically isolated from embryos of the African Claw-toed frog, *Xenopus laevis*. *Xenopus* embryos are used to elucidate genes important in morphogenesis and here provide a model for collective migration of a complex tissue. Here, we demonstrate a novel microfluidic application to etch multi-level multicellular structures in tissues. We etch the complex tissues in the *Xenopus laevis* animal cap (AC), which is composed of stratified layers of epithelial and mesenchymal cells.

To etch 3D tissues, we used our custom-designed microfluidic control system to stream a range of tissue etching reagents, adapted from decellularization protocols, over microsurgically isolated early stage embryonic tissue. Controlling the flow rate at the inner microchannel inlets individually enabled precise and dynamic chemical delivery of etchants to the 3D tissue [1]. Our system implemented a pressure-driven (nitrogen gas) approach and included two variable-volume fluid reservoirs with etchants and culture media (DFA), two microfluidic resistances and a 3-inlet converging microfluidic channel (Figure 1). The main microfluidic channels and fluidic resistance modules were fabricated with conventional soft lithography techniques from polydimethylsiloxane (PDMS). After fabrication, the main channels were cleaned and coated with fibronection. In the main T-shaped, 3-inlet microfluidic channel, the two side inlets had identical high flow rate flows, which were controlled by microfluidic resistances [2] producing a center inlet with a lower flow rate. These fluidic resistances altered the flow rates that defined the width of the etching streams in the main microfluidic channel and thus the width of the areas exposed to the etchants over the apical surface of the tissue. Inlet pressures can be dynamically changed to deliver etchant to any location over the apical surface and width of the tissue. The tissues used in these experiments were microsurgically isolated from gastrula stage *Xenopus laevis* embryo. Each tissue isolate was initially positioned inside of the outlet of the main channel and allowed to adhere to the fibronection coated glass surface for 4 hours. Once adhered, we collected time-lapse sequences of the process of etching and the response of the tissues.

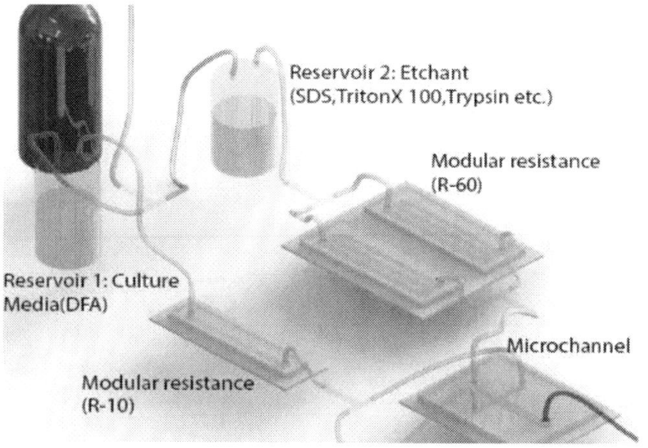

FIGURE 1: CUSTOM DESIGNED MICROFLUIDIC SYSTEM FOR 3D BIO-ETCHING

RESULTS

We first demonstrate that we can make a full cut in a local region through the entire multi-layered AC (Figure 2). We then show that we can cut a similar local region but only remove the top layer of cells of the tissue. Finally we show that we can cut different widths of cells away from defined layers of cells allowing us step-wise spatial precision for cell layer removal to examine 3D cell interactions in tissues in the future. We implemented different etching protocols including the chemicals sodium dodecyl sulfate (SDS), triton X-100, trypsin and sodium hydroxide (NaOH) to alter and shape these 3D tissues. Application of bio-etching will greatly enhance the understanding of tissue remodeling events, which will impact fields such as cancer, embryonic morphogenesis and wound repair.

Using etching we produce free-edges of epithelial cells over mesenchymal cells and free-edges of mesenchymal cells. This allows us to study multi-layer coordination of epithelial (EL) and mesenchymal (ML) cell layers and acute mechanical and behavioral response of intact epithelial and mesenchymal cell sheets for the removal of neighboring or overlying tissues.

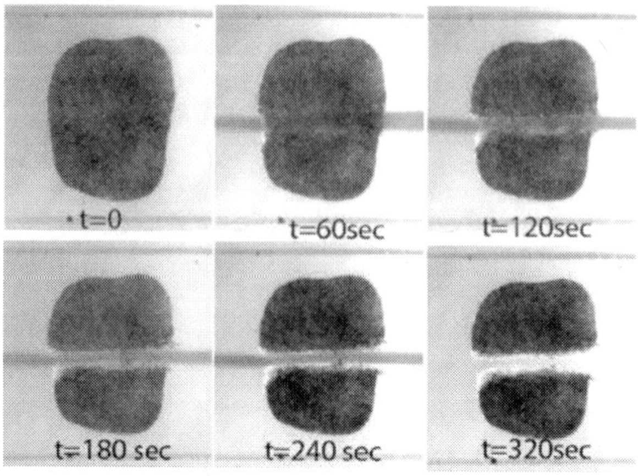

FIGURE 2: FULL CUT THROUGH MULTI-LAYER EMBRYONIC TISSUE

DISCUSSION

We demonstrated the ability to create and shape 3D tissues with subtractive manufacturing or "tissue-etching" using a microfluidics based approach. While many approaches use 2D environments with monolayers, our unique microfluidic technique provided the ability to control the 3D environment and study the interactions of individual epithelial and mesenchymal layers. The ability to control the form of multicellular tissues will be important in fields such as in tissue engineering and regeneration applications in bioengineering and medicine as well as in the synthesis of highly complex 3D integrated multicellular biosystems. Likewise, our custom microfluidic system can be configured to test a range of hypotheses concerning the control and regulation of development and cell differentiation with additional applications in tissue engineering.

REFERENCES

[1] Kim, YongTae, et al., *Journal of the American Chemical Society* 132.4 (2010): 1339-1347.

[2] Kim, YongTae, et al., *Lab on a Chip*9.17 (2009): 2603-2609.

SB³C2015
Summer Biomechanics, Bioengineering and Biotransport Conference
June 17-20, 2015,Snowbird Resort, Utah, USA
SB³C2015-356

FINITE ELEMENT MODELING OF ACTIVE TRANSMEMBRANE CELL TRANSPORT

Chieh Hou, Kelly Terlizzi, Gerard A. Ateshian

Mechanical Engineering
Columbia University
New York, NY, USA

INTRODUCTION

Multiphasic mixture theory has been applied successfully for the modeling of various biological tissues and cells, including mechano-electrochemical phenomena in tissues, osmotic loading of cells, transport of nutrients, chemical reactions, and growth and remodeling of tissues [1,2]. A finite element implementation of this framework has been introduced in the open-source finite element software FEBio, facilitating the modeling of complex biological systems [3]. The existing implementation can describe osmotic effects and passive transport (diffusion) of neutral and charged solutes. When applied to cell physiology, mixture theory can reproduce classical equations for the response of cells to osmotic loading [4], which may also be simulated in FEBio [1].

Active transport of solutes across the membrane of cells and intracellular organelles also plays a critical role in cell physiology though it may be neglected when examining certain phenomena. Recently, it was demonstrated that active transport of solutes could be incorporated into the theory of mixtures as an additional internal momentum supply in the balance of linear momentum [5]. That study showed that mixture theory could reproduce the Nernst equation for membrane potential, and also provide descriptions for the Na^+-K^+ pump, cell homeostasis, and the Na^+-Glc co-transporter. The objective of this study was to derive the general differential equations of mixture theory with active transport, implement them in FEBio, and verify the code. An analysis of the Na^+-K^+ pump-leak model and cell volume regulation [8] is also illustrated.

METHODS

Linear momentum balance for solvent and solutes α, inclusive of momentum exchange from active transport [5], is given by

$$-\operatorname{grad}\tilde{\mu}^\alpha-\frac{1}{\left(\rho^\alpha\right)^2}\mathbf{f}^\alpha\cdot\mathbf{m}^\alpha+\frac{1}{\rho^\alpha}\sum_{\beta\neq\alpha}\frac{1}{\rho^\beta}\mathbf{f}^{\alpha\beta}\cdot\mathbf{m}^\beta+\frac{1}{\rho^\alpha}\hat{\mathbf{p}}_a^\alpha=0 \quad (1)$$

where $\tilde{\mu}^\alpha$ is the mechano-electrochemical potential, ρ^α is apparent density, $\mathbf{f}^\alpha$ and $\mathbf{f}^{\alpha\beta}$ are diffusive drag tensors, $\mathbf{m}^\alpha$ is mass flux relative to the solid, and $\hat{\mathbf{p}}_a^\alpha$ is the momentum exchange resulting from active transport. Let $\mathbf{w}=\mathbf{m}^w/\rho_T^w$ and $\mathbf{j}^\alpha=\mathbf{m}^\alpha/M^\alpha$ respectively represent the solvent volume flux and solute molar flux, where ρ_T^w is solvent true density and M^α is molar mass. Expressions for $\mathbf{w}$ and $\mathbf{j}^\alpha$ may be obtained from Eq. (1) and separated into passive and active contributions. The passive flux terms were reported previously [1]; the active flux contributions are given by

$$\mathbf{w}_a=\tilde{\mathbf{k}}\cdot\sum_\alpha\frac{\mathbf{d}^\alpha}{d_0^\alpha}\cdot\tilde{\mathbf{p}}_a^\alpha,\quad \mathbf{j}_a^\alpha=\mathbf{d}^\alpha\cdot\left(\frac{c^\alpha}{d_0^\alpha}\mathbf{w}_a+\frac{\varphi^w}{R\theta}\tilde{\mathbf{p}}_a^\alpha\right) \quad (2)$$

where $\tilde{\mathbf{p}}_a^\alpha=\hat{\mathbf{p}}_a^\alpha/\varphi^w$ is the active momentum supply with units of force per fluid volume; φ^w is the solvent volume fraction; $\tilde{\mathbf{k}}$ is the effective hydraulic permeability, $\mathbf{d}^\alpha$ is the solute diffusivity in the mixture and d_0^α is its free diffusivity in solution; c^α is the solute concentration; R is the universal gas constant and θ is the absolute temperature. Internal momentum supplies cancel out.

Using our prior finite element framework [1], the statement of virtual work is now extended to include additional terms appearing in the mass balance relations for the solvent and solutes,

$$\delta A=\int_b\delta\tilde{p}\operatorname{div}\mathbf{w}_a\,dv+\sum_\alpha\int_b\delta\tilde{c}^\alpha\operatorname{div}\mathbf{j}_a^\alpha\,dv \quad (3)$$

where $\delta\tilde{p}$ is the virtual effective fluid pressure and $\delta\tilde{c}^\alpha$ is the virtual molar energy of solute α. These additional terms were implemented in FEBio, with user-specified values for the momentum supplies $\tilde{\mathbf{p}}_a^\alpha$.

RESULTS

Active neutral solute transport in 1D Consider a 1D analysis simulating active uptake of a neutral solute (e.g., via molecular motors) from an external bath into a multiphasic mixture domain of thickness h (Fig. 1a). To produce an analytical solution, solid matrix deformation is neglected, producing zero flux of solvent relative to the solid. The solute concentration in the bath is c_0, as is the initial solute concentration $c(z,t)$ at $t=0$ in the multiphasic mixture ($0 \le z \le h$). For a constant supply $\tilde{p}_a$, the solution is

$$c(z,t)=c_0+\frac{\tilde{p}_a}{R\theta}h\left(\frac{z}{h}+\frac{2}{\pi^2}\sum_{n=0}^{\infty}\frac{(-1)^{n+1}}{(n+1/2)^2}\sin\left[\left(n+\frac{1}{2}\right)\pi\frac{z}{h}\right]\exp\left[-\left(n+\frac{1}{2}\right)^2\pi^2\frac{d}{h^2}t\right]\right) \quad (4)$$

A comparison of this solution with the FEBio response is presented in Fig. 1b, showing agreement across all values of z and t.

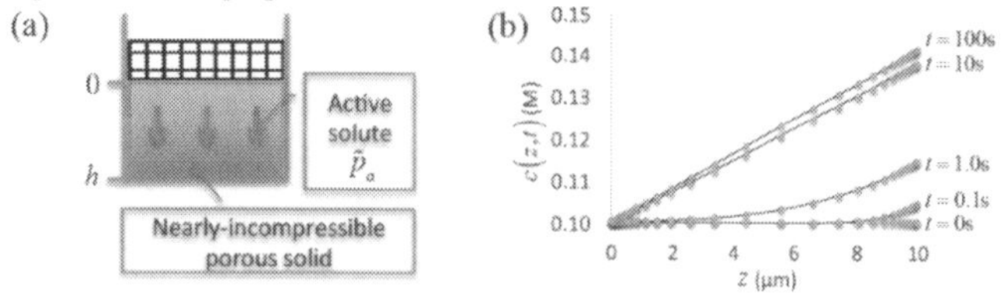

Figure 1: (a) 1D analysis of active solute transport. (b) Comparison of FEBio (symbol) and analytical (solid line) solutions at various time points; h = 10 µm, d = 10^{-5} mm²/s, $\tilde{p}_a$ = 0.01 µN/µm³.

Simplified red blood cell model with Na^+-K^+ pump For a typical red blood cell (RBC), the main transmembrane ions are Na^+, K^+ and Cl^-. Intracellular membrane-impermeant solutes include anions such as hemoglobin; small concentrations of other anions and neutral solutes are also present outside the cell [6]. For simplicity, a spherical cell model is analyzed, which incorporates the cytoplasm and membrane region. This 1D spherical model is investigated using FEBio, analytical expressions from mixture theory, and a conventional pump-leak model [6]. The cytoplasm consists of the solvent, Na^+, K^+, Cl^-, and a generic membrane-impermeant anion F^-. The membrane consists of a mixture with a neutral porous solid matrix, solvent and only Na^+, K^+ and Cl^-. Initial intracellular and fixed extracellular ion concentrations are based on literature values. The analysis examines the temporal response of intracellular concentrations for different values of the pumping supply.

From the continuity constraint for $\tilde{c}^\alpha$ described in [1], and the electroneutrality condition, ion concentrations inside the thin membrane adjacent to the cytoplasm are $c^\alpha(r_i)=B^{z^\alpha/2}c_i^\alpha$, where c_i^α is the ion concentration in the cytoplasm, $B=c_i^{Cl}/(c_i^{Cl}+c_i^F)$, and z^α is the ion charge number. Hence there exists an osmotic pressure in the membrane $\Delta p=2R\theta\left(B^{-1/2}-B^{-1}\right)c_i^{Cl}$, assuming same fluid pressure inside and outside when the membrane cannot sustain tension. Assuming uniform electric potential gradient E/h across the membrane, projecting $\mathbf{j}^\alpha$ along the radial direction r and integrating it across the membrane, ion fluxes across the membrane reduce to

$$j^\alpha=\frac{\varphi^w d^\alpha}{R\theta}\tilde{p}_a^\alpha+\varphi^w d^\alpha\frac{z^\alpha F_c E}{R\theta h}\frac{c_e^\alpha-c_i^\alpha\exp\left(z^\alpha F_c E/R\theta\right)}{1-\exp\left(z^\alpha F_c E/R\theta\right)} \quad (5)$$

where the first term on the right-hand-side is the active flux and the second term is the passive (Goldman-type) flux [6]; h is the membrane thickness and F_c is Faraday's constant. E is obtained from the zero total current condition $\Sigma_\alpha z^\alpha j^\alpha=0$. Projecting $\mathbf{w}$ along r, the solvent flux across the membrane is

$$w=\tilde{k}\frac{\Delta p}{h}+\tilde{k}\sum_\alpha\left(-\frac{R\theta}{\varphi^w d^\alpha}j^\alpha+\tilde{p}_a^\alpha\right) \quad (6)$$

In the cytoplasm, $dJ/dt=-3wJ^{2/3}/a_r$ according to the mass balance, where $J=V/V_r$ is the ratio of cell volume between current and reference states and a_r is the initial cell radius; similarly, $dn^\alpha/dt=-4\pi a_r^2J^{2/3}j^\alpha$, where $n^\alpha=c_i^\alpha V$ is the ion mole number in the cytoplasm. The cell response is calculated from the above equations and compared to the conventional pump-leak model where $J=\Sigma_\alpha n^\alpha/\left(V_r\Sigma_\alpha c_e^\alpha\right)$, from the isotonicity condition $\Sigma_\alpha c_i^\alpha=\Sigma_\alpha c_e^\alpha$.

The FEBio model includes cytoplasm and membrane regions, using multiphasic materials with only solvent and solutes for the cytoplasm and a mixture of Mooney-Rivlin solid and solution for the membrane. d^{Na} and d^K for the membrane were evaluated from homeostasis, when passive flux is balanced by pump flux: Na^+-flux $=1.2\cdot10^{-9}$ mol/m²·s for human RBC and K^+-flux is -2/3 of that magnitude [6]. d^{Cl} is 100× larger than d^{Na} [7]. $\tilde{p}_a^{Na}$ and $\tilde{p}_a^K$ are varied to maintain the 3:2 pump flux ratio. The finite element mesh uses only a small arc along the azimuth an elevation angles due to spherical symmetry (Fig. 2a); a single element is used in each region to produce homogeneous responses consistent with analytical expressions. J, E and c_i^α were evaluated from mixture theory, FEBio and the pump-leak model (Fig. 2b), with good agreement observed between mixture theory and FEBio, and small deviations from the classical pump-leak model. As expected, in the absence of pumping ($\tilde{p}_a^{Na}=0$) the cell swells continuously.

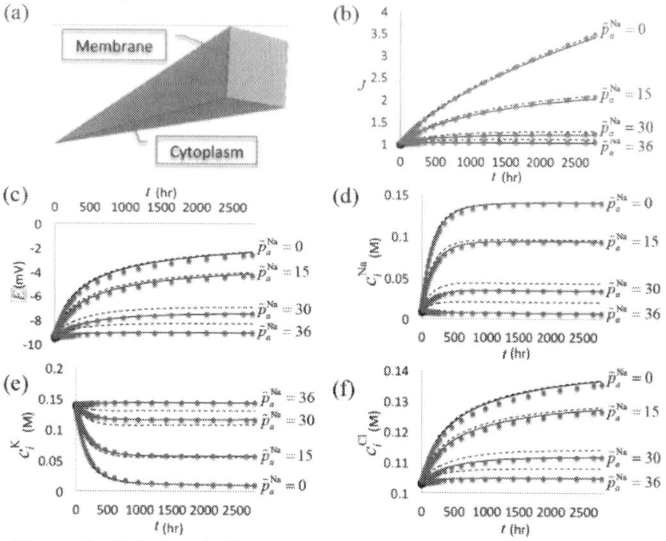

Figure 2: Cell model illustrated in (a). Cell responses, J (b), E (c), c_i^α (d-f), calculated from mixture theory (solid line), FEBio (symbols) and pump-leak model (dashed line) for various pump values $\tilde{p}_a^{Na}$ (µN/µm³). Membrane properties are $h=0.01$ µm, $\varphi_r^w=0.7$, $k=$ $2\cdot10^{-8}$ mm²/N·s, $d^{Na}=d^K=1.2\cdot10^{-13}$ mm²/ s, $a_r=5$ µm. Cytoplasm has much higher k and d^α to simulate homogeneous response.

DISCUSSION

This study describes the implementation of active solute transport in the FEBio finite element framework, using mixture theory. The code implementation is verified against analytical and numerical solutions. The modeling of a simplified RBC with passive transport and Na^+-K^+ pump shows agreement between mixture theory and FEBIO, and consistency with a more traditional pump-leak model. Results confirm that Na^+-K^+ pump is needed to maintain a constant cell volume, membrane potential and ion gradients in homeostasis [8]. Large time scales for the responses result from the very small diffusivity of ions across the RBC membrane when assuming that homeostasis is maintained only by the Na^+-K^+ pump. In reality, other pumps and co-transporters also play a role in the RBC and their contributions will be implemented in the future [8]. This study provides a novel tool for modeling active processes in cell physiology.

ACKNOWLEDGEMENTS. NIH R01 GM083925.

REFERENCES. [1] Ateshian, GA et al., *J Biomech Eng*, 135:111001, 2013. [2] Ateshian, GA et al., *Biomech Model Mechan*, 13:1105-1120, 2014. [3] Maas, S et al., *J Biomech Eng*, 134:011005, 2012. [4] Ateshian, GA et al., *J Biomech*, 39:464-475, 2006. [5] Ateshian, GA et al., *Ann Biomed Eng*, 38:1801-1814, 2010. [6] Lew,V et al., *J Membrane Biol*, 92:57-74,1986. [7]Andreoli,T.,*Membrane Physiology*, 1987. [8] Sperelakis, N., *Cell Physiology Sourcebook*, 2012.

SB^{3}C2015
Summer Biomechanics, Bioengineering and Biotransport Conference
June 17-20, 2015,Snowbird Resort, Utah, USA
SB^{3}C2015-358

RATE OF TRANSPORT OF CIRCULATING MACROMOLECULES INTO THE ARTERIAL WALL MAY DETERMINE ATHEROSCLEROTIC PLAQUE STABILITY

Zahra Mohri (1), Ethan M. Rowland (1), Lindsey A. Clarke (1), Amalia De Luca (2), Véronique Peiffer (1,2), Rob Krams (1), Spencer J. Sherwin (2), Peter D. Weinberg (1)

(1) Department of Bioengineering
Imperial College London
London, UK

(2) Department of Aeronautics
Imperial College London
London, UK

INTRODUCTION

Accumulation of lipids such as cholesterol in the inner layers of the arterial wall is thought to be an initiating event in the development of atherosclerosis. The likelihood of lipid accumulation depends not only on the concentration of lipoproteins, particularly low density lipoprotein, in plasma but also on transport properties of the wall itself: atherosclerotic lesions tend to develop in areas where the rate of transport of macromolecules from plasma into the arterial wall is high [1]. The rate of transport can be modified by altering flow [2] and by inhibiting endogenous nitric oxide (NO) synthesis [3].

Advanced atherosclerotic lesions have been divided into stable and unstable (or vulnerable) plaques. The latter are prone to rupture, which leads to formation of blood clots and acute coronary syndromes. They are characterized by a paucity of fibrous protein overlying the lesion and by a large lipid-rich core, and hence are termed thin cap fibroatheromata (TCFA). Biomechanical studies and well-established effects of lipids on inflammatory and proteolytic processes are consistent with the lipid-rich core of this lesion playing a key role in its rupture. The association of TCFA with elevated plasma cholesterol concentrations suggests that the lipid in their core derives predominantly from circulating lipoproteins.

Here we investigate the hypothesis that the fate of lesions depends, like their initiation, on the elevation of plasma macromolecule transport into the wall, with TCFA developing in regions where transport rates are particularly high. Placing a tapered, flow-restricting cuff around the carotid artery of apolipoprotein E -/- (apoE -/-) mice induces the formation of plaques upstream and downstream of the cuff; those upstream of the cuff resemble TCFA, whereas those downstream of the cuff resemble stable lesions [4]. We investigated whether, prior to the development of lesions, transport is not only elevated at both ends of the cuff but is more elevated upstream, where the TCFA-like lesions develop; we also examined the roles of flow and NO synthesis in determining rates of transport [5].

METHODS

Rigid cuffs were placed around the left common carotid artery of male wild-type (C57bl/6) mice as previously described [4]. The cuff lumen narrowed linearly from 500 to 250 μm in diameter over its length of 1.5 mm. In some animals the cuff was placed in the conventional orientation (wide end upstream) and in others it was reversed. One week later, transport properties were mapped by administering rhodamine-labelled albumin *iv*, fixing the arteries in situ 10 minutes later, and measuring tracer fluorescence in the wall of the excised carotids by en face confocal microscopy, as previously described [6] with modifications [5]. To inhibit NO production, L-NAME (50 mg/kg *ip*) was administered 30 minutes before tracer.

Flow rates in the carotid arteries of anaesthetized animals were examined by Doppler ultrasound, using a 30 MHz mechanical transducer coupled to a Vevo770 ultrasound system (VisualSonics).

Steady flow was computationally simulated with the in-house spectral/hp element solver Nektar. Geometries were obtained by microCT of vascular corrosion casts, prepared by infusing Batson's No 17 resin and maintaining it physiological pressure until it set. The images had an isotropic voxel size of 5.5 μm. They were segmented and a surface definition was extracted and smoothed. Cylindrical extensions were added to the outflows and the surface was remeshed so that the size of elements decreased as the vessel narrowed. A volume mesh of tetrahedral and prismatic boundary layer elements was generated; external faces of the elements were curved. The meshes for the forward and reverse cuff geometries contained, respectively, 25,215 and 21,912 tetrahedral elements and 7,992 and 6,834 prismatic elements.

Student's paired t-test was used to assess differences in uptake between upstream and downstream cuff margins (mean values being taken over a width of 0.7 mm centred on peak uptake). Student's unpaired t-test was used for comparisons between mice.

RESULTS

The microCT and ultrasound data demonstrated that the conventionally-oriented cuff caused a severe stenosis, restricted flow by about 50% and induced a jet downstream of the throat. Wall thickening was observed at both ends of the cuff. The reversed cuff restricted flow to the same extent but did not induce a jet.

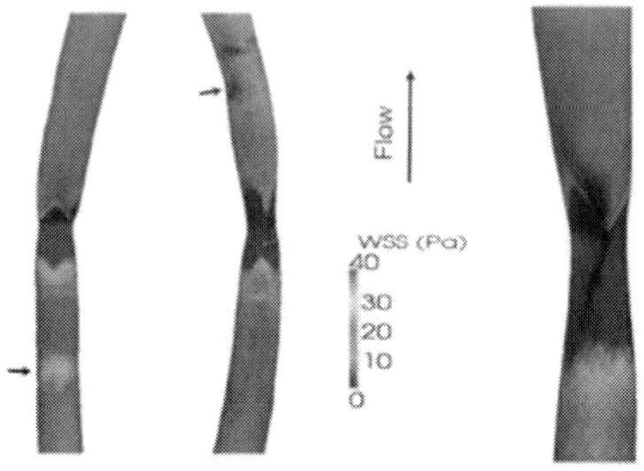

Fig 1: A. WSS with the cuff in the conventional direction (left) or reversed. The cuff throat occurs at the narrowest part of the reconstruction; the position of the wide end is indicated by an arrow. B. Detail of the throat region with the conventional cuff.

Maps of haemodynamic wall shear stress (WSS; Fig 1) indicated the presence of a small recirculation zone downstream of the conventional cuff when measured inflow velocities were increased to allow for effects of anaesthesia; no recirculation was observed with the reversed cuff. The dependence of the size of the recirculation zone on inflow velocity implies the presence of areas where the flow reverses direction during the cardiac cycle.

Uptake of tracer by the carotid artery wall of uncuffed mice was approximately uniform and the same for left and right vessels. It was elevated at both ends of the conventionally-oriented cuff, and significantly more so at the upstream end (Fig 2). Two peaks were again seen for the reversed cuff, but their heights were essentially identical (Fig 3). Two equal peaks of uptake were also seen in mice with the cuff in the conventional direction after administration of L-NAME, whilst in uncuffed mice, uptake of tracer was substantially reduced by the administration of L-NAME (data not shown).

DISCUSSION

Uptake of plasma macromolecules by the wall of the cuffed carotid of wild-type mice was elevated in regions where lesions develop in cuffed apoE -/- mice fed a Western diet. Furthermore, uptake was more elevated where lipid-rich lesions resembling TCFA occur than where more fibrous lesions are seen. The data are consistent with the longstanding view that elevated uptake leads to the development of lesions and additionally support the novel hypothesis that TCFA are triggered by an exceptionally high uptake.

The pattern of uptake was not entirely reversed by reversing the cuff orientation, showing that solid stresses cannot explain all the transport variations. Nevertheless, some features (high uptake at the upstream margin of the conventionally-oriented cuff and the downstream margin of the reversed cuff) are more easily explained by solid than fluid stresses, and coincide with thickening of the wall).

The difference between upstream and downstream peaks seen with the conventionally-oriented cuff was abolished by inhibition of endogenous NO synthesis. Further investigation is required to determine whether this reflects a direct effect on transport properties of the wall, mediated through influences of NO on endothelial cells and/or smooth muscle cells, or an influence on mechanical strain arising from a pressor effect of NOS inhibitors.

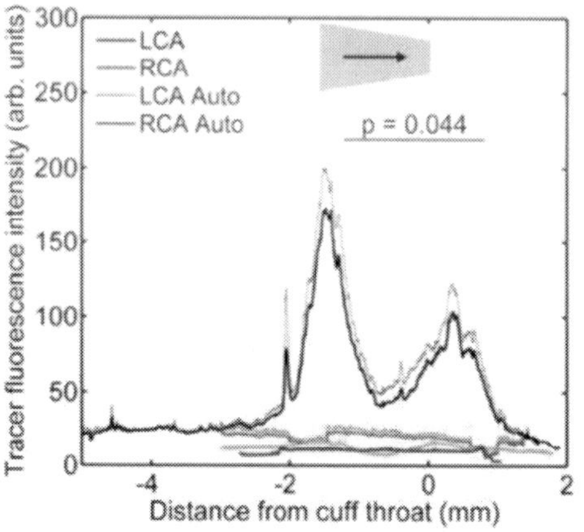

Fig 2. Fluorescence intensity for the cuffed left carotid ("LCA") and uncuffed right carotid ("RCA") of mice administered tracer or not administered tracer (indicating autofluorescence; "Auto"). Mean + 1SEM (lighter lines), n=10 (tracer), n=3 (Auto).

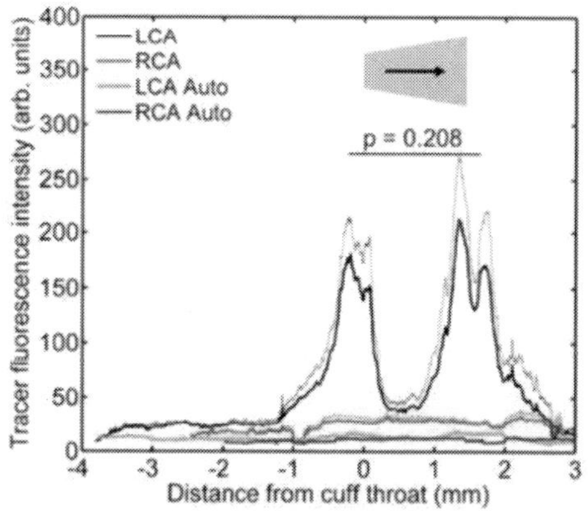

Fig 3. A plot equivalent to Fig 2 but for mice with the cuff in the reverse direction; n=8 (tracer), n=7 (Auto).

ACKNOWLEDGEMENTS

Funded by the British Heart Foundation

REFERENCES

[1] Weinberg PD, *J Vasc Res*, 41:1-17, 2004.

[2] Staughton TJ, et al., *Am J Physiol Heart Circ Physiol*, 281:H53-H59, 2001.

[3] Forster BA, et al., *Arterioscler Thromb Vasc Biol*, 17:1361-1368, 1997.

[4] Cheng C, et al., *Circulation*, 113:2744-2753, 2006.

[5] Mohri Z, et al., *PLoS One*, 9:e115728, 2014.

[6] Clarke LA, et al., *Atherosclerosis*, 224:417-425, 2012

SB³C2015
Summer Biomechanics, Bioengineering and Biotransport Conference
June 17-20, 2015,Snowbird Resort, Utah, USA
SB³C2015-359

ON THE EFFECT OF CALCIFIC CONTENT ON THE MECHANICAL BEHAVIOUR OF CAROTID PLAQUE TISSUE

Hilary E. Barrett (1), Eoghan M. Cunnane (1) and Michael T. Walsh (1)

(1) Centre for Applied Biomedical Engineering Research (CABER) Material Surface Science Institute (MSSI), Department of Mechanical, Aeronautical and Biomedical Engineering University of Limerick, Limerick, Ireland

INTRODUCTION

The success of minimally invasive endovascular stenting treatment is determined by the combined response of the pathological components in complex atherosclerotic plaque composites which control the tissue's rupture properties. A plaque lesion typically consists of a lipid core with an overlying fibrous cap accompanied by a fibrous tissue matrix and diffuse rigid calcification inclusions which together cause a severe decrease in lumen area and increase in vessel wall stresses. Specifically, the rigid calcifications inclusions have been linked to treatment failure by initiating regions of high stress concentrations under the supra-physiological loading forces causing mechanically induced plaque tissue ruptures during minimally invasive treatment. However, the mechanical behavior of the plaque and how this varies with calcification is not fully understood. Therefore, a clear need exists to use more accurate methods of characterizing the diseased tissue to better understand the effects structural components have on the mechanical behavior of the diseased tissue. The paucity of experimental mechanical data characterizing the plaque tissue in failure when under the influence of calcification has driven the need for this experimental study to characterize the tissue injury incurred when under circumferential loading conditions. In an attempt to better understand the mechanical behavior, this study will demonstrate a method of relating the biochemical and structural characteristics of carotid plaque tissue to the mechanical rupture behavior using Fourier Transform Infrared (FTIR) spectroscopy analysis and high resolution micro x-ray computed tomography (μX-CT).

METHODS

This study experimentally derives the material properties of seventeen human carotid plaque specimens, as whole structures, that were acquired from endarterectomy procedures. Uniaxial circumferential stretching tests to failure were used to examine the mechanical behavior of the diseased tissue in a manner that similarly conforms to the supra-physiological loading conditions experienced during *in vivo* stent deployment. The experimental protocol employed five preconditioning cycles, to 10% of the gauge length at a strain rate of 0.1 mm/s, to achieve a repeatable stress-stretch response and the tissue was subsequently stretched to complete failure at a displacement rate of 30% of the gauge length per second. The *in vitro* experimental tests demonstrated the mechanical response, rupture behavior and corresponding biomechanical parameters which cannot be examined through *in vivo* tissue characterization. The macroscopic failures were tracked via noncontact videography and correlated with the corresponding mechanical data to characterize the tissue in failure and obtain the peak stresses at the interface between the calcification and plaque tissue matrix. Biochemical analysis, by spectroscopic FTIR, was used to identify key pathological tissue components. The tissue content was quantified by an absorbance peak-area ratio function which is directly related to the concentration of each component. The average ratios for lipid to collagen (Lip:Col) and calcification to collagen (Calc:Col) were quantified for each plaque. A failure analysis, by high resolution μX-CT imaging and scanning electron microscopy (SEM) was performed to identify the cause of plaque failure and comprehensively classify the global internal morphology of each mechanically ruptured plaque. The calcification inclusions were quantified by volume (mm^3) and surface area (mm^2) measurements, determining the calcification's contribution to tissue biomechanics.

RESULTS

The mechanical behavior of the diseased lesions demonstrated a nonlinear heterogeneous mechanical response. The Cauchy stress and stretch data were used to quantify the stress induced on the plaque tissue structure which had an average rupture stress and stretch value (mean±std dev) of 0.414MPa ± 0.253 and 0.26 ± 0.22 respectively. A significantly large inter-specimen variation existed between the stiffness and ultimate strength values in this sample set. The biochemical analysis, by FTIR, identified the key pathological tissue components and characterized the degree of plaque progression by identifying the functional groups associated with lipid, collagen and calcification in each specimen. A high concentration of calcification was found closely adjacent to the lumen surface in both vulnerable high lipid and advanced plaque types.

The µX-CT imaging revealed the local heterogeneity of the plaques internal morphology with each specimen consisting of copious clusters of randomly dispersed micro particles throughout the tissue structure together with millimeter scale calcified masses of randomly oriented geometries throughout the cross-section of the plaques (fig.1). The quantification of the overall calcification to plaque tissue volume ratio (Calc:Plaque) shows two significantly distinct groupings exist in this sample set. The two groups were classified as lightly calcified (0>Calc:Plaque<0.20) and heavily calcified (0.3≤Calc:Plaque≤0.55). Applying this overall Calc:Plaque classification to the mechanical data demonstrates that the calcification is strongly influencing the plaques ultimate strength as the heavily calcified group contained both the stiffest and weakest mechanical properties (fig.2).

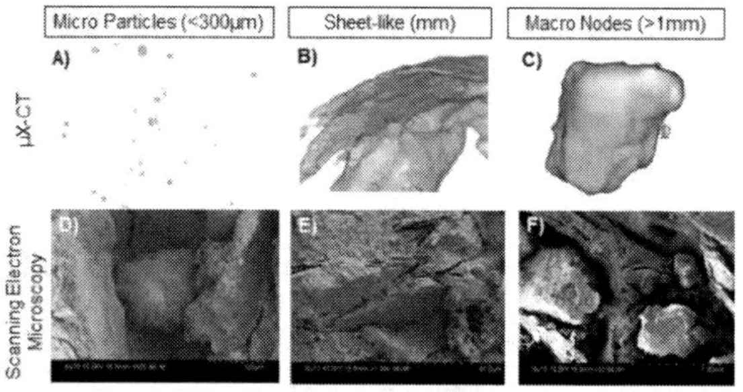

Figure 1. Calcification geometries imaged by µX-CT (A-C) & SEM (D-F).

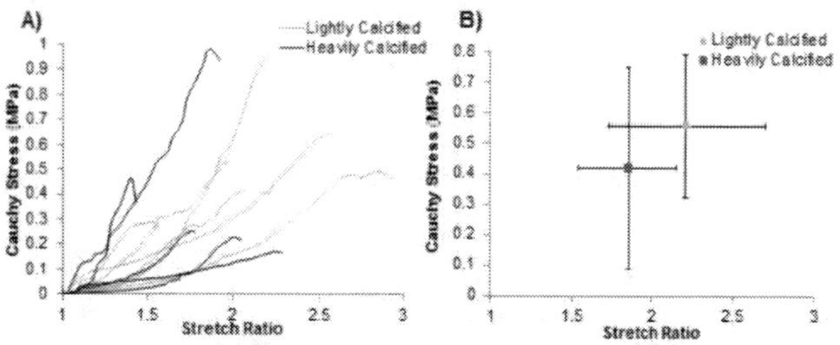

Figure 2. A) Calcification classification and B) mechanical average and standard deviation based on Calc:Plaque.

The failure analysis suggests that individual calcification nodules at the fundamental micro and macro length-scales are functionally related to the biomechanical strength properties. The analysis also revealed that ultimate plaque failure occurred along the calcification-plaque tissue interfacial boundary region for all plaques classified as heavily calcified.

DISCUSSION

This study characterizes the mechanical response, rupture behavior and corresponding biomechanical parameters of whole human carotid plaque tissue in response to supra-physiological circumferential loading conditions. Further, the mechanical rupture data was correlated with biochemical and structural measures elucidating the interaction between structural calcification inclusions and plaque tissue behavior. The FTIR analysis shows clear markers for the level of disease progression through the decreasing lipid concentrations. This decrease in lipid agrees with the AHA plaque classification that pathologically advanced plaques have lower lipid and are more fibrotic. Interestingly, calcification was found to be present independent of the progressive nature of the diseased tissue.

The µX-CT imaging shows that applying the overall calcification volume to plaque tissue volume ratio to the mechanical data may help explain the deviation in mechanical behavior between the two distinct groups. The lightly calcified group demonstrated a more linear response while undergoing a large stretch prior to reaching ultimate failure. In contrast, the increase in volume fraction in the heavily calcified group corresponded to a decrease in stretch at failure, suggesting that the calcification is limiting the deformation of the plaque. The rupture stresses fell within a wide range. This range of stresses is not easy to explain using overall volume measurements. Therefore, the individual inclusions were analyzed to assess if specific calcific inclusion geometries contributed to the experimentally derived mechanical properties. In this sample set, three distinct calcification geometries were identified. These include micro particles (<300µm), thin sheet-like structures and irregularly shaped macro nodes (>1mm). Clusters of randomly dispersed micro particles were present throughout the tissue structure. Computational studies have predicted that micro particles can cause localized stress concentrations in the tissue structure inducing a weakening effect that can diminish the ultimate strength [1]. However, the larger macro nodes of irregular shapes were the cause of ultimate failure. The weaker plaques, failed as a result of single macro calcification nodules surrounded by compliant tissue whereas the plaques consisting of agglomerated masses of calcification consuming the internal volume of the tissue required a high stress to fail. This suggests that the strength of the plaque tissue diminishes with an increasing interface area between the calcification and tissue matrix. Buffington et al. recently predicted that the geometry of calcification has a profound effect on the interfacial area stress levels and has suggested that geometry may have important prognostic value in examining the plaque [2].

This study proposes that not only calcific content but the morphology of the content is an important determinant of the plaque tissue behavior. FTIR analysis identified the biochemical composition of the key pathological components while the complementary µX-CT imaging modality revealed that the mechanical behavior of the plaque is strongly under the influence of calcification. Furthermore, plaque assessment via imaging modalities should include calcification geometrical measures and interfacial boundary area in order to accurately evaluate the risk of tissue rupture.

ACKNOWLEDGEMENTS

The authors would like to acknowledge that the source of funding for this work was from the Irish Research Council Grant no. (GOIPG/2013/1309) and the framework of the Irish Government's Programme for Research in Third Level Institutions Cycle 5, with the assistance of the European Regional Development Fund. The authors would like to thank Michael Byrne for all his help in acquiring the CT scans.

REFERENCES

[1] Maldonado, N et al., *AmJ PhysHeart Circ Phys* 303:619-28, 2012.
[3] Buffington, C et al., *CardioVas. Engineering & Technology*, 2014.

SB^{3}C2015
Summer Biomechanics, Bioengineering and Biotransport Conference
June 17-20, 2015,Snowbird Resort, Utah, USA
SB^{3}C2015-360

ACOUSTIC DETECTION OF RESPIRATORY SOUNDS IN SILICONE LUNG AIRWAY MODEL USING MICROPHONE ARRAY SYSTEM

Gabriel Pramudita Saputra (1), Kazunori Nozaki (2), Satoshi Ii (1), Chizu Habukawa (3)(1), Shigeo Wada (1)

(1) Graduate School of Engineering Science
Osaka University
Osaka, Japan

(2) Division of Dental Informatics
Osaka University Dental Hospital
Osaka, Japan

(3) Department of Pediatrics
Minami Wakayama Medical Center
Wakayama, Japan

INTRODUCTION

Acoustic assessment of a patient's lung sound was commonly used as an indicator of the healthiness of the patient's respiratory system. This method became more sophisticated especially after Laennec invented the usage of stethoscope [1]. Some conditions suspected to cause sound alteration when the disease occur are the introduction of mucus in the airway, inflammation of the airway cell, collapsing of the airway, hardening of the lung parenchyma, or the combination of these conditions. This complex situation leads to the alteration of high frequency component in the sound, appearance of crackling sound, or loud sound in a lower frequency region, which normally was not found in the normal condition [2]. However, the mechanisms of the generation of these sounds are not clearly understood.

In current study, an effort to understand the basic sound generated aeroacoustically by the airflow in the lung airway model was proposed. This model was constructed from a normal patient CT image to provide a realistic airway model. Using this model, we aimed to understand deeply the generation mechanisms, started from the location of the sound and the characteristic of the sound in the lung airway.

METHODS

To reproduce a realistic airflow inside a lung airway, a silicone airway model based on CT-image of a subject was constructed as shown in Fig. 1. The subject was an 11 years old boy in a stable condition and normal respiratory system. The silicone model size was 185×115×60 mm^3 and the trachea-end geometry was converted to a 9 mm diameter circular tube and connected to a rubber hose size of 15 mm inner diameter. In total, 39 peripheral branches can be found, and they were connected to six silicone tubes of 2 mm inner diameter and 33 silicone tubes of 1 mm inner diameter.

Air flowing through the silicone model was supplied from an air tank capacity of 120 L which flow rate was controlled to be 30 L/min by a mass flow controller (Azbil MQV0050B, Azbil Corp., Tokyo, Japan). Air was collected using an air compressor into the air tank until pressure of 0.6 MPa at first. After the pressure has been reached, the compressor will be shut down, therefore no noise from the compressor will disturb the measurement. To reduce the fluctuation of the flow inside the rubber hose, acrylic cylinders (inner diameter 80 mm) contained of urethane sponge and array of straw with inner diameter of 5 mm were constructed and located before the inlet of the silicone model. To simulate the inspiration maneuver, the air from the acrylic cylinder will go to the trachea part of the silicone model. To simulate the expiration case, the air from the acrylic cylinder will split into four acrylic cylinders. Each acrylic cylinder supplies the air to four parts of the lung model, namely upper right airways, lower right airways, upper left airways, and lower left airways.

An array consisted of five microphones (GRAS 40PH, GRAS Sound and Vibration, Holte, Denmark) was used to map the sound field on the surface of the silicone model with a distance of 2-5 mm measured from the surface of the silicone model to the tip of the microphones. The distance between each microphone was 21 mm. The microphone array recorded sound at 9 location above the surface of the silicone model with a distance set up to be 20 mm between each location as shown in Fig. 1. A sampling frequency of 20,000 Hz was used and the recorded sound was processed using 2048 points FFT in MATLAB R2012a with overlapping coefficient of 0.5. Hanning window was applied for all data sets and 50 data sets were used for calculating the ensemble averaging of the sound power spectral density.

The resulted PSD profile was used to calculate the overall sound pressure level (OASPL) of several frequency bands of interest using following equation:

$$OASPL = 20\log_{10}\left(\frac{\sqrt{\sum_{i=a}^{b}\left(P_{ref}\times 10^{PSD(f_i)/20}\right)^2}}{P_{ref}}\right) \quad (1)$$

where P_{ref} is the reference pressure equal to 2×10^{-5} Pa, $PSD(f_i)$ is the PSD value in decibels of the corresponding frequency f_i, and a and b are the indices for the lower and upper limits of the frequency of interest, respectively. Width of the bands analyzed in this study was 1000 Hz. Ten OASPL bands were calculated ranges from 1-1000 Hz (OASPL1), 1000-2000 Hz (OASPL2), 2000-3000 Hz (OASPL3), …, and 9000-10000 Hz (OASPL10). In addition, the global OASPL also calculated for frequencies between 1 and 10000 Hz. All OASPL values were subtracted to the OASPL of the background noise as reference. The relative OASPLs were shown in color graph as shown in Fig. 2.

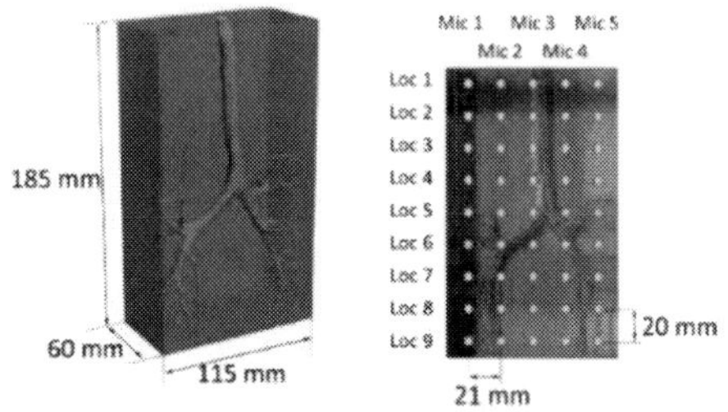

FIG 1. Left figure shows the dimensions of the silicone airway model. Yellow points on right figure show the microphone index and the location index.

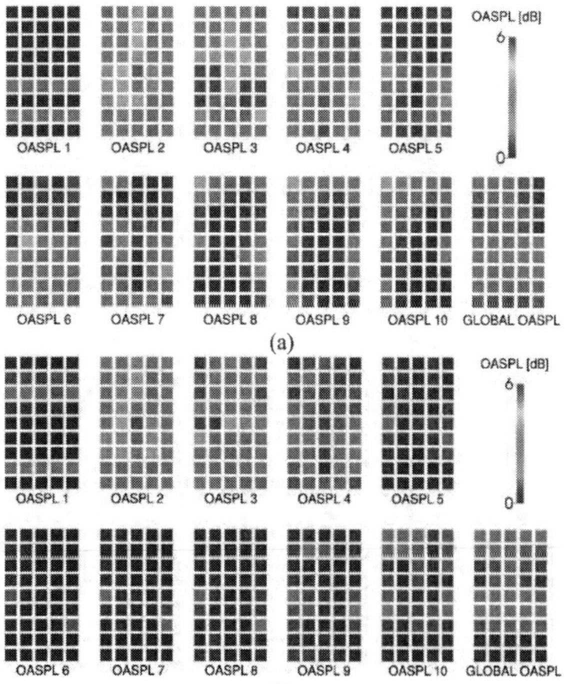

FIG 2. OASPL of measurement in (a) expiration, and (b) inspiration direction for 45 measurement points.

RESULTS

The calculated OASPLs from the measured sound on the surface of the silicone model are shown in Fig 2 for expiration and inspiration case respectively. The color-bar in the figure shows the value of the OASPL for the corresponding frequency range. For the expiration case, high OASPL value can be observed majorly at the peripheral location measured by microphone 1, 2, 4, and 5 at location 6 and 7. The highest OASPL value was found in OASPL 3 which frequency range were 2000-3000 Hz. Sound level of more than 3 dB was also observed in the OASPL 2 in the central region where the trachea was located. For the inspiration case, the highest sound level was found only on microphone 1 and 2 at location 6. However, an OASPL level of more than 3 dB was also observed for OASPL 2 on the central region of the model.

DISCUSSION

The usage of microphone array or multi-sensor measurement system was able to distinct patterns for normal and diseased subject [3], however, not much information can be gathered about how the sound was generated in the airway. Our method by using the airway geometry printed in a silicone material enables us to reduce the complexity of different material characteristics, while still consider the airflow condition that play an important role as the main sound source. Our results show that the sound generated in expiration was quite different to those generated in inspiration maneuver. The maximum value of the overall sound pressure level of the expiration maneuver majorly found in the region of peripheral airway, but in the inspiration case, the peak sound level was found in the area where the trachea and main branch was located. The size of the airway may then affect the frequency characteristics. The maximum value of the OASPL was observed to be higher in expiration maneuver than to that in inspiration maneuver. This result was contradictive in practice where medical practitioner can listen to inspiration sound easier than expiration sound. In current measurements, we found that the maximum OASPL of the sound was different between inspiration and expiration. The OASPL calculations were performed in order to calculate the integration level of the sound in a specific frequency band.

This discrepancy may come from the lack treatment of the boundaries or the connector of the silicone tubes and the model. In inspiration, existence of the throat constriction actually cannot be neglected, but in the current model, the throat was not existed. In the expiration maneuver, the silicone tube connectors may generate sound as well as a constriction.

The results presented in this report show the ability of current method to detect the original aeroacoustic sound generated by the airflow in the airway. Current simple method reveals the ability to locate the major sound source by assessing the total energy contained in the frequency spectra. In the future, more robust analysis may be performed by applying the method commonly used for the microphone array signal processing technique, such as Time Different of Arrival (TDOA), Near-field Sound Holography, and beamforming to improve the accuracy of both spatial and temporal information of the sound which then can be confirmed by the usage of Particle Image Velocimetry (PIV) to define the sound generation mechanisms for some different lung conditions.

ACKNOWLEDGEMENTS

This work was supported in part by JSPS Grants-in Aid for Scientific Research No. 24240073.

REFERENCES

[1] Dalmay, F et al., *Eur Respir J*, 8:1761-1769, 1995.
[2] Forgacs, P, *Brit J Dis Chest* 63:1-12, 1969.
[3] Kompis, M et al., *Chest*, 120(4):1309-1321, 2001.

SB³C2015
Summer Biomechanics, Bioengineering and Biotransport Conference
June 17-20, 2015, Snowbird Resort, Utah, USA
SB³C2015-361

TREATMENT OF FOCAL CARTILAGE DEFECTS USING A METAL IMPLANT: NEW BIOMECHANICAL INSIGHTS USING FINITE ELEMENT MODELING

Ashley Heuijerjans (1), Wouter Wilson (1), Keita Ito (1), Corrinus C. van Donkelaar (1)

(1) Department of Biomedical Engineering
Eindhoven University of Technology
Eindhoven, The Netherlands

INTRODUCTION

Most joint-preserving treatments for patients with focal cartilage defects rely on cartilage regeneration. For middle-aged patients with decreased regenerative potential, small metal implants can be placed to postpone total knee arthroplasty. However, the biomechanical effects of these metal implants on the adjacent and opposing cartilage are still not fully understood.

We hypothesized that elevated collagen fiber strains would be found in the superficial zone of the tissue adjacent to a defect, in particular after prolonged loading, resulting from tissue deforming into a defect. An implant might ameliorate this effect, preventing tissue from deforming. We further hypothesized that such implant might change the site with excessive strains from adjacent to the opposing cartilage, as a stiff implant may compress the opposing cartilage. This could result in increased collagen fiber strains in the superficial zone where fibers lie perpendicular to the surface. These strains might increase substantially during prolonged loading, when fluid flows out and tissue deforms. The current aim was to evaluate the above hypotheses using a finite element model of a cartilage-cartilage contact interface.

METHODS

Geometry. A plane strain 2D geometry was developed to represent simplified joint contact. The width of the cartilage was 40 mm and the thickness was 1.5 mm. The articular surfaces were assumed to be in complete contact. In the upper cartilage, a hole of width 8 mm, which is the smallest diameter for a metal implant currently available, was modeled in the defect case, while in the implant case this hole was filled with an implant. A perfect fit of the implant into the hole was assumed. Due to symmetry, only half of the geometry was simulated.

Model. Healthy articular cartilage tissue was described using the validated fibril-reinforced poroviscoelastic swelling model by Wilson et al.[1], implemented using ABAQUS V 6.14 (Simulia Inc. USA). Articular cartilage was assumed biphasic. The solid phase consisted of a non-fibrillar part, containing charged proteoglycans that develop an osmotic pressure, and a fibrillary part, representing the collagen fiber network. The metal implant, 50.000-fold stiffer than cartilage[2], was modeled as an analytical rigid part.

Boundary Conditions. Simulations were performed on equilibrated situations. Nodes on the theoretical cement lines of the upper cartilage were fixed in all directions. Nodes on the symmetry axis and on the theoretical cement lines of the lower cartilage were fixed in the horizontal direction. Pore pressure was kept zero at the free edges of the cartilage, which included the free edges in the defect in the absence of an implant.

Loads were applied on the theoretical cement line of the lower cartilage. In the fast axial loading scenario, the load was increased to 2 MPa in 1 ms. In the creep scenario, the load was increased to 0.5 MPa in 10 s followed by a constant load of 0.5 MPa for 900 s.

RESULTS

Both collagen strains and maximum principal strains were increased in presence of a cartilage defect (figure 1). Interestingly, their maxima occurred at distinct sites, which changed after sustained loading.

Fast axial loading elevated collagen strains in cartilage opposing a defect and in superficial zones adjacent to a defect. Maximum principal strains increased mainly at the edge of a defect in the deep zone. No physiologically relevant differences between intact cartilage and with implant were found.

Creep loading elevated collagen fiber strains in superficial zones adjacent to the defect to more than 10% strain. Maximum principal strains substantially increased, up to 50-fold, in the middle and deep zones of cartilage adjacent to the defect (figure 2).

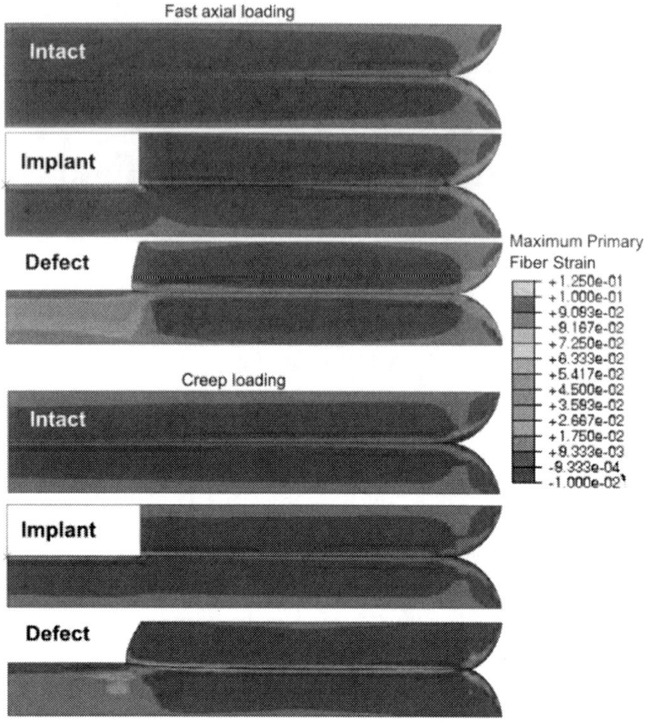

Figure 1; Maximum primary collagen fiber strains after fast axial loading (top) and creep loading (bottom) in intact, implanted and defect cartilage-cartilage contact.

DISCUSSION

Collagen fiber strains increased substantially in the superficial zones adjacent to a defect. Also, excessive maximum principal strains occurred in cartilage adjacent to a defect, with an increase up to a 50-fold compared to intact cartilage. In addition, a defect had an effect on collagen fiber strains in the opposing cartilage, caused by longitudinal deformation of tissue into the defect, which was not expected (figure 1). Similarly, the substantial increase in maximum principal strains concentrates near the edge of the defect, in the middle and deep zone, and it was caused by lateral deformation of tissue into a defect. This site correlated with locations where cartilage separation from the underlying bone, also called cartilage delamination or peel-off tearing, is known to occur clinicaly[3-5]. An implant ameliorated these lateral and longitudinal deformations, preventing high strains close to a defect. This suggested a previously undocumented supporting function of an implant. The anticipated effect on opposing cartilage was present, yet quantitatively modest.

A previous study of metal implants identified sites of elevated contact pressure on cartilage adjacent to an untreated defect[2], which are in agreement with the locations of elevated collagen strains adjacent to a defect in the present study. However, their effects on the cartilage opposing an implant were more significant than ours.

Results of animal studies with metal implants are inconsistent. Custers et al.[6-8] concluded a metal implant could not prevent degeneration, whereas others[9,10] showed positive effects. Positive clinical results have also been documented.[11,12] The reason for these discrepancies was unclear. Location in the joint or depth of implant placement likely affect the clinical result[6,10]. Studying such effects is subject of future study.

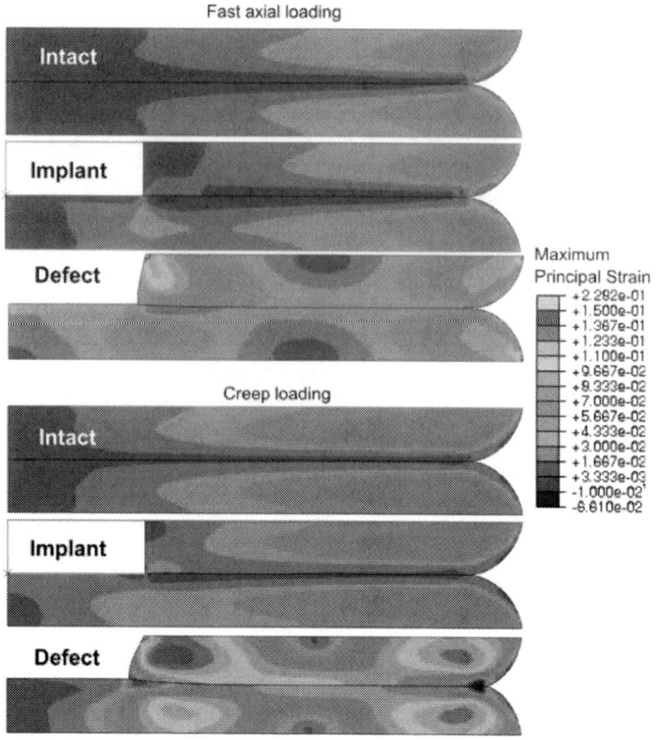

Figure 2; Maximum principal strain after fast axial loading (top) and creep loading (bottom) in intact, implanted, and defect cartilage-cartilage contact.

Due to joint incongruency and motion, it was assumed that no fluid entrapment occurs in vivo in a cartilage defect.

Limitations of the present study included 2D modeling, where a 3D model would allow studying geometrically more realistic effects such as surface congruency, joint motion and other loading modes. This could reveal potential limitations of an implant, such as adverse effects on the opposing tissue, which are not evident from the present study.

Finally, whether the computed strains may induce progressive tissue damage is subject of future study, using a damage progression model[13].

In conclusion, the present study demonstrated beneficial effects of an ideally placed metal implant as a consequence of preventing lateral deformation of adjacent tissue. With the present boundary conditions there were no significant adverse effects of the implant.

ACKNOWLEDGEMENTS

The authors would like to thank dr P. Emans, Maastricht University Medical Center, for fruitful discussions about clinical aspects.

REFERENCES

1. Wilson+, J Biomech 38:1195–204, 2005; 2. Manda+, J Biomech 44:794-801, 2011; 3. Sekiya+, Orthop 32, 2009; 4. Beaulé+, *Can J Surg* 46:463-4. 2003; 5. Beck+, *JBJS Br* 87:1012-8. 2005; 6. Custers+, OA Cart 15:937-45, 2007; 7. Custers+, OA Cart 18:377-88, 2010; 8. Custers+, JOR 28:429-35, 2010; 9. Kirker-Head+, JOR 24:1095-108; 10. Martinez-Carranza+, OA Cart 21: 739-745, 2013; 11. Becher+ Arch Orthop Trauma Surg 131:1135-43, 2011; 12. Bollars+ Knee Surg Sports Traumatol Arthrosc 10:1753-9, 2012; 13. Hosseini+ OA Cart 22:95-103, 2014

RECONSTRUCTION OF INCOMPLETE LIPID POOL GEOMETRY FOR STRESS CALCULATIONS IN ATHEROSCLEROTIC ARTERIES

Annette M. Kok, Lambert Speelman, Frank J.H. Gijsen, Jolanda J. Wentzel

Biomedical Engineering
Erasmus MC
Rotterdam, The Netherlands

INTRODUCTION

The main cause of myocardial infarction is coronary plaque rupture. Not all plaques will rupture, therefore it is of imminent importance to distinguish the rupture prone (vulnerable) plaque from the stable plaque. Vulnerable plaques are associated with expansive remodeling and a lipid rich core covered by a thin fibrous cap, better known as thin cap fibrous atheroma (TCFA).

From a biomechanical point of view, plaque rupture will occur when the wall stresses in the plaque exceed the strength of the plaque. Determining the stresses in a plaque can aid in predicting rupture. In order to calculate wall stresses, blood pressure, material properties and geometric information of the vessel wall, including the lipid pool (LP) and fibrous cap are needed.

Geometric information of the coronary wall can by high resolution intravascular imaging techniques or post mortem via histology. With intravascular ultrasound (IVUS), the lumen, calcifications as well as the transition from the media to adventitia is clearly visible, but no information on the lipid pool can be obtained. This technique has a resolution and penetration depth of 100 ± 20 µm and 5 mm, respectively. Another intravascular imaging technique is Optical Coherence Tomography (OCT) which has a high resolution (13 ± 3 µm), but a lower penetration depth of 1.25 mm. This higher resolution allows for the fibrous cap to be accurately visualized, but the lipid pools have high attenuation therefore obscuring deep structures. Hence, only the front side of the lipid pool is visible and not the backside. Therefore we don't have the full geometry of the lipid pool. Since, previous studies showed that in addition to the cap thickness also the lipid pool thickness can impact the plaque stresses [1]–[3], missing the backside of the lipid pool would potentially limit stress calculations.

The goal of this study was to evaluate whether reconstruction of the lipid pool in case of missing data affects the peak cap stress.

In this work we used histological data from an atherosclerotic vessel wall, consisting of a lipid pool and a fibrous cap. To simulate IVUS/OCT data the backside of the LP was removed and reconstructed using the cap thickness, LP angle and location of the outer vessel wall from histology data. The geometry of the reconstructed lipid pool was compared to the lipid pool obtained from histology. Wall peak cap stresses were compared in both geometries to evaluate the effect of the reconstruction on peak cap stress.

METHODS

In summary, the methods comprise five steps: 1) obtaining the vessel wall geometry based on histology, 2) determining a relationship between the known geometrical parameters and the lipid pool thickness, 3) reconstruction of the lipid pool geometry, 4) geometry assessment: comparing the reconstructed geometry to histology and 5) wall stress analysis in both geometries. All steps are explained in more detail below.

Histological geometries. Coronary plaque data was post-mortem obtained from seven patients. 13 arteries were perfusion fixed with formalin at 100 mmHg prior to the histological preparation. From these 13 arteries 54 cross sections were taken at least 200 µm apart from each other. A modified Movat pentachrome staining was used to manually delineate the lumen, lipid, intima, media and adventitia layer, see Figure 1A and 1C. When more lipid pools were present in a cross section only one per cross section was taken, in total 76 lipid pools in separate cross sections were reconstructed.

Estimate backside of the lipid pool. Based on information of the lipid pools obtained from histology, a relationship can be made between the backside of the lipid pool based on histology and known

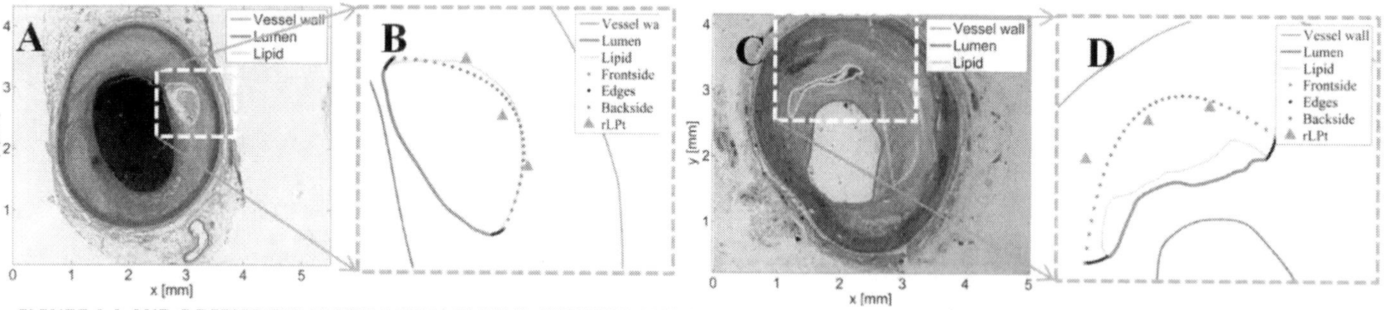

FIGURE 1 A AND C DEPICT THE HISTOLOGICAL PLAQUE, STAINING AND SEGMENTATION. FIGURE B/D SHOWS THE GEOMETRY FROM HISTOLOGY (YELLOW) AND THE RECONSTRUCTION OF THE LIPID POOL (ORANGE THE FRONT SIDE, BLACK THE EDGES AND GREEN THE BACKSIDE OF THE LIPID POOL). TO INDICATE A GOOD AND BAD RECONSTRUCTION 2 EXAMPLES ARE GIVEN OF THE HIGHEST (B) AND LOWEST (D) SIMILARITY INDEX: 0.98 AND 0.41, RESPECTIVELY.

IVUS/OCT geometrical parameters: cap thickness, lipid angle, vessel wall thickness (vessel wall thickness = distance from cap thickness to media-adventitia transition). More in detail, from these known parameters the relative lipid pool thickness (rLPt) can be determined at three positions along the lipid angle: 25, 50, 75 percentage of the lipid angle, see pink triangles in Figure 1B and 1D. Two models to estimate the rLPt were created: one at the center 50% of lipid angle ($rLPt_{50}$) and one for the 25/75 % of the lipid angle ($rLPt_{25/75}$). To correct for correlation amongst cross sections within an artery the Generalized Estimation Equation (GEE) model was used.

Reconstruction. To reconstruct the lipid pool at the edges of the frontside of the lipid pool, a part of a circle was attached with empirically determined values (radius of 140 µm *[lipid angle/mean of all lipid angles] and angle of 30°). Then, together with the three model-based rLPt, the backside is closed with a second order polynomial function, see Figure 1B and 1D.

Geometry assessment. To perform geometry assessment of the on histology based lipid pool geometry (LP_{hist}) versus the reconstructed geometry of the lipid pool (LP_{rec}) two measures were used: 1) the similarity index (SI) and 2) the relative difference in lipid pool area ($\Delta A\%=[|A_{hist}-A_{reconstr}|/A_{hist}]*100\%$).

Wall stress analysis. Contours of the vessel wall, including the lipid pool were imported and meshed in Abaqus (Dassault Systemes Simulia Corp., Providence, RI, USA) with at least three elements were over the cap thickness. A plaque consists of one lipid pool, intima, media and adventitia. All material properties were assumed to be homogeneous and incompressible. Neo-Hookean material models were used, with Young's moduli obtained from literature: 6, 1e3, 1.8e3 and 1.8e3 kPa for lipid, intima, media and adventitia, respectively. Initial stresses at 100 mmHg are calculated with the backward incremental method. Then a systolic blood pressure of 140 mmHg was applied at the lumen to mimic the potential rupture state of the plaque. In this study, von Mises stresses were calculated and reported. The maximum von Mises stress in the cap and the shoulders (15 degrees adjacent to the lipid pool) is hereafter referred to as peak (cap) stress. The peak cap stress of the LP_{hist} was compared with the peak cap stress of the LP_{rec}.

The (difference) peak cap stress, SI and ($\Delta A\%$) are not normally distributed and therefore median and interquartiles are reported.

RESULTS

To estimate the rLPt, a GEE model was created for the 50% lipid angle and 25/75 lipid angle. The GEE model estimated the $rLPt_{50}$ as: 0.056*lipid angle (rad) + 4e-4*wall thickness (µm) + -6e-4* cap thickness (µm) + 0.15 (r^2=0.54 and $p<0.05$). Similarly, $rLPt_{25/75}$ = 0.11*lipid angle (rad) + 2.8e-4 * wall thickness (µm) + -4.4e-4* cap thickness (µm) + 0.11 (r^2=0.50 and $p<0.05$). The LP_{rec} matched the LP_{hist} well with a SI and $\Delta A\%$ of 0.89 (0.85-0.92) and 13.4% (6.6-25.0), respectively. Two cases of the LP_{rec} are shown in Figure 1.

In total 72, LP geometries were analyzed for the wall stress calculations, four geometries were excluded due to technical issues. The peak stresses of LP_{rec} were highly correlated to the peak stress of LP_{hist} and are shown in Figure 2 (r^2 =0.95). The difference in peak cap stresses were 5 (1 – 12) %.

DISCUSSION

The key findings of this study are 1) a method is developed to resolve the issue of limited geometrical information of the lipid pool 2) a good agreement between the LP_{hist} and LP_{rec} geometries 3) high correlation between the peak stresses.

Even though the r^2 values of the GEE models to estimate the rLPt were low, it did not have a large effect on the reconstruction of lipid pool geometry and thus on the peak cap stress. Apparently the fibrous cap geometry plays a more important role in the peak cap stress.

In conclusion, peak cap stress can be accurately calculated in vessel wall geometries in which information on the backside of the lipid pool is absent. However information on cap geometry and wall thickness (e.g. derived from OCT and IVUS) is crucial.

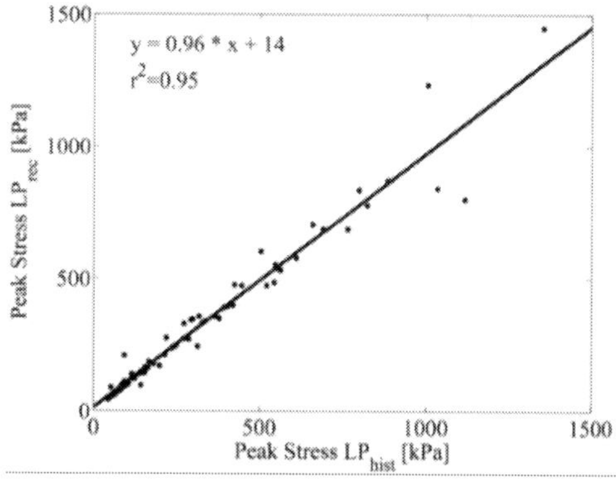

FIGURE 2: PEAK STRESS OF THE HISTOLOGY GEOMETRY PLOTTED VS. RECONSTRUCTED GEOMETRY.

ACKNOWLEDGEMENTS

The research leading to these results has received funding from the European Research Council under the European Union's Seventh Framework Programme / ERC Grant Agreement n. 310457.

REFERENCES

[1] Li, Z et al., *J Vasc Surg*, 72: 1092-1099, 2008
[2] Gao, H et al., *J Biomech*, 41: 3053-3059, 2008
[3] Akyldiz, A et al., *Biomed Eng Online*, 10: 10-25, 2011

SB3C2015
Summer Biomechanics, Bioengineering and Biotransport Conference
June 17-20, 2015, Snowbird Resort, Utah, USA
SB3C2015-363

CFD SIMULATION OF TRANSITION TO TURBULENCE FOR NEWTONIAN VS. NON-NEWTONIAN FLOW THROUGH A STENOSIS

M. Owais Khan (1), Kristian Valen-Sendstad (1,2), Dipankar Biswas (3), David M. Casey (3), Francis Loth (3), David A. Steinman (1)

(1) Mechanical & Industrial Engineering
University of Toronto
Toronto, ON, Canada

(2) Centre for Biomedical Computing
Simula Research Laboratory
Lysaker, Norway

(3) Mechanical Engineering
University of Akron
Akron, OH, U.S.A.

INTRODUCTION

Recent high-resolution CFD studies have revealed the possibility of turbulent or turbulent-like flow in the cerebrovasculature. Although studies of laminar flows have tended to conclude only a modest effect of non-Newtonian rheology on flow patterns, there are suggestions that these effects may not be negligible for predicting the onset of turbulence. For example, recent experimental studies from the Loth group have suggested that transition to turbulence in a straight pipe is delayed from the usual 2100-2300 to 2500-2700 for whole blood, and could be pushed even higher for non-Newtonian blood mimicking fluids [1]. It has also been suggested by Antiga & Steinman [2] that the continuum approximation for blood flow may break down owing to the small temporal and spatial scales of turbulent flow.

The aim of the present study was to see if these observations hold for a more physiologically relevant flow through a canonical stenosis model, with an eye towards eventually comparing CFD results against planned experimental studies using Newtonian and shear-thinning blood mimics, as well as whole blood itself.

METHODS

The model chosen for these studies was the offset cosine stenosis model proposed and studied with Direct Numerical Simulation (DNS) by Varghese et al. [3] (Figure 1). It consists of a 50% diameter reduction offset slightly in one direction to avoid numerical complexities associated with axisymmetric models. Our computational domain extended from three diameters (3D) upstream of stenosis throat to 24D downstream, in order to minimize boundary effects.

Although absolute dimensions do not matter for Newtonian simulations – Varghese et al. normalized their models to inlet diameter, and merely specified the Reynolds number (Re) – for non-Newtonian simulations it is crucial to define the model in physical units since viscosity depends on (dimensional) shear rate. A model diameter of 0.635 cm (i.e., ¼ inch) was chosen for experimental convenience, and corresponds roughly to the dimensions of a carotid or femoral artery. The viscosity of whole porcine blood was previously measured by Biswas [1] for shear rates up to 500 s^{-1}. These were fitted to a modified Cross model, having a high shear viscosity averaging 3.372 cPoise. Measured blood density averaged 1.085 g/cm^3

Transient flow simulations, with steady inflow, were performed for Re=500 to 1000 in increments of 100. Fully-developed inlet velocity boundary conditions were determined from the prescribed diameter and Re, and the experimentally-measured viscosity and density. For non-Newtonian simulations we followed the convention of defining Re based on the high shear viscosity, and this was used also for the Newtonian simulations. Resulting flow rates were 7.75-15.5 mL/s, resulting in mean inlet velocities of 24.5-48.9 cm/s

Simulations were carried using a high-performance, minimally dissipative Navier-Stokes solver [4]. No turbulence closure model was included. Meshes comprised roughly 2M linear tetrahedral elements, incorporating thin boundary layer elements. Following Varghese et al.,

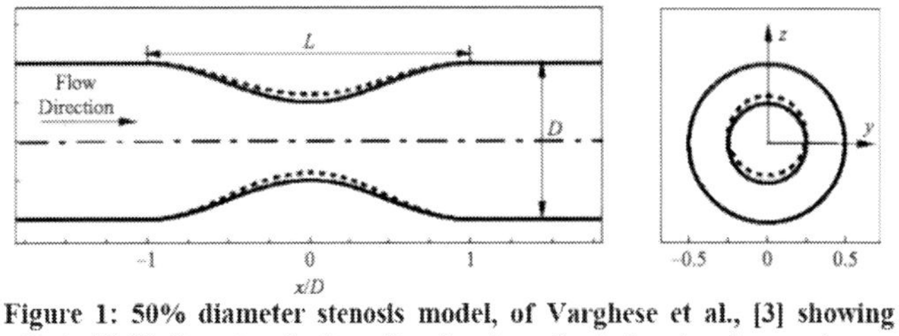

Figure 1: 50% diameter stenosis model, of Varghese et al., [3] showing offset of 0.05 diameters in the z direction in order to break axisymmetry.

meshes were densest in the region of jet between 6D and 14D, with a typical node spacing of 0.32 mm. Time-step size was set to ensure a CFL number less than 1 for all cases (e.g., 0.1 ms at Re=1000). Simulations were performed for at least 7.5 seconds to ensure washout of any initial transients. Simulations took on average about 48 hours each on 16 cores of an HPC cluster. Results were verified against Varghese et al.'s DNS simulations of Re=500 and Re=1000.

RESULTS

Instantaneous vorticity patterns shown in Figure 2 suggest that transition to turbulence occurs between Re=600-700 for the Newtonian case, but is delayed to Re=700-800 for non-Newtonian case. Notably, the non-Newtonian simulations for Re=700 suggest vortex-shedding phenomena, while the Newtonian simulation at this Re showed jet breakdown and apparent turbulence.

Mean velocity profiles shown in Figure 3 reveal a negligible effect of viscosity model except for Re-700, where the nominally laminar jet appears to persist to at least 16D for the non-Newtonian case, whereas for the Newtonian case a blunted velocity profile, characteristic of turbulence, is already evident by 12D.

RMS velocity profiles shown in Figure 4 reveal strong velocity fluctuations for both Newtonian and non-Newtonian cases at and above Re=800. At Re=700, the non-Newtonian case shows stronger RMS fluctuations than the Newtonian case, consistent with strong vortex shedding and incipient transition to turbulence as inferred from the vorticity plots. Weak RNS fluctuations at Re=600 for both cases also suggest some laminar jet instability.

DISCUSSION

Our findings indicate that transition to turbulence is delayed moderately by the non-Newtonian model, which mimics the shear-thinning behaviour of blood. The difference in critical Re appears to be roughly 15%, which is broadly consistent with that seen for Newtonian vs. blood flow in a straight tube [1]. This would seem to suggest that the shear-thinning properties of blood, as opposed to viscoelastic or other properties, are sufficient to explain the delayed transition.

As shown in Figures 2-4, excellent agreement was seen between our simulations and reference DNS solutions of Varghese et al. at Re=500 (laminar) and Re=1000 (turbulent) Newtonian cases, despite the fact that ours were carried out with 'only' 2M linear elements. Nevertheless, we intend the double the resolution by using second-order tetrahedra, in order to better resolve turbulence spectra.

The physiological or clinical relevance of these findings remains in question, as the stenosis geometry is highly idealized and only steady flow was considered. Further studies are planned to incorporate pulsatility, and more clinically relevant geometries.

ACKNOWLEDGMENTS

This work was supported by grant and salary awards from the Heart & Stroke Foundation of Canada, and a Resource Allocation from Compute Canada's SciNet HPC Consortium.

REFERENCES

[1] Biswas, D., PhD Thesis, University of Akron, 2014.
[2] Antiga, L., et al., Biorheology, 46(2), pp. 77-81, 2009.
[3] Varghese, S., et al., J Fluid Mech, 582, p. 253, 2007.
[4] Mortensen, M., et al., Comput Phys Comm, 188, pp. 177-188,
[5] Varghese, S. S., et al., J Biomech Eng, 130(1), p. 014503, 2008.

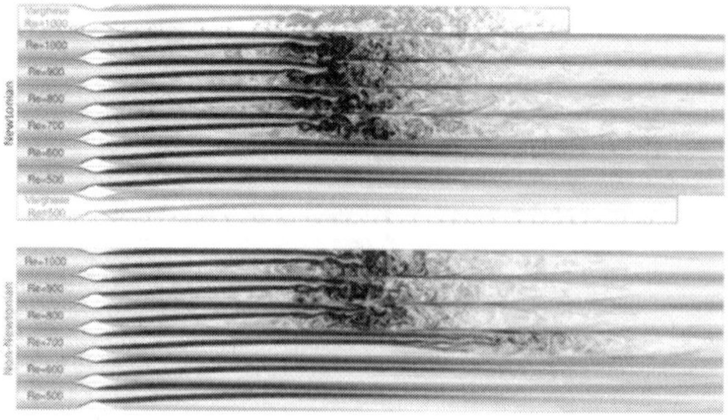

Figure 2: Instantaneous vorticity for Re=500-1000 for Newtonian (top) and Non-Newtonian (bottom) cases, showing the full extent of the CFD model domains. Shown also are Varghese et al.'s (Newtonian) DNS results for Re=500 and Re=1000, reproduced from [3] and [5], respectively.

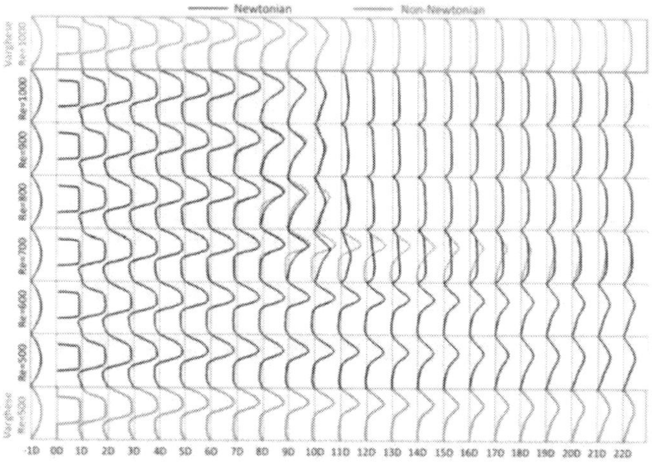

Figure 3: Velocity profiles spaced 1 diameter apart and averaged over 3 seconds of CFD data, normalized to inlet velocity, for Re=500-1000. Shown also are Varghese et al.'s (Newtonian) DNS results for Re=500 and Re=1000, reproduced from [3].

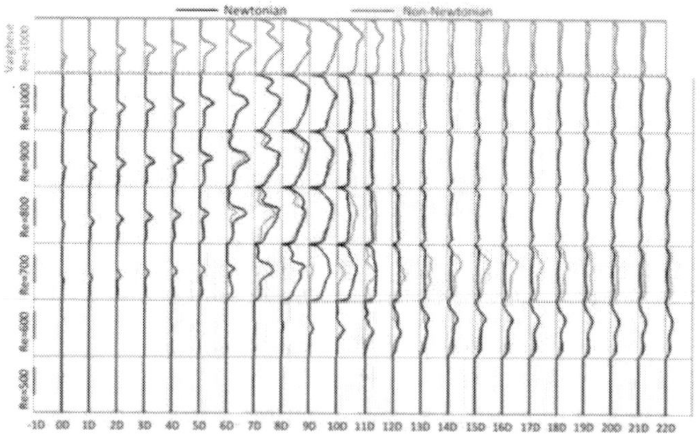

Figure 4: RMS velocities averaged over 3 seconds of CFD data and normalized to inlet velocity, for Re=500-1000. Shown also are Varghese et al.'s (Newtonian) DNS results for Re=500 and 1000, reproduced from [3].

THE IMPACT OF PARAMETER VARIATION, EXPERIMENTAL DATA AND UNCERTAINTY QUANTIFICATION FOR COMPLEX BIOMECHANICAL PROBLEMS EXAMPLIFIED FOR AAA

Wolfgang A. Wall (1), Jonas Biehler (1), Michael W. Gee (2)

(1) Institute for Computational Mechanics
Technische Universität München
Munich, Germany

(2) Mechanics & High-Performance
Computing Group
Technische Universität München
Munich, Germany

INTRODUCTION

Methodological advances in computational biomechanics in combination with the advent of relatively cheap high-performance computing platforms have led to a tremendous improvement of computational models for various diseases and mechanobiological processes. In fact, for some specific medical conditions, computational models have matured to the verge of clinical application.

The use of computational models in a clinical setting for diagnostic and predictive purposes of course requires truly patient-specific models. Unfortunately, for most, if not all applications, not all of the required patient-specific model parameters can be determined to the necessary accuracy by non-invasive measurement techniques. For example, in the context of cardiovascular structures, such as abdominal aortic aneuryms (AAA), the overall patient-specific geometry can be determined using state-of-the-art CT images. However, the patient-specific thickness of the aortic wall, which also exhibits regional variations across the AAA surface, cannot be determined from the CT data. Moreover, the exact constitutive properties cannot be determined from the image data or any other non-invasive technique. Furthermore, studies in which the aforementioned parameters have been determined by invasive experiments have shown that many of these parameters exhibit significant inter- as well as intra-patient variations.

In spite of these sometimes drastic variations, most researchers have hitherto condensed the extensive experimental findings to population averaged mean values, which are then used as input parameters for computational models. While the use of deterministic models with averaged parameters has its merits, the predictive capability of these models is often limited. We believe that the predictive capabilities of computational simulations can be enhanced if the aforementioned variations are considered in the computational assessment. This can be achieved by using a probabilistic description for the varying parameters and an efficient uncertainty quantification (UQ) framework. Thereby, the available experimental data should be used to construct the probabilistic description of the uncertain parameters. In this work we present a new approach capable of UQ for large and complex biomechanical models and exemplarily study the impact of parametric uncertainties in computational models of AAAs. Specifically, we investigate the impact of uncertainties in constitutive parameters and wall thickness on mechanical quantities typically related to AAA rupture risk, such as wall stress or strain [1]. The probabilistic description for the uncertain parameters is thereby, to a large extend, determined from extensive experimental data.

METHODS

The large nonlinear patient-specific finite element models in combination with the typically high stochastic dimension render current non-sampling based UQ approaches infeasible. In order to alleviate the computational burden that is posed even by advanced sampling based UQ methods, we use a novel approach that employs approximate models in combination with Bayesian formulations [2,3] The use of a non-parametric Bayesian regression model allows to not only establish a quantitative connection between an approximate model and an accurate and expensive high-fidelity model, but also provides confidence intervals.

With regard to the probabilistic modeling of the input parameters, we focus here on inter- and intra-patient variations of the constitutive parameters and the thickness of the aneurysmatic arterial wall, which are modeled by three dimensional, non-Gaussian random fields. The parameters of these random fields are then estimated based on our own experimental data. Two different data driven approaches to obtain the parameters of the random fields are investigated. In the first approach, these parameters are estimated based on population statistics of the measured quantities directly. The other option we pursue in this work harnesses information encoded in supplementary data using machine learning techniques. Together with experimentally obtained constitutive parameters and wall thickness measurements from more than 200 tensile tests, we have collected a set of non-invasively accessible patient-specific medical data for each of the tested samples. The data is comprised of the patients medical history and the results of biochemical blood analysis. Using this data as explanatory variables in a Bayesian regression analysis yields patient-specific, probabilistic predictions for the constitutive parameters and the wall thickness [4].

RESULTS

The flexibility and capabilities of the UQ method are demonstrated by performing uncertainty quantification with patient-specific nonlinear finite element models of AAAs [3]. We show that the computational cost for UQ with a patient-specific AAA model can be reduced to the equivalent of only a moderate number of runs of the high-fidelity model. Furthermore, this approach speeds up the computation such that results can be obtained within one day even for large forward models.

We demonstrate that the uncertainty in some of the patient-specific input parameters can have a significant impact on, e.g., wall stress. This is shown in the contour plots in Figure 1, which depicts the computed von Mises stress for six different Monte Carlo samples. In this example the wall thickness is modeled as a random field, the variance of which is determined from experimental measurements. Thus, the random field provides a realistic estimate of the variability of the wall thickness.

Moreover, we show that supplementary data can be used to create more accurate probabilistic models of uncertain mechanical input parameters.

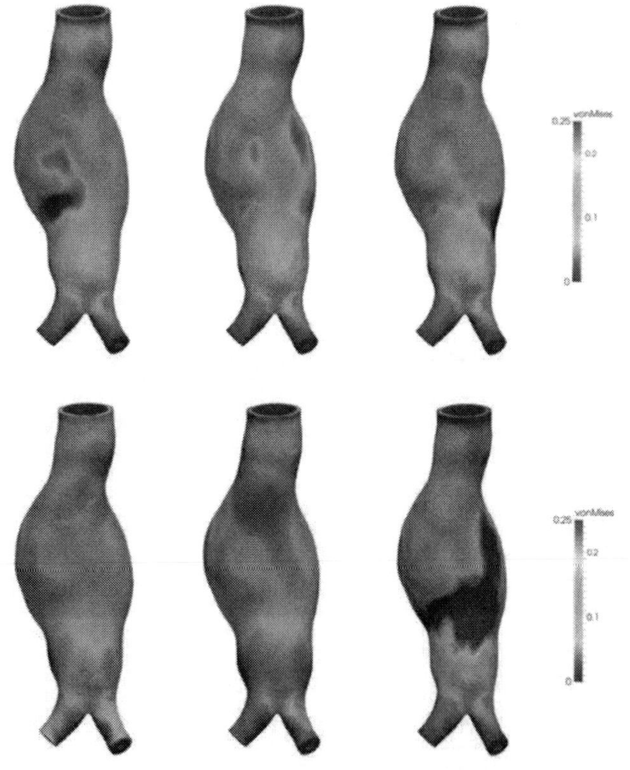

Figure 1. Contour plot of computed Von Mises stress for six Monte Carlo samples.

By harnessing information encoded in non-invasively available data, several patient-specific input parameters can be estimated more accurately. In addition, the remaining uncertainty, i.e., variance of these patient-specific input parameters can be significantly reduced. The reduced variance in the input in turn leads to reduced uncertainty in the computed mechanical output quantities, thus enhancing the predictive accuracy of the mechanical model.

DISCUSSION

In this work, we present a framework for efficient UQ in large-scale, nonlinear biomechanical problems with high stochastic dimension. This allows us, for the first time, to perform UQ using patient-specific models with respect to multiple uncertain parameters, which are modeled as random fields. Moreover, we demonstrate the utility of a data driven approach for the construction of patient-specific random fields for uncertain model parameters.

While this work exemplarily considers models of AAAs, we note that the presented approach is very general and can easily be applied to other biomechanical problems. We believe that the potentially very significant impact of uncertainties in patient-specific model input parameters on computed mechanical quantities, renders UQ a necessity when patient-specific biomechanical problems are studied.

ACKNOWLEDGEMENTS

We gratefully acknowledge the help of our clinical partners at the Department of Vascular and Endovascular Surgery, Klinikum rechts der Isar, Technischen Universität München. The countless discussion we had and their tremendous effort to collect the necessary clinical data made this study possible.

REFERENCES

[1] Maier, A., Gee, M.W., Reeps, C., Pongratz, J., Eckstein, HH., Wall, W.A., A Comparison of Diameter, Wall Stress, and Rupture Potential Index for Abdominal Aortic Aneurysm Rupture Risk Prediction, *Annals of Biomedical Engineering*, 10, 38:3124-3134, 2010.

[2] Koutsourelakis, PS., Accurate Uncertainty Quantification Using Inaccurate Models, *SIAM Journal of Scientific Computing*, 5, 33:3274-3300, 2009.

[3] Biehler, J., Gee, M.W., Wall, W.A., Towards Efficient Uncertainty Quantification in Complex and Large Scale Biomechanical Problems Using Approximate Models and Bayesian Formulations *Biomechanics and Modeling in Mechanobiology* 2014 doi:10.1007/s10237-014-0618-0

[4] Biehler, J., Kehl, S., Gee, M.W., Tanios, F., Pelisek, J., Maier, A., Reeps, C., Eckstein, HH., Wall, W.A., Non-Invasive Prediction of Wall Properties of Abdominal Aortic Aneuryms Using Bayesian Regression, 2015, in preparation

AN IN SILICO BIOMECHANICAL ANALYSIS OF THE STENT-ESOPHAGUS INTERACTION

M. Peirlinck[1], B. Verhegghe[1,2], P. Segers[1], M. De Beule[1,2]

[1]IBiTech-bioMMeda
Ghent University
Ghent, Belgium

[2]FEops bvba
Ghent University
Ghent, Belgium

INTRODUCTION

Endoscopic placement of a self-expanding stent is an established treatment for both malignant and benign esophageal strictures [1]. Over the last 20 years, esophageal stents have progressed from rigid polyvinyl plastic or rubber tubes to self-expandable metallic, plastic and biodegradable stents, whether or not semi- or fully covered. Despite these technological innovations, the association between the mechanical behavior of the stent, its design and its effect on the clinical outcome has not yet been thoroughly explored. Previous research focused on device characteristics measurable in an in-vitro experimental set-up such as radial force, chronic outward force and flexibility. It is however difficult to relate these factors to clinical outcome as more factors are involved in in-patient stent behavior.

We therefore developed a numerical framework in which a (biodegradable) polymeric braided wire stent gets virtually implanted in an esophageal anatomy.

METHODS

In order to perform this virtual stent implantation, both a finite element model of an esophageal stent as well as of the esophagus itself had to be developed.

Stent model

In this study, we focused on the *Ella-BD* stent (ELLA-CS, Czech Republic), the first bioresorbable esophageal stent introduced in Europe. The *Ella-BD* stent consists of bioresorbable polydioxanone wires braided into a flared stent structure. In the past, analytical and numerical models have been developed and validated for metallic braided wire stents [2, 3]. Due to the typical large fiber thickness of polymeric braided wire stents (PBWS), the steric interaction and friction between the stent wires cannot be neglected, as was done in the aforementioned models [4]. For that reason, a finite element model (FEM) of this PBWS was developed (Figure 1). In order to calibrate this model, in-vitro stent compression and tensile tests were performed to determine the radial strength of the stent and the elasticity of the polydioxanone wires. An in-vitro degradation study was performed, in which the full stent and separate stent wires were conserved in a phosphate buffered solution (pH=7.2) at 37°C for 10 weeks. By comparing the model's predicted evolution of radial strength though time with weekly radial compression tests of the degradation stent, the model was validated.

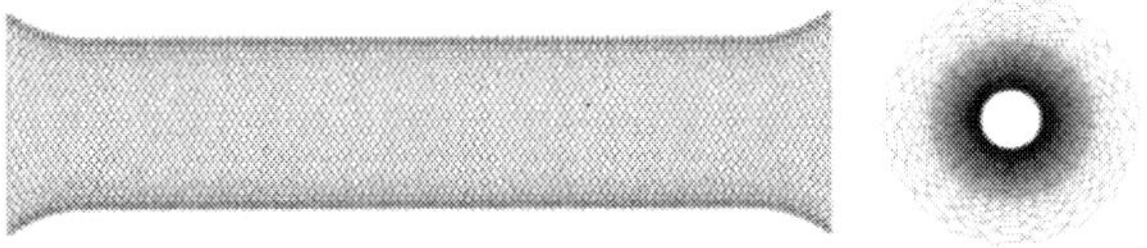

Figure 1. Stent model

Esophagus model

Analysis of the real clinical stent behavior in vivo and its influence on clinical outcome requires knowledge of the distribution of stresses and strains in the esophageal layers. We developed a simplified model of the esophagus in which we virtually implanted a PBWS. The esophageal wall was modeled as a two-layered system [5] with both the mucosal and muscle layer having their own specific material constants. We virtually crimp a 100 mm long ø 25/20/25 mm polymeric braided wire stent to the catheter, maneuver the crimped stent according to the position of an inserted guidewire and release the stent by pulling back the crimping sheet. The finite element mesh model was generated using the in-house developed geometrical and FEM pre- and postprocessor *pyFormex* (Ghent University, Belgium).

RESULTS

The in-vitro stent compression tests and tensile tests were applied to calibrate the stent model and were validated with the results from the in-vitro degradation study (Figure 2).

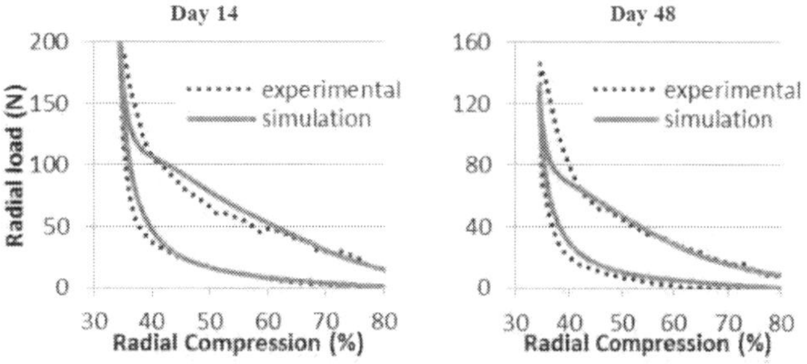

Figure 2. Validation of the stent model

The developed numerical framework gave specific insights in the in vivo mechanical behavior of the chosen PBWS's design and size on the esophageal wall. Detailed knowledge of the distribution of stresses and strains in the esophageal layers have been obtained (Figure 3).

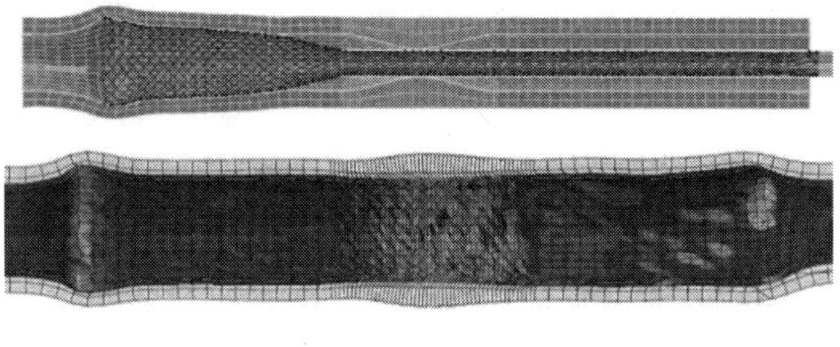

Figure 3. Virtual stent implantation procedure and pressure exerted by the stent on the esophageal wall (blue-red: 0-40mmHg)

DISCUSSION

This study is the first to study the mechanical behavior of esophageal stents in an in vivo setting. Previous research (e.g. [1]) focused on the assessment of the in vitro radial and axial force and hypothesized the influence of these experimental measurements on the actual in vivo mechanical behavior of the esophageal stent. The behavior of a stent in the esophagus can however not be simply described by radial and axial force data alone. Other factors such as the used material, the wire diameter, the cell size, the braiding angle and the presence of specific anti-migration stent end designs (e.g. flares, anti-migration collars) affect the stent's in vivo behavior.

As the stress distribution within the esophageal wall also depends on the geometry, we are now elaborating this framework with a more detailed patient-specific esophagus model. This model is based on imaging data of a 69-year-old man suffering from a malignant esophageal tumor (6cm long circumferential esophageal carcinoma starting just above the aortic arch). A thoracic CT acquisition of this patient was collected and segmented using *Mimics* software (Materialise, Belgium). The data is not detailed enough to allow separate segmentation of the inner mucosal and outer muscular layer, so we have to generalize a cross-sectional layer distribution, which can be seen in Figure 4. As the patient was not swallowing during image acquisition, an internal resting pressure of 7.5 mmHg was applied on the inner mucosal wall [6]. The developed finite element model is passive and no active muscle contraction is implemented.

We set up (and are further developing) this framework to improve our understanding of the occurrence of specific symptoms and complications following stent placement as it will provide a detailed and multi-factorial insight in the in vivo stresses and strains caused by a specific stent design in a pathological patient-specific esophageal wall. Ideally, clinicians/gastroenterologists make use of this framework in combination with high-quality randomized controlled clinical trials to explain the possible adverse events seen in specific patient populations and cases. In collaboration with technical engineers and stent manufacturers, we genuinely believe this could lead to less complications, better stent designs and better guidelines on which stent design and size are most appropriate for specific patient groups and pathologies.

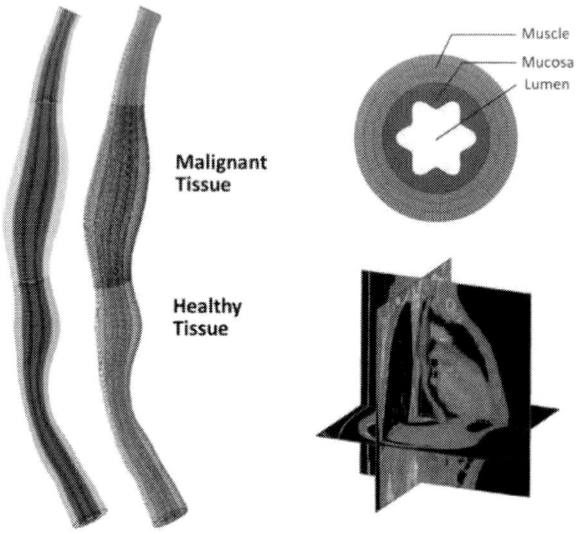

Figure 4. Patient-specific esophagus model

ACKNOWLEDGEMENTS

We thank prof. dr. Peter D. Siersema and dr. Wouter Kappelle (Department of Gastroenterology and Hepatology, University Medical Center Utrecht) for supplying the *Ella BD* stents and the imaging data required for this study.

REFERENCES

1. Hirdes, M.M., et al., *In vitro evaluation of the radial and axial force of self-expanding esophageal stents.* Endoscopy, 2013. **45**(12): p. 997-1005.
2. De Beule, M., *Finite element stent design*. 2008.
3. Jedwab, M.R. and C.O. Clerc, *A Study of the Geometrical and Mechanical-Properties of a Self-Expanding Metallic Stent Theory and Experiment.* Journal of Applied Biomaterials, 1993. **4**(1): p. 77-85.
4. Nuutinen, J.P., C. Clerc, and P. Tormala, *Theoretical and experimental evaluation of the radial force of self-expanding braided bioabsorbable stents.* Journal of Biomaterials Science-Polymer Edition, 2003. **14**(7): p. 677-687.
5. Liao, D., et al., *Two-layered quasi-3D finite element model of the oesophagus.* Med Eng Phys, 2004. **26**(7): p. 535-43.
6. Nicosia, M.A., et al., *Local longitudinal muscle shortening of the human esophagus from high-frequency ultrasonography.* Am J Physiol Gastrointest Liver Physiol, 2001. **281**(4): p. G1022-33.

SB³C2015
Summer Biomechanics, Bioengineering and Biotransport Conference
June 17-20, 2015, Snowbird Resort, Utah, USA
SB³C2015-366

APPLICATION OF GOLD PARTICLE ENHANCED CT FOR VULNERABLE PLAQUE DETECTION AND QUANTIFICATION IN ATHEROSCLEROTIC MOUSE MODELS

David De Wilde (1), Bram Trachet (1,2), Carole Van der Donckt (3), Bert Vandeghinste (1), Benedicte Descamps (1), Christian Vanhove (1), Guido R. Y. De Meyer (3), Patrick Segers (1)

(1) Department of Electronics and Information Systems
Ghent University-iMinds-IBiTech
Gent, Belgium

(2) Institute for Bioengineering
Ecole Polytechnique Fédérale de Lausanne
Lausanne, Switzerland

(3) Division of Physiopharmacology
University of Antwerp
Wilrijk, Belgium

INTRODUCTION

Atherosclerosis is a chronic inflammatory disease and often the underlying cause of myocardial infarction and stroke. One of the major clinical challenges is the need to differentiate between vulnerable plaques, which are likely to rupture and need immediate treatment, and stable plaques, which do not cause an immediate threat. The apolipoprotein E deficient ($ApoE^{-/-}$) mouse model is one of the most popular mouse models in preclinical research to study stable atherosclerotic plaques. Recently, a novel mouse model with a heterozygous mutation in the fibrillin-1 gene ($ApoE^{-/-}Fbn1^{C1039G+/-}$) has been described to develop vulnerable plaques and plaque rupture when on a Western type diet [1].

Both stable and unstable plaques have a lipid core which consists of lipid accumulating in the artery wall. Plaques are also infiltrated by macrophages and other inflammatory cells. As the presence of a high number of macrophages is one of the indicators of the unstable plaque phenotype, quantifying the presence of macrophage is an important research target for small animal imaging.

The research on molecular imaging of macrophages has focused mainly on magnetic resonance imaging, positron emission tomography and ultrasound imaging. However, the resolution of these scanners is limited for capturing the presence of atherosclerotic plaque in the smaller carotid arteries. Computed tomography (CT) imaging on the other hand is mainly used for anatomic imaging. However, in a previous pilot study we noted that AuroVist (Nanoprobes, Yaphank, NY), a gold nanoparticulate contrast agent with polyethylene glycol coating, migrated to plaque locations in these mice. Therefore, in the present study the goal was to explore the possibility of macrophage detection in murine plaques making use of the high-precision nature of micro-CT. A second goal was to employ the advantages of this technique to quantify differences in plaque morphology between $ApoE^{-/-}$ and $ApoE^{-/-}Fbn1^{C1039G+/-}$ mice.

METHODS

Experimental Protocol

Female, 6 weeks old $ApoE^{-/-}Fbn1^{C1039G+/-}$ mice (n=14) and $ApoE^{-/-}$ mice (n=9) were put on a Western type diet for 20 weeks. At weeks 10, 15 and 20 after the start of the diet, they underwent a contrast-enhanced micro-CT imaging protocol using a Triumph-II imaging system (TriFoil Imaging, Chatsworth, CA, USA). The protocol consisted of a pre-contrast injection scan at a resolution of 50 µm (figure 1-A). This was followed by an injection of 100 µl/25g body weight of AuroVist and a post-contrast injection scan (figure 1-B).

Image Processing – Quantifying Accumulation of Contrast

A rigid image registration step was performed to fit the pre-contrast image to the post-contrast image and get rid of possible movement artifacts. The common (CCA), internal (ICA) and external (ECA) carotid arteries were segmented in VMTK (www.vmtk.org). The segmented surface was then smoothed and triangulated with triangle areas ranging from 50 to 150 µm². The length of the ICA up to the branching of the stapedial artery was taken as a reference length. For every carotid bifurcation, the CCA, ECA and ICA were cut off at respectively 2, 1 and 0.95 times the reference length.

Surface triangles were categorized into (i) a group where contrast agent was present in the artery wall from a previous AuroVist injection at that position and (ii) a group where no contrast agent was present. This was done by calculating the normal at every triangle and uniformly sampling 100 points per normal in the pre-contrast image up to 300 µm from the lumen. The sampling points were assumed to be in a contrast-enhanced region if the intensity was higher than 350HU. By plotting the amount of sampling points with contrast present, a visual representation of the locations with contrast in the artery wall could be visualized on the luminal border (figure 1-C). A surface triangle was placed in the contrast agent accumulation group if more than 25% of the corresponding sampling points showed signs of contrast presence. Finally, for the three branches the contrast-enhanced area fraction (CAF [$\%_{area}$]) was calculated and a linear mixed model statistical analysis was performed to show the most significant factors influencing plaque development.

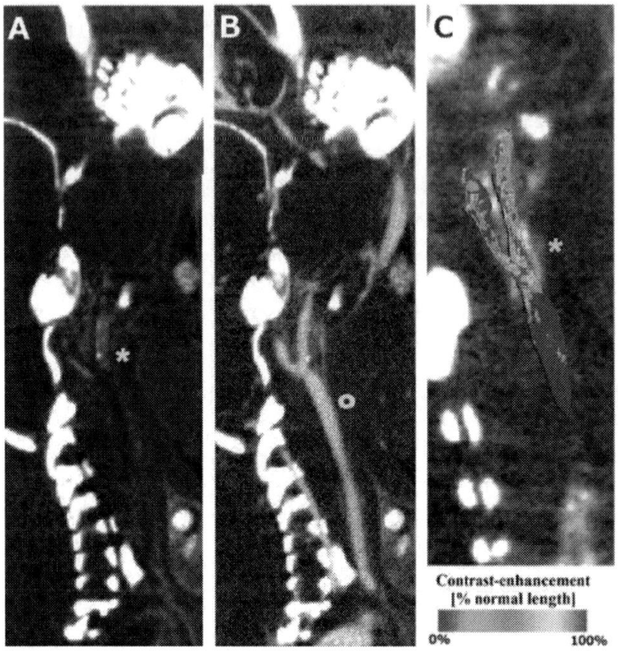

Figure 1: (A) A pre-contrast injection image at week 20. The contrast in the artery walls (*) from previous contrast injections (at week 10 and 15) is clearly visible around the carotid bifurcation. (B) The post-contrast injection image with the visible arteries (o). (C). Rendering of the lumen surface. The presence of contrast (*) in the artery wall along the surface normals is indicated in color.

RESULTS

At weeks 15 and 20 AuroVist from the previous injections was visible in the artery wall. The hypothesis that AuroVist particles are taken up by macrophages was confirmed by histology, with AuroVist particles abundantly present in macrophage rich regions. An overview of the area fractions with enhanced contrast (CAF) is given in figure 2. The highest contrast agent uptake occurred in the ECA at weeks 15 and 20 after the start of the diet. As expected, higher values were obtained for the $ApoE^{-/-}Fbn1^{C1039G+/-}$ mice compared to the $ApoE^{-/-}$ mice. Moreover, there was a clear increase in area affected by contrast accumulations over time. Statistical analysis showed that location, (CCA-ECA-ICA; $P<0.01$), time point (week15-week20; $P<0.01$) and mouse strain ($P<0.05$) were significant factors. No difference was found between the left and the right carotid arteries.

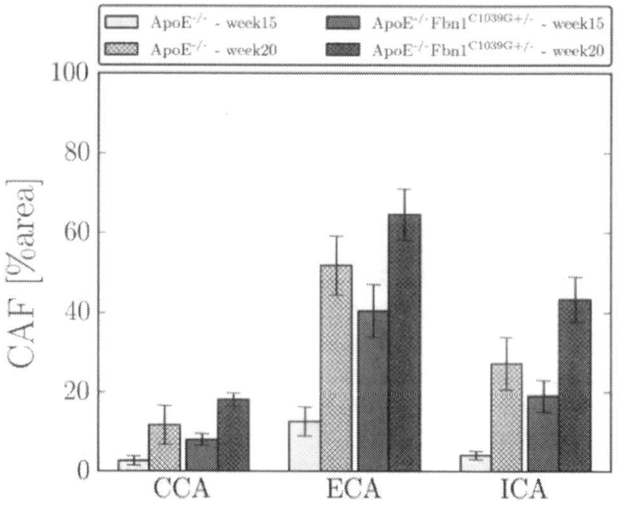

Figure 2: The contrast area fraction with contrast agent present for both strains of mice ($ApoE^{-/-}$ and $ApoE^{-/-}Fbn1^{C1039G+/-}$) at 2 time points (weeks 15 and 20) and at 3 different locations (CCA, ECA and ICA).

DISCUSSION

We have demonstrated that a technique based on successive contrast-enhanced micro-CT imaging allows to visualize and quantify AuroVist accumulations in the artery wall. Some notable differences between the branches of the carotid bifurcation were found. The regions where contrast was present are bigger in the ECA and ICA compared to the CCA. We speculate that this is linked to locally altered hemodynamics, which are known to promote macrophage infiltration. The flow distally from the bifurcation is likely to be more disturbed and this will induce more spread-out patterns in the ECA and ICA.

This technique allowed us to use high-resolution in-vivo micro-CT scans for imaging the locations of macrophage infiltration in atherosclerotic plaques. This approach has thus potential for being used in preclinical studies about plaque development. Moreover, the methodology requires only a limited amount of user interaction and this interaction has no influence on the results.

The main limitation of the proposed methodology is that animals are subjected to several micro-CT scans. This exposes them to a substantial radiation dose. The method is well-suited for mapping the zones influenced by macrophage infiltration but no depth information is preserved. Therefore it is less suited to study outward remodeling.

In conclusion, we propose a promising new tool based on in-vivo AuroVist-enhanced micro-CT to visualize macrophage infiltration. We applied the methodology in a longitudinal follow-up study in two different atherosclerotic plaque mouse models.

ACKNOWLEDGEMENTS

David De Wilde and Carolve Van der Donckt are recipients of research grants of the IWT. Christian Vanhove is supported by the GROUP-ID consortium. This research has benefitted from a statistical consult with Ghent University FIRE. Furthermore, we thank Scharon Bruneel, Francesco Iannaccone, Annette Caenen and Bieke Van der Veken for their assistance.

REFERENCES

[1] Van Herck, JL et al., *Circulation*, 120:2478-2487, 2009

SB^{3}C2015
Summer Biomechanics, Bioengineering and Biotransport Conference
June 17-20, 2015,Snowbird Resort, Utah, USA
SB^{3}C2015-368

IN VITRO CORONARY ARTERY MODEL FOR THE STUDY OF ATHEROSCLEROSIS

Elizabeth E. Antoine, Abdul I. Barakat

Hydrodynamics Laboratory (LadHyX)
Department of Mechanics
Ecole Polytechnique
Palaiseau, France

INTRODUCTION

The pathological complications of atherosclerosis, namely heart attacks and strokes, are the leading cause of mortality in the world. A common treatment for atherosclerosis is the use of endovascular stents to open blocked arteries. While broadly successful, stent expansion triggers a complex set of biological responses which leads in some cases to in-stent restenosis (ISR) or late stent thrombosis (LST).

Recent computational studies have focused on understanding and optimizing stent performance [1]. Furthermore, *in vitro* experiments have elucidated elements of the stent biological response cascade. Examples include endothelial cell (EC) dysfunction due to alterations in fluid shear stress such as found around stent struts [2] and altered phenotype of smooth muscle cells (SMCs) due to stent-induced loss of integrity of the adjacent endothelium [3]. Such studies have contributed greatly to our understanding of post-stent complications.

Elucidating the complex biological pathways involved in stent pathologies under physiological conditions remains a major challenge. For example, EC-SMC interactions in response to wounding have been studied in classical monolayer cultures; however, the similarity of such experiments to *in vivo* conditions is limited [4]. Numerical models constructed from studies of isolated phenomena *in vitro* must be validated experimentally; again, conditions similar to those found *in vivo* are necessary for meaningful results. Such validation necessitates the development of advanced experimental platforms and techniques.

In this work we present early results for a novel *in vitro* stent optimization platform which provides a scale model of the human coronary artery based on a collagen matrix, ECs and SMCs, and perfusion representative of coronary flow. The device incorporates particle image velocimetry (PIV) for direct measurement of flow at the scale of individual cells. The platform enables observation of biological response to stent insertion in real time and will be used for validation of numerical predictions as well as testing of novel stent designs.

METHODS

Collagen was acid-extracted from rat tail tendon as described previously [5] and reconstituted with 10X concentrated Dulbecco's modified Eagle's medium (DMEM) (Sigma), buffered DMEM (Sigma), sodium hydroxide (Sigma), and bovine aortic SMCs (Tebu-Bio) to obtain a solution at pH 8.4 containing 6 mg/ml collagen and 5×10^5 cells/ml. Following previous work [6], the solution was incubated at 37°C to achieve a polymerized fiber network with a mean pore diameter of 1.3 μm and compression modulus of 4600 Pa.

For long-term co-culture studies, 500 μl volumes of collagen solution were polymerized in 24-well chambers, resulting in hydrogels 15.6 mm in diameter and 2.6 mm thick. Following polymerization, bovine aortic ECs (generously provided by Georges Pompidou Hospital, Paris) were seeded on the surface of the hydrogels at confluent density (1×10^5 cells/cm^2). Hydrogels were cultured 14 days at 37°C and 5% CO_2 in bovine SMC growth medium (Tebu-Bio) which was replaced every 2-3 days. SMC-containing hydrogels without the EC monolayer were used as controls. After 14 days, cell viability was determined using a live/dead fluorescence assay (Calcein AM/Ethidium homodimer-1, Invitrogen).

For coronary platform studies, SMCs were tagged with a red fluorescent label (Vybrant CM-DiI) and suspended in a collagen solution at a density of 1×10^6 cells/ml. A length of fluorinated ethylene propylene (FEP) tubing (Cole-Parmer) with a diameter of 4.6 mm and length of 60 mm was treated with a combination of plasma, 1% polyethyleneimine, and 0.1% glutaraldehyde to render it hydrophilic. The FEP was filled with the collagen/SMC solution and fitted with end

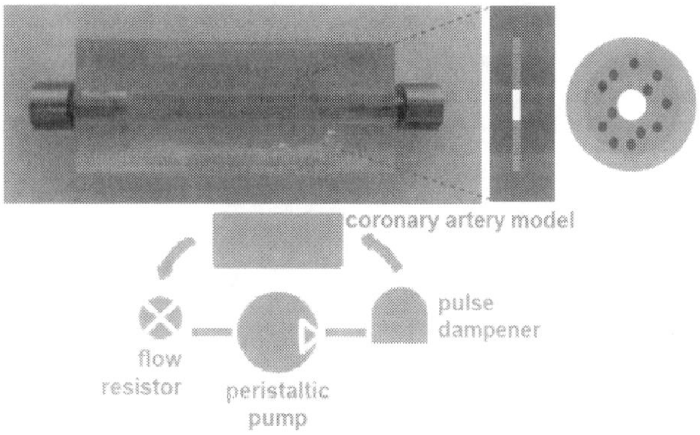

FIGURE 1: ILLUSTRATION OF DEVICE. TOP: CORONARY ARTERY MODEL IN IMAGING CHAMBER (RIGHT) WITH ENLARGED VIEW (CENTER) AND DIAGRAM (LEFT) INDICATING FEP (GRAY), HYDROGEL (ROSE) CONTAINING SMC (RED), AND CENTRAL CHANNEL (WHITE). BOTTOM: SCHEMATIC OF PULSATILE FLOW LOOP USED TO GENERATE CORONARY FLOW.

caps with a central pin. After polymerization at 37°C, the pin was removed, leaving a central channel of 1.5 mm diameter within the cell-seeded hydrogel. Imaging with an epifluorescent microscope (Ti-E, Nikon) was enabled by placing the device into an external housing and submerging it in PBS for refractive index matching across the FEP (Figure 1, top). The device was perfused with PBS at a Reynolds number of 300 and Womersley number of 3.5, conditions mimicking coronary artery flow (Figure 1, bottom). After 3 hours, cell viability was determined by incubation with Calcein AM.

To determine the feasibility of accurate flow measurement in this complex system, the circulating fluid was seeded with 1 μm red fluorescent particles (R0100, Thermo Scientific) and arbitrary low-Reynolds number steady flow was perfused through the device. Particle images were acquired at 500 ms intervals using a binned CCD camera (Retiga 1300C, Qimaging) with 4X magnification and mercury lamp illumination (Intensilight, Nikon). Following previous work, these images were analyzed using advanced PIV algorithms to quantify the instantaneous flow field [5]. Median background subtraction and high-pass filtering were applied to increase signal-to-noise ratio and reduce effects of uneven illumination. Correlation was performed using single-pass robust phase correlation with windowing and UOD validation.

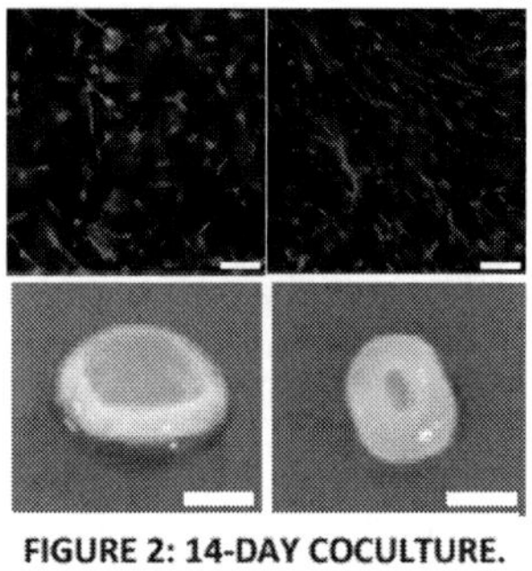

FIGURE 2: 14-DAY COCULTURE. TOP: VIABILITY (LIVE: GREEN; DEAD: RED) OF SMC CONTROL (LEFT) OR COCULTURE WITH CONFLUENT EC (RIGHT). SCALE BAR: 100 μm. BOTTOM: MACROSCALE IMAGES OF CONTROL (LEFT) AND COCULTURE (RIGHT) HYDROGELS. SCALE BAR: 5 mm.

RESULTS

SMCs were successfully cultured up to two weeks in 3D collagen hydrogels (Figure 2). Co-culture with an EC monolayer led to a phenotypic change in SMCs, identified by significant elongation of cells (Figure 2, top) as well as significant contraction of the hydrogels after 14 days (Figure 2, bottom). While 3 hours is insufficient for SMC elongation or phenotypic changes, the high colocalization of green with red signal in Figure 3 indicates similarly high SMC viability in the coronary artery model.

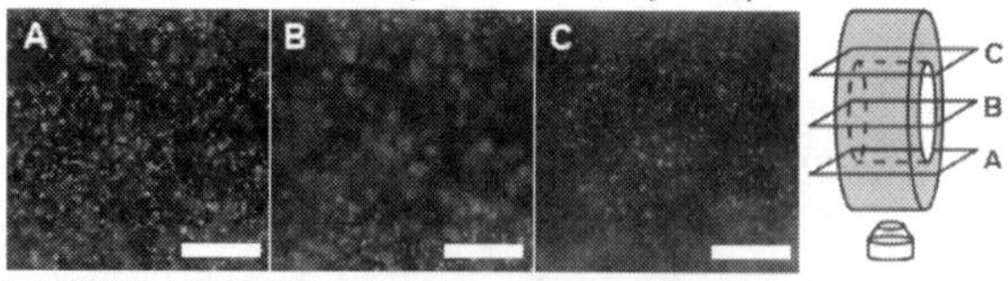

FIGURE 3: VIABILITY OF SMC IN CORONARY PLATFORM. ALL CELLS FLUORESCE RED (VYBRANT CM-DII) WHILE ONLY LIVE CELLS FLUORESCE GREEN (CALCEIN AM). IMAGES WERE ACQUIRED AT FOCAL PLANES BELOW, ABOVE, AND THROUGH THE CHANNEL LUMEN AS INDICATED IN THE SCHEMATIC. SCALE BAR: 250 μm.

Using only a single-pass PIV algorithm with minimal validation and no smoothing, instantaneous velocity fields at a resolution of 27 μm in the radial axis and 106 μm in the longitudinal axis were obtained. The resulting velocity profiles closely match the parabolic flow expected for steady low-Reynolds number pipe flow (Figure 4).

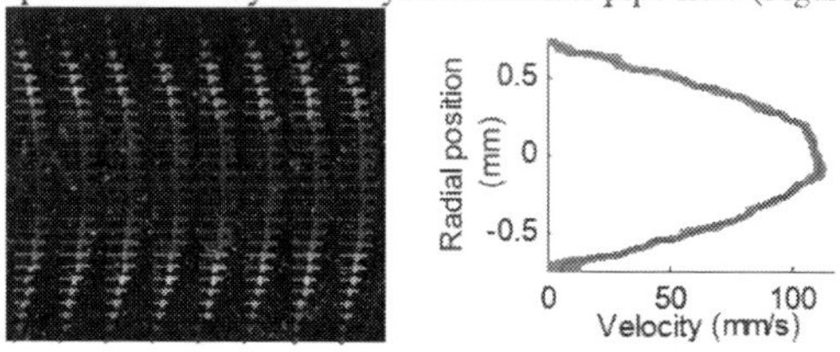

FIGURE 4: FLOW QUANTIFICATION. LEFT: PARTICLE IMAGE OVERLAID WITH INSTANTANEOUS VELOCITY FIELD. 1 IN 4 VECTORS SHOWN. RIGHT: LONGITUDINALLY-AVERAGED INSTANTANEOUS VELOCITY PROFILE (LINE) ± 1 STD DEV (SHADED).

DISCUSSION

Here we report a novel *in vitro* coronary artery platform designed for studies of biological response to stenting as well as validation of stent optimization numerical models. The 3D collagen-based device closely reproduces *in vivo* dimensions, spatial configuration, and flow parameters while enabling live cell imaging and flow quantification. Preliminary co-culture experiments indicate the potential for the device to sustain cells for several weeks while providing necessary biological cues for *in vivo*-like phenotypic changes. In future work, ECs will be introduced to the coronary artery platform and long-term studies will be performed to fully validate its performance. Furthermore, integration of a pulsed laser and high-speed camera will be completed to enable velocimetry and shear stress measurements at physiological flow rates. With these elements in place, we will use the device to study EC-SMC interactions in real time following balloon procedures and stent insertion.

ACKNOWLEDGEMENTS

E. Antoine is funded by a Whitaker Postdoctoral Fellowship. The research is partially supported by a permanent endowment in Cardiovascular Cellular Engineering from the AXA Research Fund.

REFERENCES

[1] Bozsak, F et al., *Biomech Model Mechanobiol*, 13:327-347, 2014.
[2] Fisher, A et al., *Am J Physiol Lung Cell Mol Physiol*, 281(3)L529-L533, 2001.
[3] Campbell, J & Campbell, G, *Arterioscler Thromb Vasc Biol*, 32:1784-1789, 2012.
[4] Truskey, G, *Int J High Throughput Screen*, 1:171-181, 2010.
[5] Buchanan, C et al., *Tissue Eng Part C Methods*, 20(1):64-75, 2014.
[6] Antoine, E, Doctoral dissertation, Virginia Tech, 2014.

EFFECTS OF MECHANICAL CONSTRAINTS ON CELL-GENERATED STRESS AND COLLAGEN REMODELING IN STATICALLY CULTURED MICROTISSUES

Mathieu A.J van Kelle (1,2), Sandra Loerakker (1,2), Inge A.E.W. van Loosdregt (1,2), Carlijn V.C. Bouten (1,2), Frank P.T. Baaijens (1,2)

(1) Department of Biomedical Engineering
Eindhoven University of Technology
Eindhoven, The Netherlands

(2) Institute for Complex Molecular Systems
Eindhoven University of Technology
Eindhoven, The Netherlands

INTRODUCTION

Cell-generated stress is an important factor in heart valve tissue engineering. On the one hand, cell-generated stresses that arise when the leaflets are attached to each other during heart valve *in vitro* culture, contribute to enhanced tissue formation [1]. On the other hand, after implantation and *in vivo* remodeling, retraction of the leaflet occurs due to these cell-generated stresses. A better understanding of the relationship between cell-generated stress and tissue remodeling will help to improve tissue engineering strategies, and prevent cell-mediated leaflet retraction. The aim of this study is to investigate the effect of mechanical constraints on cell-generated stress and collagen remodeling.

METHODS

In order to investigate this relationship, a microfabricated tissue gauge (μTUG) setup was used. This setup consists of 4 compliant polydimethylsiloxane (PDMS) microposts embedded in a PDMS microwell (figure 1A). Cells seeded into these wells self-assemble into a microtissue, which compacts around the microposts (figure 1C). In our setup the stiffness of the PDMS posts was varied (0.77, 1.24 and 1.70 MPa) by blending two types of commercially available PDMS (Sylgard 184 and 527) [2]. This μTUG system was used to: (1) predict cell-generated stresses using a finite element (FE) model, (2) investigate collagen remodeling using experiments, in response to the varying mechanical constraints and measure total cellular forces applied to the microposts.

The FE model predicted cellular actin stress fiber development, depending on the strain and strain-rate in these fibers [3]. These stress fibers then apply traction forces on their surroundings, which is modeled as a mixture of collagen fibers and other isotropic tissue components, attached to PDMS microposts (figure 1B). The stiffness

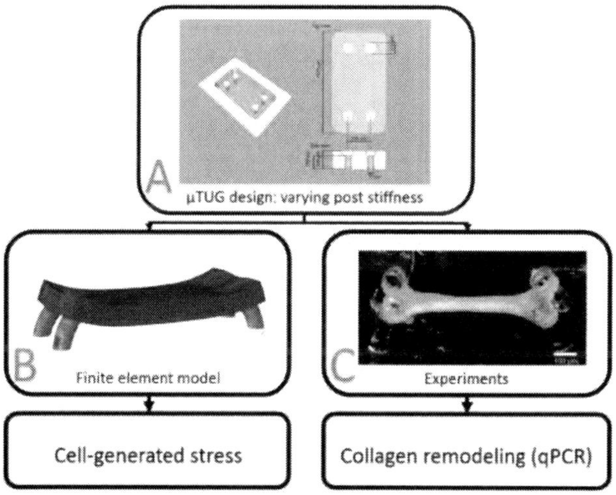

FIGURE 1: OVERVIEW OF THE METHODS
(A) μTUG SYSTEM DESIGN. [4]. (B) FINITE ELEMENT MODEL (C) HYDROGEL SEEDED CELLS SELF-ASSEMBLE INTO A MICROTISSUE, COMPACTING AROUND THE SYSTEM'S MICROPOSTS. [4]

of these microposts was varied in the model after which cell-generated stresses were predicted. In addition, experiments were performed in which tissue remodeling was investigated. Myofibroblasts were seeded in the μTUG setup (with varying micropost stiffness) using a collagen/matrigel hydrogel, and cultured for 24 and 72 hours, after which they were subjected to qPCR analysis. As collagen is the main load-bearing component of tissue engineered heart valves, the microtissues were screened for expression of important collagen remodeling governing genes: collagen type I, collagen type III, MMP-1 and MMP-2. Moreover, micropost displacement and subsequent forces on these microposts were determined to assess the total change in force exerted by the cells on their surroundings due to the change in mechanical constraints.

RESULTS

The FE model predicts that cell-generated stress is constant for every micropost stiffness in the unconstrained direction. In the constrained direction however, cell-generated stress increases with micropost stiffness (figure 3d). Experimentally determined forces on the micropost also increase with micropost stiffness (figure 3e). In fact, the relative changes in stress predicted by the FE model is in good correspondence with the relative changes in force measured in the experiments.

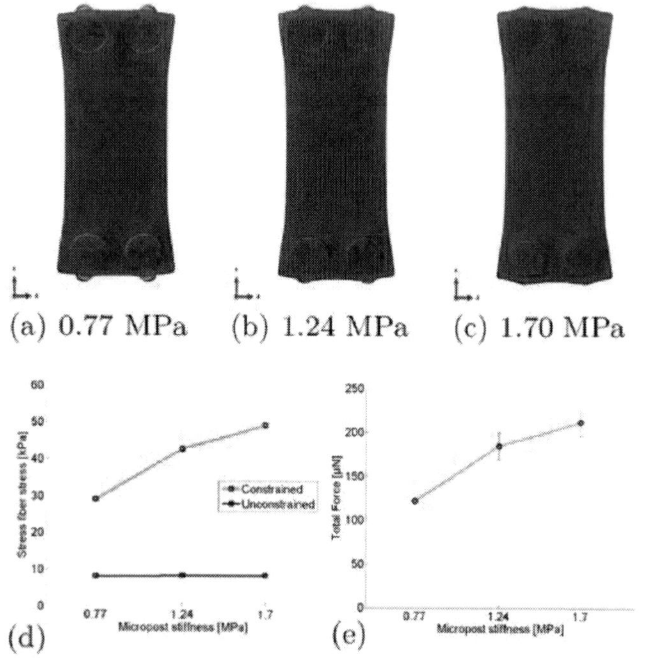

FIGURE 3: (A,B,C) FINITE ELEMENT SIMULATIONS WITH VARYING POST STIFFNESS. (D) STRESS FIBER STRESS (kPa) IN THE CONSTRAINED AND UNCONSTRAINED DIRECTION FOR INCREASING MICROPOST STIFFNESS (E) EXPERIMENTALY DETERMINED FORCE (µN) APPLIED BY MICROTISSUE ON FOUR MICROPOSTS

In contrast, gene expression levels for collagen type I and III did not change with varying micropost stiffness; there is even a slight decrease in expression compared to monolayer cultured cells (figure 4d). On the other hand, MMP-1 and MMP-2 expression is higher compared to monolayer cultured cells (figure 4d). However, again no differences in expression with micropost stiffness can be reported.

DISCUSSION

Although both the FE model and the experimental results demonstrated that cell-generated stress increased with micropost stiffness, no significant differences in gene expression were observed. Apart from MMP-1, no or small upregulation of the genes compared to unseeded cells can be observed. It is hypothesized that these low levels of gene expression were caused by the presence of high concentrations of collagen in the hydrogel, which is used to seed the cells in the microwell. As the cells were surrounded by vast amounts of collagen, they may not have been triggered to synthesize new

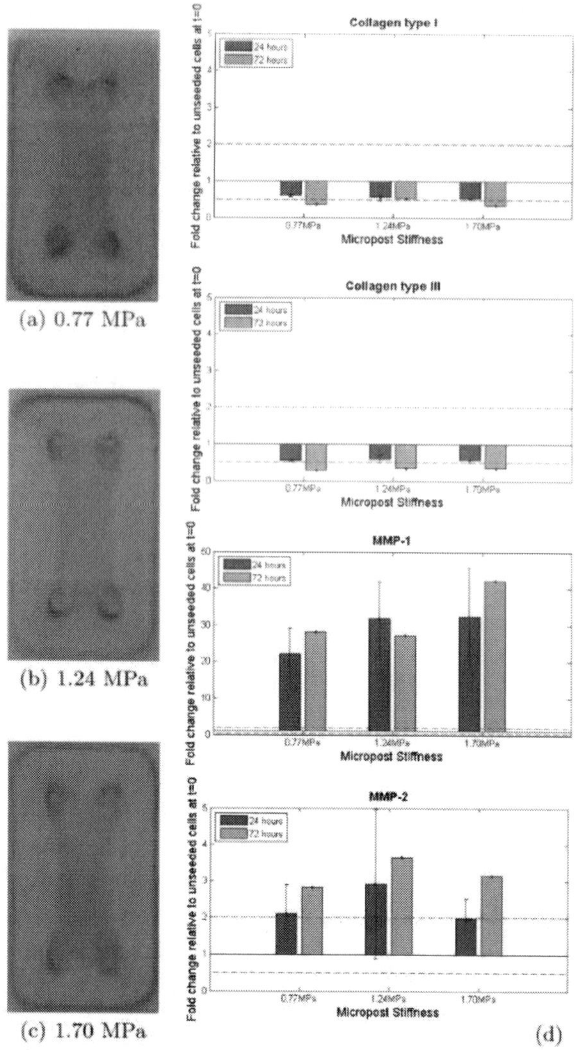

FIGURE 4: (A,B,C) MICROTISSUES ATTACHED TO POSTS WITH VARYING STIFFNESS. (D) FOLD CHANGES IN GENE EXPRESSION RELATIVE TO MONOLAYER CULTURED CELLS FOR 24 AND 72 HOURS OF CULTURE WITH INCREASING MICROPOST STIFFNESS. NOTE THAT THE SCALE FOR MMP-1 IS EXTENDED TO 50

collagen, resulting in no upregulation of the genes. Preliminary results in which we seeded the cells in a fibrin gel, rather than in a collagen-containing gel, did show upregulation of all of the previously mentioned genes. Therefore, future experiments will focus on fibrin-seeded microtissues and their response to the varying micropost stiffness. In addition, the range of micropost stiffness will be broadened by using lower or higher micropost, hence increasing or decreasing the bending stiffness of these posts.

ACKNOWLEDGEMENTS

We acknowledge the support from the Netherlands Cardiovascular Research Initiative (CVON 2012-01): The Dutch Heart Foundation, Dutch Federation of University Medical Centers, the Netherlands Organization for Health Research and Development and the Royal Netherlands Academy of Sciences.

REFERENCES

[1] Mol, A. et al., *Ann Biomed Eng,* 12:1778-1788, 2005
[2] Palchesko, R.N. et al., *PLoS One*, 7: e51499, 2012
[3] Obbink-Huizer, C. et al., *BMMB*, 1: 227-236, 2013
[4] Van Spreeuwel, A.C.C. et al., *Intergr. Biol*, 6:422-429, 2014

SB^{3}C2015
Summer Biomechanics, Bioengineering and Biotransport Conference
June 17-20, 2015,Snowbird Resort, Utah, USA
SB^{3}C2015-370

CORRELATING URETHRAL RUPTURE WITH DISTENSION OF THE URETHRA DURING THE INFLATION OF A MISPLACED TRANSURETHRAL CATHETER BALLOON

Connor V. Cunnane [1], Niall F. Davis [1, 2], Eoghan M. Cunnane [1], Rory O'C. Mooney [1], John A. Thornhill [2], Michael T. Walsh [1]

(1) Centre for Applied Biomedical Engineering Research (CABER), Department of Mechanical, Aeronautical and Biomedical Engineering, and Materials and Surface Science Institute (MSSI). University of Limerick, Ireland.

(2) Department of Urology, Tallaght Hospital, Co. Dublin, Ireland

INTRODUCTION

Urethral catheterisation (UC) is a routine procedure performed by junior doctors and nurse practitioners, with almost 25 % of hospitalised patients being catheterised during their inpatient stay [1]. The estimated incidence of iatrogenic catheter related urethral injury is 0.3 %; but this may under represent the true incidence [2-4]. Urethral injury typically occurs in men when the catheter's anchoring balloon is inadvertently inflated in the urethra [5]. Short-term complications can include pain, bleeding and acute urinary retention caused by urethral rupture. The long-term complication of urethral stricture disease is urethral reconstruction in severe cases [6].

Although iatrogenic complications from UC are well described; there are no studies demonstrating urethral strain thresholds for rupture during inadvertent inflation of a catheter's balloon in a patient's urethra. The present study aimed to demonstrate any correlation between urethral distension, urethral rupture and catheter inflation pressures during urethral catheterisation in *ex vivo* porcine models.

METHODS

Porcine urethras were acquired from male porcine models, all 7 months old, immediately after euthanasia from Fitzgerald's Abattoir (Ballylanders, Co. Limerick, Ireland) and maintained *ex vivo*. Urethral catheter inflation was performed using standard 16 French Coloplast urethral catheters in the bulbar urethra of each porcine urethra (n=21). The catheter's anchoring balloon was inflated with sterile saline using a syringe pump (Harvard Apparatus, MA, USA) at a rate of 30ml/min. The volume of saline instilled during inflation was increased in increments of 1ml and ranged from 1 to 10ml. Urethral diameter variability was mitigated by normalising the internal diametric stretch of each sample to the original diameter at the site of inflation. After inflation of the desired amount of saline into the balloon, the maximum pressure within the system was recorded using a pressure transducer (SensorTeknics UK).

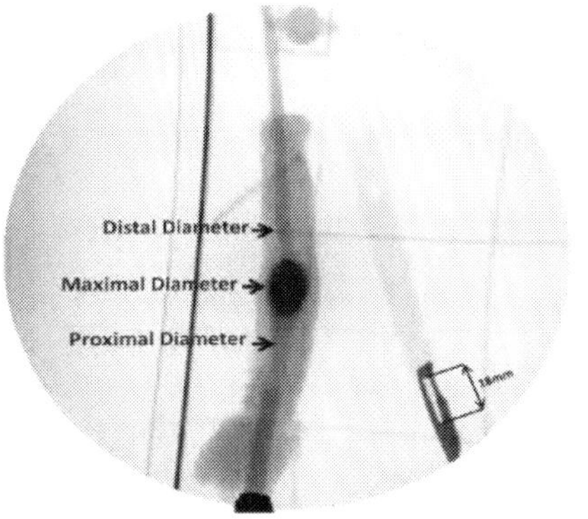

FIGURE 1: RETROGRADE URETHROGRAM ILLUSTRATING LOCATION OF DIAMETRICAL MEASUREMENTS.

Urethral trauma was then assessed with retrograde urethrography. Barium sulphide contrast was instilled into the urethral meatus using a urethral catheter via its drainage channel. A valve was ligated around the proximal end of each traumatised urethra to prevent extravasation of contrast from the proximal luminal end. The traumatised urethras were imaged with plain X-ray covering a 90° rotation in increments of 15 degrees to accurately characterise the severity of urethral trauma at each interval. The internal diametric strain was calculated by averaging the urethra's luminal diameter proximal and distal to the traumatised site as well as the maximum luminal diameter at the traumatised site from the images obtained from retrograde urethrography (Fig. 1). Contrast was instilled into a control sample to demonstrate that instillation of contrast medium during urethrography alone did not contribute to any urethral trauma; thus any trauma demonstrated at urethrography was solely as a result of balloon expansion in the urethra.

RESULTS

The variable nature of the internal luminal diameter of the porcine urethras was also measured (Fig. 2). In sample A (red line), instillation of 2ml of saline was required before contact occurred between porcine urethral tissue and the catheter's anchoring balloon. However, only 0.5ml of saline was required in sample B (green line). Consequently, urethral rupture occurred after infusing 6ml of saline for sample A compared to 5ml in sample B. Variations in internal urethral diameter prevented a reliable correlation between the volume of infused saline and urethral rupture.

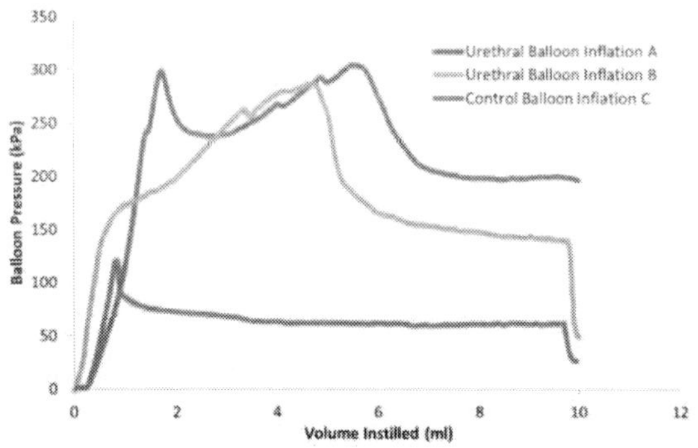

FIGURE 2: THE PRESSURE VOLUME PROFILES FOR INFLATING THE ANCHORING BALLOON IN THE PORCINE URETHRA (N=2) COMPARED TO A CONTROLLED INFLATION AT ATMOSPHERIC PRESSURE (VOLUME=10ML).

The internal diametric strain induced on each urethral specimen due to the inflation process was investigated (Fig. 3). It was discovered that urethral rupture consistently occurs at an internal diametric strain >40%. The maximum pressure recorded within the transurethral catheter's anchoring balloon during the inflation process is also represented. Findings demonstrated no evidence of urethral trauma below an inflation pressure of 150kPa, with the lowest recorded maximum inflation pressure required for urethral rupture occurring at 187kPa.

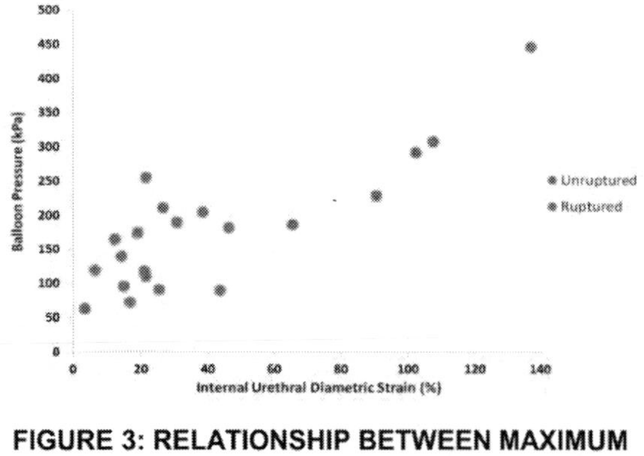

FIGURE 3: RELATIONSHIP BETWEEN MAXIMUM INFLATION PRESSURES AND INTERNAL DIAMETRIC STRAIN RECORDED FOR EACH SAMPLE TESTED (N=21).

DISCUSSION

The overall, iatrogenic complications associated with UC have decreased in recent years due to advanced training programmes that are provided by senior healthcare professionals [7]. Although supervised training programmes have led to an overall decrease in the frequency of catheter-related complications, it has been demonstrated that the anchoring balloon is still inadvertently inflated in the urethra in approximately 3% of hospitalised inpatients with resultant urethral trauma [5, 7].

In the present study we investigated the strain thresholds for urethral rupture for a transurethral catheter. Our most notable finding is that urethral rupture occurs at an internal diametric strain >40%. This important finding indicates that a misplaced anchoring balloon will cause urethral rupture when the internal urethral diameter is increased >40% of its original diameter. Diametric strain measures the change in luminal diameter due to expansion of the anchoring balloon. It is calculated as the change in diameter divided by the original diameter and is consequently expressed as a percentage (%.) In addition to this strain threshold, we also demonstrated that an anchoring balloon inflation pressure <150kPa reliably prevents urethral rupture as rupture only occurred at pressures exceeding 150kPa (Fig. 3). An important secondary finding is that the volume of saline infused into a misplaced anchoring balloon correlates poorly with the potential for urethral rupture due to internal urethral diameter variability.

CONCLUSION

Inadvertent inflation of a transurethral catheter's anchoring balloon in the urethra is a persistent cause of morbidity in urological clinical practice. The present study demonstrates that internal urethral diametric strain and threshold maximum inflation pressures are 2 important parameters that may be useful for designing a safer transurethral catheter for healthcare professionals.

ACKNOWLEDGEMENTS

The authors would like to acknowledge the Irish Research Council for funding under the EMBARK Initiative and the Irish Government under the Programme for Research in Third-Level Institutions (PRTLI) Cycle 5 and co-funded under the European Regional development fund.

REFERENCES

1. Chenoweth C, Saint S: Preventing catheter-associated urinary tract infections in the intensive care unit. Crit Care Clin 2013; 29: 19.
2. Trout S, Dattolo J and Hansbrough JF: Catheterization: how far should you go? RN 1993; 56: 52.
3. Davis NF, Mooney RO'C, O'Brien MF, et al.: Attitudes among junior doctors towards improving the transurethral catheterisation process. Ir J Med Sci. 2014 Apr 13. [Epub ahead of print]
4. Fenton AS: Anterior urethral injuries. Urology 2005.
5. Thomas AZ, Giri SK, Meagher D, et al.: Avoidable iatrogenic complications of urethral catheterization and inadequate intern training in a tertiary-care teaching hospital. BJU Int 2009; 104: 1109.
6. Incidence and Prevention of Iatrogenic Urethral Injuries Carol Kashefi, Karen Messer, Rita Barden, Carolyn Sexton and J. Kellogg Parsons.
7. Sullivan JF, Forde JC, Thomas AZ, et al.: Avoidable iatrogenic complications of male urethral catheterisation and inadequate intern training: A 4-year follow-up post implementation of an intern training programme. Surgeon 2014. Mar 6. pii: S1479 666X(14)00016-X. doi: 10.1016/j.surge.2014.02.001.

SB³C2015
Summer Biomechanics, Bioengineering and Biotransport Conference
June 17-20, 2015, Snowbird Resort, Utah, USA
SB³C2015-371

INDIVIDUAL CELL-BASED MORPHOLOGICAL ANALYSIS TO DETERMINE CHIRALITY OF EPITHELIAL MORPHOGENESIS

Michael J. Raymond, Poulomi Ray, and Leo Q. Wan

Department of Biomedical Engineering
Rensselaer Polytechnic Institute
Troy, NY, USA

INTRODUCTION

The expression of chirality in tissues is described as left-right asymmetry in cell alignment. This is a well-conserved biological property, which is seen in the arrangement of organs and tissues, the development of the GI tract and the looping of the heart. Developmental defects have been connected to changes in chiral morphogenesis, with a well-known example being *situs inversus*, a condition in which the major organs are transposed to the opposite side of the thorax [1].

Recent studies have highlighted various forms of chiral phenomena at a cellular level both in vitro and in vivo. Taniguchi et al. showed that changes in chiral biases of cells in the hindgut of *Drosophila* embryo are amongst the driving forces for directional looping. The biases were quantified by analyzing the orientation of cell edges in the epithelial sheets that comprise the hindgut [2]. Previously, we illustrated that innate chiral biases form in a variety of cells on 2D micropatterns, driven by actin. Surprisingly, we also found that cancerous counterparts of the cell lines studied express an altered chiral alignment [3]. Recently, Naganathan et al. showed that bilateral symmetry breaking in *C. elegans* is driven by active chiral torques from the actomyosin cortex. These chiral torques cause alterations in the cortical flow, which lead to symmetry breaking at the 4-cell stage [4].

It has become apparent that there is a need for analytical tools to better quantify these phenomena. Here, a Python-based image segmentation and analysis tool was developed, utilizing an individual cell-labeling scheme, to derive both geometric and chiral alignment and polarizations from cell morphology. This technique not only provides previously unattainable relationships between chirality and cell morphology, but also delivers an easy way to use the current platform for a greater scope of quantitative chiral analysis.

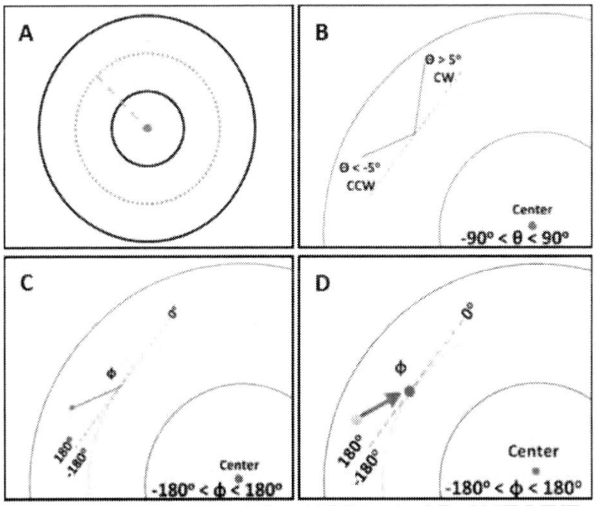

FIGURE 1: DEFINITION OF ANGLES. (A) ORANGE LINE EXTENDS TO BLUE, INDICATING FEATURES ARE WITH RESPECT TO CIRCUMFERENTIAL DIRECTION. (B) ALIGNMENT ANGLE AND (C) SHAPE POLARIZATION ANGLE DEFINITION FOR CELLS AND NUCLEI. (D) CENTROID AXIS POLARIZATION ANGLE EXTENDS FROM NUCLEAR (YELLOW) TO CELL CENTROID (RED).

METHODS

Microcontact Printing: Microcontact printing was performed using polydimethylsiloxane (PDMS) stamps and self-assembled monolayers

(SAMs) as previously described [3]. All micropatterns were in the shape of rings with an inner diameter of 100 µm and a width of 150 µm.

Cell Culture and Immunofluorescent Staining: MDCK cells were maintained in flasks with a media composition of Dulbecco's Modified Eagle Medium (DMEM) High Glucose (Life Technologies) with 10% Fetal Bovine Serum (FBS), 1mM Sodium Pyruvate, and 1% Penicillin-Streptomycin. Cells were passaged using Trypsin-EDTA, seeded onto the patterned surfaces, and cultured to confluency. Cells were fixed with 4% Paraformaldehyde (PFA) and stained for ZO-1 and nuclei (DAPI), after phase contrast imaging.

Algorithm Development and Pipeline: A Python-based algorithm was developed to segment and analyze cell edges and nuclei for the purpose of quantifying both morphological and chiral features. In brief, cell edges are skeletonized and run through a nearest neighbor protocol to find pixels with more than two neighbors, i.e. nodes. Nodes are associated with each other through pixels connecting them, which are designated edges. These are simplified to straight-edge images.

Straight-edge and nuclei images are run through a connected component protocol to produce labeled objects in the images. These objects are analyzed for basic geometric data (i.e. cell shape index, area, perimeter, aspect ratio, major and minor axes) as well as chiral markers (Figure 1), such as alignment (orientation) angle (-90° to 90°) and polarization (direction) angle (-180° to 180°), with respect to the center of the ring.

Statistical Analysis: Generated text files were converted into Excel workbooks and exported into IMB SPSS for statistical analysis. All analysis utilized a significance level of 0.05.

RESULTS

Skeletonized images were validated to determine their fidelity to the original images. The cell number, location, and basic shape were maintained for the majority of the images (Figure 2B). Cell alignment angles were shown to correspond to the orientation of the major axis of the cell and circumferential direction (Figure 2C). Visual inspection of the images confirmed shape polarization angles were directed towards the side of the cell with greater area (Figure 2D). Similarly, theses analyses were also performed on the nuclei (Figure 2E-H). In particular, the nuclear positioning relative to the centroid of the cell was also plotted (Figure 2G).

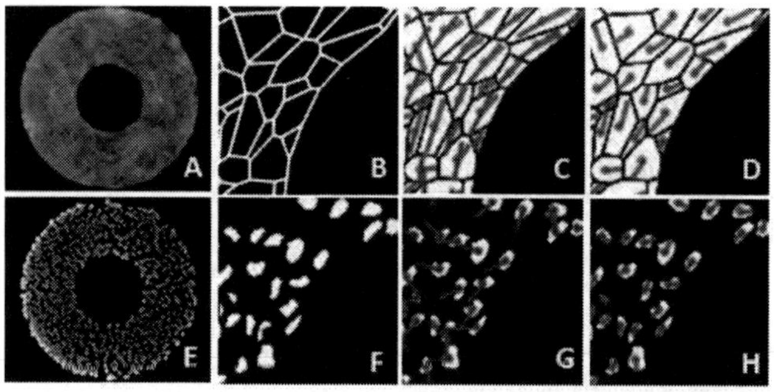

FIGURE 2: CELL-BASED CHIRAL ANALYSIS. (A) ZO-1 ANTIBODY STAINING. (B) STRAIGHT-EDGE IMAGE. (C) ALIGNMENT ANGLES (RED) RELATIVE TO THE CIRCUMFERENTIAL DIRECTION (BLUE). (D) POLARIZATION ANGLES FOR CELLS. (F) THRESHOLDED NUCLEI. (E) NUCLEI STAINING. (F) THRESHOLD IMAGING OF NUCLEI. (G) CENTROID AXIS POLARIZATION. (H) POLARIZATION ANGLES FOR NUCLEI.

Polarization, for a single ring, was further plotted into histograms for visual aid (Figure 3) and confirmed polarization angles were ±180°. Nuclear positioning is found to be more or less random (Figure 3A) with the equal chance to point to any direction. But the cell shape and nuclear shape is clearly biased to a positive angle, indicating a CW alignment based on the definition (Figure 1) and the bias of nuclear shape appears to be more significant.

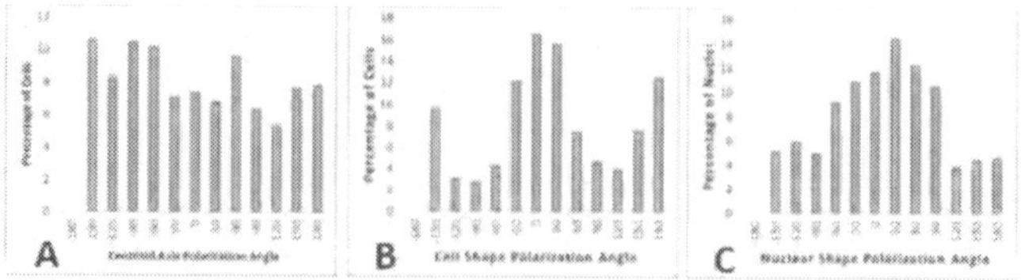

FIGURE 3: POLARIZATION ANGLE HISTOGRAMS. (A) NUCLEUS-CENTROID AXIS, (B) CELL SHAPE AND (C) NUCLEAR SHAPE POLARIZATION ANGLE.

DISCUSSION

The ability to analyze chiral morphogenesis with morphological data of individual cells in a micropatterned culture allows for a more thorough investigation of how the chiral expression develops. This also permits the usage of data mining technologies to derive previously hidden relationships between individual cells, their geometry, and location within a complex system.

Cell and nuclear shape polarization angles allow the determination of the front side of an object. It is based on the observation that migrating cells have been shown to form a teardrop shape in which that larger side is leading [5], which grants insight into the direction in which the cells are moving. The centroid axis polarization angle illustrates the angle from which the nucleus is deviating from the cell centroid; giving a second measure of cell polarization, further illustrating relations between chiral alignments and cell directionality.

Another unique aspect of this algorithm is that it has opened the door for multiple object type identification, evaluation, and correlation. Understanding how the organelles relate to the overall actions of the cell, as well as the dynamics, orientation, and localization of cytoskeletal components, will provide new understandings of why phenomena such as chirality occurs. While only the nucleus was considered, it is clear that in the future a larger number will require consideration. Both the centroid and Golgi apparatus have been implicated in chiral alignment [3] and migration alignment [5]. Other organelles may also have yet to be revealed roles.

Gaining an understanding of how edges and boundaries guide cell organization will give insight into the formation of tissues and organs during development.

ACKNOWLEDGEMENTS

The authors would like to thank NIH Director's New Innovator Award, National Science Foundation, American Heart Association, and March of Dimes for funding Support. Leo Q. Wan is a Pew Scholar in Biomedical Sciences, supported by the Pew Charitable Trusts.

REFERENCES

[1] Yokoyama, T et al., *Science*, 260:679-682, 1993.
[2] Taniguchi, K et al., *Science*, 333:339-341, 2011.
[3] Wan, L et al., *PNAS*, 10:1073-1079, 2011.
[4] Naganathan, S et al., *eLife*, 3:1-16, 2014.
[5] Ridley, A et al., *Science*, 302: 1704-1709, 2003.

TENSILE EQUILIBRIUM MATERIAL RESPONSE OF PREGNANT MOUSE CERVICAL TISSUE DURING NORMAL REMODELING

Kyoko Yoshida (1), Mala Mahendroo (2), Joy Vink (3), Ronald Wapner (3), Kristin Myers (1)

(1) Department of Mechanical Engineering
Columbia University
New York, New York, USA

(2) Department of Obstetrics and Gynecology
UT Southwestern Medical Center
Dallas, Texas, USA

(3) Department of Obstetrics and Gynecology
Columbia University Medical Center
New York, New York, USA

INTRODUCTION

Half a million babies are born prematurely each year in the United States, with rising preterm birth (PTB) rates worldwide [1]. During pregnancy, the uterine cervix acts as a mechanical barrier to keep the growing fetus inside the uterus. At the end of pregnancy, the cervix must become compliant and dilate to allow for delivery. In a normal pregnancy, the cervical extracellular matrix undergoes a dramatic remodeling process leading to changes in tissue mechanical properties, yet these relationships remain undefined. Thus, in an effort to decrease the PTB rate attributable to premature cervical remodeling, our work is focused on delineating these relationships involved in the spectrum of cervical remodeling.

To this end, the aim of this study is to measure the tensile equilibrium material response of pregnant mouse cervical tissue and to present an appropriate material model to capture our experimental data. Our previous studies show evidence of intrinsic collagen fiber remodeling during pregnancy. In a normal mouse gestation, there is a significant decrease in the mature collagen crosslink density with progressing gestation [2]. Additionally, second harmonic generation images of cervical tissue show morphological changes in the collagen structure from aligned straight fibers into dispersed undulating fibers in both mice and human pregnancy [3]. Thus, we hypothesize that the increase in cervical compliance in normal gestation is due mostly to the remodeling collagen network. Based on this hypothesis, we model the cervix as a hyperelastic fiber composite material.

The objectives of this study are to: 1. Present a material constitutive model for the cervical tissue, 2. Measure the equilibrium tensile response of the mouse cervix in normal gestation, and 3. Fit the proposed material model to the experimental data through an inverse finite element analysis (IFEA) to evaluate material parameters of mouse cervical tissue throughout gestation.

METHODS

Equilibrium tensile testing: Mechanical tests were conducted on intact cervical samples. 2-6 month old nulliparous nonpregnant (NP) and pregnant gestation day (d) 6, 12, 15, and 18 female mice were sacrificed. Upon sacrifice, reproductive tracts were isolated and stored at -80°C until testing (n=3-4 per group).

On the day of testing, the cervix was carefully isolated from the vaginal wall and the uterus. Two surgical sutures (Ethicon, Silk 2-0) were threaded through the inner canal of the cervix and attached to custom tensile grips. Samples were positioned within a universal testing machine (Instron 5948 MicroTester, 10N load cell) inside an environmental bath filled with phosphate buffered saline (PBS) + 2mM EDTA (protease inhibitor). Samples were allowed to equilibrate in the solution (about 2-4 hours) while two perpendicular CCD cameras (Point Grey GRAS-50S5M-C 75mm, f/4 lens) tracked tissue geometry (Fig 1).

After swelling equilibrium, samples were subjected to a stress-relaxation test regimen. NP, d6, and d12 samples were loaded to 0.5-3mm grip-to-grip displacements in 0.5mm increments with 30min relaxation holds. The larger d15 and d18 samples were loaded to 1-6mm grip-to-grip displacements in 0.5mm increments with 1hr relaxation holds. Throughout the test, grip-to-grip displacement, force, and tissue geometry were continuously recorded.

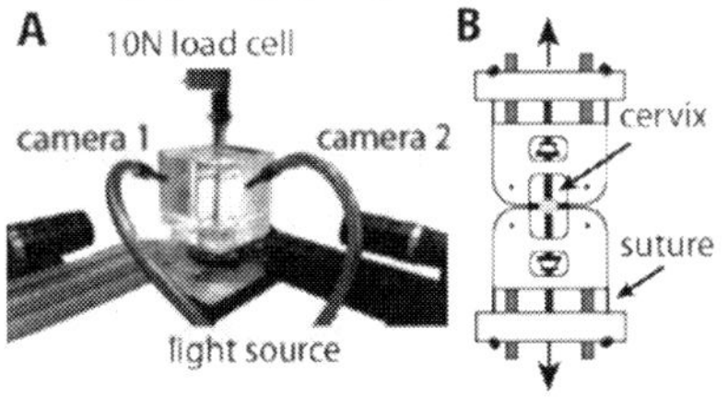

Figure 1: Experimental setup A) Universal testing machine and CCD camera positions. B) Custom tensile grips with loading direction as shown.

Because of the compliance in the sutures used, the suture-to-suture distances (defined as the cervical opening) were measured using calibrated camera images at each grip-to-grip displacement [mm] (Fig.2A). The force [N] at the end of the hold period was used to determine the cervical opening [mm] vs force [N] equilibrium relationship for each sample.

Material model for cervical tissue: In our hyperelastic fiber composite material, a randomly oriented and continuously distributed fiber network is embedded in a compressible Neo-Hookean matrix. In this first iteration of the model, we only consider the equilibrium behavior of the tissue and assume that the tissue is homogenous and isotropic in the undeformed state.

The Helmholtz free energy density of the material can be divided into a ground substance (GS) and collagen network (CN):

$$\psi(\mathbf{F}) = \psi^{GS}(\mathbf{F}) + \psi^{CN}(\mathbf{F}) \tag{1}$$

F is the deformation gradient where we assume affine motion. The ground substance is an isotropic, compressible Neo-Hookean material:

$$\psi^{GS}(\mathbf{F}) = \frac{\mu^{GS}}{2}(I_1 - 3) - \mu^{GS}\ln J + \frac{\lambda^{GS}}{2}\ln J \tag{2}$$

where $[\mu^{GS},\lambda^{GS}]$ are the Lamé constants and I_1 is the first stretch invariant. The fiber distribution is given by [4]:

$$\psi^{CN}(\mathbf{F}) = \int_0^{2\pi}\int_0^{\pi} H(I_n - 1)\psi^{fiber}(I_n)\sin\phi\, d\phi\, d\theta \tag{3}$$

We postulate an exponential power law for the collagen fiber:

$$\psi^{fiber}(I_n) = \frac{\xi}{\alpha}\exp\left(\left[\alpha(I_n - 1)^{\beta}\right] - 1\right) \tag{4}$$

Here, ξ, α, and $\beta>2$ are the fiber elastic modulus, exponential parameter, and stiffening parameter respectively. Additionally, H is the unit step function such that collagen fibers only support tension.

Inverse Finite Element Analysis (IFEA): We simplified the cervical geometry to a thick-walled cylinder with an elliptical cross section (Fig.2C). The geometrical parameters [a,b,c,t] of each sample in the undeformed state was determined using the camera images (Fig.2A,B). Based on these measurements, a 3D model of the sample was generated and meshed with hexahedral elements (Cubit, v.13.2) Using symmetry and taking 1/8 of the cylindrical length [b], the model was reduced to a 1/32 model for the finite element analysis (FEA).

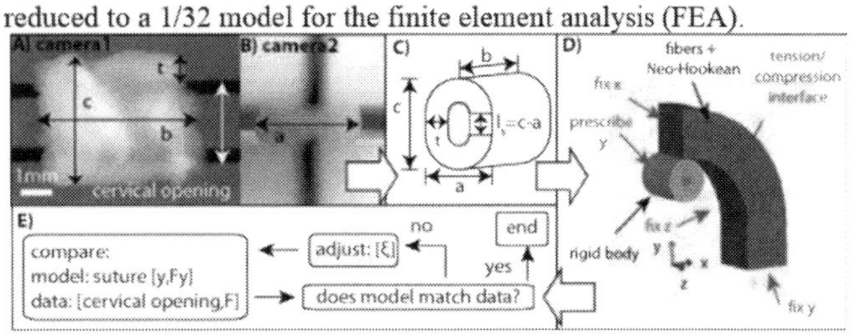

Figure 2: Overview of model analysis. A)-B) Camera images with measured geometric parameters. C) Approximation of cervical geometry. D) 1/32 model and mesh for FEA with boundary conditions (red) and materials (black). E) Flow chart for IFEA.

The generated mesh was imported into the FE software FEBio (v. 2.0), where the material model is included. The material was assigned to the generated mesh, while the sutures were modeled as cylindrical rigid bodies 0.4mm in diameter. A sliding contact interface was defined between the suture and the tissue, and boundary conditions were assigned to reflect symmetry. A displacement was prescribed to the suture to simulate the experimental tensile test. The suture position vs. reaction force curve was compared to the experimental cervical opening vs. force curve (Fig.2D,E).

Since the Neo-Hookean ground substance contributes to the compressive properties of the tissue, we based these parameters on confined compression tests conducted on human cervical tissue with E = 2 kPa and v=0.33 for all samples [5]. For this initial model fit, we kept [α=0,β=3] and only adjusted ξ by hand until the model predictions sufficiently matched the experimental data by inspection.

RESULTS

With progressing gestation, the cervical tissue increased in size while simultaneously softening. NP and d6 samples had a much stiffer response compared to d12, d15, and d18 samples. While d12 samples were consistently stiffer than d18 samples, there was a wide spread in the d15 sample response, which spanned between d12 and d18 sample response. The proposed model was able to sufficiently capture the experimental data. There was a significant decrease in ξ, the collagen fiber stiffness with progressing gestation (Fig.3).

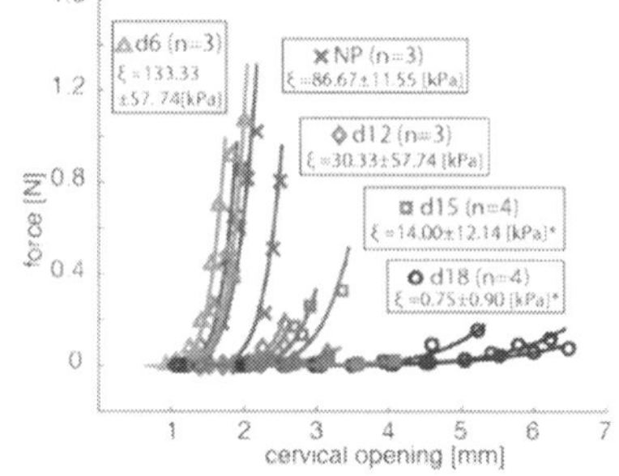

Figure 3: Tensile equilibrium data (symbols) and corresponding model fits (lines) with best-fit ξ values (average+/-standard deviation). * indicates statistically different ξ compared to NP (Student's t-test, p<0.0005)

DISCUSSION AND CONCLUSIONS

The main objective of this study was to report tensile material parameters for NP and pregnant mouse cervical tissue. Here, we present our model equations and the equilibrium tensile data of intact mouse cervical tissue. Additionally, we illustrate our method for evaluating cervical material properties, where we generate a 3D mesh of the cervix, simulate our experiment with FEA, and iterate on our model to determine the best-fit material parameters for each sample.

Our results show that increasing tissue compliance with gestation is due to a significant reduction in the fiber stiffness, ξ. This demonstrates the ability of our model to capture how intrinsic changes in collagen fiber properties lead to changes in the structural response of the intact cervix. Here, we did not adjust α and β, which may contribute to the large spread in ξ within each group. Additionally, for some of the smaller NP and d6 samples we observed an early upturn in the experimental stiffness curve. We hypothesize that these samples were pre-tensioned at the beginning of the test. In these cases, we initiated our IFEA with a level of pre-stretch to capture the data. Further investigations are needed to evaluate the contributions of reference geometry and fiber parameters to the cervical opening vs. force curve.

With the methods presented here, our future studies will focus on establishing quantitative relationships between material parameters and biochemical properties (i.e., collagen crosslinks) to develop a growth and remodeling framework for cervical tissue in pregnancy.

ACKNOWLEDGEMENTS

This study was supported in part by the NSF Graduate Research Fellowship, the Sheldon Weinig Scholars Program, and the March of Dimes Prematurity Research Initiative.

REFERENCES

[1]WHO, http://www.who.int/mediacentre/factsheets/fs363/en, 2014.
[2] Yoshida et al., *PLOS ONE*, 9(11): e112391, 2014.
[3] Zhang et al., PNAS, 109(32): 12878-83, 2012.
[4] Ateshian et al., *J. Biomech. Eng.*, 131(6):061003, 2009.
[5] Myers et al., *Acta Biomater.*, 4(1):104-16, 2008

TOWARD UNDERSTANDING THE MECHANISMS BY WHICH CARTILAGE MICROPARTICLES INDUCE SYNOVIAL INFLAMMATION IN OSTEOARTHRITIS

Amy M. Silverstein (1), Robert M. Stefani (1), Sevan R. Oungoulian (2), Eric L. Tong (1), Mukundan G. Attur (3), Steven B Abramson (3), Christopher S. Ahmad (4), James L. Cook (5), Gerard A. Ateshian (2), J. Chloe Bulinski (6), Clark T. Hung (1)

(1) Department of Biomedical Engineering
Columbia University
New York, NY, USA

(2) Department of Mechanical Engineering
Columbia University
New York, NY, USA

(3) Division of Rheumatology, Department of Medicine
New York University School of Medicine and
NYU Langone Medical Center
New York, NY, USA

(4) Department of Orthopedic Surgery
Columbia University
New York, NY, USA

(5) Department of Orthopedic Surgery
University of Missouri
Columbia, MO, USA

(6) Department of Biological Sciences
Columbia University
New York, NY, USA

INTRODUCTION

Osteoarthritis (OA) is a debilitating degenerative joint disease characterized by degradation of articular cartilage. Tissue wear naturally occurs as the result of the interaction between two distinct moving surfaces. Increased cartilage wear, leading to formation of loose particles in the joint space, has been linked to aging, overuse and obesity, often leading to OA. Clinically, cartilage and bone debris are bound to the surface or embedded deep within the synovium of patients with OA [1]. Animal models in which cartilage debris were injected into the knee, resulting in synovitis and infiltration of mononuclear cells in addition to capsular and cellular changes, have confirmed the interactive role between the synovium and cartilage wear particles in the inflammatory cascade of OA [2, 3]. However, the mechanisms mediating this interaction remain poorly understood.

Monolayer cell co-culture has previously been used to study the effects of biologic and non-biologic wear particles on cells. Thoroughly characterized, the non-biologic wear particles used for *in vitro* wear particle studies are typically 0.2-7 μm in size, as particles less than 15 μm in diameter are phagocytosed by cells of the synovial joint, regardless of composition. Compared to non-biologic wear particles, biologic counterparts are significantly more complex to study. Since biologic particles consist of extracellular matrix (ECM) components that are degradable by the cell, they may interact with the synovium beyond just being internalized. Accordingly, this study aims to characterize the effects of cartilage wear particles of a phagocytosable size on the biosynthetic activity of synovial cells.

METHODS

Synovial Cell and Cartilage Explant Harvesting: Synovial tissue and 10 mm articular cartilage explants were harvested from the knee joints of freshly slaughtered juvenile bovine calves (n=3). Synovium was digested with collagenase type IV and synovial cells were expanded in alpha-Minimum Essential media containing 10% FBS, 1% antibiotics and 5 ng/mL FGF-2 for two passages. Cartilage explants were maintained in serum free explant media [4]. Particle Generation and Preparation: Cartilage wear particles were generated from 10 mm explants (n=2) with 120 grit sandpaper (Figure 1A,B). To isolate particles capable of being phagocytosed, the resulting cartilage particles suspension was filtered with a 10 μm nylon mesh filter. Particle concentration was quantified and sub-10 μm particle sizes confirmed using a Coulter Counter [5]. Imaging: To better characterize the interaction between cartilage wear particles and synovial cells, cartilage wear particles were stained with 20 μM dichlorotriazinyl aminofluorescein (DTAF). Synovial cells were plated overnight before treatment with 250 fluorescently labeled cartilage particles per cell for 2 days. Prior to imaging, synovial cells were stained with CellTrace™

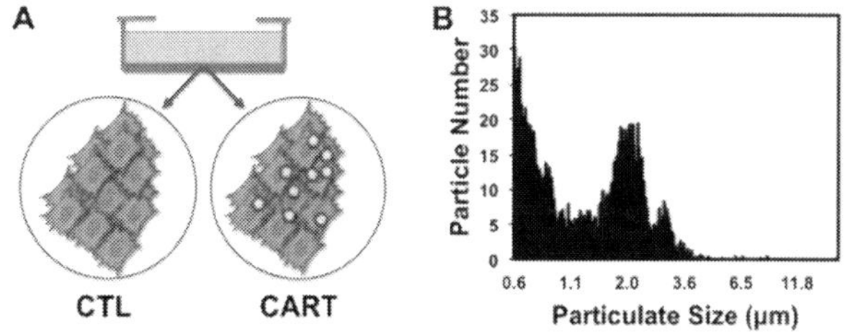

FIGURE 1. (A) SCHEMATIC OF CULTURE CONDITIONS. CTL=CONTROL. CART= CO-CULTURE WITH CARTILAGE PARTICLES. (B) COULTER COUNTER CHARACTERIZATION OF CARTILAGE PARTICULATE NUMBER AS A FUNCTION OF SIZE. HISTOGRAM REPRESENTS AVERAGE OF 4 ALIQUOTS OF DILUTED STOCK SOLUTION OF FILTERED WEAR PARTICLES.

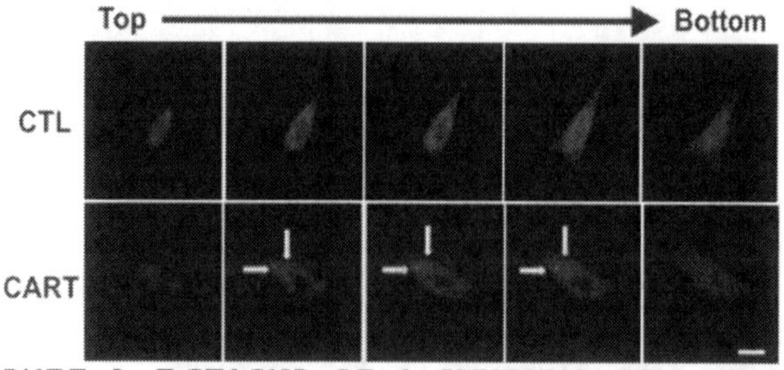

FIGURE 2. Z-STACKS OF A SYNOVIAL CELL (RED) CULTURED WITHOUT (CTL) AND WITH (CART) CARTILAGE WEAR PARTICLES (YELLOW) SHOWING INTERNALIZATION OF WEAR PARTICLES. WHITE ARROWS INDICATE PARTICLES. SCALE BAR=20 µM.

Calcein Red-Orange, AM to identify if particles were internalized or in contact with the cell surface. Z-Stack images were taken using Zeiss Confocal microscope. Co-culture of Particles with Synovial Cells: Dense monolayers of synovial cells were cultured ± cartilage particles for 5 days at a dosage of 250 particles per cell. Additionally, cartilage particles were cultured without cells as an additional control. Biochemical Analysis: DNA, glycosaminoglycan (GAG), collagen, nitric oxide (NO) and lactase dehydrogenase (LDH) were measured using the Picogreen dsDNA, DMMB, orthohydroxyproline (OHP), Griess assay and LDH assays, respectively. To determine the difference between control and wear particle treated synovial cells, DNA content as well as protein and NO synthesis by treated cells, acellular cartilage particle DNA and ECM protein content was subtracted from the measured DNA and protein content of the cells, media and debris of groups cultured with cartilage wear particles. Immunohistochemistry: Synovial monolayers were fixed in 4% paraformaldehyde after 5 days of cartilage co-culture and stained for lubricin (rabbit polyclonal primary antibody, FITC-conjugated goat anti-rabbit secondary antibody). Cell nuclei were stained with 4',6-diamidino-2-phenylindole (DAPI). Statistical Analysis: Statistical analysis was performed using Student's t-test ($p<0.05$).

RESULTS

Cartilage wear particles were generated through wearing down cartilage explants with 120 grit sandpaper. To remove particles above the desired 15 µm particulate size, the wear particle solution was filtered resulting in an average particle size of 1.06 ± 0.06 µm across 12 separate preparations. Internalization of cartilage wear particles was confirmed through confocal microscopy (Figure 2). Using a

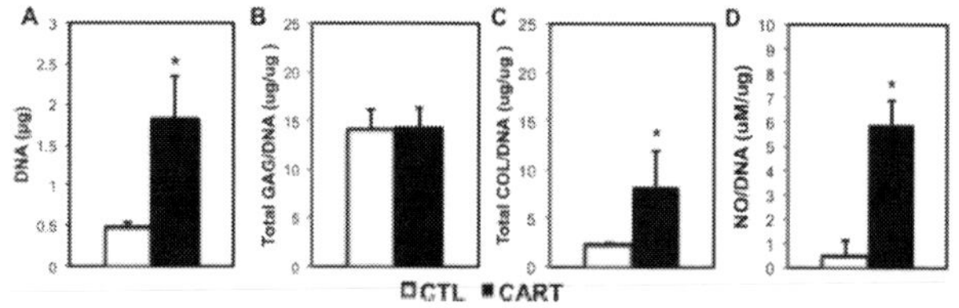

Figure 3. (A) DNA CONTENT, (B) TOTAL GAG SYNTHESIS (RETAINED BY SYNOVIAL CELL MONOLAYER AND LOST TO MEDIA), (C) TOTAL COLLAGEN (COL) SYNTHESIS, AND (D) NO PRODUCTION NORMALIZED TO DNA AFTER 5 DAYS OF CO-CULTURE. CTL = CONTROL SYNOVIAL CELLS. CART = SYNOVIAL CELLS CO-CULTURED WITH CARTILAGE WEAR PARTICLES. *P<0.05 VS CONTROL. N=5-6).

previously established dosage of 250 particles per cell, co-culture with wear particles resulted in a significant increase in cellular proliferation as evidenced by an increase in Total DNA content (Figure 3A). To evaluate if culture with cartilage wear particles induced changes in biosynthetic activity, total glycosaminoglycan (GAG), total collagen, media NO and media LDH concentrations were normalized by DNA content. Co-culture on wear particles had no significant effects on GAG synthesis (Figure 3B) or LDH (data not shown), but did significantly increase collagen synthesis (Figure 3C) and nitric oxide production (Figure 3D). Immunostaining of synovial cells cultured with cartilage wear particles showed increased lubricin staining intensity compared to control cells (Figure 4).

DISCUSSION

As cartilage tissue degrades, wear particles are released into the synovial joint and become embedded on the surface of and within the synovium. Here, we demonstrate an *in vitro* model for studying the interaction between cartilage wear particles and synovial cells. Sub-10 µm cartilage wear particles were phagocytosed by synovial cells. Co-culture of dense synovial layers with cartilage particles for 5 days resulted in increased cellular proliferation and increased production of NO. NO is a factor in the catabolic cascade of OA that regulates expression of pro-inflammatory cytokines. Accordingly, dense synovial cell monolayers showed increased expression of lubricin after co-culture with cartilage wear particles, mimicking the response of these cells to proinflammatory cytokine insult. Previous work has shown that there is increased lubricin release into the media [6] and increased lubricin staining of multilayer synovial cells [7] after exposure to interleukin. Thus, this staining may be indicative of the pro-inflammatory role of sub-10 µm cartilage wear particles.

Future work will aim to further characterize the mechanisms, such as phagocytosis and surface contact or binding, behind the interaction between the synovium and cartilage wear particles. Building upon previous work by our group, we can grow multilayer bioengineered synovium on Transwell® mesh filters [7]. This system lends itself to the development of a physiologic wear particle model of osteoarthritis that will allow us to characterize the OA-like pathophysiology in a co-culture system of synovial cells, cartilage explants, cartilage wear particles and cytokines.

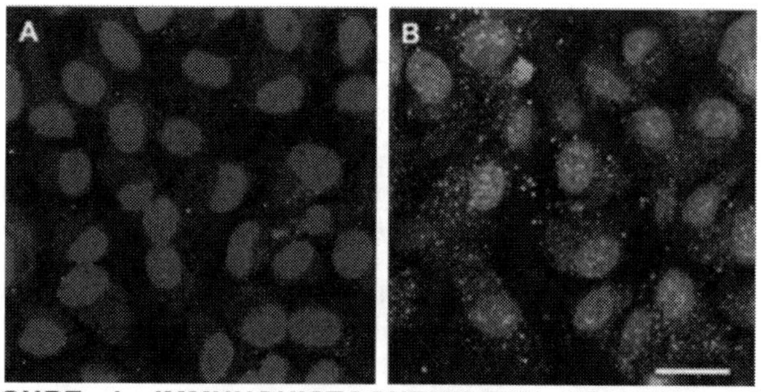

FIGURE 4. IMMUNOHISTOCHEMISTRY STAINING FOR LUBRICIN OF CONTROL (A) AND CARTILAGE PARTICLE TREATED SYNOVIAL CELLS (B). SCALE BAR=20 µM.

ACKNOWLEDGEMENTS: This work was supported in part by NIH R01AR43628 and T32AR059028.

REFERENCES: [1] Lloyd-Robers, GC, JBJS 35-B(4):627-42. [2] Chrisman, OD et al., *Yale J Bio Med*, 37:409-12,1965. [3] Evan, CH et al., *Arthritis Rheum*, 27(2):200-207, 1984. [4] Albro, MB et al., *Trasn Orthop Res Society*, 2014. [5] Oungoulian, SR et al., *J Biomech Eng*, 135(2):024501, 2013. [6] Blewis et al. *Bioeng Biotech* 106:149-160, 2010. [7] Stefani, RM et al. *Trans Orthop Res Society*, 2015.

SB³C2015
Summer Biomechanics, Bioengineering and Biotransport Conference
June 17-20, 2015, Snowbird Resort, Utah, USA
SB³C2015-375

SPECIMEN-SPECIFIC EVALUATION OF A MULTI-BODY MODEL PREDICTS INSTABILITY AND INCREASED MENISCAL LOAD IN THE ANTERIOR CRUCIATE LIGAMENT-DEFICIENT KNEE

Mohammad Kia, Kevin Schafer, Daniel Green, Andrew Pearle, Thomas Wickiewicz, Timothy Wright, Carl Imhauser

Hospital for Special Surgery
Department of Biomechanics
New York, NY, US

INTRODUCTION

Knee instability following anterior cruciate ligament (ACL) rupture is common, compromising function and causing cartilage and meniscal damage [1]. Computational modeling of the knee may be a useful tool to predict those at risk of instability and associated soft tissue damage, but this requires subject-specific synergy of models and experiments to evaluate model predictions of both knee kinematics and soft tissue loads. Such multilevel evaluations of model predictions are lacking, leaving in question the ability of models to predict clinically relevant mechanical behavior of the knee. We are developing a framework for understanding the subject-specific effect of ACL injury on knee stability and damage to secondary stabilizers such as the menisci. The aim of the current study was to assess whether a subject-specific multibody model that includes representation of discretized menisci could predict clinically relevant changes in knee function following removal of the ACL. We accomplished this through comparison of model predictions of knee kinematics and meniscal loads with a physical experiment on the same specimen that was modeled. We addressed two questions: 1) Does ACL deficiency increase coupled anterior translation and lateral meniscus load in response to combined torques known to generate knee instability? 2) Does ACL deficiency increase anterior translation and medial meniscus load in response to applied anterior loads?

METHODS

A computational model of the tibiofemoral joint of a 20 year old male cadaveric specimen consisting of subject-specific bone geometries, articular cartilage, and menisci was generated using computed tomography (CT) and incorporated into a dynamic multi-body framework (ADAMS). Ligament insertions were identified for the LCL, superficial MCL (sMCL), ACL, posterior oblique ligament (POL), posterior capsule, and meniscal horns from image data [2]. The ligaments were represented with multiple force elements covering their insertion areas; mean structural properties were described using a tension-only non-linear load-displacement relationship [3-7]. A total of 38 knee ligament fibers were included.

The lateral and medial meniscal geometries were discretized radially into 34 and 30 subelements, respectively. A linear stiffness matrix connected neighboring elements to represent a deformable geometry [8]. Tibiomeniscal coronary ligaments were modeled with a total of seven tension-only linear springs on the medial side and one fiber on the lateral side. Rigid body contact was defined between the meniscal geometries and the femoral and tibial articular surfaces as a non-linear function of penetration depth and velocity [9].

We focused on simulating clinical tests of knee stability because of their importance in diagnosing ACL injury and the relationship of the pivot shift test to clinical outcome [10]. Boundary conditions included fixing the femur rigidly and setting the tibia relative to the femur at 15° flexion, leaving it with five degrees of freedom. The pivot shift was modeled by first increasing the applied valgus moment to 8 Nm and then increasing the internal rotation moment to 4 Nm while the valgus moment remained constant. The test of anterior stability was modeled by applying a 134 N anterior load.

Model predictions of knee kinematics and soft tissue load were compared to a physical experiment using the same specimen that was modeled. The experiment was conducted using a previously described six degrees of freedom robot [9,11]. The same pivot shift and anterior stability tests as simulated in the model were performed on the cadaver with the ACL intact and after it was sectioned. All loads, boundary conditions and kinematic coordinate systems were consistent across model and experiment. Both the medial and lateral menisci were removed after identifying the kinematics of the ACL-intact and

deficient knee, and the loads borne by these structures were calculated using the principle of superposition [11]. Coordinate systems were registered across the physical and modeling environments by identifying fiducials that were rigidly fixed to the tibia and femur using a 3D digitizer (Microscribe MLX, accuracy ≤ 0.076 mm).
Model predictions of anterior stability were evaluated by one-to-one comparison with experimental data of the tibial anterior translation, and load in the medial meniscus at the maximum applied anterior load. Model predictions of the pivot shift were evaluated by identifying and comparing the increment of applied load in which the largest change in coupled anterior translation occurred. This measure was chosen because it represents the instability event that clinicians use to judge ACL integrity during this exam. Load on the lateral meniscus was also compared between the model and the experiment at the maximum applied load with the ACL present and absent.

RESULTS

The predicted anterior translation at the maximum applied anterior load in the intact and the ACL-deficient knee was 6.8 mm and 20.3 mm, respectively (Fig 1). They differed from the experiment by 0.24 mm and 1.3 mm in the ACL-intact and deficient knee, respectively. In the pivot shift of the intact knee, the largest change in coupled anterior translation was 0.35 mm occurring between 2 and 3 Nm of applied internal rotation moment in the model, and 0.26 mm occurring between 0.8 and 1.2 Nm of applied internal rotation moment in the experiment (Fig 2). In the ACL-deficient knee, the largest change in coupled anterior translation was 4.1 mm occurring between 4 and 6 Nm of applied valgus in the model, and 7.0 mm occurring between 2 and 4 Nm of applied valgus in the experiment.

During the test of anterior stability, removing the ACL caused the medial meniscus to bear 200 N and 120 N more load than with the ACL intact in the model and experiment, respectively. During the pivot shift, ACL removal caused the lateral meniscus to bear 50 N and 30 N more load in the model and experiment, respectively (Figs 3, 4).

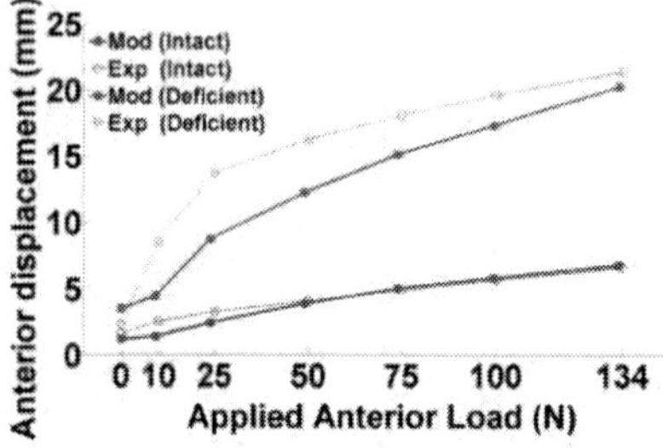

Fig 1: Anterior displacement of the tibia with respect to the femur applied anterior loads

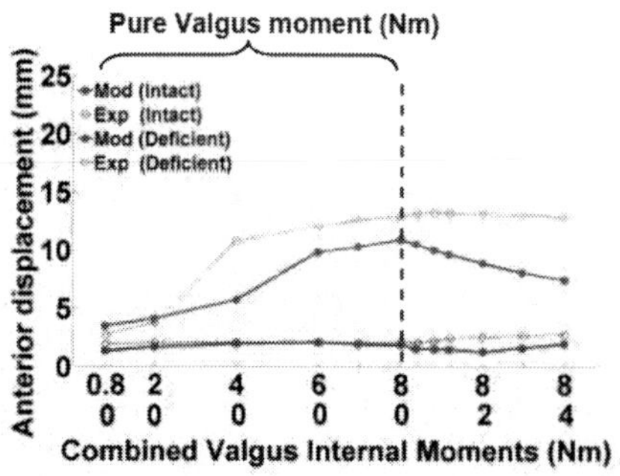

Fig 2: Anterior displacement of the tibia with respect to the femur under a combined valgus (0.8-8Nm) and internal (0-4Nm) moment

Fig 3: Tibia anterior subluxation and soft tissue loading under 134 N applied anterior load with ACL intact (left) and removed (right)

Fig 4: Tibia anterior subluxation and soft tissue loading under combined valgus and internal moments with ACL intact (left) and removed (right)

DISCUSSION

Specimen-specific experiments were used to evaluate a 5-dof multibody model of the tibiofemoral joint that included representations for major passive stabilizers including ligaments and menisci. The strength of this work is the head-to-head comparison of predicted kinematics and soft tissue loads with subject-specific experimental data. Model predictions agreed with subject-specific experimental data in terms of: (1) the magnitude of the instability event during the pivot shift; (2) the amount of anterior translation in response to an anterior load in the ACL intact and deficient knee; (3) increased load on the medial meniscus with anterior loads after removing the ACL; and (4) increased load on the lateral meniscus with the simulated pivot shift after removing the ACL.

The model predicted higher loads on the menisci than what were measured in the experiment likely due to the linear representation of meniscal stiffness properties. Non-linear properties will be implemented given their ability to better reproduce the toe-in behavior. Only population mean structural properties were used to describe the ligaments; optimization of soft tissue properties based on a subset of experimental loading conditions may improve agreement of model predictions and experimental data (Figs 1,2).

This model can be used to understand the relationship between properties of the passive stabilizers and the resulting knee kinematics, which account for many of the constituents that are surgically manipulated. The response of the model to functional loads such as those during gait should also be assessed. This tool shows promise for identifying those at risk of instability and meniscal damage following ACL injury and may help to identify those that may most benefit from ACL reconstruction.

ACKNOWLEDGEMENTS

Kirby Foundation, Institute of Sports Medicine Research, Gosnell Family

REFERENCES

[1] Kessler, T et al., Knee Surg Sports Traumatol Arthrosc. 2008.[2]. LaPrade: JBJS Am 2007.[3] Ardern, et al. Am J Sports Med, 2011.[4]. Butler, et al. J Biomech, 1992. [5]. Harner, et al. Am J Sports Med, 1995. [6]. Hauch, et al. J Biomech, 2010.[7]. Robinson, et al. J Biomech, 2005. [8]. Guess, et al. Open Biomed Eng J, 2012. [9]. Imhauser, C et al., AJSM, 2013. [10]. Leitze, et al. Clin Orthop Relat Res. 2005.[11].Woo, S et al., The Knee, 1998.

SB³C2015
Summer Biomechanics, Bioengineering and Biotransport Conference
June 17-20, 2015,Snowbird Resort, Utah, USA
SB³C2015-377

RELAXATION OF SIMULATED VISCOELASTIC FIBER NETWORKS

Rohit Y. Dhume (1), Victor H. Barocas (2)

(1) Department of Mechanical Engineering
University of Minnesota, Twin Cities
Minneapolis, MN, USA

(2) Department of Biomedical Engineering
University of Minnesota, Twin Cities
Minneapolis, MN, USA

INTRODUCTION

Structures made up of fibers are common in the biological world. Fibers and filaments of different sizes make up most materials. On a larger scale, most load-bearing soft tissues can be viewed as made up of networks of interconnected segments of, e.g., collagen. Thus, understanding the mechanics of random fiber networks can provide a deeper insight into the mechanics of various biological tissues.

Tissues are viscoelastic in nature, so the network mechanics model used should not only account for the non-linear elastic behavior but also include a viscoelastic component. This can be modelled using an appropriate viscoelastic model (like the Standard Linear Solid model) to represent the fibers in the deforming network. Adding a viscoelastic component to the fibers introduces time dependence to the stress. Under constant strain, the network relaxes and relieves stress. The relaxation time of the network must depend on the relaxation times of the individual fibers composing the network, but the dependence may not be linear even when the individual fibers are linear because the network structure and interaction between fibers might introduce some non-linearity. A material can also have multiple relaxation times due to interactions between relaxing elements.

The objective of this study was to look at how the stress relaxation behavior of a network differs from that of its constituent fibers and how it is affected by changes in the fiber relaxation times or concentrations.

METHODS

Isotropic Delaunay and Voronoi networks generated from random seed points in 3D space, were tested in simulated uniaxial extension. Fibers of the network were represented by a Standard Linear Solid (SLS) model, (Fig. 1). In order to have fibers with two different relaxation times within one network, depending on the desired concentration, fibers were selected at random, and their viscosity (η) was changed to change their relaxation time (η/E_2). Suitable values for the two fiber relaxation times were selected such that they were a factor of 2 apart. Fibers were assumed to be long, and bending deformation was neglected. At every time step, the Cauchy stress generated in the network was measured using a volume averaging approach [1].

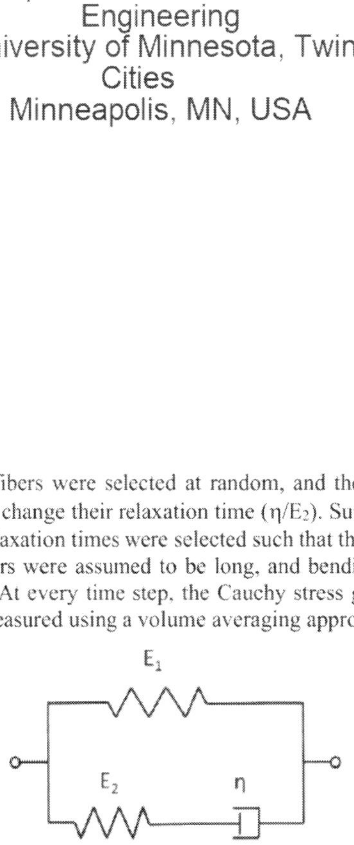

Figure 1 : Standard Linear Solid (SLS) model used to represent a fiber

The networks were given an initial stretch over a short time period and then held at constant stretch while the network stress was calculated at regular time intervals. These time intervals were chosen such that they were much smaller than the relaxation times of the network fibers thus ensuring that we did not miss any details of the stress relaxation curve. The calculated averaged Cauchy stress in the network was then fit to a Prony type series of the form

$$\sigma = A + Be^{-\frac{t}{\tau 1}} + Ce^{-\frac{t}{\tau 2}} \qquad (1)$$

First, the normalized stress data from the networks made up of only one type of fiber were fit to the curve and values for the relaxation times τ_1 and τ_2 were obtained. The stress data from the remaining networks containing both types of fibers in varying concentrations was then fit to the Prony series using τ_1 and τ_2 from above, and values for the coefficients B and C were obtained.

To check whether the network generated any new time scales, another term was added to the Prony series with a varying relaxation time, and the stress data for all the networks was fit to this new curve

$$\sigma = A + Be^{-\frac{t}{\tau 1}} + Ce^{-\frac{t}{\tau 2}} + De^{-\frac{t}{\tau 3}} \qquad (2)$$

RESULTS

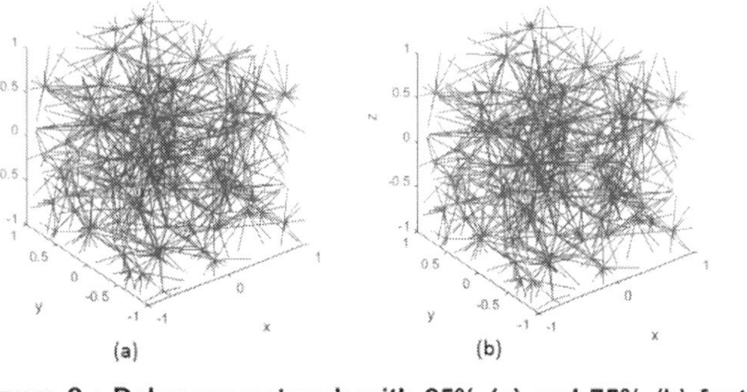

Figure 2 : Delaunay network with 25% (a) and 75% (b) fast relaxing fibers. Red fibers show fast relaxers, blue show slow relaxers

Delaunay and Voronoi networks were tested using different values of the initial stretch and relaxation times for the two fiber types. Even though the stress – strain response of the Voronoi networks was more non-linear, the two network types exhibited nearly identical stress relaxation behavior, so only Delaunay results are shown. The value of the coefficient D (equation (2)) relative to B and C, for all the networks, never more than $1e^{-6}$, which meant that no new relaxation times were present. When the network was composed of only one type of fiber, it was seen from the curve fit that there was only one effective relaxation time. It was noticed that this effective time constant was always less than the relaxation times of the two types of fibers making up the network.

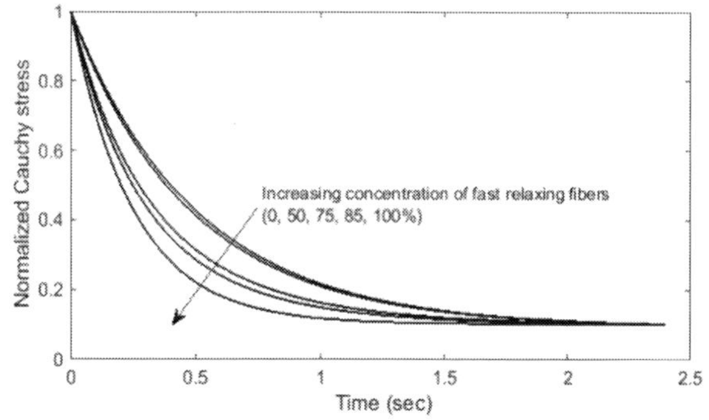

Figure 3 : Stress v/s time plot for a single network showing a significant change in the relaxation time as the fraction of fast relaxers increases above 50%. The 0% fast-relaxer and 50% fast-relaxer curves almost overlap. The fiber relaxation times used were τ_1 = 0.5 sec and τ_2 = 1 sec.

After the relaxation time values had been calculated, the normalized stress data for hybrid networks containing both types of fibers in different concentrations were fit to equation (2) as described in the previous paragraph.

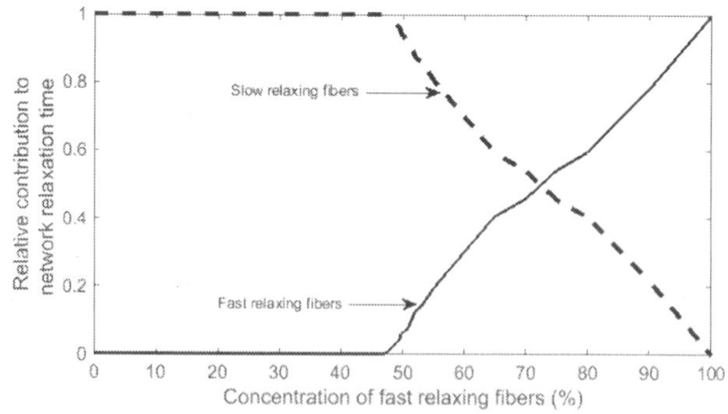

Figure 4 : Plot showing the relative contribution of the two different fiber types to the overall network relaxation behaviour for a Delaunay network (n = 7)

The fast relaxing fibers had no measurable effect on the overall network behavior at fractions below 50%. Fig. 3 shows the relative contribution of the two relaxation times as expressed as the ratio B/(B+C) and C/(B+C) from the fit of equation (2). As seen from the figure, for a fast-relaxing fiber fraction below 45-50%, the network relaxation time was equal to that of a network of all. This was found to be true regardless of the number of total fibers in the generated network or the type of network. For fast-relaxing fiber fractions between 50 and 100%, there was a roughly linear relation between fiber fraction and contribution to the overall relaxation behavior.

DISCUSSION

The effect of varying fiber relaxation times and fiber concentrations on the overall network stress relaxation behavior was investigated for isotropic Delaunay and Voronoi networks with fibers characterized by two different relaxation times.

Networks composed of a single type of fiber had only one effective relaxation time and the value of this time constant was always less than that of the individual fibers. This could be the effect of the various series/ parallel combinations of fibers between the two ends of the network being pulled.

Increasing the concentration of the fast relaxing fibers did not have any significant influence on the network relaxation time till the concentration reached around 60%. One reason for this could be that the network relaxation time decreases after the fast relaxing fibers have percolated within the network (Fig. 2). It was also seen that the network structure did not lead to any new relaxation time scales.

ACKNOWLEDGEMENTS

This work was supported by the National Institutes of Health (R01 EB005813). Computations were made possible by a resources grant from the Minnesota Supercomputing Institute.

REFERENCES

[1] Chandran, PL, Barocas, VH, 2007, J Biomech Eng 129, pp 137-147.

TOWARDS THE DEVELOPMENT OF AN ATHEROSCLEROTIC PLAQUE STRATIFICATION PARAMETER TO PREDICT ARTERIAL RESTENOTIC RESPONSE FOLLOWING ENDOVASCULAR TREATMENT

Eoghan M. Cunnane (1), Hilary E. Barrett (1), Eamon G. Kavanagh (2), Michael T. Walsh (1)

(1) Centre for Applied Biomedical Engineering Research (CABER),
Mechanical, Aeronautical and Biomedical Engineering Department (MABE),
Material Surface Science Institute (MSSI),
University of Limerick (UL),
Limerick, Ireland

(2) Department of Vascular Surgery
University Hospital Limerick (UHL)
Limerick, Ireland

INTRODUCTION

Results following endovascular treatment of atherosclerotic lesions vary according to vascular location. One of the most pronounced variances is the incidence of restenosis. Specifically, restenosis rates are approximately 4 times higher in the femoral arteries than in the carotid arteries[1,2]. This trend has necessitated high re-intervention rates for lower limb revascularisation that have driven per patient costs higher than those associated with coronary and cerebrovascular disease[3]. Numerous studies have investigated the biological and clinical factors contributing to this variance[4,5]. However, no study has compared the mechanical responses of the atherosclerotic tissue at these two locations as potential contributors to the varying restenotic response following endovascular intervention. The mechanical responses that influence restenotic response are: propensity to undergo stretch injury[6], ability to bear elevated stress levels[7], and tissue stiffness[8]. This study characterises these parameters in a sample group of excised carotid and femoral human atherosclerotic plaques to elucidate differences that may be exploited to explain the variance in restenosis rates.

Furthermore, the markers of this heightened restenosis rate remain unclear and independent predictors are required to assess patient suitability for endovascular intervention[9]. To develop such a predictor, Fourier Transform InfraRed (FTIR) spectroscopy is employed to characterise the biological composition of each plaque sample. This allows for biological parameters to be identified, correlated with mechanical properties and thus potential restenosis rates. Such a predictive parameter could be employed to pre-operatively stratify patients into different treatment strategies ensuring that patients at increased risk of developing restenosis following endovascular intervention receive open surgery or a monitoring protocol.

METHODS

Uniaxial extension testing was performed on 24 carotid and 15 femoral plaque samples gathered from consenting patients undergoing endarterectomy at the University Hospital Limerick. This method was employed to characterize the relevant mechanical responses by applying circumferential stretch in a manner comparable to endovascular device deployment. The samples are clamped in an in-house developed uniaxial tester and pre-conditioned using five loading-unloading cycles to 10% stretch at a strain rate of 0.1mm/s to facilitate for strain softening. The samples are then stretched to complete failure at a rate of 30% of gauge length per second[10]. The force and displacement values are recorded and converted to Cauchy stress-stretch ratio plots. From these plots the stretch that the plaque tissue can withstand prior to ultimate tissue failure (λ_F), the corresponding stress induced in the tissue at this point (σ_F), and also the high stiffness response (E_H) of the samples can be obtained.

Probe based attenuated total reflectance FTIR at a resolution of 2 cm^{-1} and 16 scans over the range of 4000 – 700 cm^{-1} is performed on a number of locations across the luminal surface of each sample to obtain a range of spectra (Spectrum 100, Perkin Elmer Inc., MA, USA). Following water subtraction, the areas under the key peaks are measured (Lipid: 2980–2810 cm^{-1}, Lipid Ester: 1750–1715 cm^{-1}, Collagen: 1715–1585 cm^{-1} and Calcification: 1180–900 cm^{-1}). This quantifies the ratios of collagen (Col), lipid (Li), and calcification (Ca) to the total measureable content (Tot) i.e. Col:Tot, Li:Tot and Ca:Tot.

RESULTS

Carotid and femoral plaques demonstrate significantly distinct mechanical responses. The mean λ_F, σ_F, and E_H values are all significantly lower in the femoral group compared to the carotid group. Biological characterisation reveals significantly higher levels of

Ca:Tot and Col:Tot and also significantly lower levels of Li:Tot in the femoral group compared to the carotid group, table 1.

TABLE 1: MEAN MECHANICAL RESPONSES AND BIOLOGICAL COMPOSITION OF BOTH GROUPS

Variable	Carotid	Femoral	p value
λ_F	1.89±0.36	1.62±0.28	0.02
σ_F	0.49±0.23	0.22±0.13	<0.001
E_H	0.89±0.52	0.44±0.27	<0.001
Ca:Tot	0.2±0.06	0.25±0.09	0.038
Col:Tot	0.54±0.1	0.64±0.09	0.005
Li:Tot	0.26±0.1	0.11±0.02	<0.001

Significant correlations exist between Ca:Tot and λ_F and also between Li:Tot and λ_F, and σ_F. Col:Tot does not correlate significantly with any of the measured mechanical responses, table 2.

TABLE 2: SPEARMAN'S RHO COEFFICIENTS OBTAINED FROM CORRELATING VARIABLES

	λ_F	σ_F	E_H
Ca:Tot	-0.42 (p = 0.008)	-0.3 (p = 0.061)	-0.12 (p = 0.483)
Col:Tot	-0.1 (p = 0.529)	-0.11 (p = 0.507)	-0.2 (p = 0.215)
Li:Tot	0.39 (p = 0.015)	0.41 (p = 0.009)	0.32 (p = 0.051)
Ca:Li	-0.48 (p = 0.002)	-0.49 (p= 0.002)	0.33 (p = 0.037)

Multiple regression was performed to assess the combined impact of Li:Tot and Ca:Tot on the mechanical responses. The model to predict λ_F was the only one to demonstrate a significant relationship (p=0.031). Within this model, the only terms that had significant contributions were the interaction terms i.e. (Ca:Tot)(Li:Tot). A further biological parameter was therefore defined to represent this interaction: Ca:Li. Bivariate analysis reveals significant correlations between Ca:Li and λ_F, σ_F, and E_F, table 2. Regression analysis was performed to assess the predictive power of Ca:Li. The regression models exhibit significant relationships between Ca:Li and λ_F (R^2adj=25.2% p=0.001), and σ_F (R^2adj=29.4% p=0.002), figure 1.

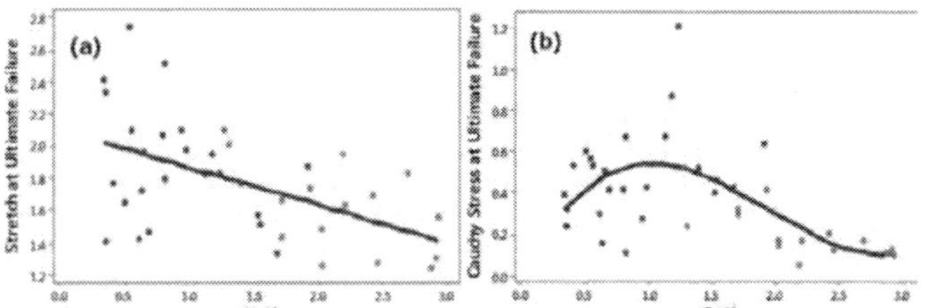

FIGURE 1: REGRESSION MODELS FOR λ_F (a) AND σ_F (b) USING Ca:Li AS A PREDICTOR

DISCUSSION

Femoral atherosclerotic tissue displays significantly lower λ_F compared to carotid plaques. This parameter has been demonstrated to positively correlate with the degree of restenosis[6]. These results suggest that the atherosclerotic intima of femoral vessels is more likely to undergo dissection during endovascular treatment with subsequent injury to the internal elastic lamina and media which triggers increased restenosis. Femoral plaques also demonstrate significantly lower σ_F compared to carotid plaques. Increased peak stresses in the artery walls are associated with increased restenosis[7]. The carotid group's ability to bear increased circumferential stress prior to ultimate failure suggests that the biomechanical environment is impacted less severely by the induced stress and that the restenotic response will be less acute than in the femoral group. Femoral plaques also display significantly lower E_F compared to carotid plaques. The stiffer atherosclerotic intima found in the carotid vessels may act as a stress-shield during treatment and allow less load to be transferred to the media. This may inhibit restenosis due to reduced disruption and decreased stress in the media. The three mechanical responses therefore offer explanation as to the distinct restenotic responses exhibited by the two vasculatures.

Despite the significant differences in mechanical responses and biological composition of the carotid and femoral human plaque tissue, the biological parameter Ca:Li has been identified as an indicator of stretch injury and stress induced during circumferential extension for both plaque groups and accounts for approximately 25% of the change in λ_F and 29% of the change in σ_F. Ca:Li correlates negatively with λ_F, figure 1a. This is expected as calcified tissue correlates negatively and lipid content correlates positively with λ_F. Calcified nodules disrupt and de-laminate fibrous stratum[10]. This decreases the load bearing capacity of the plaque tissue and reduces λ_F. Conversely, lipid rich plaques undergo increased stretch compared to fibrocalcific plaques under similar loading[11]. Therefore, expressing calcified tissue relative to lipid content exaggerates these independent relationships and creates a parameter with improved predictive power.

Ca:Li correlates negatively with σ_F, table 2. However, up to a Ca:Li value of 1, the relationship is positive, figure 1b. After this the predictive curve plateaus until to a value of 1.5 and the relationship then becomes negative. As calcified tissue levels increase within plaque, stress levels during loading also increase due to stress concentrations at interfaces between calcified and non-calcified tissue. The number of these interfaces initially increases with increasingly calcified tissue until the calcified regions begin to coalesce. Increasing calcified tissue then reduces the number of these interfaces due to coalescence[12]. This manifests as an initial rise in stress, followed by a successive decrease in stress as observed in the regression model.

It has been suggested that FTIR can be developed as a catheter based technology[13]. The parameter Ca:Li would then become a clinically available pre-operative predictor of restenotic response following endovascular intervention. Furthermore, calcified tissue and lipid content can currently be identified clinically by optical coherence tomography and magnetic resonance imaging. Employing these existing technologies in conjunction with this defined biological parameter may allow clinicians to pre-operatively predict the restenotic response of atherosclerotic plaque tissue, aid in stratifying patients that are at increased risk of developing restenosis following endovascular intervention, and optimize the treatment process thereby improving patient outcomes and reducing treatment costs.

ACKNOWLEDGEMENTS

This work was conducted under the Biomedical Engineering and Regenerative Medicine (BMERM) structured PhD programme and the Irish Government's Programme for Research in Third Level Institutions Cycle 5, with the European Regional Development Fund and the Irish Research Council EMBARK Initiative.

REFERENCES

[1] Lal BK et al, 2012. Lancet Neurol. 11, 755–763.
[2] Schillinger M et al, 2006. N. Engl. J. Med. 354, 1879–1888.
[3] Mahoney EM et al, 2008. Circ. Cardiovasc. Qual. Outco. 1, 38–45.
[4] Herisson F et al, 2011. Atherosclerosis 216, 348–354.
[5] Cassese S et al, 2014. Heart 100, 153–159.
[6] Kornowski R et al, 1998. J. Am. Coll. Cardiol. 31, 224–230.
[7] Timmins LH et al, 2011. Lab. Invest. 91, 955-967.
[8] Mahfouz RA et al, 2013. Echocardiogr. Mt. Kisco N 30, 582–587.
[9] Jones DW et al, 2013. J Am Heart Assoc. 2013;2:e000345.
[10] Mulvihill JJE et al, 2013. Acta Biomater. 9, 9027–9035.
[11] Cheng GC et al, 1993. Circulation 87, 1179–1187.
[12] Abedin M et al, 2004. Arter. Thromb. Vasc. Biol. 24, 1161–1170.
[13] Tosi G et al, 2012. Anal. Quant. Cytol. Histol. 34, 214–232.

MURINE MODEL OF BLAST TRAUMATIC BRAIN INJURY: AN IMAGING, BEHAVIORAL, AND PATHOLOGICAL ASSESSMENT

Sujith V Sajja (1), Jiangyang Zhang (1), Jeff Bulte (1), Joseph Long (2), Robert D. Stevens (1,3), Piotr Walczak (1,2), Miroslaw Janowski (1,4)

(1) Department of Radiology
Johns Hopkins School of Medicine
Baltimore, MD, USA

(2) Blast Neurotrauma Division
Walter Reed Army Insitute of Research
Silver Spring, MD, USA

(3) Anesthesiology Critical Care Medicine
Johns Hopkins School of Medcine
Baltimore, MD, USA

(4) NeuroRepair Department, MMRC,
PAS, Warsaw, Poland

INTRODUCTION:

To date, the rat has been the most widely used a pre-clinical experimental system to investigate blast traumatic brain injury (bTBI) [1,2]. However, the use of mice is very attractive due to the low cost, high-throughput, and availability of knockout and transgenic animals. The effects of explosive blast on the mouse brain remains poorly studied. The current study evaluates the effects of different level of blast intensity in a mouse model of bTBI. Moreover the study evaluates bTBI effects with the help of a multiparametric and clinically relevant readout that includes behavioral assessment, MRI, blood markers, and neuropathology.

METHODS:

Different groups of Balb/c mice (n = 8/group) were exposed to peak, static, single-blast overpressure of 17, 20, and 23 psi in addition to a repeated blast overpressure (BOP) of 17*3 and 20*2 psi with 8-9ms positive phase duration, while the control animals underwent the same procedure except for the exposure to BOP. Immediately after BOP, righting reflexes were recorded. Animals also underwent T2 and T2* MRI on day 1 to study acute gross morphological changes, using a high-resolution 11.7T MR scanner. Animal behavior was tested on day 2 and day 3 using an open-field test, and MR diffusion tensor imaging (DTI) was performed on day 4 in the 20 psi pressure group following BOP. Post-mortem brains were collected for immunohistochemistry, and plasma was collected from blood for further analysis.

RESULTS:

Survival assessment showed no deaths in the control groups, while the survival rate of the 17, 20, 23, 17*3, and 20*2 groups was 87.5%, 75%, 25%, 100%, and 11%, respectively (Figure 1). Time required for righting in the righting reflex test showed a significant correlation with overall mortality rate, but not for immediate mortality rate ($p < 0.05$). Although no gross anatomical changes were observed using T2 and T2* MRI, ventricle volumes were significantly reduced in all the groups at day 1 following BOP. Behavior assessment showed reduced overall activity in the 20, 23, and 20*2 pressure groups, with no changes in the 17 and 17*3 pressure groups ($p < 0.05$). Increased thigmotaxia (anxiety-like behavior) was observed in the 20 and 23 pressure groups. DTI data showed a significant increase in fractional anisotropy (FA) and axial diffusivity (AD) of the internal capsule (IC) and fimbria (Fi) on day 4 following BOP ($p < 0.05$) (Figure 2). Immunohistochemistry demonstrated a significant increase in apoptosis of the choroid plexus in the 17*3 group, while the analyses of other groups are currently underway ($p < 0.05$).

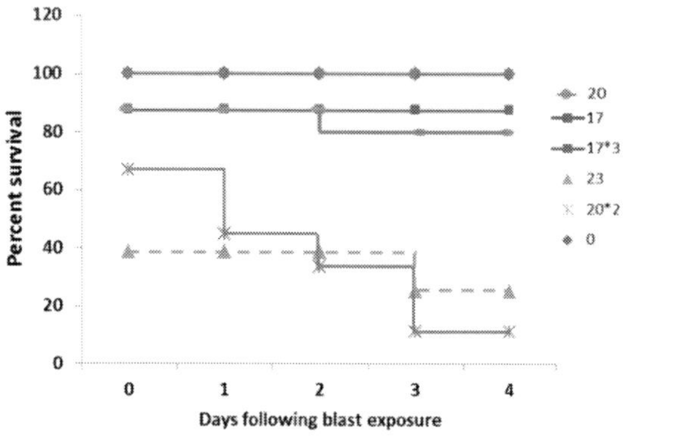

Figure 1: Time course of survival rates following blast overpressure at various different intensities.

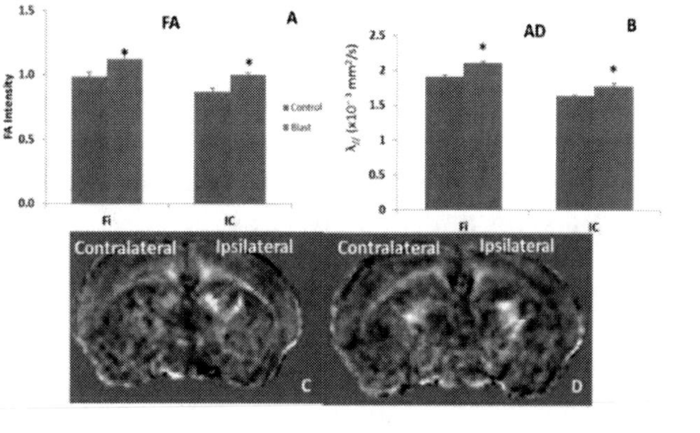

Figure 2: A significant increase in FA (A) and AD (B) was shown in Fi and IC regions of the brain at 4 days following blast overpressure exposure in the 20psi group (*$p < 0.05$). C and D are representative differential images (Avg. Blast – Avg. Control) obtained using image processing software AMIRA™.

DISCUSSION

This study successfully established a relationship between blast intensity and survival rate, as well as demonstrating specific neuroimaging and neuropathological changes in a murine model of bTBI. The reduction of ventricle volume is an important observation that could be related to increased edema. This can subsequently increase intracranial pressure, a common clinical sign following TBI, and is consistent with previous rat models of bTBI [3, 4]. Although the pathological basis for increased FA and AD on DTI are currently unknown, this data is consistent with some clinical DTI studies of bTBI. Surprisingly, we failed to observe any focal brain injury in any of the pressure groups, despite the high mortality in some pressure groups (post-mortem evaluation of animals that died was also performed). Although not studied, we suspect that non-neurological factors (e.g. lung injury) might account for some of the mortality, as has been demonstrated by others. It can be potentially related to the chest structure of mice, which warrants further research on murine biomechanics to validate the model. Collectively these results indicate a methodological framework for bTBI research in mice, opening the way for genetically modified strains to be used in further evaluating the neuropathogenesis of blast and the identification of therapeutic targets.

ACKNOWLEDGEMENTS

The research described here was supported in part by Department of Defense award number W81XWH-13-1-0389

REFERENCES

[1] Sajja, VS et al., NMR Biomed., 2012; 25(12):1331-9.

[2] Wang Y et al., J. Neurotrauma, 2011; 28(10):2171-83.

[3] Dalle Lucca JJ et al., J. Neurol Sci., 2012; 318(1-2):146-54.

[4] Dal Cengio Leonardi A, J. Biomech., 2012; 45(15):2595-602.

SB³C2015
Summer Biomechanics, Bioengineering and Biotransport Conference
June 17-20, 2015, Snowbird Resort, Utah, USA
SB³C2015-382

MOVEMENT ANALYSIS BASED ON THE SEPARATION OF ANGULAR MOMENTUM: GAIT PROGRESSION AND SEGMENT ROTATION IN PATIENTS WITH TKA

Brecca M. Gaffney (1), Will M. Johnston (1), Cory L. Christiansen (2), Jennifer E. Stevens-Lapsley (2), Kevin B. Shelburne (1), Bradley S. Davidson (1)

(1) Department of Mechanical & Materials Engineering
University of Denver
Denver, CO, USA

(2) Department of Physical Medicine & Rehabilitation
University of Colorado Denver
Aurora, CO, USA

INTRODUCTION

Patients with total knee arthroplasty (TKA) remain less functional than their healthy counterparts even up to three years following surgery [1]. This is likely caused by movement compensations that develop in the advanced stages of osteoarthritis, change again following TKA surgery, and do not fully resolve. The most common movement pattern (e.g. stiff knee strategy) is associated with unloading the involved knee and reducing demands on the quadriceps. Although this pattern is evident immediately after surgery by clinical observation, persistent long-term movement patterns can be difficult to quantify, and therefore remain untreated.

To improve the effectiveness of rehabilitation interventions, we aim to explore an interpretable dynamic framework that can be used to quantify abnormal movement patterns using the segmental contributions to whole-body angular momentum (WBAM) about the foot during gait. Previous investigations that calculated WBAM about the body center of mass (COM) demonstrated that despite large segmental contributions, the composite signal of WBAM is very small due to segmental contributions [2]. However, the individual contributions of each segment to WBAM about the body COM are not interpretable as a movement strategy to achieve forward progression.

Choosing the foot as a reference point should provide an interpretable outcome based on successful modeling of human walking as an inverted pendulum with a rotational fulcrum point at the foot [3]. To achieve forward motion, the body COM is rotated about this fulcrum in the sagittal, frontal, and transverse directions until the contralateral limb contacts the ground. Transfer to the other limb is achieved immediately after double stance, which shifts the point of reference as the gait cycle progresses.

The principle of angular momentum separation [4] allows the segmental contribution to WBAM foot to be divided into two parts and may provide additional insight into performance and strategy:

1. Angular momentum of the segment with respect to the foot.
2. Angular momentum of the segment about its own COM.

The objective of this investigation was to extend the separation of angular momentum to explain forward progression and segmental movement strategy during gait in healthy participants and patients following TKA surgery. The analysis demonstrates two dynamic concepts: 1) The contributions of segment angular momenta with respect to the foot are an indicator of forward progression and functional performance, and 2) The individual angular momenta of the segments about their COMs are an indicator of movement patterns used to achieve performance. We hypothesized that 1) momenta with respect to the foot would be similar in the healthy and TKA groups in all three planes and 2) segment momenta about their segment COMs would differ between groups and provide interpretable movement patterns.

METHODS

Six patients with unilateral TKA (2F, 4 M, age: 68.5 ± 9.6 years, BMI: 29.1±3.6 kg/m^2) 6-weeks after surgery and six healthy controls (1F, 5M, age: 67.7±7.9 years, BMI: 26.1±2.2 kg/m^2) performed three gait trials at 1.0 m/s. Motions were recorded at 100 Hz from 64 skin markers (Vicon). A 15-segment model was created in Visual 3D (C-Motion, Inc.) to obtain segment kinematics. Segment masses were calculated as a percentage of total body mass and inertial properties were determined by segment geometry [5]. A single stride from heel strike to heel strike on the involved limb (right limb for healthy controls) was analyzed for this investigation.

WBAM about the involved foot was calculated by adding the total angular momentum of each segment about the foot COM (FT). The total angular momentum is the sum of angular momentum relative to FT and the angular momentum of the segment:

$$\mathbf{h}_{FT} = \mathbf{h}_{seg/FT} + \mathbf{h}_{seg} \quad (1)$$

Angular momentum of a segment with respect to the FT is:

$$\mathbf{h}_{seg/FT} = (\mathbf{r}_{seg} - \mathbf{r}_{FT}) \times m_{seg}(\mathbf{v}_{seg} - \mathbf{v}_{FT}) \quad (2)$$

Where $\mathbf{r}_{seg}$ and $\mathbf{r}_{FT}$ are the position vectors of the segment and foot, respectively, m_{seg} is the mass of the segment, and $\mathbf{v}_{seg}$ and $\mathbf{v}_{FT}$ are velocity vectors of the segment and foot, respectively. Angular momentum of a segment about its COM is:

$$\mathbf{h}_{seg} = \mathbf{I}_{seg} \cdot \omega_{seg} \quad (3)$$

where $\mathbf{I}_{seg}$ is the moment of inertia tensor and ω_{seg} is the angular velocity of the segment. All time derivatives were observed in the inertial reference frame.

The angular momenta vectors were transformed into path coordinates, defined by the velocity vector of the body COM. Segment momenta were scaled by participant mass, body COM height, and walking speed for visualization.

RESULTS

Angular momentum of the segment with respect to the FT

For all segments and in all planes, the angular momentum of the segment with respect to the FT was 200 to 250 times larger in magnitude than the angular momentum of the segment about its COM. As a result, WBAM about the FT was primarily composed of the angular momentum of each segment with respect to the FT. The curves had a similar shape for all segments with progressively larger magnitudes with superior segments (trunk/pelvis) (Figure 1).

Comparison between the healthy and TKA groups indicated angular momentum with respect to the FT were similar in the sagittal plane (Figure 1a). In the frontal plane, patients with TKA demonstrated larger variability during unloading of the involved limb (50-60%) (Figure 1b).

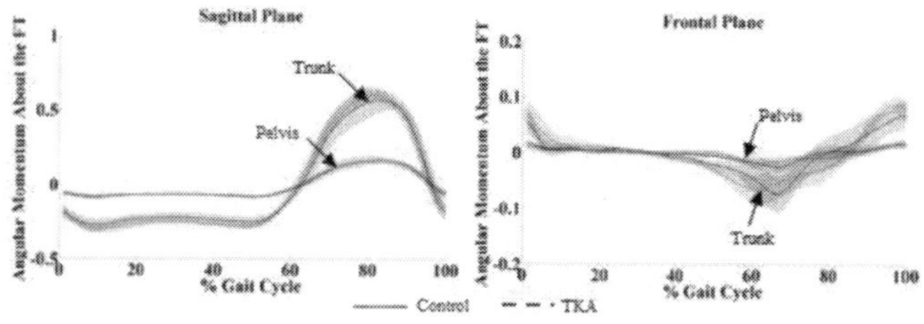

Figure 1. Segment angular momentum with respect to the foot in the (a) sagittal plane and (b) frontal plane.

Angular momentum of the segment about its COM

Sagittal Plane: Unlike angular momentum with respect to the FT, angular momenta of the segment about its COM curve shapes were very dissimilar across segments. Momenta of thigh and shank segments on both sides were much higher in magnitude than all other segments, and had similar shapes across groups (Figure 2a). Patients with TKA demonstrated smaller trunk momentum at both foot strikes compared to healthy controls (Figure 2b).

Frontal and Transverse Planes: Momenta of trunk and pelvis segments were much higher in magnitude than all other segments in both the frontal and transverse planes (Figure 2a). In the frontal plane, patients with TKA demonstrated smaller trunk rotational momentum at involved limb foot strike, and were similar to healthy controls at uninvolved limb foot strike (Figure 2c). In the transverse plane, patients with TKA demonstrated smaller trunk rotational momentum across the entire gait cycle in comparison to healthy controls (Figure 2d).

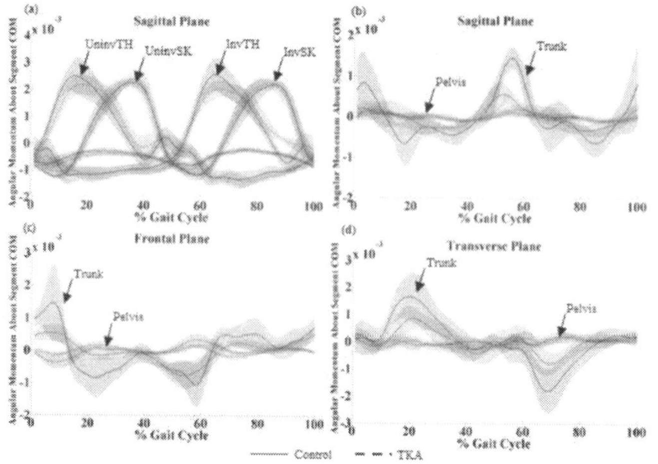

Figure 2. Segment angular momentum about each segments COM. (a) sagittal plane: uninvolved thigh (UninvTH), uninvolved shank (UninvSK), involved thigh (InvTH), and involved shank (InvSK), (b) sagittal plane: trunk and pelvis, (c) frontal plane: trunk and pelvis, (d) transverse plane: trunk and pelvis.

DISCUSSION

This investigation demonstrates differences in segmental momentum components of WBAM when taken with respect to the involved limb in patients with unilateral TKA and healthy controls during gait. This segmental separation of WBAM allows for a unique interpretation of segmental contributions to angular momentum, which can be used to interpret movement strategies.

The similar shape of segment angular momentum with respect to the FT and progressively larger magnitudes with superior segment position correspond well with the inverted pendulum model of forward progression during gait [5]. Therefore, the component of angular momentum taken about a reference point (e.g. the foot) can be interpreted as the "progressive momentum". During gait, the progressive component of angular momentum was similar across groups, which indicates that patients with TKA progressed forward similarly to healthy controls.

The differences in segment angular momentum about its COM across groups indicate unique movement strategies adopted by patients with TKA. Because this "rotational momentum" is calculated using the segment angular velocity, this term can be can be used to interpret movement strategies used to achieve forward progression during gait. Patients with TKA minimized trunk rotational momentum in all three planes, which may indicate a potential protective strategy to increase stabilization by reducing twisting and lateral rotation of the involved limb. Smaller in trunk rotational momentum in the frontal plane may indicate that patients with TKA are attempting to reduce mediolateral trunk motion and promote anterior momentum, which assists with forward progression. However, this strategy may result in adverse effects caused by asymmetric loading patterns (e.g. vertebral compression, development of hip osteoarthritis).

REFERENCES

[1] Yoshida, Y et al., *J Orthop Sports Phys Ther*, 42: 1039-49, 2012.
[2] Herr, H et al., *J Exp Biol*, 211: 467-481, 2008.
[3] Hof, A *J Biomech*, 38: 1-8, 2005.
[4] Kasdin, N et al., *Intermediate Dynamics*, 2011.
[5] Dempster, P et al. *Med Sci Sports Exerc*, 27: 1692-1697, 1995.

SB³C2015
Summer Biomechanics, Bioengineering and Biotransport Conference
June 17-20, 2015, Snowbird Resort, Utah, USA
SB³C2015-383

AN APPROACH TO INCLUDE PRE-STRESS IN A FINITE ELEMENT MODEL OF CONVECTION-ENHANCED DELIVERY WITH BACKFLOW

Gustavo A. Orozco (1), Fabián A. Urrea (1), Fernando Casanova (1), Joshua H. Smith (2), José J. García (3)

(1) Escuela de Ingeniería Mecánica
Universidad del Valle
Cali, Colombia

(2) Department of Mechanical Engineering
Lafayette College
Easton, PA, USA

(3) Escuela de Ingeniería Civil y Geomática
Universidad del Valle
Cali, Colombia

INTRODUCTION

Convection-enhanced delivery (CED) is a technique to deliver controlled amounts of drugs directly into the brain for the treatment of tumors and diseases such as Parkinson's [1]. Several CED animal studies have shown encouraging results for drug distribution [2, 3]. However, clinical trials [4] have reported low efficacy of CED, primarily due to the poor distribution of the drug, which, in many cases, has been related to backflow, a phenomena in which the infused liquid flows toward the surface of the brain in the annular gap created outside of the cannula surface.

Analytical models have complemented the clinical and experimental studies to establish the variables that reduce backflow [5]. A recent finite element model that incorporates backflow and includes material and geometrical nonlinearities was developed to predict fluid distributions under flow-controlled infusions [6]. Nevertheless, this computational model does not directly include the effect of pre-stress generated between the tissue and the cannula during insertions. This pre-stress, which has been shown experimentally to have an important influence on the backflow length, is dependent on the degree of damage during insertion, which in turn is affected by the insertion velocity [7].

Hence, the aim of this study was to include the effect of pre-stress into a finite element model in order to predict fluid flow and backflow length during flow-controlled infusions. Backflow length model predictions were compared with experimental data from agarose gel.

METHODS

Our previous axially symmetric biphasic model for infusion into the brain that includes geometric and material nonlinearities and represents backflow [6] was modified to include pre-stress. In the modified model, pre-stress was generated by applying a pre-infusion radial expansion from an initial radius to the external radius of the catheter using a rigid body. The generated pre-stress was dependent on the amount of radial expansion and the properties of the material.

After this pre-infusion radial expansion, the infusion with backflow was simulated as in our previous model. The agarose gel was represented as a biphasic material composed of a fluid and a visco-hyperelastic solid phase represented by an Ogden-type energy function and a Prony series. The hydraulic conductivity was assumed to vary with strain. The material properties of the gel were determined experimentally through unconfined and confined relaxation compression tests, as explained in a companion paper [8].

The finite element model was calibrated with results of infusions of 5 μl of 0.02% (w/v) bromophenol blue dye delivered into a 0.4% (w/v) agarose gel. Infusions were performed at three flow rates (0.5, 2.0, and 4.0 μl/min) using a 22-gauge blunt tip needle. The backflow length was visually measured as explained previously [7].

Since the insertion velocity affects both the pre-stress and the backflow length, to compare model predictions at different levels of pre-stress, three insertion velocities (0.2, 2.0, and 10 mm/s) were considered in this study. The mean backflow length at each insertion velocity was used to determine the radial expansion required in the finite element model to obtain a similar backflow length.

RESULTS

The backflow length measured in agarose gel decreased significantly ($p < 0.05$) with increasing insertion velocity and decreasing flow rate (Fig. 1). It was found that the mean experimental backflow lengths could be generated in the finite-element model using rigid-body radial displacements of 0.0090, 0.0385 and 0.1200 mm for needle insertion velocities of 0.2, 2.0, and 10 mm/s, respectively. With these radial displacements, the simulations produced pre-stresses of 623, 2662 and 8297 Pa, respectively, with maximum values near the infusion cavity and relatively constant values along the axis of the cannula (Fig. 2). The pre-stress model yielded virtually the same velocity distribution as that obtained with our previous model without pre-stress for the lowest insertion velocity, and thus for the lowest pre-stress (Fig. 3). However, for the higher pre-stress obtained at greater insertion velocities, the model with pre-stress predicted a larger fluid velocity near the infusion cavity and a smaller backflow length with respect to the model without pre-stress (Fig. 3).

DISCUSSION

In contrast with our previous finite element model of backflow [6], the present model accounts for the effect of pre-stress generated between the needle and the material during insertion. Hence, the model is capable of indirectly accounting for the effect of needle insertion velocity, which in turns affects the degree of damage in the material and the level of pre-stress. The model was able to reasonably predict the change of backflow length with insertion velocity. At a

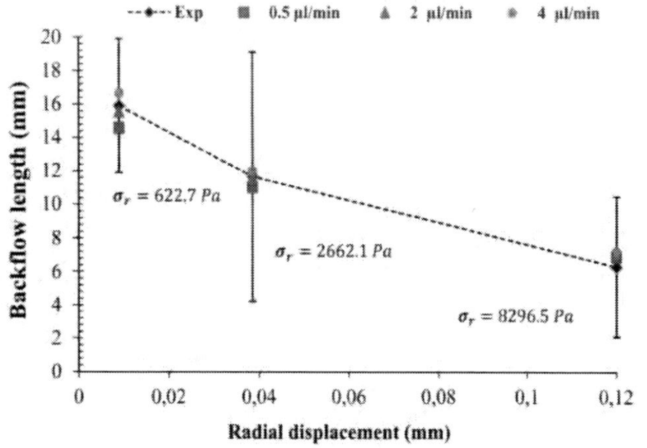

Figure 1. Experimental backflow lengths and numerical predictions for the three insertion velocities.

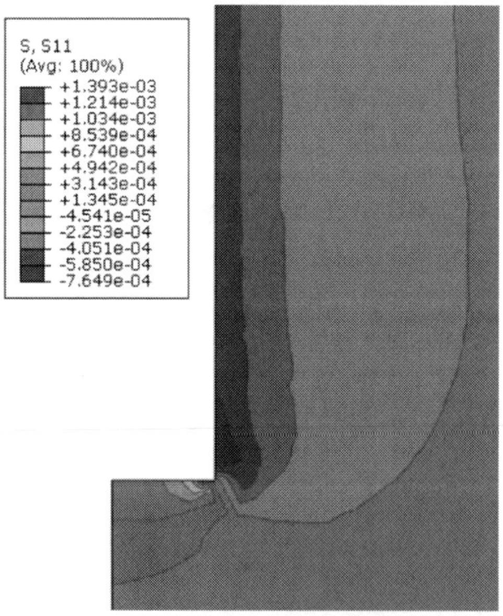

Figure 2. Pre-stress (MPa) obtained with a radial expansion of 0.009 mm, which corresponds to an insertion velocity of 0.2 mm/s.

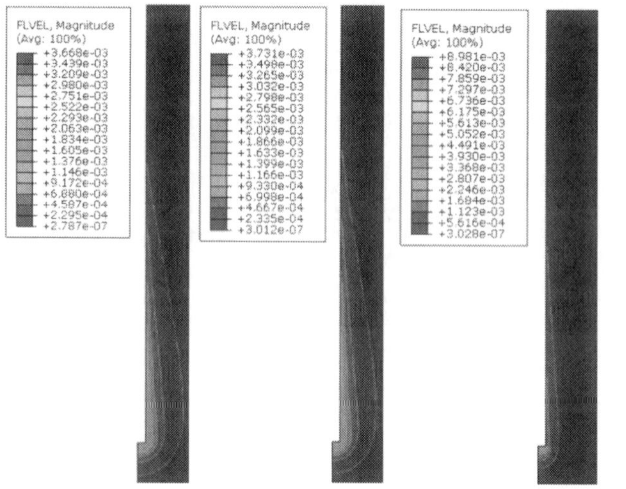

Figure 3. Fluid velocity contours in the model without pre-stress (left), the model with pre-stress of 623 Pa (center), and the model with pre-stress of 8297 Pa (right), each for a flow rate of 2 mm^3/s. Note the different color scales used.

higher pre-stress, the model showed larger fluid velocities near the tip of the catheter, which enhances drug transport to the tissue in front of the tip and not to the annular zone around the catheter.

Experimental results indicating that a higher pre-stress and lower damage were produced at higher insertion velocities are consistent with previous studies in agarose gel [7]. However, the pre-stress values were higher than those reported in the previous study [7], which may be due to the experimental and theoretical approaches used here, as well as differences in the material properties. Specifically, while intrinsic viscous effects were included in the current finite element simulations, the previous study considered the material to be hyperelastic.

Extension of the model to brain tissue will require performing animal experiments, in particular since a previous study documented that the degree of damage for brain tissue is lower at lower insertion velocities [9], opposite to that found for agarose gel.

ACKNOWLEDGEMENTS

The authors appreciate the support of the Universidad del Valle and Lafayette College to undertake this study. Thanks are also due to Colciencias program 516-2012 (contract #110656933826) "Programa Nacional de Ciencia y Tecnología de la Salud" for the financial support.

REFERENCES

[1] Bobo, R. H. et al., *Proc. Natl. Acad. Sci.*, 91:2076–2080, 1994.
[2] Brady, M. L. et al., *J. Neurosci. Methods*, 229:76–83, 2014.
[3] Rosenbluth, K. H. et al., *J. Magn. Reson. Imaging*, 36:1339–1346, 2012.
[4] Sampson, J. H., et al. *J. Neurosurg.*, 113:301–309, 2010.
[5] Raghavan, R., et al., *Phys. Med. Biol.*, 55:281–304, 2010.
[6] Orozco, G. A., et al., *Med. Biol. Eng. Comput.*, 1:1–9, 2014.
[7] Casanova, F., et. al., *J. Biomech. Eng.*, 134:041006, 2012.
[8] Cordoba, G., et al., *Proc. Sum. Biomech., Bioeng. Biotrans. Conf.*, 15-A-114-SB3C, 2015
[9] Casanova, F., et al., *Neursosc Methods.*, 237:79-89, 2014.

A COMPUTATIONAL METHOD FOR VISUALIZING FEMORAL RANGE OF MOTION FOR PATIENTS WITH SLIPPED CAPITAL FEMORAL EPIPHYSIS

Ferris M. Pfeiffer, PhD (1,2), David Tager, MD (2), Sumit Gupta, MD (2)

(1) Department of Bioengineering
University of Missouri
Columbia, MO, USA

(2) Department of Orthopaedic Surgery
University of Missouri
Columbia, MO USA

INTRODUCTION

Incidence of slipped capital femoral epiphysis typically affects adolescent patients [1]. While it is thought that obesity plays a significant factor in the development and/or progression of the disease [2], it is not entirely understood what the impact of slip severity has on the limits to normal range of motion immediately following injury, or long term with and/or without surgical intervention. It is the purpose of this study to develop and validate a computational method which can be utilized to evaluate surgical and/or non-surgical options in the treatment of chronic and/or acute slipped capital femoral epiphysis. The method is based on clinical imaging modalities (axial CT scanning) currently employed for patients presenting with symptoms consistent with the disease.

METHODS

De-identified computed tomography (CT) scans for 10 pediatric patients diagnosed at our orthopaedic clinic were acquired following IRB exemption. Each scan was manually segmented using Amira (Mercury Computers, Berlin Germany) to extract 3D surface representations of patient geometry including each ilium and both femur (Figure 1). Surface representations of each component were exported in ASCII format for further processing. A custom FORTRAN (Intel, Santa Clara, CA) routine was written to computationally transform the components in the model. In brief, the geometric center of each acetabular cup was computed from the clinical CT scan by choosing four points on the rim of the acetabulum and approximating the cup as a sphere. Once each cup center was determined, an axis was placed from center to center of the acetabulum. Anatomical landmarks were used to place a femoral neck axis originating at the greater trochanter and pointing through the center of the femoral neck toward the femoral head. Utilizing the FORTRAN routine, each femur was virtually rotated about the acetabulum center in independent planes (e.g. X-Y, X-Z, or Y-Z plane) until bony impingement was observed. Anatomical axis were set with the origin at the center of the pelvis and X axis extending from medial to lateral, Y axis extending from anterior to posterior, and Z axis extending from inferior to superior. The angle between the femoral neck vector and the anatomical axis through the pelvis was determined in each plane respectively. The difference between

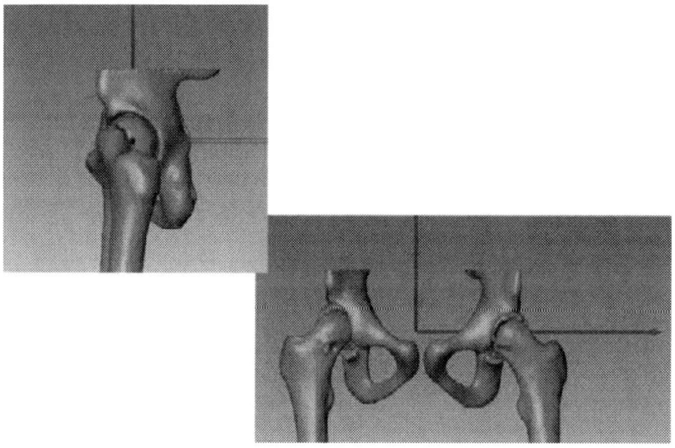

Figure 1. 3D representations of patient anatomy with anatomical axis used in the transformations

maximum positive and negative rotations was used to determine range of motion. Additionally, neutral posture was determined from the CT scan posture. The dataset of patients included 20 total joints including normal anatomy (n=6), as well as slipped stable patients (n=6) with no surgical intervention (n=4), and slipped stable patients fixed using a single fixation screw (n=4). A comparison was made between patients in the normal group (N), slipped without fixation (SnF), and slipped with fixation (SF).

RESULTS

Mean range of motions for the patients evaluated are included in figure 2.

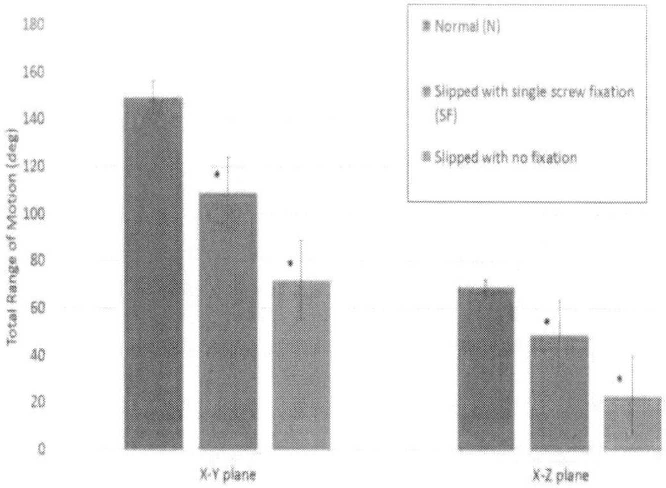

Figure 1. 3D representations of patient anatomy with anatomical axis used in the transformations

Range of motion for slipped capital epiphysis fixed with a single screw was significantly less (p=0.03) than the measured normal range of motion (N=149° vs SF=109°). Range of motion for femur with slipped capital epiphysis without supplemental fixation was significantly less (p=0.04) than the measured normal range of motion (N=149° vs SnF=71°). Range of motion without supplemental fixation demonstrated the least range of motion for any group evaluated. Rotation in the X-Y plane (lateral and medial rotation) was the largest measured (N=149 deg, SF=109 deg, SnF=71 deg).

DISCUSSION

The method described here is capable of virtually rotating patient anatomy to evaluate range of motion for slipped capital femoral epiphysis. Therefore, surgical options can be evaluated to see the effect of fixing the cap in place on range of motion as well as the effect of various surgical reductions on the clinical range of motion. It was observed that patients with slipped capital epiphysis and supplemental fixation had increased range of motion as compared to patients with slipped epiphysis and no supplemental fixation. Both slipped capital epiphysis groups had significantly less range of motion observed than patients with normal anatomy. Further studies are in the works to evaluate additional surgical options and/or osteotomies of the femoral neck to reduce deformity.

SIGNIFICANCE

This work provides a non-invasive clinical tool to evaluate the effect of various surgical and non-surgical options for patients with slipped capital femoral epiphysis. The ability to evaluate the consequences of surgical and non-surgical interventions on the range of motion of patient anatomy will provide a valuable tool for surgeons to use in surgical planning. Enhanced surgical planning is expected to improve the functional outcomes of surgical and non-surgical slipped capital femoral epiphysis patients.

ACKNOWLEDGEMENTS

This study was funded by the Comparative Orthopaedic Laboratory

REFERENCES

[1] Punnoose, AR, et al. Slipped Capital Femoral Epiphysis, **JAMA** 2013

[2] Compton, S. Childhood Obesity and Slipped Capital Epiphysis, **Radiologic Technology** 2014

SB³C2015
Summer Biomechanics, Bioengineering and Biotransport Conference
June 17-20, 2015,Snowbird Resort, Utah, USA
SB³C2015-385

OPTIMIZATION OF PARAMETERS FOR LONG-TERM STORAGE OF TISSUE ENGINEERED ARTICULAR CARTILAGE

Adam B. Nover (1), Robert M. Stefani (1), Stephanie L. Lee (1), Rebecca A. Peyser (1), Daniel R. Howard (2), Gerard A. Ateshian (1,3), Aaron M. Stoker (4), James L. Cook (4) Clark T. Hung (1)

(1) Department of Biomedical Engineering
Columbia University
New York, NY, USA

(2) Department of Orthopedic Surgery
Mount Sinai St. Luke's
New York, NY, USA

(3) Department of Mechanical Engineering
Columbia University
New York, NY, USA

(4) Department of Orthopaedic Surgery
University of Missouri
Columbia, MO, USA

INTRODUCTION

Osteochondral allografts are a typical treatment for defects in articular cartilage of >2 cm². While fresh grafts have been shown to provide a long-term solution [1], current grafts are only available for transplantation after a ~14 day mandatory disease/contamination screening [2]. Following initial harvest, grafts are clinically stored in refrigeration (4 °C) in a bovine serum-containing media [3], yet cell viability decreases over time [1,4-5], leading surgeons to generally consider 28 days (including the screening period) as the end of the implantation window [2]. Many are attempting to increase the duration of this storage window, and these protocols vary, yet a recent system published by Stoker et al. (Missouri Osteochondral Allograft Preservation System) displayed storage with day 0 viability for 63 days without serum or refrigeration [6-7]. The exclusion of serum is desirable as it reduces any concerns associated with xenogenic media components, and room temperature simplifies storage logistics.

As cartilage allografts are in insufficient supply to meet clinical demand, tissue engineering is being utilized to fabricate biomimetic grafts. To facilitate eventual clinical translation, engineered grafts would benefit from effective preservation/storage protocols. However, due to differences between native and engineered cartilage (e.g., structure, composition), it is not known if there will also be differences in their ability to be preserved. In the current study, we aim to understand how MOPS can be adopted for tissue-engineered cartilage, examining temperature dependence (Study 1) and then comparing to a clinically representative media (Study 2).

METHODS

Passage 2 adult canine knee chondrocytes were cast into 4 mm diameter, 2.3 mm thickness, 2% w/v agarose Type VII disks (30 M cells/mL). Constructs were cultured in chemically-defined media with continuous TGF-€3 supplementation (3 media changes/week) until maturity point (>200 kPa, native Young•s modulus for canine knee cartilage [8]). In Study 1, mature constructs transferred to MOPS (no TGF-€3, 1 media change/week), split into groups stored at 4, 25, and 37 °C. In Study 2, mature constructs transferred to MOPS or a protocol representative of clinical storage [3] (contains FBS, no growth factors), then groups further split into storage at 4 and 25°C. Groups referred to by media and temperature (e.g., MOPS-25). Note: MOPS-25 and Clinical-4 groups represent those in the literature, while the others are at alternative temperatures. MOPS groups in Study 2 represent a repeat of Study 1 groups. For both studies, consistent with the current clinical window, constructs were evaluated at 14 and 28 days of storage. As an extension, constructs evaluated at 42 and 56 days of storage. Constructs were evaluated for mechanical (equilibrium Young•s Modulus, E_Y, dynamic modulus, G*, 0.01 Hz), biochemical (DNA: Picogreen dsDNA kit; glycosaminoglycan, GAG: 1,9-dimethylmethylene blue dye-binding assay; collagen: orthohydroxyproline colorimetric assay), and viability (Live/Dead kit) properties. Values analyzed as absolute, and normalized to wet weight (WW), dry weight (DW), or DNA. Media also analyzed for GAG content. For both studies, 2-way ANOVA with Tukey•s HSD post-hoc (,<0.05) used to determine significant differences between individual groups and the between groups and the maturity point.

RESULTS

Study 1: Constructs reached maturity after four weeks of culture, providing a maturity point baseline of E_Y: 286±77 kPa; G*: 1.90±0.22 MPa; GAG/WW: 5.40±0.23%; and Collagen/WW: 1.57±0.32%; results are displayed in Fig. 1 as fractions of baseline values. At 28 days preserved, constructs• E_Y became significantly higher for MOPS-4 than MOPS-37. This difference was apparent by 14 days preserved in G* (not shown). GAG content followed a similar trend, with this significant difference apparent at 14 and 28 days preserved. No

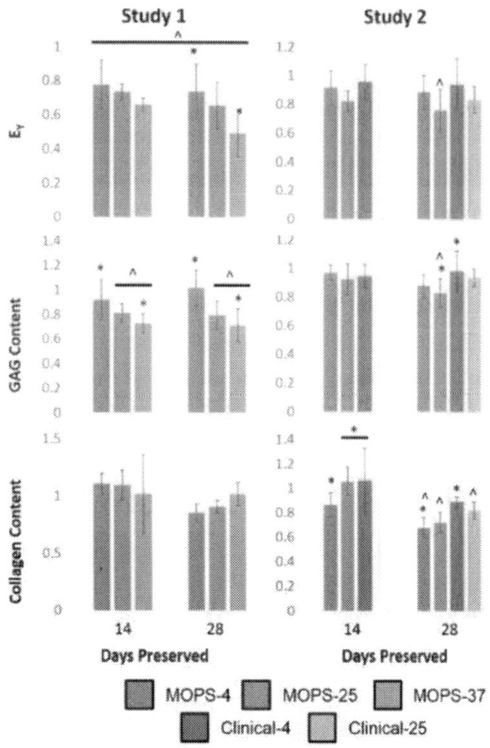

Fig. 1. E_Y, total GAG, collagen content normalized to respective maturity point values. ^ different than maturity point, * different among particular groups (p…0.05).

significant differences seen in collagen content. Compared to maturity point, E_Y and G* became significantly lower by 14 days, GAG content maintained at 4°C, but significantly lower at 25 and 37°C, and collagen maintained for the full 28 days. Viability staining showed majority live cells with no obvious differences across all groups after 28 days preserved (Fig. 2A). No obvious differences in gross morphology at 28 days preserved (Fig. 2B).

Study 2: Constructs reached maturity after four weeks of culture, providing a maturity point baseline of E_Y: 200±18 kPa; G*: 1.14±0.11 MPa; GAG/WW: 4.46±0.22%; and Collagen/WW: 1.13±0.16%. There were no significant E_Y or G* differences across groups at 14 or 28 days preserved (Fig. 1). Clinical-4 GAG content significantly higher than MOPS-25 at 28 days preserved. At 28 days preserved, collagen content significantly different between MOPS-4 and Clinical-4. At day 28, MOPS-25•s E_Y and GAG became significantly lower than the maturity point, and collagen was only maintained by the Clinical-4 group. Overall, at 28 days preserved, no significant differences between different temperatures per protocol or between different protocols per temperature.

Past 28 days preserved, in both studies, properties continued to drop over time, and differences in groups became more apparent, revealing decreasing properties with increasing storage temperature, yet no difference between MOPS and clinical protocols (E_Y shown in Fig. 3). For both studies, GAG concentration in the storage media increased significantly with increasing temperature (data not shown).

DISCUSSION

A proper storage/preservation protocol will be critical for the successful translation of tissue engineered cartilage from the lab to the clinic, giving the product a ƒshelf-life.,, MOPS is reliable for allografts and offers the benefits of being chemically-defined (no serum) and

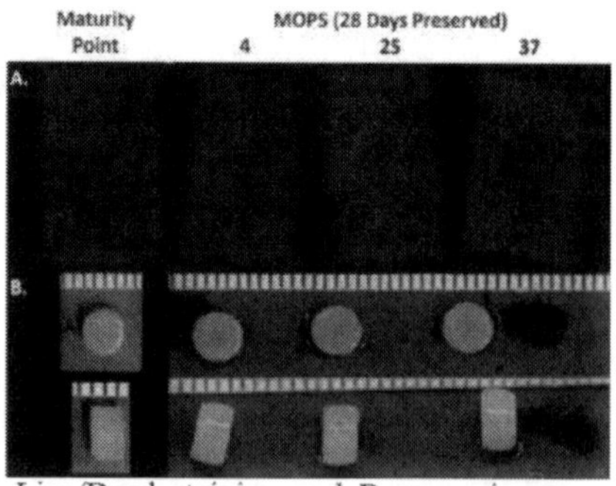

Fig. 2. A. Live/Dead staining and B. gross images at 28 days preserved for Study 1.

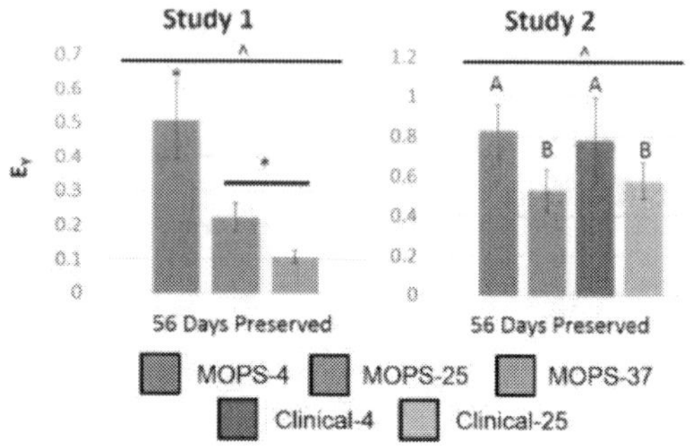

Fig. 3. E_Y values at 56 days preserved. ^ different than maturity point, * different among particular groups (p…0.05).

easy to use (weekly media change, no refrigeration or incubation), making it an appealing starting point for engineered cartilage preservation. On average between both studies, MOPS-25 maintained over 70% of key tissue properties during the current clinical window of 28 days without growth factor necessary for adult chondrocytes in culture [8-9]. Interestingly, MOPS does not extend, however, the maintenance of tissue properties beyond this window as seen with native cartilage grafts (Fig. 3). We speculate that differences in engineered cartilage arising from the *de novo* tissue growth process in the laboratory versus in vivo development, give rise to this disparate finding. In the current study, temperature appeared to play a more significant role on engineered cartilage properties than media formulation adopted (MOPS or clinical). While not significantly different at 28 days of preservation, by 56 days preserved 4°C proves better than 25°C. Temperature may affect chondrocyte metabolism as exemplified by the increased GAG in the media. Whether this media GAG reflects increased catabolism, or production of (smaller) GAGs remains to be clarified. Future work will continue to aim toward protocol optimization to increase property maintenance and extend its duration past 28 days.

ACKNOWLEDGEMENTS

This work was supported by NIH AR46568 and NIH AR060361.

REFERENCES

[1] Gross, AE et al. *Clin Orthop Relat Res* 466(8), 2008.
[2] Sherman, SL et al. *J Am Acad Orthop Surg* 22(2), 2014.
[3] Williams, RJ et al. *Am J Sports Med* 32(1), 2004.
[4] Bakay, A et al. *Int Orthop* 22(5), 1998.
[5] Beaver, RJ et al. *J Bone Joint Surg [Br]* 74(1), 1992.
[6] Stoker, AM et al. *J Knee Surg* 25(2), 2012.
[7] Cook JL, et al. *Clin Orthop Relat Res* 472(11), 2014.
[8] Ng, KW et al. *Tissue Eng* 16(3), 2010.
[9] Nover, AB et al. *Trans Orthop Res Soc* 2014.

SB³C2015
Summer Biomechanics, Bioengineering and Biotransport Conference
June 17-20, 2015,Snowbird Resort, Utah, USA
SB³C2015-386

VISCOELASTIC FOLDING INSTABILITY CONTROLS AIRWAY BRANCHING MORPHOGENESIS

Victor D. Varner (1), Jason P. Gleghorn (1), Celeste M. Nelson (1,2)

(1) Department of Chemical & Biological Engineering
Princeton University
Princeton, NJ, USA

(2) Department of Molecular Biology
Princeton University
Princeton, NJ, USA

INTRODUCTION

Space-filling, branched networks form the basic architecture for several organs in the body, including the lung, kidney and mammary gland [1]. In the early embryo, these complex structures originate as epithelial tubes — simple geometries that are subsequently molded by a sequence of branching events, patterned in both space and time. In most cases, branching involves reciprocal signaling between the branching epithelium and surrounding mesenchyme [2]. In the developing mouse lung, airway branching is highly stereotyped and regulated in part by fibroblast growth factor (FGF) signaling [3]. New epithelial branches emerge at locations adjacent to focal expression of FGF10 in the mesenchyme. This pre-pattern of FGF10 is thought to specify the locations of new branches [4]. Interestingly, however, when the mesenchyme is removed, and reciprocal signaling is effectively disrupted, isolated epithelial explants still branch in culture [5]. In the absence of a mesenchymal FGF template, it is unclear how the epithelial branching pattern is specified — a result which suggests a need for additional, non-chemically-mediated regulatory mechanisms.

Here, using a combination of ex vivo culture experiments and mathematical modeling, we show that a growth-induced mechanical instability defines the relative locations of branches within the developing murine airway epithelium in the absence of mesenchyme. Time-lapse culture of mesenchyme-free explants revealed that new epithelial branches form simultaneously, with the airway epithelium cast into a folded geometry with characteristic wavelength λ. To quantitatively investigate the mechanics of this process, we constructed a simple theoretical model for the growing airway epithelium embedded within a viscoelastic gel. Our results indicated that a viscoelastic folding instability underlies epithelial branching in the absence of mesenchyme. The dominant wavelength of the instability defines the branching pattern, and is controlled by epithelial growth rates. Taken together, these results suggest that purely physical interactions can drive the biological patterns, which underlie tissue morphogenesis in the embryo.

METHODS

Embryos were harvested from timed-pregnant CD1 mice (Charles River Laboratories) at embryonic day E12.5. Whole lung explants were dissected in ice-cold phosphate-buffered saline using fine forceps. To isolate the airway epithelium, explants were incubated in dispase (Invitrogen) on ice for 15 min, and the mesenchyme was removed manually using fine tungsten needles. Isolated epithelial explants were then embedded in three-dimensional (3D) gels of reconstituted basement membrane protein (Figure 1). Explants were cultured ex vivo at 37°C using culture medium supplemented with recombinant FGF1. Time-lapse movies were collected using a Nikon Ti-U inverted microscope customized with a spinning disk (BD Biosciences).

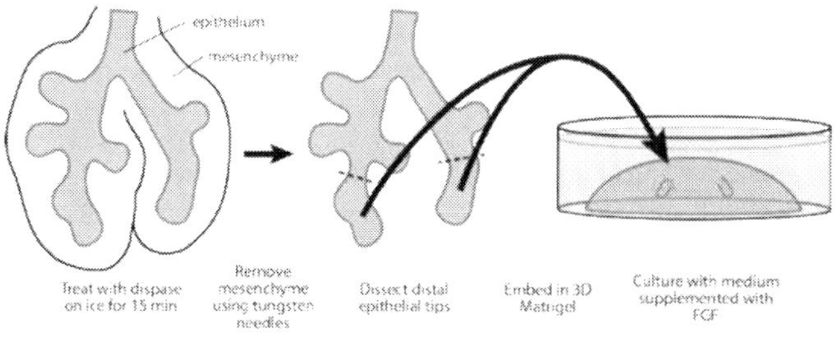

Figure 1: Mesenchyme-free culture of isolated embryonic airway epithelial explants.

Proliferating cells were detected using the Click-iT EdU Imaging Kit (Invitrogen). Briefly, mesenchyme-free explants were pulsed with 5-ethynyl-2'-deoxyuridine (EdU) for 1 hr, then fixed with 4% paraformaldehyde in PBS and washed with 0.3% Triton-X-100 in PBS. Fixed explants were then incubated with primary antibody against E-cadherin and, subsequently, AlexaFluor-conjugated secondary antibodies to simultaneously label the airway epithelium. Explants were dissected from the gel, and confocal stacks were captured. 3D reconstructions of the airway epithelium and proliferating cells were generated using Imaris (Bitplane). In other experiments, explants co-stained for EdU and Hoechst 33258 were used to quantify epithelial proliferation frequencies

To model the mechanics of mesenchyme-free branching, we constructed a theoretical model for a growing epithelial layer supported by an underlying viscoelastic foundation (Figure 2). As a first approximation, we both assumed small deformations and ignored the effects of initial epithelial curvature. Epithelial proliferation was included by decomposing the overall mechanical strain in the epithelial layer into a component due to growth (ϵ_g) and a component due to elastic deformation (ϵ^*) using the relation $\epsilon = \epsilon^* + \epsilon_g$. Thus, the governing differential equation for a growing viscoelastic layer supported by a viscoelastic foundation is given by

$$B(p)\frac{h^3}{12}\frac{d^4v}{dx^4} + N_g h\frac{d^2v}{dx^2} + \frac{\gamma B_f(p)}{2}v = -\frac{d^2M_g}{dx^2},$$

where $v(x)$ is the beam deflection, $N_g = bE\int_{-h/2}^{h/2}\epsilon_g(x,y)\,dy$ is the in-plane force due to growth, $M_g = bE\int_{-h/2}^{h/2}y\epsilon_g(x,y)\,dy$ is the growth-bending moment, and $B(p)$ and $B_f(p)$ are differential operators which define the viscoelastic properties of the growing layer and foundation, respectively [4].

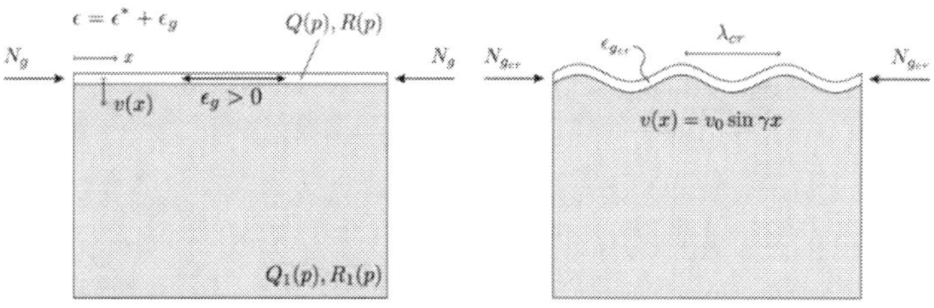

Figure 2: Schematic of model for a growing epithelial layer supported by a viscoelastic foundation.

RESULTS

We performed time-lapse imaging analysis of mesenchyme-free epithelial explants and quantified the resulting branching dynamics. In the absence of mesenchyme, new branches formed simultaneously, with the airway epithelium cast into a folded geometry with characteristic wavelength λ (Figure 3). We quantified the role of cell division during this process using EdU incorporation. Confocal stacks of immunofluorescence for E-cadherin were used to create 3D reconstructions of the epithelial surface. Mean surface curvatures were computed and compared with 3D patterns of cell proliferation in branched and unbranched explants. During this process, the airway epithelium was highly proliferative, but clear spatial patterns of proliferation did not become apparent until branches had already emerged. At this point, elevated levels of proliferation were observed

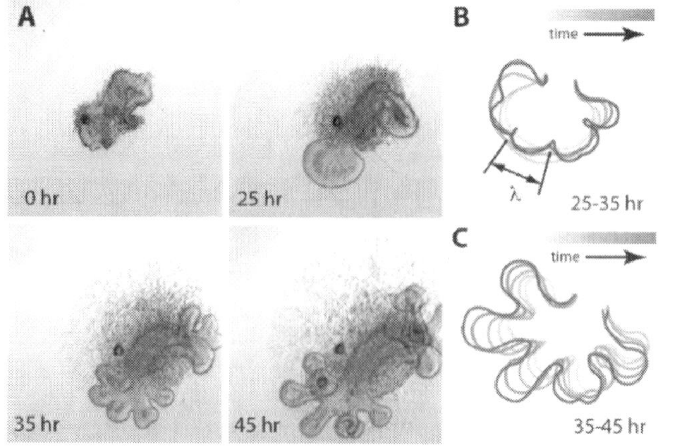

Figure 3: Dynamics of mesenchyme-free branching. (A) Time-lapse bright field images of airway epithelial explant during mesenchyme-free culture. Branches form simultaneously with a characteristic wavelength λ. (B-C) Traced contours of the airway epithelium during (B) branch initiation and (C) branch elongation.

in the tips of extending branches, where the epithelium was cast into a highly curved surface.

To investigate the mechanics of this process, we created a theoretical model for a growing epithelial layer on a viscoelastic foundation. Since spatial patterns of proliferation were not observed prior to branching, we considered the case of uniform growth, $\epsilon_g(x,y) = \epsilon_g^o$, which yielded $M_g = 0$, and $N_g = bEh\epsilon_g^o$. In this case, the growing epithelium reaches instability and folds out-of-plane with a dominant wavelength given by $\lambda_{cr} = 2\pi h\sqrt[3]{B(p)/3B_f(p)}$ (Figure 2). For the case of a viscoelastic foundation, this relation predicts shorter branch wavelengths at higher proliferation rates. To test this model prediction, we cultured mesenchyme-free explants in gels supplemented with different concentrations of FGF to modulate epithelial proliferation. These results confirmed that branch wavelengths could be tuned by varying the rate of growth in the epithelium.

DISCUSSION

During development, tissue morphogenesis is thought to be regulated primarily by biochemical signaling networks. Here, using a mesenchyme-free culture assay, we disrupted epithelial-mesenchymal signaling in the developing mouse lung and found that a mechanical instability is sufficient to drive branching morphogenesis in the absence of mesenchyme. These results suggest that mechanical cues, in addition to the well-established role of biochemical signals, need to be considered in studies of tissue morphogenesis.

ACKNOWLEDGEMENTS

This work was supported by the NIH (GM083997, HL110335, and HL118532), the David & Lucile Packard Foundation, the Alfred P. Sloan Foundation, and the Camille & Henry Dreyfus Foundation.

REFERENCES

[1] Varner VD & Nelson CM, *Development,* 141:2750-2759, 2014.
[2] Lu P & Werb Z, *Science,* 322:1506-1509, 2008.
[3] Morrisey EE & Hogan BL, *Dev. Cell,* 18:8-23, 2010.
[4] Metzger RJ & Krasnow MA, *Science,* 284:1635-1639, 1999.
[5] Nogawa H & Ito T, *Development,* 121:1015-1022, 1995.
[6] Biot MA. *Proc R Soc Lon Ser-A,* 242:444-454, 1957.

SB³C2015
Summer Biomechanics, Bioengineering and Biotransport Conference
June 17-20, 2015,Snowbird Resort, Utah, USA
SB³C2015-387

FLUID-STRUCTURE INTERACTION ANALYSIS OF MITRAL VALVE FORCES USING A COMPREHENSIVE MODEL WITH 3D CHORDAL STRUCTURE: SYNERGY OF MODELING AND EXPERIMENTS

Milan Toma (1), Morten O. Jensen (1), Daniel R. Einstein (2), Ajit P. Yoganathan (1), Richard P. Cochran (3), Karyn S. Kunzelman (3)

(1) Department of Biomedical Engineering, Georgia Institute of Technology and Emory University, Atlanta, GA, USA

(2) Computational Biology & Bioinformatics, Pacific Northwest National Laboratory, Richland, WA, USA

(3) Department of Mechanical Engineering, University of Maine, Orono ME, USA

INTRODUCTION

Intelligent design of medical devices and establishment of clinical guidelines using computational models can promote translation of fundamental concepts to help improve patient care. A current challenge for improving the accuracy of these predictions is choosing and implementing modeling boundary conditions. To optimize efficiency, current computational models of the mitral valve (MV) in literature often use a simplified geometry. For example, published studies commonly take advantage of the chordal beam-type structure to justify the use of beam elements [1-5]. However, improvement in computer technology has been exponential in recent decades, allowing the use of more detailed models without unnecessary limitations in regard to geometry and complexity.

One such application is modeling and evaluation of mitral valve dynamics. The chordal structure is an important functional part of the mitral valve and should not be omitted or simplified in computational models. Ischemic events may cause ventricular dilatation and displacement of the papillary muscle (PM) tips away from the annulus in the apical, lateral, and posterior directions [6], thereby inflicting an increased leaflet tethering and forces on the chordae tendineae [7,8]. Validating simulated results of the valve-ventricular interaction with experimental data is of paramount importance – as such, an advanced in-vitro simulator is used to validate the fluid-structure interaction (FSI) model of the mitral valve.

METHODS

Explanted ovine mitral valves were mounted in an in vitro setup, and structural data of the MV was acquired with μCT [9]. Fig. 1 illustrates the FSI model using this geometry with preserved 3D chordal structure and outline of the papillary muscles.

With this input, comprehensive computational models of the MV are used to analyze the forces exerted on the chordae tendineae from the PMs, illustrating the valvular-ventricular interaction. Each group of chords, identified by different colors in Fig. 1, causes displacement of a different part of the PM tip. The directions of the displacements are identified, e.g. the six solid black arrows in Fig. 1, and the vector average is computed by vector addition, illustrated as the dashed arrow in Fig. 1.

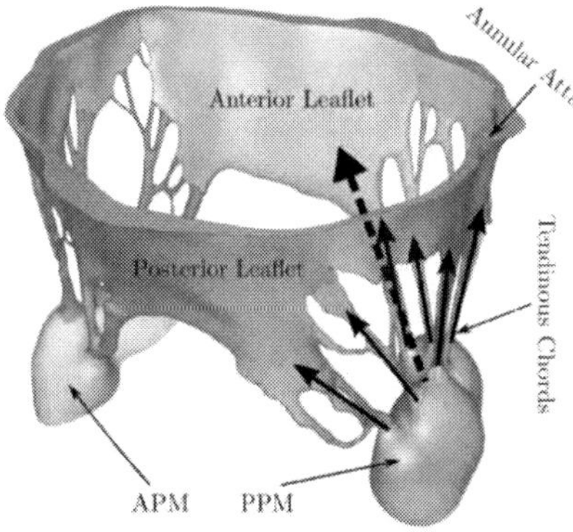

Figure 1: The mitral valve system is comprised of the annular attachment at the atrioventricular junction, two leaflets, tendinous chords and the anterior-lateral / posterior-medial papillary muscles (APM/PPM). Color-coding identifies leaflets, PMs and individual chordal branches. The valve is shown in the open configuration with the closed configuration valvular-ventricular interaction forces illustrated at individual chordal branches (solid arrows) and the resulting full PM force vector (dashed arrow).

The systolic direction of the total PM force is geometrically averaged from the individual simulated chordal force as shown in Figure 2.

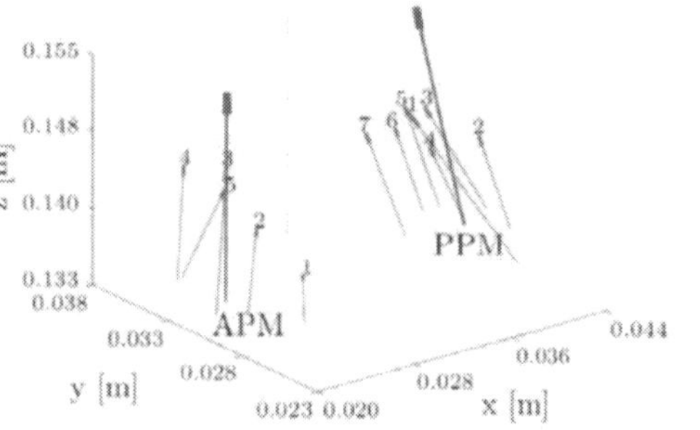

Figure 2: Geometric direction of the individual chordal and resulting total PM forces from the FSI simulation.

Experimental data from biomechanical investigations of the mitral valve were used to validate the computational model [10], Fig 3.

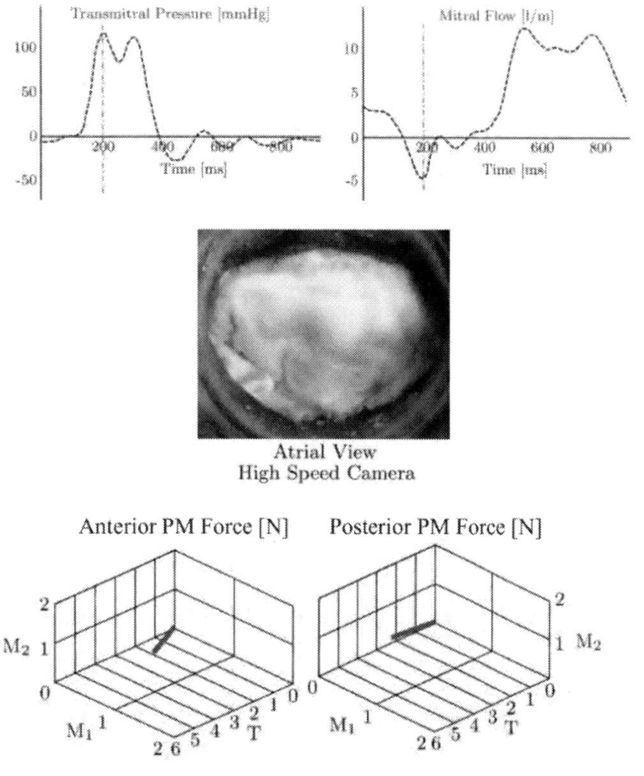

Figure 3: Three-dimensional PM force vectors of the anterolateral and posteromedial PM forces (APM and PPM, bottom) correlated with synchronized transmitral pressure and flow curves (top) and high-speed imaging (middle) [10]. "T", "M1" and "M2" denotes the experimental nomenclature for tension and momentum derived forces in the equivalent X, Y and Z directions of the FSI model. The valve is here shown in a fully closed systolic configuration (green dot-dashed line) with normal PM position.

RESULTS

As the valve closes, the hemodynamic data, leaflet dynamics and force vectors were compared between the in-vitro experiments and the FSI model. The total size scaled force magnitude of 4.1N per PM recorded in-vitro is matched by the computational model as shown in Fig 4. The direction alignment agreement between the experimental data and the fluid-structure simulation is also demonstrated in Fig 4.

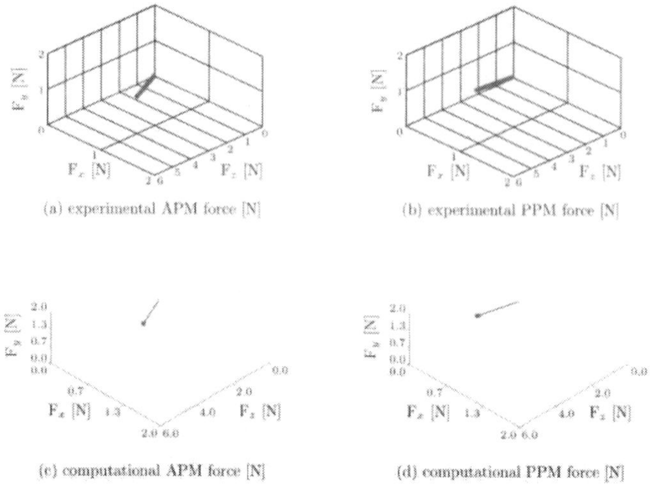

Figure 4: Validating the direction and magnitude of total papillary musle force vectors between experimental data (a+b) and fluid structure simulation (c+d) at peak systolic pressure. APM/PPM: Anterior-lateral / Posterior-medial Papillary Muscle.

DISCUSSION

In vitro and in vivo force measurements are important in validating and adjusting material parameters in computational models. A computational model based on high fidelity 3D imaging of explanted mitral valves in an in vitro setup was validated against force data recorded in this model. These comparisons will (if necessary) adjust computational model properties and provide validation. Eventually this can assist answering questions that are otherwise not possible to investigate experimentally; for example, when the force is measured to decrease in one part of the valve, it must be increased in another part to meet static equilibrium [11]. This work is important to maximize the validity of computational models of not just the mitral valve, but all biomechanical structures in the cardiovascular system.

ACKNOWLEDGEMENTS

Funding to support this work was provided by NIH grant # R01 HL52009 and R01HL092926.

REFERENCES

[1] Chandran KB, et al, *Ann Biomed Eng.* 2014 Aug 19. Epub
[2] Einstein DR, et al, *Int J for Num Methods in BME*, vol. 26, no. 3-4, pp. 348–380, 2010.
[3] Schievano S, et al, *JHVD*, vol. 18, pp. 28 34, 2009.
[4] Wenk JF, et al, *ATS*, vol. 89, no. 1546-1554, 2010.
[5] Lau KD, et al, *Med. Eng. & Physics*, vol. 32, pp. 1057–1064, 2010.
[6] Jensen H, et al, *JTCVS*, 2010;140:1312-1318
[7] He S, et al, *JHVD*, 1999;8:294-302
[8] Nielsen SL, et al, *JTCVS* 2005;129:525-31
[9] Rabbah JP, et al *ABME* 2013 Feb;41(2):305-15
[10] Jensen MOJ, et al, ABME 2001;29:406-413
[11] Rahmani A et al, JHVD, 2013;22(1):28-35

SB^{3}C2015
Summer Biomechanics, Bioengineering and Biotransport Conference
June 17-20, 2015,Snowbird Resort, Utah, USA
SB^{3}C2015-388

MICROBUBBLE VOID IMAGING – A NOVEL TECHNIQUE FOR FLOW VISUALISATION AND QUANTITATIVE ASSESSMENT OF INTRAVASCULAR MIXING IN LARGER VESSELS USING ULTRASOUND

Chee Hau Leow (1), Francesco Iori (2), Richard W. Corbett (3), Neill Duncan (3), Colin G. Caro (1), Peter E. Vincent (2), Mengxing Tang (1)

(1) Department of Bioengineering, Imperial College London, South Kensington, London, UK, SW7 2AZ

(2) Department of Aeronautics, Imperial College London, South Kensington, London, UK, SW7 2AZ

(3) Imperial College Renal and Transplant Centre, Hammersmith Hospital, London, UK, W12 0HS

OVERVIEW

Cardiovascular disease is strongly correlated with blood flow dynamics. Non-planar curvature in vascular geometry along with in-plane swirling and intraluminal mixing appear commonly present in normal arteries [1] and may be of importance in clinical interventions, including bypass grafts and arterial stents. However, there is a lack of imaging techniques capable of visualising the flow and quantifying mixing.

Microbubbles as ultrasound contrast agents have found widespread use in clinical applications [2]. Their low diffusivity and the ability to effect their controlled destruction using high acoustic pressures offers the opportunity to "inject" a void of bubbles non-invasively (yet controllably both spatially and temporally) and then observe the evolution of such a bubble void over time.

Our objective was to assess the feasibility of this novel technique -microbubble void imaging - to non-invasively visualise the flow and quantify flow-induced mixing in large vessels. This was achieved using ultrasound and controlled destruction of microbubble contrast agents, making use of high frame-rate ultrasound, the results being compared with numerical computational fluid dynamic (CFD) simulations.

Vessel mimicking phantoms with three different vessel geometries, straight, planar-curved and helical, with known effects on the flow and mixing were chosen to evaluate this technique. The flow was non-pulsatile but otherwise physiologically relevant with water as the transport medium containing suspended microbubbles. At the exit of each phantom, a transducer mounted on a Verasonics research platform was used to detect flow and microbubbles, while at a consistent distance upstream a further transducer was used to destroy microbubbles within the vessel using a brief pulse of high acoustic pressure to create a void. In parallel and using identical geometric and flow conditions, a series of CFD simulations using the STAR-CCM+ package were used to cross-validate the in-vitro findings.

In these three idealised vascular geometries, qualitative assessment of the void within the flow field was in keeping with that expected and closely mirrored by the CFD simulations (Figure 1). An entropic measure to quantify mixing [3] was applied to both the CFD and in-vitro work. Peak entropy, analogous to the maximal state of mixing during the passage of the void, correlated well between CFD and in-vitro experiments. In addition and as expected, the helical geometry exhibited the highest mixing followed by the curved and then straight geometry.

Initial in-vitro evaluation of this novel, non-invasive, high-frame rate ultrasound technique with three flow geometries in comparison with reference CFD solutions has shown the potential of this technique. While further work is required, it may have an important role in clinical care where assessment of mixing is valuable in studying fluid-wall mass transport along with the initiation and development of vascular lesions such as atherosclerosis and intimal hyperplasia.

ACKNOWLEDGEMENTS

The authors are grateful for support from the National Institute for Health Research Imperial Biomedical Research Centre, the

Garfield Weston Foundation, CD-Adapco, the British Heart Foundation and the Engineering and Physical Sciences Research Council. CHL is supported by a postgraduate scholarship from the Public Service Department of Malaysia

REFERENCES

[1] Caro CG et al., *Proc R Soc Lond A*, 452:185–197, 1996.
[2] Tang M-X et al., *Interface Focus*, 1:520–539, 2011.
[3] Cookson AN et al., *Ann Biomed Eng*, 37:710-721, 2009.

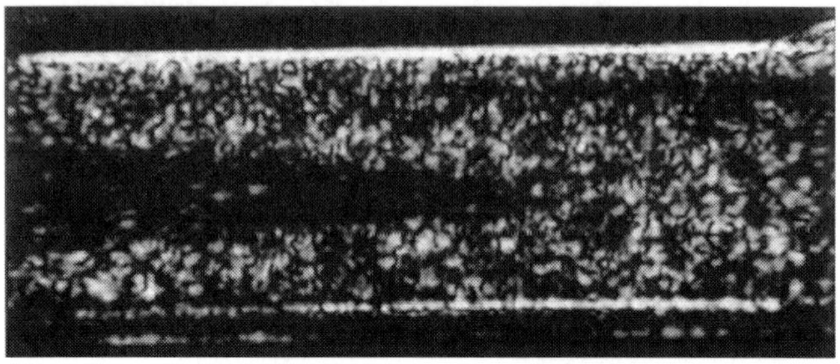

Figure 1: In vitro ultrafast contrast enhanced ultrasound images at 1000 frames per second showing an elliptic dark wave front (a bubble void created upstream) travelling in fast flow within a straight tube.

SB³C2015
Summer Biomechanics, Bioengineering and Biotransport Conference
June 17-20, 2015,Snowbird Resort, Utah, USA
SB³C2015-389

PREDICTION OF ELBOW JOINT CONTACT PRESSURES IN THE MULTIBODY FRAMEWORK

Munsur Rahman (1), Akin Cil (1, 2, 3), Antonis P. Stylianou (1)

(1) Department of Civil and Mechanical Engineering
University of Missouri-Kansas City
Kansas City, Missouri, USA

(3) Department of Orthopaedics
Truman Medical Centers
Kansas City, Missouri, USA

(2) Department of Orthopaedic Surgery
University of Missouri-Kansas City
Kansas City, Missouri, USA

INTRODUCTION

The elbow joint, recognized as the most important joint of the upper extremity serves as a fulcrum for the forearm. The articular cartilage within the elbow joint withstands repetitive mechanical forces which are about 50% body weight during activities of daily living and may reach up to 3 times body weight at about 90^0 of elbow flexion [1]. The elbow joint is the second most commonly dislocated joint in adults often resulting in significant damage to bones and ligaments. Injury induced osteoarthritis of the elbow can cause severe pain, loss of mobility, and upper limb disability [2]. Detailed knowledge of *in vivo* loading of the elbow structures is essential in understanding the biomechanical causes associated with both chronic and acute injuries and for improving design and implementation of therapeutic strategies. Currently, measuring the *in vivo* elbow joint forces and cartilage contact pressures during dynamic conditions is not possible. Computational models can enhance our understanding by providing detailed information on joint contact mechanics and kinematics in addition to tissue interactions during dynamic conditions. A review of the literature reveals few models where elbow joint cartilage contact area and contact forces have been examined under dynamic conditions.

The objective of this study was to develop an anatomically correct elbow joint model with discrete cartilage in the multi-body framework from a cadaver specimen, and evaluate the kinematics against experimental measurements. Radial collateral ligament (RCL) deficient, and lateral ulnar collateral ligament (LUCL) deficient conditions were simulated and contact pressures were compared to the intact elbow.

METHODS

Experimental measures: One fresh frozen cadaver elbow specimen (61 years old, male, left arm) was used for this study. The specimen was imaged with both CT scans and MRI imaging. Localizers were directly attached to the humerus, radius, and ulna. Each localizer was constructed out of ABS plastic using a Dimension bst1200es series 3D printer. The localizers include two orthogonal tubes filled with Vaseline that appear in the MRI images and can then be used to register the bone geometries to the experimental coordinate system. After medical imaging, the elbow was dissected by an orthopedic surgeon and mounted in a bi-axial Instron 8821 (Instron, Norwood, MA, USA) mechanical tester (Fig. 1a). The humerus head was cemented inside a cup that was attached by a hinge joint to the top ram of the mechanical tester. The intact hand was placed and secured on a slider that can slide horizontally in a single axis. The biceps and triceps tendons were sutured and attached to the 100 lbf load cells that were rigidly attached to the humerus cylinder. Four infrared rigid body markers of an Optotrak Certus system (Northern Digital, Inc., Waterloo, ON, Canada) were firmly attached to the humerus, radius, ulna, and bottom slider to record motion data. The initial position and orientation of the cadaveric elbow relative to the dynamic simulator were recorded using the Optotrak system. The elbow joints were positioned at a flexion angle of 20^0 before starting the tests. A motion profile of 190mm vertical displacement (190mm downwards from starting position and then return to initial position) was applied to the top ram of the mechanical tester. This downward displacement induced elbow flexion angles from 20° to 120°.

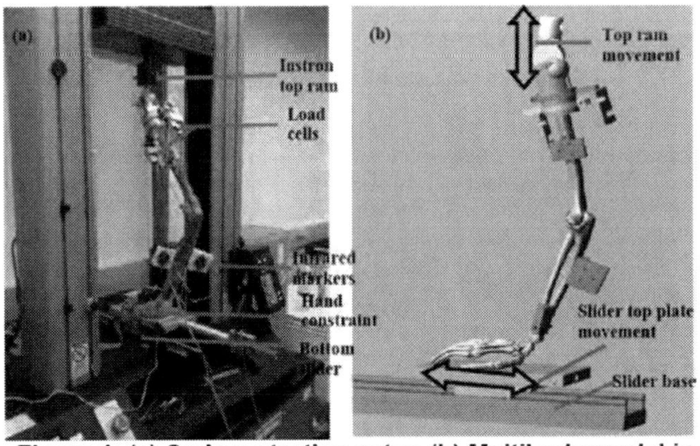

Figure 1: (a) Cadaver testing setup (b) Multibody model in Adams

Bone motion as well as biceps and triceps tendon forces were recorded during the test. After test completion, additional points were collected along the ligament insertion and origins sites using the Optotrak digitizing probe.

Computational model: Three-dimensional bone and cartilage geometries were generated from the medical images using 3D Slicer (www.slicer.org). The multibody model was created in ADAMS (MSC Software Corporation, Santa Ana, CA) by using the initial position and additional point information obtained during experimental study (Fig 1b). The bone, cartilage geometries, and ligament origin/insertion sites were aligned using the localizers in the initial position. The ligaments and interosseous membrane were modeled as non-linear springs using a piecewise function that includes the non-linear "toe" region [3]. The model included three bundles for LUCL, three bundles for RCL, three bundles for the medial collateral ligament (MCL) anterior part, three bundles for MCL posterior part, two bundles for the annular ligament, and five bundles of interosseous membrane. The LUCL and the annular ligament were wrapped around the bone to represent their anatomical physiology [4]. A custom macro was written in ADAMS to automatically divide the humerus cartilage into discrete hexahedral elements. Each element had an approximate 5 x 5 mm cross-sectional area. The macro connected each cartilage element to the humerus bone with a fixed joint located at the center of each element. The radial and ulnar cartilages were attached rigidly to the respective bone using fixed joints. The macro also defined a deformable contact constraint with no friction using modified Hertzian contact law (Eq. 1) between each humerus cartilage element with the radius and ulna cartilage geometry.

$$F_c = k_c\delta^n + B_c(\delta)\dot{\delta} \dots\dots\dots (1)$$

where k_c is the contact stiffness, δ is the interpenetration of the geometries, and $B_c(\delta)$ is a damping coefficient. The contact stiffness parameters was estimated based on simplified elastic foundation theory [5]. Local coordinate systems for the bone segments were created as described by Ferreira, King [6] and Morrey and Chao [7] to measure the ulna and radius motions relative to the humerus coordinate system. The 190 mm vertical downward/upwards motion profile of the top ram, and the force profiles collected from the biceps and triceps load cells were used as inputs to the model. Simulations were run for an intact elbow and the predicted kinematics of the ulna and radius were compared with the experimentally collected kinematics. Simulations using the same inputs were run for an LUCL deficient and an RCL deficient elbow.

RESULTS

RMS errors and correlation coefficients between the experimental and model predicted kinematics are presented in Table 1. Kinematic patterns are shown in Figure 2. Contact pressure maps for the two ligament deficient conditions and the intact elbow are shown in Figure 3.

Table 1: RMS error (deg, mm) and correlation coefficients for ulna and radius kinematics.

Kinematics description	RMS error	Correlation coefficient
Ulna internal-external rotation (Urx)	3.9	0.99
Ulna varus-valgus rotation (Ury)	5.0	0.97
Ulna flexion-extension rotation (Urz)	4.2	0.99
Radius internal-external rotation (Rrx)	5.1	0.45
Radius varus-valgus rotation (Rry)	5.9	-0.56
Radius flexion-extension rotation (Rrz)	4.3	0.99
Ulna proximal-distal position (Udx)	1.8	0.22
Ulna anterior-posterior position (Udy)	1.0	0.86
Ulna medial-lateral position (Udz)	1.4	-0.59
Radius proximal-distal position (Rdx)	5.4	0.98
Radius anterior-posterior position (Rdy)	3.3	0.59
Radius medial-lateral position (Rdz)	3.0	0.52

DISCUSSION

This study developed an anatomically correct subject specific model of the elbow joint in the multibody framework. The model was evaluated kinematically using experimental data. The model predicted ulna displacements and rotations were in agreement with the experimental data as seen in Figure 2 and the low RMS error. The largest RMS errors were observed in the radius rotations. Since the articulation between the radius and the humerus is not as geometrically constrained as the articulation between the ulna and the humerus, the higher errors in the radius kinematics were expected.

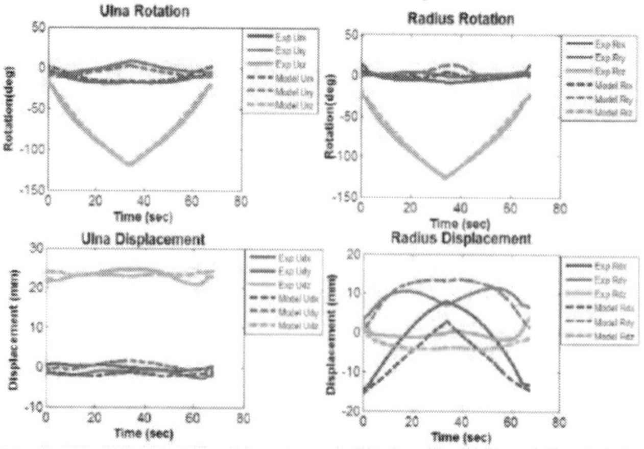

Figure 2: Comparison between experimental and model kinematics of the ulna and radius relative to humerus.

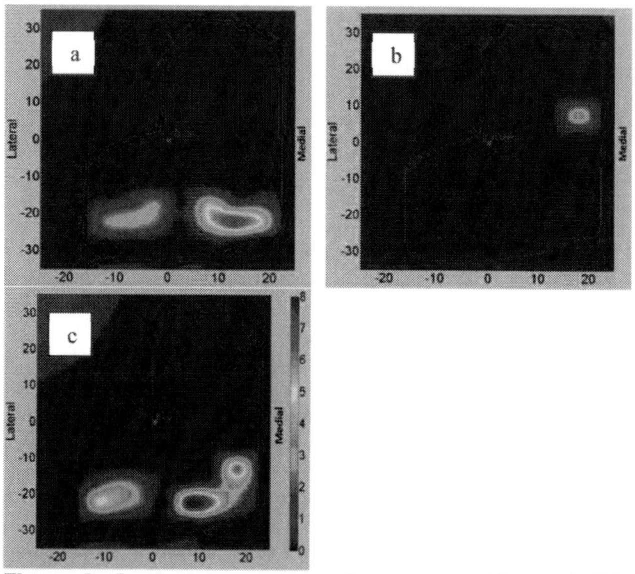

Figure 3: Contact pressure on humerus cartilage at 120° flexion for LUCL deficient (a), RCL deficient (b), and intact elbow (c).

The contact pressures on the humerus cartilage were computed for the intact elbow and two ligament deficient conditions. Removing the LUCL changed the loading pattern during deep flexion by reducing the contact between both the radius and the ulna with the humerus. The effect of removing the RCL ligament during deep flexion is very pronounced. The radius and the humerus are no longer in contact while the contact between the ulna and the humerus is greatly reduced and moves in the dorsal direction. A limitation of this study is that the model was evaluated using only kinematic data. Future studies will validate the articular contact pressure through direct measurements.

REFERENCES

[1] Safran, M. et al., Arthroscopy, 21:1381-1395, 2005
[2] Degreef I. et al.,, Minim Invasive Surg, 798084, 2011
[3] Wismans, J. et al., J Biomech, 13:677-85, 1980
[4] Rahman, M. et al., Adv. Biomecha. & App., 1:169-185, 2014
[5] Bei, Y. et al., Med Eng Phys, 26: 777-789, 2004
[6] Ferreira, L.M. et al., J Orthop Res, 29: 596-601, 2011
[7] Morrey, B.F. et al., J Bone Joint Surg Am, 58:501-8, 1976.

EVALUATION OF TWO FORMULATIONS OF POLYCAPROLACTONE FOR USE IN TISSUE SCAFFOLD APPLICATIONS

Jacob D. Harris (1), A. Sharif El-Gizawy, PhD (2), Ferris M. Pfeiffer, PhD (3)

(1) Department of Mechanical Engineering
University of Missouri
Columbia, MO, USA

(2) Department of Mechanical Engineering
University of Missouri
Columbia, MO USA

(3) Orthopaedic Surgery/Bioengineering
University of Missouri
Columbia, MO, USA

INTRODUCTION

Damage to the articular surface of knee is a fairly common injury. In one study of 31,516 knee arthroscopies, 61% of patients featured chondral lesions [1]. Another study of 1000 knee arthroscopies revealed 63% of patients to have some level of chondral lesions, 19% of those with lesions had focal chondral or osteochondral defects with an average defect size of 2.1 cm^2 [2]. Highly specialized, articular cartilage is composed of four layers which work in conjunction to provide a low-friction, tough, elastic, and flexible surface capable of bearing and transmitting joint loads [3,4]. Many surgical techniques exist which attempt to repair and restore the articular surface including marrow stimulation techniques, cell treatments and chondral plugs [5-8]. To their detriment, these repair techniques typically result in the formation of fibrocartilage, exhibiting inferior mechanical and wear properties when compared to hyaline cartilage [5, 6, 8]. Because these techniques fail to restore the surface, new techniques using 3D printing techniques and polymers to form tissue scaffolds are being widely investigated [9]. One such polymer, from the ester family, is polycaprolactone (PCL), which can be fabricated using fused deposition modeling (FDM) technology. While available in a wide variety of molecular weights with a corresponding increase in chain length, little information exists on how the molecular weight affects the mechanical performance of PCL scaffolds manufactured using the FDM process.

METHODS

Two types of FDM machines were used to print the scaffolds. The first machine, seen in Fig (1), uses a heated reservoir filled with pelletized raw material and extrusion screw to form 0.5mm dia strands in an order such as to create the 3D scaffold. A hobbyist level printer, the MendelMax 2, seen in Fig (1), utilizes polycaprolactone stock filaments to lay down smaller filaments in the desired arrangement. Needing filament stock to print, the second printer also involves a filament extruding machine, which uses a heated barrel and plunger to create the stock. Scaffolds were created using 14,000 molecular weight (MW) and 80,000 MW PCL sourced from Sigma Aldrich of St. Louis, Mo.

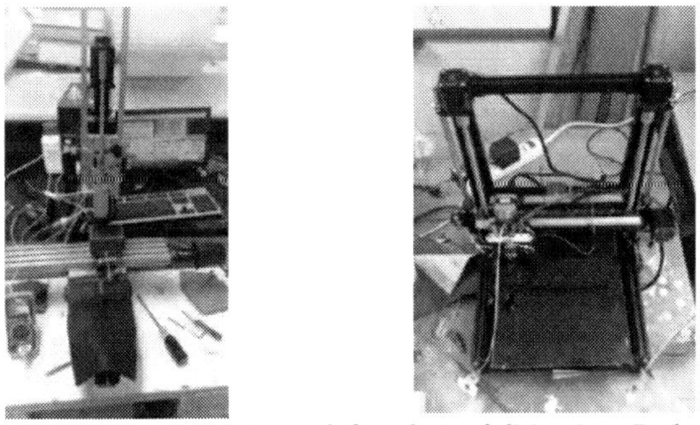

Fig (1). Reservoir printer on left and MendelMax 2 on Right

Due to the thermal properties of 14,000 MW weight material at the print temperature of 180°F, water was added to the printing well to provide additional cooling during printing. Cooling water was not necessary for the 80,000 MW material, which printed at a much higher temperature of 210°F, but had more favorable thermal properties. Completion of printing using the reservoir machine resulted in the scaffold seen in Fig (2).

After printing, the samples were trimmed and placed in a 6mm dia x 4mm deep well, seen in Fig (2), for testing using an Instron 8821S material testing machine featuring an Optotrak Certus optical tracking system from NDI Inc. Cyclic indention tests were carried out using a 2 mm plunger applying 20% strain to the samples at a frequency of 1 Hz for ten cycles.

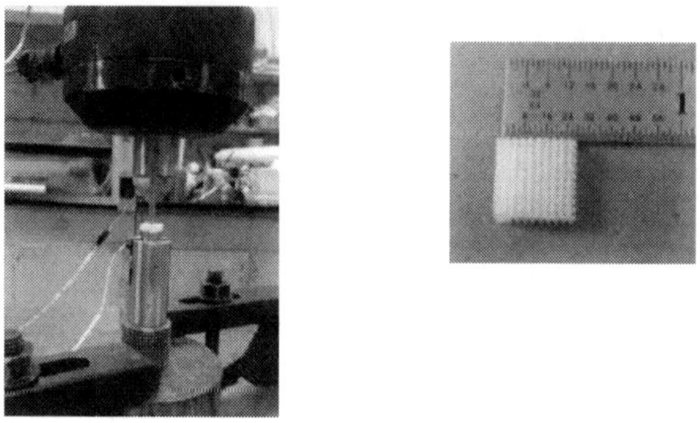

Fig (2) Indention apparatus on left and finished scaffold on right

RESULTS

Stress vs. strain data for 80,000 and 14,000 MW scaffolds is shown in figure 3. A slope taken from the first and last cycle of this data is termed the Young's Modulus and Dynamic Modulus respectively.

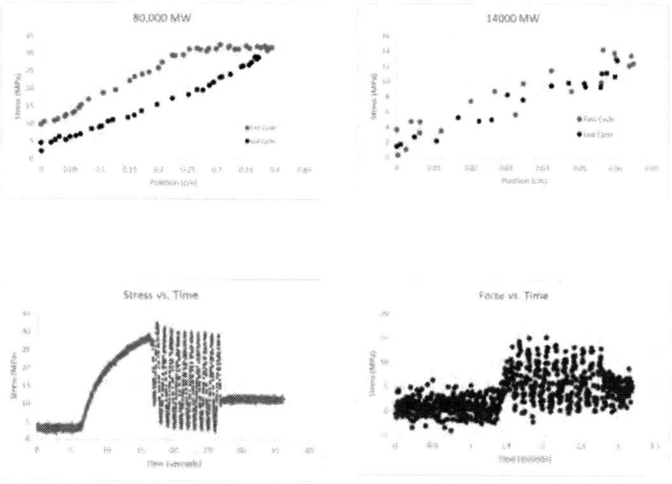

Fig. (3) Compression Test Results

Table 1. Elastic Moduli

80,000 MW		14,000 MW	
Young's Modulus (MPa)	Dynamic Modulus (MPa)	Young's Modulus (MPa)	Dynamic Modulus (MPa)
68.7825	74.285	53.62	52.86

DISCUSSION

The 80,000 MW weight featured an elastic perfect plastic region on the first cycle and by the last cycle exhibited linear behavior. Peak stress for the 80,000 MW was approximately 30 MPa. Behaving consistently through all cycles, the 14,000 MW material exhibits linear behavior and has peak stresses of approximately 14 MPa.

Uncoiling of the longer chains can help to explain the elastic perfect plastic region during the first cycle of the higher MW material. As the strain increase, the chains begin to uncoil, which requires a relatively constant force. Once completely uncoiled, the chains have aligned and continued strain requires that chains slip past one another as well as stretching of the chains themselves. Slipping past one another and stretching results in linear behavior. By the last cycle, the chains have permanently aligned, eliminating the elastic perfect plastic region. The decrease in peak stress for each subsequent cycle increased plastic deformation. Shorter chains means little uncoiling occurs, and the majority of the load is a result of chains slipping past one another and the chain stretching as strain increases. During the first cycle, the chains quickly align and subsequent cycles behave in the same way as the first cycle. No additional plastic deformation occurs after the first cycle and is indicated by the similar peak stress for all cycles.

Chain mobility offers an explanation for the lower weight bearing capacity of the lower MW material. As the chains in the lower MW material are much shorter than in the higher MW material, there is less chain entanglement. The strong bonds between molecules forming the chain can support larger loads than the weaker bonds that form between chains. When loading increases, the less entangled chains of the lower weight material rely more so on the bonds between chains to support the loads than the higher MW material does, which supports the loads through the strong bonds between the molecules forming the chain.

Differences in moduli exist between the materials, with the 80,000 MW material exhibiting a modulus in the area of 70 MPa and the 14,000 MW a modulus of roughly 53 MPa. Compared to cancellous bone with a modulus of 100-500 MPa both weights fall short of the natural bone modulus.

Only two MW and one testing regime were carried out to provide preliminary characterization of the mechanical properties of FDM printed polycaprolactone scaffolds. Continued work is necessary to ascertain the extreme limits and failure modes of the printed scaffolds. Expansion of the number of tested MW is also of critical importance, materials of 45,000 MW and potentially up to 120,000 MW are to be tested.

ACKNOWLEDGEMENTS

This study was funded by the Comparative Orthopaedic Laboratory as well as Dr. Sharif El-Gizawy

REFERENCES

[1] Curl, W et al., *Arthroscopy*, 13:456-460, 1997
[2] Hjelle, K et al.,*Arthroscopy*, 18:730-734, 2002
[3] Kheir, E et al., *Orthopedics and Trauma*, 23:450-455, 2009
[4] Pearle, A et al., *Clinics in Sports Medicine*, 24:1-12, 2005
[5] Gomoll, A et al., *Would Repair and Regeneration*, 22:30-38, 2004
[6] Ahmed, T et al., *Tissue Engineering: Part B*, 16:305-329, 2010
[7] Tetteh, E et al., *Journal of Orthopaeic & Sports Physical Therapy*, 42:243-253, 2012
[8] Moyad,T, *Cartilage*, 2:226-236, 2011
[9] Ratner, Buddy, Allan Hoffman, Frederick Schoen, and Jack Lemons. *Biomaterials Science: An Introduction to Materials in Medicine*. 3rd ed. Amsterdam: Elsevier, 2013. 1138-1150. Print.

A PERISTALTIC MECHANISM FOR CLEARANCE OF SOLUTES IN PERIARTERIAL BASEMENT MEMBRANES

M. Keith Sharp (1), Alexandra K. Diem (2), Roy O. Weller (3), Roxana O. Carare (4)

(1) Biofluid Mechanics Laboratory
Department of Mechanical Engineering
University of Louisville
Louisville, KY 40292 USA

(2) Institute for Complex Systems Simulation and Computational Engineering and Design
Faculty of Engineering and the Environment
University of Southampton, Highfield
Southampton SO17 1BJ UK

(3) Faculty of Medicine
Southampton General Hospital
University of Southampton
Southampton, Hampshire SO16 6YD UK

(4) Institute for Life Sciences
Southampton General Hospital, Clinical Neurosciences
University of Southampton
Southampton SO16 6YD UK

INTRODUCTION

Peptide amyloid beta (Aβ) is produced throughout the body, and in organs other than the brain, is eliminated through lymphatic vessels to regional lymph nodes. The brain also produces Aβ, but lacks conventional lymphatic vessels. Solutes including Aβ are transported along passages around cerebral arteries in the direction opposite that of blood flow, to arrive eventually in cervical lymph nodes [1-4]. The accumulation of Aβ in basement membranes of periarterial smooth muscle cells in Alzheimer's disease and in cerebral amyloid angiopathy [5] suggests that an age-related failure of this transport pathway may be responsible for these pathologies [6].

The mechanism driving transport in basement membranes is not clear. Because tissue pressure is relatively constant throughout the brain, longitudinal pressure differences would seem to be insufficient to directly drive flow along these spaces. Furthermore, intravascular blood pressure waves tend to cause peristaltic flow in the same direction as blood flow. Nonetheless, perivascular clearance of solutes ceases upon the death of an animal with the elimination of blood pressure pulsatility [4]. Furthermore, reduced arterial pulsatility slows perivascular transport. Thus, while it appears that pulsatility is necessary for transport, how a forward-propagating peristaltic wave results in retrograde solute transport remains to be explained. In this paper, a mechanism involving flexing fibers within the basement membrane that present greater resistance to forward flow than to reverse flow will be evaluated.

Basement membranes between endothelial cells, astrocytes, pericytes and up to several layers of smooth muscle cells form continuous channels from capillaries to large arteries entering the skull (see the hypothesized flow path in Fig. 1). Basement membranes are 100-150 nm thick and contain up to 144 different proteins (in human kidneys), the functions of most of which are poorly understood [7]. Protein concentrations are nonuniform across the membrane, and depend on the types of cells on either side of the membrane [8]. One protein (perlecan) has bean shown to regulate filtration across the glomerular basement membrane [9]. In comparison to the kidney, much fundamental information remains to be quantified about the inventory and function of the constituents of cerebral vascular basement membranes. Given the complexity being discovered in other basement membranes, it is not unreasonable to hypothesize that structures within cerebral basement membranes might be adapted to promote retrograde solute transport.

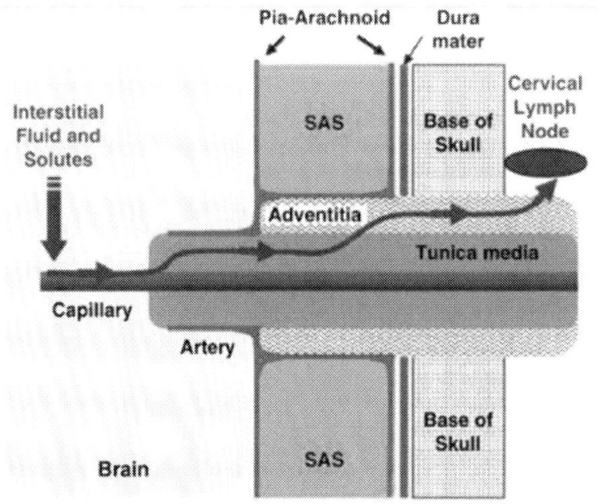

Figure 1. Hypothesized flow path from capillaries to lymph nodes [10].

METHODS

Flow through a Darcy-Brinkman medium was calculated in peristaltic flow between nominally flat plates. Symmetrical, sinusoidal waves in the walls were modeled to drive the flow at low Reynolds number and low wavenumber in channels long compared to the peristaltic wavelength [11]. The porous medium was modeled as a square array of cylinders with time-dependent orientation varying from normal to tangential to the axis of the channel. The drag coefficient for normal orientation was modeled with an empirical fit of numerical results [12], and the tangential drag was taken as half that of normal, which is exact for low cylinder concentration and conservative for high concentration. The cylinders were more or less normal to the flow during forward flow, and tangential during retrograde flow, thus promoting flow in the direction opposite that of the peristaltic wave (Fig. 2). Three schedules, or waveforms, of cylinder orientation were evaluated: 1) constant normal orientation (as a validation case), 2) switching orientation in which the cylinders were either normal or tangential to the flow, and 3) sinusoidal oscillation of orientation from normal to tangential. Switching orientation was expected to result in the greatest promotion of retrograde flow, while the latter provides a test of the sensitivity of flow to the orientation waveform. In the results to follow, longitudinal pressure gradient ΔP_λ was ignored and cylinder radius to channel half-height ratio was set to $\beta = 0.01$, based on collagen fibers (1.5 nm diameter) in a 150 nm thick channel.

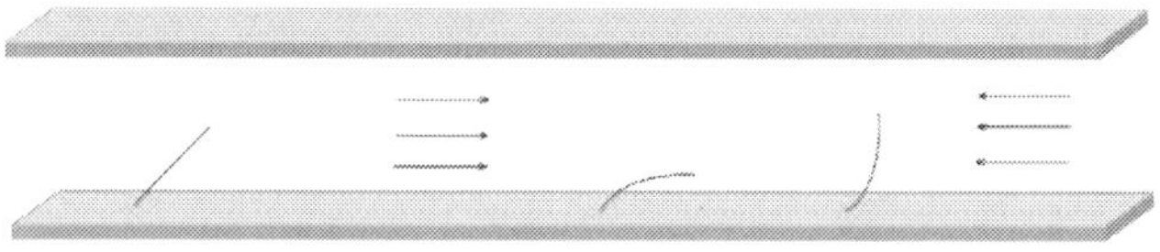

Figure 2. Orientation of cylinders to promote flow in the left to right direction. Left – cylinder orientation for zero flow. Middle – left to right flow flexes the cylinder so that it is oriented tangential to the flow for reduced drag. Right – right to left flow flexes the cylinder towards normal orientation for increased drag. For simplicity, only one cylinder is shown, and it is attached to the wall, but in the solution, cylinders are distributed throughout the channel.

RESULTS

Numerical integration with one-degree steps throughout the wave period was found to provide good convergence of the solution.

The flow rate with cylinders of constant orientation remained positive (forward), as expected, for all values of peristaltic amplitude and cylinder volume fraction. For switching orientation, negative (retrograde) flow was obtained where low amplitude ϕ/ϕ_{max}, ≤ 0.1 (where ϕ is the amplitude of wall motion normalized by the mean half-height of the channel and ϕ_{max} is the maximum normalized amplitude corresponding to cylinders touching) and high cylinder volume fraction $\varepsilon_0 \geq 0.001$ both occur (Fig. 3). (The maximum cylinder volume fraction for a square array is $\varepsilon = \pi/4$.) Over the same ranges of parameters, the increase in normalized flow rate Q/ϕ with sinusoidal versus switching orientation ranged from about 0.01 at $\varepsilon_0 = 0.0001$ to 0.05 at $\varepsilon_0 = 0.1$, which corresponded in most cases to less retrograde flow and in only a few cases to a change from retrograde to forward flow (not shown).

DISCUSSION

This model predicts dependable retrograde flow for low ϕ/ϕ_{max}, and high ε_0. Low amplitude alone causes reverse flow for cylinder concentration as low as $\varepsilon_0 = 0.0001$ when $\phi/\phi_{max} \leq 0.02$, but high concentration ε_0 is required when amplitude is greater. The direction of flow was not very sensitive to orientation waveform. Retrograde flow can be changed to forward flow with increasing peristaltic amplitude and/or decreasing cylinder concentration. Decreasing cylinder concentration might occur due to an actual reduction in proteins such as observed during ageing for collagen IV [13], or with a constant mass of proteins by dilation of the basement membrane, which has also been observed in ageing. These findings await experimental validation of the existence of flexing flow resistors, as well as measurements of flow conditions in cerebral vascular basement membranes and the changes in conditions during ageing.

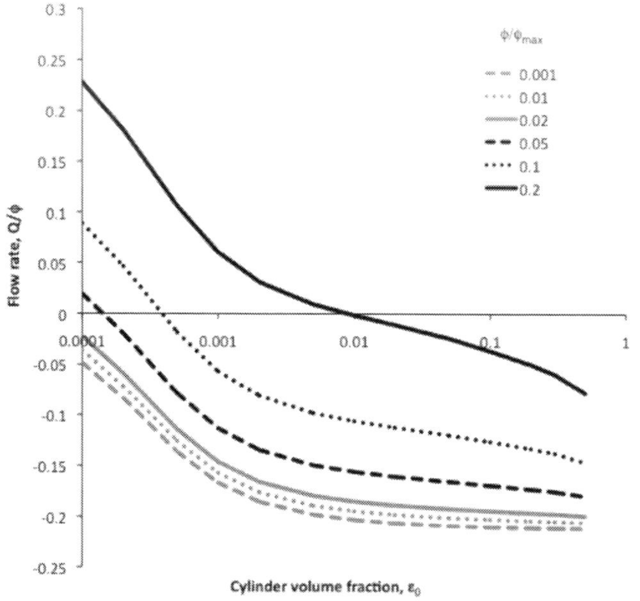

Figure 3. Flows produced by switching cylinder orientation. Retrograde flow ($Q/\phi < 0$) is promoted by large cylinder volume fraction ε_0 and low peristaltic amplitude ϕ/ϕ_{max}. $\Delta P_\lambda = 0$ and $\beta = 0.02$.

REFERENCES

[1]Szentistvanyi, *et al. Am J Physiol* 246(6 Pt 2):F835-F44, 1984.
[2] Bradbury, *et al. Am J Physiol* 240(4):F329-F36, 1981.
[3] Yamada, *et al. Am J Physiol* 261:H1197-H204, 1991.
[4] Carare, *et al. Neuropathol Appl Neurobiol* 34(2):131-44, 2008.
[5] Viswanathan & Greenberg. *Annals Neurology* 70(6):871-80, 2011.
[6] Carare, *et al. Neuropathol Appl Neurobiol* 39 (6):593-611, 2013.
[7] Lennon, *et al. J Am Soc Nephrol* 25, Online ahead of print, DOI: 10.1681/ASN.2013030233, 2014.
[8] Byron, *et al. J Am Soc Nephrol* 25 Online ahead of print, DOI: 10.1681/ASN.2013070795, 2014.
[9] Miosge. *Histology Histopathology* 16:1239-1248, 2001.
[10] Weller, et al. *Acta Neuropathol* 117:1–14, 2009.
[11] Jaffrin & Shapiro. *Ann Rev Fluid Mech* 3:13-36, 1971.
]12] Sangani & Acrivos. *Int J Multiphase Flow* 8:3:193-206, 1982.
[13] Hawkes, *et al. PLoS ONE* 7(7): e41636, 2012.

CHARACTERIZING THE CHANGE IN ANKLE CONSTRAINT FOLLOWING GRADE II AND III SPRAINS

Matthew H. Dickinson (1), Ednah G. Louie (1), Bardiya Akhbari (2), William M. Eboch (2), Sami Shalhoub (1), Lorin P. Maletsky (1, 2)

(1) Bioengineering Graduate Program
University of Kansas
Lawrence, KS, USA

(2) Department of Mechanical Engineering
University of Kansas
Lawrence, KS, USA

INTRODUCTION

Ankle sprains are among the most frequent injury to occur during physical activity.[1] It is estimated that 23,000 injuries occur each day in the United States alone.[2] Ankle sprains can lead to chronic ankle instability, which increases the risk of osteoarthritis. Understanding the kinematic changes that occur following ankle sprains will help improve the treatment of ankle injury and prevent osteoarthritis. Previous *in vitro* studies examined the effect of injuries on ankle kinematics, but focused only on select degrees of motion.[3,4]

Ankle sprains most often result from hyperinversion that causes abnormal stretching or rupturing of the anterior talofibular ligament (ATFL) and the calcaneofibular ligament (CFL).[5,6] By characterizing the ankle's full adduction/abduction (ADAB) and inversion/eversion (INEV) kinematics before and after Grade II and III sprains, a more complete picture of injury is possible. Extensive ATFL tearing indicates a Grade II sprain, and Grade III sprains occur when both the ATFL and CFL experience substantial damage. The hypotheses of this study were that the Grade II sprain would significantly decrease constraint in adduction and inversion and the Grade III sprains would reduce constraint in abduction and inversion due to the orientation of the ATFL and CFL, respectively.

METHODS

Five cadaveric ankles (4 Male) were tested with an average age of 63 years (range 49-73 years) and average BMI of 24.4 (range 18-32 BMI). All donors had no history of degenerative bone disease or ankle injury. The leg specimens had previously been frozen, thawed, and sectioned mid tibia. They were then refrozen until testing. All soft tissue more than 6.5 cm superior to the medial malleolus was stripped from the bone, and the fibula was fixed to the tibia approximately 2 cm superior to this point while the foot was in neutral position. Adhesive putty secured an aluminum fixture centered on the tibia. A loading platform was attached to the sole of the foot with screws into the calcaneus and straps around the forefoot to assist manipulations and allow a neutral position to be determined. A six-DOF tri-axial load cell (JR3, Woodland, CA) was rigidly attached to the ankle (Fig. 1).

An Optotrak Certus motion capture system (NDI, Waterloo, Canada) recorded the kinematics of the tibia and calcaneus by tracking infrared emitting bodies rigidly attached to each bone (Fig. 1). A rigid body is also attached to the talus to enable evaluations of each individual joints; however, this preliminary study reports solely on the ankle joint complex (AJC). The neutral position was determined using a bubble level for dorsiflexion/plantarflexion (DFPF) and INEV while neutral ADAB was identified by the tester's judgment.

Manual laxity manipulations were then performed with the ankle in an inverted position (foot up). The experimenter sequentially applied anterior/posterior (ANPO) load, ADAB torque, and INEV torque across the full range of DFPF while kinetic and kinematic data were recorded. This was performed slowly to avoid viscoelastic effects. Visual feedback was provided to the tester to ensure enough points at the desired loads and torques were recorded. A Grade II sprain was then simulated by severing the ATFL, and laxity measurements retaken. Subsequently, the CFL was resected and the simulated Grade III sprain was tested. Bony landmarks on the tibia and calcaneus were probed to create body-fixed coordinate systems required for the characterization of the AJC using a 3-cylindrical open-chain description of kinematics.[7]

Only the data collected while the ankle was loaded in the desired motion were used in the analysis while the unloading data was disregarded to avoid hysteresis. A unified envelope of constraint was created for each ankle to provide a basis of comparison across multiple specimens. This method has previously been described for the knee joint.[8] Briefly, a radial basis function (RBF) was trained using the experimental data selected. ANPO load, INEV and ADAB

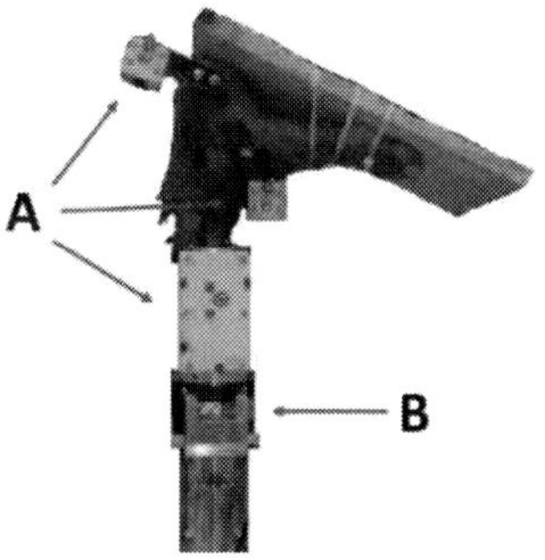

FIGURE 1. TRIAL SETUP WITH (A) RIGID BODIES AND (B) LOAD CELL

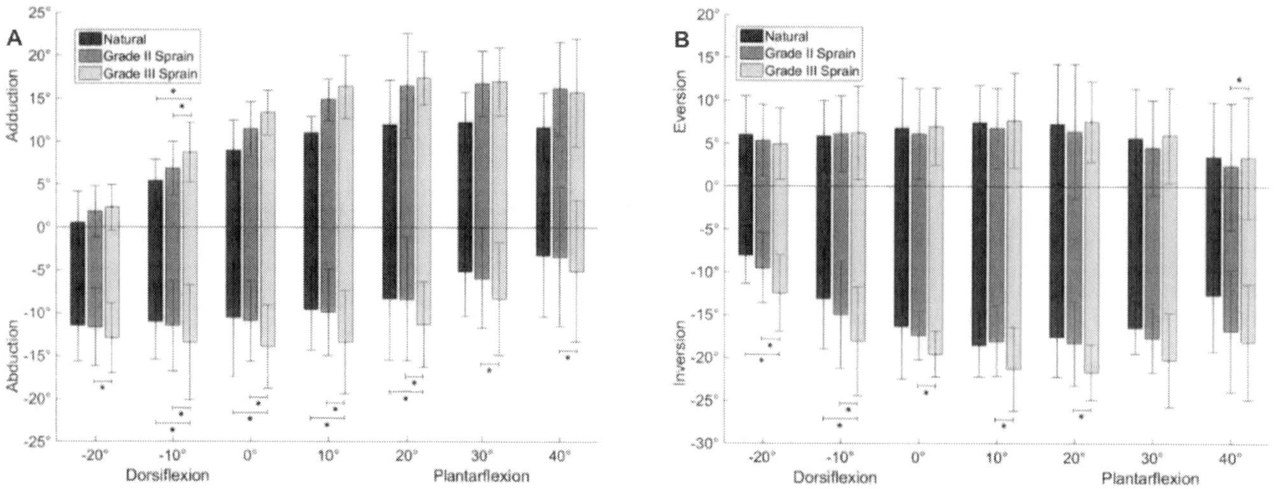

FIGURE 2. KINEMATIC CHANGES IN RESPONSE TO GRADE II AND III SPRAINS. THE MEAN KINEMATICS AND STANDARD DEVIATION IN RESPONSE TO 3 NM (A) ADDUCTION AND ABDUCTION; (B) EVERSION AND INVERSION TORQUES. ASTERISKS (*) INDICATE A SIGNIFICANT DIFFERENCE BETWEEN GROUPS (P < .05).

torques as well as the flexion angle were treated as independent variables while resulting kinematics were dependent variables. A unified envelope was created for each condition (healthy, Grade II sprain, Grade III sprain). The RBF interpolates the kinematic outcomes within the range of recorded loading conditions, which enabled the study of specific loads. Kinematics resulting from isolated 3 Nm ADAB and INEV torques were generated by the RBF and analyzed. Two-tailed paired t-tests were used to compare the three conditions of the ankles at every 10° of flexion using an alpha level of 0.05.

RESULTS

The average range of motion (ROM) in DFPF was 67.5° (range 51.2-83.1°). Given the variability in flexion ROM, a range of -20° to 40° was chosen because it reflects a typical degree of flexion, and most specimens achieved these flexion angles. No significant changes in either ADAB or INEV were observed following the simulated Grade II sprain relative to the natural ankle (Fig. 2). The Grade III sprain, however, exhibited significant differences from both the Grade II sprain and the natural ankle in abduction across all flexion angles and in adduction mid-dorsiflexion (Fig. 2A). This injury also caused considerable changes in inversion across dorsiflexion and shallow plantarflexion (Fig. 2B). The average kinematic offset was 21.9° dorsiflexion (range 2.9-37.0°), 12.6° eversion (range 6.0-26.0°), and 11.7° adduction (range 2.8-18.5°). The neutral position of the ankle determined this offset. Ankle joint ROM under isolated 3 Nm ADAB and INEV torques was on average 32.6° INEV (range 17.9-40.9°) and 32.6° ADAB (range 26.4-40.6°).

DISCUSSION

The lack of an appreciable change in ankle constraint following a Grade II sprain was unexpected; however, with a sample size of five ankles and considering normal variability between specimens, it is not entirely surprising. Additionally, the inversion data were trending toward significance in deep plantarflexion ($p = .127$ at 35° and $p = .058$ at 40°). This can be attributed to the anterior orientation of the ATFL and its elongation during plantarflexion.[9] This trend makes sense and might be significant with a larger sample size. Similarly, the ATFL is expected to resist adduction, and this trend is observed primarily in shallow plantarflexion ($p < .10$ for -5° to 20°). The large variability in flexion ROM resulted in 3 out of 5 ankles being near or at their maximal flexion at 40° while the remaining ankles still had additional mobility. This may confound the data if non-ligamentous features dominate constraint near these end points because the flexion angles of ATFL importance may not overlap between ankles.

The Grade III sprain produced significant changes in ankle constraint. The anterior/superior to posterior/inferior direction of the ligament suggested that the CFL would resist abduction as well as inversion of the ankle. Following injury, these two rotations were largely influenced across most flexion angles. An decrease in inversion constraint has clinical relevance because diagnosis of injury relies upon this partially on this measure. A slight dependence on flexion angle is also observed in the Grade III injury in regards to inversion kinematics. Rotation during dorsiflexion is more significantly altered by the CFL resection, indicating that the CFL is recruited more in dorsiflexion. Although not all of the hypotheses of the study were supported in this preliminary study, using a RBF to map out the full envelope of constraint will be beneficial once additional ankles are tested. This method allows isolated as well as specific combined loads to be studied without needing to exactly place those forces on the ankle. Additionally, a thorough characterization of ankle kinetics provides essential information for models of ankle behavior and may be important in the continued improvement of early sprain diagnosis.

ACKNOWLEDGEMENTS

This experiment was partially supported by DePuy Synthes, Inc.

REFERENCES

[1] Garrick, JG, *Am J Sports Med,* 5(6):241-2, 1977.
[2] Kelmer, E et al., *Sports Med,* 41(3):185-97, 2012.
[3] Hollis, JM et al., *Am J Sports Med,* 23(6):672-7, 1995.
[4] Kerkhoffs, GMMJ et al., *Clin Biomech,* 16(8):635-43, 2001.
[5] Wolfe, MW et al., *Am Fam Physician,* 63(1):93-104, 2001.
[6] Baumhauer, JF et al., *Am J Sports Med,* 23(5): 564-70, 1995.
[7] Wu, G et al., *J Biomech,* 35(4):543-8, 2002.
[8] Cyr, AJ et al., *Proc Inst Mech Eng H,* 228(5):494-500, 2014.
[9] Ozeki, S et al., *Foot Ankle Int,* 23(9):825-32, 2002.

PRELIMINARY FINITE ELEMENT ANALYSIS OF SUBCHONDRAL BONE CYSTS IN THE STIFLE OF THE HORSE

Lance L. Frazer (1), Kenneth J. Fischer (1), Garrett J. Noble (3), Elizabeth M. Santschi (2)

(1) Department of Mechanical Engineering
University of Kansas
Lawrence, KS, USA

(2) College of Veterinary Medicine
Kansas State University
Manhattan, KS, USA

(3) Department of Biomedical Engineering
Ohio State University
Columbus, OH, USA

INTRODUCTION

Medial femoral condylar (MFC) subchondral bone cysts (SBC) often develop in the stifle joint (knee) of young racing horses, causing lameness and limiting performance (Figure 1). Effective treatment for rapid recovery is important as delays can greatly reduce the value of the horse. Foundational research into SBC development began the early 1980's when Jeffcott et al. proposed that mechanical over-load of the MFC articular surface was the likely cause[1], however the pathogenesis is still debated.

Present treatments of equine SBCs focus on reducing bone and joint inflammation and result in mediocre success rates. Understanding the mechanics of equine SBC pathogenesis and progression should lead to more effective treatments. SBCs in the human knee are often coincident with osteoarthritis, but SBC in the young horse are without osteoarthritis and known monotonic trauma. Therefore, the equine SBC is a good model to study, without those confounding elements. Understanding equine SBCs could also lead to better understanding of and treatments for SBCs in humans.

The objective of this preliminary study was to develop a finite element model of the equine stifle joint to investigate the changes in bone and soft tissue stresses caused by a cystic lesion. We hypothesized that a cystic lesion in the medial femoral condyle will result in high stress concentrations in bone around the cyst.

METHODS

This study began by generating a finite element mesh using Simpleware software. Starting from a CT scan of a normal equine stifle joint, bone tissue was semi-automatically segmented by thresholding with manual adjustment, and the soft tissues were manually segmented (Figure 2). Each structure underwent filtering processes, such as recursive Gaussian smoothing as well as mean value filtering, to yield a smooth, anatomically sound geometry of the joint. Segmented structures include the femur, tibia, articular cartilages for both bones, ACL, PCL, MCL, LCL, lateral meniscal-femoral ligament, tibial-meniscal ligaments, as well as a cystic lesion placed in the medial femoral condyle, just deep to the articular cartilage. For this preliminary analysis with constrained loading, the four major ligaments (ACL, PCL, MCL, LCL) were removed from the model to reduce the number of elements since they were not being stretched with the simplified boundary conditions. A mesh refinement tool was used in both the area of interest (cystic area) as well as the soft tissues in order to increase the geometric accuracy and analysis accuracy. The final mesh contained 440,000 linear, tetrahedral elements with the highest concentration of elements around the cyst (Figure 3).
Bone material properties were assigned using Simpleware's grayscale feature. Apparent density of bone ranged from .007 g/cm^3 to 1.82 g/cm^3, and the equation

$$E = 6000\rho^{2.5} \qquad (1)$$

was used to relate apparent density to Young's Modulus. This equation resulted in Young's Moduli ranging from near zero to 20 GPa. The mesh and bone material properties were then imported into ABAQUS. Approximate relaxation soft-tissue material properties were assigned (Table 1) based on prior published

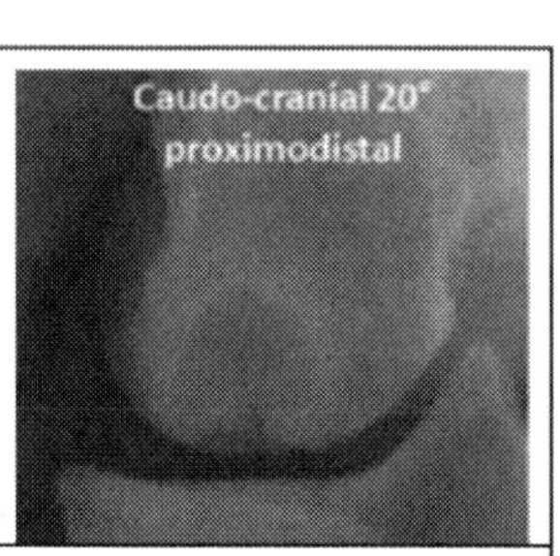

Figure 1: Radiograph of a stage-4 MFC SBC from a young Thoroughbred[2]

studies[3,4,5].

	Modulus	Poisson's Ratio
Cartilage[4]	4 MPa	0.2
Menisci[5]	20 MPa	0.2
Ligament[3]	345 MPa	0.2
Cyst	.001 MPa	0.2

Table 1: Material properties assigned to the soft tissues.

For this initial model of stance, all tissues were modeled as isotropic and linear elastic. The cyst was given relatively low material properties in order to mimic the fluid-filled structure.

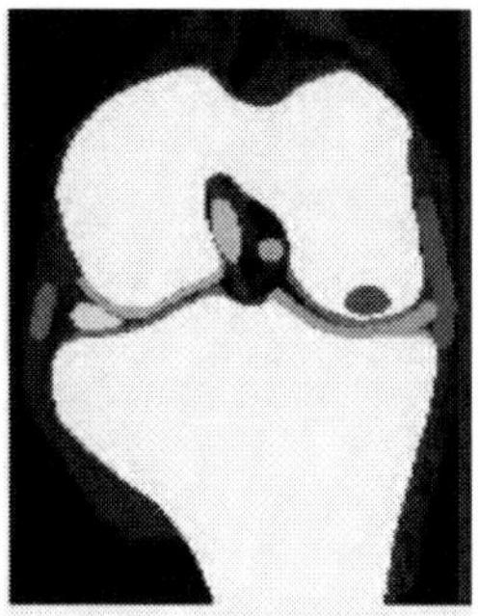

Figure 2: Frontal plane projection of the segmentation process

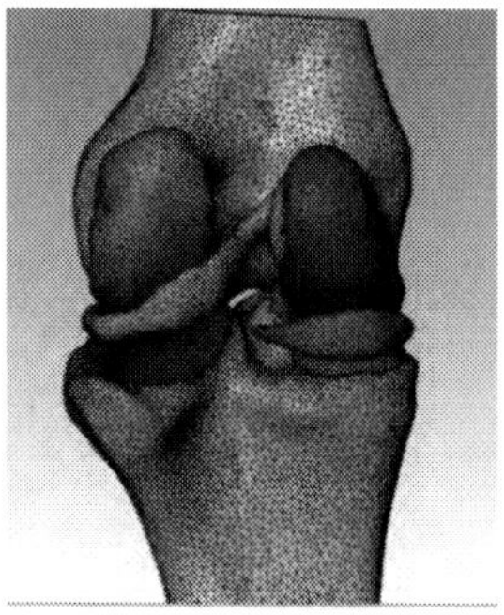

Figure 3: Shown here is the generated mesh from Simpleware

Contact surfaces were created between the two articular cartilage surfaces and for each cartilage/menisci interface, to allow normal cartilage deformation and meniscal sliding. The friction coefficient between these surfaces was set to 0.02.

To perform the analysis, displacements were fully constrained for the distal tibia. The proximal femur was constrained to move only in the vertical direction. A load of 1000 N was placed on the superior end of the femur to simulate the stance stresses of an average build horse (about 4,450 newtons, evenly distributed between four limbs, and taking into account the weight under the stifle joint). For the cyst model, the same analysis was carried out, with the properties of the cyst region changed from bone to cyst tissue.

RESULTS

Our results are consistent with our hypothesis. The cystic model resulted in tensile maximum principal stresses in the subchondral region of 1-2.5 MPa, shown in yellow and green (Figure 4, right), whereas the normal model showed no significant tensile stresses (Figure 4, left). The altered stress concentration developed most prominently in regions immediately inferior to the cyst, but also occurred posteriorly to the cyst (Figure 4, right). Approximately, the same stress distributions developed in the meniscus of each model (primarily the green), showing only alteration in the subchondral region.

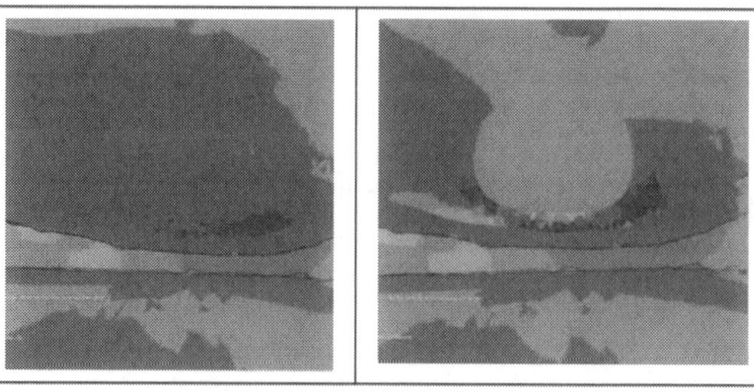

Figure 4: Shown here is a medial view of the subchondral region in the medial femoral condyle. Left is normal, and right is the cystic model. Non-zero tensile maximum principal stresses are shown as yellow (to 1 MPa) and green (to 2.5 MPa). Orange hues are near zero tensile (light) and compressive (dark) stress. Compressive stresses (to -1 MPa) are shown in red.

DISCUSSION

While this analysis is preliminary and simplistic, these results provide further insight into the mechanisms of SBC enlargements and development of peri-cystic bone thickening. The results clearly show an altered stress distribution in a MFC with a cystic lesion. Our results confirm that mechanical stresses are greatly increased around cystic lesions. Future studies will investigate these cystic lesions in much greater detail, as there is a paucity of data on the mechanics of bone cysts (we could find no prior study). The model used in this study will be enhanced by including all relevant structures such as ligaments, implementing viscoelastic properties into the soft tissues, using quadratic elements, varying the size of the cyst, adding marginal bone sclerosis, as well as creating more advanced boundary conditions in the analysis to test the joint in both extension and flexion. Our goals are to describe cyst formation as well as investigating and improving novel biomechanical treatments that have recently shown promise.[2]

ACKNOWLEDGEMENTS

The authors acknowledge the help and support from Simpleware ltd. during the segmentation process.

REFERENCES

[1] Jeffcott, LB et al., *Equine Vet J.*, 15(4):304-11, 1983.
[2] Santschi, E et al., *Vet Surg.*, 10.1111/j.1532-950X, 2014.
[3] Buttler et al., *J Biomech*, (19):425-432, 1986.
[4] Chahine et al, *J Biomech*, (37):1251-1261, 2004.
[5] Brown, Load-Bearing Role of the Human Knee Meniscus, *Doctoral Dissertation*, MIT, 1990.

SB³C2015
Summer Biomechanics, Bioengineering and Biotransport Conference
June 17-20, 2015, Snowbird Resort, Utah, USA
SB³C2015-394

INTRAOCULAR PRESSURE INCREASES IN PATIENTS WITH INTRAOCULAR GAS BUBBLES FOLLOWING A DESCENT AND SUBSEQUENT ASCENT

Lucas A. Gsellman, Rouzbeh Amini

Department of Biomedical Engineering
University of Akron
Akron, Ohio, US

INTRODUCTION

Pnuematic retinopexy is a commonly used procedure for the repair of retinal detachment. It involves the removal of a portion of vitreous humor from within the posterior eye, which is then replaced with a gas. The gas bubble presses the detached portion of the retina back into position, where it can then be reattached and left to heal. The gas bubble is eventually absorbed over a period of 1 to 8 weeks [1].

The gas bubble size changes in response to ambient pressure alteration, ocular globe deformation, and changes in ocular fluid volume. The dynamic relationship results in gauge intraocular pressure (IOP) changes. A normal gauge IOP of approximately 15 mmHg is essential for maintaining ocular shape and function. An increase in gauge IOP may lead to discomfort and temporary vision loss. It is also a risk factor for glaucoma.

It is well known that an ascent in altitude, which corresponds to a drop in ambient pressure, will cause the gas bubble to expand and lead to increased IOP [2]. Thus, it is commonplace to warn patients against large ascents, such as air travel, during their recovery time. The same consideration, however, is not always shared when the ascent in question does not bring the patient to a higher altitude than where the procedure has taken place. It was the purpose of this study to theoretically examine clinically reported cases of descents and subsequent ascents that returned patients to, without exceeding, the altitude of their surgery.

METHODS

A modified version of the mathematical model used in our previous study [2] was adapted to approximate four published patient cases reported from Shaare Zedek Medical Center, Jerusalem, Israel [3]. The model took into account aqueous flow changes due to the IOP control mechanism. The trabecular meshwork, which is the primary site of outflow was treated as a pressure dependent one-way valve. Consequently, the outflow decreased to zero when the IOP was equal to the episcleral venous pressure. Deformation of the ocular globe was approximated as linear elastic behavior as discussed previously [2]. A noted modification from the previous model was the prevention of hypotony. Ocular hypotony (i.e. negative gauge IOP) would occur when absolute IOP is less than the absolute ambient pressure. The ocular shell withstands a positive pressure gradient but would collapse under a negative pressure. Consequently, hypotony is not experienced in uninjured eyes.

The model was designed to use the input of a time elevation vector that corresponded to the travel of each patient. The path of each patient was simplified to a descent from home elevation to a low elevation at a constant rate, followed directly by a period of time at the low elevation, and lastly, a constant ascent to high elevation (Jerusalem 790m). The scenarios are representations of either a trip from the mountains to the coast with a return, or a trip through a large valley. Both are plausible scenarios due to Jerusalem's close proximity to a variety of topographical regions.

Volume lost V_{loss} and interior pressure P_{in} were solved using the system of equations below.

$$\kappa {P_{in}}^2 - \left(\kappa P_{0_{in}} + \kappa P(h) - \alpha V_0 - V_{loss}\right)P_{in} - \alpha V_0 P_{0_{in}} = 0, \quad (1)$$

$$\frac{d}{dt}V_{loss} = \mu(P_{in} - P_e) - \left[Q_{Aqu} - \lambda\left(P_{in} - P_{0_{in}} - P(h)\right)\right]. \quad (2)$$

Equations (1) an (2) were solved using Euler's Method and MatLab (The MathWorks Inc., Natick, MA). The parameters used in the equation are explained extensively in our previous work [2] and are listed in Table 1.

A parametric study was also conducted to evaluate the effects of descent rate, ascent rate, duration at low altitude, and aqueous humor flow modification. Case 1 in Table 2 was used as a base case for comparison. The model was run while all parameters of the base case were held constant except for the parameter being studied. The peak gauge IOP was calculated.

TABLE 1. PARAMETERS USED IN MODELLING EQUATIONS

Parameter	Description
P_{in_0}	Initial Absolute Intraocular Pressure (IOP)
V_0	Initial Ocular Globe Volume
α	Ratio of the Initial Bubble Size to the Globe Volume
κ	Corneoscleral Shell Compliance
μ	Aqueous Humor Outflow Facility
Q_{Aqu}	Aqueous Humor Inflow Rate
λ	Aqueous Humor Psedofacility
P_{in}	Absolute IOP
P_e	Episcleral Venous Pressure
V_{loss}	Volume Lost

TABLE 2. MODEL PARAMETERS AND PREDICTED PEAK IOP FOR THE SIMULATED CLINICALLY RELEVANT CASES. SURGERY FOR ALL FOUR PATIENTS WERE PERFORMED AT THE ELEVATION OF 790 M.

Case number	Home Elevation (m)	Lowest Elevation (m)	Descent Rate (m/min)	Ascent Rate (m/min)	Low Time (min)	Initial Bubble Size (%)	Peak Gauge IOP (mmHg)
1	790	20	40	40	60	70	34
2	790	0	40	40	60	55	36
3	395	25	40	40	60	80	46
4	240	-310	35	35	60	60	50

RESULTS

All of the simulated patient cases showed an increased peak gauge IOP (much higher than the normal value of 15 mmHg) after returning to a high altitude (Table 2). As shown in a typical simulated case (Fig. 1), the model predicted a large IOP reduction when the patient reached the lower elevation. The gas bubble volume was also reduced by ~4% (initially filling 55% of the vitreous cavity). In the case represented by Fig. 1, the peak IOP was calculated to be 36mmHg immediately following the ascent.

All parameters, used in the parametric study, had noticeable impacts on the peak gauge IOP. They are listed in increasing order of impact in Fig. 2. Modification of low time (i.e. time spent at the low

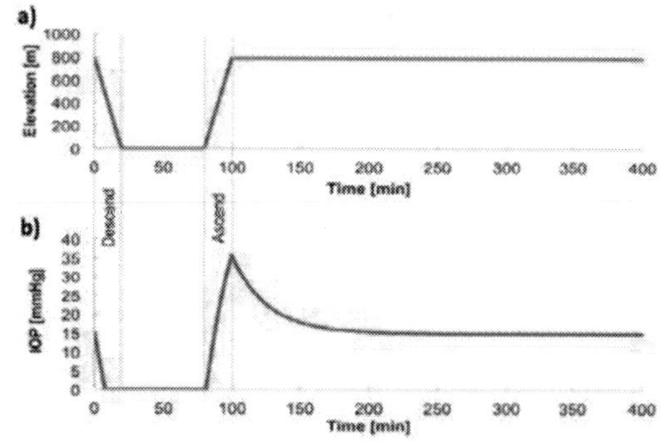

FIGURE 1. (A) ELEVATION AND (B) IOP VERSUS TIME FOR A TYPICAL SIMULATED CASE (CASE 2 IN TABLE 2)

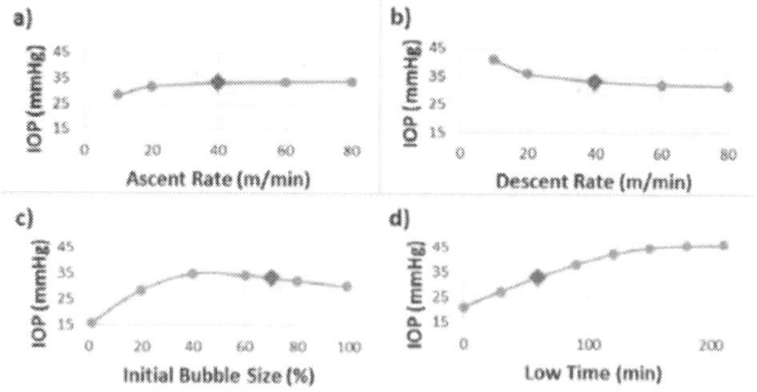

FIGURE 2. THE RESULTS OF PARAMETRIC STUDIES. CHANGES IN THE IOP VERSUS (A) ASCENT RATE, (B) DESCENT RATE, (C) INITIAL BUBBLE SIZE, AND (D) DURATION OF STAYING AT THE LOW ALTITUDE. IN ALL GRAPHS THE BASE CASE IS REPRESENTED BY THE RED DIAMOND.

altitude) resulted in the largest change in peak gauge IOP. Peak gauge IOP values ranged from 21mmHg to 46 mmHg for low times of 0min and 210min, respectively. There was less than a 3% change in peak gauge IOP for low times greater than 150min.

DISCUSSION

The longer time a patient spent at the low altitude, the higher the peak gauge IOP reached after the ascent. Such phenomenon is due to increased aqueous humor accumulation, a feedback mechanism to keep IOP at a normal level. The same mechanism also caused slower descents to result in higher max gauge IOP.

The opposite was true for the ascent; a faster ascent resulted in higher max gauge IOP. The increase in bubble size outpaces the ability for fluid to be evacuated. Increasing the outflow facility can speed up compensation during ascents. As a side note, outflow is also reduced to near zero during descents due to trabecular meshwork collapse, which mitigates the negative effect of increasing outflow. Increasing the outflow helped lower high peak gauge IOP values in all simulated cases.

A plausible worst-case scenario could be predicted, where either a slow descent, long duration at low altitude, or combination of the two has allowed time for the eye to fully adjust to the low altitude. One such case had a peak gauge IOP value of 53 mmHg with a slow descent rate of 10 m/min and a rapid ascent rate of 80 m/in. The trends predicted in our model were consistent with the recent clinical observations [3]. The discrepancies in exact values can be attributed to the simplification of the patient's paths.

Based on our simulation, patients with intraocular gas bubbles may be at high risk of elevated IOP after a descent and subsequent ascent to high altitude even without ever exceeding the initial altitude. Further, the reduction in gas bubble size at the low altitude may increase the risk of ocular hypotony and post-surgical retinal detachment. Our simulation suggests that aqueous humor flow regulating medication may help manage the risks.

ACKNOWLEDGEMENTS

The authors would like to thank Darryl Overby and Rishi Singh for the helpful discussions.

REFERENCES

[1] Chan, C et al., *Surv Ophthalmol*, 53:443–478, 2008.
[2] Amini, R et al., *Retina*, 31:1656–1663, 2011.
[3] Brosh, K et al., *Eye (Basingstoke)*, 28:892-894, 2014.

INVESTIGATING CRISPRI CELL-ENGINEERING METHODS FOR TREATMENT OF INTERVERTEBRAL DISC DEGENERATION

Niloofar Farhang (1), Jonathan M. Brunger (3,4), Joshua D. Stover (1), Pratiksha I. Thakore (3), Brandon Lawrence (2), Farshid Guilak (3,4), Charles A. Gersbach (3), Lori A. Setton (3,4), Robby D. Bowles (1, 2)

(1) Department of Bioengineering
University of Utah
SLC, UT, USA

(2) Department of Orthopaedics
University of Utah
SLC, UT, USA

(3) Department of Biomedical Engineering
Duke University
Durham, NC, USA

(4) Department of Orthopaedic Surgery
Duke University Medical Center
Durham, NC, USA

INTRODUCTION

Lower back pain (LBP) is the single leading cause of disability worldwide and has a global lifetime prevalence of 38.9% [1,2]. Degenerative disc disease (DDD) and associated pathologies are considered major contributors to LBP. The progression of DDD has been associated with an inflammatory environment that includes the presence of inflammatory cytokines (e.g. TNF-α, IL-1β) in the intervertebral disc (IVD) that are active in the degenerative process and may sensitize pain-sensing nerve fibers in the IVD. As such, therapeutic methods that antagonize the action of inflammatory cytokines in the IVD are of great interest. We are particularly interested in developing therapeutic strategies that utilize CRISPRi to modulate pro-inflammatory cytokine receptors related to the progression of DDD in both native cells and therapeutic cells delivered to the IVD.

The newly developed tool for endogenous gene regulation, Clustered Regularly Interspaced Short Palindromic Repeats interference (CRISPRi), has the potential to provide effective multiplex gene regulation for applications in DDD. It has been shown that CRISPRi can perform highly specific and effective gene knockdown in mammalian cells in a single or multiplex manner [3]. This system requires the expression of nuclease-inactive Cas9 (dCas9) by itself or fused to the Krüppel-associated box (KRAB) and the expression of short genomic loci-specific guide RNAs (gRNAs) that are complementary to the promoter of the gene [4]. We hypothesize that the CRISPRi system can be used to perform multiplex knockdown of inflammatory cytokine receptors and inhibit actions of inflammatory cytokines on cells in the IVD and retard the progression of DDD. In this work we demonstrate the modulation of tumor necrosis factor receptor type one (TNFR1) and interleukin one receptor type one (IL1R1) using CRISPRi in both Human Embryonic Kidney (HEK293T) cells and human nucleus pulpous (hNP) cells and demonstrate an ability to alter cell response to inflammatory challenge.

METHODS

CRISPRi Vector Design: Sequences for gRNAs for the TNFR1 and IL1R1 promoter region of each gene were selected with the necessary adjacent protospacer motif (crispr.mit.edu). A non-targeting gRNA was chosen as a control. Each gRNA was cloned into a plasmid under the control of the U6 promoter. To screen the gRNAs for the best knockdown, gRNA plasmids were transfected into a HEK293T cell line stably expressing dCas9-KRAB (Lipofectamine 2000, n=3 per guide). Forty-eight hours post transfection, RNA was isolated and quantitative reverse transcriptase PCR (qRT-PCR) was performed for TNFR1, IL1R1, and GAPDH. For TNFR1, the gRNA showing the greatest knockdown was cloned into two different lentiviral vectors, resulting in the co-expression of the gRNA with dCas9-T2A-GFP or dCas9-KRAB-T2A-GFP from a single lentiviral vector.

Human NP Cell CRISPRi: As a pilot study, human nucleus pulposus cells were obtained from a 51-year-old male discectomy patient and transduced with the TNFRI lentiviral CRISPRi vectors, along with a control lentiviral vector that did not express a gRNA. Polybrene was used at a concentration of 8 μg/ml to increase likelihood of transduction. Quantitative RT-PCR was then performed to identify TNFR1 knockdown in these primary human NP cells. Additionally, flow-cytometry was performed to determine transduction efficiency in human NP cells.

Functional TNFR1 Knockdown: Functional knockdown of TNFR1 was studied by quantifying TNF-α induced NF-κB activity after CRISPRi knockdown. NF-κB activity was quantified using a

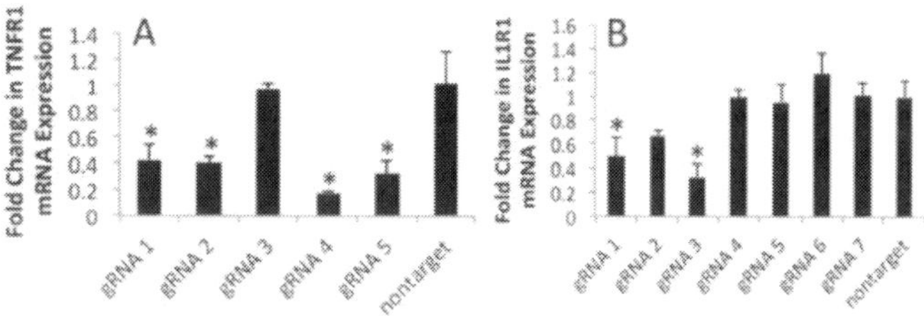

FIGURE 1: KNOCKDOWN OF (A) TNFR1 & (B) IL1R1 WITH MULTIPLE PROMOTER-TARGETING GRNAS (* = P<0.05 COMPARED TO NONTARGET CONTROL).

firefly luciferase NF-κB reporter vector that was transduced into HEK293T cells [5]. NF-κB reporter expressing HEK293T cells were also transduced with the developed TNFR1 lentiviral CRISPRi vector and cell sorted for GFP to produce a cell line expressing the TNFR1 CRISPRi system. These cells along with the non-knockdown cells were seeded at 25,000 cells/well in white opaque plates and treated with either 0 ng/ml, 150 pg/ml, 1 ng/ml, and 10 ng/ml rhTNF-α, for 24 hours (n=5 per treatment). NF-κB activity was then measured using a firefly luciferase luminescence assay (Bright-Glo, Promega). Additionally, cell number was assayed (RealTime-Glo, Promega) on a separate plate with the same experimental conditions, therefore allowing for normalization of each NF-κB luminescence signal to the number of viable cells. Luminescence for both assays was measured using a synergy HTX biotek plate reader. Activity of NF-κB is reported as a fold change relative to control and normalized by cell viability.

Statistical Analysis: To determine if knockdown of IL1R1 and TNFR1 in HEK293T cells was statistically significant a one-way ANOVA was conducted with Tukey post hoc. For statistical analysis of NF-κB activity, a two-way ANOVA with Tukey post hoc was conducted and a $p<0.05$ was considered statistically significant.

RESULTS

CRISPRi Vector Design: Quantitative reverse transcriptase PCR studies for both TNFR1 and IL1R1 in HEK293T cells show significant knockdown in four out of five guides for TNFR1 and two out of seven guides for IL1R1. Maximum knockdown in the guides screened was 82% for TNFR1 and 67% for IL1R1 (Figure 1A & B).

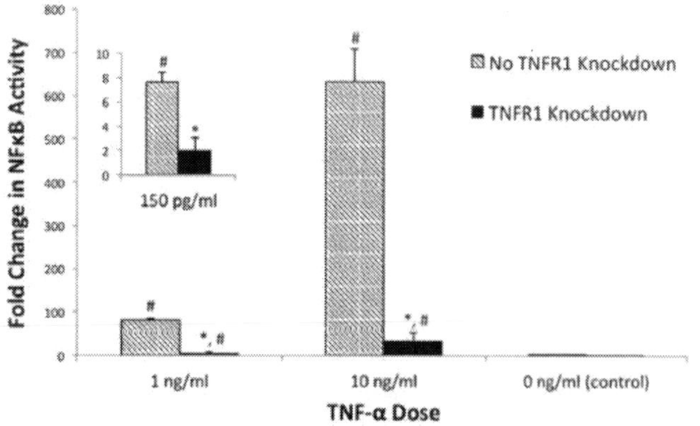

FIGURE 3: FOLD CHANGE IN NF-κB ACTIVITY BY HEK293T CELLS DOSED WITH TNF-α. DIFFERENCES BETWEEN THE KNOCKDOWN AND THE NO KNOCKDOWN CELLS ARE STATISTICALLY SIGNIFICANT (P<0.05) FOR EACH DOSE AS DENOTED BY *. ALL BUT THE 150 PG/ML KNOCKDOWN CELLS SHOWED A STATISTICALLY SIGNIFICANT DIFFERENCE IN NF-κB ACTIVITY FROM 0 NG/ML CONTROL (P<0.05), AS DENOTED BY #.

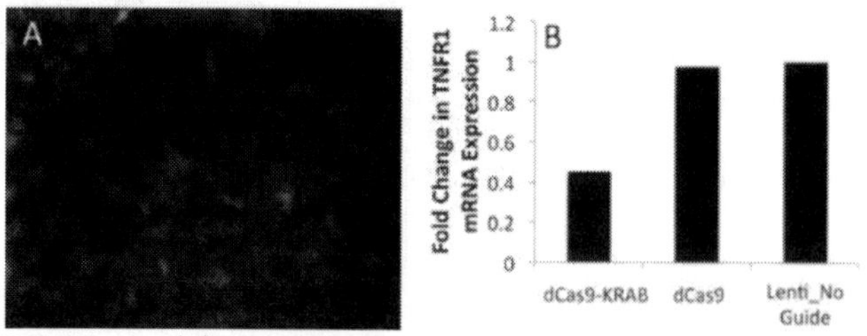

FIGURE 2: (A) TRANSDUCED PRIMARY HUMAN NP CELLS EXPRESSING CRISPRI SYSTEM AS EVIDENCED BY GFP EXPRESSION. (B) KNOCKDOWN OF TNFR1 IN NP CELLS USING TNFR1 CRISPRI LENTIVIRAL VECTOR.

Using gRNA 4 for TNFR1 with dCas9-KRAB, we saw 55% TNFR1 knockdown in NP cells but no knockdown with dCas9 alone (Figure 2B). Transduction efficiency of TNFR1 lentiviral CRISPRi vector in human NP cells was 42%.

Functional TNFR1 Knockdown: Fold change in NF-κB activity was significantly decreased in response to TNF-α after TNFR1 CRISPRi knockdown at each TNF-α dose, demonstrating a decrease in TNFR1 induced signaling (Figure 3). The decrease in activity was on the order of ~95% for the two highest doses and returned the lowest dose to baseline levels. This indicates a complete elimination of this deleterious signaling pathway at the most physiologically relevant dose.

DISCUSSION

These studies have shown great promise that CRISPRi knockdown of inflammatory receptors can be used to successfully modulate the inflammatory response relevant to disc pathology. Our results show that CRISPRi can be used to significantly reduce gene expression of TNFR1 and IL1R1 in HEK293T cells and TNFR1 gene expression in human NP cells (Figure 1 & 2). We have also shown that this measured reduction in TNFR1 gene expression also results in a loss of TNFR1 signaling, as demonstrated by the significantly lower fold change in NF-κB activity in TNFR1 knockdown vs. no knockdown cells for each TNF-α dose (Figure 3). This finding is highly relevant to disc pathology, as NF-κB signaling has been identified as a key component in the degenerative process and antagonizing NF-κB activity has shown therapeutic effects in preclinical animal models [6,7]. It is especially promising that TNFR1 CRISPRi was able to return the physiologically relevant dose (150 pg/ml) to baseline levels. Despite the 42% transduction efficiency, the CRISPRi system showed expression and activity in human primary NP cells, making their application to the disc a possibility. Future work will investigate methods to improve transduction efficiency in the primary human NP cells and investigate the ability of this system to promote viable tissue formation in the inflammatory environment in the degenerative IVD. This is a promising first step in developing and applying the CRISPR/dCas9 system to disc pathology, which holds promise both for gene therapy applications and tissue engineering applications targeting DDD.

ACKNOWLEDGEMENTS

NIH Grant F32 AR063012, Arthritis Foundation, and Nancy Taylor Foundation for Chronic Diseases.

REFERENCES

[1] Hoy, D et al., *Arthritis Rheum*, 64:2028-37, 2012. [2] Vos T et al., *Lancet*, 380:2163-96, 2012 [3] Gilbert L et al., *Cell*, 154:442-51, 2013 [4] Larson, M et al. *Nat Protoc* 8:2180-96, 2013 [5] Wilson, A et al., *Mol Ther*, 21:825-33, 2013 [6] Wuertz, K et al., *Eur Cell Mater* 23:103-20, 2012 [7] Nasto, L et. Al., *Spine* 37:1819-25, 2012

MECHANICAL CHARACTERIZATION AND A COMPUTATIONAL WEAR MODEL FOR POLYCARBONATE URETHANE AS A BEARING MATERIAL

H. Gramling (1), A. Srinivasan (1), L. Pruitt (1)

(1) Department of Mechanical Engineering
University of California, Berkeley
Berkeley, CA, USA

INTRODUCTION

Total Shoulder Replacement (TSR) devices are subjected to complex stresses *in vivo*; such loading renders traditional ultra high molecular weight polyethylene (UHMWPE) glenoids susceptible to wear, fatigue and fracture [1]. Polycarbonate urethane (PCU) is a viable orthopedic material owing to its biostability and material properties that can be tailored to mimic articular cartilage [2-5]. Bionate® 75D PCU and Bionate® 80A PCU, referring to Shore hardness, are currently used in hip replacements as a backing and bearing surface, respectively [5]. Yet, robust mechanical characterization of this polymer remains limited in scope, especially considering the high stress demands expected of orthopedic biomaterials [3,4,6]. Our work seeks to evaluate the mechanical performance of Bionate® PCU through wear assessments and corresponding computational analysis, as well as tensile behavior, with special consideration for the use of this material in TSR.

METHODS

Two formulations of PCU under the trade name Bionate® (DSM, Berkeley, CA) were tested: Bionate® 75D PCU and Bionate® 80A PCU. Tensile specimens (ISO 527-2) were obtained from injection-molded plaques by waterjet. Dogbone specimens were soaked in a sealed bath of phosphate-buffered saline (PBS) at room temperature for at least 12 days prior to testing. Tensile testing was performed on an Instron 5566 (Norwood, MA) using two protocols. First, removal time from the PBS bath was assessed for five samples at "dry times" of 1, 5 and 10 minutes. Specimens were strained at 1 mm/min until 0.25% strain, and then 500 mm/min until break (ISO 527-1). The second experiment examined the effects of strain rate on tensile properties using three samples pulled at 10, 75, and 150 mm/min, all with a dry time of 5 minutes. Raw data was processed using MATLAB (Version R2013b, Natick, MA). Uniaxial pin-on-disc wear testing was conducted on a custom tribological testing apparatus [7] using a cobalt chromium pin (radius of curvature 6.25mm) and 6 mm thick Bionate® injection molded disk, as well as a load soak control with the same materials. Disks were soaked for at least 3 days prior to testing in a bath of HyClone Alpha Calf Serum (Fisher Scientific, Hampton, NH) diluted 1:1 with deionized water, with 0.1% wt sodium azide (Fisher Chemical, Hanover Park, IL). Testing was conducted at room temperature in a bath of the same solution. Two sets of tests were conducted: the first controlled for contact stress under an average applied force of 70-100 N, the second controlled for load with an applied force of 40 N. One reciprocating cycle covered a sliding distance of 15 mm. Testing was run to either 0.5 or 1.0 million cycles, representing 6 or 12 months of use *in vivo*. Wear loss was measured gravimetrically for both test and load soak disks every 200,000 cycles.

Computational study in MATLAB is underway to compare predicted and empirical wear results. The method calculates contact stress and depth of pin penetration at each cycle based on the material properties of the pin and disk, their respective geometries, and the applied load, as per Bartel *et al.* [8]. Calculated contact stress and penetration area used to determine wear and define a new wear scar geometry. It has been shown that a power law predicts wear between polyurethane and a metal contacting surface [9]. Wear is calculated using the following formula which accounts for contact stress [10],

$$q = \alpha \cdot s \cdot A \cdot \left(\frac{p}{E}\right)^{\beta} \qquad (1)$$

where q is the worn volume [mm³], s the total sliding distance [m], A the contact area [mm²], p the contact pressure [N/mm²] calculated as described above, E the Young's Modulus of the disk [N/mm²], α and β are material-specific constants. These constants are determined by fitting Eq. (1) to data from a 5000 cycle test with gravimetric measurements taken every 1000 cycles. Wear factors for Bionate® 80A and 75D were calculated from results of one million cycle wear tests, using a combination of Archard's Law and Hertz contact theory to account for the small radius of curvature of the pin and low modulus of the disk material [11,12]. Once the constants α and β have been fitted for each material, the power law relationship in Eq. (1) may be used to compare wear predicted computationally using Hertz contact stress with empirically observed wear, and to analyze discrepancies between the Hertz and Bartel contact stress methods in predicting wear.

RESULTS

The net wear loss rate was found to be 8.01 x 10^{-7} mm³/N-m for Bionate® 75D and 66.0 x 10^{-7} mm³/N-m for Bionate® 80A. Hertz contact stresses were 9

MPa and 50.1 MPa in Bionate® 80A and 75D, respectively. Because the method of determining wear factor takes into account contact stresses, Bionate® 75D presented a smaller wear factor despite experiencing higher contact stresses at equal load relative to Bionate® 80A.

SEM imaging shows rippling of the surface indicative of adhesive wear mechanisms (Fig. 1). Ripples were about 4 µm in width on Bionate® 75D and 1.6 µm in width on Bionate® 80A.

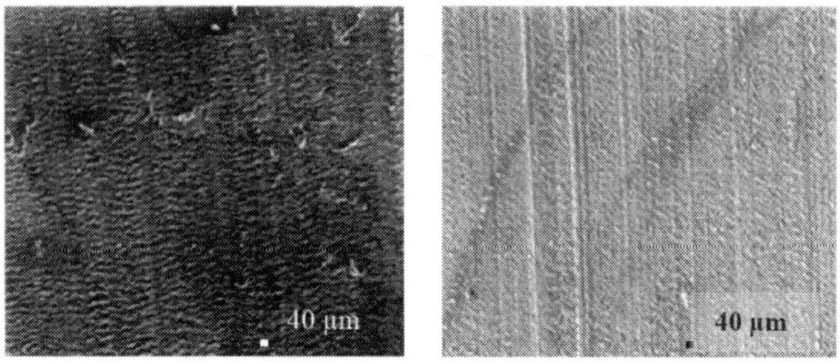

Figure 1. SEM IMAGES OF BIONATE® 75D (LEFT) AND 80A (RIGHT). RIPPLES ON THE ORDER OF 4 AND 1.6 µm ARE EVIDENCE OF ADHESIVE WEAR MECHANISMS.

Preliminary computational results indicate a "settling-in" period wherein greater contact stresses yield wear rates an order of magnitude higher over the first 100 cycles. This period corresponds with the evolution of the wear scar geometry, from initially flat to conforming to the contacting pin, and the consequent increase in contact area and decrease in contact stress to a stable "settled" value. A similar result was observed by Martinez *et al.* [10].

Table 1 summarizes the PCU material properties. Statistically significant variation in tensile material properties with dry time was observed for Bionate® 80A PCU; both the yield strength and ultimate strain varied significantly between dry times of 1-5 minutes, with $p = 0.049$ and $p = 0.012$, respectively. Bionate® 75D PCU performed consistently across all dry times. No significant differences in tensile properties were noted for strain rates under 500 mm/min. Tensile property variation between the two datasets at 5 minutes dry time may be attributed to both the large strain rate difference between 150 and 500 mm/min as well as changes in relative humidity during testing.

DISCUSSION

Bionate® 75D wear testing experienced an experimental abnormality between 600k and 800k cycles, possibly the result of third-body wear, which led to a 10x greater measurement-to-measurement mass loss rate versus previous measurements. Thus, the true wear factor for Bionate® 75D may be considerably lower than the calculated factor. Hertz contact stress is used to calculate wear factor, despite the fact that Hertz theory is not intended to handle contacting surfaces of very different moduli, such as Cobalt Chromium ($E \approx$ 200 GPa) and Bionate® PCU ($E \approx$ 1-1500 MPa). Incorporating contact stresses after Bartel *et al.* [8] will improve calculated wear factor accuracy; the method was not used in the presented results due to computational expense.

The tensile data presented compares well with that of Cipriani [2] for Bionate® 80A, however the Young's Modulus presented here for Bionate® 75D is about 150% of that found by Cipriani. The discrepancy is likely a result of testing at 25C, which is below the material's glass transition temperature, while Cipriani tested above the glass transition at 37C. Discrepancies in the yield strength of Bionate® 80A PCU with varying dry times indicates that the conditioning of devices at implantation must be optimized to minimize impact on initial performance. The high modulus, yield strength and ultimate strength of Bionate® 75D PCU provide it with favorable properties for stable fixation in vivo. Across strain rates, 80A tolerates greater strain before yield and before failure than remelted or sub-melt annealed highly-cross-linked UHMWPE. Similarly, Bionate® 75D PCU tolerates higher stresses before yield and failure than a typical cross-linked UHMWPE [13].

Future work will assess the wear response of Bionate® 80A at the same 40N load used for Bionate® 75D. Energy-dispersive X-ray spectroscopy will be used to identify the morphology of worn areas and specific wear characteristics such as ripples, to determine how hard and soft segment distribution may contribute to wear. Such results could inform manufacturing approaches and surface treatments to minimize wear of implanted devices. Future testing at use temperature (37C) is necessary to determine *in vivo* wear response. The considerable impact of contact stress on both empirical and computation results indicates a need for future ball-on-flat testing, which better approximates the *in vivo* use case of Bionate® PCU in TSR.

ACKNOWLEDGEMENTS

Authors would like to acknowledge DSM for providing Bionate® PCU samples, access to an Instron and DMA system, and several helpful discussions. Authors would also like to acknowledge helpful input from Farzana Ansari.

Funding was provided by the SMART program at UC Berkeley. This material is based upon work supported by the National Science Foundation Graduate Research Fellowship under Grant No. DGE 1106400.

REFERENCES

[1] Ansari, F, *World Congr. Biomech.*, Boston, MA, 2014.
[2] Cipriani, E, Characterization of Bionate, University of Torino.
[3] Elsner, J J et al., *Acta Biomater.*, 6:4698-707, 2010.
[4] Flannery, M et al., *Wear*, 269:331-338, 2010.
[5] Active Implants, http://www.activeimplants.com/.
[6] Bartel, D L et al., *J. Bone Joint Surg. Am.*, 68:1041-1051, 1986.
[7] Patten, E et al., *Wear*, 304:60-66, 2013.
[8] Bartel, D L et al., *Wear*, 107:193-99, 1985.
[9] Martínez, F J et al., *Wear*, 268:388-398, 2010.
[10] Martínez, F J et al., *Wear*, 284-285:52-64, 2012.
[11] Archard, J F, *J. Appl. Phys.*, 24:981, 1953.
[12] Hunter, S C, *J. Mech. Phys., Solids*, 8:219-234, 1960.
[13] Kurtz, S et al., *Biomaterials*, 23:3681-3697, 2002.

TABLE 1. TENSILE PROPERTIES OF BIONATE® PCU UNDER VARIOUS TESTING CONDITIONS.

	80A			75D			
Dry Time	**1 min**	**5 min**	**10 min**	**1 min**	**5 min**	**10 min**	
Young's Modulus (MPa)	23 ± 1	22 ± 1	21 ± 4	1560 ± 120	1538 ± 57	1602 ± 55	
Yield Strength (MPa)	0.9 ± 0.0	1.0 ± 0.1	1.2 ± 0.4	38.8 ± 3.9	39.9 ± 0.9	41.1 ± 0.9	RH: 55
Yield Strain (%)	4.1 ± 0.4	4.6 ± 0.4	6.9 ± 4.2	3.5 ± 0.2	3.4 ± 0.0	3.4 ± 0.0	
Ultimate Strength (MPa)	25 ± 1	23 ± 1	23 ± 2	38 ± 3	38 ± 1	40 ± 1	
Ultimate Strain (%)	385 ± 16	441 ± 20	427 ± 10	83 ± 10	82 ± 10	81 ± 16	
Strain Rate (Dry Time 5m)	**10 mm/ min**	**75 mm/min**	**150 mm/min**	**10 mm/min**	**75 mm/min**	**150 mm/min**	
Young's Modulus (MPa)	12 ± 1	13 ± 1	13 ± 2	335 ± 6	493 ± 16	562 ± 55	
Yield Strength (MPa)	1.6 ± 0.0	1.9 ± 0.4	2.4 ± 0.7	20.8 ± 0.9	27.4 ± 0.8	30.7 ± 0.8	RH: 63
Yield Strain (%)	13.0 ± 0.3	13.7 ± 4.2	19.3 ± 9.8	9.5 ± 0.2	6.9 ± 0.5	7.4 ± 0.8	
Ultimate Strength (MPa)	30 ± 6	30 ± 2	30 ± 0	40 ± 2	38 ± 1	36 ± 2	
Ultimate Strain (%)	399 ± 33	417 ± 13	427 ± 4	180 ± 6	183 ± 8	197 ± 12	

MICROSTRUCTURE AND DYNAMICS OF CROSSLINKED COLLAGEN GEL

Shengmao Lin (1), Linxia Gu (1,2)

(1) Department of Mechanical and Materials Engineering
University of Nebraska-Lincoln
Lincoln, NE, USA

(2) Nebraska Center for Materials and Nanoscience
Lincoln, NE, USA

INTRODUCTION

Collagen is a significant constituent of the natural ECM, and scaffolds made from collagen have been widely used in a variety of applications. The microstructure of the scaffold (porosity, mean pore size) [1, 2] and mechanical properties (Young's modulus) [3] have been shown to significantly influence cell adhesion, growth and differentiation. Therefore, controlling the microstructure and mechanical property of the collagen scaffold is essential for studying the cell-scaffold interactions and tissue synthesis.

Crosslink plays an important role in determine the mechanical property of collagen scaffold. The mechanical properties of collagen scaffolds can be altered by varying the degree of cross-linking between collagen fibers. Several studies have shown that the mechanical property of collagen scaffolds could be enhanced by introducing various chemical crosslinkers (i.e. glutaraldehyde, formaldehyde, carbodiimide) [4] or by cross-linking with physical treatment (i.e. UV irradiation, freeze-drying, heating) [5]. For example, Haugh et al. [6] has created an optimized collagen-glycosaminoglycan (CG) scaffold by DHT crosslinking which could be used in bone tissue engineering. However, the detailed relationship between the degree of cross-linking and mechanical property of collagen scaffold is missing.

Computational models have been utilized to inspect the mechanism of fiber network behaviors for fine-tuning the microenvironment of cell culture [7, 8]. Previous models have treated the crosslinks as intersection points in two-dimensional models, which were constrained either as freely rotating pin joints [9-11] or welded joints [12]. The 3D crosslinks were treated as either regular fibers [8] or torsional springs where their rotational stiffness were obtained by fitting to experimental data [7]. However, the role of crosslinks was not elucidated yet in the existing models.

In this work, the role of crosslink density on the collagen scaffold mechanical properties was investigated through computational modeling. The crosslinks with varied density corresponding to different polymerization conditions [4] were formulated in a 3D collagen network. These microscopic crosslink properties were then correlated with the macroscopic scaffold mechanics. These results could be used to guide the design of scaffold with tunable material properties.

METHODS

A representative volume element (RVE) with the side length of 40μm was used to represent collagen scaffold with a concentration of 1mg/ml, which is equivalent to the fiber volume fraction of 0.073% (Fig. 1a). 3D collagen fibers (1934 in total) were randomly distributed using the random seed algorithm [13]. The fiber is 8μm long and could be truncated to 4μm at the boundaries. The fiber diameter of 62 nm was based on the measurement for the collagen gel polymerized at 37 °C and pH 7.4 [14]. Each fiber was meshed with 2μm beam elements corresponding to the mean crosslink spacing of collagen fiber networks [15]. The crosslinks were generated between nodes when their distance is less than or equal to a certain value, referred to as crosslink threshold. Fig. 1b demonstrated 2360 crosslinks between fibers in Fig. 1a with a threshold of 800nm. The uncross-linked fibers were then removed, due that they don't contribute to the mechanics of collagen network.

The Young's modulus of collagen fiber was adopted as 50MPa [7]. Crosslinks were assumed to have the same material property as collagen fibers in the baseline model. Uniaxial tension was applied along the x-direction of the RVE. No sliding motion existed between crosslink and collagen. Nonlinear finite element models were solved using ABAQUS 6.12 (Simulia, Providence, RI).

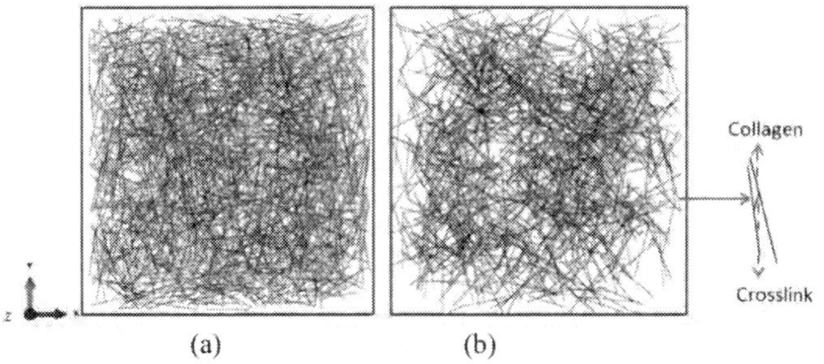

Figure 1: RVE with random distributed fibers (a) Before cross-linking; (b) Cross-linked fiber network.

RESULTS

The crosslink density was calculated as the number of crosslinks per collagen fiber. By varying the crosslink threshold from 800nm to 1600nm, the crosslink density increased from 2.09 to 4.439. As show in Fig. 2a, a plateau was clearly observed for the number of crosslinked fibers as the crosslink threshold exceeded 1200nm. This indicated a fully cross-linked collagen fiber network. A positive bilinear correlation was obtained between the crosslink density and scaffold stiffness as plotted in Fig. 2b. The rate of stiffness growth increased about four times at the turning point. The turning point is which the collagen fiber network becomes fully cross-linked, corresponding to the crosslink threshold of 1200nm.

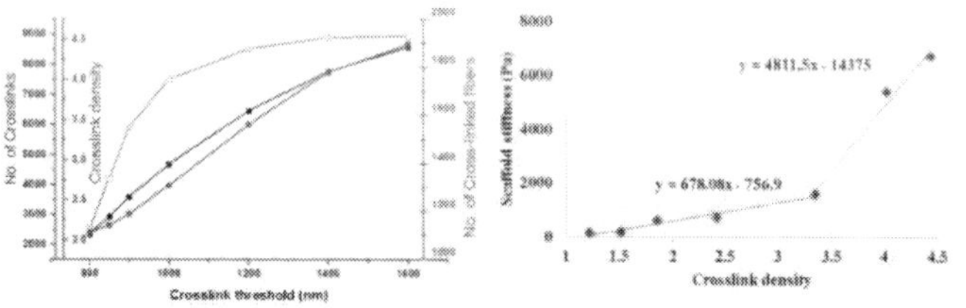

Figure 2: (a) Crosslink threshold regulated microstructure changes of collagen scaffold; (b) Correlation between crosslink density and scaffold stiffness.

DISCUSSION

The degree of cross-linking could be regulated by parameters including temperature, pH, collagen concentration and adopted polymerization agents [16, 17]. In current study, crosslink threshold, i.e., the inter-fiber nodal spacing, was used in our model to control the crosslink density. The adopted crosslink threshold from 800nm to 1600nm is adopted from Lindstrom et al. [15]. Our results have demonstrated that the scaffold stiffness increased approximately 40 times by doubling the crosslink density with the same collagen concentration. A positive bilinear correlation was found between the crosslink density and scaffold stiffness. The turning point at the crosslink threshold of 1200nm, i.e., crosslink density of 3.44, was when fibers were totally cross-linked. Further cross-linking treatment after this density point would dramatically increase the scaffold stiffness. Even a linear correlation between the crosslink density and engineered tissue stiffness was well accepted [18]. Sheu et al. has observed a dramatically increase of collagen gel stiffness after fully cross-linking [16]. Gardel et al. also demonstrated a nonlinear correlation between the crosslink density and the stiffness of actin fiber network [19]. It should be noted that their crosslink density was calculated as the crosslink mass, which included all the fibers regardless of their crosslinking status. In our work, the crosslink density was based on the amount of cross-linked fibers. However, our adopted definition of crosslink density won't change the observed bilinear correlation between crosslink density and scaffold stiffness. Moreover, no experimental evidence has ever demonstrated this bilinear correlation yet. Our observations might shed lights on future testing and experimentations. These results could improve the understanding of the mechanics of collagen scaffold for designing and regulating clinical relevant biomaterials. This work could also be extended to study how cells respond to different micromechanical environments.

ACKNOWLEDGEMENTS

This work was supported by the National Science Foundation CAREER award (CBET-1254095), and UNL Research Council Interdisciplinary Grant.

REFERENCES

1. O'Brien, F.J., et al., *The effect of pore size on cell adhesion in collagen-GAG scaffolds.* Biomaterials, 2005. **26**(4): p. 433-441.
2. Zeltinger, J., et al., *Effect of pore size and void fraction on cellular adhesion, proliferation, and matrix deposition.* Tissue Engineering, 2001. 7(5): p. 557-572.
3. Yeung, T., et al., *Effects of substrate stiffness on cell morphology, cytoskeletal structure, and adhesion.* Cell motility and the cytoskeleton, 2005. **60**(1): p. 24-34.
4. Lee, C., A. Grodzinsky, and M. Spector, *The effects of cross-linking of collagen-glycosaminoglycan. scaffolds on compressive stiffness, chondrocyte-mediated contraction, proliferation and biosynthesis.* Biomaterials, 2001. **22**(23): p. 3145-3154.
5. Schoof, H., et al., *Control of pore structure and size in freeze - dried collagen sponges.* Journal of biomedical materials research, 2001. **58**(4): p. 352-357.
6. Haugh, M.G., M.J. Jaasma, and F.J. O'Brien, *The effect of dehydrothermal treatment on the mechanical and structural properties of collagen - GAG scaffolds.* Journal of Biomedical Materials Research Part A, 2009. **89**(2): p. 363-369.
7. Stein, A.M., et al., *The micromechanics of three - dimensional collagen - I gels.* Complexity, 2011. **16**(4): p. 22-28.
8. Stylianopoulos, T. and V.H. Barocas, *Volume-averaging theory for the study of the mechanics of collagen networks.* Computer methods in applied mechanics and engineering, 2007. **196**(31): p. 2981-2990.
9. Head, D.A., A.J. Levine, and F.C. MacKintosh, *Distinct regimes of elastic response and deformation modes of cross-linked cytoskeletal and semiflexible polymer networks.* Physical Review E, 2003. **68**(6): p. 061907.
10. Heussinger, C., B. Schaefer, and E. Frey, *Nonaffine rubber elasticity for stiff polymer networks.* Physical Review E, 2007. **76**(3): p. 031906.
11. Heussinger, C. and E. Frey, *Stiff polymers, foams, and fiber networks.* Physical review letters, 2006. **96**(1): p. 017802.
12. Onck, P.R., et al., *Alternative explanation of stiffening in cross-linked semiflexible networks.* Physical review letters, 2005. **95**(17): p. 178102.
13. Digimat, A., *Software for the Linear and Nonlinear Multi-scale Modeling of Heterogeneous Materials. e-Xstream Engineering.* Louvain-la-Neuve, Belgium, 2011.
14. Raub, C.B., et al., *Noninvasive assessment of collagen gel microstructure and mechanics using multiphoton microscopy.* Biophysical journal, 2007. **92**(6): p. 2212-2222.
15. Lindström, S.B., et al., *Biopolymer network geometries: Characterization, regeneration, and elastic properties.* Physical Review E, 2010. **82**(5): p. 051905.
16. Sheu, M.-T., et al., *Characterization of collagen gel solutions and collagen matrices for cell culture.* Biomaterials, 2001. **22**(13): p. 1713-1719.
17. Charulatha, V. and A. Rajaram, *Influence of different crosslinking treatments on the physical properties of collagen membranes.* Biomaterials, 2003. **24**(5): p. 759-767.
18. Balguid, A., et al., *The role of collagen cross-links in biomechanical behavior of human aortic heart valve leaflets-relevance for tissue engineering.* Tissue engineering, 2007. **13**(7): p. 1501-1511.
19. Gardel, M.L., et al., *Elastic behavior of cross-linked and bundled actin networks.* Science, 2004. **304**(5675): p. 1301-1305.

TGF-BETA AND BMP SIGNALING PATHWAYS REGULATE CHROMATIN CONDENSATION IN MESENCHYMAL STEM CELLS IN RESPONSE TO DYNAMIC LOADING

Su-Jin Heo (1, 2), Woojin M. Han (1, 3), Tristan P. Driscoll (1, 2), Dawn M. Elliott (3), Randall L. Duncan (3), Robert L. Mauck (1, 2)

(1) Department of Bioengineering, University of Pennsylvania, Philadelphia, PA, USA

(2) Mckay Orthopaedic Research Laboratory, University of Pennsylvania, Philadelphia, PA, USA

(3) Department of Biomedical Engineering, University of Delaware, Newark, DE, USA

INTRODUCTION

Mesenchymal stem cells (MSCs) are a promising cell source for regenerative therapies given their multipotent nature [1]. Along with soluble cues, exogenous mechanical perturbations play important roles in modulating MSC lineage-specification [2]. MSC differentiation is accompanied by chromatin remodeling and changes in gene expression [3]. In previous work, we showed that dynamic tensile loading (DL) evoked chromatin remodeling and ultimately condensation [4]. Further, we demonstrated that ATP release and subsequent calcium signaling played a role in early signaling events mediating this DL-induced chromatin condensation [5]. In addition to these pathways, it is well established that transforming growth factor beta (TGF-β) and bone morphogenetic protein (BMP), members of the TGF-β superfamily, regulate cellular processes including growth and differentiation [6]. Binding of TGF-β and/or BMP to cell surface receptors initiates phosphorylation of Smad proteins and transcriptional activation of target genes [6]. Recent work has shown that Smad2/3 (pSmad2/3) signaling is activated by cyclic strain [7]. However, the role of signaling through this pathway in mechanically induced chromatin condensation in MSCs has not yet been explored. In this study, we queried whether different loading conditions (i.e, frequency and duration of applied load) regulated chromatin condensation, and ascertained whether these changes were in part mediated through the TGF-β/BMP signaling pathways.

METHODS

Aligned poly(ε-caprolactone) (PCL) nanofibrous scaffolds were fabricated via electrospinning and incubated in 20 μg mL^{-1} fibronectin in PBS overnight to enhance cell attachment [2]. MSCs isolated from juvenile bone marrow were seeded at 200,000 cells per scaffold. Constructs were cultured in a chemically defined media (CM) without exogenous growth factors [4]. Dynamic tensile loading (DL, 3%) for varying durations (30, 150, or 600 sec) and frequencies (0.2 ~ 2 Hz) was applied after 2 days of pre-culture, using a custom bioreactor [2]. To assess chromatin condensation, nuclei were stained with DAPI and scanned across their mid-sections using a confocal microscope (Zeiss, LSM 510). The chromatin condensation parameter (CCP) [8] was quantified using a gradient-based sobel edge detection algorithm implemented in MATLAB to measure edge density in nuclei. To determine the role of TGF-β superfamily signaling on DL induced condensation, pharmacologic inhibitors SB431542 (SB, 10μM, Sigma) to inhibit Smad 2/3 (TGFβ signaling) or LDN193189 (LDN, 100nM, Reagents Direct) to inhibit Smad 1/5/8 (BMP signaling) were applied for 2 hours and washed off prior to application of DL (3%, 1Hz, 600 sec). After DL, constructs were fixed and CCP quantified.

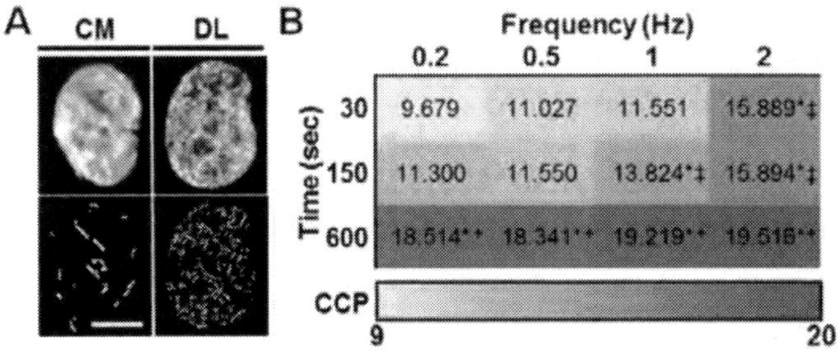

FIG 1. A) DAPI STAINED NUCLEI (TOP ROW) AND CORRESPONDING EDGE DETECTION (BOTTOM ROW) (DL: 3%, 1HZ, 600S, BAR = 3 μm), B) HEAT MAP OF CCP VALUES WITH CHANGES IN DURATION AND FREQUENCY OF DL (N = ~ 20, *: P<0.05 VS. CM CONTROL, +: P<0.05 VS. 0.2 HZ, ‡: P<0.05 VS. 0.5 HZ).

Given that our previous studies had shown that soluble ATP released to the media was a key mediator of condensation, we also collected conditioned media after each episode of DL, with and without TGF-β/BMP inhibition. This supernatant was then added to naïve (unloaded) MSC-seeded constructs. After 30 min, these constructs were fixed and CCP was determined (Fig. 2 A). To measure the concentration of ATP in the media, a luciferin/luciferase assay was performed (ATP Assay Kit, KA1661, Abnova). To determine intracellular Ca^{2+} levels with static stretch, MSCs on scaffolds were loaded with the fluorescent calcium indicator, Cal-520™ AM (15 μM, AAT Bioquest) for 1 h at 37 °C. Constructs were placed into a micro-tensile device [9] mounted on a confocal microscope. Time-series images of baseline $[Ca^{2+}]_i$ signal were recorded (every 4s for 10 min). Grip-to-grip strain of 3% was then applied (at 0.05%/s), and $[Ca^{2+}]_i$ response was measured for an additional 10 min. A custom MATLAB program was used to analyze $[Ca^{2+}]_i$ oscillations (Fig 3, A and B) [9].

RESULTS

Dynamic loading led to chromatin condensation in MSC nuclei, increasing the number of visible edges (Fig. 1A). No change in CCP was observed with 30 sec of DL, except in the case of DL at 2 Hz (Fig. 1B). DL at 1 and 2 Hz for 150 sec increased CCP, but not at 0.2 or 0.5Hz (Fig. 1B). Interestingly, with 600 sec of DL, CCP increased in all conditions, independent of loading frequency (Fig. 1B).

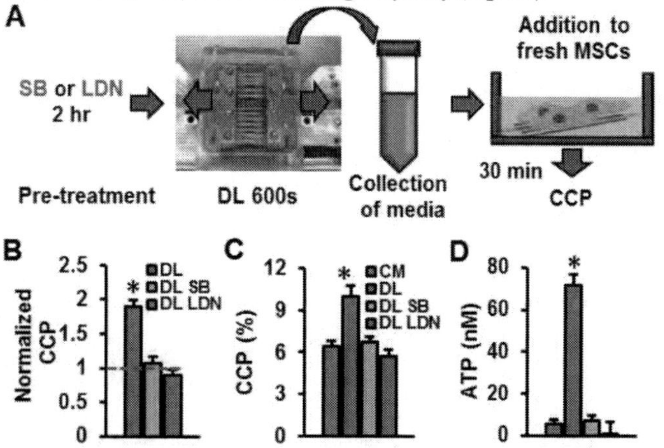

FIG 2. A) SCHEMATIC ILLUSTRATION OF CONDITIONED MEDIA STUDIES WITH INHIBITION OF TGFβ/BMP SIGNALING, B) CCP WITH 600S OF DL (3%, N = ~20, RED LINE: CM CONTROL), C) CCP IN UNLOADED SCAFFOLDS TREATED WITH DL-CONDITIONED MEDIA (N = ~20), D) ATP IN MEDIA (N = 3), *: P<0.05 VS. CM.

When DL (600 sec, 3%, 1 Hz) was applied in the context of blockade of either TGF-β or BMP signaling (SB or LDN), DL induced increases in CCP were blocked (Fig. 2B). Addition of media conditioned by DL (3%, 1Hz, 600s) increased CCP values in unloaded MSC-seeded constructs (Fig. 2C). Conversely, pretreatment of DL-constructs with SB or LDN blocked this condensation (Fig. 2C). Consistent with our previous findings, DL (3%, 1Hz, 600s) triggered ATP release into the media (Fig. 2D); both SB and LDN pretreatment blocked this ATP accumulation (Fig. 2C). This finding suggests that blockade of Smad signaling attenuates release of ATP upon DL, short-circuiting the load induced chromatin condensation events. To probe additional downstream responses, we also monitored $[Ca^{2+}]_i$ oscillations in MSCs (Fig 3A and B). With treatment with SB or LDN, there were no changes in average signal intensity, peak duration (not shown), or time to peak in any group (Fig. 3C) when constructs were exposed to grip-to-grip strain of 3%. However, changes in the number of peaks in a 10 min observation window (Fig. 3D), the time between peaks (not shown), and % of responding cells (not shown) with static stretch were all altered by inhibition of TGF or BMP signaling.

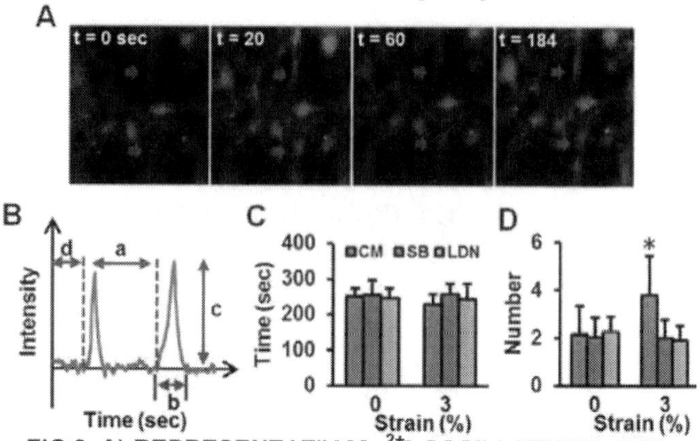

FIG 3. A) REPRESENTATIV $[Ca^{2+}]_i$ OSCILLATIONS (RED ARROWS) IN MSCS AS A FUNCTION OF TIME, B) REPRESENTATIVE $[Ca2^{+}]_i$ OSCILLATION CURVE (a: TIME BETWEEN PEAKS, b: PEAK DURATION, c: PEAK AMPLITUDE, d: TIME TO PEAK), C) TIME TO PEAK, AND D) NUMBER OF PEAKS IN 10 MIN (N = ~ 15, *: P<0.05 VS. 0%).

DISCUSSION

In this study, we showed that dynamic loading leads to rapid chromatin condensation in MSCs, and that the degree of condensation depends on both the frequency and the duration of loading. In addition, we found that blockade of either Smad 2/3 or Smad 1/5/8 signaling attenuated ATP release and subsequent alterations in Ca^{2+} signaling, and thereby limited chromatin condensation. Consistent with this finding, a recent study showed that fluid shear stress results in rapid increases in cellular contractility and ATP release [10]. Likewise, addition of either BMP or TGF-β can increase contractility, and is dependent on Smad signaling. It may be that inhibition of these pathways via pre-treatment with SB or LDN decreases cellular contractility to the point where insufficient stress is generated in the cytoskeleton and so mechanically induced ATP release with applied stretch is no longer possible. Ongoing studies are now exploring changes in cell contractility with inhibition of Smad pathways, and further delineating the manner by which these pathways interact with ATP/calcium signaling. Overall, these new data suggest that dynamic tensile loading triggers ATP release by MSCs, and that this release is modulated by both TGFβ and BMP signaling pathways.

ACKNOWLEDGEMENTS

This work was supported by the NIH (R01 EB02425 and R01 AR056624) and the Montague Research Fund from the Perelman School of Medicine.

REFERENCES

[1] Pittenger, MF et al., *Science*, 284(5411), 143-147, 1999
[2] Baker, BM et al., *Tissue Eng Part A*, 17(9-10):1445-1455, 2011.
[3] Teven, CM et al., *Stem Cells Int*, 2011(201371):1-18, 2011.
[4] Heo, SJ et al., *ORS*, 2014.
[5] Heo, SJ et al., *ORS*, 2015.
[6] Wu, MY et al., *Dev Cell*, 16(3):329-343, 2009.
[7] Saha, S et al., *Biophys J*, 94(10):4123-4133, 2008.
[8] Irianto, J et al., *Biophys J*, 104(4):759-769, 2013.
[9] Han, WM et al., *Eur Cell Mater*, 8(27):321-331.
[10] Gardinier, JD, et al., *Cell Mol Bioeng*, 7(2):266-277.

SB³C2015
Summer Biomechanics, Bioengineering and Biotransport Conference
June 17-20, 2015,Snowbird Resort, Utah, USA
SB³C2015-399

BIOMECHANICAL EFFECTS OF MENISCO-TIBIAL REPAIR

Ferris M. Pfeiffer, PhD (1,2), James P. Stannard, MD (1), James L. Cook, DVM, PhD (1), Matthew Bollier, MD (3), Patrick Smith, MD (1,4)

(1) Department of Orthopaedic Surgery
University of Missouri
Columbia, MO, USA

(2) Department of Bioengineering
University of Missouri
Columbia, MO USA

(3) Department of Orthopaedic Surgery
University of Iowa
Iowa City, IA USA

(4) Columbia Orthopaedic Group
Columbia, MO USA

INTRODUCTION

The medial meniscus, in conjunction with the medial collateral ligament (MCL), plays a crucial role in stabilizing the knee joint, particularly with consideration of anterior cruciate ligament (ACL) function [1]. Combination ACL/MCL injuries are quite common and require surgical reconstruction to restore function [2]. Most often for current clinical practice, the MCL is treated conservatively while the ACL is reconstructed. However, not all patients do well with conservative treatment of MCL injuries, and residual MCL laxity may place additional stress/strain on the ACL, meniscus, articular cartilage, and supporting structures. The goal of this study was to evaluate knee biomechanics for ACL-intact and ACL-deficient knees with concomitant peripheral medial meniscus tears or medial meniscotibial ligament rupture. We hypothesized that knee laxity would be significantly ($p=0.05$) increased and knee stiffness significantly ($p=0.05$) decreased in A-P, valgus, and external rotation in knees with peripheral medial meniscus tears or medial meniscotibial ligament rupture as compared to intact knees. We further hypothesized that surgical repair of such defects would restore knee biomechanics to pre-injury values.

METHODS

Six pairs of matched cadaveric human knees were acquired from a local tissue bank. Extraneous soft tissues were removed 10cm above and 8cm below the joint while maintaining all relevant soft tissue structures. The femur and tibia of each knee were securely potted in 1" diameter copper tubing to assist in attachment to a servo-hydraulic materials testing machine (Instron 8821s) (Fig 1). Prior to initial testing, arthroscopic assessment was used to confirm intact joint anatomy. Knees were tested in valgus (with 0°, 30°, 60°, and 90° of knee flexion), anterior drawer, and external rotation (Fig 1).

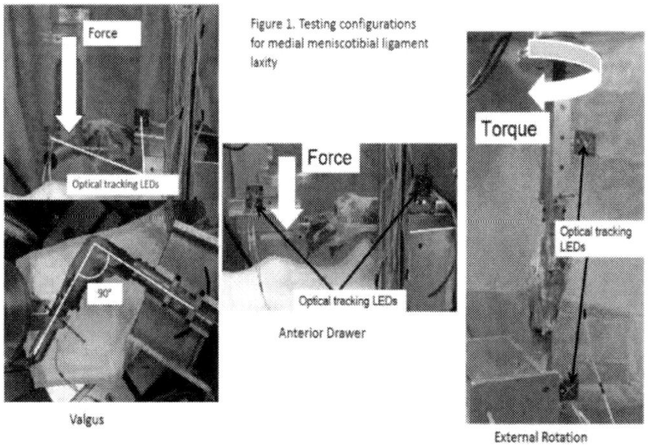

Figure 1. Testing configurations for medial meniscotibial ligament laxity

Half of the samples were tested with ACL intact while in the other half the ACL was arthroscopically transected (ACLx), additionally, either a peripheral or meniscotibial tear was created arthroscopically to produce the following groups:

1) ACL intact and peripheral tear (n=3)
2) ACL intact and meniscotibial tear (n=3)
3) ACLx and peripheral tear (n=3)
4) ACLx and meniscotibial tear (n=3)

Specimens were retested to quantify the destabilized properties. Peripheral tears were repaired with an inside-out technique using 2-0 fiber wire. Meniscotibial tears were repaired using 4 mini suture tacks and 0 Fiberwire. Specimens were tested a final time in the repaired configuration. In conjunction, finite element analysis was performed to analyze the effects of the meniscotibial ligament on knee structures using the Open Knee model. Intact geometry was modified to include either a peripheral or meniscotibial ligament tear. Simulations were performed and stress/strain predictions compared between intact and destabilized knees.

RESULTS

For all groups, the inclusion of a peripheral or meniscotibial ligament tear demonstrated significantly (p=0.05) increased laxity, as compared to controls with the ACL intact (Fig 2).

Figure 2. Results from biomechanical testing including a) valgus, b) anterior drawer with ACL intact, and c) anterior drawer with ACL transected

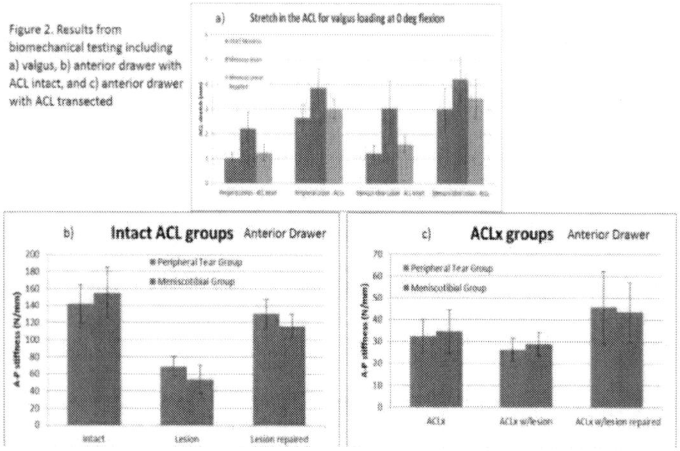

There was a trend toward significance in the ACLx groups; however, significance was not reached in any ACLx group except for testing in valgus at 30°, 60°, and 90° of flexion. Surgical repair (regardless of lesion type) significantly reduced laxity, as compared to lesion group, and was not statistically different from intact controls. There was no statistical difference between repair groups. Finite element simulations indicated a sharp increase in stress in the meniscus, cartilage, and intact ligaments in the presence of lesions regardless of type (Fig 3).

DISCUSSION

Deficiency in the ligaments of the medial compartment of the knee demonstrated a considerable effect on the stability of the joint, which was restored following surgical repair, regardless of lesion type. In specimens with the ACL transected, the addition of a lesion only marginally increased laxity versus the ACLx state. There were no significant observed or measured differences between the two lesion groups, and in effect, the peripheral tear and meniscotibial ligament tear were biomechanically similar. It was noted during finite element simulations that for the meniscotibial ligament lesion configuration, the medial meniscus was observed to "lift off" of the tibia during external rotation and valgus loading.

SIGNIFICANCE

In MCL injuries, the integrity of the medial meniscus attachment to the tibia - the meniscotibial ligament - is critical to its stabilizing function. When recognized surgically, meniscotibial ligament laxity should be surgically corrected. Medial lesions induce abnormal stresses and strains in intact anatomy. Medial meniscotibial lligament axity is biomechanically analogous to a peripheral medial meniscus tear from the capsule, which is always surgically repaired.

Figure 3. Finite element results showing a) predicted stress in the meniscus and b) predicted stress in the intact ligaments for various lesions.

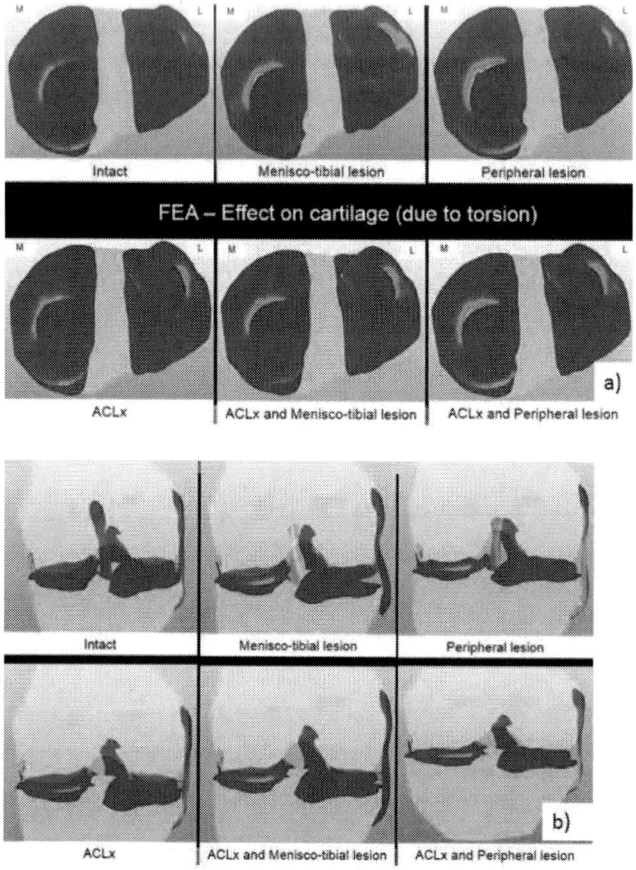

ACKNOWLEDGEMENTS

This study was funded by the Comparative Orthopaedic Laboratory and the Columbia Orthopaedic Group Research Fund

REFERENCES

[1] Bollier, M and Smith, P, Anterior Cruciate Ligament and Medial Collateral Ligament Injuries, **J of Knee Surg** 2014

[2] Marchant, MH, et. al Management of Medial-Sided Knee Injuries, Part 1: Medial Collateral Ligament, **AJSM** 2011

SB³C2015
Summer Biomechanics, Bioengineering and Biotransport Conference
June 17-20, 2015,Snowbird Resort, Utah, USA
SB³C2015-400

MODELING CELLULAR CONTRACTION ON BIOHYBRID DEVICES USING THERMAL CONTRACTION CAPABILITIES OF FINITE ELEMENT ANALYSIS TOOLS

Victoria A. Webster (1), Ozan Akkus (1,4,5), Hillel J. Chiel (2,3,4), Roger D. Quinn (1)

(1) Dept. of Mech. and Aero. Engineering
Case Western Reserve University
Cleveland, OH, USA

(2) Dept. of Biology
Case Western Reserve University
Cleveland, OH, USA

(3) Dept. of Neurosciences
Case Western Reserve University
Cleveland, OH, USA

(4) Dept. of Biomedical Engineering
Case Western Reserve University
Cleveland, OH, USA

(5) Dept. of Orthopaedics
Case Western Reserve University
Cleveland, OH, USA

INTRODUCTION

The development of biohybrid actuators and devices powered by such actuators is an area of increasing study. By culturing muscle cells on engineered, biocompatible substrates, devices can be developed that are capable of actuation of cantilevers [1] or micro-pillars [2], and even locomotion in the form of swimming [3] and walking [4], [5].

Engineering models are needed to design and simulate the behavior of such biohybrid devices. Modeling techniques often employ Finite Element Analysis (FEA) to determine the stresses generated by the cell layer based on measured device deflections. An engineering tool for predictive modeling of biohybrid devices should be computationally efficient and enable new researchers to enter the field by utilizing existing modeling technology. Many current models emulate force generation by applying individual forces or traction forces at the interface between the cell layer and the substrate. However, cell contraction can also be emulated by thermal strains, a feature that is available in most FEA packages. The challenge with this approach is that it requires the entry of thermal expansion coefficients (TEC) which will generate deformations that are equivalent to contractile forces generated by cells. The current study identified a range of TEC values that can be used to emulate cell contractions in FEA packages.

METHODS

The aim of the study was achieved by fitting published experimental data [1], [3], [5]with FEA simulations to obtain TEC variables α and ΔT. The coefficient of thermal contraction along the cell contraction direction is denoted as α and the driving temperature was denoted as ΔT. The identified TEC values were applied to an additional set of experimental studies [2], [4] to test the ability of derived TEC values to reproduce experimentally reported deformations.

Model Calibration

Many studies have developed cantilever biohybrid actuators which bend in response to cellular contraction[1], [3], [5], [6]. The deflection of such actuators can be used to approximate the stress in the cell layer using a modified form of Stoney's approximation [7],[8] to allow for the analysis of thickness ratios up to 30%. This results in the following expression for the stress in the cell layer:

$$\sigma_c = (1+\delta^3\gamma)\frac{1}{6Rt_c}\left(\frac{E_s t_s{}^3}{((1-\nu)(t_s+t_c))} + \frac{E_c t_c{}^3}{((1-\nu)(t_s+t_c))}\right) \quad (1)$$

where R is the experimentally measured radius of curvature of the beam, ν is the poisson's ratio, t_c and t_s are the thicknesses of the cell and substrate layers, respectively, E_c and E_s are the moduli of the cells and substrate, respectively, and $\gamma = E_c'/E_s'$ and $\delta = t_c/t_s$.. The σ_c value was calculated for three studies in which cantilever biohybrid devices were fabricated and tested, resulting in 7 total configurations [1], [3], [5]. For each configuration, TEC values for use in the FEA model can be determined by substituting σ_c into the linear thermal stress-strain relationship:

$$\alpha\Delta T = \frac{\sigma_c}{E_c} \quad (2)$$

This resulted in a range of parameters for further testing.

Simulation

ANSYS Mechanical APDL was used to simulate all models, using two dimensional plane strain elements. For cantilever models a fixed boundary condition was applied to one end of the model (Figure 1). In order to simulate the contraction of the cell layer, an orthotropic coefficient of thermal contraction was used. In ANSYS α and ΔT must

Table 1. Geometric and material properties for calibration, and validation studies as well as resulting $\alpha\Delta T$ values

Model	t (mm)	h (μm)	w (mm)	L (mm)	E_s (kPa)	E_c (kPa)	v	Δz (μm)	γ	αΔT
1 [5]	0.45	10	2	4	17.82	10	0.49	390	0.52	-0.4218
2 [5]	0.45	10	2	4	503	10	0.49	2	0.52	-0.0610
3 [1]	0.02	4.5	0.05	0.25	750	188	0.49	4.16	0.55	-0.0095
4 [1]	0.02	4.5	0.1	0.5	750	188	0.49	19.77	0.55	-0.0112
5 [1]	0.02	4.5	0.15	0.75	750	188	0.49	34.08	0.55	-0.0086
6 [3]	0.03	5	20	5	1500	30	0.49	1147	0.44	-0.1748
7 [3]	0.03	5	10	10	1500	30	0.49	3366	0.44	-0.1282

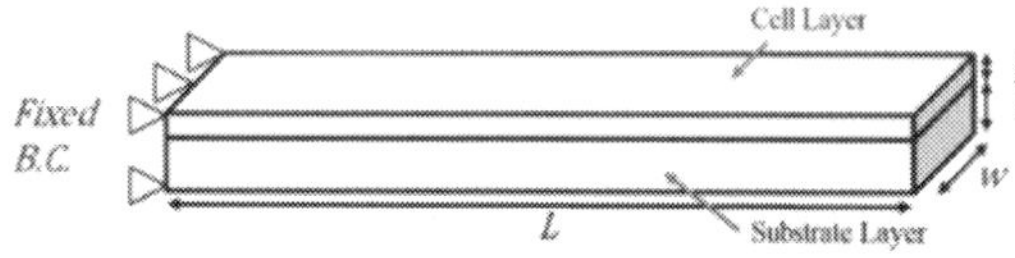

Figure 1. Schematic of cantilever models.

be entered as separate parameters. For the purposes of this study, ΔT was set at unity. The TEC along the direction of cellular alignment was set as determined by Eqs. (1) and (2) and the thermal contraction coefficients in all other directions were set to 0.

The appropriate value of γ was then calibrated experimentally in simulation. The resulting values of γ and $\alpha\Delta T$ for each of the model configurations are presented in the last two columns of Table 1.

Model Validation

In order to validate the range of $\alpha\Delta T$ determined from the model calibrations, each of the calibrating models was simulated over the range of $\alpha\Delta T$. Additionally, two non-cantilever models, a micropillar study presented by Legant et al. [2], and a biohybrid walker presented by Cvetkovic et. al [4] were simulated over the range of $\alpha\Delta T$ values.

RESULTS

Due to the wide range in moduli for both the cell and substrate layers in the studies reported in the literature, the range of $\alpha\Delta T$ values is quite large (Table 1). However this range encompasses the experimental values not only for the calibration models, but also for the additional, non-cantilever validation models. For example, when applied to the Legant et al. micropillar study, the median value of the $\alpha\Delta T$ range results in an active tension directly between the two active tensions reported for 1 Hz and 4 Hz stimulation (Figure 2).

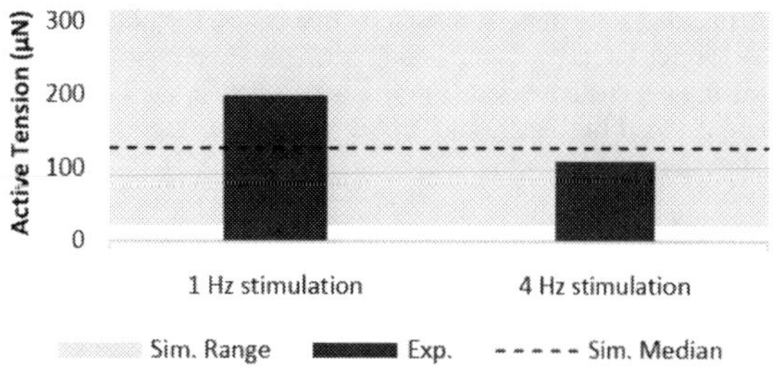

Figure 2. Comparison of experimental results [2] and simulation range. The median $\alpha\Delta T$ value results in an active tension between the experimental active tensions resulting from 1 Hz and 4 Hz stimulation.

Additionally, the use of thermal contraction is computationally faster than simulating individual cell forces. A 10 mm long cantilever model was simulated 200 times with individual contraction forces every 100 μm, or using thermal contraction. The individual force model ran in 347.5 ± 2.486 seconds whereas the thermal contraction model ran in 75.86 ± 1.318 seconds.

DISCUSSION

An earlier study attempted to obtain suitable TEC values; however, that study used one experimental study to extract TEC values [9]. The current study was a significant expansion of this previous attempt. The modeling technique presented here results in a wide range of $\alpha\Delta T$ values which can be used to simulate the actuation of biohybrid cantilevers using the thermal contraction capabilities of standard FEA tools. It is important to note that the studies used to calibrate this range have a wide range of moduli for the cell layer. Additionally, not all of these studies utilized their own measurement of the cell layer modulus, but rather relied on previously reported values. The modeling technique presented here is sensitive to the modulus of the cell layer, and as a consequence there may be some deviation from experimental results if an accurate measurement of cell layer modulus is not made. However, based on the results presented here, an initial range of TEC values of -0.061056 to -0.151514 can be used to simulate cellular contraction in most models. For models with significantly higher cell moduli, a lower TEC value is needed, whereas for models with a combination of low cell modulus and low substrate modulus, a higher TEC value is needed.

ACKNOWLEDGEMENTS

This material is based upon work supported by the National Science Foundation Graduate Research Fellowship under Grant No. DGE-0951783. This study was also funded in part by grants from the National Science Foundation (Grant Number DMR-1306665) and National Institute of Health (Grant Number R01 AR063701). Any opinions, findings, and conclusions or recommendations expressed in this material are those of the authors and do not necessarily reflect the views of the National Science Foundation.

REFERENCES

[1] J. Park et al. *Anal. Chem.*, vol. 77, pp. 6571–80, Oct. 2005.
[2] W. Legant et al. *PNAS*, vol. 106, no. 25, pp. 10097–102, 2009.
[3] A. Feinberg et al. *Science*, vol. 317, pp. 1366–70, Sep. 2007.
[4] C. Cvetkovic et al. *PNAS*, vol. 111, no. 28, pp. 10125–30, Jun. 2014.
[5] V. Chan et al. *Lab Chip*, vol. 12, pp. 88–98, Jan. 2012.
[6] K. Wilson et al. *PLoS One*, vol. 5, no. 6, p. e11042, Jan. 2010.
[7] A. Atkinson. *Br. Ceram. Proc.*, vol. 54, no. 1, 1995.
[8] C. Klein. *J. Appl. Phys.*, vol. 88, no. 9, p. 5487, 2000.
[9] K. Na et al. *NSTI-Nanotech*, 2008, vol. 3, pp. 737–740.

SB³C2015
Summer Biomechanics, Bioengineering and Biotransport Conference
June 17-20, 2015,Snowbird Resort, Utah, USA
SB³C2015-401

QUANTIFICATION OF MECHANICAL PROPERTIES OF RAT THORACIC DUCTS FOR LONG-TERM PREDICTION OF MECHANICALLY-MEDIATED GROWTH AND REMODELING

Alexander W. Caulk (1), Zhanna V. Nepiyushchikh (1), Ryan Shaw (1), J. Brandon Dixon (1,3), Rudolph L. Gleason, Jr. (1,2,3)

(1) Department of Mechanical Engineering
Georgia Institute of Technology
Atlanta, GA, United States

(2) Department of Biomedical Engineering
Georgia Institute of Technology
Atlanta, GA, United States

(3) The Petit Institute for Bioengineering and Bioscience
Georgia Institute of Technology
Atlanta, GA, United States

INTRODUCTION

Lymphatic function is directly dependent on the ability of lymphatic vessels to spontaneously and rhythmically contract to transport fluid against an unfavorable pressure gradient, a function which, if impaired, leads to lymphedema. It is well known that the mechanical environment contributes significantly to their overall contractile function, leading many to study the relationship between intrinsic contractile function and mechanical cues such as magnitude and direction of flow, inlet and outlet pressure, or pressure gradient. Also directly related to contractile function is the ability of the vessel to distend appropriately in response to mechanical loading, and there have consequently been many studies characterizing the active and passive mechanical behavior of lymphatic vessels.

Lymphatic vessels exhibit a nonlinear, pseudo-elastic, heterogeneous, anisotropic mechanical response, and experience large deformation. Several fluid-solid interaction (FSI) models have been proposed that quantify lymph propulsion through the lymphatic vasculature, most of which either neglect the nonlinear, anisotropic behavior of the tissue or employ empirical pressure-diameter relations. There is a pressing need for a predictive microstructurally-motivated mathematical model that captures the key characteristics of the mechanical response, in the framework of finite elasticity. Such a model represents a key step in exploiting important biomechanical frameworks for FSI and soft tissue growth and remodeling and holds the potential to advance our understanding of lymphatic function, dysfunction, and disease progression.

METHODS

Biomechanical testing. Thoracic ducts (n=6) from male Sprague-Dawley rats (300 – 350 g) were carefully excised and placed in physiologic saline solution (PSS) with or without calcium. Vessels were first mounted on a custom computer-controlled biaxial testing device (*1*) in PSS in a temperature controlled incubator at 37°C. Vessels were visualized using an inverted microscope while recording the transmural pressure, outer diameter, axial length, and axial force measurements. Vessels were axially stretched to 30% beyond the unloaded length (λ = 1.3), pressurized to 3 cmH_2O, and given 30-60 minutes to equilibrate to 37°C in order to establish contractility. Once contractility was established, isobaric contractile function was assessed at transmural pressures of 3, 6 and 9 cmH_2O. At each transmural pressure, the vessel was given a period of 5-10 minutes for equilibration. The same procedure was repeated for axial stretches of λ = 1.4 and 1.5. Data for analysis was taken only after the minimum and maximum diameters reached steady-state conditions.

After isobaric contractile function was assessed, the PSS in the bath and the system was replaced with Ca^{2+}-free PSS to inhibit vessel contractility and quantify the 'passive' mechanical response. Vessels were preconditioned by ramping the luminal pressure from 0 to 15 cmH_2O at axial stretch values of λ = 1.3 – 1.6 at an interval of 0.05. Vessels were then subjected to three loading cycles of pressure values from 0 to 15 cmH_2O at axial stretches identical to those during preconditioning.

Constitutive Modeling. We modeled the lymphatic vessels as an incompressible, thin-walled cylinder in the framework of finite elasticity such that the Cauchy stress can be defined by Equation (1),

$$\mathbf{T} = -p\mathbf{I} + 2\mathbf{F}\frac{\partial W}{\partial \mathbf{C}}\mathbf{F}^{\mathrm{T}} \tag{1}$$

where $\mathbf{T}$ is the Cauchy stress, p is the Lagrange multiplier that enforces incompressibility, $\mathbf{F}$ is the deformation gradient, W is the strain energy function, and $\mathbf{C}$ is the right Cauchy-Green strain tensors.

We considered the four-fiber family strain energy density function of Baek et al. (*2*) which was adapted from the original function described by Holzapfel et al. (*3*); namely,

$$W = b(I_C - 3) + \sum_{k=1,2,3,4} \frac{b_1^k}{4b_2^k}\{exp[b_2^k((\lambda^k)^2 - 1)^2] - 1\} \quad (2)$$

where b, b_1^k, and b_2^k are material parameters, $I_C = tr(\mathbf{C}) = C_{rr} + C_{\theta\theta} + C_{zz}$ is the first invariant of $\mathbf{C}$, $(\lambda^k)^2 = C_{\theta\theta}\sin^2(\alpha^k) + 2C_{\theta z}\sin(\alpha^k)\cos(\alpha^k) + C_{zz}\cos^2(\alpha^k)$ is the stretch of the kth fiber family, and α^k is the angle between the axial and fiber directions. We assume material symmetry such that, $\alpha^3 = -\alpha^4 \equiv \alpha$ and $b_1^3 = b_1^4$ and $b_2^3 = b_2^4$. Thus, equation (6) contains seven material parameters (b, b_1^k and b_2^k with $k = 1,2,3,4$) and one structural parameter (α).

We employed the model of Rachev and Hayashi (*4*) for the active stress due to muscle cell contraction; namely,

$$\hat{T}_{\theta,act} = T_{act}\lambda_\theta\left[1 - \left(\frac{\lambda_M - \lambda_\theta}{\lambda_M - \lambda_0}\right)^2\right] \quad (3)$$

where T_{act} is a parameter dependent on the activation of the muscle cells, λ_M is the stretch at which the muscle contraction is maximum, and λ_0 is the stretch at which muscle-induced force generation ceases. Due to both tonic and rhythmic contraction of lymphatic muscle cells, the parameter T_{act} was calculated for conditions describing the basal tone as well as the systolic contraction of the vessel, resulting in separate curves depicting maximum and minimum states of active contraction. All parameter values are reported for individual samples as well as in the form of the mean ± standard error of the mean (SEM).

RESULTS

Passive and active mechanical response. Contractility was established in isolated thoracic ducts, and isobaric contractile function was evaluated by monitoring time-dependent diameter changes at fixed values of transmural pressure (Fig. 1). For a single value of axial stretch, the contractile amplitude decreased (i.e. the vessel becomes more like a conduit) as transmural pressure increased. Contractile amplitude was markedly diminished upon increasing axial stretch from λ = 1.3 to 1.5. Therefore, a stretch value of 1.3 was the only value considered for the modeling portion of this study. Passive mechanical behavior was observed and showed a stiffening response that occurred at progressively lower diameters as axial stretch increased.

Material parameter calculation. Regression yielded material parameters that provided a good fit of the passive mechanical data (errors of 8.9 ± 0.8%). In general, the majority of the error in the passive fits occured either in the toe region of the plots where collagen fibers become engaged or in the higher values of axial stretch in which collagen fibers appear to be constantly engaged. Regression yielded active parameters that provided a good fit of the contractile data (Figure 1). Errors associated with the active parameters T_{act}^{max} and T_{act}^{bas} were 6.6 ± 2.5% and 6.6 ± 3.4%, respectively.

DISCUSSION

In this study, we have successfully captured both active (i.e. contractile pumping) and passive mechanical behavior of rat thoracic ducts using a microstructurally-motivated mathematical model. To date, this is the first model of lymphatic biomechanics that provides predictive capabilities while also accounting for the physical structure of the vessel in the context of finite elasticity. While this is an important step forward, there are multiple limitations of the study. First, we observed active behavior at three pressures for three axial stretches, totaling nine testing conditions. While this provided a suitable amount of data for estimation in modeling, the minutiae of stretch-dependent contractile behavior were likely lost. Also, the parameters λ_M and λ_0 were prescribed based on maximum and minimum stretch values achieved during passive and active testing, respectively. These values should be experimentally determined by including active biaxial testing such as what we have done here for an increased number of axial stretch values and isobaric conditions. In general, these simplifications were considered acceptable for the purpose of model development.

In closure, it is well known that biological tissue remodels in response to its mechanical environment. Thus, the ability of this model to quantify the internal mechanical environment of the lymphatic vessel wall provides an excellent framework for the prediction of long-term mechanically-mediated growth and remodeling. Ongoing studies include the incorporation of this model into currently available FSI models (*5*) as well as expansion into a model that supports geometric adaptation in response to mechanical perturbations in order to achieve target stress values (*6*).

ACKNOWLEDGEMENTS

We gratefully acknowledge that this work was funded by a grant from the National Institutes of Health (R01-HL113061).

REFERENCES

1. R. L. Gleason, S. P. Gray, E. Wilson, J. D. Humphrey, *J Biomech Eng* **126**, 787 (Dec, 2004).
2. S. Baek, R. L. Gleason, K. R. Rajagopal, J. D. Humphrey, *Computer Methods in Applied Mechanics and Engineering* **196**, 3070 (2007).
3. G. Holzapfel, T. Gasser, R. Ogden, *Journal of Elasticity* **61**, 1 (2000/07/01, 2000).
4. A. Rachev, K. Hayashi, *Ann Biomed Eng* **27**, 459 (Jul-Aug, 1999).
5. C. D. Bertram, C. Macaskill, J. E. Moore, Jr., *Comput Methods Biomech Biomed Engin*, (Feb 6, 2013).
6. L. A. Taber, *J Biomech Eng* **120**, 348 (Jun, 1998).

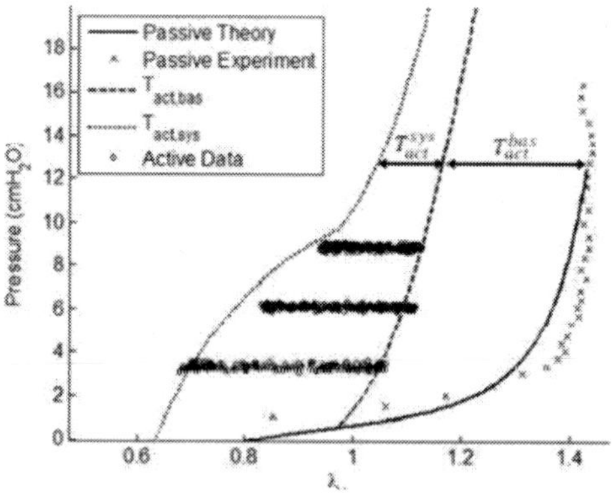

Fig. 1. Experimental data and model predictions of pressure/stretch behavior for a representative thoracic duct. Passive mechanical (x) and isobaric active contractile pumping (o) data were used to determine passive (solid line) and basal (dashed line) and maximum pumping (dotted line) material parameters for the proposed constitutive model.

NETWORK STIFFENING OF NANOFIBER SCAFFOLDS BY MINERAL

Justin Lipner (1,2), John Boyle (1,2), Victor Birman (3), Guy Genin (4) Stavros Thomopoulos (1,2,4)

(1) Biomedical Engineering
Washington University
Saint Louis, MO, USA

(2) Orthopaedic Surgery
Washington University
Saint Louis, MO, USA

(3) Engineering Education Center
Missouri University of Science and Technology
Saint Louis, MO, USA

(4) Mechanical Engineering & Materials Science, Washington University
Saint Louis, MO, USA

INTRODUCTION

Polymer based nanofibrous scaffolds have potential utility in a number of musculoskeletal applications. Electrospinning can be used to synthesize networks of fibers with controllable fiber diameters, alignments, and porosities. These scaffolds have been applied to engineer fibrous tissue such as tendon, ligament, and meniscus, but have had limited utility for mineralized tissues [1]. These limited efforts have produced a surprising contradiction: precipitation of calcium phosphate mineral onto nanofiber scaffolds stiffens the scaffolds [2-4] without stiffening the individual nanofibers [5]. We therefore performed a study to understand how this stiffening occurs with the aim of identifying design principles for mineralized nanofiber scaffolds. We hypothesized that mineral stiffened nanofiber scaffolds by enhancing fiber-to-fiber links that adhere poorly to individual nanofibers and therefore have little stiffening effect on the fibers themselves.

Two experiments were designed to test this hypothesis and uncover the mechanisms of stiffening. First, aligned nanofiber scaffolds were tested in tension at a number of off-axis angles. Here, we predicted that the stiffening effect of mineral would be more pronounced in the transverse directions (i.e., perpendicular to the predominant fiber orientation) than in the longitudinal direction (i.e., along the predominant fiber direction). Second, scaffolds were layered, mineralized, and pulled apart in a peel test. Here, we predicted that mineral would have a significant and mineral concentration-dependent effect on adhesion strength, with the fracture energy required to propagate a crack during the peel test would scaling with the spatial concentration of mineral bridges between nanofibers.

METHODS

Scaffold synthesis: Electrospinning was used to deposit 85:15 poly (lactic-co-glycolic acid) (PLGA) nanofibers onto either rotating mandrel collectors (off-axis tests) or gapped collectors (peel tests) to produce aligned mats, as described previously [4]. After plasma treatment to increase hydrophillicity, samples were either submerged in water for one hour (unmineralized groups), or mineralized in ten times simulated body fluid (10SBF) for various lengths of time. Samples were washed with deionized water for one hour and allowed to air dry prior to mechanical testing.

Off-axis tests: Test strips of 3 mm x 15 mm were cut at various angles from the predominant fiber direction (0, 15, 30, 45 and 90 degrees), as described by Nerurkar et al. [6], and tested in uniaxial tension under displacement control at a constant strain rate of 0.5% per sec (Figure 1).

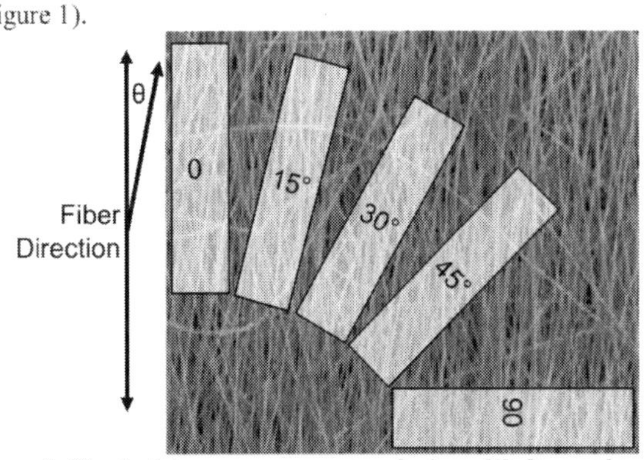

Figure 1: Test strips were prepared at multiple angles for uniaxial tensile testing.

Peel tests: Peel tests were performed by pulling the top piece from the bottom piece at an initial angle of 90 degrees at a rate of 0.05mm/sec (Figure 2). The angle between the two pieces was determined from videos of the tests. The Kendall model was used to analyze peel tests [7], according to:

$$R=\left(\frac{F}{b}\right)^2\frac{1}{2dE}+\left(\frac{F}{b}\right)(1-\cos(\Theta))$$

where R is the energy required to peel the two pieces apart, F is the force applied along the mat 1 (Figure 2), d, b, and E are the thickness, width, and modulus of the samples, respectively, and Θ is the angle between the samples.

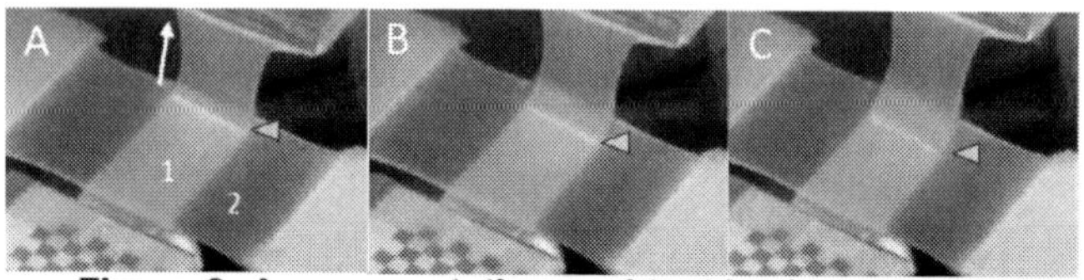

Figure 2: A representative peel test is shown at the beginning (A), middle (B), and end (C) of a test (direction of displacement is shown by white arrow). The nanofiber mats are labeled 1 and 2, and the yellow arrowhead indicates the crack front.

Statistics: To compare mechanical property outcomes, a two-way (off-axis tests) or one-way (peel tests) analysis of variance was performed, followed by post-hoc Fisher's least squared difference test when appropriate.

RESULTS

The results of both experiments supported the hypothesis. Both testing angle and mineralization affected the strength (ultimate stress) and modulus significantly ($p<0.005$), with transverse values being lower than longitudinal values in both cases, and mineralization leading to higher stiffness at all angles, and higher strength in all but the 0 degree group (Figure 3). Intermediate angles (15-45 degrees) showed strong mineralization effects on both strength and modulus.

Peel test experiments demonstrated that mineralization significantly increased adhesion energy in a dose-dependent manner (Figure 4). Between groups, only 0 vs. 5 and 5 vs. 15 comparisons were not statistically significant. The linear fit was statistically significant compared to a slope of zero ($p<0.005$).

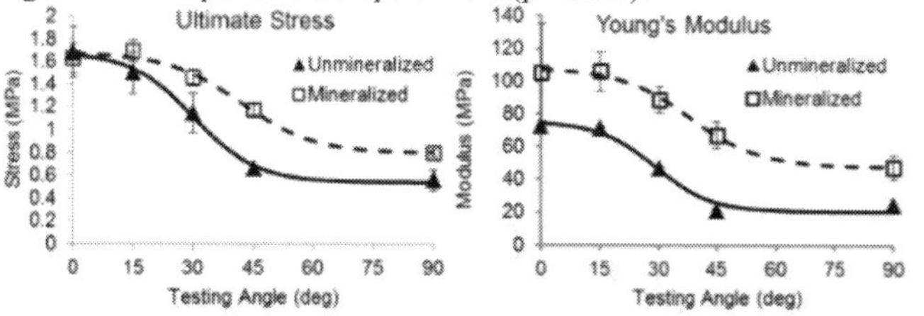

Figure 3: Results of multi-angle testing demonstrate that mineralization increases strength and modulus at angles above 30 degrees, with the maximum effect at 45 degrees. (data: mean ± SD)

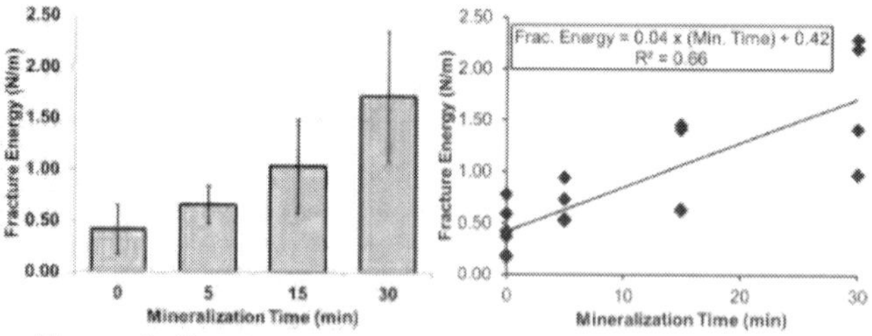

Figure 4: Fracture energy measurements suggest that a small amount of mineralization did not affect adhesion, but that mineralization had a significant and dose-dependent effect (data: mean ± SD in left figure, and all points are shown in the right figure)

DISCUSSION

The current studies supported our hypothesis that mineral stiffening of nanofiber scaffolds occurs via increased connections between fibers. Previously, unmineralized aligned nanofiber scaffolds were tested at various angles [6], demonstrating similar outcomes to the current study. Here, we showed that mineralization stiffened the scaffolds in all testing directions, and most strongly at 45 degrees (in relative terms). At this testing angle, fibers are neither aligned with the direction of load (which would provide the highest resistance) nor perpendicular to the direction of load (which would provide the least resistance). The observed effect of mineral at this testing angle suggests that it provides resistance to fiber realignment.

Peel tests of layered, mineralized, nanofiber scaffolds showed that even a relatively low level of mineral significantly ($p<0.005$) increased bonding between nanofibers. Furthermore, the effect was dose-dependent, with increased mineralization leading to higher fracture energy. The strong effect of mineral on adhesive strength between fibers supports our hypothesis that mineral forms bridges, bonding adjacent fibers together.

Current approaches for stiffening PLGA nanofiber scaffolds produce materials that are an order-of-magnitude weaker than necessary for bone and bony interface applications. The results of the current experiments provide insight into the stiffening effects of mineral on nanofiber scaffolds, and methods to test these effects.

ACKNOWLEDGEMENTS: Funding was provided by NIH R01 AR060820.

REFERENCES

[1] Liu, W. et al., *Advanced Healthcare Materials*, 1(1): 10-25, 2012.
[2] Tas, A.C. et al., *Journal of Materials Research*, 19(9): 2742-9, 2004.
[3] Liu, W. et al., *Langmuir*, 9088-9093, 2011.
[4] Lipner, J. et al., *Journal of the Mechanical Behavior of Biomedical Materials*, 40(0): 59-68, 2014.
[5] Kolluru, P.V. et al., *Acta Biomateriala*, 9(12): 9442-50, 2013.
[6] Nerurkar, N.L. et al., *Journal of Orthopaedic Research*, 25(8): 1018-1028, 2007.
[7] Kendall, K. et al., *Journal of Physics D: Applied Physics*, 8(13): 1449, 1975.

SB³C2015
Summer Biomechanics, Bioengineering and Biotransport Conference
June 17-20, 2015,Snowbird Resort, Utah, USA
SB³C2015-403

THE EFECT OF ANATOMICAL VARIABILITY ON TEMPOROMANDIBULAR JOINT MECHANICS

Jessica S. Coogan (1), Travis D. Eliason (1), Mark E. Wong (2), Daniel P. Nicolella (1)

(1) Mechanical Engineering Division
Materials Engineering Department
Musculoskeletal Biomechanics Section
Southwest Research Institute

(2) Department of Oral & Maxillofacial Surgery
The University of Texas School of Dentistry
at Houston
Houston, TX, USA

INTRODUCTION

Disorders of the temporomandibular joint (TMJ) are widespread, afflicting between 5% and 15% of the adult population [1]. Temporomandibular joint disorder (TMD) encompasses several conditions that lead to pain and dysfunction of the temporomandibular joint and the muscles of mastication. Common symptoms of TMD include orofacial pain, restricted mandibular movement, locked jaws, and sounds from the jaw joint, including clicking and popping. While the causes of TMD are not completely understood, it is thought that alterations or individual variability in joint mechanics plays a key role in the initiation and progression of TMD. However, the effect of anatomical variability on the dynamic mechanical environment within the TMJ during jaw motion is not well characterized. The objective of this investigation was to use shape modeling-based dynamic finite element modeling to investigate how variability in anatomy and muscle forces affect TMJ mechanics.

METHODS

Finite element models were constructed from cone beam CT scans of 4 individuals with a history of TMD. From the CT scans, the mandible and temporal bone were semi-automatically segmented (Seg3D, The Center for Integrative Biomedical Computing, University of Utah, Salt Lake City, UT) and meshed (TetGen v. 1.5, http://www.tetgen.org). Shape modeling methods were utilized in order to incorporate differences in anatomy between the individuals [2]. Briefly, one individual was arbitrarily chosen as the template mesh, which was registered to each of the other individuals using a coherence point drift algorithm [3]. The resulting surfaces were then defined by the same surface mesh definition and the vertices of the mesh were positioned at corresponding anatomic locations between individuals. An average model was defined by averaging vertex positions across the individuals. In addition, shape modeling methods determine independent shape modes that represent the major modes of variation within the population of included patients [2]. Models were also constructed representing +1 SD of each shape mode (Fig. 1).

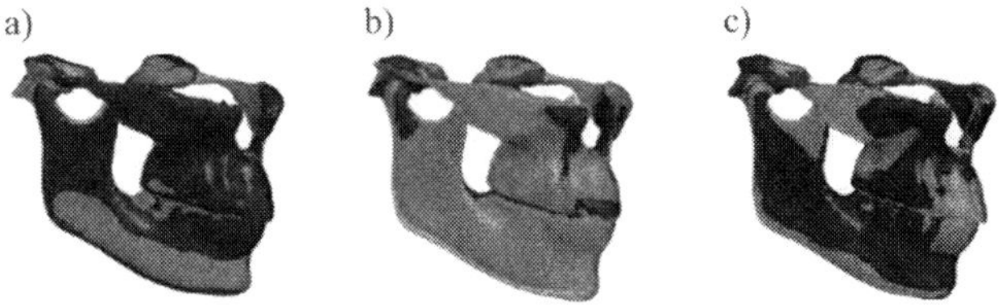

Figure 1. Comparison of average geometry and shape mode a) 1, b) 2, and c) 3. The average geometry is shown in red, and the shape mode is shown in white.

The articular disc and cartilage were created by space filling the area between the condyle and the temporal bone fossa, assuming the cartilage has a thickness of 0.4 mm [4]. Discal ligament attachments were created by projecting the outer surface of the articular disc along the condylar and fossa contours. A generic skull surface was rigidly constrained to the temporal bone in order to serve as attachment points for the TMJ muscles and ligaments. The TMJ ligament and temporalis, master, lateral pterygoid, medial pterygoid, and digastric muscles were modeled based on anatomic texts (Fig. 2).

The skull and mandible were modeled as rigid bodies. A hyperelastic material model with active contraction continuum elements was used to represent the muscles [5]. Force is generated along a locally defined fiber axis direction, and the material model parameters are taken from literature [6]. The articular disc and discal ligaments were represented by a hyperelastic material model with model parameters taken from the literature [7-8] (Table 1). The TMJ ligaments and TMJ cartilage were assumed to have the same properties as the stiffest discal ligament. All FE simulations were performed using FEBio [5].

a) b)

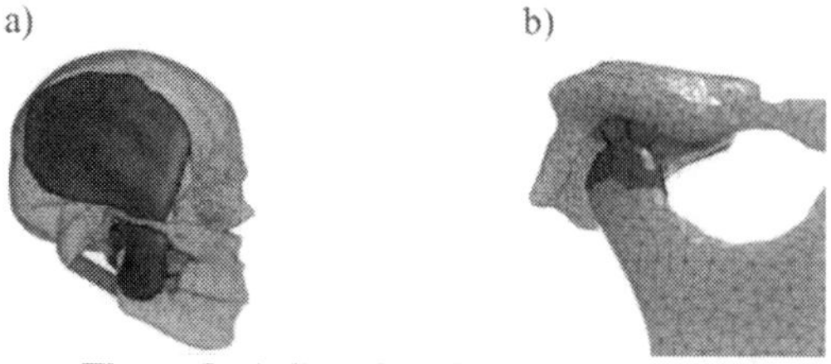

Figure 2. a) Complete finite element model of the average geometry. Ligaments and muscles are shown in red. b) Close-up of the TMJ. Cartilage is shown in blue, disc in red, and discal ligaments in translucent white.

Table 1. Material properties for the model

Material	Young's Modulus (MPa)
Muscle	0.185[6]
Articular Disc	3.9 [7]
Discal Ligaments	1.4-7.5 [8]

The muscle activations were determined using a proportional-integrative-derivative (PID) controller incorporated directly into FEBio. The reference signal was chosen to be the time varying position of the mandible and the controller varied the activation level of the muscles in order to achieve the reference kinematics. Separate PID controllers were used for different muscles. The first controller modified the activation of the left and right digastric muscle while the second controller modified the activation of the left and right lateral pterygoid muscle. Thus, motion of the mandible relative to the skull was produced by the coordinated active, time-varying contraction of the muscles and was constrained by the articular contact at the TMJ (Fig. 2b) and contact between the upper and lower dentition.

RESULTS

The PID controller was able to successfully match the prescribed displacement of the mandible (Figure 3). Each shape mode had a variable effect on the muscle activation levels required to achieve the target displacement (Fig. 4). For all models, the pterygoid muscle required higher activations than the digastric muscle. Variations in TMJ anatomy, particularly the shape of the mandibular condyle and fossa, significantly affected the pattern of stress within the disc during functional jaw motion (Fig. 5 and 6).

a) b)

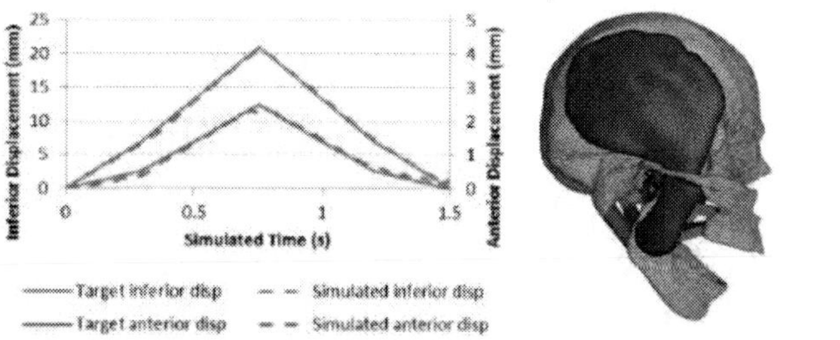

Figure 3. a) Comparison of target displacements and simulated displacements of the mandible for the average model. b) Average model shown in full opening position.

DISCUSSION

This work utilizes shape modeling methods and finite element modeling in order to create a model of the temporomandibular joint that accounts for differences in anatomy among individuals. A PID controller determined the muscle activations required to achieve target mandibular kinematics. Using the PID controller, mouth opening and closing were simulated. Differences in muscle activations and TMJ disc stress were seen across the shape modes. This work indicates that variably in TMJ anatomy results in significant difference in tissue level TMJ mechanics, which may contribute to the initiation and progression of TMD.

a) b)

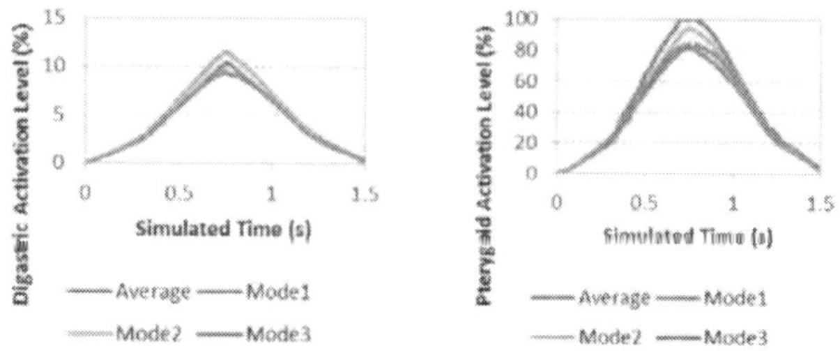

Figure 4. Comparison of the a) digastric activation level and b) lateral pterygoid activation level for the different shape modes.

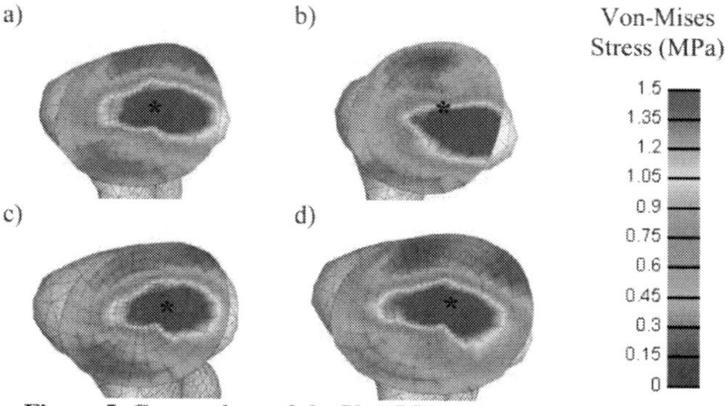

Figure 5. Comparison of the Von-Mises stress in the right TMJ disc for a) Average, b) Mode 1, c) Mode 2, and d) Mode 3

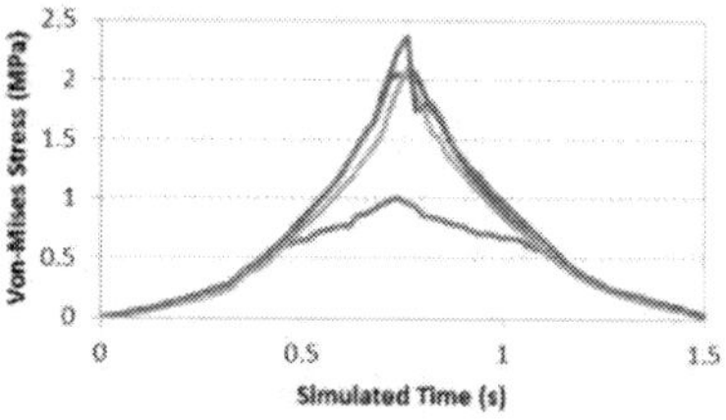

Figure 6. Anatomical differences described by independent shape modes result in significant variation in TMJ disc stresses. Stresses determined at element indicated with * in Fig. 5a above.

ACKNOWLEDGEMENTS

This work supported by the Southwest Research Institute Advisory Committee for Research IR&D project R8386.

REFERENCES

[1] Johansson, A et al., *J Orofac Pain*, 17(1): 29-35, 2003.
[2] Bredbenner, TL et al. *J Biomech*, 43(9): 1870-1786, 2010.
[3] Myronenko, A et al., *IEEE Trans. Pattern Anal. Mach. Intell*, 32(12): 2262- 2275, 2010.
[4] Singh, M et al., *J Biomech*, 42: 405-417, 2009.
[5] Maas, SA et al., *J. Biomech Eng*, 134(1), 011005-1-10, 2012.
[6] Chaudry, H et al., *J Am Osteopath Assoc*, 108(12): 680-688, 2008.
[7] Del Palomar, AP et al. *J Biomech Eng*. 128(2): 242-249, 2006.
[8] Murphy, MK et al., *J Dent Res* 92(8):753-758, 2013.

ELASTIN PRODUCTION SLOWS ANEURYSM ENLARGEMENT IN A CONSTRAINED MIXTURE MODEL OF ANEURYSM GROWTH AND REMODELLING

Kory J. Blose (1), Arturo Valentin (2), Justin S. Weinbaum (1,3), Anne M. Robertson (2), David A. Vorp (1,3,4,5,6)
1 Department of Bioengineering, University of Pittsburgh, Pittsburgh, PA, United States
2 Department of Mechanical Engineering and Materials Science, University of Pittsburgh, Pittsburgh, PA, United States
3 McGowan Institute for Regenerative Medicine, University of Pittsburgh, Pittsburgh, PA, United States
4 Department of Cardiothoracic Surgery, University of Pittsburgh, Pittsburgh, PA, United States
5 Department of Surgery, University of Pittsburgh, Pittsburgh, PA, United States
6 Center for Vascular Remodeling and Regeneration, University of Pittsburgh, Pittsburgh, PA, United States

INTRODUCTION

Abdominal aortic aneurysm (AAA) is a serious disease in need of a new treatment paradigm. Intervention, including open surgical repair and endovascular repair, involves placement of a synthetic graft to physically exclude the aneurysmal aorta, and is recommended only after the aorta reaches a critical diameter (5.5 cm). Surgical intervention does not benefit small AAAs [1], and experimental use of pharmaceutical treatment to alter the progression of small AAAs has to date not proven effective [2-5].

A potential alternative intervention for AAA would be a regenerative medicine approach; i.e., focusing on regeneration (both cells and matrix) as well as inhibiting matrix degradation [6]. The processes that initiate and expand a AAA are heavily based on cellular activity; therefore mesenchymal stem cell (MSC)-based therapy could be one approach. MSCs secrete growth factors [7, 8] that suppress inflammation and MMP activity while stimulating elastin and collagen production. MSCs can also differentiate, thus potentially providing a means to replace lost smooth muscle cells. Furthermore, MSCs have already shown promise as a treatment for AAAs in animal models when delivered by our lab to the adventitia [9], or by others systemically [10] or directly into the aortic wall [11]. MSC treatment could meet the two criteria of the ideal AAA treatment: regeneration (both cells and matrix) and inhibiting matrix degradation.

However, the properties that make MSCs attractive candidates for therapy are the same properties that make researchers and clinicians cautious about their therapeutic use. Treatments should be designed to use the minimum number of MSCs as will be needed to achieve the desired effect, which is elastin production in a mechanically stimulated smooth muscle cell. However, the question of how much elastin production is necessary remains. Here we attempted to answer this question through computational modeling. Computational models of AAA growth and remodeling [12] have been successful at capturing aspects of AAA progression including aneurysm dilation and wall thinning. The starting point of these simulations is the loss of a critical amount of elastin within the arterial wall. This loss of elastin creates a maladaptive mechanical environment which leads to the enlargement of the aneurysm [13]. Restoration of the lost elastin fibers should return the artery to a more favorable mechanical environment, possibly halting and/or reversing the progression of the AAA. Using this established computational tool, we sought to understand the effects of elastin production which will guide future experiments when determining the proper dosage of MSC for the treatment of AAA.

METHODS

We modified existing computational models of AAA growth and remodeling [12] to allow for elastin production. The current computational models do not allow for elastin production as adult cells do not normally produce elastin in its mechanically functional fiber form. In order to relieve the arterial wall from its pathologic mechanical environment characteristic of these models and actual AAAs, we explored a range of elastin mass production rates (2-16x the magnitude of natural elastin degradation rates [40 year half-life] [14]) within the growth and remodeling simulations while tracking the rate of aneurysm growth. We also calculated the enlargement rate of a AAA for each elastin production rate. Lastly, elastin production was turned on at various points in the growth history of the AAA in order to understand the effects of intervention timing.

RESULTS

Within the computational model of aneurysm growth and remodeling the production of elastin reduces the aneurysm enlargement rate (**Fig. 1**). Two key characteristics of the natural evolution of the aorta are captured by the computational model and shown in Fig. 1A; include aneurysmal enlargement (outer diameter increases) and wall thinning (the difference in outer and inner diameter decreases). When the model is altered to allow for elastin production, we find that the rate of enlargement is inversely proportional to the amount of elastin produced within the model, and the effect is exaggerated at earlier intervention time points (**Fig. 1 B, C, D**). The rate of aneurysmal enlargement at the end of the nominal model run (no elastin production) is 5.340 mm/year. Intervention at 30% growth of the vessel caused the aneurysm to begin to contract (i.e. residual tension begins to pull the artery back towards its original size) in most elastin production rates (**Table 1**).

DISCUSSION

The results of this study show that from a mechanical standpoint, elastin production within an aneurysm could relieve the maladaptive mechanical environment to an extent that slows aneurysmal enlargement. We have also shown that early intervention can reduce the enlargement rate by more than 50%. We are particularly interested in this reduction value because of its clinical significance. A 50% reduction in the enlargement rate of a small AAA could delay the need for surgical intervention by 10 years [15]. Halting the enlargement could delay the need for intervention even longer, and contracting the AAA would mean at least partial restoration of the normal physiology.

The constrained mixture models used in this study are not without limitations. Although biological phenomena are the basis of the governing equations within the model, not all biological aspects are captured such as the inflammatory response which is pronounced in AAA disease. That is not meant to discount the value of this study as it was and is still able to provide important insights to the mechanopathological nature of AAA.

These exciting results can be used to guide future experiments that study elastin production. They can also be used as a starting point for optimizing treatment protocols for AAA therapies such as MSC therapy. These findings also confirm our thoughts that regeneration of functional elastic fibers in a AAA can help slow the progression of the disease.

ACKNOWLEDGEMENTS

The authors would like to disclose funding support by the National Institutes of Health (Cardiovascular Bioengineering Training Program T32 HL076124 and Biomechanics in Regenerative Medicine T32 EB000392 to KJ Blose).

REFERENCES

[1] Greenhalgh R et al., Eur J Vasc Endovasc Surg, 16(6):462, 1998
[2] Baxter BT, Am J Surg, 188(6):628-32, 2004
[3] Baxter BT et al., J Vasc Surg, 36(1):1-12, 2002
[4] Mosorin M et al., J Vasc Surg, 34(4):606-10, 2001
[5] Thompson R et al., Adv Dent Res, 12(1):159-65, 1998
[6] Bashur CA et al., Stem Cells Trans Med, 2(6):401-8, 2013
[7] Kinnaird T et al., Circ Res, 94(5):678-85, 2004
[8] Rehman J et al., Circulation, 109(10):1292-8, 2004
[9] Blose KJ et al., Regen Med, 9(6):733-41, 2014
[10] Sharma AK et al., Circulation, 126(11 suppl 1):S38-S45, 2012
[11] Turnbull IC et al., J Surg Res, 170(1):e179, 2011
[12] Valentín A et al., Int J Numer Met BME, 29(8):822-49, 2013
[13] Humphrey J et al., J Biomech, 45(5):805-14, 2012
[14] Arribas, SM et al., Pharmacol Ther 111.3: 771-791, 2006
[15] Klink A et al., Nat Rev Cardiol, 8(6):338-47, 2011

		Rate of Aneurysm growth (mm/year)			
		Point of Intervention			
		30%	40%	45%	50%
Elastin Production Factor	2x	3.338	3.110	3.786	4.711
	3x	-0.692	3.172	3.453	4.274
	4x	-15.480	2.356	3.058	3.791
	8x	-13.302	1.352	2.236	3.302
No Intervention		5.340			

Table 1: The rate of aneurysm growth slows with increased elastin production and intervention at earlier time points.

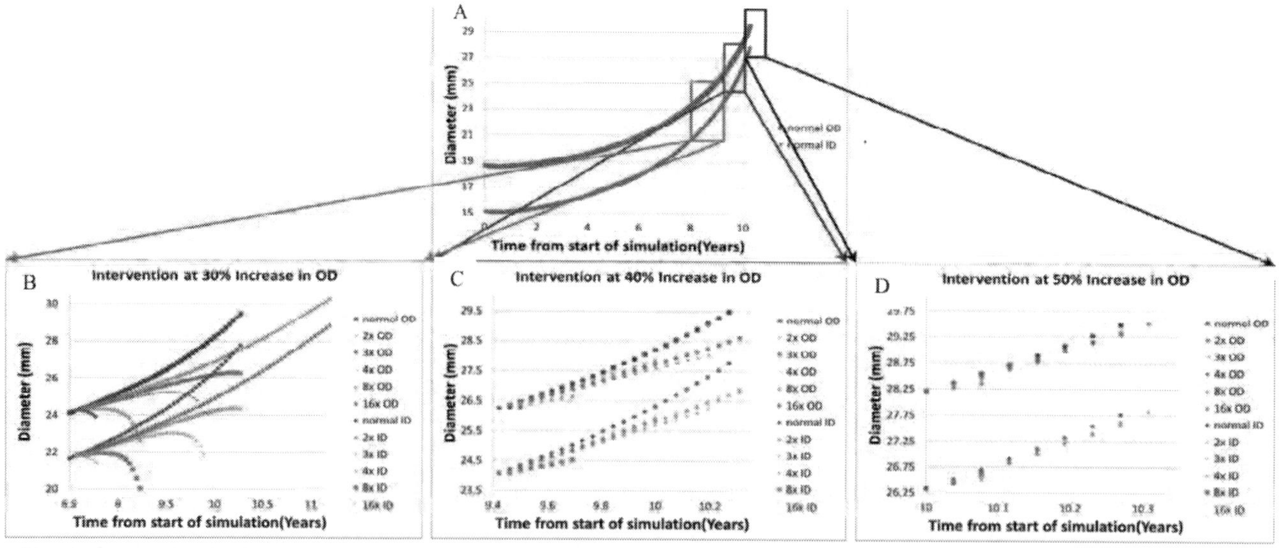

Fig. 1: Constrained mixture models show growth of aneurysmal diameter and thinning of the aortic wall (A). Intervention at 30% (B) diameter increase shows a more dramatic change in growth than later interventions (40%, C and 50%, D).

SB³C2015
Summer Biomechanics, Bioengineering and Biotransport Conference
June 17-20, 2015,Snowbird Resort, Utah, USA
SB³C2015-405

MMP-2 EXPRESSION AND THE TISSUE MECHANICS OF HUMAN ASCENDING THORACIC AORTIC ANEURYSMS

Alexander Emmott (1,2), Nastaran Shahmansouri (2), Mohammed Alreshidan (3), Stefanie Pohlod (1), Rosaire Mongrain (2), Raymond Cartier (2), Kevin Lachapelle (3), Richard Leask (1,2)

(1) Department of Chemical Engineering
McGill University
Montreal, QC, Canada

(2) Research Centre
Montreal Heart Institute
Montreal, QC, Canada

(3) Department of Cardiac Surgery
Royal Victoria Hospital
Montreal, QC, Canada

INTRODUCTION

The healthy ascending thoracic aorta (ATA) is a compliant vessel that, by the Windkessel effect, redistributes the energy of the cardiac cycle to supply blood to the arterial circulatory system. Functionally, the ATA is dominated by its medial layer, which is comprised of organized lamellae of elastin, smooth muscle cells and collagen. ATA aneurysms (ATAAs) develop as the normal mechanical properties of the vessel degenerate as a result of age, disease or congenital disorder (*e.g.*, bicuspid aortic valve). Although poorly understood, the dynamics of matrix metalloproteinase (MMP) expression has been implicated in collagen and elastin turnover in aneurysmal tissue [1,2]. In this work, we investigate the role of vessel MMP-2 expression as a co-variant to altered tissue mechanics and aortic valve type.

METHODS

Excised aneurysmal tissue was collected from patients undergoing aortic resection surgery at the Royal Victoria Hospital and Montreal Heart Institute. Biaxial tensile testing was performed within 24 hours of surgery on four 1.5 x 1.5 cm squares of the aortic ring. Each square underwent biaxial stretching (cyclic loading and unloading) to a maximum of 60 % strain using an EnduraTEC™ (Bose, Minnesota USA) tensile tester to obtain the tissue's stress-strain curves. Viscous energy loss and apparent elastic moduli were used as surrogates of the overall mechanical properties of the tissue. Samples from each quadrant tested were either fixed in 10 % formalin or frozen at -80 °C for histology and protein biochemistry, respectively.

Histological analysis using MOVAT pentachrome staining was used to assess the proportion of collagen and elastin in the tissue. Immunohistology was performed on paraffin-embedded sections of the tissue using a mouse anti-MMP-2 monoclonal antibody (MAb13431). Anti-mouse HRP-secondary and DAB substrate were used to determine the preferential localization of MMP-2 in the tissue. Overall MMP-2 expression was identified using western blot analysis on 4-12% Bis-Tris gels. Nitrocellulose membranes were immuno-blotted with rabbit polyclonal anti-MMP-2 and anti-glyceraldehyde phosphate dehydrogenase (GAPDH) antibodies. MMP-2 expression was assessed by band densitometry using ImageJ™ and normalized to GAPDH expression in the tissue.

RESULTS

Our lab has recently reported that ATAA tissue showed a marked increase in stiffness and viscous energy loss over a non-diseased control population [3]. Using similar mechanical indices, we report medial MMP-2 expression as a covariant to tissue stiffness, energy loss and aortic valve type. To date, 12 patients have been recruited for protein biochemistry work (N_{TAV}=6; N_{BAV}=6).

Quadrant-averaged axial stiffness (at 15% strain) increases with medial MMP-2 expression (Figure 1, R^2=0.667, P=0.001). While axial energy loss fraction appears to decrease with increasing MMP-2 expression, the correlation is not significant (Figure 2, R^2=0.257, P=0.092). We suggest that the occurrence of outlying values can be explained by comparative vessel pathology. In Figure 2, two patients with similar MMP-2 expression show markedly different energy loss which can be explained by a severe loss of elastin in the TAV patient compared to the BAV patient.

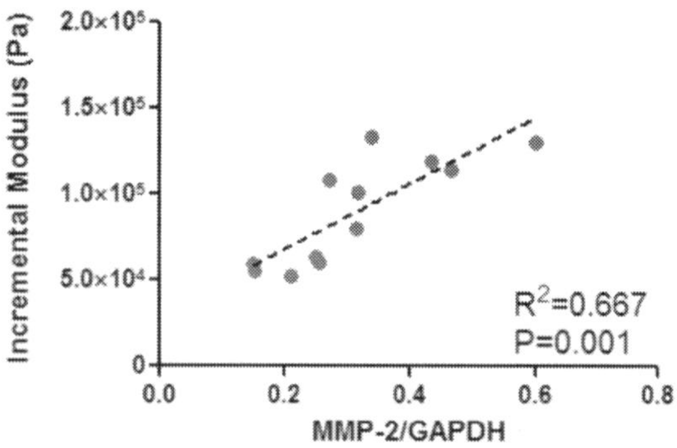

Figure 1. Incremental modulus of elasticity at 15% strain (axial direction) as a function of normalized medial MMP-2 expression.

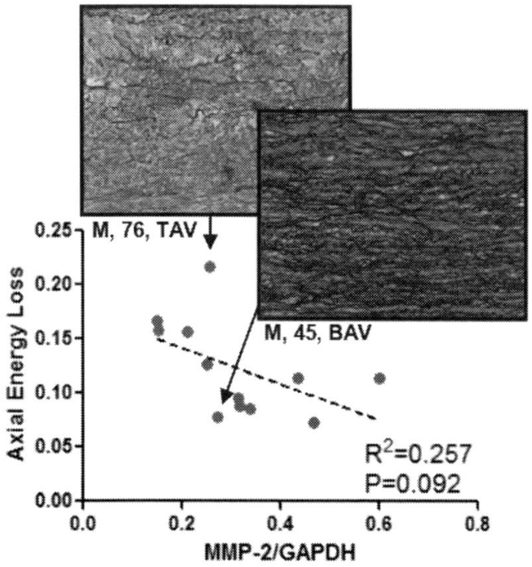

Figure 2. Energy loss fraction (axial direction) as a function of normalized medial MMP-2 expression.

Here, we confirm reports [1,4] that aneurysmal BAV patients exhibit significantly higher medial MMP-2 levels than do TAV patients (Figure 3). This increase in MMP-2 was found to occur for the inactive enzyme containing the propeptide as well as the mature, active enzyme.

Early immuno-histological results suggest (data not shown) that MMP-2 expression predominantly favors the outer medial layer (towards the adventitia) and we believe that this could be a result of focal remodeling in the tissue.

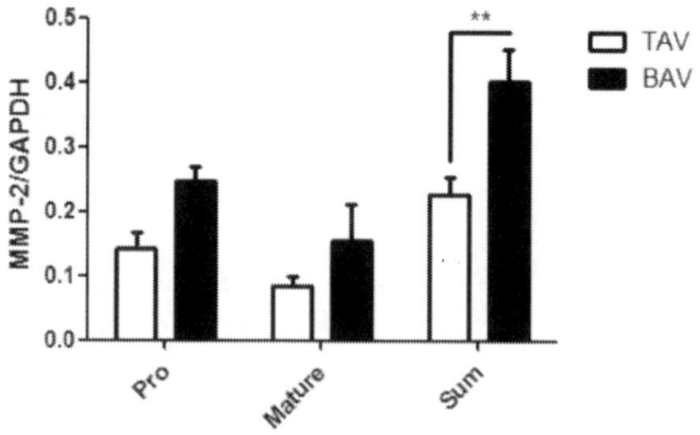

Figure 3. Medial MMP-2 expression in aneurysmal patients with tricuspid (TAV) or bicuspid (BAV) aortic valves. Expression levels are categorized by the form of the enzyme: pro-enzyme, mature enzyme or the sum of the two. Error bars are SEM and the statistical analysis is a 2-way ANOVA with a Bonferonni post-test ($P<0.01$, **).

DISCUSSION

Our preliminary study links the expression of medial MMP-2 levels with altered mechanical indices in ATAA patients. We believe that MMP-2 is central in the remodeling process and our data suggests a link between expression and stiffness and an inverse relation with expression and viscous energy loss. Our data also confirms the observation that ATAA patients with BAV have significantly more medial MMP-2 suggesting a differential pathology.

ACKNOWLEDGEMENTS

The authors acknowledge the financial support of NSERC and The Heart and Stroke Foundation. Alexander Emmott is supported by a McGill Engineering Doctoral Award (MEDA).

REFERENCES

[1] Fedak, PW et al., *J Thorac Cardiovasc Surg*, 126(3): 797-806, 2003.

[2] Collins, MJ et al., *J Clin Pathol*, 61(4): 519-523, 2008.

[3] Chung, J et al., *J Thorac Cardiovasc Surg*, 148(3), 1082-1089, 2014.

[4] Lemaire, SA et al., *J Surg Res*, 123 (1), 40-48, 2005.

SB³C2015
Summer Biomechanics, Bioengineering and Biotransport Conference
June 17-20, 2015,Snowbird Resort, Utah, USA
SB³C2015-406

STRESS ANGLE DEVICE: A NOVEL SYSTEM FOR REPRODUCING THE MECHANICAL CONDITIONS ASSOCIATED WITH REGIONS MOST SUSCEPTIBLE TO VASCULAR DISEASE

Steve Zambrano (1), Michael J. Draper (1), Brendan L. Swain (1), Caleb A. Davis (1) Michael R. Moreno (1,2)

(1) Biomedical Engineering
Texas A&M University
College Station, Texas, United States

(2) Mechanical Engineering
Texas A&M University
College Station, State, United States

INTRODUCTION

Cardiovascular diseases (CVDs) are the leading cause of death in the United States, with over 780 thousand deaths annually, which accounts for ~32% of all deaths. [1] The National Institute of Health has recorded over 83 million prevalent cases of cardiovascular diseases. [2] The impact of cardiovascular disease on the economy is evidenced in the total direct and indirect costs associated with these diseases, which amounted to $313 billion and 16.8 percent of all economic costs of illnesses in 2009. [2, 3]

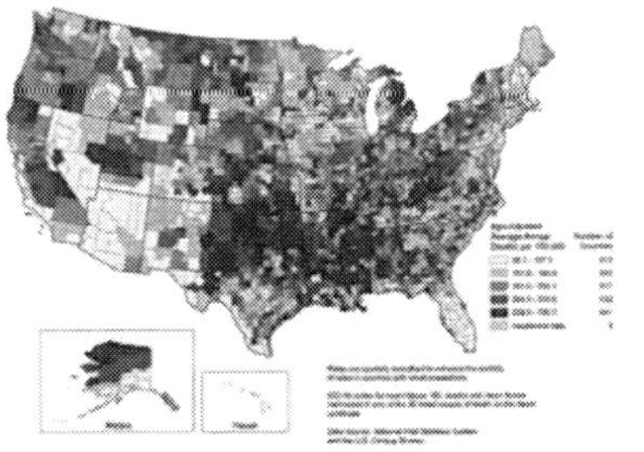

Figure 1: 2007-2009 Atherosclerosis-Related Death Rates in the United States [3]

The most common form of CVDs is atherosclerosis - the narrowing and hardening of arteries due to the buildup of plaque. Atherosclerosis is localized at branch points and curvatures in the arterial tree - where complex flow is observed. The mechanical environments of these pro-atherogenic regions differ greatly from the mechanical environment of pressure-driven flow through a straight tube, which can be used to describe the majority of the vasculature. [4] It has been widely accepted that mechanical factors such as low mean and oscillating fluid shear stress have been correlated with intimal thickening in the carotid bifurcation and the abdominal aorta. [4, 5] Thus, it has been hypothesized that mechanical stimuli may induce endothelial cell (EC) responses that are pro-atherogenic. In considering the in vivo vascular environment, hemodynamic stimulation includes flow-induced fluid shear stress (FSS) and pressure-induced cyclic circumferential strain (CS) (Figure 2). These mechanical stimuli occur naturally in the vasculature and mediate EC and smooth muscle cell (SMC) migration, proliferation, and protein expression. ECs also respond differently to different modes of shear forces (e.g. laminar, disturbed, or oscillatory flows).

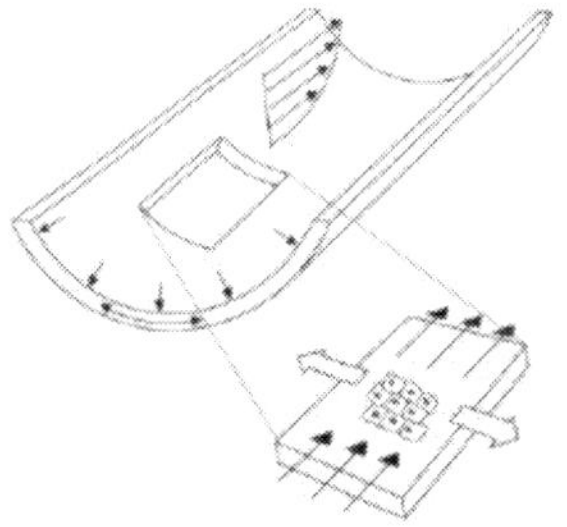

Figure 2: Fluid Shear Stress, Circumferential Strain, & Pressure in a Vessel [6]

Helical flow patterns are common in the carotid sinus, a location especially vulnerable to disease. In this region the fluid velocity vector has a substantial circumferential component and therefore shear stresses in this region are not exclusively in the axial direction (Figure 3), as they would be in a straight tube. We define the "Stress Angle" (SA) as the spatial angle between fluid shear stress and cyclic stretch, as opposed to the temporal angle which is referred to as the "Phase Angle" (PA). [7] Given that the stress angle varies in regions where disease typically forms, it is important to develop the ability to investigate these conditions.

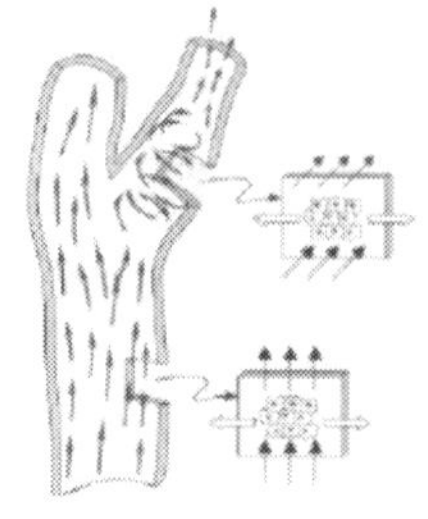

Figure 3: Carotid Bifurcation Depicting Flow Streamlines [6]

METHODS

In order to validate our hypothesis that the SA plays a definitive role in replicating the pro-arthrogenic regions seen in tortuous vessels, the following objectives will be employed: 1) Develop a device capable of subjecting ECs to uniaxial stretch and fluid shear stress simultaneously, in a manner wherein the spatial angle between fluid shear stress can be varied from perpendicular to collinear; 2) Design the system such that the PA may also be varied given evidence that the temporal angle between shear and stretch also produces significant effects on endothelial biochemical activity; and 3) Quantify the shape index and alignment angle of ECs cultured on the Stress Angle device. A cell-seeded silicon membrane will line the top plate of our modified parallel plate flow system. Both ends of the membrane are attached to a clamp, coupled to a linear shaft, and driven by a NEMA 23 stepper motor (National Instruments P70530), which cyclically applies physiologic strain to the membrane by distending the membrane across a disk. The rotatable disk and membrane can be rotated without taking the device apart, so that various stress angle states can be achieved without compromising sterility. The complete device is attached to a flow loop, which incorporates a gear pump (Ismatec BVP-Z) driven by a function generator (Goldstar 8002). A Millar (Mikro-Tip Catheter Transducer) intravascular pressure transducer is used to monitor pressure in the system. Flow rates are monitored using an ultrasonic flow probe and meter (Transonic T402) were outputs are routed to a LabVIEW-enabled laptop to acquire/control flow and cyclic stretch waveforms.

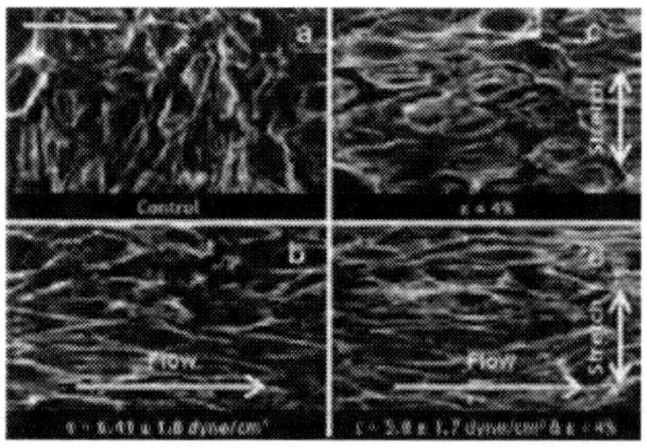

Figure 4: Bovine Aortic Endothelial Cells Exposed to Mechanical Stimuli [5]

RESULTS

Device fabrication proves that, utilizing this system, a cell-seeded membrane may be subjected to physiologically relevant mechanical conditions including controlled variation of the spatial and temporal angles between shear and stretch. The device will now be employed in studies designed for:

1) Quantifying the shape index and alignment angle of the cells at various stress angle orientations
2) Quantifying the cytoskeletal reorganization through study of cells transfected with fluorescent GFP-Actin
3) Conducting real-time PCR and additional methods of assessing gene and protein expression

These studies will accomplished through use of porcine aortic ECs under several physiological magnitudes of stretch and shear stress at specified spatial and temporal angles. These conditions will be maintained for 24 hours with images of the cells every 15 minutes.

Figure 5: Stress Angle Device Computer-Aided Design [6]

DISCUSSION

This work is the first known to investigate the effects of variations in the "stress angle" on cell behavior. In regions most susceptible to disease the "stress angle" varies widely from the perpendicular orientation characteristic of straight arteries, which are typically spared from disease. Thus, there is considerable motivation to investigate these conditions. Morphological adaptations to mechanical factors provide insight to the cell's sensing and response mechanisms. Fiber alignment is critical to the cell's transduction processes and may be key to inhibiting pro-inflammatory behaviors. Since most previous work has focused on isolating mechanical factors, there is a need to address the deficit in understanding the effects of the actual physiologic mechanical environment where cells reside. Therefore, we developed the first device capable of reconstructing the complex mechanical conditions at regions highly susceptible to disease (e.g. carotid sinus), in a manner that is controlled and straightforward to characterize. We anticipate this approach will provide invaluable data, to contribute to the understanding of the role the mechanical environment plays in disease processes and lead to potential therapeutic interventions.

ACKNOWLEDGEMENTS

No funding to be disclosed. The previous work of Robert S. Thompson is appreciated and acknowledged.

REFERENCES

[1] S. L. Murphy, *National Vital Statistics Reports,* vol. 61, no. 4, pp. 1-117, 2013.

[2] U.S. Dept. of Health and Human Services, NIH: HLBI, 2013.

[3] A. S. Go, *Heart Disease and Stroke Statistics-2013 Update,* vol. 127, pp. 6-245, 2013.

[4] S. Hammer, *Ultrasound in Med. & Biol.,* vol. 35, no. 12, pp. 2069-2083, 2009.

[5] J. E. Moore, *Annals of Biomedical Engineering,* vol. 22, pp. 416-422, 1994.

[6] C. A. Davis, S. Zambrano and P. Anumolu, *J Biomech Eng.,* 2014.

[7] Y. Qiu and J. M. Tarbell, J. Vasc. Res., vol. 37, no. 3, pp. 147-157, 2000.

SB³C2015
Summer Biomechanics, Bioengineering and Biotransport Conference
June 17-20, 2015, Snowbird Resort, Utah, USA
SB³C2015-407

ADIPOSE-DERIVED STEM CELLS FROM DIABETIC DONORS CAUSE THROMBOTIC FAILURE OF AUTOLOGOUS TISSUE ENGINEERED BLOOD VESSELS

Jeffrey T. Krawiec (1,5), Han T. Liao (3,6), Justin S. Weinbaum (1,5), Dominic J. Pezzone (1), Antonio D'Amore (1,4,5), J. Peter Rubin (3,5), William R. Wagner (1,4,5), David A. Vorp (1,2,4,5)

(1) Department of Bioengineering
University of Pittsburgh
Pittsburgh, PA, U.S.A.

(2) Department of Cardiothoracic Surgery
University of Pittsburgh
Pittsburgh, PA, U.S.A.

(3) Department of Plastic Surgery
University of Pittsburgh
Pittsburgh, PA, U.S.A.

(4) Department of Surgery
University of Pittsburgh
Pittsburgh, PA, U.S.A.

(5) McGowan Institute for Regenerative Medicine
University of Pittsburgh
Pittsburgh, PA, U.S.A.

(6) Department of Plastic and Reconstructive Surgery
Chang Gung University
Taiwan

INTRODUCTION

Many preclinical evaluations of autologous small diameter tissue-engineered blood vessels (TEBVs) utilize cells from healthy humans or animals [1]. However, these models hold minimal relevance for clinical translation of TEBV therapy, as the main targeted demographic is patients at high cardiovascular risk such as individuals with diabetes mellitus. Adipose-derived mesenchymal stem cells (AD-MSCs) represent an ideal cell type for TEBV therapy, as these can be easily and plentifully harvested from high-risk patients. However, donor demographics can negatively affect the functions of these cells [2]. Therefore, it is critically important to understand how AD-MSCs sourced from diabetic donors are affected in the context of TEBV therapy in vivo. In this study, we implanted TEBVs constructed with AD-MSCs from diabetic humans as an aortic interposition graft in a rat model. Compared to grafts using AD-MSCs from healthy controls TEBVs constructed with AD-MSCs from diabetic patients displayed a marked failure rate due to thrombosis. To investigate mechanistically we performed two follow-up studies investigating critical areas of thrombosis: platelet adhesion and fibrinolytic ability.

METHODS

Human AD-MSCs were obtained from patients who were classified into either healthy (non-diabetic, <45 years of age) or diabetic (diabetic, <45 years of age) cohorts. Human smooth muscle cells (SMC) were purchased from ATCC.

To construct TEBVs, 3 million AD-MSCs were incorporated into rat-sized (1.3 mm ID, 1 cm L) bilayered poly(ester urethane)urea scaffolds[3] using our Rotational Vacuum Seeding Device which distributes cells uniformly [4]. Following seeding, scaffolds were then incubated briefly in a spinner flask for 2 days to allow for cellular adhesion then implanted as an abdominal aortic interposition graft in a Lewis rat model for 8 weeks. Patency of the TEBVs was assessed via angiography. To explain differences seen in patency rate between diabetic and healthy AD-MSCs, we then investigated two mechanisms of thrombosis: platelet adhesion and fibrinolytic ability.

To assess the ability of diabetic or healthy AD-MSCs to bind platelets, monolayers of AD-MSCs were statically incubated with bovine platelet rich plasma anti-coagulated with a 7:1 vol/vol citrate dextrose for 30 minutes. This was followed by PBS washes to remove any unbound platelets. Platelets bound to AD-MSCs were labeled using a standard immunofluorescence protocol staining for CD41 (1:100, Kingfisher #CAPP2A) with counterstains for DAPI and F-actin (1:250, Sigma #P5282). SMCs were used as a positive control.

To assess the fibrinolytic ability of different donor AD-MSCs, conditioned media was obtained by replenishing culture media on near-confluent flasks of cells then collecting after two days in either normal (i.e. with serum) or serum-free conditions. Zymography, utilizing established methods [5] was performed using conditioned media in fibrin-based acrylamide gels (7.5% acrylamide, approximately 400 μg/mL fibrin) to provide a platform for fibrin degradation while also being able to identify active proteins (based on molecular weight) involved in this process. Following protein separation via electrophoresis, gels were incubated in a divalent cation reaction buffer (50 mM Tris HCl pH 7.4, 1 mM $CaCl_2$, 1 mM $MgCl_2$) for 7 days to allow for enzymatic degradation of the gel. Gels were then stained with Coomassie Blue to observe degradation bands.

RESULTS

While TEBVs constructed with healthy patient AD-MSCs generated patent vessels at an 8 week time point, those constructed with diabetic patient AD-MSCs displayed a marked reduced patency rate due to thrombosis: 100% (healthy) vs. 29% (diabetic) **(Figure 1)**.

Most thrombosis occurred acutely (< 1 week) and was not due to poor cell seeding of diabetic AD-MSCs which was verified to be similar to healthy AD-MSCs (data not shown).

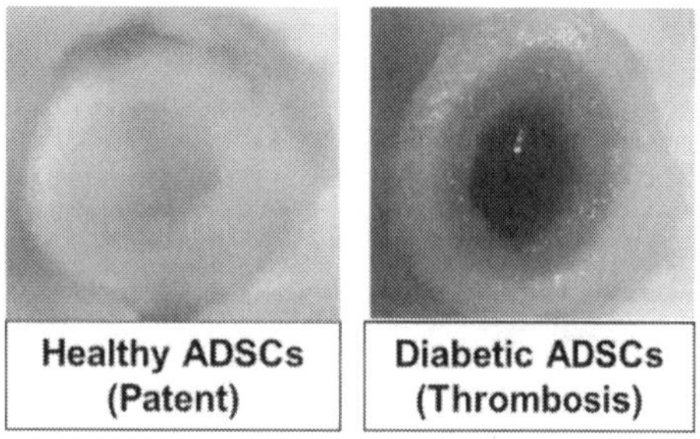

FIGURE 1. Gross morphology of explanted TEBVs created from healthy of diabetic AD-MSCs. Healthy TEBVs remained patent to the endpoint at 8 weeks (100%, n=7) while diabetic TEBVs failed acutely (< 1 week) due to thrombosis (29%, n=7).

AD-MSCs from both healthy and diabetic patient groups showed a minimal amount of bound platelets (**Figure 2A**) with no difference between the groups (**Figure 2B**). In either case the platelet binding was lower than when human SMCs were utilized as a positive control (data not shown).

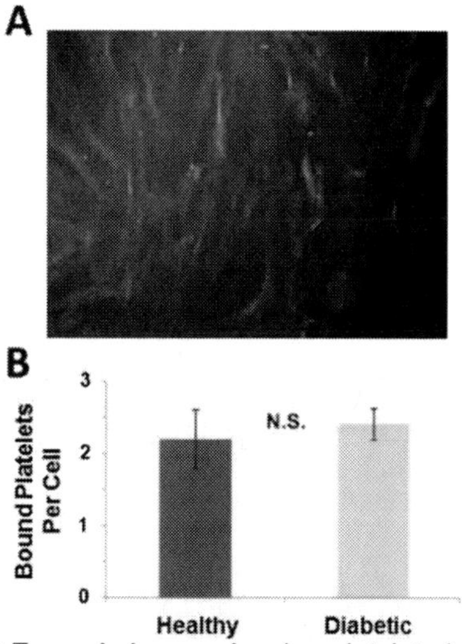

FIGURE 2. (A) Example image showing platelets (red) bound to cell bodies (body: green, nuclei: blue). (B) The number of bound platelets per cell was not statistically significant (N.S.) between healthy and diabetic ADSCs (n=3 each).

Analyzing the conditioned media (i.e. their secreted factors) of either diabetic or healthy AD-MSCs for fibrinolytic activity via zymography revealed fibrin degradation bands at ~31 kD and ~38 kD in media conditioned by AD-MSCs from healthy, but not diabetic, donors (**Figure 3**) indicating a stronger fibrinolytic ability. Diabetic AD-MSCs however do possess some activity with degradation bands present at 83 and 88 kDa (likely plasminogen) which is similar to healthy AD-MSCs (data not shown). To avoid confounding degradation bands from serum products (e.g. plasminogen) serum-free conditions were also performed which showed similar results. For this study, a 7 day time point was determined as optimal with earlier time points (1, 3, and 5 days) showing similar results but with less prominent bands.

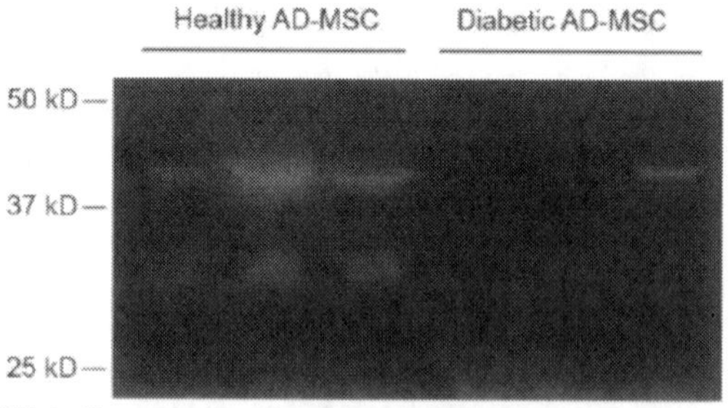

FIGURE 3. Zymogram showing fibrin degradation bands from healthy ADSC conditioned media at 38 and 31 kDa that are not present with diabetic ADSCs (n=3 each).

Currently, studies are underway utilizing fibrin-based cellular constructs to assess fibrinolysis under normal culture conditions and Western blots are being performed to identify the ~31 kD and ~38 kD bands seen during zymography.

We conclude that AD-MSCs from diabetic patients have a pro-thrombogenic phenotype which is due to a reduced fibrinolytic potential, causing failure of autologous small diameter TEBVs.

DISCUSSION

The high frequency of thrombosis seen with diabetic AD-MSC-based TEBVs highlights the importance of testing the intended donor cells – in this case a high cardiovascular risk group – when developing an autologous cell-based therapy. The reduced fibrinolytic activity seen in diabetic AD-MSCs but similar level of platelet adhesion to healthy AD-MSCs indicated that one mechanistic way cellular incorporation within TEBVs increases patency [6] occurs through the ability of those cells to efficiently break down thrombogenic material, such as fibrin. As diabetic AD-MSCs are unable to successfully perform this task, further exploration into mechanisms to enhance or supplement diabetic AD-MSCs will be imperative for the clinical translation of TEBVs.

ACKNOWLEDGEMENTS

This work was supported by the National Institutes of Health (R21 #EB016138 to DAV), the American Heart Association (#12PRE12050163 to JTK), and the Swanson School of Engineering at the University of Pittsburgh.

REFERENCES

[1] Krawiec, J et al. *Biomaterials*, 33.12:3388-3400, 2012.
[2] Krawiec, J et al. *Tissue Engineering Part A*, 2014.
[3] Soletti, L et al. *Acta biomaterialia*, 6.1:110-122, 2010.
[4] Soletti, L et al. *Biomaterials*, 27.28:4863-4870, 2006.
[5] Ahmann, K et al. *Tissue Engineering Part A*, 16.10:3261-3270, 2010.
[6] Nieponice, A et al. *Tissue Engineering Part A*, 16.4:1215-1223, 2010.

FINITE ELEMENT MODELING OF CARDIAC MUSCLE CONTRACTION

Xiaoyan Zhang (1), Kenneth S. Campbell (2), Jonathan F. Wenk (1)

(1) Department of Mechanical Engineering
University of Kentucky
Lexington, KY, USA

(2) Department of Physiology
University of Kentucky
Lexington, KY, USA

INTRODUCTION

Finite element (FE) modeling of cardiac muscle contraction is an important approach used to study the function of the heart at the organ level. Among those widely used FE models is the simple but computationally efficient time-varying "elastance" model [1]. This model, however, fails to include the force-velocity relationship of cardiac muscle contraction [2], and may misrepresent the mechanical behavior of the heart. Therefore, a more representative FE model of active contraction is needed for a better understanding of cardiac mechanics and function.

For this purpose, the present study introduces a novel FE model of cardiac muscle contraction that was developed based on the cellular model MyoSim. MyoSim is an innovative new extension of the Huxley model, which incorporates the length-dependent Ca^{2+} activation and strain rate-dependent interfilamentary effects [3]. This cellular-level model is able to capture the mechanical properties of myocytes under a wide variety of experimental conditions [3]. Incorporation of MyoSim, therefore, may improve the behavior of an organ-level FE model.

To validate the proposed FE model, the active stresses generated in a single element under different length and/or Ca^{2+} manipulations were compared to those computed by MyoSim. As a straightforward demonstration for the use at the organ level, a single cardiac cycle was simulated with a 3D FE model of the left ventricle (LV), and the pressure-volume (PV) relationship of the LV was assessed.

METHODS

A single cubic element was created with the TrueGrid (XYZ Scientific, Inc., Livermore, CA, USA). One end of the element was free to translationally move in one direction while the whole element was free to change the thickness in the other two directions.

The 3D FE model of LV was created as a truncated ellipsoid based on the geometry and size of a rat LV at early diastole. The helical fiber angles in the 3 layers of LV (i.e., epi-, mid-, and endocardium) were -60°, 0°, and 60°, respectively. The boundary conditions were set to completely constrain the displacement at the base. To simulate the passive filling and ejection of the LV, a mass flow into and out of the LV adapted from a previous study was used.

The Guccione's transversely isotropic passive material model was applied to both the single element and the LV model. The active material properties were derived from the 2-state MyoSim model of striated muscle [3]. In the proposed model, the myosin heads switch between detached and attached states, and the lengths of the cross-bridges formed were classified into 21 bins from x_1 to x_{21}. The number of myosin heads attached to the thin filament at time t with cross-bridge of length x_i, $A(x_i, t)$, was calculated by the following set of equations using a 4th-order Runge-Kutta method with explicitly fixed time step size:

$$\frac{\partial A(x_i,t)}{\partial t} = k_1(x_i)D(t) - k_{-1}(x_i)A(x_i,t) \quad (i=1, 2, \ldots, 21) \tag{1}$$

where $k_1(x_i)$ and $k_{-1}(x_i)$ are the strain-dependent rate constants for the attachment and detachment transitions, respectively. The number of myosin heads in the detached states, $D(t)$, was determined by the following equations:

$$D(t) = N(t) - N_{bound} \tag{2}$$

$$\frac{dN(t)}{dt} = a_{on}[Ca^{2+}]\left(N_{overlap} - N(t)\right) - a_{off}(N(t) - N_{bound}) + k_{plus}N_{bound} - k_{minus}(N_{overlap} - N(t)) \tag{3}$$

where a_{on} and a_{off} are rate constants, k_{plus} and k_{minus} are constants defining cooperative effects, N_{bound} is the number of sites that have already been occupied by myosin heads, and $N_{overlap}$ is the maximum number of bindings sites that heads could potentially interact with [4]. The effects of interfilamentary movement was included by using linear interpolation to displace the population distributions $A(x_i, t)$ by ½ of the half-sarcomere length change to $A_S(x_i, t)$. The final active stress generated was calculated as follows:

$$T(t) = \sum_{i=1}^{21} \rho k_{cb} A_s(x_i,t)(x_i + x_{ps}) \tag{4}$$

where ρ is the number of myosin heads in a hypothetical half-sarcomere with a cross-sectional area of 1 m^2, k_{cb} is the stiffness of cross-bridges, and x_{ps} is the length of the cross-bridge when a myosin head is bound to a directly opposed binding site.

Finally, both the passive and active material laws were implemented as a user defined subroutine in the nonlinear FE solver LS-DYNA (Livermore Software Technology Corporation, Livermore, CA, USA).

RESULTS

With a simple stretch manipulation at a constant Ca^{2+} level, the single element FE model showed good agreement with MyoSim at the same time step size (i.e., 0.001 sec) in terms of active stress development (Fig. 1); the maximum difference between the FE model and MyoSim was within 2%. Although addition of a cross-fiber stress component or the use of a smaller time step size exerted less influence on the force generation throughout the whole test, removal of global damping from FE model compromised this agreement at the later stage (Fig. 1). Moreover, increasing the complexity of the length manipulations did not affect the behavior of the proposed FE model as compared to that of MyoSim (Figs. 2 and 3). Notably, when activated by a realistic Ca^{2+} transient, the single element in the FE model generated a similar force transient to that seen in MyoSim (Fig. 4).

For the simulation of a single cardiac cycle, the 3D FE model of the LV generated a PV relationship that resembles a typical PV loop of a rat LV (Fig. 5).

DISCUSSION

The simulations run in the present study aimed to validate the proposed FE model of muscle contraction against a novel cellular-level myocyte model, MyoSim. The agreement in the force generation between the proposed FE model and MyoSim suggests that, like MyoSim, the new FE model is able to predict the mechanical behavior of a single sarcomere under various conditions. Moreover, the proposed FE model showed a potential for the future use in the study on the global function of the LV.

Importantly, due to the incorporation of MyoSim, the new FE contraction model includes not only a length-dependent Ca^{2+} activation but also a force-velocity relationship for the modeling of muscle contraction. As such, the proposed model may provide a more accurate representation of the heart. Moreover, the use of cross-bridge distribution approach provides the proposed model with a capability to simulate the LV function for different species under different conditions. Finally, since all the parameters can be adjusted to fit the cellular-level experimental data, this FE contraction model is directly linked to cellular mechanical properties and could be used to predict the effects of cellular-level changes on global heart function.

In the end, although the proposed FE model used a simple 2-state cross-bridge myosin scheme, it successfully predicted the mechanical behavior of both a single sarcomere and a rat LV in a time-efficient fashion. When used in conjunction with cellular experiments, the proposed model may be able to predict the outcomes of drug therapy for heart diseases that targets cellular function.

ACKNOWLEDGEMENTS

This study was supported by an award from the American Heart Association (14BGIA18850020) and a grant from the National Institutes of Health (R01 HL090749).

REFERENCES

1. Kerckhoffs, R.C., et al., Ann Biomed Eng, 2007. **35**(1): p. 1-18.
2. Guccione, J.M., L.K. Waldman, and A.D. McCulloch, J Biomech Eng, 1993. **115**(1): p. 82-90.
3. Campbell, K.S., J Gen Physiol, 2014. **143**(3): p. 387-99.
4. Campbell, K.S., PLoS Comput Biol, 2009. **5**(11): p. e1000560.

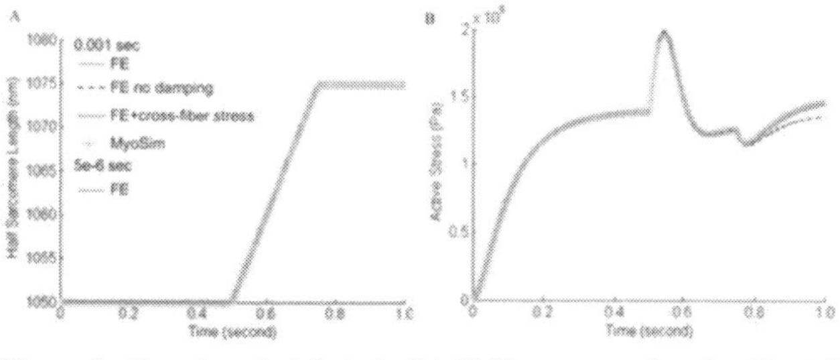

Figure 1. One-step stretch test. (A) Half-sarcomere length; (B) Active stress. pCa=4.5.

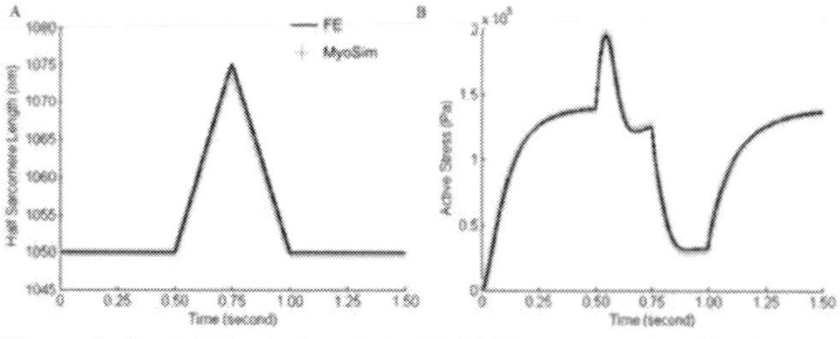

Figure 2. Stretch/shortening test. (A) Half-sarcomere length; (B) Active stress. pCa=4.5.

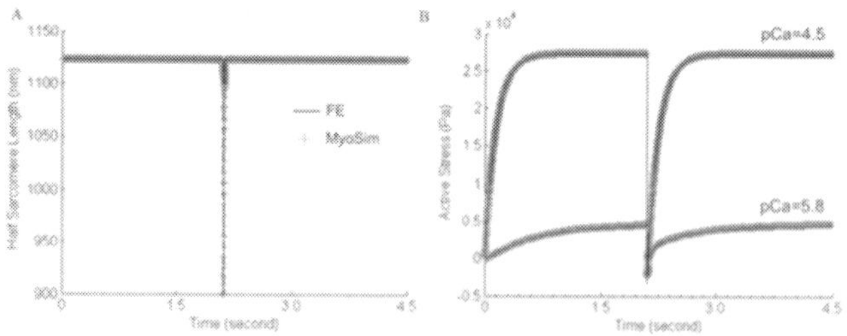

Figure 3. Tension recovery test. (A) Half-sarcomere length; (B) Active stress.

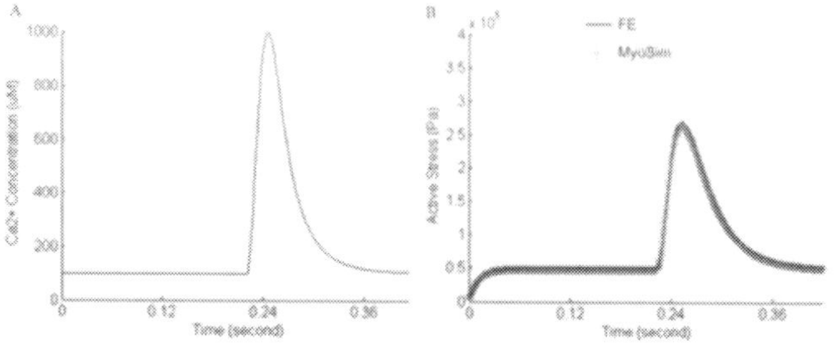

Figure 4. Isometric contraction. (A) Ca^{2+} transient; (B) Active stress.

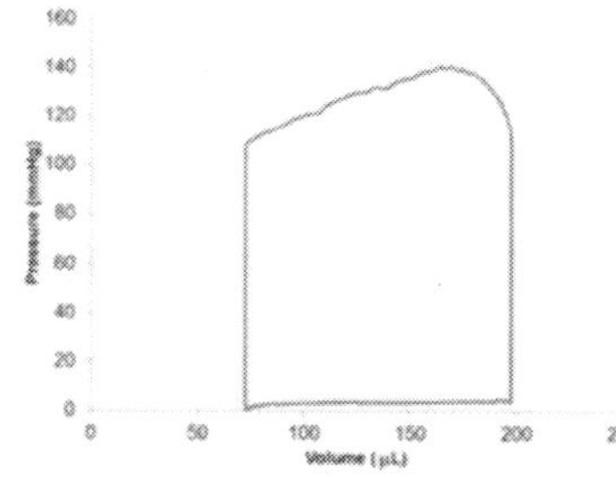

Figure 5. The PV loop of the LV FE model.

SB³C2015
Summer Biomechanics, Bioengineering and Biotransport Conference
June 17-20, 2015,Snowbird Resort, Utah, USA
SB³C2015-409

A ROBOTIC KNEE ORTHOSIS FOR LOCOMOTIVE ASSISTANCE

Saroj Thapa (1), Hao Zheng (1), Geza F. Kogler (2), Xiangrong Shen (1)

(1) Department of Mechanical Engineering
The University of Alabama
Tuscaloosa, AL, USA

(2) School of Applied Physiology
Georgia Institute of Technology
Atlanta, GA, USA

INTRODUCTION

Aging has become a major health challenge in the United States. By 2030, there will be 71 million American older adults, accounting for 20% of the U.S. population [1]. As a result, a large number of individuals suffer from lower-limb weaknesses, either from age-related degeneration of musculoskeletal structure, or pathological losses of neuromuscular functions. With impaired mobility, such individuals are more likely to live in a sedentary lifestyle, and suffer from the numerous problems associated with the sedentary lifestyle, e.g. high blood pressure, depression, obesity, etc.

To restore the mobility of these individuals, the most commonly used tool is wheelchair. Although wheelchairs are effective in transporting patients, they largely preclude the users' lower limb muscle activities and bone load-carrying. This accelerates the musculoskeletal degeneration of the user's leg. Compared with wheelchair, power-assist orthosis represents a fundamentally different, and potentially more effective approach in restoring such individuals' mobility. As a wearable device, a power-assist orthosis works in parallel with the user's limb and provides assistance on the joint level. By complementing the muscle force, a power-assist orthosis can help the user regain mobility in a natural and unobtrusive way.

Currently, the majority of powered orthotic devices are stationary or tethered devices for rehabilitative training purposes (e.g., Locomat). Self-powered mobile orthotic devices have been developed as well, including HAL-5 [2], Ekso exoskeleton, Rewalk (Argo Medical Technologies), etc. Note that these devices are usually full-body or complete lower-limb devices, and thus are difficult for users to don and doff by themselves. Unlike these existing devices, the orthosis presented in this paper is a powered knee orthosis, with the primary function of helping a user to stand up from a chair more easily and assisting the user's subsequent walking. In the literature, Pratt et al. developed a powered knee exoskeleton [3]. However, with the actuator placed behind the user, it does not allow the user to sit when fitted with the exoskeleton, and thus cannot serve the desired purpose of the power-assist orthosis in this paper.

METHODS

The robotic power-assist orthosis in this work consists of two major components, including the orthosis frame-shell assembly and an actuation unit. The former functions as the user interface that transfers the assistive torque to the human body, while the latter functions as the power source that generates the desired assistive torque.

For the design of the frame-shell assembly, there are two major requirements: 1) The frame and shells need to be able to endure significant load without excessive deformation. 2) A user should be able to don/doff the orthosis with little or no assistance from others. To meet these requirements, a unique carbon-fiber anterior-shell design is created. The use of carbon-fiber material provides the desired rigidity to resist deformation. Custom thigh and tibial shells cover the anterior of the user's leg, such that a user can don/doff the orthosis at a seated position (i.e., sliding the orthosis towards/away from the leg at the anterior). To firmly attach the orthosis to the limb, four posterior Velcro® straps coupled with chafe loops are used. To form the frame-shell assembly, two stainless steel orthotic joints (Model 1002, Becker Orthopedic, Troy, MI) are located on medial and lateral sides with aluminum uprights structurally linking the thigh and tibial shells. The shells are lined with polyethylene low-durometer foam, which can be readily added for custom-fitting the shells to accommodate a broader range of limb sizes. To resist distal migration of the orthosis, a foot piece and strut can be added to the lateral aspect of the tibial section if needed.

Due to the complex contour of human lower limb, it is difficult to mount a rotary actuator on the uneven surface. Besides, the shaft alignment would be a challenging issue. As such, a linear actuator is used as the actuation unit of the power-assist orthosis, such that it can be attached to the thigh and tibial shells with simple attachment points.

The linear actuator is formed with a light-weight DC motor (EC-4pole 30, Maxon Motors, Switzerland) coupled with a ball screw assembly (MRT 10×2 RS, Nook Industries, Cleveland, OH). The 200W DC motor is able to generate a maximum torque of 300 mNm (for repeated short-term operation). Combined with the 2 mm lead and 90% efficiency of the ball screw assembly, the actuation unit is able to generate a linear force of 848 N in both directions. To calculate the torque capacity, the actuation mechanism can be represented by a simplified crank-slider mechanism (Fig. 1a). This simplified configuration is essentially a triangle, in which a and b are the two segments with constant lengths, ϕ is the angle between these segments, and x is the length of the variable (i.e., actuated) segment. As such, the torque τ can be expressed as a function of the actuation force F and the angle ϕ:

$$\tau = F\left(\frac{dx}{d\phi}\right) = F\left(\frac{ab\sin\phi}{\sqrt{a^2+b^2-2ab\cos\phi}}\right) \quad (1)$$

As shown in (1), under a certain actuation force F, the torque capacity can be adjusted by varying the design parameters, such as a and b. To determine the design parameters, the desired actuation torque is determined according to the user's locomotive needs.

In a typical user's activities of daily life, the sit-to-stand (STS) transition is the first barrier to his/her free motion. A frail elderly person usually needs significant physical assistance in completing this motion. As such, assisting the user's STS transition is highly important for the functionality of the power-assist orthosis. After the STS transition is completed, the orthosis is expected to continue providing motion assistance in the user's ambulation until the user returns to the seated position. With this desired functionality, the primary objective for the power-assist orthosis is to provide sufficient assistance for the STS transition and ambulation, while maintaining a low weight and small volumetric profile for practical daily use. Considering the balance of torque capacity and orthosis weight/volume, the specific goal here is to provide 20% of joint torque required for the STS transition and ambulation for a 75 kg user, with a peak torque requirement of 33 N-m. To meet this torque requirement, the design parameters are chosen as a = 267mm, b = 43mm, and ϕ = 158° when the knee is straight (θ=0). Correspondingly, the orthosis is able to provide an assistive torque of 12~36 N-m, with the specific value determined by the joint angle.

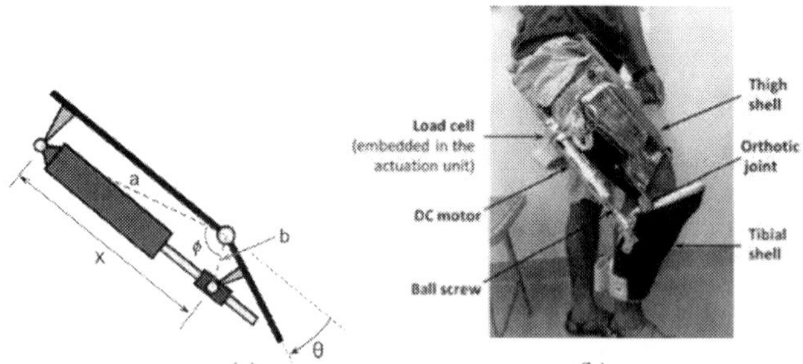

(a) (b)

Fig.1. Actuation mechanism of the powered joint (a) and a preliminary prototype of the power-assist orthosis (b).

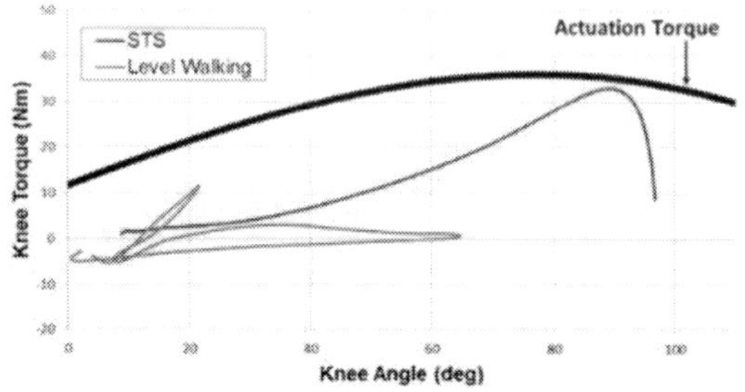

Fig. 2. Comparison of the required torque versus the torque capacity of the powered orthosis.

RESULTS

With the design above, a prototype has been fabricated (Fig. 1b). As mentioned in the previous section, a primary goal for this orthosis design is to provide 20% of the required knee torque for the STS transition and ambulation. Note that the desired assistive torque and the actuation torque capacity are both functions of the joint angle. As such, a graphical comparison on the joint angle-torque plane is conducted (Fig. 2). In this figure, the desired torque trajectories are plotted according to the related biomechanical data [4,5], with the torque magnitudes scaled down 80% to reflect the design goal. Also shown in Fig. 2 is the curve representing the actuation torque. Only extension torque is shown in this figure, since a significantly higher torque is required for the extension of the knee compared with the flexion. From the comparison of these curves, it can be clearly observed that the powered joint can provide the desired assistive torque, demonstrating the effectiveness of the orthosis design.

DISCUSSION

Presented in this paper is a new robotic knee orthosis to assist the user in his or her sit-to-stand transition and ambulation. In this orthosis design, a unique frame-shell assembly is created to provide sufficient rigidity under load and enable the user to don/doff the orthosis with little or no assistance. A motor-ball screw actuation unit is used to provide the desired joint assistance. Through kinetic calculation, the orthosis design is able to provide the desired assistive torque (20% joint torque in STS and walking for a 75 kg user), which has been demonstrated by a graphical comparison on the joint angle-torque plane.

ACKNOWLEDGEMENTS

This work is supported by the National Science Foundation under Grant #IIS-1231676.

REFERENCES

[1] Center for Disease Control and Prevention and the Merck Company Foundation, The State of Aging and Health in America 2007. The Merck Company Foundation, 2007.

[2] Tsukahara, A., Kawanishi, R., Hasegawa, Y., and Sankai, Y., 2010, "Sit-to-stand and stand-to-sit transfer support for complete paraplegic patients with robot suit HAL." Advance Robotics, vol. 24, no. 11, pp. 1615-1638.

[3] Pratt, J.E., Krupp, B.T., Morse, Morse, C.J., and Collins, S.H., 2004, "The RoboKnee: An exoskeleton for enhancing strength and endurance during walking." in Proceedings of the 2004 IEEE International Conference on Robotics and Automation, pp. 2430-2435.

[4] Schenkman, M., Berger, R., Riley, P. O., Mann, R. W., and Hodge, W. A., "Whole-body movements during rising to standing from sitting." Physical Therapy, vol. 70, no. 10, pp. 638-648.

[5] Winter, D.A., The Biomechanics and Motor Control of Human Gait: Normal, Elderly and Pathological. 2nd edition, Waterloo, ON, University of Waterloo Press, 1991.

SB³C2015
Summer Biomechanics, Bioengineering and Biotransport Conference
June 17-20, 2015,Snowbird Resort, Utah, USA
SB³C2015-410

REDUCED LYMPHATIC FUNCTION CORRELATES WITH DISEASE PROGRESSION IN A NOVEL SINGLE VESSEL LIGATION MODEL OF LYMPHEDEMA

M.J. Weiler (1,3), T.S. Nelson (2,3), J.B. Dixon (1,2,3)

(1) Wallace H. Coulter Department of Biomedical Engineering
Georgia Institute of Technology
Atlanta, GA, USA

(2) George W. Woodruff School of Mechanical Engineering
Georgia Institute of Technology
Atlanta, GA, USA

(3) Parker H. Petit Institute of Bioengineering and Bioscience
Georgia Institute of Technology
Atlanta, GA, USA

INTRODUCTION

It is estimated that more than 10 million individuals worldwide suffer from lymphedema, a chronic disease that presents with the accumulation of fluid, proteins, and adipocytes in the interstitium, resulting in a drastic enlargement of the affected limb[1]. The most common form is secondary (acquired) lymphedema, which results from the disruption of previously normal lymphatic networks, usually through injury or surgical procedure[1]. Despite the prevalence of the disease, there is very little known about the exact mechanisms of disease progression, especially as it relates to lymphatic transport function and the progressive loss thereof. Many functional lymphatics remain intact after the initial insult, yet it remains unclear how these intact vessels respond to the mechanical loading changes during the post-trauma period and how this contributes to the onset of lymphedema, which often doesn't even occur until up to 5 years after the initial trauma.

One of the biggest remaining questions with regard to lymphedema onset is how lymphatic function, especially collecting vessel transport function, changes through the disease cascade. Since lymphatic flow is driven primarily through the contractility of collecting lymphatic vessels, the ability to quantify lymphatic pump function of the remaining intact lymphatic vasculature in the context of lymphedema onset would greatly enhance therapeutic intervention strategies. We hypothesize that lymphedema results from a progressive loss of lymphatic function of the remaining intact collecting vessels, which can be visualized and subsequently quantified with NIR imaging, thus affording novel insights into the disease cascade. Therefore, we used NIR lymphatic imaging to assess changes in lymphatic transport metrics as a function of lymphedema progression. We performed this study in a newly developed rodent tail model of lymphedema, which for the first time allowed selective preservation of individual lymphatic collectors in order to track changes in vessel transport over time.

METHODS

We have modified the conventional lymphedema tail model such that a single collecting vessel will remain intact, which can then be imaged with NIR to observe changes in pump function (Figure 1). Specifically, a circumferential incision was made through the dermis to sever the dermal lymphatic vessels approximately 1 cm from the base of the animal. The incision was such that either the dominant, non-dominant, or both collecting vessels (sham) remained intact, thus

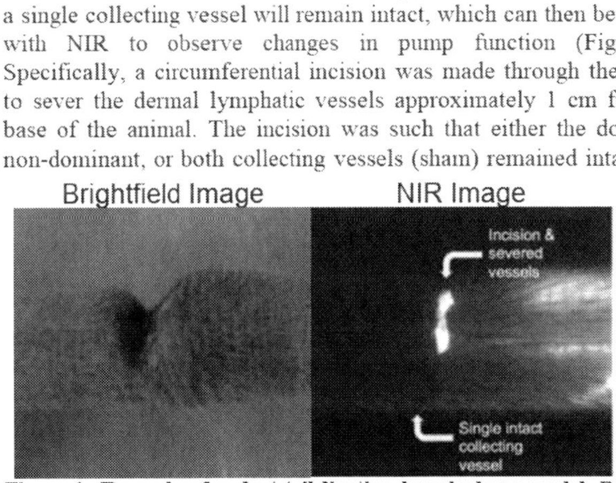

Figure 1: Example of rodent tail ligation lymphedema model. Brightfield and NIR image of tail ligation preserving single collecting lymphatic vessel.

creating three experimental groups[2]. The edges of the incision were cauterized to disrupt the relevant deeper lymphatic network and to create a final wound gap of 2–3 mm to inhibit rapid wound closure and vessel regrowth. Care was taken to maintain the integrity of the major underlying blood vessels and tendons so that the tail distal to the incision did not become necrotic.

NIR lymphatic function metrics were performed to assess the lymphatic transport capacity of the intact collectors according to previously published methods including packet transport and lymphatic pumping pressure[2]. NIR measurements were taken in one-week increments following surgery and spanning the entire duration of the disease cascade. Tail circumference measurements were recorded every three days to track changes in swelling over time. Lymph nodes were also harvested at the endpoint of the study to track nodal remodeling. Statistical significance was determined using a student's t-test with Bonferroni multiple comparison correction.

RESULTS & DISCUSSION

Swelling Cascade of Single Vessel Lymphedema Model

Tail swelling significantly increased as a result of collecting lymphatic vessel ligation compared to the sham group, and ligating the dominant vessel produced significantly more swelling than ligation of the non-dominant vessel (Fig. 2). Dominant ligation swelling peaked at 45% on day 4, while non-dominant ligation swelling reached a maximum of 28% swelling on day 14. Both ligation groups returned to normal circumference values on or soon after day 28. The sham control group experienced a mild swelling response with peak swelling of 16% on day 4 and full resolution by day 14. The timeline for tail swelling and eventual recovery follows similar trends reported in the literature for full tail ligation models, but the magnitude of swelling in our model is less than half that reported for the full ligation model. Given that we keep one collector fully intact, the magnitude difference in swelling in our model is not surprising and actually serves to test the sensitivity of NIR functional metrics in more mild cases of the disease.

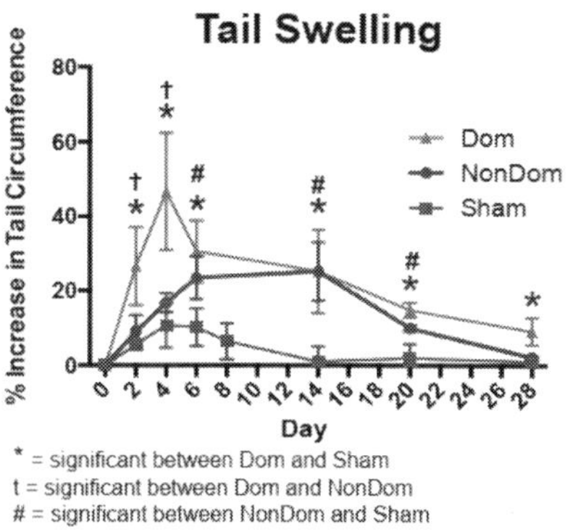

Figure 2: Tail Swelling Cascade. Plot showing swelling cascade of dominant ligation, non-dominant ligation, and sham control groups through 28 days.

Lymphatic Transport Correlates with Disease Progression

Normalized lymphatic fluorescence transport was significantly reduced in both the dominant and non-dominant ligation cases as compared to the sham control (Fig. 3). Lymphatic transport decreased to a minimum value at the 1-week time point in the dominant (8% of

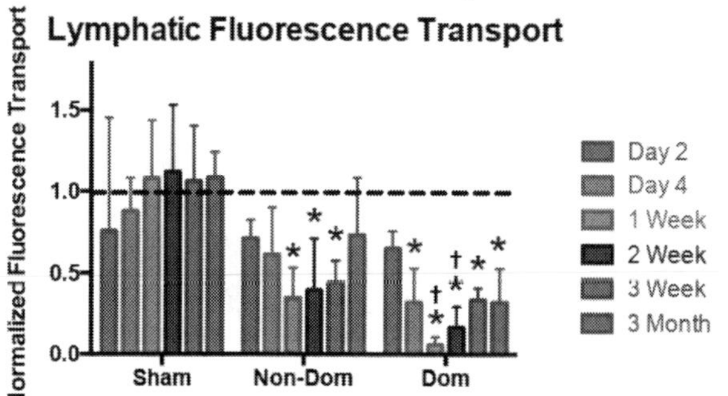

Figure 3: Lymphatic Fluorescence Transport. Plot of lymphatic fluorescence transport through the 3-month time period for all three conditions. *=significant compared to sham. †=significant compared to non-dom.

healthy transport) and non-dominant cases (35% of healthy transport) before beginning to improve at later time points. The dominant ligation resulted in significantly lower transport than the non-dominant ligation at the 1 and 2 week time points. At the three-month time point transport in the dominant ligation was still significantly reduced, while transport in the non-dominant ligation was not significantly reduced from baseline healthy values. In both ligation conditions swelling significantly correlated with reductions in lymphatic function ($p<0.001$) and no correlation existed in the sham control. These are the first results to suggest a link between swelling and reduced lymphatic function in lymphedema, and the difference in outcomes between the two ligation procedures suggests surgical planning may be an option to minimize resulting lymphedema symptoms. Furthermore, the continued transport reduction in the dominant ligation condition at the three month time point, despite the absence of swelling, suggests the intact vessel remodeled as a result of the injury and supports the theory of latent insufficiency of lymphatic collectors after injury[3].

Lymphatic Vasculature Remodels After Lymphedema

At the three-month time point swelling in all conditions completely subsided, but several factors suggest that the dominant ligation resulted in substantial remodeling in lymphatic function compared to the sham control. Percent positive fluorescent tail area, a measure of the relative distribution of lymphatic transport between the collecting vessels and the initial capillary network, was significantly increased in the dominant ligation, and effective lymphatic pumping pressure, a measure of the pumping capacity of the collecting vessels, was significantly reduced in the dominant ligation (Fig. 4). Taken together, these results suggest the lymphatic network remodeled in the dominant ligation such that the collecting vessels have a reduced transport capacity and the initial capillary network is recruited to compensate. Additionally, in the sham, the dominant lymph node was significantly larger and retained significantly more fluorescence than the non-dominant node, but this result was reversed in the dominant ligation. This result indicates that lymph transport routes were altered after the dominant ligation to favor the non-dominant node, which could have implications for immune and inflammatory responses.

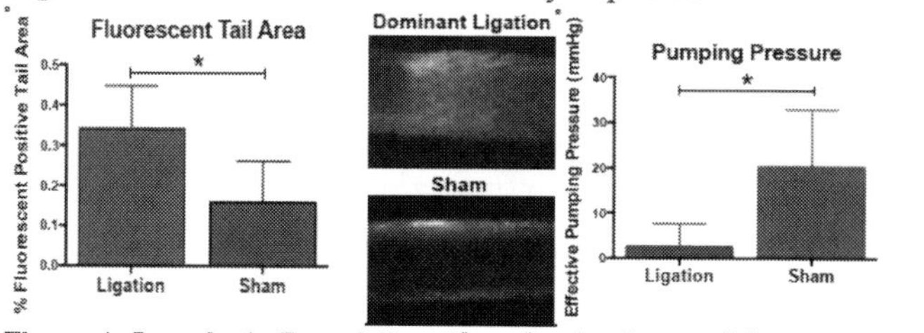

Figure 4: Lymphatic Remodeling. Plots showing increased fluorescent tail area and reduced lymphatic pumping pressure in dominant ligation condition and example images of fluorescence transport in sham and ligation conditions.

ACKNOWLEDGEMENTS

This work was supported in part by NIH R01 HL113061 and an NSF graduate research fellowship (Weiler). Dr. Dixon, Mr. Nelson, and Mr. Weiler receive research funding and they and Georgia Tech are entitled to sales royalty from LymphaTech Inc., which is developing products related to the research described in this paper. In addition, Mr. Weiler serves as the President of LymphaTech, and he and Dr. Dixon own equity in the company. This study could affect their personal financial status. The terms of this arrangement have been reviewed and approved by Georgia Tech in accordance with its conflict of interest policies.

REFERENCES

[1] Rockson, S.G. Lymphedema. *Am J Med* 110:288–295, 2001.

[2] Weiler, M. Differential transport function of lymphatic vessels in the rat tail model.... *Frontiers in Physiology* 4:1-10, 2013.

[3] Mendez, U. A chronic and latent lymphatic insufficiency follows recovery from acute... *Am J Physiol-Heart C* **303:**H1107–H1113,2012.

SB^{3}C2015
Summer Biomechanics, Bioengineering and Biotransport Conference
June 17-20, 2015,Snowbird Resort, Utah, USA
SB^{3}C2015-411

SPHERICAL SHELL MECHANICAL MODEL OF THE PACINIAN CORPUSCLE

Julia C. Quindlen (1), Henryk K. Stolarski (2), Victor H. Barocas (1)

(1) Department of Biomedical Engineering
University of Minnesota
Minneapolis, MN, USA

(2) Department of Civil Engineering
University of Minnesota
Minneapolis, MN, USA

INTRODUCTION

The Pacinian corpuscle (PC) is a cutaneous mechanoreceptor located deep in the dermis [1]. It responds to high-frequency (20-1000 Hz), low amplitude (<1 μm) vibrations on the skin surface. The PC is an ellipsoid with a single nerve fiber (neurite) along its long axis [2]. The neurite is surrounded by concentrically-aligned collagenous lamellae separated by pressurized fluid (Fig.1) [3,4]. When a stimulus is applied to the outer surface of the PC, the deformation is passed through the lamellae, distorts the neurite, opens stretch-gated ion channels, and initiates a neural response [5]. The exact mechanisms of the PC's mechano-to-neural transduction remain relatively unknown.

Loewenstein and Skalak (LS, [3]) first proposed that the interaction between fluid and lamellae causes the PC to act as a high-pass filter to vibrations. Compression of the PC is thought to cause viscous forces to develop due to fluid flow in the space between adjacent lamellae [3]. Thus, when developing a mechanical model of the PC, it is crucial to include the fluid-lamellar interactions.

LS created a mechanical model of the PC as a series of cylinders with a rigid core at the center [3]. The cylinders were joined by elastic connections with liquid in the space between adjacent layers. Biswas *et al.* (BMS, [6]) extended the LS model to incorporate details such as a viscous core and larger interlamellar spacing. Like LS, BMS used concentric cylinders to approximate the structure of the PC, but they adjusted the radius and number of lamellae to model the effect of PC size, and BMS used more realistic physical parameters.

The purpose of this study was to construct a spherical mechanical model of the PC as an alternative approximation of its ellipsoidal shape. Shell and lubrication theories were used to reduce the problem dimension. The mechanical model created in this study can be coupled to a neural model.

METHODS

The PC was modeled as an eighth of a sphere due to symmetry. Shell theory was coupled to lubrication theory to account for the fluid-lamellae interaction. The mechanics of the lamellae were modeled with the equations of equilibrium for shells of revolution derived from shell theory [7]. The behavior of the fluid was modeled by Eq.1 below, which describes fluid flow through a thin layer on a spherical surface. The solid and fluid layers were coupled through Eq.1, specifically the fluid velocity and pressure,

$$\sin\varphi\frac{\partial H}{\partial t}-\frac{1}{r^2}\cdot\frac{H^3}{12\mu}\frac{\partial}{\partial\varphi}\left(\sin\varphi\frac{\partial p}{\partial\varphi}\right)=0 \quad (1)$$

where φ is the angle measured from the vertical, H is the thickness of the fluid layer, r is the radial position of the fluid layer centerline, μ is the fluid viscosity, and p is the fluid pressure.

The PC was modeled as a series of 30 spherical shells with a lubricating fluid layer between each shell and an incompressible core located at the center (Fig. 1). Two sets of parameters were used based on previous LS and BMS models (Table 1) [3,6]. The simulations using BMS parameters had lamellar thicknesses ranging from 0.1-0.4 μm with thickness increasing with distance from the core.

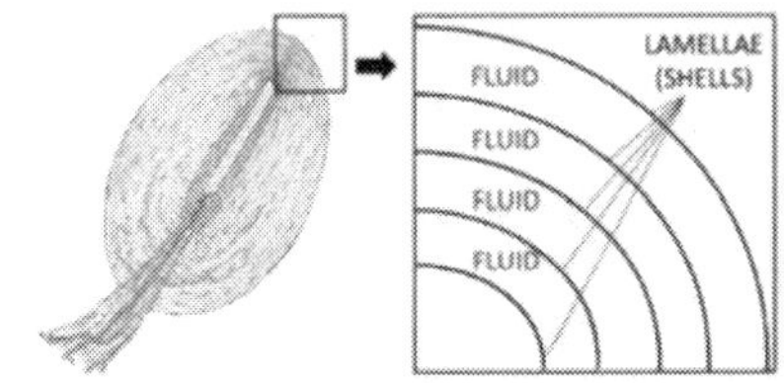

FIGURE 1: SKETCH OF A PC [4] AND REPRESENTATION OF FLUID AND LAMELLAE IN MODEL

TABLE 1: SHELL AND FLUID PARAMETERS [3,6]

	LS	BMS
Elastic Modulus (Pa)	1e3	5e5
Lamellae Thickness (μm)	1	0.1-0.4
Viscosity (mPa-s)	1	1.4, 3.5, 7

Pressure was applied to the outer shell (φ=0 to $\pi/2$) of the spherical model (Eq. 2) at various frequencies spanning from 1 Hz-10^8 Hz. Simulations were performed in MATLAB.

$$p = \frac{1}{3} + \cos(2\varphi) \tag{2}$$

RESULTS

Because the problem is linear, all results are reported as ratios and relative units. The ratio of the inner to outer shell maximum displacements at each stimulation frequency was determined (Fig. 2A). Simulations run using BMS parameters resulted in peaks in the frequency range of PC sensitivity [8], with the peak shifting to lower frequencies with increasing fluid viscosity. The sine of the phase difference between the inner and outer lamellae was calculated for various frequencies using BMS parameters at 1.4 mPa-s viscosity (Fig. 2B). At lower frequencies, the inner and outer layers were out of phase, while at higher frequencies (>10^4 Hz), they were in phase.

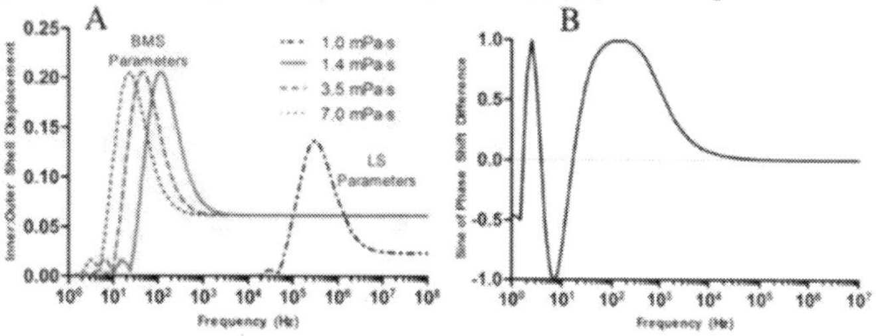

FIGURE 2: (A) RATIO OF INNER TO OUTER MAXIMUM DISPLACEMENTS VS. FREQUENCY (B) SINE OF PHASE SHIFT OF INNER AND OUTER LAMELLAE VS. FREQUENCY

The displacement of the inner and outer layers for all values of φ was calculated using BMS parameters and 1.4 mPa-s viscosity at 100 Hz stimulation (Fig. 3). These results indicate that compressions to the PC outer shell are transmitted to the neurite at smaller amplitudes.

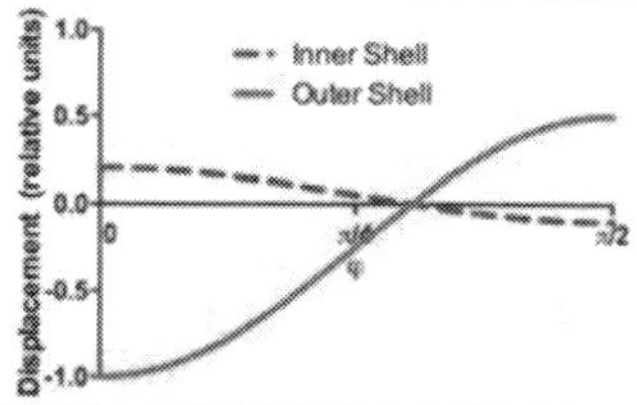

FIGURE 3: DISPLACEMENT OF INNER AND OUTER SHELLS VS. ANGLE ϕ ALONG SURFACE OF THE SHELLS

The undeformed and deformed geometries of five shells (Shells 1, 8, 16, 23, 30 where distance from the core increases with increasing shell number) were plotted for 7 mPa-s viscosity simulations (Fig. 4A). PC deformation had a larger affect on shape for layers closer to the core. The radial stretch was measured for these five layers to observe the strain differences in each shell (Fig. 4B). Deformation to the PC surface resulted in larger magnitudes of stretch in shells closer to its center. This suggests that while the structure of the PC decreases the amplitude of the displacement of layers closer to the core, it still allows the neurite to experience large stretches, thus making the fiber more likely to initiate a response from stretch-gated ion channels.

DISCUSSION

A mechanical model of the PC that couples shell theory and lubrication theory to model the fluid-lamellae interaction of the receptor at a specific stimulation frequency was created. Parameters from the literature were used to compare the results of this model to those of previous studies [3,6]. Because the PC's primary function is its role as a high-frequency vibration sensor, it was first necessary to characterize the frequency response of the model. The ratio of the maximum displacements of the inner and outer lamellae was analyzed at various frequencies to observe how displacements were transmitted to the neurite. Simulations run using BMS parameters resulted in peaks of this ratio at frequencies characteristic of PC frequency sensitivity (20-1000 Hz), while those using LS parameters peaked at much higher frequencies (Fig. 2A). This discrepancy is most likely the result of the lower lamellar thickness and higher fluid viscosity.

The results presented in Fig. 3-4 show how mechanical stimuli are transmitted through the PC layers. In Fig. 3, indentation amplitude is decreased as the stimulus is transmitted to the core of the PC, reducing the displacement experienced by the neurite. In Fig. 4, the strain experienced by shells increases for shells closer to the core. Thus, while shell displacement decreases with proximity to the core, stretch increases. The results suggest that the presence of lamellae is responsible for amplifying the strain transmitted through the PC and determining the mechanical forces experienced by the neurite. While they provide additional insights into the response of the PC, these results are in agreement with current opinions that the PC's structure plays an important role in the function of the receptor.

A strength of this study is that it reduces the dimensionality of the problem while still accounting for the coupled fluid-lamellae interactions. The model presented in this study therefore simplifies the mathematics and computations without sacrificing complexity. A limitation of this study is that the ellipsoidal geometry of the PC was approximated as a sphere. While a spherical approximation is an improvement over previous cylindrical models [3,6], it still does not account for the PC's characteristic structure and cannot address the variations in PC shape.

This study suggests that it is not only the presence of fluid and shells but more importantly the interaction between adjacent layers that allows for stretch amplification in layers closer to the core and the subsequent activation of stretch-gated ion channels on the neurite.

ACKNOWLEDGEMENTS

This research was supported by an NSF IGERT fellowship (Systems Neuroengineering, DGE-1069104).

REFERENCES

[1] Abraira, VE et al., *Neuron*, 79:618-639. 2013. [2] Pease, D et al. *J Biophys Biochem Cytol*, 3:331-342. 1957. [3] Loewenstein, W & Skalak R, *J Physiol*, 182:346-378. 1966. [4] Gray, H. Anatomy of the Human Body. 20th ed. 1918. [5] Gray, J & Sato, M, *J Physiol*, 122:594-607. 1953. [6] Biswas, A et al., *IEEE Trans Haptics*, 1412: 1-12. 2014. [7] Gould, P. Analysis of Shells and Plates. Springer-Verlag. 1988. [8] Talbot, W et al. *J Neurophys*, 31:301-334. 1968.

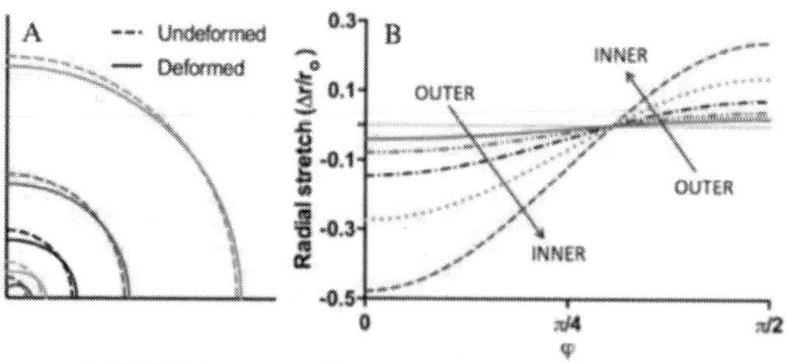

FIGURE 4: (A) UNDEFORMED AND DEFORMED GEOMETRIES OF FIVE SHELLS (B) RADIAL STRETCH FOR EACH SHELL VS. ANGLE ϕ

SB^{3}C2015
Summer Biomechanics, Bioengineering and Biotransport Conference
June 17-20, 2015,Snowbird Resort, Utah, USA
SB^{3}C2015-412

CELLULAR CHOLESTEROL CONTENT MODULATES MONOCYTE INTERACTION WITH E-SELECTIN

Amit K. Saha (1), Anand K. Ramasubramanian (2)

(1) Department of Biomedical Engineering
The University of Texas at San Antonio
San Antonio, Texas, USA

(2) Department of Biomedical Engineering
The University of Texas at San Antonio
San Antonio, Texas, USA

INTRODUCTION

The plasma membrane is segregated into lipid rafts, which are dynamic, heterogeneous regions, rich in gangliosides, cholesterol and glycosphingolipids. Lipid rafts regulate a variety of cellular processes such as protein trafficking and signal transduction [1]. Cholesterol is a major constituent of the lipid rafts and is vital in maintaining their integrity[2]. The effect of cholesterol content has been studied in a variety of cell types such as neutrophils[3], fibroblasts[4] and erythrocytes[5], among others. The impact of cholesterol on monocyte biomechanics remains understudied despite the well-established link between both cholesterol and monocytes/macrophages in atherosclerosis and other diseases. We modified the cellular cholesterol content and tested for differences in monocyte biophysical response. We altered the baseline cholesterol and studied the differences in the rolling and adhesion properties of monocytes on e-selectin, and observed significant changes in the instantaneous and short term response. As a causative agent for the observed changes, we investigated the changes in membrane lipid rift distribution and expression levels, and also changes in the membrane fluidity due to changes in the cholesterol content. We also studied the consequential distribution and expression of the receptors to e-selectin present on monocytes. Apart from this, we examined changes in cellular deformability as a possible contributing factor to the changes in the rolling and adhesion dynamics.

METHODS

Cell Culture. Primary human monocytes were isolated from buffy coat using monocyte isolation kit II by indirect magnetic labeling, and the purity of the monocytes was determined by flow cytometry using FITC-conjugated anti-CD14 antibody and a suitable isotype control. The monocytes were resuspended in RPMI 1640 with 10% FBS at required concentration depending on the assay.

Cholesterol treatment. Cholesterol depletion was performed using 10 mM Methyl-β-cyclodextrin (MβCD). For cholesterol enrichment, 0.5 mM MβCD –cholesterol complex was used with cholesterol concentration of 40mg/gm. In both cases, the cells were incubated for 30 mins. at 37° C and 5% CO_2. Cholesterol content was measured using Amplex Red Cholesterol Assay Kit, as per the manufacturer's protocol.

Fabrication of custom microfluidic platform. A silicon mold was made using standard photolithographic techniques. These silicon molds wre used as a master to make polymeric micro-channels in polydimethylsiloxane (PDMS), and final devices were assembled by bonding with clean glass slides using oxygen plasma treatment.

Lipid raft and receptor distribution and expression levels. Lipid rafts and e-selectin receptors were studied for their expression levels and distribution patterns using multispectral imaging flow cytometry.

Membrane fluidity. Membrane fluidity was assessed using pyrenedecanoic acid fluorescence intensity measurements.

Perfusion studies. For rolling and adhesion analysis, the micro-channel slide was coated with e-selectin assembled on top of an inverted microscope attached to a time-lapse digital camera A monocyte suspension was then perfused through the micro-channels using a syringe pump at a constant flow rate corresponding to a wall shear stress of 1 dyn/cm^2. Videos were captured at 20X magnification, followed by subsequent image analysis. For deformability studies, the microfluidic device was assembled on top of an inverted microscope attached to a high speed camera. We used a custom nitrogen based cell delivery system to deliver the cell suspension at a pressure of 2 psi. The videos were taken at a magnification of 20X using the proprietary software for the camera, followed by subsequent image analysis.

Statistics. All experiments were performed in triplicate, and each experiment was repeated at least thrice under independent conditions. The results are represented as mean ± SEM from one representative experiment. Statistical differences between treatments were evaluated using one way ANOVA with Tukey's post-hoc test, and significance was reported at α=0.05. * and ϕ represent significance as compared to cholesterol depleted and untreated monocytes. Consistent for all experiments (unless otherwise mentioned).

RESULTS

Changes in rolling and adhesion behavior due to cholesterol content. Alteration of cellular cholesterol significantly impacted the rolling and adhesion behavior of monocytes. Cholesterol depletion led to more cells rolling on the e-selectin coated surface (Fig. 1 a). However, only a small fraction was able to convert the transient rolling into sustained adhesion, as compared to cholesterol enrichment and untreated controls (Fig. 1 b). It also led to longer pause time and shorter rolling duration in cells. Cholesterol depletion led to the cells rolling in a faster (Fig. 1 c) and more erratic fashion (Fig. 1 d). Enrichment led to antithetical effects.

Changes in membrane properties. In order to have a better understanding of the changes in the rolling behavior, we investigated changes in lipid raft expression and distribution as a consequence of cholesterol treatment. We observed that cholesterol depletion results in decreased expression and a more uniform distribution. We probed into the changes in membrane fluidity due to cholesterol treatment. Cholesterol depletion significantly enhanced the plasma membrane fluidity (Fig. 2). We also observed significant changes in the expression levels of receptors of e-selectin on monocytes (CD44 and CD15) due to cholesterol treatment.

Changes in cellular deformability. We probed the changes in cellular deformability as a function of cholesterol load and also to understand possible links with the changes in rolling and adhesion dynamics. Cholesterol depletion was observed to decrease the cellular deformability (Fig. 3).

DISCUSSION

This work establishes a link between cellular cholesterol load and biophysical response in monocytes. We observed significant changes in rolling and adhesion behavior of monocytes due to altered cholesterol loads. And as probable reasons for the observed changes in rolling and adhesion dynamics, we investigated the changes in membrane properties and cellular deformability.

Since cholesterol is responsible in maintaining the integrity of the plasma membrane, disturbance in its natural balance alters the membrane fluidity and the lipid raft distribution and expression. Consequently, the receptors on the membrane also show altered expression levels and distribution. These can greatly influence the rolling and adhesion behavior of monocytes.

There is evidence linking cellular deformability to rolling, with stiffer cells having higher velocities[6]. The changes observed in rolling can hence partly be considered as an effect of the change in deformability.

It has been shown that monocytes in circulation can be considered as suitable representatives of cholesterol loaded macrophages[7]. Hence, our results can provide the foundation for an enhanced understanding of mechanistic features of diseases like atherosclerosis.

ACKNOWLEDGEMENTS

We gratefully acknowledge the help of Dr. Shatha Dallo, Dr. Chiungyu Hung, Dr. Marylene Palard (Microelectronics Research Center, UT Austin) and Shankar J. Evani. This work was supported by NIH HL112629. This work received technical and instrument support from The Immunology Defense Core at UTSA, funded by the National Institute on Minority Health and Health Disparities from the National Institutes of Health.

REFERENCES

1. Lundbaek, J.A., et al., Biophys J, 2003. **84**(3): p. 2080-9.
2. Roduit, C., et al., Biophys J, 2008. **94**(4): p. 1521-32.
3. Kannan, K.B., D. Barlos, and C.J. Hauser, Journal of Immunology, 2007. **178**(8): p. 5253-5261.
4. Kwik, J., et al., Proc Natl Acad Sci U S A, 2003. **100**(24): p. 13964-13969.
5. Forsyth, A.M., et al., Microvasc Res, 2012. **83**(3): p. 347-51.
6. Jadhav, S., et al., Biophys J, 2005. **88**(1): p. 96-104.
7. Devaraj, S. and I. Jialal, Arch Pathol Lab Med, 2008. **132**(9): p. 1432-5.

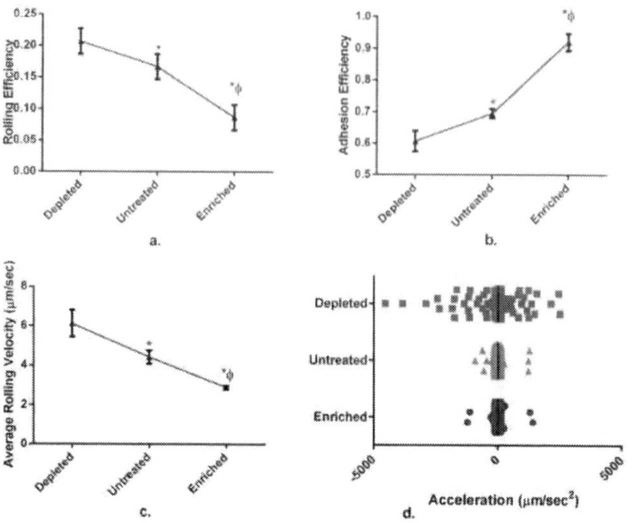

Fig. 1. Monocyte rolling and adhesion dynamics.

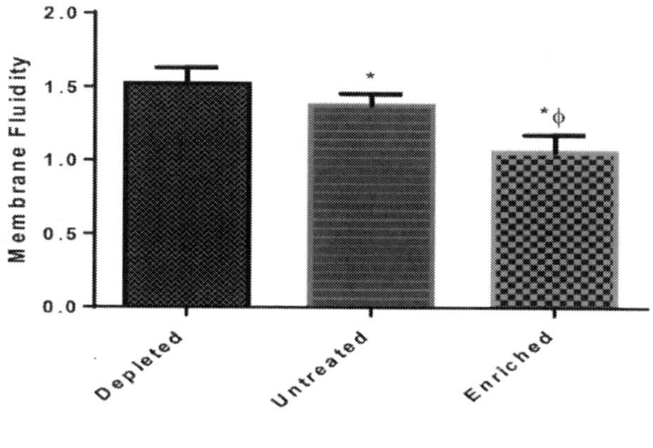

Fig. 2. Changes in membrane fluidity of human monocytes

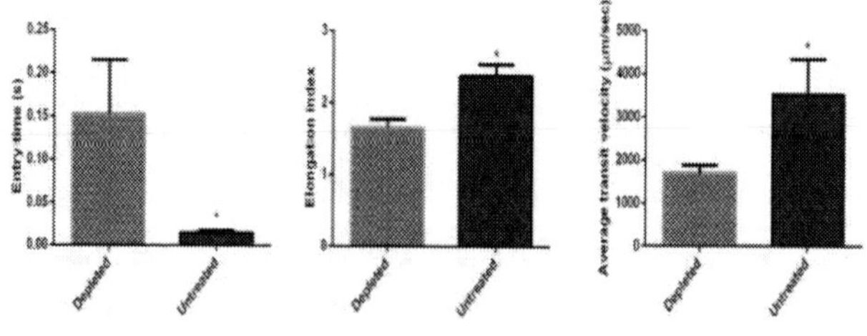

Fig. 3. Effect of cholesterol depletion on cellular deformability. Students' t-test. α = 0.05. n = 2.

SB³C2015
Summer Biomechanics, Bioengineering and Biotransport Conference
June 17-20, 2015,Snowbird Resort, Utah, USA
SB³C2015-413

SENSITIVITY STUDY OF HEAD IMPACT PARAMETERS ON INTRACRANIAL DYNAMICS

Yi Hua (1), Praveen Akula (1), Matthew Kelso (2), Linxia Gu (1, 3)

(1) Department of Mechanical & Materials Engineering
University of Nebraska-Lincoln
Lincoln, NE, United States

(2) Department of Pharmacy Practice
University of Nebraska Medical Center
Omaha, NE, Unite States

(3) Nebraska Center for materials and Nanoscience
University of Nebraska-Lincoln
Lincoln, NE, United States

INTRODUCTION

Traumatic brain injury (TBI) is a leading cause of mortality and morbidity both in civilian life and on the battlefield worldwide [1]. Survivors of TBI frequently experience long-term disabling changes in cognition, sensorimotor function, and personality [2]. Numerous rodent models, i.e., fluid percussion [3], controlled cortical impact [4], weight-drop impact [5], and closed head impact (CHI) [6], have been designed to recapitulate the various aspects of the human TBI. Among these models, the CHI model, which adapted from the controlled cortical impact model, is one of the most promising models and currently being increasingly used in many laboratories to study TBI. In a typical CHI procedure, the rat is fully anesthetized and secured in a stereotactic frame. An impact load is delivered directly to the intact skull through a pneumatically-driven impactor. Compared with other impact models, the CHI model avoids performing the craniotomy and thus the risk of operation-induced inflammation. However, various control parameters are employed by different research groups, making it difficult to compare the results. In addition, the internal brain response variables could be documented in more detail.

In this work, a series of computational experiments were conducted with the typical impact parameters (i.e., impact depth, velocity, and position) commonly found in the literature [6-8]. An anatomically detailed finite element (FE) rat head model was developed to simulate the CHI tests and to correlate the external impact parameters with the rat brain internal responses.

METHODS

A 3D FE rat head model was generated from the high-resolution magnetic resonance imaging (MRI) datasets of an adult male Sprague-Dawley rat weighing about 360 g. The image data were segmented into three main brain regions; the cerebrum, hippocampus, and cerebellum (Fig. 1). The material properties of different brain regions were adopted from reference [9]. To replicate the experimental CHI procedure, a cylindrical impactor, which connected to the bottom surface of a steel rod, was positioned perpendicular to the dorsal surface of the rat skull. A linearly ramping displacement was enforced onto the steel rod to achieve the prescribed impact depth and velocity. The interactions between the impactor and skull as well as between the brain and skull were modeled through penalty contact algorithm with tangential sliding and hard contact normal behavior. The nodes on the bottom surface of the skull were constrained in all six degrees of freedom to avoid rigid body translation. The FE model was solved using the nonlinear transient dynamic procedure Abaqus/Explicit (Dassault Systems Simulia Corp., RI, USA).

Based on the literature review [6-8], a series of computational experiments were conducted to investigate the effects of impact depth, velocity, and position on the peak maximum principal strain (MPS) responses in three different regions of the rat brain; the cerebrum, hippocampus, and cerebellum. A total of seven CHI cases were simulated, including three impact depths (1, 2, and 3 mm), four impact velocities (3, 4, 5, and 6 m/s), and two impact positions (central and lateral). The baseline model involved the use of an impact depth of 2 mm, an impact velocity of 5 m/s, and a central impact position.

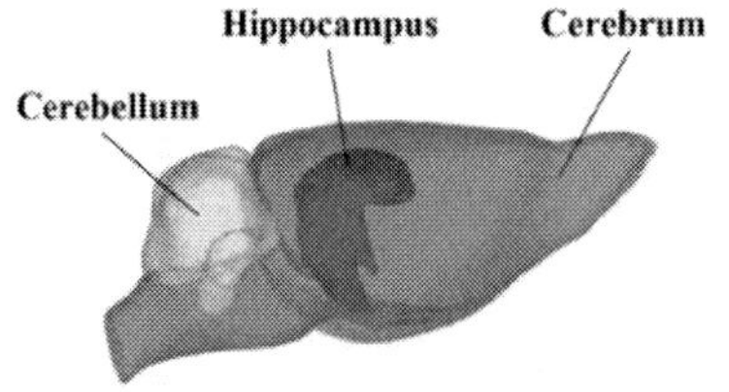

Fig. 1 An anatomically detailed finite element rat head model.

RESULTS

The effects of various impact parameters (i.e., impact depth, velocity, and position) on the peak MPS responses in the cerebrum, hippocampus, and cerebellum were shown in Fig. 2. It is observed that the peak MPS in three different regions of the rat brain increased with larger impact depth (Fig. 2a). There were 77.2%, 142.4%, and 97.1% increase of the peak MPS in the cerebrum, hippocampus, and cerebellum, respectively, when the impact depth increased from 1 to 3 mm. In contrast, impact velocity seems to have little effect on the intracranial responses (Fig. 2b). Increasing the impact velocity from 3 to 6 m/s only increased the peak MPS in the cerebrum by 18.8%, the hippocampus by 16.8%, and the cerebellum by 29.9%. Intracranial responses were also found to be sensitive to the impact position (Fig. 2c). Changing the impact position from central to lateral, the peak MPS in the cerebrum, hippocampus, and cerebellum increased by 68.1%, 57.3%, and 69.1%, respectively.

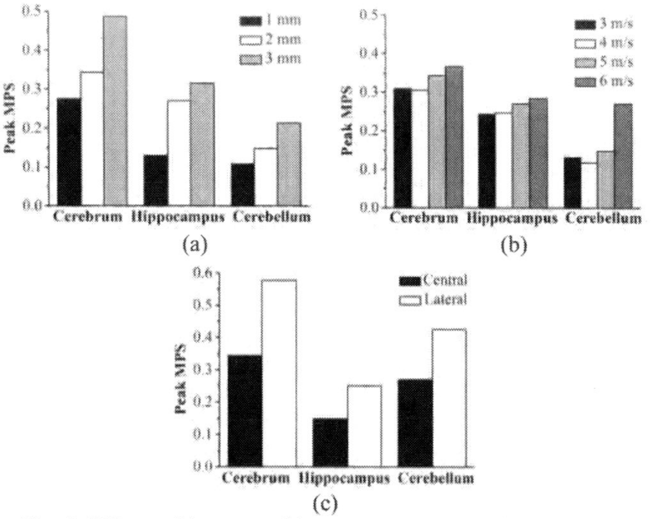

Fig. 2 Effects of the external impact parameters on the peak MPS responses in the cerebrum, hippocampus, and cerebellum: (a) impact depth, (b) impact velocity, and (c) impact position.

DISCUSSION

In this work, the effects of the external impact parameters on the intracranial biomechanical responses were systematically analyzed using the 3D FE rat head model. It is found that the sensitivities of the peak MPS in the cerebrum, hippocampus, and cerebellum with respect to the impact depth, velocity, and position are different (Fig. 2).

Impact depth is regarded as the most important factor which affecting the intracranial responses in CHI. Increasing the impact depth can cause an appreciable increase of the peak MPS in all three brain regions. This is consistent with the experimental observation that a larger impact depth could result in more severe axonal damage as well as increased permeability of blood-brain barrier in the rat brain [7]. Considering the robust effect on the variations in intracranial responses, impact depth has been used by a number of researchers to indicate brain injury severity. However, our results show that even though the impact depth was fixed at 2 mm, the model-predicted peak MPS could vary significantly through the whole brain when other impact parameters were changed. This indicates that a single impact parameter should not be used as an indicator of brain injury severity. Instead, the internal biomechanical responses of the brain such as peak MPS are more appropriate in the definition of injury severity since tissue responses are directly linked to the site-specific brain injury.

It is interesting to find that variations in impact velocity induced only small variations on peak MPS in all three brain regions. On the other hand, a wide range of impact velocities are used in different laboratories to induce desired injury levels [10]. This inconsistency could be attributed to the "overshoot" phenomenon which widely exists in the pneumatically-driven impact devices. Overshoot is defined as the maximum transient tip displacement that exceeds the user-preset penetration depth during an impact event. Brody et al. [11] reported that overshoot was strongly velocity-dependent and it tended to increase with the increase of the impact velocity. For example, the tested pneumatic device produced a 0.30 ± 0.016 mm overshoot at an impact velocity of 3.7 m/s, while at 5.2 m/s the overshoot was substantially increased to 0.61 ± 0.047 mm. This indicated that the role played by the impact velocity might be overestimated in the experiment. However, in the FE model, the impact depth is accurately defined and not affected by the velocity settings; as a result, the effect of impact velocity is found to be not so robust compared with the realistic CHI experiments.

Impact position is found to be crucial in determining the extent and location of tissue injury in rat brains. Model predictions show that lateral impact induced higher peak MPS in all three brain regions compared to the central impact. Moreover, the high strains induced by the lateral impact were mainly focused on the ipsilateral side of the brain, while those induced by the central impact were more diffusely located on both sides (Fig. 3). Our results are in consistent with the clinical observation that lateral impact inflicts primarily unilateral cortical damage, whereas central impact causes bilateral cortical alterations [7, 8].

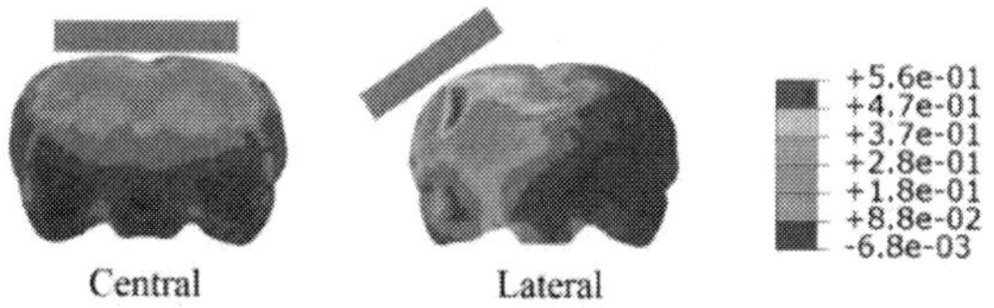

Fig. 3 Contour plots of the MPS on a coronal plane beneath the center of the impactor at the time of maximum tip displacement.

In summary, we have systematically investigated the effects of the external impact parameters on the intracranial responses using a 3D FE rat head model. Findings from this work can contribute to a deeper understanding of the individual role played by various impact parameters during an impact event, as well as enable researchers to accurately design a site- and severity-specific rodent CHI model.

ACKNOWLEDGEMENTS

The authors acknowledge Dr. Aravind Sundaramurthy for his help in segmenting different rat brain regions using Mimics®. The authors also thank Prof. Michael Boska from University of Nebraska Medical Center for proving MRI rat brain data.

REFERENCES

[1] Xiong, Y et al., *Nat Rev Neurosci*, 14: 128-142, 2013.
[2] Peskind, ER et al., *J Clin Psychiatry*, 74: 180-188, 2013.
[3] Gurkoff, GG et al., *J Neurotrauma*, 30: 292-300, 2013.
[4] Atkins, CM et al., *Neurosci Lett*, 532: 1-6, 2013.
[5] Wang, GH et al., *J Neurotrauma*, 27: 1733-1743, 2010.
[6] Maruichi, K et al., *Neuropathology*, 29: 132-139, 2009.
[7] Cernak, I et al., *Neurobiol Dis*, 17: 29-43, 2004.
[8] Laurer, HL et al., *J Neurosurg*, 95: 859-870, 2001.
[9] Mao, H et al., *J Neurotrauma*, 27: 877-888, 2010.
[10] Ren, Z et al., *J Cereb Blood Flow Metab*, 33: 834-845, 2013.
[11] Brody, DL et al., *J Neurotrauma*, 24: 657-673, 2007.

TOWARDS VERTEBRAL COMPRESSION FRACTURE PREVENTION: SIMULATING PHYSIOLOGIC FRACTURE DURING SMALL MOVEMENT ADLS

Nicole C. Corbiere (1), Stacey L. Zeigler (2), Kathleen A. Issen (1), Arthur J. Michalek (1) and Laurel Kuxhaus (1)

(1) Mechanical and Aeronautical Engineering Department
Clarkson University
Potsdam, NY, USA

(2) Physical Therapy Department
Clarkson University
Potsdam, NY, USA

INTRODUCTION

Vertebral fractures, a common sequela of osteoporosis, are under-diagnosed in North America [1]. Osteoporosis and low bone mass threaten nearly 44 million people in the US [2]. Approximately 50% of women and 25% of men will have an osteoporosis-related fracture after the age of 50 [2]. In the year 2000 it was estimated that 9 million new osteoporotic fractures occurred, and of these, 1.4 million were clinical vertebral fractures [3]. This compelling health problem causes the patient physical pain and is a financial burden on the population; therefore, it is imperative that future research works towards risk prediction and prevention of vertebral compression fractures.

It is not surprising that traumatic events, such as heavy lifting in extreme postures, can cause compression fractures. However, patients with osteoporosis often cannot recall an extreme event that could have caused a compression fracture, suggesting that fractures occur even during ordinary activities of daily living (ADLs). Evidence pertaining to the effect of the more common, lower angle motion ADLs is lacking. Thus, health practitioners have limited ability to predict the effects of potential changes in activity on fracture prevention beyond avoidance of the well-studied extremes of spinal movement. Flexion of less than 15° is the most common motion of the lumbar spine during ADLs [4]. These low-loads applied cyclically to the spine can be replicated in the laboratory by cycling an eccentric axial compressive load on a vertebral motion segment. Existing reports of vertebral fracture generation in the laboratory use fixed end conditions [5]; however, eccentric physiologic loading between healthy vertebrae is most accurately replicated using pinned end conditions (Figure 1, top right). The expected outcome is articular facet fracture. Different loading boundary conditions (such as fix-ended, Figure 1, top left) can create non-physiologic fractures. This will be confirmed by testing vertebral motion segments under two end conditions (fix-ended and pin-ended) and observing fracture mode and load. Understanding the effects of repetitive small movements on vertebral motion segment response will ultimately elucidate the mechanical initiation of vertebral compression fractures, providing clinically relevant information for fracture prevention instruction.

Towards this goal, the present work mechanically loads motion segments from cervine donors. Smit has shown that the dominant *in vivo* loading is axial compression in both quadrupeds and human bipeds [6]. Additionally, there are similarities in the mechanical response of cervine motion segments compared to a human motion segment [7-9]. This work aims to demonstrate that the cyclic loads of ADLs contribute to fracture, and that pinned end conditions are necessary to produce physiologic flexion failure.

METHODS

Vertebral motion segments were cyclically loaded in two groups with different end conditions. Nine 5-vertebra motion segments (L1-L5) were harvested from cervine donors (Nolt's Custom Meat Cutting, Lowville, NY). Six donors were male, with a mean age of 2.33 ± 0.66 years for all nine specimens. X-rays (Multix X-Ray System, Siemens, Germany) and DXA (Lunar Prodigy Advance DXA System, GE Healthcare, United Kingdom) scans were performed at Canton Potsdam Hospital prior to the removal of soft tissue. Specimens were potted in Bondo™ (Auto Body Filler, 3M, St. Paul, MN) to create end-caps to interface with the loading fixtures and were divided into two mechanical testing groups.

Specimens in Group #1 simulated flexion by loading through fixed end conditions with an eccentricity of 21.5mm (Figure 1, top left). Micro-computed tomography (μCT) scans were performed before loading and after failure. To date, two specimens were cycled at 2 Hz between displacements corresponding to upper body weight

(685N) and twice upper body weight (1370N) [5] and one specimen was cycled between the displacements corresponding to a 50N tare load and upper body weight (685N). All specimens for this group were loaded for 20,000 cycles.

The specimens in Group #2 simulated flexion by loading through pinned end conditions with an eccentricity of 10mm (Figure 1, top right). The average BMD for these specimens was 1.182 ± 0.155g/cm^2, comparable to healthy human bone. Six specimens were cyclically loaded at 2Hz between displacements corresponding to 0° and 15° flexion for 40,000 cycles. A 4-camera Qualisys (Oqus 500 cameras, Qualisys Inc., Sweden) motion capture system, with an accuracy of 0.1mm, was used to measure the flexion angle. Retroreflective markers were secured to each vertebral body with cyanoacrylate adhesive and intervertebral angles were calculated following the ISB standards [10].

After cyclic loading, all specimens in both groups were monotonically compressed to failure at a rate of 25.4mm/min [11].

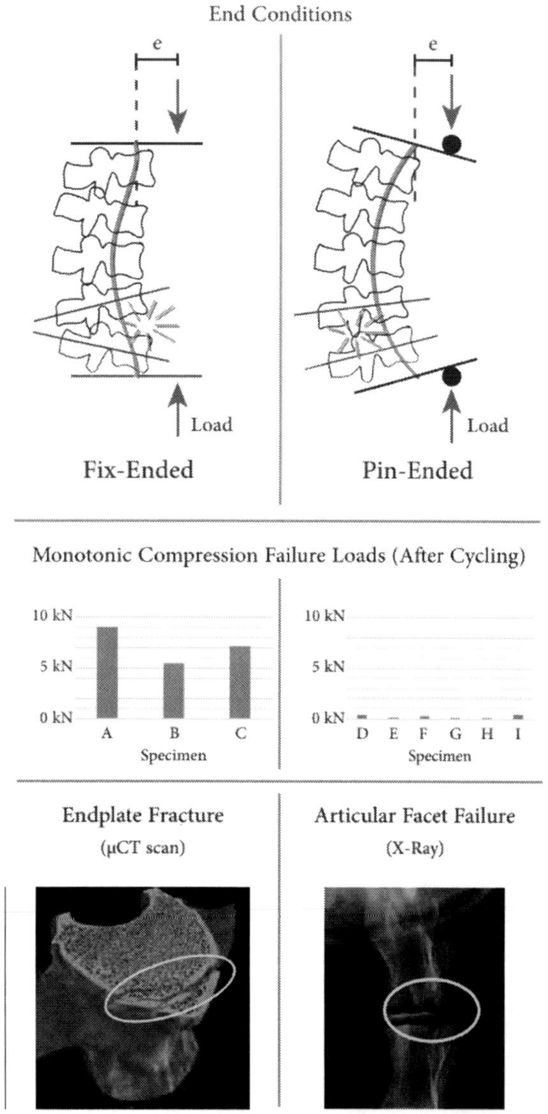

Figure 1: End conditions (top), failure loads (middle), and failure modes (bottom).

RESULTS

As expected, for all specimens, the peak load decreased with increasing cycles. The Group #1 specimens did not fail during cyclic loading. Endplate fractures, similar to those seen under the centric axial loading of trauma, were produced during the subsequent eccentric monotonic compression. These specimens experienced an average monotonic compression failure load of 7240N (Figure 1, middle left), considerably higher than the Group #2 specimens with an average failure load of 240N (Figure 1, middle right), despite loading Group #1 at a larger eccentricity. In four of the Group #2 specimens, fracture formed during cyclic loading as early as 1,400 cycles. In all nine specimens, the monotonic failure loads, after cyclic loading, were up to 52% less (Group #1) and 96% less (Group #2) than published work [6] without cyclic loading.

All specimens in Group #1 formed traumatic anterior endplate fractures (Figure 1, bottom left) due to local extension caused by fixed end conditions. Conversely, all specimens in Group #2 failed at the articular facet joints and the fracture continued to the intervertebral disc interface (Figure 1, bottom right).

DISCUSSION

These preliminary data suggest that the cyclic loading of small movement ADLs contributes to fracture formation. This work shows that pinned end conditions are necessary to create physiologic fracture. The two different groups resulted in distinct fracture modes. All specimens in Group #1 had the same failure mode (endplate fracture) as motion segments from another study which were centrically loaded with fixed end conditions [6]. These endplate fractures are not indicative of bending failure, which means that eccentric loading without proper end conditions does not induce a realistic bending failure. In addition to current research which shows that extreme motion ADLs and BMD predispose fracture, this work suggests that small movement ADLs may also make a noteworthy contribution. For example, the specimen with the lowest monotonic failure load had the highest BMD measured. The difference in monotonic failure loads between groups further demonstrates that end conditions in the lab experiments in part determine fracture mode and failure load.

Limitations of this preliminary study include using nonhuman motion segments and the small sample size. Despite these limitations, these results suggest that health care providers may need to consider small movement ADLs in fracture prevention education. Future work can further elucidate this relationship by mechanically testing more specimens (cervine and human) and a variety of ADLs. This work will ultimately provide the appropriate medical evidence for advanced fracture risk guidance and can improve rehabilitation techniques.

ACKNOWLEDGEMENTS

The authors thank Mark Hedgeland for experimental testing support and Sandra Thitten, Arlene Roach, and Canton Potsdam Hospital for the imaging. The authors have no conflicts of interest.

REFERENCES

[1] Delmas, P et al., *JBMR*, 20, 557-563, 2005.
[2] National Osteoporosis Foundation, *www.nof.org*, 2013.
[3] Johnell, O et al., *Osteo Int*, 17, 1726, 2006.
[4] Cobian, D et al., *J of ASTM Int*, 9, 2011.
[5] Nagaraja, S et al., *The Spine J*, 13, 1872-1880, 2013.
[6] Smit, T, *Eur Spine J*, 11, 137-144, 2002.
[7] Corbiere N et al., *J Biomech Eng*, 136, 2014.
[8] Kumar, N et al., *The Anat Record*, 266, 108-117, 2002.
[9] Wasinpongwanich, K et al., *J Spine*, 3, 187, 2014.
[10] Wu, G et al., *J Biomech*, 35, 543-548, 2002.
[11] McCubbrey, D et al., *J Biomech*, 28, 891-899, 1995.

OH DEER! MORPHOLOGICAL AND BIOMECHANICAL EVALAUATION OF CERVINE FEMORA

Mark J. Hedgeland (1), Morgan A. Libruk (2), Nicole C. Corbiere (1), Mario J. Ciani (2), Laurel Kuxhaus (1)

(1) Mechanical & Aeronautical Engineering
Clarkson University
Potsdam, NY, USA

(2) Physician Assistant Studies
Clarkson University
Potsdam, NY, USA

INTRODUCTION

The development of new orthopedic devices and surgical procedures depends heavily on the morphology and biomechanical behavior of the bones involved; femora are no exception. The femur, both human and animal, has been widely used in biomechanical research [1,2,3]. Animal models often provide a less expensive and/or more convenient testbed for orthopaedic device design and testing than human cadaver bones. However, before any animal model can be used as an appropriate model of human bone, the morphology and biomechanical characteristics of the bone must be documented and compared to human bone. Previously studied femora models include porcine, ovine, and canine [2,3]. Given that cervine tibiae and spine are appropriate models for human tibiae and spines, it is the goal of this study to evaluate the suitability of deer (cervine) femora as a model for human femora [4,5]. Thus the goals of this research include quantifying the morphology and mechanical properties of the cervine femur.

METHODS

Sixteen pairs (10 male and 6 female) of fresh frozen *Odocoileus virginianus* (white-tailed deer) femora were obtained from Nolt's Custom Meat Cutting (Lowville, NY). The average estimated age of each specimen was 2.1 ± 0.9 years. The 16 right femora were used for morphologic study while the 16 left femora were used for biomechanical testing. Prior to use, soft tissue structures were removed.

Ten geometric dimensions of the cervine femora were measured (Figure 1). Anteversion angle (AV), Cervico-diaphyseal angle (CD), and the Head Diameters (HDap and HPsi), were measured from photos of these specimens using ImageJ (ImageJ 1.48 nih.gov). The remaining

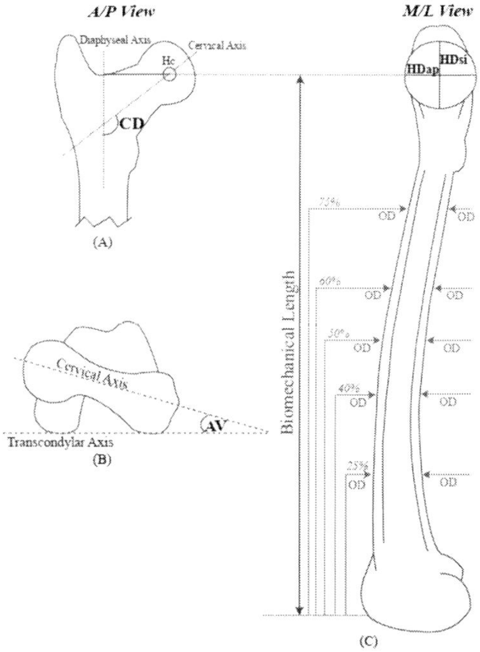

Figure 1 Measured dimensions: *(A)* A/P view of the proximal femur showing mediolateral (ML) dimension cervicodiaphyseal angle (CD) (H_C = head center) *(B)* Distal end showing angle of anteversion (AV) *(C)* M/L view of the femur showing dimensions outer diameter (OD) at percentage intervals from the distal end, head diameter in anteroposterior (AP) plane (HD_{AP}), head diameter in superoinferior (SI) plane (HD_{SI}), and biomechanical length (Lb). OD measurements were made in both the A/P and M/L planes.

dimensions were measured by hand using inside/outside calipers except for biomechanical length (L_B), which was measured using a flexible tape measure (150 cm/60 in) accurate to the nearest millimeter. Each dimension was measured three times. If by hand, three different individuals measured each dimension. The dimensions measured via photos (Fujifilm FinePix Z33WP camera) in ImageJ were measured three times by the same individual (three photos per each view of the femur) with a metric ruler included in the photo for scale. Dimensions were recorded to the nearest 0.001° or millimeter.

The sixteen remaining left femora were monotonically compressed or torqued to failure (MTS 809 Axial/Torsion Test System, Eden Prairie, MN.) First, they were potted at a 10° angle in the frontal plane using Bondo™ (Auto Body Filler, 3M, St. Paul, MN) as a rigid support material around the diaphyses of each bone [1,2]. Eight of the sixteen femurs were quasi-randomly chosen for compression testing, and the remaining eight for torsional testing. Compression was performed at a rate of 0.1 millimeters/sec and torsion at 0.2°/sec. Failure was determined by auditory and visual observation. The load-displacement data from each trial was analyzed, and the stiffness and ultimate strength of each specimen were calculated ($K_{Axial} = \Delta F/\Delta\delta$ and $K_{Torsional} = \Delta M/\Delta\theta$) using a custom MATLAB code (R2012b, Mathworks, Natick, MA).

RESULTS

A summary of the hand-measured cervine dimension is presented below in Figure 2. The average ODs of the cervine femur were found to be 23.86 and 22.79 millimeters in the A/P and M/L planes respectively. Previous research has shown that these values are comparable to the average ODs for ovine, porcine, and human femora [3]. Those results were 21.13, 25.28, and 29.91 millimeters in the A/P plane and 19.42, 25.28, and 28.53 millimeters in the M/L plane for ovine, porcine, and human femora respectively [3]. The average AV and CD for cervine femora were measured to be 11.38 and 117.03° respectively. These are comparable to average canine values of AV and CD, which were previously reported to be 27.00 and 139.80° respectively [6]. The cervine and canine femora were also found to follow a similar trend for HDap, HDsi, and Lb relative to human values [6].

Figure 3 presents the stiffnesses and ultimate strengths of the femora compared to the human values previously-reported [2]. Note that the cervine tibia has a higher torsional stiffness, and the human femur has a higher torsional ultimate strength.

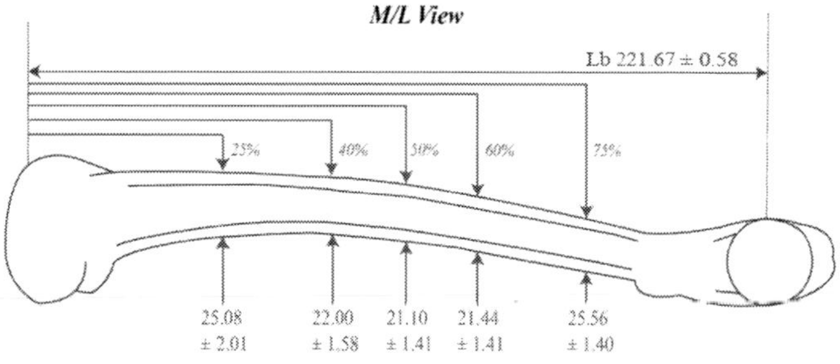

Figure 2 Average (± standard deviation) cervine OD and Lb measurements shown on a M/L view of the cervine femur. All dimensions are in millimeters.

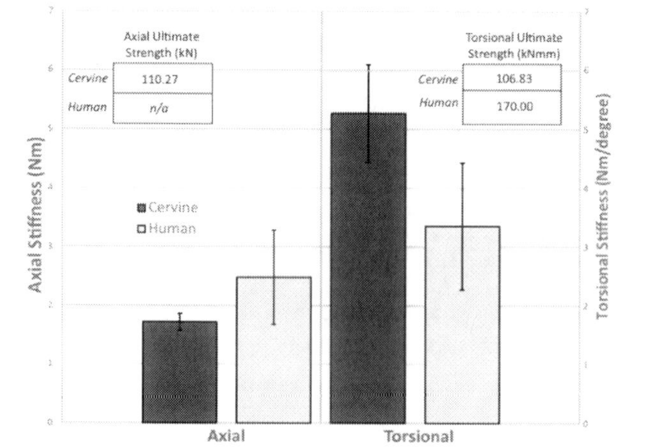

Figure 3 Average axial (left) and torsional (right) stiffness and ultimate strengths of cervine femora, compared to femora [2].

DISCUSSION

Morphologically, the cervine femur is notably shorter than the human femur (average human Lb = 431 ± 28.3 millimeters) [6]. However, cervine femur dimensions such as Lb, and average OD (in both planes) are comparable to ovine and porcine models. In comparison to human femora dimensions the cervical dimensions of AV, Lb, HDap, and HDsi here follow the same general trends as previously-reported canine dimensions (both are smaller or the same as humans for HDap, HDsi, and Lb, and larger than humans in AV) [6]. It is interesting to note that the cervine CD dimensions are smaller than the human dimension in contrast to the canine which is larger [6]. This variation away from the trend of canine dimensions may offer unique opportunities in orthopaedic device design and testing, but not so much so that the cervine dimensions are outside of the useful limits for human comparison.

The cervine femur has axial stiffness and torsion ultimate strength comparable to, and in the case of torsional stiffness, even greater than human femora. The higher torsional stiffness is not surprising given its shorter length. Thus, cervine femora could serve as biomechanical proxies for certain orthopedic research and the design of orthopedic implants, and still provide geometries of interest that other animal models cannot. The results of this study are necessarily limited by the small sample size and the stochastic nature of the harvesting process. Nonetheless, given that cervine specimens are often inexpensive and locally-abundant, the cervine femur should be considered for use in biomechanical studies.

ACKNOWLEDGEMENTS

The authors would like to thank Elizabeth Miele for her help in preparing specimens. The authors have no conflicts of interest to declare.

REFERENCES

[1] L. Cristofolini, et al., (1996), *J. Biomech.* **29**, 525-535.
[2] A. D. Heiner, T. D. Brown, (2001), *J. Biomech.* **34**, 773-781.
[3] D. C. Kieser et al *(2014)*, *Bio-Med. Mat. & Eng.*, **24**, 1693-1703.
[4] Throop, A. D. et al., (In Press), *J. Biomech. Eng.*
[5] Corbiere, N. et al., (2014), *J. Biomech. Eng.*, **136.**
[6] R. D. Bloebaum, et al., (1993), *J. Biomed. Mater. Res.* **27**, 1149-1159.

SB³C2015
Summer Biomechanics, Bioengineering and Biotransport Conference
June 17-20, 2015,Snowbird Resort, Utah, USA
SB³C2015-416

DEVELOPMENT OF GAIT SIMULATIONS USING A CONTACT FORCE FEEDBACK CONTROLLER

Jonathan P. Walter (1), Marcus G. Pandy (1)

(1) Mechanical Engineering Department
University of Melbourne
Parkville, Victoria, Australia

INTRODUCTION

Detailed analyses of healthy and pathological gait patterns computational modelling can be used to estimate muscle forces and typically involves the use of inverse dynamics or forward dynamics simulations. Knowledge of the muscle forces and tibiofemoral contact forces occurring at the knee could be particularly helpful in the design of subject-specific treatments of knee osteoarthritis. However, most lower-limb models simplify the knee as a simple hinge joint and neglect the contributions of tibiofemoral joint contact forces during muscle force estimation. The one degree-of-freedom (DOF) hinge joint allows for only one net load component, the flexion-extension moment, to be balanced with muscle forces, while the adduction-abduction (AA) moments and superior-inferior (SI) forces are ignored. Estimates of the AA contact moment and SI contact force, could be highly valuable estimates for determining the loading conditions applied to the cartilage and meniscal tissues, thus, accurate estimation of the contributions from contact and muscle forces to these net loads is necessary. Recent modelling efforts have included contact and muscle forces for multiple DOF knees during muscle force estimation, but have not included contact load information directly in their controllers[1,2] or have used inverse dynamics rather than forward dynamics[3].

This work aims to develop an improved musculoskeletal modelling algorithm that is capable of simultaneously and dynamically controlling both contact and muscle forces at the knee. The use of a feedback controller and a tibiofemoral surrogate contact model are combined in a forward dynamics simulation which balances four net load components at the knee including flexion-extension moment, adduction-abduction moment, internal-external rotation moment, and superior-inferior force. The process for creating dynamic gait simulations with feedback control and contact forces is currently in development and is described in general here.

METHODS

A musculoskeletal model of the lower limbs was developed using experimental gait measurements for three subjects. A generic model which included the trunk, pelvis, thighs, shanks and feet was scaled to each subject based on anatomical marker locations. A polygonal geometry based tibiofemoral contact model using an elastic-foundation "bed-of-springs" was placed into the scaled musculoskeletal model[4]. Tibiofemoral contact surfaces were aligned to the scaled skeletal geometries of the musculoskeletal model. Typical kinematics and contact forces during gait were determined using the contact geometries. A range of offsets to these kinematics were sampled to create a surrogate contact model. The faster surrogate contact then replaced the geometry based contact model for use during simulation[4].

A forward dynamics gait simulation algorithm using the 4-DOF knee and fast surrogate contact model was prepared in the OpenSim programming interface. The simulation was based on the computed muscle control algorithm which performs a forward dynamics simulation with continual updates to the estimates of muscle activations based on simulated and desired accelerations. For all DOF of the lower limb model, desired accelerations, a_{des}, were calculated at regular time intervals throughout the simulation based on the joint accelerations calculated from inverse kinematics, a_{exp}, and errors in joint positions, p, and velocities, v, with proportional-derivative feedback gains. Muscle forces were optimized at each time interval to satisfy the desired accelerations and minimize the sum of squares of muscle activations. Contact forces were calculated using the updated tibiofemoral knee kinematics throughout the simulation which allowed for the acceleration contributions of contact to any DOF to be calculated, a_{cont}.

Additional feedback information from the contact force contributions to the SI force and the AA moment was used to modify the desired accelerations and constrain the contact forces to reasonable estimates. A proportional feedback gain was used for the acceleration contribution from contact to the SI force, to act as a minimization of total contact force, Eqn. 1.

$$a_{des} = a_{exp} + k_p(p_{exp} - p) + k_v(v_{exp} - v) + k_C a_{cont} \quad (1)$$

A more complex term was added for the acceleration contribution from contact to the AA moment to match the experimentally measured average ratio of contact over inverse dynamic net loads, R_{CN}, using the acceleration contribution from the simulated net load a_{net}, Eqn. 2. This ratio was averaged from the stance phase of normal gait trials of five implanted knee subjects.

$$a_{des} = a_{exp} + k_p(p_{exp} - p) + k_v(v_{exp} - v) + k_{CN}(a_{cont} - R_{CN} * a_{net}) \quad (2)$$

Simulations were processed for the three subjects, one of which had an instrumented knee implant which measured tibiofemoral contact forces[5]. Simulated contact force estimates for the subject with the knee implant were compared to the measured contact forces. Simulations were repeated with modification of feedback gains to minimize errors in contact force estimates, kinematics, and to net loads.

RESULTS

Errors in estimated contact forces for the instrumented knee subject were on average 0.5 BW for the SI force and 0.3 %BW*HT for the AA moment. Errors were greater in late stance for the SI force and stayed consistent for the AA moment, see Fig. 1. Angular kinematics for all rotational joint DOFs were successfully tracked to less than 1.5 degrees RMS error.

DISCUSSION

The results presented here show that this novel force-feedback control algorithm is capable of estimating contact forces at the knee during gait while tracking joint kinematics from experimental measurements. Further work will be performed comparing simulations of the four-DOF knee with contact to simulations using a one-DOF knee without contact to determine how muscle forces differ when muscle and contact forces are estimated simultaneously. Also, further validation of the contact force estimates and refinement of the contact feedback controller is currently in development.

ACKNOWLEDGEMENTS

This work was funded by an Australian Council Discovery Projects Grant (DP120101973) and an Innovation Fellowship from the Victorian Endowment for Science and Knowledge Innovation to MGP.

REFERENCES

[1] Thelen, G et al., *J Biomech Eng*, 136:021033, 2014.
[2] Guess, T et al., *J Biomech Eng*, 136: 021032, 2014.
[3] Lin, YC et al., *J Biomech*, 43:945-952, 2010.
[4] Thelen, D, Anderson, F, *J Biomech*, 39:1107-1115, 2006.
[5] Fregly, G et al., *J Biomech*, 34:859-871, 2001.

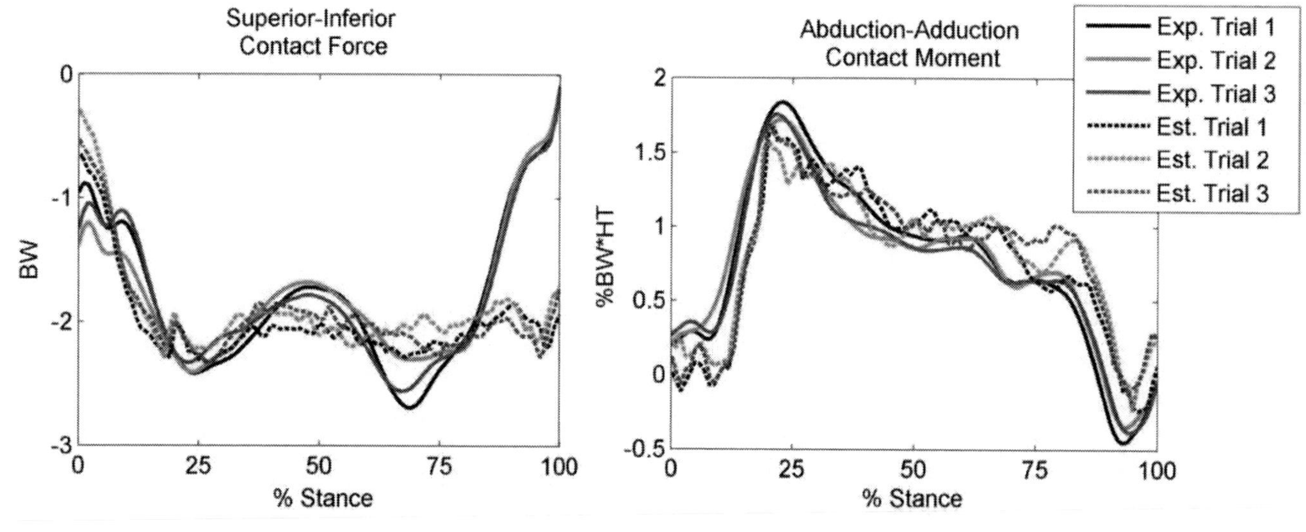

Figure 1. Estimated contact loads (dashed lines) and measured contact loads (solid lines) at the knee using a contact force feedback controller during a forward dynamics simulation of gait.

SB^{3}C2015
Summer Biomechanics, Bioengineering and Biotransport Conference
June 17-20, 2015,Snowbird Resort, Utah, USA
SB^{3}C2015-417

EXPERIMENTAL MEASUREMENT OF COLLAGEN FIBER UNCRIMPING AND RECRUITMENT WITH INCREASES IN INTRAOCULAR PRESSURE

Ian A. Sigal (1,2, 3), Ning-Jiun Jan (1,2), Huong Tran (1,2), Hiroshi Ishikawa (1,2), Katherine Davoli (1), Larry Kagemann (1,2), Joel S. Schuman (1,2,3), Gadi Wollstein (1,2,3), Kira Lathrop (1,2)

(1) Ophthalmology
University of Pittsburgh
Pittsburgh, PA, USA

(2) Bioengineering
University of Pittsburgh
Pittsburgh, PA, USA

(3) McGowan Institute
for Regenerative Medicine
Pittsburgh, PA, USA

INTRODUCTION

Abnormal ocular biomechanics are associated with several disorders that result in impaired vision or even blindness, such as myopia, glaucoma and keratoconus [1]. Ocular biomechanics are, in large extent, determined by the biomechanical properties of the connective tissues of the eye. Gradual uncrimping and recruitment of collagen fibers has been hypothesized as a mechanism underlying the nonlinear mechanical properties (stiffening) of the sclera with increased intraocular pressure [2] (Figure 1). Although some indirect evidence of recruitment exists from inverse numerical models or similarities with other tissues [3], experimental data on collagen fiber recruitment of ocular tissues remains very sparse [4].

Our goal was to measure the waviness of the collagen fiber bundles of the sclera, and the beams of the lamina cribrosa (Figure 2), and determine their uncrimping and recruitment with increases in intraocular pressure. These are two key support structures for the neural tissue in the posterior pole of the eye, and important to understand overall sensitivity to pressure.

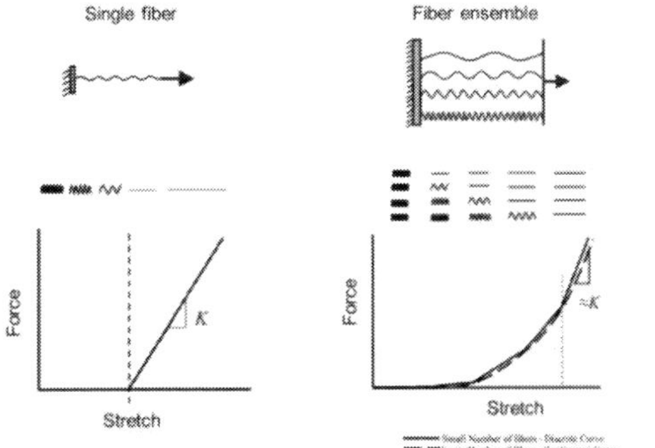

Figure 1 Model of nonlinear biomechanical behavior resulting from a process of fiber recruitment. Adapted from [3]

METHODS

23 eyes from 2 year old sheep were obtained from the slaughterhouse and processed within 8 hours of death. The eyes were fixed in 10% formalin at various intraocular pressures (0, 10, 15, 20, and 50mmHg), with at least 1 eye at each pressure. The posterior poles were separated with a trephine, cryosectioned coronally (30 μm), and at least 3 sections through the lamina cribrosa per eye were imaged using polarized light microscopy.

Images were analyzed for waviness by calculating the standard deviation of the collagen fiber orientation in manually marked straight lines longitudinal to the fiber bundles [4]. In the lamina cribrosa, the waviness of every collagen beam was measured. In the sclera, at least 900 measurements were made per eye (Figure 2).

The waviness was analyzed using mixed effect models, to account for correlations between measurements from the same sections, eyes and animals. Using the waviness at 50mmHg we defined tissue-specific thresholds to define the fraction of uncrimped fibers and a threshold waviness for a fiber to be recruited.

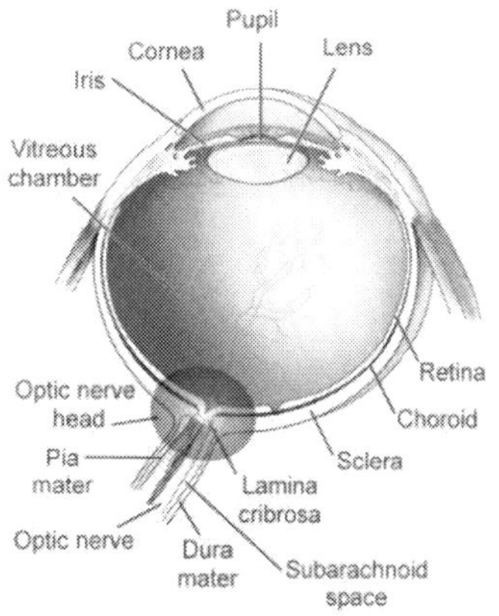

Figure 2 Parts of the eye. From [3]

RESULTS

Altogether over 21,000 manual tracings of laminar beams and sclera bundles were made, and a waviness measurement obtained for each one.

At every intraocular pressure the waviness was lower in the lamina cribrosa than in the sclera ($p<0.001$, Fig 3). As the intraocular pressure increased, waviness decreased in both the lamina cribrosa and the sclera ($p<0.001$), faster at lower pressures than at higher pressures ($p<0.001$). At all intraocular pressures, except 15mmHg, a larger fraction of sclera than LC fibers were uncrimped (Fig 4). At 15 mmHg the lamina cribrosa and sclera had the same fraction of uncrimped fibers (74%). At lower pressures, few sclera (57%), and even fewer laminar (32%) fibers were uncrimped.

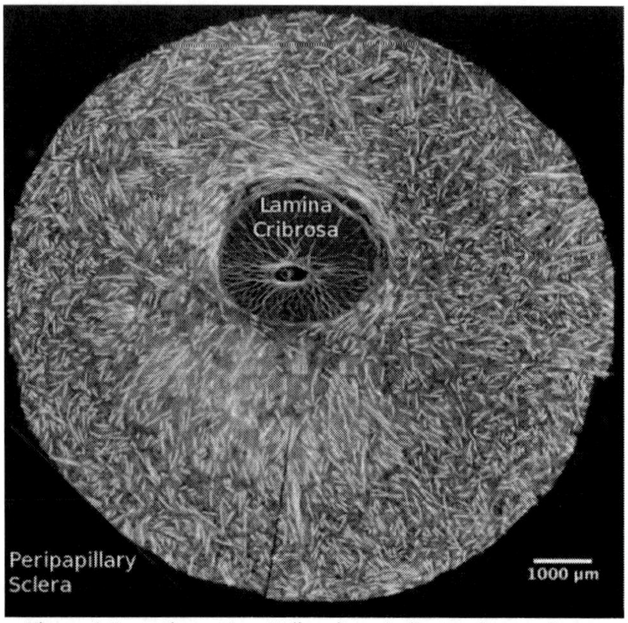

Figure 3 Example section. Yellow lines indicate where collagen waviness was measured in the sclera, and a blue line the canal border.

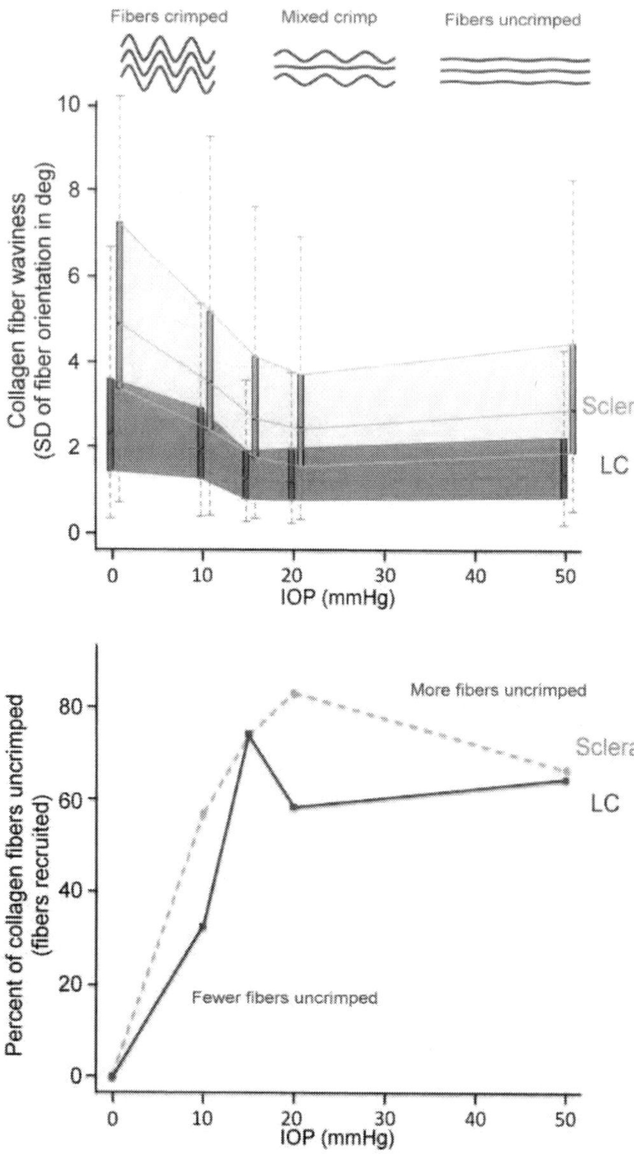

Figure 4 Effects of intraocular pressure on collagen waviness (top) and fraction of uncrimped fibers (bottom). As fibers are stretched with increased pressure, they uncrimp, decreasing in waviness. A fully uncrimped fiber bears considerably more load and is called recruited.

DISCUSSION

The decrease in waviness with intraocular pressure and associated fiber recruitment explain the macro-scale nonlinear behavior of the lamina cribrosa and sclera with pressure. Between 15 and 20 mmHg there was an increase in recruited fibers in the sclera but a decrease in recruited fibers in the lamina cribrosa. This may indicate that the sclera protects the lamina cribrosa from elevated pressures.

The causes and effects of the lower waviness of the laminar beams compared with sclera at all pressures, and the decrease in sclera recruitment between 20 and 50 mmHg remain unclear.

We have demonstrated that fiber-level stretch and tissue-level recruitment can be measured directly using waviness. Our results are consistent with those in other tissues [1-3]. Decrease in waviness indicates laminar regions under pressure-induced stretch and perhaps increased sensitivity to pressure-related tissue damage.

ACKNOWLEDGEMENTS

Financial support from NIH: R01 EY023966 (IAS), R01 EY013178 (JSS), P30 EY008098. Saundria Moed and Ryan O'Malley for help with the marking.

REFERENCES

[1] Ethier, CR, et al, *Annu Rev Biomed Eng*, 6:249-73, 2004.
[2] Grytz, R, et al., *Biomech Model Mechanobiol*, 9(2):225-35, 2010.
[3] Wognum, S, PhD Dissertation, University of Pittsburgh, 2010
[4] Sigal, IA, et al. IEEE Trans Med Imaging, 33(6):1381, 2014
[4] Sigal, IA, et al. *IOVS*, 54: E-Abstract 3158, 2013

SB^{3}C2015
Summer Biomechanics, Bioengineering and Biotransport Conference
June 17-20, 2015, Snowbird Resort, Utah, USA
SB^{3}C2015-418

TRANSCATHETER AORTIC VALVE REPLACEMENT MODEL: CRIMPING AND DEPLOYING IN PATIENT-PATHOLOGY SPECIFIC ROOTS

Matteo Bianchi (1*), Ram P. Ghosh (1*), Debapria Das (1), Gil Marom (1), Thomas E. Claiborne (1), Marvin J. Slepian (2), Danny Bluestein (1)

(1) Department of Biomedical Engineering
Stony Brook University
Stony Brook, NY, USA

(2) Sarver Heart Center
University of Arizona
Tucson, AZ, USA

(*)The first two authors have made equal contributions to this study

INTRODUCTION

Calcific aortic valve disease (CAVD) afflicts approximately 0.9% of the Unites States population with 2.8% of people over 75 years of age having moderate to severe aortic stenosis (AS) [1]. The disease is characterized by the formation of tissue similar to bone [2] on the leaflets of the AV, leading to a progressive stiffening (sclerosis) and narrowing of the aortic valve (AV) opening (stenosis). Typical symptoms such as angina and syncope tend to appear at late stage of disease, making a prompt intervention crucial for the patient [3]. Currently, replacement of the valve with a prosthetic valve is the only effective treatment for AS [4].

Transcatheter aortic valve replacement (TAVR) provides a life-saving alternative to open-heart surgical valve replacement for high-risk and inoperable patients [5]. In this type of minimally invasive intervention, a stent with a mounted bioprosthetic valve is delivered through blood vessels and deployed on the stenotic native valve. Currently, the only FDA-approved TAVR devices are the Edwards Lifesciences Sapien and Medtronic Corevalve. These TAVR valves consist of metal stent and bioprosthetic xenograft leaflets. In balloon expandable valves, such as Edwards SAPIEN, the leaflets are sutured to the stent and crimped on the delivery system prior to the procedure. As an alternative, a novel polymeric valve developed by our group [6, 7] is being commercialized by Polynova Cardiovascular, Inc. (Stony Brook, NY) and adapted to TAVR using a self-expanding stent. Polymer TAVR valves could be designed for higher tolerance to damages that may incur during TAVR crimping and deployment. Typical known intra- and postoperative TAVR complications include paravalvular leakage, heart block, thrombosis, embolization or migration of the valve in the left ventricle [8-10].

In recent years, TAVR has been studied using numerical tools and in-vitro tests to find the mechanical stress, hemodynamics and kinematics of the valve [11, 12]. However, these studies focused on the structural mechanics of the TAVR deployment and not on the postoperative conditions or their influence on the hemodynamics.

The objectives of this study are to compare the crimping mechanics of the two valves and to evaluate the effect of different TAVR deployment positions on the patient-specific hemodynamics. First, the crimping mechanics of the Polynova and Edwards SAPIEN were modeled and compared. Then, deployment of Edwards SAPIEN valve in heterogeneous patient specific models was modeled in three axial locations, in order to replicate possible surgical options during the procedure. Lastly, flow models of the deployed valve were carried out in order to determine the optimal deployment location for achieving better patient outcomes based on hemodynamics and the post deployment thrombogenic potential of the TAVR valve.

METHODS

The stents are designed in their expended state based on parametric equations [13] for an estimated crimped position (SolidWorks, DS Corp., Waltham, MA). The materials of the Polynova and SAPIEN stents were super-elastic nitinol and 316L stainless steel, respectively. The Polynova stent incorporated leaflets comprised of xSIBS (Styrene-block-IsoButylene-block-Styrene) material properties derived from our mechanical testing, whereas the SAPIEN stent had bovine pericardium material properties [14]. Hyperelastic material models were used for each study. A hollow cylinder replicated the stent crimper. Predefined radial displacement of 8 and 9 mm was applied to the inner surface of the crimper for both novel Polynova and SAPIEN valves, respectively. Precise contact modeling was included to account for the contact among the crimper, stent and leaflets, and to capture the complex leaflets' folding. The model was solved in ANSYS Explicit dynamics (Canonsburg, PA).

The CTA scans in systolic phase of two TAVR patients were obtained in the Heart Institute of Stony Brook University Hospital (SBUH) from consented patients, following an IRB protocol approved by Stony Brook University. Automatic 3D segmentation was performed in ITK-SNAP. Region competition snakes approach was employed in order to accurately extract the lumen of the aortic root and the calcifications boundaries. The outer surface of the root was then obtained in MATLAB (Mathworks Inc, Natick, MA) by extruding the lumen surface with a constant thickness for the sinus and a variable-thickness for the AV native leaflets. A solid model was then created in ANSYS Design Modeler (Canonsburg, PA) in which the calcifications were embedded in the soft tissue.

Combining the two models, an implicit finite element analysis of the TAVR procedure was performed, in which the location of the TAVR valve with respect to the AV annulus was parameterized, to replicate different choice of the surgeon during the deployment (Figure 1). The native leaflets soft tissue was modeled with an anisotropic hyperelastic material calibrated with tensile-test experimental data on excised human AV leaflets, obtained from the SBUH tissue bank. The calcium material model was calibrated performing nano-indentation measurements after μCT-scanning the calcified part of the same AV leaflets specimens. Computational Fluid Dynamics (CFD) analyses were performed for the optimal deployed configurations, to estimate the paravalvular leakage volume in diastolic phase.

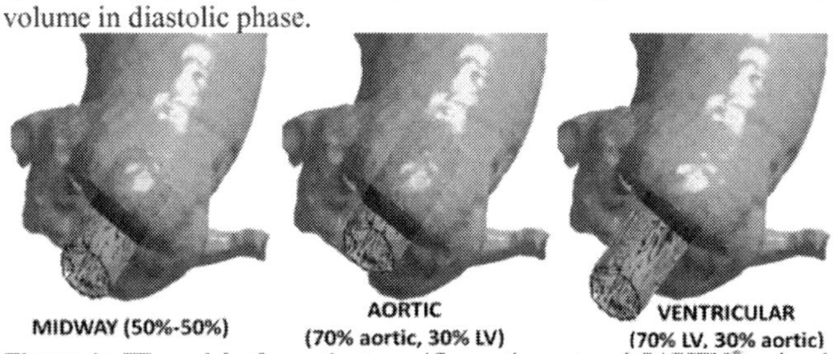

Figure 1: FE model of a patient-specific aortic root and SAPIEN® valve in midway position before deployment.

RESULTS

The maximum stress levels in the crimped valves and stents (both the absolute level, and the stress levels normalized with respect to the ultimate yield stress of the corresponding materials) were higher for the SAPIEN valve (Figure 2). The normalized value of the peak von Mises stress magnitudes to the ultimate stress levels are 9.1% and 24.7% for the novel Polynova and SAPIEN valve respectively. The maximum stresses in the leaflets were observed in the subcommisiural triangles near the leaflets bonding with the stents in both valves. The minimum stresses for both valves were observed at the belly region of the leaflets. The peak stress magnitude of the stent alone in its crimped state was also higher in SAPIEN than in Polynova. The highest stress levels of Polynova stents were observed in the aortic side of struts where

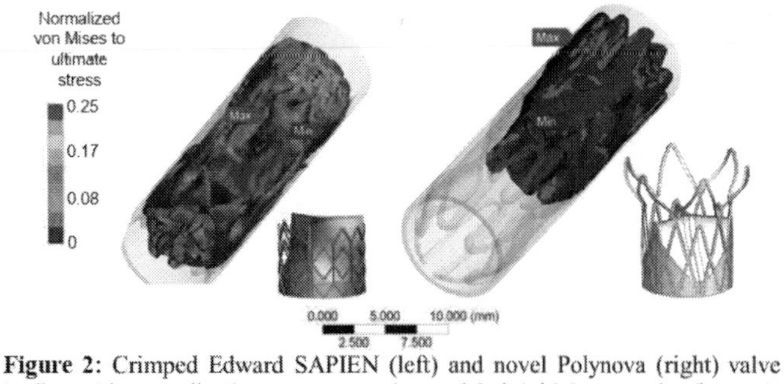

Figure 2: Crimped Edward SAPIEN (left) and novel Polynova (right) valve leaflets with normalized stress contour plots and their initial geometries (bottom).

buckling occurs whereas for SAPIEN valve the highest stresses were observed on the strut joints.

Anchorage of the valve was assessed for three different axial configurations: 70% toward the aortic side, 70% toward the ventricular side and midway (Figure 1). Stresses within the wall and interaction of calcium-soft tissue were investigated during the deployment. Higher stress levels were found in the calcium deposits than in the soft tissue (Figure 3). For both patients, who experienced valve migration during the procedure, incomplete expansion of the valve occurred and paravalvular leakage was confirmed in the correspondent CFD simulations. The risk-minimizing deployment configuration was assessed for both patients.

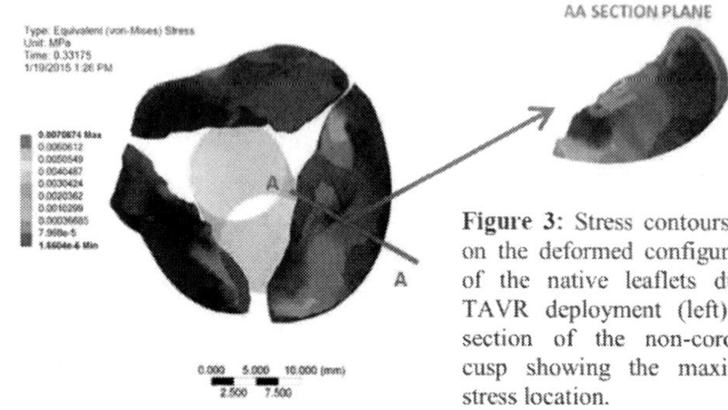

Figure 3: Stress contours plot on the deformed configuration of the native leaflets during TAVR deployment (left) and section of the non-coronary cusp showing the maximum stress location.

DISCUSSION

This study compared two TAVR valves and three deployment locations based on the crimping stresses and the deployed valves hemodynamics, respectively. The stress magnitude in the crimped configurations of Polynova and SAPIEN leaflets had small differences. However, the much lower stresses in the crimped state strongly indicate that the polymer valve may withstand damage much better than the tissue valve. Accurate patient-specific AR models including the vessel wall thickness and calcification deposits represent a useful tool to guide TAVR and to help in preventing the migration of the valve during the deployment procedure. Additionally, optimizing the positioning of the valve during deployment, as well as alternative approaches to TAVR valves tailored to patient's specific pathology, e.g. polymeric valves, may offer better procedural outcome. Ongoing studies in the CFD models include calculation of thrombogenic footprints based on stress accumulations along platelets trajectories. These models are validated with *in vitro* hemodynamics and their thrombogenicity correlated by measuring platelet activity using our platelet activity state (PAS) assay.

ACKNOWLEDGEMENTS

This study is supported by the NIH-NIBIB Quantum Award Phase II-1U01EB012487 (DB). ANSYS, Inc. is in an academic partnership with Stony Brook University.

REFERENCES

[1] Go, AS, et al., *Circulation*, 127:e6-e245, 2013.
[2] Mohler, ER, et al., *Circulation*, 103:1522-8, 2001.
[3] Kurtz, CE, and Otto, CM, *Medicine*, 89:349-79, 2010.
[4] Otto, CM, *N Engl J Med*, 359:1395-8, 2008.
[5] Cao, C, et al., *Ann Cardiothorac Surg*, 2:10-23, 2013.
[6] Claiborne, TE, et al., *Expert Rev Med Devices*, 9:577-94, 2012.
[7] Claiborne, TE, et al., *ASAIO journal*, 59:275-83, 2013.
[8] Vasa-Nicotera, M, et al., *JACC Cardiovasc Interv*, 5:858-65, 2012.
[9] Greason, KL, et al., *J Card Surg*, 28:264-6, 2013.
[10] Lauten, A, et al., *Eur Heart J*, 34:2509, 2013.
[11] Capelli, C, et al., *Med Biol Eng Comput*, 50:183-92, 2012.
[12] Wang, Q, et al., *Biomech Model Mechanobiol*, 14:29-38, 2015.
[13] Bonsignore, C, *Open Stent Design*, pp. 17-44, 2012.
[14] Sacks, MS, and Chuong, CJ, *Ann Biomed Eng*, 26:892-902, 1998.

SB³C2015
Summer Biomechanics, Bioengineering and Biotransport Conference
June 17-20, 2015,Snowbird Resort, Utah, USA
SB³C2015-419

PREDICTING REGIONS OF LOW WALL SHEAR STRESS IN THE CAROTID ARTERY BIFURCATION

Haley M. King (1), Emma J. Lichtenfels (1), Christian R. Fahrenbruck (1), Brandon L. Moore (1), Lakshmi P. Dasi (1,2)

(1) Mechanical Engineering Department
Colorado State University
Fort Collins, Colorado, USA

(2) School of Biomedical Engineering
Colorado State University
Fort Collins, Colorado, USA

INTRODUCTION

Atherosclerosis is a condition in which plaque builds up on arterial walls in the human vasculature. It is estimated that 80-90% of people over the age of 30 have some degree of atherosclerosis in their arteries, which can lead to cardiovascular disease, the leading cause of death in the U.S. [1]. The propensity of plaque to build up in a certain area in the artery is based on the particular geometry of arterial regions. Understanding, through fluids mechanics principles, why and where this plaque builds up has heavy clinical applications in preventing and treating cardiovascular disease.

One of the specific regions of vasculature in which plaque has a tendency to build up is in the carotid artery bifurcation [2,3,4]. This is the region in which the common carotid artery (CCA) splits into the internal carotid artery (ICA) and the external carotid artery (ECA). The geometry of this bifurcation contributes to the flow separation that is thought to cause regions of plaque buildup, and even subtle changes in the carotid geometry can affect the flow field significantly [5]. Immediately after the split, the ICA has a region called the internal carotid artery sinus. This sinus typically has a larger diameter than both the ICA and the ECA. These dimensional changes contribute to areas of flow separation, and in turn, low wall shear stress [6]. This low wall shear stress results in stagnant areas with low or even negative velocities which are likely regions of plaque buildup. The objective of this study was to develop a model to predict regions of low wall shear stresses with respect to the ICA-ECA flow split and the flow rate.

METHODS

To analyze fluid flow through the carotid artery, a professional glassblower was commissioned to create a scaled three-dimensional model of the bifurcation. Dimensions were taken from [7,8] in which measurements of the carotid artery were obtained in patients. The length and diameters of the common, internal, and external carotid, as well as the magnitude of curvature of the sinus in the external branch were also measured. These measurements were used to determine the scale factor of the model. Based on the available dimensions of glass tubing, a standard scale factor of 2.5 was chosen for the 3D model. Glass pipes were heated and blown to correct proportions. The model was then fixed into a flow system.

There were two variables among the tests, consistent with the research objective: flow rate (for diastole and systole) and percent of flow through each branch of the model. To account for the pulsatile flow in vivo, two flow rates were calculated and scaled to the model based on measured velocities in the carotid artery during diastole and systole. A clamp-on flow meter (Transonic Inc.) was used to verify volumetric flow rate was correct. Flow split was controlled using a downstream resistance in the ICA. The syntax of the split used in this experiment is "% flow through ICA – % flow through ECA."

Table 1: Experiments Performed

High Flow (Systole)	(1) 50% -50% split	(2) 66%-33% split	(3) 33%-66% split
Low Flow (Diastole)	(4) 50%-50% split	(5) 75%-25% split	(6) 25%-75% split

Data was collected in a single camera, 2-D PIV scenario, through the use of a high-powered laser and a high-resolution video camera. Lavision software by DaVis translated and stored the video at a rate of 1000 frames per second. For data analysis using DaVis Software, ten points within the bifurcation were chosen for velocity measurements. Point locations were strategically placed based on the typical locations of plaque buildup and areas where flow separation was occurring based on video observations. Other points were chosen as controls to compare to the velocities at points where more flow separation was expected (all points within the sinus). The velocity components at the ten points were obtained from the software at two different time points.

RESULTS

Results indicated that overall, as the flow through the ICA decreased, the streamwise velocity in the ICA sinus decreased to negative values (Figure 1) whereas velocities in the CCA and ECA increased and were more uniform (Figure 2).

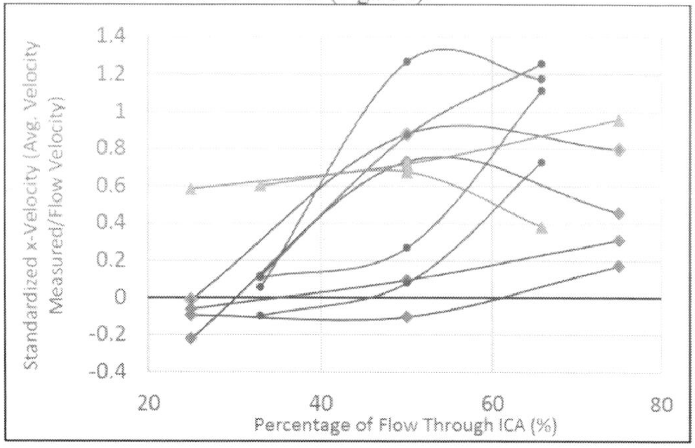

Figure 1: *Points in ICA with low or negative streamwise velocities (red= high flow, blue= low flow)*

The points in the ICA located on the opposite side of the split experienced extremely low, and sometimes negative, velocities, indicating flow separation and low wall shear stress. One point in the ICA sinus however, located right next to the bifurcation split, experienced no significantly low streamwise velocities, indicating that no flow separation is occurring here (green lines in Figure 1).

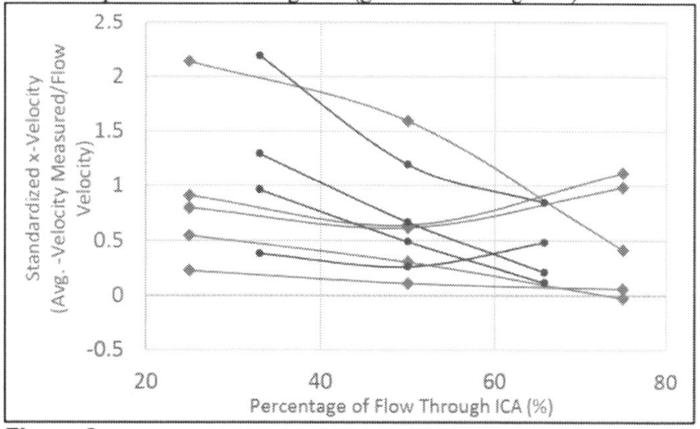

Figure 2: *Points in CCA and ECA with Positive x-velocities (red= high flow, blue= low flow)*

Although the points outside of the sinus did experience lower velocities as the flow to the ICA was increased (Figure 2), no significant negative x-velocities were observed, and the velocities were still higher than the ICA sinus values at lower flow through the ICA.

The video observations corroborated measured velocities and flow separation was visible along model's wall, but particularly significant on the wall of the ICA sinus opposite of the branch (Figure 3).

This supports the negative and zero velocities measured in this same area. There was also significant turbulence seen in the videos, indicating that assuming a stationary laminar flow is not accurate.

DISCUSSION

The results of this study indicate that flow separation, and therefore low wall shear stress that contributes to plaque buildup, is

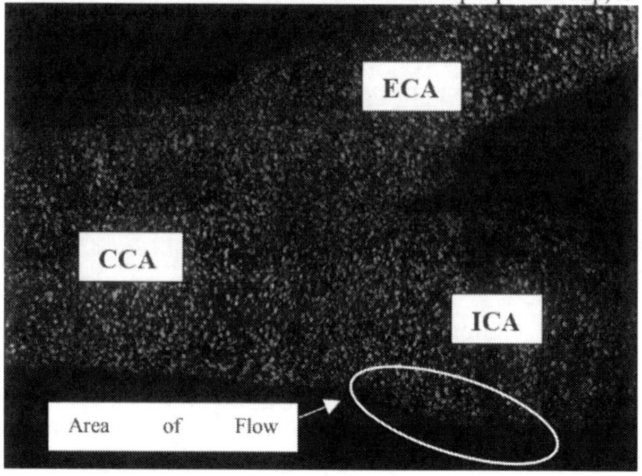

Figure 3: *3D Glass Model with Fluid Particles Video*

most common in the ICA sinus, particularly along the wall opposite of the branch. This is consistent with clinical research into the most common areas of plaque buildup [2,3,4].

Flow rate does not seem to cause flow separation as no significant differences were seen among the two flow rates tested, meaning that if low wall shear stress is present, it is present regardless of whether measurements are made during systole or diastole. The flow separation did seem to be a function of the ICA-ECA flow split (Figure 1). There were noticeable areas of flow separation, both through measured velocities and video observations, in the ICA sinus as the percentage of flow through the ICA decreased. It was also found that of the points in the ICA sinus on the wall opposite the branching, the ones with the most flow separation were closer to the initial widening of the sinus, whereas points farther along the wall, past the slight increase in diameter, showed less flow separation. This suggests that the risk of plaque buildup decreases with distance from the CCA to ICA sinus transition.

The results of this experiment indicate that plaque buildup due to low wall shear stress occurs most frequently where the ICA sinus first branches off from the CCA and on the wall opposite the flow divider, and the low wall shear stress is not necessarily a function of flow rate.

Although the glass model was created based on dimensions measured from previous research, the carotid bifurcation geometry is significantly different among individuals, so more models with varying geometries should likely be tested (currently underway) since the flow is so dependent on geometry [5]. Another limitation was neglecting the acceleration and deceleration of the fluid that is seen during systole and diastole; this experiment worked with two constant flow rates.

ACKNOWLEDGEMENTS

We would like to thank Dr. Michael Olson, the glassblower at Colorado State University, who modeled the carotid bifurcation.

REFERENCES

[1] Lopez, A.D., *The Lancet, 349*(9064):1498-1504, 1997.
[2] Zhao, S. Z., et al. *Journal of biomechanics*, 33.8:975-984, 2000.
[3] Nicholls, S.C., et al *Stroke*, 20.2: 175-182, 1989.
[4] Ku, D.N., et al *Arteriosclerosis, Thrombosis, and Vascular Biology*, 5.3:293-302, 1985.
[5] Motomiya, M. et al. *Stroke* 15.1:50-56, 1984.
[6] Zarins, K., et al. *Circulation Research,* 53.4: 502-514, 1983.
[7] Reneman, R.S., *Circulation*, 71:500-9, 1985.
[8] Ku, D.N., *J. of Biomechanics*, 20:407-21, 1987.

ANISOTROPIC HYDRAULIC PERMEABILITY IN ARTICULAR CARTILAGE AND MENISCI: A DIRECT MEASUREMENT USING BIOPSY PUNCHES

Hiromichi Fujie (1) and Ryosuke Nakamura (1)

(1) Graduate School of System Design
Tokyo Metropolitan University
Tokyo, Japan

Introduction

Articular cartilage is known to be a biphasic material; not only the solid phase of cartilage matrix but also the liquid phase of interstitial fluid generate reaction force in response to external loads. This biphasic nature was originally found by McCutchen [1] and was formulated to the biphasic theory by Mow [2]. Since then the nature of articular cartilage has been extensively determined in previous biomechanical studies. In those studies, it was important to determine the permeability of articular cartilage. One of previous methods for permeability measurement is an indirect method using confined or unconfined compression test and indentation test. With constitutive laws of articular cartilage along with Darcy's law, the permeability can be calculated. Another approach is a direct method in which fluid permeates cartilage specimens in a flow channel. In this approach, permeability is calculated from the pressure gradient and volumetric flow rate based on Darcy's law. However, it was problematic that all the methods previously developed were not suitable for the permeability measurement of a limited portion of biological tissues in varied directions.

Meanwhile, articular cartilage and menisci are known to be inhomogeneous and anisotropically structured material. We found in an analytical study that the lubrication property of articular cartilage is enhanced by its anisotropic permeability at surface [3]. This may be more significant in menisci because the inhomogeneity and structure-anisotropy are generally more significant in menisci than in articular cartilage. It is, therefore, important to determine the anisotropic property of permeability in these soft tissues for better and more precise understanding of the lubrication mechanism. Therefore, the objectives of the present study were 1) to develop a novel, confined, and direct measurement method to determine the permeability of small amount of a soft tissue using a biopsy punch and 2) to apply the developed method to the determination of the anisotropic property of the hydraulic permeability in porcine articular cartilage and menisci.

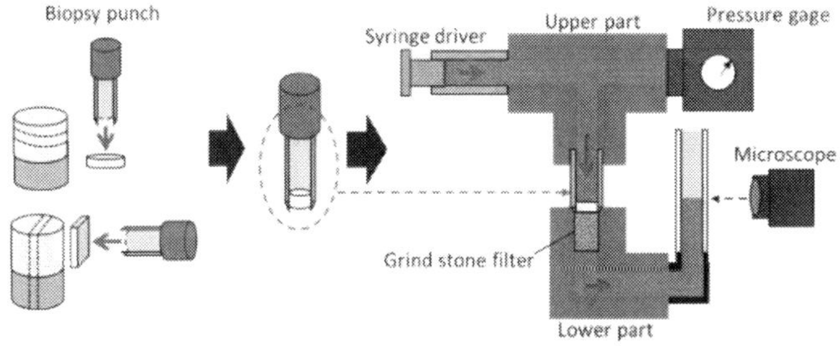

Figure 1: Preparation of articular cartilage specimens (sC/N and C/N (upper), and C/T (lower)) for permeability measurement using a permeability tester developed in the present study

METHODS

The confined permeability tester developed in the present study utilizes a biopsy punch to isolate the sample and provide the fluid flow channel for the tester (Fig.1). The tester consists of aluminum alloy-made upper and lower parts both of which have flow channels of 2.5 mm in diameter.

Cartilage-bone plug specimens (n=10) were harvested from the load bearing area of the distal femur of 6-month-old immature porcine knees. The surface layers of the plug specimens (n=5) were sliced in parallel with cartilage surface to approximately 400 μm in thickness using a microtome (Fig. 1). The middle portions of the plug specimens were also sliced in parallel with (n=5) or normal to (n=5) cartilage surface to approximately 400 μm in thickness. The cylinder of 2.5 mm in diameter taken apart from a biopsy punch was then used on the slices at the center to reduce the sample diameter. The cartilage sample was maintained inside the bottom of the biopsy punch cylinder, and they were named surface-cartilage/normal (sC/N), cartilage/normal (C/N), and cartilage/tangential (C/T) based on the flow direction in upcoming permeability tests. The medial menisci (n=9) were also harvested from the pigs. The posterior portions of the menisci were sliced along the circumferential direction (n=5) or radial direction (n=4) to approximately 900 μm in thickness using the microtome. In the same way as the preparation for cartilage sample described-above, the meniscus samples were prepared and named meniscus/radial (M/R) or meniscus/circumferential (M/C) based on the flow direction in upcoming permeability tests.

The top end of the biopsy cylinder was fixed in the channel of the upper part of the tester, while the bottom end of the cylinder (the punch and sample) was fixed in the channel of the lower part of the tester so that the lower side of the cartilage sample was in direct contact with a permeable ground stone filter. The seal between the biopsy punch with the punched sample and the channels were maintained using small o-rings. While monitored using a pressure gage (PGS-50KA, Kyowa Electronic Instruments, Japan), flow pressure was increased to 70 kPa and held at the value for 2,400s through a manual control of a syringe driver (KDS-101, KD Scientific, USA). After 1,800s, flow volume permeating the sample for 600s was monitored with a microscope (VHX-1000, Keyence, Japan) to determine the volumetric flow rate. The thickness of cartilage and menisci were measured using the microscope after the permeability test. The permeability of the samples were calculated from the flow rate and pressure gradient based on Darcy's law.

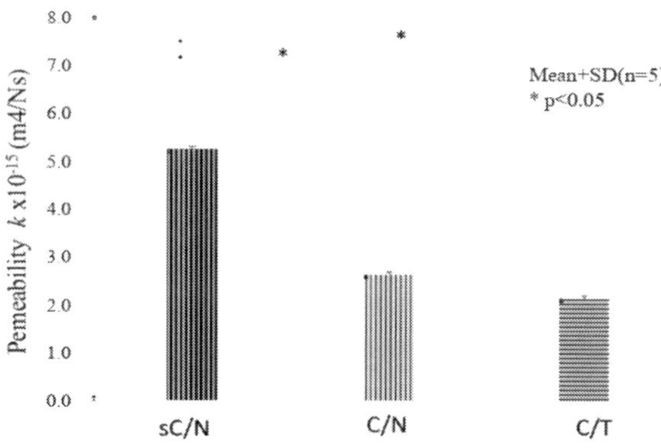

Figure 2 Hydraulic permeability of articular cartilage in sC/N, C/N, and C/T groups

RESULTS

In the permeability tests for the cartilage and meniscus samples, fluid pressure was successfully held at 70 kPa for 2,400s with a simple manual handling of the syringe driver. Results revealed that the permeability of C/N and C/T samples were (2.60 ± 0.81) x 10^{-15} m^4/Ns and (2.12 ± 0.71) x 10^{-15} m^4/Ns, respectively, with no significant difference observed between two groups (Fig. 2). In contrast, the permeability of sC/N sample was (5.23 ± 1.70) x 10^{-15} m^4/Ns and significantly higher than those of C/N and C/T samples. Meanwhile, the permeability of M/R and M/C samples were (20.24 ± 1.37) x 10^{-15} m^4/Ns and (0.77 ± 0.19) x 10^{-15} m^4/Ns, respectively. The value for M/R sample was approximately 20 times higher than that for M/C sample, with a significant difference between two sample groups. As compared with the permeability of cartilage mid-portion, the permeability of menisci was significantly higher in redial direction and lower in circumferential direction.

DISCUSSION AND SUMMURY

It is believed that the developed confined permeability – biopsy punch tester may provide advantages over existing confined and unconfined testers in terms of handling biological samples and fluid sealing. In previous permeability tests, biological tissues were cut using some surgical tools and the obtained samples were then mounted in some mechanical parts, such as Tygon tubes, consisting of a part of flow channels of permeability testers. Risk was high to break the samples due to the process of cutting and subsequent mounting, which may have harmful influences on fluid sealing during permeability tests. With the method using biopsy punches developed in the present study, the risk is low because the sample preparation is simple; just a punch to tissues. Another advantage is that it is possible to perform permeability measurement for biological tissues of small size along any direction. This enabled us to measure the hydraulic permeability of articular cartilage and menisci along different directions in the present study. Moreover, it is possible with the method to measure the permeability of cartilage surface, which may provide useful information for better understanding of cartilage lubrication. Obtained results indicated no significant difference between the permeability of C/N samples determined in the present study and that of corresponding samples determined using an unconfined permeability tester in our previous study [4], which validates the developed permeability measurement method.

Experimental results from immature pigs revealed that the permeability of articular cartilage under no deformation is isotropic. This result agrees with a finding by Reynaud and Quinn [5]; they determined normal and tangential permeability of articular cartilage through a direct measurement combined with a finite element analysis. We also found that the permeability of cartilage is significantly higher at surface than in middle layer. Results also reveled that the permeability of menisci is significantly higher in radial direction and lower in circumferential direction as compared with that of articular cartilage. It is suggested that these features as regard with the permeability of articular cartilage and menisci and obtained quantitative results are all useful for better understanding of their mechanics and lubrication.

ACKNOWLEDGEMENTS

The present study was financially supported by the Grant-In-Aid for Scientific Research B (#25252134) both from the MEXT, Japan, and the Priority Research Found from Tokyo Metropolitan University.

REFERENCES

1) McCutchun, *Nature* 184, 1959
2) Mow et al, *J Biomech Eng* 102, 1980
3) Imade et al, *Proc SBC(ASME)*, 2013
4) Susa et al, *Proc SBC(ASME)*, 2011
5) Reynaud and Quinn, *J Biomech* 39, 2006

SB^{3}C2015
Summer Biomechanics, Bioengineering and Biotransport Conference
June 17-20, 2015,Snowbird Resort, Utah, USA
SB^{3}C2015-421

THE EFFECT OF RESOLUTION ON VISCOUS DISSIPATION MEASURED WITH 4D FLOW MRI IN PATIENTS WITH FONTAN CIRCULATION: EVALUATION USING COMPUTATIONAL FLUID DYNAMICS

M. Cibis (1) K. Jarvis (2, 3), M. Markl (2,3), M. Rose(2, 4), C. Rigsby (2, 3), A. Barker (2), J. Wentzel (1)

(1) Biomedical Engineering Department
Erasmus MC
Rotterdam, The Netherlands

(2) Radiology Department
Northwestern University
Chicago, Illinois, USA

(3) Biomedical Engineering Department
Northwestern University
Chicago, Illinois, USA

(4) Medical Imaging Department
Ann& Robert H Lurie Children's Hospital of Chicago
Chicago, Illinois, USA

INTRODUCTION

Infants born with hypoplastic heart syndrome with only one fully functioning ventricle undergo consecutive interventions leading to Fontan circulation. By these interventions, the cardiovascular system is reconstructed and the systemic venous return is supplied directly to the lungs through the pulmonary arteries bypassing the right ventricle [1]. The outcome after interventions varies greatly between patients suggesting that the interventions are far from being optimal. There is growing evidence that complex flow patterns and particularly high viscous dissipation inside Fontan circulation are associated with the limited exercise performance of Fontan patients [2]. There has been therefore increasing interest in viscous dissipation inside Fontan circulation. Viscous dissipation is generally calculated by using mechanical energy balance equation [3]. This equation requires information on velocity and pressure inside the arteries. Pressure inside arteries can be obtained either by invasive measurements or by computational fluid dynamics (CFD) simulations. Viscous dissipation can also be calculated using viscous dissipation term of the Navier-Stokes equations [4]. The latter approach does not require information on pressure but only on velocities which can be measured in-vivo with 4D-flow MRI but the results might be dependent on the spatial resolution of velocities and the noise level of the measurements. In this study we aim to evaluate the influence of spatial resolution of velocity and noise on viscous dissipation calculations. We calculated viscous dissipations using CFD at different spatial resolutions and noise levels and compared them with those based on in vivo 4D-flow MRI.

METHODS

To validate the agreement between the approaches based on the mechanical energy balance equation and the viscous dissipation term of the Navier Stokes equation in viscous dissipation calculation and to analyze the effect of spatial resolution systematically, we first calculated viscous dissipation for a simulated steady flow in a straight tube with a diameter of 6.5 mm, a length of 2 cm and inflow of 8.15 mL/s. The pressure and velocities inside the straight tube were obtained by performing CFD simulation. The density and the viscosity were set as 1.06 g/cm3 and 3.5 cP (Re =242). The velocities obtained by CFD were then down-sampled to mimic MRI data at different spatial resolutions [5].

Afterwards, in vivo viscous dissipation was calculated inside the Fontan circuit for 6 patients who underwent 4D-flow MRI with the coverage of heart and the great arteries (spatial resolution: 1.9-2.5 x 1.9-2.5 x 2.2-3.3 mm^3 , temporal resolution: 38.4-41.6ms, venc: 100-150 cm/s, TR: 2.36-2.72 ms, TE: 38.4-41.6 ms, flip angle=15°). Subject-specific, time-dependent CFD simulations were performed for all subjects. MRI based velocities measured at inferior vena cava (IVC), superior vena cava(SVC) and right pulmonary artery (RPA) were used as flow boundary conditions. Left pulmonary artery (LPA) was left as stress free. At the time point with highest inflow, the CFD-based velocities were interpolated to a grid with an isotropic resolution of 0.2 mm. These velocities were then down-sampled to a grid at MRI resolution and subject-specific noise was added onto velocities, mimicking in vivo 4d-flow MRI data. Viscous dissipation was calculated using the velocities obtained from 1) CFD, 2) down-sampled CFD, 3) down-sampled CFD with noise added and 4) 4d-flow MRI measurements.

RESULTS

The viscous dissipation in the straight tube was found 6.59 µW by using the mechanical energy balance equation and 6.58 µW by using the viscous dissipation term in Navier Stokes at resolution of 0.02 mm. At lower spatial resolutions of velocities, viscous dissipation was underestimated with the latter approach (Figure-1). At the resolution of 2mm, which is representative of the spatial resolution of 4d-flow MR images in clinical routine, viscous dissipation was estimated 44% lower (2.88 µW). Figure 1 shows the viscous dissipation in the straight tube for decreasing resolution.

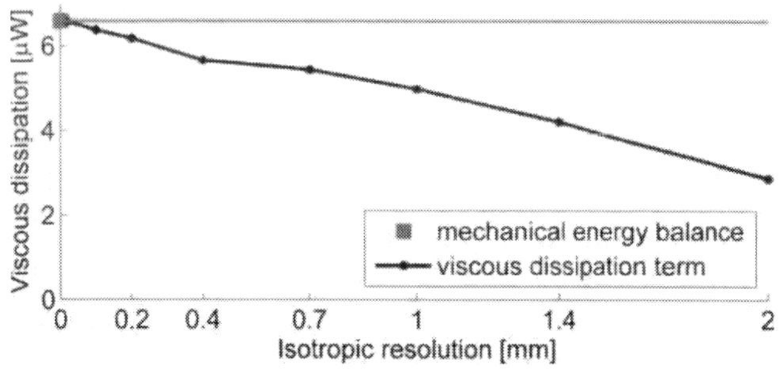

Figure-1: Viscous dissipation in the straight tube. The red line shows is based on the energy balance equation and the black data points are based the viscous dissipation term of the Navier Stokes equation at different spatial resolutions.

Maximum intensity projection (MIP) of the in vivo velocity and viscous dissipation fields for one representative case is shown in Figure-2. The MIP's of velocity and the viscous dissipation were qualitatively in good agreement between CFD, down-sampled CFD with noise added and 4D-flow MRI based calculations. The velocities and the viscous dissipation were generally larger at the T-junction. In regions containing uniform flow, such as IVC and SVC, the viscous losses were low.

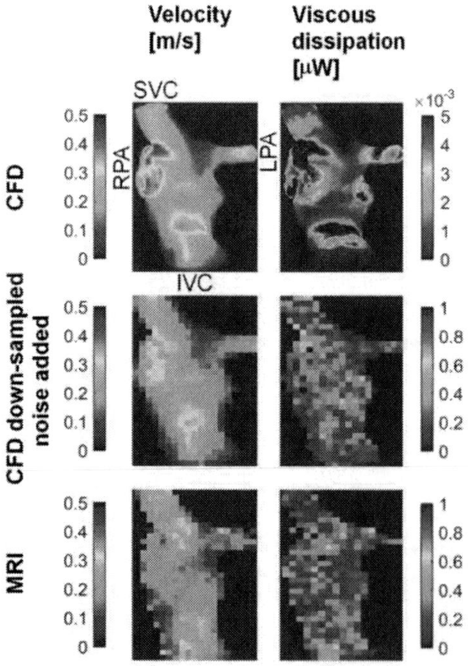

Figure-2: MIP of velocity (left) and viscous dissipation (right) obtained by, CFD, down-sampled CFD with noise added and 4d-flow MRI for one representative case.

The 4D-flow MRI based velocities (15.9±3.8 cm/s) were similar to the CFD based velocities (14.0±4.6 cm/s, p=0.09), the down-sampled CFD based velocities (13.4±4.5 cm/s, p=0.06) and the down-sampled CFD velocities with noise added (13.8±4.2 cm/s, p=0.06). Although not significant, MRI-based velocities were higher in 5 out of 6 cases (Figure-3). Similar to the study based on the straight tube, in vivo viscous dissipation was underestimated at lower resolution. Viscous dissipation based on CFD velocities (0.75±0.52 mW) was higher than the down-sampled CFD velocities (0.17±0.14 mW, p=0.03). The noise added increased viscous dissipation (0.25±0.15 mW), but the values remained lower than those based on 4D-flow MRI (0.30±0.16 mW, p=0.03). Relative viscous dissipation between subjects remained the same in all cases except for case 1 (Figure-3).

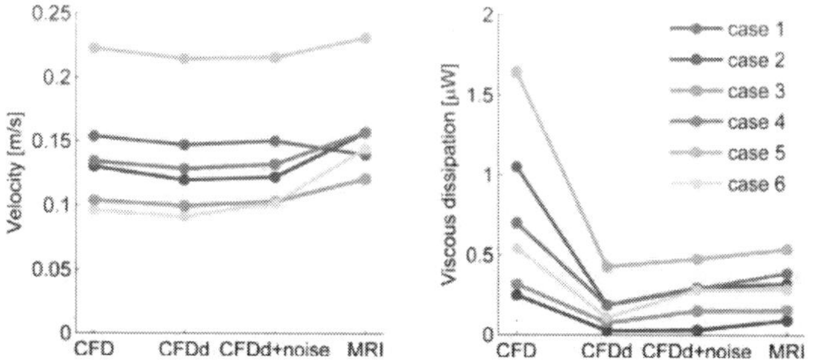

Figure-3: Mean velocity [cm/s] (left) and viscous dissipation [mW] (right) by using velocities of CFD, down-sampled CFD velocities without noise (CFDd), down-sampled CFD velocities with added noise (CFDd+noise) and MRI measurements. Each color represents different case.

DISCUSSION

This study showed the importance of spatial resolution of phase contrast MRI based velocities in calculating viscous dissipation by using the viscous dissipation term of the Navier Stokes equation. Lower spatial resolution of velocities results in underestimation of viscous dissipation compared to CFD based viscous dissipation. Nevertheless, relative viscous dissipation between different subjects was maintained irrespective of resolution and noise. We suggest that the comparison of viscous dissipation between different patient groups to detect the cases with elevated viscous dissipation is still possible, and velocities based on 4D-flow MRI with the same resolution can therefore be potentially used to calculate relative viscous dissipation among a group of patients.

ACKNOWLEDGEMENTS

This study is funded by Dutch Technology Foundation STW [Carisma 11629], NIH NHLBI grant R01HL115828; NIH grant K25HL119608

REFERENCES

[1] Gewillig M, et al. *Heart.* 2005; 91:839-846
[2] Khiabani RH, et al. *Heart.* 2014 Sep 2. pii: heartjnl-2014-306337.
[3] Bossers SM, et al. *Heart.*2014; 100:9 696-701
[4] Barker AJ, et al. *Magn Reson Med.* 2014 Sep;72(3):620-8
[5] Cibis M, et al. *NMR in Biomed.* 2014 Jul; 27(7):826-34

SB³C2015
Summer Biomechanics, Bioengineering and Biotransport Conference
June 17-20, 2015, Snowbird Resort, Utah, USA
SB³C2015-422

A FINITE ELEMENT INVESTIGATION OF FRACTURE HEALING UNDER SIMULATED MICROGRAVITY LOADING CONDITIONS

Benjamin C. Gadomski (1), Zachary F. Lerner (1), Raymond C. Browning (1,2), Christian M. Puttlitz (1,3)

(1) School of Biomedical Engineering
Orthopaedic Bioengineering Research Laboratory
Colorado State University
Fort Collins, CO
USA

(2) Department of Health and Exercise Science
Physical Activity Laboratory
Colorado State University
Fort Collins, CO
USA

(3) Department of Mechanical Engineering
Orthopaedic Bioengineering Research Laboratory
Colorado State University
Fort Collins, CO
USA

INTRODUCTION

The literature is deficient with regard to how the localized mechanical environment of skeletal tissue is altered during microgravity unloading and how these alterations affect bone remodeling and healing. Limited research has been performed to investigate the direct role of these reduced gravitational forces as they relate to fracture healing. The few *in vivo* studies that have been performed have consistently demonstrated that weight-bearing maintains skeletal integrity of healthy bones and ultimately accelerates the healing of long bone fractures by promoting rapid callus formation, while the lack of mechanical loading experienced during weightlessness leads to the inhibition of fracture healing. Alterations in the localized mechanical environment within mineralized tissues due to microgravity unloading remain inadequately described due to the experimental limitations associated with such tasks. However, the use of computational techniques may aid in elucidating the mechanical underpinnings of skeletal adaptation and healing in microgravity environments. Thus, the purpose of this study was to computationally characterize the local mechanical environment responsible for the inhibited fracture healing observed under experimental simulated microgravity conditions.

METHODS

The effects of simulated microgravity on bone remodeling and fracture healing were previously investigated in two animal studies using a large animal (sheep) model [1,2]. Animal use approval was granted by the Colorado State University Animal Care and Use Committee (Approval #11-2938A). In the first study, a trans-biarticular fixator was applied to the hindlimb of five skeletally mature sheep for 8 weeks (ExFix group). This unloading technique was shown to simulate a 0.25g environment, or a 75% reduction in loading and a loss of metaphyseal bone mineral density of 25% per month. The second experiment investigated fracture healing using this simulated microgravity model. Following a 21 day simulated microgravity period and a resultant loss of bone mineral density of approximately 18%, a 3.0mm mid-metatarsal ostectomy was performed and stabilized with an orthopaedic locking plate instrumented with a rosette strain gage. An Earth gravity (Control, n=5) group was included in which an ostectomy was created, plated, and casted, allowing full loading to be transmitted through the bone. Both groups were euthanized after 28-days and post-sacrifice mechanical, micro-computed tomography (μCT), and histomorphometric analyses were performed to quantify fracture healing.

A high fidelity FE model of the ovine hindlimb extending from the tibia to the proximal phalanges was prepared from CT imagery data of a fully mature ewe. Each bone was segmented based on Hounsfield attenuation values to generate surface representations, and 8-noded hexahedral elements were morphed to the surface geometry of each bone to create a mesh consisting of 215,000 elements. Articular cartilage was modeled as a 0.5mm, three-element thick layer extruded from the osteochondral surfaces. Transversely isotropic, linearly elastic material properties were assigned to the cortical and cancellous bone, and a hyperelastic Mooney-Rivlin material definition was assigned to the articular cartilage. A total of 8 ligaments of the metatarsophalangeal and hock joints were represented via spring elements. The external fixation components were then added to the model to create an ExFix group configuration while the base model was utilized to simulate the Control group.

To determine the appropriate model mesh resolution, strain energy density predictions from the model's hard and soft tissue constituents were compared between low, medium, and high mesh density models under identical loading conditions wherein the number of elements for each model was approximately doubled to create the next highest resolution model to ensure that the appropriate number of elements was utilized. Mode validation was performed by comparing metatarsal surface strain predictions to *in vitro* and *in vivo* experimentally-obtained strain measurements. Model predictions typically fell within one standard deviation of experimentally-derived data while the parametric convergence study demonstrated medium resolution model predictions to be within 5% of those predicted by the high resolution model. Therefore, the medium resolution model (215,000 elements) was used for the subsequent analysis.

A 3mm mid-diaphyseal ostectomy was created in the ExFix and Control models (Figure 1). Callus dimensions for each model were taken from histological data. The callus material was modeled as a linearly elastic and isotropic, and a parametric FE analysis was performed to calibrate the model four-point bending stiffnesses to the experimentally-derived results. Each model was then loaded with muscle forces from a previously-developed musculoskeletal model [3] corresponding to 100N, 200N, and 300N GRF standing loads as well as a gait speed of 0.75m/s (corresponding to the maximum speed of the housed animals, or 600N). The local strain components and hydrostatic pressure within the fracture gap and periosteal callus predicted by each model were then compared with histological results.

Figure 1. The (right) ExFix and (left) Control FE fracture models were generated by creating a 3mm ostectomy and callus (red insets) at the mid-diaphysis of the metatarsus.

RESULTS

The *in vivo* study demonstrated inhibited healing in animals exposed to simulated microgravity as compared to those that healed in a 1g Earth gravitational environment. μCT results indicated decreased callus bone volume in simulated microgravity specimens versus 1g specimens. 1g specimens routinely displayed endochondral ossification bone formation in the periosteal callus as well as lower levels of intramembranous bone formation around the periosteal callus perimeter, while the simulated microgravity specimens appeared to heal directly through intramembranous bone formation without evidence of a cartilage intermediary (endochondral ossification, Figure 2).

As expected, model hydrostatic pressure and strain predictions were greatest for a GRF of 600N (0.75m/s gait speed) in both FE models and decreased as a function of GRF. Both models predicted peak hydrostatic pressures and strains within the cortices of the fracture gap (Figure 3) contralateral to the orthopaedic fixation plate, with both parameters decreasing radially toward the callus periphery.

DISCUSSION

The mechanical unloading experienced during simulated microgravity induced inhibited fracture healing via fundamental changes to the bone formation sequelae. Previous studies have elucidated specific envelopes of pressure and strain that lead to either endochondral or intramembranous ossification (Figure 3).

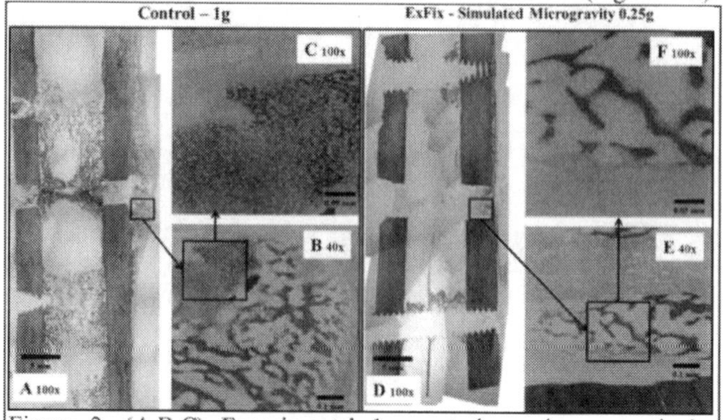

Figure 2. (A,B,C) Experimental 1g control specimens routinely displayed endochondral ossification bone formation in the periosteal callus as well as lower levels of intramembranous bone formation around the callus perimeter. (D,E,F) The simulated microgravity specimens appeared to heal directly through intramembranous bone formation.

The predictive data of this FE study suggests that both hydrostatic pressure and strain of the healing fracture contributed to alterations in the healing process, with animals exposed to a simulated microgravity environment subsequently healing via intramembranous bone formation rather than the typical endochondral ossification process experienced by animals healing in an Earth gravitational environment. These findings will help direct future countermeasures for enhancing bone healing in microgravity environments.

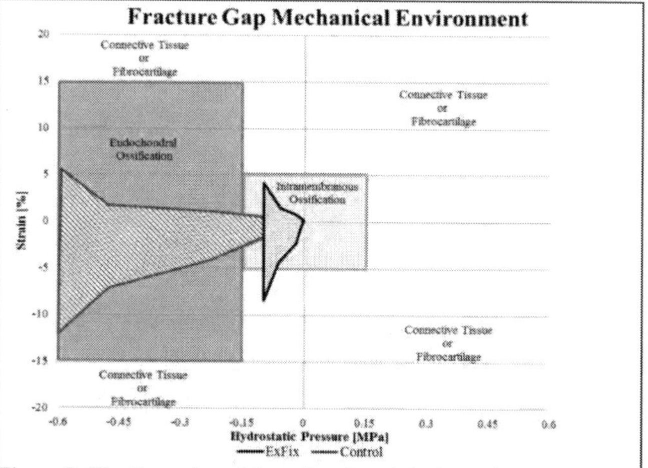

Figure 3. The Control model predicted peak hydrostatic pressure and strain within the endochondral ossification envelope for the range of GRFs while all hydrostatic pressure and strain predictions fell within the intramembranous ossification zone for the ExFix model.

ACKNOWLEDGEMENTS

Funding provided by the National Aeronautics and Space Administration (NNX11AQ81G).

REFERENCES

[1] Gadomski, B et al., *J Biomech Eng*, 136:021021-021029, 2014.
[2] Gadomski, B et al., *J Biomech*, 47:2836-2842, 2014.
[3] Lerner, Z et al., JOR, In Press.

SB³C2015
Summer Biomechanics, Bioengineering and Biotransport Conference
June 17–20, 2015, Snowbird Resort, Utah, USA
SB³C2015-423

COHERENT WALL SHEAR STRESS STRUCTURES DETERMINE THE NEAR WALL TRANSPORT IN ANEURYSMS

Amirhossein Arzani (1), Guoning Chen (2), Alberto M. Gambaruto (3), Shawn C. Shadden (1)

(1) Department of Mechanical Engineering
University of California, Berkeley
Berkeley, CA, USA

(2) Department of Computer Science
University of Houston
Houston, TX, USA

(3) Department of CASE
Barcelona Supercomputing Center
Barcelona, Spain

INTRODUCTION

Wall shear stress (WSS) is the tangential component of traction on the wall. It is one of the most extensively studied parameters used to correlate hemodynamics to various aspects of cardiovascular disease. The major motivation behind these studies is the concept of disturbed flow and how it influences chemical and physical processes at the lumen/wall. Regions of low and oscillatory WSS are often regarded as disturbed flow. However, WSS also conveys information about near wall transport [1], which is often not considered. The pulsatile nature of blood flow, and the complex anatomies in diseased arteries introduce complex temporal and spatial WSS patterns. It is not clearly understood how these unsteady WSS patterns can influence the near wall flow features. This becomes particularly important in thrombus formation studies where the transport of biochemicals near the wall are highly influenced by the WSS patterns. This study aims to characterize the near wall flow structures induced by unsteady WSS vector fields.

METHODS

Abdominal aortic aneurysms (AAA) were chosen in this study due to their complex spatiotemporal WSS patterns, and the importance of thrombus deposition in their pathology. Wall shear stress data was obtained from patient-specific computational fluid dynamics studies carried out in prior work [2]. Briefly, the models started from the supraceliac aorta and ended with the iliac arteries, including major side branches. Patient specific boundary conditions were prescribed. The computational mesh consisted of linear tetrahedral finite elements, with a maximum edge size of 0.075 cm in the interior, and five layers of boundary layer meshing with next-to-wall edge size of 250 microns.

The wall shear stress vectors were used to study the near-wall flow topology using a Lagrangian framework. Using an intrinsic orthonormal frame $(\hat{\mathbf{t}}, \hat{\mathbf{b}}, \hat{\mathbf{n}})$ one can represent the near wall velocity up to first order [1]:

$$u_t \approx \frac{\mathrm{WSS}_t \delta n}{\mu}$$
$$u_b \approx \frac{\mathrm{WSS}_b \delta n}{\mu}, \qquad (1)$$

where t and b refer to the tangential and binormal components, μ is the dynamic viscosity, and δn is a small normal to the wall distance. The methods used in [3] to compute streamlines on surface vector fields were extended to computation of time-dependent surface trajectories. Tracers were seeded on the entire aneurysm surface and advected in forward and backward time on the surface using the velocity field defined above. WSS Exposure time for each surface element was computed as the accumulated amount of time of all the tracers visiting that element divided by the square root of the element area. WSS Residence time was computed as the amount of time that each tracer takes to leave the aneurysm region, and was mapped to the tracer's initial position.

RESULTS

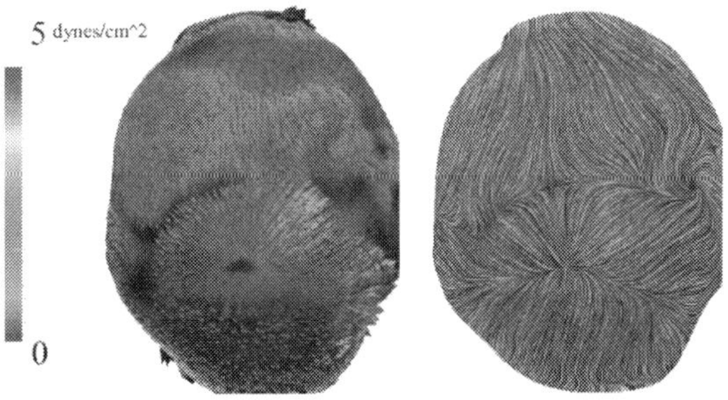

FIGURE 1: AN EXAMPLE OF INSTANTANEOUS WSS VECTOR FIELDS AND THE CORRESPONDING STREAMLINES.

Figure (1) shows an example of the complex WSS vectorial patterns observed in AAAs. The fixed points of the WSS vector fields and their stability type determines the patterns observed in steady flows. Sources, sinks and saddle points in the WSS vector field represent flow impingement, upwelling motion, and separation, respectively. The situation is more complicated in unsteady flows. Figure (2) shows the unsteady WSS tracer advection results plotted with exposure time and residence time contours. In forward time, the tracers form attracting coherent structure [4]. This structure remains confined to the region of high exposure time. In

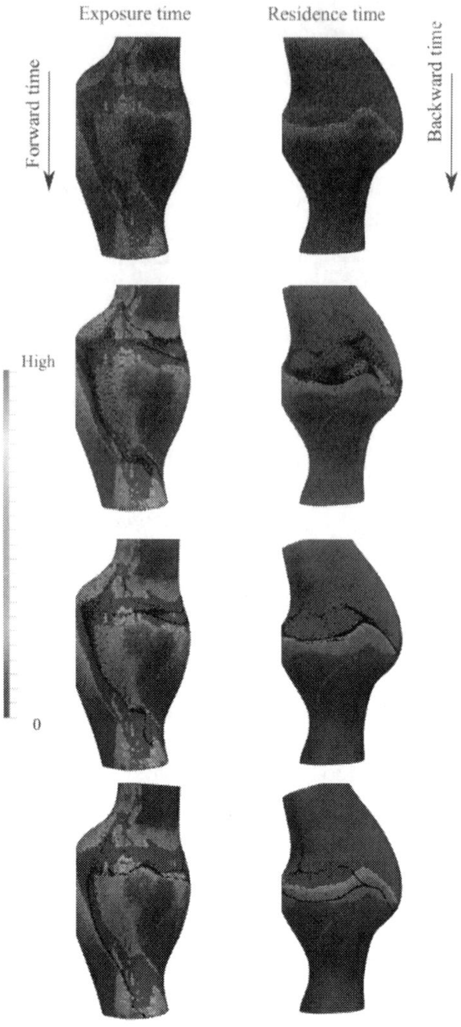

FIGURE 2: CONTOURS OF WSS EXPOSURE TIME AND WSS RESIDENCE TIME. THE FORWARD TIME AND BACKWARD TIME ADVECTION OF SURFACE TRACERS ARE SHOWN ON TOP OF THE EXPOSURE TIME AND RESIDENCE TIME CONTOURS, RESPECTIVELY.

backward time, the tracers form a repelling coherent structure. This structures determines the boundary of low residence time tracers that get flushed out of the domain, and high residence time tracers that remain inside the aneurysm for a very long time. These high residence time tracers are eventually attracted to the attracting structure in the high exposure time region.

Figure (3) shows comparison of WSS exposure time and oscillatory shear index (OSI) and time average of WSS magnitude (TAWSS). Although there seems to be a positive correlation between OSI and WSS exposure time in some regions, there are regions that do not conform to this trend.

DISCUSSION

The near wall transport topology in a AAA model was investigated. The WSS vector fields give valuable information about the near wall flow structures. The unsteady WSS vector fields form a persistent attracting coherent structure that continuously remains in the region of high exposure time. A repelling coherent structure is also formed that marks the boundary of the basin of attraction for

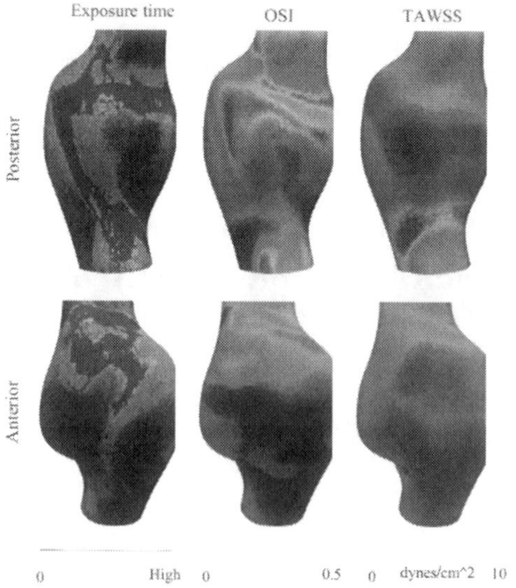

FIGURE 3: COMPARISON OF WSS EXPOSURE TIME WITH OSI AND TAWSS. ANTERIOR AND POSTERIOR VIEWS ARE SHOWN.

near wall transport. These results give valuable insight regarding transport of thrombogenic materials from the aneurysm wall. The high WSS exposure time region shows the region where these materials will be likely advected by the near wall flow, and the high WSS residence time region shows the region where the incoming materials from the wall are likely to remain inside the aneurysm for long times (many cycles). Different values for δn were chosen in the range of 0.0014–0.04 cm , but the qualitative results remained robust. The results were also robust with respect to the time of release of the tracers, due to the persistence of the repelling and attracting WSS structures, which is revealed using the Lagrangian framework.

In conclusion, for the first time we have demonstrated the existence of coherent wall shear stress structures that can potentially influence the unsteady transport of thrombogenic materials emanating from the aneurysm wall.

ACKNOWLEDGEMENTS
This work was supported by the NHLBI and an NSF CAREER award.

REFERENCES

[1] Gambaruto A, Doorly D, Yamaguchi T. Wall shear stress and near-wall convective transport: Comparisons with vascular remodelling in a peripheral graft anastomosis. *Journal of Computational Physics*, **229**(14), 5339–5356, 2010.

[2] Arzani A, Suh GY, Dalman RL, Shadden SC. A longitudinal comparison of hemodynamics and intraluminal thrombus deposition in abdominal aortic aneurysms. *American Journal of Physiology-Heart and Circulatory Physiology* **307**(12), H1786–H1795, 2014

[3] Zhang E, Mischaikow K, Turk G. Vector field design on surfaces. *ACM Transactions on Graphics*, **25**(4), 1294–1326, 2006.

[4] Shadden SC, Taylor CA. Characterization of coherent structures in the cardiovascular system. *Annals of Biomedical Engineering*, **36**(7), 1152–1162, 2008.

SB³C2015
Summer Biomechanics, Bioengineering and Biotransport Conference
June 17-20, 2015, Snowbird Resort,Utah, USA
SB³C2015-424

DETERMINATION OF THE MECHANICAL PROPERTIES OF THE IRIS USING INVERSE FINITE ELEMENT MODELING

Anup D. Pant, Rouzbeh Amini

Biomedical Engineering
The University of Akron
Akron, Ohio, USA

INTRODUCTION

Glaucoma is one of the leading causes of irreversible blindness worldwide. It is estimated that 79.6 million people will suffer from glaucoma by 2020.In the same year, 5.3 million people are estimated to be bilaterally blind due to angle closure glaucoma [1]. Angle closure glaucoma, a main category of glaucoma, is characterized by a narrow or closed anterior chamber angle, the angle between the iris root and cornea. The iris contour is a key element in angle closure pathophysiology. For example, an anterior bulging of the iris could narrow or close the anterior chamber angle.

Clinical studies have identified numerous glaucoma risk factors associated with the iris biometrics such as the curvature, thickness, and chord length. Recent investigations have also shown correlation between the iris tissue properties and angle closure glaucoma. For example, iris compressibility was significantly different when normal eyes were compared to glaucomatous ones during dynamic phenomena such as dilation [2]. Quantifying iris mechanical properties are thus of great importance in identifying glaucoma risk factors. In this study, we have used an inverse finite model to estimate iris mechanical properties assuming that the iris deformation is available beforehand.

METHODS

An axisymmetric finite element model of the iris contraction during dilation was developed previously [3]. The model assumptions were based on pharmacologically induced deformation of porcine irides in the absence of aqueous humor flow. The iris was modeled as compressible neo-Hookean solid governed by the stress balance equation:

$$\nabla.\boldsymbol{\sigma} = 0, \qquad (1)$$

where $\boldsymbol{\sigma}$ represented the Cauchy stress tensor. The Cauchy stress tensor was defined by neo-Hookean, $\boldsymbol{\sigma}_{\mathrm{NH}}$, and active dilator, $\boldsymbol{\sigma}_{\mathrm{D}}$ stress tensors as follows:

$$\boldsymbol{\sigma} = \boldsymbol{\sigma}_{\mathrm{NH}} + \boldsymbol{\sigma}_{\mathrm{D}} \qquad (2)$$

The neo-Hookean stress was defined as

$$\boldsymbol{\sigma}_{\mathrm{NH}} = \frac{G}{\det \mathbf{F}}(\mathbf{B} - \mathbf{I}) + \frac{2G\nu}{(1-2\nu)\det \mathbf{F}} \ln(\det \mathbf{F})\mathbf{I}, \qquad (3)$$

where G was the shear modulus, ν was the Poisson's ratio, **I** was the identity tensor, **F** was the deformation gradient tensor, and **B**was the left Cauchy- Green deformation tensor. **F** and **B** were defined as

$$\mathbf{F} = \frac{d\mathbf{x}}{d\mathbf{X}}, \qquad (4)$$

$$\mathbf{B} = \mathbf{F}\mathbf{F}^{\mathrm{T}}, \qquad (5)$$

where **x** was the current position of a material point and **X** was its resting position. The dilator stress $\boldsymbol{\sigma}_{\mathrm{D}}$ was applied in the dilator region (represented by the color pink in Fig 1). The dilator stress was defined by σ_{Act}, the scalar active muscle contraction stress, and $\mathbf{e}_s$, the unit vector representing the direction of the deformed dilator muscle:

$$\boldsymbol{\sigma}_{\mathrm{D}} = \sigma_{\mathrm{Act}}\mathbf{e}_s \otimes \mathbf{e}_s \qquad (6)$$

The finite element mesh was generated based on porcine iris geometry as defined in our previous work [3].

The finite element model of the iris contraction during dilation was combined with an optimization technique based on genetic algorithm [4] to calculate the iris shear modulus G and Poisson's ratio ν, assuming that the deformed configuration was known beforehand (Fig 1). In clinical setups, the iris displacement during pupil constriction or dilation can be obtained from imaging methods such as ultrasound biomicroscopy and/or anterior segment optical coherence tomography. In order to reliably calculate mechanical properties from such data, we verified our method using two sets of pseudo-experimental displacement data. In particular, we generated the deformed iris data from the predefined values of 9 kPa for G and 0.35 for ν using our forward finite element model. To access the sensitivity of the optimization method to potential experimental errors, a second set of data was generated by adding random noise within a ±%8 range of the original pseudo-experimental displacement values.

We then used our inverse modeling approach to calculate both G and $\nu$from the pseudo-experimental data sets. For every case, the objective function was chosen as the root of the sum of squared error (SSE) of the deformed configuration, i.e. the norm of the differences between the pseudo-experimental deformed position vectorsx_{exp} and the genetically driven finite element solutions vectors x_{sim} :

$$SSE = \sqrt{\Sigma\left(x_{exp} - x_{sim}\right)^2}. \quad (7)$$

The initial guesses were chosen between 5-13 kPa and 0.05-0.5 for G and ν, respectively. Following convergence of the optimization technique, SEE was calculated for a set of G (8.5-9.5 kPa) and ν (0.3-0.4) using the forward finite element model to identify whether the solutions were unique in the parameter space or not. In all case, simulations were performed using an HP Intel Xeon machine at the Ohio Supercomputing Center (Columbus, OH).

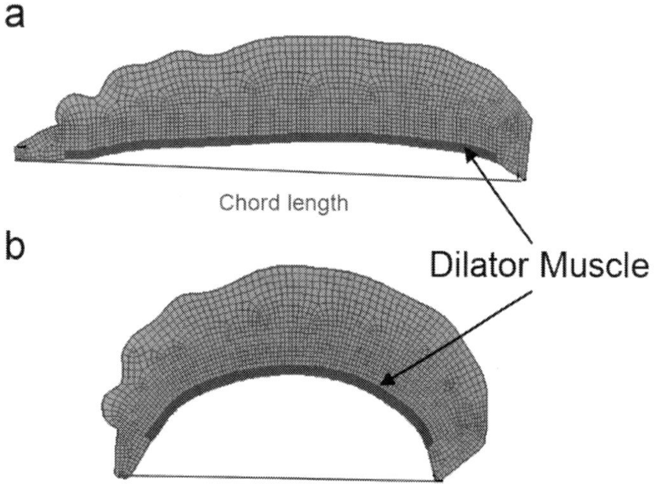

FIGURE 1: FINITE ELEMENT MODEL OF (A) THE UNDEFORMED AND (B) DEFORMED IRIS SHOWING CHANGES IN THE IRIS CHORD LENGTH BEFORE AND AFTER DILATION

RESULTS

In both noise-free and noise-added cases, the optimization technique accurately calculated the values of 9 kPa for G and 0.35 for ν. The optimization convergence was independent of our initial guess choices and broader ranges of initial guesses only increased the computational time. In each case, simulation took approximately 150 minutes (~ 100 numbers of generations) to converge. The minimum error for the noise-free data was 3.63E-05mm while for the noise-added case the error was larger (2.50E-01 mm).

When the values of G and ν were perturbated in the parameter space, no other minima were observed for the objective function (Fig. 2).As shown in Fig 2a and 2b, the SSE had only one minimum over the selected range of G and ν for the noise-free data. Similarly, only one minimum existed for the noise-added data (Fig 2c and 2d). While the addition of noise changed the value and shape of the error surface, in both cases, the error decreased as we approached the exact values of G and ν.

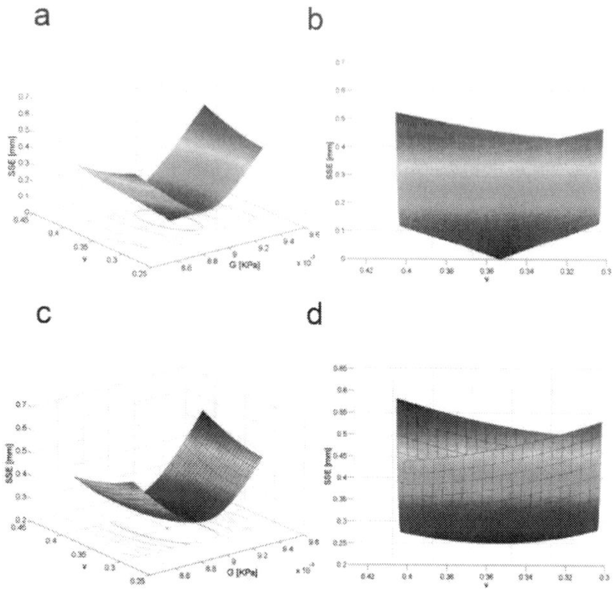

FIGURE 2: ROOT OF THE SUM OF SQUARED ERROR (SSE) PLOTTED IN (A) AND (C) VERSUS SHEAR MODULUS AND POISSON'S RATIO, ANDIN (B) AND (D)VERSUS POISSON'S RATIOFOR (A) AND (B) THE NOISE-FREE CASE AND (C) AND (D) THE NOISE-ADDED CASE.

DISCUSSION

We employed an inverse finite element method to determine the porcine iris shear modulus and Poisson's ratio. Although we have applied the method only to our pseudo-experimental data, the technique could be applied to experimental data obtained from none-invasive clinical imaging modalities such as ultrasound biomicroscopy or anterior segment optical coherence tomography.

For our future investigations, we will expand our method to obtain iris mechanical properties from deformation data acquired via ex vivo and/or in vivo experiments [3]. Quantification of such properties using clinically relevant data will enable us to identify new risk factors associated with glaucoma and help physicians detect this mysterious disease at its earlier stages.

ACKNOWLEDGEMENT

Computations were facilitated by a supercomputing resource grant from the Ohio Supercomputer Center.

REFERENCES

[1] Quigley, H.A et al., *Br J Opthalmol*, 90:262-7, 2006
[2] Quigley, H.A., *Ophthalmology*, 117:1–2, 2010
[3] Amini, R et al., *Invest Ophthalmol Vis Sci*, 53:1188-1194, 2012
[4] Storn R, Price K., *J. Global. Optim*, 11: 341–359,1997

SB³C2015
Summer Biomechanics, Bioengineering and Biotransport Conference
June 17-20, 2015,Snowbird Resort, Utah, USA
SB³C2015-425

MORPHOLOGICAL CHARACTERIZATION OF COLLAGEN FIBERS AT THE CRACK INITIATION SITES IN BIAXIALLY STRETCHED PORCINE THORACIC AORTAS TOWARD CLARIFICATION OF ANEURYSM RUPTURE MECHANISM

Shukei Sugita and Takeo Matsumoto

Department of Mechanical Engineering, Graduate School of Engineering,
Nagoya Institute of Technology,
Nagoya, Aichi, Japan

INTRODUCTION

Aortic aneurysms tend to expand gradually and sometimes result in a fatal rupture. The number of death caused by its rupture has been increasing in Japan year by year, possibly because of the rapid aging in this country. Degradation of elastin in aneurysm walls is well known and resulted in the increase in the volume fraction of collagen fibers [1]. Since elastic modulus of collagen fibers is higher than other major constituents in the aorta [2], it is reasonable that aneurysmal wall has higher elastic modulus [1]. Since tensile strength of collagen fibers is also higher than the other constituents [2], tensile strength of aneurysmal tissues should be higher, while it was similar to [3] or smaller than that of normal aortic tissues [4].

Higher heterogeneity was observed in aneurysmal tissues than normal ones [5]. Such heterogeneity may cause an increase in the degree and amount of defects in collagen fiber network in the aneurysmal walls. Such local defects may weaken the aneurysmal tissue because they easily become crack initiation sites, resulting in the rupture of aneurysms. To test this hypothesis, we developed a biaxial stretch tester which enabled us to observe crack initiation sites under a microscope [6].

In this study, morphological characteristics at the crack initiation sites were investigated in porcine thoracic aortas especially focusing on the local volume and alignment consistency of collagen fibers.

METHODS

Specimens were prepared according to our previous study [6]. Briefly, rectangular specimens of 15 mm × 20 mm of porcine thoracic aortas obtained from a local slaughterhouse were compressed 40% in the radial direction with two metal plates which have a hole at their center. After frozen at -80°C for 10 min, the specimens were sectioned perpendicularly to the radial direction into 50-μm-thick slices with a cryostat (CM3050SIV; Leica Microsystems). After being thawed at room temperature, the specimens had thin area at their center region. Only the specimens obtained in middle one-third part in the radial direction of the aorta were used for the following experiments.

Image of second harmonic generation (SHG) light of the specimens was captured with a multiphoton microscope (FV1200MPE, Olympus) to obtain distribution of collagen fibers in the specimens. Self-mode-locked Ti:sapphire femtosecond laser pulses with a pulse width of 100 fs, a wavelength of 800 nm, a repetition rate of 80 MHz, and a power of ~3 mW were used. Generated SHG light was detected using a photomultiplier tube through a 10× objective (UMPLFLN10XW, Olympus), a dichroic mirror (485 nm), and a band-pass filter (400 ± 5 nm). Images of multiple focal planes (10-μm intervals) were captured.

The specimens were then stretched until failure with a biaxial stretch tester as stated previously [6]. Briefly, each specimen was adhered to a metal frame which had a hole 10 mm in diameter at the center. A metal cylinder 6 and 8 mm in inner and outer diameters, respectively, was placed over the center of the specimen. This tester was placed under an inverted microscope (IX-71, Olympus). Images of retardance, a phase shift induced by a birefringent material, were acquired with a birefringence imaging system (Abrio-LS, CRi) through an objective (PLAPON2×, Olympus) with a light with a wavelength of 546 nm [7]. The metal frame was moved toward the cylinder stepwise by 0.5 mm to stretch the specimen biaxially. The images of the specimen during the loading process were captured with a high-speed camera (Exilim EX-F1; Casio Computer) at the rate of 300 frame/s to determine the crack initiation position. Capturing images and moving the metal frame were repeated until specimen failure.

After specimen failure, SHG images of the specimen were captured again as stated above.

All image analysis was performed with image analysis software (ImageJ 1.47v, National Institutes of Health). Crack initiation position n was determined from the images captured with the high-speed camera, and its surrounding (250×250 μm^2) area was chosen as a crack initiation area. Data were analyzed for each 50×50 μm^2 area according to the previous study [8]. Retardance in each area was evaluated in the image just before specimen failure. Since the retardance is affected not only by collagen density but also by alignment consistency of collagen fibers, SHG images were also analyzed to separate the two factors. Intensity in SHG images captured before the stretch test was summed over the entire Z-stack and was evaluated as an index of local volume of collagen fibers. To analyze alignment consistency of collagen fibers, 2-dimensional Fast Fourier Transform (2D-FFT) of the SHG images was computed. Distribution of collagen fiber alignment was obtained in a frequency domain window. An average intensity around the center calculated for every 10° was used as the distribution. Since this distribution was similar to the normal distribution, it was fitted with the Gaussian function as follows:

$$P = A + (B - A)\exp\left\{\frac{-(\alpha - C)^2}{2D^2}\right\}, \quad (1)$$

where α is the angle in the SHG image (0° corresponds to the circumferential direction), P the possibility of collagen fiber distribution, and A, B, C, and D the fitting constants. Since the standard deviation D of the Gaussian function indicates randomness of collagen fiber alignment, the parameter D was used as an index of alignment inconsistency of collagen fibers.

All data were shown in mean ± SD.

RESULTS and DISCUSSION

A typical retardance image just before specimen failure is shown in Fig. 1. Distribution of retardance was heterogeneous as shown in the previous study [8]. Retardance value at the crack initiation sites tended to be lower than that at other areas, indicating that low density and/or low alignment consistency of birefringent materials. Because retardance in the aortas was mainly caused by collagen fibers [7], this result indicates paucity and/or poor alignment of collagen fibers at the crack initiation sites.

A typical image of SHG light is shown in Fig. 2. The image clearly shows the distribution of collagen fibers. The SHG light intensity at the crack initiation sites (65 ± 9, A.U.) was significantly lower than that in other sites (72 ± 17, A.U.), indicating that density of collagen fibers was lower at the crack initiation sites.

The index of alignment inconsistency D at the crack initiation sites (31° ± 23°) was significantly higher than that in other sites (19° ± 12°), indicating that collagen fibers aligned in more random directions at the crack initiation sites.

From retardance and SHG images, both density and alignment consistency of collagen fibers were lower at the crack initiation sites than other sites. Because collagen fibers have larger tensile strength in the aortas [2], it is reasonable that density of collagen fibers was low at the crack initiation sites. It is unclear at this stage whether there was a cause-effect relationship between low alignment consistency of collagen fibers and position of the crack initiation sites. Binding force among collagen fibers might be higher when collagen fibers were aligned in one direction. If this is the case, specimens with lower alignment consistency of collagen fibers may have lower binding force. Such lower binding force might cause higher probability for crack initiation.

Crack initiation might be affected by both local volume of collagen fibers and binding force between the fibers.

ACKNOWLEDGEMENTS

This work was supported in part by KAKENHI (No. 26709002) from Japan Society for the Promotion of Science and the Science and Technology Promotion Adjustment Cost Program from Japan Science and Technology Agency.

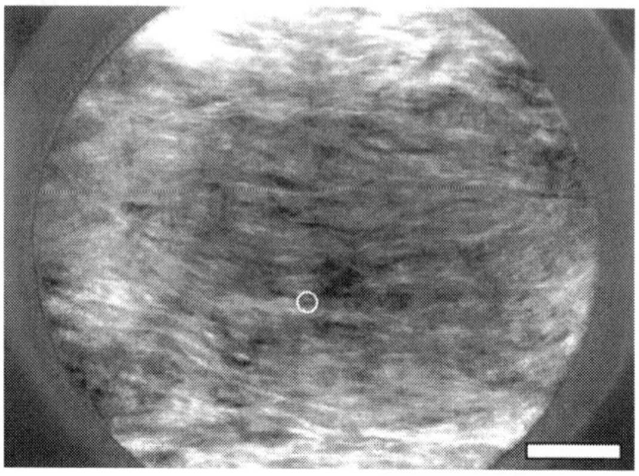

Fig. 1. A typical image of retardance distribution in a thin-sliced porcine thoracic aortic specimen. Bar = 1 mm. The white circle shows the crack initiation position.

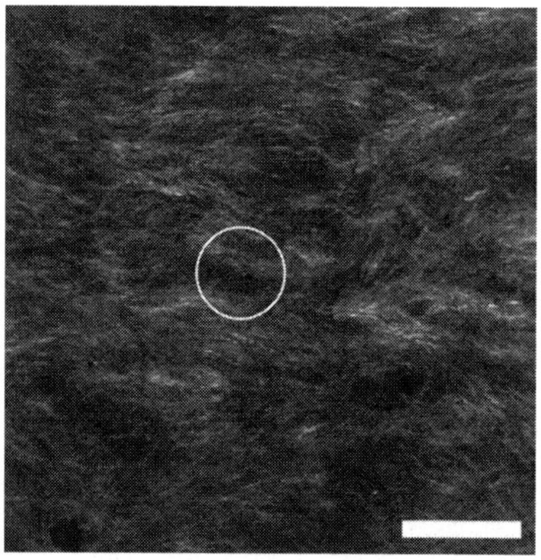

Fig. 2. A SHG light image of the specimen shown in Fig. 1. The crack initiation position is in the center of the white circle. Bar = 250 μm.

REFERENCES

[1] He CM and Roach MR, *J Vasc Surg*, 20:6-13, 1994
[2] Fung YC, Biomechanics: mechanical properties of living tissues. 2nd ed. New York: Springer-Verlag; 1981
[3] Iliopoulos DC et al, *J Thorac Cardiovasc Surg*, 137:101-109, 2009
[4] Raghaven ML et al, *Ann Biomed Eng*, 24:573-582, 1996
[5] Sugita S et al, *Cardiovasc Eng Technol*, 3:41-51, 2011
[6] Sugita S and Matsumoto T, *Biomed Eng Online*, 12:3, 2013
[7] Sugita S and Matsumoto T, *J Biomech*, 46:1403-1407, 2013
[8] Sugita S and Matsumoto T, *Biomed Mat Eng*, 23:447-461, 2013

PHENOTYPIC DIFFERENCES IN CORONARY ARTERY DISEASE PROGRESSION IN THE CLINICAL SETTING AND DEPENDENCE ON A FOCAL OSCILLATORY HEMODYNAMIC ENVIRONMENT

Lucas H. Timmins (1,2), David S. Molony (1), Parham Eshtehardi (3), Michael C. McDaniel (2), John N. Oshinski (1,4), Habib Samady (2), Don P. Giddens (1)

(1) Wallace H. Coulter Department of Biomedical Engineering
Georgia Institute of Technology and Emory University School of Medicine
Atlanta, GA, USA

(2) Division of Cardiology
Department of Medicine
Emory University School of Medicine
Atlanta, GA, USA

(3) Department of Medicine
Jacobi Medical Center
Albert Einstein College of Medicine
Bronx, NY, USA

(4) Department of Radiology and Imaging Sciences
Emory University School of Medicine
Atlanta, GA, USA

INTRODUCTION

Computational hemodynamics have been investigated as a clinical tool to aid in identifying rapidly progressing coronary artery disease (CAD) and the potential for lesions to acquire vulnerable characteristics. Prospective clinical investigations have evaluated the prognostic value of wall shear stress (WSS) in determining the characteristics of CAD progression [1,2]. We recently developed a framework that allows for examination of the focal association between WSS and virtual histology-intravascular ultrasound (VH-IVUS) defined CAD progression [3]. The investigation noted intra-patient differences in the association when data were examined in focal regions as compared to averaging around the artery circumference, highlighting the importance of examining these spatially heterogeneous variables at a regional level. Similar to previous investigations, our study did not investigate the influence of temporal oscillations in WSS vectors in combination with low time-averaged magnitude. Indeed, experimental data indicate that regions of low WSS versus low and oscillatory WSS lead to the formation of phenotypically distinct unstable and stable lesions, respectively [4].

Therefore, the aim of this investigation was to extend our developed patient-specific computational framework to a larger patient cohort and examine the association between oscillatory WSS in combination with time-averaged WSS (TAWSS) magnitude and VH-IVUS defined disease progression in the clinical setting of CAD. We present data from patients with moderate CAD and note phenotypic differences in the progression of lesions based on the presence or absence of a baseline oscillatory hemodynamic environment.

METHODS

Clinical Data All patients (n = 10) presented to the cardiac catheterization laboratory with an abnormal noninvasive stress test or stable angina syndromes and determined to have non-obstructive CAD. At the time of catheterization, biplane coronary angiography, end-diastolic gated VH-IVUS images, and Doppler derived velocity data in the left main coronary artery were collected. Patients returned at 6-months for catheterization and repeat VH-IVUS imaging data were collected. All patients provided written informed consent, and the Emory University Institutional Review Board approved the study.

Computational Modeling The patient-specific computational fluid dynamics (CFD) modeling techniques employed have been described in detail elsewhere [1,4]. Briefly, the baseline 3-dimensional lumen geometry was reconstructed from the biplane angiographic data and segmented VH-IVUS images. A volume was created from the reconstruction (Geomagic Studio 11, Geomagic, Inc.), discretized with an unstructured tetrahedral mesh (ICEM CFD 15.0, ANSYS, Inc.), and imported into a commercial solver (Fluent 15, ANSYS, Inc.). Patient-specific pulsatile velocity values were applied as a series of uniform profiles at the inlet, all outlets were assumed pressure-free, and a no-slip boundary condition was applied at the rigid wall. Following simulation convergence, nodal WSS vectors were extracted across the cardiac cycle. Vector magnitudes were averaged across the cardiac cycle yielding TAWSS values. In addition, the WSS angle deviation (WSSAD, ϕ), which quantifies the temporal angular variation between instantaneous WSS vectors and the TAWSS vector at each node, was quantified at time points (n) throughout the cardiac cycle as,

$$\phi_n = \arccos\left(\frac{\vec{\tau}_i \cdot \vec{\tau}_j}{\left|\vec{\tau}_i\right| \cdot \left|\vec{\tau}_j\right|}\right) \quad (1)$$

where τ_i and τ_j are the instantaneous and TAWSS vectors, respectively. The maximum WSSAD value in the cardiac cycle was determined to characterize the oscillatory hemodynamic environment.

Plaque Progression Quantification Changes in total area and VH-IVUS defined plaque constituent areas (fibrous, FT; fibro-fatty, FF; necrotic core, NC; dense calcium, DC) were determined by comparison of baseline and 6-month follow-up image data. With the aid of fiduciary anatomical landmarks and a developed multivariate cross-correlation algorithm [5], images were co-registered in the axial and circumferential directions, which allowed for quantification of changes in plaque areas in focal regions (i.e., sectors; Fig. 1A,B).

Data Analysis Baseline hemodynamic data were extracted at the spatial location of each VH-IVUS images in the computational domain (Fig. 1C), and nodes were divided into corresponding sectors. Within each sector, TAWSS and WSSAD data were spatially averaged. Thus, there were values for changes in VH-IVUS defined plaque areas and sector averaged hemodynamic values in each sector (Fig. 1C).

We classified TAWSS values into three groups: low (< 10 dynes/cm^2), intermediate (10-25 dynes/cm^2), and high (> 25 dynes/cm^2). Similarly, WSSAD values were classified as low (< 45°), intermediate (45°-90°), or high (>90°) oscillation. Finally, sectors were identified that exhibited both low TAWSS < 10 dynes/cm^2 and high WSSAD > 90°, herein low and oscillatory WSS.

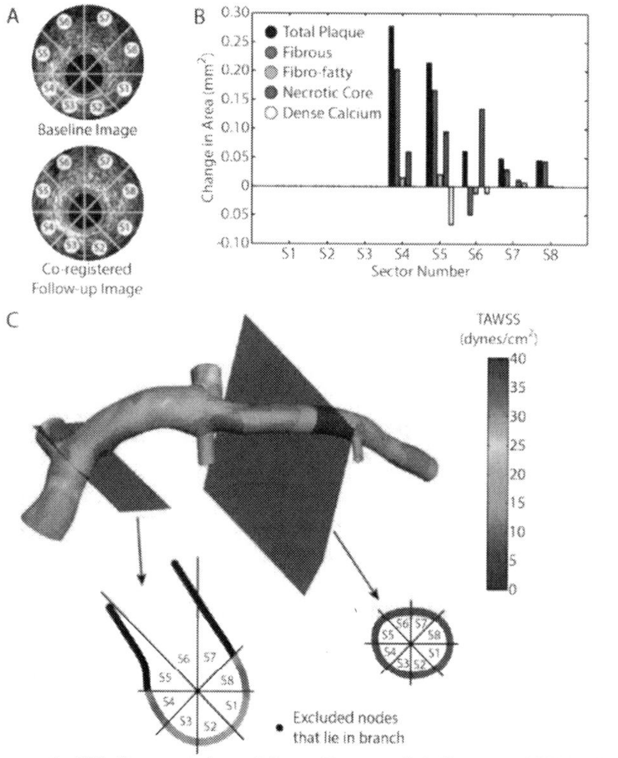

Figure 1. (A) Co-registered baseline and follow-up VH-IVUS images. (B) Quantification of changes in constituent areas in each sector. (C) Extraction of baseline TAWSS values from VH-IVUS image location and division into sectors.

RESULTS

Significant spatial and temporal heterogeneities were observed in the computed hemodynamic environment. Across the 10 patients, there were 2,352 (30.1%), 2,903 (37.1%), and 2,567 (32.8%) sectors in the low, intermediate, and high TAWSS classifications, respectively (Fig. 2A). Furthermore, temporal variations in instantaneous WSS vectors resulted in 6,234 (79.7%), 648 (8.3%), and 940 (12.0%) sectors in the low, intermediate, and high WSSAD classifications. Co-localization of low and oscillatory WSS occurred in 698 (8.9%) sectors, while low and non-oscillatory WSS occurred in 1,344 (17.2%) sectors (Fig. 2B). Heterogeneities in the baseline hemodynamic environment resulted in differences in the association between hemodynamics and 6-month VH-IVUS defined plaque progression. For example, regions of low TAWSS had an increase in plaque area, while regions of intermediate and high TAWSS exhibited decreases in plaque area (Fig. 2A). Differences were also observed when examining changes in plaque constituents. Most notably, we observed a phenotypic difference in plaque progression in regions of low TAWSS or low and oscillatory WSS (Fig. 2B). In regions of low TAWSS, total plaque, FT, and NC area increased (stable lesion), while in regions of low and oscillatory WSS total plaque and FT area decreased, but NC area increased (unstable lesion).

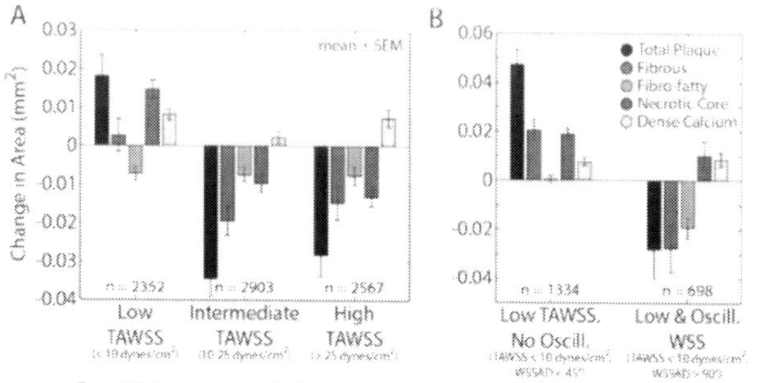

Figure 2. (A) Association between focal TAWSS and changes in total plaque area and composition. (B) Comparison of regions of low WSS or low and oscillatory WSS and changes in plaque area and composition. The numbers of sectors (*n*) in each classification is denoted.

DISCUSSION

Our investigation presents a robust methodology that allows for examination of the association between focal hemodynamics and CAD progression. We observed that phenotypic differences in progression exist depending on the presence or absence of temporal WSS vector oscillation in a low TAWSS baseline environment. It should be noted that because complex flow patterns occur in localized regions (e.g., opposite wall from side branch), only high spatially resolved analysis techniques could capture these disturbances. In fact, employing previous methods that spatially average the hemodynamic data result in very few regions of low and oscillatory WSS [1,2]. We acknowledge that data from this study disagrees with experimental data on the role of low and oscillatory WSS in atherosclerosis progression [4]; however, these data were from a surgical induced animal model of atherosclerosis and hemodynamics were analyzed on a relative, as opposed to absolute, scale. Ultimately, the clinical utility and prognostic value of WSS in identifying vulnerable lesions will require application of this framework in larger clinical studies.

ACKNOWLEDGEMENTS

This research was supported by the Georgia Research Alliance (LT, DPG), American Heart Association (DSM), Pfizer Pharmaceuticals (HS), and Volcano Corp. (HS).

REFERENCES

[1] Samady, H et al., *Circulation*, 124:779-788, 2011.
[2] Stone, PH et al., *Circulation*, 126:172-181, 2012.
[3] Timmins, LH et al., *Ann Biomed Eng*, 43:94-106, 2015.
[4] Cheng, C et al., *Circulation*, 113:2744-2753, 2006.
[5] Timmins, LH et al., *IEEE Trans Med Imag*, 32:1989-1996, 2013.

SB³C2015
Summer Biomechanics, Bioengineering and Biotransport Conference
June 17-20, 2015,Snowbird Resort, Utah, USA
SB³C2015-427

EFFECT OF HYDROXYAPATITE NANOPARTICLES ON CRYOPRESERVATION OF HUVECS

Yuanyuan Zheng (1), Jianye Wang (1), Gang Zhao* (1, 2), Tao Wang (1)

(1) Centre for Biomedical Engineering,
Department of Electronic Science & Technology,
University of Science and Technology of China,
Hefei 230027, China

(2) Anhui Provincial Engineering
Technology Research Center for
Biopreservation and Artificial Organs, Hefei,
Anhui 230027, China

* Corresponding author.
E-mail address: ZhaoG@ustc.edu.cn

INTRODUCTION

Vascular endothelial cell (EC) not only is an important target cell for rejection, but also through its antigen presenting function, phenotype change and cytokine synthesis, actively participate in the process of rejection. Human umbilical cord is an excellent source, convenient, and immunogenicity of its endothelial cells is relatively low[1,2], importantly, after cryopreservation of tissues or cells, its immunogenicity become lower according to some reports[3]. Because of these advantages, human umbilical vein endothelial cell (HUVEC) is an ideal material for surgical experiment and clinical application of vascular. In order to make use of HUVECs more effectively, the cryopreservation of HUVECs has become a very important topic in the biomedical engineering area. There are typically two approaches to achieve cryopreservation of HUVECs: conventional slow freezing and vitrification[4-7]. Slow freezing has many disadvantages while vitrification is a promising novel and simple procedure that requires less time and may become safer and more effective than slow freezing[6,8]. However, the critical cooling rates for vitrification solution are relatively high. It is reported that nanoparticles in solution offer unique thermal properties due to their physical presence and interaction with their state of dispersion[9,10]. Therefore, this thesis focuses on the effect of Hydroxyapatite Nanoparticles on the cryopreservation of HUVECs.

METHODS

HUVEC was maintained in culture medium (90% RPMI-1640) supplemented with 10% FBS. The cells were continually passaged in 75 cm² T-flasks in 5% CO_2, at 37℃. When the cells reached ~90% confluence, the cells were washed once with 1× phosphate-buffered saline (PBS). Then cells were digested in 0.25 wt% trypsin for 5 min. HUVECs were centrifuged (100 × g) for 5 min and re-suspended in fresh culture medium for further experiments.

The cryoprotectants used included 2 M 1, 2-propanediol (PROH) 2 M ethylene glycol (EG), and 0.5 M extracellular trehalose (disaccharide). Stable and single-dispersed hydroxyapatite (HA) nanoparticles used in this study (Nanjing Emperor Nano Material Co., Ltd, China) had a non-spherical shape and its diameter was 20 nm. 0.01 wt% sodium citrate and 0.005 wt% sodium hexametaphosphate were used as the dispersant. The highly distribution suspensions of HA-NPs were obtained by ultrasonic-assisted method and were characterized by dynamic light scattering. Varying concentrations of HA nanoparticles were incorporated into cryoprotectant solutions and HUVECs were vitrified by plunging HUVECs directly into liquid nitrogen. The effect of HA nanoparticles on the immediate cell mortality rates and proliferation rates post cryopreservation of HUVECs was investigated by cell count test and CCK-8 assay. The cell count test was achieved by using standard live/dead assay kit of fluorescent probes: Annexin V- EGFP and Propidium Iodide (PI). In order to determine the effect of dispersant on cell mortality rates, the highest concentration of dispersant was used as the control group first. Then, corresponding solutions without NPs while containing the same dispersant will be used as control groups.

The immunogenicity of fresh or cryopreserved HUVECs was also determined by using flow cytometry. Both HLA-A、B、C and HLA-DR antigen expression on HUVECs were determined to check the immunogenicity. After that, the mechanism of HA function was studied through a high-speed video cryomicroscopy system as illustrated in Fig. 1. The cryomicroscopy system includes the Olympus BX53 optical microscope installed with a temperature-controllable FDCS196 cryostage. What's more, the melting enthalpy of cryoprotectants was determined by DSC and the critical cooling rates of cryoprotectants were measured through ultra-fast temperature measurement system.

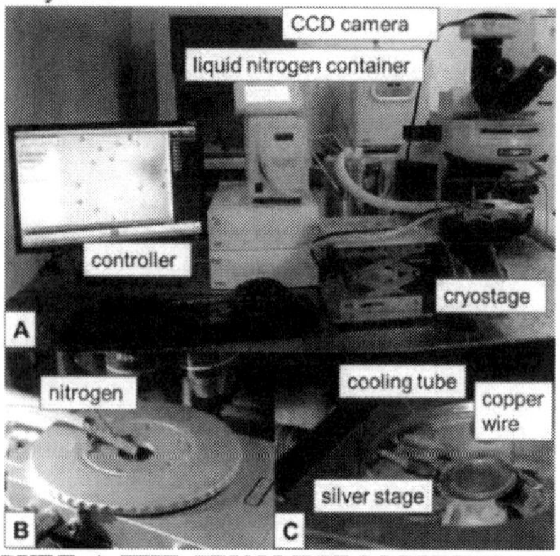

FIGURE. 1. THE CRYOMICROSCOPY SYSTEM

RESULTS

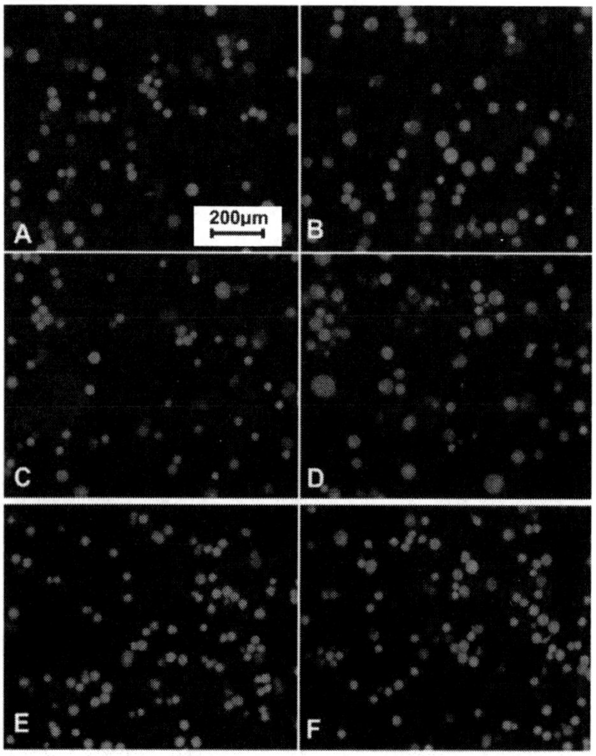

FIGURE. 2. FLUORESCENCE MICROSCOPE OBSERVATION OF HUVECS CRYOPRESERVED USING DIFFERENT CPAS

The fluorescence intensity was observed under a fluorescence microscope as shown in Figure. 2. A-F pictures represent one fluorescence microscope of vitrification solution (VS), VS+dispersant, VS+dispersant+0.01%HA,VS+dispersant+0.05%HA,VS+dispersant+0.1%HA,and VS+dispersant+0.5%HA,respevtively. Analyzing all these data with statistics software SPSS, the data obtained was expressed as average value ± standard deviation (n=5). Comparison between groups of data using t test of independent samples and $p<0.05$ was statistically significant difference. The data was indicated in Figure. 3. The results showed that HA nanoparticles reduced the mortality rate of cryopreserved HUVECs. And the proliferation rates were similar between fresh or cryopreserved groups. The results also showed adding of nanoparitcles cannot increase the cooling rate of cryoprotectant, but can decrease the amount of ice crystals during freezing and warming, therefore, the mechanical injury within and outside cells might be effectively reduced. The immunogenicity of cryopreserved HUVECs with or without HA may be different.

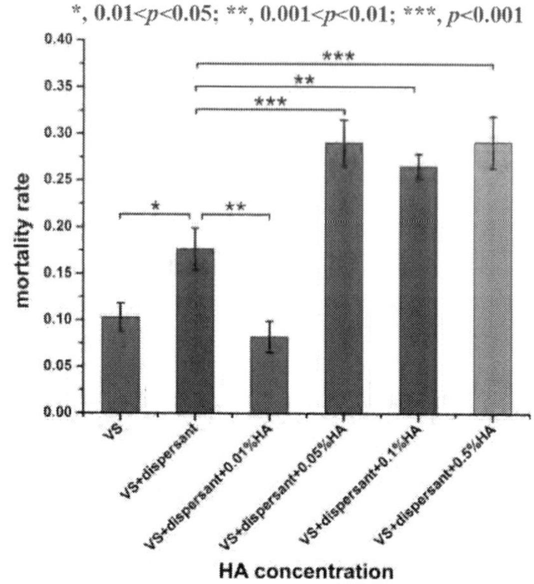

FIGURE. 3. EFFECT OF HA CONCENTRATION ON THE MORTALITY RATE OF CRYOPRESERVED HUVECS

DISCUSSION

We characterized the effect of HA on the cryopreservation of HUVECs using cell culture assays, IIF analysis and solution analysis. The solution of HA nanoparticles has an apparent effect on the mortality rate of HUVECs. The HUVECs proliferated normally and may have the reduced immunogenicity post vitrification using the combination of varying concentrations of HA nanoparticles with 2 M PROH, 2 M EG, 0.5 M extracellular trehalose.

ACKNOWLEDGEMENTS

This work was supported by the National Natural Science Foundation of China (Grant Nos. 51276179, 51476160), and the Fundamental Research Funds for the Central Universities (Grant no. WK 2100230009).

REFERENCES

[1] Rinker, B et al., Ann Plast Surg, 2(R)7, 58(6): 656-660.
[2] Cui X, et al., Cryobiology, 2007, 55(3): 295-304.
[3] Yokomise, H et al., J Thorac Cardiovasc Surg, 1996, 111(5): 930-934.
[4] Fahy, G et al., Cryobiology 21 (1984) 407-426.
[5] Fowler, A et al., Annals of the New York Academy of Sciences 1066 (2005) 119-135.
[6] Karlsson, J et al., Biomaterials 17 (1996) 243-256.
[7] Mazur, P et al. American Journal of Physiology 247 (1984) 125-142.
[8] Rall, W et al., Nature 313 (1985) 573-575.
[9] Cao, X Y et al., Joural of Wuhan University of Technology-Mater. Sci. Ed, 2003, 3: 66-68.
[10] Wang Y F, et al., Chain, 2004, 14(1): 29-32.

SCHAWNN CELL PROLIFERATION IN SCAFFOLDS WITH DECOUPLED MECHANICAL AND BIOCHEMICAL PROPERTIES

J. Stukel (1), W. Zhou (1), R. Willits (1)

(1) Biomedical Engineering
The University of Akron
Akron, OH, USA

INTRODUCTION

The peripheral nervous system (PNS) is a complex environment with multiple cell types and tissue of varying stiffness. Designing a biomaterial-based tissue engineering scaffold that promotes nerve regeneration while mimicking the *in vivo* environment remains challenging, however, because the two main cells in the PNS, neurons and Schwann cells, prefer different mechanical environments, G*~0.03-0.3 kPa [1] and ~2.5 kPa shear modulus [2], respectively. To study nerve regeneration, our lab has designed a three-dimensional (3D) microgel-based scaffold with decoupled biochemical and mechanical properties [3], within the range of the preferred mechanical environment of neurons. To form the scaffold, microgels, which are spherical hydrogels ~1μm in diameter, are functionalized with surface peptides or other molecules of interest [4] and crosslinked *in situ*. Scaffolds are formed under mild conditions, leading to high cell viability. In this project, our goal is to utilize the unique properties of the microgel-based scaffold to (1) expand the range of moduli over an order of magnitude and (2) manipulate the mechanosensing between the cells and the scaffold to improve Schwann cell proliferation, viability, and growth factor expression in scaffolds of low modulus, which are generally preferred by neurons. The first steps in this work are presented here, where the scaffold modulus range is expanded and directly correlated to Schwann cell response.

METHODS

Polyethylene glycol (PEG)-thiol microgels were fabricated via a precipitation reaction; microgels were characterized for consistent diameter, swelling, and density [4]. Scaffolds were formed by linking the microgels with 5-40 mg/mL 8-arm PEG vinyl sulfone (PEG-VS). Collagen was incorporated into the scaffold during fabrication to provide adhesion points for cells. The resulting shear modulus was measured via oscillatory shear rheometry with 5% strain and at 10 rad/second.

Cellular response within the microgel-based scaffolds was examined using primary rat Schwann cells. Cell number over 5 days was measured using Hoechst 33342-labeled DNA quantification. Viability was examined using a Live/Dead fluorescent assay and morphology was examined by fluorescently labeling the nuclei and F-actin. Gene expression of neurotrophins, specifically investigating glial growth factor, nerve growth factor, and their receptors, will be examined. Differences in Schwann cell response will be determined by a 2-way ANOVA analysis with $p<0.05$ considered significant.

RESULTS

Microgels had an average diameter of 1.28 ± 0.31 μm and polydispersity of 1.27. The 8-arm PEG-VS concentration directly correlated to the microgel-based scaffold modulus (Figure 1). The number of Schwann cells increased over the 5 days regardless of scaffold modulus, however, a peak cell number was found in scaffolds of 564 Pa after 5 days, which is lower than the previously reported optimal modulus for Schwann cells (Figure 2) [2]. Schwann cells were found to have good viability after encapsulation in the scaffolds with collagen, confirming our previous report that the encapsulation procedure was mild [3]. Based on current results, Schwann cell morphology within the scaffolds does not vary over the modulus range. Fibroblasts served as a model to compare Schwann cell morphology because previous data in our lab confirmed that fibroblasts stretch within the microgel-based scaffolds (Figure 3). Finally, growth factor gene expression from 3D scaffolds is still being collected; however, a preliminary study in 2D showed no significant differences in gene expression of nerve growth factor or glial derived growth factor with modulus.

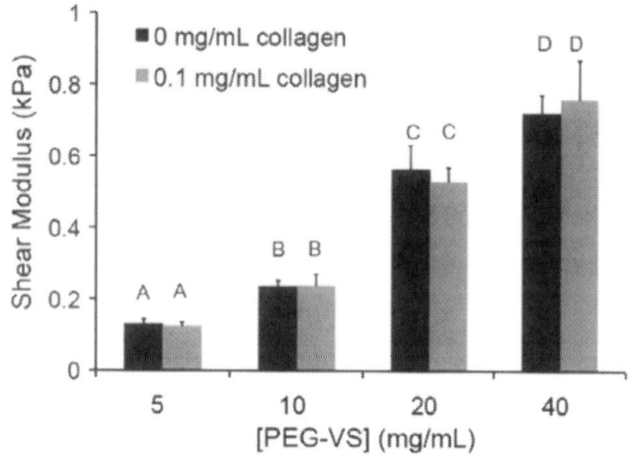

Figure 1: Effect of crosslinker concentration on the shear modulus of scaffolds with or without collagen. Statistical differences were observed between each crosslinker concentration ($p<0.05$), with letters representing different statistical groups.

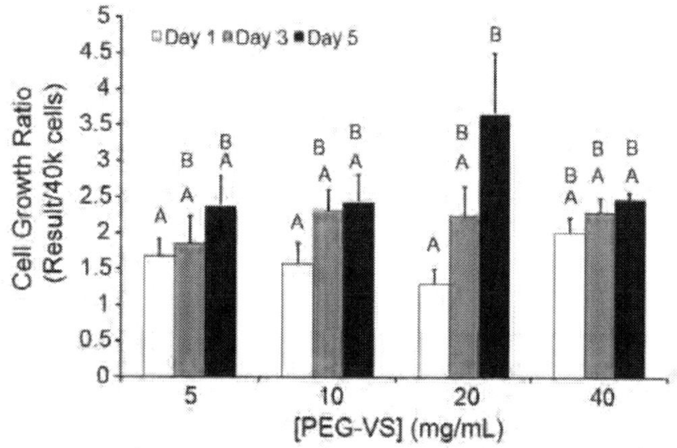

Figure 2: Preliminary data for Schwann cell proliferation in scaffolds crosslinked with varying 8-arm PEG-VS concentrations. Significant differences were observed between 20 mg/mL 8-arm PEG-VS at day 5 with 5 mg/mL 8-arm PEG-VS at days 1, 3, and 5 ($p<0.05$), with letters representing different statistical groups.

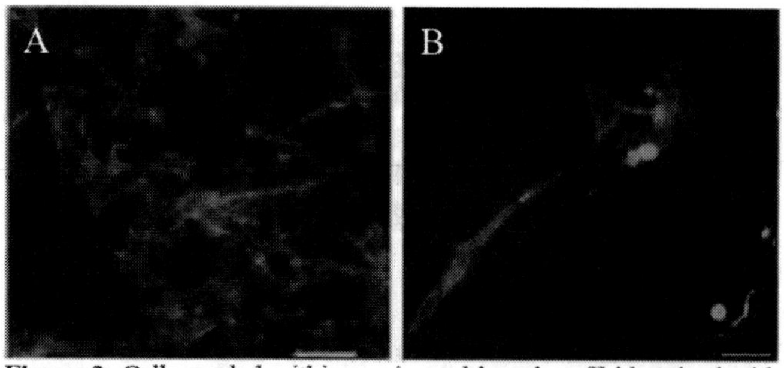

Figure 3: Cells seeded within a microgel-based scaffold stained with Hoechst 33342 and Phalloidin staining nuclei and F-actin, respectively. (A) Fibroblasts (scale bar = 100 μm) and (B) Schwann cells (scale bar = 20 μm).

DISCUSSION

Previous reports of Schwann cell morphology in 3D scaffolds are limited. Two studies of interest focus on categorizing Schwann cells in an active or quiescent state [5,6]. First, the Schmidt group showed an increase in nerve growth factor and brain derived neurotrophic factor secreted from Schwann cells when seeded in 3D scaffolds at a high density ($8x10^6$ cells/mL) with extracellular matrix proteins, collagen and laminin. Cell number was also correlated to increased proliferation and cell spreading. They hypothesized that this result is due to greater cell-cell communication because similar trends were not observed in scaffolds with a lower cell density ($2x10^6$ cells/mL) [5]. In the second study, the Tranquillo group reported an altered neural-cell adhesion molecule (NCAM) response in round Schwann cells, which they considered quiescent [6]. As NCAM is involved in neuron-glial cell interactions, its expression in Schwann cells may be an important measure the success of the cellular response within a scaffold. We are currently working to determine if we observe a similar correlation between cell number, proliferation, cell spreading, and NCAM expression in Schwann cells. By studying scaffolds of various stiffness, particularly within the stiffness range preferred by neurons, we aim to contribute to the previous work investigating Schwann cell response within 3D scaffolds.

We have developed a 3D scaffold that can be used to study how the stiffness can be manipulated in a range relevant to neural cells while maintaining the biochemical nature of the gels. Ultimately, the fully characterized Schwann cell response will be correlated to the modulus to determine a baseline for the microgel-based scaffolds. These results will then be extended to examine if local cell:scaffold binding will alter the response.

AKNOWLEDGEMENTS

This work was supported by the Margaret F. Donovan Endowment for Women in Engineering.

REFERENCES

1. Engler, et al. *Cell*, **124**(4): 677-689, 2006.
2. Gu, et al.. *Biomaterials*, **33**(28): 6672-6681, 2012.
3. Scott, et al. *Acta Biomaterialia*, **7**(11): 3841-3849, 2011.
4. Thompson, et al. *Jove,* 2013.
5. Suri and Schmidt. *Tissue Engineering: Part A*, **15**(5): 1703-1716, 2010.
6. Rosner, et al. *Experimental Neurology*, **195**: 81-91, 2005.

An Assessment of the Native Species in Articular Cartilage and Synovial Fluid as Potential Absorptive Barriers to UV-Initiated Scaffold Polymerization

A.J. Finch,[1] P.E. Donnelly,[1] P.A. Torzilli[1]

(1) Laboratory for Soft Tissue Research
Tissue Engineering, Regeneration, and
Repair Program
Hospital for Special Surgery

INTRODUCTION

When articular cartilage is damaged, its matrix is degraded and the biochemical linkages between matrix components are destroyed. Furthermore, articular chondrocytes lack the capacity to effectively synthesize new extracellular matrix components for repair. For this reason, artificial repair techniques are currently under investigation which can quickly and effectively treat damaged cartilage. One such tissue engineering technique - "scaffolding" - involves the development of a polymer lattice within cartilage that reinforces the damaged matrix. This lattice is assembled by diffusing or injecting monomers or uncrosslinked polymers into the cartilage matrix, adding a UV photosensitizer, and then exposing the tissue to UV light until a scaffold is developed. This scaffold then restores some of the structural and mechanical characteristics of the cartilage and facilitates the healing process [1,2,3].

It has been shown previously that articular cartilage, which is comprised of various organic compounds suspended in what is primarily a collagen hydrogel, absorbs UV radiation [4,5]. This attenuation of UV light is an important consideration for cartilage scaffolding: UV-initiated scaffold assembly is dependent on the effective absorption of photons by photosensitizers seeded within the scaffold matrix. If other absorbing species interfere with the absorption of UV photons by these photosensitizers, the effectiveness of the procedure will be reduced. Higher doses of UV radiation will be required to compensate for the light absorbed by articular tissue - this is problematic because higher UV doses will place healthy tissue at greater risk of becoming damaged. However if the absorptive properties of the major species native to articular cartilage are characterized, scaffolding procedures can be optimized to ensure sufficient scaffold polymerization at a minimum UV dose.

The objective of this study was to determine whether any of the major biochemical species present in cartilage are significant absorbers of UV light at 365 nm, the working wavelength of some commonly used photosensitizers. We collected and analyzed the UV-Vis absorbance spectra for type I collagen, chondroitin sulfate, sodium hyaluronate, fibronectin, synovial fluid, and proteoglycan. For each of these species the molar extinction coefficient was calculated at 365 nm. The absorption coefficient was then calculated at 365 nm and evaluated for each species at physiological concentrations. This process was then repeated for two popular photosensitizers: riboflavin and Irgacure 2529. These particular compounds are among the most cytocompatible photosensitizers used in scaffolding procedures [6,7]. The absorptive properties of the native species were then compared with those of the photosensitizers.

METHODS

Collagen type I, chondroitin sulfate, fibronectin, riboflavin, and Irgacure 2959 were purchased from Sigma (Sigma-Aldrich, St. Louis, MO), sodium hyaluronate was purchased from Lifecore (Lifecore Biomedical, Chaska, MN), and synovial fluid was extracted from normal immature bovine knee joints. Quartz cuvettes (Perkin Elmer) were filled with a range of concentrations of each sample, including those found *in vivo*, and analyzed using a Perkin Elmer UV-Vis Lambda 12 Spectrophotometer; data was collected from 200 to 750 nm (Waltham, MA). The absorbance spectra were analyzed using Sigmaplot (Systat Software, San Jose, CA). All data collected was linear, satisfying Beer's Law.

Molar extinction coefficient was calculated from Beer's Law

$$A = \varepsilon l c \quad (1)$$

Where A is absorbance, ε is the extinction coefficient, l is the path length (or the distance light travels through the sample), c is the concentration of the analyte (expressed in molarity). The extinction coefficient is generally used only to evaluate the absorptive behavior of a species in liquid solution, and is a function of wavelength [8,9].

Absorption coefficient, α, can be calculated from a modified form of Beer's Law

$$\alpha = \frac{A}{d} \qquad (2)$$

Where d is the distance light penetrates into, or through, a material. Unlike ε, α can be used to describe the light intensity attenuated by both liquid and solid solutions, and it is both a function of wavelength and concentration [8,9]. In this study, l from Eqn 1 and d from Eqn 2 are equal: both represent the width of the cuvette (1 cm).

Since the cartilage hydrogel is neither a liquid nor a solid, we evaluated it as both a liquid solution and as a solid solution by assessing both ε and α. The extinction coefficient was calculated at 365 nm for all species from the slope of a linear regression applied to absorbance versus concentration data. The absorption coefficient was evaluated for native species at concentrations in normal articular cartilage, and for photosensitizers near maximum solubility in deionized water [10].

RESULTS

Summarized absorptive properties are given in Table 1.

Table 1: Absorptive Properties of Evaluated Species

Species	$\varepsilon\ (\frac{1}{M*cm})$	$\alpha(\frac{1}{m})$
Hyaluronate	0	0
Fibronectin	0	0
Irgacure 2959	2.26	5.04
Type I Collagen	239	7.41
Chondroitin Sulfate	139 - 247	11.7
Riboflavin	13000	16.6
Proteoglycan	10000 - 14500	20.5
Synovial Fluid	*Not calculated*	52.2

The ε values for chondroitin sulfate and proteoglycan are given as ranges because neither of these compounds has a specific molecular weight: chondroitin sulfate is a polymer whose molecular weight depends on the length of the polymer chain, and there are numerous proteoglycans with various molecular weights [11].

Sodium hyaluronate and fibronectin both had absorbance values of 0 at 365 nm; no extinction coefficients or absorption coefficients were calculated at this wavelength for these species. Note that synovial fluid is an aqueous solution of proteins and other biomolecules; a molar extinction coefficient for this species was not calculated because a mole of synovial fluid is difficult to define and may be a misleading quantity.

DISCUSSION

The absorptive behavior of some of these species studied (namely collagen, riboflavin, and Irgacure 2959) have previously been evaluated [7,12] and our findings were in agreement with the existing literature. Our results indicate that some of the native species in cartilage will interfere with the absorption of UV light at 365 nm by riboflavin and Irgacure 2959.

Although ε for riboflavin is two orders of magnitude larger than that for most native species, it is the same order of magnitude as proteoglycan. Furthermore, α for riboflavin at its maximum solubility is below that of both proteoglycan and synovial fluid at physiological concentrations. In the case of Irgacure 2959, both ε and α for this photosensitizer were among the lowest quantities obtained in this study. Therefore proteoglycan and synovial fluid could significantly interfere with the absorption of light at 365 nm by riboflavin, and all of the native species evaluated, save hyaluronate and fibronectin, could significantly interfere with the absorption of this light by Irgacure 2959. These findings are consistent regardless of whether the cartilage hydrogel is evaluated as a liquid solution or a solid solution.

Riboflavin is a better absorber of UV light at 365 nm than Irgacure 2959 because it has a larger conjugated aromatic system; (see Figure 1). Single aromatic rings absorb UV light around 280 nm, however more complex conjugated systems also absorb at higher wavelengths.

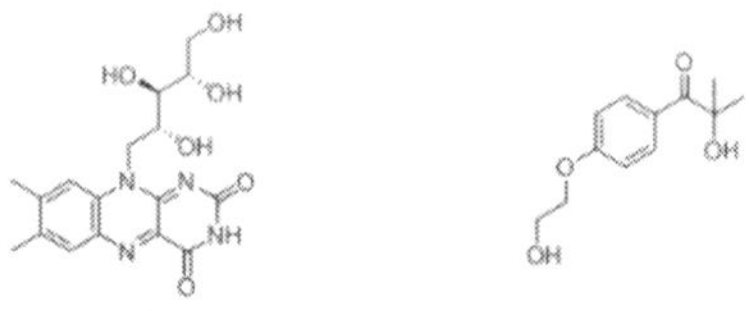

A B

Figure 1: Structures of riboflavin (A) and Irgacure 2959 (B)

These findings can be used to optimize photoinitiated polymer scaffolding procedures in articular cartilage and to minimize the risk of damage to surrounding tissue through overexposure to UV light. In articular tissue, riboflavin will be a much more efficient absorber of 365 nm UV light than Irgacure 2959 will be, based on our findings. Riboflavin is less soluble in water than Irgacure 2959, so utilizing a more soluble, substituted form of riboflavin such as riboflavin-5'-phosphate may be helpful.

A limitation of this study is that these data apply to each species in an isolated environment; each may interact differently with the others *in vivo*. This study also did not investigate the effect of free radical scavengers on the ability of photosensitizers to initiate scaffolding, which could be significant *in vivo*.

ACKNOWLEDGEMENTS

Support for this study was made possible by Grant Number R21-AR059203 (PAT) and AR007281 (PED) from the National Institutes of Health – National Institute of Arthritis and Musculoskeletal and Skin Diseases. Its contents are solely the responsibility of the authors and do not necessarily represent the official views of the NIH-NIAMS.

REFERENCES

[1] Ng, KW et al., *Tissue Eng Part A*, 18:1273-1281, 2012.
[2] Wennink, JWH, *Biodegradable Hydrogels by Physical and Enzymatic Crosslinking of Biomacromolecules* 9-41, Thesis, 2013.
[3] Nguyen, AK et al., *Regen Med*, 8:725-738, 2013.
[4] Bayliss, MT et al., *J Biol Chem*, 22:15892-15900, 1999.
[5] Ebert, DW et al., *J Biomed Opt*, 3:326-333, 1998.
[6] Bryant, SJ et al., *J Biomater Sci Polymer Edn*, 11:439-457, 2000.
[7] Metreveli, NO et al., *Exotocol Environ Saf*, 73:448-455, 2010.
[8] Denney, RC, Dict of Spec, 2nd ed. (Wiley, 1982) pp 119-120.
[9] Handbook of Vibrational Spectroscopy, v. 5 pp 3760,3777
[10] Bollett, AJ et al., *J Clin Invest*, 42:853-859, 1963.
[11] Buckwalter, JA et al., *Electron Microsc Rev* 1:87-112, 1988.
[12] Schumacher,S et al., *J Refract Surg*, 28:91-92, 2012.

SB³C2015
Summer Biomechanics, Bioengineering and Biotransport Conference
June 17-20, 2015,Snowbird Resort, Utah, USA
SB³C2015-430

IN VITRO MULTI-SCALE PATIENT SPECIFIC STUDY OF THE EFFECTS OF COARCTATION IN THE NORWOOD CIRCULATION

Tianqi Hang[1], Alessandro Giardini [2],Giovanni Biglino[2], Richard Figliola[1]

(1) Mechanical Engineering and Bioengineering
Clemson University
Clemson, SC USA

(2) Great Ormond Street Hospital for Children
London, UK

INTRODUCTION

Hypoplastic left heart syndrome (HLHS) is a congenital heart defect in which the left ventricle is severely underdeveloped. The Norwood procedure is the first stage surgical treatment for the HLHS patient. The goal is to make an unrestrictive systemic blood flow and at the same time balance it with the pulmonary flow. This is done by constructing a neo-aorta using the pulmonary artery root and the autologous aorta, and then installing a shunt to the pulmonary artery [1]. Variations of the Norwood surgery include using a modified Blalock-Taussig (mBT) shunt, which diverts blood from innominate artery to pulmonary artery (PA), or a right ventricle to pulmonary artery shunt (RVS), which diverts blood from right ventricle to PA.

Recurrent neo-aortic coarctation (NAO) is a frequent complication of the Norwood procedure. It causes changes in flow rates and hypertension in the aortic arch [2]. Conventionally, the value of a coarctation index (CoI) is used in choosing interventions to treat NAO. However, recent clinical findings within our group suggest that aortic arch morphology of Norwood patients could be a factor of hemodynamic response to NAO.

The objective of this study is to better understand the impact of coarctation on the Norwood patients of differing aortic arch morphology so as to provide the clinician supporting information in determining a course of treatment.

METHODS

Five patient-specific cases were selected, each case having a different extent of aortic arch dilation, qualitatively defined as: "very tubular", "tubular", "normal coarctation", "moderately dilated", and "highly dilated" (Fig. 1). For each patient, a three-dimensional (3D) virtual rendering of the anatomy was developed from magnetic resonance imaging and a 3D test section was printed using rapid manufacture methods [3,4]. To study a potential progression of coarctation severity in each patient, the virtual native arch anatomy was morphed by controlling the amount of narrowing of aortic isthmus, while keeping the original patient-specific aortic geometry intact. Separate models for each patient were created to provide for a range of CoI from 1.0 (none) to 0.3 (severe), where CoI is the diameter ratio between the coarctation narrowing and the descending aorta at the diaphragm. Fig. 2 shows an arch with NAO, the portion that is adjusted (highlighted) and the resulting test section.

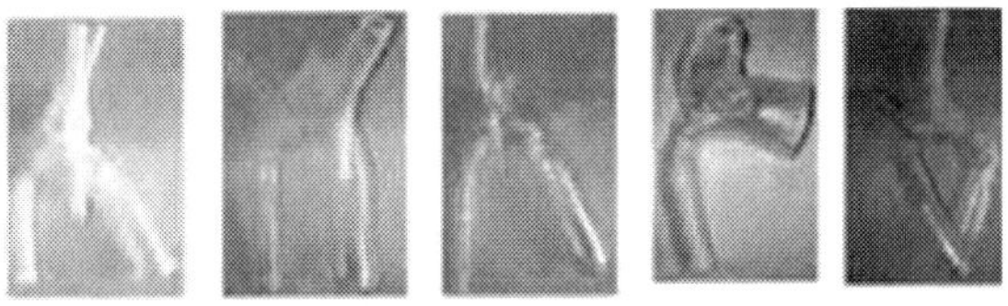

Figure 1. Range of patient arch morphologies. left to right: highly tubular, tubular, normal, moderately dilated, highly dilated.

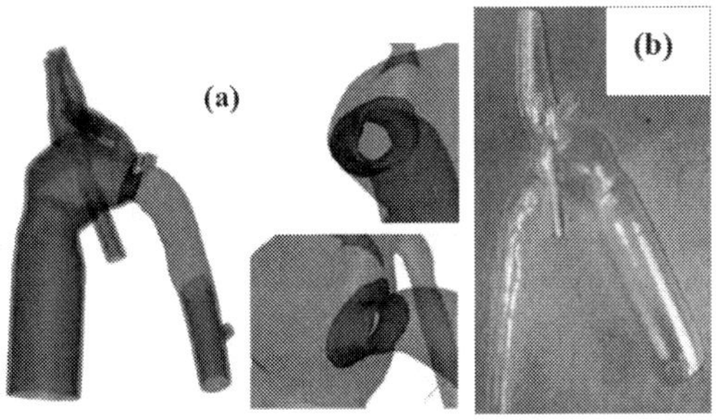

Figure 2. (a) Arch with coarctation area (highlighted). (b) Resulting aortic test section.

A multi-scale mock circulatory system (MCS) was used to assess the systems-level hemodynamic impact of progressive coarctation severity [3,5]. The MCS couples a lumped parameter network (LP) model of the circulation (Fig. 3) with the 3D test section. A ventricular assist device (VAD) (Excor, 30 cc; Berlin, Germany) serves as the univentricular heart. The MCS was tuned to patient specific conditions based on the clinical measurements. Locations for time-based pressure and flow rate measurements are indicated in Fig. 3. In addition, pressure and flow rate measurements into and out of the VAD are obtained to calculate ventricular power. The MCS function was verified and validated against the clinical data available for each of the five patients.

As an example, we compare results of increasing CoI for patients with a tubular and a dilated arch (Fig. 4). Cardiac output was held constant throughout these tests. A smaller value of CoI indicates a more severe coarctation.

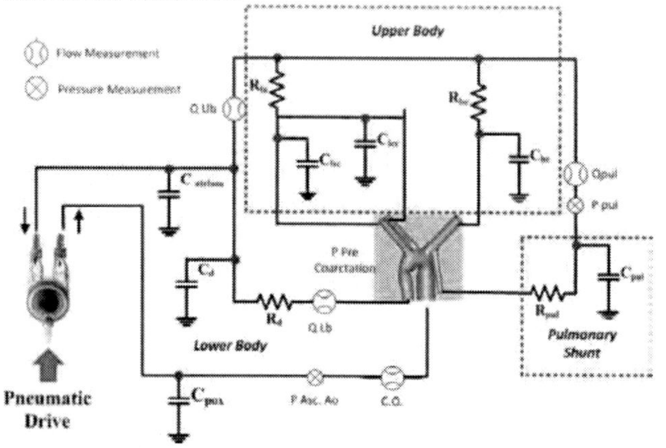

Figure 3. Mock circulatory system showing lumped parameter network coupled with 3D aorta and measurement points (Q and P). R, resistance; C, compliance; P, pressure; Q, flow rate; CO, Cardiac output; Asc.Ao, Ascending aorta; pox, proximal; pul, pulmonary shunt; d, descending aorta; atrium, atri; bc, brachiocephalic artery; lcc, left common carotid artery; lsc, left subclavian artery.

RESULTS

Fig. 4(a) and 4(b) show mean and peak pressure drop across the coarctation as measured both by the clinical catheter with pull-back method and by wall mounted pressure taps located both upstream of the coarctation and 10 diameters downstream in the descending aorta. The measured variation is nearly linear with CoI for a dilated arch, while the response of the tubular arch is very nonlinear reflecting low sensitivity to changes in modest coarctation (high CoI) and high sensitivity to changes in severe coarctation (low CoI). This behavior is even more obvious in the single ventricle power (SVP) response (Fig. 4(c) and Fig.4 (d)) where severe coarctation for tubular arch patients associates with a large increase in ventricular power.

DISCUSSION

Preliminary results show that this *in vitro* multi-scale is suitable for studying the progressive coarctation in Norwood Patients. Time-based experimental measurements of flow rate and pressures recreate the clinical measurements closely.

The impact of coarctation on the Norwood patients of differing aortic arch morphology was studied. In addition to the results presented in Fig. 4, system-level flow rates were also found to respond to CoI change. For extreme morphologies (either tubular or dilated), progressive CoI was found to lead to decreases in both the lower body and upper body flow rates while the mBT shunt flow rate increased. As coarctation develops, the shunt appears to steal a share of flow from the upper body. Depending on how much flow the shunt steals from the brachiocephalic artery, upper body flow will change with coarctation severity. Flow measurements indicate that most of the shunt steal occurs during diastole. However, for normal morphology, upper body flow can slightly increase or remain unchanged with progressive CoI. The results suggest that patient impedance is strongly associated with system-level response behavior.

The results show that patients having a dilated arch tend to have comparatively small change in upper body flow with increasing CoI compared with tubular and normal arch patients. This could be attributed to the dilated arch acting like a reservoir, supplying the upper body arteries. Or it could be that shunt stealing effect in dilated cases is less compared with other morphologies. The time-based flow and pressure results will be presented.

CONCLUSION

The results show that in vitro hemodynamic responses to progressive NAO narrowing are physiological. The results reveal the sensitivity of effects with respect to CoI, which is impossible to control and vary *in vivo*.

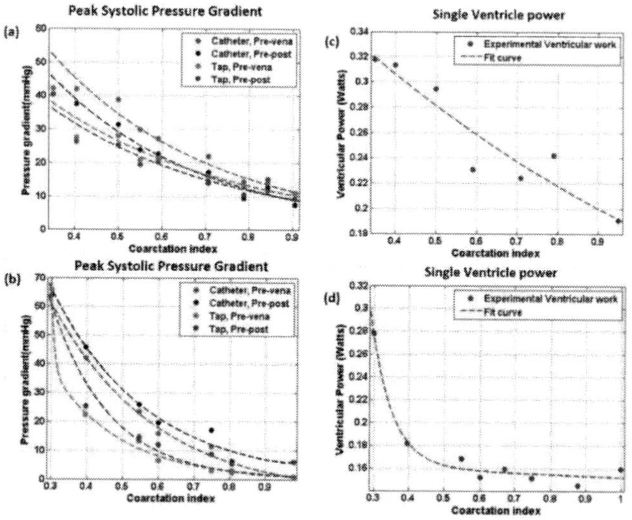

Figure 4: Hemodynamic response to CoI. Systolic pressure gradient with (a) dilated arch and (b) tubular arch. Single ventricle power per cycle with (c) dilated arch and (d) tubular arch.

ACKNOWLEDGEMENTS

This study is supported under a Transatlantic Network of Excellence grant from the Leducq Foundation, Paris. The authors are grateful for the collaboration of the MOCHA investigators.

REFERENCES

[1] Norwood W. American J Cardiology, 45(1): 87–91, 1980.
[2] Corsini, C et al., Comput Method Biomech. 17:*1572-1589, 2013*.
[3] Biglino, G et al, ASAIO J, 58(1): 25-31, 2012.
[4] Figliola R et al., Prog Ped Cardio, 30: 71-81, 2011.
[5] Biglino, G et al. ASAIO J. 59(1): 86-89, 2013.

ACCURATE PREDICTION OF COLLAGEN FIBER DISTRIBUTION USING FFT: A VALIDATION STUDY

Erica E. Morrill, Christina J. Stender, Roshani Lamichhane, Raquel J. Brown, Trevor J. Lujan

Department of Mechanical and Biomedical Engineering
Boise State University
Boise, ID, USA

INTRODUCTION

The functional behavior of soft connective tissue is strongly related to the underlying collagen fiber architecture [1]. Characterizing the fiber architecture requires objective and accurate image processing techniques to extract meaningful structural parameters [2]. A parameter often used to quantify tissue anisotropy is fiber angle distribution [2]. Accurate quantification of fiber angle distribution can advance a mechanistic understanding of collagen fiber dispersion and clarify the functional impact of tissue remodeling and repair. Furthermore, fiber angle distribution can be incorporated into constitutive models to predict mechanical behavior based directly on tissue microstructure.

A robust and easy method to quantify fiber networks is applying a fast Fourier transformation (FFT) to two- or three-dimensional images [3-5]. An FFT of imaged data will yield a spectrum that can be analyzed to acquire a parameter called the alignment index, a measure of fiber dispersion. While the alignment index provides a helpful metric to quantify fiber anisotropy, there is not a clear physical connection between the alignment index and fiber angle distribution [2]. A challenge in making a physical connection between the alignment index and fiber angle distribution is the inherent noise from FFT, and the lack of a validated method to remove noise while preserving structural information. The objective of this study is to overcome this challenge and develop an FFT protocol to accurately predict fiber angle distribution from confocal images of collagen networks.

METHODS

Overview. A custom Matlab program was developed to quantify fiber networks by fitting a semicircular von Mises distribution to the FFT spectrum. The von Mises distribution contains two structural parameters to define a fiber network: k defines the fiber angle dispersion and μ defines the preferred fiber orientation. Test patterns were developed with known k values to identify an optimal noise removal method for the FFT program. The accuracy of the optimal noise removal method was tested at variable values of k and μ. This method was then applied to measure the structural parameters of collagen fibrils in bovine ligament, imaged using confocal microscopy.

FFT Program. A two-dimensional fast Fourier transformation converted images into Fourier space. A radial sum approach defined intensity at each angle to create an angular distribution histogram of the fiber network [3]. The intensity data was squared and normalized to an area of 1 [5]. Preferred fiber orientation was estimated by fitting the data to an ellipse, where μ was defined by the semi-major axis [6]. The k parameter was calculated by fitting a semicircular von Mises distribution to the histogram data, using least squares. Prior to fitting k, the histogram data was centered about the preferred orientation.

Noise Reduction. Due to inherent noise in FFT, a noise reduction process was applied to the spectrum. One established noise reduction method is to remove all data in the angular distribution histogram below the minimum value (the minimum method); this effectively zeroes the minimal intensity value [7]. A newly proposed method was to remove all histogram data below an optimal threshold after normalization (the optimized method).

Selecting an optimal threshold required a sensitivity study of test patterns with known k values. These test patterns were automatically generated using a custom matlab program. Briefly, a binary image was populated with lines based on a von Mises probability density function for user-specified values of μ and k. The seed point for each line was randomized and the line was grown incrementally until the number of pixels exceeded the number of pixels defined at that angle by the von Mises distribution. For the sensitivity study, three 1074×1074 binary test patterns were generated for four different k values with $\mu = 90°$ (Fig. 1).

Verification of Accuracy. An additional 27 test patterns were generated for user-specified k values (k_{actual}) to quantify the error of using the FFT program to predict k (k_{pred}). The benefit of all noise removal methods was evaluated at thirteen different k_{actual} values (0.2 – 5.0). An additional analysis was conducted to determine the effect of image rotation on error at set k values (k_{actual}= 1, 5).

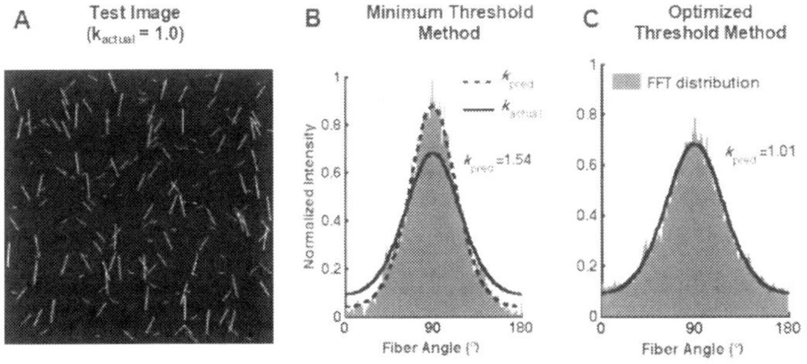

FIGURE 1. (A) FABRICATED TEST PATTERN, (B-C) FFT RESULTS FOR VARYING NOISE REMOVAL METHODS.

Validation using Confocal Images. The FFT program was evaluated for use with confocal images. Collagen fibril networks in nine bovine ligament specimens were autofluoresced at 63× using the 405 nm laser of a confocal microscope and ZEN imaging software [8]. Prior to FFT analysis, confocal images were filtered (band-pass), converted to binary images, skeletonized and dilated using ImageJ. This set of images accounted for the non-linearity and variability of the experimentally measured collagen networks. For comparison, a second set of images was created from the original confocal images by manually tracing the start and end points of the fibrils. This set of images replicated the linear topology used to generate the binary test patterns. Both image sets were run through the FFT program to calculate k and μ for the skeletonized confocal images ($k_{confocal}$, $\mu_{confocal}$) and the manually generated images (k_{manual} , μ_{manual}). This validation strategy assessed if the topographical assumptions used to generate the test patterns were valid when measuring collagen fibrils in confocal images.

Statistics. The effect of noise removal method on error was determined using a one-way ANOVA with a Games-Howell post hoc test. Effect of image rotation on error was determined using an ANOVA with a Bonferroni post hoc test. The coefficient of determination compared the structural parameters of the confocal images with the corresponding structural parameters of the manual images.

RESULTS

The sensitivity study determined an optimal removal threshold of 0.13 from the normalized intensity. The optimized threshold method had a lower error than the minimum method ($p<0.03$; Fig. 2A), and had on average 10x less error than if no noise removal method was applied ($p<0.001$; Fig. 2A).

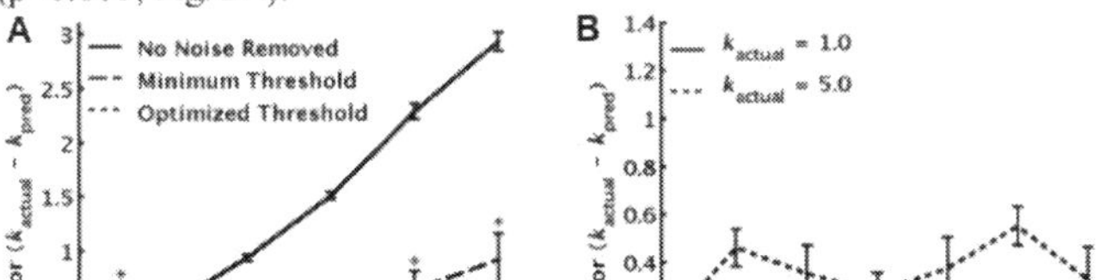

FIGURE 2. (A) EFFECT OF NOISE REMOVAL METHOD ON *k* ERROR, (B) EFFECT OF IMAGE ROTATION ON *k* ERROR (* = SIGNIFICANT DIFFERENCE BETWEEN MINIMUM AND OPTIMIZED THRESHOLD METHODS).

When using the optimized threshold method, the error in predicting k using FFT was greater for test patterns generated at k=5 than k=1 (Fig. 2B, $p < 0.001$). As test patterns were rotated, the error for predicting μ using FFT was greater for lower k values (0.94 ± 0.7° for k_{actual}=1; and 0.3 ± 0.2° for k_{actual}= 5).

A strong correlation existed when FFT was used to measure k and μ in the confocal image and corresponding manual image (Fig. 3).

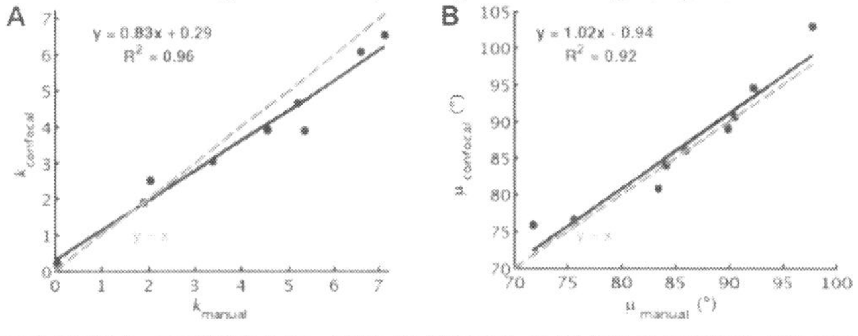

FIGURE 3. CONFOCAL AND MANUAL IMAGE CORRELATION FOR (A) *k* AND (B) *μ*.

DISCUSSION

This study successfully developed and validated an FFT program to accurately measure fiber angle distribution in confocal images of ligament microstructure. The dispersion term, k of the von Mises distribution, was used to physically link the FFT intensity histogram to fiber angle distribution. Test patterns were used to optimize a noise removal method that predicted k with a low error.

The optimization and evaluation of the FFT program required a novel testing strategy. Binary test patterns were generated with known k values using a von Mises distribution function. This enabled the error to be reliably determined when measuring k using FFT. Although a similar approach has been applied to quantify error when using FFT to measure preferred fiber orientation [9], we believe this is the first time this approach has been applied toward fiber angle distribution. In addition, the fitting of k included a procedure to center the intensity histogram about μ. We did not find prior studies that reported this procedure, but it was critical for accurate k measurement.

Limitations exist with this study. The fiber network was assumed to follow a semi-circular von Mises distribution. If different fiber distributions were observed, this fitting method would need to be modified. Secondly, this study did not investigate the effect of fiber thickness variations on the predicted von Mises parameters. However, for confocal images of collagen fibrils, the fibril thickness is relatively consistent. Finally, the optimized threshold value was only tested for measuring collagen fibrils with confocal microscopy, and has not been validated for use with other imaging modalities or tissues.

In summary, the proposed FFT protocol can serve as an analytical tool to acquire image-based statistical measurements of microstructure in confocal images. For example, this method can support the validation of structural constitutive models that embed parameters that account for fiber dispersion [10]. The verification strategy developed in this study can be applied to determine the accuracy of other FFT noise reduction methods, such as applying band-pass filters in the frequency domain [11].

ACKNOWLEDGEMENTS

This project was supported by the Biomolecular Research Center at Boise State University and funded by NIH NIGMS P20GM109095.

REFERENCES

[1] Holzapfel, G., *Biomech Preprint Series*, No. 7, 2000. **[2]** D'Amore, A. et al., *Biomaterials*, 31:5345-5354, 2010. **[3]** Ayres, C. et al., *J Biomater Sci Polym Ed*, 19:603-621, 2008. **[4]** Sander E.A., et al., *J Biomed Mater Res*, 88A:322-331, 2009. **[5]** Polzer, S. et al., Microsco Microanal, 19:1395-1404, 2013. **[6]** Sereysky, J. et al., *J Biomech*, 43:2641-2644, 2010. **[7]** Ayres, C. et al., Biomaterials, 27:5524-5534, 2006. **[8]** Monici, M, *Biotech Annu Rev*, 11:227-56, 2005. **[9]** Schriefl, A. et al., *J R Soc Interface*, 9:3081-3093, 2012. **[10]** Girard, M. et al., *J Biomech Eng*, 131, 2009. **[11]** Marquez, J.P., *Int J Solids Struct*, 43:6413-6423, 2006.

SB^{3}C2015
Summer Biomechanics, Bioengineering and Biotransport Conference
June 17-20, 2015,Snowbird Resort, Utah, USA
SB^{3}C2015-432

PERFUSION DECELLULARIZED SKELETAL MUSCLE AS A SCAFFOLD FOR THE REPAIR OF VOLUMETRIC MUSCLE LOSS

B. Kasukonis (1), J. Kim (1), T. Washington (2), J. Wolchok (1)

(1) Department of Biomedical Engineering
University of Arkansas
Fayetteville, AR, USA

(2) Department of Kinesiology
University of Arkansas
Fayetteville, AR, USA

INTRODUCTION

When provided with cues from the extracellular matrix (ECM) muscle has a strong capacity for self repair. For example, when muscle damage is mild (ex. strains, contusions, and lacerations) the underlying ECM is largely intact and repair is robust [1, 2]. However, when significant muscle volume is lost (trauma or surgical resection) the physical and chemical cues provided by the ECM are missing and the defect site is replaced with scar tissue [3]. The current standard of care to replace lost muscle is the transfer of autologous muscle flaps. Yet, the transfer of autologous muscle tissue is an invasive procedure that causes significant donor site morbidity. We suggest that the use of an ECM scaffold derived from decellularized skeletal muscle is an attractive alternative to autologous transplantation.

To create decellularized tissue scaffolds, whole tissues are typically soaked in aqueous solutions that rinse away the cellular debris. The most common decellularization solution is the detergent sodium dodecyl sulfate (SDS). The typical decellularization process relies on the slow diffusion of decellularization solutions into the tissue and cellular components out from the tissue. For thicker tissues, like skeletal muscle, the process can take two weeks or more. To accelerate the decellularization process we are interested in exploring perfusion delivery of decellularization solution. Towards this end our group has developed a benchtop perfusion decellularization system. Additionally, the system utilizes the color shift that skeletal muscle undergoes during decellularization (red to white) as a means to monitor the process optically. This abstract describes the development and preliminary in-vivo muscle repair studies using this novel system.

METHODS

Design: The design goal was to combine continuous tissue perfusion and real time optical monitoring into a single bench top system. The primary components of the system are the muscle perfusion chambers and optical detection collars with integrated data collection hardware and software. The perfusion units, optical monitoring collars, and data collection hardware are housed within and mounted onto a custom fabricated enclosure (**Figure 1**). The enclosure was designed to accommodate four side-by-side decellularization units each capable of accommodating a single muscle tissue sample. Each of the four modular perfusion units consists of a rigid acrylic cylinder (ID = 10mm; length = 40mm) that is capped a both ends by watertight ports fashioned with Leur lock style fitting and fabricated in acrylic with the aid of a 3-D printer (Objet 30, Objet, Isreal). The inlet port incorporates a removable 26-ga stainless steel hypodermic needle (length = 20 mm) that delivers decellularization solution deep into the interior of individual muscle samples. The flow of fluid into each decellularization unit is controlled via a commercial syringe pump.

The optical collar incorporates a single light emitting diode (LED) producing a peak emmitance centered at 450nm. Peak LED wavelength selection was motivated by the results of pilot study muscle optical property characterization. Located opposite to the LED, is a light dependent resistor (LDR). The LDR detector converts light intensity in photons to resistance in ohms. During decellularization, the collar is positioned over the muscle sample, such that the muscle tissue impedes the light path from the LED emitter to the LDR detector. Analog voltages collected from the LDR are converted to digital values using an 8-bit analog-to-digital converter (ADC) (Adafruit). A miniaturized single board computer (Raspberry Pi Foundation, United Kingdom) was used to power the LEDs and to control collection of the digitized LDR voltage signal from the ADC.

Characterization: To explore the influence of key perfusion parameters (flow rate and decellularization solution concentration) on device performance, muscle samples were perfused at flow rates of

0.1, 1, and 10mL/hr using SDS solutions at concentrations of 0.2 and 1%. A total of 6 parameter combinations were explored (n= 4/ parameter condition). The parametric results were compared using a two way ANOVA and Tukey's HSD post hoc analysis.

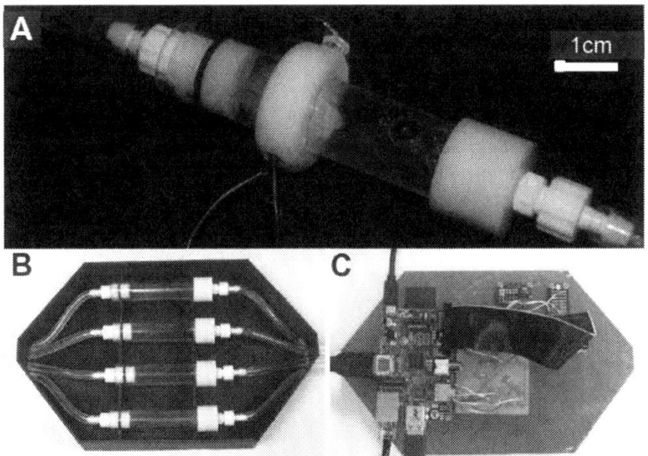

Figure 1: A. Monitoring collar on assembled decellularization chamber with muscle inserted on needle. B. Bottom device enclosure housing four decellularization chambers. C. Top device enclosure with Raspberry Pi, analog-to-digital converter and custom circuitry.

To determine whether perfusion decellularization influenced the retention of key ECM molecules within treated muscle samples the sulfated glycosaminoglycans (sGAG) and collagen content for control and perfusion samples (n=4/group) was quantified using published techniques. GAG and collagen content for each parameter was compared to diffusion decellularized samples. Statistical significance was tested using a student's t-test.

In Vivo Testing: The regenerative performance of perfusion prepared scaffolds was evaluated in a rat model. Mature Fisher 344 rats (300g) were purchased commercially (Harlan, Indianapolis, IN). A partial thickness tibialis anterior (TA) muscle defect (approximately 20% of the TA muscle) was created using a 8mm biopsy punch. A perfusion decellularized TA scaffold was cut to size and implanted into the defect. Unrepaired defects served as negative controls. The untreated contra-lateral limb served as a comparative control. Functional recovery was assessed 3 months post-surgery. The foot of each rat was secured to a force transducer and electrodes were used to stimulate the peroneal nerve. Peak TA contractile force was recorded for five contractions. The average peak force, as a percentage of the unaffected contralateral limb, was calculated for both scaffold repaired and unrepaired control groups

RESULTS

Perfusion treated samples were thoroughly decellularized at times that were well below the two week durations required for diffusion treated samples. Specifically, intracellular myoglobin, actin, and nuclei were visible is histological sections obtained from untreated skeletal muscle samples, but were absent in all samples following perfusion treatment. The decellularization solution flow rate and concentration each had a statistically significant influence on decellularization time. Generally, increasing the solution concentration and flow rates decreased the decellularization time. Specifically, increasing the flow rate from 0.1 to 10ml/hr decreased the decellularization time by 90% (from 142 to 9.9 hrs.). Increasing the SDS concentration had a more mild effect on decellularization time. Increasing the SDS concentration from 0.2% to 1% at a flow rate of 10ml/hr reduced the decellularization time from to 9.9 to 7.2 hours, a reduction of approximately 20%. When taken as a whole, the application of perfusion reduced the decellularization time from nearly two weeks for diffusion treated samples to less than a day.

We were initially concerned that perfusion treatment would strip away additional ECM molecules when compared to diffusion treated samples. However, the biochemical analysis suggests that perfusion decellularization did not significantly affect collagen or sGAG concentration when compared to diffusion decellularized control samples. For all flow rates and SDS concentrations, no significant reductions to either collagen or sGAG concentration were measured.

All animals tolerated the surgery and gained weight throughout the implantation period. The force produced by each TA muscle was normalized to total animal mass and averaged for each group. The scaffold repair and unrepaired groups produced average peak TA forces that were 72% and 54% of the unaffected contralateral limb respectively (**Figure 2**). On average, scaffold repair improved TA force generation by 31% when compared to unrepaired limbs.

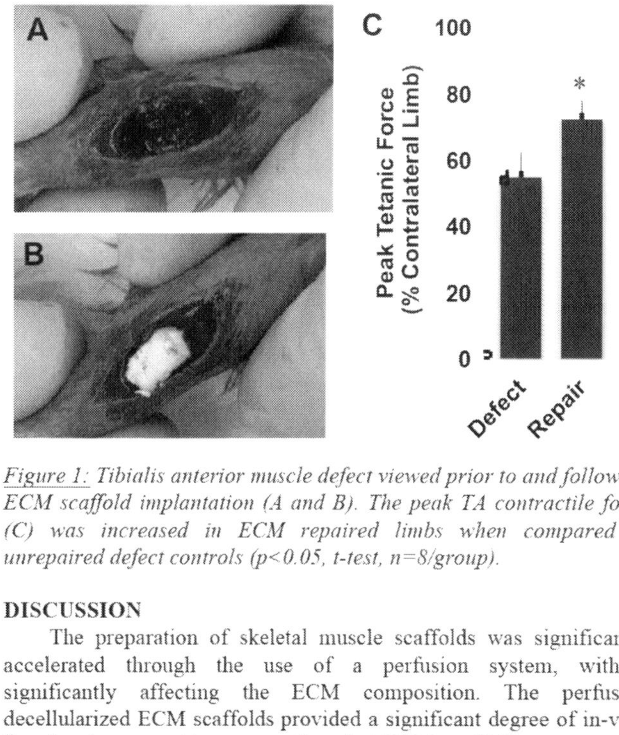

Figure 1: Tibialis anterior muscle defect viewed prior to and following ECM scaffold implantation (A and B). The peak TA contractile force (C) was increased in ECM repaired limbs when compared to unrepaired defect controls (p<0.05, t-test, n=8/group).

DISCUSSION

The preparation of skeletal muscle scaffolds was significantly accelerated through the use of a perfusion system, without significantly affecting the ECM composition. The perfusion decellularized ECM scaffolds provided a significant degree of in-vivo functional regeneration, suggesting that ECM scaffolds can provide the necessary regenerative cues needed for successful muscle repair. Current work is exploring the influence of progenitor cell co-delivery on regenerative performance, with the goal of further improving force production by ECM repaired muscles.

ACKNOWLEDGEMENTS

Funding provided through the National Institute of Arthritis and Musculoskeletal and Skin Diseases (NIH Award #: R15AR064481)

REFERENCES

[1] Hill M et al., *J Anat* 203:89-99, 2003.
[2] Mauro A et al., *J Biophys Biochem Cytol* 9:493-5, 1961.
[3] Terada N et al., *Scand J Plast Reconstr Surg Hand Surg* 35:233-8, 2001.

DETERMINATION OF THE CONVECTIVE HEAT TRANSFER COEFFICIENTS ON THE OUTER SURFACES OF THE CRYOVIVALS PLUNGED IN LIQUID NITROGEN AND 37 °C WATER BATH

Tao. Wang (1), Gang. Zhao* (1,2)

(1) Centre for Biomedical Engineering,
Department of Electronic Science & Technology,
University of Science and Technology of China,
Hefei 230027, China

(2) Anhui Provincial Engineering Technology
Research Center for Biopreservation and Artificial
Organs, Hefei, Anhui 230027, China

* Corresponding author.
E-mail address: ZhaoG@ustc.edu.cn

INTRODUCTION

Cryopreserve biological tissue and biological material plays an increasingly important role in clinical application and academic research[1]. A successful cryopreservation is depended on two major procedures means the cooling process and the rewarming process. The tissues or cells are immersed into a cryovial that filled with selected cryoprotective agents (CPA)[2,3]. The cryovial is put into liquid nitrogen quickly and cooling down to -196 °C during the cooling process. Then the tissues or cells are stored in the cryovial and keeping vitrifaction for a long time. Furthermore, the frozen tissues or cells in the cryovial must be rewarmed before use. Usually, the cryopreserved sample in the cryovial is put into 37 °C water bath during the rewarming process. Then both the cooling process and the rewarming process can be modeled as the convective heat transfer process. So, the convective heat transfer coefficient is crucial for the sample to avoid the thermal stress damage and devitrification during the cooling and the rewarming process[4,5].

Over the years, major attention has been focused on studying the effects of different CPAs. Only a few studies that investigated the convective heat transfer coefficient during the cooling process in part because the lack of readily accessible computational or experimental approaches[3,4,5]. Here, the convective heat transfer coefficient during the cooling and the rewarming process is simulated and measured based on the inverse problem methods.

METHODS

The cryovial equipped with a thermocouple was put into the liquid nitrogen after it had been filled with EC2 solution during the cooling process. The EC2 solution was composed of water, DMSO, NaCl and trehalose, and the molar ratio in EC2 solution was 0.9556, 0.0401, 0.001 and 0.0033, respectively. The cryovial with initial temperature of -196 °C was put into the 37 °C water bath during the rewarming process. The rewarming process will be suspended when the temperature of any point in the cryovial was above -1.8 °C. The temperature of the point into the EC2 solution was measured using the thermocouple. Schematic diagram of the whole experimental system during the cooling process and the rewarming process is shown in Figure 1. The whole experimental system is composed of temperature monitor, thermocouple, cryovial, EC2 solution, ice water mixture, liquid nitrogen container, 37 °C water bath, power source, computer, electric wire and data line.

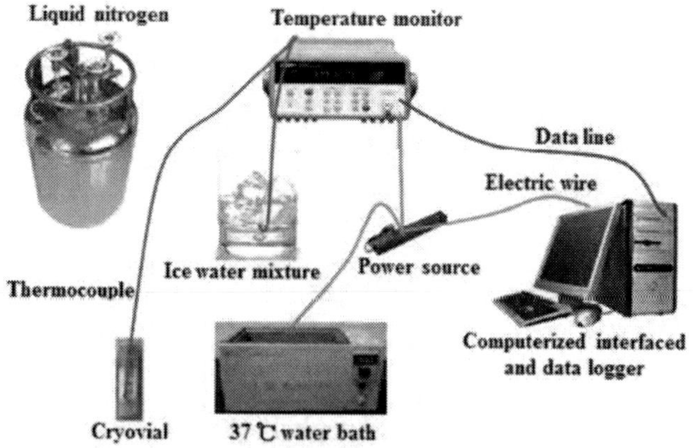

FIGURE 1. SCHEMATIC DIAGRAM OF THE WHOLE EXPERIMENTAL SYSTEM.

Heat transfer during the cooling process and the rewarming process of cryovial can be accurately described by the modified biological heat transfer equation:

$$\rho C\frac{\partial T}{\partial t}=\nabla\cdot(k\nabla T) \qquad (1)$$

where T is the temperature, t is the time, ρ is the density of the solution, C the effective thermal capacity of the solution, and k the effective thermal conductivity. The physical properties of the EC2 solution are shown in Figure 2.

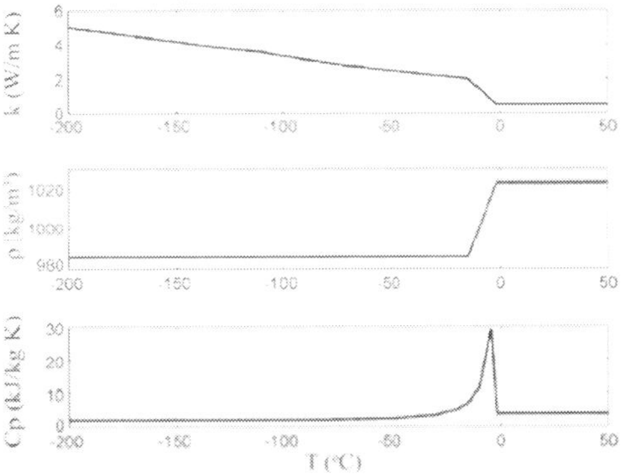

FIGURE 2. TEMPERATURE DEPENDENT THERMAL PROPERTIES OF THE EC2 SOLUTION[2].

The boundary condition during the cooling process is:

$$k\frac{\partial T}{\partial n}=h_1(T-T_c) \qquad (2)$$

where n is the normal direction heat transfer surface, T_c is the temperature of cold source, and h_1 is the convective heat transfer coefficient during the cooling process.

The boundary condition during the rewarming process is:

$$k\frac{\partial T}{\partial n}=h_2(T_g-T) \qquad (3)$$

where T_g is the temperature of cold source, and h_2 is the convective heat transfer coefficient during the cooling process.

The initial value of the convective heat transfer coefficient during the cooling and the rewarming process is supposed and the temperature of the same points A, B and C is simulated based on the Eqs. (1-3). When the temperature profiles of the three points is closest approach the experimental result, the value of the convective heat transfer coefficient is obtained due to the inverse problem methods.

RESULTS

The cryovial filled with 1.5 mL EC2 solution was put into the liquid nitrogen container during the cooling process. Zero temperature point is calibrated using ice water mixture through the whole measurement process. The temperature profile of the three points marked as A, B and C was measured during the cooling process. It took about 120 sec during the cooling process. The convective heat transfer coefficient during the cooling process is set to be 1000 W/(m^2 K) according to the previous studies[3,4]. Here, the temperature profile of the three points was simulated with h_1=300 W/(m^2 K) in this study. And the simulated results are compared with the experimental results as is shown in Figure 3.

The frozen cryovial with initial temperature of -196 °C is put into the 37 °C water bath during the rewarming process. It took about 160 sec when any point in the cryovial is above -1.8 °C. The convective heat transfer coefficient during the cooling process is set to be h_2=210 W/(m^2 K). And the simulated results of the three points are compared with the experimental results as is shown in Figure 4. Since the cryovial is cylindrically symmetric about the axis, a close-up view of the points A, B and C on the half section is shown at the bottom right of the Figure 3 and Figure 4.

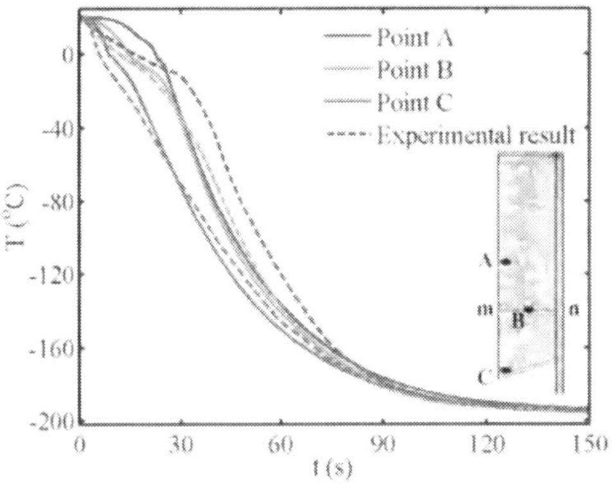

FIGURE 3. THE MEASURED TEMPERATURE PROFILES OF POINTS A, B AND C VERSUS THE SIMULATED RESULTS WHEN h_1=300 W/(M^2 K).

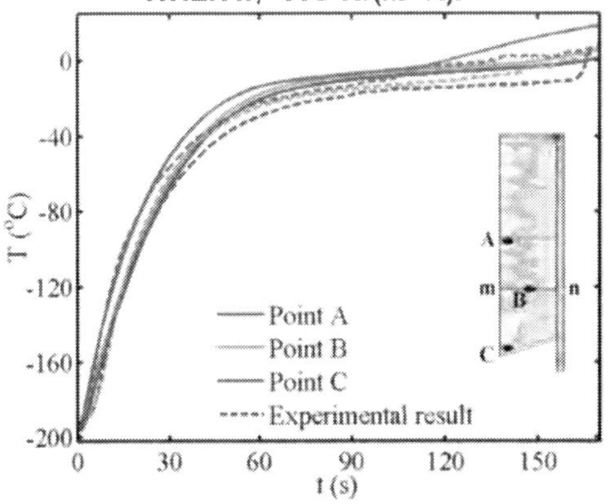

FIGURE 4. THE MEASURED TEMPERATURE PROFILES OF POINTS A, B AND C VERSUS THE SIMULATED RESULTS WHEN h_2=210 W/(M^2 K).

DISCUSSION

This study is focused on the heat transfer of the CPA solution instead of the cells since 80%-90% of the cells are composed of the water. Highly agreement was obtained between the experimental and the simulated temperature profiles when the heat transfer coefficients during cooling process and rewarming process is 300 W/m^2 K and 210 W/m^2 K, respectively. Since the convective heat transfer coefficient is mainly decided by the environment around the fluid, phase change of the fluid, the geometry surface and the physical properties of the fluid. An improved design of the cryovial will be helpful to avoid the thermal stress damage and devitrification during the cryopreservation process based on this work, which will lead to successfully cryopreserve of biological tissues and biological materials.

ACKNOWLEDGEMENTS

This work was supported by the National Natural Science Foundation of China (Grant nos. 51276179, 51476160), and the Fundamental Research Funds for the Central Universities (Grant no. WK 2100230009).

REFERENCES

[1] Wang, T et al., *Cryobiology*, 68:234-243, 2014.

[2] Luo, D.W., *Electromagnetic rewarming of cryopreserved biomaterials*, University of Kentucky (Ph.D. Thesis): 71-123, 2005.

[3] Santos, M.V et al., *Cryobiology*, 67:30-37, 2013.

[4] Santos, M.V et al., *Cryoletters*, 33:31-40, 2012.

[5] Santos, M.V et al., *Cryobiology*, 68:488-495, 2014.

DETERMINATION OF THE BIOPHYSICAL PARAMETERS OF HUVECS AND THEIR APPLICATION IN OPTIMIZATION OF THE ADDITION AND REMOVAL OF CRYOPROTECTIVE AGENTS

Yuntian Zhang (1), Dan Niu (1), Gang Zhao* (1,2), Tao Wang (1)

(1) Centre for Biomedical Engineering,
Department of Electronic Science & Technology,
University of Science and Technology of China,
Hefei 230027, China

(2) Anhui Provincial Engineering Technology
Research Center for Biopreservation and Artificial
Organs, Hefei, Anhui 230027, China

* Corresponding author.
E-mail address: ZhaoG@ustc.edu.cn

INTRODUCTION

Cryopreservation is an important resource of a variety of cells and biological tissues. While cells could be freezed for long time preservation, intracellular ice may form during preservation procedure, causing damage to cells. Special cryoprotectants are usually added into cryopreservation solutions to reduce such damage. However, some cryoprotectants are harmful to human body at room temperature. Thus it's necessary to remove them after rewarming. The adding and removing procedure will lead cells to shrink or expand. Cells will lose their biological activity if the volume changes over the limit. It is necessary to optimize the adding and removing procedure so that the cell volume changes are within certain limits. In this paper, we first studied the survival rate of HUVEC under different osmotic conditions and decided the limits of cell volume change. Then we optimized the adding and removing procedure of cryoprotectant using relative parameters, making sure that the volume changes in these procedures do not exceed given limits.

METHODS

The two parameter (2-P) formalism is now commonly used to describe volumetric responses of cells in the presence and obsence of permeable or impermeable solutes. This model is developed from the work of Jacobs and Stewart[1-3]. We could take it intactly to simulate our cryopreservation procedure. In our case, it uses L_p and P_s to describe the permeability of cells to water and cryoprotectant, respectively:

$$\frac{dV_w}{dt} = -L_p ART\left(\pi^e - \pi^i\right) \tag{1}$$

$$\frac{dN_s}{dt} = P_s A\left(C_s^e - C_s^i\right) \tag{2}$$

where superscript "e" and "i" represents extracellular and intracellular environment, respectively, subscript "s" represents cryoprotectant, A is the cell surface area, R is the universal gas constant, T is absolute temperature, π is the solution osmolarity, C is solute osmolarity.

Relation between and C can be described as:

$$\pi = C_s + C_n \tag{3}$$

where subscript "n" represents NaCl.

L_p and P_s are obtained from Arrhenius equation[4,5]:

$$L_p = L_{pg} \exp\left[-\frac{E_{Lp}}{R}\left(\frac{1}{T} - \frac{1}{T_R}\right)\right] \tag{4}$$

$$P_s = P_{sg} \exp\left[-\frac{E_{Ps}}{R}\left(\frac{1}{T} - \frac{1}{T_R}\right)\right] \tag{5}$$

where T_R is the reference temperature, L_{pg} and P_{sg} are membrane permembility to water and cryoprotectant, respectively. E_{Lp} and E_{Ps} are membrane permeability activation energy.

When loaded into anisotonic solutions of different osmolarity, cells will shrink or expand before reaching an equilibrium state, which is commonly called the osmotic fragility experiment. If solutes in extracellular solution are nonpermeable, the final equillibrium cell volume could be take as the limiting value under that extracellular osmolarity.

In order to decide the survival rate of cells under different extracellular osmotic conditions, we used PI and Calcein-AM fluorescent staining.

As for the determination of cell parameters, a detailed method was described in a previous paper[6].

RESULTS

Figure 1 shows the result of osmotic fragility experiment, where the horizontal axis represents extracellular osmolarity and the vertical axis represents cell volume. Detailed data is in Table 1. While we used impermeable solute (NaCl) in extracellular solutions, the final equillibrium volume is the limiting value under that extracellular osmolarity.

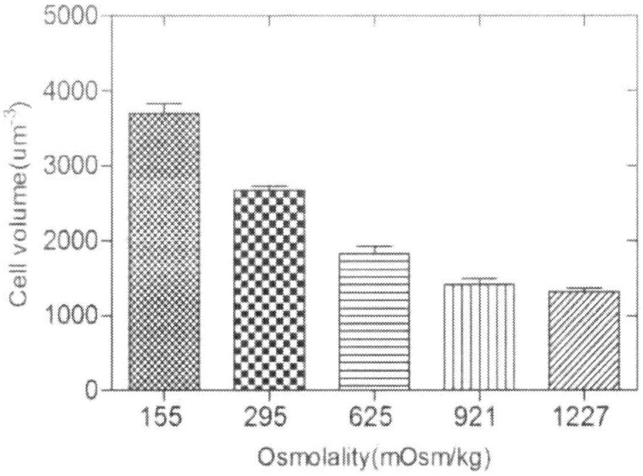

FIGURE 1. HISTOGRAM OF HUVEC VOLUME RESPONSES UNDER DIFFERENT EXTRACELLULAR OSMOLARITY.

TABLE 1. EQUILLIBRIUM VOLUME OF HUVEC UNDER DIFFERENT EXTRACELLULAR OSMOLARITY.

Osmolality (mOsm/kg)	155	295	625	921	1227
Mean	3693	2663	1826	1410	1291
Std. Deviation	139.42	58.05	99.07	77.65	23.67

Results of fluorescent staining experiment are shown in Figure 2 and Table 2. Assuming the survival rate limit to be 90%, we chose a normalized volume limit of 1.39 and 0.53.

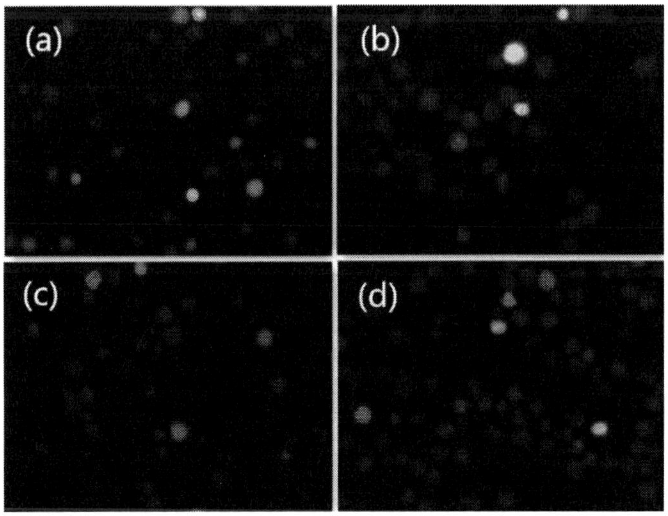

FIGURE 2. FLUORESCENT STAINING RESULTS OF HUVEC UNDER DIFFERENT EXTRACELLULAR OSMOLARITY.

TABEL 2. NORMALIZED HUVEC CELL VOLUME AND CORRESPONDING SURVIVAL RATE.

Osmolality (mOsm/kg)	155	295	921
Normalized Volume	1.39	1	0.53
Survival Rate (%)	0.98	1	0.95

According to the 2-P formalism and corresponding HUVEC parameters, we optimized the adding and removing procedure of cryoprotectant in HUVEC cells at different temperatures. Dimethyl sulfoxide (DMSO) is selected as the cryoprotectant, simulated temperatures are 5°C. The results are shown in Figure 3. It could be found that the cell volume shrinks within the limits with one-step adding (left), so it is unnecessary to optimize. However, in the removing procedure cell volume exceeds the upper limit (right, red line). Thus in an optimization procedure we remove DMSO by two steps to avoid osmotic damage. As shown in Figure3, cell volume is safe all the time in the optimized procedure (right, green line).

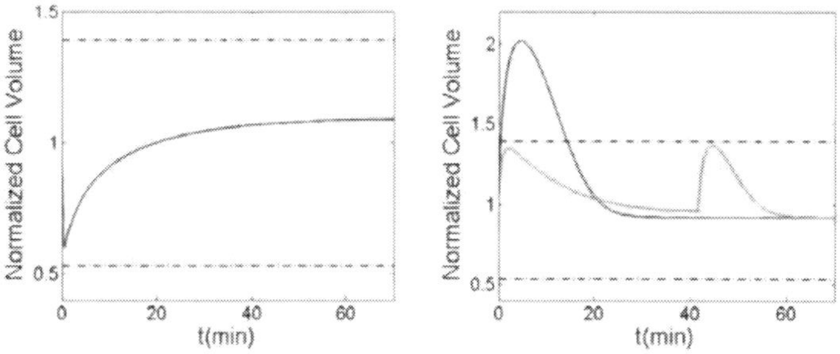

FIGURE 3. SIMULATED VOLUME RESPONSES OF HUVECS DURING ADDITION AND REMOVAL OF CRYOPROTECTANT AT 5°C

DISCUSSION

In this work, we focused on the determination of primary osmotic parameters of HUVEC and their use in optimization of the adding and removing procedure of cryoprotectant. We measured cell volume responses under different osmotic conditions and corresponding survival rate of cells. According to the survival rate we determined the limit values of cell volume change. Then we programed to simulate the adding and removing procedure of cryoprotectant. Multi-step removing was simulated to reduce osmotic damage, which would be helpful to carry out experiments for successful preservation of HUVEC.

ACKNOWLEDGEMENTS

This work was supported by the National Natural Science Foundation of China (Grant Nos. 51276179, 51476160), and the Fundamental Research Funds for the Central Universities (Grant no. WK 2100230009).

REFERENCES

[1] Jacobs, D. R. Stewart, *J. Cell. Comp. Physiol*, 1:71-82, 1932.
[2] Jacobs, *J. Cell. Comp. Physiol*, 2:427-444, 1933.
[3] Kedem, O., Katchalsky, *Biochim. Biophys.* 27:229-246,1958.
[4] J. Hunter, A. Bernard. et al., *Cryobiology*, 29:240-249, 1992.
[5] S. Leibo, *The Journal of membrane biology*, 53:179-188, 1980.
[6] Jianye Wang. et al., *Cryobiology*, 69(2):291-298, 2014.

Exploring the Response of Astrocytes to Traumatic Brain Injuries Using a Novel Bench Top Crash Tester

J. Wyatt (1), A. Walker (1), K. Balachandran (1), J. Wolchok (1)

(1) Department of Biomedical Engineering
University of Arkansas
Fayetteville, AR

INTRODUCTION

In this study we report the results of our experiments exploring astrocyte mechanobiology. Astrocytes are a specialized central nervous system (CNS) cell that plays an important role in brain function. We are interested in exploring the mechanobiology of astrocytes with the broad goal of shedding light on the cellular mechanisms that underlie the response of astrocytes to the mechanical stimuli experienced during traumatic brain injury (TBI). In particular, our group is proposing investigation of the role astrocytes play in the maintenance of CNS extracellular matrix (ECM) [1,2]. We believe a fundamental understanding of the mechanobiology of astrocyte ECM regulation could ultimately provide important insights into the mechanics of tissue degeneration and preservation following TBI.

METHODS

Our group has developed and previously reported the performance of a bench top "cellular crash test" bioreactor capable of delivering TBI mimetic impacts (deceleration and strain) to cells in culture [3]. The bioreactor was designed to deliver impact decelerations up to 200G's and substrate strains up to 50% to multiple cell culture samples. The three major components of the bioreactor are a freely moving drop shuttle whose motion is constrained along a vertical rod that terminates into an impact base (Figure 1). The shuttle is designed to carry six independent cell culture inserts (ID = 17 mm) arranged in a circle (Figure 1B). The flexible base of each cell culture insert is formed by a medical grade polyurethane (PU) scaffold t (thickness = 1mm). The highly porous scaffolds facilitate the attachment and dense cultivation of astrocytes and key to the goals of this research the scaffold also makes possible the accumulation and collection of various ECM molecules secreted by the astrocytes during growth in culture. The scaffolds are used in a manner that is analogous to the use of honeycombs to collect the honey secreted by bees. The PU scaffolds were fabricated using a sugar templating method familiar to our lab [3].

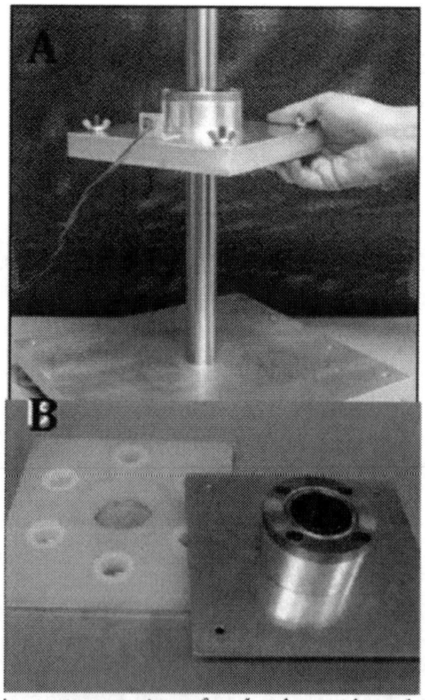

Figure 1: TBI bioreactor consists of a shuttle, guide rod, and impact base (A). Impact magnitude is sensitive to the drop height. Six cell culture inserts are housed within the shuttle seen disassembled (B).

Mouse astrocytes were purchased commercially (ATCC) and expanded following supplier guidelines. To promote cell adhesion, the PU scaffolds were incubated in fibronectin (20 ug/mL) overnight at 4°C. Scaffolds were seeded at a concentration of 2 million cells/cm^3 and impact conditioned (48 hours after seeding) using one of three conditioning schemes. No impact control

1) Impact only (peak deceleration = 150G)
2) Combined Impact + Strain (150G + 20% strain).

Impact conditioning strain and peak decellularization magnitudes were motivated by published head impact measures. Three hours after mechanical conditioning, cells were isolated from conditioned and no impact control scaffolds (n=3/group). Cellular RNA was extracted using the RNAeasy Mini Plus Kit (QIAGEN) following the manufacturer's protocol. Using 100ng of RNA, cDNA was synthesized using iScript SuperMix (BioRad). PCR analysis was performed in triplicate using CFX96 real time PCR detection system (BioRad) and SYBR Green Master Mix (BioRad). Expression levels for the key brain ECM constituents, neurocan, versican, and collagen type1 were measured relative to the housekeeping gene GAPDH. Manufacturer's software (BioRad) was used to analyze the expression data.

Longer term conditioning experiments were conducted to explore the accumulation of ECM and ECM related signaling cytokines within cell seeded scaffolds and conditioned media respectively. Astrocyte seeded scaffolds received impact only, impact + strain, and no impact conditioning three times a week (every other day) over a three week test period. Media was exchanged every 2-3 days. Conditioned media was collected from all samples at 48 hours, 1 week, and 3 weeks after first impact (n=4 / sample group). From the collected media, the concentration of key ECM regulatory cytokines (TGFB1, TIMP1, FGF and IL1-Beta) were measured using commercial enzyme linked immunosorbent assay kits (R&D Biosystems) following manufacturer guidelines.

At the conclusion of the three-week conditioning period (total impacts=9), astrocyte seeded scaffolds were immuno-stained (n=3/stain/test group) for collagen type1 co-stained with laminin or fibronectin co-stained with chondroitin sulfate proteoglycans (CSPGs). Prior to staining, cultures were fixed in 4% paraformaldehyde, permeabilized with Triton X-100, and blocked in 4% goat serum. Samples were immuno-stained using monoclonal antibodies directed against the above mentioned ECM proteins followed by the appropriate fluorescently labeled secondary antibodies (AlexaFluor). Samples were counterstained with the nuclear dye DAPI and microscopically imaged (100X) at multiple regions within the scaffold. From these images, the average fluorescent intensity for each ECM constituent, as well as astrocyte cell count, was measured for each field using image analysis software (ImageJ). Fluorescent intensity was used as an estimate of collagen, CSPG, laminin, and fibronectin accumulation.

RESULTS

Gene expression was normalized to the house keeping gene (GAPDH), as well as no impact control. Three hours after the application of a single mechanical conditioning treatment, collagen and versican gene expression were iaffected by both impact only and combined impact + strain conditioning. Specifically, collagen tyop1 gene expression was significantly decreased, while versican expression was increased in response to mechanical conditioning.

At three weeks, fluorescent intensity in cultures receiving impacts only and impacts + strain were decreased by 25% and 20% respectively when compared to controls, suggesting decreased CSPG accumulation. The relative intensity levels for fibronectin, collagen and laminin were all decreased in response to mechanical stimulation, but the decreases were not statistically significant. The astrocyte density was not significantly affected by either treatment condition.

Media levels for TGF Beta1 were decreased at all timepoints (48 hours, 1 week, 3 weeks) in response to combined impact + strain conditioning but not impact only conditioning (Figure 2). TIMP1 was decreased in response to both impact only and impact + strain conditioning at early timepoints (48 hours) but was equivalent to controls by 1 and 3 weeks. FGF and IL1-Beta concentration remained below detection limits (10pg/ml) at all timepoints points and conditions tested.

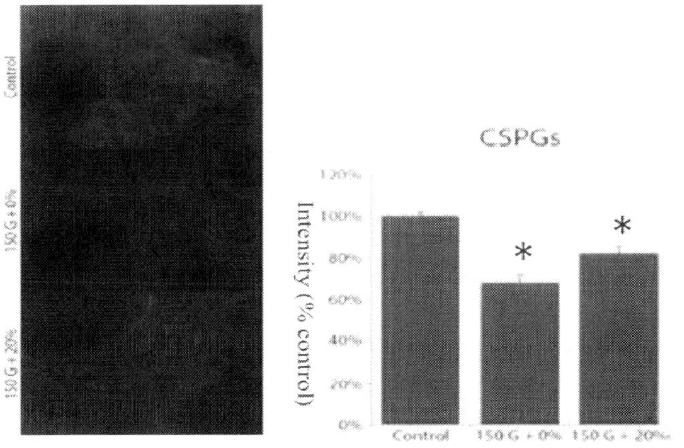

Figure 2:Representative CSPG immuno-reactivity images obtained from control, impact only, and impact + strain conditioned samples. Mechanical conditioning significantly decreased CSPG immmuno-reactivity (fluorescent intensity) when compared to controls.

DISCUSSION

Previously, our lab has reported on the development of a novel in vitro bioreactor capable of delivering both inertial forces and substrate strains to cells in culture. The bioreactor is intended to mimic the mechanic stimuli experienced by astrocytes during TBI. In this study we explored the influence of mechanical stimulation on the extracellular environment produced by astrocytes. The results from this study broadly suggest two key findings.

1) Mechanical conditioning generally down regulated the production of ECM and ECM related signaling proteins by astrocytes.
2) Astrocytes were sensitive to isolated impact only conditioning, suggesting cell mechano-sensitivity to inertial loading.

The clinical impact of these ECM changes is unclear. Current studies examining neuronal outgrowth on ECM materials collected from mechanically stimulated astrocytes may shed light on the relationship between ECM composition and neuronal health.

ACKNOWLEDGEMENTS

Financial support for this research was provided by the National Science Foundation. (Grant# CMMI-1404716)

REFERENCES

[1] Toy and Namgung, *Exp Neurobiol* 22:68-76, 2013.
[2] Heller et al., *Cellular and Molecular Biology* 7:521-31, 2014
[3] Wolchok and Tresco, *Biomaterials* 31:9595-603, 2010

BIOMECHANICS OF HUMAN TIBIA AND FIBULA FRACTURE CAUSED BY A MIXED MARTIAL ARTS KICK

Andrew Lamont (1,2), Robbin Bertucci (1,2), Youssef Hammi (2), M.F. Horstemeyer (2), Jun Liao (1,2), Hongjoo Rhee (2), Lakiesha Williams (1,2), R. Prabhu (1,2)

(1) Agricultural and Biological Engineering
Mississippi State University
Starkville, MS, United States

(2) Center for Advanced Vehicular Systems
Mississippi State University
Starkville, MS, United States

INTRODUCTION

Mixed Martial Arts (MMA) is a sport in which two athletes engage in combat with their hands and feet, each seeking to defeat the other by knock out, submission or by the decision of judges if the fight is completed without a knock out or submission. On December 28, 2013 during the UFC 168 event at the MGM Garden Arena in Las Vegas, Nevada, MMA and UFC athlete Anderson Silva suffered a compound fracture of his left tibia and fibula bones in a location above the ankle[1]. While it is common for mixed martial artists to sustain lower extremity injuries[2,3], such extent of a leg injury does not occur often on the professional level. Studies of the sport have concluded that damage to the lower extremities are fairly common, occurring in roughly 30.4% of the studied cases[4]. More specifically, damage to the lower leg was determined to occur in 4.3% of the studied cases[4].

The objective of this study was to first create a low fidelity, macro-scale rigid body simulation of the kick and to ultimately incorporate the boundary conditions (BCs) from the model into a meso- to macro-scale finite element analysis (FEA) of the incident to acquire a biomechanical explanation for why the injury occurred. Although the severity of this injury is an extreme example of the high-impact nature of MMA, it is a useful example that can be studied to benefit current and future fighters of all skill levels. A biomechanical understanding of the injury will provide MMA athletes with further insight on the reason for the severe injury and to adapt their technique if necessary in order to maximize their safety.

METHODS

GIMP[5] and Windows Live Movie Maker (Microsoft Inc., WA, USA) were first implemented to perform an image analysis of the injury incident. A video was obtained that offered both a front and side view of the kick[6] and images were collected in a frame by frame manner at both viewing angles. The displacement, velocity, and acceleration of various portions of the athletes' legs were measured and calculated from the collected images in order to obtain an initial estimation of the force of the kick.

Next, Adams/View™ rigid body dynamics software by MSC was used to design a low-fidelity macro-scale rigid body model of the kick. After measuring the length of the legs and hips of Anderson Silva and the initial locations of his ankle, knee, and hip joints, a model was able to be developed representing Silva's lower body. Additionally, the defending leg of Chris Weidman was measured and added to the rigid body model. Each joint of each athlete model was given a rotational motion function that represented the rotational displacement and velocities that had been calculated in the image analysis. After changing the material description of the model to bone and simulating it, Adams/Post Processor was used to produce graphical data to determine the maximum force of the kick. This kicking force was then compared to kicking forces found in other studies to determine its relative strength.

Abaqus/CAE™ finite element analysis software by Dassault Systemes was implemented to perform a meso-to macro-scale FEA on the injury. A lower extremity orphan mesh was imported for both Silva and Weidman. After properly placing the meshes and defining the material characteristics for skin, muscle, and bone as elastic materials [7,8,9,10,11] rotational velocity BCs were produced according to the data collected in the image analysis and rigid body dynamics model.

Animations of the FEA results were created and analyzed to determine which areas of Silva's bone experienced the greatest amounts of Von Mises stress and how that stress was produced. We developed failure causation conclusions by comparing the maximum values produced in the simulations to published data on the biomechanical properties of human bone. Additionally, the results were compared to an X-ray similar to Anderson Silva's to verify the location of the maximum stresses in his bones.

RESULTS

The initial image analysis, using GIMP, produced data that was necessary for the remainder of the project. The translational distance of each of Silva's joints was collected to determine the rotational velocities for the rigid body and FEA models. Table (1) depicts the average rotational velocity values of Silva that were implemented in the rigid body model. Additionally, the velocity of the defender's leg was initialized with a translational y-velocity of 1.68 m/s.

TABLE 1: AVERAGE ROTATIONAL VELOCITY OF ROTATIONAL JOINTS IN THE RIGID BODY MODEL

Part Rotated	Rotational Joint	Rot.Velocity (m/s)
Right knee	Right ankle	1.21
Right hip	Center of hips	-2.00
Left hip	Center of hips	2.00
Left knee	Left hip	8.77
Left foot	Left knee	19.75

These rotational velocities produced a resultant maximum force of 6573 N.

The rotational velocity that was used in the FEA model was different than the average rotational velocity used in the rigid body model. Since the model was initialized to a position fewer than 3 mm from contact, the velocity was calculated using the displacement values collected in the final frames before contact. This rotational velocity was determined to be 12.63 m/s, and helped to produce data with higher fidelity than the data produced in the rigid body model.

The FEA simulations revealed information regarding the location and development of stress in Anderson Silva's tibia and fibula. A maximum stress of roughly 250 MPa was found in the tibia while a maximum stress of roughly 220 MPa was calculated in the fibula. Both of these values exceed 132 – 181 MPa[11], which is the determined range for the ultimate stress of human bone under a bending load. As indicated in image (a) of figure (1), the maximum stress experienced by the tibia was in a location above the ankle, between the initial point of contact and the ankle. In image (b), the maximum stress in the fibula can be seen just above its connection to the foot. Both of these simulated areas of stress are coincident with the locations depicted in image (c).

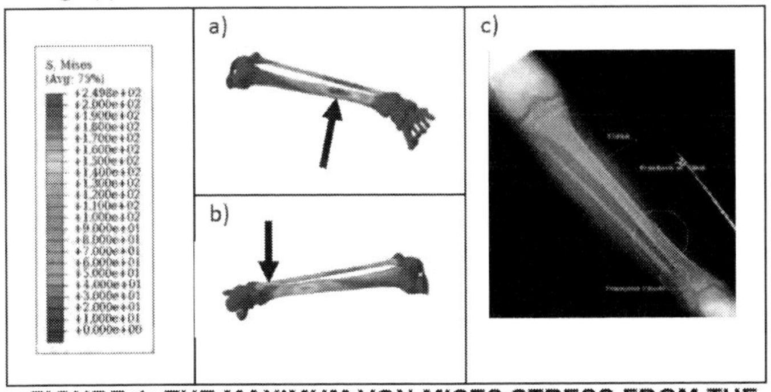

FIGURE 1: THE MAXIMUM VON MISES STRESS FROM THE FEA IS DEPICTED IN MPA. IN (A) AND (B) A FRONT AND REAR VIEW OF THE TIBIA AN FIBULA ARE DEPICTED. THE IMAGE IN (C) IS AN X-RAY IMAGE COMPARABLE TO ANDERSON SILVA'S COMPOUND FRACTURE[12].

DISCUSSION

The force that was calculated by the rigid body model simulation revealed that the kick was not excessively strong, as a study from Laurence Gray concluded the force of roundhouse kicks to range from 2759 – 9712 N[13]. However, expert analysis shows evidence that Silva may have used improper technique to strike with the flat inside ridge of his tibia, making him more susceptible to injury[14].

The FEA simulations offer data to support the above conclusion. After the initial contact of the legs, Silva's bone began to bend as his foot continued to rotate. The bending of the tibia concentrated stress at the apex of the bend which ultimately caused the bone to yield and fracture. The continuation of motion in the foot caused the fibula to bend just above its connection to the foot. Like the tibia, this bending caused the fibula to fail.

Currently, simulations are being conducted in which bone damage and failure properties have been implemented. These simulations will reveal a location and propagation pattern that may be compared to other tibial and fibular fracture images. This data will be able to further validate the conclusions already drawn by the FEA. It is planned to perform simulations with sensitivity analysis of the parameters of BCs and the surface-to-surface interactions properties, so that more true and verified data can be reported.

Results from the simulations that have been performed do, however, provide a clear explanation for why the fractures occurred. By striking his leg on the flat inside ridge at the front of his tibia and on a bony portion of Chris Weidman's knee, Silva induced an acute stress on his leg that caused it to break. These findings can be used to help present and future MMA athletes in their training so that they may avoid such a gruesome injury.

ACKNOWLEDGEMENTS

The above material is based on work supported by the Department of Agricultural and Biological Engineering and the Center for Advanced Vehicular Systems at Mississippi State University through a subcontract funded by the US Army. This material is based upon work supported by the U.S. Army TACOM Life Cycle Command under Contract No. W56HZV-08-C-0236, through a subcontract with Mississippi State University, and was performed for the Simulation Based Reliability and Safety (SimBRS) research program. Any opinions, findings and conclusions or recommendations expressed in this material are those of the author(s) and do not necessarily reflect the views of the U.S. Army TACOM Life Cycle Command.

REFERENCES

[1] Schlinsky, A. http://fansided.com/2014/01/01/ufc-168-dr-sanders-explains-anderson-silvas-surgery/#!BnwLo (3/25/2014).
[2] Bledsoe et al., *J. of Sports Sci. and Med.*, 5:136-142, 2006.
[3] Ngai et al., *Br. J. of Sports Med.*, 42(8):686-689, 2008.
[4] Rainey et al., *N. Am. J. of Sports Phys. Ther.*, 4(4):190-199, 2009.
[5] GNU Image Manipulation Program, http://www.gimp.org/ (accessed 3/27/2014)
[6] Paki Fighter., https://www.youtube.com/watch?v=ddWOhKWhsXk (accessed 2/15/2014).
[7] Diridollou et al., *Skin Research & Tech.*, 6(4): 214-221, 2000
[8] Li et al., *J. of the Royal Soc. Interface*, 9(70): 831-841, 2012
[9] Levinson et al., *J. of Biomech.*, 28(10): 1145-1154, 1995
[10] Kim et al., *BioMed Research International*, ID 197479, 2013
[11] B. LaVeau, *Biomech. of Human Motion*, 3, 1992
[12] Lalaji, A. http://insideinjuries.com/anderson-silva-broken-tibia-fibula/ (accessed 4/29/2014).
[13] Gray, L., *J. of Biomech.*, 12(8): 636-637, 1979
[14]Rutten, B. http://www.bjpenn.com/video-bas-rutten-breaks-down-anderson-silvas-leg-breaking-mistake-in-technique/ (accessed 3/27/2014)

ARTIFICIAL NEURAL NETWORKS FOR THE OPTIMIZATION OF LIGAMENT STIFFNESSES IN A COMPUTATIONAL FOOT/ANKLE MODEL

Ruchi D. Chande (1), Norma Ortiz-Robinson (2), Jennifer S. Wayne (1)

(1) Orthopaedic Research Laboratory
Departments of Biomedical Engineering
and of Orthopaedic Surgery
Virginia Commonwealth University
Richmond, Virginia, USA

(2) Department of Mathematics & Applied Mathematics
Virginia Commonwealth University
Richmond, Virginia, USA

INTRODUCTION

Computational models of human joints provide valuable insight into the joints' biomechanical function. Therefore, supplying these models with appropriate inputs is vital to their accurate, predictive ability. In modeling joint kinematics, it is important to understand soft tissue properties, such as ligament stiffnesses, as these aid in elucidating the structure-function relationship within a joint. This is particularly crucial when generating patient-specific models as properties vary from patient to patient. However, soft tissue properties specific to a patient can be difficult if not impossible to obtain and so another means to determine this information must be sought.

Artificial neural networks (ANNs) are mathematical models comprised of interconnected computational units, or neurons [1]. Through "learning," ANNs can determine relationships among data that are generally unknown [2-3]. This process involves applying a learning algorithm to known sets of input-output pairs [2-3] and iteratively adjusting system weights in order to achieve a target output(s) [1]. Once a relationship between the inputs and outputs is determined, the ANN is able to apply this knowledge to similar, never-before-seen data in order to predict the corresponding outputs [1, 3]. Because of their learning capabilities, ANNs are suitable for applications such as pattern or image recognition, classification, and curve-fitting [1, 3-7]. In the current work, ANNs were utilized to determine optimized ligament stiffness values for a foot/ankle computer model to improve its predictive ability.

METHODS

For this research, a patient-specific computational model of the lower leg in Adult Acquired Flatfoot Deformity (AAFD) was used. This model was generated in SolidWorks 2007 (SolidWorks Corp., Concord, MA) during previous work to predict function of the leg in single leg stance [8]. AAFD, a gross foot deformity resulting in misalignment of the joint, occurs in a patient secondarily to dysfunction of the posterior tibial tendon (PTT). In addition to the PTT, ligaments including the deltoid, talo-calcaneal interosseous, plantar fascia, and spring, tissues are implicated in the progression of AAFD. The degradation of such tissues was captured in the patient-specific computational model by assigning graded stiffness values to each of the aforementioned ligaments; extent of degradation was determined via clinician evaluation. Following designation of ligament stiffnesses, along with muscle loading and patient-specific bony anatomy, into the computational model, biomechanical response of the model in single-leg stance loading was simulated via COSMOSMOTION (Structural Research & Analysis Corp., Santa Monica, CA). The resulting kinematics were validated against patient radiographs and shown to have good correlation [8]. Optimizing ligament stiffnesses to improve predictive ability is warranted.

The optimization of ligament stiffnesses for this foot/ankle model can be described as a curve-fitting task, and so, an ANN to accomplish this was created. A feedforward network with a single hidden layer and 10 hidden neurons was developed using the neural network toolbox in MATLAB R2010b (The MathWorks, Inc., Natick, MA). In order to generate known input-output pairs for the ANN, the ligament stiffnesses of those structures implicated in AAFD progression (comprised of 32 linear elements in the AAFD model and divided into three tissue groupings labeled medial, plantar, and spring) were varied up or down by a given percentage and the corresponding kinematics from the model recorded. Each ligament was varied independently to produce a total of 41 datasets of kinematic measures. Specifically, the stiffness vector included values for the ligament elements while the output vector included four kinematic measures. The measures were navicular and 1st cuneiform heights (mm) and talo-1st metatarsal and talo-navicular angles (degrees) [8].

These stiffness-kinematic pairings were then fed in reverse to the ANN for training, i.e. the input constituted the kinematic measures while the output the stiffnesses. Datasets were assigned randomly to training, validation, and test groups within MATLAB, and adequate training was assessed via mean square error and regression plots. Finally, the original computational model's kinematic measures were supplied to the ANN as an unseen dataset, and the ANN's predicted stiffnesses compared to the original ligament stiffness values.

RESULTS

The training, validation, and test groups all followed decreasing trends and had similar shapes (Figure 1). No drastic increases in MSE were evident.

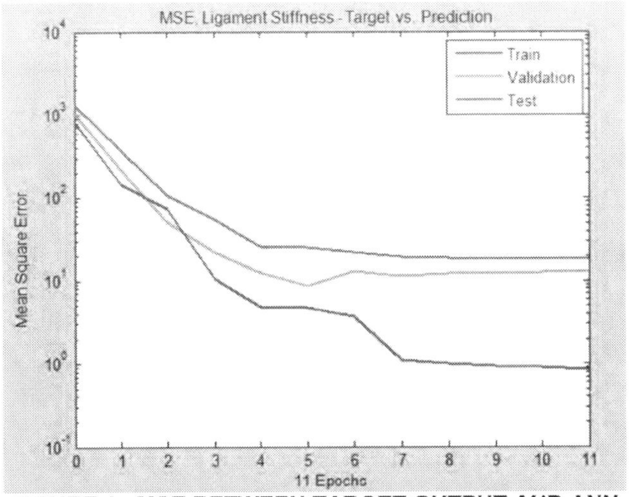

FIGURE 1: MSE BETWEEN TARGET OUTPUT AND ANN PREDICTION OF LIGAMENT STIFFNESSES.

Correlation between target outputs and ANN predictions were observed for training, validation, and test groups; all had R values ≥ 0.99 (Figure 2). Finally, following presentation of the original kinematic measures to the ANN, the maximum percent variation in the medial, plantar, and spring ligament groupings were approximately 7.5%, 14.1%, and 22.7%.

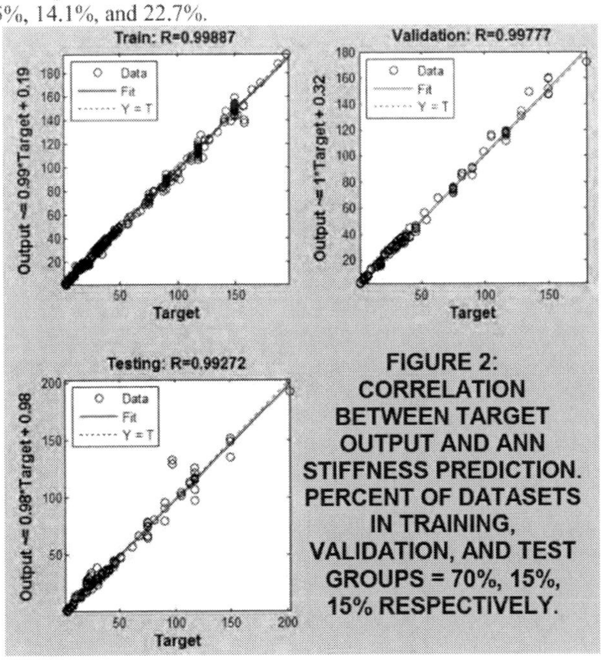

FIGURE 2: CORRELATION BETWEEN TARGET OUTPUT AND ANN STIFFNESS PREDICTION. PERCENT OF DATASETS IN TRAINING, VALIDATION, AND TEST GROUPS = 70%, 15%, 15% RESPECTIVELY.

DISCUSSION

Per the MSE plot, error decreased through the number of epochs, as expected, with the training group resulting in smaller error than the validation and test groups. Validation and test group errors were closer to one another. The difference in errors between the first and latter two groups is likely due to the data division within the network. The training group contained 70% of the 41 datasets, while the other two groups contained fewer datasets thereby allowing for higher calculated error. However, the trends of the curves follow similar, expected trends with no drastic increases in error thus indicating adequate network performance. Currently, additional datasets are being generated to provide the network with more learning scenarios and aid the network in further minimizing error during learning.

With regard to the correlation plots, excellent performance was observed among training, validation, and testing. Based on both qualitative and quantitative observation of the plots, little deviation occurred between the ANN predictions of stiffness and the target stiffness values.

Based on the good results obtained from both the MSE and correlation results, it was determined that the ANN was fit for use with additional data. A "never-before-seen" dataset, specifically the original kinematic measures from the computational model, was fed to the ANN and predictions were compared to the original ligament values. It was noted that the medial group contained the closest stiffness predictions, followed by the plantar, and finally the spring group. While the absolute ligament values did not vary much among those ligament components with the highest percent changes, their relatively smaller magnitudes resulted in the larger differences. In an attempt to remedy this in the future, multiple networks—one for each tissue grouping—will be examined so as to provide each network with fewer outputs and potentially smaller stiffness ranges during training.

From the research presented here, it was determined that an artificial neural network can be a useful means for predicting ligament stiffness values. Further data is currently being generated to supplement network training with the ultimate goal that presentation of the ANN's predicted parameters to the foot/ankle computer model will allow the latter to better reproduce radiographic measures.

REFERENCES

[1] Haykin S. *Neural Networks, A Comprehensive Foundation*, 2nd ed. Upper Saddle River, NJ: Prentice Hall, 1999.
[2] Hassoun MH. *Fundamentals of Artificial Neural Networks*, Cambridge, MA: The MIT Press, 1995.
[3] Agatonovic-Kustrin, S et al., *J Pharm Biomed Anal*, 22:717-727, 2000.
[4] Ahmed, FE, *Mol Cancer*, 4:29, 2005.
[5] Eskinazi, I et al., *Proc ASME 2013 Summer Bioengrg Conf*, 2013.
[6] Kaufman, JJ et al., *IEEE Eng Med Biol Mag*, 9:23-30, 1990.
[7] Lippman, R, *IEEE ASSP Mag*, 3:4-22, 1987.
[8]Spratley, EM et al., *J Orthop Res*, 31:1861-1868, 2013.

SB^{3}C2015
Summer Biomechanics, Bioengineering and Biotransport Conference
June 17-20, 2015,Snowbird Resort, Utah, USA
SB^{3}C2015-439

RETROSPECTIVE ANALYSIS OF ECHOCARDIOGRAPHY IN THOSE SUSPECTED OF PULMONARY HYPERTENSION WITH COMPARISON TO RIGHT HEART CATHETERIZATION

Travis B. Eason (1), Melinda V. Pyler (2), Lavanya Alapati, M.D. (1) Sanjay Mehra, M.D. (3), John M. Cahill, M.D. (1), Stephanie M. George, PhD (2)

(1) Brody School of Medicine
East Carolina University
Greenville, NC, USA

(2) Department of Engineering
East Carolina University
Greenville, NC, USA

(3) Vidant Medical Center
Greenville, NC, USA

INTRODUCTION

Pulmonary hypertension (PH) is the condition of elevated pressure within the pulmonary artery that can be caused through many different mechanisms. If left untreated, PH is life threatening within a couple of years [1]. PH can be further classified into pulmonary arterial hypertension (Group I) with elevated arterial pressure and normal venous pressure, and pulmonary venous hypertension (Group II) secondary to left heart disease, with both arterial and venous pressures elevated [2]. Proper diagnosis is essential in determining the best therapy for a patient.

The current protocol of pulmonary hypertension diagnosis begins with an echocardiogram to look for tricuspid regurgitation (TR). A TR velocity greater than 2.5 m/s is a strong indicator of elevated mean and systolic pulmonary arterial pressure (mPAP, sPAP) [1,2]. If pulmonary hypertension is suspected, patients undergo further testing with a right heart catheterization (RHC), which is considered the gold standard for diagnosis of PH. This invasive procedure obtains a direct measurement of pulmonary arterial pressure and estimates the pulmonary venous pressure through pulmonary capillary wedge pressure (PCWP). A mPAP $\geq$ 25 mmHg confirms diagnosis of PH. If the PCWP is $\geq$ 15 mmHg then the PH is considered Group II and $<$ 15 mmHg it is considered Group I [2].

Doppler echocardiography allows for the measurement of flow throughout the cardiac cycle; this data is useful in predicting pulmonary arterial pressures non-invasively and is the primary screening technique for PH. However, the method has a low specificity [3,4]. The objective of this study is to examine novel parameters derived from the TR echo signal with correlation to RHC measured pressures to improve non-invasive screening of PH.

METHODS

This project was approved by Institutional Review Boards at East Carolina University and Vidant Medical Center. Patients were selected from a chart review (2012-2014) by identifying those who had undergone both a transthoracic doppler echocardiogram and a RHC within 60 days of each other. Patients with mitral or aortic valve disease were excluded from the analysis. To date we have identified 14 patients that met the requirements. Pressure measurements (mPAP, sPAP, PCWP) were recorded from RHC and the maximum TR velocity was recorded from the echo. The DICOM image depicting the TR velocity was identified for each patient and imported into an in-house Matlab program (MathWorks Inc., Natick, MA). The Matlab program was used to extract the velocity data, velocity spectral density, and electrocardiogram parameters from the image. Scale was accounted for through the input of the time and velocity scales found in the image. The maximum velocity at each time point was found with the use of the envelope of the negative velocity signal, the regurgitation velocity.

The flow density through the observation axis is quantified by the brightness of the velocity spectrum in the Doppler signal. The signal magnitude is dependent on the gain, therefore the signal was corrected by multiplying by the inverse of the signal gain. The sum of flow densities, the number of frames with flow at a specific velocity, was calculated for each velocity over one cardiac cycle. This method identifies how much fluid is traveling at each velocity and allows one to determine if the majority of the fluid is traveling at higher or lower velocities.

Lastly, a Fast Fourier Transform (FFT) was done on the original velocity data to detect any signal frequency patterns characteristic of pulmonary hypertension. The data from the image and signal analysis were then exported into Excel for statistical and correlation analysis.

RESULTS

Study results found distinct differences in the hemodynamics between Group I and Group II pulmonary hypertension. The patient population consisted of 14 patients: 5 patients with Group I PH, 6 with Group II PH, 2 with borderline PH (20mmHg < mPAP < 25mmHg), and 1 with no PH.

It was observed that patients with Group I PH have a short peak TR, whereas Group II PH tended to have a longer ejection with higher velocity, therefore a greater flow density at higher velocities. This is further visualized when the flow density over one cardiac cycle is graphed vs. velocity (Figure 1). Patients with Group I PH exhibited a high flow density at lower velocities with the density decreasing as the velocity increased compared to Group II PH which exhibited higher flow densities across all velocities. Borderline patients had varying outcomes; one was similar to Group I PH patients and the other closely resembled the Group II PH patients.

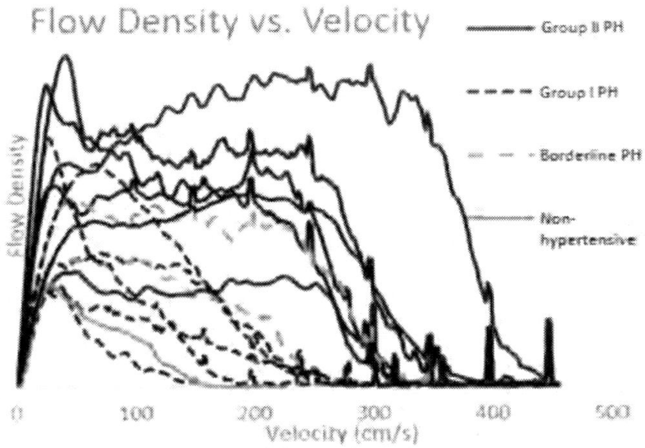

Figure 1: Flow density sum vs. velocity

Correlation analysis of TR maximum versus mPAP and sPAP was conducted as well as corresponding correlation analysis for Group I and Group II subpopulations (Chart 1). Strong correlation coefficients were found between the maximum TR and the mean and systolic PAP. There was a weaker correlation using the FFT signal for the correlation.

	TR vs. mPAP	TR vs. sPAP	FFT vs. mPAP
All patients	0.7747	0.7416	0.5284
Group I PH	0.7808	0.8086	0.4266
Group II PH	0.5957	0.6416	0.2876

Chart 1: Correlation coefficients (R) for TR vs. mPAP, TR vs. sPAP, and FFT vs. mPAP

Averages were compared between Group I and Group II PH patients (Chart 2). Significant differences in averages was found using a t-test for the average flow density and the FFT signal between Group I and Group II PH populations using a 95% CI. The difference in the maximum TR velocity was not significant between the two populations.

	Avg. Max. TR velocity (cm/s)	Avg. Flow Density	FFT (0 Hz + 2 Hz)
Group I PH	293.93	892670	222.82
Group II PH	347.87	1939919	346.72
p-value	0.210	0.0277	0.0459

Chart 2: T-Tests comparing Group I and Group II PH averages.

DISCUSSION

The objective of this study is to retrospectively analyze echo spectral data with comparison to RHC measures. Echocardiography has long been the preferred screening technique for PH, however many patients are found to have normal pressures based on RHC. We hypothesized that there was additional information in the echo data that may be of value in the diagnosis and differentiation of PH.

In a previous study, tricuspid regurgitation velocity is a predictor of systolic pulmonary arterial pressure (R=0.88, n=56) [5]. Our study confirmed that TR maximum is a good predictor of the mean and systolic pressures (R=0.7747, 0.7416, n=14). Group I PH patients had better R values in all categories compared with Group II PH patients, showing that the regurgitation velocity is more predictable with Group I PH. This may be due to the chronic progression of left heart failure which leads to Group II PH. Hemodynamics may change depending on the stage of disease, causing the weaker correlation.

The comparison of flow density vs. velocity curves was effective in distinguishing the group of pulmonary hypertension; Group II patients showed a characteristic high density signal over both low and high velocities, whereas Group I patients show a very high density signals over low velocities, but lower density at higher velocities. Non-hypertensive patients showed low density at low velocities and no signal at higher velocities. This technique may distinguish the differing etiologies in these patients before the RHC.

Interesting results were found from evaluating the TR signals in the frequency domain. By summing 0 and 2 Hz signals, a correlation was found for mPAP and sPAP, but this was weaker than the current method using maximum TR. Error in viewing the data in the frequency domain can come from the frequencies used in the ultrasound as well as how the data is recorded. Echocardiographic signals have not been extensively studied in the frequency domain, but show promise as a method to be further investigated.

We have also identified two new promising echo measures, average flow density and a FFT of the TR velocity signal which when compared between Group I and Group II patients resulted in statistically significant differences (p-values, 0.0277 and 0.0459 respectively), with both being higher in Group II PH. These parameters may be helpful ways in determining the etiology of a patient's PH.

One limitation of this study is the time in between the echo and RHC. It is reasonable to expect some delay between the procedures but the same cardiovascular environment cannot be guaranteed. In addition, echocardiographic data from this study was extracted from images, but future studies should utilize raw data from the ultrasound machines. Other limitations include variability in echocardiography signals based on the technician's skill and the machine used in the study. In summary, we have investigated novel echo measures to identify and differentiate patients with pulmonary hypertension. Two of these parameters (Average flow density and FFT (0+2 Hz) show promise in distinguishing between Group I and Group II PH. Additional patients are needed to confirm these findings.

ACKNOWLEDGEMENTS

The work was supported by the College of Engineering and Technology and the Brody Summer Scholars Research Program.

REFERENCES

[1] Klings, E et al. *Am J Respir Crit Care Med* 189.6:727-40. 2014.
[2] Simonneau G et al. *J Am Coll Cardiol*. 62(25):D34-41, 2013.
[3] Bhave NM, Ward RP. *Curr Cardiol Rep*. 13(3):258-64, 2011.
[4] Finkelhor RS et al. *Echocardiogr*. DOI:10.1111/echo.12594.2014.
[5] Selimovic N, et al. *J Heart Lung Transpl*. 26:9:927-934. 2007.

SB[3]C2015
Summer Biomechanics, Bioengineering and Biotransport Conference
June 17-20, 2015,Snowbird Resort, Utah, USA
SB³C2015-440

A PHENOMENOLOGICAL MODEL TO DESCRIBE THE VISCOELASTIC BEHAVIOR IN MULTIPLE LOADING

R. B. Babaei (1), S. D. Abramowitch (2), E. L. Elson (1,3) S. Thomopoulos (1,2), G. M. Genin (1)

(1) Mechanical Engineering & Materials Science, School of Engineering & Applied Science, Washington University in St. Louis

(2) Musculoskeletal Research Center, Department of Bioengineering, University of Pittsburgh

(3) Biochemistry & Molecular Biophysics, (4) Orthopaedic Surgery, and (5) Neurological Surgery, Washington University School of Medicine

INTRODUCTION

The Fung quasi-linear viscoelastic (QLV) model [1] is a standard tool for characterizing biological materials, and is effective for fitting stress-relaxation data. However, the model's fundamental assumption of a particular form of relaxation response has proven difficult to assess in structural biological materials. Here, we present a simple and robust spectral solution to the problem of identifying and fitting Fung QLV materials. The core of the approach is a finite series that reduces to Fung QLV model when fit to stress-relaxation data for a Fung QLV material. The output is a spectrum that enables simple graphical model identification. After validating the approach, we applied it to data for the medial collateral ligament and identified ranges over which the Fung QLV model provides accurate constitutive characterization.

BACKGROUND

The Fung QLV Model is a nonlinear history- and time-dependent viscoelastic representation for soft biological tissues, developed by Fung (1972). For a material that has never been loaded until time $t = 0$, the stress $\sigma(t)$ at time *t* is calculated from the strain history $\varepsilon(t)$ according to:

$$\sigma(t) = \int_0^t G(t-\tau)\frac{\partial\sigma^{(e)}}{\partial\varepsilon}\frac{\partial\varepsilon}{\partial\tau}d\tau \qquad (1)$$

Fung proposed the following form for the "reduced relaxation function" G:

$$G(t) = \frac{1+\int_0^\infty S(\tau)e^{\frac{-t}{\tau}}d\tau}{1+\int_0^\infty S(\tau)d\tau} \qquad (2)$$

where:

$$S(\tau) = \begin{cases} \frac{C}{\tau}, & \tau_1 < \tau < \tau_2 \\ 0, & otherwise \end{cases} \qquad (3)$$

Graphically, this represents a set of nonlinear viscoelastic elements in series:

Figure 1. Rheological feature of the QLV model.

DQLV MODEL AND SPECTRUM

We developed a discrete QLV (DQLV) representation that can be used to generate graphical relaxation spectra to from viscoelastic relaxation tests. The DQLV series reduces to the Fung QLV model in special cases. The analysis framework is analogous to the standard spectral analysis using the generalized Maxwell [2] model. The DQLV reduced relaxation function is:

$$G_{DQLV}(t) = \frac{1+T\sum_{i=1}^{n}H_i\,e^{\frac{-t}{\tau_i}}}{1+T\sum_{i=1}^{n}H_i} \qquad (4)$$

An analog exists for Fung's $S(\tau)$:

$$S_{DQLV} = \begin{cases} \frac{H_i}{\tau_i} & \tau_1 \le \tau_i \le \tau_n \\ 0 & otherwise \end{cases} \qquad (5)$$

Graphically this is represented by Figure 2, with arrows representing nonlinear springs.

Figure 2. Rheological feature of the DQLV model.

MODEL VALIDATION

Simulated and Real Experimental Data were analyzed using DQLV analysis to ensure that generalized Maxwell and Fung QLV viscoelastic materials could be identified (Figure 3)

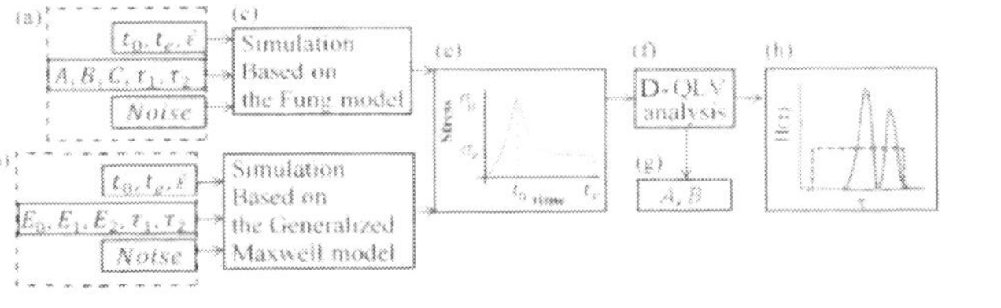

Figure 3. Study design for simulated stress-relaxation experiments.

MATERIALS AND METHODS

Rabbit MCL specimens were dissected and cut free at the insertion sites (N=6 knees). Specimens were affixed to custom-made soft tissue clamps. Cross-sectional area was determined with a laser micrometer at 3 locations. Specimens were mounted to an Enduratec (Elf 3200) testing machine. They were aligned along the loading axis using an x-y table left unloaded for 30 minutes to acclimate. The Specimens elongated to a predefined level of strain (1.25%, 2.5% and 5%). They were held for 35 minutes to reach equilibrium.

RESULTS AND DICCUSSION

Simulated Experiments

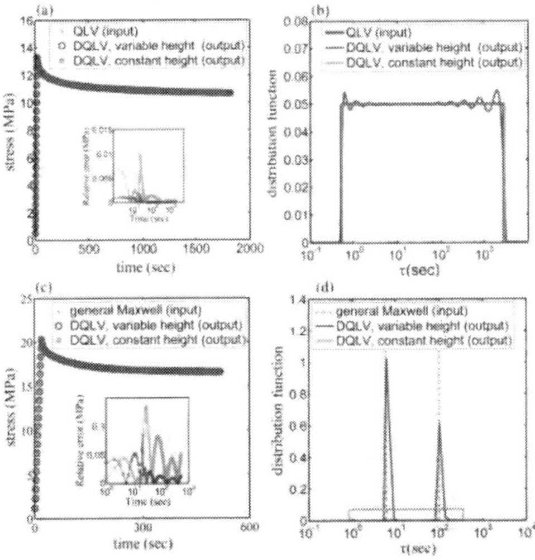

Figure 4. DQLV and QLV approaches to fit the simulated stress-relaxation data.

The Fung QLV model can fit nearly any relaxation data, including inappropriate data. Errors associated with applying the Fung QLV model become evident only at certain loading frequencies. DQLV spectral analysis can identity appropriate models, if appropriate regularization is used (Figure 4).

Rabbit MCL Experimental Data

The improvement in fit to the relaxation data using DQLV is relatively minor. However, the qualitative response identified by DQLV analysis differs fundamentally from Fung spectra. Dynamic shear testing performed by Abramowitch and Woo [3] showed that the Fung QLV model predicted MCL mechanics incorrectly. The DQLV model identifies the correct spectrum at a particular strain level from a single relaxation test, without dynamic testing (Figures 5-6).

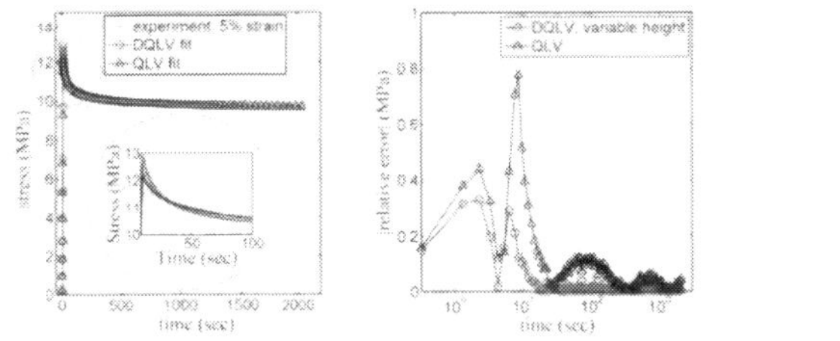

Figure 5. Stress-relaxation data of rabbit MCL for 5% strain fitted by the DQLV and Fung QLV models. (b): Relative error.

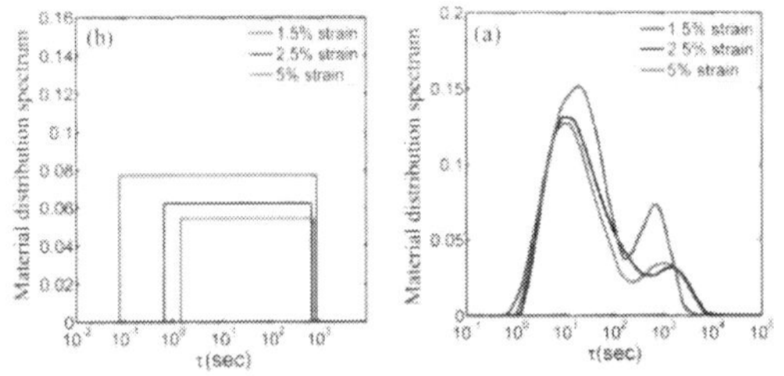

Figure 6. Determined relaxation spectra for rabbit MCL at 1.25%, 2.5% and 5% strains (a) by the DQLV model and (b) by the Fung QLV model.

CONCLUSIONS

The DQLV model enables simple, graphical analysis of stress relaxation data. The results of Figure 4 show how simple goodness-of-fit metrics are inadequate to ensure that a Fung QLV fitting is adequate for a particular material. The challenge is that the Fung model is capable of modeling the relaxation of a completely inappropriate material like a Maxwell material. This is not an issue if the mechanical behavior of interest is only relaxation, but would lead to a poor prediction if one attempted to predict the dynamic response of a Maxwell model to a Fung fitting.

For the Rabbit MCL data, the DQLV fitting identified to peaks in the relaxation spectrum. Again, a Fung QLG fit is adequate for reproducing relaxation under defined conditions, but the DQLV spectrum captures the entire relaxation spectrum and its evolution with strain and loading frequency.

ACKNOWLEDGEMENTS

This work was supported in part by the **NIH** through grants R01AR06082001 and R01HL109505, and jointly by the **NIH** and **NSF** through grant U01EB016422.

REFERENCES

[1] Fung, Y.C., *Biomechanics: Mechanical Properties of Living Tissues*, Second Ed, 1993.

[2] Roylance, D., *Engineering Viscoelasticity*., Cambridge, MA 02139: Massachusetts Institute of Technology. p8–11, 2001.

[3] Abramowitch, S.D. and Woo, S.L-Y., *SB3C*, 2005.

A TRANSVERSELY ISOTROPIC CONSTITUTIVE MODEL DESCRIBES AND PREDICTS THE CONTRIBUTION OF ELASTIN TO THE MULTIAXIAL MECHANICS OF LIGAMENT

Heath B. Henninger (1,2), Sara A. Scott (1), Benjamin J. Ellis (1), Jeffrey A. Weiss (1,2)

(1) Department of Bioengineering
University of Utah
Salt Lake City, UT USA

(2) Department of Orthopaedics
University of Utah
Salt Lake City, UT USA

INTRODUCTION

Elastin is a biopolymer network composed of a microfibril scaffold, tropoelastin monomers and covalent crosslinks between monomers. In ligament and tendon, elastin aggregates along and between collagen fibers. In our recent studies, we demonstrated that elastin contributes significantly to resisting multiaxial deformation in ligament [1, 2]. While elastin supports 20-30% of stress in ligament under uniaxial extension [1], it contributes up to 70% of load support and increases tissue stiffness when subjected to transverse deformation perpendicular to, and shear deformation along, the primary collagen axis [2]. Thus, elastin provides a significant mechanical contribution for a molecule that constitutes only 4% of the tissue dry weight [1].

The objective of this study was to develop a constitutive framework that could describe and predict the contribution of elastin to the multiaxial material response of ligament as determined by our experimental studies [1, 2]. Since elastin is localized between and along collagen fibers, we hypothesized that elastin could be represented as a transversely isotropic contribution to a solid mixture. In addition, we hypothesized that an additive decomposition of the stress into contributions from collagen, elastin and other ground substance components could describe and predict the continuum level response of the composite tissue if the individual components were derived from data targeting the mechanics of those populations.

METHODS

Experimental. Measurements of the longitudinal and transverse tensile material behavior were performed on native porcine medial collateral ligaments (MCLs) and MCLs subjected to elastase digestion. Longitudinal uniaxial tensile tests were performed using a repeated measures protocol to test specimens aligned with the primary collagen axis (N=8) [1]. A strip biaxial tensile protocol was used to test pairs of specimens taken from the same MCL, transverse to the primary collagen axis (N=8 pairs) [2]. These experiments demonstrated that elastin contributes significantly to the stress-strain response under each loading condition, but the percent contribution under transverse loading was much larger than for uniaxial extension along the collagen (Fig. 1).

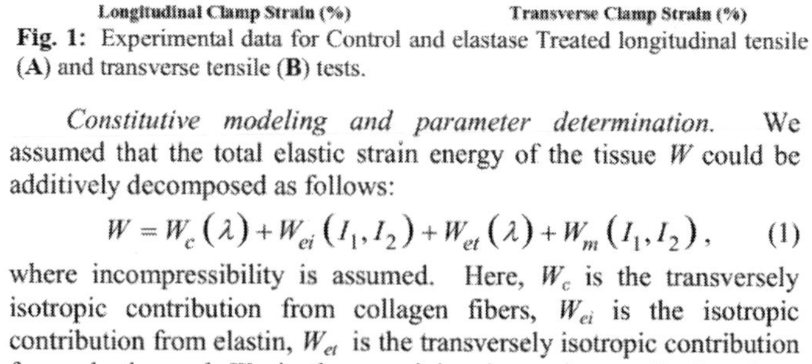

Fig. 1: Experimental data for Control and elastase Treated longitudinal tensile (**A**) and transverse tensile (**B**) tests.

Constitutive modeling and parameter determination. We assumed that the total elastic strain energy of the tissue W could be additively decomposed as follows:

$$W = W_c(\lambda) + W_{ei}(I_1, I_2) + W_{et}(\lambda) + W_m(I_1, I_2), \quad (1)$$

where incompressibility is assumed. Here, W_c is the transversely isotropic contribution from collagen fibers, W_{ei} is the isotropic contribution from elastin, W_{et} is the transversely isotropic contribution from elastin, and W_m is the remaining isotropic contribution from noncollagenous, non-elastin matrix components. I_1 and I_2 are invariants of the right deformation tensor $\boldsymbol{C}$, and λ is the stretch along the collagen fiber direction. W_m and W_{ei} were represented with an isotropic hyperelastic Veronda-Westmann model [3]:

$$W_m = W_{ei} = C_1\left[e^{(C_2(I_1-3))} - 1\right] - (C_1C_2/2)(I_1-3)\,. \qquad (2)$$

W_c and W_{et} were represented with the following exponential-linear hyperelastic model [4]:

$$\begin{aligned} \lambda W_c &= 0, && \lambda \le 1 \\ \lambda W_c &= C_3\left(e^{C_4(\lambda-1)} - 1\right), && 1 < \lambda < \lambda^* \\ \lambda W_c &= C_5\lambda + C_6, && \lambda \ge \lambda^* \end{aligned} \qquad (3)$$

Here, λ^* is the stretch at the transition from exponential toe region to linear region, C_3 scales the stress, C_4 controls rate of rise, C_5 is the slope of the linear region, and C_6 is determined from the other coefficients to ensure C^0 continuity.

Since transverse tensile tests produced an inhomogeneous deformation, the material parameter estimation capabilities of the FEBio finite element (FE) software [5] (www.febio.org) were used to determine best-fit material coefficients. Since the contributions from W_c and W_{et} in Eq. (3) are zero in compression, the material response under transverse tensile testing is only from (W_{ei} + W_m). Optimizations were performed for each sample in Control (W_{ei} + W_m) and Treated (W_m) experimental groups. The coefficients for (W_{ei}) during transverse tensile tests were determined from a fit of the difference between the control and treated samples (Fig. 2). Convergence studies were conducted to determine optimal mesh densities.

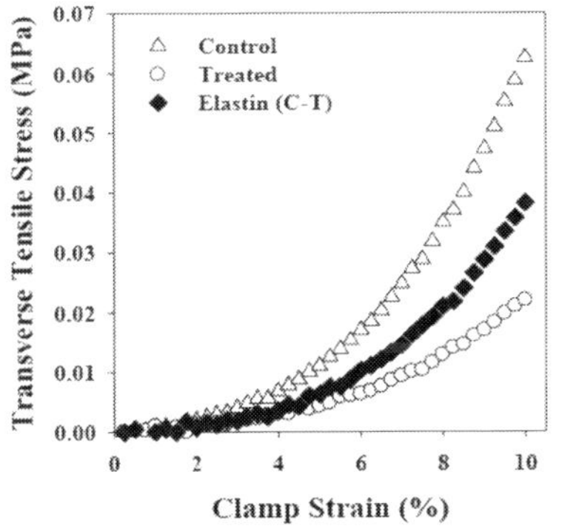

Fig. 2: Data from a transverse tensile test. The difference between Control and elastase Treated curves was assumed to represent the contribution of Elastin.

Since the longitudinal tensile experiments resulted in a homogeneous deformation at the center of the specimen gauge length, a nonlinear least squares method was used with the analytical stress-stretch equations to determine coefficients λ^*,C_3, C_4 and C_5 for W_c and W_{et}. To derive the coefficients for W_c, the Treated longitudinal response was fit assuming that only W_m and W_c were active due to elastase removal of the contributions from W_{ei} and W_{et}. Once coefficients for W_m, W_c, and W_{ei} were available, parameters for W_{et} were optimized from the Elastin response of longitudinal tests (Control – Treated). To test the predictive value of the material coefficients, a single element homogeneous deformation FE model was used with the best-fit coefficients to predict the stress of Control longitudinal tensile experiments from Eqn. (1).

RESULTS

All datasets for transverse tensile tests were fit well by the additive Veronda-Westmann models (Table 1). The C_1 coefficient, scaling the magnitude of the stress/strain curve, was lower for the treated curves versus controls (p=0.014). There was also a significant difference between the C_1 coefficients for the treated and elastin groups (p=0.045). There was no significant difference in the C_2 coefficient, scaling the rate of rise of the curve, between any of the transverse tensile groups.

Similarly, longitudinal experimental data for Treated and Elastin components were fit well with the respective fiber components of the transversely isotropic hyperelastic model (Table 1). Both the λ^* and C_5 coefficients were decreased for the Elastin material versus Treated ($p \le 0.001$). No differences were detected for C_3 and C_4 coefficients.

The best-fit coefficients provided an excellent prediction of the longitudinal stress-strain behavior for the Control group (Fig. 3, solid line).

Table 1: Material coefficients from fits of experimental data.

	Transverse Tensile			
	C_1 (MPa)	C_2		
Control	0.0014±0.0011	30.4±7.6		
Treated (W_m)	0.0003±0.0003	39.2±10.0		
Elastin (W_{ei})	0.0014±0.0014	28.2±9.5		
	Longitudinal Tensile			
	λ^*	C_3 (MPa)	C_4	C_5 (MPa)
Treated (W_c)	1.082±0.000	0.0098±0.0063	66.7±7.1	108.7±51.0
Elastin (W_{et})	1.052±0.004	0.0126±0.0080	75.1±24.4	23.9±11.1

DISCUSSION

By representing the mechanical contribution of elastin as the difference between Control and Treated experimental data, we were able to model the response of the Treated tissue devoid of elastin. Our hypothesis that the multiaxial contribution of elastin could be modeled as transversely isotropic was confirmed. These results supported the hypothesis that additive decomposition of individual material responses would sum to the continuum level tensile response of the tissue, where transverse tensile data provided inputs for the longitudinal response.

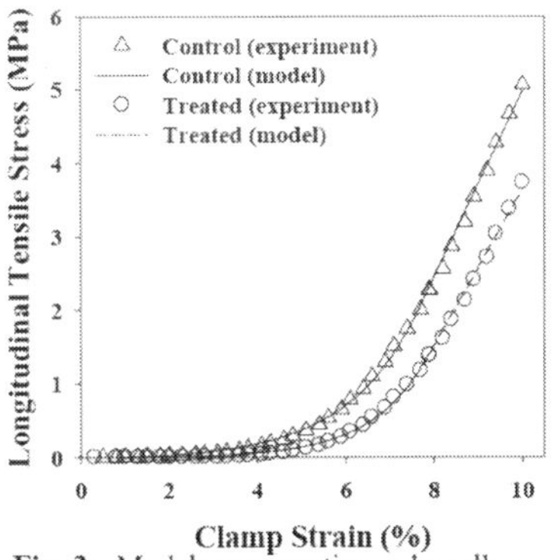

Fig. 3: Models representing only collagen fibers and matrix devoid of elastin (- -) were in good agreement with experimental data for elastase Treated tissue (O). Prediction of Control response (△) using the best-fit coefficients (–) provided excellent agreement with experiment.

Pilot trials determined that linear elastic and neo-Hookean material models could not accurately predict the Control response, where the toe region was consistently overestimated and the linear region was consistently underestimated. Given the disparity in the magnitude of force contributions from elastin between longitudinal and transverse tensile tests (~2 MPa versus ~0.02 MPa, respectively), and the excellent agreement between the models and experimental data, we conclude that the contribution of elastin to the multiaxial mechanics of ligament can be represented by an additive transversely isotropic material contribution consisting of a matrix and fiber term.

ACKNOWLEDGEMENTS

Support from NIH #AR047369 is gratefully acknowledged.

REFERENCES

[1] Henninger, HB, *et al.*, 2013, J Orthop Res, 31(8), pp. 1226-1233.
[2] Henninger, HB, *et al.*, 2014, "Elastin dominates extracellular matrix mechanics in ligament," Proc Annual Meeting of the BMES, San Antonio, TX.
[3] Veronda, DR, *et al.*, 1970, J Biomech, 3(1), pp. 111-124.
[4] Quapp, KM, *et al.*, 1998, J Biomech Eng, 120(6), pp. 757-763.
[5] Maas, SA, *et al.*, 2012, J Biomech Eng, 134(1), p. 011005.

SB³C2015
Summer Biomechanics, Bioengineering and Biotransport Conference
June 17-20, 2015,Snowbird Resort, Utah, USA
SB³C2015-442

BETTER ASSESSMENT OF ARTERIOVENOUS FISTULA PATENCY USING FUNCTIONAL DIAGNOSTIC ENDPOINTS

Krishna Subramony Anantha[1], Ehsan Rajabi-Jaghargh[1], Rupak K. Banerjee[1]

[1]Department of Mechanical and Materials
Engineering, University of Cincinnati
Cincinnati, Ohio, USA

INTRODUCTION

Arteriovenous fistula (AVF) is the widely recommended method of vascular access (VA) for hemodialysis; however, failure of AVFs due to stenosis is one of the major concerns in the dialysis population [1]. A significant stenosis (> 75% area reduction) results in considerable reduction in venous flow rate (< 500 ml/min) and ultimately, failure of the VA for supporting dialysis. Therefore, early diagnosis of stenosis can provide an opportunity to intervene in a timely manner to either assist the maturation process or avoid thrombosis. Currently, venous flow rate (Q_a) and the ratio of venous pressure to arterial pressure (VAPR) are used to detect stenosis in AVFs [2]. A $Q_a < 500$ ml/min and a VAPR > 0.55 are considered as signs of a significant stenosis. However, Q_a has only shown to be a good predictor of an inflow stenosis (a constriction at anastomosis area) in AVFs [2]. The Q_a can diagnose a significant outflow stenosis (a constriction at the outflow tract of the vein) only at the late stages when the stenosis has adversely affected the hemodynamics or resulted in thrombosis. AVFs can maintain a high Q_a at initial stage of an outflow stenosis due to formation of collateral pathways, causing diagnostics complexities. Alternatively, VAPR was developed to detect outflow stenosis in grafts but, it has shown to be a poor predictor of stenosis in AVFs.

The major criticism of the current diagnostic endpoints is that they are either based on flow or pressure, while both of these parameters change in the presence of a stenosis. We have recently showed in a pig model [2] that pressure drop coefficient (C_p: ratio of pressure drop to dynamic pressure at proximal artery) and resistance index (R: ratio of pressure drop to velocity at proximal artery) can distinguish between the adverse and favorable changes in hemodynamics and thus, can better diagnose stenosis in AVFs. Here, our *primary objective* is to assess the variation of C_p and R with changes in flow rate and stenosis severity in an idealized baseline model, developed from realistic geometries of AVF. This will allow evaluation of the efficacy of these parameters for the accurate diagnosis of stenosis.

METHODS

Geometry. The 3D geometry of the AVF and its corresponding schematic are shown in Figures 1A and 1B, respectively. The baseline geometry of the AVF was adopted from the average geometrical features of the AVFs that were created in porcine model [2]. The arterial segment of the AVF consisted of the proximal and distal arteries, while venous segment was comprised of anastomosis, S-shaped, and outflow regions. The diameters of the proximal and distal arteries and

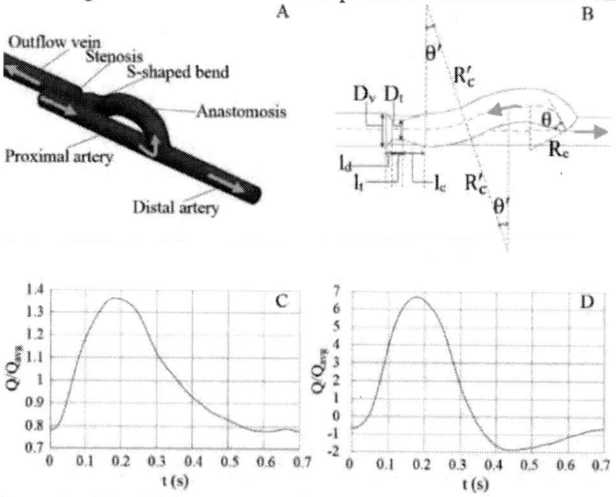

Figure 1. (A) AVF geometry, (B) schematic of AVF, normalized flow rate pulses at (C) proximal artery and (D) distal artery

the vein were 8.2, 7.0, and 9.4 mm, respectively. The anastomotic angle (θ) was 67°, while the radius of curvature (R_c) of anastomosis was 12 mm. The radius of curvature (R'_c) and the subtended angle (θ') for the successive bends of the S-shaped region were 29.9 mm and 28.7°, respectively. The stenosis was located at the end of the S-shaped region. Since there was not much information available on the stenosis profile in AVFs, the stenosis geometry was considered to be axisymmetric with trapezoidal profile, representing the shape of idealized stenotic plaques [3]. The lengths of the converging (l_c), throat (l_t) and diverging (l_d) sections were 6.0, 3.0, and 1.5 mm for all cases, respectively. D_v and D_t (Figure 1B) are the diameters of the vein and stenosis throat, respectively. The analysis was conducted for four cases with percentage area stenosis ($\%AS = \{D_v^2 - D_t^2\}/D_v^2$): 0%, 65%, 75%, and 85%. D_t were 5.6, 4.7, and 3.7 mm for geometries with 65%, 75% and, 85% AS, respectively.

Numerical Analysis. The AVF geometry was meshed using prism elements near the wall and tetrahedral elements at the core. In the numerical analysis, the vessel walls were assumed to be rigid and the blood flow was considered to be unsteady, non-Newtonian, incompressible and laminar. The analysis was performed using Fluent (version 14.5, ANSYS Inc.). The pulsatile velocity pulses were applied to the proximal artery inlet and distal artery outlet, whereas a stress free boundary condition was applied at the vein outlet. The flow rate (Q) pulses [4] at proximal and distal arteries normalized by the corresponding average Q are shown in Figures 1C and 1D, respectively. For the present study, three different average Q were considered at the proximal artery including 600, 1000, and 1600 ml/min, while the average Q at the distal artery had a constant level of 108 ml/min for all cases. These values were based on our flow measurements for AVFs in the porcine model [2].

Functional Diagnostic Endpoints. Resistance index (R) and pressure drop coefficient (C_p) were defined as follows:

$$R = \Delta p / v \quad (1)$$

$$C_p = \Delta p / (0.5 \rho v^2) \quad (2)$$

where Δp is the time-averaged pressure drop between the proximal artery inlet and vein outlet over one cardiac cycle, v is the time-averaged velocity at the proximal artery inlet, and ρ is the blood density (= 1050 kg/m^3). The R was obtained by linearly scaling the pressure drop with flow (velocity at the proximal artery), while in C_p, the Δp was normalized non-linearly with velocity (square of velocity at proximal artery).

RESULTS

Variations of Δp, R, and C_p with respect to %AS for the three different flow rates (Q) are shown in Figures 2, 3A, and 3B, respectively. Also, for more clarity, the corresponding values of Δp, R, and C_p at different %AS and Q are reported in Tables 1, 2, and 3, respectively. The Δp, R and C_p are elevated with the increase in stenosis severity (%AS) for all Q values. Also, for any %AS, the Δp and R increased with the rise in Q. However, for all %AS severities, increase

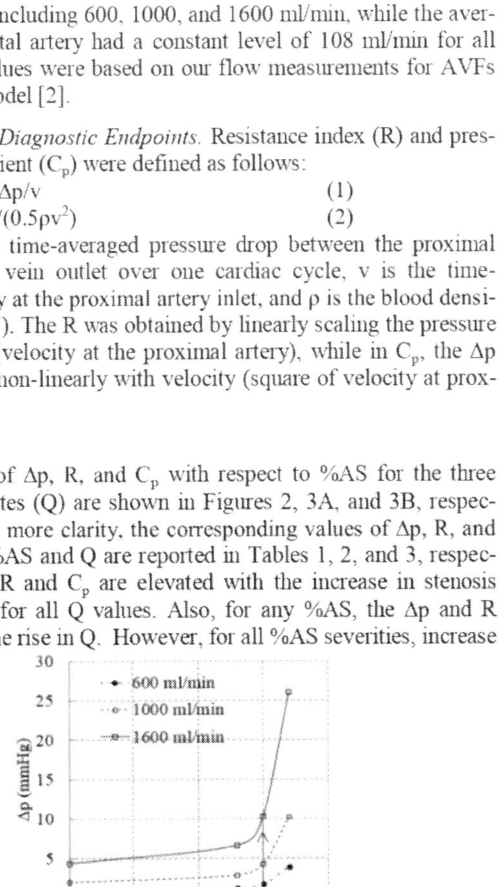

Figure 2. Variation of Δp with %AS for three different Q

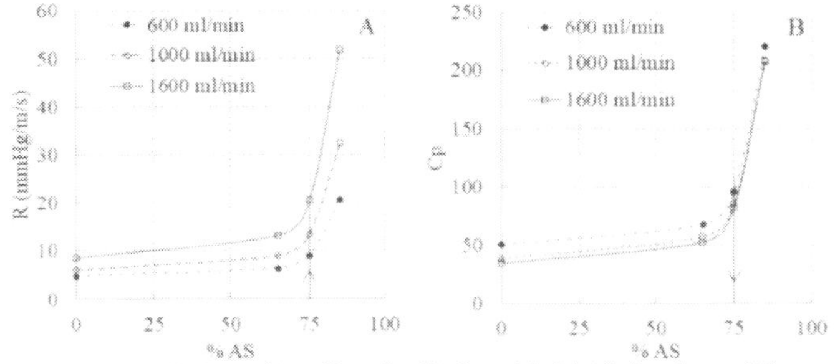

Figure 3. Variation of (A) R and, (B) C_p with %AS for three different flow rates.

in Q resulted in the decrease in C_p values. For instance, for a 75% AS as the Q decreased from 1600 ml/min to 600 ml/min, the R reduced from 20.5 to 9 mmHg/m/s, while C_p increased from 81.9 to 96.2. Here, the reduction in Q represents adverse hemodynamic condition in the AVF caused by the stenosis. Therefore, any increase in C_p can be either associated with the elevation of %AS or the decrease in Q or a combination of the two in AVFs. Thus, C_p can better diagnose the stenosis severity and its corresponding adverse effects on hemodynamics of AVFs as compared to R.

Table 1. Variation of Δp with respect to %AS for three different Q

		Δp (mmHg)			
	%AS	0	65	75	85
Q (ml/min)	600	0.9	1.2	1.7	3.9
	1000	1.9	2.8	4.2	10.2
	1600	4.3	6.6	10.3	26

Table 2. Variation of R with respect to %AS for three different Q

		R (mmHg/m/s)			
	%AS	0	65	75	85
Q (ml/min)	600	4.8	6.4	9	20.7
	1000	6.1	8.9	13.4	32.5
	1600	8.6	13.2	20.5	51.9

Table 3. Variation of C_p with respect to %AS for three different Q

		C_p			
	%AS	0	65	75	85
Q (ml/min)	600	50.9	67.9	96.2	220.7
	1000	38.7	57	85.5	207.7
	1600	34.2	52.5	81.9	206.8

DISCUSSION

Both C_p and R increased with rise in %AS, which was consistent with the findings of our recent study on porcine model [2]. In this study we determined that at all levels of %AS, R decreased with the reduction in Q, while C_p increased. The former can be explained by the fact that any reduction in Q is associated with a more pronounced decrease in Δp. The decrease in R due to reduction in Q can be considered as one of the shortcomings of this parameter as a diagnostic endpoint. However, due to non-linear relation of velocity (square term in denominator) with C_p, the index C_p has a higher resolving power for detecting the adverse effects of stenosis on the hemodynamics of AVFs. Therefore, *C_p can better distinguish the reduction in Q due to formation of stenosis as compared to R, and thus, can be used as a diagnostic endpoint to assess the functionality and patency of AVFs.*

REFERENCES

1. Dixon, BS, *Kidney International*; 70:1413–1422, 2006.
2. Rajabi-Jaghargh, E et al., *Artificial Organs*, 2014.
3. Wilson, RF et al., *Circulation*, 77:873-885, 1988.
4. Krishnamoorthy, MK et al. *Kidney Int.*, 74:1410-1419, 2008.

ELEVATED WALL SHEAR STRESS PREDICTS BRANCH GRAFT FAILURE FOLLOWING CHIMNEY ENDOVASCULAR AORTIC ANEURYSM REPAIR

R. Tricarico (1), H. Yong (2,3), A. Beck (3), S. Scali (2,3), R. Tran-Son-Tay (1,4), S. Berceli (1,2,3)

(1) Department of Biomedical Engineering
University of Florida
Gainesville, FL

(2) North Florida/South Georgia Veterans
Health System
Gainesville, FL

(3) Department of Surgery
University of Florida
Gainesville, FL

(4) Department of Mechanical and Aerospace
Engineering
University of Florida
Gainesville, FL

INTRODUCTION

The Aortic aneurysm (AA) is a degenerative vascular disorder with multifactorial pathogenesis. The 3-9% of AAs go undetected until rupture and the mortality from ruptured aneurysms is 90%. The abdominal aortic aneurysm represents the 15th most common cause of death in the world population over 65 years old.[1,2]

Approximately 20-30% of patients are unsuitable anatomic candidates for standard endovascular aortic aneurysm repair (EVAR). Nowadays, many techniques have been developed to extend landing zones for EVAR including custom fenestrated/branched grafts and surgeon-modified devices as well as chimney, periscope, and sandwich EVAR techniques.[3] The chimney graft is a stent placed parallel to the aortic stent graft to preserve flow to a vital aortic branch. The chimney graft technique can be used as a planned operation but also as a rescue procedure to salvage a side branch unintentionally covered during EVAR.[4]

The chimney technique (chEVAR) has been increasingly used to treat the branched portions of the abdominal aorta. Although successful in the short-tem, long-term durability has been questioned due to concerns about derangements in branch vessel anatomy.[5,6]

This study was performed to determine the anatomic and hemodynamic changes that result from chEVAR and to define the critical parameters that correlate to stent graft thrombosis. Specifically, we hypothesize that alterations in cross-sectional area, wall shear stress, and intra-luminal pressure precede branch graft failure and can be used as a predictive tool triggering pre-emptive remediation.

METHODS

Five patients between 65-83 years old, who had suprarenal chEVAR were chosen for analysis, with a follow-up of 0-45 months.

Pre- and post-operative CT scans were used to construct patient-specific models that recapitulate the 3D-geometry of the paravisceral aorta, through segmentation in Amira. These models were smoothed, extended, and meshed and finally they served for the computational fluid dynamic (CFD) simulations in Ansys Fluent. Boundary conditions of a stress free outflow, stationary and no-slip conditions at the walls, and uniform inlet velocity profile have been implemented to obtain the solution of the transient simulations with hypothesis of pulsatile blood flow (Figure 1). The outflow boundary conditions have been fixed as follow: 18.5% for each renal artery, 19% for the superior mesenteric artery (SMA) and 44% for the main aorta outlet in case of no occlusion; these approximations come from Taylor's curve.[7] Zero pressure was set at the aorta inlet.

Temporal changes in cross-sectional areas (CSA), intraluminal pressure, and wall shear stress (WSS) of SMA and renal artery following chEVAR have been analyzed and their correlations with vessel thrombotic failures have been evaluated with Fisher's exact test. Data were represented as mean±SD.

RESULTS

Among the five patients, 2 vessels were intentionally occluded and 2 more did not need stent graft implantation; 3 of the 11 implanted stent grafts failed during the period of follow-up due to thrombosis occlusion (Table 1).

The conformational changes in the superior mesenteric artery, which has a natural caudal orientation, were usually modest and resulted in only modest changes in local flow. On the other hand, more dramatic changes in the 3D configuration of the often perpendicular renal arteries were observed, with the stents forcing the vessels into a caudal orientation. These changes resulted in significant perturbations in flow dynamics, with elevated local intra-luminal pressure and wall shear stress (Figure 2). In particular, the superior mesenteric artery showed an increase from the pre-op situation to the first month of follow-up of maximum intraluminal pressure of 303±183 Pa; on the other hand, the values on the renal artery were 2621±3012 Pa. The superior mesenteric arteries and the renal arteries showed value of maximum wall shear stress during the first post-op month as 2.3±0.4 fold (p-value 0.052) and 2.2±1.5 fold (p-value 0.238) of pre-op, respectively.

An analysis of the hemodynamic factors associated with graft thrombosis (Table 2) demonstrates that value of the maximum shear stress of the postop greater than 35 Pa (350 dynes/cm²) showed some correlations with branch grafts occlusion. The spatial distribution of wall shear, both pre-operatively and following stent graft placement, for the three patients who had chimney graft occlusion (Figure 3) demonstrated a markedly elevated shear stress in the time interval preceding chimney graft failure. Change in maximum value of intraluminal pressure greater than 1700 Pa (15 mm Hg) or minimum lumen areas less than 20 mm^2 were not associated with graft thrombotic failure.

DISCUSSION

The placement of intravascular stents in aortic branch vessels often leads to an unavoidable change in vessel conformation, causing evident modifications in local hemodynamics. While changes in cross-sectional area, intraluminal pressure, and wall shear stress are commonplace, marked elevations in wall shear appear to be predictive of impending chimney graft failure. These findings suggest a potential role for CFD analysis in the follow-up evaluation of complex aortic endograft repairs.

ACKNOWLEDGEMENTS

Kind acknowledgement to Politecnico of Milan and Atlantis CRISP program for assistance and granting.

REFERENCES

[1] Kuivaniemi, H et al., *Circulation* 117, 242–252, 2008. [2] Celi, S et al., *InTech*, 2012. [3] Scali, S et al., *J Vasc Surg*, 1-10, 2014. [4] Ohrlander, T *et al., J. Endovasc. Ther. Off. J. Int. Soc. Endovasc. Spec.* 15, 427–432, 2008. [5] Donas, KP et al., *J Vasc Surg*, 55:659-65, 2012. [6] Schlosser, FJ et al., *J Endovasc Ther*, 20:7-12, 2013. [7] Taylor, CA et al., *Comput. Methods Appl. Mech. Eng.* 158, 155–196, 1998.

TABLE 1. OUTCOMES OF THE 5 PATIENTS ON THE SUPERIOR MESENTERIC (SMA), RIGHT AND LEFT RENAL ARTERY (RRA, LRA).

Patient #	SMA	RRA	LRA	Last CT	PostOp CT #
1	patent	patent	failed	13months	4
2	failed	occluded	patent	7months	3
3	patent	patent	failed	13months	2
4	no stent	no stent	patent	7months	3
5	patent	patent	occluded	45months	5

TABLE 2. FISCHER'S EXACT TEST ON MAXIMUM VALUE OF WALL SHEAR STRESS, INTRA-LUMEN PRESSURE MAXIMUM CHANGE AND MINIMUM CROSS SECTIONAL AREAS ON THE ANALYZED BRANCHES.

	Branch graft status		
	Failed	Patent	P-value
Max wall shear stress			
> 35 Pa	3 (60%)	2 (40%)	0.06
< 35 Pa	0 (0%)	6 (100%)	
Intraluminal press. max change			
> 1700 Pa	2 (29%)	5 (71%)	1.0
< 1700 Pa	1 (25%)	3 (75%)	
Min lumen area			
> 20 mm^2	2 (22%)	7 (78%)	1.0
< 20 mm^2	1 (50%)	1 (50%)	

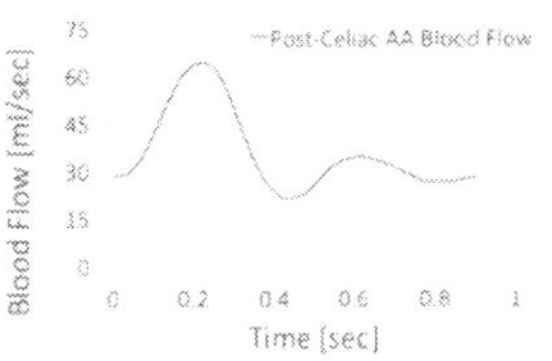

FIGURE 1. FLOW BOUNDARY CONDITION AT THE INLET; THE CURVE HAS BEEN EXTRAPOLATED FROM LITERATURE.[7]

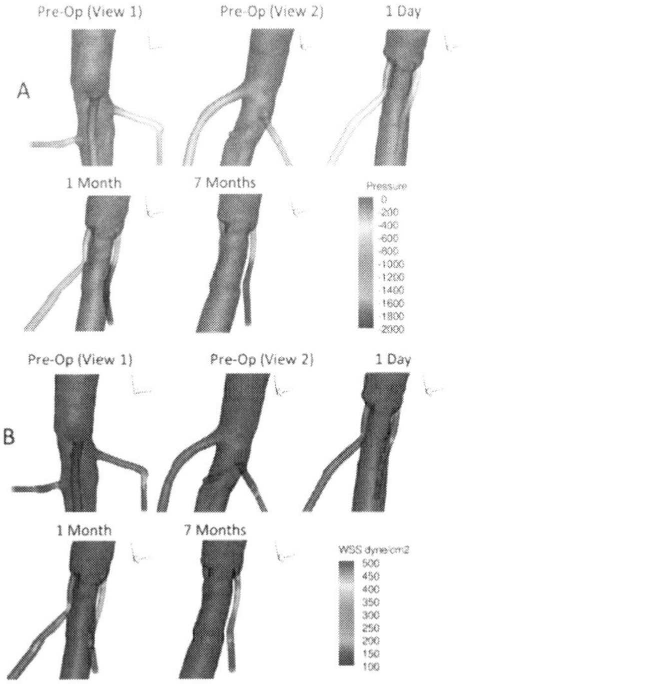

FIGURE 2. INTRALUMINAL PRESSURE (A) AND WALL SHEAR STRESS (B) ON THE WALL OF THE ABDOMINAL AORTA OF PATIENT 2 DURING THE FOLLOW UP.

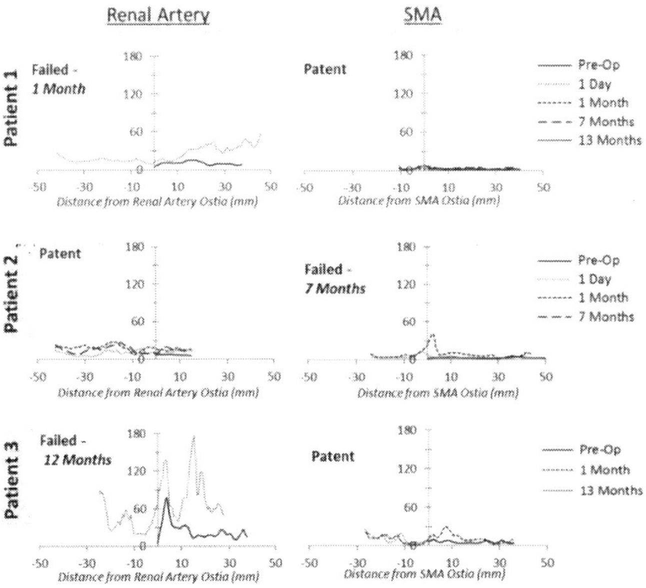

FIGURE 3. COMPARISON OF WALL SHEAR STRESS ON THE FAILED AND PATENT STENT GRAFTS.

THREE-DIMENTIONAL STRAIN DISTRIBUTION IN THE ANTERIOR CRUCIATE LIGAMENT DURING ANTERIOR TRANSLATION OF THE KNEE

Satoshi Yamakawa (1), Richard E. Debski (2), Hiromichi Fujie (1)

(1) Graduate School of System Design
Tokyo Metropolitan University
Tokyo, Japan

(2) Department of Bioengineering
University of Pittsburgh
Pittsburgh, PA, USA

INTRODUCTION

Quantitative biomechanics data on the behavior of the anterior cruciate ligament (ACL) under knee loads are necessary for better understanding of ACL injuries and improvement of ACL reconstruction. For this purpose, deformation and strains in the ACL were determined in previous studies[1-3]. Although important data were obtained in the studies it was problematic that physiological knee motions and loads were only partially simulated because of experimental limitations. Moreover, detailed data of deformation and/or strain in the attachment area of the ACL to the femur and tibia was not obtained, even though such data would be useful for the assessment and improvement of ACL reconstruction techniques. We have developed a 6-degree of freedom (DOF) robotic system for knee joint biomechanics tests in which physiological 6-DOF knee motions and loads could be simulated, and more importantly, the knee motions, once recorded could be reproduced. Recently, the image correlation method has been developed for motion tracking of arbitrary points in which detailed data of deformation and strain of an object could be obtained with a high resolution. Therefore, the objective of the present study was to apply both the robotic system and a motion tracking system based on the image correlation method to the determination of site-dependent strain in the ACL, specifically in the attachment area to the femur and tibia, during anterior translation of the knee.

METHODS

A 6-DOF robotic system consisting of a custom made 6-axis manipulator with a 6-DOF universal force/moment sensor (UFS)[4] was used. In the system, a LabView-based control program runs on a windows PC to control the displacement of, and force/moment applied to, the cadaveric knee joints with respect to the knee joint coordinate system. Porcine knee joints (n=5) were dissected down to the joint capsule and fixed to the robotic system. The anterior drawer test was performed with 50 N applied at full extension. During the test, the anterior-posterior DOF was translated under displacement control at a rate of 0.1 mm/s while all the DOFs except the AP and flexion-extension DOFs were set under force control with prescribed force/moment at zero. After the lateral condyle was removed for the observation of the ACL the intact knee motions were reproduced. The medial view of the ACL surface during the test was recorded with a CCD camera. Using a motion tracking system (VW-9000, KEYENCE, Osaka, Japan), the surface deformation of the ACL including the attachment area to the femur and tibia was analyzed. Note that, with the system, it was possible to track arbitrary points up to 50 without physical markers. In the recorded image, the ACL was transversely divided into 5 portions; 0-10% , 10-20%, 80-90% and 90-100% in the length from the femoral attachment, and was longitudinally divided to 5 portions from anterior to posterior fibers, as shown in Fig.1. Multiple gage markers were placed at the anterior and posterior edges in each portion on the recorded image, and length change between the markers were calculated to analyze local strains. Then, the ACL surface was sliced and removed, so that the femoral and tibial attachment areas were approximately 1/3 of entire attachment area. The medial view of the middle layer surface of the ACL was recorded during the reproduction of anterior translation of the intact knee. Local strains on the middle layer of the ACL were determined through the experimental manner identical to that for the intact state. This procedure was repeated again after the removal of the middle layer to determine local strains on the deep layer of the ACL during the reproduction of anterior translation of the intact knee. Note that zero strains (no deformations) were defined for the knee located at full extension with 1.0 Nm of extension moment. In a preliminary experiment, we made sure that there was almost no load interaction between the layers.

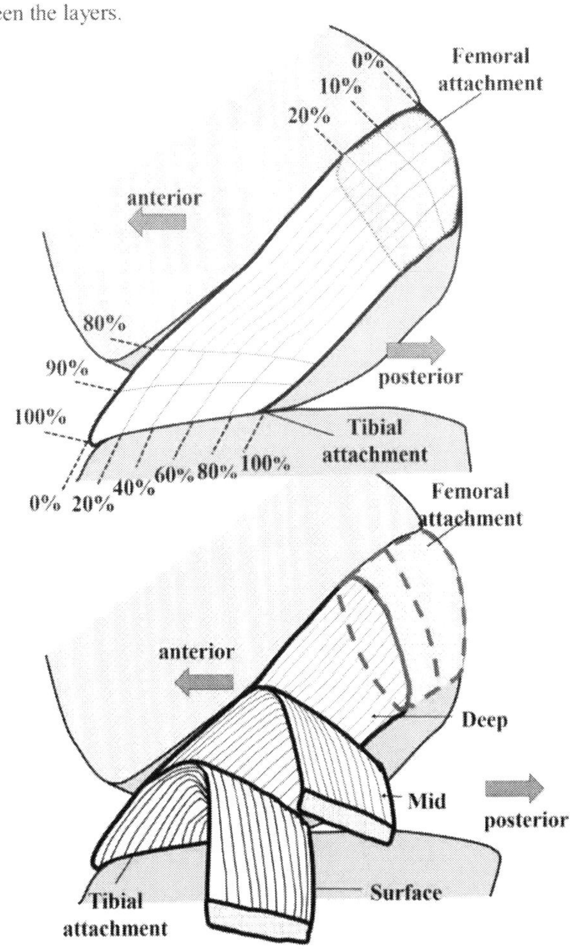

Fig.1 Twenty five portions on the ACL surface layer for the calculation of strain distribution and the sequential removal of the surface and middle layers for the measurement of 3-D strain distribution

RESULTS

Strain distribution in the ACL in response to 50 N of anterior drawer force to the knee is indicated in Table1-3. Strains in the ACL attachments (0-20% and 80-100%) were larger than those in middle portion (20-80%) in all layers. In the surface layer, the largest strain was 9.3% observed in the tibial attachment on the posterior fiber. In the middle layer, the largest strain was 7.7% observed near femoral attachment on the most posterior fiber. In the deep layer, the largest strain was 12.8% observed in the femoral attachment on the most anterior fiber.

Table.1 Strain distribution in the surface layer in response to 50 N of anterior force to the knee (percent strain)

Width / Length	Anterior 0-20%	20-40%	Mid 40-60%	60-80%	Posterior 80-100%
0-10%	2.5	0.5	0.8	3.4	0.8
10-20%	3.3	1.2	2.5	2.0	4.8
20-80%	1.1	0.7	0.8	0.3	0.1
80-90%	1.7	3.7	7.9	2.1	2.0
90-100%	4.3	3.4	6.6	9.3	6.9

Table.2 Strain distribution in the middle layer in response to 50 N of anterior force to the knee (percent strain)

Width / Length	Anterior 0-20%	20-40%	Mid 40-60%	60-80%	Posterior 80-100%
0-10%	5.3	2.4	1.1	1.9	1.7
10-20%	3.1	1.1	3.2	2.1	7.7
20-80%	0.9	0.8	0.8	0.8	2.7
80-90%	0.8	5.9	5.0	4.4	3.9
90-100%	3.6	7.1	6.5	6.7	5.9

Table.3 Strain distribution in the deep layer in response to 50 N of anterior force to the knee (percent strain)

Width / Length	Anterior 0-20%	20-40%	Mid 40-60%	60-80%	Posterior 80-100%
0-10%	12.8	1.6	1.9	2.2	2.1
10-20%	0.6	1.4	2.5	3.1	4.0
20-80%	1.3	0.8	0.9	1.6	1.8
80-90%	0.8	3.3	3.7	1.9	3.0
90-100%	3.3	1.8	3.5	3.5	5.0

DISCUSSION AND SUMMURY

The 3-D strain distribution in the ACL in response to 50 N of anterior drawer force applied to the knee was determined in the present study using the 6-DOF robotic system and the motion tracker based on image correlation method. We succeeded to obtain a detailed map of site-dependent strain data in the ACL fiber bundles.

In previous studies, video tracking systems were often involved to measure deformation and strain in soft tissues. Although measurement with high resolution was possible with the system, there was a limitation in terms of position and number of markers. It was also problematic in the previous systems that marker motion relative to tissues might affect experimental results. These problems are solved in the present study by using new motion tracking method. Since it is possible with the method to select up to 50 arbitrary points for tracking after video recording, detail and precise analysis of local strain in a wide area is possible.

Results revealed that strain in the ACL was site-dependent. Strain was significantly higher in the attachment area than those in middle portion (20-80%) in all layers. Yamamoto et al. reported that large strain occurred in the anterior fibers in the human ACL in full extension[2]. The present result suggests that the local strain in the attachments area is more complicated and more site-dependent in porcine knees. Future studies will be performed to determine the 3-Dstrain distribution of the human natural ACL and/or reconstructed ACL in response to various knee loadings using the present methodology.

ACKNOWLEDGEMENTS

The present study was financially supported by the Grant-In-Aid for Scientific Research B (#25252134) both from the MEXT, Japan, and the Priority Research Found from Tokyo Metropolitan University.

REFERENCES

1) G. Cerulli, et al., KSSTA Vol.11 p307-311, 2003
2) K. Yamamoto, et al., Medical Engineering & Physics, 1998
3) J.P. Spalazzi, et al., Wiley InterScience, 2006
4) H. Fujie, et al., J. Biomech. Eng. (ASME), 126, pp. 54–61., 2004

EQUILIBRIUM SHAPE OF THE AQUEOUS HUMOR-VITREOUS SUBSTITUTE INTERFACE

Krystyna Isakova (1), Jan O. Pralits (1), Rodolfo Repetto (1)

(1) Department of Civil, Chemical end Environmental Engineering
University of Genoa
Genoa, Italy

INTRODUCTION

Vitrectomy is a common surgical way to treat retinal detachment, a pathology that might occur with ageing or be caused by trauma, diabetes or other eye diseases. During the surgery the vitreous humor is removed from the eye and it is replaced by a tamponade fluid. The role of this fluid is to close retinal breaks and hold the retina in place against retinal pigment epithelium. The filling of the vitreous chamber with the vitreous substitute might be partial or (almost) complete, depending on the case. Currently the most commonly used fluids during the surgery are silicone oils. Owing to their hydrophobic properties a sharp interface forms between the silicone oil and the aqueous humor. The amount of area of the retina tamponated by the oil is strongly affected by the density of the oil, the surface tension at the interface and contact angle between the interface and the retina. Eames et al.[1] have proposed a theoretical model that predicts the shape of the interface between two immiscible fluids within a spherical domain. It is of clinical interest to extend the authors' work to study of the equilibrium configuration taken by the aqueous humor-tamponade fluid interface in a realistic eye geometry.

METHODS

We introduce mathematical model that predicts the shape of the interface between aqueous humor and vitreous substitute, based on the mechanical properties of both fluids. A quasi-analytical solution is found for the spherical domain, which is analogous to that produced by Eames et al.[1] and which is used to validate the numerical code[2]. We then consider a more complicated geometry that gives a good approximation of the real eye shape[3]. In this case a fully numerical analysis, conducted with *OpenFoam*, is performed, using the VOF (volume of fluid) method. The fluid under consideration is silicone oil with density of 980 kg/m^3 and the contact angle is taken to be equal to 17^0[4].

RESULTS

The results obtained employing the mathematical model and the VOF method are in a good agreement in the case of the spherical domain. This is shown in fig. 1, where we plot the shape of the interface for three different degrees of filling. In the figure the oil occupies the top part of the domain and the aqueous the bottom one. The interfaces shown on the left hand side of the plot with thin lines are obtained with the theoretical model whereas those on the right hand side (thick lines) have been obtained through the fully numerical simulations.

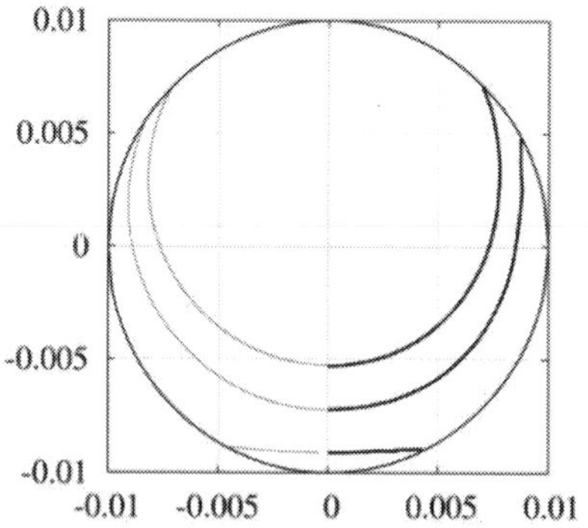

Figure 1. Shape of the interface in the spherical domain computed by mathematical model (left) and by the numerical model (right) for different volumes of tamponade fluid.

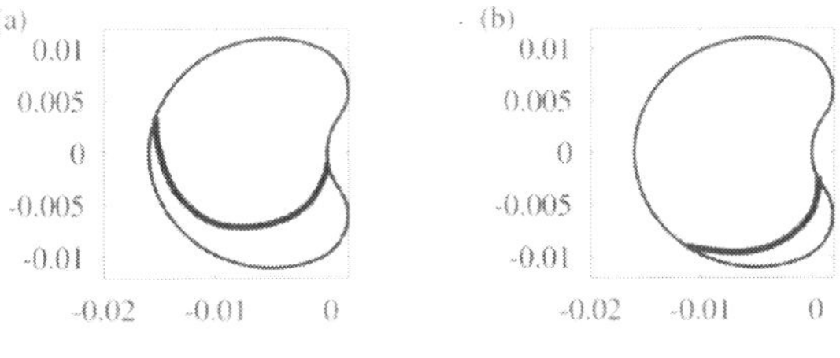

Figure 2: computed position of the interface on the plane of symmetry of the domain for two different degrees of filling.

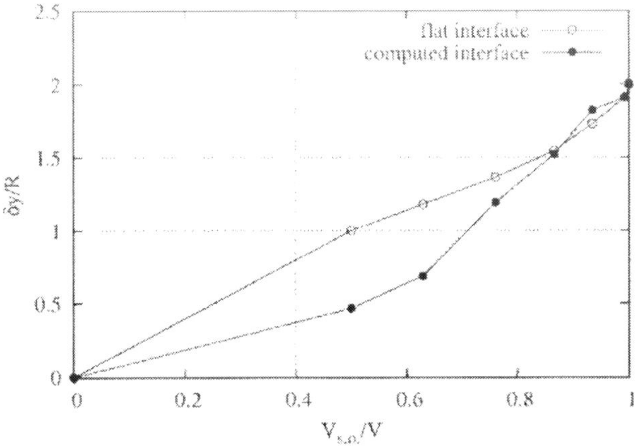

Figure 3: Distance from the top of the domain to the interface elevation on the back of the eye, scaled with the eye radius against filling ration $V_{s.o.}/V$. The black solid dots correspond to the computed shape of the interface and the empty dots are obtained assuming a flat horizontal configuration of the interface.

The shape and the position of the interface for the real-eye domain are not axisymmetric, as shown in fig.2. Fig. 2(a) corresponds to a filling ratio (volume of oil divided by the total volume of the domain $V_{s.o.}/V$) of 0.623 and fig. 2(b) to a filling ratio of 0.935.

In fig. 3 we plot the distance δy between the top of the domain and the interface along the back of the eye, normalized with the distance R between the north pole and the horizontal plane of symmetry of the domain. In the figure we also plot the corresponding values for the case of a flat horizontal interface. Each point in the figure corresponds to a different filling ratio $V_{s.o.}/V$. This shows if that a significant underestimation of the region of the retina tamponated by the oil would be obtained if the surface was supposed to be flat.

We also plot in fig. 4 the tamponated retinal area S_{total} (normalized with the total retinal surface S) as a function of the volume ratio. In analogy to what is done in fig. 3, we also plot in the figure the corresponding curve relative to the case of a flat horizontal surface. Again, it appears that the two curves are significantly different from each other.

DISCUSSION

We have analyzed the equilibrium position of the interface between aqueous humor and a tamponade fluid within a domain which reproduces realistically the shape of the vitreous chamber. The results were verified against the predictions of a theoretical model in the case of spherical domain.

The equilibrium shape of the interface is governed by the densities of the two fluids, by the surface tension at the interface and by the contact angle at the triple line of contact of the interface with the solid wall. The model allows us to estimate the surface of the retina tamponated by the oil, for a given value of the filling ratio. This is an information of practical clinical relevance since, after surgery, the patient is asked to maintain a certain position of the head for some time, to ensure that the damaged portion of the retina is covered by the tamponade fluid. Proper positioning depends on the shape taken by the interface between the oil and the aqueous. The interface assumes a highly curved and asymmetric shape, which cannot be predicted without the aid of a mathematical model. The results show that estimating the area of retina that is tamponated by the oil assuming that the surface is horizontal would lead to a significant underestimation of such a quantity.

REFERENCES

[1] Ian Eames et al. A Theoretical Model for the Predicting Interfacial Relationships for Retinal Tamponade. Investigative Ophtalmology and Visual Science, April 2010, Vol 51.

[2] C. Pozrikidis. Fluid Dynamics Theory Computation and Numerical simulation. Kluwer Academic Publishers. Londor, 2001.

[3] David A. Atchison. Optics of the Human Eye, Buterworth-Heinemann, 2000.

[4] Fawcett I, William RL, Wong D. Contact angles of substances used for internal tamponade in retinal detachment surgery. Graefes Arcb Clin Exp Ophtalmol. 1994; 232: 438-444.

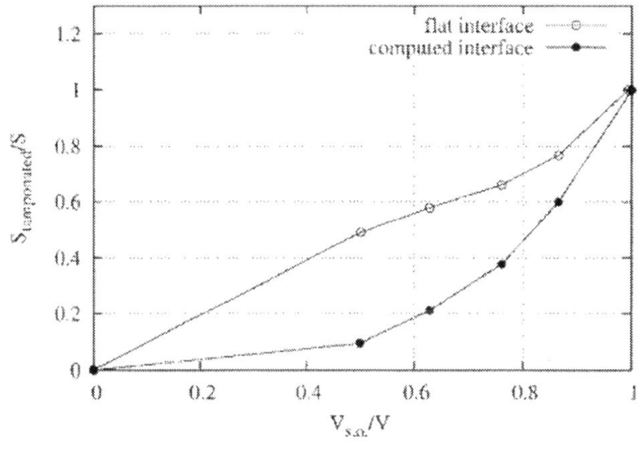

Figure 4: Surface of the tamponated retina normalized with the total retinal surface versus the filling ratio filling ration $V_{s.o.}/V$. The black solid dots correspond to the computed shape of the interface and the empty dots are obtained assuming a flat horizontal configuration of the interface.

SB^{3}C2015
Summer Biomechanics, Bioengineering and Biotransport Conference
June 17-20, 2015,Snowbird Resort, Utah, USA
SB^{3}C2015-446

DEVELOPMENT OF 3D PRINTED PATIENT SPECIFIC ASCENDING AORTIC TRAINING MODELS FOR CARDIAC SURGERY

Z. Yang (1), J. Garcia (1), K. Lachapelle (2), R. Mongrain (1), R. L. Leask (3),

(1) Mechanical Engineering
McGill
Montreal, Quebec, Canada

(2) Royal Victoria Hospital
McGill
Montreal, Quebec, Canada

(3) Chemical Engineering
McGill
Montreal, Quebec, Canada

INTRODUCTION

Cardiovascular disease is a leading cause of death in Canada and the Unites-States. Over a hundred thousand cardiac surgeries are performed every year in North America. The training of proficient cardiac surgeons requires both time and practice. Although various substitutes, such as mannequins, artificial tissues and animal organs, are available for surgical training, practice on actual patients is still an inescapable part of the learning process in becoming a seasoned surgeon. Realistic physical surgical simulators are therefore needed to lessen the reliance of practice on actual patients.

The goal of this project is to develop a protocol to capture and then 3D print patient specific surgical training ascending aortic models from clinical datasets.

METHODS

Model Creation

Gated cardiac CT scans were obtained under informed consent from the Royal Victoria Hospital in the standard Digital Imaging and Communications in Medicine (DICOM) format. Reading and segmentation of the DICOM files were done in 3D-Doctor (Able Software, Lexington, MA, USA) through a mix of manual and automatic segmentation. The lumen of the aortic arch, aortic valve and branches were segmented. Once the segmentation was completed, the lumen was then exported as a point cloud into SolidWorks Premium (Dassault System, Waltham, MA, USA). In SolidWorks extraneous points were removed from the point cloud before converting the data into a surface mesh which was then smoothed, and exported as an Stereolithography file (.stl). Using Netfabb (Netfabb, Parsberg, Germany), a wall of fixed thickness was extrapolated from the lumen.

3D printing

A series of different printers: Connex500™ (Stratasys, MN, USA), Form 1+ (fromlabs, MA, USA), Afinia H480 (Afinia 3D, MN, USA); as well as, materials: ABS (Afinia 3D, MN, USA), PLA (Afinia 3D, MN, USA), Ninjaflex (Fenner Drives, PA, USA), Vero (Stratasys, MN, USA), TangoPlus (Stratasys, MN, USA) were evaluated for producing the final models. Measurement taken form the aortic arch as well as the branch of the model were compared to the dimensions taken directly from the medical imaging and surgical dataset.

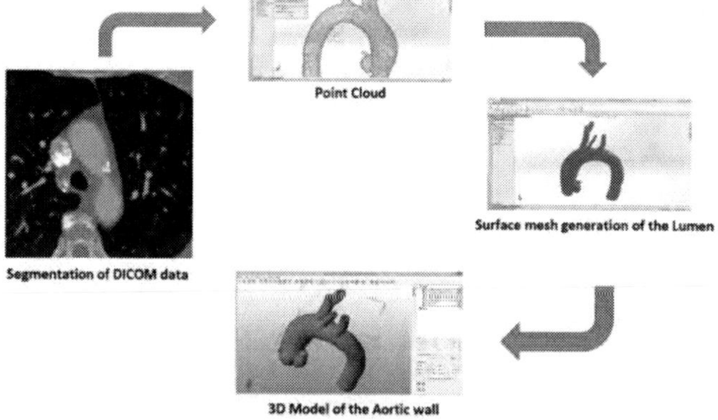

Figure 1 Model Creation

Material Testing

In attempt to create a 3D printed model matching the ascending aorta in material properties, a series of composite were printed using the Connex500™ printer. Rigid sinusoidal fibers made of VeroWhite were embedded in an elastic material, TangoPlus. A series of samples with variance in fiber geometry are created. The samples were printed in standard cruciform shape [1]. The fibers inside the cruciform were sinusoidal and placed in a 45-deg cross pattern. Fibers had a 100um x 100 um square cross-sectional profile. The fiber parameters varied were the amplitude and the angular frequency of the sine curve, as well as the vertical and horizontal spacing between fibers (Figure 2). The matrix and the fiber were created as different parts in SolidWorks. In order to avoid errors during the 3D printing process, Boolean operations were performed in order to remove all overlaps.

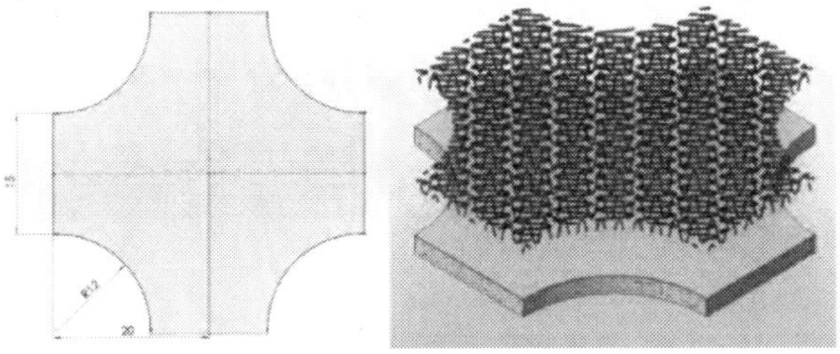

Figure 2 Standard Cruciform

RESULTS

Aortic Model

In the first stage, we printed rigid aortic model which proved to be highly accurate and are representative of in vivo ascending aorta pathological geometries provided by our surgeons. By comparing the printed model with measurements taken from the dataset, the resulting aortic models had differences in diameter of less than 1mm at the level of the aortic arch. A full statistical comparison at multiple levels of the models is now underway.

Table 1 Measurements taken from CT and VeroWhite printed model

	CT Datasets(mm)			Printed model(mm)		
Descending Aorta	20.2	21.1	21.2	20.5	20.5	21
Ascending Aorta	30	32	27.8	28	31	27.5
Brachiocephalic artery	11.1	10.79	10.9	11	11	11
Left Common carotid artery	6.6	5.3	4.4	6	5	4
Left Subclavian artery	9.3	7.3	6.1	9	8	6

The clarity of the dataset does not affect the accuracy of the resulting model. Low quality image will, however, greatly impact the time required to complete the segmentation process.

Due to high contrast with the surrounding tissues, calcifications were segmented, and included into the aortic wall. In the future these lesions will be printed with a different material.

Removal of the supporting wax used to create the tubular ascending aortic models was challenging in the rigid models. Sonic and heat removal were attempted, but have proven ineffective in total removal. Manual scraping was the only solutions.

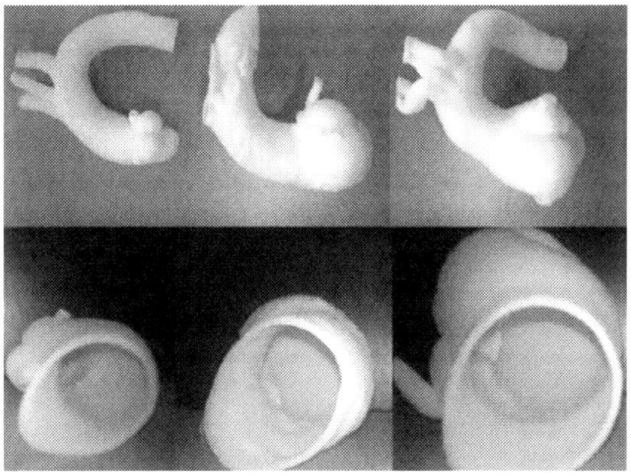

Figure 3 Printed Models in VeroWhite using Connex500™

Cruciform

A series of cruciform with different pattern of fiber have been printed and biaxial mechanical testing (Enduratec ELF 3200 BOSE MN, USA) as well as suture retention and scalpel resistance have been qualitatively evaluated. The fiber amplitude, frequency and number affect the stress-strain response of the composite materials. Suturing and cutting tests show that the printed material exhibits acceptable resistance to tearing while retaining the malleability of the TangoPlus. We will present our most up to date results of the biaxial tensile testing and micro indenter scratch testing (M3, Nanovea).

DISCUSSION

A protocol has been established for the 3D reconstruction of the human ascending aorta form patient clinical imaging. The resulting aortic models are representative of in vivo lumenal geometry.

A two material composite has been easily created in solid work then subsequently printed; allowing the control of material properties via variation in fiber pattern. The fiber will eventually have to be implemented into the 3D structure of the ascending aorta. The sinusoidal fibers will have to follow a helical pathway that roughly matches the general direction of the collagen fibers, spread out in a way that replicated the material properties tested in the cruciform. As none of the available engineering software possessed the adequate tools to accomplish this task, a Matlab™ code is currently being developed in order to place helical sinusoidal fibers onto the tubular models.

ACKNOWLEDGEMENTS

This work has been supported by Heart and Stroke Canada.

REFERENCES

[1] Escárpita, D. A. A., et al. (2012). Biaxial Tensile Strength Characterization of Textile Composite Materials, in Composites and their properties, ISBN 978-953-51-0711-8, Intech.

MECHANICAL ANALYSIS OF BONE TISSUE AS MINERAL AND ORGANIC COMPOSITE BY RAMAN SPECTROSCOPY

Masahiro Todoh (1), Shigeru Tadano (1)

(1) Faculty of Engineering
Hokkaido University
Sapporo, Hokkaido, Japan

INTRODUCTION

Bone has a hierarchical structure at different structural levels and is often regarded as a composite material consisting of mineral particles of apatite crystals ($Ca_{10}(PO_4)_6(OH)_2$) and organic matrix of Type I collagen in microscopic scale [1]. The mechanical properties of bone at macroscopic scale depend on the structural organization and properties of constituents in the microscopic scale. Recent studies found that osteoporosis is not just a simple loss of bone mass or bone mineral content, but involves significant changes in the biochemical and physical properties of the collagen matrix.

In recent years, vibrational spectroscopy has been used as a technique for studying bone properties. Infrared and Raman spectroscopies have been used to study normal and osteoporotic tissue, as well as to investigate compositional changes associated with various bone diseases [2-4]. Raman spectra can also be collected with micron-scale spatial resolution and give simultaneous information about the mineral and organic matrix components of bone. Raman spectra can also be collected over a two-dimensional area to give spatially relevant image information. With the use of multivariate statistics, small local variations in bone composition in the area surveyed.

This study investigated the relationship between the Raman shift and stress state and its effect of the anisotropic properties and degree of aging. Raman microscope system was used for the analysis of mechanical behavior of micro components of apatite crystals and collagen matrix in bone tissue with laser excitation. The bovine cortical bone specimens were loaded by the custom-made micro-tensile device attachable to Raman microscope. The relationship between the Raman shift and tensile stress state and its effect of the anisotropy properties and degree of ageing were investigated in tensile loading.

METHODS

In the experiments, Raman microscope system (inVia Reflex, Renishaw, UK) was used for the analysis of mechanical behavior of bone tissue with laser excitation (785 nm; J150GS-11, Showa Optronics, Japan). The specimens were loaded by the custom-made micro-tensile device attachable to Raman microscope, as shown in Fig. 1. This device had a load cell (LUX-B-200N-ID-P, Kyowa Electric Instruments Co., Ltd., Japan; Capacity 200 N) and uni-axial stage (TSDM40-15, Sigma Koki Co., Ltd., Japan). The specimens were kept in saline bound between two glass plates during Raman spectroscopy. The specimens were loaded from 0 to 10 N at each 2 N and then unloaded from 10 N to 0 N. The Raman shifts were determined by calculating the center of gravity of spectrum peak profile. The relationship between the Raman shift and strain state and its effect of the mechanical anisotropy and aging were investigated.

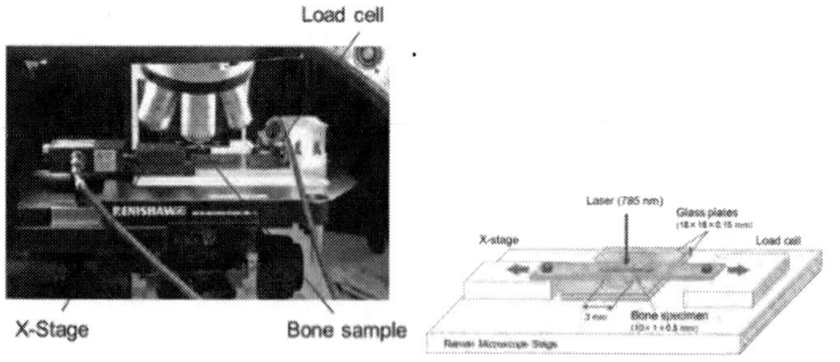

Fig. 1 Raman microscope system with micro-tensile loading device for cortical bone specimen.

Twenty four cortical bone specimens of bovine femoral diaphysis (Age: 23 months (n=12) and 9 years old (n=12)) were prepared with a size of 10.0×1.0×0.5 mm. Two types of specimens of each age were arranged with their longer sides bone-axially and circumferentially. Rough cuts were done by using a band saw and fine cuts by using a low-speed diamond saw (SBT650, South Bay Technology, Inc. USA). Both ends in 3.5mm length of specimen were bonded to metal plates for fixation to the micro-tensile device by the adhesive.

RESULTS

Figure 2 shows the Raman spectrum of bovine cortical bone. By using Raman spectroscopy, both spectrum of apatite crystals and collagen matrix could be observed as the phosphate v_1 band at approximately 960 cm^{-1} and the bands associated with collagen (Amide III at 1250 cm^{-1}, CH_2 wag at 1450 cm^{-1}, and Amide I at 1665 cm^{-1}) The most intense peak in the bone spectrum is phosphate v_1 ($v_1PO_4^{3-}$).

Figures 3 show the changes in the Raman shift of $v_1PO_4^{3-}$ by applied tissue strain. Upper and middle figures show the changes of 23-month old bone with tensile loading at bone axial (n=6) and circumferential (n=6) directions, respectively. The changes in Raman shift against bone axial stress was relatively larger than that against circumferential strain. Lower figure shows the changes of 9-year old bone with bone axial loading (n=6). The changes in Raman shift of 9-year old against tensile loading significantly decreased compared with 23-month old bone. Figure 4 show the ratio of changes of Raman shift with applied tensile stress of different aged and loading orientation of 23 months and 9 years old, respectively. From all results, the Raman shifts of mineral crystal to lower wave numbers were observed with increase of tensile strain.

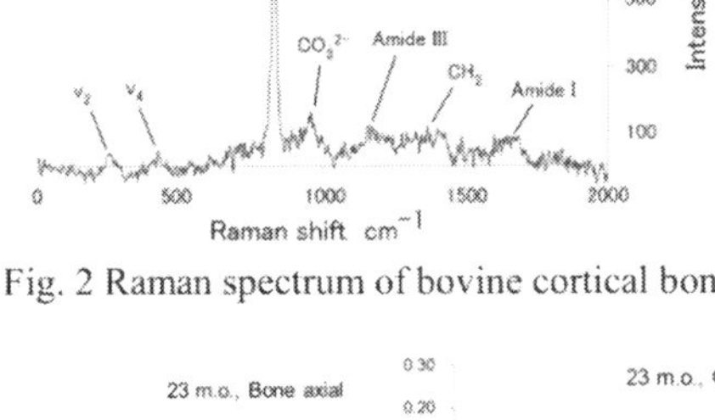

Fig. 2 Raman spectrum of bovine cortical bone

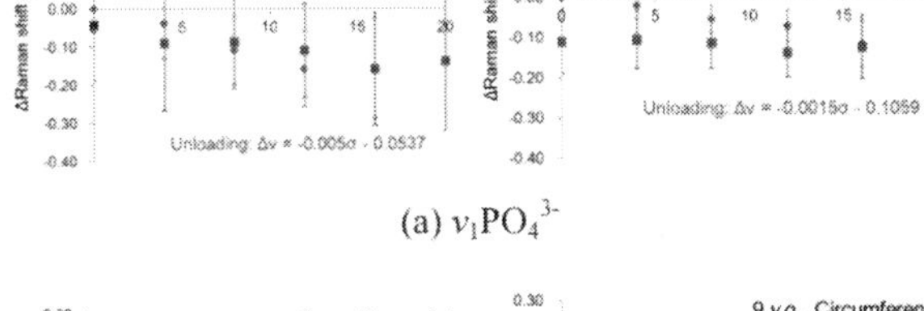

(a) $v_1PO_4^{3-}$

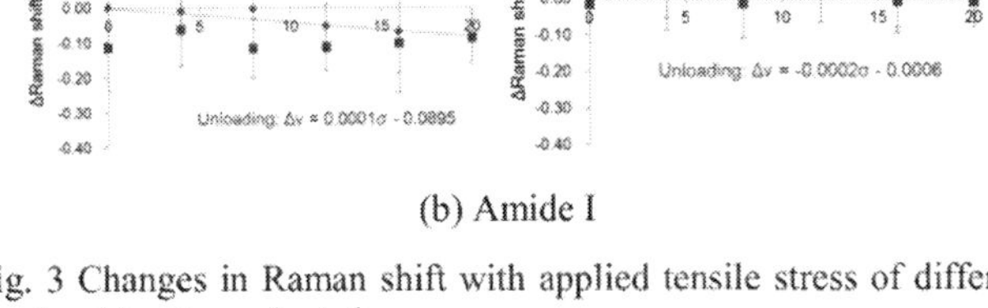

(b) Amide I

Fig. 3 Changes in Raman shift with applied tensile stress of different aged and loading orientation

DISCUSSION

This study investigated the relationship between the Ra-man shift and strain state and its effect of the mechanical anisotropy and aging. The Raman responses of mineral and collagen phase in bone tissue was well related to applied stress state. It was confirmed that this method is sufficiently useful to evaluate the stress or strain state based on the microscopic response of constituents in bone tissue. In addition, those responses of Raman shift were dependent to the mechanical properties related to structural anisotropy and degree of aging. It also suggests that the measurement of mechanical response of Raman spectrum might be useful for the bone quality evaluation as diagnosis of osteoporosis.

ACKNOWLEDGEMENTS

This research was partially supported by Grant-in Aid for Young Scientists (B) (No. 23760080) from the Ministry of Education, Culture, Sports, Science and Technology of Japan and by scholar project program from Toyota Physical & Chemical Research Institute.

REFERENCES

[1] Rho JY et al., *Med Eng & Phys*, 20:92–102, 1988.
[2] Carden A et al., *Calcif Tissue Int*, 72:166–175, 2003
[3] De Carmejane O et al., *Med Eng & Phys*, 76:207-213, 2005.
[4] Dooley KA et al., *J Biomed Optics*, 14:044018-1-8, 2009.

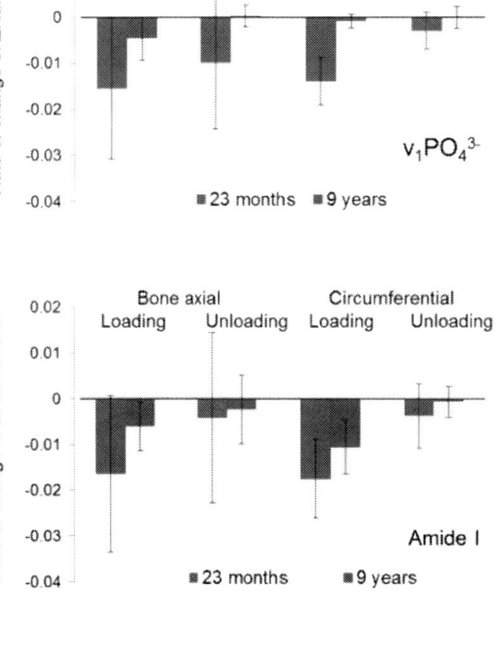

Fig. 4 Rate of changes of Raman shift of apatite crystals with applied tensile stress of different aged and loading orientation

TENSILE PROPERTIES OF STEM CELL-BASED SELF-ASSEMBLED TISSUE (scSAT) BIOSYNTHESIZED ON NANOPERIODIC STRUCTURED SUBSTRATE

K. Oya (1), Y. Tani (2), K. Koizumi (3), N. Sugita (3), K. Suzuki (4), N. Nakamura (3), H. Fujie (2)

(1) Department of Mechanical Engineering, School of Engineering, Tokai University Hiratsuka, Kanagawa, Japan

(2) Division of Human Mechatronics Systems, Graduate School of System Design, Tokyo Metropolitan University Hino, Tokyo, Japan

(3) Department of Orthopaedic Surgery, Graduate School of Medicine, Osaka University Suita, Osaka, Japan

(4) Department of Mechanical Systems Engineering, Faculty of Engineering, Kogakuin University Hachioji, Tokyo, Japan

INTRODUCTION

Stem cell-based therapies have great potentials for regeneration of tissues or organs lost by any reasons, such as accidents or diseases. A novel tissue-engineering technique for the tissues which involves a stem cell-based self-assembled tissue (scSAT) biosynthesized from synovium-derived cells has been developed [1]. It is expected as a novel material for repairing ligaments and tendons. As the scSAT is a scaffold-free construct composed of mesenchymal stem cells (MSCs) with their native extracellular matrix (ECM), it is free from concern regarding long-term immunological effects. However, the mechanical properties of the scSAT were insufficient for clinical application. In addition, the structural properties of the scSAT are isotropic while those of ligaments and tendons are anisotropic. To solve these problems, it will be effective to promote the generation of ECM in the scSAT by means of a special culture on a nanostructured surface. Our previous study indicated that a material having a certain roughness promotes cell adhesiveness and that a material having a nanoperiodic-pattern increases anisotropic property of cells [2]. In this study, a nanoperiodic structure was processed on a poly(dimethylsiloxane) (PDMS) surface via femtosecond laser processing, and the scSAT was biosynthesized on the substrate. Then, the tensile properties and histological characteristics of the scSAT were determined.

METHODS

Preparation of nanoperiodic structured PDMS substrate

Commercially available pure titanium disks with grade 2 (19 mmϕ x 1.5 mm in thickness) were wet-polished with SiC papers of grit number of 320, 600, 800, and 1,000 in the increasing order. Then, they were ultrasonically rinsed in acetone for 10 min and dried with air using an air pump.

The nanoperiodic structure on the titanium surface was patterned using a femtosecond laser apparatus (IFRIT, Cyber Laser Inc., Japan) with a seed laser (Femtolite, IMRA America Inc., Ann Arbor, MI). The processing conditions were as follows: irradiation intensity of 700 mW, pulse repetition frequency of 1 kHz, pulse width of 190 fs, wavelength of 780 nm, beam diameter of 6 mm, focal distance of 60 mm, and scanning speed of 10 mm s^{-1}. Then, the nanoperiodic structure processed on the titanium surface was transferred to PDMS substrate. The PDMS substrate subjected to nanoperiodic processing are named "Nano-PDMS". As a control, a PDMS substrate whose surface is smooth was also prepared ("PDMS"). The surface configurations of the substrates were determined using an atomic force microscope (AFM).

Biosynthesis of scSAT

MSCs obtained from the synovial membrane of human knee joints were grown in a cultivation medium (Dulbecco's modified Eagle medium (DMEM) supplemented with 10% fetal bovine serum, 100 U mL^{-1} penicillin and 100 µg mL^{-1} streptomycin). The cells were passaged four times after primary culturing. Then the cells were seeded on the PDMS substrates in a cultivation medium supplemented with 0.2 mmol L^{-1} ascorbic acid 2-phosphate to promote the biosynthesis of ECM. As a control, the cells were also seeded on the tissue culture treated polystyrene substrate ("PS"). After ECM maturation, synthesized matrices were detached from the culturing substrates and allowed to undergo natural construction for 1 h to develop scSATs.

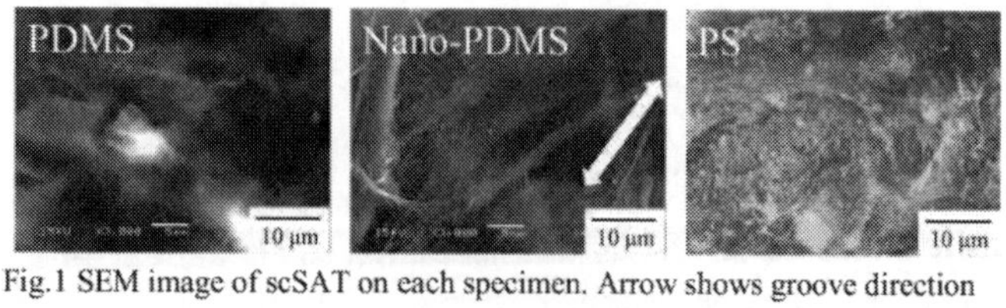

Fig.1 SEM image of scSAT on each specimen. Arrow shows groove direction

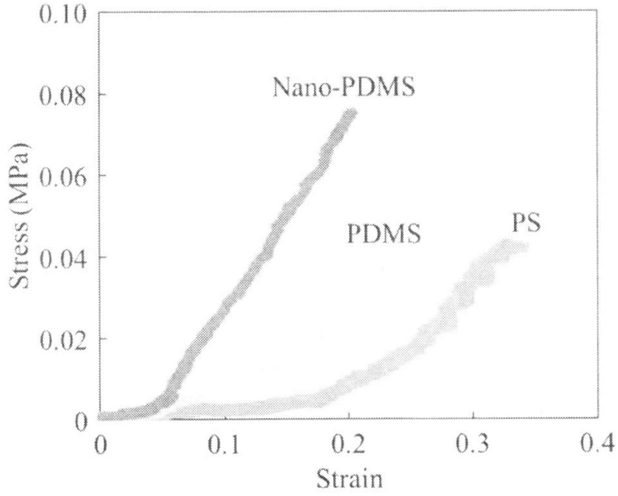

Fig.2 Stress-strain curve of scSAT on each specimen

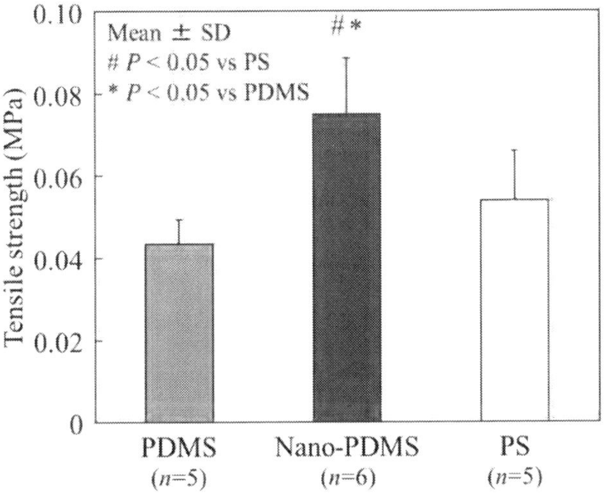

Fig.3 Tensile strength of scSAT on each specimen

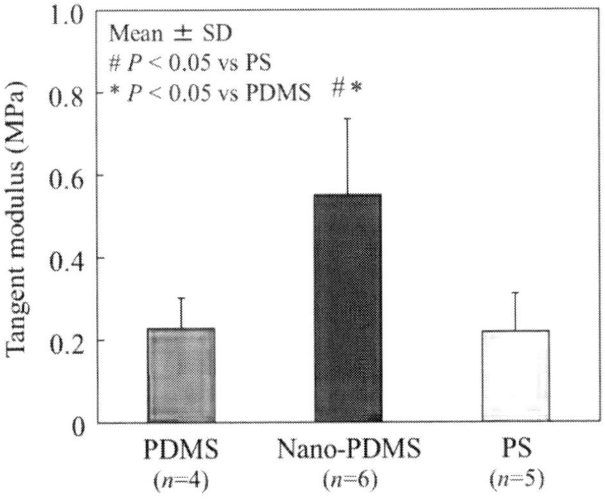

Fig.4 Tangent modulus of scSAT on each specimen

Tensile test and histological observation of scSAT

The scSAT specimens were subjected to tensile testing at a rate of 0.05 mm s^{-1} in PBS(-) solution at 37°C using a custom-made micro tensile tester [3]. The tensile load was applied in the direction parallel to the groove direction. In addition, histological observation was also performed using a scanning electron microscope (SEM) for the surface of the scSAT biosynthesized on each substrate.

RESULTS

Surface characterization of Nano-PDMS

The pitch, depth, and roughness (R_a) of the Nano-PDMS surface were 522 ± 9.1 nm, 48.1 ±23 nm, and 16.9 ± 4.7 nm, respectively. On the other hand, R_a of PDMS surface was 0.7 ± 0.0 nm.

Histological observation of scSAT

The SEM observation revealed that the densities of the scSATs biosynthesized on the PDMS and Nano-PDMS were higher than that on the PS substrate (Fig.1). No obvious difference was observed between the scSATs on PDMS and Nano-PDMS. The matrix of the scSAT on Nano-PDMS was oriented along the nanogroove direction.

Tensile properties of scSAT

Typical curves of stress-strain relationship of the scSATs produced on the substrates are indicated in Fig.2. The increase of stress occurred at lower strain in Nano-PDMS group than in other groups.

The tensile strength of the scSAT biosyntheseized on each substrate is summarized in Fig.3. The values of PDMS, Nano-PDMS, and PS were 0.043 ± 0.006 MPa, 0.075 ± 0.014 MPa, and 0.054±0.012 MPa, respectively. The tensile strength of the scSAT on Nano-PDMS was significantly larger than those on PDMS and PS.

The tangent moduli at the linear parts of the scSATs were also determined from the stress-strain curves. The values of PDMS, Nano-PDMS, and PS were 0.23±0.07 MPa, 0.55±0.18 MPa, and 0.22± 0.09 MPa, respectively (Fig.4). Tangent modulus of the scSAT on Nano-PDMS was also significantly larger than those on PDMS and PS.

DISCUSSION

Tensile properties of the scSAT biosynthesized on the Nano-PDMS substrate were enhanced compared to the scSAT on the smooth substrates. Moreover, the matrix density of the scSAT on Nano-PDMS was larger than that on PS. It has been reported that the amount of collagen content synthesized from MSCs depends on the surface roughness of the culture substrate: The rough surface enhances the synthesis of the collagen content [4]. Our results support this report. In addition, the nanopriodic structure enhanced the tissue orientation of the scSAT. Our previous study revealed that the nanogroove affects the cellular orientation [2]. It is considered that the cellular orientation affects the orientation of the scSAT. From these results, it is suggested that a nanoperiodic structured substrate processed on a PDMS surface via femtosecond laser processing enhances the tensile properties of the scSAT because it enhances ECM synthesis and regulates cellular orientation.

ACKNOWLEDGEMENTS

This study was financially supported by the Grants-In-Aid for Scientific Research B and Young Scientists B from the MEXT, Japan, and the Priority Research Fund from Tokyo Metropolitan University.

REFERENCES

[1] Ando, W et al., *Biomaterials*, 28:5462-5470, 2007.
[2] Oya, K et al., *Jpn J Appl Phys*, 51:125203, 2012.
[3] Nagai, et al, *Jpn J Clin Biomech*, 27:89–93, 2006.
[4] Dongwoo, K et al., *Biomaterials*, 33:5997-6007, 2012.

DYNAMIC VISCOELASTIC PROPERTIES OF PORCINE PATELLAR TENDON TISSUE: A STUDY OF REGIONAL VARIATION AND FREQUENCY DEPENDENT BEHAVIOUR

Sourav S. Patnaik (1,2), Taylor Szasz (1,2), R. Prabhu (1,2), Hongjoo Rhee (2), M. F. Horstemeyer (2), Jun Liao (1,2), Lakiesha Williams (1,2)

(1) Agricultural and Biological Engineering
Mississippi State University
Mississippi State, MS, USA

(2) Center for Advanced Vehicular Systems
Mississippi State University
Mississippi State, MS, USA

INTRODUCTION

Biological tissues such as tendons are viscoelastic in nature, which is a direct function of its internal composition, microstructural arrangement, and water content. Tendon stores almost 93% of its elastic energy, while ~7% is dissipated as heat energy (1). In order to quantify the exact relationship between the structure and viscoelastic behavior of the biological tissues, rheological measurements are often performed in material's linear-viscoelastic range (LVE). Up to now, dynamic mechanical testing of patellar tendon tissue has been primarily confined to tensile mode. However, some compressive loading does take place when tendon crosses over a bone or joint and thus, resulting in loss of energy. In the present study, we investigate the effect of frequency on the viscoelastic parameters (i.e. E', E" and tan δ) of porcine patellar tendon tissues in the compressive mode using a Dynamic Mechanical Analyzer (DMA). Regional variations as well as the effects of loading direction (i.e. parallel to the fibers (longitudinal direction) and perpendicular to the fiber (transverse direction)) on these viscoelastic parameters were evaluated accordingly.

METHODS

Fresh Porcine patellar tendon (n = 8 per group; animal age = 6-8 months; animal weight = 150-200 lbs.) was obtained from the local abattoir. Tendons were cleaned and excess fat/connective tissues were removed. Tendon specimens were then divided into three different regions: Central Proximal (CP), Central Center (CC) and Lateral Distal (LD) (with CP being closest to the patella, followed by CC and LD, respectively) (Fig. 1 (A)) (2). Further, these tendon specimens were divided into either longitudinal loading (Long) or transverse loading (Trans) groups (Fig. 1 (A)). Hence, the six groups were as follows: CP Long, CP Trans, CC Long, CC Trans, LD Long, and LD Trans.

For each group, samples were rectangular in cross-section (~7 x 7 mm) and with a thickness of ~5 mm. The stress distribution on the cross sectional area of the tissue was assumed to be uniform. Tests were performed within 6-7 hours of sample procurement.

Compression mode of the DMA (DMA Q 800; TA Instruments, New Castle, DE, USA) was used to evaluate specific viscoelastic parameters (E', E" and tan δ). Samples were held isothermally between two parallel plates (discs of 15mm diameter) (Fig. 1(B)) with a pre-load force of 0.05N. Samples were allowed to equilibrate (resting period of 1 min) and then subjected to a constant oscillatory strain of 3% (determined via preliminary testing – data not shown here) which is within the tissue's linear-viscoelastic range (LVE). All tests were performed at room temperature (25-27°C) with tissues in a hydrated status. Figure 1 (B) shows an example of the loss of energy between the applied strain curve and resultant stress curve of a material. The viscoelastic parameters, i.e., E', E" and tan δ values, were recorded in quadruplicate over a range of frequencies (i.e. 0.1, 1, 10, 100 Hz).

Statistical analyses were performed using SigmaStat software program version 3.1 (SPSS Inc., Chicago, IL, USA). One way ANOVA was used to find differences in the viscoelastic parameters across various frequencies and loading directions. Data was considered to be significantly different at $p < 0.05$.

RESULTS

The storage modulus and loss modulus values were found to be significantly different ($p<0.05$) across the groups. The tan δ, as expected, was in a similar range across the groups. All parameters, except tan δ, were also found to be significantly different across the frequency range (Fig 2 (A, B & C)). E', E" and tan δ for the CC region (both "Long" and "Trans" groups) were higher in magnitude as compared to the other groups. Both storage and loss modulus (for all groups) exhibited frequency dependent behavior; tan delta showed similar trend up to 10Hz, but not from 10 to 100Hz (Fig 2 (A, B & C)).

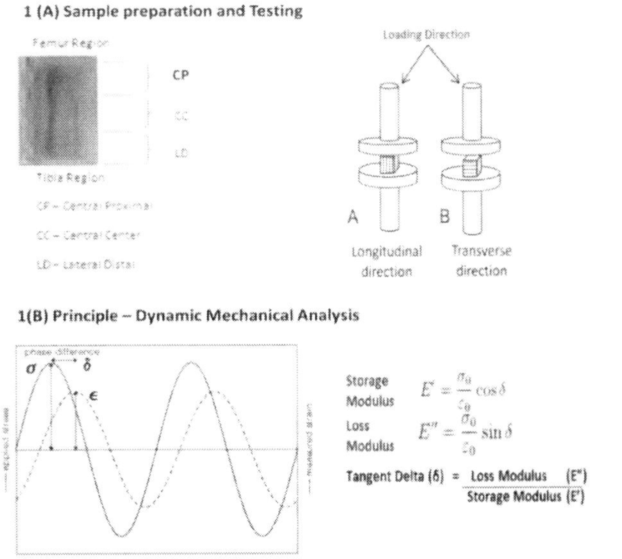

Figure 1. Dynamic mechanical testing of porcine patellar tendon – principle and sample preparation. (A) Samples were divided according to their loading directions (longitudinal or transverse) and their location as per previously established protocol (2). The groups for this study were: CP Long, CP Trans, CC Long, CC Trans, LD Long and LD Trans. (B) Principle behind calculation of viscoelastic parameters (storage modulus (E'), loss modulus (E") and phase lag (tan δ)) are shown here.

DISCUSSION

The delay between applied force and material response constitutes the phase angle or tangent delta (tan δ) of the material, a viscoelastic indicator. In general, a tan δ value less than 1.0 indicates an elastic material, while a value of more than 1.0 indicates a viscous material behavior (larger value = softer material). With the obtained range of tan δ values (0.1 – 0.47), porcine patellar tendon tissue falls in the viscoelastic range (Fig 2 (C, F & I)).

Viscoelastic parameters of tendons varied from the top proximal to the distal end. Viscoelastic parameters of CC region exhibited the highest magnitude (~40% higher) (Fig.2 (D-F)) as compared to the other groups (Fig 2 (A-C, G-I)). All viscoelastic parameters were positively correlated with increasing frequency (except tan δ at 100 Hz). This increase in storage and loss modulus with frequency indicates that the tendon tissue has the ability to absorb stress and dissipate a part of this stored energy. This synchronized cycle of energy absorption and dissipation may prevent damage to the tissues, and thus, responsible for its higher load bearing capacity.

Storage modulus values were almost three times in magnitude of their corresponding loss modulus for all the testing frequencies (Fig.2 A-B, D-E and G-H). In terms of anisotropy, storage and loss modulus values were higher in magnitude along the longitudinal direction (parallel to the fiber) than the transverse direction (perpendicular to the fiber) (Fig 2 (A-C, D-F & G-I)).

Viscoelastic properties of a material are based on the virtue of its structure-property relationship. In the same manner, microstructural arrangement of tendon fibers dictates its intrinsic viscoelastic properties (2, 3). The central region of the patellar tendon showed the highest magnitude in all of the viscoelastic parameters ((Fig 2 (D-F), as compared to the other groups. Similar observations were also reported for the central region of rabbit patellar tendon (2). Despite having differences in structural and mechanical characteristics, it was interesting to observe that the tangent delta values were consistent across the groups (regions and loading directions). The intrinsic properties of accommodating the energy change may provide further insight on the tendon's ability to withstand high loads and dissipate energy throughout. Further microstructural analysis of the tissue would provide stronger evidence for these observations.

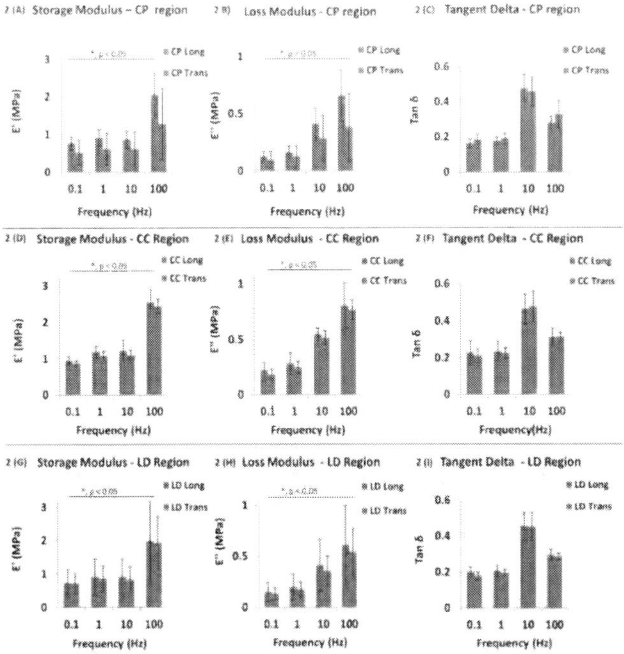

Figure 2. Viscoelastic parameters (E', E" & tan δ) obtained from dynamic mechanical testing of porcine patellar tendon are shown. Fig. 2(A-C) represent E', E" & tan δ of the CP regions, respectively, 2 (D-F) represent E', E" & tan δ of CC regions, respectively, and 2(G-I) represents E', E" & tan δ of LD regions, respectively. Longitudinal groups are shown in red histograms, while transverse groups are shown in blue color histograms. Storage and loss modulus values are significantly different across the frequencies ($p<0.05$); tan

ACKNOWLEDGEMENTS

- NSF EPS-0903787.
- Internal Funding, Agricultural and Biological Engineering, Mississippi State University.

REFERENCES

[1] Kerr, RF et al. *J. Exp. Biol*, 93: 283-302, 1981
[2] Williams, LN et al., *Ann Anat*, *190*(5), 442-451, 2008.
[3] James et al., *J Hand Surg*, 2008; 33A:102–112, 2008

LABEL-FREE PROTEIN PROFILING FOR FUNCTIONAL CHARACTERIZATION OF ENGINEERED CARTILAGE FOLLOWING INFLAMMATORY CYTOKINE EXPOSURE

Andrea R. Tan (1), Shujuan Tao (2), David Chen (2), Lewis M. Brown (2), Clark T. Hung (1)

(1) Department of Biomedical Engineering
Columbia University
New York, NY, 10027

(2) Comparative Proteomics Center
Columbia University
New York, NY, 10027

INTRODUCTION

Articular cartilage is the specialized connective tissue that lines our mobile joints with primary function in load bearing and lubrication. When damaged, cartilage exhibits a poor healing response to injury due to its avascular nature and limited cellularity [1]. Potent catabolic mediators such as interleukin-1 (IL-1) are present following joint injury [2]. Successful cartilage repair strategies that adopt engineered cartilage constructs must account for the mechano-chemical environment of the joint post-implantation. To characterize the response of fledgling tissues to chemical insult, we adopted a comparative proteomic analysis to characterize the regulation of proteins following IL-1 exposure. Such approaches afford detection of proteins with post-translational modifications and have recently been applied to studies with mesenchymal stem cells (MSCs) [3-5]..

METHODS

Cell isolation and expansion Human bMSCs were isolated from fresh unprocessed bone marrow (Lonza) of a 22 year-old male donor. Following separation via Percoll gradient, mononucleated cells were plated and adhered MSCs expanded for 4 passages in Dulbecco's modified Eagle's medium (DMEM) supplemented with 10% fetal bovine serum, 100 U/mL each of penicillin, streptomycin, and fungizone, and a growth factor cocktail (1 ng/mL TGF-β1 and 5 ng/mL bFGF). At confluence, cells were trypsinized and counted; 1.0 mL each of a 1 x10^6 cell suspension was aliquotted into 1.5 mL sterile screw-top tubes. The tubes were spun in a microcentrifuge at 37°C and 2500 rpm for 20 minutes to form a visible cell pellet at the base of the tube. Micropellets were stored in an incubator maintained at 37°C and 5% CO_2 for 28 days. Media was prepared from high-glucose DMEM with the addition of 100 nM dexamethasone (Sigma), 100 µg/mL sodium pyruvate (Sigma), 50 µg/mL L-proline (Sigma), 1% ITS+ premix (Becton Dickinson), and 1% antibiotic-antimycotic (Invitrogen). 50 µg/mL ascorbic acid (Sigma) and 10 ng/mL TGF-β3 (R&D Systems) was added fresh to the media on each media change day. Interleukin Exposure After 21 days, pellets yielded significant

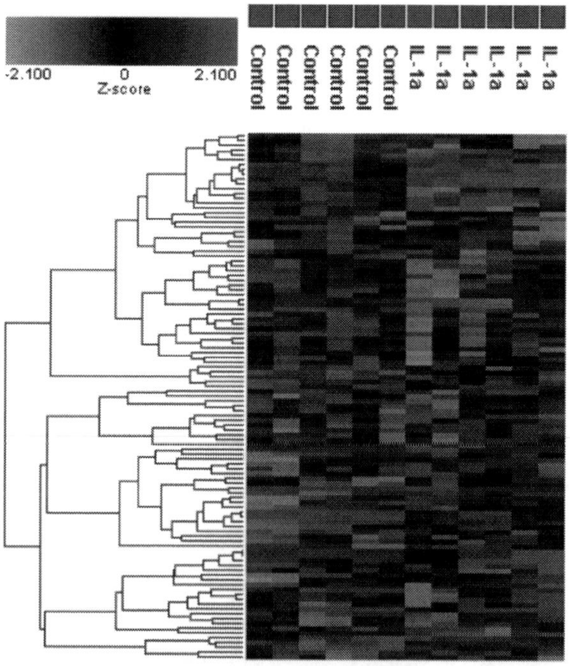

Figure 1. Hierarchical cluster analysis of the proteomic data for control and interleukin-1α (IL-1a) treated pellets.

extracellular matrix deposition, detectable by the increase in stiffness and opacity of the tissue. An insult dose of IL-1α (10 ng/mL) was then introduced into the culture system for one subset of pellets and dexamethasone was removed from the culture media from all groups, as corticosteroids have been shown to suppress the catabolic activities of pro inflammatory cytokines (reference). Proteomics Analysis After 1 week of IL-1 exposure, cell pellets were washed with ice cold PBS and physically disrupted via glass-bead beating in 0.3% SDS, TRIS-buffered saline with 1% Protease Inhibitor Cocktail. The protein fraction was precipitated using a methanol chloroform extraction and dissolved in 0.4% RapiGest SF detergent containing 50 mM ammonium bicarbonate. Dithiothreitol was added and the solution was sonicated and heated for 5 minutes to further extract the protein. A label-free protein profiling technique for mass spectrometry-based shotgun proteomics was performed, with changes in protein expression monitored using data-independent acquisition with traveling wave ion mobility spectrometry (TWIMS). Three independent biological replicates were analyzed for each group; for each replicate, two chromatograms were recorded, yielding a total of 12 chromatograms. Data were searched against a human protein database with ProteinLynx Global Server (PLGS) v2.5RC9 (Waters Corp.), and post-processed in Elucidator Protein Expression Data Analysis System (Ver. 4.0.0.2.29, Ceiba Solutions).

RESULTS

Proteomic analysis focused on proteins from which two or more peptides were detected and criteria for protein selection were based on p-value and peptide count. Cyokine exposure to human bMSC pellets resulted in differential expression of proteins (Figure 1, Table 1); specifically, proteins associated with extracellular matrix development such as fibronectin (FN1, Figure 2) and collagens (COL1A1, COL1A2, COL6A1, COL6A2, COL6A3) were down-regulated compared to control samples, supporting previous reports of matrix degradation in the presence of catabolic factors (REFS). The resulting decrease in matrix components may also influence the phenotype of the cells; for example, annexin isoforms, which have been implicated in stem cell differentiation toward a chondrocytic phenotype, were down-regulated in cytokine-treated pellets, suggesting a loss of functional components and properties. Stress-responsive proteins were found to be significantly upregulated in IL-1 treated pellets, in particular, metallothioneines (MTs), a group of low-molecular-weight, cysteine-rich, heavy metal-binding proteins (13.02 fold increase).

Table 1. Most prominently expressed proteins found in proteomics analysis of engineered cartilage pellets in the absence (control) and presence of cytokine (10 ng/mL IL-1α).

Accession Number	Protein Name	Ratio [10 ng/mL IL-1a vs control]	P-value
A0A024R462	FN1	0.61	1.1E-17
P02452	COL1A1	0.68	5.0E-06
P08123	COL1A2	0.64	1.6E-07
P12109	COL6A1	0.51	2.1E-05
P12110	COL6A2	0.43	8.5E-12
D9ZGF2	COL6A3	0.41	1.5E-22
ADA024R527	Annexin	0.75	3.2E-07
P02795	Metallothioneine-2	13.02	4.8E-15

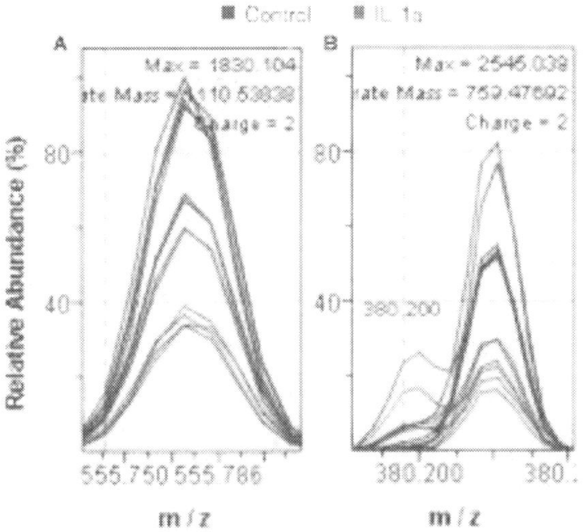

Figure 2. Differential expression of fibronectin 1. One isotopic signal from the mass spectrum of (A) peptide STTPDITGYR and (B) peptide LTVGLTR: control cells exhibit greater accumulation versus IL-1α-treated cells.

DISCUSSION

Alterations to the cell proteome following cytokine exposure were assessed using a label-free MS proteomics technique in scaffold-free, engineered cartilage constructs (micropellets). Down-regulation of extracellular matrix proteins was confirmed here, paralleling previous findings of catabolic behavior by IL-1 treated cells. More interestingly, this proteomic approach identified the upregulation of MT-2_A, previously reported to be regulated by interleukin-1 exposure in hepatoma and promyelocytic cell lines and human fibroblasts [6]. Analysis of the MT-2_A gene has previously revealed that it contains a 9 bp sequence that is recognized by the activating protein-1 (AP-1) complex of transcription factors encoded by the fos/jun family of proto-oncogenes [7] responsible for the transcriptional regulation of the collagenase gene. Furthermore, in rat spinal cord models, metallothioneines have been implicated in the pathogenesis of pain, particularly in the initiation, rather than the maintenance, of inflammatory and neuropathic pain [8]. Our findings here represent the first identification of the gene and it's potential role in mediating and initiating the downstream response to IL-1 for engineered cartilage derived from human bMSCs and may suggest a potential role of the protein in regulating not only the catabolic activity but also the pain response following knee injury. Further work investigating the differential temporal expression of MT-2_A may provide further insight into potential targets for controlling the response of cartilage-like tissue to cytokine exposure.

ACKNOWLEDGEMENTS

This work was supported in part by NYSTEM contract #C029159 and The Coulter Foundation and Columbia Technology Ventures (WHCF CU12-0369).

REFERENCES

[1] Mow, VC et al., *Cell Mech*, 345-379, 1994; [2] Pelletier, JP et al., *Semin Arthritis Rheum*, 20(6 Suppl 2): 12-25, 1991; [3] Wang, D et al., *J Biol Chem*, 279(42):43725-43734, 2004; [4] Foster LJ et al., *Stem Cells*, 23(9):1367-1377, 2005; [5] Oswald ES et al., *J Proteome Res*, 10(7):3050-3059, 2011; [6] Karin M et al., *Mol Cell Bio*, 5(10):2866-2869, 1985; [7] Angel P et al., *Cell*, 49(6):729-739, 1987; [8] Kwon A et al., *Brain Res*, 1523:37-48, 2013.

SB3C2015
Summer Biomechanics, Bioengineering and Biotransport Conference
June 17-20, 2015,Snowbird Resort, Utah, USA
SB3C2015-451

INTIMA HETEROGENEITY IN ATHEROSCLEROTIC PLAQUE STRESS CALCULATIONS

Lambert Speelman (1), Bas van Velzen (2), Anton FW van der Steen (1,2), Jolanda J Wentzel (1), Frank JH Gijsen (1)

(1) Department of Biomedical Engineering
Erasmus Medical Center
Rotterdam, the Netherlands

(2) Department Mechanical Engineering
Delft University of Technology
Delft, the Netherlands

INTRODUCTION

From a biomechanical perspective, atherosclerotic plaque rupture occurs when the stresses in the plaque (and specifically in the fibrous cap), exceed the strength of the tissue. Computational biomechanics to evaluate stresses in plaques is therefore a growing field of research. Plaque cap stresses can be calculated based on the plaque geometry, including the components, the blood pressure, and the material properties of the components. One of the unevaluated assumptions in computational plaque mechanics is that each of the components (lipid core, intima, media, and calcifications) has a homogeneous mechanical behavior. However, especially the intima has a very heterogeneous composition, including collagen fibers, smooth muscle cells, inflammatory cells, fatty material, and extracellular matrix.

In this study, we evaluated the influence of material properties heterogeneity of the intima on stress calculations of atherosclerotic plaque sections. Peak cap stresses are calculated with homogeneous and heterogeneous intima material properties and the impact of the heterogeneity on the peak cap stress was evaluated.

METHODS

To create **homogeneous intima models**, pentachrome stained histology sections of 12 plaques, pressure fixed at 100 mmHg, were used. Manual delineation of the lumen, lipid cores, intima, and wall were done to create homogeneous plaque finite element (FE) models [1]. Neo-Hookean material models were used for all components. An average Young's modulus of 445 kPa was assigned to the intima, which is the average intima stiffness, as tested with plaque indentation tests [2]. The wall (adventitia and media layer) and lipid core were assigned 1200 kPa and 5 kPa, respectively.

To create **heterogeneous intima models**, the intima section of the homogeneous models was subdivided in 4 clusters by means of a k-means clustering algorithm (fig. 1). The 4 clusters in the intima are derived based on similar histology image intensity and therefore do not directly reflect the constituents of the intima, but are used as a representative segmentation with appropriate structure dimensions.

As it is unknown which stiffness each cluster has, all clusters will be assigned a range of Young's moduli and a Latin hypercube sampling method was used to generate 100 combinations with different stiffness values for the clusters. The clusters were assigned Young's moduli ranging from 6 to 891 kPa, which is the full range of stiffness values of the intima found by plaque indentation tests [2]. A 2D Gaussian smoothing is thereafter applied to avoid sharp stiffness transitions in the intima. To create heterogeneous intima FE models, each of the FE elements of the homogeneous FE model is assigned a material stiffness according to the smoothed distribution (fig. 1).

The stresses in the homogeneous and 100 heterogeneous FE models were computed with an intraluminal pressure of 120 mmHg. Initial stresses up to 100 mmHg were accounted for [1]. The cap of the plaque was identified in the models and the 95 percentile (95-p) cap stresses were determined and compared for the homogeneous and heterogeneous models. The variation in cap stress in the heterogeneous models was related to the stiffness of the individual intima clusters.

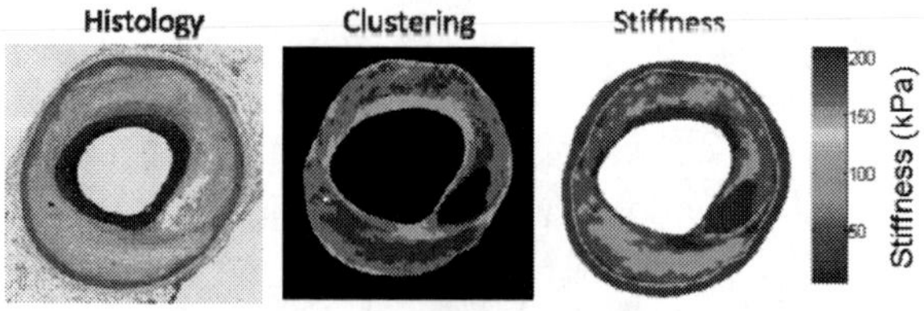

Figure 1: Histology cross-section, automatic clustering and an example of a stiffness map.

RESULTS

For the 12 plaques, all 100 simulations resulted in a 95-p cap stress (fig. 2). The relative standard deviation (STD) of the 95-p cap stress of the heterogeneous models ranged from 14% (plaque 7) to 75% (plaque 5), with an average of 32%. For most plaque models (10 of 12), the 95-p cap stress of the homogeneous model was located within the interquartile range (IQR) 95-p cap stress of the heterogeneous model (fig. 2).

There was no significant relation between the median 95-p cap stress and the 95-p cap stress variation due to the intima heterogeneity. Compared to the homogeneous models, the lowest and highest 95-p cap stress simulations resulted in an average (±STD) 95-p cap stress change of -60% (±38%) and +87% (±48%), respectively.

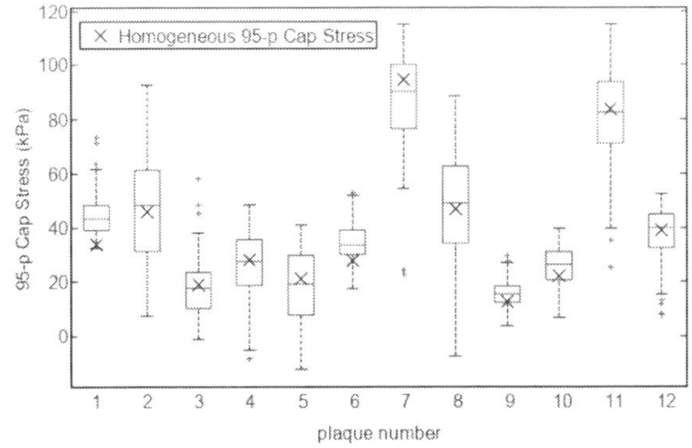

Figure 2: Bar plot of the 95-p cap stress [kPa] for the 12 plaques with the homogeneous cap stress indicated with red X. The range of the bar plots indicate the cap stresses found by the heterogeneous intima stiffness models, using the complete range of the 100 stiffness combinations.

Plaque 2 showed a change in 95-p cap stress of -60% and +102% for the lowest and highest 95-p cap stress simulations, respectively (fig. 3). The location of the peak cap stress shifted for the heterogeneous model with the highest cap stress (arrows in fig. 3).

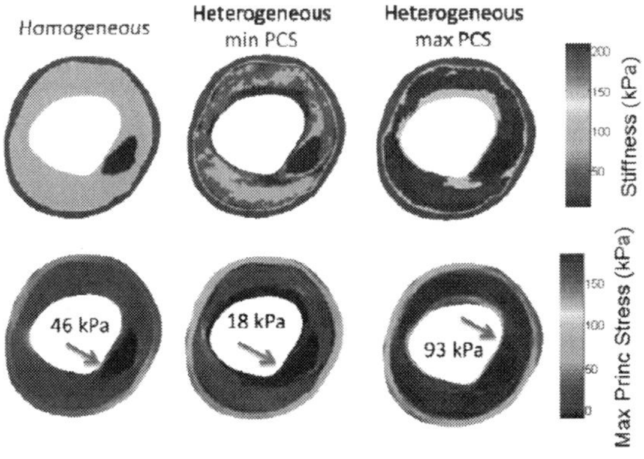

Figure 3: Stiffness (top row) and stress maps (bottom row) of the homogeneous and heterogeneous intima models with the lowest and highest 95-p cap stress of plaque 2.

The dataset was used to study the effect of the individual cluster stiffness values on the 95-p cap stress. For most plaques, the stiffness of 1 or 2 clusters showed a strong positive relation ($r>0.6$, $p<0.01$) with the 95-p cap stress, especially when the (thin) cap mainly consisted of this cluster (fig. 4, green cluster). In some cases, a negative relation was found: frequently when the area surrounding the lipid core was dominated by this cluster (fig. 4, red cluster).

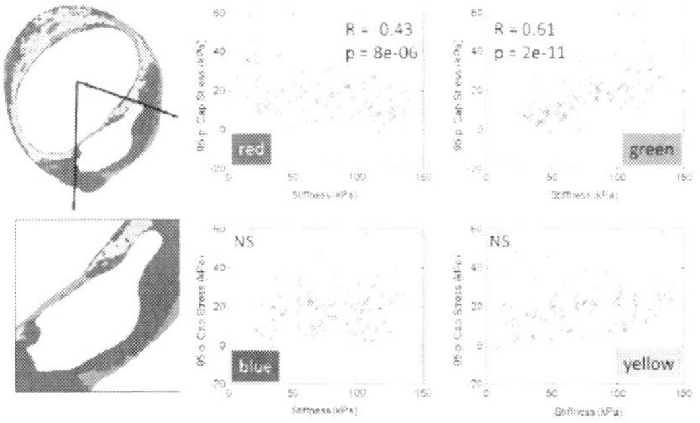

Figure 4: plaque 3 distribution of the red, green, blue, and yellow clusters (left) and the relation between the cluster stiffness and the 95-p cap stress. The stiffness of the red and green clusters show a relation with the 95-p cap stress.

DISCUSSION

Computing stresses in atherosclerotic plaques always done under the assumption of a homogeneous stiffness distribution of the intima. However, from histology it is clear that the intima has a very heterogeneous composition, and therefore stiffness distribution. In this study, the intima of 12 plaques was modeled as a heterogeneous material, based on structures visible in histology images, resulting in a representative segmentation with appropriate structure dimensions.

In most plaques, the homogeneous assumption resulted in a 95-p cap stress within the IQR of the heterogeneous 95-p cap stress. The ultimate changes in 95-p cap stress due to the intima heterogeneity were substantial, however, the relative standard deviation of the heterogeneous 95-p cap stresses was moderate (32% on average). This indicates that a large part of the combinations of cluster stiffness values resulted in 95-p cap stresses that are relatively close to the homogeneous results.

The variation in heterogeneous 95-p cap stress was not correlated to the median 95-p cap stress, so an expected cap stress variation cannot be based on the level of the cap stress. Preliminary analyses indicate that, depending on their size and distribution, not all clusters influence the 95-p cap stress. Clusters located in the cap dominate the cap stress and accurate knowledge of the cap properties is therefore important when computing 95-p cap stresses.

By assigning a large range of Young's moduli to each cluster, we created a 'worst-case scenario' for the variation in peak cap stress. The stiffness data was based on results from indentation plaque tests [2] and reflects data from 8 different plaques. A lower intra-plaque stiffness variation or an otherwise derived lower range in cluster stiffness may contribute to a reduced variation in 95-p cap stress.

As far as we know, this is the first study to include intima stiffness heterogeneity in plaque stress analysis. This dataset will provide valuable insight in the uncertainty and variation of the calculated cap stress. With this approach we will evaluate why some plaques show larger variation in peak cap stress than others, and which stress parameters are less affected by material stiffness variations.

REFERENCES

[1] Speelman et al. J Biomech. 2011 Sep 2;44(13):2376-82.
[2] Chai et al. J Biomech. 2013 Jun 21;46(10) :1759-66.

SB^{3}C2015
Summer Biomechanics, Bioengineering and Biotransport Conference
June 17-20, 2015,Snowbird Resort, Utah, USA
SB^{3}C2015-452

QUANTITATIVE ANALYSIS OF RAMAN SPECTRA FOR ASSESSMENT OF CROSSLINK CONCENTRATIONS TOWARD DIAGNOSING EARLY OSTEOARTHRITIS

Chao Wang (1), Krista M. Durney (1), Jonathan L. Kuo (2), Gregory Fomovsky (1), Jack R. Norton(2), Gerard A. Ateshian (1), Sinisa Vukelic (1)

(1) Department of Mechanical Engineering
Columbia University
New York, NY, U.S.A.

(2) Department of Chemistry
Columbia University
New York, NY, U.S.A.

INTRODUCTION

Osteoarthritis (OA) is a progressive, complex, multi-tissue joint disease with degenerative changes in the articular cartilage and subchondral bone [1], with a long asymptomatic early development and debilitating late stages. Currently there are no broadly accepted quantitative diagnostic tools for the early stages of the disease.

While OA is viewed today as a disease of a joint as an organ, with inflammation, injury, and changes in bone, articular cartilage, and synovial fluid as potential driving forces, the degradation of the extracellular matrix (ECM) components of cartilage are key to the progression of the disease [2]. The key structural components of cartilage ECM are collagens (COL) and proteoglycans (PGs). While highly hydrophilic PGs provide hydrostatic pressure, the interwoven crosslinked COL fiber network provides cartilage tensile strength and counteracts the swelling pressure of PGs. The disruption of COL network leads to its inability to withstand the swelling pressure of PGs. Along with swelling, an increased proteolytic activity shifts the ECM homeostasis towards degradation of matrix components, progressively leading to cartilage degeneration [3].

The onset of OA is characterized by changes to the structure, though not necessarily the content, of COL matrix in articular cartilage. Crosslinks (CxLs) stabilize the COL fiber network, and their disruption leads to the loss of tensile strength and structural integrity [4]. In cartilage, hydroxylysyl pyridinoline (PYR) are the major mature CxLs and pentosidine (PSD) are the senescent ones. The contribution of crosslinking to tensile strength and stability of cartilage COL matrix has been extensively demonstrated in tissue engineering studies [4] and animal models of OA [5]. In patients, increased levels of PYR and PSD in urine correlated with the severity of OA and other degenerative joint diseases [6].

Current diagnostic tools to assess the severity of OA typically detect the more advanced stages of joint degeneration. A variety of biomarkers have been proposed, including crosslinks and collagens degradation components in urine, serum, or synovial fluid of OA patients [7]. However, these methods evaluate systemic levels of damage and do not necessarily point out the specific joint affected by OA. Raman spectroscopy, in contrast, is a spatially resolved method that can pinpoint the site and severity of damage in a specific joint, and arthroscopic Raman spectroscopy probes are currently being developed [8]. Fourier Transform Infrared Spectroscopy (FTIR), a complementary technique to Raman spectroscopy, has been extensively used for normal and OA cartilage characterization in vitro [9, 10, 11]. However, the main challenge for FTIR as an in vivo diagnostic tool is its sensitivity to water, which often masks important vibrational modes. Raman spectroscopy has been previously used to characterize changes that occur in OA [8]. To date however, the method of analysis of Raman spectra has been intrinsically qualitative. Recently, we developed a quantitative method to analyze Raman spectra [12]. In this study, we have presented the application of the proposed method onto COL thin film, which serves as a preliminary and simplified model of extracellular matrix.

METHODS

The collagen solution (Advanced Biomatrix) was neutralized to PH 7.2 to 7.6 with 10× Dulbecco's phosphate buffered saline (DPBS), 0.1 M NaOH, and stored at 4 °C to minimize polymerization. Thin films of Type I collagen were prepared on alkanethiol treated gold coated silica wafers from the neutralized solution. The substrate was coated with a 5 nm layer of chromium and 18~20 nm layer of gold. Gold-coated wafers were immersed in 1-hexadecanethiol (0.5 mM; Sigma Aldrich) for 8 h. The wafers were then rinsed with ethanol and dried with filtered nitrogen. Neutralized collagen type I (COL I)

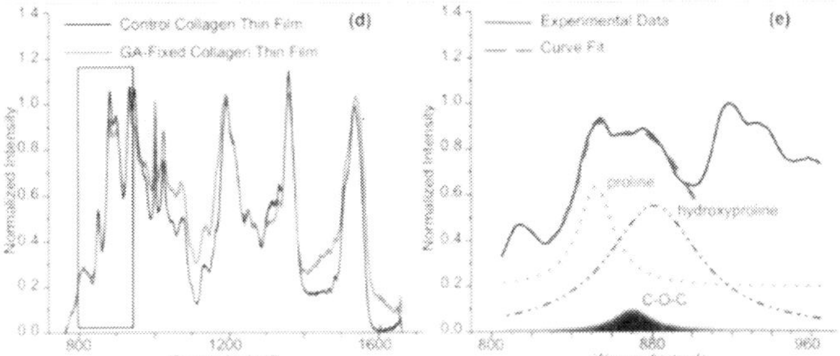

Figure 1. Raman spectra of control and GA-fixed collagen thin films. Boxed region of GA-fixed sample is decomposed, so that the Raman bands associated with the proline, hydroxyproline and ether-type, crosslink (C-O-C skeletal mode vibrational band) between the collagen can be assessed.

solution was applied onto the coated wafers and incubated at 37 °C for 12 h. After incubation, thin films were rinsed with 1× DPBS and deionized H_2O to remove the bulk salt and gels from the surface.

Collagen type I thin films were treated with glutaraldehyde (GA) solution in phosphate buffered saline (PBS) (0.2 M, PH 7.4 Sigma Aldrich). Three concentrations were used 0.05%, 0.1% and 0.2%. Each thin film group was treated for 1, 2 or 3 h, respectively. After the treatment cross-linked thin films were desorbed in PBS for 48 h.

Raman spectra were acquired with a confocal micro-spectrometer (Renishaw InVia), equipped with 1800 gr/mm. Incident laser excitation was provided by helium-neon laser with 632.8 nm wavelength, delivered by 100x objective. Raman signal was acquired by 30 accumulations, each lasting 10 s. Spatially resolved characterization was performed to generate spectral maps.

RESULTS

COL I thin films were used as simplified model to validate proposed method for quantification of the concentration of the cross-links in articular cartilage.

Three types of cross-links are introduced when COL is treated with GA [13]: (a) ether-type (C-O-C) bond (Fig. 1b); (b) a secondary amide type bond; (c) Pyridinium-type crosslinks induced through the interaction of GA molecules with a primary amine of the protein resulting in formation of the 3,4,5-substituted pyridine ring characterized by the changes in Raman band located at 1032 cm^{-1}. Pyridine ring is central chemical structure in hydroxylysyl pyridinoline (PYR), which is, along with pentosidine (PSD), responsible for tensile strength and structural integrity of ECM in articular cartilage [14].

The Raman effect represents the distortion of charge distribution by the primary incident electric field, induced by the monochromatic light source. The charge distortion results in the induction of molecular dipole moments, uniquely determined by the specific chemical bonds in the irradiated material. The sum of all induced dipole moments acts as a macroscopic polarization, a scattered light of a secondary electric field, determining the intensity of the Raman lines. Therefore, Raman spectra are a signature of the material of interest: relative contributions of individual chemical compounds to the signal can be decomposed via curve fitting and quantified. To determine concentration of the pyridinium-type crosslinks, the Raman band with a peak at 1032 cm^{-1} was isolated from the spectra and modeled as a Lorentz function. The fit was optimized via trust region reflective algorithm [15], and normalized with the Raman band associated with hydroxyproline to account for the sample heterogeneity. The integrated intensity of the Raman band was extracted from the optimization model for each measurement point and spatially resolved spectral maps were constructed (Fig. 2a). The increase in the GA concentration as well as the increase of the exposure time to GA results in the increase of the pyridinium-type crosslinks, as anticipated. The results obtained from spectral maps were averaged and summarized in Fig. 2b.

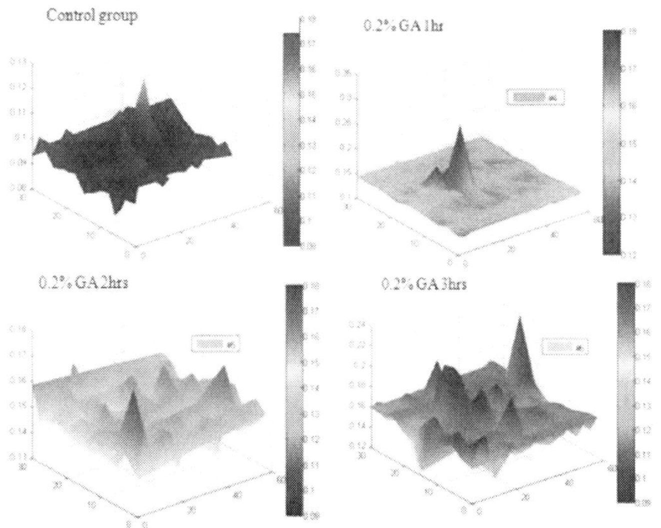

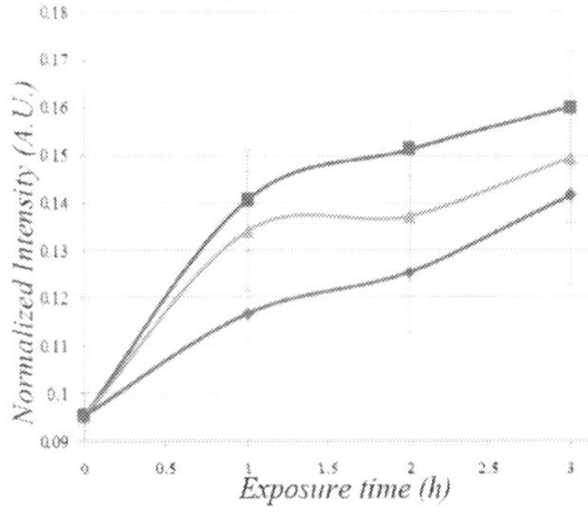

Figure 1. (a) Spatially resolved spectral maps of the normalized integrated intensity (concentration) of the 1032 cm^{-1} Raman band associated with the pyridinium-like crosslinks. Thin films were exposed to 0.2% GA solution for different time periods. (b) Averaged pyridinium-like crosslink concentrations for various percentages of GA solutions.

DISCUSSION

A novel method to measure the concentration of hydroxylysyl pyridinoline (PYR) cross-links based on the quantitative analysis of Raman spectra was proposed. The method was tested on collagen type I thin films cross-linked with GA. This is an ongoing study which aims to validate the results of the proposed method; in ongoing investigations, these results will be compared against the concentrations of PYR cross-links measured using high-performance liquid chromatography (HPLC). Further, in order to validate the method's capability to detect known changes in collagen crosslinks in the physiologic range for articular cartilage, the method will be applied onto the immature bovine articular cartilage explant model [4].

REFERENCES

[1] Ashkavand Z, et al., *J. Pharmacy Research*.7(1):132-8, 2013.
[2] Dijkgraaf LC, et al., *J. Oral. Maxill. Surg.*, 53(10):1182-92, 1995
[3] Dijkgraaf LC, et al., *J. Oral. Maxill. Surg.*, 53(8):924-9, 1995
[4] Asanbaeva A, et al., *Osteoarthritis and Cartilage.*;16:1-11, 2008
[5] Uebelhart D, et al., *Osteoarthr. and Cartilage*, 1(3):185-92, 1993.
[6] Takashi M, et al., *Arthritis & Rheumatism*, 37(5):724-8, 1994.
[7] Deberg M et al., *Osteoarthritis and Cartilage.* 2005;13(3):258-65.
[8] Esmonde-White KA, et al., *J. of biom. optics.* 2009;14(3):034013.
[9] Camacho NP et al., *Biopolymers.* 2001;62(1):1-8
[10] Saarakkala S et al., *Catrilage*, 1(4):262-9, 2010
[11] Bi X, et al., *Osteoarthritis and Cartilage*,13:1050-8, 2005
[12].Vukelić et al., *J. of Manuf. Sci. and Eng.* 132(6):061013, 2010.
[13] Jastrzebska M, et al., *J Bio: Sci Polymer Edn.*14(2):185-97, 2003.
[14] Eyre DR, et al., *Annual Reviews, Biochem.*,53:717-48, 1984.
[15] Alexabdriv NM, et al., *Structural Optimization*,15:16-23, 1998.

SB^{3}C2015
Summer Biomechanics, Bioengineering and Biotransport Conference
June 17-20, 2015,Snowbird Resort, Utah, USA
SB^{3}C2015-453

BIOMECHANICAL TESTING TO IMPROVE CONSTITUTIVE MODELS OF THE TWO-LAYERED CAROTID ARTERY MEDIA

Caleb A. Davis (1,2), Ashish Pandya (2), Stephen E. Greenwald (2)

(1) Department of Biomedical Engineering
Texas A&M University
College Station, TX, USA

(2) Blizard Institute, Barts and the London
School of Medicine and Dentistry
Queen Mary University of London
London, UK

INTRODUCTION

The arterial wall is composed of three layers: the *tunica intima*, *tunica media*, and *tunica adventitia*. However, in some mammals (though not humans), recent work has demonstrated that the media of conduit arteries such as the carotid actually consists of two distinct layers (Figure 1). On examining dozens of different mammalian species, we have found that the two-layered carotid is common to all ungulates (hooved mammals and whales), species which belong to the mammalian orders cetartiodactyla and perissodactyla. Moreover, the two-layered carotid has so far not been found in any non-ungulates, although there are many species still to be studied [1]. It is therefore currently hypothesized that this trait evolved sometime around the 65 mya extinction event and has been conserved as the mammals destined to become ungulates split into various families shortly after that time (Figure 1) [1].

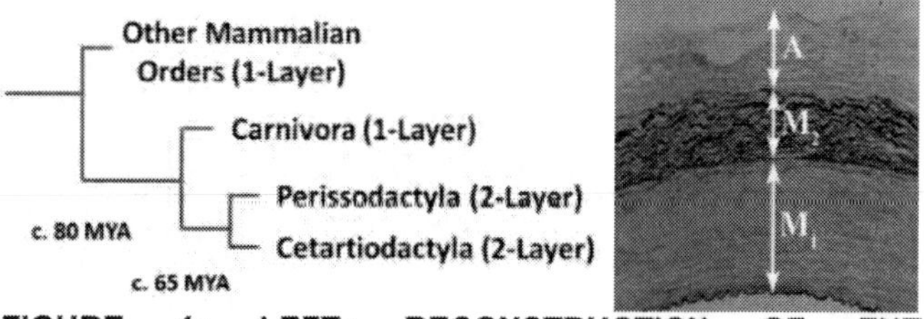

FIGURE 1: LEFT: RECONSTRUCTION OF THE EVOLUTIONARY TREE HIGHLIGHTING THE FAMILIES WITH TWO-LAYER CAROTID MEDIA [1,3]. RIGHT: CROSS SECTION OF A BOVINE CAROTID TREATED WITH MILLER'S ELASTIC STAIN SHOWING TWO-LAYERED MEDIA [2].

The two medial layers are different in structure and composition, leading researchers to speculate how each might contribute to the mechanical behavior of the vessel wall [2,4]. Because the two-layered carotid appears to have been conserved among the ungulates, it is possible that it provides some advantage in the vessel's response to mechanical forces, or at least is not disadvantageous. To study why this might be the case, we have developed *in vitro* biomechanical testing methods to produce and assess a constitutive model of the carotid artery wall which takes into account the two-layered media. We believe this will provide insight into the more general problem of the relationship between the complex structure and mechanical properties of conduit arteries.

METHODS

We developed a pressure-inflation testing protocol to separate the relative contributions of each medial layer. Biaxial vessel pressure-inflation tests were modeled after the work of von Maltzahn, et al., who considered the separate contributions of the media and adventitia and incorporated both into a constitutive model [5]. As in Timmins, et al. [2], we perform the tests on intact vessels and those with the outer medial layer removed. However, unlike that protocol, the vessels are held at various predetermined levels of axial stretch during testing.

We constructed a custom vessel bath which allows a bovine carotid artery segment to be held at the fixed axial stretch and inflated via a syringe pump (Figure 2). A pressure catheter and a force transducer measure the intraluminal pressure in the vessel and the axial tension, respectively, which are recorded at a sampling rate of up to 140 Hz. At the same time, a high-resolution camera captures the diameter change during testing at a fixed frame rate equal to the sampling rate. A custom edge-detection algorithm is used to process the video and measure the change in diameter during testing.

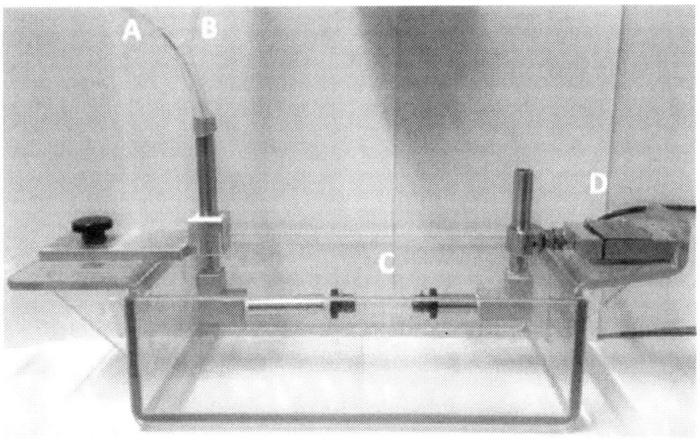

FIGURE 2: CUSTOM VESSEL BATH SET UP FOR TESTING A) CATHETER TIP MANOMETER B) FLUID INLET C) ADAPTERS FOR MOUNTING VESSEL D) FORCE TRANSDUCER

Quasi-static pressure-inflation tests consist of pre-conditioning followed by slow increase of pressure to 200 mmHg. Axial force, intraluminal pressure, and vessel diameter are measured synchronously throughout the test. To determine the individual contributions of each layer, multiple tests are performed on each vessel segment. First, the intact artery is tested at various levels of axial stretch (0%, 10%, 20%, 40%) to mimic conditions within the body and provide the necessary data for parameter fitting. Then, the artery is frozen onto a mandrel of diameter closely matching that of the vessel lumen, in a dry ice/acetone mixture (-70°C) and the outer medial layer is lathed away before it thaws; this is made possible by the visual distinction between the two layers [6]. The inflation tests are then repeated with the outer medial layer removed.

Once the data is collected, a custom MATLAB program is used to fit the parameters of a strain-energy function to the pressure, force, and diameter data. This is based on the methods employed by Timmins, et al. [2], where the form of the strain-energy function is the one developed by Holzapfel, et al. [7], modified for a two-layer model. The fiber angles in the two layers are measured from histological analysis, and the total strain energy function of the two layers is as follows:

$$\Psi_{total} = \Psi_{M_1} + \Psi_{M_2} \quad (1)$$

$$\Psi_{M_1} = \frac{c_{M_1}}{2}(I_1 - 3) + \frac{k_{1,M_1}}{2k_{2,M_1}}\left(e^{k_{2,M_1}(\lambda_\theta^2 - 1)^2} - 1\right) \quad (2)$$

$$\Psi_{M_2} = \frac{c_{M_2}}{2}(I_1 - 3) + \frac{k_{1,M_2}}{2k_{2,M_2}}\left(e^{k_{2,M_2}(\lambda_z^2 - 1)^2} - 1\right) \quad (3)$$

In these equations, λ_θ is the circumferential stretch ratio, λ_z is the axial stretch ratio, I_1 is the first invariant of the right Cauchy-Green tensor and c_{M1}, c_{M2}, k_{1M1}, k_{2M1}, k_{1M2}, k_{2M2} are material constants and the subscripts $_{M1}$ and $_{M2}$ refer to the inner and outer medial layers. The MATLAB program, based on modifications of the one used by Timmins, et al., fits values of the material constants using a multi-step nonlinear regression to take into account the properties of both layers. In the first step, data from testing the lathed specimens (consisting of only the inner layer) is used to determine the parameter values of a single layer model (Eq. 2). Then, by substituting those values into Eq. 2, the best fit for the intact specimens can be determined (Eq. 1). Finally, the constants for the outer layer (M_2) can be estimated using Eqs. 1 and 2 [2].

RESULTS

Figure 3 shows preliminary results from pressure-inflation testing of an intact bovine carotid artery segment. The features of the curves demonstrate good synchronization between the pressure, force, and diameter measurements.

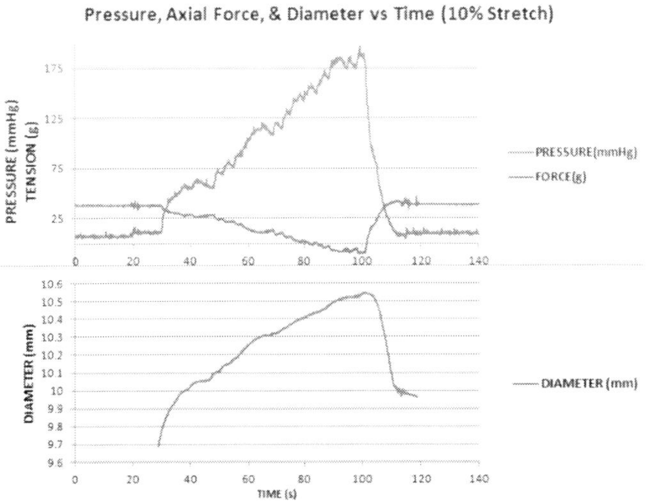

FIGURE 3: INTRALUMINAL PRESSURE, AXIAL FORCE, AND DIAMETER MEASUREMENTS FROM PRELIMINARY CAROTID ARTERY TESTING.

DISCUSSION

This work demonstrates a proof-of-concept for a new method of *in vitro* biomechanical testing of carotid arteries. The advantage of this method, based on the work done by von Maltzahn, is that it considers the individual contributions of the inner and outer medial layers. The data generated by these tests will allow for the development of a constitutive model of the arterial wall which takes the two-layered media into account. This will go a long way toward determining what, if any, mechanical advantage this structure gives the arterial wall and may help answer the question of why the two-layered carotid is observed in ungulates, but not in other mammals.

ACKNOWLEDGEMENTS

Student funding was provided by a Whitaker Fellowship from the Institute of International Education. The authors gratefully acknowledge the previous work and contributions of Talha Atcha and Dr. Lucas Timmins. We also acknowledge Guy's Hospital, London, for assistance in constructing the vessel bath used in this study.

REFERENCES

[1] Atcha, Talha, unpublished, 2014.
[2] Timmins, L et al., *Proc ASME SBC2008,* SBC2008-192480.
[3] Bininda-Emonds, O et al., *Nature*, 446:507-512.
[4] Timmins, L et al., *Am J Physiol Heart Circ Physiol*, 298:H1537-H1545, 2010.
[5] von Maltzahn, W et al., *J Biomech*, 17:839-847, 1984.
[6] Greenwald, S et al., *J Biomech Eng*, 119:438-444, 1997.
[7] Holzapfel, G et al., *J Elasticity*, 61:1-48, 2000.

INTEGRATING MORPHOLOGIC, BIOMECHANIC, BIOLOGICAL AND CLINICAL RISK FACTORS TO IMPROVE DECISION MAKING IN THE MANAGEMENT OF ABDOMINAL AORTIC ANEURYSM DISEASE

Eleni Metaxa (1), Nikolaos Kontopodis (2), Christos Ioannou (2), Yannis Papaharilaou (1)

(1) Institute of Applied and Computational Mathematics
Foundation for Research and Technology
Heraklion, Crete, Greece

(2) Department of Vascular Surgery
University of Crete
Heraklion, Crete, Greece

INTRODUCTION

In an effort to improve abdominal aortic aneurysm (AAA) rupture risk assessment several studies have identified demographic, morphometric, thrombus-related, biomechanic, and biological characteristics that associate with rupture risk. However, it is unlikely that parameters from either the biomechanics or the biology aspect of AAA rupture alone can capture the multifactorial pathology of AAA rupture, accurately predict the rupture risk and thus be trusted by surgeons for rupture risk assessment. Therefore, a paradigm shift to our approach is needed that will better capture the multifactorial pathology of AAA.

The structure of a decision tree is one that is appealing intuitively and congruent with methods of decision-making that a physician already uses on many occasions [1]. Towards such an approach, the purpose of this study was to build a decision tree that classifies AAAs into high versus low growth rate groups, based on demographic, morphometric, thrombus-based, biomechanic, and biologic characteristics that can all be measured from 3D imaging data and routine blood sample analysis.

METHODS

Thirty-four patients with diagnosed AAA, with at least one follow-up examination, were included in the study. In total, 24 characteristics from 5 categories (morphometric, thrombus related, biomechanic, biological, and demographic) were analyzed. Among those, the recently introduced asymmetry thrombus deposition index (ADTI) was also included, since a posterior thrombus deposition was found to associate with low growth rate with a specificity of ~90% [2]. Blood analysis results were available for 17 patients.

The growth rate was linearly extrapolated from the difference in maximum AAA diameter between the first and last CT scans.

The cases were divided into high and low growth rate based on the cohort's median growth rate (3.2 mm/year).

Morphometric	***Thrombus***	***Biomechanical***
1. Max Diameter (D_{max}) 2. Neck Diameter (D_{neck}) 3. Normalized diameter (D_{max}/D_{neck}) 4. Tortuosity (L/H) 5. Length of AAA (L_{AAA}) 6. Saccular Index (D_{max}/L_{AAA}) 7. Surface Area of AAA (S) 8. Volume of AAA (V) 9. Aorta-Neck Angulation 10. Neck-AAA Angulation	1. Volume of ILT (V_{ILT}) 2. Relative ILT volume (V_{ILT}/V) 3. Max ILT thickness 4. ATDI ("+1" anterior, " -1" posterior deposition)	1. Peak Wall Stress (PWS) 2. Neck vs. body PWS location 3. Anterior vs. posterior PWS location 4. Max Rupture Potential Index (RPI)
Biological	***Demographic***	
1. Creatinine 2. Platelets 3. Cholesterol	1. Gender 2. Age 3. Family History	

Table 1. List of patient and AAA characteristics. (L: length of aorta's centerline from renal arteries to aortic bifurcation, H: Distance between the two endpoints of the centerline (renal to bifurcation).

Morphometric and thrombus related markers were measured using vascular modelling toolkit (vmtk) after reconstructing the 3D image by segmenting the external and lumen surface of the aorta using ITK-snap. For the measurement of AAA length (centerline), the AAA sac was defined as the part of infra-renal aorta with diameter more than 3 cm. Aorta-neck and neck-AAA angulation were measured in the imaging software Evorad Workstation (Evorad SA, Athens, Greece).

Estimation of wall stress distribution at peak systole (120 mmHg) was performed in Workbench (*ANSYS* Inc., Berkeley, CA, *USA*), after tetrahedral mesh generation in ICEM CFD v.12.0.1 (*ANSYS* Inc., Berkeley, CA, *USA*). Hyperelastic material models for wall [3] and for ILT [4] were adopted.

Statistical analysis was performed in WEKA [5], an open source machine learning software used for data mining. The highest ranked attributes were selected with chi-square statistics with respect to growth rate class and a J48 decision tree algorithm with ten-fold cross validation was used to develop a model based on the data. A kappa statistics (κ) was also calculated.

RESULTS

The highest ranked attributes were ATDI, gender, anterior vs. posterior PWS location, neck vs. body PWS location, and relative ILT volume. The lowest ranked attribute was D_{max}. The highest ranked feature, ATDI, was also the feature with the highest information gain thus making it the first in the decision tree, which consisted of ATDI, relative ILT volume, and neck-AAA angulation (Figure 1). Twenty-six out of the 34 cases were correctly classified yielding a prediction accuracy of 76.5% with κ=0.53.

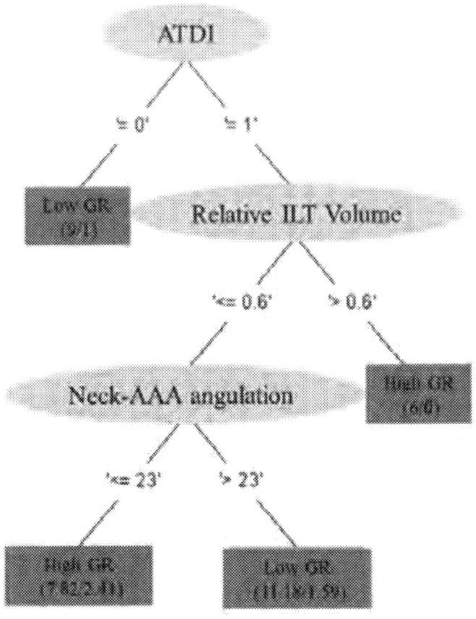

Figure 1. Model learned by J48 decision tree.

DISCUSSION

With the ongoing improvement of imaging techniques, the establishment of screening programs and the ageing population, the amount of data has greatly been increased. In order to narrow the gap between data collection and data comprehension and decision making, machine learning algorithms can build models for rupture risk assessment, that are not only appealing to physicians, but also capture the multifactorial nature of AAAs.

Shum et al. were recently the first to use a classification model to discriminate between ruptured and non-ruptured AAAs from a set of geometric features from 76 patients resulting in a decision tree model with 86.6% accuracy and κ=0.37. While the accuracy was high and the model promising, we believe that a multimodal approach, where several parameters from different categories are taken into consideration, would provide a more accurate rupture risk assessment. This is because the different categories of risk markers can potentially have a complementary role to each other. For example, biomechanics analysis provides the wall stress distribution, while blood analysis provides information about the biological activities (inflammation, thrombus formation, enzymatic activities) that may reflect the state of the degraded wall and presumably its strength. By integrating all these risk markers in one statistical model we will be able to achieve a more complete picture of the aneurysm state and predict more accurately its growth rate.

This study shows that the growth rate can be predicted with the knowledge of thrombus eccentric deposition in the sac, its relative volume, and the neck-AAA angulation with an accuracy of 76.5%. When a larger patient cohort is acquired, and more blood markers (such as C-Reactive protein, fibrinogen, D-dimer) are incorporated in the model, the decision tree may have more or different leafs and potentially increase its accuracy. A low κ (=0.53) was so far unavoidable due to the small sample size.

Although the patient cohort was small, this study presents a set of lead candidates, and demonstrates proof-of-principle for a more extensive follow-up study, providing the basis for the development of an integrated, multidisciplinary approach to the diagnosis, management and treatment of AAA.

ACKNOWLEDGEMENTS

Financially supported by the Action «Supporting Postdoctoral Researchers», co-financed by the European Social Fund (ESF) and the Greek State (LS7_2224).

REFERENCES

[1] Grobman, WA et al., *Am J Obstet Gynecol*, 194:888-894, 2006.
[2] Metaxa, E et al., *JEVT, in press*, 2015.
[3] Raghavan, ML et al., *J Biomech*, 33:475-82, 2000
[4] Vande Geest, JP et al, *J Biomech*, 39:2347-2354, 2006

SB^{3}C2015
Summer Biomechanics, Bioengineering and Biotransport Conference
June 17-20, 2015,Snowbird Resort, Utah, USA
SB^{3}C2015-455

HEMODYNAMIC PREDICTION OF THROMBUS-PRONE REGIONS IN ABDOMINAL AORTIC ANEURYSMS

Konstantinos Tzirakis (1), Eleni Metaxa (1), Nikos Kontopodis (2), Christos Ioannou (2), Yannis Papaharilaou (1)

(1) Institute of Applied and Computational Mathematics
Foundation for Research and Technology
Heraklion, Crete, Greece

(2) Department of Vascular Surgery
University of Crete
Heraklion, Crete, Greece

INTRODUCTION

Intraluminal thrombus (ILT) is present in almost all abdominal aortic aneurysms (AAAs), with several studies trying to clarify if it increases or decreases the risk of rupture: it increases local proteolytic activity and causes wall hypoxia leading to wall weakening, but it also seems to act as a mechanical buffer against wall stress. In the meantime, the prediction of its growth might be of value, as clinical studies have demonstrated that ILT growth might indicate increased AAA rupture risk [1].

Thrombus formation and growth is a complicated process, with several biological and hemodynamic factors participating in a dynamic cascade. The formation is initiated by activation of platelets, which is determined by a combination of factors including high shear stress and time exposure to such stresses. Activated platelets accumulate next in recirculation and stagnation regions of the flow field, where they attach usually to the non-endothelialized surface, where low WSS and high residence time promote their adherence. This dynamic and multifactorial process of thrombus formation and evolution makes it difficult to model and predict its deposition pattern.

In an effort to study the hemodynamic induction of ILT formation, the Lagrangian approach is usually considered for the evaluation of coherent vortex formation (that affect platelet activation), a process that requires substantial computational efforts. However, since many studies show that AAAs possess the necessary hemodynamic conditions for coagulation and activation of platelets, the AAA sac is most likely filled with activated platelets ready to attach to a thrombogenic surface if the hemodynamic conditions are favorable. Therefore, near-wall hemodynamics is most probably the major determinant of ILT deposition. Our hypothesis is that with a statistically sufficient patient cohort, a co-mapping of flow conditions at the AAA wall with the information of ILT deposition distribution will allow the construction of a statistical model that will predict thrombus-prone regions.

In the present study we compute the time averaged wall shear stress (TAWSS) distribution in two patient specific AAAs that were thrombus-free at the first scan. The luminal surface is then divided into patches. In each patch the TAWSSs and ILT thickness are averaged. A logistic regression is then performed to test the statistical significance of the association between hemodynamics and thrombus location. Although the patient cohort is small, this study demonstrates proof-of-principle for a more extensive ongoing follow-up study.

METHODS

Data Acquisition

Two CT scans of AAAs (and their follow-ups) were obtained. The 3D lumen and external wall surface of the sac were reconstructed with ITK-SNAP and smoothed in vascular modeling tool kit (vmtk). The AAA surfaces were co-registered using the aorta at the level of the renal arteries (± 5 mm) and the aortic/renal bifurcations as registration features. The ILT thickness was measured at the follow-up AAA model as the distance between lumen and external surface.

Meshing and Computational Fluid Mechanics

Flow extensions were added to the luminal surface of the first scan, and a full hexadehral mesh was constructed using ANSA (BETA CAE Systems *S.A.*, Greece) consisting of approximately 800,000 elements. In order to account for the shear thinning properties of blood, the Herschel-Bulkley viscosity model was considered. Simulations were performed using Fluent (ANSYS Inc.) with default convergence criteria set to 10^{-5}. A pulsatile profile was prescribed at the inlet with Re_{mean}=330 and Re_{peak}=1800, and results were collected after all transient effects were washed out. The TAWSS was computed and its distribution was mapped on the surface.

Surface Patching

As the shape and size of AAA change with time, a methodology was needed in order to identify the corresponding regions between follow-up images. For this purpose the "mapping and patching" algorithm implemented in vmtk was used. In brief, since the aorta is topologically equivalent to a cylinder, it can be mapped onto a rectangular parametric space that allows the comparisons between different models. The surface of the aorta is divided into a number of contiguous rectangular regions on the surface over which the quantities of interest are averaged. The important feature of the methodology is that the location of each patch is preserved despite the variability in local surface geometry, allowing for the direct comparison of each patch (and its quantities) between the first and the follow-up scan. The luminal surface of the first scan (with the hemodynamics information) and the luminal surface of the second scan (with the thrombus thickness information) were patched as seen in Figure 1B.

Statistics

For each patch, the mean TAWSS value was obtained while the subsequent ILT deposition was recorded as a dichotomous variable. Median and range of TAWSS values for the AAA regions was recorded while a Mann–Whitney test was used to compare TAWSS between thrombosed and non-thrombosed AAA regions. Moreover, a logistic regression analysis was applied to test the statistical significance of the association between TAWSS and thrombus location.

RESULTS

Thrombus deposition was observed to negatively correlate with TAWSS values. In Figure 1, the initial hemodynamics and ILT deposition at follow-up, is presented before and after surface patching. For the purpose of illustration, the inverse of time-averaged WSS is shown, which is also proportional to the relative residence time.

Median TAWSS of AAAs under evaluation was 0.14 Pa (Range: 0.02-0.42). Regarding non-thrombosed aneurysm regions median WSS was 0.14 Pa (Range: 0.03-0.42), while for those with subsequent ILT deposition respective values were 0.03 Pa (Range: 0.02-0.21). This difference was statistically significant (P-value<0.001).

The logistic regression demonstrated that there existed a significant association of low TAWSS values with subsequent ILT deposition. Specifically, there was a 0.115 (95% CI: 0.028-0.472, P-value=0.03) odds ratio of a patch to develop thrombus deposition for each 0.1 Pa decrease in TAWSS values. In other words the odds of a location developing thrombus, increased by almost 9-fold with each 0.1 Pa of TAWSS decrease.

DISCUSSION

This study investigated the relationship between a hemodynamic index and ILT deposition, and a relation was determined between areas with low WSS and thrombus deposition at a follow-up image.

While the multifactorial nature of thrombus formation has triggered a large number of researchers to study and model the hemodynamic induction of thrombus formation, there is only one recent study to our knowledge that investigated how well near-wall hemodynamics correlates with thrombus formation in patient-specific data sets [2]. However, in this study, the authors correlated the thickness change of ILT, not just the deposition of thrombus as a binary variable, with hemodynamics, and only oscillatory shear index was found to significantly correlate with thrombus growth. They conclude that for the prediction of thrombus growth, biochemical reactions with transport should also be considered besides hemodynamics. While the growth may be more difficult to model and predict, the deposition or no deposition of ILT in a thrombus-free small aneurysm can potentially be predicted by near-wall hemodynamics alone. Such a study has already been performed in intracranial aneurysms and a significant correlation was observed between regions of either low WSS or increased residence time, and thrombus deposition at follow up [3].

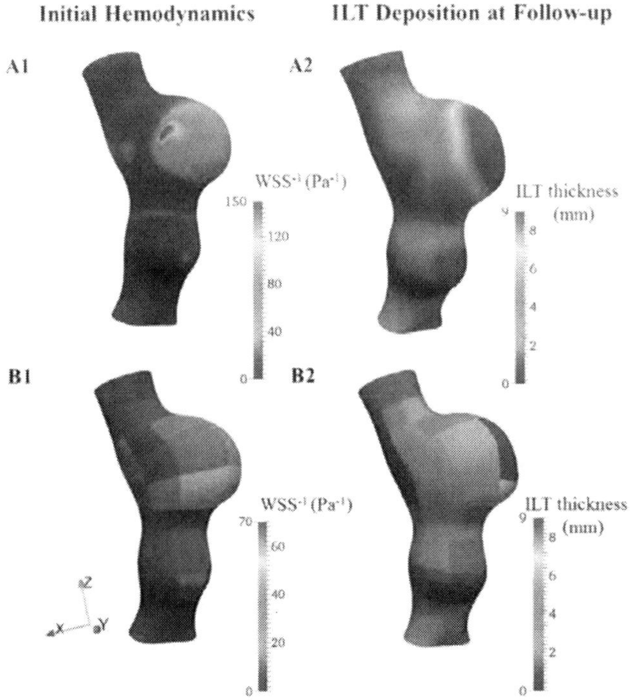

Figure 1. Initial hemodynamics (1) and ILT deposition at follow up (2) A. Before patching, B. After patching.

In this study, we are not proposing a new hemodynamic mechanism for ILT formation, but rather a convenient method for predicting it in AAAs. More patient-specific AAA cases with thrombus-free baseline sacs are currently monitored to further improve the statistical model and potentially obtain threshold values for this hemodynamic marker of thrombus deposition.

In conclusion, in spite of the complicated process of thrombus formation, WSS distribution emerges as a simple tool for identifying the possible regions of thrombus deposition in AAAs and potentially provide additional information to clinicians to help them predict in a patient-specific basis aneurysm evolution.

ACKNOWLEDGEMENTS

Financially supported by the Action «Supporting Postdoctoral Researchers», co-financed by the European Social Fund (ESF) and the Greek State (LS7_2224).

REFERENCES

[1] Stenbaek, J et al. *Eur J Vasc Endovasc Surg*, 20:466-469, 2000.
[2] Arzani, A et al. Am J Physiol Heart Circ Physiol, 307: H1786-H1795,2014.
[3] Rayz, V.L et al. *Annals of Biomedical Engineering*, 38:3058-3069.

SB[3]C2015
Summer Biomechanics, Bioengineering and Biotransport Conference
June 17-20, 2015,Snowbird Resort, Utah, USA
SB[3]C2015-458

COMPUTATIONAL MODELING OF MECHANICAL STRESSES IN STENTS: IMPLICATIONS FOR THE OPTIMIZATION OF DRUG-ELUTING STENTS

François M. Cornat, Franz Bozsak, Abdul I. Barakat

Hydrodynamics Laboratory (LadHyX)
Department of Mechanics
Ecole Polytechnique
Palaiseau, France

INTRODUCTION

Atherosclerosis is the leading cause of mortality in the world. The most common treatment for atherosclerosis today is the implantation of a stent. The earliest stents were bare metal stents (BMS); however, the unacceptably high restenosis rates associated with BMS led to the development of drug-eluting stents (DES). Although highly effective in preventing restenosis, DES appear to be associated with an increased risk of late stent thrombosis (LST) which, though rare, is often fatal when it occurs. LST following DES implantation is attributable to delayed endothelial wound healing following the massive endothelial damage induced by the high pressure deployment of a stent. The delayed endothelial healing has two primary causes: 1) the drugs eluted by DES inhibit endothelial cell proliferation, thereby slowing down vascular wound healing and 2) stents induce arterial flow disturbance which can impair endothelial cell migration [1]. We are interested in optimizing DES design and performance with the overall goal of controlling restenosis while simultaneously allowing endothelial healing. Accomplishing this goal requires a sufficiently high drug concentration in the arterial media to prevent restenosis while maintaining drug concentration at the endothelium as low as possible. The stent also needs to perturb arterial flow minimally.

We have recently developed a computational model that describes the transport within the arterial wall of drugs eluted from DES [2]. We subsequently used this model to optimize drug release from DES. To this end, we defined a cost function whose minimization allowed us to determine the optimal combination of initial drug concentration loaded onto the DES and drug release kinetics [3]. The minimization of the cost function was performed using a surrogate management framework-based gradient-free optimization algorithm. We subsequently extended the optimization to determine the optimal stent strut design that not only delivers optimal drug release but also minimizes disturbance of arterial blood flow. Not surprisingly, this optimization led to very thin struts that protrude minimally into the arterial lumen. It remained unknown, however, if these thin struts are capable of withstanding the large mechanical stresses associated with stent deployment. The purpose of the present work is to develop a computational model of the mechanical stresses in a stent as it is deployed and to evaluate the sensitivity of these stresses to stent strut shape and dimensions.

METHODS

We developed a fully parameterized three-dimensional model of a closed-cell stent design. This approach provides the ability to systematically evaluate the effect of each parameter on the mechanical stresses within the stent. To test the effect of a wide range of stent strut shapes on the computed mechanical stresses within the stent, we define the strut cross-sectional shape using a special form of the so-called "Superformula" as follows:

$$r(\beta)=\left[\left|\frac{1}{a}\cos\left(\frac{\beta}{4}\right)\right|^{n}+\left|\frac{1}{b}\sin\left(\frac{\beta}{4}\right)\right|^{n}\right]^{\frac{-1}{n}}$$

with $\beta\in\left[0,2\pi\right]$. In this equation, the parameters 'a' and 'b' are the strut semi-height and semi-width, respectively, whereas 'n' is a parameter that allows variations in stent strut cross-sectional shape ranging from diamond (n=1) to ellipse (n=2) to rectangle (n>>1).

The stent is assumed to be made of 316L stainless steel. The initial crimped stent diameter is 0.5 mm, and it is expanded to a final diameter of 4 mm, typical of a human coronary artery. The stent is expanded by applying a pressure to the inside surface of the stent. While the expansion pressure is spatially uniform at any particular time in the expansion process, its value decreases with stent radius. The total expansion time is 1 sec.

The solid mechanics study evaluates the stresses within the stent as the stent is expanded from the crimped configuration to its fully expanded state. The stent is assumed to exhibit elasto-plastic mechanical behavior with strain hardening. Different types of loads were simulated: a radial pressure applied on the interior surface of the stent simulates the balloon inflation, and a longitudinal load on the stent extremities simulates the stent compression which are the two main types of loads experienced by a stent during its lifetime [4]. The phase where the stent is implanted and subjected to arterial wall movement is also studied using an analytical model of fatigue behavior. In this case, the load was evaluated using a two-dimensional model that calculates the contact pressure between a stent strut and the arterial wall while subjected to blood pressure.

Because DES are covered with a polymer coating within which the drug is housed, we are also interested in studying the stresses within this coating during stent deployment. Several studies have demonstrated fractures in DES coating, which reduce its efficiency and perturb drug diffusion. We assume the polymer coating to be governed by a hyperelastic isotropic constitutive law. Evaluating the mechanical stress field inside the polymer coating is essential for optimizing the coating geometry and dimensions for a particular set of stent deployment conditions.

The simulations were performed using the COMSOL multi-physics software package (version 4.3a).

RESULTS

Figure 1 depicts the Von Mises stresses within a stent at a snapshot in time (t=0.65 sec) during the 1 second-long stent expansion process. The stent in this figure has 50 μm x 50 μm square cross-section struts (a = b = 25 μm and n=1000). The stresses are clearly not uniform within the stent with stress concentrations at strut junctions. These findings are consistent with those reported elsewhere [5]. The peak computed von Mises stress for this case is ~34o MPa, which clearly reaches the plastic deformation domain.

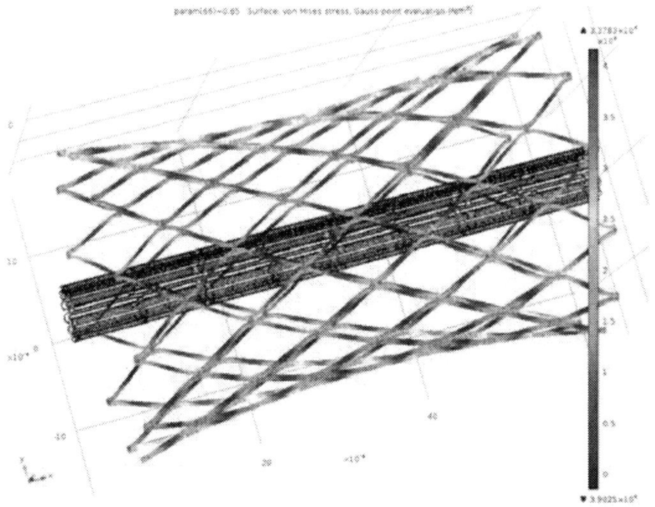

Figure 1: Von Mises stress distribution in the stent during expansion at t = 0.65 s. In gray is the initial crimped stent shape.

Performing similar simulations for the various stent strut cross-sectional shapes that result from the optimization will provide an assessment of whether or not these shapes are mechanically viable.

DISCUSSION

Mechanical evaluation of stents as performed in this study is an important consideration in the process of optimization of the design and performance of DES. It is noteworthy that while the goal of minimizing stent-induced flow disturbance in the arterial lumen pushes stent design towards thinner struts, the need for the struts to withstand the large mechanical stresses during the violent stent expansion process pushes the design towards thicker struts. Thus, the optimal strut height is expected to depend in a complex manner on stent shape, strut cross-sectional shape, stent materials, and the imposed expansion pressure profile.

All the simulations thus far have been performed on a closed-cell stent design. It would be important to expand the scope to also include open-cells stents. Furthermore, a particularly interesting extension would be the evaluation of the mechanical stresses experienced by the polymeric coating within which the DES drug is embedded. A challenge in this regard is the determination of the appropriate contact conditions to impose in the simulations.

The simulations described in this paper do not include the arterial wall against which the stent is expanded. Contact between the stent and the arterial wall would be expected to influence the mechanical stress field within the stent; therefore, future efforts will focus on incorporating this effect. Finally, some level of experimental validation of the computational results would certainly be helpful in the stent optimization process.

ACKNOWLEDGMENTS

Francois Cornat is supported by a doctoral fellowship from *Ecole Normale Superieure*. This work was supported in part by a permanent endowment in Cardiovascular Cellular Engineering from the AXA Research Fund.

REFERENCES

[1] Gojova, A, Barakat, AI. Regulation of vascular endothelial cell wound healing by shear stress. *J Appl Physiol* 98:2355-2362, 2005.

[2] Bozsak, F, Chomaz, JM, Barakat, AI. Modeling the transport of drugs eluted from stents: physical phenomena driving drug distribution in the arterial wall. *Biomech Model Mechanobiol* 13:327-347, 2014.

[3] Bozsak, F, Gonzalez-Rodriguez, D, Sternberger, Z, Belitz, P, Bewley, T, Chomaz, JM, Barakat, AI. Optimization of drug delivery by drug-eluting stents. Submitted.

[4] Pant, S, Limbert G, Curzen NP, Bressloff, NW. Multiobjective design optimisation of coronary stents. Biomaterials 32:7755e7773, 2011.

[5] Gervaso, F, Capelli, C, Petrini, L, Lattanzio, S, Di Virgilio, L, Migliavacca, F. On the effects of different strategies in modelling balloon-expandable stenting by means of finite element method. *J Biomech* 41:1206-1212, 2008.

WAVE INTENSITY ANALYSIS FROM MAGNETIC RESONANCE IMAGING: EXPERIMENTAL VALIDATION AND NEW INSIGHTS INTO SINGLE VENTRICLE PHYSIOLOGY

N. Arya (1), S. Schievano (2,3), C. Baker (2,3), T-Y Hsia (2,3), A. Giardini (3), S. Khambadkone (3), A. M. Taylor (2,3) and G. Biglino (2,3)

(1) Mathematics and Physical Sciences,
University College London, London, UK

(2) Institute of Cardiovascular Science,
University College London, London, UK

(3) Cardiorespiratory Division,
Great Ormond Street Hospital for Children,
NHS Foundation Trust, London, UK

INTRODUCTION

Wave intensity analysis has been used in the past to evaluate the working condition of the heart in relation to the rest of the vasculature (i.e. ventriculo-arterial, VA, coupling) [1]. Its applications are extensive, although limited by the invasive nature of the pressure data required to perform the analysis. Formulating wave intensity in terms of the product of simultaneous changes in velocity and fractional changes in area instead renders the analysis non-invasive, and suitable for cardiovascular magnetic resonance (CMR) imaging data [2]. CMR-derived wave intensity, however, has not been validated against pressure-based data. Hence we sought to verify the agreement between the two formulations in a CMR-compatible *in vitro* setup.
Having shown the reliability of the CMR methodology, one scenario of congenital heart disease in which insight on VA coupling is particularly desirable is hypoplastic left heart syndrome (HLHS). Following surgical palliation of HLHS, obligatory mismatched VA coupling persists for the duration of a patient's lifetime. However, volume unloading following Stage 2 surgery (i.e. bidirectional cavopulmoary connection, BCPC, or Glenn operation) may restore a more favorable coupling between the systemic right ventricle and the systemic arterial circulation. Accordingly, we sought to quantify changes to VA coupling following BCPC using CMR data alone.

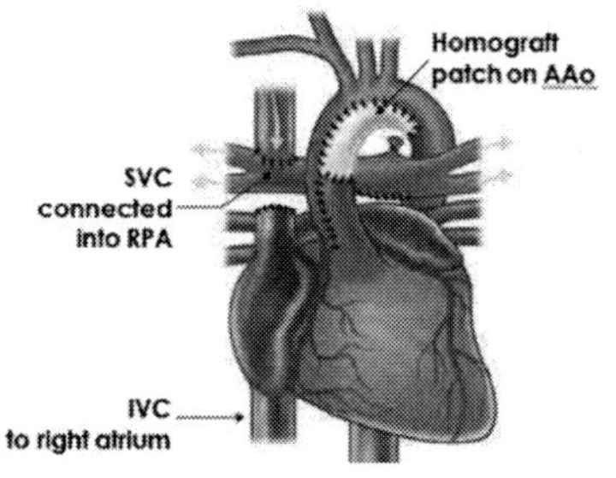

***Figure 1**: Anatomical arrangement following BCPC. SVC = superior vena cava, RPA = right pulmonary artery, IVC = inferior vena cava, AAo = ascending aorta.*

METHODS

a) *In vitro* setup

A reproducible pressure pulse was generated using a Harvard pump in a flexible arterial model (25 mm internal diameter, 2.5 mm wall thickness, 300 mm length) printed with the compliant compound Heart Print Flex® (Materialise). The model was mounted in a water-filled tank. A 6F side port was incorporated to allow for the insertion of a fibre optic pressure sensor (OpSens). Flow-velocity was measured at the inlet of the tank with an ultrasonic flow probe (9PXL Transonic). Downstream the arterial model, an air chamber allowed for regulating compliance (Figure 2). The apparatus was CMR compatible, lacking ferromagnetic components, and was placed inside the scanner (1.5T Avanto, Siemens) for acquisition of phase-contrast magnetic resonance (PCMR) data. Flow settings: stroke volume = 55 mL, heart rate = 50 bpm, diastolic time fraction = 50%. The obtained PCMR flow sequences were then analyzed using an in-house written plug-in (OsiriX) providing information on the change in area and velocity thus allowing the calculation of the wave speed and wave intensity (dI). The analysis was also performed combining pressure and velocity changes at the same location, ultimately comparing the area results (dI_A=dlnAdU) against the pressure gold standard (dI_P=dPdU).

b) *In vivo* data

CMR scans were performed on 12 patients, both prior to BCPC (age: 0.2-2.8 years, BSA: 0.28±0.04 m^2) and prior to Fontan completion (age: 0.7-3.7 years, BSA: 0.55±0.07 m^2). Functional data including ejection fraction (EF%), indexed end-diastolic and end-systolic volume (iEDV, iESV), and indexed stroke volume (iSV) were calculated from CMR data. Furthermore, wave intensity data were derived from aortic flow PCMR acquisitions using a semi-automatic house-written plug-in. Analyzing the net wave intensity patterns, peak early systolic forward compression wave (FCW) and peak end systolic forward expansion wave (FEW) were obtained as surrogates for ventricular contractility and relaxation, respectively. A ratio FCW/FEW was calculated, as a non-dimensional indicator of VA coupling. Additional coupling information was gathered using the quantity ESV/iSV, which is a surrogate for arterial elastance/ ventricular elastance (Ea/Ees). All comparisons were performed with paired t-test, $p<0.05$ indicating statistical significance.

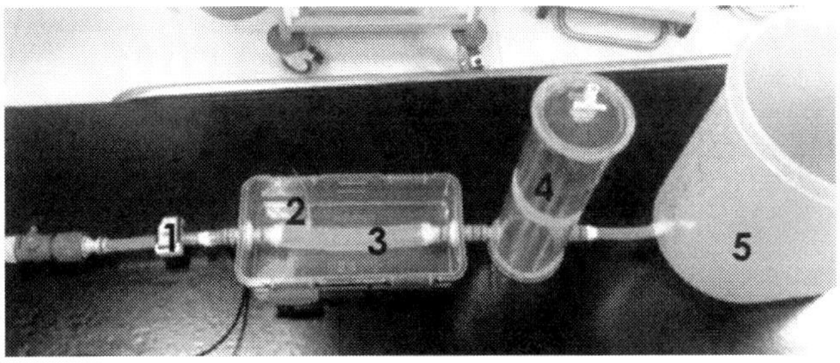

***Figure 2**: CMR-compatible experimental apparatus including 1) ultrasonic flow probe, 2) pressure measurement port, 3) compliant arterial model, 4) compliance chamber, 5) atrial chamber.*

RESULTS

Experimental data was successfully obtained and the results showed a high level of agreement between CMR and invasive data. The measured pressures and flows were extremely repeatable (mean peak values: 77.2±0.6 mmHg and 10.4±0.06 L/min, respectively). Wave speed calculations from the slopes of the Pressure-Velocity loop (7.6 m/s) and lnArea-Velocity loop (6.7 m/s) are in good agreement. Figure 3 shows the superimposition of the two sets of results, clearly highlighting extremely similar waveforms and trends.

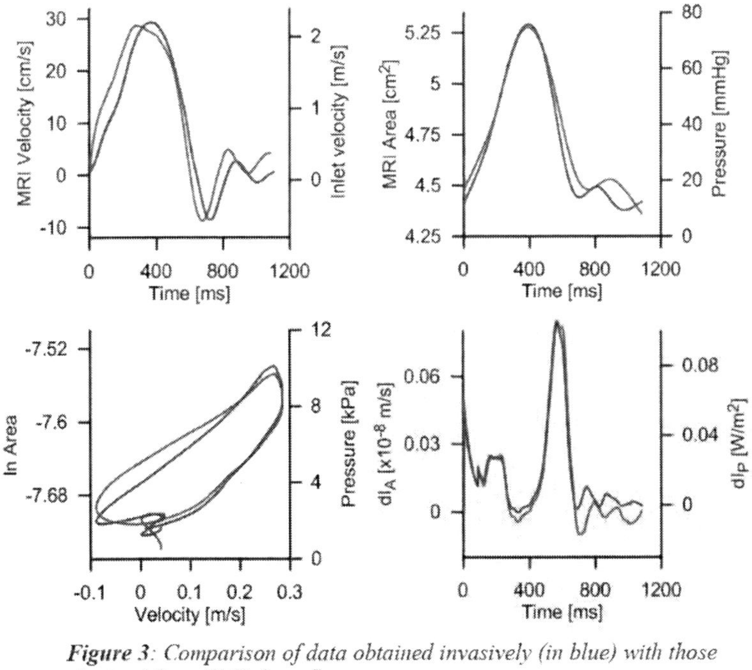

***Figure 3**: Comparison of data obtained invasively (in blue) with those obtained from CMR (in red).*

Analyzing patients' data, BCPC circulation was shown to have an overall favorable effect on VA coupling, as indicated by an increase in both FCW/FEW (3.0±1.7 vs. 8.1±5.8, p=0.006) and in Ea/Ees (0.28±0.14 vs. 0.42±0.15, p=0.006). As a result of the hemodynamic changes following BCPC, an overall improvement in EF was registered (52±13% vs. 59±7%, p=0.024). These changes are summarized in Figure 4. However, a change in volumetric data was not appreciated in this population (iEDV: 82±17 vs. 84±12 mL/m^2, p=0.9; iESV: 39±12 vs. 36±8 mL/m^2, p=0.3).

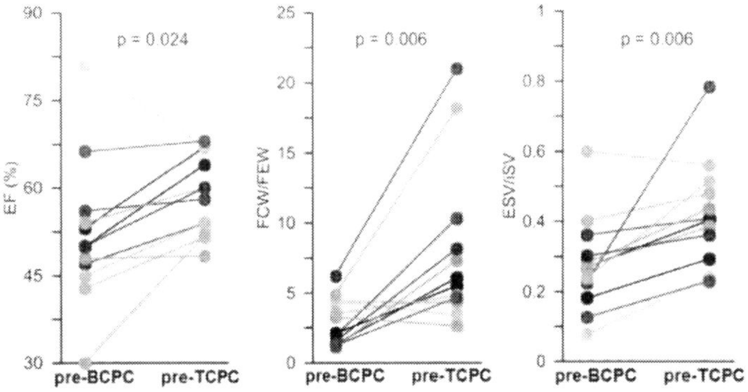

***Figure 4**: Comparison of data prior bidirectional cavopulmonary connection (BCPC) and total cavopulmonary connection (TCPC).*

DISCUSSION

In this abstract we have shown that non-invasive measurements of area and velocity show good agreement with invasive measurements of pressure and velocity. This ultimately confirms that CMR is a reliable way to obtain wave intensity information, allowing us to evaluate VA coupling in clinical scenarios where CMR data is available. This validation is further enhanced by the wave speed calculations, whereby linear fitting of the slope of the early (30 ms) portion of the loop –prior to the arrival of reflected waves– shows good agreement between the lnArea-Velocity loop and the Pressure-Velocity loop. The material of the tube provided reasonable distensibility, which ultimately led to the wave speeds being within an accepted range for arteries. This validation experiment now enhances confidence towards the use of this method to obtain and analyze coupling information. From these preliminary results we can confidently mark it as a reliable technique to calculate wave speed and take the analysis forward including wave separation and Windkessel analysis [3].

The insightfulness of the technique is confirmed by the analysis of HLHS data also presented here. Indeed, non-invasive quantification of VA coupling based on CMR data demonstrated that BCPC improves the inherent VA coupling mismatch in palliated HLHS patients, clinically intuitive but not previously quantified. It appears that rather than a specific improvement in right ventricular performance, favorable systemic vascular adaptation to the volume unloaded circulation may facilitate better VA coupling, which can lead to improved ventricular function.

REFERENCES

[1] Parker. *Med Biol Eng Comput* 2009; 47 (2): 175-88

[2] Biglino et al. *J Cardiovasc Magn Reson* 2012; 14:57

[3] Tyberg et al. *Med Biol Eng Comput* (2009) 47:221–232

SB³C2015
Summer Biomechanics, Bioengineering and Biotransport Conference
June 17-20, 2015,Snowbird Resort, Utah, USA
SB³C2015-461

CELLS ALTER TRACTION FORCE AND ORIENTATION IN RESPONSE TO LONG-TERM CYCLIC STRETCH

Heather A. Cirka (1), Qi Wen (2), Kristen L. Billiar (1)

(1) Department of Biomedical Engineering
Worcester Polytechnic Institute
Worcester, MA, USA

(2) Physics Department
Worcester Polytechnic Institute
Worcester, MA, USA

INTRODUCTION

Cell traction forces are essential to numerous cell processes such as migration and extracellular matrix assembly. Cells generate traction forces through the actin-myosin machinery of the cytoskeleton which is transmitted to the ECM via adhesion complexes. In static culture, many concomitant factors affect cell traction force generation including substrate stiffness, cell area, cell shape and presence of soluble factors. With stretch, the cell's ability to generate traction force changes rapidly and depends on magnitude [1,2], direction, and duration of stretch [1], as well as a cell's orientation with respect to stretch direction [3]. A well-documented response to cyclic uniaxial stretch of long duration (hours) is cell reorientation away from the stretch direction to the direction of zero strain [4]. However, little is known regarding the evolution of cell traction forces during reorientation. The goal of this study was to create a method for assessing changes in traction force due to long-term equibiaxial and uniaxial cyclic stretch. We hypothesized that cyclic equibiaxial stretch would decrease cell traction force (after return to the unstrained state) as consecutive stretch/release cycles have been shown to continually induce decreases in cell traction force [1]; whereas following uniaxial stretch, the cells would regain the static levels of traction force once they reorientated to the zero-stretch direction (essentially static).

METHODS

Stretch System and Compliant Culture Well Development: A custom stretch system and compliant culture well was developed to apply long-term cyclic stretch to the cells. The stretch device consists of four micro-controlled stepper motors on a low-profile aluminum base. Custom compliant culture wells were optimized for uniformity of strain field using finite element analysis. Strain field was confirmed with High-Density Mapping. Molds for the wells were designed in Solidworks and machined using a CNC machine polycarbonate sheets. Wells were cast from PDMS. **Tunable-stiffness Traction Force Substrate Preparation:** Polyacrylamide (PA) gels were attached to the PDMS substrates using protocol developed previously [5]. As soft gels are harder to attach, a stiffer gel (7.5 kPa) was first attached and then a softer gel (0.6 kPa) was layered on top. **Cell Culture:** Valvular Intersitial Cells were cultured at 10% CO2, 37°C, in standard media. Cells in passage 3-4 were used for all experiments. Cells were seeded at a density of 2,500

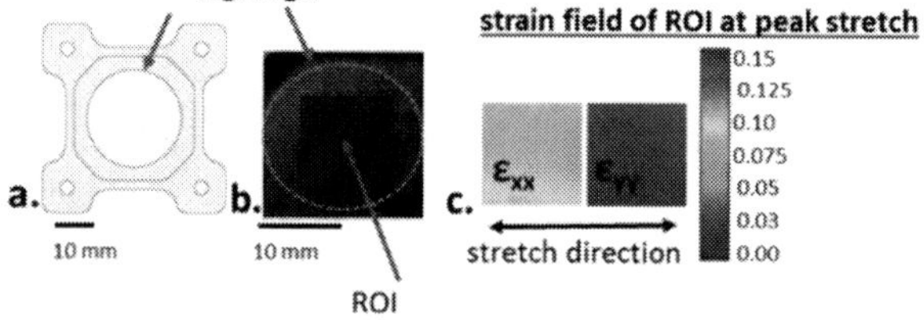

Figure 1- (a) Schematic of custom well with polyacrylamide gel attached. (b) HDM strain field analysis of region of interest on polyacrylamide gel. (c) ε_{xx} and ε_{yy} of ROI for 10% uniaxial stretch.

cells/cm^2 and allowed to attach overnight prior to stretch. TGF-β and Blebbistatin were used in two experiments to modify initial tension state of the cell. **Stretch Regimens:** Cells were stretched at 10% uniaxial or 10% equibiaxial at 1 Hz for 24 hours. **Traction Force Measurements:** Images were captured using a Zeiss inverted microscope and CCD camera and analyzed using standard methods.
Statistics: All values reported are mean ± standard deviation.
Measurements between static and stretch groups were compared using

two-tailed student t-tests assuming equal variance. A p-value less than 0.05 is considered statistically significant.

RESULTS

It has been well documented that cells exhibit and avoidance reaction with uniaxial stretch and reorient to the direction of minimal strain [9]. We expected the traction force changes with equibiaxial stretch to be more dramatic than those of uniaxial stretch as the cells would not be able to 'avoid' the strain impulse. In agreement with our hypothesis VICs that were stretched 10% equibiaxially at 1 Hz for 24 hours significantly decreased their cell generated traction force ($p<0.05$). Areas of high substrate stress, seen in static cultures under cell filipodia, were eliminated in stretched cells, and the maximum substrate stretch decreased ($p<0.05$). There was a downward trend in cell area and cells became more rounded with equibiaxial stretch ($p<0.05$). As expected, cells did not have a preferential alignment angle following equibiaxial stretch.

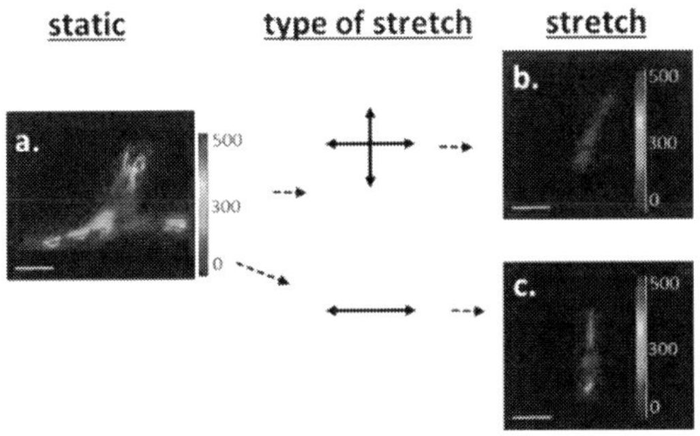

Figure 2 (a) representative figure of stress plot for a cell under static culture. High areas of substrate stress localized underneath filopodia. (b) Representative figure of a stress plot for a cell undergoing equibiaxial stretch and (c) uniaxial stretch. High areas of substrate stress are markedly reduced with stretch. Scale bars are 50 μm.

VICs that had been subjected to uniaxial stretch also had a statistically significant decrease in cell generated traction force ($p<0.05$) which was surprising as these cells are able to reorient to the direction of zero strain. This decrease in traction force (43%) was not as large as the decrease in traction force seen with equibiaxial stretch (67%). Maximum substrate stress also significantly decreased ($p<0.05$) as did cell area ($p<0.05$) although shape factor and elongation did not change.

VICs are very sensitive to presence of exogenous TGF-β in the media and differentiate into myofibroblasts, which generate more traction force compared to fibroblastic controls [10]. To modify initial tension levels in the cell, 5 ng/mL TGF-β1 was added to the media to promote differentiation to myofibroblastic phenotype. Static cells cultured for seven days in the presence of TGF-β1 generated nearly twice as much force as untreated controls. Despite the apparent shift in phenotype, cells treated with TGF-β decreased cell traction force with uniaxial stretch similar to nontreated cells. There was also a trend of decreased filopodia as quantified by an increase in shape factor ($p<0.05$), although elongation ratio did not change. In addition, TGF-β treated cells, when stretched, reoriented to a similar degree as untreated stretched cells.

Initial tension levels were decreased by culturing cells in the presence of 10 μM blebbistatin. VICs treated with blebbistatin had smaller spread area than untreated controls, and exhibited a very thin, spindle shape. There were no observed changes in cell traction force, maximum substrate stress, shape factor, or elongation, although there was a slight, yet significant increase in cell area ($p<0.05$). VICs treated with blebbistatin reoriented parallel to stretch.

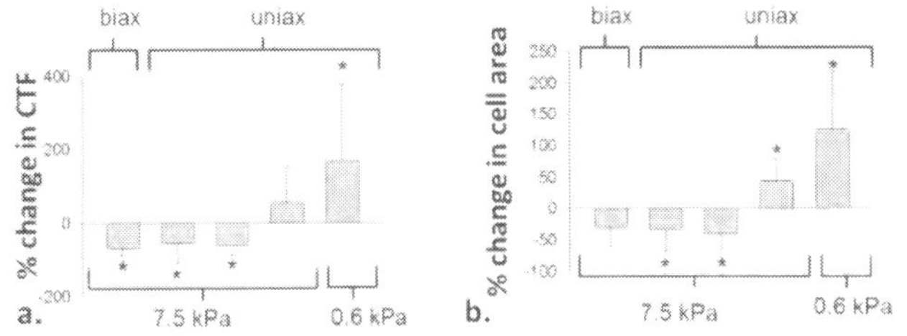

Figure 3- (a) % change in cell traction force caused by 24 hours of cyclic stretching for experiments. * indicates statistical significance (b) % change in cell area caused by 24 hours of cyclic stretch.

To decrease initial tension levels but leave cytoskeletal machinery intact, cells were cultured on 0.6 kPa gels. Static cells had a rounded morphology and very low traction force. VICs significantly increased in area, cell-generated traction forces, and elongation, and significantly decreased in shape factor, ($p<0.05$). These results indicate that cell traction force response to stretch is dependent upon initial substrate stiffness and/or tension level.

DISCUSSION

We report for the first time, changes in cell-generated traction force with long-term (24 hr) cyclic stretch on both soft and stiff substrates. Both cell types appear to alter their prestress level such that they maintain optimal stress fiber tension when stretched in agreement with theories of cell reorientation [6].

Changes in traction force were greater in equibiaxially stretched cells than uniaxially stretched cells since cells stretched equibiaxially are not able to reorient away from stretch to maintain optimal stress fiber tension. α- smooth muscle actin (αSMA) incorporation into the stress fibers with TGFβ stimulation increase basal tension but does not abrogate the decrease in traction force with stretch. With decreased myosin efficiency and thus prestress, cells orient parallel to the direction of stretch presumably to increase the stress fiber tension towards the optimal level. Surprisingly, cells on soft substrates which have similarly low prestress in static culture reorient away from stretch. This response could possibly be due to cells spreading in the direction of stretch initially, then reorienting to an angle which provides optimal stress fiber tension.

Although cell area changes have been observed with long-term cyclic stretch in previous studies [5], this study is the first to quantify cell traction force. Future work includes establishing the timecourse of changes in the prestress and evaluating protein changes indicative of phenotypic changes.

ACKNOWLEDGEMENTS

We would like to acknowledge support from American Heart Association Pre-doctoral Fellowship and Korner Family Fellowship.

REFERENCES

[1] Krishnan, R. et al., *PloS one* 4.5 (2009): e5486.
[2] Mann, J. et al. *Lab on a Chip* 12.4 (2012): 731-740.
[3] Shao, Yue, et al. Integrative Biology 6.3 (2014): 300-311.
[4] Faust, Uta, et al. PloS one 6.12 (2011): e28963.
[5] Quinlan, A. et. al *PLoS one* 6.8 (2011): e23272.
[6] Tondon, A. and Roland, K. *PloS one* 9.2 (2014): e89592.

SB³C2015
Summer Biomechanics, Bioengineering and Biotransport Conference
June 17-20, 2015,Snowbird Resort, Utah, USA
SB³C2015-462

CAPTURING DETAILED 3D MITRAL VALVE GEOMETRY FOR COMPUTATIONAL VALVE MODELING

Charles H. Bloodworth (1), Eric L. Pierce (1), Thomas F. Easley (1), Milan Toma (1), Morten O. Jensen (1), Ajit P. Yoganathan (1)

(1) Wallace H Coulter Department of Biomedical Engineering, Georgia Institute of Technology and Emory University, Atlanta, GA

INTRODUCTION

The mitral valve (MV) sits between the left atrium and ventricle of the heart and passively forces one-way flow of oxygenated blood through the left side of the heart, facilitating flow from the lungs to the rest of the body. When the left ventricle contracts to force blood through the aorta, the leaflets of the MV stretch and coapt to from a seal between the left atrium and ventricle. Insufficiencies in the MV lead to a decrease in the flow of blood to the body, an increase in work load on the ventricle, and, if left untreated, heart failure and death can eventually result. Functional Ischemic Mitral Regurgitation affects around two million people in the United States, and because prevalence of functional mitral disease increases with age, frequency of co-morbidity is increased [1]. The preferential treatment of MV disease has shifted from valve replacement to valve repair, due to a decrease in operative mortality, improved left ventricular function, and higher rates of long-term survival [2]. Along with this shift, a plethora of repair devices and solutions have entered the market, with numerous new surgical techniques.

The multitude of surgical options necessitates computational models of the MV, which would allow surgeons and engineers to evaluate repair devices, procedures and concepts before performing them, or moving on to more costly testing modalities [3]. Constructing, tuning, and validating these models relies upon extensive *in vitro* characterization of valve's structure, function and response to change. Micro-Computed Tomography (μCT) allows a high level of detail, when it comes to characterizing the geometry of the MV. We discuss here the development of a novel technique for treating MV specimens in preparation for μCT scanning. The technique provides a resulting MV geometry which is significantly more detailed, and closer to its physiological diastolic geometry.

METHODS

Experimental Setup. Ovine MVs are excised and mounted onto the annulus plate of the extensively validated Georgia Tech Left Heart Simulator (GTLHS) [4,5]. Using ultrasound imaging to view valve closure parameters under pulsatile flow, the PM positions are adjusted to achieve healthy closure. These positions are then recorded, and the chamber is dismounted from the pulse-duplicator in preparation for μCT imaging.

Bead Application for Strain Tracking. Before re-assembling the cylindrical chamber, high-contrast glass beads are applied in a grid to the leaflets of the mounted MV using a small amount of tissue adhesive (Loctite 4014, Henkel Corporation, Westlake, OH). The technique, which has been developed in parallel to the work in this study, allows for accurate location tracking of discrete points on the soft tissue between different μCT scans. Using positions of these markers, strain of the tissue can be calculated between μCT scans. This will be used to quantify the leaflet surface geometry of the MV before and after the treatment technique discussed below.

Micro-CT Imaging in the Diastolic Configuration. First generation methods developed by our group [6] have resulted in high-resolution μCT scans of the MV in an open state, however, computational simulations have since shown that these 3D data sets do not accurately represent the MV in a diastolic state. Here, we propose a method to mitigate some of the complications caused by imaging moist tissue in air. The two largest sources of error when scanning the MV in air are the hydrostatic forces of the residual moisture on the valve apparatus, and the loss of the tissue's neutral buoyancy in fluid. These phenomena lead to a 'bunching' effect on the leaflets and chordae tendineae, which leads to 3D data sets in which the leaflets appear smaller and thicker, and the chordal trees appear as bulky marginal chordae with minimal branching.

Prior to imaging in the diastolic state, the MV is lightly fixed under diastolic flow in a 0.1% Gluteraldehyde and water solution. The left heart chamber in which the MV is mounted is sealed and inserted into a closed loop with a steady flow pump (FIGURE 1). The pump maintains a peak diastolic flow of approximately 20 liters/minute through the left heart chamber and across the MV. The flow opens the valve and spreads out the chordae tendineae, as the Gluteraldehyde solution slightly stiffens and dehydrates the tissue. This works to counteract the phenomena discussed above. To produce a high resolution digital 3D geometry of the valve, the MV is then imaged by an Inveon μCT Scanner (Siemens Medical Solutions USA, Inc., Malvern, PA) set to image soft tissue (80 kV energy, 500 μA intensity, 650 ms integration time) at 43.29 μm isotropic voxel resolution.

To validate this new technique, a grid of high contrast glass beads were applied to the anterior leaflet, and the MV was imaged by μCT in two configurations before performing the fixation treatment. First, the MV was imaged in air. Next, the GTLHS was filled with water and the MV was imaged while submerged. This "submerged" state is assumed to be the ground truth geometry of the unstrained MV (note that only the beads are discernable in this image, as the MV tissue cannot be resolved in water). Finally, the valve is fixed as described above and imaged in air. The centroids of the beads were then found for each image, and the areal green strains from ground truth were calculated for the fresh and fixed states [7]. These strain fields represent the change in membrane area that the anterior leaflet experiences when taken out of water, or when fixed and taken out of water.

Data Processing. After scanning, a DICOM stack of each data set is created using Inveon Research Workplace. Segmentation and 3D reconstruction are then performed using InVesalius (InVesalius, São Paulo, Brazil). Using two different threshold levels, the surface of the valve, and the marker array are reconstructed into two separate 3D data sets. These are refined using Geomagic Studio (3D Systems, Cary, NC).

RESULTS

Preliminary results (FIGURE 2) show that the strain magnitude of the AL from ground truth to the fixed state ($-25.5 \pm 11.5\%$) is significantly lower than that to the fresh state ($-41.7 \pm 6.4\%$) for $n = 6$ points on one valve sample ($p<0.05$). The grid of six points represents the surface geometry of the AL, with negative strain indicating shrinkage from ground truth.

Additionally, more detail in the chordal tree can be seen after fixation. The smaller marginal chordae can be resolved from strut chordae from which they branch, yielding a 3D image that captures the finer detail of the subvalvular apparatus (FIGURE 3).

DISCUSSION

Computational models of the MV are useful in many ways, such as clinical diagnostics, surgical planning, device design, and device validation. The technique described here aim at providing more physiologically accurate and detailed inputs and validation points for these models. The fixation technique yields a leaflet membrane size that is significantly closer to that seen in a physiological diastolic reference state, and an unprecedented amount of detail in the chordal trees is achieved. However, our methods have certain limitations. Detail in chordal trees of the commissural leaflets is still lacking. The fixation technique provides a leaflet membrane surface area that is significantly closer to that of the ground truth state, yet still varies from a true unstrained reference state. Work is also ongoing to address these limitations and further characterize the technique on a larger sample size of mitral valves. In the coming months, results of ongoing computer simulations with data produced using these new methods are expected to show improved simulation fidelity.

ACKNOWLEDGEMENTS

Funding was provided by the NIH under grant R01 HL092926, and by the FDA under critical path grant FDA-SOL-11199843

REFERENCES

[1] Acker, MA, el al., *N Engl J Med,* 370:23–32, 2014.
[2] Milano, CA, et al., *Ann Thorac Surg,* 86:735-744, 2008.
[3] Sun, W, et al., *Annu Rev Biomed Eng,* 16:53–76, 2014.
[4] Rabbah, JPM et al., *Ann Biomed Eng,* 41:305-315, 2013.
[5] Bloodworth, CH et al., *BMES Annual Meeting,* 2014.
[6] Toma, M et al., *World Congress on Comp Mech,* 2014.
[7] Rausch, MK et al., *J Biomech,* 44:1149–1157, 2011.

FIGURE 1. SCHEMATIC OF THE STEADY FLOW SET-UP FOR VALVE FIXATION

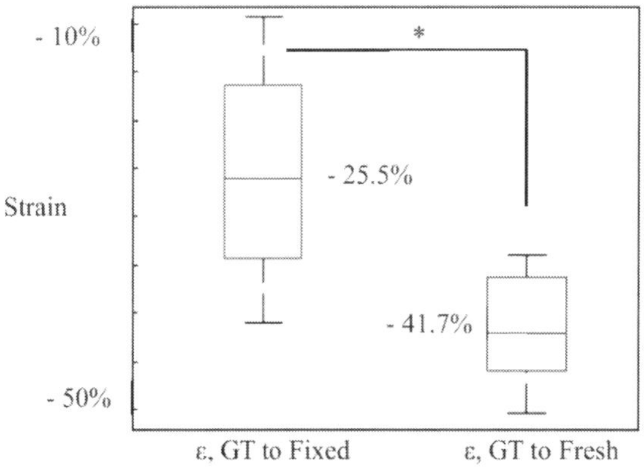

FIGURE 2. AREAL STRAIN FROM GROUND TRUTH (GT) OF THE ANTERIOR LEAFLET FOR FRESH AND FIXED STATES. STRAIN MAGNITUDE IN THE FIXED CASE IS SIGNIFICANTLY SMALLER THAN THAT IN THE FRESH CASE (* p<0.05).

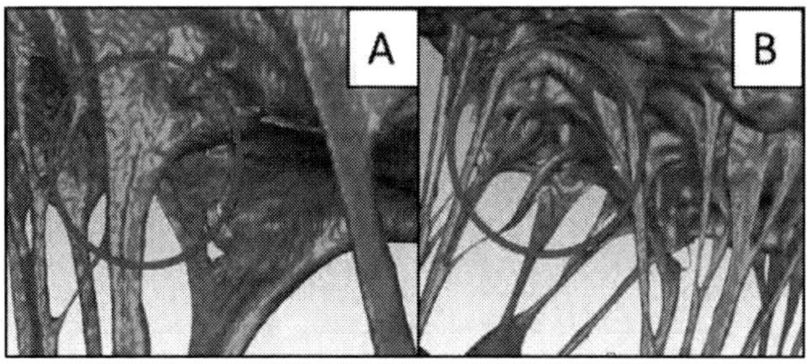

FIGURE 3. IMAGES OF THE STRUT CHORDAE AND CHORDAL TREE INSERTING INTO THE ANTERIOR LEAFLET, COMPARING CHORDAL DETAIL ACHIEVED WITH (B), AND WITHOUT (A) FIXATION

LEFT-VENTRICLE ASSIST DEVICE ANASTOMOSIS HEMODYNAMIC ANALYSIS USING DIRECT NUMERICAL SIMULATIONS (DNS) AND LARGE EDDY SIMULATIONS (LES)

Ricardo J. Bonilla-Alicea (1), W Robert Taylor (2), Alessandro Veneziani (3)

(1) Bioengineering
Georgia Institute of Technology
Atlanta, GA, USA

(2) Division of Cardiology
Department of Medicine
Emory University School of Medicine and
Atlanta VA Medical Center
Atlanta. GA. USA

(3) Mathemathics/CS
Emory University
Atlanta, GA, USA

INTRODUCTION

Left Ventricle Assist Devices (LVADS) aid the affected heart by suctioning blood from the left-ventricle and pumping it into the systemic circulation through the thoracic aorta. Destination Therapy (DT) LVAD use is ever increasing due to enhanced patient prognosis, making LVADS a viable option to heart transplantation. Unfortunately LVAD use has been associated with different pathologies such as renal dysfunction **(1)** and Von Willebrand disease **(2)**, which suggests the need of further understanding of the non-physiologic hemodynamic environment promoted inside of the thoracic aorta. One important aspect of the LVAD procedure is the location of the LVAD output graft anastomosis in the native aorta and its relationship to the resulting hemodynamics. By combining the use of medical imaging data with Computational Fluid Dynamics (CFD) we can objectively study LVAD blood flow dynamics by using patient specific aortic geometries. The geometry of the patient is obtained by performing a manual segmentation procedure on medical images obtained using Computerized Tomography (CT) scans. The resulting hemodynamics from five different LVAD output graft locations have been studied to evaluate the effect of the different positions. These results are compared with non-LVAD simulation where all the flow enters through the native aortic valve.

Boundary conditions are crucial when performing patient-specific simulations due to the limitations of obtaining in-vivo patient data. The Windkessel Lump Capacitance RCR model is used to model the outlet boundary conditions. The Windkessel model represents the cardiovascular section with an electrical circuit with resistance and capacitance elements to incorporate blood flow resistance and vessel compliance. The values for the resistance and capacitance elements have been obtained from literature data and are kept constant throughout the different simulations. The irregular shape of the patient-specific aortic geometry coupled with the high velocities produced by the LVAD output graft requires fine numerical meshes to obtain correct flow fields. In order to reduce the computational cost of performing such simulations, a Large Eddy Simulation (LES) approach has been used. Because of the clinical application of this work, low computational costs are a significant part of the project as clinicians need their results in the shortest amount of time possible.

METHODS

Patient specific geometries are obtained by performing a manual segmentation of the medical images using ITK-SNAP. The geometry is then reconstructed and smoothed by using the Vascular Modeling Toolkit (VTK) software package. The geometry includes the aortic valve, the ascending and descending aortic sections, the Innominate Artery and the Left Carotid and Subclavian Arteries. The resulting Stereolitography (STL) geometry file is then fitted with a numerical mesh using NETGEN, along with the definition of the boundary conditions.

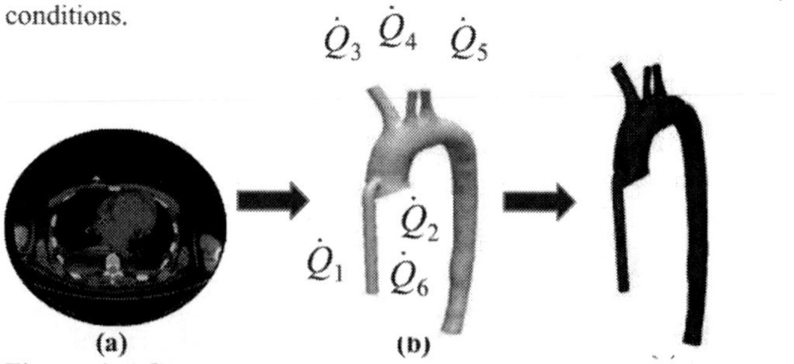

Figure 1. Geometry reconstruction process: medical image for aortic reconstruction, numerical mesh and boundary conditions

The geometry is then uploaded in the Library of Finite Elements (LIFEV) code, which solves the incompressible Navier-Stokes equations inside of the aortic geometry coupled with the inlet and outlet boundary conditions. The simulations are performed by using two different approaches, DIRECT NUMERICAL SIMULATIONS (DNS) and LARGE EDDY SIMULATIONS (LES). DNS attempts to solve all eddies up to the dissipative (Kolmogorov) scales which requires a numerical mesh with spacing h on the order of such scales. The dissipation scales are inversely proportional to the Reynolds number value (Re):

$$Re=\frac{\mu UD}{\rho} \tag{1}$$

The Re number values for the simulations presented in this work are able to be performed using a DNS approach without the need of a numerical modeling approach, such as Reynolds Averaged Navier Stokes (RANS). Underresolving of the dissipation scales results in higher convective values as the diffusion scales are not correctly solved.

In LES, the energy carrying eddies are solved while the smaller dissipative eddies are filtered. The LES model used in this simulations is based on the Leray model, where two coupled equations are solved, the NS equations with the filtered velocity term and an additional equation that solves for the unfiltered velocity term. The numerical filter used in the simulation is defined as:

$$\eta=Re^{\frac{3}{4}}L \tag{2}$$

By modeling smaller eddies the need for a fine mesh with size on the order of the dissipation scales is eliminated, allowing the use of coarser numerical meshes which significantly reduces the computational cost of the simulations.

Post-processing of the simulation results is achieved by calculating relevant hemodynamic indices that allows characterization of the flow field. The quantities with their respective definitions are the following:

1. **Wall Shear Stress (WSS)**

$$\tau=\mu\frac{du}{dy} \tag{3}$$

WSS is an important parameter when studying pro-inflammatory vessel conditions. The increased velocities produces by the LVAD device greatly increase near wall velocity gradients.

2. **Max Velocity Magnitude (V_{max})**

This is simply the maximum velocity magnitude of the blood flow inside of the geometry. The maximum velocity is expected to occur near the LVAD output graft outlet due to the reduced graft diameter when compared with a healthy native aortic valve diameter.

3. **Oscillatory Shear Index (OSI)**

The OSI measures the local recirculation of blood. High OSI values are associated with pro-inflammatory vessel conditions and it flow stagnation.

4. **Supra Aortic Volumetric Flow Splitting**

This measures the volumetric flow splitting to the upper arteries and how the position of the LVAD output graft is affecting this volumetric flow splitting. These quantities are visualized using the PARAVIEW Visualization Toolkit. Velocity and pressure contours along different planes in the geometry, in addition to velicity streamlines are calculated for each of the simulations. The volumetric flowrate values presented in tables 1 and 2 have been normalized by the inlet flowrate values of 5 liters per minute.

RESULTS

The preliminary results shown were obtained by performing a DNS simulation on a single anastomosis location. A total of five simulations were performed for this geometry. The first simulation termed Pre-Operative (Pre-Op) doesn't have the LVAD graft attached and blood flow enters through the native aortic valve. For the LVAD simulated cases, the geometry has two inlet locations, the LVAD output graft (Q_1) and the native aortic valve (Q_2). The total inlet blood flow entering through the LVAD graft and the aortic valve is kept fixed at 5 liters per minute, where the flowrate entering through the AV is expressed as a percentage of the total flowrate ω as ω=0 (0%), ω=0.1 (10%) and ω=0.2 (20%). Four simulations were performed using DNS and an additional LES simulation (ω=0) was performed using a mesh with 50% less. The simulations were ran in parallel using a total of 4 cores for a total of two heart periods (1.5s).

DNS	Q_3^*	Q_4^*	Q_5^*	Q_6^*	V_{max}	WSS_{max}
Pre-Op	0.28	0.078	0.101	0.54	33	39.5
ω=0	0.28	0.16	0.16	0.423	286	237
ω=1	0.199	0.093	0.11	0.608	202	192
ω=2	0.201	0.095	0.11	0.635	187	170

Table 1. Normalized outlet volumetric values, V_{max} and WSS_{max} for different operating conditions using DNS

LES	Q_3^*	Q_4^*	Q_5^*	Q_6^*	V_{max}	WSS_{max}
ω=0	0.25	0.13	0.15	0.49	310	251

Table 2. Normalized outlet volumetric values, V_{max} and WSS_{max} for ω=0 using LES

Δ	Q_3^*	Q_4^*	Q_5^*	Q_6^*	V_{max}	WSS_{max}
ω=0	11%	18%	6.2%	6.7%	8.4%	5.9%

Table 3. Percent difference between DNS and LES results for ω=0

Four additional anastomosis locations will be studied for the **Pre-Op, ω=0, ω=1** and **ω=2** with an **LES** approach. The **Pre-Op** simulation provides the most desirable hemodynamic flow field for a given aortic geometry.

DISCUSSION

Flow through the LVAD cannula results in higher velocity magnitudes and WSS values due to the reduced graft diameter when compared with the native aortic diameter. The maximum velocity values increase between 6-9 fold under LVAD operation, resulting in WSS values that differ greatly from normal physiologic conditions. Operation of the aortic valve reduces the amount of blood going through the LVAD graft, showing reduced velocity values as the value of ω is increased. In addition to beneficial hemodynamics, it has been shown that partial operation of the AV reduces the risk of placque formation and clotting in the AV **(3).** Volumetric splitting is not significantly affected by varying the value of ω as 0, 0.1 and 0.2.

By using the coarser mesh combined with LES, the average iteration time was reduced from 900s to 600s. Table 3 shows the normalized percentage difference between the DNS and LES results. The LES model is underdissipative when compared with the DNS model, which results in higher velocity and WSS values. Still, the results are promising as this were obtained with a reduction of approximately 33% in the iteration time. The next step is to perform LES simulations of additional patient specific aortic geometries to evaluate the effect of the different anastomosis positions on the resulting hemodynamic field.

REFERENCES

[1] Tarbell, J.M et al., *Annual Reviews of Fluid Mechanics,* 46:591-614, 2014

[2] Butler, J. et al, *Annals of Thoracic Surgery,* 81(5):1745-51, 2006

[3] John, R et al, *J of Heart and Lung Transplant,* 29(12):1321-9,2010

SB³C2015
Summer Biomechanics, Bioengineering and Biotransport Conference
June 17-20, 2015,Snowbird Resort, Utah, USA
SB³C2015-464

IMPACT OF PRE-STRETCH INDUCED SURFACE ANISOTROPY ON AXONAL REGENERATION

Chun Liu (1), Seungik Baek (2) and Christina Chan (1,3)

(3) (1) Chemical Engineering and Materials Science
Michigan State University
East Lansing, MI, USA

(2) Mechanical Engineering
Michigan State University
East Lansing, MI, USA

(3) Computer Science and Engineering
Michigan State University
East Lansing, MI, USA

INTRODUCTION

Spinal cord injury (SCI) is a leading cause of disability worldwide. In the United States, the prevalence of SCI is ~250,000 out of 300 million with 11,000 new incidences each year [1]. SCI is characterized by an acute traumatic lesion of neural elements in the spinal canal, and results in temporary or permanent deficits in sensory and motor function. After acute SCI, axons sprout briefly, but soon undergo growth arrest and retraction. The causes of failure in regeneration include lack of growth factor secreted by associated glial cells, and the lack of orientation of neurites inside the scar tissue, which prevents neurites from crossing into the site of lesion. Transplantation of Schwann cells or olfactory ensheathing cells has been shown to induce axon regeneration to some extent and partial recovery of motor and sensory functions. However, the regeneration is found to be a chaotic process due to unorganized axon alignment [2]. Therefore, to obtain sufficient regenerated axons to bridge the lesion site, properly organized axonal alignment is necessary.

The ability to use mechanical cues such as substrate stiffness, topological features, and stretch to control neural cell growth has great potential in neural regeneration post SCI [3]. In particular, biomaterials that are used to fabricate scaffolds and implantable substrates for nerve regeneration are currently intensively investigated. To elicit axon outgrowth from regenerated neurons, topological features have been reported to provide contact guidance [3]. As a result, significant efforts have been invested in developing techniques that promote axonal alignment and neural regeneration. In the case of topological guidance, aligned fibers and channels have been demonstrated to facilitate axon alignment [3]. However, reported techniques for inducing axon alignment through topological features, such as fibers, channels and patterning, were unable to promote both the length and thickness of the axons. Some scaffolds, when transplanted *in vivo*, also encounter difficulties in penetrating the axons through the channels, mainly due to a reactive cell layer that develops, thereby abrogating the regeneration process.

We previously explored surface anisotropy on a static mechanical pre-stretched poly-dimethyl-siloxane (PDMS) membrane, and predicted the effective stiffness that the cells feel in the stretched direction versus perpendicular direction [4] using a theory named "small deformation super imposed on large" [5]. We showed that surface anisotropy influenced cell alignment. Here we report that static pre-stretch not only induced surface anisotropy but also enhance axon alignment and growth. Additionally, we investigated myelination of dorsal root ganglion (DRG) neuronal axons by Schwann cells (SCs) on the pre-stretch induced anisotropic PDMS surface.

METHODS

Pre-Stretched PDMS Substrate Preparation: PDMS substrate was cured in polystyrene tissue culture dish by mixing a 10:1 solution of base and curing agent and pouring the mixture into the dish with a thickness of 1mm and subsequently plasma treated. A piece of rectangle membrane (5cm×3.5cm) was fixed on a stretching frame and stage. The membrane was evenly stretched by 10% in the longer axis.

PDMS Channel Substrate Preparation: Channel patterns of different sizes (50um, 100um, 150um and 200um) using photolithographic technique was used to fabricate micropatterned silicon wafers, which were used as molds to transfer microgrooves to PDMS substrates. PLL was added to the chamber and incubated for 2 hours before seeding with DRG neurons.

Immunocytochemistry: Immunocytochemistry was performed at room temperature on cells seeded on the stretched surface for 6, 12, and 21 days. Cells were rinsed with PBS, followed by fixation with 4.0% paraformaldehyde in PBS for 15 min, rinsed 3 times in PBS, then permeabilized with 0.1% Triton X-100 in PBS for 15 min and washed 3 times with PBS. After washing, cells were blocked in 1% BSA for 30 minutes and incubated with mouse Anti-beta III Tubulin primary antibody, or rabbit anti-P0 primary antibody, and then incubated with Alexa Fluor 488 goat anti-mouse or anti-rabbit Alexa Flour 546 secondary antibody. DAPI was used to visualize the nucleus. Confocal laser scanning microscopy (CLSM) images were obtained with Olympus FluoView1000 laser scanning confocal microscope using 20X objectives.

Quantification of Axon Orientation Distribution and Axon Thickness: Image-Pro Plus Version 4.5 was used to measure the axon orientation angles. Before measurements, the fluorescent images were converted into high contrast binary images to remove the background noises. A straight line was drawn perpendicular to the stretch direction which was used as the reference to indicate zero degree orientation. For each single axon, the orientation was measured by drawing a straight line to follow the axon pathway, and the angle between the traced line and the reference line was automatically given by the software and recorded for further analysis. All experiments were repeated at least three times. Two-sample Kolmogorov-Smirnov test were applied to evaluate the statistical significance. For axon thickness analysis, we enlarge each image to 10 times, and measured the axon thickness by pixels, which were converted into micron unit.

RESULTS

Axon alignment on pre-stretched surface: Figure 1 shows a comparison of axon orientation on stretched and unstretched PDMS substrates after 6, 12, 21 days' culture. Axons orient parallel to the stretched direction, whereas they show random orientation on the unstretched surface. The axon orientation was quantified by determining the percentage of axons that aligned at each 10 degree angle (data not shown). The axons oriented in the stretch direction and did not change significantly over time, while on the unstretched surface the axons oriented more randomly. Two-sample Kolmogorov-Smirnov test applied to evaluate the statistical significance found the distribution of cell orientations on the unstretched samples for days 6, 12, and 21 were statistically different from their respective stretched samples ($p < 0.05$).

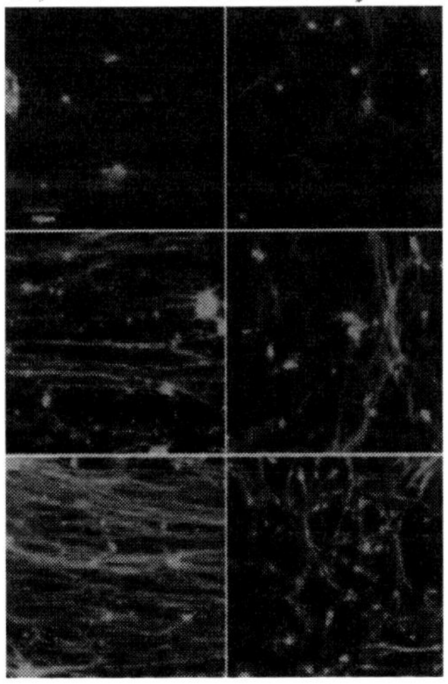

Figure 1. Axon alignment on pre-stretched and unstretched surface. Fluorescent images of DRG neurons seeded on the static pre-stretched (left) surface for 6 (top), 12 (middle), 21 (bottom) days, compared with DRG cells seeded on the un-stretched (right) surface.

Enhanced axon thickness on pre-stretched surface: In addition to axon alignment, we found that axons on the stretched surface were thicker than on the unstretched surface. Figure 2 showed the axons of DRG neurons on pre-stretched and unstretched surface after 6 days, 12days, and 21days' of culture, which were quantified (data not shown). The average thickness of the axons increased significantly on both the stretch and unstetched surfaces with time. However, the average axon thickness after 21 days culture is 1.5 times thicker on stretched than unstretched surface.

Co-culture of Schwann cells with aligned axons: Schwann cells are able to migrate into areas of damaged central nervous system and remyelinate the demyelinated axons. Since the myelination is dependent on the axon thickness, thicker axons are preferentially myelinated, thus we co-cultured SCs with aligned axons. After culturing on stretched and unstretched surfaces for 2weeks, purified SCs were added into the chamber and cultured for another week. The co-culture results were shown in Figure 3, axons were labeled by Tuj1 staining, and SCs were stained by P0. P0 is a marker of mature SCs, and provides an indication of myelination. On the stretched surface, there is a significant co-localization of green and red, suggesting SCs attached and aligned with axons on stretched surface, but on the unstretched surface, the green and red color are randomly distributed, demonstrating that the pre-stretched surface not only enhance axon thickness, but also increased the attachment of Schwann cells to the regenerated axons, which is critical for myelination to initiate.

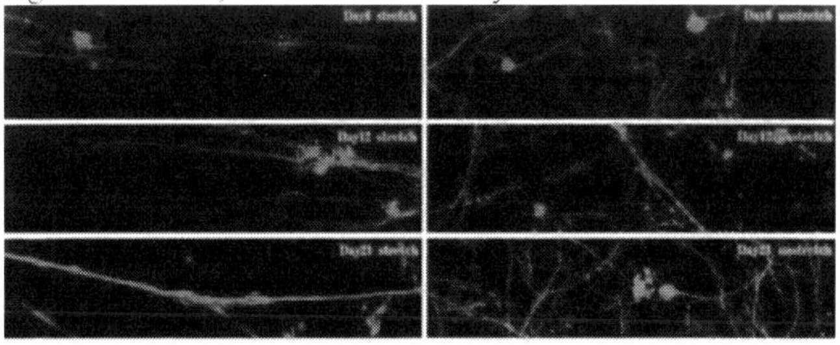

Figure 2. Fluorescent images of DRG axon thickness on stretched vs. unstretched PDMS substrate after 6, 12, and 21 days.

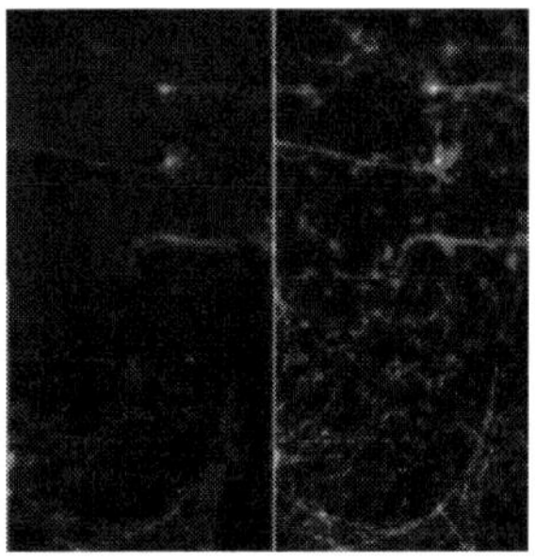

Figure 3. Co-cultured SCs with aligned axons on strreched vs. unstretched PDMS substrates. ***Top*** *(left) Tuj1 (green) staining for axons on stretched surface, (right) Overlay of Tuj1 (green), P0 (red) and DAPI (blue) staining on stretched surface.* ***Bottom*** *(left) Tuj1 (green) staining for axons on unstretched surface, (right) Overlay of Tuj1 (green), P0 (red) and DAPI (blue) staining on unstretched surface.*

DISCUSSION

Mechanical stretch has been applied in tissue engineering to guide cell behavior. It is especially important for inducing cell alignment in connective tissues such as tendons and ligaments, cardiac muscle tissues and blood vessel endothelium. Along these lines, we created, alternatively, a static pre-stretched anisotropic surface, in the absence of dynamic external mechanical stimuli that also induced cell alignment. The results of this study showed for the first time that surface anisotropy induced by pre-stretch alone induced not only cell alignment but also enhanced axon thickness, which subsequently promoted myelination of the DRG neuronal axons by SCs, critical to the regeneration process in SCI.

ACKNOWLEDGEMENTS

This study was supported in part by the NSF (CBET 0941055), the NIH (R01GM079688, R01GM089866, R01EB014986).

REFERENCES

[1] Bauchet, L. et al., *Ann Phys Rehabil Med*, 52:330-351, 2009.
[2] Li, Y. et al., *Science*, 277:2000-2002, 1997.
[3] Berns, EJ. et al., *Biomaterials*, 35:185-195, 2014.
[4] Liu, C. et al., *Cellular & Molecular Bioeng.*, 7:106-121, 2014.
[5] Baek, S. et al., *Comp. Meth. Appl. Mech.s. & Eng.*, 196:3070-78, 2007.

SB³C2015
Summer Biomechanics, Bioengineering and Biotransport Conference
June 17-20, 2015,Snowbird Resort, Utah, USA
SB³C2015-465

QUANTIFICATION OF PLAQUE SHIFT AFTER CORONARY BIFURCATION STENTING

**F. Iannaccone (1,2,*), C. Chiastra (1,3,*), F. Migliavacca (3),
F.J.H. Gijsen (1), P. Segers (2), E. Regar (4), M. De Beule (2,5), J.J. Wentzel (1)**

(1) Department of Biomedical Engineering, Thoraxcenter, Erasmus University Medical Center, Rotterdam, The Netherlands

(2) IbiTech-bioMMeda, Department of Electronics and Information Systems, Ghent University, Belgium

(3) Laboratory of Biological Structure Mechanics (LaBS), Department of Chemistry, Materials and Chemical Engineering "Giulio Natta", Politecnico di Milano, Milan, Italy

(4) Department of Interventional Cardiology, Thoraxcenter, Erasmus MC, Rotterdam, The Netherlands

(5) FEops bvba, Ghent, Belgium

INTRODUCTION

Percutaneous coronary intervention (PCI) of bifurcation lesions still represents a challenge for interventional cardiologists due to technical difficulties and associated high event rates [1]. The provisional side branch (PSB) stenting technique, which involves stent placement in the main branch (MB) with optional treatment of the side branch (SB), is nowadays the routine procedure to treat bifurcation lesions. Despite the high success rate of this technique, MB stent implantation can aggravate SB ostial stenosis, inducing SB compromise, which is considered one of the main procedural complications of PCI in bifurcations [2].

Although the mechanisms leading to SB compromise have not yet been fully clarified, the shift of the plaque or the carina towards the SB during stent expansion (i.e. plaque/carina shift) plays a fundamental role in this adverse event [2]. The aim of this study is to investigate the influence of the plaque composition and the distal angle of the bifurcation on plaque/carina shift due to stent placement using a virtual implantation environment.

METHODS

An idealized parametric model of a diseased left anterior descending (LAD) coronary artery with its diagonal branch was created using the open source software pyFormex (nongnu.org/pyformex). The bifurcation has stenoses in its proximal MB, distal MB, and SB. Particular attention was paid to the actual features of the bifurcation (Fig. 1): i) the diameter of the branches was defined according to Finet's law [3]; ii) the stenosis accounted for an increased plaque burden due to the outward growth of the plaque in the first stages of the disease [4]; iii) the plaque eccentricity; iv) a degree of stenosis suitable for stenting (60%-60%-60%); v) the curvature of the bifurcation on the heart. Material properties of the three layered arterial wall and plaque components were retrieved from *ex vivo* tests [4,5]. Three models were generated with different distal angles (45°, 57.5°, and 70°, respectively) and with 2 different plaque components: hard (i.e. fibrous) and soft (i.e. lipid) plaques, resulting in 6 different cases. A Multi-Link 8 stent (Abbott Lab., USA) was virtually implanted in all models following the PSB technique. The simulations of stent deployment (Fig. 2) were carried out using the finite element software ABAQUS (Dassault Systèmes Simulia Corp., USA) with a similar procedure as described by Mortier et al. [6].

The effect of the stenting procedure on the plaque/carina shift was investigated by virtually replicating the volumetric analysis

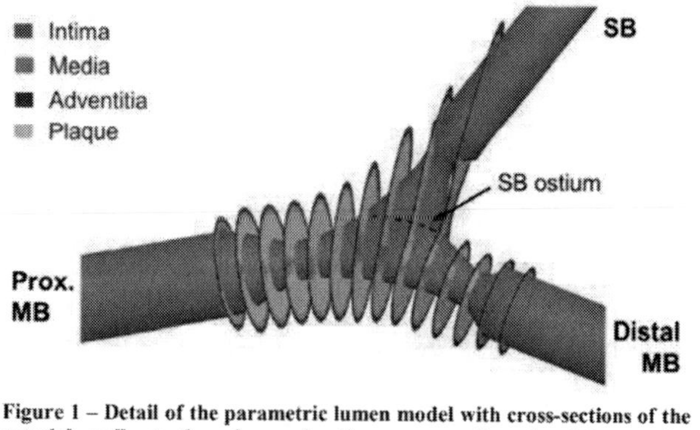

Figure 1 – Detail of the parametric lumen model with cross-sections of the arterial wall at the plaque location representing the left anterior descending coronary artery with its first diagonal branch.

* Authors equally contributed to the work.

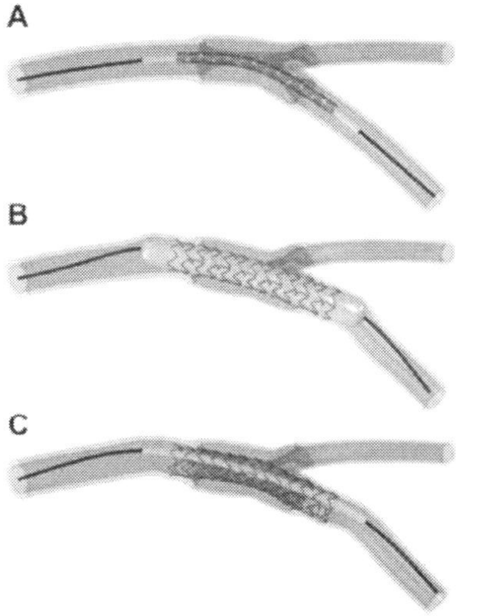

Figure 2 – Example of virtual structural simulation of the provisional side branch technique: A) Stent positioning in the main branch; B) Stent expansion by balloon inflation; C) Balloon deflation and stent recoil.

performed by Xu et al. on 44 patients' coronary bifurcations [2]. A region of the SB of 5 mm (from the SB ostium) was considered to compute SB ostial compromise (i.e. lumen volume decrease in the SB ostial segment after MB stenting), plaque shift (i.e. plaque volume increase in the SB ostial segment after MB stenting), and carina shift (i.e. vessel volume decrease in the SB ostial segment after MB stenting). In addition, the post-implantation lumen and vessel areas of 10 cross-sections in the same region were compared with the corresponding pre-implantation values.

RESULTS AND DISCUSSION

The simulations provide interesting insights in the effects of anatomical variation of the coronary bifurcation PSB stenting. The plaque/carina shift is influenced both by plaque composition and bifurcation angle.

The volumetric analysis (Fig. 3A) shows that a higher distal angle is associated with lower SB ostial compromise. In particular, a higher distal angle, i.e. a larger polygon of confluence, results in a lower plaque shift, which is minimal in the model with distal angle of 70°. Carina shift occurs only in the case with distal angle of 70°. Soft plaques magnify the effects already visible in hard plaques.

Differences between pre- and post-stenting SB cross-sectional lumen areas are negligible. However, visual inspection shows that the SB sections undergo a profound shape modification after stenting (Fig. 3C) as the device pushes the carina towards the SB resulting in an oval shape of the SB lumen. This effect might be a possible explanation of good FFR values even when the SB lumen appears very narrowed on the top view as in an angiographic-like view (Fig. 3B). The vessel projection on the angiographic images may not display the view of the minimal diameter leading to an overestimation of the outcome assessment. These results are also in agreement with previous clinical studies [2] in which the effect hereby reproduced by our numerical approach is clearly shown on intravascular ultrasound (IVUS) images of patients' coronary bifurcations.

CONCLUSIONS

The realistic 3D numerical strategy presented in this work can help to understand the mechanisms of coronary stenting in a safe, reproducible environment, and provides information that can be missed based on conventional angiography. The presented quantitative results highlight that plaque composition and anatomical differences of the LAD might be good predictors of the outcome and give insights of possible side effects due to plaque/carina shift; thus this approach can aid the physicians in defining their treatment strategy.

ACKNOWLEDGEMENTS

The authors gratefully acknowledge the ERC starters grant (BioCCora ♥ 310457) for the financial support of this research.

REFERENCES

[1] Lassen, JF et al. *EuroIntervention*, 10:545-560, 2014.
[2] Xu, JY et al. *Circ Cardiovasc Interv*, 5:657-662, 2012.
[3] Finet, M et al. *EuroIntervention*, 3:490-498, 2008.
[4] Glagov, S et al. *N Engl J Med*, 316:1371-1375, 1987.
[5] Holzapfel, GA et al. *Am J Physiol Heart Circ Physiol*, 289: H2048-2058, 2005.
[6] Loree, HM et al. *Arterioscler Thromb J Vasc Biol Am Heart Assoc*, 14:230-234, 1994.
[7] Mortier, P et al. *JACC Cardiovasc Interv*, 7:325-333, 2014.

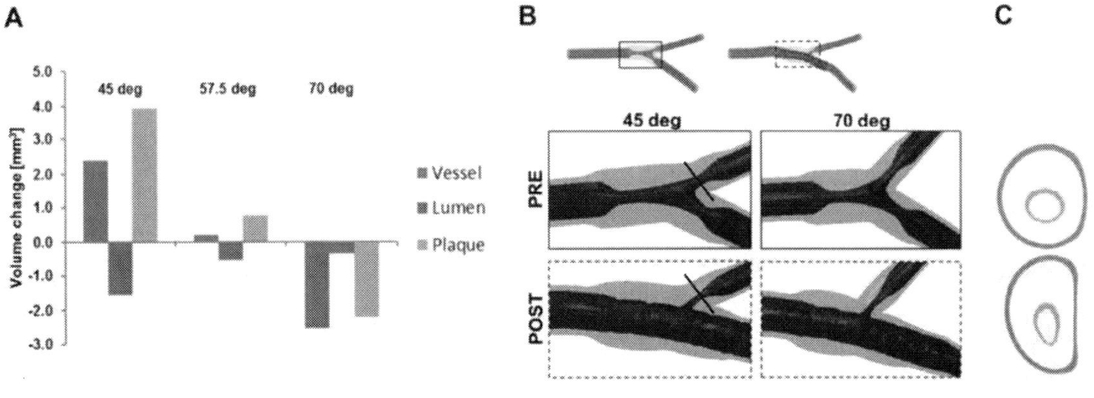

Figure 3 – Evaluation of the results: A) Volume changes of vessel, lumen, and plaque in the side branch ostial region after provisional stenting for the three cases with lipid plaque; B) Comparison of pre- and post-stenting geometries for the cases with lipid plaque with distal angles of 45° and 70°; C) Representative pre- and post-stenting side branch cross-sections at the position indicated in B), showing the lumen change after stent deployment in the main branch.

SB^{3}C2015
Summer Biomechanics, Bioengineering and Biotransport Conference
June 17-20, 2015,Snowbird Resort, Utah, USA
SB^{3}C2015-466

A STUDY ON THE INTERPLAY OF SEX AND CYCLIC STRETCH IN AORTIC VALVE CALCIFICATION

Shirin Masjedi (1), Zannatul Ferdous (1)

(1) Department of Mechanical, Aerospace and Biomedical Engineering
University of Tennessee
Knoxville, TN 37996, USA

INTRODUCTION

Aortic valve disease is a major cardiovascular disorder that affects approximately 25% of the elderly population. It has been well established that male sex is a risk factor for aortic valve disease [1][2]. However, the underlying mechanism relating male sex to increased aortic valve calcification remains unclear. A number of studies have related the sex steroid hormones and their receptors to the sex-related differences on calcification progression and have studied the effect of hormonal changes at tissue and/or cellular levels [3]. Moreover, regulatory proteins, known as the cardiovascular-selective estrogen modulators, play a key role in the sex-difference in cardiovascular disorders by governing the transcription process after activating the DNA in cardiovascular cells [4]. Similarly, a microarray analysis study on male and female porcine aortic valvular interstitial cells (VICs) identified 183 genes that are significantly different in male versus female VICs [5]. Several of these expressed genes contribute to pathways identified in calcific aortic valve disease. In addition to sex, several studies have shown a direct relationship between hypertension and aortic valve disease [6-8].

In the present study, we aim to determine if the aortic VICs isolated from male and female rats develop different amounts of calcification when cultured under representative physiological (10%) or pathological (15%) levels of cyclic stretch. The results from this study would provide crucial information at cellular level to understand if the VICs isolated from different sex demonstrate different amounts of calcification under cyclic stretching.

METHODS

9 week old Male and Female Sprague-Dawley rats were purchased from Harlan laboratory. The cell isolation protocol was approved by the Institutional Animal Care and Use Committee at the University of Tennessee-Knoxville. Briefly, the rats were euthanized using CO_2 inhalation and the aortic valves were removed from the rat hearts using a dissecting microscope. Each aortic valve was carefully cut into 5-6 pieces, then placed into a culture plate and covered with media consisting of DMEM, 10% FBS, 1% l-glutamine and 1% penicillin/streptomycin. The VICs from each sex were used for experiments at passage 7.

In order to study the effect of cyclic stretch on calcification, the VICs isolated from each sex were plated in collagen-I coated Bioflex® 6-well culture plates. First, the cells were cultured under static condition for two days for the cells to adhere to the plates. For experimental samples, calcifying culture media was then added consisting of beta-glycerophosphate, ascorbic acid and dexamethasone in regular culture medium. The culture plates were then placed into a commercial Flexcell® system (Flexcell International Inc.) and stretched under 10% and 15% cyclic stretching. The controls wells under stretch and static controls were cultured in regular media.

All the samples were collected for analysis after 7 days of stretching. Calcific nodule formation was assessed for each sample using Alizarin red staining. Image analysis was performed using Image J software for the stained images. Statistical significance was determined using one-way ANOVA. A p-value of ≤ 0.05 was considered significant. Furthermore, the samples were assessed for cell density using a Fluorometric DNA assay, for matrix metalloproteinase (MMP)-2 using gelatin zymography, and for other calcific markers using von kossa staining and alkaline phosphate (ALP) enzyme activity assay. The ALP activity was normalized to protein content measured using BCA assay (data not shown here).

RESULTS

Our results on calcific nodules measured using alizarin red staining show greater amount of calcification in male VICs than females. Figure 1 shows representative images of calcific nodules in male and female VICs cultured under 10% and 15% cyclic stretch. Figure 2a shows the percentage of area covered with calcific nodules and 2b shows their integrated density normalized to their respective controls under stretch for male and female VICs . We found that the nodule parameters were significantly greater for pathological stretch (15%) only in male VICs. Similarly, there was a significant difference in nodule parameters between male and female VICs under 15% stretch.

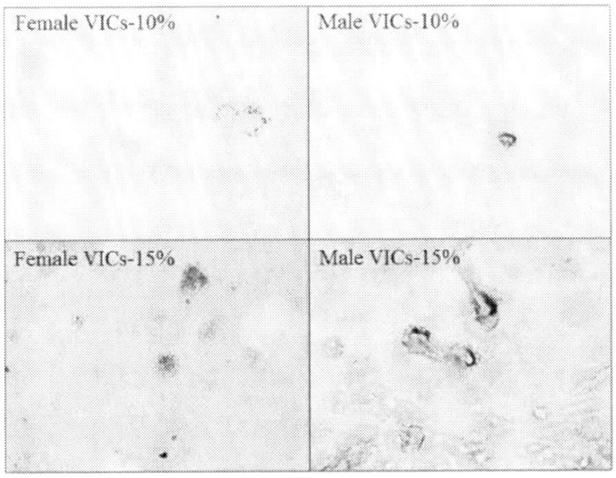

Figure 1. Calcific nodules in male and female VICs as measured using alizarin red staining. Images magnification was 20X.

a)

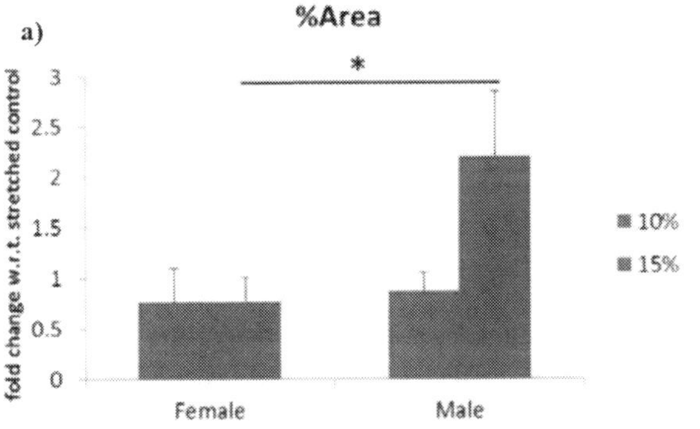

b)

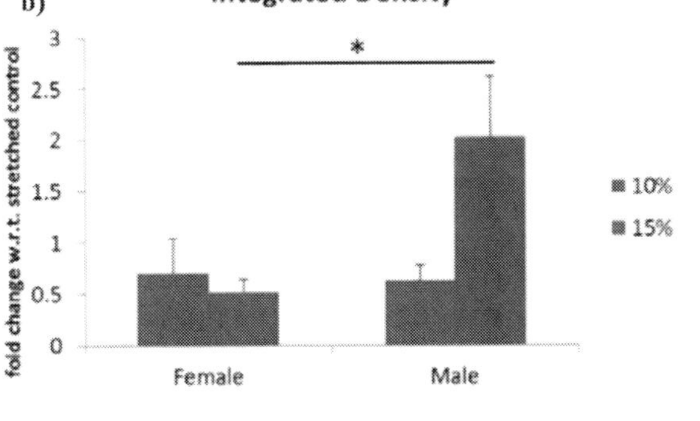

Figure 2. (a) Area percentage, (b) Integrated density of calcific nodules in male and female VICs under cyclic stretch normalized to respective stretched controls, as measured using alizarin red staining.

DISCUSSION

In this study, we observed that male VICs form greater amount of calcific nodules compared to female VICs under pathological levels of cyclic stretch. The nodule formation is an indicator of early cell accumulation, which triggers cell differentiation into diseased phenotypes. Our results support than male is at higher risk of calcification than female, and this risk is greater under pathological stretch as observed in hypertension. Further investigation on these cell types is required to fully understand the interplay between sex and cyclic stretch in aortic valve disease.

.

REFERENCES

[1] Ortlepp, JR et al., *Heart*, 89:1019-1022, 2003.
[2] Stewart, BF et al., *J Am Coll Cardiol*, 29(3):630-634, 1997.
[3] Billon-Gales, A et al., *PNAS*, 108(32):13311-13316, 2011.
[4] Mendelson, ME & Karas RH, *Science*, 308:1583-1587, 2005.
[5] McCoy, CM et al., *PloS one*, 7(7):e39980, 2012.
[6] Kostkiewicz, M et al., *Int J of Cardiol*, 71:57-61, 1999.
[7] Ferdous, Z et al., *Biomaterilas*, 32:2885-2893, 2011.
[8] Balachandran, K et al., *Cardiovasc Pathol*, 21(3):206-213, 2012.

STRAIN-DEPENDENT DEGRADATION AS A MECHANISM FOR THE PARADOXICAL EFFECTS OF MECHANICAL LOADING ON COLLAGEN FIBER ALIGNMENT IN HEALING TENDON

William J. Richardson (1,3), Stavros Thomopoulos (4), Jeffrey W. Holmes (1,2,3)

(1) Department of Biomedical Engineering
(2) Department of Medicine
(3) Robert M. Berne Cardiovascular Research Center
University of Virginia
Charlottesville, VA, USA

(4) Department of Orthopaedic Surgery
Washington University
St. Louis, MO, USA

INTRODUCTION

The structural arrangement of collagen fibers is a key determinant of the mechanical properties of many tissues, including tendon. Following injury, healing tendons typically possess lower strength and stiffness than healthy tendons, increasing the risk of re-injury. This reduction in material properties following injury and healing is in large part due to decreased alignment of collagen fibers along the primary axis of loading.

The production and organization of collagen fibers is developmentally driven by mechanical forces, which provide directional cues for cells to produce a highly aligned extracellular matrix. Paradoxically, this is not necessarily the case in healing tendon. Across a range of experimental models, increased loading consistently increased extracellular matrix content in healing tendons. However, these studies also revealed an unexpected, biphasic relationship between loading and collagen alignment: while total unloading led to poorly aligned and weak tendons as expected [1-2], exercise also led to poorly aligned and weak tendons [3-4].

One potential explanation for the lower collagen alignment in exercised tendons is damage: fibers aligned along the axis of loading experience the greatest strains and would therefore be mostly likely to rupture, while collagen fibers aligned farther from the loading axis would remain. An intriguing alternate potential mechanism is strain-dependent degradation (SDD) – the experimentally observed biphasic relationship between fiber strain and the susceptibility of collagen to enzymatic degradation (Figure 1) [5]. In the current study, we used a computational model to compare how each of these mechanisms would affect the relationship between mechanical loading and collagen structure in tendons healing under cyclic uniaxial strain, and assessed the degree to which these predictions agreed with available data.

METHODS

To predict the effects of SDD or damage on long-term collagen structure, we adapted a previously published agent-based model of collagen remodeling [6]. This model simulates many individual fibroblasts ("agents") in a collagen matrix, with each cell responding to local mechanical and structural cues that determine cell orientation, migration, collagen production, and collagen removal at each time step. Cells orient with a vector average of local cues, including principal strain direction and local fiber direction, and deposit collagen parallel to their own orientation at a rate that increases with increasing strain (linear 2-fold increase from 0% to 8% strain). Most parameters in the model are based on independent *in vitro* experiments, yet the model reproduces measured collagen fiber structures in healing myocardial infarcts in the rat [6]. In the present study, we modified the model so that collagen removal rate depended upon strain, simulating either SDD or damage at high strains (Figure 1).

We simulated a 400 μm x 400 μm region initially containing randomly oriented collagen (3% area fraction) and populated with ~500 randomly oriented fibroblasts. We predicted the time course of matrix remodeling over 8 weeks of uniaxial strain of varying magnitude (transverse strain = 0%). Fiber alignment was assessed using circular statistics and reported as the mean orientation vector length (0 = random, 1 = aligned).

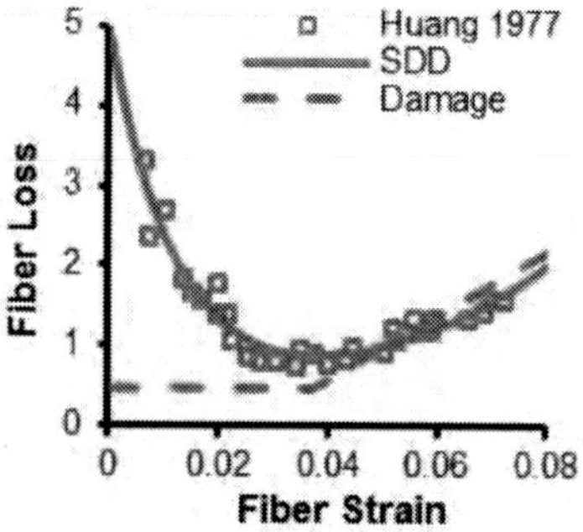

FIGURE 1. COLLAGEN FIBER DEGRADATION AND DAMAGE AS FUNCTIONS OF STRAIN.

RESULTS

Over 8 weeks of simulated healing, all levels of cyclic strain increased both collagen content and alignment compared to the completely unloaded case (i.e., zero strain) (Figure 2 A/B). In the case of strain-dependent degradation (SDD), collagen content increased with strain over most of the range simulated, reaching a plateau at high strains. Collagen alignment showed a biphasic response, sharply increasing from 0% to 4% strain, then gradually decreasing from 4% to 8% strain. The shape of this response corresponds to the inverse of the experimental degradation vs. strain curve (Figure 1).

In the case where the simulated probability of fiber damage increased at higher strains, both collagen content and collagen alignment showed biphasic responses to strain, which peaked at 4% strain corresponding to the selected threshold above which fiber damage begins to occur (Figure 2 A/B).

While both SDD and damage mechanisms produced similar biphasic trends of collagen alignment vs. strain, the two models predicted different shapes for collagen fiber orientation histograms at higher strains (Figure 2 C/D). The SDD model predicted a gradual broadening of the histogram up to 6% strain. By contrast, the damage model predicted selective loss of fibers oriented along the loading axis, producing histograms with two peaks separated by a central trough.

DISCUSSION

Using a previously published agent-based model of wound healing, we examined the effect of two mechanisms by which strain could control collagen fiber alignment. Healing was simulated for 8 weeks under cyclic uniaxial stretch at varying magnitudes. Results demonstrated that both mechanisms – strain-dependent enzymatic degradation of fibers or strain-induced physical damage of fibers – reproduced the experimentally observed biphasic relationship between mechanical loading and collagen alignment. Although these models predicted different absolute degrees of alignment, it is likely that either could be matched to measured data by adjusting model parameters. Therefore, experimental measures of average alignment alone cannot provide sufficient information to distinguish between the two potential mechanisms. However, our simulations suggest that measuring the shape of the collagen fiber orientation histogram in future studies could be useful in separating these two mechanisms. Specifically, the model predicted that SDD should result in a broadening histogram as strain increases, consistent with reports by Thomopoulos et al. in tendons subjected to exercise during healing [3,4]. By contrast, simulating loss of fibers due to damage resulted in a sharp trough within the histograms corresponding to the selective removal of fibers experiencing the highest strain.

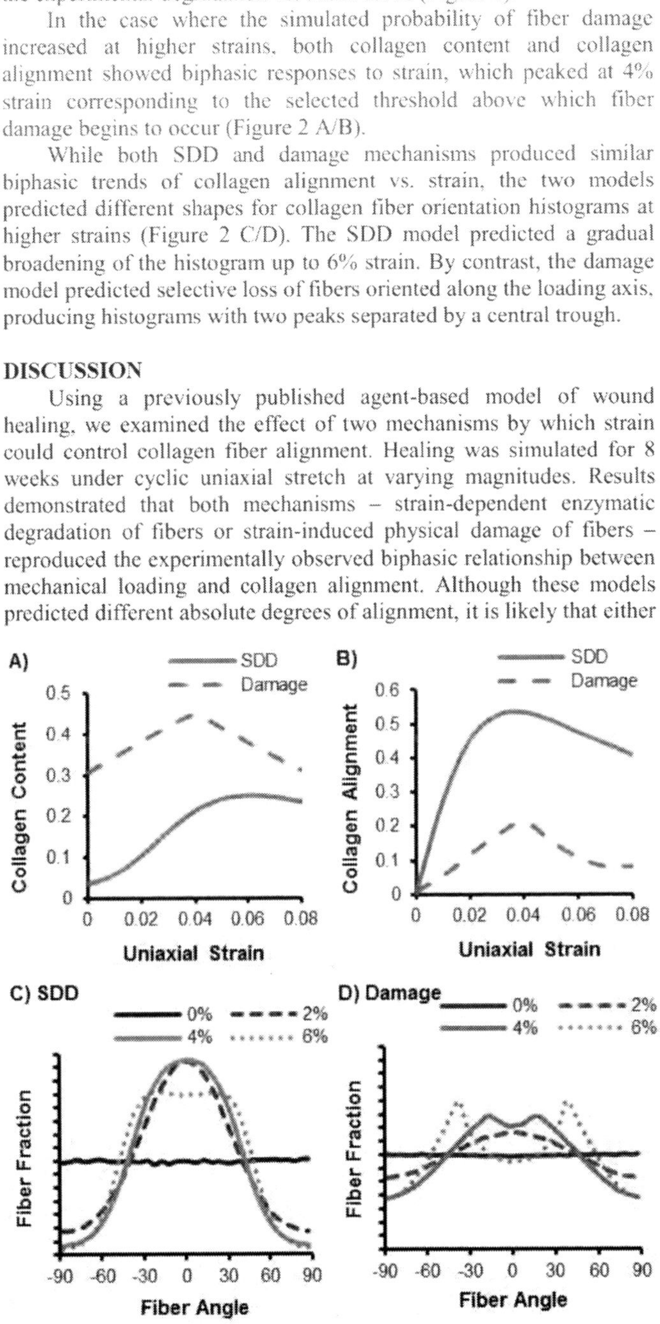

FIGURE 2. MODEL PREDICTIONS OF COLLAGEN CONTENT AND ALIGNMENT AFTER 8WKS OF HEALING UNDER VARYING LEVELS OF UNIAXIAL STRAIN.

The two models tested here also predicted very different trends in collagen content vs. strain, with the SDD model providing a better match to available data. In the SDD model, peak collagen content occurred at much higher strains than peak alignment. This is consistent with the available experimental data, which show progressive increases in collagen content across all levels of loading studied, including at levels of exercise that reduce alignment. By contrast, the damage model predicted that collagen content and alignment should decline in parallel with higher strains, a prediction inconsistent with available experimental results.

A number of important limitations of the model should be noted. First, while the strain vs. degradation rate curve employed in the SDD model was based on experimental data, the analogous curve employed in the damage model was not. Rather, thresholds and slopes were chosen for the damage curve to facilitate comparison between the two models. Accordingly, the discussion is focused on qualitative aspects of the model predictions that do not depend on the exact choice of damage parameters. Another important input to the model was the relationship between strain and collagen deposition: here, the true relationship is likely sigmoidal rather than linear, as we assumed. However, the two-fold difference in collagen production rates across the simulated strain range is consistent with multiple *in vitro* datasets [7,8]. Finally, we did not attempt to calibrate model parameters to match the measured values of collagen structure in healing tendons. While such a calibration would certainly rescale the axes of the plots presented here, it would not be expected to change the overall trends.

In summary, strain-dependent enzymatic collagen degradation was able to reproduce trends in both matrix content and alignment reported in healing tendon experiments, while physical fiber damage at high strains was not. We cannot rule out the existence of such damage, but our results suggest that damage sufficient to reduce mean alignment should be observable in the shape of fiber orientation histograms. These findings highlight the utility of agent-based models for investigating underlying mechanisms that are difficult to measure directly by analyzing their effects on global properties.

ACKNOWLEDGEMENTS

The authors acknowledge funding from the National Science Foundation (NSF/CMMI 1332530 to JWH) and the American Heart Association (14POST20460271 to WJR).

REFERENCES

[1] Galatz, L. et al., *J Shoulder Elbow Surg*, 18:669-675, 2009.
[2] Andersson, T. et al., *J Appl Physiol*, 113:1398-1402, 2012.
[3] Thomopoulos, S. et al., *J Biomech Eng*, 125:106-113, 2003.
[4] Gimbel, J. et al., *J Biomech Eng*, 129:400-404, 2007.
[5] Huang, C. et al., *J Biomed Mater Res Symp*, 8:137-154, 1977.
[6] Rouillard, A. et al., *J Physiol*, 590:4585-4602, 2012.
[7] Husse, B. et al., *AJP Reg Int Comp Phys*, 293:R1898-1907, 2007.
[8] Papakrivopoulou, J. et al. *Cardiovasc Res*, 61:736-744, 2004.

SB³C2015
Summer Biomechanics, Bioengineering and Biotransport Conference
June 17-20, 2015,Snowbird Resort, Utah, USA
SB³C2015-468

ESTIMATING AXONAL STRAIN IN WHITE MATTER USING PARANODAL PROTEINS AS FIDUCIARY MARKERS

Sagar Singh (1), Assimina A. Pelegri (2), David I. Shreiber (1)

(1) Department of Biomedical Engineering
Rutgers, The State University of New Jersey
Piscataway, NJ, USA

(2) Department of Mechanical & Aerospace Engineering
Rutgers, The State University of New Jersey
Piscataway, NJ, USA

INTRODUCTION

Traumatic injury to axons in white matter in the brain and spinal cord occurs primarily due to tensile stretch. During injury, the stress and strain experienced at the tissue-level is transferred to the axons. It is unclear how this transfer occurs, and what constituents in white matter dictate the strain experienced by axons [1]. Previous studies indicate that this transfer is influenced by axon orientation, axon geometry, and axon kinematic behavior [2]. Computational multi-scale models have been used to predict axonal strains in central nervous system (CNS) white matter tissue following tissue-level stretch [3], [4]. However, the lack of experimental data to validate model predictions remains a limiting factor, especially in predicting injury strain thresholds.

This study attempts to link axonal stretch to tissue-level by experimentally measuring length changes along the axon following macroscopic stretch. We achieve this by using contactin-associated protein (Caspr), which is expressed in axons at the paranodes [5], as a fiduciary marker to measure paranodal and internodal distances. The goals of this study were to: 1) measure internodal and paranodal length changes with tissue stretch; and 2) determine how these lengths change with development. Average changes in paranodal/internodal distances were determined for tissue at different developmental stages that was exposed to a range of tissue-level stretch. Our expectations were that paranodal and internodal distances increase with tissue-level stretch, and that the magnitude of this increase depends on the type of kinematic behavior, which reflects the stage of development and degree of myelination (Figure 1A). In our previous studies, we have demonstrated that axons display non-affine behavior early in development, when there is a lack of myelination and inter-axon coupling by glia. Axon behavior transitions to affine as myelination and glial coupling increases. We also investigated how increase in length in paranodal and internodal distances along axons differs, which would indicate non-uniform transfer of stretch. Our results show that changes in internodal and paranodal length correlate directly to tissue-level stretch at early developmental stages, where there is a lack of myelination and glial coupling. At later developmental periods, lengths continued to increase with stretch but with lower magnitude. These measurements can offer insight into how strain is transferred to axons following macroscopic stretch.

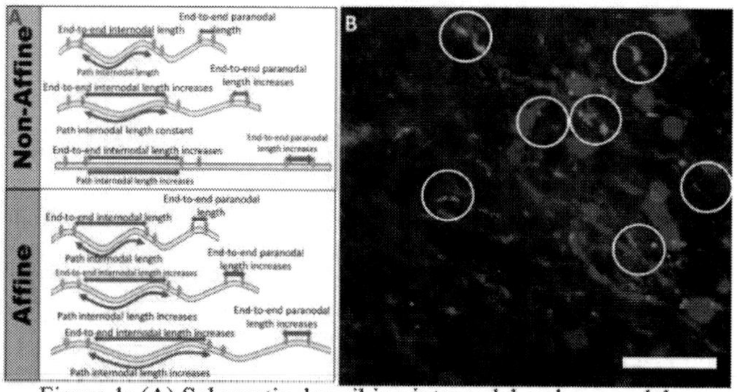

Figure 1: (A) Schematic describing internodal and paranodal lengths and differences we expect to see with kinematic behavior. (B) Immunostained section image of E18 spinal cords at 100x objective. Caspr pairs are circled. Scale bar = 20μm

METHODS

Experimental Model: Fresh chicken eggs (Charles Rivers Lab) were incubated to one of four embryonic (E) developmental stages (E12, E14, E16, and E18). Spinal cords were excised and stretched quasistatically to varying stretch levels (0%, 5%, 10%, 15%, and 20%) (N = 4 for each stretch level and developmental stage) [2]. Cords were fixed and sectioned into longitudinal 30μm thick sections using a cryostat. Sections were immunostained for Neurofilament-200kD and Caspr, and DAPI was used to label nuclei (Figure 1B).

Imaging and Analysis: Stained sections were imaged using an IX81 Inverted Olympus Microscope with a confocal spinning disc unit. Sections were imaged using an oil immersion 100x microscope lens. Image stacks were obtained in increments of 0.3μm per image through the depth of the section.

Image stacks were processed in ImageJ using the NeuronJ software plugin. Axons were traced and pairs of immunolabeled Caspr units were marked to measure paranodal distances (distance between individual Caspr stains in a pair) and internodal distances (distance between pairs of Caspr stains). Lengths were measured for axon segments that met two criteria: 1) Neurofilament staining was continuous through the length of the segment; 2) The axon length contained at least three pairs of clearly visible Caspr stains. The pathlengths and end-to-end lengths of the segments were measured.

RESULTS

Experimental Results: Internodal end-to-end length(Figure 2A), internodal pathlength (Figure 2B), and paranodal end-to-end length (Figure 2C) increased with development stage. At E12 and E14 developmental stages, internodal and paranodal distances increased linearly with stretch level ($R^2_{E12} = 0.9938$, $R^2_{E14} = 0.9930$), and the relative change in these lengths before and after stretch was roughly equal to the tissue-level stretch. At later developmental stages (E16, E18), the relative change in internodal length was consistently less than the tissue-level stretch. The mean internodal path length was constant with low-levels of stretch at E12 and E14, which is consistent with non-affine kinematics, but increased at E16 and E18. Mean paranodal distance increased with stretch level and development stage at all ages.

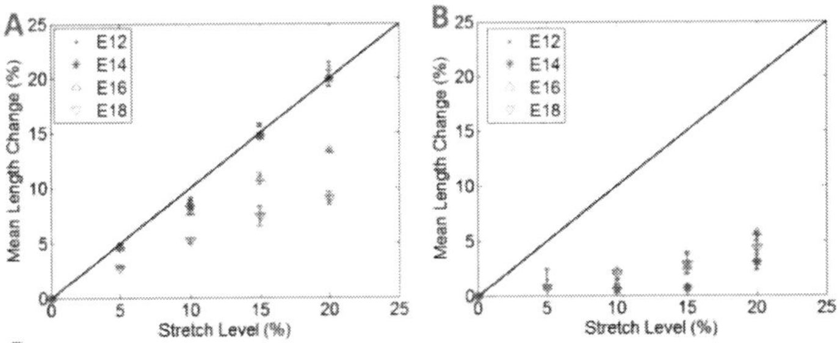

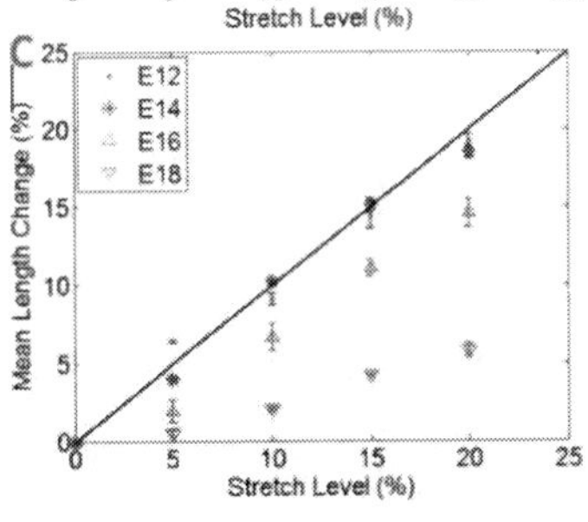

Figure 2: Mean length change was plotted against macroscopic stretch level for (A) internodal end-to-end, (B) internodal path, and (C) paranodal end-to-end distances. The black line corresponds to a 1:1 transfer of macroscopic to microscopic stretch.

DISCUSSION

This study demonstrates the use of paranodal proteins as fiduciary markers to measure how stretch is transferred from macro-to-micro elements. Caspr is expressed on axons at the Nodes of Ranvier at an early stage of development [5], making it ideal for examining changes in nodal lengths with development. We examined how paranodal and internodal distances change with stretch and development.

Based on the results of our previous studies, where axon kinematic behavior was dependent on age [2], we predicted the changes in internodal and paranodal distances differ during development. Non-affine axons first straighten before experiencing strain. Thus, we predicted that end-to-end internodal lengths would increase with stretch as the slack is taken out of the wavy axons and Caspr is displaced parallel to the direction of stretch. However, the segment pathlength would remain constant until the axon has fully straightened. Affine axons exhibit continuum-like behavior. They would experience strain while undulated, and we predicted that both internodal path and end-to-end lengths would increase at the same rate in the affine case. Measuring end-to-end and path lengths for axonal segments confirmed this, as the mean pathlength remained relatively constant at low stretch levels for E12 and E14 samples, while end-to-end length increased.

We were surprised by the divergence in change in internodal and paranodal lengths with stretch at E16 and E18. Later stages of development resulted in smaller increases of mean internodal and paranodal length than expected. One hypothesis is that the affine behavior of coupled axons leads to axons breaking at higher stretch levels at E16 and E18. To test this, we generated simple affine and non-affine finite element simulations (ABAQUS, Simulia). Greater maximum strain was predicted at the internodal regions for affine simulations, which would imply more broken axons. Depending on how broken axons behave, a greater proportion of broken axons could reduce the measured mean internodal distance. We are exploring methods to image and assess axon breakage. Paranodal distances also increased with development. The absolute amount by which these distances increased due to stretch was constant across age, which implies that kinematic behavior, and hence development, does not affect paranodal stretch.

Computational models for spinal cord and brain injury provide a method to assess trauma. The data and methods we present can potentially be used in those models to examine microscopic strain due to macroscopic events, and can provide some insight into axonal injury due to tissue-level damage.

ACKNOWLEDGEMENTS

Funding for this study was provided by the NSF CMMI Grant #1000450.

REFERENCES

[1] A. Bain and D. Meaney, "Tissue-level thresholds for axonal damage in an experimental model of central nervous system white matter injury," *J. Biomech. Eng.*, vol. 122, no.6 pp.615-22 December, 2000.

[2] H. Hao and D. I. Shreiber, "Axon kinematics change during growth and development.," *J. Biomech. Eng.*, vol. 129, no. 4, pp. 511–22, Aug. 2007.

[3] G. Karami, et al., "A micromechanical hyperelastic modeling of brain white matter under large deformation.," *J. Mech. Behav. Biomed. Mater.*, vol. 2, no. 3, pp. 243–54, Jul. 2009.

[4] Y. Pan, et al., "Finite Element Modeling of CNS White Matter Kinematics: Use of a 3D RVE to Determine Material Properties," *Front. Bioeng. Biotechnol.*, vol. 1, no. December, pp. 1–10, 2013.

[5] S. Einheber, et al., "The axonal membrane protein Caspr, a homologue of neurexin IV, is a component of the septate-like paranodal junctions that assemble during myelination," *J. cell biology*, vol. 139, no. 6, pp. 1495–1506, 1997.

SB^{3}C2015
Summer Biomechanics, Bioengineering and Biotransport Conference
June 17-20, 2015, Snowbird Resort, Utah, USA
SB^{3}C2015-469

ROLE OF SRC IN ELECTRIC FIELD-INDUCED DIRECTED CELL MIGRATION

Shun-Hao Tsao and Pen-Hsiu Grace Chao

Institute of Biomedical Engineering
National Taiwan University
Taipei, Taiwan

INTRODUCTION
Galvanotaxis, also known as electrotaxis, is a phenomenon where electric fields (EFs) direct cell migration [1]. Galvanotaxis is involved in wound healing and development [2-4], but the mechanisms are still not fully understood. Electrophoretic and electro-osmotic forces have been proposed to redistribute cell surface molecules, which lead to polarized intracellular signaling events and directed migration [5]. We previously demonstrated that lipid rafts are the primary response element of the EF-induced membrane polarization [6]. We further showed that caveolin and integrin mediate the intracellular signaling that involves polarized RhoA activation [7, 8]. Src, a non-receptor tyrosine kinase, has been shown to mediate caveolin-initiated directional migration [9] and is polarized in electrical stimulation [10]. In the current study, the hypothesis is therefore that Src mediates caveolin signaling to RhoA in EF-induced directional cell migration.

METHODS
Cell Culture
Anterior cruciate ligaments were harvested from young porcine knee joints and cut into small pieces. The explants were cultured in DMEM with 10% FBS (Invitrogen) until confluence. The isolated fibroblasts were seeded on sterile glass slides at 4×10^4 cells/cm^2 for two hours prior to migration studies. Src inhibition was performed by treating with PP2 (1 μg/ml, Calbiochem) for 2 hours. Actin disruption was achieved with cytochalasin D treatment (5 μg/ml, Cayman) for one hour.
EF Stimulation
Constant direct current (DC) EF was applied at 6 V/cm using a custom galvanotaxis system [7]. Images were taken every 15 minutes during the one-hour EF stimulation.

Fluorescence Imaging
Upon EF completion, fibroblasts were rinsed with cold DMEM and treated with 1 μg/ml Alexa-fluor 555 conjugated-cholera toxin-B (CtxB, Invitrogen) to label the GM1 gangliosides in lipid rafts, or fixed with formalin and labeled with phalloidin (Invitrogen) or antibody for caveolin-1 (Cell Signaling) and RhoA (Cytoskeleton).
Image Analysis
Cell migration were analyzed from the displacement of the centroid location of each cell. For fluorescent images, each cell was divided into four quadrants and the fluorescence intensity was measured in each region using a custom LabVIEW program (National Instruments). Asymmetry index (AI) was calculated by subtracting the intensity of the region facing anode from the region facing cathode, and normalized to the overall average intensity of the cell, where a positive value indicates anodal distribution [6].
Statistics
Statistical analyses (ANOVA with LSD post-hoc tests) were performed using SPSS (α=0.05). Results were presented as the average ± standard error of mean.

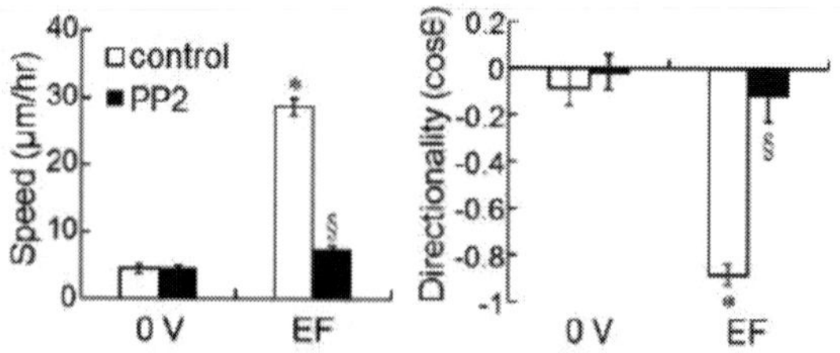

Figure 1 Effect of Src inhibition on cell migration speed and directionality (*$p<0.05$ vs 0V, §$p<0.05$ vs control, n=59-100)

RESULTS

As demonstrated previously, DC EFs stimulated directional migration toward the cathode in ligament fibroblasts (Figure 1). Lipid raft, caveolin, and RhoA all polarized toward the cathode as well (Figure 2). Inhibition of Src with PP2 abolished directed cell migration. PP2 treatment had no effect on lipid raft or caveolin polarization, but inhibited polarized RhoA distribution. EF reorganized the actin cytoskeleton to form lamellipodia in the direction of migration (cathode). Src inhibition resulted in dramatic disruption of actin and impeded cell migration. (Figure 3A). To confirm that the effect of Src on RhoA distribution was not due to cytoskeleton disruption, we also treated the cells with cytochalasin D and found only a small decrease on EF-induced RhoA redistribution (Figure 3B), suggesting that Src act upstream of RhoA signaling, independent of actin structure.

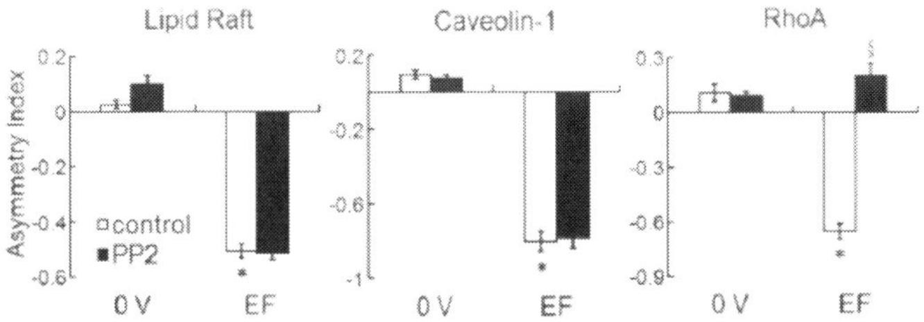

Figure 2 Effect of Src inhibition on asymmetry index of lipid raft, caveolin-1, and RhoA distribution (*p<0.05 vs 0V, §p<0.05 vs control group, n=53-146 for lipid raft, 39-93 for caveolin-1, 35-73 for RhoA)

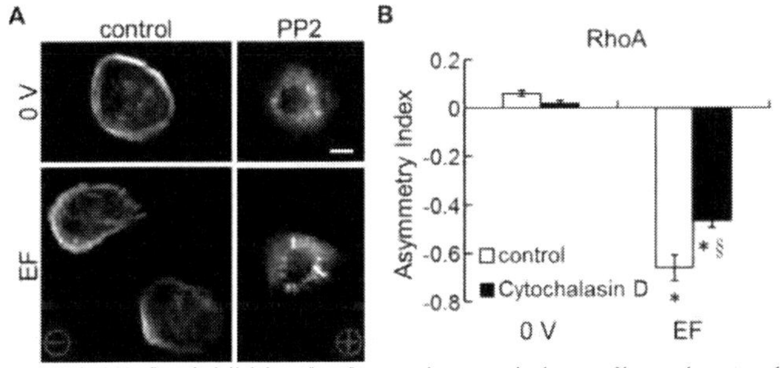

Figure 3 (A) Src inhibition lead to actin cytoskeleton disruption (scale bar = 10 μm) (B) Actin disruption with cytochalasin D had only a small effects on RhoA redistribution (n= 31-134, *p<0.05 vs 0V, p<0.05 vs control)

DISCUSSION

The goal of the current study is to investigate the role of Src in EF-induced intracellular signaling. We have demonstrated previously that EF induces lipid raft and caveolin polarization, which lead to downstream asymmetrical RhoA signaling and directed cell migration [6-8]. Using a Src-family kinase inhibitor, PP2, we found a complete elimination of directed cell migration (Figure 1). No changes in raft and caveolin polarization were found, but RhoA redistribution in response to applied EF was abolished with Src inhibition. (Figure 2). Since Src inhibition also lead to significant disruption of the actin cytoskeleton, we confirmed that actin disruption only result in a small reduction of RhoA redistribution (Figure 3). These results are consistent with our hypothesis, which is that Src acts downstream of caveolin to activate RhoA in EF-induced directional migration. Current studies are ongoing to further elucidate the role of lipid raft and caveolin on Src redistribution and activation. Our results establish a new mechanism for cell electrosensing and provide a new role in lipid raft mechanotransduction.

ACKNOWLEDGEMENTS

This work was supported by the NSC (103-2221-E-002-039) and NHRI (EX104-10411EI)

REFERENCES

REFERENCES

[1] McCaig CD, Zhao M. Physiological Electrical Fields Modify Cell Behaviour. Cell and Molecular Biology 1997;19:819-26.

[2] Guo A, Song B, Reid B, et al. Effects of Physiological Electric Fields on Migration of Human Dermal Fibroblasts. The Journal of Investigative Dermatology 2010;130:2320-7.

[3] Sun YS, Peng SW, Cheng JY. In vitro Electrical-stimulated Wound-healing Chip for Studying Electric Field-assisted Wound-healing Process. Biomicrofluidics 2012;6:34117.

[4] Song B, Zhao M, Forrester J, et al. Nerve Regeneration and Wound Healing Are Stimulated and Directed by an Endogenous Electrical Field in vivo. Journal of Cell Science 2004;117:4681-90.

[5] Allen GM, Mogilner A, Theriot JA. Electrophoresis of Cellular Membrane Components Creates the Directional Cue Guiding Keratocyte Galvanotaxis. Current Biology 2013;23:560-8.

[6] Lin BJ, Chao PH. Electric Field-induced Lipid Raft Polarization Guide Fibroblast Directional Migration. 2013 Annual Meeting of the Orthopedic Research Society.

[7] Tsai CH, Lin BJ, Chao PH. Alpha2beta1 Integrin and RhoA Mediates Electric Field-induced Ligament Fibroblast Migration Directionality. Journal of Orthopaedic Research 2013;31:322-7.

[8] Tsao SH, Lin BJ, Chao PH. Lipid Raft and Caveolin Polarization Guide Fibroblast Directional Migration in Electric Fields. Journal of orthopaedic research : 2014 Annual Meeting of the Orthopedic Research Society.

[9] Grande-Garcia A, Echarri A, de Rooij J, et al. Caveolin-1 Regulates Cell Polarization and Directional Migration Through Src Kinase and Rho GTPases. Journal of Cell Biology 2007;177:683-94.

[10] Zhao M, Song B, Pu J, et al. Electrical signals control wound healing through phosphatidylinositol-3-OH kinase-[gamma] and PTEN. Nature 2006;442:457-60.

QUANTIFICATION OF HELICAL FLOW PATTERNS IN LEFT VENTRICLES OF HEALTHY SUBJECTS AND PATIENTS WITH DILATED CARDIOMYOPATHY

Jonas Lantz (1), Carljohan Carlhäll (1,2), Tino Ebbers (1,2)

(1) Division of Cardiovascular Medicine, Department of Medical and Health Sciences, Linköping University, Linköping, Sweden.

(2) Center for Medical Image Science and Visualization (CMIV), Linköping University, Linköping, Sweden

INTRODUCTION

A fluid element that rotates about the axis of the velocity direction will give rise to a helical flow component [1]. This corkscrew pattern is usually considered physiological flow in the aorta of healthy subjects, but little is known about the helical flow inside the chambers of the beating heart.

Studies using magnetic resonance imaging (MRI) can provide time-resolved three dimensional velocity data of blood flow *in vivo* - normally called 4D flow MRI -, and quantifiable differences in blood flow patterns between healthy subjects and patients can be derived. One these quantities is helicity, which is computed as the dot product of the velocity and vorticity vector. Helicity is a pseudoscalar, meaning that it has both a magnitude and sign. Positive helicity indicate a clockwise helical pattern in the velocity direction, while a negative sign indicate a counter-clockwise flow direction.

The present study was undertaken to compare 3D helical flow patterns within left ventricles of healthy subjects and patients with dilated cardiomyopathy (DCM). The purpose was to investigate, in a quantitative manner, the flow patterns inside the left ventricle and if they change in patients with DCM.

METHODS

Ten healthy subjects and 5 patients with dilated cardiomyopathy were enrolled in the study. The patients were all classified as having a normal diastolic function at rest according to standard Doppler measurements. The study was approved by the regional ethical review board and all subjects and patients gave written informed consent before participation.

The 4D flow measurement was performed on a 1.5T MR scanner (Philips Achieva, Philips Medical Systems, Best, the Netherlands), and data were acquired during free-breathing using a retrospectively navigator-gated gradient-echo pulse sequence with asymmetric three-directional motion encoding and retrospective, vector cardiogram controlled cardiac gating. Acquisition parameters included a velocity encoding (VENC) of 100 cm/s, a relaxation time/echo time (TR/TE) of 6.3/3.7 msec, and a flip angle of 8°, parallel imaging by SENSE with a speed-up factor of 2, and a k-space segmentation factor of 2. The spatial resolution was 3x3x3 mm^3 and the field-of-view was adjusted for each subject to cover the left heart. Cine balanced steady-state free precession (bSSFP) imaging during end-expiratory breath-holds was used to acquire short-axis images in 30 timeframes across the cardiac cycle. A slice thickness of 8 mm was used and the acquired pixel size was 2.19x1.78 mm^2. Reconstructed pixel size was 1.37 mm^2 for the shortaxis images [2]. The bSSFP data were used for segmentation of the left ventricular volumes. Segmentations were performed using the freely available software Segment (v. 1.9R2046, Medviso, Lund, Sweden).

Helical flow components were calculated in Paraview 4.1 (Sandia Corporation, Kitware Inc., New Mexico, USA), as the dot product of the velocity and vorticity vectors. Helicity were computed for each voxel of the left ventricle during the entire cardiac cycle. As helicity has a sign, voxels with positive or negative helicity values were separated to investigate whether asymmetrical helical flow patterns were present.

Each cardiac cycle were split into a systolic and a diastolic part, in order to assess them separately.

RESULTS

Group averages of helicity for the healthy subjects and patients are presented in Figure 1. For the healthy subjects, overall low helicity values during systole were seen, which decrease as more and more of the blood is ejected from the left ventricle. During the early filling phase the helicity increased rapidly until diastasis, when the helicity values again began to decrease. During the late diastolic phase when the atria contracts (atrial systole) and push an additional amount of flow into the left ventricle, a small increase was seen which remained during the onset of systole. It is also interesting to see that there appeared to be a symmetrical helical flow, with equal amount of clockwise and counter-clockwise flow patterns. The standard deviation was relatively small in the healthy group, indicating a consistent flow field across the subjects.

For the patient group the standard deviation was large, indicating large variations within the group. Systole started with elevated helicity values, which decreased throughout the ejection phase. During the early filling phase helicity increased again, but not as consistent as in the healthy group. The characteristic peak of helicity observed in healthy subjects at early filling was not seen in the patient group, indicating that the flow field is altered.

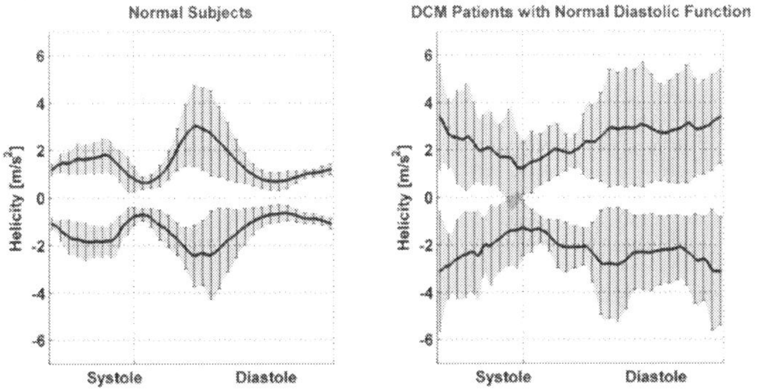

Figure 1: Group averages of helicity for the healthy subjects (left) and patients (right). Error bars indicate standard deviation within the group.

In addition, the magnitude of positive and negative helicity was investigated. Each MRI voxel that contained either a positive or negative value was counted and group averages as function of left ventricular volume are presented in Figure (2).

During a significant part of the early filling phase, the flow was mainly moving in a clockwise direction for the healthy subjects (statistical difference, $p < 0.01$). During diastasis and late diastolic filling as well as during systole there were no significant difference between clock- or counter-clockwise helical flow patterns. The patient group, on the other hand, demonstrate symmetrical flow patterns, with equal amount of clock- and counter-clockwise flow patterns.

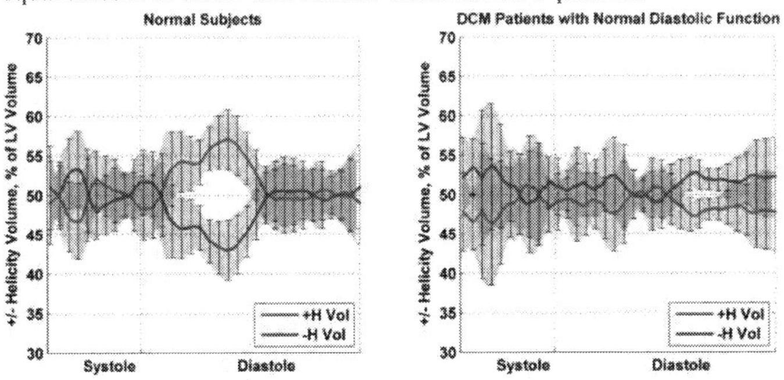

Figure 2: Group averages of portions of positive and negative helicity in the left ventricle for the healthy subjects (left) and patients (right). Error bars indicate standard deviation within the group.

In depth manual analysis of the helicity field revealed that a large region with positive helicity formed in the vicinity of the ventricular side of the anterior mitral leaflet during the early filling phase in the healthy volunteers, as shown in Figure 3,. This observation was not evident in the patient group.

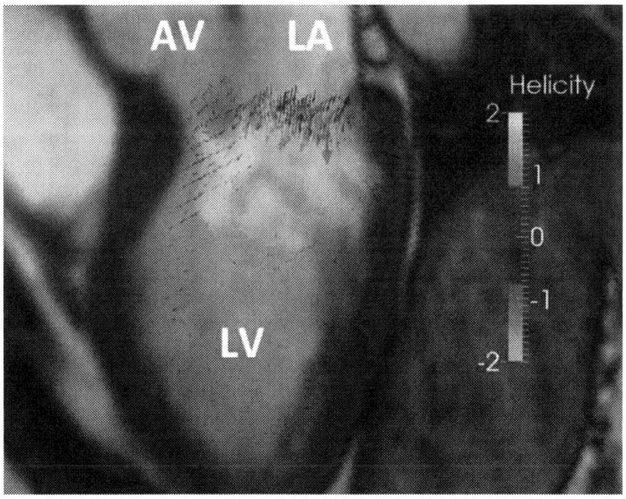

Figure 3: Velocity vectors showing velocity during early filling phase, and a volume rendering showing the resulting helical flow pattern in a healthy subject. Large clockwise flow patterns form in the vicinity of the anterior mitral leaflet, while counter-clockwise flow is present in the center of the inflow. Only helicity values |H| > 0.75 are shown for visualization purposes. LV: left ventricle, LA: left atrium, AV: aortic valve.

DISCUSSION

We studied helicity in the healthy and diseased left ventricle. The results show that for healthy volunteers the flow in the left ventricle are consistent, with helical flow forming predominantly during the filling phase. An increase of clockwise flow patterns were observed during the early filling phase, which may be explained by the formation of a large region with positive helicity in the vicinity of the ventricular side of the anterior mitral leaflet.

The patient group, on the other hand, was not as consistent in helical flow pattern, with a large variation within the group. Although the diastolic function at rest, assessed by conventional echo Doppler, was within the normal range, it is likely that same variation in active relaxation/passive stiffness of the LV myocardium as well as atrial contraction exist across the patients. Furthermore, the patients had a variation in LV remodeling which could affect the mitral valve apparatus, and hence, inflow properties.

REFERENCES

[1] Moffatt, H. K., and A. Tsinober. "Helicity in laminar and turbulent flow." Annual review of fluid mechanics 24.1 (1992): 281-312.

[2] Dyverfeldt P, Kvitting JPE, Sigfridsson A, Engvall J, Bolger AF, Ebbers T. Assessment of fluctuating velocities in disturbed cardiovascular blood flow: in vivo feasibility of generalized phasecontrast MRI. J Magn Reson Imaging 2008;28:655–663.

SB^{3}C2015
Summer Biomechanics, Bioengineering and Biotransport Conference
June 17-20, 2015,Snowbird Resort, Utah, USA
SB^{3}C2015-472

SPATIOTEMPORAL IMAGE CORRELATION SPECTROSCOPY TECHNIQUES FOR QUANTIFYING FLUID FLOW IN MICROFLUIDIC CHANNELS AND POROUS TISSUES

Brian T. Graham (1,2), Christopher Price (2)

(1) Department of Mechanical Engineering
University of Delaware
Newark, DE, USA

(2) Department of Biomedical Engineering
University of Delaware
Newark, DE, USA

INTRODUCTION

Optical observation of solute transfer and fluid flow within tissues using the focal plane sectioning ability of the confocal microscope represents an ideal method for directly and non-invasively quantifying a tissue's microfluidic environment. Fluorescent correlation spectroscopy (FCS) is an imaging approach that has been used to successfully quantify the convection of fluorescently tagged molecules in solution, but has limited applicability to both very fast and slow convective magnitudes [1]. An extension of FCS, which we term one dimensional spatiotemporal image correlation spectroscopy (1-D STICS), can be used to quantify particle convections along a user-defined flow path. 1-D STICS has been previously used to quantify the velocity of particles approximating the size of red blood cells flowing though vessels at physiological speeds [2]. Though the feasibility of this technique has been demonstrated, the range of speeds and size scales (particle and channel) over which it can be applied are unknown. Furthermore, due to limitations in the originally published correlation analysis architecture [2], application to more complex situations, such as temporal variations in flow, have been limited.

Using a commercial laser scanning confocal microscope, we have demonstrated accuracy across five-orders of magnitude of convective speeds, three-orders of magnitude of fluorophore sizes, and six orders of magnitude of dilutions. Additionally, improved analysis algorithms were implemented to increase data processing speed and to demonstrate the application of 1-D STICS to the measurement of temporally fluctuating convection through the implementation of a windowed resampling approach.

METHODS

A parametric study was used to determine the ranges and combinations of particle size, particle concentration, and flow speeds over which 1-D STICS is feasible. A syringe pump (Harvard Apparatus) was used to control the delivery of fluorescent micro-spheres (2µm, 0.2µm, or 0.02µm in diameter) through a microfluidic channel (Ibidi, Germany). Concentrations of spheres ranged from 4.6 x 10^2 to 4.6 x 10^{11} spheres/mL. Maximum flow speeds ranged from 1µm/s to 1m/s. Line scans were performed using selected combinations of imaging parameters on a Zeiss LSM780 laser scanning confocal microscope (objective, scan length, and pixel dwell time were parametrically varied). Correlations of the resultant carpet data (Figures 1A, B) were calculated using Equation (1), which, along with recursive thresholding and standard object detection techniques, was implemented in Matlab to process and interpret the 1-D STICS carpet data.

$$G(\xi,\eta)=\left|F^{-1}\left\{\left|F\left(I(x,t)\right)\cdot F^{*}\left(I(x,t)\right)\right|^{2}\right\}\right| \quad (1)$$

The recursive algorithm repeatedly adjusted the threshold then detected potential correlation streaks in the resulting binary image using geometric properties of the segmented objects. A primary correlation streak was selected from all potential streaks based on having the highest ratio of major to minor axis length and a long length relative to the other potential objects.

The binary image of the selected correlation streak was then used as a mask to isolate the original correlation. A weighted linear regression was performed on the resultant data to determine the slope of the correlation streak. This slope was then matched to a velocity map derived from the scan parameters in order to obtain the particle speed (Figure 1C). Measured particle speeds were compared to analytical solutions for rectangular channel flow [3] to determine the accuracy of the technique and to identify a threshold value, based on R^2 values of the regression analysis, which yielded the most reproducible results. Robustness of the technique was evaluated by measuring spatially varying flow profiles of laminar flow, temporal fluctuations of various time varying flow profiles, and convection within a custom capillary-mimicking hydrogel channel.

RESULTS

The range over which 1-D STICS could accurately detect particle convection in our system extended from 0.1m/s to 10µm/s, but varied with particle concentration and size (Figures 2A, B). R^2 values from the linear regression analysis correlated with accuracy and could be used to omit poor data when processing scans. The highest quality data (R^2>0.9) yielded particle speeds consistently 10-20% slower than analytical fluid predictions. Overall accuracy was demonstrated by the observation that the spatial sensitivity of 1-D STICS was high enough to quantify plug and parabolic flow profiles across channels (Figures 3A, B). Additionally, windowed resampling and analysis of data allowed detection of fast temporal variations in flow (Figures 3C, D). Finally, 1-D STICS was used to quantify convections in engineered environments where flow speed far exceeded that which is visualizable by standard confocal imaging (Figure 4A-D).

DISCUSSION

We have demonstrated the application of 1-D STICS on a commercial laser scanning confocal microscope to the quantification of particle convection and fluid flows across a large range of speeds (10μm/s-1m/s), particle sizes (2μm-0.02μm), and concentrations (4.6 x 10^2 - 10^{11} beads/mL). These length scales encompass particle sizes ranging from blood cells/platelets down to large protein complexes like lipoproteins (i.e. HDL), physiological to supra-physiological particle concentrations, and flow rates from that of aortic blood flow down to interstitial fluid convection. The systematic error observed between measured particle speed and theoretical fluid velocities can be attributed to minor losses in the pump system and inertial effects of the particles. Now that clear boundaries to 1-D STICS have been established, investigations into *in situ* fluid dynamics within tissues or engineering constructs using standard confocal techniques should be easy to implement. We believe that 1-D STICS can be implemented in the following applications: characterization of i) vascular, ii) lymphatic, and iii) interstitial flows, iv) intracellular trafficking, v) plant biology, vi) microfluidic devices, and vii) engineered tissues, amongst others. Currently, we are pursuing the application of an analogous image-based method called 2-D STICS, which has previously been used at the cellular length-scale; to map larger tissue-level 2-D transport fields in microfluidic and porous tissues using insight gained from our 1-D STICS experiments [4].

ACKNOWLEDGEMENTS:

NIH NIAMS R21 AR062738

REFERENCES:

[1] Digman M A et al., Biophysical Journal, 2005;
[2] Rossow, M et al., J Biomed. Optics, 2009;
[3] Cornish, R J, Proc. Royal Society, 1928;
[4] Herbert, B et al., Biophysical Journal, 2005

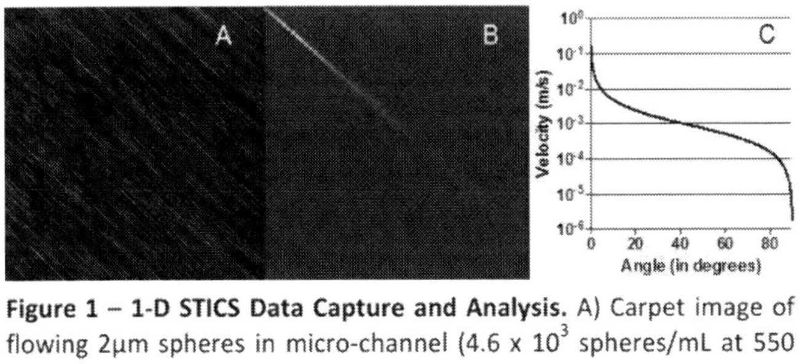

Figure 1 – 1-D STICS Data Capture and Analysis. A) Carpet image of flowing 2μm spheres in micro-channel (4.6 x 10^3 spheres/mL at 550 nL/s) captured via 1-D STICS (line = 512 pixels; 10x objective; 1.58 μs pixel dwell time). B) Spatiotemporal correlation analysis showing correlation streak, which was obtained from the carpet image. C) Relationship between correlation streak angle and particle/flow speed for the given scan parameters.

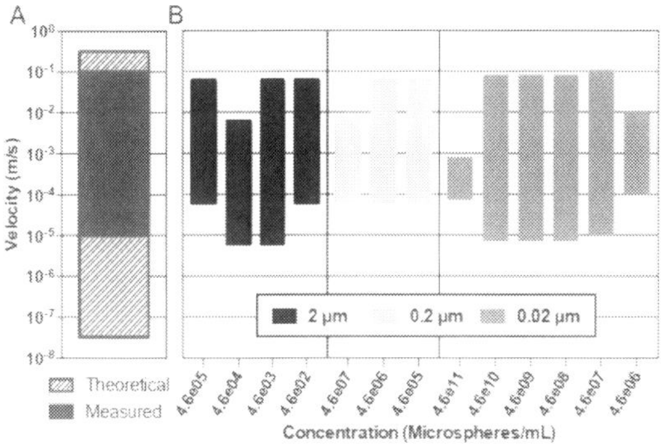

Figure 2 – Range of Detectable Speeds for 1-D STICS. A) Theoretical range of detectable particle speeds based on the limitations of our commercial laser scanning confocal microscope (Zeiss LSM780), compared to B) actual ranges of detectable velocities for three different sizes of microspheres at various concentrations.

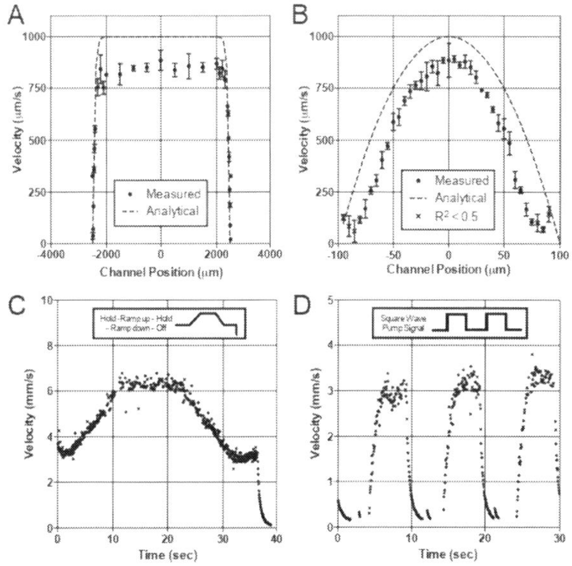

Figure 3 – Spatial and Temporal Accuracy of 1-D STICS. A) Plug flow and B) Parabolic flow profiles obtained for particle convection (0.2um spheres; 4.6x10^6 spheres/mL, at 650nL/s) through a 5mm wide x 0.2mm tall microfluidic channel. Temporal particle convection profiles captured for a C) ramp up – hold – ramp down – release and D) square-pulse flow profile. A windowed analysis protocol with window sizes of 64 lines and 32 lines were utilized, respectively.

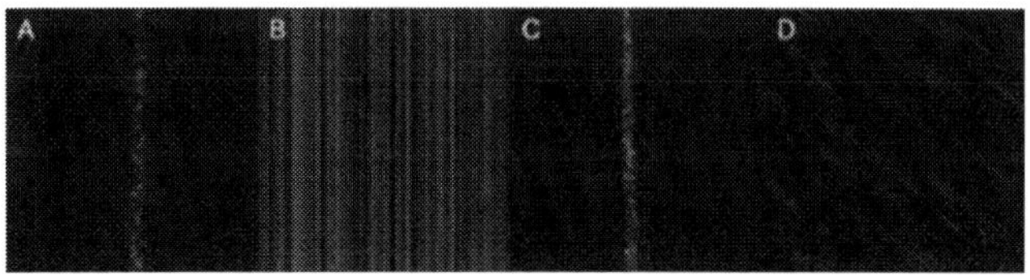

Figure 4 – Measuring microfluidic flow in engineered "capillary vessels". Image of A) stationary (non-flowing) 2um beads in a custom 10um square channel in hydrogel, designed to mimic a single capillary bed, and B) the representative carpet obtained by 1-D STICS. C) Image of 2um beads flowing through the hydrogel channel and D) a representative carpet obtained via 1-D STICS imaging.

INFLUENCE OF SULCI ON MECHANICAL RESPONSE OF THE BRAIN AND INJURY PREDICTION UNDER HIGH-RATE IMPACT

Alan C. Leung (1), Nithyanand Kota (2), Amit Bagchi (3), Siddiq M. Qidwai (3)

(1) Advanced Technology & Research Corporation
Columbia, MD, USA

(2) MCET Department
Leidos Corporation
Arlington, VA, USA

(3) Multifunctional Materials Branch
US Naval Research Laboratory
Washington, DC, USA

INTRODUCTION

The human brain contains surface topography, typically divided by the gyri for the convolutions and by the sulci (brain folds or fissure) for the grooves. Recent experimental and computational investigations into the effect of sulci on the mechanical response of the brain to high-rate impact show their influence. For example, an experimental study on porcine slices has demonstrated that for low to moderate speed translational motions of the brain, regions near sulci exhibit lesser effective stress than regions away from them [1]. Computational modeling based on reduced and idealized geometry of the sulci embedded in two-dimensional (2D) regions of the brain with boundary and loading conditions obtained from a separate three-dimensional (3D) model has shown that the maximum effective stress anywhere in the brain can be 40-90% more compared to the case where sulci are explicitly removed [2]. The differences in these and similar studies (e.g., [3]) on the effect of sulci can be attributed to the choice of models (2D slice [1] vs. 2D plane [2] vs. 3D brain [3]) and boundary and loading conditions (translation motion [1] vs. conditions derived from a 3D model [2] vs. squared sinusoidal acceleration [3]).

While these computational models are good first steps they can be improved in the areas of fidelity to sulci geometry and more realistic boundary and loading conditions, to provide more definitive information on the role of sulci in injury response, and the effect of sulci fidelity on computational prediction of injuries, especially those sustained in the battlefield. This paper addresses these concerns with specific focus on high-rate overpressure impacts. This is achieved through computational simulations of head slices in the sagittal plane with varying levels of geometric fidelity and an actual overpressure loading profile. Two specific objectives are targeted here: 1) Determine the degree of influence of sulci on brain response as a function of sulci geometry, and 2) Quantify the numerical differences due to reduction in sulci fidelity in computational models.

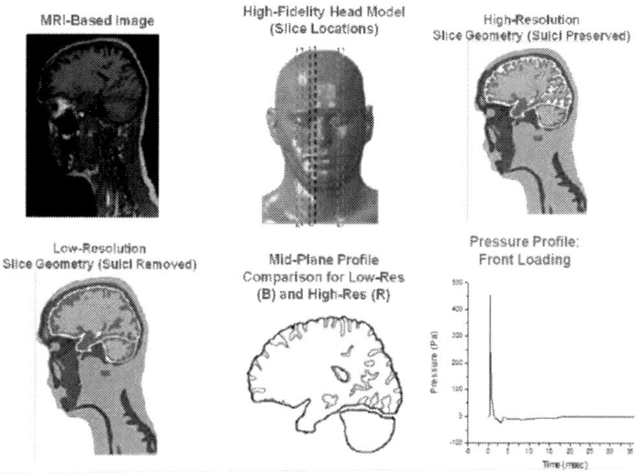

FIGURE 1. VARIOUS INGREDIENTS OF THE MODELING EFFORT: MRI SCANS, GEOMETRIC MODEL SLICES, PROFILE COMPARISON, AND LOADING PROFILE.

METHODS

To achieve these objectives, three locations (R1, R2 & L1) in the sagittal plane were chosen based on the geometric complexity of the observed sulci in both hemispheres (Figure 1). A detailed MRI scan of a 26 year old, 50th percentile male head at ~1 mm resolution was used to obtain 10 mm thick slices at these locations for simplicity. Two distinct geometric models were created from each slice and classified as low-res and high-res on the basis of fidelity to the MRI data (Figure 1). Thus, the high-res model adheres to the brain geometry as limited by the MRI scans. The low-res model is the same except for the removal of all sulci with grey matter to mimic the other extreme case.

All the major components that form these slices are accounted for in the computational models, including the brain with grey and white matter, the cerebral spinal fluid, the skull with spinal column, muscles and soft tissues, skin, etc. Brain matter is modeled as a hyper-viscoelastic material, calibrated with a broad range of low- to high-rate data. Similarly, material model parameters for all other components are calibrated through a detailed data search from open literature.

A Friedlander wave profile (Figure 1) obtained from an in-house overpressure test with a peak pressure of 450 kPa and a wavelength of 1 msec was applied to the front of the head slice models. Various out-of-plane boundary conditions were tested for their appropriateness, and based on the results the planes were fixed normally to restrict unrealistic out-of-plane deformations.

RESULTS AND DISCUSSION

The mechanical variables chosen to be highlighted in this study are most commonly used to describe brain injury, i.e, pressure, Tresca (max. shear) stress, and principal strain. Figure 2 depicts the spatial distribution of the Tresca stress and principal strain in the brain slices for both high-res (w/ sulci) and low-res (w/o sulci) models, at specific instances during the loading sequence. Both shearing and stretching are affected by the presence of sulci, but in a more significant manner compared to pressure (not shown here). However, no significant phase shift is observed in this and other data point-wise data.

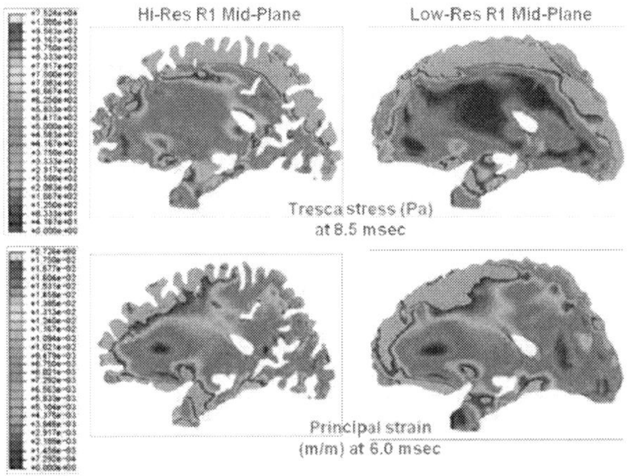

FIGURE 2. SPATIAL DISTRIBUTION OF MAXIMUM SHEAR STRESS AND STRAIN IN THE R1 SLICE MODEL.

To establish the correlation between mechanical response and the geometry of the sulci, the following process was used. As shown on the top left in Figure 4, lines were drawn emanating from the mid-point of the individual sulcus perimeter on the mid-plane of each sagittal slice. Mechanical response data were collected on these lines for all times and the maximum values were found for both types of models. Geometric parameters such as perimeter length, depth, area and total curvature were also calculated as shown on the top row of Figure 4. The maximum values of pressure, Tresca stress and principal strain for both high-res and low-res models were plotted separately, against the geometric parameters of sulci in all three slice models to discover correlations. However, the same correlations were found to be present in both types of model data (data or plots not shown here), which implied that the response was dominated by the global features of the models (geometry and material properties) and applied loading, which were similar across the models.

The data accumulated on the sulci lines were then used to compare the respective responses of the high-res and low-res models. In Figure 4, the ratio of the low-res to high-res responses is plotted for maximum Tresca stress and principal strain in the bottom two rows. R1, R2 and L1 refer to the three slices from the right and left hemispheres, whereas the subscript indicate the sulcus number in a given slice. (Over-prediction implies low-res response being larger and vice versa.) It is noted that there is a minor sulci effect on pressure: ±10%, but large effect on shear: ±35% (not shown here), and stretch: ±30%. This implies that the degree of sulci fidelity in computational models may dictate confidence in injury predictions.

In conclusion, current results show that sulci do have a high degree of influence on the mechanical response of the brain, but there is no discernible correlation with the geometric parameters of the sulci. More investigation is needed to definitely confirm this hypothesis. Secondly, the differences in the low-res model data compared to the high-res model case have strong implications on the confidence levels for brain injury prediction when fidelity is not maintained.

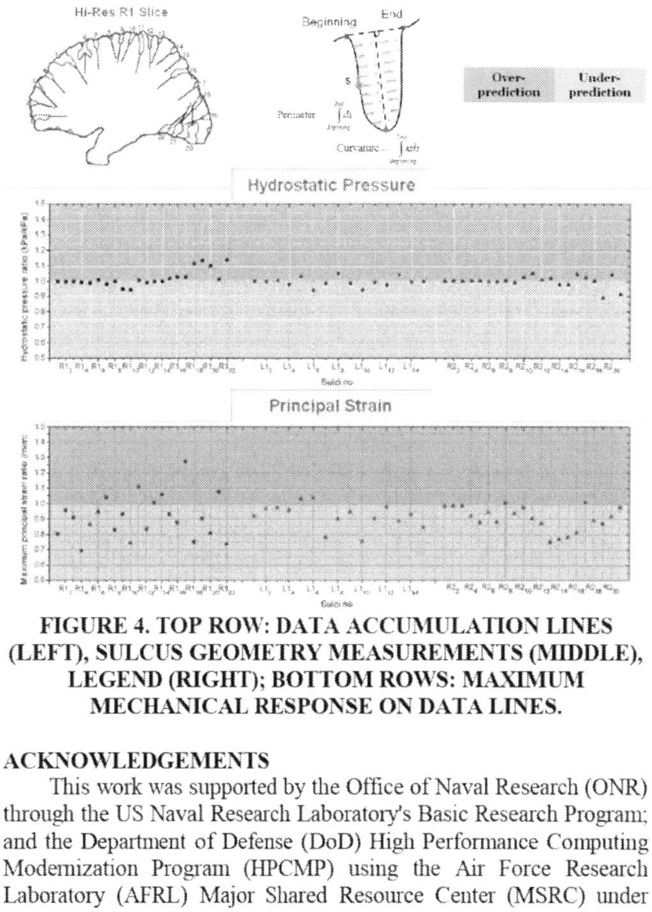

FIGURE 4. TOP ROW: DATA ACCUMULATION LINES (LEFT), SULCUS GEOMETRY MEASUREMENTS (MIDDLE), LEGEND (RIGHT); BOTTOM ROWS: MAXIMUM MECHANICAL RESPONSE ON DATA LINES.

ACKNOWLEDGEMENTS

This work was supported by the Office of Naval Research (ONR) through the US Naval Research Laboratory's Basic Research Program; and the Department of Defense (DoD) High Performance Computing Modernization Program (HPCMP) using the Air Force Research Laboratory (AFRL) Major Shared Resource Center (MSRC) under project 416, subproject 231. The authors acknowledge Dr. Ross Cotton from Simpleware® for the generation of the finite element meshes.

REFERENCES

[1] Lauret, C et al., *Med Eng & Phy*, 31:392-399, 2009.
[2] Cloots, RJH et al., *J Biomed Eng*, 36:1203-1215, 2008.
[3] Ho, J & Kleiven, S, *J Biomech*, 42: 2074-2080, 2009.

CHARACTERIZATION OF A THERMOREVERSIBLE COLLAGEN FOR FREE-FORM FABRICATION OF SCAFFOLDS

Kathryn E. Drzewiecki[1]
David I. Shreiber[1], PhD

Biomedical Engineering Department
Rutgers, The State University of New Jersey
Piscataway, New Jersey, United States[1]

INTRODUCTION

Type-I collagen is the most ubiquitous protein in the body and is easily extracted from animal sources. Extracted type-I collagen is biocompatible and biodegradable, and is widely used in regenerative medicine. Despite these advantageous biomaterial qualities, compared to synthetic polymers, it is often criticized for weak mechanical properties that are difficult to consistently control.[1] To this end, we synthesized collagen methacrylamide (CMA) by coupling methacrylic acid to the free amines of lysine residues of type-I collagen, rendering CMA photocrosslinkable.[2] CMA can be photocrosslinked in the presence of a photoinitiator and ultraviolet (UV) light to selectively stiffen the hydrogel or add bioactive, methacrylated ligands.[2] In further characterization of CMA, we discovered the material to be thermoreversible, transitioning from a suspension of triple helical fragments at low temperatures to a self-assembled fibrillar hydrogel at physiological temperature.[3] Interestingly, photocrosslinking of CMA hydrogels eliminates thermoreversibility. Using these two properties of CMA, we have established a unique method of free-form fabrication of a collagen-like material. In this paper, we will discuss the novel method of free-form fabrication of CMA and the spatial crosslinking resolution and photocrosslinking design parameters of CMA hydrogels.

METHODS

Collagen Methacrylation

As previously described, the carboxyl group on methacrylic acid was activated using ethyl-3-(3-dimethylaminopropyl) carbodiimide) and N-hydroxysuccinimide and covalently coupled to free amines on type-I collagen to produce CMA.[2] CMA was dialyzed, lyophilized, and resuspended at 3.75 mg/mL.

Free-form Fabrication

The general approach for free-form fabrication of CMA scaffolds first involves preparation of a buffered, alkalinized CMA suspension using the following formula: 20 μL HEPES, 136 μL 0.15 N NaOH, 100 μL 10X PBS, 57 μL 1X PBS, and 10 μL 10% Irgacure in neat methanol. The CMA suspension (100 μL) was plated in glass petri dishes (MatTek Corporation) and self-assembled for 30 minutes at 37 °C. Samples were then photocrosslinked in a specific spatial pattern through a photomask (Figure 1) with UV light (365 nm, 100 mW/cm^2, 50% power) for 2 minutes. The hydrogel was then placed at 4 °C for 1 hour to "cold-melt" CMA that was not exposed to UV light. The disassembled material was removed by rinsing with PBS.

Spatial Resolution

CMA hydrogels were exposed to UV light through photomasks composed of circles with diameters of 2580, 1270, 508, or 254 μm. Photopatterned CMA hydrogels were imaged using phase contrast microscopy. Spatial resolution of the images was quantified by measuring the diameter of the photopatterned circles in comparison to the photomask and calculating the percent difference (n = 7).

Design Parameters

Two design parameters were examined. First, CMA was mixed with type-I collagen to determine how much CMA is required to ensure thermoreversibility. The following volume percent mixtures of CMA:Collagen were examined: 0:100, 25:75, 50:50, 75:25, and 100:0. Hydrogels were produced as described above, self-assembled for 30 minutes at 37 °C, and cold-melted at 4 °C for 1 hour. Second, the minimum concentration of Irgacure needed to prevent thermoreversibility was investigated. The final concentration of Irgacure in the hydrogels was varied to be 0.1%, 0.05%, 0.02%,

0.01%, 0.005%, 0.001%, or 0%. Hydrogels were produced as described, self-assembled for 30 minutes at 37 °C, photocrosslinked (Blak-Ray Lamp Model UVL-21, 366 nm) for 15 minutes, and cold-melted for 1 hour at 4 °C. Thermoreversiblity was evaluated by measuring the optical density of the hydrogels before and after cold-melting and comparing the result to water.

RESULTS

Spatial Resolution

The shape of free-form fabricated CMA hydrogels (Figure 1A, left) generally matched that of the photomask (Figure 1A, right). The diameter of the hydrogel was measured from phase contrast images (Figure 1B) and compared to the three larges circles on the photomask. The smallest circle (Circle 4), while visible in the brightfield image, was not visible using phase contrast microscopy. Spatial resolution was quantified as the percent difference of the circle diameter in phase contrast images versus the photomask (Figure 1C). Photopatterned CMA circles were on average 25-40% larger in diameter than the photomask.

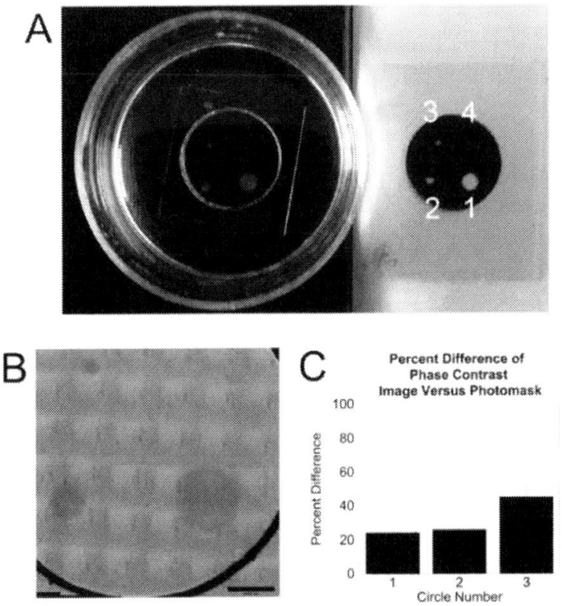

Figure 1: Brightfield and Phase Contrast Microscopy of CMA Hydrogels and Spatial Resolution of Photocrosslinking

Design Parameters

To conserve and minimize the amount of CMA used in hydrogel formulations (not yet FDA-approved), we created hybrid CMA/collagen hydrogels and determined if gels maintained thermoreversibility. The optical density of cold-melted (100%) CMA was similar to that of water (Figure 2). Composite gels demonstrated potentially graded but not fully thermoreversible behavior.

CMA is photocrosslinked using the photoinitiator Irgacure, which produces cytotoxic free radicals as a result of UV exposure.[4] Thus, we aimed to determine the minimum Irgacure concentration necessary for our free-form fabrication protocol. The optical density of CMA hydrogels with 0.1%, 0.05%, and 0.02% Irgacure were similar to that of collagen with 0.1% and 0% Irgacure, indicating lower concentrations of Irgacure were not sufficient to photocrosslink CMA and prevent thermoreversible behavior (Figure 3).

Figure 2: Optical Density of Cold-Melted Hybrid CMA/Collagen Hydrogels

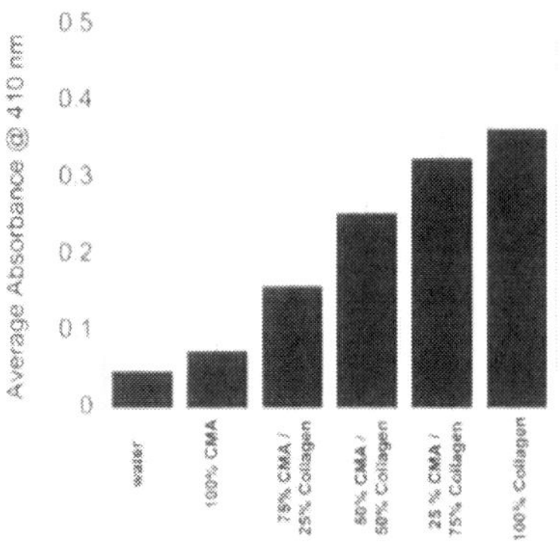

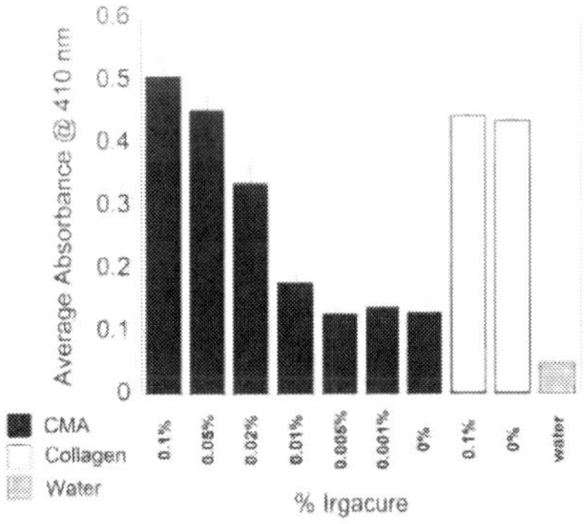

Figure 3: Optical Density of Photocrosslinked, Cold-melted CMA Hydrogels with Varied Irgacure Concentrations to Prevent Thermoreversibility

DISCUSSION

This novel method of free-form fabrication offers a simple technique to create 2D and potentially 3D structures of a collagen-like biomaterial. CMA hydrogels photocrosslinked with simple laser transparency photomasks have good fidelity to the pattern of interest. Differences in resolution often occur as a result of chemical photocrosslinkers bleeding to surrounding areas of the hydrogel, and can be tuned by minimizing UV light intensity or time of UV treatment, and by producing higher quality photomasks. Other design parameters tested, including composite hydrogels and Irgacure concentration, cannot be easily modified to maintain the needed properties of CMA for free-form fabrication. Highly concentrated (100%) CMA is needed for complete thermoreversibility of the hydrogel. Further, high concentrations of Irgacure are required to prevent thermoreversibility of photocrosslinked regions of the hydrogel. Future studies will focus on how to incorporate cells into scaffolds fabricated using this method while maintaining cell viability. Overall, this method allows for a unique method to easily rapid prototype collagen-based hydrogels for a variety of applications in the field of regenerative medicine.

ACKNOWLEDGEMENTS

The authors would like to acknowledge the following for funding support: National Science Foundation Grant ARRA-CBET0846328 and National Institutes of Health under Ruth L. Kirschstein National Research Service Award T32 GM8339 from the NIGMS.

REFERENCES

[1] Zhu, J et al., *Expert Rev Medical Devices*, 8, 5: 607–626, 2011.
[2] Gaudet, I et al., *Biointerphases*, 7, 25:1–9, 2012.
[3] Drzewiecki, K et al., *Langmuir*, 30, 37:11204–11211, 2014.
[4] Rouillard, A et al., *Tissue Eng. Part C*, 17, 2: 173–179, 2011.

SB^{3}C2015
Summer Biomechanics, Bioengineering and Biotransport Conference
June 17-20, 2015,Snowbird Resort, Utah, USA
SB^{3}C2015-475

TOWARDS THE VALIDATION OF A MULTISCALE CHEMO-ELECTRO-MECHANICAL FINITE ELEMENT MODEL USING ELECTROMYOGRAPHY

Thomas Heidlauf (1,2), Mylena Mordhorst (1,2), Sook-Yee Chong (1), Oliver Röhrle (1,2)

(1) Institute of Applied Mechanics (CE)
University of Stuttgart
Stuttgart, Germany

(2) Stuttgart Research Centre for
Simulation Technology
University of Stuttgart
Stuttgart, Germany

INTRODUCTION

Due to the fact that many physiological quantities of skeletal muscle contraction are difficult to measure in-vivo, mathematical models of muscle behavior have the potential to significantly improve the current understanding of muscle function. Most existing muscle models are single-scale models that focus on a certain aspect of the muscle contraction, such as, for example, (a part of) the cellular excitation-contraction coupling [1] or the macroscopic muscle mechanics. For a more complete description of the muscle, we developed a multiscale chemo-electro-mechanical model that combines a biophysical description of the cellular excitation-contraction coupling with a description of the propagation of electrical signals (action potentials, APs) along muscle fibers, and a continuum-mechanical model of the resulting muscle forces and deformations [2,3]. Moreover, we recently extended this modeling framework to simulate surface electromyographic (EMG) signals [4].

The EMG signal reflects the electrical activity of a skeletal muscle. The main advantage of EMG over other methods is that they can be measured in-vivo in a relatively easy and straightforward way. While previous models of the EMG [5] neglected the contraction-induced deformations of the muscle and the resulting relative movements between the source of the electrical signals (the muscle fibers) and the detection electrode, the proposed multiscale modeling framework allows in-silico predictions of the EMG under (more) realistic conditions, i.e., EMG predictions during activation-induced tissue deformations as well as for arbitrarily complex muscle geometries and muscle fiber architectures.

Predicting the muscle deformation, the resulting muscle forces, and the corresponding EMG, the multiscale modeling framework can be used to simulate isometric and non-isometric contractions and has the potential to significantly contribute to a better understanding of muscle chemo-electro-mechanics and EMG signals. Before the model, however, can be applied to physiological problems, it needs to be properly validated. While some aspects of the model have already been validated, a quantitative validation of the multiscale framework still needs to be accomplished.

This contribution reviews the key aspects of the modeling framework and presents first steps towards a quantitative validation of the multiscale model using EMG measurements.

METHODS

The proposed multiscale modeling framework consists of several parts. The membrane potential is computed using the monodomain model, where a biophysical half-sarcomere model [1] is used to describe the ionic currents crossing the muscle fiber membrane. The biophysical description has many advantages over simpler, phenomenological models, since physiological effects, such as, for example, changes in the membrane potential amplitude and the AP propagation velocity caused by membrane fatigue, result from the model rather than being prescribed as part of the model construction.

Describing the entire excitation-contraction pathway, the biophysical half-sarcomere model [1] also predicts the sarcomere-based active stresses that result from the activation-induced contraction of the half-sarcomeres. The sarcomere-based active stresses are homogenized before being included in a continuum-mechanical constitutive equation describing the relation between tissue deformation (strain) and the stress distribution. By solving the discretized governing equations of the continuum-mechanical model, one obtains the deformed geometry of the muscle as well as the resulting muscle forces.

Based on the distribution of the membrane potential, the extracellular bidomain equation is solved for the extracellular potential within the muscle and the adjacent subcutaneous tissue. The EMG signal is then the potential at the skin surface.

To summarize, the model predicts the distributions of the membrane potential and the extracellular potential (including the EMG), the contraction-induced muscle deformation and its resulting reaction forces. Moreover, due to the fact that the monodomain equation and the extracellular bidomain equation are solved on the deforming muscle geometry, the model accounts for changes in the signal propagation that might result from the muscle contraction (mechano-electric feedback). This particular feature allows the simulation of non-isometric contractions.

In addition to the model, high-density EMG signals (64 electrodes, 10mm inter-electrode distance, OT Bioelettronica, Turin, Italy) have been experimentally determined from a human biceps brachii muscle under isometric conditions.

RESULTS

Figure 1 shows the EMG signal predicted by the model for a single motor unit action potential under isometric conditions. In this example, it was assumed that the muscle fibers are straight lines that run from left to right, parallel to the surface. This mimics the parallel-fibered architecture of the biceps brachii muscle. In total, 390 muscle fibers were simulated. A 2 mm thick low-conductivity subcutaneous layer separates the muscle tissue from the surface at which the EMG is displayed.

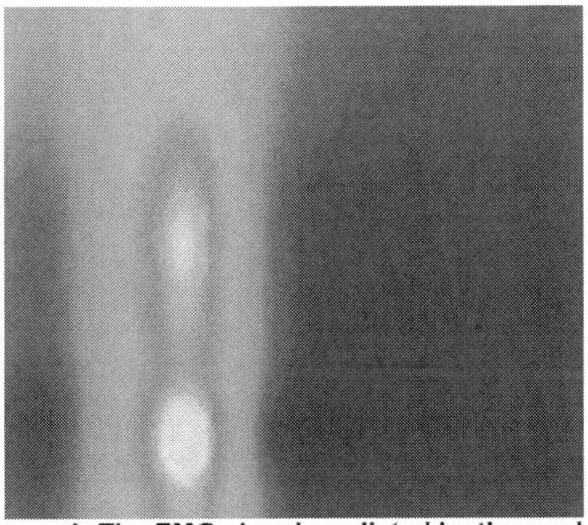

Figure 1: The EMG signal predicted by the model.

Figure 2 depicts a high-density EMG signal recorded from the skin above the human biceps brachii muscle.

DISCUSSION

In previous works, parts of the model have been validated in isolation. For example, the biophysical cell model is validated in [1], and parts of the multiscale muscle model are validated in [3]. The ability to simulate EMG signals during isometric and non-isometric contractions provides the possibility for a more comprehensive validation of the model.

A first step towards this is presented in the contribution at hand. For an isometric contraction, the EMG signal of the model was qualitatively compared to an experimentally determined signal and the results look promising. A quantitative validation has not yet been carried out but will be done in the future. This is difficult due to the fact that in the real muscle the motor unit sizes and territories (location of the muscle fibers of the active motor units) are unknown. Similarly, the thickness of the subcutaneous layer, which significantly influences the EMG amplitude, is currently unknown in the experiment but might be determined using ultrasound.

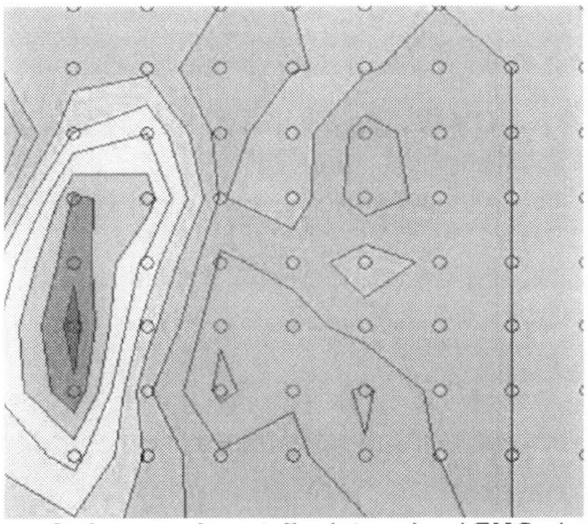

Figure 2: An experimentally determined EMG signal.

Ultrasound might also be used to measure the changes in the muscle fiber pennation angles occurring during a contraction. This can be used to validate the model-predicted deformations. Moreover, muscle forces can be measured and compared to the model's forces.

Due to the fact that the proposed model predicts, at the same time, muscle forces, muscle tissue deformations, as well as the EMG signal, one might need to appeal to a combination of different methods in order to validate the proposed multiscale model. A comprehensive validation of the model could, for example, combine simultaneous recordings of EMG, force and ultrasound.

Once the validation of the model for isometric contraction is completed, one needs to validate the model behavior under non-isometric conditions. Finally, the fully validated multiscale model can help to answer many different research questions in the field of muscle physiology.

ACKNOWLEDGEMENTS

The research leading to these results has received funding from the European Research Council under the European Union's Seventh Framework Programme (FP/2007-2013) / ERC Grant Agreement n. 306757 (LEAD).

REFERENCES

[1] Shorten, P R, et al., "A mathematical model of fatigue in skeletal muscle force contraction." *J Muscle Res Cell Motil*, 28: 293-313, 2007.

[2] Heidlauf, T and Röhrle, O, "Modeling the Chemoelectromechanical Behavior of Skeletal Muscle Using the Parallel Open-Source Software Library OpenCMISS." *Computational and Mathematical Methods in Medicine*, 2013, 2013.

[3] Heidlauf, T and Röhrle, O, "A multiscale chemo-electro-mechanical skeletal muscle model to analyze muscle contraction and force generation for different muscle fiber arrangements." *Frontiers in Physiology*, 5:498, 2014.

[4] Mordhorst, M, et al., "Predicting electromyographic signals under realistic conditions using a multiscale chemo-electro-mechanical finite element model." *J. R. Soc. Interface Focus*, accepted for publication.

[5] Mesin, L, "Volume conductor models in surface electromyography: Computational techniques." *Computers in Biology and Medicine*, 43:942-952, 2013.

SB^{3}C2015
Summer Biomechanics, Bioengineering and Biotransport Conference
June 17-20, 2015,Snowbird Resort, Utah, USA
SB^{3}C2015-476

NOVEL TECHNIQUE FOR ASSESSMENT OF MECHANICAL PROPERTIES OF CAROTID ARTERIES

Stefan N. Sanders, Frans N. van de Vosse, Marcel C.M. Rutten

Biomedical Engineering
Eindhoven University of Technology
Eindhoven, The Netherlands

INTRODUCTION

Vulnerable plaques in carotid arteries are potentially lethal or at least cause high risk of stroke when they rupture. Vulnerability research has focused on estimating geometric properties such as cap thickness and fluid dynamics, evaluating wall shear stress. Recently, the detection of plaque properties using combined imaging/numerical methods is gaining popularity. Three-dimensional methods using MRI data and fully 3D numerical models have shown the possibilities of assessing the mechanical environment of plaques, although the handling of heterogeneity is still a problem due to spatial resolution issues [1]. Higher resolution imaging methods like ultrasound cannot make a sharp distinction between plaque components. Studies on a microscopic scale, including indentation tests on human plaque samples have elucidated the highly heterogenic properties of pathological tissue [2]. Such tests, however, load the material in a direction which does not coincide with physiological loading directions, and thus may yield results that cannot be translated easily to clinical applications. A more realistic loading condition was obtained using inflation [3]. A balloon was used to deform relatively large (25-30mm) ring-shaped specimens of atherosclerotic aortic tissue. However, mechanical properties of the plaque components can only be derived indirectly, and strength assessment is not feasible with this technique. Therefore a novel technique for assessing mechanical properties of carotid arteries is developed, using experimental data and finite element models for mechanical properties estimation.

METHODS

In this study, we model a slice of artery or arterial plaque as a 2D object, and load it in a physiological way (figures 1, 2). Slices (0.3mm thickness) were cut from a fresh porcine carotid artery, obtained from the slaughterhouse.

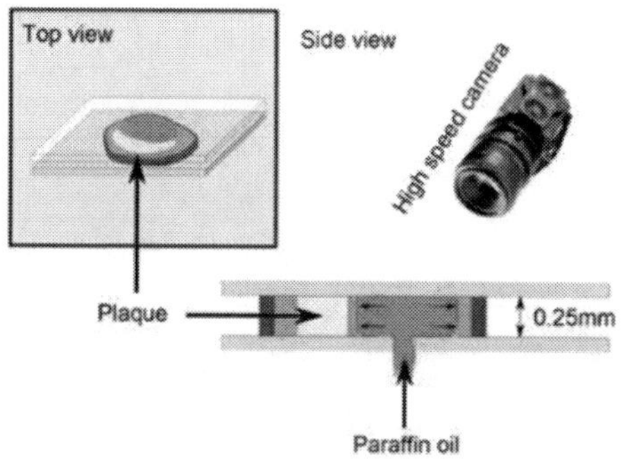

Figure 1: Schematic overview of the experiment.

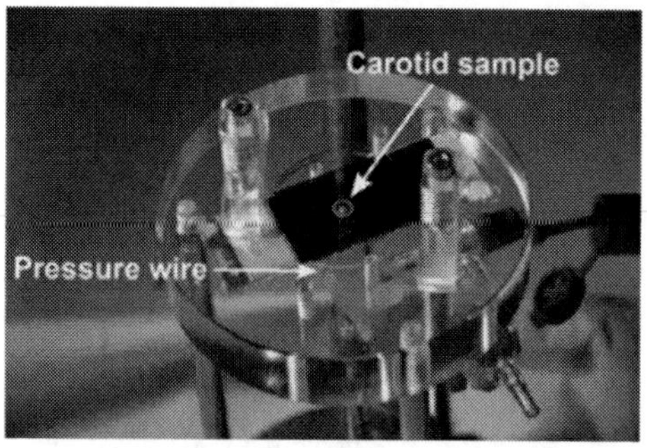

Figure 2: Experiment setup.

The slices were mounted between two horizontal microscope slides, 0.25mm apart. Through a small hole in the bottom slide, transparent paraffin oil was injected into the lumen of the artery sample, using a servo-operated pump. The oil pressurized the vascular tissue from the luminal side, and enabled the tissue to move frictionlessly.

Inflow pressure was measured using a pressure wire (St Jude Medical, Uppsala, Sweden), positioning the tip of the wire directly below the sample, in the small hole in the bottom slide. Motion of the sample was measured using a high speed video camera, synchronized to the pressure measurement at a frequency of 200Hz. The images were processed using Matlab software (R2012b) to render diameter as a function of pressure.

To assess mechanical properties, the experiment was modelled in FEBio finite element code (v2.1.0), using quadratic hexahedral elements. To start, the vascular material was modelled as being neo-Hookean. The experiment was modelled and the shear modulus of the carotid sample was assessed during pressure increase. A least-squares approach was used to compare the experiment to the model results.

RESULTS

Physiological pressures were obtained in the lumen of the sample using the pump (figure 3).

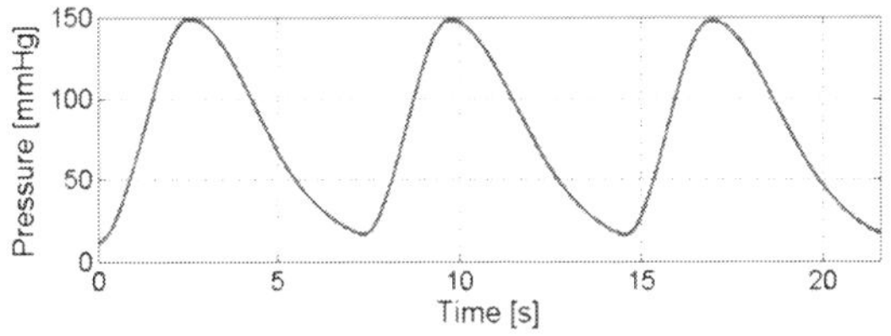

Figure 3: Pressure over time in the lumen.

The diameter of the lumen of the carotid sample changed according to the variations in pressure (figure 4).

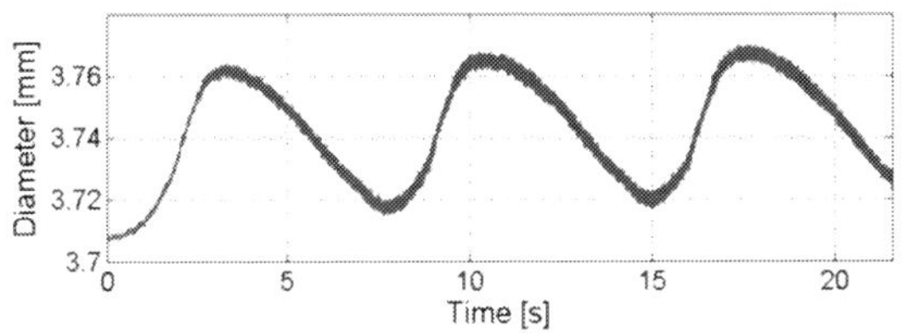

Figure 4: Lumen diameter change due to pressure.

The pressures of the first rising edge of the pressure signal were used as input for the finite element model. A shear modulus of 1MPa was found to be the closest fit to the experimental data (figure 5). It was observed that in the experiment the pressure increases faster than the diameter.

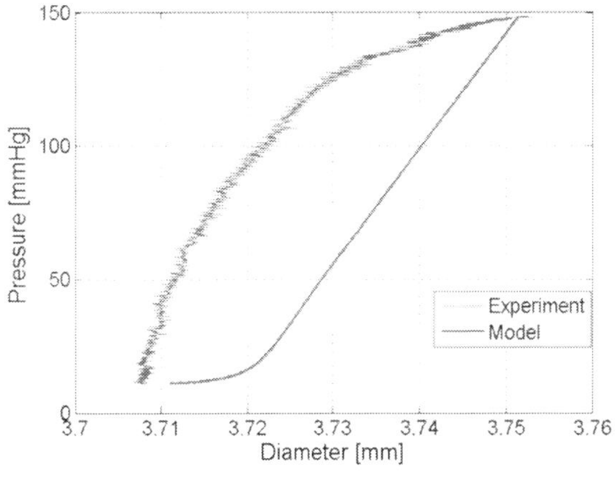

Figure 5: Experiment and model results for the diameter change due to pressure.

DISCUSSION

The first results of this novel technique for assessment of mechanical properties of carotid arteries are promising. The synchronized pressure and diameter measurement in the experiment have been successfully implemented. The carotid sample used in this experiment was found to be stiff compared to other porcine carotids, which resulted in a higher shear modulus. Paraffin oil leaks between the sample and the glass, resulting in near frictionless movement of the sample between the microscope slides. The nonlinear behavior of the tissue has been modelled with a linear material model, which may explains the difference in the pressure/diameter curves of the experiment and the model.

The main advantage of this technique is the physiological loading condition imposed in a quasi 2D setting. In future applications, heterogeneous properties, like in atherosclerotic plaque material, may be assessed as well, using vital staining techniques to distinguish different tissue components without affecting their mechanical behaviour.

ACKNOWLEDGEMENTS

This project is funded by the STW open technology program (STW #12548).

REFERENCES

[1] Lin, AP et al., *Magn Reson Med*, 60(1):8-13, 2008.
[2] Chai, CK et al., *J Biomech*, 46:1759-1766, 2013.
[3] Beattie, D et al., *J Biomech*, 120:602-607, 1998.

SB³C2015
Summer Biomechanics, Bioengineering and Biotransport Conference
June 17-20, 2015,Snowbird Resort, Utah, USA
SB³C2015-477

Temperature Control of Hydrogel Delivered Through Endoscopic Needle as Potential Treatment for Pancreatic Cancer.

Thomas L. Merrill (1,2), Jennifer E. Mitchell (2), Denise R. Merrill (2), Matthew L. Short (1,2)

(1) Department of Mechanical Engineering
Rowan University
Glassboro, NJ, USA

(2) FocalCool, LLC
Mullica Hill, NJ, USA

INTRODUCTION

Locally advanced pancreatic cancer (LAPC) has a devastating prognosis[1]. Effectively treating LAPC is challenging because lesions are hypovascular, have a high interstitial pressure, and have a dense extracellular matrix[2]. LAPC is conventionally treated with radiation and chemotherapeutic drugs given intravenously. The lesion characteristics limit systemic drug effectiveness. Localized intratumoral (IT) injection of a liquid drug is also limited by lesion characteristics causing unsafe drug leakage into surrounding tissues.

To improve outcomes, new drug delivery platforms are needed. One potential approach involves using an endoscopic ultrasound (EUS) to inject a drug-encapsulated hydrogel. EUS fine needle injection (EUS-FNI) is emerging as a promising treatment method to provide direct treatment of tumors via IT injection. Encapsulating the drug in a thermally reversible hydrogel allows the drug to be injected as a liquid (at room temperature or colder), then transition to a gel in-situ (at body temperature), creating a drug reservoir for a sustained drug release. The hydrogel will reduce drug leakage, providing drug release for weeks, limiting the number of necessary treatments.

However, hydrogel injectability is a hurdle. The long distance of travel through the endoscope (~1.2 m), the low delivery flow rate (<0.5 ml/min), and the warm temperature surrounding the endoscope (37°C), cause the phase transition to occur along the injection pathway, resulting in high delivery pressures, and clogging. To address the hydrogel injection problems, a FocalCool, LLC cooling catheter [3] (CoolGuide™) was used as a sheath around the EUS needle to control the temperature of the hydrogel during injection. Not only does the CoolGuide resolve phase transition issues, it allows a broader range of possible drug-hydrogel combinations, for treatment optimization. The CoolGuide temperature control sheath is a tri-lumen extrusion and replaces the standard circular sheath around the EUS needle. The endoscopic needle is placed in the CoolGuide central lumen, surrounded by coolant circulating through two crescent-shaped lumens. A separate cooling console provides coolant flow (Figure 1).

This study shows the temperature and pressure results of injecting hydrogel through an EUS needle with the CoolGuide sheath. CoolGuide was able to effectively control hydrogel temperature and significantly reduce required delivery pressures for the hydrogel compared to tests with a standard circular sheath.

METHODS

To characterize hydrogel delivery a custom mock loop was designed to simulate an EUS procedure accessing the pancreas. A schematic of the mock loop is shown in Figure 1. From left to right, hydrogel is pumped with a syringe pump through an EUS needle. The entire assembly was placed inside a mock EUS with a coiled heat exchanger used to maintain normal body temperature at 37°C conditions. A mock tissue section at the distal end simulated the needle penetrating tissue. The sheath did not cover the needle in the mock tissue section.

Three configurations were tested: 1) Std - a standard 19 gauge EUS needle (Cook Medical, Bloomington, IN), 2) CGTest1 - CoolGuide sheath with the standard EUS-FNI needle, and 3) CGTest2 - a CoolGuide sheath with the standard EUS-FNI needle plus an insulated needle extension. The insulated needle extension (Concept Group Inc., West Berlin, NJ) is essentially an 8 cm long microscale Thermos™ container, with a double wall separated by a vacuum layer. In all of these setups the EUS needle was placed within the mock EUS and advanced 8 cm beyond the mock EUS distal tip.

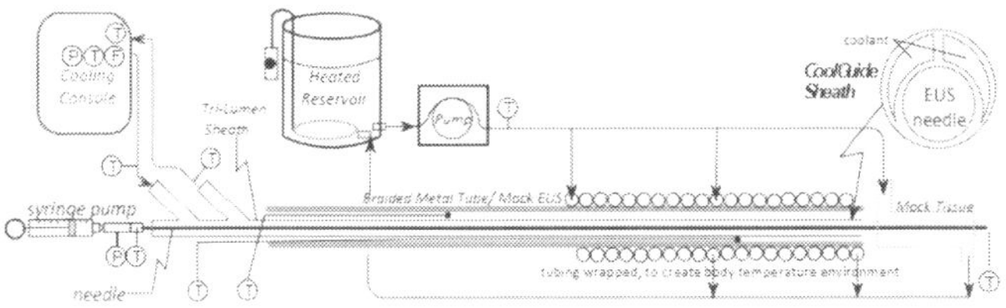

FIGURE 1: Schematic of EUS mock loop. The EUS working channel was created with a braided metal tube partially surrounded with body-temperature fluid. The cooling console provides coolant to the CoolGuide sheath surrounding the EUS needle. Temperatures and pressure measurements were recorded at designated locations.

Two fluids were tested: (1) distilled water and (2) 30% (w/v) Pluronic F127 (Sigma Aldrich, St Louis, MO). The 30% Pluronic mixture is a hydrogel with a liquid-to-gel transition temperature of 14°C. Viscosity values transition from <1 Pa-s at 4°C to 3000 Pa-s at 37°C. A 5 mL syringe of each test fluid was prepared and stored at 4°C. The fluid was pumped at 0.5 ml/min for 6 minutes using a syringe pump (Harvard Apparatus, Holliston, MA). Test conditions are described in Table 1.

Table 1: Test Conditions

Temperature Inside Endoscope Working Channel	37.5°C ± 0.5°C
Temperature of Coolant Entering CoolGuide Catheter	< 5°C
Temperature of Test Fluid Inside Syringe	< 10°C
Pump Setting	0.5 mL/min for 6 min
Each test was conducted 3 times.	

RESULTS

Testing with distilled water was completed first. These tests provided the temperature ranges that would be seen during hydrogel testing, without the concern of any phase transition.

Distilled Water: Needle exit temperatures for the standard sheath (Std) were 33.5°C and CoolGuide sheath were 30.5°C (CGTest1). However, using the CoolGuide sheath with needle insulation (CGTest2) in the mock tissue section reduced the needle exit temperature to 16°C. These results are shown in Figure 2. The pressure to pump distilled water was negligible.

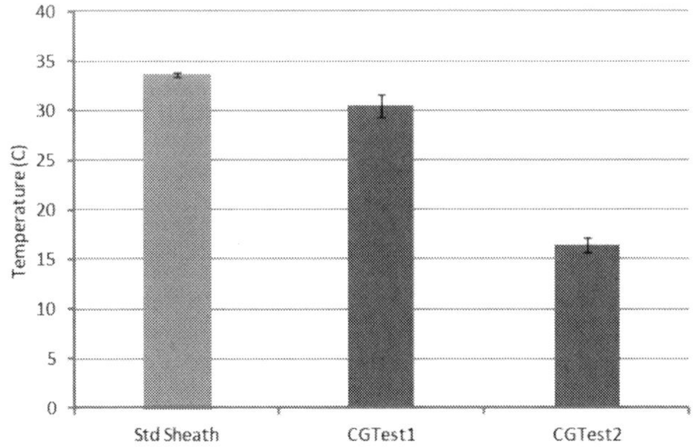

FIGURE 2: Distilled Water Delivery Temperatures. Error bars denote standard deviation of three tests. CGTest1 & CGTest2 used the CoolGuide temperature control sheath.

30% Hydrogel: Pump pressure to drive hydrogel through for the Std test exceeded the capabilities of the syringe pump (p>100 psi). This test could not be completed. Pressure to pump for CGTest1 and CGTest2 was 26.6 psi and 7.9 psi respectively. Hydrogel temperature results for CGTest1 and CGTest2 were similar to the distilled water results. Temperature and pressure data is shown in Figure 3.

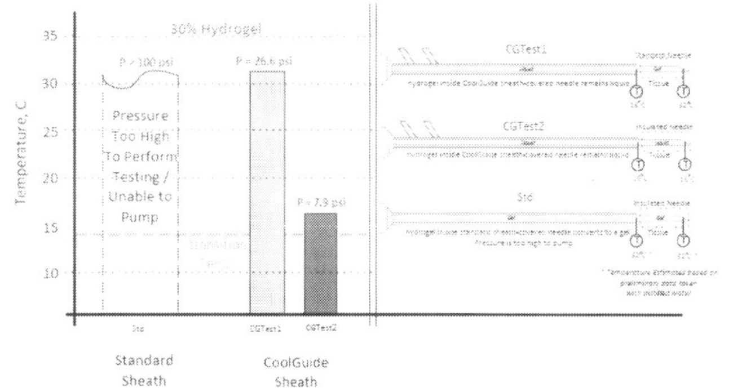

FIGURE 3: Hydrogel Delivery Temperature and Pressure. Pumping pressures are shown in red. The maximum pumping pressure for our syringe pump was 100 psi gauge pressure.

DISCUSSION

Our testing shows that temperature control is needed to deliver 30% hydrogel through an EUS needle under body-temperature conditions. The CoolGuide sheath limits the transition from liquid to gel, enabling well-controlled injection. An insulated needle extension (CGTest2) provides additional temperature control, but was not necessary to complete the test. Without temperature control the hydrogel could not be delivered.

Not only will a CoolGuide sheath allow for the delivery of hydrogels at cooler temperatures, this platform technology also will enlarge the thermal operating ranges for potential cancer fighting drug polymer combinations. Since drug release kinetics can depend highly on the constituent concentrations inside the drug reservoir, larger thermal operating ranges will provide medical researchers the opportunity to explore new approaches to destroying tumors. Finally, this delivery system may have even broader transformative potential since it can be extended to include the delivery of fiducial markers, radioactive seeds, hyperthermic nanoparticles, viral gene therapy vectors, extracellular matrix enzymes, and stem cells[4-6].

ACKNOWLEDGEMENTS

Our work was supported by NIH's National Heart, Lung and Blood Institute (NHLBI). Grant Number: 1R43HL117403-01. The content is solely the responsibility of the authors and does not necessarily represent the official view of NHLBI or NIH.

REFERENCES

[1] NCI. *Surveillance, Epidemiology, & End Results Program (SEER)* 18 2004-2010. Available at: http://seer.cancer.gov/. Accessed 05/20, 2014.
[2] Jain RK. *Sci.Am.* 310(2):46-53. Feb 2014
[3] Merrill TL. *J. Medical Devices* 2010;4(3):doi:10.1115/1.4002063.
[4] Nakai Y, Chang KJ. *Gastrointest.Endosc.Clin.N.Am.* 2012 Apr;22(2):315-24.
[5] Saftoiu A, Vilmann P. *World J Gastroenterol* 2011;17(1):42-48.
[6] American Cancer Society. Available at: http://www.cancer.org/cancer/pancreaticcancer/detailedguide/pancreatic-cancer-treating-chemotherapy. Accessed October 15, 2014.

SB^{3}C2015
Summer Biomechanics, Bioengineering and Biotransport Conference
June 17-20, 2015,Snowbird Resort, Utah, USA
SB^{3}C2015-478

MULTI-SCALE MODELS OF SKELETAL MUSCLE REVEAL THE COMPLEX EFFECTS OF MUSCULAR DYSTROPHY ON TISSUE MECHANICS AND DAMAGE SUSCEPTIBILITY

Kelley M. Virgilio (1), Kyle M. Martin (1), Shayn M. Peirce (1), Silvia S. Blemker (1,2,3)

(1) Department of Biomedical Engineering
University of Virginia
Charlottesville, VA, USA

(2) Department of Orthopaedic Surgery
University of Virginia
Charlottesville, VA, USA

(3) Department of Mechanical Engineering
University or Virginia
Charlottesville, VA, USA

INTRODUCTION

Duchenne muscular dystrophy (DMD) is a neuromuscular disorder caused by the incomplete translation of the transmembrane protein dystrophin. It is responsible for dramatically impaired muscle function and progressive muscle wasting [1]. Despite extensive experimental research there is no cure for the disease, and unanswered questions remain regarding the role of dystrophin in muscle function. How does the protein protect the cell membrane from mechanical damage? How do secondary changes in the muscle – such as fibrosis and fatty infiltration – influence the function of the muscle?

In this study, we posit that multi-scale computational models can provide a quantitative, mechanistic approach to investigate the influence of muscle disease on muscle function. To achieve this we developed a novel approach for randomly generating muscle fascicle geometries, and explored the trade-offs between mechanical function and damage susceptibility in the context of DMD-associated changes in muscle.

METHODS

We used an agent-based model (ABM) of muscle to generate new fascicle cross sections for each simulation. In order to determine the input parameters necessary to model structural changes in musculoskeletal disease, we determined the most prevalent pathological variations seen in DMD, including: density of transmembrane proteins, fiber cross-sectional area (CSA), fibrosis, and fatty infiltration [1]. Through manipulation of these input parameters, we generated fascicle geometries that accounted for both structural changes seen in disease and variability inherent in muscle tissue. The material elements from the ABM's discretized grid were then mapped onto an initialized finite-element (FE) mesh. Boundary conditions were assigned to prescribe simple shear deformation representing the shear displacement of muscle fibers and fascicles, consistent with the level of shear deformations observed in whole muscles [2].

We used a transversely isotropic, nearly incompressible, hyperelastic constitutive model to represent the muscle fibers, ECM, and boundary layer [2]. Currently there are no known measurements for the along-fiber shear modulus of muscle fibers and ECM; however, previous studies have shown that the ratio of G_{FIBER}/G_{ECM} is the critical factor in determining the contribution of structural variation on tissue-level properties [3]. Furthermore, it is unknown whether the muscle fiber is more or less stiff than the ECM, therefore G_{FIBER} was held constant during simulations, and G_{ECM} was adjusted for the two scenarios: one in which the $G_{FIBER} > G_{ECM}$ and one in which $G_{ECM} > G_{FIBER}$. The boundary layer around the muscle fascicle allowed for a unique opportunity to pre-allocate the mesh properties, boundary conditions, and applied displacements for the initialized mesh. Therefore, the boundary behavior was modeled to represent the macroscopic shear properties of a fascicle as a whole through rules of mixtures. Transmembrane proteins, represented by nonlinear elastic springs [4], were added at nodes connecting ECM and muscle fibers.

The macroscopic along-fiber shear properties of the muscle were calculated based on the displacement and shear stress of the fascicle determined from the FE simulations. The micromechanical model was first used in a parameter analysis to test the individual effect of a number of variations in the microstructure prevalent in musculoskeletal disease. Then we performed a simulated case study of DMD in which the muscle fascicle was analyzed for healthy muscle and at three stages of disease. The early stage included loss of transmembrane proteins, the middle stage included loss of proteins, fibrosis, and an increase in the variance of fiber CSA, and the late stage included loss of proteins, increased fibrosis, fatty infiltration, and large variances in fiber CSA (Fig 1A).

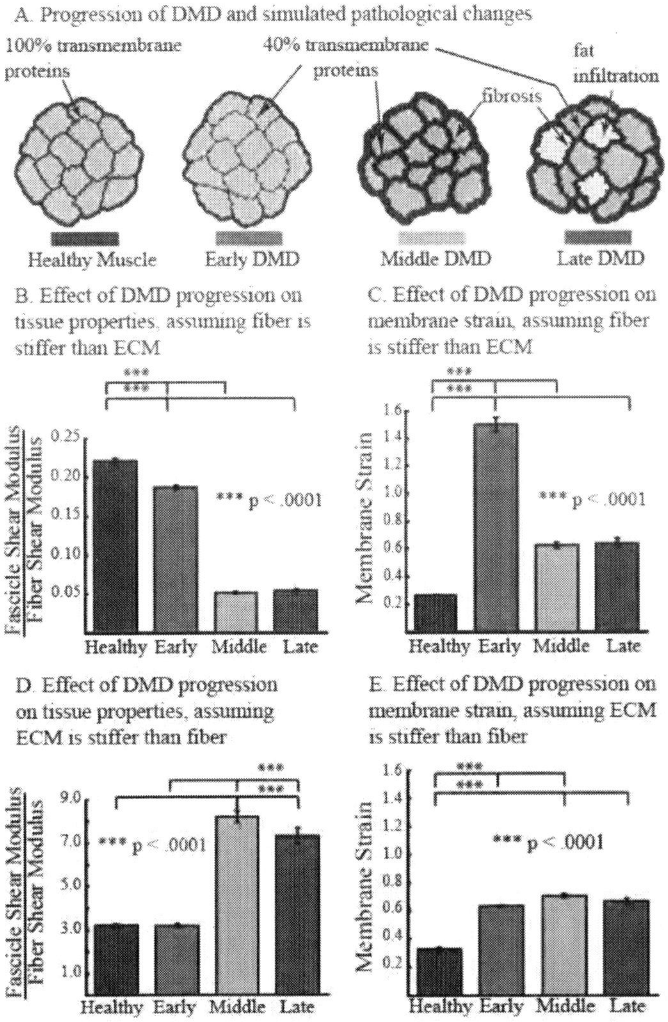

FIGURE 1. CASE STUDY REVEALS COMPLEX EFFECTS OF MUSCULAR DYSTROPHY ON FASCICLE STIFFNESS AND MEMBRANE STRAIN

RESULTS

The results of the parameter analysis differed depending on whether the ECM was stiffer than the fibers or if the fibers were stiffer than the ECM. When the muscle fibers were stiffer, both increasing the volume fraction of fibers and the density of transmembrane proteins led to a significant increase in the macroscopic shear modulus, fatty infiltration decreased the shear modulus, and variability in CSA had no effect. When the ECM was stiffer, only the volume fraction of fibers affected the tissue level properties, with an increase in volume fraction leading to a decrease in the macroscopic shear modulus. The membrane strain predictions were only affected by the density of the transmembrane proteins; and when the fiber was stiffer the effect was far more dramatic.

The case study demonstrated that DMD-associated pathological alterations influence both tissue properties and damage susceptibility. In the case where the muscle is stiffer, macroscopic shear modulus decreased significantly over time (Fig. 1B), and membrane strain initially increased (i.e. more damage susceptibility) but then decreased with fibrosis and fatty infiltration (Fig. 1C). When the ECM was stiffer, the macroscopic shear modulus increased significantly in the middle and late stages (Fig. 1D), while the membrane strain increased less dramatically but remained elevated through the middle and late stages of disease progression (Fig. 1E).

DISCUSSION

The goal of this work was to develop a computational modeling framework to investigate how disease-related changes in the muscle influence the tissue-level mechanical properties and susceptibility to membrane damage. By using an ABM to generate new geometries and mapping it to an initialized mesh, this novel modeling framework eliminates the cumbersome task of creating unique finite element meshes for each analysis and allows for an unprecedented quantity of simulations to be run in a short amount of time. Additionally, the ABM generated new fascicle cross sections with varying muscle fiber shapes and sizes for each simulation, incorporating architectural variability commonly seen *in vivo*.

Analyzing the compounding effects of microstructural variations in the DMD case study revealed the importance of understanding fibrosis in DMD. The stiffness of the ECM relative to the fiber significantly affected the degree to which disease progression influenced the fascicle stiffness and membrane strain. For example, the onset of fibrosis either alleviated the membrane strain (muscle stiffer than ECM) or increased it (ECM stiffer than muscle). This differential effect of fibrosis revealed a potential trade-off between functional preservation and protection of the membrane. The model is also consistent with the theory that transmembrane proteins protect the muscle cell from damage, because deletion of the membrane proteins resulted in increased strains in the membrane [5].

It is important to consider a number of limitations to the models presented here. First, the ECM was considered to be a homogenous structure with a constant material parameter, even within fibrosis. However, it is known that the ECM is comprised of distinct structural components, and the material properties of the ECM are likely to be altered during fibrotic disease progression. Additionally, the analyses focused on the behavior of the muscle in simple shear. We focused on shearing of the fascicle because it is a dominant mode of lateral force transmission, and the behavior in shear cannot simply be intuited from the microstructure. However, we acknowledge that simple shear does not fully represent physiological loading patterns. Finally, one of the critical challenges was the limited availability of experimental data, therefore the study focused on analyzing the relative effects of microstructural variations, given the uncertainty of the parameters.

The current study has highlighted the importance of further exploring the nature of fibrosis during disease, and the potential trade-offs between muscle function and damage susceptibility in the context of DMD. In the future, the modeling framework can be extended through a full coupling of an ABM, micromechanical model, and tissue-level model, to allow for the predictive analysis of DMD disease progression within a mechanical environment.

ACKNOWLEDGEMENTS

This study was funded by the National Science Foundation, grant # 1235224

REFERENCES

[1] Emery, A. *Lancet*, 359:687-95, 2002.
[2] Blemker, SS. *J. Biomech.*, 38:657-65, 2005.
[3] Sharafi, B., *J. Biomech.*, 43:3207-13, 2010.
[4] Zhang, C., *J Biomech*, 47:3232-6, 2014.
[5] Garcia-Pelagio, K.P., *J. Muscle Res. Cell Motil.*, 31:323-36, 2011.

A COMPARISON OF BRAIN INJURY PREDICTORS BASED ON FOUR BENCHMARK IMPACT STUDIES

Siddiq M. Qidwai (1), Nithyanand Kota (2), Amit Bagchi (1)

(1) Multifunctional Materials Branch
US Naval Research Laboratory
Washington, DC, USA

(2) MCET Department
Leidos Corporation
Arlington, VA, USA

INTRODUCTION

Traumatic brain injury (TBI caused by the propagation of a stress wave though the human head under impact is a complex phenomenon due to: complicated geometry with large number of interfaces, wide ranges of heterogeneous nonlinear material behaviors, and spatially evolving dynamic loading conditions. The obvious reasons for the lack of *in-situ*/*in-vivo* studies of human head under impact have forced researchers to rely upon computational simulations and animal testing or tissue failure data to correlate external insult and mechanical analysis variables to injury of the brain. The first method often leads to non-uniqueness because of constitutive assumptions and subjective interpretation of injury. On the other hand, the latter method tends to oversimplification by correlating brain injury under complex loading to failure under idealized conditions. As a result, approximately twenty mechanical variables have been proposed as injury predictors with specific threshold values to identify the onset of injury (e.g., see [1]).

Brain injury is broadly classified into: focal injury that occurs with immediate symptoms in a specific location and diffuse injury that occurs with delayed symptoms over a more widespread area [2]. The focal injury can be further divided into: contusion, laceration, hemorrhage, herniation, etc. Similarly, diffuse injury can be sub-classified into: diffuse axonal injury (DAI), swelling, ischemia, vascular injury, etc. Despite this hierarchy; however, most mechanical variable predictors based on point-wise, volumetric or holistic (global) measures of stress, strain, acceleration, and energy are still related to the broader two injury types only.

The use of multiple mechanical variables to predict brain injury raises the possibility that there could be overlap, redundancy or even inaccuracy associated with them. To address these concerns, in this study, we have used a validated high-fidelity computational model of the human head to compare the relative merits of these variables in predicting brain injury under impact. The point-wise variables are the focus of this study because of their more common use in predicting injury for the many kinds of impacts that are sustained in: vehicular crashes, falls and collisions, and ballistic and blast exposures.

METHODS

A validated high-fidelity geometric model of the head and neck that includes twenty five major and minor components is used for this comparison. The components include scalp, skin, muscles, eyes, cranium and facial bones, falx cerebri and tentorium cerebelli, venous and air sinuses, bridging veins, cerebro-spinal fluid (CSF), ventricles, meninges, cerebrum and cerebellum, brain stem, and cervical vertebrae, intervertebral disks, and spinal cord (Figure 1).

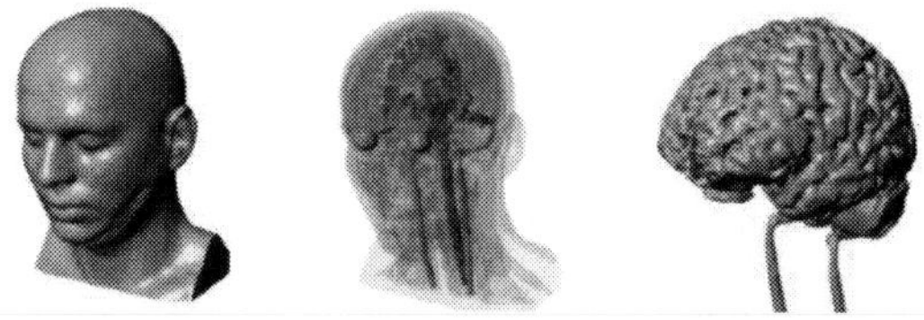

FIGURE 1: HIGH FIDELITY GEOMETRIC MODEL OF THE HUMAN HEAD AND NECK.

A high resolution finite element mesh comprising ~4.6 million tetrahedral elements-both linear and quadratic as per the numerical needs-form the core of the finite element model. Hyper-viscoelastic, hyperelastic or elastic constitutive models calibrated with the latest experimental data are used to describe responses of the solid components as appropriate. Special attention has been given to account for both low- and high-rate behaviors where data is available.

The fluidic responses of the air, blood, and CSF are modeled with equations of state and deviatoric hyperelastic models as needed.

The head model and point-wise mechanical variables (e.g., pressure, effective stress, principal strain, and shear stress) are used to predict the time, location and extent of brain injury under the same set of boundary and loading conditions as specified in four well-known experimental studies [3-6]. These four studies on automotive impacts have been chosen because of their predominant use in the validation of computational head models.

RESULTS AND DISCUSSION

The results of the simulation of an impact at 45 degrees to the head (impactor mass of 5.6 kg at 9.94 m/s) [3] are provided here as an example in Figure 2. The group of pressure, effective stress, and principal strain are used to predict focal injury, whereas the group of effective stress, principal strain and shear stress are used to predict diffuse injury. Principal strain is shown two times for the diffuse injury based on two different criteria (31% vs. 5%). The threshold values are given under each mechanical predictor.

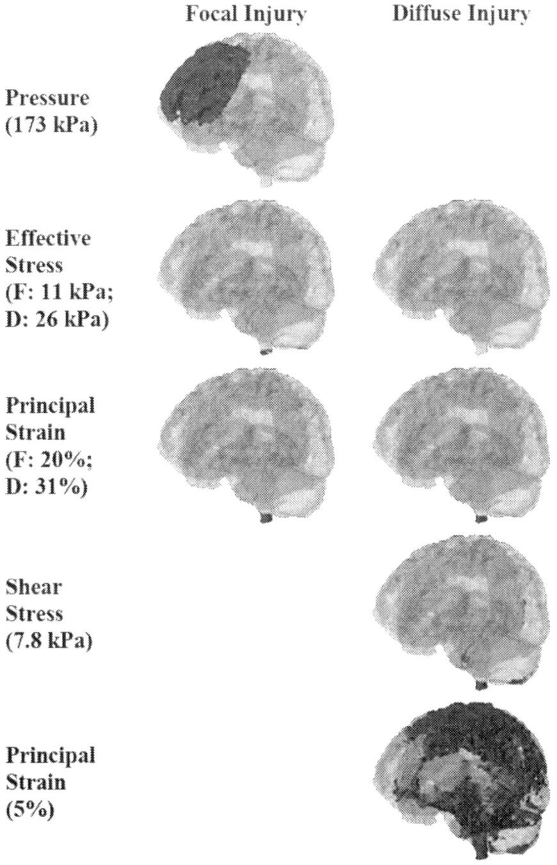

FIGURE 2: PREDICTION OF FOCAL (F) AND DIFFUSE (D) INJURIES ACCUMULATED OVER THE COURSE OF LOADING.

It can be seen for the focal injury that pressure represents a very different type of injury to that predicted by effective stress and principal strain. Similarly, effective stress, principal strain (31% threshold) and shear stress predict similar regions of diffuse injury. The obvious difference is the principal strain (5% threshold) prediction. However, one can hypothesize that the similar patterns for the first three predictors and differences with the fourth predictor may merely be due to the diverging calibration processes adopted for the variable thresholds than a fundamental difference as is seen for pressure in focal injury. One also notes that principal strain predicts both focal and diffuse injuries in the brain stem with different thresholds (20% and 31%). It leads to the question of how one region can have both types of injury without a physiological explanation.

A contrasting prediction from another simulation is shown in Figure 3. The simulation is of an experiment in which a cadaver head and neck complex in an inverted position is brought to speed and impacted against an acrylic block on the back [6]. In a departure from previous prediction, the threshold of 5% principal strain corresponds well with the thresholds of other variables, whereas as a consequence the threshold of 31% predicts relatively insignificant injury (not shown here). Clearly, threshold determination is also dependent upon the mode of loading.

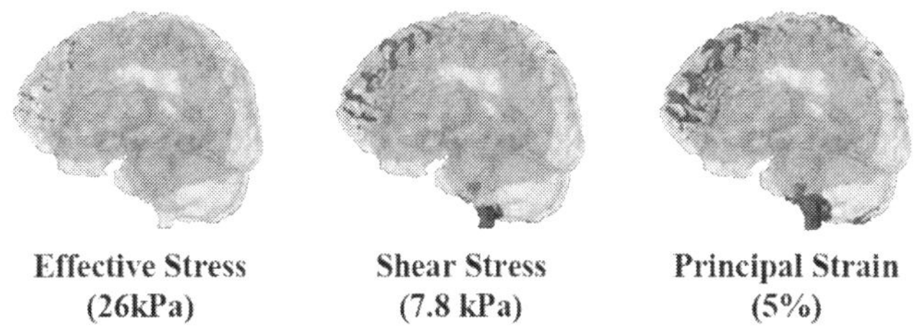

FIGURE 3. PREDICTION OF DIFFUSE INJURY ACCUMULATED OVER THE COURSE OF LOADING.

It is further interesting to note here that volumetric variables, such as the cumulative strain damage measure (CSDM) that predicts a 50% probability of DAI if 55% of the brain suffers a strain of 15% accumulated over loading time, predict no diffuse injury in these investigation which stands in stark contrast to the point-wise predictor performance.

These results and others that are not shown here demonstrate that some mechanical measures are similar thus redundant (effective stress, shear stress and principal strain) and that others do not correspond well with each other (point-wise and volumetric) and marked temporal and spatial discrepancies arise in predictions. The reasons for these observations is attributed to the non-uniqueness in the correlation process of fitting many predictors to fewer injury types and subjective interpretation of what is injury in the experiments, interrelation of variables such as effective stress, shear stress and principal strain, and choice of constitutive assumptions that are inherently subjective.

ACKNOWLEDGEMENTS

This work was supported by the Office of Naval Research (ONR) through the US Naval Research Laboratory's Basic Research Program; and the Department of Defense (DoD) High Performance Computing Modernization Program (HPCMP) using the Air Force Research Laboratory (AFRL) Major Shared Resource Center (MSRC) under project 416, subproject 231. The authors acknowledge Dr. Ross Cotton from Simpleware® for the generation of the finite element mesh.

REFERENCES

[1] Hardy, WN et al., *Int J Impact Eng*, 15:561-586, 1994.
[2] Bešenski, N, *European Radiology*, 12:1237-1252, 2002.
[3] Nahum, R et al., *Proc 21st Stapp Car Crash Conf*, 339-366, 1977.
[4] Trosseille, C et al., *Stapp Car Crash J*, 30:235-253, 1992.
[5] Yoganandan, F et al., *J Neurotrauma*, 12:659-668, 1995.
[6] Hardy,WN et al., *Stapp Car Crash J*, 51:17-80, 2007.

SB³C2015
Summer Biomechanics, Bioengineering and Biotransport Conference
June 17-20, 2015,Snowbird Resort, Utah, USA
SB³C2015-481

A COMPUTATIONAL MODEL OF BLAST LOADING TO THE EYE: A COMPARISON WITH FIELD TESTS

Shantanu Bailoor[1], Rajneesh Bhardwaj[1], Thao D Nguyen[2*]

[1]Department of Mechanical Engineering
Indian Institute of Technology Bombay,
Mumbai, 400076
India

[2]Department of Mechanical Engineering,
The Johns Hopkins University,
Baltimore, MD, 21218
USA

INTRDUCTION

Blast has become the leading cause of ocular combat trauma since the emergence of improvised explosive devices (IED). Blasts from munitions and IDEs were responsible for 80% of ocular injuries suffered in Operation Iraqi Freedom and Operation Enduring Freedom [1,2], as well as nearly 60% of all eye injuries requiring evacuation in Iraq and Afghanistan from 2003-2004 [3,4]. Studies have shown that the majority of blast injuries to the eye are caused by high-velocity fragments and by blunt force trauma from being hit by large propelled objects or from being thrown by the blast wave [1,2,4]. The incidence of primary injuries from the blast wave and the primary injury mechanisms remain unknown. Also unknown is the effectiveness of current eye armor in preventing primary blast injuries.

To investigate the effectiveness of current armor designs, Bentz and Grimm [5] subjected the Facial and Ocular Countermeasure for Safety (FOCUS) headforms to eye-level C4 air-blasts. The experiments simultaneously measured the pressure at the ocular surface of bare (control) headforms and those protected by spectacles and goggles. Results showed both spectacles and goggles significantly reduced the maximum blast pressure at the eye. The reduction was greater for larger blasts and for goggles compared to spectacles. However, goggles produced more sustained loading on the eye

We previously developed a computational model to study the blast wave loading on the eye and the effect of the stand-off distance and incident angle. We applied the model to simulate the air-blast experiments of Bentz and Grimm. Results showed excellent agreement between modeling and experiments and provides explanations for the larger reduction with larger blasts and goggles. ective equipment to reduce the effects of blast loading on the eye.

METHODS

We employed a validated fluid-structure interaction (FSI) solver described by Bhardwaj et al. [6] to solve the compressible flow problem of a propagating blast wave. The flow computations were performed on an Eulerian grid and a sharp-interface immersed boundary method [7] was used to define the boundary conditions at the fluid-structure interface. We applied a finite element model of the head developed previously by Bhardwaj et al. [6] based on the anthropometric parameters of a specific 21-year-old male, given in Weaver et al [8]. The spectacles were modeled using dimensions of standard issue spectacles designed for ballistic protection. The head and eye were assumed to be rigid for simplicity. We initially applied a Neo-Hookean model for the material behavior of the spectacles and goggles to study whether or not they were dislodged by the blast wave. Results showed negligible deformation of the eye armor and negligible effects of the deformation on the flow field. We assumed that the eye armor was rigid in subsequent simulations.

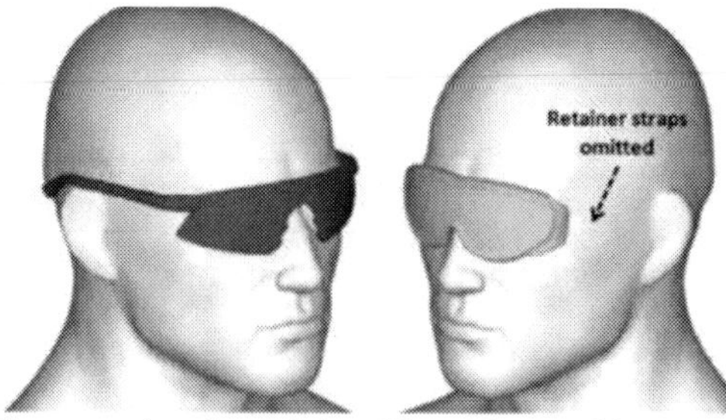

Fig 1: Computational models of headform with spectacles and goggles.

RESULTS AND DISCUSSION

Fig.2a plots the pressures computed at three locations for 0.64 kg and 6.4 kg equivalent TNT blasts. Locations 1 and 2 were in the nasal and zygomatic regions, while location 3 was at the corneal apex. Also plotted are the maximum pressure at the corneal apex measured for the the bare FOCUS headfor. For the 6.4 kg charge mass, the maximum overpressures at locations 1 and 2 were 1225 kPa and 502 kPa, which demonstrated the asymmetric distribution of the blast loading The maximum overpressures computed at the corneal apex was 750 kPa, which agreed well with the 881 kPa measured in experiments.

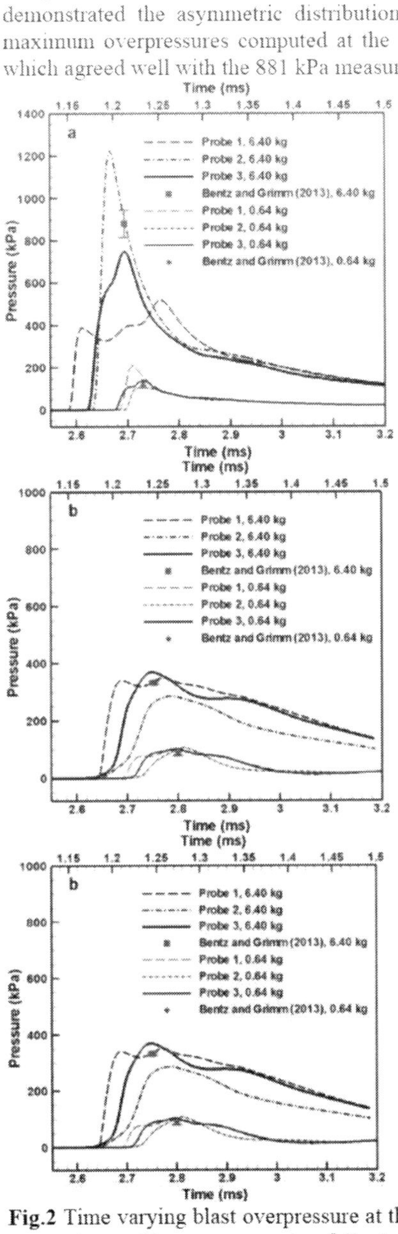

Fig.2 Time varying blast overpressure at three computed locations and comparison with measurements of Bentz and Grimm (2013) for (a) bare headforms and protected headforms employing (b) spectacles and (c) goggles for the respective 0.64 kg and 6.4 kg TNT blasts. Line plots corresponding to the 0.64 kg TNT tests refer to the lower X-axis, while those of the 6.4 kg TNT tests refer to the upper X-axis, as indicated by arrows.

Figure 2b plots the pressure computed at the 3 locations for the case of the head with spectacles. The maximum pressure measured in experiments is also plotted for comparison. The pressure computed at the three locations were significantly smaller, 348 kPa and 285 kPa for the 6.4 kg case for locations 1 and 2. This was a substantial reduction from the case of the bare headform. The pressure at the corneal apex was 369 kPa and agreed well with the 332 kPa measured by experiments

Figure 2c plots the pressure for the case of the head with goggles. The goggles provided greater reduction, achieving a 70% reduction in peak overpressures computed for the 6.4 charge mass. The pressure loading was more uniform over the eye, devoid of the sharp peaks obtained for the bare headforms, and exhibited a more sustained elevated pressure loading. The latter was caused by underwash remaining trapped in the goggles. The maximum pressure computed at the corneal apex for the 6.4 kg case was 233 kPa, which agreed well with the 204.78 kPa, measured in field test.

Overall, comparisons of the maximum blast pressure loading on the cornea between experiments and matched in the range of 3-12% for both charge masses and all 3 cases. Both the experimental and computational studies showed similar trends in reduction of peak overpressure with protective eye armor. Specifically, the goggles produced the greater reduction in maximum pressure loading on the eye. Another interesting result was that both eye armor provided a greater reduction in blast pressure loading for the larger charge mass. A larger reduction in the reflected shock front velocity was obtained for the larger charge mass, which lead to a significant decrease in the dynamic pressure of the blast wave after impact.

REFERENCES

1. Weichel ED, Colyer MH, Ludlow SE, Bower KS, Eiseman AS. Combat ocular trauma visual outcomes during operations Iraqi and Enduring Freedom. Ophthalmology 2008;115:2235-2245.
2. Mader TH, Carroll RD, Clifton SS, George RK, Ritchey P, Neville P. Ocular war injuries of the Iraqi Insurgency, January-September 2004. Ophthalmology 2006;113:97-104.
3. Ari AB. Eye injuries on the battlefields of Iraq and Afghanistan: public health implications. Optometry 2006;77:329-339.
4. Thach AB, Johnson AJ, Carroll RB, Huchun A, Ainbinder DJ, Stutzman RD, Blaydon SM, DeMartelaere SL, Mader TH, Slade CS, George RK, Ritchey JP, Barnes SD, Fannin LA. Severe Eye Injuries in the War in Iraq, 2003-2005. Ophthalmology 2008;115:377-382.
5. Bentz V and Grimm G (2013) Joint Live Fire (JLF) Final Report for Assessment of Ocular Pressure as a Result of Blast for Protected and Unprotected Eyes (Report number JLF-TR-13-01) U.S. Army Aberdeen Test Center, Aberdeen Proving Ground, MD.
6. Goroshin S, Frost DL, Levine J, Yoshinaka A, Zhang F (2006) Optical Pyrometry of Fireballs of Metalized Explosives. Propellants, Explosives, Pyrotechnics, 31: 169–181.doi: 10.1002/prep. 200600024.
7. Kingery CN, Bulmash G (1984) Airblast Parameters from TNT Spherical Air Burst and Hemispherical Surface Burst. Defence Tech Rep Report ARBL-TR-02555, U.S. Army BRL, Aberdeen Proving Ground, MD
8. Mittal R, Dong H, Bozkurttas M, Najjar FM, Vargas A, von Loebbecke A (2008) A versatile sharp interface immersed boundary method for incompressible flows with complex boundaries. Journal of Computational Physics, 2008. 227(10): p. 4825-4852.

SB^{3}C2015
Summer Biomechanics, Bioengineering and Biotransport Conference
June 17-20, 2015,Snowbird Resort, Utah, USA
SB^{3}C2015-484

TWO BUNDLES OF THE ANTERIOR CRUCIATE LIGAMENT EXHIBIT DIFFERENT MICROSTRUCTURAL PROPERTIES AND MECHANICS DURING STRESS-RELAXATION

RM Castile (1), NW Skelley (2), RH Brophy (2), SP Lake (1,2)

(1) Mechanical Eng. & Materials Science
Washington University
St. Louis, MO, USA

(2) Orthopedic Surgery
Washington University
St. Louis, MO, USA

INTRODUCTION

The double-bundle anatomic model has become widely used over the past twenty years to describe the anatomy of the anterior cruciate ligament (ACL) [1]. While cadaveric joint studies have provided supporting evidence, comprehensive studies at the tissue level are needed to determine if the material properties of the two bundles are in fact distinct or representative of a single tissue. Elucidation of these properties could help guide surgical ACL reconstruction techniques, optimize graft choice, and direct the development of novel grafts and scaffolds. Previous studies showed different macroscopic mechanical properties for the anteromedial (AM) and posterolateral (PL) bundles when tested separately [2-3]. Building upon these studies, our recent work [4] utilized a novel division-of-focal-plane quantitative polarization imaging technique to examine the material properties of samples taken from the ACL bundles during ramp-to-failure mechanical testing. We found significant differences in mechanical and microstructural properties between the two bundles; specifically, the AM bundle exhibited larger elastic moduli and failure strength compared to the PL bundle, as well as more uniformly/strongly aligned collagen fibers under load. However, the time-dependent, sub-failure properties of the human ACL have not been evaluated in a similar way, to determine if the ACL bundles have distinct tissue properties in the functional stress-strain range. Therefore, the goal of this project was to quantify the viscoelastic properties and fiber alignment of the anteromedial (AM) and posterolateral (PL) bundles in the human ACL in real-time during stress-relaxation using a novel polarization imaging technique [5]. We hypothesized that the AM and PL bundles would exhibit similar viscoelastic properties and microstructural collagen alignment during stress-relaxation.

METHODS

Sixteen donor knees with no history of knee trauma, surgery, or instability were acquired (11 male, 5 female; avg. age = 41; avg. BMI = 27). ***Sample preparation.*** All soft tissues were removed, leaving only the intact ACL and posterior cruciate ligament (PCL). An oscillating saw was used to separate the femoral condyles starting in the trochlear grove. The lateral intercondylar ridge of the femur and associated lateral bifurcate ridge between the two bundles were identified. Each bundle of the ACL was gently separated using forceps and sectioned into three pieces by making two longitudinal incisions with a number 11 scalpel, yielding a total of six samples per ligament. Free from the bony attachments, samples were sectioned to ~600 μm using a freezing-stage sliding microtome. Cross-sectional area (CSA) was measured with a 3D laser scanning system [4]. ***Mechanical Testing.*** After gluing four 0.5-mm diameter brass beads to the surface for strain analysis, samples were clamped in a tensile test machine integrated with a custom-built quantitative polarized light imaging system [5]. Using a high-resolution division-of-focal-plane polarization camera, the average direction of collagen orientation (AoP = Angle of Polarization) and the degree to which the fibers are aligned in that direction (DoLP = Degree of Linear Polarization) were quantified on a pixel-wise basis for each sample throughout the testing protocol (Fig. 1). Similar to our previous study [4], our analysis focused on average (AVG) DoLP values and standard deviation (STD) AoP values for each tested specimen, which are computed from histograms of pixelwise DoLP and AoP values extracted from the region-of-interest (ROI = tissue bounded by strain markers) of each sample. Ligament samples were preloaded to 0.1N and initial gauge length measured, Samples were preconditioned with 10 cycles of triangular wave to 3% strain, then quickly ramped to 5% strain followed by a 300-second hold to allow for stress-relaxation. Forces were acquired at 20 Hz, and digital videos were acquired at 17 fps throughout testing for optical strain analysis (using grayscale intensity images) and collagen fiber alignment analysis (i.e., DoLP and AoP).

Data Analysis. Stress was calculated as force divided by initial CSA. Peak and equilibrium stress values were computed for each curve, as well as total relaxation % (i.e., 100*(peak-equil.)/peak). For polarization analysis, thresholded clamp edges were tracked to find the exact video frame in which the sample reached the peak displacement of the ramp. For each image captured during the subsequent 300-second relaxation period, AVG DoLP and STD AoP values were extracted from a centrally-drawn ROI, and peak/equilibrium and relaxation % values were computed for these alignment parameters. Finally, relative change from peak values (i.e., Δ) were computed for stress, AVG DoLP and STD AoP, and interpolated at regular intervals throughout the relaxation period. For statistical analysis, data were averaged by bundle and *t*-tests were utilized to identify any significant differences between AM and PL samples.

RESULTS

Stress values were larger for AM samples than PL at the peak point (i.e., immediately after the applied displacement step), and remained significantly larger throughout the test until the equilibrium point (i.e., end of 300-second hold) (Fig.2A). Samples from both bundles showed significant stress relaxation (Fig.2A), which was larger for PL (36.0%) than AM (32.8%), demonstrating the viscoelastic nature of the ACL. In terms of magnitude of stress changes during relaxation, the AM bundle decreased more than the PL bundle, with statistically significant differences for the later half of the time points (Fig. 2B). In terms of collagen organization, AVG DoLP values were larger for AM than PL samples at both peak and equilibrium (Fig.3A), demonstrating more strongly (i.e., uniformly) aligned collagen in the AM bundle. During relaxation, change in AVG DoLP values was larger for AM samples, as shown by a significantly larger overall relaxation (Fig.3A) and significant changes compared to PL samples for much of the time course (Fig 3B). STD AoP values were significantly larger for PL than AM at peak and equilibrium (not shown), indicating greater variation in angles of collagen orientation. However, in contrast to other evaluated parameters, there were no differences in time-dependent ΔSTD AoP between the two bundles (not shown).

DISCUSSION

Our previous study showed that the AM bundle is mechanically stiffer/stronger than the PL bundle, and has more strongly aligned collagen, thus demonstrating different material and microstructural properties during a ramp-to-failure test [4]. The findings from the current study better define the tissue properties of the AM and PL bundles of the ACL, particularly for behavior in the lower-magnitude, functional strain range (i.e., 5%) and for time-dependent (i.e., viscoelastic) effects. The AM bundle showed significantly higher stresses compared to stresses found in the PL bundle at the same strain-level and time points, and the time-dependent changes in stresses were also different by bundle. In terms of microstructural collagen organization, the AM bundle showed more strongly aligned fibers (i.e., AVG DoLP), which were more uniformly aligned (i.e., STD AoP) in the loading direction. During the relaxation period, the change in alignment was also significantly different between bundles. Thus, while AM samples showed stronger collagen fiber alignment, they also demonstrated more change in alignment during the hold period. This may explain the larger decrease in stress for AM bundles as more fiber reorganization occurred.

These results complement our previous work [4] and continue with the pattern that the AM bundle appears to be the more "dominant" bundle, in both ramp-to-failure and stress-relaxation mechanical characterization. Several previous studies have described ACL anatomy [2,3,7,8] but there has been a relative lack of quantitative tissue level data comparing the bundles of the ACL. The findings of this study build on our previous work and provide additional insight into differences between the two bundles. With a deeper understanding of ACL properties, future investigation may be able to guide reconstruction techniques, optimize graft choice, and direct the development of novel grafts and scaffolds for ACL reconstruction.

ACKNOWLEDGEMENTS

Samples collected through the Washington University Orthopaedic Cadaver Lab, Mid-America Transplant Services Center (St. Louis, MO), and Science Care (Phoenix, AZ). Funding provided by Orthopaedic Research and Education Foundation.

REFERENCES

[1] Steckel H+ Scand J Med Sci Sports 2007. [2] Butler DL+ JBiomech 1986. [3] Butler DL+ JBiomech 1992. [4] Skelley NW+ AJSM *In press*. [5] York T+ J Biomed Opt 2014. [7] Arnoczky SP+ Clin Orthop Relat Res. 1983. [8] Odensten M+ J Bone Joint Surg Am. 1985.

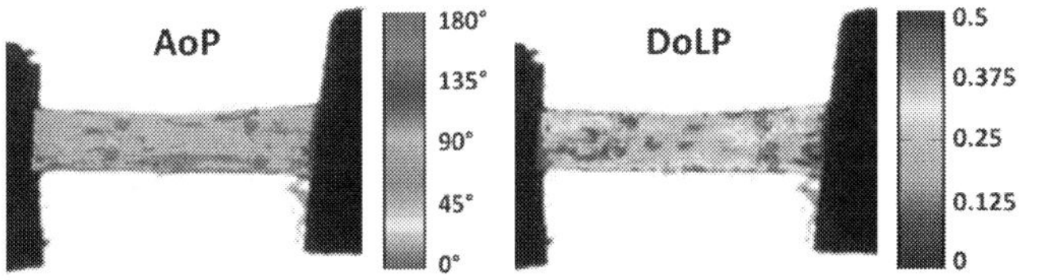

Fig. 1: A representative sample showing polarization parameters. Left: angle of polarization (AoP) represents the thickness-averaged orientation of collagen fibers (horizontal direction = 90°). Right: degree of linear polarization (DoLP) quantifies how strongly aligned fibers are in the average direction (larger numbers correspond to stronger alignment).

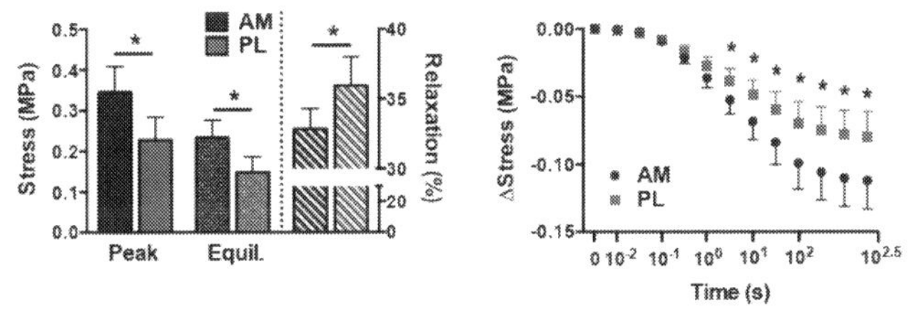

Fig. 2: (A) Stress values were larger for AM samples than PL samples at both peak and equilibrium points of the stress-relaxation test, and relaxation percentage showed significant differences between the two bundles; (B) Stress values decreased during the hold-period, with the AM bundle showing larger magnitude of changes than the PL bundle (* for $p<0.05$).

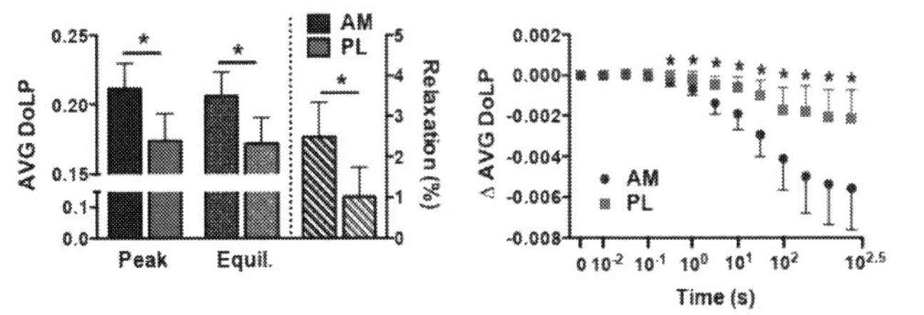

Fig. 3: (A) AVG DoLP values were larger for AM samples at both peak and equilibrium points of the stress-relaxation test, and decreased significantly more over time than PL samples; (B) ΔAVG DoLP values show that the AM samples showed more alignment change than PL over time, indicating greater changes in the strength of collagen fiber alignment during stress-relaxation. (* for $p<0.05$).

THE EFFECT OF SIZE AND LOCATION OF TEARS IN THE SUPRASPINATUS TENDON ON POTENTIAL TEAR PROPAGATION

James R. Thunes (1), Siladitya Pal (1), R. Matthew Miller(1), Richard E. Debski (1), Spandan Maiti (1,2)

(1) Department of Bioengineering
University of Pittsburgh
Pittsburgh, PA, USA

(2) Department of Chemical Engineering
University of Pittsburgh
Pittsburgh, PA, USA

INTRODUCTION

Tears in the rotator cuff are a common problem in patients over the age of 55[1]. The most common initial treatment option, physical therapy, fails in up to 50% of patients necessitating surgical intervention. One reason for failure of physical therapy is the growth of the tear, but the reasons for this phenomenon are not yet well understood. However, biomechanical stress state near a tear is thought to be the primary factor determining its propagation. In addition, cadaveric studies have revealed that supraspinatus (SSP) tendon hosts the majority of tears in the rotator cuff. Accordingly, the stress state of the SSP tendon is expected to play a major role in the propagation characteristics of the rotator cuff tear.

The objective of current study is to examine the effect of tear size and location on the stress state of the SSP tendon. Towards this end, a computational model of the supraspinatus tendon containing a tear of different sizes and at different locations was created. Principal stresses in the vicinity of the tear, while the tendon was loaded uniaxially, were enumerated. Simulations revealed that larger tears located anteriorly in the tendon experienced higher principal stress at their tips. It was also found that for majority of the tears, stress concentration was higher at their posterior tip.

METHODS

Inventor 2013 (Autodesk, Mill Valley, CA) was used to create a solid model of the supraspinatus tendon from 2D images of a human cadaveric SSP tendon (Figure 1a). The thickness of the tendon was approximated from literature[2]. A razor-sharp full-thickness tear was placed at six locations on the model tendon. The tears were oriented in the Anterior-Posterior (AP) direction and at each location, tears of three sizes (4, 8, or 16 mm) were modeled. The model was developed to represent uniaxial tensile test of the SSP tendon[3].

A custom finite element software was employed to obtain biomechanical stress state of the tendon model. The tendon was meshed with 37,000 4-noded tetrahedral elements in Trelis (csmisoft, American Fork, UT). The lateral surface of the tendon was provided with zero-displacement boundary condition. The medial surface was subjected to an applied stretch acting along the Medial-Lateral (ML) axis. The finite element model was stretched monotonically until peak principal stress at any point in the model reached the corresponding prescribed strength of the tendon material. The stretch at which the principal stress at a point reaches the corresponding material strength was defined as the critical stretch λ_C. The loading was applied over at least 1000 steps to capture the nonlinear constitutive response of the model.

A transversely orthotropic material model, based on Weiss, et al.[4], was adopted in this study. The model is composed of two parts: an isotropic, incompressible Mooney-Rivlin material representing the non-collagenous matrix and an exponential fiber model representing the collagen fibers. The strain energy function is given as

$$\Psi = \Psi_M + \Psi_f$$

$$\Psi_M = \frac{1}{2}\kappa(J-1)^2 + c_1(\bar{I}_1 - 3) + c_2(\bar{I}_2 - 3)$$

$$\Psi_f = \begin{cases} \frac{k_1}{2k_2}\left[\exp\left(k_2(\bar{I}_4 - 1)^2\right) - 1\right] & \bar{I}_4 > 1 \\ 0 & \bar{I}_4 \le 1 \end{cases}$$

where $\bar{I}_1$, $\bar{I}_2$, and $\bar{I}_4$ are the first and second invariants and the fourth pseudo-invariant of the right Cauchy-Green tensor. The material model requires four material parameters: c_1 and c_2 for the non-collagenous matrix, and k_1 and k_2 for the collagen fibers. It is assumed that the fibers are all aligned along the ML axis.

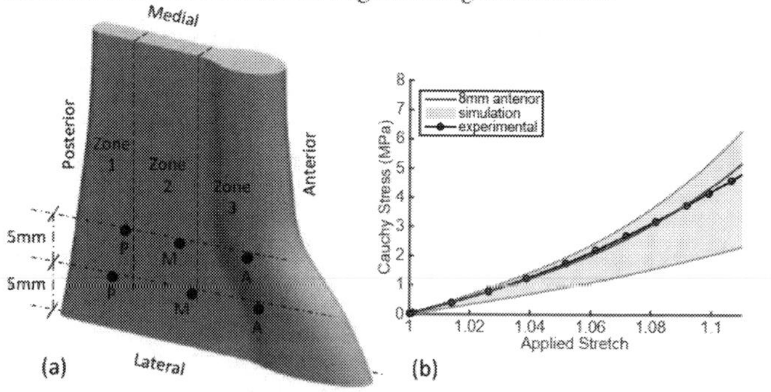

Figure 1: (a) The supraspinatus tendon and centers of the tears are shown. P, M, and A refer to tears in the posterior, middle and anterior regions, respectively. Zones for material property variation are shown by dotted lines. (b) Stress-strain responses for experimental test (black), computational simulations (red line and pink region).

Itoi, et al.[5] report a variation in the material properties along the AP direction; the elastic modulus in the anterior region was twice that in the posterior region. For this reason, the tendon model was split into three regions, each spanning 1/3rd of the tendon along the AP direction. The posterior region was designated Zone 1, the middle region Zone 2, and the anterior region Zone 3. The stiffness in Zone 3 was twice that of Zone 1 while the stiffness in Zone 2 transitioned between these values.

The material parameters were extracted by fitting computationally obtained uniaxial load-displacement curve to the experimental data using the Matlab least square optimization routine lsqcurvefit. To reduce the number of parameters to be fit, the matrix parameters were taken as constant across the tendon.

To quantify the initiation of tear propagation, it was assumed that tears would grow if the principal stress in the material reached the material strength. As with the elastic modulus, the strength increased from the posterior to anterior regions with the strength in the anterior region (20 MPa) twice that of the posterior region (10 MPa).

RESULTS

The material parameters were fit using the computational case that most closely represented the experimental test. This was the case of 8 mm tear located at the anterior location 5 mm from the lateral edge. The computational stress-strain response fit the experimental response with $r^2 = 0.997$ (Figure 1b). Material parameters for the posterior region were determined to be $c_1 = c_2 = 1$ MPa, $k_1 = 5.23$ MPa, $k_2 = 20$ MPa, while those for the anterior region were be $c_1 = c_2 = 1$ MPa, $k_1 = 5.23$ MPa, $k_2 = 30$ MPa. The stress-strain responses for all tear sizes and locations studied in this report are shown in pink. The range in stiffness is due to the variation of location and size of the tears.

Tears cause a stress concentration at the tips of the tears (Figure 2a). Due to the shape of the tendon and the difference in the material properties at the tear tips, the stress concentrations are not symmetric. This asymmetry, and additionally the variation of material strength means that the anterior and posterior tear tips reach the critical stretch at different times. In the majority of cases (14 out of 18), the posterior tip reaches the critical stretch before the anterior tip with an average decrease of 30%. There was a noticeable decrease in λ_c as tear size increased with the critical stretch decreasing between 9 and 30% depending on tear location.

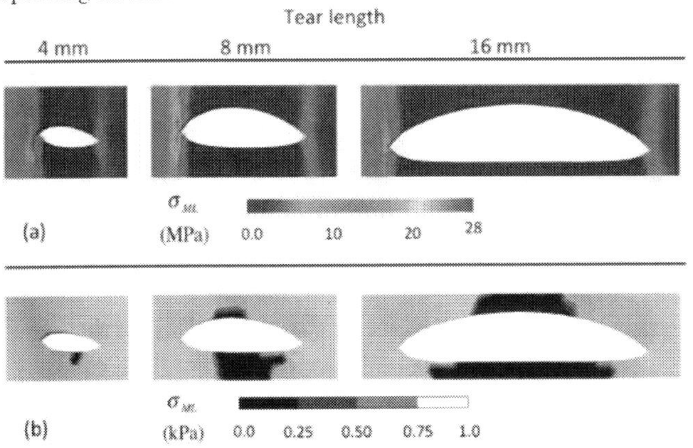

Figure 2: (a) ML stress contours in the vicinity of an 4, 8, or 16 mm tear 5 mm from the lateral edge, anterior location under $\lambda = 1.12$. (b) Stress contours of the same tear, but with the stresses scaled such that low stresses are shown in black.

In addition to stress concentrations at the tear tips, there are regions of low stress in the middle of the tear (Figure 2b). As these regions have much lower stress than the surrounding tissue (less than 1 kPa), they are considered as stress-shielded regions. There is a correlation between the size of the stress-shielded region and the size of the tear ($r^2 = 0.96$) with the region increasing with the tear size.

DISCUSSION

A computational model for human supraspinatus tendon containing a tear was constructed in this study. Uniaxial stress-strain response of the model was validated against experimental data. As opposed to other studies of the rotator cuff tendons in the literature, this model used a fiber-reinforced material that varied across the width of the tendon.

For the majority of the tears, the critical stretch for the posterior tip was lower than for the anterior tip. Thus, the tear is expected to propagate posteriorly. Similar trends have been seen in *in vivo* testing. The data suggests that the reason for this propensity for posterior growth is due to the variation in material properties across the tendon, in addition to tendon geometric features and loading conditions.

Regions of low stress were seen on the edges of the tears. These regions grew as tear size increased. *In vivo*, these regions could lead to degradation of the mechanical properties of the tendon over time.

The model used for this work is highly idealized. Specifically, it does not include the surrounding tendons, the humeral head, or localized disorganization of the collagen fibers. Additionally, the tear shapes studied were based on surgically introduced cuts instead of chronic tears. However, sufficient detail has been incorporated to allow for the study of general trends in the alteration of the stress field in the vicinity of a tear.

Finally, the propagation criterion was defined to be when the maximum principal stress reached the local material strength. This may be suitable for studying the initiation of tear growth, but it does not capture the behavior after propagation begins. For example, a tear may propagate posteriorly and then switch to propagating in the medial direction. Criteria such as the one used here are unable to capture these phenomena.

Future work will focus on high fidelity tendon geometry from CT scans for patient specific models.

ACKNOWLEDGEMENTS

Support for this work was provided by Sandia National Labs via an NPSC Graduate Fellowship and the Orthopaedic Robotics Laboratory at the University of Pittsburgh.

REFERENCES

1. Yamaguchi, K., et al., *The demographic and morphological features of rotator cuff disease. A comparison of asymptomatic and symptomatic shoulders.* J Bone Joint Surg Am, 2006. **88**(8): p. 1699-704.
2. Matsuhashi, T., et al., *Tensile properties of a morphologically split supraspinatus tendon.* Clin Anat, 2013.
3. Miller, R.M., et al., *Strain Distribution due to propagation of tears in the anterior supraspinatus tendon.* Journal of Orthopaedic Research, 2014.
4. Weiss, J.A., B.N. Maker, and S. Govindjee, *Finite element implementation of incompressible , transversely isotropic hyperelasticity.* Computer Methods in Applied Mechanics and Engineering, 1996. **135**(1-2): p. 107-128.
5. Itoi, E., et al., *Tensile properties of the supraspinatus tendon.* J Orthop Res, 1995. **13**(4): p. 578-84.

OPTIC NERVE SHEATH MECHANICS IN VIIP SYNDROME

J. Raykin (1), A. Feola (1), R. Gleason (1), L. Mulugeta (2), J. Myers (3), E. Nelson (3), B. Samuels (4), C.R. Ethier (1)

(1) Georgia Institute of Technology
Atlanta, GA

(2) Universities Space Research Association
Houston, TX

(3) NASA Glenn Research Center
Cleveland, OH

(4) University of Alabama at Birmingham
Birmingham, AL.

INTRODUCTION

Visual Impairment and Intracranial Pressure (VIIP) syndrome results in a loss of visual function and occurs in astronauts following long-duration spaceflight [1]. Understanding the mechanisms that lead to the ocular changes involved in VIIP is of critical importance for space medicine research. Although the exact mechanisms of VIIP are not yet known, it is hypothesized that microgravity-induced increases in intracranial pressures (ICP) drive the remodeling of the optic nerve sheath, leading to compression of the optic nerve which in turn may reduce visual acuity. Some astronauts present with a kink in the optic nerve after return to earth (Figure 1), suggesting that tissue remodeling in response to ICP increases may be taking place. The goal of this work is to characterize the mechanical properties of the optic nerve sheath (dura mater) to better understand its biomechanical response to increased ICP.

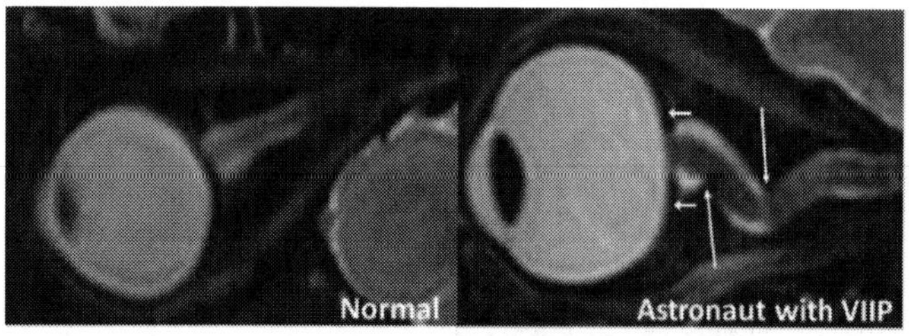

Figure 1: Tortuous optic nerve observed in an astronaut with visual disturbances following long duration space flight (taken from [2]). Long arrows: kinks in optic nerve; short arrows: posterior globe flattening.

METHODS

Porcine eyes were obtained from a local abattoir immediately following death. The optic nerve was cut away from the globe, cleaned of any periorbital tissue, and the dura was carefully peeled away from the optic nerve proper. The dura was then cannulated onto a custom built mechanical testing device that applied luminal pressure and axial stretch. Biaxial mechanical tests were performed at 37°C and in PBS. Pressure was controlled via a syringe pump, the outer diameter was recorded using a CCD camera, and the axial force was recorded using a force transducer. Pressure-diameter tests (6 cycles of 0-60 mm Hg) were performed at axial stretch ratios of λ_z =1, 1.05, 1.1, 1.15, and 1.2, where λ_z was defined as the ratio of the deformed length to the undeformed (unloaded) length. Pressures were chosen to include the range of possible physiological and pathophysiological ICP pressures. At each axial stretch, the dura was preconditioned until a repeatable pressure-diameter curve was observed. The mean circumferential and axial stresses were calculated using Laplace's law for a cylinder with closed ends.

Collagen organization at a luminal pressure of 10 mmHg was visualized using second harmonic microscopy (SHG). The mean orientation angle was obtained using a previously described fast Fourier transform (FFT) algorithm [3].

RESULTS

Biaxial testing results demonstrate that the dura exhibits a nonlinear stiffening response similar to most soft biological tissues (Figure 2). Due to the high variability in the mechanical response of the dura, a representative test is shown. We note that even though the tangent moduli differed between individual samples (up to 5x), the trends in the stress-strain curves were consistent. The tissue exhibits a large initial circumferential deformation ($\lambda_\theta \sim$ 1.29 at 10 mmHg and λ_z

= 1), followed by a marked strain-stiffening response (Figure 2c). The stiffening response occurs at progressively lower pressures with increased axial stretch. Axial stress increases with increasing pressure and axial stretch (Figure 2d). The collagen fibers of the optic nerve sheath are primarily organized in the axial direction (Figure 3). The mean angle of orientation was 0.5° from the horizontal axis in Figure 3.

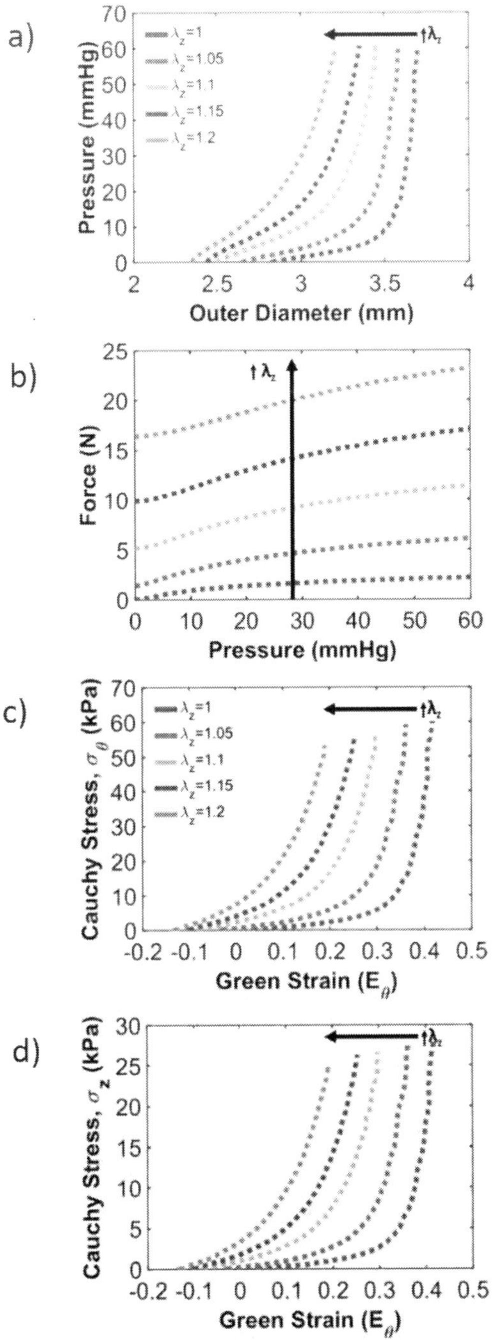

Figure 2: Representative a) pressure-diameter, b) axial force-pressure, c) circumferential stress (σ_θ) and d) axial stress (σ_z) response of the porcine dura.

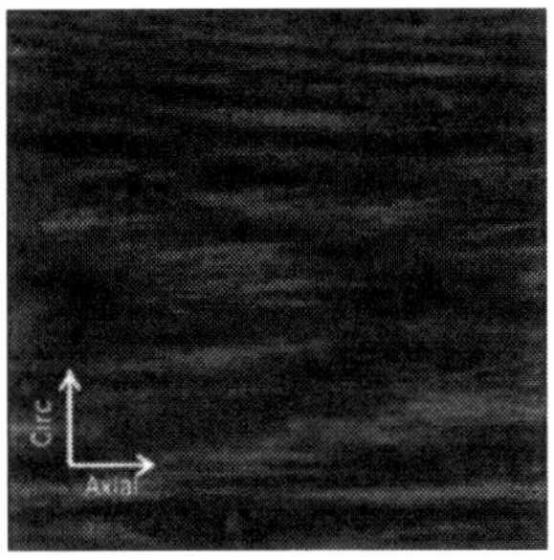

Figure 3: Representative SHG image of the porcine dura at 10 mmHg. The horizontal axis corresponds to the axis of the optic nerve.

DISCUSSION

The dura exhibits similar mechanical properties to other soft tissues. Our data indicate that large deformations occur at pressures 0-10 mmHg, which is consistent with observations in humans [4]. Compared to blood vessels, a high variability in the mechanical response was seen in the dura. The reason for this variability is currently unclear. Even though all astronauts likely experience an increase in ICP, not all of them develop the ophthalmic changes that are characteristic of VIIP. The large variation in mechanical properties that we observed could partially explain the inconsistent presentation of VIIP among astronauts. The axial orientation of the collagen fibers and a lack of circumferential cross-fibers could result in significant circumferential distension of the dura during increases in ICP. Due to Poisson effects this would lead to an axial contraction of the dura and an increase in axial forces. The increased axial tension, if sustained in the long term, could lead to cellular remodeling to lessen the axial tension, resulting in the kink seen in some of the astronauts that present with VIIP. Including these observations into computational models of the optic nerve sheath will help improve their accuracy and enable prediction of possible risk factors for VIIP.

ACKNOWLEDGEMENTS

This work was funded by NASA, grant number NNX13AP91G. Porcine eyes were generously provided by Holifield Farms (Covington, GA). We thank Devon Griffin for administrative support.

REFERENCES

1. Otto, C., *The Visual Impairment Intracranial Pressure Syndrome in Long Duration US Astronauts*. 2013.
2. Kramer, L.A., et al., *Orbital and intracranial effects of microgravity: findings at 3-T MR imaging*. Radiology, 2012. **263**(3): p. 819-827.
3. Wang, R., et al., *Differential mechanical response and microstructural organization between non-human primate femoral and carotid arteries*. Biomechanics and modeling in mechanobiology, 2014: p. 1-11.
4. Kimberly, H.H., et al., *Correlation of optic nerve sheath diameter with direct measurement of intracranial pressure*. Academic Emergency Medicine, 2008. **15**(2): p. 201-204.

FINITE ELEMENT MODEL OF CERVICAL PESSARY IN USE: EVALUATING MECHANICAL INTERVENTIONS FOR PRETERM BIRTH

Michael J. Fernandez (1), Michael D. House (2), Noelia M. Zork (3), Joy S. Vink (3), Ronald J. Wapner (3), Sachin R. Jambawalikar (4), Kristin M. Myers (1)

(1) Mechanical Engineering
Columbia University
New York, NY, USA

(2) Obstetrics and Gynecology
Tufts Medical Center
Boston, MA, USA

(3) Obstetrics and Gynecology
Columbia University Medical Center
New York, NY, USA

(4) Radiology
Columbia University Medical Center
New York, NY, USA

INTRODUCTION

Preterm birth (PTB) impacts about 15 million babies annually worldwide, including 12% of newborns in the United States [1,2]. Almost one million preterm infants die annually, and those who survive PTB can have developmental issues including problems with vision and hearing [1]. Interventions to prevent PTB are limited and efficacy is questionable [3]. Pregnant women who have a "short cervix," the condition in which the non-dilated cervical length is less than or equal to 25 mm, are more likely to deliver preterm [4]. The cervical pessary is a proposed non-invasive treatment for cases of short cervix.

The Arabin cervical pessary [5] is a ring-shaped, silicon rubber, non-surgical implant which encircles the portion of the cervix inside the vagina above the external os. It is intended to provide mechanical support to help the cervix stay closed and to correct or maintain the posterior angle of the cervical canal. The bottom edge of the pessary rests on the posterior vaginal wall, holding it in place on the cervix.

While the pessary is designed to provide mechanical support to the pregnant cervix, little work has been done to understand the mechanical stresses and strains on the cervix, beneficial or otherwise, that arise from its use. A thorough 3D biomechanical simulation can elucidate the mechanisms by which a pessary might arrest cervical shortening and enable a large breadth of comparisons between pessary design, clinical placement, and material properties.

In this study, we simulate the use of the Arabin cervical pessary in a pregnant patient with a short cervix who is at increased risk for PTB. The simulation is constructed from high resolution MRI image stacks which we segmented and used to generate a 3D finite element model (Figs. 1). This serves as a platform for research on the stress and strain states resulting from pessary use, as well as the influence of pessary size and position on cervical function.

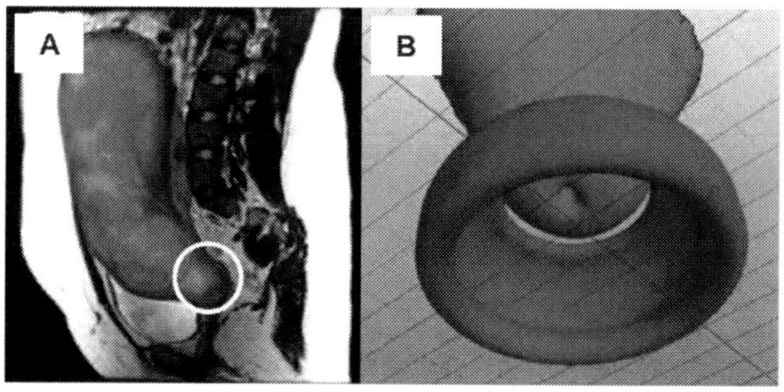

Figures 1: (A) Segmented uterus and cervix (circled in white) and (B) 3D finite element model of cervix (magenta) and pessary (blue).

METHODS

Patients were recruited and consented at Columbia University Medical Center under an IRB-approved protocol. One patient with a short cervix is featured in this study. This patient was 25 years old and was at 28 weeks of gestation at the time of the scan.

The MRI scan was performed on a 1.5 Tesla GE HDxt scanner with a 12 channel phased array torso coil. We used a Sagittal T2 Cube, 3D Fast Spin Echo pulse sequence. The slice thickness was 2.2 mm (interpolated to 1.1 mm) with a field of view (FOV) of 26 cm and image matrix of 256 x 256 for each slice. The resulting data had a resolution of 1.1 x 0.5 x 0.5 mm^3 in the sagittal, axial, and coronal directions, respectively. 90 sections were acquired in an acquisition time of approximately 6 minutes with no breath holding.

The MRI data set was imported into Simpleware ScanIP (Version 7.0, Exeter, UK) for segmentation and meshing. On appropriate image

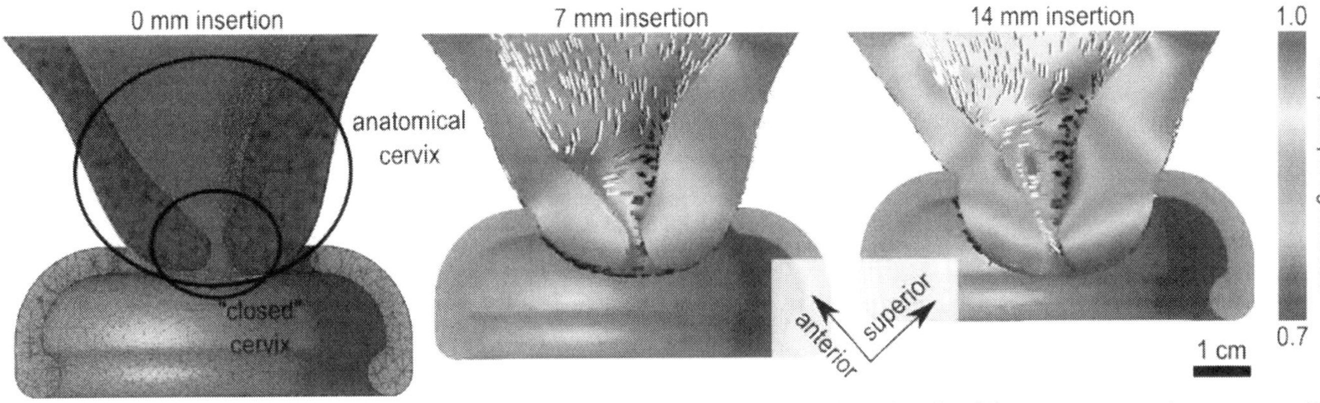

Figure 2: Insertion of cervix into pessary. Colors inside the cervix indicate highest levels of tissue compression as quantified by the 3rd principal right stretch; short line segments indicate the local direction of compression.

slices, the uterus and cervix were identified and manually segmented according to visible organ features and anatomical reference points. From the resulting 3D voxelized segmentation, a surface mesh and then a volumetric finite element mesh were generated. We created a model of a typical cervical pessary according to publically available specifications [6] using a commercial CAD package (PTC Creo 3.0), then meshed it and manually positioned it within the simulation.

The model was built in FEBio 2.2 [7]. Quadratic tetrahedral elements were used for both the cervix and the pessary in this model. The cervix contains about 23,800 elements with an average volume of 2.3 mm^3, and the pessary contains 23,900 elements with an average volume of 1.2 mm^3. We removed the uterine body from the model to reduce computation time, fixed the top edge of the cervix in 3D space, and prescribed the pessary's rigid translation spanning 0 to 14 mm with 0 mm representing negligible contact and 14 mm representing full insertion.

A cervical material model detailed in previous work by our group [8] was used for this model. Simply, it is a spherical continuous fiber distribution composite model with a neo-Hookean ground substance. In the model, fibers resist only tensile strain, and affine motion is assumed with respect to the ground substance. The pessary was defined to act as a rigid body due to its high stiffness relative to cervical tissue (The simulation was run with silicone rubber material properties then compared to the rigid body model to validate this assumption and maximum pessary strains were under 0.5%).

The sliding contact interface between the pessary and the cervix was enforced with the augmented Lagrange multiplier method in FEBio (parameters: augmentation tol.=0.2, penalty=1, facet-on-facet, search tol.=0.01, no friction).

RESULTS

Circumferentially aligned compressive strains on the order of 30% are observed in the anatomical cervix (see Fig. 2) at 14 mm insertion. This is consistent with a reduction in the diameter of the dilated cervical canal on the same order. The 3rd principal right stretch map was plotted on a cross-section view in figure 2 for three insertion depths. Also note the low levels of compression appearing in the vertical (or axial to the cervix) direction above the contact with the pessary.

Concentrations in the von Mises stress above 2 kPa are seen at the contact point between the pessary and the exterior of the cervix (Fig. 3A). Stresses up to 5 kPa are observed on the interior of the anatomical cervical canal (Fig. 3B) opposite the pessary contact area. Note that our results are based on the assumption of a zero-stress state in the reference configuration.

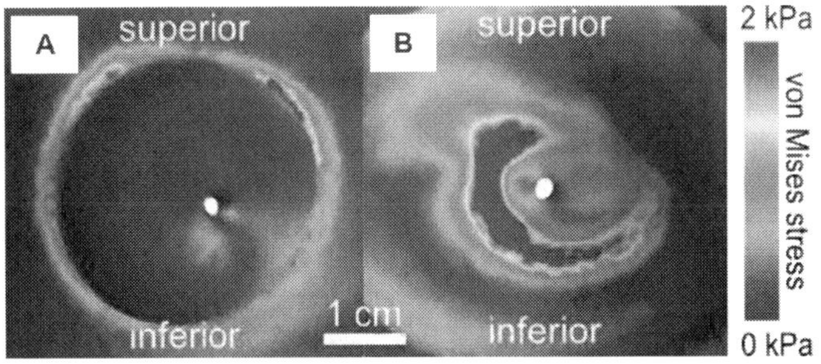

Figure 3: Effective (von Mises) stress of the exterior (A) and interior (B) of the cervix with pessary at 14 mm insertion.

DISCUSSION

Our study successfully illustrates the mechanism by which a cervical pessary may reduce dilation in patients with a short cervix. These findings are consistent with the emerging evidence that cervical pessaries can reduce the risk of preterm delivery in women with a short cervix [5]. To the authors' knowledge, this study represents the first finite element simulation of the cervical pessary's interaction with the cervix, and adds to the growing body of work related to biomechanical modeling of pregnancy [9-11].

With these tools to analyze the mechanical deformation and stresses caused by using a pessary, design changes to improve efficacy and increase patient comfort can be more readily achieved.

ACKNOWLEDGEMENTS

This work was made possible by an Imaging Pilot Award from Columbia University's Irving Institute for Clinical and Translational Research. We would also like to thank the Office of the Provost at Columbia University and our undergraduate assistant, Noah Zweben.

REFERENCES

[1] World Health Organization, "Fact sheet No. 363," 2014.
[2] Solomon, C et al., *N Engl J Med.* 370(3): 254–261, 2014.
[3] US IOM Committee on Understanding Premature Birth, 2007, http://www.ncbi.nlm.nih.gov/books/NBK11373/
[4] Iams, J et al. *N Engl J Med.* 334(9):567-72, 1996.
[5] Goya, M et al., *The Lancet*, 379(9828):1800-1806, 2012.
[6] Arabin, B et al., 2015, http://www.dr-arabin.de/e/cerclage.html
[7] Maas, S et al., *J Biomech Eng.*, 2001, http://www.febio.org
[8] Fernandez, M et al., *Comput Method Biomec*, 2015 (in review).
[9] House, M et al., *Clinical Anatomy*, 26:97–104, 2013.
[10] Jing, D et al., *J Biomech*, 45(3):455-460, 2011.
[11] Mahmoud, H et al., *J Biomech Eng*, 135(2):021009, 2013.

SB³C2015
Summer Biomechanics, Bioengineering and Biotransport Conference
June 17-20, 2015,Snowbird Resort, Utah, USA
SB³C2015-491

SPATIOTEMPORAL CELL-MATRIX INTERACTIONS DURING *EN MASSE* MIGRATION OF FIBROBLASTS ON COLLAGEN MATRICES

Altug Ozcelikkale (1), Frederick Grinnell (2), Bumsoo Han (1, 3)

(1) School of Mechanical Engineering
Purdue University
West Lafayette, IN, USA

(2) Department of Cell Biology
University of Texas Southwestern Medical Center
Dallas, TX, USA

(3) Weldon School of Biomedical Engineering
Purdue University
West Lafayette, IN, USA

INTRODUCTION

Fibroblasts play a critical role in wound healing by synthesizing and remodeling the extracellular matrix (ECM). For successful healing fibroblasts around the wound need to be activated and migrate into wound site [1]. The later stages of granulation tissue formation and wound closure depend on proper infiltration of provisional matrix by fibroblasts. Mechanical cues induced on cells by the ECM are now known to play a major role in migration [2, 3]. The regulation of single cell migration by ECM mechanical properties and microstructure are well studied [4, 5]. However it has been reported that fibroblasts are recruited to the wound site as a group, by so-called *en masse* movement, rather than as a single cell [6]. As a result traction forces generated simultaneously by multiple cells can deform the compliant ECM and result in bi-directional interactions between cells and their mechanical environment.

While active role of cells in modulating their environment during *en masse* migration have recently been recognized [7] the dynamics of this process have not been clearly understood. In order to address this gap we performed time-resolved measurements of cell movement and ECM deformation during *en masse* migration of fibroblasts on collagen matrices. Collagen densities were varied to generate different levels of hydrogel stiffness. Then migration on these compliant substrates was compared to that on collagen coated coverslips.

METHODS

Collagen matrices were polymerized from neutralized solutions of type-1 rat tail collagen. Then, early passage human dermal fibroblasts were seeded uniformly across one side of the matrix and allowed to migrate to the unpopulated region of the substrate for 24 hours (Figure 1). A promigratory growth factor medium that contained 50 ng/mL PDGF and 1% fetal bovine serum was used for all experiments. The sample was loaded onto a microscope stage incubator and fluorescence images of cell and ECM movements were acquired every 15 minutes. Time-lapse cell movement and collagen matrix deformation were quantified by single particle tracking of quantum dots (QD) internalized by the cells and particle image velocimetry of 10 μm sized fluorescent beads embedded in ECM, respectively. Deformation is reported in terms of dilatation which indicates local volumetric expansion and compression.

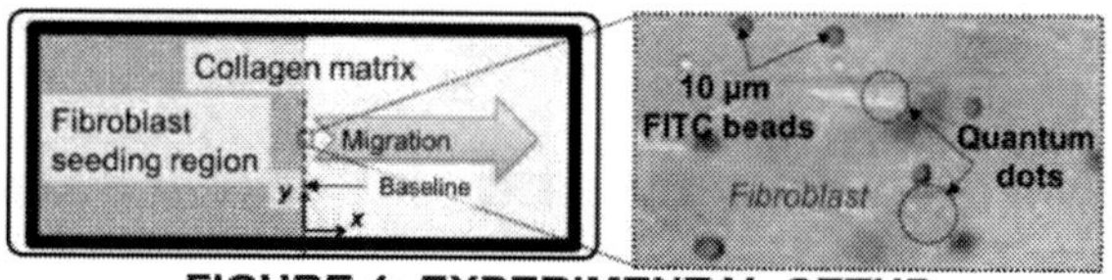

FIGURE 1: EXPERIMENTAL SETUP.

RESULTS

Migration on collagen matrices was associated with spatiotemporal matrix deformation (Figure 2). Within the first few hours, seeding region ($x < 0$) underwent gradual compaction indicated by the negative values of dilatation. The compaction is considered to be due to attachment and spreading of fibroblast. An increasing expansion in the migration region was also observed. At the onset of migration the trend was reversed and matrix in migration region was compressed. The region of compression increased as the migration front advanced (Fig. 2A). Deformation profiles indicate that after 24 hours, there was a net expansion in the seeding region and a net compression in the migration region particularly in the vicinity of the baseline. The magnitude of compression was greater and the region affected was wider for low collagen density (3 mg/mL) when compared to high collagen density (6 mg/mL) (Figure 2B).

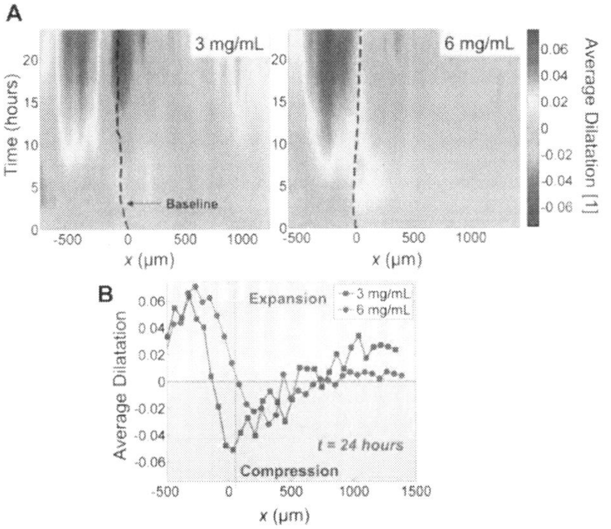

FIGURE 2: SPATIOTEMPORAL DEFORMATION OF COLLAGEN MATRICES DURING FIBROBLAST MIGRATION (B) DEFORMATION PROFILES AT 24 HOURS.

The migration behavior of cells on collagen matrices was different from that on coverslip (Figure 3A). At low collagen density, cells gradually accumulated at several locations across the baseline and migrated outwards into the unpopulated region from these focal points (yellow circles). This clustering of cells was significantly reduced in high density collagen matrix and also was not observed on coverslip. The latter cases instead showed a uniform stream of cells (yellow dashed arrow). Cell morphology was different at near and far away from the baseline. The migration field near the baseline was densely populated with cells that were partially overlapped and showed elongated morphology. Individual cells that migrated far away from baseline exhibited spreading with extended lamellipodia (Figure 3B).

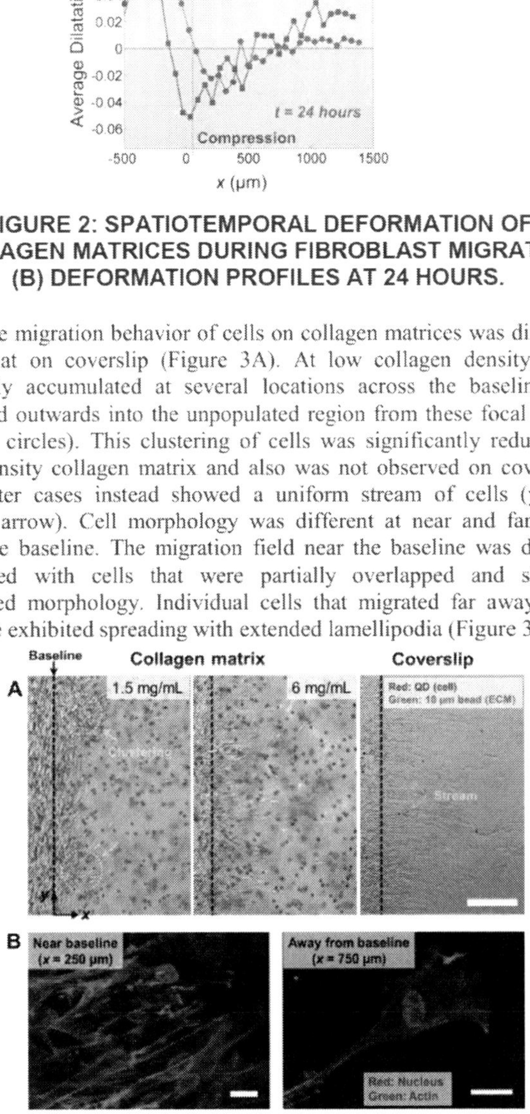

FIGURE 3: (A) FIBROBLAST MIGRATION ON COLLAGEN MATRIX VS. COVERSLIP. SCALE BAR IS 200 μm (B) CELL MORPHOLOGY NEAR VS. FAR AWAY FROM MIGRATION BASELINE. SCALE BARS ARE 20 μm.

Migration of fibroblasts on collagen matrices was hindered and delayed when compared to migration on coverslip. The number of migrating cells and the penetration depth of the cell front on collagen matrices were significantly lower than coverslip (Figure 4A). Lower collagen density resulted in less penetration. On the other hand, cell migration on collagen matrix was faster than on coverslip (Figure 4B). In all cases, the speed of the cells increased as they moved further away from the baseline. However the onset of migration was delayed on collagen matrix. It took between 6-12 hours for cells to begin migrating on collagen while migration on coverslip initiated within 4-5 hours.

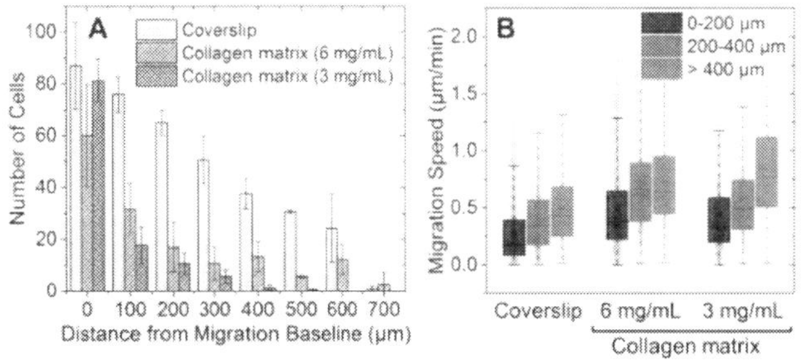

FIGURE 4: (A) DISTRIBUTION OF MIGRATING CELLS AT 24 HOURS. (B) CELL MIGRATION SPEED AT DIFFERENT DISTANCES FROM BASELINE.

DISCUSSION

Compliant substrates such as collagen matrices can represent mechanical aspects of the granulation tissue in early-phase wound healing. Our results show that *en masse* cell movement on such substrates can result in significant amount of matrix deformation. The ECM is remodeled by cells prior to migration Since migration on collagen matrices is also delayed, this initial phase may be required for initiation of *en masse* migration. The increasing expansion in the migration region prior to beginning of migration indeed hints local stiffening of the matrix, which could enable it to support traction force generated by cells and eventual *en masse* migration. The deformation induced delay in migration can explain the reason why cells on collagen matrices traveled shorter distances while having higher 'spontaneous' velocities than those of cells on coverslip.

Clustering of cells observed in low collagen density could have been caused by non-uniform deformation across the baseline. Such clustering of fibroblasts has been reported previously with serum containing growth factor environments [8]. In this study a promigratory growth factor environment with reduced serum content was used in order to alleviate the cluster formation through this route.

ACKNOWLEDGEMENTS

This work was partially supported by the B.S.F. Schaefer Award and the Discovery Park Fellowship.

REFERENCES

[1] Singer, A. J., et al., 1999, New England Journal of Medicine, 341(10), pp. 738-746.
[2] Discher, D. E., et al., 2005, Science, 310, pp. 1139–1143.
[3] Wiegand, C., et al., 2013, Wound Repair and Regeneration, 21, pp. 793–799.
[4] Hadjipanayi, E., et al., 2009, Cell Motil Cytoskeleton, 66, pp. 121–128.
[5] Miron-Mendoza, M., et al., 2010, Biomaterials, 31, pp. 6425–6435.
[6] McClain, S. A., et al., 1996, American Journal of Pathology, 149(4), pp. 1257-1270.
[7] Pan, Z., et al., 2013, Journal of Investigative Dermatology, 133, pp. 2471–2479.
[8] Rhee, S., et al., 2010, Experimental Cell Research, 316, pp. 232–244.

SB^{3}C2015
Summer Biomechanics, Bioengineering and Biotransport Conference
June 17-20, 2015, Snowbird Resort, Utah, USA
SB^{3}C2015-492

MECHANICAL PARAMETERS CHARACTERIZATION OF EMBRYONIC MOUSE HEARTS FOR STUDYING HUMAN CONGENITAL HEART DISEASE

Andres Rubiano
Mechanical and Aerospace Engineering, Gainesville, Florida, USA

Kyle G. Rowe, Dr. W. Gregory Sawyer
Mechanical and Aerospace Engineering, Gainesville, Florida, USA

Dr. Chelsey S. Simmons
Mechanical and Aerospace Engineering, Biomedical Engineering, Division of Cardiovascular Medicine
Gainesville, Florida, USA

INTRODUCTION

By the end of the second trimester of gestation, the mammalian heart goes through a compaction phase where sponge-like tissue gets remodeled into dense, powerful tissue. However, mutations in the Nkx2.5 gene produce many congenital heart defects, including a "non-compaction" phenotype. In mice that have had a human mutation of Nkx2.5 inserted into their genome, affected mouse hearts present non-compaction, and the wall of heart muscle is spongy and enlarged compared to wild type [1]. The porous appearance of these affected hearts suggests that tissue mechanics may be affected by this mutation, in addition to the traditionally characterized biophysiological defects.

Our goal was to correlate mechanical parameters to both wild type and mutated embryonic mouse hearts. Since mechanical properties can direct cell differentiation and function, differences between wild type and mutated hearts may help identify new avenues of therapy targeted at structural or remodeling proteins. Further development of this mechanical characterization tool and corresponding analysis is important because it can be implemented for any kind of excised tissue samples, from soft embryonic organs like these hearts to stiffer adult hearts or even tumors.

METHODS

Tissue Isolation and Characterization. An optical micro-tribometer (Ref. 2, Fig. 1) was used to compress the ~1mm diameter embryonic hearts. Heart tissue was generously donated by Dr. Hideko Kasahara, and all animal experiments were performed with approval from the University of Florida Institutional Animal Care and Use Committee. A small paintbrush was used to manipulate the tissue, and a razor blade was carefully used to separate the ventricles from the atria and lungs. Each isolated embryonic heart is placed inside a small well made by punching 2 mm holes through a 3 mm layer of polydimethylsiloxane (Sylgard 184, Dow Corning) on top of thick optical glass. The initial contact area of tissue with the glass bottom measured approximately 1.4 mm^2, enabling an unconfined compression test. The probe, a 1.04 mm diameter metallic rod, was programmed to compress the samples using a 100 μm/min displacement rate. Three indentation depths were attempted in sequence (300 μm, 450 μm, and 600 μm) with a 5 min hold, allowing the tissue to undergo stress relaxation, and slow retraction between each cycle.

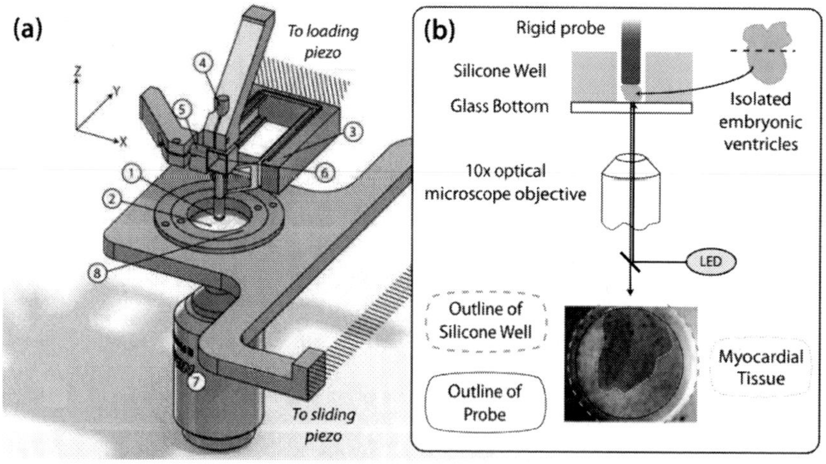

FIGURE 1. SCHEMATIC OF MECHANICAL CHARACTERIZATION TOOL
(a) A rigid probe (1) compresses tissue sample against a glass slide (2). The probe is mounted directly to a cantilevered flexure (3), and capacitance probes (4 and 5) measure the displacement of a target (6) mounted on the calibrated flexure, yielding forces. A microscope objective (7) enables viewing of the sample (8). (b) Schematic of optical pathway. A coherent light source (LED) passes through the objective and up through the transparent sample and holder assembly. The light is reflected off of the sample and probe back through the microscope objective and ultimately to a CCD camera.

Characterization Models. Two models were utilized to estimate viscoelastic properties of the embryonic hearts: (1) Stress-Strain Kelvin-Voigt Model and (2) two-element indentation Kelvin Model.

The Kelvin-Voigt model utilizes the contact area detected by the microscope objective (Fig. 1) to fit stress and strain calculations to this common model. The Kelvin model is a more traditional indentation model that uses the relaxation data, and this model was used to compare the results and qualitatively validate the stress and strain calculations from the first model. The main advantage in this model is that no additional geometric or contact models are needed.

Stress and Strain Calculations for Kelvin-Voigt Model. Cantilever-based probes typically record force-displacement values, and these data must be converted to material properties using analytical models of contact mechanics. However, because our probe diameter is actually larger than the embryonic heart diameter and we have video capture capabilities, we can use the recorded images to approximate stress-strain curves from force-displacement data. Images of the deforming heart were captured every 5 sec and used to estimate area throughout the indentation through edge-detection in *ImageJ*. Engineering stress is estimated by dividing the measured force by the calculated area. Due to the complex geometry of the embryonic hearts, volumetric strain (assuming a 1 mm starting height) was used instead of the more common linear strain. Straightforward calculations of axial strain that one might derive using Generalized Hooke's Law are not applicable to these experiments because of the poorly defined, complicated geometry.

The acquired stress, (σ(t)) and strain (ε(t)) curves were fit to a Kelvin-Voigt model to obtain the mechanical parameters of elasticity (E) and viscosity (η) [3]:

$$\sigma(t) = E \cdot \varepsilon(t) + \eta \cdot \frac{d\varepsilon}{dt}(t) \qquad (1)$$

Force-Indentation and Kelvin Model. Relaxation data was fit to a two-component Kelvin model [4]:

$$F(t) = u_0\left(k_1\, e^{-\frac{t}{\tau_1}} + k_2\, e^{-\frac{t}{\tau_2}}\right) \qquad (2)$$

This model allows us to quantify elasticity (k) and viscosity ($\mu = \tau/k$) independent from the compression's displacement, thus solving the previous depth-dependence issue. Reported coefficients correspond to the dominant component of the model, i.e. k_1 and μ_1.

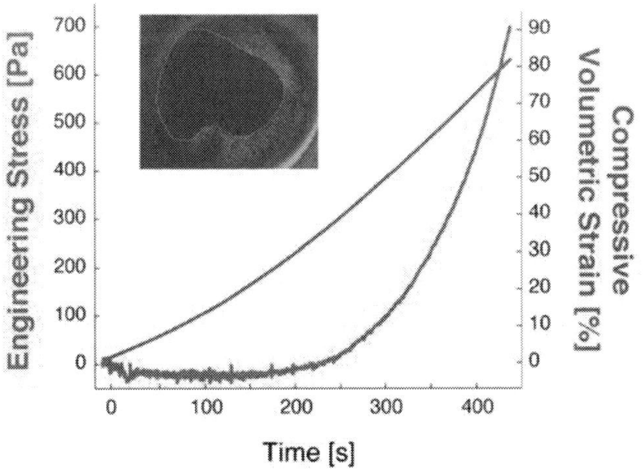

FIGURE 2. EXAMPLE DATA AND ANALYSIS FOR A SINGLE INDENTATION. Using area determined using imaging, stress and strain can be calculated during indentation. Inset shows area outlined for example heart image.

RESULTS AND DISCUSSION

Mutated hearts are softer and less viscous than wild type hearts (Fig. 3, Table 1). This project provides a baseline to test therapeutic processes in the future: correction of the mutation by genome editing, gene therapy in vitro, etc. The experimental procedure and data analysis can be utilized to characterize candidate 3D hydrogels as well as other samples of embryonic, neonatal, and even adult tissue. We found that the Kelvin-Voigt model, which uses the loading section of the curve, is dependent on indentation depth and therefore less reliable for materials where initial contact is hard to identify. Stress relaxation data seems to be more repeatable when characterizing viscoelastic biological samples, and resulting effective stiffness correlates well with elastic modulus.

ACKNOWLEDGEMENTS

The authors thank Dr. Hideko Kasahara from the Physiology Department for her donation of the embryonic hearts. The Simmons Lab acknowledges ongoing financial support from the College of Engineering and Department of Mechanical and Aerospace Engineering.

REFERENCES

[1] Ashraf H et al., *Circ Cardiovasc Genet* 7(4):423-33, 2014.
[2] Krick BA et al., *Tribol Lett* 45:185–194,2012.
[3] Levental I et al., *J Phys Condens Matter* 22:194120, 2010.
[4] Forgacs G, et al., *Biophys J* 74:2227-2234, 1998.

Kelvin-Voigt Model	E (kPa)	η (kPa·s)
Wild-type	9.8-18.9	3.1-4.2
Knock-In Mutation	2.1-5.7	444-720
Two-Component Kelvin Model	k (mN/m)	μ (N·s/m)
Wild-type	71-75	325-451
Knock-In Mutation	9.1-15.4	23.6-42.8

TABLE 1. CORRESPONDING MATERIAL PROPERTIES DERIVED FROM TWO DIFFERENT MODELS.

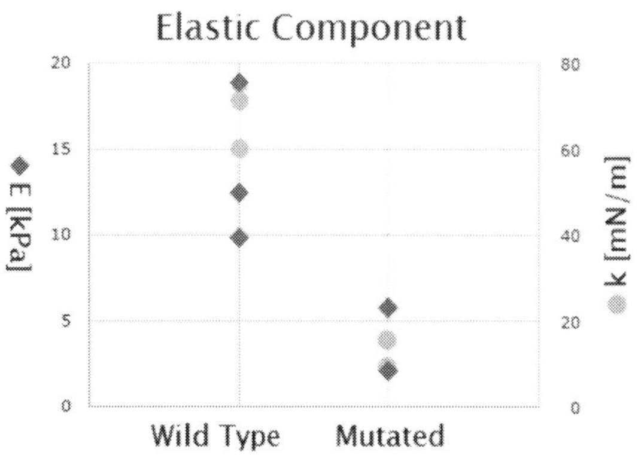

FIGURE 3. ELASTIC PROPERTIES OF MUTATED EMBRYONIC HEARTS LOWER THAN WILD-TYPE REGARDLESS OF MODEL USED.

SB³C2015
Summer Biomechanics, Bioengineering and Biotransport Conference
June 17-20, 2015,Snowbird Resort, Utah, USA
SB³C2015-493

VASO-CT FOR QUANTITATIVE MAPPING OF VESSEL WALL APPOSITION IN A FLOW-DIVERTER IMPLANT

Kajo van der Marel, Ajay K. Wakhloo, Matthew J. Gounis, Ajit S. Puri

New England Center for Stroke Research
Department of Radiology
University of Massachusetts Medical School
Worcester, MA, USA

INTRODUCTION

Flow diverter (FD) stents provide a safe and effective technique for endovascular treatment of unruptured intracranial aneurysms [1,2]. However, poor vessel wall apposition not only affects efficacy of the FD, but may also cause in-stent stenosis and late thrombosis, which increase the risks of stroke-related complications and may result in non-occlusion of the aneurysm due to so-called 'endoleak' [3].

Novel angiographic and intravascular imaging strategies enable detailed three-dimensional visualization of the FD device and the aneurysm-parent vessel complex during the procedure, which is critical to ensure proper device deployment.

First, high-resolution contrast-enhanced cone-beam CT (VasoCT) has been introduced as a superior angiographic technique for intracranial imaging of vasculature and deployed devices [4,5]. VasoCT images are reconstructed at a spatial resolution of up to 67 μm^3 after unbinned acquisition of 617 small field-of-view projection images during a rotational sweep of the C-arm.

Second, neurointerventional applications for intravascular Optical Coherence Tomography (IVOCT) are being explored. IVOCT employs near-infrared interferometry and a rotating fiberoptic wire to obtain cross-sectional images with a spatial resolution of 15-40 μm at 100-200 μm intervals [6].

However, these techniques generate large three-dimensional datasets which render identification and interpretation of poor device apposition an often qualitative and time-consuming task.

We therefore developed a semi-automatic image processing pipeline to obtain a quantitative mapping of FD device apposition to the arterial vessel wall. In this study, we evaluated this approach using VasoCT images and concurrently acquired IVOCT images in an experimental sidewall aneurysm model in canines.

METHODS

Two sidewall aneurysms were created on each common carotid artery (CCA) in female canines (n = 4) by anastomosis of sections of the resected right external jugular vein, as described before [7]. Under fluoroscopic guidance (AlluraXper FD20, Philips Healthcare, Best, The Netherlands), FD stents were placed and apposition was assessed in the right CCA using VasoCT and OCT (C7 System/Dragonfly Catheter, St Jude Medical, Westford, MA).

VasoCT images were reconstructed at 50% of the field-of-view volume at 512^3 isotropic voxels, yielding a 67 μm resolution.

Pull-back OCT image series (270 frames; 54 mm; 2.7 s) were obtained during manual flushes with viscous contrast material to generate a blood-free field. Raw cross-sectional images (504 lines, 968 pixels/line, 10 mm field-of-view) were processed and analyzed using in-house developed software to enable visualization and manual measurement of device-wall distances.

Using the Vascular Modeling Toolkit (VMTK) [8] and in-house developed software we devised a post-processing pipeline for reconstructed VasoCT images that required minimal user interaction to obtain a quantitative, two-dimensional visualization of the Euclidean distance between surface models representing the vessel wall and FD device. The pipeline comprised (1) anisotroptic image diffusion to reduce noise while preserving edges; (2) extraction of the luminal surface using level sets initialized by propagating colliding fronts from manually placed seed points; (3) parent vessel reconstruction and centerline interpolation by manual identification of the proximal, distal, and aneurysm dome points; (4) intensity-based identification of FD device end-points; (5) extraction of the FD device surface by constrained fitting of a deformable tube to image intensities representing platinum stent struts; (6) measuring device-wall apposition by calculating point-wise distances between luminal and device surface models; (7) projecting apposition measurements and distances to the centerline onto a flattened two-dimensional map, followed by 7×7 median filtering, for color-coded visualization.

RESULTS

Eight aneurysms on four vessels were analyzed. In two cases (A: subject 1, distal aneurysm; B: subject 4, proximal aneurysm), the quantitative apposition maps indicated potential malapposition of the FD device. Figure 1 shows the corresponding flattened maps of quantitative vessel wall-device apposition measurements. In each of these maps, the aneurysm is clearly visible, as well as an additional area of malapposition (dotted outlines). The maximum distance between device and vessel in the demarcated areas was 0.44 mm (A) and 0.64 mm (B). The identified area of malapposition for case A is shown on the reconstructed VasoCT image in the right panel of Figure 2.

To evaluate the accuracy of the measured malapposition, a corresponding area for each identified malapposition was manually segmented on the high-resolution cross-sectional IVOCT images (Figure 3). The maximum distance between the stent and luminal surfaces on IVOCT was 0.41 mm (case A) and 0.80 mm (B).

Perforating arteries may also appear as high-distance areas on the apposition maps (Fig. 1A, $+\pi$ distal). Therefore, similarly flattened maps of the distance between reconstructed vessel centerlines and luminal surfaces were generated (e.g., Fig. 2, left panel) that provide additional context for identification of aneurysms and side-branches (see black arrows in Fig. 2, left panel; and in Fig. 1A).

DISCUSSION

Our results suggest that quantitative maps of FD device apposition can provide the user with an intuitive representation of deployment status and can be generated with minimal user-interaction from high-resolution three-dimensional images.

High-resolution reconstructions of VasoCT images enable identification of moderate malapposition in agreement with measurements performed on IVOCT. However, careful inspection of the source images is warranted to identify potential errors in estimation of the luminal and device surfaces. Due to its low-speed rotational sweep VasoCT is sensitive to motion

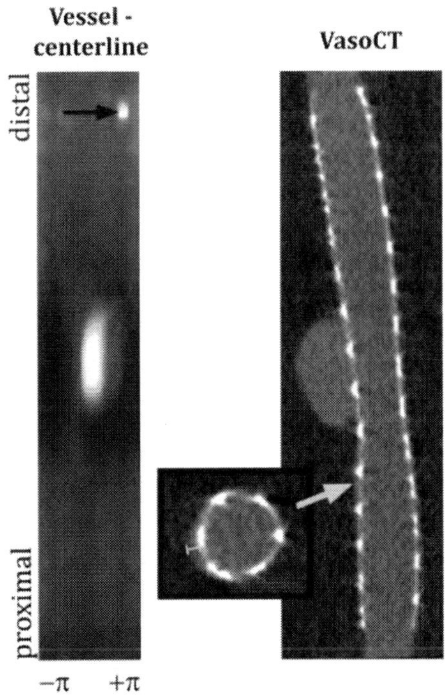

Figure 2. Left: distance from luminal surface to vessel axis (black: <1 mm; red-yellow: 1-4 mm; white: >4 mm). The black arrow indicates a perforating side-branch. Right: reconstructed VasoCT image at 67 μm spatial resolution. The yellow arrow points to the identified area of malapposition for case in Fig. 1A. The inset is the cross-sectional slice (yellow bar: 0.38 mm).

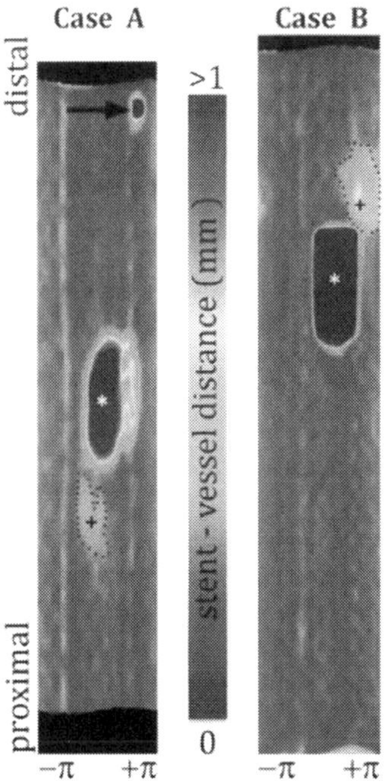

Figure 1. Quantitative apposition map based on high-resolution VasoCT. A) Subject 1, distal aneurysm. B) Subject 4, proximal aneurysm. Aneurysms (*) and areas of malapposition (+) are indicated. Color-coded areas represent vessel surface area covered by the FD device. Black arrow: side-branch.

induced by arterial pulsation. Furthermore, precise localization of the FD device is hampered by artefacts from the device's constituent metals, and may vary with the orientation relative to the rotational arc.

IVOCT's superior spatial resolution provides an unrivaled view of FD device configuration. Unfortunately, commercially available catheters for percutaneous coronary interventions are currently not compatible with the tortuous anatomy of the cerebrovascular architecture. Future work is needed to adapt the post-processing pipeline to accommodate IVOCT as a primary source of FD device apposition maps.

ACKNOWLEDGEMENTS

This study was supported in part by a grant from Philips Healthcare.

REFERENCES

[1] Wakhloo AK and Gounis MJ, *Neurosurgery* 61(2014): 111-120.
[2] Brinjikji W *et al.*, *Stroke J. Cereb. Circ.* 44(2013): 442-447.
[3] Carneiro A *et al.*, *Neuroradiology* 56(2014): 51-58.
[4] Kizilkilic O *et al*, *J. Neurosurg.* 117(2012): 45-49.
[5] Flood TF *et al.*, *J. Neurointerventional Surg* (2014).
[6] Gonzalo N *et al*, *Rev. Esp. Cardiol.* 63(2010): 893-903.
[7] German WJ and Black SP, *N. Engl. J. Med.* 250(1954): 104-106.
[8] Antiga L *et al.*, *Med. Biol. Eng. Comput.* 46(2008): 1097-1112.

Figure 3. Cross-sectional images from IVOCT. A) Subject 1, distal aneurysm (see Fig. 1A). B) Subject 4, proximal aneurysm (see Fig. 1B). Orange bars indicate FD device malapposition with a distance of 0.41 mm (A) and 0.80 mm (B), which are similar to the values calculated from VasoCT (0.44 mm and 0.64 mm, respectively).

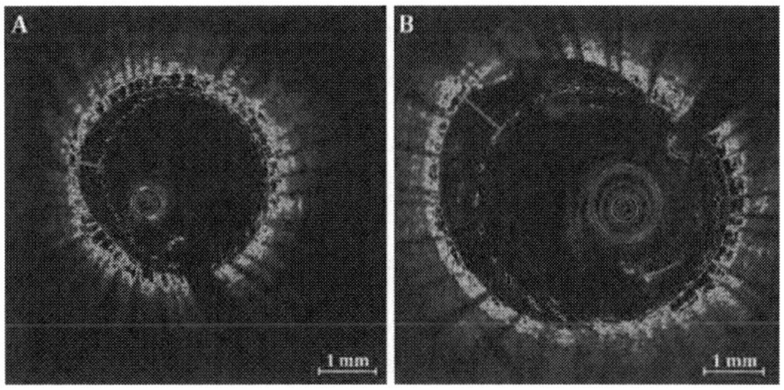

SB³C2015
Summer Biomechanics, Bioengineering and Biotransport Conference
June 17-20, 2015, Snowbird Resort, Utah, USA
SB³C2015-494

IMAGE-BASED 3D MORPHOMETRIC ANALYSIS OF THE CLAVICLE INTRAMEDULLARY CANAL IN MALE POPULATION

Jazmine R Aira (1,2), Sergio Gutierrez (1), Brandon Santoni (1,2), Mark Frankle (2,3), Peter Simon (1,2)

(1) Foundation for Orthopaedic Research and Education
Tampa, FL

(2) University of South Florida
Tampa, FL

(3) Florida Orthopaedic Institute
Tampa, FL

INTRODUCTION

Clavicle fractures predominately occur in the young adult male population [1] and are typically caused by traumatic events such as motor vehicle collisions, sports injuries, and falls [2]. The most common site for fracture is the middle third of the clavicle [1].

Current treatment for displaced mid-shaft clavicle fractures involves internal fixation with superior or anterior clavicle plating [2]. However, external plating has been associated with post-surgical complications including infection, hardware malfunction, and fracture re-occurrence [3]. In addition, exterior plating is aesthetically problematic, as the underlying hardware can be visualized and palpated from the outer skin surface.

Intramedullary (IM) fixation of mid-shaft clavicle fractures may be a feasible surgical alternative to external plate fixation. Normally, these devices are used for the treatment of long bone fractures [4]. Morphometric studies on the clavicle have been performed but are limited by a few noteworthy shortcomings. Mathieu et al. [5] evaluated the IM canal morphometry of the 10 male and 10 female clavicle CT scans. However, the methodology reported was not truly 3D as the measurements were taken on two orthogonal planes produced by CT scanner. Further, the position of clavicle was not controlled for during scanning that may potentially skew the results. King at al. [6] performed a study on 418 clavicles using CT scans, and reported on IM canal parameters. Once again, the study was not truly 3D as the measurements were taken on limited numbers of axial slices. Prior to introduction clinically, a true 3D morphometric study of the clavicular geometry is necessary to determine the requisite design features of an IM device that may be used to treat clavicle fractures. The objectives of the study were to: 1) ascertain the largest radius of the device that would fit into the clavicle canal at any location along the longitudinal axis and 2) estimate canal curvature by measuring its deviation from the clavicle as a whole, and from the fixed longitudinal axis.

METHODS

The clinical CT scans of 53 male shoulders from the patient database of the affiliated hospital were included in this study. Exclusion criteria were incomplete scans, old or acute clavicle fractures, and congenital bone anomalies. In addition, only scans with a slice thickness equal to 0.625mm were selected. A total of 39 clavicle models were segmented in MIMICS (Materialise, Leuven, Belgium) from the CT scans in DICOM format.

Following a standard segmentation protocol, only pixels representing cortical bone were included into the clavicle model. From the cortical bone mask, a second mask was obtained by filling in the cavities. Using Boolean logic, the 3D model of the canal was generated. The point cloud and volumetric models of both the clavicle and canal were exported for the study.

A custom-written automatic algorithm (MATLAB, MathWorks, Natick, MA) was utilized for the subsequent data analysis. First, principal component analysis was applied to the clavicle model geometry (represented by volumetric point-cloud of the whole clavicle) to establish a standardized coordinate system across all individuals (Fig.1a). A tightest fit box was calculated around every 3D clavicle model. This box was used as a mean of normalization to account for variability in clavicle length by assigning box length to represent 100% clavicle length. Subsequently, every box was divided into 100 sections along its length, corresponding to the longitudinal axis of the bone, at 1% increments starting from the sternal end (Fig.1b). Each clavicle slice section was further fitted with a circumscribed circle (Fig.2a), and the medullary canal of each slice section was fitted with an inscribed circle to approximate the varying shape of the cortical bone cross-section (Fig 2b).

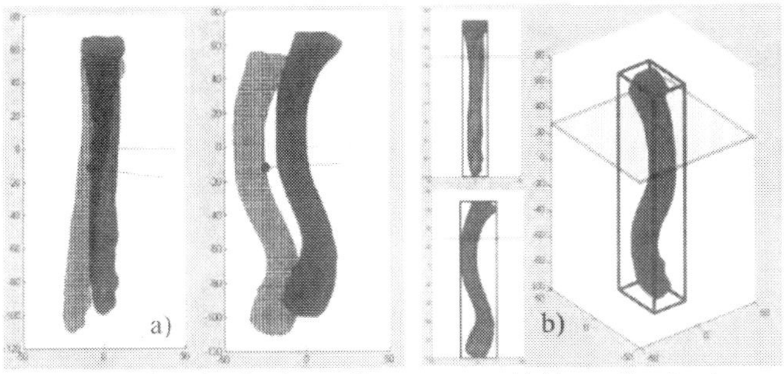

Figure 1. a) Standardized coordinate system. **b)** Normalization box.

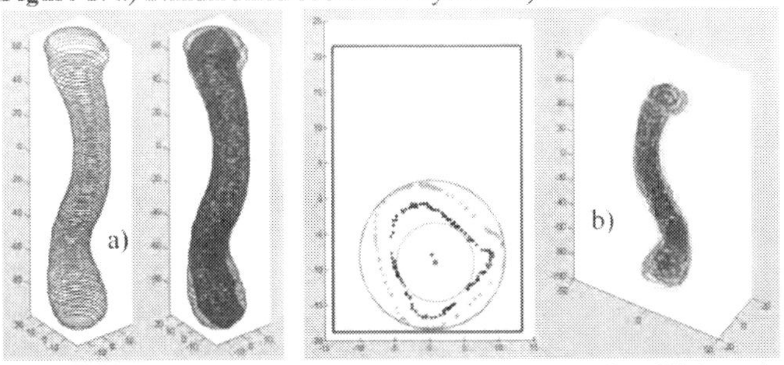

Figure 2. a) Clavicle circumscribed cir. **b)** Inscribed cir. of IM canal.

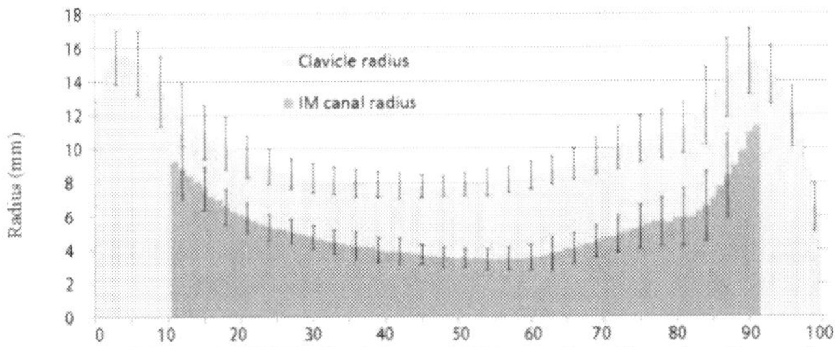

Figure 3. Mean (±STD) clavicle and IM canal radius size for each slice as a function of clavicle length (%) along the longitudinal axis.

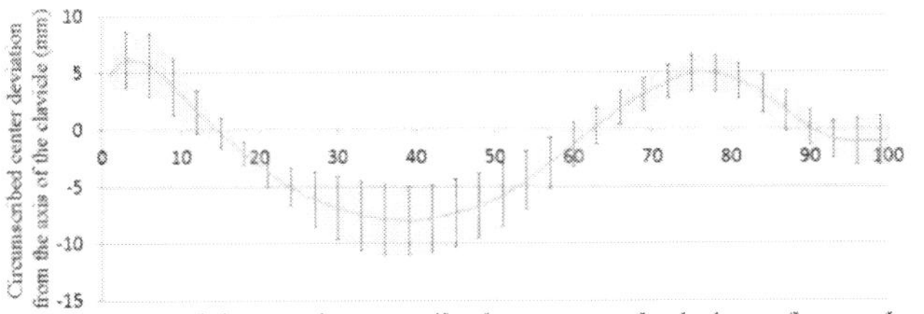

Figure 4a. Mean circumscribed center deviation from the longitudinal axis as a function of clavicle length (%).

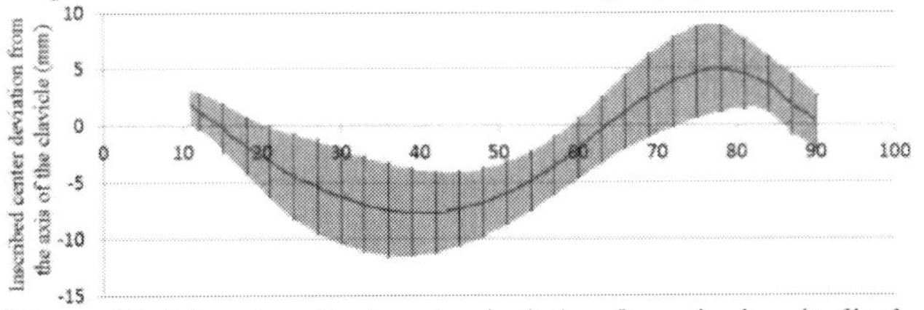

Figure 4b. Mean inscribed center deviation from the longitudinal axis as a function of clavicle length (%).

The average circumscribed and inscribed radius was taken across all patients, and the standard deviation was calculated. Given the center location in the x-y-z coordinate system for each circle, the curvature of the clavicle in the transverse plane was estimated as the anatomical anterior/posterior displacement (y-axis) as a function of longitudinal length (z-axis). Each center displacement in the y-axis was averaged across all 39 patients and reported with a standard deviation.

RESULTS

The smallest mean circumscribed radius was found to be 7.8mm at 45-50% length of the clavicle from the sternal end, while the smallest mean inscribed radius was 3.4mm at 55% clavicle length (Table 1, Fig.3). These measurements correspond to the narrowest region of the clavicle and IM canal respectively. The standard deviation was smallest at the normalized clavicle lengths corresponding to the smallest circumscribed and inscribed radii.

Disregarding the sternal and acromial 10% ends, the greatest clavicle shaft deviation from its longitudinal axis was 8.0 mm anteriorly and identified at the sternal end curvature of 40% of clavicle length. In the acromial half, the greatest deviation from the clavicle's longitudinal axis was 4.9mm at 70% of clavicle length (Fig.4a-b). The sternal curvature of the IM canal was mostly deviated anteriorly and a maximum of 7.7mm, at 42% of the clavicle length. The acromial end canal curvature was displaced 5.0mm posteriorly at 78% of the clavicle length.

Table 1. Mean (±STD) values of the clavicle and IM canal

	Circuscribed (mm)		Inscribed (mm)			Circuscribed (mm)		Inscribed (mm)	
Length	Radius	STD	Radius	STD	Length	Radius	STD	Radius	STD
10%	13.4	2.1	9.3	1.6					
15%	11.2	1.6	7.2	1.2	55%	8.0	0.8	3.4	0.6
20%	10.0	1.4	5.9	1.0	60%	8.3	0.8	3.6	0.7
25%	9.0	1.1	5.2	0.8	65%	8.8	0.8	4.1	0.8
30%	8.3	0.9	4.7	0.7	70%	9.6	1.1	4.8	1.0
35%	8.1	0.8	4.2	0.6	75%	10.4	1.4	5.5	1.1
40%	7.9	0.8	3.9	0.8	80%	10.9	1.4	5.9	1.5
45%	7.8	0.7	3.7	0.6	85%	12.5	2.3	7.8	2.1
50%	7.8	0.6	3.5	0.6	90%	15.0	2.1	11.3	2.7

DISCUSSION

The data obtained in this study presents IM canal and clavicle radius parametrized as a function of clavicle length from a 3D model of the IM canal. The use of a 3D model for morphometry study of the IM canal is a novel approach that provides us with a more accurate representation of this spatial geometry.

The smallest IM canal and clavicle radii were seen at different clavicle lengths. This suggests that, on average, one may not estimate the location of the narrowest region of canal, based on external visualization of maximum clavicle curvature alone. In addition, maximum IM canal deviation was found to be at a different clavicle distances (42% and 78%) when compared to clavicle deviation (40% and 70%). This implies that there exists an eccentricity of the IM canal center with respect to the clavicle center.

Based on this data, an IM canal device with a maximum radius of 3.4mm and fabricated to match the curvature parameters reported here may serve as a viable alternative to external plate fixation of mid-shaft clavicle fractures. This 3D data set serve as a requisite platform in the evidence-based design process of IM-based internal fixation for the treatment of clavicle fractures.

REFERENCES

[1] Postacchini F, et al. *J. Shoulder & Elbow Surg.*11:452-456. 2002.
[2] Paladini P, et al. *Translational Medicine @ UniSa* 2:47-58. 2012.
[3] Böstman O, et al. *J. Trauma.* 43(5):778–83. 1997.
[4] Shah A, et al. *J. Ortho Trauma.* 24(7):440-447. 2010.
[5] Mathieu P, et al. *Surg. & Radiologic Anat.* 36:11-15. 2014.
[6] King P.R, et al. *Eur. J. Orthop. Surg. Traumatol.* 24:37-42. 2014.

THE EFFECT OF HEAD ROTATION TO THE GEOMETRY AND HEMODYNAMICS OF HEALTHY VERTEBRAL ARTERIES

Nicolas Aristokleous (1), Ioannis Seimenis (2), Georgios C. Georgiou (3), Andreas S. Anayiotos (1)

(1) Department of Mechanical Engineering and Materials Science and Engineering, Cyprus University of Technology, Lemesos, Cyprus

(2) Laboratory of Medical Physics, Medical School, Democritus University of Thrace, Alexandroupoli, Greece

(3) Department of Mathematics and Statistics, University of Cyprus, Nicosia, Cyprus

INTRODUCTION

The vertebral arteries (VA) and the internal carotid arteries (ICA) are the four major arteries supplying the brain circulation through the circle of Willis. Significant differences in size between the left and right VA are common in healthy adults [1]. Head rotation was previously shown to influence the geometry and hemodynamics of the carotid bifurcation [2, 3]. The present study investigates if head rotation causes any changes in the geometry and hemodynamics of the extracranial and intracranial sections of the VAs, from the level of the carotid bifurcation divider to the vertebrobasilar junction, a site of frequent occurrence of atherosclerotic plaques [3, 4].

METHODS

For the geometric morphology study, a group of six presumably healthy men (mean age of 35y, range 25-50y) was used. Computational fluid dynamics (CFD) analysis was performed on a volunteer with a dominant LVA. Imaging data were acquired in the same manner as described in [2] using a 3T MRI instrument (Achieva, Philips Medical System, the Netherlands) and the study was approved by the Cyprus Bioethics Committee.

The 3D surface models were constructed for the supine (S) and prone (P) positions with rightwards head rotation using ITK-Snap (PICSL, USA) and VMTK software. For global comparison, models were clipped at two sites: a) 20mm above and below the ipsilateral carotid bifurcation and b) 20mm from both sides of the vertebrobasilar junction. The flow data was processed by Segment (Medviso AB). All blood flow rates were scaled at a constant HR of 60 bpm according:

$$Q_{scaled} = Q_{measured}(60/HR_{measured}) \quad (1)$$

CFD simulations on the reconstructed geometries at the vertebrobasilar junction were performed to calculate hemodynamic parameters such as wall shear stress (WSS), relative residence time (RRT), and localized normalized helicity (LNH).The Icem-CFD (Ansys Inc) was used to generate the two meshes of size ~$2.3 \cdot 10^6$ mixed type elements, and the numerical solution were achieved using Fluent (Ansys Inc.). On each inlet we applied the realistic pulsatile velocity waveform with Re_{avg}=125 and Re_{peak}=450 and Womersley number (α) between 2.5 and 4. The assumptions considered was: a) Newtonian fluid (μ 0.0035 kg/m·s; ρ 1050 kg/m^3); b) rigid arterial walls; c) fully developed laminar inlet flow; and d) traction-free conditions at outlet. The impact of these assumptions has been discussed elsewhere [5].

RESULTS

Table 1 shows the extracted geometric parameters at the vertebrobasilar junction for the two head postures. In the prone position, a decrease in all angles is noted.

Table 1: Average geometric parameters at the supine (S) and prone (P) head postures for the VA segment at the vertebrobasilar junction.

		S	P
Bifurcation	Angles	81.5	77.7
LVA	Angles	48.3	42.3
Planarity	Angles	13.7	1.2
Asymmetry	Angles	15.1	6.9
Curvature (1/mm)	BA	0.06	0.07
	LVA	0.11	0.12
	RVA	0.14	0.13
Tortuosity	BA	0.15	0.15
	LVA	1.33	1.18
	RVA	1.06	1.10

Fig.1 presents the reconstructed geometrical segments of the VAs at the carotid bifurcation level. As also indicated by subject-specific curvature and tortuosity data shown in Table 2, head rotation causes considerable changes in the geometric characteristics of the VAs.

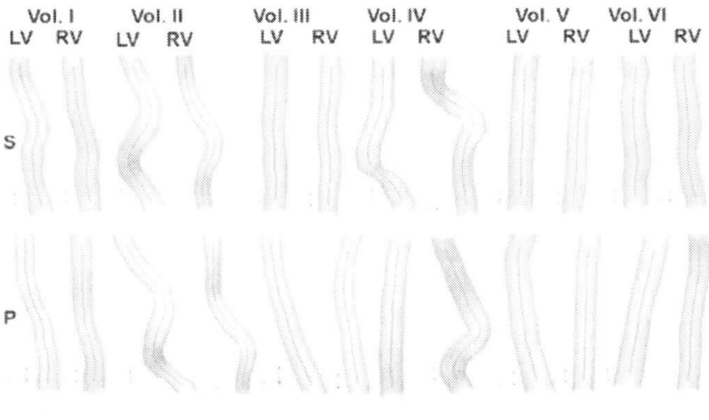

Figure 1: Reconstructed 3D models of the LVA and RVA for the six volunteers. The segments were clipped at ±20mm the carotid bifurcation level. Top: Supine head position (S) and Bottom: Prone position (P).

Fig. 2 shows the contour plots of WSS, RRT and LNH isosurface distributions at the vertebrobasilar junction. There is a notable increase in the WSS distribution in the RVA for the prone position. Large helical flow structures are observed in the area where the LVA and RVA merge into the basilar artery and continue in high intensity and in bidirectional tracts. Head position change seems to be associated with the reversal of blood rotation direction within the dominant LVA.

DISCUSSION

The results show that torsion of the neck associated with head rotation, which corresponds to the common sleeping posture in the prone position, may cause alterations in vertebral artery morphology and hemodynamics at different vessel sections. The geometric changes were clearly visible in the reconstructed geometries of the LVA and RVA, which in some cases were significantly different at the prone position (rightward head rotation) as opposed to the supine position (no head rotation).

Hemodynamic changes due to head rotation were also present in all three parameters examined (WSS, RRT, LNH) suggesting that head posture affects the developed flow fields and may influence the exposure of the VA wall to significantly different hemodynamic conditions.

Table 2: Individual geometric parameters at supine (S) and prone (P) head postures for the VA segment at the level of the carotid bifurcation divider.

		RVA		LVA	
Vol. #	Head Position	Curvature (1/mm)	Tortuosity	Curvature (1/mm)	Tortuosity
I	S	0.112	0.100	0.134	0.184
	P	0.130	0.135	0.129	0.173
II	S	0.026	0.007	0.026	0.005
	P	0.052	0.024	0.024	0.006
III	S	0.024	0.004	0.022	0.004
	P	0.034	0.010	0.029	0.002
IV	S	0.156	0.227	0.096	0.089
	P	0.149	0.233	0.030	0.011
V	S	0.033	0.005	0.040	0.005
	P	0.035	0.013	0.025	0.002
VI	S	0.036	0.005	0.031	0.002
	P	0.033	0.013	0.038	0.006

Further research is required to assess the effects of head rotation in stenotic VA patients and the role of the hemodynamic parameters examined in the development of vertebrobasilar artery stenosis.

ACKNOWLEDGEMENTS

This work was co-funded by the European Regional Development Fund and the Republic of Cyprus through the Research Promotion Foundation (Project ΥΓΕΙΑ/ΔΥΓΕΙΑ/0609(ΒΙΕ)/11).

REFERENCES

[1] J. M. Hong et al. "Vertebral artery dominance contributes to basilar artery curvature and peri-vertebrobasilar junctional infarcts," *J Neurol Neurosurg Psychiatry*, 80:1087-92, 2009.

[2] N. Aristokleous et al. "Effect of posture change on the geometric featurs of the healthy carotid bifurcation," *IEEE Trans Inf Technol Biomed*, 15(1):148-54, 2011.

[3] N. Aristokleous et al. "Impact of head rotation on the individualized common caarotid flow and carotid bifurcation hemodynamics," *IEEE J Biomed Health Inform*, 18(3):783-9, 2014.

[3] J. R. Cebral et al. "Flow-area relationship in internal carotid and vertebral arteries," *Physiol Meas*, 29:585-94, 2008.

[4] G. C. Cloud and a. H. S. Markus, "Diagnosis and management of vertebral artery stenosis," *Qjm*, 96:27-54, 2003.

[5] D. A. Steinman and C. A. Taylor, "Flow imaging and computing: large artery hemodynamics," *Ann Biomed Eng*,33:1704-9, 2005.

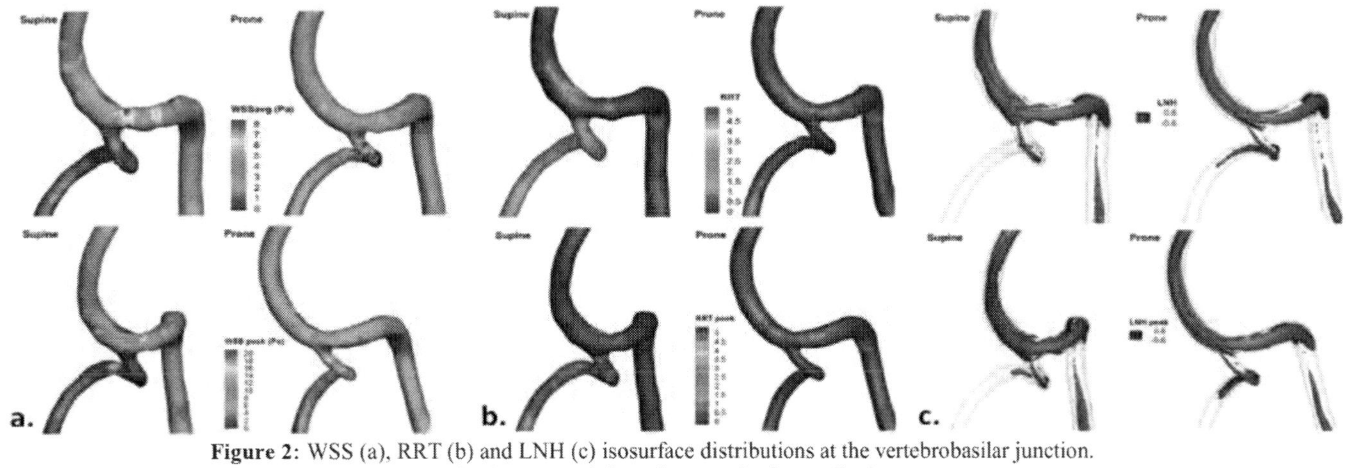

Figure 2: WSS (a), RRT (b) and LNH (c) isosurface distributions at the vertebrobasilar junction. Top: Cycle-averaged data, Bottom: Peak systolic data.

SMOOTH MUSCLE CELL ELASTIN GENERATION STIMULATED BY ADIPOSE-DERIVED MESENCHYMAL STEM CELLS

Aneesh K. Ramaswamy[1] MS, Kory J. Blose[1] BS, Justin S. Weinbaum[1,5] PhD, and David A. Vorp[1,2,3,4,5] PhD

(1) Department of Bioengineering
University of Pittsburgh
Pittsburgh, PA, USA

(2) Department of Cardiothoracic Surgery
University of Pittsburgh
Pittsburgh, PA, USA

(3) Department of Surgery
University of Pittsburgh
Pittsburgh, PA, USA

(4) Center for Vascular Remodeling and Regeneration
University of Pittsburgh
Pittsburgh, PA, USA

(5) McGowan Institute for Regenerative Medicine
University of Pittsburgh
Pittsburgh, PA, USA

INTRODUCTION

Human adipose-derived mesenchymal stem cell (hADMSC) therapy has shown a distinct effect on smooth muscle cells (SMCs) within various *in vivo* cardiovascular disease models [1, 2], including halting aortic aneurysm dilation when delivered five days after experimental insult [3]. It has been suggested that hADMSCs control aneurysmal development through regulation of interleukin-6 [4] and -17 [5], matrix metalloproteinase-2 and -9 [6, 7], and tissue necrosis factor α [8]. Given lower levels of elastin content in multiple aneurysm phenotypes [9, 10], the therapeutic effects of hADMSCs on SMCs have pointed to the induction of elastin regeneration as worthy of further examination.

Though adult human SMCs do not normally produce elastin *in vivo*, they have been shown to do so *in vitro* with external stimuli. Supernatant dosing treatments of transforming growth factor beta-1 (TGF-β1), for example, visually displayed formed elastin fibers. As hADMSCs produce TGF-β1 [11], this study hypothesized that adult SMCs co-cultured with hADMSCs in a three-dimensional fibrin-based environment would produce increased elastin.

Non-destructive measurements of SMC elastin generation within these fibrin gel constructs can provide further insight into the mechanism of vascular remodeling, achievable through the establishment of an elastin reporter gene, analogous to a collagen reporter in vascular SMCs [12]. This transcriptional elastin luciferase reporter construct, referred to as pElnLuc, could allow early and late elastin production to be distinguished during vascular therapy. We are currently using this non-destructive elastin reporter to optimize the *in vitro* hADMSC-based treatments of adult SMCs presented in this abstract. Our long-term goal is focused on using hADMSCs to promote elastin deposition within the diseased vasculature to halt the progression of unruptured aortic aneurysms.

METHODS

Fibrin gel constructs were fabricated as previously described [12, 13], using 3.7mg/mL fibrinogen, 0.21U/mL thrombin, and the cell seeding densities specified below. An elastogenic rat fetal lung fibroblast line (RFL-6) was incorporated within the gels at a seeding density of $5x10^5$ cells/mL. 200μL circular gels were fabricated in multi-well plates as previously described [12]. ε-aminocaproic acid (ACA), a fibrinolysis inhibitor, was added during media changes (8, 12, 15mM) to prevent cell-driven degradation of fibrin gel constructs.

hADMSC-SMC co-cultured constructs were made by embedding $2.5x10^5$ SMCs in fibrin gels on Flexcell Uniflex culture plates. After two days, hADMSCs (from non-diabetic adult females) embedded in fibrin gels were added on top of the constructs in seven ratios (1:15, 1:10, 1:5, 1:1, 5:1, 10:1, 15:1 hADMSC:SMC). Constructs lacking hADMSC were treated with or without TGF-β1 to serve as positive and negative controls. All constructs were subjected to cyclic uniaxial stretch (10% at 1 Hz) using a FlexCell FX-4000T. After 10 days of mechanical stimulation, constructs were imaged with an Olympus multiphoton microscope to visualize elastin via its intrinsic autofluorescence. The average intensity of images was calculated from five representative images of each sample.

To develop pElnLuc, an elastin promoter sequence was excised from a previously validated plasmid [14], with a mutation made at the start site (ATG->TTG) to allow downstream luciferase translation [15]. This ELN promoter was previously ligated into pGL4.14 (Promega) and verified via DNA sequencing. Transfection into RFL-6 and SMCs [12] is currently being completed using a pElnLuc:Fugene HD (Roche) ratio of 1ug:3μL (10μg plasmid to 30μL Fugene HD). Luminescence measurements, reflective of elastin production, will be taken by immersing gels in a 1mM Luciferin-EF (Promega) solution and quantified using a multi-mode microplate reader (Bio-Tek).

RESULTS

Endpoint, destructive modified ninhydrin assays were conducted to quantify relative elastin production by RFL-6 cells (Figure 1a) and hADMSC-SMC co-cultures (Figure 1b) within these fibrin gel constructs. RFL-6 cells showed an increase in elastin content as ACA concentrations increased, when normalized to each construct's total protein content. RFL-6-embedded fibrin gels with 12 and 15mM ACA deposited similar elastin percentage levels to native rat aorta controls.

Elastin content produced by the hADMSC-SMC fibrin gel co-cultures was quantified using the same ninhydrin assay. The 1:1 and 15:1 ratios of hADMSCs:SMCs produced statistically similar levels of elastin percentage. hADMSC-conditioned media treatments induced a statistically similar level of elastin production from SMCs as well.

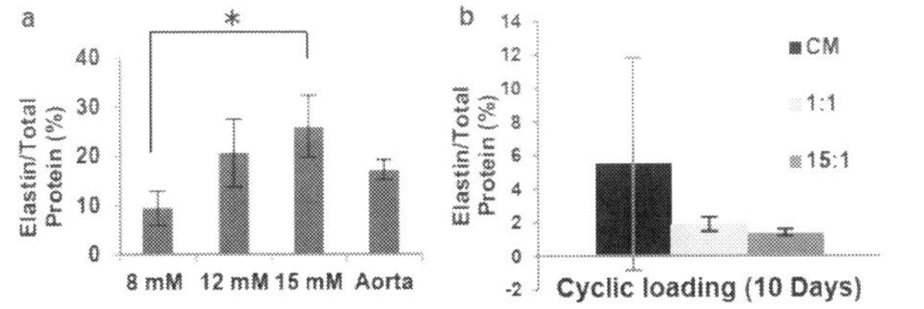

Figure 1: Relative elastin expression levels in 3D fibrin gel constructs. a) RFL-6 cells with fibrinolysis inhibitor ACA media dosing concentrations, plus rat aortic explant sections.
b) hADMSC-SMC co-cultures, after 10 days of cyclic uniaxial loading on FlexCell FX-4000. SMC cultures with hADMSC-cultured media (CM) and two co-culture cellular density ratios (1:1 & 15:1) of hADMSCs:SMCs were used. (α=0.05)

For multiphoton images of hADMSC-SMC constructs, average image intensities were greater in the hADMSC-treated constructs (7.468±1.161 for 15:1) compared to the TGF-β1-treated controls (1.996±0.164) and negative control (0.008±0.004). Figure 2a and 2b show average intensities for all groups, and Figure 2c shows a representative image from 15:1 hADMSC dose.

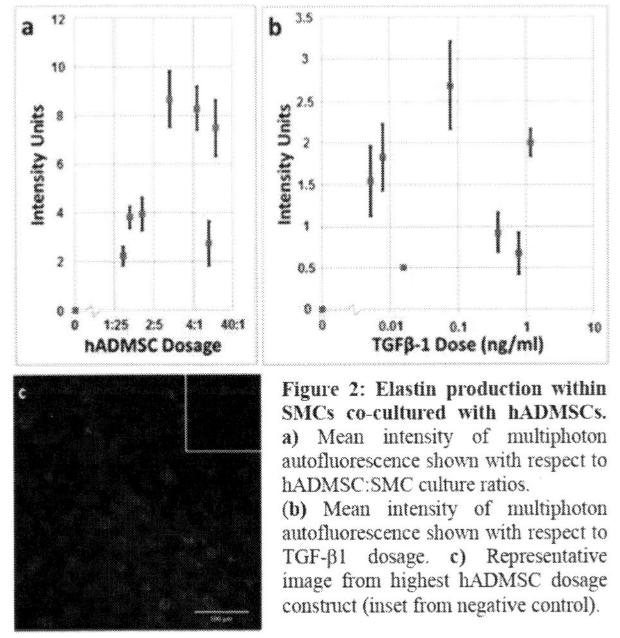

Figure 2: Elastin production within SMCs co-cultured with hADMSCs. a) Mean intensity of multiphoton autofluorescence shown with respect to hADMSC:SMC culture ratios.
(b) Mean intensity of multiphoton autofluorescence shown with respect to TGF-β1 dosage. **c)** Representative image from highest hADMSC dosage construct (inset from negative control).

DISCUSSION

Pilot studies were initially conducted with the known-elastogenic RFL-6 cells as a positive indication that cell-produced elastin could be detected in 3D fibrin gels (Figure 1a). ACA ensured that the cells did not degrade the fibrin gels, a previously-unreported property of RFL-6 cells. Statistically similar elastin percentage between 12 and 15mM and the rat aorta showed that a physiologically-relevant ratio of elastin expression is detectable using this modified ninhydrin assay.

Validation of the non-destructive transcriptional elastin reporter pElnLuc is currently being conducted using these RFL-6 cells, in order to confirm activity within a cell line that produces elastin without any stimuli needed. Studies will be done in both 2D and 3D cultures.

Multiphoton imaging of the co-culture fibrin constructs under cyclic uniaxial stimulation showed that 1:1, 5:1 and 15:1 hADMSC:SMC treatment ratios ($2x10^5$, $1.25x10^6$, and $3.75x10^6$ hADMSCs respectively) produced a significant increase in elastin deposition when compared to lower hADMSC cell densities (Figure 2a). This confirmed that that *in vitro* treatments of hADMSCs induce elastin expression by adult SMCs, and gave us the elastin-targeting therapeutic window for future hADMSC treatments on pElnLuc-transfected SMCs. Protein assays supplemented that data by showing statistically similar elastin deposition percentages at 1:1 and 15:1 hADMSC:SMC ratios (Figure 1b), indicating a possible upper limit on hADMSC dosing effectiveness. The hADMSC-SMC co-cultures produced more elastin in co-culture with hADMSCs compared to TGF-β1 treated constructs, which may be due to a synergistic effect between the many growth factors produced by hADMSCs.

Humans only produce elastin during the first few years of life. As such, elastin regeneration during disease progression has become a goal for vascular therapeutics. Our novel, continuing combination of cyclic loading and non-destructive monitoring to quantify functional elastin deposition can optimize hADMSC therapy, suitable to stimulate elastin production by mature SMCs in *in vivo* elastin-deficient vessels.

ACKNOWLEDGEMENTS

We would like to acknowledge funding through three NIH training grants awarded through the University of Pittsburgh's Department of Bioengineering: Angiopathy Training Program (T32HL094295) (AKR), Cardiovascular Bioengineering Training Program (T32HL076124) (KJB) and Biomechanics in Regenerative Medicine (T32EB003392) (KJB).

REFERENCES

1. Burks, C.A., et al., *Tissue Eng*, 2006. 12(9): p. 2591-600.
2. Zidi, M., et al., *Biomech Model Mechanobiol*, 2014.
3. Blose, K.J., et al., *Regen Med*, 2014. 9(6): p. 733-741.
4. Fu, X.M., et al., *J Transl Med*, 2013. 11: p. 175.
5. Sharma, A.K., et al., *Circulation*, 2012. 126(11 Sup 1): p. S38-45.
6. Hashizume, R., et al., *J Vasc Surg*, 2011. 54(6): p. 1743-52.
7. Park, H.S., et al., *Biochem Biophys Res Commun*, 2013. 431(2): p. 326-31.
8. Xiong, W.F., et al., *J of Immun*, 2009. 183(4): p. 2741-2746.
9. Prichard, H.L., et al., *Stem Cells*, 2008. 26(10): p. 2691-5.
10. Cao, Y., et al., *Biochem Biophys Res Commun*, 2005. 332(2): p. 370-9.
11. Melief, S.M., et al., *Stem Cell Transl Med*, 2013. 2(6): p 455-463.
12. Weinbaum, J.S., et al., *Tissue Eng Pt C*, 2010. 16(3): p. 459-67.
13. Ahmann, K.A., et al., *Tissue Eng Pt A*, 2010. 16(10): p. 3261-70.
14. Pierce, R.A., et al., *Am J Physiol-Lung Cell Mol Phys*, 2006. 291(3): p. L391-L399.
15. Quinteiro, C., et al., *J Ped Endocrin Metab*, 2002. 15(7): p. 1041-1045.

A POINT-WISE NORMALIZATION METHOD FOR DEVELOPMENT OF BIOFIDELITY RESPONSE CORRIDORS

Ian P. Marcus (1), Scott F. Gayzik (1), Kerry A. Danelson (2), Jonathan D. Rupp (3), Cameron R. Bass (4), Narayan Yoganandan (5), JiangYue Zhang (6)

(1) Biomedical Engineering
Wake Forest University
Winston Salem, NC, USA

(2) Department of Orthopedics
Wake Forest University
Winston Salem, NC, USA

(3) University of Michigan Transportation Research Institute
Ann Arbor, MI, USA

(4) Biomedical Engineering
Duke University
Durham, NC, USA

(5) Department of Neurosurgery
Medical College of Wisconsin
Milwaukee, WI, USA

(6) Applied Physics Laboratory
Johns Hopkins University
Laurel, MD, USA

INTRODUCTION

In the field of injury biomechanics, there is a continuous need for robust techniques to generate biofidelity response corridors (BRCs). BRCs provide a representative range of time-dependent responses from multiple experimental tests of a parameter from multiple biological surrogates (often cadaveric). Since BRCs ideally represent fundamental behavior of some tested population, more general than a single experimental test, they are an essential input for the design of models whether they are physical, such as anthropomorphic test devices (ATDs), or computational, such as finite element models. In the past, many BRC corridors were simply drawn by eye by bounding a set of data using a series of straight lines (e.g. [1,2]) There is a need for methods that can be consistently applied without subjective input to various loading scenarios, from traditional blunt impacts in the horizontal plane to emerging applications based on vertical loading. The desired output is a corridor with well-defined boundaries based on robust statistical principles.

Several published methods highlight BRC development for single parameter, time-history corridors. Maltese et al. use a time alignment method that minimizes the cumulative variance of the signals [3]. However, that method requires subjective selection of one signal as the standard for alignment. Recently, Nusholtz et al. published a technique that generates a representative curve (RC, essentially the mean) and evaluates repeatability within a system and reproducibility between systems [4]. Time alignment is achieved by iteratively maximizing the cross correlation of all signals within a data set, obviating the need for subjective selection of a standard signal.

The purpose of the current work is to improve the method used to calculate the Representative Curve (RC) outlined in [4] with two modifications. The proposed method limits correlation calculation to the loading portion of the curve, and, the calculation of the correlation coefficients between signals uses point-wise normalization (PWN).

METHODS

The technique begins with a set of time history signals of a response parameter from multiple tests with identical sample rates. As described in [4] the mean to mean technique is used to calculate a representative curve (RC). The heart of this approach is the calculation of a cross correlation coefficient (*CCC*) between signals, shown below in Equations 1 and 2a. For ease of comparison with [4], the notation p is for the joint cross correlation coefficient. The notation *CCC* is represents the maximized coefficient p, which occurs at time τ.

$$p(x,y) := p(x,y,\tau) \qquad 1$$

$$p(x,y,h) = \frac{\int x(t)y(t+h)dt}{\sqrt{\int x^2(t)dt \int y^2(t)dt}} \qquad 2a$$

$$p(x,y,h) = 1 - \int \frac{\sqrt{(x(t)-y(t+h))^2}}{\sqrt{x(t)^2} + \sqrt{y(t+h)^2}} dt \qquad 2b$$

The principal modification introduced in this work is that correlation coefficients are calculated using PWN (Eq. 2b), resulting in the modified correlations that are termed CCC_L. The PWN technique weighs all time points equally: each value in the resulting vector is normalized by the maximum difference that can occur between the two signals at that time point (sum of absolute values). The entire calculation is subtracted from unity because it is a sum of squares difference, and unlike Eq. 2a, better matches between curves will approach zero. To apply this method to high rate biomechanical data, it was desirable to use *only the loading portion of the curve* since time-dependent variation of unloading response generally increases with increasing strain rate The loading portion was defined as the time between the initiation of the test and the time at which the dependent mechanical variable (e.g. force or moment) reached the local maximum of the greatest magnitude. For acceleration traces, this was considered the time of peak velocity.

To demonstrate this method, biomechanical impact data originally published by Maltese et al. was used [3]. Two sets of human cadaver data (lower spine acceleration and thorax lateral force), from lateral sled impacts into a rigid wall at 8.9 m/s were downloaded from the Biomechanics Test Database maintained by the National Highway Traffic Safety Administration (NHTSA).

The data was scaled to a target body mass of 76 kg using an equal stress-equal velocity [5] method, and filtered per Maltese et al. The signals were time aligned using the PWN method. The BRC was generated using the RC (mean of the signals after optimized alignment), and a corridor that represents ± one standard deviation of the mean response. The coefficient of variation (COV) was averaged over a time window from 10 to 30 msec using the generated RC and standard deviation corridors, on both the PWN BRC and the digitized Maltase et al. BRC. Resampling, time alignment, and BRC generation was performed using MATLAB version 2014a.

RESULTS

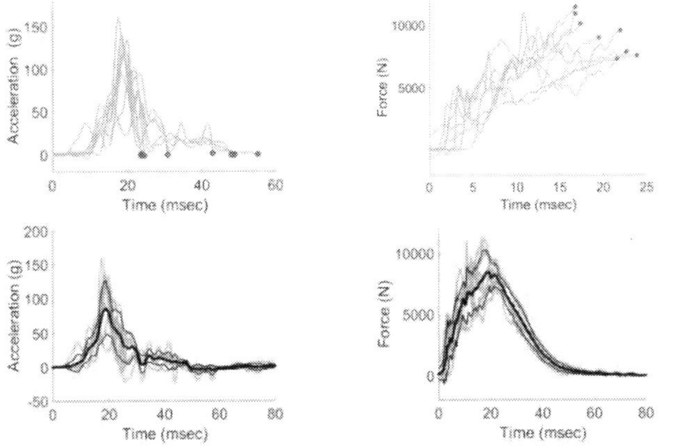

FIGURE 1. (Top Left and Right) Time shifted Lower Spine Lateral Acceleration and Thorax Lateral Force curves (cyan), respectively, with red circles representing the end of loading period. (Lower Left and Right) Time shifted Lower Spine Lateral Acceleration and Thorax Lateral Force curves (cyan), respectively, with the representative curve (thick black) and ± 1 S.D. corridor (thin black).

The average magnitude of the time shift for the lower spine and thorax force data was 4.5 msec and 5.7 msec respectfully. The algorithm ran six iterations for the acceleration based event and four iterations for the force based event before reaching the convergence criteria (no change in sum of CCC_L). The average value of the CCC_L after iterative alignment were 0.42 and 0.69 for the lateral acceleration and lateral force respectively.

DISCUSSION

The modification of the Nusholtz technique for BRCs developed in this study is a flexible framework and can be applied to time history data from various systems (i.e. ATD and/or PMHS data) and time-valued data types. One key aspect to this work is that the revised method limits the correlation calculation to the loading portion of the curve. The motivation is to intentionally weight the loading portion of the curve in the corridor generation process. Often in high rate loading tests, the boundary conditions imposed by the experimentalists play a role predominantly in the early stages of a test, and later, the response may be affected by the effects of gravity, more complex internal dynamics of the specimen, or interactions with other structures.

A brief comparison of the results in this study and those presented in Figures A1 and A5, RHF of [3] is shown below. The PWN technique (Table 1) had a lower COV than the Maltese et al method for both data types. Peak values occur at similar times despite different alignment approaches. RC peak magnitudes are similar to the previously reported values for both parameters. Acceleration shows the greatest difference but the variance in the signals is greatest at the time of peak using either method.

TABLE 1. Summary of BRCs developed using the PWN technique.

	PWN	[3]
Lower spine acceleration peak (g)	86.0	100.0
Time of peak acceleration (msec)	18.6	16.2
Thorax load wall force (kN)	8.48	8.9
Time of peak force (msec)	19.3	17.5
Average Thorax force COV %	18.5	20.2
Average Lower Spine acceleration COV %	56.8	62.5

There are several limitations to the current work. It may be difficult to apply the current approach when there is no clear loading phase due to test complexity or high frequency content. There are numerous ways of determining the peak of a signal and prospective users are encouraged to explore options to determine what best suits a given application. Also, while Nusholtz et al. stated that an RC curve test should yield CCC values of roughly 0.95 on average, this is likely the result of the ATD tests used by Nusholtz, expected to have high repeatability for a given test condition. For cadavers, we expect lower CCC values owing to specimen variation.

In conclusion, an objective, pointwise normalization method for BRC development was presented. The method is well suited for automation and both horizontal and vertical loading environments. There are a number of avenues for future work. Additional considerations will likely be necessary should this technique be used on full body cadaveric data in which phasing between body regions must be preserved. Finally, while this application is used to calculate statistical characteristics of the dataset such as mean and standard deviation BRCs, set to set comparison methods presented in [4] can be used in conjunction with the CCC_L introduced in this work to evaluate performance of ATDs vs. PMHS based data.

ACKNOWLEDGEMENTS

This work was supported under contract #N00024-13-D-6400, sponsored by the United States Army Research Lab in support of the Warrior Injury Assessment Manikin (WIAMan) program. Work was funded through subcontracts from John Hopkins University Applied Physics Lab. The authors gratefully acknowledge input from all members of the BRC Working Group including; Brian Bigler and Hattie Cutcliffe of Duke University, Joel D. Stitzel of Wake Forest University School of Medicine, Terrance Smith and Kenneth D. Wiley of Dynamic Research Incorporated, Craig Foster of TARDEC/Booz-Allen-Hamilton. The working group was organized by Andrew Merkle, Johns Hopkins University Applied Physics Lab.

REFERENCES

[1] Kent, R et al. Proc. Inst. Mech. Eng. Part D., 2006. 220(D7): p. 853-868.
[2] Neathery R. et al. Stapp Car Crash J., 1974. 18: paper: 741188.
[3] Maltese, M.R. et al., Stapp Car Crash J, 2002. 46: p. 321-51.
[4] Nusholtz, G., et al., SAE Int. J. Trans. Safety, 2013. 1(1).
[5] Eppinger, R.H. et al., SAE Technical Paper Series, 1984(840885).

SB³C2015
Summer Biomechanics, Bioengineering and Biotransport Conference
June 17-20, 2015,Snowbird Resort, Utah, USA
SB³C2015-498

INTER AND INTRA VARIATION IN THE TENSILE PROPERTIES OF THE PORCINE TEMPOROMANDIBULAR JOINT DISC

J. Lowe (1), A. Almarza (1,2,3)

(1) Department of Bioengineering
University of Pittsburgh
Pittsburgh, PA, USA

(2) McGowan Institute of Regenerative Medicine
Pittsburgh, PA, USA

(3) Department of Oral Biology
University of Pittsburgh
Pittsburgh, PA, USA

INTRODUCTION

Temporomandibular joint disorders (TMDs) encompass a variety of conditions, one being structural abnormalities of the TMJ disc. These TMDs can be caused by trauma, parafunction, unstable occlusion, functional overloading, and increased joint friction. It is estimated that roughly 10 million Americans currently have a TMD. Patients with TMD can show symptoms of persistent pain affecting normal life activities, such as malocclusion, facial asymmetry, locked jaw, and the deterioration of tissues around the joint space.

The current clinical treatment for patients with irreparable TMJ disc damage is the use of autogenous tissue sources. These tissues will inevitably resorb or create fibrous adhesions over time, making them ineffective as a long term treatment. Tissue engineering can provide implants that can remodel into tissues that possess near-native properties.[1]

Before the efficacy of such treatments can be assessed, however, the native mechanical properties of a healthy TMJ disc must be understood. Specifically, the TMJ disc seems to be better suited to withstand tension than compression. As such, several studies have investigated the tensile properties of the TMJ disc, however, the inter-disc variance in the superior-inferior direction was not studied in detail, and it could mask intra-disc variance.[2-4] It has been shown that the collagen fiber orientation in the central portion of the disc is mainly in the anterior-posterior direction, yet there is still much heterogeneity in the inferior-superior direction. Therefore, the objective of this research was to compare the variability in the tensile properties of the native TMJ disc in the inferior-superior direction both between and within samples. We hypothesize that the heterogeneous collagen fiber arrangement results in a high variance of tensile properties in the inferior-superior direction.

METHODS

Female porcine heads (n=9) from skeletally mature animals were procured from a local abattoir and the left TMJ discs were excised immediately. Anterior-posterior samples were collected from the intermediate zone of the TMJ disc, and a dogbone shape was created by punching two holes about 2 mm apart with a dermal biopsy punch. Each anterior-posterior sample was cut with a cryotome to produce 300-400 micrometer sections. Multiple sections from each disc were collected from the inferior to the superior surface, and kept in sequential order. Samples were then allowed to equilibrate in phosphate buffered saline in individual wells of a 6 well plate for at least one hour.

The samples were then tested using an Instron 5565 mechanical testing apparatus. The ends of each sample were gripped between 400 grit sandpaper in a pneumatic clamp (Figure 1). Samples were preloaded to 0.2 N. After the preload was applied, 20 cycles of 5% strain were applied at 10 mm/min for preconditioning. The samples were then pulled to failure at a strain rate of 10 mm/min. Samples were analyzed for the peak stress (stress at first rupture in the tissue), strain (strain at first rupture in the tissue), and modulus (slope of stress-strain behavior from 5-10% strain). These parameters were collected for all the inferior-superior sections of each individual disc, and averages from each disc were used to collect an aggregate value for each parameter.

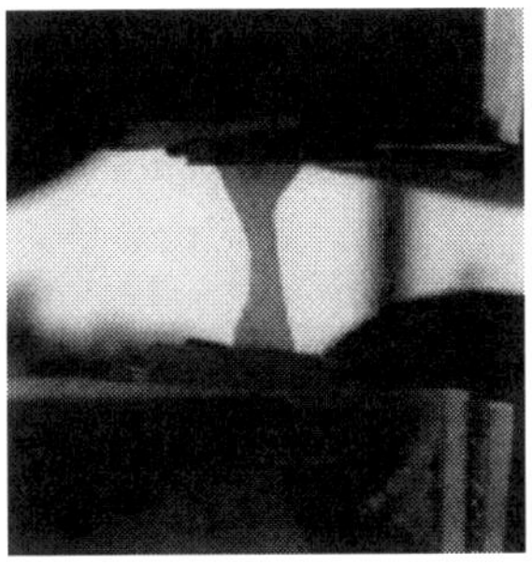

Figure 1; Dogbone sample in pneumatic grips

RESULTS

Due to the varying thicknesses of individual TMJ discs and the potential for a failed test (samples that slipped or broke at the clamp instead of the midsubstance), the number of sections for each disc fluctuated (n=2 to n=5). From these tests, the strain of all discs combined was 19.4% ± 3.9%, the peak stress was 5.1 ± 1.5 MPa, and the modulus was 38.1 ± 11.0 MPa.

The deviation within a disc was seen to be as high as 11% in strain per disc (Figure 2), 3.9 MPa in stress per disc (Figure 3), and 27.2 MPa in modulus per disc (Figure 4). This suggests that the individual variation of these parameters in a single disc from the inferior to superior direction is as high or higher than the variation across all TMJ discs.

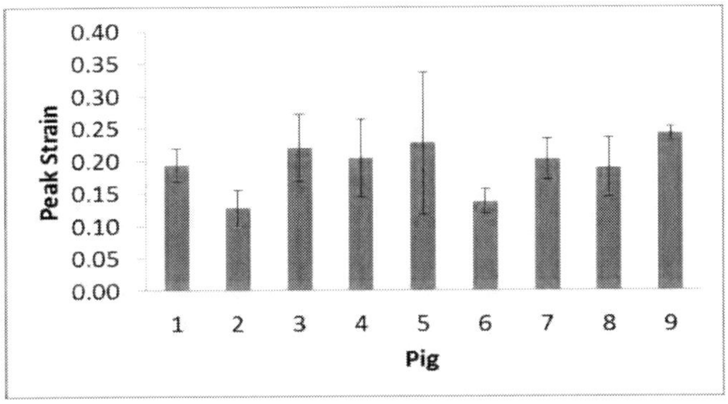

Figure 2; Strain at tissue failure for individual pigs (n=2-5). Average deviation in strain for each pig is 4.2%, compared to 3.9% across all pigs.

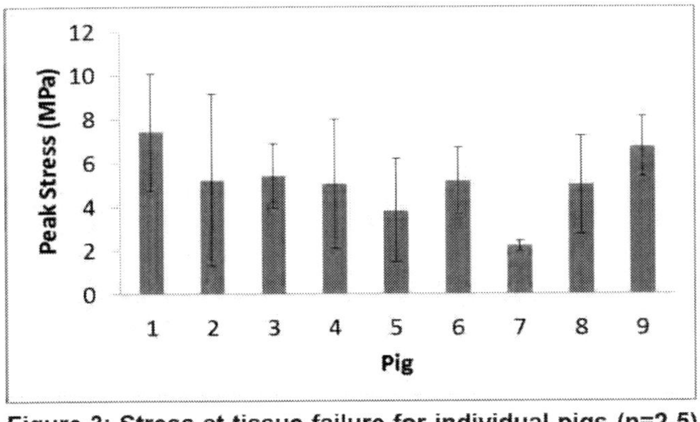

Figure 3; Stress at tissue failure for individual pigs (n=2-5). Average deviation in stress for each pig is 2.1 MPa, compared to 1.5 MPa across all pigs.

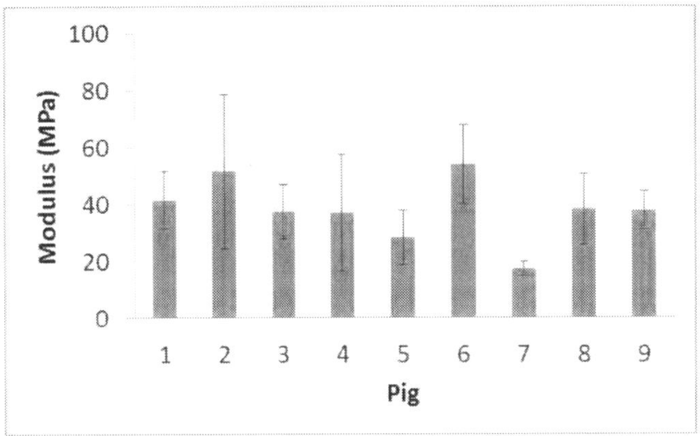

Figure 4; Modulus of the loading curve for individual pigs (n=2-5). Average deviation in modulus for each pig is 11.0 MPa, compared to 12.4 MPa across all pigs.

DISCUSSION

As shown in these results, the superior-inferior variation of the tensile properties of an individual porcine TMJ disc is either in the same range or higher than the deviations found across multiple discs. We expect this is due to the heterogeneous collagen distribution throughout the disc in the inferior-superior direction. However, since there is a preferred orientation in the anterior-posterior direction, it explains the lower variance observed between discs.

The overall values for peak stress (5.1±1.5) are within the range obtained previously (4.3 ± 2.0) tested using a similar protocol. [2] The modulus reported (37.6±12.3 MPa) is higher than the range obtained previously (23.2±10.0 MPa), and no failure strain has been previously reported using a comparable loading protocol. It was also found that sections that were greater than 400 microns would slip from the clamps. These results show that multiple inferior-superior sections for each disc must be tested in order to obtain reliable anterior-posterior tensile properties of the TMJ disc.

ACKNOWLEDGEMENTS

This research was completed with funding from the Biomechanics in Regenerative Medicine training grant from the National Institute of Health (NIH), grant #NIH T32 EB003392.

REFERENCES

[1] Brown, B et al., Inductive, Scaffold-Based, Regenerative Medicine Approach to Reconstruction of the Temporomandibular Joint Disc, *J Oral Maxillofac Surg*, 70:2656-2668, 2012

[2] Kalpacki, KN et al., An Interspecies Comparison of the Temporomandibular Joint Disc, *J Dent Res*, 90:193-198, 2010

[3] Detamore, MS, Athanasiou, KA, Tensile Properties of the Temporomandibular Joint Disc, *J Biomech*, 125:558-565, 2003.

[4] Tanaka, E et al., Shear Properties of the Temporomandibular Joint Disc in Relation to Compressive and Shear Strain, *J Dent Res*, 83:476-479

A MULTI-SCALE APPROACH IN ANALYZING FLUID/SOLUTE FLOW IN MECHANICALLY LOADED BONE

Lixia Fan (1), Shaopeng Pei (2), Xiaohan Lai (2), Xin Lu (2), Liyun Wang (2)

(1) Department of Mechanical Engineering
Nanjing University of Science and Technology
Nanjing, Jiangshu, China

(2) Mechanical Engineering Department
University of Delaware
Newark, DE, USA

INTRODUCTION

Fluid flow in the bone lacunar-canalicular system (LCS) is essential for osteocyte to sense mechanical loading, to obtain nutrients and to regulate other bone cells through transport of signaling molecules in the processes of osteoclastic targeted resorption during overuse and disuse and load-induced osteoblastic bone formation [1,2,3]. Using the fluorescence recovery after photobleaching (FRAP) approach in combination with mathematical modeling, we have quantified real time fluid flow velocity and solute transport rates in the mouse tibia LCS [4]. However, due to technical difficulties, the measurements were limited only to cortical regions 30-50 μm underneath periosteum. Transport measurements on deeper cortex and trabecular bone areas are lacking. The goal of this project was to develop a multi-scale approach combining finite element analysis (FEA) and microstructural modeling to map the tissue strain field at the whole bone level, the microscopic fluid and solute fluxes at the LCS level, and load-induced cellular stimulation in a mechanically loaded bone. This model was validated by comparing the predicted surface strains and transport enhancements against previous experimental measurements. Our preliminary results demonstrated the feasibility of this computational approach in estimating fluid flux in the LCS and the cellular stimulation forces (shear and drag forces) for osteocytes in not only cortical but also trabecular bones.

METHODS

(i) Whole Bone Model: An intact mouse tibia from an adult C57BL/6J male mouse was imaged using a Scanco μCT35 scanner at a voxel size of 20 μm. The reconstructed 998 slices were exported in DICOM format to ScanIP® (Simpleware), which meshed the entire tibia including cortical and trabecular bone into 5,112,690 tetra elements (Fig.1A). In Hypermesh® (Hyperwork, Altair), fixed constraints were imposed at the proximal tibial plateau. Similar to our experimental setup [4], a 3N cyclic compressive load was applied to the distal end of the tibia (Fig.1B). Assuming elastic material in the model, strain field was obtained using OptiStruct®, a FEA linear solver in HyperWork. The predicted strains were compared with experimental measurements to ensure that the material properties (20GPa Young's modulus and 0.33 Poisson's ratio) assigned to the FEA model were reasonable. (ii) Transport Model: For analysis of fluid and solute transport in LCS, a 3mm segment of tibial metaphysis consisting of ~ 700,000 tetra elements was selected from the whole bone FE model (20-40% distal of the proximal end, Fig.1A). The segment model was then imported to FEBio (http://www.febio.org) with the elements assigned as biphasic-solute material (20GPa modulus, 0.33 Poisson's ratio, 10% porosity, $2.8\times10^{-20}m^2$ tissue level permeability and 0.001Pa.sec viscosity). FRAP experiment [4] was simulated by setting the initial solute concentration to 0.2 at the elements around the selected photobleached element ~30μm below the anterior-medial periosteum, which was assumed to be photobleached, while the other elements were assigned 1.0. First, the FRAP transport rates (K_{diff}) were obtained by simulating non-loaded conditions with varied solute diffusion coefficients (D) in this 3D transport model. Based on the experimental results of K_{diff} for sodium fluorescein [4] a best-fit D was determined. Secondly, the loading FRAP experiment was simulated by applying the dynamic displacements, which were obtained from the whole bone model under 3N peak load, to the two cross-sectional surfaces of the transport FE model (Fig.1A) with a 0.5Hz sinusoidal waveform followed by 2 sec resting periods [4]. The boundary conditions for fluid and solute transport were listed in Fig.2A. To capture the active mixing mechanisms of canalicular flow influx into the photobleached element from surrounding elements, the increase of local solute concentration was simulated separately for the flow and reverse flow phases within one loading cycle, which both were found to enhance solute transport previously [5]. The local solute concentration during a total of 8 loading cycles (32 sec) was simulated, from which the transport enhancement (K_{load}/K_{diff}) was obtained for sodium fluorescein. The model was validated by comparing against experimentally observed transport enhancements

[4]. (iii) A Brinkman model of cellular stimulation forces: The strain and pore pressure of the region of interest, solute concentration of the photobleached element, as well as fluid and solute fluxes were obtained from the FEA model. Tissue level fluxes were scaled to the LCS levels based on porosity, assumed to be 10% [6]. The cellular stimulation forces such as fluid shearing force and drag force on transverse pericellular fibers were calculated using the Brinkman equation describing the canalicular flow [7] and measured pericellular matrix fiber spacing [8] as performed in our previous study [9]. The parameters used in these microscopic calculations were 76nm cell process radius and 160nm canalicular wall radius, measured from adult B6 mice [10], and PCM edge-to-edge spacing of 10.3nm and a fiber radius of 2 μm [8].

RESULTS

(i) Strain distribution of the whole tibia bone model under 3N compressive load on the distal end showed a tensile strain of approximately 450με on the anteromedial tibial surface with the assigned material properties matched well with the strain gage test measured data of 400με at the same location. (ii) The non-loaded FRAP FEA simulation showed that the solute diffusion coefficient D was linearly correlated with the transport rate K_{diff}. The diffusion coefficient of sodium fluorescein (376 Da) was found to be 31.8μm^2/s for experimentally observed K_{diff}=0.01709sec^{-1} in the 3D porous model (Fig.2B), which was consistent with our previous estimations after factored with 10% LCS porosity [4]. (iii) The loaded FRAP FEA simulation showed a transport enhancement of 1.24, which fell in the range of experimentally data (K_{load}/K_{diff}=1.31$\pm$ 0.24) [4]. These two results validated the FE model, which was further used to predict cellular stimulation forces at various locations. (iv) The pore fluid pressure distribution was similar to the strain distribution but with an 180° phase difference, and the pore pressure was higher at the periosteal surface than the endosteal surface. (v) The fluid flux in the model showed a heterogeneous distribution, with relatively higher flux values around endosteal surface (Fig.3A). For six selected locations, the flux varied from 0.6 to 3μm/s (Fig.3B), which resulted in much higher drag forces than shearing forces for all the locations (Fig.3C).

DISCUSSION

Using the image-based finite element analysis approach, we aimed to simulate the FRAP experiments and expand our understanding of load-induced fluid and solute transport in a large volume of bone tissue consisting both trabecular and cortical bone. In combination with our previous pericellular ultrastructural model [8], we were able to address this important issue from multiple levels, spanning from the entire bone level, LCS pore level, to the pericellular fiber level. Our results showed that this approach was feasible to estimate fluid flow and cellular stimulation forces on any locations, allowing further studies of how osteocytes' activation correlates with functional bone formation in vivo. There are several limitations in this study. The predicted canalicular flow velocity, which is sensitive to porosity, is about one order of magnitude lower than our previous estimations [4]. The porosity will be better optimized in future simulations. Sealed boundary was assumed in our model for better convergence in solute concentration simulations. However, permeability of the periosteum can introduce solute and fluid flow across the boundary and will be considered in future studies.

ACKNOWLEDGEMENTS: NIH P30GM103333; RO1AR054385.

REFERENCES

[1] Bonewald LF. Osteocytes as dynamic multifunctional cells. Ann NY Acad Sci. 2007;1116:281–290;

[2] Burger EH, Klein-Nulend J, Plas A, and Nijweide PJ. Function of osteocytes in bone-their role in mechanotransduction. J. Nutr. 125:2020S–2023S, 1995;

[3] Bloomfield SA. Cellular and molecular mechanisms for the bone response to mechanical loading. Int. J. Sport Nutr. Exerc. Metab. 11(Suppl):S128–S136, 2001;

[4] Price C, Zhou X, Li W, Wang L. Real-time measurement of solute transport within the lacunar-canalicular system of mechanically loaded bone: direct evidence for load-induced fluid flow. J Bone Miner Res. 2011;26(2):277–85;

[5] Zhou X, Novotny JE, Wang L. Modeling fluorescence recovery after photobleaching in loaded bone: potential applications in measuring fluid and solute transport in the osteocytic lacunar-canalicular system. Ann Biomed Eng. 2008;36(12):1961–77;

[6] Sharma D, Ciani C, Marin PAR, Levy JD, Doty SB, Fritton SP. Alterations in the osteocyte lacunar–canalicular microenvironment due to estrogen deficiency. Bone. 2012 51: 488–497;

[7] Weinbaum S, Cowin SC, and Zeng Y. A model for the excitation of osteocytes by mechanical loading-induced bone fluid shear stresses. J. Biomech. 27:339–360, 1994;

[8] Wang B, Zhou X, Price C, Li W, Pan J, Wang L. Quantifying load-induced solute transport and solute-matrix interaction within the osteocyte lacunar-canalicular system. J Bone Miner Res. 2013;28(5): 1075–86;

[9] Jing D, Baik AD, Lu X, Zhou B, Lai X, Wang L, Luo E, and Guo X . In situ intracellular calcium oscillations in osteocytes in intact mouse long bones under dynamic mechanical loading. 2014 FASEB J 28:1582-1592;

[10] Lai, Xiaohan; Price, Christopher; Maniscalco, Andrew; Modla, Shannon; Czymmek, Kirk; Wang, Liyun. Osteocyte Lacunar and Canalicular Cyto-Architecture in Type I Diabetic Bones, ORS 2012.

Fig. 1. (A) Whole bone FEA model of a murine tibia with the proximal end being fixed in position and load applied at distal end; A 3 mm segment was used for detailed transport modeling. (B) Loading profile consisted sinusoidal compressive load at 0.5 Hz followed by a 2 sec of resting period, similar to the one used in previous FRAP studies [1]

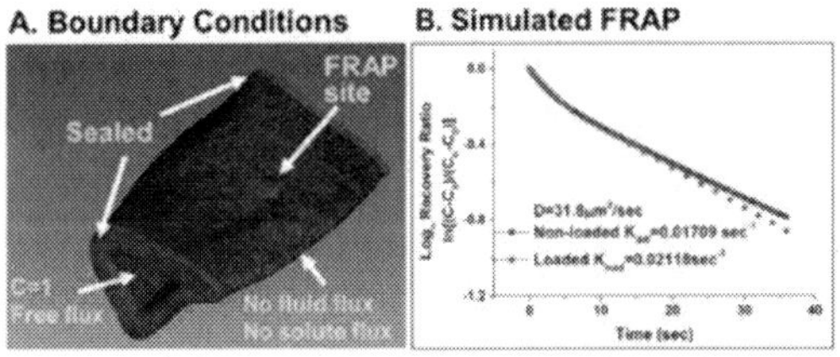

Fig. 2. (A) Boundary conditions in the transport model. (B) Simulated solute recovery was shown under non-loaded or loaded conditions at the FRAP site. An effective D was found to agree with experimentally observed K_{diff} and the K_{load} was predicted by the model.

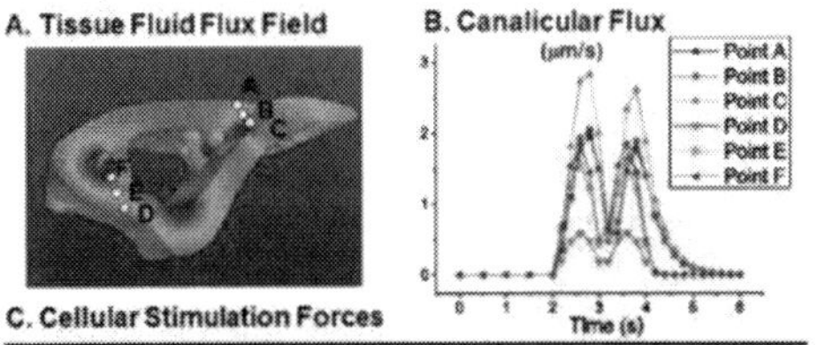

C. Cellular Stimulation Forces

Location	Canalicular Flow Velocity (μm/sec)	Shear stress (Pa)	Shearing Force (pN/μm)	Draging Force (pN/μm)
A	1.95	0.27	0.13	1.14
B	2.05	0.28	0.13	1.20
C	2.83	0.39	0.18	1.66
D	0.58	0.08	0.04	0.34
E	1.57	0.21	0.10	0.92
F	1.85	0.25	0.12	1.08

Fig. 3. (A) Fluid flux field in the 3D model (FRAP site: location A); (B) Canalicular fluid velocity profile at different locations; (C) Predicted cellular stimulation forces at the corresponding locations.

APPLICATION OF A NOVEL ROBOTICALLY SIMULATED PIVOT SHIFT FOR ANTERIOR CRUCIATE LIGAMENT RECONSTRUCTION: COMPARISON OF THE ALL-EPIPHYSEAL AND OVER-THE-TOP TECHNIQUES

Robb W Colbrunn (1), Tara F Bonner (1), Joel D Kolmodin (2), Paul M Saluan (2)

(1) Biomedical Engineering Department
Cleveland Clinic
Cleveland, OH, USA

(2) Department of Orthopaedic Surgery
Cleveland Clinic
Cleveland, OH, USA

INTRODUCTION

Anterior Cruciate Ligament (ACL) injuries increasingly affect the pediatric population. The all-epiphyseal (AE) and over-the-top (OTT) are two techniques used for reconstructing the ACL in lieu of traditional techniques so as not to disrupt the epiphyseal in growing patients. Both techniques' effectiveness in restoring intact knee kinematics has been examined previously in previous cadaveric studies [1, 2]. They concluded that both techniques provide anterior and rotational stability, yet no statistically significant conclusions were made as to which technique provided more stability. One study utilized the Rotary Loads test (5 Nm internal rotation torque, 10 Nm valgus torque) and the other used a custom mechanical pivot shift device designed to better mimic the dynamic loading condition of the clinical pivot shift exam. However, the authors of the mechanical pivot shift device paper suggest a limitation of their study was that the device was applying less than physiological loads.

The robotically simulated clinical pivot shift, a dynamic loading profile based on measurement of loads applied by a surgeon, was presented as a more robust test for detecting kinematic changes for evaluating ACL reconstructions and utilizes physiological loads [3]. This study aimed to use the simulated clinical pivot shift to:

1. Assess anterior and rotational stability provided by the AE and OTT surgical techniques
2. Compare it to alternative loading conditions

METHODS

Six fresh-frozen, cadaveric legs were transected mid-femur and mid-tibia and mounted to the Universal Musculoskeletal Simulator (Figure 1) capable of real-time force feedback using simVITRO™ software (Cleveland Clinic, Cleveland OH). Four surgical conditions were tested: Intact knee, deficient ACL, AE reconstruction and OTT reconstruction. The AE and OTT reconstruction order was randomized. For each surgical condition, three tests were performed:

1. Anterior drawer (134 N anterior, 20 N compression) at 0, 15, 30, 60, and 90° of flexion
2. Rotary loads (5 Nm internal rotation torque, 10 Nm valgus torque, 20 N compression) at 5, 15, and 30° of flexion
3. Simulated clinical pivot shift test [3]

For each specimen, the kinematics from the intact knee for each respective test was subtracted from the kinematics of the three remaining surgical conditions to acquire kinematic response relative to the intact knee. Relative anterior tibial translation (ATT) and internal tibial rotation (ITR) were reported at each tested flexion angle for the Anterior Drawer and Rotary Loads tests and for the flexion angle initiating the shift in the simulated clinical pivot shift test. This angle was at the peak ATT and corresponded to the start of the shift. Paired t-tests were used to determine significance ($p<0.05$). In one specimen the simulated clinical pivot shift could not induce a shift, even in the deficient state, and thus was excluded from the analysis.

FIGURE 1: KNEE MOUNTED IN THE ROBOTIC UNIVERSAL MUSCULOSKELETAL SIMULATOR

RESULTS

Figure 2A displays the ATT and ITR for the simulated clinical pivot shift at the flexion angle initiating the shift. Neither surgery restored intact knee kinematics. OTT exhibited 4.2 mm more ATT compared to AE ($p=0.021$) and 7.2° more ITR ($p=0.015$). OTT had no significant kinematic difference compared to the deficient ACL.

Figure 2B displays the ATT for the Anterior Drawer test. Figure 2C and 2D display the ATT and ITR (respectively) for the Rotary Loads test. Neither surgery restored intact knee kinematics nor significantly reduced ATT or ITR compared to the other surgery, agreeing with previous work with similar loading conditions [1]. The AE surgery significantly reduced ATT compared to deficient at flexion angles of 0, 15 and 30° by 5.3, 5.3 and 4.7 mm respectively and there was no significant difference between OTT and deficient, though trending was seen ($p<0.086$). (Fig 2B).

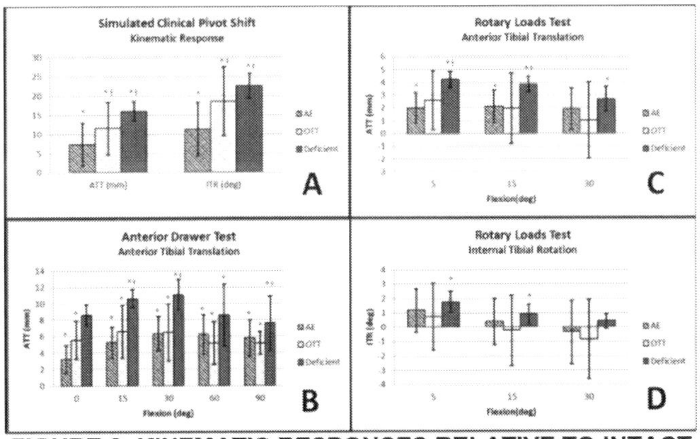

FIGURE 2: KINEMATIC RESPONSES RELATIVE TO INTACT KNEE, *P<0.05 COMPARED TO INTACT, †P<0.05 COMPARED TO AE, (NO SIGNIFICANT RESULTS BETWEEN OTT AND DEFICIENT), A) KINEMATIC RESPONSE INDUCED BY THE SIMULATED CLINICAL PIVOT SHIFT, B) ANTERIOR TIBIAL TRANSLATION (ATT) INDUCED BY THE ANTERIOR DRAWER, C) ATT INDUCED BY THE ROTARY LOADS, D) INTERNAL TIBIAL ROTATION (ITR) INDUCED BY THE ROTARY LOADS

DISCUSSION

The all-epiphyseal technique, though unable to restore native kinematics, showed significantly improved stability (4.2 mm ATT, 7.2° ITR) compared to the over-the-top technique when tested with a robotically simulated clinical pivot shift. Analysis of the Anterior Drawer and Rotary Loads tests did not detect significant differences in these surgical techniques, and are in agreement with a previous study which utilized similar loading conditions [1]. Unlike the mechanical pivot shift device [2], the loading conditions of the robotically simulated clinical pivot shift were able to identify a statistically significant difference between the two surgical techniques. It is uncertain if this difference was due to the surgical variability, loading magnitude, or type of combined loading condition that was applied.

Limitations of our study include the small sample size, surgical variability and the removal of one of the specimens. More work is needed on refining the loading profile to ensure it is able to induce a pivot shift in all specimens. While there is still room for improvement, it is currently still more effective than Rotary Loads, especially in its ability to detect rotational instabilities. Future work should include a detailed comparison of the different loading techniques to understand the benefits and drawbacks of each for elucidating knee rotational stability. These issues include the challenge of dissemination and use of this loading profile in other knee testing systems in labs around the world. The loading profile presented here requires the use of a real-time force feedback controller on a robot in order to simulate these loading conditions. However, many labs don't have access to this technology, and the rotary loads can be applied statically with effective low cost technology (e.g. deadweights and pulleys, or springs such as in the mechanical pivot shift device). In order to increase the accessibility of this loading profile to more researchers, simplified forms of it should be tested. For example, the flexion angle range can be reduced and some of the kinetics can be simplified to static loads. Only a minimal number of actuated degrees of freedom may be required and a functional testing system can be created.

The results from our study agree with previous studies [1, 3, 4] in that the Rotary Loads test does not induce large rotational changes amongst ACL surgical states. It also does not provide additional information compared to the Anterior Drawer test for measuring anterior laxity. It is our recommendation that the simulated robotic pivot shift should undergo further evaluation and refinement with an aim towards potentially replacing the Rotary Loads test in evaluating ACL reconstruction techniques.

ACKNOWLEDGEMENTS

Funding provided by Cleveland Clinic RPC grant #2012-1042

REFERENCES

[1] McCarthy MM et al., *Am J Sports Med.* 41,6: 1330-1339, 2013.
[2] Sena, M et al., *Am J Sports Med* 41: 826-836, 2013
[3] Colbrunn RW et al., *Pro ASME SBC*, SCC2013-14288, 2013.
[4] Kanamori A et al. *Arthroscopy*, 16:633-639, 2000.

SB³C2015
Summer Biomechanics, Bioengineering and Biotransport Conference
June 17-20, 2015,Snowbird Resort, Utah, USA
SB³C2015-501

DEVELOPMENT OF A TISSUE PHANTOM TO MIMIC THE THERMAL ENVIRONMENT OF A HUMAN ARM TO TEST DIGITAL THERMOMETERS

Peter Dillon (1), Oleg Vesnovsky (2), L. D. Timmie Topoleski (1,2), Jon P. Casamento (2), Laurence W. Grossman (2), Liang Zhu (1)

(1) Department of Mechanical Engineering
University of Maryland Baltimore County
Baltimore, MD, USA

(2) Center for Devices and Radiological Health
U.S. Food and Drug Administration
Silver Spring, MD, USA

INTRODUCTION

Body temperature monitoring of humans has been an important tool for helping clinicians diagnose infections, detect fever, monitor thermoregulation functions during surgical procedures, and assess post-surgery recovery. Commercially available thermometers have been developed and calibrated to their stated accuracy to predict body core temperatures using their proprietary formula, often based on the first few seconds of temperature recordings.

Our recent clinical study [1] demonstrated temperature variations from one body site to another and their deviations from the true body core temperature. The deviation was affected by environmental and physiological factors. That study was the first where temperature transients were recorded in a clinical setting by a fast responding reference thermistor bead sensor at two body sites. It is also the first time the calibrated thermistor bead sensor was placed simultaneously with a store brand digital thermometer to evaluate the digital thermometer's formula-based temperature predictions. There was a very large temperature variation during the initial 10 seconds of measurement. Unfortunately, the store brand thermometers' predictions are based on the initial few seconds of measurements, but they do not capture the initial temperature variations. Compared to the recordings from the reference thermistor bead sensor after 120 seconds, the store brand thermometers routinely overestimated or underestimated the actual temperature at the site by up to 3.7°F in both healthy and sick patients, therefore, the accuracy of those store brand thermometers is questionable.

The objective of this study was to develop a tissue-equivalent human upper arm phantom to provide a relatively realistic system to induce various thermal conditions similar to the thermal environment of a human body. Therefore, the known thermal environment can be used as a platform to evaluate accuracy of any store brand thermometers. The tissue-equivalent phantom can also be adjusted to simulate normal conditions or situations when fever occurs under various room environments.

METHODS

There are two components of this study. First, we developed a whole body model to simulate a normal thermal environment in a human body [2]. We focused our attention on the human arm in an adult model or a baby model to extract the heat generation rate needed to maintain a normal body temperature of approximately 37°C. The rationale is that the temperature at the axillary site often underestimates the true body core temperature and is often easily influenced by its environment, compared to that at the oral site. We then constructed a human phantom model, implementing the calculated heat generation rate in the whole body model. Resistors were embedded in the phantom and the temperatures at targeted sites can be adjusted via changing the current and voltage supplied to the resistors. It is expected that the maintained and known temperature at the phantom sites can be used to evaluate accuracy of store brand thermometers.

Muscle and skin surrogates were purchased from Syndaver™ Labs (skin and muscle plates, fatty, Caucasian, Syndaver™ Labs, Tampa, FL), shown in Figure 1. The muscle was rolled together to form a cylinder covered by the skin. One adult phantom human using the Syndaver™ muscle and skin is seen in Figure 1.

Based on the range of the theoretically determined heat generation rate in the upper arm [2], we identify flat resistor strips (SRFG-108/5 Silicone Rubber Heater, Omega, Stamford, CT) as suitable heating elements, since they can be easily embedded in the Syndaver™ muscle. Although there is a deviation from an actual human arm where the heating is spread over the entire arm tissue, our approach ensures that the total heat generated in the upper arm was consistent with that in realistic situations. A DC power supply

(PS300B, Hoefer, Hilliston, MA) was used to provide small incremental adjustments in voltage to the strip resistor and control the total power delivered within the range of our interest. A voltage meter was connected with the power supply to confirm the current and voltage delivered to the heating element, and therefore to calculate the total power delivered during the experiment.

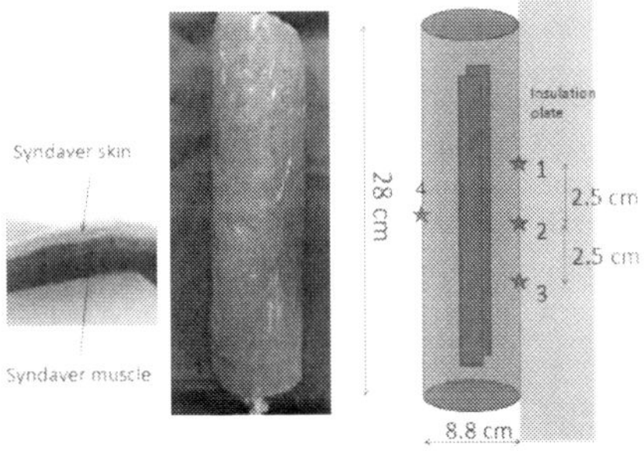

Figure 1. Syndaver™ muscle and skin plates (left image), the constructed adult arm (middle image), and a schematic diagram of the setup with dimensions. The stars in the figure represented where the thermocouples are located. The resistor strips are represented by the red plates in the schematic diagram.

Once the resistor strips were embedded inside the phantom, we wrapped the phantom with a very thin transparent food wrap to prevent any evaporation from the surface. The phantom arm was placed on an insulation plate (mimicking the human torso) so that heat transfer from that surface is minimized. The setup was designed based on temperature field simulation results, which showed very little heat transfer from the arm to the human torso [2]. Three copper-constantan thermocouples (0.1 mm in diameter) were attached to the surface of the phantom arm facing the insulation plate, while another thermocouple was attached to the other surface exposed to the room environment. Figure 1 shows the locations of the temperature monitoring sites. The air temperature of the room was also monitored and recorded. All temperature measurements were acquired and controlled via an HP control unit and an HP 6 1/2 digit voltage meter. The experiments were controlled via a LABVIEW program running in a PC. Each trial represented a fixed power supplied to the phantom arm. The steady state temperatures were recorded to show the temperature elevation achieved at the interface between the phantom arm and the insulation plate.

RESULTS

Using the COMSOL package and the Pennes bioheat equation [3], we simulated the temperature fields of an adult body and a baby body. The simulated temperature field was consistent with clinical observations in a human body. One can then calculate the total heat generation rate in the upper arm: the total heat generation rate was 1.03 W in the baby's upper arm, while it was 4.12 W in the adult's upper arm [2]. During steady state, almost all the heat generated in the upper arm escapes from the skin surface via convection to the colder ambient environment of 20°C. There is less than 5% of the heat exchanged at the interface between the upper arm and the torso.

Thermal properties of the Syndaver™ muscles were obtained using a needle probe [2]. Its thermal conductivity and thermal diffusivity are determined as 0.452±0.01 W/mK and $1.12\pm0.04\times10^{-7}$ m^2/s, respectively. For human tissues, thermal conductivity is approximately 0.5 W/mK, while the thermal diffusivity is approximately 1.32×10^{-7} m^2/s. Therefore, the tested Syndaver™ muscle had thermal properties similar to that of human tissues.

During the experiment, the ambient temperature was consistent from one trial to another, having an average and standard deviation of 21.0±0.5°C. The steady state temperatures at the four measurement sites in an adult Syndaver™ arm are presented in Figure 2. It is not a surprise to see that the highest temperature occurs at location 2 on the interface, where it is placed in the middle of the phantom surface. Temperatures at the other two locations, 2.5 cm away from location 2, are slightly smaller than (~0.5°C) that at location 2. The temperature on the surface of the Syndaver™ arm exposed to the ambient air is 9°C lower than that at the interface. A skin surface of 28°C is lower than that in a realistic situation, largely because the heating is concentrated within the heating element, rather than being distributed more uniformly as in an actual human arm.

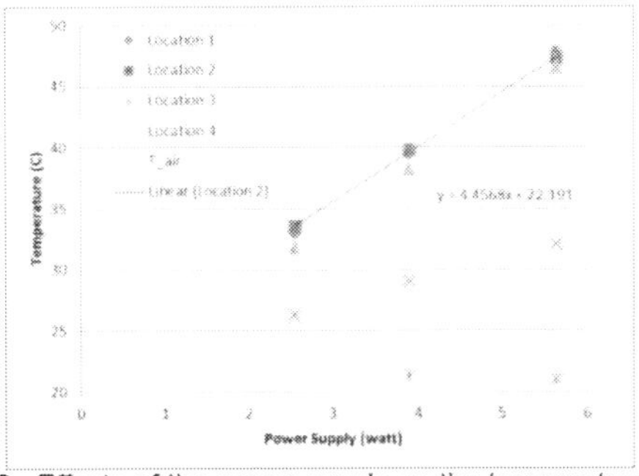

Figure 2. Effects of the power supply on the temperatures at the interface (locations 1, 2, and 3) and at the skin surface subjected to ambient air (location 4) in the Syndaver™ phantom mimicking the adult arm. The line represents the fitting of the measurement for location 2, where the maximal temperature occurs.

The effect of the power supply on the temperature elevations is also examined in Figure 2. It suggests that the total power supply should be 3.32 W to establish a steady state temperature of 37°C at the interface. Note that the measured power supply value is smaller than that predicted by the theoretical model (3.32 W vs. 4.12 W). This may also result in a relatively low temperature at the skin surface subjected to the ambient air. By adjusting the power level, one can establish various temperature rises at the interface. For example, a fever at 40°C can be achieved if the power supply is 3.9 W. There was an almost linear relationship between the power supply and the maximum temperature achieved.

In summary, a human arm phantom was constructed to mimic the thermal environment in a human arm and its thermal interaction with the torso and the colder ambient environment. We have identified material suitable for constructing artificial human arms. It has been shown that the Syndaver™ muscles have the thermal properties very similar to that of human muscle. Guided by our 3-D whole body heat transfer simulation, the total heat generated in the human upper arm was determined. Flat strip resistor heaters were imbedded into the arm phantom and a power supply connected to the heater was adjusted to establish a temperature at the arm and torso interface similar to that in *in vivo* situations. The experimentally measured heat generation rate in the arm phantom agreed well with that predicted by the whole body model. Empirical expressions were obtained to show the relationship between heating power and desirable temperatures at the interface.

ACKNOWLEDGEMENTS

This study was supported by the FDA Medical Countermeasures Initiative (MCMi).

REFERENCES

[1] Vesnovsky, O., Topoleski, L. D. T., Grossman, L. W., Casamento, J. P., and Zhu, L., 2014, *J. of Med. Dev.*, 8(3), 030905(1-3).

[2] Zhu, L. and Topoleski, L. D. T., 2014, *A report submitted to the FDA*.

[3] Pennes, H. H., 1948, *J. Appl. Phys.*, 1:93-122.

THE EFFECTS OF LOW INTENSITY VIBRATION ON BONE MINERAL DENSITY IN THE INTACT LIMB OF ANIMALS WITH A PERCUTANEOUSLY ATTACHED ENDOPROSTHESIS

Kyle A. Bodnyk (1), Garrett J. Noble (1), Matthew J. Allen (2), Noel Fitzpatrick (3), Gabriel Pagnotti (4), Richard T. Hart (1)

(1) Department of Biomedical Engineering
The Ohio State University
Columbus, OH, USA

(2) Department of Veterinary Medicine
University of Cambridge
Cambridge, UK

(3) Fitzpatrick Referrals
Surrey, UK

(3) Department of Biomedical Engineering
Stony Brook University
Stony Brook, NY, USA

INTRODUCTION

Limb amputees using traditional residual limb-to-socket prostheses can suffer from complications due to the unnatural loading applied to the skin surrounding the bone. Percutaneous bone implants are a novel way to connect prosthetic attachments to the residual limb while eliminating the prosthetic-to-skin interface, thereby increasing comfort and allowing for greater control of the prosthetic device [1]. Implants are anchored to the medullary canal of the residual bone, and a sealed skin-to-implant interface allows a portion of the implant to remain outside of the limb.

Bone implants alter the loading mechanics of the surrounding bone, which can, in turn, modify its ability to maintain a healthy skeletal structure. These outcomes may lead to increased bone resorption [2]. Thus, during post-operative rehabilitation, when weight bearing is minimal, bone resorption is more evident. Further implant loosening or even fracture can occur upon resumption of weight bearing activity.

Physical activity is necessary to maintain bone health [3]. Low intensity vibration (LIV) is a form of mechanical signal meant to induce some effects of physical activity, without the stress that may damage a fragile bone. LIV has been shown to significantly increase bone formation in several studies and animal models. These results suggest that LIV is an osteogenic signal that facilitates bone formation [4]. In a prior ovine study, hindlimb treatment with LIV induced small strains (5 $\mu\varepsilon$) on the bones and a 34.2% increase in trabecular bone in the proximal femur was observed [5]. Additionally, bone ingrowth into a turkey ulna implant treated with LIV ranged from 33-72% suggests LIV may be a beneficial means to enhance net bone formation around titanium implants [2].

During the current animal study, to non-invasively quantify bone health surrounding a percutaneous implant, dual-energy x-ray absorptiometry (DEXA) was used to measure BMD (bone mineral density) in both the implanted and intact limbs. Greater BMD may indicate denser more mature bone, while lower BMD could indicate new bone tissue that has not yet mineralized. Gait analysis, CT and histomorphometry also accompanied this study.

The objective of the present study was to investigate the effects of LIV on BMD during the initial healing of a percutaneous prosthetic attachment implant. The focus of this report is investigating the BMD in the intact limb contralateral to the implanted limb. This limb was treated with LIV but not subjected to any surgical intervention.

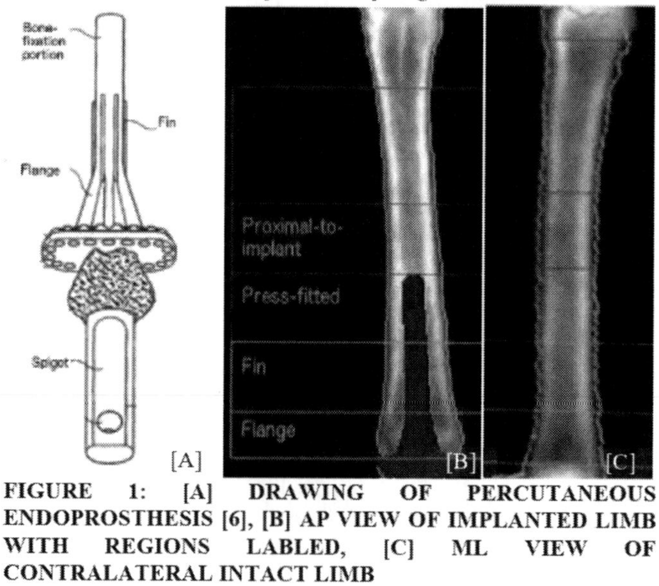

FIGURE 1: [A] DRAWING OF PERCUTANEOUS ENDOPROSTHESIS [6], [B] AP VIEW OF IMPLANTED LIMB WITH REGIONS LABLED, [C] ML VIEW OF CONTRALATERAL INTACT LIMB

METHODS

Twelve skeletally mature female Dorset sheep received custom-designed percutaneous endoprostheses press-fit within the medullary canal of metacarpal III/IV and were fitted with prosthetic feet which attached directly to the distal end of the implant. Sheep were administered LIV (0.3g, 45Hz) treatment of 10 minutes per day, 5 days a week for 12 weeks by standing on a vibration platform (modified from *Marodyne Medical*; Lakeland, FL). Controls did not receive LIV but stood on an unpowered platform next to the treatment animals (N=6/group).

Both implanted and intact contralateral limbs were scanned using DEXA (*GE Lunar Prodigy*) at weeks 4, 8, and 12 after surgery. Images were then evaluated using Encore 2008 Software (*GE Healthcare;* Little Chalfont, UK). The following four bone regions-of-interest (ROI) were specified: proximal to the implant region, press-fitted region, fin region, and flange region, as shown in Figure 1B. Scans were taken in both anteroposterior (AP) and mediolateral (ML) planes. The results were analyzed in JMP (SAS, Cary, NC) by a matched pair's analysis (time and region) and Student's t-test (control vs. treatment). As this was a preliminary study with a low sample size, we set our significance level to $\alpha<0.1$.

RESULTS

For all regions and time points, the BMD measured in the ML view was higher than that in the AP view ($p<0.05$). In the AP view, BMD was the lowest in the distal flange ROI and increased proximally at all timepoints ($p<0.003$). The ML view followed the same trend, however no differences were seen between the Press-fit and Proximal-to-Implant ROIs or the Fin and Proximal-to Implant ROIs.

In the ML flange region (Figure 2D), BMD in the LIV group was 7.3% ($p=0.06$) greater at week 12 compared to week 8 and 7.3% ($p=0.095$) greater at week 12 compared to week 4. Control BMD was 14.2% ($p=0.03$) greater than LIV BMD at week 8 while no significant differences were found in the AP view. In the AP fin region (Figure 2C), control BMD was 11.1% ($p<0.002$) greater than LIV BMD at week 4 and 10.6% ($p=0.05$) greater at week 8. Control BMD was 10.8% ($p=0.02$) greater than LIV BMD in the ML view at week 4, 10.9% ($p=0.04$) greater than LIV at week 8, and 12.3% ($p=0.09$) greater than LIV at week 12.

Analysis of the intact AP press-fitted region (Figure 2B), control BMD was 9.8% ($p=0.04$) greater than treatment BMD at week 4. For the ML view, control BMD was 7.4% ($p=0.08$) greater than treatment BMD at week 8. No significant differences were found in the intact AP or ML proximal-to-implant regions (Figure 2A).

DISCUSSION

Our results have shown that BMD measured in the ML plane is higher than that in the AP plane. This is likely due to the elliptical shape of the fused metacarpal III/IV bone. In the ML view, more bone per area is being scanned, resulting in a higher BMD reading. In all cases, BMD in the LIV group was lower than that of the control group. This appears to contradict the prior ovine study over one year that concluded LIV increases BMD over time [4], but the current study was limited to 12 weeks and animals were implanted with a percutaneous implant which affected load bearing between limbs. Baseline CT scans showed no difference in volumetric BMD between LIV and control groups, and gait was not significantly different between groups but lameness decreased over time for all animals [7]. The LIV likely caused bone adaptation, resulting in increased bone resorption followed by an increase in unmineralized osteoid causing a temporary lower BMD than the control animals for the duration of this study. This interpretation is supported by the BMD in the ML flange region (Figure 2D). Here, control and LIV group BMD were similar at week 4 but had increased at week 8 in the control group (likely due to improvement in gait) resulting in a higher BMD in the sham control group compared to the LIV group. The LIV group likely had lower BMD compared to control at week 8 due to increased remodeling activity that began to mineralize at week 12 causing the two groups to again be similar. We hypothesize treatment BMD will continue to increase and surpass control BMD beyond 12 weeks as more osteoid mineralizes. The flange region contains more trabecular bone than the other regions, which may explain why this effect was only seen in this region. Ongoing histomorphological and gait analyses are in progress to better elucidate the DEXA BMD results.

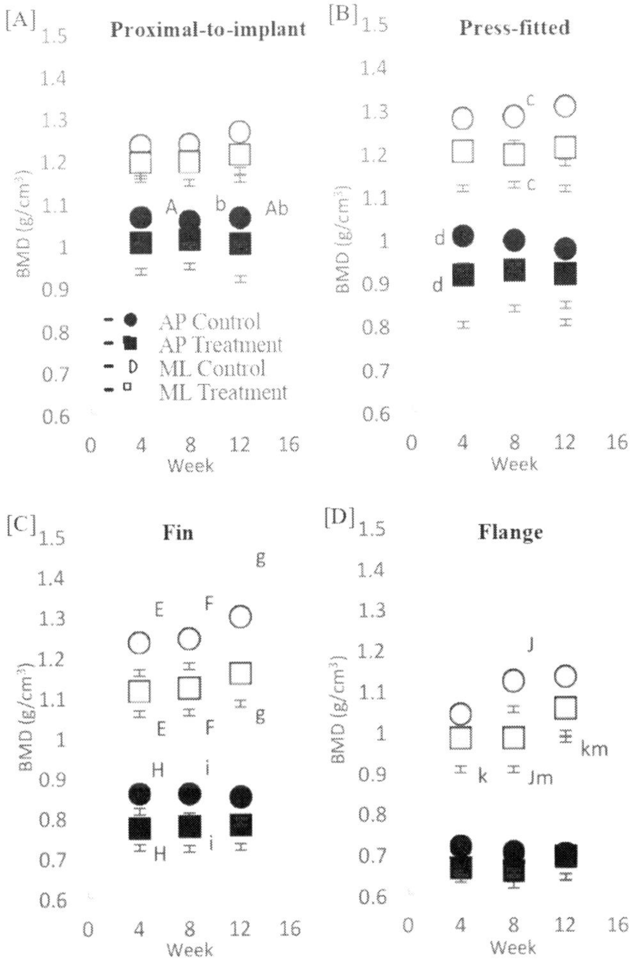

FIGURE 2: PLOTS OF BMD OVER TIME SPLIT BY VIEW IN THE [A] PROXIMAL-TO-IMPLANT ROI [B] PRESS-FITTED ROI [C] FIN ROI [D] FLANGE ROI. CAPITAL LETTERS INDICATE SIGNIFICANT DIFFERENCES (P<0.05) AND LOWERCASE LETTERS INDICATE (P<0.1)

ACKNOWLEDGEMENTS

This work is supported, in part, by funds from the Edgar C. Hendrickson Chair fund. Special thanks to Missy Matusicky, Judith Bertran, Kaleigh Peters, Brittani Jones, and Kevin Stephanoff for assistance with caring for and applying LIV to the sheep.

REFERENCES

[1] Hagberg, K. and Branemark, R. 2009. JRRD 46: 331-344
[2] Qin, Y. X., et. all 1996. JOR 14:862-870.
[3] Judex, S., & Zernicke, R. F. *J Appl Physiol*, *88*(6), 2183–2191 (2000).
[4] Luu, Y., Capilla, et. Al. JBMR 24:50–61.
[5] Rubin, C., Turner, A. S. *Nature*, *412:* 603-604, 2001
[6] Fitzpatrick, N. Percutaneous Implant. Patent WO2014064420
[7] Noble, G. J. et al. World Congress of Biomechanics, 2014 R370

SB^{3}C2015
Summer Biomechanics, Bioengineering and Biotransport Conference
June 17-20, 2015,Snowbird Resort, Utah, USA
SB^{3}C2015-503

BIOMECHANICAL CHARACTERIZATIONS OF SCAR ECM DURING THE ACUTE TO CHRONIC STAGES OF MYOCARDIAL INFARCTION

Bryn Brazile (1), J. Ryan Butler (2), Sourav S. Patnaik (1), Yanyi Xu (3), Andrew Claude (2), Raj Prabhu (1), Lakiesha N. Williams (1) Jianjun Guan (3), and Jun Liao (1)

(1) Department of Biological Engineering
Mississippi State University
Mississippi State, MS, USA

(3) Department of Clinical Sciences
College of Veterinary Medicine
Mississippi State University
Mississippi State, MS, USA

(2) Department of Materials Science and Engineering
Ohio State University
Columbus, OH, USA

INTRODUCTION

Myocardial infarction (MI) affects more than 8 million Americans, causing massive heart cell death and heart function decrease. After MI, the infarcted tissue experiences a dynamic and time dependent process, i.e., the necrotic phase, the fibrotic phase, and the remodeling phase; each phase exhibits unique structural and mechanical properties [1]. The biomechanical knowledge on the extracellular matrix (ECM) of scar tissue, mainly consisting of collagen fibers, is in much need. In this study, we took a novel approach to obtain scar ECM by applying decellularization to MI scar tissues created by permanent left coronary artery ligation (PLCAL) in a rat model. The time-dependent biomechanical and structural properties of scar ECM were then thoroughly characterized.

METHODS

Infarcted rat heart tissues were generated by a reliable and reproducible model for rat MI based on PLCAL. The infarcted hearts were harvested 15 min, 1, 2, 4, and 12 weeks after the PLCAL procedure (N = 6), to capture scar tissue development from the acute to chronic stages. Decellularized native heart tissues were used as the control. To obtain scar ECM, the infarcted hearts were subjected to 0.1% sodium dodecyl sulfate (SDS) with 0.01% trypsin, 1 mM phenylmethylsulfonylfluoride, 20 μg/ml RNase A and 0.2 mg/ml DNase at room temperature for 3 weeks. A biaxial mechanical testing system was used to characterize the scar ECM under physiologically-relevant loading conditions. Scar ECM specimens (mainly ventricle part) were trimmed into square shape, with one edge aligned with the circumferential direction and the other edge aligned with the longitudinal direction of the rat heart. Four dark markers were placed in the center region of the square sample. Samples were mounted onto the biaxial testing system via stainless steel hooks attached to 8 loops of 000 polyester suture of equal length (2 suture loops per sample edge). Each sample was subjected to 10-cycle preconditioning and then subjected to equibiaxial tension protocol of 30 N/m along the circumferential and longitudinal directions in a PBS bath. The sample extensibilities were characterized by maximum stretch along both the circumferential and longitudinal directions.

Histology was preformed for structural characterization. After decellularization, scar ECM was fixed with 10% neutral buffered formalin for 72 hours, dehydrated with graduated ethanol concentrations, and embedded in paraffin. In-plane sections of the decellularized scar ECM scaffolds were cut 5 μm thick. To elucidate changes in tissue components, Masson's trichrome and Picrosirius Red stains were used. Stained tissue sections were imaged using bright field microscopy (Nikon EC600).

RESULTS

After decellularization, a large decrease in wall thickness was observed in the native heart ECM and the scar ECM 15 minutes post infarction, which implied the collapse of cardiomyocyte lacunae (honey comb structure) after removal of heart muscle fibers. For scar ECM 1 week, 2 weeks, and 4 weeks post infarction, the decrease in wall thickness after decellularization was small. For scar ECM 12 weeks post infarction, the amount of wall thickness decrease was minimal after decellularization. The biaxial data showed that the scar ECM preserved an overall anisotropy observed in rat ventricle tissues and rat scar tissues [2], i.e., the longitudinal direction was more extensible than the circumferential direction. Scar ECM 15 minutes, 1 week, 2 weeks, and 4 weeks post infarction showed a stiffening biaxial behavior along with the time in both directions (Fig.1). The decrease of extensibility along longitudinal direction was more noticeable than the circumferential direction, which resulted in a decrease in degree of anisotropy (Fig. 1). Scar ECM 12 weeks post infarction maintained a longitudinal extensibility similar to the 4-week scar ECM, while experiencing an increased extensibility in the circumferential direction.

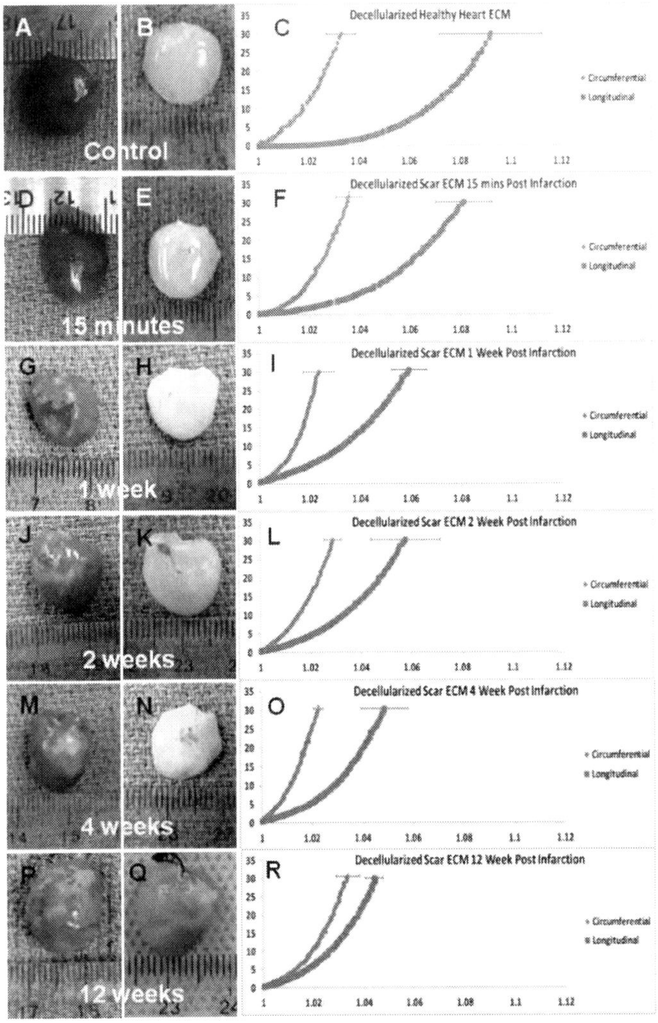

FIGURE 1: (A) NATIVE RAT HEART BEFORE DECELLULARIZATION. (B) NATIVE RAT HEART AFTER DECELLULARIZATION. (C) BIAXIAL RESULTS FOR THE DECELLULARIZED RAT HEART ECM. (D) RAT HEART 15 MINUTES POST INFARCTION. (E) 15 MINUTES POST-INFARCT HEART AFTER DECELLULARIZATION. (F) BIAXIAL RESULTS FOR THE DECELLULARIZED SCAR ECM 15 MINUTES POST INFARCTION. (G) RAT HEART 1 WEEK POST INFARCTION. (H) 1 WEEK POST-INFARCT HEART AFTER DECELLULARIZATION. (I) BIAXIAL RESULTS FOR THE DECELLULARIZED SCAR ECM 1 WEEK POST INFARCTION. (J) RAT HEART 2 WEEKS POST INFARCTION. (K) 2 WEEKS POST-INFARCT HEART AFTER DECELLULARIZATION. (L) BIAXIAL RESULTS FOR THE DECELLULARIZED SCAR ECM 2 WEEKS POST INFARCTION. (M) RAT HEART 4 WEEKS POST INFARCTION. (N) 4 WEEKS POST-INFARCT HEART AFTER DECELLULARIZATION. (O) BIAXIAL RESULTS FOR THE DECELLULARIZED SCAR ECM 4 WEEKS POST INFARCTION. (P) RAT HEART 12 WEEKS POST INFARCTION. (Q) 12 WEEKS POST-INFARCT HEART AFTER DECELLULARIZATION. (R) BIAXIAL RESULTS FOR THE DECELLULARIZED SCAR ECM 12 WEEKS POST INFARCTION.

DISCUSSION

The biaxial data shows that the scar ECM preserved an overall anisotropy observed in rat ventricle tissues, i.e., the longitudinal direction is more extensible than the circumferential direction. Scar ECM 15 minutes, 1 week, 2 weeks, 4 weeks, and 12 weeks post infarction showed a stiffening trend of biaxial behavior along with the time, in which longitudinal direction was mostly affected. The time evolving biomechanical properties of scar ECM implied the alterations of collagen fiber network from the acute to chronic stages of MI, which was verified by the histology of the scar ECM. Understanding the structure-property relationship of the scar ECM will provide insights into the disease progression, as well as the biomechanical environment of the scar tissues. Future studies will involve more post-infarction time points, as well as quantification of the fiber network in the scar ECM.

ACKNOWLEDGEMENTS

This work is supported by AHA 13GRNT17150041 and NSF EPS-0903787.

REFERENCES

[1] Holmes, J.W. et al., *Annu. Rev. Biomed. Eng*, 7:223-253, 2005.

[2] Sirry, M.S. et al., *Proceedings of 2014 World Congress of Biomechanics.*

RED BLOOD CELL DYNAMIC DEFORMABILITY AND ADHESION IN MICROSCALE FLOW DETERMINE CELLULAR HETEROGENEITY IN SICKLE CELL DISEASE

Y. Alapan (1), Y. Matsuyama (1), J. Little (2), U. Gurkan (1,3)

(1) Mechanical and Aerospace Engineering Department
Case Western Reserve University
Cleveland, OH, USA

(2) Department of Medicine
Case Western Reserve University
Cleveland, OH, USA

(3) Department of Orthopaedics
Case Western Reserve University
Cleveland, OH, USA

INTRODUCTION

Red blood cell (RBC) adhesion and deformability are the main causes leading to blood vessel occlusion (vaso-occlusion) in sickle cell disease (SCD) [1]. Here, we present a microfluidic approach to perform phenotypic and biophysical analyses, namely deformability and adhesion of RBCs at single cell level under precisely controlled microscale fluid flow [2]. We measured and compared dynamic deformability and adhesion of healthy hemoglobin A (HbAA) and homozygous sickle hemoglobin (HbSS) containing RBCs in blood samples. We introduce a new parameter to assess deformability of RBCs: the dynamic deformability index (DDI). Our results showed that DDI of HbSS RBCs were significantly lower compared to that of HbAA RBCs. Moreover, we observed subpopulations of HbSS RBCs in terms of DDI and number of adhesion sites, suggesting heterogeneity in adhesion and deformability characteristics of cells. Furthermore, we utilized the microfluidic system to perform phenotypic analyses based on number of captured RBCs to study clinical phenotypes. Here, we are presenting the results of a clinical study based on blood samples obtained and screened from 45 consenting patients under institutional review board (IRB) approval.

METHODS

The microfluidic system is composed of a Fibronectin (FN) functionalized glass surface, plastic top, and a 50 μm thick double sided adhesive film in the middle (Fig. 1a). Real-time videos of RBC deformations under fluid flow were recorded and converted to single frame images for further processing and analysis. Dimensions and cell aspect ratio (CAR) of single RBCs were determined by image processing. CAR is a calculated ratio between the vertical width and the horizontal length of a cell (Fig. 1b). Change in CAR, measured at no flow condition and at detachment instant (Fig. 1c), indicates deformability of RBCs. Adhesion sites of RBCs were determined by analyzing cell motion in response to applied flow. Number of adhered RBCs were quantified to determine clinical phenotypes.

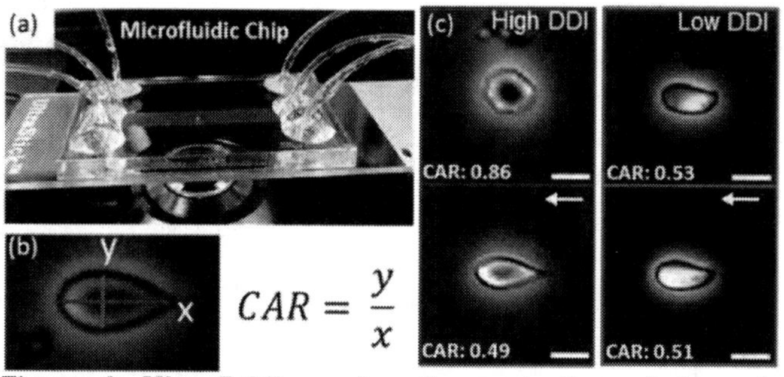

Figure 1: Microfluidic system for evaluation of dynamic deformation of RBCs in microscale physiological flow. (a) Microfluidic chip was placed on a microscope stage for live cell image recording. (b) Depiction of a RBC where the width (y) and length (x) dimensions were being measured. (c) Different types of RBCs in the absence of flow and right before detachment in response to flow. From left to right, healthy (HbAA) and HbSS RBCs. Scale bars represent 8 μm length. Arrows indicate direction of flow.

RESULTS

We observed adhesion of a morphologically heterogeneous RBC population in HbSS blood samples. We analyzed the deformability of single HbAA or HbSS RBCs by examining CAR for each individual frame that is recorded. We observed subgroups of HbSS RBCs: HbSS deformable and HbSS non-deformable. We evaluated deformability of

RBCs in time from healthy and SCD subjects using DDI (Fig. 2a). DDI was determined as the rate of aspect ratio change in time. Both HbAA and HbSS deformable RBCs exhibited dynamic deformation until reaching a deformed state, whereas HbSS non-deformable RBCs demonstrated minimal dynamic deformation (Fig. 2b).

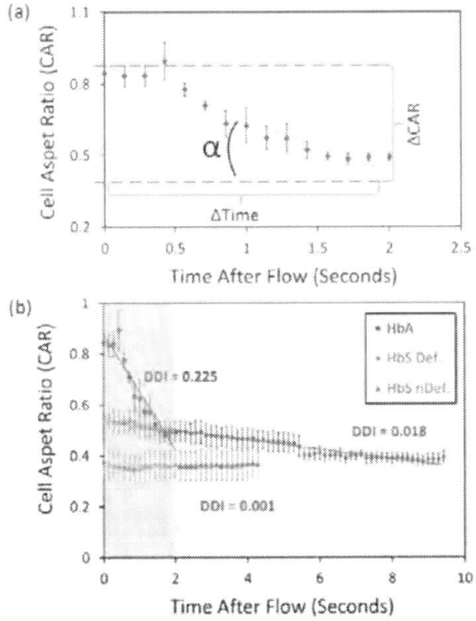

Figure 2: Heterogeneity in dynamic deformability of RBCs in SCD. (a) Dynamic Deformability Index (DDI) was determined as the rate of aspect ratio change in time. DDI = tan(α), where α is the angle between the change in CAR and the time. (b) Healthy RBCs deformed at a quicker rate to reach their deformed state than HbS RBCs for detachment flow velocities. Shown is the average curve for each group during deformation.

Deformable HbAA or HbSS RBCs displayed a single adhesion pivot point, compared to non-deformable HbSS RBCs, which showed multiple adhesion sites (Fig. 3). We observed a statistically significant difference in RBC adherence per unit area to FN, amongst RBCs from subjects with HbSS (90±15, n=28), with compound heterozygous HbSC or Sβthalassemia (22±4, n=7), or with normal hemoglobin (HbAA, 4±1, n=10, *Mann-Whitney*, $p<0.05$) (Fig. 4).

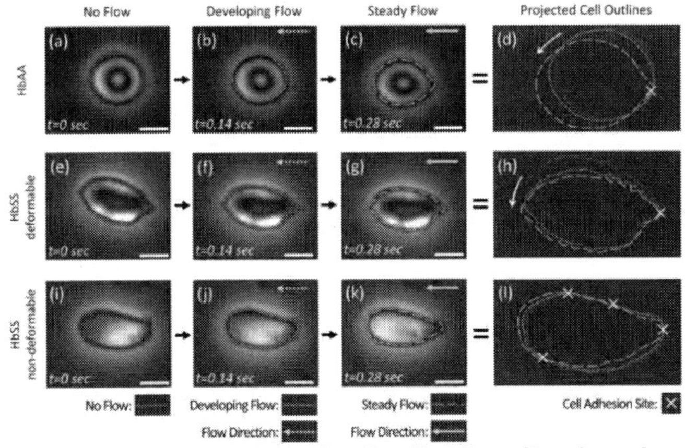

Figure 3. Determination of cell adhesion sites based on analysis of projected cell outlines at flow initiation. Outlines of individual RBCs in three consecutive frames taken over 0.28 seconds are projected to reflect the motion of the cells in response to flow initiation, for: (a-d) Hb A, (e-h) HbSS deformable, and (i-l) HbSS non-deformable RBCs. HbAA (a-d) and HbSS deformable (e-h) RBCs display cell adhesion at a single point, which acts as a pivot (d, h), resulting in rotational motion in response to fluid flow (horizontal yellow arrows). HbSS non-deformable cells display no rotational motion and show multiple sites of adhesion, as determined in projected cell outline analysis (l). Scale bars represent 5 μm length.

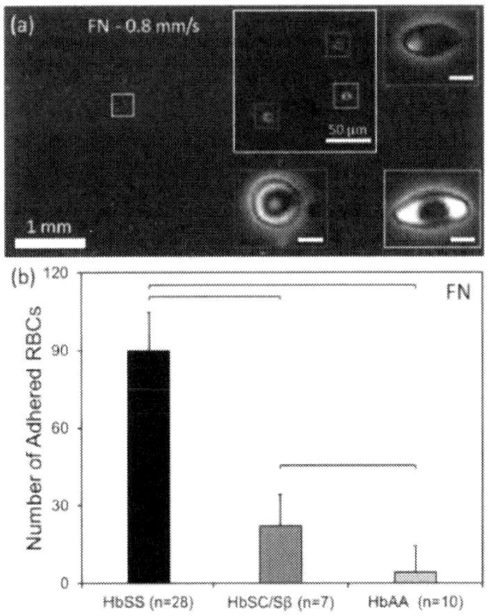

Figure 4. Determination of clinical phenotypes in sickle cell disease patients based on RBC adhesion. (a) Image of adhered RBCs on FN were obtained and quantified. Scale bars represent 5 μm length. (b) Clinical phenotypes were determined based on number of adhered RBCs on FN functionalized channels. The horizontal lines between individual groups represent statistically significant difference based on Mann-Whitney test ($p<0.05$). Error bars represent the standard error of the mean.

DISCUSSION

We present a microfluidic system to analyze RBCs for complex biophysical properties of adhesion and dynamic deformability in SCD, which reflect clinical phenotypes. Based on deformability and adhesion analyses, we report heterogeneity in HbSS RBCs, which may have significant implications for understanding vaso-occlusion and for phenotyping disease pathophysiology. Furthermore, we have shown that the microfluidic platform can be used to determine clinical phenotypes of sickle cell disease patients based on quantification of adhered RBCs on FN functionalized microchannels, in a standardized and consistent manner. Studying deformability and adhesion of cells may have implications in other multi-system diseases such as β-thalassemia, diabetes mellitus, hereditary spherocytosis, polycythemia vera, and malaria.

ACKNOWLEDGEMENTS

This work was supported by Grant # 2013126 from the Doris Duke Charitable Foundation

REFERENCES

[1] Barabino, GA et al., *Annu. Rev. Biomed. Eng.*, 12:345-367, 2010.
[2] Alapan, Y et al., *Sci. Rep.*, 4:7173, 2014.

A BOOTSTRAPPING METHOD TO ASSESS THE INFLUENCE OF AGE, OBESITY, AND GENDER ON PROBABLITY OF TRIPPING AS A FUNCTION OF OBSTACLE HEIGHT

Christina R. Garman (1), Christopher T. Franck (2), Maury A. Nussbaum (1), Michael L. Madigan (3)

(1) Biomedical Engineering and Mechanics
Virginia Tech
Blacksburg, VA, USA

(2) Statistics
Virginia Tech
Blacksburg, VA, USA

(3) Biomedical Engineering
Texas A & M University
College Station, TX, USA

INTRODUCTION

Tripping accounts for an estimated 53% of falls among older adults [1]. The most common measure for characterizing the probability of tripping while walking is the minimum foot clearance (MFC) during swing. A decrease in the central tendency (i.e. mean/median) of MFC, or an increase in MFC variability, are both associated with an increased probability of tripping [2-4]. These indirect measures of probability of tripping can lead to ambiguous results, though, when both increase or decrease simultaneously. Moreover, median MFC and MFC IQR are positively correlated [3], indicating concurrent increases or decreases in both are to be expected.

The purpose of this study was to develop a method that can determine the probability of tripping and to compare this probability between groups of interest while avoiding the noted ambiguous situations (i.e., wherein median MFC and MFC IQR concurrently increase or decrease). This method was applied to a experimental dataset to estimate differences in the probability of tripping with respect to obesity, age, and gender. These specific factors were investigated based upon reports of elevated risks of falling and sustaining a fall-related injury among obese adults [5], older adults [6] and females [7].

METHODS

Eighty participants completed the study including four gender-balanced groups comprised of 20 participants each (Table 1). Participants completed 16 trials walking along a 10 m walkway at two speeds: a self-selected gait speed and a hurried speed of 1.9 m/s. All participants wore the same brand of athletic shoes, to which were attached three reflective markers. A Vicon MX T10 motion analysis system (Vicon Motion Systems Inc., LA, CA) was used to sample marker positions at 100 Hz. MFC was determined using a method adopted from Startzell et al [2].

Table 1: Participant information (mean ± SD),

Group	Age (y)	BMI (kg/m^2))
Young Normal-Weight	24±3.3	22.3± 2.2
Young Obese	23.1±3.0	33.4±3.3
Older Normal-Weight	60.7±5.9	24.5±1.5
Older Obese	59.1±5.6	33.1±3.2

MFC values were used to create trip probability curves that indicated how the probability of tripping varied as a function of the height of a potential tripping obstacle (Figure 1). For potential tripping obstacle heights ranging from 0 - 7 cm, in increments of 2 mm, each experimental MFC value was dichotomized as either a trip (if the potential obstacle height was greater than MFC) or a non-trip (if the potential obstacle height was equal to or less than MFC). The percentages of trips at each obstacle height were then computed, serving as an estimate of the probability of tripping versus obstacle height.

A statistical bootstrapping technique was used at each potential obstacle height to determine whether the probability of tripping differed by obesity, age, or gender. The first step in this technique was to randomly reassign participant group labels (e.g., obese or normal weight when investigating obesity effects) to each participant's 16 MFC values. A probability curve was then created for each group, and the difference in trip probability between groups was calculated at each potential obstacle height. This process was repeated a large number of times (100,000 in our analysis) to obtain a distribution of differences at each potential obstacle height that would occur if group assignment was random. The distribution of these resampled differences acted as the sampling distribution of the difference under the null hypothesis that the groups had equal trip probabilities.

The second step in this technique was to examine the position, within the distribution of group differences, of the actual observed difference in probability of tripping between groups. This was done to determine if this difference was statistically significant at a level of α=0.05. A multiplicity correction was used to avoid inflation of type I errors. As such, if the difference was positioned in the outer 0.14% of the distribution, then the difference in trip probability between groups was considered statistically significant (α=0.0014). The percentage of the distribution of differences outside of the actual observed difference yielded a bootstrap *p*-value. This second step was performed for the complete set of potential obstacle heights, and between all participant groups of interest, to determine the specific heights at which the probability of tripping differed significantly between groups.

Group differences identified from this statistical bootstrapping technique were compared with group differences identified using the traditional measures of median MFC and MFC IQR. Group differences in median MFC and MFC IQR were determined using separate three-way, mixed-factor analyses of covariance (ANCOVA) with planned contrasts. Independent variables in the ANCOVAs were age, obesity, and gender, and gait speed was the covariate. Analyses were performed using JMP v7 (Cary, North Carolina, USA).

RESULTS

Obesity-related differences in the probability of tripping were not consistent between the bootstrapping technique and the ANCOVA analysis. Among older adults, the probability of tripping was higher among obese adults across a range of obstacle heights (2.4-4.2 cm), while there were no obesity effects on median MFC or MFC IQR. Among young adults there were no significant effects of obesity on the probability of tripping, nor on median MFC or MFC IQR.

Age-related differences in the probability of tripping were also not consistent between the bootstrapping technique and the ANCOVA analysis. Among normal-weight adults the probability of tripping was lower among older adults across a range of obstacle heights (2.0-4.6 cm), while age effects were not significant for either median MFC or MFC IQR. Among obese adults the probability of tripping was also lower among older adults, but across a smaller range of obstacle heights (1.2-2.4 cm), while again there were no age effects on either median MFC or MFC IQR.

With respect to gender (Figure 1), the probability of tripping was higher among females across a range of obstacle heights (0.8-4.4 cm), while both median MFC and MFC IQR were lower among females.

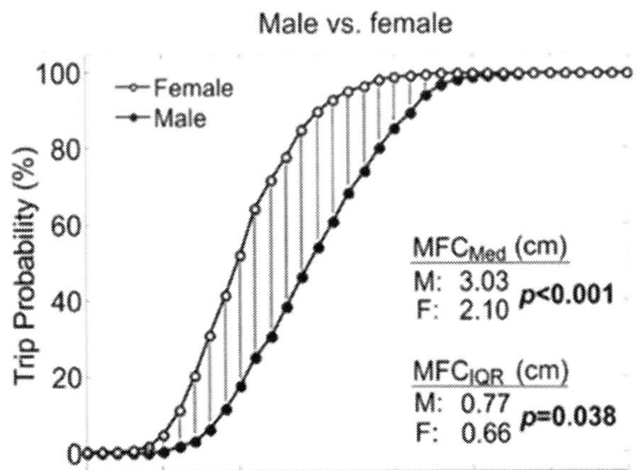

FIGURE 1: REPRESENTATIVE TRIP PROBABILITY CURVES AND MEDIAN/IQR MFC. DIFFERENCES IN TRIP PROBABILITY BETWEEN GROUPS ARE INDICATED BY A SOLID VERTICAL LINE, AND DIFFERENCES IN MEDIAN/IQR MFC ARE INDICATED BY BOLD. NOTE: M=MALE, AND F=FEMALE

With respect to gender (Figure 1), the probability of tripping was higher among females across a range of obstacle heights (0.8-4.4 cm), while both median MFC and MFC IQR were lower among females.

DISCUSSION

The technique presented here revealed differences related to obesity and age that were not identified from the more traditional approach using median MFC and MFC IQR. This technique also identified clear gender differences in the probability of tripping versus obstacle height despite ambiguous results from the ANCOVA analysis.

Furthermore, characterizing the probability of tripping as a function of obstacle height provides additional information on how individual factors may affect the probability of tripping. For example, results summarized in Figure 1a suggest that trip probability does not differ between normal-weight and obese older adults unless obstacle heights are greater than 2.4 cm. Such information may be helpful in establishing safety guidelines or standards that are more inclusive and protective for diverse populations.

Two limitations to the method presented here warrant mentioning. First, and similar to median/mean MFC and MFC IQR as commonly used to characterize probability of tripping, this method emphasizes foot clearance only at the instant that MFC occurs, even though a trip could occur at other instances during the swing phase. Second, unlike an ANOVA based upon median/mean MFC and/or MFC IQR, the current method cannot incorporate measures of covariance, or statistically control for the effects of other variables, when evaluating an independent variable of interest.

In conclusion, a pragmatic method is reported to characterize the probability of tripping as a function of obstacle height, and which can identify effects of factors not identifiable by the commonly used ANOVA analysis using MFC central tendency and variability.

ACKNOWLEDGMENTS

This work was supported by award R01OH009880 from the CDC-NIOSH, and a National Science Foundation Graduate Research Fellowship under grant DGE 0822220.

REFERENCES

1. Blake, A.J., et al., Age Ageing, 1998. **17**(6): p. 365-372.
2. Winter, D.A., Physical Therapy, 1992. **72**(1): p. 45-53.
3. Begg, R., et al., Gait & Posture, 2007. **25**(2): p. 191-198.
4. Mills, P.M., et al., Gait & Posture, 2008. **28**(1): p. 101-107.
5. Fjeldstad, C., et al., Dynamic Medicine, 2008. 7(4).
6. WHO. 2007 May 7, 2014];
7. Ambrose, A.F., et al., Maturitas, 2013. **75**(1): p. 51-61.

EVALUATING ACCURACY OF THE ALGORITHM/FORMULA USED IN STORE BRAND DIGITAL THERMOMETERS TO PREDICT BODY TEMPERATURE BASED ON THE FIRST FEW SECONDS OF THE TEMPERATURE MEASUREMENTS

Amirreza Saharkhiz (1), Oleg Vesnovsky (2), Jon P. Casamento (2), Laurence W. Grossman (2), L. D. Timmie Topoleski (1,2), Liang Zhu (1)

(1) Department of Mechanical Engineering
University of Maryland Baltimore County
Baltimore, MD, USA

(2) Center for Devices and Radiological Health
U.S. Food and Drug Administration
Silver Spring, MD, USA

INTRODUCTION

Technology has advanced to produce widely available non- or minimally-invasive home use digital thermometers that measure a temperature in 5-10 seconds. Those thermometers use an algorithm that incorporates temperature compensations or a predictive formula to project or estimate the thermal equilibrium temperature based on temperatures measured only during the initial several seconds. If the algorithm is pre-programmed into the thermometer, its predictive capability may be limited because the measurements and the path to temperature equilibrium are influenced by many environmental, physiological, and anatomical factors. For example, the amount of blood supplied to the tissue site may vary, and is not known, during a temperature measurement. Variation in the amount of blood supplied during the measurement will lead to higher or lower tissue temperatures surrounding the thermometer. Such changes in local blood perfusion are not accounted for by the temperatures measured during the initial several seconds.

The objective of this study was to evaluate how the temperature environment changes after placing a temperature sensor at a measurement site to determine whether the temperature data measured in the first few seconds are sufficient to predict the equilibrium body temperature. A specially designed thermistor bead sensor was used to measure temperature transients at the oral and axillary sites in human volunteers. A function, consisting of two exponential terms with different time constants, was fit to the recorded temperature transient data to assess possible responses of the local blood perfusion to the contact with the sensor. The data collected from this procedure can be used to understand whether a simple exponential formula can be derived to predict body core temperature and how thermoregulation in the body may affect the accuracy of temperature measurements.

METHODS

A specially designed thermistor bead temperature sensor system, with a nominal resolution of 0.001°C, was used in this study. Its response time, when immersed into a water bath without a sheath, is less than 5 seconds due to its small size [1]. The temperature measurements in the oral and axillary sites were performed on 20 healthy human volunteers (10 males and 10 females, 20-30 years old). Following normal clinical practice, the thermistor bead sensor was placed into a disposable sheath, and then the temperature sensor was inserted into the right or left sublingual pocket between the tongue and the mouth floor, due to the rich blood perfusion there, or to an axillary site under the armpit. The measurement time was 10 minutes in the mouth and 40 minutes at the axillary site, to determine the steady state temperature. Temperatures were measured and digitally recorded on a PC at a rate of 7 Hz.

Since the measurement time used in the experiment protocols is sufficiently long, we assume that a local steady state is established after either 10 minutes in the mouth or 40 minutes under the arm. The initial room temperature sensed by the thermistor bead is denoted as T_0, and the local steady state temperature at the measurement site is represented as T_{ss}. The dimensionless temperature transient data are represented by the following function:

$$\frac{T(t)-T_0}{T_{ss}-T_0} = A[1-\exp(-t/\tau_1)]+(1-A)[1-\exp(-t/\tau_2)] \qquad (1)$$

where t is time, and τ_1 and τ_2 are different time constants. The left side of Eq. 1 defines the dimensionless temperature, and it varies between zero and one. Two exponential functions on the right side of Eq. 1 are proposed, each with different time constants, τ_1 and τ_2 (where $\tau_1 < \tau_2$). A is a constant used to describe the contribution of the 1st exponential term, which describes the initial behavior of the system, while $1\text{-}A$ represents the contribution of the 2nd term, which describes the later behavior, as the temperature approaches steady state. The Solver in Microsoft Excel 2010 was used to converge the iterations for the curve fits to Eq. 1. Differences between measured temperatures, and the calculated time constants, between male and female subjects and between measurement sites, were determined using Student's t-tests.

RESULTS and DISCUSSION

The measured steady state temperatures were all approximately 37°C. The time to reach steady state temperature in the mouth, however, was inconsistent, and ranged from approximately 30-300 seconds in all the subjects on both sides of the sublingual pocket. Further, it took much longer to establish a steady state in the axillary site (700–2400 s). After 2400 seconds, the steady state temperature at the axillary site was still 0.2°C lower than that in the oral site.

The correlations between the fit curve and the experimentally measured temperature transients are excellent ($R > 0.99$). We first examine whether the body side and/or gender affect the values of the calculated parameters. Tables 1-3 give the determined values of the three parameters (A, τ_1, τ_2). Almost all the obtained p-values are larger than 0.05, indicating no significant difference between groups. The only significant differences occurred in the oral measurements. However, the mean values of either A or τ_1 do not differ that much between groups when $p<0.05$. For example, the values of τ_1 of the left side in the male and female groups are 2.81 s and 3.3 s, respectively, implying only a 15% significant difference by the gender. We conclude that neither the side of the body nor the gender affects the curve fitting parameters appreciably.

Table 1. Determined contribution of the first exponential function for the oral site and the axillary site

Gender		Male	Female	p-value
Oral	Right	97.68%±1.76%	97.82%±1.25%	*0.420*
	Left	97.14%±2.09%	96.14%±2.38%	*0.165*
p-value		*0.270*	***0.034***	
Axillary	Right	91.77%±4.16%	88.31%±7.60%	*0.114*
	Left	87.70%±6.35%	89.32%±5.98%	*0.282*
p-value		*0.055*	*0.372*	

Table 2. Determined τ_1 (s) for the oral site and the axillary site

Gender		Male	Female	p-value
Oral	Right	3.42±0.74	3.19±0.38	*0.199*
	Left	2.81±0.47	3.30±0.69	***0.041***
p-value		***0.021***	*0.341*	
Axillary	Right	6.55±2.88	5.86±2.62	*0.290*
	Left	7.34±3.13	5.52±2.32	*0.079*
p-value		*0.282*	*0.383*	

Table 3. Determined τ_2 (s) for the oral site and the axillary site

Gender		Male	Female	p-value
Oral	Right	160.18±65.71	224.18±97.94	*0.108*
	Left	204.91±100.16	180.88±73.61	*0.496*
p-value		*0.192*	*0.194*	
Axillary	Right	368.56±133.85	522.00±229.37	*0.050*
	Left	443.90±168.83	443.00±207.46	*0.490*
p-value		*0.147*	*0.220*	

We then combined all the data and evaluated whether the measuring site (oral vs. axillary) demonstrated different physiological responses during the measurements. The value of A varied from 92% to 99.7% at the oral site; among the 40 measurements, only 14 of them (35%) can be adequately fitted using only one exponential function, implying that the local blood perfusion at the oral site is sufficiently large, and only very minor increases in the local blood perfusion rate occur. In contrast, the mean value of A at the axillary site is only 89% (varying from 70% to 95%), indicating that the 2nd exponential function cannot be neglected during the fitting.

We further evaluated how the determined time constants were affected by the measurement sites. As expected, the small time constants (τ_1) for both the oral site and the axillary site are very similar (3.18 s vs. 6.32 s), shown in Figure 1. In contrast, the time constant associated with the second exponential function changed significantly, primarily affected by the measurement site (Figure 2). The average value of τ_2 was 191.64 seconds at the oral site, while it was more than doubled at the axillary site (τ_2 = 446.39 seconds). The differences in A, τ_1, and τ_s as a function of measurement site were statistically significant for the combined data ($p<0.05$).

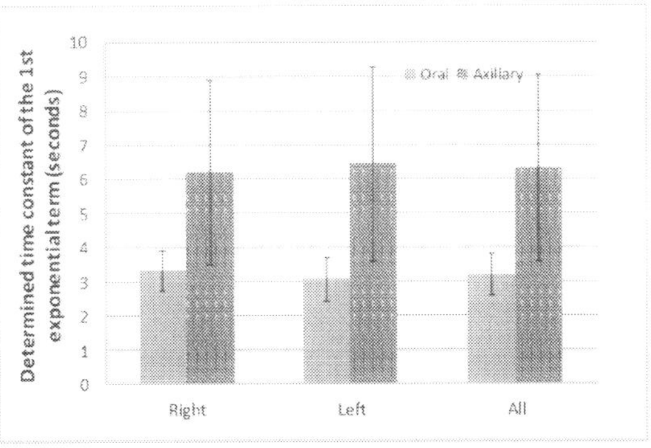

Figure 1. Determined time constant of the 1st exponential term (τ_1)

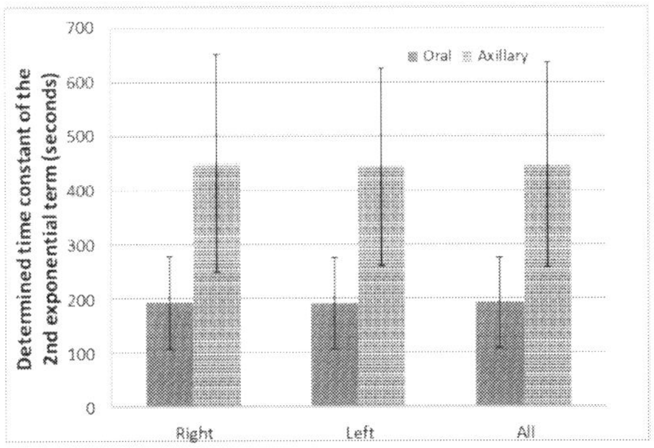

Figure 2. Determined time constant of the 2nd exponential term (τ_2)

In summary, we found that it was necessary to measure the temperatures for at least 40 minutes at the axillary site, and 10 minutes at the oral site, to establish a steady state based on a large sample of subjects. The measurement site is the dominant factor affecting the temperature transients, while neither the side of the body nor the gender has significant effects. The transient temperature measurements clearly demonstrated two phases of the thermal environment under either the oral or the axillary site. Even though at the oral site, the 1st exponential term is dominant, it would have underestimated the body temperature by up to 1°C using only the 1st function in 65% of the participants. This result is accentuated at the axillary site. If a single exponential term was used to fit the temperature transients at the axillary site, the resulting average error would have been as large as 2°C. Based on the Pennes bioheat equation [2], one can interpret that the second mechanism is possibly due to increases in the local blood perfusion rate at the measurement site. The blood perfusion rate increase is associated with a much longer transient duration, although it is relatively faster at the oral site than at the axillary site.

ACKNOWLEDGEMENTS

This study was supported by the University of Maryland Baltimore County and FDA.

REFERENCES

[1] Vesnovsky, O., Topoleski, L. D. T., Grossman, L. W., Casamento, J. P., and Zhu, L., 2014, *J. of Med. Dev.*, 8(3), 030905(1-3).
[2] Pennes, H. H., 1948, *J. Appl. Phys.*, 1:93-122.

SB3C2015
Summer Biomechanics, Bioengineering and Biotransport Conference
June 17-20, 2015, Snowbird Resort, Utah, USA
SB3C2015-508

DIET-INDUCED VASCULAR REMODELING PRODUCES A SHIFT IN COLLAGEN FIBER ANGLE DISTRIBUTION IN A MOUSE MODEL OF ATHEROSCLEROSIS

Shana R. Watson (1), Piaomu Liu (2), Edsel A. Peña (2), Michael A. Sutton (3), John F. Eberth (1), Susan M. Lessner (1)

(1) Department of Cell Biology and Anatomy
University of South Carolina
Columbia, SC, USA

(2) Department of Statistics
University of South Carolina
Columbia, SC, USA

(3) Department of Mechanical Engineering
University of South Carolina
Columbia, SC, USA

INTRODUCTION

Atherosclerotic plaque rupture is a leading cause of cardiovascular events such as myocardial infarction and stroke. Clinical complications of atherosclerosis include plaque rupture, in which rapid thrombus formation at the site of fibrous cap failure may completely occlude the vessel. In response to atherosclerotic lesion growth, arteries undergo long-lasting changes in size and composition, termed vascular remodeling. Vascular remodeling occurs whenever there is a lasting change in blood flow or blood pressure, or when the vessel wall is injured. Vessels remodel to preserve homeostatic levels of wall shear stress and circumferential tensile stress. Structural changes in vessels coincide with altered expression of matrix proteins, collagen, and elastin. With aging and atherosclerosis, the extracellular matrix (ECM) experiences changes in the fibrous structure and content of collagen and elastin [1-3]. There exists a relationship between the structural proteins and the normal function of the aorta. Therefore, investigating how the artery compensates for structural changes that occur during plaque progression will lead to a better understanding of how to prevent plaque rupture. We seek to improve our understanding of the major structural protein, collagen, and its fiber distribution and alignment to determine arterial response to mechanical loading changes during atherosclerosis-induced vascular remodeling. The central hypothesis of this study is that atherosclerotic lesion development within the aorta leads to vascular remodeling characterized by a shift in collagen fiber angles.

METHODS

We compared collagen fiber distribution in the thoracic and abdominal aorta of apolipoprotein E (apoE) knockout (KO) and apoE matrix metalloproteinase(MMP)12 DKO mouse models on control chow and high-fat Western diet (WD). The DKO mice have genetic defects in their ability to degrade and/or remodel extracellular matrix. When maintained on WD for 6-8 months, these mice develop atherosclerosis throughout the aorta. Five mice of each strain were tested for each diet. Control mice were maintained on chow diet and were approximately age-matched to WD mice. Mice were sacrificed by carbon dioxide asphyxiation and then perfused through the left ventricle with heparinized normal saline for five minutes at physiological pressure, followed by pressure perfusion fixation for 10 minutes with 10% Neutral Buffered Formalin (NBF). The aorta was opened longitudinally for en face examination of the collagen fibers. The thoracic aorta was sampled at the level of the fourth intercostal arteries, and the abdominal aorta was sampled from the infrarenal area. The samples were flattened to a slide and imaged with SHG microscopy, which can visualize collagen fibers throughout the thickness of the tissue. The SHG microscope was used to optically section each sample, allowing visualization of collagen fibers throughout the vessel wall thickness. The images of the collagen fibers were analyzed using Continuity 6.4b, a software tool that measured fiber angles relative to the circumferential direction (zero degrees) and the direction of blood flow (±90°). Image analysis of each specimen provided a 3-D histogram showing distribution of fiber angles throughout the depth of the tissue. Multivariate statistical analysis was performed in R to identify variables (mouse strain, diet, aortic position, relative depth in wall) having a significant effect ($P<0.05$) on absolute fiber angle.

RESULTS

Representative SHG images demonstrate the typical wavy arrangement of collagen fibers in the mouse aortic segments (Figure 1). 3D histograms show the distribution of collagen fiber angles throughout the wall thickness. In mice on chow diet, the average collagen fiber angle was found to be ±63° (Figure 2 & 3). However, mice on Western diet display collagen fibers oriented closer to the circumferential direction, with an average fiber angle of ±19°. For apoE KO and apoE MMP12 DKO mice, multivariate analysis showed a significant difference in average absolute fiber angle between mice on chow diet vs. Western diet. On average, fiber angle was reduced by

~43° in mice fed Western diet for 6 months compared to those on chow diet (Figure 4). There was no significant difference in average absolute fiber angle between strains. Location within the aorta and relative depth within the specimen had small but significant effects on average absolute fiber angle. Thoracic aorta specimens had average fiber angles about 0.8° lower than abdominal specimens, while the middle third of the vessel had slightly greater fiber angles (+0.3°) compared to the inner and outer layers.

DISCUSSION

We show that apoE KO and apoE MMP12 DKO mice are both subject to extracellular matrix remodeling after 6 months of Western diet feeding. Our data indicate that the aortas of both apoE KO and apoE MMP12 DKO mice fed a Western diet for 6 months undergo a significant shift in collagen fiber angle distribution. Somewhat surprisingly, there was no apparent effect of MMP12 deficiency on average absolute collagen fiber angle. MMP12 is known to play a role in elastin degradation, which can result in compensatory vessel wall stiffening. There are a few studies that associate MMP12 expression with aortic aneurysm and arterial stiffness [4]. MMP12 deficiency has also been reported to promote a "stable" atherosclerotic plaque phenotype in apoE mice fed a high-fat diet [5]. We have not yet studied the elastin content after exposure to 6 months on Western diet, but we have presented new information about changes in collagen fiber orientation in apoE KO and apoE MMP12 DKO mice fed a high-fat diet. These ECM changes coincide with the development of extensive aortic atherosclerosis, suggesting that changes in average collagen fiber angle may be part of the vascular remodeling process.

ACKNOWLEDGEMENTS

Thanks to Dr. Bruce Gao at Clemson University for use of the SHG microscope. This work was funded by NSF EPS-0903795 and CMMI-1200358.

REFERENCES

[1] Jamkhande, P et al., *Saudi Pharm J*, 22 (3):179-190, 2014.
[2] Tsamis, A et al., *J R Soc Interface*, 10 (83):20121004, 2013.
[3] Lessner, S et al., *Arterioscler Thromb Vasc Biol*, 24 (11): 2123, 2004.
[4] Ye, S. *Cardiovas Res*, 69:636-645, 2006.
[5] Johnson, J et. al., *Proc Natl Acad Sci*, 102:15575-15580, 2005.

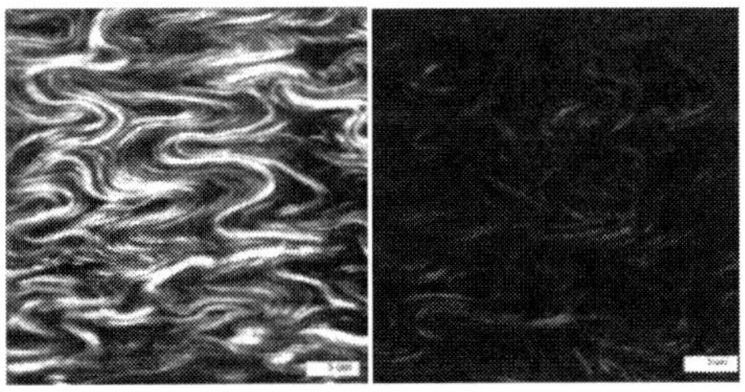

FIGURE 1. Representative SHG images of collagen fibers in aortic media of apoE KO (left) and apoE MMP12 DKO (right) mice after 6 months on Western diet. Blood flow is from top to bottom. En face; scale bar = 5 μm.

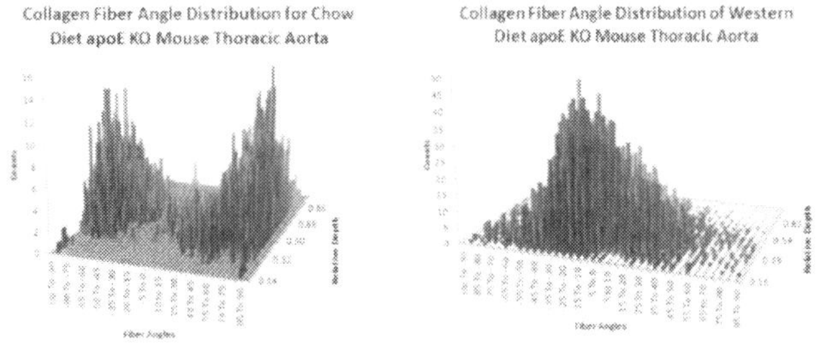

FIGURE 2. Collagen fiber angle distribution across mouse aortic wall thickness in thoracic vessel segment of apoE KO mouse on chow diet (left) and Western diet (right). Zero angle corresponds to the circumferential direction. Relative depth: endothelium = 0; outer boundary of adventitia = 1.

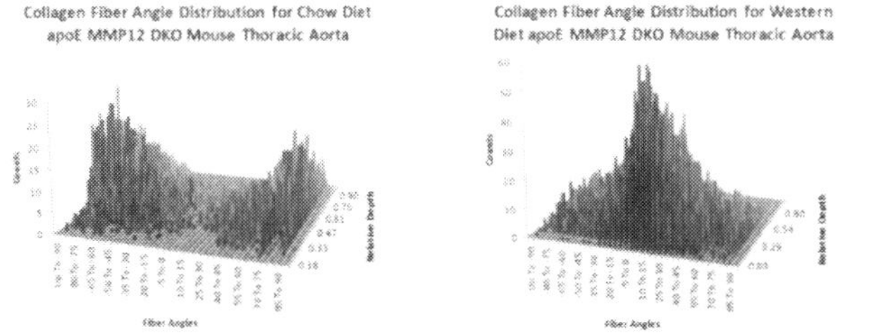

FIGURE 3. Collagen fiber angle distribution across mouse aortic wall thickness in thoracic vessel segment of apoE MMP12 DKO mouse on chow diet (left) and Western diet (right).

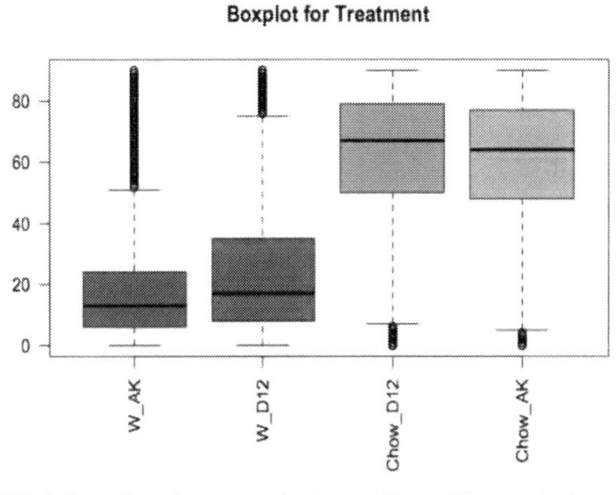

FIGURE 4. Box plot of average absolute collagen fiber angle for each treatment group (W = Western diet, Chow = chow diet, AK = apoE KO, and D12 = apoE MMP12 DKO). On average, average collagen fiber angles are significantly less for mice on Western diet vs. chow diet.

SB³C2015
Summer Biomechanics, Bioengineering and Biotransport Conference
June 17-20, 2015,Snowbird Resort, Utah, USA
SB³C2015-509

EFFECT OF OSTEOARTHRITIS ON THE MECHANICAL PROPERTIES OF HUMAN ARTICULAR CARTILAGE

Blair E. Larson (1), Kristine M. Fischenich (1), Kirk A. Kindsfater (2), Tammy L. Haut Donahue (1,3)

(1) Department of Biomedical Engineering
Colorado State University
Ft. Collins, CO, 80523
United States

(2) Orthopedic Center of the Rockies
Fort Collins, CO, 80525
United States

(3) Department of Mechanical Engineering
Colorado State University
Fort Collins, CO, 80523
United State

INTRODUCTION

Articular cartilage functions as a low friction surface that allows the joint to articulate and move freely during weight bearing activities. It can become damaged as a result of traumatic injury to the joint or with the steady progressive mechanical degradation caused by wear and tear. Destruction of cartilage and subsequent increased loading of the bone, sometimes resulting in direct tibiofemoral contact, has been shown to exasperate the development of osteoarthritis (OA).[1]

OA is the most common form of arthritis and compromises the quality of life of more than 20 million Americans.[2] Although the risk factors and symptoms are well understood, it is believed that changes within the articular cartilage originate at the nanometer level before the joint exhibits any altered function or macroscopic symptoms.[3] For this reason it is important to study the mechanical properties of cartilage and how these properties change with the progression of OA.

The purpose of this study is to experimentally determine the compressive mechanical properties of osteoarthritic articular cartilage on human tibial plateaus and compare these properties to macroscopically assessed grading of the tissue. This study will help further the understanding of the effect of osteoarthritis on articular cartilage and determine if the visual surface appearance can relates to changes in the tissue mechanical properties.

METHODS

Seven human tibial plateaus (57 ± 5 years) were used in this study. The specimens were obtained from total knee replacements performed at the Orthopedic Center of the Rockies with consent from patients and approval from Colorado State University human research institutional review board. All specimens were classified as either medial OA or advanced medial OA by the surgeon. Samples were graded according to the severity of damage to the cartilage using a modified Outerbridge scaling system. The grading ranges from a grade 0 to a grade 3. A grade 0 represents healthy, normal cartilage. A grade 1 designates destruction of the superficial cartilage such as surface fibrillation. More extensive damage that covers approximately 10%-25% of the surface of the specimen is classified as a grade 2. The most severe grade, 3, indicates a complete loss of cartilage in one or more areas covering approximately 25% of the specimen.

Indentation relaxing testing was carried out in a saline bath, with a servohydrolic material testing system (Bionic Model 370.02 MTS Corp, Eden Prairie, MN) equipped with an 8.9N load cell (Futek LSB200, Irvine, California). Tests were carried out at regions covered and uncovered by the meniscus on both the medial and lateral plateaus. Cartilage thickness was determined with a needle probe penetrating the tissue at 0.05mm/s into the cartilage.[4] Indentation relaxation testing with a 1.59 mm cylindrical, non-porous, indenter tip was carried out at 0.63 mm/s to 20% strain for 240 seconds. A Hertzian contact equation was used to determine both the instantaneous and equilibrium moduli, assuming a Poisson's ratio for human articular cartilage to be 0.5.[5]

A one way ANOVA was performed with a significance level of 0.05 using Minitab Software (Minitab15, State College, PA), to individually compare the means of the thickness, instantaneous modulus and equilibrium modulus to the assigned grade. Specific inter-group differences between mean values were determined using the post hoc Tukey test.

RESULTS

Cartilage damage was evident in all specimens and as expected the cartilage degeneration was more severe in the medial tibial plateau. The average reported grades, rounded to the nearest whole number can be seen in Table 1.

Table 1: Cartilage degradation grades

Specimen	Medial Grade	Lateral Grade
1	3	3
2	3	1
3	1	1
4	3	2
5	3	2
6	3	2
7	3	2

If regional data is combined, there was a significant decrease in the instantaneous modulus in the medial hemijoint from a grade of 1 to a grade of 3 ($p=0.008$) (Figure 1). There was a significant increase in equilibrium modulus from a grade of 2 to a grade of 3 ($p=0.02$) for the lateral hemijoint (Figure 2).

Regionally, within the medial hemijoint, there is an apparent decrease in both the instantaneous and equilibrium modulus in the covered region from grade 1 tissue to grade 3 tissues. The uncovered region of the medial hemijoint appeared to be less affected with increasing cartilage grade. The covered and uncovered regions of the lateral hemijoint had less of a pattern with increasing grade. Modulus generally declined between tissue grades of 1 and 2 but often increased between grades of 2 and 3. Due to limited sample sizes statistical analysis was not always possible at the regional level.

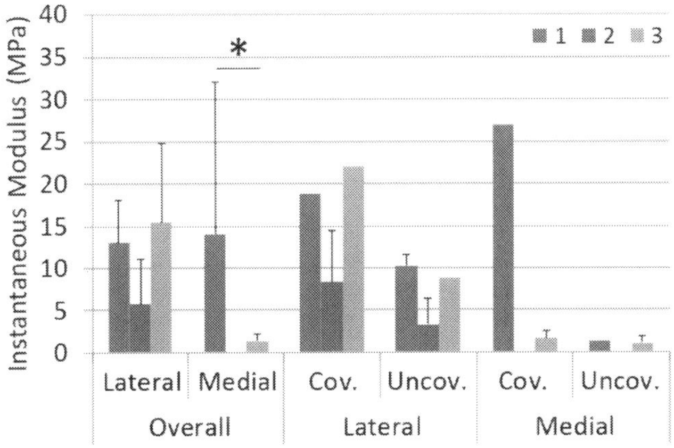

Figure 1: Instantaneous modulus for both overall hemijoint and regional levels *denotes significance

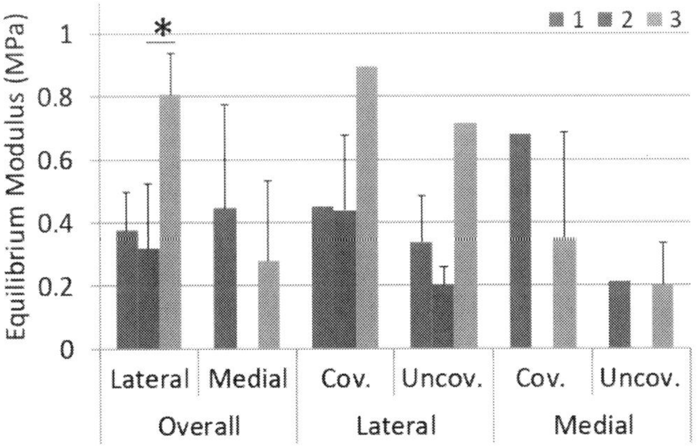

Figure 2: Equilibrium modulus for both overall hemijoint and regional levels *denotes significance

There was no statistical difference between grade and cartilage thickness in any of the hemijoints or regions. When comparing the lateral and medial hemijoints to the assigned grades there was also no statistical difference observed for cartilage thickness (Figure 3).

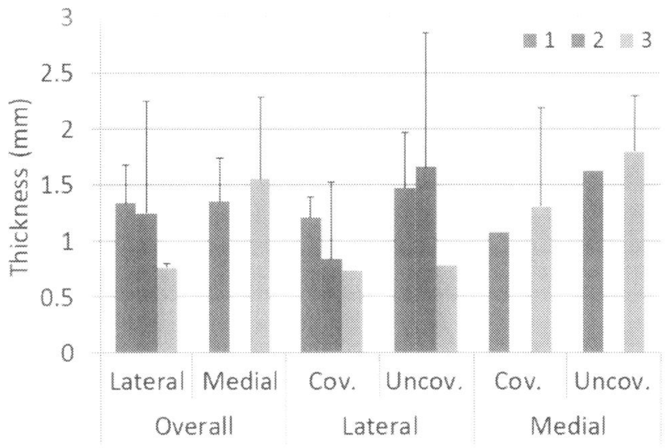

Figure 3: Thickness for both overall hemijoint and regional levels

DISCUSSION

It was determined that in the medial hemijoint there was a significant decrease in instantaneous modulus with an increase in grade. This pattern was also observed in the equilibrium modulus and the covered regions of the medial hemijoint, albeit differences were not significant. This finding is in agreement with other studies having observed a negative correlation between the mechanical behavior of articular cartilage with increasing severity of cartilage damage.[6] A previous study found that there was only a marginal decrease in the equilibrium modulus of human articular cartilage with increasing degradation[7], so a larger sample size may be necessary to observe a significant decrease.

Valgus OA, where the lateral hemijoint is predominantly affected, occurs in only about 10% of patients with OA.[8] This is why all samples used in this study were classified as medial OA and subsequently the lateral hemijoint was less damaged. The mechanical changes in compressive moduli were more variable in the lateral hemijoint. However, a study looking at osteoarthritic cartilage, irrespective of assigned damage grade, found that end-stage osteoarthritic cartilage has a decreased modulus.[9] If there were a larger sample of severely damaged lateral hemijoints, results similar to the medial hemijoint would be expected.

Osteoarthritis has been shown to cause a decrease in human articular cartilage thickness.[10] All thickness values assessed in this study were less than previously reported for healthy cartilage.[11] However, no significant differences were found between grades of individual regions. As osteoarthritis becomes more severe the increased loading in the joint leads to a significant thinning of articular cartilage[6] and it would be expected with a larger sample size to see a decrease in cartilage thickness with increasing grade.

This study showed in patients diagnosed with medial compartment OA, as the articular cartilage damage becomes more severe there is an observed decrease in the tissues compressive mechanical properties predominately in the areas covered by the meniscus. Interestingly, despite the lateral hemijoint having increased visual damage there was no obvious relation between mechanics and level of tissue damage.

REFERENCES

1.P. K. Raju *Arch. Orthop. Trauma Surg.*, 132;1267–1271, 2012. 2.J. A. Buckwalter *Clin. Orthop. Relat. Res.*, S6–S15, 2004. 3.R. Y. Hori *J. Biomech.*, 9; 259–268, 1976. 4. D. I. Isaac *J. Biomech. Eng.*, 132; 064501, 2010. 5. M. Wang *Wear*, 301;188–191, 2013. 6. R. U. Kleemann *OA&C*, 13;958–63, 2005. 7. C. G. Armstrong *J. Bone Joint Surg. Am.*, 64;88–94, 1982. 8. C.H.E Scott *Bone and Joint Journal*, 95; 436-444. 9. L. A. Setton *OA&C.*, 7; 2–14, 1999. 10. R. L. Karvonen *J. Rheumatol.*, 21;1310–1318, 1994. 11. Swann AC Br J Rheumatol., 32(1);16-25, 1993.

OBESITY AND AGE AFFECT TRIP OUTCOME AND SEVERITY FOLLOWING A LABORATORY-INDUCED TRIP

Christina R. Garman (1), Maury A. Nussbaum (1), Michael L. Madigan (2)

(1) Biomedical Engineering and Mechanics
Virginia Tech
Blacksburg, VA, USA

(2) Biomedical Engineering
Texas A & M University
College Station, Texas, USA

INTRODUCTION

The societal and economic impacts of fall-related injuries in the United States has the potential to escalate as a result of two major demographic trends. First, the prevalence of obesity has increased, from 23 % in 1988-1994 to 34.9 % in 2011-2012 [1]. Second, the number of adults over the age of 60 is projected to more than double from 2006 to 2050 [2]. These trends are problematic because individuals who are obese [3] and over the age of 65 [2] have elevated risks of falling and sustaining a fall-related injury.

Tripping accounts for an estimated 53% of falls among adults aged 65 and older [4]. Balance recovery following a trip is dependent on the successful arrest of the forward rotation of the body. Reducing the trunk angular velocity is crucial in this arrest. Moreover, larger trunk angles can inhibit recovery by interfering with lower limb recovery motions and shifting the body center of mass further from the base of support.

Trip recovery has been studied extensively in the older adult population, and it has been found that older adults successfully recover their balance without falling at a lower rate than young adults [5]. Trip recovery in the obese and older obese population has received little attention. The purpose of this study was to examine the effects of obesity and age on trip recovery outcome (fall or recovery) and trip severity quantified by peak trunk angle and angular velocity.

METHODS

Participants included 10 young (age 18-30 years) normal-weight (body mass index (BMI) of 18-24.9 kg/m^2) adults, 10 young obese adults (BMI over 30 kg/m^2), 10 older (age 60-70 years) normal-weight adults, and 10 older obese adults. Participants were tripped near mid-swing using a 7-cm-high obstacle while walking at a speed of 1.4-1.6 m/s. A safety harness was used to prevent impact with the ground in the event of an unsuccessful balance recovery. Forces applied to the harness were sampled from a load cell (Cooper Instruments and Systems, Warrenton, VA)at 1000 Hz and low-pass filtered at 20 Hz (eighth-order, zero-phase-shift Butterworth filter). Kinematic data using a modified Helen Hayes marker set was collected at 100 Hz with a 6-camera motion analysis system (MX-T10, Vicon Motion Systems Inc., L.A, CA) and low-pass filtered at 5 Hz (8th-order, zero-phase-shift Butterworth filter).

Harness load was used to classify trip recovery outcome, either as a recovery (peak load less than 30% body weight), a harness assisted recovery (peak load 30-50% body weight), or a fall (peak load exceeded 50% of body weight). Trials classified as harness assisted were removed from further analysis. Trunk angle was determined with respect to a reference angle during upright stance. Trigonometric relations among reflective markers placed on the greater trochanter and medial border of the right scapula were used in this calculation. Angular velocity was determined using a finite difference method.

Differences with respect to obesity and age on trip recovery outcome and severity (peak trunk angle and angular velocity) were analyzed using a logistic regression model and analysis of covariance, respectively. All statistical analyses were performed using JMP 10 (SAS Institute Inc., Cary, NC) with a significance level of $p \leq 0.05$.

RESULTS

Falls occurred among 50% of older normal weight adults, 80% of older obese adults, 0% of young normal weight adults, and 27.2% of young obese adults (Figure 1). The odds ratio indicated obese adults were 8.79 (C.I.: 4.62, 394.8; p=0.026) times more likely to fall compared to healthy weight adults, and older adults were 30.24 (C.I.: 1.32: 100.39; p=0.001) times more likely to fall compared to young adults.

Peak trunk angle (Figure 2) was 56.4% (p=0.046) higher among obese adults, and 78.0% (p=0.001) higher among older adults.

Peak trunk angular velocity was affected by an obesity level x age interaction (p=0.049; Figure 3). Among older adults, obese adults had 174% (p=0.012) higher angular velocity compared to normal weight adults. Among obese adults, older adults had 131.7% (p=0.011) higher angular velocity compared to younger adults.

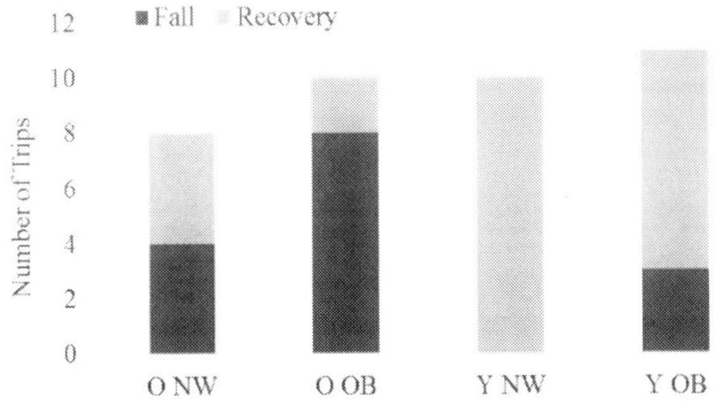

FIGURE 1: TRIP RECOVERY OUTCOME. NOTE: O=OLD, Y=YOUNG NW=NORMAL WEIGHT, AND OB=OBESE

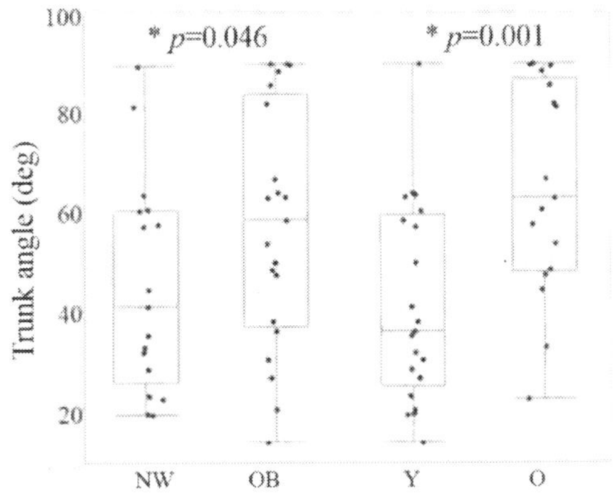

FIGURE 2: MAIN EFFECT OF OBESITY (LEFT) AND AGE (RIGHT) ON TRUNK ANGLE. * INDICATES STATISTICAL SIGNIFICANCE.

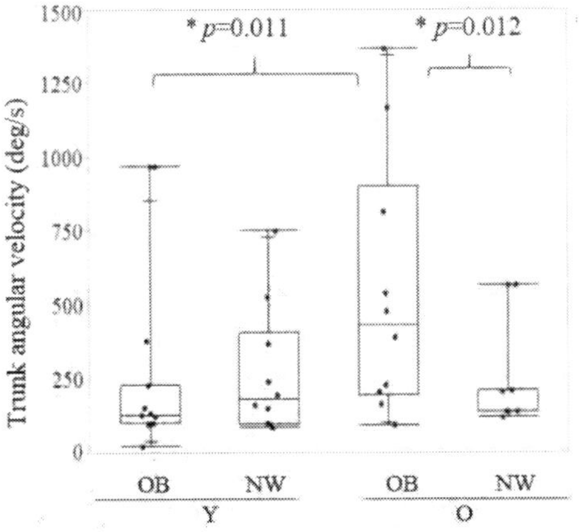

FIGURE 3: INTERACTIVE EFFECT OF OBESITY AND AGE ON TRUNK ANGULAR VELOCITY. * INDICATES STATISTICAL SIGNIFICANCE.

DISCUSSION

The purpose of this study was to examine the influences of obesity and age on trip recovery outcome and trip severity. Results indicated that following an induced trip, individuals who are obese and older adults have an increased fall risk and experience more severe trips. The results from this study, with respect to age, are consistent with those reported elsewhere [5]. With respect to obesity the only study, to the best of our knowledge, to examine this effect reported higher fall rates among obese individuals, but did not find an overall increase in fall risk [6]. The disagreement between studies may be related to differences in sample size; a larger sample size was used here, and which likely facilitated detection of a significant effect.

There are several limitation to this study that warrant discussion. First, skin movement artifacts are typically exaggerated in obese individuals, which can effect kinematic measures. However, care was taken in securing thigh and sacrum clusters to the participant to avoid such movement. Second, as with any cross-sectional study, other differences between groups besides the characteristics reported here could have contributed to the results.

This study brings to light some limiting factors for successful trip recovery. Compared to normal weight and younger counterparts, obese and older adults displayed greater trunk angles and older obese adults displayed greater angular velocity. As trunk angle and angular velocity are indicative of trip recovery outcome, the higher fall rate among obese participants found here, especially those older and obese (Figure 1), may be due in part to the higher trunk angle and angular velocity. Among obese individuals, the increased trunk angle and angular velocity may be related to an obesity-related anterior shift of the trunk center of mass [7]. This shift could act to increase the gravitational moment that rotates the body forward during a trip, and which may lead in turn to greater trunk angle and angular velocity.

This study found obese and older adults to have an increased fall risk and to experience more severe falls following a trip, and also revealed some limiting factors for successful trip recovery. The results from this study may help explain the higher rate of falls and fall related injuries among obese and older adults reported elsewhere, and can contribute to the development of intervention techniques to improve balance recovery and reduce the risk of falls.

ACKNOWLEDGMENTS

This work was supported by award R01OH009880 from the CDC-NIOSH, and a National Science Foundation Graduate Research Fellowship under grant DGE 0822220.

REFERENCES

1. Ogden, C.L., et al., JAMA 2014. **311**(8): p. 806-814.
2. WHO. 2007 May 7, 2014
3. Fjeldstad, C., et al., Dyn Med, 2008. **7**(4).
4. Blake, A.J., et al., Age Ageing, 1998. **17**(6): p. 365-372.
5. Pijnappels, M., et al., *JEK*, 2008. **18**(2): p. 188-196.
6. Rosenblatt, et al., Arch. Phys. Med., 2012. **93**(4): p. 718-722.
7. Corbeil, P., et al., IEE Trans, 2001. **9**(2): p. 126-136.

SB^{3}C2015
Summer Biomechanics, Bioengineering and Biotransport Conference
June 17-20, 2015,Snowbird Resort, Utah, USA
SB^{3}C2015-511

COMPUTATIONAL MODELING OF SYNTHETIC MESH MATERIALS: SIMULATION AND EXPERIMENTAL ASSESMENT OF MODEL PREDICTIONS

William R. Barone (1), Katrina M. Knight (1), Pamela A. Moalli (2), Steven D. Abramowitch (1,2)

(1) Department of Bioengineering
University of Pittsburgh
Pittsburgh, PA, USA

(2) Magee-Womens Research Institute
University of Pittsburgh
Pittsburgh, PA, USA

INTRODUCTION

Synthetic mesh is used in 100,000 prolapse repair surgeries per year in the United States [1]. Despite superior efficacy relative to native tissue repair, synthetic mesh is plagued by high complication rates, ranging from 11-15% of all cases [2]. These complications, including erosion and pain, are detrimental to vaginal tissue and a women's quality of life, with 20% of mesh surgeries requiring repeat operations for recurrent symptoms and complications [3].

While the exact etiology of complications is unknown, much focus has been drawn to the design of mesh products. Through trial and error, "light-weight" (>1g/cm^3), monofilament, polypropylene mesh has been found to reduce complications in the pelvic floor. Further, hernia literature has found that mesh outcomes are related to a mesh's pore size, as pores <1mm in diameter greatly enhance the inflammatory and fibrotic responses for polypropylene mesh [4]. Given the biological implications of pore size, pore deformation in response to in-vivo loads following implantation must be considered more closely.

The pelvic floor is a mechanically complex environment, with challenging geometries. Computational modeling provides a tremendous tool for optimizing mesh products to the needs of pelvic floor support. Such simulations can be used to examine the interactions between the mesh and vagina, providing details such as stress and mesh pore deformation, which are known to impact biological tissues. As a first step towards creating these simulations, the objective of this study was to develop a computational model for a specific prolapse mesh, Restorelle Directfix, and examine its response to mechanical loading. In addition, we aimed to assess model predictions by comparing pore deformation parameters including, mesh burden and minimum pore diameter, to experimental results.

METHODS

In this study a commercially available transvaginal prolapse mesh device, Restorelle DirectFix (Anterior), was considered. The device geometry was created in Solidworks (2014, DSS, Waltham, MA), such that the number of pores and their orientation were identical to the physical device (Fig 1a,b). The solid geometry was then imported into Simulation Mechanical (Autodesk, San Rafael, CA) and Gmsh (Geuzaine, Remacle) for discretization. To avoid computational bias, pores were discretized such that each was comprised of the same number and type of elements. Further, we assume each pore to be comprised of 4 fibers, creating the perimeter, and 4 knots, which joined these fibers (Fig 1d). Based on experimental observation, it was assumed that knot structures behave as pivots, allowing fiber rotation, while the fibers resist tension and buckle under compression. Thus, fibers and knots were modeled as separate Neo-Hookean materials. Material properties were obtained from matching computational

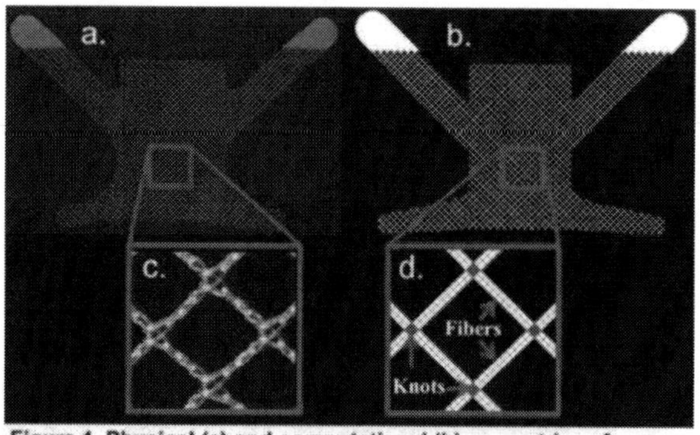

Figure 1. Physical (a) and computational (b) geometries of Restorelle DirectFix. Each pore (c) was modeled as 4 fiber and 4 knot structures (d).

simulations to experimental results from uniaxial testing of Restorelle strips (90x15mm) in 2 different orientations. Fiber behavior in tension was determined by loading the mesh with one of its two orthogonally oriented fiber groups aligned parallel to the loading axis. Knot properties were determined by orienting the pore fibers 45° from the axis of loading. For each condition, strips were loaded to 10N at 10mm/min (n=5 each). The determined material properties were: $E_{fiber} = 5000MPa$, $v_{fiber} = 0.49$, $E_{knot} = 100MPa$, and $v_{knot} = 0.01$.

For simulation and model assessment of the total transvaginal device, experimental measurement of pore deformation was made by suspending Directfix above a platform at 2 discrete locations via steel rods extending through two separate pores. The 4 arms of the device were secured in separate clamps, each with a 250g weight attached via suture. Tension was applied to the device by running the suture lines over a pulley wheel attached to posts spaced in 5° increments about the center of the device, and allowing the weight to hang freely. Two experimental scenarios were considered (n=3 each): 1) the top arms were loaded at 45° and the bottom arms were loaded at -20° 2) the top arms were loaded at 10° and the bottom arms were loaded at -45°. All loading angles are relative to the horizontal axis, with the origin at the device center. Before and during the application of load, the entire mesh was imaged from above. For computational simulation, the above experiment was recreated in FEBio, (University of Utah, MRL) using the material parameters obtained from strip testing. Traction forces were prescribed to load the device arms and the fixation rods were modeled as rigid interfaces with sliding contact. All finite element (FE) simulations were restricted to planar deformation.

The parameters chosen to assess pore deformation were mesh burden, defined as amount of mesh per unit area, and the minimum pore diameter, d_{min}. For each point on the mesh, the number of surrounding mesh points (pixels for experimental, nodes for computational) within a 2mm radius were summed, providing a local density value. Since the number and distribution of total mesh particles were dissimilar between experimental and computational models (pixels vs. nodes), each was normalized to the maximum local density of the undeformed mesh. A Bland-Altman analysis was used to compare differences between experimental and finite element (FE) results for each loading condition, with a significance of 0.05.

RESULTS

In response to loading, DirectFix experienced significant pore deformation (Fig 2). For both loading conditions, peak mesh burden increased over 230% relative to unloaded geometries and average minimum pore diameters across the entire device were decreased by at least 44%. Areas of peak deformation were observed in lower mesh arms and at the device midline, between the upper arms.

Computational modeling yielded comparable results for both normalized mesh burden and d_{min} measures (Table 1). For loading condition 1, average d_{min} values from FE simulations were 0.9% greater than experimental measures, yet average values for normalized mesh burden was ~21% greater for computational models with notable differences at the fixation rods ($p<0.05$). Results for loading condition 2 were also comparable, as average normalized mesh burden values were within 6% of experimental values, though FE average d_{min} values were 28% greater than experimental averages ($p<0.05$).

Table 1. Summary of normalized mesh burden and minimum pore diameters (d_{min}) for unloaded and loaded geometries. Experimental (Exp.) values are displayed as mean ± std, while finite element (FE) results are represented as mean values.

* indicates $p < 0.05$		Average Norm. Mesh Burden (Mesh/mm^2)	Average d_{min} Unloaded (mm)	Average d_{min} Loaded (mm)
Condition 1	Exp. (N=3)	2.60 ± 0.11*	2.14 ± 0.04	1.20 ± 0.12
	FE	3.36*	2.16	1.09
Condition 2	Exp. (N=3)	2.36 ± 0.21	2.08 ± 0.04	0.97 ± 0.03*
	FE	2.21	2.12	1.35*

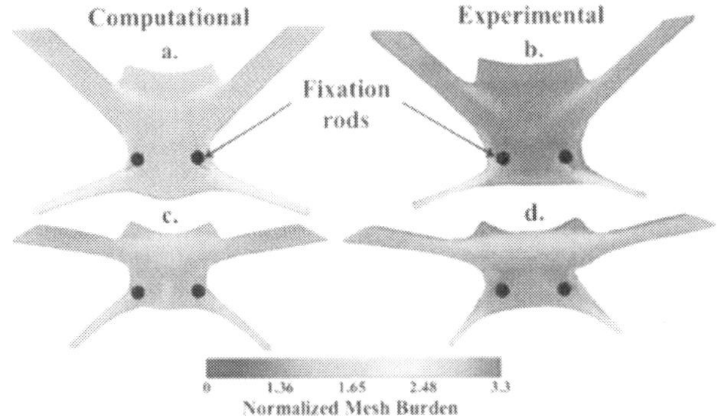

Figure 2. Contour maps of normalized mesh burden, projected on deformed geometries for loading conditions 1 (a,b) and 2 (c,d). Computational (a,c) and experimental (b,d) results show good agreement. Black circles represent fixation rod locations.

DISCUSSION

In this study a computational model for a transvaginal prolapse mesh, Restorelle DirectFix, was developed and FE predictions of pore deformation were compared with experimental findings. The proposed computational model captured the deformation of Directfix under 2 distinct loading conditions well, as the locations of increased mesh burden were similar between experimental and FE testing. Further, comparable decreases in pore diameter were observed for both computational and experimental models upon loading.

Previously, fiber reinforced continuum solids have been considered to model porous structures, however they do not allow for FE investigation of clinically relevant parameters such as pore diameter and mesh burden. Thus the approach used here is novel, allowing detailed investigation of pore geometry, providing a means by which mesh devices can be virtually examined and optimized.

When considering the results of this study is it important to note that FE simulations were restricted to planar deformations. Though loading conditions were selected to minimize non-planar deformations, self-contact and fixation created out of plane deflection experimentally. Increased vertical loading of the upper arms (condition 1) resulted in greater bunching of mesh around the fixation rods, decreasing planar measurement of mesh burden. This may explain the noticeable over prediction of mesh burden (~21%) by our model, as the FE models required all fibers to stay in plane. Predictions of mesh burden for condition 2 were similar to experimental results, as the angles of loading resulted in relatively less bunching about the rods.

Future work will include further statistical and sensitivity analysis of the current model, considering the impact of alternate constitutive models on device deformation. In addition, this model will be used to examine the mechanical interaction of prolapse mesh with patient specific pelvic floor models.

ACKNOWLEDGEMENTS

NSF Graduate Research Fellowship (NSF GRF DGE-0753293) and NIH grants (R01 HD-045590, K12HD-043441).

REFERENCES

[1] Barber, MD, et al., *JAMA*, 311(10):1023-34. 2014.
[2] Nygaard, I, et al., *JAMA*. 200(11):1311-6. 2008.
[3] Bako A., Dhar R. *IUJ*. 20:103-11. 2008.
[4] Weyhe, D, et al., *World J of Surgery*. 30(8): 1586-91. 2006.

BIOMECHANICAL CHARACTERIZATION OF PORCINE SKELETAL MUSCLE EXTRACELLULAR MATRIX

Bryn Brazile (1), Sourav S. Patnaik (1), Sallie Lin (1), Xiaodan Shi (1), Shengfa Liao (2), Raj Prabhu (1), Hongjoo Rhee (3), Lakiesha Williams, Jun Liao (1)

(1) Department of Biological Engineering
Mississippi State University
Mississippi State, MS, USA

(2) Department of Animal and Dairy Sciences
Mississippi State University
Mississippi State, MS, USA

(3) Center for Advanced Vehicular Systems
Mississippi State University
Mississippi State, MS, USA

INTRODUCTION

Skeletal muscle injury affects millions of people, causing muscle death and reduction in function. The connective tissues holding the skeletal muscle together are known as the endomysin and perimysium. The vast bulk is made of perimysium (~90%), and covers large and small bundles of muscle fibers inside the muscle. The endomysin, however, encircles each individual muscle fiber. Those connective tissues are composed of collagen fibers (Type I and III), elastin fibers, and proteoglycans with collagen fibers being the most predominant [1]. There is a great need to understand skeletal muscle injury/repairing mechanisms, especially the interaction between the skeletal muscle extracellular matrix (ECM) and the muscular/stem cells. The structural and biomechanical knowledge of skeletal muscle ECM, as well as how the ECM would affect cell behavior and destiny, is of importance of designing optimal regenerative strategy. As a baseline study, we characterize the structural and mechanical properties of decellularized skeletal muscle ECM and compare it with native skeletal muscle properties.

METHODS

Front shoulders of freshly euthanized pigs (~6 month old) were obtained from a local abattoir and transported on ice to the lab. The front shoulder was then carefully dissected to remove the Triceps Brachia long head. The triceps brachia long head muscle is the largest of the triceps group. It is a triangular shaped muscle on the dorso-lateral aspect of the forelimb. It arises from the posterior border of the scapula. The muscle was then trimmed to 30 × 10 × 3 mm strips for uniaxial testing and 20 × 20 × 3 mm squares for biaxial testing.

For skeletal muscle sample that requires decellularization, the up-right corner of the tissue was tied with a suture to mark the orientation. The muscle sample was then placed in a frame-pin supporting system that was designed to better maintain tissue macrogeometry and internal structure during decellularization. Briefly, two sides of the skeletal muscle samples were perforated with two 27G 9 31/200 BD Quincke Spinal Needles, which was then mounted onto custom made rectangular plastic frames. The skeletal muscle samples were then decellularized in a rotating bioreactor using 1% sodium dodecyl sulfate (SDS) (Sigma) with 100 U/ml penicillin and 100 μg/ml streptomycin at room temperature for 2.5 weeks. Ten-minute ultrasonic treatment (50 Hz, Branson) were applied each day; the solution was changed every day to avoid contamination and tissue deterioration.

Two different tests were used to capture the biomechanical properties of the native skeletal muscle and the decellularized skeletal muscle ECM. Uniaxial mechanical testing was performed on the tissue strips (n = 5 per group) using a Mach-1 device (Biosyntech, MN). Dog bone-shaped specimen was trimmed from the sample in a way that the testing was along muscle longitudinal direction. The samples were mounted with two stainless steel grips cushioned with emery paper. Each sample was subjected to 10 cycles of preconditioning at 10% strain, and then loaded to failure at a strain rate of 5% per second. Engineering stress was calculated by normalizing the force to the initial cross-sectional area; engineering strain was computed by normalizing the displacement to the initial grip-to-grip distance (gauge length). The maximum tensile modulus was estimated by finding the tangent value of the linear region of the stress–strain curve using linear regression. Failure stress and failure strain was also determined from the stress–strain data.

Biaxial mechanical testing was performed on the square sample (n = 5 per group) using a custom made biaxial testing machine. The biaxial loading was applied along the muscle longitudinal direction and transverse direction. Four dark markers were placed in the center region of the square sample. Samples were mounted onto the biaxial testing system via stainless steel hooks attached to 8 loops of 000 polyester suture of equal length (2 suture loops per sample edge). Each sample was subjected to 10-cycle preconditioning and then subjected to an equibiaxial tension of 60 N/m. All mechanical testing were performed in a PBS bath. The extensibilities of the native muscle and decellularized skeletal muscle ECM were characterized by the maximum stretches along the longitudinal and transverse directions.

For histological characterization, the native and decellularized samples were fixed in 2% paraformaldehyde for 2 days, dehydrated with graduated concentrations of ethanol, and embedded in paraffin. Cross-sections of the native skeletal muscle and the decellularized skeletal muscle ECM were cut to 5 μm thick for staining. To elucidate changes in tissue components, Masson's trichrome staining and Picrosirius Red staining were used to identify muscle fibers, collagen fiber network, elastin, and proteoglycans. Stained tissue sections were imaged using bright field microscopy (Nikon EC600).

RESULTS

After 2 weeks of decellularization treatment, porcine skeletal muscle tissue showed a bright white color of the typical collagenous materials, indicating the complete removal of muscle cells (confirmed by Mason's Trichrome). We also noticed that the frame-pin supporting system well preserved 3D ultrastructure, which was confirmed by the collagen network shown by histology.

Both the native skeletal muscle and skeletal muscle ECM showed a nonlinear stress-strain behavior. The native skeletal muscle was stiffer than the decellularized skeletal muscle ECM. While both groups reached over 1000 Pa, the native skeletal muscle did so at 0.7 strain; whereas, it took the decellularized skeletal muscle ECM up to 0.9 strain to reach the same stress.

For biaxial behavior, the decellularized skeletal muscle ECM showed an overall softening trend after the removal of muscle cells, resulting in increased areal strain at 60 N/m. A noticeable phenomenon is the switching of anisotropic trend. For the native skeletal muscle, the longitudinal direction was stiffer than the transverse direction. However, after decellularization the transverse direction was stiffer than the longitudinal direction. This behavior is opposite from those observed in the acellular porcine myocardial ECM [2]. The overall softening trend and the switching of anisotropy after decellularization might be correlated to either the intrinsic collagen fiber arrangement of skeletal muscle ECM or the protocol of 1% SDS treatment (higher than the 0.1% SDS treatment of the porcine myocardial ECM [2]), which warrants future thorough investigation.

DISCUSSION

In this study, we characterized the decellularized porcine skeletal muscle ECM that was obtained via a decellularization procedure. Uniaxial results showed that the native muscle is stiffer than the decellularized skeletal muscle ECM. The biaxial results revealed that, after removal of muscle cells, the skeletal muscle ECM showed an overall softening trend and the switching of anisotropy when compared with the native tissue. The extensibility of the transverse direction was found to barely change, likely implying that the skeletal muscle ECM is responsible for binding the muscle fibers together. In future studies, we will investigate the structural mechanism of the biaxial behavior of skeletal muscle ECM in depth, as well as its implications to cell-ECM interaction.

ACKNOWLEDGEMENTS

This work is supported by AHA 13GRNT17150041 and NSF EPS-0903787.

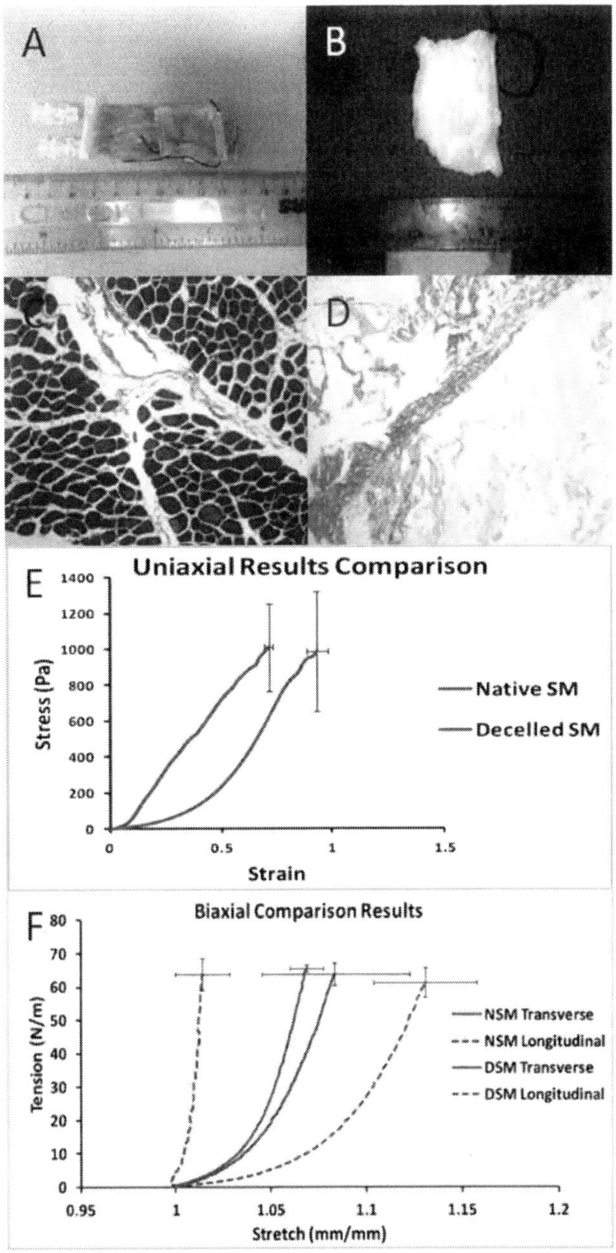

FIGURE 1: (A) SKELETAL MUSCLE BEFORE DECELLULARIZATION. (B) SKELETAL MUSCLE ECM AFTER DECELLULARIZATION. (C) HISTOLOGICAL IMAGE OF THE SKELETAL MUSCLE STAINED WITH MASSON'S TRICHROME (CROSS-SECTIONAL VIEW). (D) HISTOLOGICAL IMAGE OF THE DECELLULARIZED SKELETAL MUSCLE ECM STAINED WITH MASSON'S TRICHROME (CROSS-SECTIONAL VIEW). (E) UNIAXIAL MECHANICAL BEHAVIOR. (F) BIAXIAL MECHANICAL BEHAVIOR.

REFERENCES

[1] Purslow, P.P., *Meat science,* 70(3): 435-447, 2005.
[2] Wang, B et al., *Langmuir,* 29(35):11109-17, 2013.

EFFECT OF VARIABILITY IN ANATOMICAL LANDMARKS ON ANKLE KINEMATIC DESCRIPTIONS

Ednah G. Louie (1), Fallon G. Fitzwater (2), Matthew H. Dickinson (1), William M. Eboch (2), Bardiya Akhbari (2), and Lorin P. Maletsky (1, 2)

(1) Bioengineering Graduate Program
University of Kansas
Lawrence, Kansas, United States

(2) Mechanical Engineering Department
University of Kansas
Lawrence, Kansas, United States

INTRODUCTION

Joint kinematics provide a foundation for characterizing, assessing, and treating joint pathologies and injuries. Coordinate systems used to describe joint kinematics are defined using series of anatomical landmarks.[1] A well-defined coordinate system is critical for obtaining accurate and reliable kinematics that can be directly compared across studies. Variability in identification of anatomical landmark locations can lead to unreliable kinematic descriptions. While previous studies have been conducted to determine the effect of anatomical landmark identification on knee kinematic descriptions,[2] these effect have yet to be assessed on ankle kinematics.

This study aims to quantify the effect of uncertainty in anatomical landmark identification on reported ankle kinematics.

METHODS

The ankle specimen (73-year-old male, BMI 27) used in this study was previously frozen and thawed and refrozen before being thawed at room temperature for 12-14 hours. All soft tissue 6.5 cm proximal to the medial malleolus was removed and the fibula fixed to the tibia. The tibia was potted into an aluminum fixture and the ankle was mounted onto a custom rig in an inverted position. The tibialis anterior was loaded with a dead weight of 9 kg while the extensor digitorum longus was loaded with 4 kg of dead weight. A stepper motor used to flex the ankle was connected to the Achilles tendon using a strap and clamp. Kinematic data were collected with an Optotrak Certus optical system (NDI, Waterloo, Canada) and rigid bodies fixed to the tibia and calcaneus. After all kinematic data was acquired the ankle was disarticulated. A rigid-body probe was used to collect nine anatomical landmarks from the tibia and calcaneus (Figure 1).

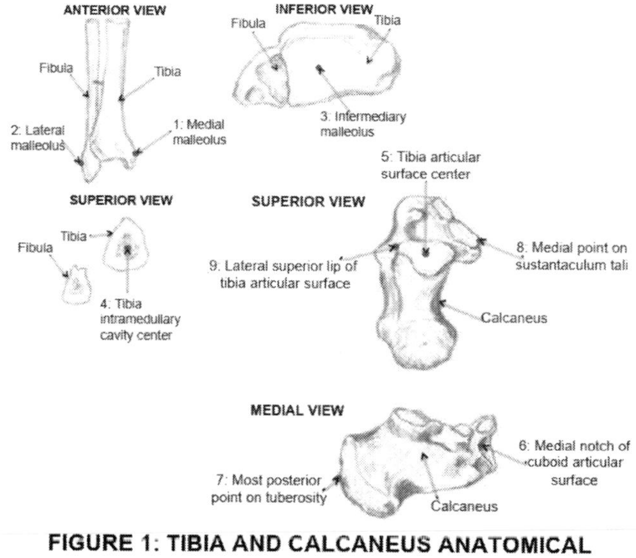

FIGURE 1: TIBIA AND CALCANEUS ANATOMICAL LANDMARKS USED TO CONSTRUCT JOINT COORDINATE SYSTEM

The nine anatomical landmarks established a customized joint coordinate system based on the three-cylindrical open-chain system used by Wu, et al. for ankle kinematic descriptions.[3] On the tibia, landmarks 1 and 2 created an initial medial/lateral (ML) axis; landmark 3 served as the origin, and landmarks 3 and 4 created the superior/inferior (SI) axis. The ML and SI axes were crossed to create the anterior/posterior (AP) axis, which was re-crossed with the SI axis to create the final ML axis. On the calcaneus, landmarks 8 and 9 created an initial ML axis; landmark 5 served as the origin, and landmarks 6 and 7 created the AP axis. The AP axis was crossed with the ML axis to form the SI axis. The SI axis was re-crossed with the AP axis to create the final ML axis.

A 30,000-cycle Monte Carlo simulation was conducted in MATLAB to determine the effect of overall variances in probed anatomical landmarks on kinematic descriptions. Three researchers collected five sets of probed anatomical landmarks. The mean and standard deviation of the five sets (summarized in Table 1) served as the deterministic input, and the 1st, 50th, and 99th percentiles of the output kinematics were plotted against plantarflexion/dorsiflexion (PF/DF) angle for the 30,000 cycles. A 20,000-cycle Monte Carlo simulation was conducted in MATLAB to verify convergence.

TABLE 1: MEAN AND STANDARD DEVIATION OF FIVE SETS OF PROBED ANATOMICAL LANDMARKS

Anatomical Landmark	Standard Deviation (mm)		
	x	y	z
1	2.8	1.5	0.3
2	1.7	3.7	0.7
3	1.1	0.2	0.7
4	0.3	0.2	0.3
5	0.1	0.4	1.1
6	1.2	1.6	0.4
7	1.4	0.9	2.5
8	0.3	1.4	0.3
9	0.8	2.7	3.4

RESULTS

Resultant kinematics were calculated for all variations of anatomical landmarks obtained from the 30,000-cycle Monte Carlo simulation. The 1st, 50th, and 99th percentile kinematics were calculated and are shown in Figure 2. The range between the 1st and 99th percentile of AP kinematics widened slightly as the ankle moved from dorsiflexion toward plantarflexion, while the range of SI kinematics narrowed along the same path. Adduction/abduction (AD/AB) and inversion/eversion (IN/EV) showed an overall shifted kinematic description of approximately ± 6° and ± 7°, respectively. ML, AP, and SI kinematics showed an overall shift of approximately ± 3 mm, ± 2 mm, and ± 2 mm, respectively.

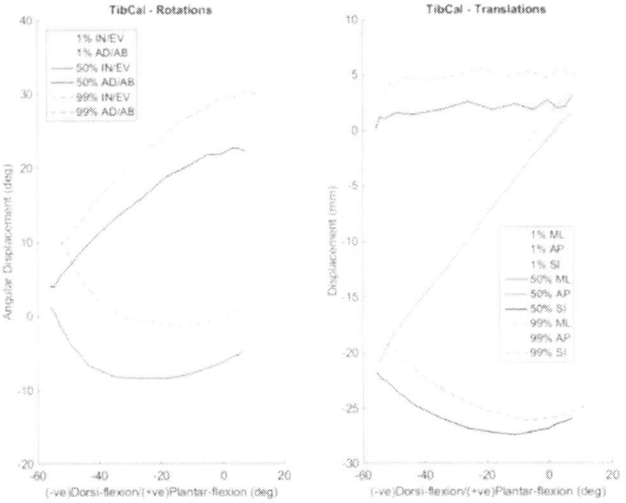

FIGURE 2: 1ST, 50TH, AND 99TH PERCENTILE PLOTS OF ROTATIONAL AND TRANSLATIONAL KINEMATICS AGAINST PF/DF

DISCUSSION

Understanding variability in reported ankle kinematics due to uncertainty in identifying anatomical landmarks is of great interest to biomechanical communities. This study varied all the anatomical landmarks simultaneously to simulate *in vitro* experimental procedures. This study found overall variances led to AP and SI variations consistent across PF/DF while variations in AD/AB and IN/EV rotation and ML translation varied across PF/DF. The large envelope for absolute kinematics indicates the kinematic to be more susceptible to changes due to variances in anatomical landmark locations. Due to the large envelope, the reliability and accuracy of both the IN/EV and AD/AB kinematic descriptions are suspect. Kinematic descriptions with a smaller range of motion were more affected by variations in anatomical landmark locations. While AD/AB, IN/EV, and ML ankle kinematics are susceptible to shifts due to identification of anatomical landmarks used to create joint coordinate systems, AP and SI are relatively less susceptible due to their smaller envelope.

The large envelope of motions due to variability in anatomical landmark locations indicate a need for improved methods of identifying anatomical landmarks. Future studies will aim to determine the sensitivity of each individual point on all kinematics.

ACKNOWLEDGEMENTS

This study was funded in part by DePuy Synthes.

REFERENCES

1. Grood ES, et al, *J Biomech Eng*, 105(2):136-44, 1983.
2. Morton NA, et al, *J Orthop Res*, 25(9):1221-30, 2007.
3. Wu G, et al, *J Biomech*, 35(4):543-8, 2002.

ISOLATION OF SUBARACHNOID HEMHORRAGE FACTORS ON A SINGLE CHIP FOR UNDERSTANDING CEREBRAL VASOSPASM PROGRESSION

Eric S. Hald (1), Kerianne E. Steucke (1), Connor Timm (1), Patrick W. Alford (1)

(1) Department of Biomedical Engineering
University of Minnesota-Twin Cities
Minneapolis, MN, USA

INTRODUCTION

Cerebral vasospasm **(CV)** is a potentially lethal disease characterized by acute hypercontractility followed by prolonged remodeling, lumen occlusion, and ischemia, typically occurring 3-7 days post-injury [1]. One common precursor implicated in CV development is subarachnoid hemorrhage **(SAH)**, which results in the release of growth and clotting factors (e.g. PDGF, TGF-β, thrombin) in the subarachnoid space. These factors are thought to act on vascular smooth muscle **(VSM)** contractile tone via complex biochemical pathways [2]. However, it is not clear which, and to what extent, individual factors (or their combinatorial effects) induce CV.

Due to the chronic nature of vascular disease, extension of culture times for *in vitro* models is desired. Recently, we developed a patterned protein deposition technique for extending culture of VSM cells on PDMS with genipin. We combined this microfluidic protein deposition technique for engineering vascular tissues with a previously developed muscular thin film **(MTF)** technique, allowing for chronic measurement of vascular contractility [3]. Here, we add a novel isolation chamber device designed to allow isolated delivery of SAH-borne factors to engineered vascular tissues on a single chip. We employ a dose-response experimental strategy to investigate the critical SAH-borne factors responsible for pathological CV vascular mechanics.

METHODS

Tissue Fabrication and Isolation

Photolithographic techniques were used to fabricate microfluidic protein deposition devices with alternating lines of 10-μm width with 10-μm pitch in a single array of five blocks of 2-mm width and 6-mm length to yield an array of five engineered arterial lamella (Fig. 1A(i)). Genipin (5 mg/ml) and fibronectin (50 μg/ml) were sequentially drawn through the devices applied to glass cover slips coated in Sylgard 184 PDMS as in [3]. Microfluidic devices were removed and acrylic isolation chamber devices were placed on the patterned substrates. These devices were fabricated prior to experimentation using a laser cutter, yielding a single array of reservoirs of 2-mm width, 10-mm length, and 6-mm height (Fig. 1A(ii)). Human umbilical artery VSM cells were then seeded into each isolation chamber at a cell seeding density of 80,000 cells/cm^2. After 24h, confluent, aligned arterial lamellae mimics formed within each chamber (Fig. 1A(iii)) and were then treated with serum-free medium to induce a contractile phenotype [4]. After 24h in serum-free medium, each chamber was treated with one of five solutions: control (serum-free medium), 10^{-12} M, 10^{-11} M, 10^{-10} M, and 10^{-9} M TGF-beta1 (or other SAH factor of interest) in serum-free medium (Fig. 1A(iv)). To simplify handling of samples and improve maintenance of isolation chamber seal, four chips were fabricated in a single acrylic clamping system (Fig. 1B). Dose-response treatment solutions were refreshed every 24h for up to seven days of treatment.

Stress Measurement

Contraction of independently-treated microtissues was measured using a previously developed MTF model [5]. At a treatment time point of interest, each of five tissues on a single chip was cut and released via the dissolution of poly(N-iso-propylacrylamide) below 32°C to form a single MTF (Fig. 1C). Stress generation in the cell layer causes the released MTF to curl. The radius of curvature can be used to calculate contractile stress, as in [5]. We determine two important stresses: 1) induced contractility via stimulation with the vasoconstrictor endothelin-1 **(ET-1)** and 2) basal tone via tissue relaxation with HA-1077, a rho-kinase inhibitor (Fig. 1D).

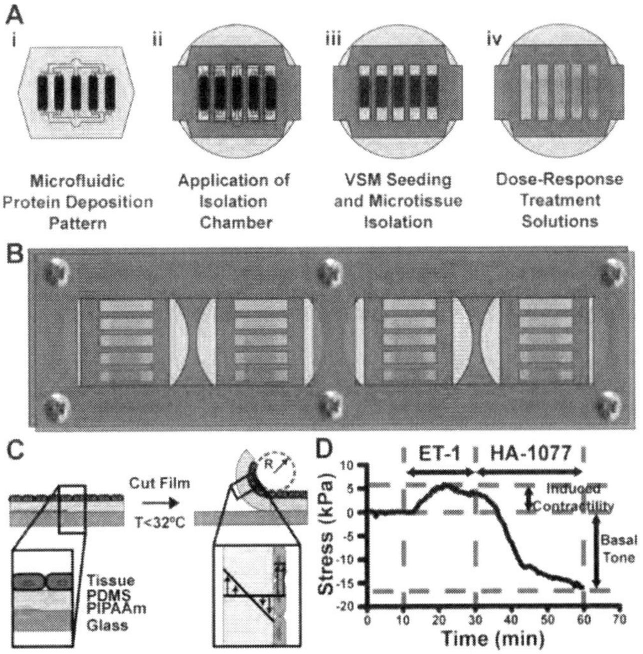

Figure 1. Tissue fabrication and isolation and MTF contractility protocols (A) (i) Branching microfluidic protein deposition device **(ii)** Application of acrylic isolation chamber **(iii)** VSM cell seeding and formation of isolated, engineered vascular tissues **(iv)** Application of variable concentrations of SAH factor(s) of interest for full dose-response. **(B)** Clamping assembly for long-term culture of isolated vascular tissues **(C)** Schematic of MTF measurement of contractile stress in the VSM cell layer **(D)** Representative temporal stress curve with serial stimulation with endothelin-1 (50 nM) and HA-1077 (100 μM) yielding measurement of basal tone and induced contractility [3].

RESULTS

Functional output for all treatments were simultaneously measured (Fig. 2A) for up to seven days of treatment. At 24h treatment, we found increasing raw stress values with increasing TGF-β concentration (Fig. 2B). We found a 4-fold increase in both normalized basal tone and normalized ET-1-induced contractility in tissues treated with 10^{-11} and 10^{-10} M TGF-β (Fig. 2C).

DISCUSSION

Here, we present an *in vitro* assay for isolated treatment and measurement of microtissue stress generation over CV-relevant time course. Our design allows dose-response experimentation with other water-soluble SAH-borne factors of interest. Investigation of the effects of PDGF and thrombin, in addition to results for TGF-β (Fig. 2), provides a platform to investigate the direct effect of these factors on vascular mechanics. We can employ a factorial experimental design to determine factor-level combinations that induce acute hyper-contractility followed by a shift to a synthetic remodeling VSM phenotype experiment. We will use this design to assess the relative importance of these SAH-borne factors, both individually and in combination, yielding further understanding of the underlying disease mechanisms and possibly illuminating desirable treatment strategies to combat the pathological vascular mechanical response in CV. Further, the ability to recapitulate CV progression in an on-a-chip system may allow future development of our technique as a pharmaceutical assay.

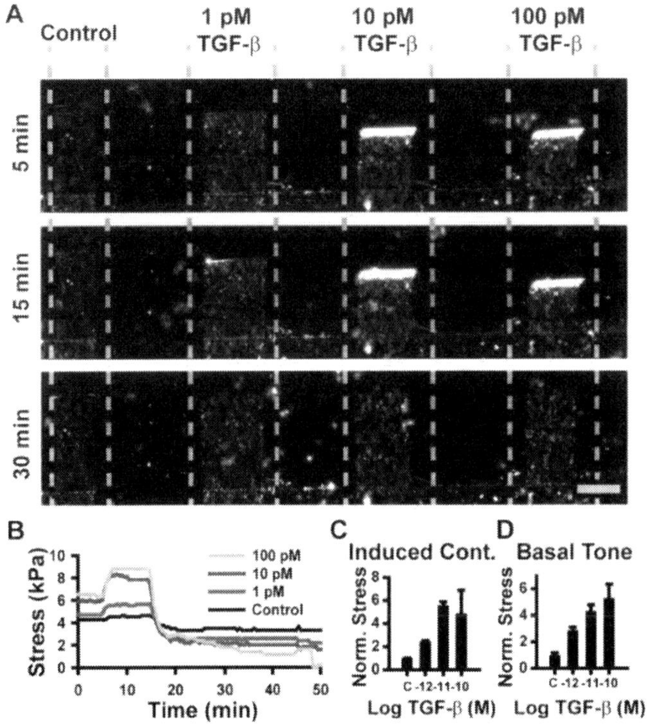

Figure 2. TGF-β dose-response contractility assay (A) Representative experimental images of MTFs treated with log concentrations of TGF-β for full dose-response on a single chip (scale bar: 1 mm) **(B)** Representative temporal stress curves for tissues shown **(C)** Normalized ET-1-induced contractility and basal tone for dose-response contractility assay (n=4, error bars: standard error).

Another key feature of our design is our ability to conduct experiments over disease-relevant time courses. Previous microcontact printing methods for engineering arterial lamella showed loss of tissue integrity as early as four days in culture [3]. We can now perform these assays up to seven days by utilizing genipin to extend culture of VSM cells [3]. Further, the microfluidic tissue fabrication and isolation methods employed here may also be used to engineer tissues using a variety of different cell types. Further work can extend our system to investigate other disease mechanisms attributed to biochemical pathways associated with water-soluble factors.

ACKNOWLEDGEMENTS

We acknowledge financial support from the American Heart Association Scientist Development Grant, 13SDG14670062 (PWA) and the University of Minnesota Doctoral Dissertation Fellowship (ESH). We also acknowledge the microfabrication resources of the Minnesota Nano Center (MNC). Parts of this work were carried out in the Characterization Facility, University of Minnesota, which receives partial support from NSF through the MRSEC program.

REFERENCES

[1] Humphrey, JD et al., *Ann Biomed Eng*, 35:1485-1497, 2007.
[2] Owens, GK et al., *Physiol Rev*, 84:767-801, 2004.
[3] Hald, ES et al., *Biofabrication*, 6:045005, 2014.
[4] Han, M et al., *Am J Physiol Cell Physiol*, 291:C50-C58, 2006.
[5] Grosberg, A et al., *Lab Chip*, 11:4165-4173, 2011.

SB3C2015
Summer Biomechanics, Bioengineering and Biotransport Conference
June 17-20, 2015,Snowbird Resort, Utah, USA
SB3C2015-515

3D STRAINS IN POSTERIOR SCLERA USING ULTRASOUND SPECKLE TRACKING

Elias R. Pavlatos[1], Benjamin Cruz-Perez[1], Hugh J. Morris[1], Hong Chen[1], Richard T. Hart[1], Jun Liu[1,2]

[1]Department of Biomedical Engineering
The Ohio State University
Columbus, OH 43210
USA

[2]Department of Ophthalmology
The Ohio State University
Columbus, OH 43210
USA

INTRODUCTION

Glaucoma is an ocular pathology associated with vision loss. Clinically, glaucoma is characterized by damage to the optic nerve, which is responsible for transmitting visual signals to the brain. It is known that such damage initiates at the optic nerve head (ONH), where the nerve axons exit the eye through the scleral canal. However, the detailed mechanisms responsible for the optic nerve damage remain largely unknown.

The peripapillary sclera, which surrounds the scleral canal, is a region of particular interest in glaucoma research. It is hypothesized that intraocular pressure (IOP) induced strains and stresses in the ONH and peripapillary sclera may play a role in the progression of glaucoma. Using inflation testing and optical surface tracking, the strains of the peripapillary sclera have been shown to be dependent on location [1], age [2], and collagen microstructure [3].

Our laboratory has developed a high-resolution ultrasound speckle tracking technique to measure through-thickness strains in ocular tissue [4]. In this study, we investigated the 3D deformation of the sclera surrounding the scleral canal during physiological IOP increase. The objective of this study was to experimentally characterize the 3D strain states near the ONH, which may help us better understand the biomechanical pathways of optic nerve damage.

METHODS

Nine eyes from human donors (age: 55.2 ± 16.8 yo) were tested within 24 hours postmortem. The anterior segment was removed using a 9-mm trephine, and the intraocular structures were removed as well. The ocular shell was mounted using a custom-built pressurization chamber [5]. The chamber was filled with 0.9% saline and connected to a programmable syringe pump (PHD Ultra, Harvard Apparatus) and a pressure sensor (P75, Harvard Apparatus) to control and monitor IOP.

Each globe was preconditioned using five pressure cycles from 5 to 30 mmHg. After preconditioning, the globe was equilibrated at a reference pressure of 10 mmHg for 30 minutes. Inflation tests were then performed with four pressure steps: 11.5, 13, 15, and 19 mmHg. After a 7 minute equilibration period at each pressure level, a 55 MHz ultrasound probe (Vevo 660, VisualSonics Inc.) was used to acquire a volume of ultrasound radiofrequency data either on the nasal side of the ONH or the temporal side, at a randomized order. The total scanned volume formed a dome centered at the ONH.

The 3D volume was created by stacking consecutive 2D frames of radiofrequency data. An automated actuator (431 Series, Newport) was used to move the ultrasound probe in the inferior to superior or superior to inferior direction (randomized for each eye). The actuator step size was 14 microns, and a 2D scan was obtained at each incremental step. The data acquisition and actuator were controlled by a customized LabView program (National Instruments).

The distributive displacements within the scleral volume were calculated using a 3D ultrasound speckle tracking algorithm that was developed based on our existing 2D algorithm [4]. The strains were estimated using a 3D least squares strain estimator [6] to fit the displacement gradients to the measured displacement data as follows:

$$\begin{aligned} u &= \frac{\partial u}{\partial x}x + \frac{\partial u}{\partial y}y + \frac{\partial u}{\partial z}z + C_1 \\ v &= \frac{\partial v}{\partial x}x + \frac{\partial v}{\partial y}y + \frac{\partial v}{\partial z}z + C_2 \\ w &= \frac{\partial w}{\partial x}x + \frac{\partial w}{\partial y}y + \frac{\partial w}{\partial z}z + C_3 \end{aligned} \quad (1)$$

where u, v, and w are the displacements obtained from ultrasound speckle tracking, x, y, and z are the kernel coordinates, and the C_i's are local intercept constants used during fitting. The Cauchy strain tensor for each kernel was then calculated, and the eigenvalues and eigenvectors of the strain tensor yielded the principal strains and vectors. Using the principal strains, the maximum shear and volume ratio were calculated as follows [7]:

$$\gamma = \varepsilon_{max} - \varepsilon_{min} \quad (2)$$

$$VR = (1+\varepsilon_1)(1+\varepsilon_2)(1+\varepsilon_3) \quad (3)$$

where γ is the maximum shear strain, VR is the volume ratio, and ε is the principal strain. The strains were visualized using the Paraview software (Kitware Inc. & Los Alamos Labs).

RESULTS

The average principal strain magnitudes within the measured scleral volume were calculated for each eye (Figure 1). Consistent in all eyes, the average 1st principal strains were negative and largest in magnitude of the 3 principal strains. The 3rd principal strains were positive and slightly lower in magnitude compared to 1st principal strains. The average 2nd principal strains were close to zero. This trend in strain magnitudes was observed at all pressure levels, while the strains increased in magnitude with increasing IOP (Figure 2, left). The average maximum shear strains were larger than each of the three principal strains in all eyes.

The principal and shear strain vectors were visualized in Paraview. For each eye, the 2nd principal strain vectors had a predominantly in-plane circumferential arrangement around the ONH (Figure 2, right). The 1st and 3rd principal strain vectors were largely oriented in the through-thickness and in-plane meridional directions respectively. As defined, the maximum shear vector is oriented at a 45-degree angle between the maximum (i.e., meridional) and minimum (i.e., through-thickness) principal strain vectors.

The volume ratio was also calculated for each eye to evaluate the compressibility of the scleral tissue. The average volume ratios at 19 mmHg were slightly less than 1 for all eyes and ranged from 0.984 ± 0.017 to 0.999 ± 0.008.

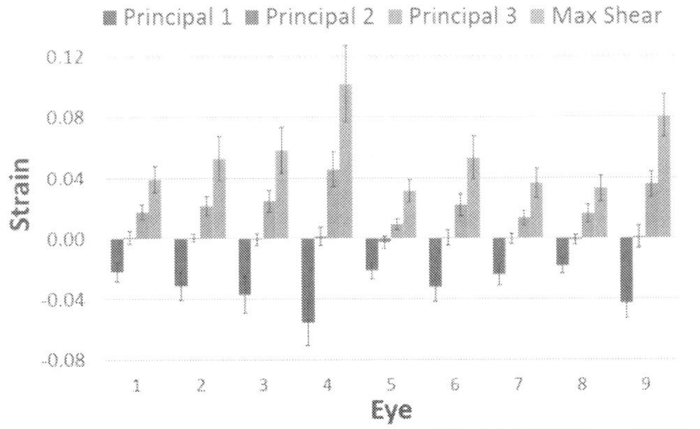

FIGURE 1. AVERAGE PRINCIPAL AND MAXIMUM SHEAR STRAINS IN HUMAN PERIPAPILLARY SCLERA AT 19 MMHG.

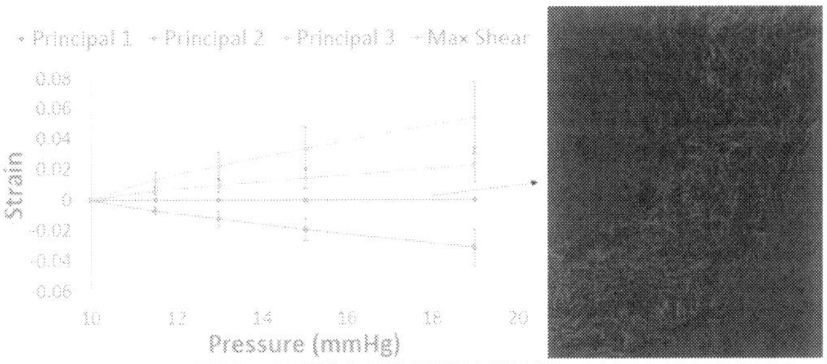

FIGURE 2. LEFT: AVERAGE PRINCIPAL AND MAXIMUM SHEAR STRAINS IN HUMAN PERIPAPILLARY SCLERA AT EACH PRESSURE LEVEL (N=9). RIGHT: 3D PLOT OF 2ND PRINCIPAL STRAIN VECTOR DIRECTIONS.

DISCUSSION

The 3D strain data showed that the peripapillary sclera immediately adjacent to the scleral canal experiences through-thickness compression (largest principal strain in magnitude) and tensile stretch of almost equal magnitude predominantly along the meridional direction during IOP increase within the physiological range (10 to 19 mmHg). This results in a maximum shear that is almost twice in magnitude compared to the largest principal normal strains. The 2nd principal strain was minimal in magnitude and largely along the circumferential direction, consistent with the preferred collagen alignment along that direction in the peripapillary sclera forming a reinforcing annulus to prevent scleral canal expansion during increases of IOP.

The volume ratios showed an overall nearly incompressible behavior, with a value approaching 1. The slightly lower than 1 volume ratio suggests a small overall volume loss during inflation, but this outcome may have been exaggerated by the ex vivo condition of the tissue (i.e., some extent of postmortem swelling and loss of water during inflation).

One limitation of the current study was that some portions of the scanned volume were not tracked well enough for strain calculations, which may or may not have significant influence on the overall outcome. Future work will improve the alignment of the ultrasound probe as well as the tracking algorithm, and investigate regional variations. In summary, our 3D strain data suggested that there is significant shear deformation in the peripapillary sclera, highlighting the need to characterize shear properties of the tissue in this region. The peripapillary sclera's ability to handle IOP-related deformation may have important implications to the ONH's homeostasis, and weaknesses in dealing with shear, compression, and stretch may all contribute to glaucomatous susceptibility.

ACKNOWLEDGEMENTS

NIH Grant R01EY020929

REFERENCES

[1] Fazio MA, et al. 2012. *IOVS* 53(9): 5326–5333
[2] Coudrillier B, et al. 2012. *IOVS* 53: 1714-28
[3] Girard MJA, et al. 2009a. *Journal of Biomechanical Engineering* 131(5): 051012
[4] Tang J, Liu J. 2012. *Journal of Biomechanical Engineering* 134(9): 091007
[5] Morris HJ, et al. 2013. *Invest Ophthalmol Vis Sci* 54: 7215-22
[6] Lopata RG, et al. 2009. *Ultrasonic imaging* 31: 1-16
[7] Malvern LE. Introduction to the Mechanics of a Continuous Medium. Englewood Cliffs: Prentice Hall Inc., 1969.

SB³C2015
Summer Biomechanics, Bioengineering and Biotransport Conference
June 17-20, 2015,Snowbird Resort, Utah, USA
SB³C2015-516

THE EFFECT OF TRABECULAE CARNAE ON THE PASSIVE COMPLIANCE OF LEFT VENTRICLE

Arnav Sanyal (1), David Halaney (2), Marc D. Feldman (2, 3), Hai-Chao Han (1, 3)

(1) Department of Mechanical Engineering
University of Texas at San Antonio
(UTSA)
San Antonio, TX, USA

(2) Department of Medicine, Cardiology Division
University of Texas Health Science Center at San Antonio
(UTHSCSA)
San Antonio, TX, USA

(3) Biomedical Engineering Program, UTSA-UTHSCSA
San Antonio, TX, USA

INTRODUCTION

Trabeculae carnae are irregular networks of cardiac muscle projecting from the endocardial surface near the mid-wall and apex of the ventricles of the heart. Patient-specific computational models of the heart, which have been used to explore the effects of various heart diseases and clinical interventions on the mechanical functioning of the heart, neglect the presence of trabecula carnae [1]. This is due to the poor spatial resolution of available imaging modalities and trabecula carnae being considered as embryologic remnants with no functional role. However, trabeculae carnae could account for a significant proportion of the ventricular mass and it was shown recently that they may play a role in ventricular functioning and adaptation [2].

Heart Failure patients with preserved ejection fraction (HFpEF) are diagnosed with impaired left ventricle (LV) diastolic relaxation and reduced LV diastolic compliance. The main cause of diastolic dysfunction is thought to be due to hypertrophy of the ventricular wall but dysfunction may be also associated with hypertrophy of the trabeculae carnae [3]. Reducing trabeculae carnae networks may provide a new approach to reduce LV stiffness and increase LV compliance, just like the loss of trabecular connections in cancellous bone can lead to reduction in bone stiffness and strength [4]. Towards this end, we investigated the role of trabeculae carnae on the diastolic compliance of the LV using both experiments and finite element simulations.

METHODS

Fresh human *ex vivo* hearts were obtained from six male organ donors who died suddenly in South Texas within 24 hours of death. Each heart was brought to 37°C, the left atrium was opened, and a condom was inserted through the cap into the LV and secured using PTFE tape. To measure baseline compliance, saline was injected through the cap into the condom, and the LV pressure was recorded to determine the pressure-volume compliance curves of these hearts up to a pressure of 50 mmHg. Following baseline compliance measurements, custom made tools were used to sever all visible free-running trabeculae in these human hearts and the compliance curve was measured again.

To explore the role of trabeculae carnae on the LV compliance and wall stress distribution, a finite element (FE) study was performed. A truncated ellipsoidal model of LV was created using Solidworks® and FE analysis was performed using ABAQUS®. The myocardium wall was assumed to behave as a Fung's anisotropic material with the following strain energy function W:

$$W = \frac{C}{2}(e^{Q} - 1)$$

$$Q = b_t E_{rr}^2 + b_f E_{ff}^2 + b_t E_{cc}^2 + 2b_t E_{rc}^2 + 2b_{fs} E_{fc}^2 + 2b_{fs} E_{rf}^2$$

where b_t, b_f and b_{fs} are material constants and E_{ij} $(i,j = r,c,f)$ are strains and subscripts r, c and f correspond to radial, fiber and cross-fiber directions. The material constants were obtained from a previous study on dog heart [5] ($b_f = 26.7$, $b_t = 2$, $b_{fs} = 14.7$, $C = 1.2$ kPa). The local alignment of the myocytes (fiber) in the myocardium with respect to the circumferential direction was assumed to vary from -37° in the epicardium to 98° in the endocardium [5]. Three layers of free-running radially oriented network of trabeculae carnae were added at the apical region of the ventricle. The trabeculae carnae were assumed to behave as Fung's isotropic material ($b_f = b_t = b_{fs} = 14.7$).

The passive end-diastolic response of the LV was obtained by applying an internal pressure of 50 mmHg on the endocardium wall. The pressure volume compliance curve of the LV was measured. Three simulations were run for comparison - first model with three layers of intact trabeculae, second model with only one layer of intact trabeculae near the midwall and apex (i.e. two layers of trabeculae most distant from the apex near the mid-ventricle region were cut off from the endocardium wall), and third model without trabeculae (i.e. all the layers of trabeculae were cut off from the endocardium wall). The end-diastolic pressure (P_{ED}) and volume (V_{ED}) relation was fit to the following equation:

$$P_{ED} = A\left[e^{K(V_{ED}-V_{ED,0})} - 1\right]$$

where A and K are fitting parameters ($V_{ED,0}$=30 ml, initial volume). The parameter K is inversely proportional to the compliance of the LV.

RESULTS

Experimental data revealed that there was an acute increase in the passive LV diastolic compliance following trabecular cutting for all six hearts (Fig. 1).

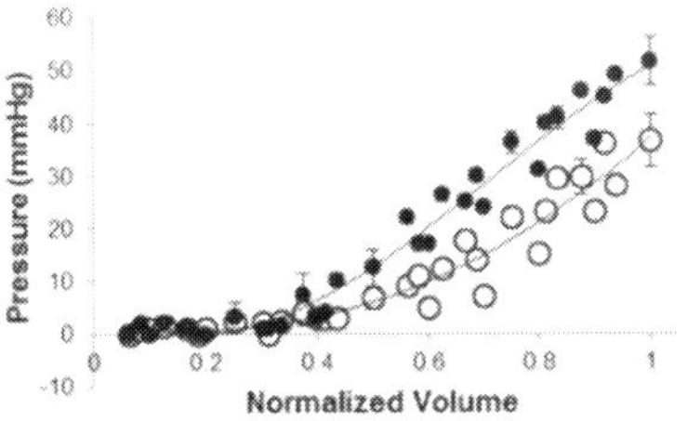

Figure 1: Passive compliance curve for six specimens of human heart before (solid circles) and after (hollow circles) trabecular cutting. The volume was normalized by the maximum volume for each specimen, $p<0.001$

The finite element simulations also demonstrated similar trends with a steeper LV compliance curve for the model with three intact layers of trabeculae (K=0.065) compared to the model without trabeculae (K=0.057) (Fig. 2). The LV compliance curve was similar for the model with only one intact layer of trabeculae (K=0.058) compared to the model without trabeculae (K=0.057), demonstrating that a similar increase in compliance of LV could be achieved by severing the trabecular network near the mid-ventricle but retaining a few trabeculae near the apex (Fig. 2).

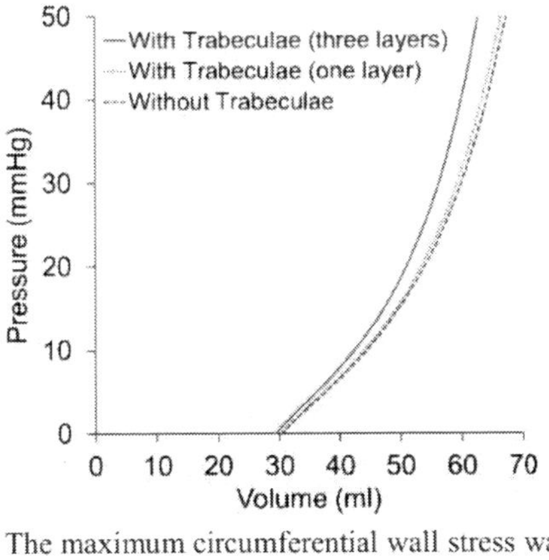

Figure 2: Passive compliance curve from FE analysis of the LV models without trabeculae, with three layers of intact trabeculae, and with one layer of intact trabeculae.

The maximum circumferential wall stress was evaluated at the apex in the endocardium and was lower for the model with three layers of intact trabeculae (37 kPa) compared to the model without the trabeculae (90 kPa) (Fig. 3). However, the maximum circumferential wall stress at the apex was comparable for the two trabeculae models with three layers and one layer of intact trabeculae (37 kPa versus 44 kPa). Further, the presence of trabeculae redistributed the wall stress from the apex more evenly to the location of trabeculae (Fig. 3). Finally, the circumferential stress in the mid-ventricle region did not show an appreciable difference with or without trabeculae (~16 kPa).

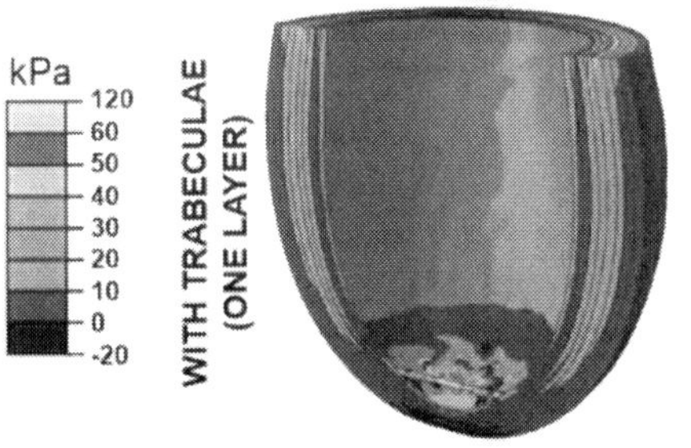

Figure 3: Circumferential stress distribution in LV at 50 mmHg pressure without trabeculae, with three layers of normal trabeculae and one layer of normal trabeculae

DISCUSSION

This is the first study to explore the role of trabeculae carnae on the diastolic compliance of left ventricle. Our results demonstrate that severing the network of free-running trabecular carnae can improve the LV diastolic compliance. While severing all the trabeculae may lead to large stresses near the apex, it can be mitigated by retaining some of the trabeculae near the apex to prevent adverse remodeling near the apex. The trabeculae near the apex have a negligible influence on the LV diastolic compliance but its presence leads to a redistribution of the wall stress from the apex more evenly into the trabeculae. Further investigation is needed to study the role of trabecula carnae on the systolic function of the heart. Clinically, our results show that selectively severing the left ventricular trabeculae carnae network, may improve the diastolic dysfunction of HFpEF patients without compromising the stress distribution in the myocardium.

ACKNOWLEDGEMENTS

This research was supported by NIH HL095852 and an Innovative Research grant from AHA award # 15IRG23320009. We also thank the Computational Systems Biology Core at UTSA for providing the computational resources.

REFERENCES

[1] Krishnamurthy, A et al., *J Comput Phys,* 244:4-21, 2013.
[2] van de Veerdonk, MC et al., *Int J Cardiovasc Imaging,* 30:357-365, 2014.
[3] Richter, GW et al., *J Cell Bio*, 18:195-206, 1963.
[4] Parfitt, AM, *Am J Med,* 82:68-72, 1987.
[5] Omens, JH et al., *J Biomech*, 26:665-676, 1993.

The Effects of the Impact of a Soccer Ball on a Human Head

Kimberly Brown (1,2), Aalaap Desai (1), Yuxiong Mao (3), M. F. Horstemeyer (1,4), Jun Liao (1,2), Lakiesha Williams (1,2), Hongjoo Rhee (1), R. Prabhu (1,2)

(1)Center for the Advanced Vehicular Systems
Mississippi State University
MS 39759

(2)Human Body Simulation Group
Dept. of Ag & Bio Eng.
Mississippi State University
MS 39762

(3) Predictive Design Technologies
GA

(4) Department of Mechanical Engineering.
Mississippi State University
MS 39762

INTRODUCTION

Much attention in recent years has been directed toward soccer-related traumatic brain injuries (TBIs). According to the American Association of Neurological Surgeons, over 24,000 head injuries in 2009 were from soccer-related incidents (1). In a study by Tysvaer et al (2), only 4.7 % of players that participated in the study complained of permanent symptoms from heading. In a more recent study of professional soccer players from Dutch soccer clubs, findings concluded that results of certain neuropsychological testing was inversely related to the number of concussions incurred in soccer and the frequency of heading the ball (3). Preliminary findings have not been significant enough to verify that heading a soccer ball leads to TBI. Another recent study concluded that heading is associated with abnormal white matter microstructure and poorer neurocognitive performance (4). While findings have directed the focus toward the possibility of heading causing TBI, some believe that it is of insignificant amount. These types of findings contribute to the ongoing debate of whether heading a soccer ball causes TBI.

Previous studies have utilized imaging techniques and other assessments involving EEG, CT, neurological exams, neuropsychological testing, and cognitive testing (5,6) While the studies have pointed toward the hypothesis of heading causing TBI, they have been unable to build upon each other thus far to create more concrete evidence. Many factors contribute to the study including gender, age, position played, and level of competition. A more controlled study would be effective in finding the impact of heading a soccer ball.

In order to eliminate external factors and obtain quantitative data, creating a finite element dynamic (explicit) model to study the pressure wave propagation and overall impact of a soccer ball on a virtual human head model would be significant. This study confirms studies and surveys previously collected from players who have headed soccer balls by showing the actual impact of heading. By using the results of this computational simulation, scientists will be able to determine the biomechanics of mild TBI. The current study is one of the first of its kind in that no previous simulations have been conducted on human head models to confirm TBI related to the pressure waves impending on the human brain upon impact of the ball.

METHODOLOGY

Finite Element Human Head Model

For this study, a finite element mesh of the male human head was developed from axial computed tomography (CT) and high resolution cryosection images from the National Library of Medicine's Visible Human Project (8). The head was segmented into four different parts including the brain, cerebral spinal fluid (CSF), skull, and the scalp (Figure 1). Using image-based meshing software (ScanFE, Simpleware, Ltd), a volumetric FE mesh was generated with tetrahedral and hexahedral elements. The FE mesh of the human head was then implemented into ABAQUS/Explicit (9) to simulate the traumatic brain injury from the heading of a soccer ball. The brain was modeled as a viscoelastic-viscoplastic material and the other components of the FE head model were treated as elastic materials as suggested by published literature (10,11,12,13).

Soccer Ball Model and Impact Scenarios

To model a soccer ball, a hollow sphere with a thin wall was created using the dimensions and pressure measurements provided by the FIFA organization (7). The diameter of the ball used is 214.6 millimeters (mm) with the wall of the ball being an included 5 mm. The soccer ball was treated as an elastic material in conjunction with properties obtained from published literature (14). Internal pressure was applied to the inner surfaces of the portioned ball at 0.08 MPa or 0.80 bar (7). The ball was positioned at a minimal distance to strike the head with a specified translational velocity. The speed of the soccer ball varied with each subsequent trial at respective velocities of 9, 14, 23, & 30 meters/second.

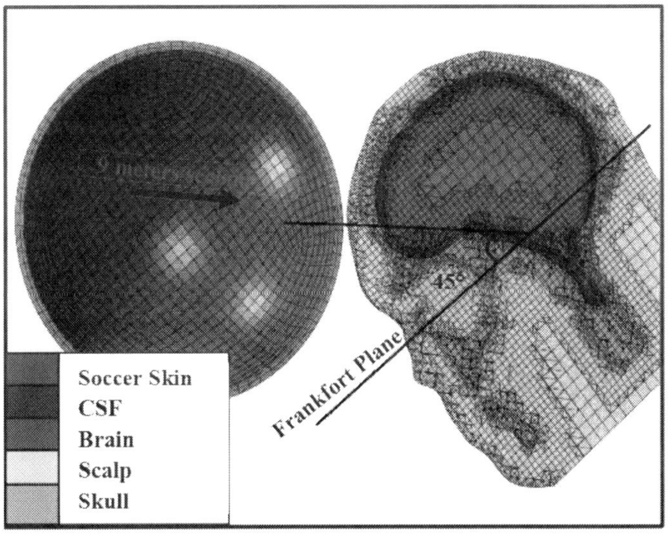

Fig. 1 A schematic of the Finite Element (FE) simulation with the ball contacting the human head 45° above the Frankfort Plane. An 817,980 element human head mesh comprising the scalp/skin, skull, cerebrospinal fluid (CSF) and brain.

RESULTS

Figure 2 (a) presents the peak impact of the soccer ball model on the human brain along the sagittal plane. Figure 2 (b) gives another contour plot of the pressure in the brain, which has been cut in the sagittal plane. These snapshots were taken in succession 1.8 ms after the impact of the ball on the human head model. By examining the pressure in the human head and the brain model (Figure 2), one can see that the peak tensile pressure occur at the counter-coup (opposite side of impact) site of brain in relation to the impact of the soccer ball.

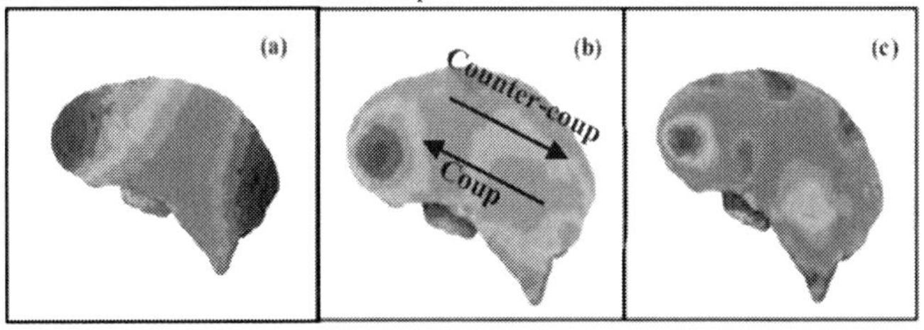

Fig. 2 Simulation of the soccer ball impacting the head at 9 meters/second in the Frankfort plane. Successive views of the pressure in the brain in the sagittal plane after ball contact.

DISCUSSION

Using Finite Element Analysis to examine the pressure effects from the impact of heading a soccer ball would be beneficial in determining the causes of mild TBIs. As shown in Figure 2(a), the peak tensile pressure occur at the counter-coup site, and persists in the same place during the majority of the soccer ball being in contact with the head. Although, the peak tensile pressure is lower in magnitude compared to American football concussions or blast-related concussions, due to the larger duration of the tensile pressure exposure, its resulting impulse is similar in magnitude and possible concussion may occur. Further analysis of different scenarios will be conducted to determine if hitting the human head from various angles, speeds, and positions effect the pressure responses in the brain.

ACKNOWLEDGEMENT

The authors would like to thank Shackouls Honors College at Mississippi State University (MSU), and the Department of Agricultural and Biological Engineering at MSU for funding this research. We express our gratitude to the Center for Advanced Vehicular Systems, at MSU, for sponsoring this research.

REFERENCES

1. American Association of Neurological Surgeons (AANS)
2. Matser JT, Kessels AG, Jordan BD, et al. Chronic traumatic brain injury in professional soccer players. Neurology 1998;51:791–6.
3. Radiology. 2013 Sep;268(3):850-7. Epub 2013 Jun 11. "Soccer heading is associated with white matter microstructural and cognitive abnormalities."
4. A. Tysvaer and O. Storli. Brit J. Sports Med. - Vol. 15. No. 3, September 1981, pp. 163-166. Association Football Injuries to the Brain: A Preliminary Report.
5. Marsha R. Zhang., Stuart D. Red., Angela H. Lin, Saumil S. Patell, Anne B. Sereno. PLOS. 2013. 8(2). Evidence of Cognitive Dysfunction after Soccer Playing with Ball Heading Using a Novel Tablet-Based Approach.
6. A Collie, D Darby, P Maruff. Br J Sports Med 2001;35:297–302. Computerised cognitive assessment of athletes with sports related head injury.
7. FIFA – Fédération Internationale de Football Association. 2012. Testing and Certification for Footballs.
8. Ackerman MJ, (1998) The Visible Human Project. Proc. IEEE 86 3: 504-511.
9. Simulia Inc. (2009) ABAQUS/Explicit User's Manual. Version 6.9 Providence, RI, 02909, USA.
10. Willinger R, Kang H-S, Diaw B (1999) Three-Dimensional Human Head Finite-Element Model Validation against Two Experimental Impacts. Ann Biomed Eng 27(3):403-410.
11. Belingardi G, Chiandussi G, Gaviglio I (2005) Development and validation of a new finite element model of human head. Int Tech Con Enhanced Safety Veh.
12. Chafi M, Karami G, Ziejewski, M (2010) Biomechanical Assessment of Brain Dynamic Responses Due to Blast Pressure Waves. Ann Biomed Eng 38(2):490-504.
13. Prabhu R, Horstemeyer MF, Tucker MT, Marin EB, Bouvard JL, Sherburn JA, Liao J-L, et al. (2011) Coupled experiment/finite element analysis on the mechanical response of porcine brain under high strain rates. J Mech Beh Biomed Matls. doi:10.1016/j.jmbbm.2011.03.015.
14. D.S. Price, R. Jones, and A.R. Harland. The Dependency of Hollow Ball Deformation on Material Properties. Sports Technology Research Group, Loughborough University, Loughborough LEICESTERSHIRE, LE11 3TU, UK

AUTHOR INDEX

AUTHOR INDEX

AUTHOR INDEX

AUTHOR INDEX

AUTHOR INDEX

AUTHOR INDEX

AUTHOR INDEX

AUTHOR INDEX

AUTHOR INDEX

AUTHOR INDEX

AUTHOR INDEX

AUTHOR INDEX

AUTHOR INDEX

AUTHOR INDEX

AUTHOR INDEX

AUTHOR INDEX

AUTHOR INDEX

AUTHOR INDEX

AUTHOR INDEX

Summer Biomechanics, Bioengineering and Biotransport Conference 2015

Snowbird, Utah, USA
17-20 June 2015

Volume 2 of 2

ISBN: 978-1-5108-2438-6

Printed from e-media with permission by:

Curran Associates, Inc.
57 Morehouse Lane
Red Hook, NY 12571

proceedings.com

Some format issues inherent in the e-media version may also appear in this print version.

Printed by Curran Associates, Inc. (2016)

For permission requests, please contact Summer Biomechanics, Bioengineering and Biotransport Organizing Committee at the address below.

Summer Biomechanics, Bioengineering and Biotransport Organizing Committee
201 Waterfront St
National Harbor, MD 20745
USA

info@sb3c.org

Additional copies of this publication are available from:

Curran Associates, Inc.
57 Morehouse Lane
Red Hook, NY 12571 USA
Phone: 845-758-0400
Fax: 845-758-2633
Email: curran@proceedings.com
Web: www.proceedings.com

WEDNESDAY, JUNE 17	3:45pm - 5:15pm

Ocular Biomechanics I — Primrose A

Session Chair: Rouzbeh Amini, *The University of Akron, Akron, OH, United States*
Session Co-Chair: Jun Liu, *Ohio State, Columbus, OH, United States*

3:45PM **Development Of A Platform For Studying Astrocyte Mechanobiology: Compression Of Astrocytes In 3D Alginate Gels** SB³C2015-655 pg. 977
John J. Mulvihill[1,2], Lisa A. Schildmeyer[1], Baptiste Coudrillier[1], Danny J. Kelly[2], C. Ross Ethier[1,3], *[1]Georgia Tech, Atlanta, GA, United States,[2]Trinity College Dublin, Dublin, Ireland,[3]Atlanta VA Medical Center, Atlanta, GA, United States*

4:00PM **Cellular Young's Modulus as a Novel Stemness Marker in the Corneal Limbus** SB³C2015-216 pg. 212
Tom Bongiorno[1], Jena Chojnowski[2], James D. Lauderdale[2], Todd Sulchek[1,3], *[1]Georgia Institute of Technology, Atlanta, GA, United States,[2]University of Georgia, Athens, GA, United States,[3]The Parker H. Petit Institute for Bioengineering and Bioscience, Atlanta, GA, United States*

4:15PM **Experimental Measurement Of Collagen Fiber Uncrimping And Recruitment With Increases In Intraocular Pressure** SB³C2015-417 pg. 543
Ian A. Sigal, Ning-Jiun Jan, Jonathan Grimm, Huong Tran, Hiroshi Ishikawa, Katherine A. Davoli, Larry Kagemann, Joel S. Schuman, Gadi Wollstein, Kira Lathrop, *University of Pittsburgh, Pittsburgh, PA, United States*

4:30PM **Phase-contrast Micro-tomography Measurements Of Intraocular Pressure-induced Deformation Of The Porcine Lamina Cribrosa** SB³C2015-147 pg.86
Baptiste Coudrillier[1], Diogo M. Geraldes[2], Nghia Vo[3], Ian C. Campbell[1], Julie Albon[4], Richard L. Abel[2], C. Ross Ethier[1], *[1]Georgia Institute of Technology, Atlanta, GA, United States,[2]Imperial College, London, United Kingdom,[3]Diamond Light Source, Didcot, United Kingdom,[4]Cardiff University, Cardiff, United Kingdom*

4:45PM **Ocular Compliance in Mice** SB³C2015-111 pg.34
Stephen A. S. Schwaner[1], Joseph M. Sherwood[2], Eric J. Snider[1], Eldon E. Geisert[3], Darryl R. Overby[4], C. Ross Ethier[1,3], *[1]Georgia Institute of Technology, Atlanta, GA, United States,[2]Imperial College London, London, United Kingdom,[3]Emory University, Atlanta, GA, United States,[4]Imperial College of London, London, United Kingdom*

5:00PM **Time And Pressure-Dependent Hydraulic Resistance Across Schlemm's Canal Endothelial Cell Layers** SB³C2015-1045 pg.1064
Alice Spenlehauer, Joseph M. Sherwood, Darryl R. Overby, *Imperial College London, London, United Kingdom*

WEDNESDAY, JUNE 17	3:45pm - 5:15pm

Design and Devices I - Vascular Disease and Therapeutic Intervention — Superior

Session Chair: James Moore, *Imperial College London, London, United Kingdom*
Session Co-Chair: Joao S. Soares, *University of Texas at Austin, Austin, TX, United States*

3:45PM **Corrosion Behavior of a Novel Biodegradable Metallic Stent** SB³C2015-240 pg.251
Jennifer Frattolin[1,2], Luca Gottellini[1,3], Olivier F. Bertrand[1,4], Rosaire Mongrain[1,2], *[1]McGill University, Montreal, QC, Canada,[2]Montreal Heart Institute, Montreal, QC, Canada,[3]Politecnico di Milano, Milan, Italy,[4]Quebec Heart and Lung Institute, Laval University, Quebec City, QC, Canada*

4:00PM **Geometric Analysis of Iliac Artery Tortuosity: Comparison to Current Clinical Practice** SB³C2015-248 pg.267
Matthew G. Doyle[1,2], Elrasheed Osman[2], Naomi Eisenberg[2], Cristina H. Amon[1], Leonard W. Tse[2], *[1]University of Toronto, Toronto, ON, Canada,[2]Toronto General Hospital, Toronto, ON, Canada*

4:15PM **In Vitro Coronary Artery Model for the Study of Atherosclerosis** SB³C2015-368 pg.453
Elizabeth E. Antoine, Abdul I. Barakat, *Ecole Polytechnique, Palaiseau, France*

4:30PM **Quantification of Plaque Shift after Coronary Bifurcation Stenting** SB³C2015-465 pg.632
Francesco Iannaccone[1,2], **Claudio Chiastra**[1,3], Francesco Migliavacca[3], Frank J. H. Gijsen[1], Patrick Segers[2], Evelyn Regar[4], Matthieu De Beule[2,5], Jolanda J. Wentzel[1], [1]*Thoraxcenter, Erasmus University Medical Center, Rotterdam, Netherlands,*[2]*Ghent University, Ghent, Belgium,*[3]*Politecnico di Milano, Milan, Italy,*[4]*Erasmus MC, Rotterdam, Netherlands,*[5]*FEops bvba, Ghent, Belgium*

4:45PM **Management of Distal Embolization during Interventional Treatment of Acute Ischemic Stroke in a Simulated Vascular Phantom** SB³C2015-589 pg.857
Juyu Chueh, Olivia W. Brooks, Ajit S. Puri, Ajay K. Wakhloo, Matthew J. Gounis, *University of Massachusetts Medical School, Worcester, MA, United States*

5:00PM **Temperature Comparison Of Catheter Ablation Between Open Irrigation And Vibration** SB³C2015-159 pg.108
Kaihong Yu[1], Tetsui Yamashita[2], Shigeaki Shingyochi[3], Kazuo Matsumoto[4], Makoto Ohta[5], [1]*Tohoku University, Sendai, Japan,*[2]*Cardiovascular Internal Medicine System Sec., JMS Co., Ltd., Tokyo, Japan,*[3]*Nidec Copal Electronics Corp., Tochigi, Japan,*[4]*Saitama Medical University, Hidaka, Japan,*[5]*Tohoku University, Sendai, Japan*

WEDNESDAY, JUNE 17	3:45pm - 5:15pm

Atherosclerosis — Wasatch

Session Chair: Michael Walsh, *University of Limerick, Ireland*
Session Co-Chair: Francis Loth, *University of Akron, OH, United States*

3:45PM **Human Coronary Plaque Morphological and Stress Vulnerability Indices Using IVUS-Based Fluid-Structure Interaction Models: A Multi-Patient Study** SB³C2015-33 pg.18
Liang Wang[1], Jie Zheng[2], Akiko Maehara[3], Chun Yang[4], Richard Bach[2], David Muccigrosso[2], Gary Mintz[5], **Dalin Tang**[1,6], [1]*WPI, Worcester, MA, United States,*[2]*Washington University, St. Louis, St. Louis, MO, United States,*[3]*Columbia University, New York, NY, United States,*[4]*China United Network Communications Co., Ltd., Beijing, China,*[5]*Columbia University, The Cardiovascular Research Foundation, New York, NY, United States,*[6]*Southeast University, Nanjing, China*

4:00PM **The Effect of Head Rotation to the Geometry and Hemodynamics of Healthy Vertebral Arteries** SB³C2015-495 pg.678
Nicolas Aristokleous[1], Ioannis Seimenis[2], Georgios C. Georgiou[3], Andreas S. Anayiotos[1], [1]*Cyprus University of Technology, Lemesos, Cyprus,*[2]*Democritus University of Thrace, Alexandroupoli, Greece,*[3]*University of Cyprus, Nicosia, Cyprus*

4:15PM **Development Of A Framework To Characterize The Role Of Wall Shear Stress In Atherosclerotic Plaque Transformation Through The Combined Use Of OCT And Vh-IVUS** SB³C2015-1064 pg.1098
David Molony[1], Lucas Timmins[1], Emad Rasoul-Arzrumly[2], Olivia Hung[2], Bill Gogas[2], Habib Samady[2], Don Giddens[1], [1]*Georgia Institute of Technology, Atlanta, GA, United States,*[2]*Emory University, Atlanta, GA, United States*

4:30PM **Phenotypic Differences in Coronary Artery Disease Progression in the Clinical Setting and Dependence on a Focal Oscillatory Hemodynamic Environment.** SB³C2015-426 pg.561
Lucas H. Timmins[1,2], David S. Molony[1,2], Parham Eshtehardi[3], Michael C. McDaniel[2], John N. Oshinski[1,2], Habib Samady[2], Don P. Giddens[1,2], [1]*Georgia Institute of Technology, Atlanta, GA, United States,*[2]*Emory University School of Medicine, Atlanta, GA, United States,*[3]*Albert Einstein College of Medicine, Bronx, NY, United States*

4:45PM **Relationship among Disturbed Shear Descriptors, the Extent of Flow Recirculation and Helicity in Carotid Bifurcation** SB³C2015-547 pg.778
Diego Gallo[1], David A. Steinman[2], Umberto Morbiducci[1], [1]*Politecnico di Torino, Turin, Italy,*[2]*University of Toronto, Toronto, ON, Canada*

5:00PM **Application of Gold Particle Enhanced CT for Vulnerable Plaque Detection and Quantification in Atherosclerotic Mouse Models** SB³C2015-366 pg.451
David De Wilde[1], Bram Trachet[1,2], Carole Van der Donckt[3], Bert Vandeghinste[1], Benedicte Descamps[1], Christian Vanhove[1], Guido R. Y. De Meyer[3], Patrick Segers[1], [1]*Ghent University, Gent, Belgium,*[2]*Institute for Bioengineering Ecole Polytechnique Fédérale de Lausanne, Lausanne, Switzerland,*[3]*University of Antwerp, Wilrijk, Belgium*

WEDNESDAY, JUNE 17	3:45pm - 5:15pm

Nano, Micro and Multiscale Mechanics — Magpie

Session Chair: Raffaella De Vita, *Virginia Tech, Blacksburg, VA, United States*
Session Co-Chair: Steven Abramowitch, *University of Pittsburgh, Pittsburgh, PA, United States*

3:45PM **Direct Measurement of Energy Landscape of Biological Interactions** SB[3]C2015-1181 pg.1307
Ahmad Haider, Daniel Potter, Todd Sulchek, *Georgia Institute of Technology, Atlanta, GA, United States*

4:00PM **Constructing Rudimentary Limit Curves For Neuronal Phospholipid Bilayer Failure And Theoretical Calcium Penetration** SB[3]C2015-1025 pg.1028
M. A. Murphy[1], Sungkwang Mun[1], M. F. Horstemeyer[1], Steven R. Gwaltney[1], Tonya W. Stone[1], Michelle C. LaPlaca[2], Jun Liao[1], Lakiesha N. Williams[1], **R. Prabhu**[1], [1]*Mississippi State, MS, United States,*[2]*Georgia Institute of Technology, GA, United States*

4:15PM **Deterioration Of Trabecular And Cortical Microarchitecture And Reduced Bone Stiffness At Distal Radius And Tibia In Postmenopausal Women With Vertebral Fractures** SB[3]C2015-231 pg.236
Ji Wang[1], Emily Stein[2], Bin Zhou[1], Kyle Nishiyama[2], Elizabeth Shane[2], X.Edward Guo[1], [1]*Columbia University, New York, NY, United States,*[2]*College of Physicians and Surgeons, Columbia University, New York, NY, United States*

4:30PM **Correlating Molecular Structure With The Mechanical Response Of Aquatic Caddisworm Silk** SB[3]C2015-325 pg.389
Nicholas N. Ashton, Russell J. Stewart, *The University of Utah, Salt Lake City, UT, United States*

4:45PM **Numerical Modeling of Human Fibrous Cap Delamination** SB[3]C2015-181 pg.148
Xiaochang Leng, Xiaomin Deng, Michael A. Sutton, Susan M. Lessner, *University of South Carolina, Columbia, SC, United States*

5:00PM **Nonlinear Viscoelastic Response of the Spinal Cord In Vivo** SB[3]C2015-150 pg.90
Snehal S. Shetye[1], Femke Streijger[2], Christopher Strickland[1], Jae H. T. Lee[2], Brian K. Kwon[2], Peter Cripton[2], Patrick D. Shipman[1], Christian M. Puttlitz[1], [1]*Colorado State University, Fort Collins, CO, United States,*[2]*University of British Columbia, Vancouver, BC, Canada*

WEDNESDAY, JUNE 17	3:45pm - 5:15pm

Organs, Morphogenesis, and Collective Cell Behavior — Maybird

Session Chair: Nandan L. Nerurkar, *Harvard Medical School, Boston, MA, United States*
Session Co-Chair: Celeste Nelson, *Princeton University, Princeton, NJ, United States*

3:45PM **Fgf8 Establishes A Contractile Gradient To Drive Directed Cell Movements In The Developing Avian Gut** SB[3]C2015-143 pg.80
Nandan L. Nerurkar, Cliff Tabin, *Harvard Medical School, Boston, MA, United States*

4:00PM **Mechanics of Early Eye Development: Interactions Between the Retina and Lens** SB[3]C2015-193 pg.170
Alina Oltean[1], Jie Huang[2], David C. Beebe[2], Larry A. Taber[1], [1]*Washington University in St. Louis, St. Louis, MO, United States,*[2]*Washington University School of Medicine, St. Louis, MO, United States*

4:15PM **Magnetic Micropillar Array To Study Epithelial Microtissue Morphogenesis And Local Mechanics** SB[3]C2015-341 pg.413
Mohammadnabi Asmani, Yan Li, Christopher Kotei, David Olsen, Zhaowei Chen, Fanjie Meng, **Ruogang Zhao**, *State University of New York at Buffalo, Buffalo, NY, United States*

4:30PM **Viscoelastic Folding Instability Controls Airway Branching Morphogenesis** SB[3]C2015-386 pg.481
Victor D. Varner, Jason P. Gleghorn, Celeste M. Nelson, *Princeton University, Princeton, NJ, United States*

4:45PM **In Silico Predictions Of Angiogenic Growth Coupled To Matrix Mechanics Within Heterogeneous Extracellular Environments** SB[3]C2015-546 pg.776
Lowell T. Edgar[1], James B. Hoying[2], Jeffrey A. Weiss[1], [1]*University of Utah, Salt Lake City, UT, United States,*[2]*University of Louisville, Louisville, KY, United States*

5:00PM **Biohybrid Swimming at Low Reynolds Number** SB^{3}C2015-1183 pg.1311
Brian Williams[1], Sandeep Anand[1], Jagannathan Rajagopalan[2], M. Taher A. Saif[1], [1]*University of Illinois at Urbana-Champaign, Urbana, IL, United States,*[2]*Arizona State University, Tempe, AZ, United States*

WEDNESDAY, JUNE 17	3:45pm - 5:15pm

Multiscale Modeling in Biotransport (Special Session) — Golden Cliff / Eagle's Nest

Session Chair: Bumsoo Han, *Purdue University, West Lafayette, IN, United States*
Session Co-Chair: Ronghui Ma, *University of Maryland Baltimore County, Baltimore, MD, United States*

3:45PM **Energetics of Water Permeation across Stretched Phospholipid/Cholesterol Bilayer: Molecular Dynamics Simulation** SB^{3}C2015-200 pg.184
Taiki Shigematsu, Kenichiro Koshiyama, Ryotaro Kurumatani, Shigeo Wada, *Osaka University, Toyonaka, Osaka, Japan*

4:00PM **Numerical Modeling of Scalp Cooling Devices for the Prevention of Chemotherapy-Induced Alopecia** SB^{3}C2015-635
Bradley Pliskow, Mehmet Kaya, Kunal Mitra, *Florida Institute of Technology, Indialantic, FL, United States* pg.939

4:15PM **Cross Capillary Mesangial Transport** SB^{3}C2015-1066 pg.1100
Sarah E. Hunt, Yoav Segal, Kevin D. Dorfman, Victor H. Barocas, *University of Minnesota, Minneapolis, MN, United States*

4:30PM **Multiscale Simulation of Shear-Induced Platelet Activation: Correlating Numerical with Experimental Results** SB^{3}C2015-302 pg.351
Peng Zhang[1], Chao Gao[1], Na Zhang[1], Seetha Pothapragada[1], Marvin J. Slepian[1,2], Yuefan Deng[1], Danny Bluestein[1], [1]*Stony Brook University, Stony Brook, NY, United States,*[2]*University of Arizona, Tucson, AZ, United States*

4:45PM **Numerical Study Of A New Thermalplasty Treatment System For Atherosclerosis** SB^{3}C2015-1110 pg.1183
Shiqing Zhao, **Aili Zhang**, Lisa X. Xu, *Shanghai Jiao Tong University, Shanghai, China*

5:00PM **An Approach to Include Pre-Stress in a Finite Element Model of Convection-Enhanced Delivery With Backflow** SB^{3}C2015-383 pg.475
Gustavo A. Orozco[1], Fabian A. Urrea[1], **Fernando Casanova**[1], Joshua H. Smith[2], Jose J. Garcia[1], [1]*Universidad del Valle, Cali, Colombia,*[2]*Lafayette College, Easton, PA, United States*

WEDNESDAY, JUNE 17	3:45pm - 5:15pm

Muscle and Joint Loading — Primrose B

Session Chair: Jun Liao, *Mississippi State, Mississippi State, MS, United States*
Session Co-Chair: Andrew Anderson, *University of Utah, Salt Lake City, UT, United States*

3:45PM **Medial and Lateral Contact Pressure Distribution Following the Implantation of a Novel Medial Meniscus Implant** SB^{3}C2015-185 pg.156
Maoz Shemesh[1], Noa Cohen[1], Eyal Zylberberg[1], Adaya Shefy-Peleg[1], Ron Arbel[2], Vincenzo Condello[3], Nogah Shabshin[4], Eran Linder-Ganz[1], **Jonathan J. Elsner**[5], [1]*Active Implants, Netanya, Israel,*[2]*Ichilov Medical Center, Tel Aviv, Israel,*[3]*Sacro Cuore Hospital, Verona, Italy,*[4]*Carmel Medical Center, Haifa, Israel,*[5]*Active Implants, Cambridge, MA, United States*

4:00PM **Effect of Meniscal Properties and Patient Variables on Knee Joint Contact Mechanics** SB^{3}C2015-213 pg.206
Hongqiang Guo[1], Thomas J. Santner[2], Po-Hsu Chen[2], Amy L. Lerner[3], Suzanne A. Maher[1], [1]*Hospital for Special Surgery, New York, NY, United States,*[2]*The Ohio State University, Columbus, OH, United States,*[3]*University of Rochester, Rochester, NY, United States*

4:15PM **Non-Invasive Method for Investigating the Relationship between Muscle Strain and Post-Exercise Blood Perfusion Using MRI** SB^{3}C2015-1175 pg.1299
Amanda K. W. Buck, Christopher P. Elder, Bruce M. Damon, *Vanderbilt University, Nashville, TN, United States*

4:30PM **Numerical Modeling of Skeletal Muscle Under High Strain and Stress Relaxation Compression Conditions** SB³C2015-247 pg.265
Benjamin B. Wheatley[1], Renee Pietsch[2], Tammy L. Haut Donahue[1], Lakiesha N. Williams[2], [1]*Colorado State University, Fort Collins, CO, United States,*[2]*Mississippi State University, Starkville, MS, United States*

4:45PM **Biomechanical Characterization of Porcine Skeletal Muscle Extracellular Matrix** SB³C2015-512 pg.710
Bryn Brazile, Sourav S. Patnaik, Sallie Lin, Xiaodan Shi, Shengfa Liao, Raj Prabhu, Hongjoo Rhee, Lakiesha N. Williams, Jun Liao, *Mississippi State University, Mississippi State, MS, United States*

5:00PM **Predicting the Stress and Intramuscular Pressure Response of Whole Skeletal Muscle Through Optimized Finite Element Analysis** SB³C2015-249 pg.269
Benjamin B. Wheatley[1], Duane A. Morrow[2], Gregory M. Odegard[3], Kenton R. Kaufman[2], Tammy L. Haut Donahue[1], [1]*Colorado State University, Fort Collins, CO, United States,*[2]*Mayo Clinic, Rochester, MN, United States,*[3]*Michigan Tech University, Houghton, MI, United States*

WEDNESDAY, JUNE 17	5:30pm - 7:00pm

Ocular Biomechanics II — Primrose A

Session Chair: Jun Liu, *Ohio State, Columbus, OH, United States*
Session Co-Chair: Ian A. Sigal, *University of Pittsburgh, Pittsburgh, PA, United States*

5:30PM **Equilibrium Shape of the Aqueous Humor-Vitreous Substitue Interface** SB³C2015-445 pg.598
Krystyna Isakova, Rodolfo Repetto, Jan Oscar Pralits, *University of Genova, Genova, Italy*

5:45PM **A Computational Model To Explore The Role Of Experimentally Determined Scleral Microstructure On Lamina Cribrosa Deformation** SB³C2015-1087 pg.1140
Avinash Ayyalasomayajula, Forest L. Danford, Jonathan P. Vande Geest, *University of Arizona, Tucson, AZ, United States*

6:00PM **Effects Of Hydration And Riboflavin/uva Collagen Crosslinking On Bovine Corneal Tensile Properties** SB³C2015-1074 pg.1114
Hamed Hatami-Marbini, *Oklahoma State University, Stillwater, OK, United States*

6:15PM **Intraocular Pressure Increases In Patients With Intraocular Gas Bubbles Following A Descent And Subsequent Ascent** SB³C2015-394 pg.497
Lucas A. Gsellman, Rouzbeh Amini, *The University of Akron, Akron, OH, United States*

6:30PM **The Effect Of Elevated Intraocular Pressure On Convective Flow In The Vitreous** SB³C2015-323 pg.385
Julie E. Whitcomb[1], Mohammad R. Kazemi[2], Michael R. Robinson[1], Mayssa Attar[1], Anita Penkova[3], Sati Sadhal[3], Susan S. Lee[1], [1]*Allergan, Irvine, CA, United States,*[2]*Consultant, San Jose, CA, United States,*[3]*University of Southern California, Los Angeles, CA, United States*

6:45PM **A Parametrized Model Of The Lamina Cribrosa For Studying Oxygen Transport** SB³C2015-312 pg.367
Fabian A. Bräu[1,2], Ian C. Campbell[1,3], Baptiste Coudrillier[1], **C. Ross Ethier**[1,3], [1]*Georgia Institute of Technology, Atlanta, GA, United States,*[2]*Technical University Munich, Munich, Germany,*[3]*Atlanta VA Medical Center, Decatur, GA, United States*

WEDNESDAY, JUNE 17	5:30pm - 7:00pm

Design and Devices II - Modeling and Simulation of Cardiovascular Therapies — Superior

Session Chair: Michael Moreno, *Texas A&M University, College Station, TX, United States*
Session Co-Chair: Lucas H. Timmins, *Georgia Institute of Technology, Atlanta, GA, United States*

5:30PM **Hemocompatibility Assessment Of Hyaluronan Enhanced Linear Low Density Polyethylene For Use In Aortic Heart Valve Leaflets** SB³C2015-1041 pg.1056
Rachael Simon-Walker, John Cavicchia, David Bark, Susan James, Lakshmi Prasad Dasi, Popat Ketul, *Colorado State University, Fort Collins, CO, United States*

5:45PM **Virtual Transplantation To Establish Donor Selection Criteria For Undersized And Complex Heart Recipients** SB³C2015-335 pg.401

Justin Ryan[1], Randy Richardson[2], Erik Ellsworth[3], John J. Nigro[3], David Frakes[1], Stephen Pophal[3], [1]*Arizona State University, Tempe, AZ, United States,*[2]*St. Joseph's Hospital and Medical Center, Phoenix, AZ, United States,*[3]*Phoenix Children's Hospital, Phoenix, AZ, United States*

6:00PM **Physical Patient Specific Simulation For Ascending Aortic Aneurysms Surgery Pre-procedural Training** SB³C2015-338 pg.407

Justine Garcia[1], ZhiLin Yang[1], Kevin Lachapelle[2], Rosaire Mongrain[1], Richard Leask[1], [1]*McGill University, Montreal, QC, Canada,*[2]*McGill University Health Network, Montreal, QC, Canada*

6:15PM **Computer Simulator of the Coil Insertion into Aneurysm: The Effect of Mechanical Property and Reference Configuration of the Coil** SB³C2015-343 pg.417

Tomohiro Otani[1], Satoshi Ii[1], Tomoyoshi Shigematsu[2], Toshiyuki Fujinaka[2], Masayuki Hirata[2], Tomohiko Ozaki[2], Shigeo Wada[1], [1]*Graduate School of Engineering Science, Osaka University, Toyonaka, Japan,*[2]*Graduate School of Medicine, Osaka University, Suita, Japan*

6:30PM **Computational Hemodynamic Assessment of a Novel Modular Anastomotic Valve Device for Hemodialysis Vascular Access** SB³C2015-1069 pg.1106

Andrew McNally[1], Philippe Sucosky[1], A. George Akingba[2], [1]*University of Notre Dame, South Bend, IN, United States,*[2]*Indiana University School of Medicine, Indianapolis, IN, United States*

6:45PM **Computational Modeling Of A Percutaneous Transvenous Mitral Annuloplasty Device Deployment Into The Coronary Sinus Vessel For Treatment Of Mitral Regurgitation: Analysis Of Anchor Sizes** SB³C2015-1068 pg.1104

Thuy Pham[1], Fanwei Kong[1], Milton Deherrera[2], Wei Sun[1], [1]*Georgia Institute of Technology, Atlanta, GA, United States,*[2]*Edwards Lifesciences, Irvine, CA, United States*

WEDNESDAY, JUNE 17	5:30pm - 7:00pm

Cardiovascular Diagnostics and Imaging — Wasatch

Session Chair: David Steinman, *University of Toronto, Toronto, ON, Canada*
Session Co-Chair: Matt Gounis, *University of Massachusetts, Worcester, MA, United States*

5:30PM **Invasive Measurement of Pulse Wave Velocity in Anesthetized Mice: a Computational-Experimental Study** SB³C2015-600 pg.875

Federica Cuomo[1], Jacopo Ferruzzi[2], Jay D. Humphrey[2], Carlos A. Figueroa[1], [1]*University of Michigan, Ann Arbor, MI, United States,*[2]*Yale University, New Haven, CT, United States*

5:45PM **Modeling Hemodynamic Effects Of Cerebral Vasospasm** SB³C2015-1049 pg.1072

Jaiyoung Ryu[1], Xiao Hu[2], Shawn C. Shadden[1], [1]*University of California, Berkeley, Berkeley, CA, United States,*[2]*University of California, San Francisco, San Francisco, CA, United States*

6:00PM **A Novel Approach to Calculating Isovolumic Pressure for Single-Beat Estimation of Right Ventricular End-Systolic Elastance** SB³C2015-544 pg.772

Alessandro Bellofiore[1], Eric Dinges[2], Sanjiv J. Shah[3], Michael J. Cuttica[3], Ranya Sweis[3], Hamorabi Mkrdichian[3], Lauren Beussink-Nelson[3], Melissa Bailey[2], James R. Runo[2], Jon G. Keevil[2], Christopher J. Francois[2], Naomi C. Chesler[2], [1]*San Jose State University, San Jose, CA, United States,*[2]*University of Wisconsin-Madison, Madison, WI, United States,*[3]*Northwestern University, Chicago, IL, United States*

6:15PM **A Method for Rapid, Accurate Calculation of Wall Shear Stress from 2D Phase Contrast Magnetic Resonance Image Data** SB³C2015-562 pg.805

Elizabeth Iffrig[1,2], William R. Taylor[1,2], John N. Oshinski[1,2], [1]*Emory University, Atlanta, GA, United States,*[2]*Georgia Institute of Technology, Atlanta, GA, United States*

6:30PM **Tracking Mass Transport Of A Drug Surrogate In Porcine Coronary Tissue Using Photoacoustic Imaging And Spectroscopy** SB³C2015-1083 pg.1132

Kenneth J. Furdella, Jonathan P. Vande Geest, Russell S. Witte, *University of Arizona, Tucson, AZ, United States*

6:45PM **Wave Intensity Analysis from Magnetic Resonance Imaging: Experimental Validation and New Insights into Single Ventricle Physiology** SB³C2015-459 pg. 622
Nikesh Arya[1], Silvia Schievano[1], Catriona Baker[1], Tain-Yen Hsia[2], Alessandro Giardini[2], Sachin Khambadkone[2], Andrew M. Taylor[1], **Giovanni Biglino**[1], *[1]University College London, London, United Kingdom,[2]Great Ormond Street Hospital for Children, London, United Kingdom*

WEDNESDAY, JUNE 17	5:30pm - 7:00pm

Multiscale Biomechanics - Coupling Muscuoskeletal, Joint, and Tissue Level Models (Special Session) — Magpie

Session Chair: Mariana Kersh, *University of Illinois, Urbana, IL, United States*
Session Co-Chair: Darryl Thelen, *University of Wisconsin-Madison, Madison, WI, United States*

5:30PM **Gait Simulation Using a Contact Force Feedback Controller** SB³C2015-416 pg.541
Jonathan P. Walter, Marcus G. Pandy, *University of Melbourne, Parkville, Australia*

5:45PM **A Multi-scale Finite Element Framework for Modeling Natural Knee Mechanics** SB³C2015-314 pg.371
Michael D. Harris, Ali Azhar, Alessandro Navacchia, Adam J. Cyr, Donald Hume, Clare K. Fitzpatrick, Paul J. Rullkoetter, Kevin B. Shelburne, *University of Denver, Denver, CO, United States*

6:00PM **Prediction Of Elbow Joint Contact Pressures In The Multibody Framework** SB³C2015-389 pg.487
Munsur Rahman[1], Akin Cil[1,2], Antonis P. Stylianou[1], *[1]University of Missouri-Kansas City, Kansas City, MO, United States,[2]Truman Medical Centers, Kansas City, MO, United States*

6:15PM **Multi-scale Models of Skeletal Muscle Reveal the Complex Effects of Muscular Dystrophy on Tissue Mechanics and Damage Susceptibility.** SB³C2015-478 pg.656
Kelley M. Virgilio, Kyle M. Martin, Shayn M. Peirce, Silvia S. Blemker, *University of Virginia, Charlottesville, VA, United States*

6:30PM **Multiscale Syntheses Of Articular Cartilage Stress In Post Aclr Model Under Physiological Loading** SB³C2015-605
Malek Adouni, Yasin Dhaher, *Rehabilitation Institute of Chicago/Biomedical Engineering, Northwestern University, Chicago, IL, United States pg.885*

6:45PM **Towards The Validation of a Multiscale Chemo-Electro-Mechanical Finite Element Model Using Electromyography** SB³C2015-475 pg.650
Thomas Heidlauf, Mylena Mordhorst, **Sook-Yee Chong**, Oliver Röhrle, *University of Stuttgart, Stuttgart, Germany*

WEDNESDAY, JUNE 17	5:30pm - 7:00pm

Musculoskeletal Tissue Engineering - Molecular, Soluble, and Mechanical Regulation of Tissue Development — Maybird

Session Chair: Danny Kelly, *Trinity College, Dublin, Ireland*
Session Co-Chair: Pen-hsiu Grace Chao, *National Taiwan University, Taipei, Taiwan*

5:30PM **Osteocyte Sclerostin Expression Is Directed By Substrate Composition And Dimensionality In Addition To Fluid Flow** SB³C2015-241 pg.253
Robert Thomas Brady[1], Andrew R. Cameron[1], David A. Hoey[2], Fergal J. O'Brien[1], *[1]Royal College of Surgeons in Ireland, Dublin, Ireland,[2]University of Limerick, Limerick, Ireland*

5:45PM **Investigating CRISPRi Cell-Engineering Methods For Treatment Of Intervertebral Disc Degeneration** SB³C2015-395
Niloofar Farhang[1], Jonathan M. Brunger[2], Joshua D. Stover[1], Pratishka I. Thakore[2], Brandon Lawrence[1], Farshid Guilak[2], Charles A. Gersbach[2], Lori A. Setton[2], Robby D. Bowles[1], *[1]University of Utah, Salt Lake City, UT, United States,[2]Duke University, Durham, NC, United States pg.499*

6:00PM **Label-Free Protein Profiling for Functional Characterization of Engineered Cartilage Following Inflammatory Cytokine Exposure** SB^{3}C2015-450 pg.608
Andrea R. Tan[1], Shujuan Tao[2], David Chen[2], Lewis Brown[2], Clark T. Hung[1], [1]*Columbia University, New York, NY, United States,*[2]*Comparative Proteomics Center, Columbia University, New York, NY, United States*

6:15PM **Bisphosphonate Gives Trauma Damaged Chondrocytes a New Life via Inhibition of Mevalonate Pathway** SB^{3}C2015-570 pg.819
Yilu Zhou, Miri Park, Liyun Wang, Enoch Cheung, X. Lucas Lu, *University of Delaware, Newark, DE, United States*

6:30PM **The Impact of Physical and Enzymatic Treatment on the Development of Tissue-Engineered Articular Cartilage Generated from Adult Human Chondrocytes** SB^{3}C2015-1099 pg.1161
Brendan L. Roach[1], Terri-Ann N. Kelly[1], Michael K. Dermksian[1], Sonia Bansal[1], Paola A. Lopez[2], Aaron M. Stoker[3], James L. Cook[3], Gerard A. Ateshian[1], Clark T. Hung[1], [1]*Columbia University, New York, NY, United States,*[2]*University of Arizona, Tucson, AZ, United States,*[3]*University of Missouri at Columbia, Columbia, MO, United States*

6:45PM **Effects Of Structural Vs. Non-structural Scaffolds On Bmp-2- And Mechanical Load-mediated Bone Regeneration** SB^{3}C2015-1180 pg.1305
Anna McDermott[1], Angela Lin[2], Robert Guldberg[2], **Joel D. Boerckel**[1], [1]*University of Notre Dame, Notre Dame, IN, United States,*[2]*Georgia Institute of Technology, Atlanta, GA, United States*

WEDNESDAY, JUNE 17	**5:30pm - 7:00pm**

Transport at the Nano- and Micro- scale — Golden Cliff / Eagle's Nest

Session Chair: Rubén Diaz-Rivera, *University of Puerto Rico at Mayagüez, PR, United States*
Session Co-Chair: Chris Rylander, *University of Texas at Austin, Austin, TX, United States*

5:30PM **Biophysical Fractionation Of Cells With Multiple Microfluidic Outlets** SB^{3}C2015-294 pg.335
Todd Sulchek, Gonghao Wang, Cory Turbyfield, Kaci Crawford, Alexander Alexeev, *Georgia Tech, Atlanta, GA, United States*

5:45PM **Pressure Drop Approach to Estimate the Apparent Slip Length in a PDMS Micro-channel** SB^{3}C2015-633 pg.935
Stephanie E. González Jiménez[1], Carlos R. Romero Peñaloza[1], Rubén E. Díaz-Rivera[1,2], [1]*University of Puerto Rico at Mayagüez, Mayagüez, PR, United States,*[2]*Purdue University, West Lafayette, IN, United States*

6:00PM **Mechano-activatable Microcapsules For Tunable Drug Delivery** SB^{3}C2015-597 pg.869
Bhavana Mohanraj, Fuquan Tu, Daeyeon Lee, George R. Dodge, Robert L. Mauck, *University of Pennsylvania, Philadelphia, PA, United States*

6:15PM **Rate of Transport of Circulating Macromolecules Into The Arterial Wall May Determine Atherosclerotic Plaque Stability** SB^{3}C2015-358 pg. 435
Peter Weinberg, **Zahra Mohri**, *Imperial College London, London, United Kingdom*

6:30PM **Gold Nanoparticle Mediated Enhanced Gene Delivery With High Selectivity For Breast Cancer Treatment** SB^{3}C2015-659 pg.985
Binita Shrestha, Liang Tang, *University of Texas at San Antonio, San Antonio, TX, United States*

6:45PM **The Effect Of Particle Deformability On The Transport Of Soft Colloids Through Porous Media** SB^{3}C2015-1148 pg.251
Eduard Benet, Louis Foucard, Franck J. Vernerey, John Pellegrino, *University of Colorado at Boulder, Boulder, CO, United States*

WEDNESDAY, JUNE 17	**5:30pm - 7:00pm**

Ligament and Tendon — Primrose B

Session Chair: Spencer Lake, *Washington University, St. Louis, MO, United States*
Session Co-Chair: Ozan Akkus, *Case Western Reserve University, Cleveland, OH, United States*

5:30PM **Two Bundles of the Anterior Cruciate Ligament Exhibit Different Microstructural Properties and Mechanics During Stress-Relaxation** SB^{3}C2015-484 pg. 662
Ryan Castile, Nathan Skelley, Robert Brophy, Spencer Lake, *Washington University, St. Louis, MO, United States*

5:45PM **Composition-dependent Mechanical Properties and Multiscale Strain Transfer in Tendon Subjected to Shear Loading** SB³C2015-173 pg.132
Fei Fang, Spencer P. Lake, *Washington University in St. Louis, St. Louis, MO, United States*

6:00PM **Evaluation of Tendon Injuries Using Shear Wave Elastography: Preliminary In-vivo Results in Human Achilles and Semitendinosus Tendons** SB³C2015-160 pg.110
Daniel H. Cortes, Stephen M. Suydam, Jennifer A. Zellers, Karin Gravare-Silbernagel, Thomas S. Buchanan, Dawn M. Elliott, *University of Delaware, Newark, DE, United States*

6:15PM **Modeling Ligament Mechanics Using Microstructural Parameters Measured from Confocal Images of Collagen Networks** SB³C2015-223 pg.222
Christina J. Stender, Evan Rust, Erica E. Morrill, Roshani Lamichhane, Raquel J. Brown, Trevor J. Lujan, *Boise State University, Boise, ID, United States*

6:30PM **A Transversely Isotropic Constitutive Model Describes And Predicts The Contribution Of Elastin To The Multiaxial Mechanics Of Ligament** SB³C2015-441 pg.590
Heath B. Henninger, Sara A. Scott, Benjamin J. Ellis, Jeffrey A. Weiss, *University of Utah, Salt Lake City, UT, United States*

6:45PM **Detection, Quantification, and Localization of Subfailure Damage in Tendon Fascicles using Collagen Mimetic Peptides** SB³C2015-628 pg.925
Jared L. Zitnay, Yang Li, Shawn P. Reese, Boi-Hoa San, S. Michael Yu, Jeffrey A. Weiss, *University of Utah, Salt Lake City, UT, United States*

THURSDAY, JUNE 18	8:00am - 9:00am

PLENARY SESSION I – Margaret Gardel — Ballrooms 1-3

THURSDAY, JUNE 18	9:15am - 10:45am

Brain Injury Biomechanics — Primrose A

Session Chair: Brittany Coats, *University of Utah, Salt Lake City, UT, United States*
Session Co-Chair: Francis Gayzik, *Wake Forest University School of Medicine, Winston-Salem, NC, United States*

9:15AM **The Effect of Facemasks on the Impact Performance of Helmets** SB³C2015-1101 pg.1165
Steven Rowson, Evan Terrell, *Virginia Tech, Blacksburg, VA, United States*

9:30AM **Estimating Axonal Strain in White Matter Using Paranodal Proteins as Fiduciary Markers** SB³C2015-468 pg.638
Sagar Singh, Assimina A. Pelegri, David I. Shreiber, *Rutgers, The State University of New Jersey, Piscataway, NJ, United States*

9:45AM **In Vivo Comparison Of Wearable Head Impact Sensors** SB³C2015-1152 pg.1259
Lyndia C. Wu, Vaibhav Nangia, Kevin Bui, Bradley Hammoor, Calvin Kuo, Fidel Hernandez, David B. Camarillo, *Stanford University, Stanford, CA, United States*

10:00AM **Image-Based Dynamic Analysis of Brain Deformation on Porcine Brain Injury Model** SB³C2015-1163 pg.1279
Arnold D. Gomez, Gregory G. Scott, Boston C. Terry, Brittany Coats, *University of Utah, Salt Lake City, UT, United States*

10:15AM **Preliminary Development and Validation of an Atlas-Based Finite Element Brain Model** SB³C2015-1145 pg.1245
Logan Miller, Jillian Urban, Elizabeth Lillie, Joel Stitzel, *Virginia Tech-Wake Forest University School of Biomedical Engineering and Sciences, Winston-Salem, NC, United States*

10:30AM **Influence of Recent Head Impact History on Biomechanics of Concussion Sustained in College Football Athletes** SB³C2015-289 pg.327
Daniel J. Sjoquist, Brian D. Stemper, Alok S. Shah, James Murtha, John R. Humm, Ashley LaRoche, Adam Pfaller, Steven Broglio, Kevin Guskiewicz, Michael McCrea, *Medical College of Wisconsin, Milwaukee, WI, United States*

THURSDAY, JUNE 18	9:15am - 10:45am

Dynamics and Rehabilitation — Superior

Session Chair: Alan Eberhardt, *University of Alabama at Birmingham, Birmingham, AL, United States*
Session Co-Chair: Tammy Bush, *Michigan State University, East Lansing, MI, United States*

9:15AM **A Support Vector Machine Based On Vertical Ground Reaction Force To Supplement Observational Gait Evaluation** SB³C2015-568 pg.817
Craig J. Simons[1], Cory L. Christiansen[2], Jennifer E. Stevens-Lapsley[2], Kevin B. Shelburne[1], Bradley S. Davidson[1], *[1]University of Denver, Denver, CO, United States,[2]University of Colorado Denver, Denver, CO, United States*

9:30AM **A Bootstrapping Method to Assess the Influence of Age, Obesity, and Gender on Probability of Tripping as a Function of Obstacle Height** SB³C2015-506 pg.698
Christina M. R. Garman[1], Christopher T. Franck[1], Maury A. Nussbaum[1], Michael L. Madigan[2], *[1]Virginia Tech, Blacksburg, VA, United States,[2]Texas A & M University, College Station, TX, United States*

9:45AM **Metabolic Consumption Using Different Repetitive Lifting Strategies** SB³C2015-1039 pg.1052
Timothy D. Craig[1], Alice E. Riley[1], Sandra A. Billinger[2], Neena K. Sharma[2], **Sara E. Wilson**[1], *[1]University of Kansas, Lawrence, KS, United States,[2]University of Kansas, Kansas City, KS, United States*

10:00AM **Design, Calibration, and Validation of a Novel In Vitro Tibial Force Sensor** SB³C2015-29 pg.10
Joshua D. Roth, Maury L. Hull, Stephen M. Howell, *University of California, Davis, Davis, CA, United States*

10:15AM **Oh Deer! Morphological and Biomechanical Evaluation of Cervine Femora** SB³C2015-415 539
Mark J. Hedgeland, Morgan A. Libruk, Nicole C. Corbiere, Mario J. Ciani, Laurel Kuxhaus, *Clarkson University, Potsdam, NY, United States*

10:30AM **The Effects of Low Intensity Vibration on Bone Mineral Density in the Intact Limb of Animals with a Percutaneously Attached Endoprosthesis.** SB³C2015-502 pg.692
Kyle Bodnyk[1], Garrett Noble[1], Matthew Allen[2], Noel Fitzpatrick[3], Gabriel Pagnotti[4], Richard Hart[1], *[1]The Ohio State University, Columbus, OH, United States,[2]University of Cambridge, Cambridge, United Kingdom,[3]Fitzpatrick Referrals, Surrey, United Kingdom,[4]Stony Brook University, New York, NY, United States*

THURSDAY, JUNE 18	9:15am - 10:45am

Cerebral and Aortic Aneurysms — Wasatch

Session Chair: Ender A. Finol, *University of Texas at San Antonio, San Antonio, TX, United States*
Session Co-Chair: Naomi Chesler, *University of Wisconsin, Madison, WI, United States*

9:15AM **Effect of Branched and Fenestrated Stent-Grafts on Renal Blood Flow.** SB³C2015-107 pg. 26
Harkamaljot S. Kandail[1], Mohamad S. Hamady[2], Xiao Y. Xu[1], *[1]Imperial College London, London, United Kingdom,[2]Imperial College Healthcare NHS Trust, London, United Kingdom*

9:30AM **Hemodynamic Changes in Treated Cerebral Aneurysms and Correlations with Long-Term Outcomes** SB³C2015-557
Michael C. Barbour[1], Patrick M. McGah[1], Michael R. Levitt[2], Kurt Sansom[1], Ryan P. Morton[2], John D. Nevra[2], Pierre D. Mourad[2], Basavaraj V. Ghodke[3], Danial K. Hallam[3], Laligam N. Sekhar[2], Louis J. Kim[2], Alberto Aliseda[1], *[1]University of Washington, Seattle, WA, United States,[2]University of Washington, Neurological Surgery, Seattle, WA, United States,[3]Dept of Radiology, Harborview Medical Center, Seattle, WA, United States pg.797*

9:45AM **Anuerysm MRI Phantoms For Direct, Ex Vivo Fluid Dynamics** SB³C2015-283 pg.319
Jeff R. Anderson[1], Orlando Diaz[2], Richard Klucznik[2], Yi J. Zhang[2], Gavin W. Britz[2], Robert G. Grossman[2], Christof Karmonik[1,2], *[1]Houston Methodist Research Institute, Houston, TX, United States,[2]Houston Methodist Hospital, Houston, TX, United States*

10:00AM **Endovascular Treatment of Intracranial Aneurysms: Finite Element Modeling of Various Intervention Strategies** SB³C2015-518 pg.723
Robert Damiano, Ding Ma, Jianping Xiang, Adnan Siddiqui, Kenneth Snyder, Hui Meng, *University at Buffalo, State University of New York, Buffalo, NY, United States*

10:15AM **Experimental Evaluation of Hemodynamics in Patient-Specific Model of Type B Aortic Dissection** SB^{3}C2015-608 pg.889
Joav Birjiniuk[1], Jean M. Ruddy[2], Mark Young[3], Lucas H. Timmins[1], Ravi K. Veeraswamy[4], David N. Ku[5], [1]*Georgia Institute of Technology and Emory University School of Medicine, Atlanta, GA, United States,*[2]*Medical University of South Carolina, Charleston, SC, United States,*[3]*Medtronic, Inc., Santa Rosa, CA, United States,*[4]*Emory University School of Medicine, Atlanta, GA, United States.*[5]*Georgia Institute of Technology, Atlanta, GA, United States*

10:30AM **Assessment And Quantification Of Transitional Flow In Intracranial Aneurysms - Highly Resolved Simulations Below The Kolmogorov Scales** SB^{3}C2015-588 pg. 855
Kartik Jain[1,2], Kristian Valen-Sendstad[2], Sabine Roller[1], Kent-Andre Mardal[2,3], [1]*University of Siegen, Siegen, Germany,*[2]*Simula Research Laboratory, Oslo, Norway,*[3]*University of Oslo, Oslo, Norway*

THURSDAY, JUNE 18	9:15am - 10:45am

Microstructure of Aneurysms — Magpie

Session Chair: Alison Marsden, *University of California San Diego, San Diego, CA, United States*
Session Co-Chair: Hai-Chao Han, *University of Texas at San Antonio, San Antonio, TX, United States*

9:15AM **Stiffer Arterial Wall Enhances Aortic Aneurysm Formation In A Mouse Model Via Elastase Infusion** SB^{3}C2015-537 pg.758
Zhijie Wang, Stephanie Morgan, Mark Golob, Zhenjie Liu, Bo Liu, Naomi C. Chesler, *University of Wisconsin - Madison, Madison, WI, United States*

9:30AM **Bayesian Calibration Of A Growth And Remodeling Computational Model Of Abdominal Aortic Aneurysms** SB^{3}C2015-564 pg.809
Seungik Baek, Liangliang Zhang, Sajjad SeyedSalehi, Jongeun Choi, Chae Young Lim, Tapabrata Maiti, *Michigan State University, East Lansing, MI, United States*

9:45AM **Role Of Aneurysm On Failure Properties Of "Radially-oriented" Collagen Fibers In Human Ascending Thoracic Aortic Media** SB^{3}C2015-572 pg.823
Siladitya Pal, Spandan Maiti, Alkiviadis Tsamis, Julie A. Phillippi, Thomas G. Gleason, David A. Vorp, *University of Pittsburgh, Pittsburgh, PA, United States*

10:00AM **Elastin Production Slows Aneurysm Enlargement in a Constrained Mixture Models of Aneurysm Growth and Remodelling** SB^{3}C2015-404 pg.517
Kory J. Blose, Justin S. Weinbaum, Anne M. Robertson, David A. Vorp, *University of Pittsburgh, Pittsburgh, PA, United States*

10:15AM **Novel Image-based Analysis Of Patient-specific Abdominal Aortic Aneurysm Wall Stress Using A Membrane Mechanics Model** SB^{3}C2015-1043 pg.1060
Mirunalini Thirugnanasambandam[1], Prahlad G. Menon[1,2], Stéphane Avril[3], Ender A. Finol[1], [1]*University of Texas at San Antonio, San Antonio, TX, United States,*[2]*Sun Yat-sen University - Carnegie Mellon University Joint Institute of Engineering, Pittsburgh, PA, United States,*[3]*Ecole Nationale Supérieure des Mines, Saint-Etienne, France*

10:30AM **Heterogeneous Elastic Properties Of Ascending Thoracic Aneurysms** SB^{3}C2015-122 pg.46
Jia Lu[1], Yuanming Luo[1], Frances M. Davis[2], Stéphane Avril[2], [1]*The University of Iowa, Iowa City, IA, United States,*[2]*Ecole Nationale Suprieure des Mines de Saint-Etienne, St. Étienne, France*

THURSDAY, JUNE 18	9:15am - 10:45am

Musculoskeletal Tissue Engineering - Matrices and Interfaces — Maybird

Session Chair: Matt Fisher, *North Carolina State University, Raleigh, NC, United States*
Session Co-Chair: Mariana Kersh, *University of Illinois, Urbana, IL, United States*

9:15AM **Perfusion Decellularized Skeletal Muscle As A Scaffold For The Repair Of Volumetric Muscle Loss** SB^{3}C2015-432 pg.574
Ben Kasukonis, John Kim, Tyrone Washington, **Jeffrey C. Wolchok**, *University of Arkansas, Fayetteville, AR, United States*

9:30AM **Fabrication Of Dense Porous Aligned Collagen Scaffolds Using 2D Plastic Compression And Porogens** SB³C2015-1034 pg.1042
Shawn Reese, Jared Zitnay, Jeffrey Weiss, *University of Utah, Salt Lake City, UT, United States*

9:45AM **Wrinkled, Wavelength-Tunable Graphene-Based Surface Topographies for Directing Cell Alignment and Morphology** SB³C2015-295 pg.337
Daniel F. Tonderys, Susan Leggett, Zhongying Wang, *Brown University, Providence, RI, United States*

10:00AM **Network Stiffening of Nanofiber Scaffolds by Mineral** SB³C2015-402 pg.513
Justin H. Lipner[1], John J. Boyle[1], Victor Birman[2], Guy M. Genin[1], Stavros Thomopoulos[1], [1]*Washington University in St. Louis, Saint Louis, MO, United States,*[2]*Engineering Education Center, Missouri University of Science and Technology, Saint Louis, MO, United States*

10:15AM **Micrometer-scale Mechanical Properties Of The Tendon-to-bone Attachment** SB³C2015-594 pg.865
Alix C. Deymier-Black[1], Yiran An[2], Andrea G. Schwartz[1], Guy M. Genin[1], Stavros Thomopoulos[1], Asa H. Barber[2], [1]*Washington University in St Louis, St Louis, MO, United States,*[2]*Queen Mary University of London, London, United Kingdom*

10:30AM **Local Microenvironment Response of Chondrocytes in Gradient Collagen Matrices** SB³C2015-639 pg.947
Tyler Novak, Benjamin Seelbinder, Celina M. Twitchell, Sherry L. Voytik-Harbin, Corey P. Neu, *Purdue University, Lafayette, IN, United States*

THURSDAY, JUNE 18	9:15am - 10:45am

Transport in Tissue and Tumor Microenvironments — Golden Cliff / Eagle's Nest

Session Chair: John Pearce, *The University of Texas at Austin, Austin, TX, United States*
Session Co-Chair: Aili Zhang, *Shanghai Jiao Tong University, Shanghai, China*

9:15AM **Temperature Control of Hydrogel Delivered Through Endoscopic Needle as Potential Treatment for Pancreatic Cancer** SB³C2015-477 pg.654
Tom M. Merrill[1], Jennifer Mitchell[2], Denise Merrill[2], Matthew Short[1], [1]*Rowan University, Glassboro, NJ, United States,*[2]*FocalCool, LLC, Mullica Hill, NJ, United States*

9:30AM **Shear Activated Nanoparticle Aggregate Delivery of Tissue Plasminogen Activator with Adjunctive Endovascular Bypass for Revascularization in Acute Ischemic Stroke** SB³C2015-303 pg.353
Matthew J. Gounis[1], Netanel Korin[2], Miklos G. Marosfoi[1], Oktay Uzun[2], Erin T. Langan[1], Anne-Laure Papa[2], Olivia W. Brooks[1], Chris Johnson[2], Ajit S. Puri[1], Deen Bhatta[2], Ajay K. Wakhloo[1], Donald E. Ingber[2], [1]*University of Massachusetts Medical School, Worcester, MA, United States,*[2]*Wyss Institute for Biologically Inspired Engineering, Boston, MA, United States*

9:45AM **Characterization Of Endothelial Cell Permeability On A Biomimetic Blood Vessel Platform** SB³C2015-217 pg.214
Yaling Liu, Antony Thomas, Christopher Uhl, *Lehigh University, Bethlehem, PA, United States*

10:00AM **Salmon Thrombin Inhibits Endothelial Barrier Disruption Through its Activation of Protein C** SB³C2015-273 pg.301
Jenell Smith, Peter Galie, David Slochower, Paul Janmey, Beth Winkelstein, *University of Pennsylvania, Philadelphia, PA, United States*

10:15AM **An In Vitro Microfluidic Tumor Model to Mimic Ductal Carcinoma In Situ** SB³C2015-673 pg.1011
Victoria Noe-Kim, Altug Ozcelikkale, Bumsoo Han, *Purdue University, West Lafayette, IN, United States*

10:30AM **Novel 3D Co-culture Assay Reveals Cancer Cell Migration Driven By Matrix-dependent Autonomous Formation Of CXCL12 Gradients** SB³C2015-532 pg.750
Christopher Moraes[1], Taisuke Kojima[2], Stephen P. Cavnar[2], Gary D. Luker[2], Shuichi Takayama[2], [1]*McGill University, Montreal, QC, Canada,*[2]*University of Michigan, Ann Arbor, MI, United States*

THURSDAY, JUNE 18	9:15am - 10:45am

Cartilage and Intervertebral Disc — Primrose B

Session Chair: Clark Hung, *Columbia University, New York, NY, United States*
Session Co-Chair: Corey Neu, *Purdue University, West Lafayette, IN, United States*

9:15AM **Quantitative Analysis Of Raman Spectra For Assessment Of Crosslink Concentrations Toward Diagnosing Early Osteoarthritis** SB³C2015-452 pg.612
Chao Wang, **Krista M. Durney**, Jonathan L. Kuo, Jack R. Norton, Gerard A. Ateshian, Sinisa Vukelic, *Columbia University, New York, NY, United States*

9:30AM **An In-situ Cartilage Characterization Method With Reduced Time And Computation Demands: Comparison Against Linear And Nonlinear Biphasic Theories** SB³C2015-226 pg.228
David L. Burris, Axel C. Moore, Xingyu Chen, Brandon K. Zimmerman, X. Lucas Lu, *University of Delaware, Newark, DE, United States*

9:45AM **Maintenance of F-Spondin (Spon1) Knockout Phenotype in 3D Cultured Murine Chondrocytes** SB³C2015-1050 pg.1074
Eben G. Estell[1], Qing Yang[2], Andrea R. Tan[1], Mukundan Attur[2], Steven B. Abramson[2], Clark T. Hung[1], [1]*Columbia University, New York, NY, United States,* [2]*New York University School of Medicine and Langone Medical Center, New York, NY, United States*

10:00AM **Impact-induced Fissuring In Articular Cartilage: Bulk And Tissue-scale Mechanics** SB³C2015-663 pg. 993
Corinne R. Henak, Lena R. Bartell, Itai Cohen, Lawrence J. Bonassar, *Cornell University, Ithaca, NY, United States*

10:15AM **Modeling The Intervertebral Discs Repair With Cellular Therapies** SB³C2015-563 pg.807
Qiaoqiao Zhu, Xin Gao, Weiyong Gu, *University of Miami, Coral Gables, FL, United States*

10:30AM **The Contribution of Collagen Fibers in Human Cartilage Endplate Mechanics** SB³C2015-660 pg.987
John DeLucca[1], Mary Boggs[1], Edward Vresilovic[2], Randall Duncan[1], Dawn Elliott[1], [1]*University of Delaware, Newark, DE, United States,* [2]*Pennsylvania State University, Hershey, PA, United States*

THURSDAY, JUNE 18	11:00am - 12:30pm

Biomechanics of Microcirculation: From Cells to Networks and Changes During Disease States (joint with JSME) — Primrose A

Session Chair: Ken-ichi Tsubota, *Chiba University, Chiba, Japan*
Session Co-Chair: Walter Lee Murfee, *Tulane University, New Orleans, LA, United States*

11:00AM **Molecular Dynamics Simulations of Lipid Bilayer Failure under Mechanical Stresses: Toward Understanding Mechanical Hemolysis** SB³C2015-214 pg.208
Kenichiro Koshiyama, *Osaka University, Toyonaka, Japan*

11:15AM **A Computational Study Of Red Blood Cells In Stokes Flow: From Single Cellular Mechanics To Suspension Rheology** SB³C2015-256 pg.277
Toshihiro Omori, Yohsuke Imai, Takami Yamaguchi, Takuji Ishikawa, *Tohoku University, Sendai, Japan*

11:30AM **Microsystems Engineering for Bridging Angiogenesis with Mechanopathology** SB³C2015-28 pg.8
Jonathan W. Song, *The Ohio State University, Columbus, OH, United States*

11:45AM **Autodigestion And Its Manifestations In The Microcirculation.** SB³C2015-292 pg.331
Geert W. Schmid-Schoenbein, Frank De Iano, Angelina Altshuler, Erik Kistler, Michael Richter, *University of California, San Diego, La Jolla, CA, United States*

12:00PM **Quantifying Hemodynamics And Wall Mechanics In Patient-specific Coronary Artery Bypass Grafts** SB³C2015-1119
Abhay B. Ramachandra, Andrew Kahn, Alison L. Marsden, *University of California San Diego, La Jolla, CA, United States pg.1201*

12:15PM **The Effects of Aging on Microvascular Network Resistance and Flow Heterogeneity in the Rat Mesentery** SB³C2015-644 pg.957
David C. Sloas, Scott A. Stewart, Richard S. Sweat, Walter L. Murfee, *Tulane University, New Orleans, LA, United States*

THURSDAY, JUNE 18	11:00am - 12:30pm

Human Dynamics — Superior

Session Chair: **Rita M. Patterson**, *University of North Texas Health Science Center, Fort Worth, TX, United States*
Session Co-Chair: **Tom Gardner**, *Columbia University, New York, NY, United States*

11:00AM **Lumbar-pelvic Coordination Differences Between Novice And Experienced Lifters In Repetitive Lifting** SB³C2015-1029 pg.1034
Timothy D. Craig[1], Alice E. Riley[1], Sandra A. Billinger[2], Neena K. Sharma[2], **Sara E. Wilson**[1], [1]*University of Kansas, Lawrence, KS, United States,*[2]*University of Kansas, Kansas City, KS, United States*

11:15AM **Design and Simulation of an Ankle Foot Simulator** SB³C2015-1171 pg.1291
Jonathan J. Miller, *University of Utah, Salt Lake City, UT, United States*

11:30AM **Improved Detection Of Dynamic Balance Using Normalized COM-COP Inclination Angular Jerk During A Golf Swing** SB³C2015-23 pg.3
Ahnryul Choi[1], Joung Hwan Mun[2], [1]*The University of Texas Health Science Center at Houston, Houston, TX, United States,*[2]*Sungkyunkwan University, Suwon, Korea*

11:45AM **Investigating the Effect of Ball Impact Location** SB³C2015-117 pg.42
Ainhoa Iglesias-Díaz, **Martin L. Tanaka**, Paul M. Yanik, Aaron K. Ball, *Western Carolina University, Cullowhee, NC, United States*

12:00PM **Designing Biomechanical Models of the Ankle: How Many Degrees of Freedom are Necessary to Reflect In Vivo Joint Kinematics?** SB³C2015-237 pg.246
Jennifer A. Nichols, Koren E. Roach, Charles L. Saltzman, Andrew E. Anderson, *University of Utah, Salt Lake City, UT, United States*

12:15PM **Obesity and Age Affect Trip Outcome and Severity Following a Laboratory-Induced Trip** SB³C2015-510 pg.706
Christina M. R. Garman[1], Maury A. Nussbaum[1], Michael L. Madigan[2], [1]*Virginia Tech, Blacksburg, VA, United States,*[2]*Texas A & M University, College Station, TX, United States*

THURSDAY, JUNE 18	11:00am - 12:30pm

Thrombus Prediction — Wasatch

Session Co-Chair: **Shawn Shadden**, *UC Berkeley, Berkeley, CA, United States*
Session Co-Chair: **Dalin Tang**, *WPI, Worcester, MA, United States*

11:00AM **Activated Expression And Shedding Of Platelet Glycoprotein IIb/IIIa Under Non-physiological Shear Stress: A Paradoxical Effect** SB³C2015-280 pg.313
Zengsheng Chen, Jun Ding, Zhongjun J Wu, *University of Louisville School of Medicine, Louisville, KY, United States*

11:15AM **Development of A Predictive Computational Model For Device-Induced Thrombosis** SB³C2015-243 pg.257
Joshua O. Taylor[1,2], Richard S. Meyer[2], Steven Deutsch[2], **Keefe Manning**[1,3], [1]*The Pennsylvania State University, University Park, PA, United States,*[2]*Applied Research Laboratory, State College, PA, United States,*[3]*Penn State Hershey Medical Center, Hershey, PA, United States*

11:30AM **Thrombotic Risk Assessment Using Transluminal Attenuation Gradient and Computational Modelling in Kawasaki Disease Patients with Coronary Artery Aneurysms** SB³C2015-1082 pg.1130
Noelia Grande Gutierrez[1], Andrew M. Kahn[2], Olga Shirinsky[3], Nina V. Gagarina[3], Galina A. Lyskina[3], Ryuji Fukazawa[4], Shunichi Ogawa[4], Jane C. Burns[2], Alison L. Marsden[1], [1]*University of California San Diego, La Jolla, CA, United States,*[2]*School of Medicine, University of California San Diego, La Jolla, CA, United States,*[3]*Sechenov First Moscow State University, Moscow, Russian Federation,*[4]*Nippon Medical School Hospital, Tokyo, Japan*

11:45AM **Hemodynamic Prediction of Thrombus Prone Regions in Abdominal Aortic Aneurysms** SB³C2015-455 pg.618
Konstantinos Tzirakis[1], **Eleni Metaxa**[1], Nikolaos Kontopodis[2], Christos V. Ioannou[2], Yannis Papaharilaou[1], [1]*Foundation of Research and Technology - Hellas, Heraklion, Greece,*[2]*University of Crete, Heraklion, Greece*

12:00PM **Thrombogenic Properties Of A Superhydrophobic Surface For Use As A Prosthetic Heart Valve Material** SB³C2015-652 pg.971
David L. Bark, Jr., Sanli Movafaghi, Brandon L. Moore, Arun K. Kota, Ketul C. Popat, Lakshmi P. Dasi, *Colorado State University, Fort Collins, CO, United States*

12:15PM **The Impact of VAD Surgical Implantation Configurations on its Thrombogenic Potential** SB³C2015-548 pg.780
Wei C. Chiu[1], Allison J. McLarty[1], Shmuel Einav[1], Marvin J. Slepian[2], Danny Bluestein[1], [1]*Stony Brook University, Stony Brook, NY, United States,*[2]*Sarver Heart Center, Tucson, AZ, United States*

THURSDAY, JUNE 18	11:00am - 12:30pm

Heart Valve Structure and Function — Magpie

Session Chair: Michael Sacks, *University of Texas, Austin, TX, United States*
Session Co-Chair: Shamik Bhattacharya, *St.Mary's University, San Antonio, TX, United States*

11:00AM **Age-dependent Changes In Stress And Strain In The Human Native Heart Valve And Their Relation With Collagen Remodeling** SB³C2015-523 pg.731
Pim J. A. Oomen[1], Sandra Loerakker[1], Daphne van Geemen[1], Jan Neggers[1], Marie-José T. H. Goumans[2], Antoon J. van den Bogaerdt[3], Ad J. J. C. Bogers[3], Carlijn V. C. Bouten[1], Frank P. T. Baaijens[1], [1]*Eindhoven University of Technology, Eindhoven, Netherlands,*[2]*Leiden University Medical Center, Leiden, Netherlands,*[3]*Erasmus Medical Center, Rotterdam, Netherlands*

11:15AM **Effects Of Leaflet Microstructure And Constitutive Model On The Closing Behavior Of The Mitral Valve** SB³C2015-339 pg.409
Chung-Hao Lee[1], Jean-Pierre Rabbah[2], Ajit P. Yoganathan[2], Robert C. Gorman[3], Joseph H. Gorman[3], Michael S. Sacks[1], [1]*The University of Texas at Austin, Austin, TX, United States,*[2]*Georgia Institute of Technology, Atlanta, GA, United States,*[3]*University of Pennsylvania, Philadelphia, PA, United States*

11:30AM **Capturing Detailed 3D Mitral Valve Geometry For Computational Valve Modeling** SB³C2015-462 pg.626
Charles H. Bloodworth, Eric L. Pierce, Thomas F. Easley, Milan Toma, Morten O. Jensen, Ajit P. Yoganathan, *Wallace H Coulter Department of Biomedical Engineering, Georgia Tech and Emory University, Atlanta, GA, United States*

11:45AM **True 3D Stresses in Heart Valve Leaflets** SB³C2015-1032 pg.1038
Bruno V. Rego, Michael S. Sacks, *The University of Texas at Austin, Austin, TX, United States*

12:00PM **A Framework For Parameter Estimation Of Heart Valves Using Inverse-Modeling Approach** SB³C2015-603 pg.881
Ankush Aggarwal, Michael S. Sacks, *University of Texas at Austin, Austin, TX, United States*

12:15PM **Immersogeometric Fluid-Structure Interaction Analysis of Bioprosthetic Heart Valves: Realistic Material Modeling and Experimental Validation** SB³C2015-1150 pg.1255
David Kamensky[1], Ming-Chen Hsu[2], Michael S. Sacks[1], Thomas J. R. Hughes[1], [1]*University of Texas at Austin, Austin, TX, United States,*[2]*Iowa State University, Ames, IA, United States*

THURSDAY, JUNE 18	11:00am - 12:30pm

Nano-, Micro-, and Multi-Scale Mechanics of Cells and Tissues — Maybird

Session Chair: Nadeen Chahine, *Feinstein Institute of Medical Research, Manhasset, NY, United States*
Session Co-Chair: Grace D. O'Connell, *University of California, Berkeley, Berkeley, CA, United States*

11:00AM **Micromechanical Heterogeneity And Anisotropy Of The Meniscus Extracellular Matrix** SB³C2015-128 pg.58
Qing Li[1], Feini Qu[2,3], Biao Han[1], Robert L. Mauck[2,3], **Lin Han**[1], [1]*Drexel University, Philadelphia, PA, United States,*[2]*University of Pennsylvania, Philadelphia, PA, United States,*[3]*Philadelphia Veterans Administration Medical Center, Philadelphia, PA, United States*

11:15AM **A Micromechanical Viscoelastic Computational Model Incorporating Progressive Detachment Of The Tau Proteins Predicting Microtubule Breaking Following Axonal Injury.** SB[3]C2015-153 pg. 96
Hossein Ahmadzadeh[1], Douglas H. Smith[2], Vivek B. Shenoy[1], [1]*Department of Materials Science and Engineering, University of Pennsylvania, Philadelphia, PA, United States,*[2]*Penn Center for Brain Injury and Repair and Department of Neurosurgery, University of Pennsylvania, Philadelphia, PA, United States*

11:30AM **Layer-specific Mitral Valve Interstitial Cell Deformations Under Physiological Loading** SB[3]C2015-340 pg. 411
Chung-Hao Lee[1], Christopher A. Carruthers[2], Ayoub Salma[1], Robert C. Gorman[3], Joseph H. Gorman[3], Michael S. Sacks[1], [1]*The University of Texas at Austin, Austin, TX, United States,*[2]*Medtronic, Minneapolis, MN, United States,*[3]*University of Pennsylvania, Philadelphia, PA, United States*

11:45AM **Relaxation of Simulated Viscoelastic Fiber Networks** SB[3]C2015-377 pg.467
Rohit Y. Dhume, Victor H. Barocas, *University of Minnesota, Minneapolis, MN, United States*

12:00PM **A Multi-Scale Approach in Analyzing Fluid/Solute Flow in Mechanically Loaded Bone** SB[3]C2015-499 pg.686
Lixia Fan[1], Shaopeng Pei[2], Xiaohan Lai[2], Xin Lu[2], **Liyun Wang**[2], [1]*Nanjing University of Science and Technology, Nanjing, China,*[2]*University of Delaware, Newark, DE, United States*

12:15PM **Blocking Inflammation Protects Against Alterations In Cell Biomecahnics, Morphology And Cytoskeleton** SB[3]C2015-1155 pg.1265
Timothy Jacobsen, Paula Hernandez, Victoria Wei, Nadeen O. Chahine, *Feinstein Institute for Medical Research, Manhasset, NY, United States*

THURSDAY, JUNE 18	11:00am - 12:30pm

Cryotherapy and Hyperthermia: 70+ Birthday Celebration for Prof. Avraham Shitzer

Golden Cliff / Eagle's Nest

Session Chair: John Bischof, *University of Minnesota at Minneapolis, Minneapolis, MN, United States*
Session Co-Chair: Liang Z. Zhu, *University of Maryland Baltimore County, Baltimore, MD, United States*

11:00AM **A Novel Method For Positioning And Operating Cryo-surgical Probes In 2-D Convex Target Areas Based On Super-positioning Of Unit Circles** SB[3]C2015-25 pg. 7
Avraham Shitzer, *Technion, Isreal Inst. Tech., Haifa, Israel*

11:15AM **Meeting The Need For Improved Monitoring Of Cryotherapy In Thin Cardiovascular Tissues With A New Micro-thermal Sensor** SB[3]C2015-1137 pg.1233
Harishankar Natesan[1], Jeunghwan Choi[1], Wyatt Hodges[2], Sean Lubner[2], Chris Dames[2], John Bischof[1], [1]*University of Minnesota, Minneapolis, MN, United States,*[2]*University of California, Berkeley, CA, United States*

11:45AM **Determination of the Convective Heat Transfer Coefficients on the Outer Surfaces of the Cryovivals Plunged in Liquid Nitrogen and 37°C Water Bath** SB[3]C2015-433 pg. 576
Tao Wang, Gang Zhao, *University of Science and Technology of China, Hefei, China*

11:45AM **Correcting Arrhenius Models for Application to Cell Death Processes Involving Intrinsic Protein Cascades** SB[3]C2015-142 pg.78
John Pearce, *The University of Texas at Austin, Austin, TX, United States*

12:00AM **Tumor Engineering to Elucidate the Effect of Mild Hyperthermia on the Transport of Single Walled Carbon Nanohorns in the Tumor Microenvironment** SB[3]C2015-586 pg.851
Matthew R. DeWitt[1], M. Nichole Rylander[2], [1]*Virginia Tech, Blacksburg, VA, United States,*[2]*The University of Texas at Austin, Austin, TX, United States*

12:15AM **Strategies Of Injecting Ferrofluid Into Tumors To Achieve Repeatable Nanoparticle Deposition And Thermal Dosage In Mangetic Nanoparticle Hyperthermia** SB[3]C2015-286 pg.323
Alexander LeBrun, Charles Bieberich, Ronghui Ma, Liang Zhu, *University of Maryland Baltimore County, Baltimore, MD, United States*

THURSDAY, JUNE 18	11:00am - 12:30pm

Cartilage Mechanics and Repair (joint with JSME) — Primrose B

Session Chair: **Hiromichi Fujie**, *Tokyo Metropolitan University, Tokyo, Japan*
Session Co-Chair: **Jennifer Wayne**, *Virginia Commonwealth University, Richmond, VA, United States*

11:00AM **Anisotropic Hydraulic Permeability In Articular Cartilage And Menisci: A Direct Measurement Using Biopsy Punches** SB^{3}C2015-420 pg.549
Hiromichi Fujie, Ryosuke Nakamura, *Tokyo Metropolitan University, Tokyo, Japan*

11:15AM **The Influence of Cartilage Morphology and Elasticity on Tibiofemoral Contact Pressures During Walking** SB^{3}C2015-667 pg.1001
Colin R. Smith, Rachel L. Lenhart, Mike F. Vignos, Jarred Kaiser, Darryl G. Thelen, *University of Wisconsin-Madison, Madison, WI, United States*

11:30AM **The Effect of Impact and Genipin Crosslinking on the Friction and Wear of Articular Cartilage** SB^{3}C2015-24 pg.5
Craig M. Bonitsky[1], Michael Selep[1], Megan E. McGann[1], Timothy C. Ovaert[1], Stephen B. Trippel[2], Diane R. Wagner[1], [1]*University of Notre Dame, Notre Dame, IN, United States,*[2]*Indiana University, Indianapolis, IN, United States*

11:45AM **Development Of Artificial Cartilage Using Two Phase Collagen/Scaffold With Mesenchymal Stem Cells** SB^{3}C2015-268 pg.297
Mitsugu Todo[1], Yusuke Nakamuta[1], Takaaki Arahira[2], [1]*Kyushu University, Kasuga, Japan,*[2]*Fukuoka Dental College, Fukuoka, Japan*

12:00PM **Evaluation Of Friction And Wear Properties Of Poly(vinyl Alcohol) Hydrogels As Artificial Articular Cartilage** SB^{3}C2015-1078 pg.1122
Seido Yarimitsu[1], Ayumi Yoshida[1], Kazuhiro Nakashima[1], Teruo Murakami[1], Saori Sasaki[2], Atsushi Suzuki[2], [1]*Kyushu University, Fukuoka, Japan,*[2]*Yokohama National University, Yokohama, Japan*

12:15PM **Relationship Of Wear Particles Of Poly(vinyl Alcohol) Hydrogel And Immune Response Of Macrophage** SB^{3}C2015-1079 pg.1124
Seiji Omata, Yoshinori Sawae, Teruo Murakami, *Kyushu University, Fukuoka, Japan*

THURSDAY, JUNE 18	12:30pm - 2:30pm

Poster Session I — BS Competition - Biofluid Mechanics — Event Center Tent

1 **Predicting Regions Of Low Wall Shear Stress In The Carotid Artery Bifurcation** SB^{3}C2015-419 pg.547
Haley M. King, Emma J. Lichtenfels, Christian R. Fahrenbruck, Brandon Moore, Lakshmi P. Dasi, *Colorado State University, Fort Collins, CO, United States*

THURSDAY, JUNE 18	12:30pm - 2:30pm

Poster Session I — BS Competition - Biotransport — Event Center Tent

2 **Generating Tumor Models From MicroCT Scan Images For Simulating Temperature Elevation During Magnetic Nanoparticle Hyperthermia** SB^{3}C2015-300 pg.347
Andrew S. Lee, Alexander LeBrun, Ronghui Ma, Liang Zhu, *University of Maryland Baltimore County, Baltimore, MD, United States*

3 **Evaluating Accuracy Of The Algorithm/formula Used In Store Brand Digital Thermometers To Predict Body Temperature Based On The First Few Seconds of The Temperature Measurements** SB^{3}C2015-507 pg.700
Amirreza Saharkhiz[1], Oleg Vesnovsky[2], Jon P. Casamento[2], Laurence W. Grossman[2], L. D. Timmie Topoleski[1], Liang Zhu[1], [1]*University of Maryland Baltimore County, Baltimore, MD, United States,*[2]*U.S. Food and Drug Administration, Silver Spring, MD, United States*

4 **Development Of A Tissue Phantom To Mimic The Thermal Environment Of A Human Arm To Test Digital Thermometers** SB³C2015-501 pg.690
Peter Dillon[1], Oleg Vesnovsky[2], L. D. Timmie Topoleski[1], Jon P. Casamento[2], Laurence W. Grossman[2], Liang Zhu[1], [1]*University of Maryland Baltimore County, Baltimore, MD, United States,*[2]*U.S. Food and Drug Administration, Silver Spring, MD, United States*

5 **Quantitative Visualization of Drug Response of Breast Cancer Cells Within a Three-Dimensional Extracellular Matrix** SB³C2015-234 pg.240
Brett S. Klosterhoff[1], Kyeonggon Shin[1], J. Craig Dutton[2], Bumsoo Han[1], [1]*Purdue University, West Lafayette, IN, United States,*[2]*University of Illinois at Urbana-Champaign, Urbana, IL, United States*

THURSDAY, JUNE 18	12:30pm - 2:30pm

Poster Session I **BS Competition - Cellular and Tissue Engineering** **Event Center Tent**

6 **Fabrication and Evaluation of Biodegradable Tissue Scaffolds for Osteochondral Defect Repair** SB³C2015-1154 pg.1263
Carly R. Garrow, Andrew J. Polk, Ferris M. Pfeiffer, *University of Missouri - Columbia, Columbia, MO, United States*

THURSDAY, JUNE 18	12:30pm - 2:30pm

Poster Session I **BS Competition - Design and Devices** **Event Center Tent**

7 **Wear Simulator for Canine Total Hip Replacements** SB³C2015-1174 pg.1297
Kevin J. Warburton[1], Evan Rust[1], Jeff Brourman[2], Trevor J. Lujan[1], [1]*Boise State University, Boise, ID, United States,*[2]*Veterinary Surgeon WestVet Animal Emergency & Specialty Center, Garden City, ID, United States*

8 **Material Property Testing of Carboxymethylated Hyaluronic Acid Hydrogel Polymer** SB³C2015-1186 pg.1315
McKenna Drysdale[1], Hee-Kyoung Lee[1,2], Barbara Wirostko[2], Brittany Coats[1], [1]*University of Utah, Salt Lake City, UT, United States,*[2]*Jade Therapeutics, Salt Lake City, UT, United States*

9 **Design of a 3-D Bioreactor for Simulation of Cerebrospinal Fluid Flow in the Third Ventricle and Aqueduct of Sylvius** SB³C2015-1164 pg.1281
Michael J. Majcher, Matthew R. Dailey, David P. Lemmer, Joseph T. Havrilak, Nic Leipzig, Bryn A. Martin, *University of Akron, Akron, OH, United States*

11 **A New Method For Determining Cross-sectional Shape And Area Of Soft Tissues Using 3d Laser Scanning** SB³C2015-176 pg.138
Yaniv Michaeli[1,2], Richard E. Debski[2], [1]*ORT Braude College, Nahariya, Israel,*[2]*University of Pittsburgh, Pittsburgh, PA, United States*

12 **Rapid Quantification Of Femoral Head Geometry From Magnetic Resonance Imaging Of Femoroacetabular Impingement** SB³C2015-643 pg. 955
Haley Ehlers[1], Michael Roberts[1], Garry Gold[2], Saikat Pal[1], [1]*California Polytechnic State University, San Luis Obispo, CA, United States,*[2]*Stanford University, Stanford, CA, United States*

13 **Design and Implementation of an Instrumented Pedal for Cycling Biomechanics Research** SB³C2015-1033 pg.1040
Luke I. Kraemer, Juan D. Gutierrez-Franco, Jake E. Deschamps, Karim C. Dudum, Eshan M. Dandekar, Scott J. Hazelwood, Hemanth V. Porumamilla, Stephen M. Klisch, *California Polytechnic State University, San Luis Obispo, CA, United States*

THURSDAY, JUNE 18	12:30pm - 2:30pm

Poster Session I | **BS Competition - Human Dynamics** | **Event Center Tent**

14 **Can Measured Muscle Synergies Reconstruct Unmeasured Muscle Excitations?** SB³C2015-573 pg.825
Nicholas A. Bianco, Carolynn Patten, Benjamin J. Fregly, *University of Florida, Gainesville, FL, United States*

15 **Using OpenSim to Predict Knee Joint Moments During Cycling** SB³C2015-1023 pg.1024
Karim C. Dudum, Jake E. Deschamps, Juan D. Gutierrez-Franco, Luke I. Kraemer, Alejandro M. Gonzalez-Smith, Eshan M. Dandekar, Scott J. Hazelwood, Stephen M. Klisch, *California Polytechnic State University, San Luis Obispo, CA, United States*

THURSDAY, JUNE 18	12:30pm - 2:30pm

Poster Session I | **BS Competition - Injury** | **Event Center Tent**

16 **The Temporal Change in Protein Biomarkers of Vitreous Humor following Blast Trauma of the Visual System** SB³C2015-1098 pg.1159
Justin A. Jones, Daniel F. Shedd, Brittany Coats, *University of Utah, Salt Lake City, UT, United States*

17 **Effect of Sclerostin Antibody on Tendon-to-Bone Healing in a Rotator Cuff Animal Model** SB³C2015-1051 pg.1076
Shivam A. Shah[1], Ioannis Kormpakis[1], Necat Havlioglu[2], Michael S. Ominsky[3], Leesa M. Galatz[1], Stavros Thomopoulos[1], [1]*Washington University of St. Louis, Saint Louis, MO, United States,*[2]*John Cochran VA Medical Center, Saint Louis, MO, United States,*[3]*Amgen Inc., Thousand Oaks, CA, United States*

THURSDAY, JUNE 18	12:30pm - 2:30pm

Poster Session I | **BS Competition - Tissue Mechanics** | **Event Center Tent**

18 **Comparison of Pulmonary Artery Material Properties between a Congenital Heart Disease Patient and a Normal Subject using Cardiac Magnetic Resonance: A Feasibility Study** SB³C2015-606 pg.887
Rajit Banerjee[1], Gavin A. D'Souza[2], Namheon Lee[3], Michael D. Taylor[3], [1]*University of Toledo, Toledo, OH, United States,*[2]*University of Cincinnati, Cincinnati, OH, United States,*[3]*Cincinnati Children's Hospital Medical Center, Cincinnati, OH, United States*

19 **Finite Element Analysis of the Biomechanical Consequences of Schmorl's Nodes** SB³C2015-1018 pg.1014
Boston B. Barham, Anton E. Bowden, *Brigham Young University, Provo, UT, United States*

20 **A Mechanical Model For Cortical Folding During Brain Development** SB³C2015-575 pg.829
Nithya Vijayakumar, Maria Holland, Ellen Kuhl, *Stanford University, Stanford, CA, United States*

THURSDAY, JUNE 18	12:30pm - 2:30pm

Poster Session I | **Multi-scale Modeling in Biotransport** | **Event Center Tent**

21 **Comparison Of MRI-measured Tracer Transport In The Rat Brain With Computational Model Prediction** SB³C2015-1149 pg.1253
Wei Dai, Garrett W. Astary, Aditya K. Kasinadhuni, Paul R. Carney, Thomas H. Mareci, Malisa Sarntinoranont, *University of Florida, Gainesville, FL, United States*

22 **A Reduced-Dimensional Transport Model for Thrombogenic Species in Large Arteries** SB³C2015-195 pg.174
Kirk B. Hansen, Shawn C. Shadden, *University of California, Berkeley, Berkeley, CA, United States*

23 **Low-Concentration Salmonella Detection Using Orbiting Magnetic Microbeads in a Continuous-Flow Microfluidic Device** SB³C2015-593 pg.863
Matthew S. Ballard, Drew L. Owen, Zachary G. Mills, Srinivas K. G. Hanasoge, Peter J. Hesketh, Alexander Alexeev, *Georgia Institute of Technology, Atlanta, GA, United States*

24 **Computational Multiphysics Model of Clot Lysis in a Completely Occluded Stenotic Artery** SB³C2015-139 pg.74
Andris Piebalgs, Xiao Y. Xu, *Imperial College London, London, United Kingdom*

25 **Effect of Intervertebral Disc Size on Nutrient Distributions: A Finite Element Analysis with Implications for In Vivo Models** SB³C2015-1077 pg.1120
Alicia R. Jackson, *University of Miami, Coral Gables, FL, United States*

26 **Reaction/Diffusion Modelling of AtzA Biocatalyst Encapsulated in a Silica Gel Matrix** SB³C2015-175 pg.136
Baris Ragip Mutlu, *University of Minnesota, Minneapolis, MN, United States*

27 **Finite Element Model of Mixed Porohyperelastic Transport in an Axisymmetric Porcine Coronary Artery** SB³C2015-184 pg.154
Michelle A. H. Armstrong, Bruce R. Simon, Jonathan P. Vande Geest, *University of Arizona, Tucson, AZ, United States*

THURSDAY, JUNE 18	12:30pm - 2:30pm

Poster Session I — Design, Dynamics and Rehab — Event Center Tent

28 **Mechanical Design of a Customizable Self-Expanding Endovascular Stent** SB³C2015-1189 pg.1321
Joel C. R. Scott, Clifton R. Johnston, Darrel A. Doman, *Dalhousie University, Halifax, NS, Canada*

29 **Which Of Four Commercially-Available Laser Scanner Systems Generates The Most Accurate Bone Model?** SB³C2015-152 pg.94
Valentina Campanelli[1,2], Stephen M. Howell[1], Maury Hull[1], *[1]UC Davis, Davis, CA, United States,[2]University of Verona, Verona, Italy*

30 **Deformation and Flow of Arterial Stenosis Model Regarding the Change in Curvature of Coronary Artery for Percutaneous Transluminal Coronary Angioplasty** SB³C2015-1142 pg.1239
Shunichi Kobayashi, *Shinshu University, Ueda, Japan*

31 **Computational Modeling Of Mechanical Stresses In Stents: Implications For The Optimization Of Drug-eluting Stents** SB³C2015-458 pg.620
Francois P. M. Cornat, Franz Bozsak, Abdul I. Barakat, *Ecole Polytechnique, Palaiseau cedex, France*

32 **Damage Analysis of Retrieved Oxidized Zirconium Femoral Components for TKA: Can Wear Breach the Oxide Layer?** SB³C2015-1135 pg.1229
Noah Bonnheim[1], Michael Ries[2], Sanjai Shukla[2], Lisa Pruitt[1], *[1]University of California, Berkeley, Berkeley, CA, United States,[2]Tahoe Fracture and Orthopaedic Clinic, Carson City, NV, United States*

33 **Design and Optimization of an In-Vitro Emboli Detector for Flow-Induced Thrombogenicity Evaluation** SB³C2015-344 pg.419
Ram Shtoltz, *Tel Aviv University, Tel Aviv, Israel*

34 **Development of Micro-scale Ultrasound Imaging for Tissue Characterization.** SB³C2015-1165 pg.1283
Jeremy Stromer, Leila Ladani, *University of Connecticut, Storrs, CT, United States*

35 **Bearing Surface Damage Analysis Of Total Shoulder Replacement Retrievals Across Fixation Designs And UHMWPE Composition** SB³C2015-531 pg.748
Louis G. Malito[1], Farzana Ansari[1], Lulu Li[1], Taylor Lee[1], Helen Park[1], Steve Gunther[2], Tom Norris[3], Mike Ries[4], Lisa Pruitt[1], *[1]University of California, Berkeley, Berkeley, CA, United States,[2]Martha Jefferson Hospital, Charlottesville, VA, United States,[3]San Francisco Shoulder, Elbow & Hand Clinic, San Francisco, CA, United States,[4]Tahoe Fracture and Orthopaedic Clinic, Carson City, NV, United States*

36 **Stress Angle Device: A Novel System For Reproducing The Mechanical Conditions Associated With Regions Most Susceptible To Vascular Disease** SB³C2015-406 pg.521
Steve Zambrano, Michael J. Draper, Brendan L. Swain, Caleb A. Davis, Michael R. Moreno, *Texas A&M University, Bryan, TX, United States*

37 **On The Characterization Of Mitral Valve Geometry And Development Of A Population-averaged Model** SB³C2015-1022 pg.1022
Amir Khalighi[1], Andrew Drach[1], Fleur ter Huurne[2], Chung-Hao Lee[1], Charles Bloodworth[3], Eric Pierce[3], Morten Jensen[3], Ajit Yoganathan[3], Michael Sacks[1], [1]*University of Texas at Austin, Austin, TX, United States,*[2]*Eindhoven University, Eindhoven, Netherlands,*[3]*Georgia Institute of Technology, Atlanta, GA, United States*

38 **Modeling the Mechanical Behavior of Polymeric Bioresorbable Stents: a Finite Element Approach** SB³C2015-658 pg.983
Nic Debusschere, Patrick Segers, Peter Dubruel, Benedict Verhegghe, Matthieu De Beule, *Ghent University, Ghent, Belgium*

THURSDAY, JUNE 18	12:30pm - 2:30pm

Poster Session I — Neuromuscular Control and Motion Analysis — Event Center Tent

39 **Artificial Neural Networks for the Optimization of Ligament Stiffnesses in a Computational Foot/Ankle Model** SB³C2015-438 pg.584
Ruchi Chande, Norma Ortiz-Robinson, Jennifer Wayne, *Virginia Commonwealth University, Richmond, VA, United States*

40 **Movement Analysis Based on the Separation of Angular Momentum: Gait Progression and Segment Rotation in Patients with TKA** SB³C2015-382 pg.473
Brecca M. Gaffney[1], Will M. Johnston[1], Cory L. Christiansen[2], Jennifer E. Stevens-Lapsley[2], Kevin B. Shelburne[1], Bradley S. Davidson[1], [1]*University of Denver, Denver, CO, United States,*[2]*University of Colorado Denver, Aurora, CO, United States*

41 **A Description of Segmental Angular Momentum Synergies using Independent Component Analysis During Gait** SB³C2015-543 pg.770
Brecca M. Gaffney[1], Cory L. Christiansen[2], Kevin B. Shelburne[1], Bradley S. Davidson[1], [1]*University of Denver, Denver, CO, United States,*[2]*University of Colorado Denver, Aurora, CO, United States*

42 **Joint Angle and Muscle Activity during Hippotherapy: A Case Study Using Motion Capture Analysis and EMG** SB³C2015-632 pg.933
Mary C. Baker, Stephen Wester, Timothy Monday, *Texas Tech University, Lubbock, TX, United States*

43 **Changes In The Range Of Motion Envelope Of The Lumbar Spine With Repetitive Lifting** SB³C2015-618 pg.907
Muhammad I. Gul[1], Timothy D. Craig[1], Neena K. Sharma[2], **Sara E. Wilson**[1], [1]*University of Kansas, Lawrence, KS, United States,*[2]*University of Kansas, Kansas City, KS, United States*

THURSDAY, JUNE 18	12:30pm - 2:30pm

Poster Session I — Best Practices in Biomechanics, Bioengineering, and Biotransport Education — Event Center Tent

44 **Best Practices in Teaching Biomechanics: Integrating Reflective Learning Activities** SB³C2015-138 pg.72
Laurel Kuxhaus, *Clarkson University, Potsdam, NY, United States*

45 **Assessment of a Bioengineering Innovation Program for Middle School Girls** SB³C2015-577 pg.833
Kristen Billiar[1], Amanda Reidinger[1], Jeanne Hubelbank[2], Helen Vassallo[1], [1]*Worcester Polytechnic Institute, Worcester, MA, United States,*[2]*Independent Program Evaluation Consultant, Sudbury, MA, United States*

46 **Teaching Undergraduate Design: An Approach Based on Industrial Experience** SB³C2015-118 pg.44
Martin L. Tanaka, *Western Carolina University, Cullowhee, NC, United States*

47 **A Progress Report From a Multi-disciplinary Capstone Experience Involving Engineering and Business Students** SB³C2015-110 pg.32
Alan Eberhardt, Joel Dobbs, *University of Alabama at Birmingham, Birmingham, AL, United States*

THURSDAY, JUNE 18	12:30pm - 2:30pm

Poster Session I — Cardiovascular Devices and Imaging in Fluid Mechanics — Event Center Tent

48 **Retrospective Analysis of Echocardiography in Those Suspected of Pulmonary Hypertension with Comparison to Right Heart Catheterization** SB³C2015-439 pg.586
Travis B. Eason[1], Melinda V. Pyler[1], Lavanya Alapati[1], Sanjay Mehra[2], John M. Cahill[1], Stephanie M. George[1], [1]*East Carolina University, Greenville, NC, United States,*[2]*Vidant Medical Center, Greenville, NC, United States*

49 **The Influence Of Sub-Optimal Acquisition Delay In A C-arm Cone-Beam CT Perfusion Study** SB³C2015-1067 pg.1102
Antonius M. de Korte, Kajo van der Marel, Juyu Chueh, Olivia W. Brooks, Ajit S. Puri, Ajay K. Wakhloo, Matthew J. Gounis, *New England Center for Stroke Research, University of Massachusetts Medical School, Worcester, MA, United States*

50 **The Relationship Of Wall Shear Stress With Clinically Relevant Metrics In Pulmonary Arterial Hypertension** SB³C2015-648 pg.963
Alifer D. Bordones[1,2], Vitaly O. Kheyfets[3], Ender A. Finol[1], [1]*University of Texas at San Antonio, San Antonio, TX, United States,*[2]*University of Texas Health Science Center at San Antonio, San Antonio, TX, United States,*[3]*University of Colorado, Denver, CO, United States*

51 **Quantitative Coronary Angiography Based Reconstructions for Wall Shear Stress Calculations in Bifurcations** SB³C2015-545 pg. 774
Jelle T. C. Schrauwen[1], Antonios Karanasos[1], Nienke S. Ditzhuijzen[1], Jean-Paul Aben[2], Jolanda J. Wentzel[1], Antonius F. W. van der Steen[1,3], Frank J. H. Gijsen[1], [1]*Thoraxcenter, Erasmus Medical Center, Rotterdam, Netherlands,*[2]*Pie Medical Imaging, Maastricht, Netherlands,*[3]*Delft University of Technology, Netherlands*

52 **Swirling Flows in Arterial Hemodynamics** SB³C2015-1072 pg. 1112
Kartik Bulusu[1], Christopher Elkins[2], John Eaton[2], **Michael Plesniak**[1], [1]*George Washington University, Washington, DC, United States,*[2]*Stanford University, Stanford, CA, United States*

53 **Integrating an Open Source Meshing Alternative into SimVascular 2.0** SB³C2015-1062 pg.1094
Adam R. Updegrove[1], Nathan M. Wilson[2], Shawn C. Shadden[1], [1]*University of California, Berkeley, Berkeley, CA, United States,*[2]*Open Source Medical Software Corporation, Santa Monica, CA, United States*

54 **Microbubble Void Imaging - A Novel Technique For Flow Visualisation and Quantitative Assessment of Intravascular Mixing in Larger Vessels Using Ultrasound** SB³C2015-388 pg.485
Chee Hau Leow, **Francesco Iori**, Richard W. Corbett, Neill Duncan, Colin G. Caro, Peter E. Vincent, Mengxing Tang, *Imperial College London, London, United Kingdom*

55 **In-vivo Validation Of The In Silico Predicted Pressure Drop Across An Arteriovenous Fistula** SB³C2015-536 pg.756
Leonard Browne[1], Khalid Bashar[2], Philip Griffin[1], Eamon Kavanagh[2], Michael Walsh[1], [1]*University of Limerick, Limerick, Ireland,*[2]*Univeristy Hospital Limerick, Limerick, Ireland*

56 **Triangulated Surface Boolean Operations for Combining 2-D and 3-D Image Segmentation for Patient-Specific Blood Flow Analysis** SB³C2015-1052 pg.1078
Adam R. Updegrove[1], Nathan M. Wilson[2], Shawn C. Shadden[1], [1]*University of California, Berkeley, Berkeley, CA, United States,*[2]*Open Source Medical Software Corporation, Santa Monica, CA, United States*

57 **Role of Wall Thickness and Tethering in the Assessment of Arterial Stiffness** SB³C2015-156 pg.102
Simona Hodis[1], Mair Zamir[2], [1]*Texas A&M University - Kingsville, Kingsville, TX, United States,*[2]*University of Western Ontario, London, ON, Canada*

58 **SimVascular 2.0: an Open Source Pipeline for Cardiovascular Modeling and Simulation** SB³C2015-656 pg.979
Hongzhi Lan[1], Nathan M. Wilson[2], Daniele Schiavazzi[1], Jameson Merkow[1], Adam Updegrove[3], Shawn C. Shadden[3], Alison L. Marsden[1], [1]*University of California - San Diego, La Jolla, CA, United States,*[2]*Open Source Medical Software Corporation, Los Angeles, CA, United States,*[3]*University of California - Berkeley, Berkeley, CA, United States*

59 **Nt-proBNP Expression Originating From The RV Myocardium In Pediatric Pulmonary Hypertension Patients Is Correlated With Both The Reactive And Resistive Components Of Vascular Impedance** SB³C2015-1130 pg.1219
Vitaly O. Kheyfets[1], Jamie Dunning[1], Uyen Truong[2], Kendall Hunter[1], Dunbar Ivy[2], Robin Shandas[1], [1]*University of Colorado, Aurora, CO, United States,*[2]*Children's Hospital Colorado, Aurora, CO, United States*

THURSDAY, JUNE 18	12:30pm - 2:30pm

Poster Session I **Embryonic, Pediatric Cardiology, and Other Fluid Mechanics** **Event Center Tent**

60 **Valveless Flow in a Thick Elastic Tube** SB[3]C2015-582 pg. 843
Pavel Kozlovsky[1], Moshe Rosenfeld[1], Ariel Jaffa[2], **David Elad**[1], [1]*Tel Aviv Univesity, Tel Aviv, Israel,*[2]*Tel Aviv Medical Center, Tel Aviv, Israel*

61 **Prediction of Downstream Velocity Waveforms for In Vitro Aortic Flow Experiments** SB[3]C2015-229 pg.234
Rafeed A. Chaudhury, Justin R. Ryan, David H. Frakes, Ronald J. Adrian, *Arizona State University, Tempe, AZ, United States*

62 **In Vitro Optimization Of The Nozzle Used In Assisted Bidirectional Glenn Procedure For Single Ventricle Stage 1 Palliation** SB[3]C2015-626 pg.921
Jian Zhou[1], Mahdi Esmaily-Moghadam[2], Timothy Conover[1], Tain-Yen Hsia[3], Alison Marsden[4], **Richard Figliola**[1], [1]*Clemson University, Clemson, SC, United States,*[2]*Stanford University, Palo Alto, CA, United States,*[3]*Great Ormond Street Hospital, London, United Kingdom,*[4]*University of California, San Diego, La Jolla, CA, United States*

63 **Mean Flow Umbilical Doppler Indices Perform Better Than Pulsatility Indices In Determining Small For Gestational Age Pregnancies** SB[3]C2015-201 pg.186
Shier Nee Saw[1], Citra Nurfarah Zaini Mattar[2], Seow Heong Yeo[3], Shu-E Soh[4,5], Yap-Seng Chong[2,6], Peter David Gluckman[6,7], Keith Godfrey[8], Seang Mei Saw[5,9], Arijit Biswas[2], **Choon Hwai Yap**[1], [1]*National University of Singapore, Singapore,*[2]*Yong Loo Lin School of Medicine, NUS & NUHS, Singapore,*[3]*KK Women's and Children's Hospital, Singapore,*[4]*Yong Loo Lin School of Medicine, NUS, Singapore,*[5]*Saw Swee Hock School of Public Health, NUS, Singapore,*[6]*Singapore Institute for Clinical Sciences, A*STAR, Singapore,*[7]*Liggins Institute, University of Auckland, Auckland, New Zealand,*[8]*University of Southampton & University Hospital Southampton NHS Foundation Trust, Southampton, United Kingdom,*[9]*Singapore Eye Research Institute, Singapore*

64 **Accounting For Clinical Data Uncertainty In Multiscale Numerical Simulation Of Single Ventricle Palliation Surgery** SB[3]C2015-309 pg.363
Daniele E. Schiavazzi[1], Tain-Yen Hsia[2], Alison L. Marsden[1], [1]*University of California at San Diego, San Diego, CA, United States,*[2]*Great Ormond Street Hospital for Children and UCL Institute of Cardiovascular Science, London, United Kingdom*

65 **Automated Tuning for Parameter Identification in Multi-scale Coronary Simulations** SB[3]C2015-541 pg. 766
Justin S. Tran, Daniele Schiavazzi, Abhay Ramachandra, Andrew Kahn, Alison Marsden, *University of California, San Diego, La Jolla, CA, United States*

66 **Investigating Environmental Causes of Congenital Heart Diseases: A Subject-Specific Computational Fluid Dynamics Study** SB[3]C2015-204 pg.192
Venkat Keshav Chivukula[1], Sevan Goenezen[2], Sandra Rugonyi[1], [1]*Oregon Health and Sciences University, Portland, OR, United States,*[2]*Texas A&M University, College Station, TX, United States*

67 **Characterization of Transition to Turbulence for Blood in an S-Shaped Pipe Under Steady Flow Conditions** SB[3]C2015-1053 pg.1080
Dipankar Biswas[1], David M. Casey[1], Douglas C. Crowder[1], Kristian Valen-Sendstad[2], David A. Steinman[2], Yang H. Yun[1], Francis Loth[1], [1]*The University of Akron, Akron, OH, United States,*[2]*University of Toronto, Toronto, ON, Canada*

68 **Quantitative Analysis Of Heart Function In Embryonic Zebrafish: Retrograde Flow In The Atrioventricular Junction** SB[3]C2015-661 pg.989
Alexander T. Bulk, David L. Bark Jr., Brennan Johnson, Deborah Gogarty, **Lakshmi P. D. Dasi**, *Colorado State University, Fort Collins, CO, United States*

69 **The Impact of Parameter Variation, Experimental Data and Uncertainty Quantification for Complex Biomechanical Problems Examplified for AAA** SB[3]C2015-364 pg. 447
Wolfgang A. Wall, Jonas Biehler, Michael W. Gee, *Technische Universität München, Garching b. München, Germany*

70 **Novel Method For 3D Reconstruction Of The Chick Embryo Cardiovascular Anatomy From Non-invasive Ultrasound Scans For Longitudinal Studies** SB[3]C2015-109 pg.30
Germaine X. Y. Tan[1], Muhammad Jamil[1], Nicole G. Z. Tee[2], Liang Zhong[1,2], **Choon Hwai Yap**[1], [1]*National University of Singapore, Singapore,*[2]*National Heart Centre Singapore, Singapore*

71 **Investigating The Two-phase Nature Of Blood Flow** SB³C2015-1102 pg.1167
Joseph M. Sherwood[1], Xuejin Li[2], George Karniadakis[2], Stavroula Balabani[3], [1]*Imperial College London, London, United Kingdom,*[2]*Brown University, Providence, RI, United States,*[3]*University College London, London, United Kingdom*

72 **Tissue Hypoxia and Murray's Law of Minimum Work Control Neovascular Growth and Remodeling** SB³C2015-1176
Sean Moore, David Hoelzle, **Joel D. Boerckel**, *University of Notre Dame, Notre Dame, IN, United States* *pg.1301*

THURSDAY, JUNE 18	12:30pm - 2:30pm

Poster Session I | **MS Competition - Biofluid Mechanics** | **Event Center Tent**

73 **Elevated Wall Shear Stress Predicts Branch Graft Failure Following Chimney Endovascular Aortic Aneurysm Repair** SB³C2015-443 pg.594
Rosamaria Tricarico[1], He Yong[1,2], Adam Beck[1], Salvatore Scali[1,2], Roger Tran-Son-Tay[1], Scott Berceli[1,2], [1]*University of Florida, Gainesville, FL, United States,*[2]*North Florida/South Georgia Veterans Health System, Gainesville, FL, United States*

74 **Better Assessment Of Arteriovenous Fistula Patency Using Functional Diagnostic Endpoints** SB³C2015-442 pg.592
Krishna Subramony Anantha, Ehsan Rajabi-Jaghargh, Rupak Banerjee, *University of Cincinnati, Cincinnati, OH, United States*

75 **Establishing In Vivo Hemodynamic Baseline In A Normotensive Rat Model** SB³C2015-1059 pg.1090
Daniela Velez-Rendon, Erica R. Pursell, Daniela Valdez-Jasso, *University of Illinois at Chicago, Chicago, IL, United States*

THURSDAY, JUNE 18	12:30pm - 2:30pm

Poster Session I | **MS Competition - Biotransport** | **Event Center Tent**

76 **Imaging Processing Algorithms For Detecting Subtle Morphological Changes In Mole Images Over Time** SB³C2015-235 pg. 242
Alireza Chamani[1], Neera Nathan[2], Thomas Hornyak[3], Liang Zhu[1], [1]*University of Maryland Baltimore County, Baltimore, MD, United States,*[2]*National Cancer Institute / NIH, Bethesda, MD, United States,*[3]*University of Maryland at Baltimore, Baltimore, MD, United States*

77 **Evaluating the Influence of Tissue Properties on the Core Temperature Using a 3D Whole Body Model** SB³C2015-1037 pg.1048
Robins T. Kalathil[1], Swarup Zachariah[1], Amit Bhattacharya[2], Rupak Banerjee[1], [1]*University of Cincinnati, Cincinnati, OH, United States,*[2]*University of Cincinnati College of Medicine, Cincinnati, OH, United States*

THURSDAY, JUNE 18	12:30pm - 2:30pm

Poster Session I | **MS Competition - Cellular and Tissue Engineering** | **Event Center Tent**

78 **Individual Cell-Based Morphological Analysis to Determine Chirality of Epithelial Morphogenesis** SB³C2015-371 pg.459
Michael J. Raymond, Poulomi Ray, Leo Wan, *Rensselaer Polytechnic Institute, Troy, NY, United States*

THURSDAY, JUNE 18	12:30pm - 2:30pm

Poster Session I **MS Competition - Design and Devices** **Event Center Tent**

79 **Fetal Monitoring with Silicone Nanocomposite Strain Gauges** SB³C2015-334 pg.339
Daniel Baradoy, Anton Bowden, David Fullwood, *Brigham Young University, Provo, UT, United States*

80 **Development of 3D Printed Patient Specific Ascending Aortic Training Models for Cardiac Surgery** SB³C2015-446 pg. 600
ZhiLin Yang[1], Justine Garcia[1], Kevin Lachapelle[2], Rosaire Mongrain[1], Richard L. Leask[1], [1]*McGill University, Montreal, QC, Canada,* [2]*Royal Victoria Hospital, Montreal, QC, Canada*

81 **Image-based 3D Morphometric Analysis of the Clavicle Intramedullary Canal in Male Population** SB³C2015-494 pg.676
Jazmine Aira[1,2], Sergio Gutierrez[2], Brandon G. Santoni[1,2], Mark A. Frankle[1,3], Peter Simon[1,2], [1]*University of South Florida, Tampa, FL, United States,* [2]*Foundation for Orthopaedic Research and Education, Tampa, FL, United States,* [3]*Florida Orthopaedic Institute, Tampa, FL, United States*

THURSDAY, JUNE 18	12:30pm - 2:30pm

Poster Session I **MS Competition - Human Dynamics** **Event Center Tent**

82 **Pseudo-Rigid Body Method for Reducing Soft Tissue Artifact: Validation and Application to Gait** SB³C2015-1147 pg.1249
Jake Deschamps, Karim Dudum, Eshan Dandekar, Scott Hazelwood, Stephen Klisch, *California Polytechnic State University, San Luis Obispo, CA, United States*

THURSDAY, JUNE 18	12:30pm - 2:30pm

Poster Session I **MS Competition - Injury** **Event Center Tent**

83 **Assessing Head Impact Sensor Validity in the Laboratory** SB³C2015-631 pg.931
Abigail M. Zadnik, Steven Rowson, Stefan M. Duma, *Virginia Tech, Blacksburg, VA, United States*

84 **Acute Failure Modes in Small Diameter Spinal Segments** SB³C2015-282 pg.317
Aubrie L. Taylor, Cassandra Bell, Anton E. Bowden, *Brigham Young University, Provo, UT, United States*

85 **Human Knee Joint Finite Element Model Using a Two Bundle Anterior Cruciate Ligament: Validation and Gait Analysis** SB³C2015-610 pg.893
Nicholas A. Czapla[1], Meghan K. Sylvia[1], Zachary F. Lerner[2], David J. Tuttle[3], Otto S. Schueckler[4], Scott J. Hazelwood[1], Stephen M. Klisch[1], [1]*California Polytechnic State University at San Luis Obispo, San Luis Obispos, CA, United States,* [2]*Colorado State University at Fort Collins, Fort Collins, CO, United States,* [3]*Radiology Associates, Inc., San Luis Obispos, CA, United States,* [4]*Central Coast Orthopedic Medical Group, San Luis Obispos, CA, United States*

86 **Quantifying Head Impact Exposure in Collegiate Women's Soccer** SB³C2015-555 pg. 793
Jaclyn Press, Steven Rowson, *Virginia Tech, Blacksburg, VA, United States*

THURSDAY, JUNE 18	12:30pm - 2:30pm

Poster Session I | **MS Competition - Tissue Mechanics** | **Event Center Tent**

87 **Development of a Human Knee Joint Finite Element Model to Investigate Cartilage Stress During Walking in Obese and Normal Weight Adults** SB³C2015-611 pg. 895
Meghan Sylvia[1], Nicholas Czapla[1], Zachary Lerner[2], David Tuttle[3], Otto Schueckler[4], Scott Hazelwood[1], Stephen Klisch[1], [1]*California Polytechnic State University, San Luis Obispo, San Luis Obispo, CA, United States,*[2]*Colorado State University, Fort Collins, Fort Collins, CO, United States,*[3]*Radiology Associates, Inc., San Luis Obispo, CA, United States,*[4]*Central Coast Orthopedic Medical Group, San Luis Obispo, CA, United States*

88 **Accurate Prediction of Collagen Fiber Distribution using FFT: A Validation Study** SB³C2015-431 pg. 572
Erica E. Morrill, Christina Stender, Roshani Lamichhane, Raquel Brown, Trevor Lujan, *Boise State University, Boise, ID, United States*

THURSDAY, JUNE 18	12:30pm - 2:30pm

Poster Session I | **Joint and Spine Mechanics** | **Event Center Tent**

89 **Influence of Bone Density and Insertion Angle on Fixation Strength of Pedicle Screws** SB³C2015-186 pg.158
Giovanni Solitro, **Farid Amirouche**, Brenden Magnan, *University of Illinois at Chicago, Chicago, IL, United States*

90 **An Open-source Toolbox for Surrogate Modeling of Joint Contact Mechanics** SB³C2015-578 pg.835
Ilan Eskinazi, Benjamin J. Fregly, *University of Florida, Gainesville, FL, United States*

91 **Effect Of Axial Compression Preload On Intervertebral Disc Torsional Mechanics** SB³C2015-127 pg.56
Semih E. Bezci, **Grace D. O'Connell**, *University of California, Berkeley, Berkeley, CA, United States*

92 **Using Dynamic Community Detection to Map Collagen Fiber Network Reorganization During Tensile Loading of the Human Facet Capsular Ligament** SB³C2015-115 pg.40
Sijia Zhang, Danielle Bassett, Beth Winkelstein, *University of Pennsylvania, Philadelphia, PA, United States*

93 **Application of a Novel Robotically Simulated Pivot Shift for Anterior Cruciate Ligament Reconstruction: Comparison of the All-Epiphyseal and Over-The-Top Techniques** SB³C2015-500 pg.688
Robb Colbrunn, Tara Bonner, Joel Kolmodin, Paul Saluan, *Cleveland Clinic, Cleveland, OH, United States*

94 **Effect Of Sacroiliac Joint Fixation On Segmental Kinematics Of Lumbar Spine: A Finite Element Analysis** SB³C2015-1146 pg.1247
Ali Kiapour[1], Derek Lindsey[2], Scott Yerby[2], Vijay Goel[1], [1]*ECORE, Toledo, OH, United States,*[2]*Si-Bone Inc, San Jose, CA, United States*

95 **Effect of Variability in Anatomical Landmarks on Ankle Kinematic Descriptions** SB³C2015-513 pg. 712
Ednah G. Louie, Fallon G. Fitzwater, Matthew H. Dickinson, William M. Eboch, Bardiya Akhbari, Lorin P. Maletsky, *University of Kansas, Lawrence, KS, United States*

96 **Biomechanical Evaluation of Proximal Junctional Failure and Kyphosis in Lumbar Spinal Fusion Surgery** SB³C2015-155 pg.100
Won Man Park[1], Kyungsoo Kim[1], Yongjung J. Kim[2], **Yoon Hyuk Kim**[1], [1]*Kyung Hee University, Yongin, Korea, Republic of,*[2]*Columbia University College of Physicians and Surgeons, New York, NY, United States*

97 **Biomechanical Comparison Between Facet Sparing Laminectomy and Laminectomy with Facetectomy in Lumbar Spine: A Finite Element Analysis** SB³C2015-653 pg.973
Shady Elmasry[1], Shihab Asfour[1], Joseph Gjolaj[2], Loren Latta[2], Frank Eismont[2], **Francesco Travascio**[1], [1]*University of Miami, Coral Gables, FL, United States,*[2]*University of Miami, Miami, FL, United States*

98 **Open Knee(s): Comprehensive Tibiofemoral Joint Testing For Specimen-specific Next Generation Knee Models** SB³C2015-530 pg.746
Tara F. Bonner, Robb W. Colbrunn, Snehal Chokhandre, Craig Bennetts, Ahmet Erdemir, *Cleveland Clinic, Cleveland, OH, United States*

THURSDAY, JUNE 18	12:30pm - 2:30pm

Poster Session I **Tissue Mechanics - General** **Event Center Tent**

112 **Extracting Mechanical Properties Of The Corneoscleral Shell From Whole Eye Perfusions** SB^{3}C2015-629 pg.927
Joseph M. Sherwood[1], Ester Reina-Torres[1], Jacques Bertrand[1], C. Ross Ethier[2], Darryl Overby[1], [1]*Imperial College London, London, United Kingdom,*[2]*Georgia Institute of Technology/Emory, Atlanta, GA, United States*

113 **Strain Rate Dependency Of The Intracellular Calcium Ion Concentration During Neuronal Membrane Mechanoporation** SB^{3}C2015-621 pg.911
Amirhamed Bakhtiarydavijani[1], Anna E. Florence[1], Michael A. Murphy[1], Sungkwang Mun[1], Jun Liao[1], Lakiesha N. Williams[1], M. F. Horstemeyer[1], Michelle C. LaPlaca[2], Raj Prabhu[1], [1]*Mississipi State University, Starkville, MS, United States,*[2]*Georgia Institute of Technology, Atlanta, GA, United States*

114 **Quasilinear and Non-Quasilinear Viscoelastic Behavior of Collagen Gels During Stress Relaxation** SB^{3}C2015-551 pg.784
Christopher E. Korenczuk[1], Victor K. Lai[2], Victor H. Barocas[1], [1]*University of Minnesota, Minneapolis, MN, United States,*[2]*University of Minnesota Duluth, Duluth, MN, United States*

115 **Direct Estimation of Three-dimensional Deformation Gradient Tensors from Volumetric Ultrasound Data** SB^{3}C2015-671 pg.1007
John J. Boyle[1], Roger Rowe[1], Frederick Damen[2], Arvin Soepriatna[2], Robert B. Pless[1], Craig Goergen[2], Stavros Thomopoulos[1], Guy M. Genin[1], [1]*Washington University in St Louis, Saint Louis, MO, United States,*[2]*Purdue University, West Lafayette, IN, United States*

116 **A Structural Constitutive Model for the Active and Passive Behavior of Biological Tissues** SB^{3}C2015-330 pg.397
Ting Tan, **Raffaella De Vita**, *Virginia Tech, Blacksburg, VA, United States*

117 **Computational Modeling Of Synthetic Mesh Materials: Simulation And Experimental Assesment Of Model Predictions** SB^{3}C2015-511 pg.708
William R. Barone[1], Katrina M. Knight[1], Pamela A. Moalli[2], Steven D. Abramowitch[1], [1]*University of Pittsburgh, Pittsburgh, PA, United States,*[2]*Magee-Womens Research Institute, University of Pittsburgh, Pittsburgh, PA, United States*

118 **Tough, Self-recovering Hydrogels Inspired By Caddisfly Silk** SB^{3}C2015-337 pg.405
Dwight D. Lane, G. Mahika Weerasekare, Sarbjit Kaur, Russell J. Stewart, *University of Utah, Salt Lake City, UT, United States*

119 **Determining The Compressive Modulus Of Mouse Trabecular Meshwork With Atomic Force Microscopy** SB^{3}C2015-1047 pg.1068
Ke Wang, Todd Sulchek, C. Ross Ethier, *Georgia Institute of Technology, Atlanta, GA, United States*

120 **Development of Shoe Sole Design Using Material Natural Frequencies** SB^{3}C2015-251 pg.271
Brett D. Steineman[1], Ted Barber[2], Tammy L. Haut Donahue[1], [1]*Colorado State University, Fort Collins, CO, United States,*[2]*Pearl Izumi, Inc., Louisville, CO, United States*

121 **The Need for Validation in Soft Tissue Constitutive Models** SB^{3}C2015-1058 pg.1088
Sandeep Madireddy, Kumar Vemaganti, *University of Cincinnati, Cincinnati, OH, United States*

122 **A Computational Study of A Simple, Transversely Isotropic Model of Soft Tissue, with Shear and Tensile Anisotropy, in Large Strain** SB^{3}C2015-202 pg.188
Yuan Feng[1], Ruth J. Okamoto[2], Guy M. Genin[2], Larry A. Taber[2], Philip V. Bayly[2], [1]*Soochow University, Suzhou, China,*[2]*Washington University in St. Louis, St. Louis, MO, United States*

123 **Development Of A Clinical Ultrasound Technique for Analysis of Protein Content Within Hydrogel** SB^{3}C2015-1109
Jessica Stukel[1], Monika Goss[2], Agata Exner[2], Rebecca Willits[1], [1]*The University of Akron, Akron, OH, United States,*[2]*Case Western Reserve University, Cleveland, OH, United States* *pg.1181*

124 **Validation Of High Rate Strip Biaxial Tension Deformations Of The Neuronal Phospholipid Bilayer Using Empirical Data** SB^{3}C2015-1120 pg.1203
M. A. Murphy[1], M. F. Horstemeyer[1], Steven R. Gwaltney[1], Tonya W. Stone[1], Michelle C. LaPlaca[2], Jun Liao[1], Lakiesha N. Williams[1], **R. Prabhu**[1], [1]*Mississippi State, MS, United States,*[2]*Georgia Institute of Technology, Atlanta, GA, United States*

125 **Collective Chiral Rotation of Epithelial Microtissues Within a Three-Dimensional Matrigel System** SB^{3}C2015-232
Amanda S. Chin, Kathryn E. Worley, Leo Q. Wan, *Rensselaer Polytechnic Institute, Troy, NY, United States* *pg.238*

126 **Finite Element Model of Cervical Pessary in Use: Evaluating Mechanical Interventions for Preterm Birth** SB^{3}C2015-490 pg.668
Michael J. Fernandez[1], Michael D. House[2], Noelia M. Zork[3], Joy S. Vink[3], Ronald J. Wapner[3], Sachin R. Jambawalikar[3], Kristin M. Myers[1], [1]*Columbia University, New York, NY, United States,*[2]*Tufts Medical Center, Boston, MA, United States,*[3]*Columbia University Medical Center, New York, NY, United States*

127 **Impact of Urinary Bladder Matrix on Vaginal Smooth Muscle Function and Structure in the Nonhuman Primate Model** SB^{3}C2015-164 pg.116
Katrina Knight[1], Zegbeh Jallah[1], Rui Liang[2], Stacy Palcsey[2], Pamela Moalli[1,2], Steven Abramowitch[1,2], [1]*Musculoskeletal Research Center, University of Pittsburgh, Pittsburgh, PA, United States,*[2]*Magee-Womens Research Institute, Magee-Womens Hospital, Pittsburgh, PA, United States*

128 **Biaxial Creep of Swine Cardinal and Uterosacral Ligaments** SB^{3}C2015-299 pg. 345
Ting Tan, Nathan M. Cholewa, Scott W. Case, **Raffaella De Vita**, *Virginia Tech, Blacksburg, VA, United States*

129 **Characterization of Soft Tissue Microstructure via Transmural SALS** SB^{3}C2015-1044 pg.1062
John G. Lesicko, Kristen R. Feaver, Michael S. Sacks, *University of Texas at Austin, Austin, TX, United States*

130 **Finite Element Modeling of the Posterior Eye in Microgravity** SB^{3}C2015-348 pg.423
Andrew Feola[1], Julia Raykin[1], Lealem Mulugeta[2], Rudolph Gleason[1], Jerry G. Myers[3], Emily S. Nelson[3], Brian Samuels[4], Ross Ethier[1], [1]*Georgia Institute of Technology, Atlanta, GA, United States,*[2]*Universities Space Research Association, Houston, TX, United States,*[3]*NASA Glenn Research Center, Cleveland, OH, United States,*[4]*University of Alabama at Birmingham, Birmingham, AL, United States*

131 **Posterior Scleral Stiffening: How is Scleral Canal Expansion Affected by the Size of Stiffening Region?** SB^{3}C2015-1048 pg.1070
Ian C. Campbell[1,2], Baptiste Coudrillier[1], C. Ross Ethier[1,2], [1]*Georgia Institute of Technology/Emory University, Atlanta, GA, United States,*[2]*Atlanta VA Medical Center, Decatur, GA, United States*

132 **Biomechanical Behavior Of Cornea When Subjected To Tension And Compression Loads** SB^{3}C2015-1076 pg.1118
Hamed Hatami-Marbini, *Oklahoma State University, Stillwater, OK, United States*

133 **Correlating Urethral Rupture with Distension of the Urethra during the Inflation of a Misplaced Transurethral Catheter Balloon** SB^{3}C2015-370 pg. 457
Connor V. Cunnane, *University of Limerick, Limerick, Ireland*

134 **Tuning Silk Fibroin Hydrogels: Genipin Crosslinking Pre-gelation Decreases Time-dependent Properties** SB^{3}C2015-1125 pg.1211
Winston H. Elliott[1], Walter Bonani[2,3], Devid Maniglio[2,3], Antonella Motta[2,3], Wei Tan[1], Claudio Migliaresi[2,3], [1]*University of Colorado- Boulder, Boulder, CO, United States,*[2]*University of Trento, Trento, Italy,*[3]*European Institute of Excellence on Tissue Engineering and Regenerative Medicine, and INSTM Trento Research Unit, Trento, Italy*

135 **Spherical Shell Mechanical Model of the Pacinian Corpuscle** SB^{3}C2015-411 pg.531
Julia C. Quindlen, Henryk K. Stolarski, Victor H. Barocas, *University of Minnesota, Minneapolis, MN, United States*

136 **Measuring Tortuosity Changes due to Central Retinal Vein Occlusion** SB^{3}C2015-1182 pg.1309
Kendall McMillan[1], Shaun Evans[1], Gil Binenbaum[2], Brittany Coats[1], [1]*University of Utah, Salt Lake City, UT, United States,*[2]*The Children's Hospital of Philadelphia, Philadelphia, PA, United States*

138 **Mechanical Properties Of Human Placenta In Normal Pregnancies And During Intrauterine Growth Restriction** SB^{3}C2015-206 pg.196
Jeanette Shifen Lau, Shier Nee Saw, Martin Lindsay Buist, **Choon Hwai Yap**, *National University of Singapore, Singapore*

139 **Determination of the Mechanical Properties of the Iris Using Inverse Finite Element Modeling** SB^{3}C2015-424 pg.557
Anup D. Pant, Rouzbeh Amini, *The University of Akron, Akron, OH, United States*

140 **A Poroelastic High Fidelity Finite Element Model Of The Osteochondral Unit To Evaluate Changes In Permeability With Osteoarthritis** SB^{3}C2015-144 pg.82
Michael E. Stender, Richard A. Regueiro, Virginia L. Ferguson, *University of Colorado, Boulder, Boulder, CO, United States*

141 **The Effects Of Stress State On The Mechanical Response And Failure Of The Neuronal Phospholipid Bilayer: A Molecular Dynamics Study** SB³C2015-1132 pg.1223
M. A. Murphy[1], M. F. Horstemeyer[1], Steven R. Gwaltney[1], Tonya W. Stone[1], Michelle C. LaPlaca[2], Jun Liao[1], Lakiesha N. Williams[1], **R. Prabhu**[1], *[1]Mississippi State, MS, United States,[2]Georgia Institute of Technology, Atlanta, GA, United States*

142 **Racial Differences In The Load-dependent Area Of The Lamina Cribrosa** SB³C2015-1172 pg.1293
Stephen J. Howerton, Forest L. Danford, Jonathan P. Vande Geest, Avinash Ayyalasomayajula, *The University of Arizona, Tucson, AZ, United States*

143 **Optic Nerve Sheath Mechanics in VIIP Syndrome** SB³C2015-488 pg.666
Julia Raykin[1], Andrew Feola[1], Rudy Gleason[1], Lealem Mulugeta[2], Jerry Myers[3], Emily Nelson[3], Brian Samuels[4], C. Ross Ethier[1], *[1]Georgia Institute of Technology, Atlanta, GA, United States,[2]Universities Space Research Association, Houston, TX, United States,[3]NASA Glenn Research Center, Cleveland, OH, United States,[4]University of Alabama at Birmingham, Birmingham, AL, United States*

144 **Modeling the Biaxial Mechanics of Brain White Matter** SB³C2015-177 pg.140
Kevin M. Labus[1], José J. García[2], Christian M. Puttlitz[1], *[1]Colorado State University, Fort Collins, CO, United States,[2]Universidad del Valle, Cali, Colombia*

145 **An In Silico Biomechanical Analysis of the Stent-Esophagus Interaction.** SB³C2015-365 pg. 449
Mathias Peirlinck[1], Benedict Verhegghe[1,2], Patrick Segers[1], Matthieu De Beule[1,2], *[1]Ghent University, Ghent, Belgium,[2]FEops bvba, Ghent, Belgium*

THURSDAY, JUNE 18	12:30pm - 2:30pm

Poster Session I — Bone Biomechanics — Event Center Tent

146 **Fabrication And Characterization Of Artificial Bone-cartilage Tissue Construction Using Mesenchymal Stem Cells** SB³C2015-207 pg.198
Takaaki Arahira, *Fukuoka Dental College, Fukuoka, Japan*

147 **Mechanical Analysis of Bone Tissue as Mineral and Organic Composite by Raman Spectroscopy** SB³C2015-447 pg.602
Masahiro Todoh, Shigeru Tadano, *Hokkaido University, Sapporo, Japan*

148 **The Effect of Pore Size on Bone Strain in the Proximal Femur** SB³C2015-1167 pg.1287
Mariana Kersh[1], Afrodite Zendelli[2], Yohann Bala[2], Ali Ghasem-Zadeh[2], Ego Seeman[2], Roger Zebaze[2], *[1]University of Illinois, Urbana, IL, United States,[2]Austin Health, Heidelberg, Australia*

149 **Multimodal Assessment of Bone Quality of the Human Rib** SB³C2015-38 pg.20
Lauren M. Mangano[1], Jean-Paul Roux[2], François Duboeuf[2], Delphine Farlay[2], David Mitton[3], Hélène Follet[2], *[1]Boston University, Brookline, MA, United States,[2]INSERM, Université Lyon 1, Lyon, France,[3]IFSTTAR, Université Lyon 1, Bron, France*

150 **The Characterization Of The Bone Marrow Mechanical Environment Using Poroelastic Finite Element Models** SB³C2015-1116 pg.1195
Joshua Gargac, Thomas Metzger, Tyler Kreipke, Hansel Weihs, Glen Niebur, *University of Notre Dame, South Bend, IN, United States*

151 **Spring Mediated Cranioplasty: a Patient Specific Numerical Model** SB³C2015-210 pg.202
Alessandro Borghi[1], Silvia Schievano[1], William Rodgers[1], Freida Angullia[2], Allan Ponniah[2], David Dunaway[2], Owase Jeelani[2], *[1]University College London, London, United Kingdom,[2]Great Ormond Street Hospital, London, United Kingdom*

152 **Preliminary Finite Element Analysis of Subchondral Bone Cysts in the Stifle of the Horse** SB³C2015-393 pg.495
Lance F. Frazer[1], Kenneth Fischer[1], Garrett Noble[2], Elizabeth Santschi[3], *[1]University of Kansas, Lawrence, KS, United States,[2]Ohio State University, Columbus, OH, United States,[3]Kansas State University, Manhattan, KS, United States*

THURSDAY, JUNE 18	12:30pm - 2:30pm

Poster Session I | **Fabrication and Manipulation of the Cellular Microenvironment** | **Event Center Tent**

153 **Mechanics of Interstitial Growth in Enzymatically Degradable Hydrogels: Characterization of Degradation Front** SB^{3}C2015-1124 pg.1209
Umut Akalp, Stephanie J. Bryant, Stacey C. Skaalure, Franck J. Vernerey, *University of Colorado at Boulder, Boulder, CO, United States*

154 **Microgeometry and Microenvironment of Mitral Valve Interstitial Cells Under Physiological Loads** SB^{3}C2015-287 pg.325
Salma Ayoub, Chung-Hao Lee, Michael S. Sacks, *The University of Texas at Austin, Austin, TX, United States*

155 **Understanding and Optimizing Electrical Stimulation of Neurons for Improving Regeneration: A Finite Element Simulation with Experimental Verification** SB^{3}C2015-520 pg.727
Robert D. Adams[1], Rebecca K. Willits[2], Amy B. Harkins[1], [1]*Saint Louis University, Saint Louis, MO, United States,*[2]*University of Akron, Akron, OH, United States*

156 **Schwann Cell Proliferation in Scaffolds with Decoupled Mechanical and Biochemical Properties** SB^{3}C2015-428 pg.565
Jessica Stukel, Wenda Zhou, Rebecca Willits, *The University of Akron, Akron, OH, United States*

157 **Wrinkled, Wavelength-Tunable Graphene-Based Surface Topographies for Directing Cell Alignment and Morphology** SB^{3}C2015-297 pg.341
Daniel F. Tonderys, *Brown University, Providence, RI, United States*

158 **Biomimetic Ex Vivo Model for Tracking Stem Cells During Microvascular Network Growth** SB^{3}C2015-1100 pg.1163
Mohammad S. Azimi, Amy L. Strong, Theresa B. Phamduy, Douglas B. Chrisey, Bruce A. Bunnell, Walter L. Murfee, *Tulane University, New Orleans, LA, United States*

THURSDAY, JUNE 18	12:30pm - 2:30pm

Poster Session I | **Multi-Scale Mechanics in Cell and Tissue Engineering** | **Event Center Tent**

159 **Optimization of Test Methods and Burst Property Characterization of Alginate Hydrogel Lung Sealants** SB^{3}C2015-630 pg.929
Patrick N. Charron, Spencer L. Fenn, Rachael A. Oldinski, *University of Vermont, Burlington, VT, United States*

160 **Numerical Simulations of Fibrous Biomaterial with Randomly Distributed Fiber Network** SB^{3}C2015-130 pg.62
Tao Jin, Ilinca Stanciulescu, *Rice University, Houston, TX, United States*

161 **Using Simulations with Realistic Fibrous Network Geometry to Find the Achievable Range of Mechanical Behaviors of Elastomeric Scaffolds** SB^{3}C2015-535 pg.754
James B. Carleton, Gregory J. Rodin, Michael S. Sacks, *University of Texas at Austin, Austin, TX, United States*

162 **A Strain Based Approach To Quantify Non-affine Behavior Of Three Dimensional Random Network** SB^{3}C2015-1080
Hamed Hatami-Marbini, *Oklahoma State University, Stillwater, OK, United States pg.1126*

163 **Empirically Determined Vascular Smooth Muscle Mechano-adaptation Laws** SB^{3}C2015-275 pg.305
Kerianne E. Steucke, Patrick W. Alford, *University of Minnesota, Minneapolis, MN, United States*

164 **Microstructure And Dynamics Of Crosslinked Collagen Gel** SB^{3}C2015-397 pg.503
Shengmao Lin[1], Linxia Gu[1,2], [1]*University of Nebraska-Lincoln, Lincoln, NE, United States,*[2]*Nebraska Center for Materials and Nanoscience, Lincoln, NE, United States*

165 **Probing Of A Complex Multi-layer Embryonic Tissue Through Novel 3D Bio-etching** SB^{3}C2015-352 pg.431
Melis Hazar[1], YongTae Kim[2], Philip R. LeDuc[1], William C. Messner[3], Lance A. Davidson[4], [1]*Carnegie Mellon University, Pittsburgh, PA, United States,*[2]*Georgia Institute of Technology, Atlanta, GA, United States,*[3]*Tufts University, Medford, MA, United States,*[4]*University of Pittsburgh, Pittsburgh, PA, United States*

166 **Numerical Simulation Of Effects Of Membrane Surface Viscosity On Tank-treading Motion Of Red Blood Cell** SB^{3}C2015-203 pg.190
Ken-ichi Tsubota, *Chiba University, Chiba, Japan*

167 **Wnt/beta-catenin Signaling Pathways Contributes To Dynamic Fluid Flow Loading Induced In Situ Osteocytic Calcium Oscillations In An Intact Mouse Femur** SB^{3}C2015-1084 pg.1134
Minyi Hu, Guowei Tian, Yi-Xian Qin, *Stony Brook University, Stony Brook, NY, United States*

168 **Nonlinear Anisotropic Mechanical Properties of Vascular Smooth Muscle Cells** SB^{3}C2015-650 pg.967
Zaw Win, Patrick W. Alford, *University of Minnesota, Minneapolis, MN, United States*

169 **Dynamic Fluid Flow Loading Induced In Situ Osteocytic Calcium Oscillations In An Intact Mouse Femur** SB^{3}C2015-1081 pg.1128
Minyi Hu, Guowei Tian, Yi-Xian Qin, *Stony Brook University, Stony Brook, NY, United States*

FRIDAY, JUNE 19	8:00am - 9:30am

Workshop: Mow and Fung Lectures - Dawn M. Elliott, Adam J. Engler — Primrose A/B

Session Chair: Louis Soslowsky, *University of Pennsylvania, Philadelphia, PA, United States*
Session Co-Chair: Guy Genin, *Washington University, St. Louis, MO, United States*

FRIDAY, JUNE 19	8:00am - 9:30am

Workshop: Problem-based Learning in Biomechanics — Superior

Session Chair: Alisa Morss Clyne, *Drexel University, Philadelphia, PA, United States*
Session Co-Chair: Kristen Billiar, *Worcester Polytechnic Institute, Worcester, MA, United States*

FRIDAY, JUNE 19	8:00am - 9:30am

Workshop: CFD Challenge 2015 — Wasatch

Session Chair: Kenichi Kono, *Wakayama Rosai Hospital, Japan*
Session Co-Chair: Kristian Valen-Sendstad, *University of Toronto, Toronto, ON, Canada*

FRIDAY, JUNE 19	8:00am - 9:30am

Workshop: Mentee-mentor Matching Mixer and Best Practices in Mentoring — Magpie

Session Chair: Naomi Chesler, *University of Wisconsin, Madison, WI, United States*
Session Co-Chair: Lakiesha Williams, *Mississippi State University, Mississippi State, MS, United States*
Session Co-Chair: Victor Barocas, *University of Minnesota, Minneapolis, MN, United States*

FRIDAY, JUNE 19	8:00am - 9:30am

Workshop: Strategies for a Successful Postdoctoral Experience — Maybird

Organizers: ASME Bioengineering Division Student Leadership Committee (special thanks to Kathryn Drzewiecki, Samira Jamalian, Paola Jaramillo, and Samantha Schoell)

FRIDAY, JUNE 19	9:45am - 10:45am

PLENARY SESSION II – Andrew McCulloch **Ballrooms 1-3**

FRIDAY, JUNE 19	11:00am - 12:30pm

PhD Competition - Biofluid Mechanics — Primrose A

Session Chair: Stephanie George, *East Carolina University, Greenville, NC, United States*
Session Co-Chair: Brandon Dixon, *Georgia Institute of Technology, Atlanta, GA, United States*

11:00AM **Increased Red Blood Cell Stiffness Increases Pulmonary Vascular Resistance And Pulmonary Arterial Pressure**
SB[3]C2015-587 pg.853
David A. Schreier, Omid Forouzan, Timothy Hacker, John Sheehan, Naomi C. Chesler, *University of Wisconsin-Madison, Madison, WI, United States*

11:15AM **Effect of Bicuspid Aortic Valve Morphotype on Ascending Aorta Hemodynamics: a Computational Study**
SB[3]C2015-304 pg.355
Kai Cao, Philippe Sucosky, *University of Notre Dame, Notre Dame, IN, United States*

11:30AM **Hemodynamic Characterization of Different Basilar Tip Aneurysm Templates Using Computational Fluid Dynamics**
SB[3]C2015-1114 pg.1191
Priya Nair[1], Brian W. Chong[2], David Frakes[1], *[1]Arizona State University, Tempe, AZ, United States,[2]Mayo Clinic Hospital, Phoenix, AZ, United States*

11:45AM **Modulation of Lymph Node Resistance during Inflammation: Experimental Measurement and Computational Modeling** SB[3]C2015-566 pg.813
Mohammad Jafarnejad[1], Joshua P. Scallan[2], Joseph M. Sherwood[1], Darryl R. Overby[1], David C. Zawieja[3], Michael J. Davis[2], James E. Moore[1], *[1]Imperial College London, London, United Kingdom,[2]University of Missouri, Columbia, MO, United States,[3]Texas A&M Health Science Center, Temple, TX, United States*

12:00PM **Inlet Flow Rate Variation and Onset of Flow Instabilities in the Carotid Siphon** SB[3]C2015-276 pg.307
Resmi KrishnankuttyRema[1], Kristian Valen-Sendstad[1,2], David Steinman[1], *[1]University of Toronto, Toronto, ON, Canada,[2]Simula Research Laboratory, Lysaker, Norway*

12:15PM **Suction Effect Produced by Active Contraction of Collecting Lymphatic Vessels Facilitates Lymphatic Filling**
SB[3]C2015-539 pg.762
Samira Jamalian[1], Mohammad Jafarnejad[1], Christopher D. Bertram[2], David C. Zawieja[3], J. Davis[4], James E. Moore[1], *[1]Imperial College London, London, United Kingdom,[2]University of Sydney, New South Wales, Australia,[3]Texas A&M Health Science Center, Temple, TX, United States,[4]University of Missouri School of Medicine, Columbia, MO, United States*

FRIDAY, JUNE 19	11:00am - 12:30pm

PhD Competition - Biotransport and Simulation — Superior

Session Chair: Victor Varner, *Princeton University, Princeton, NJ, United States*
Session Co-Chair: Alisa Morss, *Drexel University, Philadelphia, PA, United States*

11:00AM **Lymphatic Disease Phenotyping With Near-Infrared Imaging** SB[3]C2015-1104 pg.1171
Tyler S. Nelson[1], Michael J. Weiler[1], Ira L. Savetsky[2], Xiaolei Liu[3], Babak J. Mehrara[2], Hong Chen[4], J. Brandon Dixon[1], *[1]Georgia Institute of Technology, Atlanta, GA, United States,[2]Memorial Sloan Kettering Cancer Center, New York, NY, United States,[3]St. Jude Children's Research Hospital, Memphis, TN, United States,[4]Oklahoma Medical Research Foundation, Oklahoma City, OK, United States*

11:15AM **Red Blood Cell Dynamic Deformability and Adhesion in Microscale Flow Determine Cellular Heterogeneity in Sickle Cell Disease** SB³C2015-504 pg.696
Yunus Alapan[1], Yumi Matsuyama[1], Jane Little[1,2], Umut A. Gurkan[1,3], [1]*Case Western Reserve University, Cleveland, OH, United States,*[2]*University Hospitals, Cleveland, OH, United States,*[3]*Louis Stokes Cleveland Veterans Affairs Medical Center, Cleveland, OH, United States*

11:30AM **Reduced Lymphatic Function Correlates With Disease Progression In a Novel Single Vessel Ligation Model of Lymphedema** SB³C2015-410 pg.529
Michael J. Weiler, Tyler S. Nelson, J. Brandon Dixon, *Georgia Tech, Atlanta, GA, United States*

11:45AM **Finite Element Modeling Of Active Transmembrane Cell Transport** SB³C2015-356 pg.433
Chieh Hou, Kelly Terlizzi, Gerard A. Ateshian, *Columbia University, New York, NY, United States*

12:00PM **Combined Experimental and Finite Element Analysis to Determine the Diffusion Coefficient Within and Between Human Skin Layers** SB³C2015-274 pg.303
Anne M. Römgens[1], Dan L. Bader[1,2], Frank P. T. Baaijens[1], Cees W. J. Oomens[1], [1]*Eindhoven University of Technology, Eindhoven, Netherlands,*[2]*University of Southampton, Southampton, United Kingdom*

12:15PM **CFD Simulation of Transition to Turbulence for Newtonian vs. Non-Newtonian Flow Through a Stenosis** SB³C2015-363 pg.445
M. Owais Khan[1], Kristian Valen-Sendstad[1,2], Dipankar Biswas[3], David M. Casey[3], Francis Loth[3], David Steinman[1], [1]*University of Toronto, Toronto, ON, Canada,*[2]*Simula Research Laboratory, Lysaker, Norway,*[3]*University of Akron, Akron, OH, United States*

FRIDAY, JUNE 19	**11:00am - 12:30pm**

PhD Competition - Cellular and Tissue Engineering — Wasatch

Session Chair: Victor Lai, *University of Minnesota, Duluth, MN, United States*
Session Co-Chair: Colleen Witzenburg, *University of Virginia, Charlottesville, VA, United States*

11:00AM **Interstitial Cell Migration in Dense Connective Tissues is Modulated by Matrix Microstructure and Micromechanics** SB³C2015-129 pg. 60
Feini Qu[1,2], Miltiadis H. Zgonis[1,2], Robert L. Mauck[1,2], [1]*University of Pennsylvania, Philadelphia, PA, United States,*[2]*Philadelphia VA Medical Center, Philadelphia, PA, United States*

11:15AM **Cells Alter Traction Force and Orientation in Response to Long-term Cyclic Stretch** SB³C2015-461 pg.624
Heather A. Cirka, Qi Wen, Kristen L. Billiar, *Worcester Polytechnic Institute, Worcester, MA, United States*

11:30AM **Dedifferentiaton of Chondrocytes Influences Strain Transfer Measured by Deformable Image Registration** SB³C2015-561 pg.803
Jonathan T. Henderson[1], Benjamin Seelbinder[1], Alexander Veress[2], Corey Neu[1], [1]*Purdue University, West Lafayette, IN, United States,*[2]*University of Washington, Seattle, WA, United States*

11:45AM **An Active Contraction Model Of Valvular Interstitial Cells** SB³C2015-614 pg.901
Yusuke Sakamoto, Michael Sacks, *The University of Texas at Austin, Austin, TX, United States*

12:00PM **A Predictive 3D High-Content/High Throughput Screening Platform to Elucidate and Enhance Multilineage Stem Cell Differentiation** SB³C2015-601 pg.877
Amit Paul, Bo Chen, Elise DeBruyn, Michael Cho, *University of Illinois at Chicago, Chicago, IL, United States*

12:15PM **In Vitro Growth Trajectory And In Vivo Implantation Of Cell-seeded Disc-like Angle Ply Structures For Total Disc Replacement** SB³C2015-131 pg.64
John T. Martin[1,2], Dong Hwa Kim[1,2], Kensuke Ikuta[1,2], Christian G. Pfeifer[1,2], Lachlan J. Smith[1,2], Dawn M. Elliott[3], Harvey E. Smith[1,2], Robert L. Mauck[1,2], [1]*University of Pennsylvania, Philadelphia, PA, United States,*[2]*Philadelphia VA Medical Center, Philadelphia, PA, United States,*[3]*University of Delaware, Newark, DE, United States*

FRIDAY, JUNE 19	11:00am - 12:30pm

PhD Competition - Mechanics and Rehabilitation — Magpie

Session Chair: Bradley Davidson, *University of Denver, Denver, CO, United States*
Session Co-Chair: Darryl Thelen, *University of Wisconsin-Madison, Madison, WI, United States*

11:00AM **Blocking Blood-Spinal Cord Barrier Breakdown Prevents the Development of Pain Following Nerve Root Compression Injury** SB³C2015-319 pg.379
Jenell Smith, Paul Janmey, Beth Winkelstein, *University of Pennsylvania, Philadelphia, PA, United States*

11:15AM **Defining Collagen Fiber Mechanics in Neuron-Collagen Constructs Under Stretch Using Integrated Experimental & Modeling Approaches** SB³C2015-281 pg.315
Sijia Zhang, Xuan Cao, Vivek Shenoy, Beth Winkelstein, *University of Pennsylvania, Philadelphia, PA, United States*

11:30AM **Redistribution of Knee Loads Using Auditory Feedback from Pressure Detecting Shoe Insoles** SB³C2015-245 pg.261
Christopher F. Ferrigno, Ina S. Stoller, Laura E. Thorp, Najia Shakoor, Markus M. Wimmer, *Rush University, Chicago, IL, United States*

11:45AM **Pre-Clinical Assesment of a Percutaneous Leaflet Resection Device for Treatment of Degenerative Mitral Valve Disease** SB³C2015-559 pg.799
Steven Boronyak, Brett Byram, Joseph Fredi, Michael Young, W. David Merryman, *Vanderbilt University, Nashville, TN, United States*

12:00PM **Subject-specific Calibration Of Geometric Neuromusculoskeltal Models** SB³C2015-585 pg.849
Andrew J. Meyer, Carolynn Patten, Benjamin J. Fregly, *University of Florida, Gainesville, FL, United States*

12:15PM **In-Vivo Dynamic Measurement of Tibiotalar and Subtalar Joint Kinematics Using Dual Fluoroscopy: A Framework for Studying OA.** SB³C2015-527 pg.739
Koren E. Roach, Bibo Wang, Ashley L. Kapron, Niccolo M. Fiorentino, Charles L. Saltzman, Madeline Singer, Andrew E. Anderson, *University of Utah, Salt Lake City, UT, United States*

FRIDAY, JUNE 19	11:00am - 12:30pm

PhD Competition - Characterization of Tissue Mechanics — Maybird

Session Chair: Sarah Kieweg, *University of Kansas, Lawrence, KS, United States*
Session Co-Chair: Ian A. Sigal, *University of Pittsburgh, Pittsburgh, PA, United States*

11:00AM **Regional Contraction Shapes the Three-Dimensional Morphogenesis of the Embryonic Forebrain** SB³C2015-1030 pg.1036
Kara E. Garcia, Philip V. Bayly, Larry A. Taber, *Washington University in St. Louis, St. Louis, MO, United States*

11:15AM **The Dynamic Mechanical Response is Severely Altered in Collagen V Deficient Mouse Supraspinatus Tendons** SB³C2015-104 pg.24
Brianne K. Connizzo[1], Mei Sun[2], David E. Birk[2], Louis J. Soslowsky[1], *[1]University of Pennsylvania, Philadelphia, PA, United States,[2]University of South Florida, Tampa, FL, United States*

11:30AM **Evidence that Interfibrillar Load Transfer in Tendon is Supported by a Network of Small Diameter Collagen Fibrils** SB³C2015-132 pg.66
Spencer E. Szczesny[1], Kristen L. Fetchko[2], Jeffrey L. Caplan[3], Pal Pedersen[4], Dawn M. Elliott[2], *[1]University of Pennsylvania, Philadelphia, PA, United States,[2]University of Delaware, Newark, DE, United States,[3]Delaware Biotechnology Institute, Newark, DE, United States,[4]Carl Zeiss Microscopy, Thornwood, NY, United States*

11:45AM **3D Strains in Posterior Sclera Using Ultrasound Speckle Tracking** SB³C2015-515 pg.716
Elias R. Pavlatos, Benjamin Cruz-Perez, Hugh J. Morris, Hong Chen, Richard T. Hart, Jun Liu, *The Ohio State University, Columbus, OH, United States*

12:00PM **The Collagen Directionality and Dispersion and Mechanical Indentation Response in Nonpregnant Human Cervical Tissue** SB³C2015-1090 pg.1144
Wang Yao[1], Yu Gan[1], Christine Hendon[1], Joy Vink[2], Ronald Wapner[2], Kristin Myers[1], *[1]Columbia University, New York, NY, United States,[2]Columbia University Medical Center, New York, NY, United States*

12:15PM **Tribological Rehydration: Maintaining and Rebuilding Interstitial Fluid Pressure in Cartilage** SB³C2015-133 pg.68
Axel C. Moore, David L. Burris, *University of Delaware, Newark, DE, United States*

FRIDAY, JUNE 19	11:00am - 12:30pm

Undergraduate Design Competition — Golden Cliff / Eagle's Nest

Session Co-Chair: Martin L. Tanaka, *Western Carolina University, Cullowhee, NC, United States*
Session Co-Chair: Sara Roccabianca, *Michigan State University, East Lansing, MI, United States*

11:00AM **BioView: A Wearable Sensor Array For Rehabilitation Biofeedback** SB³C2015-1420 pg.1329
Tyler Maydew, Brett Donnermeyer, Kathryn Thompson, Alwyn Johnson, Bradley Davidson, *University of Denver, Denver, CO, United States*

11:15AM **The Tension Assisted Device: An Orthotic Alternative to High Tone Plantar Flexion** SB³C2015-298 pg. 343
Ron V. Perrone, Elizabeth Duncan, Cory Jeanes, Mike Martorano, Gary Bowlin, John Williams, Susan Anderson, *University of Memphis, Memphis, TN, United States*

11:30AM **DermaShift: Diagnostic Device For Pressure Ulcer Formation** SB³C2015-1421 pg.1331
Francisca Acosta[1], Hope Atina[1], Kim Le[1], Andrea Pinto[1], William Wilson[1], Erice Richardson[1], Catherine Ambrose[2], Lex Frieden[2], [1]*Rice University, Houston, TX, United States,*[2]*Texas Health Science Center, Houston, TX, United States*

11:45AM **Smartboot: An Instrumented Clinical Walking Boot for Partial Weight Bearing Training** SB³C2015-328 pg.395
Dustyn Roberts, Tim West, Michael Schenk, David Schnall, Margaret O'Brien, Melissa Groome, Brian Knarr, Jill Higginson, Anita Singh, *University of Delaware, Newark, DE, United States*

12:00PM **Soft Ankle-foot Orthotic** SB³C2015-187 pg.160
Adam Podolec, Megan Erhart, Noah Schadt, Geni Giannotti, Jared Green, Tyler Leichenberger, *Rochester Institute of Technology, Rochester, NY, United States*

12:15PM **Design of a Low-cost Haptic Assistive Handwriting Device** SB³C2015-225 pg.226
Eamon Campolettano, Alexander Croft, Kevin Fasano, Stephen Hodge, Kevin Myers, Brian Pinkard, Jessica Ross, Allison Scoular, Thomas S. Todd, Alexander A. Brown, *Lafayette College, Easton, PA, United States*

FRIDAY, JUNE 19	11:00am - 12:30pm

PhD Competition - Mechanics of Injury and Repair — Primrose B

Session Chair: Ramesh Ragupathy, *Drexel University, Philadelphia, PA, United States*
Session Co-Chair: Corey Neu, *Purdue University, West Lafayette, IN, United States*

11:00AM **Thermoresponsive, Redox-crosslinked Cellulosic Hydrogels Undergo In Situ Gelation And Restore Nucleus Pulposus Biomechanical Properties Post Nucleotomy** SB³C2015-263 pg.287
Devika M. Varma[1], Huizi A. Lin[1], Rose G. Long[2], Carine Rognon[3], Andrew C. Hecht[2], James C. Iatridis[2], Steven B. Nicoll[1], [1]*The City College of New York, CUNY, New York, NY, United States,*[2]*Icahn School of Medicine at Mount Sinai, New York, NY, United States,*[3]*Swiss Federal Institute of Technology of Zurich (ETHZ), Zurich, Switzerland*

11:15AM **A Multigenerational Collagen Damage Model Explains Engineered Cartilage Growth and Remodeling Phenomena** SB³C2015-642 pg.953
Robert J. Nims, Alexander D. Cigan, Brian K. Jones, Krista M. Durney, Clark T. Hung, Gerard A. Ateshian, *Columbia University, New York, NY, United States*

11:30AM **Influence of Intracortical Porosity on the Fracture Susceptibility of Human Cortical Bone** SB³C2015-672 pg.1009
Andrew P. Baumann, Travis L. Turnbull, Glen L. Niebur, Ryan K. Roeder, *University of Notre Dame, Notre Dame, IN, United States*

11:45AM **Effective Remodeling in Cerebral Aneurysm: a Case Study** SB³C2015-1159 pg.1271
Xinjie Duan[1], Bong Jae Chung[2], Juan R. Cebral[2], Khaled Aziz[3], Anne M. Robertson[1], [1]*University of Pittsburgh, Pittsburgh, PA, United States,*[2]*George Mason University, Fairfax, VA, United States,*[3]*Allegheny General Hospital, Pittsburgh, PA, United States*

12:00PM **Cartilage Wear Initiated by Fatigue Damage Under Physiologic Loading when Fluid Load Support and Boundary Lubrication are Compromised** SB^{3}C2015-1160 pg.1273
Krista M. Durney, Sevan R. Oungoulian, Brian K. Jones, Jason T. Suh, Clark T. Hung, Gerard A. Ateshian, *Columbia University, New York, NY, United States*

12:15PM **Cornea Damage Progression following Blast Exposure** SB^{3}C2015-1169 pg.1289
Dan F. Shedd[1], Justin A. Jones[1], Brian Zaugg[2], Brittany Coats[1], *[1]University of Utah, Salt Lake City, UT, United States,[2]John A. Moran Eye Center, Salt Lake City, UT, United States*

FRIDAY, JUNE 19	12:30pm - 3:00pm

Poster Session II **Thermal Effects and Nanoparticles** **Event Center Tent**

170 **Determination of the Biophysical Parameters of HUVECs and Their Application in Optimization of the Addition and Removal of Cryoprotective Agents** SB^{3}C2015-434 pg.578
Yuntian Zhang, Dan Niu, Gang Zhao, *University of Science and Technology of China, Hefei, China*

171 **Nonlinear Derating of High-Intensity Focused Ultrasound using Hydrophone Measurements in Water** SB^{3}C2015-542
Seyed Ahmad Reza Dibaji[1], Yunbo Liu[2], Joshua E. Soneson[2], Rupak K. Banerjee[1], Matthew R. Myers[2], *[1]University of Cincinnati, Cincinnati, OH, United States,[2]US Food and Drug Administration, Silver Spring, MD, United States* pg.768

172 **Effect of Hydroxyapatite Nanoparticles on Cryopreservation of HUVECs** SB^{3}C2015-427 *pg.*563
Yuanyuan Zheng, Jianye Wang, **Gang Zhao**, Tao Wang, *University of Science and Technology of China, Hefei, China*

173 **Feasibility of Utilizing Thermal Images for Melanoma Screening** SB^{3}C2015-126 *pg.*54
Alexander LeBrun, **Liang Zhu**, *University of Maryland Baltimore County, Baltimore, MD, United States*

174 **Application Of Mesoporous Silica Nanoparticle At Drug Delivery System** SB^{3}C2015-1372 *pg.*1327
M. Titirinli, Sevil Yücel, B. Karakuzu, Y. Basarab, *Yildiz Technical University, Istanbul, Turkey*

FRIDAY, JUNE 19	12:30pm - 3:00pm

Poster Session II **Transport at the Cell and Tissue Level** **Event Center Tent**

176 **Cancer-Associated Fibroblasts Promote Vascularization in Collagen and Fibrin Matrices** SB^{3}C2015-199 *pg.*182
M.K. Sewell-Loftin, Samantha van Hove, Gregory Longmore, Steven George, *Washington University in St. Louis, St. Louis, MO, United States*

177 **Insights Into the Hemodynamic Factors Affecting Embolus Transport for Stroke** SB^{3}C2015-584 *pg.*847
Debanjan Mukherjee, Shawn C. Shadden, *U.C. Berkeley, Berkeley, CA, United States*

178 **A Peristaltic Mechanism For Clearance Of Solutes In Periarterial Basement Membranes** SB^{3}C2015-391 *pg.* 491
M Keith Sharp[1], Alexandra Keith Diem[2], Roy O. Weller[2], Roxana O. Carare[3], *[1]University of Louisville, Louisville, KY, United States,[2]University of Southampton, Southampton, United Kingdom,[3]University of Southampton, Suthampton, United Kingdom*

179 **Trapping of Tumor Cells Using Rapid Electrokinetic Patterning (REP)** SB^{3}C2015-163 *pg.*114
Katherine N. Clayton, Seungman Park, Steven Wereley, Bumsoo Han, *Purdue University, West Lafayette, IN, United States*

180 **Visco-Hyperelastic and Biphasic Properties of a Brain Phantom Agarose Gel** SB^{3}C2015-114 *pg.*38
Gerson Cordoba[1], Gustavo Orozco[1], **Fernando Casanova**[1], Joshua H. Smith[2], Jose J. Garcia[1], *[1]Universidad del Valle, Cali, Colombia,[2]Lafayette College, Easton, PA, United States*

FRIDAY, JUNE 19	12:30pm - 3:00pm

Poster Session II — Design, Dynamics and Rehab — Event Center Tent

181 **The Effect Of Bone Defect Size And Position On Cementless Acetabular Cup Stability : A Finite Element Analysis** SB3C2015-1091 *pg.*1146
Mark H. Gonzalez, Farid Amirouche, Gianfranco Solitro, *University of Illinois at Chicago, Chicago, IL, United States*

182 **Simulation-Based Design of a Hip Actuator for Running a Mile Sprint** SB3C2015-641 *pg.* 951
John R. Rogers, Julie E. Dillon, Cameron I. McDonald, Gabriela C. Barrera-Gutierez, *United States Military Academy, West Point, NY, United States*

183 **Comparison of Human Walking Backward and Forward Using Optimization Method** SB3C2015-351 *pg.*429
Yujiang Xiang[1], Hyun-Jung Kwon[2], *[1]University of Alaska Fairbanks, Fairbanks, AK, United States,[2]The Ohio State University, Columbus, OH, United States*

184 **Beginning Braille Learning Device** SB3C2015-252 *pg.*273
Kelton Gubler, Jason Castillo, Adam Daly, Austin Eastman, Kay B. Freckleton, Andrew Silotti, *University of Utah, Salt Lake City, UT, United States*

185 **Ocular Bobbing Compensation System** SB3C2015-3803 *pg.*1335
Yucong Gu, Yuchen Yan, You Chen, Tai Kim, Ken Fischer, *University of Kansas, Lawrence, KS, United States*

186 **A Bioengineering Solution To Cure Spinal Cord Injury** SB3C2015-1161 *pg.*1275
Anita Singh[1], Jacklyn Witko[1], Brittany King[1], Alexender Herman[1], Andrea Vernengo[1], Babitha Tom[2], *[1]Rowan University, Glassboro, NJ, United States,[2]Widener University, Chester, PA, United States*

187 **High Performance Luxury 4WD (or All Terrain) Wheel Chair** SB3C2015-3731 *pg.*1333
Jason McCurry, Philip Stykes, Alex Wilfong, Martin Tanaka, *Western Carolina University, Cullowhee, NC, United States*

188 **Dynamic Balance Using The COM And COP Inclination Angle During A Golf Swing** SB3C2015-22 *pg.*1
Ahnryul Choi[1], Joung Hwan Mun[2], *[1]The University of Texas Health Science Center at Houston, Houston, TX, United States,[2]Sungkyunkwan University, Suwon, Korea, Republic of*

FRIDAY, JUNE 19	12:30pm - 3:00pm

Poster Session II — Fluid Mechanics of Atherosclerosis and Aneurysms — Event Center Tent

189 **FSI Simulations for the Hemodynamic Assessment of the Carotid Bifurcation in an Atherosclerotic Mouse Model** SB3C2015-528 *pg.*741
David De Wilde[1], Bram Trachet[1,2], Nic Debusschere[1], Francesco Iannaccone[1], Abigaïl Swillens[1], Joris Degroote[1], Jan Vierendeels[1], Guido R. Y. De Meyer[3], Patrick Segers[1], *[1]Ghent University, Gent, Belgium,[2]Ecole Polytechnique Fédérale de Lausanne, Lausanne, Switzerland,[3]University of Antwerp, Wilrijk, Belgium*

190 **Integrating Morphologic, Biomechanic, Biological And Clinical Risk Factors To Improve Decision Making In The Management Of Abdominal Aortic Aneurysm Disease** SB3C2015-454 *pg.*616
Eleni Metaxa[1], Nikolaos Kontopodis[2], Christos V. Ioannou[2], Yannis Papaharilaou[1], *[1]FORTH, Heraklion, Greece,[2]University of Crete, Heraklion, Greece*

191 **Biomechanical Response And Fiber Microstructure Coupled With Localized Protease Activity And Inhibition In The Angiotensin Ii Infused Apoe-/- Mouse Model Of Aneurysm** SB3C2015-188 *pg.*162
Darren G. Haskett[1], Tyler S. Smith[1], D. Catalina Ardilia[1], Tom C. Doetschman[1], Oliver J. McIntyre[2], Dominic V. McGrath[1], Urs Utzinger[1], Jonathan P. Vande Geest[1], *[1]University of Arizona, Tucson, AZ, United States,[2]Vanderbilt University, Nashville, TN, United States*

192 **Geometric and Hemodynamics Implications of Moyamoya Disease on Carotid Siphon** SB3C2015-264 *pg.*289
Muhammad Jamil[1], Mehnaz Haq[1], Heidi Kang[2], Zhi Rui Lee[2], Phua Hwee Tang[3], **Choon Hwai Yap**[1], *[1]National University of Singapore, Singapore,[2]Hwa Chong Institution, Singapore,[3]KKH Women's and Children's Hospital, Singapore*

193 **Coherent Wall Shear Stress Structures Determine the Near Wall Transport in Aneurysms.** SB^{3}C2015-423 *pg.*555
Amirhossein Arzani[1], Guoning Chen[2], Alberto M. Gambaruto[3], Shawn C. Shadden[1], [1]*University of California, Berkeley, Berkeley, CA, United States,*[2]*University of Houston, Houston, TX, United States,*[3]*Barcelona Supercomputing Center, Barcelona, Spain*

194 **RhoA Mediated Effects of Statin Therapy on Endothelial Cells: A Comparison to Fasudil and Latrunculin A** SB^{3}C2015-538 *pg.*760
Melissa L. Dick[1,2], Katherine N. MacDonald[1], Jean-Claude Tardif[2], Richard L. Leask[1,2], [1]*McGill University, Montreal, QC, Canada,*[2]*Montreal Heart Institute, Montreal, QC, Canada*

195 **In Vitro And Computational Fluid Dynamics Comparison Of The Flow Diversion Efficacy Of Five Commercial Stents** SB^{3}C2015-636 *pg.*941
Ronak J. Dholakia[1], Andrew Pagano[1], Fotis Drakopoulos[2], Ari Kappel[1], Chander Sadasivan[1], Xiangmin Jiao[1], David J. Fiorella[1], Nikos Chrisochoides[2], Henry H. Woo[1], Baruch B. Lieber[1], [1]*Stony Brook University, Stony Brook, NY, United States,*[2]*Old Dominion University, Norfolk, VA, United States*

196 **Effect of Red Blood Cells on Endothelial Cell Shear Stress Studied Using Discrete-Nature Blood Flow Simulations** SB^{3}C2015-1024 *pg.*1026
Brenna E. Hogan[1], Zaiyi Shen[2], Chaouqi Misbah[2], Abdul Barakat[1], [1]*Ecole Polytechnique, Palaiseau Cedex, France,*[2]*Universite Joseph Fourier, Grenoble, France*

197 **FRED versus Pipeline: An In-vitro Comparison of Cerebral Aneurysm Hemodynamics Using Particle Image Velocimetry** SB^{3}C2015-1108 *pg.*1179
Priya Nair[1], Brian W. Chong[2], Haithem Babiker[1], Justin Ryan[1], L. Fernando Gonzalez[3], David Frakes[1], [1]*Arizona State University, Tempe, AZ, United States,*[2]*Mayo Clinic Hospital, Phoenix, AZ, United States,*[3]*Duke University School of Medicine, Durham, NC, United States*

198 **In Vitro Validation of Endovascular Doppler-derived Flow Rates in Cerebral Vessels** SB^{3}C2015-526 *pg.*737
Patrick McGah, John Nerva, Ryan Morton, Michael Barbour, Pierre Mourad, Michael Levitt, Louis Kim, Alberto Aliseda, *University of Washington, Seattle, WA, United States*

199 **The Role of the Glycocalyx in Leukocyte Adhesion to the Endothelium** SB^{3}C2015-612 *pg.* 897
Karli K. McDonald, Scott E. Cooper, Richard L. Leask, *McGill University, Montreal, QC, Canada*

200 **Validation Of CFD Solver Of A Clinical Tool Using PIV On A Patient-Specific Intracranial Aneurysm.** SB^{3}C2015-623
Nikhil Paliwal, Christopher Martensen, Nicole Varble, Robert Damiano, Adnan Siddiqui, Elad Levy, Jianping Xiang, Hui Meng, *University at Buffalo, State University of New York, Buffalo, NY, United States pg.915*

201 **The Quantification of Blood Flow Patterns Induced by Endovascular Stent Grafts Using a Non-Newtonian Blood Analog** SB^{3}C2015-637 *pg.*943
Amanda Colella Centazzo, Clifton R. Johnston, *Dalhousie University, Halifax, NS, Canada*

FRIDAY, JUNE 19	12:30pm - 3:00pm

Poster Session II — Heart Valves, Mechanical Circulatory Support, Thrombosis, and Hemolysis — Event Center Tent

202 **Modulation of Platelet Microtubule Function Alters Platelet Stiffness and Mechanotransductive Responsiveness to Shear Stress** SB^{3}C2015-1036 *pg.*1046
Siu Ling Leung[1], Yi Lu[1], Danny Bluestein[2], Marvin J. Slepian[1,2], [1]*the University of Arizona, Tucson, AZ, United States,*[2]*Stony Brook University, Stony Brook, NY, United States*

203 **Convective Leakage Makes Heparin Locking of Central Venous Catheters Ineffective within Seconds: Experimental Measurements in an In Vitro Model of the Pediatric Superior Vena Cava** SB^{3}C2015-1042 *pg.*1058
Alberto Aliseda[1], Michael Barbour[1], Patrick M. McGah[1], Kurt R. Sansom[1], Kenneth Gow[2], [1]*University of Washington, Seattle, WA, United States,*[2]*Seattle Children's Hospital, Seattle, WA, United States*

204 **Aortic Flow Waveform Shape Regulates Valvular Hemodynamics And Energy Losses** SB^{3}C2015-1063 *pg.*1096
Brandon L. Moore, Lakshmi P. Dasi, *Colorado State University, Fort Collins, CO, United States*

205 **Analyzing The Effect Of Stent Geometry For Polymeric Aortic Valves** SB³C2015-1071 *pg.*1110
Atieh Yousefi Koupaei, Brandon L. Moore, David L. Bark Jr., David Prawel, Lakshmi P. Dasi, *Colorado State University, Fort Collins, CO, United States*

206 **Mathematical Modeling of Thrombus Formation in Idealized Models of Aortic Dissection** SB³C2015-170 *pg.*126
Claudia Menichini, Xiao Y. Xu, *Imperial College London, London, United Kingdom*

207 **Characterizing The Inflammatory And Coagulation Response To Trauma And Resuscitation** SB³C2015-579 *pg.*837
Elaheh Rahbar[1], Jessica C. Cardenas[2], Nena Matijevic[2], Deborah J. del Junco[2], Jeanette M. Podbielski[2], Mitchell J. Cohen[3], Bryan A. Cotton[2], John B. Holcomb[2], Charles E. Wade[2], [1]*Wake Forest University, Winston-Salem, NC, United States,*[2]*University of Texas Health Science Center, Houston, TX, United States,*[3]*University of California San Francisco, San Francisco, CA, United States*

208 **Valve Interstitial Cell Shape Regulates Cell Function And Phenotype.** SB³C2015-619 *pg.*909
Atefeh Razavi, Rachel Reynolds, Kartik Balachandran, *University of Arkansas, Fayetteville, AR, United States*

209 **Aortic Valve Calcification: Geometric And Biomechanical Analysis** SB³C2015-1055 *pg.*1084
Banafsheh Zebhi[1], Brandon L. Moore[1], Gary Luckasen[2], Lakshmi P. Dasi[1], [1]*Colorado State University, Fort Collins, CO, United States,*[2]*Medical Center of Rockies, Loveland, CO, United States*

211 **Effect Of Assymetric Deployment On The Function And Leaflet Mechanics Of Bioprosthetic TAVI Valves** SB³C2015-524 *pg.*733
Michael B. Gogarty[1], Pablo Maureira[2], Lakshmi P. Dasi[1], [1]*Colorado State University, Fort Collins, CO, United States,*[2]*Lorraine University Hospital of Nancy, Nancy, France*

212 **Vascular Compliance Effect On Endothelial-to-mesenchymal Transdifferentiation Under Highly Pulsatile Fluid Shear** SB³C2015-1111 *pg.*1185
Winston H. Elliott[1], Yan Tan[2], Nancy Tseng[2], Wei Tan[1,2], [1]*University of Colorado- Boulder, Boulder, CO, United States,*[2]*University of Colorado- Denver, Denver, CO, United States*

213 **Intimal Hyperplasia and its Implications in the Cephalic Arch: A Numerical Study of Non-Physiological Hemodynamics in Patients with Brachiocephalic Fistulae** SB³C2015-315 *pg.* 373
Seyed Mohammad Javid Mahmoudzadeh Akherat[1], Michael E. Boghosian[1], Kevin W. Cassel[1], Mary S. Hammes[2], [1]*Illinois Institute of Technology, Chicago, IL, United States,*[2]*University of Chicago, Chicago, IL, United States*

214 **In Vitro Pulsatile Flow Loop Using Human Blood To Mimic Physiological Flow Conditions** SB³C2015-664 *pg.*995
Ryan W. Oba, David Bark, Ketul Popat, **Lakshmi P. D. Dasi**, *Colorado State University, Fort Collins, CO, United States*

FRIDAY, JUNE 19	**12:30pm - 3:00pm**

Poster Session II — Respiratory and Cerebrospinal Fluid Motion — Event Center Tent

215 **Effects of Inhalation Transience on Flow Structures During Numerical Simulation of Airflow Through a CT-based Airway Geometry** SB³C2015-649 *pg.*965
Richard R. Gruetzemacher, *The University of Tennessee at Chattanooga, Chattanooga, TN, United States*

216 **Bariatric Surgery Improvements For OSA Patients** SB³C2015-151 *pg.*92
Ahmed M. Al-Jumaily, *Auckland University of Technology, Auckland, New Zealand*

217 **Acoustic Detection of Respiratory Sounds in Silicone Lung Airway Model Using Microphone Array System** SB³C2015-360 *pg.*439
Gabriel Pramudita Saputra[1], Kazunori Nozaki[2], Satoshi Ii[1], Chizu Habukawa[1,3], Shigeo Wada[1], [1]*Osaka University, Toyonaka, Osaka, Japan,*[2]*Osaka University Dental Hospital, Suita, Osaka, Japan,*[3]*Minami Wakayama Medical Center, Tanabe, Wakayama, Japan*

218 **Pressure Modulation Improves OSA Patient Therapy** SB³C2015-246 *pg.*263
Ahmed M. Al-Jumaily, *Auckland University of Technology, Auckland, New Zealand*

219 **Multicenter Comparison Of 4D Phase Contrast MRI Measurement Of Cerebrospinal Fluid Dynamics In The Cervical Spine** SB3C2015-583 *pg.*845
Suraj Thyagaraj[1], Daniel Giese[2], Francesco Santini[3], Eleonora Fornari[4], Alexander C. Bunck[2], Francis Loth[1], Bryn A. Martin[1], [1]*University of Akron, Akron, OH, United States,*[2]*University of Cologne, Cologne, Germany,*[3]*University of Basel, Basel, Switzerland,*[4]*Lausanne University Hospital, Lausanne, Switzerland*

220 **Integration Of A Spontaneous Respiratory Driver With Blood Gas Feedback Into Biogears, An Open-source, Whole-body Physiology Model** SB3C2015-327 *pg.*393
Yeshitila Gebremichael, Rachel Clipp, Jeffrey Webb, Aaron Bray, Cameron Thames, Zack Swarm, Jennifer Carter, Jeremiah Heneghan, *Applied Research Associates, Inc., Raleigh, NC, United States*

221 **A Patient-Specific Computational Model to Characterize The Impact of Neural Tissue Motion on Cerebrospinal Fluid Dynamics at the Cervical-Medullary Junction** SB3C2015-166 *pg.*120
Soroush Heidari Pahlavian[1,2], Francis Loth[1,2], Mark Luciano[3], Bryn Martin[1,2], [1]*The University of Akron, Akron, OH, United States,*[2]*Conquer Chiari Research Center, Akron, OH, United States,*[3]*Cleveland Clinic Foundation, Cleveland, OH, United States*

222 **Complexity Of The Cerebrospinal Fluid Flow In Patients Suffering From Chiari Malformation Type I - A Computational Study** SB3C2015-209 *pg.*200
Kartik Jain[1,2], Kent-Andre Mardal[2,3], [1]*University of Siegen, Siegen, Germany,*[2]*Simula Research Laboratory, Oslo, Norway,*[3]*University of Oslo, Oslo, Norway*

223 **Peclet Number Of Ciliary Transport On The Surface Of The Tracheal Lumen** SB3C2015-157 *pg.*104
Kouki Kiyota[1], Hironori Ueno[2], Keiko Numayama-Tsuruta[1], Yohsuke Imai[1], Takami Yamaguchi[1], **Takuji Ishikawa**[1], [1]*Tohoku University, Sendai, Japan,*[2]*Aichi University of Education, Aichi, Japan*

FRIDAY, JUNE 19	12:30pm - 3:00pm

Poster Session II — Musculoskeletal Soft Tissue Mechanics — Event Center Tent

224 **A Phenomenological Model to Describe The Viscoelastic Behavior in Multiple Loading** SB3C2015-440 *pg.*588
Behzad R. Babaei, *Washington University in St. Louis, St. Louis, MO, United States*

225 **Effect of Osteoarthritis on the Mechanical Properties of Human Articular Cartilage** SB3C2015-509 *pg.*704
Blair E. Larson, Kristine M. Fischenich, Kirk A. Kindsfater, Tammy L. Haut Donahue, *Colorado State University, Fort Collins, CO, United States*

226 **Conditioned Media from Degenerative Vertebral Discs Sensitizes Dorsal Root Ganglion Neurons to Heat Stimuli** SB3C2015-1040 *pg.*1054
Joshua D. Stover[1], Ibrahima Bah[2], Alexander Kotelsky[2], Mark R. Buckley[2], Brandon Lawrence[1], Robert Bowles[1], [1]*University of Utah, Salt Lake City, UT, United States,*[2]*University of Rochester, Rochester, NY, United States*

227 **Inter and Intra Variation in the Tensile Properties of the Porcine Temporomandibular Joint Disc** SB3C2015-498 *pg.*684
Jesse Lowe[1], Alejandro Almarza[1,2], [1]*University of Pittsburgh, Pittsburgh, PA, United States,*[2]*McGowan Institute of Regenerative Medicine, Pittsburgh, PA, United States,*[3]

228 **A Quantitative Evaluation of the Role of Cell Senescence in Intervertebral Disc Degeneration** SB3C2015-640 *pg.*949
Shady Elmasry, Shihab Asfour, Juan Pablo de Rivero Vaccari, **Francesco Travascio**, *University of Miami, Coral Gables, FL, United States*

229 **Characterizing the Change in Ankle Constraint Following Grade II and III Sprains** SB3C2015-392 *pg.*493
Matthew H. Dickinson, Ednah G. Louie, Bardiya Akhbari, William M. Eboch, Sami Shalhoub, Lorin P. Maletsky, *University of Kansas, Lawrence, KS, United States*

230 **Three-dimentional Strain Distribution In The Anterior Cruciate Ligament During Anterior Translation Of The Knee** SB3C2015-444 *pg.*596
Satoshi Yamakawa[1], Richard Debski[2], Hiromichi Fujie[1], [1]*Tokyo Metropolitan University, Hino, Japan,*[2]*University of Pittsburgh, Pittsburgh, PA, United States*

231 **A Finite Element Model to Evaluate the Role of the Medial Meniscotibial Attachment in Knee Biomechanics** SB^3C2015-1153 *pg.*1261
Andrew J. Polk, Ferris M. Pfeiffer, James L. Cook, James P. Stannard, Patrick A. Smith, *University of Missouri, Columbia, MO, United States*

232 **Characterization of Fatigue Failure in Bovine Meniscus.** SB^3C2015-1187 *pg.*1317
Jaremy J. Creechley[1], Trevor J. Lujan[2], [1]*Materials Science and Engineering, Boise State University, Boise, ID, United States,*[2]*Boise State University, Boise, ID, United States*

233 **Quantifying Skeletal Muscle Deformation in Three Dimensions** SB^3C2015-666 *pg.*999
Elisabeth Jensen, Kenton Kaufman, Duane Morrow, Joel Felmlee, *Mayo Clinic, Rochester, MN, United States*

234 **Treatment of Focal Cartilage Defects Using a Metal Implant: New Biomechanical Insights Using Finite Element Modeling** SB^3C2015-361 *pg.*441
Ashley Heuijerjans, Wouter Wilson, Keita Ito, Corrinus C. van Donkelaar, *Eindhoven University of Technology, Eindhoven, Netherlands*

235 **Progressive Changes In Cervical Spine Intervertebral Disc Properties During Cyclic Compressive Fatigue Loading** SB^3C2015-665 *pg.*997
Sagar Umale[1], Brian Stemper[1,2], Mingxin Zheng[3], Aidin Masoudi[3], Daniel Fama[1,2], Narayan Yoganandan[1,2], Brian Snyder[3,4,5], [1]*Medical College of Wisconsin, Milwaukee, WI, United States,*[2]*Clement J. Zablocki VA Medical Center, Milwaukee, WI, United States,*[3]*Beth Israel Deaconess Medical Centre, Boston, MA, United States,*[4]*Harvard Medical School, Boston, MA, United States,*[5]*Cerebral Palsy Clinic Children's Hospital, Boston, MA, United States*

236 **Structural Inhomogeneity Enhances Interstitial Fluid Pressurization in TMJ Condylar Cartilage** SB^3C2015-1107 *pg.*1177
Brandon Zimmerman[1], Leonardo Ruggiero[1,2], Miri Park[1], Lin Han[3], Liyun Wang[1], David L. Burris[1], Xin L. Lu[1], [1]*University of Delaware, Newark, DE, United States,*[2]*Vrije Universiteit Brussel, Brussels, Belgium,*[3]*Drexel University, Philadelphia, PA, United States*

237 **A Model To Study Articular Cartilage Mechanical And Biological Responses To Rolling And Sliding Loads** SB^3C2015-162 *pg.112*
Oliver R. Schätti[1,2,3], Luigi M. Gallo[2], Peter A. Torzilli[1], [1]*Laboratory for Soft Tissue Research, Hospital for Special Surgery, New York City, NY, United States,*[2]*Center for Dental Medicine, University of Zürich, Zürich, Switzerland,*[3]*Institute for Biomechanics, Swiss Federal Institute of Technology, ETH, Zurich, Switzerland*

238 **Ibuprofen Does Not Adversely Affect Supraspinatus Tendon Mechanical Adaptations in a Rat Model of Exercise** SB^3C2015-269 *pg.*299
Sarah I. Rooney, Rachel Baskin, Andrew F. Kuntz, Louis J. Soslowsky, *University of Pennsylvania, Philadelphia, PA, United States*

239 **Toward Understanding the Mechanisms by Which Microparticles Induce Synovial Inflammation in Osteoarthritis** SB^3C2015-374 *pg.*463
Amy M. Silverstein[1], Robert M. Stefani[1], Sevan R. Oungoulian[1], Eric L. Tong[1], Mukundan G. Attur[2], Steven B. Abramson[2], Christopher S. Ahmad[1], James L. Cook[3], Gerard A. Ateshian[1], J. Chloe Bulinski[1], Clark T. Hung[1], [1]*Columbia University, New York, NY, United States,*[2]*New York University, New York, NY, United States,*[3]*University of Missouri, Columbia, MO, United States*

240 **Regional Mechanical Properties of the Long Head of the Biceps Tendon** SB^3C2015-148 *pg.*88
Christopher W. Kolz[1], Thomas Suter[1,2], Heath B. Henninger[1], [1]*University of Utah, Salt Lake City, UT, United States,*[2]*Clinic of Orthopaedic Surgery, Kantonsspital Baselland, Liestal, Switzerland*

241 **Biomechanical Effects of Menisco-Tibial Repair** SB^3C2015-399 *pg.*507
Ferris Pfeiffer[1], James Stannard[1], James Cook[1], Matthew Bollier[2], Patrick Smith[3], [1]*University of Missouri, Columbia, MO, United States,*[2]*University of Iowa, Iowa City, IA, United States,*[3]*Columbia Orthopaedic Group, Columbia, MO, United States*

242 **Mechanical And Adhesive Properties Of Hydrogels In Tension And Shear** SB^3C2015-1097 *pg.*1157
Jennifer Kadlowec, Daniel Collins, Patrick Myers, Thomas Christiani, Jennifer Vernengo, *Rowan University, Glassboro, NJ, United States*

243 **Dynamic Viscoelastic Properties of Porcine Patellar Tendon Tissue: A Study of Regional Variation and Frequency Dependent Behaviour** SB³C2015-449 *pg.*606
Sourav S. Patnaik[1,2], Taylor Szasz[1,2], **Raj Prabhu**[1,2], Hongjoo Rhee[2], Mark F. Horstemeyer[2], Jun Liao[1,2], Lakiesha Williams[1,2], [1]*Mississippi State University, Mississippi State, MS, United States,*[2]*Center for Advanced Vehicular Systems, Mississippi State, MS, United States*

244 **Targeting Collagen Strands by Triple Helix Hybridization** SB³C2015-194 *pg.*172
Michael Yu, **Yang Li**, *University of Utah, Salt Lake City, UT, United States*

245 **Functional Tensile Properties of a Split Quadriceps Graft for Double-Bundle ACL Reconstruction** SB³C2015-180 *pg.*146
Robert Matthew Miller, Amir Ata Rahnemai-Azar, Todd Jasinski, Fabio V. Arilla, Levent Surer, Freddie H. Fu, Richard E. Debski, Volker Musahl, *University of Pittsburgh, Pittsburgh, PA, United States*

246 **A New Method for Measuring Stress Concentrations in Finite Element Analysis of Fibrocartilage Predicts Greater Fracture Risk for Angled Center Cracks** SB³C2015-326 *pg.*391
John M. Peloquin[1], Dawn M. Elliott[2], [1]*University of Pennsylvania, Philadelphia, PA, United States,*[2]*University of Delaware, Newark, DE, United States*

247 **The Effect of Anatomical Variability on Temporomandibular Joint Mechanics.** SB³C2015-403 *pg.*515
Jessica Coogan[1], Travis Eliason[1], Mark Wong[2], **Daniel Nicolella**[1], [1]*Southwest Research Institute, San Antonio, TX, United States,*[2]*The University of Texas School of Dentistry at Houston, Houston, TX, United States*

FRIDAY, JUNE 19	12:30pm - 3:00pm

Poster Session II — Injury Mechanics — Event Center Tent

248 **Sensitivity Study of Head Impact Parameters on Intracranial Dynamics** SB³C2015-413 *pg.*535
Yi Hua[1], Praveen Akula[1], Matthew Kelso[2], Linxia Gu[1,3], [1]*University of Nebraska-Lincoln, Lincoln, NE, United States,*[2]*University of Nebraska Medical Center, Omaha, NE, United States,*[3]*Nebraska Center for Materials and Nanoscience, Lincoln, NE, United States*

249 **Improving Brain-Skull Interface Through Application of Mesh Smoothing Algorithm** SB³C2015-1093 *pg.*1149
Mireille Kelley[1,2], Logan Miller[1,2], Jillian Urban[1,2], Joel Stitzel[1,2], [1]*Wake Forest University School of Medicine, Winston-Salem, NC, United States,*[2]*Virginia Tech - Wake Forest University, Winston-Salem, NC, United States*

250 **Mouse Model Of Blast Traumatic Brain Injury: An Imaging, Behavior And Pathological Assessment Study** SB³C2015-380 *pg.*471
Sujith Sajja[1], Jiangyang Zhang[1], Jeff Bulte[1], Joseph Long[2], Robert Stevens[1], Piotr Walczak[1], Miroslaw Janowski[1,3], [1]*Johns Hopkins School of Medicine, Baltimore, MD, United States,*[2]*Walter Reed Army Institute of Reseatch, Silver Spring, MD, United States,*[3]*NeuroRepair Department, MMRC, PAS, Warsaw, Poland*

251 **The Effects of the Impact of a Soccer Ball on a Human Head** SB³C2015-517 *pg.*720
Kimberly A. Brown, Aalaap Desai, Yuxiong Mao, Mark Horstemeyer, Jun Liao, Lakiesha Williams, Hongjoo Rhee, Raj Prabhu, *Mississippi State University, Mississippi State, MS, United States*

252 **Biomechanics of Human Tibia and Fibula Fracture Caused by a Mixed Martial Arts Kick** SB³C2015-436 *pg.* 582
Andrew Lamont[1,2], Robbin Bertucci[1,3], Youssef Hammi[2], Mark Horstemeyer[2], Jun Liao[1,2], Hongjoo Rhee[3], Lakiesha Williams[1,2], Rajkumar Prabhu[1,2], [1]*Agricultural and Biological Engineering, Mississippi State University, Starkville, MS, United States,*[2]*Center for Advanced Vehicular Systems, Starkville, MS, United States,*[3]*Center for Advanced Vehicular Systems, Mississippi State, MS, United States*

253 **Influence Of Sulci On Mechanical Response Of The Brain And Injury Prediction Under High-rate Impact** SB³C2015-473 *pg.*646
Alan Leung[1], Nithyanand Kota[2], Amit Bagchi[3], **Siddiq Qidwai**[3], [1]*Advanced Technology & Research Corporation, Columbia, MD, United States,*[2]*Leidos Corporation, Arlington, VA, United States,*[3]*US Naval Research Laboratory, Washington, DC, United States*

254 **Finite Element Analysis of Lower Extremity Military Boot Protection at Blast Conditions** SB³C2015-1143 *pg.*1241
Robbin Bertucci, **R. Prabhu**, M. F. Horstemeyer, Jun Liao, Lakiesha N. Williams, *Mississippi State University, Starkville, MS, United States*

255 **Development of a Computationally Efficient Full Human Body Finite Element Model** SB³C2015-638 *pg.*945
Doron Schwartz[1,2], Berkan Guleyupoglu[1,2], Bharath Koya[1,2], Joel D. Stitzel[1,2], F. Scott Gayzik[1,2], *[1]Wake Forest School of Medicine, Winston Salem, NC, United States,[2]Virginia Tech – Wake Forest University Center for Injury Biomechanics, Winston Salem, NC, United States*

256 **An Efficient and Reliable Biomechanical Testing Device to Perform Torsion Testing in Long Bones with Locking Compression Plates** SB³C2015-574 *pg.*827
Joseph P. Loftus[1], **Anita Singh**[1], Lindsay Stoy[1], Douglas J. Patterson[2], *[1]Widener University, Chester, PA, United States,[2]Christiana Care, Newark, DE, United States*

257 **A Computational Model Of Blast Loading To The Eye: A Comparison With Field Tests** SB³C2015-481 *pg.* 660
Thao D. Nguyen[1], Rajneesh Bhardwaj[2], Shantanu Bailoor[2], *[1]Johns Hopkins University, Baltimore, MD, United States,[2]Indian Institute of Technology Bombay, Mumbai, India*

258 **Supine to Prone Thoraco-abdominal Deformation and Organ Migration in a Set of Healthy Young Adults** SB³C2015-1027 *pg.*1030
Berkan Guleyupoglu, Josh C. Tan, Craig A. Hamilton, F. Scott Gayzik, *Wake Forest University School of Medicine, Winston Salem, NC, United States*

259 **Design of a Novel Shock Tube System for Blast Induced Traumatic Brain Injury** SB³C2015-598 *pg.*871
Andrew B. Robbins[1], Raoul Van Loon[2], Ashok K. Shetty[3], Michael R. Moreno[1], *[1]Texas A&M University, College Station, TX, United States,[2]Swansea University, Swansea, United Kingdom,[3]Texas A&M Health Science Center, Temple, TX, United States*

261 **A Computational Head Model Validated Against Pressure Responses Only May Not Be Used To Estimate Brain Strain Responses** SB³C2015-662 *pg.*991
Wei Zhao, Songbai Ji, *Dartmouth College, Hanover, NH, United States*

262 **Head Impact Response Resulting from Forceful Impact with Toy Swords by Pediatric Males** SB³C2015-604 *pg.*883
Stephanie M. Beeman, Steven Rowson, Stefan M. Duma, *Virginia Tech, Blacksburg, VA, United States*

263 **Behavioral And Inflammatory Consequences Of Cerebrovascual Dysfunction In Primary Blast Injury** SB³C2015 *pg.*1319
Stewart Yeoh, Kenneth L. Monson, *University of Utah, Salt Lake City, UT, United States*

264 **Pelvic Response of a Total Human Body Finite Element (FE) Model During Simulated Under Body Blast (UBB) Impacts** SB³C2015-1028 *pg.*1032
Caitlin M. Weaver[1,2], Randolph S. Coates[2], Andrew S. Merkle[3], Joel D. Stitzel[1], *[1]Wake Forest University, Winston-Salem, NC, United States,[2]US Army Research Laboratory, Aberdeen Proving Ground, MD, United States,[3]Johns Hopkins Applied Physics Lab, Laurel, MD, United States*

265 **The Effect of Pre-Crash Velocity Reduction on Occupant Response Using a Finite Element Model** SB³C2015-301 *pg.*349
Nicholas A. Vavalle[1,2], Berkan Guleyupoglu[1,2], Jeremy M. Schap[1,2], Kristofer D. Kusano[3,4], F. Scott Gayzik[1,2], *[1]Virginia Tech – Wake Forest Center for Injury Biomechanics, Winston-Salem, NC, United States,[2]Wake Forest School of Medicine, Winston-Salem, NC, United States,[3]Virginia Tech – Wake Forest Center for Injury Biomechanics, Blacksburg, VA, United States,[4]Virginia Tech, Blacksburg, VA, United States*

266 **Alteration And Failure Of Cerebral Artery Internal Elastic Lamina Following Axial Overstretch** SB³C2015-1140 *pg.*1235
Matthew I. Converse, Tessa Sommer, Kenneth L. Monson, *University of Utah, Salt Lake City, UT, United States*

268 **Prediction Of Extra-axial Injury Location From Real World Motor Vehicle Crash And Occupant Data** SB³C2015-1086
Jillian E. Urban[1], Sarah Lynch[1], Ervin Lowther[1,2], Christopher Whitlow[1,2,3], Joel Stitzel[1], *[1]Wake Forest University, Winston-Salem, NC, United States,[2]Wake Forest School of Medicine, Winston-Salem, NC, United States,[3]Translational Science Institute, Winston-Salem, NC, United States pg.1138*

269 **Effect of Neck Cable Tension During Helmeted Head Impact** SB³C2015-1113 *pg.*1189
Bethany Rowson, Steven Rowson, Stefan M. Duma, *Virginia Tech, Blacksburg, VA, United States*

270 **Traumatic Brain Injury Resulted in Increased Aquaporin-4 Expression - Relevance to Post Injury Edema** SB³C2015-647 *pg.* 961
Nasya Sturdivant, Jeffrey Wolchok, Kartik Balachandran, *University of Arkansas, Fayetteville, AR, United States*

FRIDAY, JUNE 19	12:30pm - 3:00pm

Poster Session II — Mechanics of Growth, Remodeling and Repair — Event Center Tent

271 **Damage Accumulation Modeling and Rate Dependency of Spinal Dura Mater** SB³C2015-178 *pg.* 142
Nicole Ramo, Snehal S. Shetye, Christian M. Puttlitz, *Colorado State University, Fort Collins, CO, United States*

272 **Analytical Approximation for Predicting Stress Fiber Remodeling Due to Dynamic Mechanical Stimuli** SB³C2015-215
Tommaso Ristori[1,2], [1]*Eindhoven University of Technology, Eindhoven, Netherlands,*[2]*Institute for Complex Molecular Systems, Eindhoven, Netherlands* *pg.*210

273 **Multiscale Model of Strain-Dependent Glomerular Basement Membrane Remodeling** SB³C2015-624 *pg.*917
Lazarina Gyoneva[1], Yoav Segal[1,2], Kevin D. Dorfman[1], Victor H. Barocas[1], [1]*University of Minnesota, Minneapolis, MN, United States,*[2]*VA Medical Center, Minneapolis, MN, United States*

274 **Quantification Of Transient Temperature And Thermal Damage In An Established Burn Model** SB³C2015-1166 *pg.*1285
Stephanie Lindow, **F. Scott Gayzik**, *Wake Forest University School of Medicine, Winston-Salem, NC, United States*

275 **Simulated Collagen Network Remodeling in Response to Stress.** SB³C2015-591 *pg.*859
Carley B. Hovell, *University of Minnesota, Twin Cities, Minneapolis, MN, United States*

276 **Isolation of Subarachnoid Hemhorrage Factors on a Single Chip for Understanding Cerebral Vasospasm Progression** SB³C2015-514 *pg.*714
Eric S. Hald, Kerianne E. Steucke, Connor Timm, Patrick W. Alford, *University of Minnesota, Minneapolis, MN, United States*

277 **A Finite Element Investigation of Fracture Healing Under Simulated Microgravity Loading Conditions** SB³C2015-422
Benjamin C. Gadomski, Zachary F. Lerner, Raymond C. Browning, Christian M. Puttlitz, *Colorado State University, Fort Collins, CO, United States* *pg.*553

FRIDAY, JUNE 19	12:30pm - 3:00pm

Poster Session II — Cardiovascular Tissue Mechanics — Event Center Tent

278 **Novel Technique for Assessment of Mechanical Properties of Carotid Arteries** SB³C2015-476 *pg.*652
Stefan Sanders, Frans van de Vosse, Marcel Rutten, *Eindhoven University of Technology, Eindhoven, Netherlands*

279 **Morphological Characterization Of Collagen Fibers At The Crack Initiation Sites In Biaxially Stretched Porcine Thoracic Aortas Toward Clarification Of Aneurysm Rupture Mechanism** SB³C2015-425 *pg.*559
Shukei Sugita, Takeo Matsumoto, *Nagoya Institute of Technology, Nagoya, Japan*

280 **Biaxial Mechanical Characterization of Non-Uniform Thermal Shrinkage Deformations to Guide Ablative Therapy** SB³C2015-565 *pg.*811
Steven Boronyak, W. David Merryman, *Vanderbilt University, Nashville, TN, United States*

281 **Characterization of the Fatigue Life, Dynamic Creep and Modes of Damage Accumulation within Mitral Valve Chordae Tendineae** SB³C2015-533 *pg.*752
Gillian M. Gunning, Bruce P. Murphy, *Trinity College Dublin, Dublin, Ireland*

282 **Finite Element Modeling Of Cardiac Muscle Contraction** SB³C2015-408 *pg.* 525
Xiaoyan Zhang, Kenneth S. Campbell, **Jonathan F. Wenk**, *University of Kentucky, Lexington, KY, United States*

283 **Novel Micro-Computed Tomography Technique for Soft Tissue Deformation Tracking - Application to the Mitral Valve** SB³C2015-182 *pg.*150
Eric L. Pierce[1], **Charles H. Bloodworth**[2], Ajay Naran[2], Thomas F. Easley[2], Morten O. Jensen[1], Ajit P. Yoganathan[1], [1]*Georgia Institute of Technology and Emory University, Atlanta, GA, United States,*[2]*Georgia Institute of Technology, Atlanta, GA, United States*

284 **Identification and Quantification of Extracellular Matrix Proteins at the Plaque - Internal Elastic Lamina Interface in a Mouse Model of Atherosclerosis** SB³C2015-291 *pg.329*
Lindsey Davis, Susan Lessner, *University of South Carolina, Columbia, SC, United States*

285 **The Number of Lesions Does Not Govern the Functionality of Coronary Bifurcation Lesions: A Study of the Effect of Relative Stenosis Severity** SB³C2015-224 *pg.224*
Catherine Pagiatakis[1,2], Jean-Claude Tardif[2,3], Philippe L. L'Allier[2,3], **Jennifer Frattolin**[1], Rosaire Mongrain[1,2], [1]*McGill University, Montreal, QC, Canada,*[2]*Montreal Heart Institute, Montreal, QC, Canada,*[3]*University of Montreal, Montreal, QC, Canada*

286 **Force Required to Cinch the Tricuspid Annulus: An Ex Vivo Study** SB³C2015-137 *pg.70*
Amy N. Adkins[1], Jesus Aleman[1], Edward Sako[2], Lori Boies[1], Shamik Bhattacharya[1], [1]*St.Mary's University, San Antonio, TX, United States,*[2]*University of Texas Health Science Center San Antonio, San Antonio, TX, United States*

287 **Infarcted Left Ventricles Have Stiffer Material Properties and Lower Stiffness Variation: 3D Echo-Based Modeling to Quantify In Vivo Ventricle Material Properties** SB³C2015-32 *pg.16*
Longling Fan[1], Jing Yao[2], Chun Yang[3], Di Xu[2], **Dalin Tang**[1,4], [1]*Southeast University, Nanjing, China,*[2]*Nanjing Medical University, Nanjing, China,*[3]*China United Network Communications Co., Ltd., Beijing, China,*[4]*Worcester Polytechnic Institute, Worcester, MA, United States*

288 **Development Of An Estimation Method Of Blood Vessel Configuration At No Load State For FE Patient-specific Simulation** SB³C2015-1133 *pg.1225*
Ming Yu[1], Sota Yamamoto[1], Mrie Oshima[2], [1]*Shibaura Institute of technology, Tokyo, Japan,*[2]*The University of Tokyo, Tokyo, Japan*

289 **Post Endovascular Stent Repair Of Descending Aorta: Side Effects And Development Of Aneurysm In The Ascending Aorta** SB³C2015-1095 *pg.1153*
Giampaolo Martufi, **Manal Altamimi**, Raied Aburashed, Cyrus Fiori, Jehangir J. Appoo, Elena S. Di Martino, *University of Calgary, Calgary, AB, Canada*

290 **Bicuspid Aortic Valve Hemodynamics Induce Acute Asymmetric Remodeling Of Porcine Ascending Aortas: An Ex Vivo Study** SB³C2015-1112 *pg.1187*
Samantha K. Atkins, Philippe Sucosky, *University of Notre Dame, Notre Dame, IN, United States*

291 **The Effect of Vascular Curvature on Blood Flow and Oxygen Transport in Arterio-Venous Fistulae** SB³C2015-807
Francesco Iori, Lorenza Grechy, Richard W. Corbett, Wladyslaw Gedroyc, Neill Duncan, Colin G. Caro, **Peter E. Vincent**, *Imperial College London, London, United Kingdom pg. 1013*

292 **Measuring Vessel Wall Displacement and Circumferential Strain Using Displacement Encoded with Stimulated Echo (DENSE) MRI Sequence** SB³C2015-592 *pg.861*
Elizabeth Iffrig[1,2], Xiaodong Zhong[1,3], William R. Taylor[1,2], John N. Oshinski[1,2], [1]*Emory University, Atlanta, GA, United States,*[2]*Georgia Institute of Technology, Atlanta, GA, United States,*[3]*Siemens Medical Solutions, Malvern, PA, United States*

293 **Computational Analysis Of The Effect Of Sequential Coiling On The Wall Stress Of Cerebral Aneurysms** SB³C2015-1190 *pg.1323*
Joseph E. Pichamuthu, Brian T. Jankowitz, David A. Vorp, *University of Pittsburgh, Philadelphia, PA, United States*

294 **Smooth Muscle Cell Elastin Generation Stimulated by Adipose-Derived Mesenchymal Stem Cells** SB³C2015-496
Aneesh Ramaswamy, Kory Blose, Justin Weinbaum, David Vorp, *University of Pittsburgh, Pittsburgh, PA, United States pg.680*

295 **Computational Modeling Of Passive Myocardium:A Comparison Between Two Constitutive Models** SB³C2015-347
Amir Nikou[1], Shauna M. Dorsey[2], Jeremy R. McGarvey[3], Joseph H. Gorman III[2], Jason A. Burdick[2], James J. Pilla[2], Robert C. Gorman[2], Jonathan F. Wenk[1], [1]*University of Kentucky, Lexington, KY, United States,*[2]*University of Pennsylvania, Philadelphia, PA, United States,* [3]*Gorman Cardiovascular Research Group, University of Pennsylvania, Philadelphia, PA, United States pg.421*

296 **Toward an Experimentally Validated Immersed Boundary Model of Left Ventricular Fluid Dynamics Using In Vitro Experiments** SB³C2015-1123 *pg.1207*
Boyce E. Griffith[1], Jae Ho Lee[1], Pritam Mekala[2], Arvind Santhanakrishnan[2], [1]*University of North Carolina at Chapel Hill, Chapel Hill, NC, United States,*[2]*Oklahoma State University, Stillwater, OK, United States*

297 **Role Of Cyclic Strain On Calcific Nodule Formation Among Aortic Heart Valve Cusps** SB³C2015-1085 *pg.1136*
Ying Lei, Zannatul Ferdous, *The University of Tennessee, Knoxville, TN, United States*

298 **Impact of Partial Intraluminal Thrombus Attachment on Peak Stresses on Abdominal Aortic Aneurysm Wall**
SB[3]C2015-277 *pg.309*
Juan S. Stockle, David A. Romero, Cristina H. Amon, *University of Toronto, Toronto, ON, Canada*

299 **Stability Analysis of the Continuum Constrained Mixture Model for Vascular Growth and Remodeling**
SB[3]C2015-1096 *pg.1155*
Jiacheng Wu, Shawn C. Shadden, *University of California, Berkeley, Berkeley, CA, United States*

300 **Ultrasound Monitoring of Abdominal Aortic Aneurysm Progression in a Murine Model** SB[3]C2015-1046 *pg.1066*
Arvin H. Soepriatna[1], Gurneet S. Sangha[1], Amelia R. Adelsperger[1], Evan H. Phillips[1], Clifford M. Babbey[2], Michael P. Murphy[2], Pavlos P. Vlachos[1], Craig J. Goergen[1], [1]*Purdue University, West Lafayette, IN, United States,*[2]*Indiana University - Purdue University Indianapolis, Indianapolis, IN, United States*

301 **Aortic Peak Stress Induced By Antihypertensive Medications In Aortic Dissection Patients** SB[3]C2015-552 *pg.786*
Vittoria Flamini, *New York University, Brooklyn, NY, United States*

302 **Mechanical Parameters Characterization of Embryonic Mouse Hearts for Studying Human Congenital Heart Disease**
SB[3]C2015-492 *pg.672*
Andres Rubiano, Kyle G. Rowe, W. Gregory Sawyer, Chelsey S. Simmons, *University of Florida, Gainesville, FL, United States*

303 **Machine Learning Based Structured Edge Detection for Cardiovascular Modeling** SB[3]C2015-1021 *pg.1020*
Jameson T. Merkow[1], Zhouwen Tu[1], David Kriegman[1], Nathan Wilson[2], Alison L. Marsden[1], [1]*University of California San Diego, San Diego, CA, United States,*[2]*Open Source Medical Software Corporation, San Diego, CA, United States*

304 **A Study On The Interplay Of Sex And Cyclic Stretch In Aortic Valve Calcification** SB[3]C2015-466 *pg.634*
Shirin Masjedi, **Ying Lei**, Zannatul Ferdous, *University of Tennessee, Knoxville, TN, United States*

FRIDAY, JUNE 19	12:30pm - 3:00pm

Poster Session II — Tissue Engineering and Regenerative Medicine: Materials and Interfaces — Event Center Tent

306 **Genipin Cross-linking Silk Fibroin Post-gelation Increases Gel Mechanical Stiffness** SB[3]C2015-1134 *pg.1227*
Winston H. Elliott[1], Walter Bonani[2,3], Devid Maniglio[2,3], Antonella Motta[2,3], Wei Tan[1], Claudio Migliaresi[2,3], [1]*University of Colorado- Boulder, Boulder, CO, United States,*[2]*University of Trento, Trento, Italy,*[3]*European Institute of Excellence on Tissue Engineering and Regenerative Medicine, and INSTM Trento Research Unit, Trento, Italy*

307 **Silica-Collagen Hydrogel for Corneal Replacement** SB[3]C2015-259 *pg.281*
Michael DiVito, *University of Minnesota, Minneapolis, MN, United States*

308 **In Situ Polymerization Of Thiol-acrylate Nanocomposite Foam For Bone Defects** SB[3]C2015-253 *pg.275*
Anoosha Forghani, *Louisiana State University, Baton Rouge, LA, United States*

309 **Fabrication Of Scaffolds From Different Silica Based Bioactive Glass And Investigation Of Bioactivity And Biodegredation Behaviors** SB[3]C2015-1371 *pg.1325*
Sevil Yücel[1], A. C. Ozarslan[1], B. S. Oduncu[1], P. Terzioglu[2], [1]*Yildiz Technical University, Istanbul, Turkey,*[2]*Muğla Sıtkı Koçman University, Muğla, Turkey*

310 **Evaluation Of Two Formulations Of Polycaprolactone For Use In Tissue Scaffold Applications** SB[3]C2015-390 *pg.489*
Jacob D. Harris, A. Sharif El-Gizawy, Ferris M. Pfeiffer, *University of Missouri, Columbia, MO, United States*

311 **Aqueous Biphasic Micro-printing Of Tumor Spheroids** SB[3]C2015-654 *pg.975*
Stephanie Lemmo Ham, Ehsan Atefi, Hossein Tavana, *University of Akron, Akron, OH, United States*

312 **Using Finite Element Analysis to Study the Mechanical Advantages of a Turtle's Shell** SB[3]C2015-1105 *pg.1173*
John W. Wood, Rajkumar Prabhu, *Mississippi State University, Starkville, MS, United States*

313 **Characterization of a Thermoreversible Collagen for Free-Form Fabrication of Scaffolds** SB³C2015-474 *pg.*648
Kathryn E. Drzewiecki, David I. Shreiber, *Rutgers University, Piscataway, NJ, United States*

314 **3D Printing Of B. Mori Silk Proteins For Implantable Devices** SB³C2015-320 *pg.*381
Tom M. Merrill, Maria Torculas, Jethro Medina, Brenton Boszczuk, Kyle Meehan, Ian Miller, Xiao Hu, Wei Xue, *Rowan University, Glassboro, NJ, United States*

315 **Influence of Nano- and Micro-Scale Structure of Aligned Electrospun Scaffolds on Mechanical Properties and Cell Response** SB³C2015-141 *pg.*76
Hannah M. Pauly[1], Ketul C. Popat[1], Daniel J. Kelly[2], Tammy L. Haut Donahue[1], [1]*Colorado State University, Fort Collins, CO, United States,*[2]*Trinity College Dublin, Dublin, Ireland*

316 **An Assessment of the Native Species in Articular Cartilage and Synovial Fluid as Potential Absorptive Barriers to UV-Initiated Scaffold Polymerization** SB³C2015-429 *pg.*567
Anthony Finch, Patrick Donnelly, Peter Torzilli, *Hospital for Special Surgery, New York, NY, United States*

FRIDAY, JUNE 19	**12:30pm - 3:00pm**

Poster Session II — Tissue Engineering and Regenerative Medicine: Cells, Constructs, Culture Systems, and Regeneration — Event Center Tent

317 **Effects of Mechanical Constraints on Cell-Generated Stress and Collagen Remodeling in Statically Cultured Microtissues** SB³C2015-369 *pg.*455
Mathieu A. J. van Kelle, Sandra Loerakker, Inge A. E. W. van Loosdregt, Carlijn V. C. Bouten, Frank P. T. Baaijens, *Eindhoven University of Technology, Eindhoven, Netherlands*

318 **Optimization of Parameters For Long-Term Storage of Tissue Engineered Articular Cartilage** SB³C2015-385 *pg.*479
Adam B. Nover[1], Robert M. Stefani[1], Stephanie L. Lee[1], Rebecca A. Peyser[1], Daniel R. Howard[2], Gerard A. Ateshian[1], Aaron M. Stoker[3], James L. Cook[3], Clark T. Hung[1], [1]*Columbia University, New York, NY, United States,*[2]*Mount Sinai St. Luke's, New York, NY, United States,*[3]*University of Missouri, Columbia, MO, United States*

319 **Maturation Of Human Stem Cell-derived Cardiomyocytes In 3D Tissues Through Increasing Collagen Concentrations** SB³C2015-1131 *pg.*1221
Aric Q. Pahnke, *University of Toronto, Toronto, ON, Canada*

320 **Use Of Kartogenin To Augment The Tendon-bone Tunnel Healing** SB³C2015-310 *pg.*365
Yiqin Zhou, Jianying Zhang, Guangyi Zhao, **James H-C. Wang**, *University of Pittsburgh, Pittsburgh, PA, United States*

321 **Modular Tissue Engineered Cartilage Surfaces** SB³C2015-324 *pg.*387
Audrey C. Ford, Kayla Wolf, Aditya Nandy, Anne Y. Zeng, Grace D. O'Connell, *University of California Berkeley, Berkeley, CA, United States*

322 **Tensile Properties of Stem Cell-Based Self-Assembled Tissue (scSAT) Biosynthesized on Nanoperiodic Structured Substrate** SB³C2015-448 *pg.*604
Kei Oya[1], Yuki Tani[2], Kota Koizumi[3], Norihiko Sugita[3], Kenji Suzuki[4], Norimasa Nakamura[3], Hiromichi Fujie[2], [1]*Tokai University, Hiratsuka, Kanagawa, Japan,*[2]*Tokyo Metropolitan University, Hino, Tokyo, Japan,*[3]*Osaka University, Suita, Osaka, Japan,*[4]*Kogakuin University, Hachioji, Tokyo, Japan*

323 **Ultrasound Assisted Human Mesenchymal Stem Cell Chondrogenesis: Engineering Large-scale Cartilage Grafts** SB³C2015-669 *pg.*1003
Anu Subramanian, Sanjukta Guha Thakurta, Neety Sahu, Abdul Qadir Chama, Hendrik J. Viljoen, *University of Nebraska, Lincoln, NE, United States*

324 **Osteogenic Induction Of Human Adipose Derived Stem Cells Cultured On Poly (l-lactic Acid) Scaffolds Prepared By Thermally Induced Phase Separation Method** SB³C2015-172 *pg.*130
Harish Chinnasami, Ram Devireddy, Dan Hayes, *Louisiana State University, Baton Rouge, LA, United States*

325 **Investigating the Role of Osteoactivin in Muscle Regeneration** SB³C2015-625 *pg.* 919
Jinjin Ma[1], Bing Yu[2], Andrew Baker[1], Min-Ho Kim[2], Anthony Calabro[1], Fayez Safadi[3], Christopher Malcuit[2], Kathleen Derwin[1], [1]*Cleveland Clinic, Cleveland, OH, United States,*[2]*Kent State University, Kent, OH, United States,*[3]*Northeast Ohio Medical University, Rootstown, OH, United States*

326 **Effect Of Heat Shock On Cryopreservation Of Adipose Tissue Derived Stem Cells** SB³C2015-165 *pg.*118
Mulla S. Shaik[1], Jeffrey M. Gimble[2], Ram Devireddy[1], *[1]Louisiana State University, Baton Rouge, LA, United States,[2]LaCell Incorporation, New Orleans, LA, United States*

327 **Evaluating the Consistency of Cardiomyocyte Self-assembly** SB³C2015-316 *pg.*375
Nancy K. Drew, Danny B. Baldo, Jason Q. Core, Anna Grosberg, *University of California, Irvine, Irvine, CA, United States*

FRIDAY, JUNE 19	12:30pm - 3:00pm

Poster Session II Mechanotransduction and Sub-Cellular Biophysics Event Center Tent

328 **Comparison Between Nonlinear Material Models And Fiber Network Models Reveals Importance Of Fiber Interactions In Cell Mechanosensing On Fibrous Substrates** SB³C2015-609 *pg.* 891
Maziar Aghvami[1], Kristen L. Billiar[2], Edward A. Sander[1], *[1]University of Iowa, Iowa City, IA, United States,[2]Worcester Polytechnic Institute, Worcester, MA, United States*

329 **Role of SRC in Electric Field-Induced Directed Cell Migration** SB³C2015-469 *pg.*640
Shun-Hao Tsao, Pen-Hsiu Grace Chao, *National Taiwan University, Taipei, Taiwan*

330 **Exploring the Response of Astrocytes to Traumatic Brain Injuries Using a Novel Bench Top Crash Tester** SB³C2015-435 *pg.* 580
Joe Wyatt, Addison Walker, Kartik Balachandran, **Jeff Wolchok**, *University of Arkansas, Fayetteville, AR, United States*

331 **Endoplasmic Reticulum Calcium Dynamics in Osteocyte Mechanobiology** SB³C2015-179 *pg.*144
Genevieve N. Brown, Prajesh Desai, X. Edward Guo, *Columbia University, New York, NY, United States*

333 **Movement In Engineered Valvular Tissues In Relation To Regional Flow Physics And Nutrient Transport** SB³C2015-279 *pg.*311
Manuel Salinas, *Florida International University, Davie, FL, United States*

334 **Cellular Cholesterol Content Modulates Monocyte Interaction With E-selectin** SB³C2015-412 *pg.*533
Amit K. Saha, Anand K. Ramasubramanian, *University of Texas at San Antonio, San antonio, TX, United States*

335 **Time Evolution of Contractility in Fibroblasts as a Measure of Cell Photodamage** SB³C2015-1177 *pg.*1303
Samantha G. Knoll[1], Wylie W. Ahmed[2], Taher A. Saif[1], *[1]University of Illinois at Urbana-Champaign, Urbana, IL, United States,[2]Institut Curie, Paris, France*

336 **Topology of Prestin Expressed in the CHO Cell Membrane -Atomic Force Microscopy Study-** SB³C2015-221 *pg.*218
Michio Murakoshi[1,2], Hiroshi Wada[3], *[1]Kagoshima University, Kagoshima, Japan,[2]JST, Kawaguchi, Japan,[3]Tohoku Bunka Gakuen University, Sendai, Japan*

337 **Role of Cell-Cell Interaction in Tensional Homeostasis** SB³C2015-1118 *pg.*1199
Alicia Zollinger[1], Elizabeth Canovic[2], Michael Smith[1], Dimitrije Stamenovic[1], *[1]Boston University, Boston, MA, United States,[2]Massachusetts Institute of Technology, Cambridge, MA, United States*

338 **Computational And Experimental Analysis Of Intracellular Motion, Forces, And Structure** SB³C2015-1184 *pg.*1313
Michael Mak[1], Taeyoon Kim[2], Muhammad H. Zaman[3], Roger D. Kamm[1], *[1]Massachusetts Institute of Technology, Cambridge, MA, United States,[2]Purdue University, West Lafayette, IN, United States,[3]Boston University, Boston, MA, United States*

339 **Epithelial to Mesenchymal Transition Alters Cellular Chiral Behavior** SB³C2015-238 *pg.*249
Kathryn E. Worley, Andrew K. Watrobski, David Shieh, Leo Q. Wan, *Rensselaer Polytechnic Institute, Troy, NY, United States*

340 **Modeling Cellular Contraction on Biohybrid Devices Using Thermal Contraction Capabilities of Finite Element Analysis Tools** SB³C2015-400 *pg.*509
Victoria A. Webster, Ozan Akkus, Hillel J. Chiel, Roger D. Quinn, *Case Western Reserve University, Cleveland, OH, United States*

FRIDAY, JUNE 19	3:00pm - 4:30pm

Bone Structure, Mechanics, and Function — Primrose A

Session Chair: Amy Wagoner Johnson, *University of illinois at Urbana-Champaign, Urbana, IL, United States*
Session Co-Chair: Shigeo Tanaka, *Kanazawa University, Kanazawa, Japan*

3:00PM **Collagen Bound Water Is A Strong Correlate Of Bone's Toughness** SB³C2015-262 *pg.*285
Mustafa Unal, Ozan Akkus, *Case Western Reserve University, Cleveland, OH, United States*

3:15PM **Sequential Bisphosphonate and Parathyroid Hormone Treatment Decouples Bone Remodeling in Favor of Bone Formation** SB³C2015-521 *pg.*729
Allison R. Altman, Carina Lott, Chantal M. de Bakker, Wei-Ju Tseng, Ling Qin, **X. Sherry Liu**, *University of Pennsylvania, Philadelphia, PA, United States*

3:30PM **Tissue Mineral Density Dependent Mechanical Properties of Individual Trabecula Plates and Rods Do Not Differ in Anatomic Directions but Individual Trabecular Directions** SB³C2015-242 *pg.*255
Y. Eric Yu, Ji Wang, Bin Zhou, X. Edward Guo, *Columbia University, new york, NY, United States*

3:45PM **Post-yield Damage Denatures Bone's Collagen As Determined By A Novel Molecular Spectroscopic Biomarker** SB³C2015-261 *pg.* 283
Mustafa Unal, Hyungjin Jung, Ozan Akkus, *Case Western Reserve University, Cleveland, OH, United States*

4:00PM **Measurement of Bone Mineral Density in Motor Vehicle Crash Occupants and Correlation with Age and Fracture Incidence** SB³C2015-236 *pg.*244
Ashley A. Weaver, Richarlette C. Hightower, Sarah K. Lynch, Kristen M. Beavers, Anna N. Miller, Joel D. Stitzel, *Wake Forest University, Winston-Salem, NC, United States*

4:15PM **Proximal Femoral Cortical Bone Thickness in Patients with Femoroacetabular Impingement and Normal Hips Analyzed using Statistical Shape Modeling** SB³C2015-1115 *pg.*1193
Penny R. Atkins, Prateep Mukherjee, Shireen Y. Elhabian, Sumedha Singla, Michael D. Harris, Jeffrey A. Weiss, Ross T. Whitaker, Andrew E. Anderson, *University of Utah, Salt Lake City, UT, United States*

FRIDAY, JUNE 19	3:00pm - 4:30pm

Joint Motion and Rehabilitation — Superior

Session Chair: Laurel Kuxhaus, *Clarkson University, Potsdam, NY, United States*
Session Co-Chair: Bradley Davidson, *University of Denver, Denver, CO, United States*

3:00PM **A Long-Term Simulated Degradation Study of a Synthetic Meniscus Implant: Material and Functional Properties** SB³C2015-191 *pg.*166
Maoz Shemesh[1], Adaya Shefy-Peleg[1], Eyal Zylberberg[1], Eran Linder-Ganz[1], **Jonathan J. Elsner**[2], *[1]Active Implants, Netanya, Israel,[2]Active Implants, Cambridge, MA, United States*

3:15PM **A Robotic Knee Orthosis for Locomotive Assistance** SB³C2015-409 *pg.*527
Saroj Thapa[1], Hao Zheng[1], Geza Kogler[2], Xiangrong Shen[1], *[1]University of Alabama, Tuscaloosa, AL, United States,[2]Georgia Institute of Georgia, Atlanta, GA, United States*

3:30PM **Towards Vertebral Compression Fracture Prevention: Simulating Physiologic Fracture During Small Movement ADLs.** SB³C2015-414 *pg.*537
Nicole C. Corbiere, Stacey L. Zeigler, Kathleen A. Issen, Arthur J. Michalek, Laurel Kuxhaus, *Clarkson University, Potsdam, NY, United States*

3:45PM **Effect of Arm Posture on Voluntary Activation and Moments Generated by Individuals with Tendon Transfer and Quadriplegia** SB³C2015-1151 *pg.*1257
Carrie L. Peterson[1,2,3], Michael S. Bednar[4], Anne M. Bryden[5,6], Michael W. Keith[5,6], Eric J. Perreault[1,2], Wendy M. Murray[1,2,3], *[1]Rehabilitation Institute of Chicago, Chicago, IL, United States,[2]Northwestern University, Chicago, IL, United States,[3]Edward Hines Jr., VA Hospital, Hines, IL, United States,[4]Loyola University, Maywood, IL, United States,[5]MetroHealth Medical Center, Cleveland, OH, United States,[6]Case Western Reserve University, Cleveland, OH, United States*

4:00PM **Use of a Torque Range of Motion Device to Teach Evaluation of Joint Dysfunction** SB³C2015-296 *pg.*339
Rita Patterson[1], Jeongsik Shin[2], Aditya Das[2], Vanneise Collins[1], Carol Kominski[1], Katelyn Rockenbach[1], Robert Longnecker[1], David Mason[1], [1]*University of North Texas Health Science Center, Fort Worth, TX, United States,*[2]*University of Texas at Arlington Research Institute, Fort Worth, TX, United States*

4:15PM **Development of Virtual Environment Navigation Options for Individuals Post-stroke or With Cerebral Palsy Using Kinect** SB³C2015-124 *pg.*50
Alan Eberhardt[1], Sean Pool[1], Scott Bickel[1], Gerald McGwin[1], James Rimmer[1], Laurie Malone[2], [1]*University of Alabama Birmingham, Birmingham, AL, United States,*[2]*Lakeshore Foundation, Birmingham, AL, United States*

FRIDAY, JUNE 19	3:00pm - 4:30pm

Cardiovascular Imaging — Wasatch

Session Chair: Frank Gijsen, *University of Rotterdam, Netherlands*
Session Co-Chair: Christof Karmonik, *Houston Methodist Research Institute, Houston, TX, United States*

3:00PM **Differential Hemodynamic Changes And Lumen Remodeling In The Artery And Vein Of Porcine Arteriovenous Graft And Fistula** SB³C2015-123 *pg.*48
Daniel B. Pike[1], Yong He[2], Christi M. Terry[1], Alfred K. Cheung[1,3], Yan-Ting Shiu[1], [1]*University of Utah, Salt Lake City, UT, United States,*[2]*University of Florida, Gainesville, FL, United States,*[3]*VA SLC Health Care System, Salt Lake City, UT, United States*

3:15PM **Does Aortic Open Distal and Hemi-arch Procedure Remove All Tissue Suspected for Progression of Bicuspid Valve Aortopathy?** SB³C2015-322 *pg.* 383
Alex J. Barker[1], Pim van Ooij[1], David Guzzardi[2], Emilie Bollache[1], S. Chris Malaisrie[1], Patrick M. McCarthy[1], James Carr[1], Jeremy Collins[1], Michael Markl[1], Paul W. M. Fedak[1,2], [1]*Northwestern University, Chicago, IL, United States,*[2]*University of Calgary, Calgary, AB, Canada*

3:30PM **Quantification of Helical Flow Patterns in Left Ventricles of Healthy Subjects and Patients with Dilated Cardiomyopathy** SB³C2015-471 *pg.*642
Jonas Lantz, Carljohan Carlhäll, Tino Ebbers, *Linköping University, Linköping, Sweden*

3:45PM **Preparation of a Hydrogel Phantom of Human Athrosclerotic Plaque for Medical Simulation and Imaging** SB³C2015-622 *pg.*913
Juyu Chueh[1], Tanya N. Turan[2], Truman R. Brown[2], Todd LeMatty[2], Hui Mao[3], Olivia W. Brooks[1], Matthew J. Gounis[1], [1]*University of Massachusetts Medical School, Worcester, MA, United States,*[2]*Medical University of South Carolina, Charleston, SC, United States,*[3]*Emory University School of Medicine, Atlanta, GA, United States*

4:00PM **Microcirculation-Induced MRI Signal Anisotropy in Organized Tissues: Finite Element Modeling and Validation on Perfused Hearts** SB³C2015-1162 *pg.*1277
Osama Abdullah[1], Arnold David Gomez[1], Adam Schmidt[1], Edward Hsu[2], [1]*University of Utah, Salt Lake City, UT, United States,*[2]*University of Utah, Salt Lake City, UT, United States*

4:15PM **Functional and Anatomical Measures for Outflow Boundary Conditions in Atherosclerotic Coronary Bifurcations** SB³C2015-549 *pg.*782
Jelle Schrauwen[1], Adriaan Coenen[1], Akira Kurata[1], Jolanda J. Wentzel[1], Antonius F. W. van der Steen[1,2], Koen Nieman[1], Frank J. H. Gijsen[1], [1]*Thoraxcenter, Erasmus Medical Center, Rotterdam, Netherlands,*[2]*Delft University of Technology, Delft, Netherlands*

FRIDAY, JUNE 19	3:00pm - 4:30pm

Vascular Remodeling and Stented Flow — Magpie

Session Chair: Lucas H. Timmins, *Georgia Institute of Technology, Atlanta, GA, United States*
Session Co-Chair: Zhijie Wang, *University of Wisconsin - Madison, Madison, WI, United States*

3:00PM **Effects Of Estrogen On Pulmonary Wave Reflection And Energy Transmission In Pulmonary Arterial Hypertension** SB³C2015-627 *pg.*923
Aiping Liu, Naomi Chesler, *University of Wisconsin-Madison, Madison, WI, United States*

3:15PM **Biomechanical Comparison of Glutaraldehyde-crosslinked Gelatin/Fibrinogen Electrospun Cylindrical Scaffolds to Porcine Native Vascular Tissue** SB³C2015-1128 *pg.*1217
Ehab Tamimi, Catalina D. Ardila, Darren G. Haskett, Thomas Doetschman, Jonathan P. Vande Geest, *University of Arizona, Tucson, AZ, United States*

3:30PM **Hemodynamics of Healthy Vs. Pathological Venous Valve - Fluid-Structure Interaction Computational Model** SB³C2015-212 *pg.*204
Elina Soifer[1], Dar Weiss[1], Oren Rotman[1], Uri Zaretsky[1], Shmuel Einav[1,2], [1]*Tel Aviv University, Tel Aviv, Israel,*[2]*Stony Brook University, Stony Brook, NY, United States*

3:45PM **Patient-specific Treatment Of Intracranial Aneurysms: An Automatic CFD-based Flow-diverter Optimization Principle** SB³C2015-265 *pg.*291
Philipp Berg[1], László Daróczy[1], Oliver Beuing[2], Gábor Janiga[1], [1]*University of Magdeburg, Magdeburg, Germany,*[2]*University Hospital Magdeburg, Magdeburg, Germany*

4:00PM **Vaso-CT for Quantitative Mapping of Vessel Wall Apposition in a Flow-Diverter Implant** SB³C2015-493 *pg.*674
Kajo van der Marel, Ajay K. Wakhloo, Matthew J. Gounis, Ajit S. Puri, *University of Massachusetts Medical School, Worcester, MA, United States*

4:15PM **Biomechanical Testing to Improve Constitutive Models of the Two-Layered Carotid Artery Media** SB³C2015-453 *pg.*614
Caleb Davis[1,2], Ashish Pandya[2], Stephen E. Greenwald[2], [1]*Texas A&M University, College Station, TX, United States,*[2]*Queen Mary University of London, London, United Kingdom*

FRIDAY, JUNE 19	3:00pm - 4:30pm

Cardiovascular Tissue Engineering — Maybird

Session Chair: Alisa Morss, *Drexel University, Philadelphia, PA, United States*
Session Co-Chair: Pat Alford, *University of Minnesota, Minneapolis, MN, United States*

3:00PM **Spatiotemporal Cell-matrix Interactions During En Masse Migration of Fibroblasts on Collagen Matrices** SB³C2015-491 *pg.*670
Altug Ozcelikkale[1], Frederick Grinnell[2], Bumsoo Han[1], [1]*Purdue University, West Lafayette, IN, United States,*[2]*University of Texas Southwestern Medical Center, Dallas, TX, United States*

3:15PM **Modeling the Enhancement of Extracellular Matrix Quantity and Quality in Large-Deformation Mechanically-Conditioned Heart Valve Tissue Engineering** SB³C2015-169 *pg.*124
Joao S. Soares[1], John A. Stella[2], Antonio D'Amore[2], Will Zhang[1], William R. Wagner[2], John E. Mayer[3], Michael S. Sacks[1], [1]*University of Texas at Austin, Austin, TX, United States,*[2]*University of Pittsburgh, Pittsburgh, PA, United States,*[3]*Harvard Medical School, Boston, MA, United States*

3:30PM **Tubular Pediatric Pulmonary Valves By Suturing Decellularized Engineered Tissue Tubes** SB³C2015-183 *pg.*152
Jay Reimer, Zeeshan Syedain, Bee Haynie, **Robert Tranquillo**, *University of Minnesota, Minneapolis, MN, United States*

3:45PM **Adipose-Derived Stem Cells From Diabetic Donors Cause Thrombotic Failure of Autologous Tissue Engineered Blood Vessels** SB³C2015-407 *pg.*523
Jeffrey T. Krawiec[1], Han T. Liao[1,2], Justin S. Weinbaum[1], Dominic J. Pezzone[1], Antonio D'Amore[1], J. P. Rubin[1], William R. Wagner[1], David A. Vorp[1], [1]*University of Pittsburgh, Pittsburgh, PA, United States,*[2]*Chang Gung University, Guishan District, Taiwan*

4:00PM **Biomechanical Characterizations of Scar ECM During the Acute to Chronic Stages of Myocardial Infarction** SB³C2015-503 *pg.*694
Bryn Brazile[1], Ryan Butler[1], Sourav S. Patnaik[1], Yanyi Xu[2], Andrew Claude[1], Raj Prabhu[1], Lakiesha N. Williams[1], Jianjun Guan[2], Jun Liao[1], [1]*Mississippi State University, Mississippi State, MS, United States,*[2]*Ohio State University, Columbus, OH, United States*

4:15PM **Development of a Induced Pluripotent Stem Cell Derived Cardiomyocyte Seeded Fibrin Suture for Cardiac Regeneration** SB³C2015-599 *pg.*873
Katrina J. Hansen[1], Michael A. Laflamme[2], Glenn R. Gaudette[1], [1]*Worcester Polytechnic Institute, Worcester, MA, United States,*[2]*University of Washington, Seattle, WA, United States*

FRIDAY, JUNE 19	3:00pm - 4:30pm

The Cellular Microenvironment (joint with JSME) — Golden Cliff / Eagle's Nest

Session Chair: Wei Tan, *University of Colorado, Boulder, CO, United States*
Session Co-Chair: Brendon Baker, *Boston University, Boston, MA, United States*

3:00PM **MicroRNAs and Related Tissue Remodeling Genes in Rotator Cuff With Delayed Repair in A Rat Model** SB³C2015-1089 *pg.*1142
Christopher T. Chen, Fuxin Wei, Erik Contreras, Lucas M. Chen, Zachary Shirley, William Shelton, Michael Khazzam, *UT Southwestern Medical Center, Dallas, TX, United States*

3:15PM **Soft 3-Dimensional Neotissue Microarrays as High-throughput Platforms for Interrogating Stem Cell Instructional Microenvironments** SB³C2015-1173 *pg.*1295
Michael Floren, Wei Tan, *University of Colorado, Boulder, CO, United States*

3:30PM **Long-range Communication Between Cells in Fibrous Matrices Enabled by Tension-driven Alignment of Fibers** SB³C2015-113 *pg.36*
Hailong Wang[1], Abhilash Nair[1], Brendon M. Baker[2], Britta Trappmann[2], Christopher S. Chen[2], Rebecca G. Wells[1], Vivek B. Shenoy[1], *[1]University of Pennsylvania, Philadelphia, PA, United States,[2]Boston University, Boston, MA, United States*

3:45PM **Engineered Cardiac Model System Reveals Fibroblast Threshold for Synchronized Beating** SB³C2015-158 *pg.*106
Ariane C. C. van Spreeuwel[1], Noortje A. M. Bax[1], Christopher S. Chen[2], Carlijn V. C. Bouten[1], *[1]Eindhoven University of Technology, Eindhoven, Netherlands,[2]Boston University, Boston, MA, United States*

4:00PM **Cadherin-Specific Extracellular Interactions Alter Stem Cell Sensation and Interpretation of Soft Tissue Microenvironments** SB³C2015-190 *pg.*164
Brian D. Cosgrove[1,2], Kush D. Mehta[1], Tristan P. Driscoll[1,2], Jason A. Burdick[1,2], Robert L. Mauck[1,2], *[1]University of Pennsylvania, Philadelphia, PA, United States,[2]Philadelphia VA Medical Center, Philadelphia, PA, United States*

4:15PM **Microscopic Heterogeneity in the Aortic Wall: Correlation between Mechanical Environment and Protein Expression** SB³C2015-567 *pg.*815
Takeo Matsumoto[1], Yohei Uno[1], Shintaro Iijima[1], Yoshitaka Moriyama[1], Shukei Sugita[1], Kazuaki Nagayama[1,2], Akio Matsumoto[3], *[1]Nagoya Institute of Technology, Nagoya, Japan,[2]Ibaraki University (present), Hitachi, Japan,[3]Chiba University, Chiba, Japan*

FRIDAY, JUNE 19	3:00pm - 4:30pm

Soft Tissue Mechanics — Primrose B

Session Chair: Heath Henninger, *University of Utah, Salt Lake City, UT, United States*
Session Co-Chair: Trevor Lujan, *Boise State University, Boise, ID, United States*

3:00PM **Computational Framework For Application Of Residual Stress When The Stress-free Configuration Is Unknown.** SB³C2015-1075 *pg.*1116
Steve Maas[1], Ahmet Erdemir[2], Jason P. Halloran[3], Jeffrey A. Weiss[1], *[1]University of Utah, Salt Lake City, UT, United States,[2]Cleveland Clinic, Cleveland, OH, United States,[3]Cleveland State University, Cleveland, OH, United States*

3:15PM **Using Slow and Fast Shear Waves to Estimate Shear and Tensile Moduli: Results from Simulations of Anisotropic Tissue.** SB³C2015-317 *pg.*377
Dennis J. Tweten, John L. Schmidt, Ruth J. Okamoto, Philip V. Bayly, *Washington University, St. Louis, MO, United States*

3:30PM **Experimental Measurement of Shear and Tensile Moduli In Anisotropic Tissue Using Magnetic Resonance Elastography** SB³C2015-244 *pg.*259
John L. Schmidt[1], Dennis J. Tweten[1], Maisie M. Mahoney[1], Tally Portnoi[2], Ruth J. Okamoto[1], Joel R. Garbow[1], Philip V. Bayly[1], *[1]Washington University, St. Louis, MO, United States,[2]Massachusetts Institute of Technology, Cambridge, MA, United States*

3:45PM **A Nonlinear Anisotropic Inverse Mechanics Method for Computational Dissection of Inhomogeneous Planar Soft Tissues** SB³C2015-171 *pg.*128
Colleen M. Witzenburg[1], Victor H. Barocas[2], [1]*University of Virginia, Charlottesville, VA, United States,*[2]*University of Minnesota, Minneapolis, MN, United States*

4:00PM **Structural Properties of the Anterolateral Structures of the Knee** SB³C2015-284 *pg.*321
Amir Ata Rahnemai-Azar, **R Matthew Miller**, Daniel Guenther, Freddie H. Fu, Bryson P. Lesniak, Volker Musahl, Richard E. Debski, *University of Pittsburgh, Pittsburgh, PA, United States*

4:15PM **Stiffness Characterization in Biological Materials Based on Deformation Imaging and Topology Optimization** SB³C2015-556 *pg.*795
Luyao Cai[1], Claus Pedersen[2], Ross Mclendon[2], Manuel Biedermann[2], Gergana Dimitrova[2], Jiang Yao[2], Corey P. Neu[1], [1]*Purdue University, West Lafayette, IN, United States,*[2]*Dassault Systèmes, Johnston, RI, United States*

SATURDAY, JUNE 20	11:30am - 1:00pm

Workshop: Critical Steps in Composing a Successful Mentorship Plan — Primrose A

Session Chair: Sara E. Wilson, *University of Kansas, Lawrence, KS, United States*
Session Co-Chair: Rouzbeh Amini, *The University of Akron, Akron, OH, United States*

SATURDAY, JUNE 20	11:30am - 1:00pm

Workshop: Teaching Undergraduate Design — Superior

Session Chair: Martin Tanaka, *Western Carolina University, Cullowee, NC, United States*
Session Co-Chair: Ken Fischer, *University of Kansas, Lawrence, KS, United States*

SATURDAY, JUNE 20	11:30am - 1:00pm

Workshop: SimVascular Workshop and New User Training — Wasatch

Session Chair: Alison Marsden, *University of California, San Diego, CA, United States*
Session Co-Chair: Shawn Shadden, *UC Berkeley, Berkeley, CA, United States*
Session Co-Chair: Nathan Wilson, *Open Source Medical Software Corporation, CA, United States*

SATURDAY, JUNE 20	11:30am - 1:00pm

Workshop: FEBio Workshop and Discussion — Magpie

Session Chair: Jeffrey A. Weiss, *University of Utah, Salt Lake City, UT, United States*
Session Co-Chair: Gerard A. Ateshian, *Columbia University, New York, NY, United States*
Session Co-Chair: Steve Maas, *University of Utah, Salt Lake City, UT, United States*

SATURDAY, JUNE 20	11:30am - 1:00pm

Workshop: Experimental and Computational Frameworks for Biotransport in Tumors — Maybird

Session Chair: M. Nichole Rylander, *University of Texas at Austin, Austin, TX, United States*
Session Co-Chair: Malisa Sarntinoranont, *University of Florida, Gainesville, FL, United States*

SATURDAY, JUNE 20	11:30am - 1:00pm

Workshop: Taking the Guesswork out of the Interview Process — Golden Cliff / Eagle's Nest

Organizers: ASME Bioengineering Division Student Leadership Committee (special thanks to Justine Garcia, Elizabeth Iffrig, and Corinne Riggin)

SATURDAY, JUNE 20	11:30am - 1:00pm

Workshop: Robotic Testing Systems to Study Joint and Tissue Function — Primrose B

Session Chair: Hiromichi Fujie, *Tokyo Metropolitan University, Tokyo, Japan*
Session Chair: Richard Debski, *University of Pittsburgh, Pittsburgh, PA, United States*

SATURDAY, JUNE 20	1:30pm - 3:00pm

Heart Valves and Cardiovascular Devices — Primrose A

Session Chair: Sarah Vigmostad, *University of Iowa, Iowa City, IA, United States*
Session Co-Chair: Keefe Manning, *Pennsylvania State University, State College, PA, United States*

1:30PM **Fluid-Structure Interaction Analysis of Mitral Valve Forces using a Comprehensive Model with 3D Chordal Structure: Synergy of Modeling and Experiments** SB³C2015-387 *pg.*483
Milan Toma[1], **Morten O. Jensen**[1], Daniel R. Einstein[2], Ajit P. Yoganathan[1], Richard P. Cochran[3], Karyn S. Kunzelman[3], *[1]Georgia Institute of Technology, Atlanta, GA, United States,[2]Pacific Northwest National Laboratory, Richland, WA, United States,[3]University of Maine, Orono, ME, United States*

1:45PM **Transcatheter Aortic Valve Replacement Model: Crimping And Deploying In Patient-pathology Specific Roots** SB³C2015-418 *pg.*545
Matteo Bianchi[1], Ram P. Ghosh[1], Debapria Das[1], Gil Marom[1], Thomas Claiborne[1], Marvin Slepian[2], Danny Bluestein[1], *[1]Stony Brook University, Stony Brook, NY, United States,[2]Sarver Heart Center,University of Arizona, Tucson, AZ, United States*

2:00PM **Computational Assessment of Hemodynamics in Tricuspid and Bicuspid Aortic Valves** SB³C2015-305 *pg.*357
Kai Cao, Philippe Sucosky, *University of Notre Dame, Notre Dame, IN, United States*

2:15PM **Effects Of Coronary Flow On Sinus Hemodynamics In The Presence Of A Transcatheter Aortic Valve Implantation** SB³C2015-1060 *pg.*1092
Brandon Moore[1], Pablo Maureira[2], **Lakshmi P. Dasi**[1], *[1]Colorado State University, Fort Collins, CO, United States,[2]Lorraine University Hospital of Nancy, Nancy, France*

2:30PM **Left Ventricle Assist Device Anastomosis Hemodynamic Analysis Using Direct Numerical Simulations (DNS) And Large Eddy Simulations (LES)** SB³C2015-463 *pg.*628
Ricardo J. Bonilla-Alicea, *Georgia Institute of Technology, Atlanta, GA, United States*

2:45PM **Microfluidic Facsimile Of Ventricular Assist Device Shear Stress Patterns: Towards Point-of-care Devices To Monitor Patient Thrombotic Risk** SB³C2015-1126 *pg.*1213
Annalisa Dimasi[1], Filippo Consolo[1], Lorenzo Valerio[1], Marco Rasponi[1], Danny Bluestein[2], Gianfranco B. Fiore[1], Alberto Redaelli[1], Marvin Slepian[3], *[1]Politecnico di Milano, Milano, Italy,[2]Stony Brook University, Stony Brook, NY, United States,[3]University of Arizona, Tucson, AZ, United States*

SATURDAY, JUNE 20	1:30pm - 3:00pm

Growth, Remodeling and Repair — Superior

Session Chair: Kristin Miller, *Tulane University, New Orleans, LA, United States*
Session Co-Chair: C. Alberto Figueroa, *University of Michigan, Ann Arbor, MI, United States*

1:30PM **Cell-mediated Compaction and Collagen Remodeling in Tissue-engineered Heart Valves due to Dynamic Loading Conditions** SB³C2015-219 *pg.*216
Sandra Loerakker, Tommaso Ristori, Frank P. T. Baaijens, *Eindhoven University of Technology, Eindhoven, Netherlands*

1:45PM **Non-Linear Optical Characterization of the Extracellular Matrix Changes Following Myocardial Infarction Predicts Alterations in Mechanical Properties** SB³C2015-1127 *pg.*1215
Kyle P. Quinn[1], Kelly E. Sullivan[1], Carlo A. Alonzo[1], Zachary Ballard[1], Irene Georgakoudi[1], **Lauren D. Black**[1,2], [1]*Tufts University, Medford, MA, United States,*[2]*Tufts University School of Medicine, Boston, MA, United States*

2:00PM **Tensile Equilibrium Material Response of Pregnant Mouse Cervical Tissue During Normal Remodeling** SB³C2015-372 *pg.*461
Kyoko Yoshida[1], Mala Mahendroo[2], Joy Vink[3], Ronald Wapner[3], Kristin Myers[1], [1]*Columbia University, New York, NY, United States,*[2]*UT Southwestern Medical Center, Dallas, TX, United States,*[3]*Columbia University Medical Center, New York, NY, United States*

2:15PM **Strain-dependent Degradation as a Mechanism for the Paradoxical Effects of Mechanical Loading on Collagen Fiber Alignment in Healing Tendon** SB³C2015-467 *pg.*636
William J. Richardson[1], Stavros Thomopoulos[2], Jeffrey W. Holmes[1], [1]*University of Virginia, Charlottesville, VA, United States,*[2]*Washington University, St. Louis, MO, United States*

2:30PM **A Computational Growth and Remodelling Approach to Study Adaptation and Tortuosity in a Buckled Artery** SB³C2015-580 *pg.*839
Mehdi Farsad[1], Qin Liu[2], Hai-Chao Han[2], Seungik Baek[1], [1]*Michigan State University, East Lansing, MI, United States,*[2]*University of Texas at San Antonio, San Antonio, TX, United States*

2:45PM **Quantification Of Mechanical Properties Of Rat Thoracic Ducts For Long-term Prediction Of Mechanically-mediated Growth And Remodeling** SB³C2015-401 *pg.* 511
Alexander W. Caulk, Zhanna Nepiyushchikh, Ryan Shaw, J. Brandon Dixon, Rudolph L. Gleason, *Georgia Institute of Technology, Atlanta, GA, United States*

SATURDAY, JUNE 20	1:30pm - 3:00pm

Pediatric/Embryonic Hemodynamics — Wasatch

Session Chair: Morbiducci Umberto, *Politecnico di Torino, Turin, Italy*
Session Co-Chair: Anayiotos Andreas, *Cyprus University of Technology, Cyprus*

1:30PM **In Vitro Multi-scale Patient Specific Study of the Effects of Coarctation in the Norwood Circulation** SB³C2015-430
Tianqi Hang[1], Alessandro Giardini[2], Giovanni Biglino[3], **Richard Figliola**[1], [1]*Clemson University, Clemson, SC, United States,*[2]*Great Ormond Street Hospital for Children, London, United Kingdom,*[3]*Great Ormond Street Hospital, London, United Kingdom* *pg.*569

1:45PM **Improved Post-operative Flow Modeling For Complex Peripheral Pulmonary Artery Stenosis** SB³C2015-1103
Weiguang Yang[1], Jeffrey A. Feinstein[1], Alison L. Marsden[2], Frank L. Hanley[1], Frandics P. Chan[1], Lisa W. Faberowski[1], Irene E. Vignon-Clementel[3], [1]*Stanford University, Palo Alto, CA, United States,*[2]*University of California, San Diego, La Jolla, CA, United States,*[3]*INRIA Paris-Rocquencourt, Paris, France* *pg.*1169

2:00PM **The Effect of Resolution on Viscous Dissipation Measured with 4D Flow MRI in Patients with Fontan Circulation: Evaluation Using Computational Fluid Dynamics** SB³C2015-421 *pg.*551
Merih Cibis[1], Kelly Jarvis[2], Michael Markl[2], Michael Rose[2], Cynthia Rigsby[2], Alex Barker[2], Jolanda Wentzel[1], [1]*Erasmus MC, Rotterdam, Netherlands,*[2]*Northwestern University, Chicago, IL, United States*

2:15PM **Modeling Blood Flow in Embryo-Specific Geometry of the Zebrafish Heart** SB^{3}C2015-266 *pg.*293
Pavel Kozlovsky[1], Robert Bryson-Richardson[2], Moshe Rosenfeld[1], Ariel Jaffa[3], **David Elad**[1], *[1]Tel Aviv Univesity, Tel Aviv, Israel,[2]Monash University, Melbourne, Australia,[3]Tel Aviv Medical Center, Tel Aviv, Israel*

2:30PM **Biomechanical Role For Cardiac Jelly In Pumping Mechanics Of Developing Heart During Looping** SB^{3}C2015-657
David L. Bark, Jr., Brennan M. Johnson, Bryce W. Schroder, Deborah M. Garrity, Diego Krapf, Lakshmi P. D. Dasi, *Colorado State University, Fort Collins, CO, United States pg.981*

2:45PM **Developmental Hemodynamics in the Embryonic Heart Outflow Tract** SB^{3}C2015-205 *pg.*194
Venkat Keshav Chivukula, Madeline Midgett, Sandra Rugonyi, *Oregon Health and Sciences University, Portland, OR, United States*

SATURDAY, JUNE 20	1:30pm - 3:00pm

Biomechanics in Treatment of Heart Disease — Magpie

Session Chair: Jonathan Wenk, *University of Kentucky, Lexington, KY, United States*
Session Co-Chair: Lik Chuan Lee, *Michigan State University, East Lansing, MI, United States*

1:30PM **Cardiac Reversible Growth & Remodeling Model: Predicting and Understanding the Chronic Effects of Bioinjection Therapy** SB^{3}C2015-31 *pg.*14
Lik Chuan Lee[1], Martin Genet[2], Jonathan Wenk[3], Joakim Sundnes[4], Samuel Wall[4], *[1]Michigan State University, East Lansing, MI, United States,[2]ETH Zurich, Zurich, Switzerland,[3]University of Kentucky, Lexington, KY, United States,[4]Simula Research Laboratory, Oslo, Norway*

1:45PM **The Effect Of Trabeculae Carnae On The Passive Compliance Of Left Ventricle** SB^{3}C2015-516 *pg.*718
Arnav Sanyal[1], David Halaney[2], Marc D. Feldman[2], Hai-Chao Han[1], *[1]University of Texas at San Antonio, San Antonio, TX, United States,[2]University of Texas Health Science Center at San Antonio, San Antonio, TX, United States*

2:00PM **Effect of Scar Compaction on Therapeutic Efficacy of Anisotropic Reinforcement Following Myocardial Infarction** SB^{3}C2015-615 *pg.*903
Samantha A. Clarke, Gorav Ailawadi, Jeffrey W. Holmes, *University of Virginia, Charlottesville, VA, United States*

2:15PM **Impaired Collagen Degradation Prevents Right Ventricular Hypertrophy And Dysfunction With Development Of Pulmonary Arterial Hypertension** SB^{3}C2015-645 *pg.*959
Mark Golob, Zhijie Wang, Anthony Prostrollo, Timothy Hacker, Gaoussou Diarra, Naomi Chesler, *University of Wisconsin-Madison, Madison, WI, United States*

2:30PM **The Degree Of Outflow Tract Banding Predicts Cardiac Remodeling In Chicken Embryos** SB^{3}C2015-108 *pg.*28
Madeline Midgett, Sandra Rugonyi, *Oregon Health & Science University, Portland, OR, United States*

2:45PM **Mechanical Decoupling of Nuclei from the Cytoskeleton Indicate Mechanosensitivity in a Myocardiocyte Pathology Model** SB^{3}C2015-540 *pg.*764
Benjamin Seelbinder, Sarah Calve, Corey P. Neu, *Purdue University, West Lafayette, IN, United States*

SATURDAY, JUNE 20	1:30pm - 3:00pm

Mechanotransduction I - Cellular and Sub-Cellular Biophysics (joint with JSME) — Maybird

Session Chair: Leo Wan, *Rensselaer Polytechnic Institute, Troy, NY, United States*
Session Co-Chair: James H. Wang, *University of Pittsburgh, Pittsburgh, PA, United States*

1:30PM **TGF-beta and BMP Signaling Pathways Regulate Chromatin Condensation in Mesenchymal Stem Cells in Response to Dynamic Loading** SB^{3}C2015-398 *pg.*505
Su-Jin Heo[1], Woojin M. Han[1,2], Tristan P. Driscoll[1], Dawn M. Elliott[2], Randall L. Duncan[2], Robert L. Mauck[1], *[1]University of Pennsylvania, Philadelphia, PA, United States,[2]University of Delaware, Newark, DE, United States*

1:45PM **Temperature Rise Causes Upregulation Of Tenocyte Catabolism And Enhances Gap Junctional Intercellular Communications** SB^{3}C2015-342 *pg.*415
Masataka Tashiro, Eijiro Maeda, **Toshiro Ohashi**, *Hokkaido University, Sapporo, Japan*

2:00PM **Changes In Mechanosensitivity Of Children Cells: An Approach To Investigate Bone Repair And Bone Developmental Disease** SB^{3}C2015-553 *pg.*789
Sara Barreto[1,2,3], Andrew R. Cameron[1,2,3], Dylan Murray[4], Fergal J. O'Brien[1,2,3], [1]*Tissue Engineering Research Group, Royal College of Surgeons in Ireland, Dublin, Ireland,*[2]*Trinity Centre for Bioengineering, Trinity College Dublin, Dublin, Ireland,*[3]*Advanced Materials and Bio-Engineering Research (AMBER) Centre, Ireland, Dublin, Ireland,*[4]*National Paediatric Craniofacial Center, Children's University Hospital, Temple Street, Dublin, Ireland*

2:15PM **Notch1 Mutation Leads to Valvular Calcification Through Enhanced Cadherin-11 Mechanotransduction** SB^{3}C2015-651 *pg.*969
Joseph Chen[1], Larisa Ryzhova[1], M.K. Sewell-Loftin[1], Christopher Brown[1], Stacey Huppert[2], H. Scott Baldwin[1], W. David Merryman[1], [1]*Vanderbilt University, Nashville, TN, United States,*[2]*Cincinnati Children's Hospital, Cincinnati, OH, United States*

2:30PM **Vinculin-network Mediated Cytoskeletal Remodeling Regulates Contractile Function In The Aging Heart** SB^{3}C2015-1057 *pg.*1086
Gaurav Kaushik[1], Anthony Cammarato[2], **Adam J. Engler**[1], [1]*UC San Diego, La Jolla, CA, United States,*[2]*Johns Hopkins University, Baltimore, MD, United States*

2:45PM **Acto myosin Catch Bonds And Mechano-sensitivity In Non-muscle Cells** SB^{3}C2015-1144 *pg.*1243
Franck J. Vernerey, Umut Akalp, *University of Colorado at Boulder, Boulder, CO, United States*

SATURDAY, JUNE 20	1:30pm - 3:00pm

Injury Biomechanics I: Spine, Military, Modeling — Golden Cliff / Eagle's Nest

Session Chair: Liming Voo, *Johns Hopkins University Applied Physics Laboratory, Laurel, MD, United States*
Session Co-Chair: Reuben H. Kraft, *The Pennsylvania State University, State College, PA, United States*

1:30PM **A Point-Wise Normalization Method for Development of Biofidelity Response Corridors** SB^{3}C2015-497 *pg.* 682
Ian Marcus[1], Scott Gayzik[1], Kerry Danelson[1], Jonathan Rupp[2], Cameron Bass[3], Narayan Yoganandan[4], JiangYue Zhang[5], [1]*Wake Forest Univeristy, Winston Salem, NC, United States,*[2]*University of Michigan Transportation Research Institute, Ann Arbor, MI, United States,*[3]*Duke University, Durham, NC, United States,*[4]*Medical College of Wisconsin, Milwaukee, WI, United States,*[5]*Applied Physics Laboratory Johns Hopkins Univeristy, Laurel, MD, United States*

1:45PM **A Comparison Of Brain Injury Predictors Based On Four Benchmark Impact Studies** SB^{3}C2015-479 *pg.*658
Siddiq Qidwai, Nithyanand Kota, Amit Bagchi, *US Naval Research Laboratory, Washington, DC, United States*

2:00PM **Effect Of Geometric And Material Property Changes In The Thoracic Skeleton For An Older Occupant Finite Element Model** SB^{3}C2015-125 *pg.*52
Samantha Schoell[1,2], Ashley Weaver[1,2], Nicholas Vavalle[1,2], Joel Stitzel[1,2], [1]*Virginia Tech- Wake Forest University, Winston-Salem, NC, United States,*[2]*Wake Forest School of Medicine, Winston-Salem, NC, United States*

2:15PM **Collagen Mimetic Peptide as a Marker of Mechanical Damage in Lamb Middle Cerebral Arteries** SB^{3}C2015-1020 *pg.*1018
Raymond G. Walther, Matthew I. Converse, Kenneth L. Monson, *University of Utah, Salt Lake City, UT, United States*

2:30PM **Injuries And Failure Biomechanical Responses Of Artificial Discs In The Cervical Spine: Potential Applications To Military Environments** SB^{3}C2015-519 *pg.*725
Narayan Yoganandan[1], Frank A. Pintar[1], Jamie L. Baisden[1], Joseph B. McEntire[2], Valeta Carol Chancey[2], [1]*Medical College of Wisconsin, Milwaukee, WI, United States,*[2]*U.S. Army Aeromedical Research Laboratory, Fort Rucker, AL, United States*

2:45PM **Investigation Of Possible Correlation Between Brain Tissue Response And Head Kinematics For Blast-induced Brain Injury** SB^{3}C2015-1136 *pg.*1231
Hesam Sarvghad-Moghaddam, Mariusz Ziejewski, Ghodrat Karami, *North Dakota State University, Fargo, ND, United States*

SATURDAY, JUNE 20	1:30pm - 3:00pm

Shoulder Mechanics — Primrose B

Session Chair: Richard Debski, *University of Pittsburgh, Pittsburgh, PA, United States*
Session Co-Chair: Antonis Stylianou, *University of Missouri-Kansas City, Kansas City, MO, United States*

1:30PM **The Effect Of Size And Location Of Tears In The Supraspinatus Tendon On Potential Tear Propagation** SB^{3}C2015-486 *pg.664*
James R. Thunes, Siladitya Pal, R. M. Miller, Richard E. Debski, Spandan Maiti, *University of Pittsburgh, Pittsburgh, PA, United States*

1:45PM **3D Quantification of Osteophyte Distribution on the Humeral Head** SB^{3}C2015-576 *pg.831*
Shea K. Taylor[1,2], Brandon G. Santoni[1,2], Mark A. Frankle[1,3], Peter Simon[1,2], *[1]University of South Florida, Tampa, FL, United States,[2]Foundation for Orthopaedic Research and Education, Tampa, FL, United States,[3]Florida Orthopaedic Institute, Tampa, FL, United States*

2:00PM **Surgical Accuracy of Traditional Humeral Head Osteotomy in Shoulder Arthroplasty** SB^{3}C2015-146 *pg.84*
Thomas Suter[1,2], **Christopher W. Kolz**[1], Sean T. Tagge[1], Robert Z. Tashjian[1], Ariane Gerber Popp[2], Heath B. Henninger[1], *[1]University of Utah, Salt Lake City, UT, United States,[2]Clinic of Orthopaedic Surgery, Kantonsspital Baselland, Liestal, Switzerland*

2:15PM **Validation of a Subject-Specific Computer Model of Glenohumeral Instability and Capsular Plication** SB^{3}C2015-616
Charlie Yongpravat, David Kovacevic, T Sean Lynch, Charles M. Jobin, William N. Levine, Gerard A. Ateshian, Thomas R. Gardner, Christopher S. Ahmad, *Columbia University, New York, NY, United States* *pg.905*

2:30PM **Design Considerations for Glenoid Components: A Computational Stress Analysis of Translational Motion in Shoulder Replacements** SB^{3}C2015-258 *pg.279*
Christopher Berthelet[1], Farzana Ansari[1], Lisa Pruitt[1], Tom Norris[2], Steve Gunther[3], Michael Ries[4], *[1]University of California, Berkeley, Berkeley, CA, United States,[2]San Francisco Shoulder, Elbow & Hand Clinic, San Francisco, CA, United States,[3]Martha Jefferson Hospital, Charlottesville, VA, United States,[4]Tahoe Fracture and Orthopedic Clinic, Carson City, NV, United States*

2:45PM **Damage Analysis Of Metallic And Polymeric Bearings Used In Reverse Total Shoulder Arthroplasty** SB^{3}C2015-192
Suzanne Chou[1], Isabel Yang[1], **Noah Bonnheim**[1], Farzana Ansari[1], Steve Gunther[2], Tom Norris[3], Michael Ries[4], Lisa Pruitt[1], *[1]University of California, Berkeley, Berkeley, CA, United States,[2]Martha Jefferson Hospital, Charlottesville, VA, United States,[3]San Francisco Shoulder, Elbow & Hand Clinic, San Francisco, CA, United States,[4]Tahoe Fracture and Orthopaedic Clinic, Carson City, NV, United States* *pg.168*

SATURDAY, JUNE 20	3:15pm - 4:45pm

Bone Tissue Engineering (joint with JSME) — Primrose A

Session Chair: Shigeo Tanaka, *Kanazawa University, Kanazawa, Japan*
Session Co-Chair: Ryan K. Roeder, *University of Notre Dame, Notre Dame, IN, United States*

3:15PM **Electromagnetic Field Stimulation Enhances Mechanical Properties Of Tissue-engineered Bone Constructed With Calcined Bovine Trabecular Bone Scaffold** SB^{3}C2015-336 *pg. 403*
Shigeo Tanaka, Yuki Yamashita, *Kanazawa University, Kanazawa, Japan*

3:30PM **Development And Characterization Of Novel Bone Substitutes Using Composite Scaffolds With Mesenchymal Stem Cells** SB^{3}C2015-267 *pg. 295*
Mitsugu Todo[1], Phanny Yos[2], *[1]Kyushu University, Kasuga, Japan,[2]Institute of Technology of Cambodia, Phnom Penh, Cambodia*

3:45PM **Hydroxyapatite Reinforced Collagen Scaffolds Designed for Improved Architecture, Mechanical Properties, and Tunable Growth Factor Delivery** SB^{3}C2015-174 *pg.134*
Ryan K. Roeder, Matthew J. Meagher, Holly E. Weiss-Bilka, Diane R. Wagner, Robert J. Kane, *University of Notre Dame, Notre Dame, IN, United States*

4:00PM **Microporosity Dominates Growth for Large and Small Macropores in BCP Scaffolds** SB³C2015-670 *pg.*1005
Amy Wagoner Johnson[1], Laurie Rustom[1], David Hoelzle[2], Mark Markel[3], Brett Nemke[3], Yan Lu[3], [1]*University of Illinois at Urbana-Champaign, Urbana, IL, United States,*[2]*University of Notre Dame, Notre Dame, IN, United States,*[3]*University of Wisconsin, Madison, WI, United States*

4:15PM **Fabrication Of Collagen Based 3-d Complex Constructs With Electrochemical Compaction Method** SB³C2015-154
Mousa Younesi[1], Vipuil Kishore[2], **Ozan Akkus**[1], [1]*Case Western Reserve University, Cleveland, OH, United States,*[2]*Florida institute of technology, Melborne, FL, United States* *pg.98*

SATURDAY, JUNE 20	3:15pm - 4:45pm

Spine Mechanics — Superior

Session Chair: Brian D. Stemper, *Medical College of Wisconsin, Milwaukee, WI, United States*
Session Co-Chair: Farid Amirouch, *University of Illinois, Chicago, IL, United States*

3:15PM **SEM-based Multi-scale Mechanical Modeling of Lumbar Spine Facet Capsular Ligament During Biaxial Extension** SB³C2015-1054 *pg.* 1082
Vahhab Zarei, Amy A. Claeson, Victor H. Barocas, *University of Minnesota, Minneapolis, MN, United States*

3:30PM **Simulation-Directed Design of Planar Biaxial Tests on the Lumbar Facet Capsular Ligament** SB³C2015-168 *pg.*122
Amy A. Claeson, Victor H. Barocas, *University of Minnesota, Minneapolis, MN, United States*

3:45PM **Effect Of Implant Length And Orientation On Biomechanics Of Sacroiliac Joint Stabilization: A Finite Element Analysis** SB³C2015-1141 *pg.*1237
Ali Kiapour[1], Derek Lindsey[2], Scott Yerby[2], Vijay Goel[1], [1]*ECORE, Toledo, OH, United States,*[2]*Si-Bone Inc, San Jose, CA, United States*

4:00PM **Internal Disc Strain Template Highlights Regions of High Local Strain During Compression Loading and Validates a Finite Element Model** SB³C2015-1019 *pg.*1016
Brent L. Showalter[1], John F. DeLucca[2], John M. Peloquin[1], Daniel H. Cortes[2], Alexander C. Wright[1], James C. Gee[1], Edward J. Vresilovic[3], Dawn M. Elliott[2], [1]*University of Pennsylvania, Philadelphia, PA, United States,*[2]*University of Delaware, Newark, DE, United States,*[3]*Pennsylvania State University, Hershey, PA, United States*

4:15PM **Accuracy of QCT-Based Finite Element Predictions of Vertebral Fracture When Boundary Conditions are Based on Intradiscal Pressure Profiles** SB³C2015-313 *pg.*369
Timothy Jackman, Alexander DelMonaco, Elise Morgan, *Boston University, Boston, MA, United States*

4:30PM **Differences In The Intradiscal Pressure In The L4-L5 And L5-S1 Lumbar Segments** SB³C2015-196 *pg.* 176
Hector E. Jaramillo[1], Christian M. Puttlitz[2], Jose J. Garcia[3], Kirk McGilvray[2], [1]*Universidad Autonoma de Occidente, Cali, Colombia,*[2]*Colorado State University, Fort Collins, CO, United States,*[3]*Universidad del Valle, Cali, Colombia*

SATURDAY, JUNE 20	3:15pm - 4:45pm

Biological Flows in the Interstitium and Lymphatics — Wasatch

Session Chair: Brandon Dixon, *Georgia Institute of Technology, Atlanta, GA, United States*
Session Co-Chair: Walter Lee Murfee, *Tulane University, New Orleans, LA,United States*

3:15PM **A Poroelastic Fluid/Structure-Interaction Model of the Spinal Cord and Surrounding Structures with a Cord Syrinx and Associated Stenosis of the Subarachnoid Space** SB³C2015-197 *pg.*178
Christopher D. Bertram[1], Matthias Heil[2], [1]*University of Sydney, New South Wales, Australia,*[2]*University of Manchester, Manchester, United Kingdom*

3:30PM **Neural Tissue Deformation And Cerebrospinal Fluid Flow Impedance Are Positevely Correlated At The Craniocervical Junction** SB³C2015-634 *pg.*937
Bryn A. Martin[1], Nicholas Shaffer[1], John N. Oshinski[2], Mark Luciano[3], Francis Loth[1], [1]*The University of Akron, Akron, OH, United States,*[2]*Emory University, Atlanta, GA, United States,*[3]*Cleveland Clinic Foundation, Cleveland, OH, United States*

3:45PM **In Silico and In Vitro Modelling of Flow Behaviour in Lymphatic Vessels.** SB³C2015-222 *pg.*220
Sinéad T. Morley[1,2,3], David T. Newport[1,4], Michael T. Walsh[1,2,3], [1]*University of Limerick, Limerick, Ireland,*[2]*Centre for Applied Biomedical Engineering Research, Limerick, Ireland,*[3]*Materials and Surface Science Institute, Limerick, Ireland,*[4]*Stokes Institute, Limerick, Ireland*

4:00PM **Quantification of Lymphatic Valve Resistance to Forward Flow Using Computational Fluid and Solid Modeling** SB³C2015-560 *pg.*801
John T. Wilson[1], Raoul van Loon[2], James E. Moore[1], [1]*Imperial College London, London, United Kingdom,*[2]*Swansea University, Swansea, United Kingdom*

4:15PM **A New Paradigm for the Contribution of Active Tension to the Constitutive Relation in Small Lymphatic Vessels** SB³C2015-198 *pg.*180
Christopher D. Bertram[1], Charlie Macaskill[1], Michael J. Davis[2], James E. Moore[3], [1]*University of Sydney, New South Wales, Australia,*[2]*University of Missouri School of Medicine, Columbia, MO, United States,*[3]*Imperial College, London, United Kingdom*

4:30PM **Spatiotemporal Image Correlation Spectroscopy Techniques for Quantifying Fluid Flow in Microfluidic Channels and Porous Tissues** SB³C2015-472 *pg.*644
Brian T. Graham, Christopher Price, *University of Delaware, Newark, DE, United States*

SATURDAY, JUNE 20	3:15pm - 4:45pm

Micromechanics of Atherosclerosis — Magpie

Session Chair: Dalin Tang, *Worcester Polytechnic Institute, Worcester, MA, United States*
Session Co-Chair: Susan Lessner, *University of South Carolina, Columbia, SC, United States*

3:15PM **On the Effect of Calcific Content on the Mechanical Behaviour of Carotid Plaque Tissue** SB³C2015-359 *pg.*437
Hilary E. Barrett, *University of Limerick, Limerick, Ireland*

3:30PM **Intima Heterogeneity In Atherosclerotic Plaque Stress Calculations** SB³C2015-451 *pg.* 610
Lambert Speelman[1], Bas Van Velzen[2], Anton F. W. Van der Steen[1,2], Jolanda J. Wentzel[1], Frank J. H. Gijsen[1], [1]*Erasmus MC, Rotterdam, Netherlands,*[2]*Delft University of Technology, Delft, Netherlands*

3:45PM **Diet-induced Vascular Remodeling Produces a Shift in Collagen Fiber Angle Distribution in a Mouse Model of Atherosclerosis** SB³C2015-508 *pg.*702
Shana R. Watson, Piaomu Liu, Edsel A. Pena, Michael A. Sutton, John F. Eberth, Susan M. Lessner, *University of South Carolina, Columbia, SC, United States*

4:00PM **Towards the Development of an Atherosclerotic Plaque Stratification Parameter to Predict Arterial Restenotic Response following Endovascular Treatment** SB³C2015-379 *pg.*469
Eoghan M. Cunnane[1], Hilary E. Barrett[1], Eamon G. Kavanagh[2], Michael T. Walsh[1], [1]*University of Limerick, Limerick, Ireland,*[2]*University Hospital Limerick, Limerick, Ireland*

4:15PM **Reconstruction of Incomplete Lipid Pool Geometry for Stress Calculations in Atherosclerotic Arteries** SB³C2015-362
Annette M. Kok, Lambert Speelman, Frank J. H. Gijsen, Jolanda J. Wentzel, *Erasmus MC, Rotterdam, Netherlands pg.443*

4:30PM **MMP-2 Expression and the Tissue Mechanics of Human Ascending Thoracic Aortic Aneurysms** SB³C2015-405
Alexander A. Emmott[1,2], Nastaran Shahmansouri[1,2], Mohammed Alreshidan[3], Stefanie Pohlod[1], Rosaire Mongrain[1], Raymond Cartier[2], Kevin Lachapelle[3], Richard Leask[1,2], [1]*McGill University, Montreal, QC, Canada,*[2]*Montreal Heart Institute, Montreal, QC, Canada,*[3]*Royal Victoria Hospital, Montreal, QC, Canada pg.519*

SATURDAY, JUNE 20	3:15pm - 4:45pm

Mechanotransduction II - Interactions between Cells and Their Environment (joint with JSME) — Maybird

Session Chair: Toshiro Ohashi, *Hokkaido University, Sapporo, Japan*
Session Co-Chair: Chelsey Simmons, *University of Florida, Gainesville, FL, United States*

3:15PM **A Microtissue Array Device To Screen The Lung Fibrogenic Potential Of Carbon Nanotubes** SB³C2015-350 *pg.* 427
Zhaowei Chen, Qixin Wang, Mohammadnabi Asmani, Yan Li, Yun Wu, **Ruogang Zhao**, *State University of New York at Buffalo, Buffalo, NY, United States*

3:30PM **The Impact of Pre-stretch Induced Surface Anisotropy on Axonal Regeneration** SB³C2015-464 *pg.*630
Chun Liu, Seungik Baek, **Christina Chan**, *Michigan State University, East Lansing, MI, United States*

3:45PM **Excessive Mechanical Loading Causes Aberrant Differentiation Of Tendon Stem Cells (TSCs) That Leads To The Development Of Degenerative Tendinopathy** SB³C2015-227 *pg.* 230
Jianying Zhang, **James H-C. Wang**, *University of Pittsburgh, Pittsburgh, PA, United States*

4:00PM **Both Nuclear And Extracellular Matrix Rigidity Determine The Stable Size Of Focal Adhesion Plaques** SB³C2015-228
Xuan Cao, Tristan P. Driscoll, Robert L. Mauck, Vivek B. Shenoy, *University of Pennsylvania, Philadelphia, PA, United States* *pg.232*

4:15PM **Role Of Cell Tension On Anisotropic Mechanosensing** SB³C2015-1106 *pg.*1175
Shin Min Wen, Pen Hsiu Grace Chao, *National Taiwan university, TAIPEI, Taiwan*

4:30PM **Osteocyte Mechanotransduction During Low Magnitude Mechanical Stimulation** SB³C2015-1121 *pg.*1205
Thomas R. Coughlin, Tyler C. Kreipke, Glen L. Niebur, *University of Notre Dame, Notre Dame, IN, United States*

SATURDAY, JUNE 20	3:15pm - 4:45pm

Injury Biomechanics II - Head to Foot, Modeling, Risk — Golden Cliff / Eagle's Nest

Session Chair: Steven Rowson, *Virginia Tech, Blacksburg, VA, United States*
Session Co-Chair: Ken Monson, *University of Utah, Salt Lake City, UT, United States*

3:15PM **The Effect of Grade II and Grade III Ankle Injury on the Ankle Joint Complex Kinematics and Achilles Load: A Cadaveric Study** SB³C2015-529 *pg.*743
Bardiya Akhbari, Matthew H. Dickinson, Ednah G. Louie, Sami Shalhoub, Lorin P. Maletsky, *University of Kansas, Lawrence, KS, United States*

3:30PM **Driver Injury Risk Sensitivity in Finite Element Model Reconstructions of Real World Motor Vehicle Crashes** SB³C2015-602 *pg.* 879
James P. Gaewsky, Ashley A. Weaver, Bharath Koya, Joel D. Stitzel, *Wake Forest University, Winston-Salem, NC, United States*

3:45PM **Head-neck Relative Posture Affects Fracture Outcome of the Basilar Skull and Upper Cervical Spine under Helmeted Head Crown Impact** SB³C2015-571 *pg.*821
Liming Voo, Kyle Ott, Christopher Dooley, Andrew Merkle, *Johns Hopkins University Applied Physics Laboratory, Laurel, MD, United States*

4:00PM **Investigation of Head Rotational Impulse Characteristics on Brain Strain Responses** SB³C2015-39 *pg.*22
Wei Zhao, Songbai Ji, *Dartmouth College, Hanover, NH, United States*

4:15PM **Development of a 5th Percentile Female Finite Element Model Using a Multi-Modality Image Dataset** SB³C2015-1038
Matthew L. Davis, Bharath Koya, Jeremy M. Schap, F. Scott Gayzik, *Virginia Tech-Wake Forest University, Winston Salem, NC, United States* *pg.* 1050

4:30PM **Embedded Finite Elements For Modeling Traumatic Axonal Injury** SB³C2015-1117 *pg.*1197
Reuben H. Kraft, Harsha T. Garimella, *The Pennsylvania State University, University Park, PA, United States*

SATURDAY, JUNE 20	3:15pm - 4:45pm

Lower Extremity Mechanics — Primrose B

Session Chair: Lorin Maletsky, *University of Kansas, Lawrence, KS, United States*
Session Co-Chair: Ferris Pfeiffer, *University of Missouri, Columbia, MO, United States*

3:15PM **Understanding the Mechanics of Focal Chondral Defects in the Hip: a Framework to Advance Treatment Options** SB³C2015-596 *pg.867*
Brenden J. Klennert, Benjamin J. Ellis, Travis G. Maak, Ashley Kapron, Tyler O. Kaiser, Jeffrey A. Weiss, *University of Utah, Salt Lake City, UT, United States*

3:30PM **Specimen-Specific Evaluation of a Multi-Body Model Predicts Instability and Increased Meniscal Load in the Anterior Cruciate Ligament-Deficient Knee** SB³C2015-375 *pg.465*
Mohammad Kia, Kevin Schafer, Daniel Green, Andrew Pearle, Thomas Wickiewicz, Timothy Wright, Carl Imhauser, *Hospital for Special Surgery, New York, NY, United States*

3:45PM **In-Vivo Kinematics of the Asymptomatic Hip during Dynamic Pivoting: Foundations for the Evaluation of Femoroacetabular Impingement** SB³C2015-306 *pg.359*
Penny R. Atkins, Niccolo M. Fiorentino, Michael J. Kutschke, Sara J. Fauver, Ashley L. Kapron, Christopher L. Peters, Stephen K. Aoki, Andrew E. Anderson, *University of Utah, Salt Lake City, UT, United States*

4:00PM **Open Knee(s): Magnetic Resonance Imaging for Specimen-specific Next Generation Knee Models** SB³C2015-581
Craig Bennetts[1], Snehal Chokhandre[1], Shannon Donnola[2], Chris Flask[2], Tara Bonner[1], Robb Colbrunn[1], Ahmet Erdemir[1], *[1]Cleveland Clinic, Cleveland, OH, United States,[2]Case Western Reserve University, Cleveland, OH, United States* *pg.841*

4:15PM **A New Joint Coordinate System for Robotic Testing of Cadaveric Knee Specimens** SB³C2015-554 *pg.791*
Daniel Boguszewski, Nirav Joshi, Edward Cheung, Paul Yang, Keith Markolf, David McAllister, *UCLA, Los Angeles, CA, United States*

4:30PM **Do The Laxities Of The Normal Knee At 0° And 90° Of Flexion Support The Goal Of Gap-balancing A Total Knee Arthroplasty?** SB³C2015-30 *pg.12*
Joshua D. Roth, Stephen M. Howell, Maury L. Hull, *University of California, Davis, Davis, CA, United States*

SATURDAY, JUNE 20	5:00pm - 6:00pm

LISSNER LECTURE – James A. Ashton-Miller — Ballrooms 1-3

SB^{3}C2015
Summer Biomechanics, Bioengineering and Biotransport Conference
June 17-20, 2015,Snowbird Resort, Utah, USA
SB^{3}C2015-518

ENDOVASCULAR TREATMENT OF INTRACRANIAL ANEURYSMS: FINITE ELEMENT MODELING OF VARIOUS INTERVENTION STRATEGIES

Robert J. Damiano (1,3), Ding Ma (1,3), Jianping Xiang (1,3,4), Adnan H. Siddiqui (3,4,5), Kenneth V. Snyder (3,4,5), Hui Meng (1,2,3,4)

(1) Department of Mechanical and Aerospace Engineering
State University of New York at Buffalo
Buffalo, NY

(2) Department of Biomedical Engineering
State University of New York at Buffalo
Buffalo, NY

(3) Toshiba Stroke and Vascular Research Center
State University of New York at Buffalo, Buffalo, NY

(4) Department of Neurosurgery
State University of New York at Buffalo, Buffalo, NY

(5) Department of Radiology
State University of New York at Buffalo, Buffalo, NY

INTRODUCTION

Endovascular intervention using coil embolization and flow diversion are the predominant means for treating intracranial aneurysms (IAs). Coils exclude the aneurysm from the circulation by causing thrombotic occlusion of the aneurysm sac. Furthermore, flow diverters (FDs) reduce blood inflow and gradually reconstruct the parent vessel through endothelialization, thereby obliteraterating the IA. Despite the success of these treatment strategies, poor treatment outcomes have been reported, such as aneurysm recurrence [2] and incomplete occlusion [3]. Since these intervention strategies primarily work by modifying aneurysmal flow, post-treatment hemodynamics may be a major factor in treatment outcome. Computational tools that model clinical interventions and hemodynamics in patient-specific IAs may help identify hemodynamic factors that affect treatment outcome for various intervention strategies. In order to investigate a broader range of endovascular intervention strategies, we developed a new FEM-based coiling technique for patient-specific IAs. In conjunction with our previously established HiFiVS [1] method for FD deployment and CFD, we evaluated the hemodynamic impact of four intervention strategies in a patinet-specific IA: coiling (1-8 coils), single FD, FD with adjunctive coils (1-8 coils), and overlapping FDs (1-3 FDs).

METHODS

Aneurysm Model A patient-specific internal carotid artery IA (Figure 1) was reconstructed from angiographic imaging and was virtually treated with coils (1-8), a single FD, a single FD with adjunctive coils (1-8), and overlapping FDs (1-3).

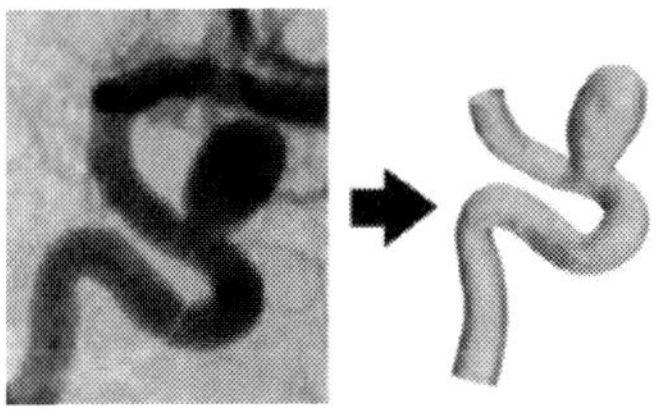

FIGURE 1. ANGIOGRAPHIC AND RECONSTRUCTED IMAGES OF THE PATIENT-SPECIFIC ANEURYSM TEST CASE.

FEM Modeling of Coil and FD Deployments Our new FEM-based technique for coil deployment explicitly models pre-shaped coils and their deployment mechanics including device packaging into a microcatheter. During pre-processing in the deployment workflow (Figure 2), we model a "pre-shaped" coil geometry [4] and isolate the IA sac from the parent vessel. The geometry is imported into the FEM software Abaqus/Explicit 6.12 (SIMULIA, Providence, RI) and discretized using Euler-Bernoulli beam elements. The coil is packaged and deployed into the IA sac via a microcatheter. After completion of the FEM simulation, the deployed discretized coil is swept with a circular profile to convert it into a 3D model for subsequent CFD simulations. For all FD deployments, we used our previously established HiFiVS method. HiFiVS incorporates all critical steps that affect the final deployed FD configuration.

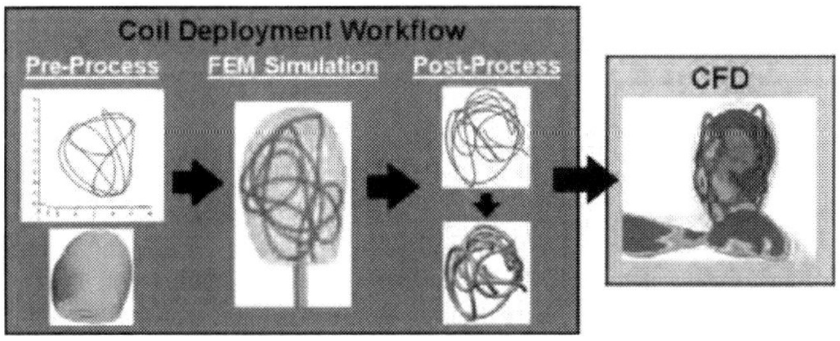

FIGURE 2. SCHEMATIC OF WORKFLOW FOR VIRTUAL COIL DEPLOYMENT.

CFD Simulations Hemodynamic modifications for the four intervention strategies were evaluated via CFD using the finite-volume based software Star CCM+ v.8 (CD-adapco, Melville, NY). The unstructured polyhedral meshing algorithm was used to generate the computational grids for each scenario. A physiologic velocity inlet condition was adopted from literature. The incompressible Navier-Stokes equations were solved under the assumptions of steady-state and laminar flow. The vessel wall and endovascular devices were assumed rigid. Blood was modeled as a Newtonian fluid with a viscosity of 3.5 cP and density of 1056 kg/m^3. Hemodynamic characteristics investigated include blood velocity streamlines, inflow rate, intra-aneurysmal velocity magnitude, WSS distributions, aneurysm-averaged velocity, and aneurysm-averaged WSS.

RESULTS AND DISCUSSION

Device Deployment Results The new FEM-based coil deployment method, in conjunction with HiFiVS, was used to deploy coils and FDs for each intervention strategy (Figure 3). The packing densities (percentage of the IA sac volume occupied by coils) for 1-8 coils were 3.7-29.6%. The deployed coils (1-8) were used for both the coiling and FD with adjunctive coil strategies. The single FD was also used in the FD with adjunctive coils and overlapping FDs strategies.

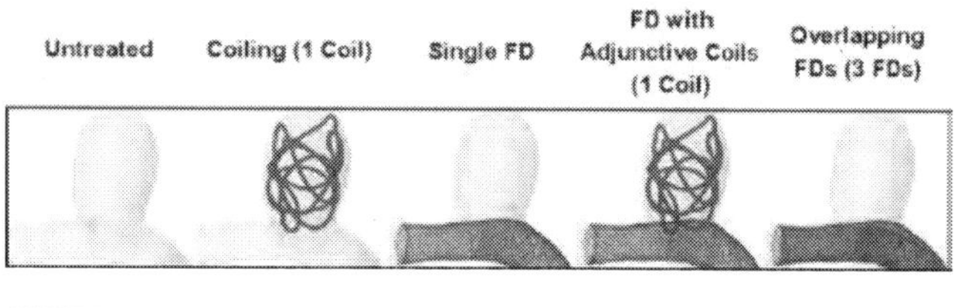

FIGURE 3. EXAMPLE DEPLOYMENT RESULTS FOR COILING (1-8 COILS), SINGLE FD, FD WITH ADJUNCTIVE COILS (1-8 COILS), AND OVERLAPPING FDS (1-3 FDS). EACH DEVICE IS UNIQUELY COLORED FOR CLARITY.

Hemodynamic Comparison of Intervention Strategies CFD results (Figure 4) revealed an impingement jet on the distal side of the untreated IA and a vortex-like flow pattern. Coils (alone and in conjunction with FD) disrupted the impingement jet and decreased flow velocity in the IA sac. The single FD and overlapping FDs did not disrupt the vortex flow pattern but decreased the strength of the impingement jet. In all intervention strategies, the devices decreased WSS magnitude in the IA sac compared to the untreated scenario. Relative hemodynamic parameters (Figure 5) showed a reduction in aneurysm-averaged velocity and WSS for all coiling scenarios compared to the untreated case. When compared with the single FD, it was clear that the majority of the inflow reduction was due to the FD and not the coils. However, additional reduction in flow parameters was seen when adding coils to the FD (adjunctive coils). A continual decrease in flow parameters was observed for successive overlapping FDs (Figure 6), but the decrease was marginal between the 2nd and 3rd FDs deployed. These results suggest coils primarily create stasis in the IA sac while FDs mainly divert inflow.

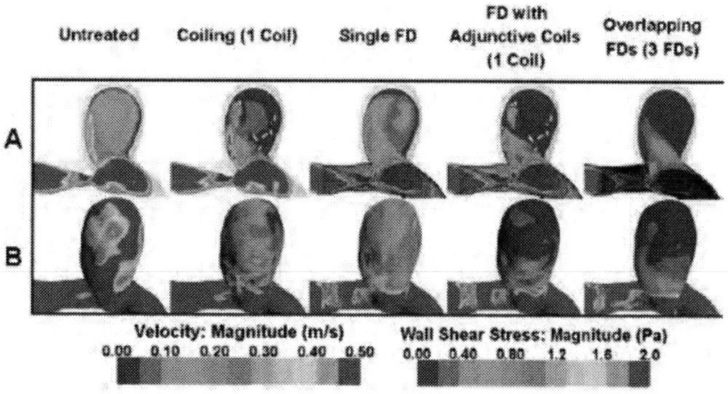

FIGURE 4. A. INTRA-ANEURYSMAL VELOCITY MAGNITUDE AND B. WSS DISTRIBUTION EXAMPLES FOR THE UNTREATED, COILING, SINGLE FD, FD WITH ADJUNCTIVE COILS, AND OVERLAPPING FDS SCENARIOS. FLOW WAS FROM RIGHT TO LEFT IN EACH PICTURE.

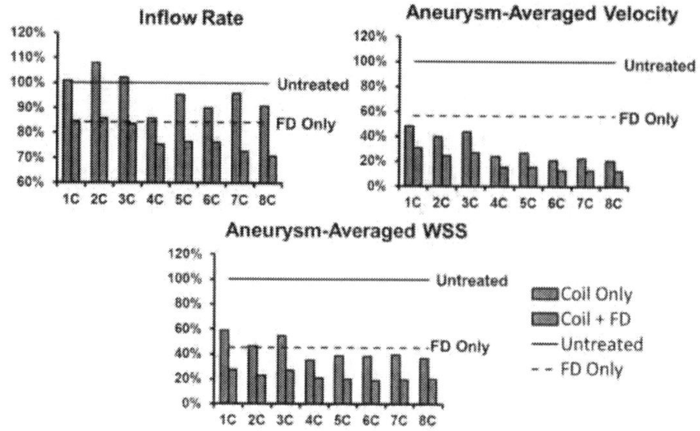

FIGURE 5. RELATIVE HEMODYNAMIC PARAMETERS FOR COILS ONLY, FD ONLY, AND COILS PLUS FD. RESULTS FOR EACH SCENARIO ARE COMPARED RELATIVE TO THE UNTREATED SCENARIO (100% BASELINE). 1C-8C CORRESPONDS TO 1-8 COILS WITH PACKING DENSITY RANGING FROM 3.7-29.6%.

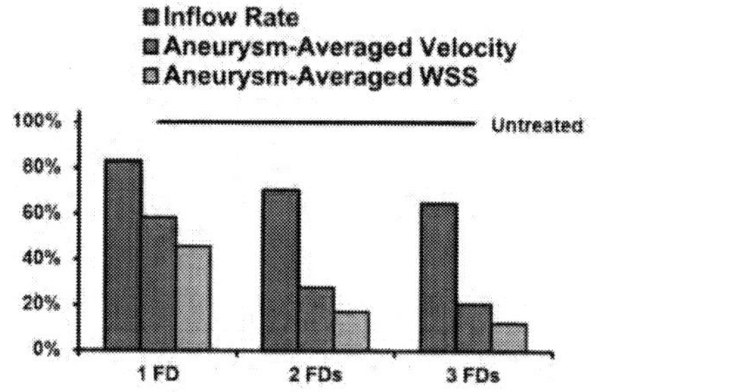

FIGURE 6. RELATIVE HEMODYNAMIC PARAMETERS FOR OVERLAPPING FDS (1-3 FDS).

CONCLUSIONS

We developed a new FEM-based coil deployment technique and combined it with our previously established HiFiVS method for FD deployment to simulate several endovascular intervention strategies including coiling, single FD, FD with adjunctive coils, and overlapping FDs. In conjunction with CFD, our modeling techniques have allowed us to investigate the hemodynamic effects of these intervention strategies in a patient-specific IA.

ACKNOWLEDGEMENTS

This abstract is based upon work supported by Covidien and computational resources were partially provided by the Center for Computational Research (CCR) at the University at Buffalo.

REFERENCES

[1] Ma D, Dumont TM, Kosukegawa H, Ohta M, Yang X, Siddiqui AH, Meng H. High Fidelity Virtual Stenting (HiFiVS) for Intracranial Aneurysm Flow Diversion: In Vitro and In Silico. Annals of biomedical engineering 41.10: 2143-2156, 2013.

[2] Raymond, J., Guilbert, F., Weill, A., Georganos, S.A., Juravsky, L., Lambert, A., Lamoureux, J., Chagnon, M., Roy, D. Long-term angiographic recurrences after selective endovascular treatment of aneurysms with detachable coils. Stroke 34, 1398-1403, 2003.

[3] Brinjikji, W., Murad, M.H., Lanzino, G., Cloft, H.J., Kallmes, D.F. Endovascular treatment of intracranial aneurysms with flow diverters a meta-analysis. Stroke 44, 442-447, 2013.

[4] Babiker MH, Chong B,Gonzalez LF, Cheema S, Frakes DH. Finite element modeling of embolic coil deployment: Multifactor characterization of treatment effects on cerebral aneurysm hemodynamics. Journal of biomechanics 46.16: 2809-2816, 2013.

SB^{3}C2015
Summer Biomechanics, Bioengineering and Biotransport Conference
June 17-20, 2015,Snowbird Resort, Utah, USA
SB^{3}C2015-519

Injuries and Failure Biomechanical Responses of Artificial Discs in the Cervical Spine: Potential Applications to Military Environments

Narayan Yoganandan[1], Frank A. Pintar[1], Jamie L. Baisden[1], B. Joseph McEntire[2], Valeta Carol Chancey[2]

(1) Department of Neurosurgery
Medical College of Wisconsin
VA Medical Center
Milwaukee, WI USA

(2) U.S. Army Aeromedical Research
Laboratory,
Fort Rucker, AL, USA

INTRODUCTION

More than one million Service members have served in conflicts in the past decade [1]. Approximately one-third of wounded warriors had head, face or neck injuries [2-4]. Approximately one-third of all spinal injuries occur to the neck. A 2009 army injury prevention report stated that nonfatal outcomes result in significant lost-duty time and decreased operational readiness; non-battle injuries account for 49% of spinal injuries. Incidence rates of degenerative disc disease are the highest in Army personnel compared to other agencies [5]. Surgical interventions include the use of conventional bone grafts and artificial discs. The later intervention is preferred in some cases due to donor site issues and early return-to-duty, RTD [6]. As the Federal Drug Administration (FDA) continues to approve artificial discs and as they are primarily developed for civilians, their efficacy in military environments needs study. Heterotopic ossification occurs around the implant in civilians [7]. Such bone growths are likely to be greater and have greater influence on the load-carrying capacity of the spine in military personnel. This group of the population sustains extraordinary physiological stresses secondary to activities such as piloting high-performance aircraft, parachute jumps, high-impact water entries, and prolonged runs with heavy loads. The increased demand places additional loads on the structural components of the cervical spine. Because of the increased loading, it is important to understand the biomechanics of these constructs at higher loads, including failure. Thus, the objectives of the present study were to examine biomechanical injury responses of FDA-approved artificial discs and conventional anterior discectomy and fusion (ACDF) compared with the intact cervical spine.

METHODS

Caprine in vivo and ex vivo animals were included in the study. The experimental protocol was approved by the U.S. Army Medical Research and Materiel Command Animal Care and Use Review Office and the Clement J. Zablocki Veterans Affairs Medical Center Animal Care and Use Committee. The animals were obtained and monitored for a short period in the veterinary unit of the VA hospital. Cervical spine radiographs were obtained before and during surgery with construct placement. The animals were anesthetized and discectomies were performed at the C3-C4 level. Bryan disc (BD), ProDisc-C (PC) or plating for the ACDF condition was used to replace the excised intervertebral disc space. Following surgery, animals were maintained in the veterinary unit for six months. Periodical x-rays were obtained to follow the recovery process. At the end of the study period, animals were euthanized, osteoligamentous cervical columns were excised, CT scans were obtained and specimens were stored at -20 degree Celsius. The control group of animals was not subjected to surgery. Those columns were also excised according to same protocols.

The C2-C5 columns were fixed at the superior and inferior ends using poly-methyl-methacrylate allowing normal curvature. Retro-reflective targets were attached to vertebral landmarks. A six-axis load cell was fixed at the distal end of the preparation to record forces and moments transmitted through the column. The loading protocol included sub-failure tests in flexion, extension and lateral bending, combined with compression. The failure test was conducted after this series in compression-flexion under a dynamic mode at a velocity of 1,000 mm/s. The specimen was placed on the platform of custom electrohydraulic testing device with an x-y cross table used for pre-positioning. The preparation was loaded with the piston of the testing device at an anterior eccentricity of 80 mm from the center of the posterior longitudinal ligament at the mid-vertebral level to induce flexion. A custom fixture was used to ensure compression-flexion loading, i.e., minimizing shear loads. Briefly, a grooved spherical plate with its curvature approximating the bending arc of the spine was used such that the roller fixture attached to the inferior end of the piston delivered the described compression-flexion loading. Off-axis loads were monitored by the six-axis load cell to ensure the intended load application. All specimens underwent radiography, CT scanning and dissection following the failure test. Temporal force and bending moment data were used to extract peak magnitudes of these variables. Ligamentous injuries and vertebral fractures were assessed based on spinal level, type and association with the type of construct, BD and PC, or fusion (ACDF). Injuries and peak failure forces and moments were compared with similar data obtained from intact/control group of specimens, i.e., animals that did not undergo surgery.

RESULTS

A total of 14 animals were used with four each for PC and BD, and three each for intact/control and ACDF conditions. Figure 1 shows intra-operative x-rays and CT scans demonstrating recovery.

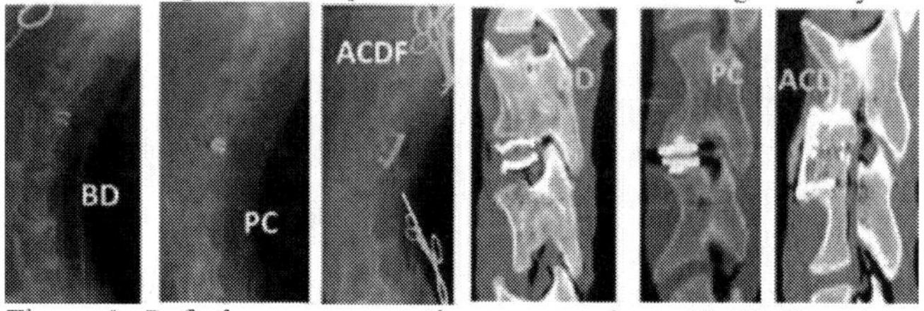

Figure 1: Left three representative x-rayss show C3-C4 discectomy and introduction of constructs. The right three sagittal images show representative CT scans of the three implant specimens used for failure tests under dynamic combined compression-flexion loading.

Force and moment histories had similar response morphologies for all conditions. Mean peak failure forces were the greatest for the ACDF (1240 ± 110 N), followed by PC (855 ± 98 N), intact (800 ± 642 N), and BD (715 ± 220 N). Mean peak moments were lowest for the intact (40.8 ± 12.4 Nm), followed by PC (71.6 ± 10.0 Nm), BD (72.8 ± 36.1 Nm) and ACDF (77.9 ± 14.2 Nm). Posterior ligamentous injuries occurred to all surgical specimens which included facet joint involvements. Facet fractures occurred in one case each with BD and PC constructs. Although posterior ligament injuries occurred in all intact specimens, fracture of the C5 vertebra occurred in one case. Injuries were more distributed in intact specimens.

DISCUSSION

Although artificial discs are evaluated in biomedical engineering and clinical studies, their purpose and outcomes are primarily focused on civilian populations and low level physiological loads (e.g., pure moments). As stated in the Introduction, these devices are being used in the military wherein subjects are healthier, younger and expected to carry increased loads compared to the civilian. While cyclic loading tests of the physical/individual artificial disc constructs are conducted, failure properties and injuries associated with combined loading and at high rates are less understood. These constructs must be evaluated under compressive loading instead of pure moments to be applicable to military populations as head supported mass in the form of helmets and night vision goggles add to the physiological mass of the head. The present experimental design study incorporated these aspects.

Because the introduction of BD/PC involves healing and recovery over time, traditional human cadaver models are less effective as they can only determine immediate post-operative properties. Thus, an in vivo animal model was used in this study. The Caprine model was selected because its head-neck alignment and physiological loading is similar to the axial alignment configuration of the human head-neck [8, 9]. Also, following surgery, for the entire six-month survival time, all animals were monitored for any unforeseen activities. Intermittent x-rays further ensured the integrity of the construct (BD, PC and ACDF) and assisted in the evaluation of the recovery and heterotopic ossifications. Thus, the animal model and excised spines from the matured/healed constructs are appropriate to study the failure properties (force and moment) and injuries at dynamic rates of loading, under combined compression flexion mode.

Failure forces in both PC and BD suggest the suitability of either artificial disc from ultimate load carrying perspective. However, failure force for ACDF were greater than these constructs and intact, suggesting that anterior plating, not present in BD or PC, provides added strength, likely reinforcing the anterior column more under this traditional surgical approach [10]. A compromising effect would be the reduction in the range of motion due to ACDF compared to BD and PC. Significantly greater moments (1.8 to 1.9 times) in PC, BD and ACDF than intact indicate that surgery alters the load-path from primarily axial mode in intact to more bending-related mechanism of internal force transfer. This changing mechanism due to the surgical intervention may have long-term implications, especially with military populations, for reasons described earlier.

Injuries in intact specimens were somewhat distributed at different levels. This is in contrast to injuries occurring at the C3-C4 level where the construct was placed (75% in PC) or at the inferior level of the construct (25% in PC and all cases in BD). In either case, this phenomenon suggests a fulcrum effect with the implant, regardless of type and nature. The fulcrum effect can also contribute to the presence of the bending-mode, discussed above. Failure of the spine at the level of introduction of the artificial disc may have important clinical and operational (return-to-duty, RTD) implications to military populations. While this is the only head-to-head study comparing two artificial discs with ACDF and intact using an accepted in vivo animal model mimicking human head-neck structures and alignment, incorporating healing/fusion and evaluating failure forces and moments at high loading rates applicable to military populations, additional tests are needed to reinforce the present conclusions.

ACKNOWLEDGEMENTS

This study was supported in part by Cooperative Agreement W81XWH-10-2-0065 with the US Army Aeromedical Research Laboratory and Department of Veterans Affairs. Views expressed here are those of the authors and do not reflect official policy/position of the funding agencies.

REFERENCES

1. Gubata, M.E., et al., Trends in the epidemiology of disability related to traumatic brain injury in the US Army and Marine Corps: 2005 to 2010. J Head Trauma Rehabil, 2014. **29**: p. 65-75.
2. Okie, S., Traumatic brain injury in the war zone. N Engl J Med, 2005. **352**: p. 2043-7.
3. Warden, D., Military TBI during the Iraq and Afghanistan wars. J Head Trauma Rehabil, 2006. **21**: p. 398-402.
4. Owens, B.D., et al., Combat wounds in operation Iraqi Freedom and operation Enduring Freedom. J Trauma, 2008. **64**: p. 295-9.
5. Mydlarz, D., Degenerative disc disease, active component, U.S. Armed Forces, 2001-2011. MSMR, 2012. **19**: p. 10-13.
6. Cochran, J., et al., Effects of Treatment for Cervical Disc Degenerative Disease in Military Populations. ASME, 2011.
7. Quan, G.M., et al., Eight-year clinical and radiological follow-up of the Bryan cervical disc arthroplasty. Spine, 2011. **36**: p 639-46.
8. Zhou, C., et al., Biomechanical evaluation of immediate stability of biodegradable multi-amino acid copolymer/tri-calcium phosphate composite interbody Cages in a goat cervical spine model. 2011. **28**: p. 63-6.
9. Pintar, F.A., et al., Fusion rate and biomechanical stiffness of hydroxylapatite versus autogenous bone grafts for anterior discectomy. An in vivo animal study. Spine, 1994. **19**: p. 2524-8.
10. Denis, F., Spinal instability as defined by the three-column spine concept in acute spinal trauma. Clin Orthop Relat Res, 1984 (189): p. 65-76.

UNDERSTANDING AND OPTIMIZING ELECTRICAL STIMULATION OF NEURONS FOR IMPROVING REGENERATION: A FINITE ELEMENT SIMULATION WITH EXPERIMENTAL VERIFICATION

Robert D. Adams (1), Rebecca K. Willits (2), Amy B. Harkins (1,3)

(1) Pharmacological and Physiological Science
Saint Louis University
St. Louis, Missouri, USA

(2) Biomedical Engineering
University of Akron
Akron, Ohio, USA

(3) Biomedical Engineering
Saint Louis University
St. Louis, Missouri, USA

INTRODUCTION

In many instances of extensive nerve damage, the injured nerve never adequately heals and a person is left with a debilitating lack of function [1]. Electrical stimulation (ES) has been shown to improve nerve regeneration *in vivo* and is a promising therapy [2, 3]. *In vitro*, ES has been shown to improve the rate of neurite growth, and in some cases to orient neurite growth [4-6]. However, it is currently not well understood how ES improves the rate of neurite growth and affects the direction of neurite growth. To complicate matters, in most neuron ES experiments only one type of ES is used (i.e. DC, AC, or pulses of a specific duration and intensity). Recently, our lab as well as others have hypothesized that the mechanism in which ES improves neurite growth is via the opening of voltage dependent Ca^{2+} channels (VDCCs) leading to a subsequent rise in intracellular Ca^{2+} ($[Ca^{2+}]_i$) [7]. However, the effect of the ES parameters, such as the intensity and duration of an electrical pulse, on how the pulse raises $[Ca^{2+}]_i$ has not been examined. Using both a computer model and *in vitro* experimental observations, our study aims to understand the mechanism in which extracellular ES increases neuronal $[Ca^{2+}]_i$ in order to optimize ES parameters for potentially improving neuronal growth.

METHODS

In our current study, using a newly developed sensory neuron finite element model, we aim to predict membrane depolarization and calcium flux caused by extracellular electrical fields, in order to better understand the effect of ES on neuronal $[Ca^{2+}]_i$. Furthermore, the model is verified by measuring the change in sensory neurons of $[Ca^{2+}]_i$ in response to extracellular electrical pulses of varying intensity and duration.

The computational model of a sensory neuron was developed in COMSOL Multiphysics. To the best of our knowledge, this is the first simulation to use the geometry of a small sensory neuron along with the two sodium channels, one potassium channel, and three calcium channels expressed on mammalian small sensory neurons. All of the channels are modeled in the style of Hodgkin and Huxley with open probability and opening time constants controlled by membrane voltage [8-11]. Furthermore, this simulation estimates the $[Ca^{2+}]_i$ of the neurons incorporating intracellular stores and Ca^{2+} pumps [12]. Three neuronal geometries were tested: 1) a neuron body with no axons or dendrites, 2) a neuron body with an extending axon, and 3) an axon with the growth cone at the end. Each of these geometries was simulated with pulses ranging from durations of 1 ns to constant DC, and from intensities of 1 V/m to $1x10^6$ V/m.

The finite element model was verified by measuring the $[Ca^{2+}]_i$ of sensory neurons using the Ca^{2+} indicator Fura-2. Dorsal root ganglia (DRG), containing only sensory neuron bodies and support cells, were dissected and dissociated from embryonic chickens and suspended in 3-D collagen scaffolds. After 48 hours of incubation, the neurons were loaded with Fura-2 and imaged using an epifluorescent microscope. After a stable baseline recording of $[Ca^{2+}]_i$ was established, neurons were stimulated with varying stimulation paradigms consisting of biphasic pulses of varying duration (10 µs to 100 ms) and varying intensity (10 V/m to 100,000 V/m), delivered at 3 Hz. The rate of rise of $[Ca^{2+}]_i$ was used to determine the nM of Ca^{2+} entry per pulse, which is directly comparable to the simulation results with COMSOL.

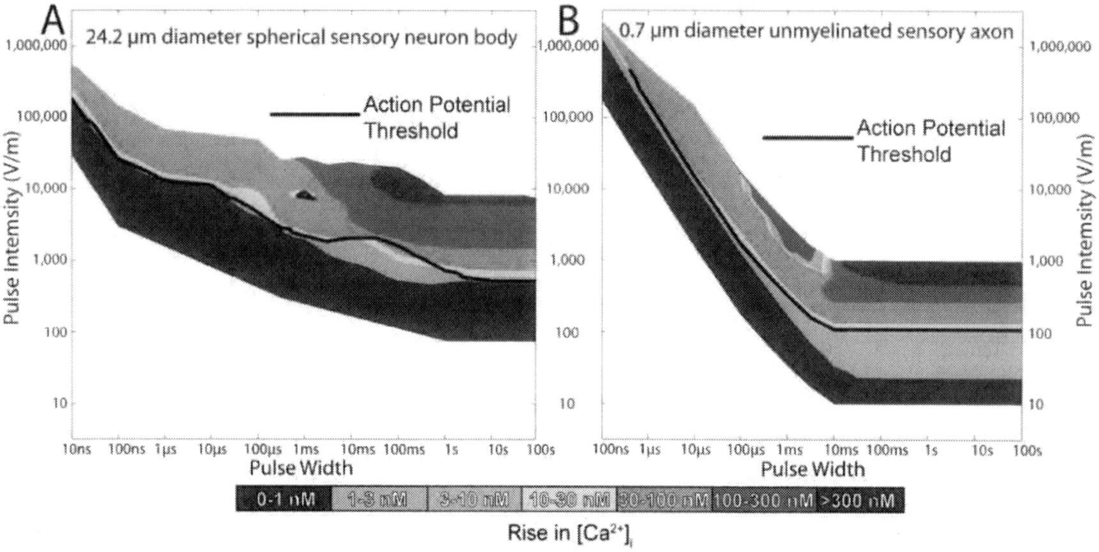

Figure 1: Simulation results show that neuronal $[Ca^{2+}]_i$ rises in response to ES. In general pulses above the action potential threshold (black line) cause the greatest rise in $[Ca^{2+}]_i$. **A**) For pulses shorter than 30 µs, the neuron body can have greater $[Ca^{2+}]_i$ rise from lower intensity stimulation. **B**) For pulses longer than 100 µs the growth cone at the tip of an axon is more sensitive to electrical fields.

RESULTS

In this study, three neuron geometries were tested facing either toward the positive or negative electrode (the cell body without any axons or dendrites is the same for positive and negative facing) and both monophasic and biphasic pulses were simulated. This resulted in 10 independent simulations that yielded numerous beneficial observations. Fig. 1 shows the simulation results from monophasic pulses on (A) a neuron body with no axons or dendrites, and from (B) a growth cone facing a positive electrode. From these simulations we found that supra-action potential stimulation causes the greatest rise in $[Ca^{2+}]_i$, but in specific areas of the graph, sub-action potential stimulation can cause a greater than 5 nM rise in $[Ca^{2+}]_i$. Additionally, from comparing the two graphs in Fig. 1, we found that growth cones are more sensitive to longer duration pulses that are greater than 100 µs in duration, and neuron bodies are more sensitive to short duration pulses that are shorter than 30 µs in duration. Neuronal measurements using Fura-2, the $[Ca^{2+}]_i$ indicator, produce similar results to the simulation in Fig. 1A.

DISCUSSION

To the best of our knowledge, this is the first neuronal simulation which describes the relationship of extracellular electrical fields to the rise in $[Ca^{2+}]_i$. Recently, simulations have been used to study extracellular electrical fields and membrane polarization [13]. These results are similar to our findings for the action potential threshold. Our simulations build on the knowledge of such models. However, by focusing on $[Ca^{2+}]_i$, they are more applicable to the nerve regeneration field because Ca^{2+} is a likely mechanism for how ES improves neuron growth. Our findings indicate that ES paradigms of varying parameters produce elevations in $[Ca^{2+}]_i$. Furthermore, this new understanding will allow ES paradigms to be designed to minimize energy usage for implanted electrodes, or conversely, to maximize Ca^{2+} entry in growth cones with sub-action potential pulses that do not cause Ca^{2+} entry to propagate throughout the entirety of a neuron.

ACKNOWLEDGEMENTS

We are grateful for our funding from the NIH (Training Grant #5T 32 GM008306-24 awarded to RDA).

REFERENCES

1. Zochodne, D.W., The challenges and beauty of peripheral nerve regrowth. *J Peripher Nerv Syst*, 17:1-18, 2012.
2. Al-Majed, A.A., et al., Brief electrical stimulation promotes the speed and accuracy of motor axonal regeneration. *J Neurosci*, 20:2602-8, 2000.
3. English, A.W., et al., Electrical stimulation promotes peripheral axon regeneration by enhanced neuronal neurotrophin signaling. *Dev Neurobiol*, 67:158-72, 2007.
4. Macias, M.Y., et al., Directed and enhanced neurite growth with pulsed magnetic field stimulation. *Bioelectromagnetics*, 21:272-86, 2000.
5. Wood, M., et al., Short-duration, DC electrical stimulation increases chick embryo DRG neurite outgrowth. *Bioelectromagnetics*, 27:328-31, 2006.
6. Adams, R.D., et al., Electrical and neurotrophin enhancement of neurite outgrowth within a 3D collagen scaffold. *Ann Biomed Eng*, 42:1282-91, 2014.
7. Yan, X., et al., Electrical stimulation induces calcium-dependent neurite outgrowth and immediate early genes expressions of dorsal root ganglion neurons. *Neurochem Res*, 39:129-41, 2014.
8. Hodgkin, A.L., et al., A quantitative description of membrane current and its application to conduction and excitation in nerve. *J Physiol*, 117:500-44, 1952.
9. Scroggs, R.S., et al., Multiple Ca2+ currents elicited by action potential waveforms in acutely isolated adult rat dorsal root ganglion neurons. *J Neurosci*, 12:1789-801, 1992.
10. Herzog, R.I., et al., Persistent TTX-resistant Na+ current affects resting potential and response to depolarization in simulated spinal sensory neurons. *J Neurophysiol*, 86:1351-64, 2001.
11. Vasylyev, D.V., et al., Membrane properties and electrogenesis in the distal axons of small dorsal root ganglion neurons in vitro. *J Neurophysiol*, 108:729-40, 2012.
12. Benison, G., et al., Modeling temporal behavior of postnatal cat retinal ganglion cells. *J Theor Biol*, 210:187-99, 2001.
13. Boinagrov, D., et al., Strength-duration relationship for extracellular neural stimulation: numerical and analytical models. *J Neurophysiol*, 104:2236-48, 2010.

SB³C2015
Summer Biomechanics, Bioengineering and Biotransport Conference
June 17-20, 2015, Snowbird Resort, Utah, USA
SB³C2015-521

SEQUENTIAL BISPHOSPHONATE AND PARATHYROID HORMONE TREATMENT DECOUPLES BONE REMODELING IN FAVOR OF BONE FORMATION

Allison R. Altman, Carina Lott, Chantal M. de Bakker, Wei-Ju Tseng, Ling Qin, X. Sherry Liu

McKay Orthopaedic Research Laboratory,
University of Pennsylvania School of
Medicine, Philadelphia, PA

INTRODUCTION:

Bisphosphonate is an osteoporosis treatment that acts to prevent bone loss by inhibiting bone resorption. However, long term bisphosphonate use may increase the risk of atypical femoral fractures [1]. Intermittent parathyroid hormone (PTH) is the only FDA approved anabolic agent which promotes bone formation. Early studies showed conflicting results regarding the efficacy of combined bisphosphonate and PTH treatment, which questioned whether PTH's anabolic effect may depend on prior resorption to initiate new bone formation [2,3]. However, recent studies have shown a positive effect of combined or tandem bisphosphonate and PTH therapies [4-7], attesting to PTH's anabolic effect in the absence of bone resorption. Therefore, we hypothesized that PTH is able to act through a resorption-independent pathway to promote modeling-based bone formation. In this study, the efficacy of PTH following 12-weeks of alendronate (ALN, a bisphosphonate) treatment was tested in ovariectomized (OVX) rats. An in vivo imaging technique was used to assess bone formation and resorption rate simultaneously and longitudinally, allowing for an evaluation of coupling between bone resorption and formation.

METHODS

Animals: 30 female SD rats (n=6/group) received surgery at 4-month of age: 24 received a bilateral OVX, and 6 received a sham OVX surgery. The study began when all rats were 6-month-old, after a 2-month development of osteoporosis in the OVX rats. The treatment plan consisted of 2 phases (P1: 0-12 wks, and P2: 12-16 wks). The OVX rats were assigned to 4 groups: (1) a Veh group treated with saline for P1&2; (2) an ALN group treated with ALN (28μg/kg 2x/wk) for P1&2, (3) a Veh+PTH group treated with saline for P1, then switched to PTH (40 μg/kg 5x/wk) for P2, (4) an ALN+PTH group treated with ALN for P1, then PTH for P2 (Fig 1). ***In Vivo* μCT Scans:** The right proximal tibia of all rats were scanned (Scanco VivaCT40, 10.5 μm) at wk0 (2 mo post surgery), then weekly from wks 11-16 corresponding with P2 (Fig 1). Trabecular microstructure of subsequent scan images was precisely aligned to baseline using an iterative registration method [8] to identify the same volume of interest (VOI). Bone microstructure was analyzed for the same VOI for wks 0, 12, 14, and 16. ***In Vivo* Dynamic Histomorphometry:** A bone sub-volume ($1.575 \times 1.575 \times 1.05 mm^3$) of each week in P2 was subtracted from the registered sub-volume of the previous week (wks 11-15) to identify the newly formed bone voxels (green) and resorbed voxels

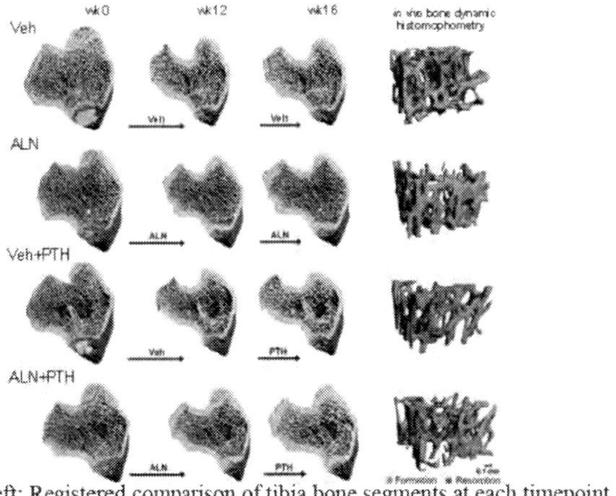

Fig 1. Left: Registered comparison of tibia bone segments at each timepoint with treatment indicated by the arrows between timepoints. Right: μCT-based in vivo bone dynamic histomorphometry, green indicates areas of new bone (formation), and red indicates lost bone (resorption) during the final week of treatment.

(red) during each week (Fig 1 Right). New bone voxels were used to calculate the bone formation rate (BFR/BS), and lost voxels to calculate bone resorption rate (BRR/BS) weekly over P2 [9]. **Femur 3-Point Bending:** The right femoral midshaft was scanned *ex vivo* for evaluation of cortical thickness and polar moment of inertia. Then the femur was subjected to a 3-point bending test for evaluation of stiffness and elastic modulus. **Spine:** Lumbar vertebra, L2 was scanned *ex vivo* (Scanco µCT35, 3.5 µm) for microstructural and tissue mineral density (TMD) analysis. TMD was calculated for bone tissue at trabecular surface layers (sTMD), and central bone tissue (cTMD). **Finite Element (FE) Analysis:** The trabecular bone sub-volume of wk 12 and 16 were converted to voxel-based FE models to estimate axial stiffness under compression. **Statistical Analysis:** Longitudinal measures were compared over time and between groups using repeated measures ANCOVA adjusted for baseline, and cross-sectional measures were compared using ANOVA.

RESULTS

Tibia Microstructure: 2-month after surgery, OVX resulted in an average BV/TV of 0.13, in contrast to 0.50 in the SHAM group at wk 0. BV/TV continued to drop in the Veh and Veh+PTH to 0.06 by wk 12 (Fig 2). ALN treatment effectively stabilized BV/TV for ALN and ALN+PTH groups. Switching to PTH resulted in a dramatic increase in BV/TV by wk 16 compared to wk 12 (40% and 42%) driven by increased Tb.Th (33% and 25%) in both Veh+PTH and ALN+PTH groups (Fig 2). ***In Vivo* Dynamic Histomorphometry:** During P2, the Veh+PTH group had a dramatic increase in BFR/BS by wk 14, which began to stabilize by wk 16. The ALN+PTH group had a similar increase in BFR/BS which remained elevated throughout P2 (Fig 3a). BRR/BS was low in all treatment groups after the onset of PTH therapy. This was further confirmed by TRAP serum ELISA (Fig 3b). Fig 3c showed that SHAM and ALN treated groups had highly coupled remodeling, with similar BFR/BS and BRR/BS. Resorption outpaced formation for the Veh group. Bone resorption and formation were decoupled in both Veh+PTH and ALN+PTH groups as shown by a substantially greater BFR/BS than BRR/BS. **Femur:** No difference was found in cortical structure or mechanical parameters between groups. **Spine Microstructure:** Compared to the tibia, OVX resulted in less reduced vertebral BV/TV (0.21 in Veh *vs.* 0.33 in SHAM). All 3 treatments groups resulted in BV/TV that was not different from SHAM. The Veh+PTH had less Tb.N but greater Tb.Th than SHAM. Compared to SHAM, Tb.N was no different from the ALN and ALN+PTH groups, while Tb.Th was greater in the ALN+PTH group. **Spine TMD:** cTMD was higher in the SHAM than all treatment groups. cTMD of the Veh+PTH group was 7% and 4% lower compared to SHAM and Veh. sTMD was significantly reduced in the Veh+PTH (6%) and ALN+PTH (4%) compared to ALN. **Tibia Stiffness:** The trabecular bone stiffness did not change during P2 in the SHAM, Veh or ALN groups. Stiffness improved in both Veh+PTH (97%) and ALN+PTH (114%) groups beyond that of all other groups.

DISCUSSION

The results of this study clearly demonstrate the efficacy of PTH following bisphosphonate therapy for stimulating new bone formation. In healthy bone, resorption and formation are coupled and balanced to sustain bone mass. Their uncoupling due to OVX results in resorption outpacing formation, and subsequent trabecular bone loss which compromises the mechanical competence of the bone. ALN treatment effectively re-couples this balance, preventing additional bone loss. PTH, however, uncouples the balance in favor of formation, resulting in thickened bone and greater structural stiffness. Resorption activities after ALN, Veh+PTH, and ALN+PTH treatments were not different and remained significantly lower than those of Veh and Sham. While we may have expected a higher BRR/BS or TRAP in the Veh compared to Sham group, the resorption-formation balance of OVX had been re-adjusted to the slower formation rate 6 months after OVX. In summary, our investigation across multiple skeletal sites suggests that ALN followed by PTH is a viable treatment strategy to maintain and improve bone quality by stimulating substantial new bone formation. The novel *in vivo* dynamic histomorphometry analysis provides direct evidence for PTH's anabolic effect in the absence of bone resorption, further confirming its capacity for modeling-based bone formation.

ACKNOWLEDGEMENTS

Penn Center for Musculoskeletal Diseases (PCMD) NIH/NIAMSP30AR050950, NIH/NIAMS T32-AR007132, University of Pennsylvania Institute on Aging (IOA) Pilot Award.

REFERENCES

[1] Seraphim A, *et al.*, 2012 J Orthopaed Traumatol, 13(4):171-7; [2] Black DM, et al., 2003 N Engl J Med, 349:1207-15; [3] Wu X, *et al.*, 2010 Cell Stem Cell, 7:571-80; [4] Cosman F, *et al.*, 2011 JBMR, 26:503-11; [5] Muschitz C, *et al.*, 2013 JBMR, 28:196-205; [6] Schafer AL, *et al.*, 2013 Osteoporosis Int, 24(10):2591-601; [7] Altman AR, et al., 2014 Bone, 61:149-57; [8] Lan S, *et al.*, 2013 Bone, 56(1):83-90; [9] Schulte FA, *et al.*, 2011 Bone 48(3):433-42.

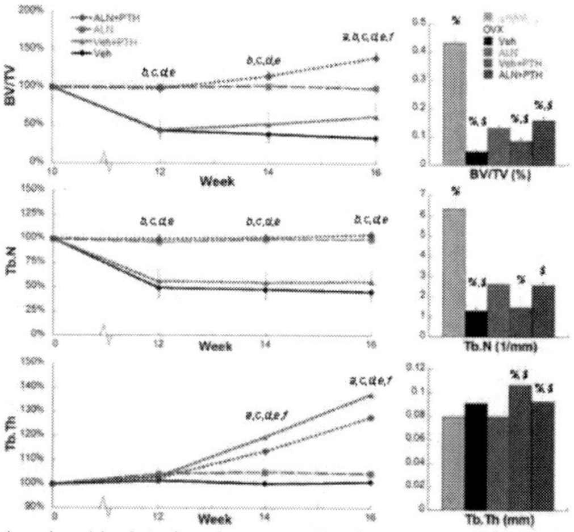

Fig 2. *In vivo* tibial trabecular microstructure parameters. Left panel: percent change from baseline; right panel: value at the study endpoint. Letters (left) indicate significant differences between groups: *a*: ALN to ALN+PTH, *b*: ALN to OVX, *c*: ALN to Veh+PTH, *d*: ALN+PTH to OVX, *e*: ALN+PTH to Veh+PTH, *f*: OVX to Veh+PTH. Symbols (right) indicate significant change from baseline (%) or week 12 ($).

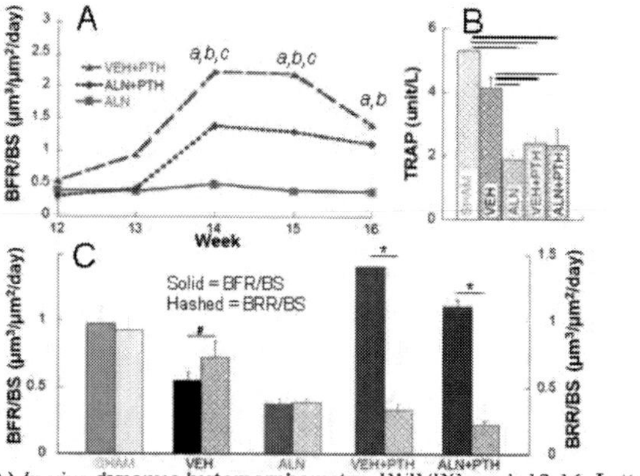

Fig 3. (A) *In vivo* dynamic histomorphometry, BFR/BS week 12-16. Letters indicate differences between groups: *a*: ALN to ALN+PTH, *b*: ALN to Veh+PTH, *c*: ALN+PTH to Veh+PTH. (B) Serum TRAP concentration, bars indicate differences between groups. (C) *In vivo* dynamic histomorphometry at week 16. Left axis: BFR/BS (solid), right axis: BRR/BS (hashed). * indicates a trend difference ($p<0.1$).

SB³C2015
Summer Biomechanics, Bioengineering and Biotransport Conference
June 17-20, 2015, Snowbird Resort, Utah, USA
SB³C2015-523

AGE-DEPENDENT CHANGES OF STRESS AND STRAIN IN THE HUMAN NATIVE HEART VALVE AND THEIR RELATION WITH COLLAGEN REMODELING.

Pim J.A. Oomen (1), Sandra Loerakker (1), Daphne van Geemen (1), Jan Neggers (2), Marie-José T.H. Goumans (3), Antoon J. van den Bogaerdt (4), Ad J.J.C. Bogers (4), Carlijn V.C. Bouten (1) & Frank P.T. Baaijens (1)

(1) Department of Biomedical Engineering & Institute for Complex Molecular Systems, Eindhoven University of Technology, Eindhoven, Netherlands

(2) Department of Mechanical Engineering, Eindhoven University of Technology, Eindhoven, Netherlands

(3) Department of Molecular Cell Biology, Leiden University Medical Center, Leiden, Netherlands

(4) Department of Cardiothoracic Surgery and Heart Valve Bank, Erasmus Medical Center, Rotterdam, Netherlands

INTRODUCTION

Collagen is the main load-bearing component of many soft tissues, including heart valves. Therefore it is important to understand its remodeling process, for only then it is possible to create tissue-engineered heart valves with long-term functionality. Tissue remodeling has been hypothesized to occur to maintain a certain mechanical homeostasis [1]. Yet, despite enormous effort of the field, the driving force of collagen remodeling remains unidentified.

Knowledge of the evolutionary changes with age in stress and strain status of native heart valves might explain which of these factors determines mechanical homeostasis. In this study, we determined the collagen architecture as well as geometric and mechanical properties of human native semilunar heart valves of fetal to adult age. These properties were used to predict age-dependent changes in stress and strain in the heart valves via finite element modeling.

METHODS

Eight sets of paired cryopreserved healthy human aortic and pulmonary valves of different age, plus one fetal valve, were obtained from postmortem donors, given permission for transplantation-oriented research. All valves were structurally and mechanically intact.

The collagen architecture of the fibrosa was visualized using confocal microscopy. Collagen fiber orientation and dispersity in the belly region were determined using a multi-scale analysis script [2].

Mechanical behavior was characterized using a custom-made micro-indentation device mounted onto the confocal microscope (Figure 1A) [3]. A global digital image correlation algorithm was used to quantify tissue deformation (Figure 1B) [4]. Material properties were estimated using inverse finite element analysis (Figure 1C,D) [5].

The mechanical and geometrical data that were obtained in this study were used as input for a finite element model (Figure 1E) [6,7]. Each heart valve was shaped in Abaqus by using the obtained age-related valve radius and leaflet thickness. The tissue was modeled as an isotropic matrix part and an anisotropic collagen fiber part. The former was modeled as a Neo-Hookean material, while for the latter an adapted Holzapfel model was used [6,7]. In each simulation, the age-dependent diastolic blood pressure was gradually applied and the circumferential and radial stretch and Cauchy stress were obtained upon full pressure application.

RESULTS

High-resolution confocal microscopy of the collagen architecture revealed that the typical circumferential collagen organization was already present in the fetal valve leaflets. On the macroscopic scale, dense fiber bundles were found in both aortic and pulmonary postnatal

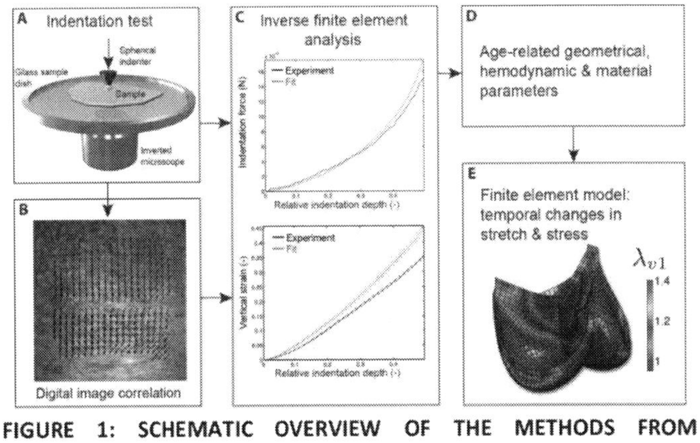

FIGURE 1: SCHEMATIC OVERVIEW OF THE METHODS FROM MECHANICAL CHARACTERIZATION TO FINITE ELEMENT SIMULATIONS.

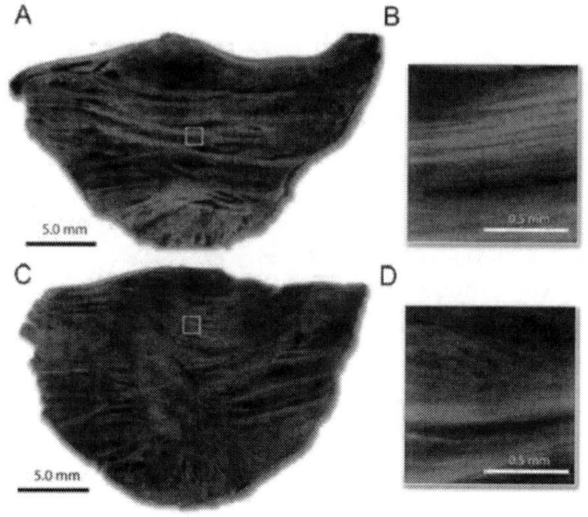

FIGURE 2. CONFOCAL IMAGES ON THE MACROSCALE AND MICROSCALE OF THE COLLAGEN ARCHITECTURE IN THE FIBROSA LAYER OF AN ADULT AORTIC (A-B) AND PULMONARY VALVE (C-D).

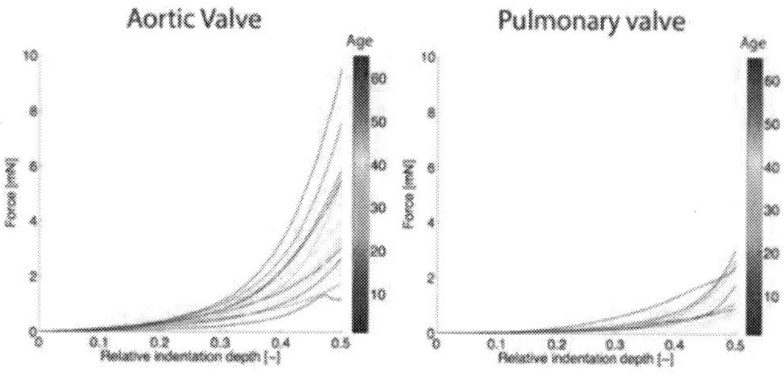

FIGURE 3. FORCE-INDENTATION CURVES OF AORTIC (A) AND PULMONARY (B) VALVES OF DIFFERENT AGES.

valves (Figure 2A,C). On the microscopic scale, the dispersity was lower in the aortic valve than in the pulmonary valve (Figure 2B,D).

Valve stiffness of the aortic valve increased as a function of age (Figure 3). In addition, the material parameters showed that circumferential stiffness of the aortic valves was higher than that of the pulmonary valves. No changes were observed for the radial direction.

Finite element simulations using age-dependent geometries, material properties, and hemodynamic conditions predicted that the rapid changes occur in aortic leaflet stretch at young age (<4 years), after which stretch homeostasis is reached in both the circumferential and radial direction (Figure 4A,C). No homeostasis was predicted for the circumferential Cauchy stress in the aortic valve (Figure 4B), with stress increasing as a function of age. The radial stress (Figure 4D) remained much smaller than the circumferential stress.

DISCUSSION

In the present study, we revealed gradual age-dependent changes in tissue collagen architecture and mechanics. We believe that these changes are related to collagen remodeling in response to changes in hemodynamics in order to maintain a mechanical homeostasis.

Age-dependent finite element simulations predicted that this homeostasis is stretch-driven for both the aortic and pulmonary valve (Figure 4A,C), whereas no such homeostasis was found for the Cauchy stress (Figure 4B,D). Moreover, the increase in tissue stress as a consequence of a rising blood pressure was not counteracted by

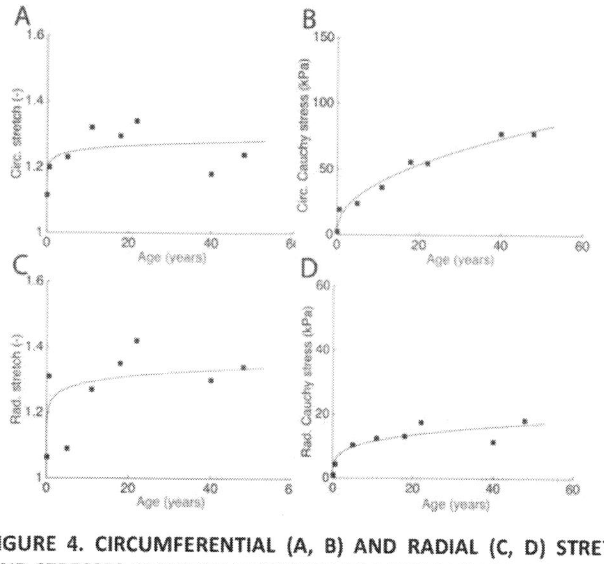

FIGURE 4. CIRCUMFERENTIAL (A, B) AND RADIAL (C, D) STRETCH AND STRESSES IN THE BELLY REGION OF AORTIC HEART VALVES.

growth and remodeling, for example by increasing the leaflet thickness to reduce the stress.

Since postnatal aortic transvalvular pressure is much higher than pulmonary transvalvular pressure, the aortic valve stretch is higher. Consequently, collagen remodeling in the aortic valve leaflets will lead to a lower collagen dispersity. Fiber bundle formation, however, occurs in both the aortic and pulmonary valve. Therefore, it appears to be a result of tissue maturation rather than of remodeling, for it cannot be explained by changes in mechanical demand.

As a consequence of changes in collagen alignment, the valve stiffness changes. Since it is known that the valvular collagen content remains unchanged after childhood [8], the age-dependent increase in stiffness in human heart valves can only be explained from collagen remodeling and maturation.

In conclusion, an extensive data set of age-dependent heart valve geometrical and material properties was created to study human heart valve biomechanics and collagen remodeling. We employed structural and mechanical changes of the valves with age to predict the evolution of leaflet stretch and stress via finite element modeling. Based on the relatively small differences in stretch with time, we suggest that collagen remodeling in the human heart valve maintains a stretch-driven homeostasis.

ACKNOWLEDGEMENTS

We gratefully acknowledge the support from the European Union's Seventh Framework Program, grant agreements no. 242008 and 604514, and the Dutch Technology Foundation (STW), Applied Science Division of NWO and the Technology Program of the Dutch Ministry of Economic Affairs, for supporting this research.

REFERENCES

1. Humphrey JD, *Cell Biochem biophys,* 50:53–78, 2008.
2. Frangi AF et al., *MICCAI,* 1496:130–137, 1998.
3. Cox MAJ et al., *J Biomech* 41:422–429, 2008.
4. Neggers J et al., *Procedia IUTAM,* 4:73–81, 2012.
5. Cox MAJ et al., *J Biomech Eng*, 128:428–436, 2006.
6. Loerakker S et al., *J Biomech*, 46:1792–1800, 2013.
7. Driessen NJB et al., *J Biomech* 40:325–334, 2007
8. Aikawa E et al., *Circulation* 113:1344–1352, 2006

SB³C2015
Summer Biomechanics, Bioengineering and Biotransport Conference
June 17-20, 2015,Snowbird Resort, Utah, USA
SB³C2015-524

EFFECT OF ASSYMETRIC DEPLOYMENT ON THE FUNCTION AND LEAFLET MECHANICS OF BIOPROSTHETIC TAVI VALVES

Michael B. Gogarty (1), Pablo Maureira (3), Lakshmi P. Dasi (1,2)

(1) School of Biomedical Engineering
Colorado State University
Fort Collins, Colorado, United States

(2) Department of Cardiovascular Surgery
CHU de Nancy
Nancy, France

(3) Mechanical Engineering Department
Colorado State University
Fort Collins, Colorado, United States

INTRODUCTION

Heart disease is the number one cause of death in the United States and continues to rise throughout the world. Aortic Stenosis (AS) due to calcification is a leading contributor to the mortality rate.

The gold standard for many years to treat this condition has been Aortic Valve Replacement (AVR)[1]. However, due to its invasive nature and long recovery times minimally invasive techniques such as Transcatheter Aortic Valve Implantation (TAVI) are becoming increasingly popular and much research has gone into developing both the techniques and tools necessary to reduce the mortality levels to be comparable to AVR.

TAVI relies on valve leaflets made from porcine pericardium attached to a mesh stent. The stent is then compressed and inserted into the femoral artery until it reaches the diseased aortic valve. The stent is then expanded over the diseased leaflets compressing them and replacing them. This method only compresses the leaflets and although it is less invasive than AVR fewer people qualify for it and it suffers from a higher mortality rate than AVR. It is also less effective than AVR since there are several key steps performed during AVR which are not duplicated with TAVI.

One of the main reasons to replace an aortic valve is due to calcification or the formation of calcium nodules in the leaflets and around the annulus. During AVR the surgeon prepares the site for a new valve by removing the diseased leaflets and tissue around the annulus. Removal of the diseased tissue and suturing of the valve are not done during the TAVI procedure. While expansion of the stent may keep the native leaflets from moving by compressing them the calcium nodules do not compress. The result can be several types of regurgitation[2][3].

The objective of this research is to study the effects of calcium nodules on TAVI valve implantations to help understand the TAVI biomechanics under realistic conditions involving non-circular deployment, as well as progressive calcification post TAVI implantation.

METHODS

FIGURE 1 shows a schematic of the chamber that was designed and built to hold a 26mm Medtronic CoreValve deployed in various radial orientations. The chamber is made of clear polycarbonate tubing allowing the valve to be observed from any angle during testing. There is an inlet and outlet port along with a viewing port allowing a camera to record the valve leaflet kinematics during testing. The chamber is connected to a left heart pulse duplicator, thereby loading the valve to physiological aortic flow and pressures (normal and hypertensive).

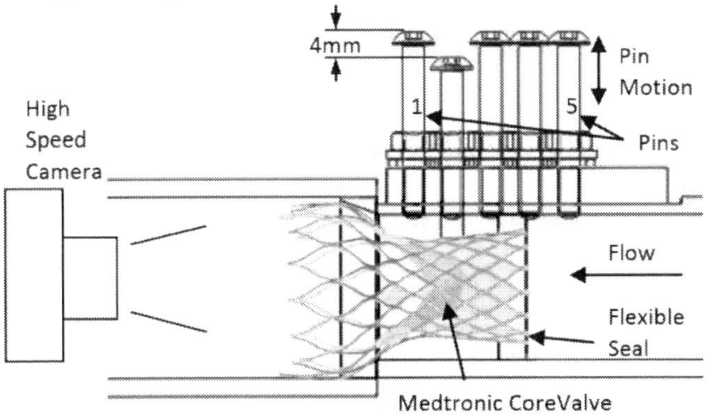

FIGURE 1: Diagram of the valve test chamber with pin 2 down 4mm

The effects of asymmetric calcium nodules were simulated by a series of screws at various locations axially along the length of the valve, FIGURE 1. These were moved radially inward to contact the stent and simulate different size nodules at different locations as point loads.

The chamber containing the Medtronic CoreValve was then attached to a beating heart simulator. This simulator allows the control of HR, BP and Flow Rate (FR) in order to realistically simulate physiological conditions. In addition, to the NI DAQ equipment used to record BP, HR and FR a LaVision (LaVision GmbH, Goettingen, Germany) high speed camera was used to collect high speed images of the valve function during testing.

Once the system was adjusted to the desired test condition one or more screws were turned to provide a 'nodule' of a desired size, such as 4, 5 and 6mm. Which is the distance the screw protrudes from the inside wall of the test chamber toward the valve.

Video at 500 Hz for each condition recorded the valve leaflet kinematics while aortic/ventricular pressure and flow waveform were recorded via two Validyne transducers (Validyne Engineering, Northridge, CA) and a Transonic flow meter (Transonic Systems Inc. Ithaca, NY).

TABLE 1 shows the conditions for the Medtronic Core Valve with one commissure rotated 60° relative to the line of screws, FIGURE 2a. Tests were also conducted with one commissure at 0° relative to the line of screws FIGURE 2b.

		Screw #	1	2	3	4	5
BP 120/80 HR 60 bpm Valve at 60°	Screw Position (mm)						
	Baseline		0	0	0	0	0
	Run 1		0	4	0	0	0
	Run 2		0	5	0	0	0
	Run 3		0	6	0	0	0

TABLE 1: Test conditions for Medtronic Core Valve

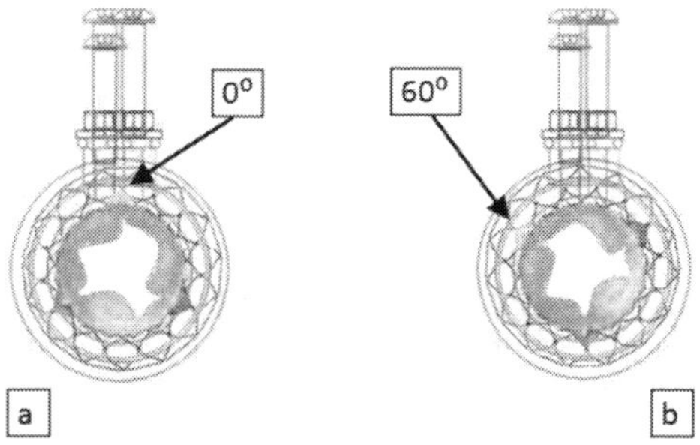

FIGURE 2: Radial Valve Orientations, a) One Commissure at 0° b) One Commissure at 60°

Each type of valve such as and Edwards Sapien valve will be tested in the chamber with the same conditions as those used for the Medtronic CoreValve.

High speed images at each condition were compared to baseline images with all of the screws retracted. Examination of the images allowed for qualitative comparison i.e., does the valve change shape, do the leaflets move differently and more quantitative comparison of time to open or close the valve. The recorded BP and flow rate values were examined to determine if regurgitation increased during each test condition.

RESULTS AND DISCUSSION

FIGURE 3 compares the flow rate vs. time for the 4 pin positions.

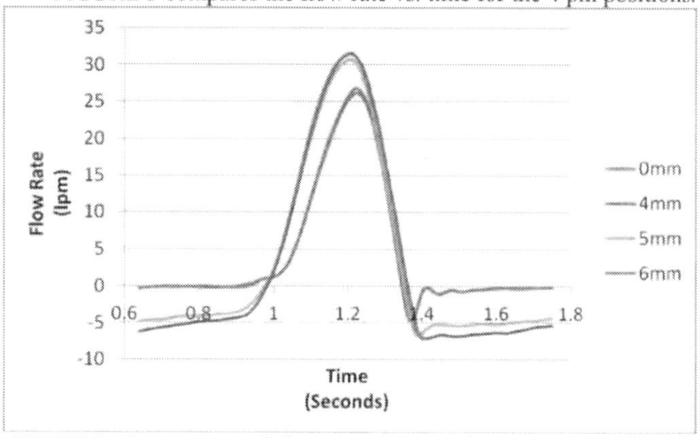

FIGURE 3: Comparison of Flow Rate vs. Time for each test condition

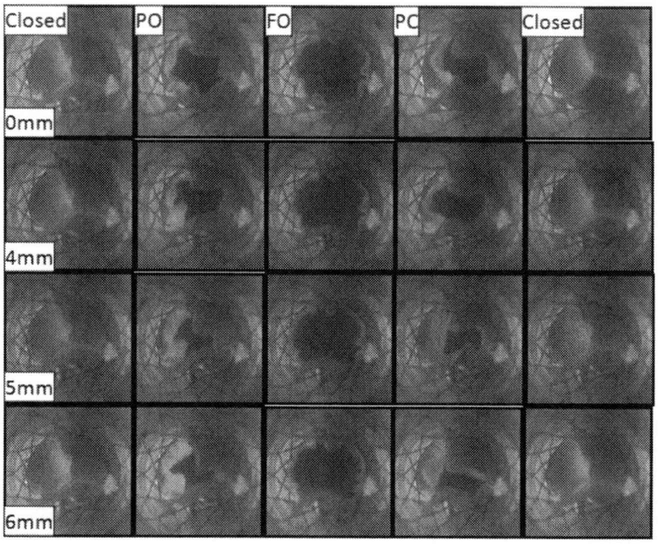

FIGURE 4: Complete valve sequence for each test condition; Closed, Partially Open (PO), Fully Open (FO), Partially Closed (PC) and Closed

FIGURE 3 shows that the peak flow rate increases to maintain physiological conditions to make up for the significant regurgitation which is a result of the pin position of 5mm. The Mean Aortic Pressure (MAP) also showed a significant drop which is consistent with regurgitation. The higher regurgitation and peak flow rates shown are the result of peravalvular leakage or flow bypassing the valve between the valve and chamber wall. The images in FIGURE 4 which show that even though the valve leaflets are affected by the pin position the leaflets seal and any regurgitation must be bypass flow.

The data demonstrates what has been suspected by cardiac researchers and surgeons alike, that calcification can adversely affect implanted TAVI valves a cause for continued study and evaluation to determine how best to alleviate these negative effects[4].

REFERENCES

[1] Zierer, A et al., *J Thor Cardiovasc Surg*, 138(5): 1067–72, 2009.

[2] Stähli, B E et al., *Cardiovasc. Diagn. Ther.*, 3(15): 15–22, 2013.

[3] Koh, E Y et al, *Interact Cardiovasc Thorac Surg.*, 1–6, 2014.

[4] Généreux, P et al., *J. Am. Coll. Cardiol.*, 61(11): 1125–36, 2013.

SB³C2015
Summer Biomechanics, Bioengineering and Biotransport Conference
June 17-20, 2015, Snowbird Resort, Utah, USA
SB³C2015-525

THE EFFECT OF MUSCLE LOADING ON ANKLE JOINT COMPLEX KINEMATICS AND ACHILLES LOAD: A CADAVERIC STUDY

Bardiya Akhbari (1), Matthew H. Dickinson (2), Ednah G. Louie (2) , Sami Shalhoub (2), Lorin P. Maletsky (1, 2)

(1) Department of Mechanical Engineering
University of Kansas
Lawrence, Kansas, United States

(2) Bioengineering Graduate Program
University of Kansas
Lawrence, Kansas, United States

INTRODUCTION

Injuries to the ankle joint are one of the most common sports injuries.[1] Inversion and adduction motion (together termed supination) injuries account for 80% of ankle's injuries.[2] Changes in the muscles' firing capability were seen after ankle sprain,[3] thus muscles activity in the gastrocnemius and soleus, tibialis anterior and extensor digitorum longus (antagonistic/agonistic muscles) will be changed after injury.[4]

Quantifying the effects of the tibialis anterior and extensor digitorum longus on the ankle joint complex will aid in muscle function analysis after injury by determining the effects of muscle weakness on the ankle joint kinematics. The purpose of this study was to measure the effects of variable tibialis anterior (TA) and extensor digitorum longus (EDL) muscle loading configurations on the ankle joint complex kinematics, and Achilles load.

METHODS

Five fresh frozen cadaveric ankles were thawed at room temperature. The tibia was sectioned 15 cm superior to the calcaneus and potted in aluminum fixtures. All soft tissue above 15 cm of the joint line was removed, except the TA, EDL, and Achilles tendon which were left intact. The ankle was then mounted onto an open chain muscle loading rig with the tibia rigidly attached in an inverted position and the foot free to move (Figure 1). The TA and EDL and Achilles tendons were isolated and clamped individually. The ankle was flexed four times over the entire range of flexion using a stepper motor attached to the Achilles tendon. Static loads were used to load the TA and the EDL. Three loading conditions were simulated: physiological loading defined as 90 N on the TA and 45 N on the EDL based on their physiological cross sectional area (PCSAs)[5], the second load set was defined as Weak EDL which reduced the EDL load from the physiological set to 50% , and the third load set defined as Weak TA which reduced the TA load from the physiological cycle to 75% of that (Table 1). Total static load on the tendons stayed 135 N during the load sets. The physiological load set was used as the baseline for comparison.

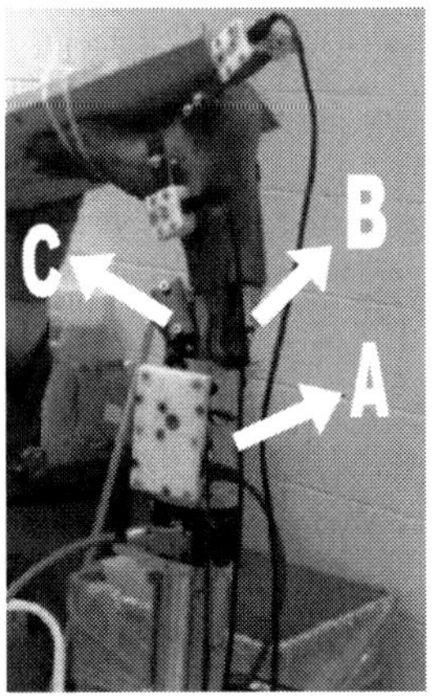

FIGURE 1: TESTING SETUP CONSISTING OF A) TIBIA FIXURE, B) MOTOR CABLE, C) MUSCLES CABLE

TABLE 1. TIBIALIS ANTERIOR AND EXTENSOR DIGITORUM LONGUS LOAD MAGNITUDES FOR THE THREE LOADING SETS

Configuration	TA	EDL
Physiological	90 N	45 N
Weak TA	67.5 N	67.5 N
Weak EDL	112.5 N	22.5 N

The kinematics of each bone were measured using IRED markers rigidly attached to the calcaneus and the tibia using an Optotrak Certus motion capture system (Northern Digital, Ontario). The loads applied by the motor onto the Achilles tendon to dorsiflex/plantarflex the ankle were measured using a 1200 N load cell (Transducer Technique, location). The difference in inversion/eversion (InEv), adduction-/abduction (AdAb) rotation, and Achilles load were calculated for the two loading cases relative to the baseline for the third motor's cycle in the -20° to 10° range of motion. A student t-tests was performed to find significant differences ($p<0.05$) between the physiological configuration and each of the two load sets individually.

RESULTS

The average kinematics of five ankles displayed significant increase in inversion throughout the whole flexion cycle (Figure 2.A). Different configuration displayed adverse effects on the ankles. The Weak EDL increased the inversion; however, the Weak TA resulted in significant and constant reduction (≈3°) in inversion (Figure 2.B). The maximum inversion difference for the Weak TA was 8.0°, although it was -3.3° for the Weak EDL. Investigation of average adduction/abduction rotation in the physiological set demonstrated changes in abduction from approximately 4° to the 5.3° of adduction over the whole range (Figure 2.C). More abduction in the Weak TA and more adduction in the Weak EDL with respect to the physiological load set throughout the full cycle were observed (Figure 2.D). The maximum relative differences were -2.1° and 0.9° for the Weak TA and Weak EDL at the 10° flexion. The physiological load set exhibited an increase in average Achilles load at early flexion and a decrease of the Achilles load in the rest of the cycle (Figure 2.E). Achilles load was measured for both Weak EDL and Weak TA across the entire cycle and it was lower than physiological Achilles load mainly (Figure 2.F). In comparison with the physiological load set, the Achilles load first decreased and then increased for the Weak TA while it initially increased and then decreased for the Weak EDL. Maximum load variance occurred at 20° of dorsiflexion and was 6.7 N and -21.4 N, respectively for the Weak TA and Weak EDL against the physiological load set. Significant differences were observed amongst the three sets for inversion/eversion motion (Figure 2).

DISCUSSION

According to the literature,[4] muscle weakness of the EDL or TA has an effect on the kinematics and kinetics. A change in the loading configuration within the TA and EDL muscles affects the ankle inversion and adduction at differing angles during the flexion range. In this study the ankle joint complex kinematics of both the Weak TA and the Weak EDL loading configuration deviated from the physiological one, with the largest dissimilarity observed in the inversion. The inversion amount was significant and increased during flexion for the Weak TA, which makes sense since the TA is the main invertor in the ankle. However, the Weak EDL inversion extent was not changed since the EDL is not as strong as the TA,[5] and the TA has its peak activity during the physiological loading set. The constant deviation from physiological configuration for the two loading cycle through the flexion range for AdAb can be explained by the coupling of motions between AdAb and InEv which means the various configuration make an offset in AdAb, but they change the InEv gradually. The ankles inversion and adduction behavior during plantar flexion was similar to Siegler[6] which studied the ankle three dimensional kinematics during natural motion without any loads attached to the muscles. The inversion amount in this study was considerably higher which can be understood by the attached tibialis anterior static load.

To keep the ankle stable, Achilles tendon has to resist or assist the dorsiflexion or plantarflexion motion and it is changed by the TA and EDL muscles strength. Thus, the Achilles tendon has to resist more loads to make the flexion possible when the TA load is lower at the early flexion; however, when the TA having a greater load capacity (Weak EDL), Achilles' load reduces since the resistance force is lower. The constant load difference in the mid-range of flexion was observed because of the transition state of Achilles' fibrils. Then, since the tendon is in flexion and was stabled the rest of load difference is going to increase.

The finding of this study can aid in verifying computational models, and enhance our understanding of muscle functions during flexion. There are a few limitations to this study. First the muscle loads used were relatively small compared to the physiological loads in the body. Low number of cadaveric specimen (five cadavers) was used and in order to improve the results, more specimens are needed. Tendons exhibit hysteresis while during repetitively motions; however, in this study the hysteresis effects decreased by setting up the experiment after third cycles. Moreover, variation in recognizing of the anatomical landmarks causes difference in the kinematic measurements for a single ankle; however the results for comparing between sets will not be changed.[7]

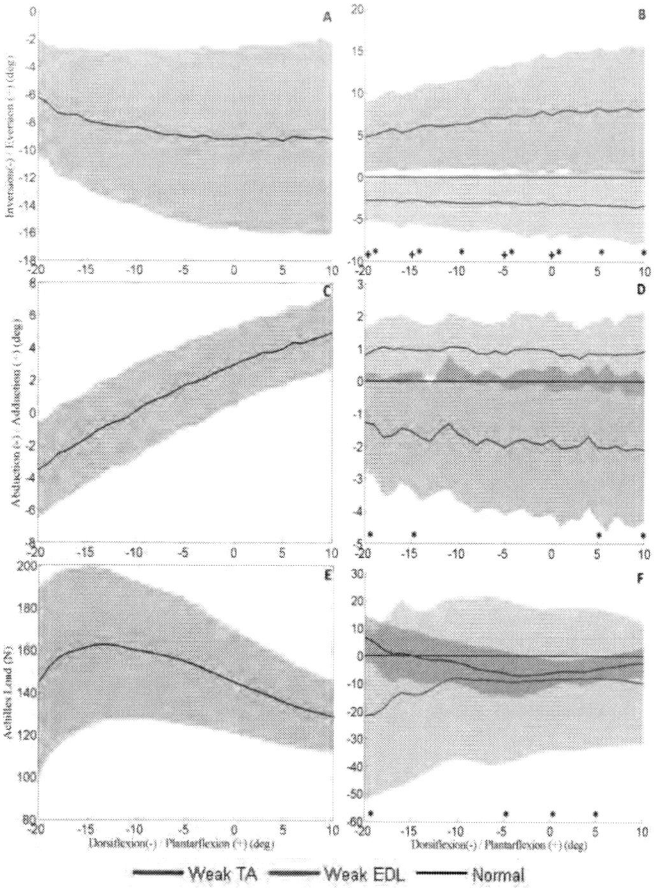

FIGURE 2: INEV (A), ADAB (C), AND LOAD (E) FOR ONE ANKLE, MEAN DIFFERENCE IN INEV (B), ADAB (D), AND LOAD (F) FOR ALL ANKLES BETWEEN PHYSIOLOGICAL AND THE OTHER TWO SETS (SIGNIFICANCE SHOWED BY * FOR WEAK TA AND + FOR WEAK EDL)

ACKNOWLEDGEMENTS

This experiment was partially funded by DePuy Synthes, Inc.

REFERENCES

1. Payne KA, et al. J Athl Train, 32(3):221-5, 1997.
2. Slimmon D, Brukner P, 39(1-2):18-22, 2010.
3. Beckman SM, et al. Arch Phys Med Rehabil, 76(12):1138-43, 1995.
4. a hilly ller . hiladelphia aunders i 566 p. p., 1983.
5. Arnold EM, et al. Ann Biomed Eng, 38(2):269-79., 2010.
6. Siegler S, et al.. J Biomech Eng, 110(4):364-73, 1988.
7. Morton NA, et al. J of Ortho Research, 25(9):1221-1230, 2007.

SB³C2015
Summer Biomechanics, Bioengineering and Biotransport Conference
June 17-20, 2015, Snowbird Resort, Utah, USA
SB³C2015-526

IN VITRO VALIDATION OF ENDOVASCULAR DOPPLER-DERIVED FLOW RATES IN CEREBRAL VESSELS

Patrick M. McGah (1), John D. Nerva (2), Ryan P. Morton (2), Michael C. Barbour (1), Pierre D. Mourad (2,3), Michael R. Levitt (2), Louis J. Kim (2), Alberto Aliseda (1)

(1) Mechanical Engineering
University of Washington
Seattle, Washington, US

(2) Neurological Surgery
Harborview Medical Center
Seattle, Washington, US

(3) Engineering and Mathematics
University of Washington Bothell
Bothell, Washington, US

INTRODUCTION

The hemodynamic environment of cerebral blood vessels is widely considered an important influence on the formation, growth, and rupture of cerebral aneurysms [1]. Much activity has been recently focused on using computational fluid dynamics to calculate hemodynamics in the intra- and- peri-aneurysmal environment [2]. One challenge to CFD of the cerebral circulation is the prescription of physiologically realistic boundary conditions such as the flow rate imposed at a model inlet.

CFD studies often rely on stereotypical boundary conditions derived from averages of healthy volunteers since patient-specific measurements are difficult to collect during assessment or treatment. However, CFD based on stereotypical boundary conditions are subject to potential uncertainty due to strong patient-to-patient variability [3]. The measurement of patient-specific cerebral blood flow rates for use as boundary conditions has the potential to improve the accuracy and reliability of CFD for cerebral aneurysm risk assessment and treatment planning.

We have previously [4,5] utilized a dual-sensor pressure and Doppler velocity endovascular guidewire (Combowire, Volcano Corp.) to obtain patient-specific blood pressure and flow rates in situ in cerebral vessels. We used this information both as boundary conditions in CFD simulations and as validation of the CFD with additional velocity and pressure measurements not used in the model setup. Good agreement was found between the CFD-based and wire-based velocities and pressure drops. However, considerable scatter, 20-40%, did exist between the two datasets.

The purpose of this study is to validate the endovascular wire-derived time-resolved volumetric flow rates in an *in vitro* setting inside models of cerebral vessels. The results establish confidence intervals for random and systematic errors associated with the wire-derived flow rates.

METHODS

A pulsatile flow loop with a blood mimicking fluid (40% glycerol/water mixture) is used to emulate the hemodynamic conditions of the cerebral vasculature. The loop pumps fluid at flow rates typical for the cerebral circulation mean 3-6 cc/s and max 5-10 cc/s. The endovascular wire acquires flow velocities inside the flow loop in both segments of straight vessels that represents the "ideal" conditions and in patient-derived models of cerebral vessels at 1:1 scale.

The Doppler wire measures the streamwise component of the flow velocity at the location of interest. The temporal resolution of the device is 10 ms. The measured wire velocity is taken to be the instantaneous vessel centerline velocity. Womersley's theory for pulsatile flow is utilized to fit a velocity profile to the Doppler velocity signal and to derive a volumetric flow rate from the integration of this axi-symmetric profile across the vessels cross-section. The in situ flow rates previously reported in [4,5] relied on Womersley's theory. Thus, the current measurements in the *in vitro* models are aimed to assess the robustness of this assumption. Measurements are typically acquired for 10-20 cardiac cycles.

The cerebral vessel models are derived from 3D rotational angiography of patients with cerebral aneurysms undergoing endovascular treatment. The models are first created with a 3D printer using ABS (acrylonitrile butadiene styrene). The "positive" model is cast in a silicone block to create a "negative." The silicone mold is then used to create a "positive" model from jewelers' wax. The wax model is cast in a polyester resin. After the resin cures, the wax is melted away leaving the final "negative" model of the aneurysm and parent vessel. Three separate cerebral vessel models are created: one is a basilar artery sidewall aneurysm, and the remaining two are internal carotid artery saccular aneurysm. The basilar artery model is shown in Fig. 1.

The wire-derived flow rates are compared against two separate and independent techniques: particle image velocimetry (PIV) and ultrasound transit time flow rate metrology. The Bland-Altman method is used to statistically assess possible bias (i.e. systematic error) and the limits of agreement (i.e. random error) between the wire-derived flow rates and either the PIV or the ultrasonic flow meter. The t-test is used to assess the statistical significance of the Bland-Altman biases.

The PIV measurements use an argon ion laser to illuminate a 2D plane parallel to the axial direction of the vessel. The flow is seeded with small, ≈20 μm diameter, lycopodium particles as flow tracers. A high-speed camera records images of the light scattered by the flow tracers at a rate of 1000 frames/s for 10 cardiac cycles. PIV measurements are collected in the straight vessel segments, and not inside the aneurysmal models. We use a commercially available ultrasound transit-time flow meter (Atrato, Titan Enterprises) to provide additional time-resolved flow rates (50 ms resolution).

RESULTS

Good agreement is found between the PIV and the wire-derived flow rates. There are no detectable biases in either the time-averaged or peak flow rates (p>.10). The coefficients of variation are 13.2% and 12.6% for the time-averaged and peak values, respectively, for the differences between the wire-based and PIV-based flow rates.

Satisfactory agreement is found between the flow meter and wire-derived flow rates. A total of 15 separate wire measurements are collected for the time-averaged and peak flow rates inside the *in vitro* models (5 per model). There is a small bias in the time-averaged flow rates, 0.568 cc/s, p<.0001, 95% CI [0.397, 0.757] cc/s, with the Doppler wire-derived values being lower than the flow meter measurements. There is no detectable bias, p>.50, in the peak flow rates between the wire-derived values and flow meter values inside the *in vitro* models. The one standard deviation limits of agreement are 0.312 cc/s and 0.747 cc/s for the time-averaged and peak values, respectively, which corresponds to relative random errors of 9.3% and 10.5%, respectively. The Bland-Altman plot for the wire-based and flow meter –based peak flow rates is shown in Fig 2.

DISCUSSION

The endovascular Doppler ultrasound velocimetry-based technique to measure time-resolved volumetric flow rates can provide reasonable, albeit imperfect, results for *in vitro* models of cerebral vessels. The random scatter is about 10%. The wire-based measurements show no systematic bias with respect to the temporal maximum flow rates. However, the wire-based values show a systematic error with respect to the time-averaged flow rates, with the relative error being about 17% too low.

The negative bias in the time-averaged flow rates does not preclude the technique's utility in obtaining patient-specific boundary conditions for CFD studies. For example, a correction factor could be applied to the time-averaged flow rates to remove the bias. As the peak values are accurate in this study, they would likely not require a correction *in vivo*.

Additionally, the current technique does have advantages over other modalities such as transcranial Doppler or phase-contrast magnetic resonance imaging. The guidewire, while invasive, can be employed in an anatomically-specific manner in any major blood vessel of the cerebral vasculature. It also integrates easily into a neurosurgeon's angiographic workflow and provides real-time blood flow measurements. Magnetic resonance-based measurements must be obtained outside of the normal angiographic workflow, and are both time-consuming and expensive.

FIGURE 1. In vitro model of a basilar artery sidewall aneurysm (Model C).

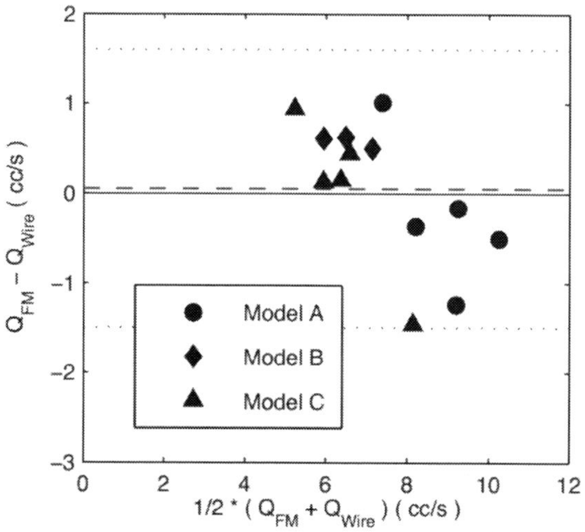

FIGURE 2. Bland-Altman plot for flow meter vs. wire-derived peak flow rates in the aneurysm models. Symbols are data points, dashed line is the mean difference, and dotted lines are ±2SD of the difference.

ACKNOWLEDGEMENTS

This work has been financially supported by NIH/NINDS grants 5R03NS078539 and 1R01NS088072. We also thank Professors Mark Ganter and Duane Storti for assistance with the 3D printing.

REFERENCES

[1] Meng, H et al., *Am J Neuroradiol*, 35: 1254-1262, 2014.
[2] Cebral, J R et al., *Am J Neuroradiol*, 32: 264-270, 2011.
[3] Marzo, A et al., *Ann Biomed Eng*, 39:884-896, 2010.
[4] Levitt, M R et al., *Am J Neuroradiol,* 35:143-148, 2014.
[5] McGah, P M et al., *Ann Biomed Eng*, 42:503-514, 2014.

IN-VIVO DYNAMIC MEASUREMENT OF TIBIOTALAR AND SUBTALAR JOINT KINEMATICS USING DUAL FLUOROSCOPY: A FRAMEWORK FOR STUDYING OA.

Koren E. Roach (1,2), Bibo Wang (1), Ashley L. Kapron (1), Niccolo M. Fiorentino (1), Charles L. Saltzman, (1), Madeline Singer (3), Andrew E. Anderson (1-4)

(1) Department of Orthopaedics
University of Utah
Salt Lake City, UT, USA

(2) Department of Bioengineering
University of Utah
Salt Lake City, UT, USA

(3) Department of Physical Therapy
University of Utah
Salt Lake City, UT, USA

(4) Scientific Computing and Imaging Institute
University of Utah
Salt Lake City, UT, USA

INTRODUCTION

Together, the tibiotalar and subtalar joints provide near six-degree-of-freedom motion at the ankle. Ankle sprains have a very high incidence at 25,000/day. One study reported approximately 33% of patients with sprains developed mechanical ankle instability, with about 40% of those patients going on to develop functional instability within one year of the injury [1]. Instability, and other biomechanical pathologies of the ankle, may cause osteoarthritis (OA) [2]. Although the mechanisms responsible for OA at the subtalar and tibiotalar joints are not well-understood, it is hypothesized that deleterious loading and kinematics initiate cartilage damage [3].

Using skin-markers, it is assumed that the talus and calcaneus move as a single rigid body, prohibiting quantification of subtalar kinematics. To better-understand mechanics of pathological ankles, the primary goal of this study was to establish kinematics of healthy, non-diseased subtalar and tibiotalar joints using dual-fluoroscopy (DF). The secondary goal was to determine the kinematic role of the subtalar joint relative to gross motion between the tibia and calcaneus.

METHODS

Ten volunteers (5 male, 5 female; mean age: 30.5±7) without a history of foot or ankle pain or pathology were enrolled in this study under ethics board approval and informed consent. An orthopaedic foot and ankle surgeon and musculoskeletal radiologist screened radiographs, not finding any instances of fractures, osseous deformities or OA. A previously described DF system [4] was arranged around a dual-belted instrumented treadmill. Volunteers ambulated on the treadmill at 0.5 m/s and 1.0 m/s and completed an unassisted balanced heelrise while DF images were acquired. Two trials were captured of each activity. A CT scan was obtained of the mid-tibia through toe-tips. The tibia, talus and calcaneus were segmented into 3D surfaces using Amira (V5.5, Visage Imaging, San Diego, CA). Model-based markerless tracking of the tibia, talus and calcaneus and calculations of joint angles and translations followed previous work [4]. Range of motion (ROM) was determined for each volunteer and averaged across trials. Bone-to-bone distance was calculated between the opposing articulating surfaces of the tibiotalar and subtalar joints in PostView [5] and plotted as color fringe plots for qualitative comparisons between subjects and activities. The minimum bone-to-bone distance was determined for each trial and averaged for each activity.

RESULTS

During gait, the tibiotalar joint experienced mean ROM of 9.5±4.4°, 3.3±1.9° and 4.8±2.4° for dorsi-/plantar-flexion, inversion/eversion and internal/external rotation, respectively (Fig. 1).

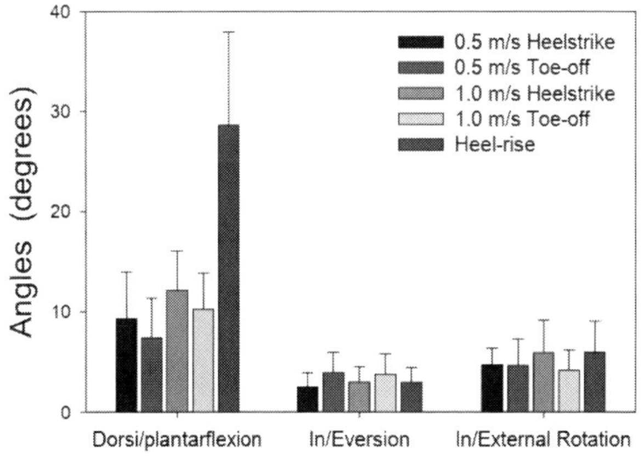

Figure 1: Tibiotalar joint angles by activity (bars = SD).

Through gait, the subtalar joint had mean ROM for dorsi-/plantar flexion, inversion/eversion and internal/external rotation of 4.5±2.6°, 9.8±4.6°, 5.8±2.8°, respectively (Fig. 2). Joint translations approached 2 mm, and appeared to be larger in the subtalar joint (Fig. 3).

The average minimum tibiotalar bone-to-bone distance during heelstrike was 0.28±0.20 mm and 0.23±0.18 mm during toe-off. The average minimum bone-to-bone distance of the subtalar joint during heelstrike and toe-off was 0.30±0.29 mm and 0.14±0.15 mm, respectively. Qualitative analysis of the bone-to-bone distance plots demonstrated the region of minimum bone-to-bone distance moving posteriorly in both the tibiotalar and subtalar joints (Fig. 4).

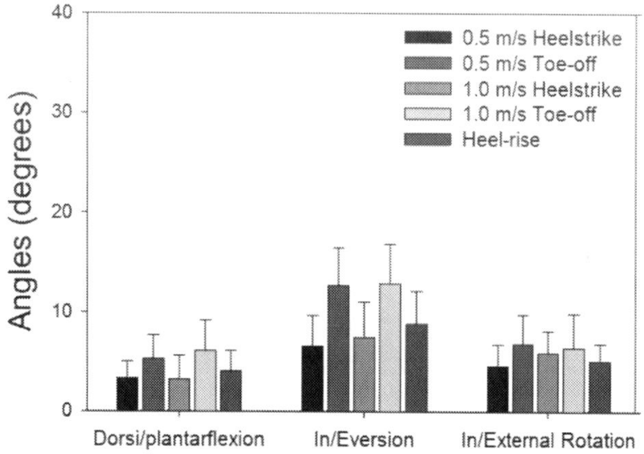

Figure 2: Subtalar joint angles by activity (bars = SD).

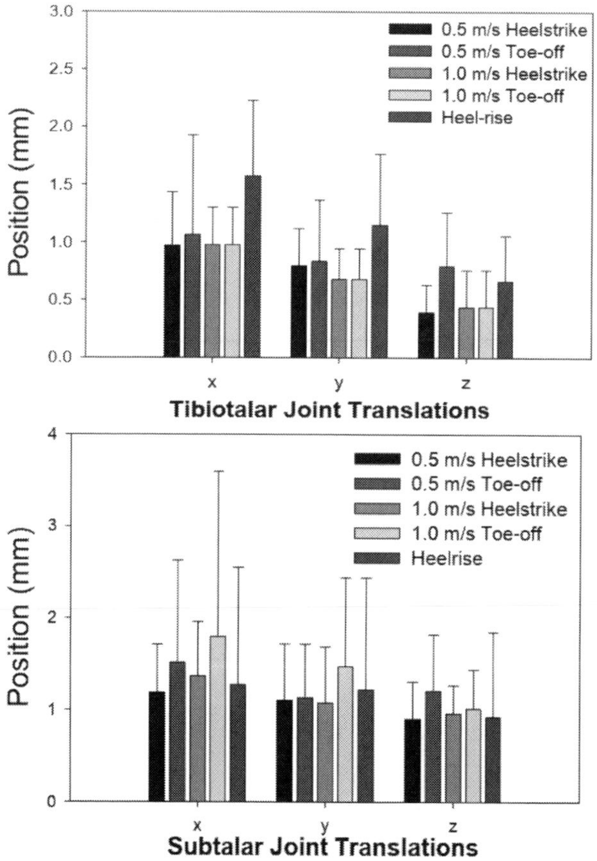

Figure 3: Joint translations by activity (bars = SD).

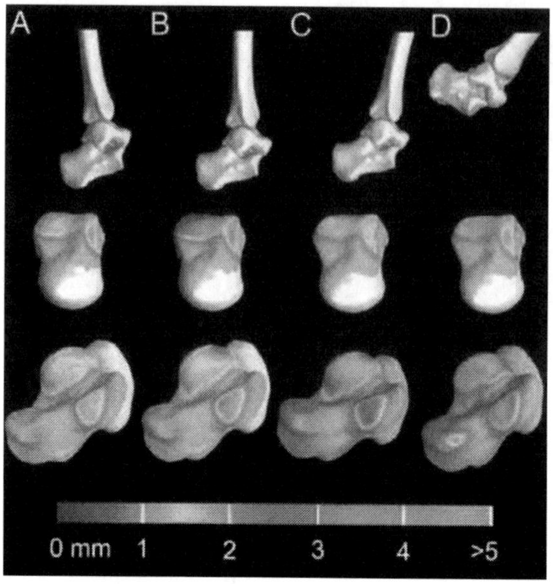

Figure 4: 3D models of the tibia, talus and calcaneus (top row) and the corresponding bone-to-bone distance fringe plots for the tibiotalar (middle row) and subtalar joints (bottom row) at heel-strike (A), early midstance (B), late midstance (C) and toe-off (D).

DISCUSSION

This is the first study to dynamically measure in-vivo subtalar and tibiotalar kinematics and bone-to-bone distance during treadmill gait. ROM results compared well to the other DF studies [6, 7]. Minimum bone-to-bone distance did not appear to vary between heelstrike and toe-off, possibly as a result of both activities involving a portion of midstance when both joints would likely be compressed. Our results demonstrated that the subtalar joint plays a secondary role in flexion at the ankle; the majority of the flexion was derived at the tibiotalar joint (Fig. 1 and 2). Conversely, the subtalar joint was primarily responsible for inversion/eversion. Although statistical analyses were not performed, the subtalar joint appeared to have more inversion/eversion during toe-off than heelstrike. Despite the differences in speed, only the tibiotalar joint flexion angles seemed to vary between the two gait speeds. The subtalar joint also appeared to have more translational movement than the tibiotalar joint. This could explain why subtalar arthrodesis often results in OA to the tibiotalar joint. Similarly, restriction of the subtalar joint may cause deleterious motions at the tibiotalar joint as a necessary compensatory mechanism, possibly predisposing the tibiotalar joint to OA.

ACKNOWLEDGEMENTS

Funding from the LS-Peery Discovery Program in Musculoskeletal Restoration and the American Orthopaedic Foot & Ankle Society is gratefully acknowledged.

REFERENCES

[1] Freeman MA, *J Bone Joint Surg Br*, 47:669-77 1965.
[2] Valderrabano V et al., *Clin Orthop Relat Res*, 467:1800-6 2009.
[3] Golditz T et al., *Osteoarthritis and cartilage / OARS, Osteoarthritis Research Society*, 2014.
[4] Kapron AL et al., *J Appl Biomech*, 30:461-70 2014.
[5] Maas SA et al., *J Biomech Eng*, 134:011005 2012.
[6] de Asla RJ et al., *Journal of orthopaedic research : official publication of the Orthopaedic Research Society*, 24:1019-27 2006.
[7] Peltz CD et al., *J Biomech*, 47:2647-53 2014.

SB³C2015
Summer Biomechanics, Bioengineering and Biotransport Conference
June 17-20, 2015, Snowbird Resort, Utah, USA
SB³C2015-528

FSI SIMULATIONS FOR THE HEMODYNAMIC ASSESSMENT OF THE CAROTID BIFURCATION IN AN ATHEROSCLEROTIC MOUSE MODEL

David De Wilde (1), Bram Trachet (1,2), Nic Debusschere (1), Francesco Iannaccone (1), Abigaïl Swillens (1), Joris Degroote (3), Jan Vierendeels (3), Guido R. Y. De Meyer (4), Patrick Segers (1)

(1) Department of Electronics and Information Systems
Ghent University-iMinds-IBiTech
Gent, Belgium

(2) Institute for Bioengineering
Ecole Polytechnique Fédérale de Lausanne
Lausanne, Switzerland

(3) Department of Flow, Heat and Combustion Mechanics
Ghent University
Gent, Belgium

(4) Division of Physiopharmacology
University of Antwerp
Wilrijk, Belgium

INTRODUCTION

Atherosclerosis preferentially occurs at locations of disturbed blood flow characterized by low and oscillatory wall shear stress (WSS). It is challenging to estimate the WSS in-vivo in mice, due to their small size and high heart rate. Computational fluid dynamics (CFD) simulations allow to quantify and visualize the WSS but rely on a rigid wall assumption that is not consistent with the buffering capacity of the carotid bifurcation. Fluid structure interaction (FSI) simulations on the other hand result in a more reliable representation of the flow but are more complex and time consuming. In addition, FSI simulations require more boundary conditions to set up the computational problem in a physiologically relevant way. In this study, a numerical FSI model of the mouse carotid bifurcation is presented. The model is based on an in-vivo geometry measured with micro CT and mouse-specific boundary conditions measured with high-frequency ultrasound. The goal of the study was to setup realistic mouse-specific FSI simulations to determine the WSS and derived parameters in the carotid bifurcation of a frequently used mouse model for atherosclerosis.

METHODS

Imaging protocol

Four female ApoE$^{-/-}$ mice underwent an imaging protocol at the age of 16 weeks. First both the left and right carotid bifurcation were examined with high resolution in-vivo ultrasound using a Vevo 2100 (VisualSonics, Toronto, Canada). Pulsed Doppler (PD) measurements were used to measure the blood velocity in the common (CCA), external (ECA) and internal (ICA) carotid arteries. At each branch the distension was measured by means of RF (radio-frequency) M-mode ultrasound. Subsequently, the mice underwent a contrast-enhanced µCT scan (Triumph-II imaging system, TriFoil Imaging, Chatsworth, USA) at an isotropic resolution of 50µm. The contrast agent, Aurovist (Nanoprobes, Yaphank, NY) was administered in a dose of one micro liter/25g body weight and made it possible to visualize the vasculature in-vivo.

Processing the imaging data

The carotid bifurcation was segmented using Mimics (Materialise, Leuven, Belgium). To get to the diastolic geometry, the segmentation was shrunk radially based on the diastolic diameter from the M-mode measurements. Furthermore, the PD-measurements were tracked and provided the instantaneous maximum velocity at the three branches.

The intrusiveness of putting a catheter in the mouse made it difficult to provide in-vivo pressure. Therefore the pressure curve was approximated by rescaling the diameter distension waveform to the in-vivo pressure range from 87 to 119 mmHg that has previously been reported for anesthetized mice at a similar age [1].

Setting up the FSI model

For the structural analysis of the FSI model, we used Abaqus/ Standard (Simulia, Johnston, US). For the arterial wall an Arruda-Boyce material model was chosen. This material model only requires two parameters, which were determined such that the simulations matched the measured distension. The behavior of the surrounding tissue was modeled using linear spring elements in Abaqus [2]. We then calculated the in-vivo initial wall stress at diastolic pressure using a backward incremental method [3]. These stresses were afterwards applied as initial conditions at the start of the FSI simulations in order to get realistic wall behavior.

For the flow analysis of the FSI model, we used Fluent (Ansys, Canonsburg, PA). A first step was to assure that the flow balance was satisfied over an entire cardiac cycle. The flow was calculated based on the measured velocity profiles and the boundary area that was obtained from a structural computation in which the artery was inflated with the approximated pressure curve. For the fluid part of the FSI simulations we applied the mass flow of the CCA as a mass flow-inlet. For the ECA we rescaled the mass flow curve such that the flow split of the ECA/ICA was 20/80%, which was the mean value measured over the 4 mice. This rescaled mass flow was then imposed at the ECA outlet. At the ICA outlet, a 3 element windkessel model was applied. This model was fitted to the rescaled flow and the pressure curve.

A strong coupling between the flow and structural solvers was realized using the in-house developed coupling code Tango (using the IQN-ILS algorithm). Three cardiac cycles were calculated to avoid transient startup phenomena. For every heart cycle 100 time increments have been used.

Postprocessing

For correct computation of the WSS, the movement of the wall should be taken into account. Therefore, the WSS was expressed in a local axis, which moved together with the arterial wall. Based on this local WSS, the time averaged WSS (TAWSS), the oscillatory shear index (OSI) and the relative residence time (RRT) were calculated.

RESULTS

The average diameter distension over the three branches in the simulations was 8.24% while an average value of 10.23% was measured in-vivo. At the ICA outlet, the resulting systolic/diastolic pressure was 117.3/87.5 mmHg, which matched very well with the target pressure. Pressure was a little higher at the inlet of the CCA (118.5/88.4) because of the small pressure drop over the carotid segment. Figure 1a displays the comparison between the FSI model pressure and the diameter distension waveform rescaled curve at the outlet of the ICA.

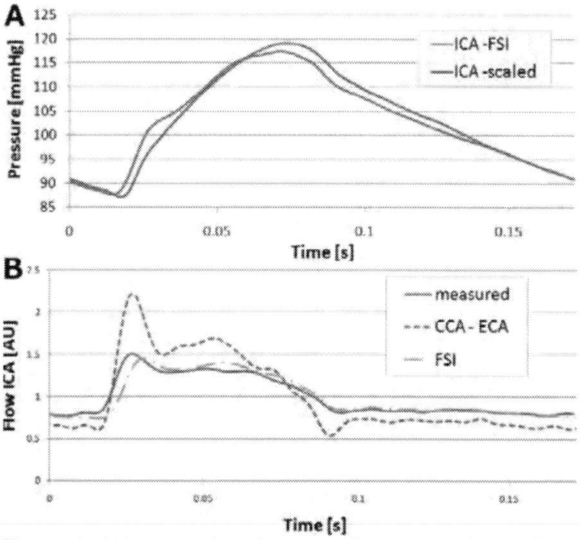

Figure 1: (A) comparison between the pressure in the FSI-model (ICA-FSI) and the pressure curve based on a rescaling of the diameter distension curve (ICA – scaled) for the ICA. (B) Comparison between the measured flow profile, the ECA flow subtracted from the CCA flow (CFD boundary condition) and the FSI flow profile at the ICA in arbitrary units.

Flows were imposed at inlet and outlet of the CCA and ECA, respectively, so their waveforms matched perfectly. The ICA flow waveform, however, is a result of the simulations. There was an excellent agreement between calculated and measured flow waveforms, with a correlation of 0.980 (Figure 1b).

The calculated TAWSS (figure 2-A) showed a zone with lower WSS at the outer bend of the ECA. For the ICA the TAWSS is clearly elevated compared to the CCA. Furthermore, for the RRT there is a zone with a clear elevation at the same location of the ECA (figure 2-B). In the ICA on the other hand, the RRT values are lower compared to the CCA.

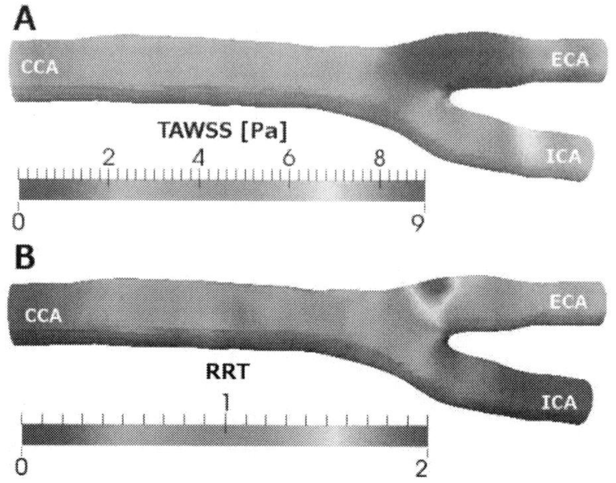

Figure 2: WSS related parameters based on the FSI results, plotted on the diastolic geometry: (A) the TAWSS and (B) the RRT.

DISCUSSION

Although it was impossible to measure the in-vivo pressure accurately enough to use it in a modeling approach, the proposed methodology (rescaling the M-mode measurements to fit the windkessel at the ICA outlet) provided consistent results. The arterial pulsatility following from the FSI model simulations closely mimics reality. Importantly, the resulting flow waveform at the ICA shows a good correspondence with measured flow velocity signals. Interestingly, Figure 1b also displays the waveform that is obtained by subtracting the flow at the ECA from the inlet flow (i.e., the flow that would be imposed at the ICA in CFD simulations). Our data thus demonstrate that it is important to include the distensibility of the carotid artery into the model to retrieve physiological flow waveforms.

The WSS related parameters highlight the zones that are prone to the development of atherosclerosis. In this case the most important location was the outer bend of the ECA close to the bifurcation. Some of our unpublished results showed that these are the same regions where the actual atherosclerosis develops first.

In conclusion, a methodology to apply mouse-specific FSI, based on in-vivo measurements, in the carotid bifurcation of mice has been shown. Afterwards, the TAWSS and RRT were calculated to highlight regions which are prone to plaque development.

ACKNOWLEDGEMENTS

David De Wilde received a research grants of the IWT. Furthermore, we thank Scharon Bruneel, Benedicte Descamps, Christian Vanhove and Mathias Peirlinck for their assistance.

REFERENCES

[1] Van Herck, JL et al., *Circulation*, 120:2478-2487, 2009
[2] Moireau, P et al., *Biomech ModelMechanobiol,* 11:1-18
[3] De Putter, S et al., *J Biomech,* 34:859-871

THE EFFECT OF GRADE II AND GRADE III ANKLE INJURY ON THE ANKLE JOINT COMPLEX KINEMATICS AND ACHILLES LOAD: A CADAVERIC STUDY

Bardiya Akhbari (1), Matthew H. Dickinson (2), Ednah G. Louie (2) , Sami Shalhoub (2)
Lorin P. Maletsky (1, 2)

(1) Department of Mechanical Engineering
University of Kansas
Lawrence, Kansas, United States

(2) Bioengineering Graduate Program
University of Kansas
Lawrence, Kansas, United States

INTRODUCTION

Inversion injuries of the ankle caused by rupture or sprain of collateral ligaments are very common and lead to deviations in joint stability.[1,2] *In-vivo* studies of ankle injuries are invasive due to a need to section the ligaments; hence *in-vitro* studies are preferred for understanding the variation in properties of injured and intact ankles. Previous in-vitro experiments studied the effects of ligaments sprain on the ankle joint complex kinematic without loading the muscles.[3] Understanding ankle joint complex (AJC) kinematics and kinetics allows for quantitative assessment of the talar tilt test and different mid-flexion assessments to be made in a total ankle replacement to create better outcomes for patients. The purpose of this study is to measure the effects of Grade II and Grade III ankle injury on the Achilles load, inversion/eversion, and adduction-/abduction of the AJC.

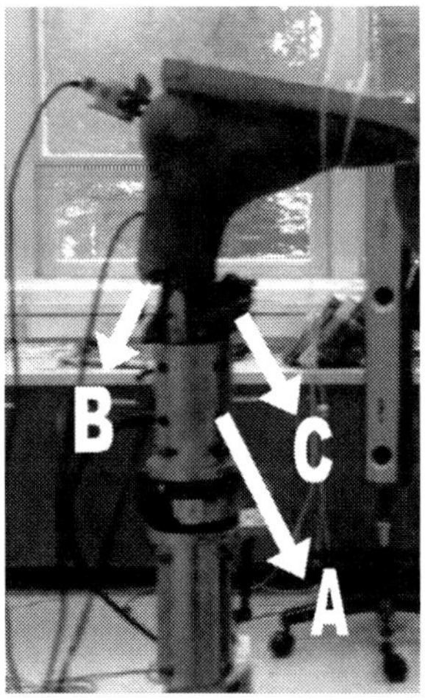

FIGURE 1: TESTING SETUP CONSISTING OF A) TIBIA FIXURE, B) MOTOR CABLE, C) MUSCLES CABLE

METHODS

Five fresh frozen cadaveric ankles were thawed at room temperature. The tibia was sectioned approximately 15cm superior to the calcaneus and potted in an aluminum fixture. The soft tissue above 15cm of the AJC joint line was removed except for the tibialis anterior (TA), extensor digitorum longus (EDL), and Achilles tendons, which were left intact. The ankle was then attached to a muscle-loading rig with the tibia rigidly fixed in an inverted position with the foot free to move (Figure 1). The TA and EDL and Achilles tendons were isolated and clamped individually. The ankle was flexed four times through its range of motion using a stepper motor attached to the Achilles tendon. Static loads were used to load the TA and the EDL. The physiological loading set was defined as 90N on the TA and 45N on the EDL based on their physiological cross sectional areas (PCSAs)[4]. Following literature,[5] a Grade II injury was simulated by resecting the anterior talofibular ligament (ATFL) followed by a resection of calcaneofibular ligament (CFL) to simulate a Grade III injury. After simulation of the injury and dissection of the ligaments, the skin and muscle were sutured to restrict excess movement of the ankle joint complex. The intact ankle experiment was used as the baseline kinematics and kinetics cycle during analysis.

The kinematics of the tibia and calcaneus were recorded using IRED markers rigidly attached to the bone and an Optotrak Certus motion-capture system (Northern Digital, Ontario). The loads applied by the motor on the Achilles tendon to dorsiflex/plantarflex the ankle were measured using a 1200N load cell (transducer technique). The difference in inversion/eversion (InEv), adduction/abduction (AdAb) rotation, and motor load (Load) were calculated for the two loading profiles relative to the baseline for the third motor's cycle. T-tests were performed to find significant differences ($p<0.05$) between the physiological configuration and each of the injured set individually.

RESULTS

The average kinematics for the five ankles displayed significant increase in inversion throughout the whole flexion cycle regardless of their injured or uninjured state (Figure 2.A). The average AdAb rotation for the physiological load demonstrated changes in the abduction angle from approximately 4° to 5.3° of adduction during the whole cycle (Figure 2.C). The physiological load exhibited an increase in average Achilles load during early flexion and a decrease in the rest of the cycle (Figure 2.E). Moreover, increasing Achilles load was seen in early dorsiflexion, which decreased through the rest of the cycle. Both the Grade II and III injuries increased the inversion and Achilles' load throughout the entire cycle. Relative to the intact ankle, AdAb rotation remained the same for the Grade II injury and increased slightly in the Grade III injury during the entire cycle (Figure 2). The Grade II and III injuries resulted in an increase in ankle inversion throughout the cycle with an average of 0.8° and 1.6°, respectively (Figure 2.B). No change was seen in ankle adduction for the Grade II injury; however, with the Grade III injury the ankle was significantly more adducted than the intact ankle (Figure 2.D). The Achilles load was begun by 6.7N compensation and decreased during the entire cycle and finished by -0.4N. The load was more and significantly different for grade III which started from 11.6N and finished at 2.7N (Figure 2.F). The results displayed no statistical difference between the Grade II injured ankle and the intact ankle, nonetheless the Grade III injury presented a statistical difference of $p<0.05$ for inversion and the Achilles loads.

DISCUSSION

The effect of the collateral ligament rupture on the AJC kinematic was displayed by previous studies without additional muscles loads.[3] In this study the tibialis anterior and extensor digitorum longus muscles were loaded by physiological statics weight to exhibit the natural motion of AJC. The ankles inversion and adduction behavior during plantar flexion was similar to (Siegler, 1988) which studied the ankle three dimensional kinematics during natural motion without any loads attached to the muscles.[6] The inversion amount in this study was considerably higher which can be understood by the attached tibialis anterior static load. Moreover, Grade II and Grade III injuries were similar to what have been previously reported in the literature.[7]

Ruptures in the ATFL and CFL affect the ankle inversion at differing angles during the flexion range. In this study the AJC kinematics of both the injury Grade II and the injury Grade III deviated from the intact ankle, with the main variation observed in the inversion and loading. The higher inversion in the injuries was caused by absence of the collateral ligaments which have a main role in the rotation regulating of AJC. Rupture in the ATFL caused increase in the inversion difference during the whole cycle, while the rupture in both ligaments caused the decrease in the difference which verifies the fact that CFL has a large effect on the kinematic in the dorsiflexion.[5]

The insignificant dissimilarity in the AdAb motion in the Grade II injury and constant difference in this motion in Grade III injury displays the CFL role in stabilizing of ankle joint complex. In addition, ATFL role in the AdAb motion was minor which shows the minor change in AdAb follows by Grade II injury. Both ligaments displayed significant effects on the Achilles load. The Achilles has to generate more loads to stable the ankle during early flexion angles after the injury, and the difference decreases with flexion. The decrease in the ankle load can be explained by Achilles' fascicle length drops in the plantarflexion. The shorter fascicle length results in the lower force generation of the Achilles tendon. Other studies displayed that patients feel pain and sooner tiredness according to the CFL injury.[5]

The finding of this study has effect on the rehabilitation of the ankle movement, verifying the computational models, improving our understanding of the ankle injury and collateral ligaments function. There are a few limitations to this study; first the muscle loads used were relatively small compared to the physiological loads in the body. Low number of cadaveric specimen (five cadavers) was used and in order to improve the results, more specimens are needed. Simulation of injuries was to completely rupture the ligaments, which is the worst case of injury and is not reflective of partial ruptures in real-life sprains. Future research is needed to understand various ruptures. Moreover, variation in recognizing of the anatomical landmarks causes difference in the kinematic measurements for a single ankle; however the results for comparing between sets will not be changed.[8]

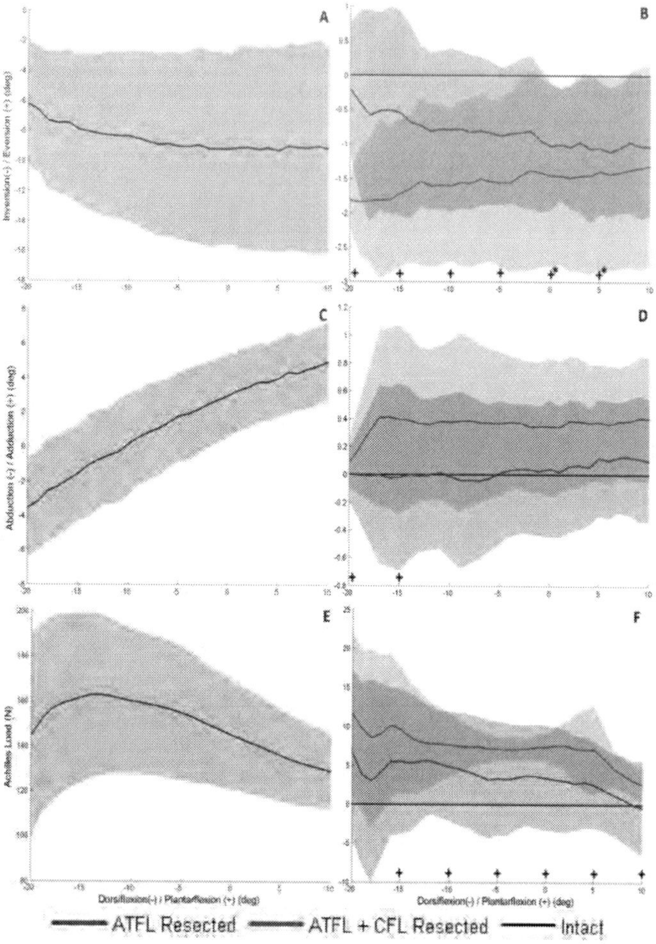

FIGURE 2: INEV (A), ADAB (C), AND LOAD (E) FOR ONE ANKLE, MEAN DIFFERENCE IN INEV (B), ADAB (D), AND LOAD (F) FOR ALL ANKLES BETWEEN PHYSIOLOGICAL AND INJURED SETS

(SIGNIFICANCE SHOWED BY * FOR GRADE II AND + FOR GRADE III INJURY)

ACKNOWLEDGEMENTS

This experiment was partially funded by DePuy Synthes, Inc.

REFERENCES

1.Brostroem L. Acta Chir Scand,128:483-95, 1964.
2.Brooks SC, et al. Br Med J (Clin Res Ed), 282(6264):606-7,1981.
3.West JE, University of Kansas. p 143 p, 2011.
4.Arnold EM, et al. Ann Biomed Eng,38(2):269-79, 2010.
5.de Asla RJ, et al. J Orthop Surg Res,4:7 , 2009.
6.Siegler S, et al.. J Biomech Eng, 110(4):364-73, 1988.
7.Petersen W, et al. Arch Orthop Trauma Surg, 133(8):1129-41, 2013.
8.Morton NA, et al. J of Orthop Research, 25(9):1221-1230, 2007.

OPEN KNEE(S): COMPREHENSIVE TIBIOFEMORAL JOINT TESTING FOR SPECIMEN-SPECIFIC NEXT GENERATION KNEE MODELS

Tara F Bonner (1,2), Robb W Colbrunn (1,2), Snehal Chokhandre (1,3), Craig J Bennetts (1,3), Ahmet Erdemir (1,3)

(1) Department of Biomedical Engineering
Cleveland Clinic
Cleveland, Ohio, USA

(2) BioRobotics and Mechanical Testing Core
Cleveland Clinic
Cleveland, Ohio, USA

(3) Computational Biomodeling (CoBi) Core
Cleveland Clinic
Cleveland, Ohio, USA

INTRODUCTION

In the development of virtual knees for clinical and scientific simulations, evaluating the predictive capacity of the model to represent joint kinematic-kinetic response is a necessary step. Obtaining accurate and comprehensive specimen-specific mechanics of the tibiofemoral joint is required to establish authenticity of the model. Historically, in knee modeling, tibiofemoral joint testing data used to evaluate a model's performance have been collected from different specimens than that of which model components were based, e.g. bone geometry and mechanical properties of tissue. Even in recent knee models, where specimen-specific joint mechanics data complements specimen-specific imaging [1], issues related to registration of anatomical images to experimentation coordinate systems (CSs) remain unclear. Without complete specimen-specific information, the source of discrepancies between model predictions and experimentation cannot be ascertained.

Recently, the Open Knee(s) project has been launched to mitigate these uncertainties with the aim to build completely specimen-specific (geometry and material) three-dimensional finite element representations of knee joints from different populations, varying gender, age and grades of osteoarthritis. Within this project's framework, specimen-specific tibiofemoral joint mechanics is measured such that specimen-specific evaluation of model performance can be conducted in an elaborate manner. The goal of this document is to provide the specific details of the tibiofemoral testing within the framework of the Open Knee(s) project.

METHODS

Five leg specimens (hip to ankle) were obtained (Table 1). All specimens were absent of knee injury, surgeries or inflammatory arthritis.

Table 1: Specimen characteristics.

Specimen #	Gender	Age	Height	Weight	BMI
oks001	Male	71	1.83 m	77.1 kg	23.1 kg/m^2
oks002	Female	67	1.55 m	45.3 kg	18.9 kg/m^2
oks003	Female	25	1.73 m	68.0 kg	22.8 kg/m^2
oks004	Female	46	1.58 m	54.4 kg	21.9 kg/m^2
oks006	Female	71	1.52 m	49.4 kg	21.3 kg/m^2

A specimen preparation protocol allowed placement of optoelectronic measurement sensors (Optotrak, Northern Digital Inc., Waterloo, Ontario, Canada) and spherical registration markers (to relate mechanics data to imaging) on the tibia, femur, and patella. Anatomical landmarks and registration spheres were digitized relative to the optoelectronic sensors, respectively, on the tibia or femur. A knee joint CS based on Grood and Suntay [2] was defined. This CS was later updated to minimize off-axis translations and rotations during passive flexion. This optimized CS was used to report joint kinematics, and the tibia CS was used to report joint kinetics. Various additional CSs were established to relate joint CSs to registration CSs and therefore geometry reconstructed from imaging. The specimen was then secured to the Universal Musculoskeletal Simulator [3], capable of 6 degrees-of-freedom and real-time force feedback using simVITRO™ software (Cleveland Clinic, Cleveland, OH). The tibia was secured to a 6-axis force-moment sensor (SI-1500-240, ATI Industrial Automation, Apex, NC) rigidly attached to the frame and the femur was secured to a custom stage rigidly attached to the robot (Rotopod R2000, Parallel Robotics System, Hampton, NH). Preconditioning was performed prior to joint testing.

Joint mechanics were characterized by laxity and combined loads tests that were executed at 0, 30, 60 and 90° flexion. Laxity testing consisted of: 1) Internal/external (IE) rotation: 0 to ±5 Nm in steps of 1 Nm. 2) Varus/valgus (VV) rotation: 0 to ±10 Nm in steps of 2.5 Nm. and 3) Anterior/posterior (AP) translation: 0 to ±100 N in steps of 10 N. Combined loads testing consisted of permutations of IE rotation moments of -5, 0, 5 Nm, VV moments of -10, 0, 10 Nm, and AP drawer force of -100, 0, 100 N. Throughout testing, a 20 N compression load was applied to the joint.

A repeatability test was performed three times (beginning, middle and end of all testing) to ensure no substantial damage to the specimen and to quantify small changes in joint laxity due to repetitive load. AP loads were applied from 0 to ±100 N in steps of 10 N at 30° of flexion.

RESULTS

An example of kinematics and kinetics for the combined loads test of specimen oks006 at 30° flexion is shown in Figure 1.

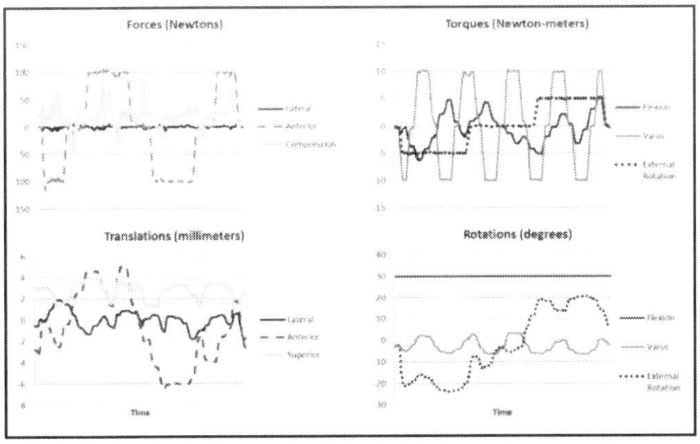

Figure 1: Kinetics and kinematics under combined loading for oks006.

Figure 2 displays the range of motion (ROM) measured from the laxity tests for the five specimens at each flexion angle. The ROM for each loading condition includes the range of kinematic responses due to the most extreme loads. It should be noted that due to load control errors during the AP laxity tests for specimen oks004, ROM was not included for 30, 60 and 90° flexion angles. Figure 2 also displays example laxity curves for specimen oks006.

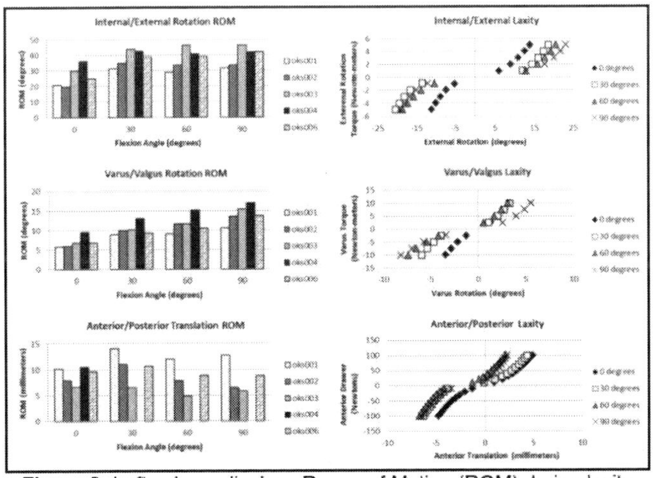

Figure 2: Left column displays Range of Motion (ROM) during laxity tests for all specimens. Right column displays laxity plots for oks006. AP ROM data is not provided for oks004 at flexion angles of 30, 60 and 90 due to loading errors.

AP ROM data from the repeatability tests are shown in Table 2. Results from oks004 are not displayed due to loading errors.

Table 2: AP ROM due to applied loads of ±100 N of anterior drawer.

	Anterior/Posterior ROM (mm)			
	oks001	oks001	oks001	oks001
Beginning	13.54	10.36	5.86	10.02
Middle	14.55	11.45	6.71	11.08
End	14.57	11.40	6.61	11.14

DISCUSSION

In order to build virtual knees authentic to specimen-specific joint mechanics, predicted kinematics-kinetic responses should match that of the same specimen for which geometry and material properties are collected and the model is developed. This document outlined the specific details on how tibiofemoral testing should be completed within the framework of the Open Knee(s) project. Tibiofemoral mechanics data was collected from five different knee specimens. Because specimen-specific geometry, tissue material properties and registration of anatomical images to experimentation CSs have been collected or will be collected for each specimen, the evaluation of the predictive capacity of each completely specimen-specific model against specimen-specific joint kinematic-kinetic response can occur.

The alternative to using specimen-specific tibiofemoral joint mechanics data is to use more generic data from the literature or data from a different specimen. It is evident in Figure 2 that kinematic-kinetic responses vary amongst specimens. If tibiofemoral joint mechanics data from specimen oks001 (or population data) were used to evaluate the predictive capacity of a model developed based on the geometry of specimen oks004, disagreements would likely occur. The likelihood of such uncertainties would only increase if material property data are gathered from literature or adapted from other specimens. As a result, it would be difficult to ascertain the source of disagreements.

Some potential limitations in robotics testing include estimates made to accommodate specimen weight during measurements of loads. It also appears that there might be a conditioning effect throughout experimentation (Table 2); more intensive preconditioning may alleviate this issue.

The presented specimens have already undergone magnetic resonance imaging [4] and tissue characterization is in progress. Five additional specimens will be tested and imaged, further diversifying the population. Specimen-specific knee models will be developed based on imaging data and mechanical tissue properties and they will be evaluated using the kinematic-kinetic data collected in this study for each respective specimen. This significant amount of heterogeneous data is acquired openly and will be made available to the community through the Open Knee(s) project.

ACKNOWLEDGEMENTS

This study was funded by NIGMS, NIH (R01GM104139, PI: Erdemir). Assistance from Katie Stemmer, Dylan Beckler, Sam Doerle and Erin Merico is appreciated. Open Knee(s) is an open development modeling project; specifications, models, and data can be accessed at http://wiki.simtk.org/openknee.

REFERENCES

[1] Mootanah, R et al., *CMBBE*, 17:13, 1502-17, 2014.
[2] Grood, E.S. & Suntay, W.J., *J. Biomech. Eng*,.105:136–144, 1983.
[3] Noble, L.D. et. al, *J. Biomech. Eng.*, 132(2), 2010.
[4] Bennetts, CJ et al., *SB*^{3}*C, 2015*.

BEARING SURFACE DAMAGE ANALYSIS OF TOTAL SHOULDER REPLACEMENT RETRIEVALS ACROSS FIXATION DESIGNS AND UHMWPE COMPOSITION

L. Malito, MS (1), F. Ansari, MS (1), L. Li (1), T. Lee (1), H. Park (1), S. Gunther, MD (2), T. Norris, MD (3), M. Ries, MD (4), L. Pruitt, PhD (1)

(1) Department of Mechanical Engineering
University of California, Berkeley
Berkeley, CA, USA

(2) Department of Orthopedic Surgery
Martha Jefferson Hospital
Charlottesville, VA, USA

(2) San Francisco Shoulder, Elbow & Hand Clinic
San Francisco, CA, USA

(4) Department of Orthopedic Surgery
University of California, San Francisco
San Francisco, VA, USA

INTRODUCTION

In vivo surface damage of total joint replacements can have significant clinical consequences as ultrahigh molecular weight polyethylene (UHMWPE) wear debris can initiate osteolysis and implant loosening [**1**]. The majority of retrieval studies have focused on polymer bearing damage in total hip replacements (THR) and total knee replacements (TKR) [**2,3**]. Yet, little attention has been given to total shoulder replacements (TSR), which experience modes of loading spanned by THR and TKR [**4**]. Few retrievals studies have examined trends in surface damage for different UHMWPE bearing formulations and different bearing fixation designs. Our methodology classifies the mode, geometric extent, and severity of surface damage in TSR retrievals of varying designs spanning a 20 year design history. We quantify trends in damage of different UHMWPE glenoid materials and fixation designs to provide insight on the origin, evolution and consequences of bearing damage *in vivo*.

METHODS

Twenty-one retrieved glenoid components with different fixation designs and UHMWPE formulations were analyzed. Fixation designs analyzed were all-polyethylene backed or metal-backed. The UHMPWE glenoid formulations present in the study consisted of Hylamer, conventional polyethylene 25 Mrad gamma irradiated in air, and finally modern cross-linked formulations. Glenoids used in the study were implanted from a period of 1987 to 2011, spanning more than 20 years. Average age at implantation was 61 years of age for the 21 implant study with an average *in vivo* time of 6.1 years. (**TABLE 1**).

A grid placed on glenoid components divided the bearing surface into 17 regions of equal area. Each region was photographed at 249 dpi (16-20x) using a Nikon 1 J1 camera and Infinivar Video Microscope lens under an AmScope LED ring light. Observers noted the presence, severity and extent of 13 damage modes (**FIGURE 1**) in each region using the measurement toolbox in ImageJ (version, location). The damage modes were chosen from polyethylene implant classification methods based on the Hood method [**2,3,5**]. Each implant was scored by at least 2 observers. Inter-observer agreement was evaluated by calculating the average percent agreement between two observers for damage mode identification and the average percent agreement with the mean for geometric extent and severity classification.

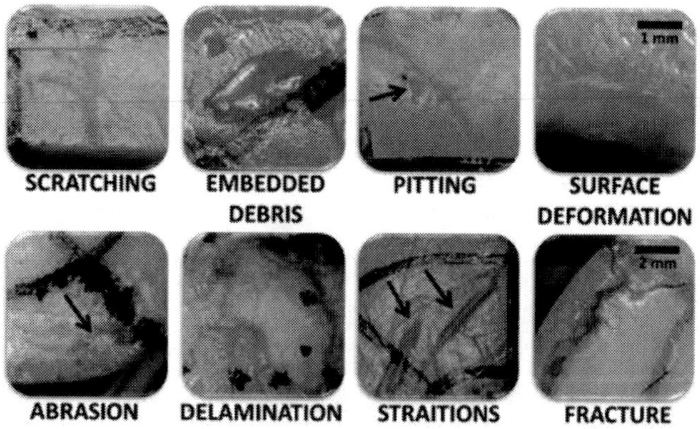

FIGURE 1: EIGHT OF THE 13 PRIMARY DAMAGE MODES OBSERVED ON UHMWPE GLENOIDS. DELAMINATION, EMBEDDED DEBRIS, RIM EROSION, ADHESION, WEAR THROUGH, SUBSURFACE CRACKING, AND BURNISHING NOT SHOWN.

RESULTS

All glenoid fixation designs and material compositions demonstrated instances of scratching, striation, pitting and surface deformation over a wide range of *in vivo* implantation times and on all samples (**TABLE 2**). Fracture was seen across all fixation designs and UHMWPE formulations in at least one instance. The group of cross-linked polyethylene glenoids had the shortest time *in vivo* and exhibited scratching, striation, pitting, surface deformation, and abrasion in the majority of samples on par with the different other UHMWPE material compositions (**TABLE 2**). Metal-backed glenoids had the longest time *in vivo* and exhibited the highest instance of embedded debris compared with the other glenoid samples (**TABLES 2**). The majority of metal-backed, Hylamer, and conventional polyethylene glenoids were fractured compared to the other groups (**TABLE 2**).

TABLE 1: SUMMARY OF CLINICAL INFORMATION

Clinical Information on Scored Glenoids						
Clinical Information	All Glenoids (n=21)	Poly-Backed Glenoids (n=16)	Metal-Backed Glenoids (n=5)	Hylamer Glenoids (n=7)	Conv. Poly 25 Mrad in Air Glenoids (n=6)	Cross-Linked Glenoids (n=7)
Age at Surgery	61 ± 10.5 Years	61.9 ± 10.1 Years	58.4 ± 12.4 Years	63 ± 10.3 Years	60.8 ± 12.6 Years	57 ± 9.2 Years
In vivo Time	6.1 ± 4.8 Years	4.2 ± 3.3 Years	13.3 ± 1.6 Years	6.9 ± 2.7 Years	10.8 ± 5.7 Years	2.0 ± 1.1 Years

TABLE 2: SUMMARY OF DAMAGE MODES PRESENT ACROSS FIXATION DESIGNS AND UHMWPE MATERIAL COMPOSITION

Glenoids With Each Damage Mode Present						
Damage Mode	All Glenoids (n=21)	Poly-Backed Glenoids (n=16)	Metal-Backed Glenoids (n=5)	Hylamer Glenoids (n=7)	Conv. Poly 25 Mrad in Air Glenoids (n=6)	Cross-Linked Glenoids (n=7)
Scratching	21 (100%)	16 (100%)	5 (100%)	7 (100%)	6 (100%)	7 (100%)
Striation	21 (100%)	16 (100%)	5 (100%)	7 (100%)	6 (100%)	7 (100%)
Pitting	21 (100%)	16 (100%)	5 (100%)	7 (100%)	6 (100%)	7 (100%)
Surface Deformation	21 (100%)	16 (100%)	5 (100%)	7 (100%)	6 (100%)	7 (100%)
Abrasion	16 (77%)	11 (69%)	5 (100%)	5 (70%)	6 (100%)	5 (67%)
Delamination	15 (71%)	11 (68%)	4 (80%)	6 (86%)	5 (83%)	3 (48%)
Embedded Debris	6 (29%)	3 (19%)	3 (60%)	2 (30%)	3 (50%)	1 (14%)
Fracture	13 (63%)	9 (57%)	4 (80%)	6 (81%)	5 (83%)	2 (36%)

DISCUSSION

To our knowledge this is the only study that assesses the presence of damage modes on TSR glenoids with different fixation designs and UHMWPE compositions. Our study encompasses more than 20 years of glenoid design methodology from metal-backed glenoids gamma sterilized in air to Hylamer glenoids and finally to modern cross-linked glenoids. The degree and extent of damage seen in shoulders raises concern that changes in fixation design and material composition have not prevented against damage to the bearing surface. A large unsolved issue is the presence of radiolucent lines after TSR implantation [**6**]. The occurrence of radiolucent lines in TSRs at the glenoid implantation site has been reported to range from 30% to 90% of implanted shoulders depending on the study [7]. The link between radiolucent lines and revision due to glenoid fixation failure has been reported to range from 0% of cases to 7% of cases in the literature [**8**]. To solve this issue, metal-backed glenoids were introduced to improve fixation, prevent against radiolucent lines, and stop potential implant loosening. However, poor ingrowth due to eccentric loading lead to early stage loosening of certain metal-backed glenoids [**9**]. The latest advent of using cross-linked UHMWPE in glenoids to reduce wear has not prevented against several damage modes already seen in glenoids. This is especially true for fracture, which can be attributed to crosslinked UHMWPE's reduced reduced fatigue crack growth resistance. One crosslinked implant evaluated in this study was retrieved due to fracture after less than 3 years *in vivo* [**10**]. Additionally damage modes such as scratching, striations, pitting, and surface deformation are present on retrieved glenoids with short *in vivo* lives, suggesting that damage can start shortly after implantation. Overall, changes in glenoid fixation design and UHMWPE material composition, including the addition of cross-linked polyethylene have not prevented against certain damage modes in TSR.

ACKNOWLEDGEMENTS

We would like to acknowledge Gerald Garcia, Suzanne Chou, Isabel Yang, and Jessica Sov for their help with this study.

REFERENCES

[1] Willert HG. *J Bio Mat Res*. 1977; 11:157-46;
[2] Hood RW. *J Bio Mat Res*. 1983; 17:829-42;
[3] McKellop HA. *Biomat*. 2007; 28: 5049-57;
[4] Terrier A. *J Shoulder Elbow Surg*. 2006; Jul/Aug 515-20;
[5] Harman, M, et al. *J Mat Sci-Mater M*, 22(5), 1137-1146, 2011.
[6] Harris W. *Clinical Ortho & related research*. 2001; 393:66-70
[7] Williams, G. and Abboud, *JSES*. 2005; 14(1):122-128
[8] Armstrong, A. and Lewis, G *JBJS Reviews*. 2013; 14(1)
[9] Boileau et al. *JSES*. 2002; 11(4):351-359
[10] Atwood S. *JMBBM* 2011; Oct 1033-1045

NOVEL 3D CO-CULTURE ASSAY REVEALS CANCER CELL MIGRATION DRIVEN BY MATRIX-DEPENDENT FORMATION OF CXCL12 GRADIENTS

Taisuke Kojima (1), Christopher Moraes (2), Stephen P. Cavnar (1), Gary D. Luker (3), Shuichi Takayama (1)

(1) Biointerfaces Institute
University of Michigan
Ann Arbor, MI, USA

(2) Dept. of Chemical Engineering
McGill University
Montreal, QC, Canada

(3) Center for Molecular Imaging, Dept. of Radiology
University of Michigan
Ann Arbor, MI, USA

INTRODUCTION

Breast cancer cells are known to migrate in response to the formation of CXCL12 gradients. Previous work in our labs have demonstrated a microfluidic source-sink model in which patterned cells that secrete ligands (source) and cells that scavenge ligands (sink) set up a gradient on cells culture on a two-dimensional surface, and migrate in response to that gradient [1]. However, this simple assay may not be sufficient to capture the complexity of the in vivo microenvironment, particularly regarding the number of ligand-binding sites available on the device surfaces. In this work, we develop a novel patterning system to pattern cells within an extracellular matrix, creating a model tissue-like environment for studies of directional cell migration.

METHODS

The fabrication process is outlined in Figure 1. 12 mm diameter glass coverslips were coated with a thin layer of polydimethylsiloxane (PDMS: Dow Corning) and cured. The PDMS surfaces were coated with a thin layer of polyacrylamide, following a previous protocol [2]. A thin slab of PDMS is prepared as a stencil, by casting it against an SU-8 mold fabricated according to standard protocols. The stencil is placed in conformal contact with the prepared substrate and oxidized at 100W for 10 min in a plasma oxidizer. Water contact angles were measured for each of the substrate conditions to identify the best processing parameters. To characterize cell adhesion directly to these surfaces, MDA-MB-231 cells were plated on each of the surfaces, and incubated for 24 hours. Devices were washed, and the number of adherent cells counted.

MDA-MB-231 breast cancer cells modified to overexpress CXCR4 ligand-scavenging receptors (CXCR4+), and to secrete Gaussia luciferase (GL), CXCR12-α-GL and CXCR12-β-GL and mixed into pre-polymerized, neutralized collagen solutions (2.5 mg/mL). These hydrogel mixtures were then manually pipetted onto the oxidized regions of the PAA-coated PDMS. The high contact angle prevented mixing. Devices were allowed to incubate for 30

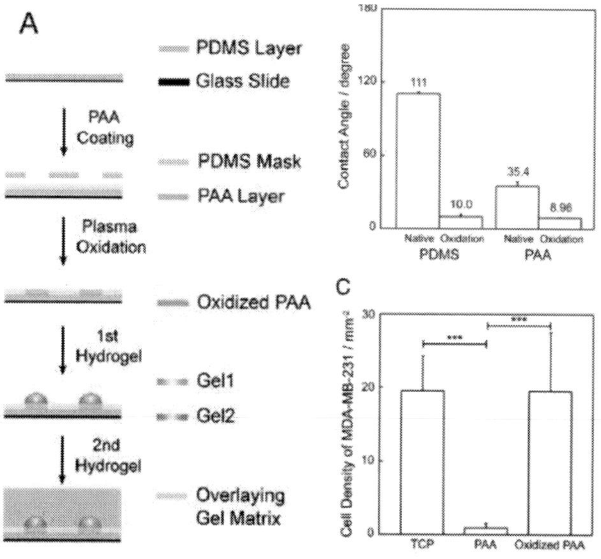

Figure 1. (A) Schematic view of 3D patterning approach. (B) Contact angle changes as a result of surface treatments are used to allow patterning of hydrogels. (C) Cell attachment, and hence, migration, is prevented on unoxidized PAA-coated materials.

minutes in a humidified, 37°C chamber to polymerize the collagen matrix. A second hydrogel layer (either collagen or collagen-matrigel mixture) was then backfilled over the polymerized gels, to create a 3D migration assay, in which cells are unable to migrate along the PAA-coated surfaces. Migration was then monitored over 24 hours.

Binding of the CXCL12 isoforms to the different matrix coatings was assessed by measuring GL photon flux from substrates that had been exposed to the ligands for 24 hours and washed thoroughly. Finite element modelling was conducted in COMSOL to assess the formation of gradients as a result of ligand binding interactions.

RESULTS

Representative images of the migration assay at Days 0 and 1 are shown in Figure 2. Green (CXCR4+) cells migrate into the separating matrix, closing the gap between the two cell populations. Quantifying these migration patterns indicates that migration after 24 hours depends on the isoform of CXCL12 present in the system, and the presence or absence of matrigel in the separating matrix (Figure 3), with matrigel significantly increasing the rates of CXCL12-driven migration, particularly in the case of the β-isoform. This is likely related to the differences in ligand-binding of the isoforms to each of

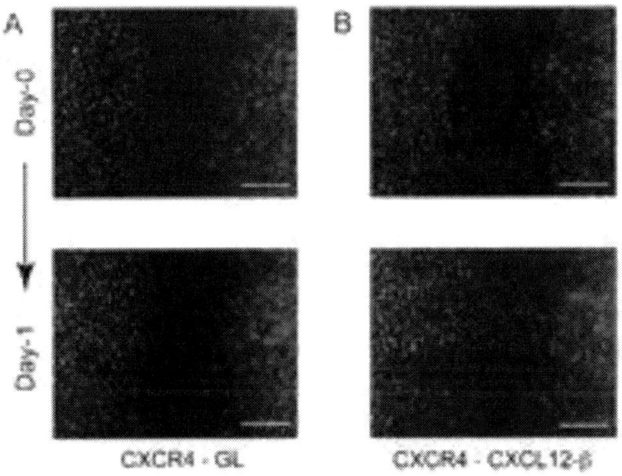

Figure 2. Representative data collected using 3D patterning techniques. After 1 day in culture, CXCR4+ cells (green) migrate through the separating matrix in response to CXCL12 secretions (red; GL = Gaussia luciferase control cells)

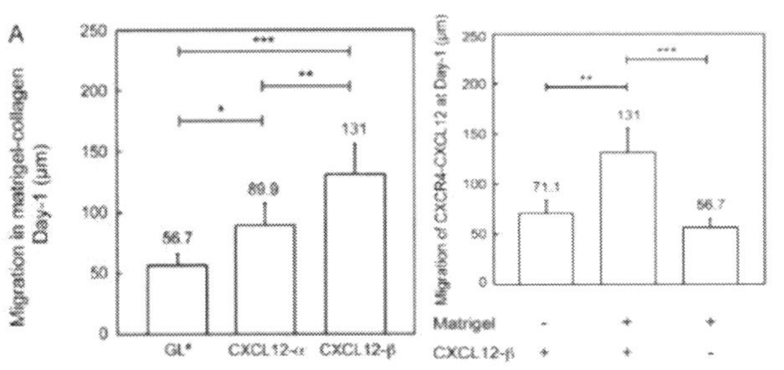

Figure 3. (A) Migration distances of CXCR4+ cells against a control and α and β isoforms of CXCL12 in a collagen-matrigel matrix. (B) Comparison of migration distances of CXCR4+ cells with and without matrigel-supplemented collagen. (* $p < 0.05$, ** $p < 0.01$, *** $p < 0.001$).

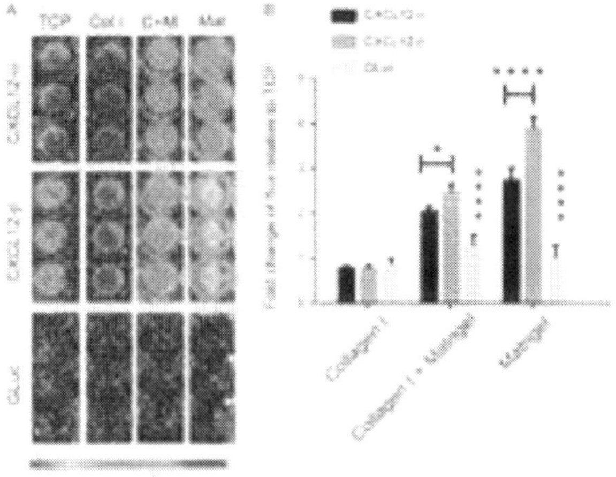

Figure 4. Ligand-binding assay indicates that the addition of matrigel significantly increases the degree of binding, in both α- and β- isoforms of CXCL12.

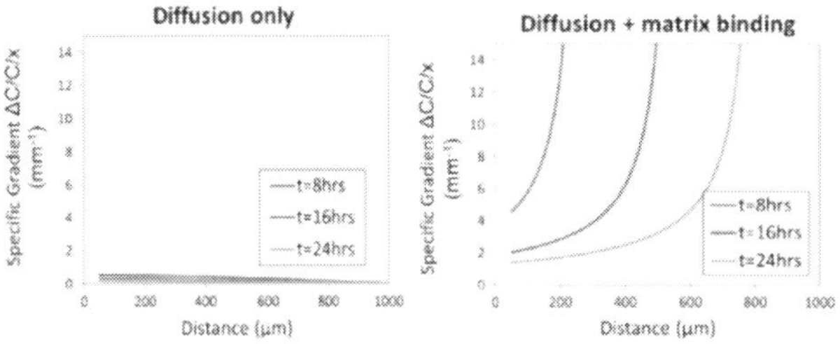

Figure 5. COMSOL Finite element simulations calculating the specific gradient created under (A) Diffusion only, and (B) ligand-binding matrix conditions.

the surfaces (Figure 4). FEA modeling indicates that a dynamic soluble gradient is formed as a result of binding interactions , with a steep specific gradient that evolves over time.

DISCUSSION

The patterning system presented here has two core advantages over existing technologies: (i) broad applicability to a variety of ECM compositions and cell types; and (ii) facile patterning of multiple cell types in defined geometric locations. Using this platform we demonstrated and modeled migration of cancer cells in a physiologically relevant matrix. Although mechanical perturbations in the ECM are a hallmark of cancer, the impact of changes in ECM content on migration are not as well recognized. This work demonstrates how 3D ECM composition may direct the formation of chemotactic gradients that underlie metastasis.

ACKNOWLEDGEMENTS

We gratefully acknowledge: Project funding (NIH, CA136829 & CA170198) and personal funding: Yoshida Scholarship to T.K.; Banting postdoctoral fellowship programs to C.M.; Advanced Proteome Informatics Training Grant No. T32 CA140044 and an NSF Predoctoral Fellowship Project Grant No. F031543 to S.P.C..

REFERENCES

[1] Torisawa et al., *Int.Biol.*,2:680-686, 2010.
[2] Fiddes et al., *Biomaterials,* 31: 315-320, 2010

SB^{3}C2015
Summer Biomechanics, Bioengineering and Biotransport Conference
June 17-20, 2015, Snowbird Resort, Utah, USA
SB^{3}C2015-533

CHARACTERIZATION OF THE FATIGUE LIFE, DYNAMIC CREEP AND MODES OF DAMAGE ACCUMULATION WITHIN MITRAL VALVE CHORDAE TENDINEAE

Gillian M. Gunning (1)(2), Bruce P. Murphy (1)(2)(3)

(1) Trinity Centre for Bioengineering
Trinity Biomedical Sciences Institute
Trinity College Dublin
Dublin 2, Ireland

(2) Department of Mechanical and Manufacturing Engineering
Trinity College Dublin
Dublin 2, Ireland

(3) Advanced Materials and Bioengineering Research Centre (AMBER)
Trinity College Dublin
Dublin 2, Ireland

INTRODUCTION

The mitral valve (MV) is a complex and finely tuned system, composed of four parts working in unison to achieve good valve function: leaflets, annulus, papillary muscles (PMs) and chordae tendineae (CT). The CT act as tethers, preventing the leaflets from inverting during systole. When one or more CT ruptures, or elongates, the leaflets can flail, resulting in mitral regurgitation (MR). MR is the most common form of valvular disease in the US, affecting 1.7% of the population, and 9.3% of people over 75 [1]. CT rupture (CTR) is one of the primary causes of severe MR [2]. In cases of CTR, regurgitation may be sudden and severe, requiring immediate emergency surgery [3]. More than half the reported cases of CTR are defined as 'spontaneous', where there is no prior cause, nor abnormality behind the sudden breaking of the CT [3, 4]. The internal structure of the CT has been investigated in the past; the core consists of a dense pattern of wavy collagen with a few elastin fibres around the edges, surrounded by a mesh of elastin, which itself is encapsulated in an endothelial layer [5]. Previous studies have investigated the mechanical properties of the CT, including: the tensile properties of human CT [6], the role of the elastin mesh in the mechanical response of CT [7], the difference in mechanical response of different types of CT under uniaxial tension [8], and an *in vitro* investigation in which physiological load and strain values were assessed [9]. However, to date there has been no investigation into the relationship between mechanical fatigue and elongation/rupture of CT. As cyclic loading is the principal loading mode of CT *in vivo,* we proposed that fatigue is a source of CTR and elongation. Consequently, the objective of this study was to characterize the relationship between applied stress, fatigue failure and dynamic creep of CT, with a view to ascertaining if mechanical fatigue is the cause of clinically significant CTR.

METHODS

The hearts of 13 pigs were obtained from a local abattoir, and the marginal CT were dissected and frozen in PBS until the day of testing. The CT were dissected such that a large chunk of PM was attached at one end, and a large area of leaflet tissue at the other, so as to facilitate gripping (fig. 1(A)). Smaller CT branches were trimmed from the main chord a short distance away from it, so as not to damage the structure itself. Customized grips were designed for the purpose of fatigue testing these delicate soft tissue samples (fig 1(B)). This enabled us to achieve chordal rupture in regions away from the grip.

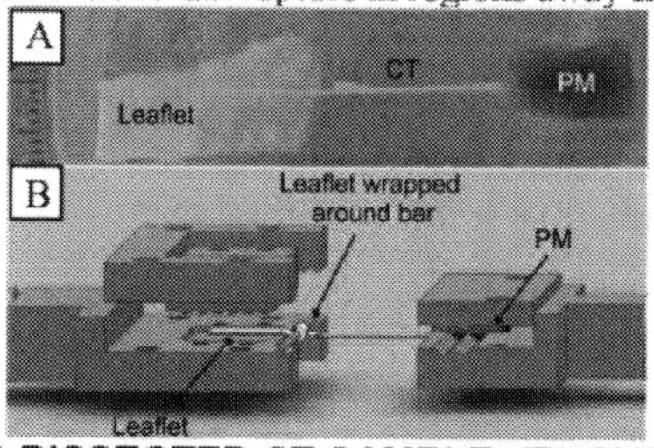

FIGURE 1 (A) DISSECTED CT SAMPLE. (B) CUSTOM GRIPS DESIGNED FOR FATIGUE TESTING SOFT TISSUE.

Both uniaxial tensile tests and fatigue tests were conducted using an M.T.S. Tytron testing machine. Samples were immersed in a water bath at 37°C. Tensile tests (*n=12)* were conducted to failure at 3mm/min. Fatigue testing was conducted at a range of peak stresses (*n*=12 at 15MPa, 10MPa, 7.5MPa and *n*=6 at 5MPa) at a frequency of 1Hz. After testing, specimens were preserved (frozen in PBS) for imaging. Some samples were prepared for histology (Leica TP 1020 tissue processor, wax embedding, and slicing to 6μm thickness). The mounted slices were treated with two different stains: Picrosirius red, and Hematoxylin and Eosin (H&E). Slides were observed under ordinary microscopy; picrosirius red slides were also observed under polarized light microscopy in order to observe collagen fibre orientation. Other samples were prepared for SEM (fixing in 2.5% gluataraldehyde, grading through alcohol, followed by drying). Results were calculated as mean ± standard deviation (SD). Statistical significance between two groups was determined using the Student's t-test, and for more than two groups using a one-way ANOVA followed by a Bonferroni post-test (significant for $p<0.05$).

RESULTS

The ultimate tensile stress of the CT was calculated and was found to be 63.33 ± 36.49MPa. The cycles to failure were plotted against the peak stress of the test, and the stress/number of cycles (SN) curve was drawn (fig 2(A)). The correlation of this relationship was determined using the Pearson coefficient, and found to be highly significant ($p<0.0005$). In order to establish the probability of failure for different peak stress groups, a Weibull plot was constructed for each stress level, describing the probability of failure and the number of cycles for each stress group. Using this data we constructed a plot of the cycles at 50% chance of failure for each stress group (fig 2(B)). This plot was fitted with a power law curve, and extrapolated to give a fatigue limit of 2.43MPa at 10 million cycles, and 0.69MPa at 2.5 x 10^9 cycles (roughly 80 years). The dynamic creep of samples was also assessed. In half of the samples, dynamic creep occurred gradually over the course of the fatigue life. In the other half, there was more than one distinct phase of dynamic creep, separated by a sudden increase in length (fig 2(C)). We measured the slopes of these curves in order to determine the rate of dynamic creep, and plotted the rate against peak stress, finding an exponential relationship between the two. H&E staining revealed a decrease in cellular content in samples which had been tested at low stress. When the picrosirius red stain was observed under polarized light, we found there to be a regular crimp in the collagen (across all groups, the average crimp period was measured as 16.1 ± 1.7 μm). However in the low stress groups there were regions of the CT where no crimp was evident, only straightened or disorganised fibres. SEM examination of the rupture tips revealed several features of rupture, including: longitudinal splitting, micro-fractures (fig 2(D)), tapered tips, blunt tips, splitting into strands, fraying, separation of layers, both crimped and straightened fibres.

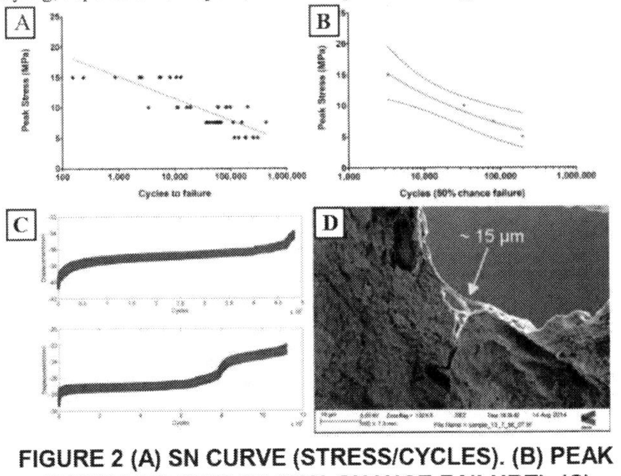

FIGURE 2 (A) SN CURVE (STRESS/CYCLES). (B) PEAK STRESS VS CYCLES (50% CHANCE FAILURE). (C) DISPLACEMENT VS CYCLES (DYNAMIC CREEP PROFILE). (D) MICRO-CRACK VIEWED UNDER SEM.

DISCUSSION

In this study, for the first time, we document the fatigue and dynamic creep properties of CT. Performing these experiments on soft tissue was possible as a new custom made grip was designed that eliminated slippage and premature failure caused by clamping effects. From our data we were able to extrapolate a fatigue limit at 10 million cycles of 2.43 MPa. To place these stresses in context, a finite element study by Rim et al. [10] of an MV with CTR, determined the stress in CT far from the rupture zone to be between 0.3 and 1MPa. However the stresses in CT close to the rupture zone were as high as 3.1MPa. In a finite element study by Prot et al. [11], the stress in healthy marginal CT was found to be ~0.7 ± 0.34 MPa, increasing to ~1.21 ± 0.47 MPa in the pathological state. According to our model the fatigue life of the CT in Prot et al.'s pathological case would be 2 x 10^6 cycles, versus their healthy case of 2.3 x 10^9 cycles, demonstrating that our model will predict significant reductions in fatigue life of diseased MV's. We found that the rate of elongation increases exponentially with increasing peak stress. Again, considering the studies mentioned above, this result would indicate that in diseased MVs where CTs are subjected to higher stresses, prolapsing leaflets may result after a short period of time. Our histological observations showed that in some regions of the low stress CT, the collagen fibres were disorganised or straightened, resembling the observations of Millington-Sanders et al. in elderly patients[5]. This suggests that the dynamic creep of the CT did not happen uniformly over the length of the CT, but rather that damage accumulated in specific regions, gradually developing into the overall elongation of the CT. We observed the tips of the ruptured CT under SEM in order to further understand the failure mechanisms. The rupture tips of human specimens have been described in the literature previously; Scott-Jupp et al. [12] describe the CT of patients that suffered spontaneous CTR, and Icardo et al. [13] describe tips of ruptured diseased CT. The descriptions from both studies are very similar to our findings of fatigue failed specimens. Micro-fractures were also found under SEM observations in our fatigue specimens, the accumulation of which may contribute to the dynamic creep observed. In summary we can derive the following observations from this study: **1.** The probability of CTR may be predicted by applied peak stress, potentially allowing us to predict CTR in patients with pre-existing conditions which increase CT stress levels. **2.** The rate of dynamic creep increases exponentially with increasing applied stress. **3.** We observed that the collagen core loses its structure and elongates in random patches along the length of the CT. This is evidence of damage accumulation, contributing to elongation and eventual CTR. **4.** Evidence of damage accumulation can also be seen in the creep profiles, where creep occurs in more than one distinct phase. **5.** From SEM analysis we have been able to identify modes of failure in fatigue which closely match clinical observations of CTR. In conclusion, we believe that the results we present here strongly suggest that spontaneous CTR and chordal elongation *in vivo* can be caused by mechanical fatigue, and that risk of rupture or elongation becomes much greater as stress in the CT increases.

ACKNOWLEDGEMENTS

The Graduate Research Education Programme and Mr Peter O'Reilly.

REFERENCES

[1] Go, AS et al., *Circulation*, 129:e28-e292, 2014. [2] Enriquez-Sarano, M et al., *Lancet*, 373:1382-94, 2009. [3] Gabbay, U et al., *Int J Cardiol*, 143:113-8, 2010. [4] Oliveira, DB et al., *Br Heart J*, 50:312-7, 1983. [5] Millington-Sanders, C et al., *J Anat*, 192 (Pt 4):573-81, 1998. [6] Lim, KO et al., *Can J Physiol Pharmacol*, 53:330-9, 1975. [7] Lim, KO et al., *Circ Res*, 39:580-5, 1976. [8] Kunzelman, KS et al., *ASAIO Trans*, 36:M405-8, 1990. [9] Ritchie, J et al., *J Biomech*, 39:1129-35, 2006. [10] Rim, Y et al., *Ann Biomed Eng*, 42:139-48, 2014. [11] Prot, V et al., *Biomech Model Mechanobiol*, 8:43-55, 2009. [12] Scott-Jupp, W et al., *J Pathol*, 133:185-201, 1981. [13] Icardo, JM et al., *Int J Cardiol*, 167:1603-9, 2013.

USING SIMULATIONS WITH REALISTIC FIBROUS NETWORK GEOMETRY TO FIND THE ACHIEVABLE RANGE OF MECHANICAL BEHAVIORS OF ELASTOMERIC SCAFFOLDS

James B. Carleton (1), Gregory J. Rodin (1,2), Michael S. Sacks (1,3)

(1) Center for Cardiovascular Simulation, Institute for Computational Engineering and Sciences
University of Texas at Austin
Austin,TX, USA

(2) Department of Aerospace Engineering and Engineering Mechanics
University of Texas at Austin
Austin, TX, USA

(3) Department of Biomedical Engineering
University of Texas at Austin
Austin, TX, USA

INTRODUCTION

Tissue engineers seek to produce tissues that function biologically and mechanically as well as the native tissues they are built to replace. One important step in achieving this goal is designing and manufacturing tissues that mimic the anisotropic, non-linear, large deformation mechanical behavior of the native tissues. This macroscopic behavior is strongly influenced by the evolution of the complex microstructural geometry of the underlying scaffold. Elastomeric scaffolds may be manufactured by the electrospinning process, which produces layers of long-fiber networks (**Fig. 1**).

FIGURE 1. SEM IMAGE: ES PEUU SCAFFOLD [1]

Characterizing this complex network geometry and exploring the effects on mechanical anisotropy has been the focus of recent work [2,3]. It has been found that knowledge of the fiber properties, fiber density, and orientation density function (ODF) are sufficient to predict the bulk mechanical behavior [2]. However, the micro- and meso-scale mechanisms responsible for the behavior remain unclear. Initial simulations based on experimentally derived geometry suggest that the load induced emergence of long fiber paths may be a mechanism contributing to the observed bulk mechanical behavior [3].

The purpose of this study is to develop and utilize computational models, based on realistic fiber geometry, to help understand the mechanisms that translate scaffold fiber network structure into tissue function. We also explore the range of macroscopic material behaviors that are achievable from the domain of producible microstructural geometries and elastomeric fiber properties. Both of these goals are essential first steps in developing the ability to design engineered tissues that mimic native tissue behavior.

METHODS

In the present work, we perform 3-D, non-linear micromechanical simulations of layers of long-fiber networks, which we have generated using a random walk procedure that mimics the electrospinning process (**Fig. 2**).

FIGURE 2. SIMULATED FIBER NETWORK GEOMETRY WITH INTERSECTIONS IN RED (A) TORTUOUS FIBERS. (B) STRAIGHT FIBERS, ISOTROPIC. (C) ALIGNED

This realistic geometric representation is an improvement over currently-used straight fiber representations, since fiber connections and curvature are accurately represented, and the resulting finite element mesh has a similar appearance to real scaffolds. As shown in **Fig. 2**, both the amount of tortuosity and the amount of fiber alignment can be controlled. The networks are generated so that key geometric statistics match those of the scaffold the model is meant to simulate [4]. The generated geometry is used as the basis of a beam element mesh for non-linear, large-deformation, micromechanical simulations.

We use a custom-built parallel finite element code to simulate networks containing large numbers of fibers undergoing large rotations.

RESULTS

The mechanical simulations capture important effects, such as fiber straightening and rotation, and the formation of long-range structures extending into the mesoscale (**Fig. 3**).

FIGURE 3. (A) THE INITIAL SCAFFOLD GEOMETRY. (B) THE SCAFFOLD IS LOADED IN SHEAR

Initially curved fibers in tension straighten and rotate toward the direction of maximum principal strain, forming long bundles of straight fibers that can extend into the mesoscale. These effects cause the slope of the stress-strain curve to increase. The other fibers are in compression and buckle as expected. After the initial transient phase of straightening with large rotations, the fiber geometry approaches a self-similar steady state. The time history of this load-induced anisotropy is further quantified by plotting the evolving probability density function of fiber orientation. Additionally, we demonstrate that a wide variety of behaviors are achievable, from very soft to very stiff, isotropic to highly anisotropic, and linear to strongly non-linear, all solely from alterations in fiber geometry with no change in intrinsic fiber behavior (**Fig. 4**).

As expected, the shape of the macroscopic stress strain curve can be tailored by varying the volume fraction and geometric anisotropy. The initial geometric anisotropy and the initial tortuosity can also be varied independently (**Fig. 5**). It is seen that both of these variables have an effect on the degree of nonlinearity of the stress-strain curve. As expected, more tortuous networks produce more non-linear curves. However, non-linear curves also result from networks of highly aligned but perfectly straight fibers.

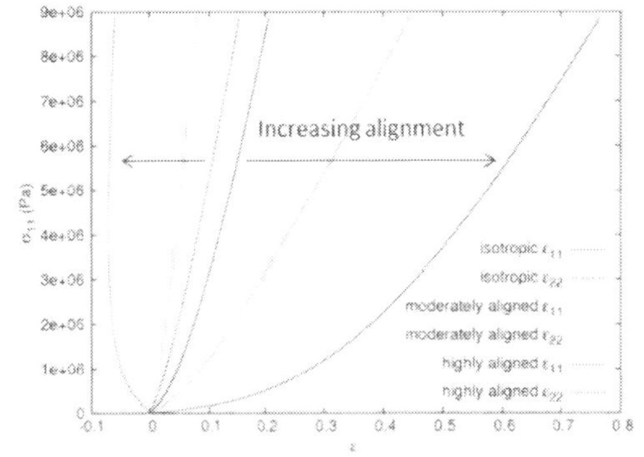

FIGURE 4. STRESS STRAIN CURVES OF SCAFFOLDS WITH DIFFERENT LEVELS OF GEOMETRIC ANISOTROPY.

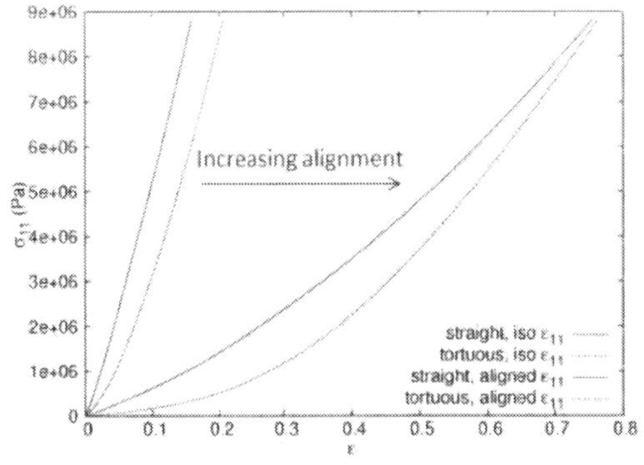

FIGURE 5. STRESS STRAIN CURVES, INDEPENDENTLY VARYING TORTUOSITY AND GEOMETRIC ANISOTROPY.

DISCUSSION

Our mechanical simulations have begun to reveal clues about how the evolution of the loaded microstructure produces the observed macroscopic mechanical behavior. Using realistic fiber network geometry, we are able to investigate the mechanisms responsible for the shape of the curve in ways not possible previously. For example, by independently varying the initial geometric anisotropy and the initial tortuosity, we can separate the effects of each of these variables on the nonlinearity of the stress-strain curve. By simulating a large variety of geometries and loadings, we see that the ODF of the fiber network approaches a steady state that is determined mostly by the loading, rather than the initial geometry of the network. Insights such as these inform macroscale material models that can be used to guide the design of scaffolds and the selection of manufacturing parameters so that the resulting engineered tissues mimic the non-linear mechanical behavior of the native tissues

ACKNOWLEDGEMENTS

The authors would like to acknowledge the funding sources for this project, National Institutes of Health NIH R01 HL089750 and NIH R01 HL68816.

REFERENCES

[1] D'Amore, A et al., *Biomaterials*, 31:5345-5354, 2010.
[2] Courtney, T et al., *Biomaterials*, 27:3631-3638, 2006.
[3] D'Amore, A et al., *JMBBM*, 39:146-161, 2014.
[4] Carleton, J et al., *Acta Biomaterialia*, 12:93-101, 2015.

IN-VIVO VALIDATION OF THE IN SILICO PREDICTED PRESSURE DROP ACROSS AN ARTERIOVENOUS FISTULA

Leonard D. Browne (1), Khalid. Bashar (1, 2), Philip. Griffin (1), Eamon G. Kavanagh (2) and Michael T. Walsh (1)

(1) Centre for Applied Biomedical Engineering Research (CABER), Department of Mechanical, Aeronautical and Biomedical Engineering, Materials

(2) Department of Vascular Surgery, Limerick University Hospital, Dooradoyle, Limerick, Ireland

INTRODUCTION

Hemodialysis is a common treatment for patients at the late stages of chronic kidney disease. In order to optimize this vital filtering process a vascular access capable of supplying high flow rates greater than 600 ml/min is needed. An arteriovenous fistula (AVF) is the access modality of choice but despite this preference it suffers from drastically poor primary patency rates ranging from 40-66% which necessitates urgent improvement.[1, 2]

Inadequate dilation or outward remodelling and venous stenosis which is characterized by aggressive intimal hyperplasia are the main contributors to these poor patency rates. Both types of remodelling are strongly linked to the abnormal hemodynamics arising from fistula creation. Yet, the specific factors which elicit remodelling events have not been unambiguously determined. This is partly due to methodological constraints which make decoupling aspects of the mechanical environment difficult within the *in-vivo* setting as the anastomosis of an AVF is known to exhibit unstable transitional to turbulent behaviour. Therefore, numerical and experimental models are often utilised and favoured to study the interaction between hemodynamic profiles and remodelling.

Numerical studies such as computational fluid dynamics (CFD) are beneficial as they can derive complex hemodynamic profiles such as shear stress and particle distributions from MRI or CT acquisitions of geometry and flow. However, the findings of these numerical models are inextricably linked to their prescribed boundary conditions. The accuracy of these numerical models are heavily reliant on *in-vivo* measurements of flow and pressure and these models will only be deemed relevant if they are able to replicate these conditions and important phenomena of the flow field.

This study serves to document the baseline conditions of the hemodynamic environment within an arteriovenous fistula by providing a quantitative measure of the pressure distribution at the time of surgery. This distribution will also pose a unique challenge to numerical models as it presents an important flow phenomenon and benchmark for numerical models to replicate.

Identifying the range of mechanical forces experienced by cells within this conduit will help determine which hemodynamic factors contribute to vascular remodeling and maturation of the venous segment of an arteriovenous fistula and could potentially be used to assist in clinical risk assessment of maturation or failure

METHODS

Two patients referred to vascular surgery for fistula creation were recruited for this study under the institutional review board approval. Informed consent was obtained from each patient. Pressure measurements were obtained and basic geometric parameters were measured. A representative model of a fistula was created and *in-vivo* measurements and numerical predictions were compared.

***In-vivo* Measurements:** The subject was in the supine position and investigations were conducted upon release of flow and prior to the full closure of the anastomotic site. A pressure catheter is introduced at the anastomosis junction and advanced into the venous segment of the AVF. The catheter is retracted in increments of 1cm as the probe returns toward the junction. Blood pressure was measured over a period of 5 cycles at each increment. Once the catheter is fully retracted it is repositioned and advanced into the proximal artery and the procedure is repeated, likewise for the distal arterial segment as shown in. All values are expressed as means ± SEM for blood pressure values. Student t-tests (two tailed) were used to determine the statistical significance ($P<0.05$).

Numerical Model: The model geometry is taken from a parametric study by Pustjens et al.,[3] whom characterized the geometry of radiocephalic and brachiocephalic fistulae 6 weeks post operatively. The diameter ratio of the parametric model was set equal to 1 and anastomosis angle of 35° to reflect the patient cases. This geometry was discretized with a structured hexahedral mesh with the package ICEM CFD (Ansys). The hexahedral mesh had a characteristic edge length of 0.15mm at the core of the mesh. The number of computational cells within the numerical model of this study was approximately 1.7 million cells

The unsteady 3D incompressible Navier-Stokes equations were solved using the segregated solver in Star CCM+ (CD Adapco) with 2nd order accuracy in space and time, a time step of $5x10^{-4}$ was applied. Rigid boundaries and the no-slip condition are enforced at walls. Blood is modelled as an incompressible Newtonian fluid with a density of 1050 kg/m^3 and a dynamic viscosity of 3.5 mPas. An inlet Reynolds number of 1000 based on mean velocity is applied at the inlet. A flow division of 100:0 was specified at the venous outlet and distal artery outlet respectively. The measurement locations are defined along the central axis of the vessels and are described by a curvilinear abscissa ranging from -6 to +6 cm.

RESULTS

A large pressure drop was measured across the fistulae and coincided with the presence of a palpable thrill (perivascular vessel vibration) in both. The systolic pressure in the vein is 3-4 times lower than that of the arterial segments as shown in Figure 1. The pulse pressure within the vein is also significantly reduced due to this irreversible pressure drop at the anastomosis. Despite the significant reduction in blood pressure across the anastomosis, the level within the vein is elevated beyond its normal physiological range, an inter subject mean of systolic pressure of 37.75 mmHg at 2cm is reported from the anastomosis.

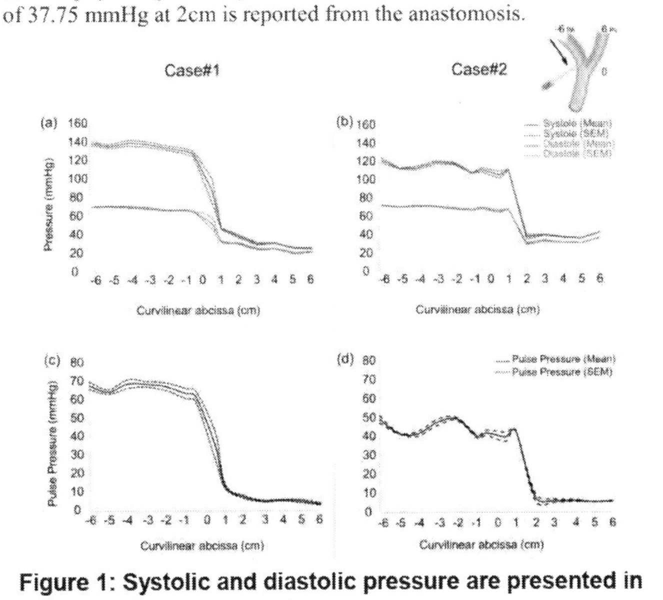

Figure 1: Systolic and diastolic pressure are presented in (a, b) and the corresponding pulse pressures are shown in (c, d) as a function of the curvilinear abscissa

The large pressure drop across the anastomosis coincides with the onset of transitional to turbulent flow and a significant increase in kinetic energy. This flow regime results in high frequency fluctuations of velocity and pressure from the numerical analysis as shown in Figure 2. Large oscillations dominate the pressure drop signal in Figure 2 (b) and can be attributed to the instabilities that arise at the anastomosis and are advected downstream into the vein.

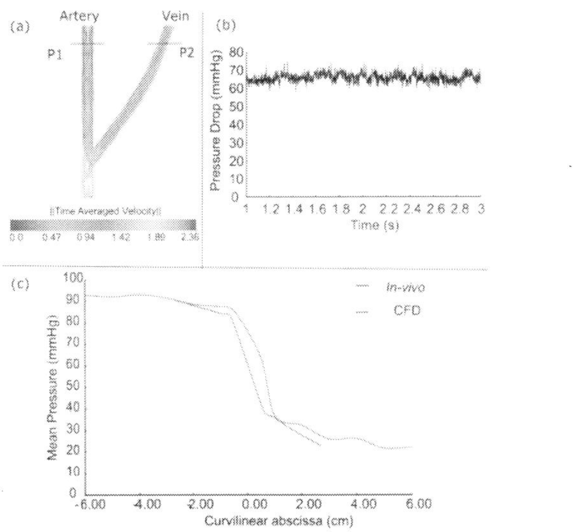

Figure 2: Large oscillations in the pressure drop between P1 and P2 (b) are observed for a Re=1000. Mean pressure is plotted as a function of curvilinear abscissa and compared to *in-vivo* results (c).

DISCUSSION

This study highlighted that the pressure distribution within the fistula of the numerical model was consistent with the distribution obtained from direct intra-operative measurements as shown in Figure 2 (c), with a relative error of 10% between the numerical and *in-vivo* pressure drop of patient case 1 between -2cm and +2 cm. The relative error between the measured and simulated results is expected due to differences between the geometry and flow conditions.

The *in-vivo* measurements of the pressure drop are much higher compared to other studies which incorporate resistance-capacitance (RC) boundary conditions.[4] This poses a question in relation to the validity of such solutions and their subsequent findings. However, discrepancies may be in part attributed to the geometrical differences and the stage of maturation.

We acknowledge the small sample size of this present study as a limitation. Larger sample sizes are needed to observe trends amongst the population of patients with fistulae and are needed to ensure a robust validation of these numerical models.

In conclusion this study has shown that important hemodynamic features can be replicated by CFD and if coupled with non-invasive acquisitions of flow and geometry such as non-contrast and cine phase MRA, critical linkages between remodeling and hemodynamics can be identified if this strategy is applied to a longitudinal analysis of an AVF as it matures. This is an aspect which we are currently addressing and aim to couple with quantifiable measures of vessel calibre. We intend for this knowledge to be directly translated to the clinical setting to improve preoperative and postoperative assessments of maturation along with the overall management of a vascular access.

ACKNOWLEDGEMENTS

The author would like to acknowledge The Irish Research Council for Science Engineering and Technology (IRCSET).

REFERENCES

1 Shenoy, S *J. Vasc. Access* 10:223–32, 2009.
2 Tordoir *et al. Nephrol. Dial. Transplant 22 Suppl 2 88–117, 2007.*
3 Pustjens *et al.* (Masters Thesis) Eindhoven: TU/e, 49 pp , 2013
4 Decorato *et al. Int. j. numer. method. biomed. eng.* 30:143–59, 2014.

STIFFER ARTERIAL WALL ENHANCES AORTIC ANEURYSM FORMATION IN A MOUSE MODEL VIA ELASTASE INFUSION

Zhijie Wang (1), Stephanie Morgan (2), Mark Golob (1), Zhenjie Liu (2), Bo Liu (2), Naomi C. Chesler (1)

(1) Department of Biomedical Engineering, College of Engineering, University of Wisconsin - Madison Madison, WI, USA

(2) Division of Vascular Surgery, School of Medicine and Public Health, University of Wisconsin - Madison Madison, WI, USA

INTRODUCTION

Abdominal aortic aneurysm (AAA) is a progressive vascular disease with high lethality that preferentially affects the elderly[1]. It remains unclear why the elderly are more susceptible to AAA, but age-related changes of the vascular extracellular matrix (ECM) likely play a role. With natural aging, elastin is degraded, collagen fibers become stiffer and less organized and smooth muscle cells (SMC) become depleted through apoptosis. Here we hypothesized that stiffened ECM associated with aging enhances AAA formation.

To test this hypothesis, we employ a transgenic mouse strain that has defects in collagen degradation. The homozygous mutant mouse ($Col1a1^{R/R}$) exhibits excessive arterial collagen accumulation, arterial stiffening and accelerated aging[2, 3]. In this study, we induced AAAs in mutant ($Col1a1^{R/R}$) and wildtype ($Col1a1^{+/+}$) mice by elastase infusion. Mechanical and biological changes in the aorta prior to and after AAA induction were measured. We found that AAA formation was enhanced and inflammatory cell infiltration was more pronounced in the mutant mice, which had stiffer aortas before AAA induction.

METHODS

Animal preparation. All procedures were approved by the University of Wisconsin – Madison Institutional Animal Care and Use Committee. Breeding pairs of $Col1a1^{tmJae}$ mice on the B6/129 background were obtained from Jackson Laboratory (Bar Harbor, ME). Genotyping by polymerase chain reaction was performed on DNA extracted from tail biopsies, as previously described[3].

AAA induction. 7-8 mo. old Col1a1 homozygous mutant mice ($Col1a1^{R/R}$) and homozygous wild-type mice ($Col1a1^{+/+}$) were randomized and subjected to elastase-mediated aneurysm induction as described previously[4]. Fourteen days post infusion, animals were euthanized and aneurysms were harvested. Age-matched, intact mice were used as baseline controls. To quantify AAA progression, the maximal external diameter of the manipulated segment of infrarenal aorta was measured by a digital caliber at the time of sacrifice.

Isolated vessel mechanical testing. After harvest, the aorta was tested in a mechanical testing system as previous reported[3]. Each aorta was mounted on 410-μm-diameter glass cannulas in a vessel chamber [Living System Instrumentation (LSI), Burlington, VT]. Care was taken to avoid vessel leaking during harvest and mounting. Then, the aorta was stretched axially 150%, at the approximate in vivo stretch ratio, to prevent buckling at higher pressures and tested at this fixed length. Calcium- and magnesium-free PBS was used for both perfusate and superfusate to obtain arterial mechanical properties in a SMC passive state. The perfusate was supplied via a steady flow pump (LSI; Burlington, VT) to achieve static pressurization at 90-120 mmHg. The vessels were allowed 30 minutes to equilibrate and then preconditioned for 10 cycles at 0.01 Hz before data recording. Superfusate was continuously circulated and maintained at 37°C, with pH = 7.4. Pressure transducers were placed in-line next to the cannulas to record pressures on both sides of the vessel. The average pressure, P, was used for all material and mechanical property calculations.

Pressure-outer diameter (P-OD) data were recorded simultaneously by IonWizard software (Version 6.0, IonOptix, Milton, MA) using in-line pressure transducers at an acquisition frequency of 1Hz; and a CCD camera (IonOptix, Milton, MA) connected to an inverted microscope (Olympus, Center Valley, PA) to measure OD at 4× magnification with an acquisition frequency of 1.2Hz. Pressure stair steps of 90, 95, 100, 105, 110, 115 and 120 mmHg were applied for 30 seconds each and P-OD data were recorded. Vessel plastic deformation was checked at a final 5 mmHg step; no plastic deformation was observed.

Mechanical properties calculation. The arteries were assumed to be at a fixed length, homogeneous, and incompressible[5]. The overall stiffness was represented by the slope of the P-stretch curve, where circumferential stretch (λ) at each pressure was calculated as OD/OD_5, where OD_5 is the OD measured at 5 mmHg.

Assuming conservation of mass and no axial extension, the wall thickness (h) as a function of OD, was calculated as:

$$h=\frac{1}{2}\left(OD-\sqrt{OD^2-OD_{120}^{\ 2}+\left(OD_{120}-2h_{120}\right)^2}\right)$$

where OD_{120} and h_{120} are the OD and h measured optically at 120 mmHg. Then elastic modulus was derived from the slope of the Piola-Kirchhoff stress – Green strain (S-ε) curve[5].

Finally, we calculated the axial stretch (λ_{axial}) of the aorta as ($L_{in\ vivo}$-L_{free})/ L_{free}, where $L_{in\ vivo}$ is the in vivo length and L_{free} is the free length after the harvest of aorta.

Histology and immunohistochemistry. After mechanical testing, aortas were fixed in 10% formalin and preserved in 70% ethanol. The vessels were then embedded in agar gel, sectioned, and stained with Picro Sirius red, Van Gieson Verhoeff and H&E. The distribution of collagen subtypes in the aorta was quantified by polarized microscopic image of the Picro Sirius red staining as described previously[5].

Immunohistochemistry was performed as described previously. The macrophage infiltration was detected using primary antibodies to mouse MOMA2 (Abcam, Cambridge, MA)[6].

Statistics. All results are presented as mean±SE. Unpaired student t test was used. All p-values were two-sided and $P<0.05$ was considered as statistically significant.

RESULTS

After elastase infusion, mutant mice (Col1a1$^{R/R}$) developed more severe aneurysms than wild-type (Col1a1$^{+/+}$) mice, which is evidenced by significantly increased AAA maximal diameter ($p<0.05$, Fig. 1A&B). MOMA2 staining for monocytes and macrophages demonstrated that the mutant mice were more inflamed ($p<0.05$, Fig. 1C&D).

From the mechanical measurement, we found that at baseline before aneurysm induction, the overall stiffness of the aorta was significantly higher in the Col1a1$^{R/R}$ mice ($p<0.05$, Fig. 2). AAA development led to a trend of increased stiffness in both mouse strains; we did not observe different stiffness in the aneurysm wall between the mouse strains (Fig. 2). Elastic moduli calculation showed the same trends of changes between the groups as found in the overall stiffness. Moreover, the axial stretch (λ_{axial}) was not different at the baseline. AAA development led to a decrease in λ_{axial} but the decrease was less in the aneurysm of Col1a1$^{R/R}$ mice ($p<0.05$, Fig. 3).

We also examined changes in ECM. We observed marked elastin degradation in the aneurysm wall of both mouse strains ($p<0.05$) as expected. Prior to AAA induction, type I collagen percentage was higher and type III collagen percentage was lower in the Col1a1$^{R/R}$ aorta ($p<0.05$). However, such differences were absent between the AAAs of these mouse strains.

DISCUSSION

In this study, we report for the first time that exacerbated AAA development was associated with stiffer arterial wall prior to the AAA induction by elastase infusion. The use of a mouse model with collagen abnormality that mimics the effects of aging[2] suggests that aging-associated ECM changes may be responsible for accelerated AAA progression and enhanced inflammation response.

Our results indicate that the stiffer aorta in the mutant mice may be a consequence of increased type I and decreased type III collagen, because type I collagen fiber is thicker and stiffer than the type III fiber. Similarly, the lack of collagen subtype difference after AAA formation explains the lack of difference in AAA wall stiffness between the strains. In addition, we also observed different longitudinal remodeling between the strains as evidenced by different λ_{axial} after AAA formation.

In summary, we found that aging-associated stiffer arterial wall enhanced AAA formation in an elastase-infusion model of AAA in the mouse. Further investigation of the role of ECM changes in inflammatory cell infiltration will deepen our understanding of aging-associated AAA formation.

ACKNOWLEDGEMENTS

The study is supported by NIH 1R21HL121493 (NCC).

REFERENCES

[1] Thompson RW et al., *Curr Problems in Surg.* 39(110), 2002.
[2] Vafaie F et al., *Aging cell.* 2013.
[3] Wang Z et al., *Biomech Model Mechanobiol.* 11(279), 2011.
[4] Morgan S et al., *ATVB.* 32(2493), 2012.
[5] Wang Z et al., *PloS one.* 8:e78569, 2013.
[6] Yamanouchi D et al., *J Vasc Surg.* 56(455), 2012.

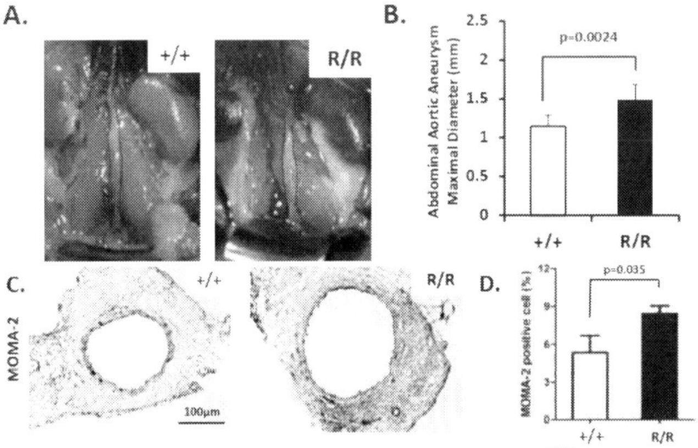

Figure 1. (A,B) AAA induction is exacerbated in Col1a1$^{R/R}$ mice. (C,D) MOMA-2 staining is more intense in the Col1a1$^{R/R}$ mice.

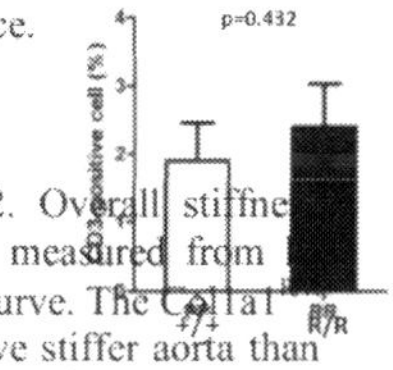

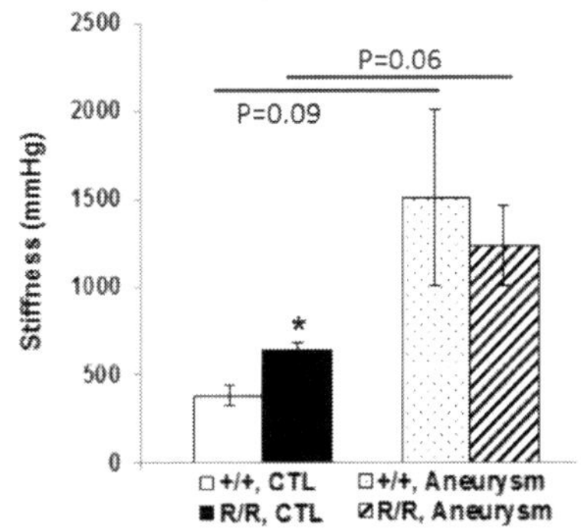

Figure 2. Overall stiffness of aorta measured from P-stretch curve. The Col1a1$^{R/R}$ mice have stiffer aorta than the Col1a1$^{+/+}$ mice before AAA induction. However, there is no difference in the aorta stiffness after AAA development. *$p<0.05$ vs. Col1a1$^{+/+}$ CTL.

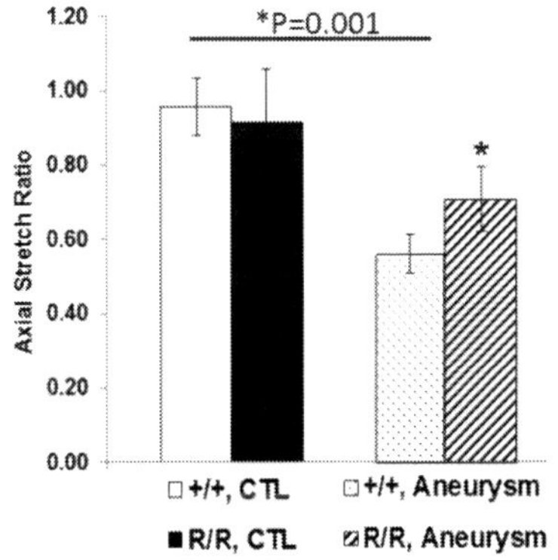

Figure 3. Changes in the axial stretch (λ_{axial}) of the aorta. There is no difference in λ_{axial} before AAA induction. However, after AAA development, the Col1a1$^{R/R}$ mice have large λ_{axial} than the Col1a1$^{+/+}$ mice. *$p<0.05$ vs. Col1a1$^{+/+}$ Aneurysm.

RHOA MEDIATED EFFECTS OF STATIN THERAPY ON ENDOTHELIAL CELLS: A COMPARISON TO FASUDIL AND LATRUNCULIN A

Melissa L. Dick (1,2), Katherine N. MacDonald (1), Jean-Claude Tardif (2), Richard L. Leask (1,2)

(1) Department of Chemical Engineering
McGill University
Montreal, QC, Canada

(2) Research Center
Montreal Heart Institute
Montreal, QC, Canada

INTRODUCTION

Statin drugs are one of the most widely prescribed drugs worldwide due to their cholesterol lowering capabilities. Statins block a key enzyme in the cholesterol biosynthesis pathway, 3-hydroxy-3-methylglutaryl coenzyme A reductase, inhibiting cholesterol formation and positively benefitting patients. However, although other available pharmaceuticals can lower patient cholesterol levels by comparable amounts, the patients taking the statin drugs fair better overall. This has led researchers to believe that statins exhibit pleiotropic effects; that is, unintentional positive effects.

Many important cellular processes and functions, such as cell shape regulation and motility, are regulated within the cholesterol biosynthesis pathway and therefore may be influenced by statin therapy. Indeed, research has shown that statins have a positive effect on endothelial cells (ECs) [1-3]. ECs line the entire cardiovascular system in a confluent monolayer and are exposed to a constant wall shear stress from blood flow. In addition, our lab has shown that statin therapy influences both EC shape and cytoskeletal structure, leading to a rounded morphology and fragmented cytoskeleton [4].

The goal of this research is to better understand how statins affect ECs through the cholesterol biosynthesis pathway by comparing their effects with other known cytoskeletal disruptors. Fasudil is a known cytoskeletal disruptor that works through direct inhibition of Rho-Kinase, a downstream effector of RhoA. Latrunculin A is a second cytoskeletal disruptor that instead works by preventing actin fiber reassembly. F-actin polymerization is halted by latrunculin A binding to G-actin monomer units and preventing their addition to the long polymeric chain. By comparing the effects of all three treatments on EC shape, cytoskeletal structure, and protein expression, we will gain further insight into the mechanism of statin drugs and their pleiotropic effects on ECs.

METHODS

Human abdominal aortic endothelial cells (HAAECs) were grown using basal PromoCell media with SupplementMix, 10% fetal bovine serum, and 1% penicillin-streptomycin. The ECs were seeded onto gelatin-coated 24-well plates at passage 5, with a concentration of 1.25 x 10^6 cells/mL. Cells were allowed to grow for 24 hours to obtain a single confluent monolayer.

Once confluent, EC media was replaced with fresh media spiked with either 10μM simvastatin, 10μM fasudil, or 0.025 μg/mL latrunculin A and allowed to grow for a further 24 hours, at which point the ECs were fixed with 1% paraformaldehyde and stored at 4°C until analysis.

EC shape index, a method to quantify the shape of a cell, was determined by staining ECs with a 4% crystal violet solution and imaging with an inverted light microscope. Images were analyzed using a MATLAB® program to determine the cell area, perimeter, and ultimately, shape index.

Cytoskeletal arrangement, F-actin in particular, was analyzed by staining the ECs with phalloidin-conjugated Alexa Fluor® 647 and was then imaged using confocal microscopy. Confocal microscopy was also used to analyze relative protein intensity by applying primary antibodies for both RhoA and LIM-Kinase to the fixed ECs. Using ImageJ, fluorescence intensity was measured as an average over the area of the cell for consistency.

RESULTS

Simvastatin causes a decrease in EC RhoA and LIM-Kinase protein expression. After 24 hours, treatment with 10μM simvastatin led to a visual decrease in the fluorescence intensity of both RhoA and LIM-Kinase, **Figure 1**. Statistical analysis shows that there is a significant decrease in the protein expression of RhoA and LIM-Kinase after 24 hours, **Figure 2**.

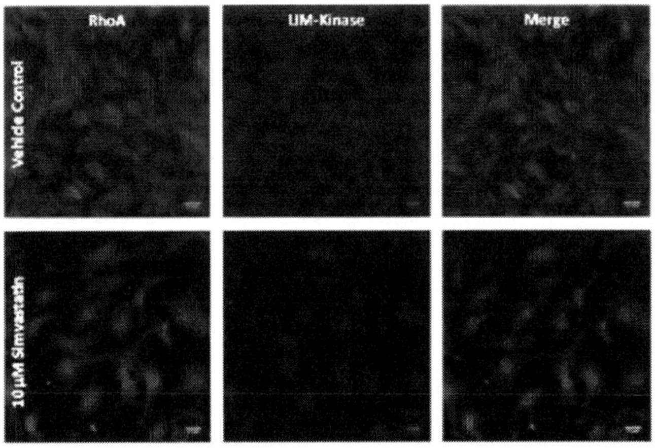

Figure 1. Endothelial cells fluorescently labeled for RhoA (green) and LIM-kinase (blue) and imaged using confocal microscopy at 20x magnification. Scale bars represent 25 μm. Images are of endothelial cells grown in static culture on glass coverslips.

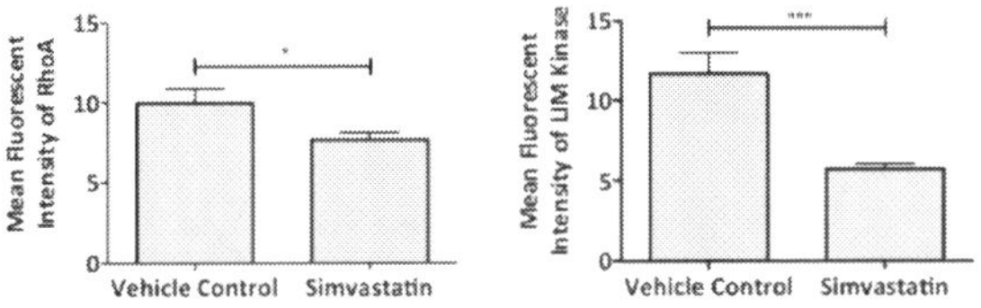

Figure 2. Fluorescence intensity measurements of endothelial cells labeled with RhoA (left) and LIM-kinase (right). * $P < 0.05$, ***$P < 0.001$ (two-tailed t-test). $N = 3$; 3 representative images captured from each condition for all 3 repeats. These values are reported as the mean ± standard error.

Fasudil and Latrunculin A influence EC RhoA and LIM-Kinase protein expression. After treatment with fasudil for 24 hours, EC protein expression, on average, was lower than the control for both RhoA and LIM-Kinase. Oppositely, treatment with latrunculin A for 24 hours caused an increase in protein expression for both RhoA and LIM-Kinase.

Simvastatin-induced EC cytoskeletal reorganization and morphology change closely resembles that of Fasudil. Visually, after 24 hours of fasudil treatment, ECs appeared rounded, similar to what was observed for simvastatin treated ECs. Indeed, shape index analysis showed a significant increase in the EC shape index when compared to an untreated control, consistent with a rounded morphology.

When fasudil-treated ECs stained for F-actin were analyzed using confocal microscopy, the cytoskeleton appeared to be fragmented and disorganized.

DISCUSSION

Our results suggest that statin therapy influences ECs through RhoA, one of many proteins implicated in the cholesterol biosynthesis pathway. Fasudil, a known inhibitor of Rho-Kinase, a downstream effector of RhoA, causes ECs to become rounded with a fragmented cytoskeleton. We have previously shown in our lab that simvastatin causes a similar rounding and cytoskeletal disorganization of ECs [4].

We did not observe any similarities between the response of ECs to simvastatin and the response with latrunculin A; in fact, the opposite trends were observed. Latrunculin A is a known actin disruptor, but instead of blocking RhoA in the cholesterol biosynthesis pathway like fasudil, it instead inhibits F-actin reassembly by binding to G-actin monomer units and preventing their addition to the long polymeric F-actin chain. The opposite trend observed suggests that statins and latrunculin A do not function along the same pathway.

We are currently investigating the effect of pretreatment with inflammatory cytokines for all 3 of the drugs. This will simulate an unhealthy, atheroprone EC layer likely to develop an atherosclerotic lesion to help us better understand the impact of statin drugs on those already suffering from cardiovascular disease. In addition, due to the constant exposure of ECs to wall shear stress from blood flow, these studies will be repeated in three-dimensional flow chambers to better represent the hemodynamics felt by ECs *in vivo*.

ACKNOWLEDGEMENTS

The authors would like to thank Paul Jonak for designing the MATLAB® shape index program, Lisa Danielczak for helping with cell culture, and the rest of the Leask Lab Group for their suggestions and support.

Financial support for this project has been graciously received from the National Sciences and Engineering Research Council of Canada (NSERC RGPIN-261938-08), the Canadian Institutes of Health Research (CIHR MOP-119292), the McGill William Dawson Fund, the McGill Engineering Doctoral Award (MEDA), the Eugenie Ulmer Lamothe Fund (EUL), the Emil Nenniger Fellowship and the Vadasz Doctoral Fellowship.

REFERENCES

[1] Rossi, J., Jonak, P., Rouleau, L., Danielczak, L., Tardif, J.C., and Leask, R.L. (2011). Ann Biomed Eng *39*, 402-413.

[2] Rossi, J., Rouleau, L., Emmott, A., Tardif, J.C., and Leask, R.L. (2010). Eur J Pharmacol *649*, 268-276.

[3] Rossi, J., Rouleau, L., Tardif, J.C., and Leask, R.L. (2010). Life Sci *87*, 92-99.

[4] Dick, M., Jonak, P., and Leask, R.L. (2013). Life Sci *92*, 859-865.

SUCTION EFFECT PRODUCED BY ACTIVE CONTRACTION OF COLLECTING LYMPHATIC VESSELS FACILITATES LYMPHATIC FILLING

Samira Jamalian (1), Mohammad Jafarnejad (1), Christopher D. Bertram (2), David C. Zawieja (3) Michael J. Davis (4), James E. Moore Jr. (1)

(1) Department of Bioengineering
Imperial College London
London, United Kingdom

(2) School of Mathematics and Statistics
University of Sydney
New South Wales, Australia

(3) Department of Systems Biology and Translational Medicine
Texas A&M Health Science Center
Temple, TX, USA

(4) Department of Pharmacology and Physiology
University of Missouri School of Medicine
Columbia, MO, USA

INTRODUCTION

Lymph formation occurs when interstitial fluid enters the lymphatic capillaries. These non-contractile vessels (also known as initial lymphatics) have very low pressures in the range 0−1 cmH_2O. Lymph then is transported from initial lymphatics to higher-pressure collecting lymphatics (10 cmH_2O in rat mesentery) and eventually reaches larger lymphatic ducts.

In contrast to the blood circulation system there is no central pump in the lymphatic system. Thus, transport of lymph (a viscous fluid) against gravity and pressure difference is accomplished by contraction of lymphatic muscle cells embedded in the walls of contractile lymphatic vessels, with the help of bi-leaflet valves throughout the lymphatic network that prevent backflow. Movements outside the lymphatic vessel such as skeletal muscle contraction help this process as well.

Observations in bat wings suggest that contraction of collecting lymphatic vessels can facilitate lymph formation by generating a suction pressure [1, 2]. In this study we measured the suction pressure in isolated collecting lymphatic vessels from rat mesentery. We define the suction pressure as the pressure inside the lymphangion when it transiently drops below the inlet pressure during muscle relaxation. This results in lymphangion filling via suction through the upstream valve during diastole (the suction effect). We also looked at changes in suction pressure caused by the resistance of non-contracting elements upstream of the collecting lymphatic vessel (similar to initial lymphatics upstream of collecting lymphatic vessels), by including a high resistance capillary tubing in the flow loop. We then used our computational model of lymphangions in series to explain how different flow parameters affect the degree to which pressure transiently falls below p_{in}.

METHODS

Mesenteric lymphatic vessels with four to five lymphangions were isolated from male Sprague-Dawley rats and cannulated with glass micropipettes. The cannulated vessels were then transported to an inverted microscope. The pressures p_{in} and p_{out} of the reservoirs beyond the pipettes were controlled with servo controllers. Vessels were imaged continuously throughout the experiment for diameter tracking. Pressure downstream of the first valve was measured using a servo-null pipette, as previously described [3]. Animal protocols were approved by the Institutional Animal Care and Use Committee at the University of Missouri-Columbia and conducted in accordance with the NIH Guide for the Care and Use of Laboratory Animals. A schematic view of the experimental setting and the equivalent lumped-parameter model are shown in Figure 1.

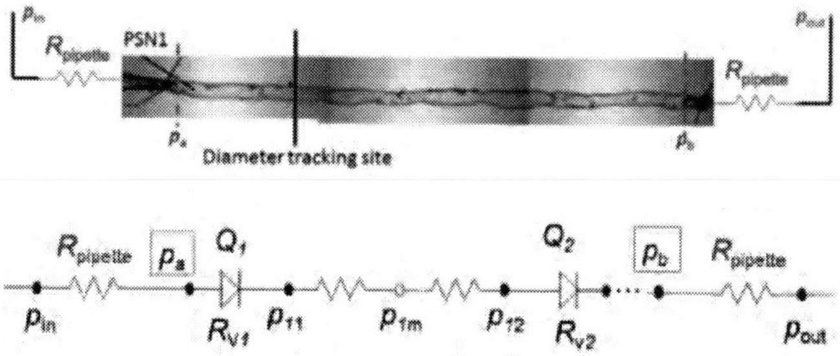

Figure 1. Top: cannulated lymphatic vessel segment with six valves. PSN1: servo-null pressure measurement site. $R_{pipette}$: resistances of upstream and downstream cannulating pipettes (4.0). p_a: pressure downstream of the inlet cannulating pipette. p_b: pressure of the outlet cannulating pipette. p_a and p_b can be calculated using our computational model. Bottom: schematic view of the equivalent lumped-parameter model (only the first lymphangion is fully represented). Q: flow rate. R_v: valve resistance. p_{11}, p_{1m}, and p_{12}: pressure calculation points in one lymphangion.

In the computational model of lymphangions in series [4], the response of each lymphangion to pressure load is determined by its active contraction as well as passive behavior of the vessel wall. The active contraction is implemented via a time-dependent and diameter-dependent active tension profile, with periods of no contraction between two subsequent contractions. The passive behavior is modeled using a non-linear pressure-diameter curve obtained experimentally [5]. The valves are modeled using a sigmoidal function [4]. The model accounts for the pressure drops due to the resistance of inlet and outlet cannulating pipettes ($= 4.0\times10^6$ g/cm^4 s), including $R_{capillary}$ ($= 4.8\times10^7$ g/cm^4 s). This enables us to calculate the pressures upstream of the first valve (p_a), and downstream of the last valve (p_b). In the cases presented here, all the lymphangions contracted synchronously (as observed in our series of experiments, though asynchronous contractions are biologically possible as well).

RESULTS

We quantified the suction pressure amplitude (suction AMP: pressure difference between baseline pressure and minimum servo-null pressure during diastole) from experimental measurements. For the six-valve segment presented in Figure 1, following addition of $R_{capillary}$, suction AMP increased from 0.85 cmH_2O to 2.4 cmH_2O but peak pressure during contraction remained almost unchanged (less than 3% decrease). Minimum and maximum diameter (and thus amplitude of diameter change) were not affected by the presence of $R_{capillary}$ (< 0.5% change). The period of diameter decrease did not change, but diastolic filling took longer when $R_{capillary}$ was included. Specifically, the time required for pressure to rise by half of the suction AMP from its minimum value increased 3.3×. The vessel took 1.6× longer to expand by half of the diameter amplitude from its minimum diameter (Figure 2). Experimental measurements further demonstrated that the effect leading to increased suction AMP was reversible; that is, when $R_{capillary}$ was removed, suction AMP decreased to its initial values.

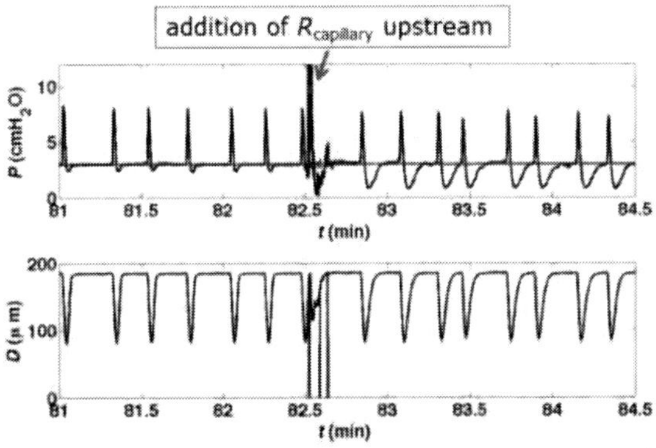

Figure 2. Pressure and diameter tracings of the vessel in figure 1. Capillary tubing with high resistance ($R_{capillary}$) was added at t = 82.5 min. p_{in} and p_{out} are constant and equal (red). Peak pressure did not change with addition of $R_{capillary}$. However, pressure dipped lower below the inlet pressure when $R_{capillary}$ was added, increasing suction AMP.

Model predictions for a chain of four lymphangions in series demonstrated that suction occurs during vessel expansion, where the positive transmural pressure ($\Delta p_{tm} > 0$) pops the vessel back open, providing the energy required for the suction effect. Diameter increases during the muscle relaxation phase and the lowering of pressure in the first lymphangion ($p < p_a$) opens the first valve. Under positive transmural pressure, vessel expansion continues during the refractory period as suction continues to result in diastolic filling. For p_{in} = p_{out} = 5 cmH_2O, when $R_{capillary}$ was included, p_a and thus lymphangion pressures dipped considerably lower below p_{in} during relaxation, creating a 1.7× larger suction pressure amplitude (suction AMP) to overcome the additional resistance upstream. Additionally, the recovery times from minimum pressure to baseline values were longer because of the additional impedance to diastolic filling (Figure 3). These findings are in agreement with our experimental measurements presented in Figure 2.

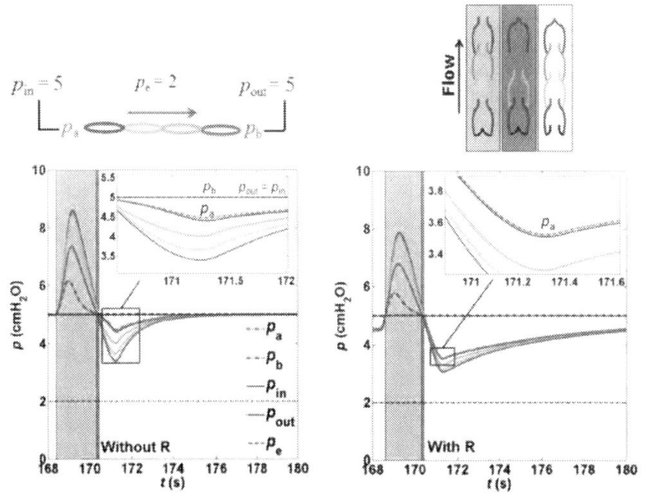

Figure 3. Model predictions of pressures in a chain of four lymphangions in series when $p_{in} = p_{out}$ = 5 cmH_2O, external pressure p_e = 2 cmH_2O. Maximum muscle tension M_0 = 100 dyn/cm during one contractile cycle. The plots are color-coded as presented in the schematic of four lymphangions in the top left corner. Three different valve states were observed during each contractile cycle; these states are shown in the top right corner and correspond to the background color in the pressure plots.

DISCUSSION

Pressure measurements in isolated lymphatic vessels from rat mesentery together with predictions of our computational model of lymphangions in series demonstrated that lymphangion filling occurs via suction during early diastole. The required energy for this effect is provided by vessel expansion occurring due to positive transmural pressure. What makes the suction effect special is that prior to it occurring, pressure in the first lymphangion is higher, and then is required to dip down below the baseline level to open the first valve and allow diastolic filling. Suction AMP decreased when the vessel pumped against an adverse pressure difference. Suction also diminished under negative transmural pressures, where vessel expansion is more difficult. Pressure measurements in segments further downstream confirmed that the suction effect is present in all the lymphangions along the vessel. A considerable increase in suction AMP following addition of a high-resistance capillary tubing upstream, suggested that suction can facilitate lymph flow from low-pressure initial lymphatics to high-pressure collecting lymphatics [6].

ACKNOWLEDGEMENTS

This research was funded by NIH grants R01-HL094269, U01-HL123420, and R01-120867. JEM gratefully acknowledges the support of The Royal Society and The Royal Academy of Engineering.

REFERENCES

[1] Hogan, RD. *The initial lymphatics,* Williams & Wilkins, 1981.
[2] Nicoll, PA et al., *Microvascular Research*, 15:257-258, 1978.
[3] Davis, MJ et al., *Am J Physiol*, 301(1):H48-60, 2011.
[4] Jamalian, S et al., *Am J Physiol*, 305(12):H1709-H1717, 2013.
[5] Bertram, CD et al., *Comput Methods Biomech Biomed Eng*, 17(14):1519-1534, 2014.
[6] Guyton, AC et al., *Lymphology*, 13(4):173-176, 1980.

MECHANICAL DECOUPLING OF NUCLEI FROM THE CYTOSKELETON INDICATE MECHANOSENSITIVITY IN A MYOCARDIOCYTE PATHOLOGY MODEL

Benjamin Seelbinder (1), Sarah Calve (1), Corey P. Neu (1)

(1) Biomedical Engineering
Purdue University
West Lafayette, Indiana, USA

INTRODUCTION

Cardiomyocytes (CMs) are known to be sensitive to their mechanical environment. The most beneficial parameters in terms of functional CM integrity are achieved around the native heart stiffness (~10 kPa) with deviations on both softer and stiffer substrates [1]. This phenomenon plays an important role in disease. In post-infarct (PI) patients, ischemic tissue can stiffen due to fibrosis causing fatal conduction deficits and arrhythmias. This pathology has been linked to the down-regulation of the sarcoplasmic reticulum (SR) Ca^{2+}-ATPase gene, *Serca2*. The coded protein pumps Ca^{2+} back in to the SR after contractions [2]. Reduced *Serca2* levels result into prolonged Ca^{2+} peaks that perturb normal beating activity. Arrhythmias can also be observed in patients carrying mutations in genes of structural proteins, including lamin A/C [3], desmin [4] and nesprins [5], which are thought to transfer mechanical cues from the cell periphery into the nucleus.

Manifestations of similar pathologies in stiffened heart tissues of PI patients and severed connections to the nucleus together suggest a role of the nucleus as a mechanosensitive mediator in CMs. However, the pathways involved remains elusive. Here, we investigated the role of specific connections from the cytoskeleton to the nucleus in mechanosensitive processes. We targeted nesprins which span through the outer nuclear membrane and bind to SUN proteins through their KASH domains, which in turn span through the inner nuclear membrane and bind to the nucleoskeleton [6]. Four different genes, *Syne1-4*, are known to code for a variety of nesprin isotypes that bind to different cytoskeleton components. To reach complete nuclear decoupling, all 4 nesprin genes were targeted for simultaneously siRNA-mediated knock down (KD). An *in vitro* CM model was developed that recapitulates arrhythmias and *Serca2* down-regulation in dependents of substrate stiffness and the effect of combined nesprin KD on gene expression and beating activity was observed. The expression of the lamin A/C coding gene *Lmna* served as another biomarker as it has been shown to scale with tissue stiffness [7].

METHODS

CMs were isolated from embryonic mice 16.5 days post conception using cold trypsin (0.25%) digestion and cultured on polydimethylsiloxane (PDMS). Thin layers of PDMS (500 µm) were poured to enable imaging using either Sylgard® 184 or 527 formulation to reach intended substrate stiffness (Table 1). Formulation per se was not responsible for observed cell behavior as stiffness-mediated effects were observed on Sylgard® 527 as well (54.7 kPa ± 5.8, ratio 1:3), however, in a lesser extent due to lower stiffness. Stiffness was determined via AFM (Table 1) using a spherical borosilicate glass tip (d=10 µm, stiffness=0.85 N/m), and Young's modulus E was calculated assuming a Hertz contact model. Substrates were coated with Matrigel (0.1 mg/ml) to enhance cell adhesion. AFM measurements of Matrigel-coated PDMS substrates confirmed that coating has no effect on substrate stiffness (not shown).

For gene expression analysis, total RNA was isolated 1, 2 and 4 days after plating on either stiff or soft substrates. RNA was reverse transcribed into cDNA and quantitative PCR was performed with SYBR® Green. Results were normalized to *Gapdh* expression. For nesprin KD, cells were transfected with 10 nM Dicer-substrate siRNA 1 day after plating using Lipofectamine®3000 and RNA was isolated 3 days after or at the day of transfection as control (Figure 1A). siRNA was designed against the KASH domain of each Nesprin gene to target all isotypes relevant for mechanical coupling. PCR primers for verification were specific for the same introns targeted by siRNA.

To analyze beating activity, cells were stained with the Ca^{2+}-sensitive dye Fura2 (5 µM) for 30 min, followed by medium change and 1 h incubation to normalize beating patterns after medium change.

Table 1. PDMS Composition and stiffness.

Referred substrate	*Formulation*	*Ratio*	*E by AFM (kPa)*
soft (native)	Sylgard® 527	1:1	11.7 ± 5.4
stiff (pathologic)	Sylgard® 184	1:10	434.3 ± 54.4

Clusters of CMs were imaged at 40× for 10 s at 9 fps using fluorescence microscopy (Nikon Eclipse-Ti equipped with Andor iXon™+ EMCCD camera) and Fura2 intensity of single cells was tracked and processed with a custom written MATLAB code.

RESULTS

CMs cultured on stiff substrates revealed distinct down-regulation of *Serca2* by 14.28 fold (93%) and up-regulation of *Lmna* by 1.65 fold after 2 days, compared to 1 day after plating (Figure 1B), and developed distinct arrhythmia with deteriorated Ca^{2+} waves and long plateau phases (Figure 1D). In contrast, almost no difference in *Serca2* and *Lmna* expression (1.02 and 1.11 fold increase, respectively) and regular, frequent (~1.6 Hz) beating patterns could be observed for cells cultured on substrates with native stiffness. siRNA-mediated KD of *Syne1-4* was successful with 99%, 96%, 75% and 80% reduction, respectively, and inversed the effect of *Serca2* and *Lmna* gene expression change on stiff substrates (Figure 1C). Non-target (mock) and non-transfected control cells showed the expected down-regulation of *Serca2* and up-regulation of *Lmna* (mock: 4.8 and 1.63 fold). Surprisingly, nesprin KD resulted in near complete depletion of calcium wave activity 3 days post transfection on stiff (Figure 1D) but also on soft substrates (not shown).

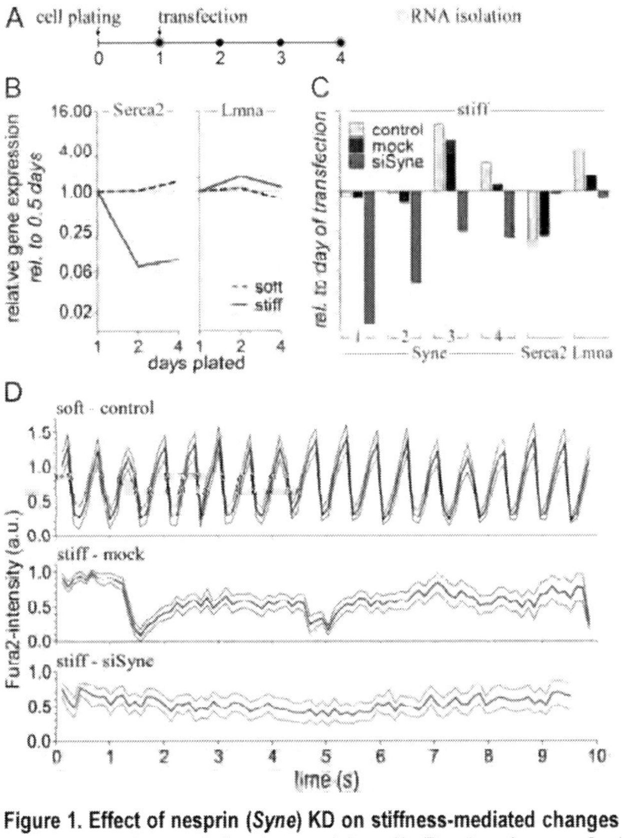

Figure 1. Effect of nesprin (*Syne*) KD on stiffness-mediated changes of gene expression and beating activity. A) Timeline for transfection experiments. **B)** Gene expression analysis of *Serca2* and *Lmna* from CMs cultured on soft or stiff substrates over 4 days. **C)** Gene expression analysis of CMs cultured on stiff substrates 3 days after transfection with siRNA against *Syne1-4* (siSyne). Control cells were either transfected with non-target siRNA (mock) or not transfected at all (control); same scales for B and C. **D)** Ca2+ wave dynamics of single cells in cluster incubated on soft or stiff substrates 4 days after plating or 3 days after transfection, respectively, using Fura2 imaging; thin lines indicate STD; n=50 cells.

DISCUSSION

We investigated to role of nuclear connections to the cytoskeleton in mechanosensitive processes of CMs. Nesprin proteins that connect the nucleus to the cytoskeleton were down-regulated and gene expression of *Serca2* and *Lmna*, and beating activity, was analyzed. *Serca2* is known to be decreased in stiffened heart tissue [2] and *Lmna* has been shown to scale with tissue stiffness [7]. The *in vitro* cell model we used showed the anticipated responses of *Lmna* up-regulation and *Serca2* down-regulation with increased substrate stiffness. siRNA transfection showed high KD efficiency of nesprin genes. The lower efficiency of *Syne3* and *4* KD might be due to the 10 and 2.5 fold increase, respectively, observed in control cells (Figure 1C). Stiffness-mediated change of *Serca2* and *Lmna* gene expression was eliminated after KD of *Syne1-4* on stiff substrates, indicating an importance of the nuclear connections in mechano-sensitive gene expression. Supporting this, a recent study revealed a correlation between intranuclear strain patterns and RNA synthesis [8].

Further, Ca^{2+} wave activity, which is directly related to beating activity, was perturbed on stiff substrates, as seen in patients. Persistent Ca^{2+} peaks indicated compromised Ca^{2+} restoration into the SR which supports previous findings for reduced *Serca2* expression as a cause of arrhythmia. Ca^{2+} wave activity was further impaired after nesprin KD. Even though normalized beating activity through restored *Serca2* expression was expected, perturbed beating goes in concert with heart pathologies reported from patients that carry mutation in lamin A/C, desmin or nesprin genes. To date, 464 laminopathies and 60 desminopathies have been found, most of them related to heart and/or general muscle pathology [3,4]. Recently, a missense mutation in nesprin 1α has been described in a patient which resulted in the need for a heart transplantation at age 26 [5]. Nesprins are at least partially redundant, which might account for the few pathologic mutations reported for Nesprin genes. This is supported by the fact that nesprin 1 and 2 double knock-out mice showed significantly elevated heart pathology compared to either single knock-out strain [9].

After additional confirmation of these preliminary experiments, further investigations will clarify if restored gene expression of *Serca2*, *Lmna*, and possibly other biomarkers is a subsequent effect of impaired beating activity, or if both result from impaired nucleo-mechanosensitive pathways. Either way, the results indicate a distinct role of nucleo-peripheral connections on mechanosensitive processes, including the manifestation of heart pathology, and support the hypothesis that the nucleus is involved in mechanical cue processing.

ACKNOWLEDGEMENTS

The authors acknowledge funding from NSF CAREER 1349735.

REFERENCES

1. Engler, A. J., *J. Cell Sci.*, 121:3794–802, 2008.
2. Fargnoli, A. S., *Ann. Thorac. Surg.*, 96:586–95, 2013.
3. Davidson, P. M., *Trends Cell Biol.*, 24:247–256, 2013.
4. Clemen, C. S., *Acta Neuropathol.*, 125:47–75, 2013.
5. Puckelwartz, M. J., *J. Mol. Cell. Cardiol.*, 48:600–8, 2010.
6. Stroud, M. J., *Circ. Res.*, 114:538–48, 2014.
7. Swift, J., *Science*, 341:1240104, 2013.
8. Henderson, J. T., *Biophys. J.*, 105:2252–61, 2013.
9. Banerjee, I., *PLoS Genet.*, 10:e1004114, 2014.

SB³C2015
Summer Biomechanics, Bioengineering and Biotransport Conference
June 17-20, 2015,Snowbird Resort, Utah, USA
SB³C2015-541

AUTOMATED TUNING FOR PARAMETER IDENTIFICATION IN MULTI-SCALE CORONARY SIMULATIONS

Justin S. Tran, Daniele Schiavazzi, Abhay B. Ramachandra, Andrew M. Kahn, Alison Marsden

Department of Mechanical and Aerospace Engineering
University of California, San Diego
La Jolla, California, United State of America

INTRODUCTION

Computational simulations of coronary artery flow can provide non-invasive information on hemodynamics and wall mechanics that can aid in treatment planning and improve understanding of disease progression. These tools have potential for clinical impact in identifying optimal geometries for coronary artery bypass grafts (CABG), in determining risk of re-stenosis in saphenous vein grafts and in determining thrombotic risk in Kawasaki disease. In this study, patient-specific geometry of the aorta and coronary arteries is constructed from CT scans and combined with finite element flow simulations using the open source software package SimVascular (simvascular.org).

Lumped parameter networks (LPN), consisting of 0D circuit element representations of hemodynamic behavior, are used to represent the heart and distal circulation that are not included in the 3D model. These networks provide a set of ODEs that can be solved for global hemodynamic quantities and used as coupled boundary conditions for the finite element flow solver [1]. The parameters of the LPN (e.g. vessel resistance, compliance, inductance, etc.) are typically tuned to match a patient's clinical data, including blood pressure, stroke volume, and ejection fraction. However, the myriad of parameters in these systems are subject to manual tuning which is time consuming and produces operator-specific parameter values that do not account for the uncertainty in the clinical measurements. This laborious process makes application to large clinical cohorts challenging. We thus propose a Bayesian registration framework combining preliminary analysis of the Fisher information matrix rank and eigenvectors with adaptive Markov Chain Monte Carlo sampling.

METHODS

A Bayesian parameter estimation approach is developed with circulation model parameters described by random variables, while a multivariate Gaussian distribution with diagonal covariance is assumed for the clinical data. The data we use to estimate the parameter values of the LPN consist of patient-specific echocardiography data, clinical data, and literature data. Echocardiography abd clinical data includes stroke volume, ejection fraction, blood pressure, and other quantities. Literature data consist of Doppler flow wire measurements of flow in the coronaries, which include the ratio of flow in systole versus diastole in the left and right coronary arteries [2]. The literature data is needed to constrain the parameters associated with flow in the coronaries, since this information is not collected for our patients and is otherwise difficult to obtain non-invasively.

Our simulations contain 35 uncertain parameters and 22 data targets. With this set of data, there is an overparameterization in the coronary LPN (i.e. large number of model parameters and limited availability of clinical measurements). This problem can be mitigated by an analysis of the Fisher information matrix (FIM) rank and eigenvectors. The FIM is a square matrix of dimension equal to the number of parameters in a system, and it considers the local changes in the output quantities given a small change in the input parameters. In order for a system to not be overparametrized, the FIM must be full rank. We calculated the FIM for each subsystem in the LPN (i.e. each of the four chambers of the heart, and each distal circulation) and found that each subsystem had a singular FIM. We then eliminated parameters from the parameter estimation based on the eigenvectors associated with zero eigenvalues. Often these eigenvectors will have a dominant entry which indicates that a change in that parameter will produce a negligible change in the outputs and can thus be held constant and not considered in the tuning procedure.

After a preliminary parameter reduction through the analysis of the FIM, their joint probability is formulated through Bayesian conjunction of likelihood and prior knowledge. This relationship is known as Bayes rule:

$$P(\theta|d) \propto P(d|\theta)P(\theta) \qquad (1)$$

where θ represents the model parameters we are interested in estimating, d represents the clinical data which we provide to constrain the parameters. $P(\theta|d)$ is the *posterior probability distribution* of the model parameters conditioned by the target data, $P(d|\theta)$ is the *likelihood function* which measures how well simulation outputs match the target data given a particular parameter set, and $P(\theta)$ is the *prior distribution* which represents our prior knowledge about the model parameters. The goal is to sample from and calculate optimal parameter values from the posterior distribution. This can be done with Markov Chain Monte Carlo (MCMC). In short, MCMC sequentially makes point-wise samples of the posterior distribution and mostly accepts parameter sets that produce highly probable posteriors. The samples are then collected to infer probability distributions for the model parameters.

Closed-loop circulation models for coronary blood flow are typically non-linear due to heart contraction and unidirectional flow in valves. Samples of non-linear circulation model parameters will be, in general, non-Gaussian and advanced MCMC should be adopted to improve convergence. Moreover, non-identifiable combinations of parameters still remain after preliminary FIM analysis and might induce multimodality in the posterior distribution leading to slow convergence of MCMC. We deal with this problem by specifying prior knowledge about the parameters through properly selected ranges.

For the prior distribution for each parameter, we assume a uniform distribution defined on a certain range of admissible parameters. For example, we disallow any of the parameters from being negative. The upper bounds for each of the parameters was chosen based on prior experience from manual tuning.

RESULTS

Using the methodologies developed in the previous section we aim to improve the ability to obtain robust statistical estimates for LPN parameters describing coronary circulation. As an initial step, we needed to detect non-identifiable parameter combinations that could not be clearly identified from the analysis of the eigenvectors and rank of the FIM. We first tested the ability of the MCMC algorithm to reproduce results of manual tuning. We tuned the coronary LPN manually until clinically acceptable outputs were obtained, then used these outputs as average target measurements with a 5% standard deviation. We then used adaptive MCMC to re-determine their distributions and provide optimal maximum a posteriori (MAP) estimates. MCMC was able to closely reproduce the target outputs, but few parameters were consistent with the expected, manually tuned, values (Table 1).

Table 1: Initial MCMC Results

Parameter	Manually tuned	MCMC
E_{lv}	7.5	5.0
R_{rv}	10.8	24.4
L_{ra}	0.025	0.009
R_{ra}	3.5	12.8
L_{la}	0.0017	0.018
R_{la}	18.6	34.5
C_{ao}	1.08	1.08

Resulting marginal distributions for many parameters were nearly uniform or characterized by multimodality, confirming that non-identifiable parameters still remain in the model formulation and additional constraints or improved prior knowledge must be specified to improve the quality of the parameter estimates. However, some parameters, such as aortic compliance, matched the expected values with limited variance.

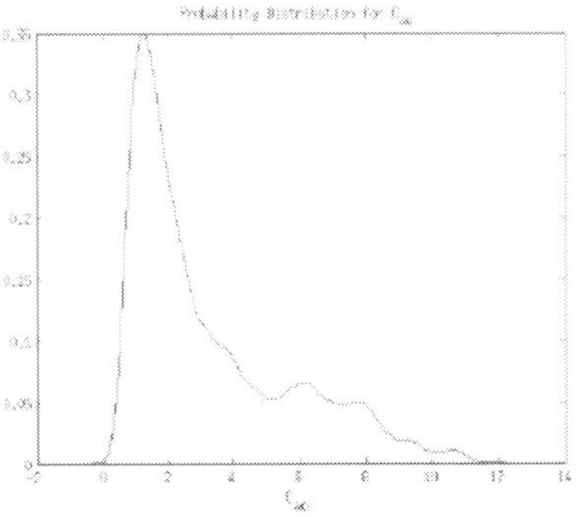

Figure 1: Distribution of C_{ao}

DISCUSSION

LPN models simulating coronary flow are characterized by a large number of parameters and relatively scarce available clinical data. In this context, optimization-based parameter estimation is difficult due to identifiability issues, and a drastic reduction in the number of parameters is often applied in practice, based on more or less justifiable assumptions. Moreover, optimization-based techniques do not account for the uncertainty typically associated with the clinical data aquisition. We show that use of a Bayesian registration approach provides numerous tools that help to detect non-identifiable combinations of parameters and provide insight on which parameters can be effectively estimated and those that are characterized by large variance. Our tests on a lumped model for the coronary circulation show signs of structural non-identifiability, where manually tuned parameter sets could not be reproduced even while exactly matching the prescribed targets. Possible improvements could be obtained by including additional clinical data or by constraining the parameters or, in a Bayesian context, the prior knowledge associated with those parameters.

Unfortunately, as a result of an extensive review of the clinical data discussed in the literature and available from echocardiography for coronary physiology, collection of additional clinical targets appears unlikely. We therefore focus current efforts on parameter reduction. This process is carried out using analysis of the samples resulting from MCMC, leading to improved understanding of which parameters are strongly correlated or poorly estimated (e.g. characterized by flat posteriors). These activities are a necesary prerequisite to application of automatic tuning procedures using real patient data.

ACKNOWLEDGEMENTS

This work was supported by an NIH RO1-RHL123689A, Burroughs Wellcome Fund CASI Award, and NSF CAREER Award.

REFERENCES

[1] Sankaran, S et al., *Annals of Biomedical Engineering*, 20:2228-2242, 2012.

[2] Ofili, E et al., *American Journal of cardiology*, 71.14:D3-D9, 1993.

[3] Schiavazzi, D et al, Uncertainty Quantification in virtual surgery hemodynamics predictions for single ventricle palliation, submitted for publication, 2015.

NONLINEAR DERATING OF HIGH-INTENSITY FOCUSED ULTRASOUND USING HYDROPHONE MEASUREMENTS IN WATER

Seyed Ahmad Reza Dibaji[1], Yunbo Liu[2], Joshua E Soneson[2], Rupak K Banerjee[1], Matthew R Myers[2]

[1]Mechanical Engineering Program, Department of Mechanical and Materials Engineering, University of Cincinnati, Cincinnati, Ohio, USA

[2]Division of Applied Mechanics, Center for Devices and Radiological Health, US Food and Drug Administration, Silver Spring, Maryland, USA

INTRODUCTION

In the design and testing of high-intensity focused ultrasound (HIFU) devices, it is important to know the HIFU-induced temperature rise in the tissue of interest. In order to reduce the required number of animal and human experiments for assessing the influence of HIFU, semi-empirical mathematical models can be used for predicting the thermal effects. However, these models require values of the pressure amplitude in the tissue, which can be difficult to obtain experimentally. Pressure measurements can be much more readily performed in water. Therefore, techniques for predicting the pressure field in tissue using measured pressures in water, i.e., derating techniques, can be valuable for characterizing HIFU systems. At low ultrasound intensities, pressure amplitudes in tissue can be estimated by measuring pressure in water and reducing it by a factor based upon typical tissue attenuation [1] (linear derating). At higher ultrasound intensities, harmonics of the fundamental frequency are generated due to nonlinear propagation effects. Higher harmonics are attenuated differently in water and tissue [2]. Energy is also converted into heat at rates that depends on the frequency of the harmonic modes. Techniques for nonlinearly transforming measured pressures in water to values appropriate for tissue are defined as *nonlinear derating*. These techniques are desirable when bioeffects of higher intensity exposure are being studied.

In this research we present a nonlinear derating method [3] that uses pressure measurements obtained in water to model the nonlinear interaction between higher harmonics in tissue. To account for the differences in modal amplitudes between tissue and water, in the modeling the water harmonics are scaled down with the aid of contours derived from numerical simulations. The results obtained from the derating procedure are compared with direct measurement of the focal pressure in the tissue phantom using hydrophone.

METHODS

Propagation of HIFU beams are often modeled using the nonlinear Khokhlov-Zabolotskaya-Kuznetsov (KZK) equation [2]. Modeling the time-harmonic pressure field as a sum of modes with a Gaussian transverse profile, the KZK equation may be reduced to a simpler system of ordinary differential equations for the mode amplitudes. In the context of the nonlinear derating technique discussed here, the system takes the following form [3]:

$$\frac{d\tilde{a}_n}{d\zeta}+\left[\gamma_n-\frac{G+i}{G-(G+i)\zeta}\right]\tilde{a}_n=\frac{inN}{4}\sum_{m=1}^{\infty}a_m\left(a_{n-m}+2^{*}_{m-n}\right) \qquad (1)$$

$$\text{where: } \gamma(f)=d\alpha(f)-\frac{2id}{\pi}\alpha(f)\log\left(\frac{f}{f_0}\right) \qquad (2)$$

Here $\tilde{a}_n$ and a_n are the scaled complex pressure amplitudes at $\rho = 0$ (where ρ is the radial coordinate scaled by the transducer radius) corresponding to the nth harmonic in tissue and water, respectively. The '*' denotes complex conjugation, n = 1, 2, ... is the harmonic number, d is the transducer focal length, $\zeta = z/d$ is the scaled axial coordinate, G is the linear pressure gain, f is the transducer frequency, and f_0 is the reference frequency usually 1 MHz. $N = 2\pi p_0\beta df/(\rho_0 c_0^3)$ is the coefficient of nonlinearity [3], where β is the nonlinear parameter for the propagation medium, p_0 is the pressure at the transducer surface, ρ_0 is the mass density and c_0 the small-signal sound speed. N in eq. (1) is corresponding to the coefficient of nonlinearity in tissue. The relationship between absorption and frequency is given by $\alpha(f) = \alpha_0\,(f/f_0)^{\eta}$, where η = 1for tissue, η = 2 for water and α_0 is the acoustic absorption at the reference frequency f_0.

A HIFU transducer with focal length of 6.26 cm and diameter of 6.4 cm was used to produce the pressure field in water. The operating frequency and the linear pressure gain were 1.05 MHz and 34.17, respectively. Material parameters included c_0 = 1482 m/s, ρ_0 = 1000 kg/m^3, β = 3.5 in water, and c_0 = 1579 m/s, ρ_0 = 1030 kg/m^3, β = 4.5 in tissue phantom. In order to account for the difference in attenuation between water and tissue, two types of pressure amplitude reductions were performed. The first, labeled "source scaling", involves performing the water measurements at a lower source pressure in water than the value for tissue. This results in a lower coefficient of nonlinearity in water than tissue (N_w/N_t < 1). The second, labeled "endpoint scaling", reduces the measured amplitudes of the water harmonics by the factor exp($-\varepsilon n\alpha_t z$), where $0 \leq \varepsilon \leq 1$ and α_t is the acoustic absorption for tissue phantom. The factor ε is required because amplitude reduction using the full attenuation leads to excessive attenuation when combined with source scaling. In our previous study [3], a set of pressure error contour plots was provided as a function of N_w/N_t and ε. The factors (N_w/N_t) = 0.82 and ε = 0.01 used in this study were chosen from those error contour plots.

Pressures were measured along the beam axis in water, using a hydrophone. The experiments in water were performed at the source-scaled pressure of 0.24 MPa. The acquired pressure trace was Fourier transformed to obtain the modal amplitudes a_n, which were inserted into the right side of eq. (1), after endpoint scaling was performed. The resulting linear, first-order equations were solved to obtain the modal amplitudes in tissue phantom ($\tilde{a}_n$). The estimated modal amplitudes were then used to construct the focal pressure waveform in tissue phantom.

In addition, a gelrite-based tissue mimicking material (TMM) was constructed based on the King et al. [4] protocol. The transducer and the tissue phantom resided in a tank of degassed water. The pressure trace at the transducer focus, right behind the tissue phantom, was recorded using a hydrophone. The source pressure at the transducer surface was p_{0t} = 0.28 MPa.

The pressure waveform at the focus as determined using the derating process was compared with direct measurement in tissue phantom. The error was calculated as:

$$pressure\ error = \frac{\left|p_{+,\,direct} - p_{+,\,derating}\right| + \left|p_{-,\,direct} - p_{-,\,derating}\right|}{p_{+,\,direct} + \left|p_{-,\,direct}\right|} \quad (3)$$

Here p_+ and p_- denote the peak positive and negative pressures, respectively.

RESULTS

Comparison between the derated modal amplitudes, $\tilde{a}_n$, at the focus and direct measurement in the tissue phantom (Fourier transform of the recorded focal pressure in TMM) is provided in Figure 1. The difference between the derated modal amplitudes and direct measurement in TMM for the first, second and third harmonic is 7%, 31% and 4%.

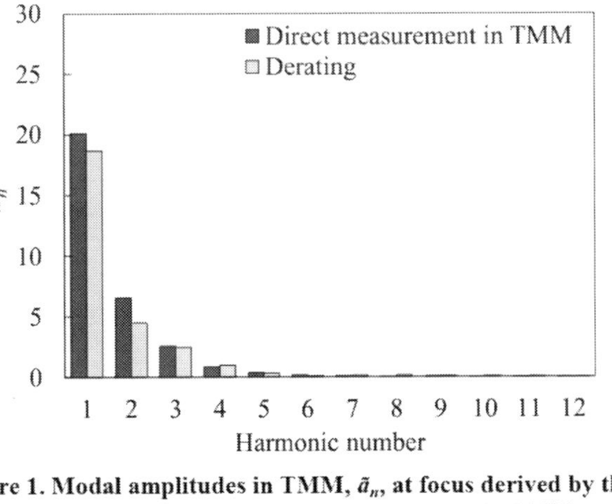

Figure 1. Modal amplitudes in TMM, $\tilde{a}_n$, at focus derived by the derating technique and direct measurement in TMM

Figure 2 shows the focal pressure waveforms obtained by derating and direct measurement in TMM. The peak positive pressure by derating and direct measurement in TMM is 6.59MPa and 8.25 MPa, respectively. The derated peak negative pressure (4.36MPa) is quite similar to the direct measurement in TMM (4.64 MPa). The focal pressure predicted by derating agrees with direct measurement in TMM within 15%.

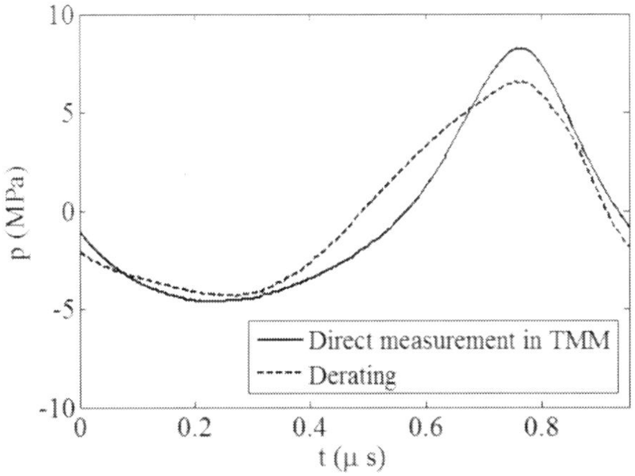

Figure 2. Focal pressure waveform in TMM, derived by the derating technique and direct measurement in TMM

DISCUSSION

The nonlinear derating technique was applied to estimate the focal pressure in TMM, using the measured axial pressures in water by hydrophone. The applied combination of source and endpoint scaling resulted in an estimation of focal pressure in TMM within 15% error. The error may be reduced by increasing the amount of source pressure in water. These experiments are underway. Finally, the technique requires experimental validation for temperature rise. The estimated temperature rise by this nonlinear derating technique will be compared with the direct measurement of the temperature rise inside the TMM using thin wire thermocouples. The technique has the potential to convert pressure measurements in water to pressure and temperature measurements in any tissue of interest, at clinically relevant power levels.

ACKNOWLEDGEMENTS

Financial supports from NSF CBET-1137166, NSF CBET-1403356 and Center for Devices and Radiological Health-US FDA are gratefully acknowledged.

REFERENCES

[1] Center for Devices and Radiological Health, "510(k) guidance document information for manufacturers seeking marketing clearance of diagnostic ultrasound systems and transducers" Report, U. S. FDA, Silver Spring, MD, 2008.

[2] Hamilton, MF et al., "Nonlinear Acoustics," Academic, San Diego, 1998.

[3] Dibaji, SAR et al., " Nonlinear derating of high intensity focused ultrasound beams using Gaussian modal sums", *J Acoust Soc Am*, 134(5): 3435-45, 2013.

[4] King, RL et al., "Development and characterization of a tissue-mimicking material for high-intensity focused ultrasound", *IEEE T Ultrason Ferr*, 58(7): 1397-1405, 2011.

A DESCRIPTION OF SEGMENTAL ANGULAR MOMENTUM SYNERGIES USING INDEPENDENT COMPONENT ANALYSIS DURING GAIT

Brecca M. Gaffney (1), Cory L. Christiansen (2), Kevin B. Shelburne (1), Bradley S. Davidson (1)

(1) Department of Mechanical & Materials Engineering
University of Denver
Denver, CO, USA

(2) Department of Physical Medicine & Rehabilitation
University of Colorado Denver
Aurora, CO, USA

INTRODUCTION

The overall goal in rehabilitation following injury/surgery is to return the patient to a functional level at or better than prior to injury. The most common orthopaedic rehabilitation strategies are strength training and movement retraining. A better understanding of movement patterns used in healthy movement and patients can provide a clearer picture of how best to target movement retraining.

A description of segment momentum coordination using synergy analysis may provide clinical insight into gait patterns; however, typical patterns have not emerged. Two previous investigations performed synergy analysis on the total angular momentum of each segment about the body center of mass (COM) [1, 2]. Synergies from these investigations were not clear and interpretable, and may be related to the chosen variable (total angular momentum about the body COM) and mathematical tool (principal component analysis).

The angular momentum of each segment about its own COM has potential to provide value in a synergy analysis. Based on an inverted pendulum model with a rotational fulcrum point about the stance foot [4], we previously analyzed two contributions to the total angular momentum about the foot during gait. Using separation of total angular momentum, two distinct components are available: 1) Angular momentum of the segment with respect to the foot, which describes the forward progression and 2) Angular momentum of the segment about its own COM, which describes the rotational momentum of each segment. These rotational angular momenta are coordinated to achieve the progressive momentum. We successfully interpreted simple features in the rotational momenta, but the large number of dimensions warrants an approach that considers all the data, such as synergy analysis.

We perform the synergy analysis in this investigation using independent component analysis (ICA), which assumes that several recorded time series input variables can be decomposed into a linear combination of statistically independent components (synergies) [3]. ICA has two distinct advantages when used for angular momentum over the two most common methods of synergy analysis, principal component analysis (PCA) and non-negative matrix factorization (NNMF). Unlike PCA, ICA results in time-varying output variables, which can be used to extract the timing of the momentum synergies. Unlike NNMF, ICA allows for negative input variables, which are an important feature in angular momentum.

The objective of this investigation was to examine patterns of segment coordination in healthy participants using synergy analysis of segment angular momentum about its respective COM (rotational momentum). We hypothesized that the synergies would be interpretable, and that a clear pattern of "normal" movement would be shared across the participants.

METHODS

Five healthy participants (1F, 4M, age: 67.2±8.7 years, BMI: 26.3±2.4 kg/m^2) performed a gait trial at 1.0 m/s. Motions were recorded at 100 Hz from 64 skin markers (Vicon). A 15-segment model was created in Visual 3D (C-Motion, Inc.) to calculate angular momentum. Segmental rotational angular momentum was calculated:

$$\mathbf{h}_{seg}=\mathbf{I}_{seg}\cdot\boldsymbol{\omega}_{seg} \quad (1)$$

where $\mathbf{I}_{seg}$ is the moment of inertia tensor and $\boldsymbol{\omega}_{seg}$ is the angular velocity of the segment. All time derivatives were observed from an inertial reference frame. Angular momenta vectors were transformed into path coordinates defined by the velocity vector of the body COM. Only sagittal plane momentum was used in this analysis.

Rotational segmental synergies in the sagittal plane were obtained across the right limb gait cycle for each participant using ICA, which performs a gradient-based optimization to maximize the independence between each component (synergy). The linear expression of ICA is:

$$\mathbf{x} = \mathbf{A}\mathbf{s} \qquad (2)$$

where $\mathbf{x}$ is the observed signal (rotational momenta), $\mathbf{A}$ is the weighitng matrix, and $\mathbf{s}$ is the independent component matrix. Therefore:

$$\mathbf{s} = \mathbf{W}\mathbf{x} \qquad (3)$$

where $\mathbf{W}$ (unweighting matrix) is the inverse of $\mathbf{A}$. We calculated $\mathbf{A}$ and $\mathbf{s}$ with the FastICA algorithm [3]:

1. Center the data by removing the mean.

$$\mathbf{x}' = \mathbf{x} - \bar{\mathbf{x}} \qquad (4)$$

2. Whiten by eigenvalue decomposition of the covariance matrix, which removes correlation between the observed data.

$$\tilde{\mathbf{x}} = \mathbf{E}\mathbf{D}^{-\frac{1}{2}}\mathbf{E}^{T}\mathbf{x}' \qquad (5)$$

where $\mathbf{E}$ is the orthogonal matrix of the eigenvectors and $\mathbf{D}$ is the diagonal matrix of the eigenvalues of the covariance matrix.

3. Choose an initial weighting vector $\mathbf{A}$.
4. Find $\mathbf{W}$ that maximizes negentropy (non-Gaussianity) of $\mathbf{W}^{T}\mathbf{x}$.
5. Solve Eq. 2 for independent components.

Based on equation (2), the product of the weighting matrix and timing curves (independent components) yields the original input signal. Within the path frame, a positive value (**As**) indicates a counterclockwise rotation and a negative value (**As**) indicates a clockwise rotation. Synergies were compared across participants using Pearson product moment correlation coefficient (r) of the weighting matrix $\mathbf{W}$. An $r \geq 0.75$ was chosen as the threshold to identify shared synergies across participants because it indicates a strong relationship between synergies [5]. If $r \geq 0.75$, synergies were averaged across participants.

RESULTS

Three common segmental rotational synergies were extracted from all participants ($r \geq 0.75$). Sagittal plane rotational momentum is largely composed of contributions from the thighs, shanks, and trunk (Fig. 1). Synergy #1 is a "Contralateral Limb Swing Synergy", and was composed of both thigh and shank momenta during the swing period of the swing limb (30% and 80%). Synergy #2 is a "Terminal Swing Synergy", and was composed of shank momenta and trunk momentum in the opposite direction. This synergy was active in a phasic pattern that corresponded to single limb and double limb support. Synergy #3 is a "Stabilization Synergy", and was composed of thigh, shank, and trunk rotational momentum and peaked during double limb support (40-60% and 90-10%).

DISCUSSION

These results indicate that similar segmental rotational momentum synergies surrounding rotational segmental momentum are shared within healthy adults. These synergies indicate that the thigh and shank are the primary contributors along with smaller contributions by the trunk. Rotational segmental synergies are symmetric in both weighting and timing throughout the gait cycle, which is expected in healthy movement patterns.

The shared synergies identified in healthy participants demonstrate that unique movement strategies are used to achieve forward progression. The Contralateral Limb Swing Synergy is likely used to generate momentum about the foot, and is primarily used to achieve forward progression [3]. The Terminal Swing Synergy is likely a shared synergy between shank segmental rotation and trunk segmental rotation used for arresting forward momentum. The Stabilization Synergy is primarily active during the double support phase and likely uses trunk, thigh, and shank segmental rotations to

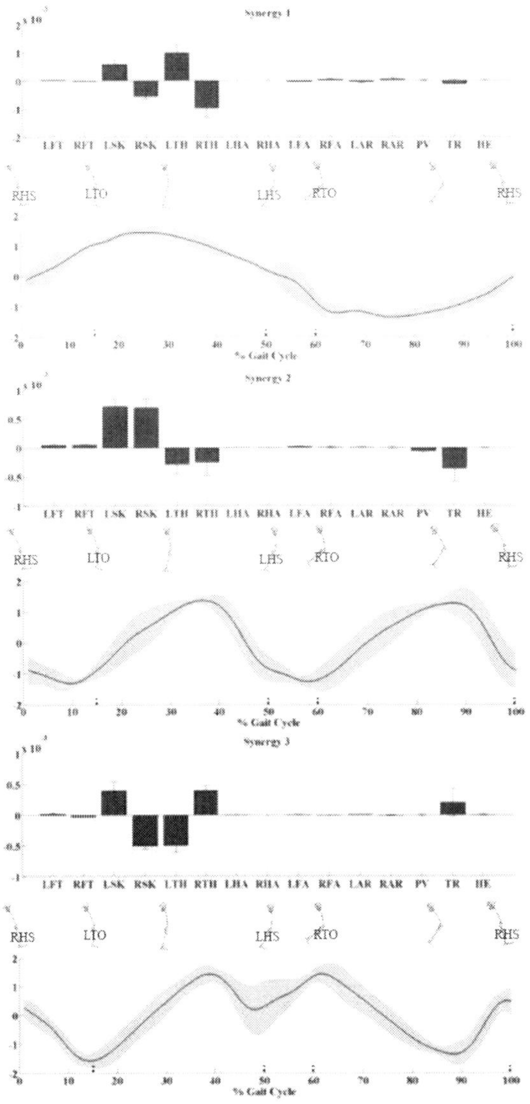

Figure 1. Three segmental rotational synergies were extracted from 5 healthy participants during gait using ICA.

stabilize the body and generate forward momentum about a fulcrum point that transition between limbs.

In summary, synergy analysis of segment rotational angular momentum, revealed interpretable patterns during gait. We will apply this analysis in two patient populations with which we commonly work (patients with total knee arthroplasty or dysvascular amputation) to determine how synergistic segmental control is altered. We anticipate that patients with pathologies will adopt asymmetric segmental rotational weighting coefficients and synergy timings. Synergies in the frontal and transverse planes will be analyzed, which will allow insight into important multi-planar activities such as turning.

REFERENCES

[1] Herr, H et al., *J Exp Biol*, 211: 467-481, 2008.
[2] Robert, T et al. *Exp Brain Res*, 197: 185-197, 2009.
[3] Hyvärinen, A et al. *Neural Networks*, 13: 411-430, 2000.
[4] Hof, A. *J Biomech*, 38: 1-8, 2005.
[5] Portney, L et al., *Foundations of Clinical Resarch*, 2000.

SB^{3}C2015
Summer Biomechanics, Bioengineering and Biotransport Conference
June 17-20, 2015, Snowbird Resort, Utah, USA
SB^{3}C2015-544

A NOVEL APPROACH TO CALCULATING ISOVOLUMIC PRESSURE FOR SINGLE-BEAT ESTIMATION OF RIGHT VENTRICULAR END-SYSTOLIC ELASTANCE

A. Bellofiore (1), E. Dinges (2), S.J Shah (5), M.J. Cuttica (6), R. Sweis (5), H. Mkrdichian (5), L. Beussink-Nelson (5), M. Bailey (3), J.R. Runo (3), J.G. Keevil (3,4), C.J. Francois (3,4), N.C. Chesler (2,4)

(1) Department of Biomedical, Chemical and Materials Engineering
San Jose State University
San Jose, CA, US

Department of (2) Biomedical Engineering, (3) Medicine, (4) Radiology,
University of Wisconsin-Madison
Madison, WI, US

Division of (5) Cardiology, (6) Pulmonary and Critical Care Medicine, Department of Medicine, Feinberg School of Medicine, (7)
Northwestern University, Chicago, IL, US

INTRODUCTION

Assessment of right ventricular (RV) function is crucial in the clinical management of a number of conditions, including pulmonary arterial hypertension (PAH). In particular, end-systolic elastance (E_{es}) has been extensively used as a measure of RV contractility. Clinical approaches to assess E_{es} are limited by the difficulty of simultaneously measuring RV pressure and volume over a number of heartbeats at varying preload, as required for the reference multiple beat method [1]. For the past decade a single-beat method has been used as a clinical research tool to estimate E_{es} [2-4]. This method is based on the interpolation of the isovolumic contraction and relaxation ranges of the pressure waveform with a sine function. The peak value of the sine function, P_{iso}, is the maximum pressure that can be reached by an isovolumic heartbeat, and is critical to calculate E_{es}.

In principle, the isovolumic ranges are determined by automatic routines evaluating the maxima and minima of the first derivative of the pressure waveform. While the single-beat method has been validated in the left ventricle [5], in the RV the isovolumic periods are shorter, resulting in a higher uncertainty of the fitted curves generated by the automatic routines [6]. In practice, the results of the automatic routines are visually checked and, if necessary, manually corrected [4].

In a previous study we showed that changes in the fitting ranges may result in large variations of P_{iso} and E_{es} [7]. Here we propose a novel method to determine the fitting ranges for evaluating P_{iso} in the RV. Our method is based on the evaluation of the second derivative of the RV pressure waveform. We hypothesize that our semiautomatic method results in reduced interobserver variability compared to the manually corrected method based on the first derivative.

METHODS

In our prospective study, subjects with diagnosed or suspected PAH underwent right heart catheterization (RHC) at Northwestern Memorial Hospital or University of Wisconsin-Hospital. All subjects gave written informed consent. The study was HIPAA-compliant and approved by the Institutional Review Board at both institutions. The RHC was performed from the internal jugular vein approach under fluoroscopic guidance. RV pressure traces were obtained using a high-fidelity solid-state micromanometer-tipped catheter (Millar Instruments, Houston, TX) at a sampling rate of 1 kHz.

RV pressure was collected over 10-30 heartbeats and analyzed using custom routines developed in LabVIEW 2013 (National Instruments, Austin, TX). Beat-averaged pressure waveforms were smoothed with a standard low-pass filter and subsequently analyzed by three observers (O1, O2, O3) using two different methods: the standard approach based on the 1st derivative of the pressure waveform, and a novel approach based on the 2nd derivative.

In the 1st derivative approach (Figure 1, *upper panels*), the fitting contraction range was set from the onset of isovolumic contraction (IC) to the point of maximum dP/dt, while the fitting relaxation range was set from the point of minimum dP/dt to the end of isovolumic relaxation (IR) [2]. The onset of the IC (end of the IR) was set at 20% of the maximum (minimum) dP/dt. The fitting ranges were interpolated using a sine wave in order to estimate P_{iso}. The observer had an option to manually correct the limits of the fitting ranges.

In the novel method proposed here, the end of the fitting contraction range and onset of fitting relaxation range were set as minimum points of the 2nd derivative of the pressure waveform (Figure 1, *lower panels*). In reality, the 2nd derivative may exhibit multiple local minima in each isovolumic period; therefore the observer selected the minimum points that ensured the best fit of the pressure waveform. The onset of the IC (end of the IR) was set at 20% of the minimum points of the 2nd derivative. All results are presented as mean ± standard deviation (SD). A paired T-test was used to compare the interobserver variability of the two methods. A p-value < 0.001 was considered evidence of statistical significance.

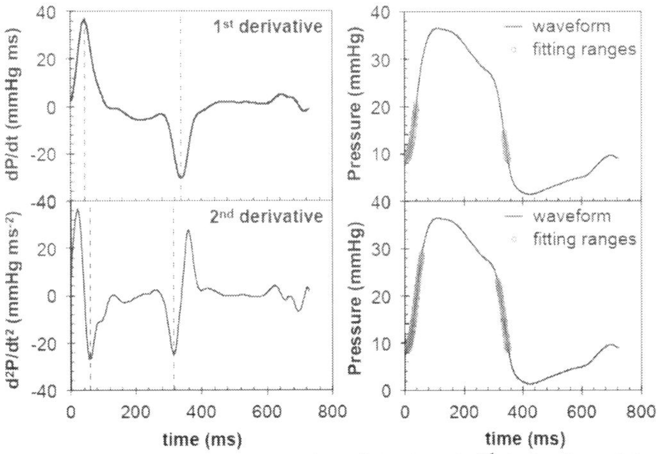

Figure 1. *Left panel.* Representative 1st (*top*) and 2nd derivative of the RV pressure of a PAH subject (*bottom*), with the relevant minimum and maximum points highlighted. *Right panel.* Fitting ranges from the automatic routines for the 1st (*top*) and 2nd derivative (*bottom*).

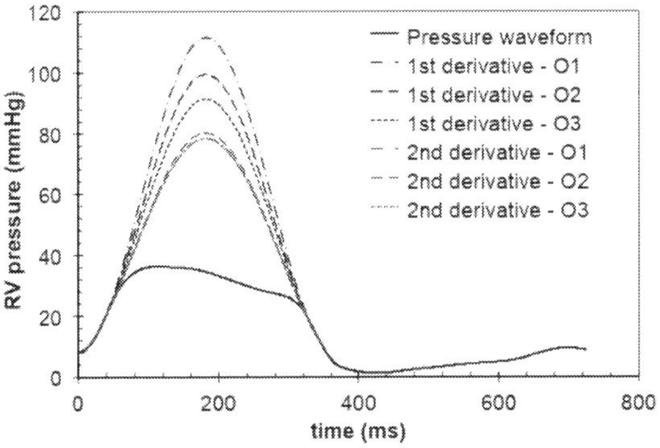

Figure 2. Sine wave interpolations of a RV pressure waveform (same subject as in Figure 1). Interobserver variability for the 1st derivative method (*in black*) is broader than the 2nd derivative method (*in red*).

RESULTS

The study group included 24 subjects (14 women, age 53.4 ± 11.3). The final diagnosis was PAH for 21. Mean pulmonary artery pressure was 43.0 ± 17.3 mmHg for the PAH subjects, and 22.7 ± 1.7 mmHg for the subjects without PAH. The representative case shown in Figure 2 (a subject with PAH) illustrates that the 1st derivative approach resulted in a broader variability of the fitting sine function compared to the 2nd derivative method. Figure 3 compares all the P_{iso} data obtained using the two methods and averaged for the three observers. The average values for the two methods exhibited a good linear agreement ($R^2 = 0.74$). The 2nd derivative method consistently produced lower values of P_{iso} (106 ± 36 *vs* 125 ± 44 mmHg, $p < 0.001$). The SD of the data from the 2nd derivative method (vertical error bars in Figure 3) were smaller than the 1st derivative method (horizontal error bars). The average SD of measurements from the 3 observers was 15.07 ± 16.12 mmHg for the 1st derivative method, and 2.54 ± 3.59 mmHg for the 2nd derivative method ($p < 0.001$).

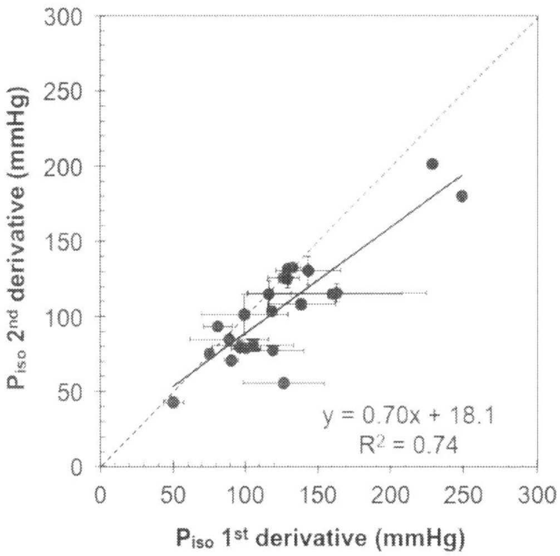

Figure 3. Comparison of the P_{iso} values for all the subjects estimated using the 1st and 2nd derivative methods. The error bars indicate the SD of the data obtained from different observers.

DISCUSSION

In this study we demonstrated a novel approach to estimate the isovolumic pressure P_{iso}, which represents the only currently viable option to assess RV contractility in the clinical setting using a single-beat method. The approach proposed here (based on the 2nd derivative of the pressure waveform) produced results that correlate well with the standard 1st derivative approach, with the advantage of a significantly lower interobserver variability. The main reason for the reduced variability is that the 2nd derivative method limits the options for manual correction to, at most, a small discrete set (minimum points of the 2nd derivative).

Another difference between the two methods is that the P_{iso} values estimated with the 2nd derivative method are significantly and consistently lower than the standard method. This is to be expected, as the 2nd derivative method considers portions of the isovolumic periods with a negative curvature, which are not included in the standard 1st derivative method. Here the advantage is that the 2nd derivative method uses more information from the pressure waveform to interpolate the sine wave (see Figure 1, *right panels*), which can be particularly important in the RV where the isovolumic periods are shorter than in the left ventricle [6].

One limitation of this study was the lack of validation of our novel method with a gold standard. However, it is worth noting that the standard 1st derivative approach also has not been validated in the human RV. The promising results presented here warrant further investigation of the 2nd derivative approach, including a comparison with the gold standard multiple beat method.

ACKNOWLEDGEMENTS

This study was supported by NIH (1R01HL105598 to NCC).

REFERENCES

[1] Latus, H et al., *J Thorac Cardiovasc Surg*, 146(6):1366-72, 2013.
[2] Brimioulle S et al., *Am J Physiol Heart Circ Physiol*, 284:H1625-H1630, 2003.
[3] Rain S et al., *Circulation*, 128(18):2016-25, 2013.
[4] Trip P. et al., *J Heart Lung Transplant*, 32(1):50-5, 2013.
[5] ten Brinke et al., *Acta Physiol*, 198, 37–46, 2010.
[6] Lambermont B. et al., *Crit Care Med*, 32(9):1886-90, 2004.
[7] Bellofiore A. et al., *Proc. BMES Annual Meeting*, September 2013.

SB³C2015
Summer Biomechanics, Bioengineering and Biotransport Conference
June 17-20, 2015,Snowbird Resort, Utah, USA
SB³C2015-545

QUANTITATIVE CORONARY ANGIOGRAPHY BASED RECONSTRUCTIONS FOR WALL SHEAR STRESS CALCULATIONS IN BIFURCATIONS

Jelle T.C. Schrauwen (1), Antonios Karanasos (2), Nienke S. van Ditzhuijzen (2), Jean-Paul Aben (3), Jolanda J. Wentzel (1), Antonius F.W. van der Steen (1,4), Frank J.H. Gijsen (1)

(1) Department of Biomedical Engineering, Thoraxcenter, Erasmus Medical Center, Rotterdam, the Netherlands

(2) Department of Cardiology, Thoraxcenter, Erasmus Medical Center, Rotterdam, the Netherlands

(3) Pie Medical Imaging, Maastricht, the Netherlands

(4) Department of Imaging Physics, Delft University of Technology, Delft, the Netherlands

INTRODUCTION

Atherosclerosis is a chronic disease of the arterial system that leads to plaque formation. One factor that plays a key-role in the progression of atherosclerosis is the wall shear stress (WSS) [1]. WSS can be determined by computational fluid dynamics (CFD). To perform CFD computations the geometry of a coronary artery is required, together with boundary conditions for the inlet and outlets. Based on bi-plane angiography recordings it is possible to reconstruct the geometry of a coronary artery [2].

Aim - For this study we investigated the effect of angiography-based segmentation and reconstructions on WSS computations in coronary bifurcations. For this we computed the WSS in phantoms of which the exact geometry was known. These WSS results were compared with the WSS results obtained from computations in angiography-based reconstructions of those phantoms.

METHODS

Phantoms & Reconstruction - Four phantoms having the geometrical characteristics of diseased coronary bifurcations were previously created [3], with representative dimensions, degree of stenosis and bifurcation angles (Fig 1A). The phantoms were imaged with bi-plane angiography following clinical protocol (Fig 1B). Two angiography images per phantom were segmented by two experienced observers blind to the original geometry. With the segmentations a 3D volume of a region of interest (ROI) was reconstructed using validated software (Fig 1C, CAAS v5.11, Pie Medical Imaging, Maastricht, the Netherlands, [2]).

CFD - Based on the two 3D reconstructions surface meshes were created using a newly developed module within the CAAS software package. Subsequently, a third surface mesh of each phantom was build using the original geometry directly. With the three surface meshes per phantom, volume meshes were created with tetrahedral cells and a five element prism layer at the wall, resulting in a mesh size of typically $3.5 \cdot 10^6$ cells (ICEM-CFD 14.5, Ansys Inc, Canonsburg, US).

Steady-state CFD computations were performed using standard numerical techniques (Fluent 14.5, Ansys Inc, Canonsburg, US). Poiseuille profiles were defined at each inlet with physiologically representative peak velocity of 15 cm/s. The outlet conditions were set as a flow ratio over the outlets of the bifurcations with a diameter-

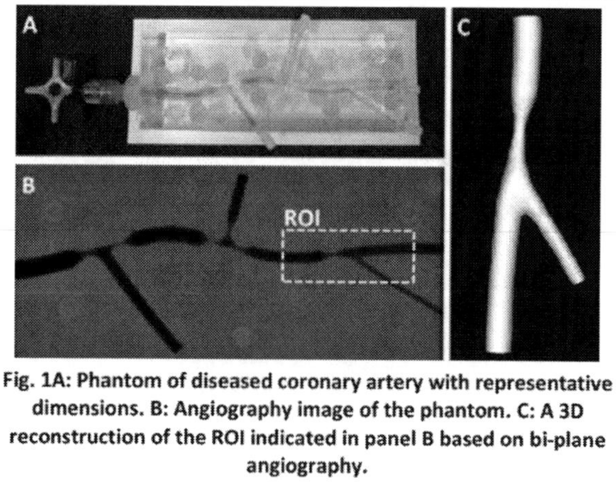

Fig. 1A: Phantom of diseased coronary artery with representative dimensions. B: Angiography image of the phantom. C: A 3D reconstruction of the ROI indicated in panel B based on bi-plane angiography.

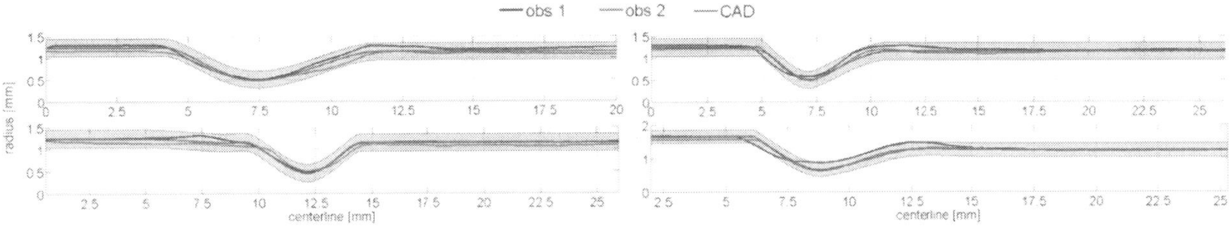

Fig 2: The centerlines of the 4 phantoms and the reconstructions based on the segmentations of the 2 observers. The deviations in the reconstructions are below the typical resolution of angiography (0.2mm) indicated by the blue zone.

based scaling law [4]. The blood was modeled as a non-Newtonian fluid with a Carreau model and a density of 1060 kg/m3.

Post-processing - From the CFD computations the 3D WSS magnitude at the wall was extracted and mapped to a 2D representation. The lengths of the 2D WSS maps were clipped in order to match the areas of the three models per phantom. The differences between the phantom and observers were quantified in three regions: proximal and distal to the stenosis, and the stenosis itself. In these regions the average WSS in circumferential direction was computed with an axial sampling of 0.45 mm.

RESULTS

The radii along the centerline of the reconstructed geometries from the two observers matched the radii from the four original phantoms (Fig. 2). The deviation of the radii was less than the typical resolution of angiography of 0.2 mm.

Fig. 3 shows the velocity results from the steady computations at 15cm/s for one of the phantoms. In all cases a jet is formed at the neck of the stenosis. Downstream of that a recirculation zone can be observed, with the direction of the jet pointed more towards the right wall. Qualitatively there is a good agreement between the velocity results in the phantom and the reconstructions from the observers, with the main differences being the strength and the direction of the jet.

The maps in Fig. 4 show the 2D WSS representation corresponding to the results shown in Fig. 3. The maps are depicted in a range of 0 to 2 Pa in order to analyze areas of low WSS areas. The areas proximal and at the stenosis agree nicely. Distal of the bifurcation all maps show the same crescent-shaped low WSS area. This is the result of recirculation zones, caused by jet formation in the stenosis. The area of low WSS from observer 1 is a bit skewed, due to a small off-set in the reconstruction of the side branch. The WSS map from observer 2 shows irregularities directly distal of the stenosis, caused by deviations in the reconstruction. The area distal of the stenosis was underestimated by observer 2, resulting in a smaller recirculation zone.

The right graph in Fig. 4 shows the mean WSS in circumferential direction distal of the stenosis for all cases. The mean absolute error between the phantom and observer 1 is 0.17±0.12Pa, and 0.22±0.15 Pa for observer 2.

DISCUSSION

Our results suggest that representative WSS results were computed with angiography-based reconstructions for coronary bifurcation. Although similar patterns of low WSS areas were found distal from the stenosis, the size and magnitude of the low WSS zones differed. Subtle deviations in the segmentation procedure near the stenosis and the side branch, influenced size and shape of the jet in the stenosis. This in turn affected the low WSS zone. Intravascular imaging techniques might be required to provide the resolution needed to reconstruct the stenosis within the necessary accuracy when jets are present.

ACKNOWLEDGEMENTS

This research was performed within a framework of the CARISMA-program of STW (Stichting Technische Wetenschappen).

REFERENCES

[1] Slager, C. et al., *Nat. Clin. Pract. Cardiovasc. Med.* 2:401–7, 2005.
[2] Girasis, C. et al., *EuroPCR*, 8(12):1451-60, 2013
[3] Girasis, C. et al., *Cath. Cardiovasc. Interv.*, 77(6):790-7, 2011
[4] van der Giessen, A. G., *J. Biom.*, 44(6):1089-95, 2011

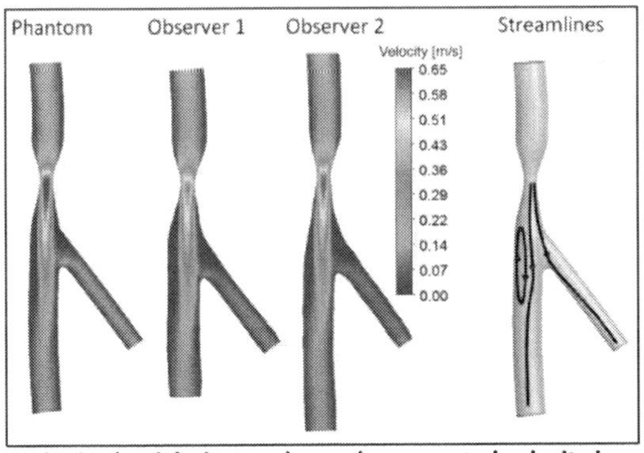

Fig. 3: The right image shows the computed velocity in a plane in the phantom, and the two middle images for the geometries from observer 1 and observer 2. The cartoon on the left depicts the general behavior of the streamlines downstream of the stenosis.

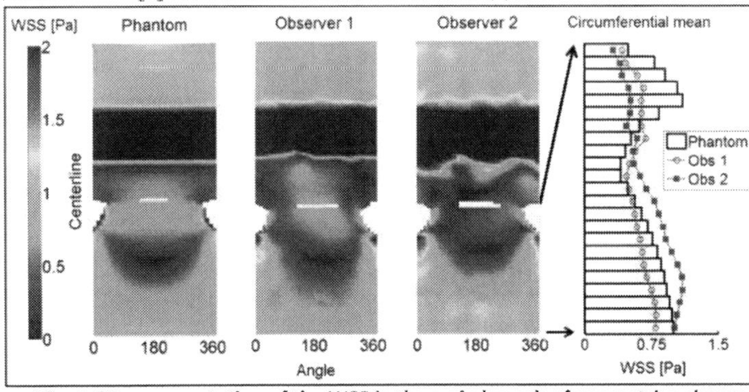

Fig. 4: 2D representation of the WSS in the main branch of geometries shown in Fig. 3. In order to represent regions of low WSS the range was set to 0-2 Pa. All maps show a similar crescent-shaped area of low WSS distal of the stenosis. The graph on the right shows the mean WSS in circumferential direction in the distal region.

SB[3]C2015
Summer Biomechanics, Bioengineering and Biotransport Conference
June 17-20, 2015,Snowbird Resort, Utah, USA
SB³C2015-546

IN SILICO PREDICTIONS OF ANGIOGENIC GROWTH COUPLED TO MATRIX MECHANICS WITHIN HETEROGENEOUS EXTRACELLULAR ENVIRONMENTS

Lowell T. Edgar (1), James B. Hoying (2), Jeffrey A. Weiss (1)

(1) Department of Bioengineering and
Scientific Computing and Imaging Institute
University of Utah
Salt Lake City, Utah, USA

(2) Division of Cardiovascular Therapeutics,
Cardiovascular Innovation Institute
University of Louisville
Louisville, Kentucky, USA

INTRODUCTION

Sprouting angiogenesis, the formation of new blood vessels from an existing vasculature, is a key process in neovascularization and leads to the expansion of an existing vascular network. Mechanical interactions and the mechanical properties of the extracellular matrix (ECM) stromal environment are important regulators of growth and neovascularization. We have previously demonstrated mechanical regulation of angiogenesis using an organ culture model based on isolated microvessel fragments in a type I collagen gel [1, 2]. However, experimental models often only permit the study of simplified environments and manipulation of conditions at the global level, and as a result these techniques often do not provide sufficient means to study the outcome of angiogenesis in more complex microenvironments such as those found in vivo.

Computational modeling facilitates the systematic testing of scenarios and hypotheses that are beyond the limitations of experimental techniques. For these purposes, we designed, implemented, and validated a coupled computational framework for modeling angiogenesis (AngioFE) in which a discrete growth model of angiogenic microvessels is coupled to a continuum finite element (FE) representation of ECM mechanics [3]. This implementation generated active stress fields within for each sprout in the ECM based solely on microvessel position and orientation. However, experiments have demonstrated that local matrix density modulates force generation and active stresses during cellular migration (e.g., [4]). The objective of this study was to extend AngioFE to allow simulation of growth within heterogeneous density environments by scaling active stress generation based on local ECM density. We then used the new capability of the model to simulate angiogenesis under homogeneous density conditions and three different mechanical microenvironments than resembled different stromal conditions that occur in vivo and cannot be easily studied using experimental techniques.

METHODS

In AngioFE, ECM properties such as fibril orientation and density are stored at the nodes of the FE mesh and interpolated at any point within the mesh using the shape functions. Microvessels are represented as a discrete collection of end-to-end line segments and initial microvessel fragments are seeded with a random position and orientation. Neovessel elongation (i.e., growth) is modeled by the addition of new line segments at each active sprout location based on local ECM information as previously described [2,3]. After each growth step, an active stress representing the stress in the matrix generated by a neovessel sprout was assigned to each sprout location. The nonlinear FE software FEBio [5] was then called to solve for the deformation based on the constitutive model, applied loading, and boundary conditions as previously described [3]. The kinematic solution predicted by FEBio was then used to updated microvessel position and ECM properties into the current configuration.

In the formulation developed for this study, the active stress field σ_s at a position $\boldsymbol{x}$ due to a particular sprout location at $\boldsymbol{x}_s$ was calculated based on the following equations,

$$\boldsymbol{r}(\boldsymbol{x}) = \|\boldsymbol{r}\|\hat{\boldsymbol{r}} = \boldsymbol{x}_s - \boldsymbol{x}, \quad (1)$$

$$\boldsymbol{\sigma}_s(\boldsymbol{x}) = \nu(\rho)\, a e^{-b\|\boldsymbol{r}\|} \cos^N\left(\frac{\psi}{2}\right) \hat{\boldsymbol{r}} \otimes \hat{\boldsymbol{r}}, \quad (2)$$

where $\boldsymbol{r}$ is the distance vector between the two points. The angle ψ was measured between the sprout orientation vector $\boldsymbol{v}$ and $\boldsymbol{r}$ (Fig. 1A). The parameters a, b, and N were selected as described previously [3]. A density scaling function $\nu(\rho)$ previously determined from experimental data was used to scale sprout stress based on local collagen density interpolated at $\boldsymbol{x}$ [2] (Fig. 1B).

We designed four numerical experiments to exercise the new features. All simulations took place within a FE meshed domain measuring 7.5 × 5.0 × 0.5 mm with an element spacing of 250 μm in each direction, similar to previous studies [2]. The mesh was seeded at 50k fragments/ml and growth was simulated for 6 days. Numerical experiments were as follows: a homogeneous density field (ρ = 3.0 mg/ml), a wound closure model consisting a narrow gap of 3.0 mg/ml collagen surrounding by two discrete regions of high density 6.0 mg/ml collagen, an ECM density gradient running from 8.0 mg/ml on the left to 3.0 mg/ml on the right, and finally a preconditioning experiment in which periodic loading was applied to the right of the mesh. In the final case, growth was simulated for 7 days and traction applied as a ramp during Day 3-4 and Day 6-7.

RESULTS

Predicted growth in the homogeneous density experiment was consistent with what we have reported in previous efforts (data not shown) [3]. Active stress created by neovessel sprouts drove the mesh to contract inward towards its geometric center, reaching Green-Lagrange strain values of roughly -0.2 in the x and y direction and -0.1 in the z direction. The density field became heterogeneous during contraction, reaching a maximum value of 5.31 mg/ml. In the wound closure simulation, microvessels vascularized the low density gap region but did not grow into the high density regions. The contraction of the low density matrix caused the gap to close, pulling the two stiffer regions closer together and narrowing the maximum width of the gap from 900 to 734 μm. The 3rd principal strain was oriented across the gap and reached a value of -0.22 (Fig. 1 C).

In the density gradient simulation, matrix contraction varied along the gradient, causing the mesh to contract inward along y and z directions where density was low. However, the density gradient along x prevented contraction along this direction, causing the narrowing of the construct to decrease while moving up the gradient forming a cap-like structure (Fig. 1 D). The contraction of the low-density regions increased matrix density up to 1.51 mg/ml. Vascular density was high in the low density regions and decreased along x due to the density gradient, and the contraction of the mesh further increased vascular density by condensing vessels together. In the final experiment, the periodic loading led to distinct changes in microvessel alignment (Fig. 1 E). Vessels began randomly orientated at Day 0 but became slightly aligned along x after the first loading period at Day 4. During the relaxation period between Day 4 and Day 6, the construct continued to contract laterally, further aligning microvessels, and the second loading period applied between Day 6 and Day 7 induced even more alignment.

DISCUSSION

Using density-dependent cellular traction forces allowed us to model matrix stiffening in vascularized constructs as the matrix is contracted as matrix density increases while the matrix contracts. Additionally, we also provide the means to simulate growth in a heterogeneous matrix density environment. Without density-sensitive scaling, matrix deformation in these simulations would be similar even though vascular growth would appear very different in each case. The wound closure, density gradient and preconditioning simulations demonstrate how heterogeneity and complexity of the matrix environment can lead to formation of different vascular topologies and deformed geometry during angiogenesis. This new approach results in better fidelity to in vivo environments for angiogenesis.

ACKNOWLEDGEMENTS:

Financial support from NIH #R01HL077683, R01GM083925 and R01EB015133 is gratefully acknowledged.

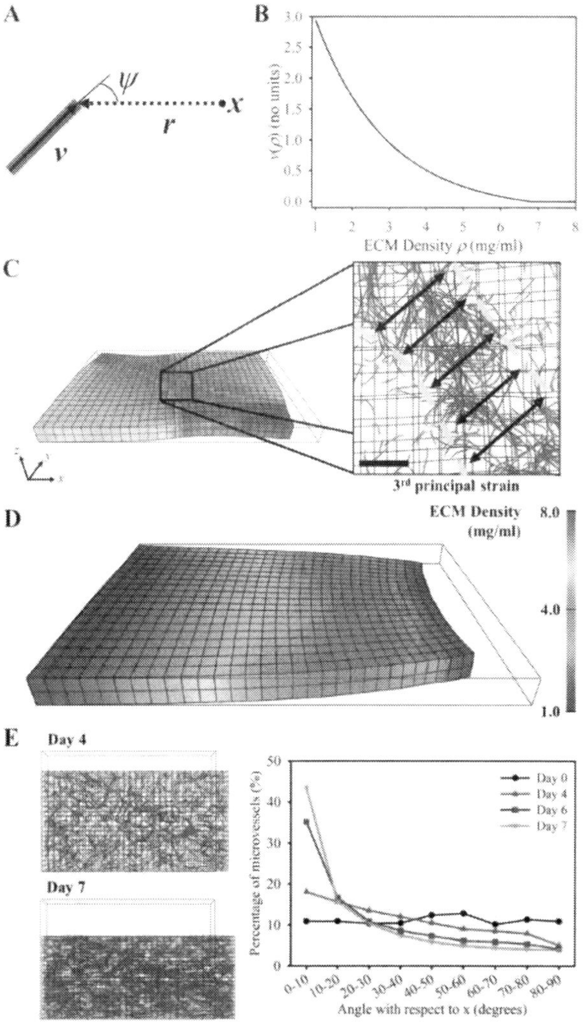

Figure 1. (A) Active sprout stresses were calculated based on distance away from the sprout (||r||) and the angle formed with respect to the sprout ψ. (B) The function ν(ρ) used to scale active stress with respect to local ECM density, ρ. (C) Wound closure simulation. Microvessels contracted the low density region causing the gap to close. Black arrows indicate 3rd principal strain. Scale bar 500 μm. (D) Density gradient simulation. The mesh narrowed the most at the right lowest-density region of the construct, resulting in a cap-like structure. (E) Preconditioning simulation. The construct elongated along x and contracted laterally after each applied loading period (Left). (Right) Distribution of microvessels angle with respect to x-axis at Day 0 (black), the end of the first loading period at Day 4 (red), the start of the second loading period at Day 6 (blue), and at the end of the second loading period and simulation at Day 7 (green).

REFERENCES

[1] Underwood, C et al, *Am J Physiol Heart Circ Physiol,* 307:H152-64, 2014, [2] Edgar, L et al, *PLOS One*, 9:e85178, 2014, [3] Edgar, L et al, *Biomech Model Mechanobiol*, 2014, [4] Kniazeva, E et al, *Am J Physiol Cell Physiol*, 297:C179-87, 2009, [5] Maas, S et al, *J Biomech Engng*, 134:011005, 2012.

SB³C2015
Summer Biomechanics, Bioengineering and Biotransport Conference
June 17-20, 2015,Snowbird Resort, Utah, USA
SB³C2015-547

RELATIONSHIP AMONG DISTURBED SHEAR DESCRIPTORS, THE EXTENT OF FLOW RECIRCULATION AND HELICITY IN CAROTID BIFURCATION

D. Gallo (1), D.A. Steinman (2), U. Morbiducci (1)

(1) Department of Mechanical and Aerospace Engineering
Politecnico di Torino
Turin, Italy

(2) Department of Mechanical & Industrial Engineering
University of Toronto
Toronto, ON, Canada

INTRODUCTION

There is clear evidence that hemodynamics plays a role in vascular disease. The carotid bifurcation particularly has been widely studied because of the preferential development of atherosclerosis, attributed to the proatherogenic role played by low and oscillatory wall shear stress (**WSS**) on endothelial cells (ECs). This theory underlies most current research on localizing factors of vascular disease, presumably oversimplifying the complex biomechanical milieu the ECs are exposed to. In this sense, recent contributions complemented **WSS** analysis with a characterization of intravascular fluid structures, in an endeavor to integrate and extend the information given by WSS-based description. In detail, we previously highlighted the role played by helical flow patterns, demonstrating that helical flow is involved in flow stability, suppressing flow disturbances and leading to a relative uniformity of WSS in the bifurcation region [1]. Additionally, Martorell et al. [2] investigated the ECs response to flow recirculation in a vessel-like *in vitro* carotid bifurcation model. Their results showed the existence of a relationship between the extent of flow recirculation and the expression of atherogenic and thrombotic biomarkers.

Moving from these observations, we calculate for the first time the extent of flow recirculation in 3 geometrically accurate image-based computational models of normal carotid bifurcation in which the full common carotid artery (CCA) is included, as it has been demonstrate to affect the hemodynamic at the bifurcation [3]. In detail, the exposure to low/oscillatory **WSS**, helical flow content and the extent of flow recirculation have been calculated and the existence of a relationship among these hemodynamic descriptors is investigated. We hypothesize that flow recirculation (1) is reflected in **WSS** directionality, that recent evidences reported to influence ECs response [4], and (2) is influenced by the amount of helical flow, which plays a stabilizing effect [1,3]. With this study, we aim to provide (although indirectly) connections between peculiar hemodynamic features and the expression of biomarkers of arterial disease.

METHODS

Computational Hemodynamics

Data from this study are a subset of the Vascular Aging – The Link That Bridges Age to Atherosclerosis (VALIDATE) study. The lumen geometry of 3 healthy carotid arteries was reconstructed from its brachiocephalic origin to well above the bifurcation from contrast-enhanced angiography. CFD simulations were carried out using a previously validated in-house solver. Following the identification of branch origin, sections located 3, 5 and 2 spheres away from the branch origin were defined respectively for CCA, internal (ICA) and external branch (ECA). These sections, denominated CCA3, ICA5 and ECA2 (Fig. 1), delimit each branch and define the zones where the below descriptors were calculated.

Hemodynamic Descriptors

Disturbed shear was evaluated in terms of relative residence time (RRT) distribution, a marker of low and oscillatory WSS. Disturbed shear was quantified as the surface area (SA) exposed to RRT above a threshold value representing the 80th percentile of distributions of RRT as previously identified [1]. To factor out the influence of vessel size, a relative exposure RRT80 was defined by scaling the SA to the surface area of the branch. Based on previous observations [1], here we use helicity to characterize intravascular flow. Considering the definition of kinetic helicity of a fluid confined to a domain of volume V, the previously introduced descriptor h_2 quantifying helicity intensity is adopted [1]:

$$h_2 = \frac{1}{\mathrm{T}\,\mathrm{V}} \int_{\mathrm{T}} \int_{\mathrm{V}} |\mathbf{v}(\mathbf{x},t) \cdot \boldsymbol{\omega}(\mathbf{x},t)| \, dVdt \qquad (1)$$

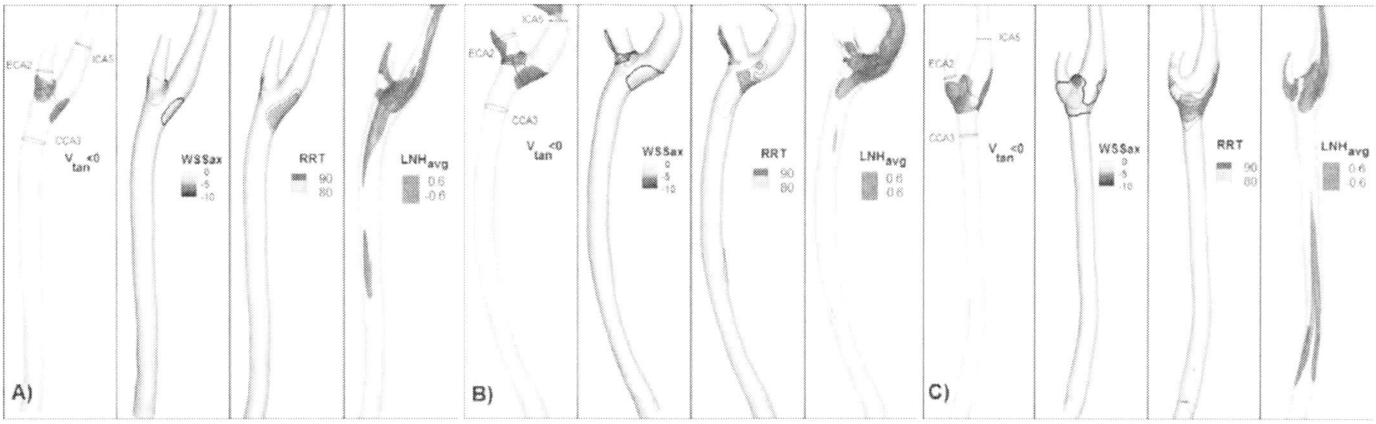

Figure 1. From left to right: isosurface of negative V_{tan} to highlight flow recirculation; contour maps of negative WSS_{ax}; contour maps of RRT above 80th or 90th percentile value; isosurface of LNH_{avg}.

where $\mathbf{v}(\mathbf{x},t)$ and $\boldsymbol{\omega}(\mathbf{x},t)$ are the velocity and the vorticity vector, respectively, and T is the duration of the cardiac cycle. The quantity local normalized helicity (LNH, $-1 \leq LNH \leq 1$), defined as the normalized dot product between local velocity and vorticity vectors, is used for the visualization of complex flow patterns. Its sign indicates the direction of rotation of fluid structures.

The definition of flow recirculation is not trivial for realistic complex vascular geometries. Based on an essential geometric attribute of the vessel, its centerline, we define the velocity V_{tan} as the component of the velocity vector aligned with the tangent to the vessel's centerline (assumed as the main direction of the flow). In this way, local flow recirculation is characterized by a negative V_{tan} value. The extent of flow recirculation was then quantified according to the descriptor proposed by Martorell *et al.* [2]. Starting from the calculation of the volume of flow recirculation (VolRec) in CCA, ECA and ICA, they introduced the descriptor normalized V_{tan}·VolRec (NVV) by multiplying V_{tan} by VolRec, dividing it by the volume of the branch and normalizing it to the CCA NVV value, i.e. the inlet segment. The same method has been adopted to define the axial component of the **WSS** (WSS_{ax}) and to quantify the relative area where the cycle averaged direction of WSS vector is opposed to the centerline direction (i.e., negative WSS_{ax}). The hemodynamic descriptors were calculated in the ECA and ICA, bounded by branch origin and ECA2/ICA5 sections.

RESULTS

An overall visualization of the hemodynamic descriptors is presented in Fig. 1. Flow recirculation (Fig. 1, $V_{tan} < 0$ isosurfaces) is evident at the bifurcation in all models, mainly in the near-wall region. Contours of negative WSS_{ax} (Fig. 1, $WSS_{ax} < 0$ region) match closely the region of flow recirculation in all models, confirming that the SA exposed to negative WSS_{ax} is an indirect measure of the extent of flow recirculation. The RRT distribution highlights that the outer walls of ECA and ICA at the bifurcation are exposed to low and oscillatory **WSS**, while isosurfaces of cycle-averaged LNH (LNH_{avg}) show the presence of large helical counter-rotating fluid structures developing in the CCA, being amplified by the bifurcation and persisting in the daughter branches. Quantitative results are reported in Fig. 2. In detail, ECA always showed higher NVV values and larger areas exposed to negative WSS_{ax} or high RRT, with respect to the ICA. On the contrary, helicity intensity is higher in the ICA than in the ECA. The variability in the quantitative hemodynamic description reflects the wide variability in the carotid bifurcation geometry.

DISCUSSION

Our findings reinforce the idea of the beneficial role of helical flow at the bifurcation, ensuring stability and reducing flow recirculation. It is also confirmed that the SA exposed to negative WSS_{ax} represents the footprint at the vessel's wall of the extent of flow recirculation, and consequently it can be a useful indicator of the impact of flow recirculation on ECs inflammatory biomarkers. At this preliminary stage of investigation, some points deserve attention. In particular, we found that NVV is always higher in ECA than in ICA, opposed to Martorell *et al.* [2]. This discrepancy may be ascribed to the level of idealization of the geometry used in their study, to their assumption about a truncated and straight CCA (the CCA length affects the bifurcation hemodynamics [3]), and to the strategy applied to prescribe condition at outflow boundaries. In order to enforce the observed links among intravascular and WSS-based indicators of disturbed flow and to better clarify some discrepancies with *in vitro* observations, in the future the study will be extended to a larger cohort.

ACKNOWLEDGEMENTS

This study has been partially granted by the Joint Project for the Internationalization of Research 2014 from Compagnia di San Paolo Foundation and Politecnico di Torino.

REFERENCES

[1] Gallo, D et al., *J Biomech*, 45:2398-2404, 2012.
[2] Martorell, J et al., *Cardiovasc Res*, 103(1):37-46, 2014.
[3] Gallo, D et al., *Ann Biomed Eng*, 43(1):68-81, 2015.
[4] Wang, C et al., *Arterioscler Thromb Vasc Biol*, 33:2130-6, 2013.

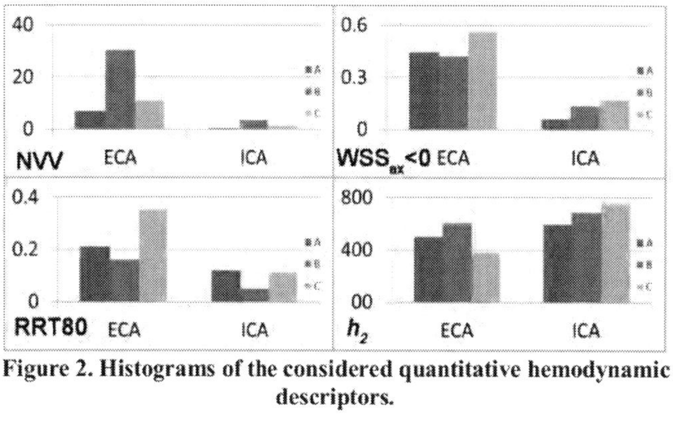

Figure 2. Histograms of the considered quantitative hemodynamic descriptors.

SB³C2015
Summer Biomechanics, Bioengineering and Biotransport Conference
June 17-20, 2015,Snowbird Resort, Utah, USA
SB³C2015-548

THE IMPACT OF VAD SURGICAL IMPLANTATION CONFIGURATIONS ON ITS THROMBOGENIC POTENTIAL

Wei C. Chiu (1), Allison J. McLarty (2), Shmuel Einav (1), Marvin J. Slepian (1,3), Danny Bluestein (1)

(1) Dept. of Biomedical Engineering
Stony Brook University
Stony Brook, New York, USA

(2) Dept. of Surgery
Stony Brook University
Stony Brook, New York, USA

(3) Sarver Heart Center
The University of Arizona
Tucson, Arizona, USA

INTRODUCTION

Continuous-flow ventricular assist device (VAD) has becom of the standard therapies for advanced stage heart failure. Several V are currently approved by the FDA for Bridge-to-transplant a destination therapy. With the steadily growing population of device recipients, various post-implant complications were reported, e.g., thromboembolic events, excess bleeding episodes, device thrombosis and device malfunction, which may lead to device replacement surgery or sudden death. Most of those complications are associated with the hyper-shear generated by VADs that increases its thrombogenicity.

To address these complications, studies tried to optimize the device design to reduce the risk of shear-induced blood damage, e.g., hemolysis or platelet activation. Despite the improved device thrombogenic performance that can be achieved by the device design optimization, remarkably few studies have considered the thrombogenic impact of the surgical implantation configuration, and its potential deleterious effect on the optimized device performance. Furthermore, with the extended functional use of VADs for destination therapy, it is essential to optimize the systemic thrombogenicity of the implanted device as part of the post-implant circulation system.

Two axial continuous-flow VADs were studied: Thoratec HeartMate II (HMII) and MicroMed HeartAssist 5 (HA5). Both share similar design concept; a three-fin (120° radially apart) inflow stator, a three-fin and six-fin outflow stators for HMII and HA5, respectively, and a rotating impeller. In their implanted configuration, both VADs require several components: elbowed inflow/outflow cannulae and an outflow graft, for connecting the VAD to the left ventricle and the ascending aorta. Thus, surgical variance occur while connecting these components together. For instance, the radial orientations of the inflow and outflow cannulae with respect to the VAD and the outflow graft anastomosis angle with respect to the ascending aorta may widely vary.

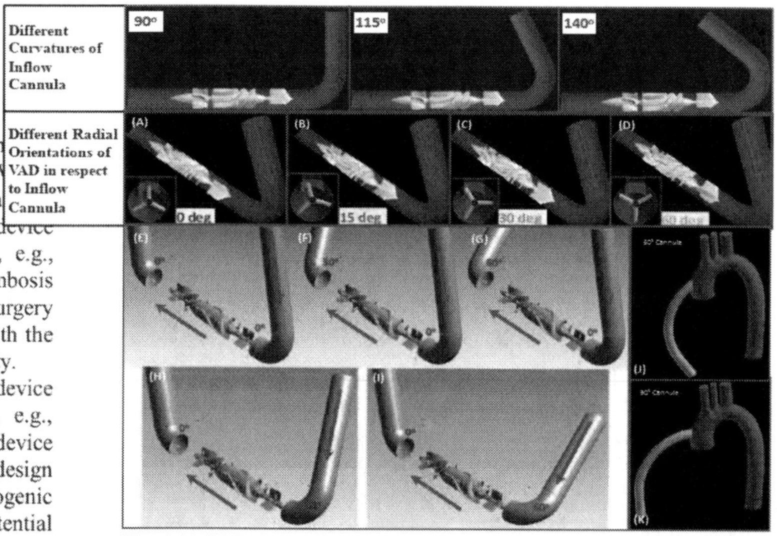

Fig. 1 (A-D) Radial orientations, 0°, 15°, 30° and 60° of HA5 inflow cannula and (E-I) Radial orientations (0°, 30° and 60°) of the HMII inflow and outflow cannulae with respect to the inflow and outflow stators, respectively. (J&K) 60° and 90° anastomosis angles, respectively, of the outflow graft with respect to the ascending aorta. The red arrows in A-I indicate the flow direction.

This study aims to quantify the effect those have on the thrombogenic potential and to determine the optimal configurations.

METHODS

Preparation of Aorta and VADs Geometries: Human aorta geometry was obtained from consented patients (approved by Stony Brook University IRB). ITK-snap and Gambit (Ansys Inc., Lebanon, NH) were utilized for geometry reconstruction (Figure 1 J&K: 60°&90° anastomosis angle, respectively). The geometries of both VADs and their associated components were prepared using Ansys DesignModeler (Fig. 1 A-D: HA5 inflow cannula radial orientations (0°, 15°, 30° and

60°); E-I: HMII inflow and outflow cannulae radial orientaitons (0°, 30° and 60°)).

Numerical Simulations of the Thrombogenic Potential: Fluid structure interaction (FSI) simulations were conducted in both VADs using the Fluent CFD package (Ansys Inc., Lebanon, NH). Blood was modeled as turbulent two phase Newtonian fluid (ρ = 1,060 kg/m^3, μ = 0.0035 kg/m-s, Temp = 37 °C) at flow rates of 5 and 4 L/min for HMII and HA5, respectively, and impeller speeds of 9,000 rpm. Approximately 30,000 platelets were seeded at the upstream of both VADs' inflow cannulae and the corresponding stress accumulation (SA) was computed along the platelet trajectories by incorporating the combined effect of stress and exposure time. Probability density function (PDF) was employed to statistically represent the distribution of the SA of the multiple trajectories.

The outflow condition of the HA5 simulation was extracted and employed as the inlet boundary condition (BC) of the outflow graft. Pressure waveforms from literatures were downscaled to represent the pathological condition and utilized as the inlet BC at the aortic root and outlet BCs at the descending aorta, brachiocephalic trunk, left common carotid artery and left subclavian artery. Approx. 5,000 platelets were released from the outflow graft inlet.

RESULTS

Radial orientation of VADs inflow/outflow cannulae: PDF results of HA5 inflow cannula showed that 0° and 60° configurations have the highest and lowest thrombogenic potentials, respectively (Fig. 2). Steady state stimulation results of the HMII inflow and outflow cannulae configuration showed velocity flow field across HMII and its associated components is affected by different configurations, especially on the downstream of the elbows and stators regions (Fig. 3)

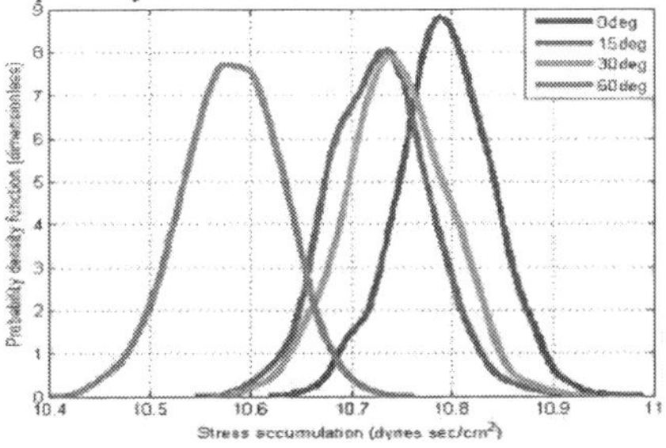

Fig. 2 PDF of HA5 inflow cannula radial orientations. The 60° and 90° orientations have the lowest and highest thrombogenic potential as compared to the 15° and 30° orientations. The 60° and 0° orientations had a mean SA of approx. 10.6 dynes-sec/cm^2 and 10.8 dyne-sec/cm^2, respectively.

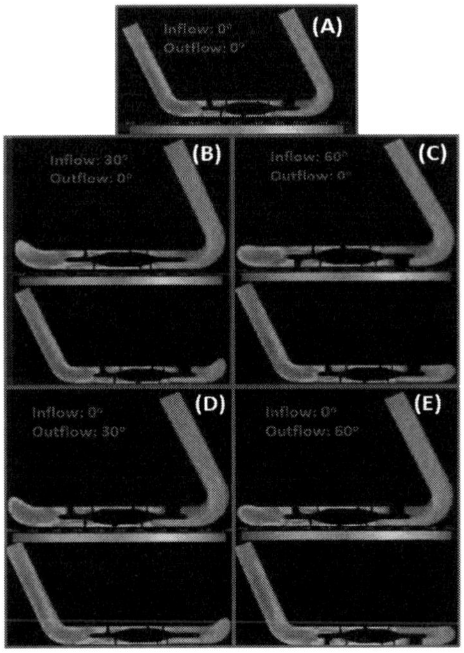

Fig. 3 Velocity flow filed of the HMII inflow and outflow cannulae radial orientations. (A) Both inflow and outflow cannulae have 0° radial orientations with respect to the VAD. (B&C) The inflow cannulae have 30° and 60° orientations, respectively, while both outflow cannulae have 0° orientation. (D&E) Both inflow cannulae have 0° radial orientation, and the outflow cannulae have 30° and 60° orientation. Differences were observed at the downstream of the inflow stator with the inflow cannula orientation varied.

Anastomosis angles of outflow graft respect to ascending aorta: The PDF of the outflow graft anastomosis angles showed that the main modes of both anastomosis angles located at the similar SA (approx. 2 dyne-sec/cm^2); however, the 90° anastomosis angle PDF had a secondary mode at its tail region (range from 50-100 dyne-sec/cm^2) (Fig. 4).

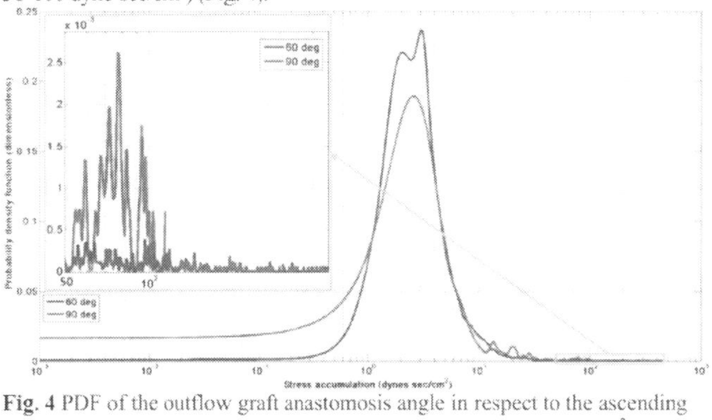

Fig. 4 PDF of the outflow graft anastomosis angle in respect to the ascending aorta. Main modes of both 60° and 90° were at approx. 2 dyne-sec/cm^2, with the 90° having a secondary mode at the tail region (50-100 dyne-sec/cm^2).

DISCUSSION

Our numerical simulations indicated that the VAD thrombogenic potential may significantly be altered while connecting the associated components differently. In our simulated conditions of HA5, the lowest thrombogenic potential occurred while the inflow cannula is 60° radially aligned with the three-fin inflow stator. This appears to be associated with secondary flow patterns formed at the elbowed pipe. When the elbowed cannula is 0° radially aligned with the stator, the stator fin is faced directly toward slower flow area downstream of the elbow; thus, platelets flowing through this region have longer exposure to elevated shear in this region.

In the HMII VAD simulations, we observed a larger slower flow region at the downstream of the stator fin when the inflow cannula was radially orientated at 60°. This appears to correlate with clinically reported pump thrombosis, since most thrombus is found in this area.

From the study of the outflow graft anastomosis angle, we found that the 60° anastomosis angle has better thrombogenic performance than the 90°. Clearly, when the blood flow is reintroduced to the aorta via 60° anastomotic outflow graft, the strong swirling jet flow generated by the VAD does not directly impinge onto the aorta wall as the 90° anastomotic outflow graft does.

In the present study, we demonstrated the importance of the VAD surgical implantation configuration, and its potential impact on the systemic thrombogenicity of the implanted VAD, which appears to affect directly the clinically reported complications. This study may be used for guidance of surgical configurations for yielding the optimal device performance. More studies with various conditions are planned.

ACKNOWLEDGEMENTS

Funding for the project was provided by the NIH: NIBBIB Quantum Award Phase II, U011EB01248 (DB).

REFERENCES

[1] Go, A.S. et al., *Circulation*, 127: e6-e245, 2013
[2] Girdhar, G et al., *PLoS One*, 7(3):e32463, 2012
[3] Chiu, W-C et al., *J. Biomech.*, 136(2): 021014, 2014
[4] Kar, B et al., *Tex. Heart I. J.*, 32: 294-298, 2005
[5] Griffith, B.P. et al., *Ann. Thorac. Surg.*, 71: S116-420, 2010
[6] Sheriff, J et al., *Ann Biomed Eng.*, 38(4): 1442-50, 2010
[7] Bhamidipati, C.M. et al., *J. Thorac. Cardiovasc. Surg.*, 140:e7-8, 2010
[8] Chiu, W-C et al., *ASAIO J.*, 60 (2): 237-40, 2014

SB³C2015
Summer Biomechanics, Bioengineering and Biotransport Conference
June 17-20, 2015, Snowbird Resort, Utah, USA
SB³C2015-549

FUNCTIONAL AND ANATOMICAL MEASURES FOR OUTFLOW BOUNDARY CONDITIONS IN ATHEROSCLEROTIC CORONARY BIFURCATIONS

Jelle T.C. Schrauwen (1), Adriaan Coenen (2,3), Akira Kurata (2), Jolanda J. Wentzel (1), Antonius F.W. van der Steen (1,4), Koen Nieman (2,3), Frank J.H. Gijsen (1)

(1) Department of Biomedical Engineering, Thoraxcenter, Erasmus Medical Center, Rotterdam, the Netherlands

(2) Department of Radiology, Erasmus Medical Center, Rotterdam, the Netherlands

(3) Department of Cardiology, Thoraxcenter, Erasmus Medical Center, Rotterdam, the Netherlands

(4) Department of Imaging Physics, Delft University of Technology, Delft, the Netherlands

INTRODUCTION

Atherosclerosis is a chronic disease of the arterial system that leads to plaque formation. A well-established hemodynamic parameter that is of interest when studying atherosclerosis in coronary arteries is fractional flow reserve (FFR). In clinical practice FFR is used to direct treatment of coronary artery disease. Another hemodynamic parameter is time-averaged wall shear stress (WSS), for which there is evidence that it plays a key role in atherosclerotic disease progression. Both WSS and FFR can be determined by computational fluid dynamics (CFD). To perform CFD computations the geometry of a coronary artery is required, together with boundary conditions for the inlet and outlets of that geometry.

Aim - The aim of this research was to study the effect of the outflow conditions on FFR and WSS in coronary bifurcations. A functional measure to set the outflow conditions was derived with computed tomography perfusion (CTP), which is an emerging non-invasive flow imaging technique. Two anatomical measures were derived from computed tomography angiography (CTA) data: Myocardial volume and coronary diameters.

METHODS

Imaging - Ten bifurcations were selected from eight patients with coronary artery disease. All patients underwent a CTA and an adenosine stress CTP [1]. CTP measures the myocardial blood flow based on the distribution of contrast medium. The concentration of contrast agent as a function of time is measured and a time-attenuation curve was constructed. From this curve the myocardial blood flow was extracted.

CFD - Segmentation of the CTA data was done with a validated software tool [2]. The two centerlines of the bifurcation were manually selected and lumen contours were segmented perpendicular to the centerlines. With the resulting contours a volume was created and converted into a mesh with a typical size $3 \cdot 10^6$ to $5 \cdot 10^6$ cells. Steady-state CFD computations were performed using standard numerical techniques (Fluent 14.5, Ansys Inc, Canonsburg, US). Poiseuille profiles were defined at each inlet based on a scaling-law [3]. The blood was modeled as a non-Newtonian fluid with a Carreau model and a density of 1060 kg/m³.

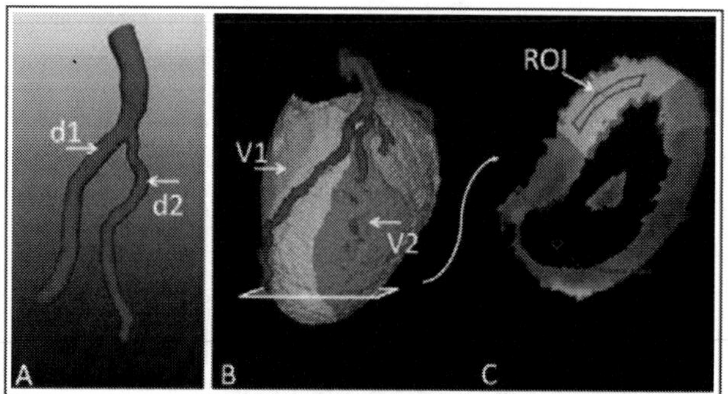

Fig. 1A: 3D reconstruction of a coronary bifurcation. From the diameters of the daughter branches (d1, d2) a flow ratio was determined. B: with the myocardial volumes (V1, V2) derived from CTA data a second set of outflow ratios was found. C: the CTP and CTA data were superimposed. The mean flow in both volumes was obtained from multiple ROI's, resulting in the third set of outflow ratios.

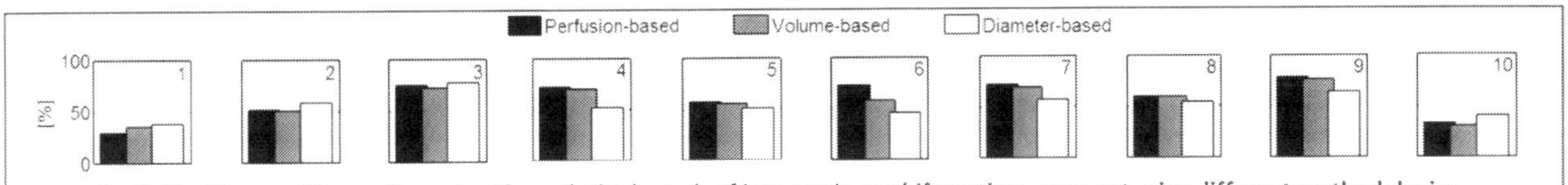

Fig 2: The three outflow ratios going through the branch of interest in ten bifurcations were set using different methodologies.

Outflow boundary conditions - In all computations the outlet boundary conditions were set as a flow ratio over the two outlets of a bifurcation. Per bifurcation three sets of outflow ratios were determined based on different measures.

1) **Diameter-based**: The diameters in the outlets were measured by a trained observer at a healthy section close to bifurcation (Fig. 1A, white arrow). The outflow ratio was determined based on the diameters in combination with a scaling law [3].
2) **Volume-based**: The myocardial volume, corresponding to each outlet of the bifurcation, was segmented using a validated Voronoi-algorithm (AZE Ltd., Tokyo, Japan) [4]. The two resulting volumes were directly scaled to set the outflow ratio over a bifurcation (Fig. 1B).
3) **Perfusion-based**: The perfusion data from CTP was analyzed by a trained observer within both the segmented volumes by sampling multiple regions of interest. The mean flow rate of each volume was multiplied with the volume itself, and the outcomes were directly scaled to set the outflow ratio (Fig. 1C) over a bifurcation.

Post processing - From the CFD results, the pressure drop and 3D WSS magnitude were extracted. The computed FFR was defined as the ratio between the pressure at the outlet and at the inlet. All 3D WSS results were mapped to a 2D representation. To quantify the influence of the different boundary condition we compared the area of overlap of WSS above the 75th percentile.

The results from the anatomical measures (the diameter-based and volume-based approaches) were compared to results from the functional measure (perfusion-based approach).

RESULTS

The flows in the two volumes in ten cases were evenly distributed. While flows where observed ranging from 37 ml/100ml/min to 169.1 ml/100ml/min, the flow difference between two adjacent volumes was not more than 15 ml/100ml/min in eight out of ten cases.

The differences between the outflow ratios found with the perfusion-based approach and the volume-based approach were small: -2% [-2%:1%]. The differences between the perfusion-based and diameter-based results were larger and showed more spread: -6% [-15%:7%] (Fig. 2).

The pressure-drop differences between the perfusion-based and the volume-based approach were small (1% [0%:4%]), whereas the differences between the perfusion-based and the diameter-based computations were more pronounced (7% [0%:18%]). By using the pressure drops, the FFRs were computed. The FFRs computed with the volume-based and diameter-based approach matched the FFR's from the perfusion-based approach extremely well (Fig. 3).

In general the locations with high WSS agreed, but the sizes of the regions differed (Fig. 4). For all cases, the difference in overlap of the high WSS areas was computed. Between the perfusion-based and the volume-based approach this was 3.0% [2.6%:16.6%]. Using the diameter-based approach larger differences with the perfusion-based approach were observed: 20.2% [10.7%:31.7%].

DISCUSSION

In this study CTP data was used for the first time to set patient-specific outflow boundary conditions in CFD computations of atherosclerotic coronary bifurcations. Currently anatomy-based measures, in combination with a scaling law, are often used to set outflow boundary conditions. However, our study showed that distinct differences were found when a functional measure is employed. One difference in particular was seen between the ratios derived from CTP data or outlet diameters; having pressure-drop differences of as much as 40%. Interestingly, the FFR's were relatively unaffected by the differences in outflow ratios, indicating that a more easily obtainable anatomical measure might suffice for these computations. The WSS on the other hand, showed local deviations due to the different outflow ratios. And while the potential of volume-based approach was demonstrated, a functional measure of the outflow ratio might be a necessity to accurately determine the WSS locally in this type of vessels.

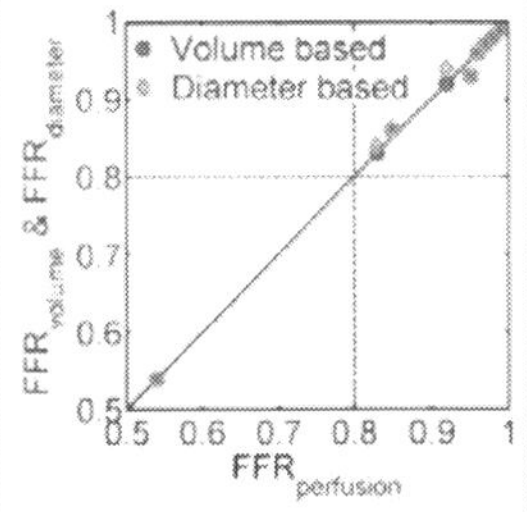

Fig. 3: The FFR values from the 3 different approaches show an excellent match.

ACKNOWLEDGEMENTS

This research was performed within a framework of the CARISMA-program of STW (Stichting Technische Wetenschappen).

REFERENCES

[1] Rossi. et al., *Euro. Heart J.*, 15(1): 85-86, 2013.
[2] Kirişli, H. a et al., *Med. Image Anal.* 17:859–76, 2013
[3] van der Giessen, A. et al., *J. Biom.* 44(6):1089-95, 2011
[4] Kurata, A. et al., *Euro. Radiol.*, 25(1): 49-57, 2015

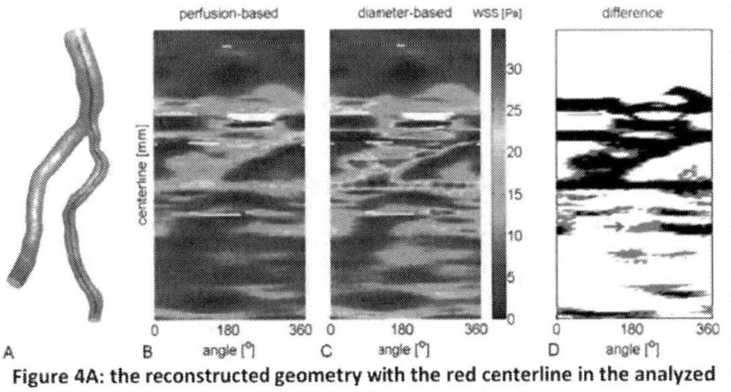

Figure 4A: the reconstructed geometry with the red centerline in the analyzed branch. B,C: 2D WSS map obtained with the perfusion-based approach and diameter-based approach. D: The difference of the 2 WSS maps with regard to high WSS. The black areas represent where both cases had high WSS, the gray areas where the two approaches differed.

SB³C2015
Summer Biomechanics, Bioengineering and Biotransport Conference
June 17-20, 2015, Snowbird Resort, Utah, USA
SB³C2015-551

QUASILINEAR AND NON-QUASILINEAR VISCOELASTIC BEHAVIOR OF COLLAGEN GELS DURING STRESS RELAXATION

Christopher E. Korenczuk (1), Victor K. Lai (2), and Victor H. Barocas (1)

(1) Department of Biomedical Engineering
University of Minnesota
Minneapolis, MN, USA

(2) Department of Chemical Engineering
University of Minnesota Duluth
Duluth, MN, USA

INTRODUCTION

Collagen is an essential structural protein found throughout the connective tissue of the human body, providing tensile strength for tendons, ligaments, vessels, and other connective tissues. In light of collagen's prevalence, understanding its mechanical behavior – including nonlinearity and viscoelasticity – is necessary to accurately quantify tissue mechanics. Extensive work has been done to analyze tissues such as tendons, ligaments, and extraocular muscle [1,2,3], but our fundamental understanding of collagen's viscoelastic response is still limited. Some work has been done to quantify collagen fiber mechanics [4,5], but much is still unknown.

A common model to quantify the nonlinear response of soft tissues is the quasilinear viscoelastic (QLV) model suggested by Y.C. Fung [6]. The QLV model considers a separable stress function dependent on both time and stretch. Both an elastic and a relaxation function are used in order to represent the response based on stretch and time, respectively. The relaxation function, G(t), is often reduced, and expressed as sum of normalized exponential functions in the form

$$G(t) = \sum B_i e^{-t/c_i} \quad (1)$$

where B_i is a material constant fitting parameter, t is the time, and c_i is a specified time constant.

Separability of terms and the ease of use make QLV attractive, but it is not clear that tissues are, in fact, quasilinear. Therefore, the validity of the QLV model remains an open question (see, e.g., [2]). Additionally, native tissues are not ideal in assessing the QLV model because reproducibility is difficult given the variation among individuals, effect of acquisition and storage conditions, changes with injury or disease, multiple cell and protein types within a tissue. We chose to examine the stretch- and time-dependent response of collagen gels as a reproducible and controllable model of tissue [7].

METHODS

Collagen gel solutions were prepared containing 14.2% 1M HEPES, 2.6% 1M NaOH, 10% 10x MEM, 6% fetal bovine serum (FBS), 0.1% penicillin/streptomycin (P/S), 0.1% fungizone, 1% L-glutamine, and 66% Invitrogen rat-tail Collagen I 3mg/mL (Gibco, Life Technologies) to obtain a final collagen concentration of 2mg/mL. The solution was prepared on ice to prevent gelation prior to incubation and was casted into dog-bone molds (Figure 1A). Each gel (n=6) contained between 6-8 mL of collagen gel solution. Porous scouring pads were cut to size and fit into the ends of the mold to allow for gel anchoring and clamp attachment points during testing. Casted gels were then incubated at 37° C and 5% CO_2 for 1-2 hours, and stored at 4° C until testing 24 hours later.

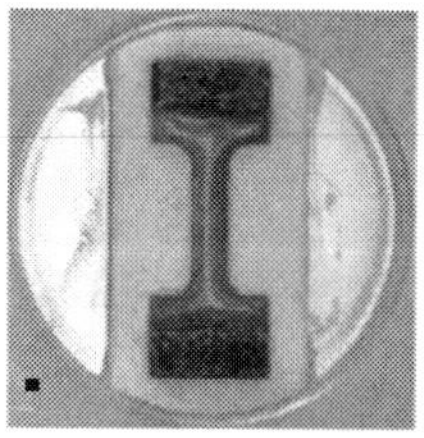
A.
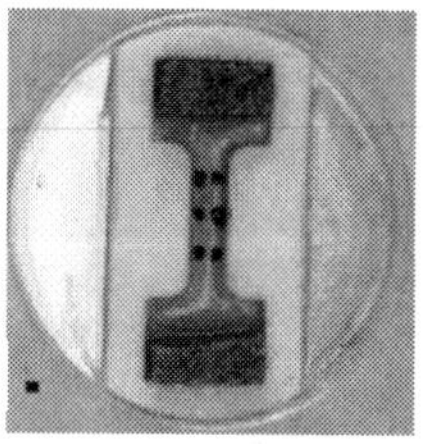
B.

FIGURE 1. (A) COLLAGEN GEL SOLUTION CASTED INTO A DOG-BONE SHAPE MOLD WITH POROUS SCOURING PAD ENDINGS. (B) 6-DOT VERHOEFF'S STAIN TRACKING PATTERN. SCALE BAR: 10 mm

Finalized collagen dog-bone gels were marked on the neck region with Verhoeff's stain in a 6-dot pattern to track neck strain throughout testing (Figure 1B). Samples were placed in a 1x PBS solution bath and tested on an Instron biaxial testing system in uniaxial mode equipped with 5N load cells. Stress relaxation experiments were performed at strains of 5% (n=2), 10% (n=2), and 15% (n=2) for 30 minutes. An initial strain ramp of 20%/s was used. Morphological measurements of the gels were taken from images and measured in ImageJ.

Experimental load and stretch measurements were analyzed in Matlab. Load cell force measurements were converted to the 1st Piola-Kirchhoff (PK1) stress by using the initial width and thickness of each sample. Stress was also normalized to the peak stress in order to fit a QLV reduced relaxation function of the form

$$G(t) = B_1 + B_2 e^{-t/2000} + B_3 e^{-t/200} + B_4 e^{-t/20} + B_5 e^{-t/2} \quad (2)$$

where the B values are parameters determined by fitting the experimental data with equation (2) and t is in seconds. B_1 represents the solid component (i.e., infinite time constant), and each following time constant is spaced in decades. The largest time constant was set to 2000 s, corresponding roughly to the duration of the stress relaxation experiment (1800 s).

RESULTS

Long-time solid like behavior was observed during stress relaxation for all collagen gels. Collagen gels at lower strains (5%) experienced slower relaxation rates compared to those at higher strains (15%) (Figure 2). The average peak stresses were larger for higher strains, with higher variability for the 10% strain case (Figure 2 inset).

When the QLV reduced relaxation function of equation (2) was fit to experimental data, it showed a mean R^2 value of 0.9958 (5% strain), 0.9863 (10% strain), and 0.9923 (15% strain), indicating excellent goodness of fit. Varying model parameters (B values) were obtained from the different strain cases. On average, the fitting parameter (B_5) associated with the lowest time constant (2 s) was higher for larger strains, while the fitting parameters (B_1 and B_2) associated with the larger time constants (∞ and 2000 s) were higher for lower strains (Figure 3).

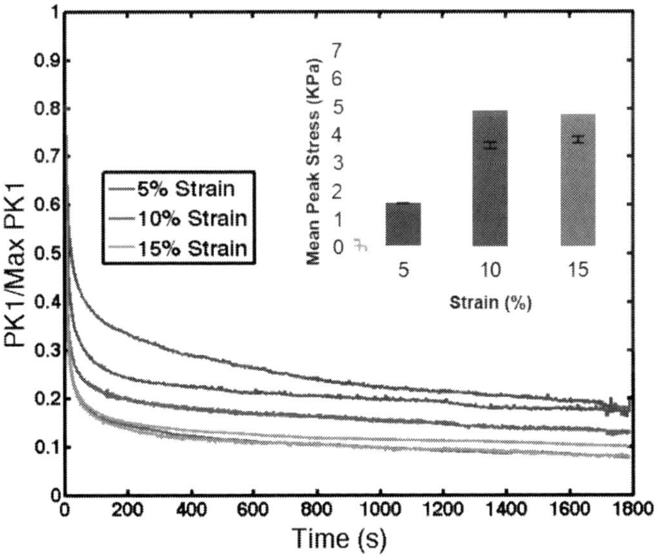

FIGURE 2. NORMALIZED STRESSES FOR EACH EXPERIMENT. (INSET) AVERAGE PEAK STRESS FOR EACH STRAIN (ERROR BARS = ST. DEV.).

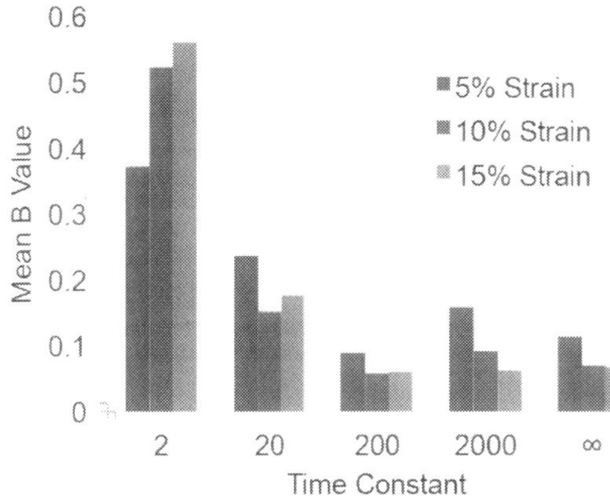

FIGURE 3. AVERAGE B VALUES FROM FITTING EQUATION (2) TO EXPERIMENTAL DATA.

DISCUSSION

An observed increase in relaxation rate with increasing strain (Figure 2) shows that collagen gel stress relaxation is both stretch- and time-dependent in a non-separable way. The QLV reduced relaxation function fit individual experiments well, but varying fit parameters were seen throughout the sample population for different strains. A greater fitting parameter associated with a smaller time constant (2 s) for larger strains indicates an increase in relaxation rate for high strains (Figure 3). On the other hand, a larger fitting parameter associated with longer time constants (∞ and 2000 s) in the smaller strain case (5%) implies a slower relaxation rate at low strains. These results confirm the observed experimental response (Figure 2).

The reduced relaxation function presents an applicable model for stress relaxation of individual collagen gels, but separability of the elastic (strain-dependent) and relaxation (time-dependent) function in the QLV model is not present. Collagen gels exhibit both a strain-dependent and time-dependent response, negating the possibility of utilizing a separable QLV model. These results confirm previous work performed on collagen-based native tissues [1], and suggest that a different model must be used to characterize the nonlinear viscoelastic response of collagen.

ACKNOWLEDGEMENTS

This work was supported by the National Institutes of Health (R01 EB005813).

REFERENCES

[1] Provenzano, P et al., *Annals of Biomed Eng*, 29:908-914, 2001. [2] Duenwald, SE et al., *Annals of Biomed Eng*, 37:1131-1140, 2009. [3] Yoo, L et al., *Investigative Ophthalmology & Visual Sci*, 50:3721-3728, 2009. [4] Shen, ZL et al., *Biophysical J*, 100:3008-3015, 2011. [5] Haut, RC et al., *J Biomech*, 5:423-430, 1972. [6] Fung, YC, "Biomechanics: material properties of living tissues." *Springer*, 1993. [7] Roeder, BA et al., *J Biomech Eng*, 124:214-222, 2002.

SB³C2015
Summer Biomechanics, Bioengineering and Blotransport Conference
June 17-20, 2015,Snowbird Resort, Utah, USA
SB³C2015-552

AORTIC PEAK STRESS INDUCED BY ANTIHYPERTENSIVE MEDICATIONS IN AORTIC DISSECTION PATIENTS

Vittoria Flamini (1), Simon Gallot Lavallée (2), Zed P. Lucienne (1), Scott Maddalo (3), Abe DeAnda (3), and Boyce E. Griffith (4)

(1) Polytechnic School of Engineering
New York University
New York, NY, United States

(2) Department of Mechanical Engineering
Imperial College London
London, UK

(3) NYU Langone Medical Center
New York, NY, United States

(4) Department of Applied Mathematics
University of North Carolina at Chapel Hill
Chapel Hill, NC, United States

INTRODUCTION

Aortic dissection occurs when a tear in the aortic wall propagates into the media to form a false lumen within the vessel wall. Rupture of the false lumen and collapse of the true lumen both carry a high risk of morbidity and mortality. Surgical treatment consists of either replacement of a portion of the aorta, or endovascular stent implantation to cover the affected segment. Both approaches carry significant risks, and it is difficult to determine the optimal timing of the intervention.

This work aims at developing computational tools to assist clinicians in their decision making process, by creating finite element (FE) simulations of the response to medications of aortic dissection patients. In fact, chronic aortic dissection patients are often treated with antihypertensive medications, as these medications have been reported to lower the mean arterial pressure MAP. However, these drugs alter the blood pressure waveform as a whole, and may result in high systolic pressures.

We expect that patient-specific models of the effect of medications on aortic dissection patients will enable the determination of areas of maximum stress concentration in the aorta, data that are not available via clinical imaging, and thus help to select the appropriate treatment, which will lower the likelihood of further dissection or rupture.

METHODS

A retrospective, bibliographic search was conducted on pubmed to identify the changes that occur to blood pressure (BP) in healthy individuals when treated with a regime of antihypertensive medications. Medications were classified in five groups depending on their class, i.e. alpha and beta-blockers, beta blockers, Ca-channel inhibitor, diuretics, etc., and one medication from each group selected for the simulations. Simultaneously, patient-specific aortic dissection geometries were segmented using Simpleware from CT-scans of aortic dissection patients, and exported in an ABAQUS format (Dassault Systémes, Providence, RI).

To obtain reliable stress distributions in patient-specific models, we need to account for initial stresses generated in the blood vessel by the baseline diastolic pressure. To do so, we implemented an iterative solution to the inverse elastostatic problem developed by Bols et al. [1], called the backward displacement method. The constitutive law used in the study is an isotropic, hyperelastic exponential law, often used for arterial wall [2]. The material properties are obtained by fitting aortic biaxial test data available in literature [3]. The inverse elastostatic routine was implemented using custom code written in Matlab (The Mathworks, Natick, MA) and Python to control finite element analyses executed in ABAQUS.

RESULTS

The analysis of the effect of the antihypertensive medications on BP, showed that although all the medications considered could lower MAP, some of them, such as nicardipine, and trichlormethiazide, reached systolic pressures higher than the physiological systolic peak, thus exposing aortic dissection patients to increased loading, see Figure 1.A.

The backward displacement method applied to this complex geometry resulted in the inclusion of the initial stress by estimating the unloaded aortic configuration, as shown in

Figure 1.B-C. The deflated, unloaded aorta can be successively loaded with the diastolic pressure to regain the same configuration as the segmented geometry, as shown in Figure 1.D.

Stress distributions in the aorta at systolic BP with the inclusion of initial stress were found for all medications analyzed; see Figure 2. Von Mises peak stress varied proportionally with systolic BP, but reached extremely high values around the aortic dissection fenestration. Compared with yield stress measured on aortic tissue, the medication-induced peak stresses in a dissected aorta, especially around the fenestration margin, were close to the ultimate stress for aortic tissue reported in literature [4]. For two medications, namely for the nicardipine and trichlormethiazide systolic loading, Von Mises stress computed were beyond the limits reported in literature, see Figure 2.

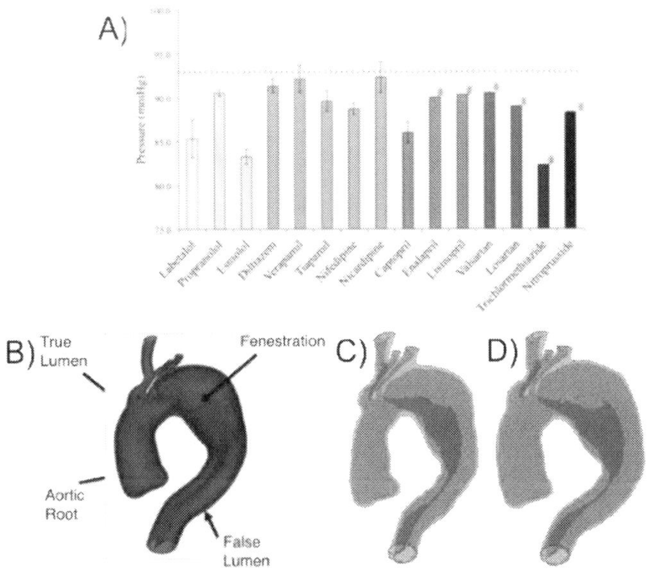

FIGURE 1: A) EFFECT OF THE SELECTED ANTIHYPERTENSIVE MEDICATIONS ON MAP; B) THE SEGMENTED GEOMETRY OF AN AORTIC DISSECTION PATIENT; C) SAME GEOMETRY UNLOADED FROM DIASTOLIC LOADING AS A RESULT OF THE BACKWARD DISPLACEMENT METHOD; D) CONFIGURATION OF THE GEOMETRY AT DIASTOLIC LOADING.

DISCUSSION

The framework implemented in this project for the evaluation of stress distribution and peak stress in patient-specific aortic geometries was successful in recovering the loaded configuration of the image-based model, see Figure 1.A, and in assigning an estimate of the stress that varies locally along the structure and identifies "weak spots", see Figure 2.

The patient-specific stress distribution obtained could be employed to evaluate the likelihood of further dissection in response to the treatment selected, and could be used to support clinical decision making.

Future work include the implementation of fluid-structure interaction simulations for the definition of aortic dissection risk parameters based on diastolic pre-stress and wall-shear stress.

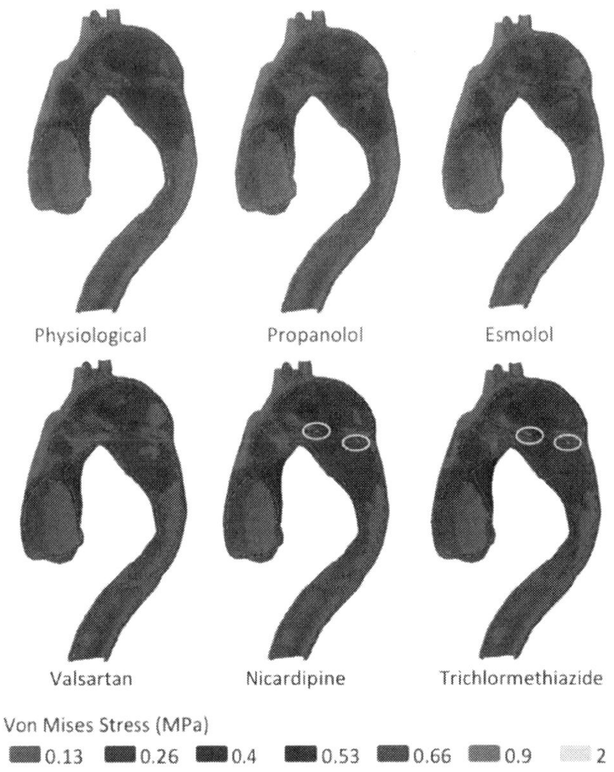

FIGURE 2: CROSS-SECTION OF THE SEGMENTED AORTIC DISSECTION GEOMETRY LOADED WITH THE SYSTOLIC PRESSURE INDUCED BY SELECTED ANTIHYPERTENSIVE MEDICATIONS AFTER INITIAL STRESS WERE ESTIMATED THROUGH THE BACKWARD DISPALCEMENT ROUTINE. THE YELLOW CIRLES INDICATE THE ELEMENTS AROUND THE FENESTRATION IN WHICH THE STRESS IS HIGHER THAN THE ULTIMATE STRESS FOR AORTIC TISSUE REPORTED IN LITERATURE.

ACKNO WLEDGEMENTS

We gratefully acknowledge support from AHA award 10SDG4320049, NIH awards GM071558 and HL117063, and NSF awards DMS-1016554 and OCI-1047734.

REFERENCES

[1] Bols, J., *J Comp App Math*, 2013, 246:10-17;
[2] Gasser, T.C., *J R Soc*, 2006, 3(6):15-35;
[3] Azadani, A.N., *Ann Thorac Surg*, 2012 93(1):87-94.
[4] Holzapfel, G. A. *Biomechanical Modelling at the Molecular, Cellular and Tissue Levels*. Springer, 2009, 259-334.

CHANGES IN MECHANOSENSITIVITY OF CHILDREN CELLS: AN APPROACH TO INVESTIGATE BONE REPAIR AND BONE DEVELOPMENTAL DISEASE

S. Barreto (1,2,3), A. R. Cameron (1,2,3), Dylan Murray (4), F.J.O'Brien (1,2,3)

(1) Tissue Engineering Research Group, Royal College of Surgeons in Ireland, Dublin Ireland

(2) Trinity Centre for Bioengineering, Trinity College Dublin, Ireland

(3) Advanced Materials and Bio-Engineering Research (AMBER) Centre, Ireland

(4) National Paediatric Craniofacial Center, Children's University Hospital, Temple Street, Dublin, Ireland

INTRODUCTION

Children have a remarkable capacity for bone healing compared to adults due to their enhanced ability to regenerate tissue. However, this enhanced potential for tissue formation can sometimes be altered in developmental diseases, such as craniosynostosis (CS), where patent sutures within the skull of a child prematurely ossify and form fused sutures, thereby changing the growth pattern of the skull and potentially restricting growth of the brain. In CS, the premature fusion of the sutures may be attributed to genetic or environmental conditions that enhance the potential for accelerated bone formation.

Since bone generation is mediated by signaling mechanisms that include stimuli from the surrounding environment [1, 2], pathological changes in the physiology of the sutures of CS children may be associated with changes in both the stimuli provided by the extracellular environment, as well as the functional capacity of cells within the sutures to sense and respond to this stimuli.

The goal of this study is to use 2D polyacrylamide (PA) gels with a range of stiffness to understand the specific mechanotransductive processes that underpin the responses of cells isolated from patent and fused sutures (PCs and FCs respectively) of CS patients. We firstly aim to characterize the PCs, FCs and cells derived from normal calvarial bone (NCs) in terms of their phenotype and potential for osteogenic differentiation. We secondly aim to identify the changes in the mechanosensitivity of cells associated with CS by comparing how the different cells respond to biophysical cues provided by changes in stiffness of PA gels. Ultimately, the identification of the molecular mechanisms that accelerate bone formation in CS patients and govern children's enhanced regenerative capacities could be used to develop advanced next generation therapeutics to promote bone repair in adults as well as children.

METHODS

Bone samples from patent sutures, fused sutures and normal (temporal-parietal) calvarial bone of children with CS were collected in PBS under ethical approval from Children's Hospital Temple Street, Dublin. Cells were extracted using a digestion buffer consisting of 300 Units/ml of collagenase Type F and 0.25% trypsin (Sigma-Aldrich, Ireland). Five sequential digestions were performed at 37°C for 10, 20, 30, 50 and 70 min, respectively. Cells from digestion III-V were pooled together and seeded into T25 flasks. Cells from passage 2-4 were characterized and used in mechanotransductive studies.

Cell characterization study: PCs, FCs, NCs and adult mesenchymal stem cells (MSCs) were cultured at a density of 3000 cells/cm^2 in 6-well plates to analyze their osteogenic potential. Alkaline phosphatase (ALP) activity and calcium deposition was measured after 7 and 14 days in culture in both growth medium (GM) and osteogenic medium (OM).

Effect of substrate stiffness on CS cells: In order to analyze the changes in the mechanosensitivity of cells from CS patients, PA gels were produced by mixing different ratios of acrylamide and bis-acrylamide monomer concentrations, and inducing free radical polymerization using ammonium persulfate (APS) and tetramethylethylenediamine [3]. Polymerization was performed between glass slides in an oxygen-free environment, using 25mm glass coverslips. Collagen type I was covalently bound to the surface of the gels using sulfo-SANPAH, before seeding with PCs, FCs and NCs at a 3000 cells/cm^2 density.

The effect of stiffness on cell morphology was evaluated by measuring cell-spread area on 1, 10, 100 and 300 kPa PA gels after 48 hours in culture. The effect of stiffness on osteogenic differentiation of cells from CS patients was analyzed using a calcium mineralization assay (Cambridge Bioscience Ltd) after 14 days in culture.

RESULTS

PCs expressed similar levels of calcium (Fig.1A) and ALP activity (Fig.1B) to MSCs after 7 and 14 days in growth medium, while FCs and NCs expressed similar values of calcium and ALP activity, which were higher than MSCs.

FCs and NCs showed higher calcium on days 7 and 14 in osteo medium compared to MSCs and PCs. ALP activity of FCs, NCs and MSCs was similar at day 7 in OM but higher than in PCs, whereas ALP activity was similar for all the groups at day 14.

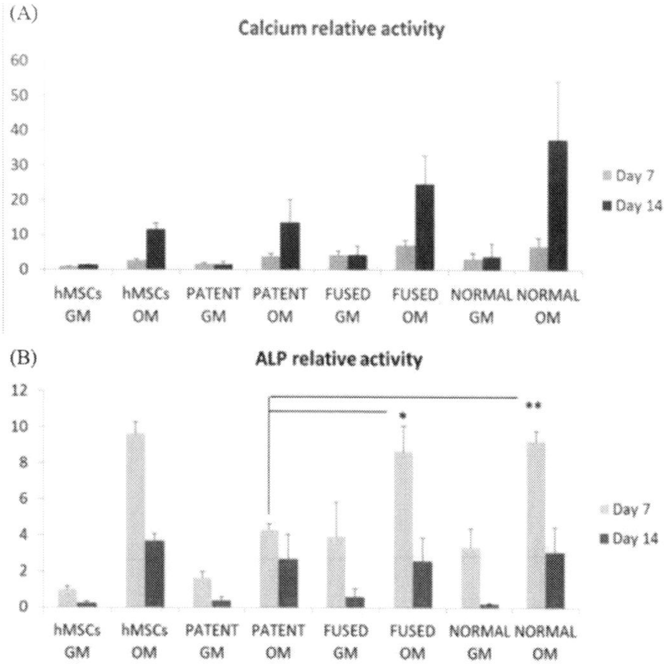

*Figure 1: (A) Calcium and (B) alkaline phosphatase activity of MSCs and cells from CS patients in GM and OM after 7 and 14 days in culture. Errors are standard deviation and significance was determined for * $p < 0.05$ and ** $p < 0.01$ (n=3).*

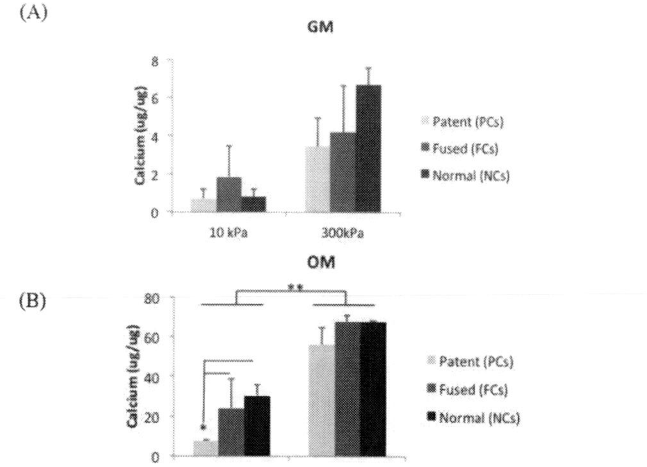

Figure 2: Calcium activity of PCs, FCs and NCs in: (A) GM and (B) OM after 14 days in culture on 10 and 300 kPa PA gels. Errors are standard deviation and significance was determined for $p < 0.05$ and ** $p < 0.01$.*

The effect of stiffness on cell morphology showed a 2.5-fold increase in the spread area of PCs cultured on 300 kPa PA gels compared to those cultured on 1 kPa gels ($p < 0.05$), while a 1.7-fold increase in spread area was seen in PCs cultured on 300 kPa PA gels compared to 10 kPa gels ($p < 0.01$). The 10 and 300 kPa PA gels were further used for the analysis of the mechanosensitivity of CS cells.

In both GM and OM conditions, calcium production of the three groups of CS cells was higher when cultured on 300 kPa PA gels compared to those cultured on 10 kPa gels (Fig.2). In OM conditions, PCs cultured on 10 kPa PA gels (Fig.2B) had a lower calcium production compared to FCs and NCs (3.7- and 4.7-fold decreases respectively). In contrast, the calcium production by PCs was similar to the levels of FCs and NCs when cultured on 300 kPa gels (Fig.2B). Also, an 8.7-fold increase in calcium production was observed in PCs cultured on 300 kPa compared to those cultured on 10 kPa gels (Fig.2B).

DISCUSSION

Initial characterization studies revealed that calcium expression in the CS cells cultured in GM was similar when compared to adult MSCs and the enhanced osteogenic potential of CS cells was not observed until cultured in OM. This indicates that CS cells are not mature osteoblasts but rather a population of osteo-progenitor cells. Despite CS cells appearing to be a progenitor population, there were differences in the level of maturation between the different CS groups, with greater expression of osteogenic markers in cells from fused and normal calvarial tissue compared to those derived from patent sutures. This suggests a more progressed bone formation in fused versus patent sutures, similar to the phenotype of cells of the mature bone located in normal calvarial tissue, as previous histomorphological studies have shown [4]. Also, higher levels of ALP were observed in cells derived from prematurely fused sutures when compare with those from patent sutures and has been demonstrated in a previous *in vitro* model of craniosynostosis [5].

Similar to the ability of CS cells to respond to biochemical osteogenic cues (i.e. soluble factors in OM), we also showed that these cells respond to biophysical osteogenic cues, with all three groups of CS cells increasing calcium expression with increasing substrate stiffness under GM and OM conditions. One of the most notable findings from these studies was that when cultured in OM, cells derived from patent sutures had the greatest increase in osteogenic differentiation with increasing substrate stiffness. This indicates a change in the mechano-responsiveness of patent sutures towards craniosynostotic fused patterns. Also, the fact that levels of osteogenesis were equivalent between the different cell types on the stiff substrates suggests that mechanotransductive processes may affect CS pathology. This is in line with a previous study that found cyclical load - another type of mechanical stimulus - also resulted in histological evidence of premature fusion in murine sutures [5]. Taken together, these findings highlight the mechanosensitivity of CS cells to biophysical cues and suggest a role for mechanotransduction in accelerating bone formation.

ACKNOWLEDGEMENTS

Funding support was provided by the European Research Council (grant agreement 239685) and the Irish Research Council.

REFERENCES

[1] Engler et al., *Cell*, 126(4):677-89, 2006.
[2] Wen et al., *Nat Mat*, 13:979-987, 2014.
[3] Justin et al., *Curr Pro Cell Bio*, 47:10.16.1-16, 2010.
[4] Milovanovic et al., *Eur Cell Mater*, 24:441-458, 2012.
[5] Oppenheimer et al., Plast Reconstr Surg, 124(6): 1840–1848, 2009.

SB^{3}C2015
Summer Biomechanics, Bioengineering and Biotransport Conference
June 17-20, 2015,Snowbird Resort, Utah, USA
SB^{3}C2015-554

A NEW JOINT COORDINATE SYSTEM FOR ROBOTIC TESTING OF CADAVERIC KNEE SPECIMENS

**Daniel V. Boguszewski (1), Nirav B. Joshi (1), Edward C. Cheung (1), Paul R. Yang (1)
Keith L. Markolf (1), David R. McAllister (1)**

(1) Department of Orthopaedic Surgery
University of California, Los Angeles
Los Angeles, CA, USA

INTRODUCTION

The joint coordinate system (JCS) first described by Grood et al. [1] has long been the standard for performing knee biomechanical studies. In this system, tibiofemoral kinematics were reported in anatomical terms about an origin midway between the tibial spines of the tibial plateau. The flexion-extension axis (FEA) was near the joint line, with the internal-external axis (IEA) of tibial rotation oriented from the Grood JCS origin between the tibial spines distally to the center of the ankle. The Grood JCS has been commonly referenced in biomechanical studies, especially those performed using robotic systems to examine knee joint kinematics and kinetics [2-5]. In fact, it is the basis for the International Society of Biomechanics standard on reporting data related to kinematics of the knee [6].

However, to date, robotic studies using the Grood JCS as the reference coordinate system have been performed at fixed flexion angles. Grood et al. [1] noted that active flexion-extension of the knee about their JCS produced significant coupled posterior tibial translation and joint distraction. While useful to convey knee kinematics in anatomic terms, it was not designed as a guide for active knee flexion during biomechanics testing. Any coupled tibial motion during active flexion of the knee can mask the true relative tibiofemoral motion. As shown by Rudy et al. [5], the magnitude of coupled tibial motion can be significantly affected by shifting the origin or point application of the JCS [5].

In a parallel study, a new JCS has been established by accurately defining the optimal FEA of the knee and the natural IEA of tibial rotation. Briefly, we identified the FEA that produced minimal coupled tibial motion during active knee flexion-extension. This axis was located 20.0 ± 4.2 mm proximal to the joint line, posterior and inferior to the transepicondylar axis. The IEA was then found by applying ± 2 Nm of internal and external tibial torque as the knee was flexed about the optimal FEA. The IEA was found to intersect the tibial plateau just medial and posterior to the plateau center, and was oriented slightly varus and tilted slightly anteriorly as it extended toward the ankle. The intersection of the FEA and IEA was the origin of the new JCS, and the cross product of the FEA and IEA formed the third axis of the new JCS. The orthogonality of the new JCS was 0.99 ± 0.01.

The rationale for establishing a new JCS was to minimize coupled tibial motions during active flexion of the knee. The objective of this study was to examine the kinematic effects when controlling knee specimens using the Grood JCS compared to our new JCS. The hypothesis was that the new JCS will reduce coupled tibial motions during active flexion of the knee.

METHODS

Thirty-eight fresh-frozen cadaveric knees were used for this study (16 males, 22 females). The mean age was 34.6 years for males (range of 19 to 45 years) and 28.4 years for females (range 16 to 42 years). Specimens were potted in PMMA for fixation. The tibia was clamped to a force-moment sensor (Omega Model Industrial Automation Load Cell, ATI Industrial Automation, Apex, NC) mounted on the end of a six degree-of-freedom (DOF) robot (KR210; KUKA Robotics Corp., Clinton Township, MI) with the femur secured to a baseplate (Figure 1). A three-dimensional digitizer (Faro Gage, FARO Technologies Inc., Lake Mary, FL) was used to reference each JCS and to align the knee with the sensor axes.

All knee specimens were secured in the robotic testing system at full extension. Using combined displacement and load control, the robot was programmed to move each tibia relative to the femur through a single cycle of active flexion from 0° to 50° while simultaneously seeking target levels of zero force and zero moment

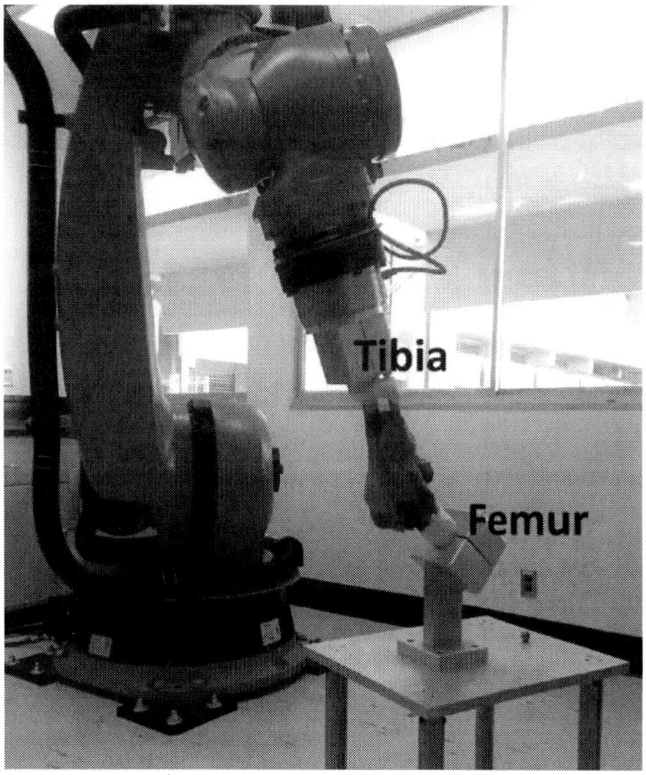

FIGURE 1. ROBOTIC TESTING SYSTEM

along the remaining anatomic DOFs. As the robot flexed the knee at a rate of 1° per second, the forces and moments of the remaining 5 DOFs were simultaneously minimized to less than 1 N and 0.25 Nm, respectively. The resulting tibial translations and rotations to maintain the target level of zero forces and moments during active flexion were recorded in real-time along each DOF.

Statistical analysis was performed using a paired two-sample t-test. Comparisons were made every 1° from 0° to 50° flexion for each anatomic DOF and for resultant tibial translation. The significance level was $p < 0.05$.

RESULTS

During active knee flexion, resultant coupled tibial motion was significantly reduced when knee joint kinematics were controlled according to the new JCS compared to the Grood JCS throughout 0° to 50° of active flexion ($p < 0.01$). At 50° flexion in the Grood system, resultant tibial translation of 21.3 ± 2.6 mm (mean ± SD) occurred in order to maintain the target level of zero forces and moments, compared to a resultant of 5.6 ± 2.8 mm of tibial translation in the new JCS (Figure 2). Posterior tibial translation was the largest contributor to the overall resultant coupled tibial translation. At 50° flexion, with the Grood JCS the tibia translated posteriorly 20.8 ± 2.6 mm compared to only 3.7 ± 3.2 mm in the new JCS.

The remaining DOFs showed only minor differences in coupled tibial motion. Medial-lateral tibial translation had a maximum difference between the two systems of 1.5 ± 1.7 mm at 50° flexion, compression-distraction tibial translation had a maximum difference of 1.6 ± 1.7 mm at 23° flexion, internal-external tibial rotation had a maximum difference of 2.0° ± 4.0° at 44° flexion, and varus-valgus tibial rotation had a maximum difference of 1.2° ± 2.5° at 23° flexion.

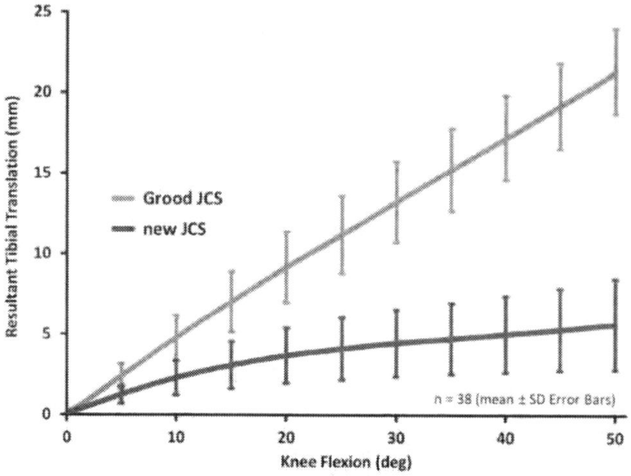

FIGURE 2. RESULTANT TIBIAL TRANSLATION

DISCUSSION

In this study, coupled tibial motion during active flexion was significantly reduced when utilizing the new JCS compared to the Grood JCS. The flexion axis in the Grood system lies near the joint line, so as the tibia was flexed about the femoral condyles, significant posterior tibial translation occurred in order to maintain zero forces and moments in the remaining anatomic DOFs. When the JCS was repositioned such that its axes were in line with the optimal FEA and IEA, coupled posterior tibial translation was reduced by as much as 74%, and the overall resultant tibial translation was reduced by as much as 82%. This is important, as coupled tibial motion due to a JCS not optimized for testing during active knee flexion can potentially mask the true tibiofemoral kinematics of the knee. Thus, the definition of the JCS can have a significant effect on knee biomechanical studies, especially those examining knee kinematics during active flexion.

This work hopes to provide an alternate JCS for robotic studies, and may prove useful in future biomechanical research investigating active knee motion. Establishing a JCS that minimizes coupled tibial motion during active knee flexion could better reflect the true tibiofemoral kinematics of the knee.

ACKNOWLEDGEMENTS

We thank Katherine Moses for her contributions to the study. The study was supported in part by the H H Lee Surgical Research Program and the Musculoskeletal Transplant Foundation (MTF # 20130216).

REFERENCES

[1] Grood et al. J Biomech Eng. 1983. 105(2):136-44.
[2] Fujie et al. J Biomech. 1996. 29(12):1577-85.
[3] Fujie et al. J Biomech Eng. 2004. 126(1):54-61.
[4] Noble et al. J Biomech Eng. 2010. 132(2):025001.
[5] Rudy et al. J Biomech. 2000. 33(9):1147-52.
[6] Woo et al. Knee. 1998. 5(4):267-88.
[7] Wu et al. J Biomech. 1995. 28(10):1257-61.

QUANTIFYING HEAD IMPACT EXPOSURE IN COLLEGIATE WOMEN'S SOCCER

Jaclyn Press (1) and Steven Rowson (1)

(1) Department of Biomedical Engineering and Mechanics
Virginia Tech
Blacksburg, Virginia, United States

INTRODUCTION

Over 265 million people worldwide are actively involved in the game of soccer and there is ongoing growth in the sport's popularity.[1] This consistent increase in the sport's prominence has led to higher awareness of potential injuries, including brain injuries. In a retrospective study done by Delaney et al. which looked at Canadian university athletes, it was found that 62.7% of soccer players had suffered concussion symptoms in a given year. Furthermore, being a female was associated with 2.6 times greater odds of having suffered a concussion.[2-3] With a concussion injury rate of 0.41 per 1000 athlete-exposures, women's soccer ranks among the top five sports with the highest rate of concussion when compared to 14 other popular men's and women's collegiate sports including football and ice hockey.[4]

In order to better understand concussive injuries in female soccer players, we aim to quantify the head impact exposure in collegiate women's soccer through on-field measurement of head impacts. While there have been studies in the past that have looked at soccer head impacts, most have been restricted to recreating impact situations in a laboratory setting in order to estimate head accelerations. Recent technological advancements in head impact sensors, however, have now made on-field measurements possible. Though as new sensors emerge, many have not been properly validated and caution must be used when interpreting the data collected. With this in mind, this study uses both visual analysis and sensor-recorded measurements when quantifying head impact exposure.

METHODS

A total of 26 players from the Virginia Tech Women's Soccer Team consented to participate in this study and IRB approval was obtained prior to testing. The subjects had an average height of 5'7" ± 2"and an average weight of 143.3 ± 15.4 lbs.

xPatch impact sensors (X2 Biosystems, Seattle, WA) were placed behind the ear of every participating player before each game and practice to record linear and rotational head accelerations (Fig. 1). Each of these sensors contains three linear accelerometers and three angular rate sensors. Data are sampled at 1000 Hz. When any event exceeded an accelerometer reading of 10g, the sensor triggered and recorded 100ms of data (10ms before and 90ms after the event). Data were downloaded from each sensor following each game and practice. Results collected were then compared to video recordings of each event in order to manually verify each impact sustained as well as categorize the type of impact that occurred. Two games were excluded from this process due to poor video quality.

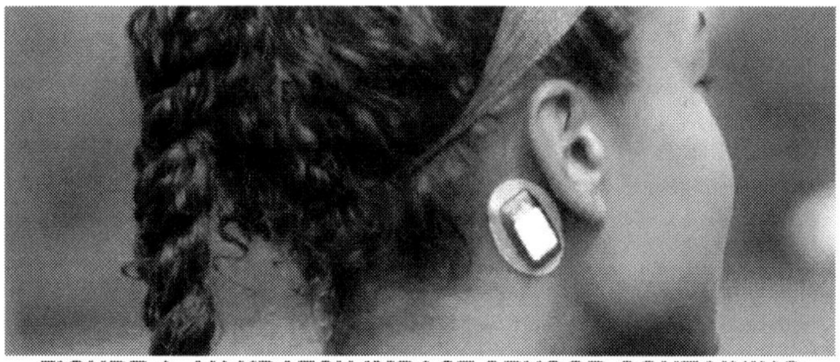

FIGURE 1. AN XPATCH IMPACT SENSOR CONTAINING THREE LINEAR ACCELEROMETERS AND THREE ANGULAR RATE SENSORS PLACED BEHIND THE EAR OF A PLAYER.

The xPatch comes equipped with a filter used to determine the validity of each impact recorded. For the purpose of this study, impacts were classified as shown in Table 1.

TABLE 1. CLASSIFICATION OF TRUE POSITIVE, FALSE POSITIVE, TRUE NEGATIVE, AND FALSE NEGATIVE IMPACTS FOR THE PURPOSE OF THIS STUDY.

	Visually identified	Sensor filter marked as real impact
True Positive (TP)	Yes	Yes
False Positive (FP)	No	Yes
True Negative (TN)	No	No
False Negative (FN)	Yes	No

These parameters were identified at various imposed linear acceleration thresholds and used to construct a positive predictive value (PPV) curve which indicated the percentage of positive recorded impacts that are true positive results (Eqn. 1).

$$PPV = \frac{TP}{TP+FP} \quad (1)$$

RESULTS

Out of a total 26 practices and 20 games, 1703 impacts were identified through video analysis. Of these impacts, 89.6% were headers. The remaining impacts were a mixture of head to head, head to body, head to ground, and unintentional ball to head contact. The average number of impacts per player per practice was 1.66 ± 0.91 and the average number of impacts per player per game was 1.63 ± 1.86. Impact totals varied by position (Fig. 2). The linear acceleration values for these impacts ranged from 6g to 113g with an average of 25 ± 16g. The average rotational acceleration spanned from 380 rad/s^2 to 26,222 rad/s^2 and had an average of 5626 ± 4223 rad/s^2.

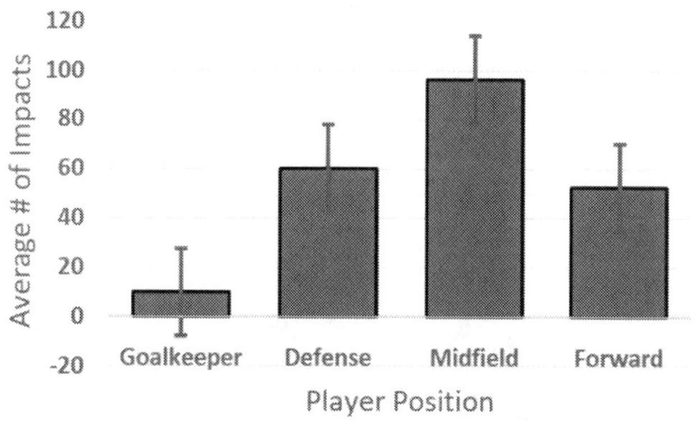

FIGURE 2. AVERAGE NUMBER OF IMPACTS SUSTAINED PER PLAYER OVER THE COURSE OF A SEASON DIVIVED INTO PLAYER POSITIONS.

The sensors reported a total of 17,865 impacts, 8999 of which were classified as real. Impact classifications (Table 1) were used to construct a PPV curve (Fig. 3). This curve suggests that the optimal linear acceleration threshold for the sensors is 34g, where 65.8% of positive recorded impacts are true positive results. At this threshold there were 383 true positive impacts, 199 false positives impacts, 15,963 true negatives impacts, and 1320 false negative impacts.

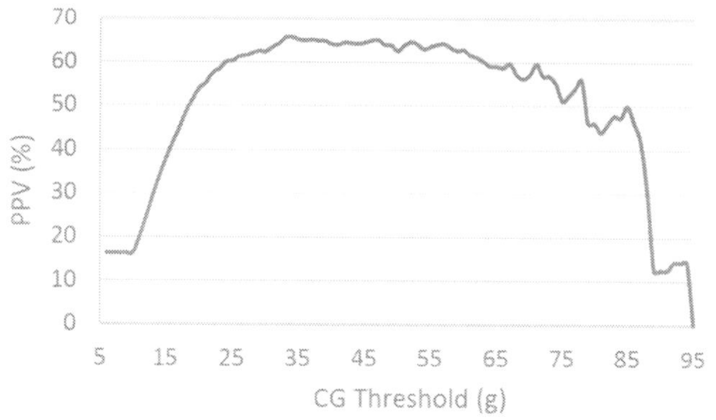

FIGURE 3. POSITIVE PREDICTIVE VALUE (PPV) ACROSS A SPAN OF LINEAR ACCELERATION THRESHOLDS.

DISCUSSION

As was expected, impact counts varied by position due to the different nature of play. Large discrepancies were found between the number of impacts actually sustained by the players and those that were recorded by the sensors. If the sensor alone was used for this study, it would have reported that the players experienced 8999 impacts in the season while we counted only 1703.

While concussive injuries among soccer players have been studied in the past, the depth of knowledge on the topic remains shallow. Most studies that have reported head acceleration values have done so under laboratory conditions. One study that has reported on-field head acceleration data was done by Hanlon and Bir, where 24 girls' youth soccer players were fitted with a wireless head acceleration measurement system implanted into a soccer headband. The girls were then asked to participate in a scrimmage and head acceleration measurements were recorded.[5] On average, the acceleration values collected in the Hanlon and Bir study were lower than those measured in this study.

It should be noted that the head acceleration measurements collected during this study can only be as accurate as the results reported by the xPatch sensors. That being said, the recorded head accelerations are higher than expected, especially rotational. The nominal risk of concussive injury from a head impact is 50% at 6383 rad/s^2 and 90% at 7483 rad/s^2.[6] The xPatch sensors recorded 217 impacts visually confirmed impacts above 50% risk and 165 impacts above 90% risk, though no head injuries were sustained by any player throughout the course of the season. The effect of skin motion on head acceleration measurements remains unknown, and may be a possible cause of error here.

While this study quantified head impact exposure in collegiate women's soccer, it also identified some of the challenges in using sensors to measure head impacts from athletes on the field. The results of this study should be considered by others when using impact sensors.

REFERENCES

[1] Kunz, M (2007, July). Big Count. *FIFA Magazine*, 10-15.
[2] Delaney, J et al., *Clin J of Sport Med*, 12:331-338, 2002.
[3] Fuller, C et al., *Br J Sports Med*, 39(Suppl I):i3-i9, 2005.
[4] Hootman, J et al., *J Athl Train*, 42(2):311-319, 2007.
[5] Hanlon, E et al, *Med Sci Sports Exerc*, 44(6):1102-1108, 2012.
[6] Rowson, S et al., *Ann Biomed Eng*, 40(1):1-13, 2012.

SB³C2015
Summer Biomechanics, Bioengineering and Biotransport Conference
June 17-20, 2015, Snowbird Resort, Utah, USA
SB³C2015-556

STIFFNESS CHARACTERIZATION IN BIOLOGICAL MATERIALS BASED ON DEFORMATION IMAGING AND TOPOLOGY OPTIMIZATION

L. Cai (1), C. Pedersen (2), R. Mclendon (2), M. Biedermann (2), G. Dimitrova (2), J. Yao (2), C.P. Neu (1)

(1) Biomedical Engineering Department
Purdue University
West Lafayette, IN, US

(2) Dassault Systèmes
Johnston, RI, US

INTRODUCTION

Medical imaging modalities enable the nondestructive measurement of deformation and internal tissue mechanics. Magnetic resonance imaging (MRI), in particular, is a preeminent technology to visualize the internal tissue structure, in addition to other physical phenomena like flow and diffusion. One specialized MRI technique, termed displacements under applied loading by MRI (dualMRI), was developed to measure displacements and strain in musculoskeletal tissues, hydrogels, and engineered constructs [1]. However, displacement and strain information does not directly describe spatial distributions of tissue stiffness, which is critical to the understanding of disease progression [2, 3]. In order to achieve stiffness mapping nondestructively in explanted tissues or even in living animals or humans *in vivo*, here we proposed an inverse modeling approach to map the internal stiffness from image-based displacements measured in engineered agarose constructs as an idealized system. In this abstract, forward simulations were performed on materials with known properties and boundary conditions. With different level of noise added, the error associated with the relative stiffness mapping technique was studied. Additionally, experimental data from two agarose plugs with different mechanical properties were analyzed and their absolute stiffness value was derived. The significance and potential of this approach was highlighted for the description of tissue degeneration, repair, and complex material properties.

METHODS

In the present study, topology optimization, which is usually used to design stiff, durable and light-weight structures, was applied to determine the stiffness distribution of agarose materials. Specifically, the design variable was the elemental relative density (ρ) of each individual finite element in the design space. The elastic modulus of the element was described as $E = \rho \times E_0$, where E and E_0 were the moduli for the current model and the base model respectively. Other than traditional topology optimizations, which use Solid Isotropic Material with Penalization (SIMP) or Rational Approximation of Material Properties (RAMP), the current work is designed to minimize the maximum absolute difference of the displacement (U) at N discrete points between the simulation and the experimental measurements:

$$Min : Max_{i=1}^{N}\left(\left|U_{exp}\left(P_i\right) - U_{sim}\left(P_i\right)\right|\right) \quad (1)$$

The optimization formulation was solved using an iterative design. In each iteration, the equilibrium of the model was determined using Abaqus (Dassault Systèmes) [4], and the sensitivities with respect to the elemental relative densities were computed for the objective function and constraints analytically [5, 6]. The elemental relative densities were updated along with the iteration until a stop criterion was satisfied.

In order to validate the approach, a forward simulation of a cylindrical model was created (Figure 1). To test whether our approach could differentiate between material regions of different stiffness, a simple bi-layer structure [7] was introduced, with a higher modulus (E=10 MPa, ν=0.49) in the bottom and lower modulus (E=5 MPa, ν=0.49) in the top. With the bottom edge of the model fully constrained, the top edge was indented to 15% of the thickness. Displacements of this loaded model were calculated by Abaqus and passed to TOSCA (Dassault Systèmes) as design reference. To inversely calculate the stiffness map, optimization was initiated using a model with linear elastic (Young's modulus E=8 MPa) properties. The difference between displacements of a target and reference model was optimized (Equation 1). To evaluate the sensitivity and stability of the method, uniform symmetric random noise was added to the reference displacement data prior to optimization. Different noise levels in Figure 1 presents the ratio of the noise amplitude to the indentation depth. The included noise levels were 0, 0.05%, 1%, 10% and 50%.

The inverse modeling approach was further tested on experimental data of engineered agarose constructs [7] (Figure 2). In the setup, two different uniform agarose constructs of 2% and 4% w/v were prepared of cylindrical shape with a 10 *mm* diameter and 6 *mm* depth. The samples were cyclically loaded by a horizontally-oriented cylindrical indenter, and loading actions were synchronized with a 14.1 *T* MRI system (Bruker Medical GMBH), in which a dualMRI (DENSE-FISP) imaging sequence [1] with a displacement encoding gradient 2.13 π/mm was programed. The displacement data measured was reconstructed to the undeformed state, and then sampled evenly as the reference displacement. To simplify the problem, plane strain was assumed in the middle portion of the imaging section and a 2D model was created. Since the cyclic load was measured, it was feasible to scale the relative stiffness map to derive the real stiffness map by comparing the supporting force in the target model with the known load applied experimentally.

RESULTS

The validation and stability study showed that the proposed inverse algorithm produced the original stiffness pattern and worked robustly within certain levels of applied noise. In the case of ideal (no noise) data (Figure 1A), the inverse modeling produced a distinctive bilayer configuration and the averaged stiffness ratio between two layers was 0.49997, which was very close to the original 0.5 setting. When 1% noise was introduced, the ratio became 0.54. Though there were several out-of-range elements near the boundary, the bilayer configuration remained. For noise levels of 10% and 50%, the optimization process stopped at iteration number 5, while the other cases with less noise continue to converge toward zero (Figure 1B; iterations after 7 not shown). The relative stiffness for all noise levels showed a stiffer bottom layer (Figure 1C).

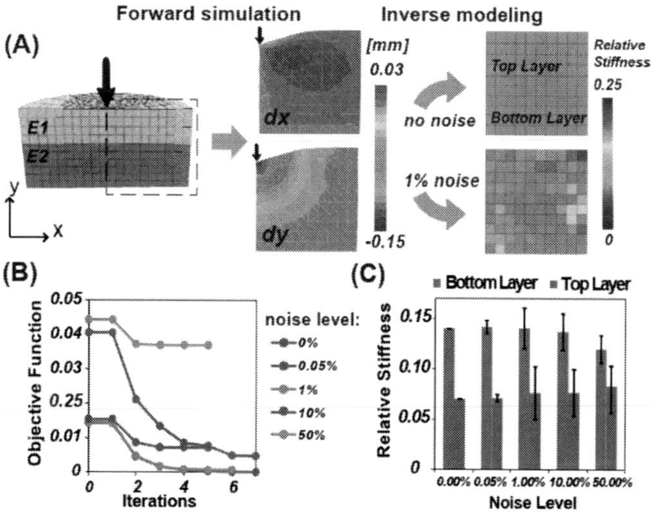

FIGURE 1. Workflow of the inverse modeling using the displacement data from forward simulation of a two layer cylinder (with E2=2×E1) (A). The relative stiffness result was presented for cases with and without added noise. The relationship between objective function value and iteration numbers (B). Averaged relative stiffness of top and bottom layer when different noise level was added (C).

Distributions of the absolute stiffness in agarose constructs were derived from MRI-based displacement data and topology optimization (Figure 2). The stiffness maps showed a distinct difference between two materials while the averaged value for 2% and 4% agarose was respectively 5.4 *kPa* and 15.7 *kPa*.. However, the stiffness patterns were not homogeneous while large elements accumulate at the boundaries. This boundary issue was also seen in Figure 1A.

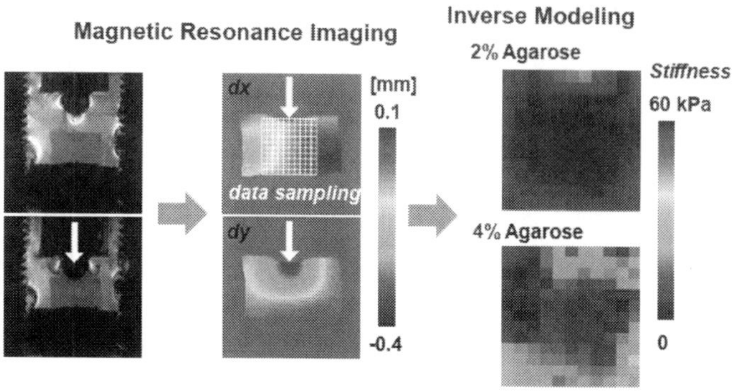

FIGURE 2. A workflow of inverse modeling for stiffness maps using MRI-based displacement data and topology optimization.

DISCUSSION

The results showed that combined deformation imaging and topology optimization is robust in the bilayer case but requires accurate input of boundary conditions. The boundary issues observed (Figure 1A, 2) might be related with the over-constricted boundary nodes, since they have smallest displacements and should be most sensitive to noise. In the agarose study (Figure 2), the calculated stiffness differs with the value measured in [7], but the ratio of stiffness between 2% and 4% is 0.35, which is between the measured instantaneous modulus ratio 0.43 and equilibrium modulus ratio 0.20. It is also important to note that cyclic loading [7] differs with the one-time loading, especially when plastic deformation was observed. Interestingly, it was also found in the stability study that more complex stiffness patterns, like higher stiffness ratio between two layers, tends to decrease the stability of this optimization, making it more vulnerable to noise. Our future endeavor will focus on different boundary settings and the analysis of bilayer samples.

In summary, our inverse modeling approach, which was built on the displacement-encoded MRI, provided more intuitive result of the stiffness distribution inside the tissue. The nondestructive nature and the various types of elemental type in Abaqus extended the potential application of this method. This approach may be useful to analyze softening observed in the timecourse of tissue degeneration, for example in diseases including osteoarthritis. It may also be possible to monitor repair success using stiffness mapping as a unique imaging biomarker for tissue function.

ACKNOWLEDGEMENTS

The authors would like to acknowledge funding from NIH R01 AR063712 and NSF CAREER 1349735.

REFERENCES

[1] Chan, D. D. and Neu, C. P., *PloS ONE*, 7:33463, 2012.
[2] Kempson, G. E., et al. *J Biomech* 4.6: 597-609, 1971.
[3] Neu, C. P. *Osteoarthr Cartilage* 22.10: 1349-1359, 2014.
[4] Abaqus Manual, Dassult Systemes, SIMULIA, 6.14
[5] M. P. Bendsøe and O. Sigmund, *Springer*, 2003.
[6] Tosca Manual, Dassult Systemes, SIMULIA, 8.1.0
[7] Griebel, A. et al., *J Biomech*, 47: 2149-2156, 2014.

SB³C2015
Summer Biomechanics, Bioengineering and Biotransport Conference
June 17-20, 2015, Snowbird Resort, Utah, USA
SB³C2015-557

HEMODYNAMIC CHANGES IN TREATED CEREBRAL ANEURYSMS AND CORRELATIONS WITH LONG-TERM OUTCOMES

Michael C. Barbour (1), Patrick M. McGah (1), Kurt Sansom(1), Michael R. Levitt (2), Ryan P. Morton (2), John D. Nevra (2), Pierre D. Mourad (2), Basavaraj V. Ghodke (3), Danial K. Hallam (3), Laligam N Sekhar (2), Louis J. Kim (2), Alberto Aliseda (1)

(1) Mechanical Engineering
University of Washington
Seattle, Wa, USA

(2) Neurological Surgery
University of Washington
Seattle, WA, USA

(3) Dept. of Radiology
Harborview Medical Center
Seattle, WA, USA

INTRODUCTION

Endovascular cerebral aneurysm treatment methods, eg. flow-diverting stents or coil embolization, are believed to reduce flow, circulation and stresses in the aneurysm sac, promoting flow stasis and thrombosis. While these methods are an improvement on previously used, invasive surgical techniques, 15-30% of aneurysms treated with a flow diverting stent or coil remain patent at midterm angiographic follow-up[1]. Treatment failure is the largest obstacle facing endovascular treatment methods, as re-treatment is often necessary and costly.

The hemodynamic environment in patients undergoing endovascular treatment is considered an important factor in long-term treatment outcome and computational simulations can be used to study changes in aneurysm hemodynamics immediately following treatment[2-3]. However, it remains unclear if and how hemodynamic changes can be directly correlated to cerebral aneurysm treatment outcome in a clinically relevant way.

In this study, patient specific computational simulations are used to investigate changes in pre- and post-treatment hemodynamics in order to ascertain the effect of treatment on metrics such as intra-aneurysm flow, wall shear stress (WSS) and wall shear stress gradient (WSSG). Clinical six-month follow up data is then used to correlate these hemodynamic changes to treatment success or failure and determine what metrics can be used to accurately predict treatment outcomes.

METHODS

Seven patients with unruptured cerebral aneurysm (9 aneurysms) were included in this study, with IRB approval and informed consent from each patient before the procedure. Each patient was treated with either an endovascular flow-diverting stent or coil embolization. Three-dimensional rotational angiography was obtained pre-treatment and contrast enhanced flat-panel CT was obtained post-treatment to determine the precise location of stent or coil. Blood velocity and pressure measurements were obtained with a dual-sensor pressure and Doppler velocity guidewire and analysis workstation (ComboWire and ComboMap, Volcano Corp, San Diego, CA). Pressure and velocity measurements were taken every 5 ms at four peri-aneurysmal locations. These measurements were made pre- and post- treatment and used to establish boundary conditions in the computational simulations[4]. It is important to note that this additional guidewire measurements fit seamlessly into the clinician workflow. Patients are brought in at six-month intervals following treatment for angiographic follow-up, at which point our clinical collaborators assess treatment outcome.

Three-dimensional vessel reconstruction is done using the Vascular Modeling Toolkit (VMTK v1.2). For patients who received a pipeline stent, a saddle-shaped surface is manually placed into each reconstruction at the location of the stent boundary determined from visual inspection of the post-treatment radiograms[5]. The stent is modeled as a thin porous surface with inertial pressure drop[6]. For those patients who received a coil embolization, the vessel domain is split into two parts, again via visual inspection of angiographic data, and the portion of the aneurysm sac packed with coils is modeled as a porous volume by adding a sink term to the momentum equation[7]. Tetrahedral volume meshes are generated using the package ANSYS GAMBIT (Release 2.4, ANSYS, Inc.). Finite-volume fluid simulations are performed using ANSYS FLUENT (Release 14.1, ANSYS, Inc.). Blood flow is assumed to be incompressible and Newtonian. At the inlet boundary, the time-dependent Womersley velocity profile based on *in situ,* wire-measured, centerline velocity is imposed. At distal outlets, pressures are prescribed using the unsteady *in situ* wire pressure measurements.

RESULTS

Simulations of the unsteady hemodynamics are performed in all 7 patients for both pre- and post- treatment cases. The CFD derived intra-aneurysmal flow rate and energy dissipation as well as WSS and WSSG on the aneurysm dome are compared for pre- and post-treatment cases. Results indicate a reduction in time-averaged and peak intra-aneurysmal flow, WSS, WSSG, and energy dissipation for all patients following treatment. Fig 1 shows time-averaged WSS pre- and post- treatment, as a representative case. Time-averaged and peak WSS were reduced by an average of 34% and 45% respectively.

The patient population is then separated by the success or failure of the treatment outcome and the changes in CFD derived variables are compared. Fig 2 shows changes in peak WSS for successful vs. unsuccessful treatment in the nine aneurysms. The median reduction in peak WSS is approximately 20% greater in patients with successful treatment than in those with incomplete sac occlusion. However, changes in intra-aneurysmal flow rate at the neck are not significant when comparing the success of treatment and both sample populations have a median reduction of 30 %.

DISCUSSION

Endovascular aneurysm treatments are designed to reduce flow into the aneurysm sac and promote intra-aneurysmal thrombosis. The results from this study confirm that flow-diverting stents and coils do reduce intra-aneurysmal flow, as well as WSS, WSSG, and energy dissipation within the aneurysm dome. By then comparing the changes in hemodynamic variables to six-month treatment outcome, we are able to determine which changes in hemodynamic variables, as well as what extent of these changes, are indicative of treatment outcome. Preliminary results suggest that changes in WSS are more indicative of treatment outcome than intra-aneurysm flow rate. While we acknowledge that the patient population is still too low to draw clinically relevant conclusions, the study will grow to include 45 patients over the next two years.

The results and methods of this study suggest that patient-specific CFD can be used to quantify a treatment's probability of success by computing the change in pre- and post- treatment hemodynamics in cerebral aneurysms. Prospectively modeling hemodynamic changes will allow clinicians to decided which treatment modality is most likely to result in complete aneurysm thrombosis – minimizing the failure rates of endovascular treatment methods.

ACKNOWLEDGEMENTS

This work has been financially supported by the NIH/NINDS (grant 1R03NS078539 and 1R01NS088072) and by an unrestricted grant from Volcano Corporation, San Diego CA.

REFERENCES

[1] Crobeddu et all, *AJNR* 34:266-270, 2013
[2] Zhang, Y et al. *Med Eng Phy* 35:608-615,2013
[3] Zulcsar, Z et al. *Acta Neurochir* 33:2010;2016, 2012
[4] McGah PM et al. *Ann. Biomed Engr.* 42:503-514
[5] Levitt, MR et al. *AJNR* 35:143-148 2013
[6] Augsburger, L et al, *Ann. Biomed. Engr.* 39:850-863, 2012
[7] Muschenborm, A et al. *Biomed Engr. Online* 12:103-116, 2013

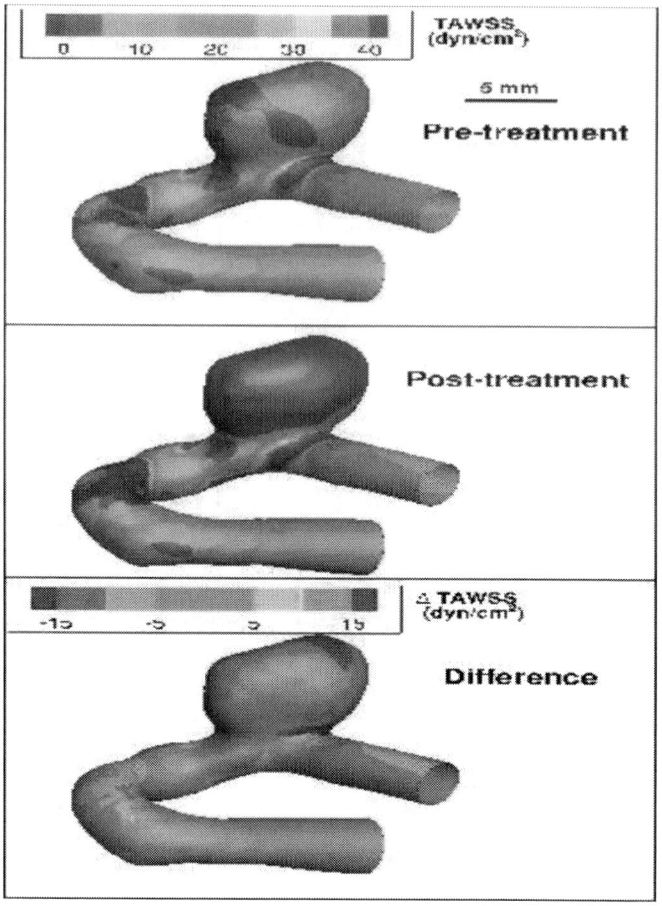

Figure 1: Comparison of pre- and post- treatment time-averaged wall shear stress

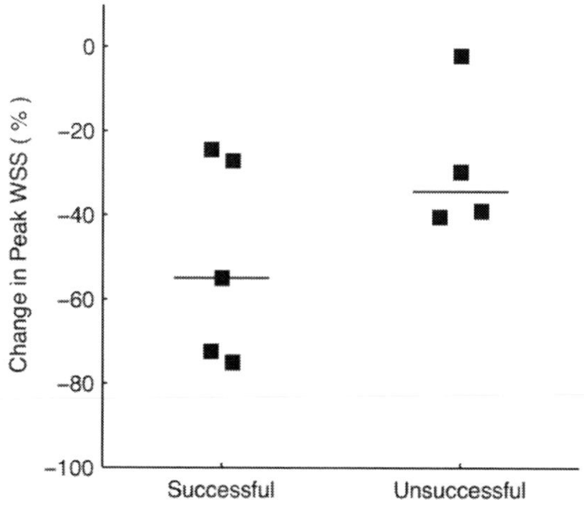

Figure 2: Percent change in maximum wall shear stress in aneurysm dome for successful and unsuccessful treatment outcomes. Bar represents median of the samples

SB^{3}C2015
Summer Biomechanics, Bioengineering and Biotransport Conference
June 17-20, 2015,Snowbird Resort, Utah, USA
SB^{3}C2015-559

PRE-CLINICAL ASSESSMENT OF A PERCUTANEOUS LEAFLET RESECTION DEVICE FOR TREATMENT OF DEGENERATIVE MITRAL VALVE DISEASE

Steven M. Boronyak (1), Brett C. Byram (1), Joseph L. Fredi (2), Michael N. Young (2), W. David Merryman (1)

(1) Department of Biomedical Engineering
Vanderbilt University
Nashville, TN, USA

(2) Division of Cardiovascular Medicine,
Department of Medicine
Vanderbilt University Medical Center
Nashville, TN, USA

INTRODUCTION

Current treatment for moderate and severe mitral regurgitation (MR) due to degenerative mitral valve (MV) disease is open-chest surgical repair or replacement, as there are no known therapies that directly affect the disease process [1]. Percutaneous approaches to MV repair have received a great deal of interest; however, there are currently no percutaneous techniques to perform leaflet resection, which is used in 70-80% of all surgical MV repairs [2-3]. Thus, there is a need for a percutaneous leaflet resection technique in the expanding MV repair armamentarium [4].

Radiofrequency (RF) ablation forms thermal lesions by resistive heating, and the primary structural component of MV leaflets is collagen, which contracts when heated to >65°C. As a result, RF ablation can be used to reduce the size of enlarged myxomatous MV leaflets (Fig. 1) [5-6]. However, application of RF ablation to MV leaflets in a dynamic hemodynamic environment is difficult without additional catheter stability. Thus, we have developed cryo-anchoring, a technique similar to cryoablation that uses pressurized nitrous oxide to provide cooling of a catheter tip, which provides reversible adhesion to moving MV leaflets [7]. The objective of this work was to translate a combined RF ablation and cryo-anchoring (RFC) catheter for treatment of degenerative MV disease from the lab bench to in vivo animal models. In order to guide and optimize RFC catheter treatment, directional leaflet shrinkage and efficacy criteria for large animal studies were quantified using ultrasound imaging in a left heart flow simulator, and catheter guidance and cryo-anchoring techniques were developed in healthy swine.

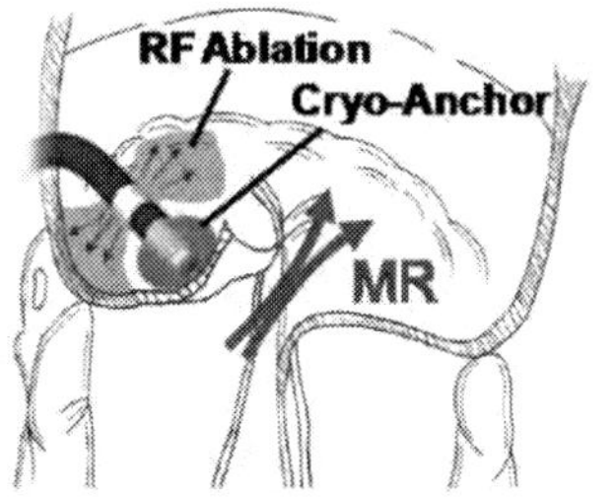

FIG. 1 – RFC catheter treatment strategy.

METHODS

Ex Vivo – Ultrasound Image Measurements: Four tracking markers were sutured onto the atrial surface of the anterior leaflet of intact excised porcine MVs with ~1 cm spacing. The intact valves, including the annulus, chords, and papillary muscles, were placed in a left heart flow loop capable of reproducing physiologic pressure and flow waveforms. Ultrasound imaging was used to track the markers in an intercommissural plane (Figs. 2A-B), and leaflet morphology was captured in the A2-P2 plane (Figs. 2A and 2C). The anterior leaflet was treated several times with the RFC catheter (each treatment: 50 W for 90 s), and MV leaflet size was quantified over the course of 5 ablations by tracking the markers in 2D views to estimate directional leaflet size reductions. Total anterior leaflet length, viewed in the A2-P2 plane, was quantified in order to provide a measure of treatment efficacy that is easily obtainable in a clinical setting.

In Vivo – Cryo-Anchor Evaluation: To evaluate cryo-anchoring ability in vivo, a Vanderbilt University IACUC-approved study was conducted in healthy swine (45-50 kg). A median sternotomy was performed and a 13 Fr sheath was placed at the apex of the heart to provide catheter access. Epicardial echocardiography was used to guide the RFC catheter to the MV apparatus using a pseudo-parasternal long-axis view. Once the RFC catheter tip was located in the vicinity of the MV leaflets, cryo-anchoring was achieved by intra-luminal delivery of nitrous oxide to maintain the cryo-anchor temperature between -20°C and -30°C for 60 s. Cryo-anchoring was obtained and verified for each leaflet independently, and animal hemodynamics were monitored for adverse consequences.

RESULTS

Maximal areal leaflet shrinkage (37 ± 3.0%) was achieved after 3-4 ablations of 50 W for 90 s (Fig. 2D). The majority of leaflet shrinkage occurred in the circumferential direction (29 ± 2.5% versus 12 ± 2.6% in the radial direction) (Fig. 2E), which is similar to previous findings and is due to collagen being preferentially aligned with this direction. Marker length shrinkage in the radial leaflet direction was comparable to the total leaflet length reduction measured in the A2-P2 imaging plane (9.1 ± 2.6%) (Fig. 2E). Fig. 2E suggests that when the leaflet length has been reduced by 10%, the total leaflet area has shrunk by approximately 40%, which provides a quantitative target for determining RFC catheter shrinkage efficacy in a large animal model of healthy swine.

The RFC catheter was successfully able to adhere to both the posterior and anterior leaflets independently of each other and is shown for the posterior leaflet in Fig. 2F. Cryo-anchoring was maintained for 60 s without negatively affecting hemodynamics as measured by blood pressures and heart rate.

DISCUSSION

Directional Leaflet Shrinkage Deformation: Early investigation into combined RF ablation and cryo-anchoring treatment has indicated that shrinkage of enlarged MV leaflets is feasible [5, 7]. In order to optimize the treatment strategy, leaflet shrinkage dynamics and maximum obtainable leaflet shrinkage must be quantified in a physiologic setting. The results from this study in a left heart flow simulator suggests that the leaflet shrinks relatively linearly with the number of ablations, and it takes 3-4 ablations to maximally shrink a 1 cm x 1 cm region of tissue by approximately 40% (Figs. 2D and 2E). Additionally, consistent with previous studies, the majority of leaflet shrinkage occurs in the circumferential direction, due to this being the direction of collagen fiber alignment. Thus, the total leaflet length, which is aligned radially, shrinks minimally (about 10%). The directional shrinkage response of myxomatous, diseased MV leaflets is currently unknown and warrants investigation.

Establishment of Pre-Clinical Shrinkage Efficacy Criteria: A further purpose of this study was to establish treatment efficacy criteria for RFC catheter testing in healthy animals without MR. As standard 2D ultrasound cannot measure the circumferentially-aligned leaflet width, and therefore the leaflet area, the shrinkage relationship between the circumferential and radial directions can be used to estimate total leaflet shrinkage from leaflet length. Leaflet length in the anterior-posterior imaging planes (A2-P2, etc.) is a more easily obtainable measurement in vivo (Fig. 2F). Given the agreement in anterior leaflet length and radial marker length (Fig. 2E), leaflet length reductions measured on clinical echocardiograms can be used to estimate total leaflet area shrinkage without the addition of tracking markers.

Cryo-Anchor Specificity: Selective administration of thermal shrinkage to each MV leaflet is essential in the RFC catheter treatment strategy. Depending on the severity and presentation of degenerative MV disease, enlarged and myxomatous leaflet tissue could be confined to a single segment of a single leaflet or be present throughout all of both leaflets. Therefore, we tested the ability of the RFC catheter to selectively cryo-anchor to each leaflet in vivo using swine (Fig. 2F). The RFC catheter successfully adhered to each leaflet independently and allowed for free motion of the un-targeted leaflet, maintaining blood pressure and heart rate at or near baseline levels.

The results from this study demonstrate that RF ablation, with cryo-anchoring stability, reduces the size of MV leaflet tissue in a physiologic flow environment by approximately 40%, providing a possible percutaneous alternative to surgical leaflet resection. Additionally, the RFC catheter cryo-anchor can selectively target each leaflet individually, allowing for RF delivery to specific enlarged leaflet segments. Further work will be completed using the leaflet shrinkage criteria, established in this study, to test RFC catheter safety and efficacy in large animal models.

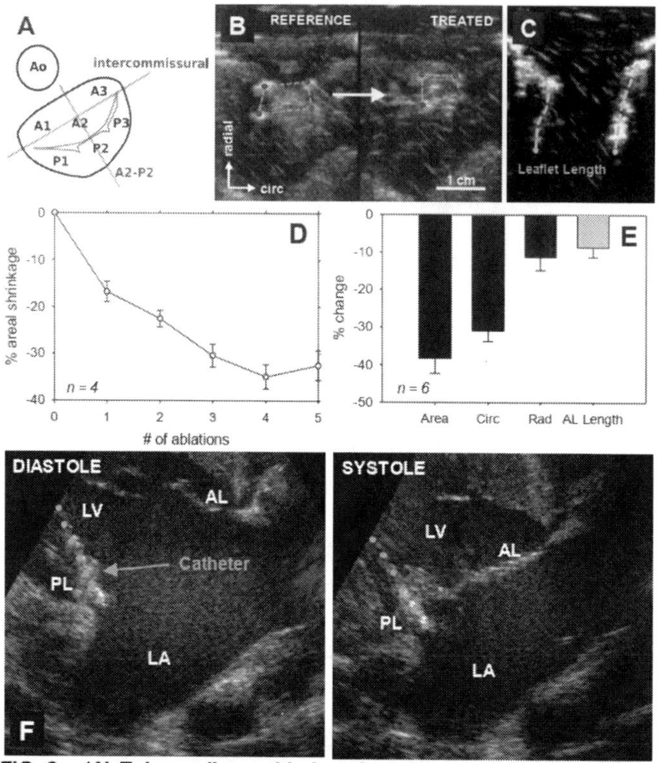

FIG. 2 – (A) Echocardiographic imaging planes used in this study, surgical view. (B) Suture markers on the atrial leaflet surface in the intercommissural view allow for quantification of directional leaflet size changes. (C) Leaflet length in the A2-P2 plane provides a measure that is more easily obtainable in the clinic. (D) Reduction in leaflet marker area over the course of several ablations. (E) Final leaflet marker size reductions (black bars, intercommissural plane) compared to final leaflet length (gray bar, A2-P2 plane). (F) RFC catheter cryo-anchored to the posterior MV leaflet, guided by epicardial echocardiography, while the anterior leaflet moves freely, in a porcine animal model. AL: anterior leaflet, LA: left atrium, LV: left ventricle, PL: posterior leaflet

ACKNOWLEDGEMENTS

This work was funded by the Wallace H. Coulter Foundation and and an American Heart Association pre-doctoral fellowship (13PRE16340018).

REFERENCES

[1] Rabkin, E., et al., *Circulation*, 104:2525-32, 2001.
[2] Flameng, W., et al., *J Thorac Cardiovasc Surg*, 135:274-82, 2008.
[3] David, T.E., *Semin Thorac Cardiovasc Surg*, 19:116-20, 2007.
[4] Boronyak, S.M., et al., *Future Cardiol*, 8:779-93, 2012.
[5] Boronyak, S.M., et al., *Ann Biomed Eng*, 40:1971-81, 2012.
[6] Williams, J.L., et al., *J Interv Cardiol*, 21:547-54, 2008.
[7] Boronyak, S.M., et al., *J Biomech*, 47:973-80, 2014.

QUANTIFICATION OF LYMPHATIC VALVE RESISTANCE TO FORWARD FLOW USING COMPUTATIONAL FLUID AND SOLID MODELING

John T. Wilson (1), Raoul van Loon (2), James E. Moore Jr. (1)

(1) Department of Bioengineering
Imperial College London
South Kensington Campus
London, United Kingdom

(2) College of Engineering
Swansea University
Singleton Park
Swansea, United Kingdom

INTRODUCTION

The lymphatic system plays a vital role in physiologic fluid and solute homeostasis as well as immune cell transport. It is responsible for the uptake of fluid and solutes from the interstitial spaces and their subsequent return to the venous system. Its dysfunction could result in a number of pathologies, including lymphedema, or swelling of interstitial fluid (IF) that could lead to chronic inflammation and/or tissue fibrosis [1,2]. Two types of valves are present within the lymphatic vasculature, primary and secondary, and both play a crucial role in maintaining net forward lymph flow. At the mesenteric level, portions of the lymphatic system consist of tubular vessels segmented by bulbous sinus regions encapsulating bi-leaflet, secondary valves that serve to prevent retrograde flow.

Lumped modeling of lymphatic pumping has shown the valve resistance to flow to be one of the most important parameters in determining pumping efficiency. Jamalian et al. showed altering the minimum valve resistance to flow resulted in orders-of-magnitude changes in the flow rate [3]. Further, Bertram et al. used experimental data to estimate open valve resistance and found it to be 0.6 x 10^6 dyn/cm^2 [4]. However, this was just an estimate and more refined values for resistance have been challenging to capture experimentally due to the difficulty in handling lymphatic vessels because of their small size (~100 µm diameter) and low flow rates (on the order of µL/hr). While computational studies incorporating a 3D geometry have been performed, these either excluded the valve [5] or did not include dynamic movement of the vessel [6].

The present study uses an uncoupled approach of finite element (FE) analysis and computational fluid dynamics (CFD) to model the opening behavior of an idealized mesenteric lymphatic valve to ultimately calculate the valve resistance to forward flow and determine how this value changes based on differences in key valve dimensions.

METHODS

Idealized Geometry. A 3D representation of a bicuspid lymphatic valve (Fig. 1) was designed based on confocal image-based data from isolated rat mesenteric lymphatic vessels: more thorough descriptions of these image acquisition experiments are detailed elsewhere [6].

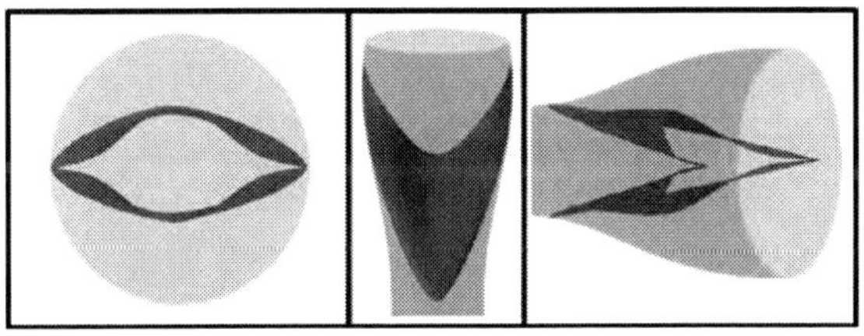

FIGURE 1. Idealized mesenteric sinus and valve. The images are positioned at different angles of view to illustrate the geometry of the idealized lymphatic valve.

Key geometric parameters were identified and the ratio of the maximum diameter of the sinus to that of the root diameter was chosen as a computational parameter of interest due to its expected high degree of influence on the resulting fluid dynamics and valve deflections. The ratio is defined by the following equation:

$$STR = \frac{D_{max}}{D_r}, \tag{1}$$

where D_{max} is the maximum diameter of the sinus, D_r is the root diameter, and STR (sinus-to-root) is the ratio between the two. It should be noted STR was varied from 1-3 for the valve geometries investigated. Parametric equations representing the idealized valve were coded in Matlab (R2014a, MA, USA) and surface meshes of the valve leaflets were generated and exported in *.stl format. The remaining computational workflow can be divided into three steps.

Step 1: Determination of Pressure Loading Conditions. To determine loading conditions for the subsequent FE analysis, CFD simulations were performed with the commercial CFD software Star-CCM+ v. 9.02.007 (CD-adapco, Melville, NY) when the valve was in a, "barely open," state with a 2 μm distance between the trailing edges of each leaflet. Steady simulations were performed and a plug-flow velocity was implemented at the inlet with an average velocity of 1.5 mm/s [Reynolds number of 0.17], which is in the physiologic range of velocities observed during *in situ* experiments [7]. The outlet boundary condition was defined to have a pressure of zero. Trans-leaflet average pressure values, P_{avg}, were calculated by subtracting surface-averaged pressure values from adjacent sides of each leaflet.

Step 2: FE Analysis. Leaflets were imported into ANSYS Mechanical (v15.0, ANSYS, Cecil Township, PA) and a Neo-Hookean material model was employed as the leaflet material property. The initial shear modulus was set as 45kPa, which is in the range values experimentally determined for arterial elastin. The annuli of the valve leaflets attached to the wall of the sinus were set as a fixed supports to prevent movement or deflection. P_{avg} was applied to the leaflets as a linear pressure load. Upon simulation convergence, the deflected leaflet geometries were exported in parasolid binary format for incorporation into subsequent CFD simulations.

Step 3: CFD Analysis. Simulations were performed using the fully deflected valve leaflets using identical boundary conditions as prescribed in Step 1.

RESULTS

Resistance was found to decrease with elevations in STR while orifice area, A_p, increased (Fig. 2).

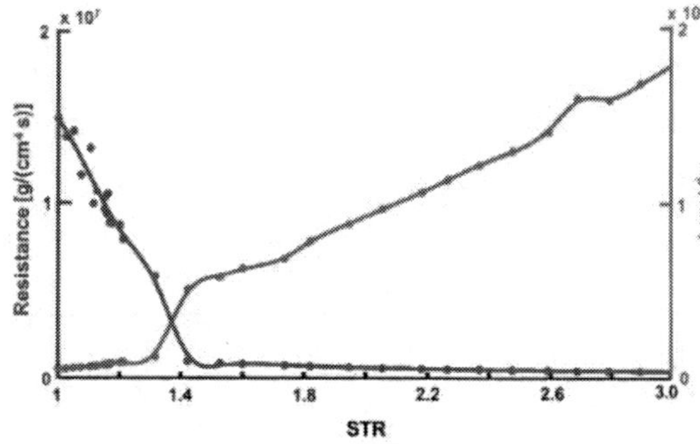

FIGURE 2. Resistance and projected orifice area, A_p, versus STR. The average resistance to flow is approximately 0.62 x10^6 [g/(cm4 s)] for STR≥1.42.

The valve resistance and A_p as a function of STR can each be divided into two separate linear regions. The range of STRs that differentiates these two separate regions occurs between STR=1.2 and STR=1.42. At low STR values, the slope of resistance vs STR is -3.0 x 10^{-7} [g/(cm^4 s)] but then decreases by two orders of magnitude beyond STR=1.42. Conversely, the slope of A_p versus STR increases from 2.0 x 10^{-5} cm^2 at low STR values to 8.0 x 10^{-5} cm^2 at higher values of STR.

CFD simulations using the deflected valve geometries showed higher velocities in simulations with smaller STR values (Fig. 2), which can be attributed to a smaller orifice area. However, it should be noted that valve leaflets did not reach the fully open configuration using STR values less than approximately 1.42. Velocity vectors indicated reverse flow into the sinus regions adjacent to the valve leaflets (Fig. 3), however the magnitude of these vectors is negligible compared to the velocity magnitude of central lymph flow through the valve orifice (-0.06 mm/s axial velocity in maximum reverse flow and 3.6 mm/s axial velocity in maximum forward flow for STR=1.42).

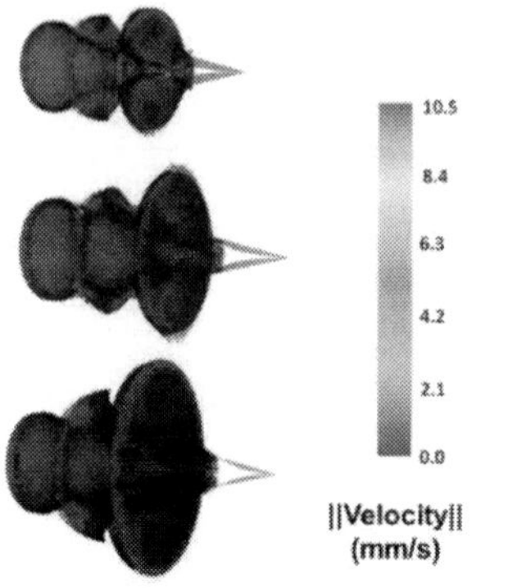

FIGURE 3. Cross-sections of velocity (vectors and contours show velocity direction and magnitude, respectively). Top) STR=1.2, Middle) STR=1.42, Bottom) STR=2.05. Note that velocity vectors are prescribed as constant length and should therefore only be used as an indication of the direction of flow.

DISCUSSION

The study presented herein investigated the effect of sinus and valve geometry on the resulting opening mechanics of the valve leaflets as well as the fluid dynamics associated with the deflected valve configurations. To the authors' knowledge, this the first study to investigate the opening behavior and mechanics of a mesenteric lymphatic valve and quantify valve resistance to forward flow. By varying STR, we were able to understand how pressure distributions due to fluid flow differ from one geometry to the next and how these pressures in turn affect the movement of the valve leaflet and subsequent resistance to flow after deflection. We have shown that the average value of resistance to forward flow (STR≥1.42) was 0.62 x 10^6 [g/(cm4 s)], a value similar to the resistance estimated by Bertram et al. of 0.6 x 10^6 [g/(cm^4 s)] [6].

Valves with high resistance occur in situations where STR is small. However, through experiments with isolated rat mesenteric lymphatic vessels, we have shown most values of STR are in the higher region (greater than approximately 1.6). This suggests physiologically that valves have developed to reach an STR value that minimizes the resistance to forward flow. It is expected that the fully dynamic behavior of the valve leaflet observed during fully-coupled simulations will influence the resistance to forward flow over time. This will be investigated in future studies.

ACKNOWLEDGEMENTS

We would like to acknowledge NIH for, in part, funding this study (Grant R01 HL094269 and U01 HL123420) and the Royal Academy of Engineering and Royal Society for funding J.E. Moore Jr.

REFERENCES

[1] Bazigou, E. et al., *Microvascular Research*, 96:38-45, 2014.
[2] Mortimer, PS et al., *J. of Clinical Invest.*, 124:915-921, 2014.
[3] Jamalian S et al., *AJP Heart and Circ. Phys.*, 305:H1709-H1717.
[4] Bertram, C et al., Biomech. Model. Mechanobiol., 13:401-416.
[5] Rahbar, E et al., *J.of Biomech.*, 44:1001-1007.
[6] Wilson, JT et al., *J. of Biomech. Eng.*, 135:051005.
[7] Dixon JB et al., Microcirculation, 13:597-610.

DEDIFFERENTIATION OF CHONDROCYTES INFLUENCES INTRANUCLEAR STRAIN TRANSFER MEASURED BY DEFORMABLE IMAGE REGISTRATION

Jonathan T. Henderson (1), Benjamin Seelbinder (1), Alexander I. Veress (2), Corey P. Neu (1)

(1) Biomedical Engineering
Purdue University
West Lafayette, Indiana, USA

(2) Mechanical Engineering
University of Washington
Seattle, Washington, USA

INTRODUCTION

Mechanical cues are important for cellular processes and in regulating gene expression by influencing chromatin compaction. Cell (e.g. chondrocyte) shape, size and gene expression are known to change as they are passaged in two-dimentional (2D) cultures [1]. Previous experiments of cells on deformable substrates have quantified minimum strain transfer to the nucleus using bulk measurement techniques, including measures of nuclear area and aspect ratios [2]. However, it is not clear how strain is transferred to the intranuclear structures because previous measurement techniques have utilized relatively low resolution strain estimates of the organelle interior. Recently, deformable image registration has been shown to provide high resolution intranuclear strain maps that can spatially correlate to chromatin domains [3]. In this study, we utilized passage zero (P0) and dedifferentiated passage four (P4) chondrocytes subjected to equibiaxial stretch of untreated and type I collagen-treated substrates to determine how these factors influence the mechanical strain transfer from substrate to nuclear domains.

METHODS

P0 or P4 chondrocytes were plated on untreated or type I collagen-treated plastic membrane substrates (Flexcell International Corp.) 24 h prior to the stretching experiments. DNA was stained with Hoechst prior to substrate deformation. Two-photon microscopy (740 nm wavelength and 250 mW power) acquired three z stack images, each three minutes apart, prior to loading using a 40× water immersion objective. Two channels were captured to visualize cells (transmitted light) and nuclei (Hoechst stain). After stretching, the undeformed cells were relocated (within 5 to 10 minutes of the applied stretch) and three more images are taken, as described previously.

Intranuclear strains were measured using deformable image registration known as hyperelastic warping [4]. To measure the nuclear strain fields, z projections of selected nuclei were used in the analyses. Briefly, with this type deformable image registration the local, subnuclear deformations are determined using images that visualize nuclei in undeformed and deformed states, and a minimization of parameters to match images.

For each combination of the experimental factors (cell passage number and substrate coating type), at least 5 substrates were stretched with 2-9 cells visualized and analyzed. A total of 110 nuclei were studied for P0 (n=20) and P4 (n=22) cells on untreated substrates, and P0 (n=34) and P4 (n=34) cells on collagen-treated substrates. Strain transfer ratio (STR) was calculated as the average intranuclear hydrostatic strain divided by substrate equibiaxial strain, assessed separately using the motion of nuclei centroids. STR was calculated for the total nuclear area (yellow lines), in addition to interchromatin (pink) and chromatin (green) regions defined by image intensity (Fig 1). A two-way ANOVA was used for statistical analysis.

Quantitative PCR was performed to confirm phenotypic dedifferentiation of chondrocyte (P0) and fibroblast-like (P4) cells using markers for extracellular matrix proteins (e.g. types I and II collagen, aggrecan; data not presented). Expression of nucleoskeleton proteins was also studied to better interpret how mechanical links to the extracellular matrix change between P0 and P4 cells.

RESULTS

Strain distributions were heterogeneous in P0 and P4 cells with observed tensile and compressive regions of the nucleus interior (Fig 1). Strain distributions for P0 and P4 nuclei revealed regional variations in chromatin domains that were further segmented for clarity. There was a significant increase ($p < 0.05$) in STR for all the measured regions between P0 and P4 cells (Fig 2). Observed trends in nuclear/cell area and aspect ratio, gene expression of nucleoskeleton components and STR are summarized in (Fig 3).

A general decrease in nucleoskeletal gene expression was observed (Fig 1C). Nuclear envelope proteins that provide nuclear stability like lamin A/C, obscurin, MAN1 and titin show a high down regulation of their respective genes from 63% (LMNA) to 95% (LEMd3), suggesting a softening of the nucleus toward P4. Similarly, LINC-complex proteins Sun1 and Nesprins show significant decrease in gene expression from 77-86%.

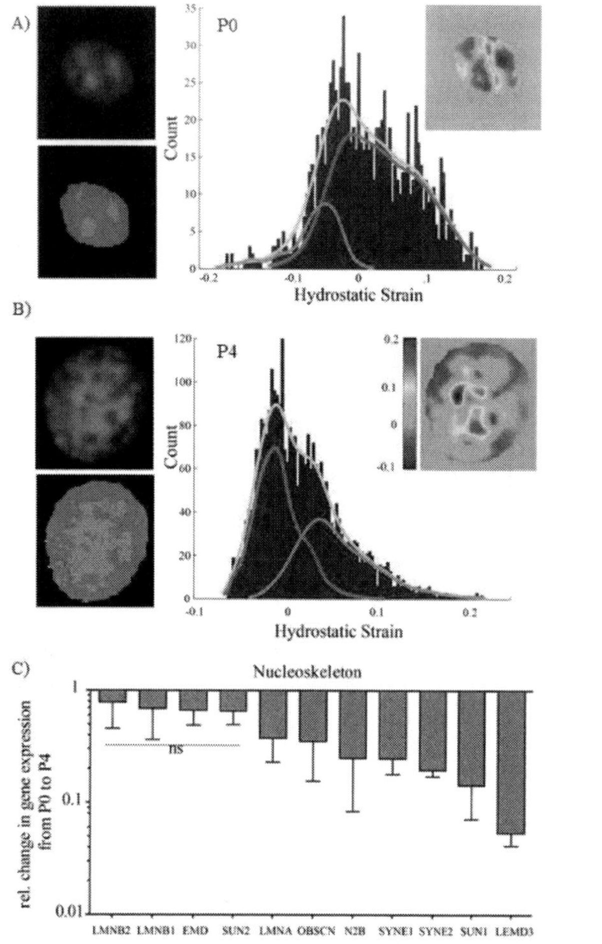

Figure 1. There were observed differences from P0 to P4 cells in intranuclear hydrostatic strain distributions, with total nuclear (yellow) and interchromatin (pink) and chromatin (green) regions noted. Changes in gene expression of P4 chondrocytes relative to fresh isolated (P0) cells for nucleoskeleton components were also observed.

DISCUSSION

The strength using hyperelastic warping is that high spatial resolution strain fields provide a detailed description of the nucleus interior compared to bulk measurements. In a similar cell substrate stretching experiment, texture correlation was used to calculate a bulk STR to the nucleus by only measuring a few points in the nucleus [2]. Warping derived measurements found areas of 10-30% tensile or compressive strains which were not presented in previous publications. Increase in STR from P0 to P4 is influenced by changes seen in the cell and nucleus during dedifferentiation. Phenotypic changes include increased cell and nucleus area and actin organization, and decreased chromatin condensation and expression of nucleoskeleton genes (Fig 3). The significance of this work will be to spatially correlate intranuclear strain and gene expression to better understand nuclear mechanobiology.

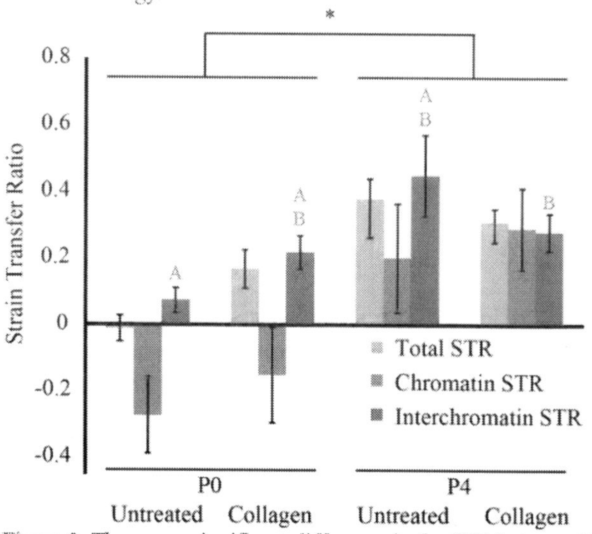

Figure 2. There was significant difference in the STR between P0 and P4 cells. We also found a significant interaction between substrate and passage for interchromatin STR ($p < 0.05$).

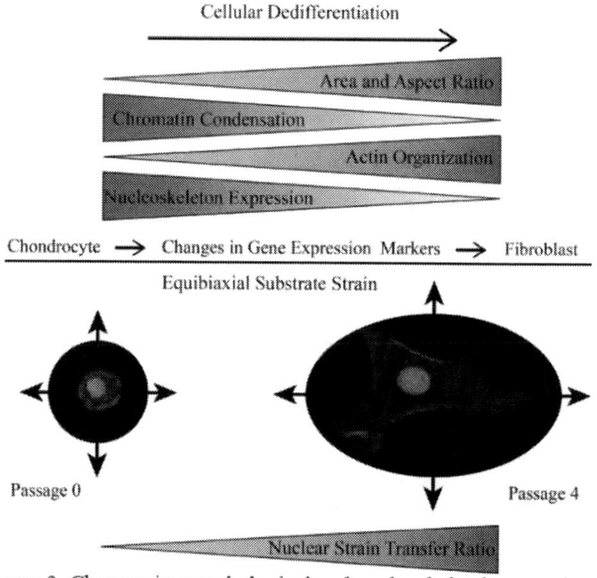

Figure 3. Changes in morphological and nucleoskeletal expressions from P0-P4 influences the STR.

ACKNOWLEDGEMENTS

The authors acknowledge funding from an Andrews Fellowship and NSF CAREER 1349735.

REFERENCES

[1] Schnabel, M. et al., *Osteoarthritis Cartilage*, 10:62-70, 2002.
[2] Gilchrist, C. et al., *J Biomech*, 40:786-794, 2007.
[3] Henderson, J. et al., *Biophysical Journal* 105, 2252-2261, 2013.
[4] Veress, A.I., et al., *Int J Biomed Imag, 2013*

SB³C2015
Summer Biomechanics, Bioengineering and Biotransport Conference
June 17-20, 2015,Snowbird Resort, Utah, USA
SB³C2015-562

A METHOD FOR RAPID, ACCURATE CALCULATION OF WALL SHEAR STRESS FROM 2D PHASE CONTRAST MAGNETIC RESONANCE IMAGE DATA

Elizabeth M. Iffrig (1), W. Robert Taylor (1,2), John N. Oshinski (1,3)

(1) Department of Biomedical Engineering
Georgia Institute of Technology
Atlanta, GA, US

(2) Division of Cardiology
Emory University & Atlanta VAMC
Atlanta, GA, US

(3) Department of Radiology
Emory University
Atlanta, GA, US

INTRODUCTION

In vascular diseases with an underlying inflammatory pathophysiology, the relationship with the frictional force the fluid exerts on the vascular wall, wall shear stress (WSS), has been studied extensively.[1] Numerous biological and computational models have demonstrated the co-localization of low and oscillatory WSS conditions with the induction of pro-inflammatory factors and early vascular disease phenotypes.[2] This body of work suggests the importance of WSS as a predictive factor in evaluating an individual's risk of developing vascular disease.

While many computational methods exist to calculate high spatially and temporally resolved WSS profiles within a vessel, these methods suffer from long computational times. If WSS is to be considered a potential 'biomarker' for assessing risk of disease, then faster reliable methods will need to be developed to determine WSS on a patient-specific basis. Herein we propose a method of measuring WSS using a an analytical solution to the Navier Stokes Equations combined with phase contrast magnetic resonance (PCMR) derived flow waveforms.[3] While this solution was generated in 1955 as a means of measuring WSS from fully developed flow in a rigid cylindrical tube with circumferentially uniformity, we have created a method built upon this solution that takes into account the increased complexity of actual blood flow seen in the abdominal aorta of healthy individuals. In this abstract we propose the development of a rapid method to calculate WSS from a PCMR image set and to apply this method in 5 healthy volunteers. This method will measure of the degree of oscillatory shear stress and could provide a clinical method for estimating the risk of aortic disease based on hemodynamic factors.

METHODS

Our method builds upon a previously published solution to the Navier Stokes equation that begins with a flow waveform and the vessel radius. We modified the solution in order to capture the circumferential variability of the shear profiles by considering flow waves from sectors instead of from the entirety of the cross section. We used these shear profiles to calculate the OSI and compared our results with those previously published. The specifics of our method are described below.

We first acquired 2D axial PCMR image sets in the abdominal aorta, encoded to measure through plane velocity (Venc 180cm/s) with retrospective ECG or peripheral pulse gating and reconstructing 30 times points in the cardiac cycle. Images were acquired in 5 healthy volunteers, all of whom were between the ages of 32 & 45 and were free from cardiovascular disease. Slice location was approximately half the distance between the renal arteries and the aortic bifurcation

The contour of the vessel lumen was determined using Segment (Medvisio). At each time point the contour was fitted to a circle and the unique radius was determined. Beginning with a radius drawn between the anterior and posterior portions of the left side of the vessel, ¼ of the lumen's cross sectional area was delineated. Flow was calculated for each time point by summing over the individual voxels within this contour (determined from the velocity value and the voxel's area) and multiplied by 4 to create a "full" vessel flow verus time wave form.

To employ the Womersley solution to this flow wave we took the Fourier transform and using the equations shown in Figure 1, determined the first 10 harmonic components for both the pressure wave and the shear stress. Summing over these harmonics at each of the time points allowed us to create our shear function for the described quadrant. The oscillatory shear index (OSI) was determined as shown in Figure 1 using a previously described equation.[4] We also calculated the angle of deviation from true posterior (90°) to the site of maximal OSI. To evaluate the range of OSI in our volunteers we computed the absolute difference in maximum and minimum values.

While this method could give us quadrant information if the entire vessel were divided into 4 equal sectors, we sought to increase our spatial resolution. We rotated the original sector described above to produce 80 unique sectors. We found that this rotation produced a distinct flow wave yielding a unique shear stress function and OSI value. We assigned this OSI value to the point in the center of the arc and calculated OSI to this point. This method yields 80 points around the circumference of the vessel and demonstrates the variability of oscillatory shear and is summarized in Figure 1.

RESULTS

The resultant OSI maps of the five volunteers are shown below in Figure 2. The OSI values ranged between 0.29 and 0.48. Peak OSI values were consistently seen on the posterior half the vessel wall, however this peak value was not always centered and deviated either left or right at average of 20.7°±11.2°. The absolute relative difference between the maximum and minimum points of OSI across the volunteers was .057±.01. Additionally, some of the volunteers had areas of increased OSI near the location of branch vessels from the abdominal aorta. The numbers obtained are consistent with values previously reported in literature for the abdominal aorta.[2]

DISCUSSION

The primary goal for this project was to develop a computationally inexpensive method using phase contrast MR to calculate WSS and OSI in the abdominal aorta. The values obtained are similar to what others have previously reported with different methods. In light of suggestions that WSS be used as a clinical indicator for risk of cardiovascular disease, this method offers an excellent tool for calculating the OSI in a healthy population. Ideally, this could be used as an early diagnostic in conjunction with other preventative measures to monitor efficacy of interventions (such as exercise and diet) known to lower risk of cardiovascular disease. However the proposed method in this project still requires validation. Additionally, the method suffers if the geometry of the vessel deviates too significantly from that of a healthy individual. In the future we plan to use this method to evaluate and compare the WSS in the abdominal aortas of men and women.

ACKNOWLEDGEMENTS

Work in this abstract was supported by an AHA Southeastern PreDoctoral Fellowship.

REFERENCES

[1] Dhawan SS et al., *Expert Rev Caridiovasc Ther*, 8(4):545-56, 2010

[2] Pedersen EM et al, *Eur J Vasc Surg*, 18:328-333, 1999

[3] Womersley JR, *J Physio*, 127:553-562, 1955

[4] Ku DN et al., *Arteriosclerosis*, 5(3):293-302, 1985

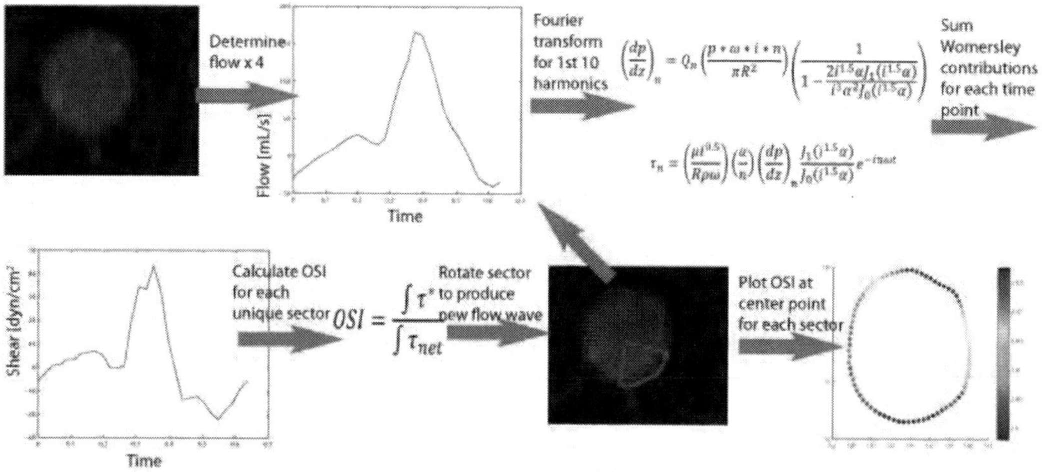

Figure 1. Summary of Method to Calculate OSI. The flow wave is calculated from the phase contrast image set within a given sector representing ¼ of the total vessel. The harmonic components of the wave are calculated and each contribution to the pressure wave and the shear are determined using the radial information from the contour as well as from the measured length of the cardiac cycle. The shear profile the OSI are determined by the summing the total shear in the proximal direction ($\dot{\tau}$) normalized to the total shear in both directions. The OSI is then plotted at the center point of each sector.

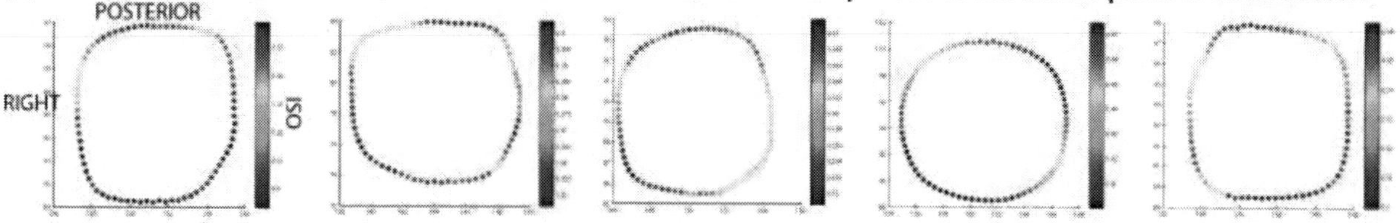

Figure 2. OSI Maps from 5 Volunteers. For each of the maps, the orientation shown in the first volunteer's map on the far left is applied. The values obtained from each of the five volunteers are similar to those previously reported for the abdominal aorta however the overall shape differs between the volunteers, suggesting the importance of capturing this variability.

SB³C2015
Summer Biomechanics, Bioengineering and Biotransport Conference
June 17-20, 2015,Snowbird Resort, Utah, USA
SB³C2015-563

MODELING THE INTERVERTEBRAL DISCS REPAIR WITH CELLULAR THERAPIES

Qiaoqiao Zhu (1), Xin Gao (2), Weiyong Gu (1, 2)

(1) Department of Biomedical Engineering
University of Miami
Coral Gables, Florida, U.S.A.

(2) Department of Mechanical & Aerospace Engineering
University of Miami
Coral Gables, Florida, U.S.A.

INTRODUCTION

While cellular therapies (e.g., cell implantation, enhancement in matrix synthesis, inhibition of matrix break down, etc.) have been shown a great potential in repairing a degenerated intervertebral disc (IVD) in vitro and in vivo experiments [1], many critical issues remain to be addressed before these experimental methods be translated into human clinical trials. One of them is that it may have significantly different outcomes in human IVDs with the similar treatment, due to the big differences in the disc size, nutritional environment, and loading patterns between human and animal discs [2]. Besides, some issues related to the cell therapy have not being resolved yet, such as the cell number needed for the implantation, timing for cellular therapies, nutrition supply needed for viability and function of the newly implanted or existing cells, and long-term outcomes of the treatment. Quantitative information on these issues is of significant importance for the successful repair of the degenerated discs. Thus, the objective of this study was to quantitatively assess the long-term efficacy of the cell therapies for IVD repair using a numerical method.

METHODS

The disc is modeled as an inhomogeneous, porous, mixture consisting of a charged solid phase [with cells and proteoglycan (PG) fixed to the solid phase], a fluid (i.e., water) phase, and a solute phase (sodium ion, chloride ion, glucose, oxygen, and lactate). The theoretical model is extended from the cell-activity coupled mechano-electrochemical model [3]. In the current model, the PG content is estimated by the chondroitin sulfated glycosaminoglycan (GAG) content. The conservation of mass for GAG is:

$$\frac{\partial\left(C^{GAG}\right)}{\partial t}+\nabla\cdot(C^{GAG}\mathbf{v}^s)=Q^{GAG}\quad, \tag{1}$$

where C^{GAG} is the molar concentration (per tissue volume) of GAG, $\mathbf{v}^s$ is the velocity of solid matrix, and Q^{GAG} is the synthetic (or degradation) rate of GAG (per tissue volume). The rate of GAG synthesis (or degradation) is assumed to be related to the cell density and GAG content by:

$$Q^{GAG}=\lambda_1\rho^{cell}-\lambda_2 C^{GAG}\quad, \tag{2}$$

where λ_1 is the GAG synthesis rate (per cell) and λ_2 is the degradation rate. The value of λ_2 is related to the half-life of GAG turnover [6]. Q^{GAG} is assumed to be zero in discs at healthy condition. Thus,

$$\lambda_1=\lambda_2\, C_0^{GAG}/\rho_0^{cell}\quad, \tag{3}$$

where C_0^{GAG} and ρ_0^{cell} are the GAG content and cell density at healthy state prior to the degeneration process.

An in silico model for the degenerated human lumbar disc has been developed based on the finite element method [3-5]. In this study, three types of cell therapies, i.e., increasing cell numbers (to simulate cell implantation) (Case I), increasing glycosaminoglycan (GAG) synthesis rate (to simulate an increase in matrix synthesis) (Case II), and decreasing GAG degradation rate (to simulate inhibition of matrix break down) (Case III), were applied to the nucleus pulposus (NP) region of a degenerated disc. For comparative purposes, a control case was also simulated in which no treatment was applied.

For Case I, the cell density was increased to 100% (Case I-a) or 200% (Case I-b) of the value in a healthy disc (i.e., 4000 cells/mm³ in NP [7]) to simulate cell implantation. In Case II, the GAG synthesis rate was increased by 100%, i.e., from 0.911×10^{-5} mmol/g dry weight/hr to 1.822×10^{-5} mmol/g dry weight/hr. In Case III, the relative GAG degradation rate was decreased by 50%, i.e., from 1.997×10^{-9}/s to 0.9986×10^{-9}/s.

Therapeutic interventions were introduced at an early stage of disc degeneration (corresponding to a mildly degenerative disc with a 7% reduction of the GAG content in the NP region and 2.2% disc height reduction, as it has been reported that the regenerative treatment is most beneficial at early stages of disc degeneration.

The loading condition on the disc was assumed to remain the same before and after treatments and the nutrition supply to disc was also assumed to be normal. Normal nutritional supply (details can be seen in [4]) was provided to all the treatment cases.

Only one upper-right quarter of the disc was simulated due to symmetry (Figure 1). The biological, chemical, electrical, mechanical, morphological changes in the disc under the various treatment cases were investigated. Results related to changes in GAG content, water content, and disc height under the treatments were presented.

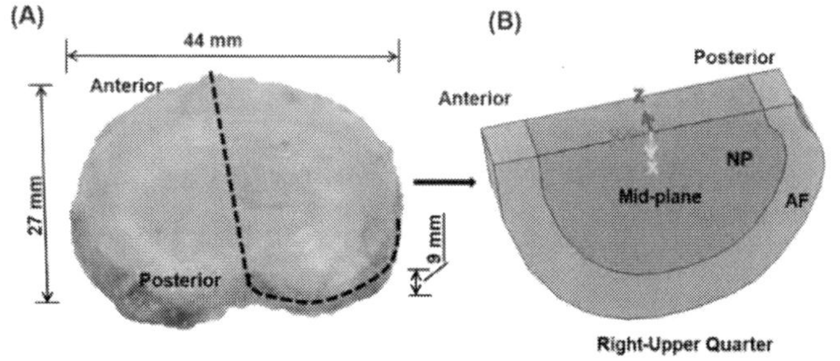

FIGURE 1: (A) GEOMETRY OF THE DISC FROM HUMAN LUMBAR SPINE (L2-3, MALE, NON-DEGENERATED) [3, 8]. (B) SCHEMATIC OF THE RIGHT-UPPER QUARTER OF THE DISC USED IN THE SIMULATIONS.

RESULTS

The changes of GAG content, water content, and disc height (all relative to the values at healthy state) were listed in Table 1. A 3D distribution of the matrix components (take GAG content as an example) in the degenerated disc with time following the three different types of treatment was shown in Figure 2. For Case I, the GAG and water contents increased (Table 1 and Figures 2C and D), and the repair process depended on the number of cells implanted. In Case I-a, the height of the degenerated disc almost returned to the value at healthy state in about 10 years. The disc height was recovered in about 1.0 year in Case I-b; at the end of 10 years, the height was 9.5% higher than that in the healthy state. For Case II, the GAG and water contents did not show significant increases in the severely degenerated region (viable cell density<50%, colored as dark-blue), but ~~increased~~ in the non-severely degenerated regions (non-dark-blue regions) (Figure 2D). The disc height was recovered in about 1.39 years (data not shown), and it was 6.0% higher than that in the healthy state at the end of treatment in 10 years (Table 1). For Case III, the GAG content, water content and the disc height (Table 1) continued to decrease with time after treatment. However, the rate of disc degeneration was reduced compared to that in the control case where no treatment was applied.

DISCUSSION

The effects of three types of cellular therapies (i.e., increasing cell density, increasing GAG synthesis rate, and inhibiting GAG degradation rate) on disc repair were simulated using the computer model developed and validated recently [4]. Our results indicate that treatment with increasing cell density (to simulate cell implantation) is effective for disc repair with a normal nutritional supply. Increasing the anabolic processes may have effects in increasing disc height, however, may not be effective for tissue repair in the regions with low viable cell density. Thus, it may not be suitable for the severely degenerated disc where most of the cells are non-viable. Inhibiting the catabolic processes in the disc can effectively slow down the degenerative process. However, this approach has no effect on the recovery of the tissue composition or disc height. The numerical model can be used to guide and optimize potential treatment therapies and has a great impact on developing new strategies to treat degenerated discs. If combined with patient-specific MRI data, this numerical tool can be used for the design and pre-clinical predictions of the treatment outcomes on a patient-specific basis.

TABLE 1: CHANGES OF GAG CONTENT, WATER CONTENT, AND DISC HEIGHT UNDER DIFFERENT THERAPIES AT THE END OF 1-YEAR (1yr), 5-YEAR (5yr), AND 10-YEAR (10yr) TREATMENT. ALL THE CHANGES WERE RELATIVE TO THE VALUES AT HEALTHY STATE.

Cases		No treatment	Case I-a	Case I-b	Case II	Case III
GAG content change (%)	1yr	-7.4	-6.6	-1.1	-2.8	-7.2
	5yr	-9.0	-5.2	+18.1	+11.2	-8.1
	10yr	-10.4	-3.9	+35.2	+23.7	-9.0
Water content change (%)	1yr	-3.2	-2.7	+0.1	-0.7	-3.0
	5yr	-4.2	-1.9	+10.1	+6.3	-3.6
	10yr	-5.1	-1.2	+18.6	+12.3	-4.1
Height change (%)	1yr	-1.6	-1.3	+0.1	-0.4	-1.5
	5yr	-2.1	-0.9	+5.1	+3.1	-1.8
	10yr	-2.6	-0.5	+9.5	+6.0	-2.2

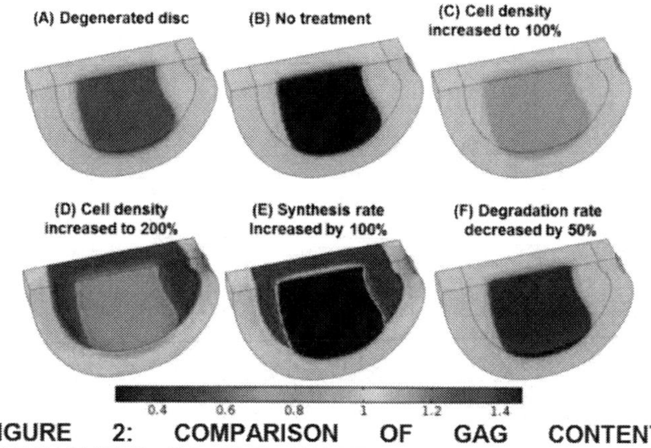

FIGURE 2: COMPARISON OF GAG CONTENT (NORMALIZED TO THE VALUE AT HEALTHY STATE) DISTRIBUTION IN THE DEGENERATED DISC BEFORE TREATMENT (A) AND WITH VARIOUS THERAPIES AT THE END OF 10-YEAR TREATMENT: (B) NO TREATMENT, (C) Case I-a, (D) Case I-b, (E) Case II, and (F) Case III.

ACKNOWLEDGEMENTS

This study was supported in part by a research grant from NIH/NIBIB (EB008653).

REFERENCES

[1] Benneker, LM et al., *Eur Cells Mater*, 27: 5-11, 2014. [2] Alini, M et al., *Eur Spine J*, 17: 2-19, 2008. [3] Zhu, Q et al., *J Biomech*, 45: 2769-2777, 2012. [4] Gu, W et al., *Spine*, 39: E1411-7, 2014. [5] Zhu, Q et al., *J Biomech*, 47: 3734-3743, 2014. [6] Sivan, SS et al., *J Biol Chem*, 281: 13009-14, 2006. [7] Maroudas A et al., *J Anat*, 120: 113-130. [8] Jackson, AR et al, *J Biomech Eng*, 133: 091006-091013.

SB³C2015
Summer Biomechanics, Bioengineering and Biotransport Conference
June 17-20, 2015,Snowbird Resort, Utah, USA
SB³C2015-564

BAYESIAN CALIBRATION OF A GROWTH AND REMODELING COMPUTATIONAL MODEL OF ABDOMINAL AORTIC ANEURYSM

Seungik Baek (1), Liangliang Zhang (2), Sajjad Seyedsalehi (1), Jongeun Choi (1,3), Chae Young Lim (2), Tapabrata Maiti (2)

(1) Department of Mechanical Engineering
Michigan State University
East Lansing, MI, USA

(2) Department of Statistics & Probability,
Michigan State University
East Lansing, MI, USA

(3) Department of Electrical and Computer Engineering
Michigan State University
East Lansing, MI, USA

INTRODUCTION

In this study, we apply Bayesian calibration for a computational growth and remodeling (G&R) model of abdominal aortic aneurysm (AAA) to predict the future behavior of the AAA by incorporating a set of longitudinal CT data. The computational model coupled with a statistic model is then validated in terms of prediction, showing the potential of our approach to the computation science and medical communities in aiding decision making. Due to the intra-patient variations, the prior distribution of model parameters has a strong influence on computational results for arterial mechanics. The first aim is to present a new systematic method to estimate the prior distribution of aortic material parameters in a constrained mixture model using previous biaxial tests. The aims are to develop a prediction framework that calibrates the physical AAA G&R model using patient-specific longitudinal computed tomography (CT)-scan images, predict the expansion of an AAA in future time, and analyze the associated uncertainty in the prediction. The Bayesian calibration is formulated for our AAA G&R computation model, which takes into account model inadequacy, the prior distribution of model parameters, measurement errors, and patient-specific longitudinal CT-scan images. First with the synthesized data, the performance and computational time are studied under different cases in sampling of model output data. Finally, we apply our Bayesian calibration for retrospective patient data from longitudinal CT scan images that illustrates its potential for the AAA management.

METHODS

Prior distribution estimation of model parameters

In the first step, we develop a statistical method to estimate the prior distribution of parameters in a constrained mixture model using previous biaxial tests of healthy abdominal aortas from 21 aortic specimens [1]. The best-fit parameters were obtained by minimizing the difference between the experimentally measured and theoretically calculated Cauchy stress for each specimen with the addition of penalty terms for constituents' homeostatic stress values and aortic wall thickness [2]. Although associated with other individual characteristics, gender and age are considered to be the basic but most relevant available variables for arterial mechanics. Using the experimental data, we studied the correlation between the estimated parameters with age and gender. We assumed that the best-fitted values we found for each parameter were the replications of a random variable with an unknown distribution. Our goal was to find the best distribution that can fit the estimated parameters. We divided our 11 parameters into two groups: Group-I with parameters that showed correlation with age or gender, and Group-II with parameters having no correlation with age and gender. For Group-I a regression between parameter values and age was used. For simplicity of regression, and by investigating the estimated parameters we chose two terms age and $1/age$. For Group-II, we investigated different distributions and transformations in order to find the best distribution for each parameter.

Bayesian calibration of a G&R AAA computation model

For a G&R AAA expansion model, a shape of elastin damage in the aortic wall initiates the growth of the aneurysm. For developing the framework, we use a damage shape with 2 parameters (θ_1 and θ_2) for the G&R computation model for the AAA expansion. Damage parameters serve as calibration parameters, which are patient specific for the AAA growth. For the statistical model, the process (represented by the diameter along axial location) is defined as $z_i = r(x_i, \theta) + \delta(x_i) + \varepsilon_i$, where both computer model $r(x_i, \theta)$ and model error $\delta(x_i)$ are assumed to be Gaussian processes. Observational error ε_i is normal distributed. Let ψ_1 and ψ_2 be the hyper-parameters in the covariance functions of $r(x_i,\theta)$ and $\delta(x_i)$ respectively, while β_1 and β_2 are the hyper-parameters in the mean functions of $r(x_i,\theta)$ and $\delta(x_i)$, respectively. In a Bayesian framework, we need to specify the prior distribution for calibration parameter θ. The statistical model then includes Gaussian process priors for both the G&R computational model and the model error, which require implementing Bayesian analysis for calibration. To investigate the prior distribution of the damage shape and difference choices of sampling times, we first use the AAA model to generate synthetic observation to produce computer model outputs. β can be integrated out easily but ψ is not. Numerical integration may not give an accurate result since it is the multi-dimensional integration. Thus, instead of integrating out ψ, we use the conditional posterior distribution and a plausible estimate of ψ plugged in as proposed by Kennedy and O'Hagan [3]. The plausible estimate of ψ is obtained in a two step approach. First, we estimate ψ_1 by maximizing $p(\psi_1| y)$ (Stage 1). Then, we estimate ψ_2 by maximizing $p(\psi_2/d, \psi_1)$ after plugging in the estimated ψ_1 (Stage 2).

RESULTS

Prior distribution estimation of model parameters

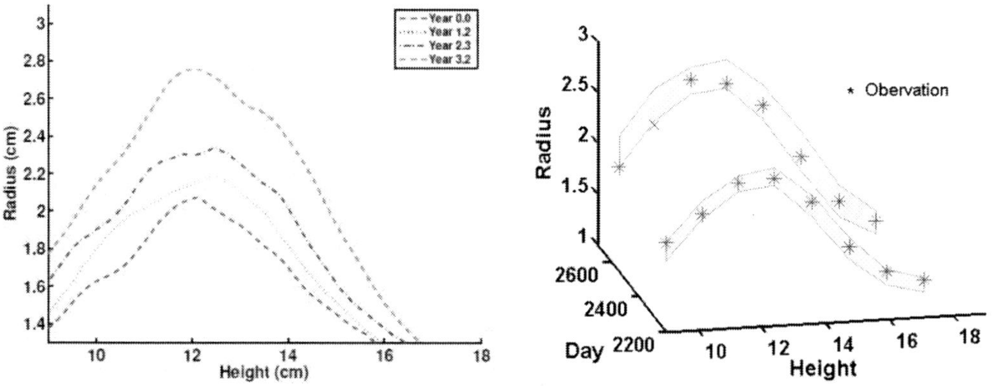

Fig. 1. Left: Radius v.s. height along the centerline measured from CT scans of patient over the surveillant time. First two scan data uses as the observation and 3rd and 4th aneurysms' shape are predicted. Right: 95% confidence band of predictions. The blue starts denote the observations lying inside the confidence band.

We investigate the correlation between the estimated parameters for each constituent and the patient age and gender; however, the results indicate that the parameters are correlated with age only. The parameters are classified into two groups: Group-I, in which the parameters $c_e, c_{k_2}, c_{k_2}, c_{m_2}, G_h^c$, and ϕ^e are correlated with age, and Group-II, in which the parameters $c_{m_2}, G_h^m, G_1^e, G_2^e$, and a are not correlated with age.

Bayesian calibration of a G&R AAA computation model

The Bayesian calibration uses a longitudinal data that we hold out the last two year observations from the calibration procedure in order to validate our approach by comparing predictions on the last two years at 2.3 and 3.3 years. There are some larger discrepancies in the lower part of predictions at 3.2 years, which can also be shown by the corresponding larger relative errors. Most of the relative errors are less than 5%. Additionally, from Fig. 1 (right), we find that only one true value lies outside of the confidence bands. The results from the simulation case study strongly suggest that Bayesian calibration can be used to combine the computer model, prior and uncertainty models, and real observations to successfully predict the QoI at future times or at unobserved locations. Information from the computer model provides a deterministic trend of the expansion process (modeled by a Gaussian process) while the model error can be treated as another Gaussian process. When we have more information from the computer model (in the form of finer grids for inputs X_c), we achieve lower prediction errors, and the posteriors of calibration parameters θ_1 and θ_2 are more likely to concentrate on their true values as illustrated in our simulation results. We also find that posteriors and predictions from our approach are robust to the selection of priors for θ_1 and θ_2.

DISCUSSION

The estimations derived from the posterior distribution are calibrated parameter values obtained by fitting the computer to the real data observed in the physical systems. We then insert the calibrated parameter values of the computer model again to predict the growth of the AAA by using the predictive distribution in the Bayesian model. Therefore, from the computer model we learn the distribution of priors for parameters, then in Bayesian Calibration from the posteriors we update the distribution of parameters according to the real observations. In this way, we tried to get more accurate information for parameters in order to improve predictions. As future work, we plan to consider complicated damage shapes, hemodynamics, and stress-mediated kinetic models with more parameters in a 3D geometry. Additionally, linearity limitation on the mean functions for Gaussian processes will be relaxed to be flexible for better prediction performance. More importantly, we will investigate how to deal with high dimensional calibration parameters when a flexible and complex damage function is used.

ACKNOWLEDGEMENTS

Authors gratefully acknowledge the support, in part, by the NIH (R01HL115185) and NSF CAREER award (CMMI-1150376).

REFERENCES

[1] J. P. Vande Geest et al., *J Biomech Eng*, 39: 1324–1334, 2006.
[2] S. Zeinali-Davarani et al., *Comp Meth Biomech Biomed*, 9: 803-817, 2011
[3] M. Kennedy and A. O'Hagan, *J. Roy Statist Soc*, 63:425-464, 2001

BIAXIAL MECHANICAL CHARACTERIZATION OF NON-UNIFORM THERMAL SHRINKAGE DEFORMATIONS TO GUIDE ABLATIVE THERAPY

Steven M. Boronyak and W. David Merryman

Department of Biomedical Engineering
Vanderbilt University
Nashville, TN, USA

INTRODUCTION

Thermal ablation devices have been used to shrink collagen-rich tissues using resistive heating to treat a variety of conditions such as glenohumeral joint laxity [1], sleep apnea [2], and anterior cruciate ligament damage [3]. Recently, thermal shrinkage of mitral valve (MV) leaflets has been proposed as a percutaneous alternative to surgical leaflet resection using an RF ablation and cryo-anchoring (RFC) catheter to shrink enlarged MV leaflets [4-5]. In order to optimize therapy and to predict the thermal shrinkage response, shrinkage deformations must be characterized for a given device. Previous studies on thermal shrinkage of collagenous tissues have used uniaxial and biaxial testing devices to characterize shrinkage deformations under a variety of loading and temperature conditions [6-7]. However, previous biaxial characterization of thermally-induced deformations have used a standard 2x2 marker tracking array, which is unable to characterize non-uniform shrinkage deformations that may arise due to ablation device configuration and tissue anisotropy. The objective of this work was (1) to develop a biaxial mechanical testing method that more completely characterizes non-uniform tissue deformations that arise in thermal shrinkage of collagenous tissues and (2) to apply this method to characterize the shrinkage response from RFC catheter treatment of MV leaflet tissue. Additionally, planar collagenous tissues that have been treated with heat have been shown to dilate slightly within the first hour after treatment [6], and we sought to (3) investigate the spatial dependence of post-treatment dilation.

METHODS

Sample Preparation: Healthy porcine MV anterior leaflets were obtained from the local abattoir (Hampton Meats, Hopkinsville, KY). Anterior leaflets were chosen for this study due to their large and continuous surface area. The center portions of the leaflets were cut into rectangular sections and 16 markers were fixed to the atrial side of the tissue samples to allow optical tracking for strain calculations during mechanical loading. Four hooks where inserted into each side of a tissue sample and attached to sutures for mechanical loading. The circumferential (direction of collagen fiber alignment) and radial leaflet directions were aligned with the loading axes (Fig. 1A). Samples were tested in biaxial load control mode to a membrane tension comparable to that experienced by healthy MV leaflets during systole (120 mmHg ≈ 90 N/m). All testing was performed with samples fully submerged in phosphate buffered saline.

Mechanical Testing and RF Ablation Treatment: Samples were initially examined to pre-condition the tissue and to quantify biaxial compliance by loading 10 times to 90 N/m of equibiaxial tension. Load was removed from the samples, and the marker reference configuration was acquired. Samples were then treated with the RFC catheter, while submerged in phosphate buffered saline, at a power of 50 W for 60 s while cryo-anchored between -40°C and -50°C (Fig. 1A-C). Samples were then pre-conditioned by loading 10 times to 90 N/m of equibiaxial tension, unloaded, and the treated configuration was acquired. Pre-conditioning and marker position acquisition was repeated again at one hour post-treatment to examine any re-dilation effects.

Deformation Characterization: Bilinear interpolation schemes are traditionally used to determine the deformation gradient from 2x2 marker arrays in biaxial mechanical testing [8]. In order to better characterize the non-uniform shrinkage deformations induced by the RFC catheter, a 4x4 marker array was used to form an isoparametric, curved-sided quadrilateral element. A bicubic interpolation, with basis functions reported from [9], was used to determine the deformation gradient tensor, **F**, within the 4x4 marker quadrilateral. Bicubic interpolation has the advantage of being C^1 continuous to avoid discontinuities when determining the displacement gradient, and the curved sides enhance interpolation fidelity. **F** was interpolated everywhere within the enclosed 4x4 marker region, and directional deformations (circumferential – F11, radial – F22) and areal deformations (determinant of the deformation gradient tensor) are reported with respect to the pre-treatment condition.

RESULTS

Representative areal deformations resulting from RFC catheter treatment are reported in the reference (pre-treatment) and post-treatment configurations (Fig. 2A-B). Deformations are greatest in the central region and along the circumferential (X11) axis. Cryo-anchor placement inhibits the formation of large shrinkage deformations along the border zones in the top and bottom of the radial (X22) axis.

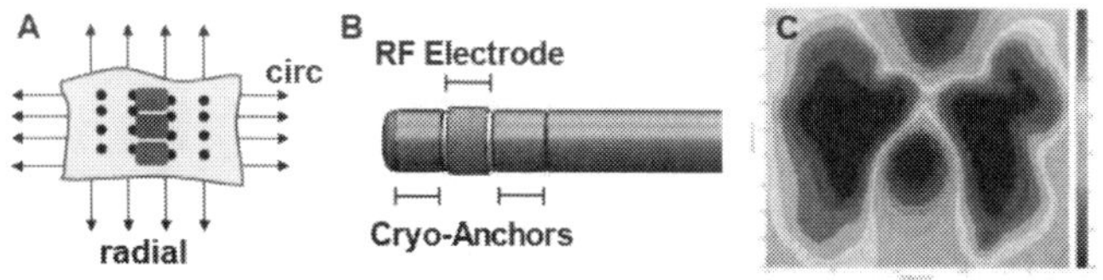

FIG. 1 – (A) 4x4 marker array shown with catheter orientation. Red: RF ablation electrode. Blue: cryo-anchors. (B) Arrangement of RF electrode and cryo-anchors on the catheter distal tip. (C) Typical thermal contours resulting from treatment with the RFC catheter. Scale from -20°C (blue) to 100°C (red).

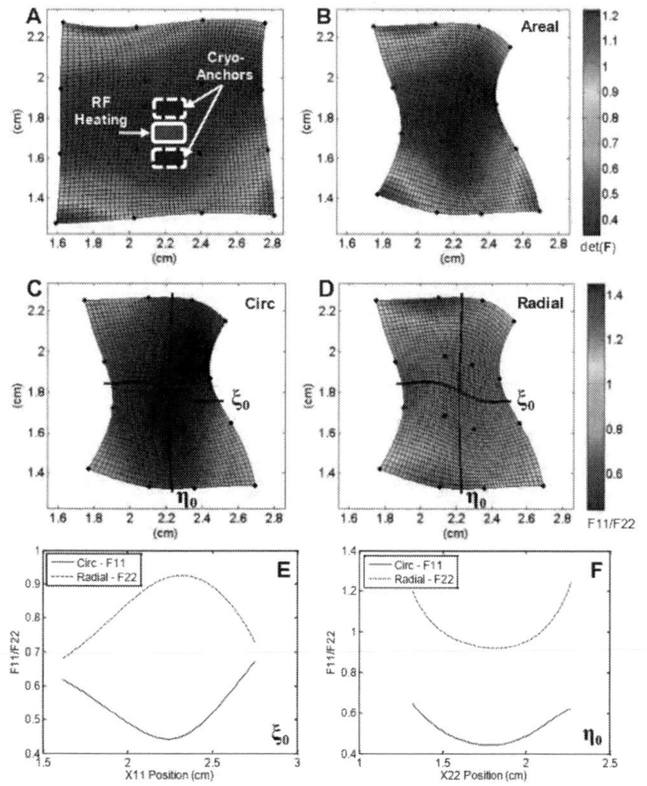

FIG. 2 – Areal deformation in the reference (A) and treated (B) configurations immediately after ablation treatment. Relative positions of RF electrode (heating) and cryo-anchors (cooling) are shown. Circumferential deformation (C) and radial deformation (D) immediately post-treatment. Circumferential and radial deformation along local coordinate system lines ξ=0 (E) and η=0 (F).

Directional deformations immediately post-treatment are reported in Fig. 2C-D, and deformation magnitude is greater in the circumferential axis due to collagen fiber alignment. The effects of cryo-anchor placement can be seen in Fig. 2D, where deformations are largest to the areas immediately left and right of the RF electrode. Circumferential and radial deformations are shown along lines $\xi=0$ and $\eta=0$ in Fig. 2E-F, in which large spatial and direction variation in deformation can be appreciated.

Tissue dilation measured one hour post-treatment was characterized by well demarcated regions of enlargement that received the greatest shrinkage deformation during treatment (Fig. 3A-B).

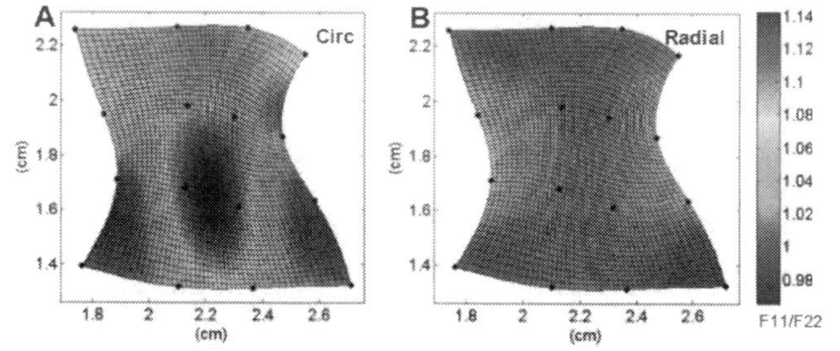

FIG. 3 – Unloaded deformations one hour post-treatment: circumferential deformation (A) and radial deformation (B).

DISCUSSION

Characterization of device-specific thermal shrinkage deformations is an important step in developing computational models and optimizing therapy for pre-procedure planning. However, most analyses of thermal shrinkage response in planar collagenous tissues have failed to capture the non-uniform deformations that occur in a device-specific and tissue-dependent manner. In this study, we develop a bicubic interpolation method to more fully characterize the non-uniform deformation response that results from the RFC catheter – a specialized ablation device that consists of an RF electrode surrounded on two sides by cryo-anchoring elements that inhibit thermal shrinkage. We found deformations that varied strongly depending on the direction of collagen fiber alignment and device alignment with the tissue. Additionally, analysis of dilation one hour post-treatment showed that dilation was greatest in areas of greatest shrinkage deformations. Further results from this analysis will be used to build computational models to guide treatment therapy and to better understand spatially-varying tissue shrinkage dynamics.

ACKNOWLEDGEMENTS

This work was funded by the Wallace H. Coulter Foundation and and an American Heart Association pre-doctoral fellowship (13PRE16340018).

REFERENCES

[1] Hayashi, K., et al., *Arthroscopy*, 12:474-81, 1996.
[2] Pang, K.P., et al., *Otolaryngol Head Neck Surg*, 136:823-6, 2007.
[3] Ma, H.L., et al., *Knee Surg Sports Traumatol Arthrosc*, 13:619-24, 2005.
[4] Boronyak, S.M., et al., *J Biomech*, 47:973-80, 2014.
[5] Boronyak, S.M., et al., *Ann Biomed Eng*, 40:1971-81, 2012.
[6] Wells, P.B., et al., *J Biomech Eng*, 126:492-7, 2004.
[7] Harris, J.L., et al., *IEEE Trans Biomed Eng*, 51:371-9, 2004.
[8] Humphrey, J.D., *Cardiovascular Solid Mechanics : Cells, Tissues, and Organs*. 2002, New York: Springer. xvi, 757 p.
[9] Lynch, D.R., *Numerical partial differential equations for environmental scientists and engineers : a first practical course*. 2005, New York: Springer. xxiv, 388 p.

SB³C2015
Summer Biomechanics, Bioengineering and Biotransport Conference
June 17-20, 2015,Snowbird Resort, Utah, USA
SB³C2015-566

MODULATION OF LYMPH NODE RESISTANCE DURING INFLAMMATION: EXPERIMENTAL MEASUREMENT AND COMPUTATIONAL MODELING

Mohammad Jafarnejad (1), Joshua P. Scallan (2), Joseph M. Sherwood (1), Darryl R. Overby (1), David C. Zawieja (3), Michael J. Davis (2), James E. Moore Jr. (1)

(1) Bioengineering Department
Imperial College London
London, United Kingdom

(2) Department of Medical Pharmacology and Physiology
University of Missouri at Columbia
Columbia, MO, United States

(3) Department of Systems Biology and Translational Medicine
Texas A&M Health Science Center
Temple, TX, United States

INTRODUCTION

Lymph nodes (LNs) are located in strategic locations throughout the body. Interstitial fluid is absorbed by initial lymphatics where it is then transported via pre-collector lymphatic vessels to collecting vessels that propel it to LNs [1,2]. Collected interstitial fluid from tissues and organs in the body passes through one and up to several nodes before being returned to the blood circulatory system [3]. This fluid, known as lymph brings in tissue specific antigens, pathogens, cytokines and chemokines as well as the immune cells (most importantly dendritic cells (DCs)) to the downstream LN that can trigger adaptive immune response by either humoral or cellular means [4,5]. LNs are where the lymphatic system and blood circulatory system intersect and closely interact with each other. LNs comprise several distinct cellular regions contained in a fibrous capsule where afferent and efferent lymphatic vessels serve as the inlets and outlets for lymph transport. It is generally thought that lymph is propelled to the subcapsular sinus (SS), passing primarily through the medulla, and exiting via efferent lymphatic vessel(s) [3,6]. Alternatively, lymph can percolate through the SS floor to B cell follicles and T cell cortex, and bathe the LN resident immune cells[7]. Furthermore, the conduit system of the LNs allows lymph and small antigens (<70 kDa) to be delivered to medullary sinuses (allowing interaction with blood vessels) to initiate the remotely controlled immune response [8-11]. Antigen carried by lymph in the LN is shown to localize in the interfollicular regions, where it is taken up by LN resident DCs that will in turn initiate the immune response [12]. The mechanism by which lymph is localized in parts of the node still remains to be elucidated.

The structure of the LNs has been shown to dramatically change after inflammation in the draining tissue. The volume of the node is sometimes increased up to five-fold within 7 days. However it is not known how expansion affects the fluid flow inside the node and if the expansion is a result of altered hemodynamic forces on the compliant LN. The swelling of the node changes the structure, which is expected to vary the resistance of the LNs to fluid flow. Despite the critical role of lymph as the carrier for molecules and cells to the LNs, little is known about its transport patterns and modification while passing through the node. This work aims to measure the resistance of the node before and after inflammation and to construct the first ever three-dimensional computational model of lymph flow in the LN.

METHODS

Animals and Tissue Preparation

Animal protocols were approved by the Institutional Animal Care and Use Committee (IACUC) at the University of Missouri-Columbia and conducted in accordance with the National Institutes of Health's *Guide for the Care and Use of Laboratory Animals*. C57BL/6 mice were anesthetized and the popliteal LNs were dissected along with its afferent and efferent lymphatic vessels. One afferent and efferent vessel were cannulated on glass pipettes of sizes 50-70 µm in diameter. Extra afferent or efferent vessels were closed off by sutures. Animals were injected with 10 ng lipopolysaccharides (LPS) and 10 ng polyinosinic:polycytidylic acid (Poly I:C) in each footpad three days before the experiments to induce inflammation.

Resistance Measurement Setup

The *iPerfusion* resistance measurement system including a differential pressure transducer (Omegadyne, USA), a sensitive flowmeter (Sensirion, Switzerland), and a linear actuator was used to measure the flow and pressure difference across cannulated lymph nodes (Figure 2A) for a number if applied pressures. Additionally, the resistance of the glass pipettes was measured before the experiments.

Computational Fluid Dynamics

An idealized geometry of the lymph node was constructed in Solidworks 2012 SP4.0 (Dassault Systèmes, France) based on the images captured in this study as well as available images in the literature (Figure 1A). The whole node is modeled as an oblate ellipsoid with similar sizes to mouse popliteal node (large diameter of 1 mm and small diameter of 0.7 mm). The final geometry was comprised several regions, each of which was either defined as a fluidic or porous region depending on the structure (Figure 1). The height of the SS is about 10μm as reported by other studies [3,6].

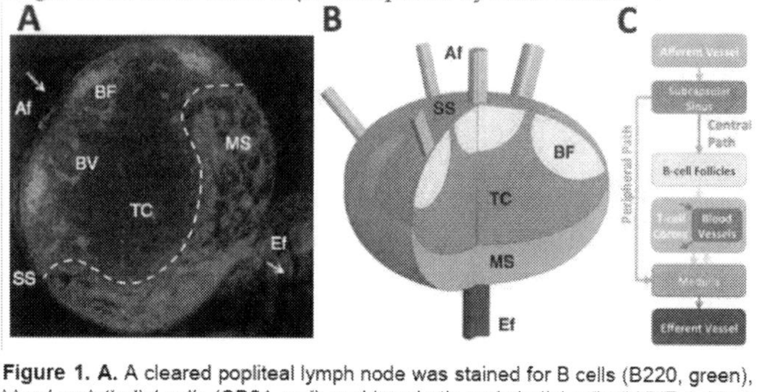

Figure 1. A. A cleared popliteal lymph node was stained for B cells (B220, green), blood endothelial cells (CD31, red), and lymphatic endothelial cells (LYVE-1, blue) and was imaged using confocal microscopy. Lymph enters the node through afferent lymphatic vessel(s) (Af). It then moves to subcapsular sinus (SS), which can then go to either B cell follicles (BF) and T cell cortex (TC) through a central path, or directly to medullary sinuses (MS) through a peripheral path. Lymph, if not absorbed by blood vessels (BV), will leave the node through efferent vessel(s) (Ef). **B.** An idealized geometry of the lymph node was made to contain the main regions of the node. **C.** A flowchart of the lymph movement through regions of the node.

Star-CCM+ (CD-adapco, London), a commercially available finite volume based solver, was used to model fluid flow inside the LN. The idealized geometry was imported, meshed and the Navier-Stokes equations were solved in the fluidic region as well as a continuity equation and Darcy's law with Brinkman's term in the porous regions. Starling's law was used to model the fluid transport between lymphatics and blood in the node. Constant inlet velocity and efferent pressure were assumed as the boundary conditions. Resistance was calculated as the pressure difference divided by the average flow going into the node. Hydraulic conductivity of medulla ($K_{medulla}$) was changed relative to that of the T cell cortex ($K_{T\ cell\ cortex}$) as the main determinant of the LN resistance.

RESULTS

The median of the resistance of healthy nodes was measured to be 0.74 mmHg/(μL/min). The median of resistance was significantly higher (two-tailed t-test, $p = 0.0086$), 2.47 mmHg/(μL/min), after inflammation (Figure 2B).

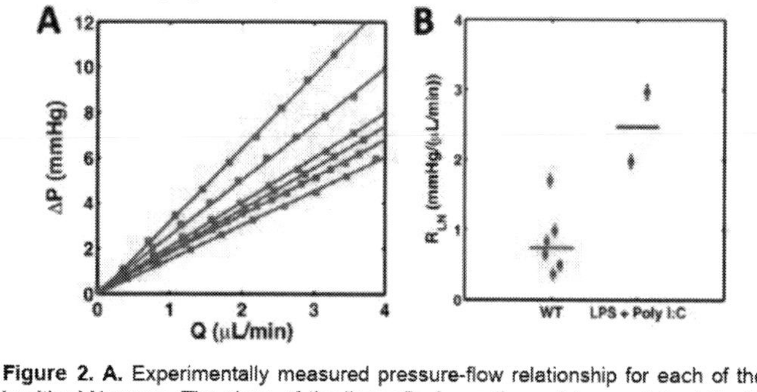

Figure 2. A. Experimentally measured pressure-flow relationship for each of the healthy LN cases. The slope of the linear fit shows the resistance of the LNs. **B.** Comparision of the resistance of healthy and inflamed LNs. Each case is shown with its 95% confidence interval (mean±2SD) and the medians for WT and inflamed LNs.

The computational model showed at high $K_{medulla}$, the changes in Q_{BV} were negligible (Figure 3A). However for the lower values of $K_{medulla}$, Q_{BV} increased to values as high as 44% of the afferent flow rate for $K_{medulla}/K_{T\ cell\ cortex}$ of 0.1 for the baseline *LA* value. Furthermore, $K_{medulla}$ was shown to be a major determinant of the apparent resistance of the whole node (Figure 3B). A decrease in $K_{medulla}$ resulted in a steep increase in the apparent resistance; for instance, an order-of-magnitude decrease in $K_{medulla}$ increased the apparent resistance of the node from 1.7 to 6.8 mmHg/(μL/min), or about a 4-fold increase.

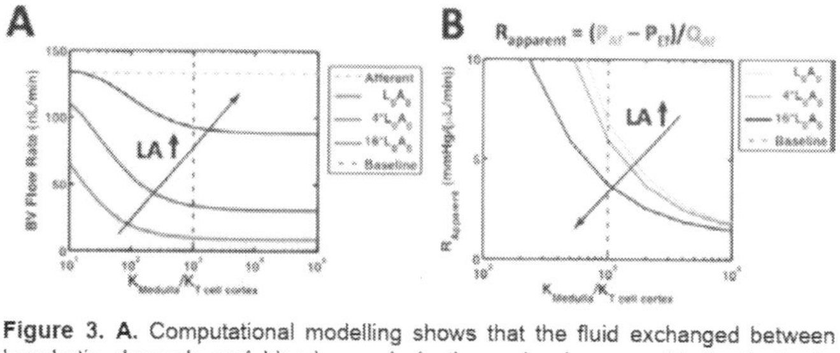

Figure 3. A. Computational modelling shows that the fluid exchanged between lymphatic channels and blood vessels in the node changes with the hydraulic conductivity of the medulla. The effect is more prominent when the surface area (*A*) and/or hydraulic conductivity of the blood vessels (*L*) increases (L_0A_0 is the baseline value of *LA*) **B.** Apparent resistance of the lymph node increases as the hydraulic conductivity of the medulla decreases.

DISCUSSION

Experimental measurements showed that lymph node resistance increases during inflammation. Computational simulations demonstrated how the resistance is dependent on the hydraulic conductivity of the medulla. We speculate that immune cell trafficking through the node during inflammation restricts the passageways for lymph transport and hence contributes to increases in the total lymph node resistance. More detailed studies are required to investigate the role of changes in lymph node resistance on lymph formation in the upstream tissue as well as implications of this resistance change for proper adaptive immune response during inflammation and vaccination.

ACKNOWLEDGEMENTS

J. E. Moore, Jr. acknowledges the support of the Bagrit Trust, the Royal Academy of Engineering, and the Royal Society-Wolfson Research Merit Award. This research is generously supported by NIH grant numbers R01 HL094269, R01 120867 (MJD) and U01 HL123420-01 (JEM & DCZ). The *iPerfusion* system was developed under EPSRC grant EP/J010499/1 (DRO).

REFERENCES

1 Zawieja, D. C. *Lymphat Res Biol* **7**, 87-96, (2009).
2 Swartz, M. A. *Adv Drug Deliv Rev* **50**, 3-20, (2001).
3 Girard, J. P. *et al. Nat Rev Immunol* **12**, (2012).
4 Lammermann, T. & Sixt, M. *Immunological reviews* **221**, 26-43, (2008).
5 Randolph, G. J. *et al. Nat Rev Immunol* **5**, (2005).
6 Ohtani, O. & Ohtani, Y. *Archives of Histology and Cytology* **71**, 69-76 (2008).
7 Gonzalez, S. F. *et al. Annual review of immunology* **29**, 215-233, (2011).
8 Gretz, J. E., *et al. J Exp Med* **192**, 1425-1440 (2000).
9 Roozendaal, R. *et al. Immunity* **30**, 264-276, (2009).
10 Gretz, J. E. *et al. Journal of immunology* **157**, 495-499 (1996).
11 Roozendaal, *et al. Intl. immunology* **20**, 1483-1487, (2008).
12 Woodruff, M. C. *et al. J Exp Med* **211**, 1611-1621, (2014).

SB³C2015
Summer Biomechanics, Bioengineering and Biotransport Conference
June 17-20, 2015,Snowbird Resort, Utah, USA
SB³C2015-567

MICROSCOPIC HETEROGENEITY IN THE AORTIC WALL: CORRELATION BETWEEN MECHANICAL ENVIRONMENT AND PROTEIN EXPRESSION

Takeo Matsumoto (1), Yohei Uno (1), Shintaro Iijima (1), Yoshitaka Moriyama (1), Shukei Sugita (1), Kazuaki Nagayama (1, 2), Akio Matsumoto (3)

(1) Department of Mechanical Engineering
Nagoya Institute of Technology
Nagoya, Aichi, Japan

(2) Department of Intelligent Systems Engineering, Ibaraki University
Hitachi, Ibaraki, Japan (present)

(3) Department of Pharmacology
Graduate School of Medicine
Chiba University
Chiba, Chiba, Japan

INTRODUCTION

The media of elastic arteries such as the aortas has a concentric layered structure composed of elastic lamina (EL) and smooth muscle-rich layer (SML), which is called a lamellar unit after Wolinsky and Glagov [1]. It is well known that the ELs are wavy in the histological sections of unloaded specimens. Recently, we found that such waviness of the ELs in the unloaded wall is released and the ELs become straight upon isolation from surrounding tissues [2]. This supports our previous observation that the ELs bear compressive residual stress and the SMLs tensile in the circumferential direction in the unloaded state [3]. We also noticed that the waviness of the ELs vary widely even between the two ELs adjacent to each other (Fig. 1) [4]. This may indicate that straight ELs bear high tensile stress in a physiological condition for they are readily stretched, while wavy ELs bear relatively low tensile stress because they might not bear tensile stress until they become straight. To test this hypothesis, we performed laser ablation experiment in this study on thin-sliced specimens of porcine thoracic aorta stretched in the circumferential direction mimicking physiological condition. As an index of the amount of tension, we measured the width of the gap caused by the ablation, and found that the local waviness of the ELs had a negative correlation with .the gap as we hypothesized.

If the mechanical environment is heterogeneous at microscopic level, protein expression of the smooth muscle cells (SMCs) may also be heterogeneous depending on the force applied to each cell. To identify proteins whose expression levels are responsive to mechanical environment, it may be worth comparing gene expression levels in the aortic tissues subjected to different mechanical environment. Recently, it is becoming clear that the wall of thoracic aorta is more compliant in the ventral side than the dorsal in human [5], rabbits [6], and pigs [7]. Taking advantage of the dorsal-ventral difference in the aorta, we have performed a DNA microarray analysis to identify the genes expressed differently between the dorsal and ventral sides. Base on this information, we examined the relationship between the local protein expression levels of the SMCs and the local waviness of the ELs adjacent to the cells, and found that some proteins had significant correlations with the waviness of the ELs.

METHODS

Preparation of thin-sliced specimen

Fifteen-µm-thick sections perpendicular to the longitudinal direction were obtained from the porcine thoracic aortas with a cryostat (CM3050SIV, Leica Microsystems, Germany). The nuclei of the SMCs were stained with Hoechst33342 (Invitrogen, USA) as fiducial markers for the SMLs. Characteristic points on the ELs were identified under a transmitted light microscope and used as fiducial markers for the deformation of the ELs. To examine the effects of freezing and thin slicing, 200-µm-thick sections were also used. Transmitted light image cannot be obtained for the thick slices, the ELs were stained with anti-elastin antibody (BA-4, Santa Cruz, USA) and Alexa Fluor 488 (Invitrogen, USA) for 200-µm-thick specimens.

Observation of microscopic deformation of aortic walls during circumferential stretch

To mimic the inflation of the aorta, the thin slices were then stretched in the circumferential direction with a laboratory-made tensile tester under an upright fluorescent microscope (BX-61, Olympus, Japan) equipped with a CCD camera (DP-70, Olympus, Japan) in an aerated Krebs-Henseleit solution at 37°C. Local elongations of the ELs and SMLs were obtained by measuring the displacement of the markers.

Laser ablation test to confirm heterogeneity in tension

To confirm heterogeneity in the circumferential stress in a physiological state, the thin slices were stretched in the circumferential direction by 35% as stated above and various points on the slices were ablated for 100 ms by a UV pulse laser (Awave355-700mW20K, Advanced Optowave, USA) at the power of 775mW through a 20x objective to make a gap both in the ELs and SMLs. The width of the gap in the circumferential direction was used as an index of tension in that region.

DNA microarray analysis

To identify the proteins responsive to mechanical environment, dorsal-ventral difference in gene expression of SMCs in rabbit thoracic aortas was analyzed with a microarray. Dorsal and ventral quarters of aortic segments were analyzed with a Rabbit Gene Expression Microarray (G2519F, 020908, Agilent, USA) based on the manufacturer's standard protocols. Microarray expression data was validated by real-time PCR. Proteins that show dorsal-ventral difference in their expression in aortic tissue were found among genes validated by real-time PCR by immunohistochemical staining.

Immunohistochemical analysis

Unloaded wall of rabbit thoracic aortas were stained for proteins identified in the previous section and correlation between protein expression in the SMCs and waviness of ELs adjacent to the cells was analyzed.

RESULTS AND DISCUSSION

The thin-slices of the aortic tissues showed highly complicated deformation at microscopic level: Rotation and shear deformation was observed in the SMCs, and the ELs became straightened first without significant lengthening and then become elongated, indicating that ELs with lower waviness bear higher tension in a physiological state. Furthermore, we found that stretch ratio was also highly heterogeneous at a microscopic level even in a single EL and SML: Stretch ratio in the ELs and the SMLs was 1.1-1.7 and 1.1-1.8, respectively, at the mean circumferential stretch ratio of 1.3 (Fig. 2). Similar results were obtained for 200-µm-thick slices.

The heterogeneity in the tension in the physiological state was confirmed by laser ablation: The gap between the ablated ends of an EL had a significant negative correlation with the waviness. Similar results were obtained for the SMLs: SMLs adjacent to ELs with lower waviness had a wider gap following ablation.

In the microarray analysis, we found 392 genes of 43,603 expressed more than twice in the ventral side as much transcript as in the dorsal side, and 296 genes expressed less than half as well. Then, we confirmed several genes with dorsal-ventral difference by real-time PCR, such as osteopontin with higher expression in the dorsal side and actin with higher expression in the ventral. It has been reported that osteopontin stimulates DNA synthesis of SMCs [8], and actin plays a pivotal role in cell contraction. Thus, SMCs in the dorsal and ventral side may have more synthetic and contractile phenotype, respectively.

Immunohistochemical analysis showed that expression of F-actin in SMCs had a significant negative correlation with waviness of ELs adjacent to the cells (Fig. 3), for example. The cells near ELs with lower waviness may bear higher tension in a physiological condition. Thus, cells subjected to higher tension may be abundant in actin filaments, although it is not clear whether the accumulation of actin filaments is a cause or an effect of higher tension.

The reason for the heterogeneity of the microscopic mechanical environment remains unclear at this stage. This might be caused by cellular activities such as remodeling and cell division.

ACKNOWLEDGEMENTS

This work was supported in part by KAKENHI (nos. 22127008 and 22240055) from Japan Society for the Promotion of Science and the Science to TM.

REFERENCES

[1] Wolinsky H, Glagov S, *Circ Res*, 20:99-111, 1967.

[2] Matsumoto T et al, Proc ASME 2008 Summer Bioeng Conf, 192450, 2008.

[3] Matsumoto T et al, *J Biomech*, 37:807-815, 2004.

[4] Matsumoto T et al, *JSME Int J Ser A*, 47:341-8, 2004.

[5] Draney MT et al., *Ann Biomed Engng*, 30:1033-45, 2002.

[6] Sugita S et al, *Trans JSME, Ser A*, 69:43-8, 2003.

[7] Kim J and Baek S, *J Biomech*, 44:1941-7, 2011.

[8] Takemoto et al, *Arterioscler Thromb Vasc Biol*, 20:624-8, 2000.

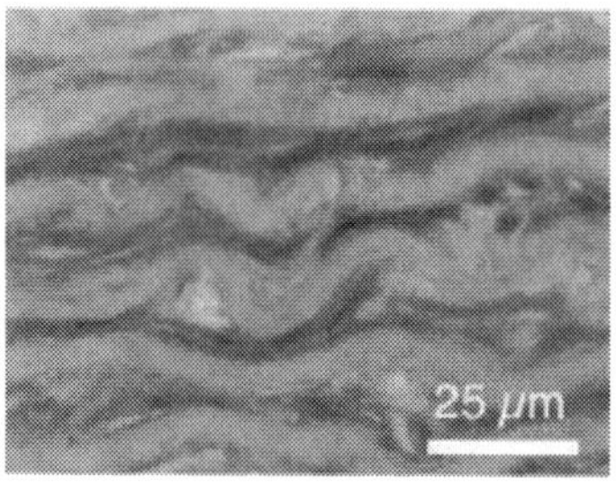

Fig. 1 A histological section of porcine thoracic aorta [4].

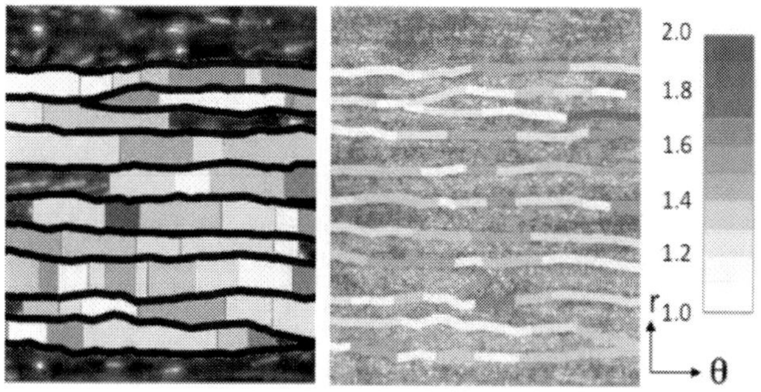

Fig. 2 Distribution of stretch ratio in SMLs (left) and ELs (right) of a thin-sliced porcine thoracic aorta stretched in the circumferential direction by 1.3.

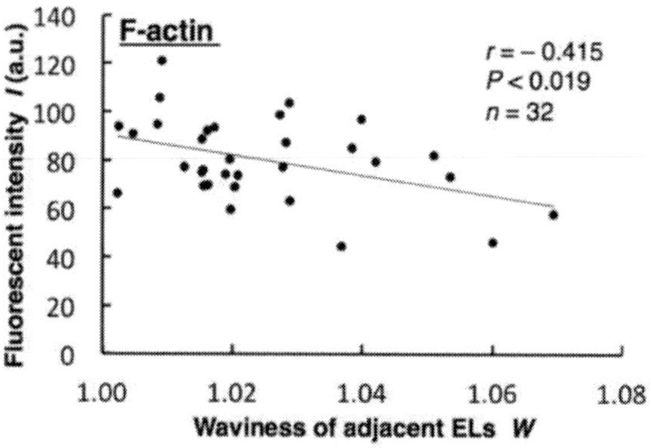

Fig. 3 Relationship between protein expression and waviness of ELs.

SB³C2015
Summer Biomechanics, Bioengineering and Biotransport Conference
June 17-20, 2015, Snowbird Resort, Utah, USA
SB³C2015-568

A SUPPORT VECTOR MACHINE BASED ON VERTICAL GROUND REACTION FORCE TO SUPPLEMENT OBSERVATIONAL GAIT EVALUATION

Craig J. Simons (1), Cory L. Christiansen (2), Jennifer E. Stevens-Lapsley (2), Kevin B. Shelburne (1), Bradley S. Davidson (1)

(1) Department of Mechanical & Materials Engineering
University of Denver
Denver, CO, USA

(2) Department of Physical Medicine & Rehabilitation
University of Colorado Denver
Denver, CO, USA

INTRODUCTION

During traditional gait analyses, the observer must evaluate a complex array of factors that include the coordinated timing and trajectories of body segments. Observation is often the only tool used by rehabilitation therapists to detect and evaluate aberrant movement patterns that occur in patients during ambulation. This is evident in clinical gait analyses performed to identify and assess conditions with motor symptoms such as Parkinson's disease, Huntington's disease, cerebral palsy, stroke, and dementia, and for evaluating progress during rehabilitation from surgery or injury. However, this qualitative approach is also susceptible to misidentification resulting from human error (incomplete or incorrect observation) and vague diagnostic criteria. [1].

An untold number of gait analyses have been published based on instrumented quantitative data to assess muscle activity, kinematics, and kinetics during ambulation [2, 3]. Although these metrics offer quantitative information with high precision and sensitivity, the instrumentation is expensive and the characteristic gait features are not available without considerable post-processing and analysis. Clinicians cite the analytical burden of instrumented gait analysis as a primary limitation preventing translation of these tools into a patient-specific clinical diagnostic [4].

Reduction in the size and price of instrumentation combined with increasing computational speed and power provides an opportunity for engineers to bridge the translational gap between observational and instrumented gait analysis. A promising computational technique based on "supervised learning" that is well suited to quantitative gait analysis is the support vector machine (SVM). This class of machine learning provides categorical outputs based on any number of continuous quantitative inputs. When implemented for gait analysis, it may be possible to use quantitative data from simple instrumentation to achieve a high level of inference on gait patterns, but without the processing and analysis bottleneck that is common in traditional instrumented analysis.

The objective of this investigation was to use inexpensive and easily attainable quantitative data from gait trials to generate an accurate classifier for two groups (healthy participants, patients one year after total knee arthroplasty surgery). Though the test data set is limited, the classifier was designed and evaluated in the context of viability as a supplemental tool to assist with clinical diagnoses. This case study provides insight into the performance of gait classification using vertical ground reaction force waveform data and the effects of algorithm penalty parameter, kernel selection, and the number of extracted features used.

METHODS

Vertical ground reaction forces (vGRF) from a single footstrike during gait were collected from 20 participants (14 healthy, 6 patients with TKA) using a floor-mounted force platform (Bertec, Columbus, OH). Each raw vGRF was down-sampled from the capture frequency (1000Hz) to form a row vector with 100 components from heel strike to toe off. This process normalized the temporal length of each individual stance period. The classification features are defined as the 100 components comprising the vGRF waveform vector. Each component of the feature vector was scaled [0, 1] relative to the full range of the training data (max vGRF amplitude) for that feature (i.e. percent of stance period). This scaling procedure prevents disproportionate weighting of features that have larger amplitude and variance. Each waveform was assigned a classification label; the healthy group was labeled as +1 and the patient group as -1.

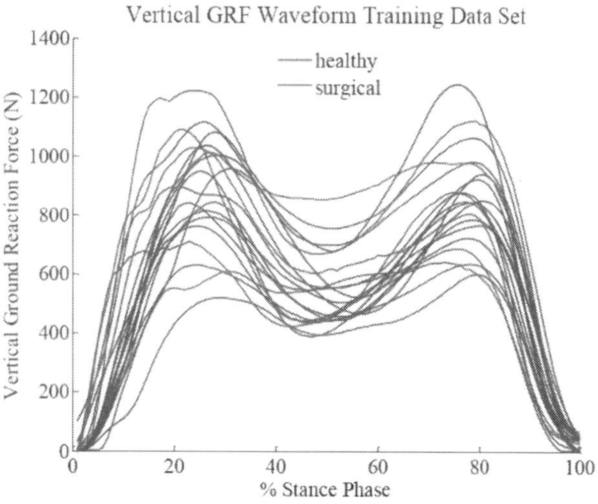

FIGURE 1: THE SVM TRAINING DATA SET INCLUDED 14 HEALTHY (BLUE) AND 6 SURGICAL TKA (RED) SAMPLES. EACH WAVEFORM WAS DISCRETIZED INTO A 100 COMPONENT FEATURE VECTOR USED TO TRAIN THE SVM.

The classification task creates a separating hyper-plane defined in the high-dimensional (100-D) feature space to use as the decision boundary between classes. For this investigation, classification was performed using a soft-margin *C*-SVM to identify if the gait waveform was produced by a healthy subject (+1) or a surgical subject (-1). The algorithm was implemented in Matlab using the LibSVM library for support vector machines [5]. A comparison of SVM classification accuracy was performed using two kernel functions (linear and radial basis function (RBF)). Kernel functions characterize the topology of the separating hyperplane.

Kernel function parameters were chosen to maximize classification accuracy. The parameter selection process used a grid search method to scan the parameter space, which includes the C parameter for the linear kernel and the C and γ parameters for the RBF kernel. The parameter grid was evaluated with the training data set using leave-one-out cross validation.

After training and testing the SVM with the full feature space, principal component analysis (PCA) was performed to identify a reduced feature space that maintains most of the original feature variance. This process may allow for extraction of the gait feature components that most distinctly characterize the vGRF waveforms [6]. To evaluate the performance of a PCA-reduced feature space, feature vectors were identified that accounted for 98% and 99.99% of the original feature space variance. The SVM was then trained and tested again with the PCA-reduced feature vectors using the parameter selection procedure described above for the full feature vector.

RESULTS

Computation time was negligible, requiring only a few seconds in every case. The accuracy of the SVM classification varied depending on the kernel (RBF or linear), and the feature vectors used (Table 1). Highest classification accuracy (95%) was achieved using the linear kernel with the full feature vector. The lowest classification accuracy (70%) was achieved with the linear kernel and a feature vector space comprising 99.99% data variance.

TABLE 1: COMPARISON OF CLASSIFICATION ACCURACY USING LINEAR AND RBF KERNELS WITH FULL AND PCA-REDUCED FEATURE VECTORS

Kernel	Classification accuracy with PCA-reduced feature vectors		
	Complete	99.99% variance	98.97% variance
Linear	95%	70%	80%
RBF	90%	85%	85%

PCA revealed that almost all feature variance could be represented using a reduced feature space: 6 principal components accounted for 98.97% of the total feature variance and 15 principal components accounted for 99.99% variance. However, the PCA-reduced feature vectors yielded lower classification accuracy than classification performed with the full feature space.

Although the highest accuracy was achieved using the linear kernel with the full feature set, the RBF kernel is less sensitive to feature selection, demonstrating comparable accuracy when using complete and PCA-reduced feature vectors. By comparison, linear kernel accuracy decreased more when using the PCA-reduced feature vectors and misclassification errors tended to classify surgical subjects as healthy.

DISCUSSION

Patients with TKA and healthy control participants had distinct gait patterns that could be categorized by SVM using only input of the vertical GRF. This case study was performed as a proof of concept in which data that are easily obtained (vertical ground reaction force) in a clinical evaluation are used within an SVM to generate an accurate classifier. The fast computation with SVM makes this a viable method for including more input waveforms and outputs that are clinically meaningful. Although the classification (healthy vs patient) is an oversimplification, the highly accurate performance in classification indicates that a simple and easily attainable signal such as the vertical GRF contains subtle information that may not be discernable through traditional clinical biomechanics analyses. Achieving high classification accuracy with a small training data set and limited features indicates that the SVM is sensitive to subtle features and suggests that classification performance should improve with additional training data that more fully characterizes the feature space. This approach can be expanded to include many additional classification categories for identifying a range of clinically relevant conditions.

To successfully deliver clinically viable diagnostic tools, engineers must consider methods to efficiently measure quantitative data and immediately translate the results into clinically interpretable information. Techniques such as SVMs may provide a method for associating high-dimensional metrics with clinically relevant diagnostic classifications and based on other inexpensive instrumentation.

REFERENCES

[1] Morgan, D et al., *Canadian Journal on Aging,* 26:19-32, 2007.
[2] Lord, S et al., *Clinical Rehabilitation,* 2155:107-119, 1998.
[3] Ferrarello, F et al., *Physical Therapy,* 93:1673-85, 2013.
[4] Toro, B et al., *Physiotherapy Theory and Practice,* 19:137-149, 2003.
[5] Chang, C, and Lin, C, *ACM Transactions on Intelligent Systems and Technology (TIST)*, 2:Article #27, 2011.
[6] Wu, J and Wang, J. *Journal of Applied Biomechanics,* 24:83-87, 2008.

BISPHOSPHONATE GIVES TRAUMA DAMAGED CHONDROCYTES A NEW LIFE VIA INHIBITION OF MEVALONATE PATHWAY

Yilu Zhou, Miri Park, Enoch Cheung, Liyun Wang, X. Lucas Lu (1)

(1) Department of Mechanical Engineering,
University of Delaware, Newark, DE, U.S.

INTRODUCTION

Zoledronic acid (ZA), an FDA approved drug for the treatment of osteoporosis, belongs to the bisphosphonate family that targets the mevalonate pathway of osteoclasts [1]. Previously, we found that systematic injection of ZA could suppress the development of post-traumatic osteoarthritis (PTOA) in the DMM (destabilization of medial meniscus) mouse model [2]. It remains unknown about how ZA could give a new life for chondrocytes after traumatic damage. The aim of this study is to explore the cellular and molecular mechanisms underlying ZA's chondro-protective effects. In specific, we will test the following hypothesis: 1) ZA can rescue the cartilage from traumatic damage during *in vitro* culture without the presence of subchondral bone; 2) ZA promotes the beneficial effects of mechanical stimulation on cartilage explants during *in vitro* culture; 3) ZA's chondro-protective effect is associated with the intracellular calcium ($[Ca^{2+}]_i$) signaling of chondrocytes; and 4) ZA regulates the mevalonate pathway of chondrocytes.

METHODS

Rescuing Chondrocytes from Traumatic Damage Cartilage explants from 3-month old calf knee joints were cultured in serum supplemented medium (DMEM, 10% FBS, 1% P/S) for 1 week to simulate traumatic damage on cartilage caused by joint bleeding. Our previous studies showed that the mechanical integrity of cartilage explants was significantly compromised after long-term exposure to serum [3-4]. The samples were further divided into two groups and cultured in chemically defined medium [4] for 4 weeks with or without the supplement of 1 μM ZA. Chemically defined medium includes DMEM, 1% ITS, 50 μg/mL L-proline, 0.9 mM sodium pyruvate, 0.1 μM dexamethasone, and 50 μg/mL ascorbate 2-phosphate. Longitudinal mechanical properties of the explants were measured, followed by GAG and collagen assay at the end of the 4-week culture. Expressions of aggrecan (AGN), type I collagen (COL1), type II collagen (COL2), ADAMTS5, and MMP-13 genes were measured using qRT-PCR after 2 weeks of rescue. At days 2, 8, and 15, five extra cartilage explants were cut into two identical halves and stained with Fluo-8 AM. The spontaneous $[Ca^{2+}]_i$ signaling of *in situ* chondrocytes in cartilage were recorded using a confocal microscope (Zeiss LSM510), and calcium videos were analyzed as previously described [5].

Promoting the Benefits of Mechanical Stimulation After harvesting, cartilage explants were cultured in chemically defined medium with or without 1 μM ZA for 4 weeks. During the culture, samples were stimulated with daily half-hour mechanical loading (10% preload followed by ±3% dynamic loading). Mechanical properties, biochemical contents, spontaneous $[Ca^{2+}]_i$ signaling, gene expressions were measured as in the first study.

Investigating Mevalonate Pathways After harvesting, cartilage explants were separated into 3 groups to check the effects of mevalonate derivatives: (1) chemically defined medium with 10 μM ZA (control); (2) control medium supplemented with 5 μM farnesol (FOH); (3) control medium with 5 μM geranylgeraniol (GGOH). Both FOH and GGOH are derivatives of mevalonate, the presence of which may reverse the ZA-induced inhibition of mevalonate pathway. After 48 hours treatment, spontaneous $[Ca^{2+}]_i$ signaling of chondrocytes was recorded. Another batch of cartilage explants was exposed to serum for 1 week to induce traumatic damage, and then separated into two groups and cultured in 1 μM ZA supplemented medium with or without FOH and GGOH (5 μM for both). Longitudinal mechanical properties of explants were measured for up to 6 weeks.

RESULTS

ZA rescued cartilage explant from traumatic damage

For cartilage explants damaged by serum exposure, the Young's modulus of the ZA rescued group increased more than 200% after 4 weeks of culture (Fig. 1A), while the control group showed a 56% increase. The responsive rate of $[Ca^{2+}]_i$ signaling of chondrocytes (Fig. 1C) is significantly higher in ZA treated group at days 8 ($p<.001$) and 15 ($p<.01$) than those in the non-ZA group. GAG content of the ZA rescued group (Fig. 1D) was significantly higher than that in the non-ZA group, while no significant difference was detected in collagen content (Fig. 1E). Anabolic gene expressions of type I, II collagens in the ZA group were significantly higher than the control group (Fig. 1F-G), while the expressions of catabolic genes, ADAMTS5 and MMP-13 (Fig. 1I-J) were significantly lower in ZA rescued samples.

ZA promoted the beneficial effects of mechanical stimulation on cartilage explants

ZA treatment significantly increased the equilibrium Young's modulus (Fig. 2A) and dynamic modulus (Fig. 2B) after 2 weeks when compared to the control group, although no difference was detected in GAG (Fig. 2D) or collagen content (Fig 2E). The responsive rate of $[Ca^{2+}]_i$ signaling of chondrocytes (Fig. 2C) was higher in ZA treated group at D2 ($p<.05$), D8 ($p<.01$), and D15 ($p<.001$). Expressions of type I, II collagen and aggrecan genes were significantly higher in the ZA treated group (Fig. 2F-H), and expressions of ADAMTS5 and MMP-13 genes (Fig. 2I-J) were significantly lower in the ZA group.

ZA regulates the mevalonate pathway of chondrocytes

After 48-hour treatment, the calcium responsive percentage of chondrocytes in the ZA+FOH (mevalonate derivative) group (25.0%) was lower ($p<.001$) than that of ZA control group (33.7%), while no significant difference was detected between ZA+GGOH (mevalonate derivative) (33.2%) and ZA control groups (Fig. 3A). After 2 weeks *in vitro* culture of cartilage explants, mechanical modulus in ZA+FOH+GGOH (mevalonate derivatives) group is lower ($p<.05$) than ZA control group (Fig. 3B-C), which indicates that mevalonate derivatives (FOH, GGOH) can reverse the beneficial effects of ZA on cartilage explants. Further mechanical testing showed that GGOH significantly dampened the ZA's chondro-protective effect (data not shown).

DISCUSSION

A number of studies have shown that bisphosphonate can suppress the development of PTOA *in vivo*, using DMM, ACL transection and carrageenan injection animal models. Most studies conjectured that the effect of bisphosphonates on PTOA is related to its strong capability in the prevention of bone loss. Our *in vitro* cartilage explant model without the presence of subchondral bone showed that, for the first time, the bisphosphonate can exert direct chondro-protective action on chondrocytes. This action can rescue the chondrocytes from traumatic damage and promote the beneficial effects of mechanical stimulation during cartilage explant culture. The $[Ca^{2+}]_i$ signaling of *in situ* chondrocytes is also significantly regulated by ZA, proving the direct effect of ZA on cells. Supplement of mevalonate derivatives reversed the chondro-protective effects of ZA, which suggests that ZA may regulate the metabolic activities through the mevalonate pathway in chondrocytes, in specific, protein geranylgeranylation. Geranylgeranylation is required for the release of MMP-13, a major proteinase responsible for cartilage degeneration. This geranylgeranylation-produced degradation enzyme could also boost the $[Ca^{2+}]_i$ activities of chondrocyte, which explains short-term treatment of GGOH didn't dampen the $[Ca^{2+}]_i$ activities. But in long-term culture as the ECM is degenerated by the enzymes, mevalonate derivatives reversed the chondro-protective action of ZA, in terms of mechanical integrity. In summary, inhibiting mevalonate pathway could serve as a potential therapeutic target to prevent the initiation of PTOA.

ACKNOWLEDGEMENTS

This work is supported by DOD W81XWH-13-1-0148 and NIH P30GM103333-01.

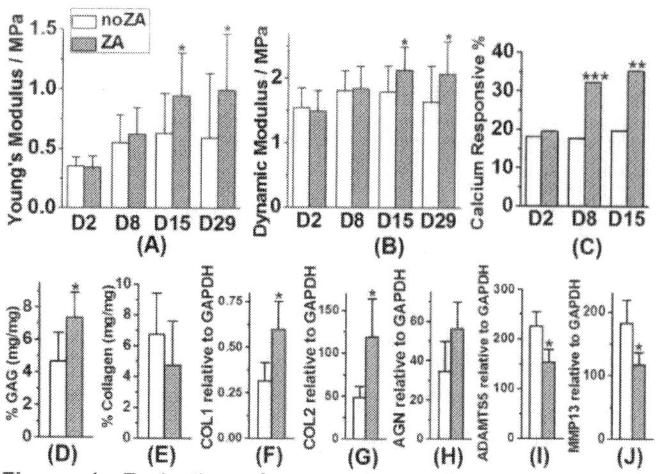

Figure 1: Evaluation of serum damaged samples that are rescued by ZA treatment. (A) Equilibrium Young's modulus; (B) Dynamic modulus; (C) GAG and (D) Collagen content after 4 weeks (*: $p<.05$). (E) Responsive percentage of the spontaneous $[Ca^{2+}]_i$ signaling of *in-situ* chondrocytes (*: $p<.05$, **: $p<.01$, ***: $p<.001$). Gene expression levels from qRT-PCR after 2-week rescue: (F) Collagen I, (G) Collagen II, (H) Aggrecan, (I) ADAMTS5, (J) MMP-13 (*: $p<.01$).

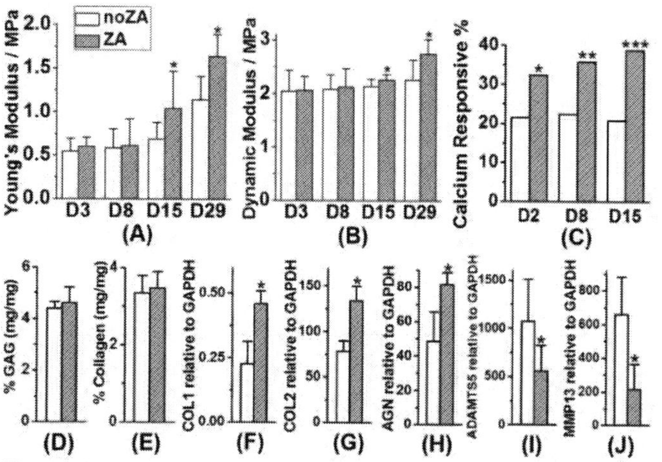

Figure 2: ZA treatment on mechanical stimulated explants: (A) Equilibrium Young's modulus; (B) Dynamic modulus; (C) GAG and (D) Collagen content after 4 weeks (*: $p<.05$). (E) Responsive percentage of the spontaneous $[Ca^{2+}]_i$ signaling of *in-situ* chondrocytes (*: $p<.05$, **: $p<.01$, ***: $p<.001$). Gene expression after 2-week culture: (F) Collagen I, (G) Collagen II, (H) Aggrecan, (I) ADAMTS5, (J) MMP-13 (*: $p<.01$).

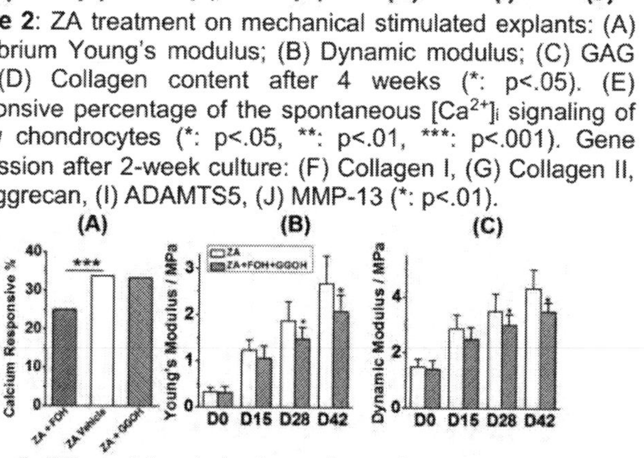

Figure 3: Effect of the derivatives of mevalonate pathway (FOH and GGOH): (A) Responsive percentage of the spontaneous $[Ca^{2+}]_i$ signaling after 48 hours treatment with ZA and the derivatives of mevalonate pathway; (B) Equilibrium Young's modulus, (C) Dynamic modulus. (*: $p<.05$, **: $p<.01$, ***: $p<.001$)

REFERENCES

[1] Ebetino, Frank H., et al. *Bone* 49.1:20-33, 2011; [2] Price, C, et al., *ORS proceedings*, 2011; [3] Zhou Y., et al. *ORS proceedings*, 2013; [4] Bian L., et al., *J Biomech*, 41.6: 1153-1159, 2008. [5] Lu X.L., et al., *JBMR*, 27.3: 563-574, 2012.

SB^{3}C2015
Summer Biomechanics, Bioengineering and Biotransport Conference
June 17-20, 2015, Snowbird Resort, Utah, USA
SB^{3}C2015-571

HEAD-NECK RELATIVE POSTURE AFFECTS FRACTURE OUTCOME OF THE BASILAR SKULL AND UPPER CERVICAL SPINE UNDER HELMETED HEAD CROWN IMPACT

Liming M. Voo, Kyle A. Ott, Chris J. Dooley, Andrew C. Merkle

Research and Exploratory Development Department
Johns Hopkins University Applied Physics Laboratory
Laurel, Maryland, USA

INTRODUCTION

Head impacts to the crown of the skull can generate compressive forces large enough to cause bony fracture of the skull and cervical spine. However, the location and severity of an injury depend on the characteristics of the impacting surface, head-neck posture, and the magnitude of the impact, among other factors. Additionally, protective equipment, such as helmets, may protect the anatomy closest to the impact site, in this case the skull vault, but convey the forces to anatomy further from impact, including the basilar skull and cervical spine [1, 2]. Fractures to the basilar skull or the atlas (C1 vertebra) and axis (C2 vertebra) of the cervical spine have high risk of fatality or quadriplegia [3]. However, their injury mechanisms and risk probability data remain largely unknown in the literature [4].

Previously, we introduced an experimental model that included the postmortem human surrogate (PMHS) head and occipito-atlanto-axial (OC-C2) joint to understand the injury mechanisms and biomechanical correlation metrics [5]. Drop-mass impact was applied to the helmeted head to simulate the scenarios of head impact with the roof structure of a military vehicle that may occur in an under-body blast (UBB) event. This unique experimental model has the advantage of preserving the essential load transfer structures from head through C1-C2 vertebrae while addressing the measurement limitations and variability of the prior PMHS models [4]. This model is capable of measuring the compressive force response across the vertebra and aligning with fracture timing captured by acoustic sensor signals. In this paper, we report how altering head-neck posture results in fracture to different bones, underscoring the complexity in the determination of head and neck injury criteria and the design of an anthropomorphic test dummy (ATD) experiencing crown impacts.

METHODS

The anatomic structures of the specimens used in this study include intact fresh PMHS head and the ligamentous cervical spine disarticulated at the C4-C5 joint. The C3-C4 vertebrae were potted with Polymethyl methacrylate (PMMA) leaving the C2-C3 joint unconstrained. Strain gages and acoustic sensors were attached to the occipital bone to detect potential basilar skull fracture events. An additional acoustic sensor was attached to the anterior surface of the C2 vertebra to capture the fracture timing of the C1/C2 vertebrae. A six-axis load cell was fixed beneath the specimen on a rigid platform (**Figure 1**). A military helmet was fitted on the specimen head following a standardized fitting procedure. A drop tower impactor with a flat metal surface applied blunt impact load to each helmeted head. Load cells on the impactor and below the specimen recorded the impact force and the force transmitted through the OC-C2 joint, respectively. High-speed digital video cameras operating at 7500 frames per second recorded the event in sagittal and coronal views. The specimen head was positioned to align the Frankfort plane horizontally or slightly flexed forward prior to the blunt impact event. The C2 vertebra orientation also varied to simulate the head-neck posture to study their effect on fracture outcome and injury mechanisms. Those posture positions were determined and verified with the aid of a mobile digital X-ray system. Each specimen was tested only once. Bone fracture was determined via post-test X-ray imaging, palpation, autopsy, and microCT imaging.

RESULTS

Three specimens were tested, each specimen tested resulted in a different type of injury. In this preliminary study, under predominantly compressive loading conditions through the head-neck junctions, the following injuries were produced: a linear fracture to the skull base, a burst fracture to the C1 vertebra, and a transverse fracture to the C2 vertebral body. Burst fracture of the C1 vertebra resulted from a test with a specimen having a forward inclined C3 vertebra and slightly forwardly inclined Frankfort plane (approximately 10 degrees from horizontal) with the OC-C2 joint approximately in the middle of its sagittal range of motion. When the head Frankfort plane was set in horizontal while C2/C3 vertebrae were aligned vertically, the crown impact resulted in basilar skull fracture in one specimen and C2 vertebral body transverse fracture in another (**Figure 2**).

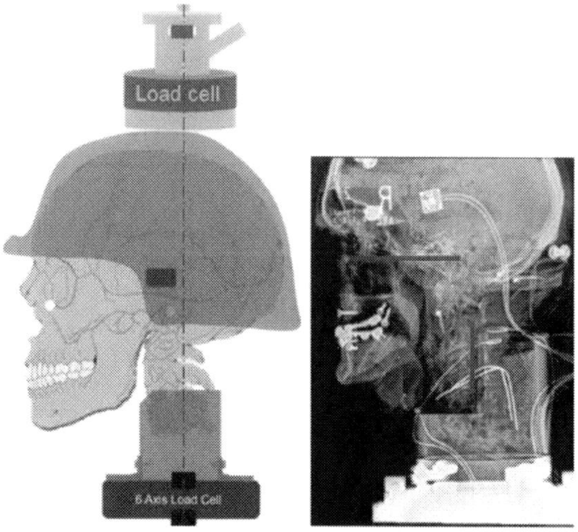

FIGURE 1. TEST SETUP AND HEAD-NECK POSTURE VERIFICATION

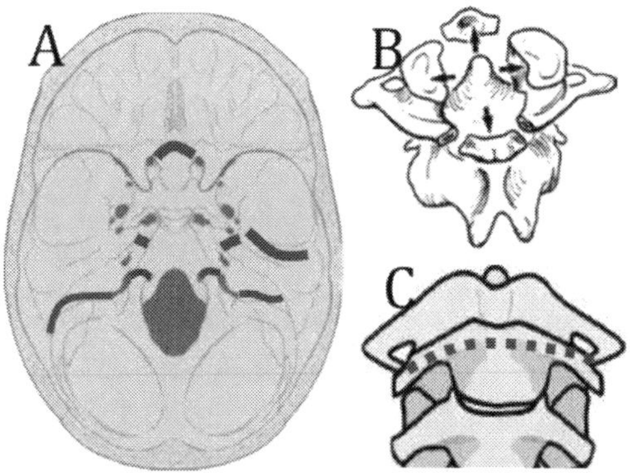

FIGURE 2. FRACTURE OUTCOMES FROM THE THREE PMHS HEAD-C2 SPECIMENS: (A) BASILAR SKULL FRACTURE (HDC2 03), (B) C1 BURST FRACTURE (HDC2 01), AND (C) C2 VERTEBRAL BODY FRACTURE (HDC2 02)

The peak compressive forces across the OC-C2 joint also differed by injury outcome (**Figure 3**). The peak forces were similar between the basilar skull fracture (4.5kN) and C2 body fracture (5.2kN) while the force creating the C1 burst fracture was considerably less (1.7kN).

DISCUSSION

The head-neck junction involves primarily three boney bodies: the basilar skull, the atlas, and the axis. Many different fracture outcomes could result from the crown impact event depending on the load transfer conditions across this complex joint. Although prior studies seem to agree that C1 burst fracture is primarily due to compressive loading, the fractures to the C2 vertebra could have many different mechanisms [4]. The replication of the fracture to the basilar skull from crown impact is particularly elusive. We propose that the precisely controlling the head-neck posture relative to impact vector is critical to controlling the injury outcome. Achieving this level of control is extremely difficult if the entire cervical spine is included (such as an entire PMHS head-neck complex and a whole body specimen PMHS). Hence our experimental model eliminates the other intervertebral joints of the cervical spine while maintaining freedom at and around the OC-C2 junction critical to this study.

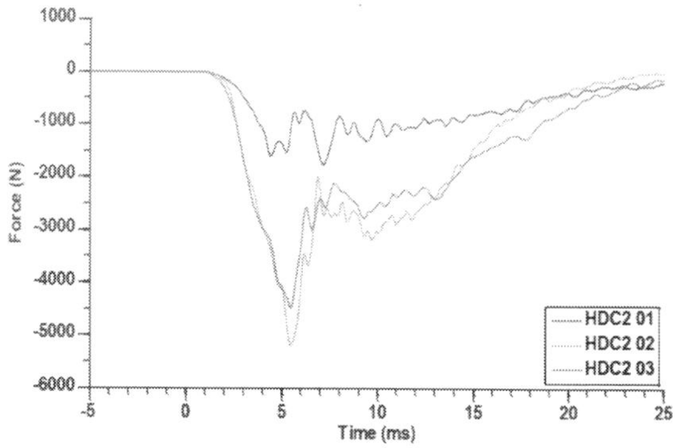

FIGURE 3. COMPRESSIVE FORCE ACROSS THE HEAD-NECK JUNCTION FROM THE THREE PMHS HEAD-C2 SPECIMENS: HDC2-01, 02 AND 03 HAD C1, C2, AND BASILAR SKULL FRACTURES, RESPECTIVELY

The pre-impact postures and the resulting compressive force magnitudes were similar between the basilar skull fracture and C2 fracture cases. It is clear that the load transfer conditions in those two cases were different than that of the C1 burst fracture. However, what sets those two conditions apart is not yet clear and warrants further investigation. The posture alignment and the vertical compressive force do not appear to be sufficient to differentiate the loading conditions in those two injury mechanisms that resulted in fractures to the basilar skull and the axis respectively.

The preliminary results of this study have shown the capability of this experimental model to replicate different types of injury mechanisms at the head-neck junction with better control of its initial posture and alignment, although only three specimens have been tested to date. They do point out the importance of the initial alignment of the head and the upper cervical vertebrae in determining the injury location. Further study will allow us to delineate how such factors control the type and probability of the injury risk at this anatomic region.

ACKNOWLEDGEMENTS

This effort was funded by contract #N00024-13-D-6400, U.S. Army Research, Development and Engineering Command. The content included in this work does not necessarily reflect the position or policy of the U.S. government.

REFERENCES

[1] Torg J, Vegso J et al, Am J Sports Med, 18:50-57, 1990.
[2] Schoenfeld A et al, J Trauma Acute Care Surg, 74:1112-, 2013.
[3] Yoganandan N, Pintar F et al, SAE Tech Paper #892438, 1989
[4] Myers B and Winkelstein B, CRC Review Biomed Eng, 23:307-209, 1995.
[5] Voo L, Ott K, Dooley C, Merkle A, World Congress Biomech, 2014.

SB³C2015
Summer Biomechanics, Bioengineering and Biotransport Conference
June 17-20, 2015,Snowbird Resort, Utah, USA
SB³C2015-572

ROLE OF ANEURYSM ON FAILURE PROPERTIES OF "RADIALLY-ORIENTED" COLLAGEN FIBERS IN HUMAN ASCENDING THORACIC AORTIC MEDIA

Siladitya Pal[1], Spandan Maiti[1], Alkiviadis Tsamis[1,4,5], Julie A. Phillippi[1,2,5], Thomas G. Gleason[1,2,4,5], David A. Vorp[1,2,3,4,5]

1. Department of Bioengineering
University of Pittsburgh
Pittsburgh, Pennsylvania, 15219
United States

2. Department of Cardiothoracic Surgery
University of Pittsburgh
Pittsburgh, Pennsylvania, 15213
United States

3. Department of Surgery
University of Pittsburgh
Pittsburgh, Pennsylvania, 15213
United States

4. Center for Vascular Remodeling and Regeneration
University of Pittsburgh
Pittsburgh, Pennsylvania, 15219
United States

5. McGowan Institute for Regenerative Medicine
University of Pittsburgh
Pittsburgh, Pennsylvania, 15219
United States

INTRODUCTION

Delamination strength (S_d) of human ascending thoracic aorta (ATA) is decreased in aneurysmal patients, and likely contributes to the risk of aortic dissection (AoD) (**Fig. 1(a)**) [1]. S_d may be influenced by local micro-architectural features of load bearing components of ATA wall tissue, namely the elastin and collagen fibers that are oriented in the radial (RAD) direction (**Fig. 1(b)**). Therefore, quantifying the biomechanical properties of RAD-oriented fibers in the wall of ascending thoracic aortic aneurysm (ATAA) may improve our understanding of the pathophysiology of AoD. A recently developed mechanistic model related the number (N) of RAD-oriented collagen fiber bridges per unit area (**Fig. 1(b)**) in the aortic media and the energy U_f required to fail a fiber bridge to the S_d of human non-aneurysmal (CTRL) ATA [2]. Possible failure modes of collagen fiber bridges in radial direction for AoD are demonstrated in **Fig. 1(c)**. This project tested thehypothesis that U_f is a key biomechanical parameter that can relate to predictable changes in the delamination strength of ATA in health and disease. Our approach took recourse of the aforementioned mechanistic model to extract the parameter U_f for non-aneurysmal as well as aneurysmal aortic specimens from patients with bicuspid aortic valve (BAV) or tricuspid aortic valve (TAV) morphology. We found that fiber bridge failure energy U_f for the aneurysmal patients of both the phenotypes is lower than that for CTRL cases.

METHODS

All human ATA experimentally-dissected specimens were gathered from our previous study [1]. Parameter N for RAD-oriented collagen fibers was quantified for all the investigated specimens (CTRL-ATA n=4, BAV-ATAA n=7, TAV-ATAA n=4) from prior multi-photon microscopy image analysis [3]. The latter provided results in both Z-RAD and Θ-RAD planes of aortic media within a small (±5μm) zone of from the experimentatlly-dissected interface. The analytical model

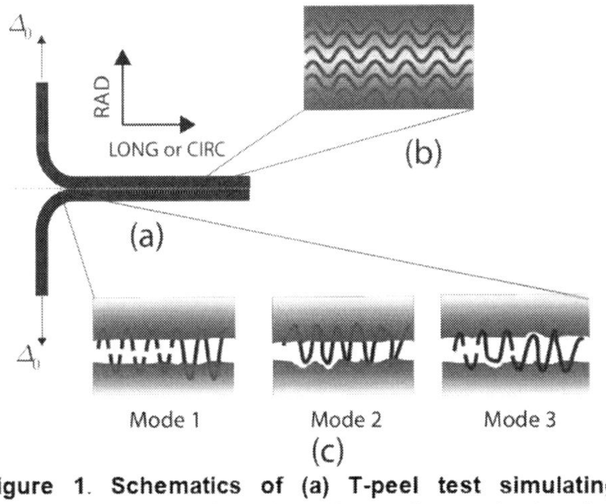

Figure 1. Schematics of (a) T-peel test simulating conditions of aortic dissection where Δ denotes separation, (b) fiber bridges in the Z-RAD or Θ-RAD planes, and (c) various modes of fiber bridge failure in the RAD-direction.

[2] was used to determine the U_f of the RAD-oriented collagen fiber in terms of S_d and N. Finally, we obtained the U_f using the S_d [1] and the

N of RAD-oriented collagen fibers for CTRL-ATA, BAV-ATAA and TAV-ATAA.

A schematic flowchart presented in **Figure 2** depicts the summary of experimental and analytical procedures undertaken to estimate U_f. To

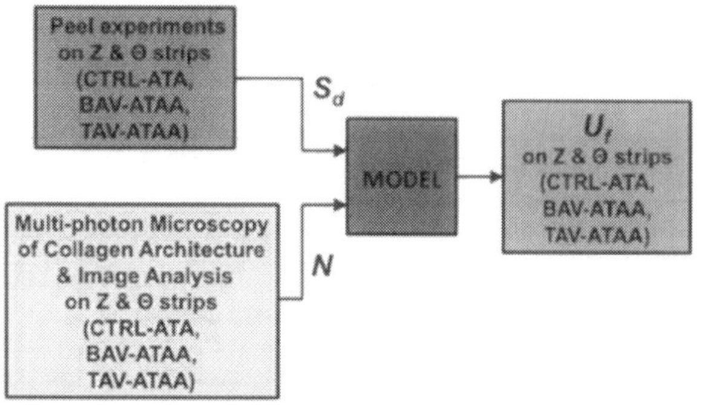

Figure 2: A flow-chart showing the procedure to estimate the fiber bridge failure U_f.

perform an assessment of the parameters N and U_f among BAV-ATAA, TAV-ATAA and CTRL-ATA and in directions Z and Θ, detailed statistical analysis was performed. Student's unpaired t-test was performed using SigmaPlot software (SYSTAT Software, Chicago, Ill) to determine significant differences in N among groups (BAV-ATAA n=7, TAV-ATAA n=4, CTRL-ATA n=4) in the media of the Θ-RAD and Z-RAD planes. One-way ANOVA followed by Tukey post-hoc test (unequal samples) was performed using the same statistics software to determine significant differences in the predicted U_f among the three patient cohorts in the Θ-RAD and Z-RAD planes. A p value of <0.05 was considered significant. Data are presented as mean ± standard error mean.

RESULTS

Number densities of RAD-oriented collagen fiber N among CTRL-ATA, BAV-ATAA and TAV-ATAA in either Θ-RAD or Z-RAD medial planes were equal (**Figure 3(a)**). There was anisotropic distribution of RAD-oriented collagen fiber bridges in the media of CTRL-ATA, while this anisotropy was not observed in the media of ATAA.

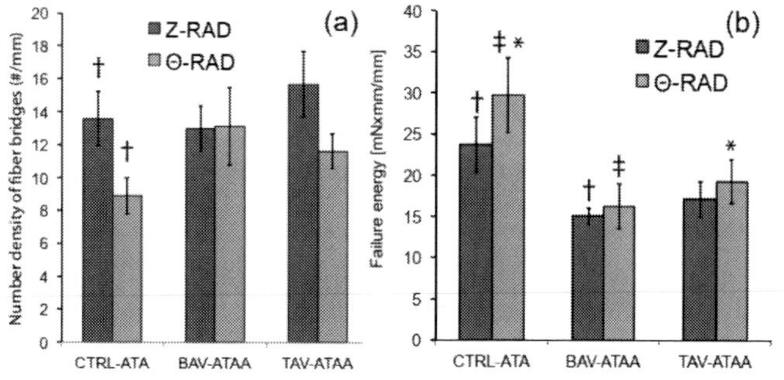

Figure 3: (a) N and (b) U_f of RAD-oriented collagen fiber bridges. Results are presented as mean ± standard error of the mean. †, ‡ and * indicate significant differences ($p<0.05$).

Extracted fiber failure energies according to the analytical model [2] are presented in **Figure 3(b)**. Our statistical analysis revealed that U_f is lower in BAV-ATAA compared to CTRL-ATA in either Z-RAD or Θ-RAD plane, and was lower in TAV-ATAA compared to CTRL-ATA in the Θ-RAD plane (**Figure 3(b)**). Also, U_f did not change between the Θ-RAD and Z-RAD planes for all three ATA phenotypes examined.

DISCUSSION

The increased number of fiber bridges in the media of Z-RAD plane of CTRL-ATA might improve resistance to delamination in that direction in accordance with the higher S_d determined by Pasta *et al.*, [1] in the Z-strips when compared with the Θ-strips of CTRL-ATA tissue. The present study also revealed U_f the energy required to fail each radially-running fiber bridge as a vital parameter influencing S_d. Failure of a fiber bridge can be realized in three separate modes: Fracture of the collagen fiber itself (**Figure 1-c,** Mode 1), extraction of the fiber from within non-fibrous extracellular matrix (**Figure 1-c,** Mode 2), and/or a combination of these two cases (**Figure 1-c,** Mode 3). U_f can be estimated from the area under the experimental load vs. displacement curve for a single collagen fiber bridge until complete failure [2]. Our analysis revealed that U_f was lower in BAV-ATAA compared to CTRL-ATA in either Z-RAD or Θ-RAD plane, and was lower in TAV-ATAA compared to CTRL-ATA in the Θ-RAD plane only. The lower U_f in BAV-ATAA may point towards a less favorable biomechanical state for the collagen fiber bridges in the diseased aorta, and thus would contribute to diminished resistance to delamination. Furthermore, despite a prominent influence of aneurysm on U_f evidenced by unchanged U_f between BAV-ATAA and TAV-ATAA in both Θ-RAD and Z-RAD planes, a notable difference in BAV-ATAA vs. TAV-ATAA aortic valve morphology was revealed since U_f of only BAV-ATAA was significantly different in the Z-RAD plane.

Reduced U_f in both BAV-ATAA and TAV-ATAA when compared with CTRL-ATA helps to explain the reduced delamination strength of ATAA measured by Pasta *et al.*, [1]. The failure energy of RAD matrix components may be a consequence of medial degeneration known to occur in BAV-ATAA and TAV-ATAA alike, despite the probable alternative upstream effector molecules that initiate this final common pathway of aneurysm formation in the two aortic valve phenotypes. Our analysis will give rise to future investigations, in which biomarkers specific to the modes of failure of a collagen fiber bridge could be clinically identified and measured. In this way, we aspire to ultimately provide insight into the pathophysiology of aortic dissection, improve risk assessment for aortic catastrophe and help guide clinical decision-making and care of patients with aortic disease.

ACKNOWLEDGEMENTS

The authors would like to acknowledge funding support of this work by the Swiss National Science Foundation Fellowships for Advanced Researcher #'s PA00P2_139684 and PA00P2_145399 (Dr. Tsamis), by the NIH Award #HL109132 (Drs. Gleason, Phillippi and Vorp), and by the University of Pittsburgh's Department of Cardiothoracic Surgery (Drs. Gleason and Vorp).

REFERENCES

[1]_Pasta, S. *et al.*, The Journal of Thoracic and Cardiovascular Surgery, 2012, 143(2): 460-467.

[2] Pal, S. *et al.*, Journal of Biomechanics, 2014, 47(5): 981-988.

[3] Tsamis, A. *et al.*, Journal of Biomechanics, 2013, 46(16): 2787-2794.

SB^{3}C2015
Summer Biomechanics, Bioengineering and Biotransport Conference
June 17-20, 2015,Snowbird Resort, Utah, USA
SB^{3}C2015-573

CAN MEASURED MUSCLE SYNERGIES RECONSTRUCT UNMEASURED MUSCLE EXCITATIONS?

N.A. Bianco (1), C. Patten (2), B.J. Fregly (1,3)

(1) Department of Mechanical & Aerospace Engineering
University of Florida
Gainesville, FL, USA

(2) Department of Physical Therapy
University of Florida
Gainesville, FL, USA

(3) Department of Biomedical Engineering
University of Florida
Gainesville, FL, USA

INTRODUCTION

Predicting muscle and joint contact forces reliably during human movement could be useful for developing improved treatments for movement disorders. However, muscle redundancy makes it impossible to calculate unique muscle force solutions without resorting to reduction methods [1]. Optimization methods can provide unique solutions, but they are dependent on subjective cost functions. While EMG-driven methods make use of experimental EMG data, they do not provide a way to predict unknown muscle excitations with confidence, relying mainly on optimization methods to calculate missing excitations. Consequently, predicted excitations may not be consistent with unmeasured experimental excitations.

Recent studies suggest that muscle synergies may provide an avenue for increasing the uniqueness of muscle force solutions. A recent optimization study improved prediction of knee contact forces by replacing a high-dimensional set of 44 independent muscle excitations with a lower-dimensional set of 5 synergy excitations [2]. The synergy excitations were extracted from 13 experimental EMG signals and were used to construct 44 interdependent muscle excitations. However, no evidence exists that experimental synergy information extrapolated from a small subset of muscles can reconstruct unmeasured excitations reliably for other muscles.

This study explores the feasibility of using muscle synergy information collected by surface EMG from 8 muscles to accurately reconstruct additional muscle excitations collected from up to 16 other muscles using surface and fine-wire EMG. The influence of the choice of 8 muscles on reconstruction accuracy is also examined.

METHODS

Experimental EMG data were collected from one healthy subject and one stroke subject. Institutional review board approval and subject informed consent were obtained. Eleven treadmill walking trials ranging from 0.4-1.4 m/s were collected for the healthy subject, and seven trials ranging from 0.2-0.8 m/s were collected for the stroke subject. For each trial, 10 consecutive gait cycles were chosen for analysis. A total of 24 EMG signals (17 surface, 7 fine wire) were collected from lower extremity muscles of the healthy subject's dominant leg, while a total of 16 EMG signals (12 surface, 4 fine wire) were collected from each leg of the stroke subject. EMG signals were band-pass filtered (zero-lag fourth-order Butterworth) from 10 to 400 Hz for surface EMG and 20 to 400 Hz for fine wire EMG to account for additional tissue motion artifacts. These data were then demeaned, full-wave rectified, and low-pass filtered (zero-lag fourth-order Butterworth) based on gait cycle period to account for speed-related changes in motion artifacts. The resulting processed signals defined experimental muscle excitation profiles. Finally, each muscle excitation was normalized to its maximum value across all walking trials.

Muscle synergy analysis was performed on subsets of 8 surface excitation profiles using non-negative matrix factorization (NMF) to obtain a low-dimensional set of time-varying signals and a corresponding set of weighting vectors that specify contributions to individual excitations [3, 4]. We define a time-varying signal as a "neural command," a weighting vector as a "synergy vector," and a neural command with its accompanying synergy vector as a "muscle synergy." Synergy vectors were allowed to vary between gait cycles, and therefore synergy analyses were performed on each gait cycle separately. Prior to synergy analysis, we normalized each excitation profile to have unit variance to allow the variations in each muscle to be considered with equal importance in the NMF algorithm [3]. The resulting neural commands were normalized to a maximum value of 1, consistent with the maximum value of muscle excitations used in muscle force optimizations. Corresponding synergy vectors were scaled in the opposite direction to keep reconstructed muscle excitations consistent.

Using the neural commands obtained from synergy analysis, we performed a least-squares regression analysis to obtain a set of synergy vectors that would reconstruct the remaining surface and fine wire excitation profiles not included in the synergy analysis. This method was termed "synergy extrapolation." Variance accounted for (VAF), a measure of goodness-of-fit that accounts for magnitude and shape, was utilized to evaluate all extrapolation solutions since the number of synergies must be provided to the NMF algorithm *a priori* [3]. Synergy analysis was performed on every possible combination of 8 surface EMG signals to provide a comprehensive set of extrapolation solutions. Each data set (healthy subject, stroke subject non-paretic leg, stroke subject paretic leg) was analyzed separately.

VAF values calculated from the individual reconstructed muscles through synergy extrapolation were averaged across gait cycles and then averaged together to create a single Extrapolation VAF for each 8-muscle combination. To gauge extrapolation ability, we recorded the number of combinations that exceeded 90% Extrapolation VAF and the Mean Extrapolation VAF of the Top 10% of Combinations for each walking speed.

RESULTS

TABLE I: HEALTHY SUBJECT

Walking Speed (m/s)	Number of Combinations at 90% Extrapolation VAF*	Mean Extrapolation VAF of Top 10% Combinations
0.4	19950	93.4%
0.5	20898	93.9%
0.6	20342	93.4%
0.7	20095	93.5%
0.8	16918	92.8%
0.9	12116	91.8%
1.0	6695	91.0%
1.1	4532	91.0%
1.2	841	89.9%
1.3	7876	91.5%
1.4	3804	90.6%

*24310 possible combinations.

TABLE II: STROKE SUBJECT

Walking Speed (m/s)	Number of Combinations at 90% Extrapolation VAF*		Mean Extrapolation VAF of Top 10% Combinations	
	Non-Paretic	Paretic	Non-Paretic	Paretic
0.2	97	2	90.7%	89.2%
0.3	167	258	91.8%	92.9%
0.4	222	0	92.7%	88.0%
0.5	276	148	93.3%	91.6%
0.6	214	177	91.5%	91.7%
0.7	284	81	93.2%	90.9%
0.8	279	0	92.3%	87.8%

*495 possible combinations.

Overall, synergy excitations extracted from muscle subsets were able to reconstruct remaining muscle excitations with high reproducibility. Small variations in the number of combinations to reach 90% Extrapolation VAF were observed across walking speeds. These variations were also observed in the Mean Extrapolation VAF of the Top 10% of Combinations, although not as acutely. For the stroke subject, there was a noticeably reduced extrapolation accuracy for the paretic leg compared to the non-paretic leg, including two walking speeds that did not produce any acceptable muscle combinations.

DISCUSSION

Though walking speed had an effect on the ability to extrapolate for both subjects, the number of acceptable combinations was sensitive to small reductions in the VAF cutoff value. For example, when the VAF cutoff was reduced to 85% for the paretic leg of the stroke subject, more than 200 muscle combinations met the cutoff for all walking speeds. Reduced extrapolation VAF could be explained by a few muscles with poorly reconstructed excitations (i.e. flexor digitorum longus, which generally only reached 85% VAF). Poor reconstructions were noted independent of signal sources.

For the stroke subject, standard synergy analysis revealed that the non-paretic leg achieved a higher VAF compared to the paretic leg at a given number of synergies, which is the opposite of results reported in Clark et al. 2010 [5]. However, using the surface EMG signals selected in that study, we found that the number of synergies required to achieve a given VAF cutoff was the same for the both legs. The fact that we did not use fine wire EMG signals when calculating neural commands may also play a role in these differences. In our synergy analyses, the number and choice of muscles had a significant influence on the solutions, as other studies have suggested [6]. In addition, analysis of EMG data from a single stroke subject is a significant limitation. Regardless, the goal of this study was only to demonstrate the feasibility of extrapolating 5 experimentally derived neural commands to other muscle excitations, not to explore the number of signals required to study stroke or to describe the general behavior of synergy solutions across large subject groups. It should be noted that 5 synergies were chosen since this number provided favorable extrapolation results across all data sets, and not based on the VAF to reconstruct the muscles in each subset. The relationship between reconstruction VAF and extrapolation VAF has yet to be investigated.

Even with large numbers of acceptable muscle combinations, there is still the possibility that a highly redundant, and therefore poorly extrapolating, muscle combination could be chosen. However, with further investigation, we have found that if a muscle combination is chosen that adequately represents all primary lower extremity functions during walking (i.e. a hip extensor, a hip adductor, a knee flexor, a knee extensor, either head of biarticular gastrocnemius, soleus, and tibialis anterior), there is a high likelihood for successful synergy extrapolation. In conclusion, synergy excitations extracted from 8 surface EMG signals were able to reconstruct other surface and fine wire EMG signals with high reproducibility.

ACKNOWLEDGEMENTS

This study was funded by NSF REU Fellowship 116104.

REFERENCES

[1] L. H. Ting, et al., *Int. J. Numer. Method. Biomed. Eng.* 2012; 28:1003–1014.
[2] J. P. Walter, et al., *J. Biomech. Eng.* 2014; 136:021031.
[3] L. H. Ting, et al., *Mot. Control Theor. Exp. Appl.* 2010; 102v-138
[4] D. D. Lee and H. S. Seung, *Letters to Nature* 1999; 401:788-791
[5] D. J. Clark, et al., *J. Neurophysiol.* 2010; 103:844–857.
[6] K. M. Steele, et al., *Front. Comput. Neurosci.* 2013; 7:105.

SB^{3}C2015
Summer Biomechanics, Bioengineering and Biotransport Conference
June 17-20, 2015,Snowbird Resort, Utah, USA
SB^{3}C2015-574

AN EFFICIENT AND RELIABLE BIOMECHANICAL TESTING DEVICE TO PERFORM TORSION TESTING IN LONG BONES WITH LOCKING COMPRESSION PLATES

Joseph P. Loftus (1), Lindsay Stoy (2), Douglas J Patterson (3), Anita Singh (4)

(1) Biomedical Engineering
Widener University
Chester, PA, USA

(2) Biomedical Engineering
Widener University
Chester, PA, USA

(3) Chief, Hand Section,
Department of Orthopaedics
Christiana Hospital, Newark DE

(4) Biomedical Engineering
Widener University
Chester, PA, USA

INTRODUCTION

The purpose of this study was to develop an economical, consistent, and repeatable torsion testing technique, including the device, for long bones affixed with compression plates. Compression plates are among the most commonly used fixation equipment in orthopedics. When focusing on forearm long bones, fracture of the radius and ulna are extremely common. It is estimated that these fractures of the radius and ulna make up 44% of hand/forearm fractures[1]. In pediatrics, it is estimated that 25% of all fractures are distal radius fractures[2]. A significant drawback in advancements of these orthopedic implants is a lack of industrial testing standard for biomechanical testing of the new iteration of these implants. Commonly used biomechanical tests include torsion, bending and combination to replicate real world scenario and the setup used for these tests costs hundreds of dollars. If a standard were to be adopted with an economical and reliable testing device, then time and money can be saved in the development and future research of the medical implants. This study details an economical and reliable custom-built torsion device with a torsion testing protocol that can be used to validate the efficacy of a newly designed forearm compression plate with various plate length, thicknesses and screw holes. These are the goals of our future projects.

METHODS

Fabrication of Torsion Testing Apparatus-The testing apparatus consists of two parts: the motor (driving side) and the sensor (reaction side). In order to perform single cycle test, the driving side must to have the capability to twist the specimen at a constant strain rate. A stepper motor (HT34-487, Omega Engineering) matched the criteria and was paired with a 15:1 gear reduction to reach a maximum torque output of 1725 in-lbs. The stepper motor was controlled by a driver (ST10-Si, Applied Motion Products), which accepts voltages from 24-80 volts. An incremental encoder was implemented in order to prevent motor stalling and ensure that all tests were run at the specified parameters. The reaction side of the apparatus was responsible for data acquisition. A reaction torque sensor (FSH00648, Futek) with a 1300 in-lb capacity interfaced with a MyDAQ controller, which send a voltage reading to LabView software. This voltage reading was calibrated in order to convert the value to torque. All design work prior to fabrication was done in SolidWorks. Strong aluminum plates varying from 0.25-1" were used for fabrication. Roller bearings were used on both the driving side and the reaction side in order to eliminate bending, ensuring the specimen was subjected to purely torsion. The base plates were designed to mount to an optical breadboard table with 1"x1" hole pattern and 1/4-20 holes. The base plate of the reaction side featured slotted holes to allow for testing of specimens of varying sizes. In order to grip the specimens, square brackets were used which allowed a tight sliding fit accepting 2" square ends. New base plates and brackets could be fabricated in order to mount to different breadboard tables and accept different size specimens. Figure 1 shows the fully assembled apparatus and schematic flow chart.

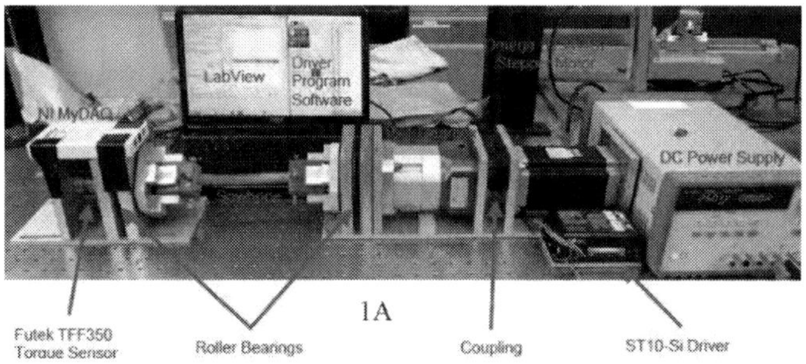

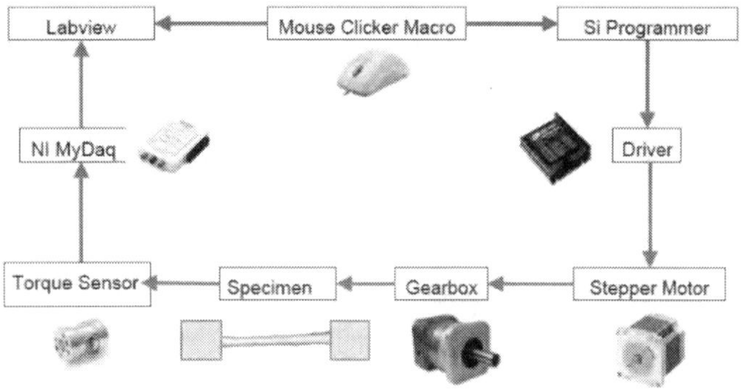

1B

Figure 1: 1A: Picture of Torsion Apparatus, 1B: Schematic flowchart detailing the apparatus components and software integration.

Potting-In order to assure repeatability of torsion testing relative to the bone orientation, each specimen must be affixed to the testing apparatus the same every time. Variation in placement can produce inconsistent results. To address this problem, a unique potting apparatus was built. ¼" acrylic was cut to fit PVC squares and then cut in half. Semicircular profiles were cut in the acrylic to match the necessary profile of the specimen. Set screws were then used to lock the specimen in place with the endcap so that the cement and plaster of paris mixture could be added. The commonly used PMMA used for potting was replicated by making a 50:50 mixture or Portland cement and plaster of paris and then mixed 3:1 with water. This PMMA substitute was then poured into the endcaps to pot each bone. Figure 2 details the potting apparatus.

Figure 2: Full potting setup with spacers

Torsion Testing-A strain rate of 0.5 degrees was chosen for validating the torsion device. For the single cycle test, the specimens were attached to the testing apparatus and tested until failure (Figure 3A). Tests were performed on 8 normal sawbones (Sawbones Worldwide) and then repeated on these broken bones with compression plates fixation performed by an orthopedic surgeon. Details of the Torque-Rotation graph obtained from actual tests and summary of bone stiffness's obtained from these single cycle failure test, as observed in the two groups, are shown in Figure 3B and Table 2.

RESULTS

Table 1: Total Bill of Materials

Item	Part Number	Cost	Seller	Quantity	Subtotal
StepperMotor	OMHT34-487	$474.00	Omega	1	$474.00
Driver	ST10-Si	$580.00	Omega	1	$580.00
Torque Cell	FSH00648	$495.00	Futek	1	$495.00
Roller Bearing	5709K33	$31.51	McMaster	2	$63.02
Bearing Ring	5709K69	$18.53	McMaster	2	$37.06
Gearbox	MMAA-GBPH-0902-NP-015-AA341	$667.00	Anaheim Automation	1	$667.00
				Total	$2,316.08

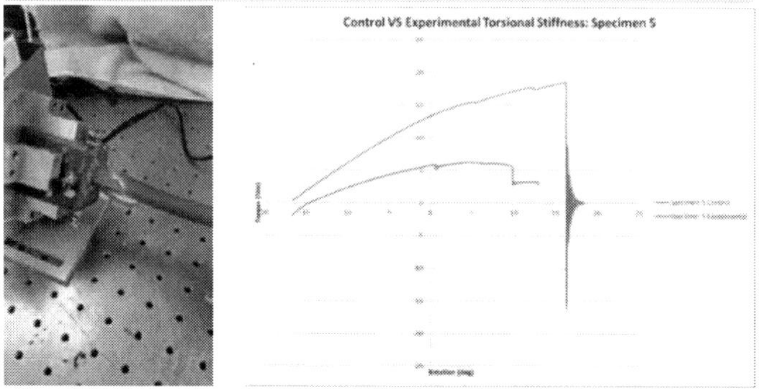

3A 3B

Figures 3: 3A: Left Demonstrated Results, 3B: Right Graphed Results

Table 2: Quantified Results

Test	N	Mean Stiffness	SD	Std Error Mean
Control (Unplated)	8	0.6594 Nm/deg	0.10634	0.0376
Experimental (Plated)	4	0.3577 Nm/deg	0.10545	0.05272

There was a 46% difference between groups where P=0.003

DISCUSSION

Torsional testing is typically performed using a biaxial servohydraulic testing machine or similar sophisticated electromechanical equipment. Due to its high cost, many researchers do not have such equipment at their disposal. In this study, we outline a highly inexpensive torsion testing device that can be used to evaluate the mechanical properties of long bones after various interventions to improve fracture healing.

ACKNOWLEDGEMENTS

Authors would like to acknowledge the contribution of Euver Joaquin Castillo, Christopher Guns, Nicholas Laverick and Thomas Senia in designing and building this device. This study was supported by Chairs Leadership Discretionary Fund offered by the Christiana Care Health Systems, Delaware.

REFERENCES

[1] Chung KC, Spilson SV. The frequency and epidemiology of hand and forearm fractures in the United States. J Hand Surg Am. 2001;26:908–915

[2] Nellans KW, et al. The Epidemiology of Distal Radius Fractures. Hand Clin. May 2012: 28(2):113-125

SB^{3}C2015
Summer Biomechanics, Bioengineering and Biotransport Conference
June 17-20, 2015,Snowbird Resort, Utah, USA
SB^{3}C2015-575

A MECHANICAL MODEL FOR CORTICAL FOLDING DURING BRAIN DEVELOPMENT

N. Vijayakumar (1), M. Holland (1), E. Kuhl (1,2)

(1) Department of Mechanical Engineering
Stanford University
Stanford, CA, USA

(2) Department of Bioengineering
Stanford University
Stanford, CA, USA

INTRODUCTION

The brain is a complex organ with a characteristic folding pattern. This folding allows for increased surface area of the brain, and accommodates a large number of neurons within the limited volume of the skull. In humans, the primary folding pattern is relatively constant, whereas secondary and tertiary folding may vary significantly across individuals [1]. From a mechanical point of view, cortical folding is the result of an instability problem as the brain develops and grows. Cortical malformations are important indicators of developmental disorders which may have severe neurological consequences. Typical examples are lissencephaly, characterized by a thicker cortex and reduced gyrification, and polymicrogiria, characterized by a thin cortex and increased gyrification [1]. Understanding the mechanisms of cortical folding may provide insight into the evolution of neurological pathologies.

The prevailing hypothesis to explain cortical folding relies on the theory of differential growth between the cortical and subcortical layers of the brain [2]. The cortex is a thin top layer of neurons, housing primarily neuron cell bodies. The subcortex is thicker, and composed largely of axons. Since the cortex expands rapidly during the second trimester of human development, folds starts to emerge around week 23 post gestation [1]. In individuals with autism spectrum disorders, the cortical growth rate starts faster than in healthy individuals, then slows down to below the norm [3]. This change in growth rate results in an increased cortical thickness and a larger number of folds. As another example, pre-term infants demonstrate slower post-natal cortical growth. This is evidenced by decreased gyrification at term when compared to full-term infants [4]. Here we explore the effects of alterations in the cortical growth rate on cortical folding and pattern formation.

METHODS

To simulate cortical folding during brain development, we model a two-dimensional plane strain section of the brain, as illustrated in Figure 1. Using the theory of finite growth, similar to plasticity, we separate the deformation into a reversible elastic part $\mathbf{F}^e$ and irreversible growth part $\mathbf{F}^g$. Both the cortex and subcortex grow isotropically within the model; the cortical growth rate is prescribed while the subcortex responds to stretch induced by cortical growth [5].

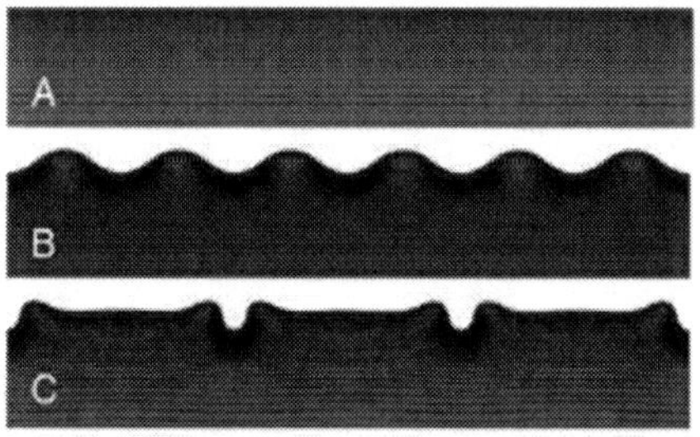

FIGURE 1. THE CORTEX, A THIN LAYER OF NEURONAL CELL BODIES, SITS ON THE SUBCORTEX, A THICK LAYER OF AXONS (A). GEOMETRICALLY CONSTRAINED GROWTH OF THE CORTEX INDUCES STRETCH, WHICH ULTIMATELY RESULTS IN SURFACE BUCKLING (B,C).

Cortical growth in previous models [2, 6] has been represented as

$$\vartheta^c_g = 1 + G_c\, t\,,\ (1)$$

where ϑ^c_g represents the growth ratio, which characterizes the increase in cortical volume,

$$\vartheta^c_g = J = V/V_0\,.\ (2)$$

To model rapid cortical growth which gradually slows down over time, we now use a hyperbolic growth model with

$$\vartheta^c_g = 1 + (G_c\, t)^\gamma \;. \quad (3)$$

This type of growth closely mimics the characteristic features of growth in autism [3]. The subcortex grows in response to cortical growth, driven by the elastic volume stretch, J^e, at the subcortical growth rate G_s,

$$\frac{d}{dt}\vartheta^s_g = G_s\,(J^e - 1) \;. \quad (4)$$

The value of the subcortical growth is determined at every time step on the integration point level using a Newton-Rhapson iterative method. This is done through a user material (UMAT) subroutine in the commercial finite element package Abaqus. A Neo-Hookean material model was used, with $G_c = 0.5$ and $G_s = 10$. Stresses are calculated from the elastic deformation,

$$\mathbf{F}^e = (1/\vartheta_g)\,\mathbf{F} \quad (5)$$

and the simulations iterate until equilibrium is reached at each point in time. The buckling patterns were compared, as well as the volume growth of specific cells in the cortex and subcortex.

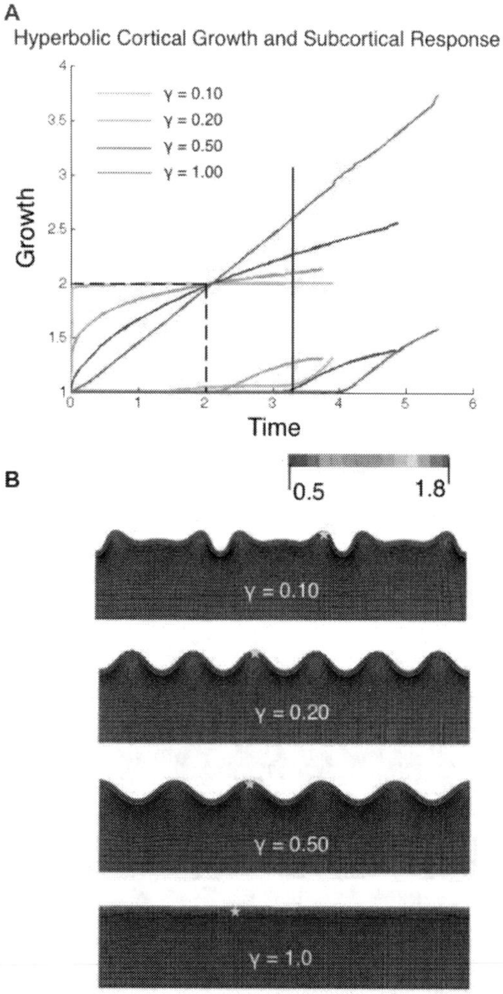

FIGURE 2. A) EVOLUTION OF CORTICAL AND SUBCORTICAL GROWTH FOR VARYING GROWTH EXPONENTS. SOLID BLACK LINE INDICATES TIME AT WHICH BUCKLING IMAGES WERE CAPTURED; DOTTED LINE INDICATES CONSTANT GROWTH LEVEL AND CORRESPONDING SUBCORTICAL GROWTH. B) SENSITIVITY OF FOLDING PATTERN WITH RESPECT TO GROWTH EXPONENT γ.

RESULTS

Figure 2 shows the sensitivity analysis for different growth exponents γ. The graph (Fig. 2A) summarizes the evolution of cortical and subcortical growth for growth exponents. Subcortical growth generally follows the pattern of cortical growth, with delayed response time. As cortical growth increases more quickly, subcortical growth has a faster response time and steeper increase. As the overstretch of the subcortex is compensated by growth, subcortical growth decays.

Linear cortical growth ($\gamma = 1$) leads to nearly linear subcortical growth, while hyperbolic cortical growth leads to hyperbolic subcortical growth (Fig. 2A). The buckling patterns vary for varying exponents, indicative of the fact that the models with smaller values of γ reached higher levels of cortical growth more quickly. Accordingly, the buckling pattern is most developed for the model with $\gamma = 0.1$, where post-buckling growth can be seen (Fig. 2B). When $\gamma = 1.0$, growth took longer and buckling was not visible at the time of image capture (Fig. 2B). Similarly, even when the models are compared at the same time and the same level of cortical growth ($t=2$, $\vartheta^c_g = 2$) steeper growth leads to more developed buckling patterns; in fact, only $\gamma = 0.1$ has begun to develop folds at that point.

DISCUSSION

The results of this study indicate that cortical growth rate determines subcortical growth rate and response time. A greater cortical growth rate leads to quicker subcortical growth. Further, subcortical growing time plays a larger role in buckling patterns than cortical growth levels alone. Models with quicker cortical growth generate shorter gyral wavelength and more folds. This is in agreement with analytical predictions using the Föppl-von Karman theory [6]. Further studies could help to elucidate whether these results are in agreement with gyrification patterns found in normal and autistic brain development. Our model is conceptually general and could, with negative growth multipliers, potentially also capture the effects of cortical shrinkage, as exhibited in disease conditions such as stroke and dementia.

ACKNOWLEDGEMENTS

Nithya Vijayakumar was supported by the Stanford Undergraduate Research Institute, Department of Mechanical Engineering, Stanford University.

REFERENCES

[1] Richman DP, Stewart RM, Hutchinson JW, and Caviness VS Jr. (1975). Mechanical model of brain convolutional development. *Science*, 189(4196), 18–21.

[2] Bayly PV, Okamoto RJ, Xu G, Shi Y, and Taber LA. (2013). A cortical folding model incorporating stress-dependent growth explains gyral wavelengths and stress patterns in the developing brain. *Physical Biology*, 10(1), 16005.

[3] Wallace GL, Robustelli B, Dankner N, Kenworthy L, Giedd JN, and Matin A. "Increased Gyrification, but Comparable Surface Area in Adolescents with Autism Spectrum Disorders." *Brain* 136.6 (2013): 1956-967.

[4] Clouchoux C, Guizard N, Evans AC, Du Plessis AJ, and Limperopoulos C. "Normative Fetal Brain Growth by Quantitative in Vivo Magnetic Resonance Imaging." *American Journal of Obstetrics and Gynecology* 206.2 (2012): 173.e1-73.e8.

[5] Budday S, Raybaud C, and Kuhl E. "A mechanical model predicts morphological abnormalities in the developing human brain." *Scientific Reports* 4 (2014): 5644.

[6] Budday S, Steinmann P, and Kuhl E. "The role of mechanics during brain development." *Journal of the Mechanics and Physics of Solids* 72 (2014): 75-92.

SB³C2015
Summer Biomechanics, Bioengineering and Biotransport Conference
June 17-20, 2015,Snowbird Resort, Utah, USA
SB³C2015-576

3D QUANTIFICATION OF OSTEOPHYTE DISTRIBUTION ON THE HUMERAL HEAD

Shea K. Taylor (1,2), Brandon G. Santoni (1,2), Mark A. Frankle (2,3), Peter Simon (1, 2)

(1) Phillip Spiegel Orthopaedic Research Laboratory
Foundation for Orthopaedic Research and Education
Tampa, FL, USA

(2) Morsani School of Medicine
University of South Florida
Tampa, FL, USA

(2) Dept. Shoulder and Elbow Surgery
Florida Orthopaedic Institute
Tampa, FL, USA

INTRODUCTION

Glenohumeral osteoarthritis (GH-OA) is a physically debilitating condition affecting up to 20% of the aged population[1]. Despite being such a high impact disorder, the clinical diagnosis of GH-OA remains subjective. Currently, diagnosis of GH-OA is done by evaluation of clinical radiographs. The following radiographic parameters and their extend assist in defining the severity of GH-OA: extent of osteophyte formations, joint space narrowing, cyst formation, and subchondral bone sclerosis. In order to eliminate the subjectivity introduced by clinicians and to better understand the etiology of the disease and its progression, it is essential to develop methods that will objectively quantify GH-OA parameters. In this study we focus on osteophyte formations, as they represent a particularly understudied aspect of GH-OA diagnosis.

Osteophytes were one of the first findings associated with osteoarthritis[2]. Nonetheless, the patho-anatomical cause underlying osteophyte development in GH-OA remains a topic of debate. Proponents of biomechanical and biochemical etiologies often clash with little data to support one view over the other. In all reality it is likely both which contribute to the formation of osteophytes; however, it is clear that a deficiency of research exists in this field. The primary objective of this study is to fill this void and enhance our understanding of osteophyte formation in end-stage degenerative disease.

In the current study, the system to quantify osteophyte distribution within anatomically relevant zones of the humeral head was developed. Following quantification, we investigated osteophyte distribution differences on a gender-specific basis. We hypothesized that a significant pattern of osteophyte distribution exists and that men would exhibit a greater volume of osteophyte development than women. We further normalized our findings relative to humeral head size, allowing us to determine if our findings are truly indicative of differences between men and women or simply an elusive correlate to bone size.

METHODS

A total of 99 patients were included in this study (56M, 43F). Inclusion criteria were established to screen the patient population from the affiliated hospital to include only patients who underwent total shoulder arthroplasty with a diagnosis of end-stage GH-OA. Patients with a congenital deformity, fracture, or secondary diagnosis (rheumatoid arthritis, avascular necrosis, etc.) were excluded. The age of all patients ranged from 60-70 years.

Pre-operatve CT scans (1.25 slice thickness, ~0.45 pixel size) were collected and imported into MIMICS (Materialize, Leuven, Belgium). A set of masks were created based on cortical bone threshold value extracting the geometry of the humerus. These included the humerus with osteophyte, humerus without osteophyte, model of osteophyte, and the intramedullary canal model (**Fig. 1b**). Particular care was taken to precisely distinguish the osteophyte formation from the humerus using the sclerotic line between them (**Fig. 1a**). All masks were exported in point-cloud and volumetric format.

A humeral coordinate system was established as a linear combination of the principal eigen vector describing the humeral canal's primary direction and the normal of the plane forming the humeral osteotomy (**Fig. 2b**). This plane of osteotomy was fitted to the set of points (> 50) manually selected on the humeral model rendering at the margin of the articular surface (**Fig. 2a**). A custom-written program was used for further analysis (MATLAB). First, the zoning algorithm was established with five anatomically relevant zones on the model of the humeral head without osteophytes (**Fig. 3a**). Using spherical coordinates, the surface of the head was automatically divided into five zones. Second, this zoning was applied to the model of the osteophyte, subdividing it into zones based on its location on the humeral head (**Fig. 3b**). Third, the pixel value of each patient, together with slice increment, was used to calculate the size of the voxel forming each data-point representing zoned osteophytes. Total volume and volume per each zone are reported. Fourth, the best fit circle was calculated based on the points that lay in the vicinity of the humeral osteotomy plane (**Fig. 2c**). Radius is reported for each patient and used to normalize volume results. Statistical analysis included the use of ANOVA and t-tests for group comparisons.

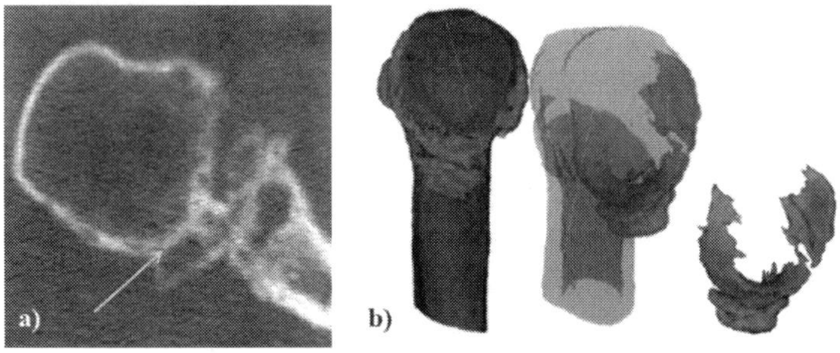

Figure 1: a) sclerotic line definition; b) input models

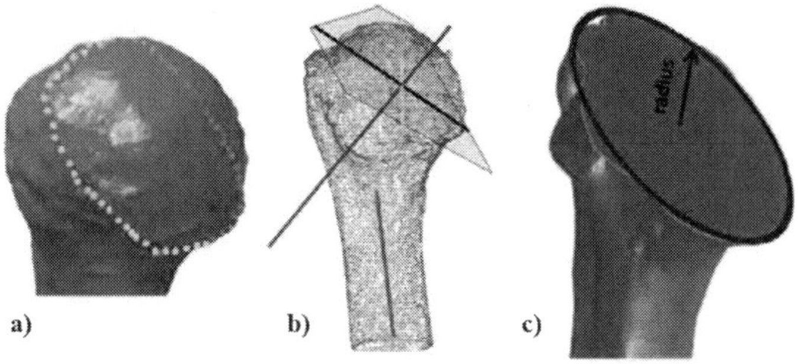

Figure 2: a) Points on articular margin; b) coordinate system; c) radius of humeral head measurement

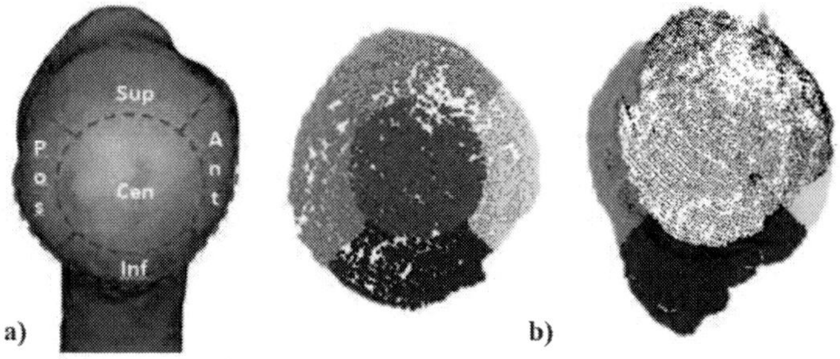

Figure 3: a) zoning algorithm on humerus; b) zoning of humerus and osteophyte

Table 1: Descriptive statistics

		Total Volume		Normalized Volume	
	Percentage	Male	Female	Male	Female
Superior	12.1%	470.5	328.0	19.3	15.7
Posterior	38.1%	1203.7	838.4	47.7	40.2
Central	1.9%	44.9	27.0	3.2	1.3
Anterior	7.8%	197.0	203.1	9.6	9.6
Inferior	40.1%	1138.6	908.0	47.9	44.8
Total	100.0%	3054.6	2304.5	127.7	111.6

RESULTS

Total osteophyte volume was significantly greater in males (p=0.022). On average, male volume of osteophyte was 3054.6±1972.9mm^3 and female volume of osteophyte was 2304.5±867.6mm^3. Zoning of the humeral head showed significant differences between zones (p<.0001)[except between posterior and inferior zones (p=.673)] in absolute osteophyte volume distribution (**Fig. 4**) and between the genders (p<.0004). Females exhibited less osteophyte formations in superior, posterior, central, and inferior zones with significance shown between genders in the posterior zone (p<.014).

Normalized osteophyte volume was calculated in order to control for subject size. Total osteophyte volume was normalized to the humeral head radius calculated for every patient at the plane of osteotomy. On average, males had larger humeral head radii than females (p<.0001). Following normalization, the total volume of osteophyte formation was not significantly different between genders (p=.2856). Zonal analysis of normalized volumes showed differences similar to non-normalized volumes between individual anatomical zones (**Tab.1**), however there were no differences observed between genders (p=0.0786).

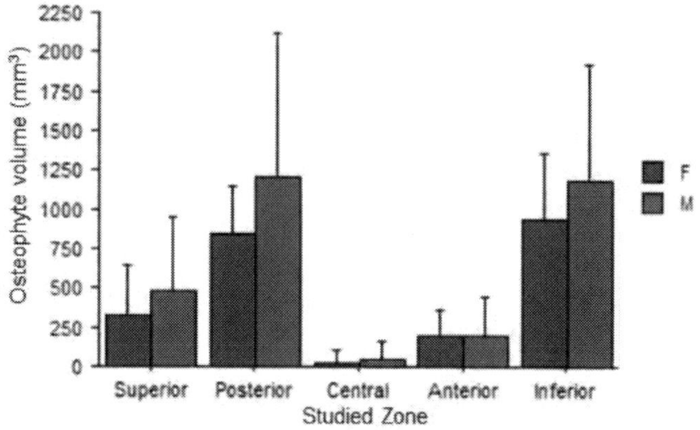

Figure 4: Zonal distribution of total osteophyte volume.

DISCUSSION

In the current study we aimed to evaluate the 3D distribution of osteophyte formation in the population of patients diagnosed with end-stage OA who underwent total shoulder arthroplasty. To our knowledge, this is the first work on the quantification of osteophyte formation in the shoulder. Novel methods for automatic, image-based calculation of the volumes representing the bony formations and automatic zoning algorithm were developed. Our results showed that men have significantly more osteophyte volume when compared to women and significantly more osteophyte formed in the posterior zone. However, when we normalized the measured volume to the size of the humeral radius (in the plane of osteotomy), there were no longer significant differences between genders. These results indicate that the formation of osteophytes in end-stage OA is dependent on the size of the individual (humeral head of male is significantly larger than in female; p<.0001) and may be independent of ones gender (p=0.2856).

For future work we plan to extend our analysis to evaluation of joint space width, subchondral bone density and cyst formation distribution. The quantified knowledge of those parameters and their relationships can be clinically beneficial to diagnosis, pre-surgical planning, and better understanding the etiology of osteoarthritis.

REFERENCES

[1] Matsen F, et al., *In: The shoulder*. p. 914. ISBN0721601480, 2004.
[2] Nichols EH, et al., *J. Medical Research*, 21(2):149-222.51, 1909.

SB^{3}C2015
Summer Biomechanics, Bioengineering and Biotransport Conference
June 17-20, 2015, Snowbird Resort, Utah, USA
SB^{3}C2015-577

ASSESSMENT OF A BIOENGINEERING INNOVATION PROGRAM FOR MIDDLE SCHOOL GIRLS

Kristen L. Billiar (1), Amanda (Zoe) Reidinger (1), Jeanne Hubelbank (2), Helen Vassallo (3)

(1) Department of Biomedical Engineering
Worcester Polytechnic Institute
Worcester, MA, USA

(2) Independent Program Evaluation
Consultant
Sudbury, MA

(3) Department of Management
Worcester Polytechnic Institute
Worcester, MA, USA

INTRODUCTION

Students' attitudes towards engineering are formed at an early age, and it has been proposed that early adolescence is a particularly appropriate time for students to explore science careers [1]. For minorities and females in particular, 1) their self-esteem may begin to decline during the middle-school years which can affect their academic performance and choice of classes in school; 2) there is a circular relationship between their affection for science, their self-esteem, and their career plans; and 3) without pursuing science and math courses in middle school, they will be left behind in high school and unable to pursue college degrees in engineering.

Studies have shown that exposure to live role models increases participants' knowledge of and interest in science careers, even if only a 2-day workshop with female collegiate science majors (trained in workshop techniques) [2]. Longer exposure to women science career role models as part of science instruction also positively affects early adolescent female (and male) students' attitude toward women in science [3].

As a *unique feature* of our NSF-funded REU program in Bioengineering, the undergraduate participants themselves participate in outreach by becoming mentors to middle school girls ("mentees") in a one-week Bioengineering Innovation Program. This activity is aimed at encouraging middle school students to remain in math and science courses by demonstrating the excitement of engineering and providing an opportunity for them to work closely with positive role models. We hypothesized that this outreach activity will also instill in the REU students a desire to support the next generation of young scientists and engineers, and will teach them mentoring and team work skills which are highly valued in the engineering workplace.

METHODS

Participants

Female students from underrepresented minority populations entering 8th or 9th grade, were recruited from Worcester-area Public Middle Schools in Massachusetts from 120 applicants. Ten girls were accepted and attended. The girls were rising 8th graders (50%) and 9th graders (50%). Half of the girls represented minorities: 20% Black, 20% Hispanic, and 10% Asian.

Activities

Each middle school girl was paired with one of the 10 REU participants. In the mornings, the REU and her middle school partner worked on an independent research project and experienced active science and engineering demonstrations facilitated by the REUs. REUs also took turns creating and directing snack activities that were related to science and engineering, e.g., cell cookies and writing a set of directions to build a structure out of fruit and candy. The girls created posters of their independent research and presented them to their parents and other visitors on the last day of the program.

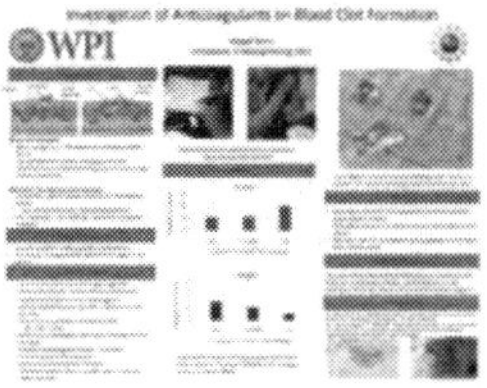

In the afternoons, the girls engaged in activities with a focus on creativity and innovation. Whole group activities addressed creativity, communication, cooperation, and presentation skills, but the girls mainly worked together in two teams of five girls on each project. The teams each selected a science heroine of their choice, researched the woman, and made a poster depicting the woman's major accomplishments. Teams also wrote and practiced a play on a topic of their choosing. They presented their posters and the play to their parents and visitors on the last day.

RESULTS

Middle School Girls' Reactions

The middle schools girls rated the week highly. The ten girls rated iBME as either "excellent" (60%) or "good" (40%). Sixty percent of the girls mentioned research with their mentors specifically as what they enjoyed most. The girls' ratings of research and creativity/leadership activities are given in the table below.

Table 1. Middle school girls' ratings of research and creativity activities

Overall Rating of Activity Session	Research in labs, Demonstrations		Play, Heroine, Posters, Games	
	#	%	#	%
I liked it a lot. (Excellent)	6	*60%*	4	*40%*
I liked it. (Good)	4	*40%*	4	*40%*
They were okay. (Satisfactory)	-	-	2	*20%*
I liked them a little bit. (Fair)	-	-	-	-
I didn't like them at all. (Poor)	-	-	-	-
Total	10	*100%*	10	*100%*

The middle school participants rated their interaction with their undergraduate mentors highly. Seventy percent of the girls indicated that their interaction was "excellent" and thirty percent of the girls checked that their interaction was "good." The girls wrote that they enjoyed their mentors because they not only "learned a lot" from them, but also found them to be "fun," "sweet," and "nice."

Impact on Participants

The participants completed a post survey that included questions about how they viewed science before and after the program. Most (90%) of the participants reported that they were more interested in science after attending the program than they were before the program. The girls reported an increase in confidence in science abilities. Almost all the girls either "agreed" (70%) or "strongly agreed" (20%) that, as a result of the week, they "feel more confident that I will be able to work in the science field."

Effects on undergraduate mentors

One of the expected outcomes of the Innovation Week was that REUs would improve overall proficiency in mentoring and their skills in mentoring middle school students in particular. Although there was a small improvement in the REUs' self-ratings of their mentoring skills from the beginning to the end of the program, the difference was not statistically significant due to their high initial ratings of their skills in the pre-program assessment (Table 2). There was no statistically significant difference between their ratings at the beginning and end of the program.

Table 2. REUs' Ratings of Agreement: I Have the Skills to Mentor a Middle School Student

Skills to mentor a middle school student	Percentage of REUs Agreeing	
	Rating at Start N = 10	Rating at End N = 10
Strongly agree	40%	30%
Agree	50%	50%
Neutral	10%	20%

Wilcoxon Signed Ranks Z =1.00, $p = .317$

DISCUSSION

The Innovations in Bioengineering Week was successfully implemented and well received. Ten middle-school girls attended. Participants were enthusiastic about the week, rating it overall as excellent or good. Morning research activities were more highly rated than afternoon creativity/innovation activities. Overall, the girls reported that the program made them more interested in science.

When the undergraduate REU participants started the program, they rated themselves as having an ability to mentor middle school students and this rating did not change significantly at the end of the program.

Based on our findings, we believe that these mentored project experience and activities will encourage the middle school participants to keep taking math and science courses so that engineering will be a possible choice for them in high school, college, and beyond. The undergraduates in the program had the opportunity to influence the educational and career plans of the younger students, which we believe will make them more likely to engage in mentorship and outreach activities in their future careers as professional engineers.

ACKNOWLEDGEMENTS

This work was funded by the NSF (REU EEC-1156821).

REFERENCES

[1] Hurd, P. D., 1978, "Early adolescence: Perspectives and recommendations (National Science Foundation Monograph SE-78-75)," Washington D.C.

[2] Oakland, T., and Young, B., 1980, "The role of models in math, science, and engineering project," UCLA Women's Resource Center, Los Angeles.

[3] Smith, W. S., and Erb, T. O., 1986, "Effect of women science career role models on early adolescents' attitudes toward scientists and women in science," *Journal of Research in Science Teaching,* 23, pp. 667-676.

SB³C2015
Summer Biomechanics, Bioengineering and Biotransport Conference
June 17-20, 2015,Snowbird Resort, Utah, USA
SB³C2015-578

AN OPEN-SOURCE TOOLBOX FOR SURROGATE MODELING OF JOINT CONTACT MECHANICS

Ilan Eskinazi (1), Benjamin J. Fregly (1,2,3)

(1) Department of Mechanical and Aerospace Engineering
University of Florida
Gainesville, FL, USA

(2) Department of Biomedical Engineering
University of Florida
Gainesville, FL, USA

(3) Department of Orthopaedics and Rehabilitation
University of Florida
Gainesville, FL, USA

INTRODUCTION

Contact models that compute the loads arising at the interface of two touching surfaces allow modelers to create realistic mechanical representations of human joints. Contact models are embedded into multi-body dynamic simulations when the complexity of pin joints, spherical joints, parallel mechanisms, and other simplified joints is not sufficient to capture variables of interest such as contact forces and ligament forces. A limitation of contact models is that they tend to be computationally expensive and hence impractical for use within iterative frameworks such as optimizations and dynamics simulations. This limitation has been tackled recently by surrogate modeling approaches that fit or interpolate previously gathered input-output data similar to a look-up table [1]–[3]. However, the process of generating, testing, and deploying these surrogate contact models remains a challenging and time consuming task. We have developed a freely available open-source program called Surrogate Contact Modeling Toolbox (SCMT) to streamline the entire surrogate contact modeling process and are making it available to the biomechanics research community at large.

SCMT consists of an application programming interface (API) and a graphical user interface (GUI) containing tools to design, generate, and test surrogate contact models. The GUI allows users to follow the surrogate modeling process without having to write any code. SCMT interacts with finite element software FEBio [4] and with Matlab which respectively perform contact simulations and train neural networks. SCMT also includes a plugin for the musculoskeletal simulation software OpenSim [5] that allows users to embed surrogate contact models generated with SCMT into OpenSim models.

METHODS

SCMT is an application that builds artificial neural network-based surrogate contact models from finite element simulations. SCMT is written in C++ and the GUI was designed with Qt. SCMT requires Windows as the operating system, and makes extensive use of third party tools and libraries such as Boost serialization libraries [6], Hammersley sequence libraries by John Burkardt [7], and the OpenMP API [8] for multithreading. Moreover, SCMT interfaces with FEBio and Matlab's neural network and code generation toolboxes.

SCMT is composed of the following tools:

1) Sample Point Generator: Creates sequences of sample points populated with inputs. The sample point generator allows the user to define six-dimensional domains to be sampled. Large sparse domains may be defined using a bounding box option while small dense domains may be defined by clustering the sample points around user-defined trajectories.

2) Sampler: Executes FEBio static analyses in a multithreaded manner.

3) Filter: Removes sample points exceeding user-defined thresholds from sets of collected sample points.

4) Out-of-Contact (OOC): Generates sample points corresponding to out-of-contact configurations. The sample points generated can be either fully out-of-contact where there are no contact patches and all contact loads are zero, or partly out-of-contact where one condyle is in contact and the other is not.

5) Surrogate Designer: Lets the user define the overall structure of the surrogate model. The structure includes the inputs and outputs to the surrogate model and to each of its stages. This tool also lets the user define the artificial neural network architectures and stopping criteria. The Surrogate Designer also interfaces with Matlab and initiates the neural network training process.

6) Tester: Evaluates the root-mean-square (RMS) and maximum absolute error between the surrogate model and test datasets. This

evaluation can be performed on a per-domain basis or for all domains together.

7) Utilities: Allows users to change the sample point data format. It can convert binary to XML and vice-versa. It can also transform sample points into comma separated values (CSV files) so that they may be explored as a spreadsheet. Another utility allows users to join different sets of sample points into a single binary file.

In addition to the GUI, we developed a plugin for OpenSim that allows surrogate contact models created with SCMT to be loaded as "Force" model components similar to springs or bushings. When the OpenSim model is loaded, the plugin dynamically loads the artificial neural networks, and when forces are computed it applies the net contact loads to the user-specified pair of rigid bodies.

As an example, we used SCMT to create surrogate contact models of tibiofemoral (TF) and patellofemoral (PF) artificial joints and evaluated their computational speed and accuracy. As a speed test, we measured the average computation time for 10,000 consecutive surrogate model evaluations on a 3.4 GHz processor. As an accuracy test, we measured the RMS errors for each of the domains calculated with test data not used for training.

RESULTS

The computational speed test resulted in a mean surrogate contact model evaluation time of 88.3 μs for the TF model and 57.6 μs for the PF model. Errors for the surrogate contact models created (Table 1) showed average force/torque errors of 13 N / 340 N-mm and 10 N / 66 N-mm for the TF and PF models, respectively. The domains corresponding to out-of-contact configurations showed RMS errors below 16 N / 370 N-mm.

TABLE 1. DESCRIPTION OF SAMPLED DOMAINS WITH NUMBER OF SAMPLE POINTS USED FOR TRAINING AND TESTING. AVERAGE RMS ERRORS FOR FORCES AND TORQUES ARE SHOWN FOR EACH DOMAIN.

Domain Name	Physical meaning	N° of sample points		RMS errors	
		Train	Test	Forces (N)	Torques (N-mm)
T1	Normal gait	23431	977	10	237
T2	All gait	20381	828	15	383
T3	Bounding box	19046	782	14	390
T4	Contact boundary	199	100	14	383
T5	Fully OOC	597	300	3	134
T6	Partly OOC	18000	1200	15	366
ALL TF	-	81654	4187	13	340
P1	Bounding box	41517	2070	9	68
P2	Contact boundary	1162	117	12	77
P3	Fully OOC	3483	234	2	18
ALL PF	-	46162	2421	10	66

DISCUSSION

Our software framework simplifies the process of creating, testing, and implementing surrogate models of joint contact. Together with the plugin, SCMT bridges the gap between contact analysis in FEBio and musculoskeletal simulation in OpenSim while at the same time radically increasing the computational speed of the contact evaluations. Contact models generated with SCMT could be used to calculate muscle and contact forces simultaneously or to calibrate ligament parameters within an optimization.

The surrogate contact models for the TF and PF joints demonstrate SCMT's capability to sample and fit distinct contact configurations. Gathering the data and training the neural networks for each of these models took two to three days for each model using a 3.4 GHz computer while using 13 concurrent threads for sampling and 12 for fitting. The initial investment however pays off by generating surrogate contact models about 15,000 times faster than the original FE simulations with relatively small discrepancies in the output loads.

Apart from the seamless integration between FEBio, Matlab, and OpenSim, the methods employed by the toolbox have several benefits. One benefit is that out-of-contact configurations can be sampled and fit to allow for the prediction of liftoff during simulation. Another benefit is that different domains of input space can be approximated with varying degrees of accuracy by adjusting the number of sample points and the domain size. Moreover, the feed-forward neural networks at the heart of the surrogate models can be trained with tens of thousands of sample points without the computer running out of memory.

SCMT has several limitations. First, the neural network-based surrogate contact models do not provide error estimates. Second, gathering the required input-output data from FE simulations can take a considerable amount of time, especially for finite element meshes containing many nodes. Training the surrogate contact models can also take a considerable amount of computation time. However, both the sampling time and training time can be reduced considerably by taking advantage of the provided parallelization options. Third, friction cannot be modeled with this approach because contact loads are assumed to be a function of pose only. Surrogate models of quantities such as contact area, center of pressure, or maximum pressure are not implemented but could be added in the future. If these measures were added, both friction and wear could potentially be incorporated, assuming that wear does not significantly affect the contact geometry.

We will eventually release SCMT to the public domain with a journal publication via simtk.org so that researchers across the biomechanics community can create and use custom surrogate contact models in their own work. We hope that this tool will lead to the widespread use of surrogate contact models and ultimately to the advancement of the field of musculoskeletal simulation.

ACKNOWLEDGEMENTS

Funding for this study was provided by NIH grant R01EB009351 and by the University of Florida.

REFERENCES

[1] Y.-C. Lin et al., Med. Eng. Phys, 32:584–94, 2010.
[2] J. P. Halloran et al., J. Biomech. Eng., 131:011014, 2009.
[3] M. Mishra et al., Biomed. Signal Process. Control, 6:164–174, 2011.
[4] S. a Maas et al., J. Biomech. Eng., 134:011005, 2012.
[5] S. L. Delp et al., IEEE Trans. Biomed. Eng., 54:1940–50, 2007.
[6] http://www.boost.org/
[7] http://people.sc.fsu.edu/~jburkardt/m_src/hammersley/hammersley.html
[8] http://openmp.org

SB³C2015
Summer Biomechanics, Bioengineering and Biotransport Conference
June 17-20, 2015,Snowbird Resort, Utah, USA
SB³C2015-579

CHARACTERIZING THE INFLAMMATORY AND COAGULATION RESPONSE TO TRAUMA AND RESUSCITATION

Elaheh Rahbar (1), Jessica C. Cardenas (2), Nena Matijevic (2), Deborah J. del Junco (2), Jeanette M. Podbielski (2), Mitchell J. Cohen (3), Bryan A. Cotton (2), John B. Holcomb (2), and Charles E. Wade (2) on behalf of The Early Whole Blood Investigators

(1) Department of Biomedical Engineering
Wake Forest University
Winston-Salem, NC, USA

(2) Department of Surgery, CeTIR
University of Texas Health Science Center
Houston, TX, USA

(3) Department of Surgery
University of California San Francisco
San Francisco, CA, USA

INTRODUCTION

Trauma is the leading cause of death in people under the age of 45 years with hemorrhagic shock accounting for nearly 50% of these mortalities [1]. Death from hemorrhage occurs very quickly usually within the first 1-3 hours following injury. It is believed that hemorrhagic shock is potentially preventable with appropriate resuscitation. Historically whole blood has been the gold standard for resuscitating severely injured patients [1]. However, as blood banking and transfusion medicine has evolved, whole blood from donors is separated into its three primary components: red blood cells (RBC's), plasma and platelets. There is an on-going debate on what the ideal ratio of these blood components is to achieve optimal patient outcomes. As a result, there is a lot of heterogeneity in resuscitation and transfusion practices across hospitals worldwide. There is a need for better understanding the underlying mechanisms of hemorrhagic shock and resuscitation.

One of the more common strategies for effective resuscitation of severely bleeding trauma patients is based on achieving primary hemostasis faster using platelets and other supplemental clotting factors. Extensive research has been done in trying to characterize the coagulopathy following an injury, termed trauma-induced coagulopathy (TIC). But our understanding of the effect of resuscitation or transfusions on coagulopathy remains fairly rudimentary. Some have reported improved survival with early transfusion of platelets in combination with RBC and plasma components [2]. But many of these studies are plagued with survival bias since a patient must survive to receive blood transfusions. Further there is limited high quality longitudinal data on trauma patients, due to the difficulty in conducting and acquiring real-time data and blood samples from this critically injured population.

The Center for Translational Injury Research (CeTIR) in Houston, TX has been instrumental to the advancement of trauma research by participating and leading a number of key and foundational studies in understanding hemorrhagic shock and resuscitation by collecting serial clinical and laboratory data points over the first 24 hours of a patient's hospital stay . Of note, a pilot randomized controlled trial called "The Early Whole Blood" (EWB) study revealed that modified whole blood (mWB) resuscitation reduced the overall blood transfusions compared to blood component therapy (COMP) in patients without traumatic brain injury [3]. However the EWB study's primary focus was on transfusion volumes, not the underlying inflammatory or coagulation response. The current study leveraged data from the EWB trial to explore the effects of mWB vs. COMP transfusions on the coagulation and inflammatory responses over time. We hypothesized that patients receiving mWB would display improved inflammatory and coagulation status compared to those receiving COMP therapy.

METHODS

The EWB study was a single-center randomized controlled pilot trial, approved by the local IRB (University of Texas Health Science Center, Houston, TX). Adult patients at highest-level trauma activations were randomized into mWB or COMP groups. The EWB study design and trial has been previously described [3].

Blood samples (~20 ml) were collected from enrolled patients in citrated and ethylenediaminetetra-acetic acid (EDTA) tubes at five time points: time 0 (i.e. hospital admission), 3, 6, 12 and 24 hours post-admission. Fresh whole blood samples collected in citrated tubes were used for thrombelastography (TEG) and platelet aggregometry assays. Remaining blood was then centrifuged to isolate plasma.

Plasma samples were then aliquoted and stored at -80°C for future analysis.

Coagulation status was evaluated using TEG, platelet aggregometry and calibrated automated thrombograms. Frozen plasma samples were thawed and evaluated for inflammatory and clotting factors. Specifically IL-1β, IL-4, IL-6, IL-10, IFN-γ, TNF-α, fibrinogen, D-dimer and activity levels of procoagulant factors II, V, VII, VIII, IX, and X were measured. Endogenous anticoagulants antithrombin (ATIII) and protein C activity levels were also measured. All coagulation proteins were measured with a Stago Compact Coagulation Analyzer (Diagnostica Stago Inc., Parsippany, NJ) following standard manufacturer protocols [4]. All inflammatory markers were measured using commercial enzyme-linked immunoassays (ELISAs).

Longitudinal statistical analyses with generalized estimating equations (GEE) were used to evaluate the effects of randomized group, time, and the interactions between group and time on changes in the measured coagulation and inflammatory markers over time. All statistics were performed with STATA (v12.1, College Station, TX).

RESULTS

A total of 107 patients were enrolled in the EWB study, 81 which adhered to "per-protocol" rules and we had complete blood samples and data from 59 patients, 34 in the COMP group and 25 in the mWB group. Patients were predominantly male with a median age of 40 years, with no significant demographic differences between groups. Patients receiving mWB and COMP transfusions appeared to uniformly display a drop in clot strength as measured by the TEG assay. TEG parameters (e.g. α-angle, MA, and G) dropped during the first 3-6 hours followed by a gradual rebound over the next 12-24 hrs. While there was a trend for greater improvement in the rate of clot formation and clot strength over the first 24 hours in the mWB group compared to those receiving COMP therapy, group effects were not statistically significant. Patients receiving mWB had improved thrombin generating potential over the first 24 hours. Fibrinogen levels at t=24hrs was significantly higher in the mWB group. Platelet function declined over time in both mWB and COMP groups along with platelet count; however, platelet function was significantly better in the mWB group particularly at t=12 hrs based on aggregometry results (Figure 1). Platelet aggregation in response to collagen and ristocetin were significantly higher in the mWB group at t=12 hrs (Figure 1). There were no distinguishing effects on the inflammatory response between the two groups (Figure 1).

The longitudinal GEE statistical model revealed significant group-time interaction effects, suggesting that timing and volume of blood product transfusions have an effect on the coagulation and inflammatory response. In particular, we observed that early administration of platelet transfusions was associated with improvements in coagulation parameters.

DISCUSSION

In this paper, we evaluated the effects of time and transfusion types and volumes on inflammatory and coagulation markers in a previously published randomized control pilot trial of severely injured trauma patients receiving mWB or COMP therapy.

We observed significant interactive group-time effects between mWB and COMP, with evidence suggesting that mWB may be superior to COMP in improving the patient's coagulation status within 12-24hrs of hospital admission. Despite observing a general decline in platelet count, similar to that reported by Kutcher et al. [5], platelet function in the mWB group was significantly better compared to the COMP group. These results further emphasize the importance of platelet function rather than platelet count in severely injured trauma patients. The inflammatory response was not remarkably different between groups.

The results from this study illustrate the highly dynamic nature of trauma-related hemorrhage and resuscitation. The interactions between RBC, plasma and platelet transfusions over time, undoubtedly, play a critical role in the patient's inflammatory and coagulation status. Further investigation is needed to understand this dynamic interaction to improve resuscitation outcomes, possibly using multi-scale modeling approaches.

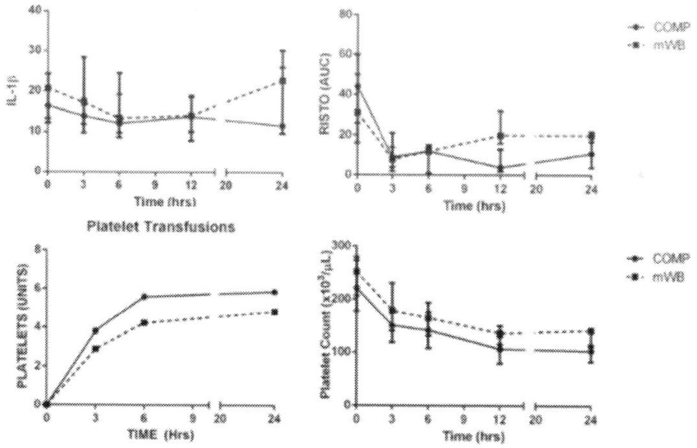

Figure 1. Inflammatory and coagulation response to trauma and resuscitation. Solid lines represent the COMP group, Dotted lines represent the mWB group. Median values with IQR are illustrated. Despite receiving platelet transfusions, platelet count dropped over time in both groups.

ACKNOWLEDGEMENTS

This study was funded by the Department of Defense; grants# W81XWH-08-C-0712 and W81XWH-07-1-0229.

Disclaimers: The views and opinions expressed in this manuscript are those of the authors and do not reflect the official policy or position of the Department of Defense, or the United States Government.

REFERENCES

[1] Murdock, A.D., et al., *Shock*, 41 Suppl 1: 62-9, 2014.

[2] del Junco, D.J., et al., *J Trauma Acute Care Surg*, 75 (1 Suppl 1): S24-30, 2013.

[3] Cotton, B.A., et al., *Ann Surg*, 258(4): 527-33, 2013.

[4] Kutcher, M.E., et al., *Ann Surg*, 260(6): 1103-11, 2014.

[5] Kutcher, M.E., et al., *J Trauma Acute Care Surg*,73(1): 13-9, 2012.

SB³C2015
Summer Biomechanics, Bioengineering and Biotransport Conference
June 17-20, 2015, Snowbird Resort, Utah, USA
SB³C2015-580

A COMPUTATIONAL GROWTH AND REMODELLING APPROACH TO STUDY ADAPTATION AND TORTUOSITY IN A BUCKLED ARTERY

Mehdi Farsad (1), Qin Liu (2), Hai-Chao Han (2), Seungik Baek (1)

(1) Department of Mechanical Engineering
Michigan State University
East Lansing, MI, USA

(2) Department of Mechanical Engineering
University of Texas at San Antonio
San Antonio, TX, USA

INTRODUCTION

In 2007, Han proposed a mathematical model that explains the physiological conditions of an artery, such as elevated blood pressure and reduced axial strain, causing buckling that is losing the stability of the artery [1]. The presented theoretical and numerical models have been tested and validated by *in vitro* experiments. Although the buckled artery may be different from the tortuous one, artery buckling finds much similarity with artery tortuosity: buckled arteries appear tortuous in shape; buckling occurs due to high blood pressure, reduced axial strain, and/or weakened wall, that all are associated with artery tortuosity. Hence, we propose testing multiple mechanisms that lead from an artery buckling to artery tortuosity.

The objective of this study was to simulate the process of adaptation in an artery from the buckled state using the computational G&R framework. We propose that the changes in the artery's wall thickness and stress are uneven because of the buckled geometry of the artery. Furthermore, the kinetic parameter in the G&R model that controls the sensitivity of G&R to the changes in the wall stress is studied to find a reasonable range for it. Finally, we illustrate that the tortuosity index increases in the artery as the result of a periodic stimulation (stretching, for instance).

METHODS

A G&R computational model developed in [2] was employed to model a buckled artery. The buckled artery's geometry is adopted from an experimental study performed in [3]. The artery's initial radius and wall thickness are $R=0.4\ mm$ and $h=0.08\ mm$, respectively. In addition, the blood pressure and the buckled artery's length are $P=21\ kPa$ and $L=15\ mm$, respectively. In the G&R model, it is assumed that the artery's wall consists of three constituents: elastin, smooth muscle cells, and four families of collagen fibers (anisotropically distributed in the artery's wall). The content of smooth muscle cells and collagen fibers can change during the process of G&R by a stress mediated turnover procedure. The formation of constituent i is given by

$$m_R^i(t)=\frac{M_R^i(t)}{M_R^i(0)}m_b^i\left[K^i\left(\frac{\sigma^i(t)}{\sigma_h^i}-1\right)+1\right]$$

where K^i and $m^i{}_b$ are sensitivity parameter and basal rate of production, respectively, and $\sigma^i(t)$ and $\sigma^i{}_k$ are the scalar function of stress and the homeostatic stress for constituent i, respectively. In addition, the survival function $q^i(\tau,t)$ is given by

$$q^i(\tau,t)=\begin{cases}\exp\left(-\int_\tau^t \kappa^i(s)\,ds\right) & t-\tau\le a_{max}^i\\ 0 & t-\tau> a_{max}^i\end{cases}$$

where K^i is a function of mechanical contribution of artery's wall such as wall shear stress or circumferential stress accounting for the rate of degradation of constituent i. Also, $a^i{}_{max}$ represents the maximum time that constituent i may survive. Accordingly, the strain energy per unit volume in the reference configuration is assumed to be

$$W^i(t)=M_R^i(0)Q^i(t)\Psi^i(0)+\int_0^t m_R^i(\tau)q^i(\tau,t)\Psi^i(\tau)\,d\tau$$

where $\Psi^i(\tau)$ is energy per unit mass of constituent i at time τ.

All the simulations start from an already buckled geometry using 2002 triangular elements in a FEM based framework. The buckled artery is left for 300 days without any changes. Then 5% increase in axial stretch or in lumen pressure is gradually applied to the artery during 30 days, and the artery is left again for another 300 days for adaptation. The artery's wall thickness and circumferential stress at the middle of the inner and outer curves are studied at each time point. In addition, the changes in the artery's center line's tortuosity index (*TI*) is studied during time such that

$$TI = \frac{\text{length of center line}}{\text{length of center line in straight condition}}$$

RESULTS

Fig. 1 shows the changes of wall thickness and circumferential stress in the buckled artery's wall as the result of *5%* stretching. The figure also shows that for different values of the sensitivity parameter (K) the artery's wall adaption procedure is different. For instance, there is no significant adaptation for $K=0.05$, while the artery's wall adapts with the stimulation very fast for $K=1.0$. As will be discussed in Discussion, the value $K=0.5$ seems to be more realistic and, thus, is chosen for the rest of results.

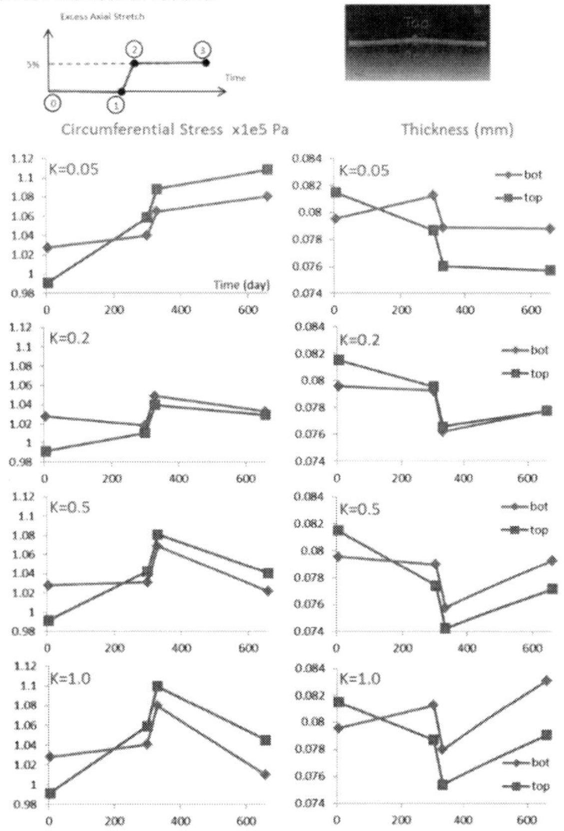

Fig. 1: The changes of the buckled artery's wall thickness and circumferential stress during one period of stimulation for different kinematic parameter.

Furthermore, Fig. 2 proposes the changes of wall thickness and circumferential stress in the same artery, as mentioned above, but as the result of 5% increase in the lumen pressure for $K=0.5$. It shows that the adaptation process is different for the inner and outer curves due to their difference in the initial wall stress and wall thickness (calculated right after buckling).

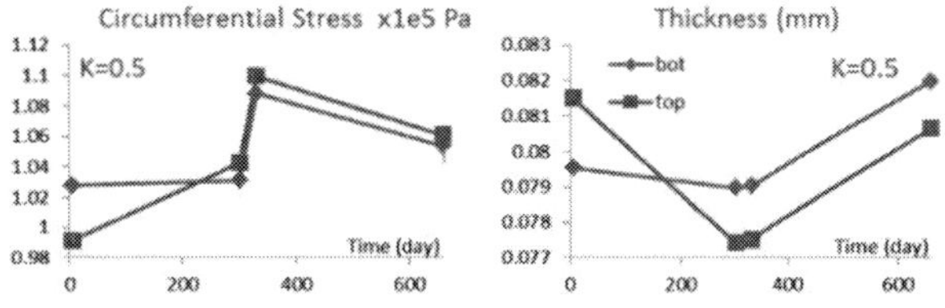

Fig. 2- Adaption process in the buckled artery due to one period of change in the lumen pressure.

Finally, Fig. 3 depicts the changes of center line tortuosity index (TI) in a buckled artery as the result of periodic stimulations (stretching or changing the lumen pressure). It is shown in the figure that TI increases after each period of applying and omitting the stimulation.

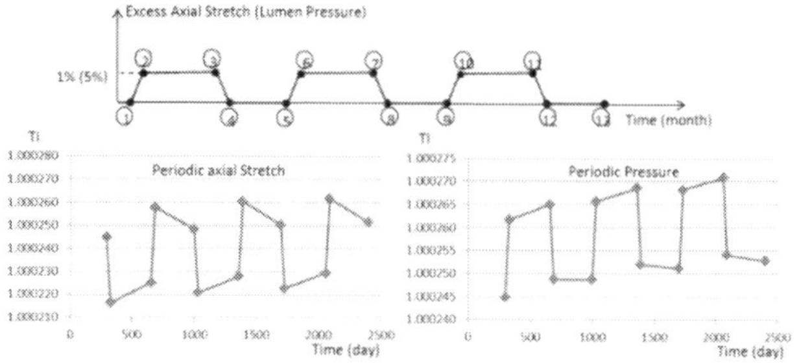

Fig3. Changes in the buckled artery's tortuosity due to periodic stimulations.

DISCUSSION

The distribution of stress is uneven in a buckled artery due to its bent geometry that leads to uneven stress/strain mediated G&R. Our simulations show that initially the wall circumferential stress in both the inner and outer curves is less than the homeostatic value of *105 kPa*. An *in vitro* experimental study on rat's carotid artery [3] suggested that stress decreases in the inner curve after buckling, but increases in the outer curve due to its higher stretch after buckling. After using the parameter estimation, we propose the sensitivity parameter K chosen around $K=[0.2\text{-}0.5]$. It is expected that the artery's wall thickness decreases during stretching and increase during releasing the stretch due the effect of *Poisson*'s ratio. This process results in elevation in the amount of wall stress. Thus, the artery's wall gets thinner by degradation of its collagen fiber and smooth muscle content in order to decrease its elevated stress. Fig. 1 shows that it is only $K=0.5$ that satisfies the expectations.

Fig. 2 shows that the tortuosity index slightly increases as the result of periodic loading and unloading in the buckled artery. In spite of buckling, which is returnable, tortuosity gradually increases artery's curvature, which may be associated with many vascular diseases in the vascular system. It is also hypothesized in [3] that buckling may be a starting point for tortuosity. Furthermore, previous research has proposed that cyclic changes in lumen pressure or cyclic stretch in the arteries (for example in cyclers) may result in higher level of tortuosity in body's arteries [4]. Our numerical simulations seem to be in-line with above research results.

While illustrating the process of adaption and the increase in the tortuosity of a buckled artery, we only use a sensitivity parameter to the wall stress using the *in intro* experiments. Thus, the future study will use the G&R simulation of arterial adaptation as a function of blood flow and wall stress such that the results are quantitatively compared with *in vivo* measurement of variables (e.g., tortuosity index, flow, pressure) as well as changes of collagen content, elastin content, number of cells and activated MMPs using histology image analysis.

ACKNOWLEDGEMENTS

Authors acknowledge the NIH under R01HL095852 and R01HL115185 and the NSF under CAREER CMMI-1150376.

REFERENCES

[1] Han, H.C., *J Biomech*, 40(16):3672-8, 2007.
[2] Zeinali-Davarani, et al., *Com Meth Biomech*, 14(9):803-817, 2011.
[3] Zhang, J., et al., *Ann Biomed Eng*, 42(8):1658-1667, 2014.
[4] Van Epps, J.S., et al., *J Biomech Eng*, 131:101005-1-11, 2009.

OPEN KNEE(S): MAGNETIC RESONANCE IMAGING FOR SPECIMEN-SPECIFIC NEXT GENERATION KNEE MODELS

C.J. Bennetts (1,2), S. Chokhandre (1,2), S.B. Donnola (3), C.A. Flask (3,4,5,6),
T.F. Bonner (1,7), R.W. Colbrunn (1,7), A. Erdemir (1,2)

(1) Department of Biomedical Engineering
(2) Computational Biomodeling (CoBi) Core
(7) BioRobotics and Mechanical Testing Core
Lerner Research Institute
Cleveland Clinic
Cleveland, Ohio, USA

(3) Department of Radiology
(4) Department of Biomedical Engineering
(5) Department of Pediatrics
(6) Case Center for Imaging Research
Case Western Reserve University
Cleveland, Ohio, USA

INTRODUCTION

Magnetic resonance (MR) imaging is a modality that is ideally suited for computational modeling of musculoskeletal joints due to its ability to non-invasively represent the three-dimensional morphology of various osseous and soft tissue structures. When using cadaver specimens, the joints are left intact allowing mechanical testing after imaging for experimental validation. Therefore, it is not surprising that magnetic resonance imaging (MRI) has been commonly used in many prior knee joint modeling and simulation studies. The knee contains several different soft tissue structures that are essential to represent anatomically and mechanically in order to accurately understand knee biomechanics using finite element (FE) analysis. These structures, including bone, cartilage, tendons, ligaments and menisci, have varying molecular compositions that influence their contrast in MR imaging. MRI is a complex imaging modality, which must be properly tuned in order to adequately identify these various tissues of interest. Currently, optimal MRI protocols for knee modeling studies have not been thoroughly evaluated. The most comprehensive prior knee modeling studies [1,2] have typically included limited sets of MR scanning settings, most likely due to convenience or other constraints, which may compromise the appropriate acquisition and representation of tissue morphology, and therefore the overall biomechanical response of the knee joint as predicted by computational modeling.

This study is part of a more comprehensive project, called Open Knee(s), which aims to generate a set of fully specimen-specific knee models, including morphology, tissue-specific material properties, and experimental joint kinematics-kinetics responses, for a sample of male and female, young and elderly, and healthy and osteoarthritic cadaveric samples. The specific goal of this study was to identify and utilize a set of MR imaging protocols on Open Knee(s) specimens that are better suited for the development of the next generation of knee models.

METHODS

The Open Knee(s) project aims to include a minimum of ten knee specimens in order to describe a cross-sectional sample of the population, consisting of male and female, young (18-35 years) with healthy cartilage, elderly (65-80 years) with healthy cartilage, and elderly with osteoarthritic cartilage (moderate, based on Kellgren-Lawrence grading [3,4]). The other requirements of the individuals from which specimens were obtained include: height (1.5-1.8 m), weight (45-90 kg), and Body Mass Index (BMI) (18.5-24.9 considered the normal range). In the current work, five specimen were imaged; four matching the aforementioned specifications and another with an age of 46 years.

The hard and soft tissue structures that are necessary to incorporate in an FE model in order to accurately simulate the mechanical response of the knee joint include the bones (femur, tibia, patella), their cartilage, the quadriceps tendon, major ligaments and menisci. The underlying shape of the bones provides the contours through which loads on the joint are transferred and distributed through contact between the articulating surfaces of the cartilage in the tibiofemoral and patellofemoral joints of the knee. The cartilage exhibits a multiphasic, rate-dependent mechanical response governed by interactions between its solid, fluid, and ionic phases, which act to resist, absorb, and distribute joint loads and reduces friction within the joint. The ligaments (anterior and posterior cruciate, medial and lateral collateral, and patellar) provide joint stability through tensile engagement and act to limit the range of motion for different degrees of freedom within the joint. The menisci conform to the opposing surfaces of the articular cartilage of the femur and tibia to provide stability during knee flexion. Geometric reconstruction of all these tissues is therefore important to develop mechanically authentic models. The tissue types of interest have varying molecular compositions, each with distinct electromagnetic resonance signatures that influence the strength of the signal intensity with respect to the particular MRI settings. Therefore, different MR scanning protocols are required to best delineate the boundaries of each tissue type within the knee.

Initial MR settings to highlight each tissue type were obtained from a study through the osteoarthitis initiative (OAI), sponsored by the National Institutes of Health (NIH) [5]. A series of imaging trials were undertaken to refine these initial protocols to best delineate the desired knee structures. The goals in tuning the MRI parameters were to increase resolution and tissue contrast, to accommodate a total image acquisition time of less than 2 hours, and to adapt to the particular MRI machine being used. A resulting set of three different MRI protocols were developed with the intention to identify: 1) all structures simultaneously for general-purpose reference, 2) cartilage and possibly menisci, and 3) ligaments, tendons, and possibly menisci. The general-purpose imaging protocol was a 3D T1-weighted sequence *without* fat suppression having an isotropic resolution of 0.5 mm, echo time (TE) of 6.01 ms, and repetition time (TR) of 20 ms. The cartilage imaging protocol was a 3D T1-weighted sequence *with* fat suppression having an anisotropic resolution of 0.35×0.35×0.7 mm, TE of 5.34 ms, and TR of 29 ms. The ligament and tendon imaging protocol was a proton-density, turbo spin echo sequence with an anisotropic resolution of 0.35×0.35×2.8 mm, TE of 9.7 ms, and TR of 10,000 ms. The ligament and tendon imaging protocol was acquired in three orientations, with the high-resolution in-plane dimension lying in the axial, sagittal and coronal planes of the knee. The reference coordinate systems remained the same for all three scanning protocols, which will allow the components of the knee model to be defined from the associated image sequence and easily recombined. Knee specimens were transported and imaged in a custom designed, cylindrical acrylic tube, supported along the axis with foam discs with tension applied to the quadriceps tendon to ensure it was lightly engaged (not slack). MR images were acquired using a Siemens Magnetom Skyra 3 Tesla MRI machine in the Department of Radiology at the Case Western Reserve University, Cleveland, Ohio.

Another aspect of the imaging procedure that was necessary for modeling and simulation involves appropriately orienting the knee joint geometry, derived from the MR images, with respect to the follow-up experimentation on joint kinematics-kinetics [6]. To accomplish this, three spherical, MRI-opaque markers were rigidly fixed to the femur and tibia. For the patella, a custom 3-D printed fixture (with three spherical, MRI-opaque markers) was fixed to the bone. The centers of these spheres can be calculated from surface coordinates of the spherical geometry from MRI, and from mechanical testing, to accurately align imaging and mechanical testing coordinate systems.

RESULTS

Figure 1 shows a mid-sagittal slice from the isotropic general-purpose imaging sequence for the currently scanned set of knee specimens, highlighting the variations in morphology that exist in our sample population.

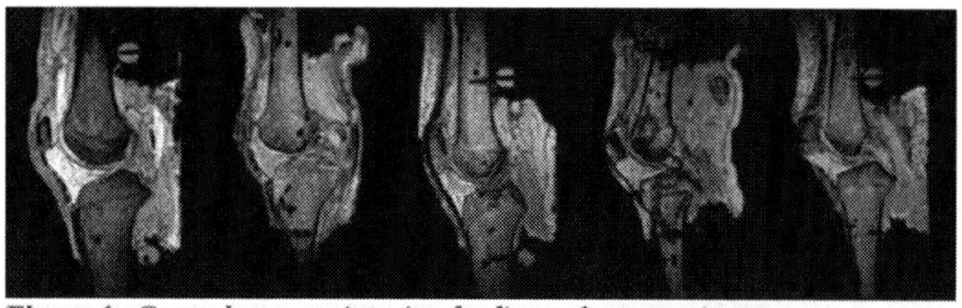

Figure 1. General purpose imaging for five cadaver specimens.

To illustrate the potential role that the selected MR protocols have on delineating the different soft tissue structures within the knee, Figure 2 includes the complete set of scans for a single specimen, identifying the corresponding structures of interest for which the particular sequence was performed. Figure 2 also shows the boundary of a spherical, MRI-opaque markers as it appears in the isotropic, general purpose MR image.

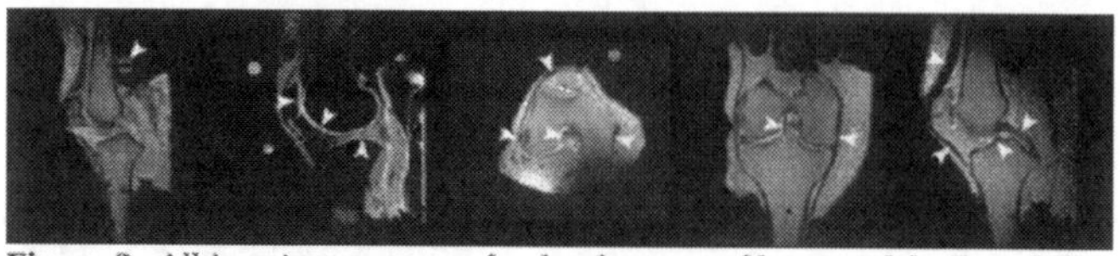

Figure 2. All imaging sequences for development of knee models. From left to right: Panel 1 – general purpose imaging; registration marker is highlighted. Panel 2 – cartilage imaging; femoral, tibial, and patellar cartilage are highlighted. Panels 3-5 – ligament and tendon imaging; cruciate, collateral, and patellar ligaments are identified along with their insertions.

DISCUSSION

A comprehensive set of MRI protocols were devised in order to represent various structures of the knee with particular emphasis on their incorporation in FE models. The imaging data acquired using this comprehensive set of refined MR protocols will likely allow a more accurate morphological representation of the associated soft-tissue structures, thereby reducing the potential error in the biomechanical response of their prospective models.

Ideally, the higher the resolution of the MR images, the more accurately the geometric boundaries of the tissue structures can be represented. Given the MR scanner and protocols that were used, the resolutions of the images obtained were as high as possible. However, using other MR scanners or through advances in MR imaging technology, higher fidelity geometric models may be possible.

The presented specimens have already undergone joint kinematics-kinetics testing [6] and tissue characterization is in progress. In the near future, the remainder of the Open Knee(s) specimens will be imaged and experimentally tested. Once image data has been acquired for all specimens, the associated knee models will be generated, which includes segmenting the tissue structures of interest from the MR images to identify their boundaries, and then meshing this geometry to determine an adequate mesh density to ensure mechanical convergence. Once the finite element models are obtained, they can be combined with experimental data describing the relative movements of the joints [6], resulting from various conditions prescribed during experimental testing, in order to evaluate the predictive capacity of the models. This significant amount of heterogeneous data is acquired openly and will be made available to the community through the Open Knee(s) project. Utilization of these data to develop and extend the next generation of fully specimen-specific models will provide unique opportunities not only for us, but for others as well, for yet-unknown applications in the study of healthy and osteoarthritic knee joint function.

ACKNOWLEDGEMENTS

This study was funded by NIGMS, NIH (R01GM104139, PI: Erdemir). Advice by Carl Winalski, MD on MRI is appreciated. Open Knee(s) is an open development modeling project; specifications, models, and data can be accessed at http://wiki.simtk.org/openknee.

REFERENCES

[1] Mootanah, R et al., *CMBBE*, 17:13, 1502-17, 2014.
[2] Kiapour, A et al., *J Biomech Eng*, 011002-1-14, 2014.
[3] Berger, MJ et al., *Arthritis Care Res*, 64(5):729-34, 2012.
[4] Braun, HJ et al., *Bone*, 51(2):278-88, 2012.
[5] Peterfy, CG et al., *OA & Cart*, 16:1433-41, 2008.
[6] Bonner, TF et al., *SB^{3}C, 2015.*

SB³C2015
Summer Biomechanics, Bioengineering and Biotransport Conference
June 17-20, 2015,Snowbird Resort, Utah, USA
SB³C2015-582

VALVELESS FLOW IN A THICK ELASTIC TUBE

Pavel Kozlovsky (1), Moshe Rosenfeld (2), Ariel Jaffa (3), David Elad (1)

(1) Department of Biomedical Engineering
Tel Aviv University
Tel Aviv, Israel

(2) School of Mechanical Engineering
Tel Aviv University
Tel Aviv, Israel

(3) Ultrasound in Obstetrics and Gynecology
Lis Maternity Hospital, Tel Aviv Medical Center
Tel Aviv, Israel

INTRODUCTION

The phrase "valveless pumping" is used to define periodic unidirectional net flow per excitation period (V_{cyc}) in a fluid-filled compliant tube, which is subjected to periodic external excitation forces at an asymmetric longitudinal location along its axis. In the past two decades many experimental and numerical studies of valveless pumping with open or closed systems were conducted as reviewed in some recent publications [1, 2]. All previous studies assumed a uniform cylindrical compliant tube, and in most cases it was a thin-walled tube (i.e., wall thickness was of the order of 10% or less of the inner diameter). However, biological conduits in which valveless pumping are naturally performed are not uniform tubes and have a relatively thick wall, for example, the tubular embryonic heart that delivers blood to the embryo body [3].

In order to allow for a comprehensive and physically realistic investigation of valveless pumping, we developed a non-uniform thick wall elastic tube model with varying cross sectional area, including local cavities and tapering, that is subjected to pressure excitations at a longitudinally asymmetric location. This work presents a comprehensive dimensionless analysis of the functional performance of the valveless pump for a wide range of system and excitation parameters.

METHODS

The physical domain of the valveless pumping through a fluid-filled thick-wall elastic tube is depicted in Fig. 1. The fluid flow was governed by the following one-dimensional (1D) dimensionless continuity and Navier-Stokes equations:

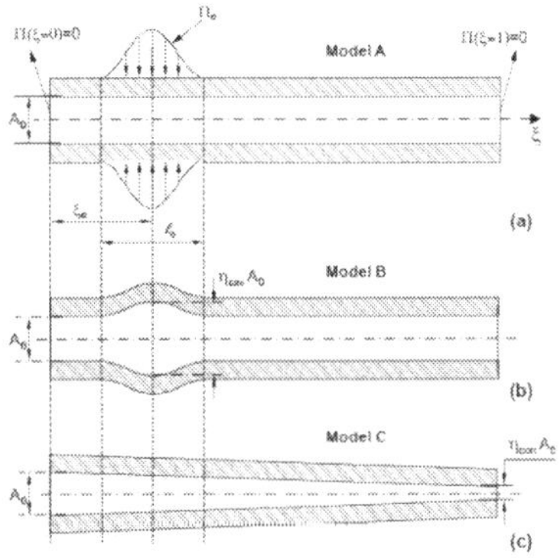

Figure 1. ζ_e and ℓ_e are the location and the width of the external pressure load, respectively; A_0 is the unstressed inner cross-section of the tube at $\xi=0$. **(a)** Uniform tube; **(b)** Tube with sinus-like cavity: $\eta_{cav}=A_0(\xi=\xi_e)/A_0(\xi=0)=$ $=1.05,\ 1.10,\ 1.15,\ 1.20$; **(c)** Tapered tube: $\eta_{con}=$ $=A_0(\xi=1)/\ A_0(\xi=0)=0.95,$ $0.90,\ 0.85,\ 0.80.$

$$\frac{\partial \mathbf{G}}{\partial \tau}+\frac{\partial \mathbf{H}}{\partial \xi}+\mathbf{B}=0 \tag{1}$$

$$\mathbf{G}=\begin{pmatrix}\alpha A_0\\ u\end{pmatrix}\qquad \mathbf{H}=\begin{pmatrix}u\alpha A_0\\ \dfrac{u^2}{2}+\dfrac{K_p}{\rho c_0^2}\left(\Pi+\Pi_e\right)\end{pmatrix}\qquad \mathbf{B}=\begin{pmatrix}0\\ f_L\dfrac{u}{\alpha}\end{pmatrix} \tag{2}$$

where τ and ξ are dimensionless time and axial coordinate, u is dimensionless axial velocity, Π and Π_e are dimensionless internal and external pressures, f_L is a dimensionless friction parameter, K_p is a flexural rigidity of the tube, $\alpha=A/A_0$ is dimensionless cross sectional area, A and A_0 are the varying and unstressed inner dimensional cross sectional areas, ρ is fluid's density, c_0 is the characteristic velocity.

The mechanical performance of the tube was governed by the following general tube law, which describes realistic collapsible patterns of the tube and also accounts for the physical effect of the wall thickness (Kozlovsky et al., 2014),

$$\Pi = m_0\left(\alpha^{m_1(\gamma)} - \alpha^{m_2(\gamma)}\right) + \beta_2 \frac{\partial\alpha}{\partial\tau} \qquad (3)$$

where m_0=3.1, $m_1=60/\gamma^{0.5}-65$, $m_2=0.7\gamma-0.4$ are tube law constants. $\gamma=h/R_0$ is the wall thickness ratio, h is the wall thickness and R_0 is the inner unstressed radius of the tube and β_2 is numerical damping parameter.

The external excitation pressure applied at an asymmetric position along the tube axis (Fig. 1a) was chosen to be the following sinusoidal function to ensure continuity in the external pressure,

$$\Pi_e(\xi,\tau) = \begin{cases} \Pi_{max}\sin^2\left(\frac{\pi(\xi-\xi_e+0.5\cdot\ell_e)}{\ell_e}\right)\sin^2\left(\frac{Str}{2\cdot DC}\tau\right), & \xi_e-0.5\cdot\ell_e \le \xi \le \xi_e+0.5\cdot\ell_e \\ & k\cdot\tau_{cyc} \le \tau \le (k+DC)\cdot\tau_{cyc} \\ 0, & \text{otherwise} \end{cases} \qquad (4)$$

where Π_{max} is the external pressure magnitude, τ_{cyc} is the time period, Str is the Strouhal number, k is the cycle number and DC is the duty cycle.

The tube ends were kept at a constant zero pressure during the simulations, hence,

$$\Pi(\xi=0,\tau)=0$$
$$\Pi(\xi=1,\tau)=0 \qquad (5)$$

All simulations began at a state of rest and were continued for k=20 pumping cycles, when a complete steady periodic state was always achieved.

The set of governing equations (Eqs. 1-3) represents a system of hyperbolic partial differential equations and it was solved using the two-step predictor-corrector MacCormack numerical scheme.

RESULTS

A comprehensive sensitivity analysis was conducted for valveless pumping with the uniform tube (Fig. 2) in order to explore the dependence of the net outflow V_{cyc} on the model input parameters as compared to the reference net outflow, $V_{cyc,ref}$, obtained for the reference parameters. The Str number had a very prominent nonlinear effect on the net outflow volume. Generally, the V_{cyc} decreased with f_L for all the tested values of Str. The peak positive outflow (i.e., V_{cyc} > 0), which is defined as a flow in positive direction of ξ, was always obtained in the vicinity of the natural Strouhal number $Str_n \approx \pi$. For f_L < 1.28 and Str in the approximate range of 4.2 < Str < 4.8, the outflow reversed its direction and V_{cyc} became negative with the maximal negative outflow occurring in the vicinity of Str ≈ 4.5.

The effect of wall-thickness ratio on the outflow was examined by changing the wall thickness ratio γ from 0.20 to 0.50 (Fig. 2). The magnitude of both positive and negative outflows increased with γ.

In valveless pumping problem in a thin wall tube under resonance conditions that the passive part of the tube outside the excitation region acts as a pump, since the total work done by this passive portion of the tube is greater than zero during the excitation cycle [1, 4]. Similarly, we have found that the work done by the passive portion of the thick-wall valveless pump was positive near the natural Str number. Moreover, the positive pumping work increased with wall thickness ratio γ for all simulated f_L parameters. In the vicinity of Str ≈ 4.5, where maximal negative outflow occurred for low f_L cases, the pumping work was also positive and increased with γ.

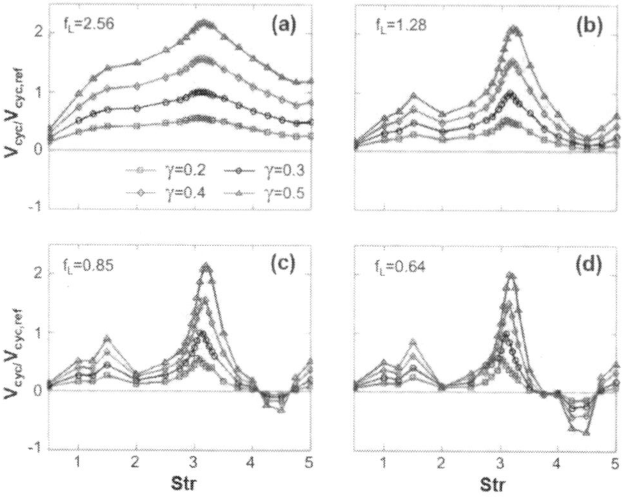

Figure 2. The effects of the wall thickness ratio γ and Strouhal number on the normalized volume of fluid displaced per excitation cycle, $V_{cyc}/V_{cyc,ref}$, for: **(a)** f_L=2.56; **(b)** f_L=1.28; **(c)** f_L=0.85; **(d)** f_L=0.64. The calculations were performed for uniform tubes. The results were normalized by the value of $V_{cyc,ref}$ obtained for the uniform tube with γ=0.30 and Str_n=3.1 for each f_L.

A similar analysis of the model with varying cavity diameters revealed an increase in the magnitude of both positive and negative outflows as the cavity diameter increased. The magnitude of the positive outflow decreased for tapered tubes. The reduction increased with decrease of the tapering ratio. However, the magnitude of the negative normalized outflow increased as the tapering ratio decreased.

DISCUSSION

The present study explored the role of non-uniformities in the tube geometry (e.g., local bulging and tapering) and wall thickness for a wide range of the friction parameter and Strouhal number.

Comprehensive sensitivity analysis of the models in this study revealed the following major characteristics of valveless pumping. Reversals of the outflow occurred only for relatively low f_L parameters. The sinus-like cavity located at an asymmetric longitudinal position, where external excitation pressure was applied, generated an additional suction effect that increased the maximal internal pressure during the excitation cycle. Accordingly, both positive and negative outflows increased with the dimensions of the cavity for all simulated f_L parameters. In tapered tubes, the magnitude of the positive outflow decreased for all f_L parameters as the tapering ratio decreased, due to higher resistance to fluid flow.

The most interesting finding of the study was the dependence of the outflow on the wall thickness ratio γ. We found that both positive and negative outflows increased with γ for all simulated f_L parameters, which means that between two valveless pumps consisting of two different elastic tubes with the same flexural rigidity K_p, the pump with a more flexible tube and thicker wall will produce a greater outflow due to its increased compliance during collapse.

REFERENCES

[1] Avrahami and Gharib, *J. Fluid. Mech.*, 608, 2008.
[2] Manopoulos et al., Phys. Fluids, 18: 017106, 2006.
[3] Forouhar et al., *Science*, 312: 751-753, 2006.
[4] Loumes et al., *Phys. Fluids*, 20: 023103, 2008.

SB³C2015
Summer Biomechanics, Bioengineering and Biotransport Conference
June 17-20, 2015,Snowbird Resort, Utah, USA
SB³C2015-583

MULTICENTER COMPARISON OF 4D PHASE CONTRAST MRI MEASUREMENT OF CEREBROSPINAL FLUID DYNAMICS IN THE CERVICAL SPINE

Suraj Thyagaraj (1), Daniel Giese (2), Francesco Santini (3), Eleonora Fornari (4), Alexander C. Bunck (2), Francis Loth (1), Bryn A. Martin (1)

(1) Conquer Chiari Research Center
Department of Mechanical Engineering
University of Akron
Akron, OH, USA

(2) Institute of Radiological Diagnostics
University of Cologne
Cologne, Germany

(3) Institute of Radiology
University of Basel
Basel, Switzerland

(4) Biomedical Imaging Center
Lausanne University Hospital
Lausanne, Switzerland

INTRODUCTION

Cerebrospinal fluid (CSF) hydrodynamics are thought to play a role in cranio-spinal disorders such as Chiari Malformation (CMI). Researchers have applied 2D phase-contrast MRI (2D PCMRI) to quantify alterations in CSF dynamics at the craniocervical junction in CMI. Results from these studies show that CSF alterations are highly spatially dependent. A detailed quantification of these alterations would necessitate acquisition of many 2D PCMRI slices with three-directional velocity encoding and require an impractical amount of MR imaging time [1]. Thus, researchers have applied 4D phase-contrast MRI (4D PCMRI) to assess these alterations in a clinically practical time frame (~10 minute acquisition). These measurements found significantly elevated peak CSF velocities in CMI patients versus controls and revealed details about CSF flow features such as vortical structures and flow jets that can be present near the craniocervical junction [2, 3]. Although these measurements show promise, the 4D PCMRI protocol remains under development by MR manufacturers and has not yet been tested for accuracy and inter- and intra-scanner reliability for detection of CSF flows (~10 to 100X smaller velocities than arterial). *The goal of this study is to investigate the reliability of 4D PCMRI assessment of CSF dynamics by construction and multi- MRI center testing of a subject-specific in vitro model of the upper cervical spine based on a patient with CMI.*

METHODS

A set of high-resolution T2-weighted MRI images (0.8 mm isotropic resolution), from a 5-year old CMI patient with 5.8 mm tonsillar descent, were manually segmented to create a geometric model of the upper cervical spine from the foramen magnum (FM) to ~5 cm caudal to the C7 vertebrae (**Figure 1**). Realistic 3D modeled spinal cord nerve roots, with ~1 mm thickness, were manually added to the model based on measurements in the literature [4]. The final model was printed by 3D stereolithography with 75μm resolution. A subject specific CSF flow waveform was measured from 4D PCMRI [3]. A pulsatile computer controlled pump was constructed and

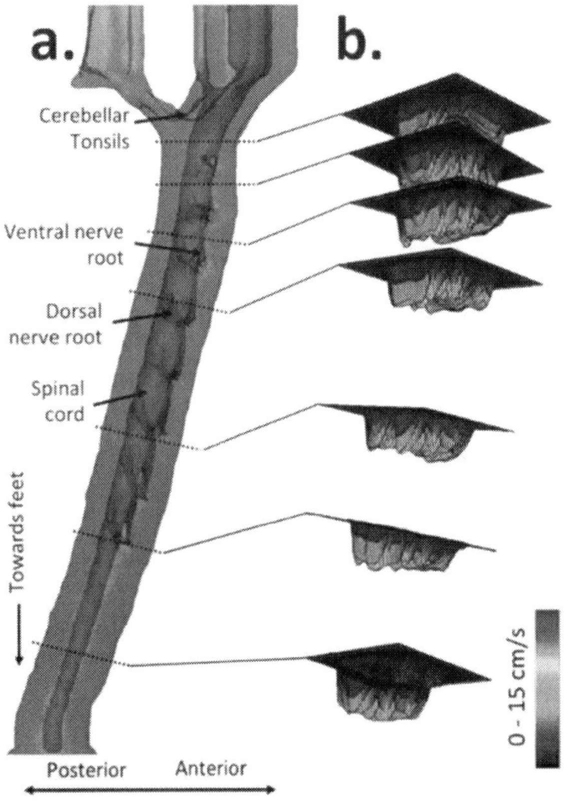

Figure 1: a) 3D reconstruction of the CSF space where flow was analyzed in this study. b) Example of 4D PCMRI measurement of CSF velocity profiles along the model axis.

programmed to deliver the CSF waveform to the model via a 25 ft length of rigid plastic tubing during the MRI experiments. The flow pump waveform was recorded by a quadrature encoder.

A consistent 4D PCMRI protocol was used to collect images, three times at four locations: 1) University Hospital, Cologne, Germany, – Philips, Achieva, 2) University of Basel, Switzerland – Siemens, Prisma, 3) CHUV, Lausanne, Switzerland – Siemens, Skyra) 4) CHUV, Lausanne, Switzerland – Siemens, Prisma. In brief, 4D PCMR images were collected with 3D velocity encoding (VENC = 15 cm/s), prospective gating, 16 phases per cardiac cycle with 30 ms time-steps, TR=7.5 s, TE=4.6 s, FA=5°, 1.5 mm isotropic resolution. Total imaging time was ~10 minutes (clinically oriented protocol). 4D PCMRI post-processing was completed using GTFlow software (Version 2.2.4, Gyrotools Inc, Switzerland). MR images were processed to determine overall velocity patterns, peak and means at C4. For visualization of velocity contours, three-directional velocity information was exported from GTFlow and re-constructed in Matlab (Version R2013b, Mathworks Corp, Natick, MA) using a custom code. CSF flow waveforms were measured by integrating pixel velocities across the area of the contour.

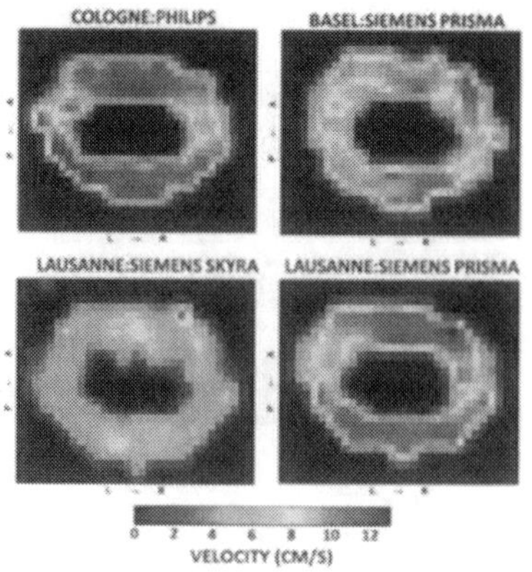

Figure 2: Peak Systolic Velocity Contours from 4D PCMRI images at the four scanning centers (reconstructed using Matlab)

RESULTS

Qualitative comparison between the 4D PCMRI measurements obtained at the four centers at the C4 vertebral junction showed markedly different velocity profiles on visual inspection (**Figure 2**). These differences were most pronounced on the images obtained from the Siemens, Skyra at Lausanne (Figure 4). The exact location of peak CSF velocity in the profile varied to a great degree for each center; albeit, all centers showed lower CSF velocities on each side of the spinal cord where nerve roots were present. In addition, regions of CSF flow jets and vortices were noticeable for images at all centers.

Quantitative comparison of peak systolic velocities had a mean value of 11.6 cm/s with an 8% standard deviation between the four centers at C4. The spatially averaged CSF velocity at peak systole had a mean of 3.5 cm/s with a standard deviation of 13%. Pulsatile flow waveforms were similar in magnitude and shape with the exception of Lausanne, Prisma (**Figure 3**). These waveforms were ~30% greater amplitude than the waveform measured at the flow pump by the quadrature encoder. The maximum difference in pump encoder position repeatability between experiments was <5%.

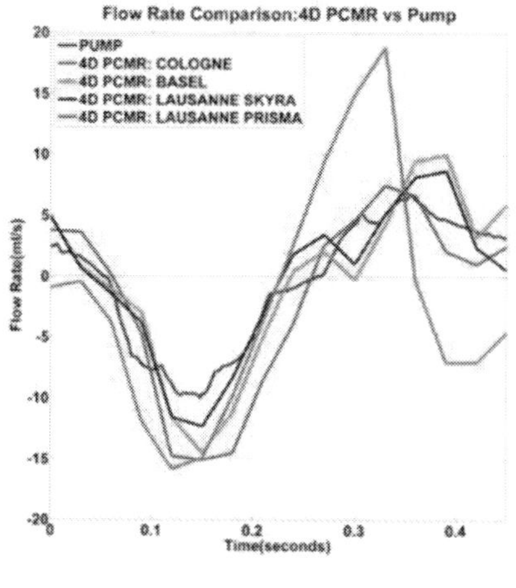

Figure 3: Flow rate measured from 4D PCMRI images at the different centers and comparison to pump measured flow

DISCUSSION

The constructed in vitro model of the upper cervical spine was useful to help understand the reliability of 4D PCMRI measurement of CSF velocities. This model was transported via airplane to multiple countries. Preliminary analysis at the C4 level shows notable quantitative and qualitative differences in CSF velocity assessment among the centers. However, an important factor for clinical use of 4D PCMRI is the level of measurement error *in comparison* to the differences that need to be detected in the CSF velocities in disease states compared to controls (e.g. CMI). Studies have shown that CSF velocities in CMI are elevated by ~100 to 350% [2]. Thus, although the measurement variability in the present study is significant, the results do not necessarily make void the possible utility of 4D PCMRI for clinical use.

Limitations. The presented results were assessed at one axial location along the spine; more locations are needed for comparison. The flow pump was connected to the model by a length of rigid tubing. However, the flow waveform may have been altered to some degree due to unavoidable tubing compliance. Our approach was to use identical tubing for all experiments and carefully tape the tubing to structures to limit any motion. In addition, tiny bubbles may have been present in the tubing length during the experiments that altered the flow waveform. To help mitigate bubbles, water used for the experiments was degassed in open air prior to conducting experiments and any noticeable bubbles were removed from the model.

ACKNOWLEDGEMENTS

This project is funded by a grant from the American Syringomyelia and Chiari Alliance Project and the Swiss National Science Foundation, International Co-operation Short Visit Award (Grant No. IZK0Z2_152766). Thanks to Soroush Pahlavian, Theresia Yiallorou, Morteza Vatani, and Dr. Jae-Won Choi for model construction assistance. The authors also thank Gerard Crelier, GyroTools LLC for support with visualization and analysis of the 4D PCMRI data presented in this study.

REFERENCES

[1] Pahlavian S.H., Martin B.A. et al., *JBME*, In submission, 2015.
[2] Bunck, A.C., et al., *Eur Radiol*, 21(8): p. 1788-1796, 2011.
[3] Yiallourou, T.I. et al.,. *PLoS One*, 7(12), 2012.
[4] Pahlavian S.H., Martin B.A. et al., *PLoS One*, 9(4), 2014.

SB^{3}C2015
Summer Biomechanics, Bioengineering and Biotransport Conference
June 17–20, 2015, Snowbird Resort, Utah, USA
SB^{3}C2015-584

INSIGHTS INTO THE HEMODYNAMIC FACTORS AFFECTING EMBOLUS TRANSPORT FOR STROKE

Debanjan Mukherjee and Shawn C. Shadden

Department of Mechanical Engineering
University of California, Berkeley
Berkeley, CA, USA

INTRODUCTION

Stroke constitutes one of the leading causes of death and disability worldwide - with an estimated 795,000 people experiencing a new or recurrent stroke in the United States every year. Roughly 1/3rd of all stroke occurrences are due to an embolic particle - a small agglomerate of cellular material - traveling to a cerebral artery and blocking blood flow to the brain. The hemodynamic factors governing the transport of these embolus particles are not yet well understood, despite its importance in stroke etiology. Additionally, extending computational methods to understand the physics of particle-laden hemodynamics in a patient-specific setting is a valuable, yet challenging, task.

Experimental studies, using either in-vitro models of vasculature segments or idealized bifurcation geometries, have mainly focused on quantifying the distribution ratio of particles across various branch vessels in comparison with the volumetric blood flow into these vessels. Most data reflect a size-dependent bias of particle transport into a branching vessel that volumetric flow division alone cannot capture.

With the recent, rapid advances in medical imaging and computational technology, patient-specific computational fluid dynamics (CFD) has gained prominence as an effective probe into arterial hemodynamics. Detailed fluid velocity data obtained from CFD simulations, along with simulations of embolic particle trajectories, may provide critical information regarding the mechanics of embolus interactions with blood flow and vessel walls. Simulation studies on embolus transport are relatively fewer and recent. They have elucidated important particle size dependent biases in cerebral vasculature, as well as a consistent preference of mid-sized particles to migrate to the brain [1]. However, the role played by the mechanics of particle interactions with fully three-dimensional flow is still to be investigated. For particle trajectory simulations, incorporation of a fully-coupled fluid-particle interaction is most accurate, but involves substantial computation costs when compared to the more common one-way coupling. Treatment of these aspects in cardiovascular simulations is limited. The objective of this research is to present an image-based CFD simulation framework to better understand the mechanics of embolus interactions with hemodynamics.

METHODS

A computer model of a carotid bifurcation was reconstructed from a CT angiography. This was discretized into a computational mesh comprised of an unstructured grid of tetrahedral elements. Blood is assumed to be a Newtonian fluid with density 1025 kg/m^3, and viscosity 0.004 Pa.s. A finite volume based numerical scheme was employed to solve the continuity and Navier-Stokes equations, and compute the velocity and pressure fields. A pulsatile, analytical, volumetric waveform was specified at the inlet. Parameters for the profile are chosen to match volumetric data in the literature. The outlet boundary conditions were set using resistances whose values were computed by assuming a flow division across branching vessels in proportion to their cross-sectional area, and a mean brachial pressure of about 93 mm/Hg.

Individual emboli were modeled as spherical particles with a nominal density of 1100.0 kg/m^3. The particle motion was modeled using a modified form of the Maxey-Riley equation[2]

$$m_p \frac{d\mathbf{v}_p}{dt} = \mathbf{F}_p = \frac{1}{2}\rho_f C_D \left(\frac{\pi D_p^2}{4}\right) |\mathbf{u} - \mathbf{v}_p| (\mathbf{u} - \mathbf{v}_p) + \mathcal{V}_p (-\nabla p + \nabla \cdot \tau) + \frac{C_a \rho_f \mathcal{V}_p}{2} \left[\frac{D\mathbf{u}}{Dt} - \frac{d\mathbf{v_p}}{dt}\right] + \mathcal{I}(\mathcal{P}, \mathcal{W}) \mathbf{F}_{contact} \,. \quad (1)$$

In Eq.1, $\mathbf{u}$ is the blood flow velocity, $\mathbf{v}_p$ is the particle velocity, ρ_f is blood density, ρ_p is the embolus density, $D_p, \mathcal{V}_p$ are the particle diameter and volume respectively, and $-\nabla p + \nabla \cdot \tau$ denote the divergence of the total fluid stress. The drag coefficient C_D is estimated using an empirical correlation based on the particle slip-velocity Reynolds number $Re_p = \rho_f \parallel \mathbf{u} - \mathbf{v}_p \parallel D_p/\mu_f$. The added mass coefficient C_a is assumed to be 0.5.

Embolus interaction with the vessel wall is incorporated through the last term in Eq. 1. The function $\mathcal{I}(\mathcal{P}, \mathcal{W})$ is an indicator function for contact between embolus particle $\mathcal{P}$ and the wall $\mathcal{W}$. For the computational mesh, this function is unity when the particle contacts one or more triangular elements on the wall, and zero otherwise. The actual form of the contact force $\mathbf{F}_{contact}$ is derived from the balance of linear and angular momenta for the particle-wall contact.

For resolving the particle-fluid two-way coupling, a simple momentum-source method is employed. Specifically, the Navier-Stokes equation is augmented by a momentum source term, for each computational cell, where the magnitude of the source is given as $\mathcal{S}_p = -(1/\mathcal{V}) \sum_p N_p \mathbf{F}_p$, where $\mathcal{V}$ is the volume of the computational cell, and $\mathbf{F}_p$ is as defined in Eq. 1. For ensuring a smooth distribution of this momentum transfer over mesh nodes of neighboring cells, a kernel averaging technique is employed, wherein the source term at $\mathbf{x}_n$ is given as $\mathcal{S}_{np} = \Phi(\mathbf{x}_n - \mathbf{x}_p)\mathcal{S}_p$, and $\Phi(\cdot)$ is a Gaussian kernel function. While alternative numerical schemes using fictitious domain type methods are also being investigated to better resolve the disturbed flow around the fluid, this momentum-source method is a reasonable first-order approximation owing to its low computational cost.

RESULTS

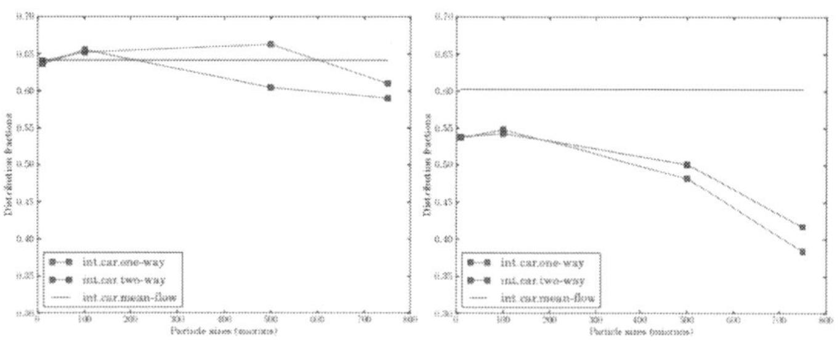

FIGURE 1: THE SIZE DEPENDENT VARIATION OF DISTRIBUTION FRACTION OF PARTICLES REACHING THE ICA FOR AN IDEALIZED (RIGHT) AND AN ANATOMICAL (LEFT) CAROTID BIFURCATION MODEL

Results for embolus distribution ratio have been compiled for the case of a human carotid bifurcation in Figure 1. The particle distribution ratio between one-way and two-way coupled models, for particle sizes that are up to about 18% of the internal carotid artery diameter (small to moderate sized emboli), were found to have a maximum difference of $\approx 5.3\%$ for anatomically realistic bifurcation model extracted from CT images. Simulations with idealized geometries of carotid bifurcations of similar dimensions show that the choice of coupling model causes a maximum difference in distribution ratios of $\approx 15\%$. Additionally, the particle distribution is seen to have a marked bias towards one branch vessel over the other, for the experiments with the anatomical bifurcation model. This is better understood by analysing the particle path lines for various release times and particle sizes. A notable, swirling flow of the embolic particles into the bifurcation is observed. Comparison with the local flow structures can be aided by considering flow helicity as an appropriate hemodynamic descriptor. The local normalized helicity is defined as [3]:

$$H_n(\mathbf{x}, t) = \frac{\mathbf{u}(\mathbf{x}, t) \cdot [\nabla \times \mathbf{u}(\mathbf{x}, t)]}{\| \mathbf{u}(\mathbf{x}, t) \cdot [\nabla \times \mathbf{u}(\mathbf{x}, t)] \|} . \quad (2)$$

The parameter $H_n \in [-1, 1]$, and denotes the angle between local velocity and vorticity, thereby providing a normalized indicator of the local swirling structure of the flow. A corresponding particle variable to characterize the swirling kinematics, is also devised. By defining an orthonormal triad $[\mathbf{e}_z, \mathbf{e}_r, \mathbf{e}_a]$ to be tangential, normal, and azimuthal to the centerline at every cross section of the vessel respectively, a swirling ratio is defined by decomposing the particle velocity as $sw = \mathbf{v}_p \cdot \mathbf{e}_a / \mathbf{v}_p \cdot \mathbf{e}_z$. The spatial extent of high negative H_n (indicating strong counter-clockwise helical flow) correlates well with that of high positive particle swirl number sw, as shown in Figure 2. Upon being introduced into a swirling flow, the particle trajectory will be determined by the strength and spatial extent of the swirling flow, the momentum with which the particle approaches this flow, and the particle's momentum response time $\tau_p = \rho_p D_p^2 / 18\mu_f$. This mechanism causes smaller and larger particles to follow differently tortuous trajectories, and in the case of our observations with the carotid bifurcation model, it preferentially directs particles towards one branch. The interplay of particle inertia and flow helicity is therefore a potentially critical mechanism that can be further quantified across multiple vasculature geometries.

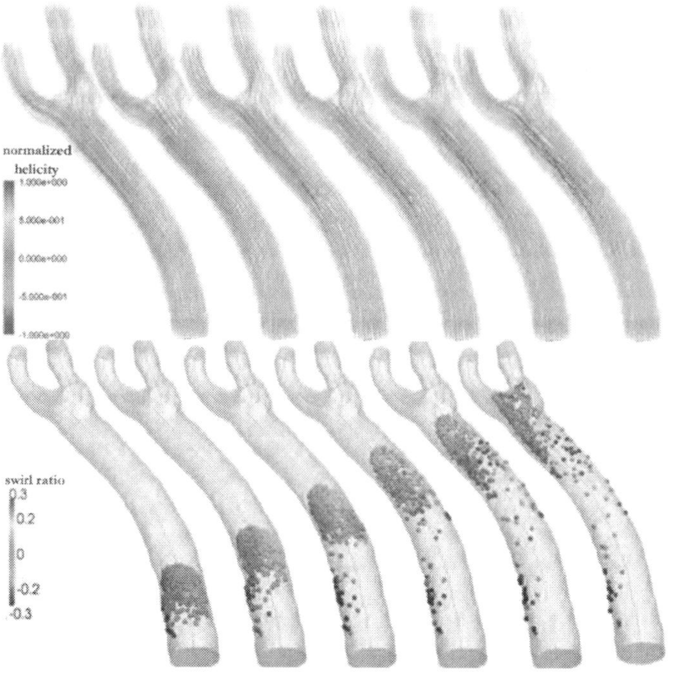

FIGURE 2: SUCCESSIVE SNAPSHOTS OF FLOW STREAMLINES COLORED BY NORMALIZED HELICITY $\mathrm{H_N}$ (TOP), AND AN ENSEMBLE OF EMBOLIC PARTICLES (750 μm) COLORED BY PARTICLE SWIRL RATIO sw (BOTTOM).

DISCUSSION

Observations from the numerical experiments with various particle sizes and release times indicate that for anatomically realistic vasculature, the curvature of vessel geometry, and dominant swirling flow play a leading order effect when compared to two-way fluid-particle momentum exchange for transporting small to moderate embolic particles. Distribution fraction estimates for idealized bifurcation geometries were therefore observed to be more sensitive to the choice of fluid-particle coupling model. Additionally, for small to moderate sized particles, in the presence of prominent circulating flow structures, one-way coupling is found to be a reasonable approximation to the fluid-particle interaction. This is of relevance for stroke simulations, since added computational costs due to two-way coupling could be detrimental in simulating ensembles of many thousands of particles for obtaining distribution statistics. Of course, for large particle sizes relative to vessel diameter, and for understanding the mechanics of embolic occlusion of an artery, the detailed resolution of fluid-particle and particle-vessel wall interactions is crucial. Hence, further investigations towards development of appropriate numerical methods are currently a focus area for the authors. Swirling, helical flow is a common feature of the hemodynamics in the major arteries [4]. The insights gained from the study on the role of flow helicity as a prominent hemodynamic descriptor is of great relevance for understanding stroke etiology. A broad range of existing studies quantify embolus distribution fraction in terms of volumetric flow distribution alone. In the context of the observations made from the study, this may not be sufficient in capturing the detailed mechanics of embolic particles with three-dimensional flow structures. Extending the study presented here to multiple patients will advance our understanding of the importance of inertial embolic particle interaction with helical flow - which is also a continuing focus area of research for the authors.

ACKNOWLEDGMENTS

This research was supported by the American Heart Association grant titled *'Embolic Stroke: Anatomic and Physiologic Insights from Image-Based CFD'*

REFERENCES

[1] Carr, I.A. et al., *Am J Physiol Heart Circ Physiol*, 305:732-739, 2013
[2] Maxey, M.R., Riley, J.J., *Phys. Fluids*, 26(4):883-889, 1983
[3] Grigioni, M. et al., *J Biomech.*, 38(7):1375-1386, 2005
[4] Lorenz, R. et al., *Magn Reson Med.*, 71(4):1542-1553, 2014

SUBJECT-SPECIFIC CALIBRATION OF GEOMETRIC NEUROMUSCULOSKELETAL MODELS

Andrew J. Meyer (1), Carolynn Patten (2,3), Benjamin J. Fregly (1)

(1) Department of Mechanical & Aerospace Engineering, University of Florida, Gainesville, FL

(2) Department of Physical Therapy, University of Florida, Gainesville, FL

(3) Neural Control of Movement Lab, Malcom Randall VA Medical Center, Gainesville, FL

INTRODUCTION

EMG-driven simulations of human locomotion using patient specific neuromusculoskeletal models may give important insights into treatments for movement related disorders. Unfortunately, creating these models is arduous and time consuming. Because of this difficulty, many researchers use scaled versions of generic models to represent their subjects. However, scaled models may be largely different from the subject being studied [1]. Differences between the scaled model and the subject may result in incorrect conclusions in some studies. Therefore, better methods are needed for expedient creation of patient-specific neuromusculoskeletal models.

This study presents a method for calibrating a scaled generic neuromusculoskeletal model to match a specific subject. Muscle moment arm, muscle-tendon (MT) length, and MT velocity, as well as other muscle parameters, were adjusted within a scaled generic OpenSim [2] model [3] such that net joint moments from muscles within an EMG-driven simulation closely matched those estimated from inverse dynamics. The most novel aspect of this calibration method is that it maintains physiological consistency between the moment arms, MT lengths, and MT velocities using a surrogate modeling method.

METHODS

Motion capture, ground reaction, and electromyography (EMG) data were collected from a high functioning hemiparetic male subject. Institutional review board approval and subject informed consent were obtained. The subject walked for several minutes on an instrumented treadmill at four speeds: 0.3, 0.4, 0.5, and 0.6 m/s. Ten gait cycles at each speed were selected for use in this study. A modified Cleveland clinic marker set was used with additional markers added to the foot. EMG data were collected from 16 muscles on each leg using a combination of surface and fine-wire electrodes. Surface EMG data were recorded for: gluteus maximus and medius, semimembranosus, biceps femoris long-head, rectus femoris, vastus medialis and lateralis, medial gastrocnemius, tibialis anterior, peroneus longus, and soleus. Fine-wire EMG data were collected for adductor longus, iliopsoas, tibialis posterior, flexor digitorum longus and extensor digitorum longus. EMG signals were high-pass filtered at 30 Hz using a 4th order Butterworth filter, demeaned, rectified, and low pass filtered at 7/T Hz, where T is the period of the gait cycle being analyzed [4]. EMG signals were scaled by their maximum values from all trials.

A generic OpenSim model [3] was calibrated to match static and kinematic data using a custom scaling technique [5]. The generic OpenSim model was first scaled to match a static trial using OpenSim's scaling algorithm. The model was then calibrated through an optimization that adjusted joint center locations, joint axis orientations, and marker locations within the segments such that inverse kinematic marker errors were minimized. Inverse kinematic calibration was performed on a single 0.5 m/s gait trial using MATLAB's nonlinear least squares optimizer. Joint parameters were adjusted only for the lower extremities, while marker parameters were adjusted for all segments. Using the calibrated model, we performed inverse kinematic and dynamic analyses for each gait trial. In addition, muscle moment arms, MT lengths, and MT velocities were calculated for 35 muscles in the leg for each gait trial using the muscle analysis tool.

A polynomial response surface model of the moment arms and MT lengths and velocities as a function of joint angles and velocities was created for use in later optimizations [6]. For each muscle that actuates a single degree of freedom (DOF), the MT length was fit using the function:

$$l_{mt} = b_0 + b_1\theta + b_2\theta^2 + b_3\theta^3 \quad (1)$$

where l_{mt} is the MT length, θ is the joint angle, and b_0 to b_3 are coefficients fitted using least squares regression. The MT velocity was the time derivative of equation (1):

$$v_{mt} = b_1\dot{\theta} + 2b_2\dot{\theta}\theta + 3b_3\dot{\theta}\theta^2 \quad (2)$$

where v_{mt} is the MT velocity and $\dot{\theta}$ is the joint angular velocity.

Similarly, moment arms were represented by taking the derivative of equation (1) with respect to θ:

$$r_{ma} = b_1 + 2b_2\theta + 3b_3\theta^2 \quad (3)$$

where r_{ma} is the muscle moment arm. For muscles that actuate multiple DOFs, more complex fits were performed using angles from other DOFs. l_{mt}, v_{mt}, and r_{ma} were fit simultaneously for each muscle. These fits were created for 35 muscles in the non-paretic (left) leg.

The muscle parameters calculated from the surrogate model were used in combination with a Hill-type muscle model to calculate the moment produced by each muscle using the following relationship:

$$M = r_{ma}F_{max}(d_{scale}a(t-\tau)\cdot f(\tilde{l}_m)\cdot f(\tilde{v}_m)+f_p(\tilde{l}_m))\cos\alpha \quad (4)$$

where M is a muscle's contribution to a joint moment, F_{max} is its maximum isometric force, a is the muscle's activation, d_{scale} is an activation scale factor included because of uncertainty in EMG scaling, t is time, τ is an optimized time delay, $\tilde{l}_m$ is the normalized muscle-fiber length, $\tilde{v}_m$ is the normalized fiber velocity, and α is the muscle pennation angle. $f(\tilde{l}_m)$, $f(\tilde{v}_m)$, and $f_p(\tilde{l}_m)$ are the normalized force-length, force-velocity, and passive force components of the Hill-type muscle model. $\tilde{l}_m$ and $\tilde{v}_m$ are calculated from MT length using the equation:

$$\tilde{l}_m = \frac{l_{mt}-l_{ts}}{l_m^o\cos\alpha}\ ,\quad \tilde{v}_m = \frac{v_m}{10\cdot l_m^o} \quad (5)$$

where l_m^o is the optimal fiber length and l_{ts} is the tendon slack length.

Muscle excitation was calculated from EMG data using the first order differential equation [7]:

$$\frac{du(t)}{dt} = (c_1e(t)+c_2)(e(t)-a(t)) \quad (6)$$

where $e(t)$ is the time varying normalized EMG signal and c_1 and c_2 are muscle-specific constants calculated using the equations

$$c_2 = \frac{1}{\tau_{deact}} \qquad c_1 = \frac{1}{\tau_{act}} - c_2 \quad (7)$$

where τ_{deact} and τ_{act} are time constants set to 60 and 15 ms for all muscles. The activation to excitation dynamics were modeled using the function:

$$a(t) = d_{nonlin}\sqrt{u(t)} + (1-d_{nonlin})u(t) \quad (8)$$

where d_{nonlin} is an activation nonlinearity parameter [8]. Initial values for l_m^o, l_{ts}, and α were taken from the literature [3]. F_{max} values were twice those taken from literature. Initial estimates for d_{scale}, d_{nonlin}, and τ were set to 1, 0, and 0 respectively. For muscles without EMG data, signals from related muscles were used to represent their activity [9].

Two EMG-driven models were calibrated using different approaches: the first adjusted only Hill-type model parameters (model 1), while the second also adjusted MT length, velocity, and moment arm values (model 2). The models were calibrated so that the net moments generated by muscles would closely match inverse dynamics loads for five degrees of freedom in the left leg. The five joint torques being matched were the: hip flexion, hip abduction, knee extension, ankle plantar flexion, and ankle eversion moments. For calibration of the first model, l_m^o, l_{ts}, d_{act}, d_{nonlin} and τ were adjusted, while calibration of the second model also varied surrogate model parameters b_0 to b_3 for each muscle (additional parameters were varied for muscles that produce moments about multiple DOFs). Moment arms were constrained to stay within 2 cm of their initial values at each time frame, and the sign of a moment arm was not allowed to change. The parameters l_m^o and l_{ts} were constrained to vary between 75% and 125% of their initial values. d_{scale} was constrained to vary between 0.1 and 1. d_{nonlin} was allowed to vary between 0 and 0.75, and τ ranged between 0 and 160 ms. All parameter values were penalized for changes away from their initial values. Constraints were used to ensure model properties such as $\tilde{l}_m$ stayed within their physiological ranges during all time frames. Model calibration optimizations were performed using MATLAB's constrained nonlinear optimizer function using an sqp algorithm. Each model was calibrated to data from 30 trials at the three slower walking speeds and used to predict joint moments for 10 trials at the fastest speed.

RESULTS

Model 2, with consistently adjusted MT lengths, velocities, and moment arms generally produced lower moment errors than did model 1 (Table 1). The moment error reductions from the geometric adjustments ranged from 1.60% for the ankle PF moment at 0.4 m/s to 56.16% for the ankle IE moment at 0.3 m/s. In general, the Ankle IE moment had the largest improvements in moment errors, while the knee FE and ankle PF moment only had small improvements of 8.05% and 1.60% at 0.6 and 0.4 m/s respectively. However, all other moment errors improved by more than 20% after geometric adjustments.

TABLE 1: Overall mean absolute errors in moment predictions for each gait speed from the model without (model 1) and with (model 2) implicit geometry adjustments.

Moment	Model Type	Calibration Errors (N-m)			Prediction Errors (N-m)
		0.3 m/s	0.4 m/s	0.5 m/s	0.6 m/s
Hip FE	1	7.07	5.81	6.23	8.99
	2	4.41	4.20	4.24	5.59
Hip AA	1	9.05	8.23	8.66	9.74
	2	5.16	4.24	5.24	4.59
Knee FE	1	6.45	5.87	5.08	5.84
	2	4.71	3.97	3.80	5.37
Ankle PF	1	6.43	5.00	7.32	6.79
	2	4.35	4.92	5.58	4.38
Ankle IE	1	5.68	5.11	4.91	5.12
	2	2.49	2.44	2.73	2.86

DISCUSSION

This study demonstrates that even small changes in musculoskeletal geometry can have a significant influence on EMG-driven moment calculations. This finding was especially true for some DOFs when performing model predictions using gait trials withheld from calibration. Therefore, muscle force optimizations or simulations that attempt to predict novel gait motions may have larger errors than their calibration trials would suggest. Additionally, these EMG-driven simulations compare favorably with previous EMG-driven model studies which reported moment errors ranging from 5 to 28 N-m [9], [10] for various joints and activities. Because of these issues, calibrating muscle moment arms using a method as described here is highly advisable when performing predictive gait optimizations.

Our proposed calibration approach involves a number of limitations. Studies that lack EMG data would be unable to perform implicit geometric adjustments. Even labs that normally work with EMG data may not have access to fine-wire electrodes and therefore would lack critical muscles such as the iliopsoas. Also, the use of time delays up to 160 ms is unconventional, though in vivo measurements made in cats suggest that this delay is reasonable [11]. Overall, the methods described here present a novel way to improve EMG-driven model moment estimates while maintaining physiological consistency.

ACKNOWLEDGEMENTS

This study was funded by NSF grant CBET 1159735.

REFERENCES

[1] Correa, TA et al., *J. Biomech.*, 44: 2096–105, 2011.
[2] Delp, SL et al., *IEEE Trans. Biomed. Eng.*, 54:1940–50, 2007.
[3] Arnold, EM et al., *Ann. Biomed. Eng.*, 38:269–79, 2010.
[4] Lloyd, DG et al., *J. Biomech.*, 36:765–776, 2003.
[5] Reinbolt, JA et al., *J. Biomech.*, 38:621–626, 2005.
[6] Menegaldo, LL et al., *J. Biomech.*, 37:1447–53, 2004.
[7] He, J et al., *IEEE Trans. Autom. Control*, 36:322–332, 1991.
[8] Buchanan, TS et al., *J. Appl. Biomech.*, 20:367–395, 2004.
[9] Doorenbosch, CAM et al, *J Electromyogr Kinesiol*, 15:429-435, 2005.
[10] Sartori, M et al, *PLoS ONE*, doi:10.1371/journal.pone.0052618, 2012.
[11] Guimaraes, AC et al., *J. Exp. Biol.*, 198:975–987, 1995.

SB³C2015
Summer Biomechanics, Bioengineering and Biotransport Conference
June 17-20, 2015,Snowbird Resort, Utah, USA
SB³C2015-586

TUMOR ENGINEERING TO ELUCIDATE THE EFFECT OF MILD HYPERTHERMIA ON THE TRANSPORT OF SINGLE WALLED CARBON NANOHORNS IN THE TUMOR MICROENVIRONMENT

Matthew R. DeWitt(1), M. Nichole Rylander (2)

(1)Virginia Tech-Wake Forest SBES
Virginia Tech
Blacksburg, VA, US

(2) Mechanical Engineering
University of Texas at Austin
Austin, TX, US

INTRODUCTION

Therapeutic use of mild hyperthermia(41-43°C) has extensively been studied for enhancement of a wide range of chemotherapeutic drugs, including the antineoplastic agent Cisplatin. While *in vitro* and *in vivo* studies have shown the combination of mild hyperthermia with certain drugs results in higher efficacy of treatment, the practicality of simultaneous application of the two modalities in the tumor environment can be difficult to achieve, limiting the overall utility of the combination. Nanotechnology can provide the means to effectively deliver both the drug of choice and the necessary thermal energy to result in enhancement. Nanoparticle-assisted photothermal therapies enable highly localized and highly selective heating through the use of nanoparticles capable of generating heat when exposed to an external stimulus such as a NIR laser. Targeted delivery of drugs with these particles can overcome issues involving regional heating and non-specific delivery of drug payload.

Single-walled carbon Nanohorns (SWNHs) are a promising nanoparticle choice for multimodal therapies for the delivery of heat and chemotherapeutic agent to produce this desired synergistic effect. We have previously shown that SWNHs can be easily loaded with Cisplatin, acting as a drug delivery vehicle, while retaining the ability to act as a photoabsorbers in the highly depth penetrating NIR region (1064nm) to produce the necessary heat to result in thermal enhancement [1].

Current studies to evaluate nanoparticle mediated thermal enhancement therapies are primarily executed in a traditional 2D cell culture system in which the particles are applied directly over seeded cells resulting in possibly exaggerated effects of synergy. These cell monolayers are also incapable of studying the complex process for a nanoparticle to reach its target (e.g. interaction with vasculature and diffusion in tumors) that can be highly influenced by hyperthermia. While animal models provide physiologic fidelity, they can be highly variable leading to confounding determination of the response to nanoparticles and leading to significant expense. Therefore, there is a need for an intermediate 3D tumor platform to better understand nanoparticle transport and to better understand the influence of hyperthermia to optimize nanoparticle mediated photothermal enhancement therapies.

In this study we have developed a novel perfusable tissue engineered model of the tumor microenvironment which is comprised of an internal endothelialized microchannel surrounded by a collagen tissue containing human breast cancer cells. We successfully validated and characterized nanoparticle transport within the system, utilizing fluorescent polystyrene nanoparticles. In addition we utilized a 3D disk co-culture model in a hanging well (Fig 1B) to understand the role of hyperthermia in the tissue level transport of SWNHs which will be utilized to develop novel nanoparticle formulations in the future to optimize nanoparticle mediated photothermal enhancement therapies.

METHODS

Human breast adenocarcinoma cells, MDA-MB-231 (ATCC) and telomerase immortalized human microvascular endothelial cells (TIME) which stably express GFP or RFP were all used in this study.

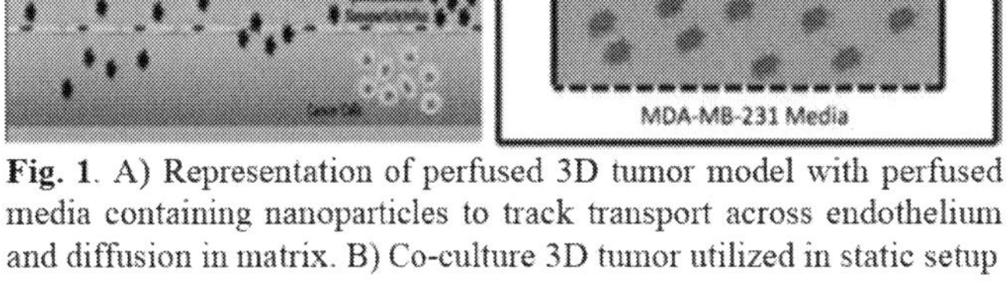

Fig. 1. A) Representation of perfused 3D tumor model with perfused media containing nanoparticles to track transport across endothelium and diffusion in matrix. B) Co-culture 3D tumor utilized in static setup

Collagen was extracted following Buchanan et al, where rat tail tendons were dissolved in 10mM HCl overnight at 4C. Following centrifugation at 30,000g for 30 min, the supernatant is collected and lyophilized. The 3D microfluidic model (Fig 1a) was prepared by mixing a 16mg/ml collagen stock solution with 10% 10x DMEM, sufficient 1N NaOH to neutralize the HCl, supplemented with DMEM containing $1x10^6$ MDA-MB-231 cells to bring the final solution to 8mg/ml collagen. Next ,the mixture was injected into cylindrical FEP tubing with a 22 or 26G needle inserted to create a vessel lumen. TIME cells were then added to the lumen and a 72 h preconditioning shear stress was applied prior to nanoparticle transport studies to allow for elongated and aligned endothelial cells [2]. A similar protocol was utilized to form the disks with the hanging well acting as a molding device and with a thin layer of acellular collagen between the cancer cells and endothelial cells to enhance endothelial viability(Fig 1B).

Fluorescent polystyrene spherical nanoparticles (40-200nm diameter) were then introduced with flow (0.05mg/ml) and an incubated live imaging microscope was utilized to measure fluorescent intensity of the lumen and collagen matrix to measure penetration as a method of validation for the system. A variety of different shear rates (10-25 dynes/cm2), matrix properties, and cell conditions were studied to determine their effect on nanoparticle transport. Following validation with the polystyrene nanoparticles, QD-tagged SWNHs were generated following protocol outlined in [3]. The temperature of the live imaging incubation was then changed from 37°C up to 43°C and similar tracking of the nanoparticles was accomplished.

The 3D culture system described in Figure 1B was utilized to study uptake by the MDA-MB-231 cells in a co-culture setup with a monolayer of TIME cells acting as a confluent endothelial barrier without the presence of a convective follow, elucidating the effect of mild hyperthermia on diffusion and uptake by the cells. The temperature was varied between 42°C and 37°C and a nanoparticle solution was administered in the endothelial media cavity for up to 6h. Digestion of the tumor construct was accomplished with collagenase over 2h and FACS (BD FACSARIA) was used to examine the cellular uptake of SWNH-QD in both the TIME Cells and MDA-MB-231 cells that were discriminated based on stable production of GFP.

RESULTS

To demonstrate the capability of the system to measure nanoparticle transport, the setup was first tested with fluorescent polystyrene nanoparticles (40-200nm diameter). After 72h of preconditioning to produce a confluent, elongated endothelium, nanoparticles (0.05mg/ml) were introduced with flow into the system and fluorescent images were captured every 5 minutes for 6 h in a live imaging incubator system (Figure 2A) and a custom matlab code utilizing ImageJ (Figure 2B) was developed to plot the radial fluorescent intensity as a function of time. Representative data is shown for 100nm particles in a system with and without the inclusion of a confluent monolayer. Data from these experiments show the size dependence transport of nanoparticles across the endothelium with relatively greater extravasation of the smaller particles (40-100nm) compared with the larger nanoparticles(200nm) which highlights the need for careful consideration of nanoparticle size when formulating novel nanomaterials for delivery in the tumor microenvironment.

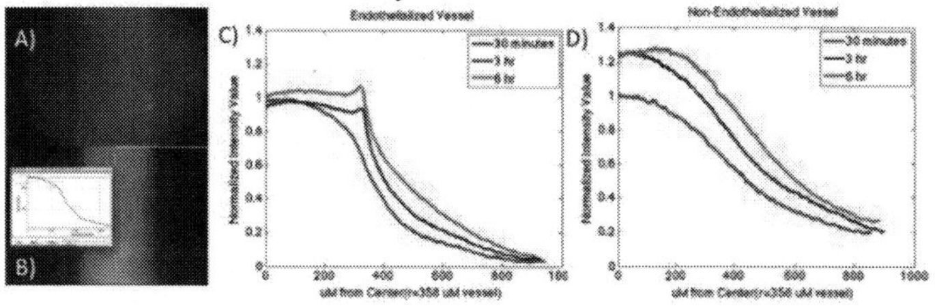

Figure 2: A & B) Fluorescent image of lumen surrounded with collagen hydrogel, and image analysis of red channel (nanoparticle) C-D) radial profiles of fluorescent intensity for endothelialized(C) and non-endothelialized channels (D)

Following characterization of the 3D microfluidic model, SWNH-QD (Figure 3B) were created and tested in the system under normothermia and mild hyperthermia (Figure 3A). Radial fluorescent intensity from the vessel wall is plotted and are shown in pairs for 30 m (blue set) 3 h (black set) and 6 h (red set) representing time after introduction of particles for both temperatures. A significant change in fluorescent intensity can be seen for the approximately 100nm particles at the 3 h time point, highlighting the effect of mild hyperthermia on extravasation and diffusion within the matrix.

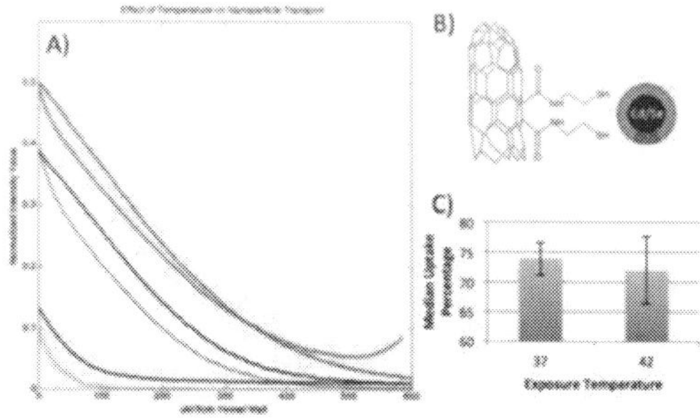

Figure 3: A) Radial intensity values of QD-SWNH fluorescence at 37 and 42°C at 30m, 3h, and 6h. B) Depiction of SWNH-QD C) FACS median uptake of MDA-MB-231 within matrix at 37 and 42°C

In addition to the convective flow model, a static disk model was utilized (Figure 1B) to determine if flow, which is commonly neglected variable in in vitro experiments, augments the effect of mild hyperthermia on the particle transport. FACS results shows insignificant difference between uptake at the two temperatures in the MDA-MB-231 cells when cultured in the co-culture setup, differing from results from 2D cell culture [1]. The hypothesized difference seen between the two models results from the production of a more aligned endothelium in the microfluidic model which would be less porous to particles at normothermia and thus be more likely to be effected by hyperthermia which is hypothesized to cause a leakier vasculature [4].

DISCUSSION

We have demonstrated that tumor engineering models can be utilized to study the transport of SWNHs within the tumor microenvironment. Two tissue engineered models, a cylindrical model with the introduction of convective flow and a static flow disk model were developed and validated to understand the effect of hyperthermia on nanoparticle transport which is crucial when developing novel photothermal enhancement therapies. A more comprehensive and controlled understanding of the transport of SWNHs at the tumor tissue level and the effects of mild hyperthermia on key transport events such as extravasation, diffusion within the matrix, and cellular uptake will allow for the optimization of such dual nanoparticle-heat delivery vehicles to result in greater efficacy in treatment prior to the use of costly animal models. Future studies will involve the addition of drugs to the nanoparticles to understand not only the transport of the particles, but also the resulting efficacy of treatment in these multimodal therapies.

ACKNOWLEDGEMENTS

Funding was granted by NSF Early CAREER Award CBET 0955072 and NIH R21 CA156078 and CA135230 and the NSF-GRFP

REFERENCES

[1] Dewitt et al., *J Biomech*, 136:021003, 2014.
[2] Buchanan et al., *Tissue Eng C*, 20:64-75, 2013.
[3] Zimmermann et al, *J. Nanopart. Res.*, 01/2013
[4] Kong et al, *Cancer Res*, 61:3027, 2001

SB³C2015
Summer Biomechanics, Bioengineering and Biotransport Conference
June 17-20, 2015, Snowbird Resort, Utah, USA
SB³C2015-587

INCREASED RED BLOOD CELL STIFFNESS INCREASES PULMONARY VASCULAR RESISTANCE AND PULMONARY ARTERIAL PRESSURE

David A. Schreier (1), Omid Forouzan (1), Timothy Hacker (2), John Sheehan (2), and Naomi C Chesler (1,2)

(1) Biomedical Engineering
University of Wisconsin-Madison
Madison, WI, United States of America

(2) Department of Medicine
University of Wisconsin-Madison
Madison, WI, United States of America

INTRODUCTION

Patients with sickle cell disease (SCD)-related pulmonary hypertension (PH) have a substantially increased risk of sudden death compared to those with SCD alone, as high as a 50% within 2 years [1, 2]. The prevalence of PH in adults with SCD is high--about 30% [3, 4]. Higher still is the percentage of SCD patients with evidence of PH via lung histology at the time of death--75% [5]. Sickled red blood cells (RBC) are more dense, rigid and frequently trapped in the microcirculation, thereby interrupting blood flow and causing ischemia, thrombotic events, increased oxidative stress, and reperfusion injury [6]. All of these consequences of SCD can and often do have drastic effects on hemodynamics including increasing pulmonary vascular resistance (PVR) [6] and systemic arterial wave reflections [7].

In this study, we sought to investigate the impact of increased RBC rigidity alone, independent of other SCD-related RBC abnormalities, on the hemodynamic changes that ultimately cause PH and increase mortality in SCD. In particular, we hypothesized that SCD-like rigid RBCs increase PVR and wave reflections in the pulmonary circulation in otherwise healthy lungs. To test this hypothesis, we infused chemically stiffened RBCs into the jugular vein of mice and measured pulmonary vascular impedance (PVZ) *in vivo*. Our results confirmed that RBC stiffness dramatically increases both PVR and wave reflections, and is likely key to poor outcomes in SCD-related PH.

METHODS

Whole blood was drawn from donor male C57BL6 mice (14 weeks-old) to obtain RBCs. To create SCD-like rigidity, RBCs were treated with 0.016% glutaraldehyde for 30 minutes, which has been shown to stiffen RBCs without changing overall shape [8]. Using established techniques from our lab [9], additional male C57BL6 mice underwent catheterization of the main pulmonary artery (PA) to obtain pressure with simultaneous echocardiography to obtain PA flow. Once baseline measurements were recorded, 50 uL of blood (Hct~45%) was drawn from each mouse and replaced with an equal volume of a 45%-by-volume suspension of chemically stiffened RBCs in physiological saline solution. All pressure and flow measurements were repeated and the cycle was repeated 2 more times, thereby increasing the concentration of chemically stiffened RBCs in the mouse.

Simultaneously obtained pressure and flow waveforms were analyzed using both time and frequency domain techniques. Metrics of PVZ were calculated in the time domain using wave intensity analysis as previously described [10]. Total PVR (Z_0) is calculated as mPAP divided by Q averaged over the cardiac cycle. The pulmonary arterial pressure waveform is separated into forward (P_f) and backward (P_b) traveling components using the linear wave separation method [11]. The index of global wave reflections (RQ) is calculated as the ratio of the amplitude of P_b to P_f. Lastly, we measured characteristic impedance (Z_C) as the derivative of pressure over the derivate of flow at 95% max flow.

We also attained metrics of PVZ in the frequency domain, P and Q were decomposed into sinusoids at harmonics of the heart rate using a Fourier transform, which expresses the data as the sum of series of sine and cosine waves. Z_0 was obtained as the magnitude of PVZ (Z) at 0 Hz; characteristic impedance (Z_C), a metric of arterial compliance, was obtained by averaging Z from the 4th to 10th harmonics.

Just prior to euthanasia, a blood sample was again drawn and viscosity measurements were obtained at a range of shear rates to compare the rheology of the final normal plus stiffened RBC suspension to the rheology of control mouse blood available from the literature [12].

RESULTS

Infusion of increasing amounts of chemically stiffened RBCs increased time domain calculated Z_0 (Figure 1A) and mPAP (Table 1). Proportionally, the increase in Z_0 was larger than the increase in mPAP because CO decreased with stiffened RBC infusions (Table 1). Pulmonary vascular wave reflections increased (Figure 1B) and characteristic impedance Z_C decreased (Table 1). There was no change in heart rate.

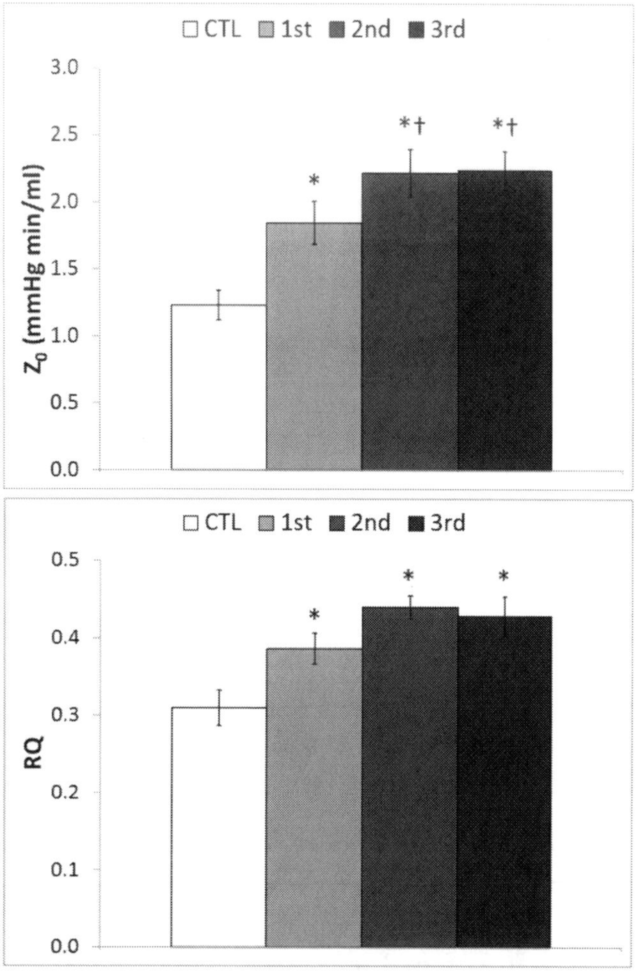

Figure 1. (A) Total pulmonary vascular resistance (Z_0) for CTL and after each injection.(B) Pulmonary arterial wave reflections (RQ). *P<0.05 versus CTL, †P<0.05 versus 1st injection. Bars represent man ±SE.

Table 1. Hemodynamic values: mean pulmonary artery pressure (mPAP: mmHg), cardiac output (CO: ml/min), heart rate (HR: beats/min), and characteristic impedance (Z_C: mmHg*min/ml).

	CTL	1st	2nd	3rd
mPAP	13.6±0.6	18.9±0.5 *	21.8±0.5 *†	23.2±0.7 *†
CO	12.0±0.7	10.7±0.6 *	10.0±0.6 *†	9.5±0.9 *†
HR	575±9	565±10	553±16	559±18
Z_C	0.26±0.03	0.22±0.01 *	0.19±0.01 *	0.19±0.01 *

Note: values are ±SE, n=13 for CTL, 1st, and 2nd, n=9 for 3rd. *P<0.05 versus CTL. †P<0.05 versus 1st injection.

There was no change in metrics of blood viscosity with three 50uL infusions of 45% suspension of chemically stiffened RBCs in saline (data not shown). We also confirmed that the chemical treatment of RBCs did not change the shape of RBCs by microscopy (data not shown).

DISCUSSION

Our results demonstrate that RBC stiffening plays a key role in altering pulmonary vascular hemodynamics and increasing right ventricular afterload. Stiffened RBCs increase pulmonary vascular resistance, evidenced by both increases in pulmonary artery pressure and decreases in cardiac output. Stiffened RBCs also acutely increase wave reflections, which is suggestive of increased arterial stiffening. The increase in mPAP in the lungs could be a possible explanation for acutely increased pulmonary arterial stiffness [13]. Additionally, increased backward wave reflection traveling to the heart (RQ) could be caused by trapping of the chemically stiffened RBCs in the lung capillaries.

The major novel finding of this study is that chemically stiffened RBCs affect both static and dynamic components of right ventricular afterload. The effects of other SCD-related RBC changes on pulmonary vascular hemodynamics and right ventricular function merit further investigation.

ACKNOWLEDGEMENTS

This work was supported by NIH grant 1R01HL086939 (NCC). The authors also would like to acknowledge Guoqing Song for his surgical expertise.

REFERENCES

1. Castro, O., M. Hoque, and B.D. Brown. Blood, 2003. **101**(4): p. 1257-61.
2. Charache, S. J Lab Clin Med, 1994. **124**(4): p. 473-4.
3. Ataga, K.I., et al. Am J Med, 2004. **117**(9): p. 665-9.
4. De Castro, L.M., et al. Am J Hematol, 2008. **83**(1): p. 19-25.
5. Haque, A.K., et al. Hum Pathol, 2002. **33**(10): p. 1037-43.
6. Machado, R.F. and H.W. Farber. Clin Chest Med, 2013. **34**(4): p. 739-52.
7. Aessopos, A., et al. Atherosclerosis, 2007. **191**(2): p. 427-32.
8. Morel, F.M., R.F. Baker, and H. Wayland. J Cell Biol, 1971. **48**(1): p. 91-100.
9. Tabima, D.M., et al. J Biomech, 2012. **45**(5): p. 799-804.
10. Mitchell, G.F., et al. Am J Physiol, 1994. **267**(5 Pt 2): p. H1907-15.
11. Westerhof, N., et al. Cardiovasc Res, 1972. **6**(6): p. 648-56.
12. Schreier, D.A., et al. J Appl Physiol (1985), 2014.
13. Bellofiore, A., et al. Ann Biomed Eng, 2013. **41**(1): p. 195-204.

ASSESSMENT AND QUANTIFICATION OF TRANSITIONAL FLOW IN INTRACRANIAL ANEURYSMS – HIGHLY RESOLVED SIMULATIONS BELOW THE KOLMOGOROV SCALES

Kartik Jain (1,2), Kristian Valen-Sendstad (2), Sabine Roller (1) and Kent-Andre Mardal (2,3)

(1) Simulation Techniques and Scientific Computing,
University of Siegen,
57076 - Siegen, Germany

(2) Center for Biomedical Computing,
Simula Research Laboratory,
P.O. Box 134, N-1325, Lysaker, Norway

(3) Department of Mathematics,
University of Oslo,
0316, Oslo, Norway

INTRODUCTION

An intracranial aneurysm is a pathological dilatation of the cerebral blood vessel, rupture of which is a major cause of morbidity and mortality in the modern world. The aneurysmal hemodynamics has been extensively modeled using Computational Fluid Dynamics (CFD). The low Reynolds number (typically < 500) has propelled the assumption of laminar flow in aneurysms in most of the CFD studies, but the assumption of laminar flow was recently questioned [1], which revealed high frequency flow fluctuations in aneurysms using Finite Element Method (FEM) based simulations of reasonable resolutions.

The issue of transition in aneurysms has been a prevalent subject of discussion, as the resolution and accuracy of the CFD has to be appropriately tailored in order to capture and assess a transitional flow. The present study aspired to scrutinize this issue of transition by performing highly resolved Lattice Boltzmann Method (LBM) based Direct Numerical Simulations (DNS) in the same 12 aneurysms as [1], and to adequately quantify and assess the flow characteristics with resolutions at the order of Kolmogorov micro-scales. The study mainly constituted of three parts – *Firstly*, the aneurysms studied in [1] were simulated with LBM for verification. *Secondly*, a space-time refinement study was carried out to evaluate the changes brought about by resolutions in aneurysmal hemodynamics. *Thirdly*, a detailed investigation was performed to uncover the *critical* Re at which transition occurs in this cohort of aneurysms.

METHODS

Clinical data for 12 middle cerebral artery (MCA) aneurysms was available from the University Hospital of North Norway, which was segmented to obtain 3D models in [1]. LBM-based flow simulations were conducted on this aneurysm cohort with a spatial and temporal discretization of 32μm and 1.81μs respectively (h4 in table 1) to verify against FEM simulations of [1].

Two aneurysms, one stable (Aneurysm1) and one unstable (Aneurysm2) were then chosen for a space-time refinement study. Table 1 lists the discretization and resulting number of cells for Aneurysm1 and Aneurysm2.

TABLE 1: SPATIAL AND TEMPORAL DISCRETIZATION OF ANEURYSM1 AND ANEURYSM2 AND RESULTING NUMBER OF CELLS

Ref.	δx (μm)	δt (μs)	nCells Aneurysm1	nCells Aneurysm2
h1	128	29.004	81 516	208 130
h2	64	7.25	714 128	1 751 495
h3	48	4.08	1 730 890	4 206 178
h4	32	1.81	5 967 144	14 366 575
h5	16	0.45	48 787 963	116 370 643
h6	8	0.11	394 432 233	929 916 251

The simulations were performed using the end-to-end parallel simulation tool chain contained in the open-source APES framework [2], and were executed on the *SuperMUC* petascale system installed at the Leibniz Supercomputing Center, Germany, using up to 16384 cores.

Blood was modeled as a Newtonian fluid with ρ = 1.025 g/cm^3 and ν = 0.035 Poise. A uniform velocity profile with a mean of 0.5 m/s was prescribed at the inlet and a zero pressure at the outlet. Initial conditions were set to zero velocity and pressure and all the vascular walls were assumed to be rigid. Average parent artery Re was 292 (range: 201-398). The LBM simulations were conducted for 8 seconds after the damping of initial conditions to statistically describe the flow characteristics namely the Turbulent Kinetic Energy (TKE) and the Kolmogorov micro scales. The quality of the spatial and temporal resolutions was assessed by calculating l^+ and t^+, which are quantified in terms of viscous length scales, and in turn are proportional to the Kolmogorov scales.

Simulations at Re varying in steps of $\delta Re = 10$ were performed on all 12 aneurysms with discretization h4 to explore the critical Re.

RESULTS

LBM simulations of the 12 aneurysms showed an excellent agreement with [1], i.e. the flow remained laminar and transitional in the same aneurysms; all sidewall aneurysms were stable and the 5/8 bifurcation aneurysms were unstable.

The refinement study revealed velocity errors of up to 16% in Aneurysm1 for discretization h1 relative to h6. The errors exhibited a second order convergence and dropped to below 2% for h4 and 1% for h5. For Aneurysm2, no fluctuations were detected at discretization h1 whereas minor fluctuations could be observed at h2 and h3 that grew considerably for h4. The fluctuations detected by h5 and h6 were similar, though they had a considerable phase difference i.e. the velocity peaks were shifted in time.

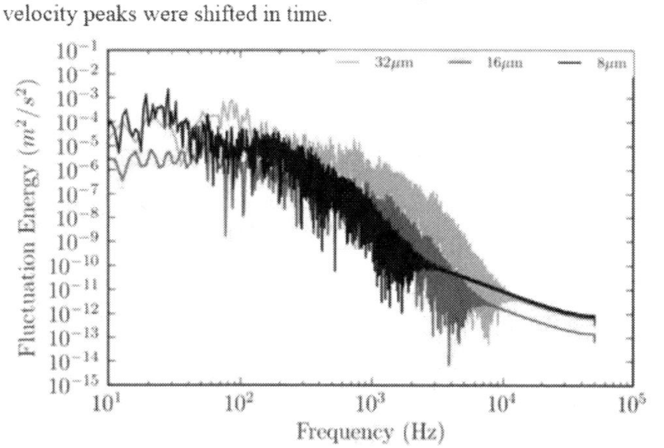

FIGURE 1: PHASE SPECTRUM OF THE TKE AT H4, H5 AND H6 FOR ANEURYSM2

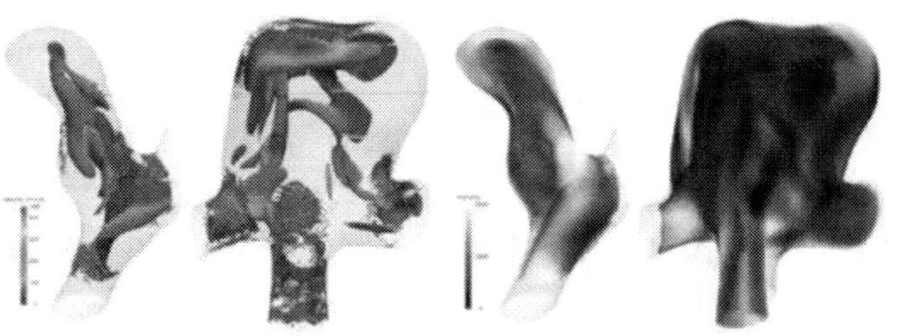

FIGURE 2: Q-ISOSURFACES (Q=0.2) COLORED BY VELOCITY FOR ANEURYSM1 AND ANEURYSM2 (LEFT HALF) AND VOLUME RENDERING OF VORTICITY (RIGHT HALF) AT H6

Power spectrum of TKE at h4, h5 and h6 is shown in figure 1 for the unstable Aneurysm2, which depicts a *revamping* in the behavior of TKE with refinements. Figure 2 shows the velocity colored Q-isosurfaces (Q=0.2), and the vorticity magnitude for A1 and A2.

The l^+ attained 1 for h4 whereas the t^+ was < 1 for all discretizations due to the intrinsically low δt in LBM.

The critical Re for each respective aneurysm with transitional flow was close to the Re at which the simulations were conducted, and fluctuations at a lower Re were not substantial. No transition was observed in the 7 aneurysms with stable flow for Re up to 550.

DISCUSSION

The present study substantiates the results of [1] that the complex anatomy of certain aneurysms promotes transition under the common modeling assumptions. It also depicts that the flow remains stable in the parent artery of these aneurysms, classifying the manifestation of aneurysm itself as an initiator of flow fluctuations.

Flow became transitional in Aneurysm2 (fig. 2) after the inflow jet collided with the upper aneurysmal wall, giving rise to coherent vortices in the bulk flow, and minuscule vortices near the walls. The non-occurrence of Aneurysm1 at bifurcation and its morphology restricted the impact of inflow jet just over the aneurysmal neck thereby inhibiting the flow from becoming transitional.

The fluctuations were isolated to bifurcation aneurysms and did not occur in the sidewall aneurysms even at considerably higher Re, a finding, which clearly characterizes the occurrence of aneurysms at bifurcation as a promoter of fluctuations in the flow.

The Re at which the present simulations are conducted is at the threshold of critical Re, suggesting the occurrence of transition at the peak systolic conditions of the MCA M1 segment. Under pulsatile conditions, the critical Re is likely to reduce, as the flow is expected to destabilize during deceleration.

The refinement study portrays that a coarser resolution is not able to detect any transition in aneurysms, and very high resolutions of up to 8μm exposes the intricate structures that appear due to transition, yet the increase in accuracy is not phenomenal. A spatial resolution of 32μm seems appropriate for LBM simulations that would allow for meticulous capture of transition, and minimal errors when the flow is laminar.

The *flow features* which are captured by such high fidelity simulations help in revealing near wall vortices, the temporal evolution of which in particular holds potential role in understanding the wall degradation resulted by transitional flow, and hence aneurysm rupture.

The highest spatial resolution of 8μm is of the order of the size of a Red Blood Cell (RBC) and the continuum hypothesis is likely to break down at this scale, as the size and concentration of RBC might obviate the formation of eddies down to the Kolmogorov scales [3]. It is however contended with the present results that a spatial resolution of 8μm allows for an improved *numerical accuracy,* as the influence of RBC interactions is anyways inconceivable when the blood is predominantly assumed as a Newtonian fluid.

ACKNOWLEDGEMENTS

We are thankful to the *Leibniz Supercomputing Center*, Munich for providing compute resources needed for this research (*PID: pr85mu*). This work has been supported by the Research Council of Norway through grant no. 209951 and a center of excellence grant awarded to the *Center for Biomedical Computing* at Simula Research Laboratory.

REFERENCES

[1] Valen-Sendstad, K., Mardal, K.A., Steinman, D.A., (2013). High-resolution CFD detects high-frequency velocity fluctuations in bifurcation, but not sidewall, aneurysms of the middle cerebral artery. *J. Of Biomechanics* 46, 402-407

[2] Roller, Sabine, et al (2013). An adaptable simulation framework based on a linearized octree. *High Performance Computing on Vector Systems*. Springer Berlin Heidelberg, 2012. 93-105.

[3] Antiga, L, and Steinman, D.A., (2009). Rethinking turbulence in blood. *Biorheology* 46.2: 77-81.

MANAGEMENT OF DISTAL EMBOLIZATION DURING INTERVENTIONAL TREATMENT OF ACUTE ISCHEMIC STROKE IN A SIMULATED VASCULAR PHANTOM

Juyu Chueh, Olivia W. Brooks, Ajit S. Puri, Ajay K. Wakhloo, Matthew J. Gounis

New England Center for Stroke Research
Department of Radiology
University of Massachusetts Medical School
Worcester, MA, USA

INTRODUCTION

Endovascular treatment of acute ischemic stroke is rapidly evolving. Recent advancements in technology, including the invention of stentretriever, have demonstrated significant improvements in clinical outcomes as compared to standard medical therapy within 6 hours of stroke onset [1]. Despite this important advancement in stroke treatment, the majority of patients remain dependent on daily care, 88% cannot return to work, and 97% continue to have a symptom from their stroke after 90 days.

A possible contributor to patient outcome may be related to clot debris generated during mechanical thrombectomy caused by variation in guide catheter utilization strategies, and as such is a potentially modifiable risk factor. Clot fragmentation can cause distal embolization in and blockage of collateral flow to the potentially salvageable tissue. The purpose of this study is to assess various catheterization strategies (type and position of the suction catheter) that can reduce the risk of distal embolization during mechanical thrombectomy.

METHODS

A vascular replica consisting of the important relevant intracranial vessels was constructed and included: 1. an internal carotid artery (ICA) that branched into the anterior cerebral artery (ACA) and middle cerebral artery (MCA), and 2. the posterior communicating artery (PComA) that represented significant collateral flow. The vascular replica was created by using a core-shell technology.[2] A lubricating coating of LSR topcoat (Momentive Performance Materials, Albany, NY) was applied to the inner surface of the silicone replica to reduce the friction between the thrombectomy device and the silicone replica. The silicone replica was connected into a flow loop with continuous monitoring of MCA flow and pressure to ensure physiologically accurate hemodynamics.[3] A small anastomosis was created between the ECA circuit and the distal MCA to simulate leptomeningeal collaterals. The PComA flow was derived from an anastomosis with the proximal ICA. By adjusting each of the Starling resistors, the baseline flow conditions were matched to normal human phase contrast MR measurements for each of the vessels creating a hemodynamically representative cerebrovascular model.[4,5]

Two clot models were used. One was the hard, inelastic clot model, which is prone to fragmentation during thrombectomy. The other clot model was soft and elastic, similar to the red and fresh clots retrieved from the stroke patients.[6] The representative clot (1cm) was injected into the flow loop into the silicone replica to form a MCA occlusion.

The Solitaire FR device was used to achieve recanalization per manufacturer's instructions in group 1, 2 and 3 (Figure 1).

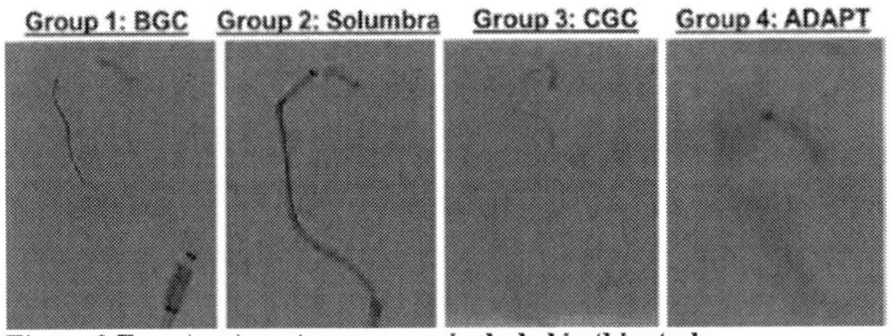

Figure 1 Four treatment groups are included in this study.

In group 1, thrombectomy was performed through an 8Fr balloon guide catheter (BGC) positioned at the cervical ICA. The Solumbra technique, thromboaspiration via a 5Fr intracranial guide catheter in the origin of the MCA, was used in group 2. In group 3, the occlusion was removed by thrombectomy through a 6Fr guide catheter (CGC) with the tip placed in the the cervical ICA. Results from mechanical thrombectomy (group 1 to 3) were compared to those from A Direct Aspiration first Pass Technique (ADAPT) using the Penumbra 5MAX catheter (group 4).

Eight experiments were carried out for each group. The maximum number of thrombectomy attempts was limited to 3. The primary endpoints were the number and size of the clot fragments. The secondary endpoints were the recanalization rate and flow restoration. Particulate analysis to both the MCA and ACA territories was performed. After thrombectomy or direct aspiration, clot fragments (>1000μm) were first separated from the collected sample by a 10mL graduated pipette. The number and the Feret diameter of these fragments were recorded. Characterization of the emboli smaller than 1000μm was conducted using the Coulter Principle (Multisizer 4 Coulter Counter, Beckman Coulter, Inc., Brea, CA). The results of particulate analysis were further grouped into 4 categories based on the fragment diameter: >200μm, between 100 and 200μm, between 50 and 100μm, and <50μm. The number of the clot fragments with size >200μm was summed to yield the size distribution due to the infrequency of the observation. For emboli with size <200μm, the average number of the clot fragments from 8 replicate experiments in each group was presented.

RESULTS

Particle analysis

In the hard clot model (Figure 2A), thrombectomy with Solumbra or BGC was most effective in reducing the rate of forming clot fragments compared to the CGC or the ADAPT technique. Thrombectomy with the BGC technique reduced total embolic particle creation in the soft clot model by at least 2-fold compared to the other techniques (Figure 2B). For emboli with size >1000μm, using CGC during mechanical thrombectomy significantly increased the risk of hard emboli formation (Figure 2C). There was no statistical difference in the creation of large particle emboli in the soft clot group between all four different access techniques (Figure 2D). In the size range between 200 to 1000μm, the BGC technique generated the most emboli, followed by the CGC, the Solumbra and the ADAPT techniques, regardless of clot type (Figure 2E and 2F). Significant difference was seen between the BGC and the ADAPT techniques in the soft clot group (Figure 2F).

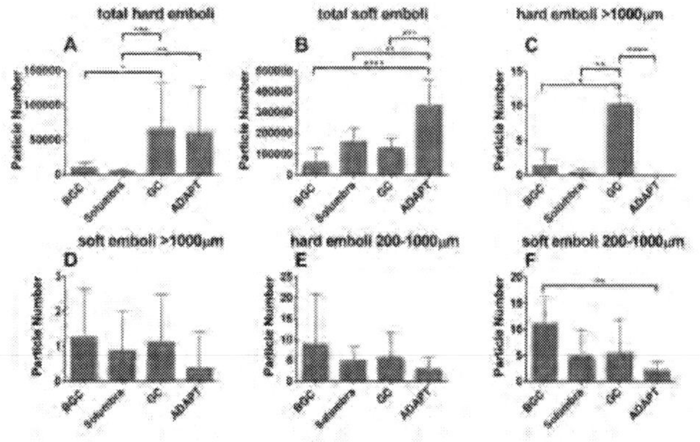

Figure 2 Number of emboli (>200μm) generated during the procedure.

The ADAPT technique led to high numbers of hard distal emboli in the size ranges 100–200μm and 50–100μm (Figure 3A and 3C). On the contrary, for soft distal emboli with size 100–200μm and 50–100μm, there was a trend for ADAPT to offer a benefit (Figure 3B and 3D). The majority of the emboli (>90%) generated during the thrombectomy procedure had a size <50μm (mean size: 12.3μm) (Figure 3E and 3F). There was a significant increase in the number of particles generated during thrombectomy using the soft clot model.

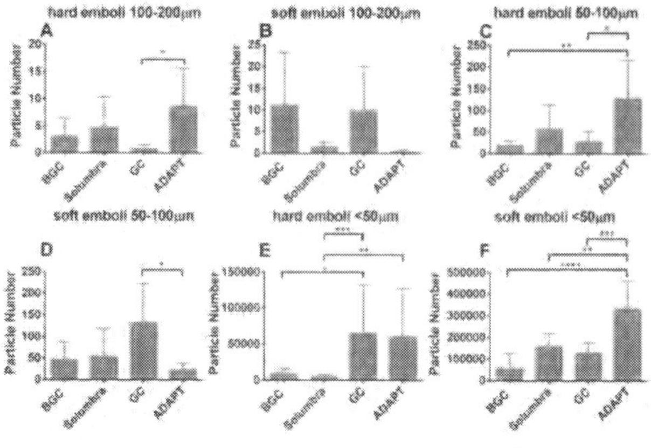

Figure 3 Number of emboli (<200μm) generated during the procedure.

Flow restoration

Complete flow restoration was achieved at the first attempt in all cases regardless of clot type in the Solumbra and ADAPT groups. The mean number of pass attempt in the BGC, Solumbra, GC, and ADAPT groups was 1.1, 1.0, 1.6, and 1.0, respectively. Deployment of Solitaire FR devices was found to be more effective for immediate MCA flow restoration in the hard clot group than in the soft clot group (49.6±19.2mL/min vs 16.9±17.6mL/min). There was no significant difference between the MCA flow before and after the balloon inflation (26.2±26.8 vs 16.6±24.2 mL/min) when the Solitaire FR devices were employed.

DISCUSSION

Distal migration of emboli during mechanical thrombectomy was observed in the previous clinical cases. In this in-vitro study, the number and size of the emboli were characterized as an indication of device safety. It was found that use of the BGC technique during a Solitaire FR thrombectomy was associated with statistically lower rates of soft distal emboli across a broad range of embolic particle sizes. The Solumbra technique may offer a benefit in several hard clot subgroups. Efficient clot retrieval observed in the Solumbra group implied that aspiration location was related to suction efficacy. On the other hand, aspiration with temporary flow arrest using the 8 Fr BGC increased suction efficacy, thereby facilitating clot removal and minimizing the chance of distal embolization.

The limitation of our model is that there was no collateral path in our system representative of the ACom. In summary, our results suggested that when encountering hard clot, use of the Solumbra or ADAPT techniques in addition to the BGC may provide an additional reduction in distal emboli and may be considered for comprehensive distal emboli reduction.

ACKNOWLEDGEMENTS

This work was supported in part from grants by eV3 Neurovascular/Covidien and Philips Healthcare.

REFERENCES

[1] Berkhemer et al. *New Engl J Med*; 372:11-20, 2015.
[2] Chueh, JY et al., *AJNR*, 30:1159-1164, 2009.
[3] Chueh, JY et al., *Stroke*, 44:1396-1401, 2013.
[4] van Ooij, P et al., *Magn Reson Med*, 69:868-876, 2013.
[5] Wahlin, A et al., *AJNR*, 34:1740-1745, 2013.
[6] Chueh, JY et al., *AJNR*, 32:1237-1244, 2011.

SIMULATED COLLAGEN NETWORK REMODELING IN RESPONSE TO STRESS

Carley B. Hovell (1), Hallie Wagner (1), Victor Lai (1,2)

(1) Department of Biomedical Engineering
University of Minnesota
Minneapolis, MN

(2) Department of Chemical Engineering
University of Minnesota – Duluth
Duluth, MN

INTRODUCTION

Human tissues routinely change in response to various environmental cues. These fundamental processes of growth and remodeling are necessary for the development of healthy tissues and the healing of tissues in the aftermath of injury or illness. The essential mechanical functionalities of many tissues are based on underlying networks of collagen fibers [1]. These collagen networks are responsible for maintaining mechanical integrity and tensile strength of the tissue [2]. Collagen network remodeling reflects on the biomechanical relationship between the extracellular matrix and the cells it supports.

In order to understand the growth and remodeling processes, the complex underlying biochemical, bioelectrical, and biomechanical interactions between cells and the extracellular matrix must also be understood. Significant strides have been made through the use of models that treat the cells and extracellular matrix as a continuum [3] [4]. These continuum models can be used to investigate the consequences of tensional homeostasis, which refers to a target stress that cells try to achieve. When this balance is disrupted by increasing the stiffness of the extracellular matrix, certain signaling pathways can be adversely affected. [5] Increased stiffness at the level of the extracellular matrix associated with increased deposition, altered composition, and alignment of the collagenous stroma can promote disruption of epithelial tissues, [6] and insufficient regulation of tensional homeostasis has been proposed to drive disorganization of tissues and promote progression of harmful diseases. [7] It is unclear, however, how tensional homeostasis is achieved at the single cell level. A simulation-based investigation was performed to explore the remodeling process by examining the behavior of a neo-Hookean cell able to degrade and/or deposit fibers in a parallel fibrillar collagen network under an imposed tissue stress.

METHODS

An aligned fibrillar collagen network was generated by Delaunay triangulation of a randomly generated point cloud within a unit cube. Fibers were defined by the Delaunay mesh edges. A constant displacement was then applied to the network. To simulate stress-dependent fiber digestion, fibers could be "turned off," or were functionally removed from the network by setting their stiffness to an extremely small value, if they were below a certain stress threshold. The cellular effects on the remodeling process were simulated by adding fibers back into the network preferentially in the orientation to the cell. The orientation of the cell, Ω_{cell}, was found via the following expression:

$$\Omega_{cell} = \frac{\Omega^k_{fibers}}{tr\left(\Omega^k_{fibers}\right)} \qquad (1)$$

where Ω was the orientation of the network and k = 2.5.[8] Fibers with an orientation value (Ω_{11}) were considered acceptable for "turning on." Once the fibers were added and removed, the stress on the network and the neo-Hookean cell stress were recalculated. The neo-Hookean contribution, σ_{cell}, was found based on the deformation of the fiber network.

The fiber forces in the network were found in this model via the following fiber force relationship:

$$F = \frac{EA}{B}\left[\exp(BE_{fiber}) - 1\right] \qquad (3)$$

where E = Young's modulus of the fiber in the limit of small strain (2.51 MPa for these collagen fibers), A = fiber area, B is a nonlinearity parameter, and E_{fiber} is the fiber Green strain.

RESULTS

Networks of increasing node densities (500-1000) were generated. When the networks were at low density, the effects of individual fibers being added and removed were more readily observed, but a minimum of 500 fibers are needed to achieve statistical homogeneity.

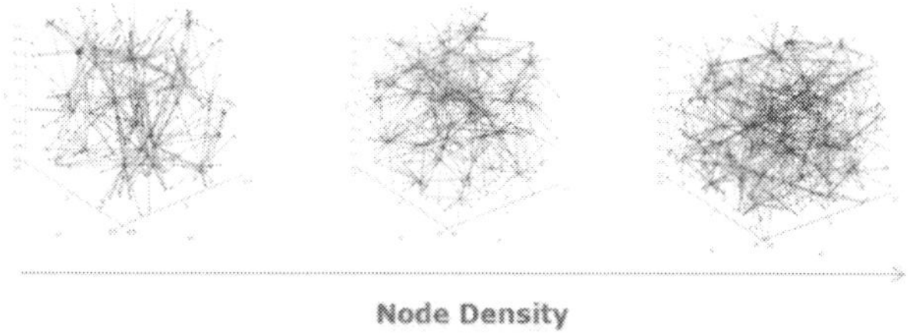

FIGURE 1. NETWORKS OF VARYING NODE DENSITIES GENERATED VIA DELAUNAY TRIANGULATION.

Thus, a network of 650 nodes was selected to undergo stretching due to its low density. The network was stretched in 50 incremental steps at 1% stretch per step for an overall deformation of 50%. The network was stretched in the X direction, and the effect of the stretching on the fibers can be seen in Figure 2.

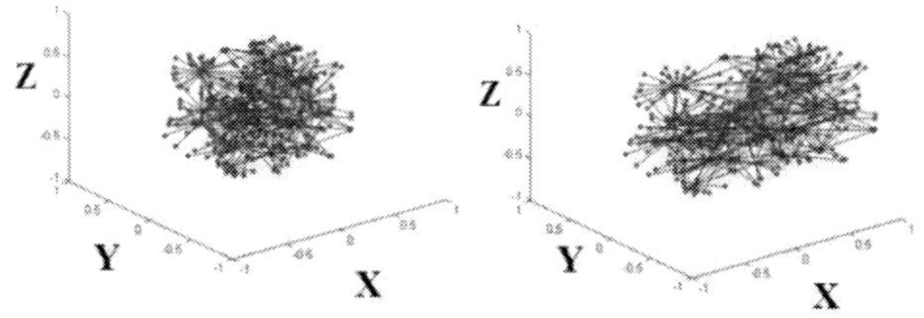

FIGURE 2. LEFT: AN UNSTRETCHED, ALIGNED DELAUNAY NETWORK. RIGHT: SAME NETWORK, STRETCHED TO 50% DEFORMATION IN X-DIRECTION.

For non-digested networks (i.e., fibers only added but not removed), remodeling, as expected, stiffened the network, as seen in Figure 3a below. The cell stress was independent of remodeling since the problem was done in controlled-stretch mode, not controlled-stress mode. Combining the data from Figure 3a and replotting them, we can see in Figure 3b that the cell stress at a given total stress is reduced by greater remodeling, as expected.

When the problem was repeated removing fibers below a critical stretch of λ = 1.05, the results (Fig. 4a) showed similar behavior in that remodeling increased the stiffness and the protection of the cell, but the total amount of material did not increase as much (Fig. 4b).

During remodeling, the overall stress was observed to increase as more and more fibers were removed and added in based on previous simulations. The parallel cell's contribution to the total stress in the system, initially negligible, increased gradually as remodeling took place. Although it was not a drastic adjustment, this result demonstrates potential application to systems, such as arteries, in which the cell stress is a major component.

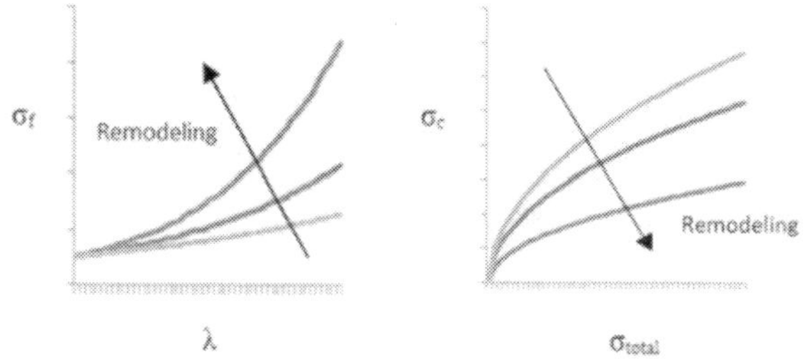

FIGURE 3. LEFT. INCREASING STRESS WITH INCREASED STRETCH AND DEPOSITION OF NEW FIBERS. RIGHT. AS TOTAL STRESS INCREASES, CELL STRESS INCREASES BUT IS SHIELDED SOMEWHAT BY THE NETWORK.

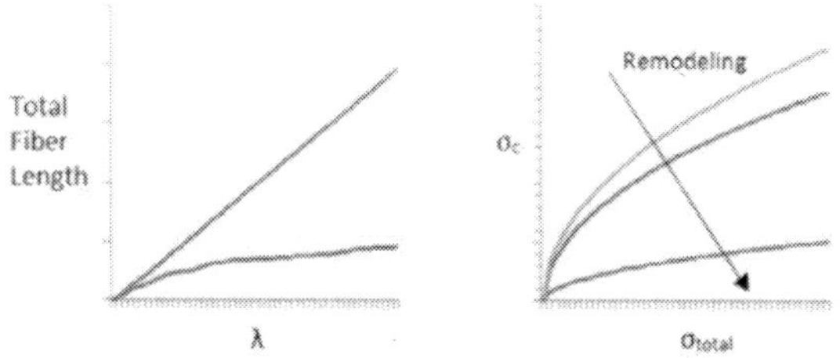

FIGURE 4. LEFT: STRESS SHIELDING IN THE REMODELING NETWORK. RIGHT: TOTAL UNLOADED FIBER LENGTH INCREASES WHEN THERE IS NO DIGESTION BUT CAN INCREASE OR DECREASE IF DIGESTION IS INCLUDED.

DISCUSSION

The resulting increase in overall stiffness of the network with more and more remodeling was as expected. The rising force component of the neo-Hookean cell in parallel with the network as remodeling was performed also yielded a logical and interesting conclusion, which was that the forces of the cell increased with the increasing stiffness of the remodeling network. This simple model did not yield a tensional homeostasis condition (i.e., σ_c did not become independent of σ_{total}), but it showed how a cell might use more matrix to shield itself from large stresses. An appropriate future expansion of this investigation would be to properly simulate the microenvironment of a cell in its entirety.

ACKNOWLEDGEMENTS

This work was supported by the National Institutes of Health (EB005813). We thank the University of Minnesota Biomedical Engineering department and the Minnesota Supercomputing Institute.

REFERENCES

[1] Y.C. Fung, Biomechanics: Mechanical Properties of Living Tissues, second ed., Springer Verlag, 1993. [2] Stylianopoulos, Triantafyllos, and Victor H. Barocas. *Computer methods in applied mechanics and engineering* 196.31 (2007): 2981-2990. [3] Ramasubramanian, A., Taber, L.A., 2008.Biomechanics and Modeling in Mechanobiology 7, 77. [4] Gasser T.C, Schulze-Bauer C.A.J, Holzapfel G.A. J. Biomech. Eng. 2002;124:355–363. [5] DuFort, C.C., Paszek, M.J., Weaver, V.M., (2011) Nat Rev Mol Cell Biol, 12:5:308-19. [6] Webster, Kevin D. et al. Biophysical Journal, 107:1:146-155. [7] Paszek, M. J., Zahir, N., Johnson, K. R., Lakins, J. N., Rozenberg, G. I., Gefen, A., Reinhart-King, C. A., Margulies, S. S., Dembo, M., Boettiger, D. et al. (2005). Cancer Cell 8, 241-254. [8] Barocas,V.H. and Tranquillo, R.T.. 119, 137-145. (1997).

SB³C2015
Summer Biomechanics, Bioengineering and Biotransport Conference
June 17-20, 2015,Snowbird Resort, Utah, USA
SB³C2015-592

MEASURING VESSEL WALL DISPLACEMENT AND CIRCUMFERENTIAL STRAIN USING DISPLACEMENT ENCODED WITH STIMULATED ECHO (DENSE) MRI SEQUENCE

Elizabeth M. Iffrig (1), Xiaodong Zhong (2), W. Robert Taylor (1,3), John N. Oshinski (1,4)

(1) Department of Biomedical Engineering
Georgia Institute of Technology
Atlanta, GA, US

(2) Siemens Healthcare Systems
Atlanta, GA, US

(3) Division of Cardiology
Emory University& Atlanta VAMC
Atlanta, GA, US

(4) Division of Radiology
Emory University
Atlanta, GA, US

INTRODUCTION

Vascular health is strongly dependent upon the vessel tissue constituents that contribute to its mechanics. In the abdominal aorta, this translates to a distensible aorta that expands to accept the pressure from the contraction of the left ventricle and then recoils to send the stored blood to the peripheral vasculature. With age, and accelerated by certain diseases, the aorta loses this capacity for distension and the risk of vascular disease, particularly atherosclerotic disease increases.[1] These changes can eventually lead to a disruptive remodeling cycle resulting in abdominal aortic aneurysm (AAA) formation.

Measuring these important mechanical properties of the vascular wall (distensibility, compliance, and strain) in a human subject is difficult because the wall thickness is very small. In the abdominal aorta, the largest vessel in the body, this thickness is only ~1.25-2 mm.[2] However, such measurements could provide a more sensitive evaluation of aortic health than the current methods (such as pulse wave velocity) of measuring compliance which are unreliable in patients with early aortic disease. Additionally, measuring wall properties in an aneurysm could provide information about risk of rupture, the ultimate often fatal outcome of AAA. In this abstract we propose a method of visualizing and measuring displacement of the abdominal aorta wall via MRI through a displacement encoded with stimulated echos (DENSE) sequence.[3] Previously DENSE imaging has been used to measure displacement and strain in the left ventricular wall. One study examined using DENSE in the ascending aorta but only two frames were acquired.[4] We sought to perform DENSE imaging in a cine mode, allowing us to measure the time resolved displacement and strain maps in the vessel wall.

METHODS

We performed DENSE imaging in 4 healthy volunteers and report the displacement fields and the circumferential strain.

The DENSE sequence is both navigator and ECG-gated and acquisition time was approximately 3 minutes. The scan produces a magnitude (or anatomical) data set, and two phase image sets one encoding phase signal as left-right motion and the other encoding anterior-posterior motion. Each of these data sets captured information at 20 time points over approximately 2/3 of the cardiac cycle.

We acquired DENSE image sets in the abdominal aorta below the origin of the inferior mesenteric artery and above the aortic bifurcation of 4 healthy individuals. Each image set was imported into an in-house Matlab program and the outer and lumen vessel walls were manually contoured. The displacement vector was determined by conversion of the phase images into displacement values using the displacement-encoded frequency (0.25 cycles/mm). Using the first frame as a reference configuration, a transformation matrix (F) was created for each subsequent frame. Using this transformation matrix, we could compute the circumferential strain (E) through Equation 1,

$$E = (F^T \cdot F - I)^{1/2} \quad (1)$$

Where F^T is the transposed deformation matrix and I is the identity matrix.

Additionally, to evaluate circumferential heterogeneity of these mechanical properties, we divided the vessel wall into four anatomical regions (left, anterior, right and posterior) and evaluated the average displacements and average circumferential strain within these regions at each time frame. We also determined the strain vs time curves for each of these regions to calculate differences in their time-dependent behavior. Figure 1 shows a representative image from one of our studies along with the data generated from the analysis described above.

RESULTS

The time-strain curves and time-displacement curves are shown as the average for the four volunteers at each frame in Figure 2. The average maximum displacement in the AP direction for the anterior sector was 2.08±1.50 mm in the anterior direction and 0.35±.17 mm in the anterior direction for posterior sector. The average maximum displacement in the LR direction for the right sector was 0.57±.06 mm to the right, while the average maximum in the left sector was .65 ± .23 mm to the left. While our strain data is beginning to show trends the inter-subject variability is still too high to perform any reliable statistical analysis.

In each patient, displacement in the anterior segment is larger than displacement in the posterior segment. This is expected as the posterior aorta is anchored to the spine and should not move significantly. In comparing the left-right displacement, the left and right sectors are similar although the variability in the right sector is significantly smaller than that of the left due to better retention of the signal in the right wall of the aorta. Overall the data appears to suffer from a high degree of variability that can be directly tied to the difficult of maintaining sufficient signal during the length of the cine scan. As this is the first time this scan has ever been performed in cine in a blood vessel, there is room for improvement.

DISCUSSION

Overall, our results are expected with respect to comparisons between the four segments and the values obtained are reasonable within a physiological context. The anterior aorta is the most unencumbered portion and should exhibit the greatest displacement. Differences between the right and left segments have yet to be uncovered if they exist due to low numbers of data sets. We plan to image more volunteers using this sequence with analysis leading to a comparison of displacement and the circumferential strain in men and women as previous reports have suggested a difference may exist in the tissue properties between the sexes.[5]

ACKNOWLEDGEMENTS

This project was supported by an AHA Pre-Doctoral Fellowship.

REFERENCES

[1] Llaurado et al, *Diabetes Care*, 35:1083-1089, 2012.
[2] Rosero et al, *J of Vasc Surg*, 53: 950-957, 2011.
[3] Spottiswoode et al, *J Magn Reson Imaging*, 27:1019-1027, 2008
[4] Haraldsson et al, *J Clin Magn Reson*, 16:6, 2014.
[5] Astran et al, *J Appl Physio*, 110:176-187, 2011.

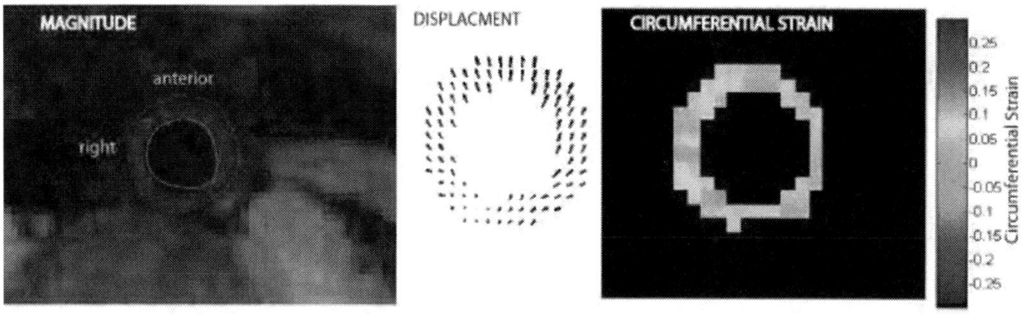

Figure 1. Representative DENSE data. The left image shows a representative magnitude image on which the vessel wall contours were manually drawn. Using the phase data described in the methods, displacement vector maps were created for each frame (center image) and the circumferential strain maps were calculated at each frame as well (left image). The orientation for all three images is shown on the far left.

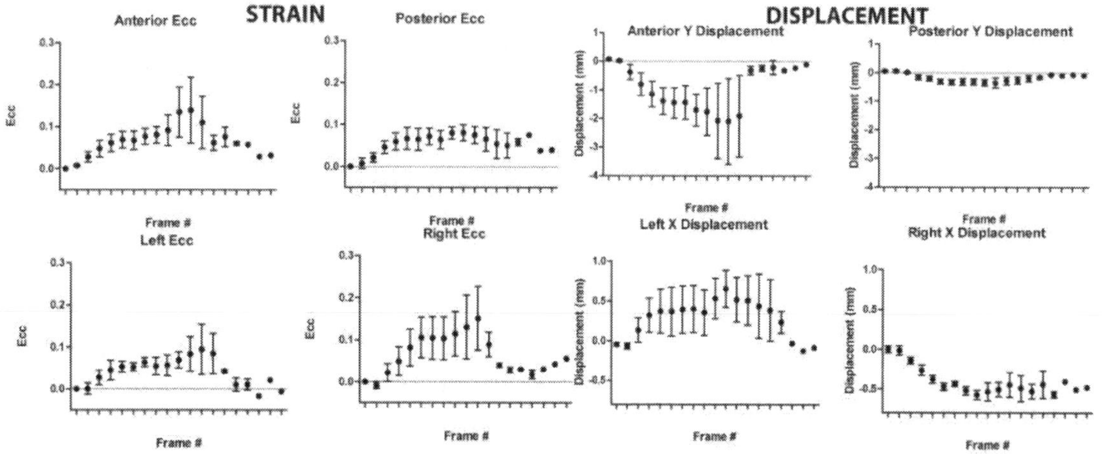

Figure 2. Circumferential Strain and Displacement. The left four plots show the average circumferential strain in our volunteers at each time frame. Because of the high variability and low numbers, no statistically significant trends emerged in comparing these four regions. However, all appear to experience similar maximum strain values. The right four figures show the average displacement. The anterior exhibits the largest degree of displacement while the right and left appear to be similar in their maximum displacement.

SB³C2015
Summer Biomechanics, Bioengineering and Biotransport Conference
June 17-20, 2015,Snowbird Resort, Utah, USA
SB³C2015-593

LOW-CONCENTRATION SALMONELLA DETECTION USING ORBITING MAGNETIC MICROBEADS IN A CONTINUOUS-FLOW MICROFLUIDIC DEVICE

Matthew S. Ballard (1), Drew L. Owen (1, 2), Zachary G. Mills (1), Srinivas K.G. Hanasoge (1),

Peter J. Hesketh (1, 2), Alexander Alexeev (1)

(1) School of Mechanical Engineering
Georgia Institute of Technology
Atlanta, Georgia, USA

(2) Institute for Bioengineering and Bioscience
Georgia Institute of Technology
Atlanta, Georgia, USA

INTRODUCTION

Every year in the United States, pathogens cause over 35 million illnesses, of which over 9 million are caused by foodborne pathogens. Among foodborne pathogens, Salmonella is the leading cause of hospitalizations and deaths [1]. There is a great need for a means of reliable and rapid detection of pathogens [2], which could help to prevent many cases of these illnesses. In order to detect the presence of such bacteria in a reliable and timely manner, a high-throughput device capable of detecting low concentrations of pathogens is required.

The traditional method of detection of pathogenic bacteria involves culture of sample bacteria [3] to raise the concentration level. Although this can result in detection of bacteria even at very low concentration levels [4], it can also be extremely time consuming, requiring an incubation time of several hours or even days.

A variety of microfluidic devices have been developed with the capability of more rapid detection of pathogens [5], but these are typically not able to detect low pathogen concentrations.

Our research is focused on the design of a continuous-flow microfluidic device that uses antibody-functionalized superparamagnetic beads orbiting around soft ferromagnetic discs to sweep through the microchannel and enable rapid capture of salmonella from the entire fluid volume [6-8]. We focus on capture of salmonella, but such a device could be functionalized for capture of other application-specific target cells.

We use numerical simulations to model the flow in the system and gain insights into how to optimize the device for salmonella capture. Monte Carlo simulations are used to obtain statistical data on the ability of the orbiting microbeads to reliably and rapidly capture salmonella from a complex fluid sample.

METHODS

We simulate the microfluidic device as a microchannel filled with a pressure-driven viscous fluid, in which suspended salmonella are modeled as cylinders with spherical caps (2 μm long and 0.5 μm in diameter). The microchannel also contains microbeads of diameter $a = 2.8$ μm orbiting around 3 μm diameter ferromagnetic discs patterned on the channel floor (fig. 1). Periodic boundary conditions in the x (flow) and y (transverse) directions allow us to simulate flow down the channel and a wide channel cross-section, respectively. The z (vertical) direction is subjected to stationary solid wall boundary conditions.

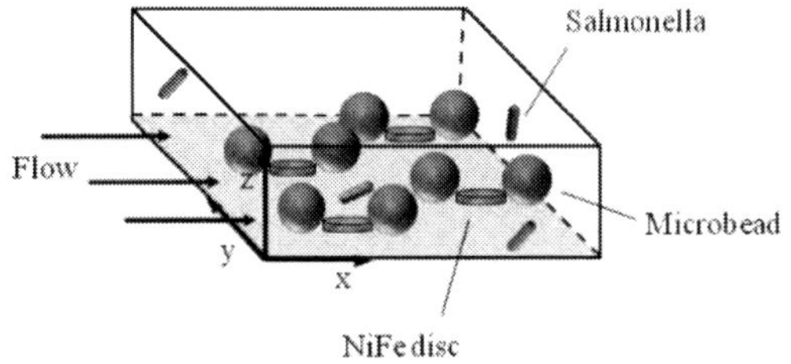

Fig 1: Schematic of section of microfluidic channel

We use a lattice Boltzmann method with immersed boundary technique [9, 10] to model the interactions between the fluid and the solid beads and salmonella. The fluid dynamics is modeled using the lattice Boltzmann method (LBM) [11, 12] with interpolated bounce-back boundary conditions [13] applied at the moving solid surfaces of the beads and salmonella based on the velocity of the solid nodes. The solid nodes experience the appropriate resulting forces from the fluid.

In order to model salmonella capture in an efficient manner, we make the following assumptions. We assume that the salmonella and beads must come into direct contact with each other for capture to occur, since the ligands and receptors are much smaller than the salmonella (tens of nanometers compared to a few microns). Based on the expected values of the association rate constant k_{on} and the dissociation constant K_D [14] and using the method of Kuo et al [15], we estimate that binding occurs much more quickly than the bead orbit time scale, and that the binding force between salmonella and beads is much greater than the hydrodynamic forces pulling them apart. This leads us to assume that salmonella that come into contact with beads are irreversibly captured.

A Monte Carlo simulation is carried out to obtain capture data on a large salmonella population. Salmonella are placed randomly in the simulation domain, and released into the system. When a salmonella bacterium is captured, the capture location is compared to its initial location to obtain a downstream distance to capture (Δx). The values of Δx for the salmonella in the population studied under specified conditions are used to create an empirical cumulative distribution function (fig. 2a), which is used to determine the downstream distance at which a desired percentage of salmonella is captured (fig. 2b).

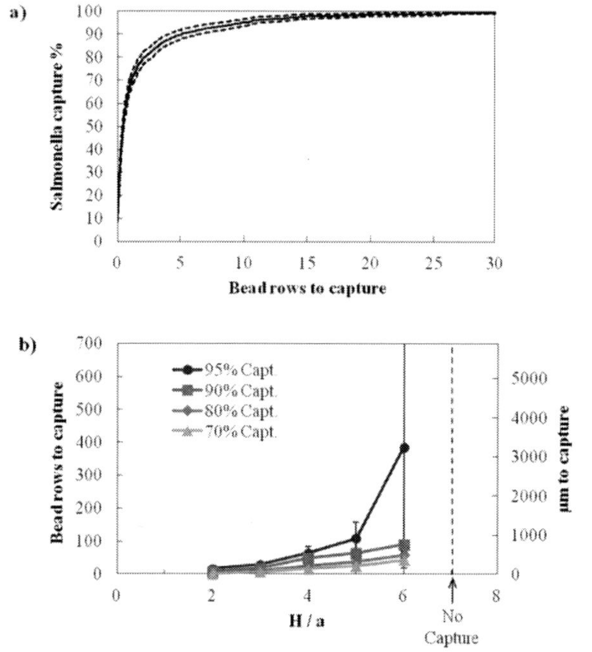

Fig 2: a) ECDF of distance to capture for salmonella population; b) Distance to capture for salmonella in microchannels of varying height to bead diameter ratio

RESULTS

As seen in fig. 2b, our Monte Carlo simulations predict that our device will capture most of the salmonella (95%) within a short distance downstream as long as the microchannel height is within limits (6a or 16.8 µm for our configuration). The distance to capture grows exponentially with channel height, varying from a few hundred microns to a few millimeters for channels of height 2a to 6a, respectively. When the channel is above 6a tall, the microbeads are no longer able to capture salmonella from all of the fluid volume and a lower percentage of salmonella will be recovered. Examination of the fluid field created by the magnetic beads helps to physically explain the relationship between channel height and capture distance. In addition to periodic horizontal flow patterns that cause the salmonella at the height of the beads to zig-zag between the moving beads until they are captured, the beads cause the fluid to move vertically upward away from the high pressure regions in front of the beads, and downward toward the low pressure regions in their wake. The downward flow component serves to draw salmonella down toward the beads, where they are efficiently captured. However, the effect of the beads on the fluid decreases with distance from the beads. Thus, salmonella in taller channels are drawn downward to the beads less, and in the case of very tall channels salmonella at the top of the channel will remain at the top of the channel.

DISCUSSION

Through numerical simulations, we demonstrated that orbiting magnetic microbeads can be used for rapid capture of a low concentration of salmonella in continuous-flow microfluidic channels of length on the order of a few millimeters. For efficient salmonella capture, the channel height should be limited to six times the bead diameter. Parallel implementation of this design on a CD platform will allow it to be a high-throughput process. Further research can be done to determine the effect of adding slanted ridges to the top of the microchannel, so as to cycle the fluid and salmonella from the top of the channel down toward the beads and thus to enable capture in a taller microchannel, which would further increase throughput. We are currently implementing our design both in a single-channel case to validate our simulation results, and in the parallel case to demonstrate high throughput processing. Implementation of this device can have a profound effect on food safety as well as on detection of other cells for which magnetic beads can be functionalized.

ACKNOWLEDGEMENTS

We gratefully acknowledge the United States Department of Agriculture (USDA) and the National Science Foundation (NSF) for financial support.

REFERENCES

[1] Scallan, E., et al., Emerg Infect Dis, 2011. **17**(1).

[2] Arora, P., et al., Biosensors and Bioelectronics, 2011. **28**(1): p. 1-12.

[3] Ivnitski, D., et al., Biosensors and Bioelectronics, 1999. **14**(7): p. 599-624.

[4] Bouguelia, S., et al., Lab Chip, 2013. **13**(20): p. 4024-4032.

[5] Boehm, D.A., P.A. Gottlieb, and S.Z. Hua, Sensors and Actuators B: Chemical, 2007. **126**(2): p. 508-514.

[6] Mao, W., et al., Bulletin of the American Physical Society, 2011. **56**.

[7] Owen, D., et al., Micromachines, 2013. **4**(1): p. 103-115.

[8] Ballard, M., et al. in *Bulletin of the American Physical Society*. 2013.

[9] Alexeev, A., R. Verberg, and A.C. Balazs, Macromolecules, 2005. **38**(24): p. 10244-10260.

[10] Ballard, M., et al., Microfluid Nanofluid, 2014: p. 1-8.

[11] Succi, S. 2001, Oxford: Oxford University Press. xvi, 288 p.

[12] Ladd, A.J.C. and R. Verberg, J. Stat. Phys., 2001. **104**(5-6): p. 1191-1251.

[13] Bouzidi, M., M. Firdaouss, and P. Lallemand, Phys. Fluids, 2001. **13**(11): p. 3452-3459.

[14] MacKenzie, C.R., et al., J Biol Chem, 1996. **271**(3): p. 1527-1533.

[15] Kuo, S.C. and D.A. Lauffenburger, Biophys. J., 1993. **65**(5): p. 2191.

MICROMETER-SCALE MECHANICAL PROPERTIES OF THE TENDON-TO-BONE ATTACHMENT

Alix C. Deymier-Black* (1), Yiran An* (2), Andrea G. Schwartz (1), Guy M. Genin (3), Stavros Thomopoulos (1), Asa H. Barber (2)
*** Contributed equally**

(1) Department of Orthopaedic Surgery
Washington University
St. Louis, MO, USA

(2) School of Engineering and Materials Science
Queen Mary University of London
London, UK

(3) Department of Mechanical Engineering and Materials Science
Washington University
St. Louis, MO, USA

INTRODUCTION

Rotator cuff injuries are extremely prevalent and are present in 50% of the population over the age of 80 [1]. Healing of these injuries is poor, with repair failure rates of up to 94% after massive tears [2]. These high failure rates are caused by the body's inability to recreate the multiscale load transfer mechanisms of the healthy tendon-to-bone attachment.

Healthy tendon-to-bone attachments exhibit a number of hierarchical structures that serve to decrease stress concentrations. At the millimeter length scale, tendon attaches to bone with a splayed geometry that increases the attachment area, decreasing stress [3]. At the micrometer length scale, an interdigitating interface is seen with gradients in mineral and collagen content, strengthening and toughening the interface [4]. At the nanometer length scale, the mineral crystal size and orientation changes along the attachment. These structural and compositional features serve to decrease stress concentrations at the interface between bone and tendon and allow for effective load transfer.

Micrometer length scales provide an opportunity to study tendon-to-bone attachments as a composite and, thus, explore the synergies between material components that define resultant mechanical function. Although millimeter-scale mechanical testing has elucidated some of the mechanisms of load transfer described above [4, 5], few efforts have attempted to examine micrometer scale mechanics. Previous work has loaded tendon-to-bone specimens in uniaxial tension and attempts have been made to measure local deformations[5]. This technique has described tissue-level mechanics, but interpretation of results has been limited by the complex attachment geometry, interference from other tissues (e.g., the tendon sheath), and difficulty in identifying the exact location of failure.

In the current study, we employed atomic force microscopy (AFM) in conjunction with scanning electron microscopy (SEM) to perform tensile tests on micrometer scale beams of tendon-to-bone attachments fabricated using focused ion beam (FIB) milling. This technique allowed for the first-ever visualization and mechanical testing of the micrometer-long transitional tissue between tendon and bone.

METHODS

The supraspinatus tendon-humeral head bone complex was collected from 56 day old CD-1 mice. The tendon-to-bone attachment was mounted in Optimal Cutting Temperature solution (Sakura Finetek Inc, Torrance, CA) and frozen at -80°C. Each tendon-to-bone attachment was then sectioned in 20 µm-thick frozen sections and mounted onto transfer tape.

In order to perform tensile tests on only the attachment tissue, samples were cut into beams with cross-sections of ~4 µm^2 (N = 4 beams from 3 mice). First, the sections were cut into 100 µm x 30 µm beams using laser capture microscopy (LCM). These beams were further trimmed via cryo-FIB to ~ 2 x 2 x 70 µm^3. These beams spanned the tendon-to-bone insertion, including unmineralized fibrocartilage, mineralized fibrocartilage, and the mineral gradient between the two tissues, as measured by energy dispersive x-ray spectroscopy (Fig 1).

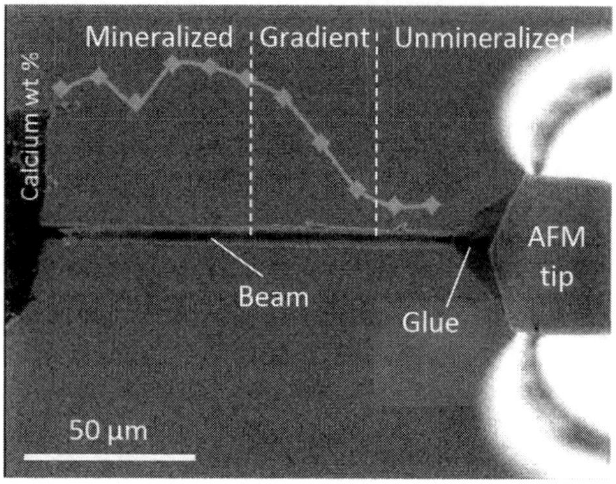

Figure 1: SEM image of a FIB-prepared beam of the tendon-to-bone attachment, connected to the AFM. The Ca wt% across the beam is plotted in orange.

The beams were mounted onto a custom-made Attocube AFM inside a low-vacuum SEM. The mineralized end of the beam was attached to an x-y stage and the unmineralized end was attached to the AFM tip using radiation sensitive glue (Fig 1). Once the glue was cured, the beam was pulled in uniaxial tension until failure. The tensile test was recorded using the SEM in backscattered electron imaging mode.

Using a silicon standard, the interferometry data from the AFM was converted into force displacement curves. The dimensions of the sample beams measured from the SEM were subsequently used to produce corresponding stress strain curves (Fig. 2). The stiffness of the beams was determined from the slope of the best fit line associated with the linear region of the stress-strain curve.

RESULTS

The average stiffness of the beams was 2.5 ± 0.7 GPa. The stiffness values varied between samples due to the variations in beam length and due to the ratios of mineralized to unmineralized tissue in each beam.

To account for these changes, a Reuss model was employed to estimate beam stiffness. The Calcium EDX signal was used to determine the amount of mineralized, unmineralized, and graded tissue in each beam. The mineralized region was assigned a stiffness of 20 GPa and the unmineralized region 100 MPa. The graded region was modelled as a linear gradient in stiffness. The average stiffness calculated from the model was 3.38±1.5 GPa, in close agreement with the experiments (Fig. 3).

The beams all failed at the beginning of the unmineralized tissue, immediately after the end of the mineral gradient.

DISCUSSION

The average stiffness of the beams was nearly 20 times higher than the values measured by tissue-level testing [5]. This difference in stiffness is likely due to scaling effects and the elimination of measurement artifacts. Effects of geometry and tissue interference in the tissue level tests may artificially decrease the stiffness.

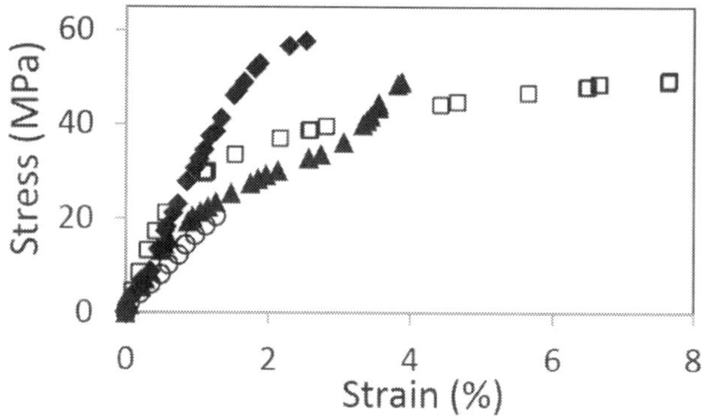

Figure 2: Plot of the stress strain curves for the 4 measured beams of tendon-to-bone attachment.

The results of the Reuss model exhibit a close match to the experimental results and suggest that the moduli of the beams measured by AFM are reasonable for these tissues. Employing such a model also indicates that the stiffness of the tendon-to-bone attachment can be accurately described as mineralized and unmineralized tissue tissues in series, with a linear gradient in between.

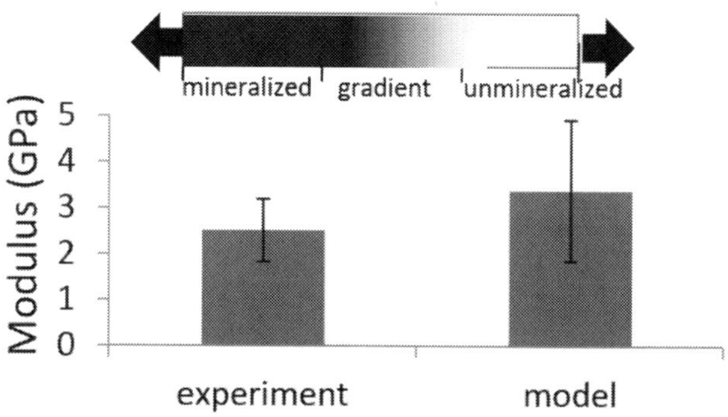

Figure 3: Plot of the average beam stiffness measured experimentally and via the Reuss model shown at top.

All of the samples failed near the beginning of the unmineralized tissues and almost immediately after the end of the mineral gradient. The region of failure coincides with the observed region of increased compliance near the end of the mineral gradient [3]. This region may serve to dissipate stresses during normal loading [6]. The results shown here suggest that this compliant region is also the point of failure during the application of excessive strain, supporting a view of the enthesis as a compliant band that shields tendon and bone from injury.

ACKNOWLEDGEMENTS

Funding was provided by the NSF CAREER 844607 and NIH U01 AR063632. Partial funding for Alix Deymier-Black was provided by NSBRI First Award Fellowship.

REFERENCES

1.Tempelhof, S., S. Rupp, and R. Seil, J. Shoulder Elb Surg. **8**(4): p. 296-299. 2.Galatz, L.M., et al., J Bone Joint Surg Am, 2004: p. 219-24. 3.Liu, Y.X., et al., Mech Mater, 2012. **44**(10): p. 005. 4.Schwartz, A.G., et al., PLoS One, 2012. **7**(11): p. 9. 5.Schwartz, A.G., et al., Bone, 2013. **55**(1): p. 44-51. 6.Liu, Y., et al., J Eng Mater Technol, 2011. **133**(1): p. 011006.

SB^{3}C2015
Summer Biomechanics, Bioengineering and Biotransport Conference
June 17-20, 2015,Snowbird Resort, Utah, USA
SB^{3}C2015-596

UNDERSTANDING THE MECHANICS OF FOCAL CHONDRAL DEFECTS IN THE HIP: A FRAMEWORK TO ADVANCE TREATMENT OPTIONS

Brenden J. Klennert (1,3), Benjamin J. Ellis (1,3), Travis G. Maak (2), Ashley Kapron (2), Tyler Kaiser (1), Jeffrey A. Weiss (1,2,3)

(1) Department of Bioengineering
University of Utah
Salt Lake City, Utah, USA

(2) Department of Orthopedics
University of Utah
Salt Lake City, Utah, USA

(3) Scientific Computing and Imaging Institute
University of Utah
Salt Lake City, Utah, USA

INTRODUCTION

With a mean incidence of 7% in the overall population, osteoarthritis (OA) of the hip is a major clinical problem and significant economic burden. Localized damage to articular cartilage, referred to herein as a focal chondral defect, occurs due to sudden mechanical overload and often leads to osteoarthritis if left untreated [1]. There is a paucity of data on hip chondral defects and therefore surgical treatment is rarely attempted. Hip joints with abnormal shape, articulation or instability are especially susceptible to this kind of injury [2]. A classic pathoanatomy of the hip that causes instability leading to chondral damage is acetabular dysplasia [3]. This is a congenital and developmental deformity characterized by a shallow acetabulum and lack of femoral head coverage [4]. In dysplastic hips, chondral defects commonly occur in the anterosuperior region of the acetabulum (Fig. 1) [5].

A clear understanding of the mechanical effects of chondral defects in the hip on the surrounding cartilage and labrum mechanics would enable us to identify the prognosis and best treatment strategy on a patient-specific basis and thus hopefully delay or prevent the onset of osteoarthritis in this patient population. The objective of this study was to assess the mechanical effects of clinically relevant chondral defects on the mechanics of the articular cartilage layers and labrum in the dysplastic hip.

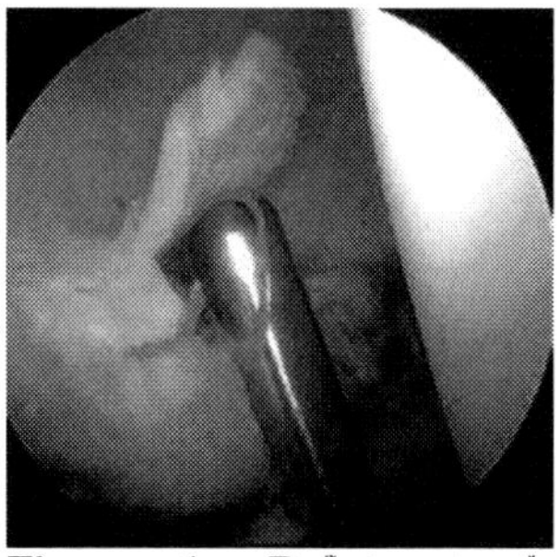

Figure 1: Defect at the chondrolabral junction of the articular surface of the acetabulum.

METHODS

A cartilage defect was simulated in a finite element model of a dysplastic hip (Fig. 2A). The FE model of a left hip from a 25 yo, 70 kg female was selected from a cohort of ten patients with acetabular dysplasia (center edge angle [CEA] 14.8 ± 4.6°, acetabular index [AI] 18.1 ± 6.9°) modeled in a previous FE study [6]. The selected hip was characterized by a CEA of 19.6° (falling just outside the first standard deviation of the dysplastic cohort) and an AI of 16.4° (well within the first standard deviation of the dysplastic cohort). The defect was modeled as a semicircular hole in the acetabular cartilage at the chondrolabral junction (Fig. 2B). Two sizes of the defect were examined: one with a diameter of 3.5-mm, and a second with a diameter of 7.0-mm. A model with no defect was also analyzed for comparison.

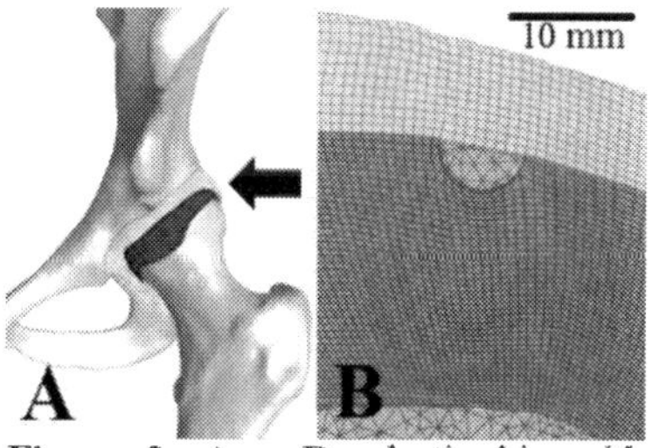

Figure 2: A – Dysplastic hip with deficient bony coverage (arrow). B – Mesh of acetabulum with large defect.

The FE models were subjected to loading and kinematics at five points representing the stance phase of the walking gait cycle: heel strike, heel strike mid stance, mid stance, mid toe off, and toe off. At

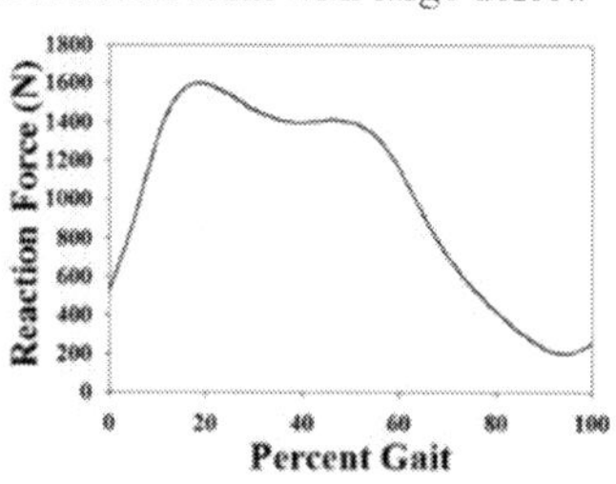

Figure 3: Gait cycle reaction forces.

each point, the joint position and joint reaction force (as % bodyweight) were defined using data from a previous study [7] (Fig. 3). The models were analyzed with the FE software FEBio (www.febio.org) [8]. Post-processing and generation of fringe plots was completed in PostView. For the five points of stance phase, the maximum shear stress, τ_{max}, and first principal Legrange strain, E_1, of each element was evaluated. Overall peak τ_{max} and E_1 on the articular and osteochondral surfaces of the acetabulum were recorded. Contact stress on the articular surface was also evaluated.

RESULTS

Peak τ_{max} increased with defect size. For the small defect and large defect models, peak τ_{max} was located on the articular surface at the start of stance phase, but as the activity progressed, the location changed to the osteochondral surface (Fig. 4 A-C).

Peak E_1 also increased with defect size (Fig. 5). For both τ_{max} and E_1, the peak values occur around the defect area, most commonly on the edge of the defect. While the addition of a defect increased maximum shear stress and 1st principal strain, the effect on contact stress on the articular surface was fairly minimal (Fig. 6). The contact stress slightly decreased in some of the undamaged regions, and slightly increased in other regions of the cartilage.

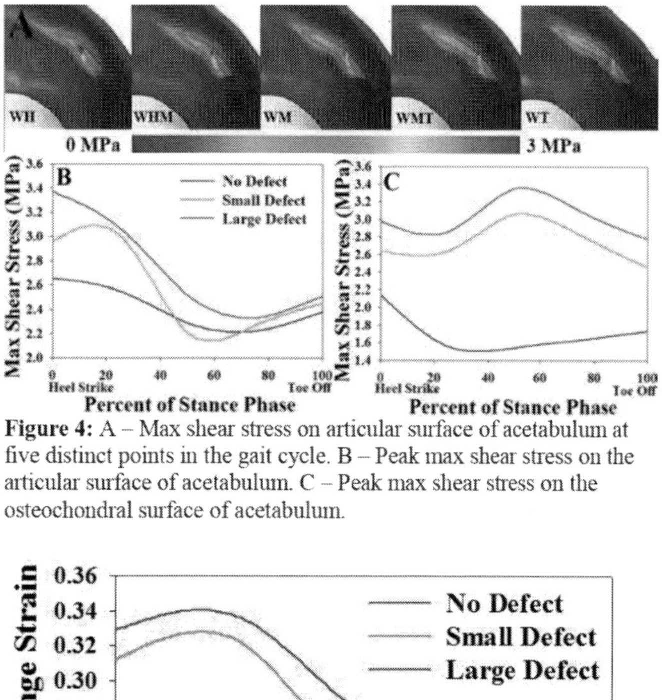

Figure 4: A – Max shear stress on articular surface of acetabulum at five distinct points in the gait cycle. B – Peak max shear stress on the articular surface of acetabulum. C – Peak max shear stress on the osteochondral surface of acetabulum.

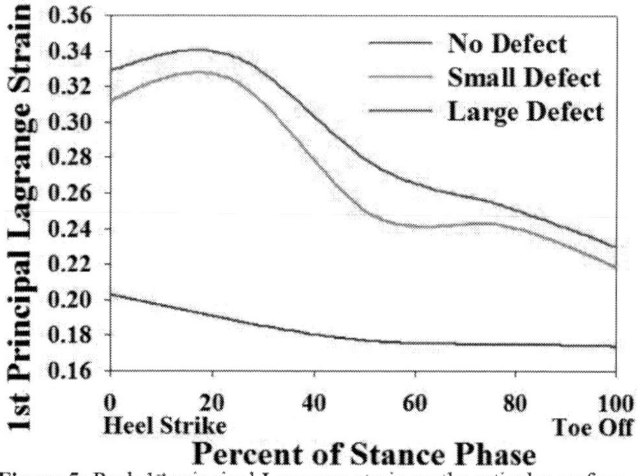

Figure 5: Peak 1st principal Lagrange strain on the articular surface of the acetabulum.

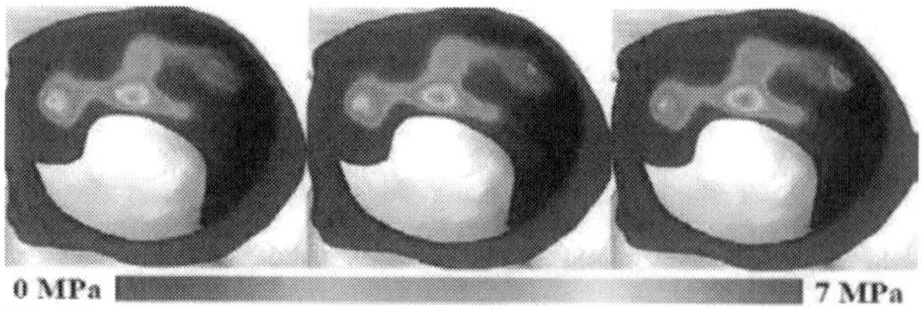

Figure 6: Contact stress distribution for heel strike for no defect (left), small defect (middle), and large defect (right) models.

DISCUSSION

In this study, the mechanical effects of clinically relevant chondral defects were simulated in a dysplastic hip using the FE method. The presence of a defect significantly increased the maximum shear stress and first principal strain of the acetabular cartilage, and this effect was amplified with larger defects. Over the stance phase of walking, the location of peak stress and strain on the acetabulum changed from the articular surface to the osteochondral boundary. Increased stress and strain in multiple regions may cause further progression of the defect and surrounding cartilage.

This study provides an important first step to the study of focal chondral defects, but further and expanded study is warranted. While the parameters (location, shape and size) of the defect simulated herein were selected to represent a defect type commonly observed in dysplastic hips, a range of defect parameters are observed clinically. Future work will extend this methodology to study the effects of defect parameters in different patient groups. Finally, only five points representing the stance phase of walking were evaluated in this study. For a comprehensive understanding of how defects affect cartilage mechanics, additional points in the gait cycle and different activities will be analyzed in the future.

Cartilage damage and osteoarthritic degeneration are thought to result from mechanical overload in the joint, such as increased shear strain at the osteochondral interface. In this study, chondral defects in a dysplastic hip were shown to cause substantial mechanical changes in the acetabular cartilage. This study demonstrates finite element modeling is a valuable tool to understand the mechanical implications of chondral defects. This study also serves as the foundation for future work investigating the relative effects of a variety of chondral defects in different hip pathologies. Further data on chondral defect mechanics may be used to improve the clinical understanding of defect prognosis and evaluate/refine cartilage repair strategies.

ACKNOWLEDGEMENTS

LS Peery Discovery Program in Musculoskeletal Restoration Investigator Grant and University of Utah Seed Grant.

REFERENCES

[1] Ding, C., et al., *Osteo and Cart*, **13**(3):198-205, 2005. [2] McCarthy, J.C., et al., *Clin Orthop*, 405:122-8, 2002. [3] Harris, W.H., *Clin Orthop*, 213:20-33, 1986. [4] Cooperman, D.R. et al., *Clin Orthop*, 175:79-85, 1983. [5] Yen, Y.M. et al., *Sports med and arthro*, **18**(2):83-9, 2010. [6] Henak, C.R., et al., *Osteo and Cart*, **22**(2):210-7, 2014. [7] Bergmann, G., et al., *J Biomech*, **34**(7):859-71, 2001. [8] Maas, S.A., et al., *J Biomech Eng*, **134**(1):011005-10, 2012.

SB³C2015
Summer Biomechanics, Bioengineering and Biotransport Conference
June 17-20, 2015, Snowbird Resort, Utah, USA
SB³C2015-597

MECHANO-ACTIVATABLE MICROCAPSULES FOR TUNABLE DRUG DELIVERY

Bhavana Mohanraj (1), Fuquan Tu (2), Daeyeon Lee (2),
George R. Dodge (3, 4), Robert L. Mauck (1, 3, 4)

(1) Department of Bioengineering
University of Pennsylvania
Philadelphia, PA, USA

(2) Department of Chemical and
Biomolecular Engineering
University of Pennsylvania
Philadelphia, PA, USA

(3) Department of Orthopaedic Surgery
University of Pennsylvania
Philadelphia, PA, USA

(4) Translational Musculoskeletal Research Center
Department of Veterans Affairs
Philadelphia, PA, USA

INTRODUCTION

A particularly desirable feature of controlled drug delivery is self-regulation, wherein physiological feedback actively regulates release kinetics. Extant systems in this domain commonly employ stimuli-responsive vehicles that rely on internal triggers to precipitate release, such as temperature, pH, or chemical reactions [1, 2]. The sensitivity of these triggers can be tuned by biomaterial composition or structure to regulate release in a given physiologic or pathophysiologic state. However, to date, such delivery vehicles have not been specifically tuned for mechanically activated release.

Tissues within the body experience mechanical perturbations across multiple force magnitudes and length scales, from mechanotransduction at the cell level [3] to the dynamics of load-bearing joints. These forces not only maintain tissue homeostasis, but can also initiate degenerative processes when applied at supra-physiologic levels [4, 5]. Given the centrality of mechanical loading in normal tissue function, our objective was to develop a mechanically activated drug delivery system to stimulate regeneration and repair in mechanically loaded musculoskeletal tissues (e.g. cartilage, muscle, bone). Towards this goal, we developed a novel class of mechano-activatable microcapsules (MAMCs, Ø~50-100μm), and illustrate here the tuning of design parameters to enable differential responses to varying applied mechanical stimuli.

METHODS

MAMC Fabrication: Mechanically activated microcapsules (MAMCs) were produced with a poly(lactic-co-glycolic) acid (PLGA) copolymer shell (doped with Nile Red) and an aqueous core containing FITC-dextran, 2 MDa, (Figure 1C). Microcapsules were fabricated using a glass-capillary microfluidic system to produce a highly monodisperse water-in-oil-in-water (W/O/W) emulsion with ~100% encapsulation efficiency [6] (Figure 1A, B). Using osmotic annealing, the wall thickness to radius ratio (t/D) was tuned by modulating the concentration of PLGA (50:50) in the middle phase, as well as the solute (NaCl) concentration in the inner and outer phases [6]. Three batches of microcapsules were fabricated, with a t/D ratio of (1) 0.008 (t~400nm, D~50μm), (2) 0.012 (t~600nm, D~50μm), and (3) 0.015 (t~1500nm, D~100μm). All capsules were maintained in 0.15M NaCl at 23°C.

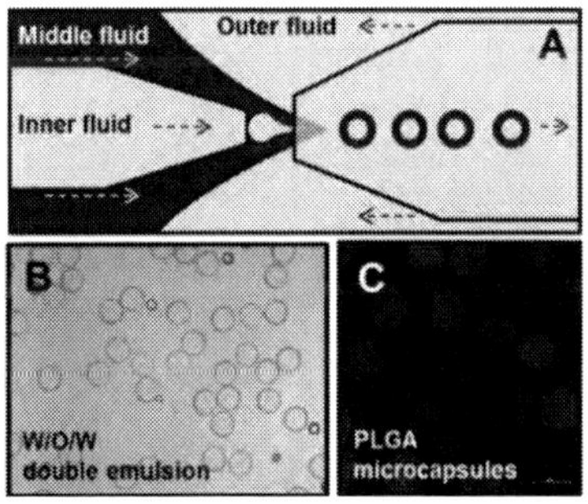

Figure 1. (A) Schematic and (B) microscopy image of W/O/W double emulsion generated from a capillary microfluidic device for fabrication of (C) PLGA microcapsules.

MAMC Compression: To demonstrate mechano-activation, a single layer of MAMCs was subjected to increasing levels of load between two glass coverslips (Figure 2). Microcapsules were loaded to ~50g, 200g, and 500g; intact, unloaded MAMCs served as a negative control while MAMCs sheared between two glass slides served as a positive control. After overnight incubation at 23°C in phosphate-buffered saline (PBS), z-stack images were collected at 20X magnification using a Nikon confocal microscope. The mid-slice was used to determine whether a single MAMC was "full" or "empty" (devoid of FITC-dextran). Fluorescence intensity of the supernatant was also assayed to measure bulk MAMC release (Ex/Em: 490/520).

MAMC Mechano-activation in 3D Hydrogels: To validate mechano-activation in a three-dimensional construct, MAMCs were embedded in a photo-crosslinked poly(ethylene glycol) diacrylate (PEGDA) hydrogel. Using a custom micromechanical compression device mounted atop a confocal microscope (Figure 3A) [7]. MAMC-laden hydrogels were compressed in unconfined compression, from 0-20% strain at steps of 4%, followed by compression until hydrogel failure. Individual MAMCs (n=2/hydrogel) were tracked through 20% strain and z-stacks collected (at 20X) at each strain level. Microcapsule deformation was quantified by measuring the elongation of the microcapsule at the mid-point for each strain step.

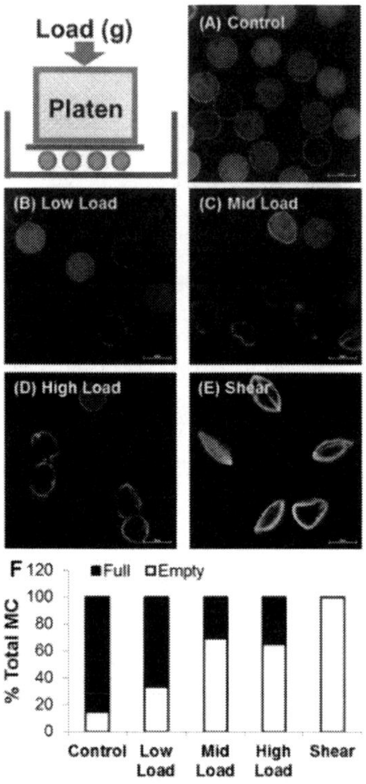

Figure 2. Mechano-activation of single MAMCs. (A-E) MAMCs containing dextran (green) with labelled shells (red) deform and fracture at higher loads. (B) Quantification of release as a function of load.

RESULTS

To assess mechano-activation, isolated MAMCs (t/D~0.015) were subjected to increasing load (Figure 2). MAMC deformation increased with applied load, resulting in a graded pattern of microcapsule rupture and release of encapsulated FITC-dextran. Sheared MAMCs served as a positive control, with ~100% of MAMCs devoid of a fluorescent signal. The fluorescence intensity of the buffer solution also correlated with applied load, confirming MAMC mechano-sensitivity (not shown).

To demonstrate mechano-activation within a 3D context, MAMCs were embedded in a 30% (v/v) PEGDA hydrogel and evaluated using a confocal-mounted compression device (Figure 3A). 30% PEGDA was chosen as an encapsulating matrix as it has a stiffness comparable to both native and mature engineered cartilage [14,15]. With compression of the hydrogel, MAMCs deformed in a dose-dependent fashion, becoming ellipsoid at 20% strain and rupturing upon hydrogel fracture (observed at ~57% strain). At t/D ratios of ~0.012 and 0.015, MAMCs underwent folding and/or sharp bending of the capsule wall, while at a t/D ratio of ~0.008, MAMCs maintained a rounded appearance with compression.

DISCUSSION

In this study, we developed a method to generate microcapsules that are mechano-sensitive and characterized the behavior of their mechano-activatable properties towards the development of a novel platform for controlled and self-regulated drug delivery. Within this MAMC population, the amount of rupture and release was readily controlled by the magnitude of the applied load. While low loads resulted in rupture of only a small fraction of MAMCs, higher loads resulted in >50% release. Within the context of a 3D environment, the deformation behavior and mechanism of MAMC rupture was in part dependent on the wall thickness-to-radius ratio (t/D) [8].

Previously, drug delivery mediated by mechanical rupture has primarily focused on mechano-chemical or thermal mechanisms to initiate release, wherein pH or temperature induced phase changes

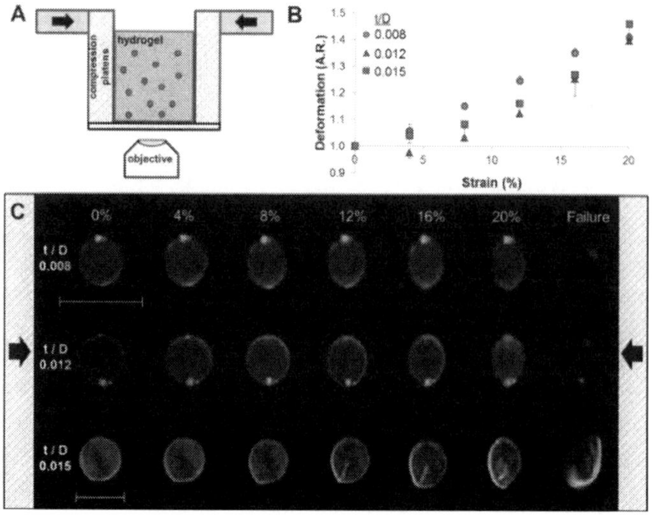

Figure 3. Mechano-activation in a 3D context. (A) Micromechanical device for imaging gel-MAMC composites with deformation. (B) Quantification of MAMC aspect ratio (A.R., n=2) as a function of fabrication parameters. (C) MAMC deformation during progressive compression of the hydrogel.

result in structural defects in the capsule shell [2,9]. Our MAMCs are designed in a manner analogous to that of self-healing polymers used in material science applications, where microcapsules containing a "healing agent" are embedded in a matrix and, upon fracture, initiate polymerization and 'repair' [10]. Building from this self-healing concept, mechano-activation of MAMCs can be designed to stimulate repair or regeneration through biologic mechanisms spurred by the release of factors from the MAMCs. Future studies will focus on tuning MAMC mechanical behavior via alteration of fabrication parameters, with the goal of designing a cohort of microcapsules with different failure thresholds in order to program sequential therapeutic delivery. Techniques such as atomic force microscopy and finite element modelling will be used to characterize MAMC strain and rate-dependent deformation, rupture, and release profiles. In addition, initial efficacy studies will measure the chondrogenic effects of growth factors (e.g. TGF- β) released from MAMCs in a dynamically loaded engineered cartilage model. This work provides an initial characterization of MAMCs, and sets the stage for the widespread application of this drug delivery system for musculoskeletal repair and regeneration.

ACKNOWLEDGEMENTS

This work was supported by the National Science Foundation, the Department of Veterans' Affairs, and the Penn Center for Musculoskeletal Disorders.

REFERENCES

[1] Kost, J. et al., *Adv Drug Deliv Rev*, 46:125-148, 2001.
[2] Datta, S.S. et al., *Adv Mater*, 26:2205-2218, 2014.
[3] Engler, A.J. et al., *Cell*, 126: 677-689, 2006.
[4] Sun, H.B., et al. *Ann NY Acad Sci*, 1211:37-50, 2010.
[5] Griffin, T.M., et al. *Exerc Sport Sci Rev*, 33:195-200, 2005.
[6] Tu, F. et al., *Langmuir*, 28:9944-9952, 2012.
[7] Farrell, M.J. et al., *Eur Cell Mater*, 23:425-440, 2012.
[8] O'Sullivan, M. et al., *Langmuir*, 25: 7962-7966, 2009.
[9] Abbaspourrad, A. et al., *Langmuir*, 29:12697-12702, 2013.
[10] White, L. et al., *Nature*, 409:794-797, 2001.

SB³C2015-598

DESIGN OF A NOVEL SHOCK TUBE SYSTEM FOR BLAST INDUCED TRAUMATIC BRAIN INJURY

Andrew B. Robbins (2), Dr. Raoul Van Loon (3), Ashok K. Shetty (4), Michael R. Moreno (1,2)

(1) Mechanical Engineering
Texas A&M University
College Station, Texas, USA

(2) Biomedical Engineering
Texas A&M University
College Station, Texas, USA

(3) College of Engineering
Swansea University
Swansea, Wales, UK

(4) Department of Molecular and Cellular Medicine
Texas A&M Health Science Center
Temple, Texas, USA

INTRODUCTION

Over the last decade, Operations Enduring Freedom (OEF) and Iraqi Freedom (OIF) have produced a dramatic increase in the prevalence of mild and moderate blast induced traumatic brain injury (bTBI) that is often undiagnosed and untreated initially. It has been estimated that 10-20% of returning OIF/OEF veterans have suffered some form of TBI. Additionally, nearly 80% of all casualties from these operations have been determined to stem from improvised explosive devices. These injury patterns have demonstrated a need for a better understanding of the mechanisms of blast induced traumatic brain injury, and there has been an uptake in research of this type. Since the 1950's shock and blast tubes have been used to study blast injury, and the last decade has seen a resurgence of this type of research. Despite this, the design of shock tubes has changed very little, and there has been relatively little effort put into replicating blast waves produced by different ordinances. We present a novel design that aims to address these issues.

METHODS

The Robbins-Moreno shock tube system was designed to improve the general design of standard shock tubes used for bTBI in several ways. Some of the design features included were:

- Tunability
  - The output wave of the device is tunable to generally match the waves produced by different ordinance.
- Mild to moderate injury
  - Must be capable of inducing mild to moderate injury in rodent models.
- MRI Compatibility
  - The does not interfere with MRI equipment and can be placed in or near an MRI.
- Portability
  - The device is easy to transport between multiple sites.

These design requirements were met using a 12 inch diameter steel tank for the driver, Mylar membranes, and a modular driven section made of up to four 5 foot sections of varying diameter polymer tubing (Figure 1).

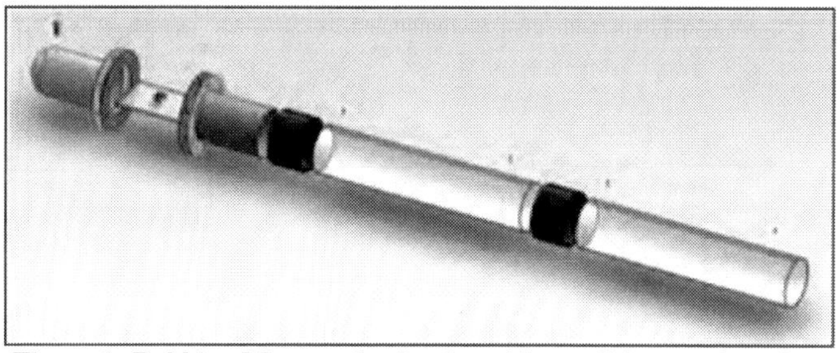

Figure 1: Robbins-Moreno shock tube with two 5 foot sections for the driven segment.

CFD analysis of the tenability of the shock tube was performed using Star CCM+. 240 simulations (axisymmetric, AUSM method) were run using different driven segment diameters (converging and diverging), driven segment lengths, working fluids, and rupture pressures. The results were analyzed and the peak overpressure, the positive impulse, and the positive phase duration were determined. These characteristics were then used to produce a set of empirical relations that can be used to tune the output wave [1].

RESULTS

Tunability

The modularity of the shock tube enables easy modification of the length and diameter of the shock tube. CFD analysis as well as limited experimental verification have shown that these parameters combined with the variable burst pressure enable the peak over pressure and the impulse (or peak over pressure and phase duration) to be changed independently of each other. Table 1 shows how modifying various parameters of the shock tube affect the characteristics of the shock wave produced.

	Peak Overpressure	Positive Phase Duration	Positive Impulse
Burst Pressure	↑	↑	↑
Tube Length	~↑	-	-
Tube Diameter	↓	-	-
Gamma	↑	↓	↓
↑ increases as input increases			
↓ decreases as input increases			
~↑ marginally increases as input increases			
Gamma: Air – 1.4 ; He – 1.66			

Table 1: Effect of changing shock tube parameters on wave characteristics.

Mild to Moderate Injury

In preliminary studies, we have induced mild injury in a mouse animal model. Twenty-five mice were exposed to blast overpressure of about 14 psi. After a six month waiting period, a number of neurobehavioral assessments were performed including the water maze test, pattern separation test, object in place test, forced swim test, novelty suppressed feeding test, elevated plus maze test. Figure 2 shows some selected results of the neurobehavioral tests that show that the mice have been impaired [2][3].

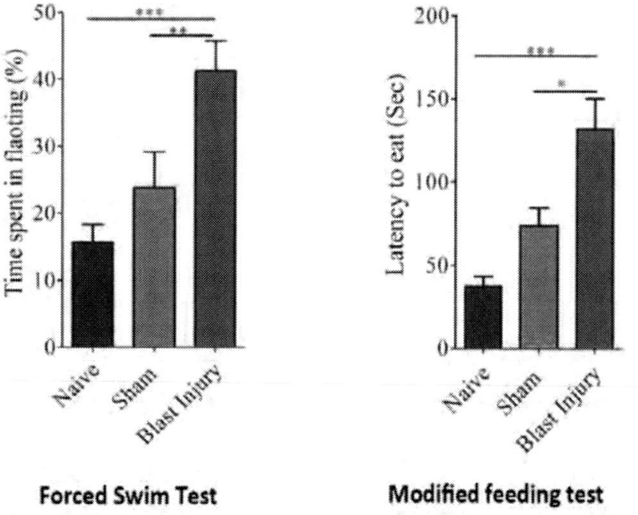

Figure 2: Depressive-like behaviour tests.

MRI Compatibility

Excluding the driver and pressure transducers, the entire construction is metal free and has a maximum length of about 23 feet, enabling the end of the shock tube to be placed into an MRI machine or near an MRI machine with the metallic components being 20 feet from the magnet. This permits imaging immediately before and after blast with essentially no time delay between injury and image acquisition. We have performed some preliminary tests of the effects of a blast wave using soft boiled eggs (figure 3).

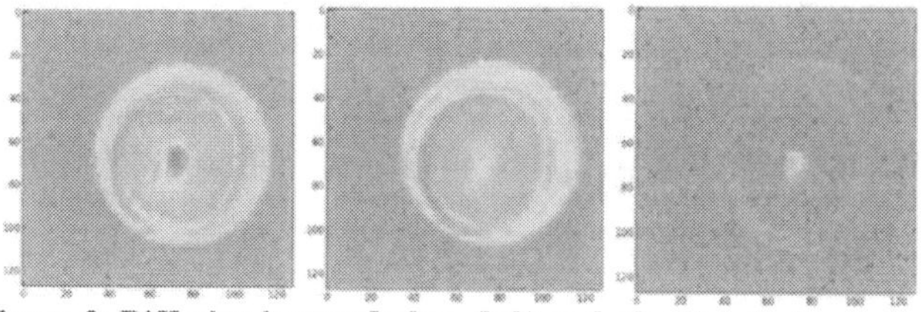

Figure 3: Diffusion images before (left) and after (center) exposure to blast waves and the difference between the images (right).

Portability

The modular polymer construction results in a shock tube that, while large, is easily disassembled and moved. Two individuals can break down the system and load it into a truck in approximately 15 to 30 minutes, and have the system back up and running in an additional 30 minutes. We routinely transport the system offsite for single day animal studies, and can tear down, setup, expose 25 animals to blast and tear down again in a single day.

DISCUSSION

The Robbins-Moreno shock tube provides the ability to study several new questions. The tunability of the device will allow us to more closely match the characteristics of the blast waves produced by ordinance. Additionally ability to independently change the over pressure and the impulse will enable us to study how these different wave characteristics contribute to injury.

The MRI compatibility of the shock tube enables rapid imaging pre and post blast, as well as novel imaging techniques such as elastography without having to remove the animal from the imaging apparatus. These techniques will enable us to obtain insight into the changes in brain tissue caused by the mechanical loading by a shock wave including changes to the mechanical properties of the tissue.

Our shock tube system has proven to be highly portable, capable of being tuned to produce different waveforms, able to induce mild to moderate bTBI, and MRI compatible. These features will enable new studies into the mechanics of bTBI, as well as the neurobehavioral consequences of this devastating injury.

For more information on the Robbins-Moreno system contact Michael R. Moreno via email: michael.moreno@tamu.edu

ACKNOWLEDGEMENTS

This research was funded in part by a NSF EAGER award.

REFERENCES

[1] Anumolu, P., Thesis, Texas A&M University, 2014.
[2] Mishra V., *Cell Transplantation*, 23:778, 2014.
[3] Shetty A. K., *Soc. Neurosci. Abstr.*. 24:522, 2014.

SB³C2015
Summer Biomechanics, Bioengineering and Biotransport Conference
June 17-20, 2015,Snowbird Resort, Utah, USA
SB³C2015-599

DEVELOPMENT OF A INDUCED PLURIPOTENT STEM CELL DERIVED CARDIOMYOCYTE SEEDED FIBRIN SUTURE FOR CARDIAC REGENERATION

Katrina J. Hansen (1), Michael A. Laflamme (2), Glenn R. Gaudette (1)
(1) Biomedical Engineering, Worcester Polytechnic Institute, Worcester, Massachusetts, USA (2) Center for Cardiovascular Biology, School of Medicine, University of Washington, Seattle, Washington, USA

INTRODUCTION

The American Heart Association reported that 7.6 million Americans survived a myocardial infarction (MI) in 2012, and 5.1 million Americans suffer from severe heart failure [1]. After an MI, infarcted tissue is negatively remodeled into non-compliant, non-contractile scar tissue instead of healthy contractile myocardium, ultimately decreasing cardiac output, which puts the patient at risk of severe heart failure. The low proliferation rate of cardiomyocytes [2] limits the hearts capability of endogenous regeneration. The current "gold standard" treatment for heart failure is a heart transplant; however, due to limited donor availability and the potential for immune rejection, a heart transplant is not an ideal treatment option [3]. Consequently, researchers have investigated other therapeutic options for heart failure, such as cellular therapy. Clinical trials have demonstrated that some cell types exhibit efficacy in increasing cardiac function after a myocardial infarction [4]. Many of these clinical trials have focused on the delivery of mesenchymal stem cells (MSC), which have been shown to have an immunomodulatory effect within the infarct region that is not capable of providing a robust restoration of mechanical function [3]. Therefore, new cell types need to be investigated for their potential to improve mechanical function after an infarct.

Delivering a contractile cell, such as an induced pluripotent stem cell derived cardiomyocyte (iPS-CM), into the infarct area may improve the mechanical function lost due to an MI [5]. The iPS-CM is not a mature cardiomyocyte; however, this cell type offers more similarities to the host tissue than a MSC, which may allow for better functional integration into the infarcted region. Although multiple cell types have shown some efficacy in improving cardiac function, all delivery methods suffer from low engraftment rates (0-10%) [6]. Previously, we have developed a fibrin microthread suture that can be used for efficient cell delivery to the myocardium [7]. We have shown that human MSCs can be seeded onto the fibrin suture and implanted into the rat heart with a 63.6±10.6% engraftment rate [7].
Previous studies have indicated that fibrin is a suitable biomaterial to support iPS-CM attachment and contraction [8]. The aim of this study is to develop and characterize an iPS-CM seeded fibrin suture for future use as a delivery platform for cardiac repair.

METHODS

Fibrin Suture Manufacturing

Fibrin microthreads were generated by co-extruding fibrinogen (70mg/mL; MP Biomedical) and thrombin (8U/mL; Sigma) into a HEPES (pH 7.4) buffered bath, then air dried to create discrete fibrin microthreads. Individual microthreads are incorporated together utilizing a suture needle to create a 2cm suture as previously described by Proulx *et al.* [9].

iPS-CM Seeding

iPS-CM were obtained cryopreserved from collaborators at the University of Washington. Cells were thawed and immediately seeded at a density of 150,000 cells/cm^2 onto either 0.67% gelatin (from bovine skin; Sigma) coated coverslips, used as a control, or onto fibronectin (10μg/mL; Sigma) coated fibrin sutures. iPS-CM seeded onto fibrin sutures were allowed to adhere for 24 hours in our custom seeding device and were transferred to un-coated 12 well plates with fresh aprotinin supplemented RPMI-B27 media (Life Technologies). Media is supplemented with aprotinin to diminish enzymatic digestion of fibrin.

Contractile Mechanics Analysis

iPS-CM were recorded using a high speed camera (Fastec HiSpec4; 60 frames per second; 1697x1710 pixels) mounted on an inverted microscope (LEICA DMIL) at days 3,7,10,14,17,21,24, and 28 post thaw. High density mapping (HDM), a phase correlation technique to measure sub-pixel displacement fields, was used to analyze iPS-CM contraction. Contraction parameters of frequency, maximum strain, and percentage of cells contracting within an iPS-CM cluster were determined.

Immunohistochemistry

On day 28 iPS-CMs were fixed with 4% paraformaldehyde. Cells were stained for F-actin (Phalloidin; Life Technologies) and sarcomeric alpha-actinin (Abcam) and counterstained with Hoechst dye. Fluorescent images were acquired using a confocal microscope.

RESULTS

iPS-CM seeded onto coverslips began contracting on day 3 with 13% of the cluster contracting. By day 10 over 90% of the cells within the cluster were contracting (Fig. 1); these cells produced a maximum contractile strain of 13.6% with a contraction frequency of 0.88Hz (Fig. 2A, B). This contraction profile was maintained with 90% or more cells in the cluster contracting through day 28.

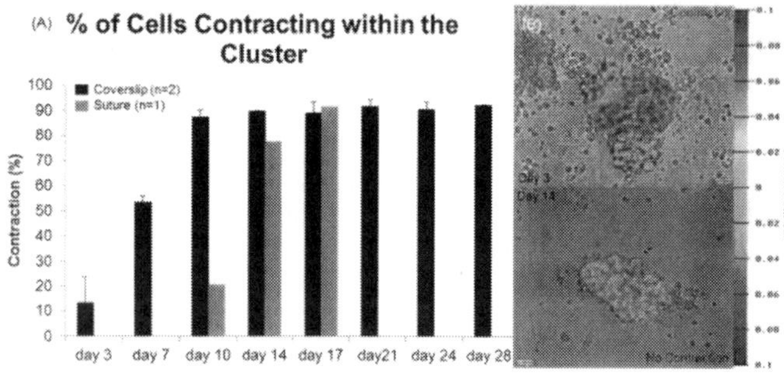

Figure 1. (A) increase of the percentage of iPS-CM within a cluster contracting from day 3 to day 28 seeded on coverslips and fibrin sutures. (B) iPS-CM cluster on day 3 with few contracting cells to day 14 with mostly contracting cells

The iPS-CM seeded onto the fibronectin coated fibrin sutures formed clusters with a similar morphology to the iPS-CM seeded onto coverslips. There were approximately ten iPS-CM clusters seeded onto the 2cm length of the suture. The cells seeded on the fibrin sutures did not begin contracting until day 7 with 20% of the cells within the cluster contracting at day 10 (Fig. 1). By day 17 over 90% of the cells within the cluster were contracting. These cells produced a maximum contractile strain of 12.3% at a frequency of 0.6Hz (Fig. 2A, B).

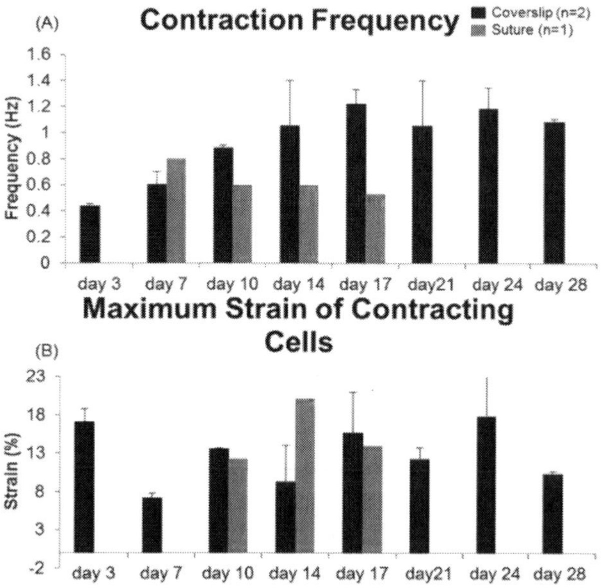

Figure 2. Contraction parameters of (A) contraction frequency and (B) maximum strain produced by contracting iPS-CM clusters from day 3 to day 28 seeded on coverslips and fibrin sutures

Immunohistochemical staining for F-actin and sarcomeric alpha-actinin resulted in positive signal within the iPS-CM clusters seeded on coverslips and fibrin sutures (Figure 3A, B). However, not all cells within the clusters stained positive for sarcomeric alpha-actinin suggesting that cardiomyocytes are not the only cell type present within the clusters. The cells that did stain positive for sarcomeric alpha-actinin did not appear to have an organized sarcomeric structure.

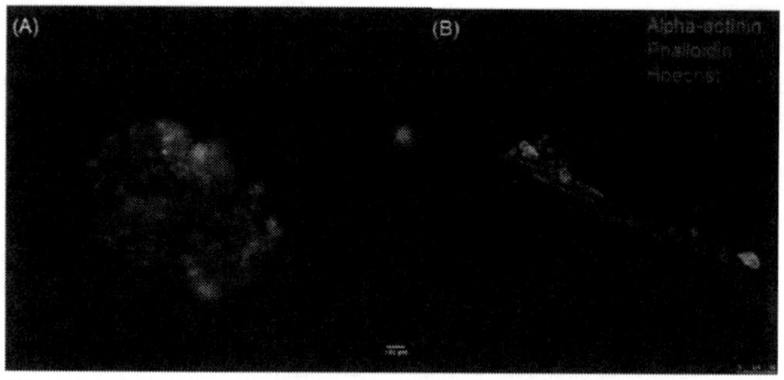

Figure 3. iPS-CM stained for Alpha-actinin, Phalloidin, and Hoechst seeded on (A) a coverslip and (B) a fibrin suture on day 28

DISCUSSION

This study has showed our ability to seed induced pluripotent stem cell derived cardiomyocytes onto a fibrin suture that will act as a novel platform for iPS-CM delivery for cardiac repair post-infarct. We have demonstrated the ability to employ high density mapping, a speckle tracking algorithm, to measure displacement fields within a cluster of contracting cardiomyocytes. The data from HDM has allowed us to quantify the number of contracting cells within an iPS-CM cluster, the maximum strain produced by the cells, and the frequency at which the cells contract. As HDM is a sterile procedure we were able to track these measurements overtime and draw conclusions about how the contraction performance of iPS-CM seeded on coverslips and fibrin sutures changes overtime.

We have demonstrated that iPS-CM seeded on coverslips begin contracting on day 3 post thaw and that the number of cells contracting increases overtime with no significant changes in contraction frequency or maximum contractile strain. The iPS-CM seeded onto the fibrin suture followed a similar pattern, however they began contracting later and did not maintain contraction through day 28.

Previous studies in our lab utilizing hMSCs have indicated a cell attachment of 20,000 cells/cm, which is significantly higher than the iPS-CM attachment found in this study (<1,000 cells/cm). Thus, further studies will investigate methods to increase iPS-CM attachment onto the fibrin suture as well as maintain contraction past day 28.

REFERENCES

[1] A. S. Go, *et al.*, *Circulation,* vol. 127, pp. e6-e245, Jan 2013.
[2] O. Bergmann, *et al.*, *Science,* vol. 324, pp. 98-102, Apr 2009.
[3] P. Jakob *et al.*, *Curr Heart Fail Rep,* vol. 10, pp. 165-76, Jun 2013.
[4] P. Menasche, *J Mol Cell Cardiol,* vol. 50, pp. 258-65, Feb 2011.
[5] A. M. Martins, *et al., Stem Cell Rev,* vol. 10, pp. 177-90, 2014.
[6] D. Hou, *et al.*, *Circulation,* vol. 112, pp. I150-6, Aug 30 2005.
[7] J. P. Guyette, *et al.*, *J Biomed Mater Res A,* vol. 101, pp. 809-18, Mar 2013.
[8] Zhang, D.,, *et al.*, *Biomaterials,* vol. 34, pp. 5813-20, 2013.
[9] M. K. Proulx, *et al.*, *J Biomed Mater Res A,* vol. 96, pp. 301-12, Feb 2011.

SB³C2015
Summer Biomechanics, Bioengineering and Biotransport Conference
June 17-20, 2015, Snowbird Resort, Utah, USA
SB³C2015-600

INVASIVE MEASUREMENTS OF PULSE WAVE VELOCITY IN ANESTHETIZED MICE: A COMPUTATIONAL-EXPERIMENTAL STUDY

F Cuomo (1), J Ferruzzi (2), J D Humphrey (2), C A Figueroa (1,3)

(1) Department of Biomedical Engineering
University of Michigan
Ann Arbor, MI, USA

(2) Department of Biomedical engineering
Yale University
New Haven, CT, USA

(3) Department of Biomedical Engineering
King's College London
London, UK

INTRODUCTION

Animal models are widely used in cardiovascular research to investigate diseases and therapies. In these models, numerous hemodynamic measurements are acquired, utilizing techniques that often times may significantly alter the baseline hemodynamic conditions of the animal. In this work we focus on how pulse wave velocity (PWV) acquired with insertion of catheters in the aortic lumen are affected by the measurement itself.

Computational tools enable the investigation of experimental conditions non-invasively. Our aims, herein, are to first reproduce the hemodynamics of a anesthetized wild-type mouse using a multi-scale computational model and rich experimental data on biomechanics and hemodynamics, and second to investigate the impact of two commercially available catheters on PWV hemodynamics.

METHODS

In vivo and in vitro experimental measurements [1,2] were performed on male wild-type mice (C57BL x 129/SvEv) between 20 and 27 weeks of age. A vascular corrosion casting was performed to build a 3D geometric model (Figure 1a). Mean blood velocity was acquired with ultrasound at ascending thoracic aorta (ATA) and infra-renal abdominal aorta (IAA) locations. Central aortic pressure was measured using a SPR-1000 pressure catheter. Lastly, a two-dimensional four-fiber family constitutive model was fitted using pressure, diameter, axial-force and axial stretch data measured in vitro at five different locations down the mouse aorta and the carotid artery.

We used the 3D simulation framework Simvascular. This is a multi-scale, fluid-structure interaction (FSI) modeling software that combines a 3D geometric model, a heart model, Windkessel models, a FSI model with regionally varying biaxial tissue properties, a posteriori gradient-based mesh adaptation method, and an external tissue support formulation [3]. The 3D geometric baseline model was obtained by segmenting the Micro-CT imaging of the vascular corrosion cast. Nine pairs of intercostal arteries were subsequently added to the model. After constructing this baseline model (Figure 1b), two additional models were created to represent typical set ups for PWV evaluation using two different commercially available catheters. The first model features two 1F diameter catheters (SPR-1000, Millar, Houston, TX) inserted in the ascending thoracic aorta and in the right iliac artery (see Figure 1c). The second model contains a 1.2F diameter dual-sensor pressure catheter (Scisense, Inc., London, Ontario, Canada) running down the descending thoracic aorta (DTA) (Figure 1d). The catheters were assumed stiff, and fixed in space. Furthermore, we assumed a complete occlusion of the vessel, at the point of catheter insertion.

A 0D lumped parameter heart model was coupled to the inlet of the 3D domain [4]. Its parameters were tuned to reproduce in vivo measurements of flow and pressure in the ATA [5]. The heart model parameters of the baseline model were kept unchanged for the simulations of the two different commercial catheters, assuming that the heart does not adapt to changes in afterload under anesthesia [6,7].

Windkessel models were coupled at each outlet of the geometric model. The Windkessel parameters were tuned to match experimentally measured pressure and flow splits [5,8]. These parameters were also kept unchanged for the catheter simulations under the assumption that anesthesia limits changes in peripheral vascular tone.

A FSI method is fundamental to capture pulse wave propagation within an arterial network. Here, a coupled momentum method was employed whereby linearized stiffness obtained from a four-fiber family model was assigned to the vessel wall of the 3D model.

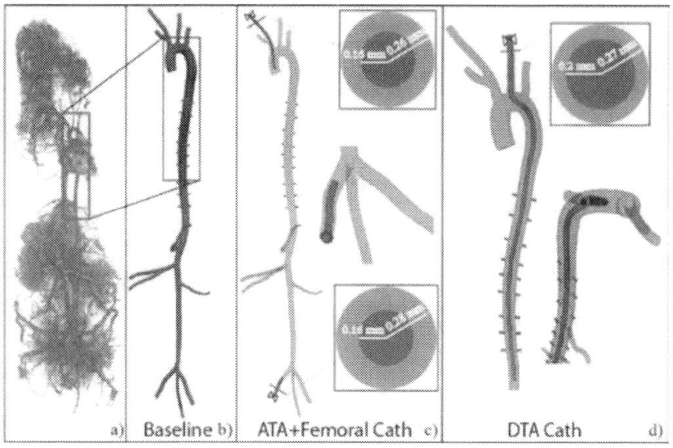

FIGURE 1: a) VASCULAR CORROSION CAST b) BASELINE COMPUTATIONAL MODEL c) ATA+FEMORAL CATHETER MODEL d) DTA CATHETER MODEL

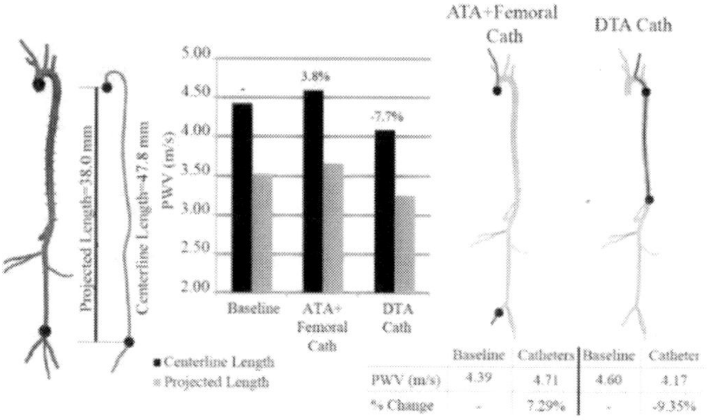

FIGURE 2: COMPARISON OF AORTO-ILIAC PWV, THE % CHANGE IS GIVEN WITH RESPECT TO THE BASELINE MODEL; COMPARISON OF PWV MEASURED BY THE CATHETER SENSORS COMPARED WITH THE BASELINE MODEL (RIGHT PART)

RESULTS

The Baseline model reproduced well the experimental data. The cardiac output was 12.03 l min^{-1}, consistent with reported values for the same strain of mice [5]; similarly, the computed ATA pressure matched well the experimental measurement. The computed transient time between pressure waveforms at the ATA and IAA locations was 10.8ms, close to the experimental transient time of 12.7ms measured between flow waveforms recorded at the same locations via ultrasound. Finally, a PWV of 4.42 m/s was obtained, similar to reported literature values for the same strain of mouse [9]. The computed distributions of flow in innominate artery, left common carotid and IAA also matched the data in literature [5,8].

The catheter simulations exhibited an increase in PWV in the ATA+Femoral Cath case, and a decrease in the DTA case. Figure 2 shows the aorto-iliac PWV computed with the actual length travelled by the pressure wave between measurement sites and with the "projected length" between the two sites. Furthermore, PWV was calculated considering the two pressure waveforms recorded at the actual sensor location for each catheter configuration, and compared to corresponding values measured in the baseline model.

An analysis of left ventricular function for the two catheter configurations showed that vessel occlusion and catheter insertion resulted in decreased stroke volume and cardiac output, with increased left ventricular/ascending aortic pressure compared to the baseline model.

Mean and peak pressure increased along the aorta in the ATA+Femoral case with respect to the baseline model; whereas mean and peak pressure decreased in the abdominal aorta in the DTA catheter case relative to the baseline model.

DISCUSSION

We were able to reproduce the hemodynamics in a healthy anesthetized mouse using a computational framework, which includes: a multi-domain method coupling a heart model with the ascending aorta and Windkessel models at each outlet of the 3D vascular model, a coupled-momentum method for FSI that included regional variation of biaxial stiffness and thickness, an external tissue boundary condition to represent perivascular support, and a mesh adaptation method. This is the first time that a combination of these methods has been used to investigate hemodynamics in murine models. Most prior works employed rigid wall assumption and/or simplified geometries [10,11], which preclude the measurement of PWV.

PWV is an important clinical metric that can be measured via insertion of pressure catheters within the arterial lumen. In humans the diameter of the device is often negligible with respect the vessel diameter, whereas in mice catheter and vessel diameters are comparable. A tacit assumption is that the changes induced by the placement of the catheters during the measurement should be negligible. One aim of this work is to investigate the degree of change in hemodynamics during murine acquisition of PWV with two different commercially available catheters. We found that the two catheter configurations affect the hemodynamics in different ways. Although both situations led to a similar decrease in cardiac output, the insertion of two Miller catheters resulted in an increase in PWV due to an overall increase in pressure.. Conversely, the dual sensor Scisense catheter decreased the PWV due to a drop in pressure caused by the bigger volume occupied by the catheter in the DTA. Often times PWV is measured using the projected length between the two locations where the pressure waveforms are recorded, instead of the actual length travelled by the wave. We calculated how this practice under-estimates the PWV.

The results presented herein are general considerations regarding catheter-based methods to measure PWV; the extents of the alterations discussed here are likely to be different depending on size, age and strain of the animal. A complete evaluation should be addressed in each specific situation.

ACKNOWLEDGEMENTS

This work was supported, by the NIH grant (R01 HL105297) and by the European Research Council Grant n. 307532.

REFERENCES

[1] Ferruzzi, J et al., *J. Biomech. Eng.*, 2015.
[2] Ferruzzi, J et al., *J. R. Soc. Interface*, 8:435-50, 2015.
[3] Xiao, N et al., *J. Comput. Phys.* 244:22-40, 2014.
[4] Lau, K and Figueroa, C, *Biomech. Model. Mechanobiol.*, 2015.
[5] Feintuch, A et al., *Am. J. Physiol. Heart Circ. Physiol.*, 292:H884-H892, 2007.
[6] Shimokawa, A et al., *J. Auton. Nerv. Syst*, 72:46-54, 1998.
[7] Usselman, C et al., *Appl. Physiol. Nutr. Metab.*, 36:376-381, 2011.
[8] Greve, J et al., *Am. J. Physiol. Heart Circ. Physiol.*, 291:H1700-H1708, 2006.
[9] Hartley, C et al., *ILAR J.*, 43:147-58,2002.
[10] Huo, Y, *Ann. Biomed. Eng.*, 36:685-699, 2008.
[11] Trachet, B, *Ann. Biomed. Eng.*, 39:2430-2444, 2011.

SB³C2015
Summer Biomechanics, Bioengineering and Biotransport Conference
June 17-20, 2015,Snowbird Resort, Utah, USA
SB³C2015-601

A PREDICTIVE 3D HIGH-CONTENT/HIGH THROUGHPUT SCREENING PLATFORM TO ELUCIDATE AND ENHANCE MULTILINEAGE STEM CELL DIFFERENTIATION

Amit Paul, Bo Chen, Elise DeBruyn, Michael Cho

Department of Bioengineering
University of Illinois at Chicago
Chicago, IL, USA

INTRODUCTION

Heterogeneity in human mesenchymal stem cell (MSC) populations remains a daunting challenge in the advancement of stem cell based therapeutics for regenerative medicine. MSCs harvested from the same donor often exhibit divergent behaviors with respect to their lineage commitment and extent of differentiation. Mounting evidence suggests a strong link between MSC heterogeneity/lineage-commitment and MSC biomechanical configurations [1-2]. The F-actin cytoskeleton of MSCs alters dramatically during multilineage differentiation [3]. Notably, these lineage-dependent changes cause MSCs to exhibit unique cytoskeletal "fingerprints" along the various stages of differentiation. There currently lacks a robust system to simultaneously measure the morphological, structural, biomechanical, and proteomic changes of multiple subcellular stem cell structures during multilineage differentiation. To address this void, and to investigate the relationship between these dynamic cytoskeletal configurations and MSC heterogeneity/lineage-commitment on a large scale, we combined high-resolution 3D microscopy and correlative atomic force microscopy (AFM) with a powerful computational algorithm recently developed in our lab. The algorithm integrates two technologies that were thought to be mutually exclusive in the past: high-content screening (HCS, detailed analysis at the single cell level) and high-throughput screening (HTS, rapid/high-volume experimentation).

METHODS

The entire HCS/HTS process was performed in an automated environment using a custom-engineered 3D cell/tissue imaging system. While undergoing multilineage differentiation (adipogenesis, osteogenesis, chondrogenesis, myogenesis, neurogenesis, vasculogenesis, and hepatogenesis), MSCs were fluorescently labeled at crucial differentiation time points for cytoskeletal and lineage-specific markers. All samples were scanned/imaged in a grid-by-grid manner via exceptionally high-resolution (100-200nm) 3D confocal/multiphoton laser-scanning microscopy (Fig. 1). All fluorescently labeled structures in each scanned grid were computationally reconstructed in 3D (Fig. 2). Through the use of a custom mosaicking tool created in MATLAB, the 3D reconstructed grids were "stitched" back together to obtain a large 12mm x 12mm 3D reconstructed mosaic of the entire sample (1000-5000 stem cells per sample). Following the 3D stitching, we modified/applied a 3D deconvolution algorithm that compensated for any irregularities in the scans. Once reconstructed, stitched, and deconvolved, the stem cells in the large 3D dataset were ready for detailed high-content 3D analysis. The first step of the analysis involved 3D segmentation. We developed a MATLAB algorithm (based on a heavily modified 3D version of a marker-based watershed segmentation technique) that accurately and consistently demarcated every stem cell in a high-volume sample (Fig. 3). Once each stem cell was segmented, the fluorescently labeled subcellular structures were also segmented in each stem cell. Each segmented stem cell/subcellular structure was then measured for over 60 individual morphological, intensity, and texture based 3D feature extractions (Fig. 3). For a true morphological analysis of dynamic cytoskeletal

Figure 1 - Magnified example: High-resolution 3D reconstruction capabilities of the HCS/HTS algorithm. (F-actin, Nuclei).

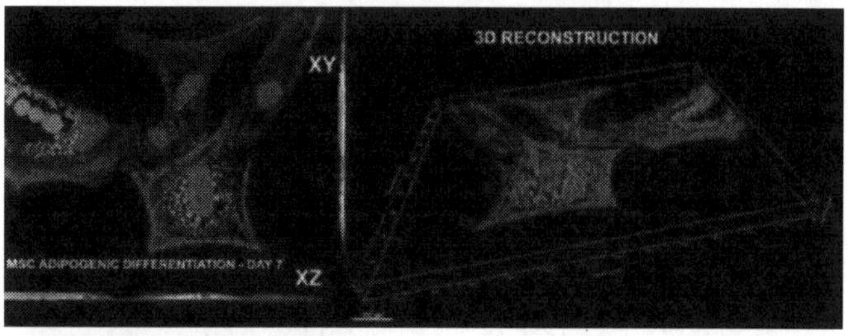

Figure 2 - 3D reconstruction of a scanned grid during MSC adipogenic differentiation (F-actin, Lipids, Nuclei).

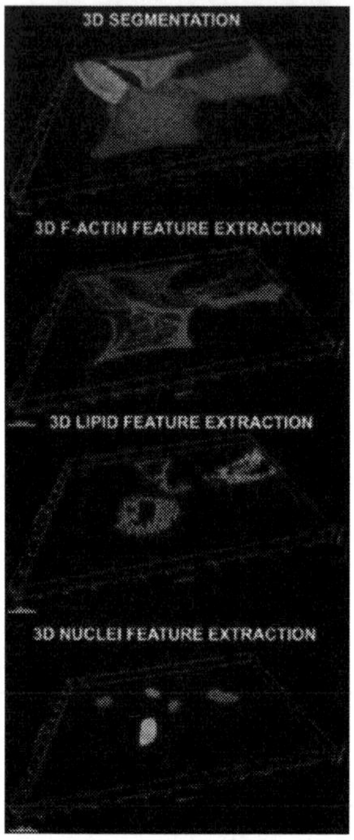

Figure 3 - Accurate 3D segmentation and feature extractions of MSCs during adipogenic differentiation.

remodeling during stem cell differentiation, we applied a novel 3D cytoskeletal fingerprinting technique (developed from open-source NIST human biometric fingerprinting algorithms). Additionally, the lineage-dependent biomechanical changes were correlated using AFM to probe multiple areas of each differentiating stem cell (Fig. 4). The extensive amount of quantitative experimental data produced from this HCS/HTS platform went through a number of customized data-mining, pattern recognition, and machine learning algorithms in order to correlate the diverse biomechanical configurations with lineage-specific changes during multilineage MSC differentiation.

RESULTS

With the abundance of quantitative experimental data, we developed a sophisticated 3D cytoskeletal fingerprinting process that accurately analyzes & predicts the mechanical transformations which are potentially required to initiate the onset of stem cell differentiation into multiple lineages (Fig. 5). With the assistance of this advanced HCS/HTS platform, many aspects of MSC differentiation were thoroughly elucidated. Perhaps most significant was a consistent pattern observed in all types of differentiation: the F-actin cytoskeleton remodeled significantly before lineage-specific cellular changes occurred (Fig.6,7). This strongly suggests that cellular mechanical transformations are a precursor to stem cell differentiation.

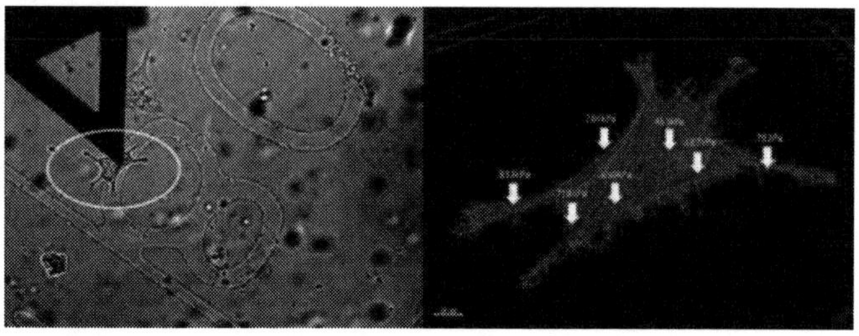

Figure 4 - Thousands of live differentiating MSCs (per sample) were probed for elasticity values (Young's Modulus) at various points of the cell. These probed cells were tracked during differentiation, and their elasticity data was correlated with their 3D biomechanical / lineage-specific imaging data. (F-actin, Nuclei).

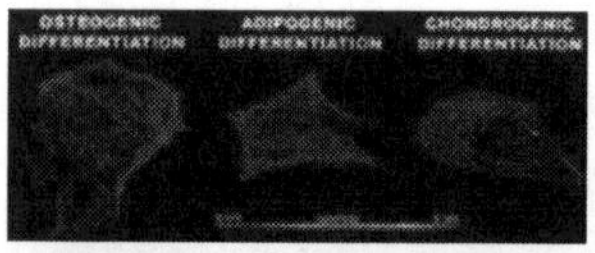

Figure 5 - As early as day 3 of MSC differentiation, the 3D cytoskeletal fingerprinting process can predict lineage commitment (between all 7 lineages) based on F-actin alone.

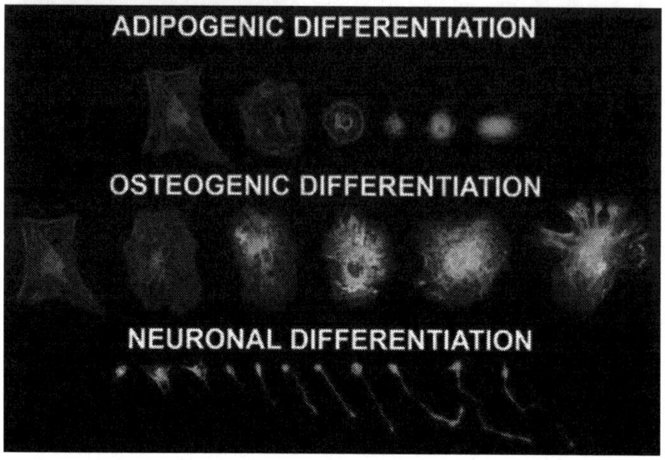

Figure 6 - Representative images of different time points during multilineage MSC differentiation (adipogenesis, osteogenesis, and neurogenesis). For each lineage, the F-actin remodels to a certain configuration before the lineage specific cellular changes occur. (F-actin, Nuclei, TOP: Lipids, MIDDLE: Osteocalcin, BOTTOM: Neuron-Specific Beta III Tubulin).

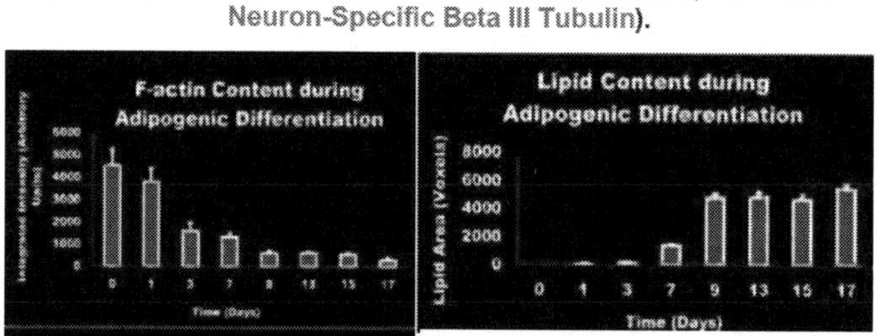

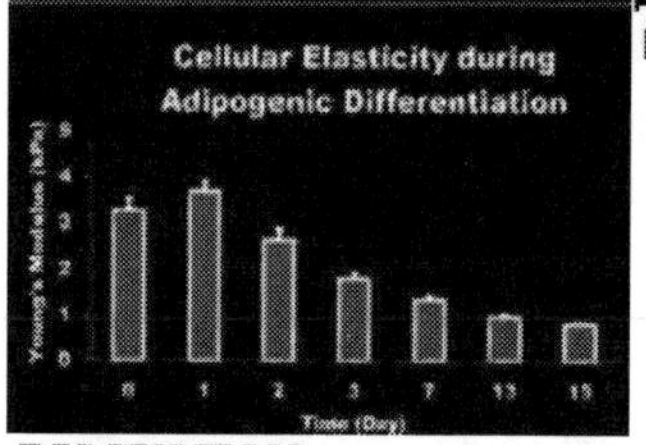

Figure 7 - Example: F-actin content dramatically decreased during MSC adipogenic differentiation. This correlated to a decrease in cellular elasticity. The lipids consistently began to form when the F-actin decreased to a certain point (between days 3 and 7).

DISCUSSION

The combinatorial approach of 3D high-resolution imaging, correlative AFM, and computational modeling provided a robust system to better understand the critical relationship between heterogeneity/lineage-commitment and cellular biomechanics during multilineage MSC differentiation. Based on the biomechanical properties and cytoskeletal configurations identified in this predictive model, we are engineering clinically relevant methods to rapidly sort heterogeneous MSC colonies into homogeneous subpopulations. The HCS/HTS system is currently being extended to induced pluripotent stem cells (iPSCs)

ACKNOWLEDGEMENTS

We would like to acknowledge and thank Dr. Peter Bajcsy for his guidance on validating the accuracy of our computational algorithms.

REFERENCES

[1] Lee, W.C. et al, *Proc Natl Acad Sci*, 111:E4409–E4418, 2014
[2] Treiser, M.D., et al, *Proc Natl Acad Sci*, 107:610-615, 2010
[3] Titushkin, I. et al., *J Biomech*, 46:521-526, 2013

SB³C2015
Summer Biomechanics, Bioengineering and Biotransport Conference
June 17-20, 2015,Snowbird Resort, Utah, USA
SB³C2015-602

DRIVER INJURY RISK SENSITVITY IN FINITE ELEMENT MODEL RECONSTRUCTIONS OF REAL WORLD MOTOR VEHICLE CRASHES

James P. Gaewsky (1,2), Ashley A. Weaver (1,2), Bharath Koya (1,2), Joel D. Stitzel (1,2)

(1) Wake Forest University School of Medicine, Winston-Salem, NC, USA

(2) Virginia Tech - Wake Forest Center for Injury Biomechanics, Winston-Salem, NC, USA

INTRODUCTION

The Crash Injury Research and Engineering Network (CIREN) surveys hundreds of motor vehicle crashes (MVCs) each year to assess occupant injury outcomes in real world crash events across the United States. Although the CIREN database contains a wealth of crash injury information including radiology reports, occupant information, and case vehicle information from event data recorders (EDR) and physical measurements, relatively few studies have computationally modeled MVCs studied by CIREN to assess occupant injuries. Previous studies have reconstructed side and oblique frontal impact crashes but were limited by the breadth and level of detail of crash injuries analyzed, the variability of vehicle models, or the number of cases analyzed [1, 2, 3].

This study developed a method to reconstruct a wide range of full frontal CIREN MVCs in LS-Dyna by tuning a generic simplified vehicle model (SVM) to perform like each specific CIREN case vehicle in a frontal crash test. The Total HUman Model for Safety (THUMS) version 4.01 was virtually instrumented to assess head, neck, chest, lung, femur and tibia responses during the crash events. For each case modeled, a sensitivity study was performed to quantify the variability in injury risk based on occupant position.

METHODS

CIREN cases were reconstructed using finite element model (FEM) simulations in a three step process to assess the variability in injury risk and severity for common frontal crash injury patterns.

Step 1: A SVM containing the floor, seat, seatbelt, dashboard, steering wheel, and airbag was developed and tuned to mimic the frontal crash response of the CIREN case vehicle for each reconstruction. Frontal New Car Assessment Program (NCAP) crash test vehicle kinematics for each case were video tracked and then incorporated into crash test reconstruction simulations with a validated model of the Hybrid III anthropomorphic test device (ATD) as shown in Figure 1~~Figure 1~~ [4].

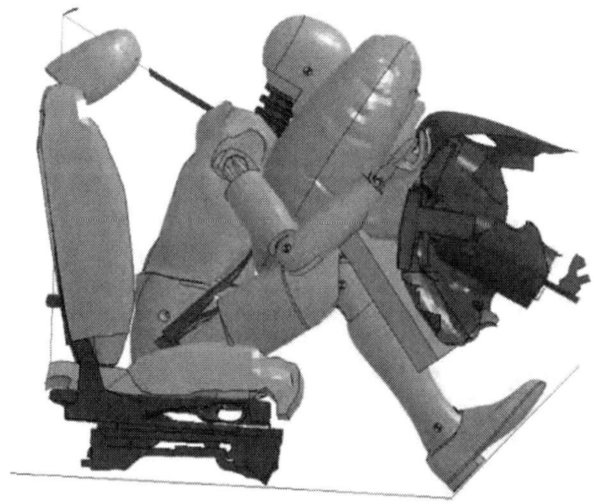

Figure 1: The SVM contains common vehicle restraint and safety systems such as seatbelts with load limiters and pretensioners, airbags, and a collapsible steering column.

For each case, ten vehicle restraint parameters (e.g. airbag inflation rate, belt pretensioner force) were varied across 240 simulations using a Latin Hypercube Design (LHD) of experiments to find an optimal set of vehicle parameters that resulted in simulated

ATD responses closely matching test data signals using the Sprague and Geers method to quantify magnitude and phase errors between corresponding data curves [5].

Step 2: The THUMS v4.01 human body model was implemented into the tuned SVM using CIREN-reported occupant positioning information, as well as a range of occupant positions by varying the seatback angle, longitudinal seat track position, D-ring height, and steering wheel position and angle using a 120 simulation LHD.

Step 3: The CIREN case's EDR velocity pulse was applied to the tuned SVM to simulate the crash for each of the 120 THUMS occupant positions. Simulated THUMS instrumentation (i.e. accelerometers, load cells, and strain measurements) was implemented and used to measure injury metrics correlated to common frontal crash injuries. AIS-based risks were estimated from the injury metrics using published injury risk curves [6, 7].

RESULTS

The SVM was successfully tuned for all frontal CIREN reconstructions. Sprague and Geers magnitude (M) and phase (P) error factors validated the FEM against the crash test ATD head, chest and pelvis accelerations (|M| < 0.23, P < 0.12) and femur and belt force (|M| < 0.12, P < 0.13) signals.

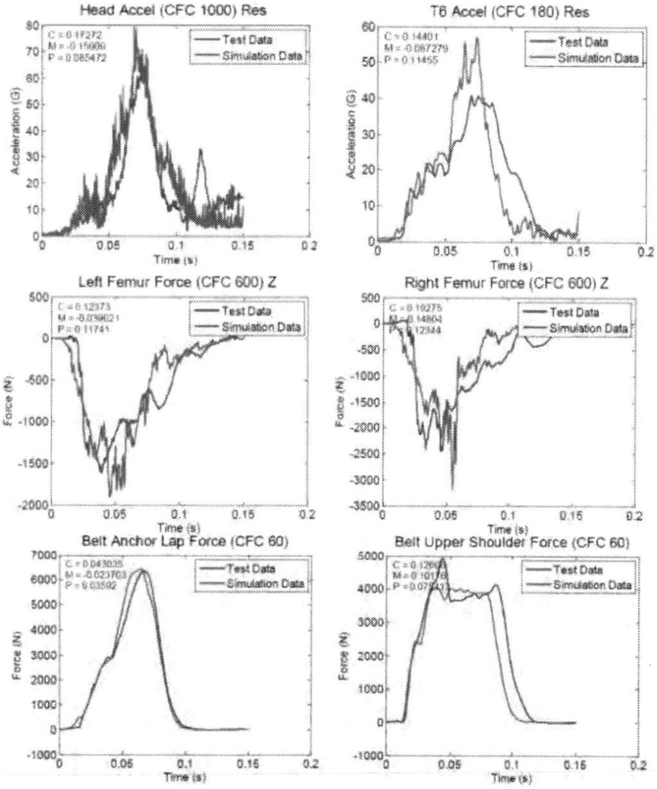

Figure 2: Head and chest accelerations, and femur and belt forces were closely matched between the real world crash test and simulated test for the CIREN case vehicle.

One reconstructed 2010 Toyota Camry occupant had no reported head injuries or knee-thigh-hip (KTH) injuries and injury risks calculated for the CIREN-reported occupant position versus the range of positions simulated were: AIS 1+, 2+, 3+ head (42%, 22-100%; 7.4%, 0-99%; 1.0%, 0-55%) as shown in Figure 3; AIS 2+, 3+ KTH (2.8%, 2-4%, 2.8%, 2-4%).

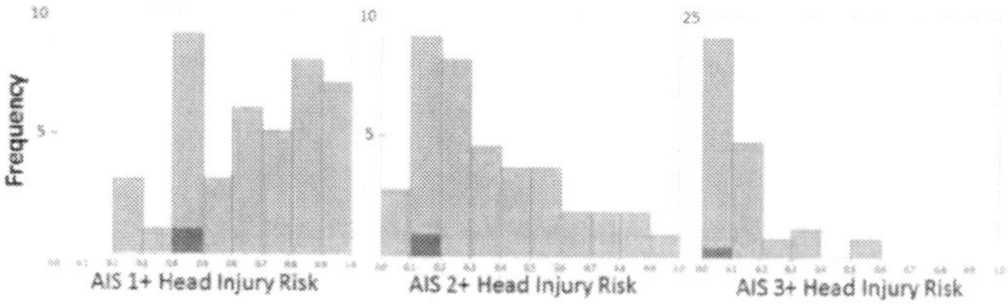

Figure 3: AIS 1+ to 3+ head injury risk across all occupant positions simulated. CIREN-reported position is dark green.

In a 2006 Chevrolet Cobalt reconstruction, the occupant's AIS 3 lung contusion (49% of lung contused: see Figure 4) was predicted since over 49% of the THUMS lung elements exceeded a 34% first principal strain threshold and AIS 2 L1/L3 vertebral fractures were predicted from THUMS' L1/L3 cross-sectional force measurements.

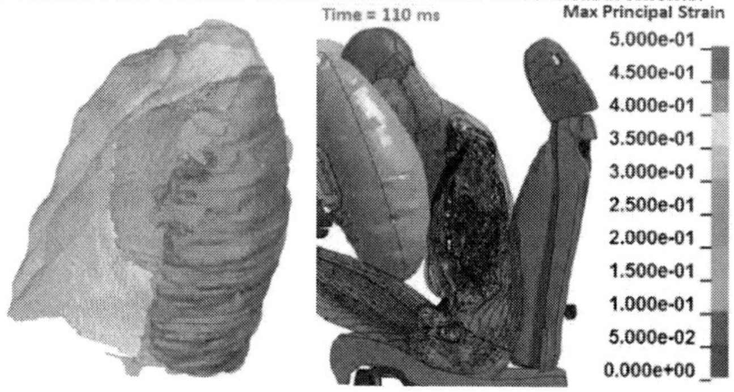

Figure 4: (Left) CIREN case occupant lung segmentation from computed tomography: healthy tissue (gray), contused tissue (pink), and pneumothorax (blue-green). (Right) Fringe plot depicting maximum principal strain in the left lung of THUMS.

DISCUSSION

Head, chest, pelvis, and femur injury risks predicted with FEMs were well correlated with documented CIREN injuries, and this methodology can be extended to reconstruct CIREN or National Automotive Sampling System (NASS) frontal crashes and potentially other crash modes. The reconstruction process allows for quantification of the sensitivity and uncertainty of injury risk predictions based on occupant position which is often uncertain in real-world crashes. THUMS injury metrics assessed include strain-based metrics for assessing organ injuries (i.e. lung) that are not easily assessed in post-mortem human subjects (PMHS) or ATDs. This information is potentially valuable to assess the effectiveness of restraint systems to prevent or mitigate injuries that are not easily studied using PMHS and ATDs.

ACKNOWLEDGEMENTS

Funding has been provided by Toyota Motor Corporation and Toyota's Collaborative Safety Research Center.

REFERENCES

[1] Danelson, KA et al., *Traffic Inj Prev*, 2015.
[2] Golman, AJ et al., *Accident Analysis & Prevention*, 64:1-8, 2014.
[3] Iraeus, J et al., *Ann Adv Automot Med*. 57:183-196, 2013
[4] *FTSS LS-DYNA Model of Hybrid III 50th Percentile Male Dummy User Manual.* May 2010: Version 7.1.3
[5] Sprague, MA et al., *Int J Numer Meth Engng*. 60:2467-2499, 2004
[6] Kuppa, S et al., *17th ESV Conference Paper,* 2001.
[7] Eppinger, R et al., *Development of Improved Injust Criteria, 1999*

SB³C2015
Summer Biomechanics, Bioengineering and Biotransport Conference
June 17-20, 2015, Snowbird Resort, Utah, USA
SB³C2015-603

A FRAMEWORK FOR PARAMETER ESTIMATION OF HEART VALVES USING INVERSE-MODELING APPROACH

Ankush Aggarwal (1) and Michael S. Sacks (1)

(1) Center for Cardiovascular Simulation
Institute for Computational Engineering & Sciences
Department of Biomedical Engineering
University of Texas at Austin
Austin, TX, USA

INTRODUCTION

Heart valves are critical components of heart because of their role in ensuring unidirectional flow of blood and opening and closing at the precise moments which should be synchronized with the contracting of myocardium. The valve leaflets are complex structural entities made from various layers with primary constituents being collagen, elastin, extracellular matrix and valve interstitial cells. Either due to congenital defects or due to the changes due to remodeling with age, the leaflets functionality is altered sometimes resulting in less efficient heart output. Such problems are hard to diagnose at an early stage and may lead to heart failure if left untreated. Therefore, a tool for determining the functional properties of heart valves from non-invasive imaging modalities will be of utmost clinical importance. Such a tool will help diagnose the valve diseases at an early stage as well as provide tools for monitoring the performance of replaced prosthetic valves. The three components of such a tool are shown in Fig. 1. In order to define the correct reference configuration, we need the pre-strain information about leaflets. For defining the unknown fiber architecture, we need the population average architecture trends that can be used to constrain parameter optimization. And the main component is development of a numerical inverse-model, which can be applied to in-vivo imaging data. In this abstract, we focus on the development of inverse model using an in-vitro dataset and briefly discuss the other components.

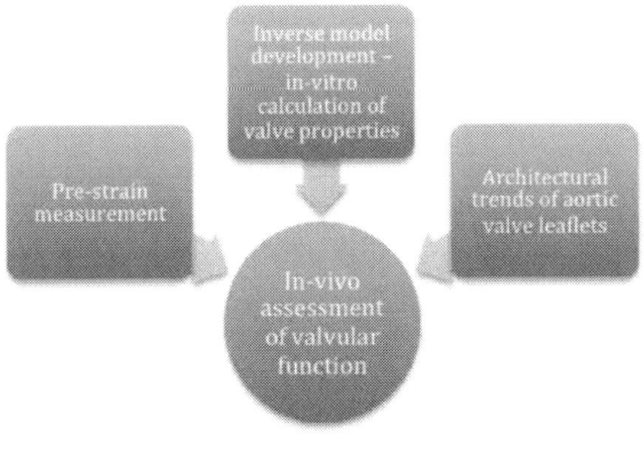

Figure 1: Three components required to build the in-vivo assessment tool for valvular function

METHODS

As a first step, we use in-vitro experimental data from pericardial bioprosthetic heart valve to develop the parameter estimation tool based upon inverse modeling. The leaflets were imaged at three different static transvalvular pressures using surface markers and dual-camera setup [1] (Fig. 2). Leaflets were then removed and their fiber architecture was determined using light scattering technique [2]. Then the leaflets were dissected and put under biaxial test to determine their stress-strain relationship. This highly comprehensive data set with high resolution, marker positions at multiple pressures, valve specific fiber architecture and biaxial data was ideal for this study – to design the inverse model, validate it and calculate it's sensitivity to various input parameters and optimization constraints. There is no existing method for inverse modeling of heart valves as per authors' knowledge. This is due to the difficulty in modeling them as well as their small size and thickness and fast movement in-vivo leading to poor quality of imaging.

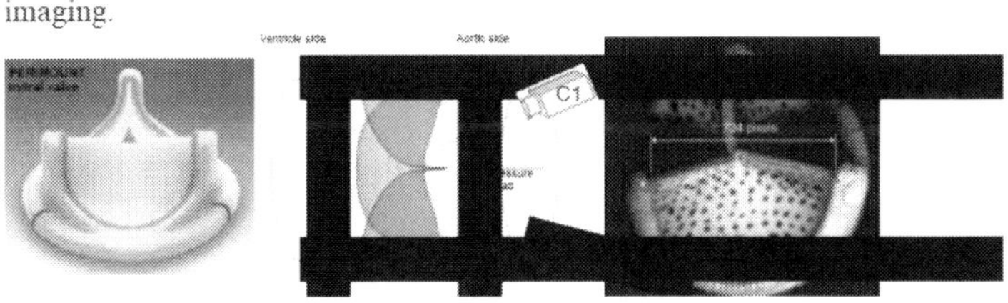

Figure 2: Experimental setup (a), the thin shell model (b) and the stresses calculated using forward model (c)

A. Forward Model

The valve leaflet was modeled as a thin shell and the contact between leaflets was treated using the augmented Lagrange multiplier method. The basal attachment was fixed as in the experiments and a normal pressure on the leaflets was applied linearly increasing to 120mm Hg. The model was solved using FEBio [3] with a quasi-static solver on single processor.

For the material model, a simplified structural model was used [4]. This allowed us to reduce the number of unknown material parameters in fiber component to only two reducing the computational time in determining them. The matrix component was modeled as a compressible neo-Hookean material and the properties were borrowed from flexural testing results previously published.

B. Algorithm

The above forward model was put in an optimization loop where the parameters were initialized with a guess and solution was iterated until convergence was achieved. For the optimization, cost function was defined as $\mathcal{F} = \sqrt{\sum_{i,\alpha}\left(x_i^\alpha - \tilde{x}_i^\alpha(c_m)\right)^2}$ where, x_i^α are the input points from experiment and $\tilde{x}_i^\alpha(c_m)$ are their projection on deformed valve surface obtained from the forward model with material parameters c_m. This is a highly non-linear function because of the contact constrained between leaflets. The non-linear least squares problem of minimizing the cost function was solved using Levenberg-Marquardt algorithm with the derivatives of the cost function calculated numerically using central difference method.

RESULTS

The cost function shape for various cases as a function of the two material parameters A and B is shown in Fig. 3. The results demonstrate the importance of using higher number of frames from in-vivo imaging for inverse modeling input. It also shows that imaging at higher pressures would give a better confidence in the results.

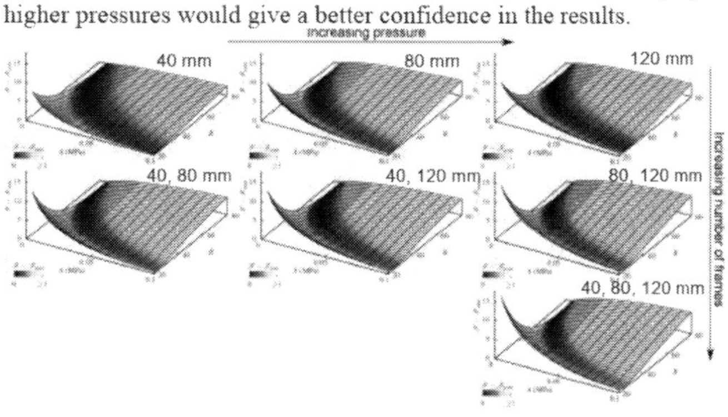

Figure 3: Resulting cost function for various pressure levels and frame numbers used

The change in cost function when average fiber architecture is utilized is shown in Fig. 4a and when it is varied the result is shown in Fig. 4b. This result demonstrates that the inverse modeling results are dependent upon the fiber architecture in an average sense.

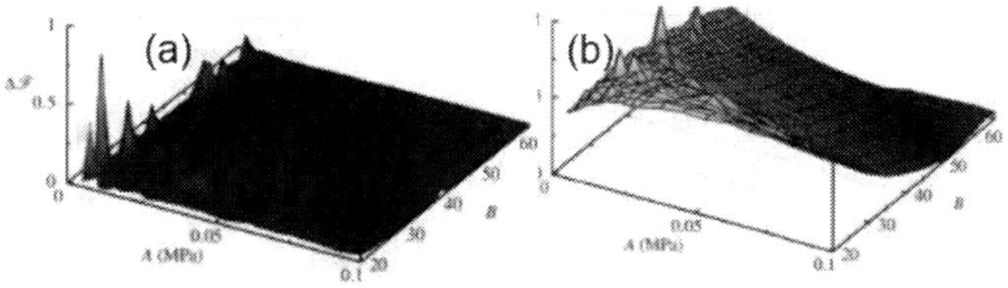

Figure 4: Change in cost function by using averaged fiber architecture, and (a) and by varying it by a small constant (b)

Table 1: Statistics of the final fit using inverse model

Pressure	Leaflet	mean(|d|)	std(|d|)	max(|d|)
40 mm	1	0.195	0.150	0.718
	2	0.238	0.196	0.977
	3	0.182	0.124	0.619
80 mm	1	0.214	0.169	0.786
	2	0.294	0.232	1.150
	3	0.199	0.140	0.574
120 mm	1	0.270	0.216	0.938
	2	0.270	0.238	1.155
	3	0.196	0.137	0.611

The results were found insensitive to the Poisson's ratio of the matrix, but highly sensitive to the noise in input point cloud from imaging. Statistics of the final fit are shown in Table 1. Comparison of predicted stress-strain relationship vs. measured are shown in Fig. 5a and the final fit of valve's deformed shape to point cloud is shown in Fig. 5b.

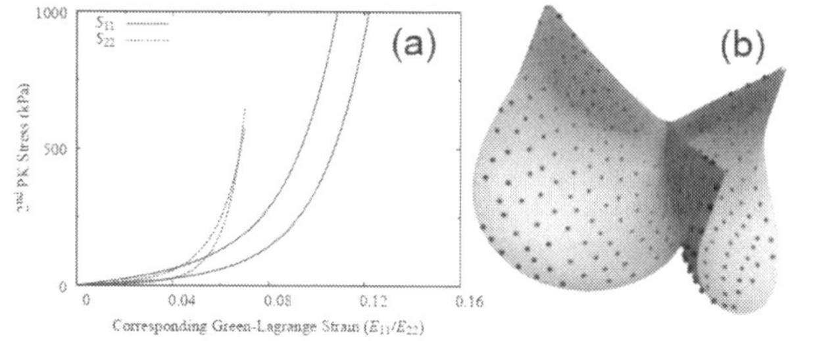

Figure 5: (a) Stress-strain curve for one leaflet predicted (blue) vs. measured (red) and (b) final valve shape fit to the point cloud from experiment at 120mm Hg

DISCUSSION

These results present the first such attempt in inverse model development for heart valves. Here each leaflet is considered to have homogenous elastic parameters. In future, the method will be extended for the parameters to vary spatially. The novelty of the present work lies in the incorporation of fiber data of each leaflet and utilizing it through simplified structural model [4]. Differences in fiber architectures of normal and pathological valve leaflets previously determined [5] can be combined with this tool to predict abnormal biomechanical behaviors. The sensitivity and convergence studies provide guidelines for creating this tool in an effective manner, e.g. the imaging resolution is a critical component of accurate determination, and higher number of frames leads to higher confidence.

To develop a method for evaluating the material parameters from in-vivo imaging, we are also working on other required components. The average fiber architecture of valve leaflets was calculated using a novel spline technique [5] and the pre-strain in leaflets was estimated by comparing the in-vivo shape and explanted leaflets' shape [6]. This information when combined with the inverse model presented in this work will lead to an in-vivo assessment tool for heart valves and help diagnose problems in the mechanical functionality at an early stage. Additionally, this approach will have the potential to serve as a general-purpose in-vivo assessment tool for heart valves – for evaluating the performance of replaced prosthetic valves as well as monitoring the progression of valve diseases. The information thus produced over time will help us understand the mechanical behavior of valve leaflets and design better surgical tools to repair and replace them.

ACKNOWLEDGEMENTS

We gratefully acknowledge the help from Will Zhang with parameter estimation and analysis of biaxial data. This work was supported by the AHA Southwest Affiliate Postdoctoral fellowship (14POST18720037 to A.A.) and NIH (R01 HL108330 to M.S.S.). The authors acknowledge the Texas Advanced Computing Center (TACC) at The University of Texas at Austin for providing HPC resources.

REFERENCES

[1] Sun, Abad, and Sacks, JBME 2005. 127(6): 905-914.
[2] Sacks, Smith, and Hiester, Ann Biomed Eng, 1997. 25(4): 678-89.
[3] Maas, Ellis, Ateshian et al., JBME 2012. 134(1)
[4] Fan and Sacks, Journal of biomechanics, 2014. 47(9), 2043-2054.
[5] Aggarwal, Ferrari, Joyce et al., Ann Biomed Eng, 2014. 42(5): p. 986-98.
[6] Aggarwal, Aguilar, Lee et al., Functional Imaging and Modeling of the Heart. 2013, Springer. p. 141-149.

HEAD IMPACT RESPONSE RESULTING FROM FORCEFUL IMPACT WITH TOY SWORDS BY PEDIATRIC MALES

Stephanie M. Beeman, Steven Rowson, Stefan M. Duma

Department of Biomedical Engineering and Mechanics
Virginia Tech
Blacksburg, VA, USA

INTRODUCTION

Injuries present a major threat to the health and welfare of children and adolescents. Play, recreational activities, and organized sports pose a risk of unintentional physical injuries, particularly since children are still gaining motor and cognitive skills. Among children 0 to 14 years of age, being struck by or against an object (stationary or moving) is the second leading external cause of TBIs (24.8%), including toys [1]. Toy swords may be categorized as war toys which have a controversial, widely debated role in child play [2]. Even while playing without intent to harm, children playing with war toys may sustain injuries due to the nature of play [2]. While extensive research in toy safety has been performed, data is unavailable with regard to toy swords. Understanding the physical capacity of children of different ages is important for the evaluation of injury risk and the design of new toys. Therefore, the purpose of this study was to quantify the ability of children to swing toy swords and the human kinematic response that could be anticipated as a result of being impacted during a forceful swinging behavior.

METHODS

A total of 36 male pediatric subjects were enrolled in this study. There were six males in the following age groups: 4, 6, 8, 10, 12, and 14 years old. Subjects were recruited throughout the community via flyers and listservs. Prior to accepting a subject for the study, the age of the potential participant was verified. The basis of the study and the experimental procedure were explained to the subjects' legal guardians and informed consent and assent were obtained. Approval to perform this study was obtained from the Virginia Tech Institutional Review Board.

Each subject completed a single trial with the same model collapsible plastic sword (Figure 1). During each trial, the subject used the toy sword to hit a head form and neck assembly of a 6-Year-Old Hybrid III Child anthropometric test device (ATD) (Figure 2). The subjects were instructed to swing the sword-like toy and hit the head form as hard as possible. Subjects were not instructed whether they were to hold the sword with one or two hands, only that they were to grasp the handle region.

FIGURE 1: COLLAPSIBLE PLASTIC TOY SWORD UTILIZED BY THE SUBJECTS DURING EACH TRIAL.

FIGURE 2: 6 Y.O. HIII ATD HEAD-NECK ASSEMBLY MOUNTED ON A LINEAR SLIDER TABLE THAT ALLOWED ANTERIOR-POSTERIOR TRANSLATION.

The head form was instrumented with a 9 accelerometer array in a 3-2-2-2 configuration which allowed linear and angular acceleration to be computed analytically. All acceleration data were sampled at 20 kHz and filtered with SAE Channel Frequency Class 1000. Resultant linear acceleration, resultant angular acceleration, and resultant angular velocity were calculated. Additionally, Head Injury Criteria (HIC) (Equation 1) and Severity Index (SI) (Equation 2) were determined. For this particular analysis, the maximum value of $(t_2 - t_1)$ in Equation 1 cannot exceed 15 ms, which refers to a HIC_{15} value.

$$HIC = \left\{(t_2 - t_1)\left[\int_{t_1}^{t_2} \frac{a(t)dt}{t_2 - t_1}\right]^{2.5}\right\}_{max} \quad (1)$$

$$SI = \int_0^t a(t)^{2.5} dt \quad (2)$$

A two-way repeated measures ANOVA was performed with post-hoc Student's t-tests. A Type I error rate of alpha=0.05 was used to assess significance.

RESULTS

Peak linear acceleration, angular acceleration, and angular velocity as well as the HIC15 and SI value were determined for each impact. Generally, the head response parameters increased with age (Table 1).

TABLE 1: HEAD RESPONSE PARAMETER VALUES WERE DETERMINED FOR ALL IMPACTS: AVERAGE (STANDARD DEVIATION) ARE REPORTED.

Age Group	Linear Accel (g)	Angular Accel (rad/s^2)	Angular Velocity (rad/s)	HIC	SI
4	8.60 (4.31)	865.30 (643.06)	2.11 (1.54)	0.28 (0.32)	0.34 (0.37)
6	12.80 (2.96)	1140.75 (309.50)	3.40 (2.33)	0.44 (0.30)	0.59 (0.38)
8	17.02 (7.40)	1998.89 (1002.30)	5.04 (2.88)	1.18 (0.89)	1.49 (1.17)
10	20.20 (7.25)	2171.98 (896.22)	4.00 (1.61)	1.29 (0.90)	1.61 (1.09)
12	19.28 (7.35)	1847.65 (469.84)	5.98 (4.67)	2.01 (1.05)	2.47 (1.32)
14	24.65 (5.88)	3194.94 (977.20)	5.45 (1.77)	2.78 (1.58)	3.16 (1.77)

Age had a significant effect on the head response parameters, per the ANOVA results ($p<0.001$). The parameters for 8-14 year olds were not significantly different, with the exception of peak angular acceleration being significantly lower for the 12 year old than the 14 year olds. The 4 year olds generated significantly lower accelerations than the 8-14 year olds, significantly lower angular velocity than the 14 year olds, and significantly lower HIC and SI values than the 10-14 year olds. The 6 year olds produced significantly lower angular acceleration than the 10-14 year olds, significantly lower linear acceleration than the 14 year olds, and significantly lower HIC and SI values than the 12-14 year olds.

DISCUSSION

This study demonstrated the range of physical capacity of male children across a broad array of ages to wield a toy sword as a striking object. Age was demonstrated to affect the resulting head response parameters, particularly when comparing the youngest age groups to the oldest age groups. Age and sword swing speed have been previously reported to exhibit a strong, positive correlation which may have influenced the impact energy [3]. The trends observed within this study are likely a result of typical pediatric and adolescent development. It is generally accepted that the strength and motor performance increases linearly throughout early childhood up to puberty, which is said to occur at approximately 13 years old in males [4, 5, 6].

Although differences were observed in physical capacity, the resulting head responses were not indicative of great likelihood for serious, concussive head injuries. The combination of maximum linear (~34g) and angular acceleration (~4800rad/s^2) corresponded to a 0.94% risk of injury [7]. Additionally, HIC and SI were used to assess potential for injury. HIC is typically associated with standards related to head injury in automotive safety testing, while SI is typically associated with standards related to sport helmet impact performance. Both provide an indication of how long the peak linear accelerations were experienced and were primarily developed to predict skill fracture. The values observed in this study are substantially lower than the injury threshold for impact trials [8, 9].

Overall, the results of this study would indicate minimal risk of head injury as a result of being struck by a particular toy sword. However, differences were observed with regard to the physical capacity of the children across the age range of the study's population. It is important to take this into consideration when designing and evaluating future models of such toys. By accounting for the physical capabilities of a specific target population, the toys can be designed with decreased inherent risks of injury.

ACKNOWLEDGEMENTS

The authors would like to thank Hasbro for providing support for this study as well as Melissa Hulse, Brock Strom, and Ryan Field for their contributions.

REFERENCES

[1] Faul, M et al., National Center for Injury Prevention and Control, 2010.

[2] Hart, JL and MT Tannock, Encyclopedia on Early Childhood Development, 1-6, 2013.

[3] Beeman, S, et al., *Biomed Sci Instrum*, 50:31-36, 2014.

[4] De Ste Croix, MB, *Journal of Sports Science and Medicine* 6:292-304, 2007.

[5] Loko, J, et al., Am J Hum Biol, 15(1):72-77, 2003.

[6] Rogol, AD, et al., *Am J Clin Nutr*, 72(2 Suppl):521S-528S, 2000.

[7] Rowson, S, and Duma, SM, *Ann Biomed Eng*, 41(5):873-882, 2013.

[8] Gadd, CW, 10th Stapp Car Crash Conference, 1966. SAE 660793.

[9] Versace, J, SAE Technical Paper Series, 1971. SAE 710881.

MULTISCALE SYNTHESES OF ARTICULAR CARTILAGE STRESS IN POST ACLR MODEL UNDER PHYSIOLOGICAL LOADING

Malek Adouni (1,2), Yasin, Dhaher (1,2)

(1) Sensory Motor Performance Program, Rehabilitation Institute of Chicago;

(2) Department of Biomedical Engineering, Northwestern University;

INTRODUCTION

Osteoarthritis (OA), which is the most common form of arthritis, afflicts 19% of women and 10% of men in North America. The annual cost of OA is > $70 billion in the United States. The knee is the site most frequently affected among lower extremity joints. Patients with ACL and/or meniscal tears were found to be at greater risk for developing osteoarthritis even after reconstruction surgery [1]. It has been proposed that injury-mediated joint instability is a primary contributor in the development and progression of join degenerative diseases. When joint stability is compromised, aberrant motion between bones arise leading to abnormal articular contact mechanics. The effect of these motions and subsequent mechanical loading to underlying cartilage during the OA process is a complex interaction that is not fully understood. We argue that the inclusion of 3D micromechanical model (atomistic molecular and continuum construct) of soft tissues (as cartilage) imbedded in a full joint model will facilitate our understanding of these interactions under isolated and functionally relevant tasks. A clear connection between micro-defects and cartilage loading, if identified, may elucidate the biochemical and mechanical underpinnings associated with the disease state. In this study we use a multi-scale constitutive model which considers the structural hierarchies of soft tissues [2]. This model creates a relationship between the aggregate mechanical behavior of the soft tissues and the basic nanoscale properties predicted by molecular simulations. The analysis reported here represent the first step of our goal since this model will be calibrated to fit the cartilage of our full scale healthy and ACL-reconstructed knee joint models (Fig.1) [3-4].

METHODS

Constitutive model: A multiplicative decomposition of the deformation gradient (F) into elastic (F^e) and plastic (F^p) parts is considered here to create the fiber reinforced composites model of cartilage. Due to the structure and the deformation mechanisms of soft tissues, the plastic flow is associated only with the uniaxial deformation of the fibrils ($F^f = F^p$). Therefore, the tissue is considered as hierarchical hyper-elasto-palstic composite material starting from the fibril level (1um) to the continuum macro tissue level (+100um). Neo-hookean generalized strain energy is used to model the fibril behavior as follow [2]

$$\psi_{fl}(\bar{I}_{1e}, \bar{I}_{4e}) = \frac{1}{2}\mu^{fl}(\bar{I}_{4e})(\bar{I}_{1e} - 3) \quad (1)$$

Where $\bar{I}_{1e}$ = tr ($\bar{C}^e = \bar{F}^e\, \bar{F}^{eT}$) and $\bar{I}_{4e}$= $n_0\, \bar{C}^e\, n_0^t$ with n_0 is the fiber direction at the initial configuration. The shear moduli, μ^{fl}, is considered as a function of the elastic fibrils deformations defined by

$$\mu^{fl}(\bar{I}_{4e}) = \mu_0\left(tanh\left(a_1(\bar{I}_{4e} - 1)\right) - a_2 exp\left(a_3(\bar{I}_{4e} - I_0)\right)\right) \quad (2)$$

This hyperbolic function used to fit the stiffness evolution of the fibrils predicted by molecular dynamic simulations during atomistic analyses [5]. Moreover, it is postulated that the yield condition (i), the plastic strain rate (ii) and the flow resistance (iii) of the tissue directly related to the effective stress of the fibrils with [2]

$$\begin{cases} \sigma_{eff} = \frac{4}{3}\bar{I}_{4e}\frac{\partial\psi_{fl}}{\partial\bar{I}_{4e}} = g_i + c\beta^2 & (i) \\ \dot{\gamma} = \dot{\gamma}_0\left|\frac{\sigma_{eff}}{\alpha(t)}\right|^{1/m} sig(\sigma_{eff}) & (ii) \\ \alpha(t) = \int h\dot{\gamma}f(\alpha) & (iii) \end{cases} \quad (3)$$

Here $g_i + c\beta^2$ represent the yield strength of the fibril which is a function of the cross link density between tropo-collagen molecules (Fig.1). This relationship serves as the basis for the multi-scale coupling between nanoscale collagen features and the aggregate materials properties of the tissue expressed at the larger scale (Fig.1).

The fiber of collagen (10um) and then the cartilage (+100um) are modeled as depth-dependent fibril reinforced composite material with incompressible neo-hookean total strain energy function as follow:

$$\psi_t(\bar{I}_{1f}, \bar{I}_4, \bar{I}_{4e}) = \frac{1}{2}\bar{\mu}_m(\bar{I}_{1f} - 3) + \frac{1}{2}\bar{\mu}_{fl}(\bar{I}_{1ef} - 3) + \frac{1}{2}\bar{\mu}_{eff}(\bar{I}_1 - \bar{I}_{1f}) \quad (4)$$

Where $\bar{I}_{1f} = \bar{I}_4 + 2\bar{I}_4^{-1/2}$, $\bar{I}_{1ef} = \bar{I}_{e4} + 2\bar{I}_{e4}^{-1/2}$, $\bar{\mu}_m = v_f v_{fm} \mu_{m\prime} + v_m \mu_m$, $\bar{\mu}_{fl} = v_f v_{fl} \mu_{fl}$. μ^{eff} is the effective shear moduli of fibril, v_f and v_{fl} are volume fraction of the fibers and fibrils, respectively. μ_m and μ_{fm} are the shear moduli of the matrix of the fiber and the matrix of the cartilage, respectively.

By satisfying both the Clausius-Duhem dissipation inequality at the macroscopic continuum level and the constraint of the incompressibility of soft tissue, the total stress σ^t can be expressed with fibril σ^f and no-fibril σ^{nf} stress tensors as follow:

$$\begin{cases} \sigma^t = \sigma^{nf} + \sigma^f & \\ \sigma^f = \frac{2}{J}\left(\bar{I}_4 \frac{\partial \psi_t}{\partial \bar{I}_4} dev(n \otimes n) + \bar{I}_{4e} \frac{\partial \psi_t}{\partial \bar{I}_{4e}} dev(n_e \otimes n_e)\right) & if\ \bar{I}_{4e} > 1 \\ \sigma^f = 0 & if\ \bar{I}_{4e} \leq 1 \end{cases} \quad (5)$$

Where $n = \bar{F} n_0 / \|\bar{F} n_0\|$

Indentation model: A cylinder with radius R=14 mm and height H=2.5 mm is considered as axisymmetric finite element model for the simulation of the indentation test proposed by [6]. The indenter is assumed rigid with a profile identical to that of the femoral condyle at the sagittal plane [6]. The depth-dependent material properties of the articular cartilage are taken based on the earlier measurements and simulations (See Fig.2a) [6]. To study the short-term response of the cartilage, 20% transient indentation are applied to the cartilage in the axial direction. The nonlinear analyses are performed using ABAQUS-Explicit 6.12 and the described material behavior is incorporated to finite elements analyses via VUMAT subroutine. The atomistic properties of the fibrils are taken based on the reported molecular simulation [2].

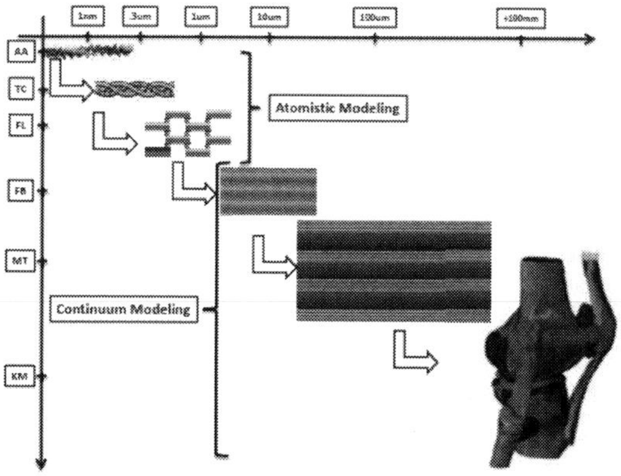

Figure.1: Schematic view of the Hierarchical features of the multi-scale model of the articular cartilage. (AA) Amino acids, (TC) tropo-collagen, (FL) Fibrils, (FB) Fibers, (MT) Macro tissue, (KM) knee model.

RESULTS

The computed reaction forces from the current model follow a similar trend as the reaction forces predicted by the earlier biphasic and hyper-elastic models of the articular cartilage (Fig.2b) [6]. The stress distribution in horizontal fibers and lateral displacement (Fig.2c) of the articular cartilage are in satisfactory agreement with values from the literature [6-7]. Increasing axial indentation to 40% (or decreasing the yield strength of the fibrils by 40%) resulted in a plastic deformation of the horizontal zone at the contact area and in the lower zone of the area initially located at distance of one fifth of the outer radius from the axis of symmetry.

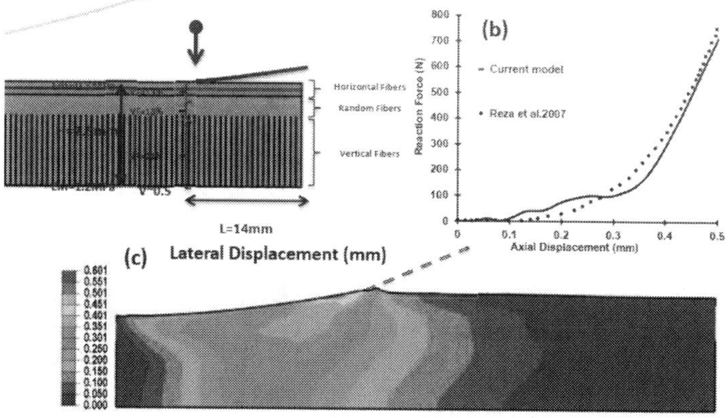

Figure.2: (a) Axisymmetric model of the articular cartilage. (b) Comparison of the predicted reaction force with poro-elastic model of Shirazi et al.2007. (C) Lateral displacement of the articular cartilage.

DISCUSSION

Our representative numerical example aims to demonstrate the basic features and numerical performance of the proposed materials model. The basic tenant of the model is a generalization of the elasto-plastic model of ligaments proposed by Tang and colleague [2] to cartilage. This work was performed as a first step to delineate the interaction between the nano- and macro-scale properties of the articular cartilage under physiological loading and during pre/post-surgical ACL replacement. From the preliminary results, predicted using an indentation test, the model showed accurate performance in the prediction of the macro elastic and inelastic stress as a function of the molecular characteristic of collagen. This multi-scale model will be calibrated to fit the articular cartilage of our full scale knee healthy and ACL-reconstruction models (Fig.1). With the creation of this cartilage model, we will have a powerful tool that accounts for the synergy between the passive nanoscale components and the active musculature of the knee joint. Therefore, the future application of the healthy and ACL-reconstruction knee models should shed further light on the initiation and the propagation of posttraumatic OA, the influence of ACL-surgical design on cartilage stress including OA prevention and the process of cartilage degradation behavior with aging [8].

ACKNOWLEDGEMENTS

Supported by: the National Institute of Health NIH.

REFERENCES

[1] Wheaton & Jensen, JPO 790:18, 2011 [2] Tang et al, ABE 1117: 30, 2009 [3] Dhaher et al, JB 3118:25, 2010 [4] Salehghaffari & Dhaher, JB 1609:17, 2013 [5] Guo et al, JMPS 1951:72, 2006 [6]Shirazi et al. JOR 609:15.2007[7]Wilson et al. JBM 357:66.2004 [8] Kempson et al, Na, 27:8, 1968.

SB^{3}C2015
Summer Biomechanics, Bioengineering and Biotransport Conference
June 17-20, 2015,Snowbird Resort, Utah, USA
SB^{3}C2015-606

COMPARISON OF PULMONARY ARTERY MATERIAL PROPERTIES BETWEEN A CONGENITAL HEART DISEASE PATIENT AND A NORMAL SUBJECT USING CARDIAC MAGNETIC RESONANCE: A FEASIBILITY STUDY

Rajit Banerjee (1), Gavin A. D'Souza (2), Namheon Lee (3), Michael D. Taylor (3)

(1) Department of Bioengineering
University of Toledo
Toledo, Ohio, USA

(2) Department of Mechanical and Materials Engineering
University of Cincinnati
Cincinnati, Ohio, USA

(3) The Heart Institute
Cincinnati Children's Hospital Medical Center
Cincinnati, Ohio, USA

INTRODUCTION

Congenital heart disease (CHD) patients with complex right heart obstructive lesions often require intervention to alleviate the obstruction in infancy. Common intervention procedures include the placement of repaired conduits, native patches, and artificial valves. Although surgical outcomes are excellent, patients are left with residual lesions including pulmonary artery regurgitation and obstruction that result in varying degrees of derangement of right ventricular-pulmonary artery (RV-PA) function [1].

Most longitudinal studies focus on evaluating either the microstructure or the pressure-strain modulus of the PA's or the RV of CHD patients independently [2, 3]. However, the evaluation of material properties of the PA's and its effect on the overall RV-PA hemodynamics is lacking. Properties such as compliance, indicated by the slope of the pressure-diameter curve, stress, and stretch of the PA wall contribute significantly to the function of the RV-PA unit. Monitoring these properties may provide valuable information on disease progression of CHD patients.

Arteries exhibit complex mechanical properties characterized by nonlinear elastic behavior under a stress condition. Modeling of the artery wall using constitutive equations coupled with *in vivo* and *in vitro* test data has been previously used to characterize these complex material properties [4]. However, application of such modeling techniques for assessing the post-operative RV-PA dysfunction in CHD patients with right heart disease has been limited due to the lack of experimental data. A previous study proposed a methodology of evaluating the arterial material properties using only *in vivo* pressure-diameter data coupled with a set of constitutive equations [5]. In this feasibility study, using pressure-diameter data, we characterized the material properties of the left PA (LPA) of a CHD patient and compared them to that of a normal subject.

METHODS

Study population. Two subjects (n=2) were included for this study. One subject (control: age = 13 years, body surface area = 1.57 m^2) had a normal RV-PA physiology and normal pulmonary valve function with no PA stenosis. The other subject (patient: age = 8 years, body surface area = 1.16 m^2) had an abnormal RV-PA function and anatomy as a consequence of the presence of a repaired RV-PA conduit placed during Ross procedure which was conducted for aortic valve stenosis.

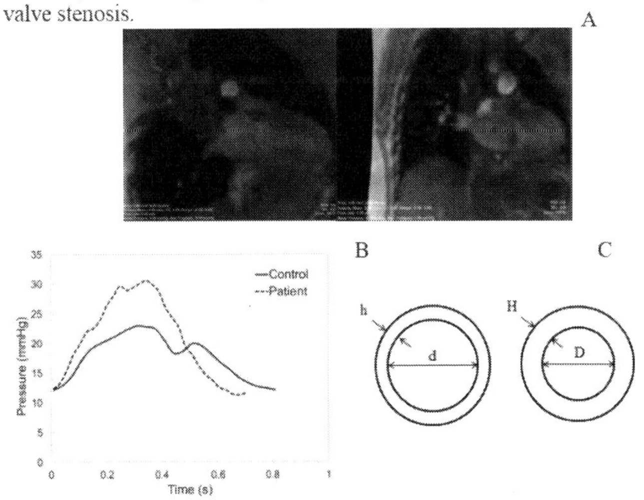

FIGURE 1. (A) MR IMAGE OF THE LPA OF THE CONTROL (LEFT PANE) AND THE PATIENT (RIGHT PANE), (B) *IN VIVO* LPA PRESSURE PULSE, (C) SCHEMATIC OF THE ACTUAL (LEFT) AND REFERENCE (RIGHT) CONFIGURATION OF THE ARTERY

Data acquisition. Cardiac MRI data was acquired using a 3.0 Tesla MRI scanner (Achieva, Philips Healthcare, Best, The Netherlands). In addition to MRI, cardiac catheterization was conducted and pressure was recorded at 30 time-points over a single cardiac pulse. CMR and cardiac catheterization were both performed in Cincinnati Children's Hospital Medical Center (CCHMC). An MRI slice showing the LPA anatomy, and the *in vivo* LPA pressure pulse are shown in Figs. 1A and 1B, respectively. The pressure and diameter data points, which were measured non-simultaneously, were synchronized using ECG gating.

Material model. A 2D hyperelastic model [5] was used to predict the nonlinear and anisotropic behavior of the LPA wall when subjected to bi-axial loading. The principal stresses were modeled using Fung's 2D strain energy function (SEF), $\psi = \frac{C}{2}(e^{Q} - 1)$

where, $Q = c_{\theta\theta}E_{\theta\theta} + c_{\theta z}E_{\theta\theta}E_{zz} + c_{zz}E_{zz}$

The principal stretches in the circumferential (θ) and axial (z) directions were defined as: $\lambda_\theta = \frac{D(A+\pi d^2)}{d(A+\pi D^2)}$ and $\lambda_z = \frac{z}{Z}$

where; D and Z represent the inner diameter and length of the vessel, respectively, in the unloaded condition (reference geometry; Fig 1C, right), and d and z represent the inner diameter and length of the vessel, respectively, in the loaded condition (actual geometry; Fig 1C, left). The principal Cauchy stresses, predicted by the model, in the circumferential and axial direction were defined as $\sigma_{\theta\theta}^{\text{mod}} = \lambda_\theta \frac{\partial \psi}{\partial \lambda_\theta}$ and $\sigma_{zz}^{\text{mod}} = \lambda_z \frac{\partial \psi}{\partial \lambda_z}$. In order to define the material model for the LPA and predict the model constants, an equilibrium condition was imposed on the artery wall and the principal theoretical stresses were defined as $\sigma_{\theta\theta} = \frac{pd^2\pi}{2A}$ and $\sigma_{zz} = \frac{d^2\pi(4F + pd^2\pi)}{4A(A + d^2\pi)}$, where p is the pressure and F is the external axial force. The four constitutive parameters, C, $c_{\theta\theta}$, $c_{\theta z}$, and c_{zz} were then evaluated using a least-square error minimization algorithm where the error was defined as the difference between the model and theoretical stresses. An in-house MATLAB® code (MATLAB Inc., Waltham, MA, USA) was developed and used for computing the material constants and further, quantifying the material properties of the LPA.

RESULTS

The values of the material model constants are specified in Table 1. Using these values in the constitutive equations, the material properties such as the pressure-diameter characteristics and the stress-stretch behavior of the LPA were computed.

TABLE 1. VALUES OF THE MATERIAL MODEL CONSTANTS

	C (kPa)	$c_{\theta\theta}$	$c_{\theta z}$	c_{zz}
Control	14.52	0.197	0.165	0.204
Patient	9.83	0.709	1.149	0.00012

Figure 2 shows the LPA pressure and corresponding diameter recorded *in vivo* during the systolic (loading) and diastolic (unloading) phases of the cardiac cycle. The close fit of the model curve to the experimental data indicates that the material model accurately predicts the behavior of the LPA. Higher pressures and consequently, larger diameter values were observed for the patient, indicating LPA dilation. Further, a greater slope of the pressure-diameter curve was observed for the patient, which indicates reduced compliance in the patient's LPA due to the presence of the repaired conduit.

Subsequently, the stress-stretch characteristics of the LPA were assessed using the material model to verify these observations. Circumferential and axial stresses in the LPA were predicted at different circumferential stretch values and at a constant axial stretch of 1.2 for the control and patient (Fig. 3).

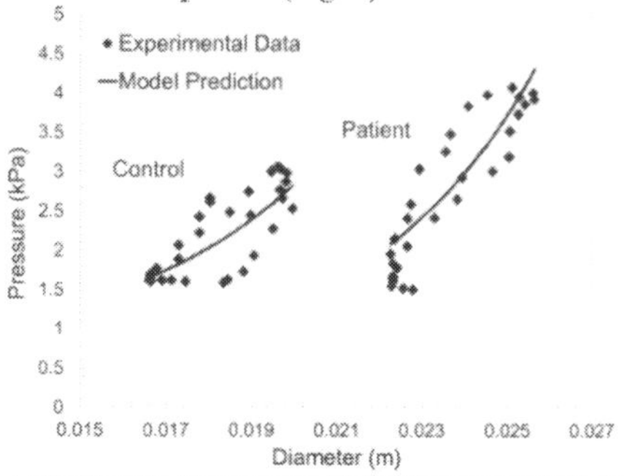

FIGURE 2. PRESSURE-DIAMETER CHARACTERISTICS OF THE LPA

As observed in Fig. 3, the stresses increased nonlinearly with the increase in circumferential stretch which showed that the model could accurately characterize the nonlinear behavior exhibited by the LPA. Higher stresses were observed for the patient in comparison to the control. Further, the circumferential stress was observed to be greater than the axial stress for both the control and the patient.

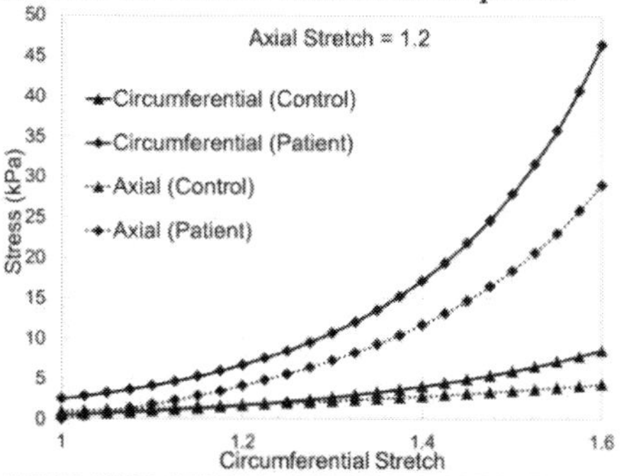

FIGURE 3. PREDICTED STRESS-STRETCH CHARACTERISTICS OF THE LPA AT A CONSTANT AXIAL STRETCH OF 1.2

DISCUSSION

This study evaluates the feasibility of using a material model in conjunction with *in vivo* pressure-diameter data to characterize the PA material properties in a CHD patient. Properties like compliance, stress, and stretch of PA wall contribute to the RV-PA function.

The pressure-diameter relationship and the stress-stretch characteristics of the patient showed that the diseased PA was less compliant compared to the normal PA. The structural changes in the PA responsible for the observed difference are likely multifactorial. The primary cause for decreased compliance in the patient's PA was the presence of a repaired conduit connecting the RV to the PA.

Thus, the material model, developed using *in vivo* pressure-diameter data, can be useful in characterizing the material properties of the PA. Further, these properties can be used in evaluating altered RV-PA hemodynamics and thus help in the longitudinal assessment of CHD patients.

REFERENCES

[1] Lee N, et al., *Biomed Eng Online*, 2014.
[2] Inuzuka R., et al., *Ann Thorac Surg*, 2013.
[3] Wang Z, et al., *Biomech Model Mechanobiol*, 2013.
[4] Kassab GS, et al., *J R Soc Interface*, 2006.
[5] Schulze-Bauer and Holzapfel, *J Biomech*, 2003.

SB³C2015
Summer Biomechanics, Bioengineering and Biotransport Conference
June 17-20, 2015,Snowbird Resort, Utah, USA
SB³C2015-608

EXPERIMENTAL EVALUATION OF HEMODYNAMICS IN PATIENT-SPECIFIC MODEL OF TYPE B AORTIC DISSECTION

Joav Birjiniuk (1), Jean Marie Ruddy (2), Mark Young (3), Lucas H. Timmins (1), Ravi K. Veeraswamy (4), David N. Ku (5)

(1) Department of Biomedical Engineering
Georgia Institute of Technology and
Emory University School of Medicine
Atlanta, GA, USA

(2) Division of Vascular Surgery, Department of Surgery
Medical University of South Carolina
Charleston, SC, USA

(3) Cardiac and Vascular Group
Medtronic, Inc.
Santa Rosa, CA, USA

(4) Division of Vascular Surgery,
Department of Surgery
Emory University School of Medicine
Atlanta, GA, USA

(5) Department of Mechanical Engineering
Georgia Institute of Technology
Atlanta, GA, USA

INTRODUCTION

Aortic dissection is a disease of the aortic wall in which the intimal layer separates from the medial and adventitial layers. Tearing in the aortic wall leads to blood flow through the original true lumen and the newly formed false lumen. Though surgical intervention is indicated in dissection involving the ascending aorta, the management of patients with descending aortic disease is less clear. While medical, endovascular, and surgical therapies have all been shown to be effective in relieving clinical symptoms, there remains no consensus as to treatment algorithms [1]. To date, only a single randomized controlled prospective clinical trial has been undertaken to study the benefit of endovascular approaches over medical therapy alone [2]. However, this and other clinical studies are obfuscated by the acuity of disease presentation and selection bias due to intervention in specific patient populations.

A proper understanding of the hemodynamic alterations due to dissection is crucial in determining the role of various treatment strategies in the restoration of normal flow. Surgical intervention can create underlying changes in aortic fluid dynamics and end-organ perfusion. Due to the limitations of translating the hemodynamics in aortae of differing animal species, an *in vitro* model of the human aorta may be more appropriate for repeated, controlled studies of the fluid dynamics in dissected aorta. Previously reported *in vitro* models, however, do not accurately reflect human anatomy [3,4,5] or lack intimal flaps [6], and are limited in clinical applicability. We demonstrate the development of an experimental model of aortic dissection, which includes a patient-specific silicone model and physiologic hemodynamic values. Further, we hypothesize that the true and false lumen pressures will equalize and that our model will recapitulate native aortic pulse propagation.

METHODS

Compliant models were fabricated from silicone in a two-mold process based on segmented patient CT images of the ascending, arch, and descending thoracic aorta. Branch vessels were excluded due to the complexity of fabrication with an intimal flap. A mold release was used in between molding steps to allow for release of the dissection flap from the aortic model wall. Once molded, the model was incised at flap entrance, located on the lateral wall at the level of the flexure of the lesser curvature, and exit to allow for filling and distension. These tears extended across 50% of the aortic circumference.

A Thoratec PVAD (Thoratec, Pleasanton, CA) ventricular assist device was attached to the model via tubing to generate pulsatile flow in the model. The tubing, model, and VAD were all filled with water, with care being taken to eliminate air in the system. Average flow rates of 3.9 L/min were achieved at a central aortic pressure of 120/80 mm Hg, 60 beats per minute, with an ejection phase lasting 300 ms. The model was catheterized in both true and false lumen using multipurpose MPA catheters, and an angioscope (Applied Vascular Devices, Laguna Hills, CA) was introduced for confirmation of catheter placement and visualization of flap motion. Pressure was measured via catheter on a pressure transducer (Harvard Apparatus, Holliston, MA), and subsequently collected using a custom LabView (National Instruments, Austin, TX) controller, sampling at 100 Hz. Pressure traces were imported to Matlab (The MathWorks, Natick, MA) for further analysis.

A lumped-parameter model of dissected aorta was developed to provide a simple prediction of the filtering characteristics of the silicone aorta. Individual segments of the aorta were considered as parallel RC circuits to account for vascular resistance and compliance. Two parallel branches with Windkessel elements were used to represent the true and false lumina of the dissected segment (Figure 1).

Luminal resistances were calculated using the dimensions of the aorta at given segments [7], whereas capacitances were determined from aortic compliance. Compliance was assumed to have minimal variation along the aorta. RC values for distal segments were treated as reported previously [8].

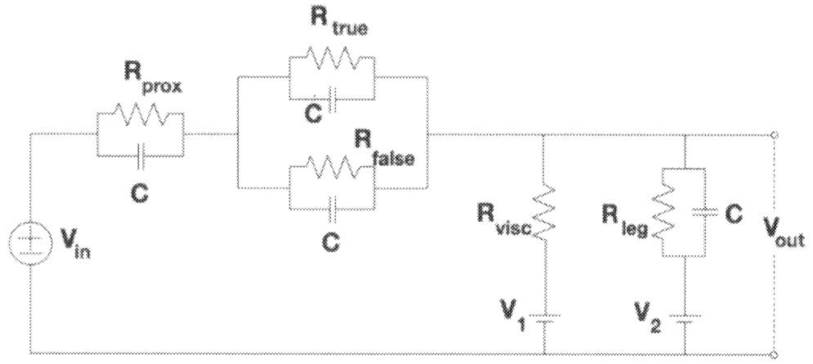

FIGURE 1. LUMPED-PARAMETER MODEL OF AORTA.

RESULTS

The fabricated model maintained its integrity under static pressure greater than 200 mm Hg and during cardiac pulsation. Distension of the intimal flap into the true lumen was observed, leading to flow in both true and false lumina. Furthermore, flap motion and model pulsation were seen during VAD pumping.

Pulse waveforms in the model aorta were found to be similar to physiologic aortic waveforms. Outlet pressures were 95 +/- 24 mm Hg, while inlet and lumped-parameter model pressure traces both were 96 +/- 28 mm Hg (Figure 2). Pressure differences between inlet and outlet did not exceed 6 mm Hg. Delays ranging from -0.04 ms and 0.02 ms were observed between inlet and outlet. True and false lumen pressures were measured at 96 +/- 25 mm Hg and 95 +/- 24 mm Hg respectively (Figure 3), with maximal pressure differences less than 4 mm Hg. Maximal delays of 0.02 ms between true and false lumina fall on the limit of detectability given a Nyquist frequency of 50 Hz.

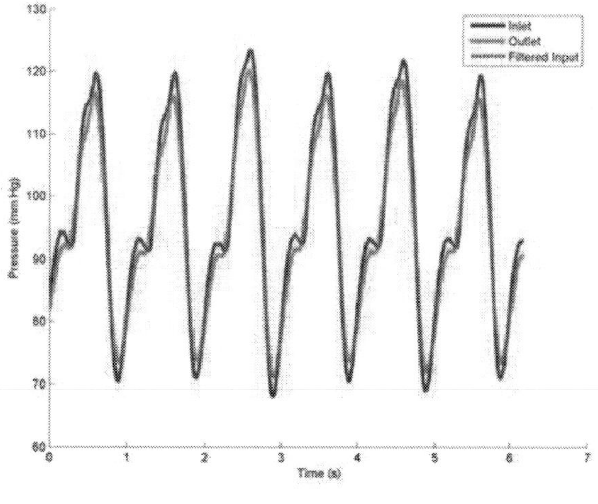

FIGURE 2. LUMPED-PARAMETER MODEL PRESSURE (FILTERED INPUT) MATCHES CATHETER (OUTLET) PRESSURE.

DISCUSSION

We demonstrate an *in vitro* model of aortic dissection with

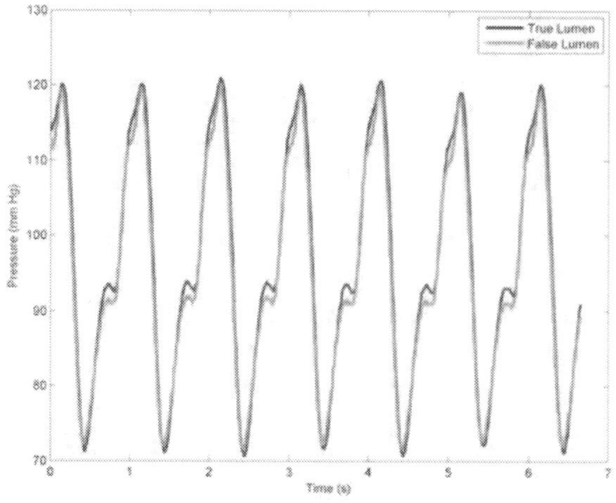

FIGURE 3. CATHETER-MEASURED TRUE AND FALSE LUMEN PRESSURES.

realistic anatomy, intimal flap, and physiologic pumping characteristics. The elastic model includes aortic compliance, providing a realistic pulsation and transfer function. The use of physiologic flow input with the VAD affords the ability to examine variations in pulse transmission and end-organ perfusion in true and false lumen. Unlike previous work in the field [3,4,5,6], this model possesses realistic aortic geometry, physiologic pulsation, and the ability to vary tear parameters as necessary.

In addition, we have conducted initial studies to apply this experimental strategy and comment on the dynamics of the aorta when dissected. We intend to correlate these results with flow variations in the dissected versus normal aorta, as variations in pressure waveforms may affect flow to different vascular branches, as well as the velocity profile at various locations throughout the aorta. These findings may inform upon treatment strategies designed to maximize end-organ perfusion.

Future refinements to the model include the addition of branch vessels as well as variation of tear geometries and locations. As branch vessels are of high clinical importance in disease management, we will examine flow into branch vessels of differing locations supplied by either the true or false lumen. These data will allow for specific investigation into the effectiveness of available interventions. Patients with aortic dissection present with heterogeneous dissection morphologies, thus we aim to investigate the role of varying dissection shape in the alteration of aortic hemodynamics.

ACKNOWLEDGEMENTS

Supported by a grant from Medtronic, Inc. We thank Kevin McTeague of Emory University Hospital for providing the VAD.

REFERENCES

[1] Fattori, R et al., *J Amer Coll Cardio*, 61:1661-1678, 2013.
[2] Nienaber, CA et al., *Circulation*, 120:2519-2528, 2009.
[3] Chung, JW et al., *Radiology*, 214:99-106.
[4] Tsai, TT et al., *J Vasc Surg*, 47:844-851, 2008.
[5] Rudenick, PA et al., *J Vasc Surg*, 57:464-474, 2013.
[6] Wen, C et al., *Annals of Biomed Eng* 38:391-402.
[7] Schoder, M et al., *Ann Thor Surg*, 83:1059-1066, 2007.
[8] Holenstein, R, Ku, D, *Biorheology*, 25:835-842, 1988.

SB³C2015
Summer Biomechanics, Bioengineering and Biotransport Conference
June 17-20, 2015,Snowbird Resort, Utah, USA
SB³C2015-609

COMPARISON BETWEEN NONLINEAR MATERIAL MODELS AND FIBER NETWORK MODELS REVEALS IMPORTANCE OF FIBER INTERACTIONS IN CELL MECHANOSENSING ON FIBROUS SUBSTRATES

Maziar Aghvami[1], Kristen L. Billiar[2], Edward A. Sander[1]

[1]Department of Biomedical Engineering,
University of Iowa, Iowa City, IA 52242
[2]Department of Biomedical Engineering
WPI, Worcester, MA 01609

INTRODUCTION

The mechanical environment of a cell regulates its behavior and plays an important part in directing large-scale processes, such as tissue remodeling and repair. This mechanical environment is complex and presents compositional, structural, and organizational cues that provide passive and active mechanical feedback to the cell. Several gel systems have been used to study how various components of the mechanical environment affect cell behavior. The most commonly used system consists of polyacrylamide (PA), which has been used with great effect to show that cells sense and respond to substrate stiffness and other mechanical factors [1, 2]. Most cells, however, reside in a heterogeneous and anisotropic extracellular matrix. *In vitro* experiments on fibrous gels have noted enhanced mechanosensing on fibrous materials compared to PA gels [3], but the underlying basis for this enhancement remains unclear. In this work, we compared a nonlinear continuum finite element (FE) model with two different multi-scale FE models consisting of a collection of distinct fiber networks (either growth networks or Delaunay networks) in order to better understand if enhanced mechanosensing and distance perception in fibrous gels arises from inherent material non-linearity or from properties unique to the fiber networks.

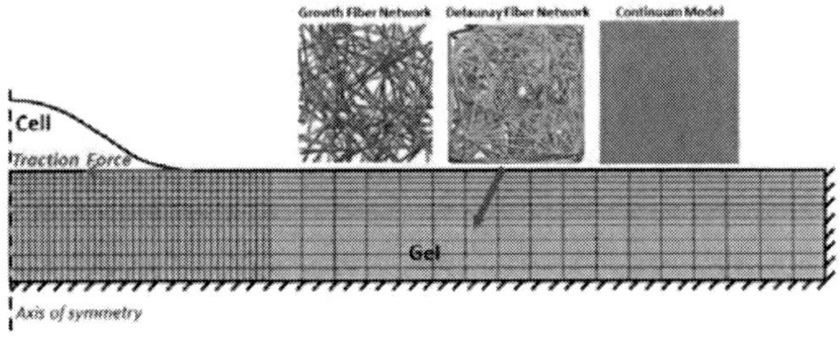

Fig. 1: Schematic of the FE model. Cell traction forces are applied to the focal adhesion area on the FE top surface. The model consists of either 3D *Growth* or *Delaunay* fiber networks or a continuum model.

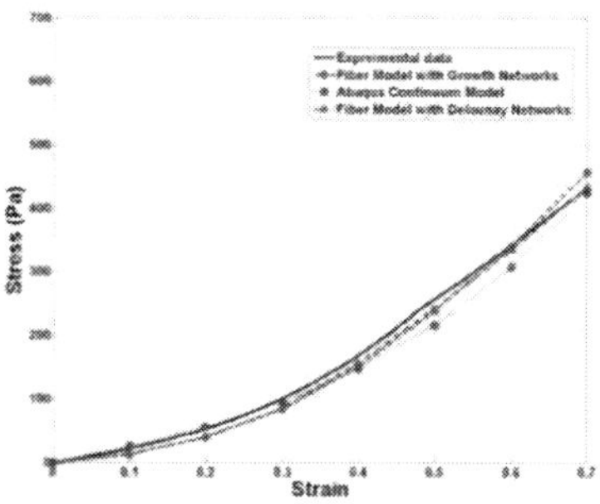

Fig. 2: Model parameters were selected to match uniaxial experimental data.

METHODS

The effect of substrate thickness (from 10 μm to 200 μm) on effective stiffness was studied with three models:

<u>*Continuum Model*</u>: A nonlinear continuum FE model was constructed in ABAQUS using a third-order (cubic) reduced polynomial strain energy function (Yeoh form) in ABAQUS.

<u>*Fiber-Based Models*</u>: Fiber-based models were constructed using a multi-scale modeling framework in which the macroscopic FE domain of the material in linked to a set of microscopic fiber networks through volume averaging [4]. Two types of fiber networks, termed *Growth Networks* [4, 5] and *Delaunay Networks* [6], were investigated in order to assess the role of network structure. The main topographical distinction between the two networks is the degree of connectivity: *Growth Networks* have a maximum of three fibers connected at a

node, whereas *Delaunay Networks* have many more. For each fiber-based model, a set of 200 unique and nominally isotropic fiber networks was created. Each network was assigned randomly to an FE element. An additive and non-interacting stress at the continuum level from a compressible neo-Hookean material was also included in the stress balance.

Parameter Selection: In order to facilitate comparisons between models, the material properties of each model were chosen so that the uniaxial stress-strain response of the model matched the data recorded for that of a rectangular 6.7 mg/mL fibrin gel.

Case Studies: Rectangular FE meshes were created that represent a spatial scale of 75 μm x 4 μm (length, width) with different depths (thicknesses):

- Case 1: 10 μm depth with 684 tri-linear hexahedral elements and 1508 nodes;
- Case 2: 30 μm depth with 1140 tri-linear hexahedral elements and 2436 nodes;
- Case 3: 50 μm depth with 1596 tri-linear hexahedral elements and 3364 nodes;
- Case 4: 200 μm depth with 5016 tri-linear hexahedral elements and 10324 nodes (for effective stiffness comparisons).

These meshes were used to examine the effect of thickness on both the continuum FE model and the two fiber-based models. Symmetry boundary conditions were applied to the left, front, and back surfaces, and fixed boundary conditions were applied to the bottom and right surfaces. A cell traction force of 1.1×10^{-10} N was distributed amongst and applied tangentially to the 22 nodes in the focal adhesion (FA) area on the top boundary over 50 equally spaced steps (Fig. 1). Differences in cell mechanosensitivity were determined by calculating the effective stiffness, which is inversely proportional to the mean displacement of the nodes in the focal adhesion (FA) area.

RESULTS AND DISCUSSION

The displacement fields for 10 μm thick gels are presented in Fig. 3 for all three models. As expected, the maximum displacements occurred in the focal adhesion (FA) area, but both the magnitudes and patterns of displacement differed noticeably with model type. Average FA displacement increased with increasing gel thickness for all models, with the largest displacements occurring in the continuum model (data not shown). Because effective stiffness varies inversely with FA displacement, a higher effective stiffness (Fig. 4) was calculated for the fiber-based models compared to the continuum model, with the *Delaunay Networks* producing a substantially higher effective stiffness compared to the *Growth Networks*. The effective

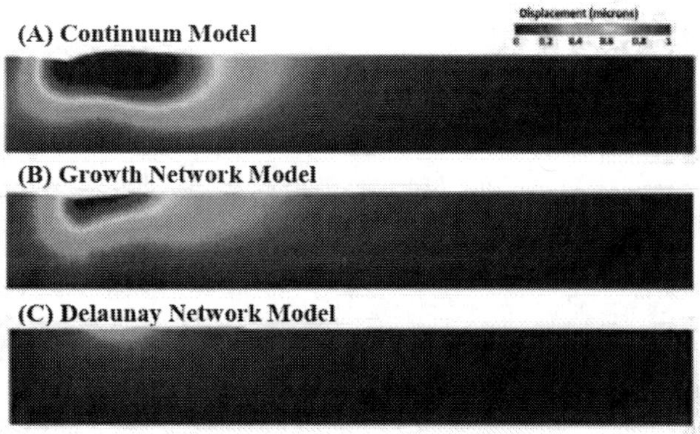

Fig. 3: Displacement contours for 10 μm thick gels for (A) the continuum model, (B) the *Growth Network* model, and (C) the *Delaunay Network* model.

stiffness was also more sensitive to gel thickness for the fiber-based models than for the continuum model, with the influence of the rigid lower boundary decaying faster for the continuum model than the fiber models (i.e., the rigid boundary can be ignored for thinner continuum gels than fiber-based model gels). The fiber-based models predict fiber realignment in response to the applied cell tractions, with greater realignment for the *Growth Networks* compared to the *Delaunay Networks* (data not shown), which might explain the difference in long-range mechanical sensing observed experimentally in fibrous gels compared to PA gels, as opposed to differences in material nonlinearity. These results help illuminate how fiber-fiber interactions and matrix microstructure might contribute to cell mechanosensing of the local mechanical environment in fibrous materials.

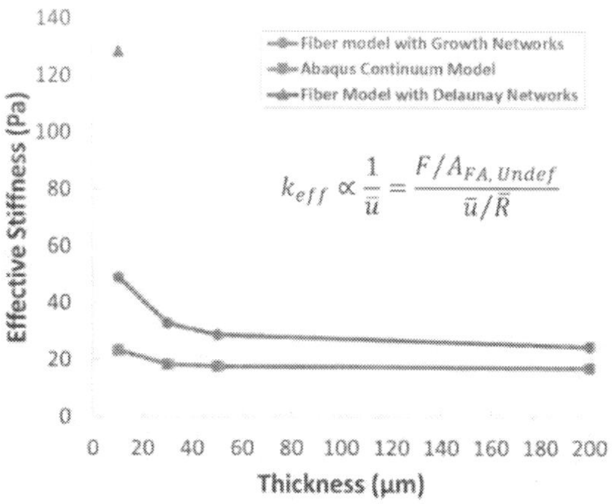

Fig. 4: Effective stiffness as a function of gel thickness; showing greater resistance of the fiber model to the cell traction forces than the continuum model with similar uniaxial behavior. Note only the 10 mm case is shown for Delaunay Networks (green).

ACKNOWLEDGEMENTS

Support for this work was provided partly by NIAMS R03 AR063967 and NSF CAREER Grant No. 1452728. Computational Resources were provided by the University of Iowa ITS Helium Cluster.

REFERENCES

[1] Pelham, R.J., Wang, Y., *Cell locomotion and focal adhesions are regulated by substrate flexibility*. PNAS 1997. 94: pp13661-13665.

[2] Engler, A.J., Sen, S., Sweeney, H.L., Discher, D.E., *Matrix elasticity irects stem cell lineage specfiication*. Cell 2006. 126: pp 645-647.

[3] Rudnicki, M.S., Cirka, H.A., Aghvami, M., Sander, E.A., Wen, Q., Billiar, K.L., *Nonlinear strain stiffening is not sufficient to explain how far cells can feel on fibrous protein gels*. Biophysical Journal 2013. 105: pp 11-20.

[4] Sander, E.A., Stylianopoulos, T., Tranquillo, R.T., Barocas, V.H., *Image-based multiscale modeling predicts tissue-level and network-level fiber reorganization in stretched cell-compacted collagen gels*. PNAS 2009. 106: pp. 17675-17680.

[5] Aghvami, M., Barocas, V.H., Sander, E.A., *Multiscale Mechanical Simulations of Cell Compacted Collagen Gels*. J Biomech Eng 2013. Vol. 135.

[6] Hadi, M.F., Sander, E.A., Barocas, V.H., *Multiscale model predicts tissue-level failure from fiber-level damage*. J Biomech Eng 2012. 134: 091005.

SB³C2015
Summer Biomechanics, Bioengineering and Biotransport Conference
June 17-20, 2015,Snowbird Resort, Utah, USA
SB³C2015-610

HUMAN KNEE JOINT FINITE ELEMENT MODEL USING A TWO BUNDLE ANTERIOR CRUCIATE LIGAMENT: VALIDATION AND GAIT ANALYSIS

Nicholas A. Czapla (1), Meghan K. Sylvia (1), Zachary F. Lerner (3), David J. Tuttle (4), Otto J. Schueckler (5), Scott J. Hazelwood (1,2), Stephen M. Klisch (1,2)

(1) Mechanical and (2) Biomedical Engineering Department
California Polytechnic State University
San Luis Obispo, CA, USA

(3) Department of Biomedical Enginering
Colorado State University
Fort Collins, CO, USA

(4) Radiology Associates, Inc.
San Luis Obispo, CA, USA

(5) Central Coast Orthopedic Medical Group
San Luis Obispo, CA, USA

INTRODUCTION

Anterior cruciate ligament (ACL) deficient individuals are at a much higher risk of developing osteoarthritis (OA) compared to those with intact ACLs, likely due to altered biomechanical loading [1]. Research indicates the ACL is comprised of two "bundles", the anteromedial (AM) and posterolateral (PL) bundles [2]. Although the function of both bundles is to restrain anterior tibial translation (ATT), each bundle has their own distinct range of knee flexion where they are most effective [3].

Articular cartilage contact stress measurements are difficult to measure in vivo. An alternative approach is to use knee joint finite element models (FEMs) to predict soft tissue stresses and strains throughout the knee. Initial and boundary conditions for these FEMs may be determined from knee joint kinematics estimated from motion analysis experiments. However, there is a lack of knee joint FEMs which include both AM and PL bundles to predict changes to articular cartilage contact pressures resulting from ACL injuries. The purpose of this study is to develop and validate a knee joint FEM using both AM and PL bundles and subsequently perform a gait analysis of varying ACL injuries.

METHODS

FEM Development. An FEM of a right knee joint was built from sagittal plane magnetic resonance images (MRIs) (GE Medical Systems, Ideal GRE, TR=7.428ms, TE=4.16ms, slice spacing=1.5mm, flip angle=45°, pixel spacing=.3156) of a healthy, 33 year old male with no prior history of injuries. FEM tissue structures modeled included: femur and tibia bone; medial and lateral menisci; femoral and tibial articular cartilage; ACL, posterior cruciate (PCL), medial collateral (MCL) and lateral collateral (LCL) ligaments.

The 3-D solids of the knee structures were created from the MR images using Mimics (Materialise NV, Leuven, Belgium) and were smoothed to remove any imperfections, before importing into SolidWorks (Dassault Systemes, Velizy-Villacoublay, France) to remove any residual overlap between structures. The ACL was divided into AM and PL bundles in SolidWorks based on their reported femoral and tibial attachment sites [4]. Soft tissue structures were meshed in TrueGrid (XYZ Scientific Applications, Inc., Livermore, California, USA) using linear, hexahedral elements. Bones were modeled as rigid bodies with 2-D shell elements. Each mesh was imported into Abaqus (Dessault Systemes, Velizy-Villacoublay, France) for FEM analyses (Figure 1).

The articular cartilage and ligaments were attached to bone using tie constraints. The distal portion of the LCL had three sets of spring elements attached, acting in the longitudinal and transverse directions to mimic tension in this ligament [5,6]. The menisci were constrained to the tibia using four sets of spring elements [7]. Articular cartilage (E=15MPa, ν=0.475 [8]) and menisci (E=59MPa, ν=0.49 [9]) were modeled as a linear elastic, homogenous, isotropic materials. Ligaments were modeled as linear elastic, transversely isotropic, homogeneous materials (Table1) [10,11,12].

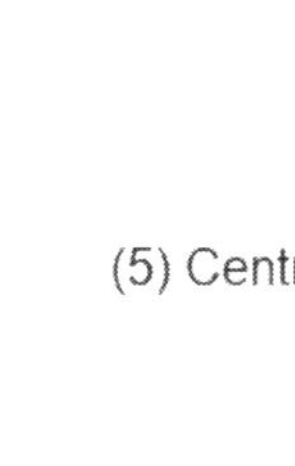

Figure 1: Right knee FEM

Table 1: Ligament material properties.

	E_L (MPa)	E_T (MPa)	ν_{12}, ν_{13}	ν_{23}	G_{12}, G_{13} (MPa)	G_{23} (MPa)
PCL,LCL,MCL	153.7	5.1	1.4	.3	1.72	1.9
AM	212.23	7.07	1.4	.3	1.72	1.9
PL	115.55	3.85	1.4	.3	1.72	1.9

FEM Validation. To validate the FEM, three experiments were simulated and FEM predictions of articular cartilage contact pressure, ATT and/or ACL strains were compared to experimental results: 1)

Axial compressive load (1000 N) applied to the tibia of intact ACL cadaver knees [13]; 2) Anterior tibial load (134 N) applied to an intact ACL and AM deficient cadaver knees [14]; 3) Posterior femoral load (130 N) applied to intact ACL cadaver knees [15]. Only cases with a flexion angle of 0° were used with appropriate boundary and loading conditions applied to the tibia and femur to replicate the three different experimental protocols. The FEM was considered validated if the predicted results were within one standard deviation of reported mean values.

Gait Analysis. A parameter study was performed on gait to predict the effects of ACL injury during gait: 1) intact ACL; 2) AM deficient ACL; 3) PL deficient ACL; 4) complete ACL rupture. The FEM simulated the 15% and 52% phases of gait by fixing the tibia and rotating the femur to the proper flexion angles and releasing all other degrees of freedom. Three individuals' gait analysis data [16] were imported into OpenSim's Joint Reaction Analysis to obtain the necessary kinematic and kinetic data to determine joint forces and moments applied at the joint center.

RESULTS

FEM Validation.

Tibial cartilage contact pressures (Figure 2A) and the ATT of an intact ACL knee (Figure 2B) deviated from the reported experimental means by less than one standard deviation. The predicted ATT of an AM deficient knee was greater than one standard deviation less than the experimental value (Figure 2B), but the predicted proportional increase in ATT from intact ACL to AM deficient (32.8%) matched very well with that observed (32.5%) [14]. FEM predicted AM strain matched the experimental results well (Figure 2C).

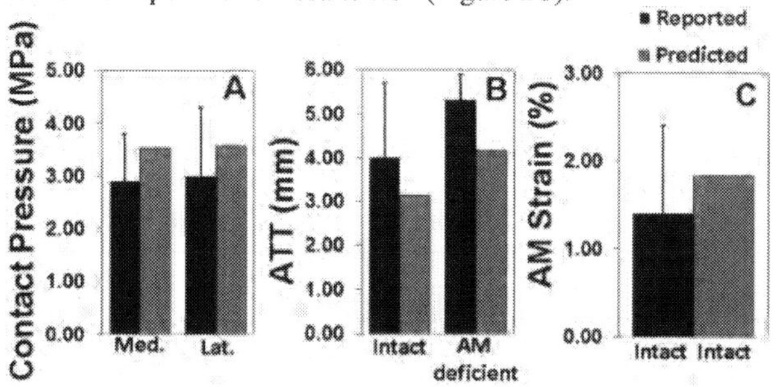

Figure 2: FEM predicted values vs reported.

Gait Analysis.

Maximum lateral pressure increased at 15% and decreased at 52% for increasing levels of ACL injury (Tables 2 and 3). ATT decreased at 15% and increased at 52% for increasing levels of ACL injury. PL deficient knees had a larger change in ATT and maximum lateral pressure at 15% gait, compared to AM deficient knees. AM deficient knees had a larger change in ATT and maximum lateral pressure at 52% gait, compared to PL deficient knees. There was a noticeable posterior shift in medial contact pressure from intact ACL knees to ACL ruptured (Figure 3)

Table 2: FEM predicted values at 15% of gait.

	Maximum medial pressure (MPa)	Maximum lateral pressure (MPa)	ATT (mm)
Intact	8.41	5.31	3.59
AM Deficient	8.45	5.39	3.41
PL Deficient	8.39	5.43	2.83
Complete Rupture	8.40	5.48	2.44

Table 3: FEM predicted values at 52% of gait.

	Maximum medial pressure (MPa)	Maximum lateral pressure (MPa)	ATT (mm)
Intact	5.78	3.94	7.17
AM Deficient	5.67	3.55	8.24
PL Deficient	5.76	3.68	7.68
Complete Rupture	5.96	3.41	8.26

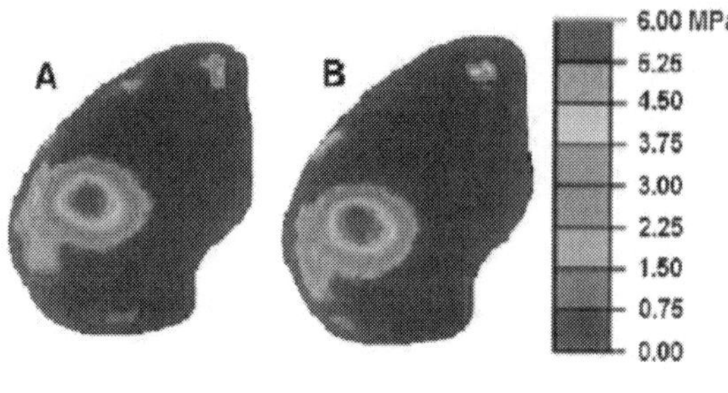

Figure 3: Predicted medial contact pressure at 52% gait: A) intact; B) ACL rupture.

DISCUSSION

FEM Validation. The FEM was considered to be validated based on the three experiments simulated. Almost all predicted values were similar to experimental results, with the exception being the predicted ATT in an AM deficient knee. Still, the predicted proportional increase in ATT from intact to AM deficient (32.8%) matched the reported proportional ATT increase (32.5%). It should be noted that the FEM has not yet been validated under combined loading. Validation for this study was performed at 0° because much of gait remains at low angles (< 20°). Reported ATT were higher than predicted values likely because experimental results were from older knees (ages 53-71) [14], consistent with findings indicating that older ligaments have demonstrated a Young's modulus reduction of 41% [6].

Gait Analysis. The 15% and 52% gait phases were chosen for their associated peak in compressive and anterior tibial joint force, respectively. Even though both phases of gait analyzed were at small knee flexion angles (<30°), predicted results suggest that combined loading has a more prominent role on bundle activation than knee flexion angle. Predicted results show that when the adduction moment and compressive load dominate (15% gait) the PL bundle supports more load than the AM bundle. FEM results also show that when the internal tibial torque, knee extension moment and anterior tibial load are more prominent (52% gait) the AM bundle supports more of the load. The decrease in ATT at 15% gait was expected due to the applied posterior direction of the tibial contact load.

ACKNOWLEDGEMENTS

Funding received from Quartus Engineering, Inc. and Cal Poly's Donald E. Bently and STRIDE Centers. The authors acknowledge Dr. Saikat Pal for assistance in FEM development.

REFERENCES

[1] Sabeka, N et al., *Current Rheumatology Reports*, 8:7-15, 2006. [2] Kopf, S et al., *KSSTA*, 17:213-219, 2009. [3] Zantop, T et al., *AJSM*, 35:223-227, 2007. [4] Mall, N et al., *Oper Tech Sports Med*, 21:2-9, 2013. [5] Noyes, F et al., *JBJS*, 58:1074-1082, 1976. [6] Chandrashekar, N et al., *J Biomech*, 39:2943-2950, 2006. [7] Hauch, K et al., *J Biomech*, 43:463-468, 2009. [8] Haut Donahue, T et al., *J Biomed Eng*, 124:273-280, 2002. [9] Guo, Y et al., *ACTA Mech Sinica*, 22:347-351, 2009. [10] Weiss, J et al., *J Biomech*, 35:943-950, 2002. [11] Butler, D et al., *J Biomech*, 19:425-432, 1986. [12] Quapp, K et al., *J Biomed Eng*, 120:757-763, 1998. [13] Seitz, A et al., *J Orthopaed Res*, 30:934-942, 2012. [14] Zantop, T et al., *AM J Sport Med*, 35:223-227, 2007. [15] Berns, G et al., *J Orthopaed Res*, 10:167-176, 1992. [16] Haight, D., *J Orthopaed Res*, 32:324-330, 2014.

SB³C2015
Summer Biomechanics, Bioengineering and Biotransport Conference
June 17-20, 2015,Snowbird Resort, Utah, USA
SB³C2015-611

DEVELOPMENT OF A HUMAN KNEE JOINT FINITE ELEMENT MODEL TO INVESTIGATE CARTILAGE STRESS DURING WALKING IN OBESE AND NORMAL WEIGHT ADULTS

Meghan K. Sylvia (1), Nicholas A. Czapla (1), Zachary F. Lerner (3), David J. Tuttle (4), Otto J. Schueckler (5), Scott J. Hazelwood (1,2), Stephen M. Klisch (1,2)

(1) Mechanical and (2) Biomedical Engineering Departments
California Polytechnic State University
San Luis Obispo, CA, USA

(3) Department of Biomedical Engineering
Colorado State University
Fort Collins, CO, USA

(4) Radiology Associates, Inc.
San Luis Obispo, CA, USA

(5) Central Coast Orthopedic Medical Group
San Luis Obispo, CA, USA

INTRODUCTION

Osteoarthritis (OA) is a degenerative condition characterized by the breakdown and loss of joint articular cartilage. While the cause of OA is not precisely known, obesity is a known risk factor [1]. Particular effort has gone towards understanding the relationship between obesity and knee OA because obesity is more strongly linked to OA at the knee than at any other lower extremity joint [2]. Although the relationship between obesity and knee OA is well established, the mechanism of pathogenesis is less understood. Excess body weight generates greater joint contact forces at the knee. However, obese individuals alter their gait, resulting in increased joint contact forces that are not proportional to body mass [3]. In this study, a partially validated knee joint finite element (FE) model was developed to predict cartilage loading during walking across individuals of varying adiposity. The model was used with kinematic and kinetic gait data to address the following hypotheses: 1) increased loading due to obesity will produce greater cartilage stress compared to the normal weight control; and 2) altered gait kinematics of obese individuals will alter the distribution of stress on the surface of the tibial cartilage.

METHODS

FE Solid Model Generation

A solid model geometry was constructed from 1.5 mm sagittal plane magnetic resonance images (MRIs) from a 33 year old healthy male with no known knee conditions. Mimics (Materialise, NV, Leuven, Belgium) was used to segment MRIs and construct 3D geometry of each bone and tissue structure. Mimic's Gaussian smoothing algorithm was used to smooth irregularities. The model included: the distal femur and proximal tibia bones; the medial collateral (MCL), lateral collateral (LCL), anterior cruciate (ACL), and posterior cruciate (PCL) ligaments; the medial and lateral menisci; and articulating cartilages of the tibiofemoral joint (Figure 1).

SolidWorks (Dassault Systemes, Velizy-Villacoublay, France) was used to assemble the model and remove any overlap between structures. FE meshes of individual structures were generated in TrueGrid Software (XYZ Scientific Applications, Inc., Livermore, California, USA) and exported to Abaqus (Dassault Systemes) for FE analysis. The femur and tibia were modeled as rigid, non-deformable shell elements due to their relatively high rigidity. All other soft tissue structures were modeled with linear 3D hexahedral elements. The ligaments were treated as transversely isotropic materials and assigned physiological material properties [4-6]. The cartilage (E=15 MPa, ν=0.475) and menisci (E=59 MPa, ν=0.49) were modeled as linear isotropic elastic materials [7,8].

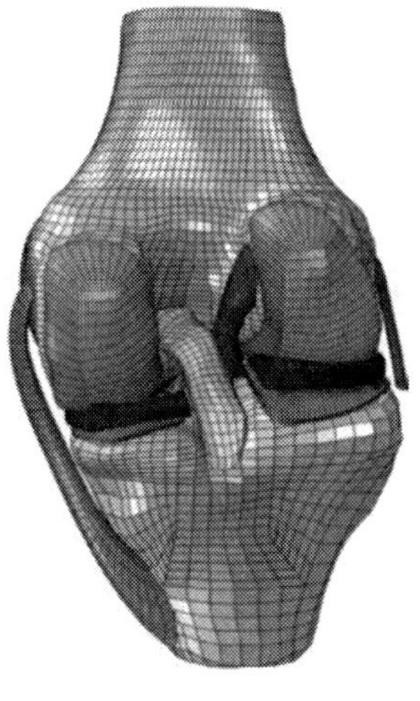
Figure 1: FE mesh of the knee.

Surface tie constraints were used to anchor cartilage and ligament structures to appropriate bony landmarks. Frictionless contact interactions were defined between the articulating surfaces of the femoral and tibial cartilages. Since the fibula is not included in this model, the distal face of the LCL was secured with linear spring elements with stiffnesses of 2.66 and 0.086 N/mm in the longitudinal and transverse directions, respectively. Spring elements were used to attach the anterior and posterior horns of the menisci to the tibial condyles. The stiffness of the meniscal springs varied from 3.2 to 5.4 N/mm [9].

Joint loads and moments were

applied to a reference node on the femur mesh approximately midway between the femoral condyles. Femur rotation in the sagittal plane was fixed at the desired flexion angle with all other degrees of freedom unconstrained. The tibia was fixed with zero degrees of freedom.

Motion Analysis and Joint Reaction Analysis

Kinematic and kinetic data was collected from normal weight (NW) and obese (OB) participants during treadmill walking at 1.25m•s^{1} [10]. Data from three NW and three OB individuals were averaged for this study. A weighted static optimization approach was used to predict muscle forces [10]. The resultant forces and moments were calculated using OpenSim's Joint Reaction Analysis and represent the knee joint contact force in the FE model.

FE Model Analysis and Output

Contact pressures at 25%, 50%, and 75% of stance were analyzed to compare NW and OB cartilage stress and regions of increased loading during the gait cycle.

RESULTS

The greatest contact pressure was observed at 25% and 75% stance for OB and NW knee conditions, respectively (Figure 2). Maximum pressure reached 9.9 MPa in the OB knee at 25% stance compared to 7.4 MPa in the NW knee at 75% stance. Both OB and NW knee conditions produced the lowest contact pressure at 50% stance. At 50% stance, minimum pressure was 3.5 MPa in the OB knee and 3.1 MPa in the NW knee.

Contact pressure was greatest in the medial cartilage at all simulated phases of stance (Figure 2). The most notable difference in peak contact pressure between OB and NW knee conditions occurred at 75% stance, where medial cartilage contact pressure was greater in the NW knee than the OB knee. The OB knee condition produced more equitable contact pressure at 75% stance, with a difference of less than 1 MPa between the lateral and medial cartilage.

Contour plots show that the medial cartilage experienced the greatest contact pressure in its central region during 25% and 50% stance (Figure 3). There was increased pressure distribution along the outer edge of the medial cartilage at 75% stance. The lateral cartilage had less varied pressure distribution. Lateral cartilage contact pressure was predominately located in the central region during all simulated phases of gait. Similar trends in contact pressure distribution were observed for OB and NW knee conditions.

DISCUSSION

There were more similarities in contact pressure distribution between OB and NW knee conditions than anticipated. Experimental gait data indicated that OB individuals had smaller flexion angles, implying a more extended knee during stance. It was hypothesized that this kinematic difference would produce altered stress patterns;

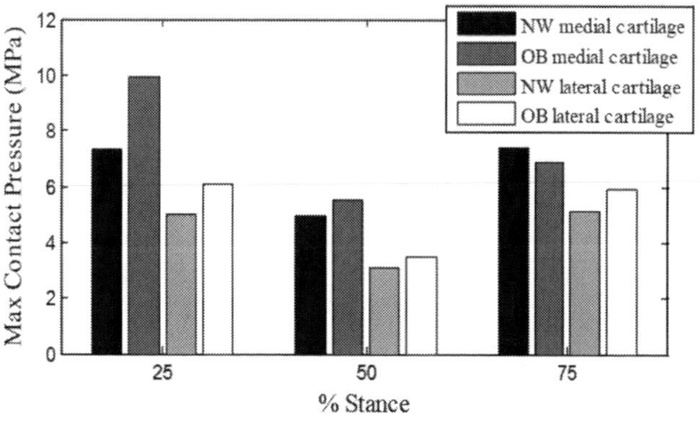

Figure 2: Maximum contact pressure on the medial and lateral tibial cartilage.

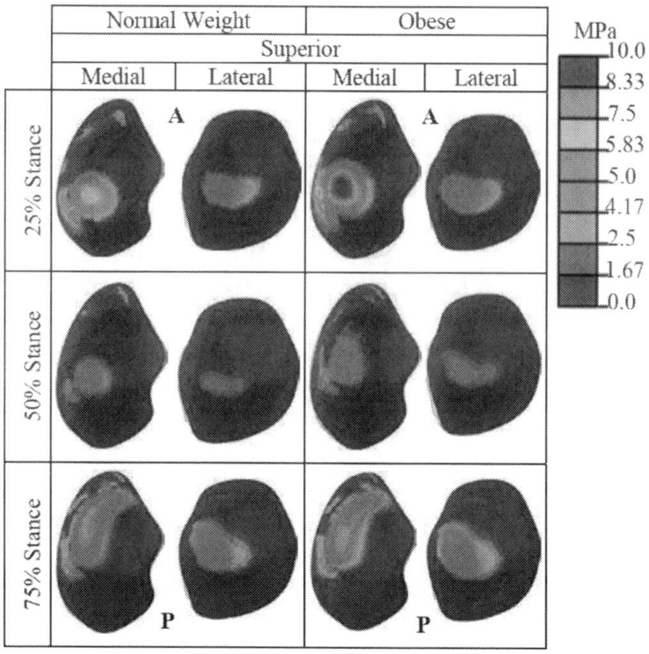

Figure 3: Contour plots of contact pressure on the lateral and medial tibial cartilage for obese and normal weight knees (A=anterior, P=posterior).

however, study results show that regions of high pressure were consistent for NW and OB knee conditions. It may be that smaller flexion angles of OB individuals result from a compensatory mechanism against larger contact forces in order to retain normal loading patterns.

Contact pressure was greater overall in the OB knee, likely due to the increase in absolute contact forces. High levels of contact stress may be associated with the development of OA; therefore, the susceptibility of obese individuals to greater than normal cartilage pressure is of concern [11]. In particular, the relative increase in contact pressure in the medial cartilage of the obese knee at 25% is indicative of a large, concentrated force acting on the cartilage surface.

The results of this study present an interesting, but limited, prediction of cartilage pressure during key points of gait; the primary limitation being that averaged data from multiple subjects was applied to a single subject specific model. Nevertheless, the results demonstrate how the use of FE modeling in combination with experimental gait data can increase the understanding of OA risk for obese patients.

ACKNOWLEDGEMENTS

Funding received from Quartus Engineering, Inc. and Cal Poly's Donald E. Bently and STRIDE Centers. The authors acknowledge Dr. Saikat Pal, for his invaluable contributions to this project.

REFERENCES

[1] Fransen, M et al., *Int J Rheum Dis*, 14:113-121, 2011. [2] Sharma, L et al., *JAMA*, 286:188-195, 2001. [3] Messier, SP et al, *Arthritis Rheum*, 52:2026-2032, 2005. [4] Weiss, J et al., *J Biomech*, 35:943-950, 2002. [5] Butler, D et al., *J Biomech*, 19:425-432, 1986. [6] Quapp, K et al., *J Biomed Eng*, 120:757-763, 1998. [7] Wangerin, S, Master's Thesis, Cal Poly, 2013 [8] Li, LP et al., *Proc Inst Mech Eng H*, 225:888-896, 2011. [9] Karen, H et al., *J Biomech*, 43:462-468, 2010. [10] Haight, D., *J Orthopaed Res*, 32:324-330, 2014. [11] Segal, N et al., *J Orthop Res*, 27:1562-1568, 2010.

SB³C2015
Summer Biomechanics, Bioengineering and Biotransport Conference
June 17-20, 2015,Snowbird Resort, Utah, USA
SB³C2015-612

THE ROLE OF THE GLYCOCALYX IN LEUKOCYTE ADHESION TO THE ENDOTHELIUM

Karli K. McDonald (1), Scott E. Cooper (2), Richard L. Leask (1,2)

(1) Department of Biomedical Engineering
McGill University
Montreal, Qc, Canada

(2) Department of Chemical Engineering
McGill University
Montreal, Qc, Canada

INTRODUCTION

Cardiovascular disease is the leading cause of death of both men and women in North America[1]. Nearly three-fourths of all deaths from cardiovascular disease are due to heart attack or stroke caused by atherosclerosis. Atherosclerosis is an inflammatory disease of the arterial wall, characterized by lipid-laden lesions. Leukocyte adhesion to the inflamed endothelium, lining arterial walls, is an important early event in disease initiation. Studies on the mechanisms underlying leukocyte adhesion have demonstrated that there is a direct relationship between wall shear stress (WSS) and the amount of adhesion and expression levels of adhesion molecules.

The apical surface of ECs, the glycocalyx, has been studied for its role in the EC mediated inflammatory response as a physical barrier to leukocyte attachment and as a mechanotransducer of WSS capable of creating an inflammatory EC phenotype. The role of the glycocalyx in leukocyte attachment is not well defined and studies thus far present conflicting conclusions[2,3].

In vitro studies are crucial in isolating the effects of hemodynamic forces on glycocalyx biology. Studies have shown EC response to steady WSS is altered when the glycocalyx is chemically degraded. In this study we seek to identify the role of the glycocalyx in leukocyte adhesion to an inflamed endothelium exposed to flow. Further, it will begin to uncover possible mechanisms for shear-mediated adhesion by examining the structure of the glycocalyx with flow and the expression of key adhesion molecules and pathways with glycocalyx degradation. Therefore, the general objective of our research is to identify the role of the glycocalyx in leukocyte adhesion. Further, we aim to determine the role of the glycocalyx as a physical barrier to attachment and its role in WSS-dependent adhesion responses by ECs.

METHODS

To examine the role of the glycocalyx in adhesion, we performed experiments with and without an intact glycocalyx using human abdominal aortic endothelial cells (ATCC, CRL-2472). Glycocalyx disruption was achieved with selective degradation of its main component, heparan sulfate (HS). The level of degradation was determined via the fluorescence intensity of images from the immunofluorescence staining of HS and confocal microscopy. Adhesion assays with and without glycocalyx degradation were conducted to quantify the number of attached leukocytes. These studies were performed under static and flow conditions. Three-dimensional cell culture models, developed by our laboratory[4] were used to expose the cells to steady WSS of 10 dynes/cm^2 (Figure 1). This was done with and without glycocalyx degradation. Following the preshearing, attachment studies with NB4 cells were conducted at 1 dyne/ cm^2 for 1 hour. The protein expression of VCAM-1, ICAM-1 and eNOS with degradation and flow are being evaluated using western blots and NF-κB translocation will be assessed using immunofluorescence staining. Two-way ANOVAs were conducted for statistical analysis followed by Tukey's multiple comparisons post-test.

RESULTS

Our results show a 33% reduction in glycocalyx thickness using enzymatic degradation, resulting in a change in cell morphology consistent with other published studies (n=3, $p<0.05$). Further, we have shown that under control conditions (no degradation), there is a significant decrease in leukocyte adhesion with pre-shearing (10 dynes/cm^2, 24 hours) compared to no pre-shearing as expected. With degradation however, both the no pre-shearing and pre-shearing conditions show firm adhesion comparable to control (n=3, $p<0.05$).

Therefore, glycocalyx disruption increases leukocyte adhesion under flow conditions. Our results also show that as expected, under control conditions, ICAM-1 is up regulated in response to steady WSS, while VCAM-1 is down regulated. With degradation however, we find that under static conditions, there is a slight decrease in ICAM-1 expression, while with flow, there is a significant increase compared to control ($n=3$, $p<0.05$). We did not find any significant change in VCAM-1 expression. Preliminary results indicate that the de-regulation of ICAM-1 expression under flow is not due to changes in NF-kB translocation, but rather through the inhibition of nitric oxide (NO) expression.

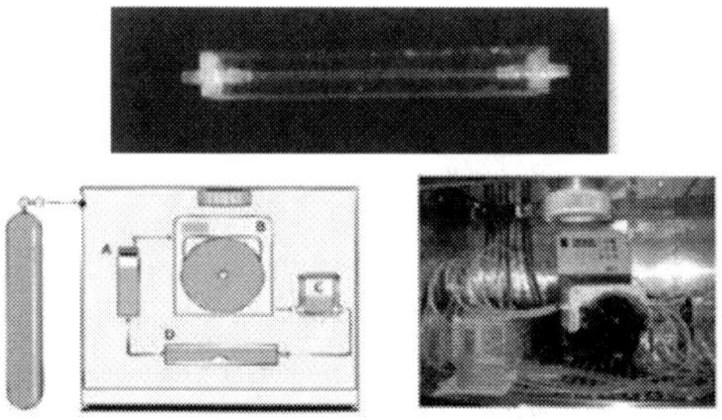

Figure 1. Three-dimensional cell culture model (Top), flow loop schematic (Bottom, Left) and image (Bottom, Right) used for perfusion experiments. In schematic drawing: A) media reservoir, B) Peristaltic pump, C) Flow dampener and D) *In vitro* model.

DISCUSSION

Our results indicate that there is an increase in leukocyte adhesion to the endothelium with degradation under flow conditions. This is in agreement with the study performed by Constantinescu et al who has shown that degradation of the glycocalyx *in vivo*, stimulates the immobilization of leukocytes to the endothelium[5]. Next, we aimed to determine whether increases in adhesion were due to the increased expression of two key firm adhesion molecules, ICAM-1 and VCAM-1, following degradation. The glycocalyx has been shown to mediate other EC intracellular pathways such as NO production and cytoskeletal reorganization in response to flow[6-8]. Whether the glycocalyx plays a role in the regulation of adhesion molecules has not been previously determined. We show that with degradation, there is a significant increase in ICAM-1 protein expression compared to control under flow. These results correlate with our adhesion assay results in that there is a significant increase in ICAM-1 expression and leukocyte adhesion under the same conditions. These results provide new insight into the role of the glycocalyx in adhesion, suggesting that it may play more than a structural role in inhibiting leukocyte adhesion. ICAM-1 expression is regulated upstream through the binding of the transcription factor NF-κB to the ICAM-1 promoter. Therefore, we sought to determine whether the changes in ICAM-1 expression were occurring through a NF-kB dependent pathway. Our preliminary results indicate that under flow and degradation conditions, NF-κB translocation to the nucleus is similar to control conditions. Therefore, glycocalyx degradation does not affect ICAM-1 expression by altering NF-κB activity. Previous studies have shown that glycocalyx degradation leads to an inhibition of NO production[7]. NO works in a negative feedback loop to inhibit NF-κB activity, thus decreasing ICAM-1 expression under flow. Therefore, we hypothesize that glycocalyx degradation impedes NO production and therefore disrupts the negative feedback loop, resulting in the de-regulation of ICAM-1 expression under flow. Experiments are currently being conducted to address this hypothesis.

Taken together, this work aims to clarify the role of the glycocalyx in adhesion and identify underlying mechanisms for such a role. We show that the glycocalyx is an important mediator of leukocyte adhesion under physiologically relevant hemodynamic flow conditions and that this EC surface structure may play a key role in modulating WSS adhesion molecule pathways. Understanding the mechanisms by which leukocytes adhere to ECs has important implications for many vascular pathologies, namely atherosclerosis. Further, identifying the role of the glycocalyx in adhesion could provide new potential areas for therapeutic intervention and insight into regenerative medicine.

ACKNOWLEDGEMENTS

We would like to acknowledge the CIHR (RN121751 – 245590) for their contribution in supporting this research. Karli McDonald is a holder of a CIHR Doctoral Scholarship.

REFERENCES

[1] Lopez, A. D et al., *Lancet* 367, 1747-1757, 2006.
[2] Mulivor, A. W et al., *Am J Physiol Heart Circ Physiol* 283, H1282-1291, 2002.
[3] Giuffre, L. et al., *J Cell Biol* 136, 945-956, 1997.
[4] Rouleau, L et al., *J Biomech Eng* 132, 071015, 2010.
[5] Constantinescu, A. A. et al., *Arterioscler Thromb Vasc Biol* 23, 1541-1547, 2003.
[6] Florian, J. A. et al. *Circ Res* 93, e136-142, 2003.
[7] Thi, M. M et al., *Proc Natl Acad Sci U S A* 101, 16483-16488, 2004.
[8] Pahakis, M. Y et al., *Biochemical and biophysical research communications* 355, 228-233, 2007.

APPLYING MEAN SOFT TISSUE PROPERTIES TO A SUBPOPULATION OF KNEE MODELS REVEALS INABILITY TO CAPTURE VARIATIONS IN KNEE STABILITY

Kevin Schafer, Mohammad Kia, Daniel Green, Andrew Pearle, Thomas Wickiewicz, Timothy Wright, Carl Imhauser

Hospital for Special Surgery
Department of Biomechanics
New York, NY, USA

INTRODUCTION

Knee stability is highly variable, being influenced by the interaction of patient-specific variables including ligamentous laxity and joint morphology. Treatments for anterior cruciate ligament (ACL) rupture such as ACL reconstruction fail to account for this heterogeneity. Thus, surgeons have little guidance on how treatment should be tailored to meet the stability profile of an individual patient. This may contribute to suboptimal outcome following treatment of ACL injury including reduced function and 5% revision rates[1, 2].

Synergy of computational models and subject-specific physical experiments provides an opportunity to identify factors driving variations in knee function including those that play an important role in treatment of ACL injury. However, current computational models typically employ one specimen or limited sample sizes of ≤ 3 and assign population mean soft tissue properties[3-5]. It is not known whether this modeling approach can capture variations in knee stability across a larger subpopulation of knees.

Factors driving passive knee function could be divided into three categories: 1) morphological features consisting of the size and shape of the articular surfaces and the bones; 2) structural properties of the ligaments and the articular surfaces; and 3) the slack in the ligaments. We have developed an approach to explore how these factors drive the passive load-displacement response of the knee using subject-specific modeling of a larger group of knee specimens integrated with direct comparison of physical experiments from those same models.

Our initial research question asked whether a computational model that includes subject-specific representations of the articular surfaces and mean structural properties of both the supporting ligaments and articular surfaces adequately captures the variations in load-displacement response observed across a subpopulation of knees.

METHODS

Models from ten knees (8 male, mean age: 53.3±12.5 years, range: 29-65 years) were developed from CT and MRI images of the bones, articular cartilage, and menisci using standard protocols with ≤ 1 mm slice thicknesses. Bones were instrumented with radiopaque fiducial markers that enabled registration of the image data to subsequent experiments on the same specimen, using a 3D digitizer (Microscribe G2X, accuracy: 0.18 mm).

Ligaments and horn attachments of the mensici were identified from the image data and represented with multiple tension-only force elements (Fig 1). Wrapping of the superficial MCL (sMCL) around the medial tibia was included. Ligaments were described using mean load-displacement properties, including a non-linear toe region and a subsequent linear region[6-10]. Ligament slack lengths were identified for each model using the optimization requirement that, at full extension, the net force on each ligament was 10 N. The force developed between contacting surfaces of the tibia, femur, and menisci was defined using rigid body contact[3, 11]. Motions of the menisci relative to the tibia were modeled using a planar constraint oriented parallel to each compartment. The equations governing model behavior including contact were generated and solved using ADAMS[11].

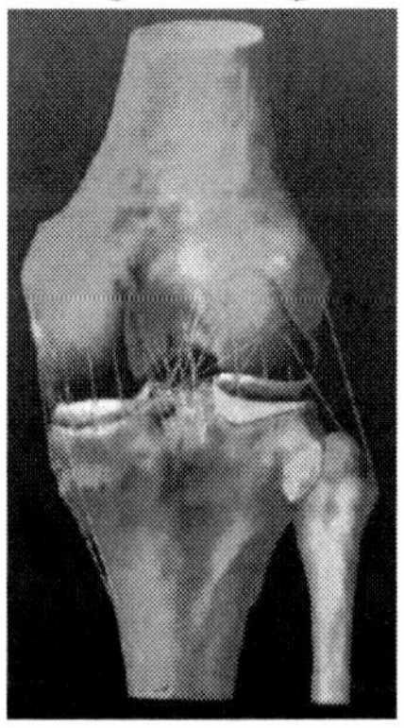

Fig 1: Multi-body model of the knee including major ligaments and menisci. Models of 10 knees were developed.

The pivot shift exam was modeled since it identifies instability, predicts clinical outcome, and can diagnose ACL rupture[12].

Boundary conditions included fixing the femur rigidly and setting the tibia relative to the femur at 15° flexion, which left it with five degrees of freedom (5-dof). The pivot shift was modeled by first increasing the applied valgus moment to 8 Nm and then increasing the internal rotation moment to 4 Nm with the valgus moment held constant. The 3D kinematics that satisfied these loading conditions was recorded.

Each of the ten knee models was evaluated using subject-specific comparisons with experimental data from the same specimen that was modeled. The physical experiment was conducted using a previously described six degrees of freedom robot[13, 14]. The same test of the pivot shift described above was performed on each of the ten knee specimens with the ACL intact. In the physical experiment the valgus moment was increased to 8 Nm in seven increments (0, 0.8, 2, 4, 6, 7, 8 Nm) and then the internal rotation moment was increased to 4 Nm in six increments (0.4, 0.8, 1.2, 2, 3, 4 Nm). The 5-dof kinematics that satisfied these loading conditions at each increment of applied torque was recorded. All loads, boundary conditions and kinematic coordinate systems were maintained between model and corresponding experiment using the aforementioned fiducials.

Differences between model predictions and experimental measurements of tibiofemoral position and compliance of the knee joint were calculated during the simulated pivot shift at each increment of torque that was applied in the experiment. The compliance was calculated for each successive increment of applied torque as the ratio of the change in tibiofemoral position per change in applied torque. The mean and standard deviation of the differences in position and compliance across all increments of torque applied in the experiment were used as a measure of the level of agreement between model and experiment. A mean difference of zero indicates perfect agreement. We focused on the comparison of internal/external rotation (IR) because it varied widely across knees.

RESULTS

In the physical experiment, applying moments representing the pivot shift to the ten cadaver knees yielded net internal rotation of 14.4 ± 5.7° (range: 8.6 to 26.8°). The predicted IR position differed from the experiment by as little as -0.5 ± 2.4° (Knee 1) and as much as -10.4 ± 6.7° (Knee 9) (Fig 2). Mean difference between predicted IR position and the corresponding experiment was within 1.1° in three models (Knees 1, 4, and 10) and ≥ 4.6° in magnitude in the remaining models (Fig 2).

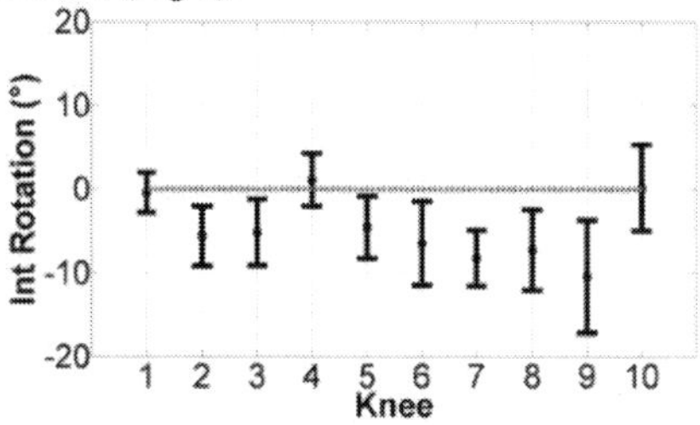

Fig 2: Difference in internal rotation position between model and experiment for ten knees (Mean ± SD). – indicates offset in external rotation

The predicted IR compliance differed from the experiment by as little as -0.3 ± 4.5 °/Nm (Knee 10) and as much as 5.4 ± 4.4 °/Nm (Knee 9) (Fig 3). Mean difference between predicted IR compliance and the corresponding experiment was within 1.6 °/Nm in five models (Knees 1, 4, 7, 8 and 10) and ≥ 2.6 °/Nm in magnitude in four models (Knees 3, 5, 6 and 9) (Fig 3).

DISCUSSION

Physical experiments on ten cadaver knees were used to evaluate 5-dof multibody models of each knee that included subject-specific articular and bony anatomy. Comparing each computational model to its corresponding physical experiment revealed that knee-specific articular morphology alone fails to capture variations in the load-displacement response measured across a subpopulation of knees. Thus, additional information is required for subject-specific modeling, which may include structural properties of the ligaments and articular cartilage, and ligament slack length.

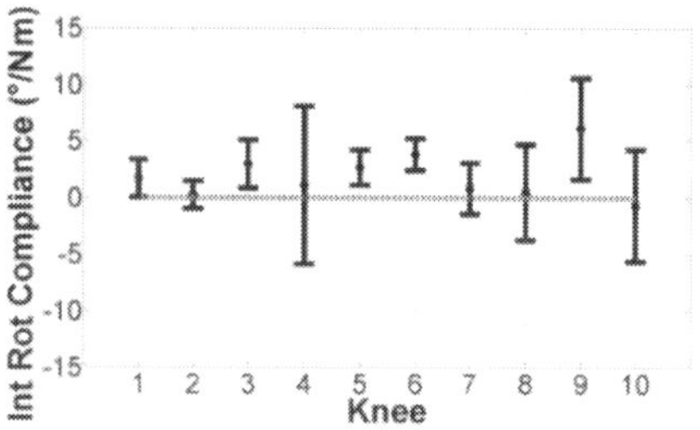

Fig 3: Difference in internal rotation compliance between model and experiment for ten knees (Mean ± SD). + indicates model is more compliant than experiment

This finding is intuitive since the articular surfaces of the tibiofemoral joint are nonconforming; however, it is important because many computational models incorporate population mean descriptions of the structural properties of the ligaments and cartilage[3-5, 15]. Such approaches enable a general understanding of knee function. However, they have limited utility in developing subject-specific treatments, since they do not capture the wide range of responses in behavior across a subpopulation. For example, internal rotation ranged from 8.6 to 26.8° in our physical experiment. Optimization of soft tissue properties based on a subset of experimental loading conditions may improve agreement of model predictions and experimental data.

Interestingly, we observed that knee models yielding the closest agreement with the corresponding physical experiment did not fall into one category in terms of their range of motion. For example, model predictions of knees 1 and 10 exhibited the best overall agreement with their experimental data, but the experimentally measured internal rotation in these knees was nearly a standard deviation below (10.4°) and above (19.4°) the mean (14.4°), respectively. This suggests that additional factors beyond the structural properties of the ligaments drive passive knee stability including ligament slack length.

The meniscal model could be enhanced beyond its current rigid body formulation. The response of the model to functional loads such as those during gait should also be assessed. This model can be used to understand the relationship between properties of the passive stabilizers and the resulting knee kinematics, which account for many of the constituents that are surgically manipulated. This integrated framework will be employed to identify factors necessary for subject-specific predictions of knee function to enhance predictability of treatment for ACL injury.

ACKNOWLEDGEMENTS

Kirby Foundation, Gosnell Family

REFERENCES

[1] Ardern CL. *Am J Sports Med*, 39: 538-43, 2011. [2] Paxton EW. *J Bone Joint Surg Am*, 92 Suppl 2: 117-32, 2010. [3] Guess TM. *Open Biomed Eng J*, 6: 23-32, 2012. [4] Kiapour A. *J Biomech Eng*, 136: 011002, 2014. [5] Shin CS. *J Biomech*, 42: 280-5, 2009. [6] Butler DL. *J Biomech*, 25: 511-8, 1992. [7] Harner CD. *Am J Sports Med*, 23: 736-45, 1995. [8] Hauch KN. *J Biomech*, 43: 463-8, 2010. [9] LaPrade RF. *Am J Sports Med*, 33: 1386-91, 2005. [10] Robinson JR. *J Biomech*, 38: 1067-74, 2005. [11] Imhauser CW. *J Biomech*, 41: 1341-9, 2008. [12] Leitze Z. *Clin Orthop Relat Res*, 229-36, 2005. [13] Imhauser C. *Am J Sports Med*, 41: 815-25, 2013. [14] Kanamori A. *Arthroscopy*, 16: 633-9, 2000. [15] Kim YH. *Proc Inst Mech Eng H*, 227: 968-75, 2013.

AN ACTIVE CONTRACTION MODEL OF VALVULAR INTERSTITIAL CELLS

Yusuke Sakamoto (1), Michael S. Sacks (1)

(1) Center for Cardiovascular Simulation
Institute for Computational Engineering and Sciences
Department of Biomedical Engineering
The University of Texas at Austin
Austin, TX, USA

INTRODUCTION

Valvular interstitial cells (VICs) play a critical role in the maintenance and pathophysiology of heart valve tissues. When activated, VICs exhibit increased levels of cytokines and extracellular matrix (ECM) synthesis and strong contraction through the expression of α-smooth muscle actin (α-SMA) fibers. However, it remains unclear how active contraction of the α-SMA fibers contribute to the overall VIC mechanical responses as well as other mechanical constituents such as nucleus, cytoskeleton, and cytosolic fluid. The objective of this study is to investigate the roles of different subcellular structures of the VICs, such as cytoskeleton, nucleus, α-SMA stress fibers (with different expression levels and contraction strengths) to the VIC mechanical responses under different mechanical loading conditions and activation states. To this end, we have developed a novel continuum-level model of VIC mechanics involving: 1) basal, non-oriented cytoskeletal network, 2) passive elastic model of the α-SMA stress fibers, and 3) active contraction model of the α-SMA stress fibers. The developed model integrates the data from micropipette aspiration (MA) and atomic force microscopy (AFM) experiments, in which the VICs are under significantly different mechanical loading conditions and activation states. Thus it enabled us to understand how the VICs function under different activation states.

METHODS

We modeled the VIC as a continuum with two main structures: the cell nucleus and cytoplasm, which occupy distinct domains within the VIC. The nucleus was modeled as an incompressible neo-Hookean material with shear modulus μ^{nuc} while the cytoplasm was modeled as a mixture of two solid phases: the basal cytoskeleton phase (represented by *cyto* superscript) and α-SMA stress fiber phase (represented by *sf* superscript), which exhibit some orientations and active contraction.

We assumed that with the absence of body and inertial forces, the conservation of linear momentum yields:

$$\nabla \cdot \mathbf{T} = \mathbf{0} \quad \text{in } \Omega_t \times (0,T) \tag{1}$$

where $\mathbf{T}$ represents the Cauchy stress tensor and Ω_t represents the domain the VIC occupies at time t. We further assumed the cytoskeleton and stress fiber phases moved together, and their stresses were additive:

$$\begin{aligned} \mathbf{u} &= \mathbf{u}^{cyto} = \mathbf{u}^{sf} \\ \mathbf{T} &= \mathbf{T}^{cyto} + \mathbf{T}^{sf} \end{aligned} \tag{2}$$

where $\mathbf{u}$ represents the displacement. The cytoskeleton phase was modeled as an incompressible neo-Hookean material, with shear modulus μ^{cyto}.

The orientation of the stress fibers at each point was described by the density function $\Gamma(\mathbf{M})$, which represented the normalized density of the fiber initially oriented along the direction $\mathbf{M}$. Also, Γ satisfies the symmetry and normalization conditions. The Cauchy stress of the fiber phase is given by integrating the contributions from every direction:

$$\mathbf{T}^{sf} = \int_0^{2\pi}\int_0^{\pi} \left(H(I_4 - 1)\mathbf{t}^p(\mathbf{M}) + \mathbf{t}^a(\mathbf{M}) \right) \sin\theta \, d\theta d\phi \tag{3}$$

where $\mathbf{t}^p$ and $\mathbf{t}^a$ represent the Cauchy stresses due to passive (elastic) fiber stretch and active fiber contraction, respectively, of the fibers initially oriented along $\mathbf{M}$. A Heaviside step function H was introduced to enforce that the passive stress only arises from fiber stretch.

We assumed that each fiber is a transversely isotropic material with the effective shear modulus μ^{sf}. Thus, we can derive the fiber passive stress from $\mathbf{t}^p$ using the fiber stretch and fiber orientations at the point. We further assumed that the effective shear modulus μ^{sf} was determined by three parameters: the intrinsic shear modulus of the stress fibers $\mu^{sf,int}$, the spatial density of the fiber ρ^{sf}, and the fiber orientation density Γ, ie.

$$\mu^{sf}(\mathbf{M}) = \mu^{sf,int}\rho^{sf}\Gamma(\mathbf{M}) \quad (4)$$

Assuming that the stress fibers contract uniformly along their directions with contraction strength f, we derived the active stress component of the stress fiber:

$$\mathbf{t}^{a} = f \cdot \Gamma_{t}(\mathbf{m})\frac{I_4}{J}\mathbf{m}\otimes\mathbf{m} \quad (5)$$

where $\mathbf{m}$ is the current direction of the fiber.

Using our VIC mechanical model, we simulated two experiments: micropipette aspiration (MA) and atomic force microscopy (AFM) of the aortic VIC (AVIC) and pulmonary VIC (PVIC) that each exhibits different expression levels of the α-SMA. In the MA experiment, VICs are in inactivated states while in AFM experiment, VICs are in activated states. Thus, using our model in conjunction with the experimental data, we investigated how the expression level and active contraction of the α-SMA fibers affect the effective mechanical responses of the VICs. We implemented our model and ran the simulations on FEBio software package [1].

In MA analysis, using the aspiration length of the VICs for different applied pressure, we determined the cytoskeleton shear modulus μ^{cyto} and the fiber effective shear modulus μ^{sf} (Figure 1). In the AFM analysis, using the indentation depth vs force curve at different indentation location, we determined the contractile strength f of the α-SMA fibers and the nuclear stiffness μ^{nuc} within the VIC (Figure 2).

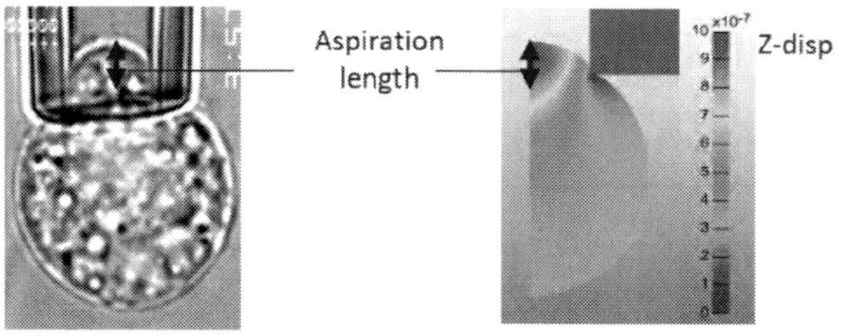

FIGURE 1: MICROPIPETTE ASPIRATION EXPERIMENT AND SIMULATION GEOMETRIES.

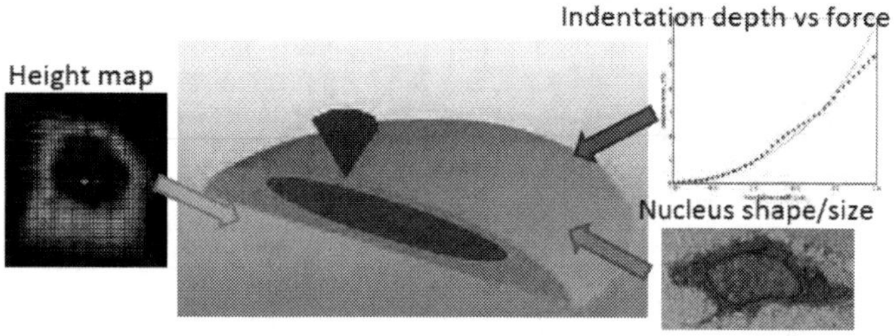

FIGURE 2: ATOMIC FORCE MICROSCOPY SIMULATION SETTINGS. THE SIMPLIFIED CELL GEOMETRY WAS USED, WITH CELL AND NUCLEUS DIMENSIONS TAKEN FROM EXPERIMENTAL DATA.

RESULTS

Using the MA experimental data of the VICs exhibiting different expression levels of α-SMA, we determined the cytoskeleton shear modulus (μ^{cyto}) and the stress fiber intrinsic modulus ($\mu^{sf,int}$) to be 7.31 Pa and 533.55 Pa, respectively. Using the intrinsic modulus and fiber expression level, we calculated the effective shear modulus μ^{sf} (in Eq. (4)) of the stress fibers within AVICs and PVICs as 533.18 Pa and 428.76 Pa, respectively. Here, we assumed that the cytoskeleton shear modulus are constant, and the VICs are inactivated states. Thus, only the amount of α-SMA within the VICs contribute to the changes in their mechanical properties.

Using the AFM experimental data, we determined that the contraction strength of AVICs and PVICs were 41.98 kPa and 3.94 kPa, respectively (Figure 3). Thus, there exists ~10-fold difference in the contractions strength between the AVICs and PVICs, implying that not only the expression level of the α-SMA fibers but also the contraction level increases from PVICs compared to AVICs. It is possible that the higher expression of the α-SMA fibers facilitates more efficient contraction. We also determined the nucleus shear modulus to be 10.69 kPa and 8.43 Pa, respectively, with no statistically significant differences (p=0.54). Thus, the nucleus' shear modulus was preserved.

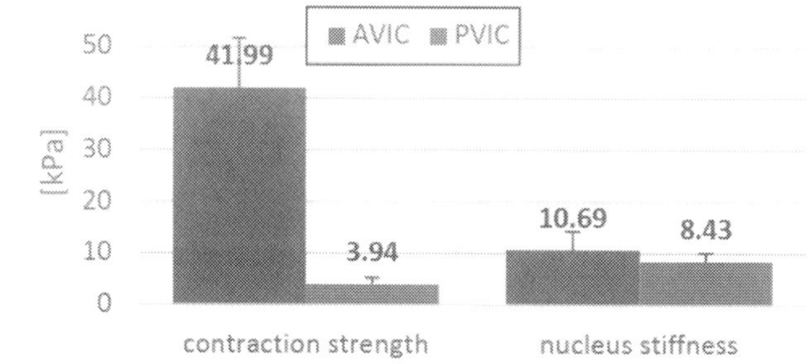

FIGURE 3: AFM SIMULATION RESULTS.

DISCUSSION

Mechanical properties of VICs have been previously measured by different techniques, such as MA [2], AFM [3], and gel contraction [3]. These studies provide useful insights that α-SMA expression upregulates the increase in stiffness of the VICs. However they did not provide the holistic view of the VIC mechanics and functions. It is particularly important because VICs undergo very dynamic and rapid mechanical changes within the native tissue. In this study, we developed the mechanical model of the VICs that is capable of capturing the mechanical responses of VICs that exhibit different expression levels of α-SMA and different contraction strengths. It has been known that VICs exhibit significantly different stiffness values for the MA and AFM measurements. Our model explains this gap between these two measurements by attributing their differences to the active contraction strength of the α-SMA fibers.

We also used our model to show that the AVICs exhibit stronger contractile activity than PVICs in activated states. In fact, it was previously observed that AVICs, seeded within a collagen gel, contracts the gel faster than the PVICs [3]. This study confirmed the previous results, and is the first to quantify the α-SMA fiber contraction strength within the VICs. It is also notable that using our model in conjunction with the AFM measurements, we can analyze the mechanical properties of the nucleus. To our knowledge, this is the first study to measure the shear modulus of the nucleus within the cell using AFM. The calibrated shear moduli of the nucleus in this study are similar to the ones measured in previous studies for different cells [4].

ACKNOWLEDGEMENTS

This study was supported by NIH Grant R01 HL-68816-01 and R01 HL119297.

REFERENCES

1. Maas, S.A., et al., J Biomech Eng-T Asme, 2012. **134**(1).
2. Merryman, W.D., et al., Am J Physiol Heart Circ Physiol, 2006. **290**(1): p. H224-31.
3. Merryman, W.D., et al., Tissue Eng, 2007. **13**(9): p. 2281-9.
4. Guilak, F., J.R. Tedrow, and R. Burgkart, Biochem Biophys Res Commun, 2000. **269**(3): p. 781-6.

SB³C2015
Summer Biomechanics, Bioengineering and Biotransport Conference
June 17-20, 2015, Snowbird Resort, Utah, USA
SB³C2015-615

EFFECT OF SCAR COMPACTION ON THERAPEUTIC EFFICACY OF ANISOTROPIC REINFORCEMENT FOLLOWING MYOCARDIAL INFARCTION

Samantha A. Clarke (1), Gorav Ailawadi (1,3), Jeffrey W. Holmes (1,2,4)

(1) Department of Biomedical Engineering
University of Virginia
Charlottesville, Virginia, USA

(2) Robert M. Berne Cardiovascular
Research Center
University of Virginia
Charlottesville, Virginia, USA

(3) Department of Surgery
University of Virginia
Charlottesville, Virginia, USA

(4) Department of Medicine
University of Virginia
Charlottesville, Virginia, USA

INTRODUCTION

Every year, almost 1 million Americans have a heart attack, or myocardial infarction (MI).[1] Due to improvements in hospital admission timing and revascularization therapies, mortality rates for acute MI have declined over the past several decades.[2] As a result, there are more MI survivors living now than ever before, making prevention of post-MI heart failure increasingly critical. Accordingly, several novel therapies – including placement immediately after infarction of surgical restraint devices originally developed for end-stage heart failure and injection of polymers directly into the infarct – have been developed to improve left ventricular function and/or prevent long-term dilation following infarction. In a combination of computational modeling and experimental studies, we showed that while increasing the stiffness of acute infarcts in all directions (isotropic reinforcement) does not improve left ventricular (LV) function, selectively stiffening large anterior infarcts in the longitudinal direction does dramatically improve LV pump function immediately after an infarct.[3,4] The objective of this study was to test whether anisotropic reinforcement provides sufficient sustained functional benefit to mitigate the post-infarction dilation of the LV that represents the first step towards post-MI heart failure.

METHODS

MI was induced in 28 canines by permanent ligation of the left anterior descending coronary artery. To generate longitudinal restraint in the reinforcement group, anisotropic Dacron patches were sutured to the epicardium over the ischemic area under longitudinal tension 45min after ligation. Previous acute sonomicrometry studies demonstrated that this procedure induces a 10% decrease in longitudinal segment lengths throughout the cardiac cycle (Figure 1, upper), without altering circumferential segment lengths. MR scans at 48hrs and 8wks post-MI were used to measure changes in infarct size (using late gadolinium enhancement) and LV geometry. Infarct and LV cavity volumes were calculated from manual contours of short axis (SA) image stacks according to Simpson's rule, with fractional slice volume adjustments at the base and apex based on long axis (LA) views. Infarct dimensions were also quantified from SA images; changes in the circumferential and radial directions were averaged across slices, and longitudinal extent was determined by the number of slices containing scar.

RESULTS

At 48hr, infarct size was 31.4±6.8% of LV mass in the control group and 33.8±5.6% in the patch group (P=NS). 48hr end-diastolic volumes (EDV) were 72.8±17.6cm^3 in the patch group vs. 92.5±20.5cm^3 in untreated controls ($P<0.05$), and end-systolic volumes (ESV) were 55.3±14.4cm^3 in patch animals, 73.3±15.2cm^3 in control ($P<0.01$). However, by 8 weeks, differences in volumes between the two groups were no longer significant (Figure 1, lower), suggesting that the previously reported acute benefits of anisotropic reinforcement are either not sufficient to mitigate LV remodeling or are not sustained over the 8-week period of the study.

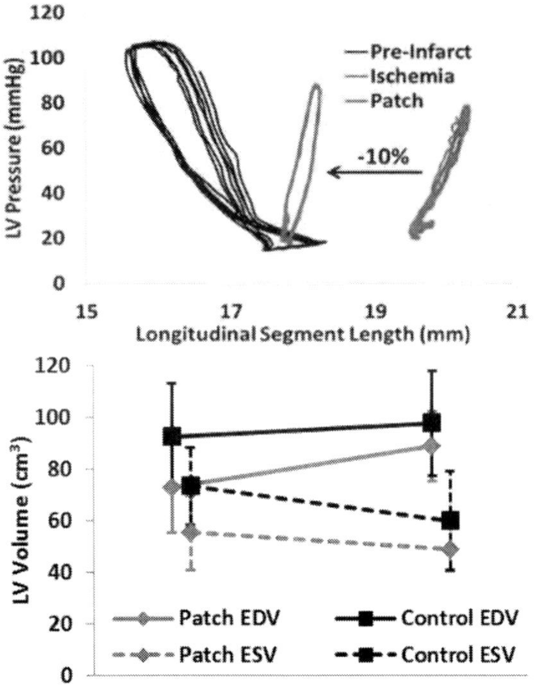

Figure 1. (Upper) Changes in acute longitudinal segment length following anisotropic patch application. (Lower) Changes in LV volume over 8 weeks in patch and control groups.

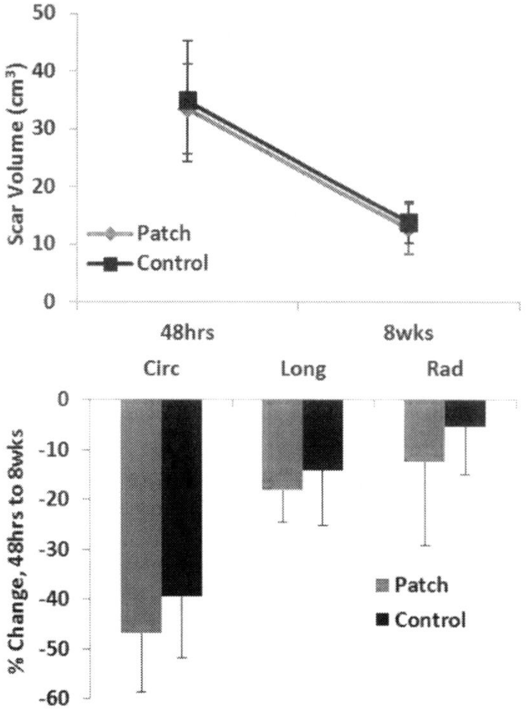

Figure 2. (Upper) Changes in scar volume and (Lower) scar dimensions over 8 weeks of infarct healing.

Analysis of changes in scar size over time revealed a potential mechanism underlying the lack of sustained benefit: surprisingly dramatic scar compaction sufficient to completely reverse the effects of anisotropic reinforcement. Infarct scar volume decreased by more than 50% in both patch and control animals (Figure 2, upper). In both groups, the majority of compaction occurred in the epicardial plane (Figure 2, lower); although dimension changes were larger in the circumferential direction, longitudinal compaction was greater than 10% on average, sufficient to reverse the dimension changes originally induced by reinforcement. Perhaps surprisingly, surgical reinforcement did not affect the degree or direction of scar compaction.

DISCUSSION

We conducted a chronic large-animal trial of anisotropic reinforcement, a strategy previously shown to dramatically improve acute LV function following an infarct, and found no improvement in LV remodeling 8 weeks following infarction compared to untreated controls. Our data suggest that compaction of the infarct scar over the 8-week study may have reversed the effect of the treatment.

Our results were surprising because the infarct healing literature typically focuses on infarct expansion – circumferential and longitudinal dilation plus radial thinning – rather than scar compaction. However, a review of studies that actually reported infarct dimensions suggests that expansion and compaction occur with similar frequency.[5] Accordingly, different therapeutic approaches may have very different efficacy in different animal models or even in different patients. We would expect surgical restraint – as in our study – to be most effective for infarcts that would normally expand if untreated. This might explain why studies that applied isotropic patches locally to prevent infarct expansion have produced much more variable results than those that employed global restraint with cardiac support devices.[6] It is less clear how polymer injection might affect scar remodeling, but designing such polymers to encourage rather than impede any naturally occurring compaction may prove important.

More broadly, our results highlight the importance of developing computational models that can accurately predict not only the acute effect of a therapy on LV mechanics, but also its long-term effects on scar and LV remodeling.

ACKNOWLEDGEMENTS

Funding for this work was provided by the National Institutes of Health (R21 HL111546) and National Science Foundation Graduate Research Fellowship (DGE-00809128).

REFERENCES

[1] Bogen, D et al., *Circ Res,* 47:728–741, 1980.
[2] Costanzo, M et al., *J. Card Fail,* 18:446–458, 2012.
[3] Fomovsky, G et al., *J Cardiovasc Transl Res,* 4:82–91, 2011.
[4] Fomovsky, G et al., *Circ Heart Fail,* 5:515–522, 2012.
[5] Richardson, W & Holmes, J. *ASME Proc [Abstract, ISBN 978-0-7918-5560-7],* 2013.
[6] Clarke, S et al., *Stud Mechanobiol Tissue Eng Biomater,* 15:169-206, 2014

SB³C2015
Summer Biomechanics, Bioengineering and Biotransport Conference
June 17-20, 2015, Snowbird Resort, Utah, USA
SB³C2015-616

VALIDATION OF A SUBJECT-SPECIFIC COMPUTER MODEL OF GLENOHUMERAL INSTABILITY AND CAPSULAR PLICATION

Charlie Yongpravat, David Kovacevic, T. Sean Lynch, Charles M. Jobin, William N. Levine, Gerard A. Ateshian[+], Thomas R. Gardner, Christopher S. Ahmad

Centers for Orthopaedic Research and Shoulder, Elbow and Sports Medicine
Department of Orthopaedic Surgery

[+]Department of Mechanical Engineering

Columbia University
New York, NY 10032

INTRODUCTION

Shoulder dislocations occur when the humeral head translates over the edge of the glenoid socket, permanently stretching the capsular ligaments. This injury of the capsular tissue results in pathological joint laxity which is a major contributor to recurrent dislocations and shoulder instability. The ideal surgical parameters to correct this pathology have yet to be established due to a lack of understanding of how shoulder kinematics and capsular mechanics are affected by different surgical capsular tightening procedures, also known as capsular plication. To address this knowledge gap, we developed subject-specific computer models of the shoulder and the glenohumeral capsule. The purpose of this study was to establish a validation methodology for our computer model of shoulder instability. To test this methodology a capsular plication repair was performed both experimentally and computationally and the measured and predicted shoulder stabilities were compared.

METHODS

Experimental Testing: One intact, fresh frozen, human cadaveric shoulder (50 y.o., female) with no signs of injury, was thawed to room temperature and dissected of all soft tissue except for the glenohumeral capsule. The scapula was rigidly mounted to a custom joint testing apparatus and the humerus was mounted to a 2 degree of freedom slide that translates in two perpendicular directions. The first axis was parallel to the glenoid face in a 30° direction inferior to the anterior axis; this corresponds to the 4 o'clock direction that is used in the clinical load and shift test [1] to evaluate the stability and laxity of the glenohumeral joint. The second axis is normal to the glenoid face, in the medial-lateral anatomic direction, and allows for the humeral head to translate over the glenoid rim during humeral head translation in the 4 o'clock direction.

To measure shoulder stability, the force-displacement curve of the cadaveric specimen was recorded. The reference orientation of the glenohumeral joint was set at 60° GH abduction via the testing apparatus (Figure 1). The humerus was translated 2 cm at a quasi-static rate of 1 mm/s in the 4 o'clock direction to the glenoid rim. A linear actuator (NLS4, Newmark, Rancho Santa Margarita, CA) with a load cell in series (Honeywell, Columbus, OH) was attached to the 2 degree of freedom slide. Custom-designed software (LabVIEW, National Instruments, Austin, TX) was used to control the linear actuator and record the output from the load cell. This shoulder stability test was performed twice, first on the native cadaveric shoulder, then again after performing a glenohumeral capsular plication.

An arthroscopic anterior capsular plication was performed by a fellowship-trained orthopaedic surgeon. The arthroscope camera (1088 HD, Stryker, Kalamazoo, MI) was inserted through a posterior viewing portal made 2 cm inferior from the posterolateral corner of the acromion. A "clock face" was defined on the glenoid surface with the superior aspect defined as the 12 o'clock position and the anterior direction corresponding with the clockwise direction. Suture anchors (Bioraptor, Smith & Nephew, London, UK) were placed into the glenoid rim at the 4:00 and 5:30 clock positions. Capsular plication was performed by using the suture that was preloaded into the anchors, passing it through the capsule 1.5 cm lateral to the glenoid rim, tying the sutures effecting a capsular plication, and securing the shortened capsule to the glenoid bony rim [2].

Subject-Specific Finite Element Modeling (Figure 2): Prior to any experimental testing, the cadaveric shoulder received an MR arthrogram and three-dimensional surface meshes of the specimen's scapula, humerus, and capsule were extracted using commercial software (Mimics, Materialize, Ann Harbor, MI).

The scapula and humerus were meshed as 5 mm rigid triangular shell elements and the capsule as 2 mm triangular shell elements (Hypermesh, Altair, Troy, MI, USA). Shoulder stability was determined by calculating the joint reaction force resulting from anteriorly translating the center of the humeral head in the 4 o'clock direction to the glenoid edge. This simulation was performed under quasi-static conditions, a nominal 10 N joint compressive load, and with no rotational degrees of freedom and one translational degree of freedom in the medial-lateral axis (Abaqus/Explicit, SIMULIA, Providence, RI).

The joint capsule was modeled as an anisotropic hyperelastic material [3]. Material parameters were obtained by an inverse finite element optimization routine. Using the load-displacement curves obtained from the native cadaveric experimental testing as the observed data and the finite element model as the model data to optimize, a sum of least squares difference cost function was minimized using the Levenberg-Marquardt algorithm (R, R Foundation for Statistical Computing, Vienna, Austria). The optimized coefficients obtained from the native capsule were then implemented into the finite element model of the plicated glenohumeral capsule to calculate the load-displacement curve and compare to the experimental cadaveric data for physical validation.

To simulate capsular plication, a "clock face" was superimposed on the glenoid surface with the superior aspect defined as the 12 o'clock position and the anterior direction corresponding with the clockwise direction. The surface mesh of the glenohumeral capsule was imported into remodeling software (Rhinoceros, Robert McNeel & Associates, Seattle, WA) that extracted a set of profile curves at the 12:30, 2, 3, 4, 5:30, 6:30, 8, 9, 10, 11:30, 12 o'clock positions. To perform a virtual anterior capsular plication, the profile lines at 4 and 5:30 were shorted by 1.5 cm to simulate the suturing and tightening of the anterior band of the IGHL (AB-IGHL). The resulting profile lines were then used to calculate a continuous surface geometry of the GH capsule [4] which in turn was used to create a surface mesh for finite element analysis.

RESULTS

Following the material parameter optimization, the computational load-displacement curve closely matched the experimental result of the native shoulder with $r^2 = 0.93$ (Figure 3). The joint reaction force when translating the humeral head increased with larger translation distances. Following capsular plication, the computational result again closely matched the experimental curve with $r^2 = 0.82$ (Figure 4) and had a higher force for a given translation as compared to the native state, indicating a tighter, and hence more stable capsule.

DISCUSSION

Cadaveric experimentation is the current standard for investigating the biomechanics of glenohumeral instability and potential effects of surgical techniques. The goal of this research is to develop a subject-specific computer model of shoulder instability from patients diagnosed with shoulder instability. This study demonstrates the accuracy of our computer model by validating the computational results against an experimentally obtained load-displacement curve. Future work will include additional cadaveric validation experiments, and evaluating different surgical techniques of capsular plication.

REFERENCES

[1] Cadet ER. Orthop Clin North Am. 2010.
[2] Peltier KE. J Shoulder Elbow Surg. 2011.
[3] Holzapfel GA. J Elasticity. 2000.
[4] Yongpravat C. WCB. 2014.

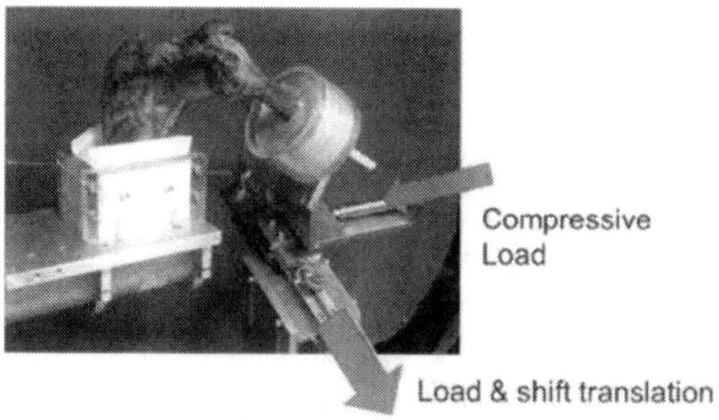

Figure 1. Experimental testing of joint stability by translating the humeral head and obtaining the load-displacement curve.

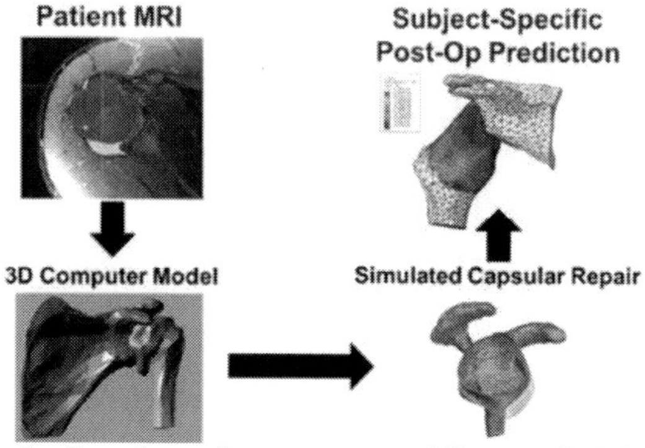

Figure 2. Overview of computer modeling methodology for subject-specific modeling. An MR arthrogram is obtained of an instability patient, a 3D computer model is extracted from the radiographic scan, several types of capsular repair can be simulated, and the resulting post-op joint mechanics can be predicted.

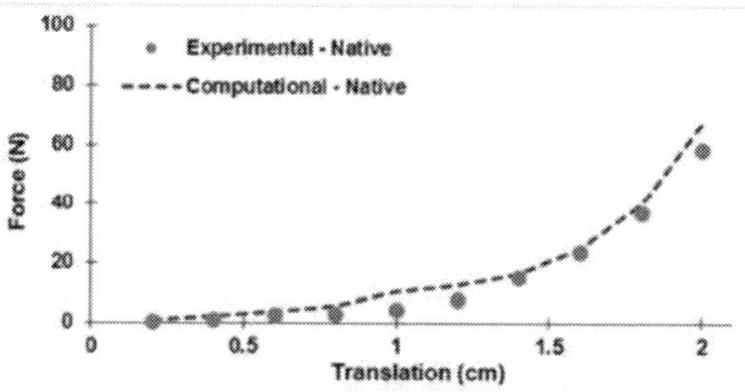

Figure 3. Load-Displacement curves of the native shoulder joint. Experimental results are plotted along with the best fit parameters for the computational model.

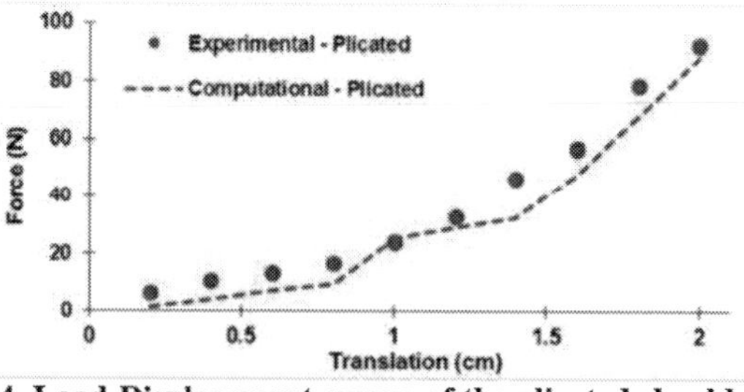

Figure 4. Load-Displacement curves of the plicated shoulder joint. Experimental results are plotted along with the prediction from the computational model.

SB³C2015
Summer Biomechanics, Bioengineering and Biotransport Conference
June 17-20, 2015,Snowbird Resort, Utah, USA
SB³C2015-618

CHANGES IN THE RANGE OF MOTION ENVELOPE OF THE LUMBAR SPINE WITH REPETITIVE LIFTING

Muhammad I Gul (1), Timothy D Craig (2), Neena K Sharma (3), Sara E Wilson (2)

(1) Chemical & Petroleum Engineering
University of Kansas
Lawrence, Kansas USA

(2) Mechanical Engineering
University of Kansas
Lawrence, Kansas USA

(3) Physical Therapy and
Rehabilitation Sciences
University of Kansas Medical Center
Kansas City, KS, USA

INTRODUCTION

Low back pain (LBP, including pain in the lumbosacral region of the spine and sciatic pain radiating to the legs) is a common and costly public and occupational health problem that has been associated with the performance of repetitive lifting tasks in the workplace [1-2]. Occupations that involve performing cyclic loading tasks have been identified to cause a 10-fold increase in low back injury [1]. Studies in animals and humans have shown that repetitive cyclic loading of ligaments of the spine can alter the range of motion (ROM) of the lumbar spine [3-5]. These studies have further shown that such changes can affect the dynamic control of spine motion, potentially increasing risk of low back injury [3]. The objective of this study was to examine effects of repetitive lifting might have on the range of motion envelope that describes the limits of lumbar motion.

A previous study done in the Human Motion and Biomechanics Laboratory at The University of Kansas identified the importance of looking at not only minimum and maximum torso flexion postures, but also the local lumbar angle limits as a function of moderate torso flexion in order to develop better understanding of the ROM envelope [6]. For each torso flexion angle, subjects are asked to flex and extend their lumbar spine while maintaining the same torso flexion to achieve a range of lumbar angles. This range is an envelope through which subjects can move during a lifting task. Since lifting tasks operate within this ROM envelope, it is important to understand this envelope and how it might change with exposure to injury risk factors such as repetitive lifting.

In this study, the specific aim is to examine change in the lumbar ROM envelope with a moderate cyclic lifting task. The hypothesis was that the upper (more kyphotic) limits of the ROM envelope would increase and the lower (more lordotic) limits of the ROM envelope would decrease after the cyclic lifting task.

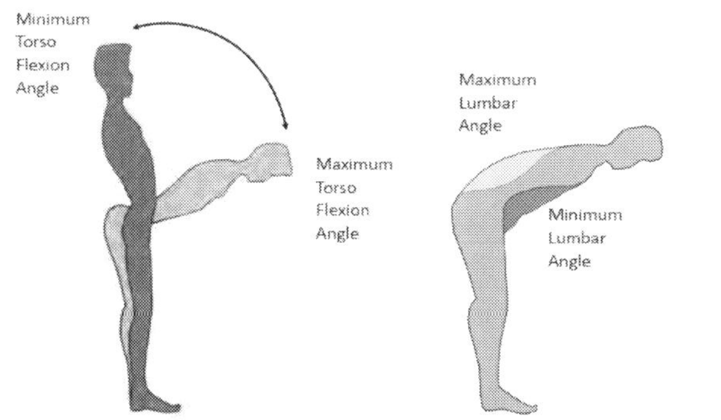

Figure 1. In this study, the range of motion envelope describes the maximum and minimum achievable lumbar angle for each torso flexion angle. Previous studies of lumbar range have focused on examining lumbar angle and torso flexion angle at maximum torso flexion.

METHODS

Nine healthy adult subjects (5 male, 4 female, 24.8±4.4 years, 1.72±.01m) were recruited for this study. Lumbar angles were recorded by attaching a set of 6-D electromagnetic motion sensors (Trakstar, Ascension Technology, Burlington, VT) to the skin at the T10 spinous process, S1 sacrum, and manubrium using a double sided adhesive tape. Each sensor returns orientation of the sensors in Euler angles (angle sequence of flexion, lateral bending, rotation) and position. This data was acquired at 100 Hz and used in a biofeedback display to allow subjects to observe their torso flexion and lumbar angle. To determine torso flexion, a line connecting the T10 and S1 sensor positions was examined and torso flexion was determined as the angle of that line from vertical in the sagittal plane. This angle represents the tilt of the trunk. Lumbar angle was determined by comparing the orientation of the T10 and S1 sensors. Specifically, the lumbar angle was the difference in the flexion Euler angle of the T10 and S2 sensors.

For the ROM envelope, the minimum and maximum lumbar angle (dependent variable) at four different flexion angles (independent variable) was measured. A biofeedback display created in Labview (National Instruments, Austin, TX) was used to display lumbar angle and torso flexion. Subjects were asked to flex their trunk to a target torso flexion and then hold that torso flexion angle while arching and flexing their low back to obtain the maximum and minimum attainable lumbar angles. Subjects were allowed to practice this movement until they were comfortable with the task. They were then asked to repeat the maximum and minimum lumbar angles three times for each of four torso flexion angles 0°, 30°, 60° and 80°.

After the ROM envelope was assessed, the subjects were asked to repetitively lift a crate with handholds using straight legs for four minutes. A weight of approximately 3% of the subject's maximum lifting force was placed in the crate. Subjects were instructed to lift at a rate of fifteen lifts per minute based on audio cues. Subjects were asked to raise the crate from floor level to waist level, pause and then lower the crate to the floor for one lifting cycle. After the 4 minute cyclic lifting task, the subjects were asked to repeat the ROM envelope assessment.

A repeated measures ANOVA was used to assess changes in the lumbar angles limits in the ROM envelope before and after the cyclic lifting task. Other independent variables in the ANOVA included trunk flexion angle and limit (maximum or minimum).

RESULTS

In this study, the lumbar range of motion at 0°, 30°, 60° and 80° of torso flexion was assessed before and after a cyclic lifting task. A repeated measures ANOVA of the lumbar angles in the range of motion envelope found the lumbar angles were significantly different between the two limits (maximum versus minimum) as expected and between the trunk flexion angles. No significant change was observed with time (pre versus post cyclic lifting tasks) ($p<0.05$). The mean range of motion limits for the lumbar angle is shown in Figure 1.

DISCUSSION

Recent studies have shown that cyclic loading of lumbar spine increases the laxity of the joints thereby increasing the range of motion of the back [3-5]. Solomonow et al. argued that such changes in lumbar spine ROM could lead to increased instability in the musculature of lower back and hence increase risk of developing LBP and other low back complications [4]. These authors observed that increased range of motion in the spine due to cyclic load in a cat model led to reduced stretch reflex activation supporting this theory. However, these past studies have only examined the extreme limits of torso flexion rather than the range of motion of the lumbar spine and low back at moderate torso flexion angles where lifting tasks might normally occur.

The results of this study reveal that performing the cyclic task in this study for four minutes did not significantly change the ROM envelope of the lower back. This finding is different from previous studies of lumbar range of motion with cyclic lifting. The first potential reason for this difference is that the ROM envelope captures primarily local lumbar flexibility rather than a global trunk flexibility (that might also incorporate characteristics of other joints such as the hip). The second potential reason is that the cyclic lifting task in this study was a fairly moderate task with a light load and moderate torso flexion excursion. To fully assess the effects of cyclic lifting on the ROM envelope, the next step of this study will be to examine how the ROM envelope changes when subjects are exposed to a longer period of cyclic lifting task, a heavier load task, and/or a higher torso flexion lifting task.

ACKNOWLEDGEMENTS

Research reported in this publication was supported by the National Institute of Arthritis and Musculoskeletal and Skin Disease of the National Institutes of Health under Award Number 1R03AR061597-01. The content is solely the responsibility of the authors and does not necessarily represent the official views of the National Institutes of Health.

REFERENCES

[1] McGill S,. *J Biomech* 30: 465–75, 1997.
[2] Bernard, B, "Musculoskeletal disorders and workplace factors," U.S. Department of Heath and Human Services, Washington, D.C. 97-141, 1997.
[3] Solomonow, M, et al., *Clin Biomech*, 18: 273-9, 2003.
[4] Solomonow, M. et al., *Spine*, 24(23): 2426-2434, 1999.
[5] McGill, S, et al., *Ergonomics*, 39: 1107-18, 1996.
[6] Maduri, A et al., *J Electromyogr Kinesiol,* 18: 807-14, 2008.

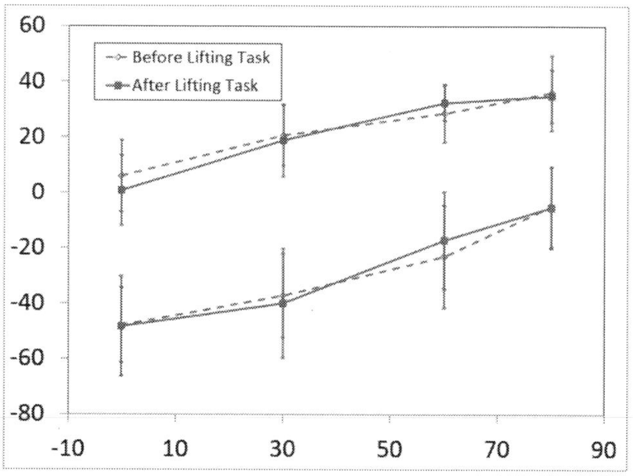

Figure 2. Subjects were asked to flex and extend their back before and after the cyclic lifting task. The figure shows an average minimum and maximum Lumbar angle plotted for 0°, 30°, 60° and 80° flexion angles. The shaded region shows standard deviation.

VALVE INTERSTITIAL CELL SHAPE REGULATES CELL FUNCTION AND PHENOTYPE

Atefeh. Razavi (1), Rachel. Reynolds (1), and Kartik. Balachandran (1)

(1)Department of Biomedical Engineering
University of Arkansas
Fayetteville, Arkansas, USA

INTRODUCTION

Valve interstitial cells play an important role in maintaining the integrity and optimum function of the valve. They are a heterogeneous population composed of mainly five different phenotypes. These phenotypes have specific functions characteristic of valve physiological and pathological conditions [1]. Activated VIC phenotype are expressed during repair processes by ECM remodeling, cell proliferation and migration and may change their phenotype via osteoblast transdifferentiation and expression of bone-like markers [2-4]. The study of this phenotypic change and pathological inducing factors and conditions contributing to cell activation and differentiation is of great importance.

Mechanical force is a factor known to regulate valve remodeling and disease progression [5-7]. Large number of mediators also has been recognized to stimulate valve microenvironment and affect calcification such as tissue growth factor–β [3]. .However the effect of cell geometry as a physical cue on VIC function and phenotype is poorly understood. Cell shape changes are shown to be correlated with various functional states of other cell types including proliferation and apoptosis[8], stem cell differentiation[9], muscle cell contractility[10] and macrophage polarization states[11].

We hypothesized that we could mimic healthy and diseased state of the VICs by controlling cellular architecture via micro-tissue engineering techniques. In this study we are particularly interested to investigate effects of VIC architecture on the function of the valve in terms of cytoskeletal structure, contractility and expression of contractile phenotype markers.

METHODS

Micro-contact printing

Fibronectin was incubated on polydimethylsiloxane (PDMS) stamps for 1 hour. Stamps had raised features composed of 10, 20, 40, 60, and 80 μm wide lines separated by 40 μm gaps. Fibronectin was microcontact printed on PDMS coverslips that had been UV-ozone treated for 15 min. Substrates then were incubated in 1% pluronic F127 for 10-15 min and 500,000 VICs were seeded per feature constrained into predefined organization guidelines.

Histochemical analysis

After 2 days of culture, cells were stained with DAPI, phalloidin and GFP-plasma membrane stain. Actin alignments were calculated by analyzing phalloidin stained images in MATLAB. Cell aspect ratio was measured through images of the live stained plasma membrane in Imagej.

Valvular Thin Film Contractility Assay (VTF)

A strip of a PIPAAm (polyN-isopropylacrylamide) was spin coated onto the glass coverslip. PDMS was then spincoated onto the PIPAAm coated coverslip and allowed to cure overnight. Cells were then seeded using micro-contact printing as before. After 2 days, cells were transferred to a petri dish containing Tyrode's buffer. Thin rectangles were cut from the PIPAAm layer. The dish was allowed to cool below 32 C dissolving the PIPAAm and releasing the PDMS and cell layer. The temperature was set back to 37C. After 2.5min, the MTFs were treated with endothelin-1 (ET-1) for 10 min followed by HA-1077 treatment for 10 min. The images were taken at 30s intervals and then analyzed using MATLAB to determine radius of curvature and cellular stress levels.[12]

Western Blot

10% polyacrylamide gel was loaded with 30 μg of total protein and was run for 1 h at 200 V. The gel then was transferred to the PVDF membrane. Membranes were incubated in primary antibody for 12 hours before being imaged with IrDye conjugated secondary antibodies. Primary antibodies were α-SMA, vimentin, SM-MHC and β-actin.

RESULTS

Study of cell architecture

We hypothesized that VIC architecture would control normal and pathological phenotype of the valve. To address this hypothesis, we constrained the geometry of the cell shape and engineered mesoscale niches by constructing extracellular matrix with varying band widths. We selected 10, 20, 40, 60, and 80 μm wide bands to provide a wide range to regulate cellular migration within the tissue. Forcing the cells into thinner patterns limited their migration leading to more elongation and alignment of cytoskeleton whereas the wider patterns allow more cells to span their width characteristic of more disorganized structure (Fig. 1A). We measured alignment of the actin filaments via analyzing the images obtained through phalloidin staining of the tissue fibers. Orientation parameter (OP) was calculated based on a fingerprint identification algorithm to quantify actin alignment[13]. Results showed that actin architecture was more than 80% aligned for all samples, and that there was a statically significant difference between thinner patterns (10 and 20) with wider ones(40, 60 and 80). There was also a trend for decreasing orientation with increasing line width

(Fig. 1B). We also calculated cell aspect ratio using live plasma membrane staining technique and measuring major and minor axes of the ellipse that best fits the stained membrane. Compared to actin alignment, the difference between pattern widths was more evident. Tissues with 10 and 20 μm width were statistically different from 40, 60 and 80 μm (Fig. 1C).

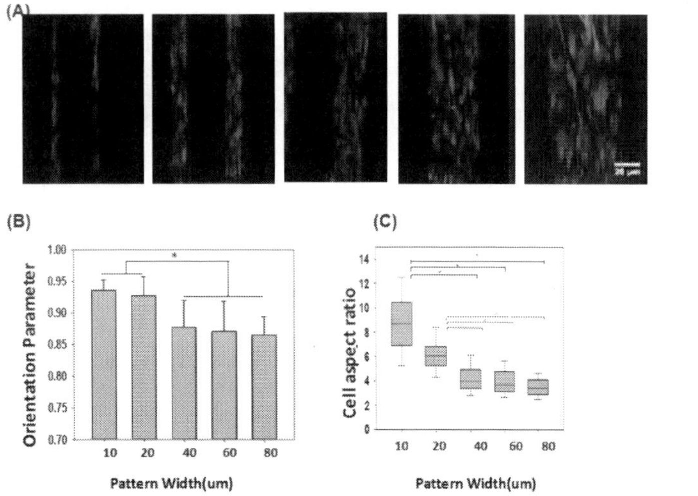

Fig1. A) Staining for actin filaments (green) and nuclei (blue), B) actin alignment parameter. C) Cell aspect ratio. (n=6; * p<0.05)

Contractile Stress

We utilized VTF assay to calculate the generated stress in various constructed tissue patterns. As depicted in Fig2, stress values were increased during the treatment with ET-1(vaso-constriction drug) and decreased after adding KA-1077(vaso-dilation drug). The difference between stress values for different band width suggests that the cell contractile output may be modulated by cell structure.

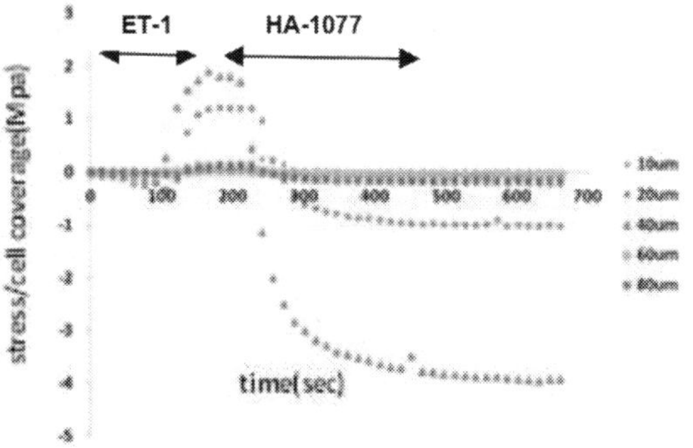

Fig2. Stress values after drug stimulation for various patterns (cell coverage=area occupied by the cells/total surface area)

Protein expression

When VICs are stimulated, they switch their phenotypes from quiescent type to activated myofibroblast. We therefore asked if the altered cell architecture in our micro-engineered models potentiated this differentiation in VICs. We measured the protein expression of α-SMA, vimentin and smooth muscle heavy chain (SMMHC), markers for contractile, quiescent and smooth muscle cell phenotype. The obtained data, presented in Fig. 3B and D shows a decreasing trend for α-SMA and an increasing trend for SMMHC expression as the tissue becomes wider. These data suggest that contractile phenotype is up-regulated for thinner patterns where the cells are more elongated. Comparing with stress data showing the maximum contraction for 40um patterns, we conclude that expected contractility for the thinner tissue pattern may be derived from factors other than change in

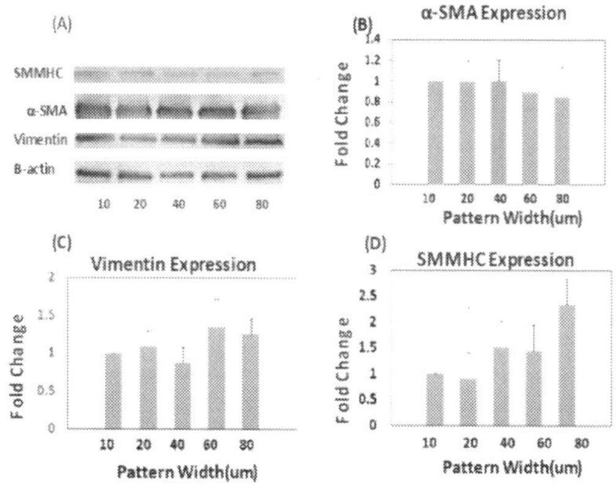

Fig3. Western blot and protein expression quantification

DISCUSSION

Studies on aortic valve disease have focused mainly on two factors contributing to aortic valve dysfunction: (I) Introduction of mechanical stress to the valve microenvironment [11] and (II) biochemical stimulation of the valve cells [5, 7]. In this study we considered the effect of valve cell geometry and structure as another factor that might affect functional behavior of the valve. Different cell architectures were designed and engineered using soft lithography and micro-contact printing techniques. The patterns constrain cells to a range of contractile, normal and disorganized structure. Cell aspect ratio and actin filaments alignments were distinctly dependent on tissue width. Contractile stress measurements showed varying stress values for the different patterns with the maximum value for 40um lines where effects of cell-cell contact and actin fiber alignment may be optimal. Contractile and smooth muscle phenotype were measured based on western blot data and demonstrated approximately opposite trends for various band widths. The correlation between contractile phenotypic changes and stress output will determine how reliable the contractile marker measurements are in defining VICs as contractile or synthetic phenotype.

ACKNOWLEDGEMENTS

We acknowledge Arkansas Biosciences Institute and Arkansas Women's Giving Circle for funding this project.

REFERENCES

[1] Liu, A.C., et al., *Am J Pathol*, 171(5):1407-1418, 2007.
[2] Osman, L., et al., *Circulation*,. 116(11 Suppl):I282-I287.
[3] Jian, B., et al., *Ann Thorac Surg*,75(2):457-466, 2003.
[4] Poggio, P., et al., *Cardiovasc Res* 98(3):402-410, 2013
[5] Balachandran, K., et al., *Am J Physiol Heart Circ Physiol*,296*(3):756-764*, 2009.
[6] Balachandran, K., et al,. *Am J Pathol*,. 177(1):49-57, 2010.
[7] Dweck, M.R., et al., *J Am Coll Cardiol*,. 60(19):1854-1863, 2012.
[8] Chen, C.S., et al., *Science*,. 276(5317):1425-1428, 1997.
[9] McBeath, R., et al., *Dev Cell*,. 6(4): 483-495, 2004.
[10] Alford, P.W., et al., *Integr Biol (Camb)*,. 3(11): 1063-1070, 2011.
[11] McWhorter, F.Y., et al., *Proc Natl Acad Sci U S A*,. 110(43):17253-17258, 2013.
[12] Grosberg, A., et al. *J Pharmacol Toxicol Methods*,. 65(3):126-135, 2012.
[13] Bray, M.A., et al., *Cell Motil Cytoskeleton*,. 65(8):641-651, 2008.

SB^{3}C2015
Summer Biomechanics, Bioengineering and Biotransport Conference
June 17-20, 2015,Snowbird Resort, Utah, USA
SB^{3}C2015-621

STRAIN RATE DEPENDENCY OF THE INTRACELLULAR CALCIUM ION CONCENTRATION DURING NEURONAL MEMBRANE MECHANOPORATION

A. Bakhtiary (1), A. E. Florence (1), M. A. Murphy (1, 2), S. Mun (2), J. Liao (1), L. N. Williams (1,2), M. F. Horstemeyer (2, 3), M. C. LaPlaca (4), R. Prabhu (1, 2)

(1) Department of Agricultural and Biological Engineering, Mississippi State University, Mississippi State, Ms, 39762.

(2) Center of Advanced Vehicular Center, Mississippi State University, Mississippi State, Ms, 39762.

(3) Department of Mechanical Engineering, Mississippi State University, Mississippi State, Ms, 39762.

(4) Department of Biomedical Engineering, Georgia Institute of Technology and Emory University, Atlanta, GA 3033.

INTRODUCTION

In the United States, 1.7 million people are estimated to sustain a traumatic brain injury (TBI) annually according to the CDC [1]. TBI Emergency Department (ED) visits for football are sited to be among the highest in sports [2]. While the football helmet designs have seemed to improve over the years, TBI related injuries seen in EDs have increased by 60% in the last ten years yielding a need for more accurate injury models [2].

During injury impact, the threshold for homeostasis in the brain can be exceeded, creating strain on the neuronal cell membrane. Once the threshold is exceeded, immediate lysis of the neuron or mechanoporation of the neuronal cell membrane may occur [3]. Literature states a strain exceeding 2-4% will cause the creation of non-specific pores in the cell membrane. The number and size of pores will increase with strain rate [3,4]. Small pores that form in the lipid bilayer membrane will reseal themselves as a poreless bilayer is energetically more favorable [3,4]. But under a high enough strain or strain rates, the resulting pores will remain open allowing non-specific molecules to move across the cell membrane down their concentration gradients [3,5]. Specifically, there is an influx of calcium ions (Ca^{2+}) into the cell resulting in damage [3,6]. Quantifying this damage will begin to link mechanical insult with clinical cellular apoptosis or damage.

Currently our research group, at Mississippi State University, is utilizing FEA brain models to design more effective protective equipment for football players [7, 8]. The effects of externally applied forces on neurons and the resulting damage needs to be quantified to improve the accuracy of these models. Once the damage is quantified at the cellular level, macro scale models will incorporate the cellular-level damage through a multiscale modeling paradigm, and hence, enable the prediction of more accurate injury metrics for the brain.

The purpose of the current research is to evaluate the strain rate sensitivity of the neuron intracellular Ca^{2+} concentration during mechanical insults to the neuronal lipid bilayer.

METHODS

In order to accurately model ion influx after mechanical insult, the fundamental laws of physics are used in combination with a real time representation of neuronal membrane pore growth in a multi-scale modeling paradigm as seen in Figure 1.

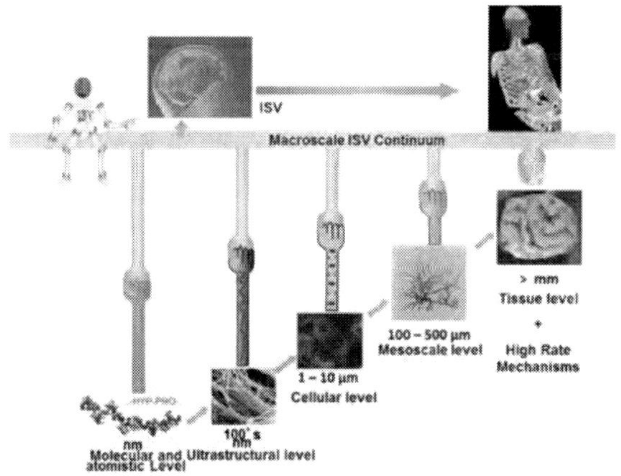

Figure 1. Multiscale modeling paradigm of the brain.

Molecular Dynamics (MD)-based equibiaxial deformations of a 72 phospholipid (POPC) bilayer (48.9 by 46.1 Å) are carried out to simulate external mechanical insults. Here, 4 rates are considered: x: 0.2 - y: 0.2 m/s, x: 0.6 - y: 0.6 m/s, x: 2.8 - y: 2.7 m/s and x: 4.0 - y: 3.8 m/s.

The data collected was then put in an in-house image analysis algorithm that counts the number of pores and calculates the average radius considering the equivalent area per pore every 0.005ns. A 5 Å cut-off pore radius is considered with respect to the Ca^{2+} and its water shell. The Ca^{2+} influx is then calculated considering a diffusion equation developed related to the number of pores and the pore size calculated from the MD simulation.

Equations: Starting with Fick's Law:

$$\frac{dn}{dt} = D\frac{A'}{\Delta x}\Delta C \quad (1)$$

where dn/dt is the rate of molecules diffusing across the membrane, D is the molecule diffusivity, A' is the effective area, Δx is the membrane thickness and ΔC is the concentration gradient across the membrane.

A concentration equation with respect to time was developed based on Equation 1:

$$[C(t)]_i = C_{eq} - \left(C_{eq} - C_o\right)e^{-\frac{Dn\pi r^2}{\Delta x V_c}\left(1-\frac{a^2}{r^2}\right)t} \quad (2)$$

where C_{eq} is the molecular concentration at equilibrium, C_0 is the molar concentration, n is the number of pores, r is the average radius of the pores, V_c is the volume of the cell, a is the radius of the molecule, and t is the time.

The DC/Dt, concentration flux, was calculated by partially deriving the concentration equation with respect to time and partially with respect to the changing pore radius, where the radius of the pores are received from the MD simulations:

$$\frac{DC}{Dt} = \frac{\partial C}{\partial t} + \frac{\partial C}{\partial r}\frac{\partial r}{\partial t} \quad (3)$$

A MATLAB code, simulating the diffusion of the Ca^{++} due to mechanoporation, was then used to calculate the new internal concentration in a step like fashion utilizing the equations developed above [9].

RESULTS

The Ca^{2+} concentration resulting from equibiaxial deformation on a lipid bilayer is shown in Figure 2 for different deformation rates on the lipid bilayer. It is shown that lower strain rates require longer times to allow for Ca^{2+} influx initiation reflecting that it takes longer for the required threshold to be reached. At the same time, higher strain rates generally result in a faster increase in Ca^{2+} concentration.

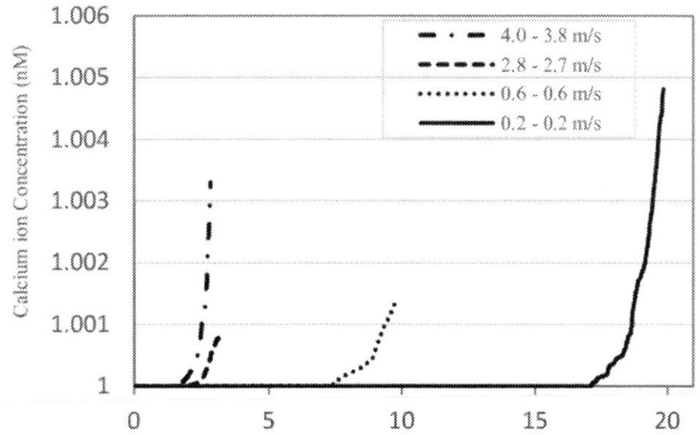

Figure 2. Ca^{++} concentration resulting from the equibiaxial deformation of the lipid bilayer at different deformation rates.

Figure 3 shows the intracellular Ca^{++} concentration with respect to the true strain for the different deformation rates. It is seen that lower strains are required for the initiation of Ca^{++} influx into the neuron when strain rates are low.

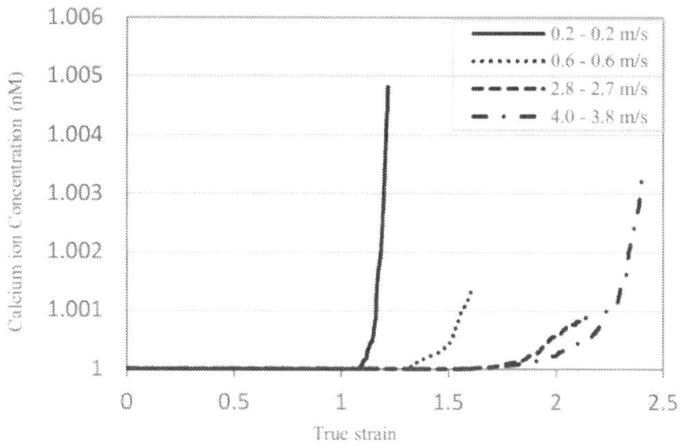

Figure 3 Ca^{++} concentration resulting from the equibiaxial deformation of the lipid bilayer at different deformation rates with respect to the true strain.

DISCUSSION

The primary goal of the current research is to evaluate the effect of deformation rate on the Ca^{2+} influx into the neuron cell.

Comparison of Figures 2 and 3 show that while higher strain rates allow for faster mechanoporation, lower strain rates require lower strain for mechanoporation and allow for the onset of Ca^{2+} diffusion. This shows that mechanoporation is a rate dependent phenomenon (Figure 2), and in general, once the Ca^{2+} influx into the neuron begins, the intracellular concentration of the Ca^{2+} increases catastrophically within a small range of subsequent strain (deformation).

Thus when designing football helmets using a multiscale model for the brain, the relevant information on mechanoporation and Ca^{2+} influx needs to be incorporated. Attention should also be given to the neuronal membrane deformation and strain rates, where the resulting threshold strains indicate the onset of catastrophic increase in the intracellular Ca^{2+} concentration.

ACKNOWLEDGEMENTS

This material is based upon work supported by the NSF through a subcontract with Mississippi State University, and was performed for the EPSCoR research program (EPS_0903787). Any opinions, findings and conclusions or recommendations expressed in this material are those of the author(s) and do not necessarily reflect the views of the NSF.

REFERENCES

[1] http://www.cdc.gov/traumaticbraininjury/pdf/BlueBook_factsheet-a.pdf
[2] Gilchrist J, et al., MMWR 60(39):1337-1342, 2011.
[3] K.A. Barbee. Annual of N.Y. Academy of Science 1066:67-84, 2005.
[4] den Ottera, W. K. The Journal of Chemical Physics 131: 205101-205109, 2009.
[5] Geddes, D. Journal of Neurotrauma 20(10): 1039-1049.
[6] LaPlaca, M.C. et al., Annals of Biomedical Engineering 25: 665-677, 1997.
[7] Prabhu R. et al. J Mech Behav Biomed Mater, 2011 Oct, 4(7):1067-80.
[8] Ladner I. S. et al., SBC2013-14649, Proceedings of the ASME Summer Bioengineering Conference, Sunriver, OR 2013.
[9] Geddes, D. M., Georgia Tech Theses and Dissertations [18332], 2005

SB³C2015
Summer Biomechanics, Bioengineering and Biotransport Conference
June 17-20, 2015,Snowbird Resort, Utah, USA
SB³C2015-622

PREPARATION OF A HYDROGEL PHANTOM OF HUMAN ATHEROSCLEROTIC PLAQUE FOR MEDICAL SIMULATION AND IMAGING

Juyu Chueh (1), Tanya N. Turan (2), Truman R. Brown (2), Todd LeMatty (2), Hui Mao (3), Olivia W. Brooks (1), Matthew J. Gounis (1)

(1) New England Center for Stroke Research
Department of Radiology
University of Massachusetts Medical School
Worcester, MA, USA

(2) Department of Neurology
Medical University of South Carolina
Charleston, SC, USA

(3) Department of Radiology and Imaging Sciences
Emory University School of Medicine
Atlanta, GA, USA

INTRODUCTION

Intracranial atherosclerotic disease (ICAD) is a common cause of stroke throughout the world. Patients with severe arterial stenosis and a recent stroke are at high risk of recurring stroke. Catheter angiography is currently the gold standard imaging technique for ICAD; however, it merely demonstrates vessel narrowing and does not provide characteristics of the atherosclerotic plaque. With less invasive angiography methods such as magnetic resonance angiography (MRA) and computed tomography angiography (CTA), the degree of stenosis is not always fully appreciated and may not disclose the underlying disease.[1]

High resolution vessel wall MRI has shown to be an excellent technique to differentiate various pathologies that may be the cause of the stenosis and allow characterization of plaque composition.[2-4] Various vessel wall imaging sequences have been proposed to suppress the signal of blood and obtain high resolution image data.[5] Currently, there is no standard protocol for MR-imaging of ICAD, nor is there a gold standard phantom to compare MR-sequences. In addition, data generated by MRI is dependent on MR technique as well as scanner hardware.

In this study, a small batch manufacturing technique is developed to create a plaque phantom that provides a platform for establishing a uniform imaging method for diagnosis of ICAD. This plaque phantom is transparent, elastic with low friction, and has uniform wall thickness. In addition, it has good compatibility with imaging modalities such as CT, MRI and three-dimensional rotational angiography. The manufacturing process and preliminary image results are described below.

METHODS

MRI imaging data from patients with intracranial atherosclerotic plaques were used to acquire the detailed structure of each plaque component. The 3D reconstruction and the conversion to STL format were performed with software packages, Mimics and Magics (Materialise, Leuven, Belgium) (Figure 1).

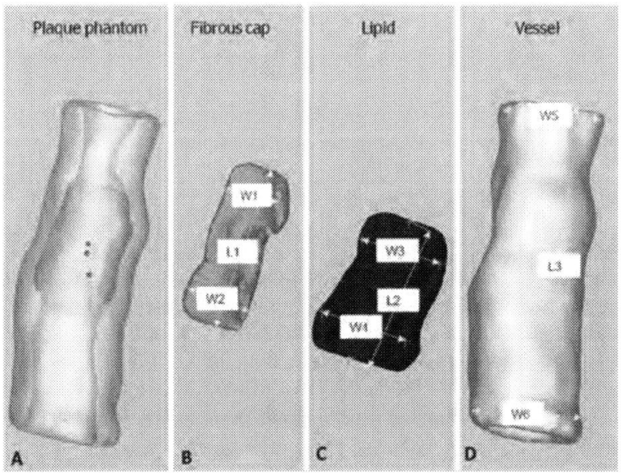

Figure 1: A phantom of atherosclerotic plaque (A) which comprises a fibrous cap (B), a lipid core (C) and a stenosed vessel lumen (D).

A stenosed basilar artery was first prepared using a core-shell mold shown in Figure 2A. Fused deposit manufacturing technique was employed to convert the virtual design into a physical model. Polyvinyl alcohol hydrogel (PVA) was infused into the ABS core-shell mold by liquid injection molding, followed by several freeze-thaw cycles for curing. The core-shell mold was immersed in xylene for ABS dissolution, yielding a PVA stenosed vessel wall (Figure 2B).

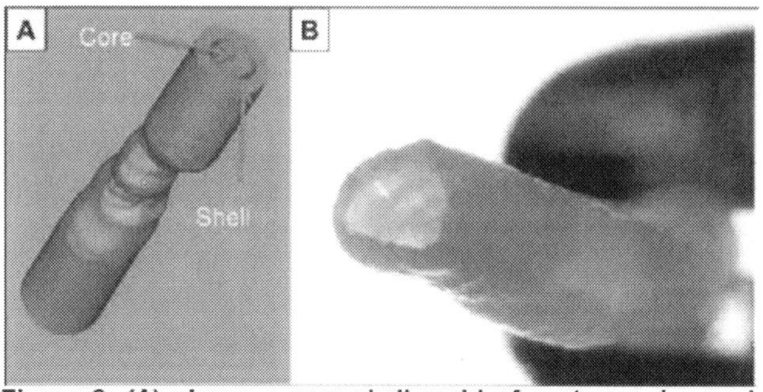

Figure 2: (A) shows a core-shell mold of a stenosed vessel. Hydrogel is infused into the core-shell model and undergoes several freeze-thaw cycles for curing. After mold dissolution in xylene, a hydrogel vascular replica is obtained (B).

A fibrous cap was constructed using a mixture of gadolinium chloride, agarose, carrageenan, sodium azide, water, and sodium chloride (T1= 2859ms, T2= 340ms). On the other hand, a lipid core was mimicked using a milk, sodium azide, and carrageenan mixture (T1= 1099ms, T2= 91ms). To precisely control the volume of each plaque component, a plaque mold made of silicone with known shape and dimension was built (Figure 3). Each mixture was infused into the silicone container, and set at -80°C for 1 hour. The silicone container was carefully cut open to release the shaped plaque component. The shaped plaque components were then glued to the PVA vascular replica of a stenosed vessel wall by adding several layers of PVA coating and performing curing process.

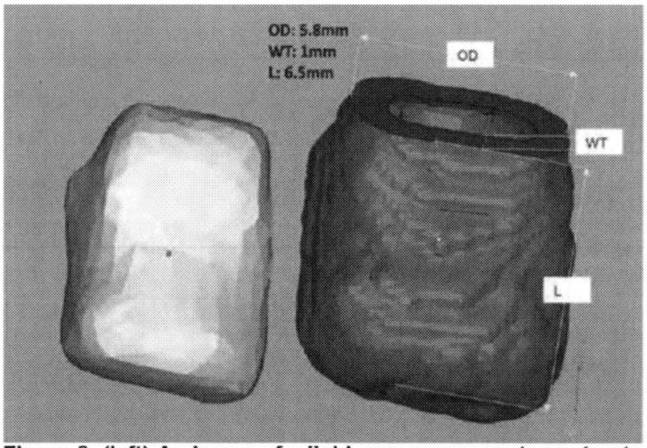

Figure 3: (left) An image of a lipid core component constructed using a mixture of gadolinium chloride, agarose carrageenan, sodium azide, water, and sodium chloride. (right) An image of a silicone mold used to form the lipid core. (OD: outer dimension, WT: wall thickness, L: length).

The plaque phantom was scanned using a Siemens 3T Trio MR scanner at 2 different sites and the results were compared.

RESULTS

Table 1 shows the dimension of each plaque component presented in Figure 1.

Table 1 The dimension of the plaque phantom

Parameter	Value (mm)	Parameter	Value (mm)
W1	2.9	L1	8.8
W2	3.4	L2	7.1
W3	4.3	L3	20.0
W4	4.4		
W5	6.1		
W6	7.0		

Variations in signal intensity between the fibrous cap, lipid core and vessel wall were observed on the T2 images acquired from both centers in Figure 4.

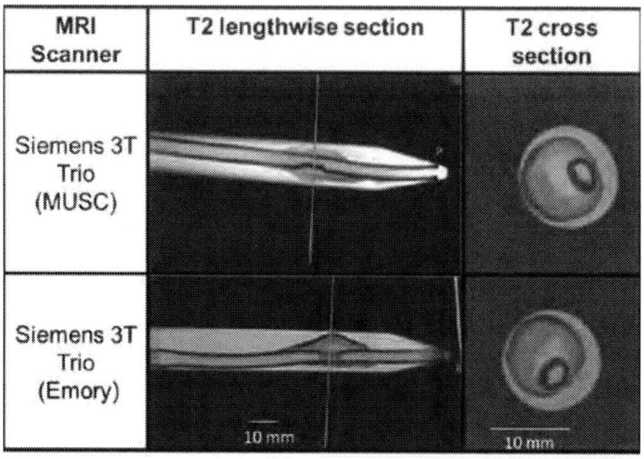

Figure 4: Sample images from basilar stenosis phantom imaged at 2 different centers (MUSC and Emory) on 2 different MRI scanners using the same sequences. The same phantom is used, but is rotated at a different angle in the lower images.

DISCUSSION

A method of making a plaque phantom based on MRI imaging data is established. The phantom described herein is composed of a stenosed vessel, a fibrous cap and a lipid core.

Unlike the silicone vascular replicas prepared in our prior work,[6] the PVA model doesn't swell in xylene after removal of the outer shell. Therefore, the time required to dissolve the ABS core is significantly increased. In our prior design,[6] the inner core was a solid piece. To solve this problem, the inner core can be modified to have a hollow structure, which allows xylene to flow into the mold and to quickly dissolve the inner core.

Our future work will focus on including the thrombotic and the calcified components of the atherosclerotic plaque into our model. The temperature effects on the phantom stability as well as the stability of the plaque phantom over time will be investigated. Finally, the plaque phantom will be connected to a flow system, and scanned under physiological flow conditions.

REFERENCES

[1] Feldmann, E et al., *Neurology,* 68(24):2099-2106, 2007.
[2] Xu, WH et al., *Atherosclerosis,* 212(2):507-511, 2010.
[3] van der Kolk, AG et al., *Stroke,* 42(9):2478-2484, 2011.
[4] Turan, TN et al., *Atherosclerosis,* 237:460-463, 2014.
[5] Degnan, A et al., *AJNR,* 33(8):1427-1435, 2012.
[6] Chueh, JY et al., *AJNR,* 30:1159-1164, 2009.

SB³C2015
Summer Biomechanics, Bioengineering and Biotransport Conference
June 17-20, 2015,Snowbird Resort, Utah, USA
SB³C2015-623

VALIDATION OF CFD SOLVER OF A CLINICAL TOOL USING PIV ON A PATIENT-SPECIFIC INTRACRANIAL ANEURYSM

Nikhil Paliwal (1, 3), Christopher Martensen (2, 3), Nicole Varble (1, 3), Robert Damiano (1, 3), Adnan Siddiqui (3, 4), Elad Levy (3, 4), Jianping Xiang (1, 3, 4), Hui Meng (1, 2, 3, 4)

(1) Department of Mechanical and Aerospace Engineering,
State University of New York at Buffalo,
Buffalo, NY

(2) Department of Biomedical Engineering,
State University of New York at Buffalo,
Buffalo, NY

(3) Toshiba Stroke and Vascular Research Center
State University of New York at Buffalo, Buffalo,
NY

(4) Department of Neurosurgery
State University of New York at Buffalo,
Buffalo, NY

INTRODUCTION

An intracranial aneurysm (IA) is a disease in the cerebral vasculature characterized by weakening and outward bulging of the arterial vessel wall. If left untreated, IAs can rupture with severe consequences. Over 50% of all ruptured IAs results in mortality from hemorrhaging in the subarachnoid space or the brain parenchyma. Hemodynamics plays a critical role in aneurysm growth and rupture[1]. Endovascular treatments such as coiling, stenting, or flow diversion aim to alter aneurysmal flow dynamics to induce thrombotic occlusion and protect the aneurysm from growth and rupture. Therefore, there is a keen need to provide a clinical computational fluid dynamics (CFD) tool to assist neurointerventionalists in treatment decision and optimize intervention for patient-specific aneurysms. To this end, AView[2], an integrated image-based vascular analysis tool has been developed, which includes modules for rapid assessment of morphological calculations, blood flow inside aneurysm, rupture-risk assessment and treatment planning in a single platform. Preliminary feedback of AView among neurointerventionalists is encouraging.

Before AView can be brought to the clinical workflow, it is important to validate it for intended use in IAs. One of the central features of AView is its flow simulation module. In the current study, we focus on the validation of the flow simulation module of AView, which uses gnuid[3], an open-source CFD solver for parallel, pulsatile and incompressible flow simulations. Gnuid solver uses a pressure-correction algorithms in solving Navier-Stokes' equation which reduces computational cost as compared to conventional CFD solvers. It has been validated for benchmark CFD cases, but has not been validated for its intentional use in IAs. We use 2D particle image velocimetry (PIV) experiments on a patient-specific aneurysm model for the validation of gnuid solver. We also use the commercial CFD solver software STAR-CCM+ (CD-Adapco, Melville, NY), as the control for our validation study.

METHODS

An 8mm internal carotid artery (ICA) sidewall aneurysm from a patient was used as the aneurysm model for this study. The 3D angiographic image (Figure 1A) was segmented in AView, to generate a surface file as shown in Figure 1B. This surface file was then used to print a 3D geometry using an Object 3D printer, Model260 V (Object-Stratasys, Inc. Eden Prairie, MN). Silicone was poured around the 3D printed geometry, and a lost-wax casting technique was employed using a low melting point metal alloy to create "soft positive" geometry. The transparent silicone phantom was obtained after melting and removing metal, as shown in Figure 1C.

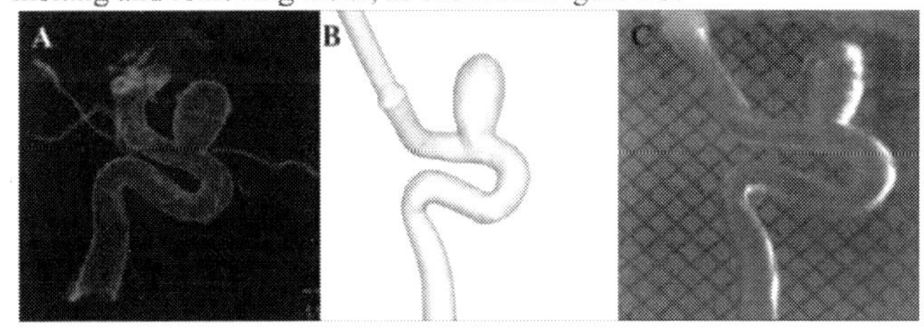

Figure 1: (A) 3D ANGIOGRAPHIC IMAGE, (B) SURFACE STL FILE, (C) SILICONE PHANTOM

PIV Setup

A flow loop was constructed that incorporated the aneurysm phantom as shown in Figure 2, where the fluid refractive index was matched to the silicone phantom (nD = 1.42) resulting in a viscosity of 10.95 cP. The flow loop was seeded with 3μm fluorescing particles. A state-of-the-art PIV system (LaVision Inc., Ypsilanti MI) including an neodymium-doped yttrium aluminum garnet (Nd:YAG) laser (LaVision, Goettingern, Germany), with a wavelength of 532nm, and a high-speed 16 bit CMOS camera at 5.5 megapixels were arranged adjacent to the flow loop aimed at the location of interest. The density of the working fluid was 1150 kg/m^3. A flow rate of 245 ml/min was used for steady state analysis, which corresponded to a Reynolds number of 260. High-speed image pairs were obtained and cross-correlated to produce 2D velocity vectors for a slice plane, at a rate of 10 Hz, and averaged over consecutive cycles.

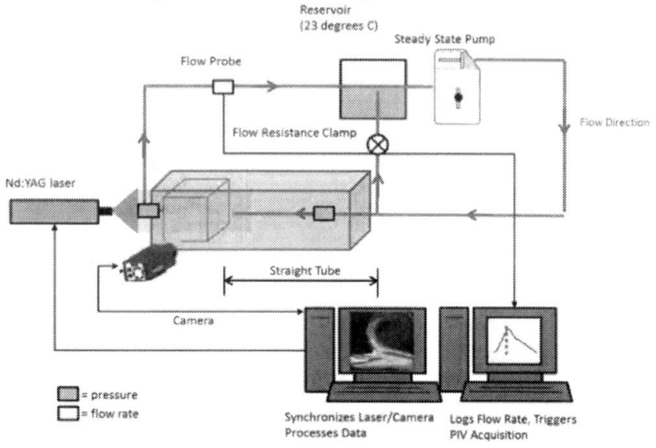

Figure 2: PIV FLOW LOOP SETUP

CFD Setup

The silicone aneurysm phantom used for PIV was imaged using 3D rotational angiography and segmented in AView to generate a precise surface mesh file for CFD. The location of the data plane for PIV was also identified by the imaged phantom to obtain the exact plane after CFD analysis. The surface file was volume meshed in AView using an edge-length of 0.25. Edge length is defined as the ratio of the maximum centerline radius to the maximum element size, thus making it relative to the geometry being meshed. Mimicking PIV setup, the fluid was considered as a Newtonian with density of 1150 kg/m^3 and a viscosity of 10.95 cP. Since gnuid solver is principally a transient CFD solver, it was given a constant inlet flow profile at the rate of 245 ml/min matching the PIV setup. The initial surface file was also used to run CFD on the commercial software STAR-CCM+ using the same flow conditions to compare with the PIV data.

RESULTS

Figure 3 shows velocity contours on a slice plane inside the aneurysm. The qualitative results for the steady state comparison suggest that the general flow profile is similar for all the 3 cases (gnuid, STAR-CCM+ and PIV) in the parent vessel. However, there seems to be a small variation in the flow pattern inside the aneurysm dome, especially in the region of low flow. Velocity contours show a better agreement between STAR-CCM+ and PIV. The boundary of velocity in PIV is choppy and does not accurately represent the boundary of the original geometry.

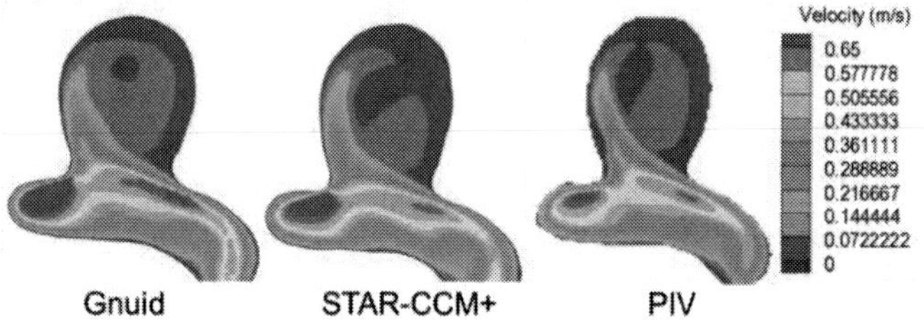

Figure 3: QUALITATIVE COMPARISON OF VELOCITY MAGNITUDE

In order to quantify the difference in the flow velocity on the plane, we used the approach proposed by Hoi et al[4]. We collected velocity magnitude from all 3 datasets along a line in the flow domain inside the aneurysm dome. Figure 4 shows the graph of the velocity magnitude on data points along the line. From the graph, it is clear that all 3 datasets were able to capture a similar flow profile along the same line. The average error in gnuid result was 19% when compared with PIV result.

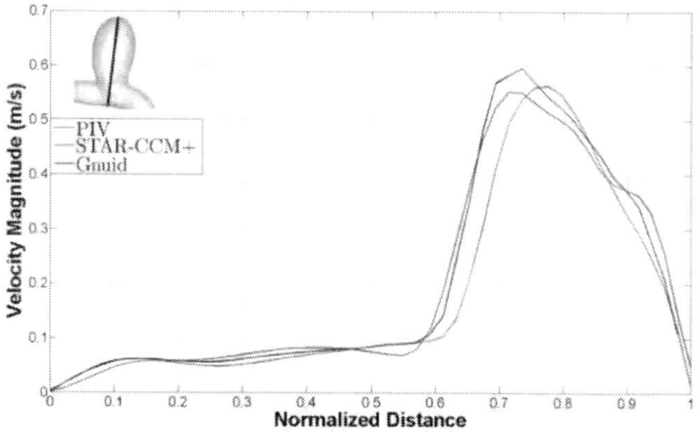

Figure 4: VARIATION IN VELOCITY MAGNITUDE ALONG A STRAIGHT LINE

DISCUSSION & FUTURE WORK

The good agreement between PIV data and STAR-CCM+ simulation result shows that our PIV setup is accurate. When validating gnuid simulation results by comparing it with PIV results, an average error of 19% was obtained. This can be attributed to two reasons; firstly, the boundary of PIV result is irregular and not encompassing the whole fluid domain. This can be eradicated by using a smaller interrogation window size at the boundaries to more accurately capture the boundary. Secondly, the gnuid solver is specifically designed for transient flow simulations, and it is expected to provide better results in case of unsteady pulsatile simulations. Hence, we plan to carry out pulsatile PIV experiments at different slice planes to obtain more accurate flow dynamics inside the aneurysms and compare it with gnuid solver on multiple patient-specific aneurysms.

ACKNOWLEDGEMENTS

We would like to acknowledge Toshiba Medical System Corporation for financial support. We would also like to thank Vincent Tutino and Dr. Ciprian Ionita (University at Buffalo) for their assistance in fabrication of the aneurysm phantom.

REFERENCES

1. Meng, H., et al., *High WSS or Low WSS? Complex Interactions of Hemodynamics with Intracranial Aneurysm Initiation, Growth, and Rupture: Toward a Unifying Hypothesis.* American Journal of Neuroradiology, 2014. **35**(7): p. 1254-1262.
2. Xiang, J., et al., *AView: A clinical tool for hemodynamic and morphological analysis of intracranial aneurysms.* Summer Bioengineering Conference, Proceedings of the ASME 2013.
3. Botti, L. and D.A. Di Pietro, *A pressure-correction scheme for convection-dominated incompressible flows with discontinuous velocity and continuous pressure.* Journal of Computational Physics, 2011. **230**(3): p. 572-585.
4. Hoi, Y., et al., *Validation of CFD simulations of cerebral aneurysms with implication of geometric variations.* Journal of biomechanical engineering, 2006. **128**(6): p. 844-851.

MULTISCALE MODEL OF STRAIN-DEPENDENT GLOMERULAR BASEMENT MEMBRANE REMODELING

Lazarina I. Gyoneva (1), Yoav Segal (2,4), Kevin D. Dorfman (3), Victor H. Barocas (1)

(1) Biomedical Engineering
University of Minnesota
Minneapolis, MN, United States

(2) Medicine
University of Minnesota
Minneapolis, MN, United States

(3) Chemical Engineering and Materials Science
University of Minnesota
Minneapolis, MN, United States

(4) VA Medical Center
Minneapolis, MN, United States

INTRODUCTION

The filtration of blood by the kidneys occurs in the glomerular capillaries, where hydrostatic and osmotic pressures direct the movement of plasma across the semipermeable capillary wall. The permeability properties of the filtration barrier must be balanced with the demands for mechanical stability of the capillaries – the wall must be stiff enough to resist the large hydrostatic pressure, yet porous enough to allow filtration. Mechanical support is provided by the glomerular basement membrane (GBM), which is mainly composed of two types of collagen IV: $[\alpha1(IV)]_2\alpha2(IV)$ and $\alpha3(IV)\alpha4(IV)\alpha5(IV)$ [1]. As part of the normal growth and adaptation of a healthy GBM, the collagen matrix is remodeled by deposition of new collagen on the epithelial (outer) surface coupled with the removal of old collagen by matrix metalloproteinases in a process that maintains the permeability and stiffness of the GBM. In some disease conditions, however, remodeling goes out of control. The genetic disease Alport syndrome is characterized by the absence of the $\alpha3(IV)\alpha4(IV)\alpha5(IV)$ collagens from the GBM. Aberrant remodeling in the Alport GBM results in thickening, local lesions, and deposition of sclerotic material. The $[\alpha1(IV)]_2\alpha2(IV)$ collagen network is thought to be more distensible [2] and less resistant to proteolytic degradation [3] than the $\alpha3(IV)\alpha4(IV)\alpha5(IV)$ network. The exact mechanism of abnormal remodeling of the Alport GBM is unclear but is suspected to be the result of entering a dangerous positive feedback loop. The goal of this study was to investigate how the structural and mechanical differences between the $[\alpha1(IV)]_2\alpha2(IV)$ and $\alpha3(IV)\alpha4(IV)\alpha5(IV)$ collagen IV networks affect remodeling of the GBM. Our focus is on strain-dependent remodeling and how the known differences between the two collagen IV networks can account for the abnormal remodeling response in Alport syndrome.

METHODS

The basement membrane of the glomerular capillary was simulated in a multiscale model (Fig. 1A) that allows us to link the nanoscale structure of the collagen IV network with the response of the capillaries on the microscale [4]. Collagen IV networks were modeled as three-dimensional networks of rods with connectivity of 4. The $[\alpha1(IV)]_2\alpha2(IV)$ and $\alpha3(IV)\alpha4(IV)\alpha5(IV)$ networks were differentiated by reconnecting the $\alpha3(IV)\alpha4(IV)\alpha5(IV)$ networks to combine collagens into structures, which we term supercoils (Fig. 1B), that had double the radius of non-combined collagens and that pre-strained the network [5]. The glomerular capillary was modeled as a ring section of a thin cylindrical tube. A hydrostatic pressure of 35 mmHg was initially applied to the inner surface of the ring and was increased by 7.5 mmHg at each step, for 5 total remodeling steps

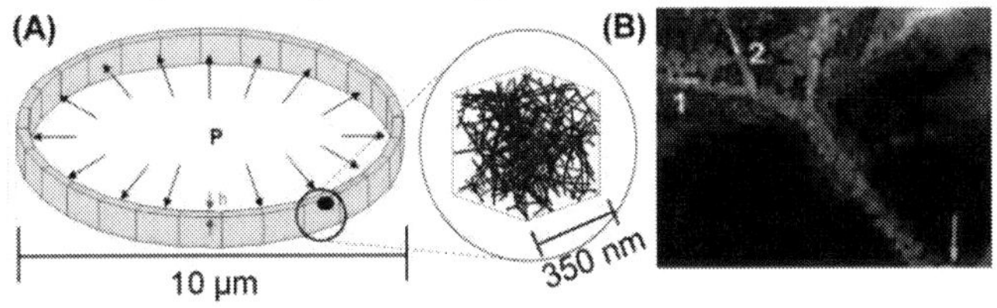

FIGURE 1. (A) SCHEMATIC OF MULTISCALE MODEL **(B)** COLLAGEN IV MOLECULES FORMING A SUPERCOIL [6]

Upon applying the load, networks were remodeled. Collagen removal was based on the work of Bhole et al., 2009 [7], for type I collagen, which suggests that more highly strained collagen fibers are degraded at a slower rate. The probability, p_r, of removing a collagen in our model was set as a function of collagen stretch, λ_c

$$p_r=\exp(-k_r(\lambda_c-1)) \quad (1)$$

where k_r was a constant parameter. Supercoils were never removed.

Fiber deposition was based on fiber orientation. A second, highly connected "phantom" network was generated based on the nodal positions of the first network but not contributing to the network mechanics or permeability. Collagen fibers were selected from the phantom network and turned "on" based on a probability, p_d, as a function of the alignment of the fibers with the circumferential direction, d_c,

$$p_d=\exp(-k_d(1-d_c)) \quad (2)$$

The Darcy permeability of the networks was calculated following our previous work [5].

RESULTS

We tested three different collagen IV networks: 1) Networks with no supercoils, representing the $[\alpha1(IV)]_2\alpha2(IV)$ networks; 2) networks with low levels of supercoiling; and 3) networks with high levels of supercoiling, the latter two representing the $\alpha3(IV)\alpha4(IV)\alpha5(IV)$ networks. For each case, N = 3. As the network was allowed to remodel based on the stretch and orientation of the collagens, the supercoiled networks had a much lower number of collagens removed and similar amounts of added collagens, compared to the non-supercoiled networks (Fig. 2).

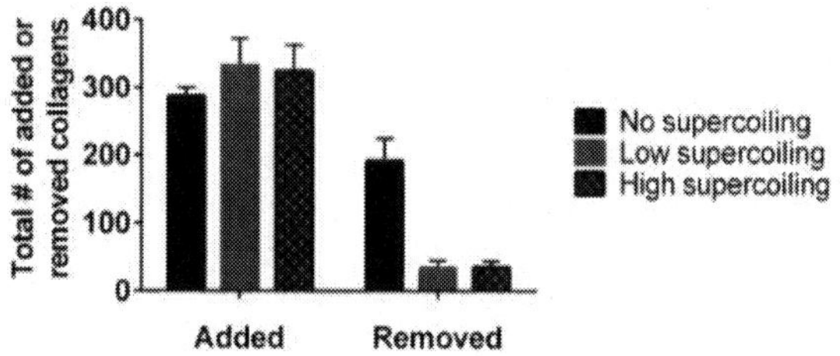

FIGURE 2. TOTAL NUMBER OF ADDED AND REMOVED COLLAGENS FOR DIFFERENT NETWORK CASES

The circumferential stress in the capillary ring was different for the three network cases throughout the remodeling steps (Fig. 3). There was a 24% increase in stress from the non-supercoiled case to the highly supercoiled case.

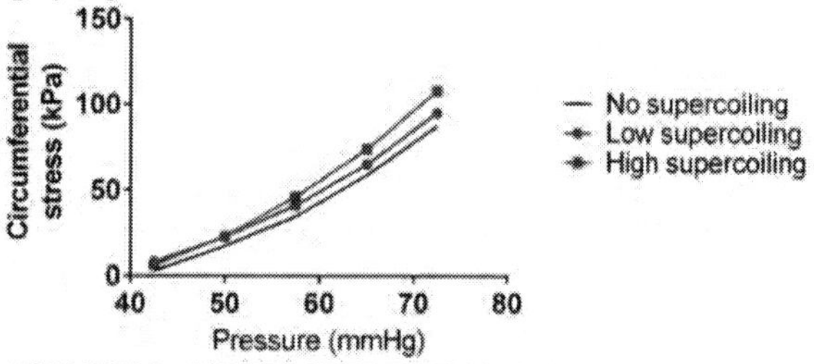

FIGURE 3. CIRCUMFERENTIAL STRESSES IN THE CAPILLARY FOR DIFFERENT NETWORK CASES

The net deposition of collagens and thickening of the networks lead to decrease in Darcy permeability (Fig. 4). The drop in permeability was larger for the supercoiled networks - 60% difference, even for low levels of supercoiling. The filtration velocity increased with remodeling (Fig. 5) due to the compression of the GBM as it stretched. That is, the drop in permeability was smaller than the drop in thickness, leading to a net increase in tissue permeability.

DISCUSSION

The fine balance between permeability and mechanics of the GBM could be easily recognized during the remodeling steps. Even small structural changes in the nanoscale collagen IV networks could cause large differences in the properties of the glomerular capillaries. From the differences in removed number of collagens, it is clear in our model that supercoiling had a protective effect on the networks. In all

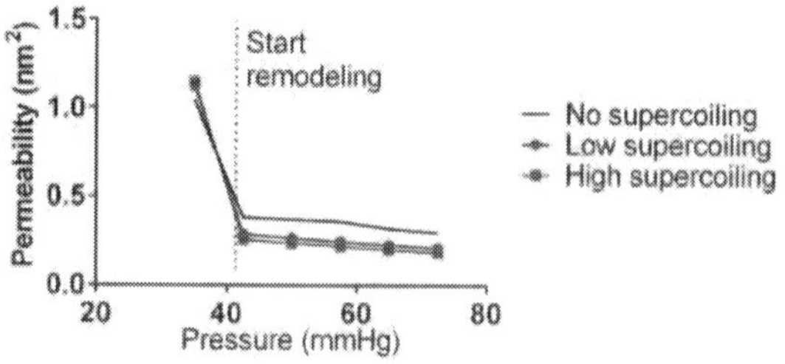

FIGURE 4. CHANGES IN DARCY PEREMABILTY DUE TO REMODELING IN DIFFERENT NETWORK CASES

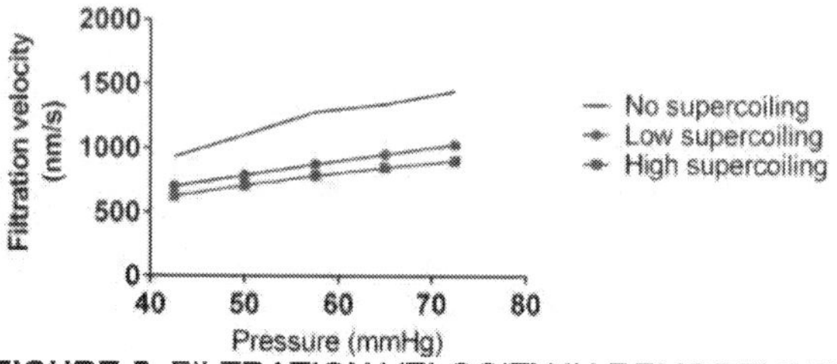

FIGURE 5. FILTRATION VELOCITY IN REMODELING COLLAGEN NETWORKS

network cases, remodeling resulted in a net increase in number of collagen molecules, as a way for the membrane to deal with the increase in pressure and stress. As the GBM thickens from overproduction of collagens to maintain mechanical stability, the conductivity of the membrane is impaired, causing increase in hydrostatic pressure. The increase in pressure, in turn, increases the stress on the capillaries, leading to a new cycle of matrix deposition and removal for mechanical reinforcement.

It should be noted that the strain-dependent remodeling presented here is pressure-controlled when in reality, the flow rate through the glomerular capillaries is the controlling factor. Future changes to the model would allow the remodeling steps to reach equilibrium so the flow rate is maintained. Also, the current representation of the multiscale model only included the contribution of the collagen IV network and did not included the epithelial cell (podocyte) layer, which plays a very important role in permeability and perhaps mechanics of the capillary wall. The cellular contribution could be addressed by adding the epithelial hydraulic conductivity to that of the GBM as explained in [8] or by adding the cellular layer to the multiscale capillary model geometry.

The values obtained for the Darcy permeability and circumferential stresses by our model are consistent with the literature: permeability is usually reported as being 1-3 nm^2 [8] and the hoop stress has been calculated to be ~50 kPa [9].

ACKNOWLEDGEMENTS

This work was carried out in part using computing resources at the University of Minnesota Supercomputing Institute and was supported by the National Science Foundation Graduate Research Fellowship and Grant No. NSF CMMI 1300649.

REFERENCES

[1] Hudson, BG et al., *J Biol Chem*, 268:26033-26036, 1993. [2] Gyoneva, L et al., *Exp Eye Res*, 113: 87-91, 2013. [3] Gunwar, S et al., *J Biol Chem*, 273: 8767-8775, 1998. [4] Chandran, PL et al., *J Biomech Eng*, 129: 137-147, 2007. [5] Gyoneva, L et al., *Ann Biomed Eng*, Nov, 2014. [6] Yurchenco, PD et al., *J Cell Biol*, 105: 2559-2568, 1987. [7] Bhole AP et al., *Phil Trans R Soc*, 367: 3339-3362, 2009. [8] Deen, W et al., *Am J Physiol Renal Physiol*, 281: F579-F596, 2001. [9] Endlich, N et al., *Semin Nephrol*, 32: 327-341, 2012.

INVESTIGATING THE ROLE OF OSTEOACTIVIN IN MUSCLE REGENERATION

Jinjin Ma (1), Bing Yu (2), Andrew Baker (1), Min-Ho Kim (2), Anthony Calabro(1), Fayez Safadi (3), Christopher Malcuit (2), Kathleen Derwin (1)

(1) Biomedical Engineering
Cleveland Clinic
Cleveland, Ohio, USA

(2) Biological Sciences
Kent State University
Kent, Ohio, USA

(3) Anatomy and Neurobiology
Northeast Ohio Medical University
Rootstown, Ohio, USA

INTRODUCTION

Wounds involving extensive damage and loss of muscle, also known as volumetric muscle loss (VML), are frequently associated with high-energy civilian trauma and combat-related extremity injuries. These injuries commonly have a high projected healthcare cost and affect the quality of many patients' life. Currently surgical reconstruction (i.e., free muscle transfer) is performed with the intent to restore partial function. However, donor site morbidity and chronic pain are frequently associated with the reconstruction. Moreover, despite the high healthcare cost, many patients eventually face or choose amputation due to the functional deficits from the injury site. Clearly, there exists a need for improved strategies to foster new muscle formation to restore the muscle function in patients with VML.

Our long-term goal is to develop effective regenerative medicine strategies for restoring musculoskeletal function in patients with injuries that involve VML. Specifically, here we will investigate a role for osteoactivin (OA) in muscle regeneration. OA (also known as GPNMB) is a transmembrane glycoprotein that is expressed in numerous cells and tissues. We have recently reported on the powerful anabolic activity of OA peptide in bone tissue [1]. Overexpression of OA in transgenic mice also showed increased muscle mass in the denervated skeletal muscle model [2]. Taken together, these data suggest that OA could play an important role in muscle regeneration, but this has not been previously investigated. Hence the objective of this study is to investigate OA in the context of healing following two types of skeletal muscle injury. First, we will measure OA expression and its relationship with MyoD1, a myogenic transcription factor, following cardiotoxin induced muscle injury. Secondly, we will investigate evidence for muscle regeneration in volumetric muscle defect with and without synthetic OA peptide delivered via a cross-linked gelatin hydrogel [3].

METHODS

Mouse Cardiotoxin Muscle Injury Model: We employed cardiotoxin injury (10μM, 100μl) in the tibialis anterior (TA) muscle in 16 week old wild-type (DBA) mice and in mice with a truncated (non-secreted) form of OA (DBA/2J) [4]. Animals were sacrificed at 3, 7 and 14 days post-injection (n=2 per group per time point), at which time the entire TA muscle was recovered. Age matched non-injured animals (n=2 per group) served as controls. RNA was extracted from a portion of the TA muscle using Trizol reagent (Life Technologies), reverse transcribed into cDNA using Quanta's qScript cDNA Supermix, and quantitative PCR was carried out using Quanta's PerfecTa Sybr Green Supermix and custom oligonucleotide primers for OA and MyoD1 on an Eppendorf RealPlex2 q-PCR system. Data is presented as fold-change from control, using GAPDH as a reference gene and the ΔΔCT method for analysis. The remainder of the TA muscle was processed routinely for histology, cryosectioned, and stained with hematoxylin and eosin (H&E).

Rat Volumetric Muscle Defect Model: Twelve male Sprague Dawley rats (250-300 g) will be used for this study. For all rats, a near full-thickness 8x4x4mm muscle defect will be surgically created bilaterally in the middle third of the TA muscle [5], corresponding to ~20% of the total muscle volume. A cross-linked gelatin plug [3] (8x4x4mm) will be delivered into one defect. A gelatin plug loaded with 0.2 or 2 μg/mg synthetic OA peptide (Anaspec), corresponding to the 18 C-terminal end amino acids of the rat OA protein [6], will be delivered into the contralateral defect. Rats will be sacrificed at 14 days and 28 days (n=3 at low and high dose OA at each time point). The whole TA muscle will be explanted and evaluated histologically for tissue morphology and neo-muscle formation in the defect using biomarkers (Developmental Studies Hybridoma Bank) specific for embryonic myosin (F1.652) and satellite cells (PAX7).

RESULTS

Mouse Cardiotoxin Muscle Injury Model: At 3 days post-injury, OA mRNA levels in the injured TA muscle of DBA mice increased by ~1700-fold (Figure 1A), together with a concomitant ~20-fold increase in the levels of MyoD1 (Figure 1B). As expected, OA levels were undetectable in the OA mutant mice (DBA/2J, Figure 1A) at 3 days, concomitant with only a ~6-fold increase in MyoD1 (Figure 1B). At days 7 and 14 days post-injury, OA levels sharply decrease in DBA mice, as do MyoD1 levels in both groups. MyoD1 levels are not different between groups at the 7 and 14-day time points. These gene expression patterns appear to correlate with a delay in muscle regeneration in OA mutant mice at 3 days post-injury (Figure 2).

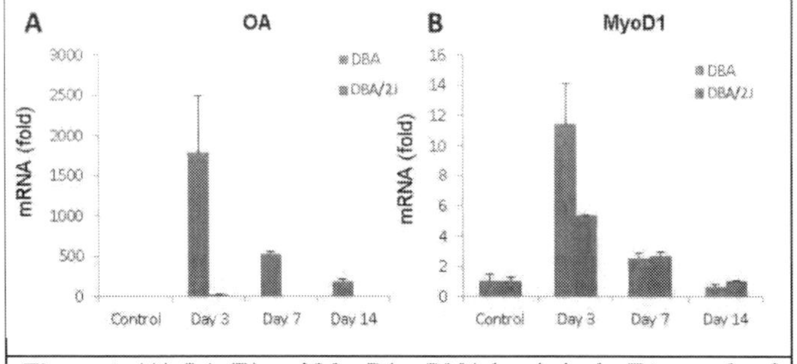

Figure 1. (A) OA (B) and MyoD1 mRNA levels in the TA muscle of wild type (DBA) and OA mutant (DBA/2J) mice at 3, 7 and 14 days after cardiotoxin induced injury. Age matched non-injured murine TA muscles served as controls (n=2 per group per time point).

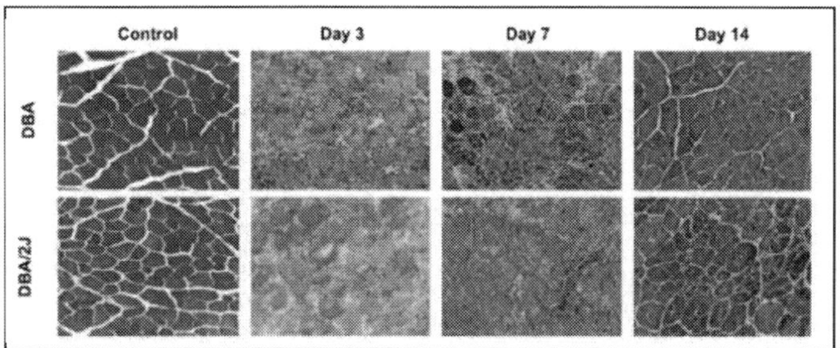

Figure 2. Representative histologic panels from the TA muscle of wild type (DBA) and OA mutant (DBA/2J) mice at 3, 7 and 14 days after cardiotoxin induced injury. Age matched non-injured murine TA muscles served as controls (H&E, 200X).

Rat Volumetric Muscle Defect Model: The rat volumetric muscle defect study is ongoing. To date we have performed a pilot study in one rat that underwent bilateral TA muscle defects filled with gelatin plugs (only) for 14 days. At 14 days the boundary of the muscle defects could be well-identified and the dimensions of the gel were approximately 8mm long by 3mm deep (Figure 3A). A dense cellular infiltrate consisting of lymphocytes, macrophages and fibroblasts was observed in the deep one-third of the gel (Figure 3B), whereas, superficial and mid-portions of the gel exhibited low to moderate cellular infiltration with regions of acellularity (Figure 3C).

DISCUSSION

Decreased OA and MyoD1 expression at 3 days in cardiotoxin injured and regenerating muscle, together with histomorphologic evidence of delayed muscle regeneration, support a possible anabolic role for OA in the acute phases of muscle repair. In a volumetric muscle defect, OA may not be able to accumulate at concentrations sufficient for functional muscle regeneration, providing the rationale for delivering exogenous OA to a muscle defect via a gelatin scaffold. Our pilot study showed that gelatin incorporated into the defect and allowed cellular infiltration. In our ongoing volumetric muscle defect study, we expect to see more newly formed muscle and less collagen deposition (fibrosis) in the defect with OA delivery. If insufficient muscle regeneration is observed in the muscle defect model with delivery of OA alone, we will consider delivering mesenchymal stem cells (MSCs) together with OA in the gelatin plug. Our previous work has demonstrated MSCs encapsulated within cross-linked gelatin and delivered in rat gastrocnemius muscle defect were associated with muscle regeneration after 8-weeks (Figure 4) [7]. Parallel work in our lab shows that recombinant OA protein promotes MSC survival, proliferation and migration (unpublished data). These results support the concept of combined OA and MSC delivery should OA alone be inadequate. Together, these efforts support our long term goal of developing effective regenerative medicine strategies for restoring musculoskeletal function in patients with injuries that involve VML.

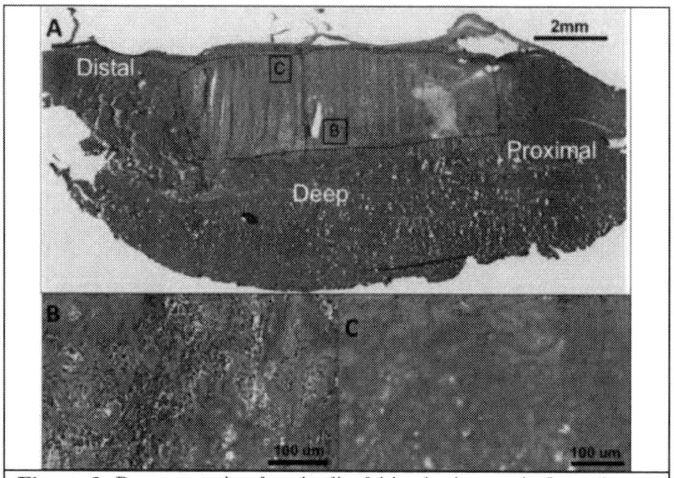

Figure 3. Representative longitudinal histologic panels from the rat TA muscle 14 days after volumetric muscle injury and filling with a gelatin plug. H&E: (A) 10X and (B, C) 200X.

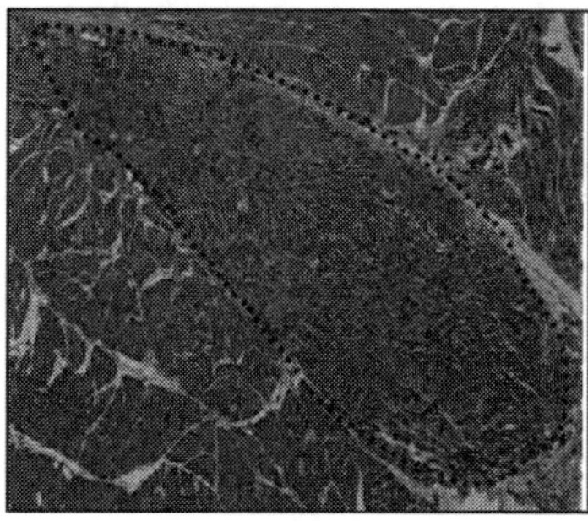

Figure 4. Representative cross section of a rat gastrocnemius muscle defect 8 weeks after volumetric muscle injury and filling with a gelatin drop encapsulated with MSCs [7]. (H&E, defect area denoted by dashed line).

ACKNOWLEDGEMENTS

This study is funded by Cleveland Clinic LRI Chair's Innovative Research Fund to JM, start-up funds from KSU to CM, and a seed award from KSU and NEOMED to MHK, FS, and CM.

REFERENCES

[1] Abdelmagid, SM et al., *Exp cell research*, 314(13):2334-51 (2008).
[2] Furochi, H et al., *J Med Invest*, 54(3-4):248-54 (2007)
[3] Bruggeman, LA et al., *Cell and Mol bioeng*, **5** (2), 194-204 (2012).
[4] Liu, N et al., *J Med Invest*, 122:2054-65 (2012).
[5] Corona, BT et al., *Biomaterials*, 34:3324-3335 (2013)
[6] Selim, AA et al., *Med Sci Monit*, 13(12):BR259-70 (2007).
[7] Guo X et al., *JW Freston Conference: Gastrointestinal Stem Cell Biology and Pathobiology*, Chicago, Illinois (2012).

In vitro optimization of the nozzle used in Assisted Bidirectional Glenn procedure for single ventricle stage 1 palliation

Jian Zhou(1) , Mahdi Esmaily-Moghadam(2), Timothy A. Conover(1) , Tain-Yen Hsia(3), Alison L. Marsden(4), Richard S. Figliola(1)

(1) Department of Mechanical Engineering
Clemson University
Clemson, SC, USA

(2) Center for Turbulence Research
Stanford University
Stanford, CA, USA

(3) Great Ormond Street Hospital
London, UK

(4) Mechanical and Aerospace Engineering Department
University of California, San Diego
La Jolla, CA, USA

INTRODUCTION

Outcomes following the Norwood and Bidirectional Glenn procedures remain unsatisfactory, with high morbidity due to unbalanced pulmonary flow. The assisted bidirectional Glenn (ABG) concept, proposed by Esmaily-Moghadam [1], is an alternative option for stage 1 surgical palliation of single ventricle defects. In the ABG, the innominate artery and the SVC are connected by a shunt with a nozzle, creating an ejector pump phenomenon that augments pulmonary flow.

The ABG has several potential advantages. First, flow from the shunt increases the pulmonary flow and systemic oxygen delivery, with potential to decrease cardiac workload and reduce cyanosis. Second, if necessary, the ABG circulation can be converted to the normal BDG circulation by blocking the shunt via a catheter device, avoiding the second stage open chest surgery. Third, the shunt flow provides pulsatility to the pulmonary blood flow, which is potentially beneficial to pulmonary vascular development [2]. The main challenge of ABG is to relieve high SVC pressure, which can cause cerebral hemorrhage and peripheral edema.

We propose that with efficient shunt design, the ABG has the potential to produce acceptable SVC pressures and avoid symptoms of SVC syndrome. Here, we study the practical realization of the ABG concept experimentally. The outcomes are compared with outcomes of the Norwood and bi-directional Glenn circulations. We use a multi-scale *in vitro* model of the ABG with a range of nozzle-to-shunt hydraulic diameter ratios β, to compare the hemodynamic performance in a typical Norwood candidate.

METHODS

A multi-scale mock circulatory system (MCS) coupling a lumped parameter network (LP) of the neonatal circulation with a three dimensional (3D) model of the anastomosis site, shown in Figure 1, was built to test the hemodynamic performance of the ABG circulation using different nozzle β ratios. The circulation is comprised of upper body, lower body, and pulmonary circulation branches. The MCS function was validated previously [3,4]. The generic 3D test section was derived from clinical imaging of 28 Norwood patients [1], and accompanying clinical pressure and flow measurements were used to tune the system values. The parametric nozzle design is shown in Figure 2. The concentric converging nozzle was situated at the SVC-shunt junction (L = 0), and the nozzle throat diameter d was varied to produce a range of nozzle ratio β = d/D from 0.24 to 0.84. Shunt angle θ was held constant at 40 degrees. The exact ligaclip shape nozzle (inset Fig. 1) proposed by Esmaily-Moghadam was also tested [1]. Parameters evaluated are: pressure and flow rate at left/right pulmonary artery (LPA/RPA) and at superior vena cava (SVC). Finally, systemic oxygen delivery (OD)

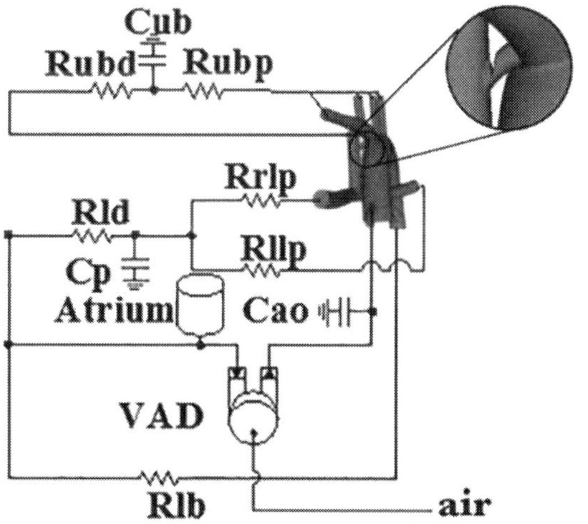

Figure 1: Schematic of Mock Circulatory System lumped parameter network coupled with the three dimensional anastomosis model for the ABG circulation; Cao, aorta compliance; Cp, lung compliance; Cub, upper body compliance; Rlb, lower body resistance; Rld, lung distal resistance; Rllp, left lung proximal resistance; Rrlp, right lung proximal resistance; Rubd, upper body distal resistance; Rubp, upper body proximal resistance; VAD, ventricular assist device. Inset: Ligaclip shunt design.

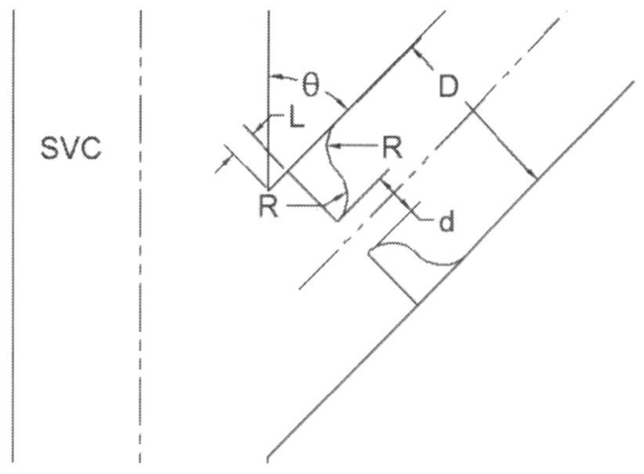

Figure 2: Dimensions of the concentric Assisted Bidirectional Glenn nozzle design used for parametric study.

is calculated from $OD = c_p Q_s - r(Q_s / Q_p)V_{O_2}$ [5] with c_p and V_{O_2} set to 0.22 mL_{O2} / mL and 0.874 mL_{O2} / mL and $r = 0.5$ based on typical patient values for a body size area of 0.33 m^2. The pulmonary and systemic flow rates are Q_p and Q_s, respectively.

RESULTS

Figure 3 shows time-resolved flow and pressure signals. Mean value and pulsatility of LPA and RPA pressure and flow all increase as β increases, as more pulsatile arterial flow is communicated via the shunt. The value of Q_p/Q_s is very sensitive to β. Q_p/Q_s increases from 0.44 to 0.72 as β increases from 0.24 to 0.84, because shunt flow increases from 0.04 LPM to 0.36 while SVC flow rate Q_{svc} stays relatively constant. Q_{svc} drops from 0.48 to 0.40 LPM as the competing shunt flow increases SVC pressure. The increase of SVC pressure is about 0.4 mmHg less than that of RPA and LPA over the range, indicating a small positive ejector pump effect of the ABG connection. Systemic oxygen delivery results are shown in Figure 4. Oxygen delivery increases from 3.11ml/s to 3.31ml/s for concentric circular nozzles as β changes from 0.24 to 0.84. However, the ligaclip design with β=0.57 provides the maximum systemic oxygen delivery of 3.34 ml/s. Lower body flow rate stayed constant at 0.66 LPM because the mean ascending aorta pressure was well-controlled at 56 mmHg.

DISCUSSION

An efficient shunt design is important for the ABG's performance. As expected, mean value and pulsatility of pulmonary flow and pressure is enhanced as β increases due to the additional pulsatile arterial flow. Though current results show a small positive pump ejector effect, the elevated SVC pressure still remains a challenge, especially for patients with high PVR. Additional nozzle optimization, including nozzle shape, angle and exact position, is needed. Systemic oxygen delivery reaches its maximum at β equals 0.57. In consideration of systemic oxygen delivery and pulmonary pressure and flow, a suggested β value lies between 0.48 and 0.60.

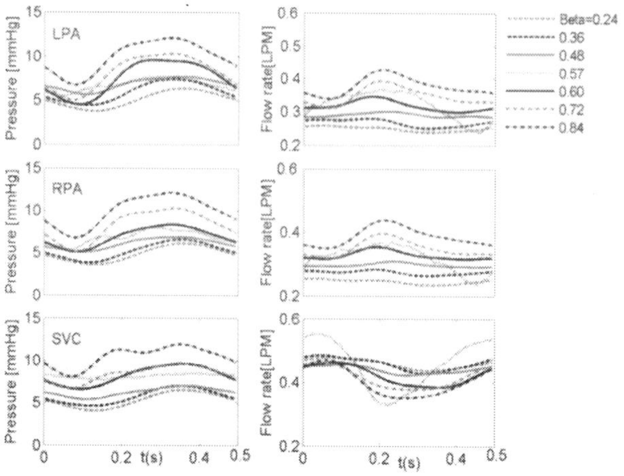

Figure 3: Time-resolved flow and pressure signals for ABG circulation with varying β ratios.

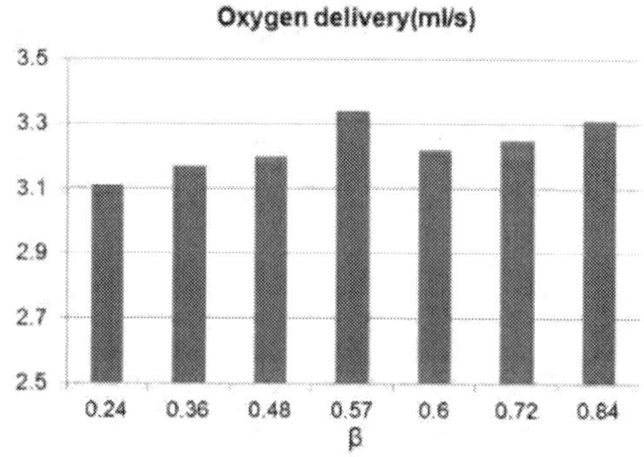

Figure 4: Comparison of experimental estimations of systemic oxygen delivery ABG circulation with varying β ratios. β=0.57 is the design by Esmaily-Moghadam[1].

ACKNOWLEDGEMENTS

This work was supported by the Leducq Foundation through a Transatlantic Network of Excellence grant. Additional support to ALM was provided from a Burroughs Wellcome Fund Career Award at the Scientific Interface, and NSF CAREER award. The authors acknowledge the collaboration of the MOCHA investigators.

REFERENCES

[1] Moghadam, ME et al, J Thorac Cardiovasc Surg. Accepted 2014.
[2] McElhinney, DB et al, Annals Thorac Surg. 66(2):668-672, 1998.
[3] Vukicevic, M et al ASAIO Journal, 59(3):253-260, 2013.
[4] Biglino, G et al, ASAIO J, 58(1): 25-31, 2012.
[5] Moghadam ME et al,. J Biomech Engr.134(5), 2012.

EFFECTS OF ESTROGEN ON PULMONARY WAVE REFLECTION AND ENERGY TRANSMISSION IN PULMONARY ARTERIAL HYPERTENSION

Aiping Liu, Naomi Chesler

Department of Biomedical Engineering, University of Wisconsin-Madison, Madison, WI

INTRODUCTION

Pulmonary artery hypertension (PAH) is a female dominant, fatal disease characterized by progressive pulmonary vascular remodeling, which increases right ventricular (RV) afterload and ultimately leads to RV failure. Despite the high incidence (the female-to-male ratio is 4:1 [1]), female PAH patients have better cardiac function and better survival rates than their male counterparts [2, 3]. Intuitively, these clinical findings suggest that the female sex hormone estrogen is an important contributor to PAH development and progression. However, the roles of estrogen in PAH remain controversial, which is referred to as "the estrogen paradox in PAH" [4].

Our recent study showed that exogenous estrogen attenuates PAH-induced arterial stiffening without affecting resistance [5]. However, how those estrogen-induced vascular changes affect pulmonary arterial (PA) wave reflection and ventricular-vascular energy transmission efficiency remain unclear. Here we hypothesized that estrogen attenuates pulmonary artery wave reflection and enhances ventricular-vascular energy transmission efficiency by preserving large artery compliance. We tested this hypothesis with *in vivo* measurements and analysis of PA pressure and flow in a mouse model of mild or early PAH.

METHODS

Animal Handling

Ovariectomized (OVX) female C57BL/6 mice, 13-16 weeks old, were purchased from a commercial vendor (Jackson Laboratory) to eliminate natural fluctuations in estrogen levels. After 1-2 weeks to ensure depletion of endogenous estrogen stores, OVX mice were implanted subcutaneously either with estradiol-17β pellets (0.1 mg per 21-day-release) to achieve a plasma estrogen level at or slightly higher than the physiological levels or with placebo pellets as control for estrogen deficiency. Immediately after pellet implantation, some mice were injected intraperitoneally with SU5416 at 20 mg/kg weekly and exposed to normobaric hypoxia at 10% O_2 (SuHx) for 21 days, which has been shown previously to create mild PAH in mice [5]. The remaining mice were kept in normoxia, were not given SU5416 injections, and were used as controls.

In Vivo Pulmonary Hemodynamics

PA pressure and flow were measured simultaneously in live mice following the procedures described by Tabima [6]. Briefly, after a mouse was anesthetized, intubated and ventilated, the chest was opened to expose the RV. A 1.2F pressure catheter (Millar Instruments) was then inserted into the RV and advanced to the main PA just distal to the pulmonary valve. The blood flow at the same location was acquired simultaneously using echocardiography (Visualsonics) with a 30 MHz probe. PA pressure and flow were recorded using a customized device (Cardiovascular Engineering).

The simultaneously acquired pressure and flow waveforms were analyzed in the time domain [7,8,9]. Total pulmonary vascular resistance Z_0, an index of distal vasculature remodeling, was calculated as the mean PA pressure divided by the cardiac output. Characteristic impedance Z_C, an index of proximal PA compliance, was calculated from the slopes of PA pressure and flow in early systole. The pulse wave reflection (Γ) is defined as $(Z_0-Z_C)/(Z_0+Z_C)$. The pulse wave velocity was calculated using the P-U method [8]. In addition, the measured pressure and flow waves are separated into forward and backward waves using the waterhammer formula [8]. The global wave reflection index R is defined as the ratio of the magnitude of backward wave to that of the forward wave (= P_b/P_f). The magnitude of wave reflection is quantified using the difference between systolic pressures of P_m and P_f (ΔSP) To assess the timing of the reflected wave, the normalized time between the beginnings of systole to the peak of reflected wave was used (Δt/T). To evaluate the ventricular-vascular energy transmission efficiency, two metrics were used: (1) the transpulmonary vascular efficiency, defined as the ratio between the total power and cardiac output (W_t/CO); and (2) the energy transmission ratio, defined as the ratio between the hydraulic power in the measured wave and hydraulic power in the forward wave (=W/W_f). Here, the hydraulic power is calculated as the averaged integral of the product of the instantaneous pressure and flow, ignoring the small kinetics terms.

Statistical Analysis

All data are presented as mean ± standard error. The significance of the changes in PA hemodynamics with estrogen treatment and SuHx exposure was assessed with the student t-test at a significance level of 0.05.

RESULTS

Estrogen Restores Proximal PA Compliance in Mild PAH

SuHx exposure increased total pulmonary vascular resistance, Z_0 (Table 1), as expected. No significant difference was found in Z_0 between the estrogen- and placebo-treated groups, indicating that estrogen has mild, if any, effects on distal PA remodeling in mild PAH. Z_C was significantly lower in the estrogen-treated group than that in the placebo-treated group. Since mean PA pressure was similar between SuHx groups, the difference in Z_C is unlikely to be due to the pressure dependence of vascular compliance, suggesting that estrogen attenuates proximal PA stiffening in mild PAH. Interestingly, the preservation of proximal PA compliance enhances the global reflection index, a measure of impedance mismatch between the proximal and distal vasculature in the estrogen-treated SuHx group.

Table 1. Parameters for pulmonary vascular remodeling

	mPAP (mmHg)	Z_0 (mmHg/μL)	Z_C (mmHg/μL)	Γ
CTL-P	16.7±0.8	1.6±0.1	0.23±0.03	0.75±0.02
CTL-E	16.7±1.2	1.5±0.1	0.24±0.02	0.72±0.03
SuHx-P	27.2±0.8*	2.9±0.3*	0.43±0.05*	0.74±0.01
SuHx-E	26.1±1.5*	2.8±0.5*	0.24±0.03#	0.84±0.02*#

* P<0.05 vs. Control; # P<0.05 vs Placebo. CTL, control; P, placebo; E, estrogen. mPAP, mean pulmonary arterial pressure.

Estrogen Attenuates Wave Reflection

SuHx significantly increased the pulse wave velocity (Table 2), a direct index of arterial stiffness, in the proximal PA. Estrogen restored the wave velocity in the PAH group, confirming that estrogen atten0.u21ates PAH-induced arterial stiffening. The reflected wave significantly increased systolic pressure in the placebo-treated SuHx group, while estrogen attenuated the effects of wave reflection on the pressure. PAH tended to increase the global reflection index (P_b/P_f) and estrogen treatment tended to attenuate the increase. The timing of wave reflection was not altered by PAH but estrogen tended to delay the arrival of reflected waves. Early wave arrival is known to adversely affect the performance of RV. These results suggest that estrogen attenuates the effects of wave reflection on RV pressure, likely through preservation of proximal PA compliance. Interestingly, this conclusion does not reconcile with the higher global reflection index in the estrogen-treated SuHx group (Table 1), suggesting that the global reflection index alone is not sufficient for assessing wave reflection. Here, the increase in the global reflection index was a result of preserved Z_C, which in fact decreased wave reflection.

Table 2. Parameters for pulmonary wave reflection

	ρc	ΔSP (mmHg)	R	Δt/T
CTL-P	1.8±0.2	1.2±0.2	0.28±0.02	0.13±0.02
CTL-E	2.0±0.2	1.1±0.1	0.27±0.02	0.15±0.02
SuHx-P	3.0±0.3*	2.5±0.4*	0.33±0.02	0.15±0.03
SuHx-E	1.7±0.3#	1.1±0.3#	0.30±0.03	0.18±0.01

Estrogen Protects Transpulmonary Vascular Efficiency

SuHx decreased the transpulmonary vascular efficiency, a measure of power necessary per unit forward flow, while estrogen attenuated the reduction in the efficiency (Fig.1A). The pulmonary transmission efficiency, a measure of the effect of wave reflection on the hydraulic power output, significantly reduced in both SuHx groups and estrogen had no effect (Fig.1B). As noted above, estrogen affects proximal PA compliance; this result suggests that estrogen improves ventricular-vascular coupling efficiency through protecting PA compliance. The insensitivity of the pulmonary transmission efficiency to the effect of estrogen suggests that this index does not reflect changes in PA compliance. .

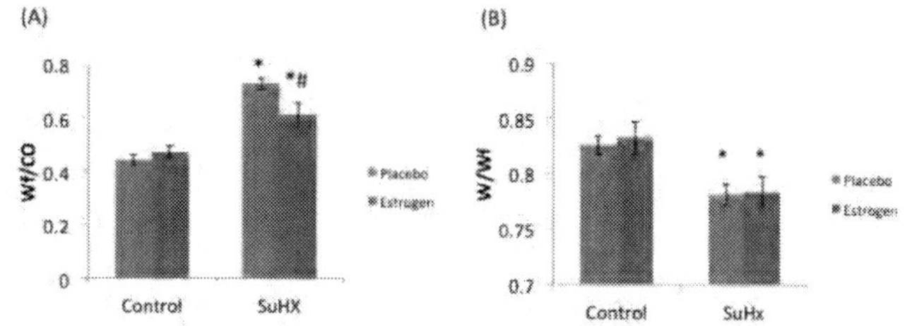

Figure 1. Effects of estrogen on ventricular-vascular energy transmission efficiency in mild PAH. (A) The transpulmonary vascular efficiency, and (B) the pulmonary transmission efficiency.

DISCUSSION

The novel finding in our study is that estrogen limits PA functional changes through preserving PA compliance in mild PAH. This protective effect of estrogen on PA wave reflection and energy transmission may have an impact on RV adaptive remodeling in PAH. To shed additional light on the estrogen paradox, future studies on the roles of estrogen in both PA and RV remodeling in severe PAH are warranted.

ACKNOWLEDGEMENTS

This study was supported by NIH R01HL086939 (NCC) and AHA postdoctoral fellowship 13POST16910091 (AL). The authors would like to thank Drs. Timothy Hacker and Guoqing Song for *in vivo* hemodynamic measurements.

REFERENCES

[1] Badesch, DB et al., *Chest*,137:376-87, 2010.

[2] Ventetuolo, CE et al., *Am J Respir Crit Care Med*, 183:659-67, 2011.

[3] Humbert, M et al., *Circulation*,122:156-63, 2010.

[4] Umar, S et al., *Am J Respir Crit Care Med*,186:125-131, 2012.

[5] Liu, A et al., *Am J Physiol Heart Circ Physiol*,307:273-283, 2014.

[6] Tabima, DM et al. *J Biomech*,45:799-804, 2012.

[7] Mitchell, GF et al. *Am J Physiol Heart Circ Physiol*, 267:1907-1915, 1994.

[8] Parker, KH. *Med Biol Eng Comput*, 47:175-188, 2009.

[9] Grignola, JC et al. *Int J Cardiol*,115:171-182, 2007.

SB³C2015
Summer Biomechanics, Bioengineering and Biotransport Conference
June 17-20, 2015,Snowbird Resort, Utah, USA
SB³C2015-628

DETECTION, QUANTIFICATION AND LOCALIZATION OF SUBFAILURE DAMAGE IN TENDON FASCICLES USING COLLAGEN MIMETIC PEPTIDES

Jared L. Zitnay (1,2), Yang Li (1), Shawn P. Reese (1,2), Boi-Hoa San (1), S. Michael Yu (1), Jeffrey A. Weiss (1,2,3)

(1) Department of Bioengineering
University of Utah
Salt Lake City, UT, USA

(2) Scientific Computing and Imaging Institute
University of Utah
Salt Lake City, UT, USA

(3) Department of Orthopedics
University of Utah
Salt Lake City, UT, USA

INTRODUCTION

Subfailure injury of dense collagenous connective tissues such as ligaments and tendons is common and results in clinically significant symptoms of pain and disability. Detection, localization and quantification of subfailure damage is extremely difficult in both clinical and laboratory settings. Medical imaging modalities such as contrast-enhanced CT and MRI can detect gross tissue failure, but these techniques do not accurately detect tissue damage occurring prior to failure [1]. In the laboratory setting, collagen structure and damage have been visualized using electron microscopy [2], X-ray diffraction [3], and atomic force microscopy [4]. However, the sample preparation requirements or inability to view collagen in its native state makes these techniques unsuitable for clinical use. Second harmonic generation microscopy has also been used to detect changes in collagen structure [5], but has not been used to localize damaged collagens.

Collagen mimetic peptides (CMP) may provide an exciting means to localize and quantify damage to the collagen triple helix at multiple physical scales. CMPs are a family of small synthetic peptides that mimic natural collagens, sharing the triple helix structural motif and a Gly-X-Y repeating sequence [6]. CMPs have the unique ability to form triple helices with denatured collagen [7, 8], providing the potential for detecting mechanically damaged collagen in tissues. This binding can be detected and quantified by the use of carboxyfluorescein (CF) tagged CMP (CF-CMP). The objective of this research was to assess the ability to detect, quantify, and localize subfailure damage in tendons using CF-CMP. We hypothesized that incrementally increasing strain during uniaxial tensile testing past the initial onset of subfailure damage would result in an increase in CF-CMP binding as evidenced by fluorescence intensity.

METHODS

RTT fascicles were dissected from Sprague-Dawley rat tails by gripping the distal end of the tail with a hemostat and holding proximally to the hemostat by hand. The hemostat was twisted around the axis of the tail until the surface skin separated and a group of fascicles could be pulled from the tail. Individual fascicles were cut from the removed tail section and intact, medium diameter fascicles (250-550 μm) were selected for tensile testing.

Fascicles were tested on a custom material test system controlled by a LabVIEW virtual interface. Separate fascicles were strained in tension to 1.0, 2.5, 5.0, 7.5, 9.0, 10.5, 12.0, 13.5, and 15.0 percent clamp strain, with 3 fascicles tested per strain group. Zero-strain length was measured as the clamp-to-clamp distance after application of a 0.03 N preload to ensure a consistent starting point for each test. Tested samples were stored in 1× phosphate buffered saline (PBS) until CMP staining. Any fascicles that ruptured prior to reaching the designated strain level were replaced with new samples.

The mechanically damaged RTT fascicles were stained using CF-CMP of sequence CF-G_3-$(GPO)_9$ [8]. Because the CMPs self-assemble into peptide homotrimers, which have little driving force to hybridize with collagen strands, the trimeric CMPs were thermally dissociated to the monomeric state before adding to the fascicle samples. Fascicles were placed in 2 ml centrifuge microtubes with 475 μl of 1x PBS. CF-CMP (150 μM) in PBS was heated to 80°C for 10 min and then quenched to room temperature by immersion in 4°C water for 15 sec. To each sample vial, 25 μl of CF-CMP solution was quickly added, resulting in a final CF-CMP concentration of 7.5 μM, and incubated overnight at 4°C. CF-CMP solution was removed and each sample was washed twice in 2 ml of 1× PBS for 30 min at room temperature. Fascicles were stored in 1× PBS at 4°C until imaging.

Each sample was analyzed for fluorescence using a Nikon

automated widefield microscope on the FITC detection channel. Fluorescence intensity of each fascicle image was analyzed using a custom Matlab script. For each sample, the section of the fascicle between the clamps was manually selected, stained regions of the sample were determined by thresholding using Otsu's method [9], and the mean image intensity within the thresholded regions was calculated. A one-way ANOVA was used to determine the effect of clamp strain on fluorescence intensity, and the Holm-Sidak method was used for post-hoc comparisons.

Transmission electron microscopy (TEM) was utilized to image the binding of CMP to damaged collagen fibers. TEM tissue sections of damaged and undamaged collagen were prepared as described previously [10]. A tissue section of 70-80 nm thickness was cut and mounted onto a carbon coated TEM grip. Purified $C(GPO)_9$ was mixed with 10-nm citrate-stabilized gold nanoparticle solution to prepare CMP conjugated gold particles (Au-CMP). A solution containing 10 nM of Au-CMP and 2% bovine serum albumin was applied to the tissue section and incubated for 15 min at room temperature. The section was washed 6 times with room temperature deionized water. In vitro binding of Au-CMP to damaged tendon collagen was confirmed by TEM observation.

RESULTS

There was a significant effect of strain level on fluorescence intensity ($p < 0.001$). Fluorescence intensity of the tested samples generally increased with increasing strain and was visible to the naked eye (Fig. 1). Intensities from 9.0 to 13.5% strain were significantly greater than intensities from 0 to 7.5% strain. Intensities from 10.5 to 13.5% strain were also significantly increased compared to 15% strain, but were not significantly different from each other. Therefore, both the intensity peak from 10.5-13.5% strain (Fig. 2A) and the intensity drop-off at 15% strain were significant.

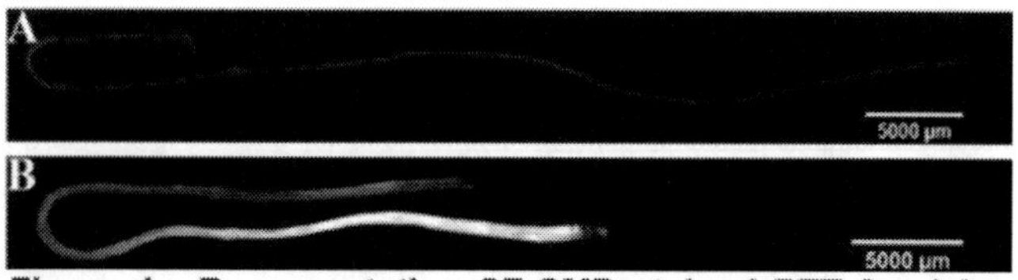

Figure 1. Representative CF-CMP stained RTT fascicles. (A) Tested to 2.5% strain, (B) tested to 12% strain.

Load curves from the mechanical test revealed that the initial onset of failure of the fascicles (i.e. decreased tangent modulus) occurred between 7.5 and 9% strain (Fig. 2B). This failure behavior continued through strains above 9%.

TEM of collagen incubated with Au-CMP confirmed CMP binding to damaged collagen fibers (Fig. 3).

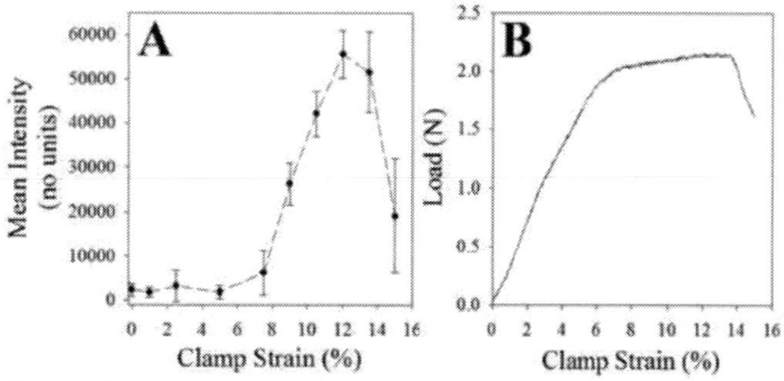

Figure 2. (A) Mean fluorescence vs. clamp strain for all RTT samples (n=3, mean std). (B) Representative load-strain curve from a 15% strain sample, showing initiation of failure at ~7% strain.

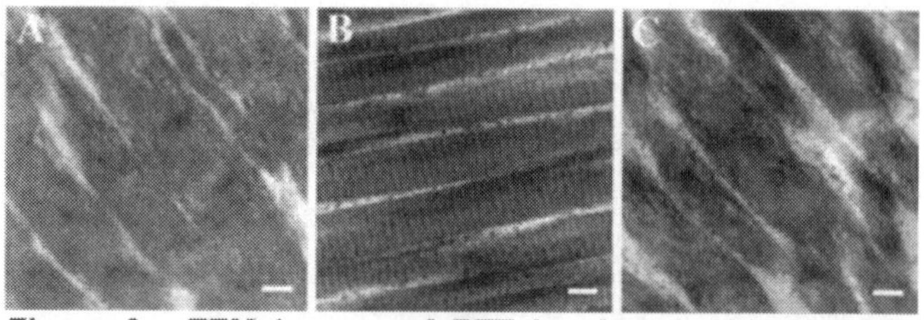

Figure 3. TEM images of RTT incubated with Au-CMP nanoparticles. (A) Damaged tendon + Au-C(GPO)9. Nanoparticles were specifically bound to the damaged areas of collagen. (B) Intact RTT + Au-C(GPO)9 shows little binding. (C) Damaged RTT + Au-CG9P9O9 (scrambled sequence) again shows little binding. Scale bars 200 nm.

DISCUSSION

We conclude that the fluorescence intensity – strain behavior is due to progressive damage of the collagen structure and CF-CMP binding. As fascicles are stretched to higher strains, more collagen triple-helices are broken and denatured, thus allowing CMP binding. This behavior was confirmed by TEM analysis, and the observations of subfailure damage are consistent with previous studies of tendon mechanics [11].

The decrease in fluorescence intensity at strains greater than 12% may be explained by areal reduction of staining intensity due to progressive sliding of fibers past each other as the fascicle failed. During the onset of initial collagen failure at approximately 7% strain, the creation of new damage still caused increased areal density of damage and thus increased mean fluorescence intensity of bound CMP. Eventually this behavior transitions so that spatial separation of damage sites results in an overall decrease of CMP fluorescence signal despite the creation of new damage. This will require further study.

The initiation of failure around 7% strain is in agreement with previous reports [12]. Other studies have captured the toe-region behavior, which can extend to 2-6% clamp strain depending on specific zero-strain length determination, and do not apply a sample preload, determining zero strain length while the fascicle is hanging under its own weight [5, 12]. Due to the 0.03N preload used in this study, fascicles were already in the linear behavior region at the zero-strain point (start of the tensile test) and failure initiation was observed at smaller strains.

ACKNOWLEDGEMENTS

Financial support from NIH R01EB015133 and R01AR047369 awarded to J.A.W and NIH R01AR060484 and R21AR065124 awarded to S.M.Y.

REFERENCES

[1] Yoganandan, N. et al., *Spine,* 26: 2443-2448, 2001.
[2] Holmes, D. F. et al., *PNAS,* 98: 7307-7312, 2001.
[3] Orgel, J. P. et al., *Structure,* 9: 1061-1069, 2001.
[4] Cisneros, D. A. et al., *J. Struct. Biol.,* 154: 232-245, 2006.
[5] Goulam Houssen, Y. et al., *J. Biomech.,* 44: 2047-2052, 2011.
[6] Li, Y. et al., *Curr. Opin. Chem. Biol.,* 17: 968-975, 2013.
[7] Li, Y. et al., *PNAS,* 109: 14767-14772, 2012.
[8] Li, Y. et al., *Bioconjug. Chem.,* 24: 9-16, 2013.
[9] Otsu, Nobuyuki, *IEEE T. Syst. Man. Cyb.,* 9: 62-66, 1979.
[10] Starborg, T. et al., *Nat Protoc,* 8: 1433-1448, 2013.
[11] Yahia, L. et al., *Matrix,* 10: 58-64, 1990.
[12] Legerlotz, K. et al., *J. Biomech.,* 43: 2274-2280, 2010.

SB³C2015
Summer Biomechanics, Bioengineering and Biotransport Conference
June 17-20, 2015,Snowbird Resort, Utah, USA
SB³C2015-629

EXTRACTING MECHANICAL PROPERTIES OF THE CORNEOSCLERAL SHELL FROM WHOLE EYE PERFUSIONS

Joseph M. Sherwood (1), Ester Reina-Torres (1), Jacques Bertrand (1), C. Ross Ethier (2), Darryl R. Overby (1)

(1) Department of Bioengineering, Imperial College London, London, United Kingdom

(2) Coulter Department of Biomedical Engineering, Georgia Institute of Technology/Emory, Atlanta, GA, USA

INTRODUCTION

Glaucoma is a leading cause of irreversible blindness. Elevated intraocular pressure (IOP) is a major risk factor in the disease, and all glaucoma treatments aim to lower IOP [1]. The main structural component of the eye, the corneoscleral shell, is compliant, and the shape of the corneoscleral shell must be maintained by IOP to ensure proper optical dimensions necessary for vision. In glaucoma, damage to retinal ganglion cells (RGCs) in the optic nerve head occurs, which may be due to elevated IOP or weakened material properties of the corneoscleral shell. Whilst the biomechanical properties of the lamina cribrosa and peripapillary sclera have been identified as important factors in optic nerve head biomechanics [2,3], the material properties of the sclera also play a role [4].

Despite only providing a lumped measurement, whole globe measurements can provide insight into relevant biomechanical properties. For example, a recent study using whole globe inflation measurements of strain showed that stiffening of the sclera increased RGC loss [5]. Additionally, analysis of the whole eye pressure-volume relationship can be used to infer lumped material properties of the corneoscleral shell [6], although such analysis has not been carried out on mice, which are increasingly being used as models for studying glaucoma.

In the present study, we used the *iPerfusion* system, developed for accurate measurement of aqueous humour outflow facility, and which simultaneously yields pressure-resolved compliance measurements, to analyse the compliance of mouse eyes. A simple mechanical model including the unloaded state of the corneoscleral shell was used to fit the data. This approach shows potential as a technique for investigating both fluid and solid aspects of ocular biomechanics in the context of glaucoma.

METHODS

Acquisition: Whole eye perfusions were carried out using the *iPerfusion* system, which comprises an automated open reservoir, a differential pressure transducer (Omegadyne, PX409) and a thermal flow sensor (Sensirion SLG64). Single eyes from 42 10-14 week old C57BL/6J mice were enucleated, fixed to a stand within a temperature regulated (35°C) PBS bath, cannulated with a 33G titanium needle and perfused with PBS + 5.5mM glucose using a nine pressure step protocol from 4 up to 21 mmHg.

Analysis: The system dynamics can be modelled using the hydraulic-electrical analogue circuit shown in Figure 1a.

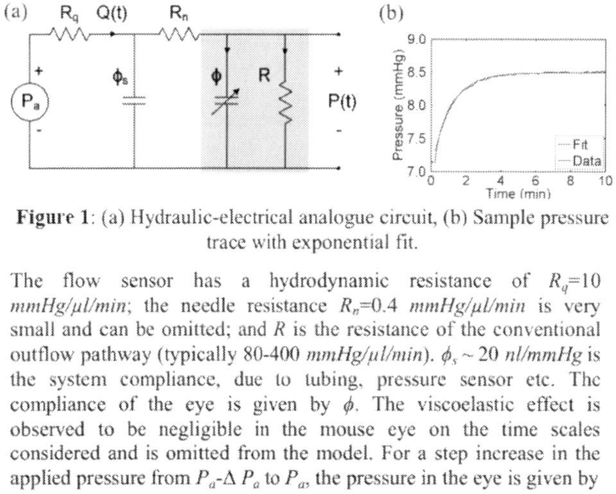

Figure 1: (a) Hydraulic-electrical analogue circuit, (b) Sample pressure trace with exponential fit.

The flow sensor has a hydrodynamic resistance of R_q=10 *mmHg/µl/min*; the needle resistance R_n=0.4 *mmHg/µl/min* is very small and can be omitted; and R is the resistance of the conventional outflow pathway (typically 80-400 *mmHg/µl/min*). $\phi_s \sim 20$ *nl/mmHg* is the system compliance, due to tubing, pressure sensor etc. The compliance of the eye is given by ϕ. The viscoelastic effect is observed to be negligible in the mouse eye on the time scales considered and is omitted from the model. For a step increase in the applied pressure from P_a-Δ P_a to P_a, the pressure in the eye is given by

$$P(t) = \left(\frac{R}{R+R_q}\right)\left(P_a - \Delta P_a e^{-t/\tau}\right) \quad (1)$$

where $\tau = RR_q(\phi_s + \phi)/(R+R_q)$. A sample trace for $P(t)$ is given in Figure 1b. The fit to Equation (1) yields an estimate for τ from which ϕ can be estimated, based on prior measurements of ϕ_s and R_q. $R = P/Q$ is calculated for each step from the last 4 minutes of data.

Model: Multi-step perfusions yield data on the compliance vs. pressure relationship for each eye, as shown by the black dots in Figure 2 (error bars are 95% confidence intervals). By fitting an appropriate model to the data, the material properties can be estimated, as follows.

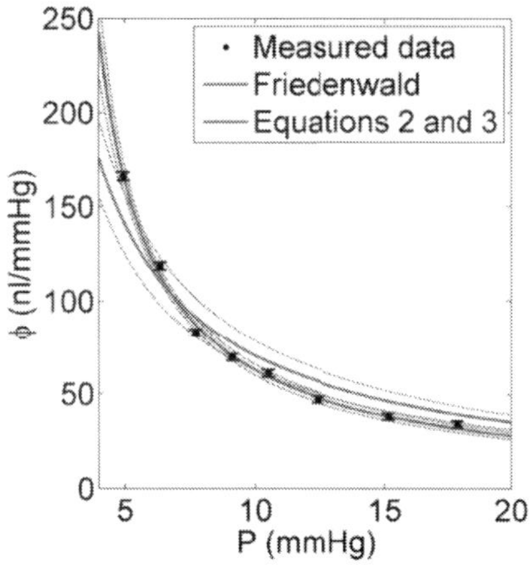

Figure 2: Compliance-pressure relationship for a typical mouse eye.

Consider the eye to be a thin walled sphere of diameter D and thickness t. For a sphere undergoing small volume changes, the hoop strain can be approximated by $\varepsilon=\Delta V/3V$. Based on the raw data, the maximum volume change (calculated using $\Delta V_i=\Sigma\ \phi_{i,j}\ \Delta P_{i,j}$ for each pressure step j and eye, i) was 1.5 μl, so the strain range is approximately 0-0.025. The corneoscleral shell is made up of predominantly collagen, and thus using a constitutive model of the form $\sigma = A(e^{\alpha\varepsilon}-1)$, Laplace's law and small strain assumptions [6,7], a model for ϕ can be derived:

$$\phi = \phi_r \left(\frac{P_r + \gamma}{P + \gamma} \right) \quad (2)$$

where $\phi_r = 3V_0/\alpha(P_r+\gamma)$ is the ocular compliance at reference pressure P_r and γ=4At/D. Note that simplifying the model for collagen to $\sigma = Ae^{\alpha\varepsilon}$ yields the Friedenwald formulation for compliance, $\phi=\phi_{rf}P_r/P$, where ϕ_{rf} is the reciprocal of the ocular rigidity. Equation 2 tacitly assumes that the unloaded state of the collagen occurs at $P = 0$. By analogy with arteries [8], it could be expected that the corneoscleral shell is pre-stressed. To account for this, the model can be altered to include the unloaded stress σ_0, which occurs at P_0. Equation 2 becomes

$$\phi = \phi_r \left(\frac{P_r - P_0 + \gamma}{P - P_0 + \gamma} \right) \quad (3)$$

where $\phi_r = 3V_0/\alpha(P_r-P_0+\gamma)$. When fitting the raw data to Equation 3, the pressure-dependent nature of the compliance means that the calculated ϕ_j does not correspond to the steady state pressure P_j, but rather some intermediate value P_ϕ. The analytical solution to the circuit with the ϕ defined according to Equation 3 is prohibitively complex, and so a numerical approach was implemented using Simulink. Considering ranges of P_0-γ = 0.5 to 2.5 $mmHg$, ϕ_r = 50 to 150 $nl/mmHg$, and R = 80-400 $mmHg/\mu l/min$ a model of the perfusion protocol was simulated. Analysis of the P_ϕ and P_j revealed that the empirical relationship

$$P_\phi = P_j - 0.56\Delta P_a \quad (4)$$

gave an RMSE of 0.08 $mmHg$. The data points in Figure 2 are plotted against P_ϕ. Note that as the reference compliance cannot be negative, ϕ_r is considered to be log-normally distributed within the population.

RESULTS

Fitting the compliance for the sample data to Equation 2 yields the red lines in Figure 2, shown with 95% confidence intervals on the fit. By comparison, the Friedenwald model (green) fails to appropriately capture the pressure dependence of the measured compliance. As P_0 and γ are both unknown constants, fitting the data to Equation 3 yields an identical fit to Equation 2, with the only difference being the interpretation of the parameters. For the population sample considered in this study, the value of the constant term (γ for Equation 2, or P_0-γ for Equation 3) was -1.85 ±1.25 $mmHg$ (mean ± 95% confidence interval (CI)). As this value is negative, Equation 2 does not have a physiologically meaningful interpretation, and Equation 3 must be used. The value of P_0 can be estimated using literature values for γ. For t, averaged values across multiple parts of the sclera in C57BL/6 yielded 47 μm [9]. Multiple authors have reported diameters for mouse eyes of ~ 3.35 mm [10,11]. Data for the mechanical properties of collagen in the mouse eye are not presently available, but based on data from human eyes, estimates of A = 12.1 kPa and α = 140 have been reported [6], which are tentatively applied here. Using these parameters, γ=5.1 $mmHg$, and thus the pressure corresponding to the unloaded state is estimated to be P_0 = 6.95 ± 1.25 $mmHg$ (mean ± 95% CI). The reference compliance (at 8 $mmHg$) was ϕ_r = 91.0 $^{\times}/$ 1.38 $nl/mmHg$ (geometric mean and 95% CI). From the derivation for Equation 3, the volume can be estimated according to $V_0 = \alpha\phi_r(P_r-P_0+\gamma)/3$ = 26.2 μl, in reasonable agreement with the volume of the eye based on globe diameter, 19.7 μl.

DISCUSSION

The *iPerfusion* system provides detailed information on the compliance response of mouse eyes with an accuracy and pressure resolution that was hitherto not possible. Analysis of the compliance-volume relationship using existing models revealed that they were not capable of fitting the data with physiologically valid values. Inclusion of a pre-stressed state in the derivation yielded physiologically reasonable parameters, e.g. the estimated volume using a value of α from the literature was close to the expected volume of a mouse eye, supporting the current findings. The pressure corresponding to the unloaded state was in the range ~5.7-8.2 $mmHg$ which is also physiologically realistic. Future experiments including anatomical measurements and spatial strain distributions will enable quantification of the material parameters of the mouse eye and thus more accurate quantification of P_0. Pre-stress in arteries has been shown to be an important mechanobiological factor for the vessel wall [8]. The present data suggests that pre-stress may be an important component in modelling the biomechanical characteristics of the corneoscleral shell. Further work is required to infer the importance of whole eye compliance on the properties of the peripapillary sclera and susceptibility to RGC loss. Nonetheless, the present approach shows promise as an accurate technique for investigating ocular biomechanics.

ACKNOWLEDGEMENTS

We gratefully acknowledge financial support from the Georgia Research Alliance (CRE) and EPSRC grant EP/J010499/1 (JMS, DRO)

REFERENCES

[1] Leske, MC et al., *Arch Opthalmol*, 121(1):48-56, 2003.
[2] Downs, CJ et al., *Exp Eye Res*, 93(2):133-140, 2011.
[3] Norman RE et al., *Exp Eye Res*, 93(1): 4-12, 2010.
[4] Sigal, IA and Ethier CR, *Exp Eye Res*, 88(4): 799-807, 2009.
[5] Kimball, EC et al., *Exp Eye Res*, 128(C):129-140, 2014.
[6] Ethier, CR, *Annu Rev Biomed Eng*, 6:249-273, 2004.
[7] McEwen, WK and St Helen, R, *J Biol Chem*, 240:2003-2010, 1965.
[8] Chuong, CJ and Fung, YC, *J Biomech Eng*, 108(2):189-192, 1986.
[9] Myers, KM et al., *Exp Eye Res*, 91(6):866-875, 2010.
[10] Nguyen, C et al. *Inv Opthalmol Vis Sci*, 54(3):1767-1780, 2013.
[11] Retmulla S and Hallett, PE, *Vis Res*, 25(1):21-31, 1985.

OPTIMIZATION OF TEST METHODS AND BURST PROPERTY CHARACTERIZATION OF ALGINATE HYDROGEL LUNG SEALANTS

Patrick N. Charron (1), Spencer L. Fenn (2), Rachael A. Oldinski (1,2)

(1) Mechanical Engineering Program
University of Vermont, School of Engineering
Burlington, Vermont, USA

(2) Bioengineering Program
University of Vermont, School of Engineering
College of Medicine
Burlington, Vermont, USA

INTRODUCTION

Lung diseases, cancers, and trauma
connective tissue lining the lung, i.e. the
to air or fluid leaking out of the lung and potential lung collapse, which is immediately life threatening. Pleural injuries may result in chronic leaking, also known as bronchopleural fistula.

To address the clinical need to seal lung leaks and prevent further tissue damage, we have devised a new method to quantitatively characterize the adhesion and burst properties of a photocrosslinked hydrogel sealant for retaining physiological lung volumes. We have fabricated methacrylated alginate (Alg-MA) and oxidized Alg-MA (Alg-MA-Ox) hydrogels utilizing visible light crosslinking of polymer solutions. The effectiveness of the hydrogels as lung sealants was evaluated and the effect of oxidation on the material's adhesive properties was determined by comparing burst pressure strengths and failure mechanism.

METHODS

Alg-MA was synthesized by reacting sodium alginate (Manugel, FMC Biopolymer) in deionized H_2O and a 20 molar excess of methacrylic anhydride (Sigma Aldrich) for 24 hours.[2] The Alg-MA product was purified by dialysis (MWCO 6-8 kg/mol, Fischerbrand) against deionized H_2O and lyophilized. To enhance adhesion[3], Alg-MA was oxidized in sodium periodate (Sigma Aldrich) solutions of varying concentrations for 24 hours to obtain Alg-MA-Ox with 10, 30, and 50% degrees of theoretical oxidation.[4] The Alg-MA-Ox was purified by dialysis against deionized H_2O (MWCO 6-8 kg/mol, Fisherbrand) and lyophilized. Alg-MA and Alg-MA-Ox were dissolved in deuterium oxide (Acros Organics) at 0.5% (w/v) and analyzed with ^{1}H-NMR spectrometry to determine the degree of methacrylation (DOM) and the degree of oxidation (DOO).[4]

To study hydrogel compositions, 3% (w/v) Alg-MA and Alg-MA-Ox aqueous solutions containing photoinitiators were formed and tested. Photocrosslinking reactants were used at the following

Y (photosensitizer, Acros Organics), 125
ır, Sigma Aldrich), and 20 mM 1-vinyl-2-
ıa Aldrich).[5]

Burst pressure and adhesion test substrates consisted of hydrated, 3 cm x 3 cm collagen squares (#28010 Collagen middles natural protein, The Sausage Maker Inc.). Rehydrated collagen squares were tested on a custom fabricated (ASTM F2392-04R10) burst strength device to ensure an intact substrate. The substrate was punctured using a 3 mm diameter biopsy punch and retested as proof of an ineffective seal (i.e., through-thickness defect).

Punctured collagen substrates were clamped between a glass slide and a Teflon mold with a 1.5 cm diameter hole centered on the puncture to prevent the solution from flowing into the burst strength device before being crosslinked. The hydrogel pre-curser solution (0.5 mL) was deposited in the mold and photocrosslinked for 10 minutes under 523 nm green visible light.[6] The substrate was loaded onto the burst pressure device (**Fig. 1**). Burst pressure tests were conducted at room temperature and an infusion rate of 75 mL/hr (Harvard Apparatus PHD 2000 Infusion). Burst strength was collected using a pressure transducer (OMEGA PX409-030AUSBH, 0-30 PSI ± 0.08% BSL).

FIGURE 1. CUSTOM FABRICATED BURST PRESSURE DEVICE CONNECTED TO PUMP (LEFT) AND PRESSURE TRANSDUCER (RIGHT).

The failure mechanism (adhesive or material failure) of the hydrogel sealant materials was also qualitatively determined. Failure mechanism was visually assessed during testing and after failure. Adhesive failure occurs when the sealant delaminates from the substrate without tearing, whereas material failure occurs when the sealant fails without delamination.

RESULTS

The methacrylation and oxidation of Alg-MA were successful and the degrees of modification were controllable. Burst pressure, degree of methacrylation, and degree of oxidation for various Alg-MA based hydrogel sealants are depicted below (**Table 1**). While oxidation decreased the likelihood of adhesive failure, the reaction also reduced the degree of methacrylation and groups available for crosslinking, as confirmed by ^{1}H-NMR. Absolute pressure values were collected during testing and plotted for comparison (**Fig. 2**) and burst pressure was calculated from absolute and atmospheric pressure. Compared to normal human peak inspiratory pressures of 25-30 cm H_2O (9.1-11.8 in H_2O), the burst pressures, at failure, for both the Alg-MA and Alg-MA-Ox blend materials exceed physiological loading conditions.[7]

TABLE 1. MATERIAL PROPERTIES OF ALGINATE BASED HYDROGEL. BLENDS REPRESENT 50:50 WEIGHT RATIOS.

Material	Burst Pressure (in H_2O)	DOM (%)	DOO (%)
Alg-MA	21.98±12.36	77	N/A
10 Alg-MA-Ox	14.20±8.03	70	10
30 Alg-MA-Ox	1.25±0.34	64	25
10 Alg-MA-Ox Alg-MA Blend	23.36±2.33	73.5	5
30 Alg-MA-Ox Alg-MA Blend	19.25±1.46	70.5	12.5
50 Alg-MA-Ox Alg-MA Blend	13.26±3.14	54	19.8

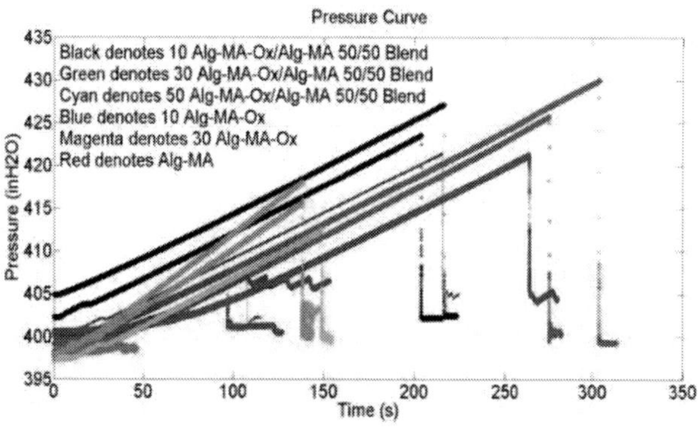

FIGURE 2. PRESSURE CURVES FOR MATERIAL TESTING; BURST PRESSURE (IN H_2O) VERSUS TIME. RED DENOTES ALG-MA, BLUE DENOTES 10 ALG-MA-OX, MAGENTA DENOTES 30 ALG-MA-OX, AND BLACK, GREEN, AND CYAN DENOTE A BLEND OF THE 10, 30, AND 50 ALG-MA-OX AND ALG-MA MATERIAL.

DISCUSSION

The basis of this study was to investigate the effect of oxidation on a methacrylate alginate-based hydrogel sealant. The oxidation reaction increases the number of available aldehyde groups available to crosslink to the collagen substrate. Unfortunately, the oxidation reaction may have resulted in material degradation due to polysaccharide chain scission. Fine tuning the degree of methacrylation and oxidation can result in a bioadhesive sealant with a range of mechanical and material properties for various applications. Increasing oxidation decreased burst pressures below the threshold of physiological pressures. Burst pressure values suggest that blending oxidized and non-oxidized materials results in strong, reliable sealants by combining the adhesive properties of the Alg-MA-Ox and the tensile strength of the non-oxidized Alg-MA.

CONCLUSION

Controlling both the degrees of methacrylation and oxidation allows for a highly tunable bioadhesive which can be adapted for a range of applications. Cytocompatibility assays with primary human pluripotent cells will be performed to determine toxicity of the Alg-MA and Alg-MA-Ox and all photocrosslinking agents. Ex vivo and in vivo respiratory mechanics data will be collected to show efficacy of Alg-MA-Ox as a lung sealant.

ACKNOWLEDGEMENTS

The authors gratefully acknowledge the College of Engineering and Mathematical Sciences, and the Office of Technology and Commercialization at UVM for financial support. This work was also supported in part by NIH grant T32 HL076122.

REFERENCES

[1] Schraufnagel, D. E., Kell, B. & Society, A. T. *Breathing in America: diseases, progress, and hope.* (American Thoracic Society, 2010).

[2] Smeds, K. A., Pfister-Serres, A., Hatchell, D. L. & Grinstaff, M. W. Synthesis of a novel polysaccharide hydrogel. *Journal of Macromolecular Science-Pure and Applied Chemistry* **36**, 981-989 (1999).

[3] Jeon, O., Samorezov, J. E. & Alsberg, E. Single and dual crosslinked oxidized methacrylated alginate/PEG hydrogels for bioadhesive applications. *Acta biomaterialia* **10**, 47-55 (2014).

[4] Jeon, O., Alt, D. S., Ahmed, S. M. & Alsberg, E. The effect of oxidation on the degradation of photocrosslinkable alginate hydrogels. *Biomaterials* **33**, 3503-3514 (2012).

[5] Park, Y. D., Tirelli, N. & Hubbell, J. A. Photopolymerized hyaluronic acid-based hydrogels and interpenetrating networks. *Biomaterials* **24**, 893-900 (2003).

[6] Fenn SL, O. R. Visible Light Crosslinking of Methacrylated Hyaluronan Hydrogels for Injectable Tissue Repair. *JBMR Part B* (2015 Under Review).

[7] Wheeless, C. Wheeless' textbook of orthopaedics. (1996).

ASSESSING HEAD IMPACT SENSOR VALIDITY IN THE LABORATORY

Abigail M. Zadnik, Steven Rowson, Stefan M. Duma

Department of Biomedical Engineering and Mechanics
Virginia Tech
Blacksburg, VA, USA

INTRODUCTION

In 2007, more than 38 million American youth (ages 5-18) participated in organized sports. Although many different injuries occurred, the most commonly reported were concussions [1]. There has been increased concern about the long-term effects of repetitive concussions [2] especially among youth athletes. Concussions in youth sports are difficult to diagnose and are typically dependent upon self-reporting of symptoms [3]. This often leads to severe under-reporting [4].

These concerns have led to the development of commercially-available head impact sensors that athletes can wear while participating in sports. These sensors alert when an impact with high risk of injury has been sustained, providing additional information to physicians and trainers in order to diagnose a concussion. Current commercially-available sensors have a variety of capabilities; some purely count impacts while others output head or helmet acceleration. As more and more sensors become available to consumers, the need for objective data to assess the capabilities of these sensors becomes more evident.

The objective of this study was to investigate laboratory methods that can be used to evaluate commercially-available impact sensors. Two such sensors were evaluated in the lab for illustrative purposes.

METHODS

A pneumatic linear impactor (Biokinetics, Ottawa, ON, Canada) was used for testing the impact sensors. The impactor face was made of ultra-high molecular weight polyethylene backed by a 35 mm of vinyl nitrile foam pad. The headform was impacted at the front, side, rear boss, and rear locations (Figure 1) at impact velocities of 3, 6, and 9 m/s. These locations and impact velocities were chosen as being most representative of potential on-field hits in football [5, 6]. Each test configuration was repeated three times. A Riddell Speed (Elyria, OH) football helmet with a facemask was worn by the headform throughout all tests.

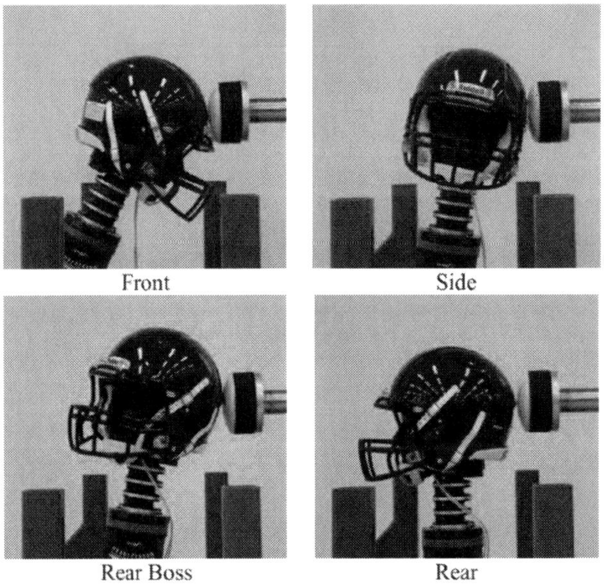

FIGURE 1: HEADFORM ORIENTATIONS FOR THE FOUR IMPACT LOCATION USED IN TESTING

Two commercially-available impact sensors were assessed under this protocol. Sensor acceleration measurements were compared to reference headform acceleration provided by a 6DX Pro 2k-18k 6DOF sensor package (DTS, Seal Beach, CA) inside a NOCSAE headform that was mounted on a Hybrid III neck. Sensor 1 had an adhesive back and was attached directly to the headform, placed behind the ear at the equivalent location of the mastoid bone. Sensor 2 was placed in a headband worn by the headform, with the sensor located in the back of the head above the nuchal line.

Linear acceleration and rotational velocity from the sensors and headform reference were processed through Microsoft Excel 2013 and MATLAB R2014a. The average and standard deviations for percent error were quantified for each test to compare sensor validity. Average percent error was the systematic error, while standard deviation was the random error. Here, negative values for systematic error corresponded to the sensor's under-predictions of reference values, whereas positive values corresponded to over-predictions.

RESULTS

A total of 72 impacts were recorded by the reference angular rate sensor in the headform. Peak linear acceleration values ranged from 29.7 to 134.0 g with an average of 72.7 g, while peak rotational velocity values ranged from 9.8 to 49.4 rad/s with an average of 28.0 rad/s.

Sensor 1 showed good correlation to reference for linear acceleration values (Figure 2). Correlation coefficients remained above 0.9 and random error remained within 10.0% except for the rear boss location which had a correlation coefficient of 0.74 and random error of 36.6%. Systematic error for each location was negligible except for the rear location whose sensor values under-predicted reference values by 20.7%.

Rotational velocity for Sensor 1 had better correlations than the linear acceleration values to reference. All correlation coefficients were above 0.96, all random error was within 10%, and all had under 10% systematic error.

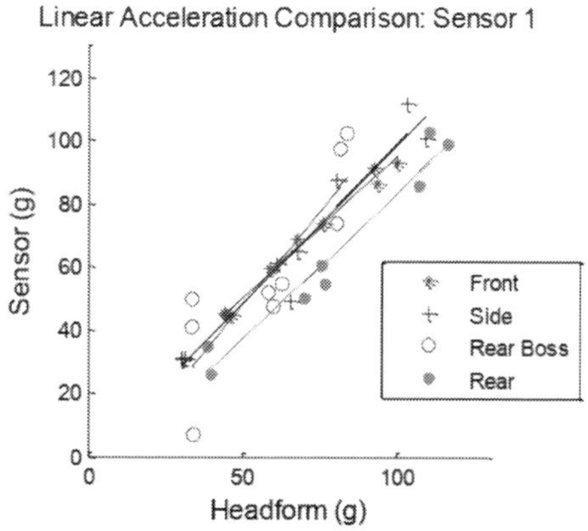

FIGURE 2: DATA AND LEAST SQUARES LINEAR FITS FOR LINEAR ACCELERATION (G) COMPARING SENSOR 1 AND HEADFORM REFERENCE. DATA ARE ORGANIZED BY IMPACT LOCATION.

Sensor 2 had large variations in linear acceleration measurements which were dependent upon impact location (Figure 3). The rear and side locations were best correlated to reference, with correlation coefficients around 0.9. The front and rear boss had correlation coefficient values of 0.39 and 0.53, respectively. All random error was above 15%, with the rear boss being as high as 72.1%. The front, side, and rear locations were systematically off by 20% to 40%, and the rear boss systematically over-predicted by 121.0%.

Rotational velocity for Sensor 2 had similar variability and over-prediction as the linear acceleration. Front and rear boss locations still had the lowest correlation coefficients of 0.58 and 0.00 respectively, and all random error was still above 15%. All locations also over-predicted by at least 25%.

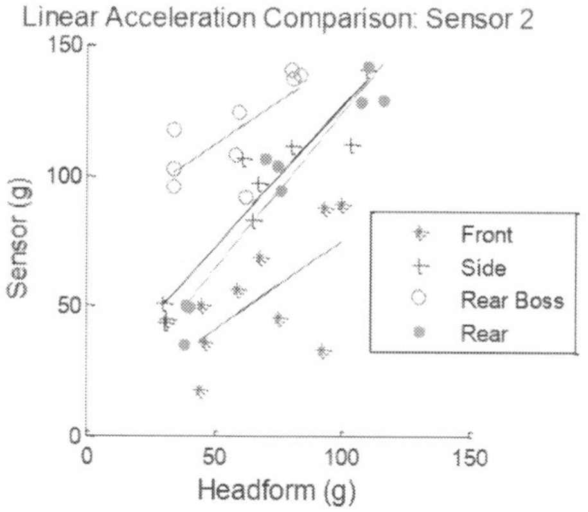

FIGURE 3: DATA AND LEAST SQUARES LINEAR FITS FOR LINEAR ACCELERATION (G) COMPARING SENSOR 2 AND HEADFORM REFERENCE. DATA ARE ORGANIZED BY IMPACT LOCATION.

DISCUSSION

The test conditions produced headform accelerations consistent with previous studies of on-field head impacts [5]. The methods were sensitive enough to determine distinct differences between sensor and reference values and between the sensors themselves.

When comparing to reference, it was shown that at certain impact locations, the sensors can be off as much as 121% from the true headform acceleration while at other locations there can be very little variation. This impact location-dependence is consistent with previous studies [7].

After comparing the two sensors, it can be seen from these results that not all commercially-available impact sensors are the same. Wide variability and large over/under-predictions from Sensor 2 indicate the lack of reliability in obtaining accurate head impact from this device. By contrast, Sensor 1 is shown to predict head impacts much more accurately in a laboratory environment.

There were several limitations of this study that must be noted. The sensors were evaluated on a dummy head which does not fully replicate the environment on the field such as hair, sweat, and skin movement relative to the skull. Also, only four distinct impact locations were tested which, although fairly representative, do not fully encompass all concussive hits.

In summary, this study used laboratory methods with similar results to on-field data in order to evaluate commercially-available impact sensors. The results suggest that caution must be executed when using these sensors as they are not all the same, and further testing is still needed before sensors can be fully assessed.

ACKNOWLEDGEMENTS

The authors gratefully acknowledge the Lewis Family Foundation for supporting this work.

REFERENCES

[1] CDC. (Oct. 28). *Heads Up: Concussion in Youth Sports Activity Report 2007-2008.*
[2] Gavett, B. E. *et al., Clin Sports Med,* 30:179-88, xi, 2011.
[3] Duhaime, A. C. *et al., J. Neurosurgery,*117:1092-1099, 2012.
[4] McCrea, M. *et al., Clin J Sport Med,* 14:13-7, 2004.
[5] Daniel, R. W. *et al., Ann Biomed Eng,* 40:976-981, 2012.
[6] Pellman, E. J. *et al., Neurosurgery,* 53:799-814, Oct 2003.
[7] Allison, M. A. *et al., Ann Biomed Eng,* Dec 5 2014.

JOINT ANGLE AND MUSCLE ACTIVITY DURING HIPPOTHERAPY: A CASE STUDY USING MOTION CAPTURE ANALYSIS AND EMG

Mary C. Baker(1), Stephen Wester(1), Timothy A. Monday(1)

(1) Texas Tech University Lubbock, TX, USA

INTRODUCTION

Hippotherapy has long been used as a therapeutic and rehabilitative technique, particularly in children with physical disabilities and/or developmental delays. Previous researchers have investigated the effects of hippotherapy sessions on various parameters, including gait kinematics and range of motion, typically assessed before and after a specified number of sessions [1-3]. While there are published data on observed or measure gait parameters before and after therapy, there are limited quantitative studies on range of motion and muscle activity during the horseback riding sessions. The goal of this case study was to measure range of motion parameters and EMG activity in relevant joint and muscle regions during horseback riding, and to determine whether there are any correlations in muscle activation and joint movement. Quantifying these measures and determining how they are correlated will allow for better justification of hippotherapy as a medically necessary and useful treatment for movement disorders in children and adults with physical disabilities.

METHODS

Data were acquired at the Windridge Equine Facility in Gilmer, TX, in a dedicated laboratory that includes an 8-camera motion capture system, custom developed for Windridge by ORBIS, Inc. A horse treadmill system is used in order to keep the horse and rider in a central position in the room to aid in obtaining motion capture data from the camera system. An inexperienced, but able-bodied, rider was used for the case study, with the treadmill speed set to 3.5 mph. The rider was equipped with 41 LED position markers, placed in regions around the head, shoulders, trunk, back, and hips, as shown in Figure 1. Motion capture data were sampled at a rate of 20 Hz.

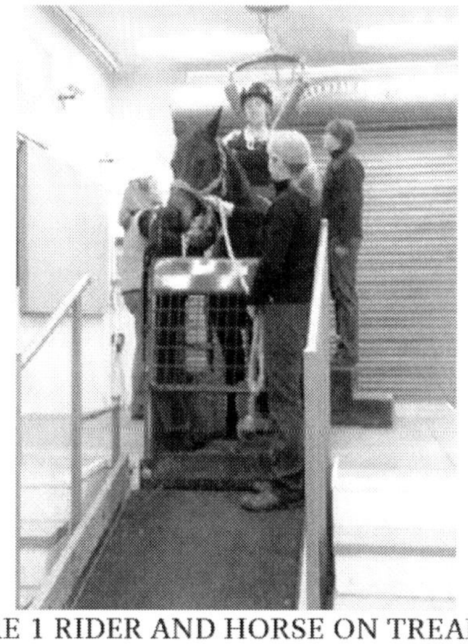

FIGURE 1 RIDER AND HORSE ON TREADMILL

EMG sensors were placed in 8 positions symmetrically left and right on the rider: neck, lower back, middle back, and abdomen. The reference electrode was placed on the bony portion of the left elbow. The EMG was acquired using the CleveMed Bioradio 150 wireless system. The wireless transmitter was placed on the rider's right hip. A Labview-based program was used to acquire the data from the EMG and to time synchronize the EMG and motion capture data using an output port from the ORBIS system. The EMG data were acquired at a 60 Hz sampling rate and then down-sampled to 20 Hz to correspond to the motion capture data record.

Motion capture data were translated to joint angle data using a proprietary, custom developed software, OBSIS. Further processing was done to eliminate dropping of camera signals by constraining to a biomechanical model. Joint angles as a function of time were obtained for the following (both left and right sides): lumbar flexion, lateral flexion, and rotation; pelvis lateral tilt, rotation, and posterior tilt; hip abduction, external rotation, and flexion; and, neck flexion, lateral tilt (right side), and rotation (right side).

A program was written in Matlab to detect local maxima and minima in the joint angle versus time data sets, and to determine an average maximum and minimum joint angle. The difference between max and min was taken to determine an angular range of motion, which was averaged across four runs. The EMG data were exported from Labview as text files that were time synchronized at t = 0 with the motion capture system. They files were also processed using Matlab, first by down-sampling at 20 Hz. They were then rectified, and further smoothing was done by using a 6 Hz filter. EMG data were compared to joint angle data using correlation – a correlation coefficient was determined between each EMG channel and each joint angle measure.

RESULTS

Table 1 illustrates the joint angle range averages and standard deviations for an average over 4 runs. Data marked with an asterisk contained one "bad" segment in a specific run, so only three of the runs were used for the averages in these data.

TABLE 1 JOINT ANGLE RANGE

Joint Angle	Average (in degrees)	St. Deviation
Lumbar Lateral Flexion	2.90	0.17
Lumbar Rotation	4.08	0.72
Lumbar Flexion	4.09	0.39
Pelvis Lateral Tilt	62.89*	5.98
Pelvis Rotation	3.00	0.26
Pelvis Posterior Tilt	61.86*	6.96
Left Hip Abduction	4.48	0.89
Left Hip External Rotation	4.74	0.57
Left Hip Flexion	3.76	0.80
Right Hip Abduction	4.80	0.68
Right Hip External Rotation	6.75	2.29
Right Hip Flexion	5.32	1.37
Right Neck Lateral Tilt	1.89	0.30
Right Neck Rotation	3.13	0.49
Neck Flexion	3.38	0.30

It can be seen from the values in the table that the largest ranges of motion are found in the pelvis lateral tilt and the pelvis posterior tilt. The standard deviation is generally about 10% of the average, with except for measure associated with right hip movement, where it is closer to 25%. A sample of joint angle versus time from Run # 4 is shown below in figures 2 and 3, for pelvis posterior tilt (PPT) and right hip flexion (RHF), respectively. The PPT angle shows a high degree of regularity, while the RHF angle appears less repeatable. This might explain the higher standard deviation in the hip measurements.

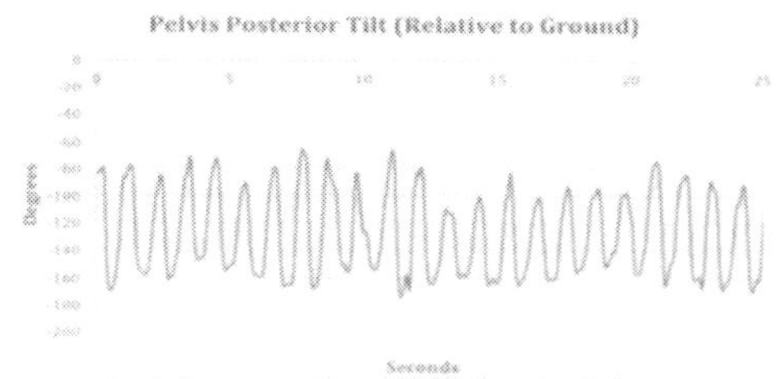

FIGURE 2 PPT ANGLE VERSUS TIME

FIGURE 3 RHF ANGLE VERSUS TIME

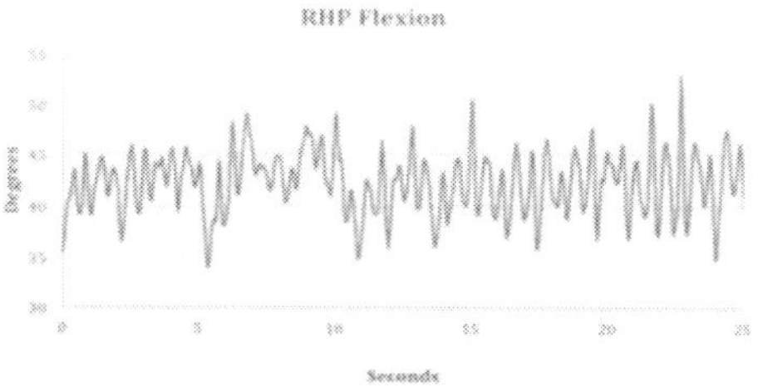

FIGURE 3 RHF ANGLE VERSUS TIME

Correlation coefficients were determined by correlating each EMG measurement with each joint angle measure. For individual runs, there were a number of correlation coefficients that appear to be significant, but no statistical testing could be done since this was a case study. An average was taken over all 4 runs, and showed that left mid-back EMG was weakly correlated with lumbar lateral flexion, pelvis lateral tilt, pelvis posterior tilt, and right hip external rotation.

DISCUSSION

The findings in this case study indicate that repeatable range of motion may be observed for multiple runs during and hippotherapy session, and that the joint angles associated with the motion are correlated in time with at least one EMG measure. Because this is a single case study, it is important not to read too much into the data except that there are clearly data points that can be identified as biomarkers and used for comparison in both 2-group and longitudinal studies of joint and muscle activity during hippotherapy.

ACKNOWLEDGEMENTS

The authors would like to acknowledge a donor gift from Tim and Nikki Winn that made this research possible. We would also like to thank Margo Dewkett, the owner of Windridge Therapeutic Equestrian Center, for use of the hippotherapy and motion capture laboratory, and Celia Bowers, the physical therapist who assisted in the data collection and trial run sessions.

REFERENCES

[5] Bergmann, G et al., *J Biomech*, 34:859-871, 2001.
[1] Benda, W et al., *The Journal of Alternative and Complimentary Medicine*, Vol.9 Num.6, p817-285, 2003.
[2] Debuse, D et al., *Physiotherapy Theory and Practice*, 25(3), p174-192, 2009.
[3] Kwon, J et al., *Arch Phys Med Rehabil*, vol.92, p774-779, 2011.

SB³C2015
Summer Biomechanics, Bioengineering and Biotransport Conference
June 17-20, 2015, Snowbird Resort, Utah, USA
SB³C2015-633

PRESSURE DROP APPROACH TO ESTIMATE THE APPARENT SLIP LENGTH IN A PDMS MICRO-CHANNEL

Stephanie E. González-Jiménez (1), Carlos R. Romero-Peñaloza (1), Rubén E. Díaz-Rivera (1,2)

(1) Mechanical Engineering Department
University of Puerto Rico at Mayagüez
Mayagüez, Puerto Rico

(2) Weldon School of Biomedical Engineering
Purdue University
West Lafayette, IN, USA

INTRODUCTION

In recent years, there has been a great interest in the manipulation and sorting of cells or particles suspended in microfluidic devices for biomedical applications [1]. Understanding the fluid mechanics at this scale is of great importance when it comes to the design of effective microfluidic devices for the hydrodynamic manipulation of single cells and micro-scale particles. For many years, scientists have been reporting a discrepancy between analytical models presuming no-slip boundary conditions and experimental results [2]. The experimental evidence points to the existence of an apparent slip near the solid surface at the micro and nano-scale. It is believed that the apparent slip occurs not only at the solid/fluid interface, but at the fluid/fluid interface as bubbles tightly bound to nano-scale pockets at the wall may provide a layer of lubrication [3]. A mathematical approximation of the apparent slip phenomena is known as partial slip. Partial slip condition is met when liquid molecules are effectively sliding on the solid surface and is characterized by the slip length (*b*) or the distance beyond the wall where the tangent of the velocity field intersects the vertical axis (Figure 1).

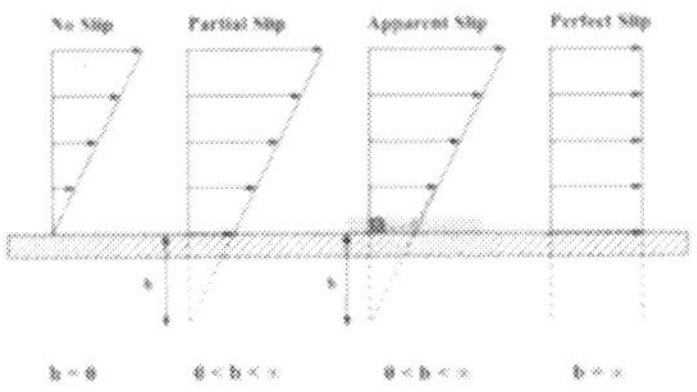

Figure 1. Types of slip boundary conditions. The arrows represent the velocity vector near the wall.

However, the mechanism behind this phenomenon still eludes us, mainly due to the difficulty of finding a reproducible experimental scheme. Instead of trying to find the mechanism behind apparent slip, in this work we take an engineering approach by presenting a correction factor for the analytical approximation of the Poiseuille-flow problem in a rectangular cross-section channel as a function of volumetric flow rate and micro-channel aspect ratio. In order to do this, we studied the pressure-drop in rectangular cross-section micro-channels under different flow conditions.

METHODS

Four rectangular cross-sectioned micro-channels where used to characterize the apparent slip. The channels had a constant height (*h*) of 26μm while the width (*w*) was varied from 100 to 800μm. The micro-channels where fabricated using photolithography to create the SU-8 masters and PDMS soft lithography to replicate the patterns. Soda lime glass was used as the bottom wall (Corning® 75x50mm μscope slides). The PDMS and the glass slide were not permanently bonded, as the natural adhesion between the two surfaces was enough to avoid leakage. However, the internal walls of the channels were exposed to air plasma for fluid priming purposes. The average thickness of the PDMS castings was 6mm in order to avoid channel deformation due to the fluid pressure [4].

For the pressure drop experiment, two solid-liquid interfaces were examined: PDMS-water interface and PDMS treated with bovine serum albumin (BSA)-water interface. BSA is typically used as a micro-channel wall coating to prevent adhesion of cells and molecules (Albumin solution 7.5%, Sigma-Aldrich). BSA was introduced in the channel following the air plasma treatment and incubated for 10min at 37°C. Afterwards, the solution was flushed from the system with DI water and thus ready for experimentation.

The pressure drop was measured for each micro-channel at different volumetric flow rates, which ranged from 50μL/hr to 750μL/hr in 50μL/hr increments. The DI water used for the experiments was distributed to the setup by a 250μL Hamilton syringe in a syringe pump (New Era Pumps Systems). The pressure drop between the channel inlet and the outlet was performed using a differential pressure sensor (PX26-005DV, Omega Engineering, see Figure 2).

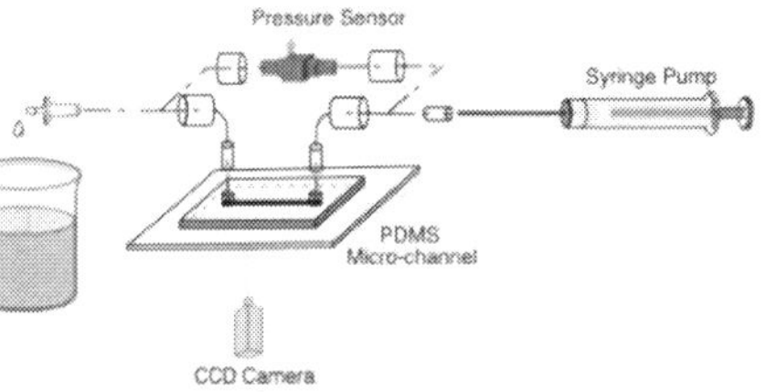

Figure 2. Pressure drop experiment setup.

Surprisingly, there is no analytical solution for the Poiseuille-flow problem with a rectangular cross-section. Even though good approximations exist, they do not take into account non-homogenous boundary conditions (BCs) such as partial slip ($u_{wall} = b\tau_{wall}/\mu$). However, in order to estimate the slip length (*b*) by using the pressure drop measurements, it was necessary to obtain an analytical approximation for the rectangular cross-section geometry considering partial slip. Therefore, we used the Poiseuille-flow solution for infinite parallel-plates with partial slip BCs as a starting point in the search of a correction factor. The hydraulic resistances for the aforementioned geometries with and without partial slip are shown in Table 1. For the no-slip case, Eqns. (1) and (2) are almost equal, except that Eqn. (2) contains a rectangular correction factor. By analogy, for the slip case, we propose to do the same with Eqns. (3) and (4) [5].

Table 1. Hydraulic resistance values for parallel plates and rectangular channels.

Parallel Plates Section (w>>h)	Rectangular Section

No Slip	$R_{\parallel} = \frac{12\mu L}{h^3 w}$ (1)	$R_{\square} \approx \frac{1}{1-0.63(h/w)} \cdot \frac{12\mu L}{h^3 w}$ (2)
Slip	$R_{s\parallel} \approx \frac{2\mu L}{h^2 w\left(\frac{h}{6}+b\right)}$ (3)	$R_{s\square} \approx c\frac{\mu L}{h^2 w\left(\frac{h}{6}+b\right)}$ (4)

Equations 2 and 4 can be substituted in the Hagen–Poiseuille law (Eqn. 5), yielding Equations 6 and 7.

$$\Delta P = R_{hyd} Q \tag{5}$$

$$\Delta P_{\square} = \frac{12\mu L}{1-0.63(h/w)} \cdot \frac{1}{h^3 w} Q \quad (6) \qquad \Delta P_{s\square} = c\frac{\mu L}{h^2 w\left(\frac{h}{6}+b\right)} Q \quad (7)$$

where ΔP is the pressure drop across the channel, R_{hyd} is the hydraulic resistance, Q is the volumetric flow rate, c is the correction factor, μ is the absolute viscosity and L is the length of the channel. In the limit that partial slip is zero ($b=0$), Equations 6 and 7 are equal, thus yielding an expression for the correction factor:

$$c = \frac{2}{1-0.63\left(\frac{h}{w}\right)} \tag{8}$$

Equation 8 only depends in the reciprocal of the microchannel aspect ratio (h/w) (Figure 3). Hence, the hydraulic resistance equation (Eqn. 4) is set for each micro-channel.

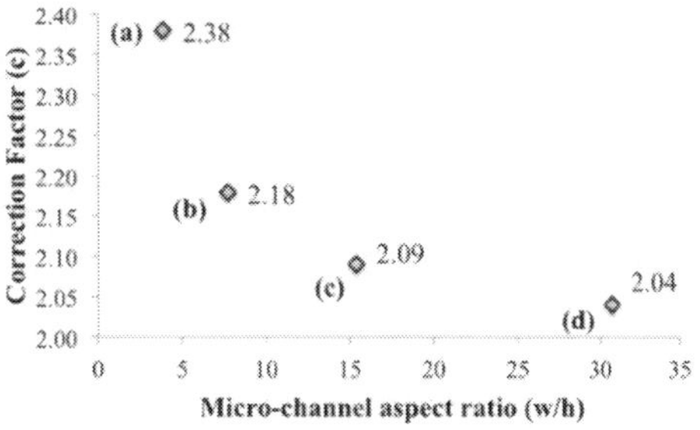

Figure 3. Correction factor versus channel aspect ratio.

RESULTS

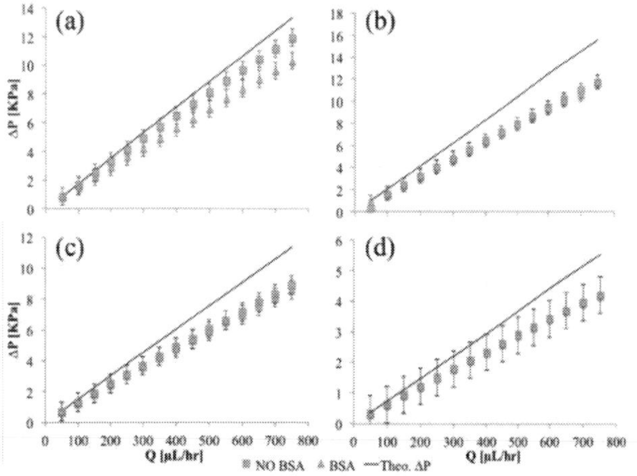

Figure 4. Experimental pressure drop as a function of volumetric flow rate. **(a)** Channel width (w) =100μm, (L)=8mm **(b)** w=200μm, (L)=20mm **(c)** w=400μm, (L)=30mm **(d)** w=800μm, (L)=30mm

Figures 4 and 5 summarize all the results of the pressure drop experiments. The theoretical (no slip, $b=0$) model was represented with a single line whereas single point data corresponds to experimental values. The dashed line in Figure 5 is the solution for Eqn. (7) considering a slip length of 1.5μm. The green dots in Figure 5 represent the solution of a finite element model (COMSOL Multiphysics) considering partial slip (b = 1.5μm).

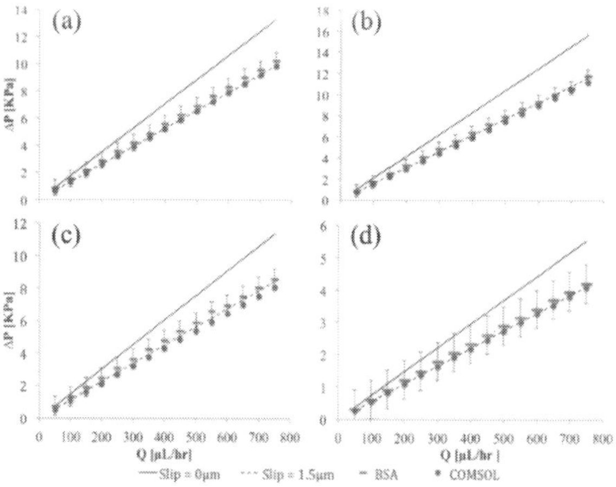

Figure 5. Pressure drop as a function of volumetric flow for PDMS/BSA-water experiments and COMSOL simulations. **(a)** w =100μm, L=8mm **(b)** w=200μm, L=20mm **(c)** w=400μm, L=30mm **(d)** w=800μm, L=30mm

DISCUSSION

For all the micro-channels and both cases under study, it was found that as the volumetric flow increased the error between the theoretical (no-slip condition) and experimental pressure drop increased (Figure 4). This means that the hydraulic resistance, which is the slope of the line, is considerably less than the theoretical one for high volumetric flows. One conclusion would be that since there is less resistance to the flow, there could be a slip velocity at the wall, if the micro-channel is not deforming or behaves as a rigid body. It can also be seen that for most of the channels (Figure 4b, c and d) there is no difference in hydraulic resistance for both BSA and No BSA interfaces, considering the error bars. It is worth to mention that for the lower end of the tested volumetric flows, less than approximately 250μL/hr, the theoretical no-slip model can be used. Once the correction factor was established for each one of the four micro-channels (Fig. 3) a hydraulic resistance equation was found for each channel. The pressure drop for each channel was plotted based on Eqn. (7), and their respective correction factor, for b = 0 (no-slip, solid line) and b = 1.5μm (dashed line) in Figure 5. This data was also plotted against the experimental pressure drop for the PDMS/BSA-water interface. Figure 5 shows that the slip length turned out almost constant, at approximately 1.5 μm, in all of the studied channels with BSA on the wall. This is to be expected to some extent since all the micro-channels walls are made with the same materials (PDMS and glass) and the walls were treated equally. This evidence suggests that the slip phenomenon is independent of the geometry of the channel (surface area) and the flow rate (wall shear stress), contrary to some of the results found the literature [3]. The results given by Eqn. 7 were verified using COMSOL. A model for each micro-channel was created and subjected to a slip length of 1.5μm in all the walls. The numerical solution paired almost exactly with the results from Eqn. 7, validating to some extend the analytical approximation. The presence of apparent slip in the studied micro-channels, supported by the experimental data, represent a reduction of 26% in the theoretical no-slip hydrodynamic resistance. This reduction is considerable and should be considered in the modeling of PDMS-glass micro-channels, especially in high flow rate applications.

ACKNOWLEDGEMENTS

This work was funded by UPRM RISE 2BEST Program, under Grant No. NIH-R25GM088023.

REFERENCES

[1] J. Nilsson et al., *Analytica Chimica Acta 649,141–157*. 2009.

[2] Lauga E. et al., *Handbook of Exp. Fluid Dyn.*, 1219 – 1240. 2007.

[3] Neto, C. et al., *Rep Prog Phys, 68:2859-2897*, 2005.

[4] Perry, C. et al. *Biomicrofluidics 6, 026501*, 2004.

[5] Bhushan, B., *Journal of Nanotechnology*, 2011.

SB³C2015
Summer Biomechanics, Bioengineering and Biotransport Conference
June 17-20, 2015,Snowbird Resort, Utah, USA
SB³C2015-634

NEURAL TISSUE DEFORMATION AND CEREBROSPINAL FLUID FLOW IMPEDANCE ARE POSITEVELY CORRELATED AT THE CRANIOCERVICAL JUNCTION

B. A. Martin (1, 2), N. Shaffer (1,2), J. Oshinski (3), M. Luciano (4), F. Loth (1, 2)

(1) Department of Mechanical Engineering,
Conquer Chiari Research Center,
The University of Akron, OH, U.S.A.

(2) Department of Biomedical Engineering,
The University of Akron,
Akron, OH, U.S.A.

(3) Department of Radiology,
Emory University,
Atlanta, GA, U.S.A.

(4) Department of Pediatric Neurosurgery,
Cleveland Clinic Foundation,
Cleveland, OH, U.S.A.

INTRODUCTION

Type 1 Chiari Malformation (CM) is a devastating neurological disorder that occurs in greater than 1 in 1000 births and is present in over 200,000 people in United States. CM has been historically characterized radiographically by a static morphometric measurement of cerebellar tonsil descent greater than 5 mm below the foramen magnum (FM, **Figure 1**). This malposition of the tonsils causes abnormal forces that act on the central nervous system (CNS) tissue and often results in painful symptoms and permanent neurologic deficit.

A better diagnostic for symptomatic CM is needed. Approximately 1-3% of the population has radiographic CM. However, only ~1 in 30 of these people are regarded as true CM patients. There is concern in the medical community that unnecessary CM operations will be performed unless better diagnostic criteria are developed. Widespread use of MRI has shown it is not unusual for someone to have no symptoms with greater than 5 mm tonsillar descent, or significant symptoms with less than 5 mm descent. The leading CM treatment is a traumatic brain surgery with little consensus for operative techniques. The goal of this study was to assess neural tissue deformation at the craniocervical junction as a diagnostic indicator of symptomatic CM and understand its relation, if any, to cerebrospinal fluid (CSF) flow impedance.

METHODS AND RESULTS

Bulk spinal cord motion (axial PC MRI). Bulk spinal cord motion in the axial direction at the FM and C2 level was assessed for symptomatic adult CM patients pre- (N=20) and post-surgery (N=16) and compared motion to healthy controls (N=14) (**Figure 2**). Spinal cord velocities were obtained by standard PC MRI measurements sensitized for CSF movement. Displacement profiles were quantified based on the integral of the average spinal cord velocities at the FM and C2. Results showed that bulk spinal cord motion at the FM for patients pre-surgery was 213% greater than healthy controls (mean 0.49 versus 0.23 mm, $p < 0.0001$) and significantly decreased to 27% after surgery (mean 0.36 mm, p=0.03).

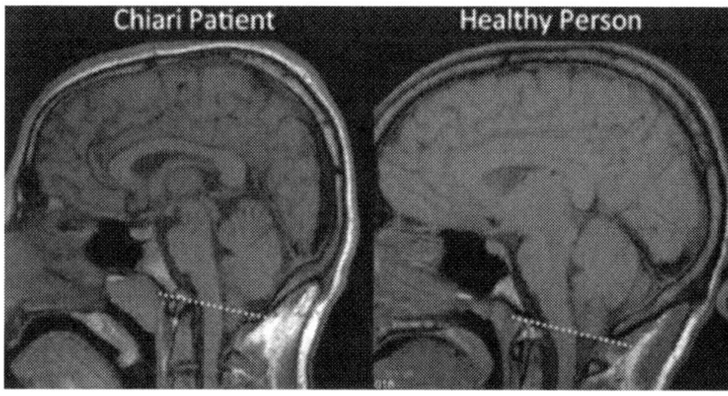

Figure 1. CM patient (left) with tonsillar descent (red) below the FM (yellow) and comparison to a healthy person (right).

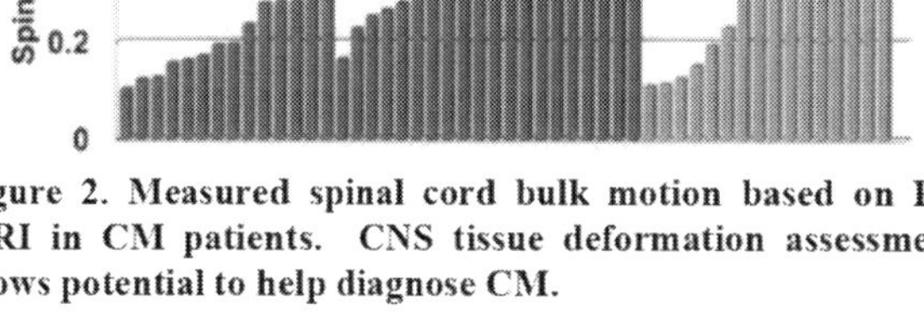

Figure 2. Measured spinal cord bulk motion based on PC MRI in CM patients. CNS tissue deformation assessment shows potential to help diagnose CM.

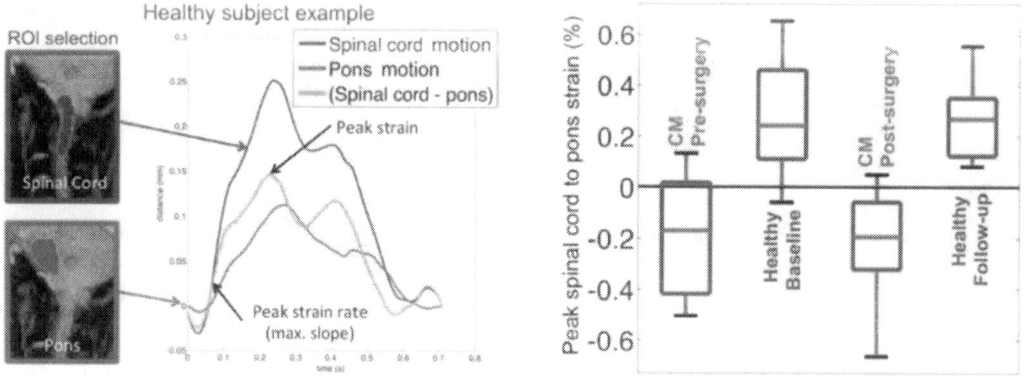

Figure 3. Quantification of strain based on PC MRI measurement of tissue motion between the spinal cord and pons (left). Peak strain varies dramatically (opposite-sign) between healthy controls and CM patients (right).

Neural tissue motion and strain (sagittal PC MRI): We quantified CNS tissue bulk motion and strain based on a high-speed sagittal phase-contrast MRI technique [1]. Deformation was assessed in controls (N=6) and CM patients pre- (N=9) and post-surgery (N=6) in terms of 1) bulk displacement of the spinal cord, pons and cerebellar tonsils and 2) maximum strain acting between the spinal cord and pons (**Figure 3**, left). Results showed a similar trend as the axial measurements; with CM patients having greater tissue deformation pre-surgery compared to controls and a decrease post-surgery. However, additional insight into the biomechanics was possible with strain quantification between the pons and spinal cord. The spinal cord was found to be compressed (peak strain = -0.20%) pre-surgery and stretched in controls (+0.32%, p=0.007) (**Figure 3**, right). Follow-up measurements showed little variation. Bulk tonsil displacement was 198% greater for patients pre-surgery than controls (p=0.047). Bulk tonsil, cord and pons motion all decreased post-surgery (from 11 to 30%), but the limited subject numbers (N=6 post-surgery) lacked statistical differences.

Correlation of CNS tissue motion and pressure dissociation. We found a positive correlation between cardiac related tonsillar bulk motion and craniospinal pressure dissociation computed non-invasively across the FM by subject-specific computational fluid dynamics modeling of CSF for each subject in this study [2,3] (**Figure 4**, top, N=26, R^2 = 0.73, P<0.001). Positive correlation was also found between CSF flow impedance at the FM and tonsillar bulk motion (R^2=0.56, P<0.001, regression not shown). For these same subjects, we also found tonsillar bulk motion to be correlated with tonsillar descent (R^2=0.51, P<0.001, **Figure 4**, bottom).

DISCUSSION

Pathophysiological importance of tissue deformation: Repetitive deformation of the CNS tissue occurs over 30 million cardiac cycles per year due to intracranial pressure differences (dissociation) caused by the pulsatile intracranial blood flow and cerebrospinal fluid (CSF) motion. Our preliminary findings support that CNS tissue deformation (tonsillar bulk motion) acts as a dynamic ball-in-cone "valve" [4-6] leading to increased pressure dissociation and plastic deformation of the tonsils (tonsillar descent). The resulting stress that acts on the CNS tissue and surrounding meninges is complex in terms of magnitude and distribution and, when abnormal, may damage tissue and result in CM symptoms. While tissue stress cannot be directly measured non-invasively, stress produces deformation that can be measured non-invasively by MRI and biomechanically quantified in terms of bulk motion, tension, compression and shear by engineering and imaging analysis. We hypothesize that symptomatic CM is correlated with MRI measurement of neural tissue dynamic deformation within specific regions of the CNS tissue. This MRI measurement may prove to be a valuable CM diagnostic indicator.

Limitations. One must use caution interpreting these results, as correlation is not causation. We do not expect tonsillar descent to correlate exactly with tonsillar bulk motion, as tonsillar bulk motion may better stratify CM patients. Future studies with greater number of subjects and usage of advanced imaging tools, such as DENSE MRI, may help to further understand the importance of cardiac related CNS tissue deformation and CSF system pathologies such as hydrocephalus and, CM, pseudotumor cerebri and others.

ACKNOWLEDGEMENTS

The authors acknowledge Conquer Chiari and NIH Grant No. 1R15NS071455-01 for support of this work.

REFERENCES

[1] Kalata, Martin, B. A. et al., 2009, Ieee T Bio-Med Eng.
[2] Martin, B. A., Kalata, W., et al. 2013, PLoS One.
[3] Shaffer, N., Martin, B. A., 2013, JBME.
[4] Oldfield, E. H., et al., 1994, J Neurosurg.
[5] Greitz, D., 2006, Neurosurgical review.
[6] Williams, B., 1981, Acta Neurochir.

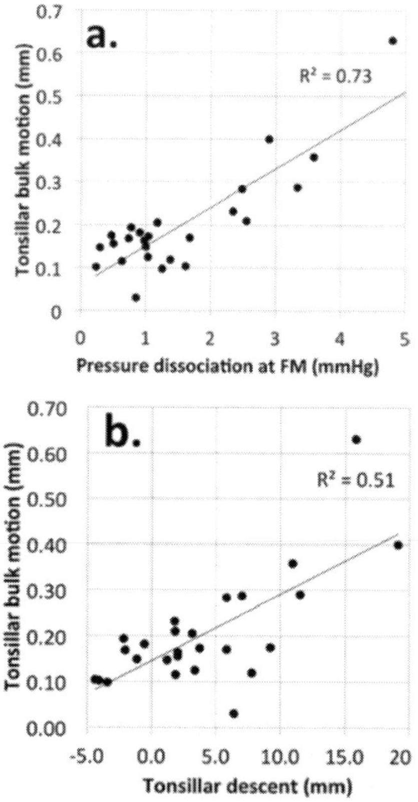

Figure 4. Relation of a) tonsillar motion and longitudinal pressure dissociation and b) tonsillar motion and descent.

SB³C2015
Summer Biomechanics, Bioengineering and Biotransport Conference
June 17-20, 2015, Snowbird Resort, Utah, USA
SB³C2015-635

NUMERICAL MODELING OF SCALP COOLING DEVICES FOR THE PREVENTION OF CHEMOTHERAPY-INDUCED ALOPECIA

Bradley L. Pliskow (1), Kunal Mitra (1), Mehmet Kaya (1)

(1) Department of Biomedical Engineering
Florida Institute of Technology
Melbourne, Florida, United States

INTRODUCTION

Alopecia continues to be one of the most distressing and traumatic side effects for patients undergoing chemotherapy treatment. The impact of hair loss on patients is regularly underestimated by health care professionals. There is a critical need to address this side-effect of chemotherapy in order to improve patient quality of life.

Scalp cooling was introduced in the 1970's as a method of reducing or preventing chemotherapy-induced alopecia. The results of scalp cooling have been very promising, providing better results than existing pharmacological methods. However, the effectiveness of scalp cooling varies considerably; implementations of scalp cooling have differed in the type of cooling device, level of hypothermia, duration of hypothermia, and administered chemotherapy drug/dose. The type or design of the cooling device for this procedure is an important factor that has not been thoroughly studied [1].

Modeling of heat transfer within the human head may provide insight into various aspects of scalp cooling device design. Previously, a numerical model was developed to study the scalp cooling procedure by quantifying the relationship between temperature and perfusion in the scalp [2,3]. However, numerical models have not been used previously to study the scalp cooling device. In the present work, a numerical model is developed to investigate factors related to a cooling device design for scalp cooling. Based upon these simulations, it is possible to study system parameters affecting performance of these devices.

METHODS

A numerical model of heat transfer is developed to simulate the interaction of a cooling cap system with the human head. A 3-D, transient, heat transfer analysis is conducted using Comsol Multiphysics software. The model geometry is discretized into two domains each described by its own set of differential equations: the head tissue and the cooling cap.

The head geometry is created as a half-ellipsoid in 3-D space based upon the head sizes from NIOSH Anthropometric Data and ISO Digital Headforms (Fig. 1) [4]. The head tissue is divided into different, homogenous layers: white matter, gray matter, skull, galea, hypodermis, dermal plexus, dermis, and hair. Each of these layers is represented using thermo-physical properties according to literature values [2,5,6,7] not presented due to brevity.

Figure 1. Head geometry modeled as a half ellipsoid in 3-D space.

Similar to previous scalp cooling models, Penne's bio-heat equation is used as a continuum approach to describe heat transfer within human tissue. Penne's bio-heat equation describes the collective effect of blood vessels as a heat sink, with all heat transfer occurring at the capillary level [8].

$$\rho c \frac{\partial T}{\partial t} = \nabla \cdot k\nabla T + \rho_b c_b \omega_b (T_A - T) + q_m , \qquad (1)$$

where T is the tissue temperature, t the time, q_m the metabolic heat production, and T_A the temperature of the main supplying arteries. Here, k, ρ and c are the thermal conductivity, density and specific heat of the tissue, respectively. When accompanied by the subscript 'b', these symbols refer to the properties of the blood, instead of the tissue.

The cooling cap geometry fits around the head and is created based upon the design being simulated. The geometry consists of the cap material and coolant. Heat transfer within the cooling cap is modeled according to the heat equation:

$$\rho c_p \frac{\partial T}{\partial t} + \rho c_p u \cdot \nabla T = \nabla \cdot (k\nabla T) + Q , \qquad (2)$$

where u is the velocity field fluid determined by solving the corresponding Navier-Stokes equation. In solids, the velocity field is zero and the equation can be simplified appropriately.

Changes in blood perfusion and metabolic heat generation due to temperature are modeled by the established Q-10 effect [2,6]. The interfaces between each of the domains are represented by boundary conditions such as convection and radiation.

First, the cooling power needed for any device to achieve therapeutic levels of scalp hypothermia is investigated. Subsequently, a liquid-cooled cap design is simulated using the model.

RESULTS

A constant, outward heat-flux is applied to the scalp (through the hair/air layer) such that heat energy is removed from the patient's head tissue. The resulting temperature profiles as a function of cooling power (Fig. 2) show that nearly 40 W of cooling power is necessary to achieve a therapeutic level of hypothermia. When at least 38.5 W of cooling power is supplied by the device, the hair follicle (at a subcutaneous depth of 2 mm) is effectively cooled to 22°C. After initiation of cooling, most of the temperature decrease occurs within the first 25-30 minutes (Fig. 3). When cooling is stopped after 3 hours, the effect of cooling is lost within the first 5 minutes (Fig. 3).

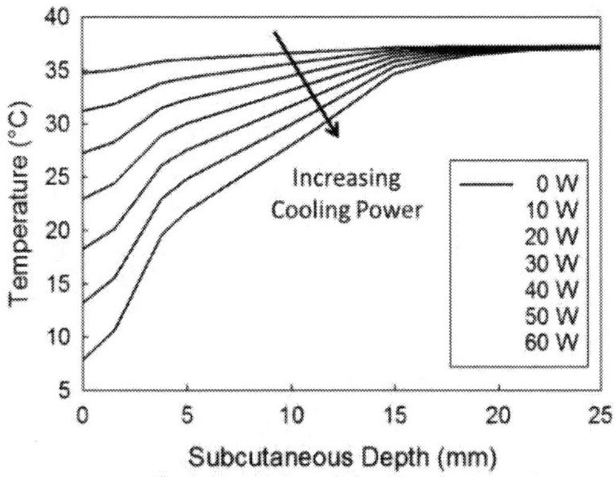

Figure 2. Steady-state temperature profile as a function of cooling power supplied by the device.

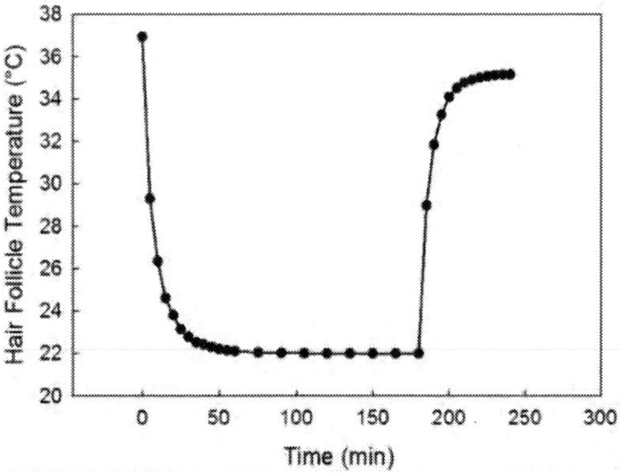

Figure 3. Hair follicle temperature as a function of time during the scalp cooling procedure (cooling power = 38.5 W).

A liquid-cooled cap design is simulated by flowing 25 gph of water at 5°C through a silicone cap that fits directly on top of the hair/air layer. Due to entrapped air within the hair layer, the hair follicle (depth of 2 mm) is only cooled to a temperature of 30°C. Increasing the thermal conductivity of the hair layer significantly improves the cooling effect of the cap (Fig. 4).

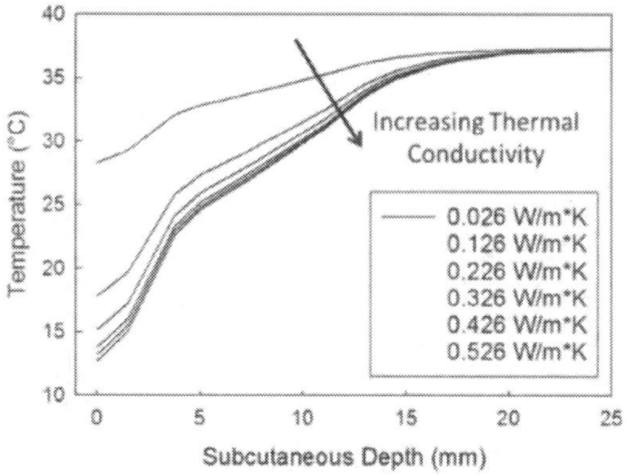

Figure 4. Steady-state temperature profile for a liquid-cooled cap design as a function of the thermal conductivity of the hair layer.

DISCUSSION

In this work, a numerical model of heat transfer is utilized to investigate parameters for a device designed specifically for the prevention of chemotherapy-induced alopecia. The model is first utilized to determine the heat transfer requirements for a cooling device that provides therapeutic hypothermia for an average patient. The transient, cooling response after initiation of cooling agrees well with previous models and is a confirmation for the need of a 30 minute pre-infusion cooling period. The transient, rewarming response shows that removal of the cap, even for a few minutes, causes significant warming of the scalp tissue. Therefore, during scalp cooling, removal of the cap for any period of time should be avoided when possible. The model is then used to simulate a sample design for a cooling cap. When a liquid-cooled cap (with coolant at 5°C) is used to cool the scalp, the thermal resistance of even a small hair layer (2 mm thick) prevents the therapy from being effective. Since a liquid-cooled cap design achieves heat transfer through conduction, it is necessary to moisten the hair with water or a water-based conditioner prior to the procedure.

The modeling tools developed here provides a powerful platform for analyzing the effect of design parameters for a scalp cooling device. Future studies using this model can consider other implementations of cooling devices, such as phase-change or convective air-cooling. Additionally, variations of patient-specific parameters can be utilized instead of parameters for an "average" patient.

REFERENCES

[1] Breed W.P., et. al., *Expert Rev. Dermatol.*, 6(1):109–125, 2011.
[2] Janssen F.E., et. al., *Phys Med Biol*, 50(17):4065– 4073, 2005.
[3]Janssen F.E., et. al., *Strojniski Vestnik/Journal of Mechanical Engineering*, 51(7-8):386-390, 2005.
[4] Zhuang Z. and Bradtmiller B., *J Occup Environ Hyg*, 2(11):567-76, 2005.
[5] van Leeuwen G.M., et. al., *Pediatr. Res.*, 48(3): 351-356, 2000.
[6] Dennis B.H., et. al., *J Biomech Eng*, 125(6):832-840, 2003.
[7]Nelson D.A. and Nunneley S.A., *Eur J Appl Physiol*, 78(4):353–359, 1998.
[8] Pennes H.H., *J Appl Physiol*, 1(2): 93-122, 1948.

SB^{3}C2015
Summer Biomechanics, Bioengineering and Biotransport Conference
June 17-20, 2015, Snowbird Resort, Utah, USA
SB^{3}C2015-636

IN VITRO AND COMPUTATIONAL FLUID DYNAMICS COMPARISON OF THE FLOW DIVERSION EFFICACY OF FIVE COMMERCIAL STENTS

Ronak J. Dholakia (1), Andrew Pagano (1), Fotis Drakopoulos (2), Ari Kappel (1), Chander Sadasivan (1), Xiangmin Jiao (3), David J. Fiorella (1), Nikos Chrisochoides (2), Henry H. Woo (1), Baruch B. Lieber (1)

(1) Department of Neurological Surgery
Stony Brook University
Stony Brook, NY, USA

(2) Department of Computer Science
Old Dominion University
Norfolk, VA, USA

(3) Department of Applied Mathematics and Statistics
Stony Brook University
Stony Brook, NY, USA

INTRODUCTION

Brain aneurysms are abnormal ballooning of intracranial arteries which when left untreated may result in fatal outcome for the patients. Neurovascular stents have been used extensively in stent-assisted coiling treatment of aneurysms. Treatment with flow diverters (low porosity, fine-mesh stents) is used for complicated aneurysm geometries not amenable to coiling. Angiographic analysis with mathematical modeling has been applied for comparison of different flow diversion designs as well as towards prediction of flow diversion efficacy [1]. High resolution computational fluid dynamics (CFD) simulations would provide the opportunity to study localized alterations in hemodynamics due to different devices. Current class of CFD studies with flow diverters in cerebral aneurysm models do not use realistic wire configurations in their models and the fine differences in porosity and pore density across the aneurysm neck might produce inaccurate results. The proposed study aims to test five commercially available neurovascular devices that have not been compared before through angiographic washout analysis and accurate microCT-based mesh reconstruction and CFD analysis in idealized sidewall aneurysm models.

METHODS

Five identical silicone elastomer replicas of an idealized geometry (Figure 1) were manufactured by dip coating. Parent artery diameter of the model was selected to be 4 mm and aneurysm diameter of 7 mm. Briefly, the geometry was printed with a rapid prototyper, the printed core was dipped in solvents to reduce surface roughness and then coated evenly with silicone using an in-house multi-axis spinner. After curing of the silicone, the core was dissolved to obtain a compliant luminal replica of the geometry. Five commercially available neurovascular stents including Pipeline (Covidien Medtronic), FRED, LVIS (Microvention Terumo), Enterprise (Codman J&J), and Neuroform (Boston Scientific Stryker) were implanted in idealized sidewall aneurysm silicone models (Figure 1).

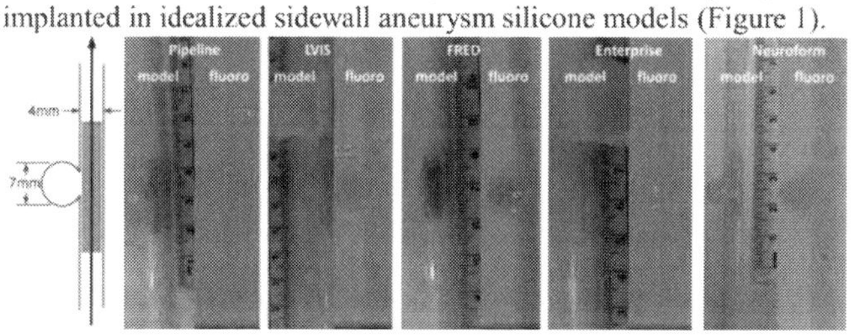

FIGURE 1: SCHEMATIC OF THE IDEALIZED SIDEWALL ANEURYSM MODEL (LEFT) AND CAMERA AND FLUOROSCOPIC SHOTS OF THE FIVE DEVICES IMPLANTED IN SILICONE MODELS

High speed angiographic washout data was acquired in all models at 15 frames/sec under steady flow through a peristaltic pump at approximately 3.5 ml/sec. As a blood replica, 50% glycerol solution was used to obtain a fluid with viscosity of 4cP at a temperature 37C. The silicone model was fit in position in the flow loop and kept immobile while acquiring the images. A 4 French injection catheter was positioned such that the catheter tip was about 50 cm away from the region of interest leaving sufficient length for contrast mixing with the working fluid. Contrast was injected through a power injector at 2.5 ml/sec for 2 seconds with a 2 second delay in injections to allow the C-arm to acquire the mask for the angiogram. Angiographic runs were acquired for a total of 6 times through each model (3 with the inlet and outlet reversed) to ensure repeatability of the data.

The aneurysm region of interest was segmented in the angiographic images to calculate the total gray scale intensity at every point in time and thus the contrast concentration-time or washout curves (Figure 2) were acquired. The contrast concentration-time curves (washout curves) within the aneurysm were then fitted using a least square approach to a previously developed mathematical model [1] that differentiates convective and diffusive flow regimes within the aneurysm.

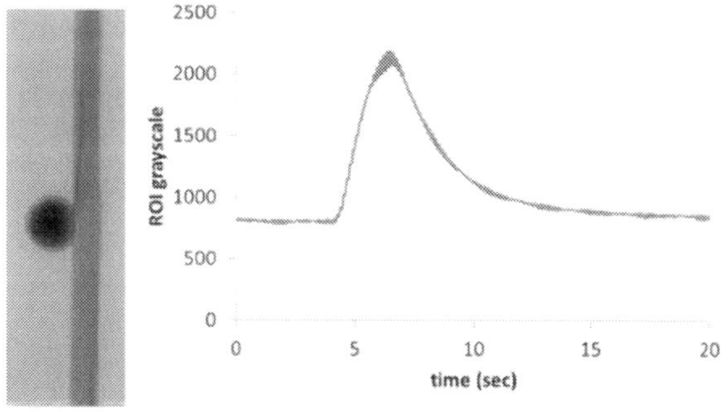

FIGURE 2: SEGMENTED ANEURYSM REGION OF INTEREST (ROI) AND CORRESPONDING CONTRAST CONCENTRATION-TIME CURVE

To obtain the exact implanted position inside the silicone model as well as geometry of the devices, the models were imaged with microCT (uCT50 Scanco Medical, Bruttisellen, Switzerland). All models except FRED were scanned at a resolution of 12microns except FRED at 6microns.

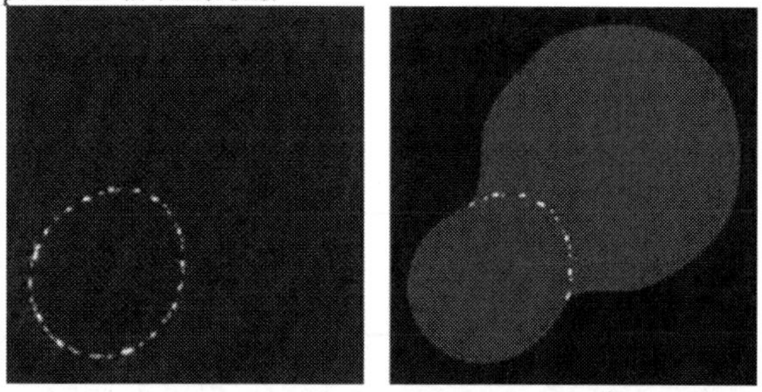

FIGURE 3: RAW MICROCT SLICE FOR PIPELINE MODEL (LEFT) AND AFTER SEGMENTATION WITH THE DEVICE WIRES CROPPED (RIGHT)

Mimics (Materialise, Leuven, Belgium) software is used to perform image segmentation on the microCT data. Two different labels are tagged to the lumen of the sidewall aneurysm model and the flow diverter. During segmentation the device is cropped to only retain the portion providing neck coverage of the aneurysm (Figure 3).

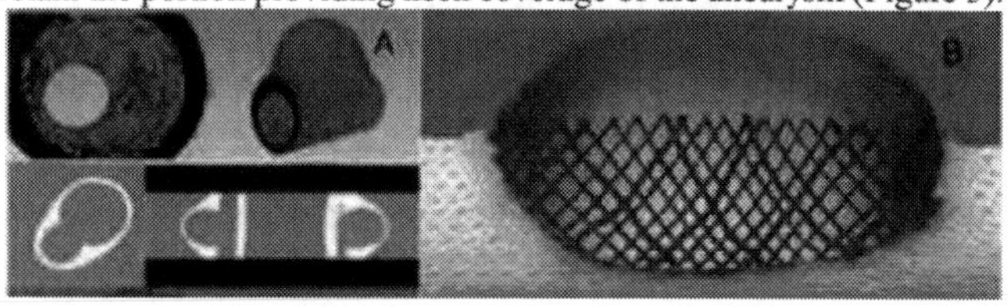

FIGURE 4: ASSEMBLY OF THE 3D OBJECT FROM 2D AXIAL SLICES (A), AND GRADATION AND THE SMOOTHNESS OF THE MESH (B)

We use an automatic and high fidelity image-to-mesh (I2M) conversion scheme that involves combination of the axial image slices using Insight toolkit's (ITK) tile image filter to stack the 2D slices and generate the volume label image that serves as an input for the creation of a Body-Centered Cubic (BCC) which creates a uniform structured BCC lattice by modifying an existing approach [2]. The BCC lattice is refined according to local feature size of labels in the 3D segmented image. The resolution of each labeled object is automatically adjusted based on the user-defined fidelity. Finally, the Mesh Compression step is used to smooth the BCC mesh in order to match the objects' boundaries (Figure 4). The mesh generated as a result is imported in ADINA 9.0 and CFD simulations in the finite element formulation will be run through each computational model.

RESULTS and DISCUSSION

Preliminary angiographic analysis (n=3 each case) indicates reduction in the amplitude of the concentration curves following implantation of the devices. Reduction in amplitude was largest for Pipeline and FRED (Figure 5). Washout parameters obtained following optimization indicate delayed convection (increase in convective time constant) for FRED as well as Pipeline as compared to the control and other neurovascular stents. The proportion of contrast transport governed by diffusion (diffusive amplitude) increased after implantation of Pipeline and FRED devices (Figure 6), with a corresponding decrease in convective amplitude.

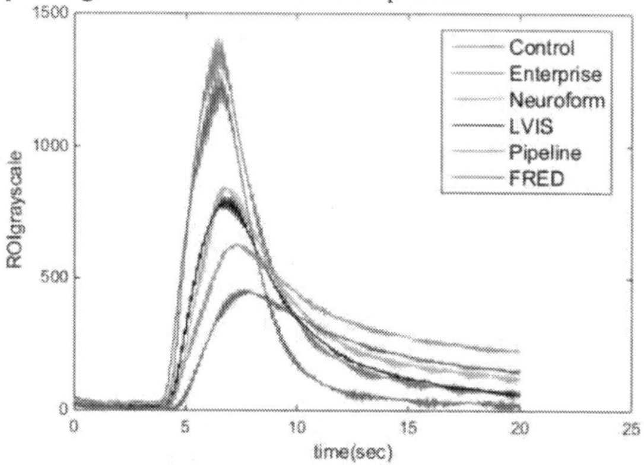

FIGURE 5: CONCENTRATION TIME CURVES FOR CONTROL MODEL WITHOUT ANY DEVICE AND THE FIVE MODELS WITH DEVICES

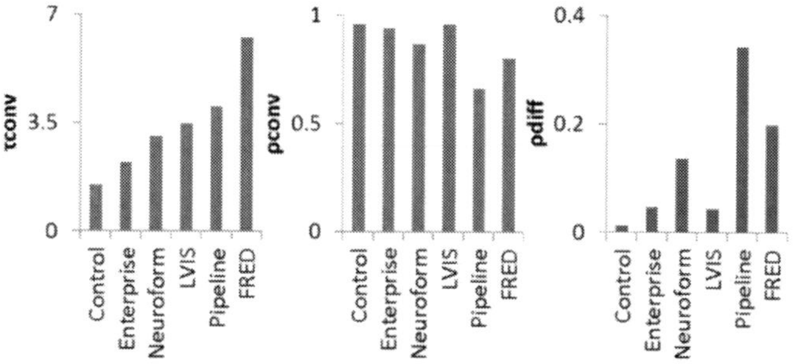

FIGURE 6: COMPARISON OF CONVECTIVE DECAY TIME CONSTANT, CONVECTIVE TRANSPORT AMPLITUDE, AND DIFFUSIVE TRANSPORT AMPLITUDE

Preliminary results show better performance of devices designed to be flow diverters. Following completion of microCT data based CFD analysis, we aim to correlate the angiographic parameters with the CFD parameters and compare device performance vis-à-vis implanted porosity and pore-density.

REFERENCES

1) Sadasivan C, et al., IEEE Transactions on Medical Imaging. 2009;28(7):1117-25
2) Y. Liu, et al., Engin. with Comp., 2012;28(4):305-318

SB³C2015
Summer Biomechanics, Bioengineering and Biotransport Conference
June 17-20, 2015,Snowbird Resort, Utah, USA
SB³C2015-637

THE QUANTIFICATION OF BLOOD FLOW PATTERNS INDUCED BY ENDOVASCULAR STENT GRAFTS USING A NON-NEWTONIAN BLOOD ANALOG

Amanda Colella-Centazzo (1), Clifton R Johnston (1)

(1) Department of Mechanical Engineering
Dalhousie University
Halifax, NS, Canada

INTRODUCTION

Endovascular stents have become the preferred treatment of many cardiovascular disorders, particularly stenotic and aneurysmal diseases. This method of treatment involves implanting a stent through the affected length of vasculature to re-establish a normal vessel diameter and restore adequate blood flow downstream. In the case of aneurysmal disease, the stent is introduced across the enlarged section to reduce the lumen and contain the flow as it passes through the aneurysm.

Despite widespread use of stents to treat aneurysms, they remain less effective long-term and lead to a significant increase in secondary interventions as compared to open repair surgery [1], while showing no significant difference in survival after 2 years [2].

Many of the failure mechanisms are associated with local blood flow disturbances, which have been shown to influence the behavior of the endothelial cells lining the arterial wall. More specifically, areas of low wall shear stress (WSS), as well as oscillatory nature of the fluid as it interacts with the vessel wall have been shown to affect the progression of atherosclerosis, the proliferation of intimal hyperplasia and the formation of thrombi [3]. These fluid dynamic considerations directly affect the long-term patency of stent grafts and have therefore prompted investigation into how the presence of the stent alters the local hemodynamic environment.

An environment representing that of the human femoral artery was constructed with optical access to the blood analog as it exits an implanted stent graft. Using Particle Image Velocimetry (PIV) the flow patterns were imaged and the changes to the peak velocity and WSS induced by stent implantation were quantified.

METHODS

A closed flow loop was designed and constructed for visualization and measurement of flow through a 10mm acrylic tube, representing a simplified geometry of the femoral artery. A common non-Newtonian blood analog representative of physiological hemodynamic parameters was circulated through the test section, comprised of a mixture of glycerol (40%), xantham gum (0.04%) and water. [4]. The mixture was exposed to a 1Hz pulsatile flow profile typical of the human femoral artery (Figure 1), with a peak velocity of 47cm/sec [5].

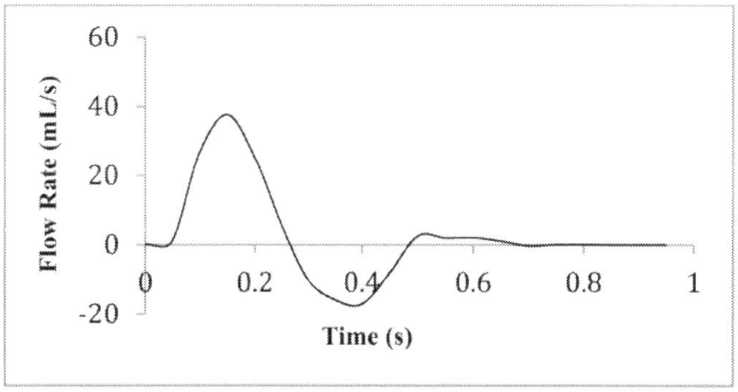

Figure 1 – 1 Hz Input waveform representing pulse profile typical of the human femoral artery.

Using PIV, cross-sectional velocity and wall shear stress immediately downstream of a Gianturco Zenith stent graft (Cook Medical, Inc., Bloomington, IN, USA) were quantified. Images were collected after deployment of a full stent, individual stent wires, as well as a baseline flow with no stent in the test section. Arrangements considered were 1 stent wire, 2 wires, 3 wires, and a full jacketed 12mm stent. All wires were from a 12 mm stent and oriented in the test section as they would be in a full stent.

The test section was embedded in an acrylic flow chamber, which was filled with 100% glycerol to correct for the refraction index differences between the circulating fluid and the surrounding acrylic tube and allow the laser sheet to reach the test area unaffected. The fluid was seeded using 10 μm silver coated hollow glass spheres, which were illuminated using a 1mm thick laser sheet.

Image frames were collected using a Photron mini UX100 at 4000fps with 20 image pairs collected per pulse, over 10 pulses. These images were then processed using PIVLab [6], using a multi-pass cross-correlation algorithm with a 50% window overlap and final interrogation window size of 32 x 32 pixels. Instantaneous velocity vector fields were averaged for each time point to obtain 20 vector maps across the pulse for each stent arrangement.

RESULTS

A 23.3% increase in centerline velocities at peak flow was observed upon introduction of the stent wire, however no difference was observed between the single, double, or triple wire configurations. Implantation of the full stent graph doubled centerline velocity at peak flow, increasing from 46.92 cm/s to 92.67 cm/s (Figure 2).

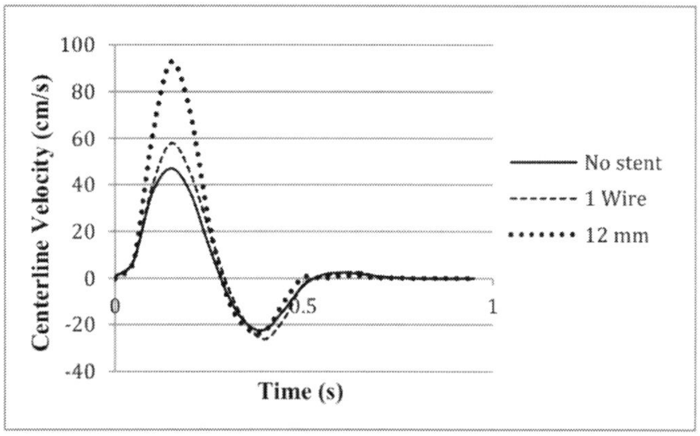

Figure 2 - Centerline velocities in stented and non-stented flow across pulse cycle at x/R=0.5 from stent outlet.

Along the near wall, the presence of a single stent wire resulted in a decrease in velocity along with a slight decrease in WSS. No discernable variations were observed in the velocity profiles when subsequent stent wires were introduced. The implantation of a full stent resulted in a substantial decrease in the velocity profile from the wall to x/R=0.66 (Figure 3). Additionally, the local WSS at peak flow fell substantially from 2.00 Pa in the non-stented case to 0.183 Pa (Table 1).

Table 1- Wall shear stress values at peak pulse measured at x/R=0.5 from stent outlet.

	No Stent	1 Wire	Full Stent
Peak WSS (Pa)	2.00	1.64	0.183

DISCUSSION

We have demonstrated that the introduction of stent grafts affects the fluid properties downstream of the stent which are known to cause disease proliferation and decrease stent patency. Furthermore, the implantation of multiple stent wires without a cover in sequence was shown to have no effect on the fluid dynamics as compared to a single wire.

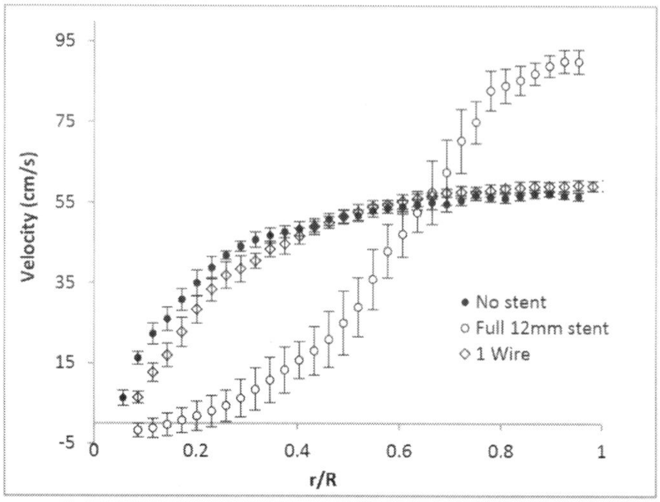

Figure 3 - Averaged velocity profiles from the near wall to the centerline at peak flow measured at x/R=0.5 from stent exit. Error bars represent standard deviation of velocity measurements.

As expected the stent struts increased friction at the wall, causing a drop in velocity in this area. However, previous studies using a 6.25 mm diameter tube, also implanted with a single stent wire had shown much larger variations in velocity profile and WSS [4]. This could be due in part to the smaller tube causing a larger ratio of stent wire to tube radius as compared to the present study.

The introduction of a full 12mm stent, the standard sizing for this vessel, caused considerable changes to the velocity profile observed at the stent outlet and induced low shear stresses at the wall. WSS in this range have been shown to encourage platelet adhesion and intimal hyperplasia [3].

REFERENCES

[1] Greenhalgh, R. (2004). Comparison of endovascular aneurysm repair with open repair in patients with abdominal aortic aneurysm (EVAR trial 1), 30-day operative mortality results: randomised controlled trial. *The Lancet*, *364* (9437), 843-848.
[2] De Bruin, J., Baas, A. F., et al. (2010). Long-Term Outcome of Open or Endovascular Repair of Abdominal Aortic Aneurysm. *The New England Journal of Medicine*, *362* (20), 1881-1889.
[3] Chiu , J.-J., & Chien, S. (2011). Effects of Disturbed Flow on Vascular Endothelium: Pathophysiological Basis and Clinical Perspectives. *Physiological Reviews*, *91* (1), 327–387.
[4] Walker M. , A., Johnston R. , C., & Rival, D. (2012). The Quantification of Hemodynamic Parameters Downstream of a Gianturco Zenith Stent Wire Using Newtonian and non-Newtonian Analogue Fluids in a Pulsatile Flow Environment . *Journal of Biomechanical Engineering*, *134* (11).
[5] Fronek, A., Coel, M., & Bernstein, E. (1976). Quantitative Ultrasonographic Studies of Lower Extremity Flow Velocities in Health and Disease. Circulation, 53(6), 957-960.
[6] Thielicke, W., & Stamhuis , E. (2014). PIVlab – Towards User-friendly, Affordable and Accurate Digital Particle Image Velocimetry in MATLAB. *Journal of Open Research Software*, *2* (1), e30.

DEVELOPMENT OF A COMPUTATIONALLY EFFICIENT FULL HUMAN BODY FINITE ELEMENT MODEL

Doron Schwartz (1,2), Berkan Guleyupoglu (1,2), Bharath Koya (1,2), Joel D. Stitzel (1,2), F. Scott Gayzik (1,2)

(1) Wake Forest School of Medicine,
Winston Salem, NC, USA

(2) Virginia Tech – Wake Forest University
Center for Injury Biomechanics,
Winston Salem, NC, USA

INTRODUCTION

Computer simulations using Finite Element Analysis (FEA) are becoming a more prominent tool in the study of motor vehicle crash safety. FEA models of the human body have the potential to greatly enhance the volume of biomechanically relevant data obtained from crash simulations. The Global Human Body Models Consortium (GHBMC) is an international consortium of industry, government, and academic research groups that have consolidated development efforts to produce advanced human body finite element models for trauma research. The GHBMC detailed 50th percentile occupant (M50-O) is an anatomically detailed finite element model with 1.3 million nodes and 2.1 million elements [1, 2]. While this complexity provides the ability to investigate the biomechanics of specific injuries, it results in a computationally expensive model. The M50-O runs a 60 ms hub impact simulation in approximately 10 hours on 48 CPU computational cluster. Since this level of detail is not always required for a given study, there is a need for a complementary, yet computationally inexpensive model.

Therefore, the goal of this study is to develop a simplified and computationally efficient human body finite element model. This model complements the M50-O by providing kinematic and kinetic data with a significantly reduced run time, using the same body habitus.

METHODS

The simplified occupant model (M50-OS) was developed using the same source geometry as the M50-O. While some meshed components were preserved, the total element count was reduced by re-meshing, homogenizing, or in some cases omitting structures that are explicitly contained in the M50-O (Figure 1).

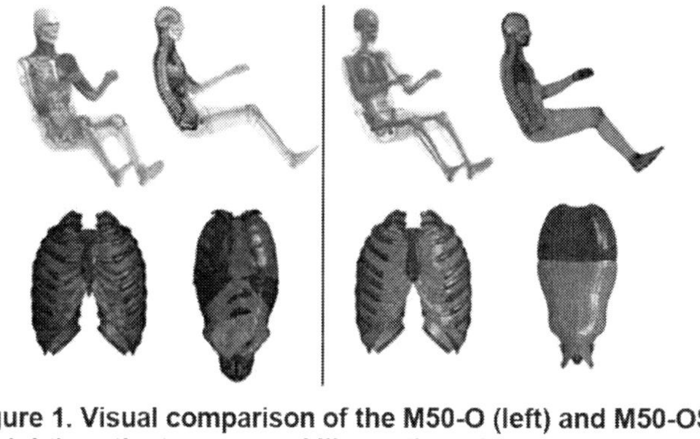

Figure 1. Visual comparison of the M50-O (left) and M50-OS (right) on the top row and illustration of the deformable thoraco-abdominal region of each model on the bottom.

Bones are included as rigid bodies, with the exception of the ribs which are deformable but were re-meshed to a coarser element density than the M50-O. Material models for all deformable components were drawn from the biomechanics literature and/or the M50-O. Kinematic joints were implemented at major articulations (shoulder, elbow, wrist, hip, knee, and ankle) and the model is designed for positioning the extremities as a pre-processing operation.

The M50-OS was simulated in multiple boundary conditions to ensure the stability and accuracy of the model. A denuded rib point-loading test, four rigid impacts, and a frontal sled pulse were completed. The results from dynamic simulations were quantitatively evaluated for magnitude and phase agreement with the experimental data using an objective comparison software package, CORrelation and Analysis (CORA v. 3.5.1).

RESULTS

The M50-OS model has 11 contacts and 354 thousand elements, 205 thousand of which are rigid and 149 thousand deformable; by contrast the M50-O model has 447 contacts and 2.2 million elements. Additional numerical comparisons are listed in Table 1.

Table 1. M50-OS element count and modeling approach in highlighted areas compared to the M50-O.

Structure	M50-O (n elements)	M50-OS (n elements)	% Change
All Deformable Structures	2,109,533	149,125	-93%
All Rigid Structures	81,273	204,694	+252%
Total	*2,190,806*	*353,819*	*-84%*

A denuded rib compression, four rigid body impacts, and one sled simulation were completed to assess the kinematic and kinetic response of the M50-OS. Denuded tests showed a good match of force vs. deflection slopes to the experimental data (Figure 2) [3]. The frontal rigid hub impact simulation the M50-OS produced a peak force and deflection within the corridor of 4.50 kN and 29.1% respectively [4]. Two rigid tests showed similar results and run times with peak forces of 4.4 kN, 7.7 kN for the abdominal bar impact (Figure 3) [5] and lateral sled impact (Figure 4) [6], respectively. The lateral plate impact at 12 m/s exhibited two peaks of roughly 20 kN (due to stiff foam used around the joints) but a more biofidelic response immediately afterward, plateauing at 9 kN at 12 ms (results not shown due to space restrictions). [7]. For all rigid impacts, the average run time was 38 times faster than its detailed counterpart. Results from a 150 millisecond frontal sled simulation [8] showed that reaction forces and kinematic trends matched experimental results well (Figure 5).

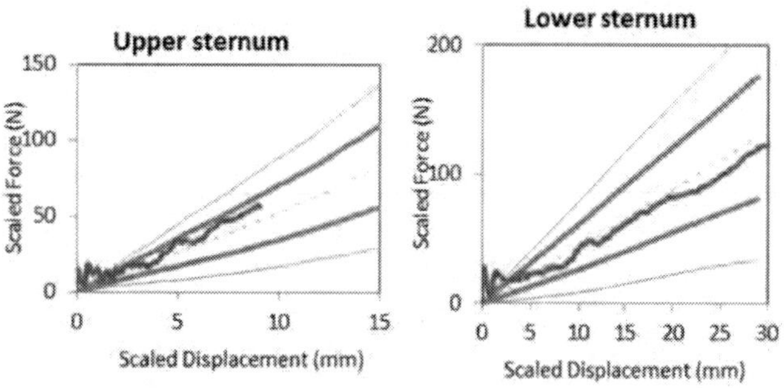

Figure 2. M50-OS force vs displacement, denuded ribs [3].

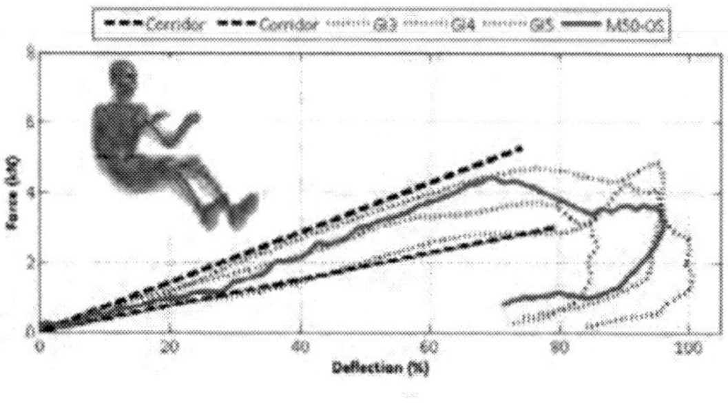

Figure 3. M50-OS force vs deflection, abdominal impact [5].

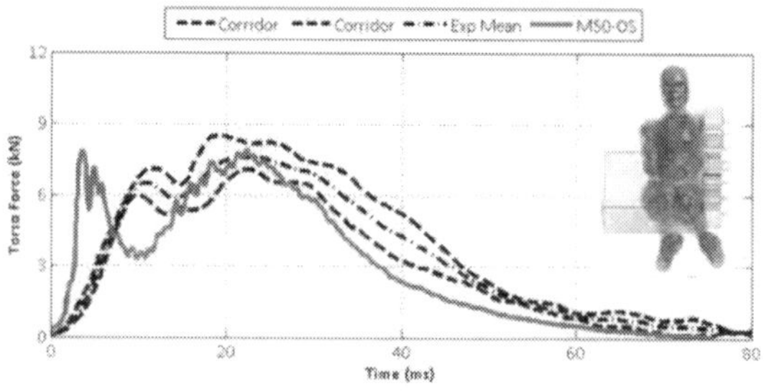

Figure 4. M50-OS force vs time, lateral sled impact [6].

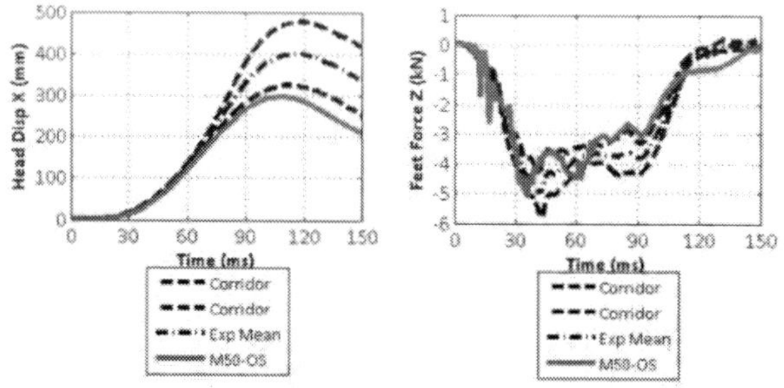

Figure 5. M50-OS head X displacement and feet Z force results, frontal sled. [8]

DISCUSSION

The global responses compared favorably with the literature force and displacement curves with peaks occurring at the correct phase in time and of an appropriate magnitude, in general. The denuded rib cage test showed that despite using a more coarse mesh the response closely matched the literature. Additional alterations to the material models utilized in the deformable components of the model may help improve the biofidelity of the M50-OS.

While local injury prediction capability is not an intended use of this model, the simplified model provides numerous benefits. It can be re-positioned rapidly, provides a potentially more biofidelic alternative to ATD models due to its anatomically based design, and can be used in parametric studies. Finally, because it has the same body habitus as the M50-O model, it can be used as a modular platform for regional biomechanics studies in the future, leveraging the detailed regional anatomy of the M50-O.

ACKNOWLEDGEMENTS

This work was funded by the Global Human Body Models Consortium (GHBMC Project No.: WFU-005). NHTSA (National Highway Traffic Safety Administration) supports the GHBMC. Support was also provided by the Wake Forest University Graduate School of Arts and Sciences. All simulations were run on the DEAC cluster at Wake Forest University.

REFERENCES

[1] Gayzik, F.S. et al., *Ann Biomed Eng*, 39(10):2568-83, 2011.
[2] Vavalle, N et al, *Ann Biomed Eng*, 41(3): 497-512, 2013.
[3] Kindig, M et al, *Comp Meth Biomech and Biomed Eng*, 1-15, 2013.
[4] Kroell, C et al, *SAE Tech Paper*, No. 741187, 1974.
[5] Hardy, W et al, *Stapp Car Crash Journal*, 45:1-32, 2001.
[6] Cavanaugh, J et al, *SAE Tech Paper*, No. 902305, 1990.
[7] Kemper, A et al, *Stapp Car Crash Journal*, 52:379-420, 2008.
[8] Shaw, G et al, *Stapp Car Crash Journal*, 53:1-48, 2009.

LOCAL MICROENVIRONMENT RESPONSE OF CHONDROCYTES IN GRADIENT COLLAGEN MATRICES

Tyler Novak (1), Benjamin Seelbinder (1), Celina M. Twitchell (1), Sherry L. Voytik-Harbin (1,2), Corey P. Neu (1)

(1) Department of Biomedical Engineering
Purdue University
West Lafayette, IN, USA

(2) Department of Basic Medical Sciences
Purdue University
West Lafayette, IN, USA

INTRODUCTION

The interaction of cells with, and their response to, the local microenvironment is critical to physiological form and function. Cells are sensitive to differences in microenvironment mechanical properties, extracellular matrix (ECM) morphology, and ligands including growth factors and ECM binding sites, among others [1, 2].

Collagen-based biomaterials are often used to investigate cell-matrix interactions due to their physiological ubiquity, biocompatibility, and ability to form a 3D matrix. Recently, efforts have been made in the development of high-density collagen matrices beyond the solubility limit, increasing density and mechanical properties toward physiological levels. To date, the most successful methods in the development of high-density collagen matrices have utilized molecular crowding (100-400 mg/mL), evaporation ($\leq$ 1000 mg/mL), and matrix compression ($\leq$230 mg/mL) [3-6]. Of these methods, only matrix compression has been shown to viably encapsulate cell populations prior to formation [3, 6, 7]. Without cellular encapsulation in the 3D engineered environment, physiologically relevant investigation into cellular response is not easily achieved, as cells that are seeded on the exterior and infiltrate to the matrix interior can lead to a degraded or remodeled ultrastructure, potentially eliminating the effects of the designed microenvironment [8]. Additionally, current high density matrices focus on the formation of sheets, and do not achieve physiologically thick dimensions. Further, current high-density solutions do not have the ability to create complex microenvironments similar to those seen in physiology. Many tissues in physiology are composed of gradients, and are often the site of injury (e.g. ligament and tendon insertion site) or specific functions (e.g. superficial zone of articular cartilage). There remains a need for development of platforms to investigate local cell-matrix interaction similar to these complex structures to gain further insight into disease etiology and regenerative medicine approaches. Here, we develop a compression regime to create collagen matrices of physiological size with spatial gradients of density, fibril morphology, and mechanical properties. We further characterized these properties and examined encapsulated primary chondrocyte gene response and protein expression over a short term culture period.

METHODS

Collagen oligomer matrices were polymerized at 5 mg/mL in rectangular molds (10×5×14, $w\times t\times h$). Matrices were densified via confined compression using a porous platen (15% porosity) over a range of strain magnitudes (0%, 50%, 90%) and strain rates (0.1%s^{-1}, 1%s^{-1}, 10%s^{-1}). Linear modulus and peak stress were measured during compression ($n=8$).

Relative collagen density was measured in the direction of compression via confocal reflectance microscopy ($n=8$) and compared to a standard curve. Local fibril properties were also assessed via cryo-SEM ($n=3$). Fibril morphology, mean fibril thickness, max fibril thickness, and fibril area fraction were quantified and compared to the control ($n=3$). Local mechanical properties were measured via AFM indentation (tip diameter=10μm) along the height of the matrix and the compressive modulus was measured from the approach curve ($n=5$).
All cell studies were performed in matrices densified at 0.1%s^{-1} using primary bovine chondrocytes. Short-term viability of cells was measured 48 h post-compression using a live/dead assay ($n=5$, <passage 5). Investigation of cellular gene expression as a result of microenvironment was investigated in a 1-week cell study for each strain magnitude ($n=5$). As positive and negative controls for chondrocytic gene expression, native cartilage explants and monolayer-plated chondrocytes, respectively, were cultured for 1-week under identical conditions. After culture, densified samples were

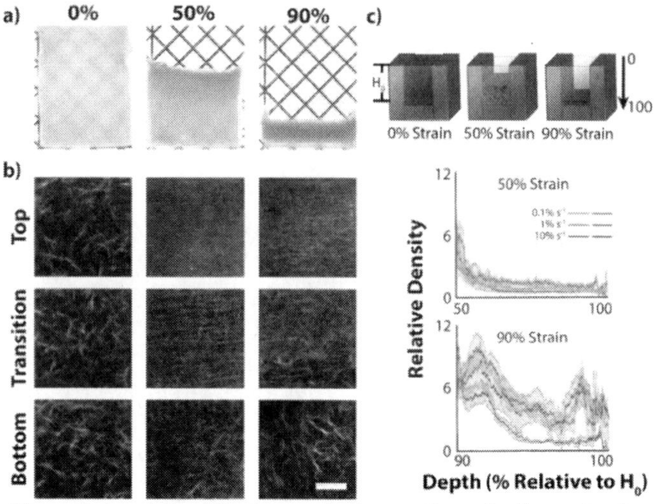

Figure 1. Local densification results in a gradient of collagen density throughout the height of the matrix. a) Transmission microscopy showed decreased opacity near the top (indented) surface (grad spacing=2.1mm). b) Confocal microscopy showed high density collagen in compressed regions (bar=100μm). c) Quantitative analysis of relative density via confocal microscopy and comparisons across strain rate (mean ± standard deviation)

cut into high and low density regions and all samples were analyzed via qPCR for chondrocytic ECM and transcription factors (*ACAN, COL1A1, COL2A2, COLA6, PRG4,* and *SOX9*). Following qPCR, we visualized expression of aggrecan (mouse monoclonal, Abcam ab3778) and type VI collagen (rabbit polyclonal, Millipore AB7821) via immunofluorescence by confocal microscopy and counterstained for f-actin (phalloidin) and DNA (Hoescht) ($n=5$). All statistics computed by ANOVA or nested ANOVA (Minitab).

RESULTS

Densification of collagen matrices resulted in the formation of a gradient in the direction of compression in collagen density, fibril morphology, and mechanical properties (Figure 1). During compression, strain rate ($p<0.001$) but not strain magnitude ($p=0.134$) increased the linear modulus of force applied. Both strain rate ($p<0.001$) and magnitude ($p<0.001$) were found to affect the peak force during compression. Confocal microscopy revealed highest collagen density near the impacted surface (up to 10-fold increase, Figure 1), decreasing along the height toward the bottom. Mean fibril thickness, maximum fibril thickness, and fibril area fraction were dependent on strain magnitude ($p=0.002$, $p=0.004$, $p<0.001$, respectively) and high vs low density region for compressed samples ($p<0.001$, $p<0.001$, $p=0.001$, respectively), where all properties increased in high density regions (Figure 2a). High compressive properties (21-fold increase) were observed in high density regions (Figure 2b) and were dependent on strain magnitude ($p<0.001$) and density region ($p<0.001$).

Cells were found to remain viable in control and densified matrices (>94% live, p=0.053 between samples). Cell density was found to increase with strain magnitude ($p<0.001$) and density region ($p<0.001$). Additionally, cells in high density, aligned fibril regions were found to be aligned where cells in low density regions were randomly oriented both at 48 h and 1 week (Figure 2c). Significant differences were found in gene expression between monolayer, cartilage explant, and aggregated collagen groups (Figure 2d) for *COL1A2, COL2A1, PRG4,* and *SOX9* ($p<0.04$ for all) but not *ACAN*

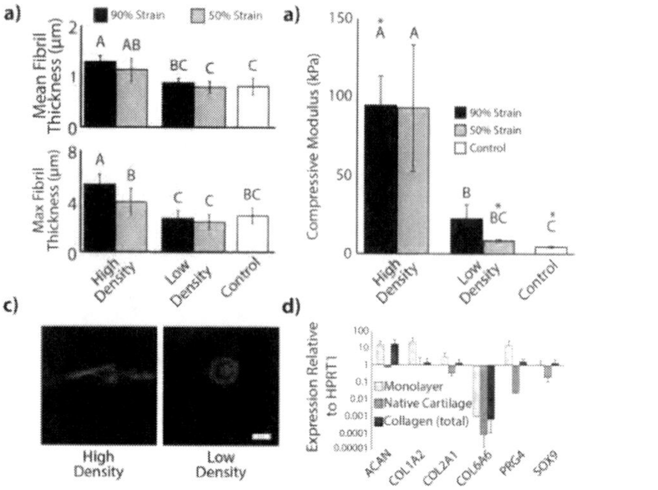

Figure 2. Densification affects (a) fibril morphology, (b) mechanical properties, and (c) cellular alignment, and exhibits (d) different gene expression from native cartilage or monolayer cultures.

and *COL6A6* ($p>0.05$ for all). No significant difference in gene expression was observed with respect to strain magnitude and density ($p>0.05$ for all). Collagen type VI and aggrecan were visualized immediately exterior to the cell body, but not in the interstitium.

DISCUSSION

This work demonstrates the utility of controlled compression modalities for the creation of unique, cellularized microenvironments of gradient density, fibril morphology, and mechanical properties. These matrices are mechanically robust and their ability to encapsulate cells allows for unique investigation of cellular response to specific and gradient microenvironments of various properties. Further, these matrices exhibited fibril ultrastructure similar to physiology, i.e. the superficial and middle zone of articular cartilage, further showing the potential for use in studies of cell response to pseudo-native microenvironments. Further, the ability to encapsulate cells and investigate their morphology and gene expression throughout a culture period shows the utility of this collagen platform in the investigation of cell response to complex microenvironments. Interestingly, we showed that chondrocytes morphologically respond to the different microenvironments contained in the gradient, however, chondrogenic gene expression was not found to be significantly influenced, over the short term, by property differences contained in high and low density regions. The range of properties seen in this platform will allow for continued investigation of cell response to high and low density collagen environments across a gradient and a range of cell types.

ACKNOWLEDGEMENTS

The authors would like to acknowledge funding from NIH R01 AR063712, NSF CAREER 1349735, Purdue OVPR Incentive Grant and NSF GRFP (TN).

REFERENCES

[1] J. R. Tse, et al., *PLoS One,* p. e15978, 2011.
[2] G. C. Reilly, et al., *J Biomech,* 43: 55-62, 2010.
[3] N. Saeidi, et al., *Biomaterials,* 33: 7366-74, 2012.
[4] P. De Sa Peixoto, et al., *Soft Matter,* 9: 11241-11248, 2013.
[5] S. Mi, et al. *J Biomed Mater Res A,* 99: 1-8, 2011.
[6] G. Mosser, et al., *Matrix Biol,* 25: 3-13, 2006.
[7] S. G. Rudisill, et al., *Acta Biomater,* 2014.
[8] S. Z. Yow, et al., *Biomaterials,* 30: 1133-42, 2009.

A QUANTITATIVE EVALUATION OF THE ROLE OF CELL SENESCENCE IN INTERVERTEBRAL DISC DEGENERATION

S. Elmasry (1), S. Asfour (1), J. P. de Rivero Vaccari (2), F. Travascio (1)

(1) Department of Industrial Engineering
University of Miami
Coral Gables, FL, USA

(2) Department of Neurological Surgery
University of Miami
Miami, FL, USA

INTRODUCTION

The etiology of intervertebral disc (IVD) degeneration is multifactorial. Among all the concomitant factors potentially contributing to disc degradation, aging is believed to play a major role [1]. Proteoglycans (PG) of disc extracellular matrix (ECM) helps maintaining tissue swelling pressure, allowing IVD to resist compressive forces [2]. The loss of PG is believed to be a major factor in age-related disc degeneration [3]. However, the mechanism of PG loss, especially in relation to aging, is still poorly understood.

Insulin-like growth factor-1 (IGF-1) is known to stimulate both cell proliferation and PG biosynthesis in a dose dependent manner [4, 5]: the upregulation of anabolic and proliferative cell activity depends on how many IGF-1 is bound to IGF cell surface receptors (IGF-1R) [6]. The IGF-1 and its binding proteins (IGFBP) are present in IVD because they are both exogenously delivered from the vascular network surrounding the IVD and endogenously secreted by disc cells [5]. In an in-vitro study on disc cells, it was found that, with aging, IGF-mediated PG biosynthesis decreased [7]. Moreover, such PG production decline concurrently occurred with a reduction of IGF-1R expression, and an increase of endogenous IGFBP production. It was hypothesized that the age-related changes in IGFBP production and IGF-1R expression may be responsible for the decline in ECM biosynthesis, which would eventually lead to IVD degeneration.

The role of IGF-1 in IVD homeostatic equilibrium is complex and involves many intertwined phenomena (i.e., competitive binding reactions with IGF-1R and IGFBP, molecular degradation, modulation of cell metabolism and viability, etc.) which are difficult to control and study during in-vivo experimental investigations. In this study, a theoretical framework describing the role of IGF-1 on ECM biosynthesis and cellularity in IVD is deployed with the objective of understanding the implications of age-related changes of cellular signals on disc homeostatic balance. More specifically, the analysis carried out in this contribution aims at quantifying to which extent the age-related decrease of IGF-1R expression and increase of endogenous IGFBP production affect IGF-mediated PG biosynthesis and cellularity in IVD.

METHODS

The analysis carried out in this study is based on previously reported theoretical frameworks describing the diffusive-reactive transport of IGF-1 in IVD, and its role in PG biosynthesis and disc cellularity [8, 9], see Figure 1. In the model, binding of IGF-1 to its cell receptor initiates the signaling cascade leading to PG production and cell proliferation. The IGF-1 availability to IGF-1R is regulated by the competitive reversible binding with IGFBPs, also present in the ECM. Since the IGF-mediated enhancement of PG biosynthesis and cellularity increases the nutritional demand of disc cells, transport and metabolism of nutrients (i.e., oxygen, glucose, and lactate) are also accounted for into the model. In addition, it is assumed that cellularity decreases as glucose levels fall below a physiological threshold [10]. Supply of IGF-1 and IGFBP to the tissue occurred via two pathways: (1) exogenously, through the vascular network surrounding IVD; (2) endogenously, secreted by disc cells.

Human lumbar IVD was schematized as a 2D axisymmetric object consisting of three anatomical regions, the nucleus pulposus (NP), the annulus fibrosus (AF), and the cartilage endplate (CEP) with dimensions and properties similar to those reported in a previous study

[8]. Model equations were solved via commercially available software for finite element analysis (COMSOL Multiphysics® 4.3, COMSOL, Inc. Burlington, MA).

The implications of age-related changes in cell signaling on the homeostatic equilibrium of IVD were investigated by comparing three hypothetical scenarios: 'young-adult', 'adult', and 'elder'. The first scenario was characterized by normal physiological expression of IGF-1R and no endogenous production of IGFBP. The other two scenarios were simulated by progressively decreasing IGF-1R expression, and increasing IGFBP endogenous production [7]. The amounts of IGF-1/IGF-1R bound complex, PG, and cell density were averaged over disc regions (NP, AF, and CEP), and their values were normalized with respect to those attained at the 'young-adult' scenario.

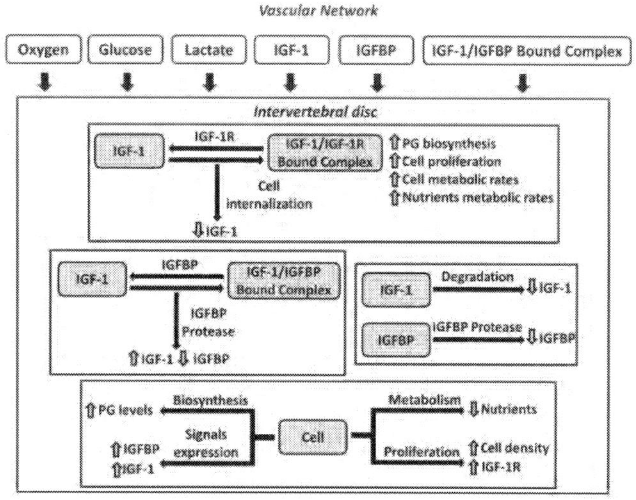

Figure 1: Schematic of a theoretical model describing IGF-1 diffusive-reactive transport in the disc and its implications on cell metabolism, viability, and biosynthetic activity [9].

RESULTS

Effects of aging on IVD homeostasis are reported in Figure 1 in terms of variation from normal physiological values. No significant changes were observed in either CEP or AF. In contrast, in NP, concentration of the IGF-1/IGF-1R bound complex reduced with aging and, consequently, also cell density and PG content decreased.

DISCUSSION

This study sheds light on age-related disruption of disc homeostasis, where aging is believed to have a significant role in symptomatic IVD degeneration [1]. We deployed a theoretical framework for IGF-mediated IVD homeostasis to investigate age-related changes in cellular signaling on cellularity and biosynthesis of the ECM in the disc. We found that aging was accompanied by decrease in both cellularity and PG concentration. However, significant differences were only observed in NP. This was because AF and CEP, being external disc regions, present higher levels of IGF-1 (due to exogenous delivery) and higher cellularity than NP. Our results are in agreement with the observed reduction of cell viability with aging [11]. Moreover, it has been shown that glycosaminoglycan concentrations in the NP of 40-60 year olds and +60 year olds were ~60% and ~30% of that for the 15-25 year-old group, while no significant differences were found in either AF or CEP [3]. Such results compare to the decrease in PG concentration reported in this study for the 'adult' and 'elder' scenarios.

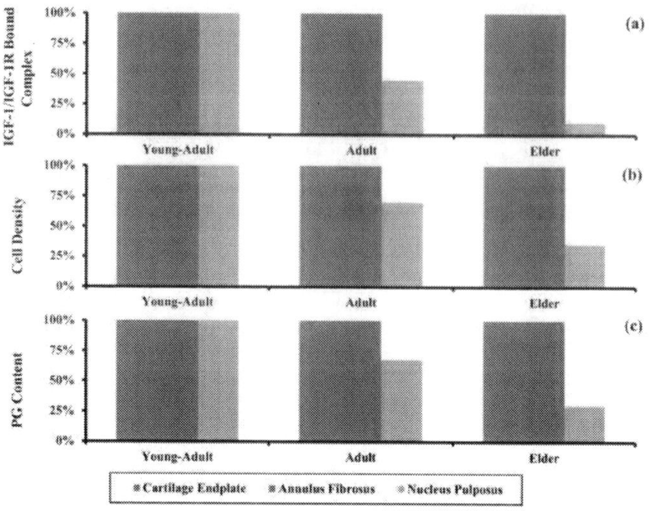

Figure 2: Aging affects ECM composition and cell viability in nucleus pulposus. (a) Content of IGF-1/IGF-1R bound complex. (b) Cell density. (c) PG Content averaged. Data were averaged over the disc regions investigated: cartilage endplate (blue), annulus fibrosus (red), and nucleus pulposus (green). Values were reported in terms of percentage variation from normal physiological values.

ACKNOWLEDGEMENTS

Study supported by funds donated to the Biomechanics Research Group of the University of Miami.

REFERENCES

[1] Roughley, Spine, 29:2691-9, 2004.
[2] Urban & McMullin, Spine, 13:179-87, 1988.
[3] Antoniou et al., J Clin Invest, 98:996-1003, 1996.
[4] Pratsinis & Kletsas, Eur Spine J, 16:1858-66, 2007.
[5] Osada et al., J Orth Res, 14:690-9, 1996.
[6] Lauffenburger & Linderman, Receptors: models for binding, trafficking, and signaling, Oxford University Press, 1993.
[7] Okuda et al., Spine, 26:2421-6, 2001.
[8] Travascio et al., J Biomech, 47:2269-76, 2014.
[9] Asfour et al., J Biomech, 48:332-39, 2014.
[10] Bibby & Urban, Eur Spine J, 13:698-701, 2004.
[11] Buckwalter, Spine, 20:1307-14, 1995.

SB³C2015
Summer Biomechanics, Bioengineering and Biotransport Conference
June 17-20, 2015, Snowbird Resort, Utah, USA
SB³C2015-641

SIMULATION-BASED DESIGN OF A HIP ACTUATOR FOR RUNNING A MILE SPRINT

**John R. Rogers (1), Julie E. Dillon (1),
Cameron I. McDonald (1), Gabriela C. Barrera-Gutierrez (1)**

(1) Department of Civil and Mechanical Engineering
United States Military Academy
West Point, NY, USA

INTRODUCTION

Exoskeleton systems that transmit force to the human body have been developed for rehabilitation [1], for load carrying [2], and for enhancing strength and endurance [3].

Kerestes et al. [4] developed a device called Air Legs that assists motion by adding energy from a phase-timed oscillation. It is a powered hip exoskeleton which assists a runner and has demonstrated reduction in metabolic cost. It uses a timed on-off control. Hollander et al. [5] describe a joint torque augmentation robot primarily for gait assistance while walking with a load.

In contrast to [4] and [5], the purpose of the device described in this paper is to run at greater speeds, and it is therefore lighter, of fewer parts, and is capable of delivering more power. Prototype hardware is being made.

Hitt et al. [6] developed a walking active ankle prosthesis which has a predetermined gait pattern based on able-bodied gait data [6] and kinetic analysis [8]. The motor controller and the system are driven by a time based function of gait [6]. Through this design, an energy efficient, powered transitibial prosthesis was successfully manufactured. In a redesign of this device for a running prosthesis, a continuous control system accounts for the variety in gait profile through defining gait parameters invariant of activity [9].

OpenSim is open source software that allows the user to create a model of the musculoskeletal structures and simulate the dynamics of human and animal movement [10]. Hamner et al. [11, 12] and Dorn [13] describe simulations of human running; simulation results include joint moments, joint angular velocities, and joint power. OpenSim can also compute metabolic rate. No physics-based musculoskeletal simulations of humans with exoskeletons exist in the literature.

It is possible to add a model of a mechanical system to the model of the human in OpenSim. The aim of this study is to document a simulation-based method of designing exoskeleton-like devices that interact with the human body; simulation is used as an aid to select mechanical components such as motors and springs.

METHODS

The authors' vision is to enable an average runner to utilize a powered hip actuation device to maintain his or her sprint pace for a mile. The device applies torque to help flex and extend the hip joints. The hip actuator comprises a motor, leg rods, and an attachment point on the thigh above the knee. The action is reciprocating: a flexion moment applied by the device to the right leg is accompanied by an equal and opposite moment (extension moment) applied to the left leg.

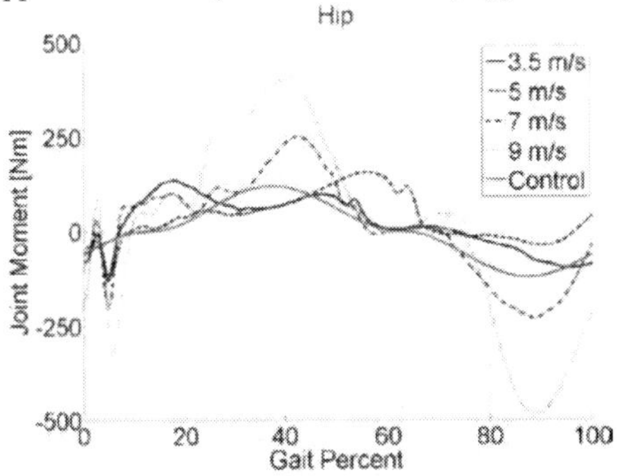

FIG 1 HIP ACTUATOR TORQUE IS TIMED TO MAXIMIZE THE CONTRIBUTION TO JOINT MOMENT.

The cylindrical electric motor is held horizontally in front (ventral) to the waist by a padded backpacking belt with connection points around the waist. The motor drives leg rods that in turn apply a force to the legs. The motor torque profile is timed to the runner's gait delivering extension and flexion moments synchronized to relieve the hip muscles of a fraction of their work. The rods apply a push/pull force on the thigh above the knee to aid leg swing. The system forms kinematic loops—one on the right side of the body, and one on the left. The link bodies include the femur, the pelvis and motor (assumed to be connected to each other), and the leg rod. The leg rods are pinned such that abduction and adduction of the legs are unrestricted, allowing natural motion of stride. The connection between the leg rod and the leg is a low-friction guide that allows for the closure of the kinematic loop. This joint allows the leg rod to move distally and proximally relative to the femur, which is required because the system is outside of the body and the mechanical joint center cannot be co-located with the center of the natural hip joint center. The controls and battery power source are located in a running backpack secured on the upper back close to the spine to minimize moment of inertia. Power and signal are carried by wires run along the body's core.

The torque profile is shown in Fig. 1 on the same plot with hip joint moments from an OpenSim analysis of over-ground running at four speeds [13]. A four minute mile is run at 6.7 m/s for comparison. The torque profile must accommodate the reciprocating action: force on the right leg must be equal and opposite to the reaction on the left leg. The peak magnitude of the torque profile shown is 120 Nm; such a torque would require a peak force of 300 N (68lb) applied to the leg of the runner. 300N is quite a large force, so gain control is used to reduce the magnitude of the torque applied. The phase of the control signal is shifted to align its positive and negative peaks as much as possible with the positive and negative peaks of the moment supplied by the hip muscles, which maximizes the contribution of the system.

The angular position profile of the system is equal to the angular position of one leg relative to the other, Fig. 2A, and the motor speed is the system speed multiplied by the motor gear ratio, Fig. 2B.

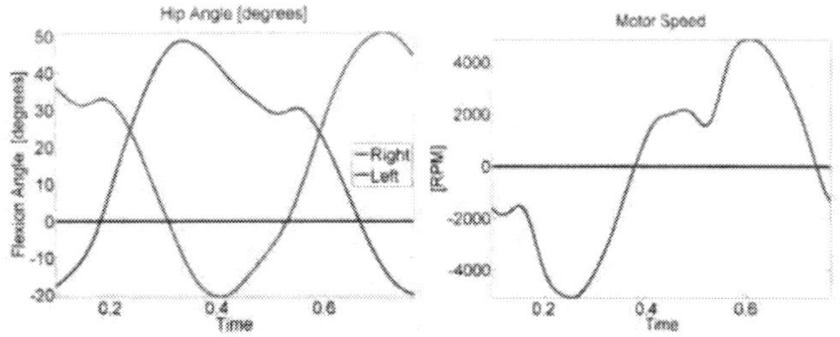

FIG. 2A HIP FLEXION ANGLES FOR RIGHT AND LEFT LEGS. FIG. 2B REQUIRED MOTOR SPEED

Motor specifications are determined from the speed and torque profiles, and the motor constants. Equations (1) through (4) are used to develop the current profile, the voltage required, and the power profile. The motor parameters are obtained from the manufacturer.

$$I = I_{NL} + \frac{T}{K_T} \tag{1}$$

$$V = \frac{N}{K_e} + R_m I \tag{2}$$

$$\omega = \frac{2\pi}{60} N \tag{3}$$

$$P = VI \tag{4}$$

T	motor torque	N	motor speed, RPM
V	motor voltage	I	motor current
P	motor input power	I_{NL}	no-load current
K_T	motor torque constant	ω	motor speed, rad/s
K_e	motor speed constant	R_m	motor winding resistance

RESULTS

The motor input power profile for one gait cycle is shown in Fig..3. This figure is for running at 4 m/s. The RMS average power is 170 W, and the peak is 320 W. A motor can be selected based on this analysis.

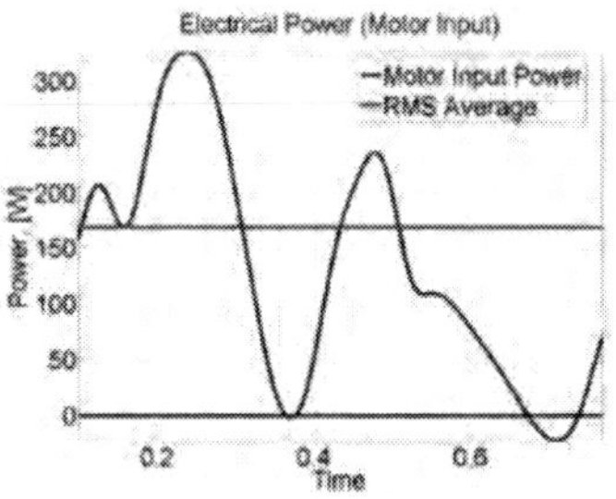

FIG. 3 MOTOR INPUT POWER OVER ONE GAIT CYCLE

DISCUSSION

An OpenSim analysis can output joint angular velocities and joint torques, the quantities that the design of wearable robotic systems depends on. In vivo measurement of muscle force and joint torque is problematic, but simulation overcomes limitations of experiments to include the estimation of forces generated by muscles and the cause-effect relationships in the complex musculoskeletal systems [10]. Simulation based prediction of the effects of machines on human metabolics and muscle force is helpful to understand the impact of this device on the user.

The next steps in this project are to run the OpenSim analysis with and without the hip actuator device. Batch scripts in Matlab will be used to programmatically operate OpenSim over the range of subjects and over a range of speeds. OpenSim will be also used to identify potential negative effects on non-hip muscles, and to estimate the metabolic effect of the device.

An important limitation of this analysis is the assumption that gait kinematics are the same with the device as without the device. This approach can answer the question "if a runner runs the same way with the device as he runs without it, will there be a metabolic savings?" If the runner adapts his or her gait to accommodate wearing the system, this analysis does not account for the change.

The control system described by Holgate [9] will be the basis for the control design of this hip actuator. Holgate [9] uses a single rate gyro to measures tibia angular velocity of an ankle prosthesis. A microcontroller derives the stride length and position in the gait cycle. In the present design, rate gyros will be placed on the thighs. In the same approach as Holgate [9], the control system will also derive similar variables but it will incorporate a different function based on the angular velocity of the thigh instead of the tibia.

Once the device is assembled and subjected to pilot testing, the authors will seek approval for a trial from the Institutional Review Board for human subject testing.

ACKNOWLEDGEMENTS

The views expressed herein are those of the authors and do not purport to reflect the position of the United States Military Academy, the Department of the Army, or the Department of Defense.

The authors wish to acknowledge the ARL-USMA Mathematical Sciences Center which funded this research in part.

REFERENCES

[1] Veneman, J et al., *Int J Robot Res* 25.3:261-281, 2006.
[2] Zoss, A et al., *Mechatronics, IEEE/ASME Trans* 11.2:128-138, 2006.
[3] Pratt, J et al., *Robot and Autom* ICRA2004.
[4] Kerestes, J et al., *ASME Int Des Eng Tech Conf* DETC2014-34405, 2014.
[5] Hollander et al., *ASME Int Des Eng Tech Conf* DETC2014-35654, 2014.
[6] Hitt, J et al., *Ind Robot: Int J*, 36.5:441-447, 2009.
[7] Whittle, M, *Gait Analysis: An Introduction*, Oxford, 1996.
[8] Hollander et al., *J Biomech Eng*, 128:788-791, 2006.
[9] Holgate, M. A. et al. *Int Conf Robot and Autom* ICRA, 2009.
[10] Delp, S et al., *Biomed Engr, IEEE Trans*, 54.11:1940-1950, 2007.
[11] Hamner, S et al., *J Biomech*, 43.14:2709-2716, 2010.
[12] Hamner, S et al., *J Biomech*, 46.44:780-787, 2013.
[13] Dorn, T et al., *J Exp Biol*, 215:1944-1956

SB³C2015
Summer Biomechanics, Bioengineering and Biotransport Conference
June 17-20, 2015, Snowbird Resort, Utah, USA
SB³C2015-642

A MULTIGENERATIONAL COLLAGEN DAMAGE MODEL EXPLAINS ENGINEERED CARTILAGE GROWTH AND REMODELING PHENOMENA

Robert J. Nims (1), Alexander D. Cigan (1), Brian K. Jones (2), Krista M. Durney (1), Clark T. Hung (1), Gerard A. Ateshian (1,2)

(1) Biomedical Engineering
Columbia University
New York, NY, USA

(2) Mechanical Engineering
Columbia University
New York, NY, USA

INTRODUCTION

Current cartilage tissue engineering (CTE) protocols successfully produce constructs matching the glycosaminoglycan (GAG) content and compressive modulus (E_Y) of native cartilage [1]. Using cell seeding densities matching native developing levels (120 million cells/mL) with adequate nutrient supply, constructs synthesize a high level of matrix leading to dramatic (9×) growth over 105 days in culture (**FIG. 1, A** vs 1× growth in **B**) [2]. Collagen content in these constructs remains below native levels (constructs ~2 % w/w vs. native explants ~10 % w/w), yet the critical mechanical role it plays becomes evident when constructs are cultured with a collagen-crosslink inhibitor (ß-aminoproprionitrile), preventing intact collagen network formation and resulting in deleterious expansion (agarose fracture, no mechanical competence) after only 28 days of culture (**FIG. 1, D** vs control in **C**) [3]. Therefore, despite the low collagen content in CTE constructs, understanding collagen deposition and network strength is critical for assessing the mechanics of engineered cartilage.

Matrix content-based mechanical models have demonstrated their utility in estimating mechanical properties in both CTE constructs and native cartilage [4-7]. Such models, however, have limited utility in describing large levels of *de novo* tissue growth as the ability for the collagen network to remodel is often neglected, or fitted to the observed growth. This study aims to develop and validate a matrix content-based mechanical model to incorporate temporal matrix deposition and predict the large degree of observed construct growth and evolving compressive modulus.

Due to the considerable swelling of growing constructs with increasing GAG deposition, we hypothesize that the continuously synthesized collagen network undergoes progressive damage with growth, compromising its ability to resist the Donnan osmotic pressure of charged GAGs. Therefore, we propose that a damage model of the collagen network will allow models to predict the extreme growth and simultaneous plateauing of mechanical properties evident in long-term CTE culture. This study uses the experimental results of [2] to inform a model relating collagen content to multigenerational fiber deposition and damage criteria. The model is then tested and validated against experimental results from a recently completed study [8].

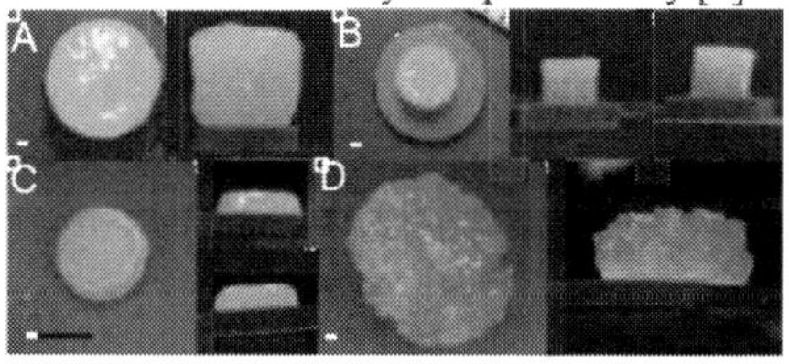

FIG. 1: HIGH GROWTH WITH NORMAL COLLAGEN DEPOSITION (DAY 105 ~ 9× GROWTH (A)) VS ORIGINAL CONSTRUCT SIZE (B) [2]. ABBERENT GROWTH AFTER DAY 28 (D) DUE TO INHIBITED COLLAGEN CROSSLINKING (CONTROL IN (C)) [3]. SCALE = 3 mm.

METHODS

Experimentally informed model: Using FEBio (www.febio.org) a homogenous material model was created to simulate the construct growth and mechanical properties of the study in [2]; in that work, small, high-cell-density constructs (Ø3 mm × 2.34 mm, 120 million cells/mL) were cultured for 14, 35, 58, 77, and 105 days at which point GAG, collagen, equilibrium compressive modulus (E_Y; 2.5 % tare and 10 % compression), and swelling ratio (SR) were measured. The material description incorporated contributions of **(1)** the agarose gel, **(2)** charged GAGs, and **(3)** fibrous collagen network. Damage was modeled in both agarose [9] and collagen following a fourth order Weibull cumulative density function of the strain energy density (SED) in each material. The damage index for each material ranges from 0 (no damage) to 1 (100% damage). The Weibull shape parameter, μ, indicates the SED threshold at which 63% of the material bonds have failed. **(1)** Agarose was modeled as a solid mixture of a neo-Hookean (E_Y =15 kPa, ν = 0) and spherical (isotropic) fiber distribution (SFD; ξ = 4.5 kPa, β = 3) materials [10]; its damage shape parameter, μ = 200 Pa, was derived from agarose tensile and failure measurements [9]. **(2)** GAG was modeled using ideal Donnan law, with bath osmolarity set at 300 mM; tissue fixed-charge density (FCD) was based on experimentally measured GAG deposition [2] assuming 2 negative charges per GAG molecule and a GAG molar mass of 513 g/mol. **(3)** New collagen generations (i) were assumed to be deposited daily based on measured synthesis rates [2]. The stress-free reference configuration of each generation was set to the current configuration of the growing construct at the start of that generation. Generational SFD fiber stiffness ξ_i and damage shape parameter μ_i were related to collagen content via

$$\xi_i = c_\xi \left(\rho_{coll,i} / \rho_{coll,sp} \right)^{\alpha_\xi}, \quad \mu_i = c_\mu \left(\rho_{coll,i} / \rho_{coll,sp} \right)^{\alpha_\mu} \qquad (1)$$

where $\rho_{coll,i}$ is the collagen content present in generation i and $\rho_{coll,sp}$ is the standard collagen content (= 1 % w/w). Power law parameters c_ξ, α_ξ, c_μ, and α_μ were fit with a custom MATLAB optimization scheme to match experimental time point E_Y and SR. A separate model lacking collagen damage (no-damage) terms was also fit.

Model validation: Fit parameters for the damage and no-damage models were used to predict results of our prior culture of constructs (Ø4 mm × 2.34 mm, 30 million cells/mL, 45 days) [8], using the experimental GAG and collagen results to evaluate the FCD and $\rho_{coll,i}$. Predicted E_Y and SR were compared to the experimental results.

Sensitivity analysis: c_ξ, α_ξ, c_μ, and α_μ were individually varied ±10% and predicted day 105 SR and E_Y changes were analyzed.

RESULTS

Fits of E_Y and SR (**TABLE 1**) were well-matched to the experimental results of [2] for the damage model. The no-damage model failed to produce the plateauing E_Y and near-linear SR experimentally seen (**FIG. 2, TOP**); estimated mechanical properties would continue to increase and SR would cease its linear rise.

Using the models to predict the results of the validation experiment [8], both SR and E_Y were well-approximated using the damage model, with good quantitative agreement against experimental results (**FIG. 2, BOTTOM**). In contrast, the no-damage model did not accurately predict construct swelling and mechanical properties.

Sensitivity analysis (**FIG. 3**) showed that SR and E_Y were most sensitive to variations in c_μ, implying the damage threshold had the greatest influence on results.

DISCUSSION

This study formulated and validated a mechanical model for predicting *de novo* engineered cartilage swelling and compressive modulus based on the temporal evolution of matrix composition. Agreement between model and experiments was best achieved when accounting for the evolving matrix damage in swelling constructs. In particular, the damage model explains why the compressive moduli of constructs plateaus around day 35, even while matrix deposition continues unabated, producing a 9-fold swelling ratio by day 105.

TABLE 1: MODEL-FIT PARAMETERS TO EXPERIMENT [2] AND NON-LINEAR CORRELATION BETWEEN MODEL AND EXPERIMENTAL E_Y.

Model	c_ξ (kPa)	α_ξ (-)	c_μ (kPa)	α_μ (-)	$R^2_{Fit,EY}$
Damage	6.22	1.37	20.15	0.63	0.94
No damage	58.90	0.14	-	-	0.89

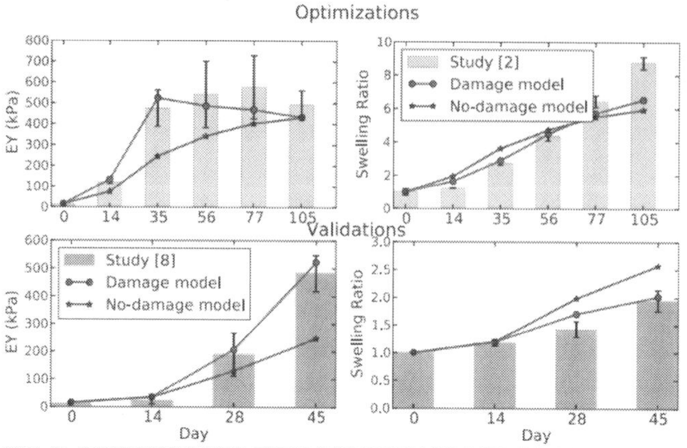

FIG. 2: OPTIMIZATIONS (TOP) BETWEEN [2] AND DAMAGE (RED) AND NO-DAMAGE (BLUE) MODELS (E_Y (LEFT) AND SR (RIGHT)). VALIDATION (BOTTOM) WITH RESULTS OF [8].

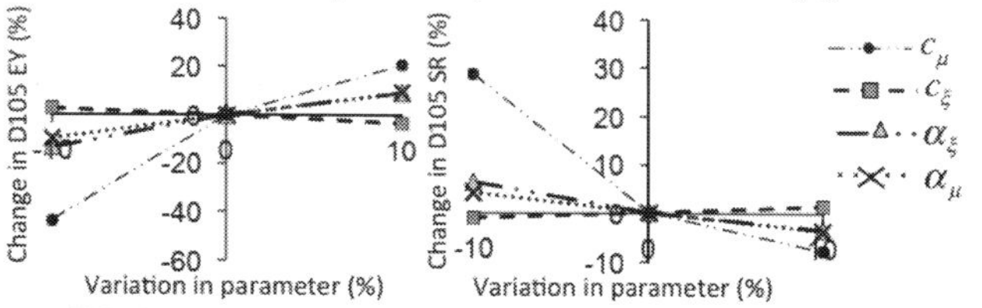

FIG. 3: OUTCOME (DAY 105 E_Y (LEFT) AND SR (RIGHT)) SENSITIVITY TO PARAMETER VARIATIONS OF ±10 %.

The Donnan swelling pressure of deposited GAGs progressively overwhelms the strength of each generation of deposited collagen matrix, producing a constant rate of swelling while maintaining a constant FCD, consistent with the observed value of E_Y. In contrast, the no-damage model is unable to fit the plateau in E_Y, and is consequently unable to predict E_Y values from the independent experiment. Prior to these investigations, the possibility that collagen might become damaged due to the rapid deposition of GAGs in CTE had not been explored. These results thus provide novel insights into this growth process, strongly suggesting that better mechanical properties are achievable if collagen synthesis precedes GAG synthesis in CTE. Similar conclusions have been drawn from models of chondroitinase (CABC) digestion [4,6], although CABC treatments dramatically attenuate collagen synthesis [2].

In future investigations, the material model presented here will be combined with nutrient supply models in an effort to predict spatial matrix deposition and mechanical properties in developing tissues. Understanding the influence of nutrient transport hindrance on tissue composition and mechanics through models such as these will prove useful in assessing appropriate culture protocols for anatomically-sized constructs. While nutrient supply models have already improved our culture of large constructs [11], implementing mechanical growth models, as presented here, will considerably improve our ability to assess optimal conditions for matrix deposition in a growing construct.

ACKNOWLEDGEMENTS

NIH AR060361, AR046568, and T32AR059038

REFERENCES [1] Byers, BA+, *Tissue Eng*, 14:1821-34, 2008. [2] Nims, RJ+, *Trans ORS*, #0670, 2012. [3] Durney, KM+, *Trans ORS*, 2015. [4] Nagel, T+ *Biomech Model Mechanobiol,* 11:325-39, 2012. [5] Driessen, NJB+, *J Biomech*, 36:1151-8, 2003. [6] Khoshgoftar, M+ *Tissue Eng A*, 20:1476-85, 2014. [7] Ateshian, GA+, *J Biomech Eng,* 131:061003-13, 2007. [8] Nims, RJ+, *J Biomech*, 47:2165-72, 2014. [9] Huang, A+, *Osteoarthritis Cartilage,* 24:29-45, 2008. [10] Cigan, AD+, *Tissue Eng*, 19:1941-8, 2013. [11] Nims, RJ+, *Tissue Eng*, 2014.

RAPID QUANTIFICATION OF FEMORAL HEAD GEOMETRY FROM MAGNETIC RESONANCE IMAGING OF FEMOROACETABULAR IMPINGEMENT

Haley Ehlers (1), Michael Roberts (2), Garry Gold (3), and Saikat Pal (1)

(1) Biomedical Engineering Department
California Polytechnic State University
San Luis Obispo, CA, USA

(2) Computer Engineering Department
California Polytechnic State University
San Luis Obispo, CA, USA

(3) Radiology Department
Stanford University
Stanford, CA, USA

INTRODUCTION

Femoroacetabular impingement (FAI) is a bone geometry abnormality common in young, athletic adults [1]. The two main types of FAI bone abnormalities are pincer and cam deformities. Pincer type deformity is when the acetabular cup covers too much of the femoral head [1]. Cam type deformity is when the femur head is not as round as it should be [1]. The current method of treatment is with arthroscopic surgery. This is when the doctor goes in with a small camera, an arthroscope, and surgical tools to clean out the damaged area by trimming the bony rim or shaving the femoral head (Figure 1).

Currently, diagnosis and treatment of FAI involves multiple computed tomography (CT) scans of the pelvic region as part of standard clinical practice. Prior to clinical intervention, CT scans are obtained to quantify the bony abnormalities and develop a surgical plan. Following the clinical intervention, CT scans are obtained to evaluate the success of the procedure and subsequent healing of the joint. Computed tomography scanning provides large contrast between bone and surrounding soft-tissue structures, enabling rapid quantification of bone geometry. However, CT scans expose patients to radiation and its side-effects, including cancer [2, 3]. Exposing the pelvic region of FAI patients to multiple CT scans is particularly alarming because this exposes the reproductive organs of young adults to radiation. As such, there is a compelling need to develop a radiation-free method for diagnosis and treatment of bony abnormalities in FAI patients.

Magnetic resonance imaging (MRI) is common. In addition to CT scanning, FAI patients currently undergo MRI scanning as part of standard clinical practice. There is no exposure to radiation from MRI, but currently there is no way to quantify rapidly bony abnormalities from MR images. The current procedure to obtain a 3D model of bony abnormalities from an MR image is for a researcher or lab technician to look at every image and manually draw a boundary around the desired structure. This manual segmentation method is a very time consuming technique and infeasible in clinics. Accordingly, the objectives of this study were – 1) to develop a computational method to quantify rapidly bone geometry from MRI of a patient diagnosed with FAI, and 2) to compare the bone geometry from our MRI-based method to bone-geometry obtained from a standard CT-based method.

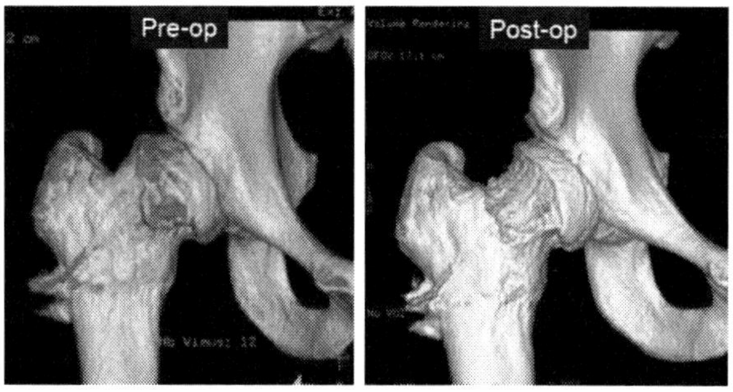

Figure 1. The femoroacetabular joint of an FAI patient prior to (left) and following a surgery (right) to remove bony abnormalities from the femoral head and neck regions.

METHODS

We acquired MRI and CT data from a 29 year old male diagnosed with FAI (Figure 2). The images were collected as part of standard clinical practice. The MRI scans were obtained from a GE Discovery M750 3T scanner with the following parameters: coronal IDEAL SPGR, in-plane resolution 0.8 mm, slice thickness 1.5 mm, repetition time 7.2 ms, echo time 3.0 ms, flip angle 45°, scan time ~12 minutes. The CT scans were obtained from a Siemens Somatom Definition AS+ scanner with the following parameters: axial in-plane resolution 0.8 mm, slice thickness 1 mm, 120 kVp, exposure time 500 ms, and tube current 317 mA.

We created a custom Matlab (Mathworks Inc., MA) code to obtain rapidly 3D geometry of the femoral head. We subtracted the water-saturated coronal MR images from the fat-saturated images. We applied an anisotropic filter [4] to delineate the edges. We used Otsu's method [5] to threshold the image such that only the pixels about a mean threshold value were displayed. The mean threshold value was determined as the mean grayscale intensity from the entire 3D image stack. This enabled us to create a black and white binary image stack while filtering out most of the background noise. We performed a connected component analysis (OpenCV project) on the binary image stack and manually selected a region-of-interest corresponding to the femoral head. We labeled the connected component with the most number of voxels within the region-of-interest; the remaining connected components were discarded. We used a canny edge filter (OpenCV project) to determine the boundary of the 3D labeled component. The 3D point cloud of the femoral head was converted to a surface model in Geomagic (3D Systems, NC). We used a standard threshold-based algorithm to create a surface model of the proximal femur from CT data. We registered the MRI and CT-based models spatially using a rigid registration algorithm and quantified the differences in surface morphology by measuring the shortest distance from a point on the CT-based model to any point on the MRI-based model.

RESULTS

We developed a custom Matlab script to quantify rapidly bony geometry from MRI of a patient diagnosed with FAI. The 3D model of the femoral head from our MRI-based method was similar to the 3D model obtained from a standard CT-based method (Figure 3); the average difference between the MRI and CT models was 0.87 mm. Our rapid quantification method required around three minutes and five mouse-clicks to create a 3D point cloud of the femoral head; in contrast, it takes around 30 minutes and several hundred mouse clicks to create a 3D point cloud using manual segmentation.

DISCUSSION

Computed tomography scanning is currently the standard for evaluating bone geometry abnormalities. However, CT scans expose patients to radiation. A method to evaluate bone geometry from MRI, an imaging technique free of radiation exposure, would eliminate the need for many CT scans.

In this study, we developed a method to quantify rapidly femoral head geometry from MRI and compared our results to a standard CT scan. Our results demonstrate a promising method for quantifying bony geometry from MRI. Our MRI-based 3D geometry of the femoral head closely matched the 3D geometry obtained from CT scanning. The average difference in surface morphology of the femoral head between the MRI and CT models was 0.87 mm. The average difference is comparable to the in-plane resolution of our MRI data; this suggests that our algorithm is likely limited by the resolution of our MRI data. The **clinical applications** of this work are to eliminate the need for CT scanning during diagnosis and treatment of bone geometry abnormalities, minimizing patients' exposure to radiation, and reducing health care costs.

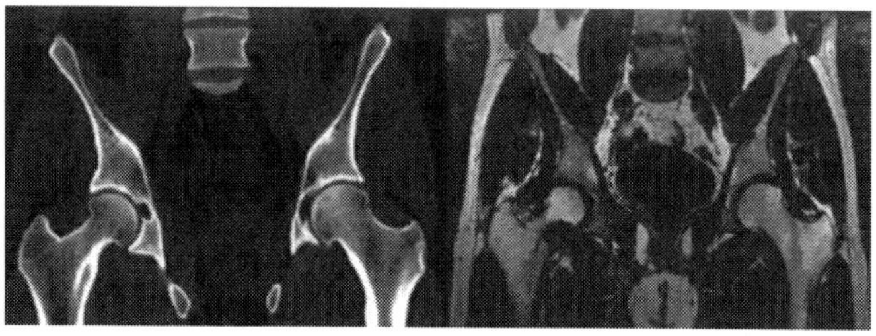

Figure 2. Computed tomography (left) and MRI (right) images of a patient diagnosed with FAI.

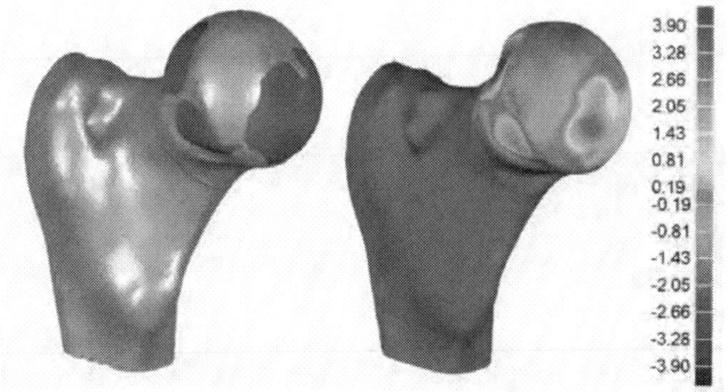

Figure 3. (left) Overlay of MRI-based femoral head geometry (blue) on CT-based proximal femur (grey) and (right) average distance (mm) in surface morphology between MRI and CT-based geometries.

ACKNOWLEDGEMENTS

We are grateful for R-IDC funding from the College of Engineering, Cal Poly San Luis Obispo.

REFERENCES

[1] Tannast, M el al., *J Reontegnology*, 188, 2007.
[2] Brenner DJ et al., *N Engl J Med*, 2007.
[3] Einstein AJ et al., *JAMA*, 2007.
[4] Perona, P et al., *IEEE*, 12:629-639, 1990.
[5] Otsu, N., *IEEE*, 9:62-66, 1979.

SB³C2015
Summer Biomechanics, Bioengineering and Biotransport Conference
June 17-20, 2015,Snowbird Resort, Utah, USA
SB³C2015-644

THE EFFECTS OF AGING ON MICROVASCULAR NETWORK RESISITANCE AND FLOW HETEROGENEITY IN THE RAT MESENTERY

David C. Sloas (1), Scott A. Stewart (1), Richard S. Sweat (1), Walter L. Murfee (1)

(1) Biomedical Engineering
Tulane University
New Orleans, LA 70118

INTRODUCTION

Aging represents the largest risk factor for cardiovascular disease, which is the leading cause of death in aged populations [1]. An essential aspect of this is the influence of aging on angiogenesis, which is clinically important in scenarios such as ischemic tissue remodeling and repair [2]. Aging has been commonly associated with impaired angiogenesis, including decreased microvascular density [3], diminished capillary growth [4], and reduced growth factor expression [5]. However, recent data in our laboratory suggests these effects may be tissue-specific and not characteristic of the rat mesentery. Aged versus adult rat mesenteries displayed larger microvascular networks with comparable vascular length density. Additionally, reports of differentially altered microvascular reactivity [6, 7] demonstrate that the effect of aging on the microcirculation is heterogeneous and more complex than impaired angiogenesis. The question of whether aged networks undergo patterning changes that might explain altered hemodynamics and impaired microvascular function remains unanswered. The objective of this study is to compare the network resistance and capillary flow heterogeneity of mesenteric microvascular networks in aged and adult rats.

METHODS

Tissue Harvest and Imaging. Mesenteric windows were harvested from adult (9 month, n=5) and aged (24 month, n=5) male Fischer 344 rats and then immunolabeled for platelet endothelial cell adhesion molecule (PECAM), an endothelial cell marker. Each tissue's entire area was imaged as a montage of 4X images on an inverted microscope (Olympus IX70, Olympus America, Inc., Melville, NY, USA).

Anatomical Network Analysis. Segment length, segment diameters, vessel types, and nodal connections were recorded per vessel segment for each network. Nodes were defined as any point of vessel bifurcation or convergence or as the terminal end of a blind capillary, feeding arteriole, or draining venule. Arteriolar versus venular phenotype was identified by relative position in the vascular network, relative vessel diameter, and endothelial cell morphology. Endothelial cells along arterioles appeared more elongated and aligned with the direction of flow when compared to venules [8]. As another measurement, network area, defined as the vascular area circumscribed by the network perimeter, was recorded for each network.

Network Model Calculations. Total network resistance per network was calculated for each group using a previously established computational model [9, 10]. The model inputs included the vessel segment lengths, segment diameters, nodal connections, and an assumed pressure drop of 65 mmHg. Network resistance was calculated by dividing the total pressure drop (65 mmHg) by total flow. The 65 mmHg pressure drop was created by assigning a pressure of 75 mmHg to inlet segments and 10 mmHg to outlet segments. These values were chosen based on previously reported values used for hemodynamic analysis of microvascular networks in the rat mesentery [11]. Segmental flow and pressure drop are calculated in the model based on conservation of mass and modified Poiseuille relationships. The apparent viscosity for each segment was updated based on empirical data that accounted for changes in hematocrit in branching vessels [9]. Resistance calculations for aged and adult networks were made for 2 cases: 1) measured diameters and 2) uniform vessel diameters. For the uniform diameter cases, diameters for each vessel in a network were set equal to 10 μm in order to delineate microvascular pattern alterations from potential diameter difference effects. In addition to total resistance, capillary flow heterogeneity was also calculated for each network. Capillary flow heterogeneity was quantified as the relative dispersion of flow through all capillaries in the network. For this calculation, capillaries were defined as all vessels with diameter less than 10 μm based on reported values in the rat mesentery [12].

RESULTS

Representative images of aged and adult rat mesenteric microvascular networks are displayed in Figure 1. For the case of measured diameters, aged and adult network resistances were comparable (aged = 2.17 ± 0.59 mmHg/(nL/min), adult = 2.21 ± 0.46 mmHg/(nL/min); p = 0.96) (Figure 2). Similarly, no significant difference was observed between aged and adult resistances for the case of constant diameters (aged = 13.70 ± 5.60 mmHg/(nL/min), adult = 18.70 ± 7.74 mmHg/(nL/min); p = 0.62). Similar to the network resistance comparison results, capillary flow heterogeneity values were not significantly different between aged and adult groups (aged = 1.67 ± 0.323, adult = 2.23 ± 0.726; p = 0.50).

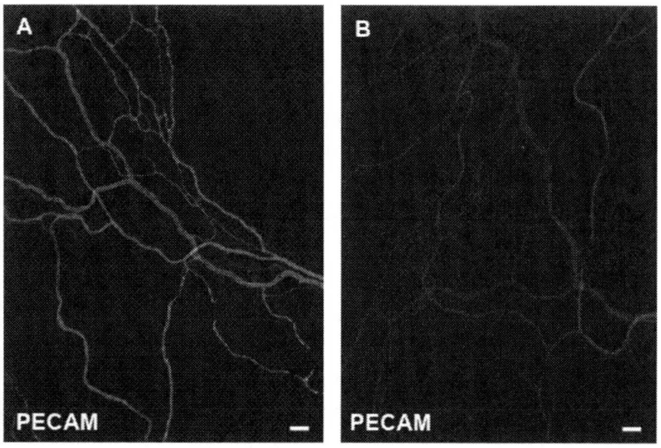

Figure 1. Representative images of adult (A) and aged (B) rat mesenteric microvascular networks. Scale bars = 100 μm.

DISCUSSION

In this study, network resistance comparisons based on intact network geometries enabled us to elucidate whether aged networks experience patterning alterations. Validation of this approach is supported by comparison of theoretical predictions and experimental measurements in the same tissue [9, 13]. Our modeling approach also enabled the isolation of potential patterning effects from diameter effects. While future measurements of direct patterning changes in aged networks (i.e. vessel tortuosity, vessel branch order) are warranted, our resistance results suggest that microvascular networks in this tissue do not undergo changes significant enough to alter local network resistance. This finding is consistent with our previous finding that hypertensive networks are characterized by resistance values similar to normotensive values [9]. For the networks analyzed in this study, no significant difference in vascular area was observed between the adult and aged networks. Based on our previous tissue level analysis suggesting that vascularized area is increased during aging, we speculate that aged tissues might contain an increased number of networks. Future analysis will be required to determine whether these networks are interconnected and how an increased number of networks influences tissue level hemodynamics.

Another indicator of network patterning and microvascular function is flow heterogeneity [14]. Similar to our resistance results, capillary flow heterogeneity was not changed in aged networks, suggesting that impaired local blood perfusion during aging is more likely controlled by local vasoreactivity changes. The topic of vasoreactivity highlights a limitation with our diameter measurements. Diameter measurements for the relative diameter case were made on vessels from fixed immunolabeled tissues. Given the evidence that aging could influence vasoreactivity [7], additional studies are required to compare our measurements with *in vivo* values.

Another limitation of this study is that our modeling approach only accounts for changes in vessel connectivity and vessel lengths. Future studies will be required to identify other possible changes such as vessel tortuosity.

In summary, network resistance and capillary flow heterogeneity, two indicators of network level hemodynamics and microvascular function, were not influenced by aging. These results suggest that microvascular dysfunction associated with aging might not be attributed to microvascular structure and the ability for vessels to remodel.

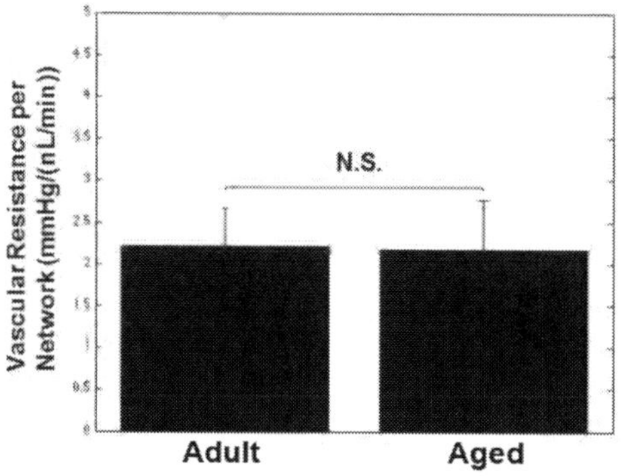

Figure 2. Quantification of network resistance for the case of relative diameters. N.S. no significant difference (p > 0.05). Values are ± SEM.

ACKNOWLEDGEMENTS

Supported by Tulane Center for Aging and NIH 5-P20GM103629-03

REFERENCES

[1] North BJ and Sinclair DA, *Circ Res*, 110:1097-1108, 2012.
[2] Lahteenvuo J and Rosenzweig A, *Circ Res*, 110:1252-1264, 2012.
[3] Sadoun E and Reed MJ, *J Histochem Cytochem*, 51:1119-1130, 2003.
[4] Yamaura H and Matsuzawa T, *Exp Gerontol*, 15:145-150, 1980.
[5] Wagatsuma A, *Exp Gerontol*, 41:49-54, 2006.
[6] Muller-Delp JM, *Microcirculation*, 13:301-314, 2006.
[7] Sinkler SY and Segal SS, *Am J Physiol Heart Circ Physiol*, 307:H830-9, 2014.
[8] Murfee WL, Rehorn MR, et al., *Microcirculation*, 13: 261-273, 2006.
[9] Yang M, Murfee WL, *Med Biol Eng Compute*, 50: 585-593, 2012.
[10] Binder KW et al., *Microcirculation*, 14: 181-192, 2007.
[11] Fenger-Gron J, Mulvany MJ et al. *J Physiol*, 488: 753-760, 1995.
[12] Lipowsky HH et al., *Amer J of Physiol*, 301: 6: H2235-H2245, 2011.
[13] Pries AR, Secomb TW, et al., *Circ Res*, 75: 904-915, 1994.
[14] Frisbee JC, Fan W, et al., *Amer J of Physiol*, 301: 4: R975-986, 2011.

IMPAIRED COLLAGEN DEGRADATION PREVENTS RIGHT VENTRICULAR HYPERTROPHY AND DYSFUNCTION WITH DEVELOPMENT OF PULMONARY ARTERIAL HYPERTENSION

Mark J. Golob (1), Zhijie Wang (1), Anthony J. Prostrollo (1), Timothy A. Hacker (3) Gaoussou Diarra (3) Naomi C. Chesler (1,3)

(1) Department of Biomedical Engineering
University of Wisconsin-Madison
Madison, WI, USA

(2) Department of Medicine
University of Wisconsin-Madison
Madison, WI, USA

INTRODUCTION

Pulmonary arterial hypertension (PAH) is a severe form of pulmonary hypertension caused by pulmonary vascular remodeling including distal pulmonary artery (PA) narrowing and proximal PA stiffening. The cause of death in PAH is typically right ventricular (RV) failure.

A common preclinical model of PAH is chronic exposure to hypoxia. To induce more severe disease in mice, chronic hypoxia can be combined with administration of a vascular endothelial growth factor (VEGF) inhibitor (SUGEN), which recapitulates characteristics of human PAH [1]. In C57BL6 mice, this combined treatment causes significant distal PA narrowing and proximal PA stiffening [2]. We have previously reported the collagen-dependence of PA biomechanical changes with development of PAH [2,3,4]; RV fibrosis is also involved in RV adaptation to PAH [5], but the role of collagen in RV remodeling is unclear.

We hypothesize that RV hypertrophy and RV dysfunction depend on collagen turnover, which is key to RV fibrosis. To test this hypothesis, we exposed transgenic mice to the combined hypoxia plus SUGEN treatment (HySu). In particular, we used the Col1a1 mutant mouse ($Col1a1^{RR}$), in which a substitution in the Col1a1 gene results in a type I collagen triple helix that is resistant to collagenase-based degradation [3,6]. The wild-type littermates ($Col1a1^{++}$) that have normal collagen degradation were used as controls. After PAH by HySu treatment, RV function was measured by *in vivo* pressure-volume relationships and various RV hypertrophy indices were examined. Our results indicate that impaired collagen degradation prevents RV hypertrophy and RV dysfunction with PAH development.

METHODS

Animal preparation. Breeding pairs of $Col1a1^{tmJae}$ mice on the B6/129 background were obtained from Jackson Laboratory (Bar Harbor, ME). Genotyping by polymerase chain reaction was performed on DNA extracted from tail biopsies, as previously described [3]. Adult female Col1a1 mice were used in this study because PAH is four more times common in women than in men [7]. Mice were grouped by genotype and treatment into four groups as shown in Table 1. HySu treatment consisted of 14 days of hypoxia (10% oxygen in an environmentally controlled chamber) and a weekly i.p. injection of SUGEN (SU5415, Sigma-Aldrich Corp) at a dose of 20 mg/kg [2]. Mice were euthanized at the age of 8-10 weeks. All procedures were approved by the University of Wisconsin-Madison Institutional Animal Care and Use Committee.

Table 1: Experimental groups (N=4-7 per group)

Group	Genotype	Treatment
1	$Col1a1^{++}$	Normoxia
2	$Col1a1^{RR}$	Normoxia
3	$Col1a1^{++}$	HySu
4	$Col1a1^{RR}$	HySu

Measures of ventricular function. Mice were anesthetized via urethane (1.6 mg/g body weight, i.p.), and the chest was opened. Synchronized pressure and volume waveforms were obtained with a 1.2 F admittance catheter (Notocord, Croissy Sur Seine, France) [4,8]. Cardiac output (CO), ejection fraction (EF), and minimum and maximum derivatives of pressure (dP/dt_{min} and dP/dt_{max}) were quantified. Analysis included at least 10 consecutive cardiac cycles.

Tissue Weights. After euthanasia the LV, RV, and septum (S), were excised and weighed. RV hypertrophy was determined using the Fulton Index (RV weight normalized by the LV+S [2]) as well as RV weight normalized to body weight [9] or tibia length [10].

Statistics. Results are reported as mean ± standard deviation. Statistical significance was assessed using a two-way ANOVA for genotype ($Col1a1^{++}/Col1a1^{RR}$) and treatment (normoxia/HySu) (R, USA, version 3.0.1). Tukey multiple comparisons were used for post hoc analysis (R). A two-sided p-value less than 0.05 was considered statistically significant.

RESULTS

RV end-systolic pressure (RVSP) was elevated to a similar degree in both sets of HySu-treated mice (Table 2), indicating the efficacy of the HySu treatment in creating PAH. Load-dependent measures of ventricular function, dP/dt_{max} and dP/dt_{min}, were elevated and lowered, respectively, in HySu-treated mice, indicating increased contractility in response to pressure overload. No differences in CO or EF were evident between HySu-treated and normoxic $Col1a1^{RR}$ mice, suggesting RV function was negligibly affected by the pressure overload (Table 2). However, EF was significantly lower in HySu-treated $Col1a1^{++}$ mice compared to normoxic $Col1a1^{++}$ mice, suggesting RV dysfunction (Table 2).

The PAH development induced by the HySu treatment also caused RV hypertrophy as evidenced by the Fulton Index (Table 2). Interestingly, HySu-treated $Col1a1^{++}$ mice showed RV hypertrophy via multiple indices, indicating both absolute and relative increases in RV mass with PAH development. However, in the $Col1a1^{RR}$ mice, there was no significant increase in the absolute or relative RV mass as represented by RV/TL and RV/BW. Since there was a significantly decreased LV+S mass with PAH development in $Col1a1^{RR}$ mice, the Fulton index increased (Table 2).

DISCUSSION

A major finding in this study is that HySu-induced PAH causes different RV remodeling responses to pressure overload in $Col1a1^{RR}$ and $Col1a1^{++}$ mice. In the $Col1a1^{++}$ mice, we observed RV hypertrophy as evidenced by an increase in all three indices, suggesting absolute and relative RV mass increase. However, only the Fulton index was significantly increased in HySu-treated $Col1a1^{RR}$ mice, and this was attributed to a decreased LV+S mass rather than an absolute RV mass increase in HySu-treated $Col1a1^{RR}$ mice (Table 2). No other relative changes in RV mass were observed in HySu-treated $Col1a1^{RR}$ mice. These results indicate RV hypertrophy was limited by the collagen degradation defect.

Consistent with RV hypertrophy, PAH led to a significantly decreased EF in HySu-treated $Col1a1^{++}$ mice. In contrast, EF was preserved in HySu-treated $Col1a1^{RR}$ mice, suggesting RV dysfunction is prevented in this genotype. Our results suggest that the collagen degradation defect prevents new collagen accumulation in the RV, which may explain the absence of increased RV mass in HySu-treated $Col1a1^{RR}$ mice. In summary, our results suggest a protective role of impaired collagen degradation for RV remodeling in PAH. The molecular mechanisms await further investigation.

Table 2: RV hypertrophy and function parameters. (*p<0.05 vs. Normoxia of the same genotype)

Strain Treatment	$Col1a1^{++}$ Normoxia	$Col1a1^{RR}$ Normoxia	$Col1a1^{++}$ HySu	$Col1a1^{RR}$ HySu
RVSP (mmHg)	30.3±4.5	33.3±5.2	49.5±16.2*	42.6±3.7*
dP/dt_{min} (mmHg/s)	-1978±339	-1932±423	-2814±673	-3123±431*
dP/dt_{max} (mmHg/s)	2629±475	2563±347	3498±331*	3793±375*
EF (%)	50.7±7.6	51.8±12.2	35.5±3.7*	48±14.7
CO (µL/min)	7738±2964	7661±3026	6014±1111	8467±1111
RV/TL (mg/mm)	1.09±0.13	1.14±0.19	1.87±0.43*	1.34±0.46
RV/BW (mg/g)	0.89±0.07	0.92±0.17	1.56±0.29*	1.17±0.36
RV/LV+S (mg/mg)	0.27±0.02	0.25±0.03	0.37±0.06*	0.34±0.11*
LV+S (mg)	70±6	77±11	80±20	60±5*
RV (mg)	18±1	19±4	29±5*	20±7

ACKNOWLEDGEMENTS

This work was supported by NIH grant 1R01HL086939 (NCC).

REFERENCES

[1] Ciuclan, L, et al. *Am. J. Respir. Crit. Care Med.* 184, 1171-1182, 2011
[2] Wang, Z et al., *PLoS One* 8 e78569, 2013.
[3] Ooi, CY et al. *Am J Physiol Heart Circ.* 299: H1823-H1831
[4] Schreier, D et al. *J. Biomech. Eng.* 135, 2013.
[5] Wang, Z et al. Physiol. Rep. 1 (7), 2013
[6] Wu, H et al. *Proc Natl Acad Sci USA.* 87 5888-92, 1990
[7] Badesch, DB et al. Chest. 137: 376-387, 2010
[8] Tabima, DM et al. *J. Physiol. Heart-Circ. Physiol.* 299, H2069-75, 2010.
[9] Borgdorff, MA et al. *J. Physiol. Heart-Circ. Physiol.* 305 H354-64, 2013.
[10] Csiszar, A et al. *Hypertension.* 54 668-75, 2009.

TRAUMATIC BRAIN INJURY RESULTED IN INCREASED AQUAPORIN-4 EXPRESSION – RELEVANCE TO POST INJURY EDEMA

Nasya M. Sturdivant (1), Jeffrey Wolchok (1), Kartik Balachandran (1)

(1) Department of Biomedical Engineering
University of Arkansas
Fayetteville, AR, United States

INTRODUCTION

Traumatic brain injury (TBI) is an extremely devastating injury experienced by soldiers in the battlefield and in full-contact sports. It has become increasingly evident that cerebral edema is one of the major factors that leads to the high mortality and morbidity of affected individuals post-TBI [1]. Cerebral edema is the swelling of brain tissue due to an increasing volume in brain water content [2]. The swelling results in the alteration of cellular physiology, biochemistry and function [1]. It is a serious condition that is difficult to treat and often results in cerebral herniation and/or death. The Aquaporin family of water channel proteins is a major route by which water crosses cell membranes and thereby, plays a critical role in the water uptake of cells following injury. Aquaporin-4 (AQP4) channels are highly expressed in astrocytes throughout the brain and spinal cord and are a key participant in the maintenance of brain water homeostasis.

Various studies have shown the upregulation of AQP4 following TBI and that cerebral edema is a major risk factor in TBI patients. However, a link between the regulation of AQP4 and cerebral edema has minimally been investigated.

We hypothesize that TBI-mediated cerebral edema occurred via an aquaporin-4 (AQP4)-mediated mechanism leading to increased intracellular fluid intake and hypertrophy. We tested our hypothesis in a three-dimensional (3D) model subjected to the strains of a traumatic brain injury. The purposes of the present study were 1) To create a cell model that successfully encapsulates and maintains viable astrocytes; 2) to effectively cause injury to the cells that resembles the injury imposed on the brain during a TBI event; 3) to determine if there is a relationship between the stress put on astrocytes during injury and the expression of AQP4; and 4) to observe the volume changes of astrocytes following a TBI.

METHODS

Development of 3D TBI model. A stretchable chamber was made by mounting a ½" tall piece of silicon tubing on top of an 8cm x 7cm polydimethylsiloxane (PDMS)-membrane. The PDMS-membrane was then braced within metal clamps. A layer of matrigel was evenly spread within the stretchable chamber and set in the 37°C incubator for 20 minutes to polymerize (Figure 1).

Cell Culture and Isolation from Matrigel. Preliminary culture of C8D1A cells was done in standard feed media (DMEM, 10% FBS, 10 mM HEPES, 1x antibiotic-antimicotic) at 37°C and 5% CO_2.

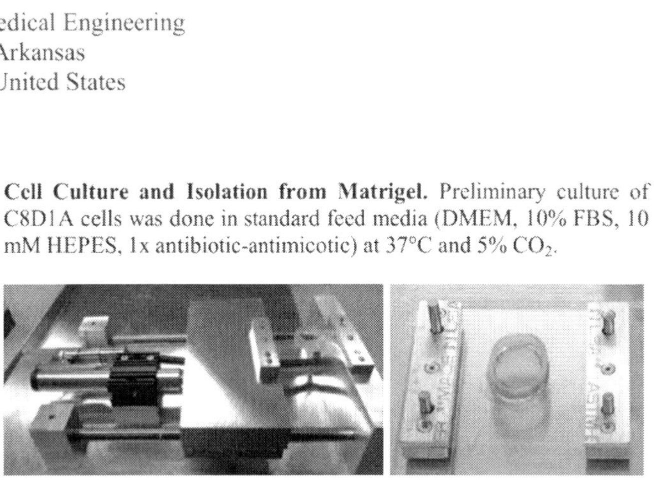

Figure 1. Stretcher device to simulate a TBI

350,000 – 500,000 cells were removed from preliminary culture, resuspended in feed media/2.5% Matrigel and seeded on top of the polymerized layer of Matrigel. Cells were given 2 hours to attach to the Matrigel before being subjected to a 5, 10, and 15% stretch at a 1000% · s^{-1} strain rate; TBI being 10% stretch at 1000% · s^{-1} [3]. 24 hours after stretch the cells/matrigel were washed 2X with PBS and scraped out of the chamber into a centrifuge tube with Corning Cell Recovery Solution. The cells sat in the Recovery Solution for 1 hour; enough time to allow the cells to separate from the Matrigel. After this period, the tubes were centrifuged at 300G for 5 mins and then the Matrigel and cell recovery solution was separated from the cells.

AQP4 Inhibition. AQP4 expression was inhibited with 20 μM Acetazolamide (AZA). Cells were in contact with AZA from the time they were seeded onto the polymerized Matrigel layer until they were completely isolated from the Matrigel and analyzed.

Western Blot. After isolation from Matrigel, the cells were lysed with lysis buffer (4M urea, 5mM EDTA, 0.5% SDS, 0.5% NP-40, 100 mM Tris pH 7.4), centrifuged at max speed 4°C and cell debris was withdrawn. A BCA protein assay was done on the supernatant. 20 μg/lane of protein was loaded onto a TGX gel and subsequently transferred onto PVDF (polyvinyl difluoride) membrane. The membrane was probed with anti-AQP4 antibody (abcam, ab46182), anti-GFAP antibody (abcam, ab7260), and anti-S100B antibody (abcam, ab4066) and visualized with a Li-Cor scanner.

Immunostains. Astrocyte/matrigel samples were stained with DAPI and AQP4. Twenty-four hours after injury, the samples were rinsed 3 times with PBS and fixed in 4% paraformaldehyde for 20 minutes, followed by blocking and immunostaining against Anti-Aquaporin-4 Antibody Cells were then incubated with Goat anti-Rabbit-594 secondary antibody and DAPI for 2 hours. The cells were then imaged using an epi-fluorescent microscope.

In Situ Calcein Imaging. Cultured cells were washed with PBS, stained with Calcein for 45 minutes. The cells were then incubated with Tyrode's buffer then stretched. Immediately after stretch, the cells were imaged under a multi-photon microscope. Images were captured every 10 minutes (Figure 2). Using MATLAB, the area, integrated density, and mean gray value of cells were measured. Then the corrected total cell fluorescence (CTCF) was measured: CTCF = Integrated Density – (Area of selected cell · Mean fluorescence of background reading). Z-stacks were acquired for 3D volume analysis.

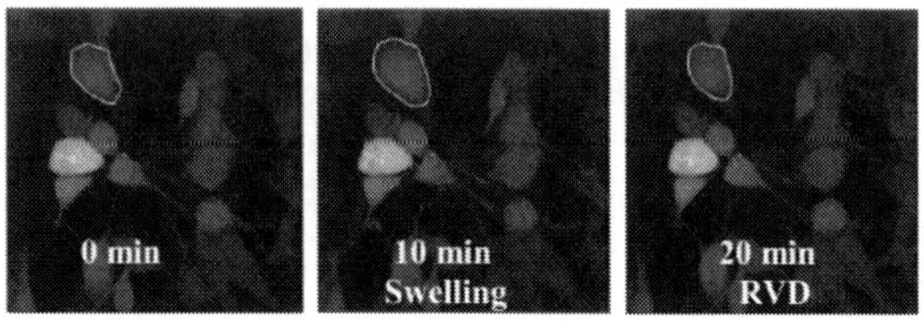

Figure 2. Calcein images immediately after stretch, 10 mins and 20 mins following stretch

Cell Vitality. Astrocyte death following TBI was assessed using fluorescent probes (LIVE/DEAD Cell-Mediated Cytoxicity Kit; Molecular Probes, Eugene OR) and flow cytometry. After 24 hours, the astrocyte/matrigel construct was rinsed 3 times with cold PBS, and the cells were then isolated from the Matrigel with the use of Corning Cell Recovery Solution. After isolation from the Matrigel each sample's cells (0, 5, 10, and 15%; with and without AZA) were resuspended in PBS and stained with 4 μL of 50 μM C12-resazurin working solution and 4 μL of 1 μM SYTOX Green stain working solution. Stained samples were then analyzed via flow cytometry.

RESULTS

TBI resulted in increased expression of AQP4 and GFAP, a marker for astrocyte activation (Figure 3), For AQP4, there is a statistically significant change (P=0.007). The addition of AZA decreased AQP4 expression in all conditions. Initial results did not show a statistically significant change in GFAP expression, but we will be increasing the sample size for further analysis. AQP4 Immunostains confirmed its increase in expression following TBI (Figure 4).

TBI resulted in increased cellular death and increasing injury

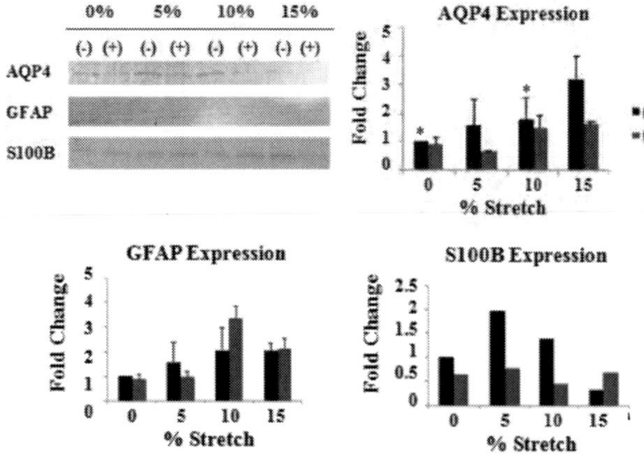

Figure 3. Representative Western blot data. * P<0.007

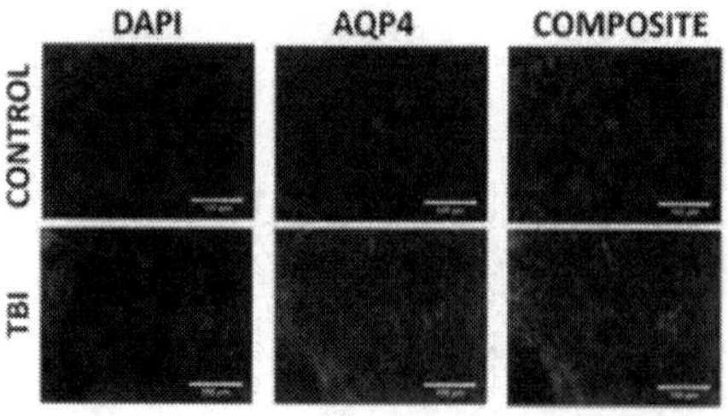

Figure 4. Unstretched astrocytes (control) in the 3D culture construct stained with DAPI and AQP4 and astrocytes in the 3D construct stretched to 10% at 1000% • s-1 strain rate for a single cycle (TBI).

from 5 to 10 to a 15% stretch. The addition of AZA significantly decreased the percentage of dead cells following a 10% and 15% stretch compared to cells without the AZA inhibitor (Figure 5).

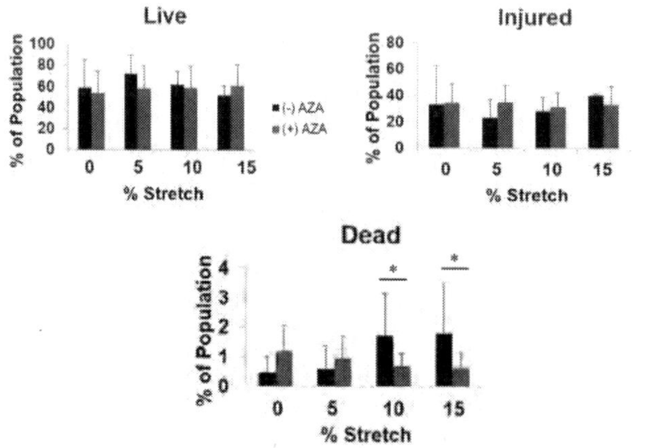

Figure 5. Cell vitality following a 5, 10, and 15% stretch. *P<0.05.

DISCUSSION

3D cell models are the most promising way to study how tissues in the body react to certain disturbances or traumas. In this study, we successfully developed an *in vitro* model to encapsulate and maintain viable astrocytes. Additionally, this 3D model has the ability to be subjected to a high speed stretch that mimics the strain experienced by cells during a TBI incident. Following injury to the 3D cell model, a noticeable increase in AQP4 expression, as seen in the fluorescent images, was observed. The AQP4 inhibitor, AZA proved to be effective in downregulating the AQP4 expression of the cells in all conditions. Continuing studies of calein expression following stretch should show the effect of stretch on cell swelling. This swelling can then be correlated with AQP4 expression following stretch. The findings in this study warrant further investigation into the mechanisms of AQP4 in the brain and the role this water channel plays in the onset and progression of cerebral edema. AQP4 could be a potential therapeutic target in the treatment of cerebral edema.

ACKNOWLEDGEMENTS

We would like to acknowledge Dr. David Zaharoff for assistance with flow cytometry. This material is based upon work supported by the NSF under grant no. CMMI-1404716 and by the NSF GRFP under grant no. DGE-1450079.

REFERENCES

[1] Donkin, JJ et al., *Curr Opin Neurol*, 23: 2293-299, 2010.
[2] Zador, Z et al., *Handb Exp Pharmacol*, 190: 159-170, 2009.
[3] LaPlaca, MC et al., *J Biomech*, 28: 1093-1105, 2005

THE RELATIONSHIP OF WALL SHEAR STRESS WITH CLINICALLY RELEVANT METRICS IN PULMONARY ARTERIAL HYPERTENSION

Alifer D. Bordones (1,2), Vitaly O. Kheyfets (3), Ender A. Finol (1)

(1) Department of Biomedical Engineering
University of Texas at San Antonio
San Antonio, TX, USA

(2) Graduate Program in Biomedical Eng.
University of Texas Health Science
Center at San Antonio
San Antonio, TX, USA

(3) Department of Biomedical Engineering
University of Colorado
Denver, CO, USA

INTRODUCTION

Pulmonary arterial hypertension (PAH) is a chronic disease that leads to a progressive rise in pulmonary vascular resistance (PVR) and stiffness, reflecting an increase in pressure in the main pulmonary artery. PAH begins with changes in the cells that line the pulmonary arteries and an increase in the right ventricle (RV) workload. Wall Shear Stress (WSS) is a known mechanism of endothelial dysfunction and studies show its relationship with the progression of PAH [2,6]. The objective of this study was to explore the relationship between physiological and hemodynamic parameters of PAH disease progression with Spatially Averaged Wall Shear Stress (SAWSS) calculated in patients with PAH through the use of computational fluid dynamics (CFD).

METHODS

With IRB approval at Allegheny General Hospital (AGH) and the University of Texas at San Antonio, 28 records of patients with PAH were reviewed retrospectively. Each patient underwent a chest computed tomography angiography (CTA) and right heart catheterization procedure to acquire images and hemodynamic data.

Patient-specific pulmonary arterial tree models were reconstructed from the CTA images using the application Mimics (Materialise, Belgium). The models were reconstructed from the main pulmonary artery until approximately the 7th generation. For the computational analysis, the inlet and outlets of the pulmonary models were extended 10 times their hydraulic diameter and the mesh was built with tetrahedral elements using ICEM (Ansys, Canonsburg, PA) following a grid convergence analysis based on WSS [1]. The protocol followed to construct each patient-specific model prior to CFD analysis is described in [2] and illustrated in Fig. 1. The finite volume numerical solver Fluent (Ansys, Canonsburg, PA) was used for the CFD simulations. Constant flow boundary conditions with a Schwartz-Christoffel mapping for the velocity profile were applied at the inlet as well as a pure resistance structured tree outflow boundary condition for each of the outlets of the geometry. The mathematical formulation and the procedure followed to generate the boundary conditions are described in our prior work in [2].

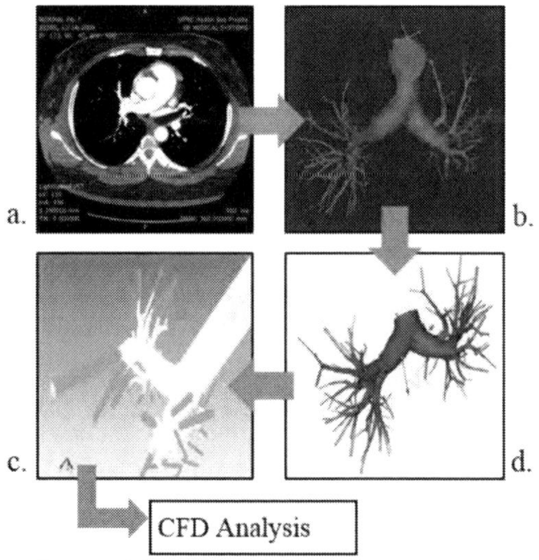

Figure 1. Protocol followed to perform patient-specific CFD analysis. The vasculature is reconstructed from CT scans (a, b). In the pre-processing step, each outlet and the inlet are extended by ten diameters (c) and a volumetric non-uniform grid is generated during the meshing step (d).

Correlation coefficients were calculated to analyze how SAWSS varies with hemodynamic and physiological variables clinically relevant for PAH. The Pearson correlation coefficient was calculated when the behavior between two variables exhibited a linear tendency, while the Spearman correlation coefficient was calculated when the relationship between the two variables was not linear but presented a monotonic tendency. Statistical significance was assessed by the p-value and 95% confidence interval (95%CI), which were calculated for every correlation coefficient. Linear and nonlinear regression analyses were performed and the goodness of fit (R^2) for each curve was calculated.

RESULTS

For each of the 28 CFD models, SAWSS was calculated and graphically correlated with pulmonary vascular resistance (PVR), arterial compliance (C), main pulmonary artery pressure (mPAP), cardiac output (CO), and main pulmonary artery diameter (mPAϕ). It was observed that there is a linear relationship between SAWSS and C (see Fig. 2), as well as cardiac output (CO), and a nonlinear relationship between SAWSS and the other three variables in this study (PVR, mPAP and mPAϕ). The Pearson correlation coefficient (r) between SAWSS and C was found to be significant, $r = 0.8117$ ($p < 0.0001$ and a 95%CI of [0.6005, 0.9171]). Before performing regression analyses, an outlier detection method was applied based on robust nonlinear regression and the false discovery rate (ROUT) [3]. Although this methodology was developed for nonlinear regression, it can also be applied for linear cases. The value of R^2 for the linear regression of SAWSS on C is 0.6548 (calculated by the sum of least squares).

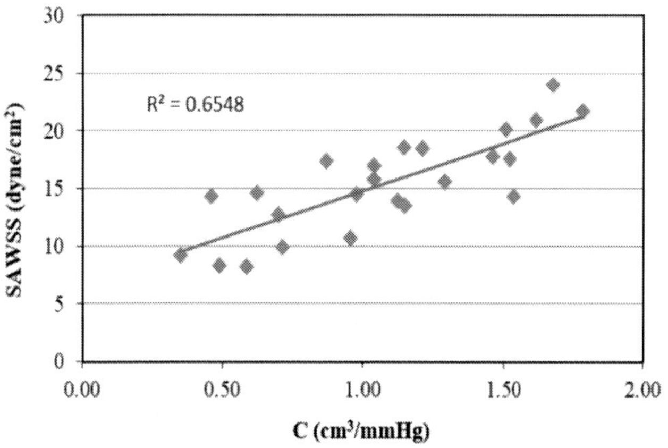

Figure 2. Linear regression of SAWSS on C ($R^2 = 0.6548$).

The relationship between SAWSS and PVR exhibits a nonlinear but monotonic tendency and a non-uniform distribution of points along the curve (see Fig. 3). Due to this nonlinear tendency, the Spearman correlation coefficient was found to be significant, namely $r = -0.6225$ ($p = 0.0015$ and 95%CI of [-0.8276, -0.2710]). A nonlinear regression analysis was performed, with R^2 calculated using weighting, a methodology designed to account for the lack of homoscedasticity of the data. Its purpose is for points anywhere on the curve to contribute equally to the sum-of-squares. It is known that random factors will yield more scatter than others, but the goal of weighting is to make those differences be entirely random and not related to the value of SAWSS [4]. The R^2 taking into account weighting is 0.5745.

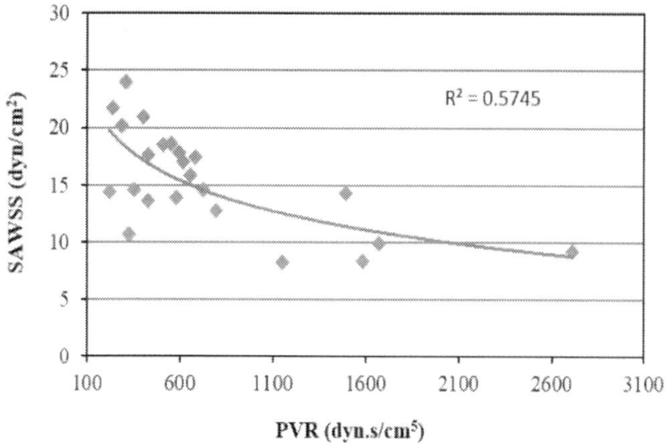

Figure 3. Nonlinear regression of SAWSS on PVR ($R^2 = 0.5745$).

The correlation between SAWSS and CO was found to be significant with $r = 0.5829$ ($p = 0.0011$ and 95%CI of [0.2682, 0.7852]). SAWSS and CO present a linear tendency and after performing an outlier check the curve obtained by linear regression yielded a $R^2 = 0.3897$. The correlation coefficients between SAWSS and mPAP, and SAWSS and mPAϕ, were found to be not significant with $r = -0.222$ ($p = 0.2561$) and $r = -0.1721$ ($p = 0.3813$), respectively.

DISCUSSION

In the systemic circulation it has been demonstrated that wall shear stress modulates endothelial function. An increase in WSS results in elevated expression of endothelial Nitric Oxide Synthase (eNOS), which is a primary modulator of pulmonary vasodilation and the target of PAH drug therapies. Therefore, it is reasonable to infer that WSS has an important effect on pulmonary endothelial health and remodeling [6]. In addition, SAWSS can be used as a quantitative metric of PAH disease progression as it correlates well with direct and indirect metrics of RV function (PVR and C) [5], and with CO. However SAWSS did not correlate well ($r < 0.5$ and $p > 0.05$) with mPAP and mPAϕ, which might indicate that the pressure and diameter of the main pulmonary artery are not good predictors of PAH progression, although more work needs to be done to understand the effects of pressure and proximal vessel geometry in PAH. Future studies will involve the use of pulsatile flow for the calculation of SAWSS and the assessment of its relationship with PVR and C in patients with PAH.

REFERENCES

[1] Prakash, S., and Ethier, C.R. J Biomech Eng, 123(2), pp. 134-144, 2001.
[2] Kheyfets, V. et al. Pulmonary Circulation, 3(1), 2015, in press.
[3] Motulsky, H. and Brown, E. BMC Bioinformatics, 7(123), 2006.
[4] Motulsky, H. and Christopoulos, A. Oxford University Press, 2004.
[5] Wang, Z., and Chesler, N.C. Pulmonary Circulation, 1(2), pp. 212-223, 2011.
[6] Tang, B. et al. Pulmonary Circulation, 2(4), pp. 470-476, 2012.

EFFECTS OF INHALATION TRANSCIENCE ON FLOW STRUCTURES DURING NUMERICAL SIMULATION OF AIRFLOW THROUGH A CT-BASED AIRWAY GEOMETRY

Richard R. Gruetzemacher (1,2), Abdollah Arabshahi (1,2), Kidambi Sreenivas (1,2)

(1) Graduate School of Computational Engineering
The University of Tennessee, Chattanooga
Chattanooga, TN, USA

(2) SimCenter: National Center for Computational Engineering
Chattanooga, TN, USA

INTRODUCTION

The human respiratory system is responsible for the continuous supply of oxygen to sustain all cells the body. This functionality also allows for the delivery of aerosolized particles, toxic or beneficial, throughout the body. A strong understanding of the fluid dynamics of the pulmonary airways, and a comprehensive knowledge of correlations between fluid phenomena in these airways and morphological features, will have a profound impact on those suffering from respiratory illness or airborne contaminants and all who may benefit from aerosolized pharmaceuticals.

Experimental studies of bifurcating, internal flows through idealized models of the airways have practical limitations. Numerical simulations, using computational fluid dynamics (CFD), have become increasingly common in recent years for studying the complex fluid phenomena in the pulmonary airways. CFD allows for easily modeling complex, realistic airways, reconstructed from medical imagery, i.e. patient-specific airways, something difficult, if possible, for *in vitro* or *in vivo* studies.

The vast majority of experimental and CFD studies of bifurcating internal flows assume that a steady solution suffices. This assumption has been adequate for the idealized models for which it was proposed. Previously, it has been practical for CFD models also. However, with decreased computational costs the validity of this assumption in CFD models of complex, realistic airways should be assessed. This study examines the complex fluid phenomena during unsteady flow through a realistic airway model generated from medical imagery to test the validity of steady state assumptions.

METHODS

A computational geometry was generated from a set of computed tomography (CT) images. Initially, Slicer3D, an open-source medical imaging package, was used for volume rendering, segmentation, and point cloud generation. The point cloud was imported to MeshLab, an open-source meshing package, for surface reconstruction via the Marching Cubes [1] algorithm. This surface was imported to Pointwise where a hybrid mesh was generated. Extensions were added to the distal airways to enable enforcement of developed flow conditions at the truncated airways not completely resolved in imaging and to aid in solution stability at the boundaries [2].

The final geometry was comprised of 25 distal airway extending as far as 9 generations. For each of the distal airways a unique boundary condition was prescribed. To determine the appropriate boundary conditions for each airway data was taken from Horsfield *et al.* [3] to approximate the typical lobar volumes. Flow rates for each lobe were computed using this data and the overall flow rates typical of breathing rates at rest. This resulted in five specific volumetric flow rates, one for each lobe, to be enforced at each lobar bronchus. These boundary conditions are more appropriate than simple uniform boundary conditions since they are based on morphological data and consider the local flow rates into the lobar bronchi.

In this study the effects of compressibility were assumed negligible. This assumption is typical for CFD studies of pulmonary airflow. *Tenasi* [4], an in-house flow solver, was used to solve the unsteady incompressible Reynolds-averaged Navier-Stokes, deemed appropriate for the flow regime of concern. *Tenasi* uses a finite volume, implicit scheme with high-resolution fluxes and was validated for three cases [5] pertaining to the flow regime of concern.

Two primary cases were completed for this study corresponding to maximum and minimum flow rates encountered in typical breathing at rest. The maximum and minimum flow rates were computed for flow rates of 7.5 and 15 breaths per minute. These breathing rates corresponded to Womersley numbers of $\alpha = 4.72$ and $\alpha = 6.68$.

RESULTS

A typical tidal volume of 0.5L was used for the two breathing rates corresponding of $\alpha = 4.72$ and $\alpha = 6.68$. Volume flux was measured at cross sections (see fig. 1) in each lobar bronchus. The volume flux at peak inspiration was within 1% of the target values for each of the five lobar bronchi.

Two steady cases were also run for comparison. These cases were run with flow rates through each lobar bronchus equal to those at peak inspiration for each Womersley number. The Reynolds numbers for these cases were 875 and 1750, respectively. The flow rates through each lobar bronchus were within 1% of the target values.

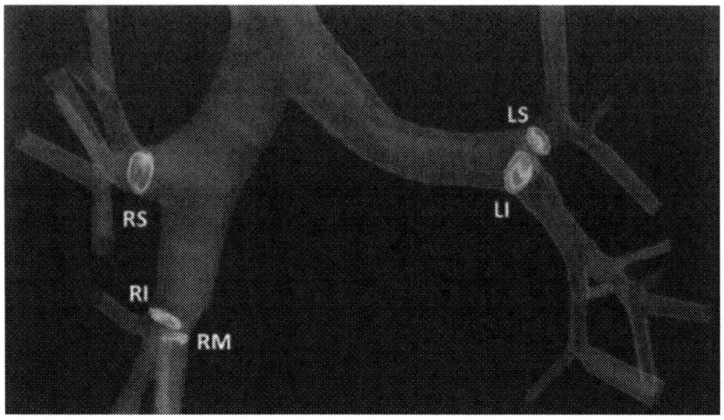

FIGURE 1: CROSS-SECTIONS ORTHOGONAL TO THE AXIAL FLOW FOR THE FIVE LOBES; RIGHT SUPERIOR (RS), RIGHT MIDDLE (RM), RIGHT INFERIOR (RI), LEFT SUPERIOR (LS), AND LEFT INFERIOR (LI).

The cross-sections (see fig. 1) where the volume flux of the lobes was measured were also used to visualize secondary flow components. Vectors in these cross sections were colored to show velocity magnitude. One cross-section image at each lobar bronchus was created showing the flow phenomena for the steady conditions. Three cross-sections were created for the unsteady cases. These cross-sections were created at phase angles over an increase from half peak inspiration to peak inspiration. The corresponding phase angles were 30°, 60°, and 90°.

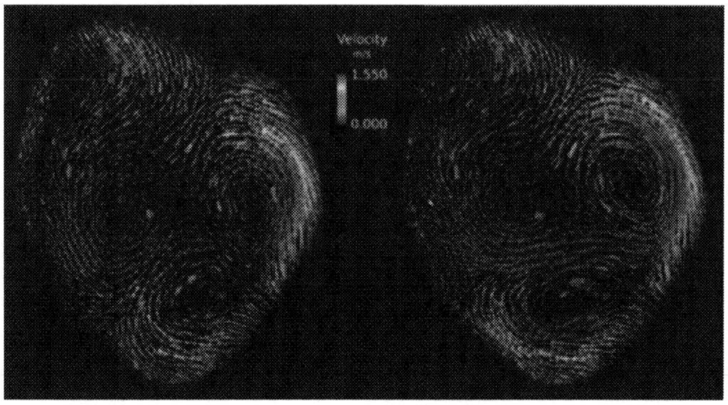

FIGURE 2: A COMPARISON OF SECONDARY FLOWS IN AN UNSTEADY SOLUTION AT PEAK INSPIRATION AND AN ANALOGOUS STEADY SOLUTION AT Re = 1750.

Comparisons were made between the steady solutions and the analogous unsteady solutions. A comparison of these cross-sections in the right superior lobe (see fig. 2), for $\alpha = 6.68$ and Re = 1750, exhibits noticeable differences between the solutions. Similar differences were present at cross-sections in two of the four remaining lobar bronchi.

Comparison of the flow development in each lobar bronchus was also made. For these comparisons the steady solution was ignored. A comparison of the increase of flow velocity between phase angles of 30° and 90° in the right superior bronchus (see fig. 3) indicates an increase in vorticity. These increases are noticeable in both the number of vorticies present and the magnitude of vorticity. Increase in the number of vorticies is noticeable in two of the four remaining lobar bronchi. Increase in the magnitude of vorticity is clearly noticeable in all of the remaining lobar bronchi.

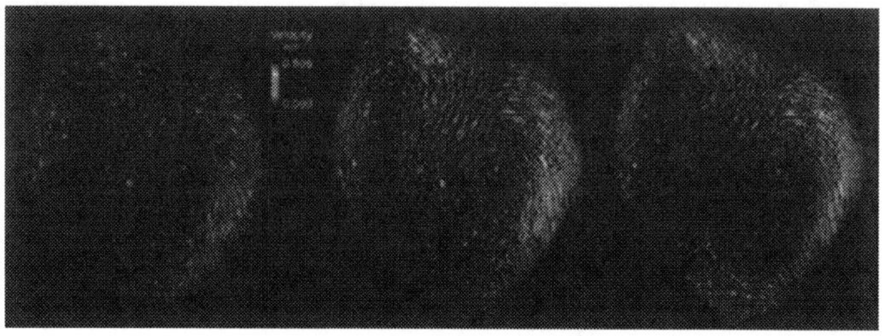

FIGURE 3: A COMPARISON OF SECONDARY FLOWS INTO THE RIGHT SUPERIOR BRONCHUS AT PHASE ANGLES OF 30°, 60°, and 90°.

Further analysis included visualization of velocity in cutting planes to identify flow features in the axial flow. Regions of separation and flow reversal were found in the high Reynolds number steady case. There were also notable differences at stagnation points of proximal bifurcations in the high Womersley number case and the analogous high Reynolds number case.

DISCUSSION

These results indicate that the assumption of airflow through CT-based airway models as steady needs further scrutiny. For the low Reynolds numbers observed at peak inspiration and the analogous steady solutions significant differences were observed from the steady solutions.

ACKNOWLEDGEMENTS

Funding for this study was made possible by the Tennessee Higher Education Commission.

REFERENCES

[1] Lorensen & Cline, *ACM Siggraph Computer Graphics*, 1987.
[2] VanErtbruggen *et al. J. appl. Physiol* 98.3 (2005): 970-980
[3] Horsfield *et al. J. appl. Physiol* 31.2 (1971): 207-217.
[4] Sreenivas *et al. AIAA Paper*, 2005, p 325.
[5] Gruetzemacher, master's thesis, 2014, scholar.utc.edu/theses/149/

SB³C2015
Summer Biomechanics, Bioengineering and Biotransport Conference
June 17-20, 2015, Snowbird Resort, Utah, USA
SB³C2015-650

NONLINEAR ANISOTROPIC MECHANICAL PROPERTIES OF VASCULAR SMOOTH MUSCLE CELLS

Z. Win (1), P. W. Alford (1)

(1) Department of Biomedical Engineering
University of Minnesota
Minneapolis, MN, USA

INTRODUCTION

Advances in 3D imaging techniques and the rise of computing power have allowed researchers to develop patient-specific models of cardiovascular mechanics that can provide physicians with an additional tool to diagnose and track vascular diseases. Non-invasive imaging is used obtain 3D images of patient's vasculature and 3D FEM models used to reconstruct fluid-solid interaction models [1]. Some current models use rule-of-mixtures approaches consider each component of tissues, including vascular smooth muscle cells (VSMCs) and extracellular matrix (ECM) components [2]. Vascular ECM mechanical properties are well studied but the isolated mechanics of VSMCs are not well understood. The Young's moduli for cells have been studied using various techniques [3] but Young's modulus is only valid for small strains. Since vascular tissues and cells undergo large strains in vivo, there is a need to determine the nonlinear mechanical properties of VSMCs. The gold standard for determining material properties is biaxial testing. We have developed a method to determine the material properties of single cells by using a microscope-mounted cell-scale biaxial stretcher.

METHODS

Cell culture. Human umbilical artery vascular smooth muscle cells were cultured at 37 °C and 5% CO_2 in growth medium consisting of Medium 199 (GenDEPOT, Baker, TX) supplemented with 10% fetal bovine serum (Gibco, Grand Island, NY), 10 mM HEPES (Gibco), 3.5 g/L glucose (Sigma-Aldrich, St. Louis, MO), 2 mg/L vitamin B12 (Sigma-Aldrich), 50 U/mL penicillin/streptomycin (Gibco), 1x MEM Non-essential amino acids (Gibco), and 2 mM L-glutamine (Gibco).

Substrate fabrication and cell seeding. Micropatterned polyacrylamide-elastomer composites were fabricated as follows. Microcontact printing was used to pattern 4000 um^2 islands of fibronectin (FN) onto a 15 mm glass coverslip. PDMS elastomer membranes were functionalized with photoinitiator benzophenone (10% w/v in 35:65 water/acetone). A 5 mL aliquot of 7.6 kPa polyacrylamide gel solution was prepared. Prior to polymerization, 50 uL acrylic acid NHS ester (1 mg/mL), 10 uL TEMED, and 25 uL APS (10% w/v) were added to the solution. 10 uL of the prepolymer solution was deposited on the functionalized elastomer and covered with the micropatterned coverslip. The gels were then exposed to UV (10.8 J/cm^2). Gels were then passivated with 4% BSA for 1 hr prior to cell seeding.

Cell stretching. Stretching experiments were conducted in serum-free media at 37 °C. Elastomer membranes were stretched uniaxially (longitudinally and transversely) to 25% (0.5%/s) strain. Brightfield and fluorescent images were obtained during stretch. Cells were then lysed with 0.1% SDS and the substrate was stretched again to obtain images of cell-free deformation. A PIV algorithm implemented in ImageJ was used to calculate bead displacements between cell and cell-free images [4]. Traction forces were calculated using an unconstrained Fast Fourier Transform Correlation algorithm [5].

RESULTS AND DISCUSSION

We have designed and built a custom biaxial stretcher with four linear actuators that can be controlled independently to perform biaxial testing at the cellular scale (**Fig. 1**). The stretcher is mounted on an inverted microscope allowing simultaneous imaging of cells and substrate deformation. An elastomer membrane is mounted into the grips of the stretcher and stretched up to 25% strain. A micropatterned soft polyacrylamide substrate doped with fluorescent microbeads allows control of the cell architecture by changing the pattern and measurement of traction forces generated by the cell.

VSMCs were micropatterned onto the substrate and stretched up to 25% uniaxially either longitudinally or transversely (**Fig. 2**). Bead displacements generated by the cells on the substrate were determined as outlined in the methods (**Fig. 2A-B**). Traction forces generated by the cells during stretch were calculated from bead displacements by comparing cell and cell-free deformation of the substrate. Traction forces were normalized to the longitudinal traction force (T_x) generated by the cell before stretch. When cells are stretched longitudinally (i.e. in the direction of primary actin alignment), there is an almost two-fold increase in traction force generated by the cells (**Fig. 2C**). When cells are stretched transversely, the traction force generated by the cells does not change (**Fig. 2D**). This result suggests that VSMCs micropatterned to mimic *in vivo* architecture have strongly anisotropic material properties.

Our previous work has shown that cellular architecture affects contractile stress generation in VSMCs [6]. However, the contribution of isolated VSMCs to the functional and mechanical properties of vascular tissues is not well understood. Here, we have developed a method to perform biaxial testing on a single cell that can be used to capture the nonlinear anisotropic mechanical properties of isolated VSMCs. We hypothesize that changing cellular architecture affects the material properties of the cell. Future work includes, developing a 3D FEM model of our experiments that will be used to calculate a strain energy density function for the cell that can be used in biomechanical models of vascular mechanics.

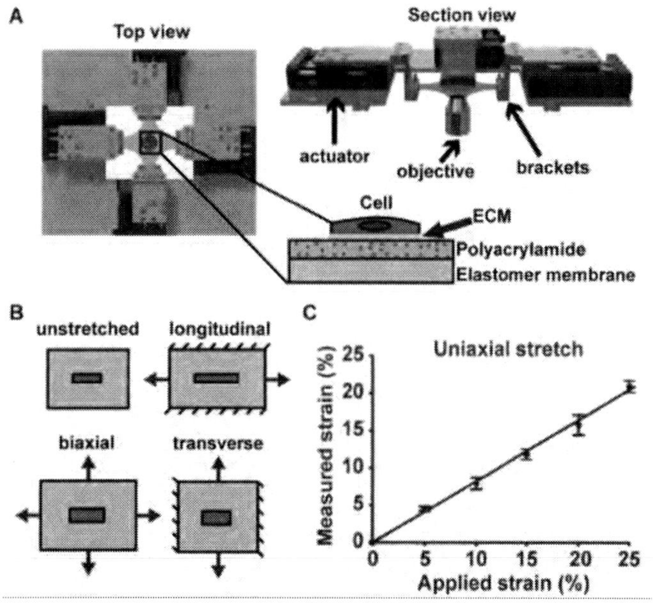

Figure 1. Cell-scale biaxial stretcher. (A) Computer renders of microscope-mountable biaxial stretcher. Inset: Cells are micropatterned onto a bead-doped polyacrylamide gel that is bonded to a stretchable elastomer membrane. (B) Elastomer membrane can be stretched uniaxially or biaxially to stretch the cell. (C) Applied uniaxial strain vs measured strain.

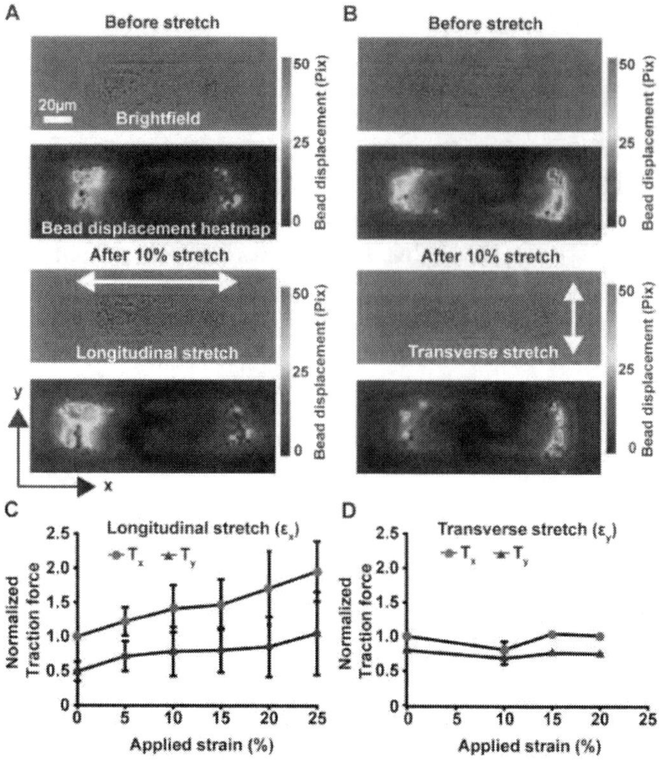

Figure 2. Cell stretching and traction force measurement. (A-B) Micropatterned cell is stretched 10% longitudinally and transversely. Brightfield images and heat maps of bead displacement magnitudes are shown. (C) Change in traction force after a cell is stretched longitudinally. Forces normalized to T_x at 0% strain. (D) Change in traction force after a cell is stretched transversely. Forces normalized to T_x at 0% strain.

ACKNOWLEDGEMENTS

We acknowledge financial support from the American Heart Association Scientist Development Grant 13SDG14670062.

REFERENCES

[1] Taylor, CA et al., *Annu Rev Biomed Eng*, 11:109-134, 2009.
[2] Alford PW et al., *Biomech and Mod in Mechanobiol*. 11:525-538, 2008.
[3] Bao G et al., *Nature Materials*. 2:715-725, 2003.
[4] Tseng Q et al., *PNAS*, 109:1506-1511, 2011.
[5] Butler JP et al., *Am J Phys*, 232:C595-605, 2002.
[6] Win, Z et al., *Integr Biol*, 6:1201-1210, 2014.

NOTCH1 MUTATION LEADS TO VALVULAR CALCIFICATION THROUGH ENHANCED CADHERIN-11 MECHANOTRANSDUCTION

Joseph Chen (1), Larisa M. Ryzhova (1), M.K. Sewell-Loftin (1), Chris B. Brown (2), Stacey S. Huppert (3), H. Scott Baldwin (2), W. David Merryman (1)

(1) Department of Biomedical Engineering
Vanderbilt University
Nashville, TN USA

(2) Department of Pediatrics
Vanderbilt University
Nashville, TN USA

(3) Department of Pediatrics
Cincinnati Children's Hospital
Cincinnati, OH USA

INTRODUCTION

Calcific aortic valve disease (CAVD) is a significant cardiovascular disorder, which results in impaired valve function and increased cardiac load. It is thought to occur in two forms, dystrophic and osteogenic, and is believed to be mediated by aortic valve interstitial cells (AVICs). Dystrophic calcification is more prevalent, found in 83% of valve explants and involves myofibroblastic activation of AVICs through enhanced smooth muscle α-actin (αSMA) and cadherin-11 leading to cell-death-mediated calcification through the nucleation of Ca and P. Alternatively, osteogenic calcification is found in 13% of calcified valves and is generated through the active secretion of osteoid matrix via osteoblast-like AVICs, which express high levels of Runx2. However, the role of dystrophic and osteogenic processes in the progression of CAVD *in vivo* is poorly understood. Identification of the contributions of dystrophic and osteogenic pathways could lead to the development of new strategies for CAVD treatment.

Genetic factors that promote CAVD provide new physiologically relevant systems that can provide deeper insights into the pathological processes occurring *in vivo*. Recent investigations have shown that Notch1 heterozygosity leads to the development of CAVD with 100% penetrance in humans [1]. Therefore, we sought to clarify the roles of dystrophic and osteogenic pathways in CAVD by assessing this genetic mutation.

In this study, we isolated Notch1 mutant AVICs from Notch1$^{+/-}$ Immortomice for the first time. Using these cells, we found that Notch1$^{+/-}$ AVICs exhibited proto-myofibroblast properties expressing low amounts of αSMA but increased amounts of the myofibroblast marker cadherin-11 and diminished osteogenic activity as evidenced by reduced Runx2 expression. Furthermore, Notch1$^{+/-}$ AVICs became fully activated myofibroblasts under mechanical stimulation and generated dystrophic calcific nodules. These data suggests that CAVD arising from the Notch1 mutation undergoes dystrophic calcification through enhanced myofibroblast mechanotransduction.

METHODS

Micromechanical properties of WT and Notch1$^{+/-}$ valves were assessed with AFM (Bruker). Murine immortalized Notch1$^{+/-}$ aortic valve interstitial cells (AVICs) were isolated and expanded *in vitro* in immorto media (DMEM, 10% FBS, 1% penicillin/streptomycin, 10 U/ml INF-γ) at 33°C and 5% CO_2. Western blotting was conducted to compare molecular signaling of wild type (WT) and Notch1$^{+/-}$ AVICs to identify changes in pathways that have been linked to CAVD – TGFβ/BMP, MAPK and PI3K/Akt. To assess AVIC mechanobiology and calcification potential, cells were seeded on Bioflex Pronectin culture plates (Flexcell International Corporation) and subject to 10% equibiaxial stretch for 24 h via the Flexcell-4000 tension system at a frequency of 0.75 Hz.

RESULTS

Tissue level characterization demonstrated that Notch1$^{+/-}$ valves have enhanced calcification, elevated cadherin-11, and increased mechanical stiffness. Von Kossa staining of the valve leaflets revealed higher uptake of stain in the Notch1$^{+/-}$ animals (Fig. 1A(a,b)). Immunohistochemistry analysis of valve tissue revealed enhanced cadherin-11 staining in the Notch1$^{+/-}$ leaflets (Fig. 1A(c,d)). Notch1$^{+/-}$ AVICs also express less αSMA and Runx2 but significantly more cadherin-11 (Fig. 1B). Further, biomechanical analysis of leaflet tissue via atomic force microscopy (AFM) indicated that Notch1$^{+/-}$ valve leaflets are significantly stiffer than WT leaflets (Fig. 1B).

We evaluated the effect of TGF-β1 and BMP2 stimulation on important signaling pathways involved in CAVD. Notably, Notch1$^{+/-}$

AVICs displayed significantly increased Akt phosphorylation at both Thr308 and Ser473 with minimal change with TGF-β1 and BMP2 stimulation (Fig. 2A,B). Using inhibitors for MAPK and PI3K/Akt signaling, we found that inhibition of Akt lead to a drastic reduction in cadherin-11 expression (Fig. 2C).

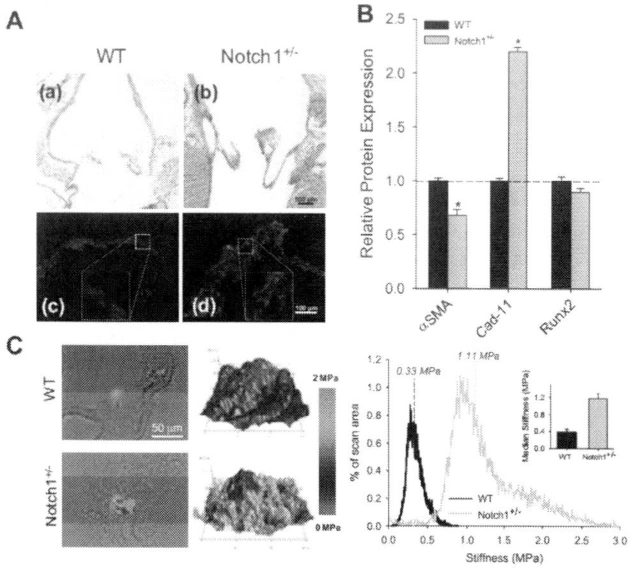

Fig. 1 – Notch1$^{+/-}$ valves reveal enhanced calcification and increased mechanical stiffness (A,C), and Notch1$^{+/-}$ AVICs exhibit upregulated cadherin-11, decreased αSMA, and reduced Runx2 expression (B).

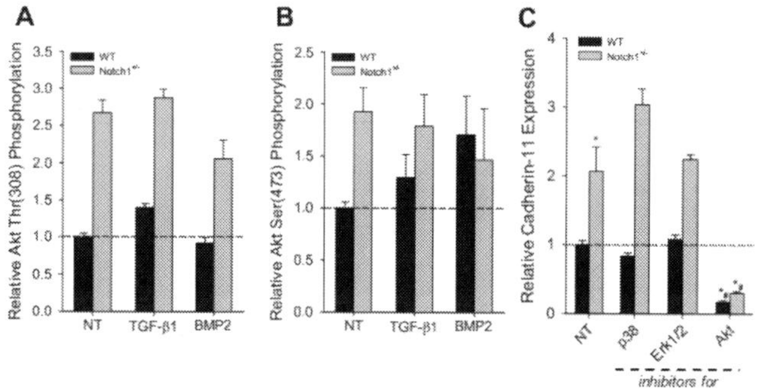

Fig. 2 – Notch1$^{+/-}$ AVICs exhibit enhanced Akt signaling at Thr(308) and Ser(473). Inhibition of Akt reveals critical role of Akt signaling in cadherin-11 expression.

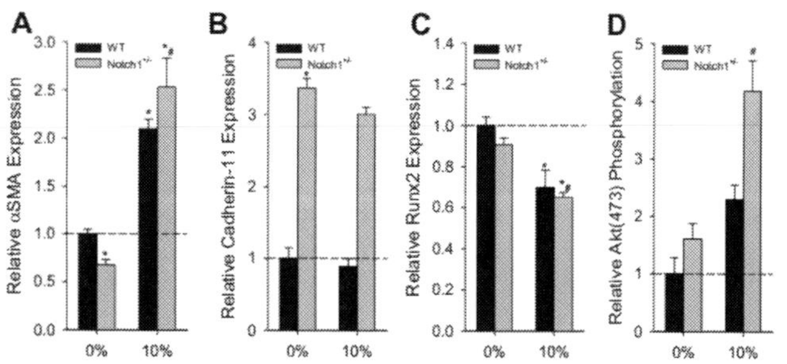

Fig. 3 – Mechanical strain increases αSMA expression but significantly reduces Runx2 expression (A,C). Cadherin-11 remains unchanged (B). Notch1$^{+/-}$ cells reveal exaggerated strain induced Akt phosphorylation (D).

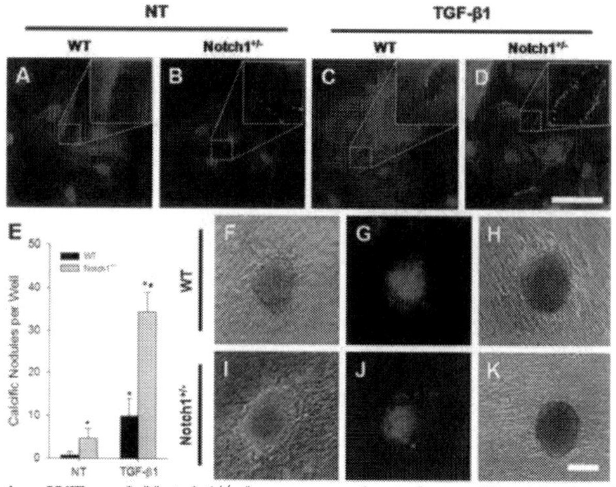

Fig.4 – WT and Notch1$^{+/-}$ become activated myofibroblasts with TGF-β1 incubation (A-D). Notch1$^{+/-}$ AVICs reveal more robust calcific nodules formation (E) and evidence of dystrophic calcification as revealed by the uptake of propidium iodide and alizarin red (G-K).

With the application 10% cyclic mechanical strain, which approximates *in vivo* diastolic loading, αSMA expression is dramatically increased, cadherin-11 expression remains unchanged, and Runx2 expression is decreased (Fig. 3A-C). Notably, unstrained Notch1$^{+/-}$ AVICs have less αSMA than WT but exceed the WT cells when strained. Similarly, phosphorylation of Akt Ser473 underwent larger increases due to strain in Notch1$^{+/-}$ AVICs (Fig. 3D).

Immunostaining revealed that Notch1$^{+/-}$ AVICs have significantly more cadherin-11; however, WT cells have more αSMA than mutant cells (Fig. 4A,B). When treated with TGF-β1, however, both WT and Notch1$^{+/-}$ AVICs revealed increases in αSMA and cadherin-11 (Fig. 4C,D). Notch1$^{+/-}$ AVICs formed significantly more calcific nodules than WT cells with and without TGF-β1; however, TGF-β1 treatment dramatically increased the number of nodules formed (Fig. 4E). Apoptosis, necrosis, and calcification stains reveal characteristics indicative of dystrophic calcification (Fig. 4F-K).

DISCUSSION

In this study, we demonstrate that Notch1$^{+/-}$ AVICs become fully activated myofibroblasts (increased αSMA and cadherin-11 expression), which leads to dystrophic calcification in a dynamic mechanical environment. Further, we demonstrate that Notch1 mutant cells suppress osteogenic activity suggesting a preference for the myofibroblast phenotype. We show that Akt signaling plays a significant role in cadherin-11 expression and that Notch1 deficiency alters AVIC mechanobiology, promoting dystrophic calcification. The findings in this report are significant because they provide paradigm shifting evidence that Notch1 mutation leads to dystrophic, but not osteogenic, CAVD. Furthermore, the clear and distinct role of increased cadherin-11 in the Notch1$^{+/-}$ AVICs is significant because it has been demonstrated as a critical component of valvular calcification, and inhibition of cadherin-11 function may be a promising approach for treating CAVD.

ACKNOWLEDGEMENTS

This work is supported by the NIH (HL115103).

REFERENCES

[1] Garg, V et al., *Nature* 437:270-274, 2005

THROMBOGENIC PROPERTIES OF A SUPERHYDROPHOBIC SURFACE FOR USE AS A PROSTHETIC HEART VALVE MATERIAL

David L. Bark Jr. (1,2), Sanli Movafaghi (1,2), Brandon L. Moore (1), Arun K. Kota (1,2), Ketul C. Popat (1,2), Lakshmi P. Dasi (1,2)

(1) Department of Mechanical Engineering
Colorado State University
Fort Collins, CO, United States

(2) School of Biomedical Engineering
Colorado State University
Fort Collins, CO, United States

INTRODUCTION

Of the numerous blood-contacting medical devices implanted, 290,000 prosthetic heart valve (PHV) replacements were performed in 2003, which is expected to increase to 850,000 by 2050 [1]. Half of these PHVs are mechanical, which are at high risk of failure through thrombosis and thromboembolism, but are more durable than the alternative bioprosthetic PHVs [2]. To mitigate the risks associated with thrombosis, patients are treated with systemic anticoagulation and antiplatelet drug regimens. These therapies exhibit limited efficacy in high flow environments, while increasing the risk for bleeding. Furthermore, systemic anticoagulation results in the most drug-related deaths occurring from adverse clinical events in the United States [3]. Therefore, less aggressive anticoagulation or antiplatelet therapy due to less thrombogenic biomaterials could provide significant benefits for the numerous patients requiring medical devices.

Superhydrophobic surfaces may provide an effective treatment that significantly reduces and in some cases eliminates thrombus formation on the material surface. These surfaces are characterized by the degree that water beads on the surface, measured by a contact angle, and exemplified by the lotus leaf in nature [4]. Here, we report on the potential for these superhydrophobic surfaces to be used as a material treatment for mechanical PHVs. Tests were conducted to quantify hemodynamic improvement, in addition to studying the immunogenic and thrombogenic responses.

METHODS

We developed a superhydrophobic coating consisting of an epoxy mixed with 20 μm perflourinated silica (perfluoro-SiO_2) particles and another version with a hierarchal mix of 10 nm and 20 μm perfluoro-SiO_2 particles. These coatings were applied to pyrolytic carbon on a 25 mm St. Jude bileaflet mechanical PHV. These coatings were also applied to a flat pyrolytic carbon surface for contact angle measurements and a glass surface to study blood cell responses.

Contact angle goniometry (Ramé-Hart) was used to assess the hydrophilicity. This technique measures the angle between the solid-liquid and liquid-vapor interface of a droplet on the surface.

Hemodynamic performance was evaluated for superhydrophobic treatments by calculating the Performance Index (PI), defined as the effective orifice area normalized with the annulus area, by performing particle imaging velocimetry (PIV), and through computations. Steady flow PI calculations were based on the ISO-5840 guidance document. PIV was performed in phase locked modes corresponding to normal physiological aortic conditions. Fluid was seeded with 1-20 μm melamine resin particles coated with Rhodamine-B. A Nd:YLF Single Cavity Diode Pumped Solid State High Repetition Rate Laser (Photonics Industries, Bohemia, NY) was used to illuminate a 0.2 mm thick measurement plane. The measurement plane was imaged with a double frame SA3 CMOS camera (Photron, Inc.), while using a calibration grid to correct image distortion. DaVis (Lavision, Inc.) was used to post-process PIV slices using an ensemble of approximately 500 phase locked measurements. Lastly, ANSYS Fluent was used to compare hemodynamic performance of a regular valve with no fluid slip at the surface and another simulation with total slip to simulate a best performance for a superhydrophobic surface. Simulations involved an unstructured three-dimensional grid (~1.8 million cells) with a half-sine wave velocity boundary condition at the inlet.

Cell responses to surfaces were evaluated by incubating materials for 1 hour with platelet rich plasma (PRP) obtained from centrifugation of Ethylenediaminetetraacetic acid anticoagulated whole blood. These samples were rinsed with Tyrodes buffer and were subsequently fixed with 1% glutaraldehyde. Samples were then imaged with scanning electron microscopy (SEM).

RESULTS

First, hemodynamic performance was tested. Fig. 1A shows the superhydrophobic valve mounted in a ring holder for flow testing with coated leaflets in the closed position. Typically water will pass through the untreated valve, but with a hierarchal treatment, water beads on the surface. This valve was placed into a testing chamber to calculate the PI in steady flow, which increased from 0.66 to 0.71 with the hierarchical treatment. Since the PI only marginally increased, a computational study of flow was performed through the valve to investigate the optimal improved performance. The pressure drop with a slip condition at the valve surface was qualitatively similar to a no-slip condition, akin to experimental measurements of the PI.

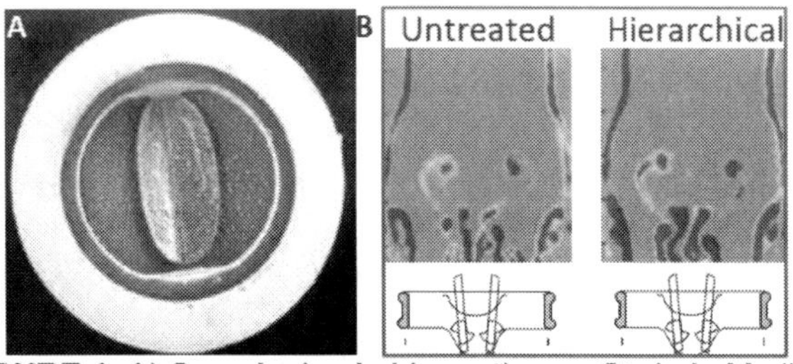

FIGURE 1: A) Superhydrophobic coating on St. Jude Medical bileaflet mechanical PHV and B) PIV results of untreated and treated PHVs.

With regards to pulsatile PIV testing, Fig. 1B shows a comparison of out-of-plane vorticity downstream of the valve during the acceleration phase for the unmodified and modified 25 mm PHV. The superhydrophobic PHV was exhibited improved flow in the central and side orifices, corresponding to a smoother vorticity profile in compared to the unmodified valve. Phase averaged turbulence was also notably reduced with the superhydrophobic coating.

Contact angle goniometry was used to assess the superhydrophobic treatment. Untreated pyrolytic carbon resulted in a contact angle of 70°. The addition of microscale perfluoro-SiO2 increased the contact angle to 134° and the hierarchical surface increased the angle to 160°, making the surface superhydrophobic, with results shown in Fig. 2.

FIGURE 2: Contact angles on pyrolytic carbon, the 20 μm particles, and the hierarchical surface (left to right).

Minimizing the thrombotic and immunogenic response is a major goal in the design of blood-contacting materials. To assess the potential, PRP was incubated on the surface with 20 μm perfluoro-SiO_2 particles and also on the hierarchical surface. A relatively large quantity of leukocytes and platelets were found on the microscale surface, as can be seen in the left side of Fig. 3. Platelets are the smaller cells, while leukocytes are the larger round cells. Many adherent platelets exhibited filopodia, demonstrating activation. Some lamellipodium could also be seen, corresponding to spreading platelets and later stages of platelet activation. Spreading was also seen in some leukocytes. Conversely, the hierarchal perfluoro-SiO_2 surface resulted in no observable platelet or leukocyte adhesion, as shown in the right side of Fig. 3. To evaluate the influence of cell and protein adhesion on the contact angle, PRP contact angles were measured before and after incubation for 1 hour. The superhydrophobic surface only changed its contact angle by 3%, while the bare surface resulted in a contact angle reduction of 67%.

FIGURE 3: SEM images of treated surfaces after incubation with PRP. A surface with 20 μm perfluoro-SiO_2 particles is shown on the left and the hierarchical surface is on the right. Inset: Human whole blood on the hierarchical surface.

DISCUSSION

The presence of a superhydrophobic surface treatment exhibits a significant, but small influence on the PI. The superhydrophobic coating is expected to result in slip at the fluid-surface interface when air pockets are present on the coating, known as the Cassie state [5]. However, even in an ideal case where there is complete slip, there is only a small change in the pressure drop across the PHV, as demonstrated by the computational results. This is likely due to the angle of the leaflets during forward flow, where flow separation leads to the rich and complex vortex structures in the wake as is seen in Fig. 1B. Therefore, the current valve design will not result in significant improvement in energy loss performance because the form drag on the valve remains high, despite the low skin friction. However, the superhydrophobic surface did result in a smoother and "less turbulent" boundary layer, which may result in a "friendlier" blood environment.

The superhydrophobic surface involving a hierarchical surface structure significantly reduced the amount of cell adhesion, indicating that these surfaces can reduce the immune and thrombotic response in a static environment. Additional studies will be performed to further evaluate cellular responses and signaling in flow.

ACKNOWLEDGEMENTS

Research reported in this abstract was partially supported by National Heart Lung and Blood Institute of the National Institutes of Health under award number R01HL119824. The content is solely the responsibility of the authors and does not necessarily represent the official views of the National Institutes of Health.

REFERENCES

[1] Yacoub, M & Takkenberg, J, *Nat Clin Pract Card*, 2:60-61, 2005.
[2] Unger, F & Ghosh, P, *Semin Thorac Cardiovasc Surg*, 14:321-323, 2002.
[3] Kutay, V et al., *J Cardiac Surg*, 21:572-577, 2006.
[4] Leslie, DC et al., *Nature Biotechnol*, 32:1134–1140, 2014.
[5] Rothstein, JP, *Annu Rev Fluid Mech*, 42:89-109, 2010.

BIOMECHANICAL COMPARISON BETWEEN FACET SPARING LAMINECTOMY AND LAMINECTOMY WITH FACETECTOMY IN LUMBAR SPINE: A FINITE ELEMENT ANALYSIS

S. Elmasry (1), S. Asfour (1), J. Gjolaj (2), L. Latta (2,3), F. Eismont (2), F. Travascio (1)

(1) Department of Industrial Engineering
University of Miami
Coral Gables, FL, USA

(2) Department of Orthopaedic Surgery
University of Miami
Miami, FL, USA

(3) Max Biedermann Institute
Mount Sinai Medical Center
Miami Beach, FL, USA

INTRODUCTION

Lumbar spinal stenosis (LSS), a disease of epidemic proportions in the United States [1], is characterized by the constriction of the spinal canal at one or multiple spine levels. The LSS is often surgically treated for decompressing the spinal nerve. However, often after surgery, stenosis propagates to adjacent spinal segments [2, 3]. Such degenerative phenomenon is known as adjacent segment disease (ASD) and its etiology, although believed to be multifactorial, is still unclear. It is hypothesized that surgeries for LSS, by altering the normal physiological spine anatomy, may change the biomechanical environment of the adjacent segments, eventually leading to ASD. Accordingly, significant research effort has been made in order to investigate and characterize the biomechanics of the spine undergoing surgical procedures for LSS. A myriad of surgical approaches for LSS are currently available, and several in-vitro and computational studies have characterized the degree of spine instability caused by surgical reductions typical of decompression surgery [4-6]. However, none of these studies has directly compared the effects of facet-sparing laminectomy and laminectomy with facetectomy at both operated and adjacent spine level.

The objective of this study was to investigate the biomechanical alterations of operated and adjacent spine levels occurring after performing facet-sparing laminectomy and laminectomy with facetectomy on the lumbar spine. More specifically, by deploying a realistic finite element model of lumbar spine, in vitro biomechanical tests were simulated and the post-operative changes in range of motion (ROM), intradiscal fluid pressure, and mechanical stress in the intervertebral discs of operated and adjacent segments were quantified.

METHODS

A three dimensional nonlinear finite elements model of the intact lumbar spine (L1- L5) was created via CT scan of a healthy male subject (Visible Human Project®). Spine vertebrae were modeled as rigid bodies, while intervertebral discs were divided in to two inhomogeneous biphasic materials. In particular, the annulus fibrosus (AF) was composed of lamellae, each of them modeled as hyperelastic fiber-reinforced biphasic media, while the nucleus pulposus (NP) was defined as non-linear isotropic biphasic media. The 7 major spine ligaments were represented as tension-only spring elements. In addition, cartilage at facet joints was modeled as isotropic elastic materials with a gap of 0.5 mm. The study was performed using noncommercial finite element software (FEBio 1.8.0, Musculoskeletal Research Laboratory, University of Utah, Salt Lake City, UT).

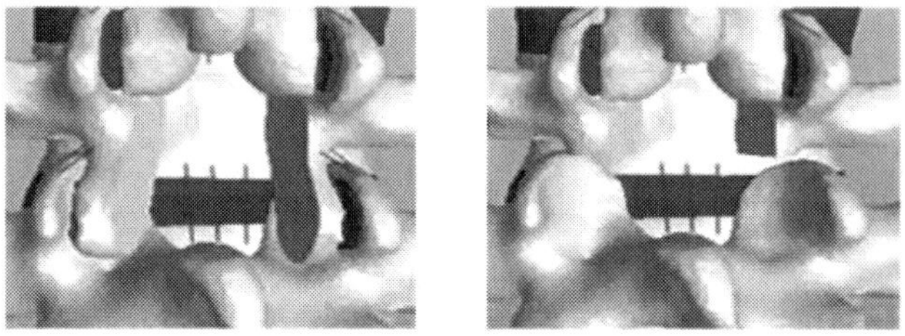

(a) Facet Sparing Laminectomy **(a) Laminectomy with Facetectomy**

Figure 1: Spine anatomical reduction for facet sparing Laminectomy and Laminectomy with Facetectomy

The model was validated by comparing its predictions to previously reported in-vitro study [6]. Subsequently, two post-operative scenarios were simulated: facet-sparing laminectomy, and laminectomy with facetectomy. For each scenario, the operated segment was L4-L5, and the alterations of spine anatomy associated to each procedure were implemented as reported in previous studies [6, 7], see Figure 1. A hybrid loading protocol was used [8]: the intact spine was loaded with a pure moment of 8 Nm in each of the main anatomic planes (flexion, extension, lateral bending, and axial rotation) together with a follower load of 400 N; operated spine models were tested with a loading conditions that yielded the same total range of motion of the intact spine. The post-operative alterations of the biomechanics of the operated and adjacent segments (L4-L5 & L3-L4) were evaluated in terms of rotational range of motion in the three anatomical planes, intradiscal pressure, and normal stress.

RESULTS

The total ranges of motion of the spine segment (L1-L5) resulting from the application of pure moment for flexion, extension, lateral bending, and axial rotation were 11.46°, 14.3°, 9.29°, and 12.24°, respectively. The post-operative changes in the intersegmental ROM of L3-L4 and L4-L5 are reported in Table 1. No major post-operative increase of ROM were found after lateral bending or extension. In contrast, during flexion, L3-L4 rotation increased up to 23% after facet-sparing laminectomy. Also, during axial rotation, L4-L5 range of motion increased up to 28% after laminectomy with facetectomy.

Table 1: Intersegmental changes in rotational range of motion

		Intact	*Facet-sparing Laminectomy*	*Laminectomy with Facetectomy*
L3-L4 (degrees)	*Flexion*	2.77	3.42	3.30
	Extension	3.05	2.87	2.84
	Lateral Rotation	2.41	2.35	2.33
	Axial Rotation	2.32	1.87	1.73
L4-L5 (degrees)	*Flexion*	4.27	4.59	5.02
	Extension	2.95	3.10	3.07
	Lateral Rotation	2.41	2.35	2.33
	Axial Rotation	0.91	0.99	1.17

The post-operative alterations of spinal segments kinematics was reflected in changes in the biomechanics of the operated and adjacent segments. More specifically, during flexion and axial rotation, intradiscal pressure raised up to 20% for both levels, see Figure 2a-b. Moreover, during flexion, the normal stresses dramatically increased up to 100% after the two procedures and at both levels, see Figure 2c. In contrast, changes in normal stress after axial rotation were minimal, see Figure 2d.

DISCUSSION

This study reports a comparative analysis on the effects of two common surgical procedures for lumbar stenosis on the biomechanics of the operated and adjacent segments: facet-sparing laminectomy and laminectomy with facetectomy. In agreement with clinical, and in vitro studies [9, 10], the model only predicted significant post-operative changes in spine kinematics during flexion and axial rotation motions. This was likely due to the fact that both procedures entail resection of interspinous and supraspinous ligaments which withstand the highest tensile forces during spine flexion [11]. In addition, removal of the facet joints in laminectomy with facetectomy reduced the spine stability during axial rotation [12].

Post-operative kinematic changes reflected in increase of intradiscal pressure and stress (Figure 2). Intradiscal pressure increase may negatively affect intervertebral disc metabolism, potentially leading to disc degeneration [13]. In this respect, facetectomy procedure mostly affected the operated level L4-L5. Conversely, facet-sparing laminectomy yielded the largest biomechanical alterations at the immediate adjacent level L3-L4. Hence, both procedures have similar potential in undermining spine health. This is despite the fact that facetectomy, involving more reductions, is often regarded as a riskier procedure for the patient in terms of spine stability.

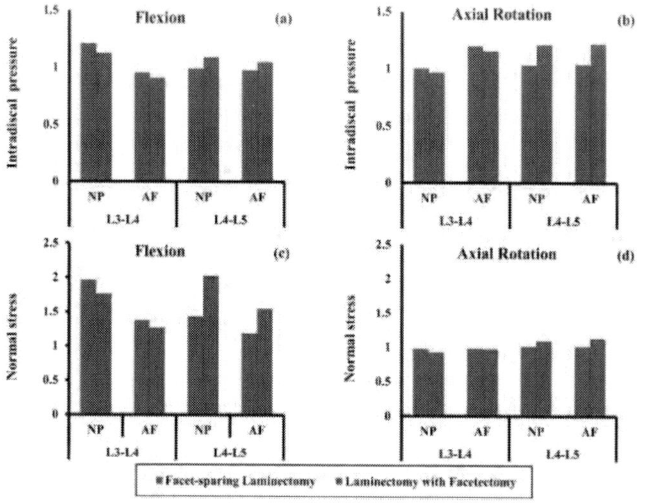

Figure 1: Intradiscal pressure and Normal stress of spine discs (L3-L4 &L4-L5) in nucleus pulposus and annulus fibrosus after flexion and axial rotation. Data are normalized with respect to 'intact spine' case.

ACKNOWLEDGEMENTS

Study supported by funds donated to the Biomechanics Research Group of the University of Miami.

REFERENCES

[1] Arbit & Pannullo, *Clin Orthop Relat R*, 384:818-825, 2001.
[2] Lee, *Spine*, 13:375-7, 1988.
[3] Rahm & Hall, *J Spinal Disord*, 9:392-400, 2008.
[4] Zander et al., *European Spine Journal*, 12:427-434, 2003.
[5] Quint et al., *European Spine Journal*, 7:229-238, 1998.
[6] Lee et al., *Spine*, 35:1789-1793, 2010.
[7] Rao et al., *The Spine Journal*, 2:320-326, 2002
[8] Panjabi et al., *Spine*, 32:959-966, 2006.
[9] Bresnahan et al., *Spine*, 34:17-23, 2009.
[10] Abumi et al., *Spine*, 15:1142-1147, 1990.
[11] Goel et al., *Spine*, 10:516-523, 1985.
[12] Adams & Hutton., *Spine*, 8.3: 327-330, 1983.
[13] Stokes & Iatridis *Spine* 29:2724-32, 2004.

SB^{3}C2015
Summer Biomechanics, Bioengineering and Biotransport Conference
June 17-20, 2015,Snowbird Resort, Utah, USA
SB^{3}C2015-654

AQUEOUS BIPHASIC MICRO-PRINTING OF TUMOR SPHEROIDS

Stephanie L. Ham (1), Ehsan Atefi (1), Hossein Tavana (1)

(1) Biomedical Engineering
The University of Akron
Akron, OH, USA

INTRODUCTION

Physiologically relevant cell-based models are important tools for the discovery of effective anti-cancer drugs. Tumor spheroids, three-dimensional compact clusters of cancer cells, imitate key characteristics of *in vivo* tumors such as poor perfusion and present a relevant method to examine drug efficacy. New techniques based on microfabricated microwells and hanging drop arrays allow high throughput spheroid culture. Nevertheless, difficulties associated with fabrication and handling of specialized devices and plates, conveniently maintaining spheroids in culture and treating them with chemical compounds, and biochemically analyzing drug response of cells using plate readers without transferring spheroids into standard plates hinder the utility of spheroids.

To address this need, we have developed a spheroid printing microtechnology using a polymeric aqueous two-phase system (ATPS) [1]. Cancer cells are suspended in an aqueous drop and printed within an immiscible, immersion aqueous phase to spontaneously self-assemble into a spheroid. Robotic liquid handling enables formation of uniformly-sized spheroids in standard 384-microwell plates with high precision and reduced labor [2]. Media exchange, drug treatment of individual spheroids, and analysis of drug response of cells -which are often challenging with other techniques- are conveniently performed by direct addition of reagents to microwells where spheroids were formed in. Histological examination of spheroids confirms expression of several key biological markers. Drug screening with biologically aggressive triple negative breast cancer (TNBC) cells and skin cancer cells show substantial differences in drug response compared to monolayer cultures [3]. This simple-to-implement technology will expedite anti-cancer drug discovery with a physiologic tumor model.

METHODS

To form an ATPS, 5%(w/w) polyethylene glycol (PEG), Mw: 35,000, 6.4%(w/w) and dextran (DEX), Mw: 500,00, are dissolved in culture media. A 384-well plate (destination plate) is filled with 30 μl of the PEG phase while one column of a 384-well plate (source plate) is filled with 10-20 μl of the DEX phase mixed with desired density of cancer cells. A robotic liquid handler aspirates 0.3 μl from each well of the source plate and dispenses it into wells of the destination plate. The cell-containing DEX phase drop settles in the immersion PEG phase and maintains the cells in close proximity to form one spheroid in each well. For histological analysis, spheroids are fixed, embedded in freezing media, cryosectioned, and immunostained for various markers including a cell proliferation marker Ki-67 and extracellular matrix proteins. A freeware, ImageJ, is used for image analysis of immunostained spheroid sections. Drug testing is performed by direct addition of various concentrations of each drug (paclitaxel and cisplatin) to spheroids in the destination plate. Spheroids of MDA-MB-157 breast cancer and A431 skin cancer cells are incubated with drugs for four days, with a drug renewal after two days of incubation. A standard PrestoBlue viability reagent protocol is adjusted for spheroid culture to determine the viability of cells in drug-treated spheroids using a microplate reader. The fluorescent signal from treated spheroids is normalized against the signal from control wells containing non-treated spheroids. Drug experiments with monolayer cultures are performed in parallel for comparison.

RESULTS

A single spheroid forms in each well of a standard 384-microplate after 24 hrs of incubation (Figure 1a). Spheroids printed with a 15×10^3 cell density show an average diameter of 235 ± 22 μm in each microplate, indicating the consistency of this technique to generate uniformly-sized spheroids (Figure 1b). This level of consistency is significant given the use of a commercial liquid handler to dispense sub-microliter volumes of the viscous DEX phase containing cancer cells. Spheroids show normal growth when maintained in culture up to one week.

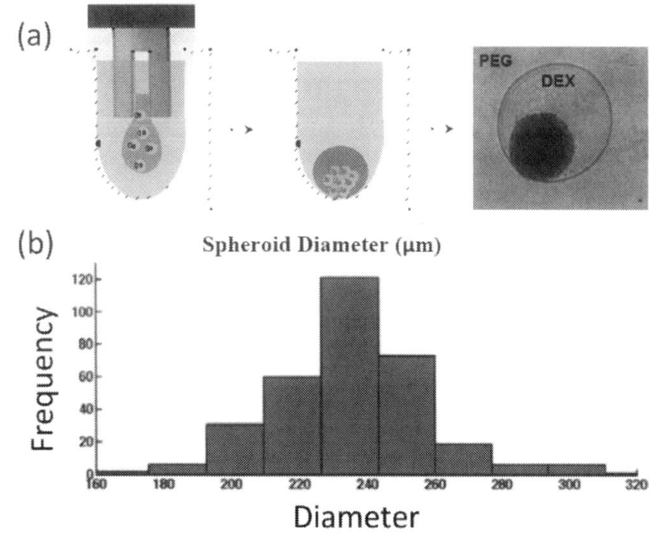

Figure 1: (a) Side-view of spheroid formation in the DEX phase drop (blue) immersed within the PEG phase solution (pink) and a top-view experimental image of spheroid. (b) Histogram of spheroids diameter from a 384-microplate shows a normal distribution. Scale bar 250 μm.

Histological immunostaining for a proliferative marker, Ki-67, reveals a homogenous distribution of proliferative cells in spheroids of 15×10^3 cell density (Figure 2, left). However, proliferative cells are distributed more at the boundary of larger spheroids made with 100×10^3 cells (Figure 2, middle). During culture, cancer cells within spheroids express key extracellular matrix proteins such as fibronectin (Figure 2, right).

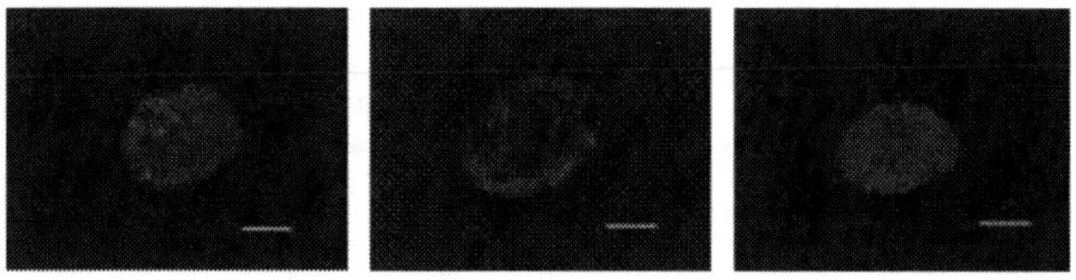

Figure 2: Immunohistochemistry of 10 μm-thick sections of MDA-MB-157 TNBC cell spheroids with Hoechst (blue) and Ki-67 (pink) for spheroids formed with 15×10^3 (left) and 100×10^3 (middle) cells. Staining of the extracellular matrix protein fibronectin (pink) in a spheroid with 15×10^3 cell density (right). Scale bar 100 μm.

Spheroids are treated with paclitaxel and cisplatin at a wide range of concentrations and their viability is biochemically analyzed using an enzymatic assay (PrestoBlue) optimized for spheroid culture [3]. Spheroids often exhibit resistance to drugs. TNBC cell spheroids exemplify this behavior and maintain a high viability after treatment with paclitaxel even at very high concentrations. On the other hand, TNBC cell monolayer cultures show a typical dose-dependent drug response (Figure 3).

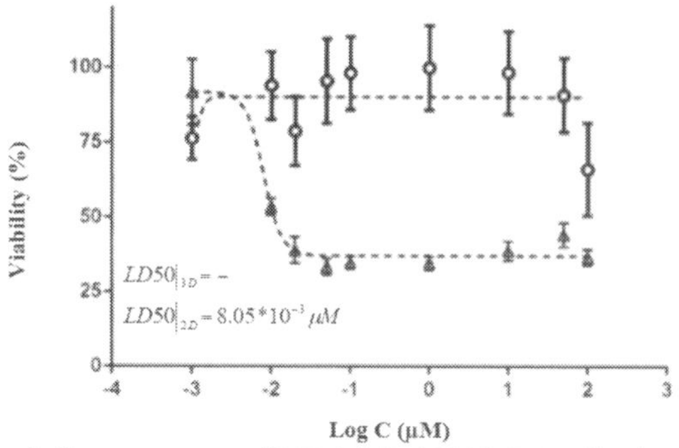

Figure 3: Drug response of MDA-MB-157 TNBC cell spheroids (circles) and monolayer (triangles) treated with paclitaxel (0-100 μM).

DISCUSSION

The aqueous two-phase tumor spheroid microtechnology generates uniform-sized spheroids in standard, commercially-available microwell plates (Figure 1). Importantly, spheroids are submerged in a culture media and individually-addressable with chemical compounds. This robotic technology significantly increases precision and reduces labor, and allows investigations of drug response of cancer cells with a relevant tumor model in a high throughput setting. The optimization of the PrestoBlue assay for spheroid cultures enables convenient analysis of effectiveness of anticancer drugs *in situ*, eradicating the need to transfer spheroids to different plates for analysis.

Staining for the proliferative marker Ki-67 in spheroids confirms that spheroids display a key feature of poorly perfused tumors. While proliferative cells are uniformly distributed in small spheroids of ~250 μm diameter, the higher cell density spheroids contain a distinct border of proliferative cells (Figure 2). This indicates that a non-proliferative, necrotic core, exists in larger spheroids. Presence of necrotic regions in tumors *in vivo* is often implicated in resistance to certain cancer therapies. Thus, this approach reproduces a key characteristic of tumors and allows testing of hypoxia targeting drugs. Expression of extracellular matrix proteins in spheroids also allows investigating cell-matrix interactions in tumor microenvironments.

Our proof-of-concept compound screening reveals significant drug resistance of spheroids compared to monolayer of cells, potentially due to the difficulty of drug diffusion into compact mass of cells (Figure 3). This reinforces the use of three-dimensional cellular models to elicit more realistic responses from cells treated with chemical compounds. Such responses often cannot be captured with monolayer of cells, leading to errors in predicting efficacy of novel chemicals. Overall, this tumor spheroid microtechnology allows quick and efficient spheroid formation and addresses complications associated with other techniques by allowing direct addition of media, drugs, and assay reagents in the same standard microplate spheroids reside in. The technology will especially be useful in research and core centers with access to robotic and screening tools.

ACKNOWLEDGEMENTS

Financial support is provided by NIH (R21CA182333).

REFERENCES

[1] H. Tavana et al., *Nature Mater.* 8:736-741, 2009.
[2] E. Atefi et al., *Adv. Func. Mater.* 24:6509-6515, 2014.
[3] S. Lemmo et al., *Cell. Mol. Bioeng.* 7:344-354, 2014.

SB³C2015
Summer Biomechanics, Bioengineering and Biotransport Conference
June 17-20, 2015,Snowbird Resort, Utah, USA
SB³C2015-655

DEVELOPMENT OF A PLATFORM FOR STUDYING ASTROCYTE MECHANOBIOLOGY: COMPRESSION OF ASTROCYTES IN 3D ALGINATE GELS

John J.E. Mulvihill (1, 3), Lisa A. Schildmeyer (1), Baptiste Coudrillier (1), Danny J. Kelly (3), C. Ross Ethier (1, 2)

(1) Wallace H. Coulter Dept. of Biomedical Engineering
Georgia Institute of Technology
Atlanta, Georgia, USA

(3) Mechanical and Manufacturing Engineering
Trinity College Dublin
Dublin, Ireland

(2) Atlanta VA Medical Center,
Decatur, Georgia, USA

INTRODUCTION

Glaucoma is the second leading cause of blindness worldwide according to the World Health Organization, affecting approximately 60 million people. As glaucoma is primarily a disease of aging, the burden is expected to increase, with the number of people affected by glaucoma set to rise to 80 million by 2020 [1].

Glaucoma is an optic neuropathy in which intraocular pressure (IOP) exceeds the threshold level that can be tolerated by the retinal ganglion cells (RGCs). IOP deforms the delicate tissues of the optic nerve head (ONH), which is thought to predispose resident ONH astrocytes to transition to a reactive phenotype; this transition is in turn believed to trigger RGC dysfunction and eventual death, resulting in vision loss. However, the details of this response, i.e. the mechanobiology of ONH astrocytes, are unclear.

Researchers have previously studied ONH astrocyte mechanobiology using 2D (monolayer) culture systems [2-4]. While 2D culture models have revealed important information, they do not mimic the 3D physiological environment of the ONH, and in fact, typically create a highly reactive environment [5]. For example, one study [7] demonstrated that most cell types, when isolated from tissues and placed into 2D cell culture, become progressively flatter, divide abnormally and lose their differentiated phenotype. However, cells can regain their physiological form and function when placed in a 3D culture environment [6]. Accordingly, our overall aim is to better understand the mechanobiology of ONH astrocytes in a physiologically relevant 3D environment. This requires a suitable 3D cell culture platform allowing mimetic biomechanical loading, e.g. compression and stretching of a 3D hydrogel with embedded cells.

In this preliminary study, we first examine morphological differences between astrocytes cultured in 2D and 3D environments, and secondly analyze the effects that mechanical loading has on astrocytes grown in 3D hydrogels.

METHODS

All studies used DI TNC1 (ATCC, CRL 2005) immortalized cell line derived from primary cultures of type 1 astrocytes from brain diencephalon tissue of 1 day old rats. Cells were maintained in Dulbecco's Modified Eagle Medium (DMEM, 10-0130CV, 4.5g/L glucose, Cellgro) with 10% FBS (Hyclone, SH30910) and 1X Pen/Strep (Hyclone, SV30010), and expanded to reach confluence.

For 3D cell culture, a 2.5% alginate (A1112, Sigma) hydrogel (0.75 g in 30 ml PBS) was prepared and mixed with a cell suspension to produce a density of 5.5 million cells/ml. This cell/alginate suspension was poured onto a 3% agarose-$CaCl_2$ gel (50ml 6% agarose-PBS, 50ml 200mM $CaCl_2$) slab in a 10 mm sterile petri dish and then covered with a second identical gel slab. The gel/cell arrangement was allowed sit 45 minutes in a sterile environment at room temperature. The diffusion of the $CaCl_2$ into the alginate caused the alginate to solidify. From the resulting solidified alginate slab, cylindrical plugs (5 mm diameter, 3mm height; Figure 1) were removed and placed in a custom made culture dish with medium. For 2D cell culture, the remaining cells were seeded in a 12 well dish.

A bioreactor system was developed to apply mechanical loads to the 3D hydrogels. The deformations generated by the system are thought to correspond to those experienced by astrocytes under normal and elevated IOPs in the lamina cribrosa (LC), the portion of the ONH where most astrocytes reside [3]. The 3D hydrogel plugs containing astrocytes were compressed (loaded), within the custom culture dish, by 10% for 24 hours at 1Hz [3] (n = 49 gels). An equal number of control hydrogels plugs (unloaded) were not subjected to compression and maintained in culture during this time.

For both 2D and 3D cell culture models, Fluorescein DiAcetate (FDA), which fluoresces only in live cells (F3677, Sigma), was used to assess cell viability. The stain was also used to highlight differences in cell morphology between the loaded and unloaded 3D environments. To compare the loaded and unloaded (control) 3D gels, confocal microscopy was used to acquire z-stacks from sections of the plugs. The 3D z-stacks were then processed, as a series of 2D images, to highlight the cells only from each 2D slice (5 µm spacing, ImageJ, NIH). The eccentricity of the resulting cells were measured (RegionProps function, Matlab, MathWorks, MA) and compared between loaded and unloaded samples by a Kruskal–Wallis test.

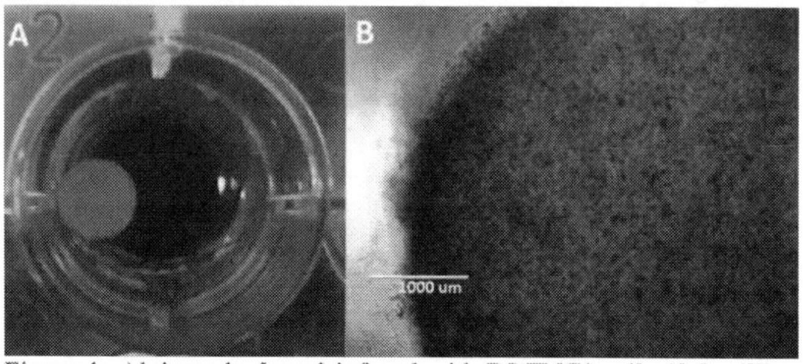

Figure 1: Alginate hydrogel infused with DI-TNC1 cells (5.5 million cells/ml). A: Overview of 3D gel, after crosslinking, in medium. B: 4x magnification of hydrogel after 3 days growth (unloaded).

RESULTS

There were clearly observable differences in astrocyte morphology between unloaded 2D and 3D cultures (Figure 2), most notably the spherical shape of cells in 3D culture versus the more elongated morphology of cells grown in 2D. FDA staining of both culture environments (Figure 3) confirmed that cells cultured in both environments were viable. Again shown in Figure 3 is the spherical shape of the cells in the 3D alginate gel, whereas the cells grown in 2D monolayer expanded, spread and clustered throughout the well.

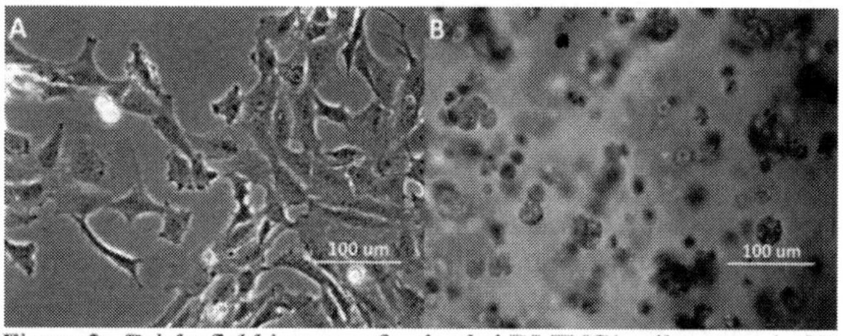

Figure 2: Bright field images of unloaded DI-TNC1 cells grown in 2D (A) and 3D (B) culture environments, 3 days after seeding. In 3D gels, cells that are out of focus can be observed.

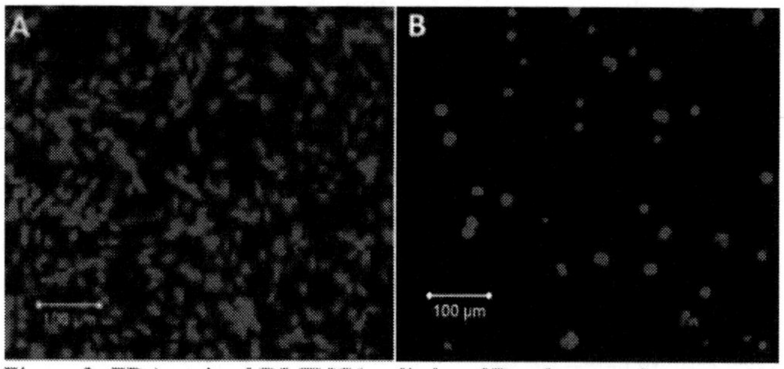

Figure 3: FDA-stained DI-TNC1 cells in a 2D culture environment (A) and in a 3D alginate hydrogel (B) (both 3 days).

Analysis of series of 2D images from the confocal z-stacks of the FDA-stained 3D hydrogels showed a clear difference ($p<0.001$) in cell eccentricity between the loaded (median eccentricity = 0.771) and unloaded (median eccentricity = 0.603) gels.

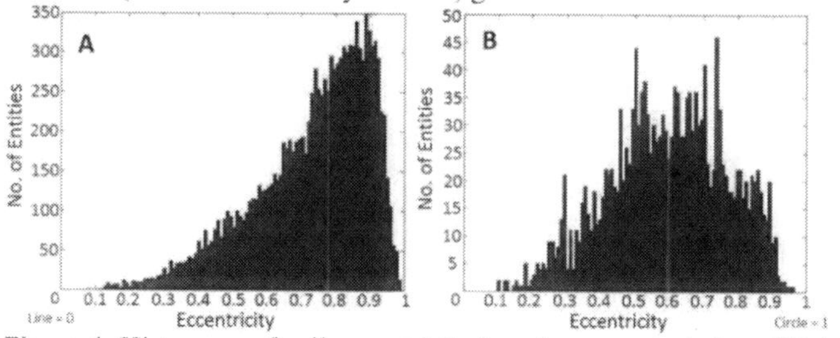

Figure 4: Histogram of cell eccentricity based on a z-stack from FDA stained hydrogels (A: unloaded, B: loaded, red line denotes median).

DISCUSSION

The main hypothesis under investigation in this study is that a 3D environment may allow for better characterization of astrocyte mechanobiology than a 2D based environment [2-4].

The findings from this study agree with previous work [2-6], in which clear cellular morphological differences between 2D and 3D culture environments were shown. Our studies showed that astrocytes were spherical in shape in the 3D alginate gel cross-linked with $CaCl_2$ (Figure 2, B) unlike the elongated shape of the cells in a 2D environment (Figure 2, A).

A previous study on astrocytes seeded in 3D alginate gels [5] also observed this spherical shape, similar to our findings. However, another study suggests that this spherical cell morphology could be due to the quick crosslinking of the alginate gel and the resulting constrained cell shape, rather than the physiological shape of the DI-TNC1 cell [8]. The dynamic switching of crosslinking methodology proposed by Gillette et al. [8] can be considered in future work to investigate whether collagen-alginate gels allow for *in vivo* based growth.

The cells exposed to load in the 3D gel exhibited more elliptical morphology versus unloaded cells (Figure 4), which we hypothesize is due to loading-induced strain. It is possible that this compressive loading (corresponding to increased IOP) can affect astrocyte morphology, which could be a factor in the transition of astrocytes to a reactive phenotype. Further work is necessary to quantitatively analyze the morphological difference between loaded and unloaded DI-TNC1 cells in 3D hydrogels.

Future work will also aim to label markers characteristic of reactive and quiescent astrocytes (e.g. GFAP, Vimentin, CSPG etc.) as well as varying cell count, percentage strain and culture duration in the bioreactor system.

ACKNOWLEDGEMENTS

ELEVATE: Irish Research Council International Career Development Fellowship – co-funded by Marie Cure Actions; Georgia Research Alliance.

REFERENCES

[1] Quigley, HA, et al., *Br J Ophthalmol*, 90(3):262-7, 2006.
[2] Yan, D et al., *Br J Ophthalmol*, 78(8):643-8, 1994.
[3] Rogers, RS et al., *Mol & Cell Proteomics*, 11(2), 2012.
[4] Kirwan, RP et al., *Mol Vis*, 11:798-810, 2005.
[5] East, E et al., *J Tissue Eng Regen Med*, 3:634-646, 2009.
[6] Baker, BM et al., *J Cell Sci*, 125(13):3015-24, 2012.
[7] Petersen, OW et al., *Pro of the Nat Acad of Sci*, 89:9064-8, 1992.
[8] Gillette, BM et al., *Advanced Materials*, 22.6:686-691, 2010.

SB³C2015
Summer Biomechanics, Bioengineering and Biotransport Conference
June 17-20, 2015, Snowbird Resort, Utah, USA
SB³C2015-656

SIMVASCULAR 2.0: AN OPEN SOURCE PIPELINE FOR CARDIOVASCULAR MODELING AND SIMULATION

Hongzhi Lan (1), Nathan M. Wilson (2), Daniele Schiavazzi (1), Jameson Merkow (3), Adam Updegrove (4), Shawn C. Shadden (4), Alison L. Marsden (1)

(1) Mechanical and Aerospace Engineering
University of California – San Diego
La Jolla, California, USA

(2) Open Source Medical Software Corporation
Los Angeles, California, USA

(3) Electrical and Computer Engineering
University of California – San Diego
La Jolla, California, USA

(4) Mechanical Engineering
University of California – Berkeley
Berkeley, California, USA

INTRODUCTION

The advancement of medical imaging techniques and computational mechanics has enabled three-dimensional patient-specific modeling and simulation of cardiovascular biomechanics. This approach can provide invaluable data to compensate for the limitations of clinical measurements and predict outcomes of alternative therapeutic and surgical interventions for individual patients [1]. The SimVascular application for image-based patient-specific blood flow simulation was originally an internal code developed in the laboratory of Prof. Charles A. Taylor (formerly of Stanford University) and has contributed to numerous advances in fundamental hemodynamics research, surgical planning, and medical device design (e.g. [2, 3]). The SimVascular 1.0 application was released as open source code in 2007 but not publicly maintained.

Recent developments in modeling, meshing, and linear/nonlinear solvers have now made it possible to develop a state-of-the-art completely open source package to provide a complete pipeline from medical image data to cardiovascular blood flow simulation results and analysis. The NSF funded the authors in 2013 to revitalize the SimVascular project to integrate the latest methods and attract new users. The goal of the project is to have SimVascular be a widely used platform for cardiovascular modeling and simulation in order to advance public health.

THE SIMVASCULAR REVITALIZATION PROJECT

When SimVascular 1.0 was publicly released in 2007, there were several major barriers preventing wider adoption by new users, large-scale application in clinical and research studies, and educational access. These included: 1) the cost and complications associated with three embedded commercial components, 2) the overall complexity of image-based blood flow simulation, 3) the lack of sustainable architecture and infrastructure, and 4) a lack of organized maintenance, documentation, and support. The NSF funded the SimVascular revitalization to overcome these barriers and foster adoption by new users.

Progress to date includes a fully open source release (i.e. explicitly removing commercial library dependences), simplifying the compilation process, including new automation tools, and an improved documentation website (simvascular.org). The documentation site includes software guides and clinical case study examples. A fully integrated test suite has also been developed with continuous testing and nightly binary creation.

The architecture of the new SimVascular 2.0 project is summarized in Fig. 1. Using CMake, the new version now supports users across all major platforms: Linux, Windows (64-bit architectures) and Mac OS X. Though we continue to support select commercial components, SimVascular 2.0 is now fully open source by integrating open source alternatives. New functionality has been introduced in the SimVascular core modules:

Image Segmentation and Anatomic Modeling Modules: These components provide more efficient algorithms to perform 2D image-based segmentation, and support for importing 3D segmentation results from external segmentation packages such as Mimics (biomedical.materialise.com/mimics) and itk-SNAP (www.itksnap.org). These modules extensively utilize the Visualization Toolkit (VTK, www.vtk.org) and the Insight Segmentation and Registration Toolkit (ITK, www.itk.org). Note that the commercial solid modeling library Parasolid is no longer publicly maintained in SimVascular due to licensing restrictions on its application programmer's interface.

Meshing Module: This module has been extended to provide open source specialized meshing techniques for vascular structures with automated tools for complex geometries using Tetrahedral Mesh Generation (TetGen, www.tetgen.org) and the Vascular Modeling Toolkit (VMTK, www.vmtk.org). **svAdapt** has been extended to support isotropic adaptive meshing using TetGen. The module continues to support meshing with the commercial package MeshSim (Simmetrix, Inc., Clifton Park, NY).

Simulation Module: This module integrates several stand-alone programs: **svPre**, **svSolver, svAdapt,** and **svPost**. **svSolver** produces efficient and stable solutions for cardiovascular models with realistic physiological conditions by implementing back-flow stabilization [4], multi-scale coupling for closed loop boundary conditions with a lumped circulation model [5], and fluid structure interaction for deformable vessel walls using the coupled momentum method [6]. A recently developed linear solver, **svLS,** with specialized preconditioners tailored to incorporate vascular resistance is invoked to handle large cardiovascular simulations and reduce solution time [7], provides an alternative to the original commercial linear solver LesLib (Altair, Inc., Troy, MI) required by SimVascular 1.0. **svPost** has been extended to directly write out VTK formatted files for visualization in ParaView (www.paraview.org).

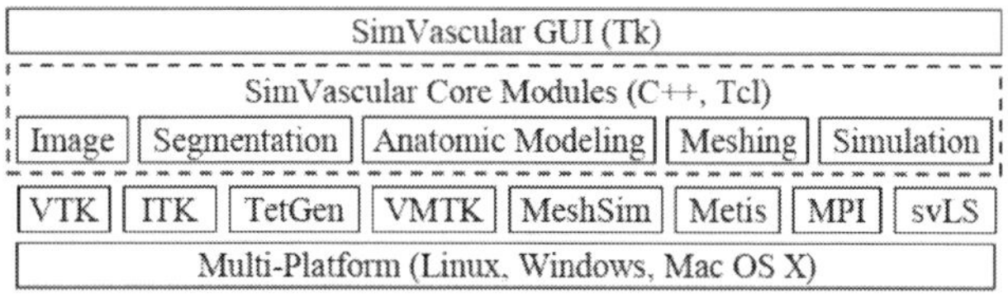

FIGURE 1. SIMVASCULAR SOFTWARE ARCHITECTURE

SIMVASCULAR MODELING PIPELINE

As a result of the revitalization project, SimVascular 2.0 now provides a complete open source pipeline for the process of cardiovascular modeling and simulation. The pipeline covers:

Image Handling: Data can be loaded from a variety of sources, including DICOM, VTK and MetaImage file formats. The **Image Module** offers a number of tools to view and manipulate the data in 2D/3D space, enabling the user to view the image data along vessel centerlines, as point clouds, or by volume rendering.

Path Planning and Image Segmentation: Users can create vessel paths (centerlines) manually. The **Image Segmentation Module** then enables creating contours of the vessel lumen along the paths via specialized segmentation tools including level set, threshold, analytic fitting, and manual segmentation. Fourier smoothing, translation, and scaling features are incorporated. Batch segmentation by level set is implemented for automated segmentation of an entire vessel with minimal user intervention. Integrated direct 3D image segmentation based on level set methods is also available.

Anatomic Modeling: The **Anatomic Modeling Module** enables a user to loft 3D solid models for vessels based on 2D segmentation and unions the vessels to create a complete 3D model of the vasculature of interest. The module can also import 3D segmentation data from other packages to build the model. User can optionally perform cleaning, smoothing, and trimming operations on the triangulated surface used to create the model as well as tag vessel surfaces as needed for simulation.

Meshing: The **Meshing Module** generates a tetrahedral unstructured mesh for the 3D anatomic model that is suitable for computational hemodynamics. To enhance mesh quality, advanced options are provided, including local refinement near selected surfaces, boundary layer meshing, spherical refinement. Mesh adaption tools utilizing *a-posteriori* error estimates are also integrated.

Blood Flow Simulation: The SimVascular pipeline enables users to perform hemodynamic simulations at realistic flow rates and pressures. Time dependent flow rates or heart models may be defined for the inlet(s), and boundary condition parameters are typically specified at the outlets by fitting available clinical data to a lumped parameter model. If the vessel wall is deformable, its mechanical properties also need to be specified. **svPre** creates binary input files for the flow solver. Additional configuration values are set in plain text solver input files, including time step control, linear solver settings, Newtonian fluid properties, etc. Users run **svSolver** to solve the 3D Navier-Stokes equations using a finite element solver with linear elements and the streamline upwind/Petrov-Galerkin (SUPG) method. Simulations can be run with a single core or with multiple cores using the Message Passing Interface (MPI). The SimVascular solver has been shown to achieve excellent scalability in prior work. **svPost** is then used to post-process results to calculate relevant hemodynamic quantities and export results to VTK visualization files.

DISCUSSION

SimVascular has become a crucial tool in cardiovascular simulation research in a small but active group of domestic and international academic labs. Modeling and simulation using SimVascular has to date formed the basis for over 300 abstracts and publications and it has been used to create a publicly available repository of over 100 clinical cases (www.vascularmodel.org) representing different parts of the vasculature.

The goal of the future work is to address the time-consuming tasks in the pipeline to enable high throughput simulations for studies with large patient cohorts. Beyond the improvements cited above in the release of SimVascular 2.0, additional efforts will likely include:

- Improving workflow and user experience with the development of a new graphical user interface (GUI) based on modern cross-platform toolkits.
- Enhancing anatomic model construction including new 3D image segmentation and machine learning methods.
- Additional modularity of the software to enable users to add their own plugins and modules (including commercial components) to extend SimVascular's functionality.
- The integration of an optimization module to enable users to automate device design and perform virtual surgical planning.

ACKNOWLEDGEMENTS

Support was provided by the NSF SSI program (Award #1339824) and an NSF CAREER Award. Software hosting is provided by simtk.org at Stanford University.

REFERENCES

[1] Taylor, CA and Figueroa, CA, *Annual Review of Biomedical Engineering*, 11:109-134, 2009.

[2] Les, AS et al., *Annals of Biomedical Engineering*, 38:1288-1313, 2010.

[3] Sankaran, S et al., *Annals of Biomedical Engineering*, 40:2228-2242, 2012.

[4] Moghadam, ME et al., *Computational Mechanics*, 48:277-291, 2011.

[5] Moghadam, ME et al., *Journal of Computational Physics*, 244:63-79, 2013.

[6] Figueroa, CA et al., *Computer Methods in Applied Mechanics and Engineering*, 195:5685-5706, 2006.

[7] Moghadam, ME et al., *Computational Mechanics*, 52:1141-1152, 2013.

SB³C2015
Summer Biomechanics, Bioengineering and Biotransport Conference
June 17-20, 2015,Snowbird Resort, Utah, USA
SB³C2015-657

BIOMECHANICAL ROLE FOR CARDIAC JELLY IN PUMPING MECHANICS OF DEVELOPING HEART DURING LOOPING

David L. Bark Jr. (1,2), Brennan M. Johnson (2), Bryce W. Schroder (2) Deborah M. Garrity (3), Diego Krapf (2,4), Lakshmi P. Dasi (1,2)

(1) Department of Mechanical Engineering
Colorado State University
Fort Collins, CO, United States

(2) School of Biomedical Engineering
Colorado State University
Fort Collins, CO, United States

(3) Department of Biology
Colorado State University
Fort Collins, CO, United Staets

(4) Department of Electrical and Computer Engineering
Colorado State University
Fort Collins, CO, United Staets

INTRODUCTION

As the first developing organ, the heart provides critical feedback to cardiovascular cells through the flow-derived stress environment corresponding to its pumping mechanism. The mechanism and blood properties, however, change as the heart transforms from a semi-straight primitive tube to the multi-chambered adult heart [1, 2]. These characteristics during the transformation are relatively poorly defined.

Zebrafish are a useful model of cardiovascular development. Their hearts develop quickly and similarly to other vertebrates. They are also transparent. During their development, a primitive heart tube structure forms by 24 hours postfertilization (hpf), which is made up of an outer myocardium and an inner endocardium, separated by an intermediate layer of cardiac jelly (CJ). Shortly after formation, unidirectional flow begins despite a lack of valves [2], followed by looping (tube bends to the right side) at approximately 36 and 48 hpf [1]. At this stage, blood appears to be forced forward through the heart via independent chamber contractions while endocardial cushions help to prevent retrograde flow, taking on the function of an atrioventricular valve, which doesn't form until ~105 hpf [1].

Here, we report on the transition of the heart pumping mechanism from early development to late stages of development. The aim is to determine how endocardial, myocardial, and CJ layers work in concert for efficient pumping. Kinematics and hemodynamics are therefore quantitated, along with direct measurements of the blood viscosity.

METHODS

Wild-type zebrafish (Danio rerio) fertilized eggs were collected, incubated, and dechorionated at approximately 24 hpf. To study kinematics, embryos were imaged at 48 hpf (cardiac looping) under a bright-field stereomicroscope (Olympus SZX12).. Embryos were placed in melted low-gelling temperature agarose gel (Sigma Aldrich, Type VII-A, A0701), positioned ventral-side up. Images were acquired at 1500 frames per second with a resolution of 1310 pixels per millimeter using a high speed camera (Photron FASTCAM SA3). Post-processing was performed using Fiji for kinematics and Matlab R2013a (The MathWorks, Inc.) to quantify flow.

Blood viscosity was calculated through force spectroscopy, using optical tweezers with a 1070-nm laser (YLR-5-1070-LP, IPG Photonics) and a high NA objective (60X water immersion, 1.2 NA; Olympus) built around an Olympus IX-71 inverted microscope.

All flow was quantified using instantaneous computed cellular speeds based on a spatiotemporal (ST) plot analysis [3]. Cells that pass along a line appear as angled streaks in the ST plot, corresponding to the speed of the cell. Measurements were ensemble averaged over a minimum of 5 cardiac cycles per sample to provide an average speed.

Lastly, the pressure gradient was derived using 2 methods. Equation (Eq.) 1 is based on the Laplace equation, while the Eq. 2 is based on flow with a low Reynolds and Womersley number:

$$\frac{\partial P}{\partial z} \propto \Delta\frac{1}{D}, \quad (Eq.1), \qquad \frac{\partial P}{\partial z} = -\frac{\mu}{D^2}\left(CU + \frac{\partial}{\partial z}\left(\frac{\partial D^2}{\partial t}\right)\right), \quad (Eq.2),$$

where D is the endocardial diameter, μ is the dynamic viscosity, U is the average velocity, and C is 16, based on the velocity profile.

RESULTS

A constricting wave closes the lumen of the heart, beginning at the sinus venosus (SV) and continues to the atrioventricular canal (AVC), Fig. 1. The wave is evaluated at 5 different locations along the heart depicted in Fig. 1A to determine how it progresses. The wave begins as the myocardial diameter decreases, nearest the SV. Simultaneously, the endocardium decreases in diameter until self-contact, Fig. 1B. Fig. 1C shows the difference in the two layers, providing an estimate of the local CJ thickness, demonstrating that the CJ is thickest when the endocardium is in contact with itself. The waveforms are qualitatively similar from the SV to 56 microns in the direction of the AVC. Similar shortening was seen lengthwise along the developing heart with 20-25% shortening of the myocardium. These waves can be seen in Fig. 1D.

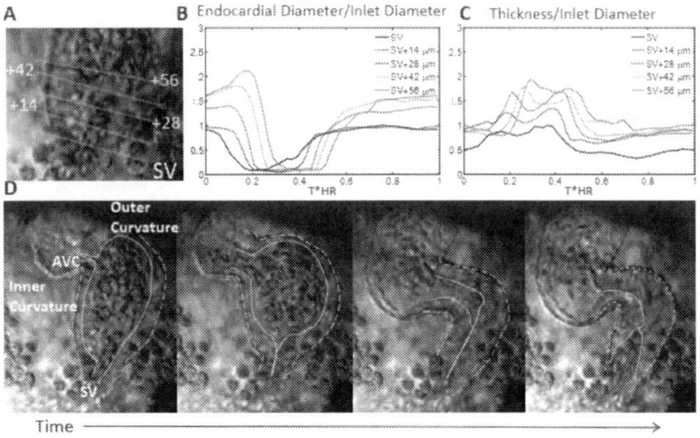

FIGURE 1: (A) Measurements are made at the SV, 14, 28, 42, and 56 μm. (B) Endocardial Diameter measurements. (C), CJ thickness. (D) Heart images during the cardiac cycle.

Calculations of the pressure differential depend on the viscosity during heart development. Therefore, viscosity was calculated by comparing force (Fig. 2A) from optical tweezers and velocity curves from ST plots at multiple points along the circulation of a single embryo, as shown in Fig 2B. The curve results in a drag coefficient of 0.4 pN*s/μm, based on a linear regression analysis, shown in Fig. 2C. The same coefficient was found by taking measurements for multiple fish at the same hpf. Assuming a cell radius of 2.5 μm and Stoke's drag ($F = 6\pi\mu r v$) for a low Reynolds number, the viscosity is 8.3 mPa*s.

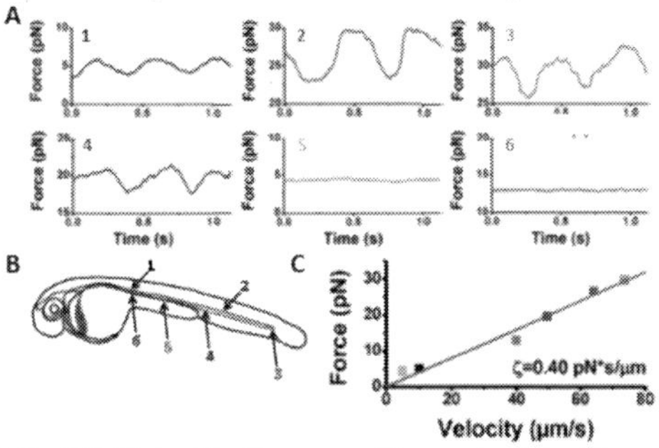

FIGURE 2: (A) Pulsatile force profiles calculated from optical tweezers at 6 points (B) along a single fish. (C) Force and velocity relationship.

Hemodynamics were evaluated during the cardiac cycle, with the centerline velocity at the SV shown in Fig 3 relative to the atrial cross sectional area of the lumen in the imaging plane, which is analogous to volume. The average Reynolds number is 0.01±0.001, while the Womersley number is 0.06±0.002. Some flow reversal is seen as the atrium begins to contract, corresponding to an interval where the SV begins closing. Very little flow passes through the SV upon closure until the end of atrial systole. Subsequently atrial diastole begins, demonstrated by and expanding area. As the atrium fills with blood, the velocity increases and then decreases. Measurements of the velocity, the diameter, and the viscosity from Fig. 1, 2 were used to estimate the pressure gradient at the SV, plotted in Fig. 3. As atrial systole begins (atrial area decreases), the pressure gradient is negative, transitioning into a positive gradient after flow reversal, as the SV opens for Eq. 2. Eq. 1 follows an opposite trend, but both equations result in a small negative pressure gradient during atrial filling.

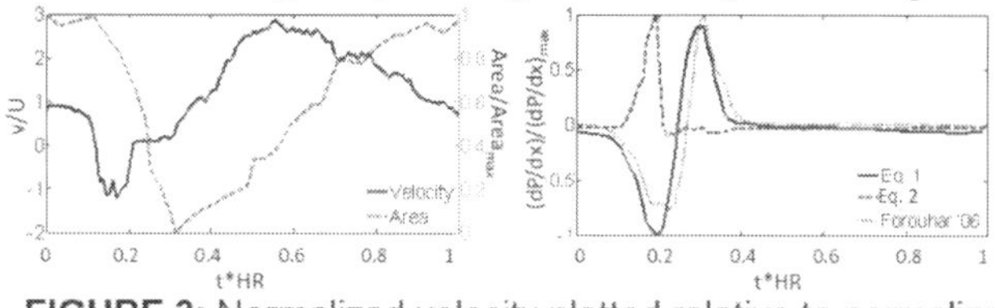

FIGURE 3: Normalized velocity plotted relative to normalized atrial area and the normalized pressure differential at the SV.

DISCUSSION

Evidence in the current study points to an active contractile pumping mechanism involving coordinated motion between cardiac layers once the heart begins looping at 48 hpf. During contraction, the CJ is leveraged to fully close the lumen of the heart through its expansion [4]. By deforming the CJ, both radial and axial myocardial length changes provide an important contribution to minimizing the necessary size of the myocardium for full closure of the endocardium, while maximizing the volume of blood expelled out of the atrium.

The pumping mechanism of the heart at 48 hpf is distinct. When observing the mechanics of the heart at 48 hpf, it is apparent that the increasing volume of the atrium during atrial diastole creates a suction for flow entering the atrium, since fluid must fill the cavity of the expanding endocardium, as seen in Fig. 1. There is also a small negative pressure gradient as the atrium expands, Fig. 3. Alternatively, during atrial systole, the blood is essentially ejected as a bolus into the ventricle, as demonstrated by the atrial area curve in Fig. 3. During this period, we assume that pressure becomes very large in the atrium, causing flow reversal at the SV, which also corresponds to a strong adverse positive pressure gradient, Eq. 2 in Fig. 3. These mechanics are distinct from those seen at in 26 hpf during initial flow [2]. At 26 hpf, the endocardial layer only momentarily contacts itself, contrary to the sustained interval of endocardial self-contact seen in the present study. Sustained self-contact allows the atrium to eject blood, while minimizing flow reversal at the SV and is important during atrial filling as the atrium expands. Pumping is also distinct from the adult heart, which contains valves. In the adult heart, the atrium plays more of a role as reservoir, as opposed to the active pumping seen at 48 hpf.

Pressure gradients derived from flow, Eq. 2 in Fig. 3, follows the opposite trend from Eq. 1 and from Forouhar et al. [3] during the contractile wave. However, the trends are similar during atrial diastole. Eq. 2 results in a negative gradient for forward flow and a positive gradient for reverse flow, as expected for the low Reynolds number reported in this study. Eq. 1 may veer from the actual pressure gradient during contraction because of variations in circumferential stress associated with active myocardial cells. Also, compression of the CJ may reverse the sign. Additionally, the previous study of Forouhar et al. is based on a simpler heart tube geometry with simpler conditions that may not be valid as the heart further develops, as shown here.

ACKNOWLEDGEMENTS

We acknowledge funding from the National Science Foundation (Award# 1235305), program director Dr. Dennis Carter.

REFERENCES

[1] Johnson, BM et al., *Cardiovasc Eng Technol*, 4:246-255, 2013a.
[2] Forouhar, AS et al., *Science*, 312:751-753, 2006.
[3] Johnson, BM et al., *J Biomech Eng*, 135:041006, 2013b.
[4] Barry, A, *The Anat Rec*, 102:289-298, 1948.

SB³C2015
Summer Biomechanics, Bioengineering and Biotransport Conference
June 17-20, 2015,Snowbird Resort, Utah, USA
SB³C2015-658

MODELING THE MECHANICAL BEHAVIOR OF POLYMERIC BIORESORBABLE STENTS: A FINITE ELEMENT APPROACH

Nic Debusschere (1), Patrick Segers (1), Peter Dubruel (2), Benedict Verhegghe (2), Matthieu De Beule (1)

(1) IBiTech – bioMMeda - iMinds
Ghent University
Ghent, Belgium

(2) PBM
Ghent University
Ghent, Belgium

INTRODUCTION

Bioresorbable stents are an emerging technological innovation within the field of cardiovascular angioplasty. Because of their temporary nature, bioresorbable stents carry the potential to overcome side effects of conventional stents such as in-stent restenosis, late stent thrombosis and fatigue induced strut fracture [1]. The gradual transfer of load from the scaffold to the blood vessel wall might facilitate vessel healing and vessel growth, which makes biodegradable stents particularly interesting for pediatric applications.

Two classes of materials are currently being used for biodegradable stents: biocorrodible metals such as magnesium alloys, and bioresorbable polymers such as polylactic acid (PLA). The Absorb bioresorbable vascular scaffold (BVS) (Abbott Vascular, Santa Clara, CA, USA) is a representative of the polymeric stents and is the most extensively investigated biodegradable scaffold at this moment. Since 2011, it has been approved with the European ce-mark.

Because relatively little is known about the mechanical behavior of these novel stents, we wanted to assess the BVS's mechanical performance via finite element analysis (FEA), which has been proven to be a useful tool for the mechanical evaluation of medical devices such as stents. In particular, we wanted to investigate the influence of the stent deployment rate during balloon expansion. Because of the high degree of viscoplasticity of the stent's polymeric backbone, the deployment rate might have an important influence on the mechanical performance and integrity of the stent.

METHODS

An implicit solution technique was chosen to model stent deployment, using the commercial finite element solver Abaqus/standard (Dassault Systems Simulia, Providence, RI, USA). To simplify the contact problem between the crimped stent and the folded balloon, the latter was modeled as a straight cylindrical tube, with anisotropic material properties, as presented by [2]. The material properties of the balloon were fit to Absorb BVS manufacturing data.

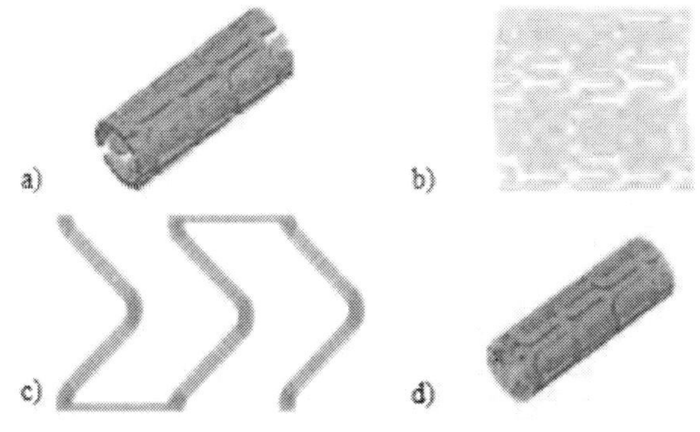

Figure 1. Reversed engineering approach to obtain the stent's finite element geometry: a) reconstruct a 3D model from micro-CT image data, b) obtain the stent's dimensions and c) a high quality mesh using pyformex, c) and perform an additional simulation to obtain the crimped stent geometry.

The geometry of the BVS stent and its finite element mesh were obtained via a reversed engineering approach, as illustrated in Figure 1. High resolution (voxel size 2.5 μm) micro-CT based image data were used as input to make a segmented 3D-reconstruction of the BVS. The 3D-geometry was subsequently imported into pyFormex (in-house developed open source software), to obtain the stent's characteristic dimensions (strut width, thickness and length etc.) and to generate a high-quality hexahedral mesh of the stent.

The Absorb BVS is made out of high molecular weight poly-L-lactic acid. An advanced material model that is sufficiently able to capture the material's viscoplastic characteristics was taken from literature [3] and implemented as a Fortran subroutine to be compatible with Abaqus/standard. The material parameters were fit by comparison to data from experimental tensile tests at different strain rates on dumbbell-shaped specimens cut out of extruded PLA-tubes (however not from Absorb tubing).

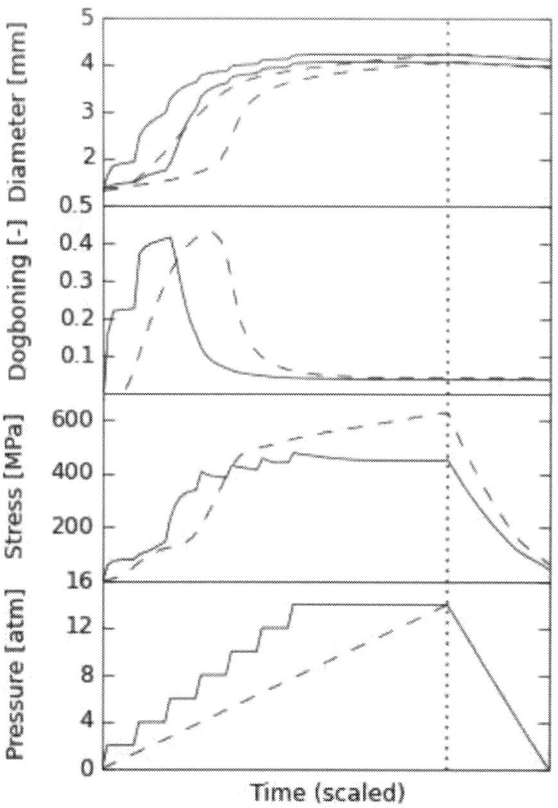

Figure 2. Maximum principal stress, dogboning and outer diameter of the BVS during balloon expansion and deflation. The dashed lines represent the outcome of the direct balloon inflation simulation, whereas the full lines display values for stepwise balloon inflation. The lowest graph shows the application of balloon pressure as a function of time. The time axis has been rescaled to be able to compare both simulations.

The generated models for balloon and stent were combined to perform two free balloon expansion simulations. During a first simulation balloon pressure was linearly increased to a value of 14 atm. In a second simulation, balloon pressure was applied according to the Absorb BVS product manual, which states that the stent needs to be deployed by stepwise pressurizing the balloon in 2 atm increments, using 5s time intervals and that pressure needs to be maintained for 30 s prior to balloon removal. The highest value of the maximum principal stress inside the stent, the degree of dogboning and the central outer diameter of the stent were recorded as a function of time for both simulations.

RESULTS

The simulation outcomes show that stent stress levels are lower and less widespread during balloon inflation in case a stepwise inflation procedure is applied. Also in the relaxed state, residual stresses are lower in case of stepwise stent deployment.

Figure 2 gives a quantitative comparison between both simulations. The lowest graph plots the application of the balloon pressure for the direct (dashed line) and the stepwise inflation procedure (full line). The second graph shows the evolution of the maximal value of the maximum principal stress over all integration points during both simulation steps. The highest values of the internal stresses are reduced with over 20% when a stepwise balloon pressure increase is used. For the relaxed configuration, reduction of stresses is less extensive. From Figure 2 we can also notice that the degree of dogboning and elastic recoil after deployment are only slightly reduced by using a stepwise inflation procedure.

DISCUSSION

The in this study presented implicit finite element solution strategy to model free balloon inflation of stents using a cylindrical balloon model and an anisotropic balloon material reflects well some of the stent's expansion characteristics such as the dogboning phenomenon. The implementation of a combined viscoplastic hyperelastic material model is required to obtain insight into to the instructions for use of the Absorb BVS stent. Although the here presented material model was fitted to very preliminary experimental data and should not be taken as representative for the actual mechanical behavior of PLLA, the material model gives a qualitative indication of how a gradual expansion procedure reduces the build-up of internal stresses and consequently the risk for strut fractures. Stepwise balloon inflation might also reduce the degree of dogboning and elastic recoil after stent deployment.

ACKNOWLEDGEMENTS

Special thanks to P. Smits (Maasstad Hospital, Rotterdam, Netherlands) and FEops (Ghent, Belgium) for the supply of the BVS samples, the Centre for X-ray Tomography of the Ghent University (UGCT) for the micro-CT scanning of the BVS stents and the company Admedes (Pforzheim, Germany) for the tensile test experiments.

REFERENCES

[1] Garg, S et al, *J Am Coll Cardiol.*, 56:S43-S78, 2010.
[2] Kiousis, D et al., *Ann Biomed Eng*, 37:315-330, 2009.
[3] Bergström, J et al, *Biomaterials*, 23:2329-2343, 2002.

SB³C2015
Summer Biomechanics, Bioengineering and Biotransport Conference
June 17-20, 2015,Snowbird Resort, Utah, USA
SB³C2015-659

GOLD NANOPARTICLE MEDIATED ENHANCED GENE DELIVERY WITH HIGH SELECTIVITY FOR BREAST CANCER TREATMENT

B.Shrestha, L.Tang

Biomedical Engineering Department
University of Texas at San Antonio
San Antonio, Texas, United States

INTRODUCTION

Cancer is the second most leading cause of death surpassed by the stroke. Breast cancer is one of the top five cancer that has low prognosis and high death rate accounting 40,000 in the year 2012[1] . Recently gene therapy for cancer treatment has emerged as a promising tool for cancer treatment. However selective targeting, effective endosomal release, and vector unpacking are the major issues. Herein we report novel nanotherapeutic system for improved selective targeting and gene delivery using gold nanoparticles for anti-cancer therapy. Gold nanoparticles were carrier of choice as these are one of the most efficient delivery platforms due to its unique optical properties, ease of functionalization, controllable size and biocompatibility[2]. Polyethyleneimine (PEI) is considered to be gold standard for gene therapy due to its proton sponge effect.PEI was hence a primary choice to load siRNA via electrostatic binding to the nanotherapeutic system[3]. In this study, we showed selective and enhanced transfection efficiency of our nanoparticles for improved anti-cancer effect in HER2 over expressing breast cancer cell line SKBR3.

METHODS

Gold nanospheres (GNS) were synthesized using standard Turkevich method. Briefly, 20 ml of 1 mM of gold chloride was heated at 100°C under magnetic stirring. Soon after boiling 2.2 ml of 1% sodium citrate was added to the solution and was stirred until the color changed to ruby red. The solution was filtered and stored at 4°C.Anti-HER2 were thiolated using trauts reagent following the vendor recommended procedure[4]. As prepared GNS were conjugated with thiolated antibody and SH-PEG-COOH polyethylene glycol (PEG, Mw: 5000).Typically, 1ml of 28 uM thiolated HER2 was added 10 ml of GNS. To this 20 ml of 150 uM PEG solution was added and stirred for 2 hours followed by centrifugation to remove unbound PEG molecule. The carboxyl group was activated using EDC and NHS in the ratio 4:1 in order to obtain negative charge on the surface.PEI was assembled on the negatively charged surface simply by adding 2 mg /ml PEI to the above resultant solution. The mixture was stirred for 2 hours and was centrifuged to remove unbound PEI. To load siRNA to the positively charged surface, siRNA at various N/P ranging from (5:1 to 20:1) ratio was added and incubated for 1 hour at 4°C.The particles were characterized by UV-Vis spectrum, zeta potential measurement, SEM, microplate reader and florescence microscope.

To determine the binding affinity of the particles, gel electrophoresis using 2% agarose gel was used. Electrophoresis was carried out at constant voltage of 120 V for 20 min.

Release from PEI-GNS: The FITC tagged nucleic acid were used. The particles were incubated with Hep2 and SKBR3 cell line for 2, 4 and 6 hours. The florescence intensity was observed at different time points.

MTS assay was used to determine cytotoxicity of the nanotherapeutic system. The experiment was performed in Human epithelial cell line Hep-2 (ATCC CCL-23) and metastatic breast cancer cell line SKBR3 (ATCC).Hep2 cells was cultured in Eagle's Essential Medium (EMEM,ATCC,USA) supplemented by 10 % Fetal Bovine Serum (FBS) and SKBR-3 cell line was cultured in Mc Coy's 5A mediun (ATCC ,USA) supplemented with 10% FBS. Both cell line were cultured in 5% CO2 environment at 37oC.Approximately $10X10^5$ cells were seeded in each well and was incubated overnight at 37oC in 5% CO2 environment. Different concentration of nanotherapeutic complexes was added to the cells in serial dilution ranging from 100% to 3.125 % and was further incubated for 24 hour. Cells with no nanoparticles were our positive control. After 24 hours, the media was replaced with 100 ul of fresh media followed by addition of 20 ul of MTS/PMS solution. After further incubation for 4 hour at 37oC and absorbance at 490 nm was measured using Synergy 2 microplate reader (BioTek,USA).

RESULTS

The GNS synthesized by Turkevich method were well dispersed with high stability (fig 1a). SEM image demonstrated that the particles were approximately 12 nm in size (fig 1b, 1c). Anti-HER2 was conjugated to GNS for selective targeting HER2 over expressing SKBR3 cells. Anti-HER2 conjugation was confirmed by UV-Vis spectrum and ELIZA result (fig 1d, 1g,). Anti-HER2 modified GNS particles were incubated with Hep2 and SKBR3 cells respectively. The florescence microscope shows successful cellular uptake of the particles in Hep2 and SKBR3 cell. However, there was significantly higher uptakes in SKBR3 cells due to HER2 overexpression (fig.1i,1j) compared to Hep2 cells (fig 1e,1f). As such, Anti-HER2 conjugation to the nanoparticles can reinforce the GNS-based targeting to breast cancer SKBR3 cell line. The zeta potentials of bare GNS, PEG GNS, PEI-PEG-GNS and siRNA were -38.0, +44, and +3 mV respectively, confirming different coatings on GNS (fig 2a, 2b). Binding affinity of the particles was determined by gel electrophoresis. Nucleic acids were loaded onto nanoparticles at different N/P ratios (1:1, 5:1, 10:1, 15:1, 20:1). Dissociation of nucleic acid was not observed for N/P ratio higher than 15:1 (fig 2c). Therefore N/P ratio 15:1 was determined to be optimal for further experiments. Cell viability testing showed excellent biocompatibility of the nanoparticles and coating on the surface (fig 2e). Scattered florescence in the cell cytoplasm of SKBR3 showed the efficient endosomal escape and release of conjugated nucleic acid. As comparison, little florescence signal was detected in Hep2 cells. These results indicated that the particles could be selectively targeted to the desired tumor cells with high transfection efficiency.

DISCUSSION

Gold nanoparticle based nanocarrier system for enhanced gene delivery in SKBR3 cell with high selectivity was developed. PEI-GNS facilitated the effective endosomal escape and vector unpacking. This design is a promising multifunctional delivery platform for combinational anti-cancer therapy, which is the current focus of our study. Future work includes further functionalization with different therapeutic agent for simultaneous delivery of gene and chemotherapeutic agent. The system will be further optimized to actively control the release profile with smart feedback system.

ACKNOWLEDGEMENTS

The authors thank Kleberg Advanced Microscopy Center, UTSA for microscope facilities. This study was partially funded by the National Institutes of Health.

REFERENCES

1. Kleinsmith, L.J., *Principles of cancer biology*. 2006: Benjamin-Cummings Pub Co.
2. Shen, J., et al., *Multifunctional Gold Nanorods for siRNA Gene Silencing and Photothermal Therapy*. Advanced healthcare materials, 2014.
3. Godbey, W., K.K. Wu, and A.G. Mikos, *Poly (ethylenimine) and its role in gene delivery*. Journal of Controlled Release, 1999. **60**(2): p. 149-160.
4. Wang, X., et al., *Gold nanorod biochip functionalization by antibody thiolation*. Talanta, 2014.

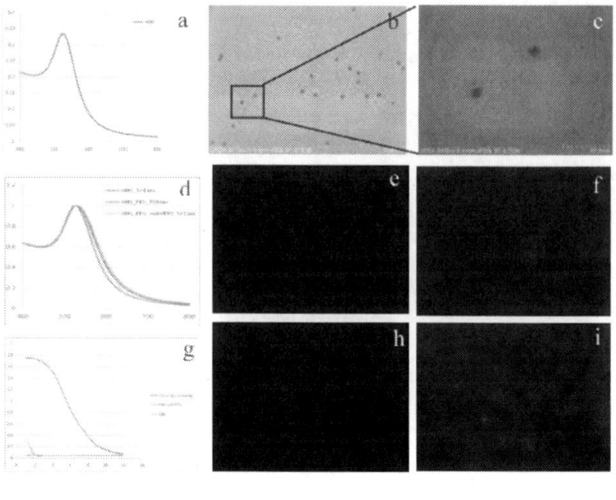

Fig 1.UV-Vis spectrum of GNS (a)before and (d) after conjugation of Anti-HER 2.(b)SEM image of GNS and (c) size analysis.(g)Eliza test. Florescence microscope images of (e, f) Hep2 and (h, i) SKBR3 cells.

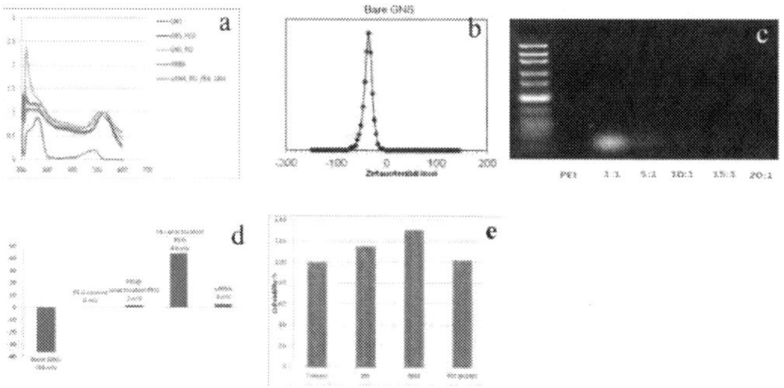

Fig 2. (a) UV-Vis spectrum of GNS after assembly of PEG, PEI and siRNA. Zeta potential measurement of (b) bare GNS (d) after assembly of PEG, PEI and siRNA. (c) Gel electrophoresis at different N/P ratio.(e) Cytotoxicity evaluation.

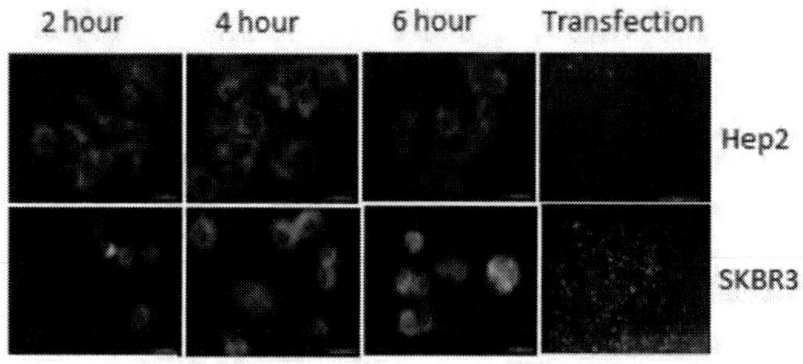

Fig 3. Cellular uptake and transfection of the nanoparticles. Time dependent release of nucleic acid in Hep2 and SKBR3 cells incubated with nanoparticles for 2, 4 and 6 hours. The cell cytoplasm was stained with cell mask and nucleic acid was tagged with FITC. Transfection efficiency of particles in (g) HEP2 cells and (h) SKBR3 cells.

SB³C2015
Summer Biomechanics, Bioengineering and Biotransport Conference
June 17-20, 2015,Snowbird Resort, Utah, USA
SB³C2015-660

THE CONTRIBUTION OF COLLAGEN FIBERS IN HUMAN CARTILAGE ENDPLATE MECHANICS

John F. DeLucca (1), Mary B. Boggs (2), Edward J. Vresilovic (3), Randall Duncan (1,2), Dawn M. Elliott (1)

(1) Department of Biomedical Engineering
University of Delaware
Newark, DE, USA

(2) Department of Biological Sciences
University of Delaware
Newark, DE, USA

(3) Department of Orthopedic Surgery
Pennsylvania State University
Hershey, PA, USA

INTRODUCTION

The cartilage endplate (CEP) is a thin layer of hyaline-like cartilage serving as the interface between the vertebral bone and the intervertebral disc. The CEP functions to control transport of solutes and water into and out of the disc and is a mechanical barrier between the nucleus pulposus and the vertebral body. These functions are influenced by structural and mechanical properties of the CEP which arise from the proteoglycan content and the collagen organization within the tissue. Previously we demonstrated that CEP permeability and extra-fibrillar matrix compressive modulus are related to the tissue's proteoglycan and water contents, respectively [1]. CEP collagen content correlates with the tensile modulus [2] however, to date the collagen organization, and the role of collagen in CEP mechanics remain unknown.

Because of the important functions the CEP plays in the disc transport cycle, the CEP is usually included in finite element models (FEM) of the disc. Critical to an accurate FEM of the disc is an appropriate choice of constitutive models and material properties which reflect the tissue's structure. Historically, most models of the CEP only incorporate mechanical properties measured in compression and neglect collagen organization and tensile mechanical properties altogether [3, 4]. Collagen is an essential component of structure-function relationships in cartilaginous tissues. In articular cartilage, for example, collagen resists lateral tensile deformation, resulting in fluid pressurization and compressive load bearing [5, 6]. Therefore, the **objective of this study was to determine the role of collagen in CEP and disc mechanics**.

METHODS

CEP Structure and Constitutive Model

To determine the appropriate fiber constitutive model, a Pfirmann grade 2 disc was processed for histology. Samples were stained with Picrosirius red to show collagen content and viewed with polarized light with 45° cross-polarizers. Collagen organization was determined qualitatively. To further assess collagen organization, a second Pfirmann grade 2 disc was selected and processed for scanning electron microscopy. From histology and SEM, fibers appeared to be oriented parallel to the vertebrae and therefore an ellipsoidal fiber distribution (EFD) was selected [6]. The EFD is defined by the collagen fiber modulus (ξ) and fiber non-linearity (α). The anterior-posterior and lateral fiber directions were assumed to produce the same fiber modulus (ξ_1), while the contribution of fibers to the axial direction of the disc was set to 10% of ξ_1. Fiber non-linearity was assumed to be independent of fiber direction. The EFD formulation was therefore:

$$\xi(\theta,\varphi) = \left(\frac{\sin^2\varphi + 100\cos^2\varphi}{\xi_1^2}\right)^{-\frac{1}{2}}$$

$$\alpha(\theta,\varphi) = \alpha_1$$

CEP fiber properties were determined by fitting the tensile response from Fields et al. [2] using an inverse finite element method in FEBio [7]. The average tensile response was fit to a biphasic-swelling model, developed from our previous work, reinforced with an ellipsoidal fiber distribution (EFD):

$$\boldsymbol{\sigma} = -(p_f + p_{osm})\mathbf{I} + \boldsymbol{\sigma}_{ISO} + \boldsymbol{\sigma}_{EFD}$$

Where σ is the applied stress, p_f and p_{osm} are the fluid and osmotic pressures, σ_{ISO} is the isotropic extrafibrillar matrix stress, and σ_{EFD} is the fiber stress. Extra-fibrillar matrix, permeability, and biochemical properties were taken from our previous work (Table 1) [1].

Finite Element Modeling of CEP Mechanics

The role of CEP fibers in disc mechanics was determined using a validated intervertebral disc finite element model (Fig. 1) loaded in axial compression [4]. Simulations were performed with and without inclusion of CEP fibers to evaluate their role in disc mechanics. The simulated disc was hydrated to equilibrium for 24 hrs with the vertebrae fixed in order to pressurize the NP. After hydration, 10% axial compression was applied over 10 seconds and held 6.5 hrs to equilibrium.

CEP axial strain (E_{ZZ}), shear strain (E_{YZ}), and fluid flux were calculated along the anterior-posterior axis of the mid-sagittal plane of the disc (*yz*-plane), and disc reaction force was calculated across the superior vertebra.

RESULTS

Structure and Fiber Constitutive Model

Picrosirius red staining demonstrated that collagen is a prominent component of the CEP matrix. Collagen fibers appeared parallel to the vertebrae and NP (Fig. 2) and are not arranged like fibers in articular cartilage which has distinct zones of varying fiber distributions. The CEP has a planar distribution of fibers and does not have fibrillar anchoring to the adjacent bone or the NP. Modeling this ellipsoidal distribution fit the experimental data well ($R^2 = 0.97$) and resulted in a fiber modulus of $\xi_l = 7.1$MPa and a fiber nonlinearity of $\alpha = 2.88$.

Finite Element Results

When fibers were excluded from the model, the CEP was predicted to experience axial tensile strains (E_{ZZ}) (Fig. 3A) and shear strains (E_{YZ}) exceeding 20% (Fig. 3B) under axial disc compression. The fiber-reinforced model produced the expected CEP compressive strain and shear strains reduced (-10% to +8%) compared to the fiber-less model (-23% to +14%). Fluid fluxes through the CEP were doubled in the fiber-reinforced model because of reduced CEP deformation (Fig. 3C). Including CEP fibers increased peak disc reaction force by 13% but did not change equilibrium reaction force (Fig. 3D).

DISCUSSION

The fiber-reinforced model successfully demonstrated that CEP fiber-reinforcement has important effects on CEP function and likely disc transport. Fiber-reinforcement reduces CEP strain and permits greater convective transport. Notably, CEP shear strains (E_{YZ}) are large (>8%) despite inclusion of fiber-reinforcement. Large shear strains may explain the prevalence of CEP delamination found in herniated disc tissue [8]. Furthermore, fiber-reinforcement may enhance convection and the diurnal transport cycle. Previous finite element studies have described the important roles of disc solute transport via diffusion and convection and how these may contribute to disc cell function [9-11]. The present study suggests that inclusion of CEP and the CEP fiber-reinforcement may be an important factor in these models. Moreover the effect of moderate degeneration and advanced degeneration with damage remains to be determined. Damage to the CEP may be related to enhanced transport in advanced degeneration [12].

In conclusion, modeling the fibrous structure of the CEP has significant implications for disc mechanics. Transport through the CEP is likely dependent on the CEP fiber distribution, as fiber-reinforcement reduces CEP deformation and increases fluid flux.

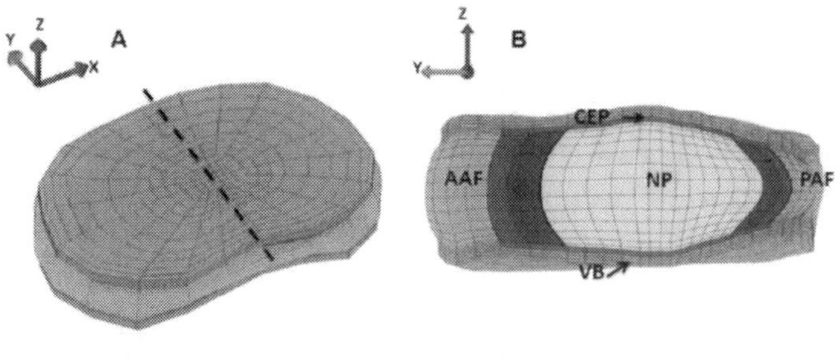

Figure 1: Intervertebral disc finite element geometry

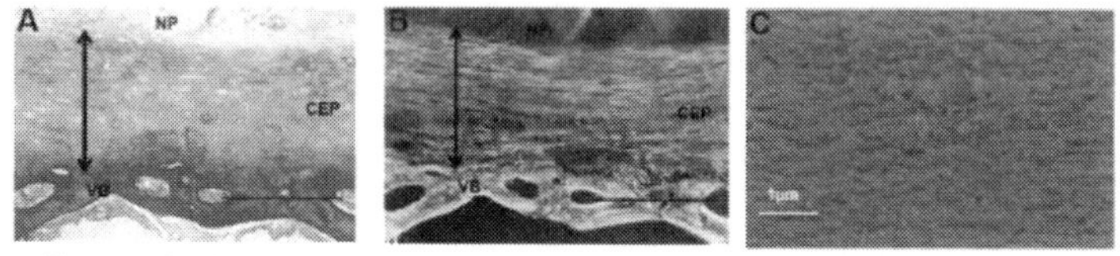

Figure 2: (A) Picrosirius red, (B) polarized light, and (C) SEM of CEP collagen

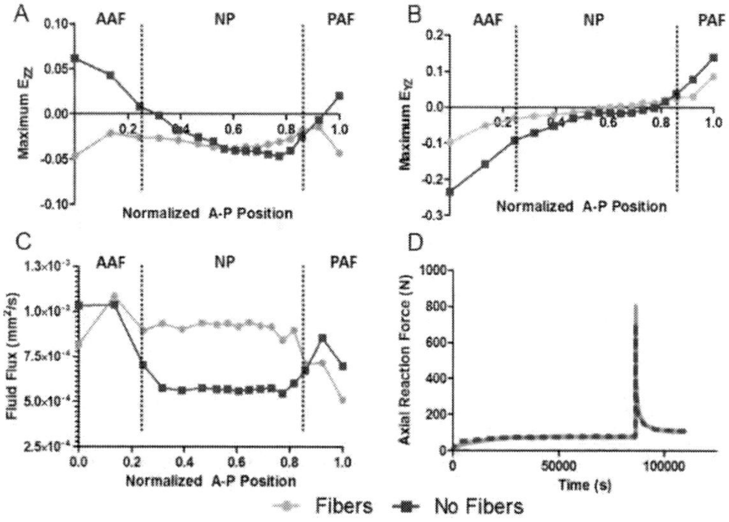

Figure 3: Effects of CEP fiber-reinforcement on CEP and disc mechanics (AAF = anterior inner annulus fibrosus, PAF = posterior inner annulus fibrosus)

E (MPa)	*v*	*β*	*k(mm⁴/Ns)*	*M*	*FCD*	Water %
0.33	0.25	1.06	0.00025	3.9	365	62.3

Table 1: CEP properties for FEA

ACKNOWLEDGEMENTS

This study was funded by NIH R01AR050052.

REFERENCES

[1]DeLucca et al., *Trans. ISSLS,* 2014. [2]Fields et al., *J Ortho Res*, 32(2):245-52, 2014. [3]Jackson et al., *CMBBE*, 142(2):195-204, 2011. [4]Jacobs et al., *J Biomech*, 47:2540-6, 2014. [5]Moore and Burris, *J Biomech*, 47(1):148-53, 2014. [6]Ateshian et al. *J Biomech Eng*, 131(6), 2009. [7]Maas et al., *J Biomech Eng*, 134(1), 2012. [8]Lama et al., *Eur Spine J*, 23(9):1869-77, 2014. [9]Ferguson et al., *J Biomech*, 37(2):213-21, 2004. [10]Zhu et al., *J Biomech Eng*, 45(16):2769-77, 2012. [11]Shirazi-Adl et al., *J Biomech*, 43(7):1330-36, 2010. [12]Rajasekaran et al., *Spine*, 29(3):2654-67, 2004.

SB³C2015
Summer Biomechanics, Bioengineering and Biotransport Conference
June 17-20, 2015, Snowbird Resort, Utah, USA
SB³C2015-661

QUANTITATIVE ANALYSIS OF HEART FUNCTION IN EMBRYONIC ZEBRAFISH: RETROGRADE FLOW IN THE ATRIOVENTRICULAR JUNCTION

Alexander T. Bulk (1), David L. Bark Jr. (1,2), Brennan M. Johnson (2), Deborah M. Garrity (3), Lakshmi P. Dasi (1,2)

(1) Department of Mechanical Engineering
Colorado State University
Fort Collins, CO, USA

(2) School of Biomedical Engineering
Colorado State University
Fort Collins, CO, USA

(3) Department of Biology
Colorado State University
Fort Collins, CO, USA

INTRODUCTION

Development of the heart is affected by mechanical factors associated with blood flow. Therefore, changes in flow can result in abnormal development, which can lead to severe congenital heart defects. To better understand the relationship between flow and congenital defects in humans, various animal models including zebrafish have been used, due to their similarity in heart development.

As the zebrafish heart begins to form, it starts as a tube prior to looping, which involves an increasingly defined atrium and ventricle. After ~105 hpf, a valve develops between the atrium and ventricle in a region called the atrioventricular junction (AVJ), which prevents blood from moving in retrograde motion. However, before the valve develops, a small fraction of retrograde flow can be found for a short duration of each cardiac cycle. Previous studies indicate that this flow is important for valve development at the AVJ [1].

Though retrograde flow through the AVJ appears to influence valve development, it is not well understood what causes retrograde flow during development. This study focuses on three factors: 1) AVJ diameter at maximum reverse flow, 2) contractility of the atrium, and 3) the phase difference between atrial and ventricular contraction. These factors are evaluated after altering the mechanical environment for 6 hours and are compared for fish from an unaltered environment.

METHODS

Zebrafish embryos are transparent, making their heart visible under bright-field microscopy. Therefore, to image the heart, we used a stereomicroscope (Olympus SZX12). A centrifuge was used to alter the mechanical environment and cardiac preload to determine the influence on the heart. Fish were spun at 70G for a 6 hour period. 4 fish were spun between 36 and 42 hpf and another 4 fish were spun between 42 and 48 hpf. These were compared with 6 control fish that were not spun. Hearts were then examined at both 48 hpf and 55 hpf.

Blood flow analysis was performed using a program (KAT) written in Matlab (Mathworks, Inc.) that utilizes spatiotemporal (ST) plots [2]. Example ST plots are shown in Fig. 1, demonstrating changes in diameter at the AVJ. These plots are also used to calculate fluid velocity through a centerline parallel to flow at the AVJ. Cells traversing along the line result in a streak that has a slope equal to the velocity of the cell. Previous studies have determined that the mean velocity is 74% of the centerline velocity, so the flowrate, Q(t) is calculated through:

$$Q(t) = \frac{\pi}{4} D(t)^2 * 0.74V(t) \tag{1}$$

where, D is the diameter, and V is the centerline velocity. Data is ensemble averaged over at least 5 cardiac cycles that are determined through periodic motion of the heart wall.

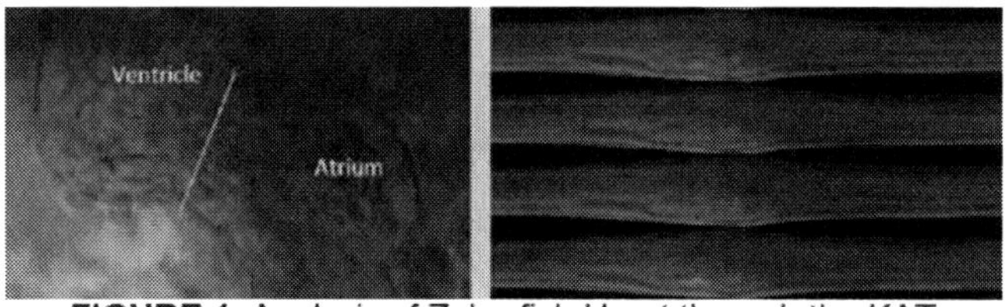

FIGURE 1: Analysis of Zebrafish Heart through the KAT program with reference line drawn across the AVJ and resultant ST plot

Retrograde flow was quantified through a parameter called the reverse flow fraction (RFF), which is defined as the sum of all negative values of Q(t) over the sum of positive Q(t) for a single cardiac cycle. ImageJ was used to determine the ejection fraction, a parameter used to assess the atrial contractility. To calculate the measure, polygons were carefully drawn around the interior of the endocardial lining of the atrium to measure the end-diastolic and end-systolic volumes (EDV and ESV). The ejection fraction was calculated by taking the difference in EDV and ESV divided by the EDV. The KAT program was used to analyze the phase difference between the atrium and ventricle's initial contraction.

The RFF values measured from the KAT program were then compared to the three measured factors using a chi-squared analysis in order to find relationships to the retrograde flow through the AVJ.

RESULTS

A box plot of the RFF measurements is shown in Fig. 2. Wildtype fish had little to no reverse flow through the AVJ, whereas groups of spun fish showed a significant change in reverse flow, except for fish spun between 36 and 42 hpf viewed at 48 hpf.

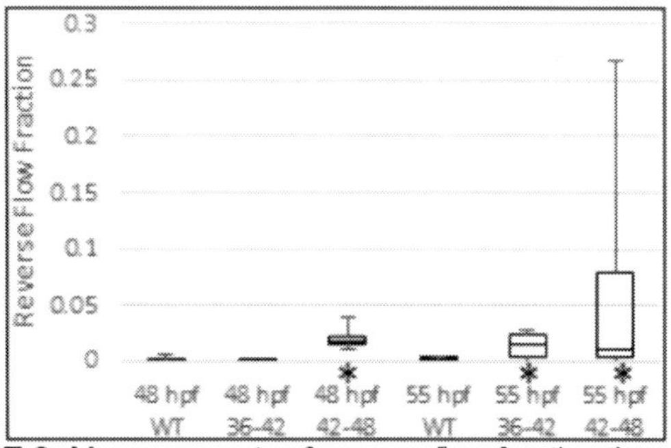

FIGURE 2: Measurements of reverse flow fraction through the AVJ (* $p < 0.05$)

A box plot of the AVJ diameter at maximum retrograde flow is shown in figure 3. Again, there was a significant change in all spun fish from those that were not spun, except for fish spun during the 36-42 hpf time point.

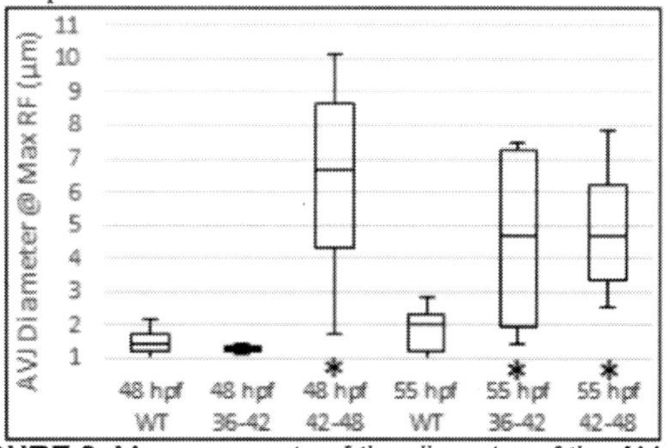

FIGURE 3: Measurements of the diameter of the AVJ at maximum retrograde flow (* $p < 0.05$)

The atrial ejection fraction is shown in Fig. 4. There was no statistical difference in atrial ejection fraction measurements for spun fish viewed at 48 hpf. However, there was some significance for fish viewed at 55 hpf, with p-values of .17 and .02 for the 36-42 and 42-48 hpf spin groups, respectively.

There was no statistical significance in atrial contraction phase differences relative to ventricular contraction between spun and wildtype fish in any group, so results were not included in the abstract.

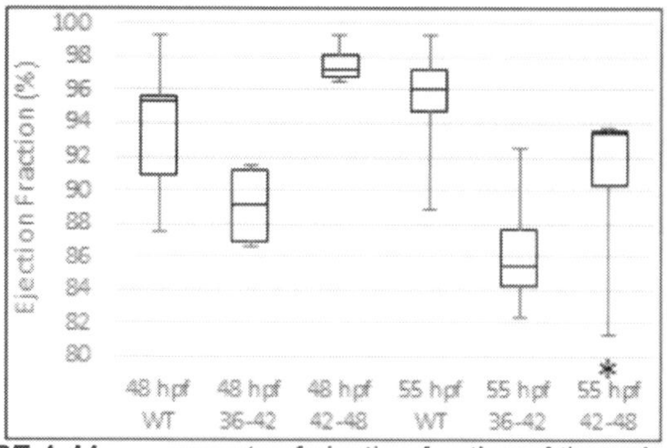

FIGURE 4: Measurements of ejection fraction of the atrium (* $p < 0.05$)

DISCUSSION

RFF significantly increased for zebrafish that were spun and then imaged at 55 hpf, indicating that the altered mechanical environment before (36- 42 hpf) or during cardiac looping (42-48 hpf) may have a sustained influence on RFF during later development, corresponding to a post-looping stage (55 hpf). However, spinning the fish only appeared to have an impact on RFF during cardiac looping (48 hpf) immediately after the spin (42-48 hpf). Therefore, it may take time for RFF to be affected from the altered mechanical environment that exists during the centrifugation.

The AVJ diameter appeared to follow similar trends to the RFF for spin groups, relative to control fish. Therefore, a relationship may exist between maximum AVJ diameter and the RFF. An explanation for this possible relationship is that the AVJ's ability to close, reflected by the RFF, may depend on the necessary change in diameter from a fully open state to a fully closed state at the AVJ.

The ejection fraction did significantly lower at 55 hpf for fish spun at 42-48 hpf. Although the difference was not striking, there may be an impact on the ability of the atrium to contract when the mechanical environment is altered during looping (42-48 hpf).

The atrioventricular contraction phase difference affects the difference in pressure between the atrium and ventricle. Therefore, it was assumed that there would be significant difference in phase difference of the spun fish groups, since the RFF was affected by spinning, especially in the fish viewed at 55 hpf. P-values for the phase difference ranged between .17 and .33 though, so the measured values did not statistically support this assumption.

In conclusion, the RFF does change during later stages of development, which may impact valve formation [1]. This change may be associated with the change in relaxed AVJ diameter. We plan to perform additional statistical comparisons on this data to further investigate the impact of the altered mechanical environment on heart development.

ACKNOWLEDGEMENTS

We acknowledge funding from the National Science Foundation (Award# 1235305), program director Dr. Dennis Carter.

REFERENCES

1. Vermot J et al., *PLoS biology*, 7:e1000246, 2009.
2. Johnson BM et al., *J Biomech Eng*, 135:041006, 2013.

A COMPUTATIONAL HEAD MODEL VALIDATED AGAINST PRESSURE RESPONSES ONLY MAY NOT BE USED TO ESTIMATE BRAIN STRAIN RESPONSES

W. Zhao (1), S. Ji (1,2)

(1) Thayer School of Engineering
Dartmouth College
Hanover, NH, USA

(2) Department of Surgery and of
Orthopaedic Surgery
Dartmouth College
Hanover, NH, USA

INTRODUCTION

Traumatic brain injury (TBI) continues to be a major socio-economical and health issue. Indisputably, TBI is initiated through mechanical events; however, the biomechanical mechanisms of TBI remain elusive to date. Linear ($\boldsymbol{a}_{lin}$) and rotational ($\boldsymbol{a}_{rot}$) accelerations are thought to be related to coup/contrecoup injuries and focal/diffuse type of brain injuries, respectively [1]. While these findings were supported by early experimental work, the kinematic variables and their variants alone do not inform tissue-level brain mechanical responses that are thought to initiate the injury in the first place.

Finite element (FE) models of the human head are increasingly employed to investigate TBI biomechanical mechanisms because of their unique capability to provide tissue-level response. Unfortunately, different models are discordant in response predictions even under the same head impact [2]. This may appear surprising because most head FE models have been validated against brain biomechanical response data drawn from the same few sets of cadaveric head impacts in terms of pressure [3,4] and/or relative brain-skull displacements [5,6]. While it is perhaps desirable to fully validate a head model against all of these experimental data, some models have only validated against brain pressure responses [7–10]. A recent dimensional analysis suggests that brain pressure responses from translational head impact is uniquely determined by $\boldsymbol{a}_{lin}$ (magnitude and direction) as well as brain size and shape, but is insensitive to brain shear stiffness due to brain's near incompressibility [11]. Therefore, we investigate whether an FE head model validated against brain pressure alone can be used to estimate brain strain responses.

METHODS

All brain response simulations were obtained using the Dartmouth Head Injury Model (DHIM), which has been successfully validated against relative brain-skull displacement and intracranial pressure from cadaveric experiments as well as strain from a live human volunteer [11–13]. In this study, we first verify that brain pressure responses are uniquely determined by $\boldsymbol{a}_{lin}$ magnitude for a given impact direction and brain model (i.e., same size and shape for the brain), as long as a near incompressibility property for the brain is maintained. Next, we verify that with sufficient brain incompressibility, the pressure responses are indeed *insensitive* to brain shear stiffness properties.

A temporal profile of $\boldsymbol{a}_{lin}$ was arbitrarily generated using a sinusoidal waveform with its peak magnitude (a_{lin}^{p}) corresponding to the 95th percentile a_{lin}^{p} in ice-hockey (i.e., 49 g, where g is 9.8 m/s²) [2] with additional oscillation based on the following function (**Fig 1**; time, *t*, in seconds):

$$a_{lin}(t) = 49\times sin^2(100\times\pi t) + 20\times sin^2(1000\times\pi t) \quad (1)$$

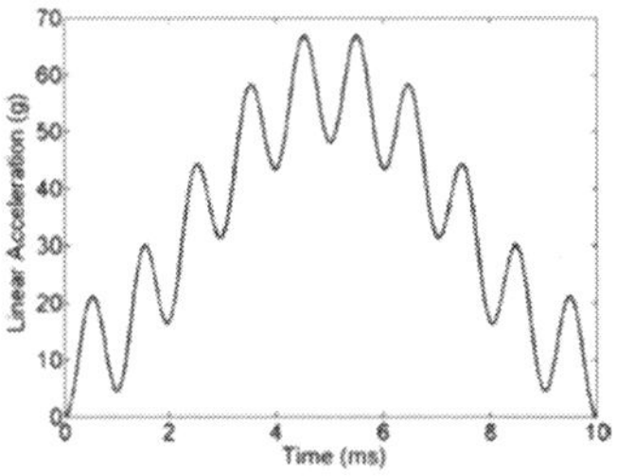

Fig. 1 An arbitrarily generated temporal profile of $\boldsymbol{a}_{lin}$ based on a sinusoidal function with an additional oscillation

The $\boldsymbol{a}_{lin}$ profile was applied to the rigid skull in the anterior–posterior direction, representing a frontal impact. Brain bulk modulus

(K) was varied by five orders of magnitude (1E-2 to 1E+2 relative to the baseline value of 219 MPa) [12] while maintaining the brain's shear stiffness properties. Second, different material models and their shear stiffness properties were adopted [14–18] while maintaining the same baseline K. To quantify the variation in pressure as a result of the change in K and shear stiffness properties, the root mean squared error (RMSE) based on the differences in pressure magnitudes relative to that from the baseline simulation was calculated.

RESULTS

When K was equal to or above the baseline value (219 MPa), the temporal shapes of coup and contrecoup pressure (P_{coup} and P_{c_coup}) correlated well with that of $\boldsymbol{a}_{lin}$ (with a slight delay of ~0.5 ms). Little change in P_{coup} was found when further increasing K above the baseline value, although changes in P_{c_coup} were somewhat more significant (**Fig. 2; Table 1**). However, the similarity in the temporal shapes faded with lower K values and eventually diminished with a much lower K value, as further confirmed by the rather large RMSE values relative to the baseline in these cases (**Table 1**).

Regardless of the brain shear stiffness properties (models and/or property values), P_{coup} and P_{c_coup} were nearly identical to their baseline counterparts (**Fig. 3;** only showing the results using the "compliant" material in [15] for illustration.), as quantitatively confirmed by the small RMSE values (**Table 2**).

These results clearly suggest that with sufficient incompressibility for the brain (i.e., high K values $\geq$ 219 MPa), brain pressure responses do not depend on brain shear stiffness properties, confirming previous observations [14]. Because different shear stiffness properties (model and/or property values) could produce substatnially different shear or strain responses of the brain regardless of their nearly identical pressure responses [19], an FE model validated agaist pressure resposnes alone may not be sufficient to ensure unique strain responses [7–10].

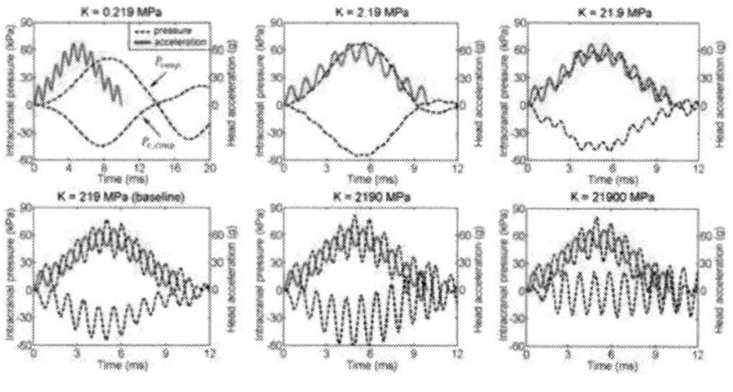

Fig. 2 Correlation between P_{coup}/P_{c_coup} and the given a_{lin} over time for a range of brain bulk moduli.

Table 1 RMSE of P_{coup} and P_{c_coup} for different K of the brain.

K (MPa)	0.219	2.19	21.9	219	2190	21900
$RMSE_{coup}$ (kPa)	25.09	12.40	10.47	0	4.25	4.74
$RMSE_{c\_coup}$ (kPa)	20.63	14.40	10.94	0	10.27	11.06

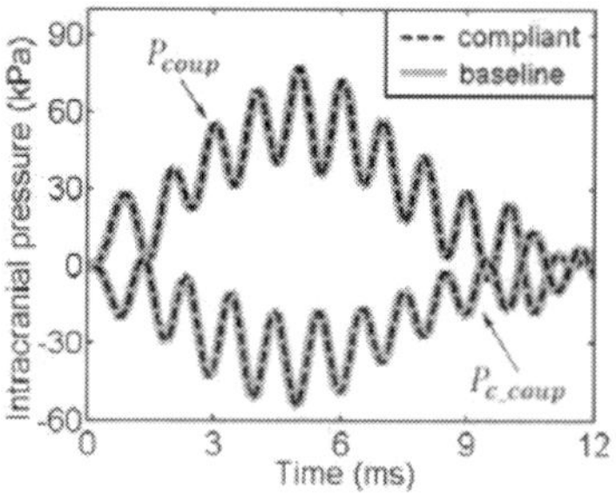

Fig. 3 P_{coup} and P_{c_coup} obtained from the "compliant" model in Kleiven (2007) [15] compared with the baseline response. Similarly, pressure responses from all other shear stiffness properties were virtually identical and are not shown here.

Table 2 RMSE of P_{coup} and P_{c_coup} for different shear stiffness properties (model and/or property values) of the brain.

	Ogden [15]			MR	Viscoelastic		
shear stiffness	a	c	s	[14]	[16]	[17]	[18]
$RMSE_{coup}$ (kPa)	0	0.048	0.086	0.030	0.296	0.058	0.164
$RMSE_{c\_coup}$ (kPa)	0	0.034	0.052	0.024	0.177	0.057	0.097

MR: Mooney-Rivlin; a: average; c: compliant; s: stiff

DISCUSSION AND CONCLUSION

With sufficient incompressibility (i.e., a high bulk modulus), brain pressure responses do not depend on brain shear stiffness properties. Therefore, an FE model only validated against brain pressures may not be used to estimate brain strain-related responses. Validations against both pressure and strain responses are necessary to ensure model biofidelity for estimating tissue-level responses.

ACKNOWLEDGEMENTS

This work was sponsored, in part, by the NIH Grant R21 NS078607 and the Dartmouth Hitchcock Foundation.

REFERENCES

[1] King et al., *IRCOBI Conferenc*, 1-12, 2003.
[2] Ji et al., *Ann. Biomed. Eng.*, 42(1): 11–24, 2014.
[3] Nahum et al., *Proc. 21st Stapp Car Crash Conf.*, 337–366, 1977.
[4] Trosseille et al., *Proc. 46st Stapp Car Crash Conf.*, 235–253, 1992.
[5] Hardy et al. *Stapp Car Crash J.*, 45: 337–368, 2001.
[6] Hardy et al. *Stapp Car Crash J.*, 51: 17–80, 2007.
[7] Kraft et al., *PLoS Comput. Biol.*, 8(8): e1002619, 2012.
[8] Wright et al., *J. Neurotrauma*, 30(2): 102–118, 2013.
[9] Morse et al., *Ann. Biomed. Eng.*, 42(12): 2501–2511, 2014.
[10] Cotton et al., *Comput. Methods Biomech. Biomed. Engin.*, 2015. DOI: 10.1080/10255842.2014.994118.
[11] Zhao et al., *Biomech. Model. Mechanobiol.*, 2014. DOI: 10.1007/s10237-014-0634-0.
[12] Ji et al., *J. Neurotrauma.* 2014. DOI:10.1089/neu.2013.3268.
[13] Ji et al., *Biomech. Model. Mechanobiol.*, 13(5), 1121–1136, 2014.
[14] Kleiven et al. *Stapp Car Crash J.*, 46: 123–144, 2002.
[15] Kleiven, *Stapp Car Crash J.*, 51: 81–114, 2007.
[16] Zhang et al., *Stapp Car Crash J.*, 45: 369–394, 2001.
[17] Takhounts et al., *Stapp Car Crash J.*, 52: 1–31, 2008.
[18] Mao et al., *J. Biomech. Eng.*, 135(11): No. 111002, 2013.
[19] Meaney et al., *J. Biomech. Eng.*, 136(2), No. 021008, 2014.

IMPACT-INDUCED FISSURING IN ARTICULAR CARTILAGE: BULK AND TISSUE-SCALE MECHANICS

C.R. Henak (1), L.R. Bartell (2), L.J. Bonassar (1,3), I. Cohen (4)

(1) Department of Biomedical Engineering
Cornell University
Ithaca, NY, USA

(2) Department of Applied & Engineering Physics
Cornell University
Ithaca, NY, USA

(3) Sibley School of Mechanical and Aerospace Engineering
Cornell University
Ithaca, NY, USA

(4) Department of Physics
Cornell University
Ithaca, NY, USA

INTRODUCTION

Post-traumatic osteoarthritis (PTOA) accounts for at least 12% of all osteoarthritis, thus causing a large economic burden [1-3]. PTOA is a unique subset of osteoarthritis that is caused by a traumatic injury. Joint-scale causes of PTOA are injuries such as intra-articular fracture and recurrent sprain, indicating that an altered mechanical environment initiates PTOA [2].

Early damage in osteoarthritis occurs at the tissue scale, and includes features such as fissuring and fibrillation [4]. Previous research has evaluated the relationship between bulk mechanics and tissue-scale damage relevant to PTOA, but has failed to evaluate both damage and mechanics on the same spatial scale. This has led to discrepancies in the mechanical thresholds which cause damage (e.g., [5,6]). These discrepancies suggest that further research is warranted to evaluate mechanics and damage on the same spatial scale.

Therefore, the objectives of this study were twofold: establish the ability to measure differences in bulk mechanics between fissured and intact cartilage following impact injury in our newly developed impact apparatus, and develop methods to evaluate tissue-scale strain during impact-induced fissuring.

METHODS

Articular cartilage was dissected from immature bovine stifle joints (1-3 days old). Cylindrical cartilage explants were removed from various locations on the tibial plateau and the femoral condyles. Prior to testing, samples were stored at -20°C. All samples were cut in half to create hemicylinders. For bulk strain measurements, samples were stained for 30 minutes in 5-dichlorotriazinylaminofluorescein, a dye that binds to amines and proteins. Following staining, samples were rinsed briefly in PBS. For tissue-scale strain measurements, polystyrene microspheres (2 μm diameter, Polysciences, Inc., Warrington, PA) were placed on the flat surface of the hemicylinder and allowed to dry briefly in order to form a pattern that could be tracked using digital image correlation (DIC).

Each cartilage sample was mounted onto a custom impact device integrated with an inverted microscope [7] (Figure 1). The impact device consists of a backplate and a spring-loaded impactor. The deep portion of each sample was attached to the backplate with cyanoacrylate, while the articular surface was impacted. Cylindrical impactors with 0.8 mm diameter and 1.2 mm diameter impact tips were used. The spring was compressed various amounts in order to deliver a range of impacts, some of which caused fissuring. During impact, a Phantom V7.1 digital camera captured images at 1,000 – 10,000 frames per second using a 5× or 10× objective. Samples were illuminated using a mercury lamp, and were immersed in PBS during testing.

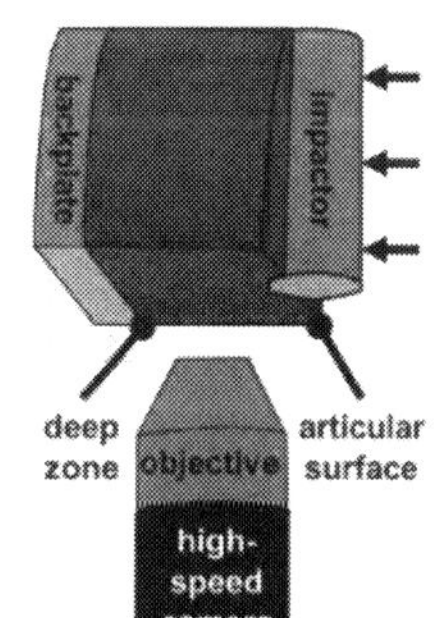

Figure 1. Impactor schematic showing sample, impactor and objective.

Bulk metrics were evaluated for 12 samples. Six samples fissured during impact and six samples remained intact. For each sample, the bulk axial strain was calculated from the thickness between the backplate and the impact tip in comparison to the undeformed thickness. Force was calculated from the movement of the backplate, which was calibrated using weights. Average axial stress was calculated using the length of the impactor contact on the sample and the diameter of the impactor. Strain rate, force rate and stress rate were calculated between subsequent images. Differences between intact and fissured samples were tested using an unequal variance t-test. *P*-values were corrected for repeated measures using Bonferroni correction. Significance was set at $p \leq 0.05$.

Tissue-scale Green-Lagrange strain was calculated using DIC with a grid with 45 μm spacing in the reference configuration [8,9]. One fissured and one intact sample were evaluated. Both samples were imaged at 10,000 frames per second. The displacement in each image was calculated relative to the previous image, due to very large displacements overall. Strain fields were evaluated, and the invariants maximum shear strain and first principal strain were calculated.

RESULTS

Fissured samples had significantly larger bulk strain, force, strain rate, force rate and stress rate than intact samples (Figure 2). Stress was nearly significantly different between the two groups. On average, the fissured samples had 1.7 times larger strain, 2.1 times large force, 2.0 times larger stress, 1.8 times larger strain rate, 2.0 times larger force rate, and 1.8 times larger stress rate than the intact samples.

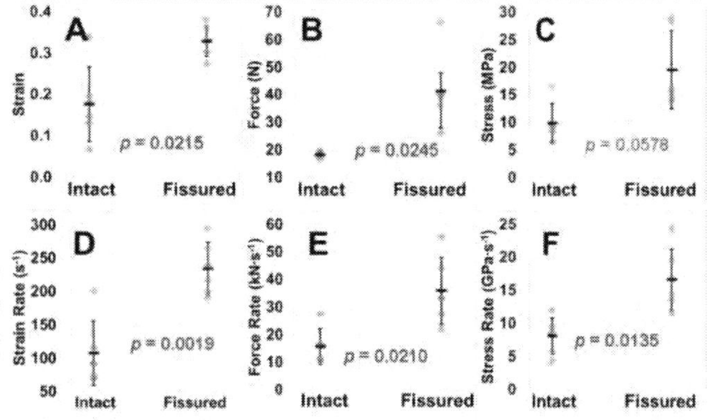

Figure 2. Bulk mechanics: A – strain, B – force, C – stress, D – strain rate, E – force rate and F – stress rate. Nearly all metrics were significantly larger in fissured samples.

Tissue-scale strain demonstrated highly heterogeneous patterns, with larger magnitudes in the fissured sample (Figure 3) than in the intact sample (Figure 4). Specifically, the peak tissue-scale first principal strain was 39.3 times larger in the fissured sample than in the intact sample, while the peak maximum shear strain was 8.5 times larger in the fissured sample than in the intact sample. In both cases, peak maximum shear strain and peak first principal strain occurred just under the impact tip.

DISCUSSION

Consistent with previous studies, most bulk mechanics were significantly different between fissured and intact samples. However, the specific magnitudes at which damage occurred vary between the present study and previous studies. For example, in the present study, the average peak stress for intact samples was 8.5 MPa, and the average peak stress for fissured samples was 16.7 MPa. Previous studies have reported average peak stresses of 30 MPa for intact samples with 50 MPa for fissured samples in the mature equine stifle [6], 3 MPa for intact samples with 9 MPa for fissured samples in the mature bovine elbow [5], and 5.6 MPa for intact samples with 6.2 MPa for fissured samples in the immature rabbit knee [10]. These discrepancies demonstrate the weakness of bulk mechanics as an indicator of tissue-scale damage.

The evaluation of tissue-scale strain demonstrated very large differences between intact and fissured samples. While bulk metrics were only approximately a factor of two larger in fissured than intact samples, tissue-scale strains were 8 to 39 times larger in fissured than intact samples. These differences suggest that tissue-level mechanics offer promise for ongoing investigation into the specific causes of PTOA-like damage in articular cartilage.

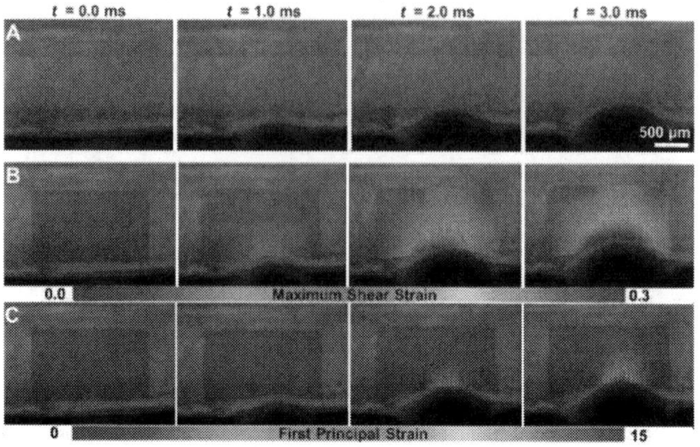

Figure 3. Tissue-scale mechanics for fissured sample: A – images during impact. B – maximum shear strain during impact. C – first principal strain during impact. Strain values were large under the impact tip. Note the large scales in comparison to Figure 4. Articular surface is down.

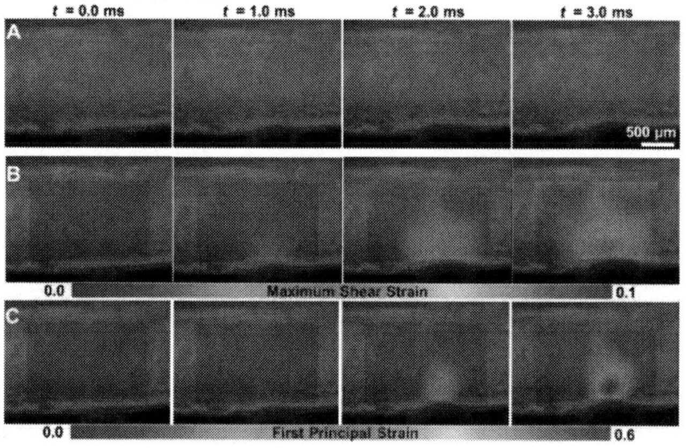

Figure 4. Tissue-scale mechanics for intact sample: A – images during impact. B – maximum shear strain during impact. C – first principal strain during impact. Note the smaller scales than in Figure 3. Articular surface is down.

ACKNOWLEDGEMENTS

Funding from NIH #AR062677. Funding from NSF GRFP (LRB).

REFERENCES

[1] Lawrence et al., *Arth Rheum*, 58(1):26-35, 2008. [2] Brown et al., *J Orthop Trauma*, 20(10):739-744, 2006. [3] Buckwalter et al., *Cartilage*, 4(4):286-294, 2013. [4] Grenier et al., *J Biomech*, 47:645-652, 2014. [5] Natoli et al., *Ann Biomed Engg*, 36(5):780-792. [6] Bolam et al., *Am J Vet Res*, 67(3):443-447. [7] Bartell et al., *Proc World Cong Biomech*, 2014. [8] Jones et al., *Exp Mech*, 54:971-985, 2014. [9] Jones, *MatLab File Exchange*, 2013. [10] Borrelli et al., *J Orthop Res*, 28:1307-1314, 2010.

SB^{3}C2015
Summer Biomechanics, Bioengineering and Biotransport Conference
June 17-20, 2015, Snowbird Resort, Utah, USA
SB^{3}C2015-664

IN VITRO PULSATILE FLOW LOOP USING HUMAN BLOOD TO MIMIC PHYSIOLOGICAL FLOW CONDITIONS

Ryan W. Oba[1], David L. Bark Jr.[1], Lakshmi P. Dasi[1,2], Ketul C. Popat[1,2]

(1) Department of Mechanical Engineering
Colorado State University
Fort Collins CO
USA

(2) School of Biomedical Engineering
Colorado State University
Fort Collins CO
USA

INTRODUCTION

According to the American Heart Association, heart disease is the leading cause of death for both men and women in the United States, resulting in over 600,000 deaths annually [1]. Over the years, marked improvements to heart valve implant designs as well as surgical techniques have resulted in better patient outcomes, yet valve replacement does not provide a definitive cure to the patient. Complications due to native valve disease are traded for those seen with prosthetic valve disease (i.e. post-operative bleeding, thrombocytopenia, hypertriglyceridemia, hyperkalemia, and hypersensitivity) [2].

The heart valve replacement market is currently dominated by two classes of valves; mechanical and bioprosthetic. While these replacement valve types prolong patient lifespans, there still exists a need to develop new valves that will overcome shortcomings such as thrombus formation or calcification in the valves, resulting in even better patient outcomes. While groups are working to improve heart valve implants, current methods used to validate these new designs require extensive animal testing, a large amount of time and money, offer little control over conditions such as flow and pressure, and ultimately fail to provide a means of correlating hemodynamic and hemo-compatibility data. Thus, there exists a need to develop a new model that can accurately predict in vivo replacement heart valve performance.

The central hypothesis of this research is that an in vitro pulsatile flow loop model can mimic physiological flow conditions seen by replacement heart valves and be used to assess replacement heart valve hemodynamics and hemo-compatibility. A specific hypothesis of this research is that the location and degree of thrombus formation in heart valve implants in vivo, can be accurately predicted using our in vitro flow loop model.

METHODS

The in vitro flow loop was designed to simultaneously subject two identical prosthetic heart valves to tunable physiological aortic flow conditions. The first component of the flow loop is the inner polycarbonate chamber which houses two oppositely oriented identical prosthetic heart valves. This chamber is ultimately filled with blood. These materials were chosen because they can be passivated to blood, are clear for visualization, and can withstand physiological pressures. Separated from the inner chamber by silicone membranes, are two outer polycarbonate chambers. The outer chambers are filled with water and connected to a piston pump whose input waveform is controlled via computer. By doing so, the volume and frequency used to change the pressure in the outer chambers can be controlled. Changes in the piston position generate a corresponding membrane response (expanding or contracting). In the case where the input waveform is a sine wave, a pulsatile flow in the inner chamber is created. Inner and outer chambers were machined using a computer numerically controlled vertical endmill (VMC Series, HAAS Automation, Oxnard, CA). The flow loop design, valves used in this study, and flow loop loaded with human blood are shown in Figure 1.

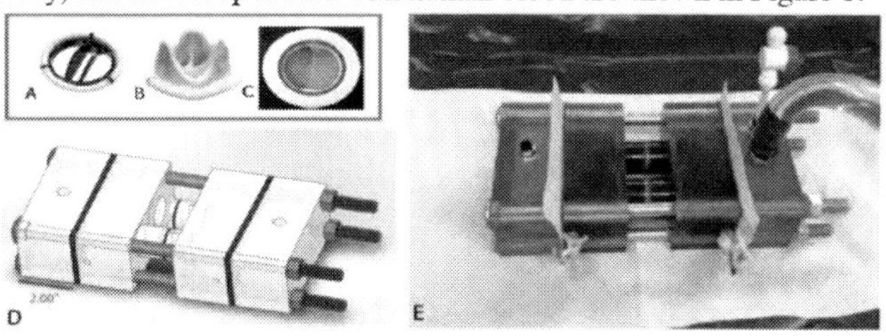

FIGURE 1: A., B., C. Valves Used **D.** Flow loop CAD model **E.** Fabricated flow loop filled with human blood.

Initial validation of flow and pressure was performed using a blood viscosity-matching water-glycerin analog as well as two types of valves; 22mm mechanical valve (St. Jude Medical, Saint Paul, MN), and 22mm bioprosthetic valves (CarboMedics Inc, Austin, TX). Pressure was measured using pressure transducers (DP15 Series, Validyne, Northridge, CA), and flow was measured using an external flow probe (PXL Series, Transonic, Ithaca, NY).

Secondary validation of the flow loop was performed through the use of particle image velocimetry (PIV). PIV results reveal details to the flow such as stagnation regions, local fluid velocity vectors, and vortex formation in the flow. After dispersing fluorescent particles (1-20μm diameter PMMA-Rhodamine B-Particles, Dantec Dynamics, Denmark) in the water-glycerin, a laser sheet was shown through the cross section of the heart valve implants. A high speed camera then captured the flow downstream of the valves. During post processing, by discretizing the high speed camera's frames into a grid, the program (DaVis, LaVision, Goettingen, Germany) can calculate fluid velocity.

Groups previously showed that 3.2% citrated blood can be recalcified and that clot formation can be controlled in vitro [3]. This citrate buffer was used in this study as well. Prior to the full scale loading of the flow loop, preliminary work with smaller scale aliquots of the 3.2% citrated blood were placed inside glass vials on a shaker plate. Then, differing amounts of calcium chloride were injected into each vial and the time points for initial clot formation were recorded.

RESULTS

Flow within the testing apparatus can be modified in two ways; by change of stroke volume of the piston pump or by the change of the frequency of the piston pump motion. This corresponds to modeling various heart sizes or modeling various heart rates, respectively. Figure 2 illustrates the changes in flow and differential pressure seen across two identical mechanical heart valves due to a change in stroke volume of the piston pump. By changing the input waveform to the piston pump, the flows seen by the valves adjust accordingly. Positive values of pressure and flow correspond to opening and closing stages of the valves, whereas the flat sections correspond to the closed state of the valves. Minor negative values in the flow illustrate that the valves used had small levels of regurgitation, or reverse flow during testing. PIV results, also shown in Figure 2, depict vortex formation originating from the tips of the mechanical heart valve leaflets and continuing to form downstream from the valves.

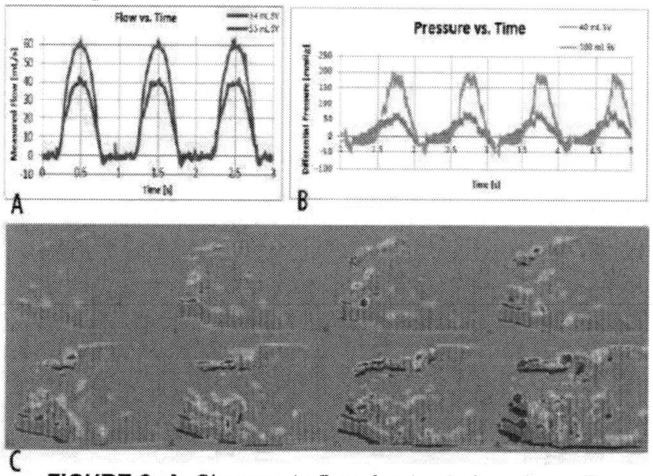

FIGURE 2: A. Changes to flow due to stroke volume **B.** Changes to pressure due to SV **C.** PIV of valve downstream.

The corresponding citrate:calcium ratio and clot times using the citrated blood are shown in Figure 3. Results published by Fallon et. al are shown alongside these as well.

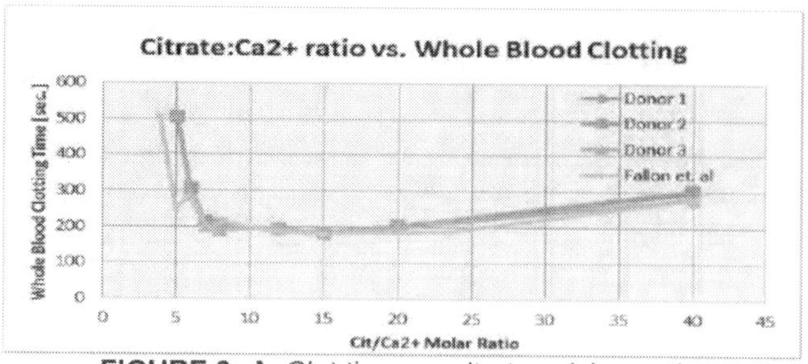

FIGURE 3: A. Clot time vs. citrate:calcium ratios

DISCUSSION

The hemodynamic results from this study show that pulsatile flows within the flow loop can be accurately adjusted to match the range of physiological levels seen in the body. Similarly, the corresponding changes to pressure due to changes in flow result in pressures which remain in the physiological regime. Because PIV results in the flow loop show similar vortex formation on MHV leaflets to those seen in vivo[4], this demonstrates that flow features are physiological. Results from the blood citrate studies show that it is possible to anticoagulate the blood prior to use, and that upon the addition of varying levels of calcium that the blood will begin to clot at a predictable time point. Results seen during this study were similar to those seen by Fallon et. al.

Advantages this flow loop model include the high degree of control over the flow conditions seen by the heart valves, the ability to pair high quality pressure and flow data with hemo-compatability results, its generation of smooth pulsatile flows, and its use of full scale-valves, as opposed to other models which only tested valve materials[3]. With this data, one major advantage of this flow loop over existing methods is. One shortcoming of this flow loop is the fact that longer term tests using blood such as those needed to assess the fatigue life of the valve, cannot be performed. To achieve this, animal testing may still be required.

For future work, quantification of blood clotting seen within the flow loop will require the use of hemoglobin assays and scanning electron microscopy (SEM); hemoglobin assays provide insight into how much of the blood clotted whereas SEM helps to quantify clot size and cellular morphology. The next step will be to show that the blood flown across the valves resulted in clot formation in similar locations (i.e. hinges on a mechanical valve), as seen in previous literature. After flowing recalcified blood in the flow loop for 2 hours, SEM images of clot formation on the valves will be taken. These results will be analyzed using ImageJ software. Once the final validation of the flow loop is complete, various heart valve geometries and materials will be tested and compared and these results will be used to help researchers create superior new heart valve designs.

For brevity, not all data is shown in this abstract. However, we intend to present additional data for validation of the flow loop at the conference, including additional hemodynamic as well as bio-compatibility results.

ACKNOWLEDGEMENTS

Research reported in this publication was supported by the National Institutes of Health under award number R01HL119824. The content is solely the responsibility of the authors and does not necessarily represent the official views of the National Institutes of Health.

REFERENCES

[1] Kochanek, KD et al., *Natl. Vital Statistics Reports*, 60(3), 2009.

[2] Dasi, LP et al., *J Biomechanical Eng.*, 132(2):021005, 2010.

[3] Fallon, AM et al., J *Thrombosis Research*, 121(3):397-406, 2007.

[4] Yoganathan, AP et al., *Ann Biomed Eng.*, 42(3):526-40, 2014.

SB^{3}C2015
Summer Biomechanics, Bioengineering and Biotransport Conference
June 17-20, 2015,Snowbird Resort, Utah, USA
SB^{3}C2015-665

PROGRESSIVE CHANGES IN CERVICAL SPNE INTERVERTEBRAL DISC PROPERTIES DURING CYCLIC COMPRESSIVE FATIGUE LOADING

Sagar Umale (1), Brian D. Stemper (1,2), Mingxin Zheng (3), Aidin Masoudi (3), Daniel Fama (1,2), Narayan Yoganandan (1,2) Brian Snyder (3,4,5)

(1) Department of Neurosurgery, Medical College of Wisconsin, Milwaukee, WI

(2) Clement J. Zablocki VA Medical Center, Milwaukee, WI Department

(3) Beth Israel Deaconess Medical Center, Boston, MA

(4) Harvard Medical School, Boston, MA

(5) Cerebral Palsy Clinic Children's Hospital, Boston, MA

INTRODUCTION

Prolonged exposure to whole body vibration in environments such as, helicopters, fast boats, and land-based vehicles can contribute to neck, back, and shoulder pain in military personnel [1,2]. Prior experimental studies have shown that repetitive loading of spinal units may cause bony injuries such as endplate and cortical fractures [3,4], and clinical hypotheses have included development of intervertebral disc herniation [2]. However, due to fatigue, changes in disc material properties can alter the mechanical response of the spinal column, change segmental load sharing resulting in increased injury risk to other tissues, and may become permanent with continued repetitive loading. To our best knowledge, progressive changes in soft tissue properties have not been reported for compressive fatigue in cervical spine segments. Therefore, it is important to characterize these changes due to vibrational environment and by progressive loading and determining the changes during fatigue and compare pre and post fatigue changes. The purpose of this study was to characterize changes in mechanical properties of cervical spine disc due to low magnitude compressive fatigue loading.

METHODS

Four cervical bone disc bone spine motion segments without posterior elements obtained from post-mortem human subjects (mean age: 58.2±3.2 yr) were fixed at cranial and caudal extents using polymethylmethacrylate. Specimens were exposed to a protocol that included fatigue loading for 5 sets of 10,000 cycles with tension compression and viscoelastic assessments prior to the first and after each fatigue set. Fatigue testing was performed with specimens submerged in a physiologic saline bath with temperature maintained at 34 deg C and consisted of repetitive compression applied using the piston of an electrohydraulic testing device (MTS Systems, Eden Prairie, MN) between 0 and 150 N at 2 Hz using a sine wave function. Tension compression tests were performed at a static speed of 0.1 mm/s to 10% strain. The viscoelastic test involved compressing the specimen to 10% strain and allowing it to relax under the constant deformation. Disc height was measured from x-rays obtained after each fatigue set and an average of three readings was measured as disc height. Axial force was measured at 100 Hz during the entire test using a uniaxial load cell attached to the piston. After each test the segment was allowed to settle for 15 minutes in the saline bath and then the next test was performed.

RESULTS

Cycle-by-cycle intervertebral disc height during each fatigue set is plotted in Figure 1. Disc height loss was greater during the first and second fatigue sets. The disc regained some height during the inter-set relaxation period. The average decrease in disc height during fatigue loading was observed to be 16.5±6.5%, 12.5±5.8%, 9.7±3.4%, 6.9±1.8% and 5.9±3.1% from fatigue set 1 to fatigue set 5 respectively, whereas about 10% height was recovered after first two fatigue sets and about 5% was recovered after later three. The average decrease in disc height was calculated as 15% for all the four segments over 50,000 cycles. The stress relaxation curves for segment 4 are plotted in Figure 2. Segments demonstrated the most flexible behavior during the first and second fatigue sets and became progressively stiffer for subsequent fatigue sets. The curves for pre fatigue and post first and second fatigue test shows similar viscoelastic behavior whereas post third, forth, and fifth fatigue the stiffness of the disc can be observed to be increased, as the stress required for same strain is increased. This observation is also illustrated in Figure 3, in which average stiffness for all the segments during fatigue is plotted. It can be observed that during first three sets there was not much change in stiffness of the disc, whereas fourth and fifth sets showed a high increase in stiffness. The tension and compression tests showed that after fatigue the disc was less stiff in tension, whereas stiffer in compression. The viscoelastic test demonstrated decrease in viscous properties of the disc. The relaxation time was observed to be lesser for disc with increase in fatigue.

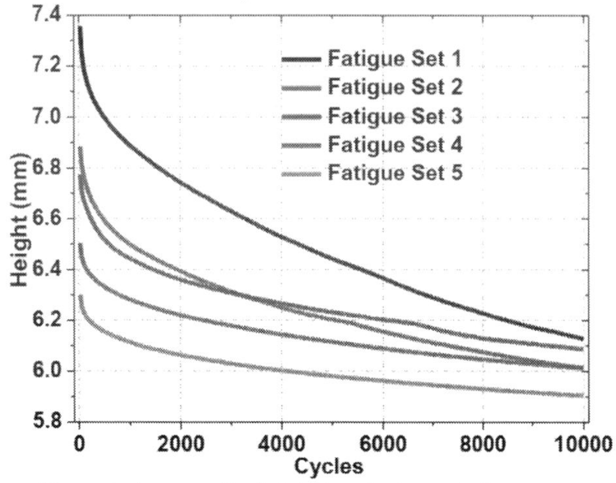

Figure 1. Average disc height loss during each fatigue set.

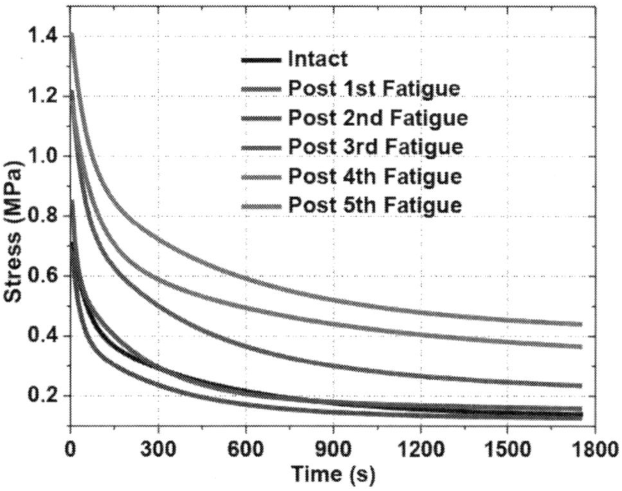

Figure 2. Stress relaxation curves for segment 4 post fatigue.

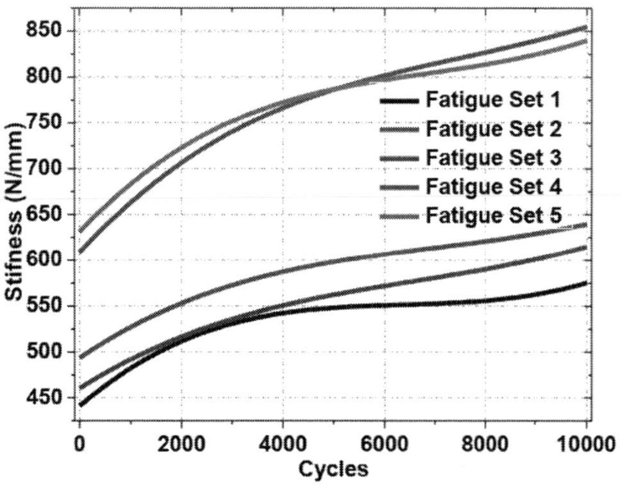

Figure 3. Average stiffness during each fatigue set.

DISCUSSION

The disc height was observed to be reduced after each fatigue set for each segment. It can also be observed from Figure 1 that the curve profile for first two fatigue sets is similar and the curve profile for next three fatigue sets is similar. It shows that for the first two fatigue sets the disc demonstrates viscoelastic properties, whereas for the next three sets the disc shows only elastic response with decrease in height. It can be observed from Figure 2 that the initial stiffness of the disc increased up to 65% between intact segment and post fifth fatigue set. Thus there was a gradual decrease in viscous response and a gradual increase in stiffness of the disc. If the curves from Figure 2 and Figure 3 are grouped during the first three fatigue sets the stiffness of the specimens was relatively less as compared to the following two fatigue sets. This behavior is observed due to decrease in disc height and loss in viscous behavior. Loss in viscosity can also lead to ore fractional losses and thus failure of the soft tissue.

The tension compression tests however showed that the stiffness of the segments was decreased in tension. Progressive disc height loss can contribute to pain-related conditions including intervertebral foraminal stenosis. These changes also contribute to altered loading sharing of the intervertebral segment that can contribute to elevated loading of adjacent structures such as the facet joints, eventually resulting in an accelerated degenerative process. Understanding these mechanisms and the transition from acute to chronic changes is a key to the development of exposure limits for repetitive axial loading environments that will be used to limit the onset of neck and low back pain symptoms for military personnel. Given inter-personal differences in spinal biomechanics based on gender, age, and loading history, among other factors, it is conceivable that these changes could be assessed using non-invasive techniques such as multiplanar ultrasound under controlled loading inputs to the cervical spine [5]. This remains an ongoing focus of the current line of investigation.

ACKNOWLEDGEMENTS

This work was supported by the U.S. Army Medical Research and Materiel Command Fort Detrick, Maryland (01026018, Contract No. W81XWH-13-1-0050), and the Department of Veterans Affairs Medical Research.

REFERENCES

1. Harrison MF, et al. (2012). A predictive logistic regression equation for neck pain in helicopter aircrew. Aviat Space Envir Med. 83, 6, 604-8.
2. Pope MH, Magnusson M, Wilder DG. (1998). Low back pain and whole body vibration. Clin Orthop Relat Res.(354):241-8.
3. Goel, V.K., et al. (1987). Response of the Ligamentous Lumbar Spine to Cyclic Bending Loads, in: Second Annual Meeting of the North American Spine Society. Banff, Alberta, Canada., pp. 294–300.
4. Adams M & Hutton W (1980). The effect of posture on the role of the apophysial joints in resisting intervertebral compressive forces. JBJS B62: 358-62.
5. Z. Mingxin, M. Aidin, S. Umale, D. Buckland, N. Yoganandan, B. Stemper, T. Szabo, and B. Snyder, "Dual Ultrasound Can Measure Kinematic Motion and Intervertebral Disc Deformation of Cervical Spine.," presented at the ORS 2015 Annual Meeting, Las Vegas, Nevada, March 2015., 2015, p. 1096.

QUANTIFYING SKELETAL MUSCLE DEFORMATION IN THREE DIMENSIONS

Elisabeth R. Jensen (1,2), Duane A. Morrow (1), Joel P. Felmlee (3), Kenton R. Kaufman (1)

(1) Division of Orthopedic Research
Mayo Clinic
Rochester, MN, USA

(2) Mayo Graduate School
Mayo Clinic
Rochester, MN, USA

(3) Department of Radiology
Mayo Clinic
Rochester, MN, USA

INTRODUCTION

Skeletal muscle health is critical to independence and well-being. Muscle strength enables mobility and interaction with the environment as well as important exercise-related health benefits. Muscle weakness can manifest itself in several ways. In cases like sarcopenia and collagen VI disorders, it is known that passive muscle structures undergo stiffness changes [1,2]. We have yet to identify how these changes alter muscle mechanics on a functional level.

Several methods have emerged in recent decades that enable the visualization and quantification of muscle deformation. Cine Phase Contrast (CPC) MRI samples spatially distributed velocity maps at finite time points throughout a motion cycle, allowing internal muscle motion to be tracked over time at a high resolution [3,4]. This technique has previously been used to quantify strain distribution in skeletal muscles. However, until now the results have been limited to two dimensions.

The objective of this study was to advance CPC MRI-based analysis of skeletal muscle strain to three dimensions and to establish the validity of the output. Specifically, we sought to compare the output of the CPC data to structural elements visible in T2-weighted MRI images.

METHODS

Subjects A female subject (24 y.o., 61 kg) with no neuromusculoskeletal health deficits was recruited to participate in this study. The design and methods were approved by the Mayo Institutional Review Board (IRB #13-002825).

Setup A custom MRI-compatible apparatus was used to passively rotate the subject's ankle while minimizing extraneous limb movement (less than 3 degrees and 6 mm of lower limb motion) as validated using a motion capture system (Cortex, Motion Analysis Corp., Santa Rosa, CA). The ankle was rotated from neutral (0 degrees) through 26 degrees of plantarflexion and back to neutral in 2 seconds. The motion cycle was triggered by a square wave (0.5 Hz) generated by a function generator (Hewlett-Packard, Palo Alto, CA).

Data collection MRI data were collected using a commercially available Fast 2D Phase Contrast sequence (50 mm/s v_{enc}, 12 time segments, 4 vps) on a 1.5T MRI system (Signa HDX 16.0, GE Medical Systems, Waukesha, WI). The sequence was gated using a cardiac simulator triggered by the same square wave described in the previous paragraph. Nine adjacent 5-mm-thick image slices were collected in the sagittal plane with a 24 cm field of view (FOV) and 192x192 in-plane resolution. The three dimensions of velocity were encoded in the frequency, phase, and through-plane directions independently. Axial T2-weighted image stacks were collected with the ankle stationary at neutral and at 26 degrees plantarflexion. Slices were arranged to cover the length of the TA (5x0 mm thickness, 20 cm FOV, 320x320 in-plane resolution).

Data processing The TA was manually segmented from axial T2-weighted images collected with the neutral ankle position (Mimics, Materialise NV, Leuven, Belgium). The segmentation was smoothed and exported as a surface mesh. Using the surface mesh to define the boundaries, a 3D hexahedral mesh with 10824 elements was manually generated (TrueGrid, XYZ Scientific, Livermore, CA) for use in Abaqus (SIMULIA, Waltham, MA). The trajectory of each mesh node was independently tracked using the CPC-generated velocity maps (smoothed and corrected according to [5]), the forward-backward integration method [6,4], and custom software (MATLAB, The MathWorks, Natick, MA). The output was a single 3D trajectory for each node throughout the experimental range of motion. The trajectories were applied as nodal boundary conditions in Abaqus to obtain the final deformed mesh model of the lengthened TA. Local element orientation was defined such that axis 1 aligned with the long axis of the TA and axis 2 aligned with the local orientation of the internal aponeurosis, as determined from T2-weighted images.

Validation testing To establish validity of the final model, several tests were employed, including the following: 1) A surface mesh of the deformed mesh model (exported from Abaqus) was overlaid on the T2-weighted images of the lengthened TA (plantarflexed ankle) to establish face validity of the model. 2) The transition line between positive and negative E13 shear strain of the mesh was compared to the location of the internal aponeurosis, visible on the T2-weighted images. It was expected that these should align because the pull on the aponeurosis during passive muscle elongation facilitates fascicle rotation, or shear (Figure 1).

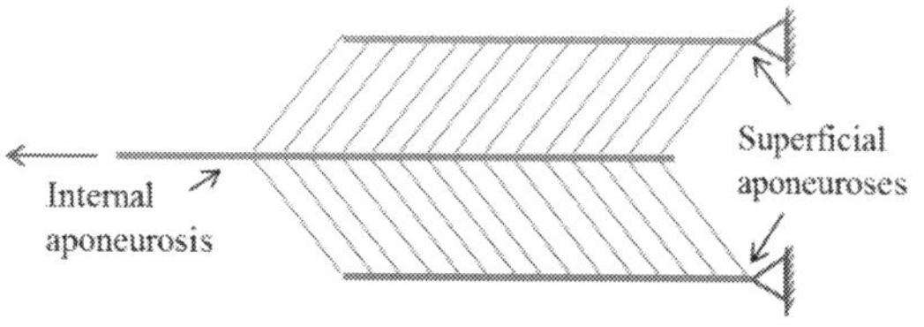

FIGURE 1: Muscle fibers on opposing sides of internal aponeurosis experience opposite shear strains with passive muscle lengthening.

RESULTS

The validation testing results were as follows: 1) The static T2-weighted images collected with the plantarflexed ankle agreed with the deformed mesh; the mesh outline matched the image-identified TA position (Figure 2a,b). 2) The transition line between positive and negative E13 shear strain aligned with the internal aponeurosis throughout the length of the muscle (example cross-section shown in Figure 2b,c).

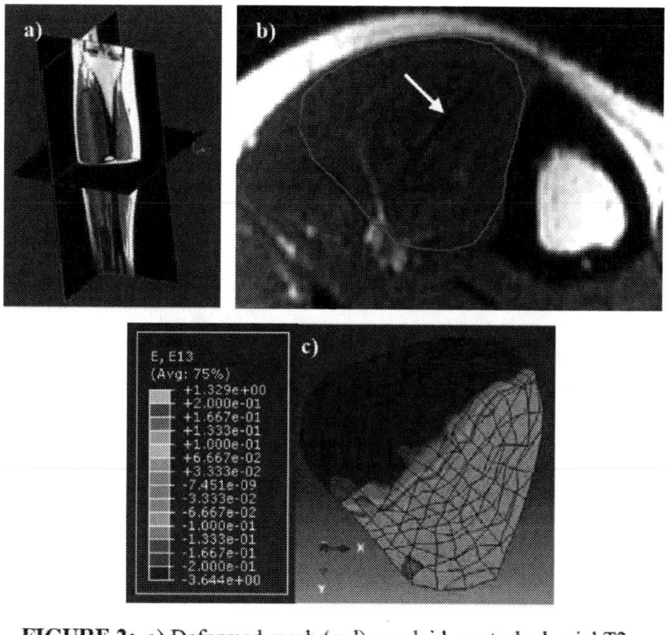

FIGURE 2: a) Deformed mesh (red) overlaid on stacked axial T2-weighted MRI images taken with plantarflexed ankle. b) Axial cross section of deformed mesh as identified in (a). Arrow points to aponeurosis. d) Corresponding cross section of deformed mesh, showing transition between positive and negative E13 shear strains.

DISCUSSION

The ability to observe and quantify the internal dynamics of skeletal muscle is an important tool in understanding functional changes that lead to weakness. Previous studies on this topic have, out of necessity, attempted to represent complex three-dimensional dynamics in two dimensions. The current study demonstrates a novel approach using CPC MRI to quantify three-dimensional dynamics of skeletal muscle.

Given the limited literature on internal skeletal muscle dynamics, validation of the current model is restricted to methods which focus on whole muscle and major tendon units. Thus the authors chose to confirm muscle position using static images and assess the shear strain differentiation across the aponeurosis. Overall, the demonstrated results suggest validity of the final mesh output.

The deformed mesh aligned appropriately with the TA when the ankle was plantarflexed, as expected. This means that the trajectories of the most superficial nodes behaved as expected. As a result it can be inferred that the motions of each of the individual nodes in the model were appropriate in both magnitude and direction.

More significantly, the positive-to-negative shear strain transition closely coincided with the aponeurosis. This provides model support on the mesh element scale. Mechanical analysis of bi-pennate muscle under passive elongation predicts opposing fascicle rotation on opposite sides of the aponeurosis, generating opposing shear strain. The ability of the model to identify the proper transition line demonstrates strong internal validity.

This work has demonstrated a novel method of quantifying three-dimensional spatially- and temporally-resolved muscle deformation. Future applications of these data include analysis of the 3D distributed principle, shear, and volumetric skeletal muscle strain, as well as detailed morphological changes. The current method focuses on two time points (ankle at 0 and 26 degrees plantarflexion), but the data allow time-resolved analysis of the entire motion, which may be useful for some applications. The current study is limited to analysis of passive elongation due to the long duration of the data collection, but future application of accelerated CPC sequences will significantly reduce data collection time and enable data collection during active contraction [7].

ACKNOWLEDGEMENTS

Funding for this study was provided by the National Institutes of Health (T32AR56950 and R01HD31476) and Mayo Graduate School.

Special thanks to John Novotny for sharing data analysis code.

REFERENCES

[1] Gajdosik, RL et al. *Journal of Orthopaedic & Sports Physical Therapy*, 29:181-190, 1999.

[2] Lampe, AK and Bushby, K. *Journal of Medical Genetics*, 42(9):673-685, 2005.

[3] Sinha, S et al. *Journal of Magnetic Resonance Imaging*, 20(6):1008-19, 2004.

[4] Zhou, H et al. *Journal of Magnetic Resonance Imaging*, 25(1):175-184, 2007.

[5] Jensen, ER et al. *Journal of Biomechanics*, 48(1):95-103, 2015.

[6] Pelc, NJ et al. *Journal of Magnetic Resonance Imaging*, 5(3):339-345, 1995.

[7] Johnson, KM et al. *Magnatic Resonance in Medicine*, 60(6): 1329-1336, 2008.

SB³C2015
Summer Biomechanics, Bioengineering and Biotransport Conference
June 17-20, 2015,Snowbird Resort, Utah, USA
SB³C2015-667

THE INFLUENCE OF CARTILAGE MORPHOLOGY AND ELASTICITY ON TIBIOFEMORAL CONTACT PRESSURES DURING WALKING

Colin R. Smith (1), Rachel L. Lenhart (2), Jarred Kaiser (1), Michael F. Vignos (1), Darryl G. Thelen (1,2,3)

(1) Department of Mechanical Engineering
University of Wisconsin-Madison
Madison, WI, USA

(2) Department of Biomedical Engineering
University of Wisconsin-Madison
Madison, WI, USA

(3) Department of Orthopedics and Rehabilitation
University of Wisconsin-Madison
Madison, WI, USA

INTRODUCTION

Articular cartilage overloading during gait is recognized as a factor affecting the development and progression of knee osteoarthritis (OA). Accordingly, the external Knee Adduction Moment (KAM) is often used as a surrogate measure of contact force on the medial compartment of the tibiofemoral joint, where OA often develops (1). However, the KAM may not be an indicator of localized tissue stress and strain which can induce damage over multiple loading cycles. Thus, it may be more relevant to estimate cartilage contact pressure and tissue deformation during gait. However, current gait models employ a highly simplified knee joint model that only allows for estimates of net joint contact loads (2). To overcome this limitation, we developed a multi-body dynamic knee model that includes representations of the major ligaments and articular surfaces of the tibiofemoral and patellofemoral joints (3). The increased degrees of freedom included in this model allow for prediction of cartilage pressures in the context of full-body walking simulations.

There are three objectives of this study. First, to introduce an Enhanced Static Optimization (ESO) simulation routine to simultaneously calculate muscle forces, cartilage pressures, ligament forces, and secondary knee kinematics from subject-specific gait measures. Second, to evaluate whether the KAM is predictive of medial cartilage contact pressures during gait. Finally, to investigate how variations in cartilage morphology and material properties can alter cartilage contact pressures and deformations.

METHODS

A 3 body, 12 DOF knee model was developed from MR images of a healthy adult female (1.65 m, 61 kg). Fourteen ligament bundles were represented by nonlinear elastic springs, with wrapping surfaces used to prevent penetration of bony geometries. Articular cartilage surfaces were segmented from the images and represented by high resolution meshes. Given the bone segment positions in space, cartilage contact pressures were computed by detecting overlap between articulating surfaces, and using an elastic foundation model to compute pressure based on the depth of penetration and an assumed elastic modulus for cartilage tissue. The predictive capacity of the model was validated by comparing simulated passive and active knee motion with *in vivo* 3D knee kinematics measured with dynamic MRI (3). The knee model was integrated into an existing lower extremity musculoskeletal model (2), which included 43 muscles acting about the hip, knee and ankle joints.

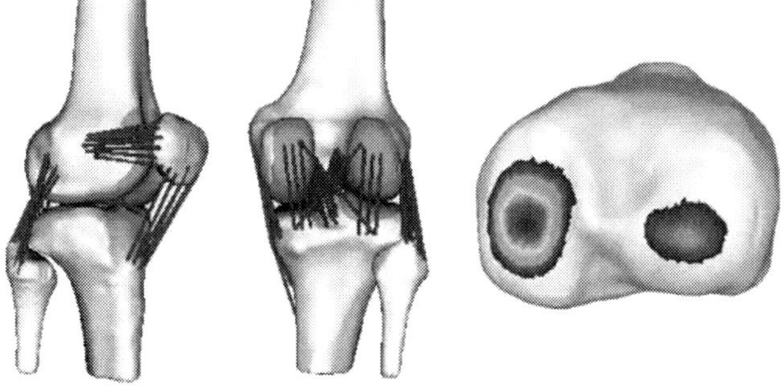

Figure 1 – Multibody 12 degree of freedom knee model and representative pressure map

To estimate tissue loading and knee kinematics during gait, we developed the ESO routine. Similar to standard static optimization, ESO uses numerical optimization to calculate active muscle forces needed to satisfy whole body equations of motion. However, ESO also simultaneously calculates the secondary joint kinematics (5 tibiofemoral and 6 patellofemoral DOF) which induce ligament forces and cartilage contact pressures that, along with the estimated muscle forces, generate the measured primary joint accelerations and zero accelerations in the secondary knee DOF. The formal cost function of the optimization minimizes the sum of the squared muscle activations, weighted by muscle volume. A regularization term, the sum of squared changes in secondary kinematics from the previous time step, is used to constrain the solution over time.

We used the model and ESO routine to estimate internal tissue loads from measured gait kinematics and ground reaction forces collected on the same subject for which we constructed the knee model. These analyses were repeated for 5 values of tibiofemoral cartilage thicknesses (combined tibial and femoral thickness ranging from 2mm-10mm) and 6 values of cartilage modulus of elasticity (5 MPa – 25 MPa). For each solution, we computed the mean contact pressure and mean percent deformation for the contacting cartilage on both the medial and lateral tibial plateau.

RESULTS

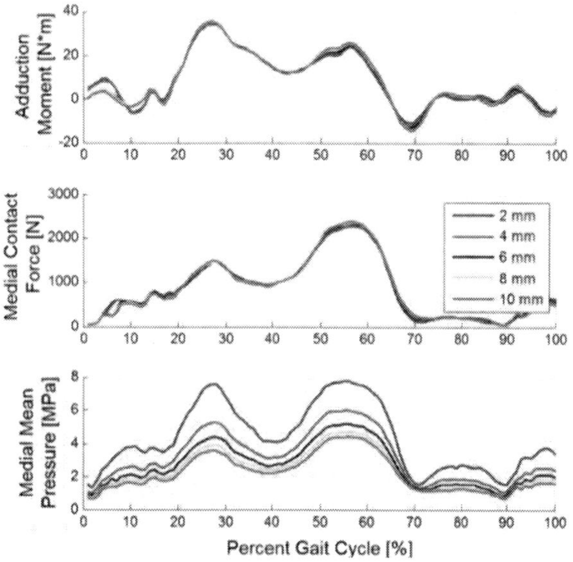

Figure 2 - Knee adduction moment, medial contact force and medial mean contact pressure over a simulated gait cycle for various cartilage thicknesses

The nominal simulation results indicate that the KAM shows a similar temporal trend over the gait cycle to the medial contact force and the mean medial cartilage pressure. However, as the thickness of the cartilage is varied from 2 mm to 10 mm the mean cartilage pressure varies by a factor of 2 while the knee adduction moment remains virtually unchanged (Figure 2). The medial contact force also shows nearly no change. The same results were seen for variations in the cartilage elastic modulus, where the mean pressures showed substantial changes over the gait cycle between elastic modulus values, but the KAM and medial contact force remained unchanged (not shown).

Solving the ESO routine with different cartilage thicknesses and elastic moduli had a relatively small effect on the frontal plane knee alignment, inducing only subtle changes in the KAM. However, cartilage properties substantially altered both the mean contact pressure and percent cartilage deformation (Figure 3).

Decreasing cartilage elastic modulus from 25 to 5 MPa reduced the mean contact pressure from 8 to 5 MPa, but increased the cartilage deformation from 9% to 22%. In contrast, decreasing the combined cartilage thickness from 10 to 2 mm resulted in increases in both the mean contact pressure (4.2 to 7.6 MPa) and the mean cartilage deformation (20% to 30%).

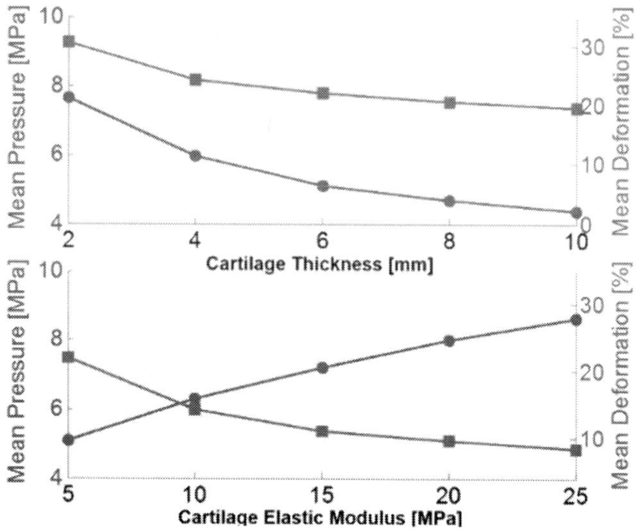

Figure 3 – The effect of cartilage thickness and elastic modulus on medial tibial cartilage mean pressures and mean deformations (depth of cartilage overlap/cartilage thickness*100%) at the second peak of tibiofemoral force during gait

DISCUSSION

We demonstrated the capability of our 12 DOF knee model and EOS simulation routine to predict cartilage contact pressures during movement. This novel approach allows calculation of localized cartilage pressures in the knee joint and has great promise for better predicting the development of OA.

Our simulation results indicate that the KAM does show the same temporal trends as the medial contact force and mean pressure over a gait cycle. However, the KAM and medial contact force are not able to predict changes in pressure that arise from changes in material properties of the cartilage. These results are similar to studies of loads measured from instrumented knee implants which showed the trend of the KAM correlates well to the medial knee contact force [4], however the KAM was less able to predict changes in the medial knee contact force associated with gait modifications [5].

The sensitivity of cartilage contact pressure and deformation to cartilage properties is relevant to consider in the context of OA. As OA progresses, cartilage becomes thinner and the collagen network is damaged leading to a decrease in elastic modulus [6]. Our results suggest these changes can alter both the localized cartilage tissue stresses and strains, potentially leading to further cartilage damage and degeneration. Further study is needed to assess how these observations vary with alternative gait patterns and different cartilage surfaces.

ACKNOWLEDGEMENTS NIH EB015410

REFERENCES

[1] Sharma, L et al., *Arthritis & Rheum.*, 41:1233-40, 1998.
[2] Arnold EM et al., *Ann of Biomed Eng*, 38:269-79.
[3] Lenhart RL et al., *WCB*, 2014.
[4] Zhao, D. et al., *J. Orthop. Res.*, 25:789-797, 2007.
[5] Walter, JP et al., *J. Orthop. Res.*, 28:1348-54, 2010.
[6] Silver, FH et al., *Matrix Biol.*, 21:129-137, 2002

ULTRASOUND ASSISTED HUMAN MESENCHYMAL STEM CELL CHONDROGENESIS: ENGINEERING LARGE-SCALE CARTILAGE GRAFTS

Anu Subramanian (1), Sanjukta Guha Thakurta (1), Neety Sahu (1), Abdul Chama (1) and Hendrik J. Viljoen (1)

(1) Department of Chemical and Biomolecular Engineering, University of Nebraska, Lincoln, NE USA

INTRODUCTION

Tissue engineered cartilage (TEC) offers an exciting modality for repairing large defects, as cell-free approaches may not be sufficient to enable the repair of large defects. Beyond the biological challenges that must be addressed to treat such joint disorders, it remains a significant engineering challenge to generate cartilage grafts with large dimensions. Central to a successful tissue engineering strategy to grow functional tissue equivalents is a bioreactor that maintains cells seeded on biodegradable polymeric scaffolds and provides essential gas and nutrient transport, as well as the mechanical stimuli necessary to promote extracellular matrix (ECM) synthesis. Previous research has shown that viability, homogeneity, and quality of engineered tissue constructs can be improved when research-sized constructs (~6mmx3mm) were cultured in perfusion and rotation bioreactors versus conventional static culture methods. Till date tissue engineering has not met the requirements for large-sized constructs. But ultra-sound may prove to be a powerful tool in our effort to produce large grafts; notably US was noted to be chondroinductive and helps maintain the chondrocyte phenotype[1-4]. However, to fully capitalize on these positive findings and address the need for high quality large grafts, it is imperative to address the question of scale: how can mechanical stimulation be transferred from a transducer to the cells in a scaffold to optimize ECM production and quality? Whilst in situ measurements of the macroscopic medium are feasible, it is not possible to probe conditions either inside the scaffold or cells, except by dissecting and staining specimens. Thus the practical approach calls for mathematical models to assist with the US bioreactor design. Thus first, a systematic approach is described to evaluate the power delivery to the bioreactor and calculate the US field in the different media that constitute the bioreactor, including the porous scaffold that is typically used to support tissue growth. A KLM model is used to relate the pressure (and velocity) at the transducer/liquid interface with the input voltage. A Fourier transform (FT) of the beam enables us to present the input field as a series of plane waves and the inverse FT of the combined result is then used to construct the complete acoustic field in the bioreactor. Several valuable insights are gained from the model: it provides a quantitative picture of the mechanical energy distribution in the reactor and therefore the amount of energy that cells are exposed to[2,3]. Finally, we propose here, a continuous method to construct large-sized TEC by first directing stem cell differentiation to chondrocytes by controlling the microenvironment and stimulation with ultrasound (US) in a bioreactor developed at UNL and secondly, maintaining hMSC-derived chondrocytes under optimal US stimulation in the bioreactor for a longer duration to allow matrix deposition.

METHODS

hMSCs were purchased from Lonza Inc (Walkersville, MD), expanded in α-MEM basal media supplemented with 10% FBS and antibiotics. Passage 4 hMSCs were seeded onto pre-wetted Biomerix™ scaffolds (~23 mmx 3mm) at $4x10^6$ cells/scaffold and cultured in the automated US-assisted bioreactor developed at UNL-Lincoln. US stimulation was at 14 kPa (5 MHz, 2.5 Vpp), 5 mins/application and at 6-applications/day. Following four study groups were adopted: Group 1: US(-),TGFβ3(-); Group 2: US(+),TGFβ3(-);Group 3: US(+),TGFβ3(+); where 10 ng/ml of TGFβ3 was included for the first two-weeks of culture and constructs were cultured for a total of 8-weeks. At indicated time points, cell-seeded scaffolds were randomly retrieved and assayed. Cellular viability was determined by Live-Dead™ assay kit, total collagen and total glycosaminoglycan (GAG) were measured via the DMMB assay and hydroxyproline assay kits and normalized to dsDNA. Viscoelastic properties of constructs were measured with BOSE Electroforce 3200, dynamic stiffness was determined from unconfined compression testing. Histological and immunohistochemical sections were processed using standard histological procedures at the Tissue Science Facility, University of Nebraska Medical Center (Omaha, NE). Both ANOVA and student t-test were used to compare all study groups.

RESULTS

While US has been employed in both in vitro and in vivo applications to modulate cell functions, we are the first group to have shown that the optimal mechanical coupling of the US signal to cellular and nuclear processes that control proliferation and differentiation occurs at the primary resonant frequency(~5.2±0.8

MHz) of the cell[2]. Through the mathematical analysis of natural joint movement, we have also shown that ultrasound effectively reproduces the mechanical stimulation experienced by chondrocytes under physiological conditions and the typical chondrocyte oscillates with a frequency close to the model predicted resonant frequency (~5.6 MHz) following impact. Taking theoretical and experimental results together suggests that US at 5 MHz is an effective modulator of chondrogenesis. To capitalize on the positive bioeffects of low-intensity-US (LIUS) and applying them to the field of cartilage tissue engineering, our laboratory has designed and developed a bioreactor configuration that uses US to stimulate chondrocytes maintained in in-vitro culture over a range of US stimulations[4]. Aspects of ultrasound that would negatively affect cells, including temperature and cavitation, were shown to be insignificant for the ultrasound protocols used covering a wide range of frequencies and pressure amplitudes, including the ones proposed here.

As the US bioreactor is central to the generation of cartilage constructs, we have undertaken a modelling analyses to understand how the acoustic pressure distribution changes when cell infiltration progresses and systematically evaluated the power delivery to the bioreactor and calculated the US field in the different media that constitute the bioreactor, including the porous scaffold that is used in this proposal. To do so, we have resorted to Biot theor. Figure 1 (Panel A) shows a color map of the pressure when the scaffold is not populated with cells – note the excellent distribution of pressure (and concomitantly mechanical energy) in the scaffold. Panel B depicts a scaffold that is almost totally infiltrated. Interestingly, the populated scaffold presents a better US conducting medium and pressure peaks are actually higher than in the surrounding water.

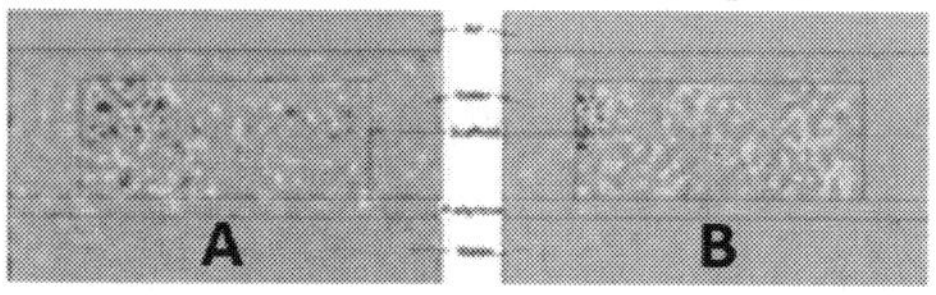

Fig: 1

A mathematical model was developed that combines a transducer model with a numerical scheme for predicting the ultrasonic field to correlate the pressure amplitude in the bioreactor with the transducer input voltage and current. In our configuration, the total distance between the transducer face and the water-air interface is approximately 20 mm. Thus, the bioreactor resides completely within the near-field of the transducer (the far-field begins at approximately a^2/λ≈130mm when f=5MHz). Several valuable insights are gained from the model: it provides a quantitative picture of the mechanical energy distribution in the reactor and the amount of energy that cells are exposed to (Fig.2). Our results show that the ultrasonic beam has very high directivity and the pressure field within the scaffold is fairly evenly distributed provided the diameter of the scaffold is larger than that of the transducer; this finding sets an obvious design condition which we will incorporate as we culture constructs with larger diameters[1].

Fig: 2

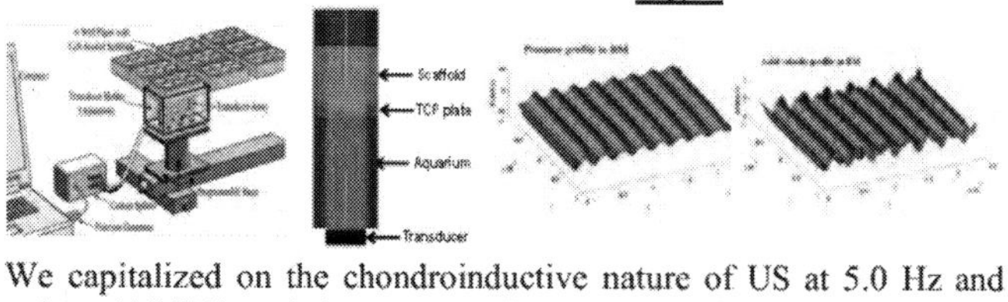

We capitalized on the chondroinductive nature of US at 5.0 Hz and cultured hMSC seeded constructs (25 x 3mm) in the US-bioreactor for 8-weeks, where non-stimulated constructs served as controls.

Fig: 3

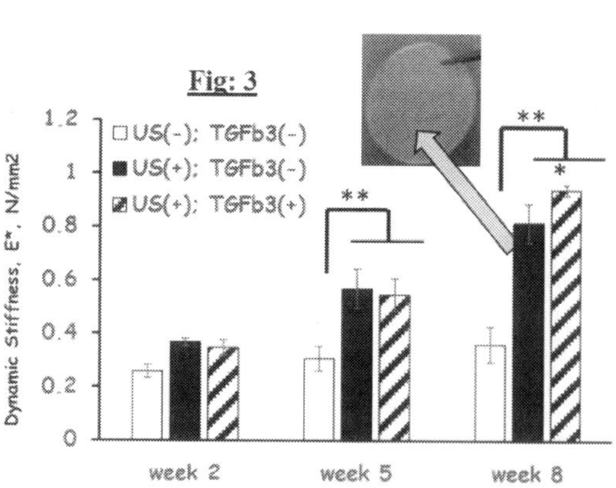

Viscoelastic properties of constructs generated were measured in fully submerged condition using BOSE electroforceTM, as detailed in task-2 and the dynamic stiffness shown in Figure 3. At week-8, constructs cultivated under US [US(+) TGF]3has a dynamic stiffness of 0.94 ± 0.02 MPa and permeability of 5.58x10-11 m4/N.s. In general, constructs form the control group [US(-) TGFB3(-)] had properties similar to wet cell-free scaffolds. Total GAG, collagen and chondroitin sulfate (CS) were also assayed and shown in figure 4. Interestingly, GAG and CS deposition was only noted in US treated constructs. Large-scale constructs could easily be handled with forceps and did not bend under their own weight (Fig.3); but could be bent when applying force with the forceps, suggesting their potential to be molded to the contours during surgical implantation.

Fig: 4

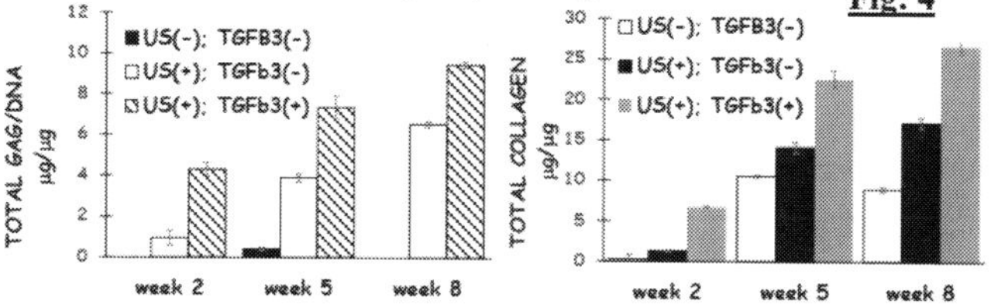

DISCUSSION

This work describes a novel paradigm for the generation of large-scale cartilage grafts starting with human mesenchymal stem cells, where the phases of cell-seeding, chondrogenesis of hMSC, culture of hMSC-dervied chondrocytes are performed directly of a 3D scaffold in an US-assisted bioreactor. Modeling was employed during bioreactor development as part of a rational design strategy, leading to US stimulation regimens in the bioreactor that facilitated uniform stimulation throughout the large scaffold volume. While generation of large-scale grafts from adult chondrocytes has been reported, to our knowledge, this is the first report of engineered human cartilage constructs generated from hMSC. The 25 mm constructs engineered in our bioreactor had 5-10% of the dynamic stiffness of native cartilage tissue. The culture conditions could be further optimized to further increase construct maturity. In future work, the further integration of cell signaling investigations within US-bioreactors will lead to a quicker realization of the promise of tissue engineering and regenerative medicine.

ACKNOWLEDGEMENTS

Support from Nebraska-Research-Initiative is sincerely acknowledged.

REFERENCES

[1] Guha Thakurta et al., *Acta Biomater,* 10: 4798–4810, 2014.
[2] Louw et al., *Ultrasound Med Biol,* 39:1303–1319, 2012.
[3] Louw et al., *Ultrasound Med Biol* 10.1016 /j.ultrasmedbio.2014.12.015.
[4]Subramanian et al., *Tissue Eng Part C Methods,19:244–255, 2013.*.

SB³C2015
Summer Biomechanics, Bioengineering and Biotransport Conference
June 17-20, 2015,Snowbird Resort, Utah, USA
SB³C2015-670

MICRO POROSITY DOMINATES BONE GROWTH FOR LARGE AND SMALL MACROPORES IN BCP SCAFFOLDS

Amy J. Wagoner Johnson (1,2), Laurie E. Rustom (2), David. J. Hoelzle (3), Mark D. Markel, Brett Nemke, Yan Lu (4)

(1) Department of Mechanical Science and Engineering
University of Illinois at Urbana-Champaign
Urbana, IL, USA

(2) Department of Bioengineering
University of Illinois at Urbana-Champaign
Urbana, IL, USA

(3) Department of Aerospace and Mechanical Engineering
University of Notre Dame
Notre Dame, IN, USA

(3) School of Veterinary Medicine
University of Wisconsin
Madison, WI, USA

INTRODUCTION

Repair of bone defects remains a major challenge despite substantial efforts over the past 25 years. Bone defects are challenging or impossible to treat for a large segment of the population and can result in significant pain, disability and disfigurement. Porous materials that are calcium phosphate (CaP)-based are well-established candidates for bone repair due to their osteoconductive, and sometimes osteoinductive, properties. Many researchers have shown that these materials are more effective as bone scaffolds if they contain both macroporosity (> 100 μm) and microporosity (< 20 μm), though the mechanism responsible for their efficacy remains under debate[1]. Our research uses biphasic calcium phosphate (BCP) scaffolds that are fabricated by 3D printing of a colloidal ink. They consist of layers of orthogonal rods, the spacing of which forms the macropores. The rods themselves can be microporous or solid. Scaffolds with microporous rods are referred to as microporous (MP) and those with solid rods are referred to as non-microporous (NMP). Within research on microporosity, we have shown that microporosity in the size range of 5-10μm that is within scaffold rods generates capillary forces that draw cells into the micropores[2]. Cells in the micropores of MP scaffolds form bone in the micropores[3], and these scaffolds show enhanced bone regeneration over NMP scaffolds[4-6]. We hypothesized that this effect would still dominate with macropores approximately twice as large as used previously. We further hypothesized that such microporosity could be used to pattern bone in scaffolds containing microporous (MP) and non-microporous (NMP) domains.

METHODS

Five types of BCP scaffolds, 8 mm in diameter, were fabricated according to established procedures[7]. These included four types of monolithic scaffolds: MP and NMP scaffolds, each with either "large" or "small" macropores. "Small" refers to rod spacing used in previous work[3-6], on the order of the rod diameter (~400 μm), while "large" refers to spacing approximately twice that. The fifth type of scaffold contained four domains, each with the macro and micropore characteristics of one of the four monolithic scaffolds (e.g. see Fig. 2). These four-domain scaffolds are referred to as "Quad" scaffolds. All scaffolds were implanted into the ramus of pig mandibles for 3 weeks ($n \geq 7$, monolithic; n=5, Quad) and then scanned using microcomputed tomography (micro-CT). A custom algorithm was used to segment tissue and scaffold, and the average bone volume fraction in the available space was quantified and compared for all scaffolds.

Bone distribution was also quantitatively compared for monolithic scaffolds using a ½ bone fraction (½BF) value. This is defined as the fraction of bone contained within half of the fractional volume of the defect, measured from the defect center. For a scaffold that has bone that is perfectly uniformly distributed, we expect this ½BF to be 50%. Bone distribution for Quad samples was visualized using heat maps. Here, 500-image stacks of data from a single sample were collapsed to generate heat-maps, with the hottest colors signifying the regions with highest bone fraction.

Student's t-tests were used to statistically compare the effects of microporosity and macropore size on bone volume fraction in the available space and the bone distribution. All MP scaffolds or domains were compared to all NMP scaffolds or domains, regardless of macropore size. Similarly, all scaffolds with small macropores were compared to all scaffolds with large macropores, regardless of microporosity.

RESULTS

Microporosity significantly influenced the fraction of bone in the available space ($p \ll 0.01$) in monolithic scaffolds, consistent with our previous work[4–6] (Fig. 1a). However, macropore size did not (p=0.21). Monolithic MP scaffolds had an average bone fraction of 38.4 +/- 8.5% compared to 25.1 +/- 5.2% for NMP scaffolds. Monolithic scaffolds with small macropores had an average bone fraction of 32.9 +/- 9.7% compared to 29.5 +/- 9.8% for those with large macropores.

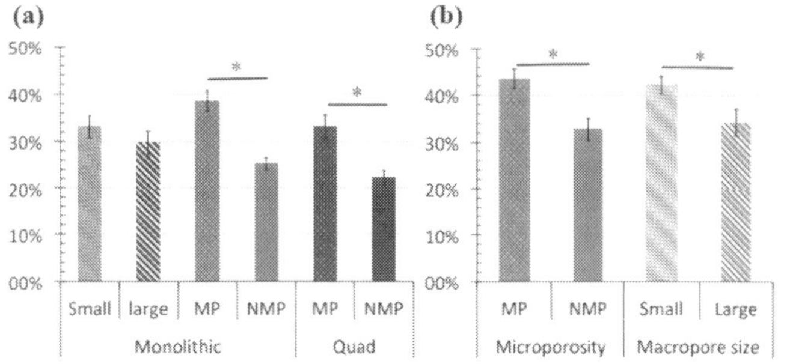

*FIG. 1: **(a) Fraction of bone in available space.** Microporosity significantly influenced the fraction of bone (p<<0.01) in monolithic scaffolds; macropore size did not (p=0.21). Bone fraction was significantly greater in MP domains as compared to NMP domains (p=0.003) in Quad scaffolds. **(b) Bone distribution in monolithic scaffolds.** Distribution was significantly more uniform in monolithic MP scaffolds (p<0.001) and in monolithic scaffolds with small macropores (p=0.01). (a,b) Bone fraction in monolithic MP and NMP scaffolds was similar to that of MP (p=0.28) and NMP domains (p=0.11) in Quad samples.*

Bone distribution was significantly more uniform in monolithic MP scaffolds than in NMP scaffolds (p<0.001, Fig. 1b). In MP scaffolds, the ½BF was 43.6 +/-8.0% as compared to 32.7 +/-9.3% in NMP scaffolds. In comparing the effect of macropore size on bone distribution in monolithic scaffolds, scaffolds with small macropores also had a more uniform bone distribution as compared to those with large macropores (p=0.01). The ½BF for scaffolds with small macropores was 42.2 +/-7.4% compared to 34.2 +/-11.4% for those with large macropores.

In Quad scaffolds (Fig. 2), MP domains had a significantly higher bone volume fraction in the available space as compared to NMP domains (p=0.003, 1a). Fig. 2c shows a representative heat map of a Quad sample indicating the density of bone throughout the scaffold. MP and NMP domains had an average bone volume fraction of 32.9 +/- 5.7% and 22.1 +/- 3.5%, respectively. Interestingly, bone fraction in available space in monolithic MP and NMP scaffolds is comparable to that in MP and NMP domains, respectively, of Quad samples (p=0.28 and p=0.11, respectively).

DISCUSSION

As observed in previous work, microporosity enhanced bone volume fraction in BCP scaffolds that also had macropores, and it also improved bone distribution. Here, this effect was observed regardless of macropore size. This confirms our hypothesis that the effect of micropores dominates even with larger macropores.

Macropore size did not influence bone volume fraction for the two macropore sizes tested. However, the "large" macropore size decreased the uniformity of bone resulting in less bone at the center. This result might indicate the existence of a cut-off macropore size above which bone is unable to fill the entire defect.

Results from Quad scaffolds indicate combining regions with different microporosities in the same scaffold allows us to direct and pattern bone growth. The similarity of results obtained in MP/NMP Quad domains to those in MP/NMP monolithic scaffolds may indicate that domains in multi-domain scaffolds do not influence growth in adjacent domains. However, the relatively low number of samples in this study may have prevented the detection of an interaction.

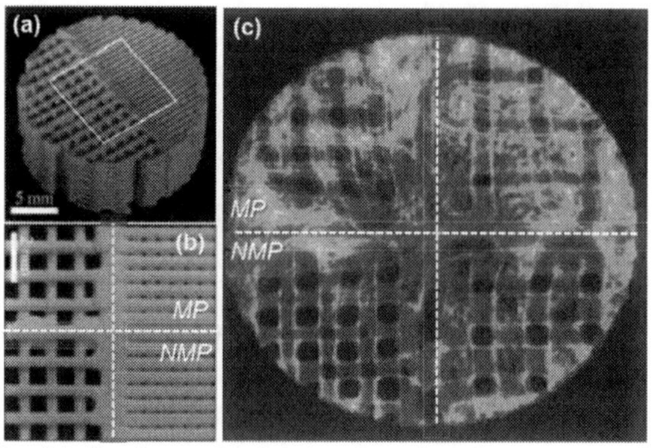

*FIG. 2: **(a) Quad sample showing four domains corresponding to the four monolithic scaffolds.** These included microporous (MP) and nonmicroporous (NMP) scaffolds, shown in the top and bottom of (b), respectively, with large and small macropores, shown on the left and right of (b), respectively. **(c) Bone distribution in the four domains.** Here, the small macropores are on the left and the large are on the right. The MP domains had significantly more bone than the NMP domains (p=0.003). (a,b) adapted from[7].*

Results from the comparison of "large" and "small" Quad domains were inconclusive, perhaps due to manufacturing defects in the Quad scaffolds. However, since results from the monolithic scaffolds suggest that scaffolds may only support a uniform growth below a certain macropore size, the incorporation of domains with different macropore sizes, below the cut-off value, in multi-domain scaffolds may allow to customize the density of the bone-scaffold composite.

CONCLUSIONS

This *in vivo* study quantified bone regeneration using micro-CT in four monolithic scaffolds, MP and NMP with large and small macropores and one scaffold with four quadrants, each corresponding to one of the monolithic scaffolds. The research demonstrates that, consistent with previous work, microporosity increases the fraction of bone and improves bone distribution in monolithic scaffolds. However, macroporosity does not significantly change bone fraction for the two sizes studied here. MP domains in Quad samples have a significantly higher bone fraction as compared to those that are NMP. The bone fraction in MP and NMP domains in Quad samples was not significantly different than in the corresponding monolithic scaffolds. Finally, microporosity may be used to direct growth in scaffolds containing MP and NMP domains. The work has important implications in scaffold design and in understanding factors that influence bone regeneration *in vivo* at both the macro and microscales.

ACKNOWLEDGEMENTS

This work was supported in part by the AO Foundation, Grant S-11-17W, Switzerland.

REFERENCES

1. Chai, Y. C. *et al. Acta Biomater.* **8,** 3876–87 (2012).
2. Polak, S. J., *et al. Acta Biomater.* **9,** 7977–7986 (2013).
3. Lan Levengood, S. K. *et al. Biomaterials* **31,** 3552–3563 (2010).
4. Woodard, J. R. *et al. Biomaterials* **28,** 45–54 (2007).
5. Lan Levengood, S. K. *et al. Acta Biomater.* **6,** 3283–3291 (2010).
6. Polak, S. J. *et al. Acta Biomater.* **7,** 1760–71 (2011).
7. Hoelzle, D. J. *et al. J Biomech Eng* **133,** 101001 (2011).

SB³C2015
Summer Biomechanics, Bioengineering and Biotransport Conference
June 17-20, 2015, Snowbird Resort, Utah, USA
SB³C2015-671

DIRECT ESTIMATION OF THREE-DIMENSIONAL DEFORMATION GRADIENT TENSORS FROM VOLUMETRIC ULTRASOUND DATA

John J. Boyle (1), Roger Rowe (2), Frederick W. Damen (3), Arvin H. Soepriatna (3), Robert B. Pless (2), Craig J. Goergen (3), Stavros Thomopoulos* (1,2), Guy M. Genin* (2) (*contributed equally)

(1) Department of Orthopaedic Surgery
Washington University in St. Louis
St Louis, MO 63110
USA

(2) School of Engineering and Applied Science
Washington University in St. Louis
St Louis, MO 63108
USA

(3) Weldon School of Biomedical Engineering
Purdue University
West Lafayette, IN 47907
USA

INTRODUCTION

Mechanical characterization of inhomogeneous and/or geometrically complex three-dimensional biological tissues requires precise and accurate determination of strain fields. Digital image correlation is a well-established technique for determining strain fields on the surfaces of deforming materials. The technique involves matching patterns between pairs of images to estimate the displacement of certain regions or features on a sample [1]. The most advanced of these techniques, although limited to two-dimensional (2D) manifolds, have proven effective for applications such as tracking deformations in dynamic magnetic resonance (MR) images of the heart and brain [2]. However, artifacts from out-of-plane motion severely limits strain tracking in 2D slices from volumes imaged using MR, ultrasound, or confocal imaging. We therefore developed the first image warping technique for full volumetric data sets with the goal of reliably determine strains within tissue volumes.

Traditionally, strain calculations are performed after digital image correlation by binding the midpoints of matched regions to form elements. Then, the initial and displaced positions of the midpoints are used to estimate the deformation gradient tensor, **F**, in each element. **F** relates a material vector $d\mathbf{X}$ in the undeformed reference configuration to the corresponding spatial vector $d\mathbf{x}$ in the deformed configuration using a least squares fit (LSF) [2]:

$$d\mathbf{x} = \mathbf{F}\,d\mathbf{X} \tag{1}$$

The Green-Lagrange strain tensor, **E**, can then be then calculated as:

$$\mathbf{E} = \tfrac{1}{2}(\mathbf{F}^T\mathbf{F} - \mathbf{I}) \tag{2}$$

where **I** is the second order identity tensor.

The method we developed improves upon the state of the art in two ways, and is an extension of our earlier 2D approach [3]. First, the algorithm uses non-rigid rather than rigid image registration. Rigid registration uses cross correlation of undeformed and deformed regions to estimate the displacements of the midpoints, which are then used to estimate **F.** We previously showed, as others have, that a non-rigid approach improves estimates of displacements [3]. Non-rigid approaches improve mapping of displacements by optimally warping the undeformed image before cross-correlation through iterative minimization of an energy function:

$$\sum[I(\mathbf{W}(\mathbf{x};\mathbf{p})) - T(\mathbf{x})]^2 \tag{3}$$

where $T(\mathbf{x})$ is a template image, and $I(\mathbf{W}(\mathbf{x};\mathbf{p}))$ is an image, I, warped by a defined warping function $\mathbf{W}(\mathbf{x};\mathbf{p})$ whose warping parameter $\mathbf{p}$ can be modulated.

The second improvement that we made in 2D was to develop a warping function analogous to the deformation gradient tensor to directly estimate the strain. Using the following warping function:

$$\mathbf{W}(\mathbf{x};\mathbf{p}) = \begin{bmatrix} (1+p_1) & p_3 & p_5 \\ p_2 & (1+p_4) & p_6 \\ 0 & 0 & 1 \end{bmatrix}\begin{bmatrix} x \\ y \\ 1 \end{bmatrix} \tag{4}$$

F for the region was directly extracted from:

$$\mathbf{F} = \begin{bmatrix} (1+p_1) & p_3 \\ p_2 & (1+p_4) \end{bmatrix} \tag{5}$$

Directly estimating strain from the warping function resulted in a ~3 order-of-magnitude improvement in both accuracy and precision compared to standard cross-correlation techniques [3].

In the following, we expanded upon these techniques to utilize a 3D warping function rather than a 2D one. The new method allows for precise characterization of 3D strains in a volumetric data by directly estimating deformation gradient tensors from the intensity data.

METHODS

Theory. Using the same energy function as (3), we replace the 2D image warping function (4), with the volumetric warping function:

$$\mathbf{W}(\mathbf{x};\mathbf{p}) = \begin{bmatrix} (1+p_1) & p_4 & p_7 & p_{10} \\ p_2 & (1+p_5) & p_8 & p_{11} \\ p_3 & p_6 & (1+p_9) & p_{12} \\ 0 & 0 & 0 & 1 \end{bmatrix} \begin{bmatrix} x \\ y \\ z \\ 1 \end{bmatrix} \quad (6)$$

The 3D deformation gradient tensor is then directly extracted from the first nine components of the warp:

$$\mathbf{F}_{3D} = \begin{bmatrix} (1+p_1) & p_4 & p_7 \\ p_2 & (1+p_5) & p_8 \\ p_3 & p_6 & (1+p_9) \end{bmatrix} \quad (7)$$

Generation of Test Data with Known Strains. The new method was tested on a series of artificial image volumes of a simulated contracting spherical cell in a linear, isotropic extracellular matrix (ECM). ECM was modeled visually as a randomly generated volume of random ellipsoids. This ECM was then strained according to Eshelby's exact solution for strains around a contracting spheroid [4]. To simulate ultrasound images and assess performance when image volumes are noisy, random noise was added to the image voxels at increasing percentage levels of the voxel intensity range.

Experimentally Measured Ultrasound Data of a Beating Mouse Heart. Ultrasound images of a beating mouse heart were captured with a high frequency ultrasound system (40 MHz, Vevo2100, VisualSonics). Briefly, a linear step motor attached to the transducer was synchronized to an electrocardiogram (EKG) of the mouse heart. Ultrasound slices were captured over the duration of the heartbeat, and then the transducer was moved to capture an adjacent ultrasound slice. After the entire capture, the EKG data was used to synchronize the slices and create a full volume of a single heartbeat [5]. The Purdue Animal Care and Use Committee approved all procedures.

RESULTS

Three-dimensional strains measured with the new methods replicated the input strains used to generate the volumetric data of the synthetic developing Eshelby inclusion (Figure 1). Because the algorithm calculates strains over a fixed-size region, strain concentrations smaller than the region size are lost to averaging. However, strains over all distances greater than the voxel size were captured to within a few percent (Figure 2).

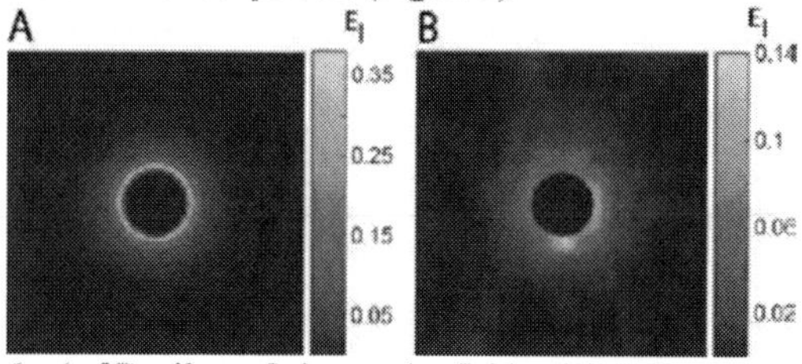

Figure 1. A 2D slice of the strain field arising from the Eshelby solution (A), and strains estimated from simulated images (B). The strain field was highly accurate over distances greater than the voxel size.

Artificially adding noise to image volumes of an Eshelby inclusion revealed that the methods were insensitive to noise (Figure 2). With 0.001 RMS noise, results were nearly identical to a noiseless image (red and blue lines, respectively). Adding 0.01 RMS noise decreased the accuracy and precision of the calculation by about twofold compared to that of a noiseless image (yellow and blue lines, respectively). Lastly, adding 0.2 RMS noise results in a three-fold decrease of accuracy and precision of calculation compared to that of a noiseless image (purple and blue lines). While the accuracy and precision of the calculation decreased by three fold from 0 to 0.2 RMS noise, the computation time with added noise increased 10-fold (data not shown).

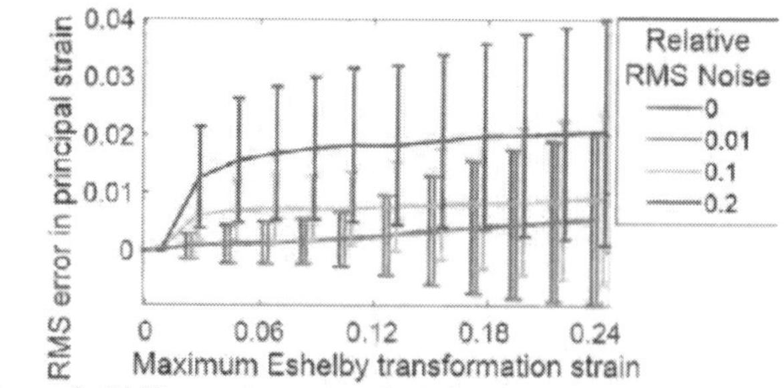

Figure 2. RMS error in strain calculation versus average strain for the Eshelby solution. Note that the maximum strain (x-axis) in the Eshelby solution surrounds the cell (Figure 1A).

Results for the ultrasound of a beating mouse heart demonstrate that measurement of strain from real ultrasound data is feasible (Figure 3). Strains measured with the new method are consistent with manually measured strains, confirming that the methods are appropriate for this type of data.

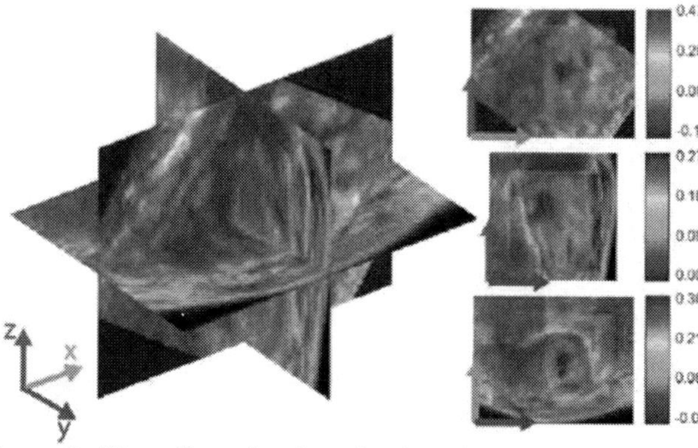

Figure 3: Three-dimensional strains in a beating heart, visualized using high frequency ultrasound.

DISCUSSION

Our results demonstrate, for the first time, a method that is fully capable of measuring complex, non-linear, three dimensional strain fields. Importantly, the insensitivity to noise of the technique is promising for use with ultrasound imaging, which is inherently noisy. We further demonstrated the robustness of the method by completing a full 3D strain analysis of a beating mouse heart captured via ultrasound. There are, however, several limitations with the method. Most importantly, it assumes that changes in strain over the timescale of the image frame-rate are small: any large sudden change in strain will result in a lack of convergence on a solution. This is particularly challenging when considering that most 3D imaging techniques are inherently slow. Our mouse heart data set overcomes this issue by capturing distinct high frame rate slices and synchronizing them to a biological signal to create a full volumetric dataset, rather than capturing the entire 3D volume at once. Future studies will evaluate the efficacy of the technique for human echocardiograms and for ultrasounds of isometrically loaded rotator cuffs.

ACKNOWLEDGEMENTS: This work was supported in part by the NIH through grants U01EB016422 and R01HL109505.

REFERENCES

]1] Schreier, H. W., Sutton, M. A. & Orteu, J.-J. Springer: 2009.
[2] Bayly, P.V.,et al. *J Biomech Eng* 126, 2004.
[3] Boyle J J, et al. *J Royal Society of the Interface*, 100, 2014.
[4] J. D. Eshelby, *Proceedings of the Royal Society A*, 241 1957.
[5] F.W. Damen et al. *World Congress of Biomechancs* 2014

SB³C2015
Summer Biomechanics, Bioengineering and Biotransport Conference
June 17-20, 2015,Snowbird Resort, Utah, USA
SB³C2015-672

INFLUENCE OF INTRACORTICAL POROSITY ON THE FRACTURE SUSCEPTIBILITY OF HUMAN CORTICAL BONE

Andrew P. Baumann, Travis L. Turnbull, Glen L. Niebur, and Ryan K. Roeder

Department of Aerospace and Mechanical Engineering
University of Notre Dame
Notre Dame, IN 46556
United States of America

INTRODUCTION

Intracortical porosity (Ct.Po) is increasingly recognized as critically important in predicting fracture risk and treating osteoporosis [1-3]. Elevated Ct.Po has been correlated with decreased elastic modulus, yield strength, and fracture toughness under static loading [4-6]. However, Ct.Po is poorly correlated with areal bone mineral density (aBMD) [7,8], suggesting that it may be able to account for gaps in fracture prediction by aBMD alone [9].

Pharmacological treatments for osteoporosis have varying effects on Ct.Po and cortical bone strength. Bisphosphonates, RANKL inhibitors, and cathepsin K inhibitors each inhibit osteoclastic resorption through different mechanisms. Bisphosphonates have been shown to preserve or decrease Ct.Po through suppressed intracortical remodeling, resulting in either maintenance or improvement of bone strength, respectively [10,11]. However, there has been debate as to whether or not long-term bisphosphonate treatment could lead to increased fracture susceptibility via increased mineralization levels and/or microdamage accumulation [12]. Thus, a more complete understanding of the relative influence of Ct.Po, mineralization, and microdamage on fracture risk is needed to improve assessment of fracture risk and the efficacy of pharmacological treatments for osteoporosis.

The objective of this study was to investigate the relative influence of Ct.Po, mineralization, and microdamage on the fracture susceptibility of human cortical bone (Fig. 1). A preliminary study using tissue from a single male donor showed that fatigue microcracks that initiated fracture were located near elevated Ct.Po but not elevated mineralization [13]. Therefore, the present study investigated tissue from multiple post-menopausal female donors and also investigated characteristics of Ct.Po that could be predictive of fracture risk.

METHODS

Ten fresh-frozen female human femora (NDRI) were evenly distributed across two age groups (50-64 and 80-94 y/o). A 4.5 cm segment of the diaphysis, centered at 75% of the total femur length, was excised from each femur, and wet-milled into paired cylindrical specimens from each anatomic quadrant, with a gauge section 3.0 mm in diameter and 10 mm in length (Fig. 1). Paired specimens were randomly divided into either a control or compressive fatigue group (n = 4/group/donor; N = 80 total). Four specimens were excluded from further analysis due to premature fracture. Specimens were hydrated in PBS at all times and stored at -20°C wrapped in gauze in airtight containers during interim periods.

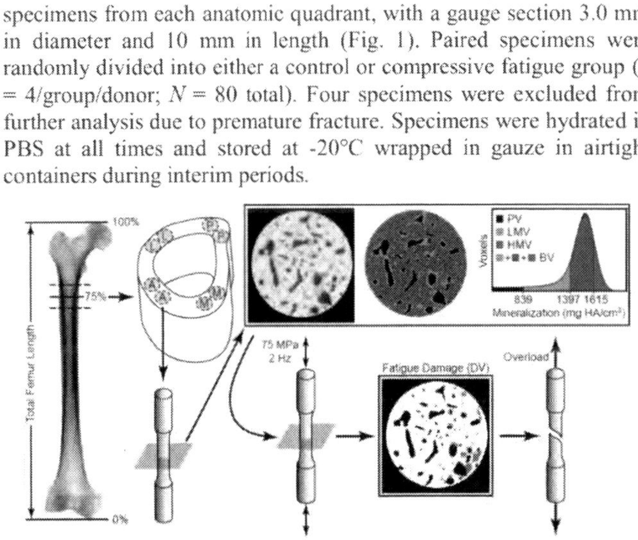

Figure 1. Paired cortical bone specimens were harvested from each anatomic quadrant of female human femora at 75% of total femur length. Specimens were sequentially imaged by micro-CT as-prepared, after fatigue loading, and after an overload to fracture. Thresholded images showed Ct.Po (PV), low mineralization (LMV), high mineralization (HMV), and fatigue microdamage (DV).

The entire gauge length of each specimen was sequentially imaged by micro-CT (Scanco µCT-80) at 10 µm resolution, 70 kVp voltage, and 113 mA current with 1000 projections at 600 ms integration time. Noise in grayscale images was reduced using a Gaussian filter (sigma = 0.8, support = 1). Pore volume (PV) was segmented from bone volume (BV) at a constant global threshold. Mineralization levels were segmented within the BV to identify the highest (HMV) and lowest (LMV) 20% of mineralization across all specimens (Fig. 1). Fatigue microdamage was created by loading (Bose ELF 3300) in cyclic uniaxial compression (75 MPa, 2 Hz, R = 0.1) at 37°C in PBS until achieving a 5% reduction in secant modulus. All specimens were stained by barium sulfate to label microdamage [13,14] and imaged again by micro-CT. Thresholded images revealed the damage volume (DV) which was registered with as-prepared images [13] (Fig. 1). Finally, all specimens were loaded in uniaxial tension at a crosshead speed of 25 mm/s to measure the elastic modulus, ultimate tensile strength, and work of fracture. Specimens were imaged again by micro-CT and registered with the previous image reconstructions to determine the fracture initiation site by the presence of a DV on the fracture surface.

Two-point correlation functions, $g(r)$, for encountering one microstructural feature (e.g., PV, LMV, HMV, etc.) as a function of the 3-D radial distance from another feature (e.g., the DV that initiated fracture, or all DVs) were calculated for each specimen [13]. Data were normalized to account for differences in the volume sampled at each radial distance, r, and the total volume of each feature of interest, such that $g(r) = 1$ for random probability.

Ct.Po characteristics were measured from as-prepared micro-CT images of specimens. The transverse pore area (Po.Ar) and centroid were used to compute an equivalent pore diameter (Po.Dia), pore spacing (Po.Sp), pore clustering (Po.Cl = coefficient of variation of Po.Sp), and radial pore location (Po.Lo).

RESULTS

The older age group exhibited greater Ct.Po, Po.Dia, and Po.Lo ($p < 0.01$, t-test), but Po.Sp and Po.Cl were not statistically significant. The fatigue group exhibited greater microdamage (DV/BV) than the control group ($p < 0.0001$, t-test), resulting in decreased elastic modulus, ultimate tensile strength, and work of fracture compared with the control group ($p < 0.005$, t-test).

Two-point correlation functions showed a greater than random probability for PV, and a random probability for HMV, located spatially adjacent to the fracture initiation site DV or all DVs (Fig. 2). However, the spatial correlation for PV was stronger for the fracture initiation site DV compared with all DVs.

The mean Po.Dia of pores associated with the fracture initiation site was significantly greater than the overall mean Po.Dia ($p < 0.0001$, t-test). Among the pooled fatigue specimens, 78% of fracture initiation sites were located adjacent to a pore with an equivalent diameter (Po.Dia) in the upper 50th percentile of the overall Po.Dia distribution (Fig. 3).

DISCUSSION

Microdamage (DV) at the fracture initiation site was spatially correlated with PV but not HMV (Fig. 2a). Therefore, fractures initiated at fatigue microcracks located near elevated Ct.Po but not elevated mineralization. This suggests that the net benefit of reduced or maintained Ct.Po due to a pharmacological treatment outweighs the net detriment of elevated mineralization. All damage sites (DVs) were also spatially correlated with PV but not HMV (Fig. 2b). This suggests that the fracture resistance of cortical bone is compromised by a critical subset of damage sites and associated pores. In fact, Po.Dia was predictive of fracture initiation sites (Fig. 3), suggesting that a larger Po.Dia was more likely to accumulate fatigue damage due to elevated local stress. Micro-CT based finite element models could be used to quantify these stress concentrations and predict fracture risk.

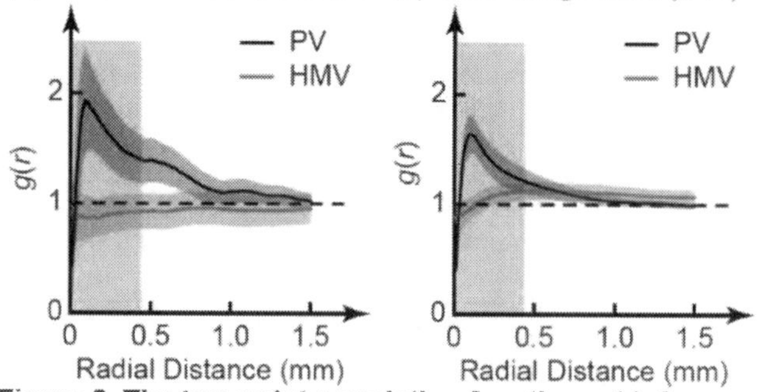

Figure 2. The two-point correlation function, $g(r)$, for pore volume (PV) and high mineral level volume (HMV) at radial distances away from (a) the fracture initiation site or (b) all damage sites (DVs) in fatigue loaded specimens showing the mean (trace) and 95% confidence interval (shading) for $g(r)$. Note that $g(r) = 1$ corresponds to random probability. A distance of 440 μm is shown (gray) as the typical distance for a microcrack to break free of microstructural barriers and propagate to macroscopic fracture [15].

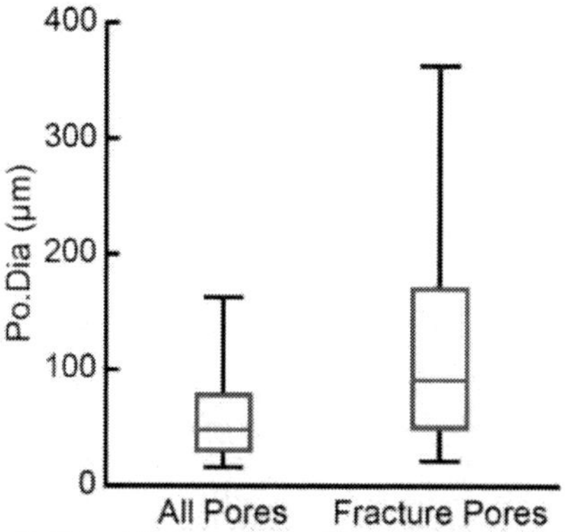

Figure 3. Box plot showing the equivalent Po.Dia distribution for all pores in fatigue loaded specimens (left) compared with pores associated with the fracture initiation site for each specimen (right). Boxes and whiskers show the 5th, 25th, 50th, 75th, and 95th percentiles.

ACKNOWLEDGMENTS

This research was partially supported by the Investigator Initiated Studies Program of Merck Sharp & Dohme Corporation.

REFERENCES

[1] Burghardt, AJ et al., *J Bone Miner Res*, 25:983-993, 2010. [2] Feik, SA et al., *J Anat*, 191:407-416, 1997. [3] Thomas, CD et al., *J Anat*, 206:115-125, 2005. [4] Schaffler, MB et al., *J Biomech*, 21:13-16, 1988. [5] Yeni, YN et al., *Bone*, 21:453-459, 1997. [6] Ural, A et al., *J Biomech Eng*, 129:625-631, 2007. [7] Zebaze, RM et al., *Lancet*, 375:1729-1736, 2010. [8] Nicks, KM et al., *J Bone Miner Res*, 27:637:644, 2012. [9] Sornay-Rendu, E et al., *J Bone Miner Res*, 22:425-433, 2007. [10] Borah, B et al., *J Bone Miner Res*, 25:41-47, 2010. [11] Burghardt, AJ et al., *J Bone Miner Res*, 25:2558:2571, 2010. [12] Allen, MR et al., *Bone*, 49:56-65, 2011. [13] Turnbull, TL et al., *J Biomech*, 47:3135-3142, 2014. [14] Landrigan, MD et al., *Bone*, 48:443-450, 2011. [15] Mohsin, S et al., *J Anat*, 208:81-89, 2006.

SB³C2015
Summer Biomechanics, Bioengineering and Biotransport Conference
June 17-20, 2015,Snowbird Resort, Utah, USA
SB³C2015-673

AN *IN VITRO* MICROFLUIDIC TUMOR MODEL TO MIMIC DUCTAL CARCINOMA *IN SITU*

Victoria Noe-Kim (1), Altug Ozcelikkale (1), Bumsoo Han (1,2)

(1) School of Mechanical Engineering
Purdue University
West Lafayette, IN, USA

(2) Weldon School of Biomedical Engineering
Purdue University
West Lafayette, IN, USA

INTRODUCTION

Ductal carcinoma *in situ* (DCIS), where malignant cells that show potential to acquire invasive properties form in the lining of the breast milk ducts, constitutes a significant part of newly diagnosed breast cancer cases. The commonly applied treatment remains to be excision of the diseased tissue that causes patient discomfort and can lead to loss of functionality [1]. Developing alternative non-invasive treatment modalities for DCIS is desired but still remains challenging due to lack of appropriate pre-clinical models that can simulate the disease's diverse pathophysiology, differential response to treatments, and patient-to-patient variations. Traditional *in vitro* models such as static cell cultured as monolayer, three-dimensional spheroids and tissue scaffolds are convenient, but insufficient for representing the physiological characteristics of DCIS tumor microenvironment, thus leading to failure in predicting clinical outcomes [2]. On the contrary, small animal models such as xenografts better represents *in vivo* tumors, but the environmental parameters are difficult to control [2, 3]. Hence, it is critical to develop a model system that can recapitulate physiologically relevant characteristics with high degree of control. In order to address this need, a new *in vitro* model was developed based on microfluidics that mimics tumor microenvironment of DCIS by forming a lumen structure with extracellular matrix (ECM) collagen gel and luminal epithelial cancer cells. The advantages of microfluidic devices are that it is reproducible, the environmental parameters are easily controlled, and it has potential in reducing cost of research if expensive reagents are utilized [2].

METHODS

Design of DCIS-on-Chip

A lumen structure to represent the milk duct in breast was generated along the microfluidic channel using a fluid dynamic phenomenon called viscous finger patterning as seen in Figure 1B. Based on this concept, as the less viscous fluid passes through the ECM hydrogel, it generates a hollow, tubular structure.

Once the lumen structure is produced, MCF-7 breast cancer cell line is seeded along the lining of the lumen. MCF-7 is selected because it is known to retain several characteristics specific to mammary epithelium [4].

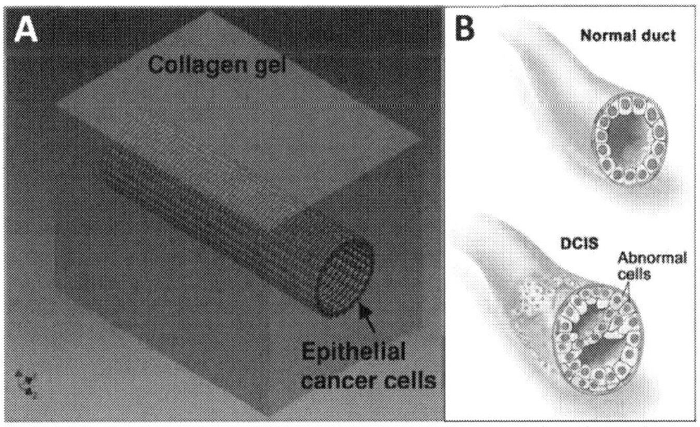

FIGURE 1: A) 3-DIMENSIONAL REPRESENTATION OF MCF-7 BREAST CANCER CELLS AT THE LINING OF LUMEN STRUC TURE FORMED THROUGH COLLAGEN GEL. B) SCHEMATIC VIEW OF MILK DUCT [5].

Device fabrication

The device consisted of a top layer with ports only and a bottom layer with the microchannel structure. It was fabricated using polydimethylsiloxane (PDMS) - curing agent mixture casted onto the mold. The PDMS-curing agent mixture was degassed and then cured on a hot plate. Using biopsy punches, the 2-mm diameter inlet port and the 3-mm outlet port were created on the top layer. Lastly, the two modified layers were bonded by treating the surface through plasma treatment. The details of the chip configuration are presented in Figure 2.

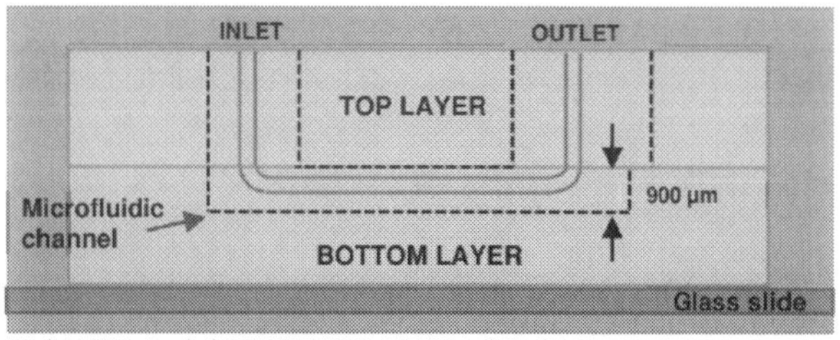

FIGURE 2: SCHEMATIC VIEW OF CHIP DEVICE FOR DCIS. LUMEN STRUCTURE REPRESNTED IN PINK.

Cell culture and hydrogel preparation

MCF-7/GFP cell line was trypsinized and prepared as cell suspension with seeding density of 20,000 cells/µL. The ECM was represented by type I collagen with final concentration of 6 mg/mL. In order to visualize ECM, the 3-µm fluorescent (Ex/Em = 543/592 nm) microbeads were mixed in the collagen solution.

Lumen patterning and cell seeding

Prior to generating the lumen, 6 mg/mL of collagen solution was added to the microchannel using a micropipette. Immediately, culture medium droplets were placed at the inlet port. The lumen formed instantly, and then the chip device is incubated at 37 °C for the collagen to polymerize and fix the structure. A droplet of cell suspension was flown through the lumen by passive pumping and let it sit in the incubator at 37 °C for 10 minutes and then another droplet was added. This time the device was rotated 90 degrees when placed in the incubator. This process was repeated two more times to cover the whole lumen surface [7].

RESULTS

A straight, tubular, and hollow structure in collagen hydrogel was consistently reproducible using passive pumping method. In order to visualize the lumen structure and MCF-7/GFP cells attached to its surface, confocal images of cancer cells and ECM hydrogel were combined producing a 3D reconstruction of the new *in vitro* model as demonstrated in Figure 3. It was evident that the lumen formed as a cylindrical, hollow structure as seen in Figures 3A and 3B. The average diameter of the lumen was approximately 350 µm. Figure 3C, a zoomed in image of Figure 3A, clearly showed a layer of MCF-7/GFP cancer cells that were seeded along the surface of the lumen. Moreover, the entire surface of the lumen was covered by MCF-7/GFP cells, as seen in Figure 3D which displays a z-projection of the cells and tubular structure in microfluidic channel.

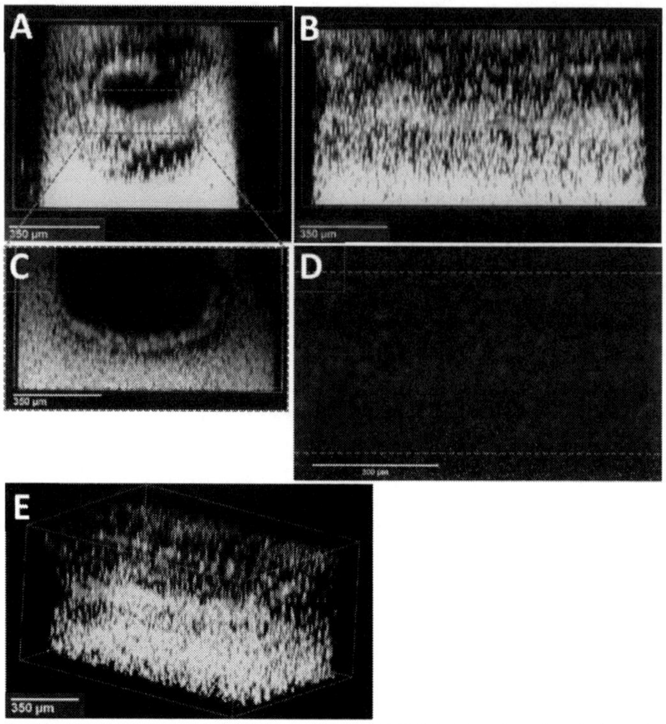

FIGURE 3: 3-DIMENSIONAL RECONSTRUCTION OF CONFO CAL IMAGES, REPRESENTING MCF-7 IN GREEN AND ECM IN ORANGE. A) FRONT, B) SIDE, C) ZOOMED IN, D) Z-PROJ ECTION, & E) ISOMETRIC VIEWS.

DISCUSSION

In conclusion, a new *in vitro* model of DCIS in microfluidic platform using viscous finger patterning was successfully developed. Furthermore, the lumen diameter, about 350 µm, in this study may be a good representation of milk ducts towards the periphery. A range of the duct diameters is measured to be 0.6 to 4.4 mm where ducts become smaller as it reaches periphery of the breast in asymptotic women [7]. Although this model may not demonstrate as accurate representation of *in vivo* models, it will effectively help bridge the gap between the traditional *in vitro* model and *in vivo* model. Finally, future studies will be directed toward co-culture system with other cell types such as stromal fibroblast to study epithelial mesenchymal transition (EMT), a process in which epithelial cells acquire invasive properties, and tumor heterogeneity.

ACKNOWLEDGEMENTS

This research was funded by CA 129287 from NIH/NCI, CTR grant from Indiana CTSI (via UL TR000006 from NIH), Grant CBET-1009465 from NSF and grants from Walther Cancer Foundation.

REFERENCES

[1] Siziopikou, K. P., *Arch. Pathol. Lab. Med.*, 137:462-466, 2013.
[2] Ozcelikkale, A. et al., *Mol. Pharm.*, 10:2111-2126, 2013.
[3] Richmond, A., and Su, Y., *Dis. Model. Mech.*, 1:78-82, 2008.
[4] Cell Biolabs, Inc., "MCF-7/GFP Cell Line," 2010.
[5] National Cancer Institute, http://www.cancer.gov, 2014.
[6] Bischel, L. L., et al., *Biomaterials*, 34:1471–1477, 2013.
[7] Geddes, D. T., 2009, *Int. Breastfeed. J.*, 4:4, 2009.

SB^{3}C2015
Summer Biomechanics, Bioengineering and Biotransport Conference
June 17-20, 2015,Snowbird Resort, Utah, USA
SB^{3}C2015-807

THE EFFECT OF VASCULAR CURVATURE ON BLOOD FLOW AND OXYGEN TRANSPORT IN ARTERIO-VENOUS FISTULAE

Francesco Iori (1), Lorenza Grechy (1), Richard W. Corbett (2), Wladyslaw Gedroyc (3), Neill Duncan (2), Colin G. Caro (4), Peter E. Vincent (1)

(1) Department of Aeronautics, Imperial College London, South Kensington, London, UK, SW7 2AZ

(2) Imperial College Renal and Transplant Centre, Hammersmith Hospital, London, UK, W12 0HS

(3) St Mary's Hospital, Praed Street, London, UK, W2 1NY

(4) Department of Bioengineering, Imperial College London, South Kensington, London, UK, SW7 2AZ

OVERVIEW

Arterio-Venous Fistulae (AVF) are arterio-venous connections commonly formed surgically in the wrist or upper arm of patients with End Stage Renal Disease (ESRD) who need long-term haemodialysis [1]. If they 'mature' (including dilating) and remain patent, blood flow through them increases substantially, thus providing suitable access for haemodialysis. Unfortunately, however, patency of AVF is often severely reduced by the inflammatory disease Intimal Hyperplasia (IH) and/or by thrombosis, causing unfavourable clinical outcomes, and significant additional costs for healthcare systems [2].

There is considerable evidence to suggest that the inherently unphysiological blood flow patterns within AVF can lead to development of IH and their ultimate failure. In this study the computational fluid dynamics package STAR-CCM+ was used to simulate unsteady blood flow and oxygen transport in various AVF configurations. Specifically, simulations were undertaken in idealized AVF configurations, as well as configurations obtained via 'virtual surgery' (based on CT scan data of parent arteries). Our objective was to understand how vascular curvature alters flow and oxygen transport patterns within AVF.

Our results show that arterial curvature has a dramatic impact on flow and oxygen transport within AVF. In particular, if the vein is connected to the outer curvature of an arterial bend then flow patterns within the artery are stabilized *c.f.* if the vein is connected to the inner curvature of an arterial bend, or a straight artery. Also, if the vein is connected to the outer curvature of an arterial bend then oxygen transport to the arterial wall is suppressed *c.f.* if the vein is connected to the inner curvature of an arterial bend, or a straight artery.

It is hoped understanding gained as a result of these studies can be used to design novel AVF configurations with 'favourable' blood flow patterns that achieve consistently high patency rates. Results may also help solve a range of other important healthcare problems; such as failure of stenting procedures, arterial bypass grafts, and organ transplants, due to unfavorable flow patterns and subsequent IH.

ACKNOWLEDGEMENTS

The authors are grateful for support from the National Institute for Health Research Imperial Biomedical Research Centre, the Imperial College Healthcare Charity, the Garfield Weston Foundation, CD-Adapco, the British Heart Foundation and the Engineering and Physical Sciences Research Council.

REFERENCES

[1] Konner, K et al., *J Am Soc Neph*, 14:1669-1680, 2003.
[2] Roy-Chaudhury, P et al., *J Am Soc Neph*, 17:1112-1127, 2006.

FINITE ELEMENT ANALYSIS OF THE BIOMECHANICAL CONSEQUENCES OF SCHMORL'S NODES

Boston B. Barham, Anton E. Bowden

Mechanical Engineering
Brigham Young University
Provo, UT, US

INTRODUCTION

Schmorl's nodes are intrusions of the disc into the vertebral endplate in the lumbar spine and are common in spinal imaging (Figure 1). They are significantly associated with the presence and severity of spinal disc degeneration [1], which is a major cause of work disability and a costly health care issue. There is a large variation in Schmorl's node topology and morphology, however they are most commonly round and located in the center of the vertebra. However, the clinical significance of Schmorl's nodes has remained unclear. The purpose of the present work was to utilize nonlinear finite element analysis (FEA) to identify the mechanical consequences of a Schmorl's node on intervertebral disc pressures and segmental kinematics.

METHODS

A previously validated and published nonlinear finite element model of the lumbar spine [2] was used as a baseline model for comparison (Figure 2). Material property data for the various components of the model were based on previously published data (Table 1). The model was previously validated against experimental and published data for quality of motion, range of motion, cortical strains, and disc pressures.

A second model was obtained by modifying the baseline FEA model using Altair Hyperworks. Utilizing the morphing function, a Schmorl's node was simulated at the top of the mesh elements of the L2 vertebra (Figure 2), patterned after a common round Schmorl's node geometry. This approach was

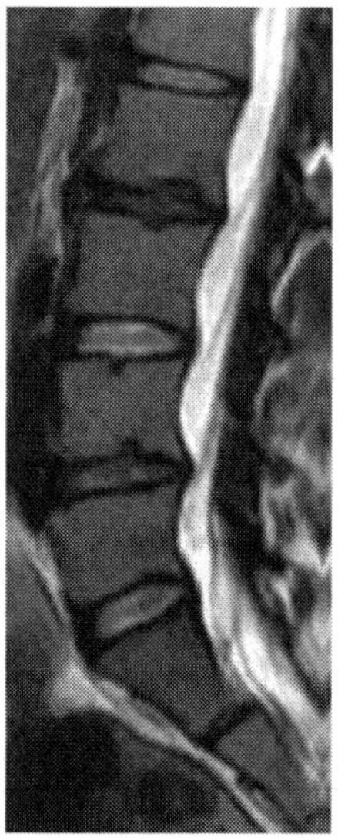

FIGURE 1: MRI SCAN SHOWING A VARIETY OF SCHMORL'S NODE MORPHOLOGY AND TOPOLOGY.

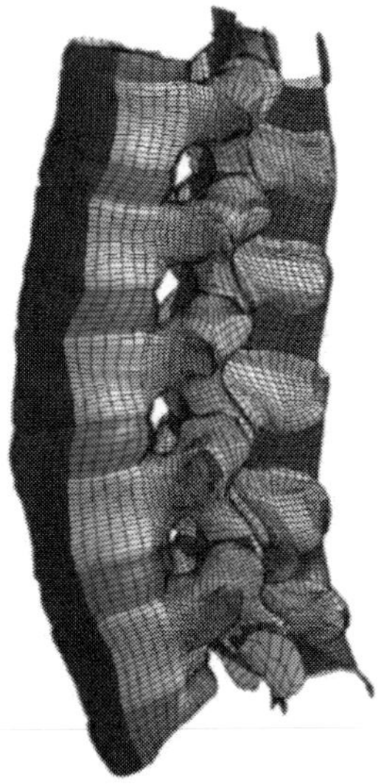

FIGURE 2: NON-LINEAR FINITE ELEMENT MODEL OF THE LUMBAR SPINE. [2]

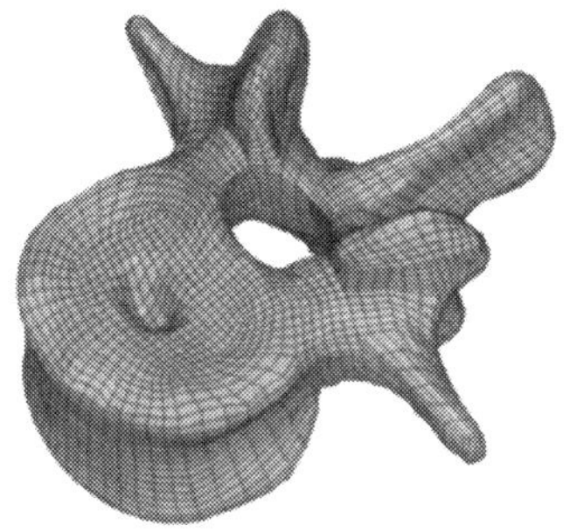

FIGURE 3: SIMULATION OF A COMMON SCHMORL'S NODE, MORPHED FROM THE CONTROL MODEL.

optimal because FE mesh topology and connectedness of elements was identically preserved between the two models with the exception of the Schmorl's node geometry. Identical material properties were also used for both models.

Boundary conditions were applied to the inferior T12 vertebral endplate and the superior S1 endplate. The S1 endplate was fixed and a 444 N compressive follower load was applied to simulate gravity loading of the upper torso as well as the muscular compressive loading [3]. The T12 endplate was then subjected to an applied extension moment of 7.5 Nm based on previously published work [2].

TABLE 1: MATERIAL PROPERTIES

Structure	Formulation	Modulus	Poisson's ratio	References
Cortical bone	Isotropic, elastic shell elements	12000	0.2	(Goel et al., 1995, Polikeit et al., 2003)
Cancellous bone	Density dependent anisotropic, elastic hex elements	E_z=4730ρ$^{1.56}$ E_x= 0.42E_z E_y= 0.29E_z	0.23, 0.4, 0.38	(Morgan et al., 2003, Ulrich et al., 1999)
Nucleus pulposus	Mooney-Rivlin hex elements	0.5, 0.05*	Incompressible	(Noailly et al., 2005, Bowden et al., 2008)
Inner annulus fibrosus	Anisotropic, elastic hex elements	5.6,0.34,0.19	1.86,0.88,0.14	(Elliott and Setton, 2001)
Outer annulus fibrosus	Anisotropic, elastic hex elements	17.45,0.27,0.19	1.77,0.33,0.14	(Elliott and Setton, 2001)

RESULTS

Disc pressures were calculated by averaging the pressures from elements located at the center of the intervertebral disc. Figure 3 shows a contour map of the pressure throughout both models at the level of the Schmorl's node. Average disc pressures where computed twice, once after application of the 444 N follower load and again after the full 7.5 Nm extension moment was applied (Table 2). Disc pressures were substantially higher in the Schmorl's node model as compared to the baseline model. For example in full extension, with the applied follower load, the average disc pressure was 46% higher in the Schmorl's node model.

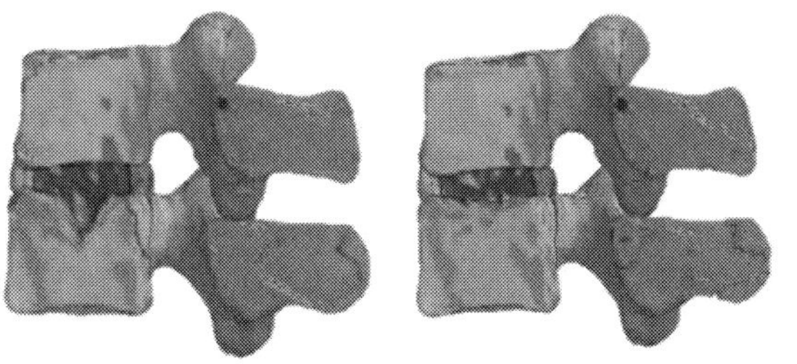

FIGURE 4: VERTICAL CUT OF CONTROL AND SCHMORL'S NODE MODELS SHOWING DISTRIBUTION OF PRESSURE.

TABLE 2: DISC PRESSURES (KPA)

Model	After Follower Load	After Applied Moment
Control	355	463
Schmorl's Node	530	676

Range of Motion was measured for two regions, first for the entire lumbar spine and second for the L1-L2 segment (Table 3). Very little change in ROM was observed as a result of the presence of the Schmorl's node.

TABLE 3: RANGE OF MOTION (DEGREES)

Model	After Follower Load	After Applied Moment
Control		
Lumbar Spine	48.1	35.2
L1-L2 Segment	4.2	5.0
Schmorl's Node		
Lumbar Spine	48.2	35.4
L1-L2 Segment	4.0	4.8

DISCUSSION

Based on our results, presence of a Schmorl's node significantly increases disc pressure with respect to the otherwise identical control model. These results are significant in that they identify that geometrical effects of Schmorl's node presence may be a precipitating factor in disc degeneration for some patients.

The present work also shows that the presence of Schmorl's node does not significantly impact the movement of the spine. Angles of deflection showed a 4.7% change in deflection when loaded with an applied moment. Thus, biomechanically, the area of influence of a Schmorl's node would be confined to the affected segment and would not likely induce adjacent level effects.

ACKNOWLEDGEMENTS

We acknowledge the use of the Fulton Supercomputing Laboratory at Brigham Young University for the use of the computational resources used for the finite element simulations.

This material is based on work supported by the National Science Foundation (CMMI-0952758). Any opinions, findings, and conclusions or recommendations expressed in this material are those of the authors and do not necessarily reflect the views of the National Science Foundation.

REFERENCES

[1]Florence P.S. Mok, et al., Spine (Phila Pa 1976). 35(21):1944-952, 2010.

[2] Von Forell, G. A., Comput. Methods Biomech. Biomed, 2013, (in press).

[3] Patwardhan, A. G., J. Orthop. Res., 21(3): 540–546, 2003.

SB^{3}C2015
Summer Biomechanics, Bioengineering and Biotransport Conference
June 17-20, 2015,Snowbird Resort, Utah, USA
SB^{3}C2015-1019

INTERNAL DISC STRAIN TEMPLATE HIGHLIGHTS REGIONS OF HIGH LOCAL STRAIN DURING COMPRESSION LOADING AND VALIDATES A FINITE ELEMENT MODEL

Brent L. Showalter (1), John F. DeLucca (2), John M. Peloquin (1), Daniel H. Cortes (2), Alexander C. Wright (3), James C. Gee (3), Edward J. Vresilovic (4), Dawn M. Elliott (2)

(1) Department of Bioengineering
University of Pennsylvania
Philadelphia, PA, USA

(2) Department of Biomedical Engineering
University of Delaware
Newark, DE, USA

(3) Department of Radiology
University of Pennsylvania
Philadelphia, PA, USA

(4) Department of Orthopaedic Surgery
Pennsylvania State University
Hershey, PA, USA

INTRODUCTION

Tissue strain is an important indicator of mechanical function in load-bearing joints such as the intervertebral disc. High local strains indicate regions that are more susceptible to damage. However, the enclosed structure of the disc hinders the measurement of internal disc strain. We recently developed a method for measuring 3D internal disc strain in discs undergoing axial compression.[1] However, subject variability made it difficult to analyze population trends in disc strain. This variability can be reduced by averaging multiple images, which is called creating image templates.[2] **Our primary study objective was to create a disc strain template for intervertebral discs undergoing axial compression.**

Because of the complexities of internal disc structure and mechanical properties, finite element models (FEMs) are valuable tools to study disc mechanics.[3, 4] Unfortunately, FEMs are often constructed with significant assumptions concerning internal disc structure and mechanical properties, and they are routinely validated using only simple, and often single, global parameters such as amount of height loss during compression. In consequence, FEM validation lacks rigor and the validity of the disc FEM results internal to the disc is of concern. **Our secondary objective was to use a disc strain template to validate the internal strains calculated by a disc finite element model.**

METHODS

To create a disc strain template, 3D internal strains were first measured for 7 Grade III human L4-L5 discs.[1] Next, an average disc shape, the disc anatomical template, was created. The transforms between the individual subject discs and the disc template were then used to map the strains from the subject discs to the template space and averaged to create the disc strain template.

Data Acquisition

In previous work,[1] bone-disc-bone segments prepared by first removing the posterior elements and then potting in PMMA. Samples were compressed to a reference (50N) and to a loaded condition (11.3 ± 3.2% axial strain). After displacement was held for 3 hours, it was fixed. The sample was moved to a 7T MRI scanner and imaged. Images with 0.3 mm isotropic resolution were acquired with a sequence designed to enhance lamellar contrast (Turbo Spin Echo, TR/TE = 3000/34 ms, ETL = 7). Strain within the annulus fibrosus was calculated by using Advanced Normalization Tools (ANTs) to register the reference and loaded images.[5] Because of decreasing signal-to-noise ratio with distance from the RF coil, we only calculated strains in two-thirds of the disc.

Disc Anatomical Template

The intervertebral discs were manually segmented from the reference images. A disc anatomical template was created using an optimization algorithm in ANTs. Anatomical template quality was assessed by comparing with an existing average disc.[6]

Disc Strain Template

The strain components of the seven subject discs were mapped to the template space and averaged. By defining strain in relation to the disc anatomy rather than Cartesian coordinates, the tensor orientation was preserved when strains were mapped. The effectiveness of the strain template at reducing individual variability was assessed by examining the first strain invariants of the subject discs and the strain template. For analysis, the annulus was divided into five regions: anterior, anterolateral, lateral, posterolateral, and posterior.

Finite Element Model

Loading was simulated in a biphasic finite element model with the same conditions used to create the strain template.[7] The FE predictions were resampled and compared with the strain template.

RESULTS

Disc Anatomical Template

The resulting disc template had dimensions within 1 mm of an independently created shape model.[6]

Disc Strain Template

The distribution of strain invariants in the subject discs varied substantially between discs (Fig. 1). The subject discs shown have different distribution and peak invariants, such as those marked by the red arrows. The peaks are eliminated in the template, but the trend of higher strain in the posterior region remains.

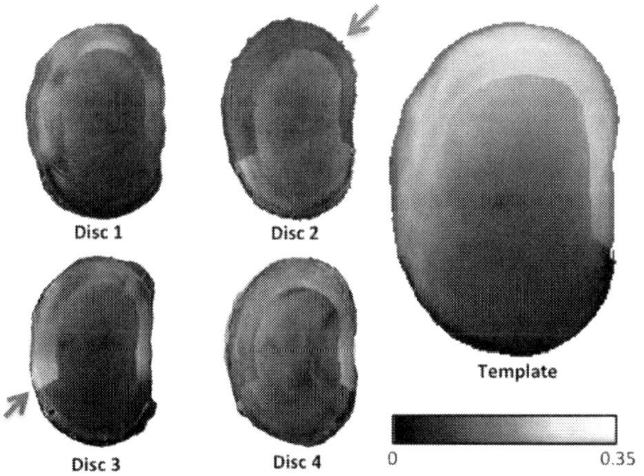

Figure 1: First strain invariant of representative discs and template.

Axial strains were of larger magnitudes than radial and circumferential strains (Fig. 2). Axial compression strains were greater in magnitude in the posterior regions of the disc than the anterior. The lateral annulus experienced the greatest circumferential strains with a corresponding minimum in radial strain. Although subject variability has been reduced in the template, there is still a wide range of local strains.

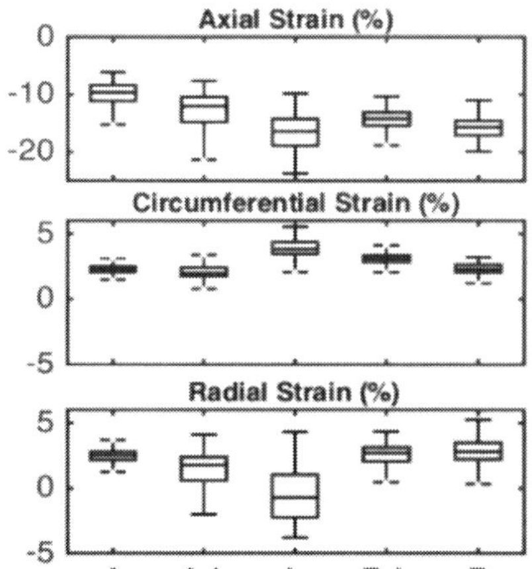

Figure 2: Percent strains in the anterior (A), anterolateral (A-L), lateral (L), posterolateral (P-L), and posterior (P) annulus fibrosus.

Finite Element Model

The magnitudes of the disc strain template and the strains predicted by the finite element model are very similar (Fig. 3), with values differing by 1.9 ± 1.6% strain. The largest difference is in the posterior region, where the finite element model had peaks in radial strains.

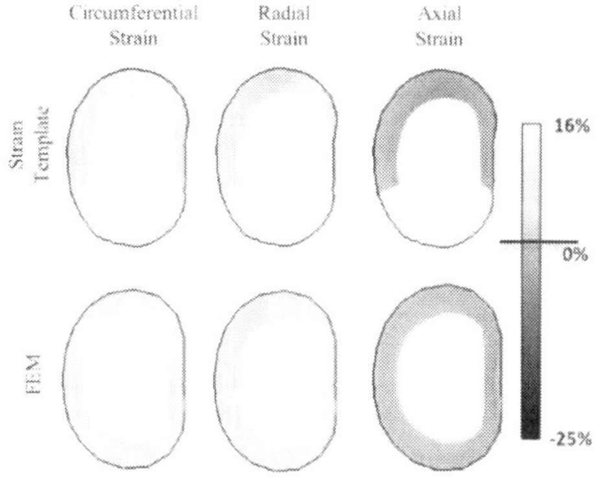

Figure 3: Comparison of Disc Strain Template and FEM

DISCUSSION

The disc strain template is a substantial advance in studying soft tissue mechanics. The 3D strain maps from which the template is derived were created non-invasively using physiologically relevant loads. The combination of 3D maps from several discs into a single template allows for the study of population trends by reducing individual subject variability (Fig. 1). While the eliminated individual variations may yield information that is important in other contexts, the substantial variation between subjects hinders the study of population trends. By reducing subject variability and preserving population trends, the disc strain template enables evaluation of the effects of variables such as loading conditions, disc pathologies, damage mechanisms, and treatments on tissue mechanics.

The strain distributions of the template are consistent with disc anatomy. Axial strain was greater in the posterior region of the disc than the anterior region, which is a result of the thinner annular wall and lower disc height in the posterior. Also, circumferential strain was higher in the lateral and posterolateral region of the disc, which is consistent with higher occurrence of radial tears in those regions.

The wide range of observed axial strain magnitudes highlights the inhomogeneity created by the disc's structure and composition. The mean measured strain is comparable to the applied strain of -11%, but the range of -25% to -5% is substantial. Thus, even in a very simple loading condition, there are regions of the discs structure that are experience high strains that may eventually lead to tissue damage. The wide range is also important to consider in the context of the disc's cellular environment.

This is the first study to compare the internal measurements of a finite element model with internal strain measurements (Fig. 3). The similarity between the two methods is a rigorous validation of the FEM.

In conclusion, the disc strain template provides an unprecedented view of the internal intervertebral disc strain. In addition, this template has been used to validate a finite element model. The technique opens the door for a wide range of studies, including the effects of disc pathologies, damage mechanisms, and treatments. Moreover, wider adaptation of the disc strain template technique is relevant to measure population-relevant 3D internal strain in other soft joints.

ACKNOWLEDGEMENTS

This study was funded by NIH R01AR050052 and NSF GFRP.

REFERENCES

[1]Yoder et al., *J Biomech Eng*, 136(11), 2014. [2]Lancaster et al., *Hum Brain Map*, 28:1194-205, 2007. [3]Schroeder, Y et al., *Eur Spine J*, 15:S361-71, 2006. [4]Natarajan, R et al., *JBJS*, 88A:36-40, 2006. [5]Avants et al., *Neuroimage*, 54:2033-44, 2011. [6]Peloquin et al., *J Biomech*, 47: 2452-9, 2014. [7]Jacobs et al., *J Biomech*, 47:2540-6, 2014.

SB³C2015
Summer Biomechanics, Bioengineering and Biotransport Conference
June 17-20, 2015, Snowbird Resort, Utah, USA
SB³C2015-1020

COLLAGEN MIMETIC PEPTIDE AS A MARKER OF MECHANICAL DAMAGE IN LAMB MIDDLE CEREBRAL ARTERIES

Raymond G. Walther (1), Matthew I. Converse (1), Kenneth L. Monson (1,2)

(1) Department of Mechanical Engineering
University of Utah
Salt Lake City, UT, USA

(2) Department of Bioengineering
University of Utah
Salt Lake City, UT, USA

INTRODUCTION

Traumatic brain injury (TBI) affects an estimated 1.7 million people annually in the United States, 52,000 of whom die as a result of TBI [1]. TBI often causes dysfunction of cerebral blood vessels [2], and has also been found to correlate with an increased risk of stroke up to 5 years after the initial injury [3, 4]. Because TBI causes damage to cerebral blood vessels, characterization of this damage presents a unique opportunity to quantify the loading and/or deformation experienced locally during an injury.

Previous work in marking mechanical damage to blood vessels has primarily been focused on SEM, TEM, or histological imaging, each of which require intensive sample preparation and which would not be practicable for an in vivo application [5, 6, 7]. Recent discoveries of the usefulness of collagen mimetic peptide (CMP) as a marker for denatured collagen suggest that it may be useful as an alternative means of characterizing collagen damage in cerebral blood vessels.

CMP is a family of small synthetic peptides that mimic natural collagens, sharing collagen's triple helix structure, as well as the Gly-X-Y triplet repeat sequence [8]. Recent work has shown that CMP can bind to denatured collagen, both in vivo and in isolated tissue [9, 10]. This binding can be visualized and quantified by CMP labeled with a carboxyfluorescein (CF) tag (CF-CMP).

As a first step toward in-vivo implementation, this study seeks to characterize the effectiveness of CF-CMP as a marker for damage due to various levels of axial overstretch in isolated lamb middle cerebral arteries.

METHODS

Dissection and Sample Preparation: The heads of two lambs were obtained from a meat packing plant where they were sacrificed with a humane electrical stunner. The heads were transported on ice to a facility where the brains could safely be removed. Within 5 hours of death, the lamb brains were removed from the head, and a slice of tissue surrounding each middle cerebral artery (MCA) was dissected out. These tissue slices were placed in PBS, and transported on ice to the dissection facility.

Sections of MCA, each between 4 and 5 mm in length, were dissected from the surrounding tissue, taking care to remove connective tissue. Side branches of the MCA segments were ligated with individual fibrils from unwound 6-0 suture, so as to hold pressure. After dissection, samples were stored in PBS at 4°C until mechanical testing. Seven MCA sections were prepared in total.

Mechanical Testing: The mechanical testing of the MCA segments was performed in a manner similar to that described previously [11, 12]. Briefly, each segment was cannulated with hypodermic needles, and secured with 6-0 suture and cyanoacrylate glue. In order to allow determination of local stretch values, microspheres were applied to the surface of the blood vessel.

After mounting the MCA segment, it was preconditioned by oscillating the luminal pressure (6.7-20 kPa) for five cycles while length was held constant at various sub-failure axial stretch values up to $\lambda_Z \approx$ 1.05 times in vivo stretch. Following preconditioning, non-control samples were subjected to an axial overstretch at one of four prescribed levels (M1: $\lambda_Z \approx 1.3$; M2: $\lambda_Z \approx 1.4$; M3: $\lambda_Z \approx 1.6$; M4: sample pulled to failure). With the exception of the samples pulled to failure, all samples were then cut perpendicular to their axis using scissors, making two cylindrical segments, one for use in this study, and one used in a separate study of incubation times.

CMP Solution Preparation and Staining: Concurrent with mechanical testing, a PBS solution containing 20 μM of CF-CMP was prepared and placed in a 150 mL water bath on a hot plate set to 70°C, taking care to minimize light exposure. The CF-CMP solution was left in the water bath for 10 minutes, then removed and allowed to cool to room temperature for 5 minutes.

After cooling, 90 µL of CF-CMP was pipetted into a well for each half vessel segment, and the vessel segments were moved from PBS to the CF-CMP solution, taking care to ensure the vessels were fully submerged. Segments were allowed to incubate in the CF-CMP solution for 18 hours ("overnight") at 4°C.

At the end of the incubation period, each segment was rinsed in PBS for a total of 30 minutes, taking care not to expose the segments to light. After rinsing, the segments were cut open by slicing the sidewalls lengthwise with scissors.

Once each segment was cut open, it was laid flat on a glass slide, with the lumen side down. The segments were then covered with fluoromount, and a coverslip was placed on top.

Microscopy Method: Each sample was imaged at 10x on an Olympus confocal microscope. Confocal stacks were taken with a step size of 2 µm from the luminal wall (defined as the slice 6 µm to the luminal side of where the internal elastic lamina was in focus) to the exterior edge of the vessel wall (defined as the slice where adventitial features reached peak brightness)..

Analysis Method: Side by side stacks of each vessel were stitched together to form a single stack covering the full circumference of the vessel using a custom Matlab script. These stacks were then analyzed using Matlab to find the number of bright pixels (the number of pixels within 10% of the maximum pixel value). In order to account for variations in wall thickness, this metric was plotted as a function of normalized wall depth (defined as the depth divided by the maximum wall depth).

RESULTS

Typical Confocal Images: Figure 1 presents a single confocal slice of each overnight incubation sample, taken at approximately 75% through the thickness of the vessel wall. Examination of Figure 1 shows that at a normalized wall depth of approximately 75%, the M3 and M4 damage level samples show streaks with a high level of marking, and lower background marking.

Matlab Quantification: Figure 2 presents the percent of each confocal slice for each image found to be "bright", that is, within 10% of the maximum pixel brightness. This metric demonstrates the difference between samples, with high percentages of bright pixels for the M3 and M4 level samples as compared to controls and samples stretched to less damaging levels. Marking is also shown both by examination of confocal stacks, and by the percent of bright pixels to be much higher in the adventitial portion of the vessel walls.

DISCUSSION

The fact that marking primarily occurs at more damaging (M3, M4) stretch levels agrees well with similar preliminary work done in ligaments in another lab at the University of Utah, which showed that CMP primarily marks collagen damage at near-failure levels. It is worth noting that the CMP binding was primarily observed as bright streaks oriented parallel to the vessel axis. This marking peaks in the adventitia, which has been shown to have a substantial component of longitudinal fibers [13].

This study shows that CMP may be a viable marker for near-failure stretch levels in isolated vessels. Such a marker could be useful in characterizing blood vessel stretch levels and their associated damage in association with in vivo injury models. In order to be useful in such applications, further work would need to include more samples, with a better characterization of the local level of overstretch, and a more refined quantification method. In addition, for in vivo applications, this method would need to be refined for marking vessels embedded in brain tissue.

The preliminary evidence that this study provides also suggests that CMP could be useful as a delivery method for treatment of damaged blood vessels, as it preferentially binds to collagen that has been exposed to extreme damage levels. Again, any such treatment would require further exploration of CMP in an in vivo setting.

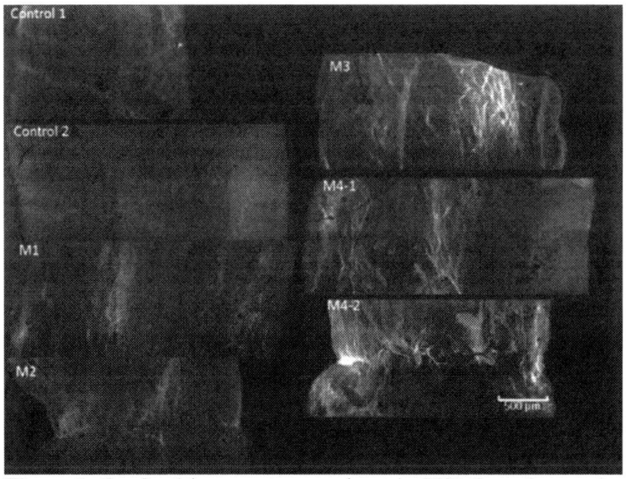

Figure 1. Confocal images at approximately 75% through vessel wall thickness for control and damaged (M1-M4) samples. Binding of CMP is most evident in samples stretched to more injurious (M3, M4) levels.

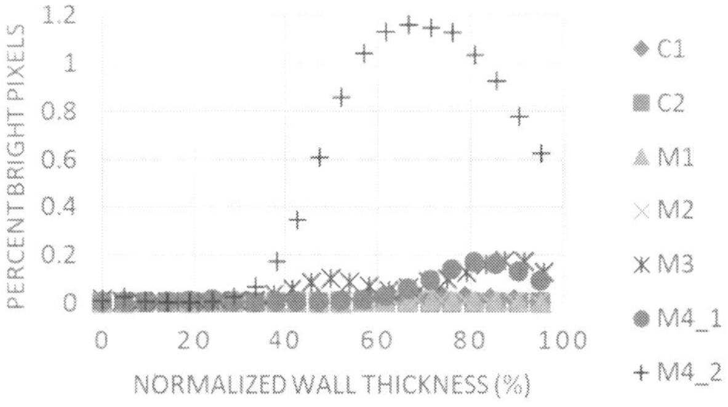

Figure 2. Percent of "bright" pixels (within 10% of maximum pixel brightness) as a function of wall thickness for control, and damaged (M1-M4) samples. This metric is most evident in samples stretched to the most injurious (M3, M4) levels.

ACKNOWLEDGEMENTS

Li Yang, S. Michael Yu for their providing of and assistance with CF-CMP solutions.

REFERENCES

[1] Langlois, JA et al, Centers for Disease Control and Prevention, National Center for Injury Prevention and Control, 2004
[2] DeWitt, DS et al, *J Neurotrauma*, 20:795-825, 2003
[3] Burke, JF et al, *Neurology*, 81:33-39, 2013
[4] Chen, Y et al, *Stroke*, 42:2733-2739, 2011
[5] Yamamoto, Y, et al, *Neurosurgery*, 30(1):1-6, 1992
[6] Ohkawa, M et al, *Radiology*, 198(1):179-184, 1996
[7] Langille, BL et al, *Arterioscler Thromb Vasc Biol*, 12:307-317, 1992
[8] Li, Y et al, *Current Opinion in Chemical Biology*, 17:968-975, 2013
[9] Li, Y et al, *Proc Natl Acad Sci U S A*, 109:14767-14772, 2012
[10] Li, Y et al, *Bioconjugate Chemistry*, 24:9-16, 2013
[11] Monson, K et al, *Ann Biomed Eng*, 36:2028-2041, 2008
[12] Bell, ED et al, *J Biomech*, 46:91-96, 2013
[13] Finaly, HM et al, *J Vasc Research*,32(5):301-312, 1995

SB³C2015
Summer Biomechanics, Bioengineering and Biotransport Conference
June 17-20, 2015, Snowbird Resort, Utah, USA
SB³C2015-1021

MACHINE LEARNING BASED STRUCTURED EDGE DETECTION FOR CARDIOVASCULAR MODELING

Jameson T. Merkow
ECE Dept., UC San Diego
San Diego, CA

Zhouwen Tu
Cog. Sci Dept., UC San Diego
San Diego, CA

David Kriegman
CSE Dept., UC San Diego
San Diego, CA

Nathan Wilson
Open Source Medical Software Corporation
Santa Monica, CA

Alison L. Marsden
MAE Dept., UC San Diego
San Diego, CA

INTRODUCTION

Patient-specific cardiovascular simulation has become increasingly necessary and effective for non-invasively obtaining hemodynamic data not available from medical imaging, and developing specialized treatments and surgeries at no risk to the patient. Recent applications include congenital heart disease, and evaluating risk in of bypass graft failure [1, 2]. Patient specific flow simulations require accurate three dimensional models to be constructed through segmentation of patient image data. Furthermore, statistically significant studies require large patient cohorts, making current segmentation methods inadequate.

Standard open source packages such as ITK-SNAP [3] rely heavily on region growers, and active contours. These methods enjoy a flexible framework that enforces edge smoothness making them attractive for constructing vascular models for CFD. However, active contours and region growers have many limitations, such as sensitivity to local minima, and they lack contextual information. Many improvements have been proposed, however they increase design complexity and require substantial manual tuning to achieve acceptable accuracy. Recently, we have begun integrating machine learning methods into the open source SimVascular project [4]. These methods learn context and appearance parameters automatically to specifically address the above shortcomings. Learning methods have shown competitive results in a wide range of medical imaging applications, such as neural segmentation, 'vessel-ness' measurement, and prostate segmentation [5–7]. However, these methods perform sub-optimally when used to construct cardiovascular models for CFD since they do not enforce smoothness.

Our approach is to improve edge localization directly using a learning based classifier that automatically leverages structural context. This approach provides a number of key advantages. First, the classifier determines which appearance features are most discriminative in context and requires no parameter tuning. This produces a flexible and reliable framework. In addition, an adaptive prior can be built directly into a learned classifier. These methods will be incorporated to enhance segmentation capabilities in the open source SimVascular project.

METHODS

In this section, we describe our classifier, input features, and sampling methodology for producing robust edge maps of vascular structures and ignore unimportant edges. To classify image patches we use the powerful structured forest classifier proposed by Dollár and Zitnick [8]. This classifier produces a voxel patch label based on a multi-channel input patch.

Classifier—Structured forests employ a randomized ensemble of decision trees to classify image patches. A decision tree produces a final classification by making a number of binary decisions, each based on a single feature.

Decision forests were long believed to be sub-optimal for edge detection because information gain is ill-defined for continuous labels. However, Dollar and Zitnick showed that decision trees could be trained by leveraging the structural nature of edges which allows a discrete mapping of label space $Y \rightarrow C$ where $c \in C, C = \{1, ..., k\}$. Using discrete label space, C, information gain can be calculated. Let $\mathbf{T}_L$ denote a training set $(\vec{x}, l)$ with input features $\vec{x} = (x_1, x_2, ...x_n)$ and label $l \in L$, we seek a mapping:

$$\Pi_\phi : Y \xrightarrow{\Pi_\phi} C \quad s.t. \quad IG(\mathbf{T}_C, i) \approx IG(\mathbf{T}_Y, i) \tag{1}$$

where i denotes the ith feature of $\vec{x}$, C is a structured discrete label space, Y is a continuous label space, and Π_ϕ denotes the mapping function parameterized by ϕ.

Finding a mapping to discrete output space C requires an intermediate mapping $Y \rightarrow Z$, where similarity can be easily measured. Traditionally, patch similarity is computed by measuring similarity between individual voxel pairs. This represents a computational challenge for voxel patches. In a $8 \times 8 \times 8$ voxel patch, there are $\binom{8 \cdot 8 \cdot 8}{2} = 130,816$ individual pairs. This is avoided by employing a two stage dimensionality reduction process: 1) only m dimensions of Z are sampled, 2) Principal Component Analysis (PCA) projection is calculated and only the k strongest components are preserved. The discrete set C is found by computing K-means clustering on Z and quantizing the top $log_2(k)$ PCA components. Membership assignment is found by computing

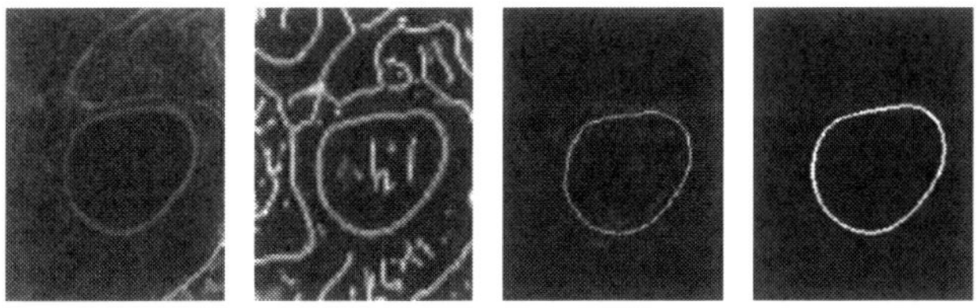

(a) (b) (c) (d)

FIGURE 1: Cross sectional comparison of edge maps. (a) Gradient magnitude. (b) Canny edge map. (c) Proposed method after thresholding and thinning. (d) Ground truth.

$z_i = \Pi_\phi(y_i)$ and selecting the z_k that maximizes the sum of similarity over all z_i.

Input Features—At each sample location a $16 \times 16 \times 16$ voxel patch is taken, feature are computed and $8 \times 8 \times 8$ ground truth is extracted. We use three image features, one atlas feature and a compound image-atlas feature. Each of the three image features are available at two resolutions, these features are voxel intensity, gradient magnitude and Canny edge response. *A priori* information is incorporated via a distance map calculated from user-generated centerlines. Third, our compound feature is the projection of the image gradient onto the gradient of the distance map. This feature captures voxels where the two gradient directions are aligned. Lastly, intra-patch difference features were computed by applying a strong point spread function, then calculating pairwise differences.

Selection of Training Samples—We used an intelligent sampling scheme to maximize responses at edge locations while minimizing responses at edge adjacent neighborhoods. Though positive regions in a patch-to-patch classifier are any patches that contain any edge voxels, we found that only using edge aligned patched gave the best results. We took negative samples from two regions: 70% of samples were taken from patches neighboring edges,the remaining 30% were taken from any negative region.

RESULTS

To evaluate our method, we constructed a dataset of ten CT scans taken from different patients. All scans are of the thoracic region and have ground truth segmentation of the ascending & descending aorta, the aortic arch and any brachiocephalic, common carotid, and subclavian arteries visible in the scan. These images were taken for clinically indicated purposes and the ground truth segmentation were constructed using manual segmentation in SimVascular by expert researchers, then verified by a cardiologist. Seven of the ten images were randomly selected for training, and the remaining three were used to calculate accuracy. All ground truth segmentations are derived from actual 3D models for use in blood flow simulation, so they contain salient and latent edges suitable for cardiovascular flow.

We compare our approach to gradient magnitude, Canny response, and a patch based classifier using the same input features.To fairly compare our method with canny response and gradient magnitude, we apply an image mask to the accuracy calculation that to disregard voxels that are greater than 15 voxels from ground truth edge. We generated ROC curves by thresholding the edge response by increasing values then calculating the overlap between the result and the ground truth labels.We see in Figure 2 that our approach easily out performs canny and gradient edge maps, and furthermore out performs the patch based classifier by a wide margin. Figure 1 show a cross sectional example comparison of the widely used edge maps and the edge map produced by our method.

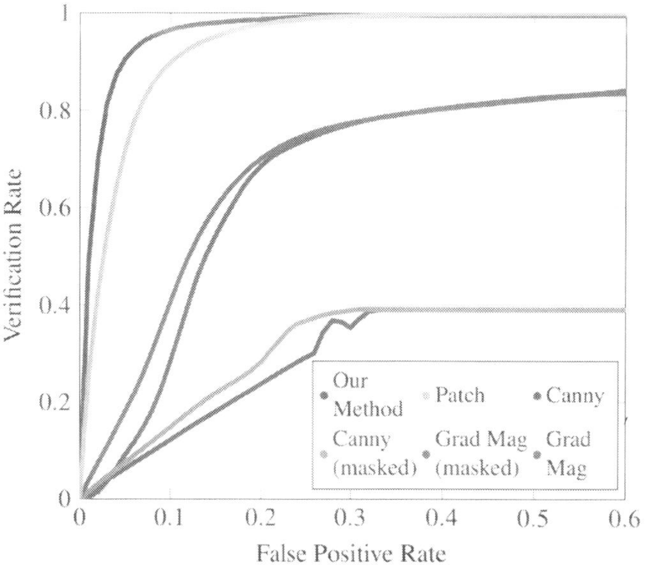

FIGURE 2: Performance comparison of proposed method with other leading methods demonstrating improved accuracy.

DISCUSSION

In this work, we demonstrate using learned edge detectors in medical imaging for cardiovascular model construction. We show that our method out performs other widely used methods, gradient magnitude and canny edge response, and other learned edge detectors. These edge maps can be used in conjunction with other methods to reduce their design complexing by removing local minima and increase *a priori* context. Incorporation of these methods will increase efficiency of the model construction process in SimVascular and other open source projects by reducing the number of parameters to tune. Implementation into SimVascular is underway and will be provided through the open source project at `http://simvascular.org`.

ACKNOWLEDGMENTS

This work was supported by NSF SSI grant through ACI (Award# 1339824), NSF CAREER OCI 1150184, and a Burroughs Wellcome Fund Career Award at the Scientific Interface. Additional support for SimVascular was provided by simtk.org at Stanford University.

REFERENCES

[1] Vignon-Clementel, I. E., Marsden, A. L., and Feinstein, J. A., 2010. *Progress in Pediatric Cardiology*, ***30***(1-2), Dec., pp. 3–13.

[2] de Zélicourt, D. A., Marsden, A. L., Fogel, M. A., and Yoganathan, A. P., 2010. *Progress in Pediatric Cardiology*, ***30***(1-2), Dec., pp. 31–44.

[3] Ho, S., Cody, H., and Gerig, G., 2003. *Submitted to MICCAI 2003*.

[4] Schmidt, J. P., Delp, S. L., Sherman, M. A., Taylor, C. A., Pande, V. S., and Altman, R. B. *Proceedings of the IEEE*, ***96***(8), pp. 1266–1280.

[5] Tu, Z., Narr, K. L., Dollar, P., and Dinov, I., 2008. *Medical Imaging*.

[6] Zheng, Y., Loziczonek, M., Georgescu, B., Zhou, Kevin, S., Vega-Higuera, F., and Comaniciu, D., 2011. *SPIE Medical Imaging*, Mar., pp. 79621K–79621K–12.

[7] Li, W., Liao, S., Feng, Q., Chen, W., and Shen, D., 2012. *Physics in medicine and biology*, ***57***(5), p. 1283.

[8] Dollar, P., and Zitnick, C. L., 2013. *ICCV*, pp. 1841–1848.

SB³C2015
Summer Biomechanics, Bioengineering and Biotransport Conference
June 17-20, 2015,Snowbird Resort, Utah, USA
SB³C2015-1022

CHARACTERIZATION OF MITRAL VALVE GEOMETRY AND DEVELOPMENT OF A POPULATION-AVERAGED MODEL

Amir H. Khalighi (1)(2), Andrew Drach (1), Fleur M. ter Huurne (3), Chung-Hao Lee (1), Charles Bloodworth (4), Eric L. Pierce (4), Morten O. Jensen (4), Ajit P. Yoganathan (4), Michael S. Sacks (1)(2)(5)

(1) Center for Cardiovascular Simulation,
Institute for Computational Engineering and Sciences,
The University of Texas at Austin
Austin, TX, USA

(2) Department of Mechanical Engineering,
The University of Texas at Austin
Austin, TX, USA

(3) Department of Biomedical Engineering,
Eindhoven University of Technology
Eindhoven, Netherlands

(4) Cardiovascular Fluid Mechanics Laboratory,
Wallace H. Coulter Department of Biomedical
Engineering, Georgia Institute of Technology
Atlanta, GA, USA

(5) Department of Biomedical Engineering,
The University of Texas at Austin
Austin, TX, USA

INTRODUCTION

The Mitral valve (MV) is the atrioventricular valve regulating blood flow between the left atrium and ventricle. The MV has a dual-flap structure comprised of the anterior and posterior leaflets, which coapt to prevent backflow of blood into atrium as left ventricle contracts during cardiac systole. The leaflets are attached to the papillary muscles through chordae tendineae giving a parachute shape to the MV apparatus. Abnormalities in MV geometry can lead to valve malfunction and further heart complications. MV regurgitation is one the most common valvular insufficiency disorders, affecting more than 2 million people in the United States [1]. Recent studies have suggested that heart valve repair is a more effective treatment compared to the prosthetic heart valve replacement [2]. However, the natural anatomical variations and pathological alterations complicate surgical planning [3], and demand a better understanding of MV geometry characteristics and patient-to-patient dissimilarities.

Below we describe the data processing pipeline we have developed for characterizing MV geometry and building a population-averaged geometric model of the MV apparatus. The proposed approach enables the analysis of the MV geometric features in a multi-scale framework suitable for geometry characterization and quantification of intra-patient variations. In the following sections, we review the steps and discuss how this method can be used to perform high-fidelity geometry modeling of MV.

METHODS

MV geometry data was acquired from an Inveon micro-computed tomography (μCT) system (Siemens AG, Germany) configured to acquire data with the following settings: 80 *keV* energy, 500 *μA* intensity, and 650 *ms* integration time, 43 *μm* resolution with isotropic voxels. All MV specimens in the present study were scanned in the fully-open state, corresponding to the left ventricular diastole.

The acquired μCT images were imported into Simpleware ScanIP (Simpleware Ltd, Exeter, UK) and MATLAB (Mathworks, United States) to extract point cloud representations of the atrial and ventricular surfaces of the valve leaflets. The point cloud datasets were then tessellated in Geomagic Studio (3D Systems, Inc., United States). Chordae tendineae structure was extracted using NURBS fitting (Centerlines module in ScanIP software environment).

The analysis of leaflet geometry was performed by decomposing geometrical features into the large- and fine-scale models. Large-scale structural information was characterized by a superquadrics surface as defined by Solina and Bajcsy in [4]. Eq. 1 denotes the form for the superquadrics surface employed

$$F(x,y,z)=\left[\left(\frac{x}{a_1}\right)^{2/\varepsilon_2}+\left(\frac{y}{a_2}\right)^{2/\varepsilon_2}\right]^{\varepsilon_2/\varepsilon_1}+\left(\frac{z}{a_3}\right)^{2/\varepsilon_1}=1 \quad (1)$$

where a_1, a_2, a_3 are the dimensional parameters while ε_1 and ε_2 regulate the shape. The fitting parameters were found by minimizing objective function of the form

$$R=\sqrt{a_1 a_2 a_3}\left(F^{\varepsilon_1}-1\right) \quad (2)$$

The residuals from superquadrics fitting were resolved by L_2-projection of the point cloud onto the finite element mesh of the superquadric surface. This residual field was then analyzed in the spectral domain to characterize fine-scale geometrical features. Conformal Fourier Transform (CFT) method [5] using Non-uniform Fourier Transform (NUFFT) [6] enabled us to quantify and filter fine-scale features to the user-specified level spatial smoothness (low-pass box filter with cutoff frequency).

Tubular topology of Chordae tendineae structure allows for simplified modeling approach to quantify the geometric features. Chordae structure was modeled by NURBS-based medial axis reconstruction with two output quantities: centerline NURBS curves and along-axis pointwise cross-sectional area variations.

Geometry characterization and study of variations among five datasets was performed for the large-scale, fine-scale, and for chordal structure. Averaging was performed at large-scale level by taking the mean of dimensionality parameters as well as shape factors. The PSD fields modeling fine-scale features were averaged in spectral domain and then mapped back to the spatial representation. Chordae structure was analyzed using structural graph theory as in [7] (for topology characterization) and empirical distribution functions (for length and number of branches).

RESULTS

The results of applying proposed method to one of the studied data-sets (Fig.1A) are shown. Fig.1B illustrates the point cloud of midsurface extracted from the μCT imaging data. The overlaid surface corresponds to the fitted superquadric surface based on minimization of L_2 residual specified by eq. (2). Fig.1C depicts the unfolded projection of the residual field (fine-scale features) onto the superquadric surface after reconstruction using a filtered spectrum. Chordae geometry analysis is shown in Fig. 1E. Extracted smooth centerlines and selected cross-sections are highlighted in red. The provided results demonstrate the capability of the proposed approach to capture, quantify and filter all of the geometric features of the MV apparatus in a unified geometric framework (Fig.1D,F)

DISCUSSION

Based on the geometry analysis pipeline presented here, geometry-characterization and population-averaging of the entire MV can be carried out in a consistent manner with adjustable level of detail. Geometry characterization provides insight into MV performance through quantification of changes in the MV geometry under healthy and pathological conditions. Population-averaged models are essential for unbiased computational simulations to unravel the true response in predictive modeling. The proposed framework allows controlling the level of details incorporated in final model through direct application of filters in spectral domain. Moreover, the final population-averaged model can be conveniently imported into the mesh generation software packages to build finite element computational models.

ACKNOWLEDGEMENTS

Research reported in this publication was supported by National Heart, Lung, and Blood Institute of the National Institutes of Health under award number R01HL119297. The content is solely the responsibility of the authors and does not necessarily represent the official views of the National Institutes of Health.

REFERENCES

[1] Enriquez-Sarano, M. et al., *Lancet*, 373(9672):1382-1394, 2009.
[2] Fedak, P. W. et al., *Circulation*, 117(7):963-974, 2008.
[3] Bothe, W. et al. *ANN THORACIC SURG*, 95(4):1475-1483, 2013.
[4] Solina, F., and Bajcsy, R., *PAMI*, 12(2):131-147, 1990.
[5] Zhu, C. et al., *PIER*, 120:165-179, 2011.
[6] Greengard, L., and Lee, J. Y., *SIAM review*, 46(3):443-454, 2004.
[7] Alhashim, I. et al. *ACM TOG*, 33(4):158, 2014.

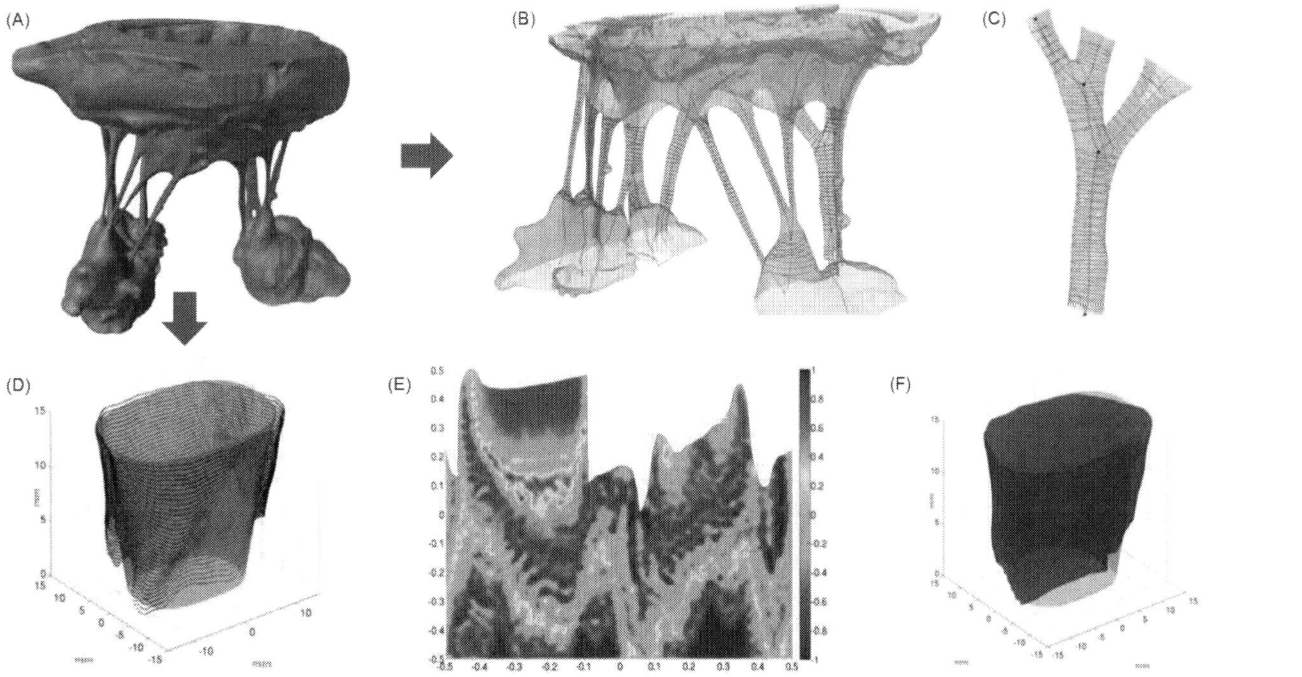

FIGURE 1. (A) CT DATA, LEAFLETS ARE SHOWN AS BLUE, CHORDAE TENDINEAE ARE SHOWN AS RED, AND GREEN REPRESENTS PAPILLARY MUSCLES. (B) CHORDAL STRUCTURE MODELED BY MEDIAL AXIS REPRESENTATION. (C) BRANCHING IN CHORDAL STRUCTURE. (D) POINT CLOUD DATA IS SHOWN AS BLACK, AND THE LARGE-SCALE (FITTED SUPERQUADRIC) MODEL IS SHOWN AS A SHADED SURFACE. (E) UNFOLDED VIEW OF A RECONSTRUCTED FINE-SCALE MODEL, LOWER BOUNDARY CORRESPONDS TO ANNULUS, UPPER BOUNDARY IS FREE EDGE, COLORMAP REPRESENTS NORMALIZED DEVIATIONS FROM THE FITTED SURFACE. (F) RECONSTRUCTED SURFACE USING THE TWO-SCALE MODEL.

SB³C2015
Summer Biomechanics, Bioengineering and Biotransport Conference
June 17-20, 2015,Snowbird Resort, Utah, USA
SB³C2015-1023

USING OPENSIM TO PREDICT KNEE JOINT MOMENTS DURING CYCLING

Karim C. Dudum (2), Jake E. Deschamps (1), Juan D. Gutierrez-Franco (1), Luke I. Kraemer (1), Alejandro M. Gonzalez-Smith (1), Eshan M. Dandekar (3), Scott J. Hazelwood (1,2), Stephen M. Klisch (1,2)

(1) Mechanical Engineering Department
California Polytechnic State University
San Luis Obispo, CA, USA

(2) Biomedical Engineering Department
California Polytechnic State University
San Luis Obispo, CA, USA

(3) Kinesiology Department
California Polytechnic State University
San Luis Obispo, CA, USA

INTRODUCTION

Cycling is a relatively low impact activity conventionally recommended as a rehabilitative or fitness sustaining exercise for patients at a high risk for knee osteoarthritis (OA) [1,2]. Expanding our understanding of knee joint loads is necessary to develop and improve evidence-based prescriptions for cycling as a rehabilitative and fitness therapy that limits the risk for knee OA.

OpenSim (www.simtk.org) is an open source biomechanical analysis software that can partition predictions of external joint loads (or net muscle moments) into muscle and joint contact loads [3]. Joint contact loads more accurately represent cartilage tissue loading and hence risk for cartilage damage and/or OA [4]. As a first step towards predicting knee joint contact loads during cycling, we hypothesized that OpenSim can predict external knee joint moments that are consistent with published data [5,6]. To address this hypothesis, we conducted cycling experiments and used OpenSim's scale tool, inverse kinematics (IK) solver, and inverse dynamics (ID) solver to model the recorded activity.

METHODS

Equipment. A LifeCycle GX stationary bike (Life Fitness, Schiller Park, IL) was used to record foot loads during the motion analysis studies. Each stock pedal was replaced with a custom pedal box containing a 6-axis load cell (AMTI, Watertown, MA) [7] to record kinetic data. An 8-camera Owl motion analysis system (Motion Analysis Corp., Santa Rosa, CA) recorded kinematic data. Cortex software (Motion Analysis Corp., Santa Rosa, CA) was used to post-process kinematic and kinetic data before usage in OpenSim.

Experimental Studies. Inclusion criteria included normal body mass index (BMI = 18.5-24.9) and no history of leg injuries. Retroreflective markers were placed on male subjects (n = 2) using a full body Helen Hayes marker set [8]; each subject first maintained a pose for 3 seconds, while a static trial was captured. Then, a dynamic capture was taken as subjects performed 2 minutes of continuously seated cycling at ~70 RPM while grasping the handlebars. Experimental protocols were approved by Cal Poly's Human Subjects Committee and were designed to minimize risk to human subjects.

Data Processing. Cortex exports kinematic data in a .trc file and kinetic data as a .kin file. The .trc file type contains 3D coordinate locations of markers relative to the origin of the capture volume, which is easily imported into OpenSim. The .kin files, which contain force vector locations, center of mass positions, and force vectors corresponding to the capture rate of the kinematics, have a file header unrecognizable by OpenSim. Thus, we used Excel to manually extract the appropriate force vectors from .kin files, 1 from each load cell, and saved the kinetic data with the appropriate .mot file header recognizable by OpenSim.

The static capture data were used to scale a full body OpenSim model [9]. Virtual markers were placed on the model in accordance with the Helen Hayes protocol [8]. The OpenSim scale tool used the .trc file from the static capture in order to correlate the experimental marker data with the virtual marker set created in the OpenSim environment, thus dimensioning model segment lengths and consequent mass distributions and further aligning model joint rotation axes with the experimental case. The IK tool, using the .trc file from the dynamic capture, solved for model specific kinematics in the form of joint angles and was formatted as a .mot. The ID tool utilized the IK output and the manually formatted .mot file representing the kinetic vectors, selectively applied as a body force to the talus of each foot, in order to quantify all model specific joint moments and residual pelvic loads. Since seat and handle bar forces were not directly measured, we used published values for these forces [6] to calculate a statically equivalent system of such loads at the pelvis. These handlebar/seat equivalents (HBSEs) were compared with OpenSim predicted pelvic loads from ID.

RESULTS

Both IK (Figs. 1, 2) and ID (Fig. 3) results were consistent between subjects. On average, knee joint angles ranged from 115.5° to 17.4° in flexion (Fig. 2). Further, knee flexion moments ranged from -14.7 N-m to 31.1 N-m (Fig. 3). Averaged over the entire crank angle, ID results for pelvic loads differed from HBSEs calculated from literature values [6] by up to 66.0 N in the superior-inferior direction and up to 33.4 N in the anterior-posterior direction (Fig. 4).

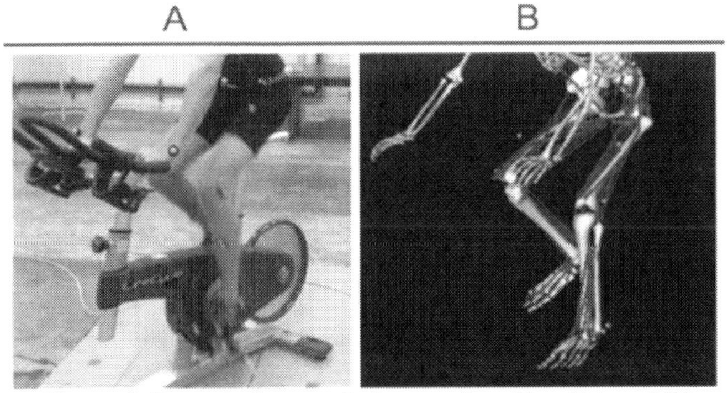

Figure 1. Sample frame of observed motion (A) and IK results (B).

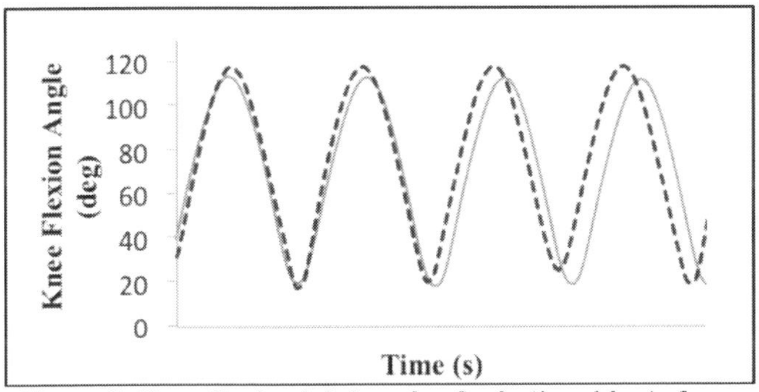

Figure 2. Knee flexion joint angles for both subjects from IK results for 2.5 revolutions of cycling.

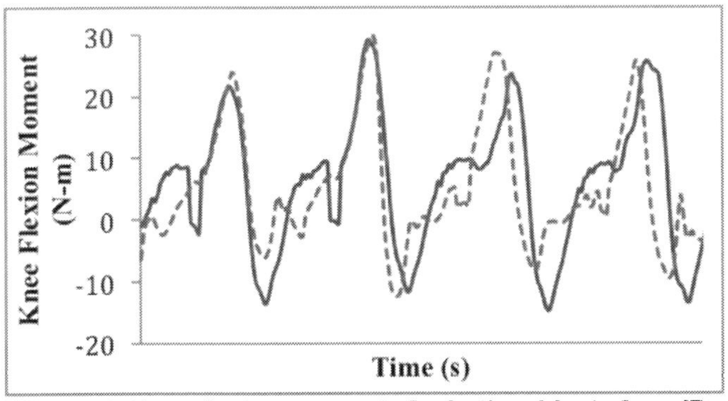

Figure 3. Knee flexion moments for both subjects from ID results for 2.5 revolutions of cycling.

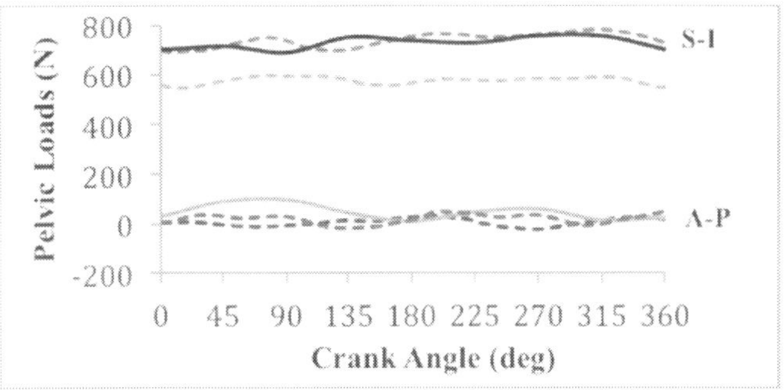

Figure 4. Pelvic loads for each subject (dotted) compared with HBSEs (solid). A-P = Anterior-Posterior, S-I = Superior-Inferior.

DISCUSSION

Our results support the hypothesis that OpenSim can effectively model cycling and predict external knee joint (i.e. net muscle) moments. Although peak values for knee flexion moments were lower than published values [6], other pilot experiment results revealed that higher peaks can be obtained by increasing machine resistance levels. The OpenSim predicted pelvic loads from ID reasonably matched the HBSEs. Interestingly, the ID results predicted the expected result that pelvic loads were dominated by the superior-inferior component; in that direction, ID results differed from HBSEs by 9.2%. Handlebar forces in the medial-lateral direction were not reported [6]; thus, appropriate HBSEs for that direction could not be computed.

In this study, kinetics were defined as body forces rather than more accurately defined as point forces, thus future work will include restructuring the method for kinetic data collection and subsequent import into OpenSim. Further, future work will include implementing a marker set more conducive to reducing soft tissue artifact. OpenSim's residual reduction algorithm (RRA) will be used in an effort to reduce modeling errors. Following RRA, further use of OpenSim tools [3] will be used to predict both muscle and joint contact loads, addressing our long-term goal of developing an improved understanding of knee joint contact loads during cycling for patients at high risk for knee OA.

ACKNOWLEDGEMENTS

Funding received from Quartus Engineering, Inc. and Cal Poly's Donald E. Bentley and STRIDE Centers.

REFERENCES

[1] McLeod, William D., and T. A. Blackburn. *The American journal of sports medicine* 8.3 (1980): 175-180.

[2] D D'Lima, Darryl, et al. *Clinical orthopaedics and related research* 440 (2005): 45-49.

[3] Dunne, James. "User's Guide." *SimTK.org*. Stanford University, 23 May 2014.

[4] Delp, Scott L., et al. *Biomedical Engineering, IEEE Transactions on* 54.11 (2007): 1940-1950.

[5] Ruby, Patricia, M. L. Hull, and David Hawkins. *Journal of biomechanics* 25.1 (1992): 41-53.

[6] Stone, Cal, and M. L. Hull. *Journal of biomechanics* 28.4 (1995): 365-375.

[7] Kraemer et al. SB^{3}C Conference Abstract (2015) (submitted)

[8] *KinTools RT For Cortex*. Santa Rosa: Motion Analysis Corporation, 2010. User's Manual.

[9] Hamner, SR, A Seth et al. Journal of Biomechanics, doi:10.1016/j.jbiomech.2010.06.025 (2010)

SB³C2015
Summer Biomechanics, Bioengineering and Biotransport Conference
June 17-20, 2015,Snowbird Resort, Utah, USA
SB³C2015-1024

EFFECT OF RED BLOOD CELLS ON ENDOTHELIAL CELL SHEAR STRESS STUDIED USING DISCRETE-NATURE BLOOD FLOW SIMULATIONS

Brenna E.Hogan (1,2), Zaiyi Shen (2), Chaouqi Misbah (2), Abdul I. Barakat (1)

(1) Laboratoire d'Hydrodynamique (LadHyX)
Ecole Polytechnique
Palaiseau, France

(2) Laboratoire Interdisciplinaire de Physique (LiPhy)
Université Joseph Fourier
Grenoble, France

INTRODUCTION

By virtue of its anatomic location, the vascular endothelium is constantly exposed to fluid mechanical shear stress due to the flow of viscous blood. In medium and large arteries, shear stress regulates endothelial structure and function and determines predisposition to atherosclerosis [1,2]. In the microvascular, shear stress on the endothelial surface plays an important role in leukocyte-endothelial interactions and is thus a critical regulator of inflammation.

Although blood is 40-45% by volume erythrocytes (red blood cells (RBCs)), the vast majority of computational studies as well as *in vitro* experimental investigations of the effect of shear stress on endothelial cells have neglected the effect of RBCs. Thus, their effect on the shear stress on the endothelial cell surface remains unknown.

In the present work, we explored the effect of various concentrations of RBCs on the flow field in the vicinity of the endothelial cell surface using a Lattice Boltzmann-Immersed Boundary Method code where red blood cells are simulated as discrete, deformable structures interacting with the flow. We also took advantage of the amenability of the Lattice Boltzmann code to complex geometries in order to more realistically simulate the endothelial surface as a wavy surface (rather than the flat surface assumed in most other computational studies).

METHODS

The study involved flow simulations using a 2-dimensional Lattice-Boltzmann scheme with immersed boundary method. The immersed boundary method is used to simulate the RBC dynamics in the blood flow. Lattice Boltzmann is built upon a theory known as lattice gas hydrodynamics which applies to mesoscopic fluid particles and has been established as a computational alternative to direct Navier-Stokes simulations. Lattice gas hydrodynamics is based on the collective dynamics of fictitious particles on the nodes of a lattice. The quantity of interest is the particle distribution function, $f_i(\vec{x},t)$, which represents the probability of finding a fluid particle at the node $\vec{x}$ at time t and travelling with speed $\vec{c_i}$. In our simulations, we use the 2-dimensional 9-speed lattice (D2Q9) such that $i = 0 \dots 8$. The evolution equation of the distribution functions in time and space is as follows:

$$f_i(\vec{x}+\vec{c_i}\Delta t, t+\Delta t) = f_i(\vec{x},t) - \frac{1}{\tau}\Delta t\left(f_i(\vec{x},t) - f_i^{eq}(\vec{x},t)\right) \quad (1)$$

The right-hand side of Equation (1) represents the effect of collisions of the fluid populations, where τ represents a relaxation time towards a local equilibrium, f_i^{eq}, which is typically taken as a second-order expansion of the Maxwell-Boltzmann distribution

$$f_i^{eq} = w_i\rho\left[1 + \frac{\vec{u}\cdot\vec{c_i}}{c_s^2} + \frac{(\vec{u}\cdot\vec{c_i})^2}{2c_s^4} - \frac{\vec{u}\cdot\vec{u}}{c_s^2}\right] \quad (2)$$

where $c_s = 1/\sqrt{3}$ is the speed of sound in the lattice and w_i are weight factors. The relaxation time, τ, is related to the kinematic viscosity of the fluid such that

$$\vartheta = \rho c_s^2 \frac{\Delta x^2}{\Delta t}\left(\tau - \frac{1}{2}\right) \quad (3)$$

where Δx and Δt are the grid spacing and time step, respectively. The macroscopic quantities including the local density and local fluid velocity are recovered from the discrete populations f_i as

$$\rho(\vec{x},t) = \sum_i f_i(\vec{x},t) \quad (4)$$

$$\vec{u}(\vec{x},t) = \frac{1}{\rho(\vec{x},t)}\sum_i f_i(\vec{x},t)\vec{c_i} \quad (5)$$

The shear stress at the endothelial cell surface can be evaluated in the Lattice Boltzmann simulation at each lattice node using the equation

$$\sigma_{xy} = -\left(1 - \frac{1}{2\tau}\right)\sum_i\left(c_{ix}c_{iy} - \frac{\delta_{xy}}{2}\vec{c_i}\vec{c_i}\right)\left(f_i(\vec{x},t) - f_i^{eq}(\vec{x},t)\right) \quad (6)$$

where δ_{xy} is the Kronecker delta [3].

Simulations were run for channels ranging in width from $10\mu m$ to $50\ \mu m$ and for RBC concentrations ranging from 0% to 45%. All the simulations were for a channel Reynolds number of one (Re=1).

RESULTS

In the two-dimensional simulations, the RBC concentration, $\emptyset$, defined as:

$$\emptyset = \frac{A_{RBC}}{A_{channel}} \tag{7}$$

where A_{RBC} and $A_{channel}$ are the RBC and channel areas, respectively. Figure 1 depicts two examples of the simulations for different channel sizes with different RBC concentrations.

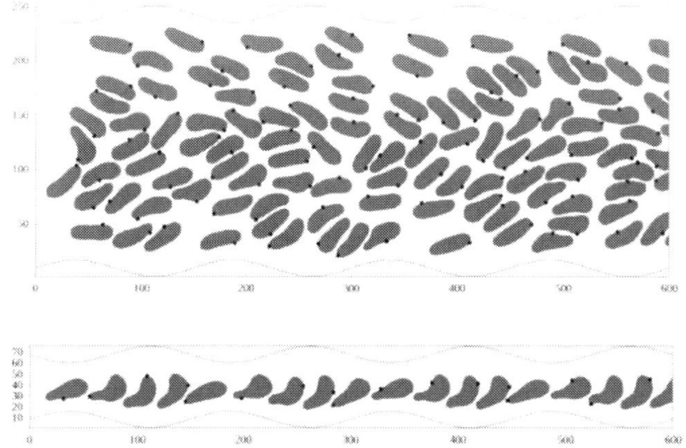

FIGURE 1: Simulations of RBCs in channels of varying width and RBC concentration. TOP: $50\mu m$ channel with $\emptyset = 0.4$. BOTTOM: $15\mu m$ channel with $\emptyset = 0.25$. (Black dots mark tracers that track rotation of RBC throughout the simulation.)

As already described above, we are interested in determining the effect of RBCs on the shear stress on the endothelial cell surface. To gain an appreciation for the magnitude of the shear stress, we evaluate the average endothelial near-wall velocity (i.e. the average over the entire channel length of the velocities of the nodes immediately above the endothelial surface). Figure 2 illustrates this average near-wall velocity as a function of RBC concentration for a $10\mu m$ channel. The results clearly demonstrate that the near-wall velocity (and hence wall shear stress) increases with RBC concentration.

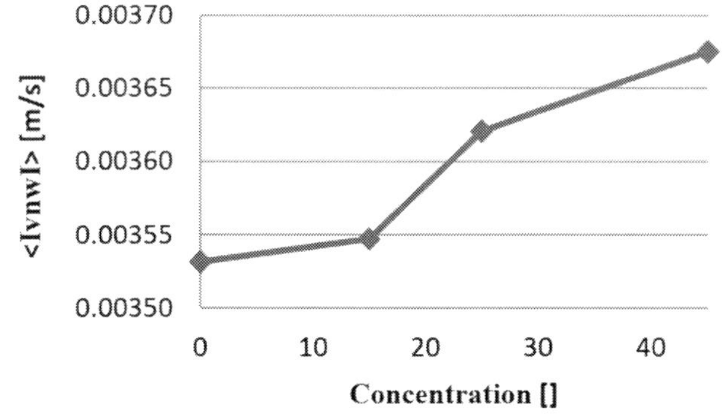

FIGURE 2: Near-wall velocity, $< |v_{nw}| >$, as a function of red blood cell concentration, $\emptyset$.

DISCUSSION

These recent results support the notion, that the presence of RBCs in blood flow modifies the flow field and thus affects the near-wall velocity at the endothelium. In future work, we will more completely explore the parameter space by studying the impact of larger channel diameters, higher flow Reynolds numbers, and different values of RBC rigidity. Furthermore, we will perform experiments to measure the effect of RBCs on the flow field near the endothelium using particle image velocimetry.

This research represents the first step in studying the effect of RBCs on flow-mediated mechanotransduction in vascular endothelial cells. The research findings are expected to identify possible pathways by which RBCs can influence endothelial cell responsiveness to flow.

ACKNOWLEDGMENTS

Brenna Hogan is supported by a doctoral fellowship from Ecole Polytechnique.

This research is funded in part by a permanent endowment in Cardiovascular Cellular Engineering from the AXA Research Fund.

REFERENCES

[1] Davies, PF. Flow-mediated endothelial mechanotransduction. *Physiol Rev*, 75:519-560, 1995.

[2] Barakat, AI. Blood flow and arterial endothelial dysfunction: mechanisms and implications. *Comptes Rendus Physique*, 14:479-496, 2013.

[3] Kruger, T., Varnik, F., Raabe, D. Shear stress in lattice Boltzmann simulations. *Phys. E*, 79: 046704, 2009.

CONSTRUCTING RUDIMENTARY LIMIT CURVES FOR NEURONAL PHOSPHOLIPID BILAYER FAILURE AND THEORETICAL CALCIUM PENETRATION

M. A. Murphy (1,2), Sungkwang Mun (2), M. F. Horstemeyer (2,3), Steven R. Gwaltney (4,5), Tonya W. Stone (2,3), Michelle C. LaPlaca (6), Jun Liao (1,2), Lakiesha N. Williams (1,2), R. Prabhu (1,2)

(1) Department of Agricultural and Biological Engineering
Mississippi State University
Mississippi State, MS, USA

(2) Center for Advanced Vehicular Systems
Mississippi State University
Mississippi State, MS, USA

(3) Department of Mechanical Engineering
Mississippi State University
Mississippi State, MS, USA

(4) Center for Computational Sciences
Mississippi State University
Mississippi State, MS, USA

(5) Department of Chemistry
Mississippi State University
Mississippi State, MS, USA

(6) Department of Biomedical Engineering
Georgia Institute of Technology
Atlanta, GA, USA

INTRODUCTION

Both human and economic losses result from the approximately 1.7 million new cases of traumatic brain injury (TBI) in the United States every year [1]. To obtain a better understanding of TBI, we must investigate the neuronal cell membrane's response to multiple stress states for incorporation into a physics-based multiscale modeling framework for the brain [2]. A good starting point is the phospholipid bilayer because many TBI cases are a result of mechanoporation in the neuronal phospholipid bilayer [3], but prior experimental [4] and *in silico* [5] studies have primarily focused on the equibiaxial stress state because the bilayer is assumed to be planar. This assumption of equibiaxial deformations is appropriate biologically, because the bilayer would likely be placed under equibiaxial stresses during typical cellular functions where pressure from cellular processes would dominate stresses placed on the cell membrane. However, the same assumption is not necessarily true for deformations related to injury mechanics, so other stress states must be accounted for. Fortunately, the assumption that the bilayer is planar leads to an obvious non-biological analogue—sheet metal. When determining strains at which sheet metal will fail, a tool known as a forming limit diagram (FLD) is often utilized to create two dimensional (2-D) strain spaces, showing the strain magnitudes where failure is likely to occur for each stress state [6]. To utilize the FLD, the major and minor strains at failure are plotted against one another to create limit lines that determine if a structure is likely to fail given the strains in that space [6]. In this study, the authors utilize molecular dynamics (MD) and the concept of FLDs to create a simple strain map and limit diagram showing limits for the initial peak stress, pore formation, water penetration, and pores large enough to allow passage to calcium ions for the equibiaxial, non-equibiaxial, strip biaxial, and uniaxial stress states.

METHODS

Seventy-two 1-palmitoyl-2-oleoylphosphatidylcholine (POPC) phospholipids and 9,070 TIP3P water molecules (original structure sourced from NIH's Laboratory of Computation Library [7] and 6,828 TIP3P water molecules were added) was equilibrated for ten nanoseconds using the program LAMMPS [8] and the CHARMM36 all-atom lipid force field [7]. The equilibrated structure was subjected to constant velocity deformations in the x and y dimensions, as determined using strain rates for each dimension (Table 1) and correspond to the von Mises strain rate ($\dot{\epsilon}_{VM}$) = 5.5 × 10^8 s^{-1}. The z dimension was allowed to relax freely during deformations. Structure failure was assumed to occur when water fully penetrated the bilayer and formed a chain of water through both bilayer leaflets.

Table 1. Stress States and Corresponding Strain Rates

Stress State	$\dot{\epsilon}_x$ (s^{-1})	$\dot{\epsilon}_y$ (s^{-1})
Equibiaxial	8.18×10^8	8.18×10^8
Non-Equibiaxial	9.45×10^8	4.72×10^8
Strip Biaxial	8.18×10^8	0.0
Uniaxial	8.18×10^8	Relax

Structure coordinates were output every 5 ps for image analysis. The structure was visualized using OVITO [9]. Periodic images were displayed, showing a 3x3x1 (a total of 9 boxes) structure. By displaying periodic images, all pores that were large enough for detection were displayed fully in at least one location when viewed from the perpendicular (top) view. All atoms were colored black and output using the default rendering options. MATLAB [10] was used to analyze these images. All pores on the exterior periodic image boundaries and duplicate pores were removed before determining the number of pores, pore areas, and time when initial pores began to appear. This analysis was performed for pores with a minimum axis diameter length greater than 0 Å to detect all pores and 10 Å to detect pores large enough for Ca^{2+} penetration. Strains were recorded for the peak stress, first detected pore, water penetration, and first detected pore large enough to accommodate Ca^{2+} diffusion.

RESULTS

Deformation of the bilayer resulted in water penetration due to pore formation (Figure 1 (A)). These pores continued to expand and many became large enough to theoretically allow passage to Ca^{2+} ions (Figure 1 (B)). Further, the equibiaxial and non-equibiaxial stress states exhibited considerably lower strains to pore formation and failure than the strip biaxial and uniaxial cases (Figure 2).

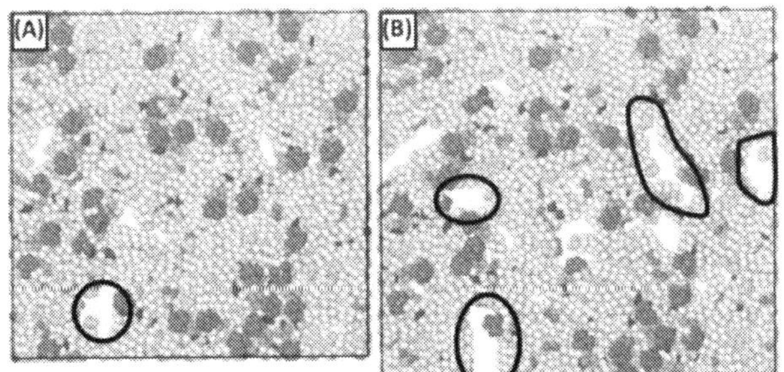

Figure 1: Perpendicular (top) view of the equibiaxial case, showing the structure poration at (A) Water Penetration and (B) when pores large enough for Ca^{2+} penetration are first detected. Water has been hidden for clarity. Regions with water penetration have been circled.

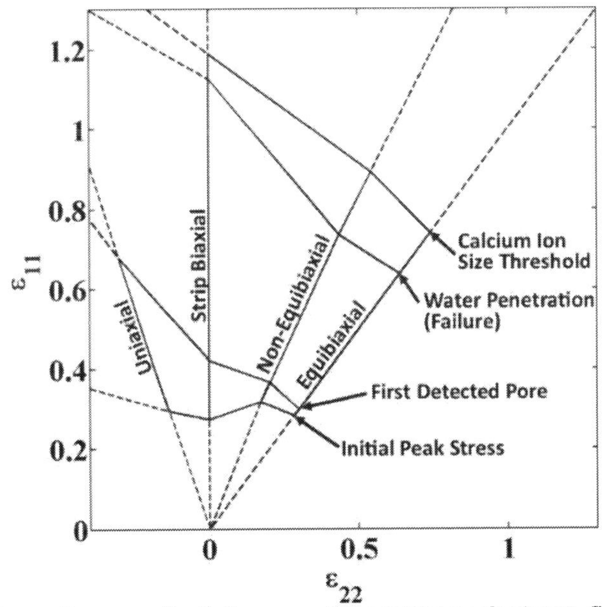

Figure 2: ϵ_{22}-ϵ_{11} limit diagram of the initial peak stress, first detected pore, water penetration, and Ca^{2+} threshold (r = 5 Å) for equivalent equibiaxial, non-equibiaxial, strip biaxial, and uniaxial tension deformations at $\dot{\epsilon}_{VM}$ = 5.5 × 10^8 s^{-1}.

DISCUSSION

The potential for developing a limit diagram for the phospholipid bilayer has been explored in the current study. While it is known that the mechanoporation of the bilayer results in chemical cascades and injury to neurons [3], it still remains unknown at what point that injury occurs. Potential methods of damage initiation can be seen clearly in MD studies, where pore formation and water penetration can be observed directly, but MD are restricted to small length and time scales. These scale restrictions make studying many aspects of TBI inaccessible, especially when studying long-term effects of penetration on cellular systems. However, it is possible to discern specific phenomena to help progress the understanding of the TBI, such as examining the effects of deformation at the nanoscale to determine failure criteria, such as when conditions are met to allow water or Ca^{2+} penetration into the cell (Figure 1).

To better understand how these nanoscale phenomena affect TBI, we must better understand them for each stress state. The limit diagram serves this purpose, showing that the equibiaxial case and non-equibiaxial stress states result in considerably lower strains during initial pore formation and failure which indicates they are more detrimental during high rate deformations than other stress states, especially the uniaxial stress state for which failure was not observed (Figure 2). Additionally, a threshold size for Ca^{2+} penetration has been implemented thorough image analysis, showing that the potential for Ca^{2+} penetration exists shortly after the penetration of water, a much smaller molecule, occurs. Because it allows for these relations to be easily observed and can potentially be used to view safe strain spaces, the limit diagram is an analysis tool that provides a method to explore deformations of the phospholipid bilayer failure, potentially allowing for a relatively simple way to determine cellular damage in the future.

ACKNOWLEDGEMENTS

This material is based upon work supported by the Department of Agricultural and Biological Engineering and the Center for Advanced Vehicular Systems at Mississippi State University. Further, it has also been supported by US Army, through a subcontract with Mississippi State University (W912HZ-13-C-0037), and was performed for the CRES-GV research program. Any opinions, findings and conclusions or recommendations expressed in this material are those of the author(s) and do not necessarily reflect the views of the U.S. Army ERDC ERC.

REFERENCES

1. Faul, M., et al., *Traumatic Brain Injury in the United States: Emergency Department Visits, Hospitalizations, and Deaths 2002-2006*, N.C.f.I.P.a. Control, Editor. 2010, Centers for Disease Control Prevention, National Center for Injury Prevention and Control: Atlanta (GA).
2. Prabhu, R., et al., J Mech Behav Biomed Mater. 4:1067-1080, 2011.
3. Farkas, O. and Povlishock, J.T., *Cellular and subcellular change evoked by diffuse traumatic brain injury: a complex web of change extending far beyond focal damage*, in *Prog Brain Res*, T.W. John and I.R.M. Andrew, Editors. 2007, Elsevier. p. 43-59.
4. Evans, E., et al., Biophys J. 85:2342-50, 2003.
5. Koshiyama, K. and Wada, S., J Biomech. 44:2053-8, 2011.
6. Horstemeyer, M.F., *Integrated Computational Materials Engineering (ICME) for Metals: Using Multiscale Modeling to Invigorate Engineering Design with Science*. 2012: Wiley.
7. Klauda, J.B., et al., J Phys Chem B. 114:7830-7843, 2010.
8. Plimpton, S., J Comput Phys. 117:1-19, 1995.
9. Stukowski, A., Modell Simul Mater Sci Eng. 18:015012, 2010.
10. MATLAB and Image Processing Toolbox Release 2013a, The MathWorks Inc., Natick, MA, 2013.

SB^{3}C2015
Summer Biomechanics, Bioengineering and Biotransport Conference
June 17-20, 2015,Snowbird Resort, Utah, USA
SB^{3}C2015-1027

SUPINE TO PRONE THORACO-ABDOMINAL DEFORMATION AND ORGAN MIGRATION IN A SET OF HEALTHY YOUNG ADULTS

Berkan Guleyupoglu (1,2), Josh C. Tan (1,2), Craig A. Hamilton (1,2), F. Scott Gayzik (1,2)

(1) Virginia Tech – Wake Forest
Center for Injury Biomechanics
Winston-Salem, NC, USA

(2) Department of Biomedical Engineering
Wake Forest School of Medicine
Winston-Salem, NC, USA

INTRODUCTION

Medical imaging is a valuable tool in the study of biomechanics. In this study, a set of MRI imaging data is used to explore abdominal organ compression and migration between two common postures: supine and prone. Imaging in different postures furthers understanding of how the body adapts and changes shape internally based on body position.

Such data can be used to inform various efforts in the study of biomechanics. Medical device design optimized by imaging data may further improve devices targeted at specific disorders or disabilities [1]. Imaging also allows for the development of computational human body models, which can be a platform by which to virtually evaluate design concepts. Data in multiple postures can inform such models of typical changes in postural adjustments [2].

However, medical image data used biomechanics studies are often retrospectively acquired, and researchers are unlikely to encounter scans of healthy individuals with the pulse sequences, resolution, and anatomy desired in the data set. The purpose of this study was to obtain a dataset and evaluate the effects of prone to supine postural adjustment, specifically analyzing thoraco-abdominal compression and organ migration.

METHODS

We prospectively acquired scans in both prone and supine postures (Figure 1), from 22 healthy young adults; M:F 1:1, with age, height, and weight of 28.8±7.0 years, 173.0±7.8 cm, 70.6±10.9 kg. The study was conducted under approval from the WFU Institutional Review Board (#IRB00018578). While prone, subjects' arms were superior to the head and supported by a foam insert. Breath-held, T1-weighted MRI scans were acquired with a Siemen's Skyra 3T, with in-plane resolution of 1.56 mm (TR: 4.10 ms, TE: 1.23 ms, thickness: 2 mm, matrix 256x256, FOV 400 mm).

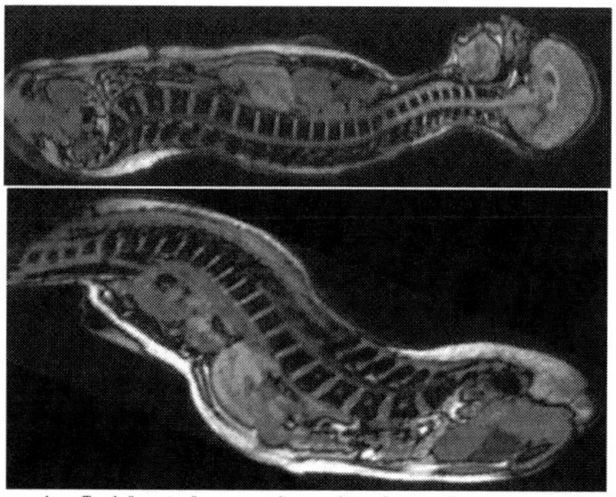

Figure 1. Subject in supine (top) and prone (bottom) postures in a sagittal view.

Thoraco-abdominal compression was taken to be the perpendicular distances from the anterior edge of L4/L5 to the table (in the case of prone) or to the exterior body surface (in the case of supine). The gross compression values were investigated for differences between postures and gender using a two-tailed, paired sampled t-test.

Organ migration and compression both required selection of an anatomical marker. The portal vein and liver interface was used as well as the splenic vein and splenic arteries interface with the spleen.

To assess compression, a sagittal slice at the location of the marker was used and a measurement from posterior to anterior organ surfaces was taken. Differences were explored using a two-tailed paired samples t-test.

Organ migration used the same anatomical markers and a local coordinate system defined at L1. The coordinate system was created by using 4 points around L1 that were centrally located on the vertebrate (in relation to the head-foot distance). Using vector math, the locations of the anatomical markers were redefined into the local coordinate system and the difference of the distances of the organs in prone and supine positions were calculated.

RESULTS

Gross thoraco-abdominal compression was found to be 9.0% (71.6±12.9 mm prone vs. 78.7±12.4 mm supine). A significant difference between postures was found using a two tail, paired samples t-test ($p=0.0013$). Gender differences were not noted. Liver compression was on average 17.7% (110.4±4.1 mm prone vs. 134.8±2.6 mm supine; Figure 3.A). Spleen compression was found to be 3.5% (53.8±1.9 mm prone vs. 57.3±2.9 mm supine; Figure 3.B). Significant differences in liver compression were found ($p<0.0001$) however, there were no significant differences noted for spleen compression.

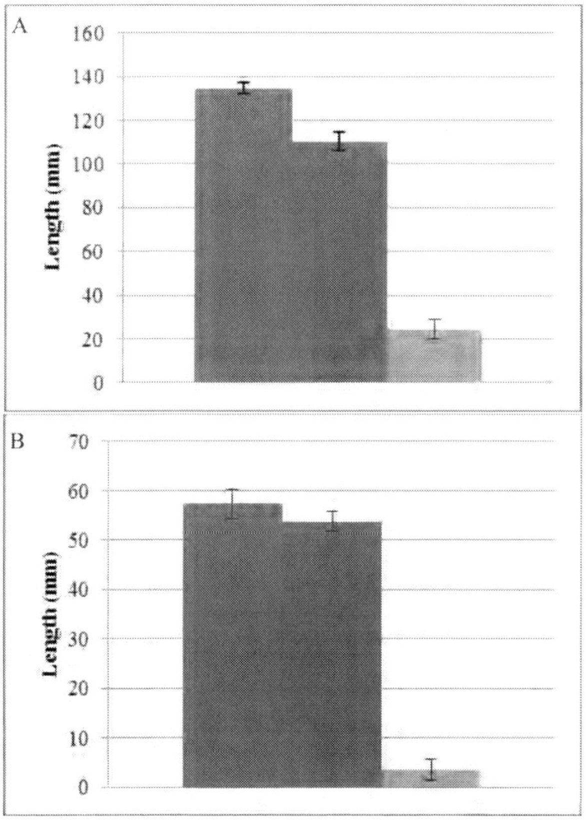

Figure 3. Average measurements in supine (left) and prone (middle) as well as their average compression in millimeters (right) for the A) Liver and B) Spleen. Error bars are representative of the standard error for each case.

Organ migration was found to occur as a result of posture. Organ migration was measured in the local coordinate system with Δx being analogous to ΔP-A (transverse plane), Δy analogous to ΔL-R (sagittal plane) and Δz analogous to ΔH-F (coronal plane). The liver was found to have an average 8.3 mm Δx (posterior migration), -11.4 mm Δy (rightward migration) and a 15.1 mm Δz (migration towards the head) migration. The spleen was found to have on average migrated -4.1 mm Δx (anterior migration), -6.1 mm Δy (rightward migration), and 1.7 mm Δz (migration towards the head; Figure 4).

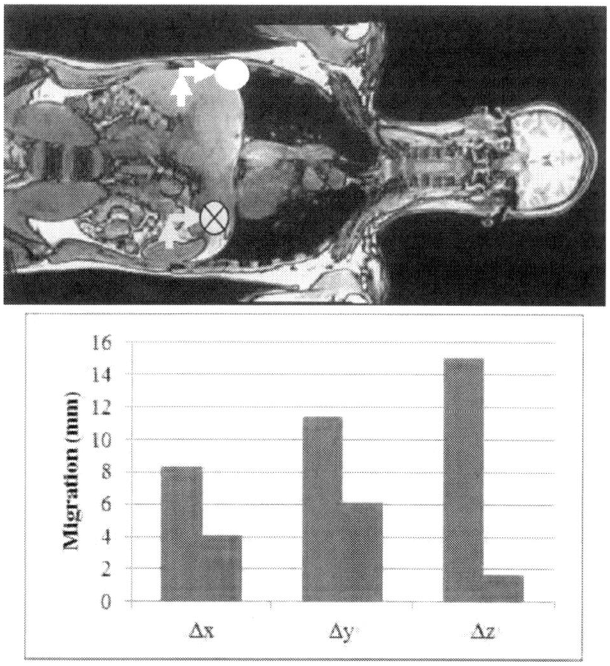

Figure 4. Coronal view of a subject. The white arrows indicate average direction of liver migration and yellow arrows indicate average direction of spleen migration (on top). Blank circle indicates migration towards the posterior while a circle containing an "X" indicated anterior migration. Migration of liver (blue) and spleen (red) are shown graphically for average 3-dimensional migration (on bottom). Chart does not indicate in-plane direction.

DISCUSSION

The purpose of the investigation was to quantify the effects of posture on gross thoraco-abdominal compression, organ compression and migration. To investigate these effects, MRI scans of 22 individuals were used in conjunction with anatomical markers. The findings are limited to healthy adults. The compression that occurs in the thoraco-abdominal region is a result of posture and likely by extension, body mass. These models could be used as baseline assessments for organ locations in human body models that mimic either the prone or supine posture.

ACKNOWLEDGEMENTS

This work was sponsored by the U.S. Army Research Lab through cooperative agreement W911NF-13-2-0039.

REFERENCES

[1] Karadimas, EJ et al, *J Spinal Disord Tech*, 19:495-500, 2006.
[2] Gayzik, FS et al, *Annals of BME*, 39:2568-2583. 2011.

SB³C2015
Summer Biomechanics, Bioengineering and Biotransport Conference
June 17-20, 2015,Snowbird Resort, Utah, USA
SB³C2015-1028

Pelvic Response of a Total Human Body Finite Element (FE) Model During Simulated Under Body Blast (UBB) Impacts

Caitlin M. Weaver (1,2), Randolph S. Coates (3), Andrew C. Merkle (4), Joel D. Stitzel (1)

(1) School of Biomedical Engineering and Sciences
Wake Forest University
Winston-Salem, NC, USA

(2) Soldier Protection Sciences Branch
US Army Research Laboratory
Aberdeen Proving Ground, MD, USA

(3) WIAMan Project Management Office (PMO)
US Army Research Laboratory
Aberdeen Proving Ground, MD, USA

(4) Biomechanics and Injury Mitigation Systems Program
Johns Hopkins University Applied Physics Lab
Laurel, MD, USA

INTRODUCTION

Injuries to the warfighter caused by explosive mechanisms have increased from 7% in World War II to 81% during Operation Iraqi Freedom (OIF) and Operation Enduring Freedom (OEF) [1]. UBB events in theater are the cause of many serious injuries sustained by the warfighter to the pelvis, spine, and lower extremities [2]. These injuries are often debilitating, resulting in increased healthcare expenses and a reduced quality of life. Injury prediction for UBB events continues to be a challenge due to the limited availability of UBB-specific test studies. UBB injury prediction methods are subject to injury criteria developed for motor vehicle crash injury prediction, often limited to automotive loading rates [3].

This study focuses on the pelvic injury response of the Global Human Body Models Consortium (GHBMC) 50th percentile seated FE human body model (v4.2) [4]. The input data used for this study was obtained from testing performed by the Biomechanics Product Team (BIO PT) for the U.S. Army Warrior Injury Assessment Manikin (WIAMan) project. The purpose of the WIAMan effort is to create an enhanced capability to assess risk to Soldiers in the UBB environment for use in live-fire test and evaluation and protection technology development. This includes the creation of a soldier-representative, biomechanically-validated anthropomorphic test device (ATD). Evaluation of GHBMC model fidelity and injury response is based on biofidelity targets (corridors) created using pelvis accelerations obtained from experimental testing of UBB-type loading using post mortem human subjects (PMHS) [5].

METHODS

The input data used for this FE study was obtained from vertically accelerative load testing performed on PMHS by the WIAMan BIO PT. Acceleration pulses applied to the FE rig were obtained from accelerometers attached to the floor and seat of experimental test vehicle rigs. These acceleration pulses were used to perform UBB FE simulations. For these simulations, the GHBMC was positioned in a FE vehicle rig seat within the measured tolerances used to position the PMHS for experimental testing (Figure 1). All simulations were performed using LS-DYNA software.

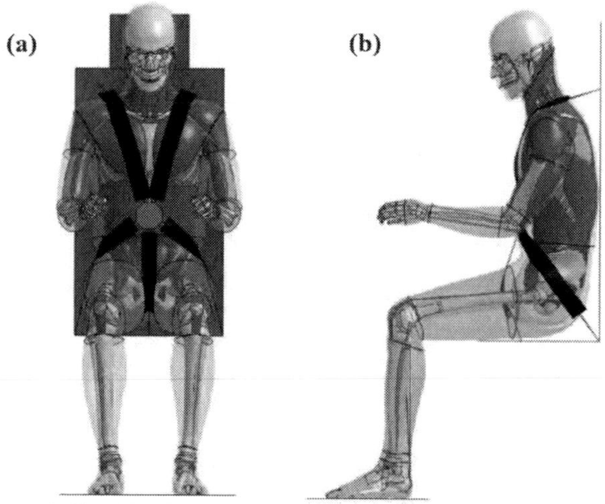

Figure 1. Test Rig Design with Belted GHBMC (a) Front View (b) and Side View.

Acceleration data from nodes in the S1 region of the pelvis of the GHBMC were extracted from the simulations. The extracted S1 acceleration data was compared to S1 data recorded from the experimental biofidelity corridors created from preliminary WIAMan experimental test data. Corridors were created using data filtered at 345, 1050, and 1750Hz. The corridors were generated using a standard approach determined by a Biofidelity Response Corridor (BRC) working group. The approach aligns non-normalized signals using the Nusholtz method, transforms signals to principal component space using eigenvectors and eigenvalues, and generates ±1 and ±2 standard deviation equivalent corridors [6].

RESULTS

The FE S1 acceleration showed good correlation with the biofidelity corridors. An analysis was performed using an objective rating method (CORrelation and Analysis, CORA) using the BRC generated corridor curves. The ±1 and ±2 SD curves were used for the inner and outer corridor limits, respectively. The average corridor curve was used as the cross-correlation reference. The CORA analysis showed good correlation (70% or higher) with an average of 70.7%, 83.5%, and 82.5% for the data filtered at 345, 1050, and 1750Hz, respectively.

Figures 2 through 4 show the CORA analysis of a FE simulation for data at all three filter levels. For these figures, the cross-correlation reference curve is represented in yellow, the inner corridor limit curves are represented in green, the outer corridor limit curves are represented in blue, and the FE simulation data curve is represented in red. The CORA rating method used for this analysis produced a CORA score of 74.6% for data filtered at 345Hz (Figure 2), 80.6% for data filtered at 1050Hz (Figure 3), and 78.2% for data filtered at 1750Hz (Figure 4).

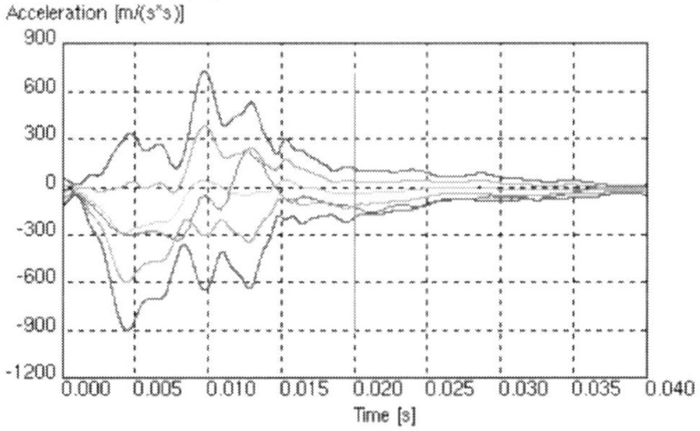

Figure 2. CORA Rating for Curve Comparison Between Biofidelity Corridors Generated with PMHS Test Data and an FE Simulation with GHBMC with CORA Score of 74.6%.

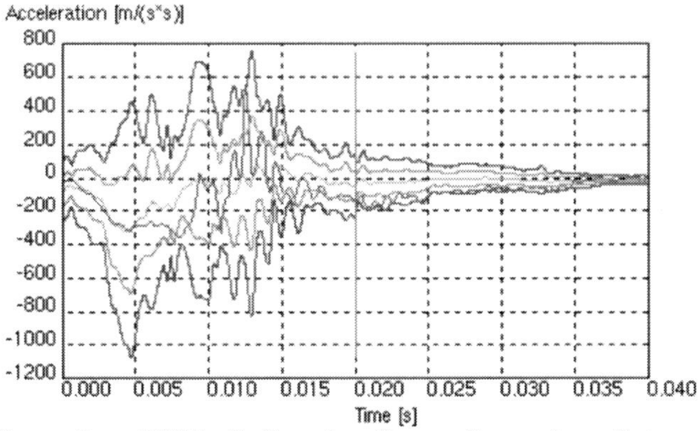

Figure 3. CORA Rating for Curve Comparison Between Biofidelity Corridors Generated with PMHS Test Data and an FE Simulation with GHBMC with CORA Score of 80.6%.

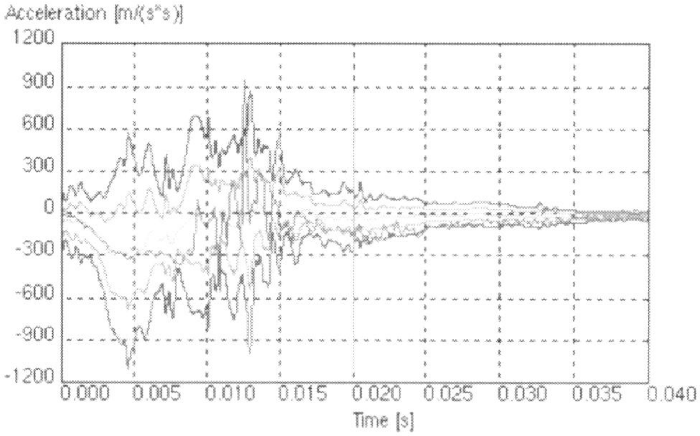

Figure 4. CORA Rating for Curve Comparison Between Biofidelity Corridors Generated with PMHS Test Data and an FE Simulation with GHBMC with CORA Score of 78.2%.

DISCUSSION

Injuries to the pelvis, spine, and lower extremities are frequently sustained as a result of UBB events. Though several studies have been conducted in an attempt to investigate the effects of UBB loading on the lower extremities, there have not been any significant research studies involving these effects on the pelvis. The current focus of biomechanical research on the pelvis has been on automotive rate loading. This focus presents an issue for UBB related injuries because common injuries such as pelvic ring and ischium fractures are not commonly caused by automotive rate loading. As a result, it is important to conduct research studies on the pelvis involving UBB type loading to understand the causation and mechanisms of these types of pelvis injuries [2].

The initial work for this FE study has shown good correlation for results comparison between PHMS experimental testing and FE human body model acceleration outputs of the S1 region of the pelvis. Further FE UBB impact simulations and additional human body model metrics will be compared to the experimental biofidelity corridors. To date, the comparison of full body UBB experimental testing to drive and compare with full body FE simulation metrics for UBB is unique. This data was acquired with the explicit purpose of enhancing vehicle and soldier survivability.

ACKNOWLEDGEMENTS

The authors would like to acknowledge the US Army Research Lab, the John Hopkins Applied Physics Lab, and the Science, Mathematics And Research for Transformation (SMART) Scholarship for Service Program for their support and collaboration.

REFERENCES

[1] Yoganandan, et. al. Clinical Biomechanics. 28 (2013) 302-609.
[2] Merkle, A. C., et. al. John Hopkins APL Technical Digest, Vol. 31. No. 4. 2013. 286-295.
[3] Bailey, A. M., et. al. IRCOBI Conference 2013. IRC-13-25. 158-170.
[4] Vavalle, N. A., et. al. Ann Biomed Eng, vol. 41, pp. 497-512, Mar 2013.
[5] Chancey, Carol, et. al. Program for Warrior Injury Assessment Manikin (WIAMan). Presented at USAMRC Industry Day, March 15-18, 2011.
[6] Nusholtz, Guy S., et al. Statistical Considerations for Evaluating Biofidelity, Repeatability, and Reproducibility of ATDs. No. 2013-01-1249. SAE Technical Paper, 2013.

LUMBAR-PELVIC COORDINATION DIFFERENCES BETWEEN NOVICE AND EXPERIENCED LIFTERS IN REPETITIVE LIFTING

Timothy D. Craig (1), Alice E. Riley (1), Sandra A. Billinger (2), Neena K. Sharma (2), Sara E. Wilson (1)

(1) Department of Mechanical Engineering Department
University of Kansas
Lawrence, KS, USA

(2) Department of Physical Therapy and Rehabilitation Science
University of Kansas Medical Center
Kansas City, KS, USA

INTRODUCTION

Low back pain (LBP) is a costly and common problem accounting for approximately 40% of all disability compensation claims for individuals under 45 years old in the United States [1]. National Institute for Occupational Safety and Health has identified several occupational conditions to be associated to LBP including repetitive lifting [2].

Previously Maduri et al. measured lumbar range of motion (ROM) to normalize lumbar motion in subjects of varying levels of flexibility [3]. This study observed that a novice group of lifters demonstrated a highly kyphotic posture relative to ROM during a cyclic lifting task and it was speculated that this coordination may predispose these lifters to injury.

The goal of this study was to examine the differences between novice and experienced lifters using the ROM method to determine if lifting experience changes the selected lifting strategy. In addition, two lumbar pelvic coordination strategies, including a highly kyphotic strategy and a neutral strategy, were examined to determine the effects of lift training on lifting strategy.

METHODS

Twenty four subjects, 14 males and 10 females, participated in this study with the approval of the Human Subjects Committee from the University of Kansas Medical Center. Subjects were divided into two groups based on experience lifting weights. Subjects were classified as experienced (EXP) if they lifted weights at least 3 times per week over the previous year. Subjects were classified as novice (NOV) if they did not regularly lift weights over the previous year. This resulted in 11 EXP lifters with 7 males and 4 females. The NOV group had 7 males and 6 females for a total of 13 subjects.

A lifting task was created where the subjects lifted a milk crate repeatedly from ground level to waist level. The lifting task was repeated continuously for four minutes with each lift cycle (ground-waist-ground) lasting four seconds based on an auditory cue to the subject. The milk crate contained weight equivalent to 3% of a single max isometric lift collected prior to the lifting trials.

Three electromagnetic sensors (MotionStar, Ascension Technology, VT) collected position and orientation data at 100Hz using MotionMonitor Software (Innsport, IL). Sensors were located at T10, S1, and the manubrium and were used to define the trunk flexion angle (FA) and lumbar angle (LA). Where the FA was the angle created between T10-S1-vertical and the LA was the difference in angular tilt of the sensors located at T10 and S1 describing the curvature of the lumbar spine.

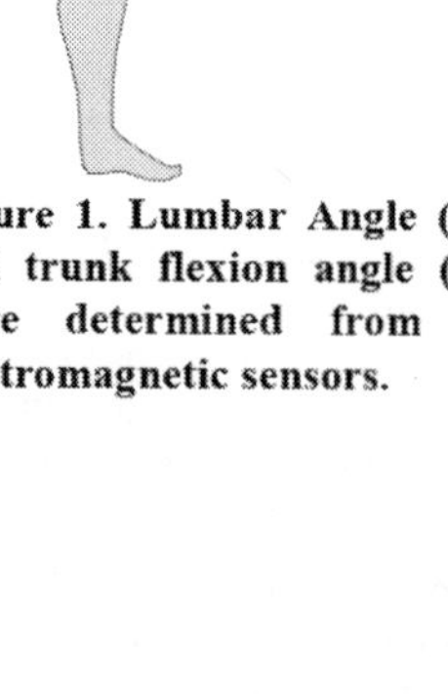

Figure 1. Lumbar Angle (LA) and trunk flexion angle (FA) were determined from the electromagnetic sensors.

The ROM of the LA was recorded at four flexion angles between 0 and 80 degrees. These data were linearly interpolated to define a ROM envelope for all FA. This ROM envelope was used in defining the trained lifting strategies by normalizing LA to individual ROM. One trained strategy required a highly kyphotic, or pelvis first (PF), strategy as shown in Figure 2. The second strategy required a neutral spine (NU).

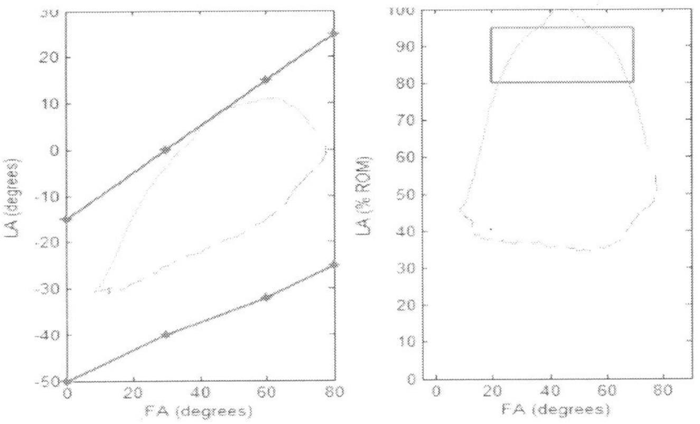

Figure 2. Normalizing the lumbar angles to subject ROM.

A biofeedback program displayed the instantaneous feedback of the relationship between the normalized LA and the FA on an xy plot along with a target box for the given strategy. The target box spanned between 20 and 70 degrees FA for both trained strategies. For the PF strategy the target ranged between 80% and 95% LA while the range for NU ranged between 42% and 58% LA. During the trained strategies subjects were instructed to make the cursor pass through the target box during the extension (lifting) phase and perform the descending phase however they choose. The self selected strategy was always performed prior to introducing the biofeedback and therefore did not have visual feedback for lifting coordination.

The trials were divided into individual lift cycles, then lift phase, and finally into quadrant ranges based on FA. The highest range of FA, quadrant four, was removed from the analysis as many subjects failed to reach the range towards the end of the trial. The normalized LA was averaged for four cycles for all data in the quadrants for a given lift phase. A repeated measures ANOVA was performed on these values with differences considered significant at $p \leq 0.05$.

RESULTS

It was observed that the NOV group self selected a strategy similar to the PF trained strategy while the EXP group self selected a strategy similar to the NU strategy. During the extension phase of lifting the self selected lifting strategy of NOV lifters was more kyphotic than EXP lifters, starting at 69.9% and 42.4% respectively.

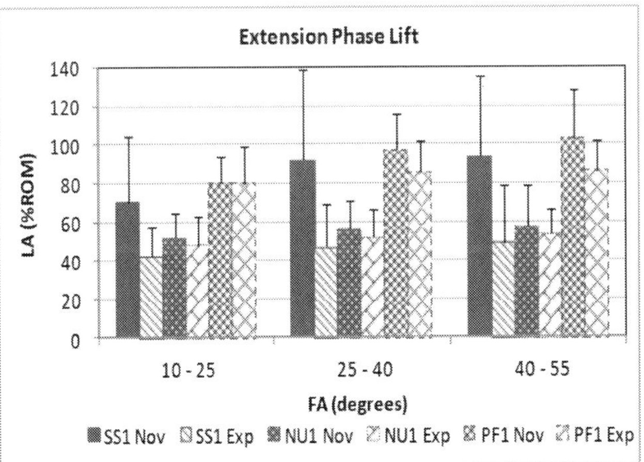

Figure 3. Normalized LA for the three ranges of FA examined.

Figure 3 demonstrates the normalized LA increased for both groups over the quadrants with the NOV group reaching a maximum LA of 93.4% compared to 49% for the EXP group. Statistical differences ($p \leq 0.05$) were observed for the three different strategies, the lift phase direction, and the quadrants.

Both groups adjusted to the trained strategies with lower LA in the neutral (NU) strategy and higher LA in the PF strategy. However, it was observed that the NOV group remained more kyphotic than the EXP group in both trained strategies. It was also observed that the NOV group increased LA from the self selected to the PF strategy even though the self selected strategy was already in the target region.

DISCUSSION

It was hypothesized that the NOV group would select a more kyphotic SS lifting strategy approaching their ROM limits while the experienced group would select a more neutral strategy, maintaining near the middle of their ROM. This hypothesis was supported by the significant difference ($p \leq 0.05$) between NOV and EXP lifters with the NOV group self selecting a strategy that resembles the PF lifting strategy while the EXP group selects a strategy more similar to the NU strategy. This result would agree with Maduri et al. where the subjects, believed to be novice lifters, were found to use highly kyphotic posture [3]. One other study has also shown that experienced lifters have reduced lumbar flexion angles during lifting tasks [4]. However, this previous study did not assess LA relative to the subjects' range, so the finding that EXP lifters used neutral LA postures is novel.

A necessary step to determining low back injury prevention training is improving the understanding of lifting strategies. To create a training method to reduce low back pain it is useful to use the observations from studies examining the differences between novice and experienced lifters [5-6]. It is believed that experienced lifters will avoid potentially injurious strategies and could provide insight to decreasing injury risk.

This current study is limited by the method of determining the ROM. The procedure of collecting LA at given FA was performed statically while the task itself was dynamic. Additionally the criteria for the EXP group was based weight lifting experience while other studies have defined experience as manual materials handling workplace experience.

Future work will examine whether training subjects and low back pain patients to use a NU strategy may reduce incidence of low back injury.

ACKNOWLEDGEMENTS

Research reported in this publication was supported by National Institute of Arthritis and Musculoskeletal and Skin Diseases of the National Institutes of Health under award number 1R03AR061597-01. The content is solely the responsibility of the authors and does not necessarily represent the official views of the National Institutes of Health.

REFERENCES

[1] Lis, A et al., Euro Spine J 16(2):283-289, 2007
[2] Bernard, B et al., USDoHaH, 97-141, 1997
[3] Maduri, A et al., J Electomyogr Kinesiol, 18(5): 807-814, 2008
[4] Plamondon, A et al., Ergonomics, 53(10): 1239-1253, 2010
[5] Authier, M et al., Int J Occup Saf Ergon, 1(3): 262-275, 1995
[6] Marras, W et al., Clin Biomech, 21(4):345-352, 2006
[7] Lee, J et al., J Biomch, 46(6):1211-1215, 2013

SB3C2015
Summer Biomechanics, Bioengineering and Biotransport Conference
June 17-20, 2015, Snowbird Resort, Utah, USA
SB3C2015-1030

REGIONAL CONTRACTION SHAPES THREE-DIMENSIONAL MORPHOGENESIS OF THE EMBRYONIC FOREBRAIN

Kara E. Garcia (1)

Philip V. Bayly (2), Larry A. Taber (1)

(1) Department of Biomedical Engineering
Washington University in Saint Louis
Saint Louis, MO, USA

(2) Department of Biomedical Engineering
Washington University in Saint Louis
Saint Louis, MO, USA

INTRODUCTION

Early embryonic structures play a critical role brain development. In humans, embryonic malformations have been directly linked to birth defects including anencephaly and holoprosencephaly, and abnormal brain structure has been associated with autism, schizophrenia, and bipolar disorder [1]. Through neurulation, the brain first emerges as a straight, cerebrospinal fluid (CSF)-filled tube [2]. Using experiments and axisymmetric models, Filas et al. (2012) showed that actomyosin on the apical side of neuroepithelium contracts to form primary boundaries between the forebrain, midbrain and hindbrain regions. However, subsequent boundary formation and propagation remains unstudied from a mechanical perspective.

Using experimental and computational models, this study examines the three-dimensional mechanics behind forebrain division at the anterior intraencephalic sulcus (AIS). Models must additionally consider complex geometry due to optic vesicles (prospective eyes) protruding laterally from the forebrain and rapid inflation due to increased CSF pressure. The first aim of this study was to determine the mechanism behind AIS formation, accomplished with perturbation experiments. The second aim was to quantify and model this mechanism, as well as potential changes during later stages of increased pressure and boundary propagation.

METHODS

Experimental Methods: Fertilized white Leghorn chicken embryos were incubated 40-72h to the desired stage then extracted using the filter paper method [1,2]. For perturbation experiments, embryos were cultured for 6h (from Hamburger and Hamilton stage 11 to stage 12+) under altered pressure, altered contraction, or control conditions. To inhibit or enhance contraction, embryos were cultured in media containing 60uM blebbistatin (myosin II inhibitor) or 30nM calyculin A (myosin II phosphatase inhibitor), respectively. To relieve CSF pressure, embryos were cultured with an open glass tube inserted to allow flow between the brain lumen and external fluid. Three-dimensional geometries were recorded using optical coherence tomography (OCT), and diameters of each brain region were measured from bright field images. To quantify actin, embryos were fixed in 3.7% formaldehyde then stained with phalloidin. Apical actin was imaged using 3D confocal microscopy or the 2D flat-mount technique described in Filas et al. (2012) and Guthrie et al. (1991). In-plane cell sheets were reconstructed from 3D confocal using ImageJ Volume Viewer, and actin orientation was determined using ImageJ Directionality, a Fourier spectrum analysis.

Computational Methods: Three-dimensional models were created using ABAQUS FEA (D. S. Simulia). Geometry was matched to OCT measurements, while nonlinear material properties and CSF pressures were taken from previous literature [4-6]. Apical contraction was approximated with membrane elements contracting to half of their original length in directions defined tangent to the surface.

RESULTS

Perturbation experiments (Figure 1) showed that actomyosin contraction significantly impacts AIS boundary shape (arrows in 1B, n=9-10 per group). Under normal development, the AIS decreased in diameter with respect to both initial and adjacent vesicle diameters (1A, D). These effects were exaggerated when contraction was enhanced with calyculin and prevented when contraction was inhibited with blebbistatin (1A-D). OCT coronal cross-sections of the AIS (1C) suggest specific regions of increased contraction, in agreement with Nicholson-Flynn et al. (1996) who found increased high molecular weight tropomyosin (TM) in similar locations of the rat embryo (1E). These locations are the roof plate and floor plate (up and down arrows, respectively) and optic stalks (arrowheads). Perturbations to relieve pressure did not prevent AIS formation (n=5, data not shown).

Confocal imaging after initial boundary formation (Figure 2B, C) revealed uniform apical cell size and strong circumferential alignment of actin in locations of TM expression. Consistent with Filas et al. (2012), alignment was also seen along the full circumference of primary boundaries but not in the adjacent vesicles (2B). As shown in Figure 2A, modeling circumferential contraction at locations of aligned actin produced realistic boundary shapes. For older embryos (Figure 3), actin staining showed smaller apical shapes in the vesicles compared to the boundary and younger stages.

It is worth noting that, for flat-mounted samples, deformation due to bending could artificially increase the observed trends in boundary orientation and decrease the observed trends in cell size. However, 3D confocal results (n=3 stage 12-13) were consistent with trends seen in flat-mounted samples (n=6 stage 11-14, n=6 stage 16-18). Flat-mounted results were also consistent with trends in Filas et al. (2012) and Guthrie et al. (1991).

DISCUSSION

Perturbation experiments support actomyosin contraction as the primary mechanism of AIS boundary formation. Furthermore, apical alignment of actin suggests greater circumferential contraction in the boundary region. This circumferential alignment in the boundary and isotropic actin in surrounding vesicles fits previous reports on the cylindrical midbrain-hindbrain boundary [1]. However, the AIS differs in that cylindrical optic stalks provide a second axis of circumferential alignment, perpendicular to the first (Figure 2A, blue). This observation raises an interesting question: Why does actin align preferentially in the circumferential direction?

Computational models in this study are limited to predefined, time-dependent contraction. However, future models will explore the effects of stress-dependent growth and contraction. The mechanism of stretch activation, in which cells contract in response to tension, has been previously studied for blood vessels and embryonic wound healing [6]. Preliminary models suggest that points of contraction, such as localizations of TM expression, could lead to primarily circumferential stress in the boundaries (not shown). Though CSF pressure could also provide tension, initial differences between boundaries and vesicles would be small, and perturbation experiments showed that embryos still initiate boundaries without pressure.

On the other hand, actin staining in older embryos revealed increased apical constriction in vesicle walls compared to boundaries. This is consistent with the observations of Guthrie et al. (1991) for rhombomeres in the hindbrain. Importantly, studies have also shown that CSF pressure and vesicle diameter increase drastically from stage 12 to 18 [2,5]. For a pressurized vessel, this difference in diameter could lead to increased vesicle stresses and potentially contraction. Future experiments and models will explore the effects of decreased pressure and contraction at later stages of boundary propagation.

One limitation to this study is that it focused on short-term contraction and did not consider growth effects due to cell division (on the order of 8-12h). Furthermore, molecular signaling pathways related to contraction and boundary formation were outside the scope of this study. Future work should address these questions as well as those related to pressure and stretch activation.

ACKNOWLEDGEMENTS

This work was funded by the National Institutes of Health Grants T32 EB018266-01 and R01 NS070918.

REFERENCES

[1] Filas, B et al. *Phys Biol.* 9(6):066007, 2012.
[2] Hamburger, V and Hamilton, H, *J Morphology,* 88:49-91, 1951.
[3] Guthrie, S et al., *J Neurobiol.*, 22(7):742-754, 1991.
[4] Xu, G et al., *J Biomech.*, 132(1):011005, 2010.
[5] Jelinek, R and Pexieder, T, *Physiol Bohemoslov*, 17:297-305, 1968.
[6] Taber L, *Phil. Trans. R. Soc. A*, 367:3555-3583, 2009.
[7] Nicholson-Flynn, K et al., *J Neurosci.*, 16(21):6853-6863, 1996.

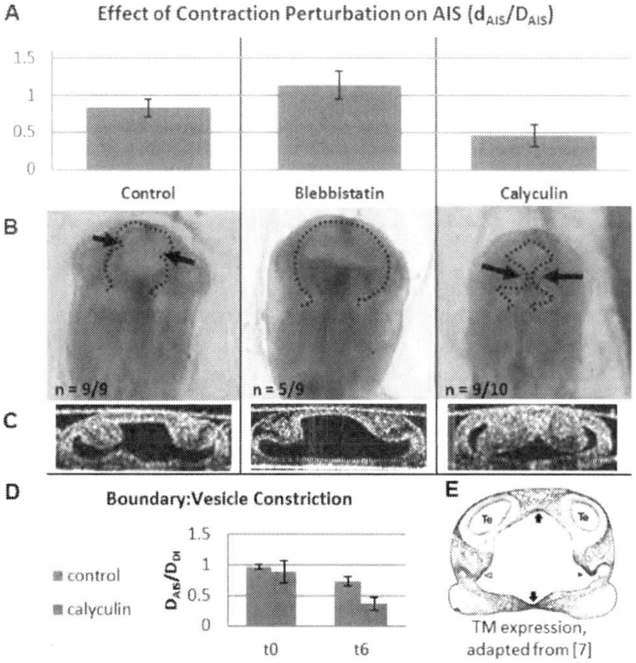

FIGURE 1. ALTERED CONTRACTION CHANGES BOUNDARY MORPHOGENESIS.

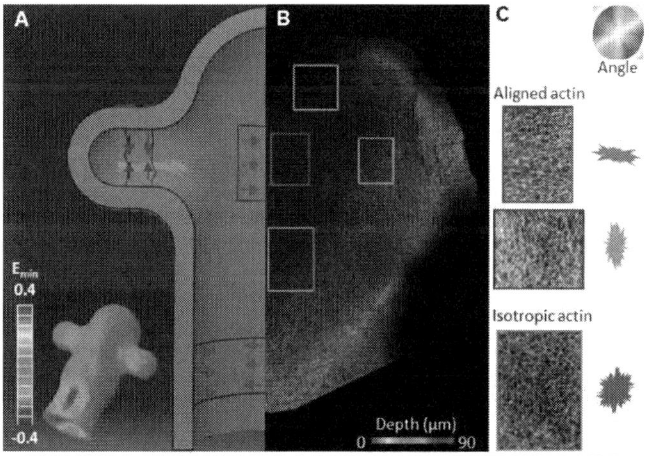

FIGURE 2. REGIONS OF ACTIN ALIGNMENT SUPPORT CIRCUMFERENTIAL CONTRACTION.
(Scale bar in B: 250µm across, 90µm color range)

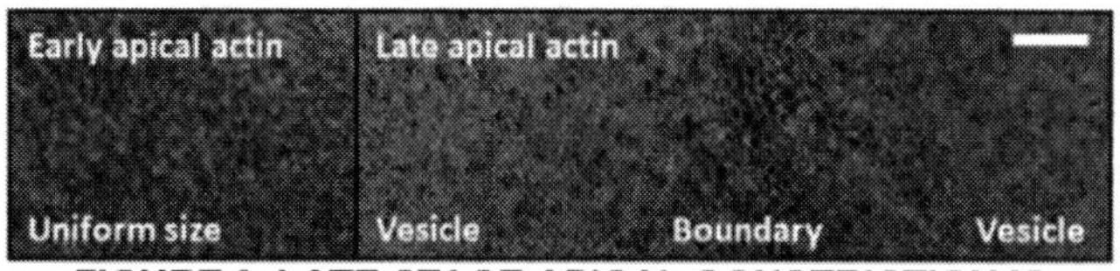

FIGURE 3. LATE STAGE APICAL CONSTRICTION IS GREATER IN VESICLES. (Scale bar: 100µm across)

TRUE 3D STRESSES IN HEART VALVE LEAFLETS

Bruno V. Rego and Michael S. Sacks

Institute for Computational Engineering and Sciences
Department of Biomedical Engineering
The University of Texas at Austin
Austin, TX, USA

INTRODUCTION

Transmural stress distributions in the valve leaflets are highly heterogeneous, as they combine the responses of the distinct load-bearing layers. While the mechanical behaviors of each layer have been previously investigated [1], fundamental questions remain as to the optimal approach for integrating these results into a model of true 3D stress in the intact valve leaflet tissues. In addition to an accurate description of stress throughout the valve tissue, a detailed consideration of structural factors such as fiber crimp and orientation distribution is also necessary to establish a clear structure-function relationship in each layer, and thus in the intact native valve leaflet. Structural information is typically obtained in a non-preconditioned, unloaded state. In contrast, mechanical testing samples first undergo a series of permanent set-like deformations, such as layer separation and preconditioning, which are time-independent and stable for the duration of mechanical testing [2]. Pre-runtime deformations thus transform a specimen kinematically into a new stress-free state where strain energy is defined, which must serve as the reference in stress analysis to satisfy Cauchy elasticity. The discrepancy between the initial free-floating and initial runtime states presents a need to formulate a structural model which can account for these effects directly. To address this issue, we have derived a constitutive relation which allows for strain energy and structural properties to be considered with respect to different referential configurations, according to the states from which each data set was obtained.

METHODS

Mechanical Data. We chose to apply the theory using AV mechanical data from a previous study, obtained per the methods previously described in detail [1]. In brief, rectangular test samples from the lower belly region of fresh porcine AVs were acquired and an array of four fiducial markers was permanently affixed to each side. All intact specimens, as well as their separate fibrosa and ventricularis layers, were subjected to tension-controlled biaxial testing under seven protocols of varying circumferential-to-radial tension ratios (0.1:1, 0.5:1, 0.75:1, 1:1, 1:0.75, 1:0.5, and 1:0.1). Each specimen was imaged in its initial free-floating state, in its post-preconditioned stress-free runtime state for each protocol, and continuously during testing, such that the fiducial marker array could be tracked throughout loading.

Kinematic Analysis. As a way of concisely accounting for and quantifying deformation between the various states β_i of each AV leaflet specimen, we introduce the notation shown in Fig. 1, which labels initial free-floating, initial runtime (post-preconditioned), and currently loaded (at time t) states both for single layers and the intact leaflet. We also define ${}^{j}_{i}\mathbf{F}_l$ as the deformation gradient between states β_i and β_j for layer l (note that we assume the leaflet behaves as a tightly bonded structure, so l is not specified between any two intact states, since the gradient applies identically to both layers).

In order to assign a single runtime reference state to each specimen, each fiducial marker's initial runtime coordinates were averaged individually ("pooled coordinate averaging") across all biaxial test protocols for that specimen. Each specimen's average initial runtime state was subsequently used as the reference configuration for all stress analysis. A custom-written Mathematica (Wolfram Research Corp.) implementation of a Nelder-Mead optimization algorithm was utilized to fit the constitutive model to the experimental data.

Theoretical Framework. Following established procedures [2,3], we consider an ensemble of fibers with varying slack strain and use a recruitment function D at the ensemble level to account for a range of slack strains, all referenced to β_0 for the case of a single layer (Fig. 1), using

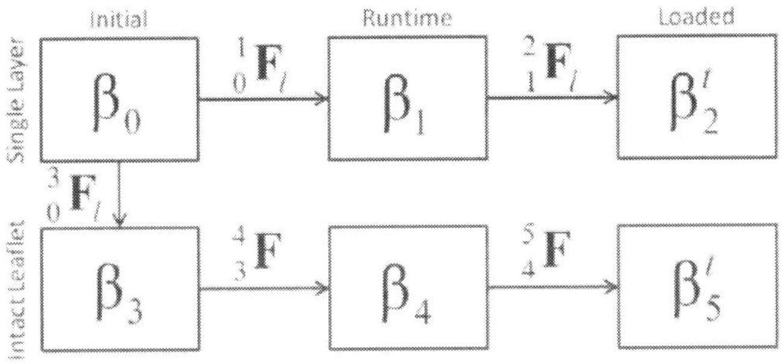

Fig. 1 Schematic diagram of the different AV states, establishing notation for specimens in initial, runtime, and externally loaded states.

$$D_0(\mu,\sigma,{}^{t}_{0}E_{ens})=\begin{cases}\dfrac{y^{\alpha-1}(1-y)^{\beta-1}}{B(\alpha,\beta)({}_{0}E_{ub}-{}_{0}E_{lb})} & y\in[0,1]\\ 0 & \text{otherwise}\end{cases}$$

$$y=\frac{{}^{t}_{0}E_{ens}-{}_{0}E_{lb}}{{}_{0}E_{ub}-{}_{0}E_{lb}},\ \bar{\mu}=\frac{\mu-{}_{0}E_{lb}}{{}_{0}E_{ub}-{}_{0}E_{lb}},\ \bar{\sigma}=\frac{\sigma}{{}_{0}E_{ub}-{}_{0}E_{lb}},\ \alpha=\frac{\bar{\mu}^2-\bar{\mu}^3-\bar{\sigma}^2\bar{\mu}}{\bar{\sigma}^2},\ \ \beta=\alpha\frac{1-\bar{\mu}}{\bar{\mu}} \quad (1),$$

where $B(\alpha,\beta)=\frac{\Gamma(\alpha)\Gamma(\beta)}{\Gamma(\alpha+\beta)}$ is the Beta function with shape constants α, β, Γ the gamma function, and ${}^{t}_{0}E_{ens}=\mathbf{N}^{T}\,{}^{t}_{0}\check{\mathbf{E}}\,\mathbf{N}$. However, the recruitment function parameters need to be recast in β_1. To do this we define D_1 from D_0 with the same mean μ and standard deviation σ defined in (1) but mapped to β_1, using

$$D_1(\mu,\sigma,{}_{0}E_{lb},{}_{0}E_{ub},{}^{t}_{0}E_{ens},{}^{t}_{1}E_{ens})=\begin{cases}\dfrac{y^{\alpha-1}(1-y)^{\beta-1}}{B(\alpha,\beta)({}_{1}E_{ub}-{}_{1}E_{lb})} & y\in[0,1]\\ 0 & \text{otherwise}\end{cases}$$

$$y=\frac{{}^{t}_{1}E_{ens}-{}_{1}E_{lb}}{{}_{1}E_{ub}-{}_{1}E_{lb}},\ \bar{\mu}=\frac{\mu-{}_{0}E_{lb}}{{}_{0}E_{ub}-{}_{0}E_{lb}},\ \bar{\sigma}=\frac{\sigma}{{}_{0}E_{ub}-{}_{0}E_{lb}},\ \alpha=\frac{\bar{\mu}^2-\bar{\mu}^3-\bar{\sigma}^2\bar{\mu}}{\bar{\sigma}^2},\ \ \beta=\alpha\frac{1-\bar{\mu}}{\bar{\mu}} \quad (2),$$

$${}^{t}_{1}E_{ens}=\frac{{}^{t}_{0}E_{ens}-{}^{1}_{0}E}{1+2\,{}^{1}_{0}E},\ {}_{1}E_{lb}=\frac{{}_{0}E_{lb}-{}^{1}_{0}E}{1+2\,{}^{1}_{0}E},\text{and }{}_{1}E_{ub}=\frac{{}_{0}E_{ub}-{}^{1}_{0}E}{1+2\,{}^{1}_{0}E}.$$

The resulting expression for the ensemble stress is thus

$${}^{t}_{1}S_{ens}\left({}^{t}_{1}E_{ens}\right)=\eta_c\int_0^{{}^{t}_{1}E_{ens}}\frac{D_1(\mu,\sigma,{}_{0}E_{lb},{}_{0}E_{ub},x)}{(2x+1)}\left(1-\frac{\sqrt{2x+1}}{\sqrt{2\,{}^{t}_{1}E_{ens}+1}}\right)dx \quad (3).$$

The matrix is a single phase using a neo-Hookean model [3]. When incompressibility is enforced at the final tissue level, the final tissue level 2nd PK stress tensor is given by modifying eqn. 3 to obtain

$${}^{t}_{1}\check{\mathbf{S}}\left({}^{t}_{1}\check{\mathbf{E}}\right)=\eta_c\int_{\phi}\Gamma(\mu_\Gamma,\sigma_\Gamma,\phi)\left\{\int_0^{{}^{t}_{1}E_{ens}}\frac{D_1(\mu_D,\sigma_D,{}_{0}E_{lb},{}_{0}E_{ub},x)}{(2x+1)}\left(1-\frac{\sqrt{2x+1}}{\sqrt{2\,{}^{t}_{1}E_{ens}+1}}\right)dx\right\}(\mathbf{n}\otimes\mathbf{n})d\phi$$

$$+\eta_m\left(\check{\mathbf{I}}-\frac{{}^{t}_{1}\check{\mathbf{C}}^{-1}}{{}^{t}_{1}C_{33}}\right),\ \text{where}$$

$${}^{t}_{1}E_{ens}=\mathbf{n}^{T}\,{}^{t}_{1}\check{\mathbf{E}}\,\mathbf{n},\ \ \Gamma_{\phi}[\mu_\Gamma,\sigma_\Gamma,\phi]=\Gamma_{\theta}(\mu_\Gamma,\sigma_\Gamma,\theta)\frac{\lambda^2}{{}^{1}_{0}J_{2D}}=\Gamma_{\theta}(\mu_\Gamma,\sigma_\Gamma,\theta)\frac{\mathbf{N}\cdot{}^{1}_{0}\check{\mathbf{C}}\cdot\mathbf{N}}{\det\left[{}^{1}_{0}\check{\mathbf{F}}_{2D}\right]},\ \theta=\theta\left({}^{1}_{0}\mathbf{F},\phi\right) \quad (4)$$

Eqn. 4 has a total of eight model parameters η_c, η_m, μ_Γ, σ_Γ, μ_D, σ_D, ${}_{0}E_{lb}$, ${}_{0}E_{ub}$, all with a physical meaning and all referred to β_0. When implementing eqn. 4, the following considerations should be noted:

1. The bounds on the outer integral are treated as $\phi\in[-\pi/2, \pi/2]$. As long as $\Gamma(\phi)$ is sufficiently narrow and of zero magnitude near $\phi=\pm\pi/2$, this should be sufficient.
2. All model parameters are understood to be determined in the β_0 configuration.
3. When examining the appropriateness of this approach, it is important to follow the assumption that all fibers remain slack during the set deformation.
4. While all model parameters are referred to β_0, the stress and strain tensors are referenced to β_1.
5. Even if $D({}_{0}E_s)$ is independent of θ in β_0, it will in general be a function of ϕ in β_1. Independence of orientation will only be maintained under isotropic set deformations with no shear.
6. The matrix phase is always referred to β_1.

RESULTS

Optimizations of the parameter set for the full constitutive model resulted in extremely good fits to experimental data for both intact and single-layer specimens ($r^2 > 0.95, p < 0.001, n = 6$). Representative samples of fitted data illustrate substantial layer-wise transmural variation of mechanical properties through the AV leaflet (Fig.2).

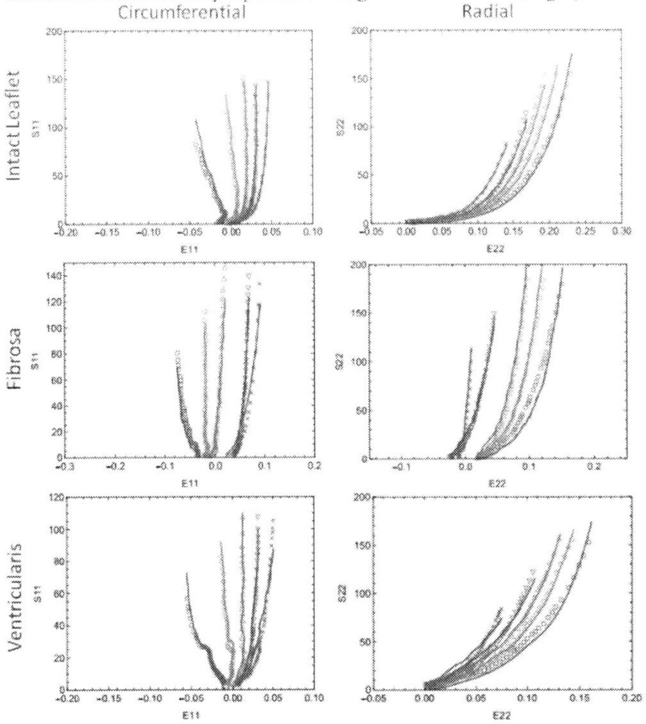

Fig. 2 Representative plots for the AV layer data.

DISCUSSION

In the current study, a constitutive model which accurately describes layer-wise AV leaflet behavior with respect to a post-preconditioned state has been attained. A framework has also been developed that allows equivalent strain states in both load-bearing layers and the intact leaflet to be compared directly, thus providing an accurate account of transmural stress distribution in the AV. This achievement also lays important groundwork for developing a more complete understanding of structure-function relationships in the AV, as structural data collected in β_0 and β_3 can be transformed into a corresponding permanent-set reference configuration through the assumption of affine fiber kinematics, and can then be integrated into future analyses of leaflet response to external stresses. In addition, the methods developed here can be easily generalized to the mechanical analysis of other planar soft tissues, and thus have the potential to apply to various endeavors in the field of biomechanics.

ACKNOWLEDGEMENTS NIH grants R01HL119297 and R01HL108330.

REFERENCES

[1] Stella, J and M Sacks, *J Biomech Eng*, 129:757-766, 2007.
[2] Lokshin, O and Y Lanir, *Biomaterials*, 30:3118-3127, 2009.
[3] Fan, R and M Sacks, *J Biomech*, 47:2043-2054, 2014.

SB[3]C2015
Summer Biomechanics, Bioengineering and Biotransport Conference
June 17-20, 2015, Snowbird Resort, Utah, USA
SB³C2015-1033

DESIGN AND IMPLEMENTATION OF AN INSTRUMENTED PEDAL FOR CYCLING BIOMECHANICS RESEARCH

Luke I. Kraemer (1), Juan D. Gutierrez-Franco (1), Jake E. Deschamps (1), Karim C. Dudum (2), Eshan M. Dandekar (3), Scott J. Hazelwood (1,2), Hemanth V. Porumamilla (1), Stephen M. Klisch (1,2)

(1) Mechanical Engineering Department
California Polytechnic State University
San Luis Obispo, CA

(2) Biomedical Engineering Department
California Polytechnic State University
San Luis Obispo, CA

(3) Kinesiology Department
California Polytechnic State University
San Luis Obispo, CA

INTRODUCTION

Cycling is a common, low-impact activity used for recreation, exercise, and rehabilitation. Knee joint loading can be predicted using inverse dynamic analyses of pedal load cell and kinematic data measured during cycling biomechanics experiments. Several studies have successfully measured foot loading at the pedals, e.g. by using custom instrumented pedal spindles outfitted with strain gauges and a potentiometer to measure crank angle [1-3]. Such designs are relatively complex, requiring difficult machining and component fabrication, and require post-processing of strain gauge data.

The long-term goal of this study is to calculate knee joint loading and, ultimately, cartilage tissue stress to provide evidence-based prescriptions for rehabilitative and fitness sustainment exercises for those who are at high risk for knee osteoarthritis. The objective of this project was to design, fabricate, and implement an instrumented pedal system using existing load cells for use in cycling biomechanics research. Measured data retrieved from the load cells during cycling experiments will be compared to similar studies to verify that this project was successful.

METHODS

Design criteria were specified as follows: 1) to replicate foot motion that occurs with a bicycle's stock pedals, 2) to easily implement the pedals on different bicycles, 3) to incorporate two existing, 6-channel load cells (AMTI AD2.5D-250, Watertown, MA), and 4) to implement with an 8-camera motion capture system (Motion Analysis Corp., Santa Rosa, CA) for tracking motion of the load cell coordinate systems.

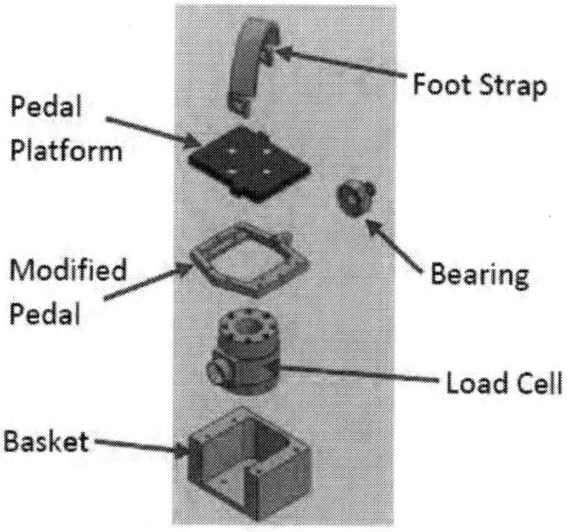

FIGURE 1. EXPLODED VIEW OF INSTRUMENTED PEDAL DESIGN.

__Design and fabrication__. The design (Fig. 1) replicates the circular foot motion in order to best estimate pedal loading that occurs with a non-instrumented pedal. This was accomplished by lowering the load cell into the body of the pedal. Such a design is difficult to achieve with standard bicycle pedals that have a 2-bearing system with a spindle that extends the entire width of the pedal and, thus, would impede force transmission between the foot and the pedal. Tioga MT-Zero pedals (Tioga, Gardena, CA) were used because their spindle-less design consequently resulted in ease of implementation with the load cells. The Tioga pedals have a 1-cm long axle that mates with one oversized bearing which threads into the crank. Thus, there is no spindle to interfere with the pedal body where the foot contacts the load cell. Additionally, the Tioga pedals are made from high quality chromoly steel, which is strong enough to allow machining without sacrificing structural integrity. To modify the Tioga pedals, the center skeleton was removed until there was enough room to drop a load cell through the modified pedals (Fig. 1). A custom basket was machined and attached underneath each modified pedal. The load cells were bolted to the baskets and positioned low enough so that the top surface of the load cells followed the standard trajectory of bicycle pedals. The design was finished by fabricating and bolting a pedal platform with a strap (Cybex, Medway, MA) on top of each load cell in an attempt to evenly distribute foot loading and assure secure, continuous contact. The pedal platforms have an aluminum plate for rigidity and a rubber layer to provide friction to enhance contact and to reduce vibrations. In summary, this design required manufacturing two relatively simple components: the pedal platforms and the load cell baskets, as well as minor modifications to the pedals.

Load cells. The existing 6-channel load cells measure forces and moments in three directions and output the data to a data acquisition box without the need of further calculations, filtering, or signal processing of the raw data.

Motion analysis system. Four reflective markers were strategically placed on each pedal (Fig. 2A). During experiments, the 8 digital cameras track marker trajectories and are post-processed to determine crank angle, as well as the origin location and axes directions of the pedal coordinate system (Fig. 2B).

Implementation. To assure that the system is working as desired, cycling experiments were performed at ~ 70 RPM cadence for 3 subjects and pedal loads were compared to published data [1-3]. Inclusion criteria were normal body mass index and no history of leg injuries. Experimental protocols were approved by Cal Poly's Human Subjects Committee and were designed to minimize risk to human subjects.

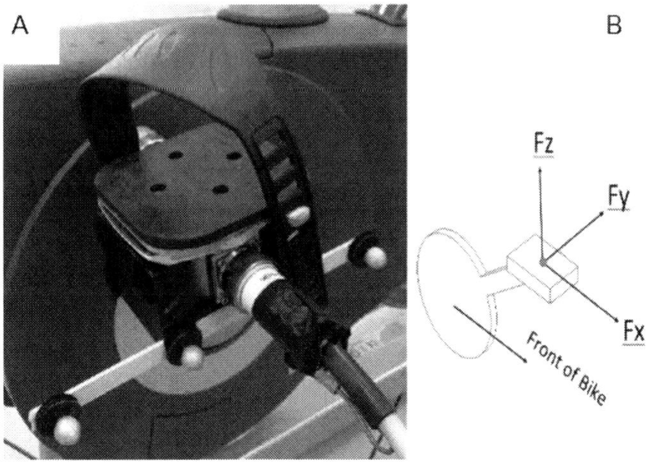

FIGURE 2. A) INSTRUMENTED PEDAL AND MARKER ASSEMBLY MOUNTED ON BICYCLE. B) COORDINATE SYSTEM USED [1].

RESULTS

Cycling experiments using the load cell instrumented pedals yielded force readings as shown below (Fig. 3). These results were compared to previous cycling studies and the range of forces agreed favorably [1-3]. A comparison of the load cell pedal readings to a strain gauge pedal system is shown below (Table 1).

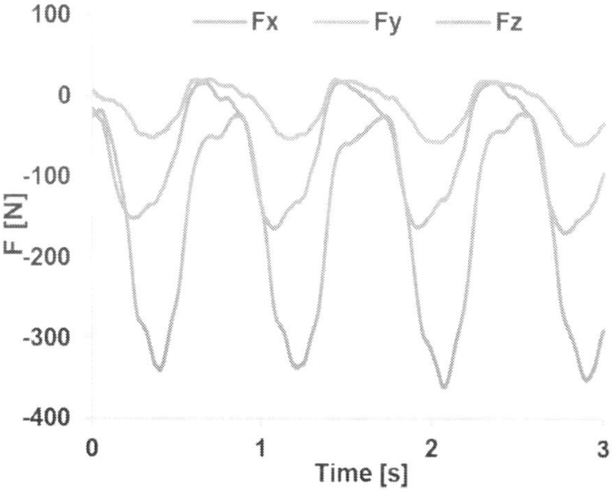

FIGURE 1. MEASURED FORCES ACTING ON THE PEDAL FROM ONE SUBJECT. USING THE COORDINATE SYSTEM SHOWN IN FIG. 2B, THE -X, -Y, AND -Z VALUES CORRESPOND TO FORCE (F) COMPONENTS IN THE POSTERIOR, LATERAL, AND DISTAL DIRECTIONS.

TABLE 1. COMPARISON OF RESULTS OF LOAD CELL INSTRUMENTED PEDAL VERSUS STRAIN GAUGE PEDAL FROM [1].

	Fz Range	Fx Range	Fy Range
Load Cell Pedal	-350N to -10N	-120N to 15N	-70N to 10N
Strain Gauge Pedal	-325N to 0N	-100N to 40N	-75N to 0N

DISCUSSION

The objective to design and fabricate an instrumented pedal for cycling biomechanics research was accomplished using load cells and by modifying a Tioga Mt-Zero Pedal. The finished product was simple, functional, and easily implemented with standard 9/16" bicycle cranks. Experiments were conducted with 3 subjects with a motion analysis system to capture the motion of the load cell origin and coordinate systems. Results for load cell data were consistent with published results using a spindle instrumented with strain gauges. Future aims are to use the instrumented pedal in cycling biomechanics research to estimate knee joint loading in subjects at high risk for knee osteoarthritis, including obese/overweight subjects, transtibial amputee prosthetic users, and ACL deficient and ACL reconstructive surgery patients.

ACKNOWLEDGEMENTS

Funding received from Quartus Engineering Inc. and Cal Poly's Donald E. Bently and STRIDE Centers.

REFERENCES

[1] Hull, M, Davis, R, *J Biomech*, 14:843-856, 1981.
[2] Hull, M, Ruby, P, Hawkins, D, *J Biomech,* 25:41-53, 1992.
[3] Hull, M, Boyd, T, Neptune, R, *J Biomech*, 30:505-511, 1997.

FABRICATION OF DENSE POROUS ALIGNED COLLAGEN SCAFFOLDS USING 2D PLASTIC COMPRESSION AND POROGENS

Shawn Reese (1,2); Jared Zitnay (1,2); Jeffrey Weiss (1,2,3)

(1) Department of Bioengineering
University of Utah
Salt Lake City, UT, USA

(2) Scientific Computing and Imaging Institute
University of Utah
Salt Lake City, UT, USA

(3) Department of Orthopedics
University of Utah
Salt Lake City, UT, USA

INTRODUCTION: Type I collagen hydrogels are commonly used as cell culture scaffolds [1]. However, the low collagen concentration (typically less than 0.5%/weight) results in a scaffold that is mechanically inferior to many native tissues, which have a significantly higher collagen content (~30%/weight for tendon) [2]. Recently, plastic compression has been used as a means of densifying collagen gels to increase their strength. This method uses 1D compression to cause fluid exudation from the scaffold while compressing the collagen fibrils [1]. Although this method is successful in enhancing mechanical properties, the resulting scaffolds lack porosity and secondary fibril organization such as aligned collagen fibrils (as seen in tendon). In this research, we developed a novel method that combines 2D plastic compression and hydrophilic porogen particles to create dense collagen scaffolds that have both aligned fibril organization and controlled porosity. The resulting porosity and collagen fibril alignment were examined using both two photon imaging and micro CT imaging.

METHODS: Hydrophilic polyurethane microsphere aggregates were manufactured using a phase inversion technique. These particles were used as a porogen and had sizes that ranged from 25-500 μm. Particles were disbursed in acid solubilized type I collagen (6 mg/ml) at a concentration of 12 mg/ml (Fig. 1). Gels were polymerized into a rectangular shape (25 × 25 × 10 mm) and compressed by 80% along the width and 60% along the thickness using a custom 2D compression device (Fig 2) at a rate of 0.5 mm/min. After compression, scaffolds were left in the 2D compression device for 2 hours in order to allow the gels to reach equilibrium. In one group of gels, a 5% formalin solution was pipetted onto the gel prior to the start of this period to crosslink the collagen and help maintain shape, while a second group did not receive formalin treatment. Compressed gels were soaked in an 80% ethanol solution for two days to dissolve the porogen particles and then rinsed overnight in phosphate buffered saline (PBS).

A sample from the formalin crosslinked group was imaged at 16X and 60X using a two photon microscope (λ=855 nm). For this scaffold, porosity was computed from thresholded 16X images in 2D by dividing the number of dark pixels by the total number of pixels. A scaffold from the uncrosslinked group was stained with osmium tetroxide and uranyl acetate [3] and imaged using micro CT with a voxel size of 10 μm. The resulting datasets were resampled to a 20 μm voxel size for image analysis and 3D reconstruction. An adaptive thresholding algorithm was used to convert the 2D image slices to binary images. Porosity was computed from these images with the same algorithm used for the two photon images.

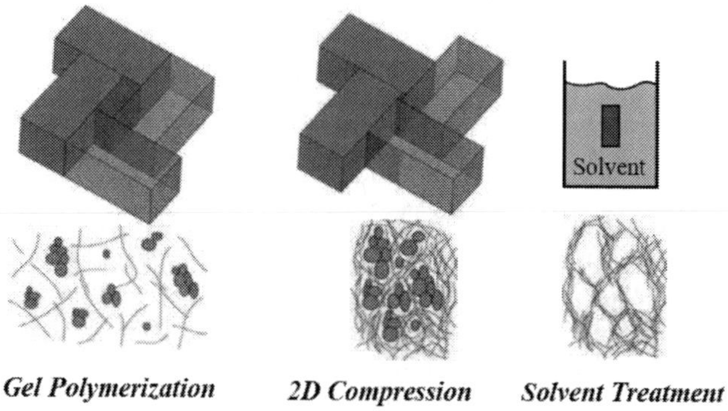

FIGURE 1. Collagen (pink) was polymerized with porogen particles (blue) into a rectangular shape (*Left*). The gels were then compressed in two dimensions, which densified the collagen gel and aligned the fibrils with the axis perpendicular to the plane of compression (*Center*). Removal of the porogen particles via a solvent left a porous scaffold with dense, aligned and interconnecting collagen structures (*Right*).

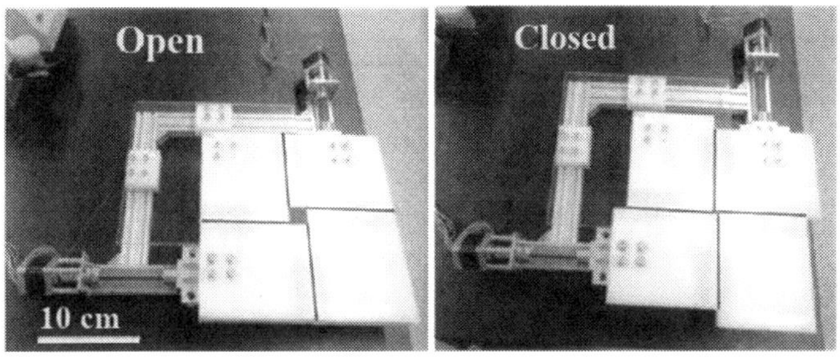

FIGURE 2. The 2D compression system is shown in an open state (*Left*) and closed state (*Right*), where the jaws are controlled by stepper motors.

RESULTS: Both groups of gels maintained a rectangular cross section after 2D compression and solvent based porogen removal. The gels that were fixed with formalin did not experience any change in dimensions after removal from the 2D compression system (Fig 3). For this group, volume was reduced by 98%, resulting in a densification of 12.5 times and a final collagen concentration of 7.5% within the scaffold (Table 1). The unfixed gels expanded by ~50% after removal from the 2D compression system and solvent extraction of the porogen, resulting in a final volume reduction of 75%, a densification of 5 times and a final collagen concentration of 2.5%.

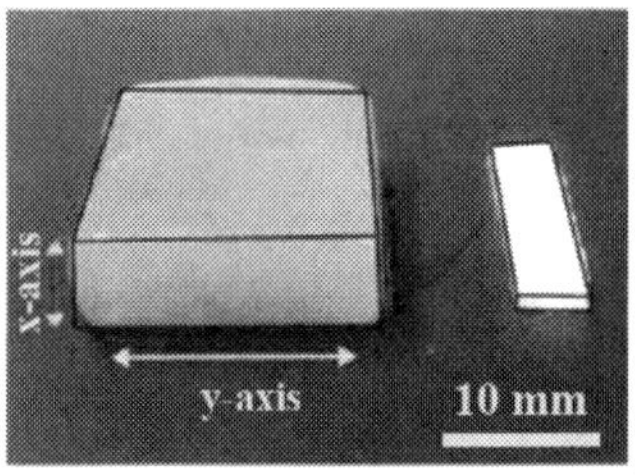

FIGURE 3. An uncompressed gel (*Left*) is shown next to a fully compressed gel that was fixed in formalin while in a compressed state. The y-axis and x-axis of compression is denoted. The pH indicator (pink) was washed from the gel after porogen removal, resulting in a white scaffold (*Right*).

Two photon imaging of the formalin fixed constructions revealed a network of pores ranging from 25-500 μm in size and embedded within a dense, aligned collagen matrix (Fig. 4), where collagen fibrils were highly aligned in the axis perpendicular to the plane of 2D compression. CT imaging of the stained, unfixed samples showed an interconnected network of pores separated by collagen (Fig. 5). The collagen appears to be organized into fibers, which are also aligned with the axis perpendicular to the applied 2D compression. Porosity computed for the formalin fixed group was 32%, while it was 37% for the unfixed group.

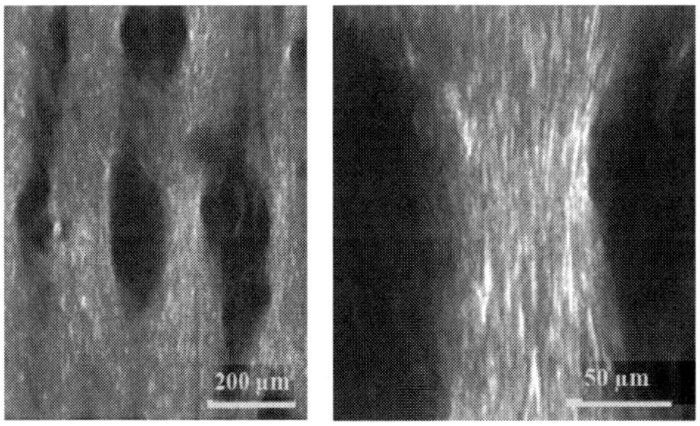

FIGURE 4. A 16X two photon image of a crosslinked sample (*Left*) shows an interconnected network of collagen fibrils with numerous pores. A 60X image of this same sample (*Right*) shows highly aligned and dense collagen fibrils.

Group	ε_x	ε_y	V/V_0	Porosity	Collagen Density
Fixed	.2	.4	.08	.32	7.5%
Unfixed	.33	.6	.25	.37	2.5%

TABLE 1. The final 2D strains after solvent based porogen removal (designated ε_x, ε_y), volume ratio (final volume divided by original volume), computed porosity and final collagen density within the processed scaffold are compared for the fixed and unfixed groups.

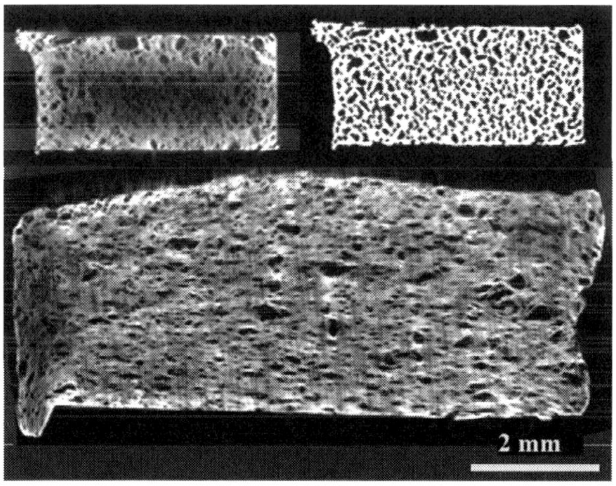

FIGURE 5. Longitudinal cross section (*Top Left*) and horizontal cross section (*Bottom*) of the CT data from an unfixed sample shows a highly porous network and aligned collagen fibers within the scaffold. An adaptive thresholding algorithm was used to binarize the cross sections (*Top Right*) so that porosity could be computed.

DISCUSSION: Application of plastic compression to collagen gels in two dimensions densified the scaffolds by a factor of 5.0 in the unfixed group and a factor of 12.5 in the fixed group, which resulted in a final collagen concentration of 2.5% and 7.5%, respectively. Two photon microscopy and micro CT image data demonstrated that the resulting scaffolds had a dense collagen network with interconnecting pores, where both the pore structure and the collagen fibrils were aligned with the axis perpendicular to the 2D compression plane. The increased density of collagen resulted in scaffolds that are noticeably more robust than the collagen gels while maintaining a porous structure, which may facilitate infiltration of cells and diffusion of nutrients.

Without cross linking of the collagen, the scaffolds expanded by ~50% after compression, however the porosity remained nearly the same. This suggests that both the pores and connecting collagen expanded at a similar rate during the porogen removal process. To achieve higher density scaffolds, a suitable cross linking method will be needed prior to solvent based extraction of the porogen. Although formalin was used in this study, a number of less cytotoxic crosslinking agents would likely yield similar results. The dense and highly aligned nature of the resulting collagen scaffolds may find use in the engineering of replacement connective tissues such as tendon and ligament. By varying the porogen concentration, magnitude of 2D compression and cross-linking method, a vast array of scaffolds with controllable porosity, stiffness and alignment may potentially be generated.

ACKNOWLEDGEMENTS: Financial support from NIH #R01EB015133.

REFERENCES: [1] Brown, Exp Cell Res, 319(16) 2013. [2] Benjamin, Histol Histopathol, 12(4) 1997. [3] Starborg et al. Nat Protoc, 8, (2013).

SB³C2015
Summer Biomechanics, Bioengineering and Biotransport Conference
June 17-20, 2015,Snowbird Resort, Utah, USA
SB³C2015-1035

EFFECT OF INTERSPINOUS DEVICE ON LUMBAR SPINE: A FINITE ELEMENT STUDY

Deniz U. Erbulut (1,2), Iman Zafarparandey (2), Chaudhry H. Raza(2), Ismail Lazoglu (2), Ali F. Ozer (2)

(1) Department of Neurosurgery
Koc University
Istanbul, Turkey

(2) Department of Mechanical Egineering
Koc University
Istanbul, Turkey

INTRODUCTION

Interspinous process (ISP) devices are utilized in surgical treatment of lumbar spine pathologies by prevention of extension motion. Several *in vitro* and finite element (FE) studies have analyzed the effect of ISP devices on the instrumented and adjacent levels under various biomechanical parameters. ISP devices are categorized as either static or dynamics based on the design mechanism. Static ISP devices, such as X-STOP (SFMT, Concord, CA, USA), ExtenSure (NuVasive, San Diego, CA, USA), and Wallis (Abbott Spine, Bordeaux, France), are incompressible and provide a set amount of space between the processes. Static devices are looser in flexion and tighter in extension.

Among static devices, X-STOP is designed to prevent extension motion while allowing undisturbed motion in other planes [1]. It was also found that the X-STOP did not affect range of motion (ROM) of the adjacent segments in other motion planes. In an *in vitro* study, it was found that intradiscal pressure (IDP) is reduced in extension after instrumentation but did not have affect in other motion directions [2]. However, majority of studies on ISP devices are *in vitro* analyses which have certain limitations. The IDP cannot be calculated correctly *in vitro* with pressure transducers [2]. Alternatively, FE studies provide the benefit of examining the inner workings of the lumbar spine under varying loading conditions.

The objective of current FE study was to evaluate the effect of an ISP device on the biomechanical parameters of an instrumented lumbar spine using hybrid testing protocol. The biomechanical parameters, including ROM, IDP, and facet loads (FL) were predicted at both the index level and the adjacent levels.

METHODS

A three-dimensional FE model of the lumbar spine, L1-L5, was developed using multi-block technique. Bony parts were meshed using hexahedral elements (C3D8). The detailed meshing procedure has been reported in our previous study [3]. Facet joints between the vertebrae were simulated using 3-dimensional gap contact elements (GAPUNI). The ligaments were simulated using 3-dimensional truss elements with non-linear motion. The complete L1-L5 model consisted of 72193 nodes and 55650 elements. The material properties of various components were derived from the literature [4]. The model was validated against published *in vitro* studies [5-8] by flexible (load control) approach under 10Nm pure moment and 400N compressive follower load. The nodes on the upper endplate of L1 were coupled to a flying node and pure moment of 10Nm was applied the flying node. A follower load of 400N was applied at each motion segment using connector elements between adjacent vertebrae.

The FE model was modified by inserting ISP device between the L3 and L4 processes. The design of the ISP device was similar to X-STOP with oval cross-section. The device had a core that could be adjusted between the processes without applying any injury to the interspinous and supraspinous ligaments. The isotropic elastic formulation was used to simulate the material properties of titanium (Young's modulus: 115 GPa, Poisson's ratio: 0.3). The intact and instrumented models were simulated in extension plane using hybrid protocol with hybrid moment of 14.5Nm. ROM, FL, IDP were investigated in instrumented model at the index and adjacent levels. Moreover, the stress at the bone-implant interface was also predicted.

RESULTS

The hybrid protocol for adjacent-level effect investigations predicted that the ISP device decreased the ROM, FL, and IDP at the index level relative to the corresponding values for the intact spine in extensions. ROM at the L3-L4 segment decreased by 80.6% after instrumentation relative to the ROM at this segment in the intact lumbar spine. Further, the ROM values in adjacent segments—L2-L3 and L4-L5—increased by 29.7% and 21%, respectively, relative to those in the corresponding segments in the intact spine. FL compared at the index and adjacent levels in both the models revealed that at the index level in the instrumented model, the facet loads decreased significantly up to 99.9% relative to the intact model. However, at the superior adjacent level (L2-L3) and the inferior adjacent level (L4-L5), the FL increased up to 60%. The IDP at the adjacent levels increased by 40% and 6.5% at L2-L3 and L4-L5, respectively, after instrumentation.

The predicted IDP at the index level in the intact model was 0.98 MPa but it dropped by 52.6% after instrumentation. In contrast, the predicted IDP at the superior adjacent level in the intact model was 0.87 MPa but it increased significantly by 40% after instrumentation. Further, the predicted IDP at the inferior adjacent level in the intact model was 0.80 MPa but it increased moderately by 6.6% after instrumentation.

DISCUSSION

ISP devices have been analyzed in the literature for their effect biomechanical behavior of the lumbar spine. Both the index and the adjacent levels are extensively analyzed in prior studies. However, no study applied the hybrid protocol to determine adjacent-segment effects. A hybrid testing protocol is appropriate for the biomechanical evaluation of spinal adjacent level [9]. Hence, we evaluated the biomechanical effect of ISP devices on the index and adjacent segments.

Previous studies have determined that the all commercially available ISP devices affect lumbar biomechanics only in extension and show no effect in other motion planes. The instrumented model showed reduction in the ROM, IDP, and FL at the implanted level in extension. Our kinematic results at the index level in extension were in agreement with the corresponding results in the literature. Although we applied the hybrid protocol, which leads to higher loading (14.5 Nm), the restriction of motion in extension at the implanted level was almost the same as that in the aforementioned studies, owing to the design of ISP devices that makes them restrict motion in extension.

However, the question is whether ISP surgery affects the kinematics of adjacent segments. According to previous studies, ROM at the adjacent segments was not affected significantly in flexion-extension motion whereas significant increase was reported for lateral bending and axial rotation. In contrast to these findings, our present FE model predicted increased motion in the superior and inferior adjacent segments after ISP implantation in extension under hybrid loading (pure moment of 14.5 Nm) and follower load of 400 N. Although an increase in motion may not lead to adjacent-segment hypermobility because of changes in extension only and is possibly not clinically significant, it may be useful to investigate the effect of motion changes on other components at adjacent levels, such as facet joints and intervertebral discs.

In this study, the predicted facet loads of our FE model for intact model at the index level were consistent with those reported in literature [4]. The facet load reduced at the index level in extension after implantation, similar to prior studies who also reported change of 10-20% in facet loads at adjacent levels [10]. In contrast to their results, our FE model predicted that the facet load increased by up to 60% in extension at both the superior and the inferior adjacent levels. This can be due to *in vitro* use of pressure-sensitive film in the facet joint investigation which has a ±15% error associated with the pressure measurement [2]. Moreover, variation in the contact area of the pressure sensor and unsuitability of the load control protocol for adjacent-level investigations raise questions about accuracy of facet load measurements *in vitro*.

The FE model predicted that the ISP device would release the IDP at the index level in extension which was in agreement with cadaver studies in literature. However, the FE model predicted that the IDP at the superior adjacent segment, L2-L3, increased by almost 40% after ISP implantation and that this IDP was very close to that in the intact case at the inferior adjacent segment, L4-L5, in extension. Unfortunately, these results disagree with those of a cadaver study by Swanson et al. The reasons for this disagreement is that Swanson et al. performed load control instead of displacement control for adjacent-level investigation, and used pressure transducer for pressure measurement.

Successful outcomes of the use of ISP devices in the lumbar treatment have been reported in the literature; however, some studies report ISP failure as well [11, 12]. Bowers et al. [13] reported the fracture of spinous processes of 3 out of 13 patients after X-STOP implantation and suggested that the degree of osteoporosis and over-distraction of the interspinous space with an oversized implant are possible causes. Stress concentration at the bone after ISP implantation is one of the main reason for occurrence of fracture.

Our FE study conclusively showed that the stress distribution in the spinous processes at the index level changed significantly after instrumentation. In the intact model, most of the stresses were concentrated at the facet joints. After instrumentation, a greater part of the stress was transferred to the instrumented region. Spinous processes may be strong enough to withstand the stresses after instrumentation [14], but ISP devices may cause gradual erosion at the bone–implant interface under constant stress and motion.

ACKNOWLEDGEMENTS

Please acknowledge any grant or other funding support or the assistance of others as appropriate.

REFERENCES

[5] Bergmann, G et al., *J Biomech*, 34:859-871, 2001.

1. Lindsey, DP et al., *Spine*.28:2192-7. (2003).
2. Wilke, H-J et al., *European spine journal*.17:1049-56. (2008).
3. Erbulut, D et al., *Medical engineering & physics*.36:915-21. (2014).
4. Kiapour, A et al., *Spine*.37:E581-E9. (2012).
5. Niosi, CA et al., *European Spine Journal*.15:913-22. (2006).
6. Schmoelz, W et al., *Journal of spinal disorders & techniques*.16:418-23. (2003).
7. Yamamoto, I et al., *Spine*.14:1256-60. (1989).
8. Schilling, C et al., *European Spine Journal*.20:297-307. (2011).
9. Panjabi, M et al., *Spine*.32:959-66. (2007).
10. Wiseman, CM et al., *Spine*.30:903-7. (2005).
11. Miller, JD et al., *Journal of Neurosurgery: Spine*.12:210-3. (2010).
12. Kim, DH et al., *The Spine Journal*.12:466-72. (2012).
13. Bowers, C et al., *Neurosurgical focus*.28:E8. (2010).
14. Shepherd, DE et al., *Spine*.25:319-23. (2000).

MODULATION OF PLATELET MICROTUBULE FUNCTION ALTERS PLATELET STIFFNESS AND MECHANOTRANSDUCTIVE RESPONSIVENESS TO SHEAR STRESS

Siu Ling Leung (1), Yi Lu (2), Danny Bluestein (3), Marvin J. Slepian (1,3)

(1) Departments of Medicine and Biomedical Engineering,
Sarver Heart Center, the University of Arizona
Tucson, AZ, USA

(2) Department of Mechanical Engineering,
the University of Arizona
Tucson, AZ, USA

(3) Biomedical Engineering Department,
Stony Brook University
Stony Brook, NY, USA

INTRODUCTION

Thrombosis of mechanical circulatory support devices, e.g. ventricular assist devices (VADs), is a major limitation of this otherwise life-saving technology. VAD thrombosis is driven largely by shear-mediated platelet activation, as a result of supraphysiologic shear ("Hypershear") experienced in passage through these pump systems. Recently it has been reported that conventional anti-platelet agents, utilized clinically to limit VAD thrombosis have little to no effect in limiting hypershear-mediated platelet activation [1]. This lack of efficacy stems largely from the fact that the majority of present agents target or interfere with biochemical, rather than physical mechanisms of platelet activation. Shear, as a physical force may induce platelet activation via direct damage or via transductive mechanisms, i.e. "mechanotransduction," a defined series of mechano-biochemical pathways and steps, which convert physical input signals into a sequence of biochemical events, including tubulin polymerization and microtubules depolymerization. Platelet shape change, a typical marker for platelet activation, and variation in platelet overall stiffness were the after effect of microtubule modifications.

Herein, the connection between microtubule integrity, platelet overall stiffness and platelet sensitivity to elevated shear were studied. We found that the overall stiffness of platelets may be a relevant variable in determining its sensitivity to shear-mediated platelet activation. In this study, we (1) describe a novel method to measure the overall stiffness of single platelets using dielectrophoresis force, (2) examined the efficacy of modulating microtubule organizations as a means of modulating platelet stiffness (3) examine the correlations between platelet stiffness and platelet responsiveness to shear.

METHODS

Platelet-rich plasma was isolated from venipuncture blood and centrifuged at 1300 rpm for 12 mins. Platelets were separated from plasma by Sepharose 2B gel filtration, using a column equilibrated in platelet buffer containing 125mM NaCl, 10mM HEPES, 25mM D-(+)glucose, 27mM KCl, 0.5mM Na2HPO4, 2mM MgCl2(6H2O), 1mM Na3-Citrate-6H2O and 0.1% BSA, pH 7.4. Gel filtrated platelets were counted with a Z2 Coulter particle counter.

Platelet stiffness was measured through a modified electrodeformation assay typically used on cancer cells or protoplasts [2, 3]. Before measurements, platelets were stained with Alexa Fluor 594 conjugated wheat germ agglutinin and diluted in isotonic buffer [3] containing 8.5%w/v sucrose and 0.3% w/v dextrose until the medium conductivity reached ~150μS/cm. 50μl of the diluted platelets sample were pipetted onto the parallel microelectrodes with gap separations of 10μm [4]. A 127μm-thick hollow polymer spacer was placed on top of the chip to create a fluid chamber and the chip was covered with a coverslip for microscopic observation. Alternating electric current in a sine wave pattern was generated across the microelectrodes by a function generator connected to a wideband power amplifier. The amplitude of applied voltage was monitored by a digital oscilloscope. The sample chip was observed under an Eclipse E800 fluorescence microscope, equipped with a 100X Plain Fluor oil objective. A CCD camera captured images of platelets deformed by electrical forces. Images were processed and quantified by NIH ImageJ software. All glass surfaces on the chip were coated with Sigmacote siliconizing reagent and rinsed with D.I. water to prevent surface-induced platelet activation

Shear-mediated platelet activation study: A hemodynamic shearing device was utilized to shear platelets at a concentration of

2x10[5] platelets/μL for 2, 5 and 10 mins, utilizing a shear stress of 70 dyne/cm[2] [5]. Platelets activation levels following elevated shear exposure were quantified by a chemically modified prothrombin-based platelet activation state (PAS) assay [6]. Basically, thrombin generation rates from samples were measured utilizing acetylated prothrombin and compared against maximum standard, sonicated platelets at 75W for 10s.

RESULTS

According to the theory of dielectophoresis, the magnitude of DEP force is proportional to the square of the applied voltages. In our experiment, platelet deformations under sequential increase of applied voltages, from 0 to 90Vpk-pk at 15V steps, were measured using the experimental setup as shown in Figure 1. Rapid deformations were observed in the meantime applied voltages were applied. Images of deformed platelets were captured in a duration ~5s. Figure 2 shows reversible deformations increase from 0.012 to 0.160 in the control group when applied voltages were increased from 15 to $90V_{pk\text{-}pk}$. For platelets treated with the microtubule stabilizer Paclitaxel (taxol), an increase in deformation only from 0.004 to 0.084 was observed with increasing applied voltages. Whereas, an increase in deformation from 0.034 to 0.226 was observed in platelets treated with the microtubule depolymerizing agent,Colchicine.

Recent studies reported that platelet activation accumulate progressively under subsequent exposure to shear [7]. Based on this, we sheared the platelets under constant rate for 2, 5 and 10mins to achieve elevated level of shear. A progressive increase in platelet activation level from 5.0%, 40.1% to 93.4% was observed respectively at 2, 5 and 10mins in the control study (Figure 3). When platelets were treated with colchicine, activation levels were dropped to 5.2%, 35.2% and 51.6% at 2, 5 and 10mins. Conversely a dramatic increase in shear response was observed for platelets treated with Taxol, with activation levels at 2, 5 and 10 mins of 32.6%, 59.7% and 98.2% respectively.

DISCUSSION

Today few limited techniques have been developed to measure the mechanical properties of single, non-surface adherent cells or cell fragments such as platelets. In the present study, we successfully demonstrate an electrodeformation assay, which can serve as a valuable tool for measuring resting platelet stiffness. Also, we modulated platelet overall stiffness through stabilizing or depolymerizing their microtubules structures using cytoskeletal drugs. Significantly, we demonstrated reducing the overall stiffness of the platelet through modulating platelet microtubule structure reduces shear-mediated activation on platelets. Thus, we shown that reducing stiffness, i.e."rubberizing" the platelet, can reduce its sensitivity to elevated shear mediated activation.

ACKNOWLEDGEMENTS

This project was funded by NIH (NIBIB Quantum U01EB012487).

REFERENCES

1. L. Valerio, et al. *Journal of Heart and Lung Transplantation*, 33: S143-S144, 2014.
2. J. Chen, et al. *2011 Ieee 24th International Conference on Micro Electro Mechanical Systems (Mems)*: 1119-1122, 2011.
3. P. K. Wong, et al. *J Biomech*, 38: 529-35, 2005.
4. Y. Lu, et al. *Anal Chem*, 85: 3971-6, 2013.
5. J. Sheriff, et al. *Ann Biomed Eng*, 38: 1442-50, 2010.
6. J. Jesty, et al. *Analytical Biochemistry*, 272: 64-70, 1999.
7. M. Nobili, et al. *ASAIO J*, 54: 64-72, 2008.

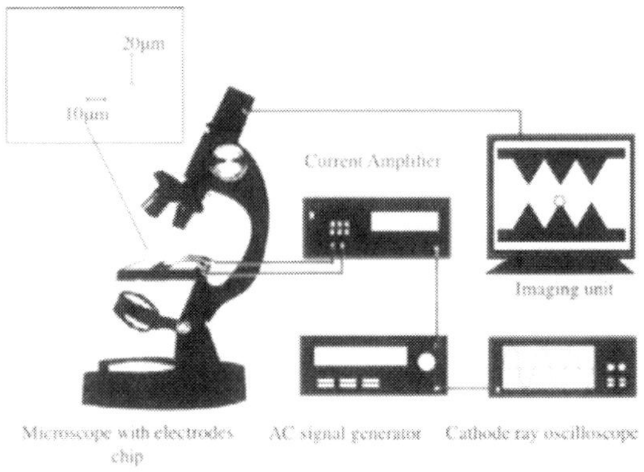

Figure 1 Schematic diagram of the experimental setup for electrodeformation-based platelet stiffness measurement. (Insert: Geometry of the Ti-Au-Ti microelectrodes)

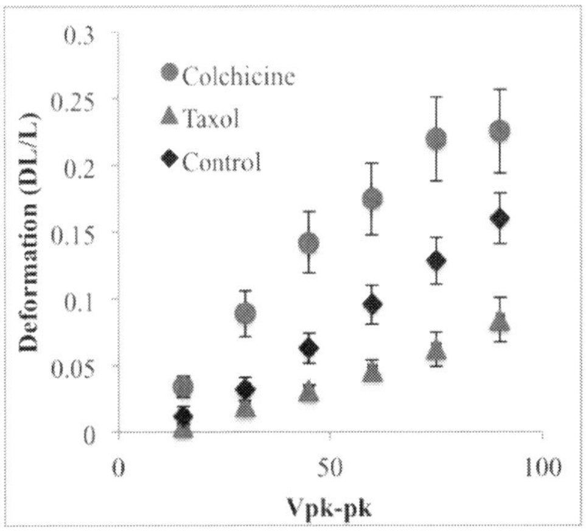

Figure 2. Stiffness responses of untreated, colchicine- and paclitaxel-treated platelets vs. applied electric field strength at 1MHz applied frequency.

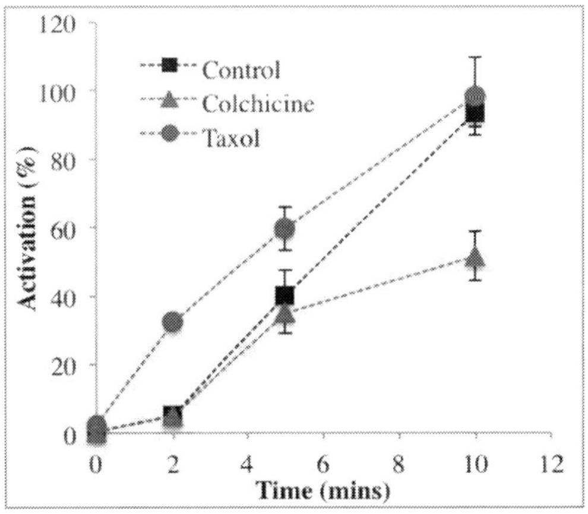

Figure 3. Platelet activation levels of untreated, colchicine- and paclitaxel-treated platelets under elevated level of shear.

SB³C2015
Summer Biomechanics, Bioengineering and Biotransport Conference
June 17-20, 2015, Snowbird Resort, Utah, USA
SB³C2015-1037

EVALUATING THE INFLUENCE OF TISSUE PROPERTIES ON THE CORE TEMPERATURE USING A 3D WHOLE BODY MODEL

Robins T. Kalathil (1), Swarup A. Zachariah (1), Amit Bhattacharya (2), Rupak K. Banerjee (1)

(1) Department of Mechanical and Materials Engineering, University of Cincinnati
Cincinnati, Ohio, USA

(2) Department of Environmental Health
University of Cincinnati College of Medicine
Cincinnati, Ohio, USA

INTRODUCTION

Accumulation of heat in the human body during firefighting activities for a sustained time period causes body temperature to reach critical levels. This increase in body temperature leads to a build-up of heat stress. Elevated heat stress can lead to adverse health effects, including unconsciousness and/or cardiac arrest [1]. Thus, accurate prediction of thermal response of the human body during firefighting scenarios is of high importance. Determination of the core body temperature (T_c) using computational method requires a reasonable estimate of tissue properties and numerical formulation to calculate thermal response of the human body. In a previous study, we have utilized a computational whole body model to determine the T_c of firefighters during firefighting training drills [1]. The study evaluated firefighting data that included periodic work (Sc) and rest (R). Tissue properties of humans at rest show significant variation among different individuals [2, 3], but the effect of variation in each property on the core body temperature (T_c) has not been quantified.

Hence, the objective of the current study is to determine the influence of variation in tissue properties on the T_c. These properties include: density (ρ), specific heat (c), thermal conductivity (k), blood perfusion (ω) and metabolic rate (q_m). McIntosh *et al.* [2] and Duck [3] have specified a range of estimated values for the tissue properties of humans for resting condition. These ranges have been utilized in determining the influence of tissue properties on T_c.

METHODS

The whole body model was developed using a combination of different geometric shapes such as cylinder, cuboid, and sphere to represent the limbs, torso, and head, respectively. The computational model was based on a human being with a height of 1.80 m, weight of 80 kg, and body surface area of 1.90 m² [1]. The model was divided into four domains: head, muscle, organ, and gut.

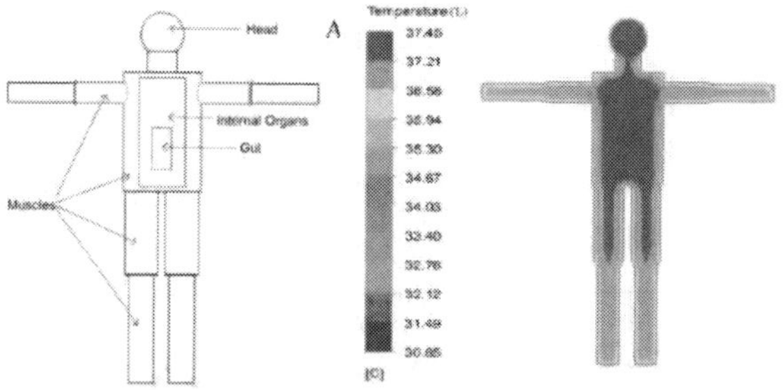

FIGURE.1: GEOMETRY OF THE WHOLE BODY MODEL (A) INITIAL TEMPERATURE FIELD IN 25°C AIR (B)

The transient thermal response of the whole body model was determined by solving two equations simultaneously: 1) Pennes bioheat equation, to evaluate the temperature distribution in the body and 2) The energy balance equation, to determine the change in blood temperature due to a transient tissue-blood heat exchange. Pennes equation is defined as:

$$\rho_t c_t \frac{dT_t}{dt} = k_t \nabla^2 T_t + q_m + \rho_{blood} c_{blood} \omega (T_{blood} - T_t) \qquad (1)$$

where ω is the blood perfusion rate, ρ_t is the density of the tissue, c_t is the specific heat of the tissue, k_t is the thermal conductivity of the tissue, T_{blood} is the blood temperature, T_t is the tissue temperature and q_m is the heat generated due to metabolism. T_{blood} in the human body was represented as a lumped system and was computed using the energy balance equation as shown below.

$$(\rho c V)_{blood} \frac{dT_{blood}}{dt} = -(\rho c \omega_{avg})_{tissue} V_{body} (T_{blood} - T_{wt}) \quad (2)$$

where V_{blood} is the total volume of blood in the body, ω_{avg} is the average blood perfusion, and T_{wt} is the weighted average tissue temperature.

The values of metabolic rate (q_m) and blood perfusion (ω) for individual tissues [2] were assigned to the four domains of the computational model. The q_m for each domain (eg. Organ) was determined by calculating the weighted average of the metabolic rates of all internal organs/tissues (eg. heart, lungs, liver etc.) representing different domains. Details of different domains are presented in Table 1. A similar approach was adopted for determining the range of ω. Further, a single weighted average value of density, specific heat, and thermal conductivity were also computed for the whole body from the available tissue properties as shown in Table 2. Other input parameters to the model, such as the heart rate time series and details of a firefighting suit, were similar to those described in our previous study [1]. T_c was computed by varying one input parameter at a time while keeping all the other tissue properties at their baseline values presented in Tables 1 and 2. Eleven numerical computations were executed: one with baseline values and the rest by varying limiting the (maxima and minima) values of tissue properties. It should be noted that only the steady-state q_m and ω were varied. The transient variations of q_m and ω were dependent on the heart rate time series and T_c, respectively [1, 4].

TABLE 1: BASELINE VALUE AND ITS VARIATION ALONG WITH VALUES FROM THE PREVIOUS STUDY OF ω AND q_m

Domain	Current study		Previous Study [1]	
	ω (1/s) [2]	q_m (W/m³) [2]	ω (1/s)	q_m (W/m³)
Head	0.00574±13%	9865.5 ± 18.4%	0.00833	9225.0
Muscle	0.00055±35%	597.5 ± 18.4%	0.00050	553.5
Organ	0.00425±30%	4529.6 ± 18.4%	0.00126	1401.5
Gut	0.01330±52%	6994.0 ± 18.4%	-	-

TABLE 2: BASELINE VALUE AND ITS VARIATION ALONG WITH VALUES FROM THE PREVIOUS STUDY OF k, ρ AND c

Property	Current Study	Previous Study [1]
k (W/m.K)	0.41 ± 7.0% [2, 3]	0.50
ρ (kg/m³)	1055 ± 4.3% [2, 3]	1060
c (J/kg.K)	2698 ± 14.8% [2, 3]	3166

RESULTS

Our previous study [1] reported a maximum difference of 0.3°C between computed and experimental T_c during firefighting activity (Figure 2). The tissue properties used in this study can be found in Table 1 and 2. Comparison of the baseline T_c obtained from the current study with the experimental T_c [1] showed a deviation of 1.6°C, or 4.1% (maximum % increase between baseline T_c and experimental T_c) as presented in Figure 3.

Temperature deviations presented in this section refer to the *maximum* deviation observed in T_c with respect to baseline T_c due to the variation in input parameters. A temperature rise of 0.5°C (a 5.5% increase with respect to experimental T_c) due to an increase in the steady-state q_m is evident in Figure 3. When *c decreased,* a noticeable increase of 0.38°C (a 5.2% increase with respect to experimental T_c) in T_c (Figure 3) was observed. An increase of 0.1°C in T_c was noted when ω was varied to its lower bound. Minimal deviations in T_c were observed due to the variations in k and ρ. A maximum variation of 0.01°C in T_c was observed when k was varied by 7%. A 4.3% change in ρ resulted in a difference of 0.01°C from the baseline T_c.

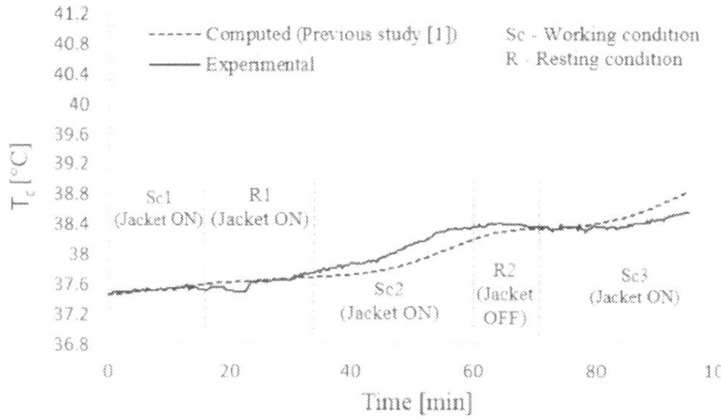

FIGURE 2: COMPUTED AND EXPERIMENTAL T_c DURING FIREFIGHTING ACTIVITY [1]

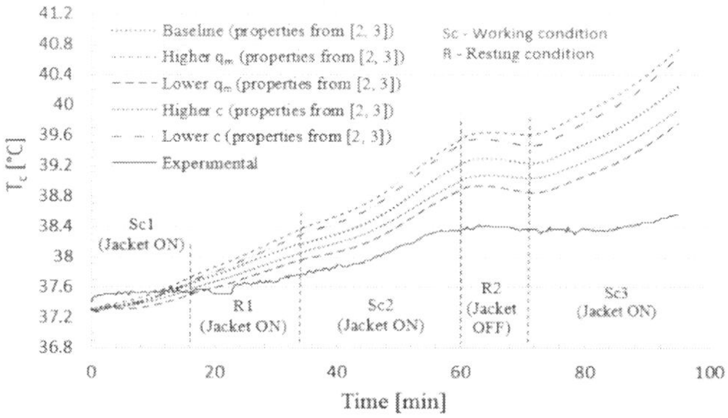

FIGURE.3: COMPUTED T_c DURING FIREFIGHTING ACTIVITY WITH BASELINE AND VARIATIONS OF q_m AND c ALONG WITH EXPERIMENTAL T_c

DISCUSSION

The effect of variations in the steady-state values of the tissue properties on T_c using a whole body model was demonstrated in this study. Effects of change in q_m and c on T_c were significant as compared to the influence of ρ, ω and k. The notable difference between T_c from the current and previous studies can be attributed to the variability in the q_m and c listed in Table 1 and 2.

Assumptions and Limitations. The property values listed in Table 1 and Table 2 shows large variability. The T_c values can be improved with better accuracy of tissue properties.

Future Scope. The current study will be extended to find the uncertainty of T_c due to combined effect of the variation in the tissue properties and input parameters.

ACKNOWLEDGEMENTS

This research study was supported by the National Institute for Occupational Safety and Health Targeted Research Training Program of the University of Cincinnati Education and Research Center Grant #T42-OH008432.

REFERENCES

[1] Zachariah S.A et al., *WCB*, 2013.
[2] McIntosh R. L et al., *Biophys. Rev. Lett.*, 5:129-151, 2010.
[3] Duck F.A., Physical Properties of Tissue, *Academic Press: London* 1990.
[4] Simmons G. H. et al., *Exp. Physiol.* 96 (9):822-828, 2011.

DEVELOPMENT OF A 5TH PERCENTILE FEMALE FINITE ELEMENT MODEL USING A MULTI-MODALITY IMAGE DATASET

Matthew L. Davis (1,2), Bharath Koya (1,2), Jeremy M. Schap (1,2), F. Scott Gayzik (1,2)

(1) Wake Forest University School of Medicine
Winston-Salem, NC, USA

(2) Virginia Tech-Wake Forest University Center for Injury Biomechanics
Winston-Salem, NC, USA

INTRODUCTION

Motor vehicle injuries and fatalities remain a leading public health concern worldwide. In 2013, the World Health Organization reported more than 1.2 million deaths as a result of motor vehicle crashes [1]. To mitigate the societal impact of these injuries, researchers are using a variety of tools, including finite element (FE) models. Such models are often developed to represent an average (50th percentile by height and weight) male occupant. In order to predict the response of a greater portion of the driving public, there is an interest in developing such models for other cohorts among the driving public. This study focuses on the female driver in the 5th percentile of height and weight. In the past, 5th percentile female models have been scaled from existing average male models, since scan data for such a specific target anthropometry is limited [2]. However, as part of the Global Human Body Models Consortium (GHBMC) project, comprehensive image and anthropometrical data of a 5th percentile female was acquired. The objective of this study is to review the data collected for this effort, including medical imaging, CAD development, and the discretization of the CAD into a finite element model.

METHODS

The subject recruitment and imaging protocol was approved by the Wake Forest School of Medicine Institutional Review Board (IRB #5705). An individual representing the 5th percentile female (F05) in terms of height (149.9cm), weight (48.0±0.63kg), and body mass index (21.4kg/m^2) was selected and imaging data was collected. Additionally, 15 external anthropomorphic measurements were taken from the subject to meet criteria published by Gordon et al [3]. Computed tomography (CT), Magnetic Resonance Imaging (MRI), Upright MRI, and external anthropometry of the individual representing the F05 were obtained in supine, seated, and standing positions with image data deemed anatomically normal after examination by a radiologist. In addition to the screening anthropometry, 52 external bony landmarks, and surface data were acquired via a 3D digitizer with the subject in a driving posture using a custom, adjustable buck (Figure 1).

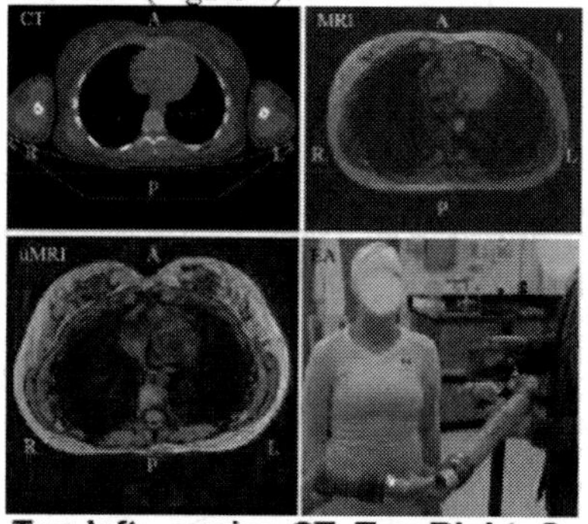

Figure 1. Top left:, supine CT, Top Right: Supine MRI, Bottom Left: upright MRI, Bottom Right: External Anthropometry. Images show transverse slices through the thorax.

Images were acquired with slice thicknesses ranging from 0.65 mm to 2 mm and fields of view ranging from 250 mm to 500 mm. The image sets were used for segmentation of bones and soft tissue. Image segmentation was completed using Mimics software with manual and semi-automatic techniques. Because the model is being developed to represent a female in the driving position, geometries were initially segmented using the higher resolution supine scan data and then transformed into the seated image space to posture-dependent placement of anatomic structures. In order to ensure appropriate inertial properties of the model throughout blunt impact scenarios, thoracoabdominal organ volumes were also compared to targets for the F05 published in the literature [4]. After developing STL data from medical imaging, Non-Uniform Rational B-Spline (NURBS) surfaces with tangential continuity were constructed over all segmented data. The outer flesh of the model was developed from the surface scan data obtained using the laser scanner.

The CAD dataset was then utilized for the mesh development of the GHBMC F05 occupant model. Meshes were primarily developed using commercially available hexahedral meshing software (TrueGrid, Beta-ANSA, Hypermesh). A limited number of structures with simple geometries (such as the radius and ulna) were morphed from meshes in the GHBMC average male model. One of the focuses of mesh development was to reduce the need for contacts between structures by increasing the number of nodally connected geometries. As an example of this approach, the mesh in the lower extremity and upper extremities were developed radially outward from the bone, with muscle properties assigned to elements as part of the structured mesh. The quadriceps and hamstring muscles were then developed by assigning new part IDs to elements in the thigh flesh using the CAD data. This element assignment technique allowed a continuous mesh from the outer surface of the model down to the bone.

RESULTS

The selected volunteer (24 years old, female) was a good fit in terms of height and weight (149.9 cm, 48.1 kg) with deviations from the target values of 0.7% and 1.9% respectively. In addition to an initial screening, the subject was also evaluated based on 15 anthropomorphic measurements from Gordon et al. For all 15 measurements, the average subject percent deviation was 4.1% (cutoff for inclusion was 5%). Following recruitment, 54 bony landmarks were recorded in both the seated and standing postures via palpation.

In total, 66 scan series were collected across all modalities for a total of 14,170 images. The CAD dataset consists of 182 individual bones. Explicit representations of 32 organs were also developed, including brain, thoracic, and abdominal tissues relevant to crash induced injuries. Major vasculature was also represented to provide the appropriate biomechanical connections in these regions. The exterior skin has a surface area of 1.46 m^2, which is within 3% of estimates in the literature for a female of the same height and body weight. The assembled CAD model can be seen in Figure 2.

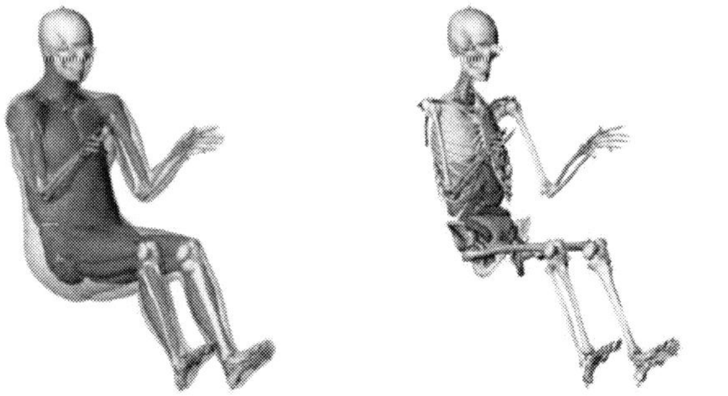

Figure 2. F05 Occupant CAD Model

The alpha version of the GHBMC F05 occupant model has been meshed and is currently undergoing regional validation. The model consists of 850 parts, 2.3 million elements, and 1.3 million nodes. The model was developed with a target average element edge length of 2 mm. Material definitions have been applied for each part based on experimental data found in the literature. The model met stringent thresholds for element quality: jacobian (>0.3 for all solid elements and >0.4 for all shells), tet-collapse (>0.2 for all elements), and a minimum time step value of 0.01 μs. The complete model can be seen in Figure 3. Also displayed in the upper right image is a cross section of the lower extremity displaying the element assignment technique for developing muscles connected nodally to the bone and flesh.

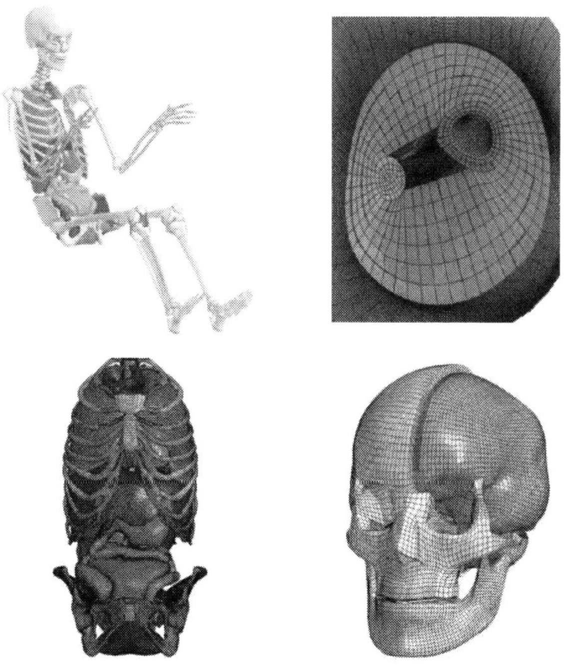

Figure 3. Mesh of the GHBMC F05 Occupant Model

DISCUSSION

This study presents a methodology and comprehensive dataset for the development of a 5th percentile female finite element model. The majority of structural components in the model were meshed using a structured hexahedral approach. While developing meshes, care was taken to not only capture important geometric features of the CAD, but also to promote node-to-node connections where appropriate. The goal of developing nodal connections between parts was to both increase robustness and significantly decrease the number of contacts required. The current GHBMC average male model devotes roughly 40% of CPU time on processing contacts, which motivated this modeling scheme to reduce the reliance on contacts.

Future work for the model will be biomechanical validation using a variety of experimental datasets. Initial validation will be focused on regional (i.e. head, neck, etc.) performance and then will be extended to whole body responses. Once validated, the data obtained from this model will be valuable for the development of new vehicle safety devices. The model will also be useful for making comparisons to existing anthropomorphic test devices currently in use, such as the Hybrid III 5th percentile female.

ACKNOWLEDGEMENTS

Funding was provided by the Global Human Body Models Consortium, LLC (GHBMC).

REFERENCES

[1] WHO, "Global status report on road safety." Geneva, Switze, 2013.
[2] Gordon C, et al. DTIC Document 1989.
[3] Vezin, P., et al, *Enhanced Safety of Vehicles Conference,* 2005.
[4] Davis M.L., et al. Traffic Inj Prev. Acc Manuscript, Dec. 2014.

SB^{3}C2015
Summer Biomechanics, Bioengineering and Biotransport Conference
June 17-20, 2015,Snowbird Resort, Utah, USA
SB^{3}C2015-1039

METABOLIC CONSUMPTION USING DIFFERENT REPETITIVE LIFTING STRATEGIES

Timothy D. Craig (1), Alice E. Riley (1), Sandra. A. Billinger (2), Neena K. Sharma (2), Sara E. Wilson (1)

(1) Department of Mechanical Engineering Department
University of Kansas
Lawrence, KS, USA

(2) Department of Physical Therapy and Rehabilitation Science
University of Kansas Medical Center
Kansas City, KS, USA

INTRODUCTION

Low back pain is a serious concern and a leading cause of disability accounting for 40% of all compensation claims in the United States for individuals under 45 years [1]. Many studies have examined back pain and the National Institutes for Occupational Safety and Health have identified risk factors, including repetitive lifting, for low back disorders [2-3]. One study identified occupations with cyclic loading tasks to have a 10 fold increase in low back injury [4].

In order to create prevention training programs to reduce risk of low back injury, it is necessary to better understand lifting strategies and how they are selected. It is believed that experienced lifters may have developed avoidance for strategies that could increase risk of injury. Previously, experienced lifters were observed to have reduced lumbar flexion angles [5]. Conversely, Maduri et al. observed that a group, believed to be novice lifters, performed a repetitive lifting task using a highly kyphotic strategy approaching their range of motion (ROM) limits [6]. It has been speculated that novice lifters might perform a more kyphotic strategy for its energetic advantage.

The goal of this study was to examine the energy expenditure of multiple lifting strategies in a repetitive lifting task completed by both novice and experienced lifters. It was hypothesized that using a highly kyphotic lifting strategy would use less energy than a strategy utilizing a more neutral spine.

METHODS

Twenty-seven subjects, 16 males and 11 females, participated in this study with the approval of the Human Subjects Committee from the University of Kansas Medical Center. The subjects were divided into two groups based on weight lifting experience. Subjects having lifted weights three times per week for the previous year were considered experienced and individuals who had not lifted weights regularly in the previous year were considered novice. The experienced group (EXP) had a total of 11 subjects with 4 females and 7 males while the novice group (NOV) had 7 females and 9 males for a total of 16 NOV lifters. Subjects performed a repetitive lifting task to examine the energetics of using three different lumbar pelvic coordination strategies. The three strategies examined included a self selected strategy, a highly kyphotic (PF or pelvis first), and a neutral (NU) strategy.

The lifting strategies were performed using a repetitive lifting task. The task involved lifting a milk crate from the ground to waist level and then returning to near ground height. The crate held a small amount of weight normalized to 3% of a single max lift. The crate was never released and held throughout the trial that lasted approximately 4 minutes. A metronome controlled the rate of lifting by beeping at 1Hz with each lift cycle (ground-waist-ground) lasting four seconds.

Oxygen consumption was measured and analyzed through the collection of expired gases using a metabolic cart (Parvo Medics Inc., Sandy UT). Gas and flowmeter calibrations were performed following the specifications of the manufacturer. Data were averaged using a 5 second window.

Two angles, lumbar angle (LA) and trunk flexion angle (FA), were measured using three electromagnetic sensors (MotionStar, Ascension Technology, VT) collecting position and orientation data at 100Hz using MotionMonitor Software (Innsport, IL). The sensors were located at T10, S1, and the manubrium. The FA was defined as the angle between T10-S1 and vertical and the LA was the difference in angular tilt of the sensors at T10 and S1 to describe the spinal curvature.

The LA was normalized using individual subjects range of motion (ROM) by collecting the maximum and minimum kyphotic postures at 0, 30, 60, and 80 degrees of flexion. A ROM envelope

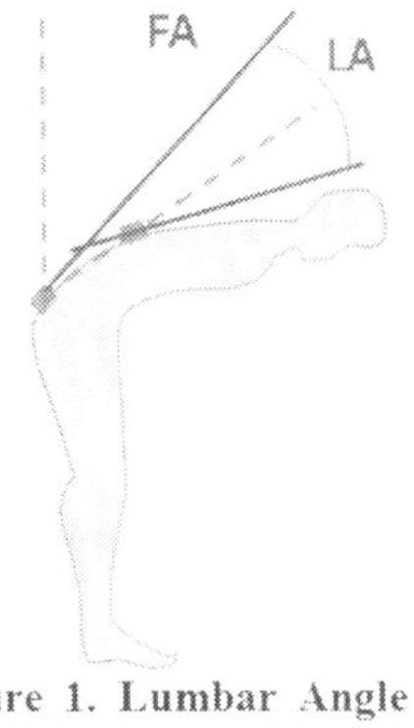

Figure 1. Lumbar Angle (LA) and trunk flexion angle (FA) were determined from the electromagnetic sensors.

expanded the ROM using linear interpolation to calculate a normalized LA at any given FA. Biofeedback displayed an XY-plot with the FA on the x-axis and normalized LA on the y-axis for visual feedback. A target box was displayed to control for the lifting strategy. The target box ranged between normalized LA of 42% and 58% ROM for the NU strategy and 80% to 95% ROM for the PF strategy. For both trained strategies, the FA range for the target was between 20 and 70 degrees. Prior to introducing the biofeedback program, subjects completed the lifting task using a self selected strategy without visual feedback. During the trained strategies, subjects were instructed to make the cursor pass through the target box during the extension (lifting) phase of the lift.

The metabolic data was divided into three 30 second regions starting at 50s after the lifting task commenced. The volumetric oxygen consumption (VO2), normalized by subject mass, was examined using an average and standard deviation per subject for each region. A repeated measures ANOVA was performed on the three time regions for the three lifting strategies using the average and standard deviation measures of VO2.

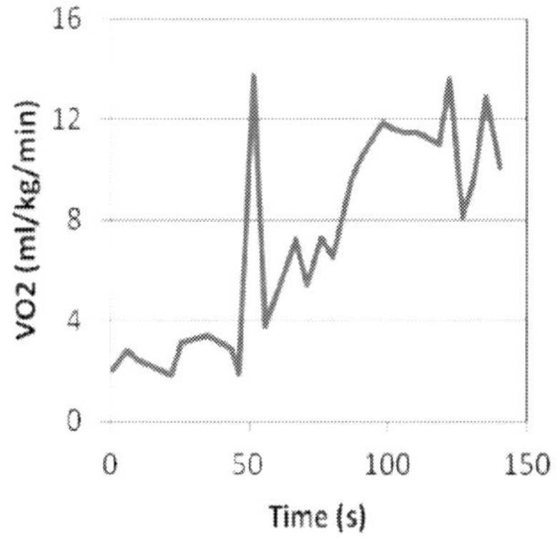

Figure 2. Sample trial of the VO2 measurement demonstrates the large jumps observed in VO2 measurements.

RESULTS

It was observed that the average VO2 demonstrated no significant difference between lifting strategies or between different levels of lifting experience. However, it was observed that the standard deviation of the VO2 measurement was large with a significant difference ($p<0.05$) in the lifting strategies. Additionally the average VO2 measurement increased significantly over time for both the NOV and EXP groups.

DISCUSSION

It was hypothesized that the NOV group would prefer a highly kyphotic lifting strategy that would provide an energy savings. The belief was that the coordination could implement a stretch-shortening cycle such as that seen in other body segments [7]. Similarly the EXP group was hypothesized to prefer a neutral strategy that would require more energy. Although the preference to specific strategies was supported, the observed energy expenditure for the strategies does not support the hypothesis.

Although the results do not prove the hypothesis, an interesting observation was made on the variability of the VO2 measure. The ANOVA examining the standard deviation of the VO2 measure demonstrated a significant difference ($p<0.05$) for the strategies. Both the NU and PF strategies had larger deviations. This difference may be a result of the increased difficulty subjects faced in performing a trained strategy rather than using a self-selected strategy. The introduction of the biofeedback is believed to have altered the neurological control loop and the variation resulting is the effect of the adjustment to the control strategy.

The current study is limited by the criteria to determine experienced lifters. This study classified individuals with weight lifting experience as experienced lifters whereas other studies define experience as manual materials handling workplace experience. Additionally the procedure of collecting the lumbar angle at a given flexion angle was performed statically while the task itself was dynamic.

Future work will involve investigating the alteration of the neurological control with the introduction of biofeedback and training. Training programs including biofeedback may have a neurological component that could affect the results. Additionally future work will examine whether training subjects to use a neutral strategy could reduce incidence of low back injury over time.

ACKNOWLEDGEMENTS

Research reported in this publication was supported by National Institute of Arthritis and Musculoskeletal and Skin Diseases of the National Institutes of Health under award number 1R03AR061597-01. The content is solely the responsibility of the authors and does not necessarily represent the official views of the National Institutes of Health.

REFERENCES

[1] Lis, A et al., *Eur Spine J*, 16(2): 283-298, 2007.
[2] Benard, B, *USDoHaH* 97-141, 1997.
[3] Health NIoOSa, *USDoHaH NIOSH* 81-122, 1981.
[4] McGill, S, *J Biomech* 30:465-475, 1997.
[5] Maduri, A et al., *J Electromyogr Kinesiol,* 18: 807-14, 2008.
[6] Plamondon A et al., *Ergonomics*, 53(10):1239-1253, 2010.
[7] Horita, T et al., *Euro J Appl Physiol*, 79:160-167, 1999.

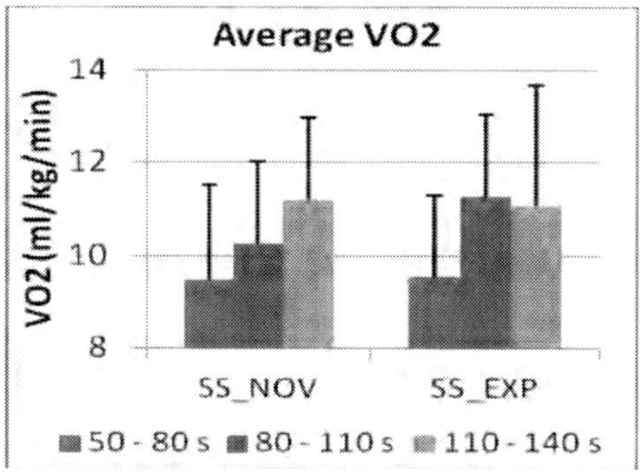

Figure 3. Average VO2 between groups demonstrate no difference between lifting experience but did have a significant difference with time.

SB³C2015-1040

CONDITIONED MEDIA FROM DEGENERATIVE INTERVERTEBRAL DISCS SENSITIZE DORSAL ROOT GANGLION NEURONS TO HEAT STIMULI

Joshua D. Stover (1), Ibrahima Bah (2), Alexander Kotelsky (2), Mark R. Buckley (2), Brandon Lawrence (3), Robert D. Bowles (1,3)

(1) Department of Bioengineering
University of Utah
Salt Lake City, UT

(2) Department of Biomedical Engineering
University of Rochester
Rochester, NY

(3) Department of Orthopaedics
University of Utah
Salt Lake City, UT

INTRODUCTION

Pro-inflammatory cytokines have been implicated as active components in many events and processes associated with back pain, including disc degeneration and intervertebral disc (IVD) herniation [1]. It has recently been hypothesized by our group and others, but not tested, that inflammatory factors released from degenerative discs can contribute to back pain by sensitization of nociceptive neurons to noxious stimuli [2]. In addition, neurons exposed to low pH can produce nociceptive sensitization. As a result, the degenerative IVD environment is primed to sensitize nociceptive neurons that may play a significant role in discogenic back pain. However, it is unclear what aspects of the biomechanical environment of the degenerative IVD interact with these pain sensing nerve fibers. The nociceptive neurons innervating the healthy and degenerative IVD are calcitonin gene-related peptide (CGRP) positive [3] and CGRP expressing peptidergic neurons have been associated with noxious heat sensitivity [4]. We hypothesize that nociceptive neurons exposed to factors produced in the degenerative disc environment become sensitized and fire at lower temperatures than those of healthy patients. In this study, we test the feasibility of this hypothesis by imaging dorsal root ganglion neuron action potentials in response to heat after exposure to human degenerative disc conditioned media.

METHODS

Degenerative Intervertebral Disc (IVD) Conditioned Media: Lumber IVD tissue was obtained from 3 patients undergoing surgical treatment (i.e. discectomy) for back pain symptoms. IVD tissue was cleaned, cut into small pieces (~3mm^2), transferred to a 75 cm^2 tissue culture flask and cultured for 48 hours at a media to tissue concentration of 3.5 ml/g (DMEM-HG, 50 µg/ml ascorbic acid, 5µg/ml gentamicin, 0.125 µg/ml fungizone). After incubation, conditioned media was collected and stored at -80°C until use.

Rat Dorsal Root Ganglion (DRG) Neuron Culture: Postnatal 1-3 day Fisher rat DRGs were explanted, placed in L-15 medium, and cleaned of anterior and posterior roots and connective tissue. Ganglia were dissociated in collagenase (1.33%) for 30 minutes, then trypsin (0.25%) for 15 minutes, followed by trituration through a fire polished Pasteur pipette in DNase. Contaminating Schwann cells were eliminated by the addition of 20 µl purified O4 antibodies to a 1 ml cell suspension in DMEM then by the addition of 50 µl purified rabbit complement. Rat DRG neurons were seeded onto laminin coated 35 mm tissue culture dishes at a density of 50,000 cells per dish and cultured in 1.5ml of SATO$^-$ medium for 4-6 days. At the time of experiments, media were replaced with fresh SATO$^-$ and 0.75 ml of DMEM or degenerative disc conditioned media and cultured for 24 hours.

Human DRG Neuron Cell Culture: Fresh human DRGs were provided by the Miami Tissue Bank. Ganglia were cleaned of anterior and posterior roots and connective tissue and digested overnight in collagenase IV and dispase. The next day, human DRGs were seeded onto laminin coated 35mm dishes and subjected to the same cultured conditions as the rat DRGs. Fewer human DRG neurons were obtained and tested due to scarcity of fresh human DRG availability.

Calcium Imaging Experiments: Cells were loaded with the calcium dye Fluo-4AM (3µM) and incubated in the dark at 37 °C for 1 hour. Fluorescent measurements of calcium were performed using a multi-photon microscope (Bruker/ Prairie View, excitation 810 nm, emission 545nm, 0.5 Hz). Neurons were incubated at 37 °C for 15 minutes to establish a baseline calcium signal and then exposed to temperatures of 38, 39, 40, 42, and 44 °C while imaging.

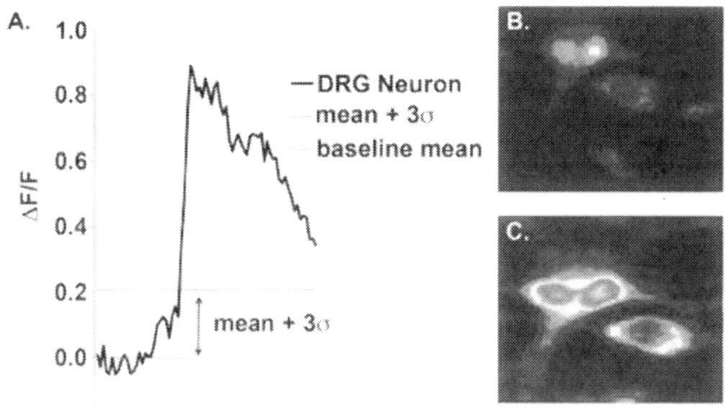

FIGURE 1. (A.) REPRESENTATIVE CALCIUM IMAGING ΔF/F PLOT DEMONSTRATING NEURON ACTION POTENTIAL AT (B.) BASELINE AND (C.) AFTER HEAT STIMULI.

Image and Statistical Analysis: Image analysis was conducted using Fiji software. The background signal was subtracted from each cell and the mean baseline ($F_0(t)$) was calculated at 37°C. ΔF/F (($F(t) - F_0(t))/F_0(t)$) was calculated for each cell across the entire experiment. The baseline mean and standard deviation of ΔF/F was calculated for all cells at 37°C. Neurons were considered to be firing in response to heat if the ΔF/F for the cell was 3 standard deviations greater than the mean baseline value at 37°C (Figure 1). A Chi-squared test with a Bonferroni correction was used to determine statistical significance.

RESULTS

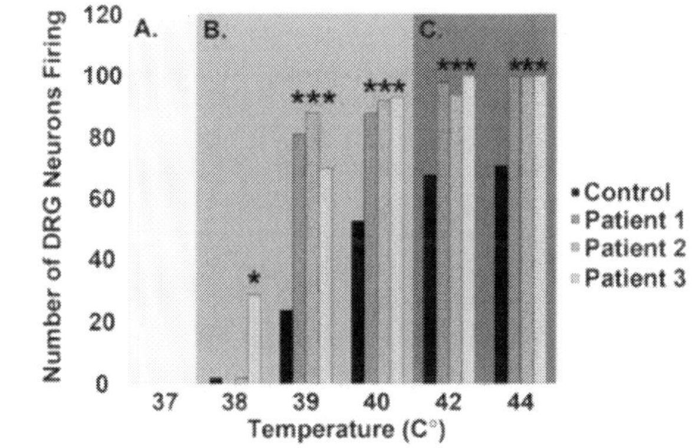

FIGURE 2. RAT DRG NEURON FIRING AFTER EXPOSURE TO DEGENERATIVE IVD CONDITIONED MEDIA (N=100, * = P<0.001 COMPARED TO CONTROL) AT ESTIMATED TEMPERATURES EXPERIENCED (A.) AT REST, (B.) DURING ACTIVITY, AND (C.) VISCOELASTIC HEATING.

Both rat and human DRG neurons were sensitized to heat after exposure to degenerative IVD condition media (Figure 2 and Table 1). The number of rat DRG neurons firing at 38 °C in cells exposed to conditioned media from patient 3 (29/100) increased significantly (P<0.01) when compared to the control group (Figure 2). A dramatic increase in DRG neuron firing occurred between 38-39°C in the degenerate IVD conditioned media exposed groups. The number of neurons firing at 39, 40, 42, and 44 ° C in rat DRG neurons exposed to conditioned media from all three patients were significantly increased (P<0.001) compared to the control group (Figure 2). For the control group, 68% and 71% of DRG neurons were heat sensitive at 42, and 44 °C, respectively. These results are in agreement with a previous study using similar methods for imaging and determination of neuron firing [5]. However, even at 44 °C, the control group firing does not reach the levels of DRG neuron firing of the conditioned media exposed neurons at lower temperatures (i.e. 39 °C).

The human DRG neurons showed similar results to the rat DRG cells when exposed to patient 2 conditioned media (Table 1). The number of human DRG neurons exposed to conditioned media was significantly greater than the number of neurons firing in the control group at all temperatures tested, as low as 38 °C. Fewer cells and patient conditions tested due to scarcity of fresh human DRGs.

	MEDIA CONDITIONS	
Temperature (C°)	**Control**	**Patient 2**
37	0 of 5	0 of 13
38	0 of 5	5 of 13*
39	0 of 5	2 of 13*
40	1 of 5	3 of 13*
42	1 of 5	5 of 13*
44	1 of 5	9 of 13*

TABLE 1. HUMAN DRG NEURON FIRING AFTER EXPOSURE TO DEGENERATIVE IVD CONDITIONED MEDIA (* =P <0.005 COMPARED TO CONTROL)

DISCUSSION

In the present study, we demonstrated that factors released by human degenerative discs sensitized neurons to elevated temperature by significantly increasing the number of neurons firing at temperatures as low as 38 °C in both rat and human DRGs. These results demonstrate the shifting of the firing threshold of DRG neurons in response to physiologically relevant heat stimuli in the IVD.

Normal body temperature fluctuations range from 36.4 – 37.7 °C [6]. The lowering of the neuron-firing threshold near this range suggests that patients with degenerative discs could experience nociceptive neuron firing with relatively small increases in temperature in the degenerative IVD due to core temperature changes, viscoelastic heating, and/or inflammation induced increases in temperature. Core body temperatures have been demonstrated to increase into a range from 38 – 40 °C during activity [7]. This temperature range would induce increased firing from all patients tested here. Additionally, temperatures above 40°C have been measured and modeled in fibrous tissue due to viscoelastic heating [8], which may be relevant to annulus fibrous tissue during loading.

Our results demonstrate that degenerative IVD produces factors capable of sensitizing neurons to heat stimuli. This may indicate that discogenic back pain may be related to interactions between the inflammatory cytokines present in the disc, nociceptive neurons innervating the disc, and heating stimuli, such as viscoelastic heating and changes in core temperature. Future work will focus on determining which factors (e.g. inflammatory cytokines) are influencing the sensitization of DRG neurons and measuring/modeling temperatures in the IVD during loading.

ACKNOWLEDGMENTS

All multiphoton microscopy was conducted at the University of Utah Cell Imaging Core facility.

REFERENCES

[1] Ribsud, M.V. et al. *Nat Rev Rheumatol.* 10: 44-56, 2014. [2] Bowles, R.D. et al. *Osteoarthritis Cartilage*, 21: S59-60, 2013. [3] Ashton. I.K. et. al. *J Orthop Res.* 12:186-192, 1994. [4] McCoy, E. S. et al. *Neuron,* 78: 138–151, 2013. [5] Greffrath, W. et al. *Neuroscience*, 104: 539-50, 2002. [6] Mackowiak, P.A., et al. *JAMA.* 268: 1578-1580, 1992 [7] Gonzalez-Alonso, J. et al. *J. Appl Physiol,* 86:1032-1039, 1999. [8] Wilson A.M. et. al. *J Biomech.*, 27: 899-905.

SB[3]C2015
Summer Biomechanics, Bioengineering and Biotransport Conference
June 17-20, 2015,Snowbird Resort, Utah, USA
SB³C2015-1041

HEMOCOMPATIBILITY ASSESSMENT OF HYALURONAN ENHANCED LINEAR LOW DENSITY POLYETHYLENE FOR USE IN AORTIC HEART VALVE LEAFLETS

Rachael L. Simon-Walker (1), John C. Cavicchia (2), David L. Bark Jr. (1,2), Susan P. James (1,2), Lakshmi P. Dasi (1, 2), Ketul C. Popat (1,2)

(1) Department of Mechanical Engineering
Colorado State University
Fort Collins, CO, USA

(2) School of Biomedical Engineering
Colorado State University
Fort Collins, CO, USA

INTRODUCTION

Bioprosthetic and mechanical heart valves have greatly reduced mortality and improved the quality of life of those suffering from heart valve disease but remain plagued with issues of durability (bioprosthetic) or require the use of life-long anticoagulation therapy (mechanical) [1, 2]. It follows that developing a heart valve that is both long-lasting and non-thrombogenic remains a critical concern. As an alternative to currently used bioprosthetic and mechanical heart valves, synthetic polymers have been sought after for their robust properties which lend themselves to enhanced longevity and the ability to be manufactured to have similar hemodynamic properties to that of the native aortic valve [3, 4]. However, as most synthetic polymers are naturally hydrophobic in nature, thrombosis is a great risk as hydrophobic surfaces promote platelet activation/adhesion and can trigger the coagulation cascade. As a result, no polymer based heart valve has been approved for the market to date [4].

To address this drawback we have developed a semi-interpenetrating polymer network using linear low density polyethylene (LLDPE) and hyaluronan (HA), a naturally occurring hydrophilic polysaccharide found in the native aortic valve leaflets, to create a highly hydrophilic and highly durable material for use in heart valves. It is hypothesized that the hydrophilic HA on the surface will combat thrombosis at the blood-material interface by mitigating the primary step in the generation of thrombosis through the reduction of protein adhesion thereby resisting platelet adhesion/activation and the coagulation cascade. The use of LLDPE was selected for the HA+LLDPE composite due to its durability and mechanical properties which can be manufactured to similar qualities to heart valve leaflets.

In preliminary studies, we have evaluated the hemocompatibility of three hydrophilic polyethylene based surfaces with varying captive bubble contact angles: LLDPE (84°), tissue cultured polystyrene (TCPS, 62°) and HA imbued LLDPE (HA+LLDPE, 30°). Results indicate that the highly hydrophilic HA+LLDPE may result in a more hemocompatible material. Immediate and future work will compare TCPS and untreated LLDPE to LLDPE imbued with varying concentrations of HA. Hemocompatibility will be determined through standard platelet activation/adhesion studies, whole blood clotting analysis, cytotoxicity studies, fibrinogen binding, contact and compliment activation, thrombin-antithrombin complex assay (TAT), and thrombin generation assays (TGA).

METHODS

Preliminary studies were performed to compare the hemocompatibility of untreated LLDPE with that of TCPS (CellStar®, Greiner Bio-One GmbH, Germany) and HA+LLDPE. 8 mm discs of each material were first disinfected through sonication in 70% ethanol for 30 min, rinsed in phosphate buffered saline (PBS) (2x), allowed to air dry in a sterile hood, and UV treated for 30 min. LLDPE and HA+LLDPE substrates were adhered onto a subset of the TCPS disks via Vetbond a commonly used surgical glue (3M, St. Paul, MN, USA) in order to keep the thin sheets of LLDPE from floating in any liquids. Platelet adhesion studies were then performed using the following techniques.

Whole blood from a healthy donor was drawn via venipuncture into 5 standard 6 mL vacuum tubes coated with the anti-coagulant ethylenediaminetetraacetic acid (EDTA), a calcium chelator. The first tube was discarded to avoid the presence of tissue factor, endothelial-derived ultra long von Willebrand Factor, and locally activated platelets acquired upon initial puncture of the skin. Tubes were centrifuged at 150 g for 15 minutes and then allowed to rest for 10 minutes. The plasma supernatant was pooled into a fresh 25 mL conical tube. 0.5 mL of plasma was pipetted onto discs of TCPS,

LLDPE, and HA+LLDPE (adhered to TCPS discs) in a 48 well plate. The well plate was placed upon a horizontal shaker plate at 100 rpm in an incubator set at 37 degrees Celsius and 5% CO_2 for 120 min. After 2 hours, unadhered cells were gently removed by aspirating the supernatant from the side of the wells and rinsing the substrates with PBS (2x). The substrates were incubated with 1 mL of a 5 μM calcein-AM solution for 20 min at room temperature. They were then rinsed with PBS (1x) and immediately imaged using a Zeiss Axioplan 2 fluorescence microscope (Zeiss, Thornwood, NY, USA). Images were taken at 5x, 10x, and 25x. Cell adhesion was determined using fluorescence microscope images and quantified using ImageJ software (National Institutes of Health, Bethesda, MD, USA).

Whole blood clotting kinetics was also analyzed using human whole blood drawn via venipuncture to ascertain the influence of the anti-thrombogenic properties of hydrophilic surfaces. The primary tube was discarded and all the blood was collected in 10 mL tubes absent anti-coagulants. 5 μL of this blood was immediately allocated to 200 μL of DI water at t=0, while the rest was placed upon the substrates within a 48 well plate. Blood was allowed to clot on the surfaces for up to 60 min and the free hemoglobin concentration was measured at regular intervals (15, 30, 60 min). After the specified clotting times, the substrates were transferred into a different 48-well plate and immersed in 0.5 mL of DI water. The samples were agitated for 30 s and left in the DI water for 5 min to release the free hemoglobin from the red blood cells. After 5 min, 200 μL of the blood-DI water solution was transferred to a 96 well plate, placed in a plate reader and analyzed at a wavelength of 540 nm.

RESULTS

Preliminary results indicate no significant differences between TCPS and LLDPE in the reduction of platelet adhesion to the surface. However, cell coverage analyzed using the ImageJ software from fluorescence images showed a nearly 45% decrease in cell coverage with plasma placed on HA+LLDPE.

FIGURE 1: Comparison of platelet adhesion on linear low density polyethylene (A;LLDPE) vs tissue cultured polystyrene (B;TCPS) and hyaluronan imbued LLDPE (C;HA+LLDPE) based on calcein labeling and imaged at 25x. Free hemoglobin was significantly reduced in the LLDPE+HA group as compared to LLDPE and TCPS (p≤.01).

Whole blood clotting results show that while there was no significant difference in reduction of free hemoglobin between TCPS and LLDPE at t=15 and t=30, a significant difference was noted at t=60 ($p \leq .05$). HA+LLDPE however greatly reduced clotting at all time points as compared to untreated LLDPE and TCPS. While free hemoglobin was reduced by approximately 68% at t=15, 76% at t=30, and 84% at t=60 for TCPS and LLDPE, HA+LLDPE had a reduction of only 30% at t=15, 54% at t=30, and 63% at t=60 in free hemoglobin ($p \leq .01$).

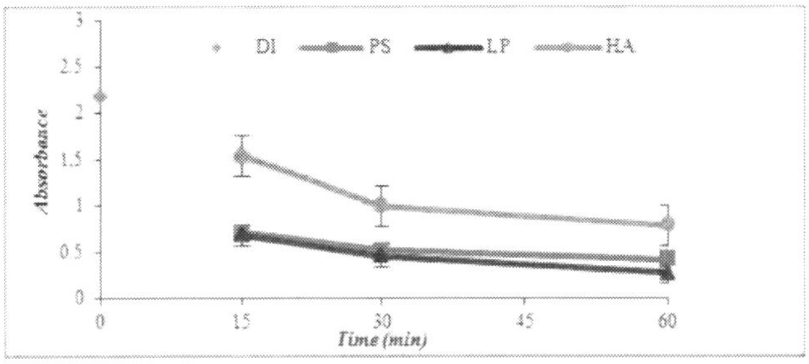

FIGURE 2: Free hemoglobin measured via absorbance. Significant changes in hemoglobin concentration were found in the HA+LLDPE group at all time points (p≤.01) .

DISCUSSION

When blood contacts a foreign material the first phase of the healing process is to initiate the transport of proteins to the surface of the material from blood flow. These proteins adhere to the surface, signaling platelet activation and more permanent adhesion. The corresponding coagulation cascade can initiate with the platelet surface as a catalyst. As proteins adsorb they can change conformation and expose hydrophobic sites present within the proteins thereby creating increased stability in the bond between the surface and protein, while also providing more binding sites for translocating platelets on the surface. Hydrophilicity thus greatly impacts the hemocompatibility of biomaterials and can be used to manipulate materials to become more bio and hemocompatible by eliminating the exposure of the hydrophobic domain in the plasma proteins. Results of this study indicate that highly hydrophilic surfaces (HA+LLDPE surfaces) result in decreased platelet coverage and decreased clotting as compared to less hydrophilic surfaces (TCPS; LLDPE). Preliminary results also suggest that hydrophilic surfaces may have a contact angle threshold which waylays the act of early clotting and overall platelet adhesion. Future work will analyze varying contact angles on whole blood clotting results over time and their comparison to platelet adhesion and activation. Overall hemocompatibility of surfaces with varying percentages of HA cross-linked with LLDPE will also be assessed. Future testing will include a more in depth look at the hemocompatibility of these surfaces through thrombin-antithrombin assays and thrombin generation assays, complement and contact activation, fibrinogen binding, and further whole blood clotting and platelet activation studies including SEM analysis of platelet activation.

Synthetic polymers have been sought out as potential candidates for manufacturing heart valve leaflets but have ultimately failed due to their hydrophobic nature. Having the ability to create hemocompatible surfaces on such materials by increasing hydrophilicity could lead to a new generation of heart valves that speak to the present pitfalls of currently used industry valves.

ACKNOWLEDGEMENTS

Research reported in this abstract was supported by the National Institutes of Health under award number R01HL119824.

REFERENCES

[1] Hammermeister, K. et al., *J Am Coll Cardiol.*, 36(4); 1152-1158. 2000.
[2] Pirabot, P. et al., *Circulation.*, 119; 1034-1048. 2009.
[3] Clairborne, T. et al., *J Biomech Eng.*, 135; 2013.
[4] Ghanbari, H. et al., *Trends Biotechnol.*, 27(6);359-67. 2009.

SB[3]C2015
Summer Biomechanics, Bioengineering and Biotransport Conference
June 17-20, 2015,Snowbird Resort, Utah, USA
SB³C2015-1042

CONVECTIVE LEAKAGE MAKES HEPARIN LOCKING OF CENTRAL VENOUS CATHETERS INEFFECTIVE WITHIN SECONDS: EXPERIMENTAL MEASUREMENTS IN AN INVITRO MODEL OF THE PEDIATRIC SUPERIOR VENA CAVA

Michael C. Barbour (1), Patrick M. McGah (1), Kurt Sansom(1), Kenneth Gow(2), Alberto Aliseda(1)

(1) Mechanical Engineering
University of Washington
Seattle, WA, USA

(2) Seattle Children's Hospital
Department of Surgery
University of Washington
Seattle, WA, USA

INTRODUCTION

Central Venous Catheters (CVCs) are commonly used to provide vascular access during hemodialysis in patients with end-stage renal disease. In recent years, the use of CVCs for hemodialysis has grown in the United States and they are used in approximately 25% of hemodialysis patients[1]. Despite their widespread use, CVCs have a high incidence rate of catheter-related infection and are associated with known elevated risk of mortality and morbidity[2]. When compared to arteriovenous fistulas, CVCs carry a 2-3 fold increase in a patient's risk of death and 5-10 fold increase in risk of infection[3]. CVCs are also prone to malfunction and 54% of CVCs remain patent for less than 90 days[4].

A common complication associated with CVCs is thrombus formation at or near the catheter tip[5]. A thrombus will occlude the catheter lumen and restrict flow rates during dialysis, requiring an intervention and occasionally a full catheter exchange. In an attempt to prevent such complications, hospitals routinely administer an anticoagulant, often a heparin/saline solution, into the catheter lumens following dialysis - a procedure known as *heparin locking*. The recommended concentration for a heparin lock is 1,000 U/mL, but some centers will use concentrations as high as 10,000 U/mL[6]. The logic behind this practice is that the locking fluid remains inside the catheter lumens for the duration of the interdialytic period, preventing clot formation.

Although it has been the standard for many years, the benefits of heparin locking are controversial, and the process seems to have been adopted without sufficient supporting evidence[7]. A number of trials have found weak or conflicting evidence for the benefit of heparin in preventing CVC thrombus formation, in comparison to pure saline[8,9,10]. Heparin locking is also associated with added risks to the patient: during the interdialytic cycle heparin will leak out into the systemic blood circulation, placing both adult[11] and pediatric[12] patients at risk of systemic anticoagulation and bleeding incidences.

METHODS

An Ash-split, 10 Fr, non-tunneled catheter with a combined luminal volume of 1.8 cc is used in the experiments. The flow conditions inside the pulsatile flow loop are designed to mimic the hemodynamic environment of a pediatric (5-7 yrs. old) Superior Vena Cava (SVC). The working fluid is a 60:40 by weight water:glycerin mixture with a viscosity of 3.2 cP at a temperature of 25°C. The density of the mixture, measured with a standard hydrometer, is 1,050 kg/m^3. The flow is driven by a pulsatile blood pump at 54 bpm, a stroke volume of 30 mL, and a 35%-65% systole-diastole waveform.

Characterization of the flow in the model SVC inside the pulsatile flow loop is performed by particle image velocimetry (PIV) starting at two vessel diameters upstream of the location where the catheter is introduced into the flow loop.

We investigate the dynamics of heparin leakage from CVCs using planar laser induced fluorescence (PLIF). These measurements, together with the velocity profile measured by PIV, allows for the quantification of the amount of locking solution that is lost from the catheter during the interdialytic cycle. Two separate experimental configurations are used in this study in order to investigate the influence of convective-diffusive flux on the leakage process. Locking fluid losses are measured from a CVC placed in both static fluid and the pulsatile flow loop, allowing for the calculation of losses caused by diffusion and convection respectively. Losses due to pure instillation are also measured from a catheter placed inside the pulsatile flow loop, but with the pump inactive, to separate the effect of instillation from convection.

RESULTS

Data from a time-resolved PLIF experiment is shown in fig 1. The RhB concentration in the measurement plane is integrated over the cross sectional area of the model SVC and plotted versus time, expressed in cardiac cycles. The concentration of RhB measured downstream of the catheter tip rapidly rises and then decays in an almost exponential manner following the onset of convection. The peak in RhB concentration and convective mass flux only lasts a few seconds (cardiac cycles), after which a negligible flux of locking solution is detected. The instantaneous mass flux, also shown in fig 1 is calculated in combination with the PIV velocity measurements.

The results from the cumulative mass flux show the mean RhB leakage from instillation alone is 28.16% of the initial RhB mass in the catheter lumen. When the catheter is subjected to pulsatile flow conditions for 10 cycles and 60 cycles, the convective losses increase to 42.96% and 43.9% respectively. The three measurements of cumulative mass fluxes obtained from mass balance in the catheter lumen (instillation, convection for 10 cardiac cycles and convection for 60 cardiac cycles) are compared using an independent T-Test and a Wilcoxan Rank-Sum test. When comparing the mass balance loses due to convection, 10 cardiac cycles vs 60 cardiac cycles, the P-values for the independent T-Test and Wilcoxan Rank-Sum test are .723 and .596, respectively. Thus, the measurements are found to be equivalent (they could be measuring the same quantity). This is consistent with the data from the time-resolved mass flux that shows only the first few cardiac cycles contribute to the cumulative mass flux. When instillation-only fluxes are compared to convective flux experiments (for 10 cardiac cycles), however, the P-values from both tests are $p<.01$, rejecting the null hypothesis and showing statistically significant differences: the leakage due to catheter instillation is distinct from those due to instillation and convection.

DISCUSSION

The results presented here, in conjunction with our numerical work published previously[13], suggest a clear two-stage mechanism for heparin leakage from CVCs. The early stage leakage, which lasts less than 10 seconds, is caused by instillation imparting momentum to the lock fluid and to a convective flux from the surrounding pulsatile blood flow. The late stage leakage, which constitutes the remainder of the interdialytic phase, is limited by the slow diffusive flux of heparin from inside the catheter lumen where the concentration is still high.

Leakage in the early phase occurs very rapidly and contributes to significant losses of up to 40% of the initial catheter locking solution. Previous studies[14] have suggested that early phase leakage lasts for 10 minutes post-instillation, but the instantaneous PLIF results presented here show that the majority of lock fluid losses occur within the first 10 cardiac cycles. The comparison of PLIF mass flux measurements of RhB leakage from a CVC subjected to pulsatile conditions for 10 vs 60 seconds show that there is no statistical difference between the two. In other words, the losses that occur after 10 seconds are statistically insignificant when compared to the losses that occur during the first 10 seconds. This new early phase time-scale of 10 seconds is in agreement with our recent numerical study[13] which estimates the time necessary to deplete an Ash-Split CVC tip to be <10 seconds.

The comparison of mass losses from experiments with and without pulsatile conditions after instillation demonstrate that the early phase leakage is a result of two separate mechanisms: instillation and convection. Previous *in vitro* studies[15] have reported instillation losses in the range of 20-30% but the influence of pulsatile flow conditions on locking solution leakage had yet to be quantitatively studied. The instillation loss rates in our study are consistent with this previous study. Further, after subtracting this instillation loss, we find that convection due to the pulsatile flow environment around the tip of the CVC inserted into the SVC or right atrium, induces a fast flux where an additional 10-15% of locking fluid is leaked from the catheter tip. We hypothesize that these additional losses are due to a convective flux of locking solution forced by the hemodynamic flow around and into the catheter tip. The time-dependent pressure gradient induced by the inertia of the pulsatile blood flow creates a force that drives locking solution out of the catheter through some of the ports at the tip. This fluid is replaced by blood flowing into the catheter from other upstream ports, and therefore the replenishment of heparin concentration is only through diffusion from the remainder of the CVC luminal volume, a slow mechanism. CFD of the near-tip region showed blood entering the catheter through the side ports and transporting heparin out of the tip and into the systemic circulation[13]. The region of the catheter with side-ports, the near tip region, constitutes approximately 10% of the catheter locking volume, limiting the convective flux for the catheter used in this study. This result implies that, unlike instillation losses that push locking solution out of the tip but replaces it with more locking solution with approximately equal heparin concentration, convective fluxes deplete the near-tip region of nearly all heparin replacing the locking solution with blood from the SVC.

ACKNOWLEDGEMENTS

This work has been financially supported by the NIH/NINDS (grant 1R03NS078539 and 1R01NS088072), by an NSF CAREER award (CBET-0748133) and by the UW Royalty Research Fund.

REFERENCES

[1] Ethier et al. *Nephrol Dial Transpl*, 23(10):3219–3226, 2008.
[2] Ravani et al. *J. Am Soc Nephrol*, 24:465–473, 2013.
[3] Rehman et al. *Clin. J. Am. Soc. Nephrol.*, 4(2):456–460, 2009.
[4] Napalkov et al. *BMC Cardiovascular Disorders*, 13(1):1–1, 2013.
[5] Beathard GA. *Catheter thrombosis*, 2001.
[6] Moran et al. *Semin Dialysis*, 21:441–445, 2001.
[7] Matthew et al. *J. Adv. Nursing*, 65(10):2007–2021, 2009.
[8] Rabe et al. *Intensive Care Medicine*, 28(8):1172–1176, 2014.
[9] Cesaro et al. J. *Clin. Oncology*, 27(12):2059–2065, 2009.
[10] Goossens et al. *Ann. Oncology*, 24(7):1892– 1899, 2013.
[11] Karlaasan et al. *Nephrol Dial Transpl,* 16:2072–2074, 2001.
[12] Moritz et al. *Pediat Nephrol*, 18:68–70, 2001.
[13] McGah et al. *ASAIO Journal*, 60(4):443–451, 2014.
[14] Markota et al. *Nephrol Dial Transpl*, 24(5):1550–1553, 2009.
[15] Sungur et al. *Nephrol Dial Transpl*, 22(12):3533–3537, 2007.

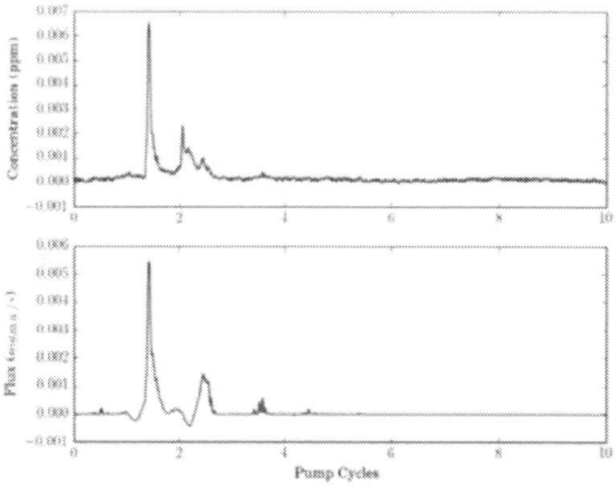

Figure 1: Concentration and Mass Flux of RhB 3 diameter downstream of catheter tip, as a function of time, over 10 cardiac cycles

NOVEL IMAGE-BASED ANALYSIS OF PATIENT-SPECIFIC ABDOMINAL AORTIC ANEURYSM WALL STRESS USING A MEMBRANE MECHANICS MODEL

Mirunalini Thirugnanasambandam (1), Prahlad G. Menon (1,2), Stephane Avril (3), Ender A. Finol (1)

(1) Department of Biomedical Engineering
University of Texas at San Antonio
San Antonio, Texas, USA

(2) Department of Electrical and Computer Engineering
Sun Yat-sen University - Carnegie Mellon University
(SYSU-CMU) Joint Institute of Engineering,
Pittsburgh, Pennsylvania, USA

(3) Center for Biomedical and Healthcare Engineering
Ecole Nationale Supérieure des Mines
Saint Etienne, France

INTRODUCTION

Abdominal aortic aneurysms (AAA) are characterized by the focal dilatation of the infrarenal region of the abdominal aorta by more than 50% of its normal, healthy diameter. The current clinical management of AAA is based on a single geometric variable - the maximum diameter of the AAA - which is not a reliable predictor of aneurysm rupture for all individuals. A maximum diameter of 5.5 cm, on average, or a growth rate of 1 cm/year is the current clinical standard for AAA elective repair. Patients who do not satisfy either of these criteria are kept under surveillance and not recommended elective repair. This does not account the small aneurysms that do rupture while on surveillance and the large aneurysms that remain stable for years. Therefore, there is need for an accurate clinical marker that can be used for decision making for elective intervention.

The objective of this study was to derive the patient-specific AAA wall stress distribution, which is an important biomechanical parameter that characterizes risk of rupture. This was done using an innovative image-based technique with the AAA wall modeled as a nonlinear elastic membrane. This aspect of using membrane mechanics helps to circumvent the need for a constitutive model to describe the material properties of the arterial wall.

METHODS

MRI and CT imaging. Contrast-enhanced computed tomography (CT) images were acquired retrospectively to reconstruct the aneurysm geometry and derive regional variations of wall thickness. ECG-gated MR images were acquired prospectively to capture wall motion over 20 phases of a cardiac cycle.

Segmentation and image fusion. The CT images of a representative AAA were segmented using an in-house segmentation and geometry quantification code (AAAVasc), to identify the lumen, inner wall and outer wall contours following the protocol described in [1]. Wall thickness was estimated at 72 points along the circumference of each CT slice. These points together constitute the vertices of a 3D surface representation of the AAA.

The CT reconstructed AAA model was encoded with regional wall thickness information. In addition to this model, ECG-gated time-resolved MR images at each phase of the cardiac cycle were segmented in axial slices (at every available time instant) using the open-source cardiac MRI suite Medviso Segment (Medviso AB, Sweden). This yielded the lumen contours by means of a semi-automatic edge-attracting active contouring function [2], following which 3D surface geometries of the lumen and the outer AAA wall were exported [3]. The centerlines of the CT and MR based 3D models at each time instant were aligned and the two representations of the same AAA were co-registered by iterative closest point based rigid-body registration (see Fig. 1). Each vertex on the reconstructed 3D CT surface was matched against the 3D MR reconstructed AAA outer wall surface at each available cardiac phase by establishing correspondences between each vertex of the CT surface and the closest surface location on each co-registered MR surface reconstruction. In this manner, the wall thickness at each vertex of the CT AAA reconstruction was warped onto the MR-derived AAA reconstruction at the 20 cardiac phases. Together, these constituted a time-series of surface models encoded with accurate CT-based wall thickness, which were pipelined as an input to a nonlinear elastic membrane model for computing AAA wall stresses.

Membrane model. When the AAA wall is modeled as a non-linear elastic membrane, the spatial distribution of intraluminal pressure is assumed uniform. A weighted finite difference method was used to solve the local equilibrium equations at each node on the AAA wall. Intraluminal blood pressures at each of the 20 phases of the cardiac cycle in which the MR images were acquired were obtained by interpolating the average AAA intraluminal pressure waveform described in Veer et al. 2008 [4] and scaling it with the patient-specific cuff SP/DP (systolic/diastolic pressures).

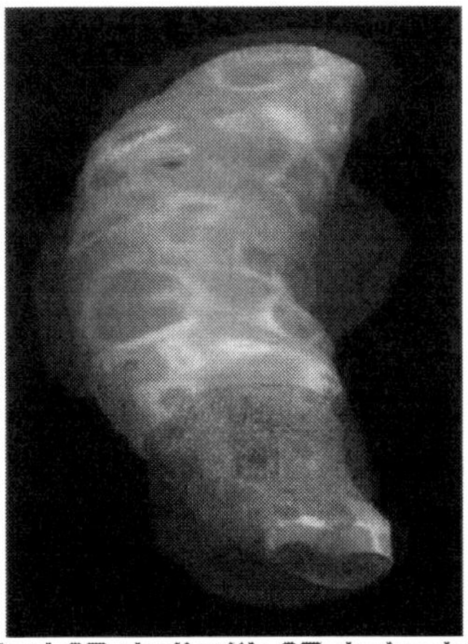

Figure 1: Original CT shell with CT-derived wall thickness co-registered with the MRI outer shell prior to warping.

Initially, the domain of interest was discretized to create a triangular finite element mesh. The unknown components of stress and their spatial variations between nodes were evaluated at each point on the AAA wall by assuming a linear shape function. The stress tensor was evaluated at each point on the pressurized membrane model of the AAA wall at each phase of the cardiac cycle. This was done by solving the boundary value problem while assigning Neumann boundary conditions at the boundaries of the AAA domain. There were three equilibrium (scalar) equations at every point and three unknowns (σ_{11}, σ_{22}, and σ_{12}), which were the three components of the stress tensor in the plane of the membrane. Thus, the tension in the vessel wall was determined without using a material model that describes the elastic properties of the wall. Using σ_{11}, σ_{22}, and σ_{12}, the principal stresses and maximum shear stress were evaluated using Equations (1) and (2):

$$\sigma_{I,II} = \frac{\sigma_{11}+\sigma_{22}}{2} \pm \sqrt{\frac{(\sigma_{11}-\sigma_{22})^2}{2} + \sigma_{12}^2} \quad (1)$$

$$\tau_{max} = \frac{(\sigma_I - \sigma_{II})}{2} \quad (2)$$

The first and second principal stresses and the maximum shear stress were evaluated on the entire surface of the AAA wall.

RESULTS AND DISCUSSION

A plot of the spatial distribution of principal stresses and shear stress derived for the membrane model using patient-specific geometry and wall thickness distribution is presented in Fig. 2, where 'A' and 'P' represent the anterior and posterior anatomical directions, respectively. The corresponding standard error box plots with outliers are also highlighted in Fig. 2. Table 1 provides the descriptive statistics of these parameters.

From Fig. 2 it can be observed that the region of maximum principal stresses and maximum shear stress is on the postero-lateral side of the aneurysm, which in turn exhibits high surface curvature. The distribution of stresses evolves as the morphology of the AAA changes over the cardiac cycle. The neck region is typically not the most likely site of AAA rupture. From [5] it can be seen that the posterior wall and the retro-peritoneal regions are more prone to rupture as compared to any other location on the aneurysm. The skewness and kurtosis values from Table 1 depict a near-normal distribution with the principal stress distributions slightly skewed. The box plots shown in Figure 2 were generated using cut-off values corresponding to the 1st quartile minus 2.2 times the inter-quartile range to the 3rd quartile plus 2.2 times the inter-quartile range. The outliers seen in the principal stress distributions depict extreme peak wall stresses that can lead to erroneous conclusions.

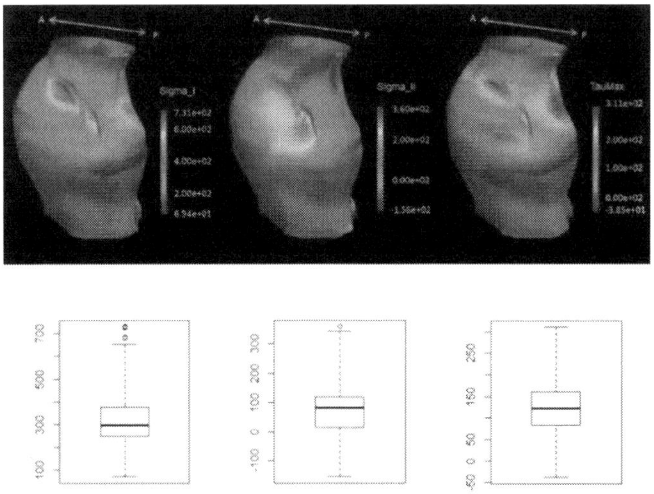

(a) (b) (c)

Figure 2: (a) Distribution of (a) first and (b) second principal stress and (c) maximum shear stress with corresponding boxplots and outliers.

Table 1: Descriptive statistics: principal stresses and maximum shear stress (kPa)

	Mean	*Median*	*SD*	*Skewness*	*Kurtosis*	*Confidence interval*	
σ_I	321.65	299.03	100.7	0.96	0.90	316.9	326.3
σ_{II}	71.93	82.65	82.91	0.09	0.23	68.09	75.77
τ_{max}	124.86	121.94	56.71	0.22	-0.13	122.2	127.5

Image-based patient-specific AAA stress analysis was performed assuming a nonlinear membrane model for the AAA wall. The primary advantage of the novel method presented herein is the fact that computationally expensive finite element analyses were not needed and no assumptions were made regarding local and global AAA morphology or material properties. This clinically relevant and promising method has potential to lead to the development of a fast and reliable tool to predict AAA rupture risk which may be impactful for timely and effective clinical management.

REFERENCES

[1] Shum, J., et al. Medical Physics 37(2010): 638-648.

[2] Heiberg, E., et al. *BMC Medical Imaging* 10.1 (2010).

[3] Menon, PG., et al. *Journal of Cardiovascular Magnetic Resonance, 16*(Suppl 1) (2014): 385.

[4] van't Veer, M., et al. *Journal of Vascular Surgery* 48.6 (2008): 1401-1407.

[5] Simão da Silva, E., et al. *Cardiovascular Surgery* 8.7 (2000): 526-532.

SB³C2015
Summer Biomechanics, Bioengineering and Biotransport Conference
June 17-20, 2015, Snowbird Resort, Utah, USA
SB³C2015-1044

CHARACTERIZATION OF SOFT TISSUE MICROSTRUCTURE VIA TRANSMURAL SALS

John G. Lesicko, Kristen R. Feaver, Michael S. Sacks

Center for Cardiovascular Simulation
Institute for Computational Engineering and Sciences
Department of Biomedical Engineering
The University of Texas at Austin
Austin, Texas, USA

INTRODUCTION

Understanding the microstructural components of soft tissues is essential in determining structure-function relationships and in guiding the construction of biomaterials for use in conjunction with other living tissues. Small angle light scattering (SALS) is a proven means of quickly acquiring sample-wide collagenous soft tissue structural information. However, traditional SALS techniques are limited: collection of depth-specific structural information is not possible due to the transmission-based nature of the technique such that location-specific microstructure data is inherently depth averaged. As health care becomes increasingly reliant upon multi-scale organ models, the need for more detailed structural fiber information rises. Techniques such as micro-CT are able to provide fiber direction information in three dimensions, yet offer no other direct means to characterize the microstructural elements of the tissue. Thus, in order to better study the depth-specific microstructure of native and engineered soft tissues, we have developed a new experimental and analytical method for acquiring *transmural* structural information via SALS, dubbed Transmural SALS (TSALS).

METHODS

The SALS technique for collecting structural properties works by raster scanning specimens through the path of a low power laser beam and collecting the location-specific angular distributions of local microstructural fiber from the spatial distribution of scattered light. This process yields large field-of-view microstructure maps that are limited in resolution by only the raster step size and the diameter of the incident beam. Data obtained using SALS techniques takes the form of a two-dimensional set of equally spaced nodes, with each node representing a location-specific orientation distribution function (ODF), $\Gamma[(\theta)]$, obtained from the angular distribution of local microstructural fibers. The ODF is the basis for all further analysis of microstructure properties.

Experimentally, TSALS data collection techniques closely match traditional SALS procedures. Yet, rather than scanning intact tissues, samples are histologically sectioned at the desired axial resolution through the depth of the tissue and each individual slice is scanned. These serial scans are registered computationally to match the original configuration of the tissue sample and result in a three dimensional set of fiber orientation distributions that can be analyzed using structure tensor methods.

To provide a compact method for both data storage and manipulation in modeling, we utilized a 4th rank structure tensor, $\tilde{\mathbf{H}}$, to represent the ODF at each point in the tissue. We start by defining a direction variable, $\hat{\mathrm{N}} = \hat{\mathrm{N}}(\theta)$, and the associated 2D ODF $\Gamma\left[\hat{\mathrm{N}}(\theta)\right]$. From that ODF, we define $\tilde{\mathbf{H}}$ as:

$$\tilde{\mathbf{H}} = \mathrm{H}_{ijkl} = \int_{-\pi/2}^{\pi/2} \Gamma[\hat{\mathrm{N}}]\mathrm{N}_i \mathrm{N}_j \mathrm{N}_k \mathrm{N}_l d\hat{\mathrm{N}} = \int_{-\pi/2}^{\pi/2} \Gamma[\hat{\mathrm{N}}(\theta)]\mathrm{N}_i(\theta)\mathrm{N}_j(\theta)\mathrm{N}_k(\theta)\mathrm{N}_l(\theta)\hat{\mathrm{N}}(\theta)d\theta$$

These structure tensors act as a concise mathematical descriptors of the fiber orientation properties, and allow for a close-fit reconstruction of the original ODF using well-defined operations, as shown in Figure 1. From $\tilde{\mathbf{H}}$ we can compute all fiber orientation-related parameters such as preferred direction (PD) and orientation index (OI), a measure of in-plane fiber splay.

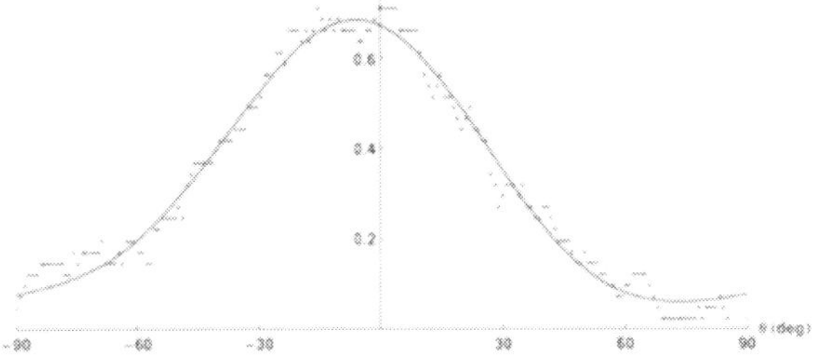

Figure 1. Recovered ODF (red) overlaid on experimental data (scatter).

RESULTS

TSALS yields a spatially accurate, three-dimensional, nodal data set from which we can glean valuable sample-specific information on spatial variation in microstructure property via gradients of scalar, vector, and tensor values throughout the transmural thickness of the tissue. Additionally, 3D reconstructed data can then be visualized using isosurfaces, vector fields, and tensor glyphs in ParaView (Kitware, Clifton Park, NY). All of the analysis methods have been validated using 3D data sets in which the ODF varies locally by spatial position (Figure 2).

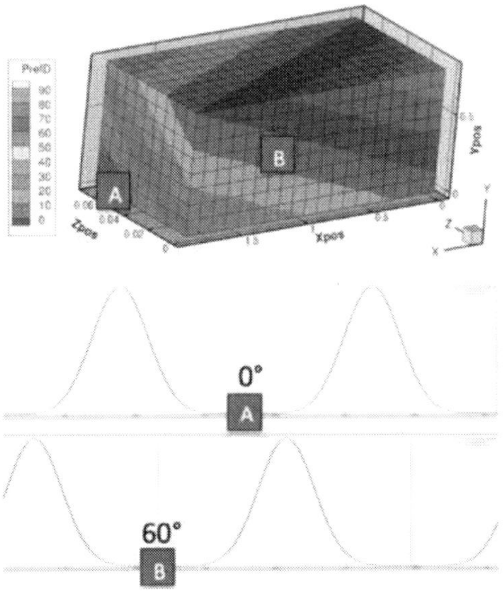

Figure 2. Graphical representation of 3D microstructural data generated for method validation: linear PD variation in all directions (top), and example ODFs demonstrating the shift in angular preferred direction throughout (bottom).

This information is used in conjunction with mechanical property databases to model and predict structure-function behavior of various tissues. Actual tissue structure information databases, such as the one presented in Figure 3, are now available for future modeling and simulation studies. To date, this method has been used to characterize the soft tissue microstructure of the following fibrous tissues: bovine tendon, porcine aortic valve leaflets, infarcted porcine myocardium, and porcine sclera.

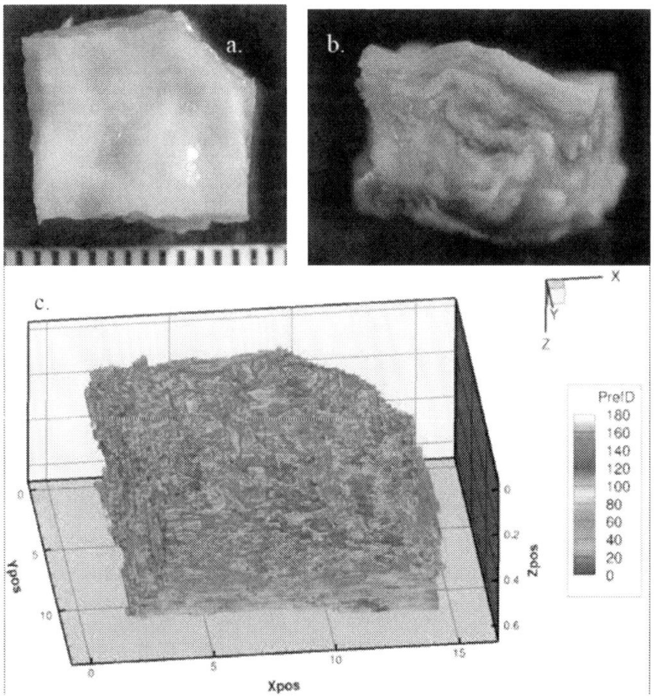

Figure 3a. Apex view of the endocardial surface of intact infarcted ovine myocardium, **b.** Transmural cross-section of tissue with endocardium at the top, **c**. Three-dimensional visualization of transmural microstructure map: shown here is preferred fiber direction.

DISCUSSION

In conclusion, we present a new technique for succinctly characterizing the depth-specific microstructure of any fibrous soft tissue or tissue-like material using experimentally proven data acquisition methods, yet utilizing novel and concise structural tensor analysis methods. The technique is a computationally efficient means of representing microstructure information obtained from ODFs, and is not necessarily limited to SALS. The resulting data is easily mapped to finite element models of biological materials, and provides a platform for previously unachievable structural analysis.

ACKNOWLEDGEMENTS

NIH grants R01HL119297 and R01HL108330.

REFERENCES

[1] Sacks, M. S. (2003). J Biomech Eng **125**(2): 280-287.
[2] Sacks, M. S., et al. (1997). Ann Biomed Eng **25:** 678-689.
[3] Advani, S.G. and Tucker, C.L.. (1987) Journal of Rheol **31**: 751-784.

TIME AND PRESSURE-DEPENDENT HYDRAULIC RESISTANCE ACROSS SCHLEMM'S CANAL ENDOTHELIAL CELL LAYERS

Alice Spenlehauer (1), Joseph M. Sherwood (1) and Darryl R. Overby (1)

(1) Department of Bioengineering
Imperial College London
London, UK

INTRODUCTION

Elevated intraocular pressure (IOP), caused by increased aqueous humor (AH) outflow resistance, is a key risk factor for glaucoma, the second leading cause of irreversible blindness in the world [1]. AH is a transparent, blood-derived fluid that bathes the anterior segment of the eye. In exiting the eye at the irido-corneal angle, the bulk of AH crosses the inner wall endothelium of Schlemm's canal (SC) in the basal-to-apical direction through micron-sized transendothelial pores. Reduction in pore density occurs in glaucoma [2, 3], suggesting that the hydraulic resistance of SC inner wall may be important for IOP regulation.

Relatively little is known about the mechanisms of pore formation. However, pores often associate with *giant vacuoles* (GVs) [4], which are large dome-like cellular outpouchings that result from the unique basal-to-apical pressure drop across the inner wall. GVs induce large biomechanical strain on SC endothelium, and strain has been observed to trigger pore formation in SC endothelial cells (SCECs) [5, 6]. *In vitro* studies have shown that GV size and pore density both increase with pressure drop [7, 8], while hydraulic resistance across SCEC layers decreases with pressure drop [7]. Furthermore, *in vitro* perfusion studies have shown that GVs are dynamic structures that form and collapse within minutes, even when the perfusion pressure is kept constant [7]. Thus, we hypothesize that the hydraulic resistance of SCEC layers exhibits a time dependence that reflects the time-dependent biomechanical response of SCECs under a fixed transendothelial pressure gradient.

In this study, we used a custom-built pressure-controlled *iPerfusion* system to measure the hydraulic resistance of SCEC monolayers seeded on a porous membrane, over a number of perfusion pressures. Time-resolved measurements enabled investigation of the temporal change in resistance for a given pressure. Additionally, by perfusing individual monolayers over more pressure steps than previously examined [7], we were able to more fully characterize the pressure-dependent hydraulic resistance of SCEC cell layers.

METHODS

Human SCECs from one ostensibly healthy donor (SC78) were isolated following previously described methods [9] and cultured in low-glucose DMEM (Gibco) supplemented with 10% fetal bovine serum (Hyclone), 1% L-glutamine (Sigma-Aldrich) and 1% penicillin streptomycin (Sigma-Aldrich). For perfusion experiments, SCECs were seeded at confluency (45,000 cells/cm^2) on filters with track-etched pores of diameter 0.4 µm (0.33 cm^2 cell growth area, 4x10^6 pores/cm^2, 3470 Transwell, Corning) using an established protocol [7]. Cells were cultured on the filter for 48 hrs prior to perfusion. Eight samples from the same cell line were included in this study.

The perfusion system consists of a manometric reservoir controlled by a linear actuator (L35, Nanotech) which is used to apply the desired pressure upstream of the system; a thermal flow meter (SLG64-0075, Sensirion) to continuously measure the flow (Q) through the filter and cell layer; and a differential pressure transducer (PX409, Omegadyne) to continuously measure the pressure drop (P) across the cell layer and filter. For each experiment, a series of five increasing 5 mmHg pressure steps lasting 10 minutes each were applied across the whole perfusion system. The cell layer was perfused in the basal-to-apical direction to mimic *in vivo* conditions. The cell layer was observed with a microscope before and after the perfusion to check that it had remained intact. Using the pressure and flow measurements acquired, the hydraulic resistance of each cell layer was computed for each pressure step ($R_p = (P*A)/Q$ where A is the filter surface area). All calculated R_p values were corrected to account for the hydraulic resistance of the filter, 0.12±0.06 mmHg.µL^{-1}.min.cm^2 (mean ± 2 standard deviations (SD)).

RESULTS

For all SCEC layers, a time-dependent decrease in R_p was observed for all pressure steps (Figure 1). The relative decrease in R_p from the beginning of steady state (once the compliant response of the system had subsided: ~ 2 minutes) to the end of the pressure step was found to range from a few percent to about 20% over approximately 10 mins. Considering each cell layer separately, the relative decrease in R_p over time was greater at lower pressure drops (Figure 2).

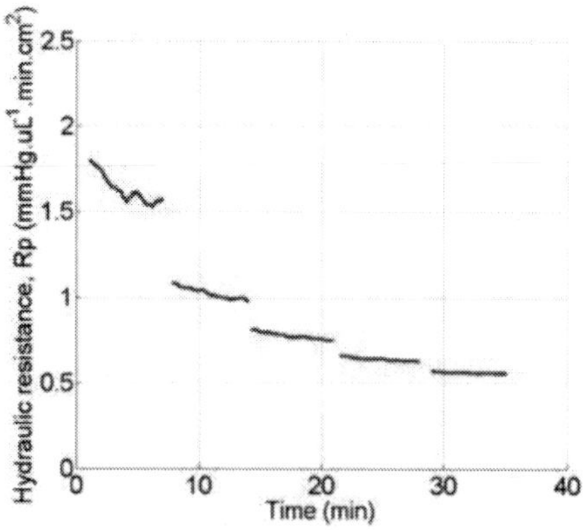

FIGURE 1 – Representative plot of R_p versus time for an individual cell layer over 5 pressure steps. There is a time-dependent decrease in R_p for each pressure step.

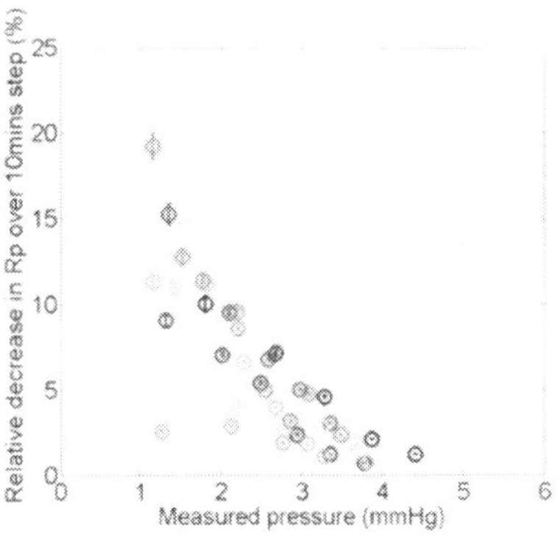

FIGURE 2 – The relative decrease in R_p over each step as a function of the pressure drop across the SCEC layer. Data from the same cell layers are shown in the same color, (mean ± 2SD).

Rp decreased significantly with increasing pressure drop for all cell layers (Figure 3). In plotting the average R_p for each step against the corresponding average pressure drop across the cell layer, it was apparent that the data could be modeled by an exponential decay:

$$R_p(P) = R_\infty\left(1 - e^{-\beta P}\right) + R_0 e^{-\beta P} \quad (1)$$

where R_∞ is the predicted hydraulic resistance at large pressures, $\beta>0$ is the decay parameter of the exponential, and R_0 is the predicted hydraulic resistance at 0 mmHg. The exponential fit predicts a nearly two-fold reduction in hydraulic resistance between 2 and 6 mmHg, which is consistent with prior measurements over the same pressure range [7]. The parameters of this model were estimated by fitting Equation 1 to the pressure-dependent behavior of each cell layer (Figure 3) using weighted non-linear regression. The decay parameter β ranged from 0.61 to 1.08 $mmHg^{-1}$; R_0 ranged from 2.35 to 8.35 $mmHg.\mu L^{-1}.min.cm^2$; and R_∞ ranged from 0.29 to 0.59 $mmHg.\mu L^{-1}.min.cm^2$. For comparison, the maximum hydraulic resistance of the SCECs *in vivo* is approximately 0.324 mmHg $mmHg.\mu L^{-1}.min.cm^2$ (calculated based on data from [10]).

Following the perfusion, all SCEC layers appeared intact without large holes or apparent areas of cell detachment.

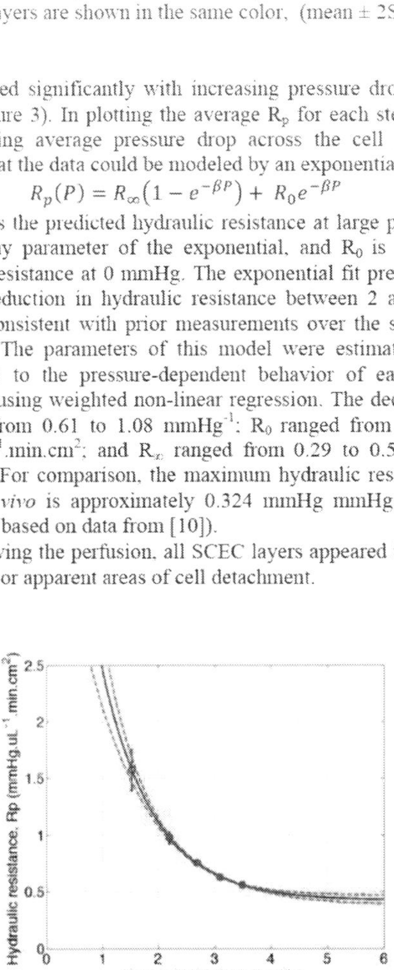

FIGURE 3 – Representative plot of average R_p value on a step against the corresponding average pressure for an individual cell layer (mean ± 2SD). The blue line is an exponential fit to the data and the magenta dashed line is the 95% confidence interval on this fit.

DISCUSSION

We measured the hydraulic resistance across *in vitro* SCEC layers using the *iPerfusion* system. An interesting finding was the significant time-dependent decrease in R_p measured within individual pressure steps. A similar behavior has been reported in bovine aortic endothelial cells [11], but not in SCECs. In addition, we show that for a given cell layer, the relative change in R_p over time decreases with increasing pressure drop. The observed changes in hydraulic resistance may be related to the dynamics of GV or pore formation that appear to occur over similar time scales [7] and may saturate at higher pressure drops. However, longer duration perfusions over a wider pressure range are necessary to more fully characterize the time- and pressure-dependent behavior of SCEC monolayers. Furthermore, by mounting the perfusion system on a microscope as done previously [7], it will be possible to simultaneously visualize GV formation directly, so as to relate changes in hydraulic resistance to the biomechanical response of SCEC layers on the cellular scale.

Previous perfusion studies of SCECs have reported that R_p decreases with increasing pressure drop [7]. Our data support this finding and reveal that an exponential decay curve is an appropriate model of the pressure-dependent decrease in R_p. The parameters extracted from the exponential fit may provide insight into the biomechanical properties (e.g. stiffness) of SCECs, which become altered in glaucoma and contribute to increased AH outflow resistance and elevated IOP [8].

A limitation of the current model is that junctions between SCECs *in vitro* are largely immature, due to low expression of cell-cell adhesion molecules such as VE-cadherin and CD31 that are down-regulated in SCEC culture [12]. Hence, the absolute values of R_p reported may not be representative of the *in vivo* cell layer. Nonetheless, the relative change in R_p with time or with pressure (as captured by the exponential decay) may reflect pore formation or widening of paracellular spaces. Ongoing studies are examining the relationship between time and pressure-dependent changes in R_p and the biomechanical properties of SCECs.

ACKNOWLEDGEMENTS

We would like to thank Kristin Perkumas (Duke University, Durham, NC) for providing human SCECs.

REFERENCES

[1] Tham, Y et al., *Ophthalmology*, 121:2081-2090, 2014
[2] Allingham, RR et al., *Invest.Ophthalmol.Vis.Sci.*, 33:1661-1669, 1992
[3] Johnson, M et al., *Invest.Ophthalmol.Vis.Sci.*, 43:2950-2955, 2002
[4] Inomata, H et al., *Am.J.Ophthalmol.*, 73:760-789, 1972
[5] Braakman, ST et al., *Exp.Eye Res.*, 2014
[6] Overby, DR et al., *Exp.Eye Res.*, 88:656-670, 2009
[7] Pedrigi, RM et al., *Exp.Eye Res.*, 92:57-66, 2011
[8] Overby, DR et al., *Proc.Natl.Acad.Sci.U.S.A.*, 111:13876-13881, 2014
[9] Stamer, WD et al., *Invest.Ophthalmol.Vis.Sci.*, 39:1804-1812, 1998
[10] Johnson, M et al., *Invest.Ophthalmol.Vis.Sci.*, 24:320-325, 1983
[11] Tarbell, JM et al., *J.Appl.Physiol.(1985)*, 87:261-268, 1999
[12] Perkumas, K et al., *Exp.Eye Res.*, 96:82-87, 2012

ULTRASOUND MONITORING OF ABDOMINAL AORTIC ANEURYSM PROGRESSION IN A MURINE MODEL

Arvin H. Soepriatna (1) Gurneet S. Sangha (1), Amelia R. Adelsperger (1), Evan H. Phillips (1), Clifford M. Babbey (2), Michael P. Murphy (2), Pavlos P. Vlachos (1, 3), Craig J. Goergen (1)

(1) Weldon School of Biomedical Engineering
Purdue University
West Lafayette, Indiana, USA

(2) Center for Aortic Disease
Indiana University-Purdue University Indianapolis
Indianapolis, Indiana, USA

(3) Department of Mechanical Engineering
Purdue University
West Lafayette, Indiana, USA

INTRODUCTION

An abdominal aortic aneurysm (AAA) is a localized enlargement of the abdominal aorta, the largest artery that supplies blood to the body. If left untreated, AAAs can continue to expand and eventually rupture, an often fatal event with a mortality rate of 90%.[1] Given that AAAs affect between 0.5% and 3.2% of the United States' population, AAA ruptures cause approximately 16,000 deaths per year, making them the 10th leading cause of death in America.[1,2,3] Even though surgical repair is an effective treatment option, patients whose aneurysms do not exceed a 5 cm diameter threshold are not recommended to undergo surgery, as it is an invasive procedure that carries surgical risk.[4] Since AAA diameter is not always a great predictor of aneurysm rupture, some patients suffer from vessel rupture without ever being recommended for surgical repair.[5] As such, there is a need to investigate an alternative and more robust predictor of aneurysm growth and rupture.

In efforts to accomplish this goal, non-invasive imaging has been used to monitor and analyze how aortic wall mechanics lead to vessel growth and remodeling prior to rupture. The current study uses ultrasound to investigate how AAA volume and circumferential cyclic strain change throughout disease progression in the apolipoprotein E-deficient/angiotensin II mouse aneurysm model.

METHODS

Animals: We used apolipoprotein E-deficient (apoE$^{-/-}$) C57BL/6 mice for this study. We induced AAAs by implanting an ALZET® osmotic pump in the dorsal, subcutaneous space of each animal. The osmotic pumps systemically released angiotensin II (AngII) at a rate of 1000 ng/kg/min for 28 days. Ultrasound monitoring showed AAA formation in 7 mice. Following euthanasia on day 28, the aortas were pressure-perfused with saline, fixed in 4% paraformaldehyde, embedded in paraffin, and sectioned at 10 μm thickness. We stained aortic cross-sections using Verhoeff Van Gieson (VVG) and Masson's Trichrome (MTC) stains and compared the images with that obtained from non-linear optical microscopy as described by Phillips et al.[6]

Data Acquisition: A Vevo2100 ultrasound system (VisualSonics) with a MS550D probe (40 MHz central frequency) was used for all image acquisitions. Before pump implantation, longitudinal and transversal B-mode cine loops, M-mode cine loops, and 3D images of the abdominal aorta were acquired. Following pump implantation, we used ultrasound to monitor the aorta for AAA formation. Once an aneurysm was identified, we acquired ultrasound images of the AAA on days 3, 7, and 14 post-diagnosis.

Volume and Strain: We used a VisualSonics software package to measure aortic volume and diameter. The AAA and true lumen volumes were determined after manually segmenting their boundaries in the acquired 3D images. The false lumen volume was calculated by subtracting the true lumen volume from the total AAA volume. These volumes were normalized using measurements for aneurysm length, and the segmented boundaries were used to generate 3D volume renderings of the AAA anatomy. We calculated the Green-Lagrange (G.L.) circumferential cyclic strain using Equation 1 and the measured systolic ($d_{systolic}$) and diastolic ($d_{diastolic}$) diameters obtained from longitudinal M-mode cine loops.[7]

$$\text{Green-Lagrange Strain} = \frac{1}{2}\left[\left(\frac{d_{systolic}}{d_{diastolic}}\right)^2 - 1\right] \qquad (1)$$

Statistics: All data are shown as mean±SE. Analysis of variance (ANOVA), followed by a Tukey-Kramer multiple comparison post-hoc analysis, was performed using SAS 9.4 statistical package to test for significance in aortic volumes and strain at different days. An alpha of 0.05 was used to determine significance.

RESULTS

The two-fold increase in total aortic volume, observed on day 3, confirmed AAA formation (shown by + in Fig. 1). AAA expansion was associated with significant increases in true and false lumen volumes (AAA: 1.9±0.2 to 2.9±0.5 mm^2, true lumen: 0.7±0.1 to 0.9±0.1 mm^2, false lumen: 1.2±0.3 to 2.0±0.5 mm^2) between days 3 and 14 (shown by * in Fig. 1). A significant increase in false lumen volume was also observed between days 7 and 14 (shown by # in Fig. 1). The increase in false lumen volume attributed to 83% of AAA expansion. The contributions of the true (red) and false lumen volumes to that of the total AAA (yellow) can be seen in a representative 3D rendering (Fig. 2A), and the aortic dissection can be seen in an example long axis B-mode (Fig. 2B).

The G.L. circumferential cyclic strain significantly decreased from 18.6±2.4% to 2.9±0.6% with AAA formation in the suprarenal aorta, after which the strain remained relatively constant with AAA expansion (*$p<0.05$; Fig. 3). VVG histological staining, showing elastin in black and collagen in red, revealed breakage in the medial layer of the aortic wall, which led to the formation of the false lumen (Fig. 4A). MTC staining, showing collagen in blue and cell nuclei in purple, revealed collagen buildup in the false lumen (Fig. 4B).

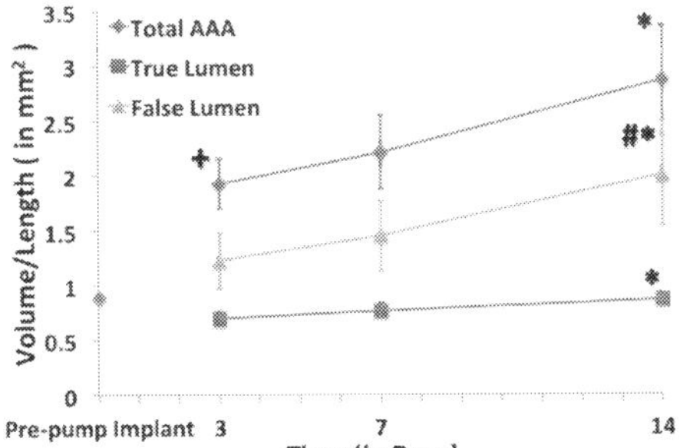

FIGURE 1: TOTAL AORTIC, TRUE LUMEN, AND FALSE LUMEN VOLUMES PER LENGTH.

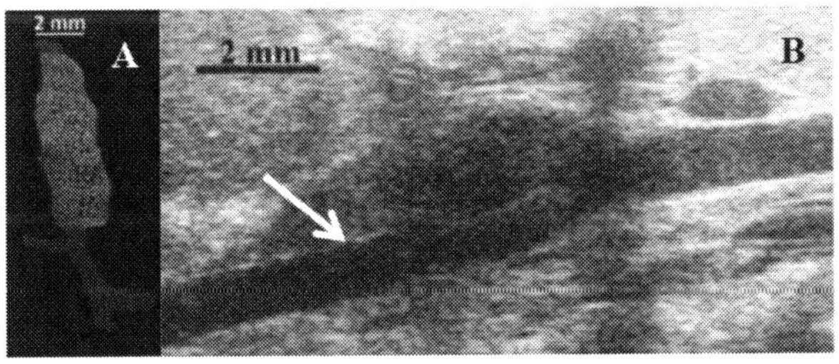

FIGURE 2: (A) 3D RENDERING OF A SEGMENTED AAA 14 DAYS POST-DIAGNOSIS. (B) B-MODE IMAGE OF A DISSECTING ANEURYSM SHOWING SEPARATION OF THE MEDIAL AND ADVENTITIAL LAYERS (ARROW).

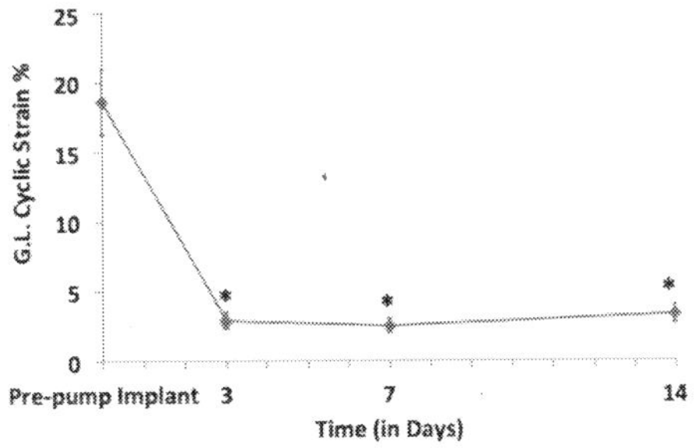

FIGURE 3: G.L. CIRCUMFERENTIAL CYCLIC STRAIN.

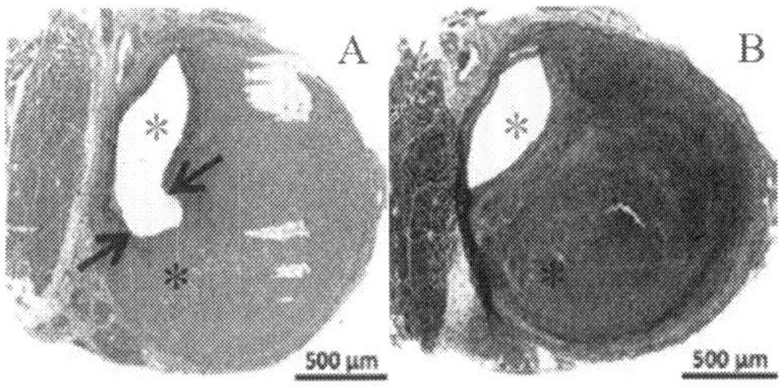

FIGURE 4: (A) VVG-STAINED SECTION OF AAA TISSUE SHOWS MEDIAL TEAR (ARROWS) SEPARATING THE TRUE AND FALSE LUMEN (RED AND BLACK ASTERISKS). (B) MTC-STAINED SECTION OF AAA TISSUE SHOWS COLLAGEN ACCUMULATION IN THE FALSE LUMEN.

DISCUSSION

The purpose of this study was to demonstrate the use of ultrasound in monitoring and characterizing AAA development. The standard errors observed are likely attributed to the inherent variability in AAA development of the mouse model. Taken together, these results suggest that AAA expansion is associated with a significant increase in false lumen volume. Rapid false lumen growth is consistent with the idea that in a high blood pressure environment, medial dissections introduce blood flow into the adventitial layer of the aorta.[8] The decrease in the G.L. circumferential cyclic strain following AAA formation suggests an increase in vessel stiffness associated with the degradation of elastin and increased collagen turnover (Fig. 4).[6]

While it is clear that ultrasound can be used to quantify false lumen volume and in vivo aortic strain, the relationship between them and aortic rupture requires further investigation.[9,10] Prior experience and literature show that AAA incidence ranges from 40% to 80% for this mouse model.[4,11,12] However, most studies investigating AAA formation only confirmed AAA formation at the end of the study.[13] As a result, these studies do not account for discrepancies resulting from varying onsets of initial aortic expansion.[13,14] For this reason, incorporating ultrasound as a noninvasive imaging tool can be beneficial. The results of this study demonstrate how ultrasound can be used to monitor how changes in the false lumen volume and aortic strain vary with AAA progression.

ACKNOWLEDGEMENTS

This work was supported by funds from the American Heart Association (14SDG18220010 to CJG) and National Institutes of Health (T32 DK101001-01).

REFERENCES

[1] Upchurch, R et al., *Am Fam Physician*, 73:1198-1204, 2006.
[2] Lederle, F et al., *Ann Intern Med*, 126:441-449, 1997.
[3] Kent, K et al., *J Vasc Surg*, 52:539-548, 2010.
[4] Daugherty, A et al., *Arterioscler Thromb Vasc Biol*, 24:429-434, 2004.
[5] Rissland, P et al., *J Biomech Eng*, 131:031001, 2008.
[6] Phillips, E et al., *Biomed Res Int*, Article ID 413189.
[7] Holzapfel, G et al., *Nonlinear Solid Mechanics: A Continuum Approach for Engineering*. Chichester: Wiley, p.455, 2000.
[8] Juang, D et al., *Circulation*, 118:e507-e510, 2008.
[9] Chang, C et al., *J Am Coll Cardiol*, 52:1170-1176, 2008.
[10] Favreau, J et al., *J Vasc Surg*, 56:462-469, 2012.
[11] Yoshida, S et al., *Circulation*, 130:A20013, 2014.
[12] Goergen, C et al., *Arterioscler Thromb Vasc Biol*, 31:270-9, 2011.
[13] Hashizume, R et al., *J Vasc Surg*, 54:1743-1752, 2011.
[14] Fu, et al., *J Transl Med*, 11:175, 2013.

SB³C2015
Summer Biomechanics, Bioengineering and Biotransport Conference
June 17-20, 2015, Snowbird Resort, Utah, USA
SB³C2015-1047

DETERMINING THE COMPRESSIVE MODULUS OF MOUSE TRABECULAR MESHWORK WITH ATOMIC FORCE MICROSCOPY

Ke Wang (1), Todd Sulchek (1), C. Ross Ethier (1)

(1) Coulter Department of Biomedical Engineering
Georgia Institute of Technology/Emory University
Atlanta, Georgia, USA

INTRODUCTION

Elevated intraocular pressure (IOP) is an important risk factor for glaucoma [1], and is usually caused by an increased resistance to the outflow of aqueous humor via the so-called conventional outflow pathway [2]. The principal site of outflow resistance in this pathway is in the trabecular meshwork (TM) and the inner wall of Schlemm's canal [3]. Stiffness (compressive modulus) of the extracellular matrix of the TM can affect TM cell behavior [4], which can in turn lead to changes in TM stiffness. It has been suggested that this interplay could result in a progressive increase in IOP [5]. Additionally, the stiffness of human trabecular meshwork, as measured using atomic force microscopy (AFM), was increased in glaucomatous eyes compared with control eyes (glaucomatous eyes: mean, 80.8 kPa, range, 0.5 – 565.3 kPa; control eyes: mean, 4.0 kPa, range, 0.5 – 10 kPa) [5]. However, it is not known if increased TM stiffness in human glaucoma is a cause of increased outflow resistance, or is an epi-phenomenon, e.g. due to glaucoma medications.

Mice are powerful tools for understanding aqueous humor dynamics in health and disease [6]. Because IOP differs between mouse strains, inter-strain comparisons provide an alternate way to investigate possible correlations between TM stiffness and IOP. TM stiffness has been measured in human and porcine eyes [7]. However, no data has been published on mouse TM stiffness, likely due to the challenge of making measurements on this small tissue. AFM is attractive for determining the mechanical properties (stiffness) of mouse TM, since it has high force sensitivity and spatial resolution [8]. This preliminary study evaluates the feasibility of making such measurements.

METHODS

Sample Preparation

Measurements were performed in eyes of adult C57BL/6J female mice (mean age = 15 weeks, range 12-17 weeks; n=5). Mice were sacrificed using CO_2, and the eyes were enucleated within 5 minutes of death and stored in Dulbecco's phosphate buffered saline (DPBS) at 4°C before use. The eyes were dissected using microscissors by making a round cut in the posterior sclera centered on the optic nerve. The lens was expelled and the iris was removed by traction. The anterior eye was divided into four quadrants, and the choroid and remaining vitreous were removed by gently scraping the sclera with forceps or a swab. One of the quadrants was attached to a Plus gold slide (Electron Microscopy Sciences, Hatfield, PA), oriented so that the inner (uveal) TM faced upwards. No glue was needed since the sample was secured by the permanent positive charges on the slide surface. AFM measurements were performed at room temperature within 24 hours of enucleation.

Instrumentation

Samples were transferred to a MFD-3D AFM (Asylum Research, Santa Barbara, CA) without drying and continuously immersed in DPBS during measurement. Silicon nitride cantilevers with an attached borosilicate sphere (radius, 5 um; nominal spring constant, 0.1 N/m; Novascan Technologies, Inc., Ames, IA) were used. Cantilevers were calibrated by measuring the thermally induced motion of the unloaded cantilever before measurements. A series of forcemaps were obtained in regions located along a line which started on the cornea and ended on the pigmented region, representing the sclera. The starting location for this line was determined visually, and thus it was impossible to ensure that the starting points were identical between samples. Each forcemap region consisted of 4 × 4 force curves over a lateral scan size of 10 μm. The center-to-center distance between adjacent forecemap regions was 10 μm. Each force curve was taken at a rate of 8 μm/s.

Data Analysis

For the 16 force curves in each forcemap region, those lacking either linear behavior or a clear cantilever contact point were not included in the analysis. The remaining force curves were fit to a Hertz model for a sphere in contact with a flat surface to determine the local compressive modulus (Igor Pro software, 6.34A). The average modulus from the valid measurements points within a forcemap region was taken as the compressive modulus of that region. These moduli from all forcemap regions in one sample were then plotted as a function of region location along the measurement line (cornea to sclera). The anterior border of the TM was putatively defined as the anterior-most location adjacent to a modulus peak and had a compressive modulus less than 10 kPa. Moving posteriorly, the first location which had modulus larger than 10 kPa was treated as the posterior border of the TM.

RESULTS

The compressive moduli profiles of the samples tested showed several broadly similar features (Figure 1). Specifically, tissue was stiff in the cornea and decreased gradually as the measurement location moved posteriorly. Most eyes then showed a local peak stiffness in between the corneal region and a region with softer modulus. However, towards the posterior end of the measurement zone, the modulus increased again and showed large variations. Based on the criteria listed above, a putative TM location was identified in each sample (red brackets; Figure 1); however, the anterior-posterior extent of the TM varied significantly from eye to eye. The compressive modulus of C57BL/6J mouse TM, as identified, ranged from 1.23 to 5.42 kPa (mean, 3.91 ± 1.73 kPa; Table 1).

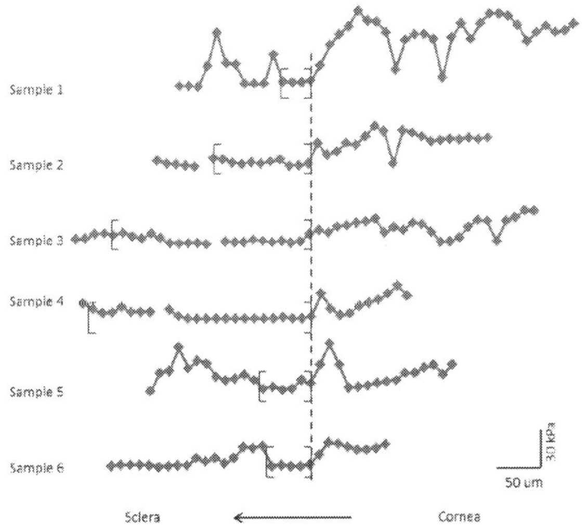

Figure 1: Compressive modulus values measured in mouse TM by AFM. Measurements were made from right to left (cornea to sclera) as described in the text. All curves have been aligned at the putative anterior border of TM (dashed line). The regions inside red brackets are considered as the TM region.

Table 1. Compressive Modulus for TM (E) in C57BL/6J mouse samples. N = number of forcemaps taken on each sample; SD = standard deviation.

Sample	Hours from Enucleation to Measurement	E ± SD (kPa)	N
1	17	4.94 ± 0.64	43
2	24	4.83 ± 2.50	36
3	4	3.37 ± 2.82	50
.4	6	2.84 ± 2.89	35
5	22	5.42 ± 1.28	33
6	3	1.23 ± 0.70	30

DISCUSSION

This study measured the local compressive modulus of mouse TM for the first time. We observed significant variations in stiffness as a function of measurement position, as expected based on known mechanical properties of the relevant tissues in other species. For example, cornea is stiffer than TM in human eyes [5, 9]. We also observed stiffness peaks between stiff and soft regions. At the putative anterior edge of the TM, this peak may be due to a transition zone between corneal endothelium and TM known to exist in both human and bovine eyes [10]. Although the exact composition of this zone remains unknown, it is very likely the place where Descemet's membrane terminates. Given the fact that Descemet's membrane is stiffer than corneal endothelium in rabbits [11], we speculate that, in the mouse eye, there is a similar (stiffer) transition zone, which may account for the modulus peak at the putative anterior edge of the TM. However, it is necessary to confirm the existence of a transition zone in mouse eyes by histology. The large variation in modulus for regions on the posterior side of the curves in Figure 1 may be due to residual iris root, ciliary body and/or choroid in the preparation.

A modulus threshold of 10 kPa was chosen to help identify the TM, since 10 kPa is the maximum elastic modulus of normal human TM. However, it must be noted that the data reported on human TM was obtained from the Schlemm's canal side of TM, while in this study, indentations were taken on the inner (uveal) side of the TM. Further, the anterior-posterior extents of the TM measured in this study were highly variable, which was unexpected. This suggests that this method of defining the anterior and posterior boundaries of the TM is insufficient and must be improved.

Nonetheless, AFM has potential to allow us to estimate the location of TM in mouse eyes and measure TM stiffness. Future experiments will combine AFM with improved TM visualization to validate AFM-based methods for TM location.

ACKNOWLEDGEMENTS

We gratefully acknowledge funding support from the Georgia Research Alliance.

REFERENCES

1. Gordon, M.O., et al., Arch Ophthalmol, 2002. **120**: p. 714-20;
2. Grant, W.M., AMA Arch Ophthalmol, 1951. **46**: p. 113-31.
3. Johnstone. M.A., et al., Am J Ophthalmol, 1973. **75**: p. 365-383.
4. Russell, P., et al., Invest Ophthalmol Vis Sci, 2008. **49**: p. 629-35.
5. Last, J.A., et al., Invest Ophthalmol Vis Sci, 2011. **52**: p. 2147-52.
6. McKinnon, S.J., et al., Exp Eye Res, 2009. **88**: p. 816-24.
7. Camras, L.J., et al.,. Invest Ophthalmol Vis Sci, 2012. **53**: p. 5242-50.
8. Last, J.A., et al., Invest Ophthalmol Vis Sci, 2010. **51**: p. 6083-94.
9. Last, J.A., et al., J Struct Biol, 2009. **167**: p. 19-24.
10. Yu, W.Y., et al., J Biomed Biotechnol, 2011. **2011**: p. 4127-43.
11. Thomasy, S.M., et al., Acta Biomater, 2014. **10**: p. 785-91.

SB³C2015
Summer Biomechanics, Bioengineering and Biotransport Conference
June 17-20, 2015,Snowbird Resort, Utah, USA
SB³C2015-1048

POSTERIOR SCLERAL STIFFENING: HOW IS SCLERAL CANAL EXPANSION AFFECTED BY THE SIZE OF STIFFENING REGION?

Ian C. Campbell (1, 2), Baptiste Coudrillier (1), C. Ross Ethier (1, 2)

(1) Wallace H. Coulter Department of Biomedical Engineering
Georgia Institute of Technology and Emory
Atlanta, GA, USA

(2) Rehabilitation Research and Development
Atlanta VA Medical Center
Decatur, GA, USA

INTRODUCTION

The posterior eye contains a complex connective tissue network surrounding delicate neural tissue essential for delivering visual information to the brain. In glaucoma, deformation of the posterior eye, particularly as result of elevated intraocular pressure acting on the optic nerve head (ONH), is thought to contribute to vision loss. However, the link between the complex biomechanics of the ONH and vision loss is poorly understood and is thus a topic of great research interest. Previous biomechanical computer modeling [1] has examined how the ONH deforms in response to intraocular pressure and has shown that the stiffness of the sclera is a key determinant of posterior eye and ONH biomechanics. Consequently, modulation of scleral stiffness has been proposed as a treatment for glaucoma [2-4].

If scleral stiffening is to be developed as a glaucoma therapy, it is important to understand how stiffening will alter ONH biomechanics. Specifically, it would be clinically important to know how much stiffening would be effective, and how the spatial pattern of scleral stiffening would affect the deformation of ONH tissues. To date, no study has directly examined these questions. Here, we use two computer models of the posterior eye, one employing a generic structure for the lamina cribrosa (the site where neural tissue passes from inside the eye through the sclera to form the optic nerve) and the other employing complex connective tissue structure derived from a micro-computed tomography scan of a pig eye. We hypothesize that stiffening the sclera immediately surrounding the optic nerve (the so-called peripapillary sclera) will have a major effect on the expansion of the scleral canal and that as the stiffened region of the posterior eye expands, additional stiffening will have a diminishing effect on scleral canal expansion. This study will help guide development of methods of delivering scleral stiffening agents *in vivo*.

METHODS

We modeled the geometry of the posterior 1/3 of the porcine eye using Gmsh 2.8.5. We specifically delineated three zones of sclera based upon connective tissue structure: the annular ring (the region immediately adjacent to the lamina cribrosa and optic nerve), the peripapillary sclera (the region around the annular ring where the scleral thickness changes), and the posterior sclera (the rest of the eye model). Each of these geometric regions of sclera was further subdivided into concentric rings (Figure 1), each of which could be virtually stiffened to parametrically simulate delivery of scleral stiffening to regions of various sizes. The annular ring was broken into 3 evenly-sized regions, the peripapillary sclera was broken into 10 rings, and the posterior sclera was broken into 50 rings. We modeled the lamina cribrosa as a cylindrical region anterior to the optic nerve, with the optic nerve being surrounded by a sheath of pia mater.

This geometry was used as input for inflation modeling using FEBio 2.2.2 in order to simulate the effects of an intraocular pressure of 30 mm Hg. All sections of sclera were modeled as a Mooney-Rivlin solid with an embedded matrix of collagen fibers following the von Mises distribution, aligned with $\kappa=0$ for the posterior sclera (planar isotropic), $\kappa=0.8$ for the peripapillary sclera (slight circumferential alignment), and $\kappa=2$ for the annular ring (highly circumferentially aligned fibers) [5]. The pia mater and optic nerve were modeled as isotropic linear elastic tissues, and the lamina cribrosa was either modeled as a generic linear elastic region made of soft neural tissue (Model A) or a subject-informed lamina cribrosa using connective tissue volume fraction and beam orientations derived from a micro-computed tomography scan of a porcine optic nerve head processed with Frangi's filter (Model B) [5].

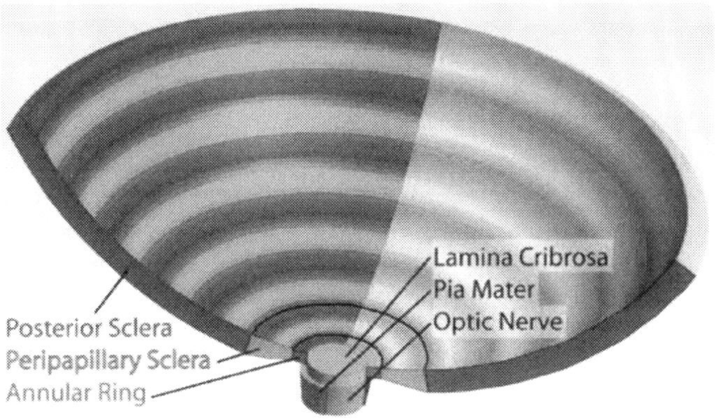

FIGURE 1: CUTAWAY OF GEOMETRY OF MODEL EYE. EACH COLORED STRIPE REPRESENTS A REGION WHERE SCLERAL STIFFNESS COULD BE INDIVIDUALLY MODULATED.

For both Model A and Model B, we simulated scleral stiffening of between zero and all 63 regions (starting adjacent to the lamina cribrosa and adding regions outward from there) stiffened by either a factor of 1.75 (Model A1, Model B1) or a factor of 2.5 (Model A2, Model B2). For each model, we computed scleral canal expansion (SCE), defined as the percent change in scleral canal circumference at the anterior surface. SCE was calculated at 30 mm Hg.

RESULTS

Scleral canal expansion as a function of the degree of scleral stiffening is plotted for Model A in Figure 2 and for Model B in Figure 3. The magnitude of total SCE is comparable to previous studies [6, 7]. An increased magnitude of scleral stiffening reduced total scleral canal expansion (Model A1 vs. Model A2, Model B1 vs. Model B2). Employing a generic, soft lamina cribrosa resulted in a greater SCE than when using a subject-informed lamina cribrosa including connective tissue volume fraction (Model A vs. Model B).

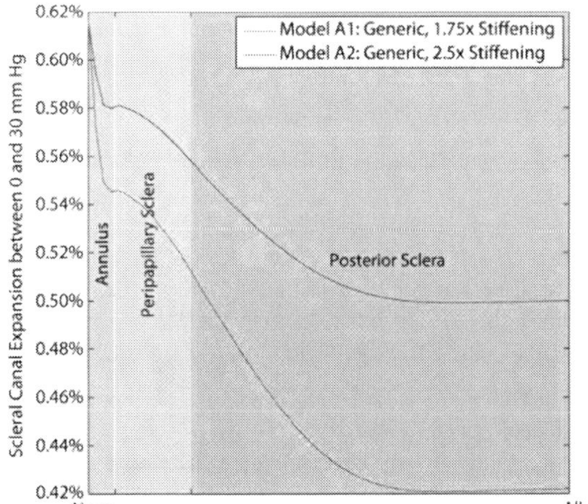

FIGURE 2: SCLERAL CANAL EXPANSION IN MODELS EMPLOYING A GENERIC LAMINA CRIBROSA

Stiffening only the annulus resulted in steeply decreased SCE in the generic lamina cribrosa model (Model A), but resulted in an increased SCE in the porcine subject-informed model (Model B). Additionally stiffening the peripapillary and posterior sclera resulted in reduced SCE in all models, and all models reached an asymptotic level of posterior eye stiffening at which point additional stiffening had essentially zero effect on SCE.

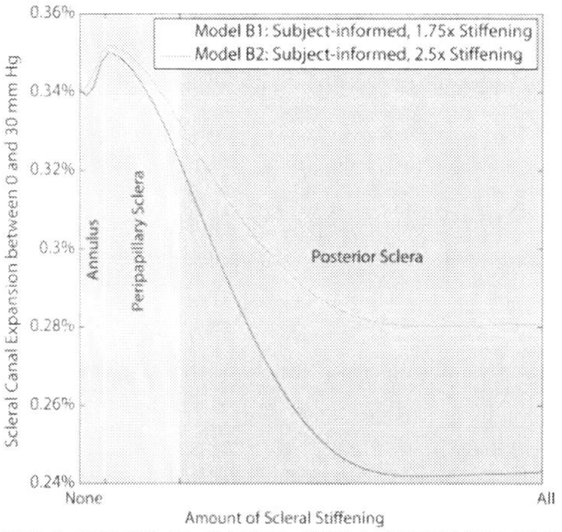

FIGURE 3: SCLERAL CANAL EXPANSION IN MODELS EMPLOYING A SUBJECT-INFORMED LAMINA CRIBROSA. NOTE DIFFERENT VERTICAL SCALE VS. FIGURE 2.

DISCUSSION

Our finding that stiffening only the annulus in the subject-informed lamina cribrosa model (Model B) is surprising, as we hypothesized that reinforcing the scleral tissue immediately surrounding the lamina cribrosa would shield the ONH from excessive tensile deformation. Our subject-informed lamina cribrosa contains complex anatomic structure without a perfectly circular perimeter and containing the porcine ventral groove, a radial protrusion of connective tissue into the center of the lamina cribrosa. Therefore, this unexpected finding may be explained by non-uniform expansion of the lamina cribrosa in this model and may not be replicated in the human lamina cribrosa. Further analysis, especially models based upon the human eye rather than the porcine eye, will be important for determining if stiffening only the annulus will increase scleral canal expansion.

Furthermore, all models agree that stiffening the peripapillary sclera and posterior sclera will reduce scleral canal expansion. Our results suggest that there is an optimal area of the posterior eye outside which stiffening had no further effect on SCE, and therefore if we assume that SCE is directly related to visual damage in glaucoma, the eye may not benefit from delivery of further scleral stiffening agents outside this optimal region.

ACKNOWLEDGEMENTS

The authors gratefully acknowledge financial support from the Georgia Research Alliance and also from the Veterans Health Administration, Rehabilitation Research and Development, under grant number 1IK1RX001791-01. We also thank Dr. Richard L. Abel (Imperial College London), whose assistance obtaining the micro-CT scan of the porcine eye was essential for deriving Model B.

REFERENCES

[1] Sigal et al. *IOVS* 2005.
[2] Wollensak & Spoerl. *J. Cataract and Refract. Surg.*, 2004.
[3] Kimball et al. *Exp Eye Res*, 2014.
[4] Gonzalez et al. *ARVO E-abstract #2173356*, 2015.
[5] Campbell et al. *J. Royal Soc. Interface*, 2015.
[6] Girard et al. *J. Biomech Eng*, 2009.
[7] Coudrillier et al. *Biomech. Mod. Mechanobiol.*, 2012.

SB³C2015
Summer Biomechanics, Bioengineering and Biotransport Conference
June 17–20, 2015, Snowbird Resort, Utah, USA
SB³C2015-1049

MODELING HEMODYNAMIC EFFECTS OF CEREBRAL VASOSPASM

Jaiyoung Ryu (1), Xiao Hu (2), Shawn C. Shadden (3)

(1) Department of Mechanical Engineering
University of California, Berkeley
Berkeley, CA, US

(2) Departments of Physiological Nursing/Neurosurgery
University of California, San Francisco
San Francisco, CA, US

INTRODUCTION

Aneurysmal subarachnoid hemorrhage (aSAH) is estimated to affect between 5 to 10 per 100,000 person per year [1]. At least a quarter of aSAH patients die, and half of survivors are left with persistent neurological deficit. Delayed cerebral ischemia (DCI) is the leading cause of the mortality and morbidity among patients who survive the initial hemorrhage. The major cause of the DCI is the cerebral vasospasm. Understanding how vasospasm affects blood flow locally and systemically is essential for the development of patient monitoring systems, diagnostic tools and treatment strategies. Indeed, factors such as blood velocity in the major cerebral arteries, as measured using transcranial Dopper, is often used to monitor for vasospasm.

A quantitative study on the hemodynamic effects of vasospasm is challenging due to the complexity of the problem, which involves the cerebral autoregulation (CA), the redistribution of blood flow by collateral vessels, and coupling with intracranial pressure. Pioneering work by Lodi and Ursino [2] incorporated a lumped parameter (LP) model for arterial blood flow and CA. They simplified the vasculature into two LP networks, representing the vasculatures with and without vasospasm, which were connected by a collateral lumped vessel model. Recently, nonlinear blood flow simulations have been employed for cerebral blood flow modeling using 1-D [3, 4] and 3-D formulations. Study of cerebrovascular constriction with non-linear blood flow modeling coupled with sophisticated CA mechanisms has not been considered to our knowledge.

In this study, we employ a 1-D nonlinear cerebral blood flow model [4] coupled with Ursino's CA model to investigate the hemodynamic effects of vasospasm at the middle cerebral artery (MCA), which is a common site. We consider a vascular model that includes important systemic arteries (aorta, subclavians, brachials) as well as the vasculature through the Circle of Willis (CoW) collateral pathways (Fig. (1)). This allows an accurate computation of pulsatile blood flow, whereby the effects of CA on systemic redistribution of the blood flow can be considered. Vasospasm is modeled by prescribing vessel area reduction in a middle portion of the MCA (marked with × in Fig. (1)). The effect of the vasospasm is studied under normal and impaired CA conditions, since CA is not always functioning properly in aSAH patients.

METHODS

The arterial tree (Fig. (1)) is modeled as a deformable 1-D pipe network, with each pipe representing a vessel segment. The 1-D blood flow governing equations are obtained by integrating flow quantities over the cross-sectional area. The conservation of the mass and momentum in the axial direction are:

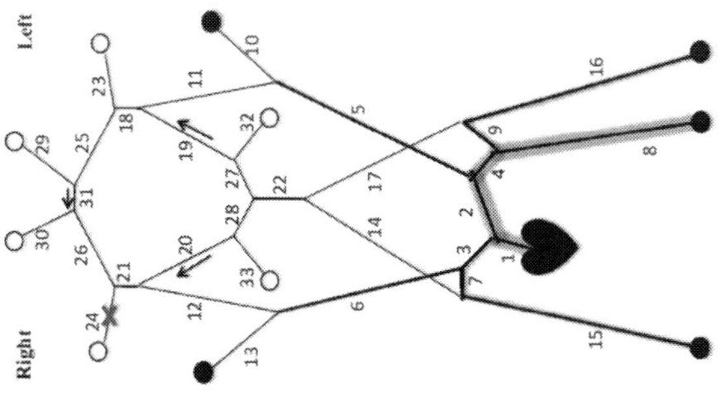

FIGURE 1: A schematic representation of the 1-D arterial network for the simulation (similar figure can be found in Alastruey et al. [3]). Outflow boundaries marked with ○ are coupled with CA model. Windkessel model is used at the boundaries marked with ●. The vasospasm location is marked with × at the right middle cerebral artery.

$$\frac{\partial A}{\partial t} + \frac{\partial AU}{\partial x} = 0, \qquad (1)$$

$$\frac{\partial U}{\partial t} + \frac{\partial}{\partial x}\left(\frac{1}{2}U^2 + \frac{p}{\rho}\right) = \frac{f}{\rho A}, \qquad (2)$$

where x is distance in the axial direction, $A(x,t)$ is the vessel's cross section area, $U(x,t)$ is the transversely averaged velocity, and $p(x,t)$ is the transmural pressure, $\rho = 1.05 \times 10^3 kg \cdot m^{-3}$ is the blood density and $f(x,t)$ is the friction force due to viscosity. f can be determined by assuming a velocity profile, which in this case is assumed to be $u = \frac{\gamma+2}{\gamma}U\left[1-\left(\frac{r}{R}\right)^{\gamma}\right]$. This leads to $\gamma = \frac{2-\alpha}{\alpha-1}$, where α is a momentum flux correction factor given by $\alpha(x,t) = \frac{2\pi\int_0^R u^2 r dr}{AU^2}$. $\gamma = 9$ $(\alpha = 1.1)$ was chosen based on previously published data. The viscous force becomes $f = -\frac{2\pi\alpha\mu U}{\alpha-1} = -22\pi\mu U$, where $\mu = 4.05 \times 10^{-3} Pa \cdot s$ is the blood viscosity. To close the system of equations (Eqs. (1) and (2)), a constitutive relation of the pressure and the cross section area of the vessel is derived using Laplace's Law:

$$p - p_0 = \frac{\beta}{A_0}\left(\sqrt{A} - \sqrt{A_0}\right), \qquad (3)$$

where A_0 is the cross sectional area of the vessel at its reference state, p_0 is the pressure outside the vessel, and $\beta = \frac{\sqrt{\pi}h_0 E}{1-\sigma^2}$ is a variable depending

TABLE 1: Flow rates and proportion of those at major arteries (right side only) at the baseline state and with MCA occlusion. Q_{inB} is the combined flow rates at the CCA and VERT at the baseline state.

Case	Q	CCA	VERT	MCA	ACA	PCA
Base-line	$Q\ (ml/s)$	5.63	1.11	2.21	1.47	1.11
	$Q/Q_{inB}\ (\%)$	83.5	16.5	32.8	21.8	16.5
Vaso-spasm	$Q\ (ml/s)$	3.91	1.06	0	1.50	1.12
	$Q/Q_{inB}\ (\%)$	58.0	15.7	0	22.3	16.6

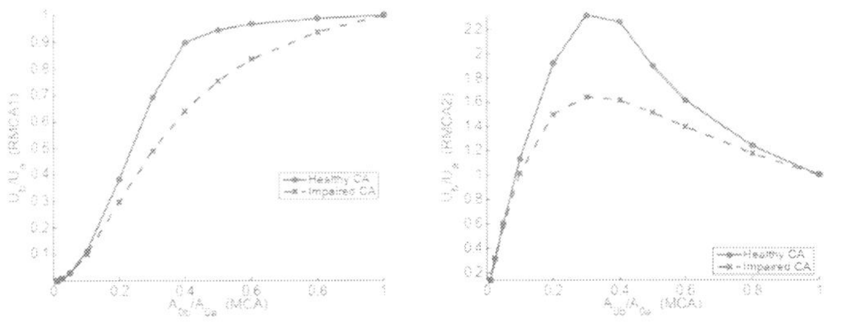

FIGURE 2: Flow velocities at the proximal non-restricted (left) and restricted (right) segments of the right MCA during the vasospasm.

on the properties of the vessel, including the Young's modulus (E), the thickness of the vessel wall (h_0), and the Poisson's ratio ($\sigma = 0.5$).

The hyperbolic system of partial differential equations is solved in each vessel segment of Fig. (1) with the inflow/outflow boundary conditions and vessel junction relations. For inlet boundary conditions, a periodic inflow rate $Q(t)$ at the ascending aorta is prescribed using a half sine wave ($\hat{Q}_{\max} = 430ml/s$) with duration $t_s = 0.3s$ (systole), and zero for the rest of the period $t_d = 0.6s$ (diastole) [3]. Outflow boundary conditions are calculated by CA outflow formulations [4, 5] for cerebral outlets (marked with ○ in Fig. (1)) and Windkessel models for systemic arteries (marked with ● in Fig. (1)). The spatially and temporally evolving flow variables and cross section area at each uniformly discretized cell are solved with a finite volume method, where 2nd order spatial scheme and two-step Adams-Bashforth temporal scheme are employed.

RESULTS

A vasospasm after aSAH was modeled by prescribing vessel area reduction, $A_{0a} \rightarrow A_{0b}$, in the middle portion of the right middle cerebral artery (MCA, vessel segment #24 in Fig. (1)), where $A_{0b} < A_{0a}$. Table (1) lists flow rates (absolute and relative) for major cerebral arteries at the baseline state and with MCA occlusion. In this limiting case, complete blockage of MCA was assumed and auto-regulation was neglected. At the baseline state, flow rate values at the middle, anterior (ACA, #30), and posterior (PCA, #33) cerebral arteries were close to the target state of the CA outflow model, which matches previous measurements. Most blood through the PCA was from the VERT, and most blood through the MCA and ACA were from the CCA, as expected. The occlusion of the MCA caused significant reduction in CCA blood flow due to increased distal resistance. Some blood from the CCA was rerouted to the PCA by the posterior communicating artery (PcoA, #20) resulting in the reduction of VERT blood flow. Flow through the ACA and PCA showed slight increase.

In Fig. (2), various levels of the MCA vasospasm (area reduction) are considered. Both healthy and impaired CA cases are compared at the non-restricted and restricted segments of the MCA. As the vasospasm progresses, blood flow is reduced consistently. At the non-restricted vessel portion, A_0 is maintained and the blood flow velocity decreases accordingly. Healthy CA can maintain the blood flow $\sim 90\%$ when MCA vessel area is reduced to 40% of the initial state. With impaired CA, blood flow is reduced to $\sim 60\%$ for the same diameter reduction. As the vessel area is reduced further, CA reaches its limit to maintain flow (M for the MCA becomes -1 in Fig. (3) left), and MCA flow velocity decrease precipitously. Interestingly, blood flow velocity at the restricted portion on the other hand demonstrates a non-monotonic variation both for normal and impaired CA cases. This is important since flow velocity in the MCA is used clinically to diagnose vasospasm.

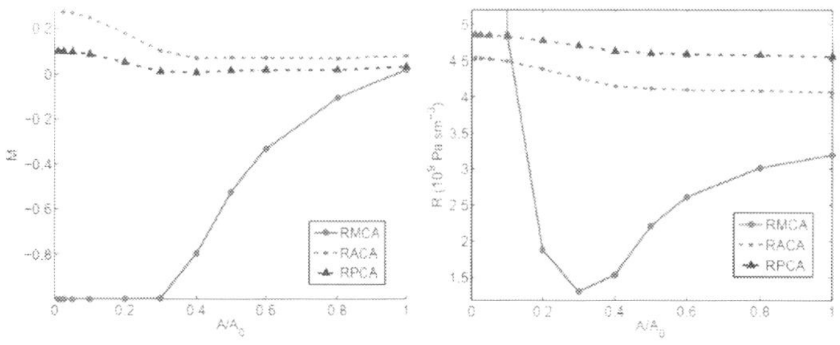

FIGURE 3: The change of the CA activation factor M (left) and the equivalent resistance of cerebral outlets (right) during the vasospasm.

As the MCA constriction progresses, blood velocity at the constricted segment (Fig. (2) right panel) increases–to almost twice the initial state for normal CA (red curve). As area is reduced further however, velocity decreases rapidly. This reduction of the blood velocity can be mis-interpreted as a sign of recovery if one assumes a constant flow flux, which leads to a false negative diagnosis. The non-monotonic velocity change is similar for both normal and impaired cases, with the normal CA case presenting a more dramatic change. Equivalent resistance, $R_{eq} = \Delta p/Q = \frac{p - p_{ic}}{Q}$, at cerebral outlets are shown in Fig. (3) right. R_{eq} at the MCA outlet decreases initially affected by the reduction of Δp, then increases rapidly with the decrease of Q. The narrowed MCA vessel becomes a major resistance to the blood flow causing systemic rerouting of flow and the observed nonlinear, non-monotonic relation.

DISCUSSION

The role of CA on the hemodynamic changes during a vasospasm following aSAH has been studied. A pulse wave propagation in a large range of compliant arterial network is considered using a multiscale approach, which involves coupling of 1-D nonlinear blood flow solver and a sophisticated CA outflow model. It was observed that healthy CA can maintain blood flow even in a challenging situations (60% MCA vessel area reduction), with approximately $\sim 30\%$ larger blood flow than impaired CA case. The evolution of cerebral vasospasm is often monitored by measuring the flow velocity in the MCA, with an increase velocity indicating vasospasm progression. However, our results indicate that a non-monotonic relationship between blood flow velocity and area reduction, which can lead to incorrect diagnosis of vasospasm progression.

ACKNOWLEDGMENTS

This work was supported by the Hellman Faculty Fellows program, NIH NS076738, and UCSF Institute of Computational Health Sciences.

REFERENCES

[1] J. Van Gijn and G. J. E. Rinkel. *Brain*, 124(2):249–278, 2001.

[2] C. A. Lodi and M. Ursino. *Annals of biomedical engineering*, 27: 257–273, 1999.

[3] J. Alastruey, K. H. Parker, J. Peiró, S. M. Byrd, and S. J. Sherwin. *Journal of biomechanics*, 40:1794–1805, 2007.

[4] M. Connolly, X. He, N. Gonzalez, P. Vespa, J. DiStefano III, and Xiao Hu. *Medical engineering & physics*, 36:354–363, 2014.

[5] X. Hu, V. Nenov, M. Bergsneider, T.C. Glenn, P. Vespa, and N. Martin. *IEEE Trans Biomed Eng*, 54:597–610, 2007.

MAINTENANCE OF F-SPONDIN (SPON1) KNOCKOUT PHENOTYPE IN 3D CULTURED MURINE CHONDROCYTES

Eben G. Estell (1), Qing Yang (2), Andrea R. Tan (1), Mukundan Attur (2), Steven B. Abramson (2), Clark T. Hung (1)

(1) Department of Biomedical Engineering
Columbia University
New York, New York, United States

(2) Division of Rheumatology
NYU Hospital for Joint Diseases.
New York University School of Medicine
NYU Langone Medical Center
New York, New York United States

INTRODUCTION

Osteoarthritis (OA) is a debilitating condition marked by the progressive degeneration of cartilage from the articulating surfaces of diarthrodial joints. This disease affects 27 million or 14% of adults in the United States and 34% of those above 65, with a projected increase of 40% over the next 25 years[1]. One estimate puts the total costs associated with treating this disease and lost productivity to be $89.1 billion annually[2].

The progressive loss of cartilage in OA is driven by phenotypic changes in the population of chondrocytes within the tissue, but the molecular mechanisms underlying these changes are still poorly understood. Elucidating these mechanisms may point to molecular targets for strategies to mitigate or prevent cartilage degeneration. To this end, our collaborators have performed extensive gene expression analysis via Affymetrix microarray on nondiseased and OA cartilage to identify over 200 molecules that are up-regulated in osteoarthritis[3].

One such molecule, F-spondin (FS), traditionally classified as a neuroregulatory protein, has been shown to be expressed at elevated levels in OA cartilage[4]. This work provides a possible pathway for regulation of cartilage degradation by FS during OA, and links this elevated expression with chondrocyte hypertrophy[4]. To further investigate this pathway, Spon1 knockout (Spon1-/-) mice have been developed and characterized; initially showing increased bone formation[5]. The current collaborative work seeks to isolate chondrocytes from Spon1-/- mice and establish optimal expansion protocols in order to yield an adequate cell number for 3D culture in tissue-engineered cartilage. This culture system will provide a better model to characterize the influence of FS on the phenotypic changes of chondrocytes during OA.

METHODS

Spon1-/- mice were previously generated in collaboration with the Texas Institute of Genomic Medicine by targeted deletion of exon 1 in C57BL/6 mice, confirmed by Southern blot analysis and PCR using probes for specific wild type (wt) and mutant loci[5]. All procedures performed in the study were approved by the Institutional Animal Care and Use Committee (IACUC) of New York University School of Medicine.

Study 1: 2D Culture. Articular cartilage was harvested from two six-month-old mice (one wt and one Spon1-/-) and chondrocytes were isolated by previously established mouse protocols[6]. 2D cultures of chondrocytes were maintained with Chondrocyte Growth Medium (Cell Applications). To establish long-term viability of cell line and retention of chondrocyte phenotype, chondrocytes were sub-passaged into 12-well plate and grown to 90% confluency for gene expression analysis at 2 and 12 weeks, performed via SYBR Green PCR (Life Technologies) to quantify relative RNA expression of typical bone and cartilage markers.

Study 2: 3D Culture. After 12 weeks of culture and validation of maintained chondrocyte phenotype, wt and Spon1-/- chondrocytes were centrifuged to create pellets of $2.5*10^5$ cells/pellet (n=3 per group). Pellets were maintained in Chondrogenic Media, DMEM containing 100 μg/ml Sodium Pyruvate, 50 μg/ml L-Proline, 1% ITS+, 1% PSAM. Fresh media was changed on a 2-2-3 day schedule and supplemented continuously with 50μg/mL ascorbic acid, 100nM dexamethasone, and 10ng/mL TGFβ-3. Pellets were cultured for 28 days and then analyzed by qPCR for typical bone and cartilage markers to evaluate maintenance of chondrocyte phenotype and influence of Spon1 knockout on chondrocytes in 3D culture. Relative expression levels of target genes were calculated after normalizing with geomean of two housekeeping genes, -GAPDH and HPRT1.

RESULTS

Study1: Wt and Spon1-/- chondrocytes remained viable in 2D culture over 12 weeks. A chondrogenic phenotype was maintained in both groups throughout culture time as evaluated by qPCR at 2 and 12 weeks, showing high levels of cartilage extracellular matrix (ECM) markers (collagen type II (COLII) and aggrecan), and low levels of bone markers (osteocalcin and COLI) (data not shown). Interestingly BSP, a typical marker for bone tissue, was expressed dramatically higher in Spon1-/- over wt chondrocytes.

Spon1-/- chondrocytes exhibited higher expression levels of cartilage markers (COLII and aggrecan) than wt, but also show moderate increases in catabolic markers MMP13 and ADAMTS4 and 5 (Table 1).

TABLE 1: BONE AND CARTILAGE MARKERS IN 2D CULTURE OF WT AND KO CHONDROCYTES

Marker	Relative mRNA Expression (x-fold increase vs. wt)
BSP	149.6
COLII	24.5
Aggrecan	3.9
ADAMTS4	1.5
ADAMTS5	4.5
COL10	1.8
MMP13	2.1

Study 2. Gene expression analysis of the pellets after 28 days of culture showed maintenance of chondrogenic phenotype and similar trends of higher expression in Spon1-/- versus wt (Figure 1). At day 0, Spon1-/- shows increased expression over wt of ECM markers COLII and aggrecan, with dramatically higher fold-expression after 28 days. Hypertrophic markers MMP13 and COL10 showed a similar trend, mirroring the increased expression of Spon1-/- over wt shown in 2D. Expression of the catabolic markers ADAMTS4 and 5 did not however follow the 2D results; showing decreased relative expression in Spon1-/- versus wt. As in the 2D cultures, BSP was found to be up-regulated in F-spondin deficient chondrocytes, as well as other molecules previously identified as potential OA targets such as periostin.

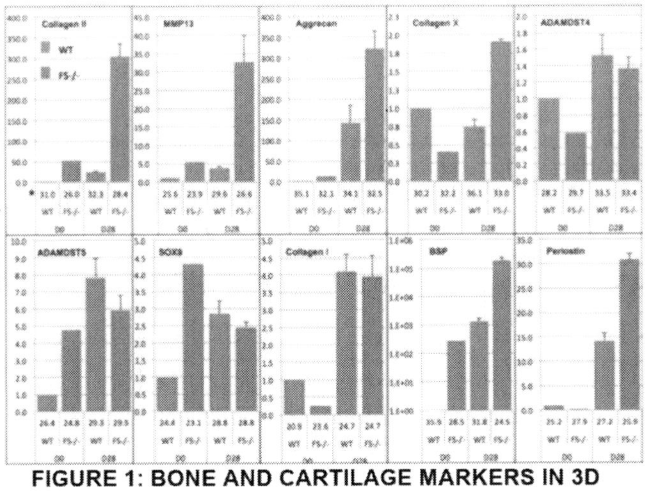

FIGURE 1: BONE AND CARTILAGE MARKERS IN 3D PELLET CULTURE OF WT AND KO CHONDROCYTES.
*** Ct values**

DISCUSSION

This study marks the first isolation of chondrocytes from the novel Spon1 (F-spondin) knockout mouse and culture in a 3D system. By successful long-term culture of Spon1 knockout chondrocytes, we have shown that F-spondin is not essential for proliferation of chondrocytes. In both 2D and 3D culture, Spon1-/- chondrocytes sustained a chondrogenic phenotype but also showed higher relative expression of chondrogenic markers than their wt counterparts. This increase was particularly marked in COL2, the major structural component of cartilage ECM, with a 24.5-fold increase over wt in 2D and 12-fold increase in 3D. Previously we have reported the association of F-Spondin with hypertrophic marker (MMP-13) in human chondrocytes. However in the current study, in 3D cultures in the presence of TGFβ-3, F-spondin deficient chondrocyte expressed elevated levels of bone marker and hypertrophic markers (MMP-13, Col X). Furthermore, in vitro studies F-spondin treatment augmented collagen and proteoglycan degradation in OA chondrocytes. Therefore, F-spondin may play an inhibiting role on cartilage ECM turnover.

Interestingly, several markers often associated with chondrocyte hypertrophy – a common cellular outcome in OA – were also up-regulated in the Spon1 deficient chondrocytes in the presence of TGFβ-3. This result combined with the dramatic increase in expression of bone marker BSP, another factor recently linked to chondrocyte hypertrophy in OA[7], suggest that F-spondin may play an inhibitory role in TGF-b3 induced hypertrophic changes promoted during 3D culture. The inclusion of TGF-b3 in our chondrogenic growth medium must be taken into account, as binding of TGFβ-3 to ALK1 receptors has been shown to increase MMP13 expression. Future work will seek to better characterize the influence of TGFβ-3 on chondrocyte phenotype be screening for ALK receptor levels and SMAD signaling. Overall, spon1 deficient chondrocytes expressed increased bone and hypertrophic markers.

Incorporation of the novel Spon1 knockout chondrocytes into a simple 3D culture system will allow us to complement the genomic analyses performed with biochemical analysis of the tissue developed by the chondrocytes as the engineered tissue matures. Future work will focus on more complex 3D scaffolds, such as the agarose scaffold currently employed in our laboratory for cartilage tissue engineering. This versatile scaffold will allow for the application of current tissue engineering techniques such as dynamic compressive loading, and the resulting experimental model will have increased control over physiological parameters and be thus better suited to elucidate the mechanisms behind OA regulating molecules like F-spondin.

ACKNOWLEDGEMENTS
Funded by NIAMS/NIH AR60361 (CTH) and AR054817 (SBA),

REFERENCES

1. Helmick, CG et al., *Arthritis Rheum*, 58:15–25, 2008.
2. Leigh, JP et al., *J Rheumatol*, 28:1647–1654, 2001.
3. Attur, MG et al., *Curr Issues Mol Biol*, 4:129–146, 2002.
4. Attur, MG et al., *FASEB J*, 23:79–89, 2009.
5. Attur, MG et al., *Osteoarthritis Cartilage*, 20:S59, 2012.
6. Gosset, M et al., *Nat Protoc, 3*:1253–1260, 2008.
7. Pesesse, L et al., *Osteoarthritis Cartilage*, 21:1913–1923, 2013.

EFFECT OF SCLEROSTIN ANTIBODY ON TENDON-TO-BONE HEALING IN A ROTATOR CUFF ANIMAL MODEL

Shivam A. Shah (1), Ioannis Kormpakis (1), Necat Havlioglu (2), Michael S. Ominsky (3), Leesa M. Galatz (1), Stavros Thomopoulos (1)

(1) Department of Orthopaedic Surgery
Washington University in St. Louis
Saint Louis, MO, USA

(2) Department of Pathology
John Cochran VA Medical Center
Saint Louis, MO, USA

(3) Metabolic Disorders Research
Amgen, Inc.
Thousand Oaks, California, USA

INTRODUCTION

Over 250,000 rotator cuff repairs are performed annually in the United States, but surgical repair is associated with a high anatomical failure rates [1]. Repair of a torn rotator cuff tendon involves re-apposing the tendon to its original attachment site on the humeral head. Bone loss at the insertion site after tendon detachment has been reported at multiple anatomic sites [2-5]. Furthermore, suppression of bone resorption through the use of bisphosphonate treatment has led to improved healing [3,4]. The goal of the current study was to enhance tendon-to-bone healing through the promotion of bone formation. Sclerostin antibody treatment has led to increased bone formation, mass, and strength in rodent models of both osteoporosis and fracture repair [6]. The use of systemic sclerostin antibody treatment for enhanced tendon-to-bone healing has not previously been evaluated. We hypothesized that treatment with sclerostin antibody (Scl-Ab) would prevent the bone loss observed after rotator cuff injury and improve tendon-to-bone healing.

METHODS

Adult Sprague-Dawley rats were used in this study, as approved by the University's Animal Studies Committee. The supraspinatus (SS) tendon was sharply elevated from the humeral head and repaired, as previously described, in both shoulders [7]. Animals were either left untreated (CTL group) or administered sclerostin antibody (Scl-Ab VI) via subcutaneous injections (25 mg/kg) at the time of injury and repair and every two weeks until sacrifice (Scl-Ab group). All animals were allowed free cage activity and sacrificed after 2, 4, or 8 weeks of healing for histology (N=18), bone morphometry (N=61), quantitative real-time PCR analysis (qRT-PCR, N=40), and biomechanics (4 and 8 weeks only, N=30). For histology, 5 μm sections were stained with hematoxylin and eosin, toluidine blue, Goldner's trichrome, or Von Kossa staining and semi-quantitatively analyzed by NH (blinded) using a tendon-to-bone maturing score from Ida *et al.* [8]. Insertion continuity, bone resorption, cell and fiber alignment, and cellularity were part of 9 factors evaluated on a scale from 1-4. A higher score was indicative of improved tendon-to-bone healing [8]. For bone morphometry, SS-humeral head samples were scanned at a resolution of 20 μm to determine bone volume (BV), bone mineral density (BMD), and trabecular bone parameters (μCT 40, Scanco). The region of interest included bone within the humeral head proximal to the growth plate. For biomechanical analysis, uniaxial tensile tests to failure were performed on a material testing frame (Instron 5866) in a saline bath at 37 °C. For gene expression, RNA was isolated from either the SS tendon or the humeral head and qRT-PCR (Fluidigm) was performed for a panel of 27 genes related to bone formation, tendon, and fibrocartilage. An ANOVA followed by Fisher's LSD post hoc tests was used to determine effects of time and treatment. Additional statistical comparisons with the non-injured group were done with 2-tailed student t-test. $p<0.05$ was considered significant.

RESULTS

__Histology__ - No repair site failures or gaps were seen at the time of dissection. Histological sections demonstrated tendons reattached to the bone via fibrovascular scar and bone loss was evident in the CTL group at later time points. Semi-quantitative, blinded analysis showed more mature tendon-to-bone healing and new bone formation in the Scl-Ab group compared to CTL groups at 4 and 8 weeks.

__Bone Morphometry__ – There were significant effects of time and treatment on bone morphometry. After 8 weeks of healing, the Scl-Ab group had significantly more bone with increased bone mineral density compared to the non-injured group (Figure 1). Scl-Ab treatment resulted in a 40.3% increase in BV/TV and 36.6% increase in BMD (data not shown) compared to CTL group after 8 weeks of healing following injury. Analysis of the trabecular bone showed that four weeks of healing following injury led to a significant decrease in trabecular number in both CTL and Scl-Ab groups compared to the non-injured group (Figure 1). At 8 weeks following injury, the CTL group had an even larger decrease in trabecular number, while trabecular number returned to a level similar to the non-injured group in the Scl-Ab group. Trabecular thickness significantly increased in both CTL and Scl-Ab groups compared to the non-injured group after 2, 4, and 8 weeks of healing (data not shown).

Biomechanics – There were significant effects of time and treatment on biomechanics. Injury caused significant decreases in maximum load (32.9%), ultimate stress (68.4%), modulus (75.9%), stiffness (48.3%), and resilience (64.5%), compared to non-injured group (Figure 2). There were no significant differences between the Scl-Ab and CTL group following injury and 4 weeks of healing. Additional work to evaluate the mechanical properties after 8 weeks of healing is in progress. In the non-injured tendon, Scl-Ab treatment caused significant increases in ultimate stress (47%) and resilience (137%) compared to the CTL group. Scl-Ab also caused significant decreases in modulus (39.6%) and stiffness (51.2%) compared to the non-injured, CTL group.

Gene Expression – In the CTL group, gene expression in the bone tissue adjacent to the repair was significantly different than uninjured bone for sclerostin (-3.1x), osteoprotegerin (-2.6x), Runx2 (+1.8x), osterix (+2.2x), DKK1 (-4.3x), DMP1 (-1.8x), and tenomodulin (-3.6x) after 8 weeks of healing (Figure 3). However, after 8 weeks of healing with Scl-Ab treatment, only Runx2 (2.4x) remained significantly different than the non-injured CTL group. In the non-injured bone tissue, Scl-Ab significantly increased Runx2 and SOST compared to CTL, consistent with increases in osteoblastogenesis and bone mass, respectively. In tendon tissue, collagen type I, collagen type III, Aggrecan, and MMP2 gene expression were significantly increased after 2 weeks of healing in both groups compared to non-injured controls (data not shown). There was no significant effect of treatment, and expression levels of these genes decreased towards non-injured group levels with longer durations of healing.

DISCUSSION

Results from the study support the hypothesis that Scl-Ab treatment could reduce the bone loss that accompanies rotator cuff injury and repair and potentially improve functional properties. Scl-Ab treatment resulted in significantly more bone and higher bone mineral density after 8 weeks of healing, though biomechanical data are not yet available. At 4 weeks, Scl-Ab treatment had not yet increased bone volume or repair strength, indicating that longer treatment with Scl-Ab may be necessary to observe improvements in mechanical parameters. In the bone tissue adjacent to the repair, Scl-Ab treatment appeared to accelerate the return of certain genes to non-injured levels. In tendon tissue, the temporal pattern of gene expression during repair was characterized, and Scl-Ab treatment did not affect this process. The bone-specificity of the Scl-Ab effect was expected based on the expression pattern of sclerostin in bone, rather than in tendon.

Interestingly, Scl-Ab treatment in healthy tendon-to-bone attachments caused an increase in ultimate stress (strength) and resilience but a significant decrease in modulus and stiffness. This result is consistent with a previous finding using bisphosphonate alendronate [3], and was reflected in increases in Runx2 and SOST expression in the bone in the current study. Increased mineralization at the tendon attachment may affect strength, modulus, and toughness differently. Biomechanical testing results from the 8 week healing group will help clarify this finding.

In summary, bone loss is observed at many tendon-to-bone repair sites. The results from this study hold promise for Scl-Ab treatment to enhance tendon-to-bone repair.

ACKNOWLEDGEMENTS: Sclerostin antibody provided by Amgen and UCB Pharma. NIH F31-AR066452 and R01-AR057836.

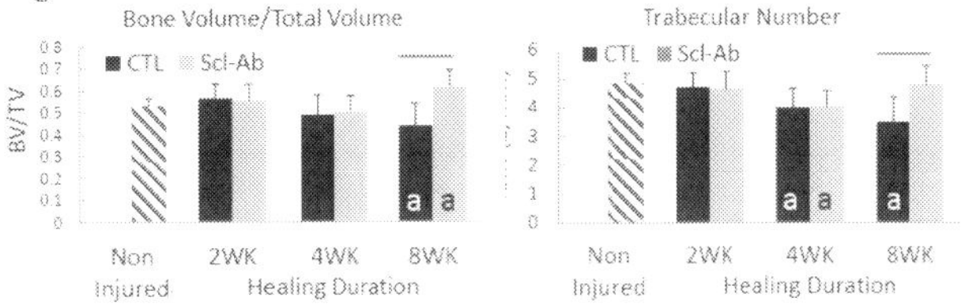

Figure 1: Scl-Ab led to increased bone formation and attenuated trabecular bone loss following injury and repair. (a: significantly different than Non-Injured CTL; line: significantly different).

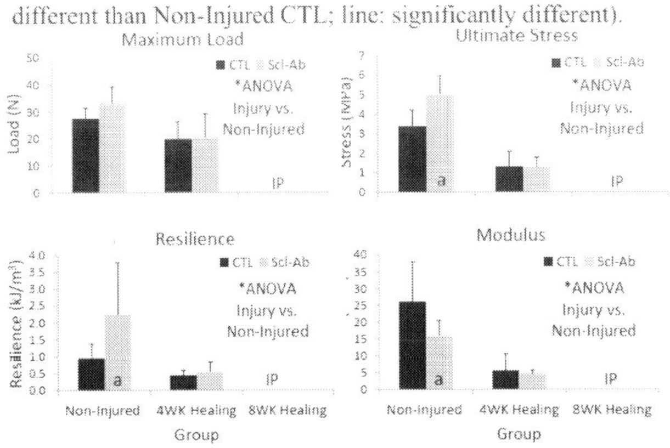

FIGURE 2: Injury caused a decrease in mechanical parameters after 4 weeks of healing. Scl-Ab led to increased ultimate stress and resilience but decreased modulus in healing tendon attachments. (a: significantly different than Non-Injured CTL; IP: In Progress).

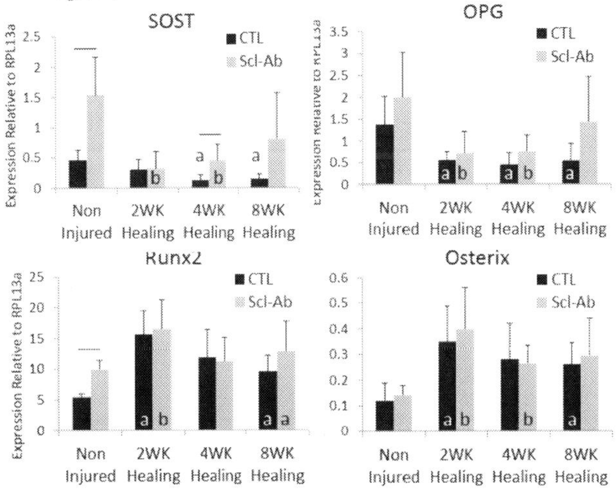

FIGURE 3: In bone tissue, Scl-Ab treatment helped accelerate return of gene expression levels to non-injured baseline. (a or b: significantly different than Non-Injured CTL and Non-Injured Scl-Ab, respectively; line: significantly different).

REFERENCES

[1] Galatz, L.M., et al., J Bone Joint Surg Am, 2004. 86-A(2):219-24.
[2] Galatz, L. M. et al. J Orthop Res, (2005) **23**, 1441-1447.
[3] Thomopoulos, S., et al. J Orthop Res, (2007) **25**, 473-479.
[4] Thomopoulos, S., et al. J Orthop Res (2008) **26**, 1611-1617.
[5] Galatz, L. M. et al. JSES (2009) **18**, 669-675.
[6] Ke, H. Z.., et al. Endo Rev (2012) **33**(5):747–783.
[7] Manning, C. N., et al. J Orthop Res (2011).
[8] Ide, J., et al. JSES (2009) **18**, 391-398.

SB^{3}C2015
Summer Biomechanics, Bioengineering and Biotransport Conference
June 17-20, 2015,Snowbird Resort, Utah, USA
SB³C2015-1052

TRIANGULATED SURFACE BOOLEAN OPERATIONS FOR COMBINING 2-D AND 3-D IMAGE SEGMENTATION FOR PATIENT-SPECIFIC BLOOD FLOW ANALYSIS

Adam R. Updegrove (1), Nathan M. Wilson (2), Shawn C. Shadden (1)

(1) Department of Mechanical Engineering
University of California, Berkeley
Berkeley, CA, USA

(2) Open Source Medical Software Corporation
Santa Monica, CA, USA

INTRODUCTION

The process to create complex 3-D patient-specific anatomic models from a series of 2-D image segmentations utilizing solid model lofting operations [1] has been used extensively for patient-specific blood flow simulation (e.g. [2]). In addition, direct 3-D image segmentation techniques are widely popular and have been frequently utilized to construct localized anatomic models for clinically relevant problems such as cerebral and abdominal aortic aneurysms. The geometric extent of the anatomic models constructed directly from 3-D segmentation techniques are often limited compared to anatomic models constructed from 2-D techniques due to factors including poor image quality and difficulty in selecting appropriate spatially-varying segmentation parameters in 3-space. Further, since 2-D segmentation techniques lead to the creation of boundary representation solid models (with vessel walls defined as NURBS) and 3-D techniques were simply spatial regions enclosed by a densely triangulated surface, it is difficult to combine models constructed using different segmentation techniques. This has led to a common accepted modeling tradeoff: more extensive and flexibly defined vascular models created by lofted 2-D segmentations, or less extensive and potentially better-conforming vascular models created by 3-D segmentation.

In this work, a solution is proposed enabling the use of 3-D segmentation for certain regions of interest (e.g. aneurysmal sac) while simultaneously using 2-D segmentations to include surrounding vasculature structure as desired. In particular, a technique to create triangulated vessels from 2-D segmentations without using solid modeling packages has been developed. In addition, a new implementation of robust Boolean operations for complex triangulated surfaces has been developed and implemented into the open source package SimVascular (www.simvascular.org).

METHODS

Creating a 3-D patient-specific anatomic model consists of three core steps: creating 3-D segmentation(s), creating "vessels" from groups of 2-D segmentations, and finally performing Boolean operations to create a unified flow domain. It should be noted that models can be constructed using only 2-D segmentations if desired. In the SimVascular application, a core benefit of the methodology described herein is that a user can utilize a fully open source software pipeline to construct and mesh the anatomic model, without need for multiple software products.

The 2-D segmentations are stored as ordered groups of segmentations along particular vessel paths of interest. A 3-D surface triangulation is created for each "vessel" (i.e. topologic cylinder). Briefly, a 2-D segmentation represents the lumen boundary as a closed and ordered piecewise linear curve defined in a plane transverse to the vessel of interest. The 2-D segmentations for a given vessel are sampled (s_1) to a uniformly distributed set of points given a user defined resolution. The points are then reordered for each segmentation such that the initial point on each curve is the closest point in 3-D space to the next adjacent curve. This leads to a series of "streaklines", or piecewise linear paths, along the length of the vessel. A Kochanek-Bartels cubic spline is then fit to the points and the path is uniformly sampled (s_2) at a user-defined spacing. The vessel wall is then triangulated by creating triangles from the s_1*s_2 points such that the outward normal of each triangle is consistent. Polygonal caps are added to the topologic cylinder and then triangulated using a VTK filter so the final vessel surface contains only triangles. Figure 1 demonstrates the process of going from segmentations to a lofted solid for a given vessel.

To create the entire patient-specific lumen, the vessels are united using Boolean Operations for triangulated surfaces. This operation consists of three distinct steps:

1. **Finding the intersection lines between two surface meshes** – The intersection of oriented bounding box trees of each surface mesh indicate finite regions in space where the triangulated surfaces may intersect. Each surface triangle lies in a distinct plane and the intersection line formed from these planes can be found and restricted to the bounding boxes. If this finite line is intersected by two edges of the triangle, a true intersection line has been found. During this process, intersected triangles are tagged as needing to be revisited for re-triangulation.

2. **Re-triangulating the intersected surfaces with the addition of the intersection lines**. The re-triangulation is done separately for each input surface. Triangles that have been identified as needing to be remeshed (this includes all marked triangles and triangles sharing an edge to these triangles), are processed and re-triangulated using both Delaunay 2D and ear clipping methods. From the re-triangulation, new cell information is added to the intersecting lines in order to subsequently complete the full Boolean process. In addition, information regarding the intersecting points and new triangles containing an intersecting line is retained for the Step 3.

3. **Creating the Boolean of the two surfaces (i.e. Union, Intersection, Difference)**. The Boolean uses an algorithm that orients the intersection lines and fills regions of each surface based on intersection lines that divide the surface into separate regions [3]. Each intersection line is connected to two cells on each input surface. When the lines are followed in a directional manner, one cell has a counter-clockwise (CCW) orientation, and the other has a clockwise (CW) orientation. The regions connected to these cells can then be filled according to the direction (CCW or CW) they have. For closed input surfaces, the intersection always contain completely closed intersection lines (i.e. the first and last point of the intersection curve are the same). However, for open input surfaces (e.g. "uncapped vessels"), it can be possible to have surfaces that do not fully intersect and thus the intersection lines form an open line rather than a closed loop. When each region of a surface has been filled with a specific orientation, appropriate combinations of the two surfaces give the different Booleans desired (Union, Intersection, Difference).

RESULTS

The creation of a lofted surface starts with a segmentation, whether it be 2-D, 3-D or both. Then we can connect these segmentations using the techniques described in the Methods section above. Figure 3 displays a lofted pulmonary artery from 2-D segmentations with two different resolutions.

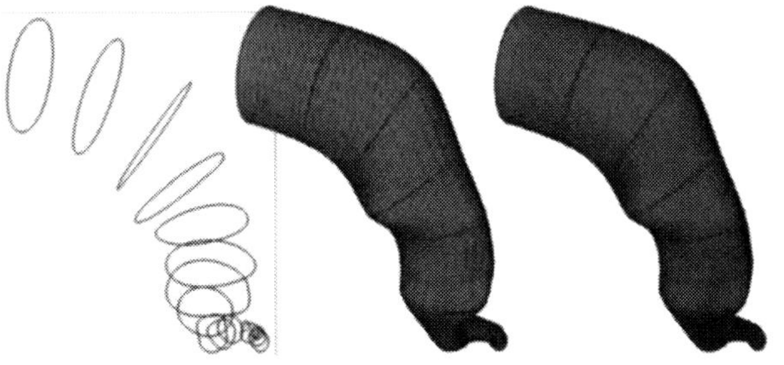

Figure 3. Solid created from lofting 2D segmentations using cubic splines; Left – segmentations; Middle – lofted solid with circumferential resolution of 30, axial resolution of 6; Right – lofted solid with circumferential resolution of 60, axial resolution of 6.

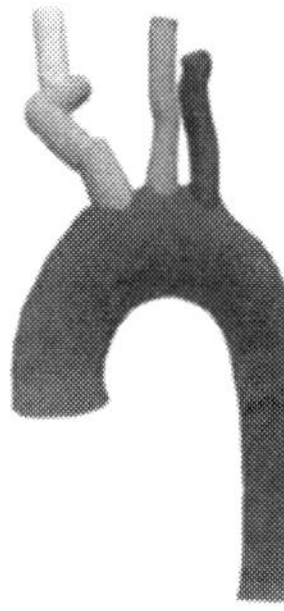

Figure 1. Combing 3D and 2D segmentations.

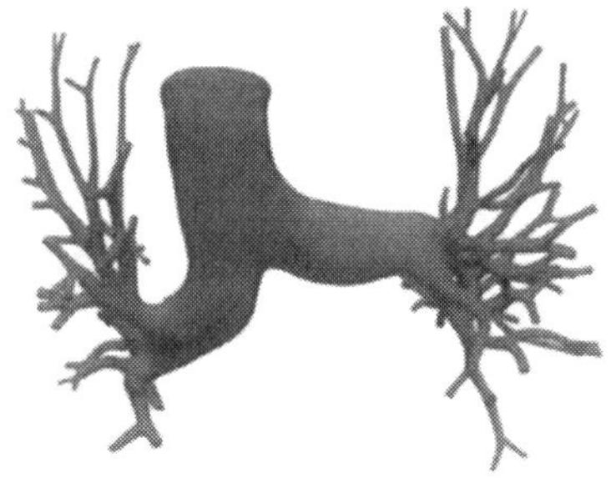

Figure 2. Pulmonary vascular model united using Boolean operations for triangulated surfaces.

After the surfaces are lofted separately, the vessels are united using the Boolean operations. To demonstrate the combination of 2-D and 3-D segmentation methods, a 3-D model of the aortic arch was created using the open source tool ITK-SNAP (www.itksnap.org). This model was united with four vessels created from lofted 2-D segmentations. The successful result of this Boolean operation is displayed in Figure 2. Combining several branches (2-D or 3-D segmented) can give an entire complex vascular network as seen in the pulmonary model displayed in Figure 3.

DISCUSSION

Developing robust Boolean operations for triangulated surfaces addressed two main technical challenges in vascular model construction. First, this enables vascular model construction using popular 2D image segmentation techniques without the need for previously utilized external commercial solid modeling packages such as Parasolid. Second, this enables a hybrid segmentation approach where one can leverage the strengths of both 3D segmentation and 2D segmentation techniques as appropriate when constructing more extensive vascular models.

Although multiple implementations of Boolean operations for triangulated surfaces exist, none was robust enough in our testing for operations of the sort described herein. For example, VTK implements triangulated surface Boolean operations; however, this implementation breaks down in many instances of combining complex patient specific geometries. The GNU Triangulated Surface Library (GTS, http://gts.sourceforge.net/) implements a more robust method than found in VTK; however, in order to include the GTS Boolean operations, the entire GLIB and GTS libraries must be added, and moreover customization needed to facilitate specification of boundary condition for subsequent simulations is difficult. Our custom package provides a robust, efficient, and lightweight solution for Boolean operations of triangulated surfaces. It was built to be seamlessly integrated into VTK, as consistent with much of SimVascular, to ensure modularity and compatibility of our implementation.

ACKNOWLEDGEMENTS

Support was provided by an NSF Graduate Fellowship (NSFGF) and the NSF SSI Program. The SimVascular code is hosted by simtk.org.

REFERENCES

[1] Wilson NM, et al. *Proc MICCAI*, 2001, pp. 449-456.
[2] Arzani A, Shadden SC. *Phys Fluids* 24(8), 081901, 2012.
[3] Mei G, Tipper JC. *arXiv preprint:1308.4434*, 2013.

SB^{3}C2015
Summer Biomechanics, Bioengineering and Biotransport Conference
June 17-20, 2015,Snowbird Resort, Utah, USA
SB^{3}C2015-1053

CHARACTERIZATION OF TRANSITION TO TURBULENCE FOR BLOOD IN AN S-SHAPED PIPE UNDER STEADY FLOW CONDITIONS

Dipankar Biswas (1), David M. Casey (1), Douglas C. Crowder (2), Kristian Valen-Sendstad (3), David A. Steinman (3), Yang H. Yun (2), Francis Loth (1,2)

(1) Department of Mechanical Engineering
The University of Akron
Akron, Ohio, USA

(2) Department of Biomedical Engineering
The University of Akron
Akron, Ohio, USA

(3) Department of Mechanical and Industrial Engineering
University of Toronto
Toronto, Ontario, Canada

INTRODUCTION

Hemodynamics have been shown to play a key role in the pathogenesis and morphogenesis of arterial disease [1]. Correct evaluation of the wall shear stress is necessary to predict the pathophysiology of the diseased vessel. Non-invasive measurement of velocity and shear stress is difficult due to the small size of the blood vessels. For this reason, researchers sometimes use computational fluid dynamic simulation of image-based models to evaluate the development, progression and diagnosis of cardiovascular disease [2]. It is well established that blood behaves like a non-Newtonian fluid at low shear rates (<200 s^{-1}) and like a Newtonian fluid for high shear rates [3]. Shear thinning blood analogs can approximate blood rheology; however, the precise variation of viscosity with shear rate in whole blood is difficult to match [4]. Most *in vitro* and *in silico* experiments assume blood to be Newtonian fluid under the presumption that the prevalent shear rates in their model are high (>200 s^{-1}) [5]. This assumption may yield accurate results depending on the flow conditions and geometry. There are cases where the non-Newtonian nature of blood altered the flow pattern and the hemodynamic forces considerably [6]. Blood flow in the human body is typically laminar; however, transitional flow has also been observed in various locations, notably diseased locations.

Stehbens et al. investigated transition to turbulence (TT) using dye injection and visual observation for 21 S-shaped geometries, varying the lengths, radii of curvature and size [7]. None of his critical Reynolds numbers (Re_{cr}) were above 1200 confirming previous researchers [8]. All these studies employed water as the test fluid. The present study examined one of these geometries using blood and a Newtonian fluid as test fluids. The goal of the current study was to compare velocity profiles for blood and a Newtonian fluid under steady flow conditions in an S-shaped geometry.

METHODS

Construction of an in-vitro flow system that employs blood as the test fluid demands special attention as factors like non-compatible materials, altered temperature, and elevated shear rates aggravate thrombus formation and altered rheology. The experimental system setup is shown in Figure 1. The flow rate was controlled to within 1.5%. The test fluid temperature was maintained at 37 ± 0.3°C. Velocity measurements were taken using a 20 MHz pulsed Doppler ultrasound probe (pulse repetition frequency, PRF = 62.5 kHz) and processed by an ultrasound module (Crystal Biotech, Northborough, MA). The S-shaped pipe was made using a vinyl tubing (6.35 mm inner diameter) that was morphed in a channel made from wood. The S-shape had a radius of curvature of ~ 5D.

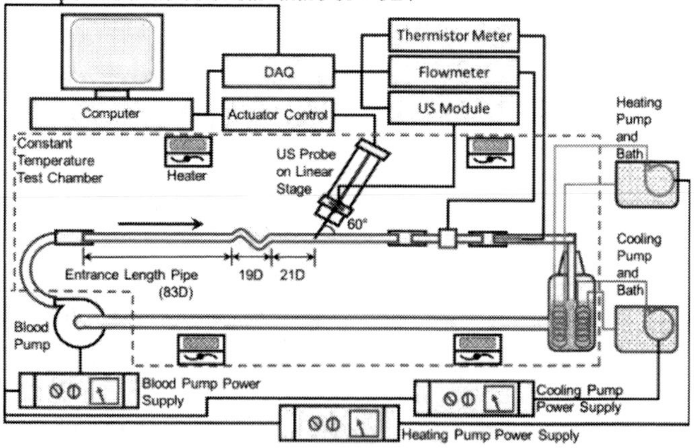

Figure 1: Schematic of flow system.

Viscosities were measured for shear rates ranging from 5 to 500 s^{-1} (AR2000ex) at 37°C using a 60 mm acrylic parallel plate with a gap size of 500 micron. The rheometer was calibrated prior to each experiment and was shown to be within 1% of a viscosity standard (400 s^{-1}) and repeatability errors were also ±1%.

The signal processing method employed is similar to that described by Thorne et al. [9] and Jones [10]. The Doppler signal, volume flowrate and temperature were recorded simultaneously using a data acquisition system (NI USB-6229, National Instruments, Austin, TX), at a rate of 250 kS/s for three seconds.

Software was developed to process the Doppler signal for mean and unsteady velocity using a fast Fourier transform (FFT, Matlab, Mathworks, Natick, MA). Mean velocity was obtained by averaging many power spectra taken over the three-second period [10]. The mean velocity was then computed using the well-known Doppler relationship [9].

$$V = \frac{c \cdot f_d}{2 \cdot f_t \cdot \cos\alpha} \qquad \text{[Eq.1]}$$

Where c is the speed of sound in the fluid being measured, f_d is the Doppler frequency shift (Hz), f_t is the transmitted beam frequency (20MHz), and α is the angle of incidence between the ultrasound beam and the direction of flow (60°).

Three samples of whole porcine blood with 10% sodium citrate were tested. In addition, three samples were tested of a Newtonian analog fluid that consisted of water-glycerin (WG) mixture with 55:45 by percent volume with a similar viscosity to that of blood. Titanium dioxide (TiO2) particles (0.1g/liter, Spectrum Chemicals, T1081, Gardena, CA) were used as ultrasound reflectors for the water glycerin mixtures.

RESULTS

Blood showed a shear thinning nature with viscosities 4.34 ± 0.46 cP, and 3.02 ± 0.14 cP at 20 and 400 s^{-1}, respectively. Percentage hematocrit were 34.6, 35.3 and 35.8% for blood samples 1, 2 and 3, respectively. Two velocity markers were utilized to determine the Re where TT first occurs for each test fluid. First, centerline velocity fluctuations demonstrate the Re at which large scale fluctuations first appear in the flow. These fluctuations can be seen as a sudden increase in turbulent kinetic energy (TKE, normalized by mean velocity squared) as Re is increased. The TKE was calculated as the mean of the TKE of seven measurement locations near the center for all Re and Re_{cr} was determined where this value was 0.07. Second, this method can be applied to velocity traces at each radial measurement location across the pipe (Figure 2) to insure that the entire flow field is following the centerline trend. S-shaped pipe TKE showed $Re_{cr} \approx 1025$ (SEM = 37) and 1148 (SEM = 61) for WG and blood, respectively. Using centerline velocity fluctuation the mean Re_{cr} of TT for WG was ~1000 (SEM <1) and for blood was ~1133 (SEM = 40). Considerable amount of "turbulent like" fluctuations were observed for WG at Re=500. From the profile shapes (i.e., blunting), the mean Re_{cr} of TT for WG was ~1150 (SEM <1) and for blood was ~1258 (SEM = 20) (Figure 3).

Quantitative comparison between the mean velocity for blood and WG showed that for laminar flow (Re <= 1100), only small differences (< 10%) were observed between blood and WG mean velocity profiles for radial locations near the pipe center (r < 1mm or 30% of radius). However, differences as high as 20% were observed at the larger Reynolds numbers in the region where TT was different for the two fluids (1100 < Re < 1300). Velocity profiles were again similar at Re>1300.

DISCUSSION

The viscosity results for our experiments show blood rheology to be shear thinning as reported in other studies in the literature [3]. Blood rheology did not change appreciably (< 2% for shear rates > 1 s^{-1}) between measurements taken before and after each experiment nor did we observe changes in RBC morphology. Re for blood for all experiments was based on the measured viscosity at high shear rate (400 s^{-1}) where viscosity no longer changes with shear rate. Distinct differences were observed between the TKE distribution in blood and WG. WG was observed to have more TKE near the walls than blood. Even though a small difference was observed between the Re_{cr} of blood and WG (~10%), it is difficult to conclude that there is a difference since it might be due to the cumulative errors of various parts and processes of the experimental system. In the S-shaped pipe, higher shear might have been predominant which caused the blood to act as a Newtonian fluid.

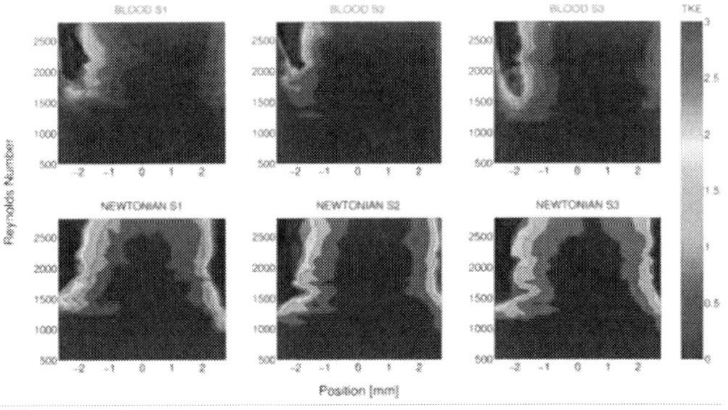

Figure 2: TKE distribution as a function of both Re and radial position (blood top row, Newtonian analog bottom row).

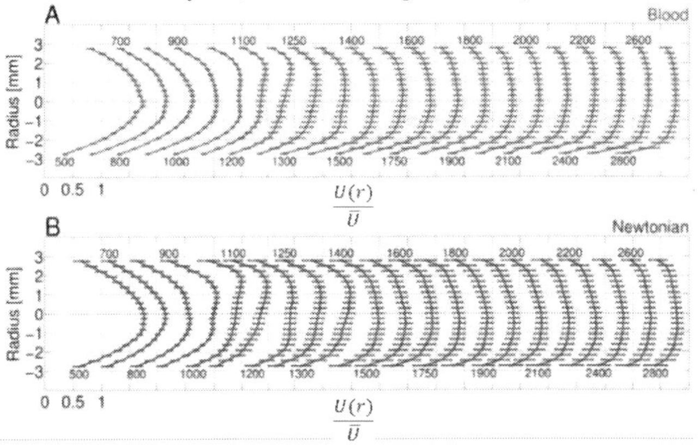

Figure 3: Normalized mean velocity profiles with RMS (error-bars).

REFERENCES

[1] Fry, D.L et al., *Circ Res*, 22(2):165-97, 1968.

[2] Steinman, D.A et al., *J Biomech Eng*, 135(2):021016, 2013.

[3] Brooks, D.E et al., *J Appl Physiol*, 28(2):172, 1970.

[4] Brookshier, K.A et al.,. *Biorheology*, 30(2):107-16, 1993.

[5] Loth, F et al., *Annual Review of Fluid Mech.*, 40:367-393, 2008.

[6] Chen, J et al., *J Biomech*, 39(11):1983-95, 2006.

[7] Stehbens, W.E et al., *J Exp Physiol*, 72(2):201-13, 1987.

[8] Stehbens, W.E et al., *Q J Exp Physiol Cogn Med Sci*, 44(1):110-7, 1959.

[9] Thorne, M.L et al., *Ultrasound Med Biol*, 34(7):1163-73, 2008.

[10] Jones, S.A., *Critical Reviews in Biomedical Engineering*, 21(5):399-483, 1993.

SB³C2015
Summer Biomechanics, Bioengineering and Biotransport Conference
June 17-20, 2015,Snowbird Resort, Utah, USA
SB³C2015-1054

SEM-BASED MULTISCALE MECHANICAL MODELING OF LUMBAR SPINE FACET CAPSULAR LIGAMENT DURING BIAXIAL EXTENSION

Vahhab Zarei (1), Amy A. Claeson (2), Victor H. Barocas (2)

(1) Mechanical Engineering
University of Minnesota, Twin Cities
Minneapolis, MN, USA

(2) Biomedical Engineering
University of Minnesota, Twin Cities
Minneapolis, MN, USA

INTRODUCTION

The facet joint is a synovial joint formed between the superior and inferior articular facets of two adjacent vertebrae [1]. The facet joint is surrounded by a ligamentous capsule containing dense, highly aligned collagen fibers; it stabilizes the joint, keeps the synovial fluid in the joint space, and restricts the motion of spine during physiological motions. The lumbar facet capsular ligament (LFCL) is a possible source of low back pain, because its collagen fibers are innervated by nociceptors (pain-sensing receptors), and it undergoes large deformations during spine motion [2]. It has been estimated that 15% to 40% of the low back pain is due to the facet joints [3].

Heterogeneity, anisotropy, nonlinearity, and time-dependency complicate the mechanical behavior of LFCL. The ability of fibers to realign under load further complicates the behavior. When the tissue is loaded, fibers initially rotate more than stretching until they orient along the load direction, at which point they stiffen the tissue and provide mechanical strength. Thus, the mechanical behavior of the LFCL is highly dependent on the spatial orientation of its collagen fibers [1]. The LFCL contains other components, including cells, water, and non-collagenous biomacromolecules that might contribute to its mechanical response.

Here we present a novel SEM-based multi-scale method accounting for the LFCL structural complexities (anisotropy and inhomogeneity) to simulate its mechanical behavior during biaxial testing.

METHODS

A multi-scale finite element technique [4] was used to simulate the biaxial tensile test. Briefly, macro-scale stresses are found based on micro-scale forces in fiber networks, which deform based on macro-scale displacements. These fibers follow the phenomenological constitutive equation [5] $F=A[\exp(BE)-1]$, where F is the fiber force, A and B are constitutive constants and E is the fiber Green strain.

In order to account for sample anisotropy and heterogeneity, the networks are generated using an SEM-based fiber orientation tensor measured for 87 different regions in a human L3-L4 LFCL sample. Fast Fourier Transform (FFT) analysis explained in [6] was used to calculate the orientation tensor for each element. Hence the fiber orientation for each point on the tissue was found based on its distance from the marked elements. Delaunay networks were generated from random 3D seed points stretched according to the localized orientation tensor. Finally, the LFCL geometry was made based on the image of the biaxial test using ABAQUS software (Fig. 1).

This geometry and networks were used in the simulation to obtain force-stretch ratio curves in lateral-medial (L-M) and superior-inferior (S-I) directions. Moreover, the same biaxial test was simulated using a model with isotropic networks instead of the SEM-based networks.

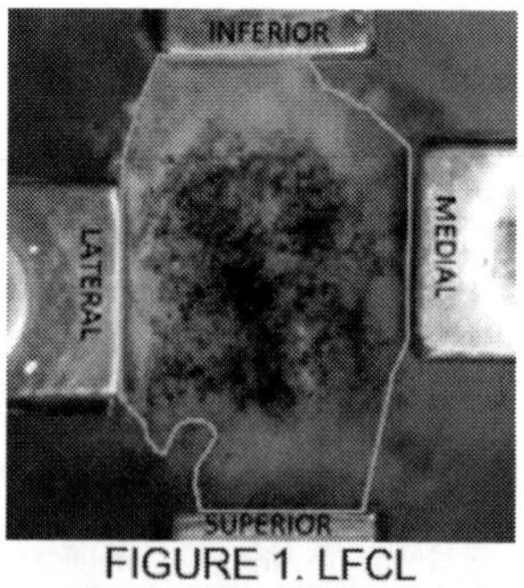

FIGURE 1. LFCL DOMAIN CREATED BASED ON TISSUE SURFACE

RESULTS

The SEM images successfully taken from the different regions on the tissue show aligned collagen fibers. The generated networks match with the actual fiber orientation for each image (Fig. 2).

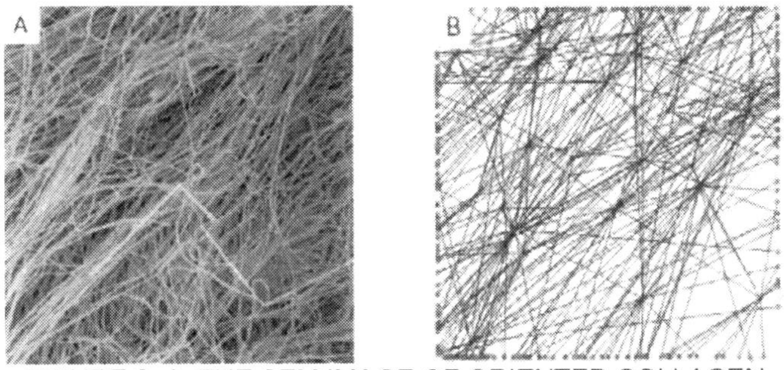

FIGURE 2. A: THE SEM IMAGE OF ORIENTED COLLAGEN FIBERS AND B: THE GENERATED NETWORK WITH THE SAME ORIENTATION (2-D PROJECTION)

L-M and S-I Force-displacement curves were plotted and compared to the experimental results (Fig. 3). The peak force for the experiment is 11.0N and 13.7N for the L-M and S-I directions respectively. The values for the aligned SEM-based networks were 11.1N for L-M and 10.4N for the S-I directions. For the isotropic fiber-based model, the peak force for the L-M and S-I directions are 9.4N and 11.6N respectively.

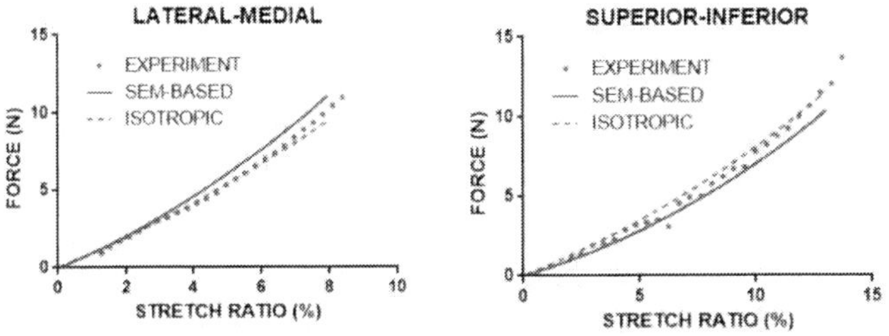

FIGURE 3. FORCE-STRETCH RATIO CURVES IN THE L-M AND S-I DIRECTIONS FOR THE EXPERIMENT, THE SIMULATION WITH SEM-BASED AND ISOTROPIC NETWORKS

Spatially varying Cauchy stresses are represented on the tissue geometry in Fig. 4. The average normal stress over the whole sample in L-M and S-I directions and the average maximum 2D shear stress are 231KPa, 212KPa and 118KPa respectively. The maximum shear stress is roughly uniform, except for a low-stress band running from superior-lateral corner towards inferior-medial one.

DISCUSSION

The mechanical response of the LFCL during biaxial tensile testing was investigated using an SEM-based multi-scale method. This method is based on the tissue microstructural features extracted from SEM. To the best of our knowledge, no study of this type has been done previously.

Other types of imaging, such as polarized light, have been used to characterize the fiber orientation within the cervical FCL [7]. SEM overcomes the problem that polarized light cannot pass through the thicker, denser LFCL, but it is not without other shortcomings. Variations in collagenous fiber diameter were observed, but could not be quantified. Moreover, the fiber orientation through the sample thickness (normal to the surface) could not be captured. Hence, we assumed that there is only a small degree of orientation in this direction, i.e. fibers are assumed to be mainly parallel to the surface of the tissue. In addition, SEM only represents the fibers on the surface. Mechanical behavior depends also on the fibers orientation throughout the thickness of the sample.

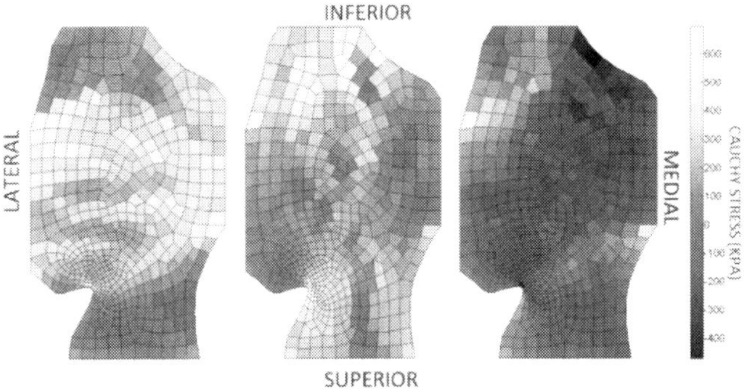

FIGURE 4. THE CAUCHY STRESS FIELD PVER THE TISSUE. NORMAL STRESS IN A: L-M, B: S-I DIRECTIONS. C: MAXIMUM PLANAR SHEAR STRESS

SEM images confirmed that the main components of LFCL are bundles of collagenous fibers with varying concentration over the heterogeneous tissue. These images also showed that in most locations, the collagen fibers are highly aligned, which highlights the anisotropy of the tissue.

In the force-stretch ratio diagrams, the model matches the experiment well, but so does the isotropic model, suggesting that the role of the microstructural alignment in this experiment may not be large. This surprising result may be due to the role of other non-modeled components, the relatively small amount of fiber rotation in the biaxial test, or errors in the measurement due to slipping or failure of the tissue during the test.

The detailed-fiber-based approach is much more demanding computationally and is thus only to be preferred over a neo-Hookean or fiber-composite model if the benefit is clear. For near-equibiaxial loading, at least based on the sample studied, it appears that a simpler constitutive model would be sufficient to predict the gross behavior of Figure 3 and can give a stress field similar to that of Figure 4, but the fine detail and heterogeneity would be lost. If one's goal is to examine the microstructural behavior of the tissue, then a microstructural model is obviously needed.

In conclusion, this work introduces a new methodology to account for microstructural features of LFCL, and it can be used for more mechanical characterization.

ACKNOWLEDGEMENTS

This work was supported by a NIH grant [1R21EB009788]. Part of this work were carried out in the Characterization Facility, University of Minnesota (UMN), which receives partial support from NSF (MRSEC program). We extend our thanks to UMN Anatomy Bequest program, UMN Supercomputing Institute, our collaborator Prof. B. Winkelstein (U of Pennsylvania), and the following individuals for their contribution: Prof. P. Provenzano, Mr. C. Frethem and Mr. J. Carruth (UMN) and especially Prof. E. Sander (U of Iowa).

REFERENCES

[1] JF Behrsin et al., *J Surg Radiol Anat* 10: 211-219, 1988 [2] L Kalichman et al., *Semin Arthritis Rheum* 37:69-80, 2007 [3] S. J. Dreyer et al., *Arch Phys Med Rehabil* 77:290-300, 1996 [4] Stylianopoulos T et al., *J Comput Methods 196:2981–2990, 2007* [5] Billiar KL et al., *J Biomech*, 34:859-871, 2001. [6] E. A. Sander et al., *J Biomed Materials* 122:327–335, 2008. [7] K. P Quinn et al,. *J Biomed.* 14(5), 054010, 2009

SB[3]C2015
Summer Biomechanics, Bioengineering and Biotransport Conference
June 17-20, 2015,Snowbird Resort, Utah, USA
SB[3]C2015-1055

AORTIC VALVE CALCIFICATION: GEOMETRIC AND BIOMECHANICAL ANALYSIS

Banafsheh Zebhi (1), Brandon L. Moore (1), Gary J. Luckasen (2), Lakshmi P. Dasi (1,3)

(1) Department of Mechanical Engineering,
Colorado State University
Fort Collins, Colorado, USA

(2) Medical Center of the Rockies,
Loveland, Colorado, USA

(3) School of Biomedical Engineering,
Colorado State University
Fort Collins, Colorado, USA

INTRODUCTION

Calcific aortic valve disease (CAVD) is a progressive disease of dysregulated mineral metabolism which has been observed in the cardiovascular system for many decades[1]. Calcification of the cardiovascular system predominantly affects the aorta, coronary arteries, peripheral arteries, and the aortic valve. Calcification is characterized by the formation of tissue similar to bone on the cusps of the aortic valve[2]. Aortic valve calcification has been studied in the literature, particularly the location and severity of aortic valve calcification[3] and pattern of calcific deposits for stenotic and regurgitant aortic valves[4].

The main objective of the present study is to reconstruct three-dimensional aortic valve calcification patterns. Geometric structure is analyzed for the right coronary cusp (RCC), left coronary cusp (LCC), non-coronary cusp (NCC), right coronary artery (RCA), and left coronary artery (LCA) in order to determine a reliable spatial pattern for calcium deposition in different areas that are exposed to different mechanical loads. This work should lead to future artificial aortic valve designs that can be optimized in a patient-specific manner.

METHODS

A total of six patients were studied (age 80 ± 7 years, 83% men). All patients had aortic valve disease and were referred to Medical Center of the Rockies (Loveland, CO, USA) for multislice computed tomography (MSCT) of the chest. The MSCT examinations were performed with a Philips Brilliance 64 channel CT scanner (Philips Healthcare, Amsterdam, Netherlands). The thickness of slices taken was 0.67 mm, the vertical spacing between the pixels was 0.33 mm and horizontal pixel spacing was 0.748 mm. Image resolution was 512x512 pixels. Basic measurements including annulus diameter, sinotubular junction diameter, distance between cusps and the aortic axis, and cusp-to-cusp distance were performed by Philips iSite PACS software version 2.6.1.

An in-house MATLAB code (2014b, MA, USA) was written to stack 2D CT images and a C code was written to generate a 3D model. The 3D model was then segmented in Tecplot 360 (2012 R1, WA, USA) by thresholding to extract the location of aortic valve, leaflets, aorta and calcium nodules. The calcium volume was calculated on each of the left coronary, right coronary, and non-coronary cusps as well as the left and right coronary arteries by integrating over each area.

Calcium volume was computed on all cusps and coronary arteries, and statistical analysis was performed to determine location and severity of calcification. All statistical analysis was performed with Kaleida Graph software version 4.1.

Additional experimentation was also performed to examine the effects of coronary flow on aortic root hemodynamics. An in vitro particle image velocimetry (PIV) protocol was developed to visualize flow patterns within both non-coronary and coronary sinuses. A Medtronic Hancock II porcine bioprosthesis was uses as the model aortic valve for these experiments. It was placed inside a clear, acrylic chamber machined to mimic the outer walls of the aortic root.

RESULTS

The segmented calcium nodules and reconstructed aortic root for six patients are presented in Fig. 1. Calcium and aortic root tissue are shown in red and yellow, respectively. Calcium was present in NCC

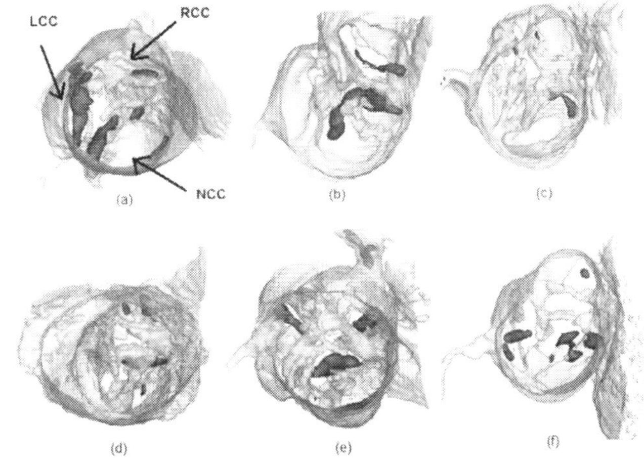

FIGURE 1. The aortic roots of 6 patients were segmented in Tecplot. Calcific nodules are colored in red while native aortic root tissue is yellow.

and RCC for all patients; however, calcification in LCC, RCA, and LCA areas did not exists in all of the patients. Fig. 1 (a), (b), (d) and (f) show calcification on RCC, LCC and NCC when RCA and LCA

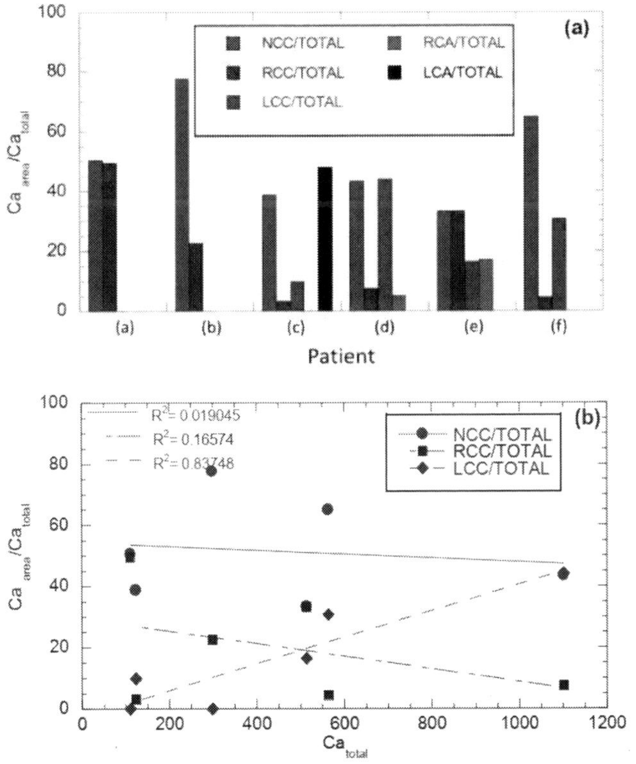

FIGURE 2. Calcium volume was computed for different regions within the aortic root. Relative volume is shown a) for each patient and b) as a function of total volume.

are not calcified. In contrast, Fig. 1 (c) and (e) show nodule locations when RCA and LCA are calcified. Distribution of calcium in different areas is shown in Fig. 2. Data in Fig. 2 is normalized with respect to the maximum calcium in each patient to compare the relative distributions between different anatomical regions. NCC is the main location for calcium deposition which contains 33% to 77% of the total calcium (Fig. 2-a). Generally, the lowest calcium was deposited on RCA (Average $Ca_{RCA}/Ca_{total} \approx 4$ %) (Fig. 2-a). The calcium deposition ratio in LCC increased as total calcium increased; however, RCC and NCC did not show a significant trend with change in total calcium (Fig. 2-b).

DISCUSSION

According to Figs. 1 and 2, which show qualitative and quantitative calcific nodule distributions, calcium deposition is generally higher on NCC than all other areas. This might be due to the absence of a coronary artery on NCC. Results from hemodynamics experiments (Fig. 3) provide support for this reasoning. Specifically, these results show significant fluid motion and shear stress near the leaflet in the coronary sinus that is not present in the non-coronary sinus. Since low shear stress and a lack of "washout" flow in the sinus have both been correlated to disease, this could help explain the findings of most calcification on NCC.

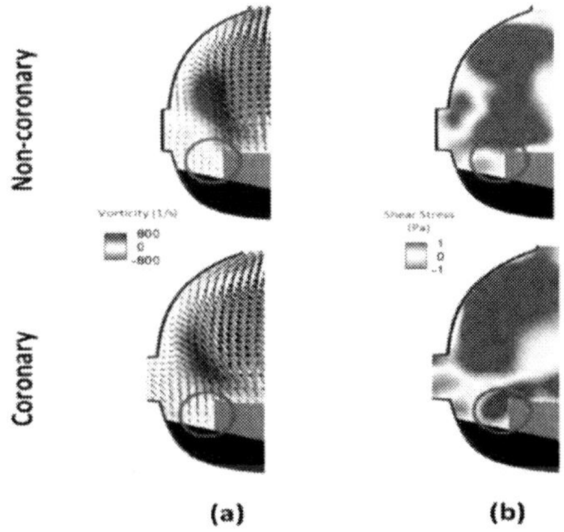

FIGURE 3. Experimental results showing a) velocity vectors and vorticity contours as well as b) fluid shear stress contours at peak systole for two sinus cases. The fluid region near the leaflet fibrosa is circled in red (black region is signal loss).

ACKNOWLEDGEMENTS

Research reported in this publication was partially supported by the National Institutes of Health under award number R01HL119824.

REFERENCES

[1] Hjortnaes, J., New, S. E., & Aikawa, E. Visualizing novel concepts of cardiovascular calcification. *Trends in cardiovascular medicine*. 2013; *23*(3), 71-79.

[2] Mohler, E. R., Gannon, F., Reynolds, C., Zimmerman, R., Keane, M. G., & Kaplan, F. S. Bone formation and inflammation in cardiac valves.*Circulation*. 2001; *103*(11), 1522-1528.

[3] Ewe SH. Location and severity of aortic valve calcium and implications for aortic regurgitation after transcatheter aortic valve implantation. *The American journal of cardiology*. 2011;108:1470

[4] Thubrikar MJ, Aouad J, Nolan SP. Patterns of calcific deposits in operatively excised stenotic or purely regurgitant aortic valves and their relation to mechanical-stress. *Am. J. Cardiol.* 1986;58:304-308

Vinculin-network mediated cytoskeletal remodeling regulates contractile function in the aging heart

Gaurav Kaushik (1), Anthony Cammarato (2), and Adam J. Engler (1,3)

Department of Bioengineering
University of California, San Diego
La Jolla, CA, USA

(2) Division of Cardiology, Department of Medicine
Johns Hopkins University
Baltimore, MD, USA

Sanford Consortium for Regenerative Medicine
La Jolla, CA, USA

INTRODUCTION

The average age worldwide is projected to increase dramatically in the coming decades, and advanced age is a risk factor for cardiac dysfunction (*1*). Treating age-related heart failure is complicated due to its heterogeneous etiologies. Identifying individual, age-associated compensatory events is made difficult because of the diversity of aging processes within and between conventional models as well as an abundance of maladaptive events. However, integrated approaches that identify conserved hallmarks of cardiac aging and verify their functional consequences can assist in identify therapeutic targets.

To assess conservation of known aging hallmarks, we analyzed the left ventricular proteomes of adult and aged simians (*Macacca mulatta*) and rats (*Rattus norvegicus*) using mass spectrometry (MS), representing the first detailed analysis of the human-like aging simian cardiac proteome. Subsequent bioinformatic investigations suggest that the cytoskeletal vinculin network has a central, conserved role during cardiac aging. Using *Drosophila melanogaster*, a rapidly aging and genetically tractable model system, we observed that cardiac overexpression of vinculin induced extensive cytoskeletal and cellular remodeling. It also correlated with preserved contractile function with age and enhanced contractility in wildtype fly hearts. Average and maximal lifespan were significantly extended in *Drosophila* with cardiac-specific vinculin-overexpression. In contrast to a perception that remodeling is predominantly maladaptive, these data suggest a beneficial role for vinculin-mediated restructuring in the aging heart.

METHODS

6 adult (6mo) and 6 aged (24mo) female F344xBN F1 rats were obtained from the National Institutes on Aging (NIA). Rat colony maintenance and all experiments were performed in accordance with UC San Diego IACUC protocol S11032. 4 adult (average: 11.63 years old, range: 8.75-14.12) and 5 aged (average: 22.4 years old, range: 18.81-25.48) male rhesus monkeys were maintained at the NIA in in accordance with NIA IACUC protocol AG000238-07 (Effects of Aging on Experimental Atherosclerosis in Nonhuman Primates). Rat and rhesus monkey samples where flash frozen for subsequent proteomic analysis.

All fly lines were obtained from the Bloomington *Drosophila* Stock Center at Indiana University or the Vienna *Drosophila* RNAi Center (VDRC). *yw* and w^{1118} strains served as the wild-type control lines. Flies were raised on standard agar-containing food at 25°C. Cardiac-specific perturbation of gene expression was achieved via the Gal4-UAS system as described previously (*2*). Virgin females expressing the transcription factor Gal4 under control of the cardiac-specific tinman (tinHE-Gal4) promoter were crossed with male flies containing the desired transgene under control of the Gal4-recognizing UAS promoter sequence. Female progeny of tinHE-Gal4 and w^{1118} served as control. Proper insertion of both constructs was validated via heart-specific gene expression analysis.

Data comparison was subjected to either a non-parametric Student's t-test with unequal variance assumption, repeated measures one-way ANOVA, one-way ANOVA with post-hoc Tukey correction, or two-way ANOVA with post-hoc Tukey correction as indicated. Pooled data are represented as mean +/- standard error of the mean unless otherwise indicated. All experiments were performed with biological triplicates of indicated sample size.

RESULTS

Proteomic quantification of adult versus aged rhesus monkey (11 vs. 22 years) and rat (6 vs. 24 months) left ventricular free wall myocardium was performed. Tryptic peptides were detected using reversed-phase liquid chromatography mass spectrometry (MS) and a total of 1,206 (monkey) and 1,086 (rat) non-redundant proteins were quantified. Of these proteins, 602 were common between species. Based on label-free quantification via normalized spectral counts, cytoskeleton proteins had the greatest increase with age, especially in monkeys and were associated with cardiac toxicological functions, such as arrhythmia, hypertrophy, and failure as assigned by Ingenuity Pathway Analysis (IPA). Cytoskeletal networks assembled by IPA centered on vinculin, which were upregulated in rhesus monkeys (+32%) and rat (+35%) with age. Histology sections in rats indicated an age-related increase in intercalated disc (ID) associated vinculin in aged rat myocardium, suggesting that increased vinculin expression may reinforce the cortical actin superstructure during aging, leading to measurable changes in cortical stiffness. To directly evaluate this, atomic force microscopy (AFM)-based nanoindentation was performed on isolated rat cardiomyocytes. However, isolated cells at these ages were poorly spread and minimally adherent, suggesting that isolation compromised cytoskeletal integrity. Thus, mechanical assessment required use of *Drosophila*, an *in situ* model system with protein homology and a tubular heart that enables one to preserve cytoskeletal structure while probing the heart mechanically.

To directly assess *Vinculin* overexpression as a conserved aging event, we examined the functional consequences of cardiac-specific *Vinculin* overexpression. Transgenes were strategically expressed in w^{1118} hearts, which did not stiffen with age, via the cardiac-specific tinHE-Gal4 driver in conjunction with UAS-inducible Vinculin (*VincHE*) or -interfering RNA (RNAi) against Vinculin (*VincKD*) (Figure 1A). Cardiac-specific qPCR verified knockdown and/or overexpression for each respective gene. Vinculin overexpression (*VincHE*) resulted in increased cardiac stiffness compared to controls, with preference at IDs, similar to 5 wk *yw* (green triangles, Figure 1B). *MhcKD* exhibited reduced cardiac stiffness across the heart (red circles, Figure 1B). However, *MhcKD+VincHE* flies were indistinguishable from control (purple diamonds, Figure 1B), suggesting that *Vinc* overexpression could restore basal cardiac stiffness in hearts with myosin deficiency. Cytochalasin D treatment reversed the stiffening phenotype in *VincHE* in a pattern matching 5 wk old *yw* genotype, suggesting remodeling of the actin superstructure by increased vinculin in *VincHE*. Increased vinculin expression and costameric localization, cortical stiffening, and similar response to drug treatments occurred in both *VincHE* and 5 wk *yw* flies, suggesting a causative role for vinculin in inducing age-related changes observed in *yw* and correlate with age-associated cytoskeletal changes observed in mammals and aging patients.

For relative force production, *VincHE* shortening velocity was significantly higher at all viscous loads and displayed less-severe load-dependent decline in fractional shortening versus all other genotypes. *MhcKD* velocities were diminished at all loads and resembled 5 wk w^{1118} flies. However, *MhcKD+VincHE* had substantially improved shortening velocity compared to *MhcKD* (Figure 1C) and significant extension in lifespan (Figure 1D). These data suggest that *VincHE* hearts produce greater contractile forces and that increased vinculin expression can rescue the diminished force production of *MhcKD*, as observed in *MhcKD+VincHE*. Computational studies have suggested a role for vinculin in regulating sarcomere lattice spacing and contractility (3). Therefore, electron micrographs of cross-sections through *VincHE* and control cardiac myofibrils were examined to investigate alterations in ultrastructure. Similar average thick filament lattice spacing (45.74 and 45.79 nm) were observed in control and *VincHE*. However, significant reduction in lattice spacing variance (3.11 nm for control vs. 0.89 nm for *VincHE;* F-test $p < 10^{-48}$) was observed for *VincHE* flies. These data suggest improved myofilament arrangement contributes to enhanced contractility.

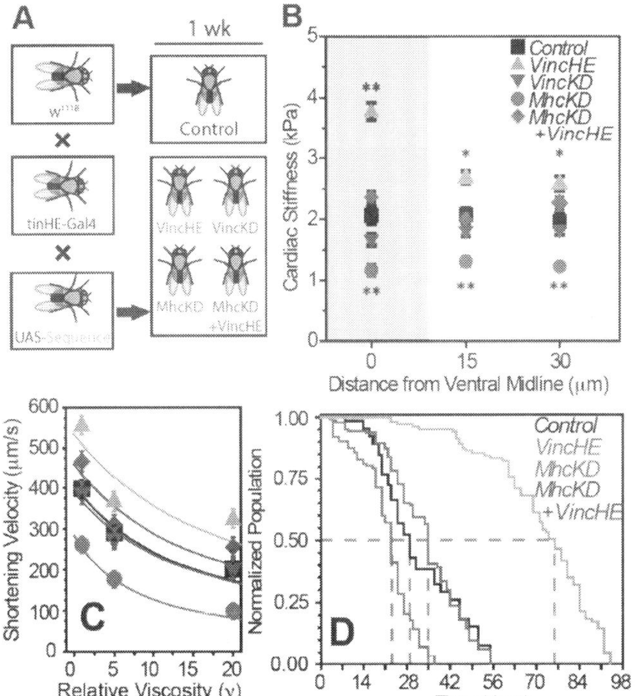

Fig. 1: (A) Schematic of transgenic flies, (B) their heart tube stiffness relative to intercalated discs (green), (C) the shortening velocity of their heart muscle vs. load, and (D) their lifespan.

DISCUSSION

Here we have outlined a cytoskeleton-based compensatory mechanism in the aging heart that is conserved across species. Although the vinculin network was primarily featured here, proteomic analysis suggests an even wider array of cytoskeletal proteins are differentially expressed with age and warrant further investigation. Other vinculin/catenin-family proteins, and the signaling molecules that regulate their translation and post-translational modifications, could play complementary roles. By presenting distinct aging proteomes and providing a proof-of-concept study of vinculin in cardiac aging, we aimed to establish a resource that will facilitate a broader examination of cardiac senescence so that additional therapeutic targets and regulators of aging heart function can be identified. We believe that our findings highlight the interplay of the molecular, cellular, and ultrastructural state of the cytoskeleton in maintenance of function in the aging heart and that these studies of mechanotransduction can directly inform clinical outcomes.

ACKNOWLEDGEMENTS

This work was supported by NIH (HHSN268201000032C, P01HL077180, T32HL105373, R01AG045428, DP02OD006460, R01GM32443, and R21HL106529), AHA (10SDG4180089 and 13PRE14410037), and American Federation for Aging Research.

REFERENCES

[1] Yancy, CW et al., *Journal of the American College of Cardiology*, 62: e147-239, 2013.
[2] Staudt, N et al. *PLoS genetics*, 1: e55, 2005.
[3] Tangney, JR et al. *Biophys J*, 104: 1623-1633, 2013.

SB3C2015
Summer Biomechanics, Bioengineering and Biotransport Conference
June 17-20, 2015, Snowbird Resort, Utah, USA
SB³C2015-1058

THE NEED FOR VALIDATION IN SOFT TISSUE CONSTITUTIVE MODELS

Sandeep Madireddy, Kumar Vemaganti

Department of Mechanical and Materials Engineering
University of Cincinnati
Cincinnati, Ohio 45221-0072

1. INTRODUCTION

Constitutive models of soft tissue mechanical response are widely used in biomedical and biomechanical applications. These constitutive models are usually calibrated using experimental data from simple laboratory tests like uniaxial tension and compression. However, the models are not always validated against more complex modes of deformation. In this study, we highlight the need for such validation studies for soft tissue mechanics.

Soft tissue is assumed to be an isotropic incompressible hyperelastic medium. The widely used Ogden model is calibrated against uniaxial tensile tests on porcine brain tissue obtained from the literature [1] for two choices of the model order ($N = 2, 3$). To account for the variability in the experimental response, the calibration is carried out using a probabilistic Bayesian framework, leading to distributions of the Ogden model parameters [2].

These distributions are used to predict the responses of the two models ($N = 2, 3$) under spherical indentation using analytical force-indentation relations from the literature [3]. We find that the responses of the two models are drastically different even though both provide very good fits to the experimental data.

2. MATERIALS & METHODS

2.1. Constitutive Model

The isotropic incompressible hyperelastic behavior of soft tissue is modeled using the N-th order Ogden strain energy density function

$$\Psi = \sum_{i=1}^{N} \frac{2\mu_i}{{\alpha_i}^2}(\lambda_1^{\alpha_i} + \lambda_2^{\alpha_i} + \lambda_3^{\alpha_i} - 3) \qquad (1)$$

where λ_i are the principal stretches, and μ_i and α_i are constitutive model parameters to be determined from experimental data. The shear modulus of the material is given by $\mu = \Sigma_{i=1}^{N} \mu_i$. The resulting relation for the Cauchy stress as a function of stretch for a uniaxial deformation is

$$\sigma_{11}(\lambda) = \sum^{N} \frac{2\mu_i}{\alpha_i}(\lambda^{\alpha_i - 1} - \lambda^{-\frac{\alpha_i}{2} - 1}) \qquad (2)$$

2.2. Spherical Indentation Response

The analytical expressions for the force-indentation response of a hyperelastic material proposed by [3] for the Ogden strain energy density function are used in this study. For the N-th order Ogden model, indentation force F is related to the indentation depth δ as

$$F(\delta) = \sum_{i=i}^{N} \left[\frac{B\pi a^2}{\alpha_i} \left[\left(1 - 0.2\frac{a}{R}\right)^{\frac{-\alpha_i}{2} - 1} - \left(1 - 0.2\frac{a}{R}\right)^{\alpha_i - 1} \right] \right] \qquad (3)$$

where E_0 is the initial Young's modulus, G_0 is the shear modulus, R is the radius of the indenter, and a is the contact radius. Furthermore, $B = \frac{40E_0}{9\pi(1-\nu^2)}$, $a = R^{0.503 - 3.97\times10^{-6}\delta}\delta^{0.498}$, and $G_0 = \frac{E_0}{3} = \sum_{i=i}^{N} \mu_i$.

2.3. Bayesian Calibration

The Bayesian approach to model calibration takes the noise in the experimental data into consideration to obtain estimates of the model parameters. In this probabilistic approach, we seek distributions of the parameters (rather than deterministic values) that maximize the likelihood that the model can predict the observed experimental data. This is accomplished using the Bayes theorem in which prior knowledge (distribution) P about the model parameters $\boldsymbol{\theta}$ is updated in light of the experimental data $\mathbf{d}$ to obtain the posterior distributions of the parameters

$$P(\boldsymbol{\theta}|\mathbf{d}) = \frac{P(\mathbf{d}|\boldsymbol{\theta})P(\boldsymbol{\theta})}{P(\mathbf{d})} \qquad (4)$$

The prior probability on the parameters, $P(\boldsymbol{\theta})$, can be assumed to be a uniform distribution on the parameter space. It is one way to represent our prior ignorance about the parameters.

The noise associated with experimental measurements can reasonably be represented as a Gaussian process based on maximum entropy considerations [4]. The likelihood function $P(\mathbf{d}|\boldsymbol{\theta})$ can now be written as

$$P(\mathbf{d}|\boldsymbol{\theta}) = \prod_{i=1}^{n} \frac{1}{\sqrt{2\pi{\Delta_i}^2}} \exp\left[-\frac{(\sigma_{11}(\boldsymbol{\theta}, \lambda_i) - D_i)^2}{2{\Delta_i}^2}\right] \qquad (5)$$

where n is the number of points at which the data D_i are measured and Δ_i is the size of the error bar at point i. A nested sampling-based algorithm

3. RESULTS & DISCUSSION

The posterior probability distribution functions for the model parameters μ_i and α_i ($N = 2, 3$) are obtained using MULTINEST. Both models provide very good fits to the experimental data using the maximum likelihood estimate (MLE): see Fig. 1(a). Using a value of 0.01m for the indentor radius, the uncertainty in the parameters is propagated to the indentation response for both models. For a given model, this results in a distribution of indentation force for each indentation depth. The maximum likelihood response is shown in Fig. 1(b), and the distributions are shown in 1(c)(d).

For both cases, the indentation responses are multimodal, and are dramatically different with the maximum likelihood value jumping between the multiple modes. Also the uncertainty in the indentation increases with the depth of indentation.

4. CONCLUSIONS

In this work, we study the importance of validating hyperelastic constitutive models of soft tissue. A Bayesian approach is used to calibrate, and estimate the probability distributions on the parameters using uniaxial tension tests done on porcine brain tissue. The resulting parameter uncertainty is propagated to estimate the indentation response for different depths of indentation.

Both models provide excellent fits to the calibration data but produce drastically different responses under indentation loading. This highlights the importance of model section ([6]) and validation of the constitutive model against other deformation modes. Which of these is the right indentation response can only be determined by carrying out further experiments. This also means that model development cannot be a linear process, where one simply fits a model after carrying out an experiment, and then uses the model to make predictions about the response of the system under study. Rather it is necessary for model development and experimentation to go hand in hand, with each informing the other in a systematic loop until one is satisfied with the predictive capability of the model.

5. ACKNOWLEDGEMENTS

We are grateful to the University of Cincinnati Simulation Center for providing financial support for this work.

References

[1] B. Rashid, M. Destrade, M. D. Gilchrist, in: 2012 IRCOBI Conference Proceedings 12-14 September 2012-Dublin (Ireland): IRC-12-49, International Research Council on the Biomechanics of Injury, pp. 405–416.

[2] S. Madireddy, B. Sista, K. Vemaganti, International Journal for Uncertainty Quantification (under review).

[3] D. C. Lin, D. I. Shreiber, E. K. Dimitriadis, F. Horkay, Biomechanics and Modeling in Mechanobiology 8 (2009) 345–358.

[4] D. S. Sivia, Data Analysis: A Bayesian Tutorial, Oxford University Press, 1996.

[5] F. Feroz, M. Hobson, M. Bridges, Monthly Notices of the Royal Astronomical Society 398 (2009) 1601–1614.

[6] S. Madireddy, B. Sista, K. Vemaganti, Computer Methods in Applied Mechanics and Engineering (Accepted 2015).

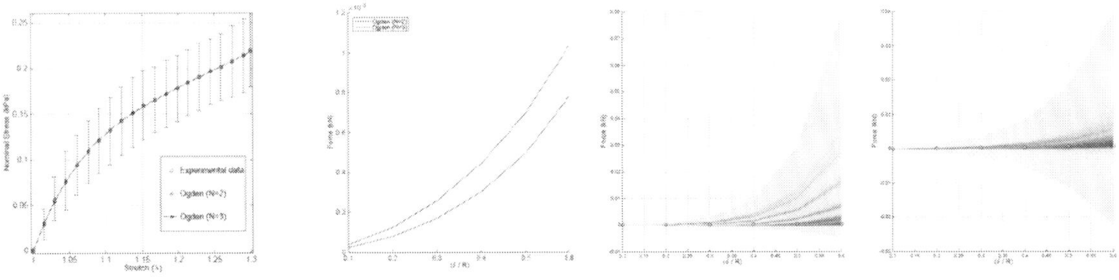

(a) (b) (c) (d)

Figure 1: (a) Result of model calibration. Both Ogden $N = 2$ and $N = 3$ provide excellent fits to data. (b) Maximum likelihood prediction for indentation force using calibrated models. (c) and (d) Distribution of indentation force response using Ogden $N = 2, 3$, respectively.

SB³C2015
Summer Biomechanics, Bioengineering and Biotransport Conference
June 17-20, 2015,Snowbird Resort, Utah, USA
SB³C2015-1059

ESTABLISHING *IN VIVO* HEMODYNAMIC BASELINE IN A NORMOTENSIVE RAT MODEL

Daniela Vélez-Rendón (1), Erica R. Pursell (1), Daniela Valdez-Jasso (1)

(1) Department of Bioengineering
University of Illinois at Chicago
Chicago, IL, USA

INTRODUCTION

Pulmonary arterial hypertension (PAH) is characterized by an abnormal high pressure in the pulmonary circulation, leading to the vascular remodeling of the arteries, changing its thickness and structural component. Likewise, it imposes a pressure overload to the right ventricle (RV). The RV has the ability of adapting when the afterload increases, by augmenting its wall thickness and contractility. However, if the increment in the afterload is much higher compared to its capacity of adaptation, it leads to its cardiac failure and death [1-3].

Currently, there is no cure for PAH, only treatments that lessen the symptoms and improve the quality of life of the patients. In a study by REVEAL Registry showed that PAH patients have a median survival rate of 7 years [4]. Nowadays, not much is known about what leads to PAH and the mechanisms underlying RV dysfunction and failure, indicating the gap in our knowledge of the right ventriculo-pulmonary vascular system and the disease [5].

In the present work, we aim to establish a baseline of the hemodynamic properties of the organs affected by PAH, namely the RV and the pulmonary vasculature, and elucidate their interaction in a normotensive group. The *in vivo* hemodynamic measurements are time-series of ventricular pressure and volume, pulmonary arterial blood flow, and simultaneous ventricular-vascular pressures, of a rat during an open chest surgery.

METHODS

In vivo hemodynamic measurements were taken from two male Sprague-Dawley rats (12-weeks-old, weighing 322.3 + 3.25 g). After five days of acclimatization, the animals were placed in an induction chamber supplied with 5% isoflurane. Once the animals were unconscious, they underwent a tracheotomy and were intubated (E-Z Anesthesia Ventilation System, Palmer, PA) and supplied with a tidal volume of 2.4 ml at a respiratory rate of 50 breath/min. To expose the heart and the great vessels a thoracotomy was subsequently performed.

The RV is identified and punctured with a 21-gauge needle to immediately place the 1.9F admittance pressure-volume catheter (Transonic Scisense, Ontario, Canada). After obtaining pressure-volume reference measurements during steady state conditions, the inferior vena cava is occluded for a few seconds to analyze how the disrupted blood venous return affects the end-systolic and end-diastolic pressure volume relationships, as previously done by *Wang et al.* [6]. Following the pressure-volume measurements, the catheter is exchanged for a 1.6F dual-pressure sensor catheter, and placed passed the pulmonic valve. Simultaneous time-series pressures are then recorded from the RV and pulmonary artery. Once reproducible data is obtained, the flow probe (Transonic Scisense, Ontario, Canada) is placed on the pulmonary artery for blood flow measurements. The data is collected and pre-processed in Labchart Software (ADInstruments Inc. Colorado Springs, CO). The hemodynamics measurements methods are summarized in Figure 1.

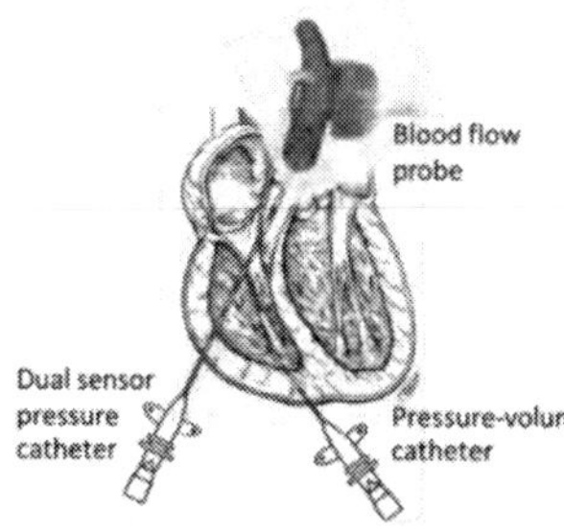

Figure 1. Schematic of the hemodynamics measurements.

At the conclusion of the hemodynamic studies, the animals were exsanguinated. Ventricular, vascular, and ventriculo-vascular performance metrics were heart rate (HR), cardiac output (CO), stroke volume (SV), dP/dt_{max} and ejection fraction (EF).

RESULTS

Surgeries lasted approximately two hours. Figures 2-3 are representative hemodynamic waveforms. Figure 2 shows one representative RV pressure-volume (PV) loop in the steady state and during inferior vena cava occlusion. As shown in the figure, during occlusion, the end-systolic volume (ESV) and end-diastolic volume (EDV) shifted to the left, and decrease in the pressure due to the reduction in the incoming blood flow.

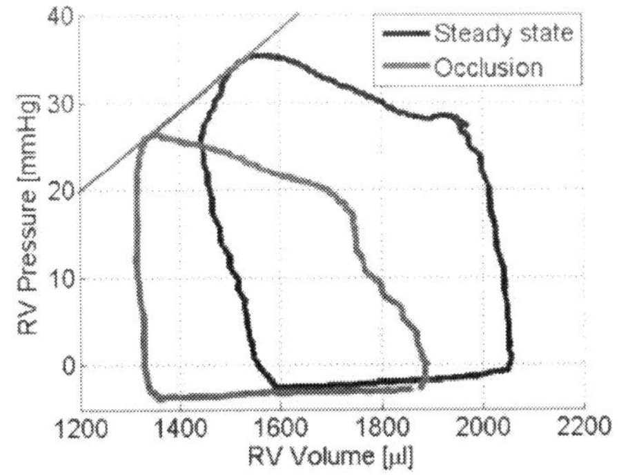

Figure 2. Representative *in vivo* pressure-volume loops during steady state (black) and inferior vena cava occlusion (red). The blue line represents the ESPV relationship (ESPVR) where we have changes in preload.

The measured blood flow through the pulmonary artery as a function of time is shown in Figure 3a. Blood pressure in the RV and PA measured simultaneously are shown as time-series in Figure 3b. As expected, the RV pressure ranges from 0 to 30 mmHg whereas the PA range is smaller and only ranging from 10 to 30 mmHg. One of the clear differences between the two pressure waveforms is the dichotic notch of the PA, which occurs when the pulmonary valve closes and the diastole begins.

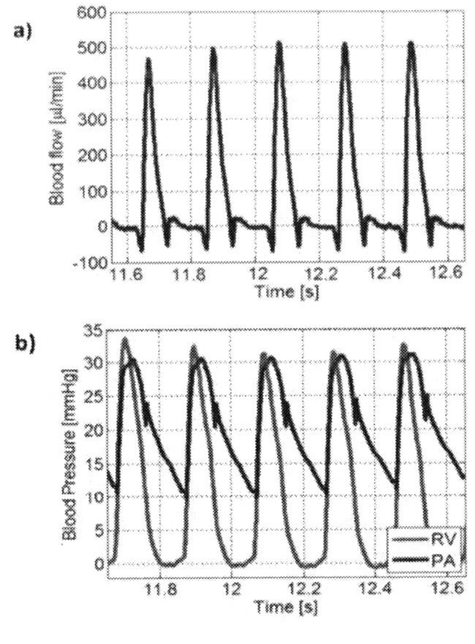

Figure 3. a) Blood flow (μL/min) acquired in the PA. b) Blood pressure (mmHg) measured in the RV (red) and the PA (blue).

Figure 4 shows the summary statistics of the hemodynamic data collected in the two animals. The parameters are in the range of previously reported works. In particular, the HR is close to when the animal is conscious, indicating that the minimal effects of anesthesia during the experiment. Equally, the mean values of dP/dt_{max}, CO, SV and EF are high and within the range of good cardiac performance and level of contractility.

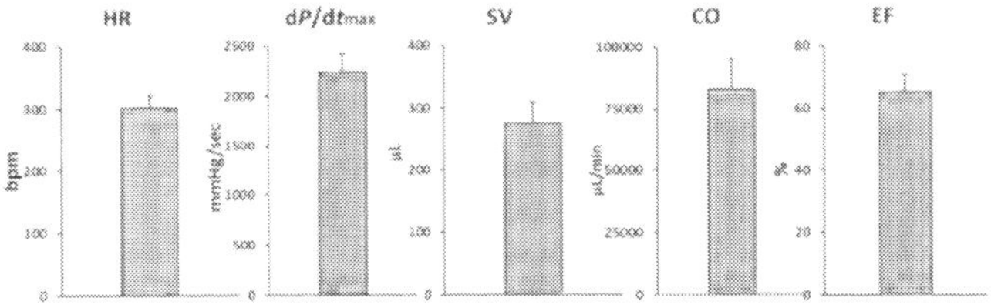

Figure 4. Summary statistics of *in vivo* hemodynamic measurements from a normotensive group. HR: heart rate; dP/dt_{max}; SV: stroke volume; CO: cardiac output; and EF: ejection fraction.

DISCUSSION

This study shows the first steps towards the establishment of the baseline of the hemodynamic properties of the organs affected by PAH. In this study we present *in vivo* measurements of three different signals namely blood flow from the PA, and pressure from the PA and RV, all measured simultaneously, which as far as we are aware of it has not been reported previously.

Our results reflect that both animals were normotensive, due to the good indices recorded of the cardiac performance as show in Figure 4, being consistent with results published by *Wang et al.* [6] for a normotensive group. Despite they performed the experiment in a mouse model, our outcomes in the mean values of dP/dt_{max}, EF and CO are within the same ranges. The blood flow time-series has the waveform as expected. It is our hypothesis that these signals will not be start changing their shape and getting off sync once PAH is induced in these animals.

ACKNOWLEDGEMENTS

We would like to acknowledge Gizzel Gomez and Jesse Gerringer for their support during the surgeries, and Dr. Filip Konecny from Transonic for his assistance during the data processing.

REFERENCES

[1] Pulido, T et al., Macitentan and Morbidity and Mortality in Pulmonary Arterial Hypertension, *N Engl J*, 13:369:809, 2013.

[2] Vonk-Noordegraaf, A et al., Right heart adaptation to pulmonary arterial hypertension, *JACC*, 62: D22–33, 2013.

[3] Wang, Z et al., Changes in Large Pulmonary Arterial Viscoelasticity in Chronic Pulmonary Hypertension. *PLOS ONE*. 8: 1-9, 2013.

[4] Benza, R et al., An Evaluation of Long-term Survival From Time of Diagnosis in Pulmonary Arterial Hypertension From the REVEAL Registry, *CHEST*, 142: 448-456, 2012.

[5] Voelkel et al., Right Ventricular Function and Failure, *Circulation*, 114: 1883-1891.

[6] Wang, Z et al., Progressive right ventricular functional and structural changes in a mouse model of pulmonary arterial hypertension, *Physiol Rep*, 1: 1-11, 2013.

SB³C2015
Summer Biomechanics, Bioengineering and Biotransport Conference
June 17-20, 2015,Snowbird Resort, Utah, USA
SB³C2015-1060

EFFECTS OF CORONARY FLOW ON SINUS HEMODYNAMICS IN THE PRESENCE OF A TRANSCATHETER AORTIC VALVE IMPLANTATION

Brandon L. Moore (1), Pablo Maureira (2), Lakshmi P. Dasi (1, 3)

(1) Mechanical Engineering Department
Colorado State University
Fort Collins, Colorado, United States

(2) Department of Cardiovascular Surgery
Lorraine University Hospital of Nancy
Nancy, France

(3) School of Biomedical Engineering
Colorado State University
Fort Collins, Colorado, United States

INTRODUCTION

Aortic valve stenosis is a highly prevalent form of heart disease in the U.S., which is often caused by calcification of the valve leaflets. Development of calcification is largely regulated by hemodynamics. Specifically, sinus flow is of interest since calcification preferentially develops on the aortic side of the valve [1]. Additionally, the non-coronary leaflet often calcifies first [2], so detailed information on flow differences between the sinuses is crucial for understanding hemodynamic influences on disease in the aortic valve.

Once calcification has progressed to the point of severe aortic stenosis and a patient has become symptomatic, valve replacement is generally recommended [3]. However, conventional open heart surgery is not a viable option for high-risk patients. In this case, transcatheter aortic valve implantation (TAVI) has been developed as an alternative means of treatment. Since this technology is still relatively new, little is known about the long-term effects of such valves. Therefore, the objective of this work is to elucidate TAVI valve sinus hemodynamics with and without coronary flow that could have long-term implications for younger (i.e. lower-risk) patients.

METHODS

Flow Loop Setup: An in vitro setup was designed to generate aortic flow and to examine resulting hemodynamic parameters and leaflet kinematics. A schematic of this flow loop setup is shown in Figure 1. The major components of this flow loop include a piston pump, aortic valve and valve chamber, compliance chamber, and resistance valve. The motion of the piston pump was prescribed to generate physiologic aortic flow waveform, while adjustable compliance and resistance ensured physiological bulk hemodynamics (BP ~ 120/80mmHg; HR ~ 60bpm; Mean Flow Rate ~ 5 L/min). A secondary flow circuit branching off of the aortic sinus was also added

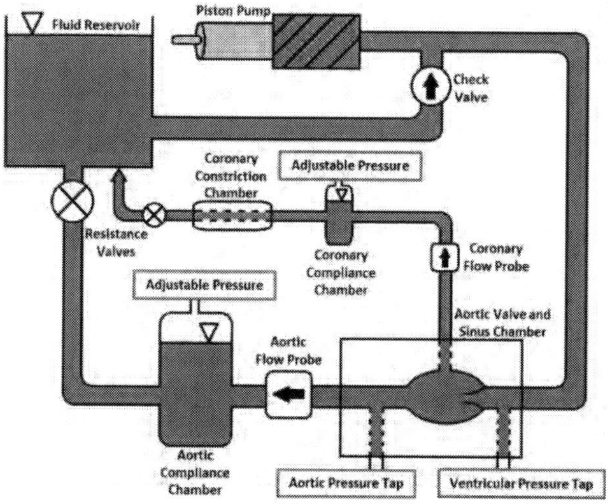

FIGURE 1: Schematic of left heart flow loop that was used to govern pressure and flow profiles across the aortic valve

to simulate coronary flow. This circuit includes compliance and resistance as well as a dynamically-controlled Windkessel chamber to add flow resistance during systole. Such systolic resistance is meant to mimic coronary vessel constriction that results from myocardial contraction. To examine non-coronary sinus flow, this coronary circuit was clamped-off just downstream of the sinus.

A 25 mm Medtronic Hancock II T505 porcine bioprosthesis valve was mounted in a clear, acrylic sinus chamber, and a Medtronic CoreValve was inserted into this model aortic valve. The chamber was machined to provide a smooth transition between the valve and the walls of the mock aortic root. Dimensions were chosen based on in vivo CT scan data.

Particle Image Velocimetry: High-resolution, time resolved particle image velocimetry (PIV) was conducted to capture velocity fields within one of the sinuses of Valsalva. Velocity vectors were calculated using advanced PIV cross correlation approaches with a 50% overlap multi-pass approach starting from one 32×32 pixel interrogation followed by one 16×16 pixel interrogation passes. Post processing was performed using adaptive median filtering. The effective spatial resolution was 29 microns per pixel with a temporal resolution of 4000 Hz.

Data Analysis: Coronary flow was tuned to match physiologic waveforms and was recorded with a flow probe. Velocity fields were generated from PIV, and vorticity and shear stress contours were computed based on these velocity vectors.

RESULTS

Figure 2 shows coronary flow waveforms for the model native valve control case as well as a TAVI valve case. Aortic flow reaches a

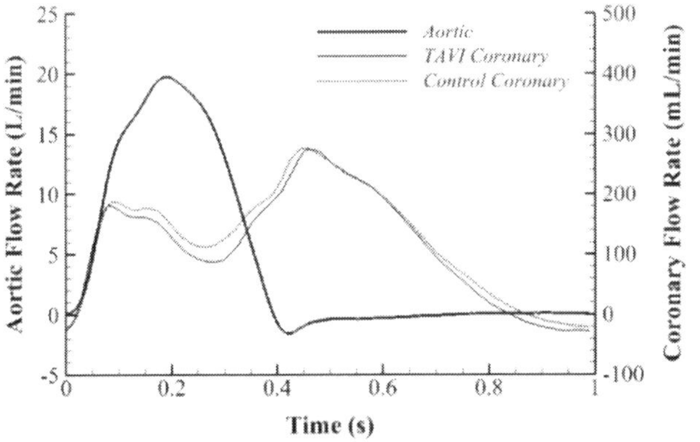

FIGURE 2: Aortic and coronary flow waveforms with and without TAVI valve

peak of about 20 L/min and forward flow persists for duration of approximately 0.4s. Peak coronary flow occurs during early diastole for each case and reaches a value of ~275 mL/min. Maximum systolic

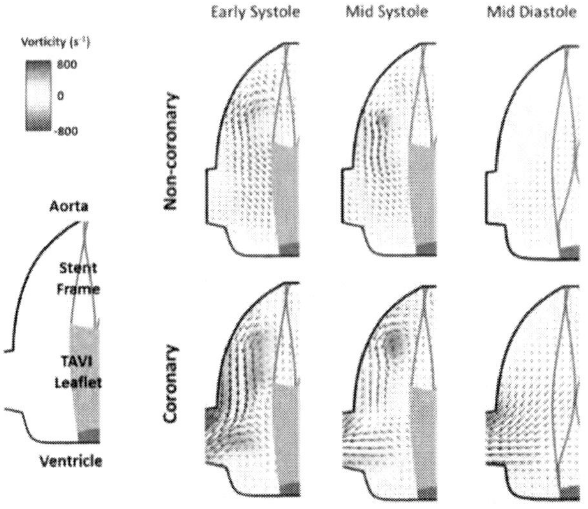

FIGURE 3: Velocity vectors and vorticity contours at three distinct time points during the cardiac cycle

coronary flow is around 175 mL/min and is generally less for the TAVI case.

Velocity vectors and vorticity contours are presented in Figure 3. Both sinus cases display some systolic counterclockwise vorticity representative of a classical sinus vortex. However, there is more fluid motion near the base of the sinus at all time points in the coronary sinus versus the non-coronary sinus. Voriticity magnitude near the leaflet is higher during systole when coronary flow is present, with peak values around 200 s^{-1} and 250 s^{-1} for non-coronary and coronary sinuses, respectively. Additionally, fluid shear stress in this region is also significantly higher for the coronary sinus (Figure 4) throughout the cardiac cycle.

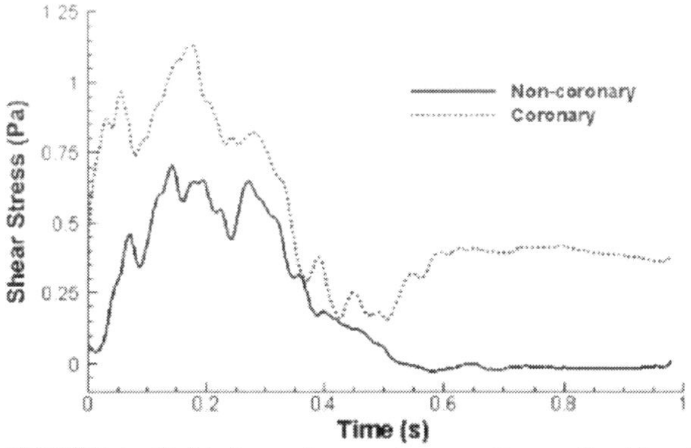

FIGURE 4: Fluid shear stress values on the aortic side of the valve leaflets over one cardiac cycle

DISCUSSION

When comparing native valve and TAVI coronary flow rates, there is a slight reduction in systolic coronary flow for the TAVI case. This could be due to lowered pressure within the sinus that is caused by a higher pressure drop across the smaller-diameter TAVI valve.

Differences also exist between the TAVI non-coronary and TAVI coronary sinuses. The increased flow in the base of the sinus that occurs in the coronary sinus could be beneficial from a disease standpoint. It provides a "washout" effect to move otherwise stagnant blood out of the sinus. This is helpful since stagnation is often associated with thrombosis [4]. Similarly, low velocity fluid causes low wall shear stress on the valve leaflets, which could exacerbate native valve calcification and/or lead to calcification of the TAVI leaflets.

ACKNOWLEDGEMENTS

Authors gratefully acknowledge funding from National Institutes of Health (NIH) under Award Number R01HL119824, and the American Heart Association under award 11SDG5170011.

REFERENCES

[1] Weinberg, E.J., et al., *Hemodynamic Environments from Opposing Sides of Human Aortic Valve Leaflets Evoke Distinct Endothelial Phenotypes In Vitro.* Cardiovascular Engineering, 2010. **10**(1): p. 5-11.

[2] Sucosky, P., et al., *Altered Shear Stress Stimulates Upregulation of Endothelial VCAM-1 and ICAM-1 in a BMP-4-and TGF-beta 1-Dependent Pathway.* Arteriosclerosis Thrombosis and Vascular Biology, 2009. **29**(2): p. 254-260.

[3] Nishimura, R., et al., *2014 AHA/ACC Guideline for the Management of Patients With Valvular Heart Disease: Executive Summary.* Journal of the American College of Cardiology, 2014. **63**(22): p. 2438-2488.

[4] Wootton, D. and D. Ku, *Fluid Mechanics of Vascular Systems, Diseases, and Thrombosis.* Annual review of biomedical engineering, 1999. **1**(1): p. 299-329.

SB^{3}C2015
Summer Biomechanics, Bioengineering and Biotransport Conference
June 17-20, 2015,Snowbird Resort, Utah, USA
SB³C2015-1062

INTEGRATING AN OPEN SOURCE MESHING ALTERATIVE INTO SIMVASCULAR 2.0

Adam R. Updegrove (1), Nathan M. Wilson (2), Shawn C. Shadden (1)

(1) Department of Mechanical Engineering
University of California, Berkeley
Berkeley, CA, USA

(2) Open Source Medical Software Corporation
Santa Monica, CA, USA

INTRODUCTION

Patient-specific anatomic models constructed from medical imaging data can be geometrically complex given the natural branching structure of vasculature as well as the potential for anatomical abnormalities due to disease. In addition, it is not uncommon to have models with a wide range of length scales. For example, anatomic models for simulating coronary blood flow can include the aortic arch, which can be an order of magnitude larger in diameter than the modeled coronary arteries.

Once a researcher has created an anatomic model of interest, the next major step in the process of patient-specific blood flow simulation is to discretize the model. Ideally the process of discretization, or mesh generation, needs to be robust, computationally efficient, and require minimal user interaction. In addition, the ability to adapt the mesh based on *a-posteriori* estimates is often desired.

SimVascular (www.simvascular.org) is an open source package to perform image-based, patient-specific blood flow simulation. While the source code of SimVascular 1.0 has been publicly available since 2007, previous releases have relied explicitly on an integrated commercial library, MeshSim (Simmetrix, Inc., Clifton Park, NY), to generate meshes. This dependence of SimVascular 1.0 on commercial libraries for meshing (as well as commercial libraries for solid modeling and solving linear equations) has limited adoption and more widespread use of this application. The goal of the work described herein was to integrate into SimVascular a robust and easy to use alternative to the commercial MeshSim library, thus removing one potential deterrent to new users in exploring the capabilities of the software.

METHODS

SimVascular 1.0 only supported one internal model data type for meshing. Specifically, the user generated a boundary representation solid model representing the desired anatomy. Internally, this was done by using the commercially available Parasolid solid modeling library (Siemens PLM Software, Plano, Texas). The Parasolid model was passed directly to the MeshSim library integrated in SimVascular 1.0 in order to discretize the domain.

In this work, an alternative representation of an anatomic model for meshing is utilized. Specifically, the flow domain of interest is defined as a single region enclosed by a triangulated lumen surface. The lumen surface consists of a set of consistently oriented triangular faces that approximate the lumen boundary.

The triangulated lumen surface may be directly created from 3-D image segmentation methods inside of SimVascular, imported from open source packages such as ITK-SNAP (itksnap.org), or imported from commercial segmentation packages such as Mimics (biomedical.materialise.com/mimics). In addition, techniques described elsewhere can be used to generate triangulated models from lofting 2-D segmentations or from a combination of 2-D and 3-D segmentation results [1].

As a first step, SimVascular 2.0 was extended to support MeshSim "Discrete Models" (DM). In particular, MeshSim provides functions to create DMs from triangulated surfaces. The current DM implementation inside of SimVascular only supports meshing a single region. Meshing parameters for a DM are specified similarly to those of a Parasolid model.

In addition to the commercial option, a fully open source capability was implemented in SimVascular 2.0. Briefly, the method begins with a triangulated lumen surface, and is completed when the region is completely filled with tetrahedrons. The user specifies a series of meshing parameters to guide boundary layer meshing, local mesh refinement, and radius based meshing (see Figure 1). Two external open source packages are utilized in the meshing process: the

Vascular Modeling ToolKit (VMTK, www.vmtk.org) and TetGen (www.tetgen.org).

In order to provide options similar to those available in the commercial meshing module of SimVascular, techniques for open-source mesh adaption were also developed. Mesh adaption enables a user to adapt or remesh their domain based on estimates of errors in the solution. That is, the Hessian is used to provide *a-posteriori* error estimates on which to set a mesh sizing function [2]. Refining and/or coarsening the mesh based on this sizing function ideally will lead to a computationally-efficient mesh that reduces the overall error in the solution while minimizing additional run time needed to simulate additional cardiac cycles for the adapted mesh.

RESULTS

To demonstrate the new methods, a simple patient-specific model of the aorta including the left and right common iliac branches was meshed using the new algorithm in SimVascular 2.0 and compared to a mesh generated using MeshSim for the same geometry. In Figure 2, a section of the mesh is extracted to show the difference of the two meshing techniques when similar mesh options and mesh sizes are applied. In particular, a boundary layer mesh was generated with three layers and a mesh edge size of 0.2 cm. Select meshing statistics are given in Table 1. As shown in the table, a similar number of elements with comparable mesh quality were generated using both meshing techniques.

Steady flow blood flow simulations were run using both meshes to compare the impact of the meshing techniques on simulation results of interest. A physiologically reasonable steady volumetric flow rate was applied to each mesh. Figure 3 displays the wall shear stress magnitude after reaching steady state. The instantaneous wall shear stress was visualized and compared favorably as shown. After running the steady simulation, each mesh was adapted to the solution using techniques described previously; this information is displayed in Table 1. Differences in mesh elements and nodes are likely due to three factors. MeshSim does not adapt within the boundary layer, TetGen only refines and does not coarsen elements, and TetGen and MeshSim differ in adaption technique. MeshSim applies an anisotropic mesh function (different based on direction); whereas, TetGen applies an isotropic mesh function (same parameter for each direction).

DISCUSSION

Results of a newly developed integrated open source mesh generation algorithm presented herein are promising. For certain classes of clinically relevant anatomy, the techniques can be utilized to create meshes of comparable density and similar quality to those generated by a best-in-class commercially available library. Techniques integrated from VMTK such as the extraction of centerlines and using the distance to centerlines to assist in specifying meshing parameters are particularly useful when meshing models of large geometric extent with significantly varying caliber of included vessels. The availability of the open source mesh generation library is a critical component in the overall project effort of SimVascular 2.0 to encourage new users to test the software. The commercial MeshSim library continues to be more robust in our testing for a wider class of problems, so it is anticipated that the SimVascular development team will continue to support it. Future research will continue to improve the open source meshing options. In particular, integration of multi-region meshing and techniques to implement anisotropic adaptation will be investigated.

ACKNOWLEDGEMENTS

Support was provided by an NSF Graduate Fellowship and NSF SSI Award #1339824. Software hosting for SimVascular is provided by simtk.org.

REFERENCES

[1] Updegrove A, et al. Triangulated Surface Boolean Operations for Combining 2-D and 3-D Image Segmentation for Patient-Specific Blood Flow Analysis: SB3C; 2015; Snowbird, Utah

[2] Sahni O, et al. *CMAME 195*, 5634-5655, 2006

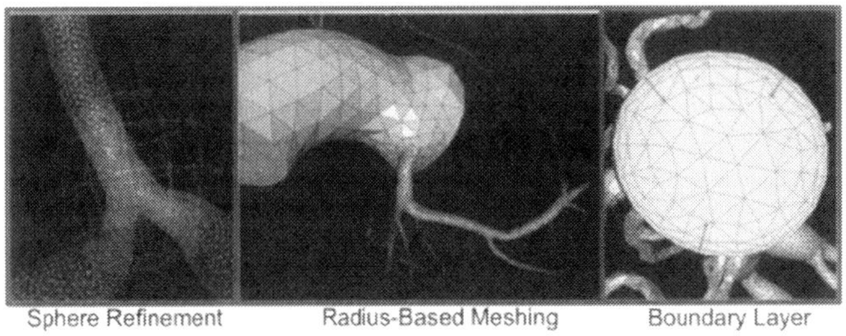

Figure 1: Types of advanced meshing provided in the open-source meshing module of SimVascular 2.0

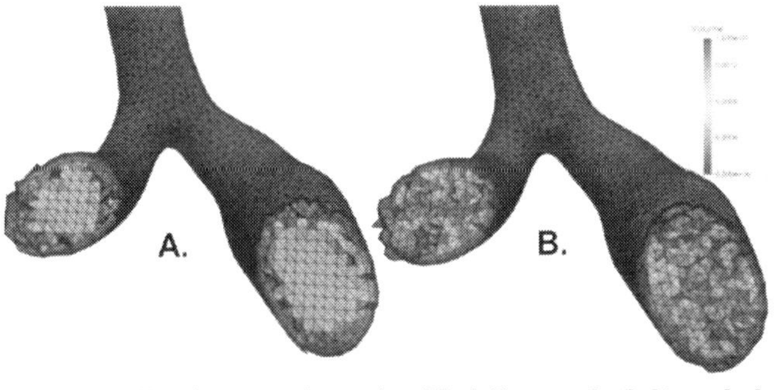

Figure 2: Visual comparison of a MeshSim mesh (left) and the TetGen mesh (right).

Table 1: Mesh Statistics

	A. MeshSim	B. TetGen
# Of Elements	436,048	456,309
# Of Nodes	82,235	81,887
Min/Max Volume (cm^3)	6.84e-6 / 2.57e-3	2.26e-5 / 1.59e-3
Min/Max Aspect Ratio	1.028 / 12.74	1.00 / 7.45
Min/Max Radius Ratio	1.00 / 21.29	1.00 / 8.37
# Of Elements after adaption	453,142	481, 472
# Of Nodes after adaption	83,220	85,880
Min/Max Volume (cm^3) after adaption	3.21-6 / 1.74e-2	6.09e-6 / 1.59e-3

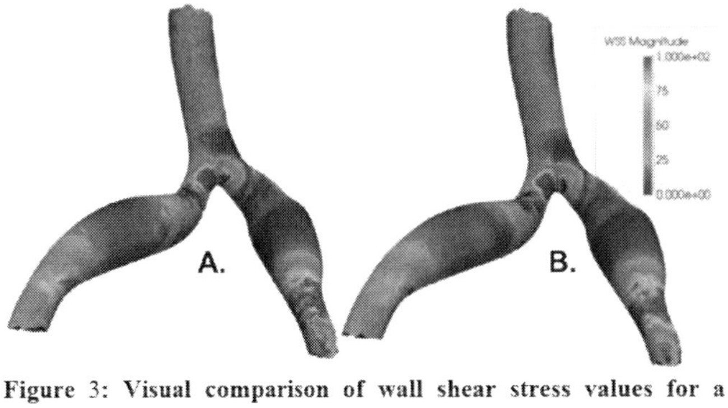

Figure 3: Visual comparison of wall shear stress values for a MeshSim mesh (left) and TetGen mesh (right) for steady flow.

AORTIC FLOW WAVEFORM SHAPE REGULATES VALVULAR HEMODYNAMICS AND ENERGY LOSSES

Brandon L. Moore (1), Lakshmi P. Dasi (1, 2)

(1) Mechanical Engineering Department
Colorado State University
Fort Collins, Colorado, United States

(2) School of Biomedical Engineering
Colorado State University
Fort Collins, Colorado, United States

INTRODUCTION

A healthy human heart supplies oxygen to the body tissues by pumping blood through the aortic valve, which opens and closes approximately 3 billion times in an average human lifespan [1]. Due to this large workload, most mechanical parameters associated with the aortic valve are optimized for maximum efficiency. Therefore, even slight alterations in fluid dynamics across this valve can have serious implications for pumping efficiency and therefore cardiac health.

One parameter that is often indicative of cardiac health is the shape of the aortic flow waveform. Often approximated as a half-sine wave, this curve typically peaks earlier than midway through the forward flow phase in vivo. However, in some cases this peak flow location can be shifted to a later time point. One cause of this shift is slowed calcium re-uptake in the myocardium, a condition which prolongs systolic contraction and is associated with a failing left ventricle [2]. While it is known that a shifted waveform can indicate adverse mechanical conditions, these specific conditions and their implications are still unclear. Therefore, this study aims to explore the valvular dynamics that are present for differing aortic flow waveforms.

METHODS

Flow Loop Setup: An in vitro setup was designed to generate multiple aortic flow waveforms and to examine resulting hemodynamic parameters and leaflet kinematics. A schematic of this flow loop setup is shown in Figure 1. The major components of this flow loop include a piston pump, aortic valve and valve chamber, compliance chamber, and resistance valve. The motion of the piston pump was manipulated to create three different aortic flow waveforms, each with a different temporal peak flow location, while adjustable compliance and resistance ensured physiological bulk hemodynamics (BP ~ 120/80mmHg; HR ~ 60bpm; Mean Flow Rate ~ 5 L/min).

A 25 mm Medtronic Hancock II T505 porcine bioprosthesis valve was mounted in a clear, acrylic sinus chamber. The chamber was machined to provide a smooth transition between the valve and the walls of the mock aortic root. Dimensions were chosen based on in vivo CT scan data.

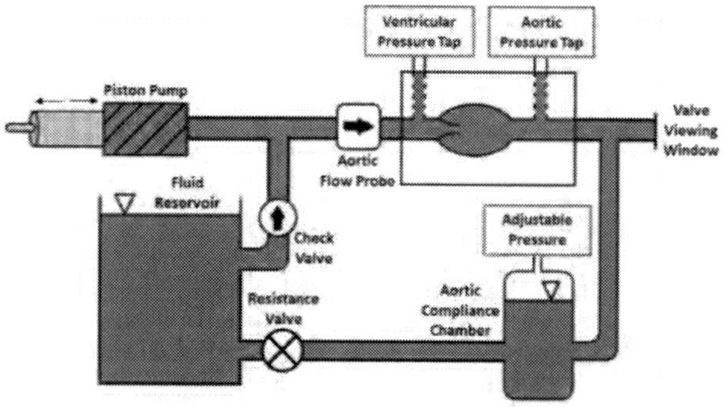

FIGURE 1: Schematic of left heart flow loop that was used to govern pressure and flow profiles across the aortic valve

Particle Image Velocimetry: High-resolution, time resolved particle image velocimetry (PIV) was conducted to capture velocity fields within one of the sinuses of Valsalva. Velocity vectors were calculated using advanced PIV cross correlation approaches with a 50% overlap multi-pass approach starting from one 32×32 pixel interrogation followed by one 16×16 pixel interrogation passes. Post processing was performed using adaptive median filtering. The effective spatial resolution was 29 microns per pixel with a temporal resolution of 5000 Hz.

Data Analysis: Leaflet tip location was tracked manually using raw camera images from PIV trials. Pumping power was also calculated from transvalvular pressure gradient and flow rate measurements. Velocity fields were generated from PIV, and vorticity contours were computed based on these velocity vectors.

RESULTS

Figure 2 shows flow rate and transvalvular pressure gradient measurements used to calculate pumping power requirements. Average flow rates for early, mid, and late flow times were 5.4, 5.4, and 4.9 L/min, with peak velocity occurring at 0.092, 0.202, and 0.324 s, respectively. Power loss across the valve over one cardiac cycle (normalized for flow rate) was 0.027, 0.030, and 0.051 W/(L/min) for early, mid, and late flow peaks, respectively.

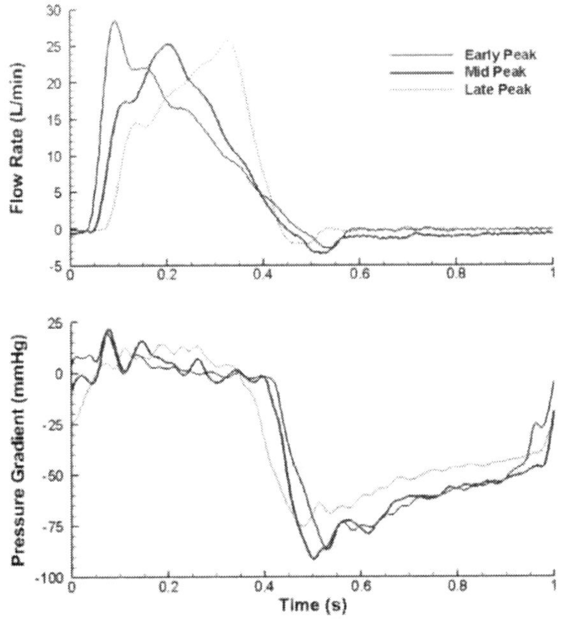

FIGURE 2: Flow rate and pressure gradient waveforms

Velocity vectors and vorticity contours are presented at times of peak flow for each case (Figure 3). Maximum vorticity magnitudes at times of peak flow rates are approximately: 920, 870, 810 s^{-1} for early, mid, and late flow peaks, respectively.

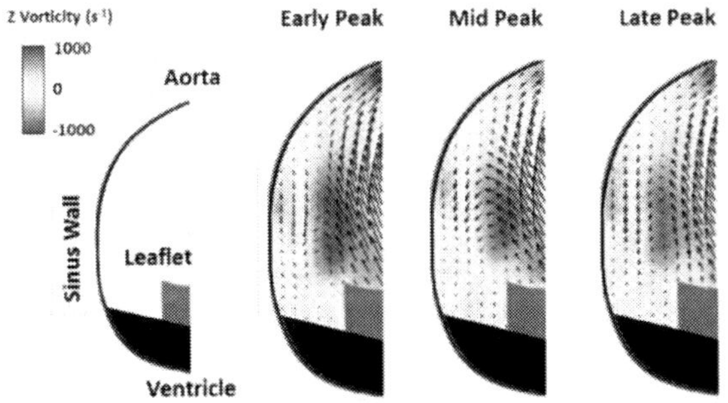

FIGURE 3: Velocity vectors and vorticity contours at peak flow

Leaflet kinematics were also tracked and results of radial leaflet tip position vs time are shown in Figure 4. Since camera images were focuses on the sinus region, the leaflet tip only became visible when the valve was partially to fully open. Leaflet tip position reached a radial position of approximately 11-12mm from the aortic axis upon maximal opening in each case, with the fullest and longest duration of opening for the early peak flow case.

DISCUSSION

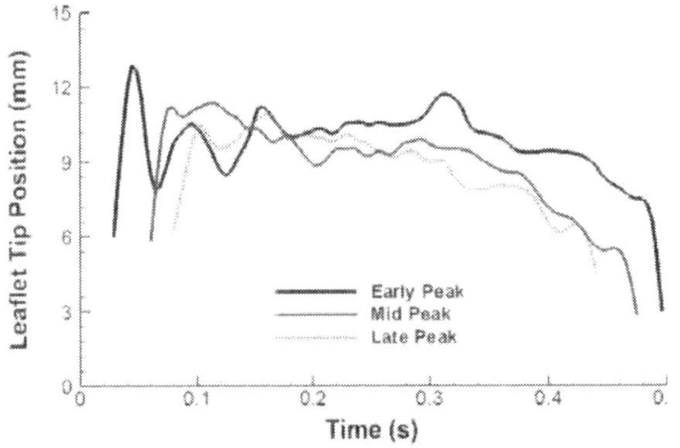

FIGURE 4: Distance from aortic axis to leaflet tip was manually tracked throughout systole for each flow case

As shown in Figure 2, a higher transvalvular pressure gradient yields higher flow acceleration. Likewise, a lower yet more sustained pressure gradient causes slower flow acceleration. The peak flow rate in all cases is similar. Gharib et al [3] demonstrated that a faster ramp yields a lower formation number, which means that maximal vortex circulation occurs at a lower stroke volume for higher flow acceleration. So an earlier flow peak could mean a stronger sinus vortex. This is supported by vorticity contours presented in Figure 3, which shows a stronger vortex for an earlier peak flow.

One of the main aspects of this study is the power required to pump fluid across the aortic valve. Consistent with clinical findings, a shifted aortic flow curve is associated with adverse cardiac mechanics. It is now clear that these adverse mechanics are a result of increased workload on the ventricle for a later peak flow time.

Leaflet kinematics tracking showed earlier peak flow time being correlated to longer valve open duration. This can be explained by examining pressure gradients for each case. A certain positive threshold must be reached to initiate opening, while a similar negative pressure gradient is required to close the valve. The opening threshold is reached earlier with more rapid flow acceleration in the early peak case, and the closing threshold is reached sooner with more rapid deceleration in the late peak case. This helps explain differences in energy losses. Rapid fluid deceleration present in the late peak case leads to more energy loss than allowing the fluid to decelerate gradually in the early and mid peak cases.

The main implication of this work is that a failing ventricle, which often lacks the ability to quickly accelerate fluid, requires more energy to pump blood through the aortic valve. Furthermore, lower vorticity that accompanies this lack of acceleration could have implications for disease due to lowered wall shear stress on the leaflet and lack of washout flow in the sinus.

ACKNOWLEDGEMENTS

Authors gratefully acknowledge funding from National Institutes of Health (NIH) under Award Number R01HL119824, and the American Heart Association under award 11SDG5170011.

REFERENCES

[1] Gould, S.T., et al., *Hemodynamic and Cellular Response Feedback in Calcific Aortic Valve Disease.* Circulation Research, 2013. **113**(2): p. 186-197.

[2] O'Rourke, M.F., A. Pauca, and X.-J. Jiang, *Pulse wave analysis.* British Journal of Clinical Pharmacology, 2001. **51**(6): p. 507-522.

[3] Gharib, M., E. Rambod, and K. Shariff, *A universal time scale for vortex ring formation.* Journal of Fluid Mechanics, 1998. **360**: p. 121-140.

SB^{3}C2015
Summer Biomechanics, Bioengineering and Biotransport Conference
June 17-20, 2015,Snowbird Resort, Utah, USA
SB^{3}C2015-1064

DEVELOPMENT OF A FRAMEWORK TO CHARACTERIZE THE ROLE OF WALL SHEAR STRESS IN ATHEROSCLEROTIC PLAQUE TRANSFORMATION THROUGH THE COMBINED USE OF OCT AND VH-IVUS

David S. Molony (1), Lucas H. Timmins (1,2), Emad Rasoul-Arzrumly (2), Olivia Y. Hung (2), Bill D. Gogas (2), Habib Samady (2), Don P. Giddens (1,2)

(1) Wallace H. Coulter Department of Biomedical Engineering
Georgia Institute of Technology and Emory University School of Medicine
Atlanta, GA, USA

(2) Division of Cardiology, Department of Medicine
Emory University School of Medicine
Atlanta, GA, USA

INTRODUCTION

Atherosclerotic plaques prone to rupture, termed vulnerable plaques, are typically characterized by a large necrotic core and a thin overlying fibrous cap. Rupture of this type of plaque, termed thin-cap fibroatheromas (TCFA), is responsible for the majority of acute coronary syndrome events, hence, an understanding of the development and progression of these plaques is critical. It has been found that hemodynamics are associated with changes in plaque size and morphology, specifically that plaques enlarge in regions of low wall shear stress (WSS) [1, 2] and transform to a more vulnerable phenotype in regions of high WSS [2].

These prospective studies of atherosclerosis were performed using intravascular ultrasound (IVUS) or virtual histology IVUS (VH-IVUS) and, as a consequence, were limited by their spatial resolution. The greater resolution of optical coherence tomography (OCT) (10μm vs 100μm) allows for the identification of a thin fibrous cap, serving as the only intravascular imaging modality that can identify TCFA. However, compared to IVUS, OCT is limited by its poor penetration depth, leading to uncertainty in true lesion size and plaque burden. IVUS on the other hand can measure plaque burden and by using radiofrequency tissue analysis, the components of the plaque (such as necrotic core) can be identified. Given the complementary strengths of these imaging modalities, their combined use can result in a greater understanding of the natural history of coronary artery disease (CAD), and in particular how hemodynamics play a role in plaque progression and transformation.

This work has 2 aims, (1) to develop a methodology to co-register serial VH-IVUS and OCT images and (2) to assess the role of hemodynamics in plaque transformation. This is achieved by correlating baseline hemodynamics and focal changes in plaque features as assessed by VH-IVUS and OCT over a 12 month period.

METHODS

Clinical Data The patient population involved in this study presented to the cardiac catheterization laboratory with an abnormal noninvasive stress test or recurrent angina pectoris. During initial catheterization, biplane coronary angiography, OCT images, VH-IVUS images, and Echo Doppler derived velocity data were collected. Patients returned at 12-months for repeat catheterization and the final cohort of patients examined in this study was determined from those with both baseline and follow-up VH-IVUS and OCT imaging (n=13). All patients provided written informed consent, and the Emory University Institutional Review Board approved the study.

Image Co-registration VH-IVUS images are automatically segmented and then manually checked and edited if required by an expert reader using EchoPlaque v4.0 (INDEC, Santa Clara, CA). OCT images were segmented automatically using the dynamic programming approach described by Wang et al. [3]. Briefly, images are transformed into the polar domain and a local cost function is generated which describes the pixel intensity difference between the lumen and arterial wall. An accumulated cost function image is then calculated and an optimal path can be determined by tracing the maximum value from the bottom to the top (figure 1). Fibrous caps, if present, are also segmented semi-automatically using a previously validated dynamic programming algorithm [4]. Here, a human user must specify a region of interest and the cap is then segmented based on a cost function describing the pixel intensity difference between the lipid and the overlying fibrous cap. The benefit of these automatic techniques is the removal of intra-operator error [4].

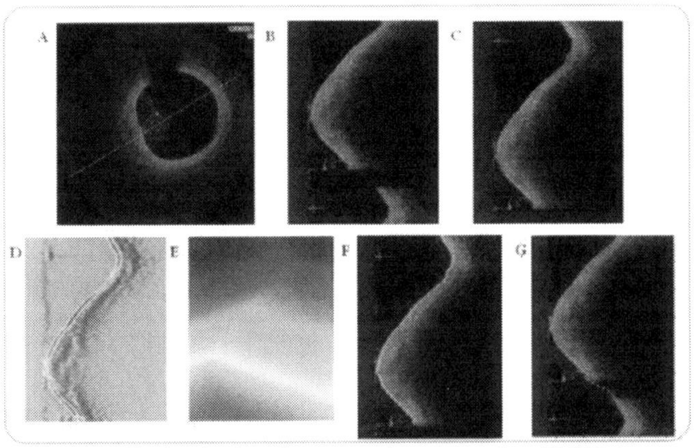

Figure 1. (A) Optical coherence tomography (OCT) image. (B) OCT image converted to polar domain. (C) Guidewire shadow artifact shifted to bottom of the image (D) Cost image (E) Cumulative cost image (F) Optimal path through cumulative cost image (G) Lumen segmentation interpolated over guidewire shadow artifact

IVUS and OCT images are co-registered using a Matlab (Mathworks, Natick, MA) based GUI (figure 2). The user identifies corresponding branches in both modalities and images are matched longitudinally and circumferentially. Longitudinal co-registration is achieved based on the known distances between corresponding branches and circumferential co-registration is based on determining the angular difference between features in longitudinally co-registered images. This difference is quantified by determining the angular difference between identical branch points in both modalities as an initial solution. This can then be refined using a cross correlation algorithm which uses the lumen centroid to artery wall distance [5] (figure 3).

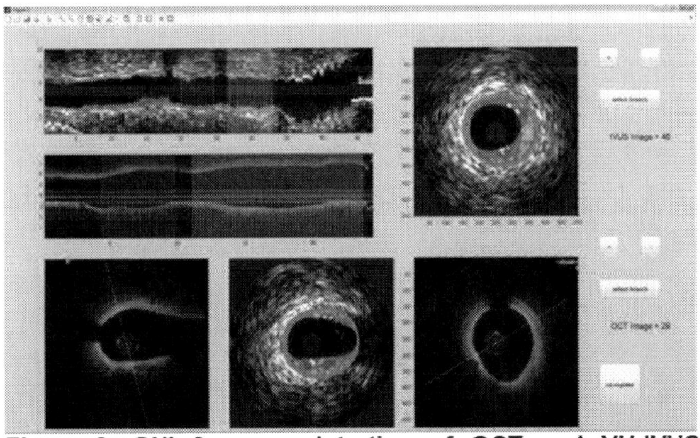

Figure 2. GUI for co-registration of OCT and VH-IVUS images. The user selects branch points which co-registers images longitudinally. Longitudinally co-registered regions of interest are displayed using the same color in the longitudinal view (top left and right panels). Co-registered images are displayed in the bottom left and right panels

Hemodynamic Modeling A 3D model of the coronary artery is generated using the ANGUS method. VH-IVUS images are stacked on a reconstruction of the IVUS catheter generating a 3D model. IVUS is preferred to OCT as these images are EKG gated and hence will not have an error generated from longitudinal movement of the catheter during pullback as experienced by OCT. Following 3D model reconstruction hemodynamic simulations are performed using patient-specific velocity waveforms. WSS is then extracted from these hemodynamic simulations and the relationship between WSS plaque changes from baseline to follow-up are characterized in focal regions based on a previously developed methodology which co-registers baseline and follow-up VH-IVUS images [5]. Follow-up OCT images are then co-registered with follow-up VH-IVUS images.

RESULTS

The new methodology is illustrated using an example patient. Two regions of interest are identified based on side-branches and these are highlighted in the longitudinal views of figure 2. An example cross-correlation of longitudinally co-registered OCT and VH-IVUS images is shown in figure 3. In the far right panel the lumen to radii distances are shown for both images as well as the co-registered OCT image. The solid blue line indicates the angle with the highest correlation with the lumen to radius distance from the VH-IVUS image. The accuracy of the algorithm can be visually noticed in the middle panels.

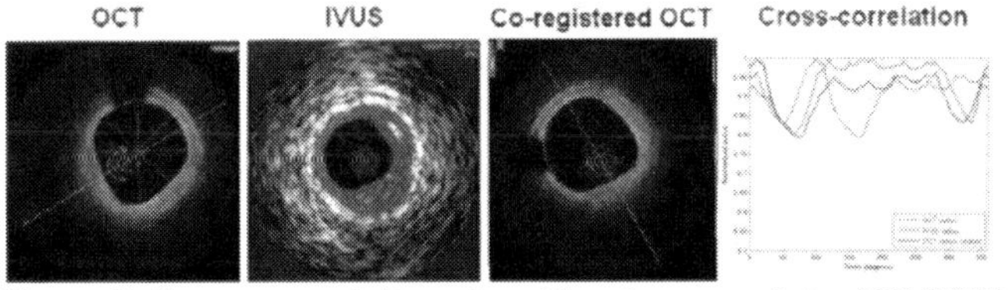

Figure 3. Cross correlation algorithm to co-register VH-IVUS and OCT images

DISCUSSION

A new methodology that co-registers VH-IVUS and OCT images has been developed, which can be used to assess plaque progression and transformation in serial studies. The method has the advantage of fusing these complementary imaging modalities so that evolution of plaque burden, plaque constituents, and the fibrotic cap can be investigated in specific subjects. Further, by combining this methodology with CFD, relationships between plaque variables and hemodynamics can be assessed. This framework is being applied to a cohort of mildly diseased patients (average plaque burden of 33%), which may help further the understanding of the role of hemodynamics in atherosclerotic plaque progression.

ACKNOWLEDGEMENTS

This research was supported by the Georgia Research Alliance (LT, DPG), American Heart Association (DSM), National Institutes of Health (OYH), and Volcano Corporation and Forest Pharmaceuticals (HS).

REFERENCES

[1] Stone, PH et al., *Circulation*, 126:172-181, 2012.
[2] Samady, H et al., *Circulation*, 124:779-788, 2011.
[3] Wang Z, et al., *Proc. SPIE*, 7889, 2011.
[4] Zahnd G, et al., *Information Processing in Computer-Assisted Interventions Lecture Notes in Computer Science*, 8498:78-89, 2014
[5] Timmins, LH et al., *Ann Biomed Eng*, 43:94-106, 2015.

SB³C2015
Summer Biomechanics, Bioengineering and Biotransport Conference
June 17-20, 2015, Snowbird Resort, Utah, USA
SB³C2015-1066

CROSS CAPILLARY MESANGIAL TRANSPORT

Sarah E. Hunt (1), Yoav Segal (2), Kevin D. Dorfman (3), Victor H. Barocas (1)

(1) Biomedical Engineering
University of Minnesota
Minneapolis, MN, USA

(2) Medicine
University of Minnesota
Minneapolis, MN, USA

(3) Chemical Engineering and Materials Science
University of Minnesota
Minneapolis, MN, USA

INTRODUCTION

For fifty years, filtration and transport in the glomerular tuft has been modeled as ocurring in either a single, long cylindrical capillary [1], or in a network of connected capillaries [2], but without transport between adjacent capillary segments (fig. 1). In reality, glomerular capillaries are connected by the mesangium, a tissue composed of smooth-muscle-like cells [3]. Although plasma fluxes through the mesangium are small compared to filtration into the urinary space [3],

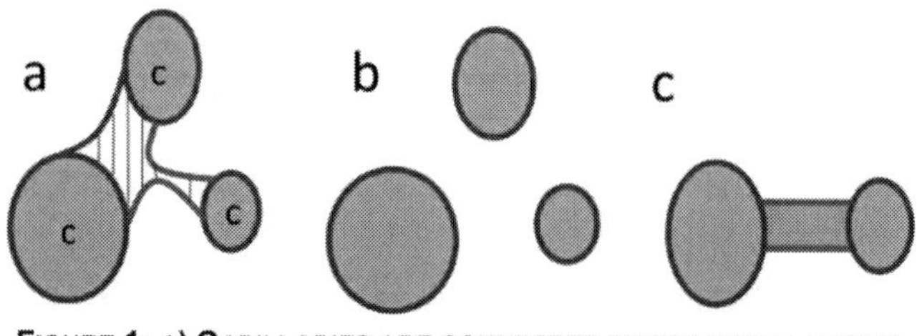

FIGURE 1: A) CAPILLARIES ARE CONNECTED BY MESANGIAL TISSUE (GRAY) B) PREVIOUS WORK MODELED ISOLATED CAPILLARIES C) CURRENT MODEL INCLUDES MESANGIAL INTER-CAPILLARY FLOW

transport of solutes between adjacent capillaries, and/or transport between capillaries at vastly different afferent to efferent locations, could significantly influence the overall flow and profile and solute distribution within the glomerular tuft. In this work, we present a model of mesangial flow between adjacent glomerular capillaries.

METHODS

Our model consists of two cylindrical glomerular capillaries connected by a rectangular mesangium (fig. 1c). Although this is still a considerable simplification of the extremely complex glomerular geometry, it represents an important first exploration of the effect of capillary-capillary communication via transmesangial flow and diffusion. We assumed that the glomerular capillary was well mixed. The mesangial matrix was modeled with Darcy's law, as previously [5], to represent flow through the extracellular filled space. Mesangial cells were assumed to be impermeable to fluid flow.

Pressure profiles along the glomerular capillaries (in the afferent-efferent direction) were set *a priori*, as in [1]. Then the concentration in the glomerular capillary was calculated using the following equation:

$$\frac{dc}{dz} = \frac{2kR}{M} P_{ref} c^2 \frac{P_{UF}}{P_{ref}} \left(\pi - \sin^{-1}\left(\frac{m}{2R} \right) \right) \tag{1}$$

Here, k is the permeability of the glomerular capillary wall, R is the radius of the capillary, M is mass flow rate of protein, and m is the width of the adjacent mesangial matrix. P_{UF} is the ultrafiltration pressure, given by [1]:

$$P_{UF} = \left(p_{cap} - P_T \right) - a_1 c^2 - a_2 c^2 \tag{2}$$

Here, P_T is the hydrostatic pressure in the urinary space, and a_1 and a_2 are coefficients given in [5] to calculate osmotic pressure. Hydrostatic pressure in the urinary space was set to 1.3 mmHg, and osmotic pressure in the urinary space was assumed to be zero. The pressure and concentration profiles thus calculated were used as boundary conditions on flow in the mesangial matrix, to calculate the mesangial matrix pressure and concentration profiles. These values were then used to calculate a flux into the capillaries, and capillary concentrations were adjusted using the following expressions:

$$\frac{d(Qc)}{dz} = h\left(D \frac{\partial c_{mes}}{\partial x} + \frac{\kappa}{\mu} \frac{\partial p_{mes}}{\partial x} c \right) \tag{3}$$

$$\frac{dQ}{dz} = 2h\frac{\kappa}{\mu}\frac{\partial p}{\partial x} - 2kRP_{ref}\frac{P_{UF}}{P_{ref}}\left(\pi - \sin^{-1}\left(\frac{m}{2R}\right)\right) \qquad (4)$$

In these equations κ is the permeability of the mesangial matrix, h is the width of the mesangial flow channel, c_{mes} and p_{mes} are concentrations and pressures in the mesangium, and Q is the volumetric flow rate. This process was iterated until the concentration values converged.

We present results for cocurrent flow in parallel capillaries of different sizes, where the radius of the right hand capillary was held constant at 7 µm, while the radius of the left hand capillary was varied to 4.5, 4, 3.5, and 3 µm. In both capillaries the permeability, pressure drop, viscosity, and length were the same, while the volumetric flow in each was determined by Poiseuille's law.

RESULTS

The small capillary (left) concentrates quickly, driving a concentration gradient into the mesangial matrix (fig. 2). (Figures 2 and 3 show results for a small capillary radius of 3 µm.) The large

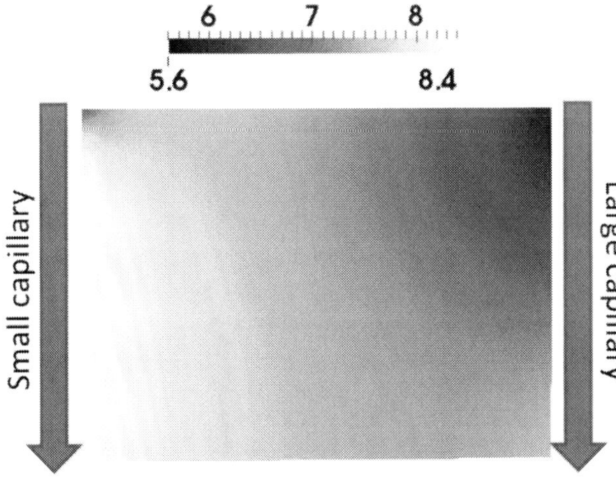

FIGURE 2: ALBUMIN CONCENTRATION IN THE MESANGIAL MATRIX (g/100 mL). BLUE ARROWS – CAPILLARY FLOW DIRECTION

capillary on the right concentrates more slowly because it has a larger volumetric flow and its surface area to volume ratio is lower. The mesangial velocity is very slow compared to capillary velocities (0.26 µm/s vs. 1.37 x 10^4 µm/s), and thus the total flux into the mesangium is small. Comparing the results of our capillary and mesangium model to those of an isolated capillary shows that the osmotic pressure reaches filtration equilibrium in the mesangial model close to when it reaches equilibrium in the model for a capillary alone (fig. 3). For the larger

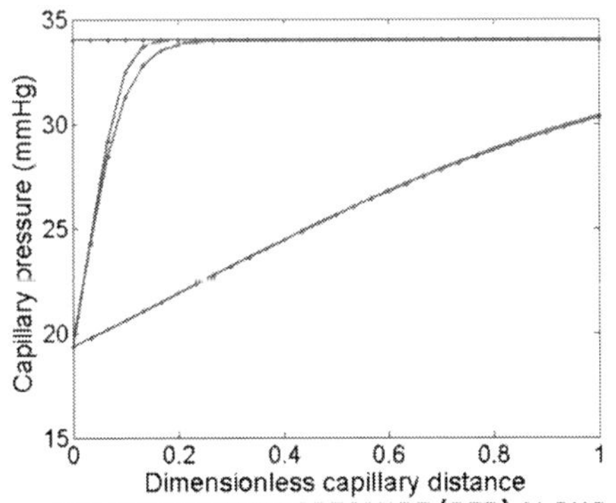

AND HYDROSTATIC PRESSURE (RED) ALONG DIMENSIONLESS CAPILLARY LENGTH. GREEN- WITHOUT MESANGIUM, BLUE – WITH A MESANGIUM. TOP LINES ARE SMALL CAP., LOWER LINES ARE THE LARGE CAPILLARY

capillary the results are essentially identical (lower lines in fig. 3). However, as the size of the left hand capillary becomes much smaller than the right hand capillary, the two models begin to diverge (fig. 4), with capillaries in the mesangial model taking longer to reach filtration equilibrium. Figure 4 shows the ratio of the equilibrium point of an isolated capillary versus that of two capillaries connected with a mesangium. Thus, even small mesangial fluxes can have an effect on overall transport.

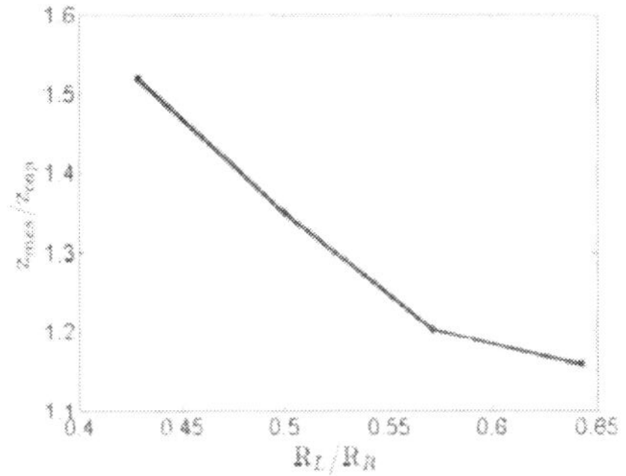

FIGURE 4: AS THE LEFT CAPILLARY, R_L, BECOMES MUCH SMALLER THAN THE RIGHT CAPILLARY, R_R, THE POINT OF FILTRATION EQUILIBRIUM SHIFTS FURTHER DOWN THE CAP. LENGTH

DISCUSSION

Including mesangial transport in calculations of glomerular capillary flow is important because it affects how quickly the capillaries reach filtration equilibrium. The loss of plasma across the capillary wall concentrates glomerular albumin, which increases the plasma osmotic pressure. This increased osmotic pressure eventually counteracts completely the trans-capillary hydrostatic pressure difference that drives filtration. Previous work [1,2] established the location of filtration equilibrium as a function of glomerular capillary permeability, volumetric flow rates, and hydrostatic pressure. Our work suggests that filtration equilibrium may be altered when capillaries of differing size can exchange plasma and ions through the mesangium (figure 4). Specifically, communication between capillaries can lead to shorter equilibration distances. Although the differences here are small, the effects may be magnified when capillaries of different concentration are in close proximity, as demonstrated in the rat [2]. This is especially the case in countercurrent flow, which occurs at least at the entrance to the glomerulus [3].

This model did not account for the effects of capillary size on apparent blood viscosity, and red blood cell flow, or the effects of those factors on concentration. It also ignores the effect of protein concentration on viscosity. These details may be included in future work to provide a more accurate picture of the role of mesangial transport.

ACKNOWLEDGEMENTS

This work was supported by NIH 1F31DK097947 to SEH.

REFERENCES

[1] Deen WM, et al. *Am J Physiol* 223:1178–83, 1972.
[2] Brenner BM, et al. *Am J Physiol* 263:F562–72, 1992.
[3] Elger M, et al. Adv Anat Embryol Cell Biol 139: 1–98, 1998
[4] Hunt SE, et al. In progress
[5] Myers BD, et al. *Circ Res* 36: 425–35, 1975.

SB³C2015
Summer Biomechanics, Bioengineering and Biotransport Conference
June 17-20, 2015,Snowbird Resort, Utah, USA
SB³C2015-1067

THE INFLUENCE OF SUB-OPTIMAL ACQUISITION DELAY IN A C-ARM CONE-BEAM CT PERFUSION STUDY

Antonius M. de Korte, Kajo van der Marel, Juyu Chueh, Olivia W. Brooks, Ajit S. Puri, Ajay K. Wakhloo, Matthew J. Gounis

New England Center for Stroke Research,
Department of Radiology
University of Massachusetts Medical School,
Worcester, MA, USA

INTRODUCTION

Stroke is the leading cause of morbidity and the fourth leading cause of death in the Unites States. Eighty-five percent of all strokes are ischemic, in which an occlusion prevents the tissue downstream from getting blood and nutrients to survive.[1] Up-to-date information on tissue infarct is crucial in the decision-making in acute stroke management and can be done by perfusion studies in CT and MRI.[2] One of these perfusion parameters is cerebral blood volume (CBV), theoretically related to cell function and has the potential ability to distinguish salvageable penumbra from already infarcted brain.[3] CBV is diminished in infarcted and ischemic areas.[4] CBV-Maps can be derived from differences in contrast enhancement between a mask and a contrast filled run. Contrast enhancement is ideally in steady-state throughout the entire acquisition, as CBV equals the area under the attenuation curve. [5]

About one-third of ischemic strokes can potentially be treated by intra-arterial intervention, for example by thrombectomy, in which extracting the embolus restores blood flow.[6] Technical improvements have enabled acquisition of CBV-Maps on C-Arm Cone-Beam CT (CBCT) systems in the intervention suites. The ability to measure CBV peri- and intra-operational can change the workflow of acute ischemic stroke significantly. Patients could be brought directly into the angiosuite where diagnosis and treatment could be given without any time lost. CBV-Maps may enable patient triage for intra-arterial treatment, as recanalization in large infarcts is futile.[7]

Acquisition of CBCT CBV-Maps requires two 200° rotational sweeps of the C-Arm, and steady-state contrast enhancement is required during the entire second acquisition. Optimal timing of the acquisition delay to match the onset of the steady-state can reduce contrast volume and radiation dose, and improve image quality. The onset time of the steady-state phase is affected by patient-specific factors and scanning parameters.[5]

The exact influence of sub-optimal timing of the acquisition delay on image quality is still unknown. Investigating this in stroke patients is unethical because of the contrast and radiation dose exposure, as well as time lost to final treatment. A phantom study gives the ability to investigate the qualitative differences of CBV-Maps with different acquisition delays in a controlled environment.

We therefore evaluated the impact of sub-optimal acquisition delays on CBV-values measured with C-Arm CBCT in a perfusion phantom.

METHODS

Perfusion Phantom

A cylindrical MRI-compatible phantom with a homogenous simulation of brain parenchyma consisting of plastic 3.4mm-spheres was designed in our lab. Phantom dimensions: 7cm x 8cm (*height x diameter*) (Fig 1). One input artery splits into 3 feeding branches into the phantom. Three inlets penetrate into the spheres and feed the parenchyma equally trough small pores. Venous outflow is set-up in a similar manner as arterial inflow.

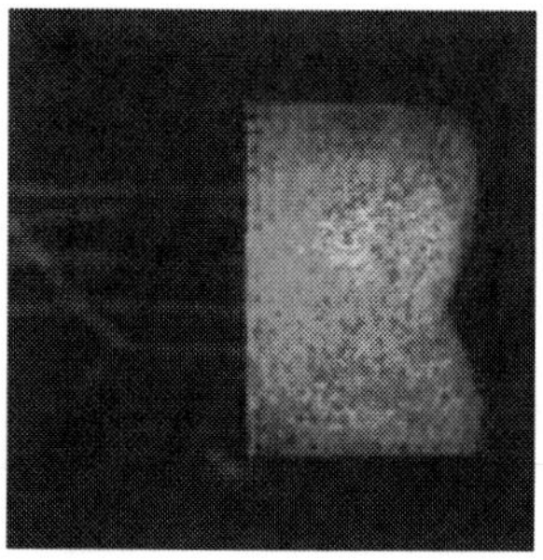

Fig 1. Lateral 2D-Perfusion, Time To Peak

Experimental Setup

The phantom was connected to a peristaltic pump (MasterFlex L/S, Thermo Fisher Scientific, USA). The flow rate was set on 600mL/min. Tap water was used as streaming fluid. A Y-connection into the feeding artery enabled injection of contrast medium (iohexol 350mgI/mL, Omnipaque; GE Healthcare) by a power injector (Mark V ProVis; Medrad, Indianola, PA, USA). Liquid was not recirculated because of the repetitive administration of contrast agent.

Image acquisition

C-Arm CBCT data were obtained on an x-ray angiography C-Arm System. (AlluraXper FD20; Philips Healthcare) Before all acquisitions, a system-specific air calibration of the C-Arm system was performed.

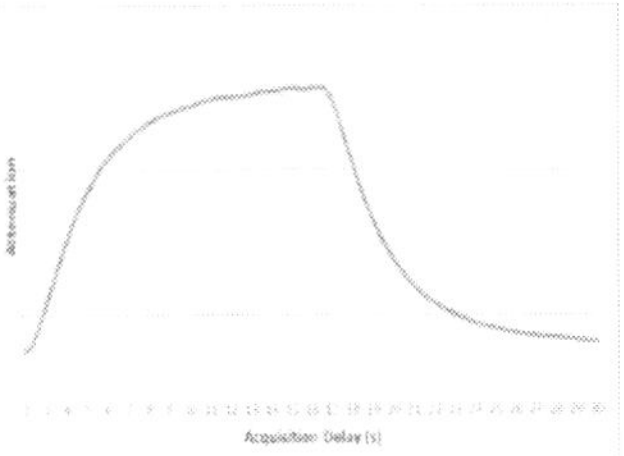

Fig 2. Time Attenuation Curve 2D-Perfusion. 3mL/s 100% contrast

First, 2D-Perfusion acquisitions at 3 frames per second for 15 s were made with different infusion rates to generate time-density curves. Infusion rates were 3 mL/s (50% contrast), 6 mL/s (50% contrast), and 2, 3 and 4 mL/s (100% contrast).

The time density curve with the best plateau phase was chosen and time to steady-state was determined (3mL/s 100% contrast).

Subsequently, a CBCT acquisition with soft tissue contrast was used to collect data for CBV-Maps. Each CBV-Map was calculated from two acquisitions. The first acquisition was obtained without contrast. The second acquisition was acquired with administration of contrast with an identical protocol as for the 2D-Perfusion. C-Arm CBCT images (FOV = 251.5 × 194.5 × 251.5 mm^3, voxel size = 0.34 mm^3) were reconstructed from 620 x-ray projections generated during a 10.4-second rotational motion of the x-ray source over approximately 200°. Acquisition delays of 3, 4, 5, 6, 7, 10, and 15 s were chosen to evaluate the influence of timing on CBV measurements. With a maximum contrast injection duration of 20 s, all acquisitions but the 15 s delay had a continuous injection of contrast during the entire run.

Image processing and data analysis

2D-Perfusion images were analyzed on a Philips XtraVision Workstation. CBV values were calculated for all voxels within a small Region of Interest (ROI) manually drawn with ITK-SNAP (Version 3.2.0). Spheres in this ROI were segmented manually on a high resolution reconstruction of the first non-contrast acquisition, and its volume was subtracted from the ROI volume to obtain the ground-truth value for CBV measurements. CBV-Maps were calculated for all different acquisition delays. Arterial Input was obtained from ROIs on two arterial inlets. CBV is expressed in mL per 100mL of phantom volume.

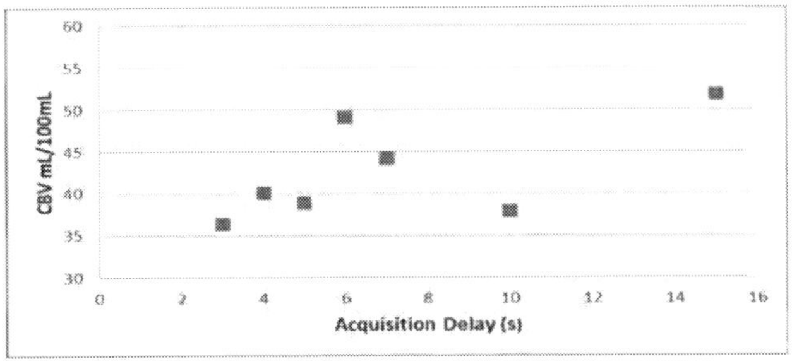

Fig 3. CBV (mL/100mL) in ROI with different acquisition delays.

RESULTS

The acquisition delay (AD) of 10 s gave the longest delay to reach steady-state. The standard deviation of the attenuation measured with 2D-Perfusion during the entire contrast filled acquisition with a 10 s acquisition delay was less than 0.01%.

The ground-truth CBV in the selected ROI was 44.88 mL/100 mL. Acquisition of CBV-Maps with an acquisition delay of 7 s produced the smallest deviation from the manually determined volume, with only a 1.18% difference (44.19 mL/100mL).

Calculated CBV in steady-state (acquisition delay = 10 s) in the ROI was 37.80 mL/100mL (-14.47%). Decreasing acquisition delays to as early as 3 s resulted in an average CBV of 41.72 ± 5.10 mL/100mL (±7.05%). A longer acquisition delay of 15 s resulted in a CBV of 51.74 mL/100mL (+14.58%). (Fig 4.)

DISCUSSION

Preliminary results of this perfusion phantom study confirmed that CBV values vary with acquisition delay. Sub-optimal timing seems to result in lower CBV values, but there also seems to be a variation in CBV even when steady-state is reached, which cannot be explained by sub-optimal timing.

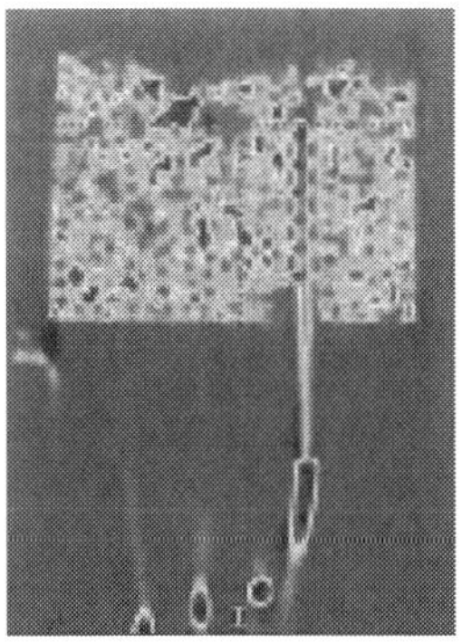

Fig 4. CBV-Map Perfusion Phantom

These are the first results with this perfusion model, and improvements will be made for further studies. Redesigning the in- and outflow structure could reduce potential recirculation and stasis of fluid. This also could increase the clearance rate of contrast, which helps reducing the circulating volume per minute so that a more physiologic situation can be created. There was some trapping of air in the phantom, which made measurements more difficult. It also prevented quantitative measurements for comparison by MRI.

However, this model gives us the ability to simulate basic brain perfusion and study the different factors that play a role in C-Arm CBCT CBV measurements. This model is the first step to evaluate the impact of acquisition delay on CBV-Map quality. The next step is to design a model that represents the heterogeneity of the brain during stroke so that we can evaluate the effect of sub-optimal timing of acquisition delay on CBV-Map quality in this model as well. This hopefully brings us closer to quantitative CBV measurements in the angiosuite.

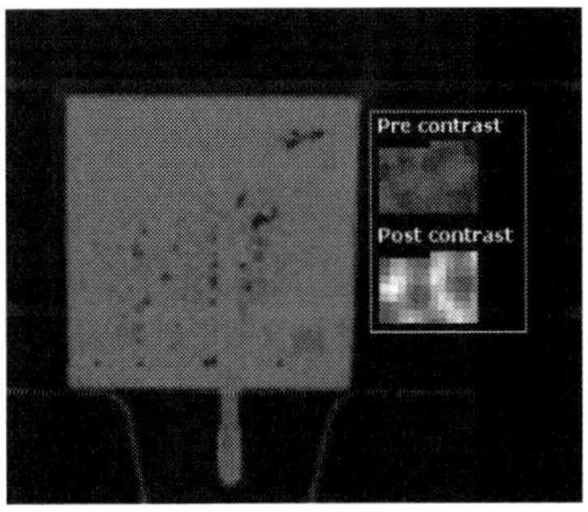

Fig 5. XperCT pre-contrast run with ROI.

ACKNOWLEDGEMENTS

This study is supported in part by a generous grant from Philips Healthcare.

REFERENCES

[1] Go, A.S., et al., *Circulation*, 127:e6–245, 2013.
[2] Schaefer, P.W., et al., *Stroke*, 39:2986–92, 2008.
[3] Srinivasan, A., et al., *Radiographics*, 26 Suppl 1, 2006.
[4] Struffert, T., et al., *AJNR*, 33:618-25, 2012.
[5] Caroff, J et al., *Neuroradiology*, 56:245–49, 2014.
[6] Heldner, M.R., et al., *Stroke*, 44:1153–57, 2013.
[7] Spiotta, A.J., et al., *JNIS*, 6:511-16, 2013.

SB[3]C2015
Summer Biomechanics, Bioengineering and Biotransport Conference
June 17-20, 2015,Snowbird Resort, Utah, USA
SB[3]C2015-1068

Computational modeling of a percutaneous transvenous mitral annuloplasty device deployment into the coronary sinus vessel for treatment of mitral regurgitation: analysis of anchor sizes

Thuy M. Pham (1), Fanwei Kong (1), Milton Deherrera (2), Wei Sun (1)

[1]WHC Department of Biomedical Engineering
Georgia Institute of Technology/ Emory University
Atlanta, GA 30313313-2412

[2]Advanced Materials/ Technology Group
Edwards Lifesciences, Irvine, CA

INTRODUCTION

In functional mitral regurgitation (MR), the mitral annulus (MA) becomes dilated due to left ventricular dysfunction thus induces regurgitation. Functional MR represents a major clinical challenge: often poorly recognized and undertreated and is associated with high mortality. Recently, minimally invasive percutaneous transvenous mitral annuloplasty (PTMA) approaches using entirely catheter-based methods have been developed to treat patients who are not suitable candidates for open-heart surgery. One of the approaches is to utilize the proximal location of the coronary sinus (CS) to the MA to percutaneously deploy a self-expandable PTMA device into CS vessel. The device consists of three components: the proximal and distal anchors and 'spring-like' bridge (Fig. 1a). The distal anchor is first deployed into the great cardiac vein (GCV) and the proximal anchor is expanded at the CS ostium. The bridge held in its open position by the biodegradable suture material (BDM), which dissolves after 1 to 2 months, allowing slow and steady shortening of the device as well as allowing the two anchors to be firmly fixed after tissue ingrowth. When the bridge section totally contracts, it indirectly reshapes the MA and decreases MR.

Although the approach has shown to be feasible in animal [1, 2], device fracture was reported in human clinical trials [3, 4]. The mechanism of failure is unknown. It is hypothesized that the biomechanical interaction between surrounding tissues and the stent plays a critical role in the outcome of the deployment and the device performance. For instance, excessive oversize of the device size (diameter) may result in CS vessel rupture, while insufficient device oversize may lead to device migration and recurrence of MR. However, there is no method to predict the outcome of PTMA device. In this study, we develop a methodology to quantitatively predict the outcome of PTMA device after implantation. Our initial approach is to determine the appropriate anchor's size that is not only without imposing high stress on vessel, but also generate adequate radial force to resist stent dislodgement in a MR patient. The optimal device size was determined based on the biomechanical interaction between the PTMA device and CS vessel—contact forces, peak stress strain on CS vessel, and stent fatigue life.

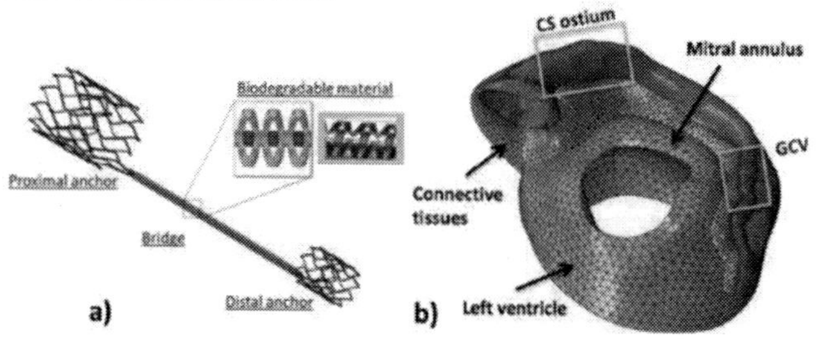

Figure 1 – a) PTMA device consists of proximal, distal anchors and bridge with biodegradable material; b) Finite Element models consists of coronary sinus (CS) and great cardiac vein (GCV), mitral annulus, connective tissues and partial left ventricle.

METHODS

Stent model. Typically, the outer diameter (OD) of the proximal and distal anchors should be 2 – 4 mm oversizing of the CS and GCV diameters. The CS and GCV have diameters of 9.0 and 4.3 mm, respectively. From the patient CS/GCV geometries, 5 proximal and 4 distal anchors of difference OD were created: proximal OD of 10, 11, 12, 14 and 15 mm; distal OD of 5, 6, 7, and 8 mm. Optimal design is achieved without imposing peak stress over tissue ultimate tensile strength and shear force not exceeding radial force. The Nitinol properties of stent is described as: E_A = 51.7MPa, ν_A = 0.3, E_M = 47.8MPa, ν_A = 0.3, εL = 0.063, σ_L^S= 600 MPa, σ_U^E= 670MPa, σ_U^S =

288MPa, σ_L^E= 254MPa, T_0 = 37°. The proximal and distal anchors were modeled using 40,077 and 21,635 C3D8I elements, respectively.

Patient model and material properties. The CT images of a patient with known MR were collected from the Hartford Hospital (Hartford, CT). The CS (7,786 C3D8I elements; 0.74 mm thickness at CS ostium and 0.36 mm thickness at GCV), partial left ventricle (LV, 15,812 C3D4 elements), and connective tissues (CT, 53,183 C3D4 elements) were reconstructed into 3D FE model using Avizo and Hypermesh (Fig. 1b). The material properties of tissues were obtained from biaxial testing of human cadaver tissues and fitted to the modified anisotropic hyperelastic Holzapfel-Gasser-Ogden material model [5] (Fig. 2): CS: ρ = 1.1 g/cm^3, C01 = 14.407, C10 = 0.137, k1 = 6.681 kPa, k2 = 44.076, κ = 0.066, θ = 11.394 degree; LV: ρ = 1.1 g/cm^3, C01 = 0.037, C10 = 0.144, k1 = 6.080 kPa, k2 = 98.367, κ = 15.388, θ = 6.79 degree; Tissue: ρ = 0.9 g/cm^3, E = 1.6 MPa, ν = 0.499. The ultimate tensile strength of the CS tissue is 0.92 $\pm$ 0.37 MPa (min=0.5MPa, max=1.4MPa) obtained from uniaxial tensile tests of CS (n = 7).

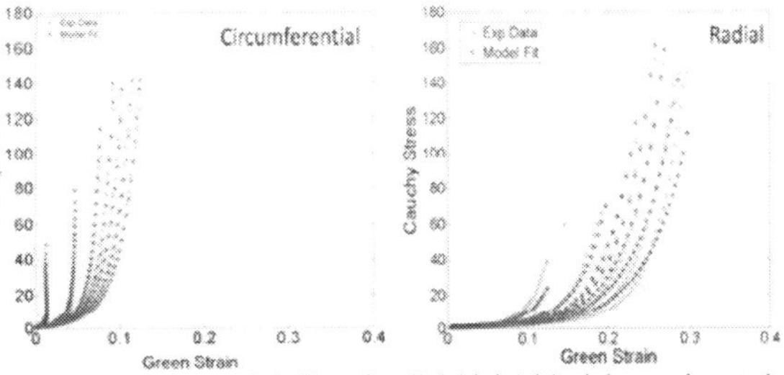

Figure 2 – Circumferential (left) and radial (right) biaxial experiment data (white circles) of human CS vessel (82 years old male) fitted to the modified Holzapfel-Gasser-Ogden material model (red circles).

FEA simulation and analysis. The simulation consists of three steps: Step 1 involves crimping of the anchor, step 2 is to release the stent as it comes in contact with the vessel, and step 3 includes inflation of the vessel at 20 mmHg physiological pressure. Each anchor was crimped to 3 mm OD using a cylindrical crimper model (M3D4R; 23,275 elements) with no friction. The friction between anchors and CS is assumed to be 0.1 in the following steps. The radial and shear forces of are the sum of the forces from all stent nodes that are in contact with the CS inner wall (*cstress,*cforce). Stent fatigue was analyzed using the Goodman-Haigh diagram: the maximum, ε_{max}, and minimum ε_{min}, strain values at all nodes on stent were obtained, where the mean strain, ε_{mean}= (ε_{max}+ε_{min})/2 and half-amplitude strain, $\Delta\varepsilon$ = (ε_{max}-ε_{min})/2 were calculated at 0 and 20 pressure levels; all points were plotted and compared with constant life curves [6]. The stent fatigue safety factor can be predicted using the equation: SF = 0.4%/$\Delta\varepsilon_{max}$. To analyze anchor migration, a simulation was performed on the entire PTMA device with the bridge segment. The bridge (73,013 C3D8R elements, Fig. 1a) was pulled 20 mm in the axial direction so that the entire device length is equivalent to the CS implantation length. The force used to stretch the bridge is the stent retraction force or the force exerting on the CS and transmitting to the MA once the BDM is dissolved, drawing both anchors together. This force was compared to the total radial forces of both anchors to determine the risk of anchor dislodgement. Each simulation was run with ABAQUS/Explicit v6.13 on a Linux cluster using 32 2.8GHz CPUs for 5 hours.

RESULTS

Fig. 3 shows the strain distribution of CS ostium and peak values after step 2. Given the same strut thickness, stent expansion decreased from PROX 13 and DIST 8 as the result of reduction in CS strain and expansion. Tables 1 tabulates the peak stress and stress, total normal and shear contact forces of CS ostium and GCV. Peak stress at CS ostium is below the UTS at DIST 3 while stress on GCV is approaching UTS at size 1. Radial force increases with increase in anchor size, however, vessel stress approached UTS at higher radial force. The shear forces are well below the normal forces for most sizes in all steps. From these values, it is determined that PROX 12 and DIST 6 are the optimal designs (Table 1 in yellow boxes). The safety factor values for the optimal designs are 13.33 and 5.00, which are higher than other designs. The reaction force used to stretch the bridge was 12.70 N.

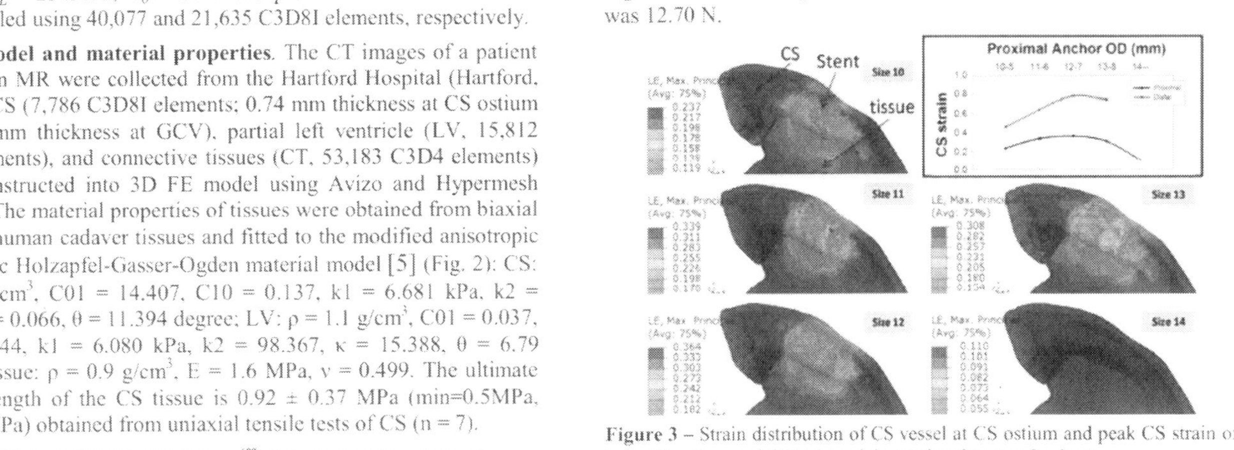

Figure 3 – Strain distribution of CS vessel at CS ostium and peak CS strain of both CS ostium and GCV (top-right graph) after step-2 release.

Table 1. Proximal and distal anchor simulation results.

	Step1: Stent Release					Step2: CS vessel inflation					
Proximal Anchor	Stress (MPa)	Strain	Shear Force(N)	Normal Force (N)	Shear/Normal Force	Stress (MPa)	Strain	Shear Force(N)	Normal Force(N)	Normal Force(N)	SF
10	0.59	0.31	0.02	0.24	0.09	0.61	0.31	0.02	0.25	0.10	0.80
11	1.16	0.35	0.16	1.69	0.10	0.75	0.36	0.11	1.12	0.10	2.00
12	1.81	0.45	0.42	4.86	0.09	1.27	0.42	0.33	3.59	0.09	13.33
13	2.42	0.33	0.63	6.62	0.10	1.67	0.30	0.40	4.98	0.08	1.33
	Step1: Stent Release					Step2: CS vessel inflation					
Distal Anchor	Stress (MPa)	Strain	Shear Force(N)	Normal Force (N)	Shear/Normal Force	Stress (MPa)	Strain	Shear Force(N)	Normal Force(N)	Normal Force(N)	SF
5	0.63	0.46	0.01	0.07	0.09	0.68	0.50	0.00	0.05	0.09	1.21
6	1.54	0.65	0.03	0.37	0.09	1.43	0.65	0.03	0.31	0.08	5.00
7	2.77	0.78	0.11	1.19	0.09	2.09	0.75	0.06	0.66	0.09	2.86
8	3.29	0.75	0.61	1.98	0.31	2.21	0.73	0.08	0.97	0.09	4.00

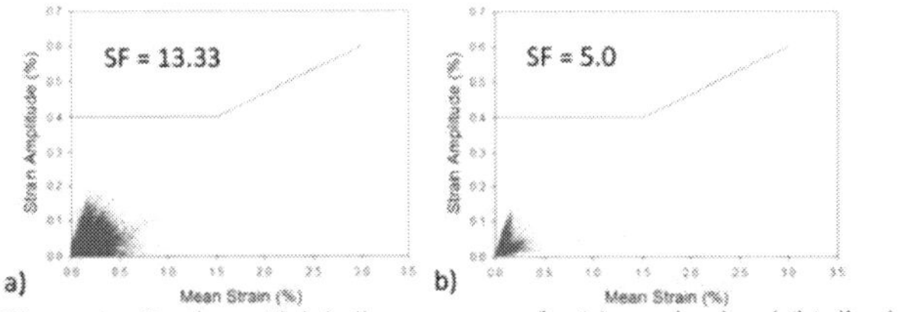

Figure 4 – Goodman-Haigh diagram curves for (a) proximal and (b) distal anchors.

DISCUSSION

Current problems in the development of PTMA device are the stent migration, fracture and coronary sinus perforation. This study evaluated optimal PTMA anchor size given the vessel stress and contact forces. The radial forces of both anchors should be equal or greater than the stent retraction force in order to prevent device dislodgement. However, the sum of radial forces for optimal designs is below the stent retraction force (5.23 vs. 12.70N). Increasing the anchor OD can improve radial force, however, increase the risk of CS perforation from high vessel stress. This difference in force, however, can be compensated by tissue-ingrowth—enables by slow absorption of BDM up to 2 months. This study represents, to our knowledge, the first computational study of the PTMA intervention using realistic patient-specific geometry and human tissue properties to enhance understanding of stent-tissue interaction. Future works include simulation of the entire stent deployment into the CS vessel and determine the outcome of PTMA deployment from MA reduction area.

ACKNOWLEDGEMENTS

This work was supported in part NIH HL104080 and HL108240 grants and a NIH NRSA Individual pre-doctoral fellowship No. HL097722.

REFERENCES

[1] Duffy, S. J., et al., Catheter Cardiovasc Interv., 68(2), pp. 205-210.
[2] Dubreuil, O., et al., Catheter Cardiovasc Interv., 69(7), pp. 1053-1061.
[3] Webb, J. G., et al., Circulation., 113(6), pp. 851-855. Epub 2006 Feb 2006.
[4] Harnek, J., et al., JACC: Cardiovascular Interventions, 4(1), pp. 115-122.
[5] Gasser, T. C., et al., J R Soc Interface, 3(6), pp. 15-35.
[6] Cavanaugh Jr., KJ, et al., Journal of ASTM International, 3(6), pp. 8-18.

COMPUTATIONAL HEMODYNAMIC ASSESSMENT OF A NOVEL MODULAR ANASTOMOTIC VALVE DEVICE FOR HEMODIALYSIS VASCULAR ACCESS

Andrew McNally (1), A. George Akingba (2), Philippe Sucosky (1)

(1) Department of Aerospace & Mechanical Engineering
University of Notre Dame
Notre Dame, IN, USA

(2) Department of Surgery
Indiana University School of Medicine
Indianapolis, IN, USA

INTRODUCTION

End-stage renal disease (ESRD) results from conditions that chronically decrease renal function and is most commonly treated with hemodialysis (HD) while patients await kidney transplant. Current methods for maintenance of HD vascular access are fraught with problems that compromise patency of these surgically created shunts. The major cause of shunt failure is thrombosis caused by occlusion of the outflow venous anastomosis and draining vein. Intimal hyperplasia (IH), i.e., a thickening of the vessel wall, initiates the pathogenesis of shunt stenosis/thrombosis and has been associated with recirculating flows and low wall shear stress (WSS), preferentially in wall regions where WSS < 0.5 Pa[1]. Therefore, one way to improve vascular access patency is through the preservation of normal venous hemodynamics in HD patients.

In pursuit of this goal, a novel vascular access technology based on a modular anastomotic valve device was developed. The device aims at isolating the graft from the systemic circulation between dialysis periods, while enabling normal AVG vascular access during dialysis periods. The device can be sewn onto the native vessels and hemostatically coupled to conventional graft material, while the modular design allows the surgeon to achieve a custom fit in each patient (Fig. 1). The MAVD features a shuttle valve system which can alternate between the open and closed positions in order to block or allow AVG access to systemic blood flow. The goal of this study was to quantify the hemodynamics of the MAVD mounted in the venous position during and between hemodialysis periods.

METHODS

3D model construction: Two geometries were created by virtually implanting the MAVD in both closed and open positions in a normal vein geometry replicating inter- and intra-dialysis periods,

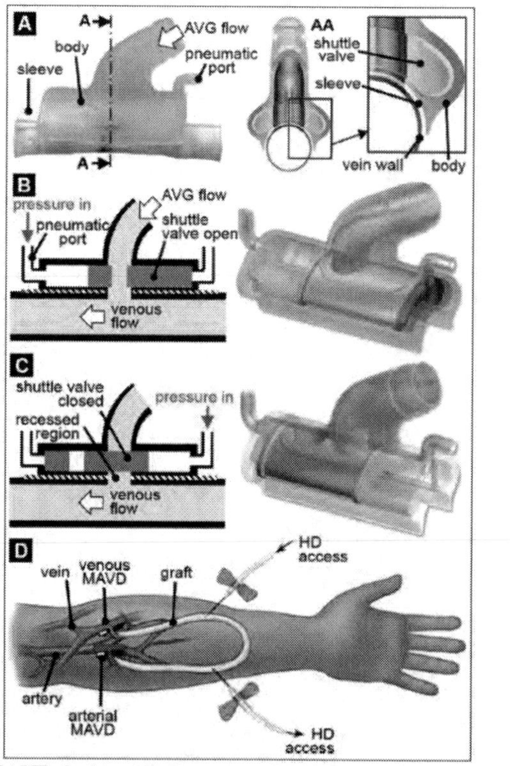

Fig. 1: MAVD design. (A) Prototype MAVD, (B) schematic of operation in open and (C) closed positions, and (D) clinical implementation.

respectively. The virtually implanted MAVD uses an anastomotic angle of 30-degrees, which has been shown to result in less abnormal hemodynamics[2]. A third geometry consisting of a native vein model without a shunt was also generated to provide a reference hemodynamic state. The vein wall was treated as an incompressible, isotropic and hyperelastic material using a previously published 5-parameter, third-order Mooney-Rivlin model calibrated with respect to biaxial vein tissue data[3]. Blood was modeled as an incompressible non-Newtonian fluid using a modified power-law model[2].

Boundary conditions: Three simulations where performed by imposing realistic flow boundary conditions at the inlets of each of the three models (i.e., open MAVD, closed MAVD and native vein). The flows through the closed MAVD configuration and the vein model were simulated using a steady venous flow rate of 320 mL/min. The unsteady flow in the open MAVD configuration was modeled by prescribing a steady flow rate of 320 mL/min at the vein inlet and a pulsatile velocity waveform at the graft inlet with a mean flow rate of 1000 mL/min, obtained from previously published graft access flow measurements[4]. The cardiac cycle simulated a physiologic heart rate of 76 beats per minute.

FSI modeling approach: The 3D fluid-structure interaction (FSI) models were created using commercial software (ANSYS Inc.). The governing equations were solved in their arbitrary Lagrangian Eulerian (ALE) forms and consisted of the flow continuity and momentum equations, the aortic wall momentum equation and three coupling conditions enforcing continuity of displacements, continuity of velocities and compatibility of traction at the fluid-structure interface.

Hemodynamic endpoints: The venous hemodynamics were assessed in terms of streamlines, velocity, WSS and pressure. The potential existence of a secondary flow in the closed MAVD model was investigated by characterizing the flow in five equidistant transverse cross-sections. Low WSS (LWSS) Regions on the vein wall where WSS fell below 0.5 Pa were identified. These regions were further characterized in terms of temporal shear magnitude (TSM), and oscillatory shear index (OSI) a measure for recirculating flow, and RRT, a robust indicator of disturbed flow.

RESULTS

The streamline, velocity and WSS fields predicted in the normal vein and closed MAVD models are shown in Fig. 2. The flow patterns captured in the distal sections indicate progressive attenuation of

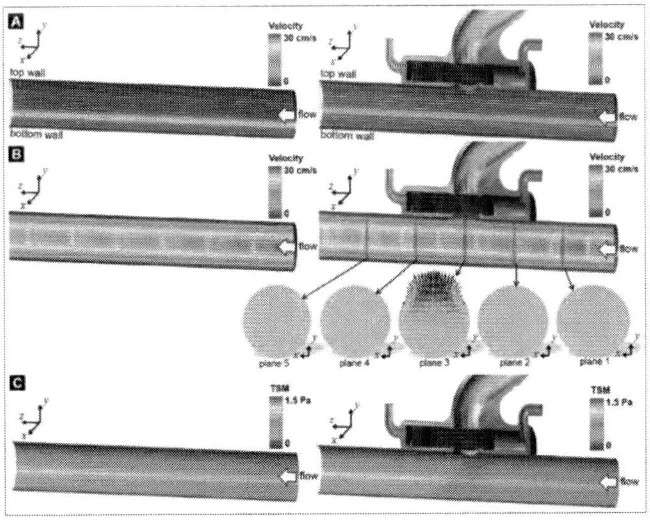

Fig. 3: Native vein and Closed MAVD (A) Streamlines, (B) velocity fields on symmetry and 5 cross sections, (C) TSM fields.

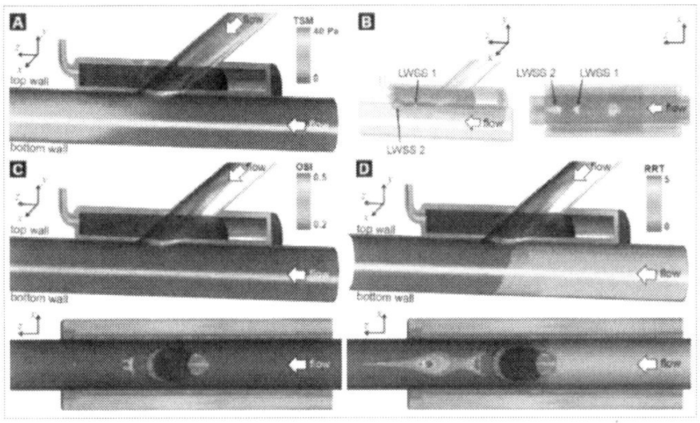

Fig. 2: Open MAVD venous WSS metrics; (A) TSM field, (B) Low WSS regions, (C) OSI field, and (D) RRT field.

secondary flow and recovery of flow alignment along the vein axis. The comparison of the flow in the vein and the closed MAVD reveals a difference of only 1.1% in maximum longitudinal velocity magnitude and 0.3% in pressure drop across the vein.

WSS predictions within the open MAVD are shown in Fig. 3. The analysis of the longitudinal WSS indicates the existence of two venous LWSS regions in the wake of the recessed region. The analysis of the longitudinal WSS metrics captured in LWSS regions 1 and 2 indicates the mild bi-directionality of the flow in those regions (OSI = 0.21 and 0.07, respectively), and fluid residence times (RRT = 2.47 and 2.89, respectively) similar to current vascular accesses.

DISCUSSION

This study suggests the capability of the MAVD to isolate the graft from the systemic circulation and to normalize flow at the vein-graft anastomosis between hemodialysis periods. The maintenance of normal WSS levels on the venous wall is expected to preserve venous homeostasis and potentially improve vascular access patency. These results demonstrate an improvement over and confirmation of previous studies by the incorporation of venous compliance.

Ongoing work includes the development of a bench-top vascular access loop to experimentally validate the simulated hemodynamics, and to provide an apparatus for the sterile conditioning of porcine vascular tissue to the WSS environment of the MAVD and current vascular access methods. These future studies will confirm the results shown here, and provide a first link to the biologic implications of this novel vascular access device.

ACKNOWLEDGEMENTS

This project was funded with support from the Indiana Clinical and Translational Sciences Institute funded, in part, by Grant Number UL1TR001108 from the National Institutes of Health, National Center for Advancing Translational Sciences, Clinical and Translational Sciences Award.

REFERENCES

*[1] Akoh, J. A. Prosthetic arteriovenous grafts for hemodialysis. J. Vasc. Access **10**, 137–47 (2009).*
[2] McNally, A. et al. Novel modular anastomotic valve device for hemodialysis vascular access: preliminary computational hemodynamic assessment. J. Vasc. Access (2014).
*[3] Rezakhaniha, R. et al. A structural model of the venous wall considering elastin anisotropy. J. Biomech. Eng. **130**, 31017 (2008).*
*[4] Tricht, I. Van et al. Experimental analysis of the hemodynamics in punctured vascular access grafts. ASAIO J. **51**, 352–9 (2005).*

SB[3]C2015
Summer Biomechanics, Bioengineering and Biotransport Conference
June 17-20, 2015,Snowbird Resort,Utah, USA
SB³C2015-1070

DEVELOPMENT OF A FINITE ELEMENT MODEL OF THE PEDIATRIC OCCIPITO-ATLANTOAXIAL COMPLEX FOR STUDYING OSODONTOIDEUM AND ATLANTO-OCCIPITAL DISLOCATION

Rinchen Phuntsok (1, 2), Marcus D. Mazur (3), Vijay M. Ravindra (3), Douglas L. Brockmeyer (3), Benjamin J. Ellis (1, 2)

(1) Department of Bioengineering
University of Utah
Salt Lake City, Utah, USA

(2) Scientific Computing and Imaging
University of Utah
Salt Lake City, Utah, USA

(3) Department of Neurosurgery
University of Utah
Salt Lake City, Utah, USA

INTRODUCTION

The pediatric occipito-atlantoaxial complex (C0-C1-C2) is a primary location of pediatric cervical spine injuries. It is estimated that 60 – 80% of spinal injuries in the pediatric population occur in this region [1]. The presence of distinct age related anatomical features, such as more compliant ligaments compared to adults, as well as ossification centers and cartilage in the vertebrae predispose children to injury [2]. Common pediatric pathological conditions include osodontoideum and atlanto-occipital dislocation [3, 4]. Due to the lack of pediatric cadavers, there have been very few experimental studies on the pediatric cervical spine and there have been no direct measurements of pediatric cervical spine ligament material properties reported in the literature. Experimental studies by Ouyang et al. and Luck et al. have provided data limited to tension and flexion-extension rotation of the cervical spine [5, 6]. Finite element (FE) modeling can be used to gain a better understanding of the biomechanics and ligament properties of the pediatric occipito-atlantoaxial complex. The one pediatric cervical spine FE model reported in the literature is a dynamic (explicit) model motivated by the need to study car crash injuries [7]. The experimental and computational studies, to date, do not address the effects of injuries, pediatric craniocervical diseases or the surgical interventions being used to address those conditions. Furthermore, previous studies have not clearly elucidated how much more compliant pediatric cervical spine ligaments are compared to adult ligaments. The overall objective of this research is to develop a FE model of the pediatric occipito-atlantoaxial complex that can be used to simulate pathological conditions, including osodontoideum and atlanto-occipital dislocation. The specific goal of this study was to determine the relative decrease in stiffness of published adult ligament material properties in order for the pediatric model flexion/extension range of motion (ROM) to match published experimental data.

METHODS

A subject with a radiographically normal cervical spine was selected for this study (IRB#00066287).The patient was a healthy 13 month old, 9 kg female, who had a diagnostic computer tomography (CT) angiography study for a soft tissue injury to her neck, which revealed no evidence of bony, ligamentous, or vascular injury. Volumetric image data was acquired using CT with a field of view encompassing 240 mm, 512 x 512 acquisition matrix and 0.5 mm slice thickness. CT data were segmented (Amira, Visage Imaging, CA) and cross sectional contours of the bones, cartilage, and ligaments were extracted. Polygonal surfaces were generated and smoothed. Polygons composing the surfaces of the boney portions of C0, C1, C2 and C3were converted directly to shell elements and represented as rigid bodies. The transverse atlantal ligament (TAL), tectorial membrane (TM), andcartilage surfaces, including ossification centers and

Table 1. Adult ligament stiffness (N/mm) and the scaled values used in the pediatric model [8].

Ligament	Adult	50%	10%
Apical	35	17.5	3.5
Alar atlantal and occipital	17.8	8.9	1.78
Anterior Atlanto Occipital	20	10	2
Posterior Atlanto Occipital	7.3	3.65	0.73
Anterior Longitudinal	38.5	19.25	3.85
Facet Capsule	29.4	14.7	2.94
Posterior atlanto-axial	4.2	2.1	0.42

cartilage in the vertebrae, were imported into ANSA (BETA CAE Systems USA, Inc.) and meshed with tetrahedral elements. A mesh convergence study was conducted during which the element edge lengths were reduced from 0.8 mm to 0.3 mm. This resulted in an increase in flexion/extension ROM of only ~1%. All other ligaments were represented with tension-only discrete elements (Table 1).For the upper cervical facet joints, closely-spaced discrete elements (54 at C0-C1 and 43 at C1-C2, per side) were used to represent the facet capsular ligaments[8].

Cartilage was modeled as an isotropic hyperelastic Mooney-Rivlin material, and the tectorial membrane (TM) and transverse atlantal ligament (TAL) were modeled as neo-Hookean materials (Table 2). Pediatric ligament properties were scaled from adult

Table 2. Soft tissue material properties used in the pediatric finite element model.

Soft Tissue	Adult Modulus (MPa)	Model Modulus (MPa)	Poisson's ratio
TAL	8.3-24.7	10	0.4 [8]
TM	4.3-12.5	5	0.4 [8]
Cartilage	10	5	0.4 [9]

properties reported in the literature [8]. Adult ligament properties were reduced to 50% and 10%.C0 and C1 were free of constraints while C2 and C3 were fixed. The sliding-tension-compression contact in FEBio was defined for the articulating cartilages, while sliding_with_gaps contact interfaces were defined between tectorial membrane and transverse ligament, and between transverse ligament and the odontoid. Both of these contact algorithms use the penalty method for enforcement. The model was subjected to ±0.1 Nm in flexion/extension, axial rotation and lateral bending. Models were preprocessed using PreView, analyzed using FEBio and post-processed using PostView [10]. C0-C2 rotations were measured for all loading conditions. The averages and standard deviations reported by Luck from pediatric flexion-extension experimental data for 12 pediatric cadavers ranging between 5-216 months was used to evaluate the flexion-extension predictions from the models[6].

RESULTS

At 10% of adult ligament stiffness the model predicted a range of motion (ROM) within one standard deviation of published [6] pediatric ROM data for flexion-extension (Fig. 1). The model behaved relatively stiffer in extension compared to flexion. Axial rotation and lateral bending (Fig. 2) results show a symmetric ROM, however there are currently no published pediatric experimental data for comparison.

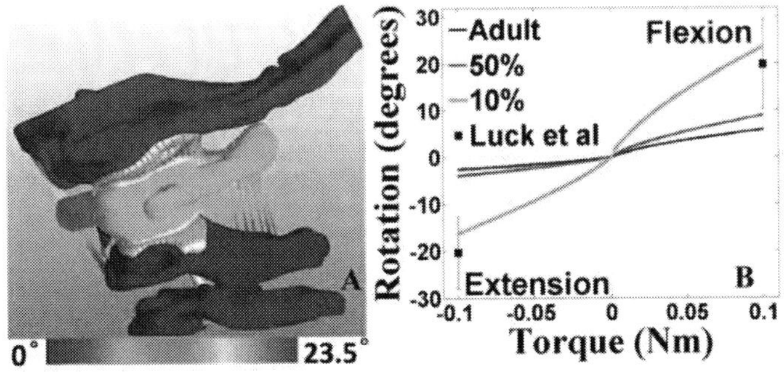

Figure 1: (A) FE model of Flexion at 10% ligament stiffness. (B) Flexion-extension ROM with decreasing ligament stiffness.

The model predicts a relatively stiffer range of motion in both axial rotation and lateral bending compared to flexion-extension. As expected, the models for each ROM showed increased laxity with the decrease in ligament stiffness.

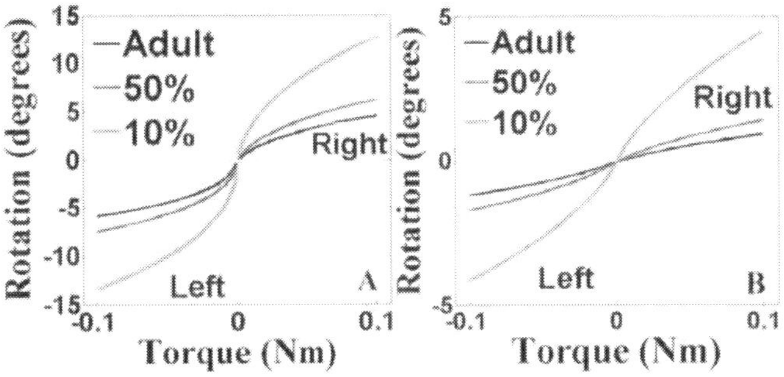

Figure 2: (A) Axial rotation and (B) lateral bending ROM with decreasing ligament stiffness.

DISCUSSION

The pediatric occipito-atlantoaxial FE model developed in this study predicted an accurate flexion-extension ROM when the ligament material properties were reduced to 10% of the adult ligament properties. This gives a reasonable prediction of pediatric cervical spine ligament stiffness and the relationship between flexion-extension ROM and ligament stiffness. Ouyang et al. also performed bending tests on specimen ranging between 2 to 12 years old [5]. Although their data is harder to interpret at the smaller torques, it appears that they reported ~ 10° ROM in flexion/extension for a ±0.1 Nm torque applied to a 2 year old specimen. The flexion-extension ROM predictions from our model were significantly higher than those reported by Ouyang et al., which is not unexpected given the large range of flexion-extension ROMs reported in both experimental studies for the pediatric population. The results from out model are corroborated by the only other pediatric FE model reported in the literature [7] The model developed by Dong et al had a ROM for flexion-extension of ~35-40° which is close to the ROM (~40°) of our model. In the current study we also simulated lateral bending and axial rotation. Although there are currently no other experimental or computational results to evaluate those predictions, we do have some confidence in those predictions because they are symmetric and reasonable compared to the differences seen between pediatric and adult flexion-extension ROMs. In the future, higher torques will be applied to the model and disease specific models of the pediatric occipito-atlantoaxial complex will be created.

ACKNOWLEDGEMENTS

The Marion L. Walker, M.D. Endowment is gratefully acknowledged for funding this research.

REFERENCES

[1] Cirak, B. et al.,*J Pediatr Surg*,39: 607-612, 2004.[2] Huelke, D. et al.,*SAE Technical Paper*, 1992.[3] Klimo, P., Jr. et al.,*Neurosurg Clin N Am*,18: 463-478, 2007.[4] Klimo, P., Jr. et al.,*Neurosurg Clin N Am*,18: 599-620, 2007.[5] Ouyang, J. et al.,*Spine (Phila Pa 1976)*,30: E716-723, 2005.[6] Luck, J.F., PhD Thesis, Duke University, 2012.[7] Dong, L. et al.,*Ann Biomed Eng*,41: 2538-2552, 2013.[8] Leahy, P.D., PhD Thesis, Colorado State University, 2012.[9] Zhang, H. et al.,*Conf Proc IEEE Eng Med Biol Soc*,6: 6165-6167, 2005.[10] Maas, S. A. et al.,*J Biomech Eng*,134: 011005, 2012.

ANALYZING THE EFFECT OF STENT GEOMETRY FOR POLYMERIC AORTIC VALVES

Atieh Yousefi Koupaei (1), Brandon L. Moore (1), David L. Bark Jr. (1, 2), David A. Prawel (1, 2), Lakshmi P. Dasi (1, 2)

(1) Department of Mechanical Engineering
Colorado State University
Fort Collins, CO, USA

(2) School of Biomedical Engineering
Colorado State University
Fort Collins, CO, USA

INTRODUCTION

Aortic valve disease (AVD) is a major cause of death around the world. Currently over 290,000 heart valve (HV) procedures are performed annually worldwide [1]. This number is expected to triple by 2050. Therefore the demand for artificial HVs is expanding at a rate of 10-12% per year [2]. Several drawbacks are associated with existing bioprosthetic and mechanical HVs such as durability and thrombosis, respectively. Polymeric aortic prosthetic HVs can overcome durability issues, may be designed to reduce antithrombotic therapy, and may reduce current production costs.

The aim of this work is to initiate the design of low-cost tri-leaflet polymeric aortic HVs, by determining geometries that result in hemodynamic performance comparable to bioprosthetic HVs. Specifically, 3 features will be investigated: 1) the aspect ratio between HV's inner diameter and its height, 2) the influence of leaflet arches, and 3) the effect of a flexible stent.

METHODS

To analyze various designs, the stents for polymeric HVs have been 3D printed (uPrint SE). The aspect ratio between stent height and diameter (H/D) was varied for three different valves using Acrylonitrile Butadiene Styrene (ABS), with ratios listed in Table 1. A fourth HV stent was created from TPU with a low profile, which adds flexibility to the stent posts. The latter valve is designed to evaluate the effect of adding a degree of freedom to the stent poles. Each of the four valves has an inner diameter of 21.5 mm.

TABLE 1: Parameters used for stent profile design

	Low Profile	Medium Profile	High Profile
H/D	0.6	0.7	0.88

The leaflets for each valve were made of Linear Low Density Polyethylene (LLDPE). A rectangular strip of LLDPE was used as the primary leaflet design. A secondary design involved additional arch shapes for each leaflet. The arch height to diameter (h/D) aspect ratio was varied (Table 2) to evaluate its effect on coaptation. All leaflet strips were wrapped around the stents to create a cylinder.

TABLE 2: Parameters used to design leaflets

	No Arch	Short Arch	Long Arch
h/D	0	0.116	0.21

Valves were placed in a clear acrylic chamber made to mimic the walls of the aorta [3]. This chamber was placed into a flow loop that simulates the performance of left heart. The fluid inside the flow loop was a mixture of 60%-40% water and glycerin to match the density and viscosity of blood. In order to obtain physiological conditions, the flow rate was adjusted between 4.5-5.2 L/min and blood pressure was controlled at 120/80 mmHg. Another set of experiments were performed to simulate severe hypertension condition, which involved a mean aortic pressure of 160 mmHg (blood pressure 180/140 mmHg).

High-speed videos were taken to analyze the overall performance of each valve and to evaluate leaflet closure. To do this, the valves were placed in a clear dog-leg chamber. Images of the valve were captured for both normal pressure and severe hypertension.

Regurgitant flow was evaluated for each of the valves by integrating flow curves obtained from a probe (Transonic) in the flow loop. Regurgitation is defined as the ratio between reverse flow to the total amount of flow in the loop multipled by 100%.

Particle Imaging Velocimetry (PIV) was performed on two valves at normal blood pressure to capture the velocity field along the centerplane of one leaflet using a LaVision system. The valves used for this study included the low profile valve with a short arch and the high profile valve with no arch. Velocity vectors were calculated using DaVis (LaVision) with a 50% overlap multi-pass approach starting from one 32×32 pixel interrogation followed by one 16×16 pixel interrogation passes. Post-processing was performed using adaptive median filtering. The effective spatial resolution was 28 microns per pixel and the temporal resolution was 2000 Hz.

RESULTS

Kinematics and regurgitation were studied for each of the valve designs. First, valves without an arch were studied to evaluate the effects of H/D aspect ratios on valve closure. Taller stent profiles resulted in more complete closure, as depicted in Fig. 1. Whereas the low profile valve resulted in incomplete closure, or incompetence.

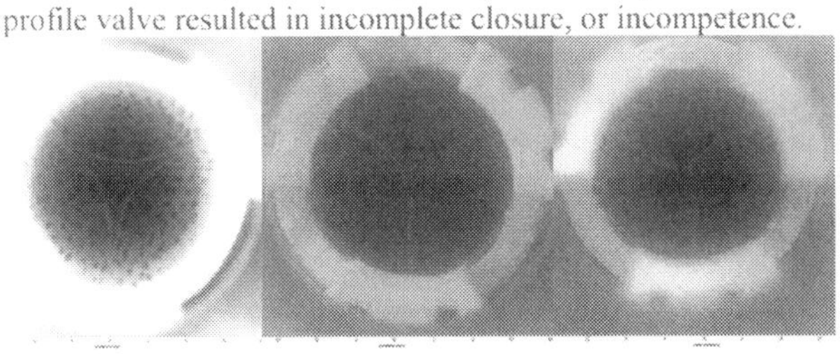

FIGURE 1: Effect of H/D on leaflet closure. H/D= 0.6 (left), H/D= 0.7 (middle), H/D = 0.88 (right).

Arches were added to leaflets to determine if they could increase the valve closure for the low profile valve. Arches did reduce the gap between leaflets, as shown in Fig. 2. However, too much material exists for the long arch, resulting in folds in the material.

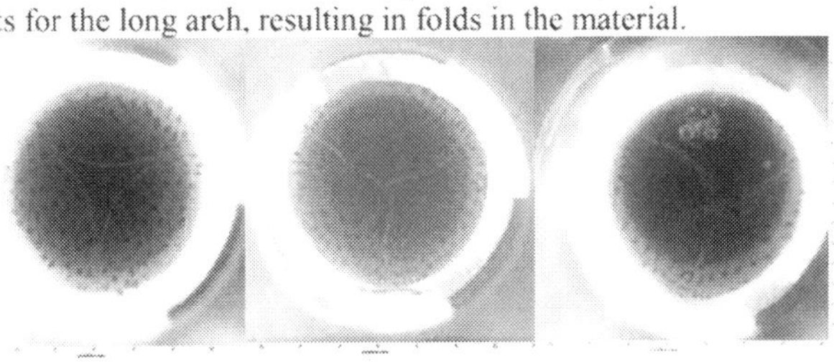

FIGURE 2: Effect of h/D on leaflet closure for a low profile HV with h/D =0 (left), h/D= 0.116 (middle), and h/D = 0.21 (right).

A flexible stent was compared with a rigid stent for a low profile valve to determine if competence could be increased. The two HVs shown in Fig. 3 resulted in similar closure.

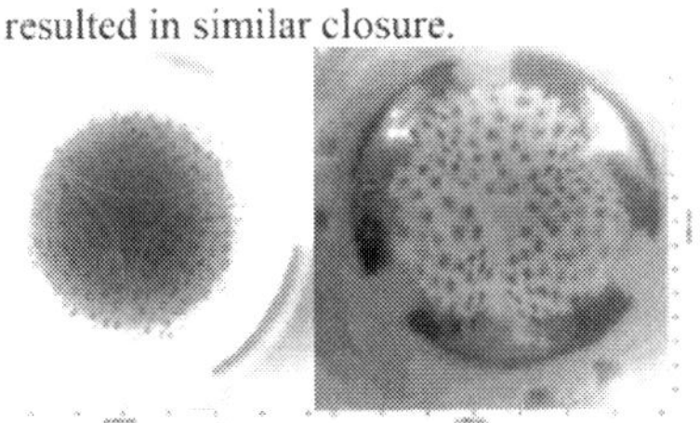

FIGURE 3: Closure of a fixed (left) and flexible stent (right).

Regurgitant flow was determined for representative valves and is presented in Table 3. Values reflect valve closure presented in Figs. 1-2, with the addition of a medium profile stent and a short arch. States of hypertension decreased regurgitation for the low and high profile configurations, but increased regurgitation for other configurations. The addition of arches to the low profile valve had the most profound impact on reducing regurgitation. Comparatively a high profile stent maintained 22.2% regurgitant flow.

TABLE 3: Regurgitation fraction percentage (%)

	Low Profile	+ Short Arch	Medium Profile	+Short Arch	High Profile
Normal Pressure	87.23	2.6	27	3.18	22.2
Hypertension	6.6	7.6	27.6	5.2	7.8

Fig. 4 shows velocity and vorticity along the centerplane of one leaflet in peak systole. A high profile HV is shown relative to a low profile HV with short arch. Overall, there appeared to be increased vorticity in the high profile stent.

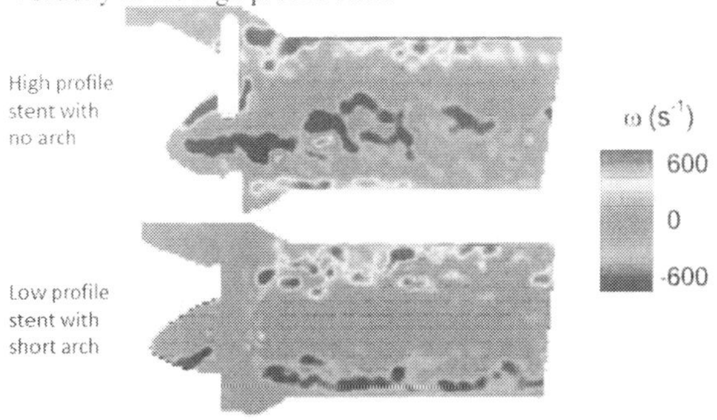

FIGURE 4: Velocity vectors and vorticity contours along the center plane of best performing HVs: a high profile stent with no arch (top) and a low profile stent with a short arch (bottom).

DISCUSSION

The comparison between flow rate for low, medium and high profile valves indicate that the high profile stent results in better closure and less regurgitant fluid flow during diastole. Since low profiles are clinically desirable, we evaluated methods to reduce regurgitant flow for the low profile design. The short arch for both low and medium profile stents reduced regurgitation and improved leaflet coaptation. The longer arch did not provide much improvement over the short arch. Therefore, 0.116 is an acceptable h/D aspect ratio for leaflets. It can be implied from data presented in table 3 that the addition of arches to the leaflets had the strongest effect in reducing regurgitation. Even though the high profile HV showed good closure and coaptation, deformation of the commissures may result in the increased regurgitation seen in Table 3. Alternatively, the flexible stent increased coaptation, reducing the gap in the valve during diastole, but to a lesser degree relative to the alternatives. Velocity profiles and vorticity contours exhibit good agreement with previous studies and demonstrate that the short valve profile may also provide more desirable hemodynamics [4]. Overall, the addition of arches was most effective at reducing regurgitation, allowing the use of a low profile valve, which has many desirable properties. Further hemodynamic analysis will be performed to characterize valve performance.

ACKNOWLEDGEMENTS

Research reported in this abstract was supported by the National Institutes of Health under award number R01HL119824.

REFERENCES

[1] Yacoub MH & Takkenberg JJM, *Nat Clin Pract Card*, 2:60-61, 2005.
[2] Black MM & Drury PJ *The Pathology of Devices*, 127-159, 1994.
[3] Yap CH et al., *Biomech Model Mechan*, 11:171–82, 2012.
[4] Moore BL & Dasi LP *ASME 2013 Summer Bioengineering Conference*, V01AT04A009-V01AT04A009, 2013.

SB³C2015
Summer Biomechanics, Bioengineering and Biotransport Conference
June 17-20, 2015,Snowbird Resort, Utah, USA
SB³C2015-1072

SWIRLING FLOWS IN ARTERIAL HEMODYNAMICS

Kartik V. Bulusu[1], Christopher J. Elkins[2], John K. Eaton[2] and Michael W. Plesniak[1]

(1) Department of Mechanical and Aerospace Engineering
The George Washington University
Washington DC USA

(2) Department of Mechanical Engineering
Stanford University
Stanford CA USA

INTRODUCTION

Secondary flow structures arise due to curvature-related centrifugal forces and pressure imbalances. Known as "spiral blood flow structures" in clinical parlance, secondary flow structures have been characterized to play a protective role in preventing aortic arterial wall damage. The lack of these structures has been linked to carotid atheromatous disease, renal artery stenosis and rapid deterioration of renal function [1]. The effect of streamwise and cross-stream centrifugal forces under pulsatile (physiological) flow conditions towards the occurrence of multi-scale, multi-strength secondary flow structures has been explored [1-2]. These forces affect the motion and stability of secondary flow structures. Under physiological inflow conditions, complicated effects, such as asymmetry and spatio-temporal distributions arise. The analytical procedures to predict secondary flow (vortical) structures are complicated, requiring Navier-Stokes equations in toroidal coordinates and perturbation techniques [3].

This study is motivated by clinical evidence of arterial secondary flow structures, observed during patient diagnosis and associated with curved arteries and their pathophysiological conditions [4-5]. The overarching goal of this study is the detection of three-dimensional shapes of secondary flow structures using magnetic resonance velocimetry (MRV) techniques with guidance obtained from high fidelity, particle image velocimetry (2D-PIV) measurements.

Experiments were performed using multi-harmonic, physiological, carotid artery-based inflow conditions (Womersley number = 4.2) with a premise that concomitant secondary flow structures demonstrate phase-dependent behavior. MRV and 2D-PIV techniques were implemented independently, on a 180-degree curved artery test section to investigate spatio-temporal secondary flow structure morphologies.

EXPERIMENTAL METHODS

A 180-degree acrylic-based, curved artery test section with curvature ratio (r/R), 1/7, was used as an idealized geometry for curved arteries (Figure 1). The main experimental assumptions made are that of arterial rigidity and Newtonian behavior of blood. The following Newtonian blood analog fluids were used for both MRV and PIV experiments: (i) Glycerol-DI Water (60-40, % by weight) (ii) 79% saturated NaI, 20% pure glycerol, and 1% water (by volume) [6]. Kinematic viscosity of both blood analog fluids were measure using an Ubbelhode viscometer at approximately 25 °C (ν = 3.55 cSt ± 2.8%).

The MRV-technique offers the advantage of three-dimensional velocity field acquisition without requiring optical access or flow markers. Accordingly, the saturated NaI-based, blood analog fluid with refractive index matched to the acrylic test section (1.45 ± 3.4%) was used only for PIV. The physiological carotid artery waveform was reconstructed from ultrasonic flowmeter measurements on the left carotid artery reported by Holdsworth *et al.* [7].

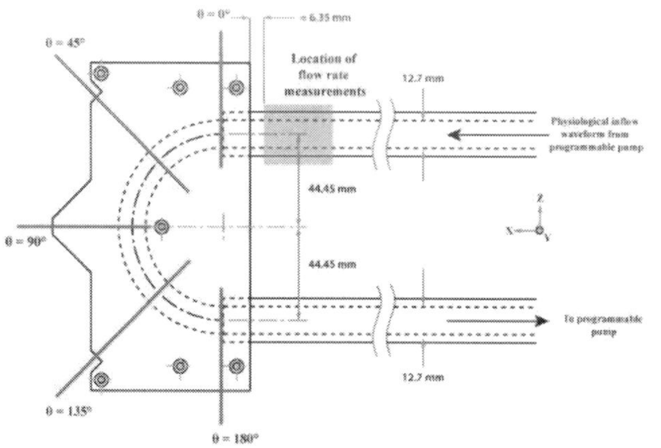

Figure 1. Schematic drawing of the 180° curved artery test section used in 2D-PIV and MRV experiments.

This waveform was digitized to discrete, evenly spaced time instances (t) spanning a period (T) of 4 seconds and supplied to the test section using a programmable pump (ISMATEC BVP-Z) controlled using a data acquisition card (NI-USB-6229). In Figure 2 the results of flow rate measurements using the carotid artery waveform are shown.

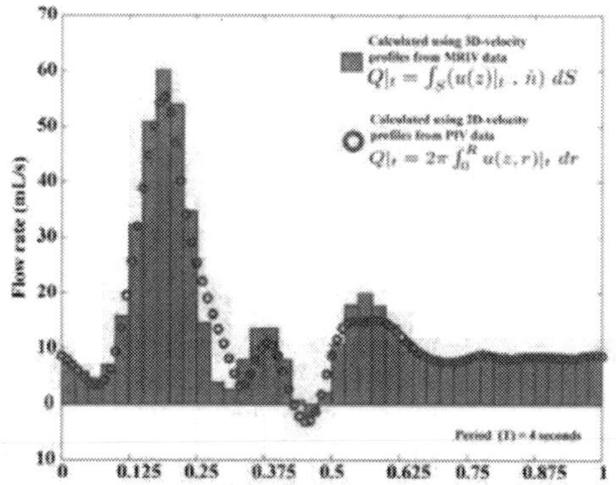

Figure 2. Flow rate measurements from carotid artery-based inflow conditions; green bars (MRV-data in X-Y plane), red circles (2D-PIV-data in X-Z plane).

The details of experimental recording parameters for PIV and MRV are presented in Tables 1 and 2. A novel wavelet decomposition-based approach to detect coherent secondary flow structures was applied to both PIV- and MRV-data, the details of which follow.

Table 1. PIV recording parameters

Flow geometry	2D-circular, cross-sections parallel to light sheet
CCD Array (pixels)	x 1376, y 1040
Final number of vectors	x 86, y 65
Spatial resolution	0.459 mm
Vector spacing (50% overlap)	0.22941 mm
Temporal resolution	100 phases in 4 seconds
Pulse delay	Δt = 600 – 3200 μs
Seeding material	Flouro-Max Red™ fluorescent polymer microspheres
Seeding particle diameter (d_p)	≈7μm
Dynamic spatial range (DSR) [1]	32:1
Dynamic velocity range (DVR) [1]	60:1 (at t/T = 0.21)

Table 2. MRV recording parameters

Flow geometry	3D - 180° curved artery test section
Imaging volume	200 mm x 116 mm x 21.6 mm
Imaging matrix	332 x 192 x 36
Spatial resolution	0.6 mm (isotropic)
Temporal resolution	40 phases in 4 seconds
Velocity encoding	R/L 80 cm/s, A/P 40 cm/s, S/I 90 cm/s

COHERENT STRUCTURE DETECTION

We envisioned secondary flow phenomena as multi-scale and multi-strength occurrences of swirling, rotation-dominated or strain-dominated vortical structures. This necessitated size-structure-strength considerations (via wavelets) for accurate coherent structure detection.

Continuous wavelet transform (CWT) was implemented using a Ricker wavelet (as the mother wavelet, ψ) to characterize the shape of two-dimensional vortical patterns. Guidelines on the choice of the mother wavelet (ψ) and the restrictions thereof, are noted herein as (i) preservation of the L^2 norm, (ii) admissibility and (iii) good localization and smoothness both in physical and spectral spaces [8-9]. The 2D Ricker wavelet is represented by equation 1,

$$\Psi(x,y) = \left\{2 - \left(\frac{x}{\sqrt{l}}\right)^2 - \left(\frac{y}{\sqrt{l\varepsilon}}\right)^2\right\} exp\left\{-\left(\frac{\left(\frac{x}{\sqrt{l}}\right)^2 - \left(\frac{y}{\sqrt{l\varepsilon}}\right)^2}{2}\right)\right\} \quad (1)$$

where l is the scale factor and ε controls the wavelet isotropy. The suitability of the 2D- Ricker wavelet toward vortical pattern characterization has been elaborately discussed [9-11]. Our algorithmic approach (PIVlet 1.2) implements wavelet transforms on vorticity data and computes Shannon entropy of the wavelet transformed vorticity ($\tilde{\omega}$) toward an optimal wavelet scale and the resolution of multi-scale secondary flow structures [1, 12]. Measurement ensembles of planar cross-sectional vorticity fields (ω) from MRV- and 2D-PIV-data were are treated with PIVlet 1.2 to generate wavelet-transformed vorticity fields ($\tilde{\omega}$) (see Figure 3).

RESULTS

The $\tilde{\omega}$-fields presented in Figure 3, are the result of an exhaustive pattern search for secondary flow structures and represent multi-scale vortical patterns. Phase-averaged, two-dimensional, PIV-data at the systolic peak (t/T = 0.18) and certain cross-sectional planes have been compared to MRV-data to facilitate the characterization of large-scale, Dean-, Lyne- and Wall-type (D-L-W) vortices (Figure 3). The similarity of these large-scale structures in both PIV- and MRV-data is reasonable given the spatio-temporal resolution of both techniques. The variation in strengths and scales is suggestive of vorticity distributions in three dimensions. Hence, the potential of generating the complete, three-dimensional geometry of the secondary flow structure is being explored with the MRV-data within the constraints of spatio-temporal resolution (see Tables 1 and 2.)

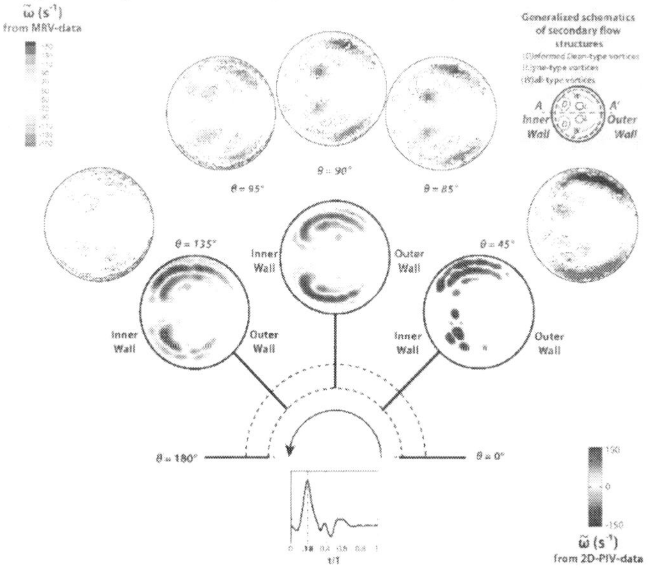

Figure 3. Comparison of 2D-PIV and MRV-data with wavelet-transformed vorticity fields ($\tilde{\omega}$).

DISCUSSION

Arterial secondary flows were investigated using two experimental techniques (MRV and 2D-PIV), with interesting morphologies of multiple pairs of secondary flow structures observed. Establishing the presence of multi-scale secondary flow structures in an idealized 180° curved artery model facilitates extension to realistic arterial model geometries using MRV. Further image processing and analysis of the MRV data will lead to elucidation of the complex spatio-temporal secondary flows that occur in the arterial vasculature.

ACKNOWLEDGEMENTS

This material is based in part upon work supported by the NSF (CBET-0828903), and GW Center for Biomimetics and Bioinspired Engineering (COBRE).

REFERENCES

[1] Bulusu, K.V. & Plesniak, M.W., *Exp Fluids*. 54(3), 1493, 2013.
[2] Webster, D.R. & Humphrey, J.A.C. *Phys Fluids* 9(2), 407, 1997.
[3] Siggers, J.H. & Walters, S.L., *J. Fluid Mech.* 600, 133- 165, 2008.
[4] Kilner et al., *Circulation* 88: 2235–2247, 1993.
[5] Sengupta, P. P *et al., Heart Failure Clin.*, 4:325–332, 2008.
[6] Deutsch, S *et al., Annu. Rev Fluid Mech.* 38, 65–86, 2006.
[7] Holdsworth, D.W *et al., Physiol. Meas.* 20(3), 219–40, 1999.
[8] Farge, M *et al., CTR, Proc. Summer Program*. 331–348, 1990.
[9] Kailas, S.V. & Narasimha, R., *Exp Fluids*. 27, 167–174, 1999.
[10] Schram, C. & Riethmuller, M.L., *Meas Sci Tech.* 12, 1413, 2001.
[11] Varun, A.V *et al., Exp Fluids*. 45, 857–868, 2008.
[12] Bulusu, K.V *et al., Exp Fluids*. 55, 1832, 2014.

SB^{3}C2015
Summer Biomechanics, Bioengineering and Biotransport Conference
June 17-20, 2015,Snowbird Resort, Utah, USA
SB^{3}C2015-1074

EFFECTS OF HYDRATION AND RIBOFLAVIN/UVA COLLAGEN CROSSLINKING ON BOVINE CORNEAL TENSILE PROPOERTIES

Hamed Hatami-Marbini (1), A. Rahimi (1), R. Maulik (1), J. siegmann (1)

(1) School of Mechanical & Aerospace Engineering
Oklahoma State University
Stillwater, OK

INTRODUCTION

Keratoconus is an eye disorder in which the cornea softens and bulges out due to the intraocular pressure. This eye disease affects almost 1-5 people in every 10,000 and is not usually diagnosed before the teenage years. While its exact cause is still unknown, it is believed that patients who wear hard contact lenses are more at risk. Previous studies have clearly showed that the progressive non-inflammatory stromal thinning in Keratoconus affects corneal biomechanical properties and its normal curvature. The changes in corneal curvature impair the vision and result in serious vision disorders such as astigmatism and myopia.

The riboflavin/UVA collagen crosslinking is a relatively new technique for improving biomechanical properties of the Keratoconus cornea and subsequently halt the progression of the disease [1]. The efficacy of this treatment option has been characterized using various mechanical testing; nevertheless, the uniaxial tests are most commonly used. Using this technique, Wollensak et al. concluded that maximum tensile stress in the riboflavin/UVA treated porcine and human corneas was increased about 70% and 300%, respectively [2]. Similarly, McCall et al. reported about 160% increase in the breaking strength of rabbit cornea due to the corneal collagen crosslinking treatment [3]. The corneal stroma is a hydrated soft tissue and its water content is expected to affect its biomechanical properties. Indeed, we have recently shown that hydration of corneal samples significantly affect the measurements of the uniaxial tensile testing method [4]. It was observed that swollen corneal samples had softer tensile properties compared with the dehydrated corneas. The present study was designed to extend this previous study and investigate possible effects of hydration on corneal stiffening due to the riboflavin/UVA collagen crosslinking treatment, as measured using the uniaxial tensile testing method.

METHODS

Fresh bovine eyes, obtained from a local slaughterhouse, were brought to the laboratory on ice. From each intact eye globe, a corneal scleral ring was dissected with a surgery scissor and a scalpel. Prior to being subjected to the UVA/riboflavin collagen crosslinking procedure, the scleral corneal rings were placed in 0.1% riboflavin with 20% dextran T500 solution until they reached an equilibrium swelling state. A Pachymeter (DGH Pachette 3, DGH Technology, Inc., Pennsylvania, USA) was used to measure the thickness of samples. At the start of the collagen crosslinking treatment procedure, the specimens were placed on a semi-sphere surface to be able to mimic their natural in-vivo curvature. They were then subjected to the UVA irradiation with an irradiance of 3 mW/cm^2 for 30 minutes while drops of riboflavin solution was added on their surface every five minutes. After the UVA/riboflavin crosslinking treatment, a double blade cutting device was used to punch a rectangular strip from the nasal temporal direction of each corneal sclera ring. The strips were prepared from nasal-temporal orientation so that we could avoid unwanted variations in the measurements due to the corneal anisotropic response.

The uniaxial tensile tests were done using a dynamic mechanical analyzer (RSA-G2, TA instrument, Delaware, USA). In order to investigate the effects of hydration on tensile properties of collagen cross-linked corneas, the cross-linked specimens were soaked in the Ophthalmic Balanced Salt Solution (OBSS) or dehydrated them in air until their thickness reached the thickness of one of the three thickness groups, i.e. 700, 1100, and 1500 μm. These strips were then secured between the clamps of the DMA machine and the immersion chamber was filled with mineral oil in order to prevent any swelling/deswelling during the experiments. All samples were subjected to a preconditioning step for creating a similar stress-strain history in them. The preconditioning procedure included five rapid loading/unloading cycles followed by three short stress-relaxation steps. The experiments were done by applying 5% axial strain with a displacement rate of 2 mm/min. During the experiments, the applied force was measured and the stress was calculated from dividing the recorded force by the initial cross-sectional area of the strips. Similar experiments were done on untreated samples.

RESULTS

Figure 1 shows the tensile stress-strain behavior of riboflavin/UVA collagen crosslinked corneal strips at different thicknesses. This plot also compares the measured mechanical behavior of collagen crosslinked specimens with that of the untreated ones. It is seen that the riboflavin/UVA collagen crosslinking procedure increased the stiffness of the bovine strips. Furthermore, the samples tested at lower thickness had significantly stiffer response while the stiffness decreased with decreasing the thickness (hydration).

DISCUSSION

The present work was performed to determine the influence of hydration on tensile properties of the riboflavin/UVA collagen cross-linked cornea. The thickness of the samples was used as a representation of their hydration. It has been shown that there is one-to-one relation between thickness and hydration of bovine cornea [5], i.e.

$$H_w=3.7\ ln(t/0.295)$$

where H_w is the hydration in mg dry tissue / mg water and t is thickness in mm. Note that although hydration of all of the strips in each thickness group is not necessarily the same, a significant difference between their hydration is still expected. Figure 1 shows that, similar to its effects on porcine, human, and rabbit corneas, the riboflavin/UVA collagen crosslinking procedure strengthened the tensile properties of the bovine cornea.

Furthermore, this plot shows that the stiffening effect of the collagen cross-linking treatment, as measured by uniaxial tensile testing method, is *hydration dependent*. This is an important conclusion and implies that, an accurate estimate of the effects of this treatment procedure on corneal biomechanics is only possible if the tensile behavior of riboflavin/UVA collagen crosslinked and untreated specimens is measured at the same level of hydration. For example, if the hydration of the specimens were not controlled and the thickness of the treated and untreated strips was 1100 and 700 um, respectively, we would have incorrectly concluded that the riboflavin/UVA had no strengthening effects on bovine cornea. Because of deturgescent effect of dextran, which is applied on the surface of the samples during the riboflavin/UVA, it is possible that some of previous reports tested the crosslinked samples at a lower level of hydration than the control group and overestimated the influence of this treatment option on biomechanical properties of the cornea. Future studies are required to address this possibility and fully characterize how this treatment option influence mechanical properties of the human and porcine corneas.

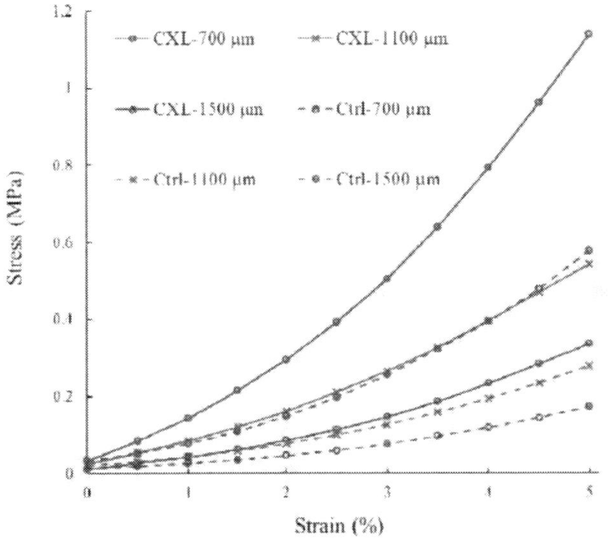

Figure 1. Stress-strain of the riboflavin/UVA collagen cross-linked bovine cornea at different levels of thickness (hydration) compared with that of the untreated samples.

ACKNOWLEDGEMENTS

This project has been funded in whole or in part with the start-up fund from Oklahoma State University.

REFERENCES

[1] Spoerl, E, Huhle, M, Seiler, T, Exp Eye Res 66:97-103, 1998.

[2] Wollensak, G, Spoerl, E, Seiler, T, Am J Ophthalmol, 135:620-627, 2003.

[3] McCall, AS, et al., Invest Ophthalmol Vis Sci 51:129-138, 2001.

[4] Hatami-Marbini, H, Rahimi, A, J Mech Behav Biomed Mater 36:90-97, 2014.

[5] Hatami-Marbini, H, Ann of Biomed Eng, 42:1740-1748, 2014.

COMPUTATIONAL FRAMEWORK FOR APPLICATION OF RESIDUAL STRESS WHEN THE STRESS-FREE CONFIGURATION IS UNKNOWN

Steve A. Maas (1,2), Ahmet Erdemir (3), Jason P. Halloran (4), Jeffrey A. Weiss (1,2)

(1) Department of Bioengineering
University of Utah
Salt Lake City, Utah, USA

(2) Scientific Computing and Imaging Institute
University of Utah
Salt Lake City, Utah, USA

(3) Computational Biomodeling (CoBi) and
Department of Biomedical Engineering
Cleveland Clinic
Cleveland, Ohio, USA

(4) Mechanical Engineering Department
Cleveland State University
Cleveland, Ohio, USA

INTRODUCTION

Many biological tissues exhibit some level of residual stress in vivo. These stresses originate from development, growth, and remodeling and are thought to be one of the main sources of discrepancy between in vivo and ex vivo material characterization. Residual stresses are usually characterized by *prestrain*, which can be *in-situ strains* (i.e. strain that exists in the absence of loading), or *residual strains* (i.e. strains that exist after the tissue was excised from the body). In-situ strains in diarthrodial joints are the in the range of 3-10% [1]. Residual strains have been observed in many studies and have been found non-negligible [2,3,4]. Consequently, it is important to take the residual stress into account in computational models.

Computational approaches generally assume that the residual stress is elastic in nature and can be defined through a prestrain [5]. More specifically, a multiplicative decomposition of the deformation gradient is assumed, and the total deformation gradient $\mathbf{F}_e$ is defined as the product of a prestrain $\mathbf{F}_p$ and a load induced deformation gradient $\mathbf{F}$:

$$\mathbf{F}_e = \mathbf{F} \cdot \mathbf{F}_p . \tag{1}$$

Several investigators have specified this prestrain by applying a (reversible) deformation to a stress-free reference configuration [3,6,7]. In this case the prestrain deformation gradient is defined as the gradient of the deformation map that takes the stress-free configuration to the unloaded in vivo state. The major limitation of this approach is the assumption of a global stress-free configuration since in general, there is no reason to assume that such a stress-free state can be found.

We developed and implemented a method that achieves prestrain without the assumption of a zero-stress reference configuration. The approach is general enough to accommodate many types of prestrain, but for illustration, we assume that the prestrain is determined by the stretching of fibers in a transversely isotropic material. This allows us to write the prestrain deformation gradient as a uniaxial stretch along the fiber direction. Thus, in order to prescribe a prestrain in our method, it is sufficient to know the local, possibly inhomogeneous, fiber stretch in the material, which can be determined experimentally. Since prescribing a deformation gradient may result in a stress field that is not compatible with the geometry and boundary conditions, we have implemented an iterative method that finds the configuration that achieves equilibrium and maintains the prescribed pre-stretch approximately.

METHODS

Our method was implemented in the open-source finite element software FEBio (www.febio.org) [8]. In addition to the multiplicative decomposition of the total deformation gradient (equation 1), we assumed a decomposition of the prestrain deformation gradient.

$$\mathbf{F}_p = \mathbf{F}_c \cdot \hat{\mathbf{F}}_p \tag{2}$$

Here, $\mathbf{F}_c$ is a correction term that is to be determined by satisfying user-defined constraints and $\hat{\mathbf{F}}_p$ is an initial guess of the prestrain deformation gradient. In our implementation $\hat{\mathbf{F}}_p$ is either prescribed fully, or defined in terms of some other strain measure. For instance, if λ_p is the fiber stretch (which can often be measured experimentally), then

$$\hat{\mathbf{F}}_p = \mathrm{diag}\left[\lambda_p, 1/\sqrt{\lambda_p}, 1/\sqrt{\lambda_p}\right] . \tag{3}$$

Currently, the constraints can be either a zero distortion of the reference mesh, or enforcing a desired level of prestrain. The method iteratively updates $\mathbf{F}_c$ until the constraint is satisfied to within a user-defined tolerance.

The method is illustrated using two examples. In the first example, a desired value of fiber strain is applied to a finite element model of the medial collateral ligament (MCL) of the knee. The MCL was modeled as a transversely-isotropic Mooney-Rivlin material. The initially applied fiber strain is incompatible with the reference geometry and results in an additional distortion of the geometry. This distortion also alters the effective fiber stretch. The proposed method was used to recover the desired fiber stretch.

The method was also applied to a computational model of the knee from the OpenKnee project [9,10]. OpenKnee is a collaborative project that provides detailed, freely available, finite element models of the knee. For this abstract a simplified OpenKnee model was used to simulate a 5 degree passive flexion and the effective stresses are compared between a residually stressed model using a homogenous 3% fiber strain of the ligaments, and a model without prestrain applied. The bones were modeled as rigid and the ligaments were assigned a transversely-isotropic Mooney-Rivlin material.

RESULTS

Without correction, applying a 3% homogenous fiber strain results in stress field that is incompatible with the reference configuration and deforms the mesh. This in turn alters the effective fiber strain (Figure 1). After one iteration the algorithm finds a slightly deformed reference configuration that approximates the desired fiber strain more closely. After five iterations the applied strain field is nearly homogenous.

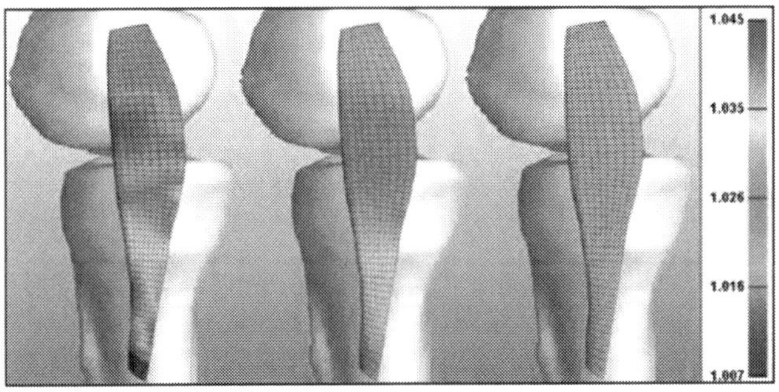

Figure 1. Contour plot of fiber stretch. Left, strain ranges between 0.76%-4.45%. Middle, after 1 iteration, strain: 2.53%-3.37%, Right, after 5 iterations, strain: 2.95%-3.04%.

In the OpenKnee model a significant difference can be observed in the effective stress magnitude and distribution in the ligaments between the residually stressed and unstressed models (Figure 2). Although the simplified model does not simulate an actual physical loading scenario, it does illustrate how even a small prestrain can have a drastic effect on the effective stress distribution in the ligaments.

DISCUSSION

In summary, we implemented a method in FEBio for applying pre-strain without the assumption of a stress-free configuration. This method is more versatile than existing methods since a complex prestrained in-vivo configuration can be achieved that would be difficult to achieve by deforming a stress-free configuration. However, with little effort, the method can also be used to solve the "traditional" approaches to prestrain where a global stress-free configuration is assumed.

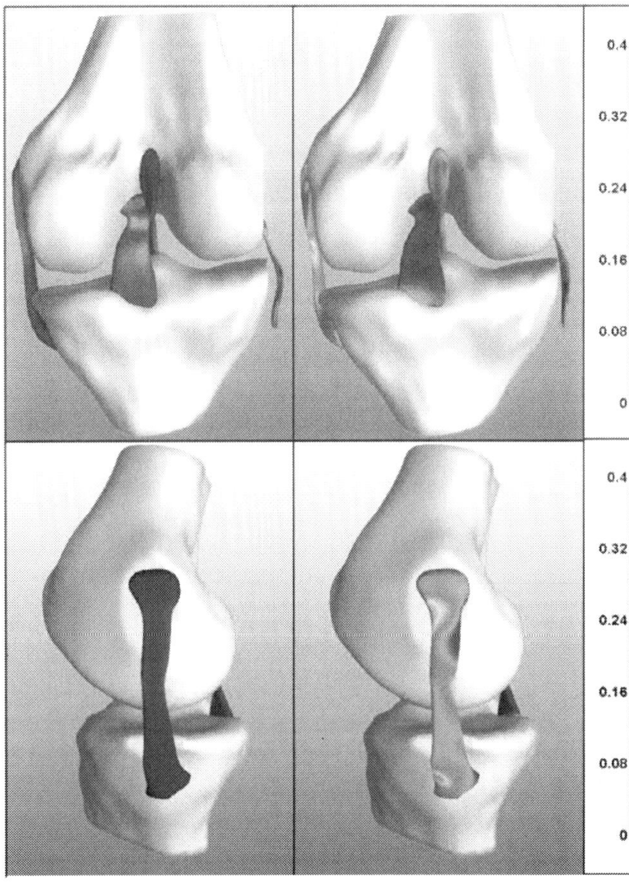

Figure 2. Effective stress of OpenKnee model (simplified) with 5 degree passive flexion. Left, without prestrain. Right, with 3% homogenous fiber strain. Top row, posterior view. Bottom row, medial view.

Another advantage of the proposed computational framework is that it can recover a deformation gradient field that is either kinematically incompatible with the reference configuration, or enforce a desired pre-strain level even when the full prestrain deformation gradient is unknown.

ACKNOWLEDGEMENTS

This work was funded by NIH R01GM083925 and NIH R01GM104139.

REFERENCES

[1] Gardiner, J.C., et.al., *Clin. Orthop. Relat. Res.*, 391:266-274, 2001
[2] Fung, *Journal of Bioemdical Engineering*, 1986
[3] Rausch, M.K., et. al., *J. Mech. Phys. Solids*, 61:1955-1969, 2013
[4] Wang H.M., et. al., *Biomech Model Mechan*, 13:99-113, 2014
[5] Johnson B.E., Hoger A., *J Elasticity*, 41:177-215, 1995
[6] Balzani D, et.al., *Acta Biomater*, 2:609-618, 2006
[7] Holzapfel G.A., Ogden R.W., *J. R. Soc. Interface*, 7:787-799, 2009
[8] Maas S.A., et.al., *J. Biomed. Engn*, 132:011005, 2012
[9] Erdemir A, Sibole S., *OpenKnee User's guide*, 2010
[10] Erdemir A, *J. Med. Device*, 7(4):0409101-4091010, 2013

BIOMECHANICAL BEHAVIOR OF CORNEA WHEN SUBJECTED TO TENSION AND COMPRESSION LOADS

Hamed Hatami-Marbini (1)

(1) School of Mechanical & Aerospace Engineering
Oklahoma State University
Stillwater, OK

INTRODUCTION

The cornea is a hydrated soft tissue at the outer surface of the eye. It transmits the incident light while acting as a resilient barrier to the external and internal forces. From the anterior to posterior, the cornea consists of epithelial cells, Bowman's zone, stroma, Descemet's membrane, and endothelial cells. The biomechanical properties of the cornea are mainly derived from the stromal layer, which forms about ninety percent of corneal thickness. The stroma is composed of micrometer thick parallel-to-the-surface collagenous lamellae. Each lamella contains thin collagen fibrils which are regularly packed in a hydrated matrix of extracellular substances such as proteoglycans and salts. There have been many different studies over past decades to predict the corneal stromal behavior in normal and diseased states. In particular, four different mechanical testing methods, i.e. inflation experiments, uniaxial strip extentiometry, indentation, and compressive experiments, have been used. Many of these previous investigations resulted in a wide range of material parameters. For example, the Young's modulus have been found to vary from 0.01 to 50 MPa [1,2]. This wide range of variation is often explained in terms of the variations in testing conditions, such as gripping techniques and presence of a preconditioning step, or intrinsic differences between different specimens such as their age or species. Nevertheless, little attention has been made to elucidate the possible effects of hydration on biomechanical properties of cornea. We hypothesized that one of the reasons why there exists wide variation in the previously reported material properties of the cornea might be the lack of control on corneal thickness (hydration). In this paper, we discuss our recent findings using different testing techniques, i.e. unconfined compression and tensile strip method, to characterize the effects of hydration on corneal biomechanics.

METHODS

Corneal scleral rings were excised from fresh bovine and porcine eyes within few hours after slaughter. Corneal disks and strips were prepared using a biopsy punch and a double blade cutting device, respectively. Corneal buttons were used for unconfined compression experiments and corneal strips were used for uniaxial tensile tests.

A DHR-2 Rheometer (TA Instruments, Delaware) and a RSA-G2 dynamic mechanical analyzer (TA instrument, Delaware) were used to perform the mechanical tests. For unconfined compression experiments, the chamber of the rheometer was first filled with the solution and the circular specimens were then placed on the center of the impermeable platen surface [3]. A set of successive unconfined compression stress relaxation experiments were performed by applying a ramp compressive strain and holding the rheometer's head still for 30-60 minutes. During the experiments, the reaction force with mN accuracy and the displacement with um accuracy were recorded. We obtained the reaction stress by dividing the recorded force by the initial area of the samples. Furthermore, the compressive strain at each step was computed by dividing the change in sample thickness by their initial thickness. For uniaxial tensile experiments, the corneal strips were soaked in the Ophthalmic Balanced Salt Solution (OBSS) or dehydrated them in air until their thickness reached a predetermined thickness [4]. Then, the samples were secured between the clamps of the testing machine and the immersion chamber was filled with mineral oil. The mineral oil was used in order to prevent any swelling/deswelling of the specimens during the tensile experiments. All samples were subjected to a preconditioning step consisting of five rapid loading/unloading cycles followed by three short stress-relaxation steps. Although preconditioning might cause microstructural changes, this procedure created a similar stress-strain history in samples and allowed us to better assess the roles of hydration on their tensile properties. After preconditioning of the strips and determining their initial length by applying a tare force, displacement-control tensile testings were performed. The experiments were done using strips from nasal-temporal and inferior-superior directions and at different displacement rate in order to determine the corneal anisotropic and rate-dependent behavior.

RESULTS

The equilibrium compressive stress as a function of thickness is plotted in figure 1. The inset of Figure 1 shows the compressive stress variation as a function of time in the stepwise unconfined compression experiments. During the ramp loading, the compressive stress increased non-linearly and decreased to an equilibrium stress during the relaxation stage. Figure 2 shows the tangent modulus at 5% strain as a function of thickness of bovine corneas. It was seen that with increasing the thickness (hydration), the tensile modulus significantly decreased.

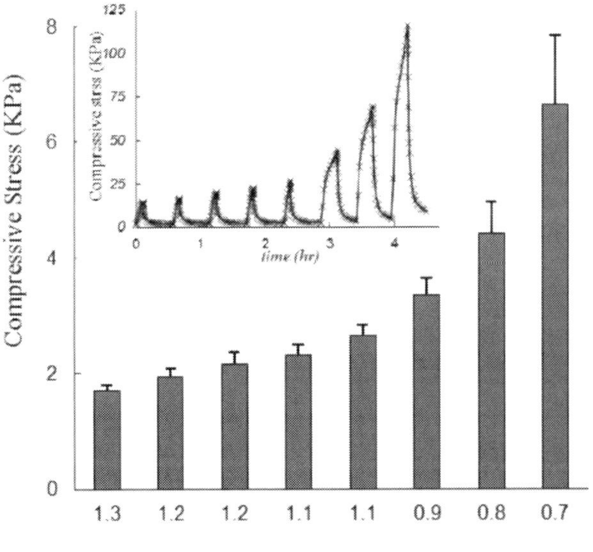

Figure 1. The equilibrium compressive stress as a function of average thickness of the corneal samples. The inset show shows the average stress-time behavior in unconfined compression tests.

DISCUSSION

The present work was performed to determine the thickness dependent biomechanical properties of the cornea. It was seen that mechanical properties of the corneal stroma strongly depend on its thickness (hydration), Figures 1-2. It is known that the negatively charged proteoglycans form an extended network between the collagen fibrils of the corneal stroma. The presence of proteoglycans creates an imbalance between the ionic concentration inside and outside of the tissue and subsequently a strong tendency for the stroma to imbibe water and swell. It is also known that as the thickness becomes smaller, the distance between negative charges become smaller and the net negative charge density increases. Therefore, it could be expected that a higher compressive force is measured with decreasing the thickness, Figure 1. On the other hand, it has been suggested that the glycosaminoglycan side chains of the proteoglycans form interfibrillar bridges between collagen fibrils [5]. With increasing the thickness, the bonds between collagen fibrils and proteoglycan side chains possibly break and this is why a softer mechanical response was observed with increasing the thickness, Figure 2. It is noted that the thickness of the samples was used as a representation of their hydration in the present study. This is because previous studies showed that there is one-to-one relation between thickness and hydration of cornea [6,7].

In summary, the present research clarified the important roles of hydration changes on compressive and tensile properties of the cornea. It was concluded that since dehydrated corneas have stronger biomechanical properties compared with the swollen ones, careful attention should be taken in controlling the hydration of samples when performing experimental studies on corneal tissue.

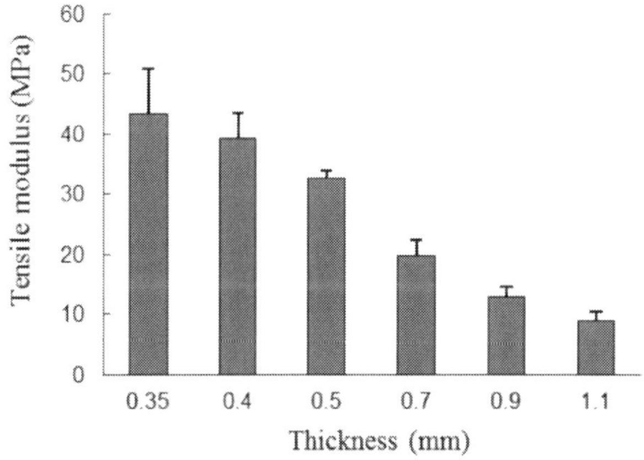

Figure 2. The effect of thickness on tensile modulus of cornea.

REFERENCES

[1] Elsheikh, A, et al., J. Refract. Surg. 23: 808-818, 2007.
[2] Hatami-Marbini, H, et al. Exp Eye Res. 120:103-108, 2014.
[3] Hatami-Marbini, H, et al., J Biomech. 10:1752-1758, 2013.
[4] Hatami-Marbini, H, et al., J Mech Behav Biomed Mater 36:90-97, 2014.
[5] Hatami-Marbini, H, et al., Mater Res Soc Symp Proc, 1239-1244, 2009.
[6] Hatami-Marbini, H, Ann of Biomed Eng, 42:1740-1748, 2014.
[7] Hedbys, BO, et al., Exp. Eye Res. 5:221-228, 1966.

EFFECT OF INTERVERTEBRAL DISC SIZE ON NUTRIENT DISTRIBUTIONS: A FINITE ELEMENT ANALYSIS WITH IMPLICATIONS FOR *IN VIVO* MODELS

Alicia R. Jackson, PhD

Department of Biomedical Engineering
University of Miami
Coral Gables, FL USA

INTRODUCTION

Low back pain is a condition resulting in billions of dollars in costs each year in the United States alone. Although the origins of the condition remain unclear, degeneration of the intervertebral discs (IVD) of the spine is strongly associated with back pain. Poor nutritional supply is believed to be main cause of disc degeneration.

Numerous animal models have been used for *in vivo* and *in vitro* studies of IVD degeneration and possible treatments strategies (e.g., intraveneous injections of growth factors), including horses, pigs, dogs, rabbits, and rats. In addition to differences in cellular activity and tissue properties, these discs vary widely in their size and geometries. Such variation can have significant effects on the chemical microenvironment around disc cells, including differences in nutrient and metabolite concentrations and pH levels.

The IVD is the largest avascular structure in the human body. As such, nutrients, such as glucose and oxygen, must diffuse from the surrounding vasculature to cells. For human discs, this distance can be as large as 7-8 mm [1], making for very low nutrient levels and acidic pH at the disc center. Such precarious conditions have key influences on the disc cells. Thus, it is imperative to employ a disc model capable of closely mimicking the particular nutritional environment in the IVD in order to translate findings to the human IVD. The objective of this study is to investigate the effect of disc size on nutrient distributions in the IVD using finite element analysis. Three disc sizes were chosen: human, pig, and rat. Mechanical loading and endplate permeability were also investigated to determine if the effects of such conditions are similar for different sized discs.

METHODS

In this study, the theoretical formulation published in our earlier work was used [2]. The theoretical framework was based on mechano-electrochemical mixture theory [3]. Nutrient levels were coupled to cellular consumption as in our previous study [2]. Strain-dependent properties for permeability, solute diffusivities, fixed charge density and tissue porosity were incorporated into the model.

In order to investigate the effect of disc size on the distribution of nutrients in the IVD, three different disc sizes were used: human, pig, and rat lumbar discs. The size of the human disc was based on that of an L2-L3 human disc (41 y.o. M, Grade I) [2]. Based on findings in the literature, the human disc was then scaled for each of the two animal discs [4,5]. For pig, the disc dimensions were scaled to 66% of the human disc, while the rat was scaled to 10%. The general shapes of the discs were all the same. Each IVD consisted of 3 distinct regions: annulus fibrosus (AF), nucleus pulposus (NP), and cartilaginous endplate (CEP). The CEP was permeable above the NP region only. In this study, only the upper right quarter of the disc was modeled due to symmetry. Tissue and cellular properties as well as solute concentrations at the disc edges were the same for all sizes.

For each disc size in this study, two physical effects were investigated: endplate permeability and mechanical loading. The endplate permeability was altered by changing the porosity of the CEP tissue, which in turn alters the permeability; the endplate was either considered normal or blocked (~80% reduction in permeability). Two loading conditions were investigated: reference, or undeformed, and a 10% static compressive strain.

RESULTS

Results for minimum glucose concentration for all 12 (= 3 sizes x 2 CEP conditions x 2 loading conditions) conditions are shown in Table 1. The minimum glucose level in the pig was 3.3x larger than in the human disc; likewise, the level was 5.3x larger for the rat IVD. Furthermore, 10% strain resulted in an 18.7% decrease in the minimum level for the human IVD, compared to a 3.6% decrease in the pig IVD and no decrease for the rat IVD. Finally, the blocked CEP led to a 46.7% decrease in the minimum glucose concentration in the human disc (compared to normal CEP, no strain), while it caused 8.9% and 0% decreases in the pig and rat IVDs, respectively.

Three dimensional concentration distributions for the three disc sizes in reference configuration (i.e., normal CEP, no strain) are shown in Figure 1. Figure 2 shows the glucose concentration through the disc radius for all 12 conditions investigated.

Table 1: Minimum Glucose Concentrations (mM) for Each Disc Size at Various Boundary Conditions

	Human	Pig	Rat
Normal CEP, no strain	0.75	2.48	4.0
Normal CEP, 10% strain	0.61	2.39	4.0
Blocked CEP, no strain	0.40	2.26	4.0
Blocked CEP, 10% strain	0.21	2.15	4.0

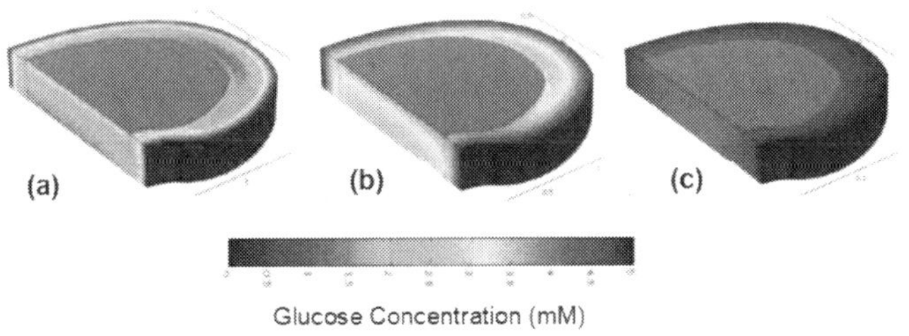

Figure 1: Glucose concentration profiles in the disc for three different disc sizes: (a) human IVD; (b) pig IVD; and (c) rat IVD. Note all discs have normal CEP permeability and no mechanical strain applied. Discs are shown the same size for clarity.

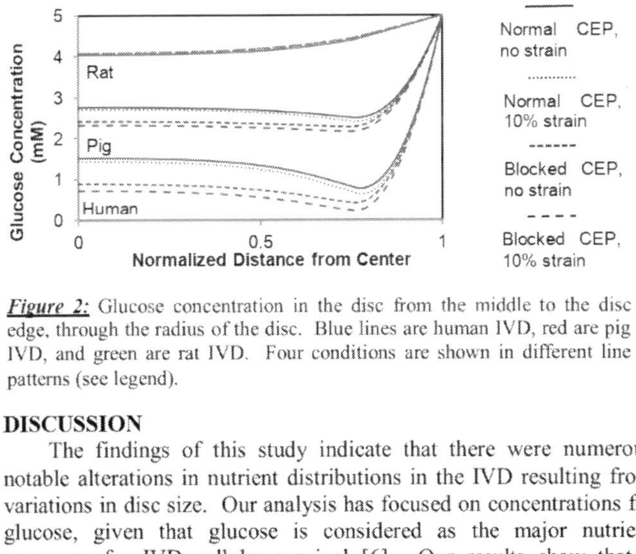

Figure 2: Glucose concentration in the disc from the middle to the disc edge, through the radius of the disc. Blue lines are human IVD, red are pig IVD, and green are rat IVD. Four conditions are shown in different line patterns (see legend).

DISCUSSION

The findings of this study indicate that there were numerous notable alterations in nutrient distributions in the IVD resulting from variations in disc size. Our analysis has focused on concentrations for glucose, given that glucose is considered as the major nutrient necessary for IVD cellular survival [6]. Our results show that a smaller disc size results in much higher concentrations of glucose in the IVD. This is notable since levels below 0.5 mM have been shown to trigger cell death [6]. In this study, only the human disc with blocked CEP showed levels reaching below this threshold. Further, steep concentration gradients are only present for the human IVD, with very little gradient for the rat disc.

It was also determined that boundary conditions such as loading and blocked CEP more strongly affected the glucose levels in human IVD as compared to the smaller discs; no changes were seen for the rat IVD for any of the conditions examined. These findings indicate that *in vivo*-like conditions, such as endplate calcification and compressive loading, have a more pronounced role in the human disc than for smaller animal models. This is similar to findings from a previous study which showed a more crucial role of CEP permeability for the transport of solutes into human IVD compared with smaller discs [7].

In vivo and organ culture models of the IVD are necessary for better understanding tissue pathology, and for developing new strategies to treat and/or prevent degeneration. However, the particular nutritional environment in the human IVD plays a key role in cellular activity and health status of the tissue, and thus should be mimicked in disc models. Our results show that, even in relatively large animal models, such as the pig, nutrient levels in the disc never reach extreme levels similar to those seen in the human disc; for small disc models, like the rat lumbar IVD, very little nutrient deprivation is apparent. Therefore, nutritional limitations that might hinder the efficacy of a treatment strategy in the human IVD would not likely be present in smaller disc animal models.

Previous studies have investigated the effect of disc size on the distribution of intravenously administered solutes in the IVD; these studies also found that alterations in disc size have a large impact on the concentration distribution of solutes in the tissue [7,8]. This study further reinforces the premise that disc size plays a key role in concentration distributions in the tissue, and is thus important for accurate representation of the *in vivo* environment.

There are several limitations in the present study that should be noted. First, tissue properties and cellular metabolic rates were considered the same for all three disc sizes investigated. In reality, the properties for each animal model would vary, which might have an effect on nutrient distributions in the disc. However, the drastically different nutrient levels determined in this study are likely indicative of the general trends for these tissues. Another weakness of this study is the use of scaled geometries, based on the human IVD geometry, rather than exact models for each animal. The scaling factors used were based on studies in the literature comparing human discs with other animal discs. The two animal discs chosen have shapes similar to that of the human IVD. Nonetheless, more accurate results would be obtained using exact animal disc geometries.

In conclusion, this study has investigated the effect of disc size on the nutrient distributions in the IVD under mechanical loading and reduced endplate permeability. Results show that smaller disc size leads to much higher nutrient levels in the disc, making it difficult to interpret the relevance of studies from animal models compared to the human disc. Thus, size consideration should be taken into account when developing *in vivo* or organ culture models of the disc to study disc degeneration changes and treatment strategies, in order to most closely mimic the human *in vivo* environment.

ACKNOWLEDGEMENTS

This study was funded by the University of Miami College of Engineering.

REFERENCES

[1] Stairmand, JW, et al., *Spine*, 16(4):444-9, 1991.
[2] Jackson, AR et al., *Comp Meth Biomech Biomed Eng*, 14(2):195-204, 2011.
[3] Lai, WM et al., *J Biomech Eng*, 113(3):245-58, 1991.
[4] O'Connell, GD et al., *Spine*, 32(3):328-33, 2007.
[5] Beckstein, JC et al., *Spine*, 33(6):E166-73, 2008.
[6] Bibby, SR & Urban, JPG, *Eur Spine J*, 13:695-701, 2004.
[7] Motaghinasab, S et al., *Eur Spine J*, 23:715-23, 2014.
[8] Motaghinasab, S et al., *J Biomech*, 45:2195-2202, 2012.

SB³C2015
Summer Biomechanics, Bioengineering and Biotransport Conference
June 17-20, 2015,Snowbird Resort, Utah, USA
SB³C2015-1078

EVALUATION OF FRICTION AND WEAR PROPERTIES OF POLY(VINYL ALCOHOL) HYDROGELS AS ARTIFICIAL ARTICULAR CARTILAGE

Seido Yarimitsu (1), Ayumi Yoshida (1), Kazuhiro Nakashima (2), Teruo Murakami (1)
Saori Sasaki (3), Atsushi Suzuki (3)

(1) Research Center for Advanced Biomechanics
Kyushu University
Fukuoka, Japan

(2) Department of Mechanical Engineering
Kyushu University
Fukuoka, Japan

(3) Faculty of Environment and Information Sciences
Yokohama National University
Yokohama, Japan

INTRODUCTION

The number of total joint replacement is increasing every year [1]. This type of surgery can relieve pain and recover the function of articular joints. The popular common bearing surface for artificial joint is a combination of an ultra-high molecular weight polyethylene (UHMWPE) and a cobalt–chromium–molybdenum (Co-Cr-Mo) alloy. The lubrication mode of artificial joints composed of UHMWPE and Co-Cr-Mo alloy appears to be mixed or boundary lubrication modes in severe loading conditions. Artificial articular cartilage has been proposed for improvement of lubrication mode in artificial joints [2,3] by actualizing the adaptive multimode lubrication [4] with extremely-low friction and minimum wear similar to natural synovial joints. In addition, artificial articular cartilage can be used as the implant material for the focal defects of natural articular cartilage. Artificial articular cartilage has great potential for the recovery of joint function.

In previous study, we evaluated the tribological behaviors of three types of poly(vinyl alcohol) (PVA) hydrogels, which are physically cross-linked by hydrogen bonding [5, 6]. We found that the friction and wear properties of hydrogels depend on the types of hydrogels. However, reduction of friction and wear of PVA hydrogels were needed for clinical application of PVA hydrogels as artificial articular cartilage.

The aim of this study is to develop the PVA hydrogels with extremely-low friction and minimum wear and to evaluate the lubrication property of PVA hydrogels.

METHODS

The Reciprocating friction tester used in this study is shown in Fig.1. Sliding pairs of a Cobalt-chromium-molybdenum (Co-Cr-Mo) alloy ball (φ26 mm) and a PVA hydrogel plate were tested in reciprocating friction test. The applied load was 5.88 N. The sliding velocity was 20 mm/s and the total sliding distance was 100 m. In this study, ultra-pure water was used as a lubricant for friction test.

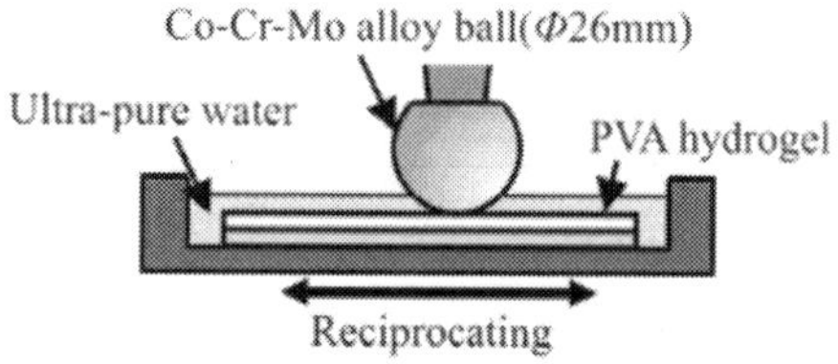

Fig.1 Schematic image of reciprocating friction tester

In this study, three types of PVA hydrogels, PVA-FT (freezing-thawing) gel, PVA-CD(cast-drying) gel and PVA-hybrid gel were developed. 15 wt% aqueous solution of PVA (polymerization degree: 1,700, saponification degree: 98.0~99.0 nol%) was used as raw material for the preparation of PVA hydrogels. PVA-FT gel was prepared by repeated freezing-thawing method, and freezing-thawing cycles were 4 times. PVA-CD gel was prepared by drying of PVA solution in the temperature and humidity controlled chamber. Temperature and relative humidity at drying process were 8 ℃ and 50 %RH, Respectively. After the drying process, the samples were swollen in pure water and PVA-CD gel was obtained. For the preparation of PVA-hybrid gel, PVA-FT gel was firstly prepared in polystyrene dish (Φ90 mm). Then, PVA solution was poured onto the PVA-FT gel and the samples were dried in the temperature and humidity controlled chamber. After the drying process, the samples were swollen in pure water and PVA-hybrid with laminar structure composed of PVA-FT and FT gels were obtained. Two types of PVA-hybrid gel were prepared in this study. First one was prepared by single drying process with temperature of 8 ℃ and relative humidity

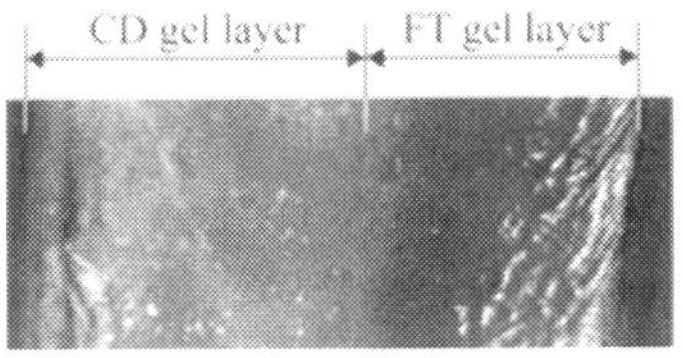

Fig.2 Cross-sectional image of PVA-hybrid gel

of 50 %RH (PVA-hybrid gel A). The second one was prepared by additional drying process with temperature of 20 ℃ and relative humidity of 40 %RH after the drying process with temperature of 8 ℃ and relative humidity of 50 %RH (PVA-hybrid gel B). Side of PVA-CD gel was used as frictional surface in friction test of PVA-hybrid gel A and B.

RESULTS

Friction coefficients at the steady state in friction test were shown in Fig.3. PVA FT gel showed high friction coefficient such as above 0.13. In contrast, PVA-CD gel, PVA hybrid A and B showed quite low friction coefficient below 0.01.

Microscopic images for intact and worn surface of PVA hydrogels were shown in Fig.4. Intact surface structure of PVA-FT gel was completely lost after friction test. Some scratches and roughing of surface were observed on the surface of PVA-CD gel. Some surface detachments were observed as pores on the surface of PVA-hybrid gel A. The intact surface structure of PVA-hybrid gel B remained after friction test and showed excellent wear resistance.

DISCUSSION

PVA-FT gel showed high friction and poor wear resistance, and PVA-CD gel showed quite lower friction than PVA-FT gel but wear was not completely suppressed. PVA-hybrid gel B obtained by two-phase drying process showed extremely low friction and minimum wear in this study. When drying processes of PVA-hybrid gel A and B were finished, residual water content in dried sample was almost the same in this test but wear resistance was different. It is considered that second drying process with higher temperature and lower relative humidity contributed to the reinforcement of surface structure of PVA-hybrid gel by forming more cross-linking points.

The structure of superficial layer of PVA-CD gel and PVA-hybrid gels was a key to the extremely low friction property, and it is considered that additional cross-linking to the surface structure is essential to obtain excellent wear resistance. In the next study, surface properties of PVA hydrogels (structure, adhesivity) were researched and preparation condition would be optimized to obtain PVA hydrogel with extremely low friction and minimum wear.

ACKNOWLEDGEMENTS

This research was supported by Grant-in-Aid for Specially Promoted Research of JSPS (23000011).

REFERENCES

[1] Kurtz, S et al., *J Bone Joint Surg Am*, 87:1487-1497, 2005.
[2] Sasada, T et al., *J Jpn Soc Biomat*, 3:151-157, 1985 (in Japanese)
[3] Unsworth, A et al., *Engng Med*, 17(3):101-104, 1988.
[4] Murakami, T et al., *Proc IMech, Part H: J Engng Med*, 212:23-35, 1998.
[5] Suzuki, A et al., *Proc World Tribology Congress 2013 Torino*, in USB, 2013.
[6] Murakami, T et al., *ORS 2014*, PS2 1177, 2014.

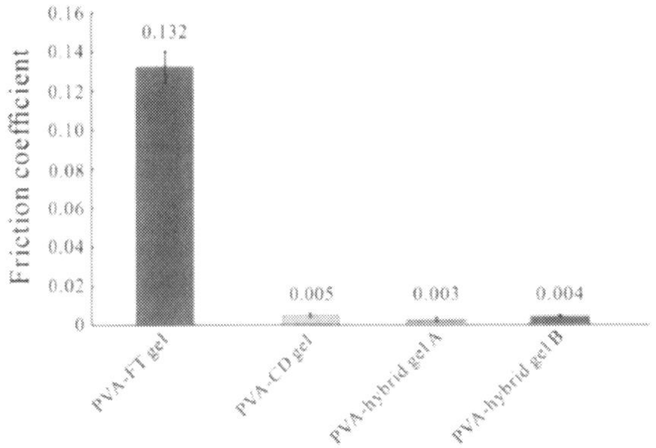

Fig.3 Friction coefficient of PVA hydrogels at the steady state in friction test

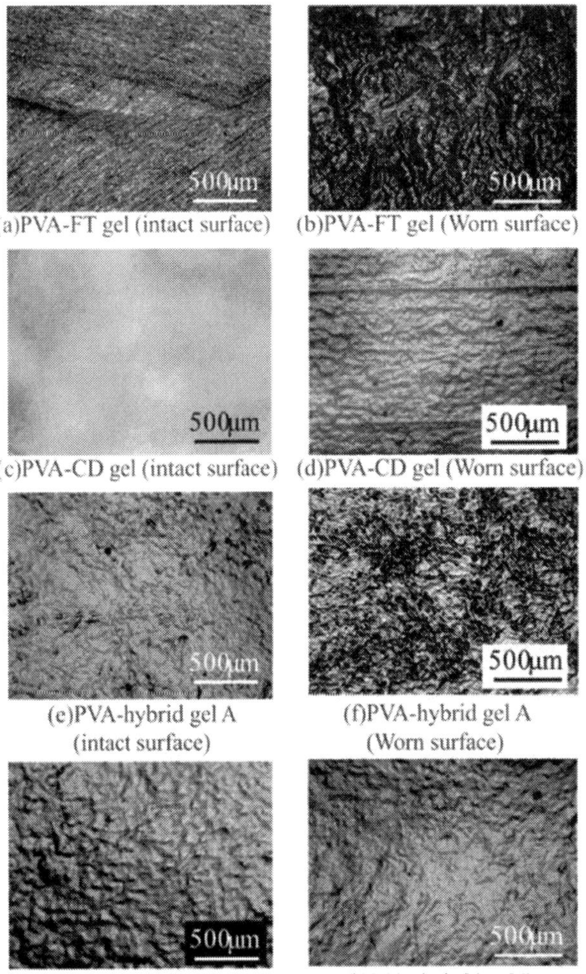

(a)PVA-FT gel (intact surface) (b)PVA-FT gel (Worn surface)
(c)PVA-CD gel (intact surface) (d)PVA-CD gel (Worn surface)
(e)PVA-hybrid gel A (intact surface) (f)PVA-hybrid gel A (Worn surface)
(g)PVA-hybrid gel B (intact surface) (h)PVA-hybrid gel B (Worn surface)

Fig.4 Microscopic images of intact and worn surface of PVA hydrogels

RELATIONSHIP OF WEAR PARTICLES OF POLY(VINYL ALCOHOL) HYDROGEL AND IMMUNE RESPONSE OF MACROPHAGE

Seiji Omata (1), Yoshinori Sawae (1,2), Teruo Murakami (1)

(1) Research Center for Advanced Biomechanics
Kyushu University
Fukuoka, Japan

(2) Department of Mechanical Engineering,
Faculty of Engineering,
Kyushu University. Fukuoka, Japan,

INTRODUCTION

Interventional surgery such as total joint replacement for treatment of osteoarthritis (OA) is proficiently induces to promptly recover quality of life. The wear resistance of joint materials has been improved in most of total joint replacements, but significant wear occurs under severe frictional condition. In such cases, the endurance life of the artificial joints is limited to 10–20 years by aseptic loosening owing to wear particle-induced osteolysis as the active bone resorption by osteoclasts associated with the wear particles.

Since biocompatible hydrogels for artificial cartilage implant (ACI) have high lubricity, several researchers have focused on modifying the surface of the artificial joint materials with soft layer to reduce the wear and extend the endurance limit of the implant. Recently, several clinical investigations have reported that the poly(vinyl alcohol) (PVA) hydrogels were used as artificial cartilage for repair of locally damaged cartilage and remained unscratched for over 2 years [1]. However, it is necessary to keep making an effort of these clinical investigations for long term in the future.

For developing a novel biotribological material for ACI, Murakami *et al.* focused on a PVA hydrogel as an artificial cartilage biomaterial with ultra-low wear and friction coefficient [2]. In this paper, to confirm the validity and safety of the gel in human joints, we observed the morphology of wear particles and assessed macrophage immune responses by applying hydrogel wear particles.

METHODS

First, we conducted tribological tests for collecting PVA wear particles. We prepared a PVA hydrogel, by repeated freezing-thawing method (PVA-FT). To collect PVA hydrogel wear particles, we use a uniaxial reciprocating ball-on-plate tribometer as shown in Fig. 1. The applied load was 4.9 N and the maximum contact pressure was approximately 0.1 MPa. Sliding speed was 50 mm/s. The reciprocating stroke was 50 mm. Total sliding distance was 3.0 km in atmospheric air at 37 °C and 100 %rh. Lubricant is ultra-pure water. To assess PVA hydrogel wear particle, we observed dried PVA wear particles using a high-resolution field emission gun scanning electron microscope (SEM; SU8000; Hitachi High-Technologies).

Second, we assessed the effect of the PVA wear particles on cytotoxicity and immune response of macrophages. To estimate the macrophage immune response to wear particles, we used the PMA-stimulated THP-1 cell line as the macrophage. We prepared two control media; the basal medium was the negative control (NegC) and culture medium with 1 μg/mL of lipopolysaccharide (LPS) was the positive control (PosC). For assaying proinflammatory cytokines in culture medium, we measured IL-1β and TNF-α levels using an enzyme-linked immune sorbent assay (ELISA) with a human IL-1β ELISA Kit.

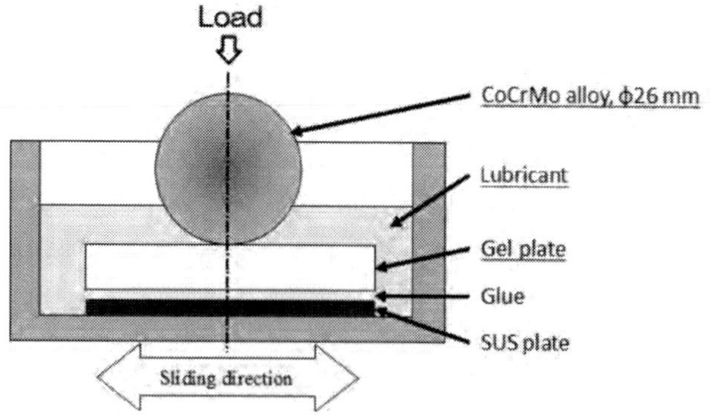

Fig. 1. Schematic drawing of a uniaxial reciprocating ball-on-plate tribometer for preparing PVA wear particles.

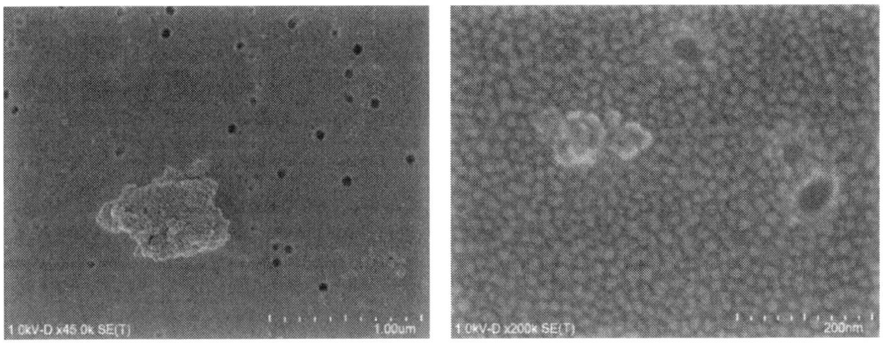

(a) (b)

Fig. 2. SEM images of large PVA wear particles of PVA-FT (WP-FT).

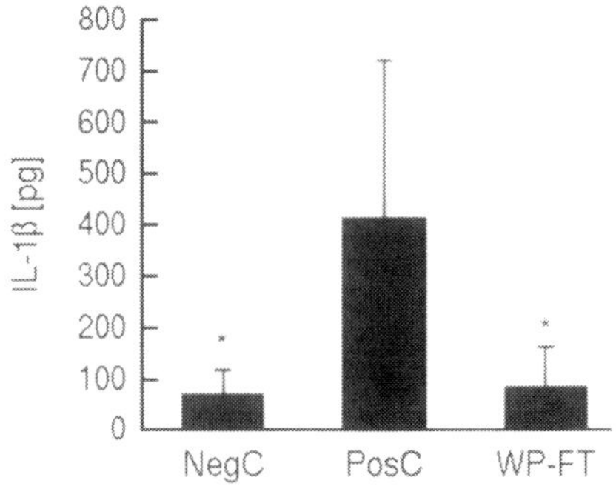

Fig. 3. Cytokine syntheses of IL-1β by applying PVA-FT wear particle (WP-FT). Asterisks (*) indicate statistical significant respective culture groups compared to positive control (PosC), *: $d > 1$, $P < 0.01$, $(1 - \beta) > 0.8$.

In statistical analysis, ELISA results were expressed as the mean ± standard deviation (*SD*). The significance of the difference between each experimental group was evaluated using the two-tailed Welch's *t*-test. If results of pairwise comparisons between two groups were $d > 1$, $P < 0.01$ and $(1 - \beta) > 0.8$ simultaneously, we judged the difference as significant in the tangent modulus, where d is Cohen's d, a kind of the effect size; P is the level of significance; and $(1 - \beta)$ is the power of the test that had been calculated using the R language.

RESULTS

We examined SEM images of dried PVA wear particles as shown in Fig. 2. Although the range of particles between 0.5 and 5 μm was chiefly observed (Fig. 2(a)), a few large-sized particles such as greater than 5 μm were found. We found a number of submicron-sized wear particles approximately 100 nm. Most micron-sized particles were spherical and elongated, and appeared to be aggregations of nano-sized PVA particles approximately 20 nm in diameter. These nano particles were attached to the surface of the 0.05 μm pore-sized filter as shown in Figs. 2(b).

We measured cytokines released into the culture medium using ELISA for assessing the influence of PVA wear particles on the immune response of macrophage. We focused on IL-1β as the most important marker cytokines of the immune response as shown in Fig. 3. Exposing macrophages to wear particles in the particle-stimulated group, compared to negative control, did not change cytokine syntheses. In contrast, positive control group was significantly increase biosynthesis of both cytokines.

DISCUSSION

To develop a novel artificial cartilage with superior tribological function, we surveyed whether PVA hydrogel wear particles were nontoxic and biocompatible. We observed the morphology of the dried wear particles using SEM for evaluating PVA wear particles. SEM observations showed that dried micron-sized PVA wear particles were formed as larger-sized particles by close packing of 20- to 50-nm-sized particles. Many of these nanoparticles homogeneously adhered to the particle-capture filter. Then, dynamic light scattering measurements confirmed that the wear particles were mostly nano-sized (data not shown). We believe that the nanoparticles occurred in association with the wear test because microcrystals, which were crosslinked due to hydrogen bonds in the PVA hydrogel, were gradually scraped off by frictional shear deformation. Although the wear particles are small during the early period, large-sized particles also were formed when the sliding distance is lengthened [3]. Thus, many of the micron-sized wear particles were likely generated during the wear test in this study.

Since proinflammatory cytokines which are synthesized by phagocytizing the wear particles are generally-cited as a major factor of osteolysis and aseptic loosening, we investigated the effect of the PVA hydrogel wear particles on macrophage immune response by assessing the IL-1β cytokine as shown in Fig. 3. At the results, few particle-stimulated macrophages were positive stained for cytokines compared to the negative control, and there were remarkably fewer positively stained macrophages than the LPS-stimulated positive control. ELISA did not show any significant increase in IL-1β synthesis in culture medium, compared to negative control. We confirmed that PVA wear particles did not induce macrophage immune response. Thus, PVA hydrogels not only as a bulk material but also its wear particles, have remarkable biocompatibility.

SUMMARY

We surveyed the effect of PVA hydrogel wear particles on the immune response of THP-1 macrophages using a freezing-thawing PVA hydrogel, PVA-FT gel, to investigate the possibility whether these particles lead to aseptic loosing of the artificial cartilage implant system. Most of the particles were nanoparticles up to 50 nm in diameter. Considering the particle volume distribution, there were considerable micron-sized particles over 5 μm in diameter. There were very few micron- and submicron-sized wear particles around 1 μm in diameter in SEM observations of dried PVA wear particles. Cytokine synthesis in the PVA-FT wear particle-stimulated group was significantly lower than that of the LPS-stimulated positive control. Thus, PVA wear particles may not affect the macrophage immune response.

ACKNOWLEDGEMENTS

The financial support was given by the Grant-in-Aid for Specially Promoted Research of Japan Society for the Promotion of Science (JSPS KAKENHI Grant Number: 23000011).

REFERENCES

[1] Baker, M.I. *et al.*, *J Biomed. Mater. Res. Part B*, 100B:47–56, 2014.
[2] Murakami, T. *et al.*, *Proc. IMechE Part J: J. Eng. Tribol.* 228 (2014) 1099–1111.
[3] Ikeda, K. *et al.*, *Tribol. Online*, 8:122–127, 2013.

SB³C2015
Summer Biomechanics, Bioengineering and Biotransport Conference
June 17-20, 2015,Snowbird Resort, Utah, USA
SB³C2015-1080

A STRAIN BASED APPROACH TO QUANTIFY NON-AFFINE BEHAVIOR OF THREE DIMENSIONAL RANDOM NETWORK

Hamed Hatami-Marbini (1)

(1) School of Mechanical & Aerospace Engineering
Oklahoma State University
Stillwater, OK

INTRODUCTION

Three Dimensional randomly cross-linked filamentous networks are the building blocks of many different biological systems such as cytoskeleton, extracellular matrix, and tissue scaffolds. For example, cytoskeleton is a complex network of F-actin, microtubules, intermediate filaments, binding proteins, and different cross-link filaments which plays numerous roles such as governing cell migration, controlling cell shape, and transmitting force [1]. There have been many different studies in which scientists investigated the mechanical properties of fibrous structures in order to understand and explain the physical origins of their complex response. A majority of these previous have focused on modelling fibrous materials by a two dimensional intertwined fibrous networks. Based on these pioneering studies, it is well accepted that the deformation of these structures is non-affine over a wide range of scales. We have recently developed a new non-affinity measure based on the gradients of the displacement field into which the probing length scale is incorporated [2]. This non-affinity measure was used to study the degree of non-affinity of two dimensional networks. In the present work, we extend this strain-based non-affinity to three dimension and investigate the influence of the characteristic length scales on non-affine behavior of three dimensional networks.

METHODS

The first and essential step in modeling biological fibrous networks is to obtain and reconstruct their true random microstructure. Here, a three dimensional fiber network is generated by randomly placing fibers of length L_0 in a cubic domain of size L, Fig. 1. The shortest distance between all fibers are determined and those closer than a critical value are assumed to be rigidly connected. The fibers are assumed to have a circular cross-section and are characterized by axial and bending stiffness. There are three characteristic length scales in the system. The fiber length, L_0, and the mean segment length, l_c, which are representative of the network structure, and the square root of the ratio of bending stiffness and axial stiffness, l_b, which denotes the relative importance of bending and axial stiffness. For the cylindrical filaments considered in the present study, l_b is equal to half of the cross-sectional radius. The total energy of the system is written as the sum of the strain energies associated with axial and bending deformation of all fibers that are present in the simulation box. The deformation of the system loaded by displacement-imposed boundary conditions is obtained via an energy minimization procedure. Once the nodal displacements are obtained, strains are computed on various probing length scales by choosing a quadruplet of network nodes such that they form approximately an equilateral tetrahedron. The displacement field u_i, $i=1,2,3$ within a tetrahedron is written in terms of its nodal displacements using the finite element interpolation functions. Then, the constant strain field ε_{ij} $i,j=1,2,3$ associated with each tetrahedron is obtained using

$$\varepsilon_{ij}=1/2(u_{i,j}+u_{j,i}).$$

The resulting gradients are an average of the underlying mechanical fields over the length scale r which is equal to the cube root of the volume of the respective tetrahedron. The three dimensional non-affinity measure is defined as the fluctuation of the actual deformation gradients relative to their affine estimates, i.e.

$$H_{ij}=<(\varepsilon_{ij}-\varepsilon_{ij}^{aff})^2>_r.$$

Note that the index r denotes the probing length scale and the symbol $<\ >_r$ represent the average of the inside quantity over all possible length scale r.

RESULTS

Figure 2 shows different components of non-affinity measure for a network with fiber number density of 800 fiber per unit volume when subjected to uniform uniaxial tensile strain in the x_1-direction. It is seen that all components of the $\boldsymbol{H}$ tensor is non-affine and follow the same scaling. The cut-off is at the large scale proportional to the cube root of the cubic simulation box.

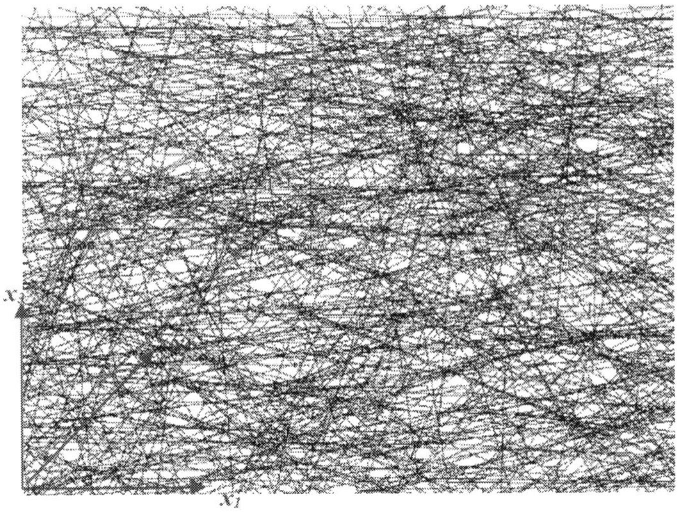

Figure 1. A representation of a three dimensional random fiber network.

DISCUSSION

The present work was performed to investigate how the non-affinity of three dimensional fiber networks depends on different characteristics of these complex systems. It is observed that all components of the non-affinity measure follow a power law scaling with r over the entire range of probed length scales, Figure. 2. It is also seen that the scaling exponent does not significantly depend on far field loading and fiber initial length L_0. Therefore, it is concluded that the elastic fields of three dimensional random fiber networks, defined as the square of fluctuations of the strain field, exhibits a power law dependence on the scale of observation. The other details of this study and a comprehensive analysis of the findings will be presented in a forthcoming publication.

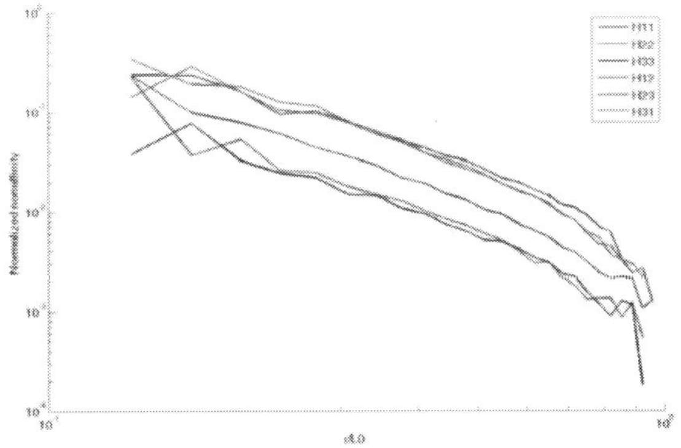

Figure 2. Different components of the normalized Non-affinity measure as a function of normalized length scale.

ACKNOWLEDGEMENTS

This project has been funded in whole or in part with by National Science Foundation grant CMMI-1351461.

REFERENCES

[1] Hatami-Marbini, H, Mofrad, M, in Complex Fluids in Biological Systems: Experiment, Theory, and Computation, 187-205, 2015

[2] Hatami-Marbini, H, Picu, RC., Phys Rev E, 062103, 2008

DYNAMIC FLUID FLOW LOADING INDUCED *IN SITU* OSTEOCYTIC CALCIUM OSCILLATIONS IN AN INTACT MOUSE FEMUR

M. Hu (1), G. Tian (2), YX. Qin (1)

(1) Department of Biomedical Engineering
Stony Brook University
Stony Brook, NY, USA

(2) CMIC-Two Photon Imaging Center
Stony Brook University
Stony Brook, NY, USA

INTRODUCTION

Current clinical evidences point to the critical role of mechanotransduction in various bone processes. The magnitude of mechanotransductory signals within bones can be influenced by distribution of intramedullary pressure (ImP) [1]. Direct dynamic fluid loading into mouse marrow cavity has shown to influence ImP and bone fluid flow (BFF) that leads to osteogenic response [2]. Although osteocytes have been suggested as the major mechanosensors in bone, how *in situ* osteocytes embedded within a mineralized bone matrix respond to the mechanical stimuli derived from direct dynamic fluid flow stimulation is still unclear. While *in vitro* osteocytes show unique calcium oscillations to fluid shear, the objective of this study was to visualize and quantify calcium responses of *in situ* osteocytes in response to dynamic fluid flow stimulation into the marrow cavity of an intact *ex vivo* mouse femur via a confocal imaging technique.

METHODS

We have recently developed a dynamic fluid flow loading system for intact mouse long bones and confocal imaging observation of instant osteocytic Ca^{2+} response to the applied stimuli of various loading frequencies. Stony Brook University IACUC approved all animal protocols. Female C57BL/6J mice of 3 months old were used to obtain fresh femur samples immediately after euthanasia by CO_2 inhalation. Before imaging, the bone samples were incubated in DMEM with 5% FBS and 1% penicilin/streptomycin. A hole was drilled into the marrow cavity from the distal end by using a 24-gauge micro drill bit. Fluo-8 AM (Santa Cruz Biotechnology, Inc., Santa Cruz, CA) was used for calcium fluorescent label for 2 hours before confocal imaging [3]. For dynamic fluid flow loading, a 24-gauge catheter was connected to a water-filled syringe pump to tightly seal the drilled hole of each femur sample. The syringe pump was controlled by a function generator that set the loading frequencies and amplitudes (Figure 1B). Loading frequencies of 1 Hz, 5 Hz, 10 Hz, and 20 Hz with 1 V of constant loading magnitude were applied to the bone samples (n = 5 per group) for 10 sec of baseline, 30 sec of loading, and 10 sec of post loading. Real time confocal imaging with 40X objectives, 488-nm laser excitation, and 2 frames/sec was performed to capture the calcium signals of the osteocytes' cell bodies within the bone that was subjected to dynamic fluid flow stimulation. For data analysis, the fluorescent intensity of each cell body was extracted as a function of time using the microscope software. By normalization to the baseline intensity, the percentage of response cells, number of calcium spikes, first spike magnitude, and the time to the first spike were quantified.

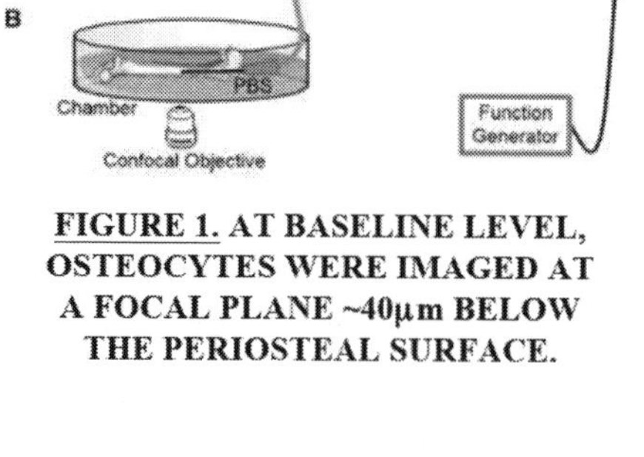

FIGURE 1. AT BASELINE LEVEL, OSTEOCYTES WERE IMAGED AT A FOCAL PLANE ~40μm BELOW THE PERIOSTEAL SURFACE.

RESULTS

At baseline level, osteocytes were imaged at a focal plane ~40mm below the periosteal surface (Figure 1A). Dynamic fluid flow stimulation at 1 Hz did not induce any calcium response within the osteocytes. However, 5 Hz, 10 Hz, and 20 Hz of loading lead to 25%, 93%, and 52% of responsive cells, respectively (Figure 2). Results of number of calcium spikes showed that loading of 5 Hz, 10 Hz, and 20 Hz lead to 1.4, 2.6, and 3.0 spikes, respectively (Figure 2). Results of peak magnitude normalized to the baseline value showed that loading of 10 Hz lead to 5% and 7% greater calcium spike magnitude compared to 5 Hz and 20 Hz of loading, respectively (Figure 2). Moreover, loading of 10 Hz exhibited about 65% and 11% shorter calcium initiation time compared to 5 Hz and 20 Hz of loading, respectively (Figure 2).

DISCUSSION

Interesting and promising data shown in this study provided direct evidence that osteocytes displayed calcium spikes in response to *in situ* dynamic fluid flow stimulation into intact mouse femur in a frequency dependent manner. The number of responsive cells, number of Ca^{2+} spikes, first spike magnitude, and initiation time were dependent on the loading frequencies, where 10 Hz loading seemed to be optimal. Using an advance approach in skeletal mechanobiology, real time observation of the Ca^{2+} signals clearly indicated the instant *in situ* osteocytic response to the mechanical signals derived from dynamic fluid flow stimulation. This study provides advances for our better understanding of bone mechanotransduction, which may lead to great insights into current clinical challenges.

ACKNOWLEDGEMENTS

This work is kindly supported by the National Institute of Health (R01 AR52379 and R01AR61821), the US Army Medical Research and Materiel Command, and the National Space Biomedical Research Institute through NASA contract NCC 9-58.

REFERENCES

[1] Qin et al., *J Biomech.* 2009 Jan 19;42(2):140-5. [2] Kwon et al., *J Bone Miner Res.* 2010 Aug;25(8):1798-807. [3] Jing et al., FASEB J. 2014 Apr;28(4):1582-92.

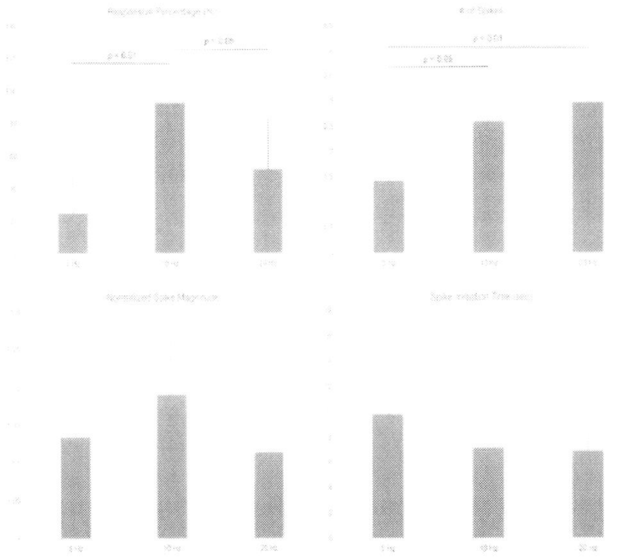

FIGURE 2. DYNAMIC FLUID FLOW STIMULATION AT 1Hz DID NOT INDUCE ANY CALCIUM RESPONSE WITHIN THE OSTEOCYTES. HOWEVER, 5Hz, 10Hz, AND 20Hz OF LOADING LEAD TO 25%, 93%, AND 52% OF RESPONSIVE CELLS, RESPECTIVELY. NUMBER OF CALCIUM SPIKES, PEAK MAGNITUDE NORMALIZED TO THE BASELINE VALUE, AND THE INITIATION TIME OF THE FIRST OBSERVED CALCIUM SIGNAL SUGGEST FREQUENCY DEPENDENT RESPONSES.

SB³C2015
Summer Biomechanics, Bioengineering and Biotransport Conference
June 17-20, 2015,Snowbird Resort, Utah, USA
SB³C2015-1082

THROMBOTIC RISK ASSESSMENT USING TRANSLUMINAL ATTENUATION GRADIENT AND COMPUTATIONAL MODELLING IN KAWASAKI DISEASE PATIENTS WITH CORONARY ARTERY ANEURYSMS

N. Grande Gutierrez (1), A. M. Kahn (2), O. Shirinsky (3), N.V. Gagarina (3), G. A. Lyskina (3), R. Fukazawa (4), S. Ogawa (4), J. C. Burns (2), A. L. Marsden (1)

(1) Mechanical and Aerospace Engineering
University of California San Diego
San Diego, California, USA

(2) School of Medicine
University of California San Diego
San Diego, California, USA

(3) Sechenov First Moscow State University
Moscow, Russian Federation

(4) Department of Pediatrics
Nippon Medical School Hospital
Tokyo, Japan

INTRODUCTION

Kawasaki Disease (KD) is the most common cause of acquired heart disease in childhood in developed countries. KD can result in coronary aneurysms in up to 25% of patients if not treated with intravenous immunoglobulin therapy within the first ten days of fever [1]. Coronary artery aneurysms (CAA) put KD patients at risk for thrombosis, myocardial infarction and sudden death. Clinical guidelines recommend CAA diameter >8mm as the criterion for initiating systemic anticoagulation [2, 3].

Patient specific modeling and hemodynamic simulations in KD patients suggest that hemodynamic data can predict regions with increased risk of thrombus formation and may be superior to simple assessment of the diameter of the aneurysm [4]. This additional flow information in CAAs may support clinical decisions and provide quantitative means to decide if KD patients should be put on systemic anticoagulation therapy. Specifically, regions with high Particle Residence Time gradients (PRTg) have correlated with regions of thrombus formation. In addition, results suggested that higher risk of thrombus formation may be linked to more fusiform shape aneurysms rather than saccular aneurysms with greater sphericity.

While patient specific modeling can provide a wealth of hemodynamic information, there are numerous challenges associated with performing these simulations for routine clinical use. Therefore, we propose Transluminal Attenuation Gradient (TAG) analysis as a potential methodology to extract flow information from image data alone in KD patients with CAA. While TAG has been previously introduced to evaluate the functional significance of stenoses in the coronary arteries, it has not previously been used for aneurysms.

The goal of this study is to analyze the correlation between TAG and geometric and hemodynamic parameters in KD patients with CAA and evaluate TAG for prediction of thrombotic risk.

METHODS

TAG is determined from the change in radiological attenuation on CT angiography (CTA) images in Hounsfield units per vessel length and reflects the kinetics of contrast agent in a coronary artery. TAG analysis has been used for characterizing coronary artery stenoses [5, 6]; however this approach has not been used previously for CAA. TAG is calculated based on intensity values at consecutive cross-sections along the vessel from the ostium to distal regions of the vessels until the cross-sectional area falls below 2 mm^2. Seven KD patients with CAA and four normal controls were enrolled in this study. A total of 10 CAA vessels (3 RCA, 6 LAD, 1 LCX) and 6 normal control vessels (3 RCA, 2 LAD, 1 LCX) were analyzed.

Patient-specific anatomic models for flow simulations were constructed from CTA image data from 7 KD and 1 normal control using the open source software package SimVascular (simvascular.org). Cardiovascular flow simulations were performed using the finite element Navier-Stokes solver in SimVascular to obtain pressure and velocity fields in the aneurysm and wall shear stress and Particle Residence Time gradients. Blood was modeled as a Newtonian fluid and walls were assumed to be rigid. Simulations used a Lumped parameter Network (LPN) to model the heart and vascular boundary conditions coupled to a linear solver. For coronary arteries we used specialized boundary conditions that account for changing coronary resistance due to heart contraction.

Aneurysms were characterized using the following geometric parameters: Maximum aneurysm diameter, Aneurysm Shape Index (ASI) and Sphericity Index. ASI was calculated as the ratio of the aneurysm length to the maximum aneurysmal diameter. Aneurysm sphericity was defined using the following equation:

$$\psi = \frac{\pi^{\frac{1}{3}}(6V)^{\frac{2}{3}}}{A} \quad (1)$$

where V is the aneurysm volume and A is the surface area.

RESULTS

TAG values for the CAA vessels were significantly different compared with the non-aneurysmal vessels (mean -29.19 vs. -11.12, p=0.016) as shown in Figure 3. A Wilcoxon rank sum test was used to determine p-values. Figure 1 and Figure 2 show representative examples of TAG measurements on a normal vessel versus a CAA vessel for right and left coronary arteries respectively.

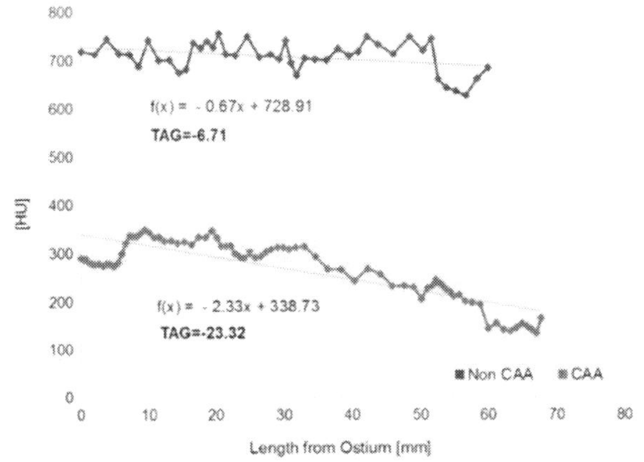

Figure 1 Representative Transluminal Attenuation Gradient measurement for a Left Anterior Descending (LAD) coronary artery.

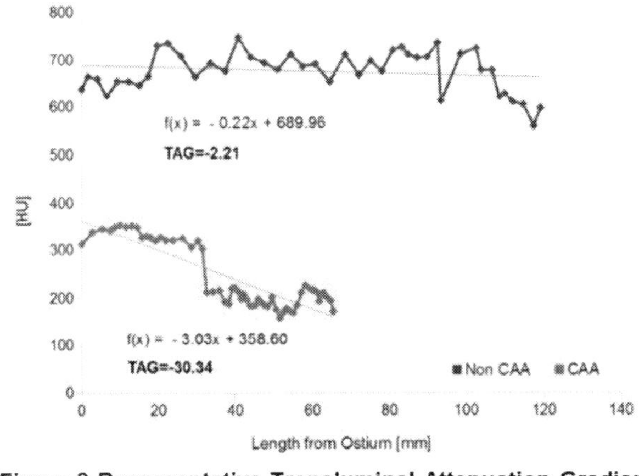

Figure 2 Representative Transluminal Attenuation Gradient measurement for a Right Coronary Artery (RCA).

In addition, TAG values showed negligible correlation (-0.15) with maximum diameter of the aneurysm while showing high correlation with ASI and Sphericity Index (0.85, -0.89). Correlation was measured using Pearson's linear correlation coefficient.

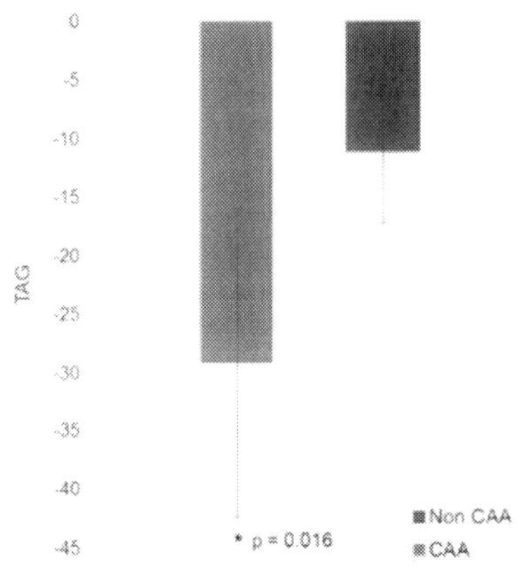

Figure 3 Comparison of TAG mean value and standard deviation between CAA vessels and control group.

TAG was compared to PRTg and other hemodynamic data obtained for each patient. In general, vessels with abnormal TAG values had regions of low wall shear stress and hemodynamic parameters in ranges previously related with increased thrombotic risk.

DISCUSSION

Our results demonstrate that TAG is significantly different in KD patients with CAA compared to the control group with no aneurysms. In agreement with previous simulation studies, TAG values for right and left coronaries in patients with aneurysms should be taken into account in addition to diameter measurements when analyzing CAA severity. However, a larger sample size is needed in order to obtain statistically significant results for CAA grouped by right vs. left coronary artery compared to normal control vessels.

For the group of patients analyzed there was very low correlation between maximum aneurysm diameter and TAG values, suggesting that quantitative metrics derived from the combination between TAG and computer cardiovascular simulations may be superior to diameter alone for thrombotic risk assessment for KD patients.

This study suggests a potential use of TAG measurements to evaluate CAA in KD patients. Thrombotic risk stratification for CAA and decisions about initiation of systemic anticoagulation in KD patients may be improved by incorporating TAG and should be evaluated in future prospective studies.

ACKNOWLEDGEMENTS

'la Caixa' Fellowship, NSF CAREER Award OCI 1150184, Burroughs Wellcome Fund Career Award at the Scientific Interface.

REFERENCES

[1] Kato, H et al., Circulation, 94(6):1379–1385, 1997.
[2] Suda, K et al., Circulation 123(17):1836–1842, 2011.
[3] Sugahara, Y et al., Pediatr Cardiol 29(2):398–401, 2008.
[4] Sengupta, D et al., Biomech Model Mechanobiol.13(6):1261-76, 2014.
[5] Steigner, M et al., Circ Cardiovasc Imaging, 3(2):179-86, 2010.
[6]Choi, J et al., JACC: Cardiovasc Imaging, 4(11):1149–1157, 2011.

SB³C2015
Summer Biomechanics, Bioengineering and Biotransport Conference
June 17-20, 2015,Snowbird Resort, Utah, USA
SB³C2015-1083

TRACKING MASS TRANSPORT OF A DRUG SURROGATE IN PORCINE CORONARY TISSUE USING PHOTOACOUSTIC IMAGING AND SPECTROSCOPY

Kenneth J. Furdella (1), Russell S. Witte (1, 2), Jonathan P. Vande Geest (1, 3)

(1) Department of Biomedical Engineering
University of Arizona
Tucson, AZ, United States

(2) Department of Medical Imaging and Optical Sciences
University of Arizona
Tucson, AZ, United States

(3) Department of Aerospace and Mechanical Engineering
University of Arizona
Tucson, AZ, United States

INTRODUCTION

The introduction of drug eluting stents (DES) has dropped restenosis rates to below 5% [1]. Statistically, each year approximately 484,000 people receive a DES. Even with a low restenosis rate of 5%, the number of reintervention cases performed per year is over 20,000 [2, 3].

Sirolimus and its derivatives are commonly used drugs loaded onto a DES. Sirolimus stops the proliferation and migration of smooth muscle cells in the coronary arteries (CAs), which helps prevent restenosis. In order to mimic sirolimus transport, a drug surrogate, DiI, was used to evaluate mass diffusion into the CA and myocardium. DiI was chosen due to its similarity to sirolimus, such as its lipophilic properties and molecular weight [4].

Intravascular imaging modalities have become the standard for restenotic imaging, but are limited in their depth of penetration, or lack contrast for tracking the drug in the tissue. Although intravascular ultrasound can achieve depths of several centimeters, it typically has inadequate contrast for tracking the delivery and distribution of a drug or drug surrogate. Alternatively, photoacoustic (PA) imaging offers sufficient depth and contrast for tracking an optical agent in the heart or blood vessel. PA is an imaging modality that combines a non-ionizing laser pulse with ultrasound technology. First, a laser pulse causes transient heating and thermal expansion of the tissue. This rapid expansion generates pressure waves (ultrasound) which are detected by ultrasound receivers, which are then used to form an image proportional to the optical absorption of the sample. Multiple wavelengths can further be used to identify compounds and estimate concentrations using PA spectroscopy [5]. Finally, by combing pulse echo ultrasound (structure/anatomy) with PA imaging (optical absorption), more information about the tissue can be obtained. In this study, we used PA imaging and spectroscopy to estimate the distribution of a drug surrogate (DiI) within the CA and a sectioned porcine myocardium at different concentrations. PA images were taken to create a relative calibration curve of DiI diffusion into the myocardium.

METHODS

The optical absorption spectrum of DiI was determined by using a spectrometer (Ocean Optics). Specifically, DiI has two notable peaks at 530 nm and 570 nm and the entire spectrum was used to determine the pixels that correlated to the DiI absorption spectrum.

For PA imaging, a wavelength tunable laser (Surelite-II OPO, Continuum) from 430 nm – 670 nm with 5 nsec pulses was coupled into a 1.5 mm diameter optical fiber (CeramOptec). Our patented setup of the PA enabling device (PED) is seen in Figure 1 allows for PA and pulse echo (PE) imaging with a clinical scanner and linear array [6].

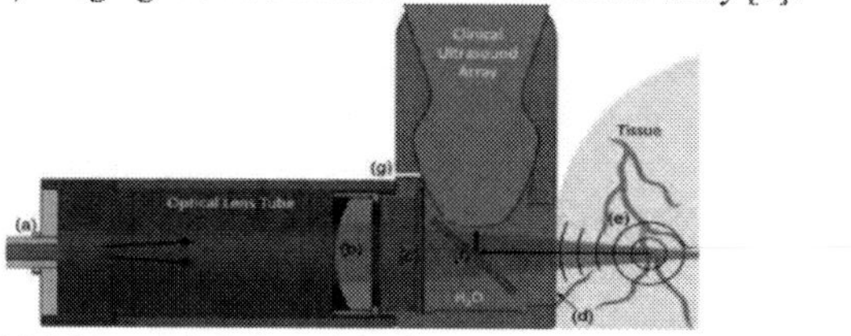

Figure 1:Light travels left to right from the optical fiber (a), to the optical lens tube (b), passes the glass window (c) , through water and transparent membrane in the PED, and into the tissue (e). The generated sound wave is scattered from the tissue back towards the PED. The 90° glass plate (f) reflects the sound wave into the ultrasound array where it is detected and processed. (d) The drain for the PED [7].

In one series of experiments, an in situ porcine CA (proximal and distal) was cut, cannulated, and injected with water, which can be seen in Figure 2. A PE image (not shown) was acquired to locate the blood vessel. A baseline PA image was taken to identify residual absorbers in the vessel. All PA and PE images were recorded with a 10 MHz clinical US linear array (Zonare, L10-5). The laser incident on the heart was approximately 2mJ for all testing. Next, the CA was injected with DiI at a concentration of 1/50, 1/10, and 1/2 of 0.12 mg/ml. DiI concentrations were injected into the vessel, and after 20 minutes the vessel was flushed with water, and PA imaged. The DiI spectrum was constructed based on the linear combination of the DiI absorption spectrum [7].

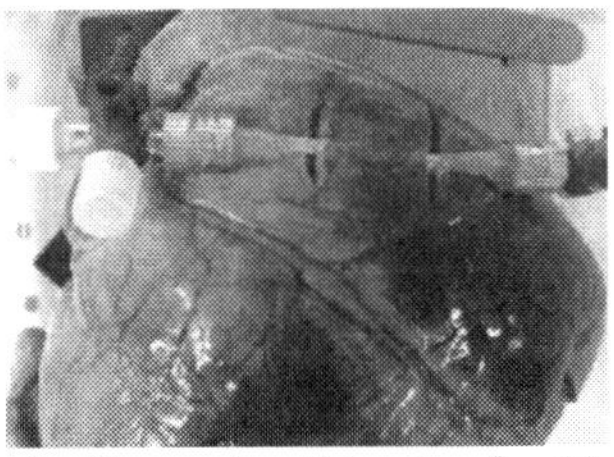

Figure 2: In situ porcine CA setup for PA imaging

In another study, porcine myocardium was excised from an intact porcine heart and sectioned into 1x1x0.5cm rectangular sections. The samples were soaked in DiI for 24 hours at concentrations of 0, 2%, 4%, 8%, 16%, 32%, 64%, and 100% concentration of 0.12 mg/ml DiI solution. The samples were removed from DiI solution and embedded in 1% agarose gel. A 3D scan of the samples were taken using B mode PA imaging with images taken every 100μm with a computer-controlled servo (Velmex). A calibration curve was generated by plotting the concentrations vs. intensity of the PA signal. After PA imaging, the samples were frozen in liquid nitrogen, cryosectioned, and imaged with fluorescence microscopy for comparison.

RESULTS

In the in situ porcine CA study, the correlation between the DiI optical absorption spectrum and PA spectrum was determined. In the images, the pixels that correlated with the DiI spectrum can be seen in Figure 3 (left). A cross section of the CA can be seen in Figure 3A (right) and is highlighted in the green box. The green arrow shows the top of the CA. Figure 3B and C (right) show a zoomed in portion of the CA at concentrations of 0.1 and 0.5, respectively. Figure 3D (right), shows the absorption spectrum at the top of the CA.

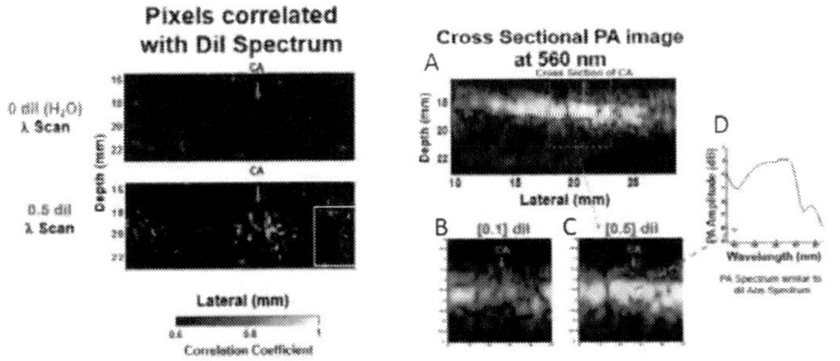

Figure 3: *Left* pixels correlated with DiI spectrum at 0% and 50% DiI. *Right* (A) Cross section of in situ CA using PA imaging. The green arrow denotes the top of the CA. (B & C) PA showing the intensity of DiI in CA at 0.1 and 0.5, respectively. (D) The photoacoustic absorption spectrum at the top of the CA for the 0.5 concentration of DiI.

Representative PA and PE images of the excised myocardium can be seen in Figure 4A. In the PA images, the brighter the pixels the higher the correlation to the DiI spectrum, using a linear combination of the DiI absorption spectrum. The calibration curve comparing the PA intensity vs. concentration can be seen in Figure 4B and plotted on a $\log_2$ scale.

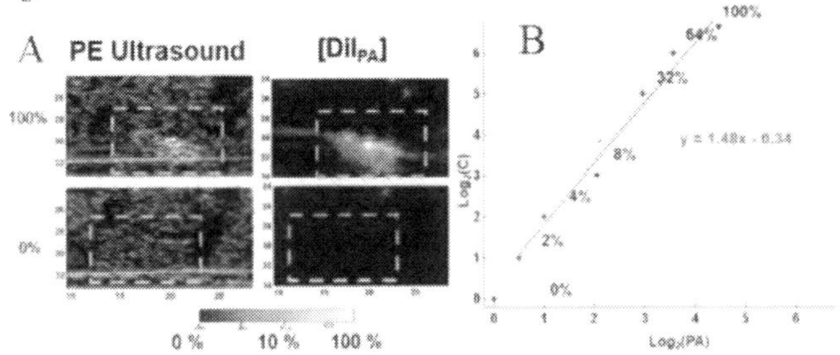

Figure 4: (A) PA and PE images of 0% and 100% DiI. (B) Calibration curve of PA intensity vs. DiI concentration and best linear fit.

DISCUSSION

Using two different preparations we have demonstrated that DiI can be detected and quantified in the porcine CA and myocardium using PA imaging and spectroscopy. At multiple wavelengths, DiI can be separated from the intrinsic background absorption of cardiac tissue. Also, the intensity of the PA signal increased linearly with the concentration of DiI in the tissue. The equation of line for the calibration curve (from Figure 4B) was found to be y=1.48x-0.34. This curve can act as a reference to compare future PA experiments. In the future, we plan on testing dyes with different hydrophobicities and evaluating diffusion of multiple dyes simultaneously in the CA.

One potential limitation is that the resolution of the PA imaging system described in this study (~300 μm) may not be sufficient for tracking absorbers and monitoring drug delivery inside the coronary vessel wall. However, this may still be achievable using high resolution intravascular photoacoustic (IVPA) systems. We anticipate that these baseline studies may lead to future *in vivo* animal and human studies for monitoring drug delivery to the CA and preventing restenosis and other cardiac abnormalities.

REFERENCES

1. Sidhu, S., et al., *A meta-analysis of trials comparing Cypher and Taxus stents in patients with obstructive coronary artery disease.* Br J Clin Pharmacol, 2006. **61**(6): p. 720-6.
2. Go, A.S., et al., *Heart Disease and Stroke Statistics--2014 Update: A Report From the American Heart Association.* Circulation, 2013.
3. Auerbach, D., Maeda, J., and Steiner, C., *Hospital Stays with Cardiac Stents, 2009.* Agency for Healthcare Research and Quality. Rockville, MD. , April 2011(HCUP Statistical Brief #128).
4. J. T. Keyes, B. R. Simon, and J. P. Vande Geest, "Location-dependent coronary artery diffusive and convective mass transport properties of a lipophilic drug surrogate measured using nonlinear microscopy.," *Pharmaceutical research*, vol. 30, no. 4, pp. 1147–60, Apr. 2013.
5. Su, J. L., et al. (2009). "Photoacoustic imaging of coronary artery stents." Opt Express **17(22): 19894-19901.**
6. L. G. Montilla, R. Olafsson, D. R. Bauer, and R. S. Witte, "Real-time photoacoustic and ultrasound imaging : a simple solution for clinical ultrasound systems with linear arrays," vol. 58, pp. 1–12, 2013.
7. Bauer, Daniel R., et al. "Simultaneous detection of multiple contrast agents with photoacoustic spectroscopy." *Ultrasonics Symposium (IUS), 2012 IEEE International.* IEEE, 2012.

SB³C2015-1084

WNT/BETA-CATENIN SIGNALING PATHWAYS CONTRIBUTES TO DYNAMIC FLUID FLOW LOADING INDUCED *IN SITU* OSTEOCYTIC CALCIUM OSCILLATIONS IN AN INTACT MOUSE FEMUR

M. Hu (1), G. Tian (2), Y-X. Qin (1)

(1) Department of Biomedical Engineering
Stony Brook University
Stony Brook, NY, USA

(2) CMIC-Two Photon Imaging Center
Stony Brook University
Stony Brook, NY, USA

INTRODUCTION

Intense research on Wnt signaling pathways in bone biology have been focused on for over a decade, which include their critical role in skeletal development, maintaining bone mass, potential therapeutic development for regenerative medicine. Wnt signaling pathways with subtle alterations in the intensity, amplitude, location, and duration would significantly affect skeletal development, as well as bone remodeling and regeneration [1]. Central focuses of the related research fall on how Wnt/Lrp5 signaling regulates osteoblasts and osteocytes, the roles of Wnt signaling players in bone development and osteoclastogenesis, translational studies for clinical therapeutics, and diagnostics centered around inhibiting Wnt pathway antaganists, etc. However, the cross-talks between Wnt/β-catenin signaling with other mechanically induced osteogenic signaling events remain unknown. Specifically, the role of Wnt/β-catenin signaling on osteocytic Ca^{2+} Oscillations, one of the earliest mechanotransduction events in bone cells, is worth the investigation, which may ultimately explain the underlying mechanism of how Wnt/β-catenin affects osteoblastogenesis. Therefore, the objective of this study was to use a confocal imaging technique to visualize and quantify calcium responses of *in situ* osteocytes under dynamic fluid flow stimulation into the marrow cavity of an intact *ex vivo* mouse femur with dynamic blocking of Wnt/β-catenin signaling.

METHODS

In the present study, ICG-001 (Selleck Chemicals, United States and Europe) was used to antagonize Wnt/β-catenin-TCF-mediated transcription. Intact mouse femur samples were collected from female C57BL/6J mice of 3 months old immediately after euthanasia by CO_2 inhalation, followed by incubation in DMEM with 5% FBS and 1% penicillin/streptomycin. A 24-gauge hole was drilled into the marrow cavity from the distal end. By separating into control and inhibitor-treated groups (n=5 per group), the samples in the inhibitor-treated group were pre-incubated with 25μM of ICG-001 for 1 hour at room temperature before Ca^{2+} dye incubation. The samples from both groups were stained with Fluo-8 AM (Santa Cruz Biotechnology, Inc., Santa Cruz, CA) was used for calcium fluorescent label for 2 hours before confocal imaging [2]. For dynamic fluid flow loading, 10Hz with 1V of constant loading magnitude was applied to the bone samples (n = 5 per group) for 10 sec of baseline, 30 sec of loading, and 10 sec of post loading. Real time confocal imaging was performed to capture the calcium signals of the osteocytes. The fluorescent intensity of each cell body was extracted as a function of time for data analysis. Statistical analysis was done by One-way ANOVA.

RESULTS

Data showed significantly reduced osteocytic Ca^{2+} responses when Wnt/β-catenin signaling pathway was blocked by ICG-001. Upon Wnt/β-catenin signaling blockage, the percent responsive cells, number of Ca^{2+} spikes, and the peak magnitude normalized to the baseline value were decreased by 83% ($p < 0.001$), 12%, and 7% ($p < 0.05$), respectively; while the initiation time of the first observed Ca^{2+} signal was increased by 19%, compared to the control (Figure 1).

DISCUSSION

Coincided with our previous study, the controls from the present study showed osteocytes that displayed calcium spikes in response to 10Hz *in situ* dynamic fluid flow stimulation. Advance confocal imaging approach allowed a real time observation of the Ca^{2+} signals, which clearly indicated the instant *in situ* osteocytic response to the mechanical signals derived from dynamic fluid flow stimulation. More impressively, using a Wnt/β-catenin signaling blocker provided a pilot finding of potential cross-talking or inter-dependency between Wnt/β-catenin signaling and Ca2+ influx signaling. This study provides advances for our better understanding of bone mechanotransduction, which may lead to great insights into current clinical challenges.

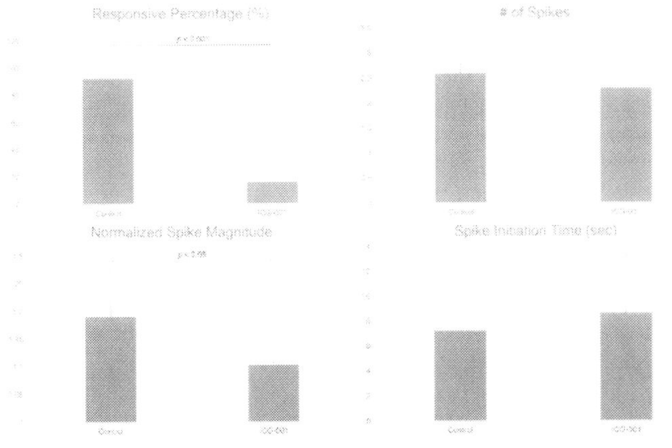

FIGURE 1. Wnt/β-catenin signaling blocker (ICG-001) significantly diminished the Dynamic fluid flow stimulation induced osteocytic Ca2+ response at 10 Hz. Upon Wnt/β-catenin signaling blockage, the percent responsive cells, number of Ca2+ spikes, and the peak magnitude normalized to the baseline value were decreased by 83%, 12%, and 7%, respectively; while the initiation time of the first observed calcium signal was increased by 19%, compared to control.

ACKNOWLEDGEMENTS

This work is kindly supported by the National Institute of Health (R01 AR52379 and R01AR61821), the US Army Medical Research and Materiel Command, and the National Space Biomedical Research Institute through NASA contract NCC 9-58.

REFERENCES

[1] Monroe et al., Gene. 2012 January 15; 492(1): 1-18. [2] Jing et al., *FESEB*. 2014.

SB3C2015
Summer Biomechanics, Bioengineering and Biotransport Conference
June 17-20, 2015,Snowbird Resort, Utah, USA
SB3C2015-1085

ROLE OF CYCLIC STRAIN ON CALCIFIC NODULE FORMATION AMONG AORTIC HEART VALVE CUSPS

Ying Lei (1), Zannatul Ferdous (1)

(1) The Department of Mechanical, Aerospace, and Biomedical Engineering
The University of Tennessee
Knoxville, Tennessee, USA

INTRODUCTION

Calcific aortic valve disease (CAVD) is the leading cause of mortality in the elderly population in developed countries [1]. The end-stage manifestation of this disease is the formation of calcific nodules, which leads to valve thickening, increased stiffness, and improper opening and closing of the valve [2]. To date, there is no pharmacological therapy for CAVD; thus surgical replacement is necessary in most cases [3].

Aortic valvular interstitial cells (VICs) are responsible for maintaining valve integrity and function [4]. They are thought to initiate CAVD due to an inflammatory response [2]. In the aortic valve, calcification predominantly occurs on the fibrosa side, which encounters more disturbed blood flow compared to the ventricularis side. The shear stress and strain field induced by the flow are thought to contribute greatly to the pathology of CAVD [5]. Although previous studies have shown that the aggregation and calcification of VICs can be affected by the strain field, only a handful of studies have assessed the influence of strain levels on nodule formation [6-10]. In addition, since calcification frequently initiates at the non-coronary cusp of aortic valves before affecting the coronary cusps[11], investigating the response of VICs from coronary and non-coronary cusps to physiological vs. pathological strain field will provide new insights into the mechanisms of CAVD.

In this study, we investigated whether VICs isolated from the coronary or the non-coronary cusps would calcify at different amounts when cultured under representative physiological (10%) or pathological (15%) cyclic strain.

METHODS

The left, right and non-coronary cusps were carefully isolated from freshly sacrificed porcine aortic valves and were digested in collagenase II solution in an incubated shaker for 4 hours at 37∘C. The cell solution from each cusp was filtered, centrifuged and then plated into tissue culture flasks with Dulbecco's Modified Eagle's Medium supplemented with 10% fetal bovine serum, 1% penicillin-streptomycin, and 1% L-glutamine.

VICs at passage 4 were seeded onto Bioflex collagen-I coated culture plates (Flexcell International Corporation) at a density of 50,000 cells/cm^2 in 5 ml of regular media. The cells were then cultured under static condition in regular media for 2 days to ensure cell attachment and confluence. At day 2, the plates were placed in the Flexcell-5000 tension system and cells were subjected to either 10% or 15% of cyclic equibiaxial strain at a frequency of 0.3 Hz for 7 days. Experimental wells were cultured in osteogenic media, which is regular media supplemented with β-glycerol phosphate, L-ascorbic acid and dexamethasone. Control wells under strain were cultured in regular media.

On Day 9, the VICs were fixed in 10% formalin and stained for alizarin red and von kossa staining. Positively stained nodules were manually counted under the microscope. The Image J software was used for image analysis to measure the areas and perimeters of nodules. The data obtained was further normalized to respective strain controls. Data represents average and standard error. Statistical significance was determined using single-factor ANOVA. A p-value of ≤0.05 was considered significant.

RESULTS

1. Larger Nodules in Groups Subjected to Bigger Level of Cyclic Strain

For all of the experimental groups (left, right or non-coronary cusps) subjected to either 10% or 15% cyclic strain, the nodules had significantly larger area than their respect static control groups (Fig. 1, $p < 0.05$). The normalized area ranged from 1.4 to 1.8. All of the experimental groups also had significantly greater average perimeter of nodules than their control groups, except the left cusp cells subjected to 15% strain. The nodules in 15% strain groups had larger average normalized area than 10% strain for all three types of cells. However, only the right cusp group subjected to 15% strain had significantly larger average normalized perimeters than 10% strain. For the left and non-coronary cusp cells, the differences between 10% strain and 15% strain groups were not significant.

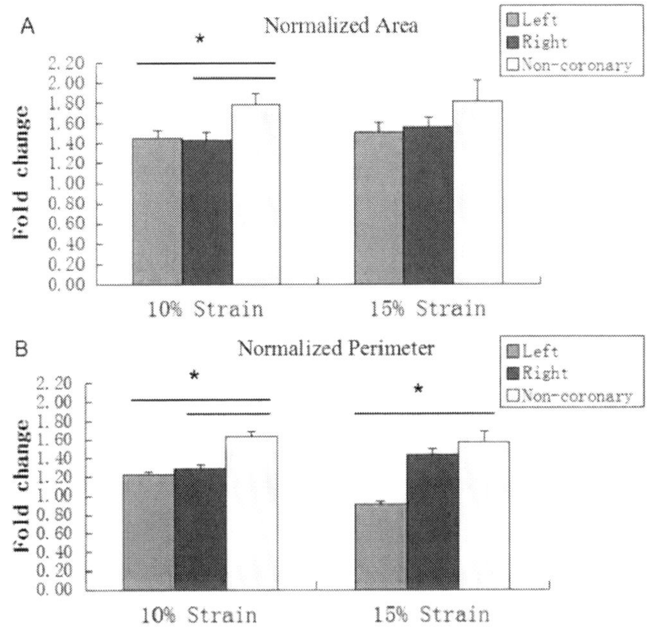

Fig 1. NORMALIZED AREA AND PERIMETER OF CALCIFIC NODULES IN CELLS ISOLATED FROM THE LEFT, RIGHT AND NON-CORONARY CUSPS (* indicates $p < 0.05$.)

2. Larger Nodules in Wells Containing VICs from Non-coronary Cusps than Coronary Cusps.

To determine the response of VICs from coronary and non-coronary cusps to varying levels of strain under calcifying condition, the areas and perimeters of nodules from the groups were measured and compared. Nodules were defined as round with sufficient stain intensity, and contained elongated cells in a radial pattern around the periphery of the nodule. In groups subjected to 10% cyclic strain, the normalized area and perimeter in the non-coronary cusp group were significantly larger than the left and right cusp groups ($p<0.05$). However, the differences between the left and right cusps were not significant. In groups subjected to 15% cyclic strain, the average normalized area and perimeter in the non-coronary cusp group were greater than the coronary cusps groups. However, significance was not achieved for normalized area among the groups under 15% strain. The normalized perimeter in the left cusp group was significantly smaller than the other two groups ($p<0.05$).

DISCUSSION

In this study, we observed that VICs from the non-coronary cusps generated larger nodules than the coronary cusps. This is similar to results from previous studies in our laboratory, where we found that VICs from the non-coronary cusps expressed greater amounts of calcific marker than the coronary cusps in static culture. The altered response of non-coronary cells to the cyclic strain under calcifying condition suggests that they might have inherently different phenotypic characteristics than the coronary cells.

REFERENCES

[1] Ferdous, Z et al., "Differences in valvular and vascular cell responses to strain in osteogenic media." Biomaterials 32.11: 2885-2893. (2011)

[2] Fisher, C et al., "Calcific nodule morphogenesis by heart valve interstitial cells is strain dependent. "Biomechanics and modeling in mechanobiology 12.1: 5-17. (2013)

[3] Lueders, C et al., "Rapid manufacturing techniques for the tissue engineering of human heart valves. "European Journal of Cardio-Thoracic Surgery: ezt510. (2014)

[4] Brody, S et al., "Approaches to heart valve tissue engineering scaffold design." Journal of Biomedical Materials Research Part B: Applied Biomaterials 83, no. 1: 16-43. (2007)

[5] Ferdous, Z et al., "Influence of cyclic strain and decorin deficiency on 3D cellularized collagen matrices." Biomaterials 29.18: 2740-2748. (2008)

[6] Robicsek F et al., "Mechanical stress as cause of aortic valve disease. Presentation of a new aortic root prosthesis." Acta Chir Belg 102(1):1 - 6 (2002)

[7] Merryman WD et al., "Synergistic effects of cyclic tension and transforming growth factor-beta1 on the aortic valve myofibroblast." Cardiovasc Pathol 16(5):268 - 276. (2007)

[8] Simmons CA et al., "Mechanical stimulation and mitogen-activated protein kinase signaling independently regulate osteogenic differentiation and mineralization by calcifying vascular cells." J Biomech 37(10):1531-1541. (2004)

[9] Balachandran K et al., "An ex vivo study of the biological properties of porcine aortic valves in response to circumferential cyclic stretch." Ann Biomed Eng 34(11):1655 - 1665. (2006)

[10] Balachandran K et al., "Elevated cyclic stretch altersmatrix remodeling in aortic valve cusps: implications for degenerative aortic valve disease." Am J Physiol Heart Circ Physiol 296(3):H756 - H764. (2009)

[11] Merryman WD "What modulates the aortic valve interstitial cell phenotype?" Future Cardiol 4(3): 247 - 252. (2008)

SB³C2015
Summer Biomechanics, Bioengineering and Biotransport Conference
June 17-20, 2015,Snowbird Resort, Utah, USA
SB³C2015-1086

PREDICTION OF EXTRA-AXIAL INJURY LOCATION FROM REAL WORLD MOTOR VEHICLE CRASH AND OCCUPANT DATA

Jillian E. Urban (1), Sarah K. Lynch (1), Ervin Lowther (2)
Christopher T. Whitlow (2,3), Joel D. Stitzel (1)

(1) Center for Injury Biomechanics
Virginia Tech-Wake Forest University
Winston Salem, NC, USA

(2) Department of Radiology, Neuroradiology
Wake Forest School of Medicine
Winston Salem, NC, USA

(3) Translational Science Institute
Wake Forest School of Medicine
Winston Salem, NC, USA

INTRODUCTION

It is estimated that 1.7 million people per year sustain a TBI [1]. The leading causes of TBI are falls, MVCs, blunt trauma (i.e., workplace or sports-related trauma), and assaults [2]. MVCs are the leading cause of TBI-related death, and adults of age 20 to 24 are at the highest risk. Of the 1.7 million people who sustain a TBI per year, only 3% of these injuries result in death. Although the number of people who die from head injury appears to be relatively small, the cognitive effects from TBI can cause devastating impacts on quality of life and may result in premature mortality [3-5]. Survival post-head injury is often predicted based on age, Glasgow coma scale (GCS) score, injury severity score (ISS), and presence of hematoma on a computed tomography (CT) scan [6-8]. It has been found that extra-axial bleeds are an independent risk factor for survival [6]. The objective of this study is to evaluate the volume and resulting extra-axial injury extent, distribution, and intracranial location from real world motor vehicle crash data and determine which variables are predictors for injury location.

METHODS

Axial non-contrast head soft tissue computed tomography (CT) data within the first 24 hours post-crash and radiology reports were collected for cases with CT data within the CIREN database demonstrating extra-axial head injuries (AIS 3+) based on 1998 [9] and 2005 [10] AIS coding. Detailed data regarding the occupant, injury, vehicle, and crash were also collected. These data included: age, gender, height, weight, and the involved physical component (IPC) in the vehicle attributed to the SDH. In the CIREN database, IPCs are the objects thought to be contacted by the occupant considered to play a role in producing an injury based on specific evidence-based coding guidelines from vehicle crash, occupant kinematics, and injury evidence. Crash type, delta-v, and barrier estimate speed (BES) were also collected for the highest severity impact. Data were collected using the CIREN SQL interface and SQL developer (Oracle, Redwood Shores, CA). All cases selected underwent a full case review with medical, engineering, and crash reconstruction specialists to determine injury causation and are "complete" in the database.

Head CT scans were segmented using manual and semi-automated techniques within Mimics v14 and v16 (Materialise, Leuven, Belgium) [12]. All brain injuries were identified from descriptions within the radiology reports. The volume of each hemorrhage was calculated from the size and number of voxels within the segmentation. Injury segmentations will be reviewed by a board certified radiologist to ensure proper identification of the injury. This analysis was specifically focused on extra-axial bleeds including subdural hematoma (SDH), subarachnoid hemorrhage (SAH), and epidural hematoma (EDH). The point of contact on the head was identified by the presence of a superficial soft tissue/scalp contusion on the CT images and on the 3D CT reconstructions of the skin. If swelling did not occur, the point of contact on the head was approximated by the IPC, known occupant kinematics, and the injury causation scenario. Cases with unidentifiable contact location were excluded from the analysis. The centroid of the contact location was computed to reduce the estimated point of contact to a single coordinate location.

A local head coordinate system was established from bony landmarks that were identified from the 3D reconstruction of the occupant skull. The landmarks were at the left and right external auditory meatus (EAM) and nasion. The axis of the local coordinate system was determined from the perpendicular intersection of the nasion landmark projection on the line between the left and right EAM (Figure 1A).

This point was established as the origin of the local head coordinate system. The local x-axis was defined as the vector between the origin and nasion and the local y-axis was defined as the vector defined between the origin and the right EAM. The z-axis was determined from the cross product between the x- and y-axis. Each occupant local head coordinate system was aligned to a global coordinate system using a transformation matrix defined by translations and rotations.

The transformation matrix between the local and global coordinate systems were applied to 3D data (point cloud) of the injury and soft tissue swelling and then converted to a spherical coordinate system. Incremental calculation of injury volume was performed at 0.05 radian by 0.05 radian increments along the azimuth and elevation axes of the spherical coordinate system. Each 0.05 radian by 0.05 radian increment is referred to as an injury sector. The injury volume was calculated from a 3D Delaunay triangulation to create a polygonal surface of the convex hull of the points within each injury sector and projected onto the surface of a unit sphere established from the normalized distance between the left and right EAM (Figure 1B). Injury distribution and extent was quantified in terms of theta (about z-axis) and phi (from the xy-plane), respectively. Lastly, the spherical coordinate system was divided into eight regions, to determine the generalized injury location (Figure 1C).

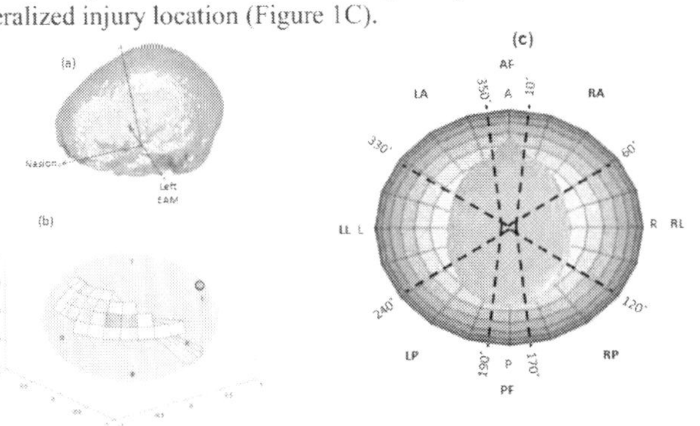

FIGURE 1. (A) LOCAL COORDINATE SYSTEM (B) EXAMPLE INJURY SECTOR (INCR =0.2 RADIANS) PROJECTION ONTO UNIT SPHERE. BLUE SQUARES SHOW INJURY LOCATION AND RED DOT SHOWS CENTROID OF THE CONTACT LOCATION. (C) INJURY LOCATION DIVISIONS BY AZIMUTH (THETA) ANGLE MEASUREMENT. ***AF = ANTERIOR FALX, RA = RIGHT ANTERIOR, RL = RIGHT LATERAL, RP = RIGHT POSTERIOR, PF = POSTERIOR FALX, LP = LEFT POSTERIOR, LL = LEFT LATERAL, LA = LEFT ANTERIOR.***

A discriminant analysis was performed to predict the injury location given known response variables. Regularized discriminant equations (λ=0.5, γ=0.5) were used to determine the canonical coefficients and receiver operator characteristic (ROC) curves were used to evaluate the success of the prediction. Response variables include age, body mass index, volume of injury, theta and phi of contact centroid, contact stiffness, maximum crush, delta-v/BES, and principal direction of force (PDOF). BES was used only if delta-v was not available from the crash reconstruction.

RESULTS

Of the occupants in the CIREN database with extra-axial head injuries, there were 134 cases with adequate soft tissue CT data collected within 24 hours post-crash demonstrating the injury. Of the 134 cases, 38 were excluded because they were not isolated extra-axial cases and 6 were excluded due to obscure/uncertain IPC coded for the respective injury. This resulted in a total of 90 scans to be included in the discriminant analysis. Each injury type of was analyzed individually for this analysis. There were 50 cases with SAH, 34 cases with SDH, and 6 cases with EDH. The computed injury volumes were right skewed and were log transformed to satisfy normality assumption.

IPC data were classified as 'Hard' (i.e., Tree, Pole), 'Compliant' (i.e., Pillars, Header), and 'Soft' (i.e., Airbag) to evaluate the IPC based on contact stiffness. 23.3% of the cases were coded as contacting hard surfaces, 73.3% were coded as compliant surfaces, and only 3% were coded as soft surfaces. The measured volume was not significantly different between the three groups. The average injury sector for each case was classified into eight locations based on the average sector location in the spherical coordinate system. The top three average extra-axial injury locations were Right Lateral (35.3%), Right Anterior (22.1%), and Left Anterior (13.2%). The least common average injury locations were Left Lateral and Left Posterior (2.94% for each).

Discriminant analysis was used to assess the injury location given known crash and occupant response variables. The prediction priors were specified such that they were proportional to the occurrence. 44% of the variation was described by the first canonical variable which was influenced most by elevation (phi) of the impact location (0.618), contact surface stiffness (-0.435), azimuth (theta) of the impact location (0.380), log of the injury volume (0.378), and PDOF of the crash (0.300). A total of 35.3% of the cases were misclassified (Table 1). The area under the ROC curve (AUROC) reveals good prediction for all injury locations (Table 2).

TABLE 1. ACTUAL ROWS BY PREDICTED COLUMNS.

	AF	LA	LP	LL	PF	RA	RP	RL
Anterior Falx	3					1		
Left Anterior		6			1	1		
Left Posterior			2					
Left Lateral				1				1
Posterior Falx		1			3			
Right Anterior		2	1			10		2
Right Posterior						1	3	3
Right Lateral	2		1			5		16

TABLE 2. NUMBER OF CASES AND AUROC.

Injury Location	Number of Cases	AUROC
Anterior Falx	4	0.8988
Left Anterior	9	0.9733
Left Posterior	2	1.000
Left Lateral	2	0.9923
Posterior Falx	5	0.9671
Right Anterior	15	0.8043
Right Posterior	7	0.9451
Right Lateral	24	0.9032

DISCUSSION AND RESULTS

Discriminant analysis is a dimension-reduction technique used to identify which combinations of variables best describe the between-class variation. In this study, injury locations were classified based on the average location of the injury volume and input responses were used to predict the resulting injury location. Elevation and azimuth of the impact location, contact surface stiffness, log of the injury volume, and PDOF were important variables in discriminating average extra-axial injury location. While this approach is limited to extra-axial injuries, this analysis may be performed on a per-injuries basis and may ultimately provide information related to the contributing factors to certain brain injuries. These data will ultimately lead to a better understanding of the biomechanics of head injury.

ACKNOWLEDGEMENTS

The authors would like to thank NHTSA for funding this project.

REFERENCES

[1] Coronado et al., Head Trauma Rehabil, 2005. [2] Faul et al., CDC, 2010. [3] Ventura et al., Arch Phys Med Rehabil, 2010. [4] Brown et al., NeuroRehabilitation, 2004. [5] Fazel et al., JAMA Psychiatry, 2014. [6] Signorini et al., J Neurol Neurosurg Psychiatry, 1999. [7] Pennings et al., Arch Surg, 1993. [8] Brooks et al., Arch Phys Med Rehabil, 2013. [9] AAAM, 2001. [10] AAAM, 2008. [12] Urban et al, J Neurotrauma 2012.

SB^{3}C2015
Summer Biomechanics, Bioengineering and Biotransport Conference
June 17-20, 2015,Snowbird Resort, Utah, USA
SB³C2015-1087

A COMPUTATIONAL MODEL TO EXPLORE THE ROLE OF EXPERIMENTALLY DETERMINED SCLERAL MICROSTRUCTURE ON LAMINA CRIBROSA DEFORMATION

Avinash Ayyalasomayajula (1), Forest L. Danford (2), Jonathan P. Vande Geest (1,2,3,4)

(1) Department of Aerospace and Mechanical Engineering, University of Arizona, Tucson, AZ, USA

(2) Department of Biomedical Engineering, University of Arizona, Tucson, AZ, USA

(3) Graduate Interdisciplinary Program of Biomedical Engineering University of Arizona Tucson, AZ, USA

(4) BIO5 Institute, University of Arizona, Tucson, AZ, USA

INTRODUCTION

Glaucoma is a family of optic neuropathies that lead to an irreversible loss of vision. Primary-open angle glaucoma (POAG) is one kind of glaucoma characterized by an increased angle at the iris, in the anterior chamber. Visual information from the eye is transmitted to the brain via retinal ganglion cells (RGCs) whose axons pass through a highly porous region in the posterior eye, specifically, the optic nerve head (ONH), called the lamina cribrosa (LC). Loss of vision in POAG results from the death of RGCs often as a consequence of a chronically elevated intraocular pressure (IOP). Previous experimental and computational studies have speculated that the variations in the mechanical properties of the tissues in and around the ONH likely influence the biomechanical response of the LC differently, and affect the viability of RGCs [1].

In an effort to characterize the influence of scleral mechanical properties on the LC, computational studies have modeled the scleral tissue as an elastic, hyperelastic, and anisotropic hyperelastic material. Pijanka et al have used inverse finite element techniques and determined the bulk orientation of collagen fibers in different regions of the posterior human sclera [2]. The orientation of collagen fiber obtained in their computational study is a depth-averaged measure of the collagen orientation. Previous studies from our laboratory have reported on a technique to estimate the fiber angle orientation at different depths along the scleral thickness in the different regions of the eye (viz. superior, inferior, nasal, and temporal) using small angle light scattering experiments (SALS), and commented on the differences in the collagen fiber orientations and fiber splay observed in eyes from different races [3, 4]. To our knowledge, the effect of variations in the microstructural organization of scleral collagen on LC strains has never been reported.

To address this, the current study utilizes a 3D finite element model of a posterior human eye, and the collagen fiber orientations from glaucomatous and non-glaucomatous eyes measured using SALS experiments, to compare their influence on LC strains. We believe that such a computational model would provide better understanding of the factors that could lead to a disproportionate incidence of POAG in different demographics.

METHODS

Geometry and FE mesh

The geometry of the posterior eye model included the prelaminar tissue, sclera, lamina cribrosa, postlaminar tissue, and pia mater. The geometry was based on the descriptions provided in previous literature [1, 5]. The geometry meshed with linear 3D elements as shown in **Figure 1**. The sclera was meshed using 8-node hexahedral elements. A mesh convergence study was performed to ensure optimal mesh density was achieved. The mesh that was used for the parametric study consisted of 530,462 elements and 651,607 nodes.

Material properties:

The prelaminar, postlaminar, LC, and piamater were assigned neohookean material properties. The Holzapfel anisotropic material model was used to simulate the response of the sclera [6]. The values of the model parameters (C_{10}, D_1, k_1, k_2) were obtained by fitting the data published previously by Coudrillier et al [7], to the Holzapfel model. This fitting yielded C_{10} = 38100Pa, D_1 = 1e-6Pa^{-1}, k_1 = 48775Pa, k_2 = 8.51e-7. The peripapillary sclera was identified as a 1mm wide annular region around the ONH. This region was assigned the same values of the Holzapfel model parameters but with collagen fiber orientation aligned with the equatorial direction.

The LC, prelaminar, postlaminar, and piamater were modeled as neohookean materials. See **Table 1** for the material property values used for these regions.

Tissue	C_{10} (Pa)	$D_1(Pa^{-1})$	Source
Lamina cribrosa	43624.2	4.6e-7	[8]
Post laminar	100671.1	2e-7	[1, 8]
Pre laminar	100671.14	2e-7	[1, 8]
Pia mater	399659.9	1.5e-7	[8, 9]

Table 1: values of the neohookean constants used in the simulations

Boundary conditions:

All regions in the model had tied surface constraints with their surrounding regions to avoid any relative motion between them. The base of the model was constrained in equatorial and axial directions. An internal pressure of 15mmHg was applied on the prelaminar surface.

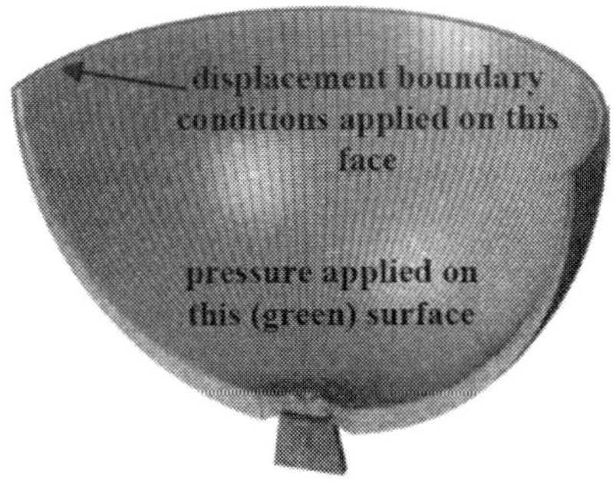

Figure 1: Geometry of the posterior eye model with boundary conditions

Two cases of scleral microstructural organization were compared: a glaucomatous eye of a 75yr old Caucasian male, and a non-glaucomatous eye from a different 75yr old Caucasian male.

Tissue	Fiber orientation (degrees)		Fiber splay	
	G	N	G	N
Superior	103.97	100.56	0.471	0.486
Inferior	87.85	78.8	0.439	0.469
Nasal	99.3	82.37	0.476	0.442
Temporal	107.58	88.7	0.441	0.434

Table 2: Scleral collagen fiber orientation obtained from SALS experiments. '**G**' and '**N**' correspond to glaucomatous and non-glaucomatous eyes respectively.

The fiber orientations obtained from the four regions of each of the two eyes were imposed on the finite element model (see **Table 2** for the average fiber orientation and the fiber splay). Within each region, the fiber orientation and fiber splay were homogeneous. The maximum principal strains in the LC from the two models are compared.

RESULTS

The maximum principal strains in the LC from the glaucomatous model were found to be approximately 3% higher compared to the maximum principal strains obtained from the non-glaucomatous eye model (see **Figure 2**). The axial displacement in the LC was also found to be higher in the glaucomatous eye compared to the non-glaucomatous eye, by the same percent magnitude.

DISCUSSION

The current study provides a computational framework that can be used to simulate multiple cases of fiber orientations measured in glaucomatous and non-glaucomatous eyes and study the effect on LC strains. It was observed that the LC in the glaucomatous eye model experienced slightly higher strains and axial displacements compared to the non-glaucomatous eye model. It should be mentioned that the same Holzapfel model parameters were used for the sclera in both models. The average orientations and splays of the scleral collagen were found to be very similar in superior and inferior regions but noticeably different in the nasal and temporal regions. This could have contributed to the difference in the LC strains observed in our simulations.

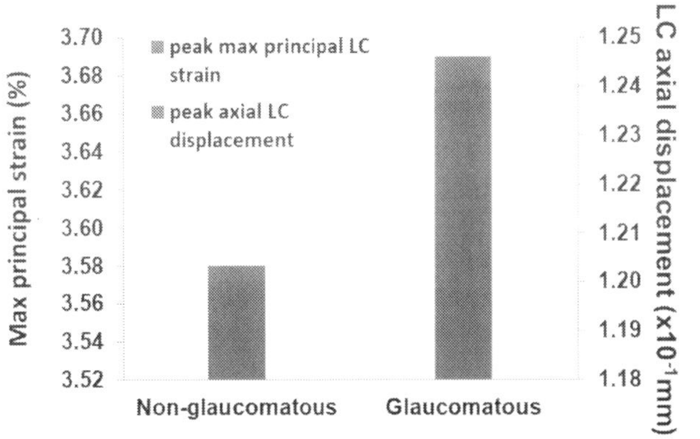

Figure 2: variation in maximum principal strain, and axial displacements in the LC with varying collagen fiber angle orientation in the sclera

Current and future work:

Ongoing work in our lab is focused on implementing depth-dependent fiber orientations (instead of region-averaged values) into the model, and investigating their influence on the mechanical response of the LC. It could be possible that at higher pressures on the prelaminar surface lead to greater differences in the LC strains and displacements. Future work involves determining the Holzapfel model parameters for sclerae from glaucomatous and non-glaucomatous eyes and investigate the differences (if any) in their material properties, which could likely result in different trends in the variation of strains in the LC.

ACKNOWLEDGEMENTS

Funding for this work was provided by the NIH grant 1R01EY020890-01A1 to JPVG.

REFERENCES

[1] Sigal, I et al., *Investigative Ophthalmology and Visual Sciences*, 45:4375-4387, 2004Pijanka, J et al., *IOVS*, 53:5258-5270, 2012

[2] Pijanka, J et al., *Investigative Ophthalmology and Visual Sciences*, 53:5258-5270, 2012

[3] Yan, D et al., *Investigative Ophthalmology and Visual Sciences*, 52:821-829, 2011

[4] Danford, F et al., *Investigative Ophthalmology and Visual Sciences*, 54:7922-7932, 2013

[5] Oyster W, *The Human Eye: Structure and Function*, Sinauer Associates, 1999

[6] Holzapfel, G A, *Journal of Elasticity*, 61:1-48, 2000

[7] Coudrillier, B et al., *Investigative Ophthalmology and Vision Sciences*, 53:1714-1728

[8] Friberg, T R, *Experimental Eye Research*, 47:429-436, 1988

[9] Uchio, E, *Clinical Science*, 83:1106-1111, 1999

SB³C2015
Summer Biomechanics, Bioengineering and Biotransport Conference
June 17-20, 2015,Snowbird Resort, Utah, USA
SB³C2015-1089

MICRORNAS AND RELATED TISSUE REMODELING GENES IN ROTATOR CUFF WITH A DELAYED REPAIR IN A RAT MODEL

Christopher T. Chen (1,2), Fuxin Wei (1) , Erik Contreras (1), Lucas M. Chen (2), Zachary Shirley (1), William Shelton (1), Michael Khazzam (1)

(1) Department of Orthopedic Surgery,
UT Southwestern Medical Center,
Dallas, TX, USA

(2) Biomedical Engineering Program,
UT Southwestern Medical Center,
Dallas, TX, USA

INTRODUCTION

The majority of rotator cuff tears seen clinically are atraumatic and chronic in nature, which can contribute to joint inflammation and tendon degeneration. Although inflammation is an integrative part of the healing process, excessive or prolong inflammation with high levels of matrix-degrading enzymes have been associated with impaired healing of the rotator cuff [1,2]. However, the underlying mechanism and tissue remodeling factors associated with chronic tear and rotator cuff repair were not completely clear [2,3].

microRNA (miR), a novel regulator of mRNA, acts in a tissue-dependent manner to suppress the translation of targeted mRNAs, but their roles in rotator cuff healing are not clear [4,5]. The purpose of this study was to evaluate the healing characteristics and miRs of a full thickness rotator cuff tear that has undergone delayed repair in a rat animal model and examine the differences seen as compared to acute repair. We hypothesized that decreased biomechanical strength as well as increased tissue inflammation and remodeling in the rotator cuffs with delayed repair as compared to the acute repaired tendon.

METHODS

Thirty eight mature male Sprague-Dawley rats were randomized to either acute or delayed rotator cuff repair (IACUC #2012-0218). Nineteen rats underwent detachment of the supraspinatus and infraspinatus tendons from the greater tuberosity followed by immediate repair (Acute) through bone tunnels [6], while 19 rats underwent detachment of the supraspinatus and infraspinatus tendons followed by delayed repair 2 weeks later using bone tunnels (Chronic). Sham surgery was performed on the contralateral shoulder and used as the control (Sham). Six weeks following repair, animals were sacrificed for histology (H&E, Saf-O/Fast green; n=5/group) and biomechanical analysis (Stiffness, Peak Load, Failure Load and Failure Energy; n=7/group). The degrees of tendon degeneration (Bonar's scores) and synovial inflammation were evaluated by two independent observers.

Rotator cuffs for mRNA expression were snap-frozen for and analyzed for COL1A1, COL3A1, TGB-β1, MMP-9, MMP-13 and IL-6 using qPCR (n=7/group). Rotator cuffs for mRNA expression were snap-frozen for and analyzed for COL1A1, COL3A1, TGB-β1, MMP-9, MMP-13 and IL-6 using qPCR (n=7/group) where selected samples (n=3) were analyzed for miRNA expression. Statistical analysis was performed using one-way ANOVA with FLSD post-hoc to determine the significance (α=0.05). The targeted mRNAs of miRNAs were predicted by MirTarget2 with a Target Score (TS) > 50 [5,7]

RESULTS

We found that the rotator cuff with delayed repair (Chronic) has a tendency for greater tendon degeneration with increased inflammatory cells (Fig. 1A-1C) as compared to the Acute (immediately repaired) group (p=0.06, Fig. 1D) where both Acute and Chronic groups have significantly higher degrees of tendon degeneration and inflammation than those of Sham group ($p<0.001$). Both the Stiffness and Peak Load in the Acute group were significant lower than those of Sham group ($p\leq0.05$, Figs. 2A and 2B). In the Chronic group, further decreases of Stiffness and Peak Load as compared to the Acute group were found (32.9% and 30.3%, respectively). There was no difference between groups in Failure Energy. The findings from biomechanical analysis were correlated well with the mRNA expression. Significant increases in the expression of IL-6, MMP-9, MMP-13, TGF-β1, Col-1, and COL-3 were found in the Acute and Chronic groups as compared to the Sham. Further increases of IL-6, MMP-13, TGF-β1, and COL-3 expression were found in the Chronic group as compared to the Acute repair group as shown in Figs. 3A and 3B.

Fifty-seven miRs showed significant differences between groups, including 14 miRs with significant differences between acute and delayed repair groups ($p<0.05$). Correlation analyses between miRs and mRNAs revealed two distinct groups: 1) miR-18a*, miR-34b*, miR-297, miR-328b-3p, and miR-493* for inflammation and matrix degradation; 2. miR-223*, miR-365, miR-365*, miR-505*, and miR-3578 for tissue repair (Table 1). IL-6 has been shown to promote increased miR-18a* expression in hepatocytes, where it acts in a positive feedback loop of IL-6 signaling pathway. Our data shows a strong correlation between IL-6 and miR-18a*. Potential mRNA targets of selected miRs were predicted by MirTarget2 with a TS >50 as shown in Table 2.

DISCUSSION

Our findings show that there is an impaired healing in the rotator cuff with a delayed repair as compared to immediate (acute) repair. Although we did not found significantly changes in synovial inflammation in histology, there was a tendency of increase in synovial fibroblasts which are associated with a greater expression of IL-6 and MMP-13. MMP-13 is the primary collagens in tissue remodeling and can effectively cleave type 1 and 3 collagen. These findings support our hypothesis that chronic tear is associated with inferior biomechanical properties and impaired repair of rotator cuffs, which are more closely replicates what is seen clinically. Future studies are needed to determine the factors and mechanism that impacts tendon to tendon and bone healing.

Our results also showed that chronic tear is associated with inferior biomechanical properties and impaired repair of rotator cuffs, which are more closely replicates what is seen clinically. The expression of miRs related to the inflammatory process is significantly altered in the injured rotator cuff compared to the Controls. Of great interest, there are two distinct groups of miRs that correlate oppositely with inflammation and degradation and tissue repair genes. This suggests that miRs work as a team to mediate the healing of rotator cuff tendons.

ACKNOWLEDGEMENTS

We thank the technical supports from UTSW RNA microarray Core facility. This study was supported by a grant from AANA.

REFERENCES

[1] Yamamoto, A. et al., J Shoulder Elbow Surg, (2010) 19:116-20.
[2] Robertson et al. Am J Sports Med (2012) 40:1993–2001.
[3] Trenkmann et al. Arthritis & Rheumatism, (2013) 65: 916–927.
[4] Van and Olson Nature Rev Drug Discovery, (2012) 11, 860-872.
[5] Wang and El Naqa Bioinformatics (2008) 24:325-332.
[6] Thomas et al., J Shoulder Elbow Surg, (2012) 21:23-31.
[7] Brock et al. J Biol Chem (2011) 286:40142–50.

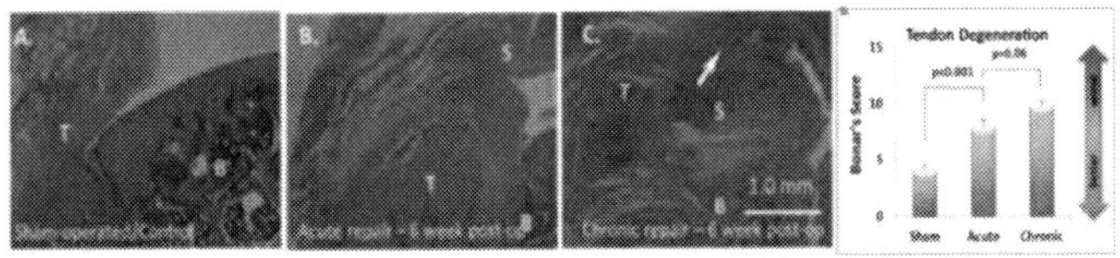

Fig. 1 H&E staining of A) sham-operated (Control), B) acutely repaired (Acute) and C) delayed repair (Chronic) rotator cuffs. D) Bonar's score for tendon degeneration showed that increased tendon degeneration in Chronic and Acute groups compared to Sham ($p<0.001$).

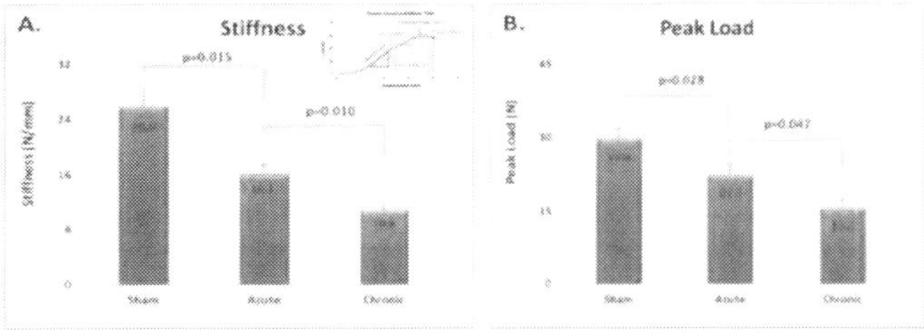

Fig. 2. The A) Stiffness and B) Peak Load) of the repaired and sham-operated rotator cuffs. Significant decreases of Stiffness and Peak Load were found in the Acute and Chronic groups compared to Shams.

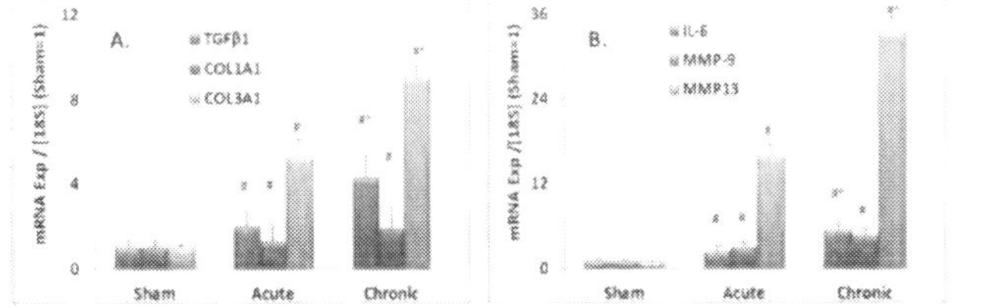

Fig. 3. The expression of tissue repair (TGB-β1, COL1A1, COL3A1) and tissue degeneration genes (IL-6, MMP-9, and MMP-13) in the repaired and sham-operated rotator cuffs. The upregulation of TGF-β1, COL3A1, IL-6 and MMP-13 was found in the Chronic group. # indicates $p<0.05$ as compared to Sham. ^ indicates $p<0.05$ between Acute and Chronic groups

miRs	IL-6	MMP-13	MMP-9	COL3	TGFβ1	COL1
miR-328b-3p	0.65	0.41	0.68	0.77	-0.18	-0.21
miR-297	0.72	0.34	0.56	0.44	-0.33	-0.48
miR-493	0.81	0.11	0.49	0.64	-0.56	-0.53
miR-18a*	0.79	0.25	0.39	0.64	-0.26	-0.53
miR-291a	0.77	-0.08	0.34	0.75	-0.56	-0.52
miR-34b*	0.72	0.07	0.25	0.75	-0.42	-0.51
miR-365	-0.56	-0.08	-0.14	-0.46	0.40	0.34
miR-365*	-0.71	0.02	-0.31	-0.89	0.40	0.46
miR-223*	-0.73	-0.34	-0.57	-0.56	0.11	0.45
miR-92a-1	-0.61	-0.34	-0.78	-0.42	0.46	0.29

Table 1. Correlation analysis between miRs and mRNAs showing two distinct groups that correlate oppositely with pro-inflammation cytokines and tissue repair genes.

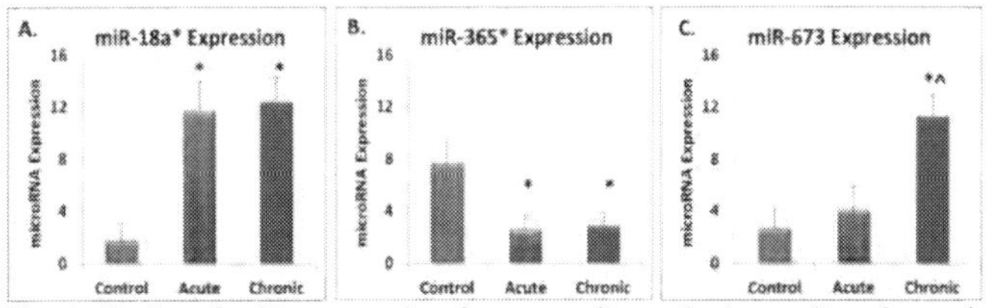

Fig 4. Ratio of miR expression between Control, Acute Repair, and Chronic Repair, shown in log base 2. * indicates $p<0.05$ as compared to Control. ^ indicates $p<0.05$ between Acute and Chronic groups

miRs	Predicted mRNA Target (Target Score)			
miR-328b-3p	BCL2 (92)	Nkx2-3 (70)		
miR-297	CD59 (100)	Rock2 (95)	Tnip3 (87)	
miR-493	KIF3B (93)			
miR-18a-star	TIMP2 (89)	CITED2 (84)	TGFBR3 (73)	SMAD5 (66)
miR-291a	Mapk9 (86)	Pak7 (70)		
miR-365	Ets1 (95)	Creb5 (77)	Mapk1ip1l (72)	
miR-365-star	Irak4 (82)	Col6a2 (78)	MMP14 (72)	Hdac1 (70)
miR-223-star	Tank (98)	Bmper (88)	Trappc10 (87)	Rhobtb3 (86)
miR-92a-1	S100a9 (90)	Elk1 (90)		

Table 2. Potential targets predicted by MirTarget2 for selected miRs.

THE COLLAGEN DIRECTIONALITY AND DISPERSION AND MECHANICAL INDENTATION RESPONSE IN NONPREGNANT HUMAN CERVICAL TISSUE

Wang Yao (1), Yu Gan (2), Christine P. Hendon (2), Joy Vink (3), Ronald J. Wapner (3), Kristin M. Myers (1)

(1) Department of Mechanical Engineering
Columbia University
New York, NY, USA

(2) Department of Electrical Engineering
Columbia University Medical Center
New York, NY, USA

(3) Department of Obstetrics and Gynecology
Columbia University Medical Center
New York, NY, USA

INTRODUCTION

The role of the cervix as a mechanical barrier is crucial during pregnancy. It must be strong and closed to maintain the fetus inside the uterus prior to term. The preterm birth (PTB) rate in the United States in 2011 was at 11.7%, affecting half million babies per year [1]. The material properties of the cervix play a key role during pregnancy. Characterizing the anisotropic material properties of the internal os, which is the end of cervix that connects to the uterus, is particularly important because this is the location of the initiation of premature cervical funneling which might lead to PTB.

Magnetic resonance imaging (MRI) [2], X-ray [3], second harmonic generation (SHG) [4, 5], and optical coherence tomography (OCT) [6] studies have shown that the collagen ultrastructure has a preferred orientation, which acts to resist cervical deformation. However, current literature lacks quantitative collagen fiber distribution data needed to integrate into a material constitutive model for the cervix. Therefore, the overall goal of this research is to construct a material constitutive model for the human cervix, which incorporates collagen ultrastructure, and to study its homogeneity within the cervix. The objectives of this study are to: use an OCT technique to determine the alignment and local distribution of the collagen fibers within the cervix, incorporate the OCT results into the ellipsoidal fiber distribution (EFD) neo-Hookean material model, and simulate the spherical indentation mechanical test using a finite element analysis (FEA).

METHODS

Tissue Collection: 18 nonpregnant (NP) and 4 term pregnant (PG) cervical tissue samples were collected from consented hysterectomy patients. Patient age ranged from 29 to 49 with various obstetric histories. Each cervix was sliced perpendicular to the inner canal. A slice from the internal os was used for OCT and mechanical tests. The thickness of the slices ranged from 2.42 mm to 6.51 mm. Collagen directionality data from 4 NP cervices are presented here in this abstract.

Optical Coherence Tomography (OCT): Details about the setup and data analysis techniques for the OCT methods are stated in [7]. Briefly, a gradient based fiber orientation algorithm was used to measure the directionality of local fibers [8]. The local orientation was calculated in every 1 mm × 1 mm area on the *en face* plane. Twelve 1 mm × 1 mm areas were selected for local fiber dispersion analysis.

Material Model: The details of EFD neo-Hookean model and parameters are stated in [9]. The ellipsoidal fiber distribution (EFD) is expressed in spherical coordinates as

$$D\left(\mathbf{n}_o, \frac{b}{a}\right) = \left(\frac{\cos^2\theta\sin^2\phi}{a^2} + \frac{\sin^2\theta\sin^2\phi}{b^2} + \frac{\cos^2\phi}{b^2}\right)^{-\frac{1}{2}}. \quad (1)$$

The local fiber distributions were converted into polar coordinates and an ellipse was fit to each distribution. The normalized major radius a and minor radius b of the ellipse was used to describe the distribution.

Here, dispersion ratio is defined as b/a where $b/a = 1$ describes randomly distributed fibers and $b/a < 1$ describes fibers with preferred directionality. A three dimensional distribution was defined as an ellipsoid, where we assume the radius in third direction is the same value of the minor radius of the ellipse.

Simulation of Spherical Indentation (FEA): A representative finite element geometry was built for a cervical slice taken at the internal os. The patient was 43 years old at the time of hysterectomy and she previously had one full term vaginal delivery. An investigation of various fiber distributions was conducted, while other material model parameters were taken from [9]. A hexahedron mesh generated by Cubit (v.13.2) was imported into FEBio (v.2.0.1) for FEA.

Clamped boundary conditions were placed on the bottom of the slice. The indenter was modeled as a rigid body sphere. Frictionless facet-to-facet sliding contact was prescribed between the tissue and the indenter.

RESULTS

Fiber Orientation Map: 69 volumes were stitched as a new volume with surface area 32 mm × 28 mm, as shown in Fig. 1A. The fiber orientation map is shown in Fig. 1B.

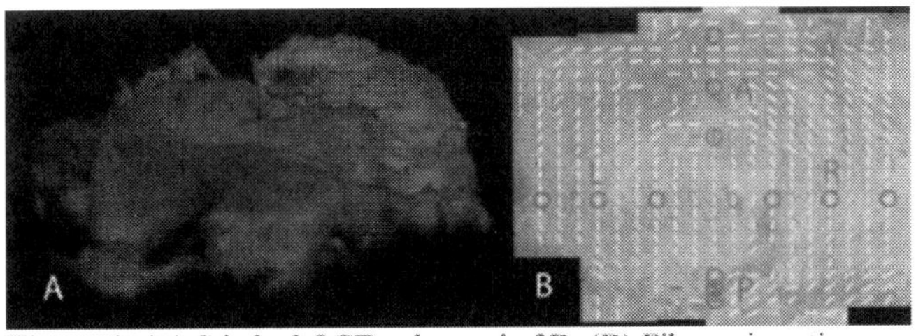

Figure 1. (A) Stitched OCT volumes in 3D. (B) Fiber orientation map on stitched OCT image. The 2D plane is parallel to the top surface with a vertical depth of 245 μm. The angle of each bar represents the orientation of local fibers where brighter color represents better aligned fibers. Red circles are the spots taken into local distribution analysis.

Local Fiber Distribution: The local fiber distribution was quantified by the dispersion ratio as shown in Table 1. Data from three other patients are included in the table. The averaged dispersion ratio is 0.76, and overall the outermost radial regions of the cervical slice had the tightest alignment when compared to the other regions of the cervix. However, this result was not statistically significant. There was no clear trend between anterior (A), posterior (P), patient left (L), and patient right (R) regions.

Table 1. Dispersion ratio b/a for 4 NP patients

n=4	A	P	L	R
Innermost	0.73±0.18	0.80±0.12	0.77±0.10	0.74±0.15
Middle	0.75±0.15	0.79±0.09	0.81±0.11	0.85±0.10
Outermost	0.73±0.10	0.69±0.30	0.73±0.12	0.72±0.08

Note: Three rows ("Innermost", "Middle", and "Outermost") represent radial locations of the mid-stroma area that are closest to the inner canal, in the middle, and closest to the outer boundary of the slice, respectively.

Finite Element Analysis: Three material models were evaluated by FEA. Model descriptions, principal strain magnitudes and directions under the indenter tip are shown in Table 2 and Fig. 2. Model I and II have a large difference in the magnitude of principal strains and reaction force on the indenter. The reaction force is twice as big for Model II than that of Model I due to the addition of the fiber network. The 1st and 2nd principal strains for Model I are much larger than those of Model II which means fibers in Model II prevent tissue deformation. The 3rd principal strain for all models is similar. There is no difference between model II(a) and II(b) in terms of strain magnitude. The 3rd principal strains for model II(a) and II(b) are both in axial direction but the 1st and 2nd principal strains have different directions. For less dispersion model (II(a), tighter circumferential alignment), 1st principal strain for a material element under the indenter tip is in radial direction and 2nd principal strain is in circumferential direction. This feature is not clear in Model II(b) where both principal strains are in the plane perpendicular to axial direction and between the radial and circumferential directions.

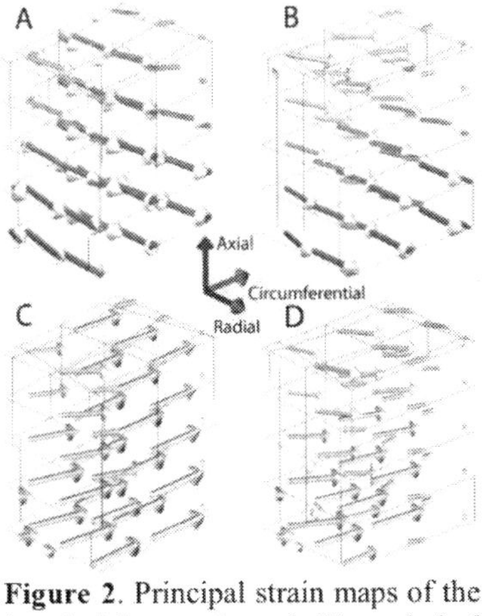

Figure 2. Principal strain maps of the 1st (subfigure A and B) and 2nd (subfigure C and D) principal strains of Model II(a) (subfigure A and C) and II(b) (subfigure B and D).

Table 2. Principal strains under indenter tip

Model	Material	b/a	1st	2nd	3rd	Fz (mN)
I	NH	-	36.6% (R)	25.4% (R/C)	-40.8% (Z)	33
II(a)	EFD NH	0.20	8.5% (R)	4.8% (C)	-43.2% (Z)	68
II(b)		0.76	8.5% (R/C)	5.1% (R/C)	-43.1% (Z)	67

Note: "NH" – neo-Hookean. "EFD NH" – EFD neo-Hookean (fiber added to Model I). "1st", "2nd", "3rd" – Three principal strains. "R", "C", "R/C", "Z" – strain's direction is "radial", "circumferential", "between radial and circumferential", or "axial". "Fz" – reaction force on indenter.

DISCUSSION

In this research, we present a flow of experiments and data processing techniques to study the ultrastructure of human cervix and incorporate the ultrastructure in finite element analysis. Same protocols will be applied to all the remaining untested specimens. In addition, indentation and tension tests will be performed in the future to determine material properties and validate our proposed equilibrium constitutive model through inverse finite element analysis (IFEA).

The research and conclusions described in this abstract has the following limitations. First, due to short of time, not all experiments have been finished. More data would bring more mathematically significant conclusions. With all specimens tested by indentation and tensile test, a sounder evaluation system of human cervix with both tensile and compressive properties will be developed. Second, the current framework makes some simplifications and assumptions that we could deliberate on. Collagen fibers have multiple types with different structures and functions. Second harmonic generation imaging, crosslink assay and other techniques would help us to differentiate their separate contributions. The collagen fiber might have preferential alignment in the longitudinal direction of the cervix so the third radius of the ellipsoid might be different than b. Slicing the cervix through the inner canal and imaging the longitudinal section will help us to understand the fiber orientation in axial direction.

REFERENCES

[1] Solomon, C., Iams, J., Prevention of Preterm Parturition, *N Engl J Med*, 370 (3): 254-261, 2014.
[2] Weiss, S. et al., *Anat Rec*, 288A: 84-90, 2006.
[3] Aspden, R. M., *Coll Relat Res*, 8: 103-112, 1988.
[4] Drezek, R. A. et al., *Cancer*, 98: 2015-2017, 2003.
[5] Myers, K. et al., *Eur J Obstet Gynecol Reprod Biol*, 144: S82-S89, 2009.
[6] Gan, Y. et al., submitted to *Biomed Opt Express*, 2014.
[7] Gan, Y. et al., *Biomed Opt Express*, 4 (10): 2150, 2013.
[8] Gan, Y., et al., *Conf Proc IEEE Eng Med Biol Soc*: 3873-3876, 2014.
[9] Myers, K. et al., submitted to *J Biomech*, 2014.
[10] Yao, W. et al., *J Mech Behav Biomed Mater* 34: 18-26, 2014.

SB^{3}C2015
Summer Biomechanics, Bioengineering and Biotransport Conference
June 17-20, 2015,Snowbird Resort, Utah, USA
SB³C2015-1091

THE EFFECT OF BONE DEFECT SIZE AND POSITION ON CEMENTLESS ACETABULAR CUP STABILITY : A FINITE ELEMENT ANALYSIS

Mark H. Gonzalez (1). Farid A. Amirouche (2), Gianfranco Solitro (3)

(1,2,3) Department of Orthopedic Surgery
University of Illinois at Chicago
Chicago, Illinois, USA.

(1,2,3) Department of Mechanical Engineering
University of Illinois at Chicago
Chicago ,Illinois, USA

INTRODUCTION

Cementless total hip arthroplasty for the treatment of osteoarthritis has enjoyed superior long term results. The acetabular component is hemispherical and has an outer shell with pores generally between 50 – 400 microns that allow bone ingrowth and incorporation. The acetabulum is underreamed by 1 milimeters and the cup in impacted into position. The impaction of the cup creates hoop stresses that create a frictional force between the cup and the bone. The frictional force holds the cup in position until bone can ingrow into the porous surface.

In situations where there are bone deficiencies of the acetabular rim , initial cup fixation may be less that ideal. When the bone rim is deficient the rim can expand with less force diminishing the compressive force. The second factor is physical contact between the cup and bone. Bone deficiencies can lead to decreased contact, which also diminishes the frictional force.

Micromotion greater than 150 -220 microns may preclude ingrowth of the bone into the porous surface of the cup.(1) In this scenario Adjunctive fixation may be necessary to stabilize the cup.

To further investigate cup stability in the face of a bone defect a mathematical model of the pelvis and specifically the cup bone interface was created.(2) A validated model was created based on a patient specific CT scan with MIMICS. The model will then be imported into a CAD program to create the bone defects. The model will then be imported into ANSYS for FEA. The Acetabular cup will be loaded with a combined profile of torque and compression to mimic the gait cycle.

The purpose of creating the validated model is to measure micromotion of a cup bone construct and the stresses involved. Two specific questions will be posed. 1. How does the size of a bone defect influence cup stability. 2: How does position of a bone defect influence cup stability.

The current research is unique in that it is an attempt as it is to understand and quantify the relative degree of micromotion associated with bone defects of different sizes and different locations. This knowledge is critical to eventually understanding how any bone defects of the acetabulum can contribute to overall arthroplasty stability. Current modeling of the acetabulum is primariuly geometric and will allow creation of a geometry based prosthesis This research will lead to methodology to optimize reconstruction based on mechanical principles..

METHODS

The acetabula of a single pelvis are dissected. On one side a bone defect is created. A Ct scan of the pelvis is then obtained. Cementless aacetabular cups are then placed in each acetabulum using a 2 mm underreaming technique. After placement of the acetabular cups a second CT scan of pelvis is obtained to document cup position.

Each acetabulum is then isolated and mounted in a custom made jig. The cup is loaded with an instron machine The maximum load setting ranged from 25-1500N in increments of 25N for a maximum load of 25-300N (i.e., a maximum of 25, 50, 75, etc.), and in increments of 50N for 300-1500N (i.e., a maximum of 300, 350, 400, etc.). The moment arm generated a maximum torque that ranged from 0.45-27 N-m.

The load profile applied to the acetabular cup was chosen to simulate a combination of the greatest forces experienced by a THA patient during walking and subluxation.(3,4)

Six LVDT position microsensors sensors are placed to circumferentially about the cup bone interface. The micromotion of the cup in relation to the bone is measured in both the superior-inferior and anterior- posterior planes.

The CT Scans of the acetabulum without defect are used to create models of the two hempelvices. (MIMICS) The models are imported into 3-Matic and based on the CT scans of the cup implants, the virtual cups are placed into the acetabular models. It was decided to bind the cup to the rim of the inverted U in an arc covering 270 degrees. (Fig.1,2) The bonded contact is created in an equatorial rim accounting for 23% of the overall surface area of the cup.

The model is validated with correlation less than 40 microns up to 800 newtons of compressive forces.

Micromotion under identical loading conditions was then measured with 1.5 x 1.5 cm defects in 8 different positions at 45 degree intervals. (Fig (3).

Peak stresses in the acetabulum were also calculated

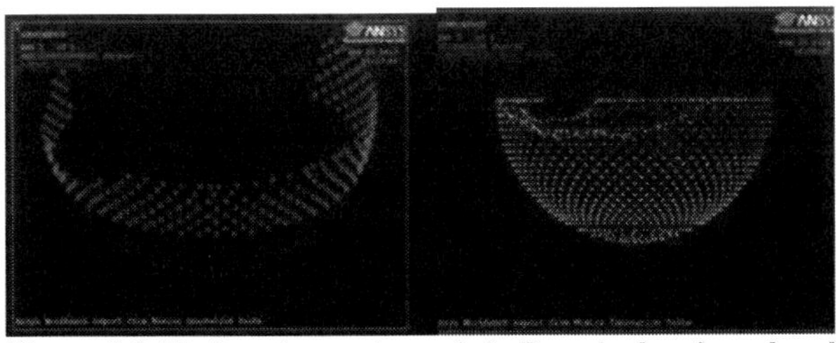

Figure1,2:Finite element model. Constrained nodes in 270 degree superiorly based arc.

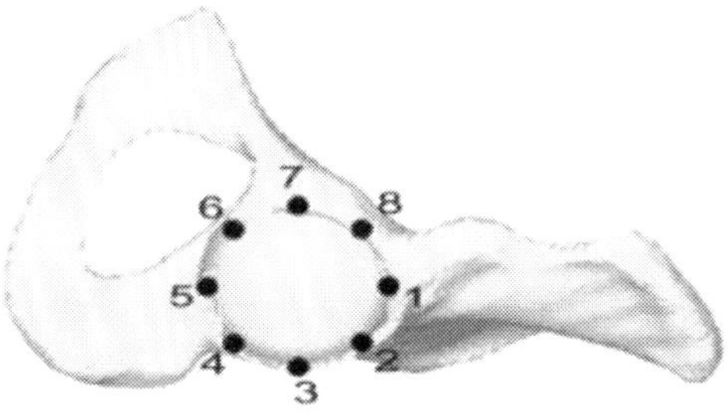

Fig. 3 Position of Defects 1-8. The defects are spaced 45 degrees apart. Defects 1 and 5 are along the superior inferior axis that connects the anterior superior iliac spine and the ischium Defects 7,8 are in the anterior column and defects 3,4 are in the posterior column. Defects 5,6 are inferior to the columns.

The second experiment compared rim defects of small .75 x.75 cm, medium 1x1 cm and 1.5 x1.5cm. at position 2 and position 7.

RESULTS

The micromotion of the cup in the intact acetabular model as well as the 8 different defects are plotted in Figure 4. Defects 3, 4 and 7 showed the most micromotion. All three of these defects were columnar and uncontained.by the contact area of the cup. At 1000 Newtons the micromotion at these locations was close to double control. Defect 1 which was in the ischium and was contained actually showed less micromotion than control. Defects 5,6 are inferior to the columns and defects in these aress showed minimal increase in micromotion. Defects 2 and 8 are both contained by the bonded area of the cup and show modest increases in micromotion

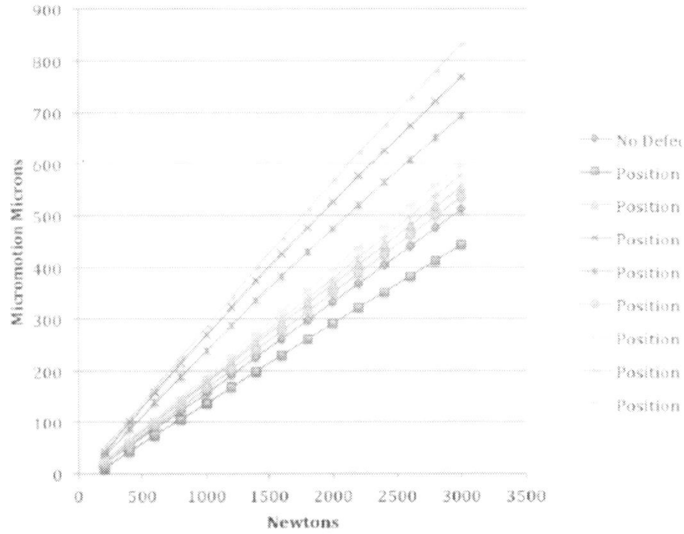

Figure 4 : Plotting compressive load vs micromotion of the acetabulum without and with defects in different positions.

Different size defects were evaluated at position 2 which is contained and defect 7 which is not contained. An increase in defect size at the contained defect showed only modest increases in micromotion. Defect size at position 7 which is not contained had a much greater effect on micromotion.

Peak stresses were greatly increased about uncontained acetabular defects.

DISCUSSION

Bone defects had a variable effect on cup micromotion depending on the position of the defect. Bone contact of an underreamed cementless cup is primarily about the rim of the acetabulum . When bone defects of the anterior or posterior columns are uncontained the micromotion of the cup under loads is much greater than defects that are contained. Defect size was much more critical in uncontained than contained defects of the acetabular columns .

In situations where a cementless acetabular prosthesis is implanted columnar defects not bridged by the prosthesis may destabilize the cup. Uncontained defects are also potential sites of crack initiation because of high peak stresses. When reconstructing an acetabulum with large uncontained defects the degree of potential micromotion may warrant the use of adjunctive fixation such as screws or a cup cage construct.

REFERENCES

(1)Pilliar, R et al.,*Clin Orthop.* 108-113,1986.

(2) Zivkovic, I et al, *Journal of Biomedical Engineering,* vol. 132, no. 4, p. 41008, 2010.

(3)Bergmann, G et al., . *J Biomech,* vol. 26, pp. 969-990, 1993.

(4)Stansfield B ey al., *Clinical Biomechanics,* vol. 17, pp. 130-139, 2002.

IMPROVING BRAIN-SKULL INTERFACE THROUGH APPLICATION OF MESH SMOOTHING ALGORITHM

Mireille E. Kelley[1,2], Logan E. Miller[1,2], Jillian E. Urban[1,2], Joel D. Stitzel[1,2]

(1) Biomedical Engineering
Wake Forest University School of Medicine
Winston-Salem, NC 27101

(2) Biomedical Engineering
Virginia Tech - Wake Forest University
Center for Injury Biomechanics
Winston-Salem, NC 27101

INTRODUCTION

With an average of 53,000 traumatic brain injury (TBI)-related deaths per year in the US, TBI is a leading cause of disability and injury-related death [1]. The mechanisms of TBI are not fully understood, but further study may help prevent, diagnose, and treat these types of injuries. Computational models are beneficial in studying TBI in that experiments can be run repeatedly and quickly, unlike animal or cadaver tests, and the models provide a large amount of information.

The complex structure of the brain plays an important role in injury. Studies have investigated the biomechanical consequences of brain features, such as sulci and gyri, in the cerebrum using finite element (FE) models. The studies found a decrease of the peak maximum principal strain within different regions of the brain compared to models with a homogeneous surface and local peak stress distributions in the sulci under loading conditions [2, 3]. Therefore, anatomically accurate sulci and gyri are important features of an FE model.

One way to create FE models is by using a voxel-based approach. During this process, image segmentation is used to identify voxels within regions of interest which are then converted to hexahedral elements in the FE model [4]. This approach results in a stair-stepped surface on each part, which may affect the mechanical response of the interface between parts. Smoothing is a way to modify the hexahedral elements to create more realistic interfaces. Taubin developed an algorithm to smooth polygonal meshes, which is an improvement upon Laplacian smoothing by including an "unshrinking" step [5]. This step aims to retain the original volume and is important because without it the mesh will iteratively lose significant volume and the elements will distort [6]. The objective of this study is to adapt the smoothing algorithm developed by Taubin for application to an FE model of the human head. Smoothing the interfaces will ultimately provide a more accurate model with an improved brain-skull interface and a more powerful tool for predicting brain injury

METHODS

Voxel-based FE model

For this study, the FE model of the brain developed by Miller et al. known as the atlas-based brain model (ABM) will be used [4]. The ABM was created from the International Consortium for Brain Mapping (ICBM) atlas [4]. The model consists of approximately 2 million nodes and elements. The model uses solid 1mm isotropic hexahedral elements for the brain (combined cerebrum and cerebellum), ventricles, and CSF. The skull is simulated by a 1mm layer of rigid shell elements generated in LS-DYNA to completely cover the CSF. Prior to mesh smoothing, the ABM was partitioned into 8 sections to improve processing time. CSF elements were then added to each section in the ±z-direction along the brain-skull interface to ensure the brain was completely encapsulated by the CSF.

Mesh smoothing

A custom MATLAB program was developed based on the smoothing algorithms proposed in Boyd et al. [5], Taubin [6, 7], and Chen et al. [8]. The mesh smoothing algorithm consists of the following steps:

- Each node is classified into categories and assigned a number to signify its order as surface (4), interface (3), interior (2), or fixed nodes (1) as seen FIGURE 1 (c). Surface nodes are associated with less than 8 elements and are only associated with the CSF part, interface nodes are associated with elements of two parts, and interior nodes are only associated with elements of the same part.
- Neighbors for each node are found based on the distance from the central node and are subject to constraints. The constraints imposed in this algorithm create non-symmetric neighborhoods for the interface nodes, which results solely in identification of other interface nodes as neighbors. The surface and interior nodes can use neighbors from any of the classifications. An optional classification is a fixed node, which does not have neighbors and therefore will not move.

- Apply Taubin smoothing to every node in the FE mesh for k iterations using the smoothing parameters λ and μ.

$$w_{ij} = \frac{1}{N_i}$$

$$\Delta x_i = \sum_{j \in i^*} w_{ij}(x_j - x_i)$$

Smoothing step

$$x_i = x_i + \lambda \Delta x_i$$

Unsmoothing step

$$x_i = x_i + \mu \Delta x_i$$

where $k = 2$, $\lambda = 0.4307$, $\mu = -0.6732$ [8], N_i is the number of neighbors for node i, w_{ij} is the inverse of the number of neighbors associated with node i, and n is the number of nodes in the mesh. The values for k, λ, and μ were chosen to maintain element quality and preserve volume while $\lambda < 0$ and $\mu < -\lambda$ [7].

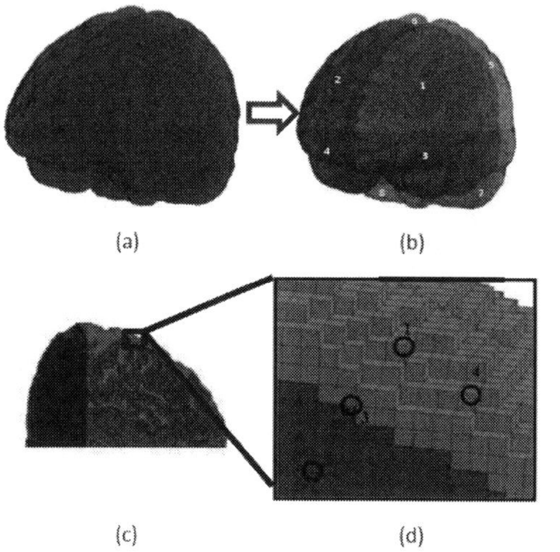

FIGURE 1: A) ORIGINAL BRAIN MODEL B) BRAIN MODEL DIVIDED INTO 8 SECTIONS. C) SECTION 1 (PART OF FRONTAL LOBE) ISOLATED D) ILLUSTRATION OF THE NODE CLASSIFICATIONS: FIXED (1), INTERIOR (2), INTERFACE (3), AND FIXED (4).

Element quality metrics, including warpage, aspect ratio, skew, and jacobian, were measured to optimize the smoothing parameters (λ and μ) and maintain good element quality. LS PrePost and HyperMesh were used to check volume of the mesh and quality of the elements. The thresholds for the element quality metrics were chosen based on Global Human Body Model Consortium (GHBMC) limits [9, 10].

RESULTS

The results of smoothing each of the eight sections of the brain model are shown in FIGURE 2, TABLE 1, and TABLE 2. An average of 0.20% failed the jacobian test. An average volume loss of 0.59% was seen between the original model and the smoothed model.

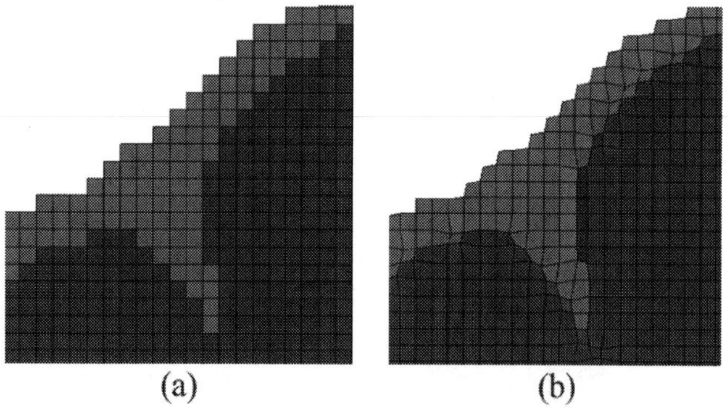

FIGURE 2: A) ORIGINAL INTERFACE BETWEEN BRAIN (RED) AND CSF (BLUE) B) INTERFACE BETWEEN BRAIN AND CSF AFTER SMOOTHING.

TABLE 1: NUMBER AND PERCENT OF ELEMENTS FAILING QUALITY TESTS.

	Acceptable Range	Measured
Number of elements	--	1993972
Jacobian (% fail)	>0.30	3991 (0.20%)
Warpage (% fail)	<50.00	44 (0.00%)
Aspect Ratio (% fail)	<8.00	0 (0.00%)
Skew (% fail)	<60.00	0 (0.00%)

TABLE 2: PERCENT VOLUME LOSS FOR EACH PART AFTER SMOOTHING.

	Original model Volume (mm^3)	Smoothed model Volume (mm^3)	Percent Loss in Volume
Total	1993988	1982249	0.59%

DISCUSSION

A smoothing algorithm was developed, using previously developed methods, to smooth the interfaces between parts of the ABM [7]. The model was sectioned into 8 parts and smoothed by the method presented in the current study. The smoothed model had less than 1% loss in volume and only 0.20% of elements failed the jacobian test. The small volume loss was due to smoothed elements on the surface, but was not large enough to affect the integrity of the model. Additionally, the few elements that failed the jacobian did not fall below 0.20. The results of smoothing the interface between the brain and CSF using non-symmetric neighborhoods can be seen in FIGURE 2. The results demonstrate that the smoothing algorithm preserved volume and maintained element quality.

Developing an anatomically accurate and high resolution FE model of the head is important for studying the mechanisms behind TBIs. Previous research has shown accurately modeling sulci and gyri in computational models is important because of their effect on stress distribution and strain measurements in the cerebrum [2, 3]. The ABM includes sulci and gyri, but the stair-stepped interfaces can be smoothed, which may improve the accuracy of the stress and strain measurements in FE simulations. Additionally, the smooth surface created using this algorithm creates interfaces where a sliding contact definition can be more easily implemented. A sliding contact definition would be an improvement over the current contact definition in ABM, which uses merged nodes.

Further work will include applying the smoothing algorithm to the entire brain in a single step. Dynamic simulations of the entire brain model will be run to evaluate stress and strain metrics at the depths of the sulci as well as throughout the brain. The smoothing algorithm presented here will ultimately lead to an improved FE model of the brain and allow more accurate brain injury prediction.

ACKNOWLEDGEMENTS

The authors would like to acknowledge Elizabeth Lillie for her help with development of the Atlas-Based Brain Model and the Wake Forest University Graduate School of Arts and Sciences for funding to conduct this research.

REFERENCES

[1] V. Coronado, *et al.*, *MMWR Surveill Summ,* vol. 60, pp. 1-32, 2011.

[2] J. Ho and S. Kleiven, *J Biomech,* vol. 42, pp. 2074-80, 2009.

[3] R. J. Cloots, *et al.*, *Ann Biomed Eng,* vol. 36, pp. 1203-15, Jul 2008.

[4] L. Miller, et al., "Development of an Atlas-Based Finite Element Head Model," ed: BMES, 2014.

[5] S. K. Boyd and R. Muller, *J Biomech,* vol. 39, pp. 1287-95, 2006.

[6] G. Taubin, ed: IBM T.J. Watson Research Center, 1996.

[7] G. Taubin, *Eurographics,* vol. 19, 2000.

[8] Y. Chen and M. Ostoja-Starzewski, *Acta Mechanica,* vol. 213, pp. 155-167, 2010.

[9] C. Untaroiu, *et al.*, *Ann Biomed Eng,* vol. 41, pp. 513-26, 2013.

[10] H. Mao, *et al.*, *J Biomech Eng,* vol. 135, p. 111002, Nov 2013.

THE HIP JOINT CENTER ESTIMATES FROM SKIN-MARKER-BASED METHODS DO NOT CORRESPOND WITH MEASUREMENTS USING DUAL FLUOROSCOPY

Niccolo M. Fiorentino (1), Penny R. Atkins (2), Michael J. Kutschke (1), Ashley L. Kapron (1), K. Bo Foreman (1,3), Andrew E. Anderson (1-4)

(1) Department of Orthopaedics
University of Utah
Salt Lake City, Utah, USA

(2) Department of Bioengineering
University of Utah
Salt Lake City, Utah, USA

(3) Department of Physical Therapy
University of Utah
Salt Lake City, Utah, USA

(4) Scientific Computing and Imaging Institute
University of Utah
Salt Lake City, Utah, USA

INTRODUCTION

Perturbations in the location of the hip joint center (HJC) can substantially alter calculations of hip joint angles, moments, and reaction forces of gait models [1]. The hip joint lies deep within the pelvis, making it impossible to directly measure the in-vivo location of the HJC with reflective skin markers. The use of pelvic landmarks and an accompanying regression equation, referred herein as "CODA", is the most common approach to estimate the HJC [2]. More recently, functional methods, referred herein as "FHIP", estimate the HJC from skin markers as the femur moves relative to the pelvis in a predefined motion [3,4]. Both FHIP and CODA assume the HJC does not move relative to the pelvic coordinate system. To date, a gold standard has not been available to quantify the accuracy of HJC measurement and to test the assumption that the HJC remains at a set position within the pelvis. We developed a high-speed dual fluoroscopy (DF) system that quantifies the in-vivo spatial position of the femur and pelvis to submillimeter accuracy [5]. In this study, we used DF as a reference standard to compare spatial positions estimated by CODA and FHIP.

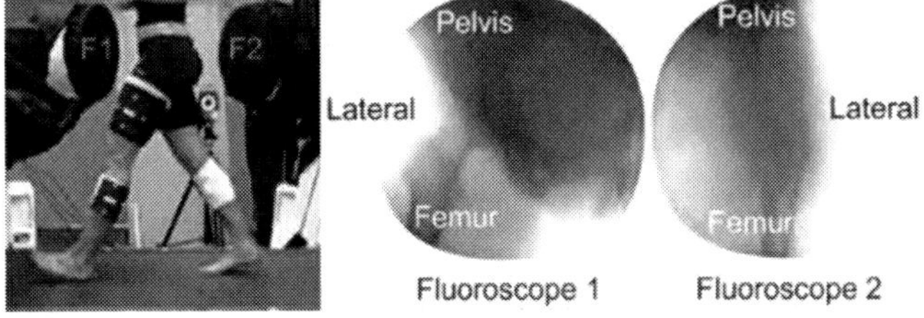

Figure 1. Experimental Setup. Subjects were imaged simultaneously with a 10-camera Vicon system and dual fluoroscopy. Subjects were imaged on a treadmill (left); example fluoroscopy images (right). F=Fluoroscope.

METHODS

Eleven young, healthy volunteers signed informed consent to participate in this IRB approved study (Age: 23+/-2.2 yr, BMI: 21.1+/-1.9). Subjects were imaged simultaneously at 100 Hz using a 10-camera Vicon system and DF (Fig. 1). Skin markers were placed on the pelvis at the anterior superior iliac spines (ASIS), posterior superior iliac spines (PSIS) and iliac crests, with a cluster of four markers on the thigh (Fig. 2).

Skin marker data were captured and processed in Vicon Nexus and imported into Visual3D for filtering (6 Hz) and HJC calculation (Fig. 2). Visual3D defined the CODA pelvis HJC based on equations in the literature [2], and FHIP was calculated as the point about which the femur rotates relative to the pelvis during a pre-defined motion of circumduction, ab/adduction and flexion/extension [3,4]. Hip joint centers were compared at a single time point while the subject stood in a neutral position (i.e., static) and while walking at a self-selected speed. CODA and FHIP were defined in the marker-based pelvis anatomical coordinate system and then transformed to the DF system.

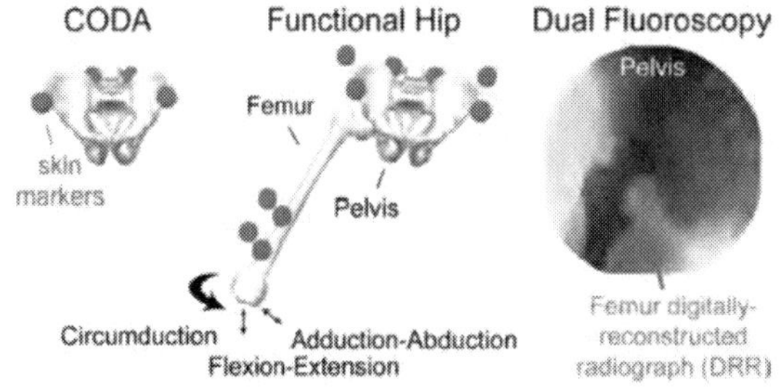

Figure 2. Hip Joint Center Measurement Methods. The hip joint center was measured with three techniques. With submillimeter accuracy [5], the DF method served as the reference standard.

Model-based tracking of the DF images measured the location of the pelvis and femur [5,6]. Briefly, computed tomography (CT) scans of each subject were segmented to generate 3D models of the pelvis and proximal femur. Digitally reconstructed radiographs (DRR) were created from the 3D CT models and aligned with DF images (Fig. 2). Transformations required to align the DRR with the DF images defined the spatial position of the bones. The DF HJC was quantified as the centroid of a best-fit sphere to the femoral head [7].

To compare the CODA, FHIP and DF HJCs in the same coordinate system, the CODA and FHIP HJCs were transformed into the DF coordinate system. The HJC locations were compared in the pelvis anatomical coordinate system, as measured by DF. HJC locations on the standing trial were compared statistically with a t-test.

RESULTS

For the static trial, the CODA HJC differed significantly from DF in the superior-inferior direction and approached significance in the anterior-posterior direction (P=0.06) (Fig. 3). The FHIP differed significantly in all three anatomical directions. During walking, the DF HJC remained relatively constant in the DF-measured pelvis anatomical coordinate system compared to the CODA and FHIP HJC (Fig. 4). The range of values for the ensemble average never exceeded 1.8 mm in any direction for DF, but the maximum range was 6.3 mm and 6.5 mm for CODA and FHIP, respectively.

DISCUSSION

To our knowledge this study is the first to directly measure the 3D position of the hip joint center under static and dynamic conditions. Our results indicate that the CODA and FHIP methods may misrepresent the location of the HJC, which has been shown to have a significant effect on biomechanical model outputs. For CODA, differences could be attributed to subjects in the current study possibly possessing different bony anatomy than the population in the literature study that developed regression equations, and/or minor differences in marker placement between study subjects and the literature study's landmark identification [2]. For FHIP, differences could arise from the FHIP's functional definition as compared to the geometric (at the femoral head center) definition for the DF method.

For walking, the assumption that the HJC mostly remains constant in the pelvis anatomical reference frame seemed valid (Fig. 4, top). However, skin markers failed to capture this phenomenon, as demonstrated by fluctuation over the gait cycle (Fig. 4, middle and bottom). Soft-tissue artifact of the skin marker relative to the underlying bony anatomy likely explains an origin of the fluctuations. Furthermore, if the FHIP and CODA landmarks were defined further from the true HJC point of rotation, they would exhibit larger movements during the gait cycle for the same joint angle changes.

Future work will incorporate the DF solutions into biomechanical models to assess the influence of HJC location and errors made by the FHIP and CODA pelvis on joint mechanics. The DF data herein may also serve to recalibrate the FHIP and CODA pelvis methods.

ACKNOWLEDGEMENTS

Funding from NIH R01 EB016701-02 and LS Peery Discovery Program in Musculoskeletal Restoration are acknowledged. We thank Justine Goebel and Madeline Singer for their contributions.

REFERENCES

[1] Lenaerts, G et al., *J Biomech*, 42:1246-1251, 2009.
[2] Bell, A et al., *J Biomech*, 23:617-621, 1990.
[3] Schwartz, MH and Rozumalski, A, *J Biomech*, 38:107-116, 2005.
[4] Camomilla, V et al., *J Biomech*, 39:1096-1106, 2006.
[5] Kapron, A et al., *J Appl Biomech*, 30:461-470, 2014.
[6] Bey, MJ et al., *J Biomech Eng*, 128:604-609, 2006.
[7] Harris et al., *Eur J Radiol*, 83: 788-796.

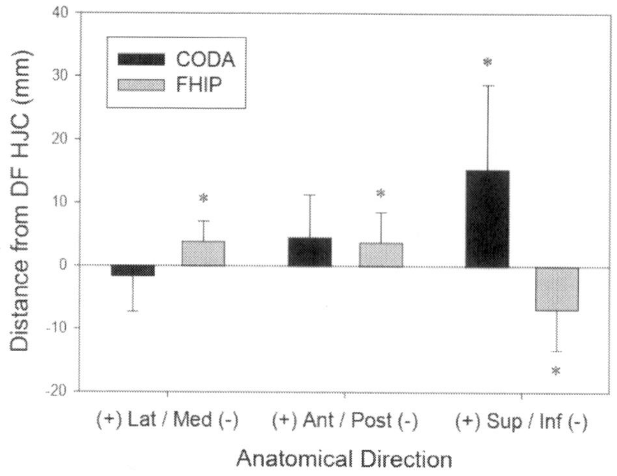

Figure 3. Static HJC Comparison. The hip joint center (HJC) locations estimated from skin markers (CODA and FHIP) were compared to the DF measurement during a standing trial. Data were averaged across all subjects. * indicates significance at P<0.05.

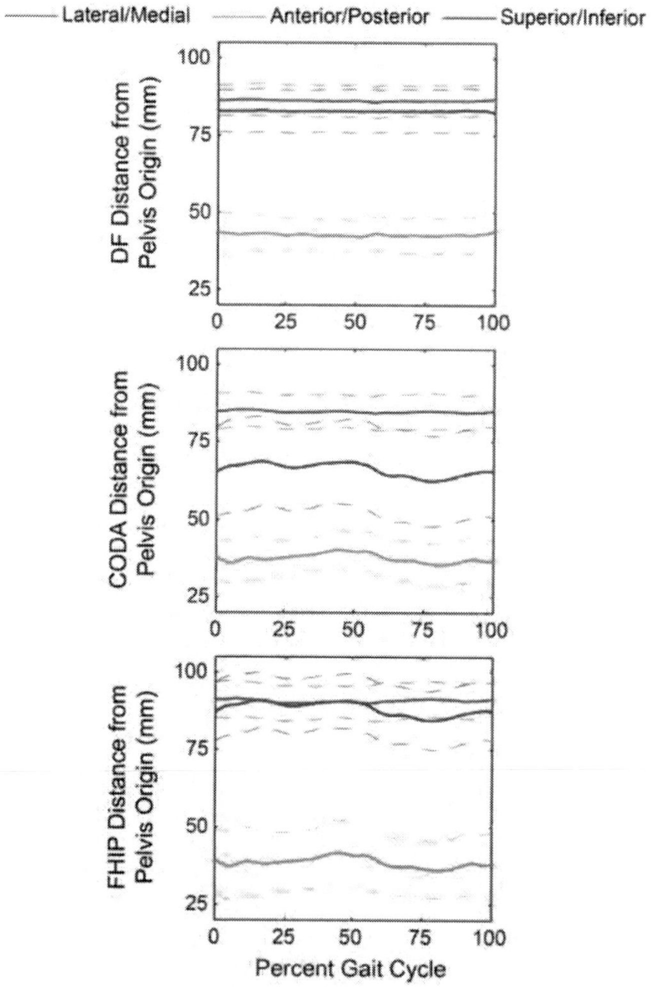

Figure 4. Dynamic HJC Locations. Ensemble average (solid line) and standard deviation (dotted line) for the HJC locations over the time-normalized gait cycle (heel strike to heel strike).

POST ENDOVASCULAR STENT REPAIR OF DESCENDING AORTA: SIDE EFFECTS AND DEVELOPMENT OF ANEURYSM IN THE ASCENDING AORTA

Giampaolo Martufi (1), Manal N. Altamimi (2), Raied Aburashed (3), Cyrus Fiori (3), Jehangir J. Appoo (4), Elena S. Di Martino (1,5)

(1) Civil Engineering
University of Calgary
Calgary, Alberta, Canada

(2) Biomedical Engineering Graduate Program
University of Calgary
Calgary, Alberta, Canada

(3) Undergraduate Engineering Program
University of Calgary
Calgary, Alberta, Canada

(4) Cardiac Sciences and Surgery
University of Calgary
Calgary, Alberta, Canada

(5) Centre for Bioengineering
University of Calgary
Calgary, Alberta, Canada

INTRODUCTION

Thoracic aortic aneurysms are an indolent but catastrophic process. As the aorta dilates, the wall is thought to weaken and risk of rupture or dissection increases. Traditional treatment of descending thoracic aneurysms involved an invasive large thoracotomy incision and replacement of the aneurysmal segment with a hand sewn Dacron graft. Over the past 2 decades, advances in technology have led to the use of covered endovascular prosthesis placed inside the aortic wall to exclude the aneurysm sac. These procedures are minimally invasive and done through a small incision in the groin crease with the aid of intraoperative fluoroscopy. In addition, Thoracic Endovascular Aortic Repair (TEVAR) has been shown to decrease perioperative morbidity and mortality [1].

The response of the arterial wall to endovascular stenting is not well understood. Additionally, the proximal and distal effects of having a nitinol or stainless steel prosthesis in the vascular system are unknown. This potential change in compliance has been postulated as a cause for a clinical phenomenon known as "retrograde aortic dissection".

Consequently, the purpose of this study was to evaluate changes in the wall mechanics in the ascending aorta proximal to where the endovascular stent graft is placed. Secondary purpose was to investigate the growth of the ascending segment after endovascular placement of a stent in the descending aorta. Stress and diameter changes in the ascending aorta were evaluated at each follow-up.

METHODS

Patient's population

From December 2005 to December 2013, 16 patients (14 males and 3 females, aged of 60 ± 10 years) with descending thoracic aortic lesions underwent endovascular repair. Institutional Review Board approval was obtained for the collection and use of anonymized data from the subjects. Patients were monitored prior to surgery and between 2 to 24 months after the surgery with repeated contrast-enhanced multislice Computed Tomography (CT).

Geometry Reconstruction

A 3D model of the thoracic aorta was obtained from CT images using the segmentation features of the image processing software ScanIP (Simpleware Ltd., Exeter, UK). A reliable smoothing procedure was applied to improve the reconstructed geometry without changing the natural details of the thoracic aorta. A mesh generating software (ScanFE; SimplewareLtd., Exeter ,UK) was used to generate three-node shell elements (quadratic, with 5 integration points, 2 mm thick) for the Finite Element (FE) analysis. The mesh was then refined by using a second mesh-processing package (Hypermesh; Altair Engineering, Inc. Troy, MI, USA). The number of shell elements for the aortic wall ranges from 50000 to 100000 dependent on the size and shape of each patient-specific geometry.

Finally, the maximum diameter was manually evaluated directly from CT images by using the existing measuring tool in Scan IP (Simpleware Ltd., Exeter, UK).

Finite Element Analysis

Non-linear isotropic constitutive descriptions of the wall as proposed previously were used for the FE analysis. In details, the arterial wall was modeled using the nonlinear hyperelastic formulation derived from Raghavan and Vorp [2], who examined the uniaxial properties of excised ascending thoracic aortic aneurysm tissue [3]. The strain energy function W used to model the aortic wall was as follows:

$$W = C_1(I_1 - 3) + C_2(I_1 - 3)^2 \quad (1)$$

where C_1 and C_2 are stress like parameter and I_1 is the first invariant of the left Cauchy-Green tensor. Population mean values of $C_1 = 80$ kPa and $C_2 = 170$ kPa were used [3].
Wall stress was calculated with the multi-purpose finite element program ABAQUS 6.13-3. The mean arterial blood pressure of 100 mmHg (13.33 kPa) was prescribed, and the nodal degrees of freedom of the distal ends of supra-aortic vessels, the descending aorta and the proximal aortic sinus were fixed in all directions. No contact with the surrounding organs was considered. The stress field was evaluated and the Maximum Wall Stress of the Ascending aorta (MWSA) was computed.

RESULTS

The maximum principal Cauchy stress distribution obtained for one representative preoperative model is plotted in Figure 1.
Areas with high stresses were located in the posterior region of the descending aortic aneurysms, slightly above the location of the maximum diameter where a maximum principal stress of 312 kPa was measured. The recorded MWSA was 200 kPa, located in the posterolateral region of the ascending aorta. The measured preoperative maximum diameter of the ascending aorta was 32 mm.
Stress and a diameter increase were observed in the postoperative analysis. In details, for the postoperative representative model the MWSA increased from 200 kPa to 212 kPa, while the ascending aorta diameter increased from 32 mm to 33.2 mm with an average growth rate of 0.2 mm/months. Additionally, not only the magnitude of MWSA was increased in the postoperative model but also the location changed over time.

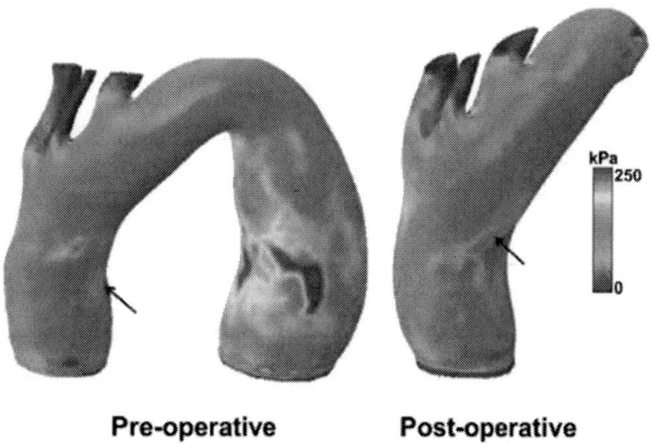

Figure 1: Changes in maximum principal Cauchy stress distribution of the aortic wall prior and after endovascular treatment of the descending aorta. The arrows point to the location of the Maximum Wall Stress of the Ascending aorta (MWSA).

In the entire cohort of patients an increase in maximum diameter of the ascending aorta from 33.5 ± 3.0 mm to 35.7 ± 2.8 mm was recorded between the pre- and post- operative assessments. This diameter increase corresponded to an average growth rate of 0.36 ± 0.39 mm/months. The MWSA rose in average from 202 ± 16.7 kPa in the preoperative model to 212.5 ± 20.0 kPa in the postoperative model.

DISCUSSION

Improved results with TEVAR have led to possible therapeutic intervention for patients that were previously not deemed candidates for open treatment. In addition, TEVAR has shown to decrease perioperative mortality compared to traditional open surgery for pathologies such as aneurysmal disease and acute type B dissection [1].
The tissue response to implantation of an endovascular device involves a complex interplay among device design, materials, and deployment technique. The interaction between stent and vascular walls has been addressed by numerous investigators [4]. Different reports examining aneurysm size after endovascular grafting show that aneurysms shrink following successful endovascular treatment and enlarge when there are persistent endoleaks [5].
However, our data underlines the sensitivity of the proximal segment of the aorta to endovascular grafting of the distal aorta and suggests a stent-mediated aortic remodeling of the ascending aorta.
The maximum diameter growth of the ascending aorta after stenting of the descending segment was about 0.36 mm/months in our cohort. As an immediate consequence of diameter increase and geometrical changes of the ascending aorta, the stress distribution changes and the maximum stress rise post stent-graft in the descending aorta. Apart from the stress increase the location of the maximum stress changed considerably over time.

Whether endovascular repair of the descending aorta causes adverse aortic remodeling of the ascending aorta with consequent aneurysm formation requires further investigation and will likely be the topic of future study.

ACKNOWLEDGEMENTS

The work was partially supported by the Natural Sciences and Engineering Research Council of Canada (NSERC) Discovery Grant, MITACS Postdoctoral Elevate Award and NSERC CREATE Postdoctoral Fellowship Award.

REFERENCES

[1] Piffaretti, G et al. *J Card Surgery,* 1:26, 2006.
[2] Raghavan ML and Vorp DA, *J Biomech*, 33: 475-482, 2000.
[3] Vorp, DA et al., *Ann Thorac Surg,* 75:1210-1214, 2003.
[4] Back, M et al., *J Vasc Surg*, 19:905-911, 1994.
[5] Matsumura, JS et al., *J Vasc Surg*, 25:113-123, 1997.

STABILITY ANALYSIS OF THE CONTINUUM CONSTRAINED MIXTURE MODEL FOR VASCULAR GROWTH AND REMODELING

Jiacheng Wu and Shawn C. Shadden

Department of Mechanical Engineering
University of California, Berkeley
Berkeley, CA, USA

INTRODUCTION

Abdominal aortic aneurysm is a significant cause of mobility and mortality in the US. It has been suggested that a better understanding of rupture risk involves the stability of aneurysm expansion due to vascular growth and remodeling (G&R), i.e., rupture might be a result of unstable aneurysm expansion. Previous research in applying the constrained mixture theory of vascular G&R to aneurysm growth has focused on *computational* investigations. However, *stability analysis* of the underlying adaptive mechanism has received less attention.

Recently, [2] defined the concept of mechanobiological stability based on a Lyapunov analysis of differential equations of mass density and vessel wall position. But these equations do not explicitly account for the dependence of the total strain energy on the history of the growth. In this work, we directly start from the more physiologic integral state equations of the continuum based constrained mixture model of G&R and derive the state equations (time-delayed differential equations) describing the progression of aneurysm expansion. By employing an exponential form of the constituent survival function, which is commonly used to describe natural decay, the delayed differential equations can be reduced to an ODE system. Linear stability analysis is subsequently applied to this ODE system and a stability criteria is obtained based on G&R kinetic parameters and material properties. This leads to improved understanding of how physiologic parameters affect long term stability of aneurysm growth.

METHODS

The purpose of this work is to derive theoretical stability criteria for G&R using a continuum-based constrained mixture model. To make this analysis tractable, we assume (1) the vascular geometry is cylindrical; (2) deformation only contains radial expansion; (3) constituents align in the circumferential directions to sustain radial expansion; (4) the material is linear and strain energy density function is $\hat{W} = \frac{1}{2}\mathbf{E}\epsilon^2$, where $\mathbf{E}$ is the material stiffness and ϵ is the strain.

Constrained mixture theory of growth and remodeling

Based on the constrained mixture model, mass density is given by

$$M(t) = M(0)Q(t) + \int_0^t m(\tau)q(t-\tau)d\tau \ . \quad (1)$$

The total strain energy is given by

$$W(t) = \frac{M(0)}{\rho}Q(t)\hat{W}(\mathbf{C}_{n(0)}(t)) + \int_0^t \frac{m(\tau)}{\rho}q(t-\tau)\hat{W}(\mathbf{C}_{n(\tau)}(t))d\tau \quad (2)$$

where

$$\hat{W}(\mathbf{C}_{n(\tau)}(t)) = \frac{1}{2}\mathbf{E}(\lambda_{n(\tau)}(t)-1)^2 = \frac{1}{2}\mathbf{E}\left(G_h\frac{r(t)}{r(\tau)}-1\right)^2 \ . \quad (3)$$

The force balance in the circumferential direction yields

$$Pr(t) = T_\theta = \frac{1}{\lambda_z}\frac{\partial W}{\partial \lambda_\theta} \ . \quad (4)$$

Therefore, the equations describing the long-term expansion of an aneurysm are

$$\begin{aligned} M(t) &= \int_0^t m(\tau)q(t-\tau)d\tau & (5)\\ Pr(t) &= R\int_0^t \frac{m(\tau)}{\rho}q(t-\tau)\mathbf{E}\left(G_h\frac{r(t)}{r(\tau)}-1\right)\frac{G_h}{r(\tau)}d\tau & (6)\\ m(t) &= M(t)\left(k_g(\sigma(t)-\sigma_h)+f_h\right) & (7)\\ \sigma(t) &= \frac{Pr(t)}{h(t)} & (8)\end{aligned}$$

where $h(t) = \frac{M(t)}{\rho J(t)}$ and J is the determinant of deformation gradient tensor. Eq. (7) defines the stress-mediated growth law based on wall tension deviation and k_g is the feedback gain.

Reduce time-delayed system to ODE system

Since the current behavior of the system depends continuously on all past time history, the system is a continuous time-delayed system. It is convenient if we can convert it to an ODE system depending only on current states. By assuming the survival function has exponential decay

$$q(t) = \exp(-\alpha t), \quad \alpha > 0 \ , \quad (9)$$

the delayed differential state equations can be reduced to the following ODE system

$$\begin{aligned} \dot{M}(t) &= M(t)k_g(\sigma(t)-\sigma_h) & (10)\\ \dot{r}(t) &= \frac{1}{k(t)}\left(\alpha r(t) - \frac{m(t)}{r(t)}k_1\right) & (11)\\ \dot{y}(t) &= k_2\frac{m(t)}{r^2(t)} - \alpha y(t) & (12)\\ m(t) &= M(t)\left(k_g(\sigma(t)-\sigma_h)+f_h\right) & (13)\\ \sigma(t) &= \frac{\rho Pr(t)^2}{M(t)R} & (14)\end{aligned}$$

where $k(t) = \left(\frac{1}{P}y(t) - 1\right)$, $k_1 = \frac{R\,\mathbf{E}\,(G_h-1)G_h}{P\rho}$, $k_2 = \frac{G_h^2\mathbf{E}R}{\rho}$ and $f_h = \alpha$, and R is vessel radius in the reference configuration. The variable $y(t)$ is an extended state generated from the reduction process. However, later it can be shown that $y(t)$ can be decoupled from the system using linearization.

Stability analysis of the ODE system

The stability of the aneurysm expansion is characterized by the stability of the ODE system we obtained above, which is obtained by studying the stability of the equilibria of the system.

Linearization around the equilibrium point

We assume the steady state value for mass density and vessel radius are M_h and r_h respectively. Therefore, based on Eq. (14), the steady value of circumferential stress σ_h is $\frac{\rho P r_h^2}{M_h R}$. Perturbing the nonlinear ODE system around the homeostatic values

$$M(t) = M_h + \Delta M(t) \quad (15)$$

$$r(t) = r_h + \Delta r(t) \quad (16)$$

$$\sigma(t) = \sigma_h + \Delta\sigma(t) \quad (17)$$

with

$$\left|\frac{\Delta M}{M_h}\right|, \left|\frac{\Delta r}{r_h}\right|, \left|\frac{\Delta \sigma}{\sigma_h}\right| \ll 1 . \quad (18)$$

gives the the linearized state equations for Δr and ΔM

$$\frac{d}{dt}\begin{bmatrix}\Delta r \\ \Delta M\end{bmatrix} = \begin{bmatrix}A & B \\ C & D\end{bmatrix} \times \begin{bmatrix}\Delta r \\ \Delta M\end{bmatrix} \quad (19)$$

where

$$A = \frac{1}{k_h}\left(2\alpha - \frac{2k_g r_h^2 \rho P}{R M_h}\right), \quad (20)$$

$$B = \frac{1}{k_h}\left(-\frac{\alpha r_h}{M_h} + \frac{k_g r_h^3 \rho P}{M_h^2 R}\right), \quad (21)$$

$$C = \frac{2k_g \rho P r_h}{R}, \quad (22)$$

$$D = -\frac{k_g \rho P r_h^2}{R M_h}. \quad (23)$$

It can be noticed that the linearized subsystem for Δr and ΔM is decoupled with the perturbation of the extended state Δy.

Stability analysis for the linearized state equations

To gain understanding about the stability of the linearized system, i.e. whether G&R results in stable aneurysm expansion or unstable expansion, we need to first find the eigenvalues of system matrix by setting the characteristic polynomial for the system be equal to zero

$$\begin{vmatrix}A-\lambda & B \\ C & D-\lambda\end{vmatrix} = \lambda^2 - (A+D)\lambda + (AD - BC) = 0 . \quad (24)$$

It can be easily verify that $AD - BC = 0$ based on Eq.(20–23). Therefore, the corresponding eigenvalues are

$$\lambda_1 = A + D, \quad \lambda_2 = 0 . \quad (25)$$

This means $[\Delta r, \Delta M] = [0, 0]$ is a **degenerate** fixed point of the linearized system.

For a degenerate fixed point, asymptotic stability of the linearized system, i.e.

$$\lim_{t\to\infty} \Delta r, \Delta M, \Delta\sigma = 0$$

cannot be obtained no matter how we choose G&R parameters(k_g, α, $\mathbf{E}$). However, neutrally stable can be obtained under certain conditions.

Theorem 0.1. (Neutrally Stable Fixed Point) *Consider an autonomous dynamical system $\dot{\mathbf{x}} = \mathbf{A}\mathbf{x}$ with system matrix*

$$\mathbf{A} = \begin{bmatrix}A & B \\ C & D\end{bmatrix} \quad (26)$$

whose eigenvalues are λ_1 and λ_2. Let $\tau = Trace(\mathbf{A}) = A + D$, $\Delta = \det(\mathbf{A}) = AD - BC$. If

$$\Delta = 0, \quad \tau < 0 \quad (27)$$

then $\lambda_1 < 0$ and $\lambda_2 = 0$ and there exists a line of neutrally stable non-isolated fixed points passing through $\mathbf{x} = \mathbf{0}$. A neutrally stable fixed point denotes a fixed point which is Lyapunov stable but not attracting.

RESULTS

Based on Theorem 0.1, the neutrally stabilizing condition for the linearized system (19) at the fixed point $[\Delta r, \Delta M] = [0, 0]$ is $A + D < 0$, i.e.

$$k_g > \frac{\alpha R M_h}{\rho P r_h^2}\left(\frac{1}{1 + \frac{\mathbf{E} M_h}{2\rho P r_h}}\right) \triangleq k_{cr} . \quad (28)$$

Eq. (28) gives the **stability criteria** for aneurysm expansion in terms of G&R parameters (k_g, α), material properties ($\mathbf{E}$, ρ), transmural pressure P and geometry (R, r_h).

By taking the time derivative of Eq. (14), we can obtain the linearized ODE for the time evolution of wall tension deviation

$$\begin{aligned}\frac{d}{dt}\Delta\sigma(t) &= \frac{\rho P}{R}\left(\frac{2r(t)}{M(t)}\dot{r}(t) - \frac{r^2(t)}{M^2(t)}\dot{M}(t)\right) \\ &\approx \left[\frac{1}{k_h}\left(2\alpha - \frac{2k_g r_h^2 \rho P}{R M_h}\right) - \frac{k_g \rho P r_h^2}{R M_h}\right]\Delta\sigma \\ &= (A + D)\Delta\sigma , \end{aligned} \quad (29)$$

which implies that stability criteria (28) also assures the exponential stability for wall tension deviation, i.e.

$$\lim_{t\to\infty} \Delta\sigma(t) = 0 . \quad (30)$$

However, for deviation of mass density ΔM and radius Δr, only neutral stability can be obtained. In conclusion, as $t \to \infty$ only $\Delta\sigma$ will converge to zero while ΔM and Δr will only stay bounded but without converging to zero.

DISCUSSION

Eq. (28) gives the stability criteria for aneurysm expansion by defining a critical value k_{cr} for the feedback constant k_g in terms of G&R parameters (k_g, α), material properties ($\mathbf{E}$, ρ), transmural pressure P and geometry (R, r_h). When $k_g > k_{cr}$, aneurysm expansion is stable while $k_g > k_{cr}$ causes instability of aneurysm expansion, i.e. unbounded aneurysm growth will be caused.

The stability criteria also indicates that for lower vessel wall stiffness $\mathbf{E}$ or higher decay rate of the constituent α (determined by constituent life span), the critical value k_{cr} becomes higher, which makes the stability criteria harder to satisfy, as expected. Therefore, these factors tend to destabilize the expansion process of aneurysm and the risk of aneurysm rupture increases.

The stability analysis also shows that even when the stability criteria is satisfied, only $\Delta\sigma$ will converge to zero as $t \to \infty$ while ΔM and Δr will only stay bounded but not necessarily converge to zero. This matches observation in [1] that in stable aneurysm expansion cases, only wall tension σ is able to recover the homeostatic value while the geometry and mass density cannot return to the original homeostatic states.

REFERENCES

[1] S Baek, KR Rajagopal, and JD Humphrey. Competition between radial expansion and thickening in the enlargement of an intracranial saccular aneurysm. *Journal of Elasticity*, 80(1-3):13–31, 2005.

[2] CJ Cyron, JS Wilson, and JD Humphrey. Mechanobiological stability: a new paradigm to understand the enlargement of aneurysms? *Journal of The Royal Society Interface*, 11(100):20140680, 2014.

SB3C2015
Summer Biomechanics, Bioengineering and Biotransport Conference
June 17-20, 2015,Snowbird Resort, Utah, USA
SB3C2015-1097

MECHANICAL AND ADHESIVE PROPERTIES OF HYDROGELS IN TENSION AND SHEAR

Daniel Collins (1), Patrick Myers (1), Thomas Christiani (2), Jennifer Vernengo (2), Jennifer A. Kadlowec (1)

(1) Mechanical Engineering
Rowan University
Glassboro, NJ 08028

(2) Chemical Engineering
Rowan University
Glassboro, NJ 08028

INTRODUCTION

As the human body ages, the intervertebral disc (IVD) decays, typically resulting in Annulus fibrosis (AF) tears and Nucleus pulposus (NP) herniation, a phenomenon which affects roughly 80% of the human population at some point in their lives.[1] This often occurs from the dehydration of the NP, and the AF then accepting more load without support from the NP.[2] While disc fusion can prevent pain associated with this occurrence, it reduces some of the range of motion in the back, thereby reducing the quality of life for the patient.[3] Discectomies offer an alternative, but are often plagued by complications such as device migrations.[4] The use of a hydrogel bioadhesive to replace the NP offers an option that allows the patient to retain full range of motion without suffering pain associated with IVD decay.[5] The objective of this study was to characterize a novel bioadhesive composed of poly(N-isopropylacrylamide) (PNIPAAm) grafted with chondroitin sulfate (CS) (PNIPAAm-g-CS) by determining the adhesive tensile and shear properties.

METHODS

Using previously reported methods, PNIPAAm-g-CS was synthesized via redox polymerization.[6] Alginate microparticles, AMPS, (142 $\pm$ 18 μm) were produced by crosslinking with calcium chloride in a water in oil emulsion based on methods described by Paques.[7]. For all samples, a 5% (w/v) solution of PNIPAAm-g-CS was made in phosphate buffered saline (PBS) at room temperature. Alginate microparticles were suspended in the solution at a concentration of 50 mg/mL. Annulus Fibrosus tissue was harvested from a porcine spine that was purchased at a local butcher.

Adhesive properties of the PNIPAAm-g-CS in contact with the interior layer of porcine AF tissue were tested for both tensile and shear mechanical behavior. Tensile testing was conducted in a manner similar to that previously reported[7] and based on a modified version of ASTM F 2258-05. For this study, two samples of porcine AF were cut to 7 mm x 7 mm and affixed horizontally parallel to an upper delrin cylinder and lower fixture mounted on a FGS-200PV E-force test stand with cyanoacrylate adhesive and warmed to 37° C. Hydrogel solution (60 μL) at 15° C was applied by pipette onto the samples and allowed to solidify for 5 minutes in a 37° C saline bath. The upper fixture was raised at a rate of 2 mm/min and force displacement data were recorded until failure.

Shear testing was conducted using a test method developed based on ASTM standard F 2255-05 and a custom designed and fabricated fixture.[8] Two tissue samples of porcine AF were cut to dimensions 5mm x 10mm, affixed to plexiglass slides with cyanoacrylate adhesive and warmed to 37°C in a saline bath. Hydrogel (60 μL) was applied to the samples by pipette and allowed to solidify for 5 minutes. The samples were displaced relative to one another using a FGS-200PV E-Force Test Stand at a rate of 5 mm/min until the hydrogel failed in shear, while force and displacement data were recorded.

Both tensile and shear stress were calculated as force divided by initial contact area, with strength being the maximum stress value. Work of adhesion was calculated by integrating force with respect to displacement using a trapezoidal approximation. A total of n=7 samples were used to calculate a mean and 95% confidence interval for each property.

RESULTS

Tensile and shear strengths with 95% confidence intervals for two different formulations of hydrogels, 5% (w/v) PNIPAAm-g-CS, with and without 142 μm alginate microparticles (AMPS), on porcine AF are presented for comparison in Table 1. Similarly, adhesive works for these two formulations are presented in Table 2.

Table 1 – Adhesive strengths for different hydrogels when tested on porcine AF tissue

Hydrogel	Shear Strength (Pa)	Tensile Strength (Pa)
5% PNIPAAM-g-CS	2800 ± 724	1615 ± 336
5% PNIPAAM-g-CS + 142 μm AMPS	10340 ± 1504	4503 ± 615

Table 2 – Work of adhesion for different hydrogels when tested on porcine AF tissue

Hydrogel	Shear Work (mJ)	Tensile Work (mJ)
5% PNIPAAM-g-CS	0.618 ± 0.186	0.667 ± 0.363
5% PNIPAAM-g-CS + 142 μm AMPS	2.454 ± 0.230	1.126 ± 0.212

Table 1 results clearly show that the adhesive strengths in both shear and tension are higher for PNIPAAM-g-CS with 142 μm alginate microparticles than the counterpart PNIPAAM-g-CS alone. Similarly, Table 2 results clearly show that the works of adhesion in both shear and tension are higher for PNIPAAM-g-CS with 142 μm alginate microparticles than the counterpart PNIPAAM-g-CS alone. Further, both hydrogel fomulations have shear strengths greater than tensile strengths; and shear and tensile works that are approximately equal for PNIPAAM-g-CS alone to even greater shear than tensile work for PNIPAAM-g-CS with alginate microparticles.

DISCUSSION

The addition of alginate particles to the hydrogel increased the adhesive strengths and works consistently compared to the hydrogel without particles. While particulate infusion has an effect on the chemistry of the hydrogel, which increases its adhesion; it also presents a matrix of solid-in-gel, which creates solid-on-solid contact, which may also have an effect of added these properties, especially in shear. These particularly promising results are important as we consider this material as an implantable replacement for NP, which may resist expulsion due to these higher strengths and works.

Consistent with our previous findings, the maximum adhesive strengths of both hydrogel formulations, with and without particles, was greater in shear than in tension,[8] with several reasons potentially explaining this. First, large tensile deformations result in necking and thus reduced cross-sectional area at the point of failure. Since initial contact area is used in calculations for adhesive strengths, these engineering values are less than actual or true strength. Assuming the hydrogel is incompressible and using large deformation theory, actual values, which would be greater, could be calculated. Meanwhile, the difference in engineering versus true strengths would be considerably less in shear. Secondly, as mentioned above, the microparticles may also provide frictional load contributing to adhesion in shear that is more pronounced than in tension.

The adhesive works for PNIPAAM-g-CS alone are similar and not reflective of strengths, both of which are proportional to force. As noted by the confidence intervals, sensitivity to experimental conditions, particularly temperature and force measurements may limit conclusions here. The adhesive work for PNIPAAM-g-CS with particles was two times greater in shear than in tension, which is similar to the strength difference for this hydrogel, which may be explained by previously mentioned issues of solid-solid interaction and friction.

This work provides results for mechanical properties of a bioadhesive hydrogel, namely PNIPAAM-g-CS, in shear and tension. For these modes of deformation, adhesive strength and work are only two properties discussed here in this study for a limited set of hydrogel formulations. Based on promising shear strength and work results of PNIPAAM-g-CS with 142 μm alginate microparticles, shear tests with other particle sizes and particle concentrations are planned. Further, in order to determine the viability of the hydrogel as suitable NP implant, further mechanical properties will be evaluated such as expulsion and compression.

ACKNOWLEDGMENTS

Funded by the NIH/NIBIB 1R15 AR 063920-01.

REFERENCES

[1] Borges, A et. al., *Acta Biomaterialia* 7:3412-3421, 2011. [2] Leone, G et al., *J. Biomedical Materials Research* 84A2:391-401, 2008. [3] Carl, A et. al., *Spine* 4:6:S325-S329, 2004. [4] Jin, D. et. al., *J. of spinal disorders & techniques* 16:4:331-337, 2003. [5] Dastjerdi, A et. al., *Acta Biomaterialia* 8(9):3349-3359, 2012. [6] Wiltsey C. J et al., *Mater. Sci. Mat. Med.* 837-847, 2013. [7] Paques JP. *Food Hydro.* 31: 428-434, 2013. [8] Mauriello et al. *BMES Annual Conf.* 2014.

The Temporal Change in Protein Biomarkers in the Vitreous Humor following Blast Trauma

Justin A. Jones (1), Daniel F. Shedd (2), Brittany Coats PhD (2)

(1) Department of Bioengineering
University of Utah
Salt Lake City, Utah, United States

(2) Department Mechanical Engineering
University of Utah
Salt Lake City, Utah, United States

INTRODUCTION

Ocular injuries due to blast exposure have increased in occurrence over the last several decades. Between 1983 and 2002, 36,110 bombings occurred in the United States, resulting in 5931 injuries [1]. The incidence of eye injury due to blast trauma with soldiers has increased from 9% to 13% since the Vietnam War [2]. Progression from ocular injury to ocular disease is not adequately characterized; and currently, indicators of injury progression to vision loss are largely unknown. Common closed globe injuries, including, retinal detachment, retinal tears, and optic nerve fiber degeneration [3], can elicit a cellular inflammatory response that releases proteins into the vitreous humor of the eye.

One prominent ocular protein biomarker is the neurofilament heavy chain (NfH). It is believed that NfH is released from degenerating retinal ganglion cells and their axons into the vitreous [4]. Other protein biomarkers of importance are cytokines. Cytokines are signaling proteins present in the inflammatory cascade. A particularly well known cytokine is vascular endothelial growth factor (VEGF) and has importance to pathologic angiogenesis [5]. Its subcomponents are said to be involved in endothelial cell migration, proliferation, survival and permeability and are typically present any time there is an inflammatory response [6].

The goal of this study was to discover if protein biomarkers known to reflect ocular injury can be used as reliable early identifiers of vision loss due to blast exposure.

METHODS

All testing procedures were approved by the University of Utah Institutional Animal Care Use Committee (IACUC). Male Long Evans rats (n = 24; 300-350g) were placed in a 6 meter long by 15.24 cm internal diameter blast tube, and exposed to a 30 psi overpressure blast with a 7 msec duration (Figure 1). Experimental animals were separated into three survival time groups: 1 day, 1 week and 4 week. Before the blast exposure was performed, each animal was weighed and anesthetized using a mix of ketamine and dexmedetomidine with a dosage of 65mg/kg and 0.14mg/kg, respectively. A visual eye examination was then performed by an ophthalmologist. The animals were exposed to a blast perpendicular to the sagittal plane from right to left. After the blast, the animal was removed and another visual eye examination was performed.

After the respective survival time was reached, the animals were sacrificed by formalin perfusion fixation through the heart using standard practices [7]. The whole globe eyes and brains were harvested at necropsy. The eyes were then eviscerated and the vitreous and lens removed. The vitreous and lens were separated by placing them into a filter centrifuge tube and spun down (10k rpm, 10 min). The separated vitreous (approx. 50 μL) was diluted with phosphate buffer saline (PBS) until a total volume of 150μL was reached. The sample was then separated into three equal tubes.

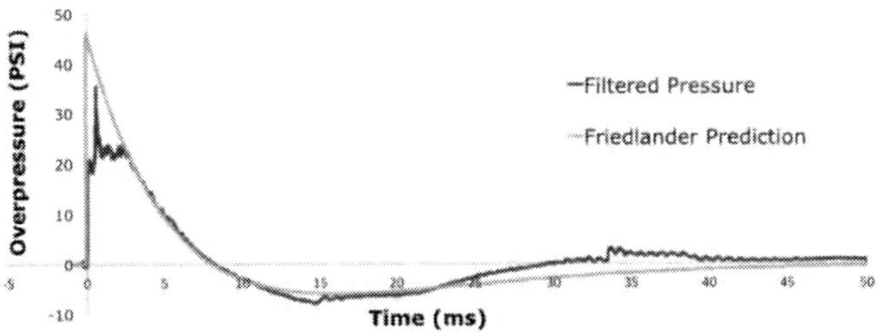

FIGURE 1. PRESSURE-TIME HISTORY AT LOCATION OF ANIMAL PLACEMENT WITHIN BLAST TUBE

To analyze the NfH content, an ELISA protocol was used according to an ELISA technique previously developed by Petzold et al [8]. The microtitre plates were coated overnight at 4°C with 100 µL of capture antibody, SMI35. The plates were then washed three times for 10 minutes using a barbitone buffer wash containing 0.1% BSA, and 0.05% Tween 20. After washing, 250µL of barbitone block with 1% BSA was added to each well and the plate was incubated at room temperature (RT) for 1 hour. After another wash cycle, 50µL of sample, standard, or negative control was added to each well of the plate in duplicate. After one hour incubation at RT the wash processes was repeated. After washing, 100µL of second antibody was added to each well of the plate and incubated for 1 hour at RT. Following a third wash cycle 100µL HRP-labeled swine anti-rabbit antibody was added to the plates and incubated for one hour at RT. After a final wash 100µL TMB substrate was added and incubated for 20 minutes in a dark room, the reaction was stopped by adding 50µL of 1 M HCL. The absorbance was then read using an ELISA plate reader at 450nm with 750nm reference wavelength.

The analysis of inflammatory cytokines was performed using a commercially available kit (RayBio® Rat Cytokine Antibody Array G). These kits tested for the 19 cytokines including VEGF, LIX and TNF-α. The methods to develop these kits were done according to the manufacturer's instructions and can be found on RayBio® Tech website. Once the glass chip was developed, the intensities were read using a GENEPIX™ 4000A microarray scanner at an excitation frequency of 532nm.

At this point in time, sample size is small (n=4 per group), so 5 one-way ANOVAs were performed. Two assessed significant changes across time points within each eye side, and three assessed significant differences between the right and left eye at each time point. Collection of control data is ongoing and is not included in this abstract. All analyses were performed using JMP® software with a p-value < 0.05 considered significant.

RESULTS

Both the left and right eyes showed a general increase in NfH concentration, but this increase was only significant in the right eye between 1 day and 4 weeks (p=0.044, Figure 2). There were no significant differences between NfH concentration of the ipsilateral (0.126±0.02 ng/ml) and contralateral (0.137±0.05 ng/ml) eyes at the 4 week time point.

Several cytokine proteins in the eye contralateral to the injury significantly decreased over time post-injury (GM-CSF, IL-1β, IL-4, IL-10, LIX ,TNF-α), but remained relatively constant in the eyes ipsilateral to injury (Figure 3). At 4 weeks after the injury, LIX and TNF-α were significantly higher in the eye ipsilateral to the injury

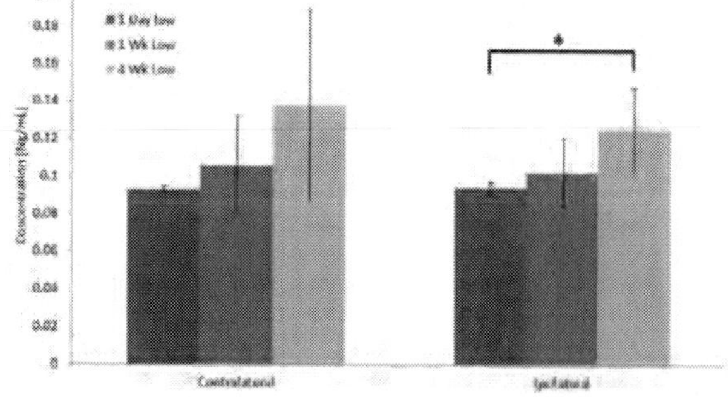

FIGURE 2. CHANGES IN NFH CONCENTRATION POST INJURY IN IPSILATERAL (RIGHT) AND CONTRALATERAL (LEFT) EYES. *p<0.05

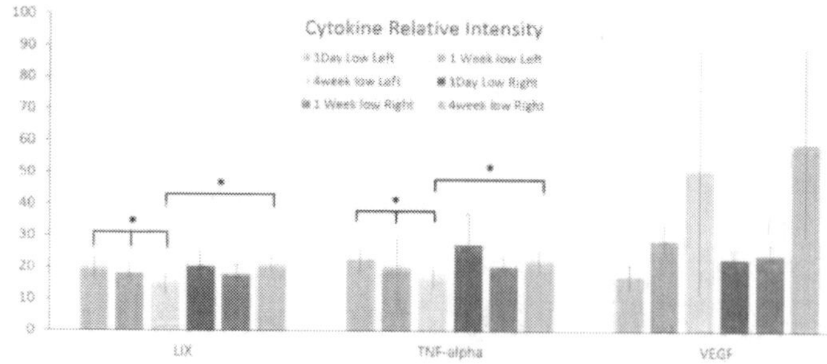

FIGURE 3. CYTOKINE RELATIVE INTENSITIES

compared to the contralateral eye (p = 0.015 and 0.043, respectively). No significant differences were found in either eye for any of the remaining proteins of the cytokine array, including VEGF (shown above).

DISCUSSION

The significant decrease in NFH and cytokine proteins with time suggests that there was an acute (1 day) inflammatory response that occurred in both eyes, and only significantly decreased in the eye contralateral to the blast pressure. NfH is hypothesized to release from degenerating retinal ganglion cells and their axons. Previous research has shown that decreased axonal transport is preceded by cytoskeletal changes and degradation of NfH [4]. The sustained elevation of NfH in the ipsilateral eyes in this study suggest that cytoskeletal changes in the retina are ongoing at 4 weeks after injury. This long-term injury may lead to future vision degradation. Longer-term analysis, vision assessment and control group evaluation needs to be completed before making this assertion.

We found no significant change in VEGF in either eye. This was surprising as significant increases in VEGF have been reported in many ocular disorders including diabetic retinopathy, diffuse macular edema, retinal vein occlusion and retinal detachment [5]. Instead, we found several changes in interleukins, TNF-α, and LIX. The significance of these proteins in the blast injury response will be explored further do determine if they are merely be a generic inflammatory response to the blast that is quickly resolved.

In summary, the blast pressure in this study appears to create some damage to the retina that is potentially recovered quickly in the contralateral eye, but not in the ipsilateral eye. Longer time points are currently being explored to determine the resolution of NfH to baseline levels post injury. Future work will include completion of the control groups as well as performing higher blast pressures to perhaps create higher levels of injury. We are also working to combine the findings of this study with changes in visual acuity and histology in the same animals.

ACKNOWLEDGEMENTS

We would like to thank USAMRMC #W81XWH-12-1-0243 for support of this project.

REFERENCES

[1] Kapur GB, et. al. *J Trauma,* 59:1436-44 2005.
[2] Weichel E., Colyer, *et al. Ophthalmology,* 115:2235-2245, 2008.
[3]Dougherty A., et al. *Brain Injury* 25:8-13, 2011.
[4] Balaratnasingam C. et al. *IOVS,* 49:986-999. 2008
[5]Cross M, et al. *TIBS,* 28:488-494, 2003.
[6]Miller J., et al. *Am J Pathol* 145:574-584, 1994.
[7]Gage G. J, et al. *J. Vis. Exp.* (65), e3564, 2012
[8]Petzold A., et al. *J Neural Transm* 116:1601-1606, 2009.

SB³C2015
Summer Biomechanics, Bioengineering and Biotransport Conference
June 17-20, 2015, Snowbird Resort, Utah, USA
SB³C2015-1099

THE IMPACT OF PHYSICAL AND ENZYMATIC TREATMENT ON THE DEVELOPMENT OF TISSUE-ENGINEERED ARTICULAR CARTILAGE GENERATED FROM ADULT HUMAN CHONDROCYTES

Brendan L. Roach (1), Terri-Ann N. Kelly (1), Michael K. Dermksian (1), Sonia Bansal (1), Paola A. Lopez (4), Aaron M. Stoker (3), James L. Cook (3), Gerard A. Ateshian (1,2), Clark T. Hung (1)

(1) Department of Biomedical Engineering
Columbia University
New York, NY, USA

(2) Department of Mechanical Engineering
Columbia University
New York, NY, USA

(3) Missouri Orthopaedic Institute
University of Missouri at Columbia
Columbia, MO, USA

(4) School of Engineering
University of Arizona
Tucson, AZ, USA

INTRODUCTION

Adult human chondrocytes as a cell source for functional tissue-engineered articular cartilage have not been fully investigated *in vitro*. Previous efforts have focused on biochemical [1-3] and/or mechanical [3,4] characterization. The studies presented here combine previous work from our group that has demonstrated a capacity to produce functional adult human chondrocyte constructs, as well as the effective enzymatic/physical treatments of immature chondrocyte tissues. Adult human chondrocytes cultured in a 3D agarose hydrogel have been shown to produce tissues with mechanical properties reaching ~150kPa [6,7]. The removal of glycosaminoglycan (GAG) via chondroitinase ABC (chABC)-treatment has been shown to significantly improve the collagen content of tissue-engineered cartilage seeded with immature cells [6,8-11]. By extension, the dynamic and tensile properties have also shown marked improvements with chABC-treatment. Further, the introduction of channels has proven effective in improving the biochemical and mechanical properties of immature tissue-engineered cartilage [12,13]. Based on these results and our early efforts with adult canine chondrocytes (data not shown), we hypothesize that, separately, chABC treatment and channel creation will yield significant improvements in tissue properties compared to their respective controls.

METHODS

To test our hypotheses, healthy human chondrocytes were isolated, passaged in expansion media (1 ng/mL TGF-β1 and 5 ng/mL FGF-2) [6,8], cast at 60×10^6 cells/mL in 2% (w/v) low-melt agarose, and cultured for 42 days in chondrogenic media (ITS+, 50 μg/mL ascorbic acid) with 10 ng/mL TGF-β3 [6,8]. For Study 1, a subset of the constructs was digested in 0.005 U/mL chABC for 7 days (Figure 1, middle panels) [11]. For Study 2, a 1-mm channel was cored in the center of the constructs (Figure 1, bottom panel). For both studies, untreated constructs were also maintained as a control (Figure 1, top panel). Unconfined compression testing was performed [6,8] to determine equilibrium compressive, tensile and dynamic compressive moduli and permeability of the constructs [14]. After testing, the constructs were assayed for GAG and collagen [6,8]. Two-way ANOVA (α=0.05) with Tukey post hoc test was performed to determine statistical significance (n=3-9).

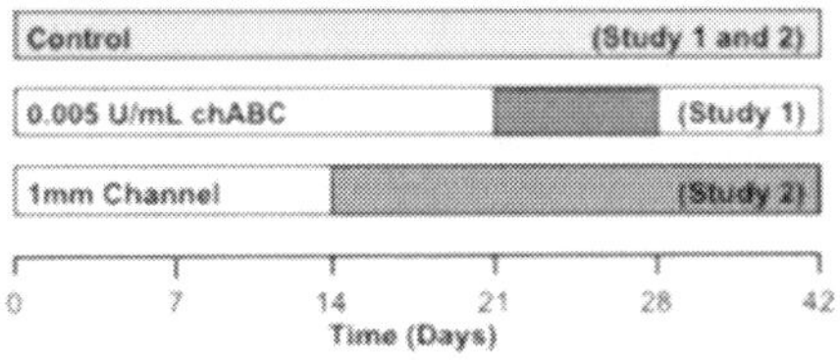

FIGURE 1. EXPERIMENTAL PROTOCOL FOR EXAMINING THE EFFECTS OF CHABC-TREATMENT (STUDY 1) AND 1-MM CHANNEL (STUDY 2) ON ADULT HUMAN TISSUE-ENGINEERED ARTICULAR CARTILAGE.

RESULTS

Study 1: The results of this study showed that chABC-treatment did not significantly improve the tissue properties of engineered articular cartilage respective to controls (Figure 2). chABC treated groups did show a significant improvement in GAG and the dynamic modulus from day 28 to 42, a trend not observed for control tissues. Interestingly, collagen showed significant increases from day 28 to 42 for both groups.

Study 2: The results of this study demonstrate a marked increase in GAG, tensile modulus, dynamic modulus, and Young's modulus

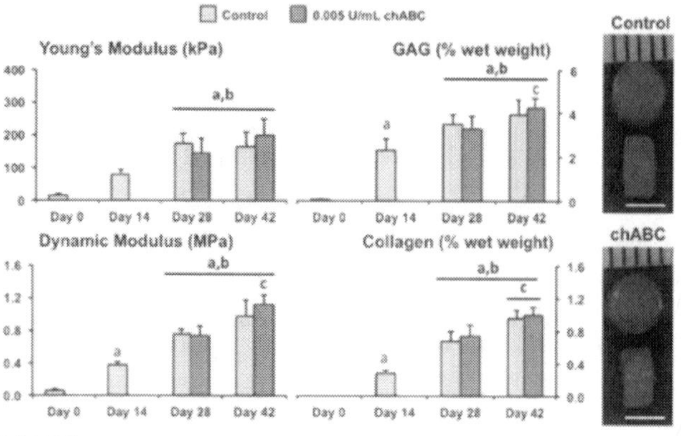

FIGURE 2. DEVELOPMENT OF ADULT HUMAN ENGINEERED ARTICULAR CARTILAGE FOLLOWING CHABC-TREATMENT (STUDY 1). (A) P<0.01 VS. DAY 0, (B) P<0.05 VS. DAY 14, (C) P<0.05 VS. DAY 28 (N=3-9). INSET (BOTTOM RIGHT) IMAGES OF TYPICAL CONSTRUCTS AT DAY 42 (SCALE BAR = 4 MM).

(Figure 3). The Young's moduli of tissues that received a 1-mm channel were statistically higher than controls. There was no difference in collagen as normalized by wet weight. There was a statistically significant decrease in permeability of tissues that received a 1-mm channel between day 14 and day 42, a change not observed to the same extent in control samples. The difference in tissue properties is illustrated by tissue opacity at day 42, with a gradient of opacity from the periphery to the center seen in the control group, but not in the channel group (Figure 3, inset).

DISCUSSION

To date, there has been limited success in engineering adult human cartilage. Protocols optimized for immature bovine cells are being challenged to accommodate a cell source with lower plasticity and metabolic rates. chABC-treatment improves the collagen content of engineered constructs created from immature cells [6,8-11], however, the current work demonstrates that previously robust findings with immature chondrocytes do not directly translate to human tissue [14]. It was observed that GAG, collagen and the dynamic modulus of chABC-treated samples significantly increased from day 28 to day 42 (Figure 2). For control tissues, despite the observed plateau in most parameters after day 28, collagen continues to be actively produced at day 42, suggesting that collagen may have an intrinsically longer timescale of development than GAG. Taken together with recent findings that earlier treatment may be appropriate [11], these findings highlight the potential impact of chABC-treatment on late-stage growth and therefore suggest that further investigation into the time course of application is warranted.

Previous studies have demonstrated that the creation of channels in tissue-engineered cartilage improves nutrient availability and produces a more homogenous tissue [12-13,15]. In Study 2, a 1-mm central channel was created at day 14. By day 42, this modification produced a significant increase in Young's modulus when compared with controls and led to corresponding increases in GAG, collagen and dynamic modulus. Further, the permeability of the channel group significantly decreased from day 14 to day 42 (Figure 3), a difference not observed in the control. This suggests the channel increased nutrient availability in the center of the tissue, improved deposition of extracellular matrix proteins, and resulted in an increase in density. This was confirmed visually (Figure 3 inset) at day 42, as channeled tissues displayed a level of opacity absent in controls, with visible shrinking of the channel. The permeability of channeled constructs, 0.00493 $mm^4/N\text{-}s$, when compared with the permeability of native articular cartilage, ~0.001-0.002 $mm^4/N\text{-}s$, suggest that the use of channels is a viable approach for improving the functional properties of engineered constructs derived from adult human chondrocytes.

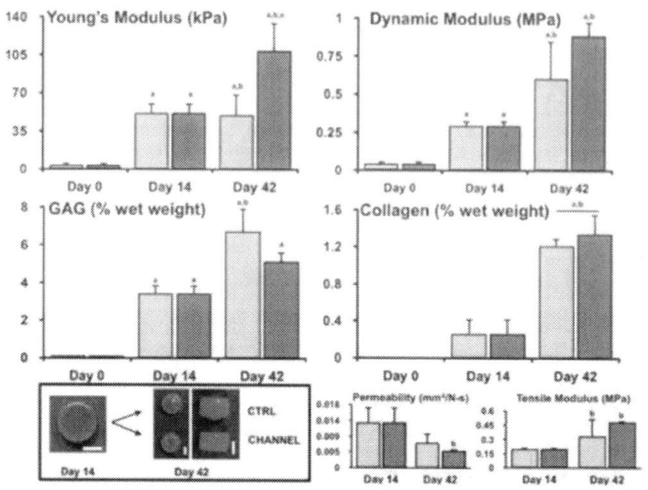

FIGURE 3. DEVELOPMENT OF ADULT HUMAN TISSUE-ENGINEERED ARTICULAR CARTILAGE FOLLOWING CREATION OF 1-MM CHANNEL. 42-DAY CULTURE PERIOD (STUDY 2). (A) P<0.01 VS. DAY 0, (B) P<0.0005 VS. DAY 14 (N=3-5). *INSET (BOTTOM LEFT)* DAY 14, OPACITY GRADIENT FROM CENTER TO PERIPHERY. AT DAY 42, SWOLLEN GEOMETRY IN CONTROL SAMPLES, WITH EVIDENT GRADIENT IN OPACITY (SCALE BAR = 2 MM).

These data highlight the difficulties associated with engineering of adult human cartilage, but suggest promising tools that can be utilized in future work. Here, low dose treatment was shown to provide a modest improvement in mechanical and biochemical properties and may prove more advantageous with the application of dynamic loading, shown to improve collagen architecture and increase nutrient availability [6,16,17]. While a 1-mm channel yielded improved tissue properties, future work must consider the use of an array of channels and the inclusion of stirring to enhance nutrient availability [13]. These techniques, in combination, may further progress the development of engineered adult human cartilage.

ACKNOWLEDGEMENTS

This work was funded by the NIH [AR46568 (CTH), R21EB014382 (CTH, TNK), 5T32AR059038 (BLR)], NSF Graduate Research Fellowship (BLR), and the Amgen-SURF program (PAL).

REFERENCES

[1] Demarteau+ *BBRC* 310:580-88, 2003. [2] Wang+ *Biomat* 27:4434-42, 2006. [3] Lehmann+ *Eur J Histo* 57:205-15, 2013. [4] Smeriglio+ *TE A* 21:147-55, 2015. [5] Pleumeekers+ *Eur Cell Mater* 27:264-80, 2014. [6] O'Connell+ *Eur Cell Mater*, in review. [7] Oswald+ *Trans ORS*, 2009. [8] Kelly+ *JOB* 46:1784-91, 2013. [9] Bian+ *TE A* 15:2065-72, 2009. [10] Ng+ *TE A* 16:1041-51, 2010. [11] O'Connell+ *Eur Cell Mater* 27:312-20, 2014. [12] Bian+ *OAC* 17:677-85, 2009. [13] Cigan+ *JOB* 47:3847-54, 2014. [14] Cigan+ *TE A* 19:1941-48, 2013. [15] Buckley+ *TE A* 15:3212-20, 2009. [16] Kelly+ *JOB* 39:1489-97, 2006. [17] Kelly+ *Biorheology* 41:223-37, 2004.

BIOMIMETIC *EX VIVO* MODEL FOR TRACKING STEM CELLS DURING MICROVASCULAR NETWORK GROWTH

Mohammad S. Azimi (1), Amy L. Strong (2), Theresa B. Phamduy (1), Douglas B. Chrisey (1,3), Bruce A. Bunnell (2), Walter L. Murfee (1)

(1) Department of Biomedical Engineering
Tulane University
New Orleans, LA, US

(2) Department of Pharmacology
Tulane University
New Orleans, LA, US

(3) Department Physics and Eng. Physics
Tulane University
New Orleans, LA, US

INTRODUCTION

The use of stem cells to manipulate angiogenesis offers a promising approach for future therapies for multiple pathological conditions, such as diabetic retinopathy, cardiac ischemia, and peripheral arterial disease [1]. While mesenchymal stem cells derived from bone marrow and adipose tissue potentially have been shown to enhance angiogenesis via paracrine signaling and direct cell incorporation, our understanding of how to manipulate them or target where they go within a tissue still remains relatively unclear. New models are needed that enable cell fate and function within intact microvascular networks. An *ex vivo* tissue culture model that enables real-time investigation of cell migration, fate, and function during microvascular network growth, including both angiogenesis and lymphangiogenesis does not currently exist.

We have recently introduced the rat mesentery culture model as a model for time-lapse investigation of mechanistic cell–cell interactions at specific locations across intact blood and lymphatic microvascular networks *ex vivo* [2]. The objective of this study was to show that the rat mesentery culture model could also be used for stem cell lineage studies by demonstrating the capability of tracking the fate of exogenously delivered human adipose-derived stem cells (hASCs).

METHODS

All experiments were performed in accordance with the guidelines of the Tulane University Institutional Animal Care and Use Committee. Adult male Wistar rats were anesthetized and prepared aseptically for surgery. Animals were then euthanized and mesenteric windows were harvested from the ileum. A mesenteric window was defined as the thin, translucent connective tissue between the artery/vein pairs that feed the small intestine. Tissues were rinsed in sterile DPBS and spread on tissue culture inserts [2]. Human adipose-derived stem cells (hASCs) with red fluorescent protein (RFP) were divided into two groups. In the first group, RFP-transfected hASCs were seeded onto tissues by dripping 200 μL of cells in suspension at $5x10^5$ cells/mL. The culture was incubated for 30 minutes, and after that the excess cell suspension was aspirated to make sure all the remaining cells are attached to the tissues. Then the inserts were placed in 6-well plates. Tissues were cultured for 3 or 5 days in cell culture media.

Additional experiments were performed for the second group to demonstrate the feasibility of bioprinting ASCs onto the rat mesentery tissues in culture. In order to combine the advantages of the rat mesentery and capabilities of laser direct-write cell printing, A UV-transparent quartz disk, termed "ribbon," was spin-coated with a thick layer of liquefied 20% w/v gelatin. A 1 mL suspension of trypsinized ASCs at $2x10^6$/mL was pipetted onto the coated ribbon surface. Cells were incubated at 37°C to partially embed cells in the liquefied gelatin layer, followed by another incubation at room temperature for the gelatin to solidify. Prepared ribbons were inverted to face the flattened mesentery tissues for cell transfer. A single pulse of 193 nm ArF excimer laser interacted with the sacrificial gelatin layer on the ribbon, forming a local vapor pocket to eject a single droplet of desired cells onto the tissue [3]. Subsequent groups of cells, targeted in real-time with the aid of an *in situ* camera, were additively deposited into computer-programmable positions to complete the array patterns and were cultured for 3 days.

Tissues were labeled with FITC conjugated BSI-lectin to identify endothelial cells along blood and lymphatic microvascular networks. Lectin labeling of endothelial cells on day 0 and later time points enabled time-lapse imaging of cells during angiogenesis. Co-cultures were imaged on day 0 to confirm the presence of hASCs on the tissues, and then at later time-points to evaluate the cells' adopted morphology, attachment, and migration.

RESULTS

Immediately after ASC seeding onto the rat mesentery tissues, cells displayed a rounded morphology. In comparison, at 1 and 3 days, the delivered ASCs exhibited an elongated morphology (Fig. 1). Multi-planar imaging also revealed that hASCs were located within the same focal plane as blood vessels. By day 3, a subset of hASCs was observed in a perivascular location typical of vascular pericytes (Fig. 2). Laser direct-write printing enabled the patterning of ASCs into distinct clusters within the cultured microvascular networks. Comparison of day 0 and day 2 images support cell migration (Fig. 3).

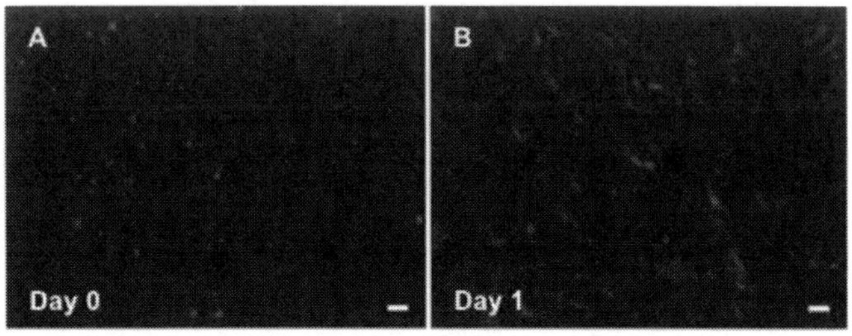

FIGURE 1. A) RFP-HASCS ON DAY 0 AFTER SEEDING. CELLS SHOWED A SPHERICAL MORPHOLOGY AS EXPECTED. B) RFP-HASCS ONE DAY AFTER SEEDING. CELLS ACQUIRED A SPINDLE-SHAPED MORPHOLOGY COMPARED TO DAY 0. SCALE BARS = 100 μM.

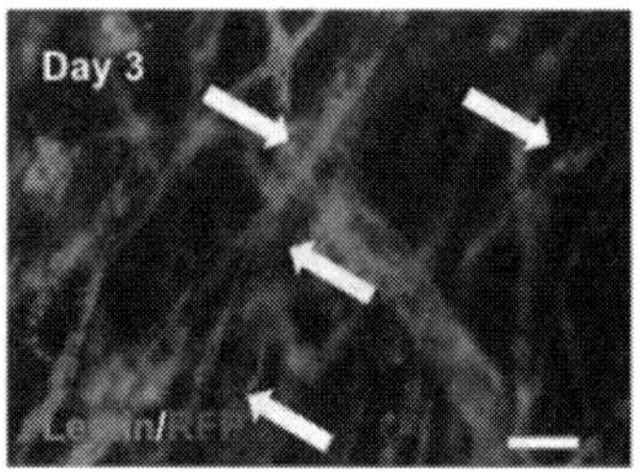

FIGURE 2. FATE OF ASCS IN RAT MESENTERIC MICROVASCULAR NETWORKS. RFP STABLY TRANSFECTED CELLS IN THE RAT MESENTERY CULTURE MODEL AFTER 3 DAYS IN CULTURE. ARROWS INDICATE CELLS IN PERICYTE LOCATION. SCALE BAR = 50 μM.

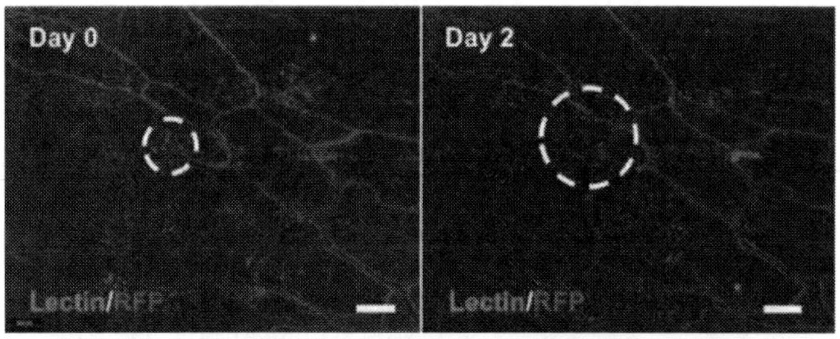

FIGURE 3. PRINTED HASCS ON MESENTERY TISSUE USING LDW TECHNIQUE. CELLS WERE PRINTED IN A 4X4 ARRAY ON DAY 0. DAY 2 IMAGES SUPPORT THE CELL MIGRATION WITHIN THE MICROVASCULAR NETWORK. SCALE BARS = 500 μM.

DISCUSSION

Our results suggest the rat mesentery culture model can be used for tracking ASC fate and function within an intact microvascular network. This capability potentially provides a novel tool for pre-clinical screening of stem cells. Compared to other common angiogenic *in vitro* assays, such as 2-D and 3-D cell cultures [4], and aortic ring assay [5], an advantage of the rat mesentery culture model is that cells can be tracked during angiogenesis and lymphangiogenesis within an intact microvascular network scenario.

The injection of stem cells *in vivo* confirms their ability to integrate within the mesentery and, more importantly, their ability to become pericytes and influence microvascular remodeling within mesenteric microvascular networks. Our results support the ability of ASCs to incorporate into mesenteric networks when seeded onto mesenteric tissues *in vitro*. The observation of ASCs in pericyte locations along capillaries is consistent with evidence from the literature supporting that ASCs can differentiate into pericytes [6].

The laser direct-write method integrates controls for environmental temperature and humidity management, which allows for optimal droplet formation, minimizes printed droplet evaporation, and maintains mesentery tissue hydration. In addition, the optical-based technique allows for real-time monitoring of cells on the ribbon to select desired cell groups, as well as single cells, for printing. More importantly, the video feed also allows for the ability to visually locate vascularized areas on the mesentery tissue and potentially position cell groups relative to desired vessels [3]. The use of laser direct-write printing offers a migration metric in which cells can be tracked from an initial position towards blood vessels or lymphatic vessels. Cell viability after printing is supported by their elongated morphology and migration.

A limitation of the current study is that ASCs can only be tracked over a 3-5 day time course. We have previously shown that cells within cultured rat mesenteric tissues remain viable for at least 7 days. However, future studies are required to determine how long the hierarchical microvascular network structure can be maintained. Conducting longer time course ASC studies will enable the tracking of cells after the initial stages of angiogenesis. However, for now our results support the use of the rat mesentery culture model for short term cell lineage studies, which would be valuable for investigating how cells migrate and what they do immediately after delivery.

ACKNOWLEDGEMENTS

This work was supported by the Tulane Center for Aging and NIH 5-P20GM103629-03 to W.L.M.

REFERENCES

[1] Kelly-Goss MR et al., *Microcirculation*, 21:345-357, 2014.
[2] Stapor PC et al., *Am J Physiol Heart Circ Physiol*, 304:H235-245, 2013.
[3] Phamduy TB et al., submitted paper.
[4] Davis GE et al., *In Vitro Cell Dev Biol Anim*, 36:513-519, 2000.
[5] Nicosia RF and Ottinetti A, *Lab Invest*, 63: 115-122, 1990.
[6] Mendel TA et al., *PLoS One*, 31:e65691, 2013.

THE EFFECT OF FACEMASKS ON THE IMPACT PERFORMANCE OF HELMETS

Steven Rowson and Evan Terrell

Department of Biomedical Engineering and Mechanics
Virginia Tech
Blacksburg, Virginia, United States

INTRODUCTION

As many as 3.8 million sports-related concussions occur in the United States annually [1]. Due to recent reports of the potential long-term neurodegenerative effects of repetitive concussions, increased efforts to reduce the number of concussions sustained by athletes have been initiated. There are three primary intervention strategies to reduce the incidence of injury: teaching athletes proper technique to reduce high risk head impacts, modifying rules to reduce high risk scenarios, and improving protective equipment. While player education and rule changes focus on reducing the number of head impacts, improved protective equipment further reduces risk when incidental head impacts occur. This paper focuses on the intervention of improving protective equipment, specifically related to the laboratory testing used to design and regulate performance of helmets.

Current helmet testing standards do not test with facemasks attached to the helmet, even though most players in contact sports are equipped with facemasks during play. The National Operating Committee on Standards for Athletic Equipment (NOCSAE) evaluates whether or not a helmet is likely to prevent catastrophic head injury. Like NOCSAE, ASTM also has a set of standards that assess a helmet's ability to prevent catastrophic head injury. The Virginia Tech Helmet Ratings supplement the existing standards by assessing the ability of helmets to reduce concussion risk. These testing protocols all currently use drop tests to evaluate helmet performance without a facemask attached. There are no data published in the literature on how the addition of a facemask may alter helmet performance. In theory, attachment of a facemask may stiffen the structural response of the helmet shell, which will interact with the helmet liner differently, and affect overall performance.

It is important for laboratory test conditions to be representative of the impact conditions experienced by athletes during play so that laboratory-assessed performance is translatable to the field. The objective of this study is to investigate whether facemasks affect the performance of the helmets during laboratory testing. These data have direct applications towards current and future helmet standards.

METHODS

A total of 120 impact tests were performed to assess the effect of facemasks on helmet performance. One hockey helmet (Reebok 11k, size Large) and one football helmet (Schutt Vengeance VTD, size Large) were evaluated. Two samples of each helmet model were tested, one helmet with and one helmet without an attached facemask (Fig.1).

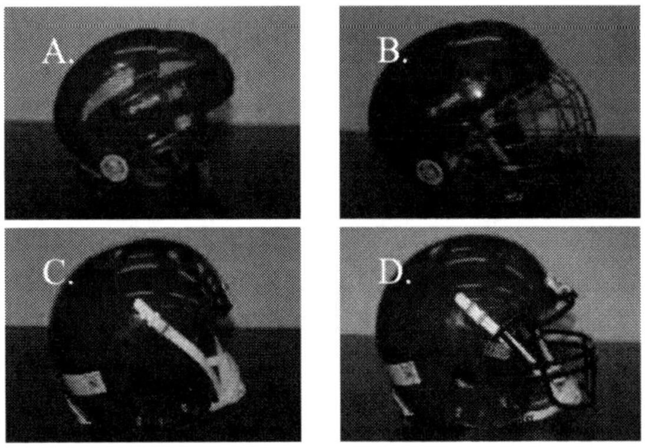

FIGURE 1. HELMETS EVALUATED: REEBOK 11K HOCKEY HELMET WITHOUT (A) AND WITH (B) FACEMASK, AND SCHUTT VENGEANCE VTD FOOTBALL HELMET WITHOUT (C) AND WITH (D) FACEMASK.

A pneumatic linear impactor (Biokinetics, Ottawa, Canada) was used to simulate impacts experienced by athletes during play. The 14 kg impactor ram was equipped with an impacting surface designed to replicate the impacting characteristics of a typical football helmet. A medium sized NOCSAE headform was mounted on a Hybrid III 50th percentile neck. The headform was instrumented with a six-degree-of-freedom sensor package (6DX Pro 2k-18k, DTS, Seal Beach, CA) consisting of three linear accelerometers and three angular rate sensors. The base of the neck was mounted to a sliding table simulating the effective mass of the torso during impact. Two impact velocities (6 m/s and 9 m/s) were tested at three impact locations (front, side, and rear boss). Five repeated trials were performed at each test configuration. For each test, data were sampled at 20k Hz.

Acceleration data were transformed to the center of gravity of the NOCSAE headform and filtered to channel frequency class (CFC) 180 using a 4 pole phaseless Butterworth low pass filter. Angular rate data were filtered at CFC 155 and differentiated to compute rotational acceleration. Statistical analyses consisted of two-factor ANOVAs and post hoc student t-tests.

RESULTS

Overall, tests with a facemask generated 1.0% higher linear head acceleration than tests without a facemask attached (Fig. 2). For the hockey helmet, the facemask did not significantly affect linear acceleration for any combination of impact velocity and location. On average, linear acceleration of the headform was 1.2% higher without a facemask than with one for the hockey helmet tests. For the football helmet, the facemask did not significantly affect linear acceleration for any combination of impact velocity and location, with the exception of the 9 m/s impact to the front location ($p = 0.023$). On average, linear acceleration was 3.2% higher for tests with a facemask than those without.

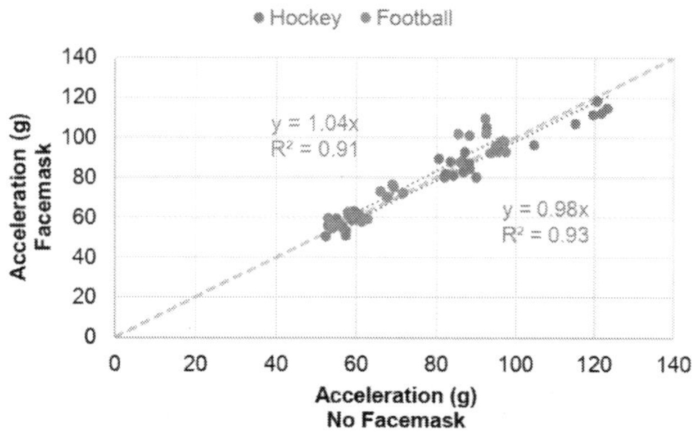

FIGURE 2. COMPARISON OF LINEAR ACCELERATIONS FOR TESTS WITH AND WITHOUT FACEMASKS BY HELMET TYPE. THE DASHED GRAY LINE ILLUSTRATES NO DIFFERENCE IN RESPONSE.

Overall, tests with a facemask generated 3.0% higher rotational head acceleration that tests without a facemask attached (Fig. 3). For the hockey helmet, the facemask did not significantly affect rotational acceleration for any combination of impact velocity and location, with the exception of the 9 m/s impact to the side location ($p = 0.002$). On average, rotational acceleration of the headform was 2.4% higher with a facemask than without one for the hockey helmet tests. For the football helmet, the facemask did not significantly affect rotational acceleration for any combination of impact velocity and location. On average, rotational acceleration was 3.6% higher for tests with a facemask than those without.

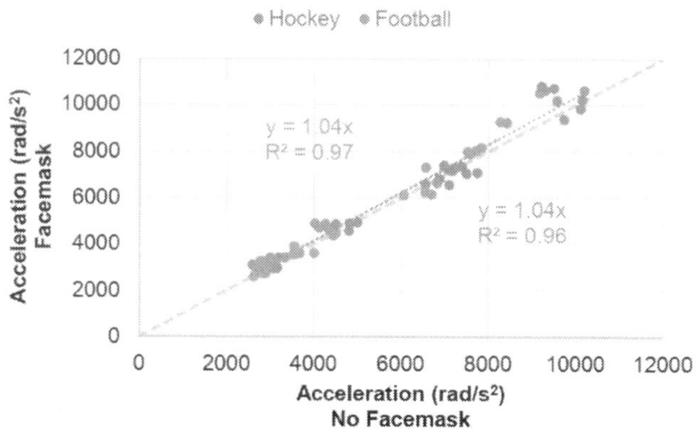

FIGURE 3. COMPARISON OF ROTATIONAL ACCELERATIONS FOR TESTS WITH AND WITHOUT FACEMASKS BY HELMET TYPE. THE DASHED GRAY LINE ILLUSTRATES NO DIFFERENCE IN RESPONSE.

DISCUSSION

The findings of this study suggest that facemasks do not greatly affect the performance of helmets during impact testing. The average effect was less than 4% for both hockey and football helmets. Previous clinical studies on the effect of facemasks on injury rates can provide context to these findings. Between 1995 and 2002, several prospective studies were performed that followed hockey teams and analyzed the head, neck, and facial injuries sustained by players wearing different types of facial protection. In 1995, one study reported no increase in head and neck injuries in leagues that mandated facemask use compared to leagues that did not [2]. Similarly, another study found no difference in concussion risk between players wearing full versus half face shields [3]. A third study reported no difference in concussion risk between players wearing full, half, or no facemasks [4]. These studies suggest that the small differences in impact performance resulting from facemask attachment likely do not translate clinically.

Adding a facemask generally affected the football helmet more than the hockey helmet. This is likely due to the differences in how the facemasks attach to each helmet. The facemask of the football helmet is attached in such a way that does not allow relative movement between the facemask and helmet shell. In contrast, the facemask of the hockey helmet is attached by a hinge at the top and straps on the side, which allows for relative motion between the mask and helmet shell. Furthermore, the football facemask had a larger gauge than the hockey facemask, resulting in roughly double the mass. The more rigid attachment of the football facemask likely explains the difference between helmet types.

In summary, testing with a facemask seemed to have a negligible effect on helmet performance when compared to testing without one. These findings should be considered when developing future helmet testing standards.

REFERENCES

[1] Langlois, JA et al, *J Head Trauma Rehabil*, 21(5):375-8, 2006.
[2] LaPrade, RF et al., *Am J Sports Med*, 23(6):773-775, 1995.
[3] Benson, BW et al., *Jama*, 282(24):2328-2332, 1999.
[4] Stuart, MJ et al., *Am J Sports Med*, 30(1):39-44, 2002.

SB^{3}C2015
Summer Biomechanics, Bioengineering and Biotransport Conference
June 17-20, 2015,Snowbird Resort, Utah, USA
SB^{3}C2015-1102

INVESTIGATING THE TWO-PHASE NATURE OF BLOOD FLOW

Joseph M. Sherwood (1,2), Xuejin Li (3), George M. Karniadakis (3), Stavroula Balabani (2).

(1) Department of Bioengineering, Imperial College London, UK

(2) Department of Mechanical Engineering, University College London, UK

(3) Division of Applied Mathematics, Brown University, Providence, RI, US

INTRODUCTION

Detailed knowledge of blood flow characteristics is of fundamental importance in understanding a wide range of physiological and pathological functions. The multiphase nature of blood, being composed of discrete cells suspended in a continuous medium, drastically alters its flow characteristics from those of a simple fluid such as water. When measured in rheometers, blood displays shear thinning properties; reported to be a result of red blood cell (RBC) aggregation at low shear rates and RBC deformation at high shear rates. Additionally, the viscosity of blood is highly dependent on the haematocrit (RBC concentration). Under flow, RBCs migrate away from the wall, forming a region of low RBC concentration commonly termed a cell-depleted layer (CDL), in long straight vessel sections which is enhanced around bifurcations [1]. An alternative approach is to consider local concentration as a continuous variable [2]. Such an assumption requires scales wherein, in a time-averaged sense, the haematocrit at a given location is reasonably represented by its mean value. In such a continuum approach, the concentration of the suspending medium varies in the opposite manner to the RBC concentration. Furthermore, the velocities of RBCs and SM have been shown to be influenced by the RBC concentration [3].

In the present study, we investigate microscale blood flow in the paradigm of a two-phase fluid, with spatially resolved measurements of velocity and concentration of both the RBCs and the suspending medium. Numerical modelling using the established dissipative particle dynamics (DPD) methodology, and a continuum approach using an empirical viscosity model, are compared to the experimental results.

METHODS

Experimental: Details of the experimental setup and methodologies used in the present study have been reported previously [2] and are described briefly below. Human blood was collected via venepuncture and mixed with 1.8 mg/ml EDTA to prevent coagulation. The samples were separated via centrifugation and the plasma and buffy coat removed. The RBCs were then washed twice in PBS and resuspended at a haematocrit of 0.25. Neutrally buoyant 1.1 μm fluorescent microparticles (Invitrogen) were added at a volume concentration of 0.1% to provide the velocity of the SM and for certain cases, Dextran 2000 at 5 *mg/ml* was added to induce RBC aggregation. A polydimethylsiloxane (PDMS) microchannel was bonded to a glass microscope slide with sequential 90° bifurcations, with 5 channel widths between them, and a 50×50μm cross-section. The pressure in each branch of the channel was regulated using a Fluigent MFCS system. The microchannel was placed on an inverted microscope with a 10× air objective and illuminated alternately with Nd:YAG laser (New Wave Solo) and microstrobe LED illumination. Images were captured on a CCD camera (Hamamatsu C8484-05CP). All acquisition was carried out using a custom system built in LabVIEW. The strobe and laser illuminated images were processed using ensemble averaged PIV algorithms and image processing methodologies to extract concentration distributions [2].

Numerical: The DPD approach is well established and has been described elsewhere [4]. The motion of RBCs flowing in a straight channel is modelled using a multiscale RBC (MS-RBC) model. In the MS-RBC model, the cell membrane is modeled by a 2D triangulated network with N_v vertices. The vertices are connected by viscoelastic bonds to impose proper membrane mechanics. Cell-cell aggregation interaction is approximated using the Morse potential. For the MS-RBC model, the Morse potential interactions are implemented between every pair of vertices of separate RBCs if they lie within a defined potential cutoff radius.

The continuum modelling approach used ANSYS CFX to model the blood as Quemada fluid, with shear- and haematocrit-dependent experimentally defined empirical parameters. The measured RBC distributions were used as an input into the simulations and boundary conditions were based on flow rates calculated from the PIV measurements.

RESULTS

Figure 1 compares the bluntness of the haematocrit and velocity profiles. Definitions for the indices have been described previously [2]. Comparing the values of B^*_H (for which 1 would be a uniform distribution of RBCs), it can be seen that although the absolute bluntness predicted by the DPD simulations is marginally lower than the experimental data, the overall trends are well replicated. For the non-aggregating samples, the bluntness decreases slightly as the flow rate increases, probably due to increase RBC migration at higher flow rates. In the presence of RBC aggregation, a large proportion of the cells lie in the channel centre, but as the flow rates increase, the aggregates break up and the profiles become blunter.

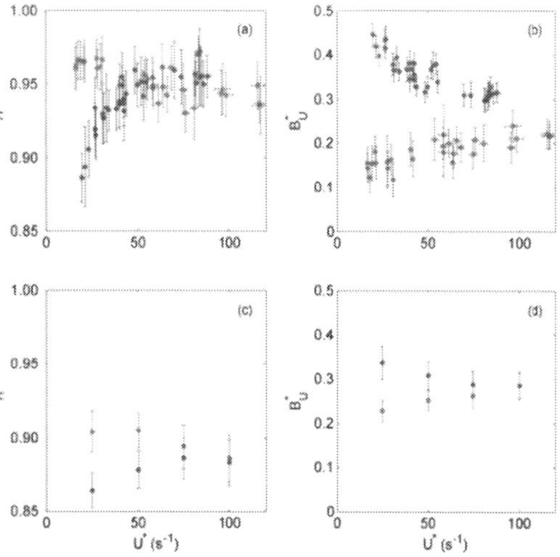

Figure 1: Bluntness indices for RBC concentration B^*_H – (a) Experimental, (c) DPD. Bluntness indices for RBC velocity B^*_U - (b) Experimental, (d) DPD.

The velocity bluntness varies inversely with the haematocrit bluntness for both experimental and computational data sets, and for both aggregating and non-aggregating blood samples. However, not only the RBC velocity is altered by the RBC concentration distributions. As can be seen in Figure 2, the suspending medium velocity is notably higher than the RBC velocity in a branch of the bifurcation, and both are skewed. The haematocrit profile for the same case is skewed in the opposite direction. However, in the other branch, the relative velocity is negligible.

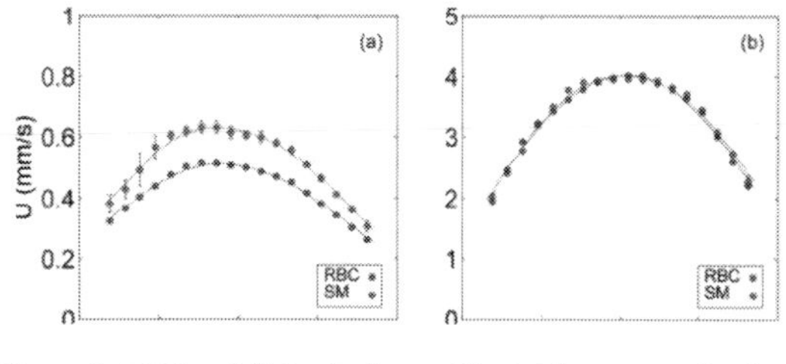

Figure 2: RBC and SM velocity profiles (a) in a perpendicular side branch and (b) in the other branch of the bifurcation for an aggregating blood sample.

These variations in relative velocity are altered by flow rate, flow ratio and RBC aggregation, and have also been observed in the DPD simulations. Work is underway to provide accurate comparison of the relative velocity using the DPD technique.

The continuum modelling approach instead calculates a 'total velocity' by weighting the individual velocities by their relative concentrations. This is shown by the red dots in Figure 3, with the grey shading area indicating uncertainty on the experimental measurements. The dark blue line shows the velocity profile calculated by the continuum model, which compares favourably with the experimental data, particularly when considered relative to the analytical solution (cyan) and the profile calculated when assuming a uniform RBC distribution (green). This technique is has low computational expense, but limited resolution and presently requires experimental data as an input.

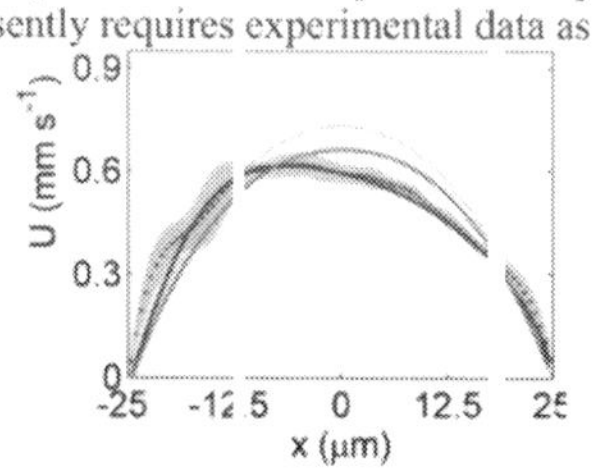

Figure 3: Comparisons of experimental (red), analytical (cyan) and numerical (green/dark blue) velocity profiles in the side branch of the bifurcation.

DISCUSSION

The present results show some of the various aspects of microscale haemodynamics that occur as a result of the specific characteristics of RBCs. Numerical modelling approaches are able to reproduce experimental data with some success. The MS-RBC model can predict the velocity, location, orientation etc. of every cell in the flow and thus yields a wealth of data. The continuum model is very efficient, and has the potential to be applied to a large number of vessels, but requires a valid approach to predict the distribution of RBCs, rather than using input from experimental data if the model is to find further utility. Existing approaches based on RBC convection/diffusion [5], were found not to be able to reproduce the experimental data (data not shown).

In addition to the necessity for better understanding of haemodynamics on all length scales, and particularly in the microvasculature, the present work also has implications for experimental acquisition. In particle image velocimetry measurements of blood flow, RBC and SM velocities are often considered to be equivalent. However, the two phase nature should be considered when assuming artificial tracers follow the flow of blood, rather than just of the suspending medium, or that the RBCs themselves are representative flow tracers. Clinical imaging techniques that use the RBCs to measure flows (e.g. ultrasound) may be similarly affected. Additionally, we hypothesise that there may be a role for these complex microscale haemodynamics in microvascular diseases such as diabetes and malaria, through altering shear stresses on the endothelium.

ACKNOWLEDGEMENTS

We would like to thank Prof. Kim Parker for his advice.

REFERENCES

[1] Ong, PK et al., *Microvas Res*, 82:118-125, 2012.
[2] Sherwood, JM et al., *PLOS One*, 9(6): e100473, 2014
[3] Sherwood, JM et al., *Biomech Mod Mechanobiol*, 13(2): 259-273, 2014.
[4] Li, X et al., *Phys Biol*, 9:026010, 2012.
[5] Bark, DL and Ku, D, Biophys J, 105:502-511, 2013

IMPROVED POST-OPERATIVE FLOW MODELING FOR COMPLEX PERIPHERAL PULMONARY ARTERY STENOSIS

Weiguang Yang (1), Jeffrey A. Feinstein (1, 2), Alison L. Marsden (3), Frank L. Hanley (4) Frandics P. Chan (5), Lisa W. Faberowski (6) and Irene E. Vignon-Clementel (7)

Departments of (1) Pediatrics (Cardiology), (2) Bioengineering, (4) Cardiothorac Surgery, (5) Radiology, (6) Anesthesia
Stanford University

(3) Department of Mechanical and Aerospace Engineering
University of California, San Diego
La Jolla, CA, USA

(7) INRIA Paris-Rocquencourt and Sorbonne Universités UPMC Univ. Paris 6, Laboratoire J.-L. Lions,
Paris, France

INTRODUCTION

Peripheral pulmonary artery stenosis (PPAS) is one of the most common cardiac abnormalities among patients with Alagille syndrome, a genetic disorder.[1,2] PPAS can result in right ventricular (RV) hypertension and pulmonary flow disparity.[1,2] For complex PPAS, surgical reconstructions are performed to relieve obstructions by patching the pulmonary artery (PA) with homografts. Even in patients with initially successful interventions, post-operative outcome predictions and optimal surgical strategies are not well established.[2] In recent years, image based computational flow simulations have been used to study hemodynamics in congenital heart diseases and simulation derived surgical designs have been translated into clinical applications for single ventricle defects.[3,4,5,6] However, to our knowledge, no flow modeling has been performed for PPAS. Moreover, in these previous studies, the influence of the three dimensional model on post-operative hemodynamics was emphasized and the lumped model parameters of the boundary conditions were generally assumed unchanged.[7,8] Although the conventional methods have provided us with some important insights into post-operative hemodynamic performance, the longer term changes in distal vasculature due to surgery were neglected. Experimental studies have demonstrated the arterial adaptation and remodeling due to flow perturbations.[9,10] Results suggested that the shear stress is maintained by vessel dilation and constriction.[9] Thus, it is necessary to take into account the pulmonary adaptation and remodeling in post-operative modeling. The purpose of this study is to improve post-operative predictions for complex PPAS by adapting outflow boundary conditions.

METHODS

Two Alagille patients (male, age range: 1.9-2.9 years) with PPAS who underwent PA plasty were identified. Using pre-operative CT data, three dimensional PA models were created using SimVascular (www.simvascular.org). Based on the pre-operative models, virtual surgeries were performed to represent the post-operative geometry. Patients' demographic data and major pre-operative evaluations are summarized in Table 1.

Patient	Age (yr)	BSA (m^2)	RVP (mmHg)	LPS (RPA/LPA)
A	2.9	0.58	75/9	86%/14%
B	1.9	0.48	74/11	80%/20%

Table 1: Major pre-operative clinical evaluation results. BSA = body surface area; RVP=right ventricular pressure. RVP is presented as systolic/diastolic pressures and lung perfusion scan (LPS) is shown as right PA (RPA)/left PA(LPA) flow split.

Blood flow in the PA with severe stenosese was simulated with a three dimensional finite element Navier-Stokes solver that accounts for wall compliance (fluid-solid interaction).[11] The inflow waveform was taken from a previous study [12] and scaled to match the patient's cardiac index that was derived during pre-operative catheterization. A three element Windkessel model was applied to each of the model outlets.[13] Boundary conditions for the pre-operative models were tuned to match the pre-operative PA pressures and pulmonary flow splits obtained from catheterization and lung perfusion scans. For each outlet, a structured tree model with the equivalent pre-operative resistance was created to represent distal small arteries.[14] Flow distribution in the tree segments was computed with Poiseulle's law.

A two-step approach was used to predict the post-operative hemodynamic performance. First, the pre-operative boundary conditions were applied to the virtual post-operative model and the flow changes due to the surgical reconstruction were computed. Then, the tree segments were dilated or constricted to restore the pre-operative shear stress. The outflow resistance for each outlet was thus updated. The final simulation was performed with these adapted outflow boundary conditions.

RESULTS

Figure 1 shows mean pressures in the pre- and post-operative models for patients A and B. A significant pressure drop, higher than 30 mmHg is found in the LPA for the pre-operative model due to the tight stenosis. Patching the LPA effectively reduces the PA pressure.

Table 2 summarizes the predictions for the pulmonary flow split given by the conventional method and adapted outflow boundary conditions. Comparing with the post-operative measurements, simulations with unchanged outflow boundary conditions predict post-operative flow splits with 10% errors. The differences between predictions and in-vivo measurements are reduced to 1-4% when the downstream PA adaptation is taken into account for the outflow boundary conditions.

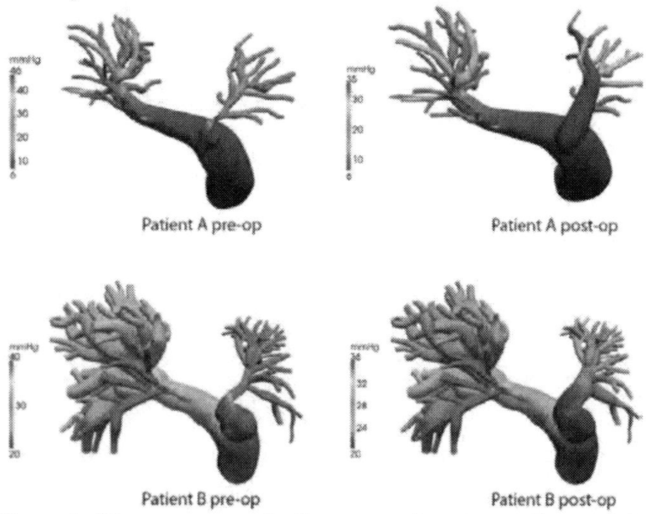

Figure 1: Mean pressures in the pre- and post-operative models for patients A and B.

Patient	Post-op LPS	unchanged outflow BCs	adapted outflow BCs
A	53/47	65/35	52/48
B	63/37	73/27	67/33

Table 2: Post-operative lung perfusion scan (LPS) results and post-operative predictions with unchanged and adapted outflow boundary conditions (BCs)

DISCUSSION

In this study, imaged based computational flow simulations and virtual surgeries were employed to study the hemodynamic performance for Alagille patients with PPAS. Generally, the downstream pulmonary vascular resistance dominates the pulmonary pressure and flow distribution because the RPA and LPA develop similarly and the resistance created by the proximal PAs is small compared to the downstream vascular bed. In contrast, the LPA is significantly hypoplastic in these Alagille patients with complex PPAS resulting in a large difference in size between the RPA and LPA. Simulations show that the tight stenosis at the origin of the main LPA plays an important role in RV hypertension and pulmonary flow disparity.

Both in-vivo measurements and simulations show that relieving obstructions reduces PA pressures and improves pulmonary flow distribution. However, the post-operative pulmonary flow distribution is not determined by the augmentation of the ostial stenosis only, but also by the adaptation and remodeling that occur in distal small vessels. This was seen in the post-operative simulations without changing the outflow boundary conditions, which underestimated the pulmonary flow in the LPA. The surgical repair created an initial flow change allowing more flow streaming to the LPA in the patients. The downstream small arteries dilated with decreased vascular resistance and further improved the perfusion in the LPA. Kamiya et al. [9] and Razavi et al. [10] also demonstrated similar arterial adaptation and remodeling in dogs and rats.

Since the conventional method did not accurately predict post-operative pulmonary flow distribution, a new method that models the pulmonary remodeling based on the constant shear stress hypothesis has been introduced. The structured tree model provides corrections for post-operative outflow boundary conditions and better agreement with in vivo measurements is obtained. However, the uncertainties in the virtual surgery and pre-operative evaluations can affect the accuracy of the prediction: the initial flow perturbation is determined by the surgical reconstruction and baseline hemodynamic values are tuned to match the pre-operative measurements.

The outflow resistance decreases or increases with increasing or decreasing flow under the constant shear stress hypothesis. Although it is known that post-operative pulmonary flow distribution varies with time, the PAs cannot adapt to flow perturbations without bounds. Future studies should incorporate the constraint factors associated with the pulmonary growth and remodeling into the prediction.

In conclusion, the adapted outflow boundary conditions improve post-operative hemodynamic modeling for complex PPAS and may provide a useful tool for identifying optimal surgical plans.

ACKNOWLEDGEMENTS

The authors would like to thank Elisa McCarthy and Michelle Ogawa, Nathan Wilson, Ana Ortiz and Ali Johnson for their assistance in obtaining clinical data and model construction. This work is funded, in part, by the Leducq Foundation (ALM and IEVC).

REFERENCES

[1] Krantz, I D et al., *J Med Genet*, 34:152-157, 1997.
[2] Monge, M C et al., *JTCVS*, 145:476-481, 2013.
[3] Marsden, A L, Physics of Fluids, 25:101303, 2013.
[4] Marsden, A L, *Annu Rev Fluid Mech*, 46: 519-546, 2014.
[5] Vignon-Clementel, I E et al., Prog Pediatr Cardiol, 30:3-13, 2010
[6] Arbia, G et al., JSR, 186:44-55, 2014
[7] Yang W. et al., JTCVS, 143:1086-1097, 2012
[8] Baretta, A et al., *Phil. Trans. R. Soc. A*, 369: 4316-4330, 2011.
[9] Kamiya A et al., Am J Physiol, 239:H14-21, 1980.
[10] Razavi, H et al., *AJP-Lung Cell Mol Physiol, 304:L17-28, 2013*
[11] Figueroa CA et al., *CMAME*, 195:5685-5706, 2006.
[12] Tang, B et al., *Ann Biomed Eng*, 39:347-358, 2011.
[13] Vignon-Clementel, I. E. et al., *CMAME*, 195: 3776-3796, 2006.
[14] Olufsen M. et al., *J Fluid Mech*, 705:280-305, 2012.

LYMPHATIC DISEASE PHENOTYPING WITH NEAR-INFRARED IMAGING

Tyler S. Nelson (1, 2), Michael J. Weiler (2,3), Ira L. Savetsky (4), Xiaolei Liu (5,6), Babak J. Mehrara (4), Hong Chen (5), J. Brandon Dixon (1,2,3)

(1) George Woodruff School of Mechanical Engineering
Georgia Institute of Technology
Atlanta, GA USA

(2) Parker H. Petit Institute for Bioengineering and Bioscience
Georgia Institute of Technology
Atlanta, GA USA

(3) Wallace H. Coulter Department of Biomedical Engineering
Georgia Institute of Technology
Atlanta, GA USA

(4) Memorial Sloan Kettering Cancer Center
New York, NY USA

(5) Oklahoma Medical Research Foundation
Oklahoma City, OK USA

(6) St. Jude Children's Research Hospital
Memphis, TN USA

INTRODUCTION

The lymphatic system consists of a network of vessels, nodes, and accessory organs that play an essential role in the body's balance of interstitial fluid volume by draining lymph from the interstitium back to the bloodstream. This system is vital for maintenance of oncotic pressure gradients, immune cell trafficking, and lipid transport. Despite the critical roles that this system performs, very little is known about the lymphatic vasculature in comparison to the blood vasculature, which can be attributed, in part, to the difficulty associated with imaging lymphatic vessels. Near-infrared (NIR) imaging has emerged in the literature as a novel lymphatic imaging modality to simultaneously provide sufficient spatial resolution to visualize small initial lymphatics and temporal resolution to capture the dynamic lymphatic pump function responsible for fluid transport.

Functional NIR imaging of the lymphatics has primarily focused on tracking the frequency of fluorescence pulsations, or "packets", of the collecting vessels. However, contractile frequency is only one factor that affects overall lymph transport, and thus its use in determining lymphatic dysfunction is limited. We recently reported on a novel system for quantifying the lymphatic pumping pressure of a collecting vessel chain that builds upon NIR imaging [1]. Here we report the utility of this method, in combination with our entire functional NIR imaging toolbox, to distinguish between various phenotypes of lymphatic failure in a variety of lymphatic-related pathologies: 1) a genetically modified mouse model with subtle abnormalities in lymphatic development, 2) a mouse model of diet-induced obesity, and 3) a mouse model of surgically induced secondary lymphedema.

METHODS

Near-infrared (NIR) imaging was used to examine functional differences in lymphatic transport between control and a variety of experimental groups of C57BL6 mice. The NIR imaging procedures were conducted on mouse tails and hind limbs following previously published protocols[1,2]. These procedures involve the intradermal injection of an appropriately sized NIR probe, which is subsequently transported from the injection site by the lymphatics. The probe is then excited with an external light source and fluorescent images of the lymphatic vessels are captured at 10 frames per second. Functional metrics were evaluated via custom image processing algorithms that track fluorescence within the vessels. In each experiment four established functional metrics were evaluated; packet frequency, pumping pressure, emptying rate (the rate at which fluorescence is cleared from a vessel during upstream occlusion), and fluorescence area (a measure reflective of the total number of initial and collecting vessel conduits recruited for fluorescence transport).

Details of the three models mentioned above are as follows: 1) Genetic model – a lymphatic endothelial cell epsin knockout (LEC-DKO) mouse that results in altered VEGF-C/VEGFR3 signaling and subsequent malformation of lymphatic valves[3]; 2) Obesity model – mice were fed a 60% fat diet for 34 weeks and compared to age and sex matched mice on a standard chow diet[4]; 3) Lymphedema model – all but one collecting lymphatic vessel on the tails were surgically ligated and allowed to undergo tail swelling, resolution, and vessel remodeling for three months. Sham surgical controls were included in which a comparable dermal wound was made but all collecting vessels were left intact.

RESULTS

Lymphatic Endothelial Cell Epsin Knockout

LEC-DKO mice show no signs of abnormal lymphatics 15 days after birth as determined by traditional histological techniques. However with our highly sensitive imaging approach, we were able to detect several abnormalities in the function of the lymphatic network. LEC-DKO mice exhibited enhanced reliance on the initial lymphatic network for fluid clearance as reflected in the representative images and quantification of total fluorescence area **(Fig. 1 A-C)**. LEC-DKO mice exhibited significant declines in lymphatic packet frequency (**Fig 1 D**). Upon upstream lymphatic flow occlusion, significant differences between the vessel emptying rates of control and LEC-DKO mice were observed (**Fig. 1 E**). In fact, KO mice exhibited a negative emptying rate, suggesting the collecting vessels experienced back flow once the effects of the upstream injection pressure were blocked off with the occlusion cuff. Because the lymphatic vessels of LEC-DKO mice did not clear their contents during occlusion, the lymphatics exhibited no detectable lymphatic pumping pressure above the resolution limit of the system (0.5 mm Hg), while controls showed normal values of effective pumping pressure (16.8 mmHg).

Figure 1

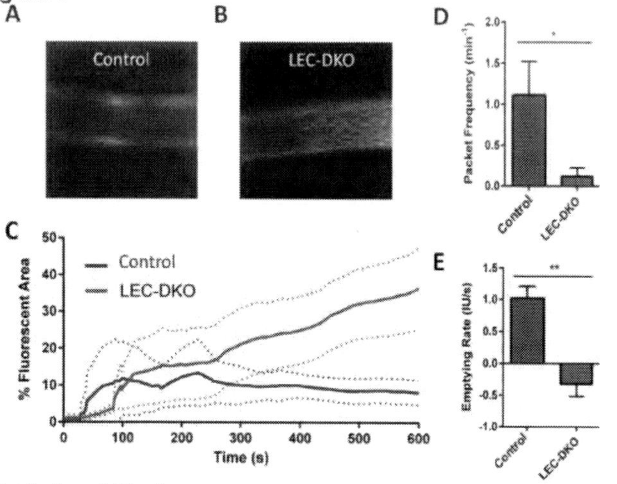

Diet-Induced Obesity

Packet frequency of lymphatic vessels in the tails and limbs was significantly decreased in obese mice (**Fig. 2 B-C**). No differences were observed in pumping pressure, emptying rate, or fluorescence area suggesting that the underlying lymphatic pump function is not severely compromised, but that differences in transport kinetics are due to some other factor such as the elevated inflammatory state in the dermis of these animals. Example fluorescence tracking plots, used to calculate packet frequency, are shown in (**Fig. 2 A**).

Figure 2

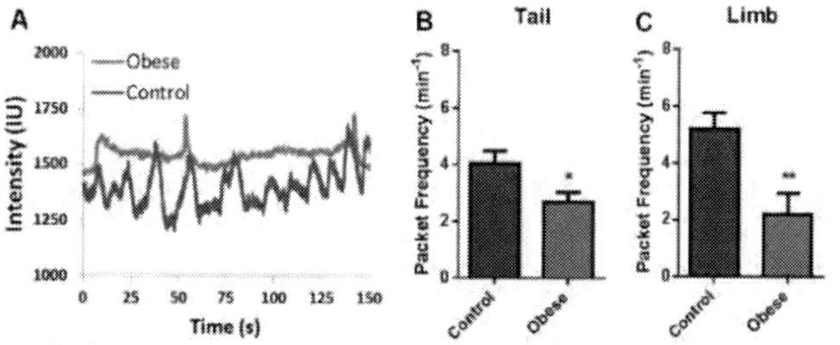

Lymphedema

While animals undergoing surgical lymphatic disruptions of some of the collectors have significant swelling and global decline in lymph transport, after 3 months the swelling is gone and the overall lymphatic architecture appears to be restored to normal (images not shown). However, unligated collecting lymphatic vessels continued to exhibit significant underlying defects in transport capacity as compared to the surgical sham controls. Specifically, pumping pressure was drastically reduced (**Fig. 3 A**), and significant increases in the total fluorescence area were observed, (**Fig. 3 B**), suggesting a compensatory reliance on the initial lymphatic vessel network to make-up of for the progressive decline in the collecting lymphatic pump over the course of lymphedema development.

Figure 3

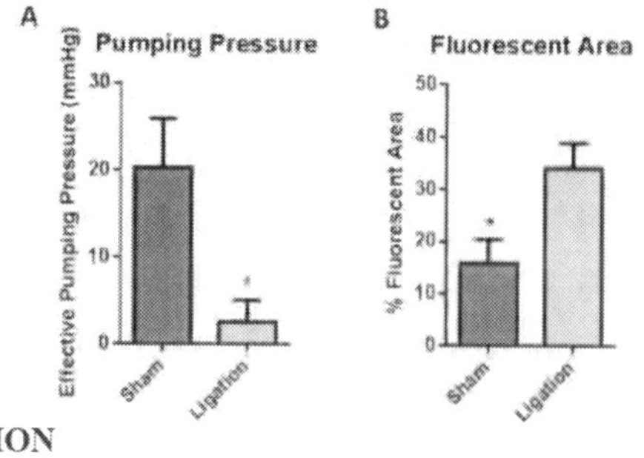

DISCUSSION

The results of these studies demonstrate not only the utility of functional NIR imaging in a variety of lymphatic pathologies, but can provide unique insight into the different phenotypes of lymphatic failure that might present during disease. The detection of these phenotypes would not be possible with gold standard histological or lymphoscintigraphy imaging approaches. Since the techniques are non-invasive, they not only provide a snapshot in time, but also can track the progressive change in lymphatic function over the course of the disease or therapy, providing direction for future hypothesis testing as it relates to the underlying mechanisms of lymphatic failure. For example, the lack of pumping pressure and negative emptying rate in LEC-DKO mice suggests that even though the valves in these mice appear morphologically normal in adults, there is an underlying defect in their functional ability to overcome an adverse pressure gradient. In the obese mice, even though transport kinetics are reduced (which has previously been shown to be linked to inflammation[4]), the capacity of the lymphatics to generate the pressure needed to move fluid against an adverse pressure gradient is normal, suggesting the lymphatics are not damaged and that deficits in transport could be improved pharmacologically. Lastly, results from the lymphedema model suggest that normal functioning lymphatics could remodel when they are mechanically overloaded in such a way that the vessels lose their ability to pump normally over time. Without the insight provided by this NIR imaging toolbox, these new research directions would not have been immediately obvious. Lymphatic dysfunction manifests as a spectrum of impaired transport modalities, and the ability to non-invasively distinguish these various modes of failure is a promising development both for the future of clinical lymphatic diagnostics, and an improved understanding of underlying disease mechanisms.

ACKNOWLEDGEMENTS

This work was supported in part by NIH R01 HL113061 (BD), NIH R01HL-093242 (HC), and an NSF graduate research fellowship (MW). BD, TN, and MW receive research funding and they and Georgia Tech are entitled to sales royalty from LymphaTech, which is developing products related to the research described in this paper. In addition, MW serves as the President of LymphaTech, and he and BD hold equity in the company. This study could affect their personal financial status. The terms of this arrangement have been reviewed and approved by Georgia Tech in accordance with its conflict of interest policies.

REFERENCES

[1] Nelson, T. S. *et al. Am. J. Physiol. Regul. Integr. Comp. Physiol.* 306, R281–90 (2014).
[2] Weiler, M.J. *et al. Biomed. Opt.* 17, 066019–1–10 (2012).
[3] Liu, X. *et al. Sci. Signal.* 7, ra97 (2014).
[4] Savetsky, I. L. *et al. Am. J. Physiol. Heart Circ. Physiol.* 307, H165–72 (2014).

USING FINITE ELEMENT ANALYSIS TO STUDY THE MECHANICAL ADVANTAGES OF A TURTLE'S SHELL

John W. Wood (1), Rajkumar Prabhu (2)

(1) Agricultural and Biological Engineering
Mississippi State University
Starkville, MS, United States

(2) Agricultural and Biological Engineering
Mississippi State University
Starkville, MS, United States

INTRODUCTION

In the past few years, research into biological materials and microstructures has grown immensely. By studying these materials and their structures, scientists can imitate nature and drastically improve a limitless number of mad-made materials (Liu and Jiang, 2011). An example of these materials being studied is the shell of a red-eared slider turtle. A recent study found that the inner structure of the shell allowed it to easily deform under small loads, but as the load increased the structure stiffened and deformation decreased (Krauss et al., 2009). This ability of the shell to stiffen is due to three distinct layers (Fig. 1) that contain voids and as a load is applied, these layers collapse and become more and more dense. This feature allows it to absorb a large amount of energy before failing.

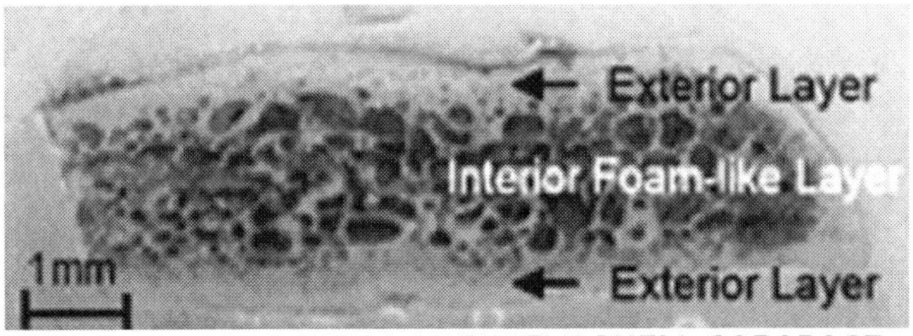

FIG. 1 – THREE LAYERS OF TURTLE SHELL CARAPACE

Seeing as how this shell can absorb so much force without breaking, another study was done to see if this biological design could be used in designing a more efficient man-made armor (Damiens et al., 2011). The researchers in that study focused mainly on the mechanical properties of the top section (carapace) of a box turtle. Through various mechanical tests, they found all of the information required to design a finite element (computer-simulated) model.

This study will use the information gathered by Damiens et al. (2011) and design a 3-D finite element model that matches the material properties of a turtle shell as close as possible. The main problem to solve will be determining the factors needed to create a model that deforms and absorbs energy in a way that closely resembles the specimens in the mechanical tests. Once this is done, the model can be used to simulate various scenarios to see if it will make a viable armor.

METHODS

Explain what was examined, developed or done to answer the research questions. Explain how it was done. Explain how the resulting data/results were analyzed.

In order to create an approximate model of the turtle shell, all of the mechanical testing data from Damiens et al. (2011) was gathered and organized into stages of production. The finite element program that was used to create the model was Abaqus. Initially, a model with three layers was created (with appropriate dimensions). Each of the layers were given initial properties including: density, Young's modulus, and Poisson's ratio. As the model was refined, more attributes were added to each layer, such as: viscosity, porosity, Bulk modulus, and permeability. After each attribute was added, the shell model was then subjected to a simulated three-point bending test (Fig. 2).

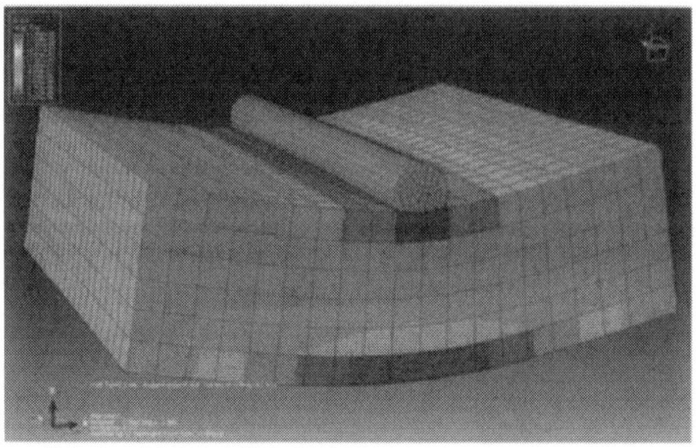

FIG. 2 – SIMULATED THREE-POINT BENDING TEST

From this, a stress vs. strain graph was created and compared to the same graph from the mechanical testing (Fig. 3) to see how closely they resembled one another.

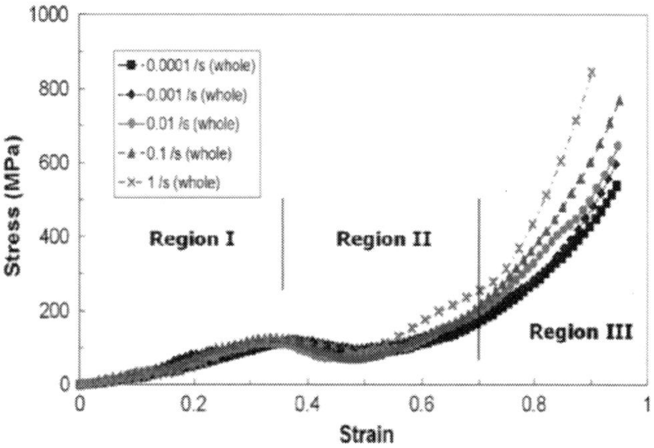

FIG. 3 – STRESS VS. STRAIN GRAPH FROM MECHNICAL TESTING

This part of the study is still being completed. Once the simulated graph matches the graph in Fig. 3, more complex simulations will be done to see if this material has enough energy-absorbing attributes to be used as a viable armor.

RESULTS

Currently, there are very few results to report. As more and more properties are added to the model, the two stress versus strain graphs become more closely related. The most crucial part of the simulation is creating a middle structure that can collapse in on itself and maintain its structural integrity. The first graphs generated were fairly linear, but more recent graphs are starting to resemble the mechanical testing graph. The largest improvement to date came from substituting a poro-elastic property with permeability. This substitution led to a more stable structure and will ultimately lead to an ideal simulated model.

DISCUSSION

As seen in Fig. 3, the stress versus strain graph are divided up into three distinct regions:

- Region 1: linear elastic behavior
- Region 2: perfectly inelastic behavior
- Region 3: densification (stiffening) behavior

Understanding what causes these behaviors is key to creating an identical model. The first region is caused by all three shell layers absorbing the subjecting force. Once this force reaches a critical point (~119 MPa) the inner, porous core begins to collapse. This is what causes the second region. As the middle shell layer begins to collapse those voids and densify, the energy from the force is absorbed, causing the inelastic plateau. If the force continues to increase, the middle core eventually cannot collapse anymore and, after a certain amount of strain, the stress begins to increase drastically until failure (~750 MPa). Creating a model that follows these behaviors exactly will be nearly impossible, but getting it close will still provide a useful tool for further investigating this material for use as a highly efficient armor.

ACKNOWLEDGEMENTS

The authors would like to acknowledge the financial support from the Department of Defense (DoD), the computer programs supplied by the Center for Advanced Vehicular Systems (CAVS), and the educational support from the Agricultural and Biological Engineering (ABE) staff at Mississippi State University.

REFERENCES

[1] Liu, Kesong, Jiang, Lei. "Bio-inspired design of multiscale structures for function integration." *Nano Today* 6.2 (2011): 155-175.

[2] Krauss, S., et al. "Mechanical Function Of A Complex Three-Dimensional Suture Joining The Bony Elements In The Shell Of The Red-Eared Slider Turtle." *Advanced Materials* 21.4 (2009): 407-412.

[3] Damiens, R., et al. "Research Paper: Compressive Behavior Of A Turtle's Shell: Experiment, Modeling, And Simulation." *Journal Of The Mechanical Behavior Of Biomedical Materials* 6. (2011): 106-112.

SB³C2015
Summer Biomechanics, Bioengineering and Biotransport Conference
June 17-20, 2015,Snowbird Resort, Utah, USA
SB³C2015-1106

ROLE OF CELL TENSION ON ANISOTROPIC MECHANOSENSING

Shin-Min Wen and Pen-Hsiu Grace Chao

Institute of Biomedical Engineering
Nation Taiwan University
Taipei, Taiwan

INTRODUCTION

Cells in situ reside in a variety of matrices and exhibit a wide variety of phenotypic morphologies. The actin cytoskeleton organization is influenced by cell morphology that results in changes in cell tension [1, 2]. As the cytoskeleton supports cellular structures and mediate numerous intracellular processes, perturbations to the cytoskeleton via deformational load can change cytoskeleton organization and influence global cellular functions, including proliferation and phenotype [3-5]. Parallel and perpendicular uniaxial loading (relative to the cell long axis) has been shown to result in differential collagen production [6]. Few studies, however, have examined the mechanism of anisotropic mechanosensing. In the current study, we investigate how cell anisotropy and uniaxial loading interact to influence the actin cytoskeleton structure and nuclear shape.

METHODS

Microfabrication PDMS membranes were treated with 2 mg/ml dopamine hydrochlode (pH=8.5) and microcontact printed with 25 μg/ml fibronectin. Uncoated regions were blocked with 1% BSA. Circle (diameter = 40 μm) and elliptical patterns were ~1250 μm² in area and the elliptical pattern had an aspect ratio of 5.

Cell Culture Human bone marrow-derived mesenchymal stem cells (MSCs) were kindly provided by Dr. Shih-Chieh Hung (YMU [7]). Cells were seeded on the patterned PDMS membranes overnight. Myosin activity was inhibited with 50 μM blebbistatin in serum-free medium for one hour. Rho-kinase (ROCK) and myosin light chain kinase (MLCK) were inhibited with 10 μM Y27632 and ML-7, respectively, in serum-free medium for 30 minutes.

Mechanical Stimulation A custom bioreactor applied 10% uniaxial strain for 60 seconds (strain rate: 0.1/sec) either parallel or perpendicular to the long axis of the cells.

Imaging and Analysis Cytoskeleton organization was examined by fluorescently labeling f-actin with Alexa Fluor 488-conjugated phalloidin (Invitrogen) and the nuclei were labeled with by DAPI. F-actin content distribution was quantified by stacking fluorescent images of individual cells and calculate the average intensity in each pixel location. Nuclear aspect ratio was analyzed using Image J.

Statistical Analysis SPSS was used to perform ANOVA analysis with post-hoc tests (LSD, α=0.05)

RESULTS

In circular cells, f-actin formed radial and circumferential structures that were lost with 10% strain for one minute (Figure 1). Myosin inhibition with Y27632, ML-7, and blebbistatin suppressed actin structures in all unloaded cells. Interestingly, 10% strain stimulated f-actin formation in the Y27632 and ML-7 groups, albeit short and disorganized. No treatment (loading or drugs) had any significant effects on nuclear shape in circular cells (data not shown). In elliptical cells, f-actin formed bundles in the direction of the long axis, which were also lost with parallel strain (Figure 2). Surprisingly, actin bundles were observed after one minute of perpendicular loading, exhibiting a direction-dependent actin structural change. When ROCK was suppressed with Y27632, parallel loading dramatically increased f-actin in the elliptical cells, while perpendicular loading did not change f-actin intensity. Suppression of myosin activity with blebbistatin had similar effects as Y27632. No apparent changes in f-actin content from mechanical strain were observed when the cells were treated with ML-7, where all groups exhibited low f-actin intensity. Nuclear shape generally followed similar trend as the f-actin content, where high f-actin levels corresponded with high nuclear aspect ratio (elongated nuclei, Figure 4). However, when f-actin increased after parallel strain with Y27632, no significant changes were found in nuclear shape. On the other hand, in the blebbistatin group, parallel and perpendicular strain both changed nuclear shape, although f-actin only increased in the parallel-loaded group.

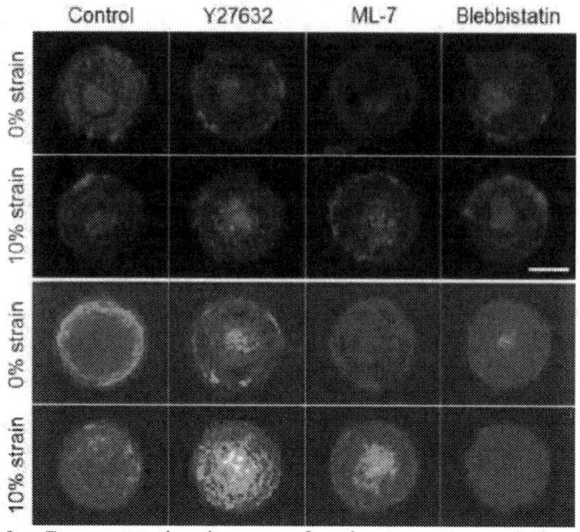

Figure 2. Representative images of actin structures and fluorescent heat maps of 10-15 cells as a quantitative measure of actin intensity distribution. Scale bar = 20 μm.

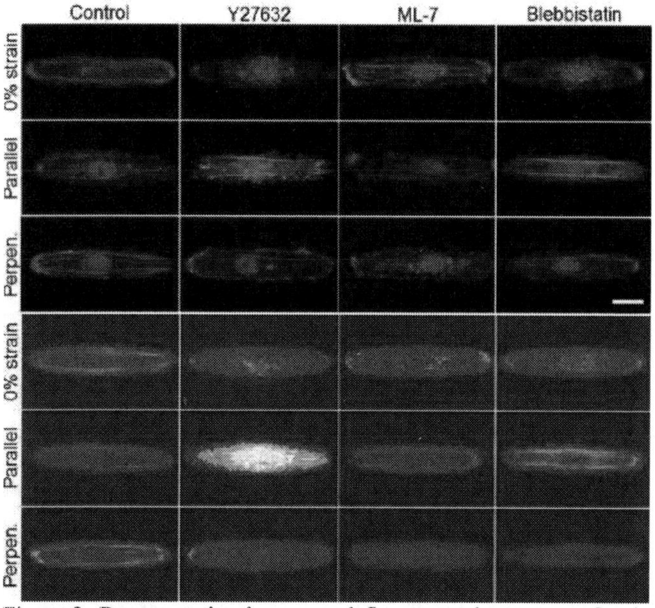

Figure 3. Representative images and fluorescent heat maps of actin structure. Scale bar = 20 μm.

DISCUSSION

In this study, we investigated the initial cytoskeletal and nuclear response to uniaxial loading and explored the role of cell tension on anisotropic mechanosensing. All treatments that interfered with myosin changed the effects of loading on f-actin content and structure, confirming our initial hypothesis. Our results suggest that actomyosin tension mediates actin severing when subjected to on-axis strain. Tension reduction also decreased nuclear aspect ratios, consistent with the role of cell tension on nuclear morphology [8]. Surprisingly, inhibition with Y27632 and blebbistatin did not suppress new f-actin formation upon parallel loading (Figure 2), suggesting that additional, myosin-independent factors mediate this adaptive response. The changes in nuclear shape from uniaxial strain with blebbistatin treatment further demonstrated the role of myosin-independent actin structures in nuclear morphology control. Moreover, differential effects from parallel and perpendicular strain may also be attributed to MLCK regulation, since ML-7 treatment did not lead to the same increase in f-actin upon parallel loading. MLCK has been shown recently to control focal adhesion in a myosin-independent manner [9]. Current studies are ongoing to investigate the involvement of myosin-independent cytoskeletal components as well as the equilibrium effects of cell tension on cytoskeletal reorganization.

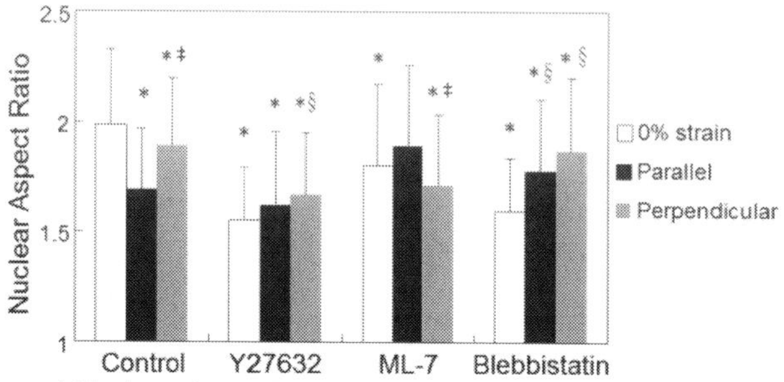

Figure 4 Nuclear shape changes in response to mechanical strain and myosin disruption (n=22-181, *$p<0.05$ vs control 0% strain, §$p<0.05$ vs 0% strain, ‡$p<0.05$ vs parallel loading)

ACKNOWLEDGEMENTS

This work was supported by the NSC (103-2221-E-002-039) and NHRI (EX104-10411EI)

REFERENCES

[1] Thery M, Pepin A, Dressaire E, et al. Cell distribution of stress fibres in response to the geometry of the adhesive environment. Cell Motil Cytoskeleton 2006;63:341-55.
[2] Kumar S, Maxwell IZ, Heisterkamp A, et al. Viscoelastic Retraction of Single Living Stress Fibers and Its Impact on Cell Shape, Cytoskeletal Organization, and Extracellular Matrix Mechanics. Biophys J 2006;90:3762-73.
[3] Altman GH, Horan RL, Martin I, et al. Cell differentiation by mechanical stress. FASEB J 2001;16:270-2.
[4] Maniotis A, Chen CS, Ingber DE. Demonstration of mechanical connections between integrins, cytoskeletal filaments, and nucleoplasm that stabilize nuclear structure. Proc Natl Acad Sci U S A 1997;94:849-54.
[5] Lee C-F, Haase C, Deguchi S, et al. Cyclic stretch-induced stress fiber dynamics – Dependence on strain rate, Rho-kinase and MLCK. Biochem Biophys Res Commun 2010;401:344-9.
[6] Lee CH, Shin HJ, Cho IH, et al. Nanofiber alignment and direction of mechanical strain affect the ECM production of human ACL fibroblast. Biomaterials 2005;26:1261-70.
[7] Hung S-C, Pochampally RR, Hsu S-C, et al. Short-Term Exposure of Multipotent Stromal Cells to Low Oxygen Increases Their Expression of CX3CR1 and CXCR4 and Their Engraftment In Vivo. PLoS ONE 2007;2:e416.
[8] Khatau SB, Hale CM, Stewart-Hutchinson PJ, et al. A perinuclear actin cap regulates nuclear shape. Proc Natl Acad Sci U S A 2009;106:19017-22.
[9] Chen C, Tao T, Wen C, et al. Myosin Light Chain Kinase (MLCK) Regulates Cell Migration in a Myosin Regulatory Light Chain Phosphorylation-independent Mechanism. J Biol Chem 2014;289:28478-88.

SB³C2015
Summer Biomechanics, Bioengineering and Biotransport Conference
June 17-20, 2015, Snowbird Resort, Utah, USA
SB³C2015-1107

STRUCTURAL INHOMOGENEITY ENHANCES INTERSTITIAL FLUID PRESSURIZATION IN TMJ CONDYLAR CARTILAGE

B. Zimmerman (1), L. Ruggiero (1,2), M. Park (1), L. Han (3), L. Wang (1), D. Burris (1), X. L. Lu (1)

(1) Mechanical Engineering
University of Delaware
Newark, DE, USA

(2) Mechanics of Materials and Constructions
Vrije Universiteit Brussel
Brussels, Belgium

(3) Biomedical Engineering
Drexel University
Philadelphia, PA, USA

INTRODUCTION

The temporomandibular joint (TMJ), composed of a fibrocartilaginous disc separating the opposing articular surfaces of the glenoid fossa and mandibular condyle, is the only moving joint in the human head [1]. Although lubrication is critical for extended use of the TMJ over a lifetime [2], the frictional properties of TMJ cartilaginous tissues have not been studied as extensively as those of regular hyaline cartilage.

Many theoretical [3,4] and experimental [5,6] studies have proved that interstitial fluid pressurization (IFP) is a critical lubrication mechanism in hyaline cartilage. In hyaline cartilage, the fluid pressure often supports over 95% of the mechanical loading under physiological conditions, minimizing the load borne by friction-causing solid-solid contacts. Our recent study [7] suggested that IFP plays an essential role in the lubrication of TMJ condylar cartilage.

Despite employing IFP for lubrication, as do typical cartilaginous tissues, TMJ condylar cartilage has a unique ultrastructure that is drastically different from regular hyaline cartilage (Fig. 1, left). Condylar cartilage is characterized by an extra fibrous superficial layer covering the three hyaline-like proliferative, mature, and hypertrophic zones (collectively termed 'middle-deep zone' herein). Large type I collagen bundles comprise the majority of this layer, running tangent to the articular surface and rooting vertically into the middle-deep zone. In hyaline cartilage, the tension-compression nonlinearity of the collagen fibers is critical to induce and maintain fluid load support, with the peak fluid load support directly related to the ratio of tensile to compressive moduli [4]. It is natural to conjecture that the presence of the superficial fibrous zone in condylar cartilage could also induce tension-compression nonlinear behaviors, affecting joint lubrication.

Due to difficulties in experimentally measuring fluid pressure inside cartilage, numerical simulations based on mixture theories are often used to analyze fluid pressure fields inside loaded cartilage. In this study, a finite element (FE) model was built to simulate compression and sliding on TMJ condylar cartilage to investigate the roles of the superficial zone in IFP. To characterize the heterogeneous, nonlinear nature of condylar cartilage, indentation and tensile testing was performed to determine its mechanical properties. This study aims to (1) determine the tensile and compressive moduli of superficial and middle-deep zones of condylar cartilage, and (2) implement a FE model, using the determined material properties, to investigate the effect of structural heterogeneity on IFP and lubrication. The results are compared with experimental trends revealed by tribology tests.

METHODS

Healthy TMJs were harvested from fresh porcine heads, chosen due to similarities to human TMJ [8]. The mandibular condyle was mounted on a previously-reported custom micro-tribometer [6] and the central region was tested under six normal loads and 7 sliding speeds in order to examine the transition region in which IFP can be detected.

To obtain samples for indentation testing, cylindrical plugs were harvested from the central region of TMJs, and the superficial zone removed by sledge microtome. A creep test was performed on both zones with a φ=1.2mm porous plane-ended indenter. Following surface alignment and equilibration under a 5mN preload, a 20mN load was applied until equilibrium. Rectangular strips of both zones were prepared similarly by microtome. Tensile testing of the strips was performed as described by [9] with an Instron 5848 Microtester. The tensile moduli were obtained from the linear portion of the relaxed stress-strain curve.

The tissue was modeled in the FE program FEBio [10] as a bilayer material composed of the superficial and middle-deep zones. According to the ultrastructure of condylar cartilage, both superficial and middle-deep zones were modeled as fiber-reinforced biphasic materials, with differing fiber reinforcements selected based on tissue ultrastructure. Determination of material parameters in both constitutive models was achieved by curve-fitting the experimental creep indentation data.

The experimental frictional testing was replicated in FEBio with a rigid impervious spherical probe (φ=6.4mm) compressed on the tissue surface with specified strain before initiation of sliding along the surface. Sliding simulations were performed at sliding speeds of 1mm/s and 2mm/s under 10% compressive strain, and for 10% and 15% compressive strain at 1mm/s sliding speed. The fluid load support ratio, defined as the ratio between the integration of fluid pressure and the total traction over the contact area, was calculated to represent the level of IFP in the loaded tissue. A physiologically-representative heterogeneous model (composed of superficial and middle-deep zones) and a homogeneous model (middle-deep zone only) were created to examine the effect of the superficial zone on IFP.

RESULTS

The mechanical properties obtained by curve-fitting with the FE model indicated vastly different properties for each zone of condylar cartilage. The superficial zone displayed a prominent tension-compression nonlinearity, with a ratio of tensile to compressive moduli of approximately 1098:1 (H_{+A}=30.73±12.97 MPa vs. H_{-A} =0.028±0.016 MPa, n=9). In contrast, the tension-compression nonlinearity of the middle-deep zone is much weaker, at 17.4:1 (H_{+A}=2.43±1.75 MPa vs. H_{-A} =0.14±0.09 MPa, n=9).

The fluid pressure, total stress, and solid matrix stress in condylar cartilage after 10 sliding cycles are shown in the heterogeneous and homogeneous models (Fig. 2). In the presence of the superficial zone (heterogeneous model), the magnitude of fluid pressure and total stress are higher than those in the homogeneous model. Despite this, the stress on the solid matrix is lower in the heterogeneous model, indicating an ability of the fluid phase to protect the solid matrix.

To quantify the contribution of fluid pressure during sliding, the fluid load support ratio is plotted over time under varying speeds and normal forces for both heterogeneous and homogeneous models (Fig. 3). The physiologically representative heterogeneous model displays a dependence of fluid load support on sliding speed, but is independent of normal strain, which agrees with experimental results (Fig. 1). In both cases, the presence of the superficial zone provides for much greater peak fluid load support and a decreased rate of exudation. Peak fluid load support is 96.7% in the heterogeneous model, compared to only 82.8% in the homogeneous model.

DISCUSSION

Previous work has demonstrated that tension-compression nonlinearity of the solid matrix contributes essentially to IFP in loaded hyaline cartilage, as the high tensile stiffness resists tissue expansion under compressive loading. Peak fluid load support is directly related to the tension-compression nonlinearity [4]. The tensile to compressive moduli ratio has been reported as 36:1 in the superficial zone and 9.4:1 in the deep zone of patellar cartilage, and 67:1 in the superficial zone and 8.5:1 in the deep zone of femoral cartilage [5]. In this study, the tensile to compressive moduli ratio in the superficial zone of condylar cartilage is extremely large (1098:1), approaching the values of fibrous tissues such as tendons [11], while the ratio in the middle-deep zone (17.4:1) is similar to the deep zone of hyaline cartilage. Therefore, the tension-compression nonlinearity of TMJ condylar cartilage should be able to significantly boost interstitial fluid pressurization during loading.

FE simulations confirmed that fluid pressure can support up to 96.7% of the applied load on TMJ condylar cartilage under 10% compressive deformation and 2mm/s sliding speed, approaching the 99% upper limit predicted in a previous study [4]. In contrast, the fluid load support in the homogeneous model in significantly lower, demonstrating the critical role played by the fibrous layer in lubrication and IFP. The heterogeneous structure also slows the depletion of fluid pressure compared to the homogeneous model. The fluid load support ratio can be maintained over 95% after 10 sliding cycles under constant 10% deformation (Fig. 3, left), suggesting fluid load support may be maintained at a very high level in cartilage *in vivo* during motion.

An increase in sliding speed from 1mm/s to 2mm/s enhances fluid load support, supporting our experimental results showing a decrease in friction with increasing sliding speed (Fig. 1, middle), occurring due to the probe migrating over the tissue before exudation can occur, thus maintaining IFP. In the heterogeneous model, fluid load support remained virtually identical at strain levels of 10% and 15%, consistent with our experiments demonstrating no correlation between friction coefficient and loading magnitude (Fig. 1, right).

These results strongly suggest that the tissue heterogeneity confers a unique nonlinear mechanical behavior to TMJ condylar cartilage, representing a functional adaptation to maximize IFP-related lubrication at the cartilage surface.

ACKNOWLEDGEMENTS

This work is supported by DOD W81XWH-13-1-0148, NIH P30GM103333-01, and NIH U54-GM104941. The authors would like to thank Dr. Gerard Ateshian for assistance with FEBio.

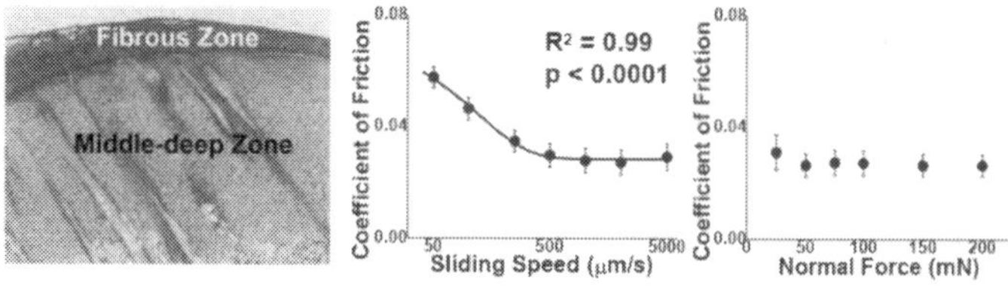

Figure 1. (Left) Histology image of condylar cartilage, stained with Sirius red for collagen. (Middle) Friction coefficient as a function of sliding speed, displaying an exponential relationship ($p<0.0001$). (Right) Friction coefficient as a function of normal loading, with no significant correlation ($p=0.37$). In experimental plots, n=6.

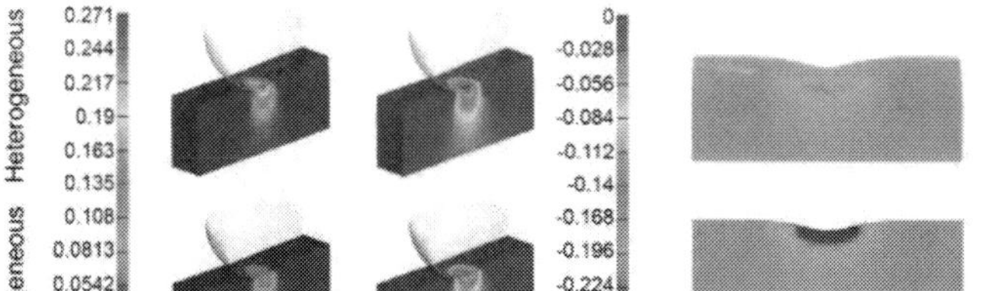

Figure 2. FE simulation of fluid pressure, total stress, and stress on the solid matrix after ten sliding cycles for the heterogeneous (top) and homogeneous (bottom) models. Solid matrix stress is the sum of fluid pressure and total stress.

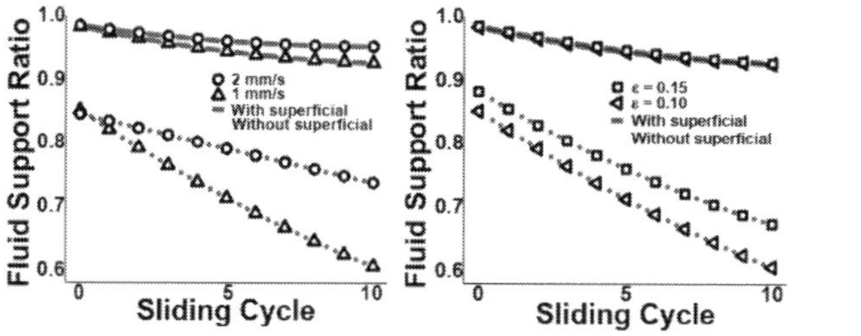

Figure 3. (Left) Fluid load support ratio as a function of time and sliding speed (normal strain=10%). (Right) Fluid load support ratio as a function of time and normal strain (sliding speed=1mm/s). In each case, the heterogeneous model (with superficial) provided greater fluid load support.

REFERENCES

[1] Kuroda, S et al., *Osteoarthr Cartilage*, 17:1408-1415, 2009. [2] Tanaka, E et al., *Ann Biomed Eng*, 36:14-29, 2008. [3] Soltz, M and Ateshian, G, *J Biomech Eng*, 122:576-586, 2000. [4] Soltz, M and Ateshian, G, *J Biomech*, 31:927-934, 1998. [5] Park, S et al., J Biomech, 36:1785-1796, 2003. [6] Bonnevie, E et al., *Tribo Lett*, 41(1):83-95, 2011. [7] Zimmerman, B et al., *J Dent Res*, 94(1):85-92, 2015. [8] Bermejo, A et al., *Oral Surg Oral Med Oral Pathol*, 75:18-23, 1993. [9] Huang, C et al., *J Biomech*, 38(4):799-809, 2005. [10] Maas, S et al., *J Biomech Eng*, 134:011005, 2012. [11] Yin, L and Elliot, D, *J Biomech*, 37:907-916, 2004.

SB^{3}C2015
Summer Biomechanics, Bioengineering and Biotransport Conference
June 17-20, 2015, Snowbird Resort, Utah, USA
SB^{3}C2015-1108

FRED VERSUS PIPELINE: AN IN-VITRO COMPARISON OF CEREBRAL ANEURYSM HEMODYNAMICS USING PARTICLE IMAGE VELOCIMETRY

Priya Nair (1), Brian W. Chong (2), Haithem Babiker (1), Justin Ryan (1), Fernando L. Gonzalez (3), David Frakes (1,4)

(1) School of Biological and Health Systems Engineering
Arizona State University
Tempe, Arizona, USA

(2) Department of Radiology
Mayo Clinic Hospital
Phoenix, Arizona, USA

(3) Department of Neurosurgery
Duke University School of Medicine
Durham, North Carolina, USA

(4) School of Electrical, Computer and Energy Engineering
Arizona State University
Tempe, Arizona, USA

INTRODUCTION

An estimated 5% of world population harbor unruptured cerebral aneurysms (CAs), a pathological out-pouching of blood vessels in the brain [1]. CA rupture is a serious condition associated with a 45% mortality rate [2]. Endovascular treatments for CA repair have gained popularity over surgical techniques due to their minimally invasive nature and shorter recovery time. Endovascular devices achieve aneurysmal occlusion by disrupting flow within the aneurysm thereby promoting thrombosis. Pipeline Embolization Device (PED) is one such low-porosity device used in the clinic. Although low-porosity devices greatly reduce intra-aneurysmal flows, there is a clinical concern that these devices may occlude perforating vessels resulting in decreased cerebral perfusion [3]. To prevent this various dual-layered stents were designed, one of which is the Flow-Redirectional Endoluminal Device (FRED). The FRED has a low-porosity region at the center of stent, functioning as a flow diverter at the aneurysmal neck, and higher-porosity region at the proximal and distal ends providing stability to the device without compromising the patency of smaller blood vessels. In this study, we compare the effects of PED and FRED on CA hemodynamics using Particle Image Velocimetry (PIV).

METHODS

A patient-specific CA was segmented and reconstructed from CT angiography dataset using Mimics (Materialise, Lueven, Belgium). The computational model was translated into an optically clear, lost-core urethane model (Figure 1) that was mounted on a flow loop as shown in Figure 2. A mixture of aqueous sodium iodide, glycerin and water, seeded with 8μm fluorescent microspheres, was chosen as the blood analog. Flow rates of 3, 4, and 5 mL/s were investigated under steady and pulsatile flow conditions.

A stereo-PIV system was used to obtain volumetric flow information through the aneurysm models. An Nd:YAG laser light sheet was used to illuminate the fluorescent particles through the model and two CCD cameras were used to capture the particle images. The cameras were fitted with low-pass filters to permit capture of particle images while blocking laser reflections from the metallic stents. PIV images were acquired across the aneurysmal volume before and after deployment of PED and FRED. A commercial cross-correlation algorithm was employed to compute velocity vectors within the aneurysmal volume, at each plane. Volumetric velocity information was then used to calculate the root-mean-squared velocities (V_{rms}) within the aneurysm for the untreated and treated cases using in-house Matlab codes. The effect of each stent in altering aneurysmal hemodynamics was analyzed by observing the percentage change in measured hemodynamic response before and after treatment.

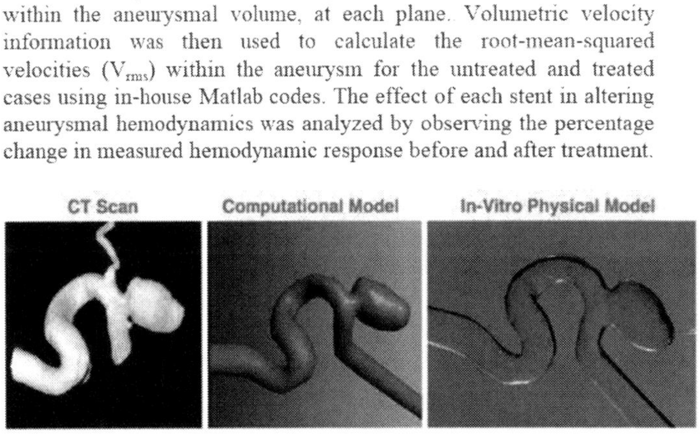

Figure 1: Manufacturing process of a lost-core urethane model

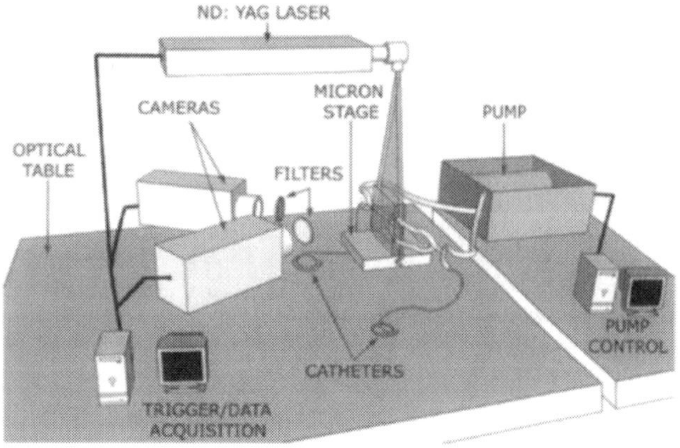

Figure 2: Schematic of the PIV system

RESULTS

Reduction in aneurysmal V_{rms} was observed following deployment of the low-porosity stents for both, steady and pulsatile, flow conditions. Figure 3 represents V_{rms} observed before and after deployment under steady flow conditions for two example flow rates. Reductions of over 55% and 65% were observed after treatment with PED and FRED respectively. The effect of the low-porosity devices on intra-aneurysmal V_{rms} under pulsatile flow conditions was similar across the three investigated flow rates (Figure 4). A qualitative comparison between the effects of PED and FRED on intra-aneurysmal velocity magnitude is also presented in Figure 5.

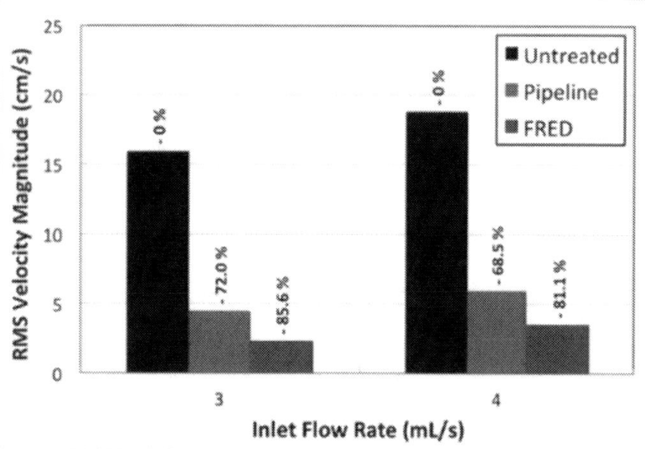

Figure 3: RMS velocity magnitude (in cm/s) within the aneurysm after treatment with PED and FRED with percentage reductions under steady flow conditions.

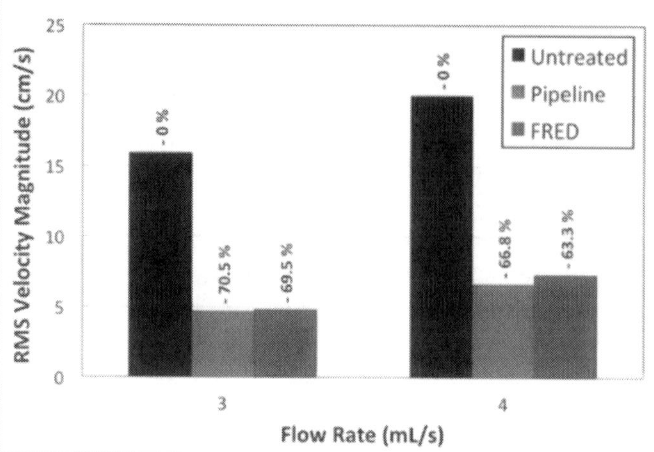

Figure 4: RMS velocity magnitude (in cm/s) within the aneurysm after treatment with PED and FRED with percentage reductions under averaged pulsatile flow conditions.

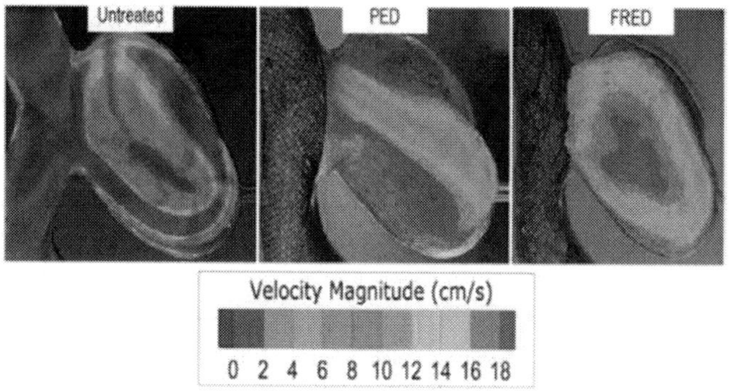

Figure 5: Center-plane vector plots color-coded by velocity magnitude (in cm/s) depicting the aneurysm before treatment (left), after treatment with PED (middle) and FRED (right).

DISCUSSION

FRED achieved greater than 50% reductions in root mean squared velocities within the aneurysms, comparable to a PED. Furthermore, uniform distribution of reduced flow velocities were observed after treatment with the FRED whereas a low velocity jet was observed after deploying a pipeline across the aneurysmal neck. Uniform reduction in velocities may promote thrombosis within the aneurysm, and over time may lead to vascular remodeling at the aneurysm neck eventually resulting in isolation of the aneurysm from circulation [4]. Furthermore, the presence of high porosity stent design on the proximal and distal end would prevent occlusion of perforating arteries around the aneurysm.

ACKNOWLEDGEMENTS

The authors acknowledge funding from the following sources: American Heart Association Beginning Grant in Aid, National Science Foundation CAREER Award, Brain Aneurysm Foundation Research Grant, and Women in Philanthropy Society Category B Grant

REFERENCES

[1] Seibert, B et al., Frontiers in Neurology, 2: 45, 2011
[2] Babiker, MH et al., J Biomech, 45:440-447, 2012
[3] Roszelle, BN et al., *J NeuroIntervent Surg*, 0:1-7, 2012
[4] Piotin, M et al., *Radiology*, 243:500-508, 2007

DEVELOPMENT OF A CLINICAL ULTRASOUND TECHNIQUE FOR ANALYSIS OF PROTEIN CONTENT WITHIN HYDROGELS

J. Stukel* (1), M. Goss* (2), A. Exner (2,3), R. Willits (1)
***These authors contributed equally to this work.**

(1) Biomedical Engineering
The University of Akron
Akron, OH, USA

(2) Biomedical Engineering
Case Western Reserve University
Cleveland, OH, USA

(3) Radiology
Case Western Reserve University
Cleveland, OH, USA

INTRODUCTION

Ultrasound is an attractive technique for quantitative characterization of tissue engineering constructs because it is nondestructive, highly sensitive, and allows for simultaneous imaging of multiple samples [1]. These attributes are ideal for imaging hydrogels because ultrasound does not require drying or coating the scaffolds, which is common in other imaging techniques. While other groups have used higher frequency transducers to study differences in collagen fibers and cell density [2], high frequency transducers are more expensive and not widely available. We aim to evaluate the use of clinical ultrasound, with a 12 MHz transducer, to determine differences in protein concentration and extent of fibrillogenesis within poly(ethylene glycol) (PEG) hydrogels. Validating that our technique can distinguish between protein concentrations and the extent of fibrillogenesis is the first step towards application for *in situ* imaging of collagen fiber formation in tissue engineering scaffolds.

METHODS

Collagen type I (0-2.5 mg/mL) was mixed into 10% (w/v) PEG-diacrylate (3300 MW) solutions containing Irgacure 2959, phosphate buffered saline (PBS), and deionized (DI) water prior to crosslinking. Two sets of gels were made, one with collagen in 0.2% acetic acid, referred to as non-fibrillar, and another with collagen suspended in PBS so fibers formed due to the increase in pH, referred to as fibrillar (n=5). The gels were crosslinked by exposure to ultraviolet light in a Teflon mold between glass to form uniform rectangular gels.

Collagen gels were imaged using a Toshiba clinical ultrasound system (Aplio500) with a linear array transducer (PLT-1204AX, centered at 12MHz) for B mode image acquisition. Five samples were placed in line in a lab-designed water-filled acrylamide phantom that was fixed atop the ultrasound transducer. Each sample was imaged three times through different cross sections. Each image represented a map of the signal intensity. To further characterize collagen content and structure within the gels, reflectance microscopy will be utilized.

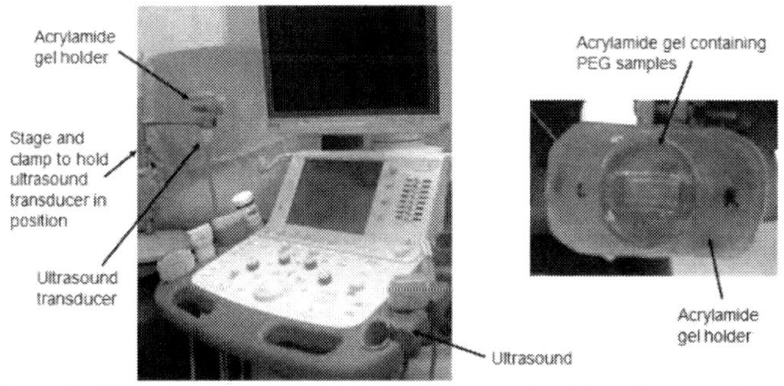

Figure 1. Ultrasound set up. The water filled acrylamide gel containing the PEG gels was secured above the ultrasound transducer.

Data analysis was performed on the ultrasound signal intensity from each sample. Three regions of interest were manually chosen per image to obtain a representative sample and avoid edge effects of the PEG gels. Then, they were analyzed to quantify the signal intensity. Intensity values across the range of collagen concentration are reported as average ± standard deviation. Statistical significance was determined using a 2-way ANOVA and Tukey post hoc test, with $p<0.05$ considered significant.

RESULTS

Visible differences were observed from the greyscale analysis with varying fibrillar collagen concentration, but not for gels with non-fibrillar collagen (Figure 1).

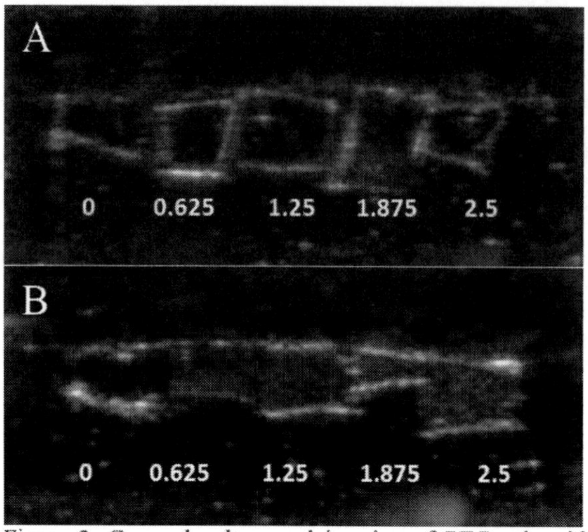

Figure 2. Greyscale ultrasound imaging of PEG gels with varying concentrations of (A) non-fibrillar and (B) fibrillar collagen. The edges of the gel are white due to the gel-water boundary. Each gel is labeled with the appropriate collagen concentration (mg/mL).

The three regions of interest were chosen within each gel from the greyscale images to calculate the ultrasound intensity, plotted as average ± standard deviation for fibrillar and non-fibrillar collagen (Figure 2).

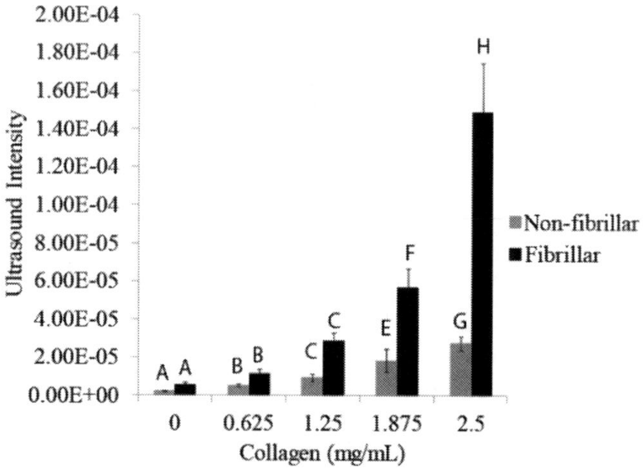

Figure 3: Ultrasound intensity from greyscale images of PEG gels with 0-2.5 mg/mL non-fibrillar or fibrillar collagen. Significant differences were noted between globular and fibrillar collagen for 1.875 and 2.5 mg/mL collagen ($p<0.05$), noted by different letters. No significant differences were noted between non-fibrillar and fibrillar collagen for groups with the same letter.

DISCUSSION

In this study, we have demonstrated the use of clinical ultrasound for the imaging of PEG hydrogels with varying collagen concentrations in two forms, non-fibrillar and fibrillar collagen. We hypothesize that fibrillar collagen has a greater signal intensity because fibers cause more light scattering than in the non-fibrillar collagen. It has been shown that thin, loosely packed collagen fibers appear brighter under ultrasound as compared to thick mature collagen [2]. Collagen in the non-fibrillar gels we tested had not formed fibers, and therefore resulted in a decreased signal intensity as compared to the fibrillar collagen. While previous work has used high frequency ultrasound to image collagen [2,3], this study demonstrates the ability of clinical ultrasound to determine changes in collagen content. The differences in signal intensity between the non-fibrillar and fibrillar collagen 1.875 and 2.5 mg/mL indicate that this ultrasound technique may eventually be used for imaging collagen integration in tissue engineering scaffolds and fibrillogenesis *in situ*.

ACKNOWLEDGEMENTS

This work was supported by the National Institute of Biomedical Imaging and Bioengineering (R01EB016960) and the Margaret F. Donovan Endowment for Women in Engineering.

REFERENCES

[1] Solorio, L, *Drug Deliv Transl Res.*, 2(1): 45-55, 2012.
[2] Anthony, P. et al., *American Heart Journal.*, 133(3): 364-368, 1997.
[3] Mercado, K.P., *Tissue Eng. Part C Methods.*, 2014 (Epub ahead of print)

SB³C2015
Summer Biomechanics, Bioengineering and Biotransport Conference
June 17-20, 2015,Snowbird Resort, Utah, USA
SB³C2015-1110

NUMERICAL STUDY OF A NEW THERMALPLASTY TREATMENT SYSTEM FOR ATHEROSCLEROSIS

S Zhao (1), A. Zhang (1,2), Lisa X. Xu (1,2)

(1) School of Biomedical Engineering
Shanghai Jiao Tong University
Shanghai, China

(2) Med-X Institute
Shanghai Jiao Tong University
Shanghai, China

INTRODUCTION

Atherosclerosis is the most common cardiovascular disease. In atherogenesis, the endothelium plays a crucial role, since endothelial injury is the initial event in the process. Besides, the synthetic phenotype of VSMCs, migration and proliferation are believed to be key factors in atherosclerosis and restenosis.[1] And recently, Tang [2] identify a new type of stem cell in the blood vessel wall, named multipotent vascular stem cells (MVSCs). In response to vascular injuries, MVSCs, instead of smooth muscle cells, become proliferative, and differentiate into smooth muscle cells and chondrogenic cells, thus contribute to the restenosis of the blood vessel after treatment.

To treat the unwanted region of the plaque while open the occluded region of the blood vessel, thermal ablation has been combined with conventional angioplasty.[3]However, during the treatment, the endothelial cell layer is usually the first victim[4]The pure mechanical force induced by the dilation of balloon, the freezing temperature on the surface of the balloon or the high temperature used cause further damage to the endothelium.

Thus, in this study, a selective treatment of the atherosclerosis is proposed, in which the endothelial cell will be protected, and the abnormally proliferation of VSMCs and MVSCs be inhibited with accurate control of thermal therapy.

A thermal angioplasty device is also designed. In order to evaluate the feasibility of accurate control of the thermal therapy, a finite element model is designed to analyze the temperature and stress distribution during the treatment.

METHODS

The designed thermal angioplasty balloon is shown in Figure 1. The agent at low temperature such as liquid nitrous oxide or other similar gas is pumped into the balloon, it expands the balloon while lowering the local temperature to a certain value. A radiofrequency (RF) heating by the flexible electrodes on the surface of the balloon is used to ablate the proliferated region. With the cooling of the flowing agent the balloon, the intima of the arteries could be spared from the RF heating.

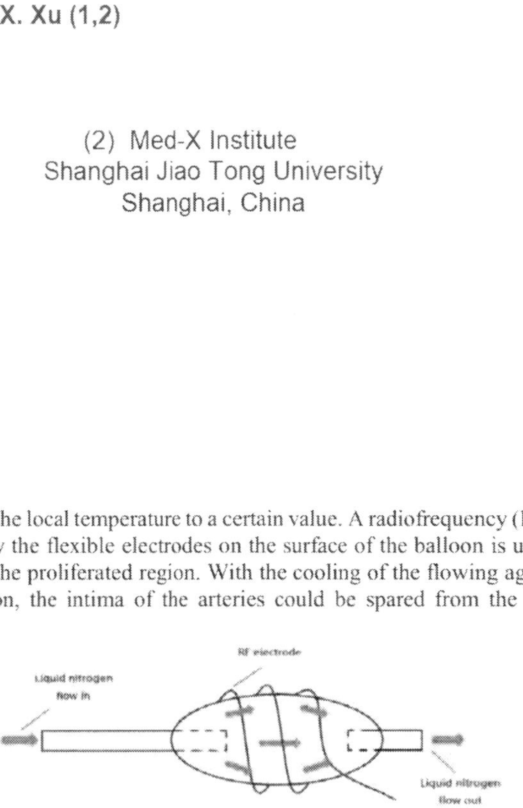

FIGURE 1 SCHEMATIC PICTURE OF THERMAL BALLOON

Aiming to evaluate the feasibility of accurate control of thermal therapy, a finite element model including solid mechanics, heat transfer and electric currents field is made by using COMSOL Multiphysics. Some of electric and thermal parameters are cited from [5], [6].

The geometry of model is shown in Figure 2. The gray one represents the artery, and the blue one represents the radiofrequency electrode.

The RF frequency used in this model is 460 kHz, which is much lower than 1 MHz, thus the quasi-static approximation of the electrical field in the tissue can be applied [7]:

$$\nabla \cdot (\sigma_t \nabla V) = 0 \qquad (1)$$

Where V is the electrical potential, and σ_t is the electrical conductivity of the artery.

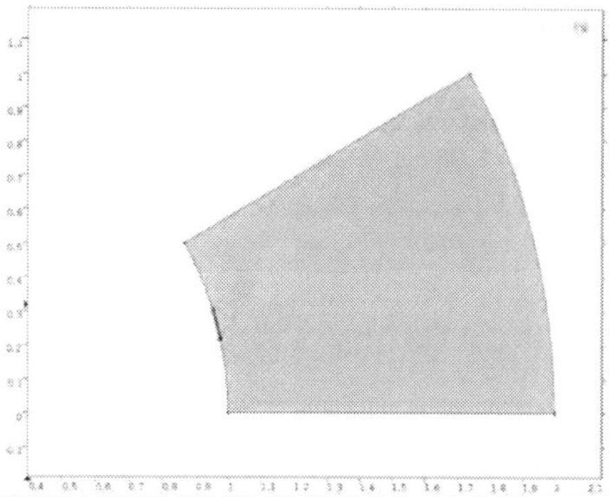

FIGURE 2 THE GEOMETRY OF ARTERY MODEL

The specific absorption rate of RF energy in the tissue Q_{SAR} (W/m^3) is:

$$Q_{SAR} = \sigma_t(\nabla V)^2 \quad (2)$$

Pennes' equation that describes the heat transfer in the tissue is [8]:

$$\rho_t c_t \frac{\partial T_t}{t} = \nabla \cdot (k_t \nabla T_t) + \rho_b w_b (T_b - T_t) + Q_{meta} + Q_{SAR} \quad (3)$$

where ρ_t, c_t , k_t stand for the density, heat capacity and thermal conductivity of tissue, respectively; T_b, w_b , ρ_b are the arterial blood temperature, blood perfusion rate, the density of blood; Q_{meta} is the metabolic rate of tissue, Q_{SAR} is the specific absorption rate of tissue which is calculated by the equation(1) and (2). And the artery obey the Hook's Law and a linear thermal expansion is applied:

$$e_{ij} = \frac{1+v}{E}\sigma_{ij} - \frac{v}{E}\sigma_{kk}\delta_{ij} + \alpha(T - T_0)\delta_{ij} \quad (4)$$

Where σ_{ij} is the stress tensor, e_{ij} is the strain tensor, v is Poisson's ratio, E is Young's modulus, the constant α is the coefficient of linear expansion. And v, E, α are input parameters in the model.

RESULTS

The voltage of radiofrequency is set as 35V, and on the inner surface of the artery, the boundary condition is set as 10 °C. After 0.2 second of RF heating, the simulated temperature contour is shown in Fig.3.

As shown in Fig.3, high temperature region appears in the middle of artery. It indicates the device can give a hyperthermia condition to the lesion of atherosclerosis. With increasing of the heating time, the highest temperature can be further increased.

Besides, the stress distribution after the RF heating is also obtained. Results shown that the Von-Mises stress on the wall of the artery is about 10% larger than that during the conventional angioplasty, and this difference may be acceptable.

DISCUSSION

The model has successfully simulated the temperature and the stress distribution of the design. The results confirm that by adjusting the voltage, the temperature of cooling agent and heating time, accurate control of thermal therapy can be realized. The stress on the vessel wall during the heating is only a litter larger than that obtained in conventional angioplasty. By further taking the real physiological anatomy of the atherosclerosis plaque, the dependence of the artery's mechanical property on temperature, the evaporation process of the circulating agent into consideration, a more accurate prediction can be made and the treatment plan better designed.

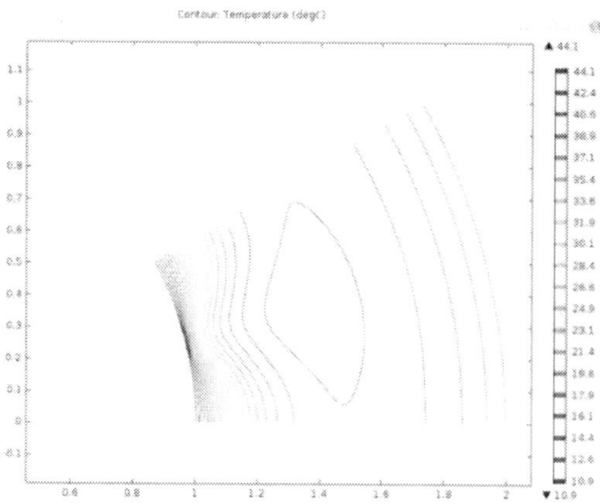

FIGURE 3 TEMPERATURE CONTOUR

ACKNOWLEDGEMENTS

This work has been supported National Natural Science Foundation of China (NSFC 51476101). The authors would like to thank Dr. Yuanyuan Shen for her support in COMSOL software.

REFERENCES

[1] R. Ross, "The pathogenesis of atherosclerosis: a perspective for the 1990s," *Nature,* vol. 362, pp. 801-9, 1993 Apr 1993.

[2] Z. Tang, A. Wang, F. Yuan, Z. Yan, B. Liu, J. S. Chu, *et al.*, "Differentiation of multipotent vascular stem cells contributes to vascular diseases," *Nature communications,* vol. 3, p. 875, 2012.

[3] E. D. Grassl and J. C. Bischof, "In vitro model systems for evaluation of smooth muscle cell response to cryoplasty," *Cryobiology,* vol. 50, pp. 162-173, Apr 2005.

[4] C. Chaabane, F. Otsuka, R. Virmani, and M. L. Bochaton-Piallat, "Biological responses in stented arteries," *Cardiovascular Research,* vol. 99, pp. 353-363, Jul 2013.

[5] M. Pop, A. Molckovsky, L. Chin, M. C. Kolios, M. A. S. Jewett, and M. Sherar, "Changes in dielectric properties at 460 kHz of kidney and fat during heating: importance for radio-frequency thermal therapy," *Physics in Medicine and Biology,* vol. 48, pp. 2509-2525, Aug 2003.

[6] A. Bhattacharya and R. L. Mahajan, "Temperature dependence of thermal conductivity of biological tissues," *Physiological Measurement,* vol. 24, pp. 769-783, Aug 2003.

[7] S. Tungjitkusolmun, S. T. Staelin, D. Haemmerich, J. Z. Tsai, H. Cao, J. G. Webster, *et al.*, "Three-dimensional finite-element analyses for radio-frequency hepatic tumor ablation," *Ieee Transactions on Biomedical Engineering,* vol. 49, pp. 3-9, Jan 2002.

[8] H. H. Pennes, "Analysis of tissue and arterial blood temperatures in the resting human forearm (Reprinted from Journal of Applied Physiology, vol 1, pg 93-122, 1948)," *Journal of Applied Physiology,* vol. 85, pp. 5-34, Jul 1998.

VASCULAR COMPLIANCE EFFECT ON ENDOTHELIAL-TO-MESENCHYMAL TRANSDIFFERENTIATION UNDER HIGHLY PULSATILE FLUID SHEAR

W.H. Elliott (1), Y. Tan (2), M. Li (2), W. Tan (1,2)

(1) Department of Mechanical Engineering
University of Colorado at Boulder
Boulder, Colorado, U.S.A.

(2) Department of Pediatrics
University of Colorado at Denver
Aurora, Colorado, U.S.A.

INTRODUCTION

Neointimal hyperplasia (NIH) exhibited excess collagen-type-I deposition on the blood vessel wall, and unmitigated proliferation of fibroblasts/myofibroblast-like cells exhibiting α-smooth muscle actin (α-SMA). Complications in NIH treatment and prevention are determined by cell type and origin. Cells present in NIH express α-SMA, which may derive from either smooth muscle cells or fibroblasts. Emerging evidence shows that these α-SMA+ cells could also be transitioned from endothelial cells (ECs), through a process called endothelial-to-mesenchymal transdifferentiation (EndMT).

EndMT involves the transition of differentiated ECs to mesenchyme-like cells. EndMT occurs in both embryonic development, but also contributes partially to pulmonary, and cardiac fibrosis, signaled by inflammatory cytokines such as transforming growth factor-β (TGF-β) [1], by physical forces such as fluid shear [2], or by EC apoptosis with subsequent TGF-β release [3]. EndMT is characterized by a mixed mesenchymal phenotype with decreased platelet endothelial cell adhesion molecule-1 (PECAM-1), increased α-SMA, [1,2], CD34 – endothelial progenitor cells (EPC) marker [5], and fibroblast specific protein-1 expression (FSP-1). [4] Additionally, FSP-1 indicates macrophage cell-type in previous EndMT study. [4]

Previously, we have shown that unidirectional, highly pulsatile flow (HPF) causes endothelial dysfunction such as pro-inflammatory effects. Inflammation may proceed cell apoptosis and cell transition for wound "healing". Herein, we examine EndMT mechanism through characterizing EC phenotypic alteration and cell apoptosis under HPF, and subsequent increases of FSP-1 and CD34 expression in remaining cells. This is performed on the stiff substrate and compliant tubes. Increased vascular compliance might rescue the EC from this EndMT, a devastating transition.

METHODS

Human pulmonary arterial ECs (HPAEC) were obtained from Lonza Inc, and kept in a CO2 incubator at 37°C and 5% CO_2. Cells were thawed as needed and grown to approximately 70-80% confluency using EGM-2 with additional growth factor and fetal bovine serum bullet kit (Lonza, Basel, Switzerland). Once confluent, cells at passages of 3-8 were seeded on two glass slides for the flow experiment.

Flow Setup and Experiments Glass slides or elastic substrates with varied compliances were chemically functionalized with 20% of aqueous sulfuric acid and plasma etching, respectively, and then coated with 6% of 3-aminopropyltriethoxysilane in acetone. After silanization, substrates were treated with 1.5% of glutaraldehyde solution. This formed an aldehyde which could initiate amine linkage to protein coating. Substrates were coated with 25 μg/ml fibronectin aqueous solution for 30-60min to provide a functional attachment layer for the HPAECs. HPAECs at a concentration of 6.0×10^5 per ml were seeded on the fibronectin-coated substrates and given one hour to attach to the surface. Cells were covered in growth media and allowed to reach confluency, before transferring to flow chamber apparatus.

Sterilization of the flow circuit was completed through perfusion of flow system with 10% hydrogen peroxide (H_2O_2) under ultraviolet light within a biosafety cabinet Type II for a minimum of 30 minutes. Circulating media used in all experiments consisted of a basal media (EBM-2, Lonza, Basel, Switzerland), medical grade dextran (MP-products Inc, Solon, OH) (7% w/w), and penicillin/streptomycin (2% v/v), to maintain similar blood viscosity and high shear stress, while reducing volumetric flow rate. A digital flow meter (Alicat Scientific Inc, Tucson, AZ) made direct measurement of flow conditions.

Two flow conditions were applied within a closed fluid circuit: HPF at 1 Hz using a pulsatile blood pump (Harvard Apparatus Inc;

Holliston, MA) and dampened flow using peristaltic pump (Cole Parmer). Both mean flow rates were equivalent, with mean surface shear stress of 1.2 Pa. Low pulsatile flow (LPF) or steady flow showed little variation in flow velocity with pulsatility index (PI=△ Velocity/Mean Velocity) less than 0.2, while the PI value of HPF was determined to be 1.7.

Western Blot Analyses were performed as per the manufacturer's suggestions (Invitrogen, Carlsbad, CA). Following electrophoresis, resulting protein was transferred to nitrocellulose membrane. Membranes were blot with PECAM-1 mouse monoclonal, TGFβ1 mouse monoclonal, and 1A4 (α-SMA) antibodies (all from Santa Cruz Biotech, Santa Cruz, CA).

Apoptosis Assay Apoptosis assay was applied to cells after exposure to flow for 24 hr. FLICA poly caspase reagent (Immunochemistry Technology LLC, Bloomington, MN) including fluorescent-labeled inhibitor of caspases, is a simple yet accurate method to measure apoptosis via caspase activity in whole cells. The assay was performed by following the manufacturer's instruction.

RESULTS

Cell Apoptosis Precedes EndMT We have shown that EC released TGF-β during EC apoptosis. These in conjunction with our findings showing EC transition to mesenchymal fibroblastic phenotype and release of TGF-β1 after 48 hr in HPF, which further induced fibroblast proliferation and migration, suggest that PAECs might show enhanced apoptotic activity prior to the EndMT process and their further influences on neighboring cells. We therefore asked whether the occurrence of endothelial transition into a pathological fibroblastic phenotype were preceded by endothelial apoptosis. To address this question, caspase assay was used to assess cell apoptosis after 24 hr flow conditioning. Results showed that HPF induced much higher percentage of apoptotic cells when compared to LPF or steady flow (Figure 1). The co-existence of apoptotic cells and non-apoptotic cells are interesting facts of the HPF-induced changes.

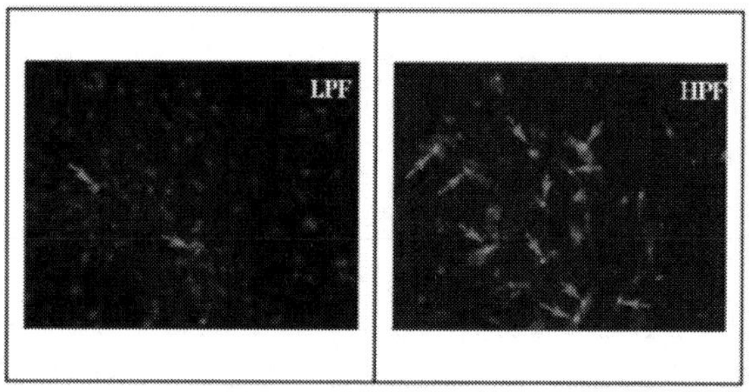

Figure 1: APOPTOSIS ASSAY SHOWS HIGHER CELL APOPTOSIS UNDER HPF. Caspase (green), Nuclei (Blue)

Cell markers additionally show changes between flow conditions. EndMT is shown to occur, with decreases in PECAM-1, and increased α-SMA expression. Expression of both CD34 and FSP-1 indicate increases, though results do not show significance in two-tailed t-tests. This may be due to EPC, and CD34 presence occurring in both populations, regardless of HPF. Additionally, more time may be necessary to allow for FSP-1 expression.

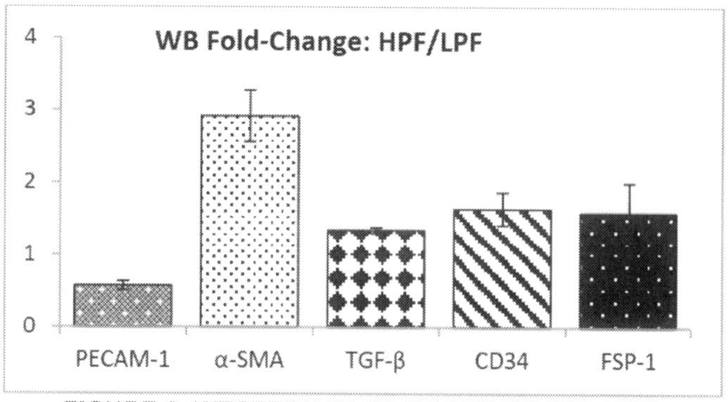

FIGURE 2: WESTERN BLOT ON ENDMT-RELATED MOLECULES. RESULTS SHOW COMPARATIVE EXPRESSION OF EC PHENOTYPIC MOLECULE (PECAM-1) IS REDUCED WHILE EXPRESSION OF MYOFIBROBLAST PHENOTYPIC MOLECULES (α-SMA, TGF-β, CD34 AND FSP-1) ARE INCREASED IN HPF, WHEN COMPARED TO LPF.

DISCUSSION

Our results indicate HPF signals EC apoptosis, subsequent TGF-β release, and resulting EndMT of ECs on a stiff substrate. ECs on the stiff substrate (no compliance) undergo apoptosis and release TGF-β [3] in HPF environment, signaling EndMT.[1,2] This was supported by significantly increased expression of caspase, α-SMA, TGF-β, and decreased PECAM-1 expression. We show that HPF may contribute to NIH, and that resulting EndMT cells may be "survivor" cells of EPC-like, myofibroblasts-like or macrophage-like cells, evidenced respectively by increased CD34, α-SMA or FSP-1 expression. [5] Increased vascular compliance might rescue the EC from this EndMT, a devastating transition.

ACKNOWLEDGEMENTS

The authors wish to acknowledge funding sources, including AHA (13GRNT16990019 to W.T) and NHLBI (HL097246 to W.T.).

REFERENCES

[1] Zeisburg, EM et al., *Nat Med*, 13: 952-961, 2007.
[2] Egorova, AD et al., *Circ Res*, 108:1093-1100, 2011.
[3] Sakao, S et al., *Am J Physiol Lung Cell Mol Physiol* 291:L362-L368, 2006
[4] Österreicher, CH et al., *PNAS*, 108: 308-311, 2011
[5] Sakao, S et al., *FASEB*, 19: 1178-1180, 2007

SB³C2015
Summer Biomechanics, Bioengineering and Biotransport Conference
June 17-20, 2015, Snowbird Resort, Utah, USA
SB³C2015-1112

BICUSPID AORTIC VALVE HEMODYNAMICS INDUCE ACUTE ASYMMETRIC REMODELING OF PORCINE ASCENDING AORTAS: AN EX VIVO STUDY

S. Atkins, P. Sucosky

Department of Aerospace and Mechanical Engineering
University of Notre Dame
Notre Dame, IN, 46556
USA

INTRODUCTION

The bicuspid aortic valve (BAV) is the most common form of inherited cardiac anomalies and differs from the normal tricuspid aortic valve (TAV) as its anatomy features two leaflets instead of three. The BAV is the first indicator for surgical valve replacement and is frequently associated with secondary aortopathy such as aortic dilation, dissection, aneurysm and rupture. In fact, BAV patients are at a 5-10 increased risk for aortic dissection [1]. Interestingly, BAV ascending aortas (AAs) dilate asymmetrically, with the site of dilation seemingly corresponding to the location of the impinging valvular orifice jet on the aortic wall. For patients with the most common type-I BAV (i.e., left-right-coronary cusp fusion), dilation preferentially localizes to the AA convexity, while the AA concavity is less likely to dilate. Matrix metalloproteinases (MMPs), enzymes involved in the remodeling and dilation process, are found both in TAV and BAV AAs of patients experiencing aortic dilation [2]. The predominant theory of BAV aortopathy to date has been the genetic theory, which states that the aortic wall is genetically weakened because of an inherited defect; however, evidence of a specific gene is lacking. The emerging hemodynamic theory states that abnormal hemodynamics experienced by the BAV aortic wall lead to endothelial cell activation and extracellular matrix (ECM) remodeling [3]. Given the particular sensitivity of vascular tissue to its surrounding hemodynamics, the fluid wall shear stress (WSS) abnormalities experienced on the wall of the ascending aorta downstream of a BAV may be involved in the pathogenesis of BAV aortopathy; however, there have been no ex vivo studies to date to confirm this hypothesis. Therefore, the goal of this study is to determine the role of regional WSS in a TAV and BAV AA on the development of acute remodeling (MMP-2 and MMP-9 expression) in ascending aortic tissue harvested from areas that seem to be prone, and conversely, protected from the development of dilation. Validation of the hemodynamic theory may provide therapeutic modalities aimed at new target molecules capable of normalizing the BAV flow and slowing disease progression in BAV patients.

METHODS

<u>WSS Characterization:</u> The local WSS in the convexity and concavity of a type-I BAV AA was obtained computationally in an idealized three-dimensional AA fluid-structure interaction model operating under physiologic flow conditions. WSS values were obtained 1 cm distal to the sinotubular junction on the convex and concave sides of the AA.

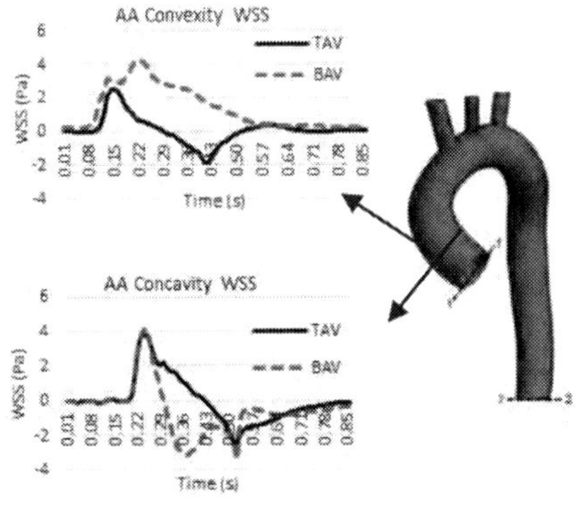

Figure 1: TAV and BAV AA WSS waveforms from the concavity and convexity. 1-1 = inlet section; 2-2 = outlet section.

<u>Experimental Groups and Tissue Conditioning:</u> Waveforms from both the convex and concave regions were then programmed into a

cone-and-plate bioreactor capable of replicating pulsatile WSS in a sterile environment (Fig. 2).

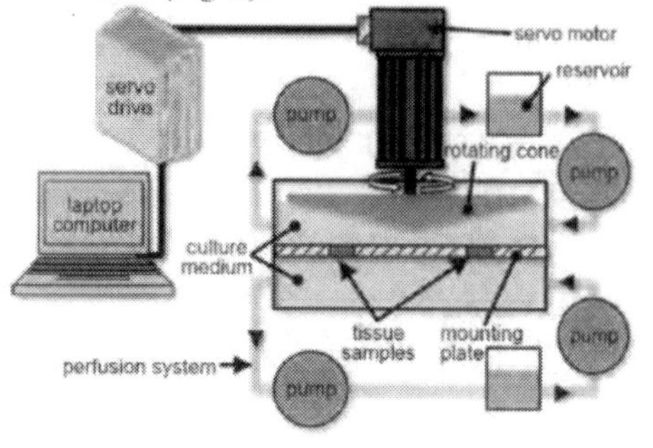

Fig. 2: Single cone-and-plate bioreactor setup.

Circular tissue specimens (7-mm diameter) were isolated from porcine AAs, in the same regions as those from which WSS was characterized computationally. Tissue harvested from the wall convexity was exposed to the WSS variations predicted in the same region of either a TAV or BAV AA, while tissue from the wall concavity was subjected to the WSS variations predicted at the same location in a TAV or BAV AA. Controls consisted of fresh tissue samples from either location. All tissue samples were conditioned for 48 hours.

Immunostaining: Slices 7 µm thick were sectioned and warmed on a heater (37 °C) for approximately 20 min. Sections were blocked for 1 h at room temperature with 200 µL of blocker [10% Goat Serum (Sigma-Aldrich), 0.2% TritonX-100 (Sigma-Aldrich), 1% dimethyl sulfoxide (DMSO; Thermo Fisher Scientific)] in PBS. Next, primary antibodies diluted in blocker were incubated overnight at 4 °C at the following dilutions: MMP-2 (1:200,EMD Millipore) and MMP- 9 (1:200, EMD Millipore). Sections were then rinsed 3 times in PBS and incubated with goat-anti-mouse secondary antibody (1:100, Santa Cruz) for 2 h at room temperature. Next, the sections were rinsed 3 times for 5 min in PBS before counterstaining with 1 4',6-Diamidino-2-phenylindole (DAPI, Sigma-Aldrich) and mounting with fluorescence mounting medium (Dako). Sections were cover-slipped and stored at 4 °C. Slides were subsequently imaged under the mercury lamp of a Nikon E600 imaging microscope using a TR/FITC/DAPI filter. Results were quantified using ImageJ and normalized to the fresh controls.

Statistical Analysis: Data was analyzed using ANOVA to determine if there was a significant contribution on measured parameters, followed by a Bonferroni post-hoc test. A p-value of 0.05 was used to measure statistical significance.

RESULTS

Enzymatic expression of MMP-2 and MMP-9 in BAV and TAV AA WSS conditioned convexity and concavity samples is shown in Fig. 3. Exposure of tissue specimens from the aortic wall convexity to BAV AA WSS from the convexity resulted in increased MMP-2 (2.3 fold, $p < 0.05$) and MMP-9 (2.5 fold, $p < 0.05$) expressions relative to the TAV AA WSS treatment and control. In contrast, exposure of tissue specimens from the aortic wall concavity to TAV AA and BAV AA WSS from the concavity showed significantly decreased MMP-9 (0.5 fold, $p < 0.05$) expression as compared to the TAV AA WSS group and control. Changes in MMP-2 expression on the aortic concavity between the BAV AA WSS group and the TAV AA WSS group as well as the control were insignificant.

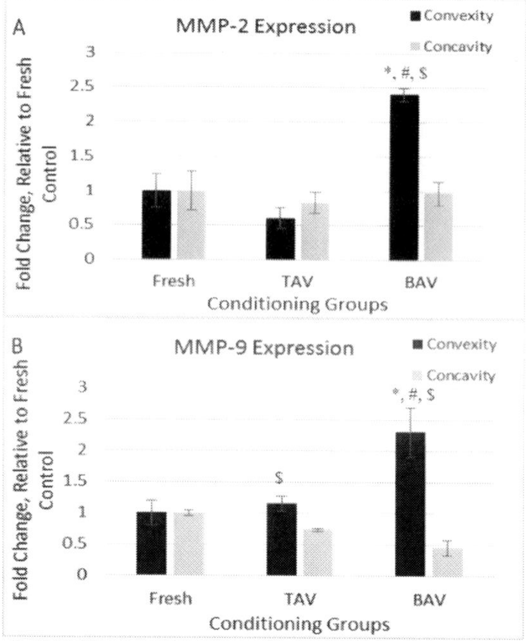

Figure 3: (A) Enzymatic expression of MMP-2 in the convexity and concavity. Mean ± SEM. * = $p < 0.05$ vs. Fresh, # = $p < 0.05$ vs. TAV, $ = $p < 0.05$ vs. concavity. (B) Enzymatic expression of MMP-9 from the convexity and concavity. Mean ± SEM. * = $p < 0.05$ vs. Fresh, # = $p < 0.05$ vs. TAV, $ = $p < 0.05$ vs. concavity.

DISCUSSION

Studies investigating the impact of the BAV anatomy on WSS and disease have been lacking to date. This study demonstrates for the first time that the local hemodynamics experienced by the BAV concavity has a protective effect on the aortic wall by down-regulating MMP-9 expression relative to the local hemodynamics experienced by the BAV convexity. Previous animal models have demonstrated that MMP-2 and MMP-9 work synergistically to promote aneurysm formation [4]. More studies are needed to elucidate the potential causality between local WSS and local biological remodeling events in BAV AAs. However, this study confirms the sensitivity of aortic tissue to WSS abnormalities and demonstrates the susceptibility of BAV hemodynamic stresses to focally mediate aortic medial degradation along the disease-prone wall convexity. The results provide compelling support for the role of hemodynamics in BAV secondary aortopathy.

REFERENCES

[1] Januzzi JL, Isselbacher EM, Fattori R et al.: Characterizing the young patient with aortic dissection: results from the International Registry of Aortic Dissection (IRAD). results from Int Regist Aortic Dissection (IRAD) 2004;43:655–659.

[2] LeMaire SA, Wang X, Wilks JA, Carter SA, Wen S, Won T, et al.: Matrix metalloproteinases in ascending aortic aneurysms: bicuspid versus trileaflet aortic valves. J Surg Res 2005 Jan;123:40–8.

[3] Atkins SK, Sucosky P: Etiology of bicuspid aortic valve disease: Focus on hemodynamics. World J Cardiol 2014 Dec 26;6:1227–33.

[4] Longo, GM, Xiong, W, Greiner, TC, Zhao, T, Fiotti, N, Baxter B: Matrix metalloproteinases 2 and 9 work in concert to produce aortic aneurysms. J Clin Invest 2002;110:625.

EFFECT OF NECK CABLE TENSION DURING HELMETED HEAD IMPACT

Bethany Rowson, Steven Rowson, and Stefan M. Duma

Biomedical Engineering and Mechanics
Virginia Tech
Blacksburg, Virginia, United States

INTRODUCTION

Although the Hybrid III Anthropomorphic Test Device (ATD) was originally designed for automotive crash tests, it has also been widely used to study sports injury biomechanics. In 2003, Pellman et al. reconstructed concussive impacts of professional football players with Hybrid III ATDs [3]. In 2010, Gwin et al. performed tests to evaluate a proposed helmet standard by the National Operating Committee for Standards in Athletic Equipment (NOCSAE) to better simulate in vivo head impacts [2]. A NOCSAE headform mounted on a Hybrid III neck was used for these tests.

For automotive safety testing, the Code of Federal Regulations, Part 572.33 specifies that the axial cable running through the center of the Hybrid III neck be tightened by torqueing the jam nut on it to 12 in-lb. Since these specifications were written for automotive applications, they may not necessarily apply to the field of sports injury biomechanics. The purpose of this study was to determine if applying torque to the jam nut on the neck cable in the Hybrid III neck has an effect on linear acceleration, angular acceleration, and angular velocity of the head.

METHODS

A medium NOCSAE head form, modified to house one triaxial linear accelerometer and one triaxial angular rate sensor mounted on a Hybrid III neck, was used for tests with a pneumatic linear impactor. The head was fitted with a Riddell Revolution Speed adult size large helmet for all tests. The head and neck were mounted on a sliding table that allowed for motion in the direction of impact. The impactor had a convex face and padding to simulate a helmeted head.

Two conditions (tension and no tension) were tested at four different locations and three speeds for each location, resulting in a total of 24 tests. For the tension tests, the neck was tightened to 12 in-lb. The four impact locations were: front, side, rear boss, and rear. The average velocities (95% CI) were 4.67 m/s (4.60, 4.73), 7.07 m/s (7.03, 7.10), and 9.32 m/s (9.28, 9.36).

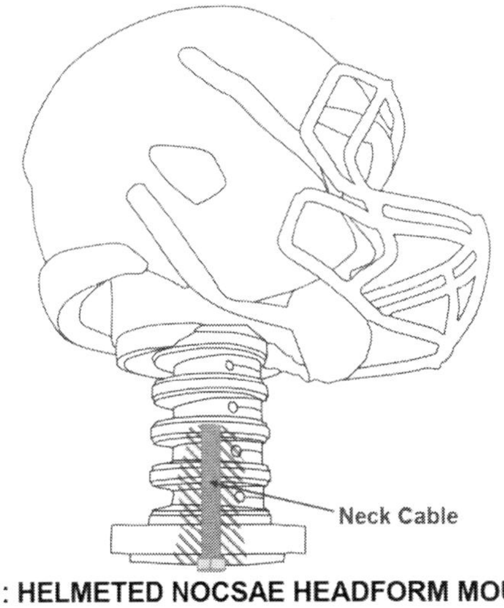

FIGURE 1: HELMETED NOCSAE HEADFORM MOUNTED ON HYBRID III NECK

All linear acceleration and angular velocity data were collected at a frequency of 20,000 Hz. The linear acceleration data were filtered to CFC 1000 in accordance with SAE J211, and the angular velocity data were filtered to CFC 155 [4]. Angular acceleration was calculated by differentiating the angular velocity data. All linear acceleration data were transformed to the CG of the headform.

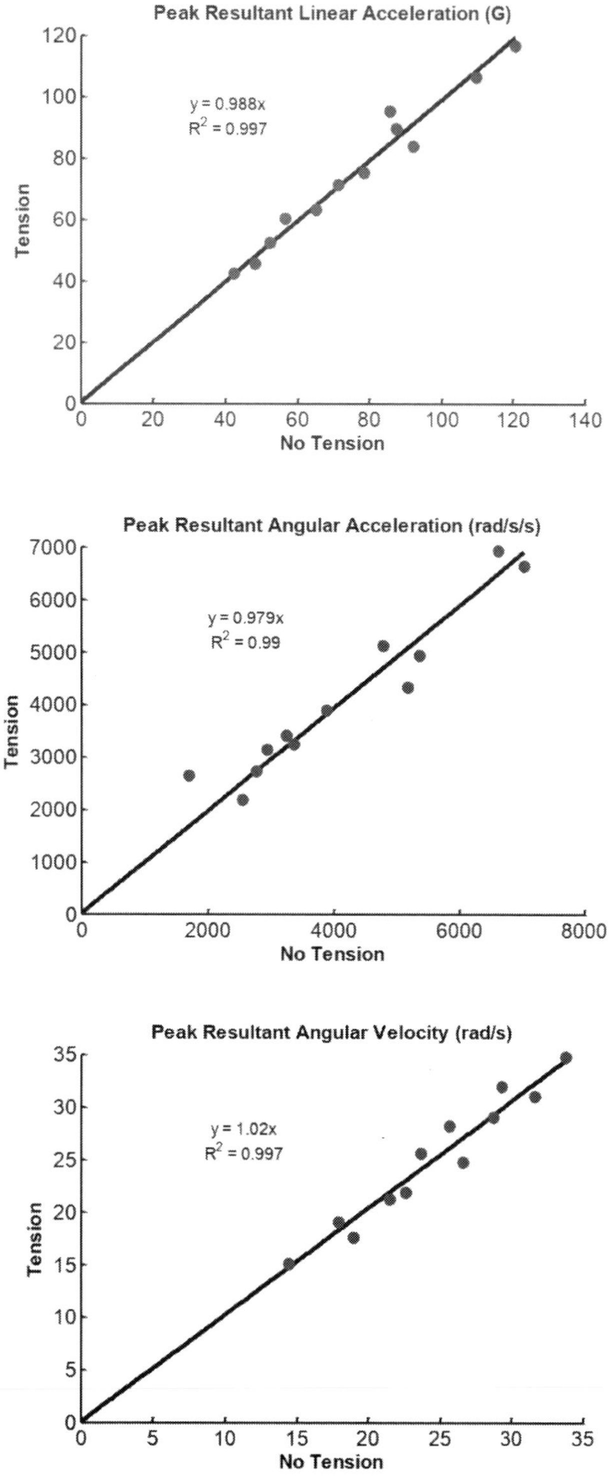

FIGURE 2: LINEAR REGRESSIONS OF TENSION VERSUS NO TENSION TESTS FOR: PEAK RESULTANT LINEAR ACCELERATION (TOP), ANGULAR ACCELERATION (MIDDLE), AND ANGULAR VELOCITY (BOTTOM)

The percent difference between each paired test condition (no tension and tension) was calculated for peak resultant linear acceleration, angular acceleration, and angular velocity (Eq. 1).

$$\frac{X_{NT} - X_T}{X_T} \times 100\% \qquad (1)$$

Where X represents either linear acceleration, angular acceleration, or angular velocity, while the subscripts denote no tension (NT) and tension (T). The results are reported as an average +/- standard deviation for each metric.

RESULTS

Two-sided paired Student's t-tests were performed for peak resultant linear acceleration, angular acceleration, and angular velocity. There were no significant differences between groups, with p-values of 0.61, 0.84, and 0.33, respectively. Linear regressions for each outcome variable are shown in Figure 2, with the associated equation of the line and R^2 value displayed. All regressions were constrained through zero at the origin.

The percent differences for paired test conditions were 0.95% +/- 5.38%, 0.03% +/- 14.15%, and -1.40% +/- 5.87% for peak resultant linear acceleration, angular acceleration, and angular velocity respectively.

DISCUSSION

These results demonstrate that there are no significant differences in peak linear acceleration, angular acceleration, or angular velocity of the head when the neck cable is tightened to the prescribed tension of 12 in-lb. In 2008, Below et al. looked at the effects of varying neck cable torque values [1]. They evaluated 8, 10, 12, and 14 in-lb, and found significant differences in peak neck rotation, peak moment about the occipital condyle, and peak Y-axis moment for different torque values. They suggested that as long as the tension is the same for every test in a series, the values can be adjusted to meet the desired response. They also noted that increasing the specified torque requirement could be one was to decrease neck variability between tests. Although there are differences in the biomechanical response of the neck with varying torque values, there does not appear to be any significant difference in peak head acceleration values, which are often used for applications such as helmet performance evaluation and standards.

In conclusion, this study evaluated resultant head kinematics with variations in neck cable tension, and found no significant differences in peak values between tests with no cable tension and those performed at the prescribed cable tension for automotive tests. These results have implications for future sports head injury tests, specifically helmet evaluations.

REFERENCES

[1] Pellman, E.J., et al., Concussion in professional football: reconstruction of game impacts and injuries. Neurosurgery, 2003. 53(4): p. 799-812; discussion 812-4.

[2] Gwin, J.T., et al., An investigation of the NOCSAE linear impactor test method based on in vivo measures of head impact acceleration in American football. J Biomech Eng, 2010. 132(1): p. 011006.

[3] Rowson, S., et al. Rotational Head Acceleration Measurement Techniques and Headform Characteristics: Implication on Helmet Impact Testing. in 41st International Workshop on Human Subjects for Biomechanical Research. 2013. Orlando FL.

[4] Below, J.D., et al., Evaluation of Neck Bracket Angles and Neck Torque Procedures in the Hybrid III Small Female Neck Flexion Test. SAE Technical Paper 2008-01-0530, 2008.

SB³C2015
Summer Biomechanics, Bioengineering and Biotransport Conference
June 17-20, 2015,Snowbird Resort, Utah, USA
SB³C2015-1114

HEMODYNAMIC CHARACTERIZATION OF DIFFERENT BASILAR TIP ANEURYSM TEMPLATES USING COMPUTATIONAL FLUID DYNAMICS

Priya Nair (1), Brian W. Chong (2), David Frakes (1,3)

(1) School of Biological and Health Systems Engineering
Arizona State University
Tempe, Arizona, USA

(2) Department of Radiology
Mayo Clinic Hospital
Phoenix, Arizona, USA

(3) School of Electrical, Computer and Energy Engineering
Arizona State University
Tempe, Arizona, USA

INTRODUCTION

Cerebral aneurysms (CAs) are localized dilations of weakened blood vessels in the brain, most commonly found at arterial bifurcations in the circle of Willis [1]. CA rupture may lead to a critical condition called subarachnoid hemorrhage (SAH) that is associated with a high degree of morbidity and mortality [2]. Effective management of CAs is therefore essential.

Hemodynamics is known to play an important role in CA initial and progression, but is still considered to a limited degree in CA evaluation and treatment planning. In the clinic, geometric features of the aneurysm such as aneurysm size and dome-to-neck ratio (DNR) are extensively characterized during CA evaluation. CA size has been found to be an important predictor of rupture where a higher rupture rate has been observed in larger aneurysms [3]. DNR, on the other hand, is often used to develop suitable treatment methodologies [4]. CA size and DNR, along with parent vessel curvature, can alter hemodynamics at the aneurysm site that may promote further growth eventually leading to rupture [5].

In this study, we quantify the effects of geometric variations on idealized basilar tip aneurysm (IBTA) hemodynamics using computational fluid dynamics (CFD) to help provide an improved geometric/hemodynamic basis for clinical decision-making. Eight unique combinations of three geometric factors: Dome size (DS), DNR, and parent-vessel contact angle (PV-CA) are developed using a 2-level, full-factorial design. Hemodynamic responses within the aneurysm and at the neck plane are investigated at three flow rates under steady and pulsatile flow conditions.

METHODS

A 2-level, full-factorial design was employed on three geometric factors: DS, DNR and PV-CA to generate eight unique combinations of CA templates, as shown in Figure 1. The corners of the cube correspond to low/high values of a particular factor. For example, **1 (4,3:2,45)** represents **IBTA-1** with **4** mm DS, **3:2** DNR and **45°** PV-CA. The edges of the cube are color-coded to represent changes in one factor keeping the other two factors constant. Red, green and blue edges represent changes in DS, DNR and PV-CA, respectively.

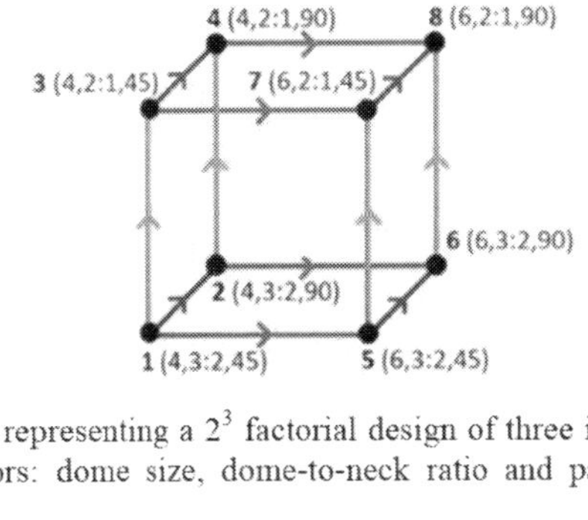

Figure 1: Cube representing a 2^3 factorial design of three investigated geometric factors: dome size, dome-to-neck ratio and parent-vessel contact-angle

The computational geometric templates were designed in SolidWorks (Dassault Systemes, Waltham, Massachusetts, USA), where the aneurysm was modeled as a sphere with DS and DNR in accordance to Figure 1. Inlet and outlets were modeled as cylinders of diameter 4 mm, with inlet angles of either 45° or 90°. The models were prepared for CFD simulations using ANSYS ICEM 12.1 (ANSYS, Inc., Canonsburg, Pennsylvania USA) where the blood volume was discretized into millions of unstructured tetrahedral elements. The meshed geometries were then imported into ANSYS

Fluent for CFD simulations. Flow rates of 2, 3, and 4 mL/s were investigated under steady and pulsatile flow conditions. Blood was assumed to incompressible and non-Newtonian, and a no-slip, rigid boundary condition was applied at the vessel wall. Blood density and viscosity were chosen to match a blood analog solution, which will be later used for experimental comparison.

The CFD data was analyzed using Tecplot 360 (Tecplot, Inc., Bellevue, Washington, USA) to calculate hemodynamic responses within the aneurysmal volume and at the neck-plane. Distal aneurysmal wall shear stress (WSS), root-mean-squared velocity magnitude (V_{rms}) within the aneurysm and at the neck, and cumulative flow entering the aneurysm per second (cross-neck flow) were the responses calculated. Statistical analysis of variance (ANOVA) was then performed on the obtained responses using JMP Pro (JMP, SAS Institute Inc., Cary, North Carolina, USA) to understand the effect of the geometric factors on hemodynamics.

RESULTS

Percentage change in hemodynamic responses was calculated by changing one geometric factor, keeping the other two factors constant. Increasing DS increased distal aneurysmal WSS across all models (except the aneurysm with low DNR and high PV-CA), aneurysmal and neck-plane $V_{rms,}$ and CNF whereas increasing DNR from 3:2 to 2:1 decreased all computed hemodynamic responses under steady and pulsatile flow conditions (Example: Figure 2). Changing PV-CA was found to affect only distal aneurysmal WSS, where the 45° templates had higher distal WSS compared to their 90° counterparts.

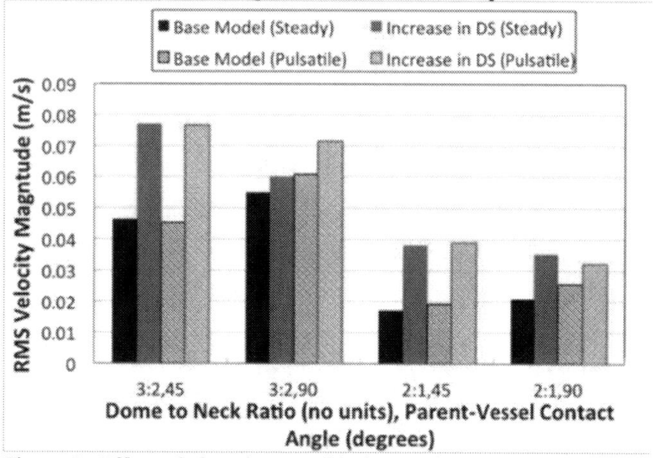

Figure 2: Effect of changing DS (black to red boxes) on aneurysmal V_{rms} at 3 mL/s

The percentage contribution of each factor on hemodynamics, along with 2- and 3-factor interactions, under steady and pulsatile flow conditions were computed using ANOVA (Table 1). Aneurysmal DNR had the greatest impact across all computed hemodynamic responses (>37%). DS had greater impact on aneurysmal and neck-plane V_{rms}, and CNF, whereas PV-CA had the most impact distal aneurysmal WSS and neck-plane V_{rms}.

DISCUSSION

The results obtained in this study, in agreement with existing literature, shows that geometric changes can significantly alter CA hemodynamics [5,6]. The three investigated factors, particularly DNR, had an impact on calculated hemodynamic responses. Aneurysm templates with lower DNR have larger neck, which may permit greater inflow of blood into the aneurysms thereby creating an undesirable hemodynamic environment. Aneurysmal DS is already established as a strong predictor of rupture. ANOVA results showed that DS had the second largest effect on calculated responses. From a clinical perspective, aneurysms with higher DS have been found to have greater risks of rupture, which correlated with our findings of increased velocities and flow conditions in larger domed aneurysm. Our results agreed with findings that showed that higher PV-CA is associated with elevated levels of hemodynamic responses that may promote aneurysmal growth or regrowth [6]. Whether high or low WSS is detrimental to a CA still remains a topic of debate. Elevated aneurysmal WSS can lead to vascular remodeling resulting in endothelium destruction, and areas of low WSS can disintegrate the vessel wall [7]. Several combinations of geometric factors also had an impact on CA hemodynamics, the most prominent of which was combinations of PV-CA with DS, DNR and both.

Future work would include extending our database to anatomical aneurysms to understand how our results can be translated to a more clinically relevant scenario. Furthermore, we would like to include some experimental methods to calculate hemodynamic responses on IBTA templates.

From our results, we observed that geometric changes, individually and in combination, altered hemodynamic responses in IBTA templates. We believe that understanding the effect of geometric variations on hemodynamics could potentially translate to the clinical setting for CA evaluation and treatment planning.

Factors	Distal Aneurysmal WSS (%)	Aneurysmal Vrms (%)	Neck-Plane Vrms (%)	Cross-Neck Flow (%)
Dome Size	8.28 (8.66)	23.07 (4.86)	21.84 (14.06)	43.94 (43.61)
Dome to Neck Ratio	46.21 (57.86)	71.95 (75.05)	53.37 (56.71)	36.93 (37.37)
Contact Angle	17.91 (15.09)	0.05 (3.95)	12.72 (18.55)	3.74 (2.59)
Dome Size * Dome to Neck Ratio	0.73 (0.01)	0.06 (0.24)	6.23 (3.97)	9.73 (11.71)
Dome Size * Contact Angle	11.93 (11.35)	3.66 (8.53)	0.13 (0.12)	0.39 (0.03)
Dome to Neck Ratio * Contact Angle	9.08 (3.75)	0.16 (4.68)	4.34 (6.51)	3.81 (4.29)
Dome Size * Dome to Neck Ratio * Contact Angle	5.86 (3.29)	1.05 (2.70)	1.36 (0.08)	1.47 (0.39)

Table 1: Percentage contribution of individual factors, and 2- and 3-factor interactions under steady (outside parenthesis) and averaged pulsatile (within parenthesis) flow conditions at 3 mL/s

ACKNOWLEDGEMENTS

The authors acknowledge funding from the following sources: American Heart Association Beginning Grant in Aid, National Science Foundation CAREER Award, Brain Aneurysm Foundation Research Grant, and Women in Philanthropy Society Category B Grant

REFERENCES

[1] Meng, H et al., *Stroke*, 38:1924-1931, 2007
[2] Connoly, ES et al., Stroke, 43:1711-1737, 2012
[3] Wiebers, DO et al., The Lancet, 362:103-110, 2003
[4] Brinjikji, W et al., Am J Neuroradiol, 30:1513-1517, 2009
[5] Sfroza, DM et al., Ann Rev Fluid Mech, 41: 91, 2009
[6] Hoi, Y et al., J Neurosurg, 101:676-681, 2004
[7] Cebral, JR et al., Am J Neuroradiol, 32:264-270, 2011

PROXIMAL FEMORAL CORTICAL BONE THICKNESS IN PATIENTS WITH FEMOROACETABULAR IMPINGEMENT AND NORMAL HIPS ANALYZED USING STATISTICAL SHAPE MODELING

Penny R. Atkins (1,2), Prateep Mukherjee (3,4), Shireen Y. Elhabian (3,4), Sumedha Singla (3,4), Michael D. Harris (1,2), Jeffrey A. Weiss (1-4), Ross T. Whitaker (1,3,4), Andrew E. Anderson (1,2,4,5)

(1) Department of Bioengineering
University of Utah
Salt Lake City, UT, USA

(2) Department of Orthopaedics
University of Utah
Salt Lake City, UT, USA

(3) School of Computing
University of Utah
Salt Lake City, UT, USA

(4) Scientific Computing and Imaging Institute
University of Utah
Salt Lake City, UT, USA

(5) Department of Physical Therapy
University of Utah
Salt Lake City, UT, USA

INTRODUCTION

Hip osteoarthritis (OA) develops in one out of four people over their lifetime [1]. Within the last decade, femoroacetabular impingement (FAI) has been implicated as a primary cause of hip OA in young adults [2]. Cam-type FAI is characterized by an aspherical femoral head and reduced head-neck offset caused by a bony lesion at the anterolateral head-neck junction. Cam FAI may initiate abnormal contact between the femur and acetabulum that results in early cartilage degeneration and bone remodeling, especially near the limits of hip flexion and internal rotation. Bone is resected in cam FAI patients to improve head-neck offset and sphericity of the femoral head. The depth and volume of resection is crucial to patient outcome and satisfaction [3]. Inadequate resection is noted as the primary reason for repeat surgery as it will not restore range of motion [4]. However, too aggressive of a resection may weaken the femoral neck, leading to an iatrogenic hip fracture [5]. The shape and depth of resection has been evaluated experimentally and computationally [6, 7]. However, prior studies did not consider the effect of the anatomical variations seen in femurs with cam FAI. As cortical bone contributes to the majority of load bearing within the hip [8], it is important to understand how cortical bone thickness is distributed throughout the femoral head in control hips and cam FAI patients, especially at the location of the lesion. The purpose of this study was to quantify and compare variation in proximal femoral head cortical bone thickness and total volume of the cortex in anatomically normal controls and patients with cam-type FAI using statistical shape modeling.

METHODS

With Institutional Review Board approval, volumetric computed tomography (CT) images of the proximal femur were acquired using a Siemens SOMATOM 128 Definition CT Scanner (120 kVp tube voltage, 512 x 512 acquisition matrix, 1.0 mm slick thickness, 0.9-1.0 pitch, 250 mAs + Care-Dose tube current, 300-400 mm field of view) on 21 cam FAI patients (20 males) and 39 control subjects (25 males). The two groups were not significantly different in age, weight, height, and body mass index (BMI), with averages and standard deviations of the patients and (controls) of 31±11 (27±8) years, 79.9±17.4 (84.3±16.9) kg, 176.4±8.8 (180.3±8.1) cm, and 25.9±5.7 (25.8±5.0) kg/m^2, respectively.

Patients were diagnosed with cam FAI using standard clinical measures [9] and were scheduled for surgery. The symptomatic side was analyzed in patients, while the analyzed side was chosen randomly for controls. CT datasets were upsampled to improve resolution. Cortical and trabecular bone layers of each proximal femur were semi-automatically segmented and reconstructed from the CT image data using Amira (v5.4, Visage Imaging, San Diego, CA). Reconstructed surfaces of the cortical and trabecular layers of bone were used to determine the thickness of the cortical bone over the surface of the femoral head and proximal shaft above the lesser trochanter. Triangular meshes representing the outer cortex were created for each femur, with cortical thickness specified at each node.

ShapeWorks [10], an open-source distribution software, allows for the generation of correspondence-based statistical models using optimization iterations to reduce the variation in correspondence placement across populations. Correspondence points (n=2048) were automatically placed on each sample surface in ShapeWorks using a gradient descent energy function to derive the mean cam and control femur shape. To effectively tease out the inherent shape variations of the proximal femur alignment of the reconstructed bones is critical. Thus, shape preprocessing included alignment using an iterative closest point algorithm and cropping of the shape by a plane placed above the lesser trochanter. Correspondences were first placed on the entire surface; the normal of each correspondence was evaluated to determine a set of the best correspondences on the set of femurs. Planes were transposed from a single femur based on this set of best correspondences. A sensitivity analysis was completed to evaluate the effect of plane transposition when compared to manually determined plane positions distinct for each femur (our prior method of choice, using the variation in cropping plane slope and offset). The resultant mean shapes were compared using Hotelling T-square tests.

From statistical shape modeling, a principle component analysis (PCA) provides information about the modes of variation between the shapes of the population and subgroups within the population. The modes from the PCA were evaluated for the cumulative percent variance. The initial modes that summed to 90% of the total variance were used for all analyses. Mean shapes for cam FAI and asymptomatic controls were compared based on their maximal thickness in the region of the cam lesion.

Due to the predominance of male patients with cam FAI, a comparison of male patients (n=20) and male controls (n=25) was also completed. Similar to the entire population, age, height, weight, and BMI were not statistically varied between the two groups of males.

RESULTS

Inspection of the mean shape in each group revealed that patients with cam-type FAI had increased cortical thickness in the anterolateral regions of the head-neck junction (Fig. 1), which is typically the location of cam lesion and subsequent bone resection during surgery. The maximum thickness of 4.96 mm in the head-neck junction in cam patients was greater than that of the control femurs at 3.63 mm. The mean cam FAI femur and the asymptomatic control mean femurs were significantly different (T^2=63.5, p=0.00001) when comparing the first seven modes of variation accounting for 90% of the total variance.

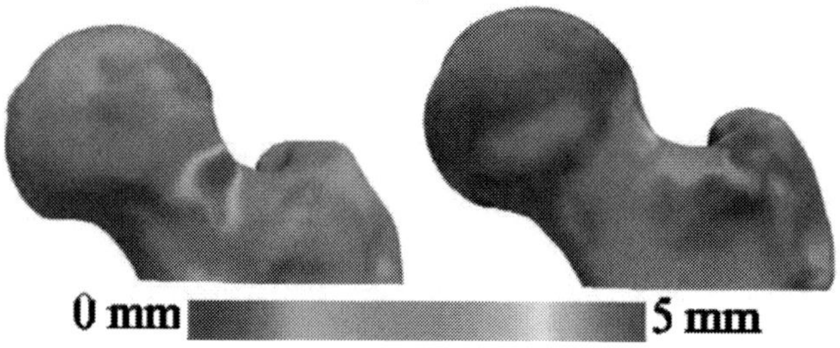

Figure 1. Mean cam (left) and control (right) shapes representing the mean cortical bone thickness for each subject cohort.

Thickness variations of the male populations were similar to the entire population with maximal thicknesses of 4.96 mm in the head-neck junction in cam patients greater than that of the male control femurs at 3.25 mm. The two shapes were also significantly different (T^2=50.9, p=0.00007) when comparing the first seven modes of variation, accounting for 90% of the total variance.

Analysis of alignment and automatic cropping planes showed that the transposed planes were within 9.9 degrees and 10.9 millimeters of the manually defined planes. In general, the transposed planes were located proximal of the manually defined plane. However, the resultant mean shapes did not appear to be affected by the plane placement other than the direct effects of variations in plane location (see Fig. 2).

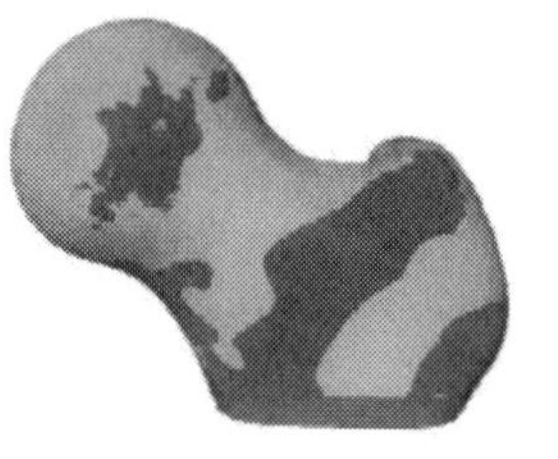

Figure 2. Overlay of semi-transparent mean shapes generated from manual planes (dark blue) and transposed planes (light blue). Note the additional length of femur (distal) from manually generated planes despite the agreement in overall shape.

DISCUSSION

In comparison to femurs of control subjects, cortical bone thickness in patients with cam FAI was increased in the anterolateral region of the femoral neck. Increased cortical bone thickness in these regions could be caused by bone remodeling in response to repetitive impingement, or aberrant bone from incomplete closure of the physeal plate. The characteristic bump observed on the mean cam shape is consistent with clinical observations. General shape variations between groups agreed with prior work [9].

Understanding the geometry of the trabecular and cortical layers of bone in patients with cam-type FAI could improve pre-operative planning by providing resection templates. The volume and distribution of cortical bone in the proximal femur of patients with cam-type FAI is crucial to understanding the implications of resection on the forces in the femoral neck. Determination of these metrics and presentation of anatomical variation using shape modeling for a larger population may reduce the need for extensive imaging prior to surgery.

The analysis approach used in this study could be extended to larger studies and other populations with hip pathology to improve our biomechanical understanding of the hip. The improvement in automatic alignment and cropping methods should streamline preprocessing steps in ShapeWorks. Automatic cropping reduces the manpower required to prepare large datasets that are typical for shape analysis. The final shape results do not seem to be affected within the accuracy of plane placement presented within. The methods used for sensitivity analysis are further extendable to evaluating the effectiveness of preprocessing methods to ensure that the dataset has been appropriately prepared for effective analyses.

ACKNOWLEDGEMENTS

NIH R01-EB01670, NIGMS 8 P41 GM103545-15 from the NIH, and LS Peery Discovery Program in Musculoskeletal Restoration.

REFERENCES

[1] LB Murphy, et al. *Osteoarthritis and Cartilage,* 18: 1372-9, 2010.
[2] R Ganz, et al. *Clinical orthopaedics and related research,* 466: 264-72, 2008.
[3] MJ Philippon, et al. *Am J Sports Med,* 35: 1918-21, 2007.
[4] AV Papavasiliou and NV Bardakos. *Bone Joint Res,* 1: 131-44, 2012.
[5] OR Ayeni, et al. *J Bone Joint Surg Am,* 93: e47, 2011.
[6] RM Mardones, et al. *J Bone Joint Surg Am,* 87: 273-9, 2005.
[7] E Rothenfluh, et al. *The American journal of sports medicine,* 40: 2002-8, 2012.
[8] G Holzer, et al. *Journal of Bone and Mineral Research,* 24: 468-74, 2009.
[9] MD Harris, et al. *J Orthop Res,* 31: 1620-6, 2013.
[10] J Cates, et al. *Inf Process Med Imaging,* 20: 333-45, 2007.

THE CHARACTERIZATION OF THE BONE MARROW MECHANICAL ENVIRONMENT USING POROELASTIC FINITE ELEMENT MODELS

Joshua A. Gargac (1), Thomas A. Metzger (1), Hansel Weihs (2), Tyler Kreipke (1), Glen L. Niebur (1,2)

(1) Bioengineering Graduate Program
University of Notre Dame
Notre Dame, IN, USA

(2) Aerospace and Mechanical Engineering Department
University of Notre Dame
Notre Dame, IN, USA

INTRODUCTION

Bone adapts its internal structure in response to external mechanical loads, allowing the body to maintain a robust, efficient skeleton. This adaptation is driven by cells that sense mechanical loads and signal adaptation cascades [1, 2]. However, the determination of the *in situ* mechanical environment of these cells is difficult, since bone has a complex geometry and the relevant features span multiple length scales. For example, little is known about the mechanical environment of the bone marrow, because it is encased in the opaque bone, making observation and measurement difficult. However, bone marrow is composed of many mechanically sensitive cells.

Finite element (FE) models can determine stress in complex geometries when generated from computed tomography (CT) images [3]. Poroelasticity theory can further be incorporated in such models to estimate the marrow mechanics during whole femur mechanical loading. The objective of this study was to investigate bone marrow mechanics during whole bone loading by (1) validating the ability of FE models to capture experimentally measured mechanical behavior of whole femora; (2) quantifying the effects of poroelasticity on whole bone mechanics, and (3) evaluating the mechanostimulatory potential of pressure and shear stress in the bone marrow.

METHODS

Five porcine femurs from five animals 6-8 months of age were procured at the time of slaughter (Martin's Custom Butchering, Wakarusa, IN). Testing was completed within 8 hours post-mortem to minimize storage effects on marrow. Miniature pressure transducers (Precision Measurements Inc., Ann Arbor, MI) were inserted in four holes drilled approximately 5 mm apart along the curvature of the femoral neck. Strain gage rosettes were fixed to the anterior and posterior aspects of the lesser trochanter for validation of finite element models. The instrumented femurs were loaded in cyclic compression, from approximately 600 to 2000 N sinusoidally at 1 Hz while submerged in 37°C water. Pressure, load, and strain data were monitored at 100 Hz (LabVIEW, National Instruments, Austin, TX). Strain gages on one femur failed during the experiment. The data were filtered using a 5 Hz third-order Butterworth filter.

Following testing, the femurs were imaged using a spiral *in vivo* μCT system (Bioscan, Paris, France) at 70 kVp, 100 mA, and reconstructed with a 0.143 mm cubic voxel size using a Hanning filter. Strain gages remained in place during scanning in order to identify their locations on the final models. The locations of the pressure transducers were identified from the drill holes in the μCT images.

The outer boundaries of the femurs were traced on each image slice using Analyze Visualization Workshop (AVW, BIR, Mayo Clinic) and imported into a general-purpose FE program (ADINA 9.0.1, Watertown, MA). A 10-node tetrahedral mesh was generated, and material properties for each element were assigned based on published modulus-density relationships for bone [3, 4]. Transverse orthotropic material properties were applied assuming the material aligns with the principal stress directions calculated from the first iteration of the FE models [5]. A poroelastic constitutive model was applied assuming the pore bone marrow is incompressible and that the pores in trabecular bone were fully saturated with marrow.

Permeability, *k*, was applied spatially based on a relationship between bone mineral density (BMD) and permeability determined from computational fluid dynamics models at three locations within the proximal femur. To account for marrow viscosity, the intrinsic permeability was divided by the measured viscosity of 85 Pa*s [6]. The sensitivity of the results to viscosity was tested in one model by applying a viscosity of 0.4 Pa*s [7]. Displacement boundary conditions were applied to mimic the experimental loading. A zero-pressure boundary was defined on the periosteal surface and at the distal end of the model. After the initial model, the density dependent elastic moduli and viscosity were scaled for each femur to match the whole bone stiffness measured experimentally [8].

The calculated pore pressures and principal strains at the transducer and gage locations were compared to the experimental results. Pressure gradients were calculated within each element of the models. Assuming pores were cylindrical, and oriented mainly in the flow direction, the drag force, f, was estimated (Eq. 1-3), where Q_A is the volumetric flow rate per a unit of cross-sectional area, v_p is pore velocity, and ϕ is the porosity. Marrow shear stress was estimated from these data by dividing drag force by trabecular spacing calculated from BMD using a relationship from μCT images of trabecular bone.

$$Q_A = \frac{k}{\mu}\frac{\partial P}{\partial x} \quad (1) \qquad v_p = \frac{Q_A}{\phi} \quad (2) \qquad f = 4\pi\mu v_p \quad (3)$$

RESULTS

The models were initially stiffer than the experiment, and were calibrated to the measured whole bone stiffness ($R^2 = 0.98$; Fig. 1A). Poroelasticity increased the stiffness of the FE models by 23% (±1.0%). The results agreed reasonably with the experimentally measured strain at the anterior surface (Fig. 1B; 1 femur excluded due to poor gage data).

Changing the viscosity did not affect the stiffness of the models, but the pressures scaled proportionally, decreasing by 99.5%. Scaling to the experimental data suggested the marrow viscosity was 8.5 Pa*s. After this calibration, the pressure was on average 1.6% (±83%) lower than the measured pressures at the pressure transducer centroids.

Across the samples at peak loading conditions, the median pore pressure in the metaphysis was 47 Pa with 20% of the volume having pressures above 200 Pa (Fig. 2). On average, the median shear stress in the metaphysis was estimated as 0.18 Pa, with 28% of the volume having shear stresses greater than 0.5 Pa.

DISCUSSION

The results of this study show the poroelastic constitutive model can reasonably calculate strain and pressure values when used in specimen specific FE modeling. The incorporation of poroelasticity not only affects the overall stiffness of the models, it allows for the characterization of the mechanical environment of the bone marrow. While pressures were below the mechanostimulatory threshold of 200 kPa for mesenchymal stem cells, the estimated shear stress was well above this threshold of 0.5 Pa [9], suggesting that physiological activity may stimulate a mechanoregulatory response in the bone marrow. The elevated shear stress magnitudes agree well with previous computational models of marrow flowing through trabecular bone [10-12], which predicted marrow shear stresses exceeding 0.5 Pa. However, these results rely on a highly simplified model.

The use of a single density-modulus relationship is often unable to predict surface strains and deformation in models of long bones [8]. The density-modulus relationship was used to create a modulus distribution, and then both modulus and viscosity were calibrated for each sample to match the loading measurements. The measured pressures suggest the viscosity of marrow is much greater than 1 Pa*s.

Previous studies have employed poroelasticity to account for interstitial fluid around osteocytes in cortical bone [13], while this study modeled the interaction of the bone marrow fluid phase within trabecular bone at the mesoscale. The development of this model can be used to provide boundary conditions to fluid structure interaction models of bone marrow flowing through trabecular bone as part of a multi-scale model of bone.

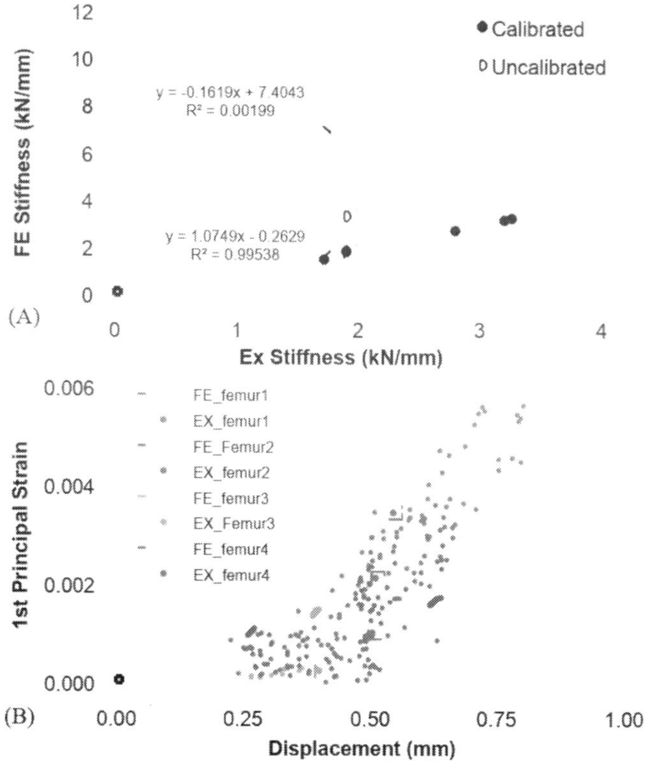

Fig. 1: FE model validations to experimental values of whole bone stiffness (A), and 1st principal strain at the gages (B)

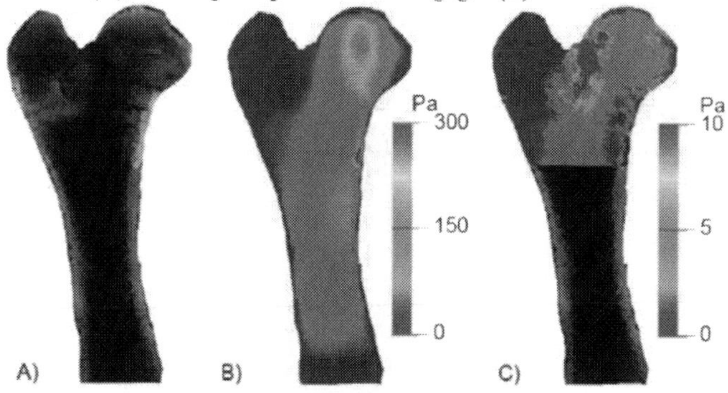

Fig. 2: (A) Tissue grayscale (B) pore pressure, and (C) shear stress.

ACKNOWLEDGEMENTS

NSF-1100207

REFERENCES

1. Duncan R et al., *Calcif Tissue Int,* 57(5):344-58, 1995.
2. Papachristou D et al., *Bioessays*, 31(7):794-804, 2009.
3. Taddei, F et al., *Med Eng Phys*, 26(1): 61-9, 2004.
4. Wirtz, D et al., *J Biomech*, 33(10): 1325-30, 2000.
5. Marangalou J et al., *Bone*, 60:213-220, 2014.
6. Metzger T et al., *JMBBM*, 40:307-13, 2014.
7. Bryant J et al., *Proc Inst Mech Eng H*, 203(2):71-5, 1989.
8. Eberle S et al., *Med Eng & Phys*, 35(7):875-883, 2013.
9. Luu Y et al., *J Bone Miner Res*, 24(1):50-61, 2009.
10. Coughlin T, et al., *J Biomech*, 45(13): p. 2222-9, 2012.
11. Metzger T et al., *J Biomech Eng*, 137(1), 2015.
12. Birmingham E et al., *Ann Biomed Eng*, 41(4):814-26, 2013.
13. Knothe Tate M et al., *J Biomech*, 33(2):247-54, 2000

EMBEDDED FINITE ELEMENTS FOR MODELING TRAUMATIC AXONAL INJURY

Harsha T. Garimella (1) and Reuben H. Kraft (1)

(1) Department of Mechanical and Nuclear Engineering
The Pennsylvania State University
University Park, PA, USA

INTRODUCTION

Many successful attempts have been made to include the anisotropy of the axonal structure of the brain, into the constitutive models for a more realistic and accurate finite element simulations [1]. But all the studies so far have used an averaging technique or axonal orientation density functions to incorporate the axonal orientations into the head finite element models and none of the models have dealt with fiber crossings. With the recent advancement of Diffusion Tensor Imaging (DTI) and Diffusion Spectrum Imaging (DSI) and other rapidly developing neuroimaging modalities, there remains a need to develop more complex anisotropic constitutive models involving intravoxel heterogeneity of axonal fibers (presence of two or more families of fibers). However, using the previously established approaches, the underlying mathematical formulation in finite element modeling increases in its complexity as more and more fibers are taken into account. In this paper, we employ the 'embedded finite element method' to represent the white matter fibers in the finite element model of the brain. This method serves two advantages. First and foremost is the use of an explicit finite element representation of tractography, which helps us obtain strains at axonal level that can be used into study DAI. Secondly, it provides a means to incorporate the anisotropic properties. This method enables the modeling of unlimited number of fibers with great potential to model structure and function with in the brain.

METHODS

Here we describe a method called 'embedded element method' to introduce anisotropic detail into the human brain explicitly that can handle fiber crossings. A suite of medical imaging and software tools are used to obtain an individual specific embedded element model. Figure 1 explains the complete process followed in the presentation. Figures 1(a) and 1(b) shows the MRI and DTI scans of the female soccer player obtained from the Penn State Center for Sports Concussion Research and Service. Figure 1(c) shows the parcellation and segmentation of the 3D volumes rendered from the 2D scans into different anatomical components. Figure 1(d) shows the diffusion tensor tractography model obtained from the DTI scans. Figures 1(e) and 1(f) shows the finite element models of human head and tractography, respectively. Figure 1(g) shows the incorporation of the tractography into the human head finite element model and the respective finite element analysis under impact loading conditions.

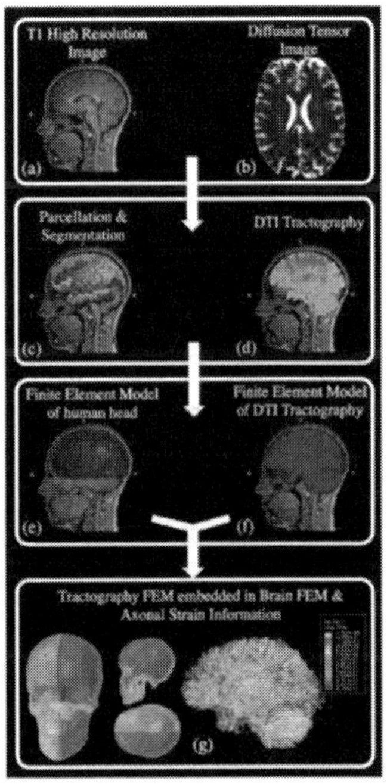

Figure 1. Process Flowchart adopted in this paper.

Over the past decade, a number of embedded representations for reinforcements have been published [2]. This approach is desirable when used to achieve advantages of a regular mesh and at the same time modeling complex reinforcement details [3]. In this model, the reinforcement is considered an axial member built into the host element in such a way that its displacements are compatible with those of surrounding elements (matrix). This approach does not constrain the number of fibers that can be embedded into a given finite element model. This also helps us eliminate the mathematical complexity that is generally involved in defining the transverse isotropic mathematical equations with multiple fibers. This technique is used to specify that group of elements is embedded in host elements. If the node of an embedded element lies within a host element, the node becomes an "embedded node". The translational degrees of freedom are constrained to the interpolated values of the corresponding degrees of freedom of host element. Here we are using an embedded truss element to represent the axonal fibers in the matrix.

RESULTS

The human head model contains the main anatomic components of scalp, skull, CSF, brain and axonal fibers. The undeformed and deformed configurations of human head as well as the axonal structure have been shown in the Figure 2. Confidence in the human head finite element model has been established by comparison of these models to standard Nahum and Hardy experiments for pressure and displacement validations, which will be discussed in the presenettion. This procedure eliminates the need for a separate attention to incorporate the orientation of axonal fibers into the human head finite element model through averaging methods.

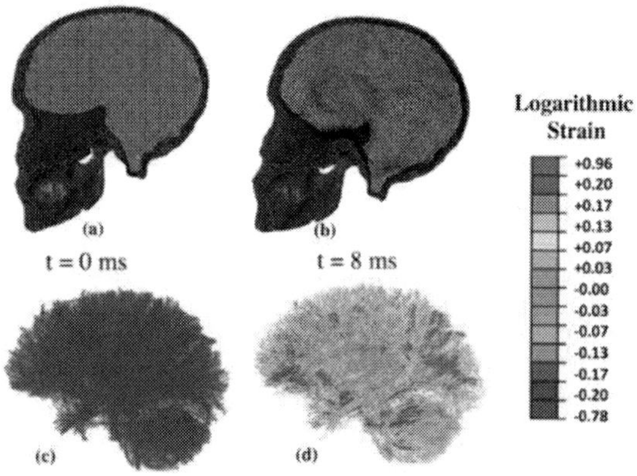

Figure 2. Figure showing the deformation in the brain and the corresponding localized axonal strains as shown in the figure.

DISCUSSION

It has been shown that axonal strain is a mechanism that leads to functional damage of neurons, which can be used as a measure for the onset of injury in computational models. The advantage of using such a measure is that it offers a method of coupling the cellular mechanisms of damage with macroscale mechanical loading. In all the studies, so far the main problem with using the axonal strain as the injury measure is the lack of a standard procedure to extract axonal strains in their respective orientations. Integrating an axonal structure with the exact orientation into a finite element model of human head has been a major problem in computational modeling. Different averaging techniques as well as orientation density functions at both model level and element level have been used previously to account for the anisotropic orientation of the axons. With the method described here, we can solve the problem of extracting the axonal strains without involving any of the mathematical complexity described in the previous literature. The main advantage of the embedded element method is that it enables the use of an explicit representation of the fiber network present in the brain and offers no constraints to the number of embedded fibers, i.e., any number of fibers can be embedded into the model, thus making this approach robust and a ideal method to implement diffusion spectrum imaging tractography (presence of fiber crossings) as well.

In the development of the multi-scale modeling framework, assumptions were made to simplify the analysis. A no slip assumption between the interface and the matrix has been made. Assuming slip between the matrix as well as the fiber would give a more accurate strains and the presence of slip also causes shearing of the axons, which is also a primary mechanism in the axonal strain injury. The white matter is assumed to be hyperelastic in nature. Viscoelasticity was not included in the finite element models. It was acknowledged that to accurately model the damage, the constitutive model should be extended to account for the time dependent behavior of the brain tissue. The exclusion of the viscoelasticity may affect the outcome of our results, leading to smaller shear strains and thus less predicted damage. Future efforts are focused on improving the mechanical description of the brain tissue. Another limitation of this analysis is that the mechanical properties as well as the injury thresholds of the axonal fibers are not known. The material properties of the brain are used to represent the axonal fibers in the model. To our knowledge, there has been no literature reporting the elastic and failure thresholds i.e., the material properties of axonal bundles. However, there are studies that attempt to estimate fiber bundle properties using magnetic resonance elastography, however these methods include artifacts from the surrounding tissue as well since the axonal tracts are not isolated. Another limitation is the use of axonal injury threshold. The thresholds for primary axotomy (the rupture or severing of axons) have been reported, as well as the induction of axonal pathology during mechanical loading, however, most studies have been performed at the single axonal level rather than at the bundle level. A comprehensive axonal strain analysis will be done in future and will be validated against the neuroimaging findings. In future, these models can be potentially used to develop structure - function models of the human brain.

ACKNOWLEDGEMENTS

All the DTI/DSI data used here is being provided by Pennsylvania State University Center for Sports Concussion Research and Service, University Park, USA. The authors thank Dr. Sam Slobounov and Dr. Brian D. Johnson for the data provided. This work was supported in part through instrumentation funded by the National Science Foundation through grant OCI– 0821527.

REFERENCES

[1] R. H. Kraft, P. J. Mckee, A. M. Dagro, and S. T. Grafton, "Combining the Finite Element Method with Structural Connectome-based Analysis for Modeling Neurotrauma: Connectome Neurotrauma Mechanics," PLoS Comput. Biol., vol. 8, no. 8, p. e1002619, Jan. 2012.

[2] D. V. Philips and O. C. Zeinkiewicz, "Finite element non-linear analysis of concrete structures," Proceeding Inst. Civ. Eng., vol. 61, no. 3, pp. 59–88, 1976.

[3] A. E. Elwi and T. M. Hrudey, "Finite Element Model for Curved Embedded Reinforcement," J. Eng. Mech., vol. 115, no. 4, pp. 740–754, Apr. 1989.

SB³C2015
Summer Biomechanics, Bioengineering and Biotransport Conference
June 17-20, 2015, Snowbird Resort, Utah, USA
SB³C2015-1118

ROLE OF CELL-CELL INTERACTION IN TENSIONAL HOMEOSTASIS

Alicia J. Zollinger (1), Elizabeth P. Canović (2), Michael L. Smith (1), Dimitrije Stamenović (1)

(1) Department of Biomedical Engineering
Boston University
Boston, Massachusetts

(2) Department of Material Science and Engineering
Massachusetts Institute of Technology
Cambridge, Massachusetts

INTRODUCTION

Past works have shown that adherent cells have a preferred level of tension, and will act on their environment to ensure that this tension is maintained at a preferred set point value [1, 2, 3]. This process was termed tensional homeostasis and is considered a biological necessity in many tissue types, particularly in the endothelium. The breakdown of tensional homeostasis has been related to the progression of cancer and athersclerosis [2, 3, 4]. It has been suggested that tensional homeostasis must exist across a broad range of length scales, spanning from the subcellular level to the tissue level [2]. Two recent studies challenged this idea, however.

In one study it has been shown that the traction field of isolated living endothelial cells have erratic, dynamic fluctuations over a long observation time [5]. An additional study showed that in response to mechanical deformation, the contractile force of isolated living fibroblasts does not return to a preferred homeostatic level [6]. Taken together, these studies sugested that isolated cells do not have the ability to maintain tensional homeostasis and that multicellular cooperation may be needed for the maintenance of tensional homeostasis in tissues.

In this study we used micropattern traction microscopy (MTM) to study dynamic fluctuations in the traction field of bovine aortic endothelial cells organized in clusters of varying sizes over a period of two hours. Our results showed that the traction fluctuations decreased with increasing cluster size and increasing strength of the traction field.

METHODS

Cell Culture Bovine aortic endothelial cells were cultured in Dulbecco's Modified Eagle media with 1 g/L glucose (Corning) supplemented with 10% bovine calf serum and 1% penicillin and streptomycin. Cells were maintained in a sterile incubator at 37°C and 5% carbon dioxide. Cells were grown to 90% confluence before trypsinization and used during passages 5-14. Cells were seeded onto micropatterned polyacrylamide gels and allowed to adhere for 18 hr before imaging.

Micropatterning and Polyacrylamide Gel Fabrication Fibronectin was isolated from human blood plasma (Valley Biomedical) and labeled with AlexaFluor 488 (Invitrogen) as previously described. Labeled fibronectin was then used to micropattern polyacrylamide gel substrates (2 μm diameter dots at 6 μm center-to-center separation) using soft lithography and an indirect patterning technique. Polyacrylamide gels were created with 0.13% bisacrylamide in order to yield an elastic modulus of 6.7 kPa and a Poisson's ratio of 0.445. Gels were ~700 μm thick [7].

Imaging All fluorescent, brightfield and DIC images were taken over the course of two hours using an Olympus IX81 microscope with a 40× water objective (1.15 NA, 0.25 mm WD) and a Hamamatsu Orca R2 camera controlled with Metamorph software. The microscope is outfitted with an environmental chamber that maintained the sample at 37°C and 5% CO_2. After traction imaging, a fluorescent nucleus stain was used to confirm cluster size.

Image Processing Image processing was done using a custom script written in Matlab. In order to calculate traction forces, the program finds the displacement of each micropatterned dot and uses gel elastic properties to calculate the total force applied to the dot (Fig 1). Each cell or cluster was imaged every 5 min for 2 hr resulting in a total of twenty-five time points. By summing the magnitude of all traction forces applied at a given instant to a given cell, we calculated the net traction and obtain timelapses of the net traction [7].

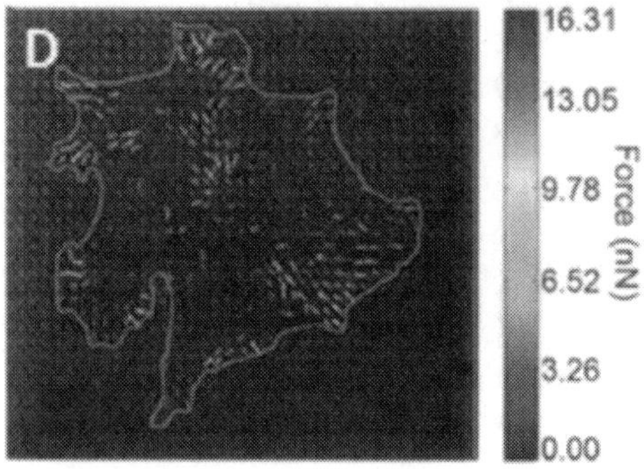

Figure 1: Tractions of a ten cell cluster using MTM. Unit of color bar is nN

Normalized Standard Deviation (NSD) To quantitate temporal fluctuation of the net traction we calculated the NSD as follows

$$NSD = \frac{1}{x_o}\sqrt{\frac{1}{n}\sum_{i=1}^{n}(x_i - \langle x \rangle)^2} \qquad (1)$$

where n is the number of data points, x is a time-varying signal, $\langle x \rangle$ is the time average of that signal, and x_0 is the initial value of the signal.

RESULTS

We observed a significant decrease in NSD of cell net traction (Fig. 1) with increasing cluster size ($p < 0.023$). These data suggested that the clusters of increasing cell numbers promoted tensional homeostasis. We also found that regardless of the cluster size, NSD decreased with increasing time average of the net traction ($p < 0.002$).

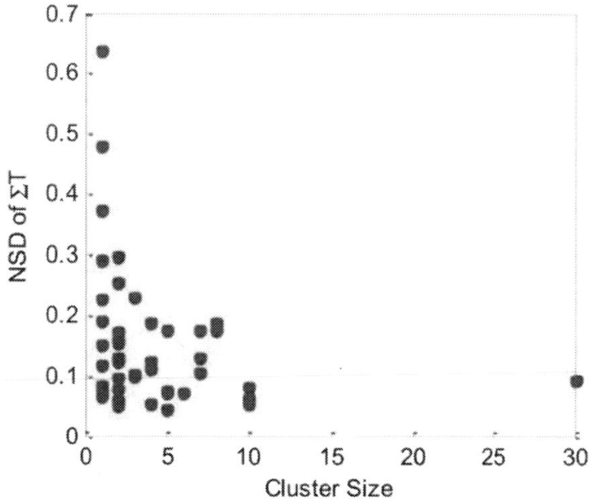

Figure 2: Normalized standard deviation (NSD) of the net traction plotted versus the number of cells in each cluster. As cluster size increased, NSD decreased.

We hypothesized that the observed decrease in NSD with increasing cluster size might be either due to statistical averaging (i.e., random fluctuation in a sample decrease with the square of the sample size), or due to cell-cell interactions. To investigate these possibilities, we combined eleven single cells into clusters of varying size and measured the NSD of the net traction of these groups of cells. We found a decrease in the NSD that was qualitatively similar to the relationship in Fig. 2. However, in the case of non-interacting cluster, the NSD vs. the average net traction had an inverse square relation, whereas interacting clusters exhibited a power-law relationship. Furthermore, we performed a Kolmogorov-Smirnov test on the NSD of the ten cell clusters and the ten grouped cells and found that the test rejected the hypothesis that the distributions came from the same population. Taken together, these results suggested that the observed decrease in NSD with increasing cluster size was not result of statistical averaging and that cell-cell coupling might be important for maintaining tensional homeostasis.

DISCUSSION

The two most significant findings of this study were i) that isolated endothelial cells did not maintain tensional homeostasis, which was consistent with previous observations in isolated fibroblasts [6], and ii) that multicellular clusters promoted homeostasis with increasing cluster size, which was a novel finding. These findings suggest that homeostasis does not exist across multiple length scales as purported previously, but that it may require a higher level of organization than that of a single cell.

We explained our findings as interplay of stochastic processes and cell-cell interactions. However, to obtain a comprehensive description of tensional homeostasis, future studies should consider the effects of mechanical interdependence between cells, cross-signaling between cell-cell and cell-matrix adhesions, rigidity sensing, and matrix composition.

ACKNOWLEDGEMENTS

This project was funded by the Training Program in Biology and Physiology as well as NSF grants CEBET-1150467 and CMMI-1362922.

REFERENCES

[1] Brown, RA et al., *J Cell Physiol*, 175:323-332, 1998
[2] Humphrey, JD et al., *Cell Biochem Biophys*, 50:53-78, 2008
[3] Chien, S et al., *Am J Physiol Heart Circ Physiol*, 292:H1209-24, 2007
[4] Paszek, MJ et al., *Cancer Cell*, 8:241-254, 2005
[5] Krishnan, R et al., *Am J Physiol Cell Physiol*, 303:C368-75, 2012
[6] Webster, KD et al., *Biophys J*, 107:146-55, 2014
[7] Polio, S et al., *Acta Biomaterialia*, 8:82-88, 2012

SB³C2015
Summer Biomechanics, Bioengineering and Biotransport Conference
June 17-20, 2015, Snowbird Resort, Utah, USA
SB³C2015-1119

QUANTIFYING HEMODYNAMICS AND WALL MECHANICS IN PATIENT-SPECIFIC CORONARY ARTERY BYPASS GRAFTS

Abhay B. Ramachandra (1), Andrew Kahn (2), Alison L. Marsden (1)

(1) Department of Mechanical and Aerospace Engineering
University of California San Diego
La Jolla, CA, USA

(2) Department of Medicine
University of California San Diego
La Jolla, CA, USA

INTRODUCTION

Coronary artery bypass graft (CABG) surgery revascularizes diseased coronary arteries and is among the most common major hospital-based procedures in the United States. Most CABG patients require multi-vessel revascularization and arterial grafts, despite having superior long-term patency, are limited in their availability. Hence approximately 70% of CABG procedures use saphenous vein grafts (SVG). Vein graft failure post CABG surgery is a critical clinical problem with failure rates of more than 50% within ten years. An increasing number of CABG patients also require repeat procedures, which carry higher morbidity and mortality rates. As a result, vein graft failure continues to pose a significant clinical and economic burden.

Hemodynamics, wall mechanics and stresses are known to contribute to cellular and structural changes in the vessel wall and are hypothesized to be integrally linked to SVG failure. Existing noninvasive clinical measurement techniques are incapable of directly measuring clinically relevant quantities such as wall shear stress (WSS), wall shear stress gradient (WSSG) and oscillatory shear index (OSI). Here, using advancements in medical imaging, patient-specific coronary modeling [1] and high performance computing, we develop a novel image-based coronary simulation framework to non-invasively characterize the mechanical stimuli acting on bypass grafts, including detailed local hemodynamics and vessel wall mechanics. We compare the biomechanical conditions between venous and arterial grafts.

Quantifying these stimuli is a necessary step to fill an unmet clinical need for early identification of patients at increased risk of developing SVG obstruction.

METHODS

We used a customized version of the open source software SimVascular (www.simvascular.org) and built 3D patient-specific models from CT images collected for clinically indicated purposes with IRB approval. We meshed the geometry using the commercial software MeshSim (Simmetrix Inc., NY).

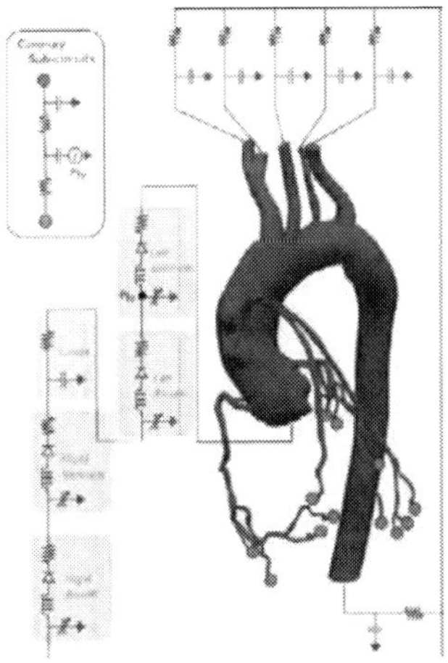

Figure 1: Schematic of multiscale lumped parameter network for closed loop simulations, including heart and coronary models.

We employed novel coupling algorithms for multiscale modeling with fluid structure interaction, in which the governing Navier-Stokes equations and Cauchy's equations are solved in the 3D domain using stabilized finite elements, and coupled to a closed loop circulatory model in the 0D domain (Figure 1) [2]. A custom linear solver was used to improve efficiency, employing a specialized preconditioning method that accounts for vascular resistance. The coupled Momentum Method (CMM) [3] was used for fluid structure interaction, and couples the equations of wall deformation at the variational level as a boundary condition for the fluid domain, keeping the computational mesh fixed during the simulation. The formulation uses a linear membrane model enhanced with transverse shear and has been previously validated against analytical and *in vitro* experiments with excellent agreement up to a radial deformation of approximately 10%. Variable wall thickness and material properties were assigned to different vessels in the CABG model using material constants of 0.27 MPa for aortic vessels, 1.15 MPa for the coronaries, 1.4 MPa for the arterial grafts and 5 MPa for the venous grafts (Figure 2).

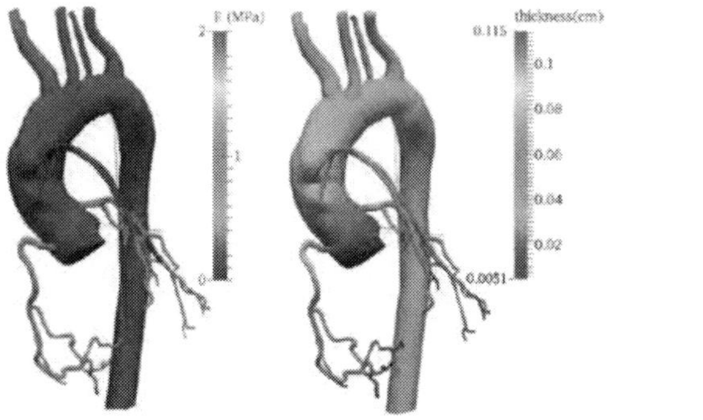

Figure 2: Varying material properties (Young's Modulus, left) and thickness (right) across vessel walls in a patient-specific CABG model.

The multiscale modeling approach allows one to obtain detailed 3D data in regions of interest (e.g. bypass graft hemodynamics) and global circulatory parameters (e.g. ventricular pressure volume loops) in a fully coupled dynamic fashion. The lumped parameter network (LPN) incorporates blocks to model the heart, upper and lower body, pulmonary vascular bed, and coronary arteries, as shown for a CABG example in Figure 1. The LPN parameters are systematically tuned to match non-invasive patient specific clinical targets, such as blood pressure, cardiac output, and ejection fraction.

RESULTS

Simulations were run for three patient specific CABG models. The framework reproduced the clinically observed behavior of coronary flow waveforms, which is out of phase with systemic flow. The flow and pressure waveforms are emergent from the coupled 3D-0D model and are an improvement over existing practices of prescribing flow and pressure at inlets and outlets of the coronary geometry. Cardiac cycle averaged WSS, calculated from closed loop rigid wall simulations for 3 patients (n_{venous}= 5, $n_{arterial}$=3) is noticeably low in the venous grafts, compared to arterial grafts. The normalized area of the graft where cycle averaged WSS is less than the pathological limit of 4 dyn/cm^2, was noticeably higher in the venous grafts. These findings are consistent with long-term adaptations observed in the clinic in which venous grafts are more prone to atherosclerosis than arterial grafts.

DISCUSSION

Results demonstrate noticeable differences in hemodynamics between venous and arterial grafts in patient-specific models. Motivated by these preliminary results, current efforts are aimed at simulating additional patients with variable wall properties and fluid-structure interaction.

With the current patient-specific framework provides a comprehensive quantification of mechanical stimuli acting on arterial and vein grafts in patient specific models, and also lays the groundwork for early identification and risk stratification of patients at increased risk of developing SVG obstruction. This framework is a necessary step towards future work linking fluid-solid interaction (FSI) simulations with vein graft growth and remodeling to create the patient-specific fluid-solid-growth (FSG) framework to predict graft failure mechanisms in CABG patients.

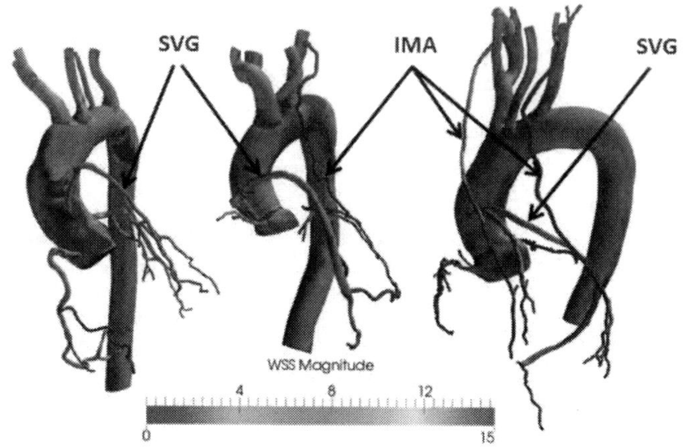

Figure 3: Preliminary results showing differences in cycle averaged WSS distribution between internal mammary artery (IMA) and saphenous vein graft (SVG) hemodynamics in a cohort of three patients simulated with rigid walls and closed loop boundary conditions.

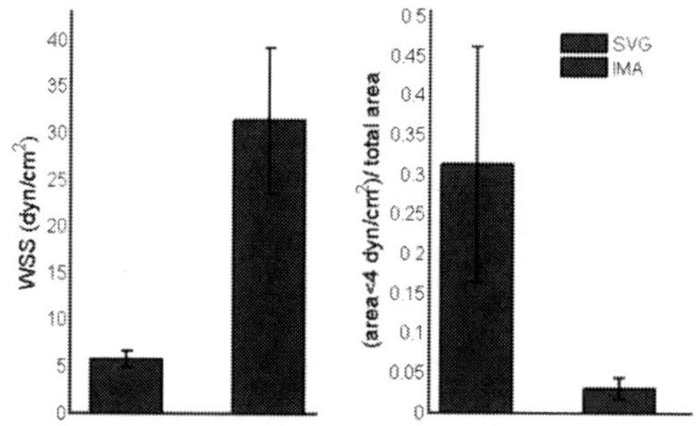

Figure 4: Cardiac cycle averaged quantities averaged across venous (n=5) and arterial (n=3) grafts. Error bars indicate +/- standard deviation.

ACKNOWLEDGEMENTS

Funding was provided by NIH R01-RHL123689A, a Burroughs Wellcome Fund Career Award at the Scientific Interface, and NSF CAREER OCI 1150184. We thank Weiguang Yang for his help with implementation of variable wall property methods into SimVascular.

REFERENCES

[1] Kim, H. J. et al, *Ann. Biomed. Engg*, 38:3195-3209, 2010.
[2] Sankaran, S et al., *Ann. Biomed. Engg*, 40:2228-2242, 2012.
[3] Figueroa, C. A. et al., *CMAME*, 195: 5685-5706, 2006.

SB³C2015
Summer Biomechanics, Bioengineering and Biotransport Conference
June 17-20, 2015,Snowbird Resort, Utah, USA
SB³C2015-1120

VALIDATION OF HIGH RATE STRIP BIAXIAL TENSION DEFORMATIONS OF THE NEURONAL PHOSPHOLIPID BILAYER USING EMPIRICAL DATA

M. A. Murphy (1,2), M. F. Horstemeyer (2,3), Steven R. Gwaltney (4,5), Tonya W. Stone (2,3), Michelle C. LaPlaca (6), Jun Liao (1,2), Lakiesha N. Williams (1,2), R. Prabhu (1,2)

(1) Department of Agricultural and Biological Engineering
Mississippi State University
Mississippi State, MS, USA

(2) Center for Advanced Vehicular Systems
Mississippi State University
Mississippi State, MS, USA

(3) Department of Mechanical Engineering
Mississippi State University
Mississippi State, MS, USA

(4) Center for Computational Sciences
Mississippi State University
Mississippi State, MS, USA

(5) Department of Chemistry
Mississippi State University
Mississippi State, MS, USA

(6) Department of Biomedical Engineering
Georgia Institute of Technology
Atlanta, GA, USA

INTRODUCTION

Approximately 1.7 million new cases of traumatic brain injury (TBI) occur every year in the United States, which result in both human and economic losses [1]. Yet, much is still unknown regarding this common injury. One key component in understanding TBI is studying the cell membrane and its primary constituent, the phospholipid bilayer. It has been shown in both experimental [2] and computational [3] studies that the phospholipid bilayer is rate dependent with regards to the stress response. This trend is simple to see, but validating results quantitatively is more difficult. Experimentally, phospholipid bilayers lyse at low surface tensions, even when stiffened with cholesterol[4], but it is often difficult to replicate real world parameters in the lab. One of these parameters often includes strain rate, with biological experiments being limited to relatively low rates when compared to testing for non-biological materials. Conversely, molecular dynamics must use extremely high strain rates, resulting in a situation where simulation and experimental results cannot be directly compared. These differences require that estimates based on experimental data and equations be utilized for comparison, such as those derived by Evans et al. [2] for estimating the surface tension based on loading rate. The current study uses a fitting method and equation previously used for experimental deformations of phospholipid bilayers to fit data from simulated bilayer systems and show that the current data follows the same quantitative trends found in experimental studies.

METHODS

Seventy-two 1-palmitoyl-2-oleoylphosphatidylcholine (POPC) phospholipids and 9,070 TIP3P water molecules (original structure sourced from NIH's Laboratory of Computation Library [5] and 6,828 TIP3P water molecules were added) was equilibrated for five nanoseconds using the program LAMMPS [6] and the CHARMM36 all-atom lipid force field [5]. The structure was then replicated by a factor of two in the x dimension, resulting in a 144 phospholipid structure. Both the 72 and 144 phospholipid structure were equilibrated for another five ns. The equilibrated structures were subjected to constant velocity deformations in the x dimension (Table 1). The z dimension was allowed to relax freely during deformations. Failure was assumed to be when full water penetration occurred, resulting in a chain of water extending through both bilayer leaflets.

Table 1. Deformation Strain Rates and Loading Rates

Case	$\dot{\epsilon_x}$ (s^{-1})	R_γ (N·m·s^{-1})
72x1.0	2.0×10^8	5.2×10^8
72x2.0	4.1×10^8	7.9×10^8
72x3.0	6.1×10^8	8.6×10^8
72x4.0	8.2×10^8	1.4×10^9
72x5.0	1.0×10^9	1.9×10^9
144x1.9	2.0×10^8	4.5×10^8
144x3.8	4.1×10^8	7.9×10^8
144x5.7	6.1×10^8	1.2×10^9
144x7.6	8.2×10^8	1.3×10^9
144x9.5	1.0×10^9	1.6×10^9

The approach described in Evans et al. [2] was followed to determine appropriate constants for Equation 1:

$$\frac{\gamma}{\gamma_\delta} \cong ln\left(\frac{\mathcal{R}_\gamma}{v_{0\delta}} * \gamma_\delta\right) \quad (1)$$

Where γ is the rupture surface tension, γ_δ (N·m^{-1}) is a tension scale, $v_{0\delta}$ (s^{-1}) is the frequency for spontaneous nucleation of defects, and $\mathcal{R}_\gamma$

($N{\cdot}m^{-1}{\cdot}s^{-1}$) is the loading rate. Specifically, $\mathcal{R}_\gamma$ for each deformation case was found using the slope from the initial portion of the surface tension vs time curves (Table 1). Similarly, γ_δ values were determined from the slope of the peak and failure surface tension values vs $\ln(\mathcal{R}_\gamma)$ curve. The relation

$$\mathcal{R}_{\gamma=0} = \nu_{0\delta} * \gamma_\delta \quad (2)$$

was then used to determine $\nu_{0\delta}$ after solving for $\mathcal{R}_{\gamma=0}$ using the equation of the linear trend line from the surface tension vs $\ln(\mathcal{R}_\gamma)$ curve [2].

RESULTS

Strip biaxial deformations of the phospholipid bilayer resulted in a sharp initial peak in surface tension. The loading rate $\mathcal{R}_\gamma$ was determined for each simulation case as shown in Figure 1 and recorded in Table 1 above. The surface tension then gradually decreased and water penetration (failure) occurred at considerably lower surface tension values. Peak and failure surface tension values presented two unique curves (Figure 2); therefore, both were fit using the procedure described by Evans et al. [2]. The resulting fits were in good agreement with the data for both the peak and failure surface tension values with R^2 values of 0.97 and 0.90 for the peak and failure fits, respectively.

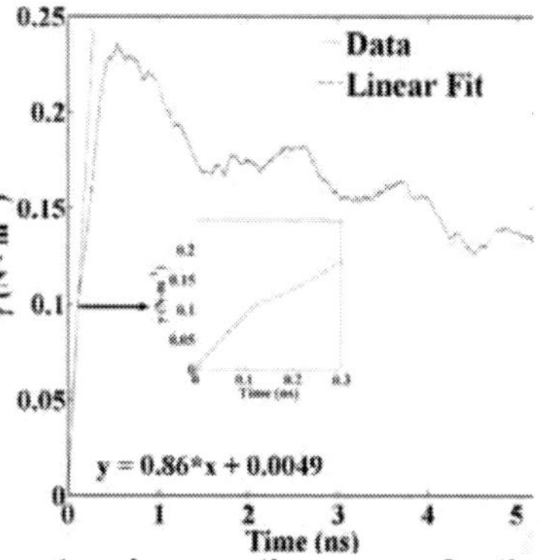

Figure 1: Surface tensions vs time curve for the 72x3.0 case with a linear fit of the linear portion of the curve (enlarged for clarity) shown for determining the loading rate $\mathcal{R}_\gamma$.

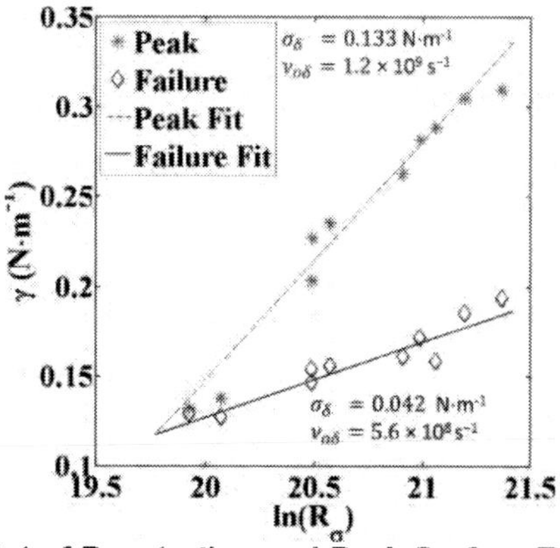

Figure 2: Plot of Penetration and Peak Surface Tensions vs $\ln(\mathcal{R}\gamma)$ and the resulting fits

DISCUSSION

We have shown in this study that the strip biaxial results agree qualitatively with previous studies and have determined appropriate constants for both the peak and failure surface tension values, showing that they are described correctly by Equation 1. While the constants found in the current study do allow for a good fit using the original equation, they differ considerably from those found in Evans et al. [2], who found that $\nu_{0\delta} = \sim 0.2\ s^{-1}$ and $\gamma_\delta = 4.0 \times 10^{-3}\ N{\cdot}m^{-1}$ for similar phospholipid types. This difference indicates that there may be more than one high rate regime than those originally tested experimentally and fit by Evans et al. [2]. However, the data still appears to follow the same trend, so there may be limited regimes of loading rates for which determined constants are valid. Physically, it is reasonable that these constants would change, because it is expected for the nucleation rate $\nu_{0\delta}$ to increase with the deformation rate. However, the reason why the constants do not change over a range of loading rates, leading to defined regimes, may not be as readily apparent. The authors believe these loading rate regimes may correspond to some physical phenomena. In this case, the constants found above may correspond to loading rates where the phospholipid bilayer cannot adequately rearrange to form the typical energetically favored form, resulting in the expansion of multiple pores that are less circular. This implies some point exists where the linear curve, on a semi-log plot, representing the deformation rate regime seen in the current study will change to another, similar linear curve with different constants until the regime (and constants) observed experimentally by Evans et al. [2] are found.

The idea of different regimes would explain the conflict between the observed trends in the surface tension curves in the current study (Figure 1) and those of other studies. For example, in Koshiyama and Wada [3], they found that during high rate equibiaxial tension the phospholipid bilayer rose to an elevated surface tension more slowly until failure occurred at the maximum surface tension, at which point the surface tension dropped quickly. Although their study also involved high rate deformations, the rates in the current study are slightly higher. In the plots for peak and failure surface tension (Figure 2), the two fits intercept at a point just under the rates currently examined. This interception may indicate that the regime change occurs in the loading ranges just under those tested currently and that the failure surface tension will be higher than the peak surface tension in this lower regime, consistent with the observations made by Koshiyama and Wada [3]. Assuming that the bilayer does behave differently in these regimes, this finding would indicate that the phospholipid bilayer, and, in effect, cell membranes, may have a much higher resilience to high rate impulse deformations than deformations performed experimentally at lower rates indicate and greatly affect current understandings of TBI.

ACKNOWLEDGEMENTS

This material is based upon work supported by the Department of Agricultural and Biological Engineering and the Center for Advanced Vehicular Systems at Mississippi State University. Further, it has also been supported by US Army, through a subcontract with Mississippi State University (W912HZ-13-C-0037), and was performed for the CRES-GV research program. Any opinions, findings and conclusions or recommendations expressed in this material are those of the author(s) and do not necessarily reflect the views of the U.S. Army ERDC ERC.

REFERENCES

1. Faul, M., et al., *Traumatic Brain Injury in the United States: Emergency Department Visits, Hospitalizations, and Deaths 2002-2006*, N.C.f.I.P.a. Control, Editor. 2010, Centers for Disease Control Prevention, National Center for Injury Prevention and Control: Atlanta (GA).
2. Evans, E., et al., Biophys J. 85:2342-50, 2003.
3. Koshiyama, K. and Wada, S., J Biomech. 44:2053-8, 2011.
4. Needham, D. and Nunn, R.S., Biophys J. 58:997-1009, 1990.
5. Klauda, J.B., et al., J Phys Chem B. 114:7830-7843, 2010.
6. Plimpton, S., J Comput Phys. 117:1-19, 1995.

OSTEOCYTE MECHANOTRANSDUCTION DURING LOW MAGNITUDE MECHANICAL STIMULATION

Thomas R. Coughlin, Tyler C. Kreipke, Glen L. Niebur

Bioengineering Graduate Program
University of Notre Dame
Notre Dame, IN, 46556
United States of America

INTRODUCTION

Physical activity affects bone mass by stimulating a cellular response. Osteocytes, the bone cells that reside within the mineralized matrix, are believed to sense bone deformations or induced fluid flow within their lacuno-canalicular network [1]. In contrast, low magnitude mechanical stimulation (LMMS) regimes, such as vibration, can increase bone formation in trabecular bone in the absence of significant matrix strain and, hence, osteocyte signaling [2, 3]. However, cells in the marrow may be responsible for mechanotransduction during LMMS, as various marrow cells are mechanosensitive [4], and vibration induces shear stresses in the marrow during vibration [5]. Alternatively, osteocytes may sense LMMS via some currently unknown mechanism.

Osteocytes regulate bone formation by multiple signaling proteins, including sclerostin. Sclerostin, encoded by the SOST gene, inhibits bone formation by antagonizing Wnt signaling and BMP expression, which are both critical for osteoblast differentiation and function [6]. Sclerostin is expressed in the absence of loading [7], while SOST downregulation is necessary for a bone formation response [8]. Mechanical loading decreases SOST expression and the number of sclerostin positive osteocytes *in vivo* [9]. Sclerostin is necessary to regulate bone formation, and mutations to the coding region of the SOST gene are the cause of high bone mass disorders such as Van Buchem disease.

The goal of this study was to determine whether sclerostin is regulated by osteocyte mechanotransduction during vibration. Bioreactors were used to culture trabecular bone explants and to subject osteocytes to LMMS. The aims of this study were to (1) determine the anabolic effects of LMMS in our culture system using micro-computed tomography (μCT) and (2) quantify the number of osteocytes expressing sclerostin after LMMS.

METHODS

Bioreactors were used to culture cylindrical trabecular bone explants. Explants were excised from porcine cervical vertebrae (Martins Meats, Wakarusa, IN) using a diamond tip core drill (Starlite Industries, PA). In the bioreactor, explants were supplied with a continuous flow of media via a peristaltic pump (89% HG DMEM, 10% FBS, 1% AB/AM, media was supplemented with 20 mM β-glycerol phosphate and 50 μM ascorbic acid-2-phosphate). LMMS was applied to treatment groups with an acceleration amplitude of 0.3 g at 30 Hz for 30 min, twice per day (Fig. 1).

To determine the effect of our culture and loading regime on explant bone adaptation, eight trabecular bone explants (sham, n=4; LMMS, n=4) were prepared from two pigs as described above. Sham and LMMS groups were cultured for four weeks as described above, 5 days/week with the SHAM group not subjected to vibration.

Explants were imaged prior to culture by μCT (Scanco μ-CT-80, Bruttisellen, Switzerland) at 20 μm isotropic voxel size, 70 kVp x-ray source at 114 mA, and 200 ms integration time in a sterile fixture. At the end of the experiment, the explants were imaged with the same protocol, and registered to the pre-culture scans to allow direct comparison between time points [10]. Architectural and volume fraction parameters were calculated for a 7.5 mm diameter cylindrical region of the explants [11]. At the end of culture, marrow was removed for tripotentiality assays.

Short-term 3D culture experiments were used to examine the effect of LMMS on sclerostin expression in osteocytes. Eighteen explants (control, n=6; sham, n=6; LMMS, n=6) were excised from two pigs. Control explants were fixed with 10% formalin on the day of slaughter for 48 hours. Sham and LMMS explants were cultured for 5 days in 3D bioreactors. LMMS explants were loaded twice/day for 30 mins for 4 days while SHAM explants were unloaded. Explants were fixed for 48 hrs after the 5th day of culture.

Fixed explants were demineralized with 10% w/v EDTA, processed, paraffin embedded, and sectioned longitudinally at 6 μm for immunohistochemistry. Sections were dried at 60°C, reacted for endogenous peroxidase activity, blocked, and incubated with anti-sclerostin primary antibody (Abcam). Secondary antibody labeling was achieved with the Vectastatin Elite ABC kit (Vector Labs, Inc) with diaminobenzidine (DAB) as the chromogen. Sections were counterstained with methyl green and coverslipped. Immuno-labeled sections were imaged at 500x and the number of sclerostin positive osteocytes was quantified.

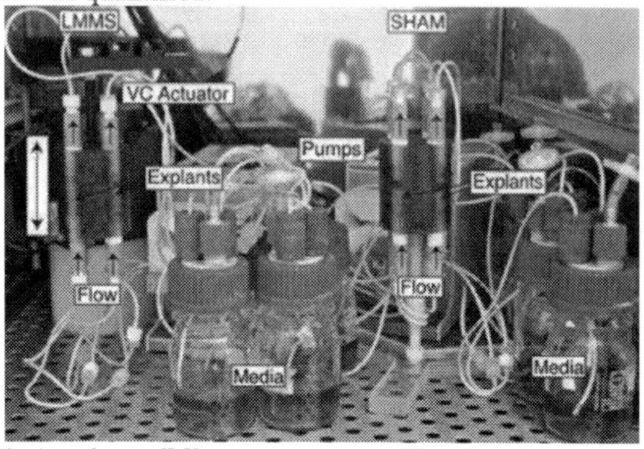

Figure 1. A voice coil linear actuator with sub-micron closed loop control applies LMMS to culture chambers. A peristaltic pump continuously circulates media from sample specific media reservoirs.

RESULTS

Cells extracted from the explants after four weeks of culture demonstrated tripotentiality in osteogenic, adipogenic, and chondrogenic assays (Fig. 2). On average, BV/TV decreased in both sham and LMMS groups (Fig. 3). All sham explants had a decrease in BV/TV while one LMMS explant had an increase in BV/TV (Fig. 3). Two LMMS explants had an increase in trabecular thickness (Tb.Th.), while all sham explants had a decrease in Tb.Th.

In the short-term study more than 60% of osteocytes were stained positive for sclerostin. There was no significant difference in the number of osteocytes expressing sclerostin in the SHAM or LMMS explants compared with controls (Fig. 4).

DISCUSSION

Osteocyte mechanotransduction due to LMMS was investigated in an explant culture model. The explants remained viable for the duration of the study indicated by viable MSCs and changes in bone turnover. The number of sclerostin positive osteocytes was similar in loaded and unloaded explants, suggesting that LMMS does not induce a mechanobiological response in osteocytes. However, we did not assess other osteocyte signaling pathways, such as RANKL and OPG. While sclerostin is a key regulator of WNT signaling, other signaling molecules are also important. Further, there was no difference in sclerostin positive osteocytes in control explants versus sham explants. Similarly, in hind limb suspended mice and control mice there was no difference in sclerostin levels [9].

One limitation of this study is that the amount of sclerostin expressed was not quantified. The number of osteocytes positive for sclerostin was measured, but the osteocytes may have a change in SOST gene expression that was not detected. For example, in a different study the number of sclerostin positive osteocytes did not demonstrate changes in SOST gene expression [9].

In contrast to previous results, where explants were cultured for 19 days [10], LMMS did not increase trabecular bone volume fraction compared with non-loaded controls. However, the duration of the study and the loading protocol were different than this study. Interestingly, our results demonstrate a large variation in bone turnover responses in loaded explants.

Strain induced fluid flow in the LCN is believed to be the primary mechanism driving osteocyte mechanotransduction. As the number of osteocytes expressing sclerostin did not change after LMMS, LMMS may be sensed by an alternate mechanism, such as the marrow, as these cells are highly mechanosensitive [4]. This study indicates that osteocytes may not play a role in mechanotransduction during LMMS.

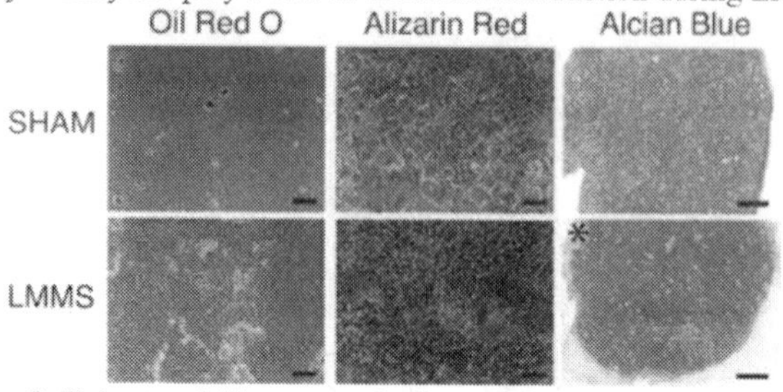

Figure 2. Tripotentiality assays indicated that MSCs were present in the marrow after four weeks in culture. * indicates that two of four samples did not stain intensely for Alcian Blue.

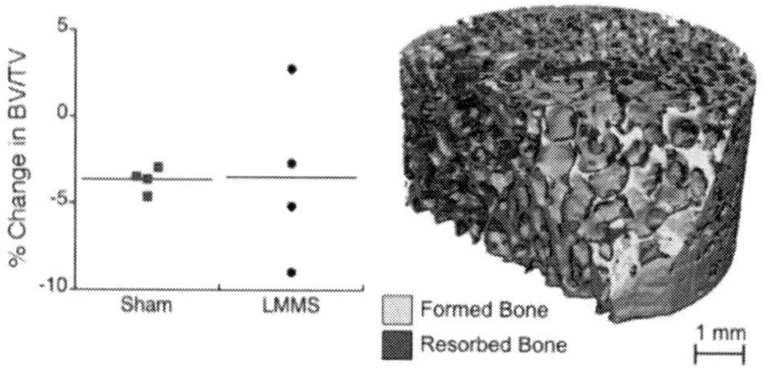

Figure 3. On average, bone volume fraction (BV/TV) decreased by 4% in both sham and LMMS groups after the 4-week experiment. Registered 3D μCT scan pre- and post-experiment.

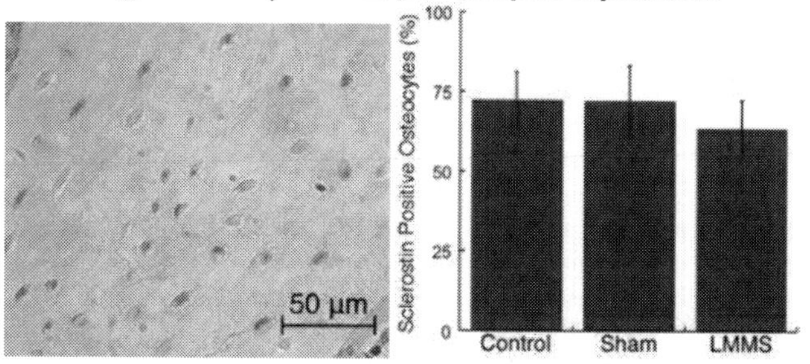

Figure 4. More than 60% of osteocytes were sclerostin positive in all groups (sclerostin=brown, nuclei=green).

ACKNOWLEDGEMENTS

NSF CMMI 1100207 and 1435467

REFERENCES

1. Weinbaum, *et al.*, 1994, J of Biomech 27(3):339-60. 2. Judex, *et al.*, 2007, J of Biomech 40(6):1333-1339. 3. Rubin, *et al.*, 2001, FASEB J. 15:2225-2229. 4. Castillo and Jacobs, 2010, Curr Osteo Reps 8(2):98-104. 5. Coughlin and Niebur, 2012, J of Biomech 45(13):2222-2229. 6. Bellido, 2013, Calcif Tissue Int. 7. Poole, *et al.*, 2005, FASEB J 19(13):1842-4. 8. Tu, *et al.*, 2012, Bone 50(1):209-17. 9. Robling, *et al.*, 2008, J Biol Chem 283(9):5866-75. 10. Birmingham, *et al.*, 2014, Ann Biomed Eng. 11. Schulte, *et al.*, 2011, Bone 48(3):433-42.

SB³C2015
Summer Biomechanics, Bioengineering and Biotransport Conference
June 17-20, 2015, Snowbird Resort, Utah, USA
SB³C2015-1123

TOWARD AN EXPERIMENTALLY VALIDATED IMMERSED BOUNDARY MODEL OF LEFT VENTRICULAR FLUID DYNAMICS USING IN VITRO EXPERIMENTS

Boyce E. Griffith (1), Jae Ho Lee (2), Pritiam Mekala (3), Arvind Santhanakrishnan (3)

(1) Departments of Mathematics and Biomedical Engineering and McAllister Heart Institute
University of North Carolina
Chapel Hill, NC, USA

(2) Department of Mathematics
University of North Carolina
Chapel Hill, NC, USA

(3) School of Mechanical and Aerospace Engineering
Oklahoma State University
Stillwell, OK, USA

INTRODUCTION

Cardiovascular disease continues to be the leading cause of death in the United States, and is now the leading cause of death worldwide. Several recent studies using clinical data [1-4] have shown the diagnostic potential of intra-ventricular fluid dynamics in characterizing cardiac health, and computational models of cardiac mechanics and fluid dynamics promise to provide physics-based approaches to patient risk stratification and treatment planning. Computer models of the heart also promise to enable cardiovascular device design, and in silico tests, which are beginning to be accepted by regulators as evidence of device efficacy and safety, also promise to accelerate device approval.

Before computational models can be widely deployed for pre-surgical or pre-interventional planning, or for medical device design or virtual testing, rigorous method verification and model validation of must be conducted [5-7]. As part of its Critical Path initiative, the US Food and Drug Administration is working to determine the role of computational fluid dynamics (CFD) in the regulation of cardiac devices. An initial FDA study compared [8] CFD predictions of flow through an idealized cardiovascular device to experimental particle image velocimetry (PIV) data of flow through the same device. These studies found large discrepancies between the predictions of computational fluid dynamics (CFD) simulations and high-fidelity experimental measurements, highlighting the importance of careful experimental validation prior to the clinical application of CFD and fluid-structure interaction (FSI) models and simulations.

This study presents initial work towards the development an experimentally validated computational model of left ventricular mechanics and fluid dynamics based on the immersed boundary (IB) method [9,10]. We use a flow phantom consisting of a flexible-walled physical model of the left ventricle (LV) to generate experimental data on intraventricular filling fluid dynamics using 2D phase-locked particle image velocimetry (PIV). A computational model corresponding to this in vivo experimental model has also been developed, and in ongoing work, we are applying an adaptive version of the IB method to simulate the dynamics of this in vitro LV model.

METHODS

Experimental An *in vitro* flow circuit provides physiological pulsatile flow through an optically clear, flexible-walled LV phantom. Particle image velocimetry (PIV) via a high-speed two-dimensional PIV system (LaVision GmbH, Goettingen, Germany) is used to obtain instantaneous and time-averaged flow data. Postprocessing uses DaVIS 8.0 (LaVision GmbH) to determine velocity fluctuations, viscous stresses, Reynolds stresses, and particle residence times, and to identify coherent flow structures.

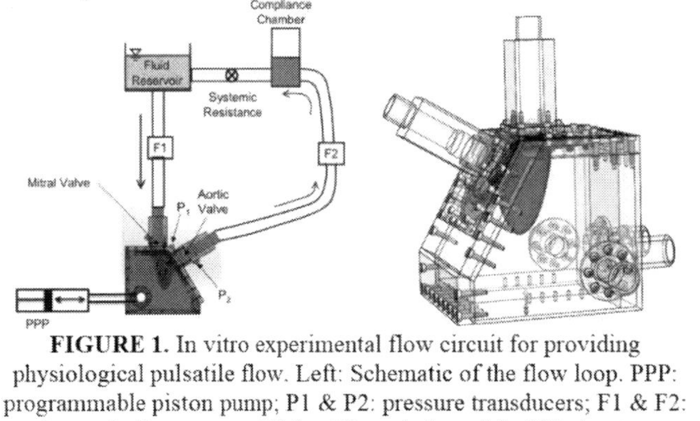

FIGURE 1. In vitro experimental flow circuit for providing physiological pulsatile flow. Left: Schematic of the flow loop. PPP: programmable piston pump; P1 & P2: pressure transducers; F1 & F2: ultrasonic flow meters. Right: 3D rendering of the LV phantom.

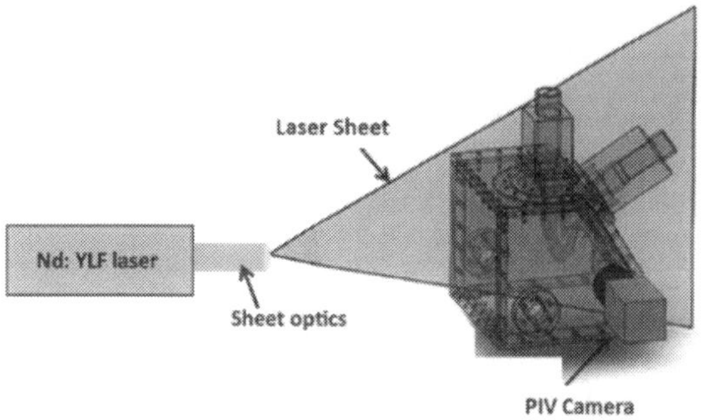

FIGURE 2. Schematic of the particle image velocimetry (PIV) setup for intraventricular flow studies. The flow field along the central LV plane (direction of flow indicated in blue) is visualized using a green laser sheet, as shown.

Computational The IB method describes the dynamics of an incompressible elastic body immersed in a viscous incompressible fluid. The IB formulation of the equations of motion uses an Eulerian description of the momentum, viscosity, and incompressibility of the coupled fluid-solid system along with a Lagrangian description of the deformation and elasticity of the structure. The equations are:

$$\rho \frac{\mathrm{D}\mathbf{u}}{\mathrm{D}t}(\mathbf{x},t) = -\nabla p(\mathbf{x},t) + \mu \nabla^2 \mathbf{u}(\mathbf{x},t) + \mathbf{f}(\mathbf{x},t),$$

$$\nabla \cdot \mathbf{u}(\mathbf{x},t) = 0,$$

$$\mathbf{f}(\mathbf{x},t) = \int_U \nabla \cdot \mathbb{P}(\mathbf{X},t)\, \delta(\mathbf{x} - \chi(\mathbf{X},t))\, \mathrm{d}\mathbf{X} - \int_{\partial U} \mathbb{P}(\mathbf{X},t)\, \mathbf{N}(\mathbf{X})\, \delta(\mathbf{x} - \chi(\mathbf{X},t))\, \mathrm{d}A,$$

$$\frac{\partial \chi}{\partial t}(\mathbf{X},t) = \int_\Omega \mathbf{u}(\mathbf{x},t)\, \delta(\mathbf{x} - \chi(\mathbf{X},t))\, \mathrm{d}\mathbf{x},$$

in which $\mathbf{x} \in \Omega \subset \mathbb{R}^3$ are physical coordinates, $\mathbf{X} \in U \subset \mathbb{R}^3$ are material coordinates, $\mathbf{N}(\mathbf{X})$ is the outward unit normal to ∂U, $\chi(\mathbf{X},t) \in \Omega$ is the physical position of material point $\mathbf{X}$ at time t, $\mathbf{u}(\mathbf{x},t)$ is the Eulerian description of the material velocity field, $p(\mathbf{x},t)$ is the Eulerian pressure, $\mathbf{f}(\mathbf{x},t)$ is the Eulerian elastic force density, $\mathbb{P}(\mathbf{X},t)$ is the first Piola-Kirchhoff stress tensor associated with the elastic body, and $\delta(\mathbf{x})$ is the three-dimensional Dirac delta function.

RESULTS

Experimental Initial experimental results from PIV studies of LV filling are shown in Figure 3.

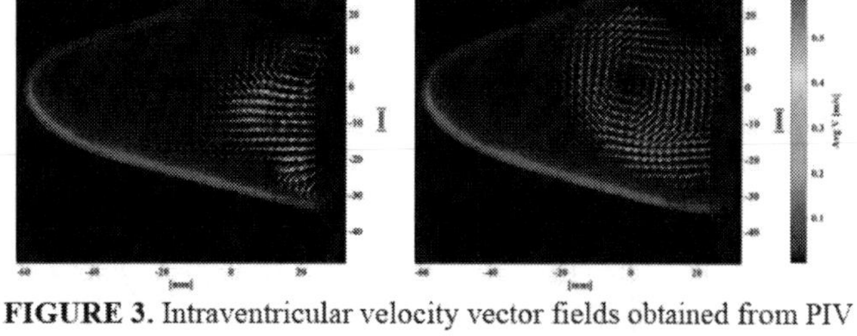

FIGURE 3. Intraventricular velocity vector fields obtained from PIV during early (left) and late (right) deceleration of early filling.

Computational At present, we have developed an initial computational model of the LV phantom, as shown in Figure 4. Simulation work using this model is ongoing, and we have performed preliminary simulations of early filling, as shown in Figure 5.

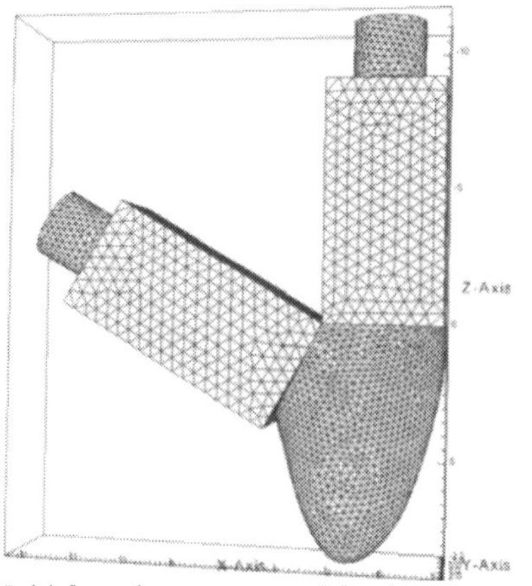

FIGURE 4. Initial meshes constructed for the IB model of the LV phantom, including inflow and outflow blocks.

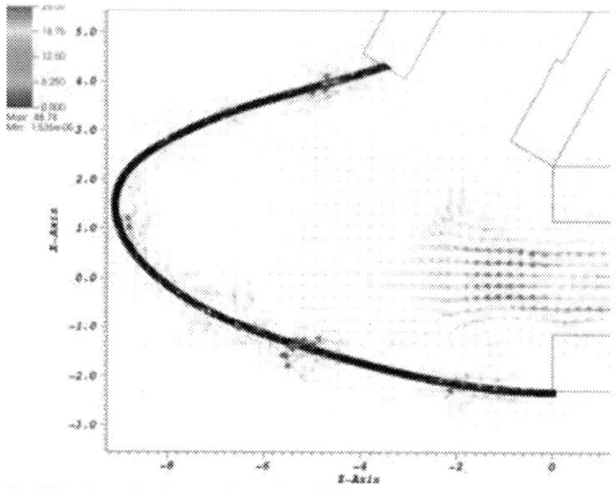

FIGURE 5. IB simulation of early filling using a flexible model of the *in vitro* experimental model of the LV and inflow and outflow blocks.

DISCUSSION

We are working to develop an experimentally validated computational model of the LV. To date, we have obtained quantitative flow data using PIV for an *in vitro* LV model, and we have developed a corresponding computational model of this *in vitro* model. Simulation work using this new computational model is ongoing.

ACKNOWLEDGEMENTS

We gratefully research funding from National Institutes of Health awards GM071558 and HL117063, National Science Foundation awards DMS 1016554 and OCI 1047734, and Oklahoma Center for the Advancement of Science and Technology award HR14-022.

REFERENCES

[1] Rodevand, O et al., *J Am Soc Echocardiogr*, 12:500-507, 1999.
[2] Hong, G et al., *JACC Cardiovasc Imaging*, 1:705-717, 2008.
[3] Eriksson, J et al., *Am J Physiol Heart Circ Physiol*, 300:H2135-2141, 2011.
[4] Poh, K et al., *Eur Heart J Cardiovasc Imaging*, 13:385-393, 2012.
[5] Taylor, C and D Steinman, *Ann Biomed Eng*, 38:1188-1203, 2010.
[6] Krittian, S et al., *Ann Biomed Eng*, 38:2676-2689, 2010.
[7] Kung, E et al., *Ann Biomed Eng*, 39:1947-1960, 2011.
[8] Stewart, S et al., *Cardiovasc Eng Technol*, 3:139-160, 2012.
[9] Peskin, C, *Acta Numer*, 11:1-39, 2002.
[10] Gao, H et al., *Int J Numer Meth Biomed Eng*, 30:1199-1222, 2014.

SB[3]C2015
Summer Biomechanics, Bioengineering and Biotransport Conference
June 17-20, 2015,Snowbird Resort, Utah, USA
SB³C2015-1124

MECHANICS OF INTERSTITIAL GROWTH IN ENZYMATICALLY DEGRADABLE HYDROGELS: CHARACTERIZATION OF DEGRADATION FRONT

Umut Akalp (1), Stacey C. Skaalure (2), Stephanie J. Bryant (2,3), Franck J. Vernerey (1)

(1) Mechanical Engineering
University of Colorado Boulder
Boulder, Colorado, USA

(2) Department of Chemical and Biological Engineering,
University of Colorado Boulder
Boulder, Colorado, USA

(3) Bio frontiers Institute, Material Science and Engineering Program
University of Colorado Boulder
Boulder, Colorado, USA

INTRODUCTION

Controlling biological growth within a cell-laden polymeric scaffold is a critical challenge in the tissue engineering community. Indeed, construct growth must often be balanced with scaffold degradation and is often coupled to varying degrees of deformation that originate from swelling, external forces and the effects of confinement. These factors have been shown to affect growth in many ways, but to date, our understanding is mostly qualitative. While cell sensing, molecular transport and scaffold/tissue interactions are believed to be important players, it will be critical to quantify, predict and control these effects in order to eventually optimize tissue growth in the laboratory. This study will introduce a mixture model in order to quantitatively understand scaffold-mediated mechanisms of transport, deposition, degradation and elasticity that affect the overall growth of a tissue subjected to finite deformations [1].

In this work, we will specifically focus on cartilage engineering in degradable PEG hydrogel scaffolds within which a population of cells (chondrocytes) releases a degrading enzyme and extra-cellular matrix (ECM). Our first objective is to investigate the role of enzymatic degradation on ECM diffusion and their impacts on two important outcomes: the extent of ECM transport (and deposition) and the evolution of the scaffold's mechanical integrity [2]. Knowing the characteristics of the degradation behavior that leads to successful tissue growth; we then incorporated the modeling and the experimental approaches in order to investigate if such a hydrogel design is feasible.

METHODS

We propose a formulation in which the macroscopic evolutions in tissue size, density as well as the appearance of residual stresses can be directly related to changes in internal composition by considering three fundamental principles: mechanical equilibrium, chemical equilibrium and molecular incompressibility [3]. Mechanical and chemical equilibrium are defined based on Flory-Huggins theory, which characterizes a hydrogel through two parameters: crosslinking density ρ_x and polymer solvent interaction parameter χ. The resulting model allows us to pay particular attention to the features that are critical to the interaction between growth and deformation: osmotic pressure and swelling, the strain mismatch between old and newly deposited material as well as the mechano-sensitive cell-mediated production.

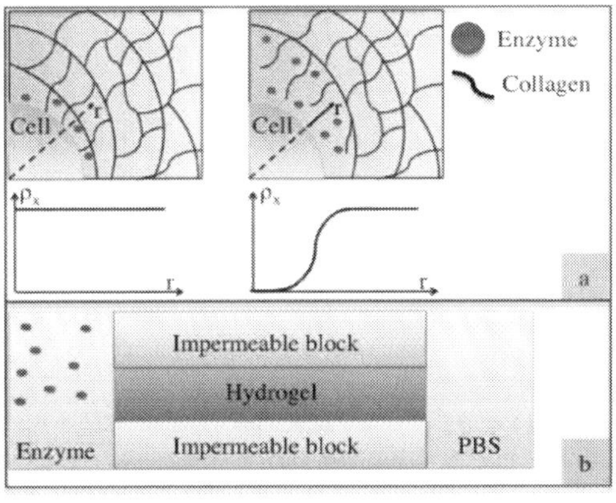

Figure – 1 (a) Cell-mediated degradation of the hydrogel results in the varying profile of crosslinking density along the radius r. (b) The experimental setup in order to investigate the degradation front characteristics.

1D experiments were also conducted in order to investigate the characteristics of degradation that is caused by the diffusion of cell-secreted enzymes in different hydrogel designs (Fig. 1 a). Diffusivity of enzyme in hydrogels was determined matching the computer simulation results with the experimental data obtained from unidirectional diffusion of fluorescence labeled collagenase in a non-degradable hydrogel specimen (Fig. 1 b). Fluorescently labeled, enzymatically degradable hydrogels were used in the same experimental setup and the degradation front, which describes the evolving crosslinking density spatially over time, was observed experimentally using the image processing. Three different hydrogel formulations (stiff, moderate and soft) were used to observe the effect of gel properties on the characteristics of the degradation front. The computer simulations were also performed to investigate the degradation front and the results were matched with the experimental results in order to determine the enzyme rate constant.

RESULTS

There are two quantities that allow us to describe the degradation front in enzymatically degradable hydrogels, that are, front speed and width. The enzymatic degradation is coupled with the diffusion of enzyme through the hydrogel, therefore, two extreme cases for degradation front are: (a) diffusion-dominated (b) degradation-dominated. In order to describe the competition between degradation and diffusion quantitatively, we introduced a dimensionless parameter [2]

$$\Phi = f(r_e, k, D_e, C_e), \quad (1)$$

which is a function of enzyme diffusivity (D_e), rate constant (k), concentration (C_e) and radius of gyration (r_e). Our aim is to use this as design parameter in order to investigate the degradation front shape and the speed in PEG hydrogels.

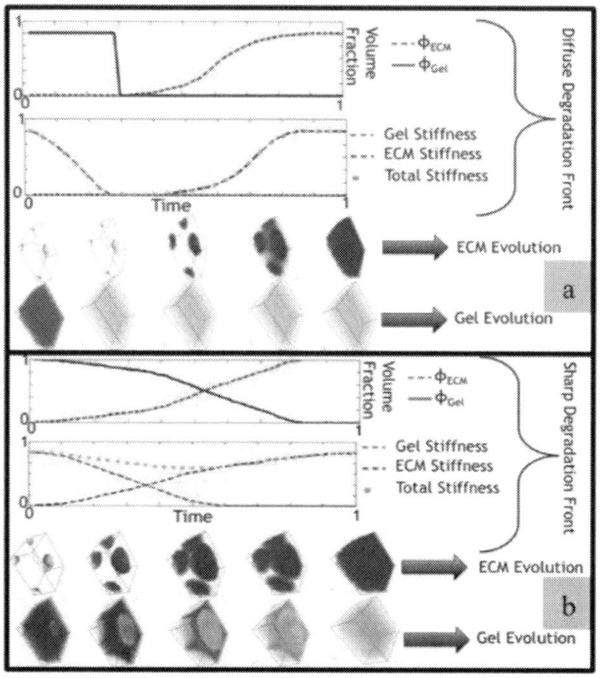

Figure – 2 3D Model results for hydrogels with different degradation characteristics. (a) Uniform degradation results in the disconnected ECM clusters, and therefore, sharp decrease in macroscopic stiffness. (b) Sharp degradation front that moves with the ECM deposition front provides mechanical integrity.

Firstly, we investigated the high values of Φ, for which the enzyme diffuses easily in the hydrogel (Fig. 2 a). Because the hydrogel degrades slowly with respect to the diffusion, the crosslinking density profile is relatively uniform. That results in a sharp decrease in the macroscopic stiffness of the structure, because the hydrogel degrades before the deposited ECM clusters are connected, which could provide the necessary mechanical support. Therefore tissue loses its mechanical integrity. In the case of low Φ (Fig. 2 b), for which the diffusion is relatively slow, the degradation front is sharp and followed by ECM deposition, which provides the additional support to the structure. In this case, the deposited ECM clusters are connected before the hydrogel is degraded completely, and therefore, the mechanical integrity is maintained. These results show that a hydrogel design that degrades with a sharp degradation front that moves with a speed that is consistent with the deposition rate enables the successful tissue growth.

We further investigated the feasibility of controlling the degradation front using three different hydrogel formulations. As it can be seen in the Fig. 3 a and b, highly cross-linked gel (low Φ) has sharp front profile that moves slowly because of the small mesh size ξ of the pores that impedes the diffusion. As expected, the loosely cross-linked gel (high Φ) has the widest degradation front with the highest speed, which may cause the cleavage of the deposited ECM clusters.

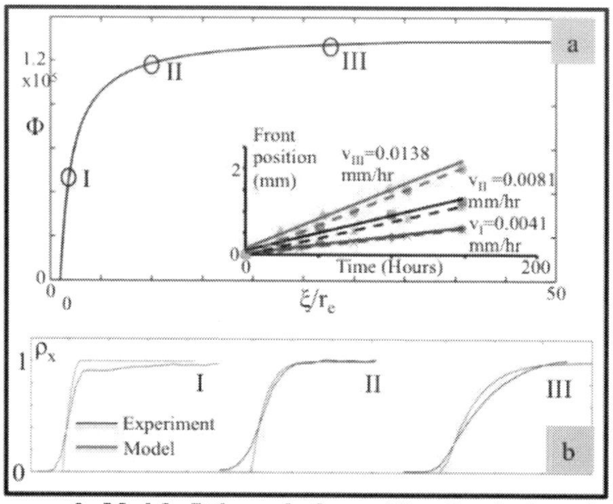

Figure – 3 Model fitting of the unidirectional degradation experiments. In (a) it is shown the corresponding Φ values for three different hydrogels and the velocities of the front (dashed lines are model predictions). In (b) the model predictions of the front shapes are compared with the experimental data.

DISCUSSION

In this study, we introduced a model in order to quantitatively understand the mechanisms that characterize the tissue growth in cell-laden hydrogel scaffolds. Numerical results based on finite element analysis show that if properly tuned, enzymatic degradation can create a degradation front that is key to maintaining scaffold stiffness while allowing ECM deposition. Incorporating the experiments and the model, we have shown that such a hydrogel design is feasible.

ACKNOWLEDGEMENTS

FJV gratefully acknowledges the support of the National Science Foundation under the CAREER award 1350090. FJV and SJB acknowledge that research reported in this publication was supported by the National Institute of Arthritis and Musculoskeletal and Skin Diseases of the National Institutes of Health under Award Number 1R01AR065441.

REFERENCES

[1] Vernerey FJ et al., *Comp. Meth. in Biomech. & Biomed. Eng.*, 15:1197-1210, 2011.
[2] Dhote V et al., *Biomechanics and Modeling in Mechanobiology*, 13:167-183, 2014.
[3] Dhote V et al., *JMBBM*, 19:61-74, 2013.

TUNING SILK FIBROIN HYDROGELS: GENIPIN CROSSLINKING PRE-GELATION DECREASES TIME-DEPENDENT PROPERTIES

W.H. Elliott (1,2), W. Bonani (2,3), D. Maniglio (2,3), A. Motta (2,3), W. Tan (1), C. Migliaresi (2,3)

(1) Department of Mechanical Engineering
University of Colorado at Boulder
Boulder, Colorado, U.S.A.

(2) Department of Industrial Engineering
and BIOtech Research Centre
University of Trento
Trento, TN, Italy

(3) European Institute of Excellence on Tissue
Engineering and Regenerative Medicine, and
INSTM Trento Research Unit
Trento, TN, Italy

INTRODUCTION

Silk fibroin derived from domesticated silk worms (Bombyx mori) is a long chain protein (~375 kDa) with repeat GAGAGAS and GAGAGAT amino acid sequences consisting of 90% of their primary structure. These repeat units form β-sheet secondary structures with hydrophobic side groups on their surface, leading to intra- and intermolecular, closely-packed crystallization structures,[1] acting as "physical" crosslinks, resulting in stable gel structures. These gels can be highly porous and exhibit excellent cell attachment and viability.[2] Fibroin gelation may occur through reduction of solution pH until the molecule isoelectric point is reached, and resulting spinodal decomposition of the solution.[3, 4] To decrease the solution pH for β-sheet crystal formation, we have previously used high-pressure carbon-dioxide treatment (PCT) to enhance gel properties.[5]

Genipin, due to its low cytotoxicity and ease of use, is a common chemical crosslinker, which bonds to separate primary amide groups.[6] Silk fibroin molecules contain only 2% amino acids with primary amide side-chains. [7] This limits genipin crosslink (GCX) formation.

Tuning of dynamic mechanical properties is important in tissue engineering application. Addition of GCX normally increases hydrogel stiffness. Herein, we show decreased stiffness at higher frequencies in SF hydrogels through GCX formation before gelation.

METHODS

Silk fibroin solutions in water were prepared according to Kim et al., [8] with limited modifications. SF solution underwent gelation without genipin (Control), or GCX formation at 37°C for either 24 (Pre24CX) or 48 hours (Pre48CX). Gelation through PCT occurred at 60 bar, and was previously described in Floren, et al. [5] Control gelation occurred at 37°C, while Pre24CX and Pre48CX gelation occurred at room temperature (RT) to reduce residual genipin activation.

Gel Filtration Chromatography Gel filtration chromatography (GFC) analysis of the fibroin samples was conducted with a Shodex SB-805HQ column (Shodex OH pak®, Showa Denko, Munich, Germany). Freeze-dried fibroin-based hydrogels and solutions were re-dissolved in 9.3 M LiBr at 65°C for 0.5 h. The obtained fibroin solutions were dialyzed against DI water at RT and diluted with eluent solution (3 M Urea, 0.02 M Tris HCl, 0.15 M NaCl, pH 7.5) to obtain a protein concentration in the range of 0.5-0.8 mg/ml. The chromatography system was operated with a flow rate of 1 ml/min at 27 ± 1 °C and elution was detected with a Jasco UV-1570 detector set (Jasco, Bouguenais, France) at 224 nm. Calibration curve was obtained with Low/High Molecular Weight Gel Filtration Calibration Kit (GE Healthcare Europe, Freiburg, Germany).

Scanning Electron Microscopy Imaging Imaging of samples was obtained using a scanning electron microscope (Quanta 200 FESEM). Prior to imaging, lyophilized cross sections were sputter coated (Biorad SC500, Hemel Hempstead, U.K.) with a thin layer of gold or platinum/palladium mixture.

Dynamic Mechanical Analysis Dynamic mechanical analysis (DMA) was performed to establish the viscous and elastic properties of the material. Frequency sweep of cyclic, uniaxial compression testing was used to determine approximate location of the glass transition frequency (ω), storage moduli (E'), and loss moduli (E'') of samples. Samples were placed between horizontal, quartz slides attached to both the base fixture and driving piston of a mechanical testing machine. (Bose, Eden Prairie, MN, U.S.A.) The driving plate was brought into contact with the gel, and DI was placed around the gel, with surface tension allowing DI to maintain its integrity for the duration of the test. Subsequently, an average strain of 25% was applied in compression and the system allowed to reach the equilibrium. Dynamic strain was then oscillated at peak-to-peak amplitude of 20% strain. Oscillation frequencies ranged from 2.202E-3 to 1 Hz. Data was recorded after 4 seconds of pre-conditioning of the sample for each frequency.

RESULTS

Results from GFC analysis showed significant increases in both number-average and weight-average molecular weights (M_N and M_W, respectively) for Pre48CX, when compared to the control gels without GCX. The Pre48CX curve illustrating the GCX formation presented a distribution more skewed towards higher molecular weights, with greater molecular weight average (M_W) presented by control gels.

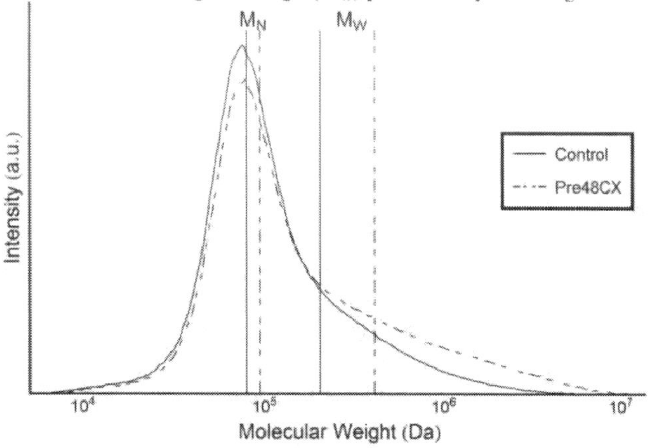

FIGURE 1: MOLECULAR WEIGHT DISTRIBUTIONS OF CONTROL AND PRE48CX

SEM images show that the gel microstructure changes with the crosslinking. Two types of morphology, were found in SEM images (Figure 2). Control gels display porous microstructure with limited connectivity between pores. In comparison, gels with GCX before PCT show more fibrous morphology and higher pore connectivity.

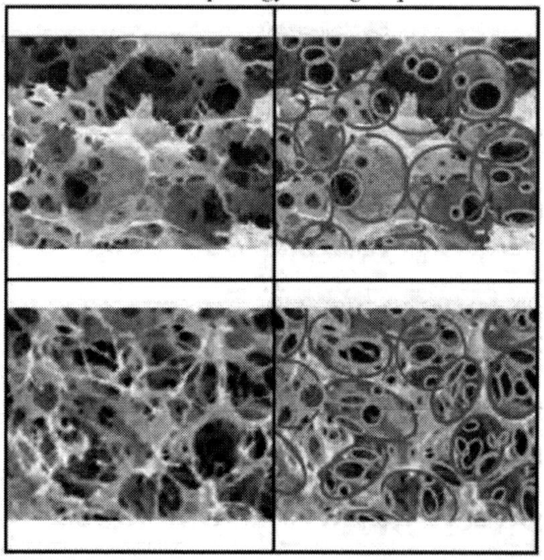

FIGURE 2: SEM IMAGES OF CONTROL (TOP) AND PRE48CX (BOTTOM) GELS WITH THE PORE STRUCTURE (RED) AND PORE CONNECTIVITY (BLUE) HIGHLIGHTED IN THE PICTURES ON THE RIGHT PANEL

E' and E'' results from DMA reveal typical viscoelastic response of gels, with lower and upper plateau regions at lower and higher frequencies, respectively, separated by a transition region. Results indicate that both Pre48CX and Pre24CX decrease in stiffness at higher frequencies. This is likely due to greater pore connectivity in these gels, as shown in Figure 2.

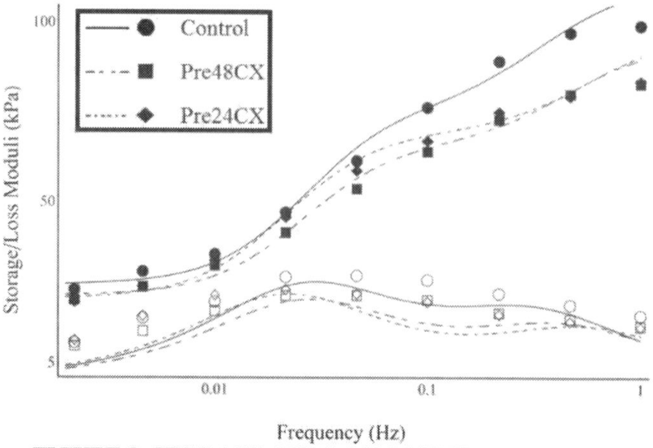

FIGURE 3: STORAGE AND LOSS MODULI AT VARIED FREQUENCY

DISCUSSION

Crosslinking of silk fibroin hydrogels before gelation caused changes in morphology, and decreased stiffness at higher frequencies. This is counterintuitive as increased GCX has previously led to stiffer response from hydrogels. Skewing of molecular weight distributions towards higher values could change gelation mechanics and rates, [4] causing more distinct gelation nodes.[10] This might lead to more fibrous morphology, and greater pore connectivity, thereby decreasing resistance to fluid expulsion from pores.

ACKNOWLEDGEMENTS

The authors wish to Lorenzo Moschini for help in High Pressure Liquid Chromatography, and Gel Permeation Chromatography data acquisition and analysis, and acknowledge funding sources, including NHLBI (HL119371 to W.T.).

REFERENCES

[1] Hu, X. et al., *Macromolecules,* 39: 6161-6170, 2006
[2] Wang, X. et al., *J Controlled Rel,* 121: 190-199, 2007
[3] Ayub Haider, Z. et al., *Biosci, Biotechnol, Biochem,* 57: 1910-1912, 1993
[4] Skripov, V.P., Skripov, A.V., *Phys-Uspekhi,* 22:389-410, 1979
[5] Floren, M. et al. *Biomacromolecules,* 13:2060-2072, 2012
[6] Butler, M.F., et al., *Polym Sci Part Polym,* 41: 3941-3953, 2003
[7] Marsh, R.E. et al., *Biochim, Biophys, Acta,* 16:1-34, 1955
[8] Kim, U.-J. et al., *Biomaterials,* 26: 2775-2785, 2005
[9] Hofland, G.W., et al., *Int Dairy J,* 13: 685-697, 2003

MICROFLUIDIC FACSIMILE OF VENTRICULAR ASSIST DEVICE SHEAR STRESS PATTERNS: TOWARDS POINT-OF-CARE DEVICES TO MONITOR PATIENT THROMBOTIC RISK

A. Dimasi (1), F. Consolo (1), L. Valerio (1), M. Rasponi (1), D. Bluestein (2), G. B. Fiore (1), A. Redaelli (1), M. Slepian (2,3)

(1) Department of Electronics, Information and Bioengineering, Politecnico di Milano, Milano, Italy

(2) Biomedical Engineering Department, Stony Brook University, Stony Brook, NY, USA

(3) Department of Medicine and Biomedical Engineering, Sarver Heart Center, University of Arizona, Tucson, AZ, USA

INTRODUCTION

Ventricular assist devices (VADs) have emerged as effective, therapeutic systems providing restoration of hemodynamics for patients with progressive, advanced and often end-stage heart failure [1]. Many of these devices have unfortunately been plagued by device-associated thrombosis [2, 3]. VAD thrombosis is a serious complication that may result in reduced function, heart failure, systemic emboli, potential pump stop and even death. For this reason, all VAD patients are treated with anti-thrombotic agents. Presently these drugs are administered largely empirically without "personalization" of the regimen for a given patient or device, often resulting in therapeutic ineffectiveness and thrombotic events. To date, few systems are available to assess the thrombotic risk of patients; moreover, most of them are not amenable to rapid, daily bedside use. Furthermore, they do not examine anti-thrombotic drug efficacy under actual flow and shear conditions associated with a given VAD. Indeed, non-physiological shear stresses have been demonstrated to be a major means of activation of platelets in blood contacting devices [4].

In recent years our group has developed a methodology that characterizes the shear stress profile and thrombogenic potential of devices known as Device Thrombogenicity Emulation (DTE) [5]. The DTE approach is based on combined *in silico – in vitro* methods: CFD analyses are used to characterize shear stress histories of platelet flight trajectories within a given device; actual platelet activation, related to individual, "hotspot" trajectories is verified following *in vitro* exposure of platelets to the computational-derived shear stress patterns utilizing a Hemodynamic Shearing Device (HSD), with activation quantified by means of a biochemical assay (the platelet activation state assay, PAS). To date this methodology has been utilized to examine the effect of shear stress on platelet activation and to iterate and improve VAD design [6-9]. More recently, it has been utilized to examine the efficacy of anti-thrombotic therapy for a given patient and a given VAD. However, the DTE testing methodology can be performed only with large laboratory-based instrumentation, e.g. the HSD, with highly trained personnel, thus it is not suitable to frequently monitor blood susceptibility in routine clinical use.

We hypothesize that the individual shear and flow characteristics of a given VAD, i.e. defined "hotspot" trajectories, may be emulated in a microfluidic channel system, i.e. "reverse engineered" to create a VAD microfluidic facsimile. This would allow development of a small, VAD-specific, personalized Point-of-Care (POC), lab-on-a chip system that may be utilized at the bedside or in an outpatient setting to assess the adequacy of anti-thrombotic drug protection.

METHODS

In the present work, we tested the feasibility of using microfluidic technologies to design a microfluidic platform able to replicate the shear-flow characteristics of VADs. We focused our attention on the HeartMateII (HMII, Thoratec Corportation, Pleasanton, CA, USA) device, a mainstay of present clinical therapy. From the work of Chiu et al [8], we extracted platelet flight trajectories and shear stresses.

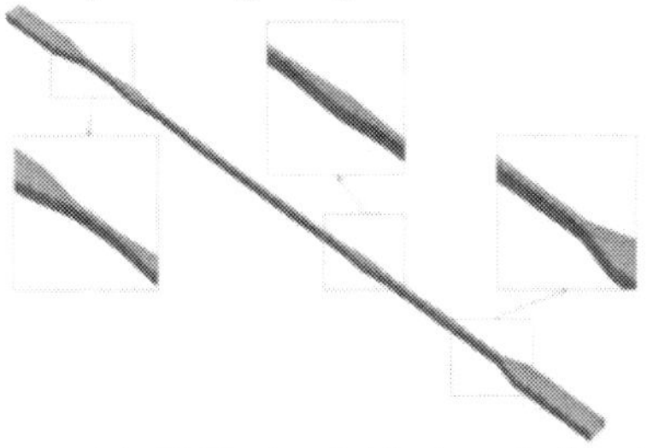

FIGURE 1. Geometry of the HMII microfluidic channel with enlargements on the shear stress generating features.

For the present study, a single waveform was taken as a reference to design the microfluidic unit (Fig. 1): the latter was made of a series of consecutive stress-generating features (a series of channel narrowings and expansions) that were sized in order to replicate the VAD-specific shear stress trajectory in terms of the number, magnitude of shear stress peaks and their duration. The device is characterized by a reference rectangular cross-section of 120 μm x 50 μm (width x height). The stress-generating features were sized in order to generate the desired shear stress peaks at a constant flow rate of 12 μl/min, assuming a fluid resembling whole blood.

CFD analyses. A transient two-phase flow simulation was performed in ANSYS Fluent® (Ansys Inc., USA): a discrete particulate phase having characteristic of human platelets (a diameter of 3 μm and the density of water, 998 Kg/m^3), was modeled to flow through the microchannel and the fluid surrounding it was modeled as incompressible with a density of 1060 Kg/m^3, while a non-Newtonian power-law model was implemented and used to model the rheological properties of blood at 37 °C. The simulation was run with a time step of 0.1 ms until all the particles escaped the fluid domain. At each time step, the scalar stress (σ) along each platelet trajectory was computed and for each trajectory, the linear stress accumulation (SA) was obtained by integrating the scalar stress over the exposure time. The probability density function (PDF) of the SA of the microfluidic device was then obtained, which represents the 'Thrombogenic Footprint' of the device. The SA distribution of the HMII microfluidic model and the scalar stress curves along the trajectories were compared to those reported by Chiu et al [8], in order to assess if the microfluidic device is able to replicate the shear-related thrombogenic potential of the actual VAD.

Preliminary *in vitro* tests. A microfluidic PDMS platform made of the designed microchannel unit was designed and fabricated using standard soft lithography techniques. Preliminary experimental tests were performed to characterize the thrombogenic potential of the platforms: the latter were perfused with ovine gel-filtered platelets (GFP) in a closed loop using a multichannel Ismatec IPC (IDEX Health & Science GmbH, Futtererstr, Wertheim, Germany) peristaltic pump. GFP was collected after predefined perfusion time (0', 15', 60', 120'), or number of passages through the single microchannel unit (0, 20, 80, 160 passages) and analyzed through the PAS assay.

TABLE 1. SA distributions of the HMII VAD and the simulated microfluidic model: the 10th, 50th and 90th percentiles of the PDFs of SA are reported.

	SA percentile distribution [Pa s]		
	10th	50th	90th
HMII micro	0.3	0.9	2.7
HMII VAD	0.3	0.8	2.9

RESULTS

CFD analyses. The PDF of the SA obtained from the simulation of the HMII microfluidic model fit well with that computed with the macroscale simulation [9], as shown in Table 1 where the 10th, 50th and 90th percentiles of SA are shown.

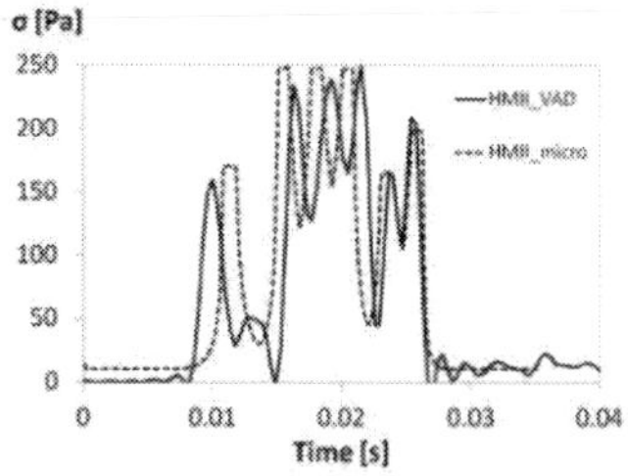

FIGURE 2. Scalar stress curves in the simulated HMII microfluidic device and in the HMII VAD [9].

A good comparison was also obtained in terms of scalar stress waveforms: in Fig. 2, the scalar stress curves corresponding to the 90th percentile of SA obtained with HMII microfluidic simulation is shown in comparison to the corresponding waveform of the macroscale VAD.

Experimental *in vitro* tests. In Fig. 3a a microscope image of the PDMS microfluidic platform is shown. In Fig. 3b the results of the *in vitro* preliminary test are reported as PAS level versus the number of passages through the microchannel unit.

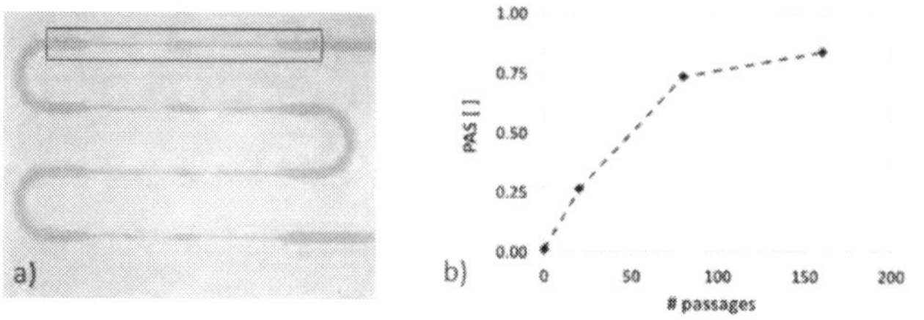

FIGURE 3. a) Optical microscope image of the microfluidic platform perfused with stained water. The red rectangle highlights the single microchannel unit. b) PAS versus the number of passages though the microfluidic units.

Our results demonstrate that the microfluidic platforms are able to induce a significant level of platelet activation, with the PAS assay capable of capturing the dynamics of this phenomenon: activation increases with repeated passages through the device, with non linear dynamics that is very rapid at the beginning and then tends to a plateau close to the maximum level of activation of the sample. Further studies will be performed to validate the platforms as a test bench for platelet activation studies under VAD-like shear stress flow conditions.

DISCUSSION

This initial study establishes the viability and basis for the development of a POC system able to examine anti-thrombotic drug efficacy of patients, using a system truly representative of the flow and shear conditions that platelets are exposed to. CFD analyses were used to characterize the shear stress patterns over a large population of platelets flowing through the device. The microfluidic platform is very flexible as it allows one to reach shear stresses unattainable with the HSD, as well as to test concurrent shear stress histories. For example, a standard 2 cm^2 microfabricated PDMS chip can host tens of parallel microfluidic patterns to provide manifold parallel experiments or to test different shear stress history patterns simultaneously.

Future work will be addressed to: 1. validation of a variety of VAD-emulation microfluidic platforms in comparison with HSD studies, 2. to develop and test more sophisticated sample handling procedures, and 3. to downscaling and on-chip integration of the platelet activation assay, in order to truly advance the system towards practical clinical POC diagnostic use.

ACKNOWLEDGEMENTS

This work was partially financially supported by Fondazione Cariplo grant no. 2241-2011 and NIH/NIBIB 5U01EB012487.

REFERENCES

[1] Stewart, GC and Givetz MM, *Circulation*, 125:1304-1315, 2012.
[2] Mehra, MR et al. *J Heart Lung Transplant,* 33(1):1–11, 2014.
[3] Starling, RC et al., *N Engl J Med*, 370(1):33–40, 2014.
[4] Girdhar, G et al., *Expert Rev Med Devices*, 5(2):167-81, 2008.
[5] Girdhar, G et al., *PLoS One*, 7(3):e32463, 2012.
[6] Nobili, M et al., *ASAIO J*, 54(1):64–72, 2008.
[7] Xenos, M et al., *J Biomech*, 43:2400–09, 2010.
[8] Chiu, W et al., *J Biomech Eng,* 136:021014, 2014.
[9] Pelosi, A et al., *Biomech Model Mechanobiol*, 13(2):349-61, 2014.

NON-LINEAR OPTICAL CHARACTERIZATION OF THE EXTRACELLULAR MATRIX CHANGES FOLLOWING MYOCARDIAL INFARCTION PREDICTS ALTERATIONS IN MECHANICAL PROPERTIES

Kyle P. Quinn (1), Kelly E. Sullivan(1), Carlo A. Alonzo (1), Zachary Ballard (1), Irene Georgakoudi (1), Lauren D. Black III (1,2)

(1) Department of Biomedical Engineering, Tufts University, Medford, MA, 02155, USA

(2) Cellular, Molecular, and Developmental Biology Program, Sackler School for Graduate Biomedical Sciences, Tufts University School of Medicine, Boston, MA 02111, USA

INTRODUCTION

Understanding the microstructural organization and mechanical function of the extracellular matrix (ECM) is critical for the development of therapeutic strategies, which regulate the wound healing process and maximize functional recovery following disease or injury. In the heart, excessive fibrosis following myocardial infarction (MI) can impair diastolic function, which can lead to maladaptive hypertrophic growth and ultimately heart failure [1]. However, the relationships between tissue composition, organization, and mechanical function that lead to maladaptive growth are challenging to elucidate in the myocardium due to rapid changes in cellularity after MI and an inability to characterize both ECM microstructure and function non-destructively.

In this study, we overcome those challenges through the use of multiphoton microscopy and whole organ perfusion decellularization to directly relate the microstructure and mechanical properties of isolated myocardial ECM in a rat model of MI. Second harmonic generation (SHG) imaging has often been utilized to identify collagen fibers and characterize tissue organization, while two-photon excited fluorescence (TPEF) emanating from collagen fibers can also provide a measure of their crosslinking [2]. Subsequent tensile mechanical testing and compositional analysis after this non-destructive imaging enabled a direct quantitative assessment of how ECM composition, fiber organization, crosslinking, and ECM stiffness relate to each other.

METHODS

Animal experiments were approved and conducted in accordance with the Tufts Institutional Animal Care and Use Committee. MI was induced in male Sprague-Dawley rats (250-275 g) by coronary artery ligation. Healthy and infarcted hearts ranging from 1 to 8 weeks post-MI were harvested and decellularized via retrograde perfusion [3]. Decellularized myocardial tissue strips were excised from the left ventricle of each heart, and images of the decellularized samples were obtained with a Leica TCS SP2 laser scanning microscope equipped with a tunable titanium-sapphire laser (Mai Tai; Spectra Physics; Mountain View, CA) and a 63x water immersion objective (NA 1.2). Image volumes were acquired through 5μm z-steps at both 740nm and 800nm excitation using two non-descanned PMT detectors with 400(±10)nm and 525(±25)nm emission filters. Image intensities were normalized by PMT gain and laser power as previously described [4]. Collagen-positive voxels were defined where the normalized SHG intensity exceeded 0.05, and the average intensity within the collagen-positive mask was computed for each of the image channels. The cumulative collagen autofluorescence and SHG intensity within each image volume was also computed by summing all intensity values within the collagen voxels and normalizing by the total image volume. Fiber orientation was determined through two-dimensional Fourier transform of the SHG images and angular sampling; mean fiber orientation and directional variance were computed from the derived angular distribution.

Following imaging, tissues were loaded in tension in the circumferential direction of the heart using a uniaxial mechanical testing machine. The elastic modulus of the tissue was computed by modeling the stress-strain data using a probabilistic model of fiber recruitment [5]. After mechanical testing, composition of the infarcted myocardium was characterized through a QuickZyme total collagen assay and LC-MS/MS-based proteomics. Correlations among multi-photon metrics, collagen content, and the tissue's elastic modulus were computed, and significant differences among time points were determined through ANOVA with post hoc Tukey HSD tests.

RESULTS

Healthy and infarcted hearts exhibited clear macroscopic (Fig. 1A) and microstructural differences that were apparent through multi-photon microscopy. Qualitative examination of the merged TPEF/SHG image volumes revealed that healthy heart tissue has a sparse arrangement of relatively large diameter collagen fibers, while the scar of infarcted samples consisted of a dense, aligned network of small collagen fibers (Fig. 1B). Autofluorescent features were apparent in most images and not only restricted to the SHG-positive collagen voxels (Fig. 1B). Infarcted hearts demonstrated a higher density of collagen-positive voxels compared to healthy hearts ($p<0.05$) (Fig. 1C). The average SHG intensity within the collagen voxels also differed among the healthy and post-infarct time points. The average SHG intensities between weeks 1-4 post-MI were significantly lower than healthy tissue ($p<0.01$). However, at 8 weeks, the average SHG intensity was greater than all earlier time points ($p<0.001$). To estimate total collagen content within the image volume, the average SHG intensity was integrated across all collagen voxels, and this cumulative SHG intensity measurement demonstrated a significant increase ($p<0.05$) over different post-MI time points. However, all post-MI time points demonstrated significantly lower average TPEF intensities within SHG-positive collagen voxels ($p<0.05$) relative to healthy hearts, particularly in the 525nm emission channel (Fig. 1D). The reduced average TPEF intensity within the collagen voxels of infarcted tissue suggests that the ECM of the scar is significantly less cross-linked within the first 2 months after MI compared to healthy tissue.

Changes in matrix composition also coincided with altered tensile mechanical properties. Compared to the mechanical response of healthy myocardium ECM, infarcted decellularized tissue exhibited a reduced elastic modulus ($p<0.05$) at weeks 2-8 post-MI (Fig 1D). Compositional analysis of the decellularized tissues demonstrated a significant increase in total hydroxyproline (HP) content at weeks 4 ($p<0.01$) and 8 ($p<0.05$) post-MI as compared to the healthy and week 2 post-MI groups. Interestingly, HP content and the elastic modulus were not correlated with each other ($R=-0.21$, $p=0.33$). However, the elastic modulus was significantly correlated with average TPEF intensities of the collagen-containing voxels ($p<0.05$), with the average TPEF intensity at 740nm excitation and 525 nm emission providing the strongest correlation ($R=0.61$; $p<0.001$). HP content, on the other hand, was significantly correlated with cumulative intensity measurements of SHG and all TPEF channels ($p<0.05$), with the strongest correlation ($R=0.46$, $p<0.01$) between HP and the cumulative TPEF intensity at 740nm excitation and 525 nm emission.

DISCUSSION

This study demonstrates a variety of microstructural and compositional changes in the myocardial ECM following infarction via multiphoton microscopy. Our findings indicate that the reduced elastic modulus of the ECM within the infarct scar is related to the lack of collagen crosslinks in newly deposited immature collagen. The substantial level of collagen deposition within the first 2 months following infarction may be initiated by fibroblasts in an effort to counteract the greater compliance of immature collagen and increase the bulk tissue strength and stiffness needed to maintain the structural integrity of the myocardium. This would suggest efforts to promote the maturation of the collagen network may help minimize the excessive deposition of weak, compliant collagen and prevent further fibrosis without impairing the integrity of the injured wall. The label-free multi-photon imaging metrics established here may serve as non-invasive biomarkers of ECM status for evaluating dynamic tissue changes during disease progression and evaluating the efficacy of novel therapies.

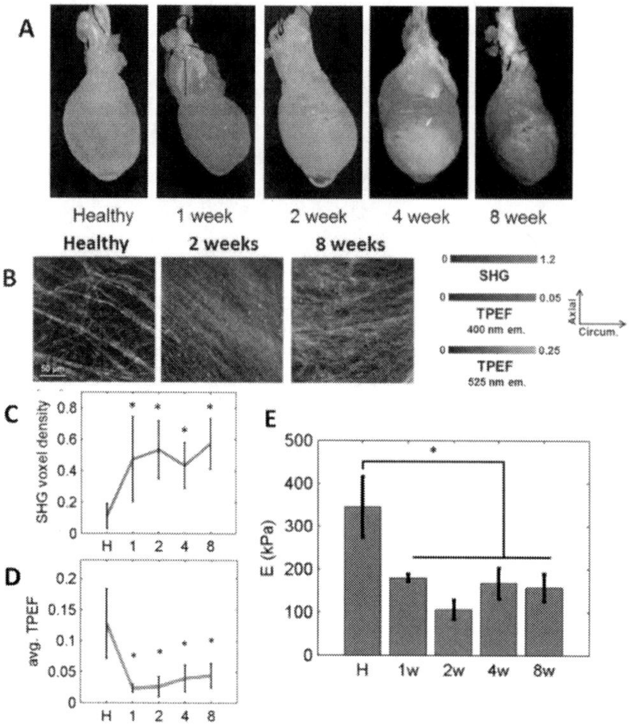

Figure 1: **(A)** Perfusion decellularized rat hearts at various times points following the induction of MI. Note the blue suture denoting the location of coronary artery ligation and the more opaque ECM present at later time points indicating significant remodeling. **(B)** Representative image volume projections from healthy ECM, 2 weeks post-MI ECM, and 8 weeks post-MI ECM. **(C)** The proportion of SHG-positive voxels increased following MI ($p<0.05$). **(D)** The average autofluorescence intensity (740nm excitation, 525nm emission) within the collagen-positive regions was significantly lower ($p<0.05$) in infarcted ECM relative to healthy samples, indicating a reduction in average fiber crosslinking upon infarct scar formation. **(E)** Average fiber elastic modulus of ECM strips at the various time points. Note the significant decrease in fiber modulus following the induction of MI ($p<0.05$).

ACKNOWLEDGEMENTS

The authors are very grateful to Mark Aronovitz in the Molecular Cardiology Research Institute of Tufts Medical Center, for his assistance in the development of the in vivo model of MI. This work was funded in part by the NIH Awards R00 HL093358 to LDB, F32 AR061933 to KPQ, and R01EB007542 to IG.

REFERENCES

[1] Mandinov, L et al Cardiovascular research 45:813-825, 2000.
[2] Zoumi, A. et al, Proc Natl Acad Sci USA 99:11014-11019, 2002.
[3] Ott, HC et al, Nat Med 14:213-221, 2008.
[4] Quinn, KP et al, Scientific Reports 3:3432, 2013.
[5] Hurschler, C et al, J Biomech Eng 125:415-422, 2003.

BIOMECHANICAL COMPARISON OF GLUTARALDEHYDE-CROSSLINKED GELATIN/FIBRINOGEN ELECTROSPUN CYLINDRICAL SCAFFOLDS TO PORCINE NATIVE VASCULAR TISSUE

Ehab Tamimi (1), Catalina D. Ardila (1), Darren G. Haskett (1), Thomas Doetschman (3,4,5), Jonathan P. Vande Geest (1,2,5,6)

(1) Graduate Interdisciplinary Program of Biomedical Engineering, The University of Arizona, Tucson, AZ, USA, 85721

(2) Department of Aerospace and Mechanical Engineering, The University of Arizona, Tucson, AZ, USA, 85721

(3) Department of Cellular and Molecular Medicine, The University of Arizona, Tucson, AZ, USA, 85721

(4) Sarver Heart Center, The University of Arizona, Tucson, AZ, USA, 85724

(5) BIO5 Institute for Biocollaborative Research, The University of Arizona, Tucson, AZ, USA, 85721

(6) Department of Biomedical Engineering, The University of Arizona, Tucson, AZ, USA, 85721

INTRODUCTION

Heart disease is the major cause of death in the United States, with an estimated 83.6 million Americans with one or more type of cardiovascular disease (CVD). Mortality data from 2010 shows that CVD is the underlying cause of death for 1 out of every 3 Americans. The highest percentage of CVD-related deaths were attributed to coronary heart disease, with 158,008 coronary artery bypass graft procedures performed in 2010 [1]. Biological grafts, such as autologous native vessels, have been the gold standard for small diameter grafts. Unfortunately, this source of grafts is limited by availability, shorter durability due to calcification [2], and inherent compliance mismatch between the native vessel and the graft which can lead to conditions such as intimal hyperplasia. There is an increasing demand for affordable, compatible and more easily accessible coronary vascular grafts.

Tissue-engineered vascular grafts (TEVGs) offer an alternative source of grafts that can be developed to be nonthrombogenic, biocompatible, nonimmunogenic, resistant to infection, and compliance-matched to the native vessel. Since compliance mismatch is a major cause of restenosis and graft failure, it's important that the mechanical properties of the TEVG be comparable to that of native vessel. The purpose of this study was to investigate the feasibility of developing an electrospun acelluar gelatin/fibrinogen TEVG that could compliance-match porcine left descending coronary arteries (LDCA) by investigating the effect of crosslinking time on the mechanical properties.

METHODS

Mechanical data from LDCAs extracted from hearts obtained from the University of Arizona Meat Laboratory post-mortem have been previously collected by our research group [3]. Gelatin/fibrinogen cylindrical constructs were fabricated by electrospinning a 10% (w/v) of gelatin/fibrinogen (80:20) in HFP onto a rotating translating mandrel. Electrospun sheets with this composition have been shown by our lab to promote porcine vascular smooth muscle (SMC) cell adherence and proliferation compared to alternative ratios [4]. The constructs were suspended inside a 25% glutaraldehyde (GLUT) chamber and exposed to the GLUT vapor phase for 2, 8 and 24 hours. Multiphoton imaging was used to determine the effect of GLUT crosslinking on construct fiber orientation distribution.

To determine the mechanical properties of the constructs, a custom-made microbiaxial optomechanical device (MOD) was used. This device has the capability of measuring circumferential and axial stress-strain responses simultaneously and has been previously used by our laboratory group to assess the mechanical properties of the LDCA [3]. For each crosslinking experimental group, circumferential and axial 2nd Piola Kirchhoff stresses were plotted against the circumferential and axial Green strains, respectively. The circumferential and axial tangential moduli were extracted from the stress-strain curves by determining the slope of the curves at different locations, compliance values were calculated and the mechanical data was fitted to the Fung strain-energy constitutive model described by Equation (1).

$$W = \frac{c}{2}(e^{Q} - 1) \qquad (1)$$

where $Q = A_1 {E_{zz}}^2 + A_2 {E_{\theta\theta}}^2 + 2A_3 E_{zz} E_{\theta\theta}$, W is the strain energy density, c, A_1, A_2, and A_3 are material constants and $E_{\theta\theta}$ and E_{zz} are the circumferential and axial strains, respectively. For each of these calculated values, a comparison was made between each experimental group and the LDCA.

RESULTS

Multiphoton images qualitatively showed that crosslinking did not have an effect on fiber orientation. Statistical analysis revealed that the effect of crosslinking time on compliance was significant (p-value=0.05). Two-tailed two-factor t-tests were performed to compare the compliance values of each experimental group to the LDCA. The constructs crosslinked for 8 hours was the only experiments group that showed compliance not statistically different than the LDCA (**Figure 1**). Circumferential tangential moduli showed less similarity between the experimental groups and the LDCA with only a few values in the 2 and 8 hour group with no statistical difference (**Figure 2**).

The Fung equation was shown to be a suitable constitutive model to predict the mechanical responses of all constructs and the LDCA with reasonable R^2 values (Figure 3). The 8-hour constructs showed the most axial mechanical similarity to the LDCA.

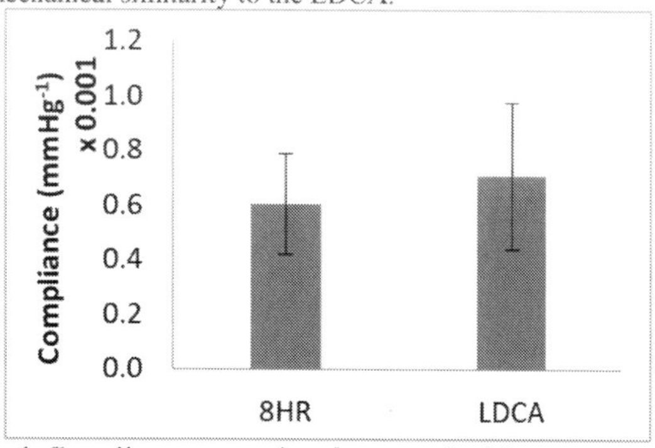

Figure 1. Compliance comparison between the 8-hour constructs and the LDCA

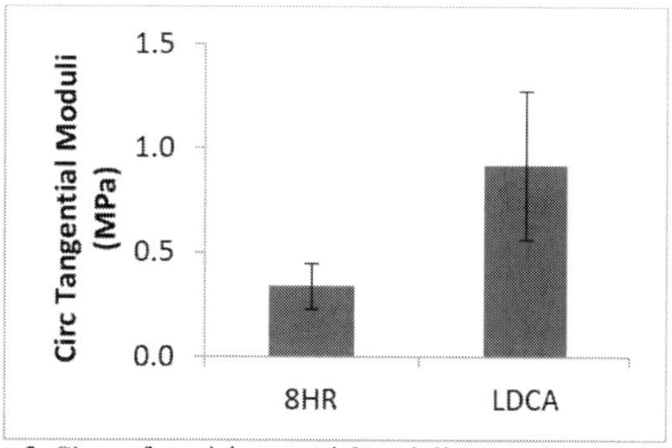

Figure 2. Circumferential tangential moduli comparison between the 8-hour constructs and the LDCA at 120 mmHg

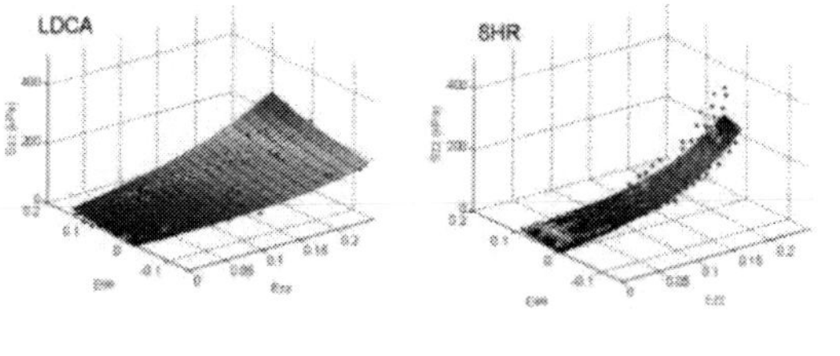

	C (kPa)	A_1	A_2	A_3	R_2
8HR	49.4 ± 4.9	16.5 ± 1.4	7.8 ± 1.45	2.6 ± 0.6	0.80±0.07
LDCA	95.8 ± 27.5	6.9 ± 2.2	3.7±1.2	1.6±0.6	0.788±0.04

Figure 3. (Top) Axial stress-strain fitted Fung equation surface plots for each experimental group plotted against data points from all three replicates (Bottom) Average Fung strain-energy equation constants and R^2 values for Fung equation surfaces.

DISCUSSION

All results showed that GLUT crosslinking has a biaxial stiffening effect on the gelatin/fibrinogen construct, showing that it may be possible to modify the mechanical properties of the constructs to meet specifications required by vascular graft transplantation through manipulating the crosslinking time. The circumferential and axial tangential moduli comparison showed that the mechanical properties are within the same order of magnitude for constructs crosslinked for 2 and 8 hours, with some values matching the LDCA with no statistically significant differences.

Multiphoton imaging could be used to evaluate the change in fiber orientation for constructs under axial and circumferential stretch to further explain the mechanical behavior shown in the stress-strain curves. Other biopolymers could be added to the construct composition to better compliance-match native vasculature while giving it more mechanical integrity to withstand higher biaxial strains. The effect of crosslinking on biocompatibility and biodegradability is still largely unknown. Future studies should include the effect of the SMCs growth on the mechanical properties and determine if SMCs remodel the fibrous constructs. In addition, studies could investigate the effect of the circumferential and axial stimulation of SMC-seeded constructs on scaffold remodeling and ECM deposition.

KNOWLEDGEMENTS

The authors acknowledge funding from the NIH (1R21HL111990 to JPVG, Co-I Doetschman and Slepian). The authors would also like to acknowledge the help of Darren Haskett, Forest Danford, Corina MacIsaac, Jamie Lynn Hernandez, and Joshua Alan Uhlorn

REFERENCES

1. Go, A.S., et al., *Heart disease and stroke statistics--2014 update: a report from the American Heart Association.* Circulation, 2014. **129**(3): p. e28-e292.

2. Kurobe, H., et al., *Concise review: tissue-engineered vascular grafts for cardiac surgery: past, present, and future.* Stem Cells Transl Med, 2012. **1**(7): p. 566-71.

3. Keyes, J.T., et al., *Deformationally dependent fluid transport properties of porcine coronary arteries based on location in the coronary vasculature.* J Mech Behav Biomed Mater, 2013. **17**: p. 296-306.

4. Ardila, D.C., et al., *TGFbeta2 differentially modulates smooth muscle cell proliferation and migration in electrospun gelatin-fibrinogen constructs.* Biomaterials, 2014. **37C**: p. 164-173.

NT-PRO BNP EXPRESSION ORIGINATING FROM THE RV MYOCARDIUM IN PEDIATRIC PULMONARY HYPERTENSION PATIENTS IS CORRELATED WITH BOTH THE REACTIVE AND RESISTIVE COMPONENTS OF VASCULAR IMPEDANCE

V. Kheyfets, J. Dunning, U. Truong, K. Hunter, D. Ivy, R. Shandas

(1) Department of Bioengineering
University of Colorado
Aurora, CO, USA

(2) The Heart Institute
Children's Hospital Colorado
Aurora, CO, USA

INTRODUCTION

Pulmonary arterial hypertension (PAH) is a progressive disease characterized by a gradual increase in pulmonary vascular impedance, which ultimately leads to right ventricular (RV) hypertrophy and dysfunction. The RV myocardium responds to pressure and volume overload by secreting a neurohormone, B-type natriuretic peptide (BNP) [1], that has relatively modest vasodilator properties but could serve as a non-invasive marker of vascular-ventricular function. NT-proBNP is a cleaved byproduct of BNP that offers numerous logistical advantages for clinical monitoring over BNP (including longer in-vivo half-life and storage life) [2]. Particularly for pediatric patients, a non-invasive analog of vascular function could replace invasive right heart catheterization. NT-proBNP is a promising candidate for the following reasons: (1) It can be measured from blood serum with relative ease and at low cost; (2) it has been shown to predict outcome; and (3) serum levels have been shown to respond to therapy. In previous work, biomarker expression has been evaluated with respect to outcome, with some attention paid to functional and hemodynamics markers in pediatric PAH [3]. In fact, studies on pediatric patients have even shown that BNP levels will decrease concurrently with therapeutic intervention in cases of mitral valve stenosis [1]. However, to our knowledge, this is the first article relating NT-proBNP expression with the resistive and reactive components of afterload in the pediatric PAH population.

The current study aims to answer two questions regarding the use of NT-proBNP as a clinical indicator of disease in pediatric PH: (1) Does NT-proBNP predict the resistive and reactive components of afterload in pediatric PH patients; and (2) does NT-proBNP originating from an otherwise structural and functionally normal LV myocardium, compromise its efficacy as a marker of afterload?

METHODS

Right heart catheterization (RHC) data was retrospectively analyzed for N = 54 datasets, belonging to 36 pediatric PAH patients at Children's Hospital Colorado (some patients underwent a bi-annual follow-up RHC, which was considered as an individual dataset). Each dataset was confirmed to have normal LV morphology and function at the time of RHC from echocardiography by a trained cardiologist. The cardiologist noted the extent of inter-ventricular septal flattening. In this study, we consider two degrees of septal flattening severity: *(1) normal to mild (N = 40); and (2) moderate to severe (N = 14).*

Serum NT-proBNP concentration was analyzed, for blood samples collected within 5-days of RHC, using an electrochemiluminescence immunoassay (Mayo Medical Laboratories ProBNP II, Roche Diagnostics, IN, USA).

The resistive and reactive components of vascular impedance were estimated from RHC data using Eq. (1) and (2), respectively.

$$PVR = \frac{mPAP - PCWP}{CO} \tag{1}$$

$$C = \frac{CO}{HR * PP} \tag{2}$$

In Eq. (1) and (2) *PVR = pulmonary vascular resistance, C = pulmonary vascular compliance, mPAP = mean PA pressure, PCWP = pulmonary capillary wedge pressure, HR = heart rate, PP = pulse pressure,* and *CO = cardiac output.*

The data presented in this study is analyzed using two statistical tests: (1) linear correlations and (2) comparing two sample means (PAH patients with flat septum vs. normal septal geometry). For both tests, each dataset was normalized using a logarithmic operation. NT-proBNP, which has been shown to vary with age in pediatric patients, was normalized by subject age at the time of RHC. When comparing the means of two samples, an F-test was done to compare variances, followed by the appropriate t-test. For both statistical tests, results within a 95% confidence interval were considered statistically significant.

RESULTS

Within the entire sample considered, BNP and NT-proBNP revealed a statistically significant correlation ($R^2 = 0.79$, $P_{two\text{-}tail} < 0.001$). The following data specifically compares clinical variables with NT-proBNP, but the strong correlation between the presence of both proteins suggests that the following conclusions could be extrapolated to BNP.

Patients that showed a moderate to severely flattened inter-ventricular septum, had a significantly higher mPAP, resistive afterload, and NT-proBNP levels, while RV CO was unchanged (see Fig. 1).

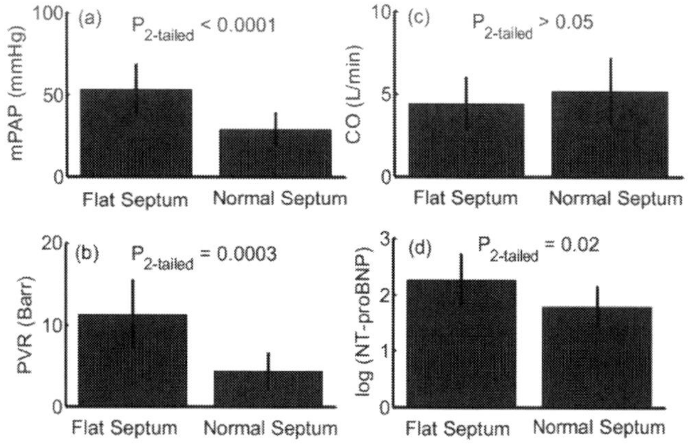

Figure 1. Bar plots (with standard deviation shown as error bars) presenting a statistical difference in mPAP (a), resistive afterload (b), ventricular pump function (c), and NT-proBNP (d) expression between patients with normal and flattened inter-ventricular septum.

The resistive component of afterload, normalized by body surface area (BSA), described 81% of the variability in NT-proBNP in patients with a normal inter-ventricular septum (see Fig. 2). However, in patients with moderate to severe septal flattening, the correlation is insignificant ($R^2 = 0.21$, $P_{two\text{-}tailes} = 0.1$). The correlation coefficient for both groups analyzed together was $R^2 = 0.75$ ($P_{two\text{-}tailes} < 0.001$).

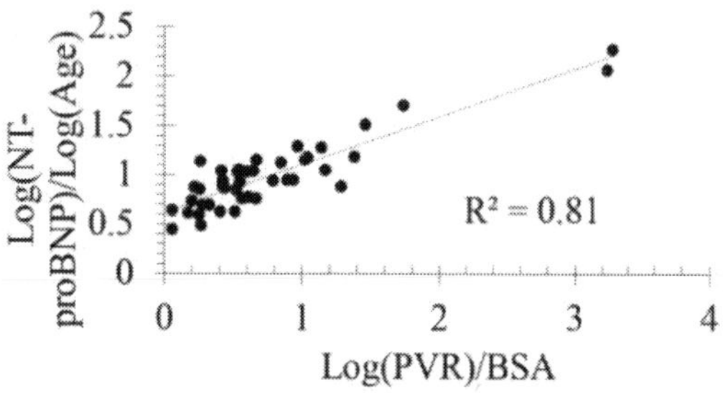

Figure 2. Log(PVR)/BSA correlated against Log(NT-proBNP)/Log(Age) ($R^2 = 0.81$, $P_{two\text{-}tailed} < 0.001$).

A decrease in compliance suggests an increase in the reactive component of impedance. The homeometric reactive component described 80% of the variability in NT-proBNP in patients with a normal inter-ventricular septum (see Fig. 3). However, again, patients with a moderate-severely flattened septum revealed no correlation between reactive afterload and NT-proBNP ($R^2 = 0.17$, $P_{two\text{-}tailed} = 0.15$). The correlation coefficient for both groups combined, analyzed together, was $R^2 = 0.69$ ($P_{two\text{-}tailed} < 0.001$).

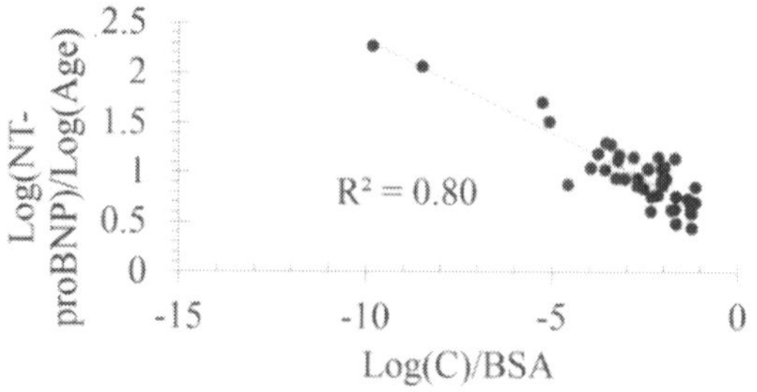

Figure 3. Log(C)/BSA correlated against Log(NT-proBNP)/Log(Age) ($R^2 = 0.80$, $P_{two\text{-}tailed} < 0.001$).

DISCUSSION

The resistive component of impedance represents dissipated hydraulic energy that must be replaced by the RV. The reactive component does not contribute to dissipated flow energy, but still offers an impediment to the flow. In this study, we found that both the resistive and reactive components of afterload are equally correlated with the release of biochemical markers indicative of myocardial distress (excessive stretching/wall stress).

Consistent with previous findings, a gradual increase in PVR eventually alters the inter-ventricular pressure gradient, therefore causing a moderate to severely flattened inter-ventricular septum. While patients presented with a flattened septum showed a statistically higher serum NT-proBNP concentration, a closer look at the correlative data suggests a concerning issue about its origin. If this study offers a framework for estimating reactive and resistive afterload from non-invasive biomarkers, we do so with a cautious note: It appears that the flattened septum, resulting from excessively high PVR, could consequently increase LV free-wall distension and put excessive stress on the LV myocardium, thus causing it to secrete NT-proBNP. Under these conditions, proteins originating from both ventricles will mix in the systemic circulation, thus compromising the specificity of this neurophoromone in predicting characteristics of pulmonary vascular impedance. Furthermore, it could jeopardize the biomarker's ability to accurately explain RV function, but should not influence the prediction of disease outcome.

An alternative explanation could be that, in scenarios of septal flattening, the RV is releasing the added NT-proBNP. Possibly due to the advanced stage of RV dysfunction, maybe as some programmed compensation, or simply as a marker of some remodeling transition.

In conclusion, we show that serum NT-proBNP concentration could provide a non-invasive estimate of pulmonary vascular impedance for pediatric PAH patients. Nevertheless, this estimate should be used with caution for patients with extremely high PVR, especially when septal flattening has been observed

ACKNOWLEDGEMENTS

This work was funded by NIH RO1 HL 114753 and NIH K24 HL081506 awards.

REFERENCES

[1] *Esteves, W. A. et al., Journal of interventional cardiology, 26(5), pp. 501-508, 2013.*
[2] *Foris, V. et al., Chest, 144(1), pp. 274-283, 2013.*
[3] *Bernus, A. et al., Chest, 135(3), pp. 745-751, 2009.*

MATURATION OF HUMAN STEM CELL-DERIVED CARDIOMYOCYTES IN 3D TISSUES THROUGH INCREASING COLLAGEN CONCENTRATIONS

Aric Q. Pahnke (1,2), Milica Radisic (1,2)

(1) Chemical Engineering and Applied Chemistry
University of Toronto
Toronto, Ontario, Canada

(2) Institute of Biomaterials and Biomedical Engineering
University of Toronto
Toronto, Ontario, Canada

INTRODUCTION

Cardiac tissue engineering has emerged as an intriguing course to mitigate the detrimental effects of cardiovascular disease. Specifically, in vitro models, like heart organoids, can replicate cellular and tissue organization seen in the native heart and can be used as platforms to explore therapies in management of cases like cardiac arrhythmias which have caused sudden cardiac death [1]. Furthermore, increasing reliability of induced pluripotent stem cell technology allows researchers to form tissues with the same cellular genome as a patient which can be used for precision medicine [2].

This field continues to innovate, advancing adept designs capable of monitoring electrical and contractile behavior of tissues while providing physiological cues for organization[3]. As these techniques aid in the analysis of the tissues, evidence has mounted that cardiac tissues do not replicate the mechanical or electrical properties of the native heart. Formed tissues also lack hallmark features of mature myocardium which includes complete sarcomeric structures and t-tubules. These properties could be essential to drug testing results and *in vitro* analysis. We hypothesize that cardiac tissue maturation state can be improved by increasing the collagen content of formed tissues. We compare the resulting electrical and mechanical effects on in vitro cardiac tissues after modification of cellular environment and find data to suggest maturation state improves with collagen density.

METHODS

Cell culture

Tissues were prepared from human induced pluripotent stem cells differentiated into cardiomyocytes using a variation of previous methods [4]. Cells were maintained in Mtesr medium and cultured 72 hours before induction. After day 10 of induction into cardiomyocytes, cells were dissociated using Collagenase type II.

Tissue Formation

Cells were then mixed into a collagen gel as previously accomplished [5] at 20 million cells per milliliter. Specific ratios of collagen were controlled by water ratio addition and were tested from 2.5 mg/ml to 3.5 mg/ml. Cell mixture was then added to a preformed mold of PDMS [Fig. 1] and incubated for 25 minutes at 37° C and 5% CO_2 to allow gel to harden. After the gel hardened media was added and formed tissues returned to incubate.

Tissue Dynamics

Electrical properties of excitation threshold and maximum capture rate of the tissues were determined through introduction of biphasic pulses of electric field stimulation as reported previously [6]. Strength of contraction was calculated from taking video of the top surface of each post [Fig. 1] according to Eq. I.

$$Wo = \delta * 24EI\left(\frac{1}{3(a+b)^4-4(a^3(a+b))+a^4+4(a+b)^3c-4a^3c}\right) \quad \text{Equation I}$$

Tissue Staining

Formed cardiac tissues were fixed in 4% paraformaldehyde, immobilized in paraffin wax, and sliced at 4 microns. Sliced samples were stained using Hematoxylin and eosin stain (H&E) and Masson's Trichrome to see gross tissue morphology and collagen deposition. Other samples were stained with cardiac tissue markers cTNT, connexin 43, and α-actinin (not shown).

Statistical Analysis

Statistical analysis was performed using Matlab. Data reported as mean ± standard error. Statistical significance was analyzed by using a one-way analysis of variance (ANOVA), followed by a post-hoc Tukey tests. An $\alpha=0.05$ was considered significant for all tests.

RESULTS

Cells began beating on day 10 of differentiation. After tissue formation, CMs became denser in the middle of the tissue where most of the tension is expected to be experienced. Also, correct stiffness of the collagen gel proved to be essential as tissues were unstable and broke in gels of lower collagen content.

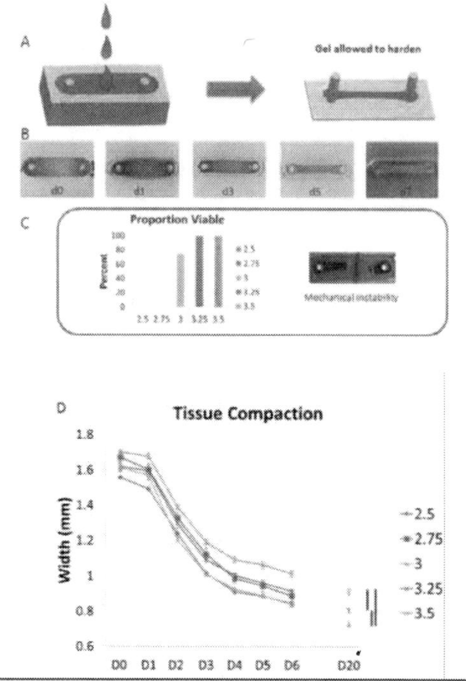

FIGURE 1: TISSUE FORMATION AND STABILITY. (A) REPRESENTATION OF TISSUE FORMATION. GEL AND CELL MIXTURE PLACED INTO PREFORMED PDMS MOLD. (B) TISSUE COMPACTION OVER 7 DAYS (C) PROPORTION OF TISSUES VIABLE AFTER 20 DAYS. (D) TISSUE WIDTHS OF DIFFERING GEL STIFFNESS OVER 20 DAYS.

The tissues caught electrical signals up to and exceeding 6 Hz whereas gels made of lower collagen concentrations were much less (~4.5) [12]. Excitation Threshold (ET) represents the lowest voltage needed to excite a tissue. A lower ET is indicative of a matured state. Higher gel concentrations have largely improved ET. The force-frequency relationship as seen in adult human myocardium is seldom seen in engineered heart tissue, much less in naïve iPSC-CMs. Many intracellular proteins are involved in calcium cycling and their expression is correlated to the force-frequency relationship. Here we see a small indication of this relationship from electrically matured tissues and would expect more with more mature tissues. Similar to native tissue forces, these tissues first trend upwards and then decrease force as frequency increases. Tissues with higher maximum capture rate (MCR) also have higher frequency at maximum force. Gel Concentration affects MCR which is the highest frequency at which tissues can be stimulated. Unlike ET, a higher value is desired. The high MCR in 3.5 gels could be caused by more developed intracellular calcium cycling signifying a mature tissue.

Hematoxylin and eosin stain (H&E) of viable groups which indicates the health of a tissue. Dense packs of cells allude to connections between CMs. A dense tissue would indicate a functional tissue able to contract synchronously and produce forces able to bend a post. Masson's Trichrome is typically used to note areas of fibrosis where collagen content is high. In terms of the heart, fibrotic tissue is unable to conduct electricity or contract. Cells seemed more spread out in higher collagen concentration tissues but, interestingly, did not show signs of fibrosis.

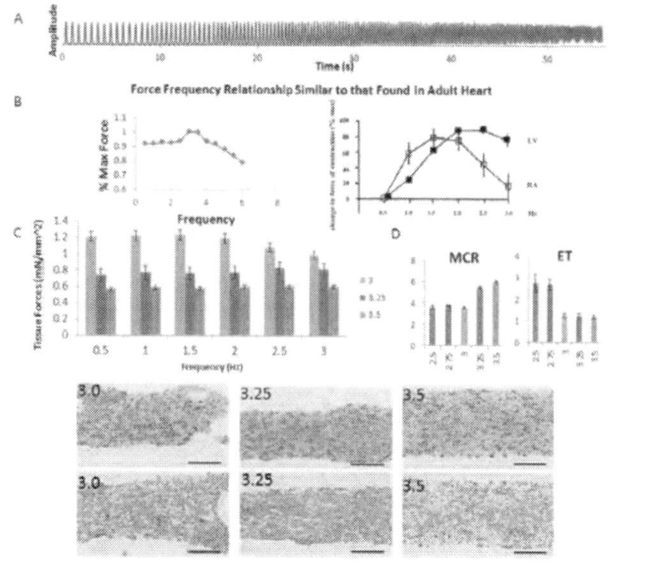

FIGURE 2: ELECTRICAL AND SPATIAL ORIENTATION OF ORGANOIDS (A) ORGANOID BEING STIMULATED IN A RANGE FROM 1.5-6HZ. (B) TISSUE COMPACTION OVER 7 DAYS (C) INCREASING GEL CONCENTRATION DECREASES FORCE OF CONTRACTION (A=0.05). (D) ELECTRICAL PROPERTIES OF TISSUES. (E) MASSON'S TRICHROME AND H&E STAINS.

DISCUSSION

Cardiac tissue formation can prove useful as a surrogate in drug testing and in precision medicine. For formed tissues to be of most use they must have hallmarks of native cardiac tissue and have similar properties. Cells within the tissues respond to environmental cues and we hypothesized that cardiac tissue maturation state can be improved by increasing the collagen content of formed tissues. Herein we show data supporting the electrical and spatial maturation of cardiac tissues. Formed tissues resembled the native myocardium in the force frequency relationship and also improved in MCR and ET measurements. Furthermore, it was shown that the tissue properties can be controlled through the cellular environment. As engineered tissues are increasingly used in clinical settings and drug discovery, replicating native tissue properties will prove necessary to see true results applicable *in vivo*. This work shows that control over the initial cellular environment is essential to forming tissues that resemble the native myocardium.

ACKNOWLEDGEMENTS

Funding for this work was generously supported by Canadian Institutes of Health Research (CIHR) and Microfluidic Applications and Training in Cardiovascular Health (MATCH).

REFERENCES

[1] Nunes, S et al. *Nat. Methods*, vol. 10, no. 8, pp. 781–787, 2013.

[2] Takahashi, K et al. *Cell*, vol. 131, no. 5, pp. 861–872, Nov. 2007.

[3] Thavandiran, N et al. *PNAS*, vol. 110, no. 49, pp. E4698–E4707, 2013.

[4] Lian, X et al. *PNAS.*, May 2012.

[5] Miklas, J et al. *Biofabrication*, 2014.

[6] Radisic, M et al. *PNAS* vol. 101, no. 52, pp. 18129–18134, 2004.

SB³C2015
Summer Biomechanics, Bioengineering and Biotransport Conference
June 17-20, 2015,Snowbird Resort, Utah, USA
SB³C2015-1132

THE EFFECTS OF STRESS STATE ON THE MECHANICAL RESPONSE AND FAILURE OF THE NEURONAL PHOSPHOLIPID BILAYER: A MOLECULAR DYNAMICS STUDY

M. A. Murphy (1,2), M. F. Horstemeyer (2,3), Steven R. Gwaltney (4,5), Tonya W. Stone (2,3), Michelle C. LaPlaca (6), Jun Liao (1,2), Lakiesha N. Williams (1,2), R. Prabhu (1,2)

(1) Department of Agricultural and Biological Engineering
Mississippi State University
Mississippi State, MS, USA

(2) Center for Advanced Vehicular Systems
Mississippi State University
Mississippi State, MS, USA

(3) Department of Mechanical Engineering
Mississippi State University
Mississippi State, MS, USA

(4) Center for Computational Sciences
Mississippi State University
Mississippi State, MS, USA

(5) Department of Chemistry
Mississippi State University
Mississippi State, MS, USA

(6) Department Biomedical Engineering
Georgia Institute of Technology
Atlanta, GA, USA

INTRODUCTION

With approximately 1.7 million new cases every year in the United States alone, traumatic brain injury (TBI) has become a great concern that results in both human and economic losses [1]. However, injury to brain is complicated and dependent on many different aspects, including the failure of the neuronal cell membrane and its primary component, the phospholipid bilayer. This failure can have devastating results for the cell, including apoptosis and necrosis [2]. Multiscale constitutive material models attempt to replicate the brain's anisotropy to obtain more accurate stress responses during finite element analysis studies of TBI scenarios [3], requiring lower length scale information and molecular dynamics (MD). It is generally accepted that multiple stress states must be considered when creating constitutive models. However, currently, deformation studies of the phospholipid bilayer have primarily focused on examining the structure equibiaxially, both *in vitro* [4] and *in silico* [5, 6], with little focus given to other stress states, such as Tomasini et al. [6] who also examined surface shear. For the development of higher fidelity multiscale models, multiple stress states must be quantified. In this study, the authors use MD to compare the von Mises (equivalent) stress response of four stress states due to deformations of a simplified phospholipid bilayer.

METHODS

Seventy-two 1-palmitoyl-2-oleoylphosphatidylcholine (POPC) phospholipids and 9,070 TIP3P water molecules (original structure sourced from NIH's Laboratory of Computation Library [7] and 6,828 TIP3P water molecules were added) was equilibrated for ten nanoseconds using the program LAMMPS [8] and the CHARMM36 all-atom lipid force field [7]. The equilibrated structure was subjected to constant velocity deformations in the x and y dimensions. The z dimension was allowed to relax freely during deformations. The examined stress states can be seen in Figure 1, where the cases 72x4.0y3.8, 72x4.6y2.2, 72x4.0y0.0, and 72x4.0yRel correspond to Figure 1 (A), (B), (C), and (D), respectively, and indicate the deformation velocities with the form 72x[x deformation velocity]y[y deformation velocity]. For example, 72x4.0yRel represents the uniaxial case where the x dimension is deformed at 4.0 ms^{-1} and the y dimension is allowed to relax freely under a pressure of 1 atm. The phospholipid bilayer was assumed to have failed when full water penetration occurred, determined visually using OVITO [9], resulting in a chain of water extending through both bilayer leaflets.

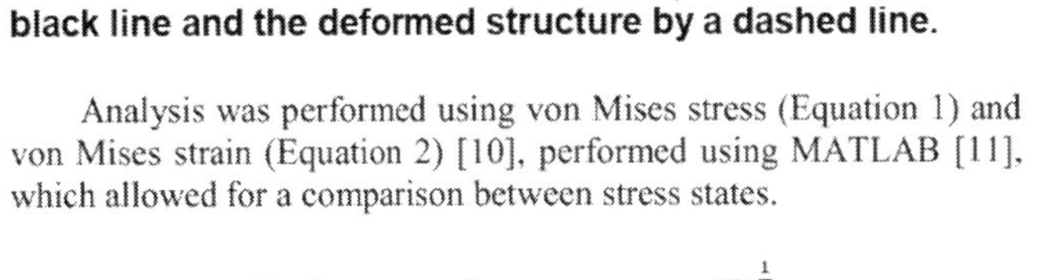

Figure 1: Representative perpendicular (top) schematic of the deformations performed for (A) Equibiaxial, (B) Non-equibiaxial, (C) Strip biaxial tension, and (D) Uniaxial tension deformations. The initial structure is represented by a solid black line and the deformed structure by a dashed line.

Analysis was performed using von Mises stress (Equation 1) and von Mises strain (Equation 2) [10], performed using MATLAB [11], which allowed for a comparison between stress states.

$$\epsilon_{VM} = \left(\frac{2}{9} * \left[\left(\epsilon_x - \epsilon_y\right)^2 + (\epsilon_x)^2 + \left(\epsilon_y\right)^2\right]\right)^{\frac{1}{2}} \quad (1)$$

$$\sigma_{VM} = \left(\frac{1}{2} * \left[\left(\sigma_x - \sigma_y\right)^2 + \left(\sigma_y - \sigma_z\right)^2 + (\sigma_x - \sigma_z)^2\right]\right)^{\frac{1}{2}} \quad (2)$$

Where σ_{VM} and ϵ_{VM} are the von Mises stress and strain. Similarly, $\sigma_x, \sigma_y, \sigma_z, \epsilon_x, \epsilon_y$, are the principle in-plane stresses (pressures) and true strains of the simulation box. Equation 1 was also used as an equivalent von Mises strain rate equation where ϵ_x and ϵ_y were replaced by their respective strain rates. The von Mises strain rate equation was used to determine $\dot{\epsilon_x}$ and $\dot{\epsilon_y}$ for each stress state correlating to the equivalent von Mises strain rate $(\dot{\epsilon}_{VM}) = 5.5\times 10^8$ s^{-1} so that equivalent stress states could be compared.

RESULTS

Deformations of the phospholipid bilayer resulted in an initial peak in stress prior to extensive deformation and then began decreasing (Figure 2). As the bilayer was deformed further, water penetration occurred and resulted in pores expanding more quickly (Figure 2). The equibiaxial and non-equibiaxial cases resulted in higher stresses than other stress states and the von Mises strain at penetration for the strip biaxial case was more than twice the strain for stress states (Figure 3). Additionally, water penetration did not occur in the uniaxial case.

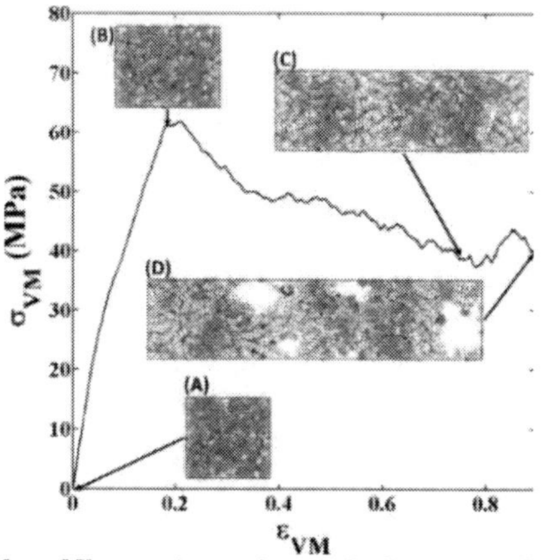

Figure 2: Von Mises stress-true strain curve for the strip biaxial case. Snapshots of the top view of the phospholipid bilayer with water hidden show the phospholipid bilayer at the (A) initial structure, (B) first peak stress, (C) full water penetration (failure), and (D) multiple fully penetrating pores.

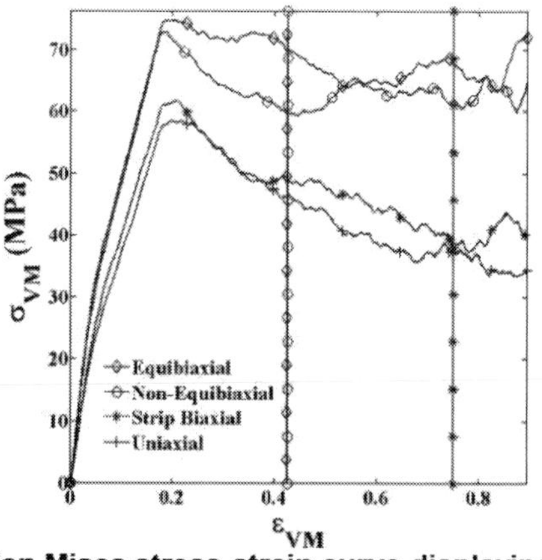

Figure 3: Von Mises stress-strain curve displaying the effect of the stress state (Equibiaxial, Non-Equibiaxial, Strip Biaxial, and Uniaxial Tension) at $\dot{\epsilon}_{VM} = 5.5\times 10^8$ s^{-1}. Vertical lines with markers matching their corresponding curve display when full water penetration occurred.

DISCUSSION

The effect of stress state on the resulting stress of the phospholipid bilayer and the point of failure when performing deformations at high rates has been examined in the current study. This additional characterization of the stress response of the phospholipid bilayer during high rate deformations will provide more data for implementation into multiscale TBI models.

The current simulations show that phospholipid bilayer stress initially increases, peaks, and decreases with failure occurring long after the stress peak (Figure 2). This trend is contrary to previous studies of the phospholipid bilayer, where the observed trend is a more gradual increase in stress that continues to increase or remains constant until failure. However, the authors believe this difference in trend to be due to exceeding the loading range in which the phospholipid bilayer can rearrange sufficiently to obtain favorable energy state, resulting in the multiple expanding pores seen in Figure 2 (D). Additionally, the bilayer experiences considerably higher stresses and a much lower von Mises strain at failure during the equibiaxial and non-equibiaxial cases than other stress states (Figure 3), while the uniaxial does not experience failure. These observations indicate the equibiaxial and non-equibiaxial stress states are much more detrimental. The uniaxial displays that the bilayer can effectively rearrange to prevent the creation of pores large enough to allow water penetration. However, while a free bilayer could potentially rearrange as long as the number of phospholipids allowed when pulled in only one direction, there are many obstructions in a cell membrane that would restrict phospholipid movements in a deformation scenario.

ACKNOWLEDGEMENTS

This material is based upon work supported by the Department of Agricultural and Biological Engineering and the Center for Advanced Vehicular Systems at Mississippi State University. Further, it has also been supported by US Army, through a subcontract with Mississippi State University (W912HZ-13-C-0037), and was performed for the CRES-GV research program. Any opinions, findings and conclusions or recommendations expressed in this material are those of the author(s) and do not necessarily reflect the views of the U.S. Army ERDC ERC.

REFERENCES

1. Faul, M., et al., *Traumatic Brain Injury in the United States: Emergency Department Visits, Hospitalizations, and Deaths 2002-2006*, N.C.f.I.P.a. Control, Editor. 2010, Centers for Disease Control Prevention, National Center for Injury Prevention and Control: Atlanta (GA).
2. Farkas, O. and Povlishock, J.T., *Cellular and subcellular change evoked by diffuse traumatic brain injury: a complex web of change extending far beyond focal damage*, in *Prog Brain Res*, T.W. John and I.R.M. Andrew, Editors. 2007, Elsevier. p. 43-59.
3. Prabhu, R., et al., J Mech Behav Biomed Mater. 4:1067-1080, 2011.
4. Evans, E., et al., Biophys J. 85:2342-50, 2003.
5. Koshiyama, K. and Wada, S., J Biomech. 44:2053-8, 2011.
6. Tomasini, M.D., Rinaldi, C., and Tomassone, M.S., Exp Biol Med (Maywood). 235:181-8, 2010.
7. Klauda, J.B., et al., J Phys Chem B. 114:7830-7843, 2010.
8. Plimpton, S., J Comput Phys. 117:1-19, 1995.
9. Stukowski, A., Modell Simul Mater Sci Eng. 18:015012, 2010.
10. Horstemeyer, M.F., *Integrated Computational Materials Engineering (ICME) for Metals: Using Multiscale Modeling to Invigorate Engineering Design with Science*. 2012: Wiley.
11. *MATLAB*. 2013, The MathWorks Inc.: Natick, MA.

DEVELOPMENT OF AN ESTIMATION METHOD OF BLOOD VESSEL CONFIGURATION AT NO LOAD SATE FOR FE PATIENT-SPECIFIC SIMULATION

M. Yu (1), S. Yamamoto (1), M. Oshima (2)

(1) Mechanical Engineering
Shibaura Institute of Technology
3-7-5 Toyosu Koto-ku, Tokyo, Japan

(2) Interfaculty initiative in Information Studies
The University of Tokyo
4-6-1 Komaba Meguro-ku, Tokyo,Japan
Country

INTRODUCTION

One of the important issues in a patient-specific modeling and simulation is to predict the hemodynamics as realistic as possible. The patient-specific model for an FEM analysis is constructed from the medical data such as CT or MRI data in order to perform an accurate simulation under patient specific geometry. Although the geometry obtained from the medical image data can be assumed to represent the geometry under a certain blood pressure ranging from 80 to 100 mmHg, the stress and stress can not be obtained from them. Since such conditions are unknown, they are assumed to be zero in a FSI or a structural analysis. Therefore, it is necessary to estimate such conditions as the initial condition,

In this study, a simple tube which is regarded as image-based geometry under 80mmHg is compressed according to the stress-strain relationship from the reverse direction such that the geometry at no load state condition can be obtained. This estimated model is used to simulate the blood flow under pulsatile pressure. The effects of the configuration with the present method is investigated by comparing the results of stress and displacement to those of the simulation within estimation, i.e. the original configuration constructed from the medical image data.

METHODS

The simple tube which is regarded as CCA under 80mmHg blood pressure is created by ICEM-CFD. The length is 25mm, the internal radius is 3mm and the thickness of the vessel wall is 1mm. The total number of elements is 27920 while the total number of nodes is 24531. The original model is designated as model A in this paper. In order to avoid the tube from bucking during compression of the wall, the soft compressive material is inserted in the luminal region of the tube. A nonlinear hyper elastic Ogden material is applied to the vessel wall. This material properties of the stress-strain relationship is obtained from the tensile test done by Yamada in 2010 as shown Fig. 1. The experiment is conducted at the range $\varepsilon > 0$. Therefore, in order to realize compression of the material, we created the negative region by taking the point symmetry curve of the stress-strain relationship of Yamada model. The configuration of the tube at no load state is estimated by loading 80mmHg pressure to the pipe surface to compress the model. Both edge surfaces of the tube are fixed. After compression, the resulting shape becomes a new analysis model, which is designated as model B. Since the difference of blood pressure between model A and model B is 80mmHg, it is necessary to add 80mmHg pressure onto model B before applying the pulsatile pressure. The pressure curve prescribed to two models in this study is described in Fig.2. All analysis are calculated by LS-DYNA.

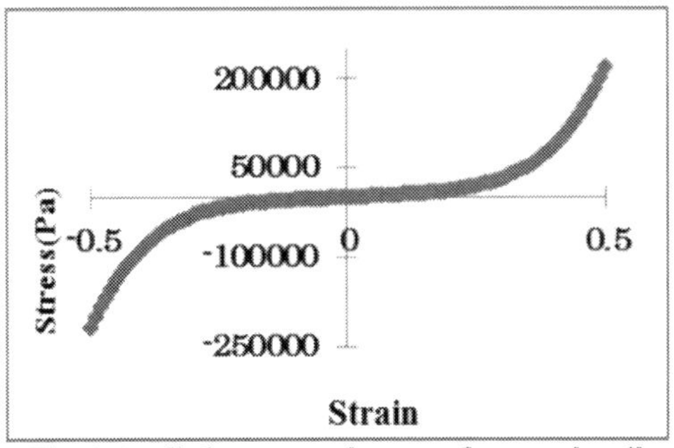

Fig. 1 Applied stress-strain curve for vessel wall

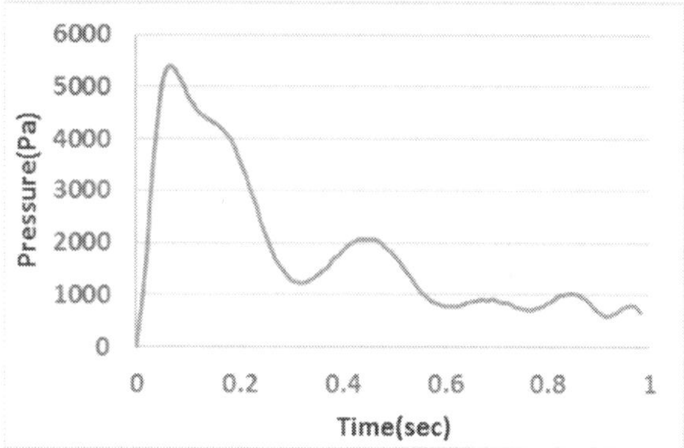

Fig. 2 Applied pulsatile pressure curve for model A model B

RESULTS

Figure 3 shows the difference in the displacement of diameter while pulsatile pressure varies from 80mmHg to120mmHg. Figure 4 shows the maximum principal stress for each model. In addition, Fig. 4 compares a relative maximum principal stress from 80mmHg for two cases. All results are referenced from the same point on internal surface. The peak of systole is at 0.06 sec, and the end of systole is at 0.3 sec.

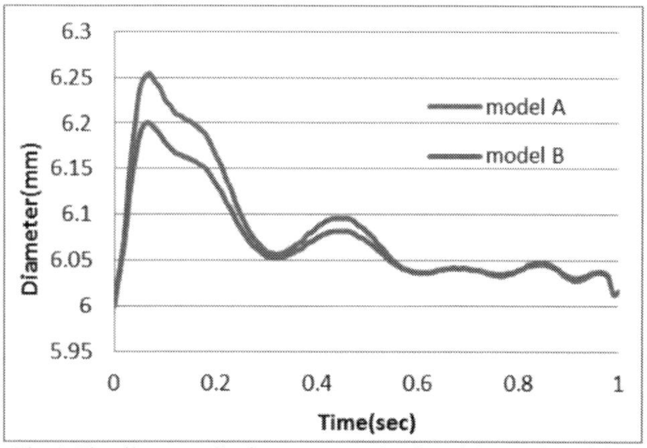

Fig. 3 The diameter displacement result of model A and model B

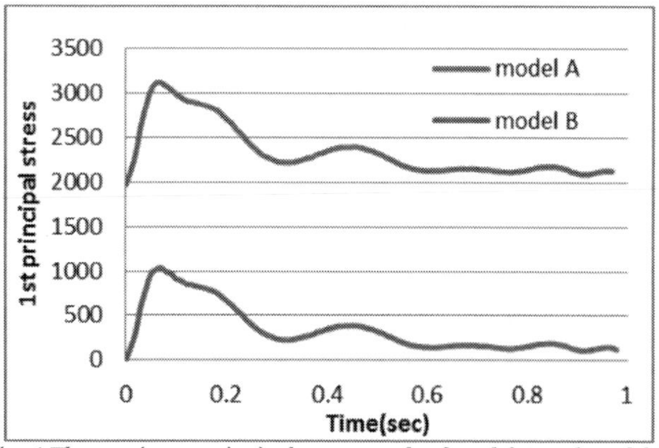

Fig. 4 The maximum principal stress result of model A and model B

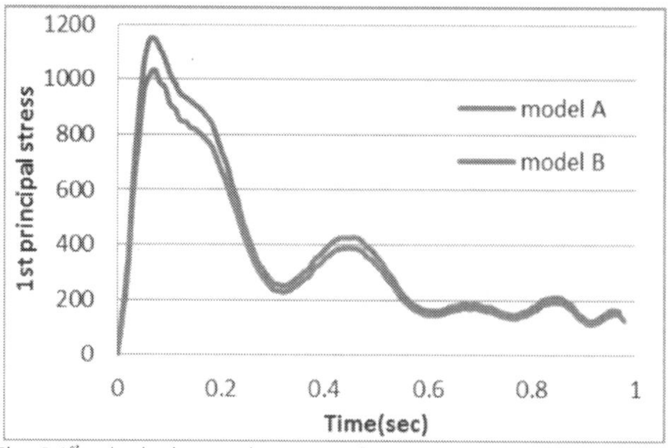

Fig. 5 1st principal stress increment from 80mmHg pressure state of model A and model B

DISCUSSION

As shown in Fig.3, the maximum difference in the diameter between two models is 5.4×10^{-3} mm which is less than 2% of the initial diameter. The result shown in Fig.4 indicates that the stress in model B is different from that in model A. This different is caused by the difference in configuration at the initial state such that model A deforms from 80mmHg state whereas model B deforms from 0mmHg state. When we discuss the stress under the realistic physiological blood pressure, it is important to consider the configuration at no load sate. The relative stress increment does not show a significant difference that the maximum difference is 132Pa at 0.06 seconds and two lines are similar at low pressure. This result from the stress-strain curve which is applied to the models. Even though the material is a non-linear hyper elastic one, it behaves as a linear material at this range of pressure. If the bigger pressure over 80mmHg is applied to the model or if different material curve are used, in which the curve has a strong non-linear characteristic, the difference between two models might become apparent.

The stress is one of important factors to evaluate the vascular rupture. Therefore appropriate analysis condition and model are necessary for the accurate estimation. As future work, we will use different thickness ratio, stress-strain curve and blood pressure.

ACHNOWLEDGEMENT

This work was supported by JSPS KAKENHI Grant Number 24300158.

We also received great advices from Professor Shu Takagi, the University of Tokyo.

REFERENCES

[1] Ryo Torii et al., International Jornal for Numerical Methods in Fluids, Vol 65, Issue 1-3, pages 324-340, 10-30 Janurary 2011.

[2] S Zeinali-Davarani et al, Computer methods in Biomechanics and Biomedical Engineering, Volume 14, Issue 9, 2011

[3] Hiroshi Yamada et al., Jornal of Biomechanical science and Engineering, Vol. 5,No. 4, 2010.

GENIPIN CROSS-LINKING SILK FIBROIN POST-GELATION INCREASES GEL MECHANICAL STIFFNESS

W.H. Elliott (1,2), W. Bonani (2,3), D. Maniglio (2,3), A. Motta (2,3), W. Tan (1), C. Migliaresi (2,3)

(1) Department of Mechanical Engineering
University of Colorado at Boulder
Boulder, Colorado, U.S.A.

(2) Department of Industrial Engineering
and BIOtech Research Centre
University of Trento
Trento, TN, Italy

(3) European Institute of Excellence on Tissue
Engineering and Regenerative Medicine, and
INSTM Trento Research Unit
Trento, TN, Italy

INTRODUCTION

Silk fibroin derived from domesticated silk worms (Bombyx mori) is a long chain protein (~375 kDa) with repeating GAGAGAS and GAGAGAT amino acid sequences consisting of 90% of their primary structure. These repeat units form β-sheet secondary structures with hydrophobic side groups on their surface, leading to intra- and inter-molecular, closely-packed crystallization structures,[1] acting as "physical" crosslinks and resulting in stable gel structures. Resulting crystal structures are embedded within amorphous entanglements by the remaining primary structure. The result is a triphasic material, with fluid-filled pores, viscoelastic amorphous entanglements, and elastic crystalline structures. These gels can be highly porous and exhibit excellent cell attachment and viability. [2]

Gel mechanical properties are especially increased by high-pressure carbon-dioxide treatment (PCT) decreasing solution pH, causing β-sheet crystal formation,[3] or by chemical crosslink. Genipin is a common chemical crosslinker due to its low cytotoxicity and ease of use. Genipin bonds to two separate primary amide groups, [4] thus interacting with fibroin molecule N-terminus, and amino acids with primary amide side-chains. These amino acids comprise 2% of silk fibroin primary structure, outside β-sheet crystal structures. [5, 6] GCX formation may be hindered after β-sheet crystallization from restricted motility of amorphous primary structure, which contain available primary amide groups. Increased genipin crosslink (GCX) increases mechanical stiffness of hydrogels. [7]

Control over the crosslinking mechanism permits the tuning of dynamic mechanical properties, which is important in tissue engineering application. Herein, we show increased stiffness in silk fibroin hydrogels through genipin crosslink (GCX) formation during or after gelation.

METHODS

Silk fibroin solutions in water were prepared according to Kim et al., [8] with limited modifications. SF solution underwent gelation without genipin (Control), or GCX formation at 37°C for 24 hours, either concurrent with gelation (CCX), or post-gelation (PostCX). Gelation through PCT occurred at 60 bar, and was previously described in Floren, et al. [5] Control and CCX gelation occurred at 37°C, while Post24CX gelation occurred at room temperature (RT) to reduce premature genipin activation.

GCX concentration by confocal microscopy Previous literature shows GCX content is fluorescent with the excitation peak at 590 nm and emission peak at 630 nm. [8] Lyophilized gels were rehydrated using DI for 24 hours. Maximum GCX efficiency was measured by Control gels immersed in 15 mM glycine solution and 1 mM genipin for 24 hours at 37°C. Samples were removed from DI and separated on a single chamber slide, partitions removed, and a second glass slide placed over the samples to ensure homogeneous depth of field. Samples were then imaged using confocal microscopy (Nikon Corporation, Chiyoda, Tokyo, Japan). The excitation and emission wavelengths were 561nm and 630 nm, respectively. For imaging, microscope aperture was maximally opened with the gain set to maximize emission intensity without saturation for the all gel samples. After maximum emission intensity was determined within the sample set, program settings were maintained as a method of comparing between samples. Max GCX allows the determination of the overall GCX efficiency. Summation of emission intensity over the given area was recorded and converted to numerical byte values for each pixel in a 64 x 64 matrix (resolution of 4.97 nm/pixel). Images were analyzed using NIS-Elements Viewer (Nikon Corporation), and emission intensities were determined with ImageJ (NIH, Bethesda, MD, USA).

Scanning Electron Microscopy Imaging Imaging of samples was obtained using a scanning electron microscope (Quanta 200 FESEM). Prior to imaging, lyophilized cross sections were sputter coated (Biorad SC500, Hemel Hempstead, U.K.) with a thin layer of gold or platinum/palladium mixture.

Dynamic Mechanical Analysis Dynamic mechanical analysis (DMA) was performed to establish the viscous and elastic properties of the material. Frequency sweep of cyclic, uniaxial compression testing was used to determine approximate location of the glass transition frequency (ω), storage moduli (E'), and loss moduli (E'') of samples. Samples were placed between horizontal, quartz slides attached to both the base fixture and driving piston of a mechanical testing machine. (Bose, Eden Prairie, MN, U.S.A.) The driving plate was brought into contact with the gel, and DI was placed around the gel, with surface tension allowing DI to maintain its integrity for the duration of the test. Subsequently, an average strain of 25% was applied in compression and the system allowed to reach the equilibrium. Dynamic strain was then oscillated at peak-to-peak amplitude of 20% strain. Oscillation frequencies ranged from 2.202E-3 to 1 Hz. Data was recorded after 4 seconds of pre-conditioning of the sample for each frequency.

RESULTS

Increased GCX efficiency is present in PostCX compared to CCX, as evidenced by confocal imaging. Both measured values showed significant differences from the control, according to paired t-test (Figure 1). Here we found the increase in stiffness was associated with increased crosslink density.

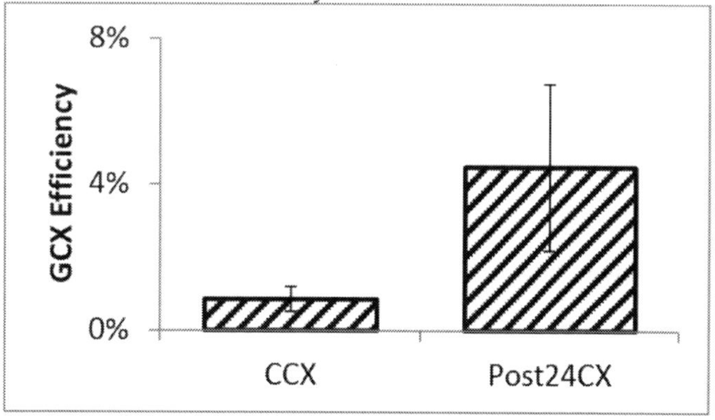

FIGURE 1- GCX EFFICIENCY FROM CONFOCAL MICROSCOPY

Gel morphology showed no perceivable changes between samples in SEM imaging. This indicates fluid evacuation from pores maintains similar response between samples.

DMA response shows stiffer response for both CCX and PostCX gels, when compared to control. CCX response shows significantly higher stiffness response at higher frequencies, while PostCX shows stiffer response at all frequencies (Figure 2).

E' and E'' results from DMA tests reveal typical viscoelastic response of gels, with lower and upper plateau regions at lower and higher frequencies, respectively, separated by a transition region. Values for E'' show similar results to E' across all samples when compared to control.

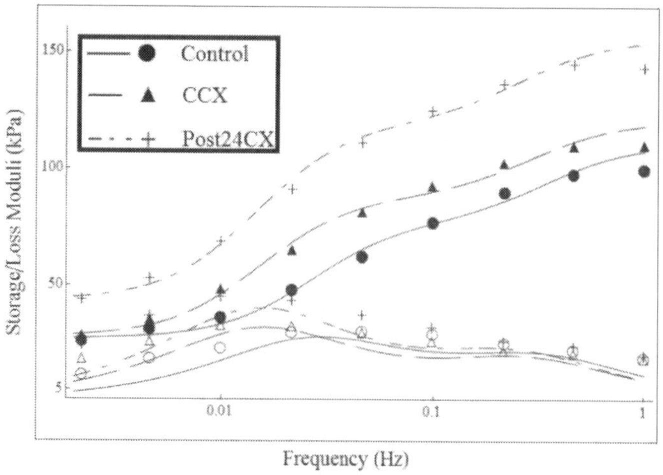

FIGURE 2- STORAGE AND LOSS MODULI

DISCUSSION

Crosslinking of silk fibroin hydrogels after gelation causes changes to molecular structure, and increased stiffness at all frequencies. High stiffness increase was seen in PostCX, likely due to increased GCX content as compared to CCX. Decreased crosslinking might occur in CCX because of increased pH of the system during GCX treatment, which was evidenced by both confocal data and stiffness response. Cross-linking of amorphous chain entanglements likely affects reptation mechanics, leading to stiffer response.

ACKNOWLEDGEMENTS

The authors wish to acknowledge funding sources, including NHLBI (HL119371 to W.T.).

REFERENCES

[1] Hu, X. et al., *Macromolecules,* 39: 6161-6170, 2006
[2] Wang, X. et al., *J Controlled Rel,* 121: 190-199, 2007
[3] Floren, M. et al. *Biomacromolecules,* 13:2060-2072, 2012
[4] Butler, M.F., et al., *Polym Sci Part Polym,* 41: 3941-3953, 2003[5] Marsh, R.E. et al., *Biochim, Biophys, Acta,* 16:1-34, 1955
[6] Ha, S.-W., et al., *Biomacromolecules,* 6: 2563-2569, 2005
[7] Madhavan, K., et al., *Acta Biomater.* 6: 1413–1422, 2010
[8] Kim, U.-J. et al., *Biomaterials,* 26: 2775-2785, 2005
[9] Sundararaghavan, H.G., et al., *J Biomed Mat Res A,* 87A: 308-320, 2008

DAMAGE ANALYSIS OF RETRIEVED OXIDIZED ZIRCONIUM FEMORAL COMPONENTS FOR TKA: CAN WEAR BREACH THE OXIDE LAYER?

Noah Bonnheim (1), Michael Ries, MD (2), Sanjai Shukla, MD (2), Lisa Pruitt, PhD (1)

(1) Department of Mechanical Engineering
University of California, Berkeley
Berkeley, CA, USA

(2) Tahoe Fracture and Orthopaedic Clinic
Carson City, NV, USA

INTRODUCTION

Oxidized zirconium was developed as an alternative material to cobalt-chromium (CoCr) for use as a bearing surface in total knee arthroplasty (TKA) and total hip arthroplasty (THA). Osteolysis from polyethylene wear continues to be a leading cause of implant failure after five years, and the development of lower friction, scratch-resistant opposing bearing materials such as oxidized zirconium to reduce adhesive and abrasive wear is of paramount interest [1].

Ceramics have demonstrated superior wear behavior and biocompatibility compared with CoCr, however their use has been restricted to femoral heads in the hip due to their brittle nature [2, 3]. Wrought zirconium alloy (Zr-2.5%Nb), when oxidized by thermal diffusion in heated air, develops a stable, low-friction ceramic oxide at the surface, which gradually transitions to the alloy substrate [4]. Oxidized zirconium offers the tribological properties of a ceramic with the mechanical ductility and fracture resilience of a metal [3]. *In vitro* studies have shown that oxidized zirconium femoral components in the knee can reduce polyethylene wear by between 42-85%, depending on test conditions, compared with CoCr [5,6].

While numerous *in vitro* studies of oxidized zirconium demonstrate a clear performance benefit, clinical studies are limited. Laskin reported a more rapid return of flexion and functional milestones with oxidized zirconium compared with CoCr, however more recent studies by George et al. and Hui et al. found no difference in clinical or radiographic outcomes [7,8,9]. Continued research into the *in vivo* performance of oxidized zirconium is essential to characterize the clinical merits of the material and further illuminate if it is a superior alternative to CoCr.

This study examined the wear damage on four retrieved oxidized zirconium femoral components for TKA using scanning electron microscopy and energy-dispersive x-ray spectroscopy. The condition of the 5μm oxide ceramic layer was assessed to determine whether it was breached *in vivo*. Penetration of wear through the oxide layer would represent a novel research finding regarding the clinical performance of oxidized zirconium.

METHODS

Four oxidized zirconium TKA femoral components (OXINIUM™, Smith & Nephew, Memphis, TN) were analyzed using scanning electron microscopy (SEM) and energy-dispersive x-ray spectroscopy (EDS). Average *in vivo* duration was 42.5 months (range: 24-84) and average patient age was 52.8 years (range: 34-62). Wear scars on the articulating surfaces of the four implants were analyzed to determine if their chemical signatures were consistent with the zirconia oxide layer or zirconium substrate, which inherently differ in their oxygen content.

Severe tool damage during retrieval exposed a significant amount of the zirconium substrate for one of the retrievals. This area, along with an undamaged area on a non-articulating portion of the implant, was analyzed as a control to ensure the EDS measurement technique could discriminate between oxidized surface layer and un-oxidized substrate.

RESULTS

The SEM/EDS measurement modality was able to successfully discriminate between the chemical signatures the oxide surface layer and the un-oxidized substrate at the site of tool damage. **Figure 1A** shows an SEM image of the boundary between the oxide surface layer (bottom of image) and the un-oxidized substrate (top of image). **Figure 1B** shows an EDS generated map of oxygen content, where oxygen is shown in light green. The visible demarcation between oxide surface layer and substrate in Figure 1A also presents as a boundary of oxygen content in Figure 1B. This is an indication of the validity of the measurement technique: measuring the chemical composition of the wear damage, specifically with respect to oxygen, can discern whether or not damage has reached the zirconium substrate.

FIGURE 1A (LEFT). SEM IMAGE OF BOUNDARY BETWEEN OXIDE SURFACE LAYER AND SUBSTRATE. FIGURE 1B (RIGHT). EDS MAP OF OXYGEN CONTENT.

There is no evidence to suggest that the oxide layer for any of the four implants was breached due to *in vivo* wear. The chemical composition of the wear scars on the articulating surface generally matched that of the undamaged oxide surface. The measured weight percent zirconium, oxygen, and niobium of the articulating surfaces of the four implants are reported in **Table 1**.

TABLE 1. WEIGH PERCENT ZIRCONIUM, OXYGEN, AND NIOBIUM OF ARTICULATING SURFACE WEAR SCARS.

	Sample 1	Sample 2	Sample 3	Sample 4
Wt % Zr	69.2	66.7	61	70.5
Wt % O	18.7	15.8	21.6	22.6
Wt % Nb	2.9	3.5	3	3.6

The weight percent oxygen for the wear scars, undamaged oxide surface, and un-oxidized substrate are reported in **Table 2**. No significant difference was found between the average oxygen content of the wear scars and the oxygen content of the unworn oxide surface ($p=0.153$). The average oxygen content of the wear scars (19.7±3.1 Wt%) was found to be significantly greater than the oxygen content of the un-oxidized substrate ($p=0.001$).

TABLE 2. AVERAGE WEIGHT PERCENT OXYGEN FOR WEAR SCARS, UNDAMAGED OXIDE SURFACE, AND UN-OXIDIZED SUBSTRATE.

	Wear Scar Average (n=4)	Undamaged Oxide Surface (n=1)	Un-Oxidized Substrate (n=1)
Wt % O	19.7	22.6	0

Two of the four implants (Samples 2 and 4) had measurable amounts of titanium on their articulating surface (1.51 and 0.84 Wt%, respectively). This likely indicates intraoperative deposition from the tibial tray during either implantation or explanation.

DISCUSSION

This investigation of the *in vivo* integrity of oxidized zirconium through SEM/EDS chemical analysis is the first of its kind to our knowledge. The retrieval analysis shows that the material appears to exhibit adequate durability for long term-implantation and the tribological properties of the ceramic layer are maintained *in vivo*. For the sample analysis, SEM/EDS data indicated that *in vivo* dynamics do not cause penetration of the oxide ceramic layer, a finding in line with *in vitro* studies.

This study is the first to report on the presence of titanium on the oxidized zirconium femoral surface, which was likely caused by contact between the femoral component and the tibial tray during either implantation or explanation. The clinical implications are not known; however, it is hypothesized that they are insignificant.

The average time *in vivo* (42.5 months) for the retrievals studied is shorter than is typical for TKA procedures. It is possible that the relatively short *in vivo* duration underrepresented normal wear damage. Investigation of retrieved implants with a longer *in vivo* duration using a similar experimental methodology presents important future work to better characterize the clinical merits of oxidized zirconium.

ACKNOWLEDGEMENTS

We would like to acknowledge Farzana Ansari and Hannah Gramling for help with these studies.

REFERENCES

[1] Gioe, T, et al. Why Are Total Knee Replacements Revised? *Clinical Orthopaedics and Related Research*, 428, pp. 100-106, 2004.
[2] Wang, S, Zhang, S, Zhao, Y. A Comparison of Polyethylene Wear Between Cobalt-Chrome Ball Heads and Alumina Ball Heads After Total Hip Arthroplasty. *Journal of Orthopaedic Surgery and Research*, 8(20), pp. 1-4, 2013.
[3] Hunter, G, Jones, W, Spector, M. Oxidized Zirconium. In: Bellemans, J, Ries, M, Victor, J Eds. *Total Knee Arthroplasty: A Guide to Get Better Performance.* Würzburg, Germany: Springer, pp. 370-377, 2005.
[4] Hunter, G, et al. Creation of Oxidized Zirconium Orthopaedic Implants. *Journal of ASTM International*, 2(7), pp. 16-20, 2005.
[5] Ezzet, K, et al. Oxidized Zirconium Femoral Components Reduce Polyethylene Wear in a Knee Wear Simulator. *Clinical Orthopedics and Related Research*, 428, pp. 120-124, 2004.
[6] Spector, M, et al. Wear Performance of Ultra-High Molecular Weight Polyethylene on Oxidized Zirconium Total Knee Femoral Components. *The Journal of Bone and Joint Surgery*, 83-A(2), pp. 80-86, 2001.
[7] Laskin, R. An Oxidized Zr Ceramic Surfaced Femoral Component for Total Knee Arthroplasty. *Clinical Orthopedics and Related Research*, 416, pp. 191-196, 2003.
[8] George, G, et al. Clinical and Radiographic Comparison of Oxidized Zirconium and Cobalt Chrome Femoral Components of a Single Design Primary Total Knee Arthroplasty. *ISRN Biomaterials*, 2013, pp. 1-7, 2013.
[9] Hui, C, et al. Five-Year Comparison of Oxidized Zirconium and Cobalt-Chromium Femoral Components in Total Knee Arthroplasty. *The Journal of Bone and Joint Surgery*, 93-A(7), pp. 624-630, 2011.

SB3C2015
Summer Biomechanics, Bioengineering and Biotransport Conference
June 17-20, 2015,Snowbird Resort, Utah, USA
SB3C2015-1136

INVESTIGATION OF POSSIBLE CORRELATION BETWEEN BRAIN TISSUE RESPONSE AND HEAD KINEMATICS FOR BLAST-INDUCED BRAIN INJURY

Hesam Sarvghad-Moghaddam (1), Mariusz Ziejewski (1), Ghodrat Karami (1)

(1) Mechanical Engineering Department, North Dakota State University, Fargo, ND, USA

INTRODUCTION

Blast Induced Traumatic Brain Injury (bTBI) occurs as a result of the impact of blast waves on the soldiers' head. Upon impingement of the shockwaves on the head, the high pressure waves propagates through the soft tissue of brain. Due to the relative displacement of the brain caused by the sudden acceleration/deceleration imposed by the waves, high pressure profiles, as well as shear stresses are formed at different lesions of the brain. They will cause damage to the tissues leading to contusion, diffuse axonal injury, and intracranial hemorrhage.

Several injury criteria have been proposed based on both tissue responses of the brain, such as shear stress, shear strain, and intracranial pressure (ICP), as well as head kinematics such as linear acceleration [1-3]. Although head acceleration is commonly used as the major brain injury predictor, several researchers have suggested the use of tissue responses as the main injury predictors. King et al. [4] evaluated the linear and angular acceleration and shear strain of the brain tissue for unprotected and helmeted heads under similar impact scenarios. Observing no change in angular acceleration upon using the helmet, they concluded that the tissue response of the head is a more viable injury predictor than the input accelerations. Zhang et al. [1] proposed an injury criterion for mild TBI based on the shear stress levels in the brain tissue. Using the angular and linear acceleration data as the input to their finite element (FE) model, they evaluated the brain response in terms of ICP and shear stress. To the authors' best knowledge, the study on the correlative relationship of brain's tissue response and the head kinematics have mainly been performed for head injuries under impact. However, as the injury mechanism for the bTBI is different from impact induced TBI, evaluation of head and brain biomechanics under blast shockwaves is of great importance.

Accordingly, in the research presented here, a computational approach was taken to investigate possible correlation between the brain's tissue response and head kinematics to find out whether the linear acceleration can be used as a sole injury predictor is defining brain injury for bTBI. The significance of such a study is that it can help improve the design and material of the current military helmets for enhanced blast protection as well as develop state of the art equipment to evaluate the risk of injury for the soldiers.

METHODS

A comprehensive finite element (FE) analysis was performed on the interaction of North Dakota State University Head Model (NDSUHM) with blast-induced shockwaves [5]. In order to comply with the threshold of lung injury, a an overpressure of approximately 520 kPa due to blast was generated around the head and it was exposed to identical boundary conditions from front, back, top, and side orientations. LS-DYNA nonlinear FE solver was employed to model the generation and propagation of blast waves, and the response of the 3D NDSUHM to the incoming waves. A numerical algorithm was implemented through a robust multi-material Arbitrary Lagrangian-Eulerian (ALE) technique to perform the fluid-solid interaction (FSI) of blast waves and the head [5].

Linear elastic material properties have been selected for all the head components except for the brain which was modeled using the hyper-viscoelastic constitutive law [5].

RESULTS

The tissue response of the brain as well as the linear acceleration of the head was evaluated in all directions and were compared against each other. Propagation and interaction of blast waves with the head and the development of the resulting maximum pressure profile at the coup site and the shear stress at the brainstem is shown in Figure 1 for front blast. The peak linear acceleration for the head as well as the highest peak values for ICP and shear stress were observed for the blast from the side.

Figure 2 show the variation of peak ICP and shear stress values in each direction with the linear acceleration. An important finding was the direction dependent response of the head to an identical loading in different directions in terms of both tissue and kinematical responses. Moreover, undergoing a higher linear acceleration did not necessarily impose higher tissue responses (blast from top). Time histories of brain ICP and head's linear acceleration showed similar patterns in all directions in terms of the peak duration while shear stress pattern varied with changing the direction. Compared to maximum shear stress, much better correlation was found between the ICP peak ICP values and the head acceleration.

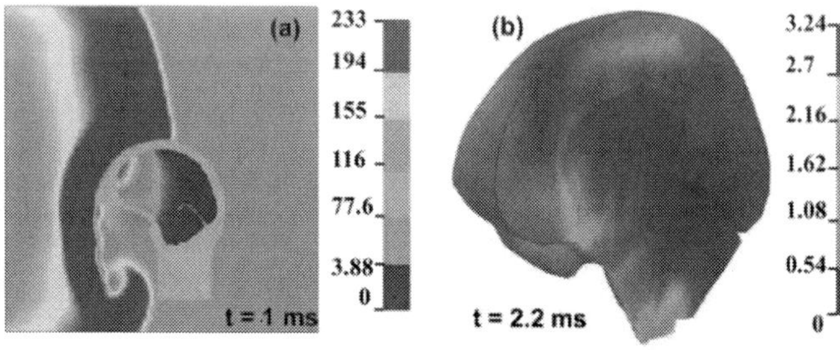

FIGURE 1. (A) BLAST WAVE PROPGATION AND ICP (KPA) DEVELOPMENT (B) MAX. SHEAR STRESS AT BRAINSTEM

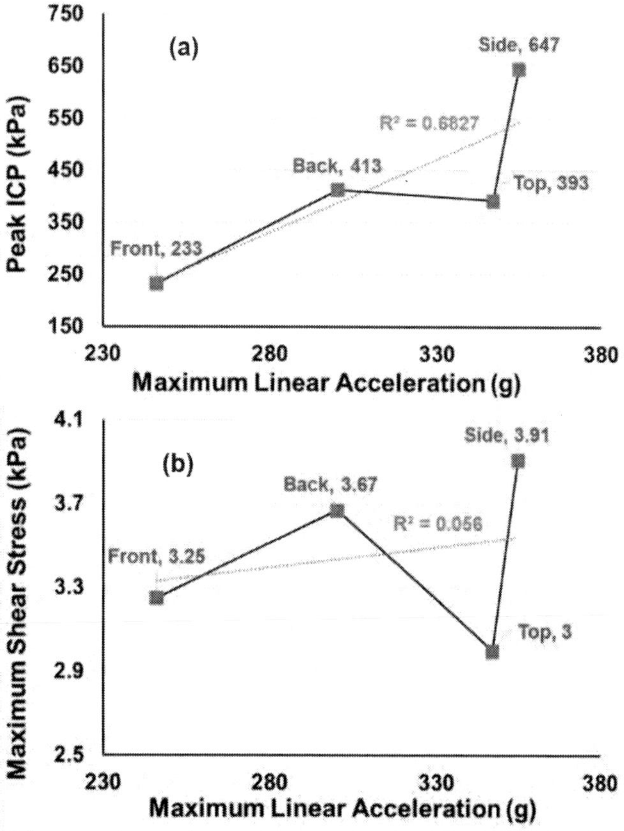

FIGURE 2. RELATIONSHIP BETWEEN ACCELERATION OF HEAD AND (A) PEAK ICP (B) MAX. SHEAR STRESS

DISCUSSION

Human head and its components-specifically brain- are composed of different constitutive parts with different shapes, functions, and materials. It mean that the response of the head would differ based on the location of impact. Accordingly, the idea behind the present research was to investigate the accuracy of considering linear acceleration as a sole injury predictor by finding the response of the head to an identical blast load from four different directions. Regarding the injury thresholds introduced by Ward et al. [2] (peak coup ICP = 235 kPa) the head was marginally in the safe zone only for the front blast (ICP = 233 kPa). However, based on Zhang et al. criterion (max. shear stress = 7.8 kPa), the brain was far from mild TBI in all scenarios. However, the range of linear acceleration of head (>200 g) puts the head in the severe and critical injury range based on [3]. These discrepancies between the head kinematics and brain's response is an indication of the lack of credibility of linear acceleration to be a sole indicator of brain injury. As shown in Figure 2a, the variation of ICP levels showed an acceptable correlation with linear acceleration, while shear stress basically showed no effective correlation with that (Figure 2b). This would be in agreement with the results of King et al. [4] relating linear acceleration variations to pressure gradient. While the acceleration of the head differed in each direction due to the directional behavior of the head, still for pretty close peak values of acceleration in top and side blasts, the brain predicted a much higher peak ICP value in side blast (647 kPa), nearly 64% more than that in top blast (393 kPa). However, the shear stress was only 30% higher in side blast and was still small (3.9 kPa). Accordingly, another disadvantage of using only the linear acceleration as the injury predictor is that in cases like the one mentioned, it would not be possible to discern different types of injury as the pressure gradients mainly relates to concussion and contusion injuries while shear stress is associated with diffuse injuries such as the diffuse axonal injury (DAI) which can be more critical than concussion.

The results of this specific case study showed that the tissue response of the brain, such as shear stress, strain, and ICP, are much more viable injury predictors compared to head acceleration. The reason lies within the fact that the brain response is an indicative of the local response of the brain tissue to the applied load. It was observed that while linear acceleration may be able to help define the generality of pressure-related injuries such as concussion, but it is unable to predict the extent and level of injury. Moreover, due to very weak correlation with shear stress variation, it might not be able to accurately predict the diffuse and shear related injuries.

The relationship between head kinematics and the brain response would be of significance in designing the protective tools such as helmet for head protection. The possibility of any correlation among these parameters would be lower as the application of helmet on the head would increase the complexity of the head response.

ACKNOWLEDGMENTS

The authors would like to acknowledge the Army Research Office (ARO) for their financial support.

REFERENCES

[1] Zhang, L et al., 2004, *J. Neurotrauma* 126(2): 226-236, 2004.
[2] Ward, C.C.et al., *Proc. 24th Stapp Car Crash Conference*, SAE Paper Number 801304: 347–360, 1980.
[3]St Clair, VJM, et al., *TRL Project Report* PPR213, 2007.
[4] King, A.I. et al. *Proc. IRCOBI, Lisbon, Portugal*, 1-12, 2003.
[5] Sarvghad-Moghaddam, H. et al. *Comput Methods Biomech Biomed Engin*, 1-10: 2014.

SB³C2015
Summer Biomechanics, Bioengineering and Biotransport Conference
June 17-20, 2015,Snowbird Resort, Utah, USA
SB³C2015-1137

MEETING THE NEED FOR IMPROVED MONITORING OF CRYOTHERAPY IN THIN CARDIOVASCULAR TISSUES WITH A NEW MICRO-THERMAL SENSOR

Harishankar Natesan (1), Jeunghwan Choi (1, 2), Wyatt Hodges (3), Sean Lubner (3), Chris Dames (3), John Bischof (1)

(1) Department of Mechanical Engineering
University of Minnesota
Minneapolis, Minnesota, USA

(2) Department of Engineering
East Carolina University
Greenville, North Carolina, USA

(3) Department of Mechanical Engineering
University of California
Berkeley, California, USA

INTRODUCTION

Atrial fibrillation (AF), which affects millions of people every year, is commonly treated by focal thermal therapy of the pulmonary vein (PV). Specifically, thermal and cryoablation for PV isolation has been used for more than 10 years in AF treatment [1, 2]. These treatments require improved monitoring of probe contact, tissue thickness, and freeze or heat completion through the wall for optimal outcomes. Unfortunately, clinical imaging such as Ultrasound (US), Magnetic Resonance Imaging (MRI), and Computed Tomography (CT) have resolutions similar in scale (~ 1 - 2 mm) to the PV thickness, and therefore are unable to adequately resolve changes at the mm and sub mm scale within the PV during treatment. Thus, alternatives are urgently needed. Here then we present a new, micro-thermal sensor based on the "3ω" technique [4] modified for thin, anisotropic and/or composite biological systems [5]. We use this sensor to measure and report the first thermal properties of PV (1-2 mm thick) and esophageal tissue (2-3 mm thick). Further, we demonstrate the ability of the sensor to discriminate between tissue vs. fluid contact (with and without flow). We also demonstrate new proof of principle that the sensor can measure tissue thickness in the 1 to 2 mm regime and the initiation and completion of the freezing process with an accuracy better than 0.1 mm. Together, these new capabilities stand to vastly improve the monitoring and therefore application of thermal probes to treat disease in thin cardiovascular tissues.

METHODS

The supported 3ω technique is used to measure thermal conductivity (k) of thin (≤ mm thicknesses) biological tissues. It uses a sensor made up of a thin gold heater line (3 mm X 30 μm X 200 nm) deposited on a glass substrate with a tissue sample on top. An AC current excites the heater line producing thermal waves traversing the

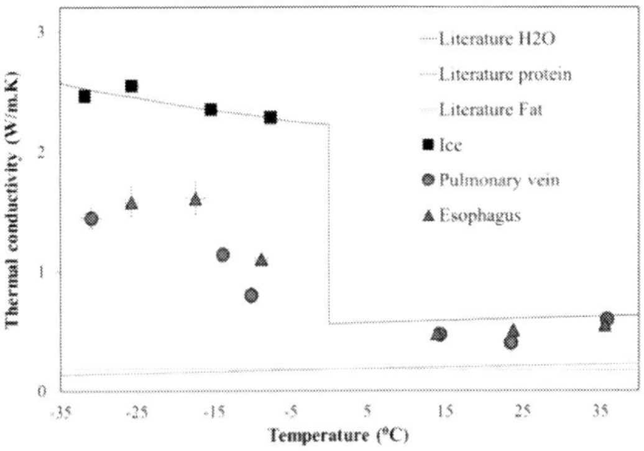

Figure 1. Supported 3ω measurements of porcine PV and porcine esophagus (N=3). The difference in thermal conductivity between frozen and unfrozen tissue is roughly a factor of 2. The straight line represents thermal conductivity of water, ice, protein and fat from literature [3]

sample and the substrate. By measuring the third harmonic voltage across the heater line and knowing the k of the substrate, the k of the sample region traversed by the thermal waves can be determined. The sampled region (i.e., thermal penetration depth (L_t)) can be controlled by changing the frequency (f) of the AC current, where $L_t \propto 1/f^{0.5}$. In this study, the sensors are embedded on a flat glass substrate in contact with cooling surfaces for preliminary proof of concept experiments. Future studies are envisioned where these sensors are integrated directly onto freezing probe or catheter surfaces.

RESULTS

Our initial experiments using this sensor demonstrate better than 5% uncertainty in k measurements for sample thicknesses down to 100 microns in water and mouse liver tissue [5]. Here we expand this data set to include samples of porcine PV < 2 mm thick and surrounding esophageal tissues <3 mm thick as shown in Fig 1. These properties can then be used directly for treatment planning, or used to inform tissue contact, thickness and freeze initiation and completion.

First, to verify our ability to sense probe contact, we used our existing supported 3ω sensor to differentiate between tissue contact and flow conditions. Fig. 2 (bottom, A) shows a constant heat conductance (h_c) as water is at rest. As water flows in Fig 2 (bottom, B), h_c rises due to convective heat loss. Fig 2 (bottom, C) shows that placing a piece of tissue (1 mm thick: here represented by bologna) on top of the sensor blocks the water flow and thereby drops h_c down to the lowest value of all cases.

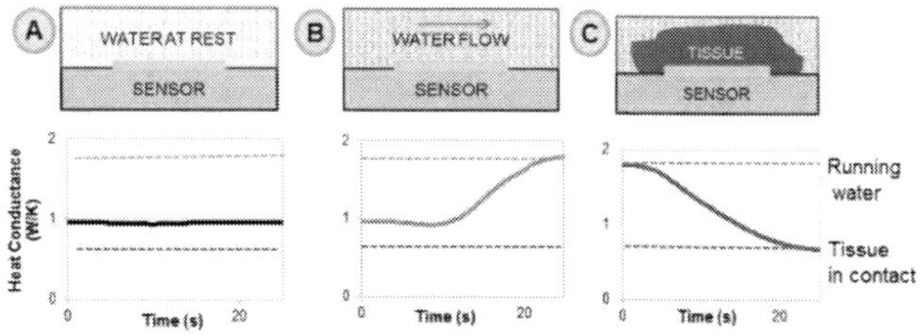

Figure 2. *Top: In-vitro experiment to identify flow vs. tissue contact: (A) water at rest, (B) water flowing, and (C) tissue in contact; Bottom: Heat conductance of flowing water > water at rest > model tissue.*

Next, to verify our ability to measure thickness, we prepared several samples of thin ice of varying thickness, L, and recorded the raw 3ω signal ($\sim 1/h_c$) in Fig. 3. The backside boundary conditions were also varied, approximating both isothermal (highly-conductive copper back) and well-insulated (low-conductivity plastic) conditions. Below a characteristic frequency ($f_c \sim \alpha/L^2$), the 3ω voltage deviates (black arrows in Fig 3) irrespective of the backside boundary condition, thereby providing a measure of the thickness.

Finally, as a preliminary demonstration of freeze-front tracking in real-time, we used the sensor to monitor a propagating ice/water freezing front, through its effect on the 3ω voltage. In this example the front surface was perpendicular to, and propagating through, the heater line, rather than the more common situation of a front parallel to, and not touching, the heater line (See Fig. 4). At the same time, we used a video camera to record the actual location of the freezing front. The preliminary comparison shows that we can measure the water/ice interface from the voltage signal alone, at better than 0.1 mm.

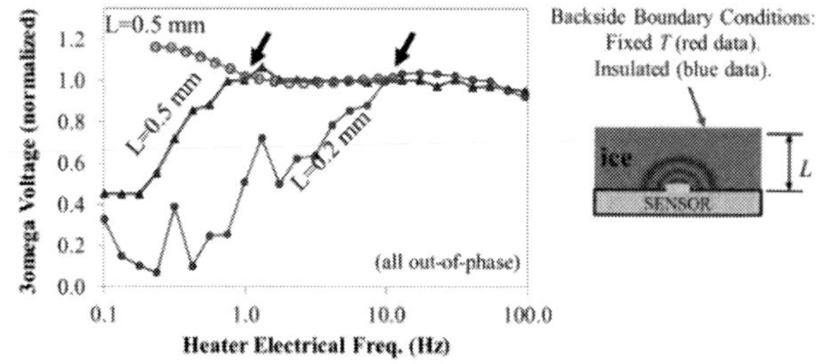

Figure 3. *Experimental confirmation of the ability to detect sample thickness through a frequency sweep of 3ω data.*

DISCUSSION

We have provided the first proof of principle that a supported 3ω sensor can monitor important dynamic aspects of cryotherapy in thin cardiovascular tissues such as PV. Specifically, we measure and report the first thermal conductivity measurements of the thin tissues involved (PV and esophagus) (Fig. 1). Further, we show the ability to sense tissue contact vs. stagnant or flowing fluid in Fig 2 to know when a probe is properly deployed. We then demonstrate the new ability to measure tissue thickness in Fig 3, which can help plan how deep to freeze. Finally, we also demonstrate the ability to measure the initiation and completion of tissue freezing with accuracy better than 0.1 mm in water in Fig 4. All of these capabilities will help adequately freeze the PV without over-freezing into adjacent structures such as the esophagus.

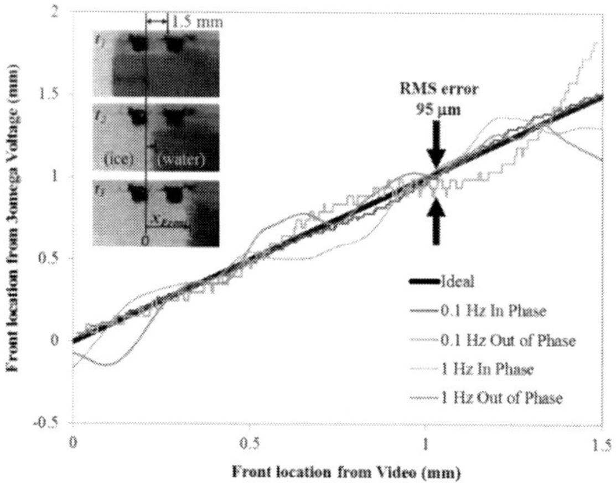

Figure 4. Using 3ω voltage to track location of a freezing front. *Inset: Images from three different times, showing how video is used to locate the true front location. The 3ω heater line is not visible but is located between the two black square marks. Plotted data: Front location deduced from 3ω voltages. The RMS error is less than 100 microns, confirming excellent tracking potential.*

In summary, this highly innovative work represents several conceptual firsts in the monitoring of thin tissue cryotherapy with a novel micro-thermal sensor. In future, this sensor can be modified to be disposable and deployed on a cryoprobe or balloon to measure thin tissue properties, contact, thickness and the initiation and completion of freezing. In addition to improving monitoring of freezing in PV as envisioned here, other monitoring of heat and cold treatments in peripheral artery disease, renal or hepatic artery (hypertension and diabetes) or even tissue fusion applications are possible.

ACKNOWLEDGEMENTS

Funding from NSF CBET 1236760 and Medtronic, Inc.

REFERENCES

1. Avitall, B., et al., Journal of Cardiovascular Electrophysiology, 2003. **14**(3): p. 281-286.
2. Sarabanda, A.V., et al., Journal of the American College of Cardiology, 2005. **46**(10): p. 1902-1912.
3. *ASHRAE Handbook.* Refrigeration, ASHRAE, New York, 2002: p. 81-830.
4. Cahill, D.G. and R.O. Pohl, Physical Review B (Condensed Matter), 1987. **35**(8): p. 4067-73.
5. Lubner, S.D., et al., Review of Scientific Instruments, 2015. **86**(1).

SB³C2015
Summer Biomechanics, Bioengineering and Blotransport Conference
June 17-20, 2015,Snowbird Resort, Utah, USA
SB³C2015-1140

ALTERATION AND FAILURE OF CEREBRAL ARTERY INTERNAL ELASTIC LAMINA FOLLOWING AXIAL OVERSTRETCH

Matthew I. Converse (1), Tessa Sommer (2), Kenneth L. Monson (1,2)

(1) Department of Mechanical Engineering
University of Utah
Salt Lake City, UT, USA

(2) Department of Bioengineering
University of Utah
Salt Lake City, UT, USA

INTRODUCTION

Traumatic brain injury (TBI) is one of the major health care problems in the United States, contributing to nearly a third (30.5%) of all injury-related deaths [1]. Although damage to brain tissue is of primary concern in TBI, nearly all significant traumatic brain injuries include some element of injury to the blood vessels, and any injury or dysfunction of the vasculature puts neural tissue at risk.

Two recent clinical studies have identified TBI as a significant risk factor for stroke [2, 3]. Both investigations excluded patients with intracranial hemorrhage at the time of TBI but still found increased risk of stroke. The reported correlation up to five years after the initial event suggests that TBI may induce unrecoverable damage to the vasculature. Little is known about how vessels are injured and how an initial injury affects blood vessel function. This knowledge gap limits new diagnostic, preventative, and treatment strategies.

We hypothesize that mechanical stretch induces permanent subfailure damage to cerebral blood vessels – specifically to the elastic structures. We test this hypothesis by examining structural damage to the internal elastic lamina (IEL) in isolated lamb middle cerebral arteries (MCA) following a single axial overstretch. We explore failure thresholds of the elastic layer as well as changes to the size and shape of fenestrae (holes) inherent to the IEL. Insights into changes in elastic structures may serve to explain, in part, this interesting link between TBI and stroke.

METHODS

Vessel dissection and cannulation

Two eight day old lambs were euthanized with an overdose of beuthanasia. Immediately after death, left and right middle cerebral arteries (MCAs) were resected and transported to the lab on ice in Hank's Buffered Saline Solution. Nine relatively straight segments (approx. 3-5 mm in length) of the MCAs were dissected, and side branches were ligated with individual fibrils of unwound 6-0 silk suture. Vessels were then cannulated to two size-matched hypodermic needles with 6-0 silk suture and cyanoacrylate glue. Microspheres were also spread over the vessel surface to allow visual tracking of deformation during testing. All testing was carried out within 48 hours of death.

Preconditioning

Prior to axial overstretch, each vessel was preconditioned by oscillating the luminal pressure (50-150 mmHg) for five cycles while the length was held constant at axial stretch values up to 1.05 times the in vivo length. Axial in vivo length was estimated as the length at which pressure cycles had a negligible effect on the tensile load [4]. The vessel was then stretched axially from a buckled length up to the maximum length seen in preconditioning while held at an internal pressure of 2 mmHg. The unloaded length was defined as the length at which axial force began to rise during this procedure.

Axial overstretch

Six of the nine arterial lengths were subjected to a single axial overstretch ($\lambda_{z\_max} = 1.29 \pm 0.08$) at 0.1 mm/s with an internal pressure of 100 mmHg. Three control segments were only preconditioned. Samples were then returned to their in vivo length and pressure (100 mmHg) and fixed in 4% PFA for 30 mins.

Confocal imaging

Strips of the arterial wall were then flattened onto glass slides for confocal imaging of the IEL. Eight 60x images were taken at different locations on each strip. The area, shape, and orientation of IEL fenestrae were quantified using a custom MATLAB program.

RESULTS

Failure of the internal elastic lamina

Of the six segments that were overstretched, one ($\lambda_{z\_max} = 1.17$) showed no sign of layer-specific failure and four ($\lambda_{z\_max} = 1.23 \pm 0.05$) had failure of the IEL, yet continued to hold pressure. A sixth segment ($\lambda_{z\_max} = 1.38$) also had failure of the IEL but lost pressure at $\lambda_z = 1.37$. In these latter five segments, the onset of IEL failure occurred at $\lambda_z = 1.29 \pm 0.02$. Failure of the IEL was observed in the images taken during the axial overstretch and was confirmed in the confocal microscope (Figure 1).

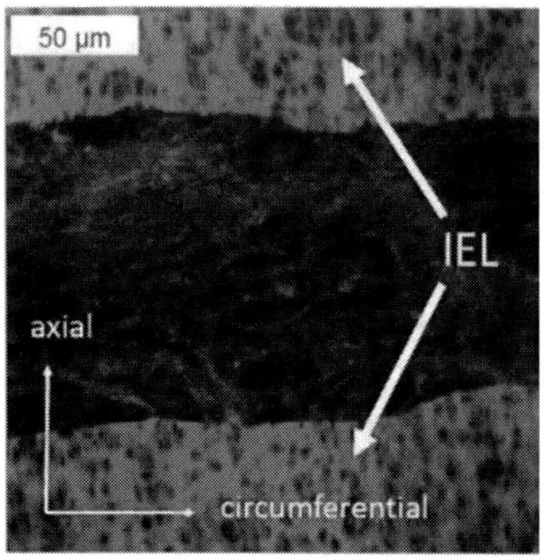

Figure 1: Confocal image showing a ruptured internal elastic lamina (IEL). Small holes (fenestrae), inherent to the IEL, can also be seen.

Changes in IEL fenestrae

Statistically significant changes in fenestrate size were found only when comparing overstretched vessels to a control from the same lamb. Two overstretched segments (λ_{z_max} = 1.17 and 1.30) had larger fenestrae ($p < 0.05$) and larger major to minor axis ratios ($p < 0.001$) than a neighboring control. The more severely stretched of these two had fenestrae with larger axis ratios than the less stretched segment ($p < 0.001$). The observed elongation in fenestrae shape was in the axial direction.

DISCUSSION

The purpose of this study was to test the hypothesis that the internal elastic lamina of cerebral arteries exhibits permanent structural damage following axial overstretch. The results confirm this hypothesis as catastrophic rupture of the IEL was observed as well as subtle changes to the IEL fenestrae.

Failure of the internal elastic lamina

To the authors' knowledge, this study is the first to report failure of the internal elastic lamina from axial loading of cerebral blood vessels. Stemper et al. reported internal layer subfailures in porcine aortas; however, the specific failed structure was not identified [5].

Relationships with collagen damage

Further investigation of the images and force signals taken during axial overstretch has shown that IEL failure typically occurs immediately after the rupture of nearby large collagenous fibers (Figure 2). These fibers likely prevent the vessel from axial over-distension [6], and, upon breaking, transfer high levels of stress to focal regions in the adjacent layers.

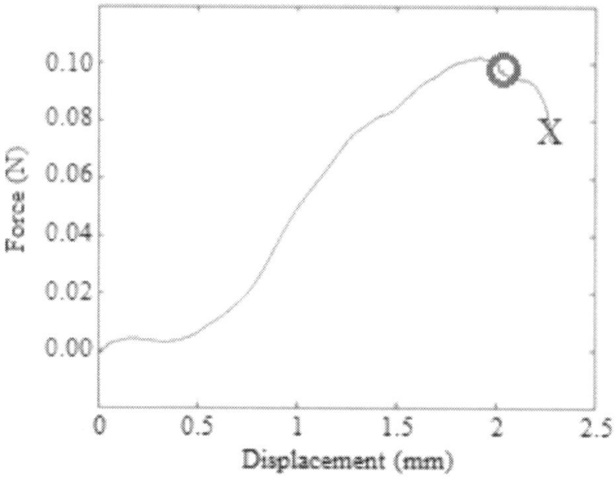

Figure 2: Force signal during the axial overstretch of one of the artery segments. The rupture of large collagen fibers (manifest as drop in force) is followed by rupture of the IEL (O), then loss of pressure (**X**).

Related work in our lab further confirms this link between collagen damage and IEL failure. A recently developed marker for mechanically damaged collagen has been used on a separate set of lamb cerebral arteries. This unpublished work shows that the onset of ruptured collagen fibers coincides with 'yielding' in the force signal. Furthermore, collagen damage is most apparent near regions of IEL failure.

Significance

These findings may, in part, explain the recently reported correlation between TBI and stroke (2, 3). First, a ruptured IEL would weaken the arterial wall, possibly making it more susceptible to hemorrhage. A weak or degraded IEL has previously been linked with human cerebral aneurysms [7, 8]. Secondly, a ruptured IEL adversely exposes arterial media to blood which may increase the risk of thrombosis. Whether or not there is a direct link between IEL failure and stroke, a ruptured IEL is likely to compromise vascular mechanics following TBI.

Unfortunately, since vessels continue to hold pressure following its rupture, IEL failure would not result in immediate hemorrhaging and therefore would go undetected in a medical scan. Therefore, future work should focus on understanding if, where, and under what conditions the IEL ruptures in vivo.

ACKNOWLEDGEMENTS

This research was supported by the Primary Children's Medical Center Foundation Integrated Science Award.

REFERENCES

[1] Faul, M, et al., *TBI in the US Report*, CDC, 2010.
[2] Burke, JF et al, Neurology, 81:33-39, 2013.
[3] Chen, Y et al, Stroke, 42:2733-2739, 2011.
[4] Van Loon, P, Pergamon Press, 14:181-201, 1977.
[5] Stemper BD, et al., Ann Biomed Eng, 35:285-291, 2007.
[6] Gasser, TC, et al., J R Soc Interface, 3:15-35, 2006.
[7] Scanarini, M, et al., Acta Neurochir Suppl, 43:171–182, 1978.
[8] Stehbens, WE, Arch Neurol, 8:272-285, 1963.

EFFECT OF IMPLANT LENGTH AND ORIENTATION ON BIOMECHANICS OF SACROILIAC JOINT STABILIZATION: A FINITE ELEMENT ANALYSIS

D. Lindsey (1), A. Kiapour (2), S. Yerby (1), V. Goel (2)

(1) SI-BONE, Inc., San Jose, CA

(2) Engineering Center for Orthopaedic Research Excellence (ECORE) Departments of Bioengineering and Orthapedic Surgery

INTRODUCTION

Sacroiliac (SI) joint fusion using minimally invasive placement of implants is an option to relieve pain resulting from SI joint disorders[1,2,3]. Hardware orientation and placement are factors that can affect fixation strength in the pelvis. Due to the density variations within the sacrum, altering implant length and placement (to allow for placement in higher bone density regions) during SI joint fusion may be beneficial. This finite element study investigated implant length and placement on SI joint stability and range of motion (ROM).

METHODS

An experimentally validated model of L1-pelvic segment (Figure 1) was used to investigate the effects of SI joint fixation[4,5]. This finite element model has been used previously to evaluate surgical fixation of the lumbar spine and the effects of leg length discrepancy. The model was comprised of the lumbar spine, pelvis, and both femurs and contained 60,126 elements and 77,540 nodes. The model was developed from a computed tomography (CT) scan of a young man's pelvis without any irregularities or severe degeneration. Material properties for the bone, ligaments and joints are based upon the previous study[5]. Heterogeneous properties were assigned to the bone of the pelvic segment.

SI joint fixation was simulated by placing three triangular TPS coated implants (iFuse Implant System®; SI-BONE, Inc., San Jose, CA) across the left joint. The implants configurations were either posterior (inline in sacral inlet view) or a trans-articular (convergent in sacral inlet view). The superior implant length was varied to end at either mid ala or mid sacrum (Figure 2).

Surgically, the implants are placed by a pin, drill, broach, and impaction procedure. Due to the TPS coating of the Ti implants a press fit is required to insert the implants; thus to simulate the insertion of each device a mating part with a longitudinal cut was created within the model. The mating part's properties were that of the surrounding pelvic model, including the SI joint gap and the bone material properties. To determine the press fit force, a separate smaller mating part (to simulate the overlap of the press fit) was modeled and the force to open the longitudinal cut such that an implant could be easily inserted was determined. This force was then applied to close the longitudinal cut of the original mating part within the model to simulate the press fit. The implants were assigned material properties of Ti6Al4V ELI ($E = 115$ GPa, $v=0.32$).

To simulate the physiological loading, a compressive follower load of 400N plus a bending moment of 10N-m moment were applied to the intact and instrumented models to create Flexion-extension (Flex-Ext), lateral bending (LB), and axial rotation (AR) motions. Range of motion (ROM) was calculated for the SI joint and each lumbar motion segment. Motions were calculated at spinal segments and the left (treated) SI joint in all models. The left and right motions for bending and rotation were combined into lateral bending and axial rotation, respectively. The peak load at contra-lateral SIJ was computed for all cases.

RESULTS

Implant placement reduced the SI joint ROM in the four models by an average of 70% (flexion), 66% (extension), 38% (LB), and 68% (AR). Lengthening the superior implant from mid ala to mid sacrum reduced the SI joint ROM an average of 10%. Trans-articular placement reduced the SI joint ROM by an average of 14% compared with the posterior placement. The most stable (mid-sacrum; trans-articular) combination was 22% stiffer than the least stable (mid ala; posterior). The load at SIJ is shown in Figure 4.

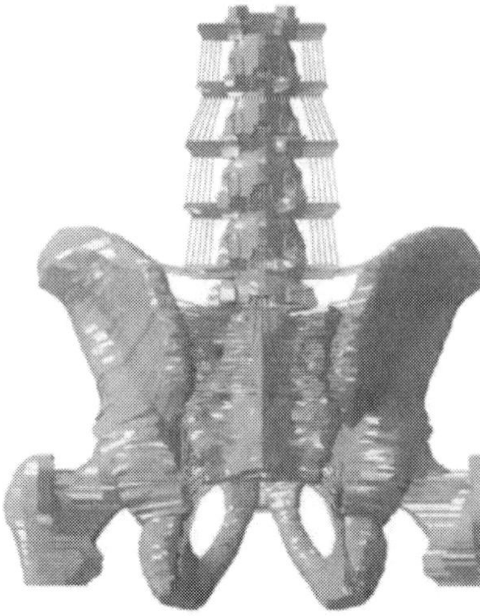

Figure 1. Finite Element model of intact L1-Pelvic segment.

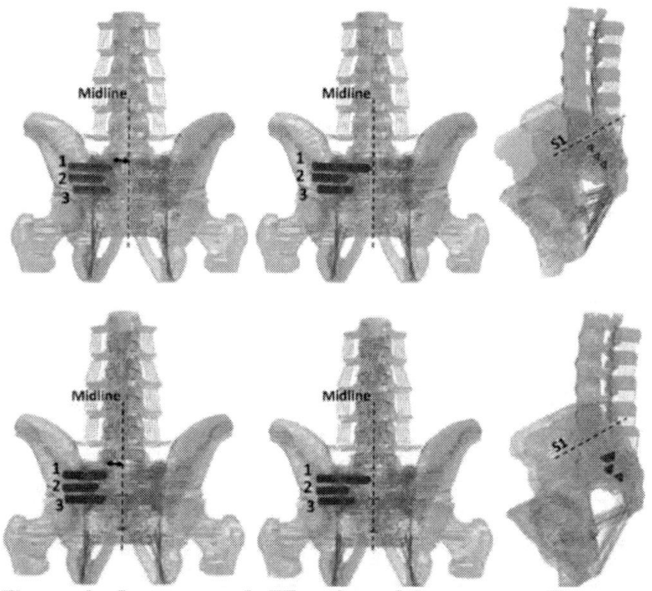

Figure 2. Instrumented FE spine-pelvic segment. Top row: posterior (inline in sacral inlet view) (Mid Ala: left; Mid Sacrum: center, sagittal view for both: right). Bottom row: trans-articular technique (convergent in sacral inlet view) (Mid Ala: left; Mid Sacrum: center, sagittal view for both: right).

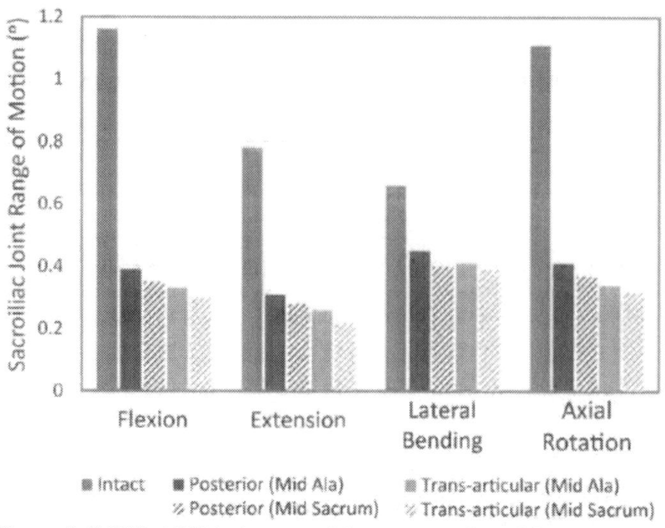

Figure 3. ROM at SIJ in intact and instrumented models.

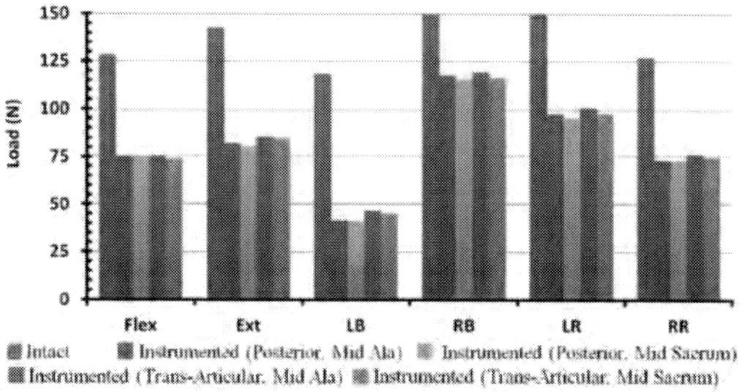

Figure 4. Max. Contact Load at Contra-lateral SI Joint (LB & RB: left and right bending, LR & RR: left and right rotation)

DISCUSSION

Implant length and placement both altered the stability of the treated SI joint. A longer superior implant and trans-articular placement increased stability. The peak load at contra-lateral SIJ decreased slightly in lateral bending and axial rotation loadings as top implant extended to Mid-Sacrum compared to Mid-Ala configuration or the implants were placed in Trans-Articular configurations. This study demonstrates that placement of the implants within higher bone density regions allows for up to 22% more SI joint stability.

ACKNOWLEDGEMENTS

This study was support by a fund from SI-BONE Inc.

REFERENCES

[1] Hoy, D et al. Ann Rheum Dis. 2014 Jun;73(6):968–74.
[2] Murray, CJL et al. Lancet. 2012 Dec 15;380(9859):2197–223.
[3] Sembrano, JN et al.Spine. 2009 Jan 1;34(1):E27–32.
[4] Ivanov, AA et al. Spine. 2009 Mar 1;34(5):E162–9.
[5] Kiapour, A et al.J Orthop Res 2012 Oct;30(10):1577–80.

SB^{3}C2015
Summer Biomechanics, Bioengineering and Biotransport Conference
June 17-20, 2015,Snowbird Resort, Utah, USA
SB^{3}C2015-1142

DEFORMATION AND FLOW OF ARTERIAL STENOSIS MODEL REGARDING THE CHANGE IN CURVATURE OF CORONARY ARTERY FOR PERCUTANEOUS TRANSLUMINAL CORONARY ANGIOPLASTY

Shunichi Kobayashi(1), Daiki Miyamoto (2)

(1) Faculty of Textile Science and Technology
Shinshu University
Ueda, Nagano, Japan

(2) Graduate School of Science and Technology
Shinshu University
Ueda, Nagano, Japan

INTRODUCTION

Percutaneous Transluminal Coronary Angioplasty (PTCA) is one of the minimally-invasive treatments for coronary disease. Detailed assessment of the deformation of plaque in a coronary disease during the dilatation of PTCA balloon catheter, and recovery of blood flow is important for the development of the balloon catheter. Thus, we developed the arterial stenosis model as a diseased blood vessel model, made of polyvinyl alcohol (PVA) hydrogel (tube portion) and vinyl acetate (plaque portion). In the experiment, pulsatile flow perfused to the stenosis model and PTCA balloon catheter with guide wire was inserted into the stenosis model. The curvature of stenosis model was changed in synchronization with pulsatile flow. The flow rate, pressure and deformation of the stenosis model were examined before the insertion, during the dilatation, and after the removal of the PTCA balloon catheter.

METHODS

Figure 1 shows the stenosis model. The inner diameter D and the thickness h of straigt tube, and the length of stenosis Ls are, 3.5 mm, 1 mm, and 8 mm, respectively. Nominal stenosis severity St and eccentricity Ec is at zero luminal pressure are

$$St = (1 - Ds / D) \times 100 \ [\%],\ Ec = e / ((D - Ds) / 2) \times 100 \ [\%] \quad (1)$$

We used the nominal stenosis models of St = 60 - 80%, and Ec = 100%. Polyvinyl alcohol (PVA) hydrogel was used for healthy straight sections (tube portion). Dimethyl sulfoxide was added to PVA solution before the gelation. The mechanical property of this PVA hydrogel was closer to that of coronary artery [1]. Vinyl acetate was used for the plaque to represent the plastic deformation after the dilatation of the PTCA balloon catheter. The Young's modulus E (the average value when the strain was from 0 to 0.5) of the vinyl acetate was set to 0.07MPa at 24 °C. Figure 2 shows the experimental setup. The working fluid used was water at 24 °C. The computer-controlled gear pump was used for the pulsatile flow. The average upstream and downstream pressures were set at 90 mmHg and 70mmHg, respectively. Flow rates and pressures upstream and downstream of the stenosis model were measured. The stenosis model was attached on the sliding surface of rotating eccentric cam to change the curvature of stenosis model. The rotation of the eccentric cam is in synchronization with pulsatile flow. Lumenal area / average curvature, change of curvature radius of the stenosis model were set at 0.08, 10mm, respectively[2]. The guide wire was inserted into the stenosis

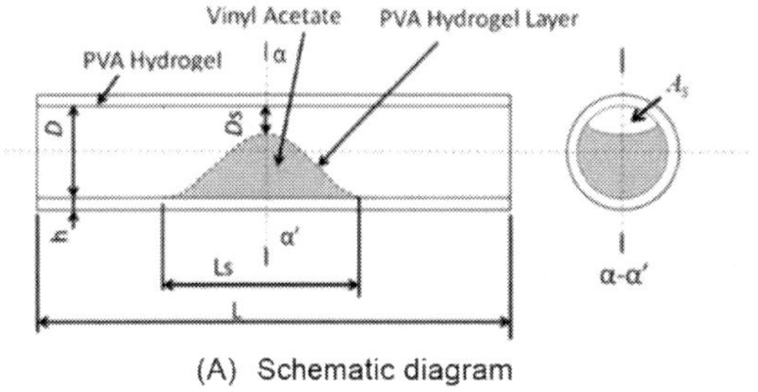

(A) Schematic diagram

(B) Photograph
Figure 1. Stenosis model

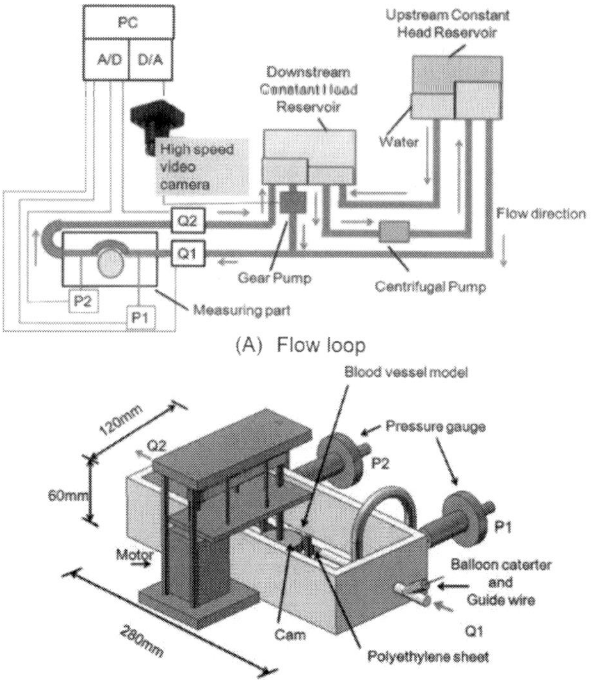

(A) Flow loop

(B) Details of measuring part

Figure 2. Experimental setup.

model and the PTCA balloon catheter was inserted along the guide wire. The PTCA Balloon catheter was pressurized and dilatated by the indeflator. Tomographic images of the stenosis model were taken by Micro CT.

RESULTS AND DISCUSSION

The vinyl acetate (plaque part) was deformed concavely and kept after catheter removal. We measured luminal area from CT images and evaluated true stenosis severity St_{ture} :

$$St_{ture} = (A_S / A) \times 100\ [\%] \tag{2}$$

where A and As are luminal areas of straight portion and apex of stenosis, respectively. Figure 3 shows the change of St_{ture}. Reduction of St_{ture} by the dilatation of balloon catheter is greater with greater nominal stenosis severity. Figure 4 shows the measured upstream and downstream flow rates. During dilatation, there was no flow but the flow rate was increased after dilatation and catheter removal. We evaluated two rates of flow rate:

$$R_{QI} = Q_{ave\_ad} / Q_{ave\_bi},\ \ R_{QR} = Q_{ave\_ad} / Q_{ave\_st0} \tag{3}$$

where Q_{ave_ad} is the average flow rate after the dilatation and removal of the catheter, Q_{ave_bi} is average flow rate before the catheter intersection, Q_{ave_st0} is average flow rate when St = 0 % as the no stenosis healthy arterial model. R_{QI} means the ratio of increasing flow by the dilatation in the stenosis model, and R_{QR} means the ratio of recovery of flow to the healthy condition. Figure 5 shows the relationship between nominal stenosis severity St and the ratios of increasing flow and recovery of flow R_{QI} and R_{QR}. R_{QI} increases with an increase in St. This is explained that the increase amount of

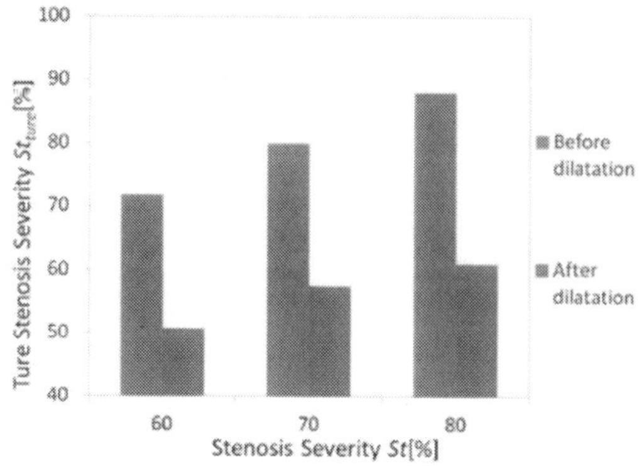

Figure 3. The change of true stenosis severity by the dilatation of balloon catheter.

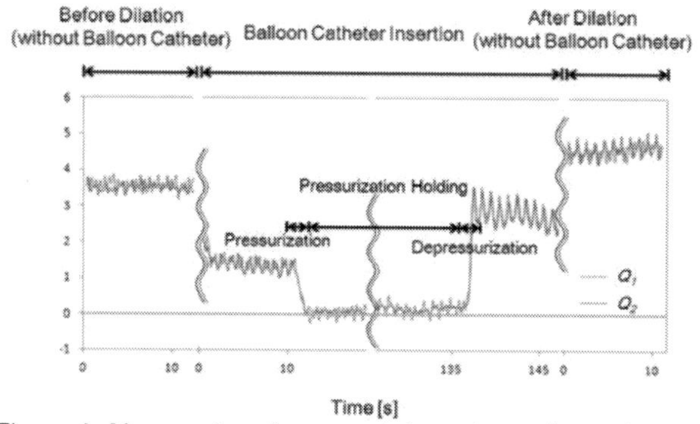

Figure 4. Measured upstream and downstream flow rates Q_1, Q2 (St = 70%)

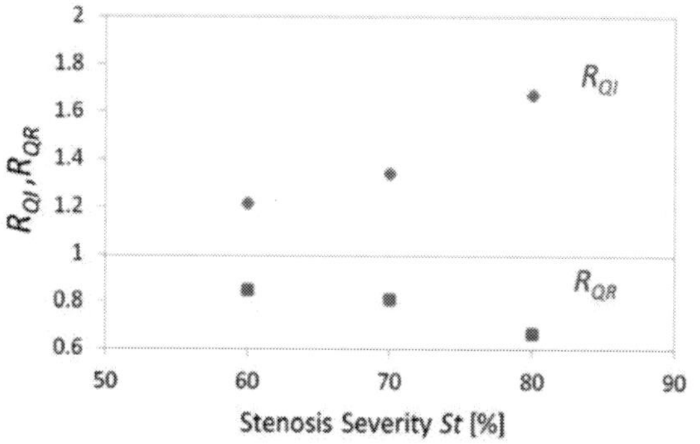

Figure 5. Relationship between nominal stenosis severity St and the rations of increasing flow and recovery of flow R_{QI}, R_{QR}.

luminal area by the dilatation at the stenosis increases with an increase in the volume of plaque. This is also explained the reduction of St_{ture} by the dilatation, which is shown in Figure 3. R_{QR} decreases with an increase in St. This is simply explained that the St_{ture} after the dilatation is greater with greater St, which is shown in Figure 3.

REFERENCES

[1] Ji, J et al., *J. Biomech. Sci. Eng.*, 6 (2):79-88, 2011.
[2] Weydahl, E. S et al., *J.Biomech*, 34 (9) , 1189-1196, 2001.

SB³C2015
Summer Biomechanics, Bioengineering and Biotransport Conference
June 17-20, 2015,Snowbird Resort, Utah, USA
SB³C2015-1143

FINITE ELEMENT ANALYSIS OF LOWER EXTREMITY MILITARY BOOT PROTECTION AT BLAST CONDITIONS

Robbin E. Bertucci (1,2), R. Prabhu (1,2), M. F. Horstemeyer (2), Jun Liao (1,2), Lakiesha N. Williams (1,2)

(1) Department of Agricultural & Biological Engineering
Mississippi State University
Starkville, MS, USA

(2) Center for Advanced Vehicular Systems
Mississippi State University
Starkville, MS, USA

INTRODUCTION

Anti-personnel (AP) landmines have become extremely prevalent in battle due to their relative low cost, maintenance, and simplicity to deploy. These land mines are often activated when a military personnel steps on the mine [1]. This threat has led to the importance of evaluation and improvement of personal protective equipment (PPE) in many countries.

Standard military boots provide minimal protection under blast loading and fail easily. Several new designs have been made to specifically counteract AP mines. These boots contain wedge designs, honeycomb energy absorbing structures, and structures increasing stand-off distance from the ground.

A variety of approaches have been used to test PPE, but each methodology has issues addressing the following goals: consistency and repeatability, transferability, ability to rank protection based on injury outcome, and correlation to level of trauma. Finite element analysis has become an increasing research area in PPE development due to its ability to provide a reproducible testing means with low cost. For numerical procedures to help with the advancement of PPE, more complex lower extremity models and material models need to be developed.

This study focuses on integrating our lower extremity mesh from previous work into a common military boot and evaluating the boot's ability to protect the lower extremity and dissipate the blast wave.

METHODS

A commonly worn military boot was obtained from the Reserve Officers' Training Corps at Mississippi State University. A surface scan of the boot was taken and imported into ScanIP, 3D image processing software. The boot was then manually and segmented into the following different materials: leather covering, outer rubber sole, foam mid-sole, and insole. The laces were excluded from the model since they have no purpose in mitigating the blast wave.

Next the boot was manually manipulated in ScanIP in order to incorporate it into our lower extremity model. The previous geometry of the lower extremity was developed from axial computed tomography scans and cryosections contained from the National Library of Medicine's Visible Human Male. Once appropriately aligned, the boot and lower extremity were meshed.

Using Abaqus/Explicit [Simulia], blast simulations were executed using the incident wave procedure. Numerous simulations were performed varying in blast overpressure and duration. An overview of the setup can be seen in Figure 1. The contact definition between materials was fixed due to our findings in previous studies that this has little effect on the results at such high rates.

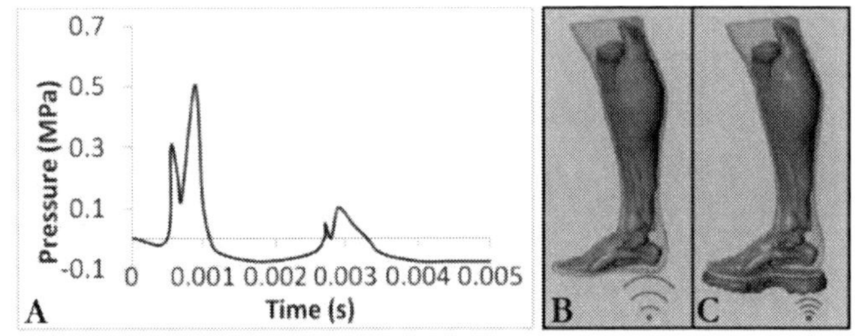

Figure 1. Basic setup for blast simulations. (A) Example air pressure profile used for the blasts, (B) setup without boot, and (C) setup with boot.

Elastic material properties from literature were used for the skin, bone, and muscle. The boot mesh was simplified to contain only the outer sole, midsole, and insole. The leather was omitted since it has little effect on the blast mitigation. Rubber, polyurethane foam, ethylene-vinyl acetate (EVA) material properties were obtained from literature and assigned to the specific parts of the boot. The rubber and EVA were modeled using the Marlow model for hyperelasticity, and the polyurethane foam was modeled using the hyperfoam model. Damage initiation and evolution were also included for the outer and midsole properties.

RESULTS

A 3D military boot model was successfully developed and incorporated into our lower extremity model. For preliminary studies, the lower extremity was meshed from the knee down and the leather portion of the boot was ignored. The mesh contained ~ 3 million elements. Pressure, stress, and strains were evaluated in the same manner as our previous studies (in each material and along the line of blast wave loading) and compared to the results from our simulations of the lower extremity without a boot.

The boot was found to absorb and dissipate the majority of the pressure prior to entering the lower extremity (Figure 2). Displacement is greatly affected by the boot. When the pressure wave enters the boot, the rubber outsole can redistribute the wave rather than it propagating directly through the lower extremity biological materials.

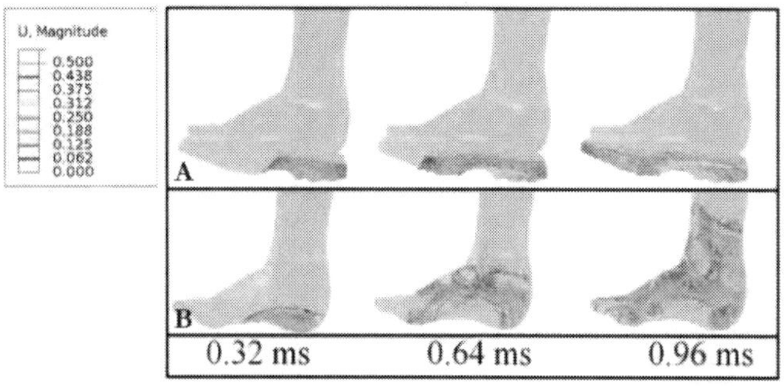

Figure 2. Displacement magnitude (mm) in all materials. (A) Simulation with boot and (B) simulation without boot.

Peak pressures in the lower extremity were observed in the tibia and foot bones (Figure 3). The pressure observed in each biological material is 100 times lower in the simulations with the boot compared to the simulations without the boot.

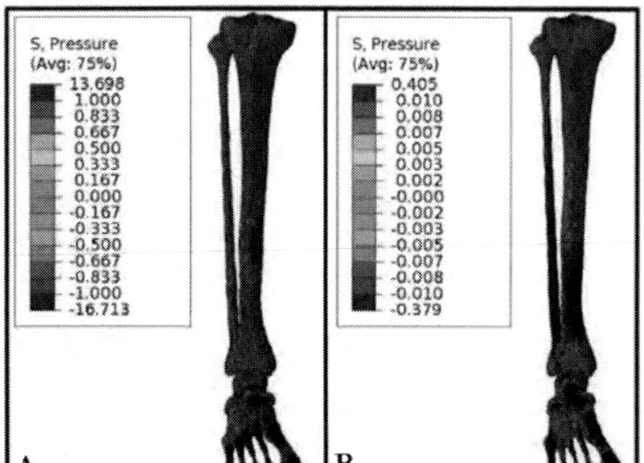

Figure 3. Pressure (MPa) in the bones at 1.52 ms. (A) Simulation without boot and (B) simulation with boot.

No material damage was observed in the first blast conditions examined. A higher peak pressure of approximately 100 MPa with a duration of 0.25 ms, showed damage to the tread of the boot (Figure 4). With such a fast and high pressure profile, the simulations had difficulty completing. These issues are being explored.

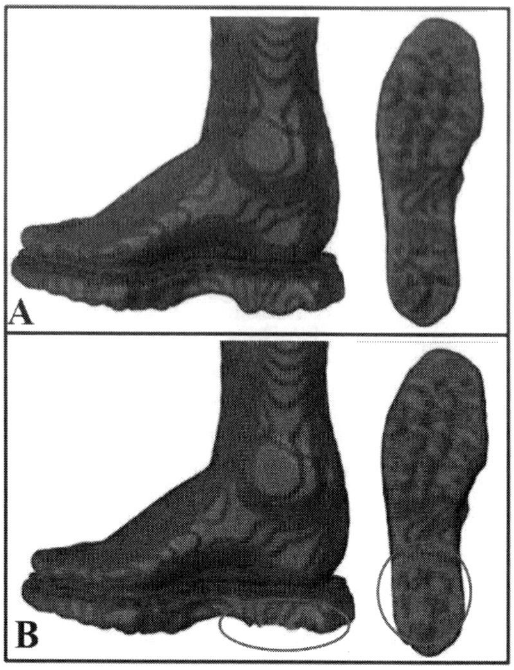

Figure 4. Material damage in high blast pressure simulation. (A) Undeformed and (B) deformed at 0.2 ms.

DISCUSSION

This is the first of many PPE studies that will be investigated using our lower extremity model. In this study we were able to successfully incorporate a commonly worn military boot into our realistic lower extremity model. The results allowed for the evaluation of how much energy is dissipated by these boots. At low blast pressures, boots have been seen to absorb the majority of the pressure [2].

Being able to evaluate how much energy is dissipated by these boots can help the military to be able to investigate the level of protection different boots provide. In the future, more realistic material models will be incorporated into the simulations which will provide a better understanding and evaluation of injury threshold criteria in the lower extremity when exposed to blast wave loading.

ACKNOWLEDGEMENTS

This material is based upon work supported by the U.S. Army TACOM Life Cycle Command under Contract No. W56HZV-08-C-0236, through a subcontract with Mississippi State University, and was performed for the SimBRS research program. Any opinions, findings and conclusions expressed in this material are those of the author(s) and do not necessarily reflect the views of the U.S. Army TACOM Life Cycle Command.

REFERENCES

[1] Chaloner, E.J. et al., J R Army Med Corps, 148: 38-43, 2002.
[2] Islam, S. et al., J Mine Action, 4.2, 2000.

SB[3]C2015
Summer Biomechanics, Bioengineering and Biotransport Conference
June 17-20, 2015,Snowbird Resort, Utah, USA
SB³C2015-1144

ACTO-MYOSIN CATCH BONDS AND MECHANO-SENSITIVITY IN NON-MUSCLE CELLS

Franck J. Vernerey and Umut Akalp

Mechanical Engineering and Material
Science and Engineering,
University of Colorado Boulder
Boulder, Colorado, USA

INTRODUCTION

The contraction of non-muscle cells is a phenomenon at the heart of many biological phenomena such as cell migration, tissue morphogenesis or wound contraction. Recently, it was also shown that mesenchymal stem cells rely on their ability to contract in order to sense the mechanical properties of their surrounding and further choose their differentiation pathway. Understanding the fundamental mechanisms underlying contraction and its dependency on factors such as applied force and external stiffness [1] is therefore becoming critical for a number of bio-medical applications. The molecular mechanisms by which non-muscle cells generate motion and forces depend on a series of highly coordinated events occurring in both the cytoskeleton and at the cell membrane. Of particular importance are long contractile elements of the cytoskeleton, made of acto-myosin filaments, often denoted as stress fibers, that anchor to the external environment through focal adhesions. These filaments assemble from a variety of proteins present in the cytosol such as globular actin and non-muscle myosin, a process that is known to be driven by the nature of a cell's mechanical environment. The aggregation of actin into filament and stress fibers is strongly influenced by mechanics; on stiff substrates, cells develop a strong acto-myosin network and display a strong contractility while on soft substrates, cells are typically devoided of stress-fibers and hardly contract [1,4]. How are these cells capable to sense stiffness and accordingly modulate its cytoskeleton and contraction is still an open question.

To better understand acto-myosin mechanics and perhaps shine a light on the cause of cell mechano-sensitivity, we build a mechanistic model of a stress-fibers, starting from the interaction between actin filaments and myosin heads (known as cross-bridges) at the molecular scale. We particularly concentrate on a particular aspect of this interaction that we denote the catch-bond behavior. Indeed, a recent experimental study [2] has shown that acto-myosin bonds are particular in that their life time is extended under the application of a moderate force (up to 6 pN), a concept that contrasts with the more conventional Bell's response in biological adhesion. The aim of this abstract is to show, through our simple model, that this molecular behavior can indeed be a critical player in the mechano-sensitivity of non-muscle cells.

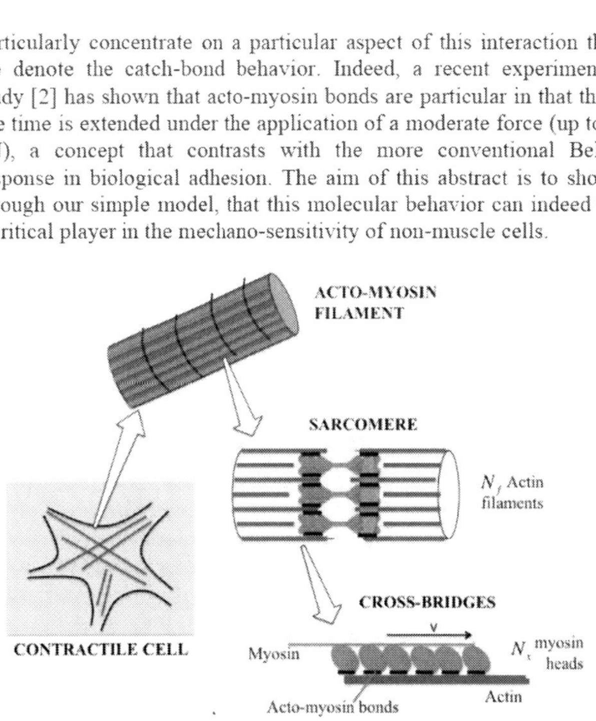

Fig. 1: Hierarchical model of a stress-fiber from the molecular scale (cross-bridges) to the micro-scale (sarcomere) to the macro-scale (filament bundle)

METHODS

In this work, we propose a mechanical analog of stress-fibers based on their hierarchical organizations into sarcomere units and cross-bridges (Fig. 1). At the micron-scale, a half-sarcomere is made of a number Nf actin filaments connected to myosin via cross-bridges. These bridges are themselves made of connections between myosin heads and binding sites on the actin filaments and can be represented as a parallel assembly of three key elements: (a) a contractile element that captures the change in sarcomere length due to the motion of myosin heads on actin filaments, (b) an elastic element whose stiffness k corresponds to that of the combined actin-myosin filaments and (c) bond elements that capture, as a whole, the adhesion between myosin heads and their corresponding actin. Let us assume that a maximum of N_x cross-bridges can be formed during contraction. As a whole, the sarcomere can thus be modeled as a nested parallel assembly of $N=N_f.N_x$ elementary units that are represented in the right of Fig. 1. The total contraction δ of the sarcomere arises from two components: the sliding δ_s of filaments and the elastic deformation δ_e of the cross-bridges. After going through classical thermodynamics arguments, one can show that the equilibrium of a sarcomere is found by solving two coupled equations interpreted as the balance of mechanical forces and the balance of chemical potentials:

$$\eta N k \delta_e - F = 0$$

$$-N E_b + N \ln\left(\frac{\eta}{1-\eta}\right) = 0$$

where η is the fraction of attached cross-bridges in a sarcomere, F is the external force and E_b is the energy stored in the acto-myosin bond. For a catch-bond, this energy may be cast in the following form [3]:

$$E_b(f) = E_b^0 + \alpha\left[1 - e^{f/f_0}\right] - f\Delta x$$

where f is the force in a single bond while the other quantities are parameters defining for instance, the critical force at which the bond stability is maximal. Importantly, the parameter α describes the catchiness of the bond. If α=0, we recover Bell's law and the bond destabilizes with force, while α>0 indicates the presence of a catchy behavior. In the following section, we aim to show that the bond catchiness (measured by α) governs the mechano-sensitivity at the sarcomere level, as well as at the cellular scale.

RESULTS

The actin cytoskeleton is known for its ability to sense the stiffness of the external environment [1]. To better understand the prediction of our model, we consider a virtual experiment in which a sarcomere is attached to a rigid support on one end and to a compliant substrate, represented by a spring of stiffness K, on its other end (Fig. 2).

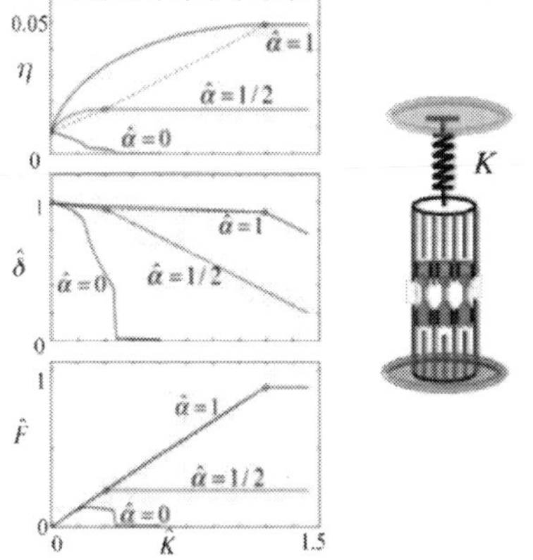

Fig. 2: **Contraction of a sarcomere against a compliant environment of stiffness K. We show (a) the stress fiber activation, the total contractile strain and normalized contractile force in terms of K for different bond catchiness.**

We observe that when a catch-bond behavior is considered, the sarcomere increases it contractility and activation with external stiffness, as observed in a number of experimental studies. A lack of bond-catchiness, however, leads to a very small contraction that is quasi independent of the external stiffness. To study whether such a behavior can be observed at the cellular scale, we built a lattice model of the acto-myosin cytoskeleton of a cell contracting on a bed of micro-needle whose stiffness could be varied through different geometric design. In these simulations, each lattice connection is described by the sarcomere model described above. Fig. 3 shows that the nature of the actin cytoskeleton and the cell contraction strengthen with needle stiffness when the catch-bond behavior is considered, consistent with experimental measurement. No contraction and a quasi-inexistent actin structure are observed otherwise.

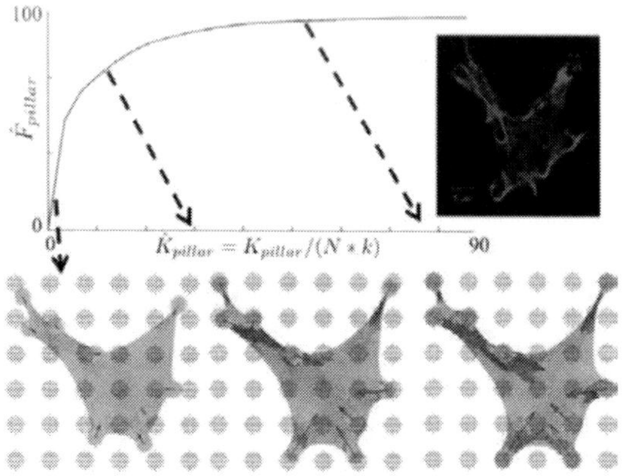

Fig. 3: Mechano-sensitivity at the cellular scale. When catch-bonds are considered, the well-known increased contraction with external stiffness is observed. If bond-catchiness is not considered, we do not observe a relationship between contraction and external stiffness.

DISCUSSION

To summarize, this abstract presents mechanistic model that shows that the catch-bond behavior of acto-myosin units in the stress fibers of non-muscle cells could be responsible for their mechano-sensitivity. The model exhibits a positive feedback between the existence of external force and contractile mechanisms occurring at the molecular scale. When a cell contracts against a stiff substrate, cross-bridges are subjected to large forces, which tends to reinforce their stability. This makes stress-fibers more active and able to generate larger forces. The model therefore suggests that the bond response of acto-myosin units could be at the heart of cell sensing, migration and differentiation.

ACKNOWLEDGEMENTS

The author would like to acknowledge the support of the National Science Foundation under the CAREER award 1350090.

REFERENCES

[1] Trichet et al, *PNAS* 109(18): 6933, 2012
[2] Guo B and Guilford W. *PNAS*, 103(26):9844
[3] Thomas W et al, *Annual review of biophysics*, 37:399
[4] Vernerey FJ and Farsad M. *Journal of the Mechanical Behavior of Biomedical Materials*, 4(8):1683, 2011

PRELIMINARY DEVELOPMENT AND VALIDATION OF AN ATLAS-BASED FINITE ELEMENT BRAIN MODEL

Logan E. Miller[1,2], Jillian E. Urban[1,2], Elizabeth M. Lillie[1,2], Joel D. Stitzel[1,2]

(1) Biomedical Engineering
Wake Forest University School of Medicine
Winston-Salem, NC, USA

(2) Center for Injury Biomechanics
Virginia Tech-Wake Forest University School of Biomedical Engineering and Sciences
Winston-Salem, NC, USA

INTRODUCTION

Each year, approximately 1.7 million people in the United States suffer from traumatic brain injury (TBI) [1]. TBI is a leading cause of disability and injury-related death, accounting for nearly one third of all injury-related deaths [2]. To prevent and treat these types of injuries, the fundamental injury mechanisms need to be well-characterized and understood. Various methods have been used over the years to study brain injury including animal tests, cadaver studies, anthropomorphic test devices (ATDs), and computational models [3]. Finite element (FE) models are powerful tools because they can provide spatial and temporal distributions of stresses and strains over the problem domain. The quality of a model's predictions, however, is dependent on the accuracy of the modelled geometry and the model's ability to describe complex material response.

Many FE brain models with varying degrees of anatomical accuracy, element size and material properties, have been developed to study brain injury over the last several decades. Early models represented the brain-skull structure with a simple fluid-filled spherical shell; subsequently, model complexity increased, and the most advanced models now include major anatomical brain structures and different material properties for each component. Among these are the Kleiven brain model, Simulated Injury Monitor (SIMon) as well as the head models of full-body FE models such as the Total Human Model for Safety (THUMS) and Global Human Body Models Consortium (GHBMC) [3][4][5][6]. The geometry of the various structures within head models are derived from medical images, such as computed tomography (CT) and magnetic resonance imaging (MRI). The model development process typically involves image segmentation to define 3D surfaces of desired structures followed by a surface-based mesh generation process. During this process, complex anatomical features, such as the intricate folds on the surface of the brain (sulci and gyri), are simplified and the brain is modeled as a 'smooth' surface, surrounded by a uniform layer of cerebrospinal fluid (CSF). This simplification and reduction in geometrical accuracy could significantly impact the model's prediction of local brain response and injury. Indeed, studies have indicated that the intricate folding patterns on the surface of the brain greatly affect both the distribution and magnitude of stress and strain throughout the brain [7]. Another limitation of previous models is the substantial amount of time required for the surface-based mesh generation process, which hinders production of patient-specific models. Patient-specific models to study head injury are desirable due to significant variations in brain morphology between individuals [4].

The objective of the current study is to create a high resolution, anatomically accurate brain model developed from the International Consortium for Brain Mapping (ICBM) brain atlas using a voxel-based mesh generation approach. This atlas-based brain model (ABM) is proposed as a new tool for analyzing brain injury that can accurately predict local stresses and strains in brain tissue by accounting for the unique folded structure of the brain. ABM material properties have been optimized and validated based on local brain motion measured from cadaveric impact test C755-T2 [8].

METHODS

The ABM FE model was developed from the ICBM atlas representing an average adult male brain [9]. The ICBM atlas contains 58 independently labeled regions of the brain including cortical gyri, white matter, internal gray matter, and CSF. An FE model was created from this image set by converting each voxel into a single element of the same size using a custom code developed in MATLAB (The

MathWorks, Natick, MA). The labeled regions for the CSF and ventricles were created and assigned to different parts, and the remaining regions were combined and assigned to a single brain part. Because there were places in the model where the outer surface of the brain is in direct contact with the skull, a single layer of CSF was added around the entire model. Finally, to simulate a skull a 1 mm thick layer of shell elements was generated in LS-DYNA to completely enclose the model.

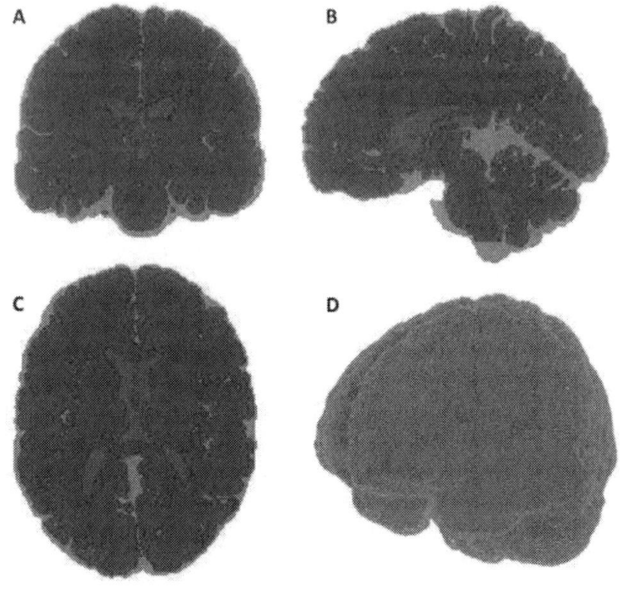

FIGURE 1: Coronal (A), sagittal (B), axial (C), and isometric (D) views of the ABM.

Material properties used in initial development of the model were similar to those used in SIMon (TABLE 1) [3]. SIMon models the brain and CSF with Kelvin-Maxwell viscoelastic formulation and the ventricles with an elastic fluid, with properties similar to water.

TABLE 1: Material Properties

Part	Material Properties
Brain	ρ=1040 kg/m^3, K=558.47 MPa, G_0=1.66 kPa, G_I=0.928 kPa, β=16.95 s^{-1}
CSF	ρ=1050 kg/m^3, K=4.966 MPa, G_0=100 kPa, G_I=20 kPa, β=0.01 s^{-1}
Ventricles	ρ=1000 kg/m^3, E=0 MPa, ν=0.5, K=2100 MPa
Skull	ρ=35200 kg/m^3, E=6900 MPa, ν=0.3

As initial validation, the ABS was subjected to the impact conditions of cadaveric test C755-T2 conducted by Hardy (2007) and local displacements at various locations within the brain were analyzed [8]. Performance of the ABM was quantified using CORA (CORrelation and Analysis) to calculate ratings for x- and z-displacement signals for each NDT. CORA evaluates the correlation of two curves by calculating a rating using a corridor and cross-correlation method. Brain material parameters (ρ, K, G_0, G_∞ and β) were optimized using a Latin Hypercube Design of Experiments with 100 simulations. CORA was used to identify the simulation with optimal material properties.

RESULTS

CORA values for each simulation were calculated, and the simulation with the highest CORA rating had the following material parameters: ρ=1031 kg/m^3, K=846.474 MPa, G_0=4.916 kPa, G_I=0.9593 kPa, and β=33.52 s^{-1}. The total CORA ratings for original and optimized ABM are 0.258 and 0.441, respectively. X- and z-displacements are reported for a1, a2, p1 and p2 below (FIGURE 2).

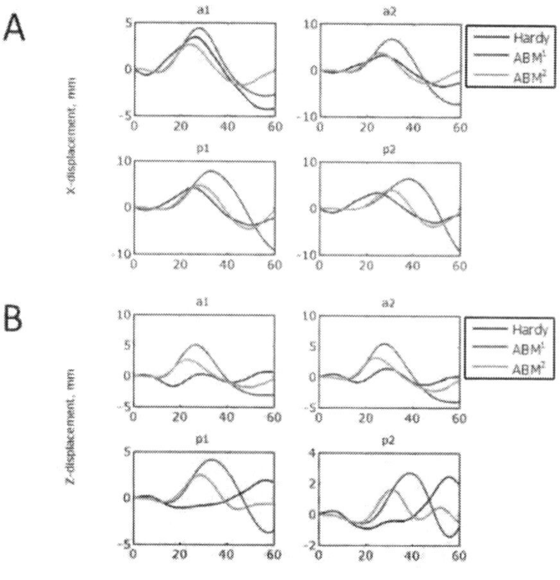

FIGURE 2: X-displacement (A) and z-displacement (B) time histories for ABM a1, a2, p1, and p2 NDTs.

DISCUSSION

A finite element model of the human brain developed from the existing ICBM atlas using a voxel-based approach was presented in this study. Validation against C755-T2 relative displacements show good agreement with experimental data for many NDTs in magnitude, phase and shape.

Future development of ABM includes the improvement of material properties by considering the response of multiple experimental tests simultaneously during the optimization process. This optimization should be conducted following the integration of brain structures, such as the falx cerebri and tentorium cerebelli. Additional improvements may also include investigating the brain-skull interface to enhance the modeling of this contact. The model developed in the current study produces an anatomically accurate brain model by employing voxel-based mesh generation. Further development of this model will lead to a better understanding of the biomechanics of head injury as well as provide a tool to predict and prevent such injuries.

ACKNOWLEDGEMENTS

Funding is provided by the National Institutes of Health (R01 NS082453). Simulations were run on the DEAC Cluster at WFU

REFERENCES

[1] Faul, M et al., CDC and Prevention, 2010.
[2] Centers for Disease Control and Prevention, MMWR, 2011
[3] Takhounts, EG et al., *Stapp Car Crash J.*, 47:107–133, Oct. 2003
[4] Kleiven, S et al., *J. Biomech.*, 35(2):153-160, 2002.
[5] Iwamoto, M et al., International IRCOBI Conference, 2002
[6] Mao, H et al., *J. Biomech. Eng.*, 135(11):111002, 2013
[7] Lauret, C et al., *Med. Eng. Phys.*, 31(3):392-399, 2009
[8] Hardy, WN, et al., *Stapp Car Crash J.*, 51:17–80, 2007
[9] Mazziotta, J et al., *Philos. Trans. R. Soc. Lond. B Biol. Sci.*, 356(1412):1293–1322, 2001

EFFECT OF SACROILIAC JOINT FIXATION ON SEGMENTAL KINEMATICS OF LUMBAR SPINE: A FINITE ELEMENT ANALYSIS

D. Lindsey (1), A. Kiapour (2), S. Yerby (1), V. Goel (2)

(1) SI-BONE, Inc., San Jose, CA

(2) Engineering Center for Orthopaedic Research Excellence (ECORE)
Departments of Bioengineering and Orthapedic Surgery

INTRODUCTION

The pain generators for low back pain include the lumbar spine, hips, and sacroiliac (SI) joints[1,2,3]. One treatment is arthrodesis to stiffen the motion segment to relieve the pain. Rigid lumbar fixation has been shown to increase biomechanical stresses at the adjacent levels. The use of rigid fixation for SI joint fixation may alter the biomechanical stresses of the adjacent lumbar segments, although the magnitude and potential consequences are currently unknown.
This finite element study investigated the impact of SI joint fixation on both the SI joint and the adjacent lumbar segment ranges of motion (ROM).

METHODS

An experimentally validated model of L1-pelvic segment (Figure 1) was used to investigate the effects of SI joint fixation[4,5]. This finite element model has been used previously to evaluate surgical fixation of the lumbar spine and the effects of leg length discrepancy. The model was comprised of the lumbar spine, pelvis, and both femurs and contained 60,126 elements and 77,540 nodes. The model was developed from a computed tomography (CT) scan of a young man's pelvis without any irregularities or severe degeneration. Material properties for the bone, ligaments and joints are based upon the previous study[5].

SI joint fixation was simulated by placing three triangular TPS coated implants (iFuse Implant System®; SI-BONE, Inc., San Jose, CA) across the left joint according to the manufacturer's recommendation (Figure 2). The implants were oriented such that they were within the central portion of the sacrum, inline in the inlet view, and parallel in the outlet view. Surgically, the implants are placed by a pin, drill, broach, and impaction procedure. Due to the TPS coating of the Ti implants a press fit is required to insert the implants; thus to simulate the insertion of each device a mating part with a longitudinal cut was created within the model. The mating part's properties were that of the surrounding pelvic model, including the SI joint gap and the bone material properties. To determine the press fit force, a separate smaller mating part (to simulate the overlap of the press fit) was modeled and the force to open the longitudinal cut such that an implant could be easily inserted was determined. This force was then applied to close the longitudinal cut of the original mating part within the model to simulate the press fit. The implants were assigned material properties of Ti6Al4V ELI ($E = 115$ GPa, $v=0.32$).

To simulate the physiological loading, a compressive follower load of 400N plus a bending moment of 10N-m moment were applied to the intact model to create Flexion-extension, lateral bending, and axial rotation motions. Range of motion (ROM) was calculated for the SI joint and each lumbar motion segment. In the instrumented models, a hybrid loading protocol which included gradually increasing the bending moment at the cranial end of the segment until the overall motion of the segment matched that of intact under 10Nm bending moment was simulated. Motions were calculated at L1-L2, L2-L3, L3-L4, L4-L5, L5-S1, and the left (treated) SI joint in all models. The left and right motions for bending and rotation were combined into lateral bending and axial rotation, respectively.

RESULTS

Loading of the intact model resulted in L1 – pelvis motions of 28.26° in flexion, 14.27° in extension, 52.84° in lateral bending, and 29.31° in axial rotation. Loading the treated model (to the intact angles) required moments of 10.26 N-m for flexion, 10.15 N-m for extension, 10.13 N-m for lateral bending, and 10.28 N-m for axial rotation. The treated model moments were increased by 2.6% in flexion, 1.5% in extension, 1.3% in lateral bending, and 2.8% in axial rotation (Figure 3). The instrumentation with iFuse implants reduced the SI joint ROM in flexion-extension (54%), lateral bending (25%), and axial rotation (50%). The ROM at the adjacent level (L5-S1) increased in flexion-extension (3%), lateral bending (1%), and axial rotation (4%).

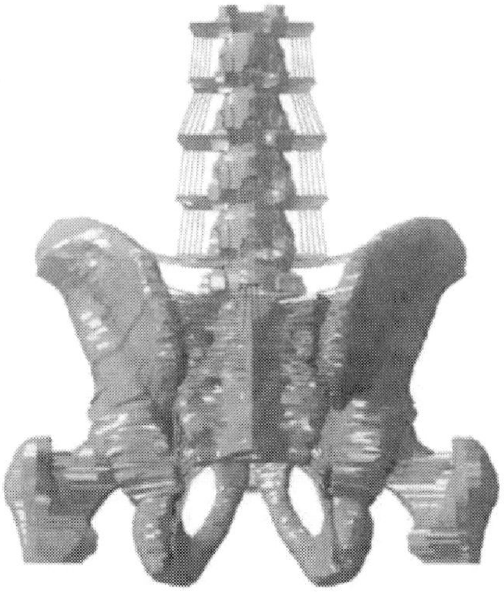

Figure 1. Finite Element model of intact L1-Pelvic segment.

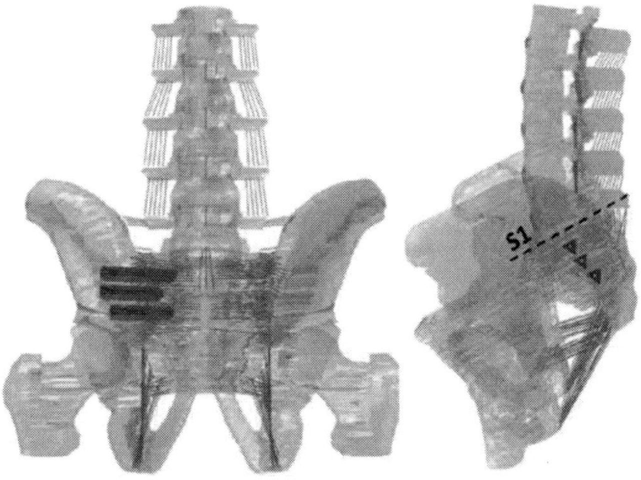

Figure 2. Posterior and sagittal views of the FE model of L1-Pelvic segment instrumented with three iFuse implants at SIJ.

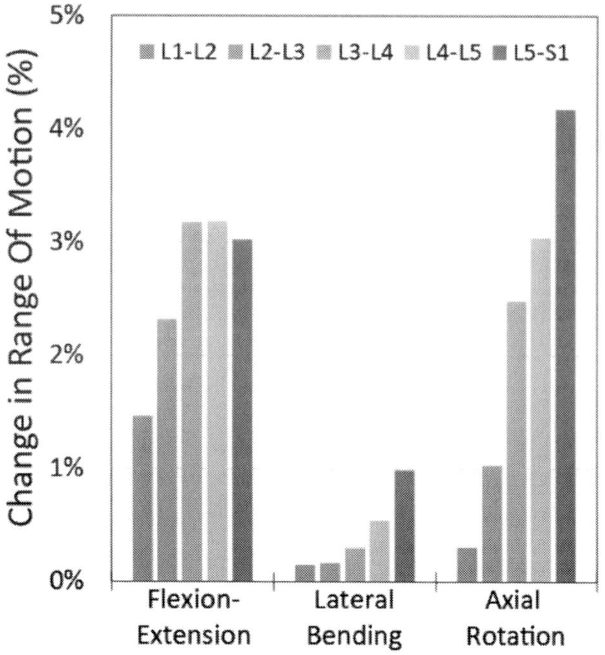

Figure 3. Lumbar ROM Change after SI Joint Fixation.

DISCUSSION

The results of the current study show that the hybrid loading protocol required slightly larger moments to take the instrumented segment to intact ROM in all planes of motion. As noted in previous studies, rigid fixation of a motion segment increases the overall construct stiffness[5,6]. Ivanov et al reported moment increases of 46 - 88% after a single level lumbar fixation. Kyaw et al reported moment increases of 122 - 469% after rigid fixation of degenerative lumbar model. The applied moments in this study were increased much smaller amounts - suggesting that fixation of the SI joint does not substantially stiffen the lumbo-pelvic construct.

The SIJ fixation reduced the SI joint ROM in flexion-extension, lateral bending, and axial rotation, consistent with a previous cadaveric experiment. Although adjacent segment ROMs were minimally increased when compared with lumbar fixation studies that report adjacent segment increases of 20-136%. Minimal increase in ROM of the adjacent lumbar levels suggest that SI joint fixation will not be deleterious to those levels.

ACKNOWLEDGEMENTS

This study was support by a fund from SI-BONE Inc.

REFERENCES

[1] Hoy, D et al. Ann Rheum Dis. 2014 Jun;73(6):968–74.
[2] Murray, CJL et al. Lancet. 2012 Dec 15;380(9859):2197–223.
[3] Sembrano, JN et al.Spine. 2009 Jan 1;34(1):E27–32.
[4] Ivanov, AA et al. Spine. 2009 Mar 1;34(5):E162–9.
[5] Kiapour, A et al.J Orthop Res 2012 Oct;30(10):1577–80.
[6]Kyaw, TA et al.Eur J Orthop Surg Traumatol Orthopédie Traumatol. 2014 Jul;24 Suppl 1:S283–7.

SB³C2015
Summer Biomechanics, Bioengineering and Biotransport Conference
June 17-20, 2015,Snowbird Resort, Utah, USA
SB³C2015-1147

PSEUDO-RIGID BODY METHOD FOR REDUCING SOFT TISSUE ARTIFACT: VALIDATION AND APPLICATION TO GAIT

Jake E. Deschamps (1), Karim C. Dudum (2), Eshan M. Dandekar (3), Scott J. Hazelwood (1,2), Stephen M. Klisch (1,2)

(1) Mechanical Engineering Department
California Polytechnic State University
San Luis Obispo, CA, USA

(2) Biomedical Engineering Department
California Polytechnic State University
San Luis Obispo, CA, USA

(3) Kinesiology Department
California Polytechnic State University
San Luis Obispo, CA, USA

INTRODUCTION

A common way to study human joint loads during various activities is to use a camera-based motion capture system to film retro-reflective markers adhered to a subject's skin. When reconstructing the underlying bone kinematics, it is often assumed that the skin moves rigidly with the bone. However, skin motion is influenced by muscles, adipose tissue, and flexing joints, introducing errors known as soft tissue artifact (STA). STA is considered the largest source of error in human movement analysis [1]. Considerable effort has been aimed at reducing STA but, even for the most innovative methods, STA persists.

Recently, [2] proposed implementing multi-marker clusters, analyzing each 3-marker subset as a triangular Cosserat point element (TCPE), and applying a filtering algorithm. In a pendulum validation test that simulated STA, this analysis resulted in lower errors than other state-of-the-art methods. However, results were reported only about a known axis and the method has not been applied to gait analysis. Here, the application of TCPEs and filtering algorithm are termed the pseudo-rigid body (PRB) method, which we propose to be a more physical descriptor as the method models each body segment as a "pseudo"-rigid deformable body.

This study aimed to: develop the PRB algorithm; recreate the STA simulation to validate the algorithm and report errors for out-of-plane angles; and contrast the PRB method's performance in gait analysis with a more traditional Helen – Hayes (HH) marker set.

METHODS

PRB Method. For the PRB method used here, seven makers were used for each body segment (Fig. 1A). At each time point, all combinations of three markers are analyzed as TCPEs. A director vector approach [2] is used to determine the deformation gradient (*F*) and Lagrangian strain (*E*) tensors for each TCPE:

$$\boldsymbol{d}_i \otimes \boldsymbol{D}^i = \boldsymbol{F} = \boldsymbol{R}\boldsymbol{U} \quad (1)$$

$$\frac{1}{2}(\boldsymbol{F}^T\boldsymbol{F} - \boldsymbol{I}) = \boldsymbol{E} \quad (2)$$

Two present configuration director vectors are formed from subtracting the three marker locations (Fig. 1B), with a third formed by normalizing the cross product of the first two, yielding $\boldsymbol{d}_i$. Reference configuration reciprocal director vectors, $\boldsymbol{D}^i$, are formed by using the previous process to obtain director vectors, and then using cross products to form reciprocals. Using *F*, *E* and rotation (*R*) tensors are calculated. The TCPEs which have the lowest strain magnitudes and most consistent rotation tensors are selected, and rotation vectors are calculated and averaged [2]. An algorithm for this process was coded in MATLAB (MathWorks, Natick, MA) to analyze fully recorded motions.

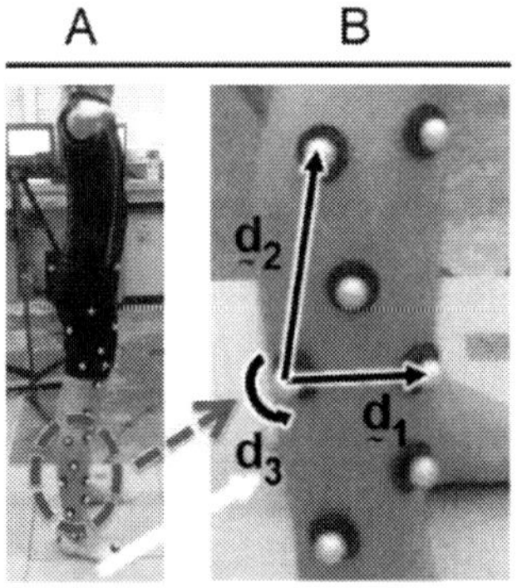

Fig. 1. Director vectors illustrated on the PRB marker set for gait analysis.

Motion Capture System. A motion capture laboratory with eight cameras and Cortex software (Motion Analysis Corp., Santa Rosa, CA) was used to collect data. A walkway and force plate (AMTI, Watertown, MA) was used during gait. Marker trajectories were filtered using a 4th-order Butterworth filter and cutoff frequency of 6 Hz.

Pendulum Validation. Similar to [2], a 300 mL silicone implant was attached to the end of an 80 cm long rigid pendulum to simulate STA. 4 markers were attached to the pendulum to estimate true RB motion.

Seven markers were attached to the implant's surface. Three "free swing" trials were conducted where the pendulum was released and allowed to swing uninterrupted. Three "impact swing" trials were conducted with a simulated impact: the pendulum was grabbed and released before returning to its starting position. The pendulum's RB motion was estimated from markers on the implant using the PRB method. Rotation angles were decomposed into in- and out-of-plane angles. The algorithm's marker usage identified the relative reliability of each marker, allowing for the formation of the most and least reliable marker sets from the three most used and the three least used markers. The performance of these marker sets gives a range of predictions of how limited marker sets compare to the PRB method. Maximum and root mean square (RMS) errors were determined.

Gait Analysis. Knee kinematics for three subjects were obtained using both HH and PRB marker sets. For the HH method, markers were attached to the skin over the ASISs, sacrum, medial and lateral femoral condyles, medial and lateral malleoli, and tops of the feet. Asymmetric offset markers were placed on each thigh and each shank. For the PRB method, seven markers were evenly distributed on both the right thigh and shank. Reference position information was gathered with each subject standing with feet roughly hip width apart. A Cortex script was used to determine the functional centers of the ankle, knee, and hip joints. After the reference trial, medial markers were removed to allow subjects to walk normally. After a warm-up period, each subject walked across the room six times at a self-selected pace. Cortex was used to determine knee kinematics for the HH marker set. The PRB algorithm was adapted to determine knee kinematics. Each trial was normalized to consist of one gait cycle, and average plots were created to present mean and standard deviation for each method across the gait cycle. Experimental protocols were approved by Cal Poly's Human Subjects Committee and were designed to minimize risk to human subjects.

RESULTS

Pendulum Validation. The PRB method's error was lower than those of the limited marker sets (Table 1), with one exception. The least reliable marker set had relatively high errors. Out-of-plane angles (Fig. 2) were low for most of the motion, but were considerably larger when the pendulum was stopped at an angle.

Table 1. RMS and maximum errors averaged across free and impact swing trials for the PRB method and two limited marker sets.

Trial	Error Type	PRB	Most Reliable	Least Reliable
Free Swing Averages	RMS Error	.68	.89	3.59
	Maximum Error	2.41	3.13	14.46
Impact Swing Averages	RMS Error	.75	2.07	6.24
	Maximum Error	5.03	14.01	25.92

Gait Analysis. Kinematics differed between the PRB and HH methods (Fig. 3). The PRB method showed larger knee flexion angles for most of the motion. Near the end of stance, varus rotations diverged for the two methods. Internal rotations differed during the entire gait cycle.

DISCUSSION

Pendulum Validation. RMS error was lower than the value reported by [2]. Although a direct comparison cannot be made, this suggests that the PRB algorithm was properly implemented. Large errors observed with limited marker sets imply that the PRB method is more accurate than limited marker sets, even when those are well devised.

Results in [2] took advantage of the known axis of rotation. This information is usually unavailable in gait analysis, as neither the knee nor hip can be modeled as a simple pin joint. Out-of-plane rotations during the pendulum test would manifest as varus or internal rotations during gait, which would have strong, erroneous connotations. The largest out-of-plane rotations occurred when the pendulum was being initially displaced. The implant was attached to the pendulum with one clamp at the top and one at the bottom, allowing the implant to rotate about the pendulum to an extent likely beyond what occurs during gait. Since skin does not terminate at single connection points, it is unlikely that the out-of-plane rotation magnitudes seen here would occur during gait.

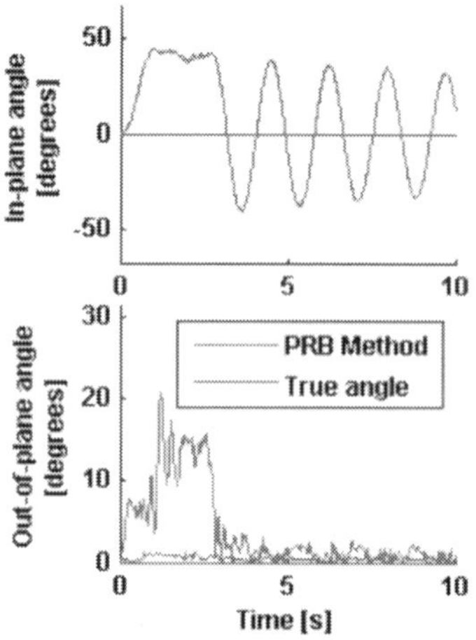

Fig. 2. True and PRB determined in- and out-of-plane rotations for a free swing trial.

Gait Analysis. Differences were observed between the two methods for all kinematic components. These kinematics are compared with those reported in studies that used invasive methods to circumvent STA. The varus rotation identified by the PRB method near the end of stance closely matches [3]. The internal rotations from the PRB method correlate with one subject examined in [4], while those identified by the HH method seem to have an unreasonably large range of motion. These comparisons must be interpreted with caution due to high subject variability in knee kinematics. The results support the development of several future aims: comparing the PRB method to the rigid body least-squared optimization method [3]; expanding the pendulum STA simulation to more accurately model the knee joint; and integrating the PRB method with inverse dynamic analysis.

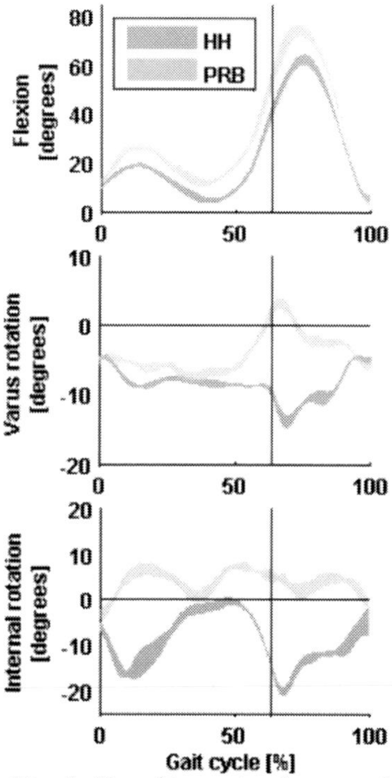

Fig. 3. Knee kinematics determined using HH and PRB methods averaged across six trials. Plot thickness indicates one standard deviation.

ACKNOWLEDGEMENTS

Funding received from Quartus Engineering, Inc. and Cal Poly's Donald E. Bently and STRIDE Centers.

REFERENCES

[1] Leardini, A. et al., *Gait Posture*, 21.2:212-225, 2005. [2] Solav, D. et al., *Int J Eng Sci*, 85:1-9, 2014. [3] Andriacchi, P. et al., *J Biomech Eng*, 120.6:743-749, 1998. [4] Reinschmidt, C. et al., *Gait Posture*, 6.2:98-109, 1997

SB[3]C2015
Summer Biomechanics, Bioengineering and Biotransport Conference
June 17-20, 2015, Snowbird Resort, Utah, USA
SB³C2015-1148

THE EFFECT OF PARTICLE DEFORMABILITY ON THE TRANSPORT OF SOFT COLLOIDS THROUGH POROUS MEDIA

Eduard Benet (1), Louis Foucard (2), Franck J. Vernerey (3), John Pellegrino (4)

(1, 3, 4) Mechanical Engineering
University of Colorado at Boulder
Boulder, Colorado, United States

(2) Civil Engineering
University of Colorado at Boulder
Boulder, Colorado, United States

INTRODUCTION

The transport of soft colloids, soft vesicles and other small deformable particles such as animal cells, or bacteria, through a porous medium, is of crucial relevance in the study of cancer metastasis, the fabrication and development of filtration membranes, microfluidic devices and other processes involved in applications related to medicine or energy. For instance, filtration is critical to the purification of residual water in food and fossil fuels industries [1]. In the pharmaceutical industry, new drug delivery techniques to transport and deliver drugs in a more efficient way use a filtration process in the fabrication and purification of liposomes [2]; small, soft vesicles formed by a lipid bilayer that encapsulate a water solution where the drugs are stored. And in a similar field, new designs of microfiltration devices are used to capture circulating tumor cells, and then be able to design better treatments [3].

Even though many of those particles can undergo very high deformations, most filtration designs and theoretical models have neglected their deformability in the estimation of permeability [4], and only recently some groups considered the hindrance factors of non-spherical particles [5]. In many application however, particle elasticity is critical for the design of new membranes, but also to enhance our ability to forecast the performance of existing materials. For example, the change of elasticity of certain cells during diseases (such as cancer) may be used as a unique differentiating feature to sort and eventually remove such cells from the blood of patients [6]. More generally, modeling tools will provide a quantitative understanding of the interaction between elastic colloids and membranes with the aim to potentially optimize and design new medical devices and technologies.

With these applications in mind, we have developed a computational multiscale approach that aims at characterizing, understanding and predicting the effect that, those deformations at the microscopic level, have on the macroscopic results of the filtration process [7]. This simulation technique will further be used to physically understand and improve the design of new membranes and measurements that specifically target the particle deformability as a relevant mechanical property. Among all the different effects that could be studied at this level, e.g., the adhesion of the particles, its interactions with fluid chemistry, etc. our interest relies on the geometry of the membrane as a primary differentiating characteristic, hence the focus in this paper is on the relationship between the elasticity of the particle and the porosity of the system, and its effects on the overall permeability of the system. Moreover, the method is not specific for filtration membranes and could be generalized to other porous media such as cell movement through tissues.

METHODS

A numerical approach based on the extended finite element is used to solve the creeping flow inside a portion of a filtration membrane or a microfluidic device (Reynolds number close to 0). The system is represented by a 2-dimensional porous flow due to a pressure gradient on the vertical direction (figure 1). The simulation uses periodic boundary conditions to represent the bulk response of tissue or membrane at larger length-scales. Within the flow, the particle-based moving interface method developed in [8] is used to track the deformation of the vesicles and its interaction with the membrane geometry. The objective of each simulation is to compute the macroscopic permeability of the system for a given type of vesicle(s) based on the results of the simulation of a microscopic portion of the membrane with dimensions H and W. The homogenization procedure [6] is as follows: first, the Stokes flow in the membrane without vesicles is computed; hence the permeability of the solvent is obtained. This is done by computing the total vertical flux of solvent, and homogenizing to obtain the flux through the whole membrane Q_y^s:

$$Q_y^s = \frac{1}{WH\Delta t}\int_{\Delta t}\left[\int_{-W/2}^{W/2}(-Hq_s\, d\xi)\right]dt \qquad (1)$$

Where Δt is the time required for a fluid element (such as a vesicle) to travel across the section, and q_s is the macroscopic solvent flow.

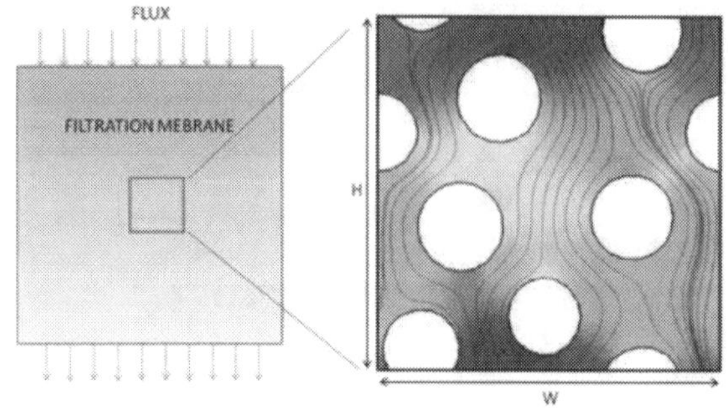

Figure 1: Stokes flow around a periodic portion of the membrane due to a constant pressure gradient across the vertical axis Δp. The white circles represent the geometry of the membrane. The black lines are the flow streamlines and the background color is the pressure.

Second, the flux of vesicles can be obtained based on their speed v_v, and after homogenizing this relation over the total membrane the flux of vesicles on the membrane is obtained:

$$Q_y^v = \frac{1}{WH\Delta t}\int_{\Delta t}\left[\int_{-W/2}^{W/2}(-Hv_v d\xi)\right]dt \qquad (2)$$

Finally, in a similar way the vertical macroscopic pressure gradient can be calculated as:

$$\nabla_y p = \frac{1}{HW\Delta t}\int_{\Delta t}\int_{\Gamma}\left(p\left(\xi,\frac{H}{2}\right) - p\left(\xi,-\frac{H}{2}\right)\right)d\xi dt \qquad (3)$$

With these variables known, the permeability of the solvent and vesicles for a given geometry can be found as:

$$\kappa_s = \frac{\mu Q_y^s}{\nabla_y p} \qquad \kappa_v = \frac{\mu Q_y^v}{\nabla_y p} \qquad (4)$$

where μ is the viscosity of the fluid. The simulations are repeated for different scenarios where the elastic properties of the vesicles are changed. And again with geometries of different porosities until enough data is collected.

RESULTS

The model is first used to investigate the permeability of a fibrous network made of circular obstacles, as shown in figure 1. After non-dimensionalizing the model, the mechanical properties of the vesicle can be characterized by the capillary number $C_a = \mu Q_y^v/\gamma$, which corresponds to rigid vesicles when it is small and highly deformable ones when it is large, (γ the surface tension of the vesicle boundary). The permeability of the network is also a dimensionless variable which uses as a reference the average permeability of the system without vesicles. Figure 2, shows the transition from very stiff vesicles (a), which get trapped at the overture, to softer ones (b), which are able to squeeze and cross the network even though they are slowed down in the process. At first glance the results indicate whether a vesicle is capable or not of crossing certain network and how the membrane flow is changed during this process. However, slow vesicle velocities during the permeation could also indicate the increasing probability of jamming and cake formation within the porous space.

Figure 3 shows how the solvent permeability of the system is affected by the presence of vesicles. In addition to reduce their own permeability, stiffer vesicles also affect the solvent permeability due to their ability to adapt and obstruct pores. Opposite to this, softer vesicles produce an effect similar to particles that are small enough to pass through the pores without getting trapped. In that case the system tends to reach the permeability of the fluid without vesicles.

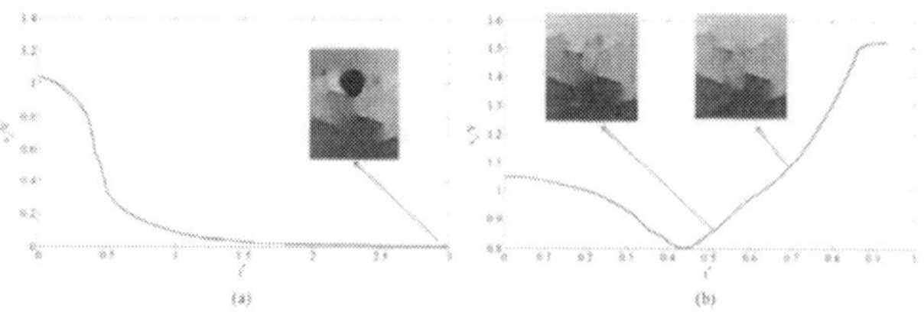

Figure 2: Vesicle speed as a function of non-dimensionalized time t* for network 1, C_a=0.04 (a) and C_a=0.2 (b).

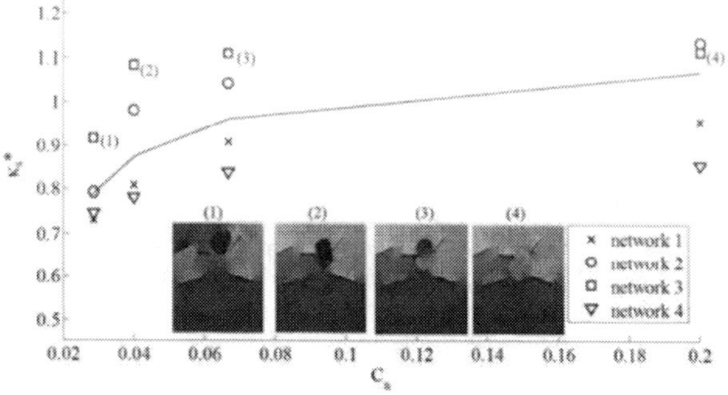

Figure 3: Relative permeability of the network as a function of the capillary number in four different networks.

DISCUSSION

The results obtained with these examples clearly state that particle deformability plays a significant role in the permeation of soft vesicles through porous networks. The method clearly captures how the mechanical properties of the vesicles affect the overall permeability of the membrane, but it also captures the importance of the network itself. For instance, one could hypothesize that, for a given overture size, a highly porous network could lead to much lower values of the vesicle permeability due to the fluid deviation. This is not true for less porous systems, where the flow can only follow certain paths. Therefore the method is able to provide a new insight on how particle deformability can be used as a critical feature in different applications.

ACKNOWLEDGEMENTS

The authors gratefully acknowledge the National Science Foundation (NSF) Industry/University Cooperative Research Center for Membrane Science, Engineering and Technology (MAST) at the University of Colorado Boulder (NSF Award IIP 1034720) for stimulating broad-based and impactful research.

REFERENCES

[1] Cheryan, M. *Membrane Technology*, 2005(2):5-7, 2005
[2] Gregoriadis et al., *Drugs*, 45(1):15-28,1993
[3] Zheng et al., *Biomed Microdevices*, 13:203-213, 2011
[4] Faibish et al., *J. of colloids and interface science*, 204(1):77-86, August 1998.
[5] Ruth, E. et al., *Industrial & Engineering Chemistry Research*, 48(5):2404-2413, 2009
[6] Hur et al., *Lab Chip*, 11, 912, 2011
[7] Foucard, L. et al., *Computer Modeling in Engineering and Science*, 98(1), pp 101-127, 2014.
[8] Foucard, L. et al., *Computer Methods in Applied Mechanics and Engineering*, 283 280–302, 2015

SB³C2015
Summer Biomechanics, Bioengineering and Biotransport Conference
June 17-20, 2015, Snowbird Resort, Utah, USA
SB³C2015-1149

COMPARISON OF MRI-MEASURED TRACER TRANSPORT IN THE RAT BRAIN WITH COMPUTATIONAL MODEL PREDICTION

Wei Dai (1), Garrett W. Astary (2), Aditya K. Kasinadhuni (3), Paul R. Carney (3,4,5,6), Thomas H. Mareci (2), Malisa Sarntinoranont (1)

(1) Department of Mechanical and Aerospace Engineering, University of Florida, Gainesville, FL, USA

(2) Department of Biochemistry and Molecular Biology, University of Florida, Gainesville, FL, USA

(3) J. Crayton Pruitt Family Department of Biomedical Engineering, University of Florida, Gainesville, FL, USA

(4) Department of Neuroscience, University of Florida, Gainesville, FL, USA

(5) Wilder Center of Excellence for Epilepsy Research, University of Florida, Gainesville, FL, USA

(6) Department of Pediatrics, Division of Pediatric Neurology, University of Florida, Gainesville, FL, USA

INTRODUCTION

To improve the treatment of neural diseases, convection-enhanced delivery (CED) is a promising method for local drug delivery in central nervous system (CNS). CED has the advantage of delivering macromolecular therapeutic agents greater distances throughout tissue than diffusion-driven mechanisms and circumvents the blood-brain-barrier (BBB). It has undergone rapid development in recent years. One of the challenges of this drug delivery method is proper targeting of a specific brain region while avoiding unintentional leakage to adjacent brain regions. Many factors, e.g. infusion site, cannula shape, duration, drug flow rate, molecular weight, and tissue properties will affect CED drug delivery. To improve targeting, voxelized 3D computational models have been developed by our group to predict flow and distribution patterns during CED infusion [1, 2]. These models are based on magnetic resonance diffusion tensor imaging data sets. Tissue transport properties are assigned on a voxel-by-voxel basis for flow and mass transport through porous media.

Structural tissue changes occur in the brain with disease progression. For example, changes in the hippocampus have been found in temporal lobe epilepsy (TLE), Alzheimer's disease, and schizophrenia. Patient-specific computational models that account for structural changes can be generated from MR-DTI data sets and used to predict therapeutic agent transport in the brain during local delivery aiding surgical planning of CED to treat neurological diseases. The hippocampus is organized complex, heterogeneous region in the brain. It has a rolled structure of densely packed layers of neurons, their projecting axons, and a hippocampal fissure. Previously in our group, computational models for CED have been developed to simulate large molecular drug delivery in the ventral hippocampus [1, 2].

With the additional utilization of generalized anisotropy (GA) [3] data from high angular resolution diffusion imaging (HARDI), we are able to improve segmentation in the dorsal hippocampus. In this study, we are able to develop CED models for the dorsal hippocampus that account for hippocampal fissures. In previous studies by our group, quantification of *in vivo* concentration patterns during CED was achieved through dynamic contrast enhanced MRI (DCE-MRI) by using measured relaxivities [4]. Using this approach, we are able to compare predicted and measured CED distributions in the dorsal and ventral hippocampus. This is the first time that we compare simulation results from the voxelized model with concentration measurements from *in vivo* experiments.

METHODS

Diffusion tensor imaging (DTI) data was collected from an excised and fixed rat brain scanned in a 17.6 Tesla, 89 mm vertical bore magnet system (Bruker NMR Instruments, Billerica, MA). Data was obtained using a recovery time (TR) of 4000 ms and echo time (TE) of 28 ms, and two diffusion weightings: b = 100 s/mm2 (number of averages (NA) = 8, 7 diffusion gradient directions) and b = 2225 s/mm2 (NA = 2, 64 diffusion gradient directions). The diffusion gradient directions were assigned using a method of electrostatic repulsion. The field of view (FOV) was 26.98 mm × 17.86 mm, with a matrix size of 142 × 94 in 63 slices of 0.19 mm thickness. Bilinear interpolation was used to obtain an image matrix of 284×188×126 with an isotropic resolution of 0.095 mm × 0.095 mm × 0.095 mm. Normalized average diffusivity (AD) and fractional anisotropy (FA) values were used to segment white matter (WM), gray matter (GM), and cerebrospinal fluid (CSF) of the whole rat brain. The hippocampus region that contains crossing fiber layers was segmented using GA data which better accounted for orientational variation in a voxel (multiple crossing fibers). The hippocampus fissure was also assigned manually on the GA map slice-by-slice using a Matlab subroutine (Matlab v.R2012a, Mathworks, Natick, MA).

DTI data was used in the voxelized modeling approach for CED in the rat hippocampus. Simulations solved for porous media transport using Darcy's law and the convection-diffusion equation [1, 2]. The porosity was set to 0.26 in tissue and 1 in CSF regions. The site of infusion was modeled as a single voxel source region with constant velocity and normalized concentration. Post-imaging processes were done in Matlab and medical imaging software ITK-SNAP.

In vivo CED infusions into dorsal and ventral rat hippocampi were performed in a previous study by our group [4]. Tissue concentration profiles were calculated using measured relaxivities with dynamic constrict enhanced MRI (DCE-MRI) through equations relating signal intensity to the contrast agent concentration [4]. All the image processing used an in-house software MAS [5] written in Interactive Data language (IDL) (IT Boulder, CO).

RESULTS AND DISCUSSION

Image segmentation of dorsal and ventral hippocampi for a fixed brain is shown in Fig. 1. By comparing with a rat brain atlas [6], the segmentation based on AD, FA, and GA maps reflect the heterogeneous structure found in the hippocampus. The segmentation of the hippocampal fissure brought greater connectivity of fluid spaces into the model.

Comparative CED distribution patterns from anterior to posterior regions are shown in Fig. 2 (a) and (b), for the dorsal and ventral hippocampus, respectively. In the dorsal hippocampus, CED infusate transports mainly in medial-lateral and anterior-posterior directions; in the ventral hippocampus, the infusate transports mainly in inferior-superior and anterior-posterior directions. With fixation, tissue shrank due to cross-linking [7]. As dimension changes between *in vivo* and fixed brains, we accounted for fixation shrinkage when comparing distribution volumes *in vivo* and fixed brain by multiplying simulated distribution volume by a factor of 1.117.

Predicted CED distribution patterns generally matched those measured with *in vivo* DCE-MRI, see Fig. 2. CED distribution volumes were also calculated using a 5% concentration threshold. DCE-MRI data were compared with computational model predictions within the same infused regions, as shown in Fig. 3. Overall, predicted CED distribution volumes in the dorsal hippocampus more closely matched experiment data than in the ventral hippocampus. Even though the brain structure can vary between animals, the simulated distribution predicted using fixed animal brains appeared to capture the major CED drug delivery patterns in the rat brain. Collection of *in vivo* DTI data with which to generate voxelized models would provide further validation on an individual rat basis.

ACKNOWLEDGEMENTS

We would like to thank Dr. Luis Perez who collected the fixed rat brain DTI data. All the MRI data in this study was obtained at the Advanced Magnetic Resonance Imaging and Spectroscopy Facility in the McKnight Brain institute and National High Magnetic Field laboratory of UF. This work was funded by NIH R01NS063360.

REFERENCES

[1] Kim, JH et al., *Med Biol Eng Comput*, 48: 203-214, 2010.
[2] Kim, JH et al., *Ann Biomed Eng*, 40: 2043-2058, 2012.
[3] Ozarslan, E et al., *Magn Reson Med*, 53: 866-876, 2005.
[4] Astary, GW, MR-guided real-time convection-enhanced delivery. University of Florida, 2011.
[5] http://marecilab.mbi.ufl.edu/software/MAS/
[6] Paxinos, G and Watson, C, The rat brain in stereotaxic coordinates. Academy Press, 2007.
[7] Ma, Y et al., *Front Neuroanat,* 2: 2008.

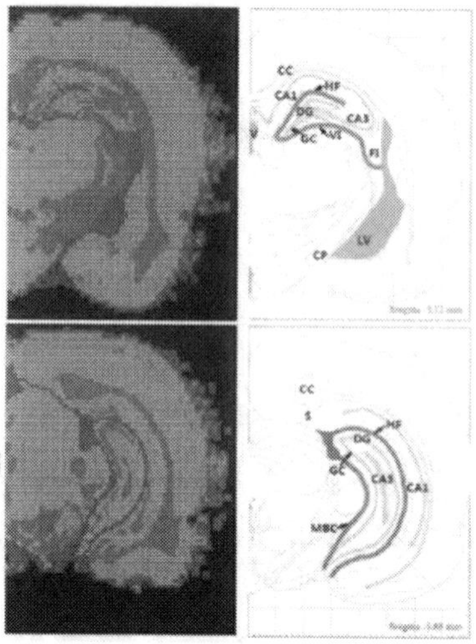

Figure 1. (Left) Segmentation of dorsal and ventral hippocampi at the infusion plane (Brown = Impermeable regions, Green = GM, Orange = WM, and Blue = CSF spaces). (Right) Rat brain atlas.

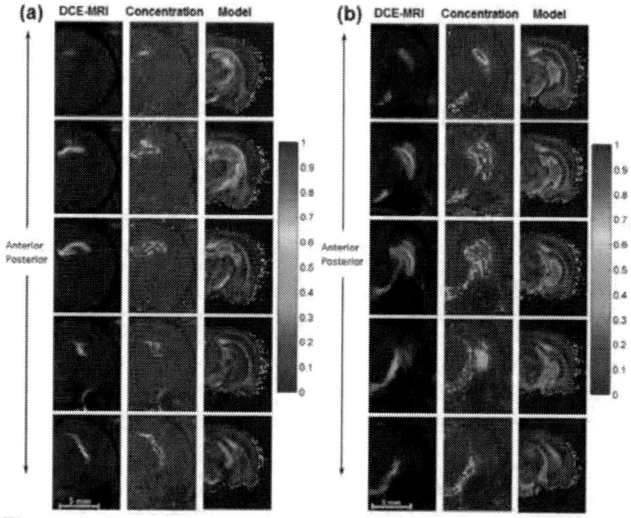

Figure 2. Comparison of experimentally measured DCE-MRI distributions with predicted CED into the dorsal (a) and ventral (b) hippocampus. 7.35 µL of tracer with infusion rate of 0.3 µL/min.

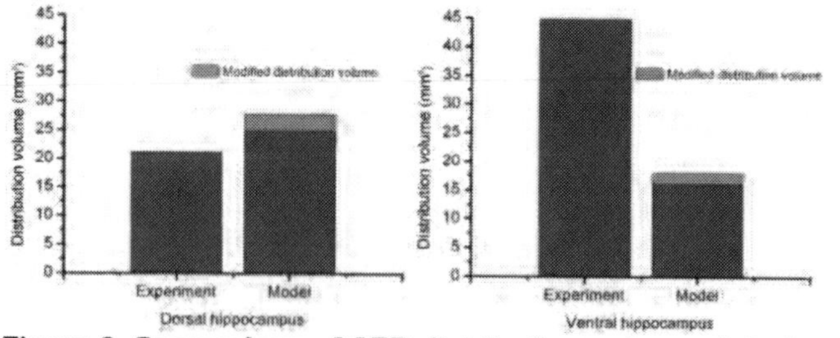

Figure 3. Comparison of CED distribution volumes into the dorsal (left) and ventral (right) hippocampus. Adjusted distribution volumes are in green (+11.7%)

SB^{3}C2015
Summer Biomechanics, Bioengineering and Biotransport Conference
June 17-20, 2015, Snowbird Resort, Utah, USA
SB^{3}C2015-1150

IMMERSOGEOMETRIC FLUID-STRUCTURE INTERACTION ANALYSIS OF BIOPROSTHETIC HEART VALVES: REALISTIC MATERIAL MODELING AND EXPERIMENTAL VALIDATION

David Kamensky (1), Ming-Chen Hsu (2), Michael S. Sacks (1), Thomas J.R. Hughes (1)

(1) Center for Cardiovascular Simulation
University of Texas at Austin
Austin, TX, USA

(2) Department of Mechanical Engineering
Iowa State University
Ames, IA, USA

INTRODUCTION

BHV design has remained relatively static over the past decade, with only small improvements in durability; BHVs frequently need to be replaced 10-15 years after being implanted [1]. The design of newer, more durable valves could be accelerated with predictive modeling. It is generally believed that mechanical stresses drive the degradation of BHVs [1-3]. The forces driving BHVs are due to the surrounding blood flow. The effect of hydrostatic forcing on a closed BHV may be modeled as a prescribed pressure load, but such a model cannot capture the transient strains in opening valves or the water hammer effect on a closing valve, both of which may contribute to long term structural fatigue. A complete mechanical model of a BHV must include FSI. The purpose of this work is to develop a robust framework for fluid-structure interaction (FSI) simulation of bioprosthetic heart valves (BHVs).

However, BHV FSI cannot yet be conveniently simulated with "off-the-shelf" technologies. Aspects that make the problem especially difficult include the similar densities of the fluid and structure, the large deformations of the structure, and changes in topology of the fluid domain as the valve leaflets contact one another. Due to these difficulties, FSI simulation of heart valves remains at the cutting edge of computational mechanics and has seen virtually no use as a predictive tool by bioengineers. Past research efforts on BHV FSI have primarily focused on the basic computational FSI challenges, with little regard to realistic modeling of BHV leaflets. In Refs. [4,5], we reviewed existing FSI technologies and developed a new numerical method for simulating BHV FSI that, in the tradition of Peskin's immersed boundary method [6], allows the structure discretization to move independently of a background fluid mesh. The present work extends this method to include a realistic soft tissue constitutive model and compares FSI simulations with experiments [7] that measured BHV deformations *in vitro*.

MATHEMATICAL MODEL AND NUMERICAL METHOD

We model blood as incompressible Newtonian fluid and BHV leaflets as hyperelastic Kirchhoff-Love shell structures. The fluid-structure interface is subject to no-slip and no-penetration boundary conditions. To allow the structure to move independently of the fluid discretization, we formulate the fluid-structure constraints weakly as an augmented Lagrangian. Summarizing Ref. [4], we obtain an effective numerical method for this problem as follows: First, we replace the fluid subproblem with its variational multiscale counterpart [8], which contains mesh-dependent stabilization terms to model subgrid turbulent flow features. We then formally eliminate the tangential portion of the fluid-structure Lagrange multiplier $\boldsymbol{\lambda}$, enforcing the no-slip condition by Nitsche's method. This models unresolved boundary layers on coarse meshes [9]. In the case of a thin structure moving through an unfitted fluid mesh, Nitsche's method reduces to a penalty method for the no-slip condition. We compute the remaining scalar Lagrange multiplier $\lambda = \boldsymbol{\lambda}\cdot\mathbf{n}$ (enforcing the no-penetration condition) by semi-implicit time integration, in which the penalty forces of the augmented Lagrangian are treated implicitly, then used to update λ explicitly in each time step. For stability analysis, this algorithm may be related to fully implicit integration of a feedback boundary condition, in the spirit of Ref. [10]. Our discretization of the structure subproblem uses technology from isogeometric analysis to directly analyze NURBS surface representations of the BHV leaflets. The smoothness of NURBS function spaces allows us to discretize the thin shell problem with Galerkin's method, as in Ref. [11]. The use of a smooth displacement solution space also improves performance of sliding contact between valve leaflets [12]. We directly capture the spline representation of the deforming structure geometry in the fluid mesh and identify our technique with the emerging class of *immersogeometric* methods [4], which aim to capture design geometries on unfitted analysis meshes.

In the present contribution, we enhance the realism of the constitutive model in the structure subproblem. In Ref. [4], we assumed a St. Venant-Kirchhoff material. This does not capture the nonlinear behavior of soft tissues. In the present work, we implement the structural constitutive model of Ref. [13], which accounts for the nonlinear stress-strain relation of the collagen fiber network reinforcing planar soft tissue. The structural model homogenizes stresses due to collagen fiber ensembles distributed among different in-plane orientations. Assuming plane stress, it gives the stress-strain relation (cf. [13, Eq. (13)])

$$\mathbf{S} = \int_{-\pi/2}^{+\pi/2} \Gamma(\theta) S_{\mathrm{ens}}[E_{\mathrm{ens}}](\mathbf{N}\otimes\mathbf{N})\,d\theta + \phi_{\mathrm{m}}\mu_{\mathrm{m}}(\mathbf{I} - C_{33}\mathbf{C}^{-1}) \quad , \qquad (1)$$

where $\mathbf{S}$ is the 2nd Piola-Kirchhoff stress, θ is an angle parameterizing in-plane material directions $\mathbf{N}(\theta)$, $\mathbf{I}$ is the identity tensor, $\mathbf{C}$ is the right Cauchy-Green deformation tensor, C_{33} is the through-thickness component of the Cauchy-Green tensor (which can be determined from in-plane components by assuming incompressibility), ϕ_m is the volume fraction of extracellular matrix (ECM), and μ_m is the stiffness of the ECM, which is modeled as an incompressible neo-Hookean material. The ECM pressure has been formally eliminated by invoking the plane stress assumption S_{33}=0. E_{ens} is the component of the Green-Lagrange strain $\mathbf{E}$ in the fiber ensemble direction $\mathbf{N}$, viz. $E_{\mathrm{ens}}=\mathbf{N}\bullet\mathbf{E}\mathbf{N}$. $\Gamma(\theta)S_{\mathrm{ens}}[E_{\mathrm{ens}}]d\theta$ is the stress contribution of collagen fibers in direction $\mathbf{N}$, where Γ is a probability distribution indicating the density of fibers in each direction and S_{ens} gives the effective tensile response of fibers stretched by E_{ens}. This tensile response can be obtained from structural considerations by homogenizing over collagen fiber slack strains [13, Eq. (6)], but, in the present work, we use a closed-form exponential response [13, Eq. (8)], for reasons of computational efficiency. The distribution Γ can be derived directly from measurements of BHV leaflet fiber architecture. Ref. [14] found that spline representations of leaflets, such as those in the present work, are well suited to mapping fiber architecture data onto valve models.

NUMERICAL RESULTS

In Refs. [4,5], we found that the numerical method outlined in the previous section can simulate cyclic FSI of BHVs in physiologically-realistic flow conditions. As we move toward more realistic constitutive models, we consider the extent to which our computed results match *in vitro* measurements. Iyengar et al. [7] measured BHV leaflet deformations throughout a cardiac cycle, simulated within a flow loop. The cited study is documented for laboratory reproducibility rather than validation of numerical methods and, short of rebuilding the apparatus to obtain further measurements, it is impossible to recreate the experiment *in silico* with sufficient precision for meaningful quantitative validation. However, we may currently make qualitative comparisons between computations and experiments which, when contrasted with the results of pure structural dynamics analysis (e.g. Ref. [15]), illustrate the value of accounting for FSI. The experiments of Ref. [7] found, e.g., that BHV leaflets open through a deformation that initiates near the attached edge, then spreads toward the free edge. This is reproduced in FSI simulations (Figure 1), but not in pressure-loaded structural analysis; a pressure load drives the free edges apart immediately.

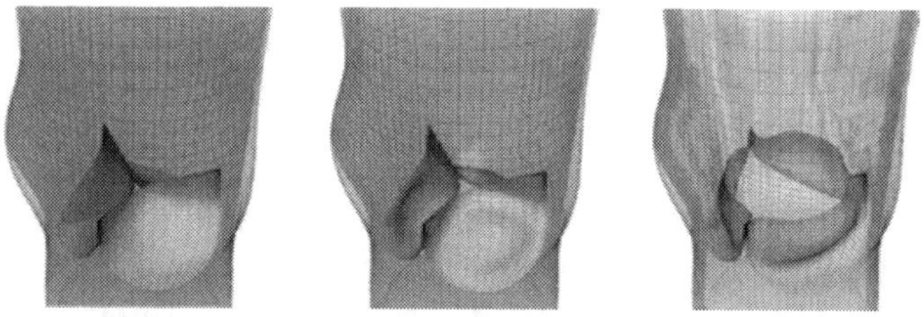

FIGURE 1: BHV OPENING IN AN FSI SIMULATION.

CONCLUSION AND OUTLOOK

This contribution extends the computational FSI methodology of Ref. [4] to include realistic material modeling of BHV leaflets and compares the results of FSI simulations with those of *in vitro* experiments. We previously published details of the numerical method and demonstrations of its accuracy and robustness [4,5]. In the present work, we combine this technology with the structural soft tissue constitutive model of Fan and Sacks [13]. We also make initial comparisons between FSI computations and experiments which, while not constituting rigorous validation, are encouraging, and we believe that useful insights into BHV function may already be gained with the current FSI technology. By relating changes in the material properties to long-term mechanical fatigue of the leaflets, we may begin to investigate the causes and effects of BHV degradation.

ACKNOWLEDGEMENTS

D. Kamensky was supported by the CSEM Graduate Fellowship at UT Austin. This research was supported by NIH grant R01 HL108330.

REFERENCES

[1] R. F. Siddiqui et al., *Histopathology*, 55:135-144, 2009.

[2] F. J. Schoen and R. J. Levy, *Annals of Thoracic Surgery*, 79(3):1072-1080, 2005.

[3] P. Pibarot and J. G. Dumesnil, *Circulation*, 119(7):1034-1048, 2009.

[4] D. Kamensky et al., *Computer Methods in Applied Mechanics and Engineering*, 284:1005-1053, 2015.

[5] M.-C. Hsu et al., *Computational Mechanics*, 54:1055-1071, 2014.

[6] C. S. Peskin, *Journal of Computational Physics*, 10(2):252-271, 1972.

[7] A. K. S. Iyengar et al., *Annals of Biomedical Engineering*, 29(11):963-973, 2001.

[8] Y. Bazilevs et al., *Computer Methods in Applied Mechanics and Engineering*, 197:173-201, 2007.

[9] Y. Bazilevs and T. J. R. Hughes, *Computers and Fluids*, 36:12-26, 2007.

[10] D. Goldstein et al., *Journal of Computational Physics*, 105(2):354-366, 1993.

[11] J. Kiendl. PhD thesis, Lehrstuhl für Statik, Technische Universität München, 2011.

[12] L. De Lorenzis et al., *International Journal for Numerical Methods in Engineering*, 87:1278-1300, 2011.

[13] R. Fan and M. S. Sacks. *Journal of Biomechanics*, 47(9):2043-2054, 2014.

[14] A. Aggarwal et al., *Annals of Biomedical Engineering*, 42(5), 2014.

[15] H. Kim et al., *Annals of Biomedical Engineering*, 36(2):262-275, 2008.

SB³C2015
Summer Biomechanics, Bioengineering and Biotransport Conference
June 17-20, 2015,Snowbird Resort, Utah, USA
SB³C2015-1151

EFFECT OF ARM POSTURE ON VOLUNTARY ACTIVATION AND MOMENTS GENERATED BY INDIVIDUALS WITH TENDON TRANSFER AND QUADRIPLEGIA

Carrie L.Peterson (1,2), Michael S. Bednar (3), Anne M. Bryden (4,5), Michael W. Keith (4,5), Eric J. Perreault (1,2,6) and Wendy M. Murray (1,2,6)

(1) Sensory Motor Performance Program
Rehabilitation Institute of Chicago
Chicago, IL, USA

(2) Research Service
Edward Hines Jr., VA Hospital
Hines, IL, USA

(3) Dept. of Orthopaedic Surgery
Loyola University Chicago
Maywood, IL, USA

(4) Cleveland FES Center
MetroHealth Medical Center
Cleveland, OH, USA

(5) Dept. of BME
Case Western Reserve University
Cleveland, OH, USA

(6) Depts. of BME and PM&R
Northwestern University
Evanston, IL, USA

INTRODUCTION

Voluntary control of elbow extension improves functional abilities for individuals with quadriplegia. When cervical spinal cord injuries occur at the C5 and C6 level, individuals with quadriplegia typically lack voluntary control of elbow extension due to complete or partial paralysis of the triceps. Fortunately, voluntary elbow extension can be improved in these individuals via tendon transfer, a surgical procedure in which the muscle-tendon path of a non-paralyzed muscle is transferred to extend the elbow. Two common procedures to enable voluntary elbow extension in individuals with C5 and C6 quadriplegia are the biceps-to-triceps and the posterior deltoid-to-triceps transfer.

Tendon transfers will provide the most benefit if individuals can both activate the muscle for its new function and generate large elbow extensor moments in functionally relevant arm postures. In nonimpaired muscle, posture affects the maximum voluntary moment a muscle can generate across a joint due to its influence on: 1) musculoskeletal moment-generating properties (e.g., moment arms, muscle force-length and force-velocity relationships) and 2) the ability of individuals to maximally activate a muscle (i.e., voluntary activation) [1,2]. Posture-dependent musculoskeletal properties are altered after tendon transfer. Specifically, moment arms are changed after biceps and deltoid transfer, and surgical tensioning influences the moment-elbow angle relationship [3]. Because individuals with quadriplegia cannot fully activate the brachioradialis transfer to restore lateral pinch [4], voluntary activation may have an important effect on the maximum elbow extensor moment that individuals with biceps or deltoid transfer can generate in different postures.

The purpose of this study was to determine the effect of arm posture on voluntary activation and maximum isometric elbow extensor moments generated by arms with biceps transfer and arms with deltoid transfer in three functionally relevant arm postures. The results of this study will inform whether a particular posture is advantageous for generating elbow extensor moments.

METHODS

Voluntary activation and maximum voluntary elbow extensor moments were assessed in seven individuals (5 men, 2 women, 34.8 ± 12 years old) with C5 or C6 spinal cord injuries who had undergone either biceps or deltoid transfer. Eleven arms (6 deltoid transfer; 5 biceps transfer) were assessed. Tendon transfer subjects were at least one year post-operative. The experimental protocol was approved by the Institutional Review Board of the Edward Hines, Jr. VA Hospital. The privacy and welfare of subjects was protected.

Subjects completed the protocol in each of three functional postures: postures to initiate a reach in the horizontal plane and overhead, and a posture consistent with initiating a pressure-relief lift. Subjects were seated in their own wheelchair and an elbow moment transducer [5] was affixed to their arm to record moments (Figure 1). First, subjects completed three maximum voluntary elbow extension trials while moments were recorded at 1000 Hz. Maximum effort was held for 5 to 7 seconds with two minutes rest between trials. The maximum moment was computed as the greatest average extensor moment maintained for 0.5 seconds. Next, subjects were asked to

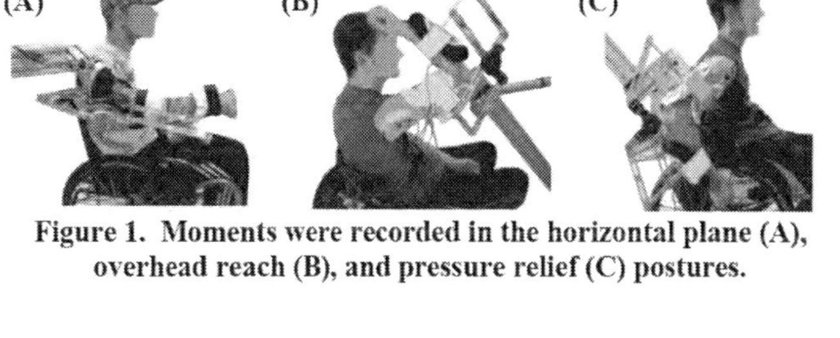

Figure 1. Moments were recorded in the horizontal plane (A), overhead reach (B), and pressure relief (C) postures.

perform moment-matching trials at 0%, 33%, 66% and 100% of their maximum voluntary elbow extension moment. An electrical stimulus event (single pulse, 100 μs duration) was delivered after the subject reached the target moment. Stimulation was delivered using a constant current stimulation unit (DS7AH; Digitimer Ltd., Welwyn Garden City, UK) connected to two disposable, self-adhesive Ag/AgCL electrodes (Noraxon, Inc., Scottsdale, AZ) placed on the transferred muscle (a cathode on the motor point and an anode 2.5 cm distal to the cathode along the muscle path). Stimulus intensity was set at 120% of the threshold intensity, above which further increases resulted in no change in the elbow extensor moment at rest. Target moments were presented in a randomized order and subjects completed at least 6 trials at each effort level.

For each moment-matching trial with stimulation superimposed, the pre-stimulus elbow moment and twitch amplitude were computed. Pre-stimulus elbow moment was computed as the average moment maintained 250 ms prior to the stimulus event. The twitch amplitude was computed as the difference between the maximum elbow extension moment occurring within 150 ms after the stimulus event, and the pre-stimulus moment. For each transferred arm and posture, a linear regression of twitch amplitudes and pre-stimulus moments was extrapolated to predict the elbow moment that could be generated if the entire motor pool innervating the transferred muscle was recruited [6]. Voluntary activation was computed as the ratio of the maximum voluntary moment to the predicted moment.

Separate linear mixed-effect models and ANOVAs were used to compare mean voluntary activation and maximum elbow extensor moments across postures within each tendon transfer group using R software (R Foundation for Statistical Computing, Vienna, Austria). Posture was treated as a fixed factor and arm as a random factor. The inverse of the group variances were included as weights in each model to account for unequal variances. Post hoc comparisons were performed when the main effect of posture was significant ($p < 0.05$).

RESULTS

Voluntary activation did not differ by posture for the biceps transfer arms ($F_{(2,38)} = 0.085$, $p = 0.92$; Fig. 2). Within the biceps transfer group, maximum voluntary moments were greater in pressure relief compared to the horizontal plane posture ($F_{(2,38)} = 13.6$, $p < .0001$; Fig. 2). Within the deltoid transfer group, voluntary activation was greater in the horizontal plane compared to the overhead reach posture ($F_{(2,34)} = 7.54$, $p = 0.002$; Fig. 2). Maximum voluntary moments did not differ by posture within the deltoid transfer group ($F_{(2,43)} = 1.20$, $p = 0.31$; Fig. 2).

DISCUSSION

In the subjects we evaluated, none of the postures we tested were particularly advantageous for generating elbow moments in the deltoid transfer arms. Arm posture did affect the ability of individuals to voluntarily activate the deltoid transfer in elbow extension. However, no differences in maximum elbow moments were seen across postures in the deltoid transfer arms, suggesting the effect of posture on activation was offset by its effects on musculoskeletal moment-generating properties. In contrast, for the biceps transfer, the pressure relief posture was advantageous for generating an elbow extension moment relative to reaching in either the horizontal plane or overhead. In these subjects, arm posture did not affect the ability of individuals to voluntarily activate the biceps transfer in elbow extension.

The biceps muscle is estimated to be longer in the pressure relief posture relative to the horizontal plane or overhead reach [7]. Thus, the increase in elbow moment with biceps length suggests the biceps

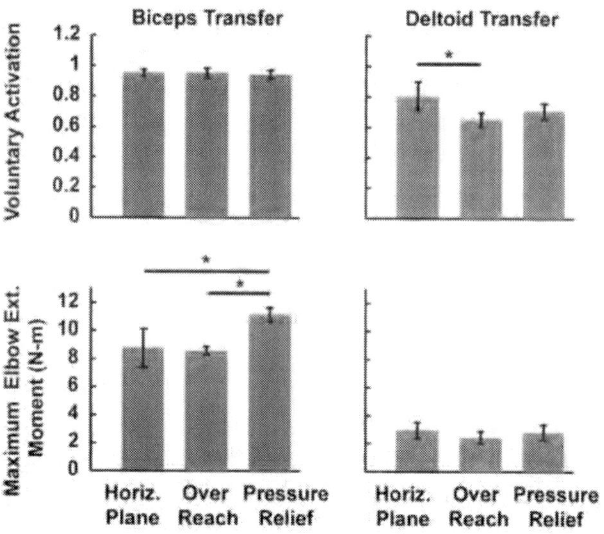

Figure 2: Mean voluntary activation and voluntary moments in each arm posture. Significant differences in the means ($p < .05$) are indicated by *. Error bars are ± one standard error.

transfer operates on the ascending portion of the force-length curve within the range of postures we tested.

Our results suggest that inter-subject differences in the maximum elbow moment and shoulder angle relationship reported previously after deltoid transfer [8] may be due to differences in voluntary activation across posture, in addition to moment-generating properties. Previous work has demonstrated the effect of deltoid tensioning during surgery and slippage of the interposition graft each on the relationship between elbow angle and elbow extensor moments [3]. Our study supports the importance of these factors affecting the moment-angle relationship of the deltoid transfer.

Activity gains after deltoid transfer are greater in reaching overhead relative to assisting with pressure relief [9]. However, we found no difference in maximum elbow moments generated by deltoid transfer arms between the overhead reach and pressure relief postures. Thus, our results suggest that the decreased ability of the deltoid transfer to assist with pressure relief relative to reaching overhead is due to differences in the task requirements (i.e., reaching overhead requires less moment than a pressure relief lift), and not due to differences in voluntary moments with arm posture.

ACKNOWLEDGEMENTS

This study was funded by the Rehabilitation Research & Development Services of the Dept. of Veterans Affairs (Merit Review Award B7515R). We thank Vikram Darbhe for his help with the data collection and experimental design.

REFERENCES

[1] Prasartwuth, O. et al., *J Physiol,* 571(1): 243–252, 2006.
[2] Becker, R. and Awiszus, F., *Muscle Nerve,* 24:667–672, 2001.
[3] Friden, J. and Lieber, R.L., *J Hand Surg Am,* 26(1):147-55, 2001.
[4] Johansen, M.E. et al., *J Hand Surg Am,* 31(5):747-753, 2006.
[5] Memberg, W.D., et al., *Clin Biomech,* 16(10):918-920, 2001.
[6] Sheild, A. and Zhou, S. *Sports Med,* 34(4):253-67, 2004.
[7] Peterson, C.L., et al., *Conf Proc Meeting of IEEE EMBS,* 2014.
[8] Kirsch, R.F. et al., *IEEE Trans Rehabil En,* 4(4):403-409, 1996.
[9] Wangdell, J. and Friden, J., *J Hand Surg Am,* 37(5):1003-10, 2012.

SB³C2015
Summer Biomechanics, Bioengineering and Biotransport Conference
June 17-20, 2015,Snowbird Resort, Utah, USA
SB³C2015-1152

IN VIVO COMPARISON OF WEARABLE HEAD IMPACT SENSORS

Lyndia C. Wu (1), Vaibhav Nangia (2), Kevin Bui (1), Bradley Hammoor (1), Calvin Kuo (2), Fidel Hernandez (2), David B. Camarillo (1,2)

(1) Bioengineering Department
Stanford University
Stanford, CA, USA

(2) Mechanical Engineering Department
Stanford University
Stanford, CA, USA

INTRODUCTION

Wearable sensors have the potential to enable precise descriptions of the cause of brain trauma by recording injury events in the field. Instrumenting humans has many challenges, including adequately coupling acceleration sensors to the structure of interest. The head in particular is covered by several layers of soft tissue and hair, and has few sites to rigidly mount sensors. Nevertheless, several types of sensors have been or are being commercialized for consumer and research use. It is imperative to quantify the coupling of these sensors to the human head and to evaluate how this might alter predictions of brain trauma. To ensure fidelity of data and research findings, devices need to be fully evaluated prior to mass deployment. In addition, evaluation of sensors *in vivo* is essential to ensure the quality of data from live human subjects.

Instrumented mouthguards have been evaluated *in vitro* using dummy validation [1]. Since the mouthguard couples to the skull through upper dentition, it is expected to have good skull coupling. In addition, skin patch and skull cap sensors have been developed, but their validation has not been reported in literature. Previous studies with skin-mounted accelerometers report artifacts due to skin dymamics [2]. It is likely that tissue dynamics would affect sensor performance, but such effects have not been quantified.

This study aims to evaluate and compare instrumented mouthguard, skin patch and skull cap sensors *in vivo*.

METHODS

We evaluated skull coupling of the sensors using high speed video tracking to visualize an ear-canal target as the skull reference (Fig. 1), since previous research confirmed skull coupling of deeply-inserted custom-fit earplugs [3]. A volunteer wore an instrumented mouthguard, skin patch, and skull cap, and underwent sagittal soccer head impacts at initial ball speed of 7 m/s (Fig. 1A), which is the average header speed in youth soccer [4]. An expandable foam earplug was inserted approximately 20mm into the ear canal, similar to the depth at which a well-coupled sensor was mounted. We took high speed stereo video at 1000 frames per second to track motion of the three sensors with respect to the ear-canal skull reference (Fig. 1B).

After evaluation of skull coupling, we compared kinematic measurements of the skin patch and skull cap sensors with mouthguard as the reference, using trials where mouthguard and ear-canal were found to have minimal relative displacement. To investigate whether sensor errors due to tissue or textile dynamics could be modeled, we also tried fitting skin patch and skull cap measurements using a second order linear model.

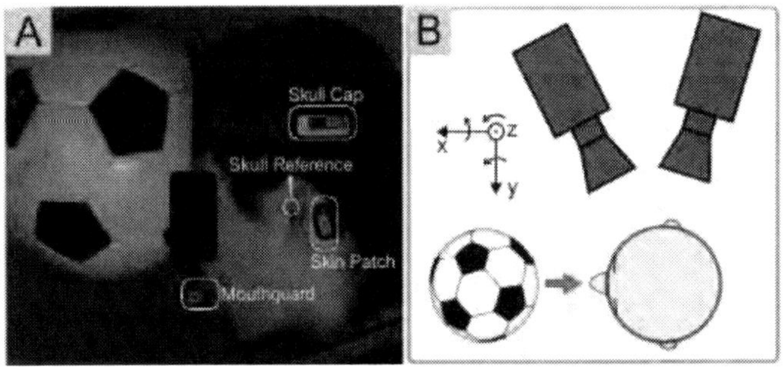

FIG. 1 – *IN VIVO* EVALUATION AND COMPARISON OF INSTRUMENTED MOUTHGUARD, SKIN PATCH, AND SKULL CAP.

RESULTS

For the mouthguard, all head impact trials (n = 16) show relative displacements from the ear-canal reference of within 1mm (mean =

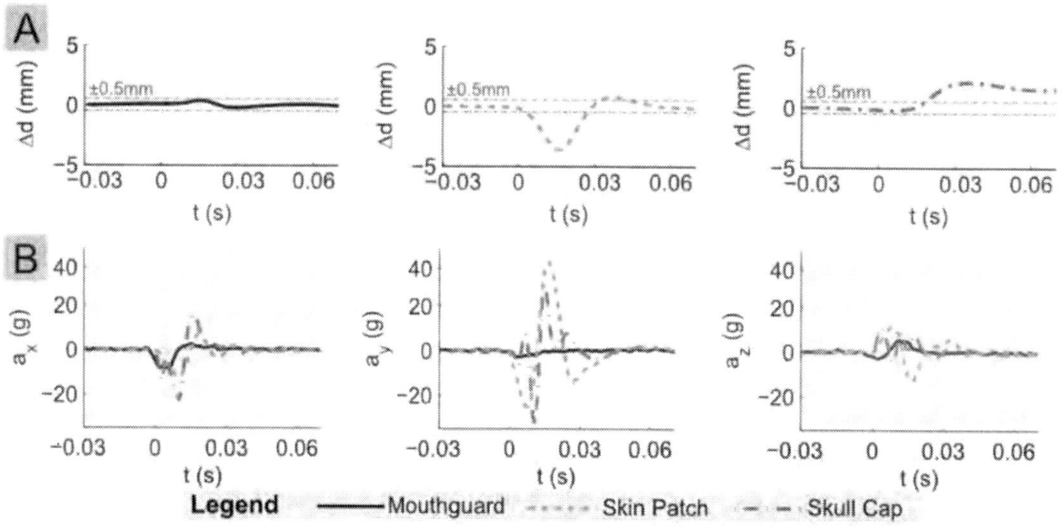

FIG. 2 – COMPARISON OF SKULL COUPLING AND SENSOR MEASUREMENTS OF THE MOUTHGUARD, SKIN PATCH, AND SKULL CAP.

0.5mm, standard deviation = 0.2mm), which is on the same order as video error. In addition, 10 of the 16 trials showed relative displacements of within 0.5mm. We used these 10 trials for subsequent sensor measurement comparisons with the mouthguard as the reference sensor, since good skull-coupling was demonstrated. On the other hand, the skin patch and skull cap sensors showed more significant movement at the moment of head impact. The skin patch displaced by 1.7 – 4.1mm, and the skull cap displaced by 2.1 – 13.2mm.

Comparison of skull coupling and sensor measurements for one sample trial are shown in Fig. 2. In this trial, the mouthguard displaced less than 0.5mm, while the skin patch and skull cap displacements were more significant (Fig. 2A). This skull decoupling resulted in sensor over-predictions by the skin patch and skull cap. As shown in Fig. 2B, these sensors not only over-predict accelerations in the sagittal plane (a_x and a_z), they also had significant over-predictions in the out-of-plane direction (a_y). In addition, the direction of peak measurements by the skin patch and skull cap were sometimes incorrect compared to the mouthguard reference.

When we modeled skin patch dynamics in anterior-posterior translation, we found an average natural frequency of 23.8 Hz (±2.3 Hz), a damping ratio of 0.27 (±0.09), and system mass of 8.5 g (±1.6 g). On the other hand, skull cap dynamics could not be explained by the model very well, showing large variations in model parameters.

DISCUSSION

Custom-formed mouthguard couples well to the head under our impact conditions with closed-jaw, while the skin patch and skull cap sensors do not. This was an expected outcome, since the mouthguard is custom-formed to couple well to the teeth. It does confirm that clenching force, similar to firm clamping on a dummy headform, is able to hold the mouthguard in place even during impact. Mouthguard coupling in the field remains to be tested. However, in anticipated field impacts, the teeth are likely clenched, resulting in similar conditions as our tested scenario. The skin patch and skull cap sensors, on the other hand, consistently show dislocations from the skull during soccer impacts.

Skull coupling issues led to sensor measurement errors in skin patch and skull cap. Confirming video coupling observations, the skin patch and skull cap sensors showed over-estimations of skull kinematics. By comparing per-axis measurements, it becomes apparent that the skin patch and skull cap would 1) predict wrong direction of peak acceleration, and in turn predict wrong directionality of impact, 2) predict significant out-of-plane motion. Such errors would lead to incorrect predictions of injury risks if the measurements were used directly. Sensor errors either need to be modeled and corrected in post-processing, or reduced by improving skull coupling.

A simple second order model approximation seems to work reasonably well at modeling skin effects, in agreement with previous studies of skin dynamics [5]. The average estimated mass of the skin patch system (8.5g) is greater than mass of the sensor itself (5g), showing that some underlying tissue mass is also activated during the impacts. On the other hand, behavior of the skull cap sensor is less predictable than the skin patch sensor, varying from impact to impact. This is likely due to the fact that it was directly impacted by the ball. Ball force and impact location were important factors for skull cap system not accounted for by the simple linear model. In addition, we did not evaluate other directions, which might have different models. And due to the observations of significant out-of-plane motion, cross-coupling of axes likely happen. Therefore, in practice, higher degree-of-freedom dynamic models may be necessary to fully predict sensor errors.

ACKNOWLEDGEMENTS

This study was funded by the National Institute of Health and the David and Lucile Packard Foundation.

REFERENCES

[1] D. B. Camarillo, et al., Ann. Biomed. Eng., 41(9):1939-49, 2013.
[2] C. Reinschmidt, et al., J Biomech., 30(1):729–732, 1997.
[3] R. S. Salzar, et al, SAE Int. J. Passeng. Cars, 1(1):1367–81, 2014.
[4] N. Shewchenko, et al., Br. J. Sports Med., 39(1):i26–32, 2005.
[5] W. Kim, et al., J. Biomech. Eng., 115(1):47–52, 1993.

SB³C2015
Summer Biomechanics, Bioengineering and Biotransport Conference
June 17-20, 2015, Snowbird Resort, Utah, USA
SB³C2015-1153

A FINITE ELEMENT MODEL TO EVALUATE THE ROLE OF THE MEDIAL MENISCOTIBIAL ATTACHMENT IN KNEE BIOMECHANICS

Andrew J. Polk (1,2), Ferris M. Pfeiffer, PhD (1,2,3), James L. Cook, PhD (1), James P. Stannard, MD (3), Patrick A. Smith, MD (3)

(1) Comparative Orthopaedic Lab
University of Missouri
Columbia, MO, United States

(2) Department of Bioengineering
University of Missouri
Columbia, MO, United States

(3) Department of Orthopaedic Surgery
University of Missouri
Columbia, MO, United States

(4) Columbia Orthpaedic Group
Columbia, MO, United States

INTRODUCTION

Peripheral tears of the medial meniscus and meniscotibial ligament insufficiency are both common clinical problems, especially in sports medicine, and are thought to place additional stress on the anterior cruciate ligament (ACL), meniscal attachments, and articular cartilage. The intent of this study is to utilize a validated finite element model of the human knee to determine the effect of peripheral medial meniscus tears or medial meniscotibial insufficiency in ACL-intact and ACL-deficient knees on knee joint biomechanics, and to evaluate the effectiveness of two surgical repair techniques.

We hypothesize that with the inclusion of medial meniscal attachment insufficiency (either by deficiency of the meniscotibial ligament or a peripheral tear), knee laxity will be significantly increased and that knee stiffness in A-P, valgus, and external rotation will be significantly reduced as compared to knees with an intact meniscotibial attachment. We further hypothesize that the stress in the remaining tissues (i.e. articular cartilage, meniscal horn, and ACL if present) will be significantly increased in knees with a deficient meniscotibial attachment.

METHODS

A 3D geometrically accurate finite element (FE) model (Figure 1) was used to evaluate stress, strain, and deformations for six different configurations including 1) intact, 2) ACL transected (ACLx), 3)ACL intact w/peripheral meniscus tear, 4) ACLx w/peripheral meniscus tear, 5) ACL intact w/meniscotibial ligament transected, 6) ACLx w/meniscotibial ligament transected, 7) peripheral meniscus tear repaired using 2-0 Fiberwire, 8) meniscotibial ligament repaired with 2-0 Fiberwire. Geometry was utilized from the Open Knee open source model [1],[2]. Non-linear viscoelastic material models were utilized and analysis of A-P shear, valgus, and external rotation was performed using FEBio non-linear finite element solver.

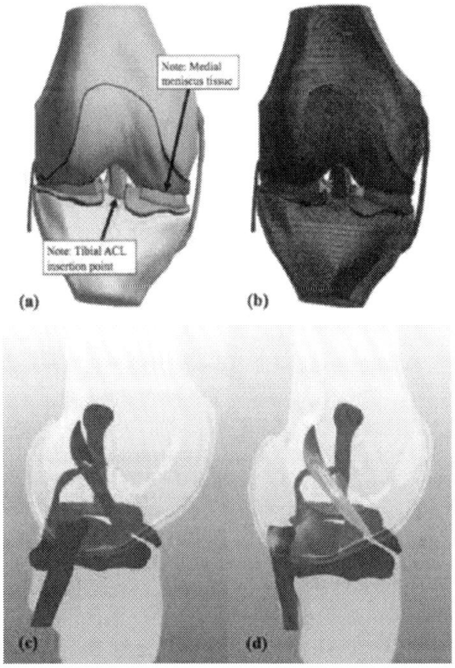

Fig 1. Finite element model of meniscotibial ligament including a) knee geometry used, b) FE mesh, c) Stress plot of ACL for intact medial attachment, and d) Stress plot of ACL for deficient medial meniscus attachment.

Biomechanical test data was used for validation of the finite element model. For this, 6 pairs of matched knees were removed of all soft tissue 10cm above and 8cm below the joint line, leaving the joint intact. All knees had parapatellar exposure with removal of the fat pad. The femur and tibia of each knee was then securely potted in copper tubing. Prior to injury, all knees were tested (using an Instron 8821s biaxial servo-hydraulic load frame) in valgus at 0°, 30°, 60°, and 90° of flexion (Figure 2), anterior laxity (Figure 3), and external rotation (Figure 4). Knees 1-3 (right) had peripheral medial meniscus tears created and knees 4-6 (right) had meniscotibial tears created with a scalpel by means of arthroscopic surgery and were retested.

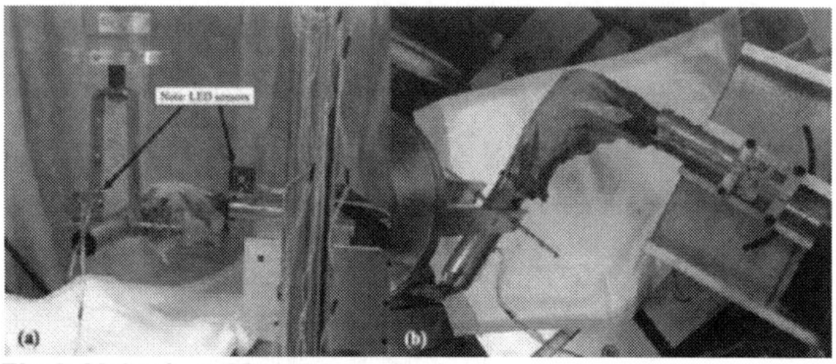

Fig 2. Valgus at 90 degrees flexion

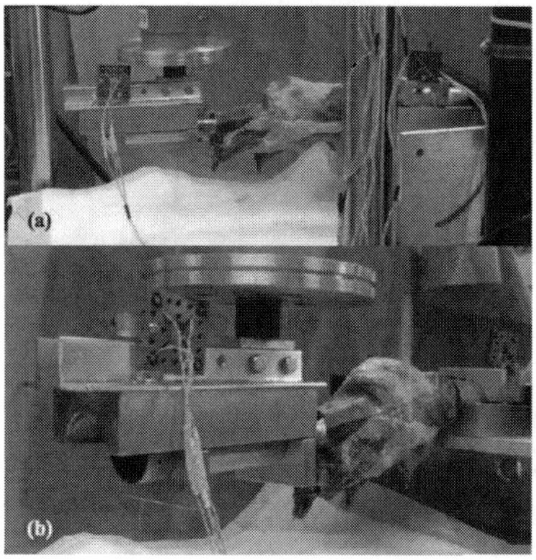

Fig 3. Anterior-posterior (AP) shear

Fig 4. External Rotation

Following post-injury testing, the two lesions were repaired and tested for a final time. The peripheral medial meniscus tear was repaired with "gold standard" inside-out suturing with 2-0 Fiberwire on long needles. The meniscotibial lesion was repaired with 4-spaced mini Arthrex suture tacks with 0-Fiberwire sutures attached to the implants. Real-time 3D motion of the femur and tibia was tracked through attached LED sensors (Figure 2a) in conjunction with an NDI Optotrak Certus motion capture system. Additionally, imaginary points were placed at the femoral and tibial ACL insertion points, as determined through arthroscopic probing, relative to an established global coordinate system.

The contralateral left knee for each pair 1-6 had the ACL transected and was tested using the same procedure. Knees 1-3 (left) had peripheral medial meniscus tears created and knees 4-6 (left) had meniscotibial tears created. All knees were tested pre-injury, post-injury, and post-repair for valgus at 0°, 30°, 60°, and 90° flexion (Figure 2), anterior laxity (Figure 3), and external rotation (Figure 4). This group of 6 left knees addressed the impact of either a peripheral medial meniscus tear or medial meniscotibial lesion on stability of the knee in an ACL-deficient knee.

RESULTS

Finite element models (Figure 1) confirmed the hypothesis that medial meniscal attachment plays a significant role in the biomechanics of the knee joint. Significant increases in the stress carried by the remaining tissues (articular cartilage, meniscal horns, and intact ACL) were predicted in the meniscal attachment deficient knee models.

Biomechanical testing (Figure 5) indicated a significant increase in joint laxity for both ACL deficient as well as both meniscal lesion groups.

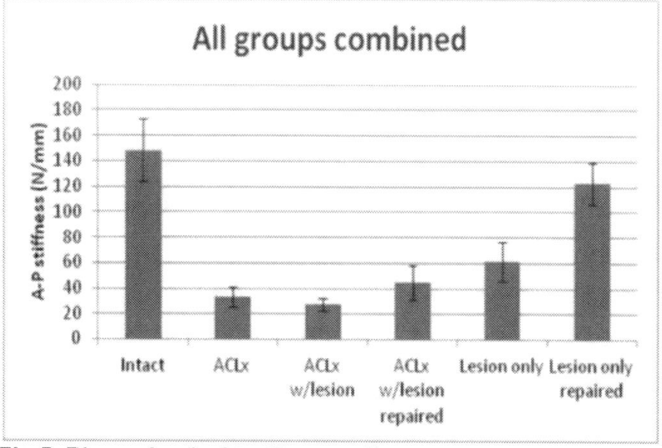

Fig 5. Biomechanical testing results

DISCUSSION

The results of this work indicate that the medial meniscal attachment significantly affects knee joint stability. Primary joint laxity was observed in A-P shear loading. Finite element models agreed well with biomechanical testing.

Based on the findings of this study, we conclude that medial meniscus attachment contributes to knee joint stability primarily in A-P shear and external rotation. We also conclude that stress in the ACL is significantly increased in samples with insufficient medial meniscus attachment. Results indicate a need to repair medial meniscal insufficiency when present.

ACKNOWLEDGEMENTS

Funding for this study was provided by the Comparative Orthopaedics Laboratory. FE models were adapted from the Open knee open source model.

REFERENCES

[1] Erdemir A. and Sibole S. Open Knee: A Three-Dimensional Finite Element Representtion of the Knee Joint, User's Guide, Version 1.0.0, December 17, 2010. (2010)

[2] Sibole, S., Bennetts, C., Borotikar, B., Maas, S., van den Bogert, A. J., Weiss, J. A. And Erdemir, A. Open knee: a 3D finite element representation of the knee joint, 34th Annual Meeting of the American Society of Biomechanics, August 18-21, 2010, Provideence, RI. (2010)

FABRICATION AND EVALUATION OF BIODEGRADABLE TISSUE SCAFFOLDS FOR OSTEOCHONDRAL DEFECT REPAIR

Carly R. Garrow (1,3), Andrew J. Polk (1,3), Ferris M. Pfeiffer, PhD (1,2,3)

(1) Department of Bioengineering
University of Missouri – Columbia
Columbia, Missouri, United States

(2) Department of Orthopaedic Surgery
University of Missouri – Columbia
Columbia, Missouri, United States

(3) Comparative Orthopaedic Laboratory
Columbia, Missouri, United States

INTRODUCTION

Degeneration of articular cartilage, caused by trauma or a variety of other issues, can lead to joint deterioration, osteoarthritis, and significant joint pain. Joint repair/replacement technologies which utilize metal and plastic have a limited lifespan and the potential for adverse reactions in the body. A biodegradable scaffold which allows the body to assist in the regrowth of articular cartilage can prevent osteoarthritis and joint pain from developing or worsening. A biodegradable tissue scaffold can be placed in the joint in the place of the degenerate articular cartilage. As the scaffold bears some of the mechanical loading of the joint, the articular cartilage repairs itself and begins growing around and through the tissue scaffold. The cartilage will continue to refill the space it had occupied previously, while the tissue scaffold slowly degrades. The purpose of this study was to determine the optimal fused deposition manufacturing (FDM) printing parameters which could generate a tissue scaffold with ideal pore size and mechanical strength to be utilized as a substitute for damaged osteochondral tissue.

METHODS

A custom 3D FDM printer (Figure 1) was created in order to print the biodegradable plastic Polycaprolactone (PCL). A 14,000 molecular weight PCL was used throughout this project. The melting point of the PCL was determined and used as a baseline temperature for the heated extruder. A small temperature range for the extruder was determined about 20 degrees above the melting point to ensure a proper extrusion rate. Prints were created in water to allow for hardening of the scaffold into the appropriate shape, creating a need to establish an optimal water temperature. Different combinations of extruder and water temperatures were tested until an appropriate scaffold pore size and mechanical strength were achieved. Two types of prints were compared for symmetry and surface flatness, both important features of a tissue scaffold. The first tissue scaffold was directly printed 6mm in diameter, while the second was a 6mm piece cut from a 20mm print. Indentation testing (Figure 2) was performed to evaluate the mechanical strength of each printed scaffold. In brief, the 6mm diameter printed scaffold was placed on an anvil which was attached to the table of an Instron 8821s servo-hydraulic load frame. Each scaffold was indented using a 4mm diameter punch at a rate of 0.1mm/s until ultimate failure of the scaffold. Applied force was recorded using the Instron data acquisition system.

Figure 1. FDM printer for manufacturing tissue scaffolds

Displacement was measured using optical tracking sensors (NDI Certus). Stress vs. strain plots were generated (Figure 3) and the elastic modulus was calculated from the slope of the stress vs. strain curve.

Figure 2. Indentation test setup

In order to evaluate the effect of filament lay down arrangements, a finite element analysis of various configurations was run in conjunction with biomechanical testing. In brief, 1/4 models of 6mm diameter by 4mm in height porous scaffolds were generated using Solidworks CAD software (Figure 4). Boundary conditions were consistent with biomechanical testing. Three configurations of lay down patterns were simulated including 1) 90-90-90 single band, 2) 90-90-90 double band, and 3) 30-60-90 single band. Stress vs. strain curves were generated for the FE simulations (Figure 3) and compared with biomechanical testing results for the 90-90-90 single band configuration to confirm accuracy of the FE simulations.

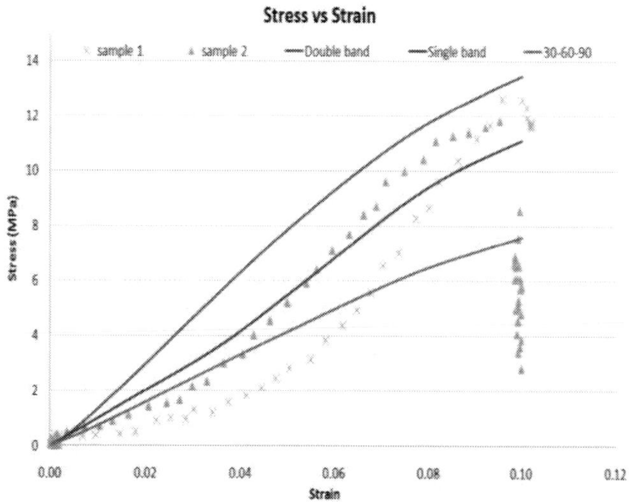

Figure 3. Stress vs. Strain plot for two scaffold tests, FE results for three different lay down patterns

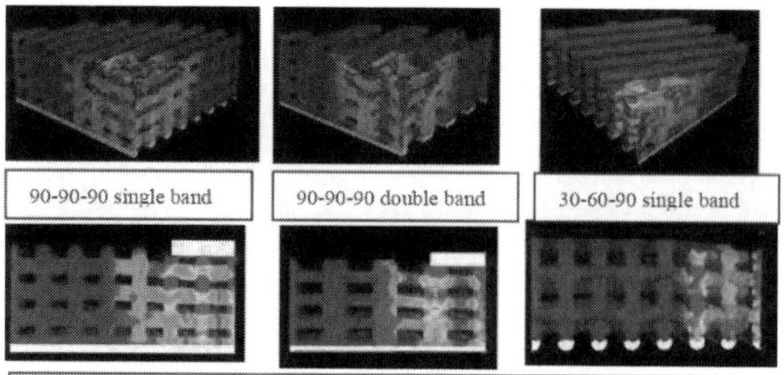

Figure 4. FE results for indentation simulations for three lay down patterns of PCL filaments

Table 1. Measured and simulated properties

	Elastic Modulus (MPa)	Yield Strength (MPa)
Physical Samples		
Cancellous bone (human)	90-150	2-20
scaffold 1	236.18	11.9
scaffold 2	178.73	11.4
FEA analysis		
Single Band 90°	132.23	11.1
Double Band 90°	147.4	13.4
30°-60°-90° layered	82.17	7.6

RESULTS

An extruder temperature ranging from 185° to 195°F and a water temperature of 75°F were established as ideal parameters for creating the optimal scaffold. The measured and simulated elastic modulus and yield strength for the various scaffold designs is shown in Table 1.

DISCUSSION

FE simulations predicted an elastic modulus and yield strength which were consistent with values measured in biomechanical tests. Simulations of various filament lay down patterns predicted a range of moduli and yield strength which will allow for future scaffold properties to be "tuned" to the specific application. The elastic modulus and yield strength of the tested and simulated scaffolds is in the range of human cancellous bone.

ACKNOWLEDGEMENTS

This study was funded by the Comparative Orthopaedic Laboratory.

BLOCKING INFLAMMATION PROTECTS AGAINST ALTERATIONS IN CELL BIOMECAHNICS, MORPHOLOGY AND CYTOSKELETON

Timothy Jacobsen (1), Paula Hernandez (1), Victoria Wei (1), Nadeen O. Chahine (1,2)

(1) Center for Autoimmune and Musculoskeletal Disease
Feinstein Institute for Medical Research
Manhasset, NY USA

(2) Department of Neurosurgery
Department of Molecular Medicine
Hofstra North Shore LIJ Health System
Manhasset, NY USA

INTRODUCTION

Intervertebral disc (IVD) degeneration (DD) is characterized by increased catabolic activity, ECM breakdown, and elevated levels of pro-inflammatory cytokines, particularly TNF-α and IL-1β [1]. DD and associated pathophysiologic loss of ECM, modulates hydrostatic pressure and biomechanical loading in the nucleus pulposus (NP) and can impact cell mechanobiology [2,3]. Stimulation of NP cells with either the inflammatory cytokine TNFα, or with lipopolysaccharide (LPS), a ligand of the toll-like-receptor-4 (TLR4) inflammatory pathway, has been shown to cause significant increases in both cell volume and membrane hydraulic permeability [4]. TAK-242 is a cell-permeable cyclohexenecarboxylate that disrupts TLR4, but not TLR1-3 or TLR5-10, through interaction with intracellular adaptor molecules TIRAP and TRAM [5]. The goal of this study is to examine the effect of a TLR4 inhibitor at mitigating biophysical changes in NP cells that are induced by inflammatory stimulation. We hypothesize that disruption of the TLR4 pathway by TAK-242 will prevent changes in biophysical properties of NP cells induced by inflammatory stimulation with LPS or TNFα.

METHODS

NP Cell Culture: NP tissue was isolated from bovine lumbar spines, minced and digested using standard enzymatic protocols.. Inflammatory Stimulation: Cells were incubated in serum-free DMEM supplemented with 0.1 μg/mL LPS or 10 ng/ml TNFα for 24 hours. Treatment Groups: Cells were treated with TAK-242 (0.1, 1, 10, 100 μM, EMD Millipore) from 1 hour pre-stimulation until completion of the 24 hour inflammatory stimulation period with LPS or TNFα. Osmotic Challenge: NP cells were seeded in a PDMS, Y-shaped micro-fluidic channel, and incubated for 30 minutes to allow for adhesion while maintaining a rounded morphology [4,6,7]. Cells were equilibrated in a 333 mOsm/L NaCl solution. A hyper-osmotic shrinking step (333 to 466 mOsm/L) followed by a hypo-osmotic swelling step (466 to 333 mOsm/L) were then applied using NaCl solutions at room temperature. Cell cross-sections were imaged during osmotic loading (0.5 Hz, 5 minutes) using DIC imaging. Cell volume-responses were curve fitted to a mixture theory model [4,7], to derive the osmotically active intracellular water content (ϕ) and membrane hydraulic permeability (Lp). Parameters were analyzed using 2-way ANOVA and Fisher LSD post-hoc test using Statistica. Biochemical Assays: Cell supernatants were collected and assayed for nitric oxide content using Griess reaction (Roche). Supernatants were also assayed for viability using LDH assay (Roche LDH detection kit). Cell Morphology: In dedicated cell preparations, DIC images of rounded NP cells were acquired at 333mOsm/L and cell radius was measured using matllab. Nuclear Morphology: Nuclei were stained with NucRed Live 647 (Molecular Probes) and incubated in 333mOsm NaCl. Cells were imaged on a confocal microscope (60x). Analysis of nuclear morphology was performed with ImageJ and CellProfiler [8] and chromatin condensation was quantified with Matlab program adapted from [9], reported as chromatin condensation parameter (%CCP). For nucleus statistics, we used unpaired t-test with $p<0.05$ as significant. Cytoskeleton: Cells were seeded in 1.2% alginate for 1 day (to maintain rounded morphology), treated with inflammatory challenge for 24 hours, then fixed and released from alginate for cytoskeletal analyses. Cells were stained with a primary antibody to detect β-tubulin (mouse anti-β tubulin, Sigma), or vimentin (mouse anti-vimentin, Abcam) with anti-mouse Alexa 488 secondary antibody. Cells were also co-stained with Rhodamine-Phalloidin (Molecular Probes) to identify F-actin. Cells were imaged using confocal microscopy and analyzed with Image J.

RESULTS

Treatment of NP cells with LPS or TNFα lead to a significant increase in cell radius ($p<0.05$ vs. untreated, Figure 1A). LPS treatment also induced a significant increase in the radius of the nucleus ($p<0.01$, Figure 1B), though negligible changes in chromatin condensation due to LPS treatment were observed (not shown). Inflammation due to LPS or TNFα treatment induced significant changes in actin cytoskeleton (Figure 2). An enrichment of punctuate aggregation of F-actin in the cortical region is observed in inflammatory treated cells vs. control (Figure 2). No significant changes were observed in microtubule structure. Some thickening of vimentin network was also observed (Figure 2). Treatment with the TAK-242 inhibitor resulted in a dose dependent inhibition of both LPS induced and TNFα induced NO release, though greater sensitivity was observed against LPS (*$p<0.05$ vs. LPS or TNFα, Figure 3). During hyper-osmotic loading step (333 to 466 mOsm/L), significant increases in hydraulic permeability (Lp) were observed in LPS or TNFα treated cells. Co-treatment with TAK-242 reduced Lp levels back to baseline (*$p<0.05$ vs. untreated, Figure 4a). No significant changes in intracellular water content were observed vs. untreated cells. Significant increases in cell radius were also observed in LPS or TNFα treatments, and are mitigated in cells co-treated with TAK-242 (*$p<0.05$, Figure 4b).

DISCUSSION

The goal of this study was to investigate the specific contribution of inflammation on cell biophysical properties and morphology of isolated NP cells. Inflammatory stimulation increases NP cell and nucleus radius, suggesting that coupled changes occur in both the cell and nucleus morphology due to inflammation. We also observed alterations in cytoskeleton elements due to inflammation in NP cells grown in 3D alginate beads, consistent with our previous studies on NP cells in 2D [4]. Loss of F-actin from the intracellular non cortical regions was observed, along with enrichment in cortical F-actin with punctuate aggregations, after inflammatory stimulation. The vimentin intermediate filament network encapsulating the perinuclear region was observed to be thicker, suggesting that the fibers are more compact in those areas. Changes in vimentin network may alter the perinuclear space, and may be related to changes observed in the nucleus radius. Inflammatory treatment also alters the biophysical properties of NP cells, as indicated by significant increases in membrane hydraulic permeability (Lp), with no effect on osmotically active water content (ϕ). The small molecule inhibitor, TAK-242, inhibited LPS- and TNF- induced pro-inflammatory signaling (nitrite release). TAK-242 was not found to increase cell death over a 24 hour treatment period (LDH assay, data no shown). TAK-242 inhibited LPS- and TNF- induced increases in Lp, returning levels to that of untreated control cells. We have previously shown that Lp is correlated with NP cell radius [4]. Protection against changes in Lp using TAK-242, was found to be associated with a coordinated inhibition in cell radius change. This study has demonstrated that inhibition of TLR4 can mitigate changes in mechanobiology of NP cells. Consequently, our findings identify a potential effect of systemic or local chemical insults on the mechanobiology of resident cells, and may represent cellular changes that trigger future cell-cell or cell-ECM changes in degeneration.

ACKNOWLEDGEMENTS: This study was funded in part by NSF CAREER Award 1151605 (NOC) and the Alison Conyngham Neisloss Collaborative Women in Science Award.

REFERENCES: [1] Le Maitre +, Arth Res Ther, 2005. 7(4). [2] Urban +, Spine, 1998. [3] Iatridis+, JOR, 1997. [4] Maidhof +, PLoS ONE, 2014. 9(6). [5] Matsunaga +, Mol Pharm, 2011. 79. [6] Chao +, J Biomech, 2005. 38. [7] Albro +, Biophys J, 2009. 97. [8] Carpenter +, Genome Biology, 2006. [9] Irianto +, Med Eng Phys 2014

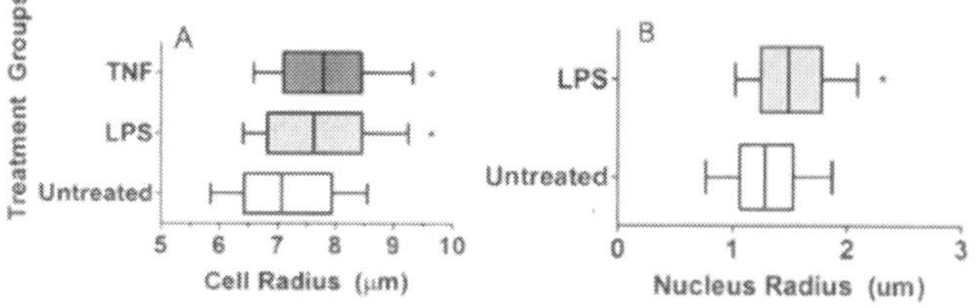

Figure 1: Treatment of NP cells with LPS or TNFα led to (A) significant increase in cell radius. (B) LPS treatment also induced a significant increase in the radius of the nucleus ($p<0.01$ vs untreated).

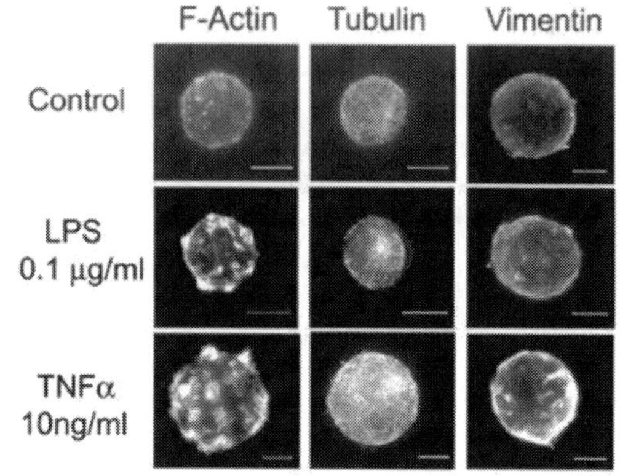

Figure 2: Inflammation induced changes in cytoskeleton, punctuate aggregation of F-actin and thickening of vimentin network in cortical regions. Scale bars: 10 μm

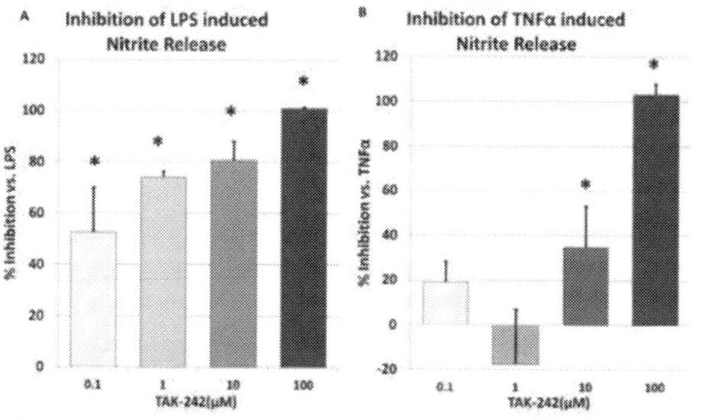

Figure 3: Treatment with TAK-242 results in a dose dependent inhibition of (A) LPS induced and (B) TNFα induced NO release (*$p<0.05$ vs. no inhibitor).

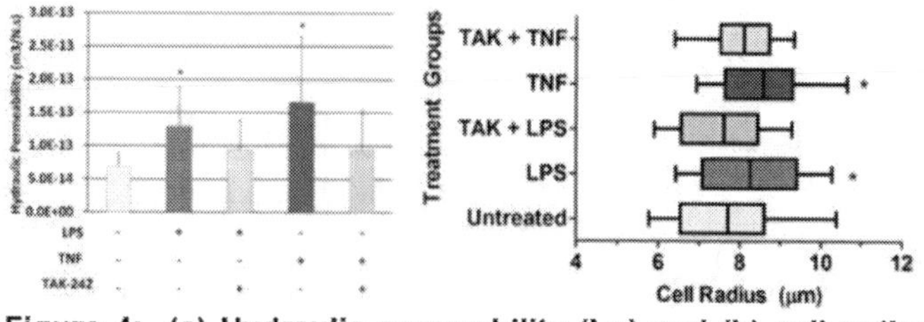

Figure 4: (a) Hydraulic permeability (Lp) and (b) cell radius changes in inflammatory stimulation and inhibitor treated groups. *$p<0.05$ vs. untreated. Data shown (mean ± SD) from hyper-osmotic loading step (333 to 466 mOsm/L).

SB³C2015
Summer Biomechanics, Bioengineering and Biotransport Conference
June 17-20, 2015,Snowbird Resort, Utah, USA
SB³C2015-1156

2D/3D REGISTRATION TO FIND HOST BONE COVERAGE OF RSA IMPLANTS

Jonathan W. Keimel (1,3), Kristi L. Krebes (2,3), Andres F. Cabezas (1,2,3), Adam Lorenzetti (4), Brandon G. Santoni (1,3), Mark A. Frankle (4), Peter Simon (3,1)

(1) Department of Chemical & Biomedical Engineering
University of South Florida
Tampa, FL, USA

(2) Department Biological Sciences
University of South Florida
Tampa, FL, USA

(3) Philip Spiegel Orthopaedic Research Laboratory
Foundation for Orthopaedic Research and Education
Tampa, FL, USA

(4) Dept. Shoulder & Elbow Surgery
Florida Orthopaedic Institute
Tampa, FL, USA

INTRODUCTION

Glenohumeral osteoarthritis (GH-OA) is the most common reason for shoulder replacement. Shoulder arthroplasty procedure provides reliable pain relief and restoration of function [1]. However, glenoid component wear and aseptic loosening are among the most common reasons for revision [1]. In the cases of severe glenoid bone loss, or revision of surgery, anatomic glenoid reconstruction is particularly challenging. Common methods utilized to reconstruct the anatomic glenoid include an asymmetric reaming of the glenoid, augmented glenoid component, and bone grafting of the glenoid. Selection of the method depends on the severity of deformity and the degree of bone loss. While asymmetric reaming may aid surgeon design the treatment for patients with mild bone loss, it can be an insufficient solution for cases with severe bone deficiencies. For those severe cases, an augmentation with either specific glenoid component implant or bone stock is generally suggested [2]. Although, augmented glenoid components represent a promising solution for severe posterior bone loss, there are currently no long-term clinical studies available. On the other hand, bone grafting provides a biologic solution in cases of substantial bone loss that prevents adequate version correction or secure seating of a glenoid component. The increase in surface contact area due to the graft allows for more baseplate coverage and in turn, may lead to better patient outcomes [2].

Our goal in this pilot experiment is to develop a technique to quantify the area of host bone baseplate coverage from the set of conventional clinical pre/post-surgical CT scans and radiographs.

METHODS

The patient selected for this pilot experiment underwent shoulder arthroplasty procedure due to severe GH-OA. In addition to insufficient bone stock (**Fig.1**) patient had irreparable rotator cuff tear. As a consequence of primary and secondary diagnoses the reverse shoulder arthroplasty (RSA) was selected. Pre-operative CT scan of the affected shoulder and post-operative x-rays (frontal, true-AP, oblique and sagittal) of the patient were acquired in DICOM format. Information on the implants (DJO RSP, DJO, Austin, TX) used during the procedure were also collected.

A pilot CT scan was imported into Mimics (Mimics 17: Materialise; Leuven, Belgium) and final patient specific 3D scapula model was segmented based on the threshold value of the cortical bone. The current research edition of Mimics software includes a specialized module capable of determining placement of surgical implant components through utilizing a pre-operative CT scan and post-operative planar radiographs. Although a minimum of two non-identical radiographs are necessary for a successful 2D-3D registration [3], visualization of the rib-cage on the radiographs pose a source of decreased accuracy. To overcome this problem, a set of four, non-identical conventional clinical radiographs (frontal, true AP, oblique and sagittal) were used. Lastly, the 3D model of the implant in STL format was imported.

The implant and scapula's outer contours are projected onto the clinical x-ray using a virtual emitter. These projections are aligned to multiple 2D x-ray images, with the contours of the 3D projections conforming to their respective outline as presented on the x-ray. The alignment is accomplished through manual and automated registration techniques in an iterative fashion. The final scapula and implant position were then exported to SolidWorks (Dassault Systemes; Waltham, MA), where the polylines representing the intersection between the implant face and glenoid body were generated. The area of the polyline was used to determine host bone coverage.

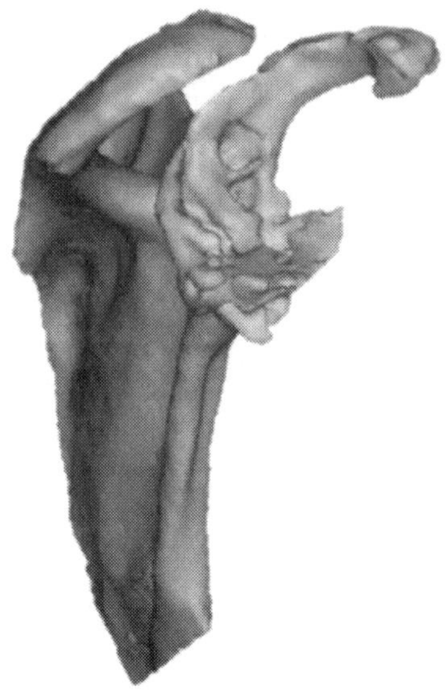

Figure 1: Scapula with severe bone loss (left)

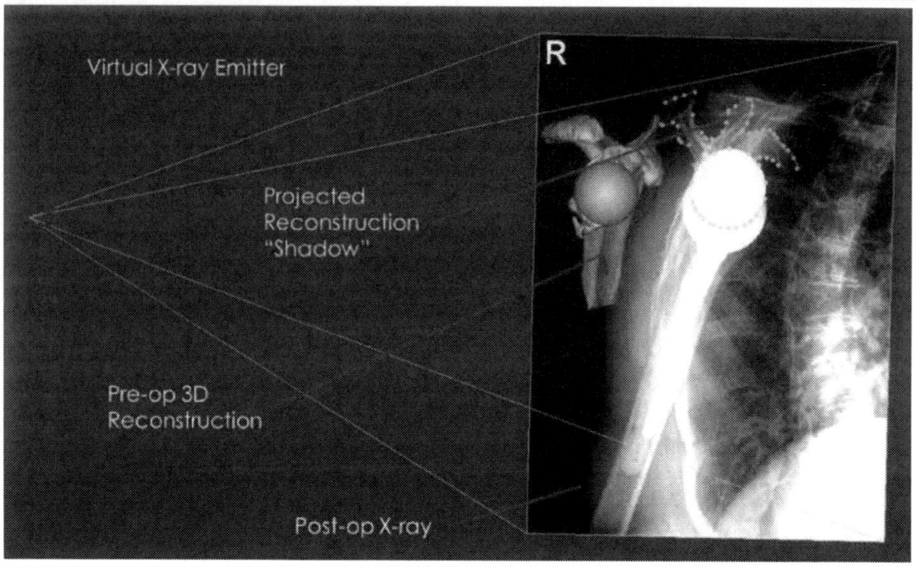

Figure 2: The 3D reconstruction projection of both the implant and scapula in aligned with a contour manually created on the x-ray. This process is repeated for multiple x-rays, until implant alignment is satisfactory (<10% error) on all x-rays. (right)

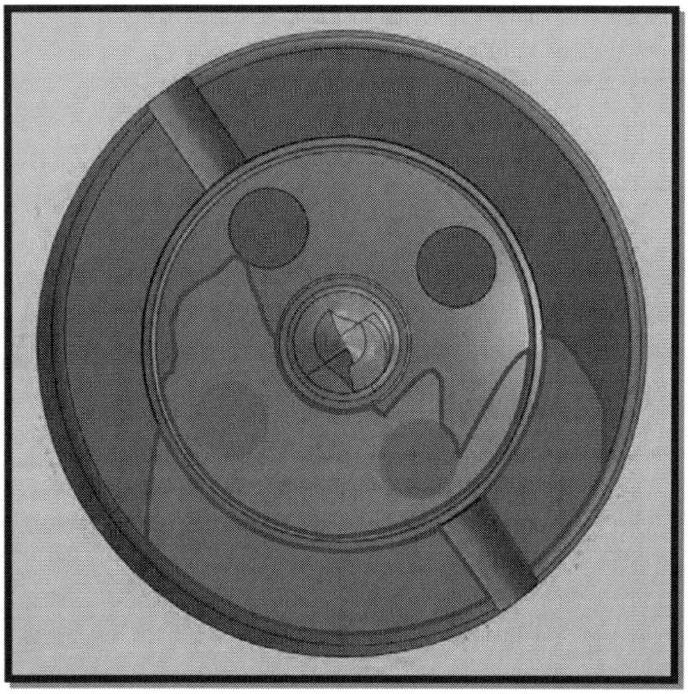

Figure 3: The area of host bone coverage on implant.

RESULTS

The area of the host bone coverage on implant was 346 mm^2 (**Fig.3**). This represents 41% of the total implant contact area. The overall precision of 2D-3D registration reported by the MIMICS program was 90%.

DISCUSSION The primary objective of this experiment was to test whether the 2D-3D registration of pre-operative CT and post-operative radiographs of shoulder arthroplasty patient is a feasible option in evaluating the glenoid component-host bone interface. A glenoid component loosening and dislocation represents most common complications seen after shoulder arthroplasty [5]. To mitigate these problems it is essential to acquire optimal host bone coverage intra-operatively. Bone grafts are a vital option in cases of severe posterior bone loss or complex revision surgeries when adequate bone stock is missing and version correction through anterior reaming would compromise the glenoid vault.

In the present abstract, the results of the pilot experiment with 2D-3D registration of the pre-operative CT scans and post-operative radiographs of the primary RSA patient are presented. We were able to reach 90% overall precision in the registration process and to estimate an area of host bone coverage on implant equal to 346 mm^2. The crucial obstacle posed by the existing overlap of clavicle and rib cage in studied radiographs was successfully surmounted. Those additional anatomical structures made it difficult to precisely outline the contours of scapula. However, radiographs taken from other angles helped significantly to increase overall accuracy of the presented model.

For the future work we plan to use this technique on the larger cohort of shoulder arthroplasty patients. In a long term, the utilization of this technique in retrospective large cohort studies will help us to better understand the role the bone graft play in successful outcome of the shoulder arthroplasty.

ACKNOWLEDGEMENTS
We would like to thank Materialise for sharing the Research Edition of Mimics as well as DJO for sharing the models of their implants.

REFERENCES

[1] Warner JP, et al. *J Shoulder and Elbow Surgery,* 2014
[2] Sears B, at al. J AAOS, 20:09-604,2012
[3] Vigneron L et al. *White Paper:Materalize*. 2014
[4] Virani N, et al. *J Shoulder Elbow Surg,* 22:179-187, 2013.
[5] Norris TR, et al. *J Shoulder Elbow Surg,* 11:130-135,2002

DO EXTERNAL LOAD MEASURES PREDICT KNEE CONTACT FORCE CHANGES DUE TO WEIGHT LOSS?

Nathan R. Sauder (1), James C. Coburn (2), Melinda K. Harman (3), Heather K. Vincent (4), Daryl D. D'Lima (5), Benjamin J. Fregly (1)

(1) Department of Mechanical & Aerospace Engineering
University of Florida
Gainesville, FL, USA

(2) Center for Devices and Radiological Health
Food and Drug Administration
Silver Spring, MD, USA

(3) Department of Bioengineering
Clemson University
Clemson, SC, USA

(4) Department of Orthopedics and Rehabilitation, Division of Research
University of Florida
Gainesville, FL, USA

(5) Shiley Center for Orthopaedic Research & Education, Scripps Clinic
La Jolla, CA, USA

INTRODUCTION

Knee osteoarthritis (OA) researchers often use external load measures (e.g., the knee adduction moment) as indicators of internal knee contact forces. External loads have been used for this purpose when studying knee OA progression [1], differing leg alignments [2], modified gait patterns [3][4], and weight loss [5]. Using instrumented knee implant data, recent studies have shown that changes in external loads can be poor indicators of changes in internal knee contact forces produced by gait modification [6][7]. No studies to date have used instrumented knee data to evaluate whether changes in external loads are good indicators of changes in internal knee contact forces produced by weight loss.

This study uses data collected from a female subject with an instrumented knee replacement to investigate the relationship between changes in external load measures and changes in internal knee contact forces resulting from weight loss. Changes in knee inverse dynamic loads, vertical ground reaction force, and the knee flexion angle are compared to measured changes in knee contact forces to determine the predictive ability of the external measures.

METHODS

Two data sets were collected from one female subject with an instrumented knee implant and neutral leg alignment. Institutional review board approval and subject informed consent were obtained. The first data set was collected when the subject was age 65 years with a weight of 873 N and the second when she was age 69 years with weight 775 N (11% weight loss). Data from the second set were released as part of the Third Knee Grand Challenge competition [8]. Weight loss was achieved by dietary modification. Both data sets include 5 trials of overground walking at self-selected speed. Data from each trial was comprised of surface marker trajectories, ground reaction loads, and six loads from the instrumented prosthesis [9].

A subject-specific OpenSim [10] model was created using geometry generated from CT scan data of the subject's pelvis and left leg and CAD models of the instrumented implant. Markers from each data set were attached to the model, and inverse kinematics was performed for each trial. One trial of fluoroscopy data was scaled to the length of each individual trial and used to approximate knee abduction-adduction and internal-external rotations as well as superior-inferior, medial-lateral, and anterior-posterior translations. Resulting motions from inverse kinematics were filtered at 6 Hz and used in inverse dynamic analyses.

Data from the instrumented implant were filtered with a fifth degree Woltring filter with a cutoff frequency of 6 Hz. Medial and lateral forces were estimated from the instrumented implant data using regression equations derived using an elastic foundation contact model of the implant [8].

Four quantities were analyzed for each of nine external and internal knee loads and the knee flexion angle. The four quantities analyzed were first peak value during stance, second peak value during stance, trough value (minimum value between the two peaks), and average value during stance. The nine variables were the knee abduction-adduction moment (KAM), knee superior-inferior force (KSF), knee flexion-extension moment (KFM), vertical ground reaction force (VGRF), knee flexion-extension angle (KFA), total knee contact force (TCF), medial knee contact force (MCF), lateral knee contact force (LCF), and medial/total knee contact force ratio (M/T). Since knee flexion angle does not have two peaks and a trough, peak 1 was defined as the maximum value up to 30% of the gait cycle, the trough was defined as the minimum value between 30 and 50%, and peak 2 was not calculated. Unitless absolute and relative ratios between changes in each quantity/variable and changes in body weight were calculated (Eq. 1 and Eq. 2).

$$\Delta_{ABS} = \frac{q - q_0}{w - w_0} \quad (1)$$

$$\Delta_{REL} = \frac{(q - q_0)/q_0}{(w - w_0)/w_0} \quad (2)$$

In these equations q_0 and q represent the pre and post weight loss values of a particular quantity/variable combination and w_0 and w are the pre and post weight loss body weight.

No statistical analyses were performed due to the lack of coupling between trials at different weights, lack data from intermediate time points, and lack of multiple subjects. For this reason, we selected ±20% as an acceptable difference between an external change ratio and a corresponding internal change ratio. Only force quantities (i.e. not moments, angles, or force ratio) were compared for absolute changes due to differences in units.

RESULTS

No absolute differences in external forces were within 20% of absolute differences in internal forces (Table 1). Six external quantities had relative differences that were within 20% of relative differences in internal quantities: stance average of VGRF with TCF (18% error), first peak of KSF with MCF (16% error), trough of KAM with LCF (13% error), and stance averages of MCF with KSF (0.1% error), VGRF (10% error), and KFM (19% error) (Table 2).

TABLE 1: ABSOLUTE CHANGE RATIOS.

	Peak 1	Peak 2	Trough	Stance Avg
KAM	--	--	--	--
KSF	0.72	0.21	1.21	0.65
VGRF	0.73	0.34	1.33	0.62
KFM	--	--	--	--
KFA	--	--	--	--
TCF	1.12	-2.36	4.22	1.16
MCF	1.75	-0.30	4.81	1.21
LCF	1.07	-0.38	0.40	-0.05
M/T	--	--	--	--

TABLE 2: RELATIVE CHANGE RATIOS.

	Peak 1	Peak 2	Trough	Stance Avg
KAM	3.64	3.36	4.59	4.48
KSF	0.82	0.24	1.48	0.99
VGRF	0.77	0.33	1.48	0.89
KFM	-10.20	1.72	56.70	1.18
KFA	-0.12	--	-0.38	0.59
TCF	0.54	-1.23	2.60	0.75
MCF	0.98	-0.19	3.12	0.99
LCF	1.60	-0.76	5.30	-0.14
M/T	0.46	0.94	0.73	0.25

DISCUSSION

This study explored the relationship between changes in measurable external quantities and changes in internal knee contact forces produced by significant weight loss. In general, changes in external quantities were poor indicators of changes in corresponding internal quantities. For example, over stance phase, a 1 N reduction in KSF was not matched by a 1 N reduction in TCF. Furthermore, in some cases (e.g., several second peak values), changes in external quantities did not even match the correct sign of changes in internal quantities. However, relative changes in a few external quantities were similar in magnitude to the relative changes in corresponding internal quantities. In particular, over stance phase, the relative change in average medial contact force was well predicted by the relative change in average knee superior-inferior force.

Six pairs of quantities had relative changes in an external quantity within 20% of the relative change in a related internal quantity. Of these six pairs, one pertained to total contact force, one to lateral contact force, and four to medial contact force. Of particular interest are the KAM-LCF trough relationship and KSF-MCF relationships. KAM is commonly used as an indicator of MCF or of the distribution of contact force between medial and lateral compartments [2]. This relationship is not generally seen in the data presented here. However, the similarity between relative changes in trough values of KAM and LCF could be due to such a relationship. For this subject, changes in MCF were matched more closely by changes in KSF than by changes in KAM, possibly due to the subject's neutral leg alignment. Relative changes in both the first peak and the average value of MCF over stance phase were similar in magnitude to the relative changes in the same quantities of KSF. The stance phase average in particular was very well predicted (~0.1% error). This result could indicate that, for certain subjects, changes in first peak and average value of MCF over stance phase may be well predicted by changes in KSF.

There are several limitations to this study. Due to the rarity of subjects with instrumented knee implants who have lost significant weight, data were available from only one subject. Subjects with different leg alignments could produce different results. In addition, since data were available at only two time points, the analysis was restricted to exploring one-to-one relationships between changes in external and internal quantities. For example, the relative change in knee adduction moment peak 1 may always be roughly 3.7 times the relative change in medial contact force peak 1, but such a relationship could not be confirmed using the available data. Though no correction was made for differences in self-selected walking speed, the difference between pre and post weight loss speeds was only 0.1 m/s. Finally, the acceptable difference of ±20% was somewhat arbitrary, though 20% might be an acceptable error for clinical application.

The results of this study reveal that one should not, in general, assume that changes in external inverse dynamic loads caused by weight loss are indicative of changes in internal knee contact forces. However, for some individuals, relative changes in specific internal quantities may be predicted by relative changes in related external quantities. Most notably, changes in the first peak and stance average of MCF may be predictable using KSF.

ACKNOWLEDGEMENTS

This research was funded by NIH grant R01EB009351 and by the University of Florida.

REFERENCES

[1] Miyazaka, T et al., *Ann Rheum Dis*, 61:617-622, 2002.
[2] Andriacchi, TP, *Orthop Clin North Am*, 25:395-403, 1994.
[3] Dowling, AV et al., *J Biomech Eng*, 132:071007, 2010.
[4] Barrios, JA et al., *J Biomech*, 43:2208-2213, 2010.
[5] Messier, SP et al., *Arthritis Rheum*, 52:2026-2032, 2005.
[6] Meyer, AJ et al., *J Orthop Res*, 31:921-929, 2013.
[7] Walter, JP et al., *J Orthop Res*, 28:1348-1354, 2010.
[8] Fregly, BJ et al., *J Orthop Res*, 30:503-513, 2012.
[9] Kirking, B et al., *J Biomech*, 39:1744-1751, 2006.
[10] Delp, SL et al., *IEEE T Bio-Mech Eng*, 54:1940-1950, 2007.

SB³C2015
Summer Biomechanics, Bioengineering and Biotransport Conference
June 17-20, 2015,Snowbird Resort, Utah, USA
SB³C2015-1159

EFFECTIVE REMODELING IN CEREBRAL ANEURYSMS: A CASE STUDY

Xinjie Duan (1), Bong Jae Chung (2), Juan R. Cebral (2), Khaled M. Aziz (3), Anne M. Robertson (1)

(1) Department of Mechanical Engineering & Materials Science, University of Pittsburgh, Pittsburgh, PA, USA

(2) Department of Bioengineering, George Mason University, Fairfax, VA, USA

(3) Department of Neurosurgery, Allegheny General Hospital, Pittsburgh, PA, USA

INTRODUCTION

Rupture of intracranial aneurysms is the leading cause of subarachnoid hemorrhage, a devastating type of stroke with high mortality and disability rates. In order to avoid rupture, an aneurysm must maintain its structural integrity even under extreme loading, for example during weight lifting. This structural integrity can be estimated, for example, through a factor of safety (FOS), defined as the structural capacity divided by the design load. It follows from a simple force balance that, as a roughly spherical aneurysm increases in size, the intramural stress will increase, even under fixed transmural pressure. Hence, the wall will need to adapt to maintain the same FOS, for example by increasing the wall thickness.

In a previous work [1], we analyzed the mechanical properties for 8 unruptured human aneurysms. Laplace's Law for a sphere was used to estimate the intramural stress and assess the structural integrity of these aneurysms. Here, we build on this work and provide a more accurate estimate of the structural integrity using patient specific biomechanical models. We chose two human cerebral aneurysms that had similar mechanical properties (including failure), but substantially different wall thicknesses and geometries. We evaluated their FOS based on stresses estimated from finite element studies of the mechanics, using patient specific geometries and material properties.

METHODS

As detailed in [1], aneurysm domes were harvested following surgical clipping and mechanically tested within 48 hours. Meridonal strips were obtained from the upper dome and tested in a custom designed uniaxial loading system compatible with multiphoton microscopy (MPM), enabling simultaneous mechanical testing and structural imaging [2]. The samples were tested to failure to obtain the mechanical response and ultimate strength. Collagen architecture was imaged using the SHG signals from MPM.

Two human cerebral aneurysms (CA25 and CA26) with similar mechanical properties and substantially different wall thicknesses and shapes were chosen for study. CA26 from the distal ACA was chosen for its unusual wall thickness of 0.45 mm, nearly 3 standard deviations above the mean of 0.22mm for the sample population (n=15), [1]. CA25, an AcomA aneurysm, was chosen for comparison. Its wall thickness of 0.24 mm was within a standard deviation of the mean. CA25 was nearly spherical in shape and CA26 was irregularly shaped.

CA25 and CA26 were similar in terms of mechanical properties. In particular, their high stress stiffnesses were 8.25 MPa and 7.37 MPa, respectively (population average: 12.4 ± 6.6 MPa) [1]. Their ultimate stresses (Cauchy) were 1.63 MPa and 1.90 MPa, respectively (population average: 1.39 ± 0.50 MPa).

For biomechanical modelling, the aneurysm wall was idealized as an isotropic, incompressible hyperplastic material with Cauchy stress,

$$\underline{T} = -p\underline{I} + C_1 e^{C_2(I_1-3)} \underline{B} \tag{1}$$

where I_1 is the first invariant of the Left Cauchy Green tensor $\underline{B}$. The patient specific material parameters C_1 and C_2 were obtained from uniaxial studies [1], Table 1. The lumen surfaces were obtained from pre-surgical 3D rotational angiography images. The wall thickness was prescribed as a constant, obtained from patient specific measurements [1]. Shell elements were used for the wall, motivated by the small ratio of thickness to equivalent diameter in thie aneurysm, Table 1. Transverse shear stiffness (K_{11}) was estimated using a linear method introduced by Ma et al. [3]. Zero displacement boundary conditions were applied at the aneurysm neck. ABAQUS/Standard 6.12 was used to perform the finite element analysis at equilibrium.

RESULTS

Results of the mechanical testing and MPM imaging for samples CA25 and CA26 are given in Figure 1. The collagen architecture for

CA25 and CA26 for both luminal and abluminal projections displayed similar features including densely packed, thin collagen fibers at the luminal side (Fig.1a & d) and disorganized, thick collagen fibers on the abluminal side (Figs.1 b & e). Differences in thickness between CA25 and CA26 are visible in Figs.1 c & f. Figure 1g shows the mechanical behaviour of the two aneurysms in uniaxial tension, displaying similarities in stiffness and failure properties.

TABLE 1. Aneurysm wall morphology and material properties used in modeling

Case	*Wall Thickness t_0 (mm)*	*Equiv. Diam. (mm)*	*K_{11} (N/mm)*	*C_1 (MPa)*	*C_2*	*R^2*
CA25	0.24	4.5	0.067	0.114	4.38	0.99
CA26	0.45	6.8	0.176	0.303	1.71	0.96

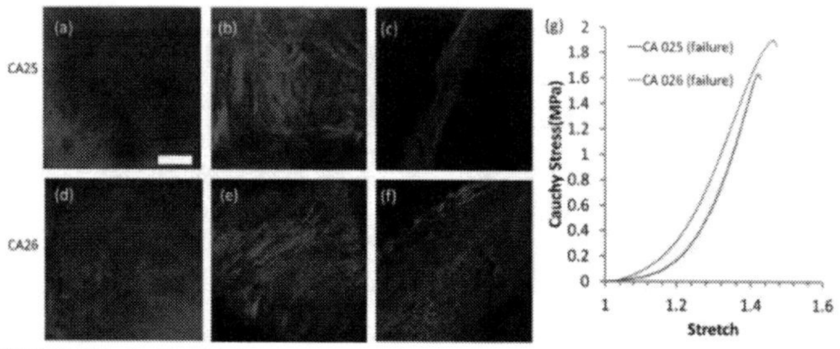

FIGURE 1. MPM images and uniaxial failure tests. Collagen fibers are visible in projected stacks of MPM images on (a) luminal side (b) abluminal side and (c) cross-section view of CA25; and (d) luminal side (e) abluminal side and (f) cross-section view of CA26. Scale bar= 100um. (g) Stress-strain curves from uniaxial failure tests.

The von Mises stress distribution in CA25 is relatively homogeneous, ranging from 38 KPa to 94 KPa while that in CA26 ranges from 12 KPa to 160 KPa, Fig. 2. Estimates using Laplace's Law predicted intramural stresses in CA25 and CA26 of 72 KPa and 58 KPa, respectively, consistent with the range found here. The larger heterogeneity in CA26 arises from its more complex shape. An FOS was calculated using measured ultimate stress and maximum von Mises stress for loading to 200 mmHg (obtained from patient specific computational studies, not shown). The FOS for both CA25 and CA26 are 8.1 and 6.8, respectively for even this extreme loading condition, suggesting effective remodelling in both cases.

DISCUSSION

This case study provides an example of the capacity of at least some aneurysms to reduce intramural stresses through effective growth and remodelling. CA26 had a larger effective diameter and more complex shape than CA25. In order to maintain a FOS comparable to that in CA25, the walls of the CA26 would either need to have stronger materials (higher ultimate stress) or thicker walls. In CA26, this was achieved by a substantial wall thickening.

Using immunofluorescence staining, it was found that the increased wall thickness in CA26 was achieved by an increased numbers of layers of type III collagen. Further these layers were endowed with a dense cell population, Fig. 3a, suggesting CA26 was capable of continued maintenance and remodelling.

The computational analysis of the intramural stresses showed that, as a result of this increased thickness, the von Mises stresses in CA26 were comparable with those in CA25, except in an isolated region of high stress in CA26. Even with this focal elevated stress, the FOS for CA26 for 200 mmHg was well above six, indicating effectively remodelling in the wall that was robust to physiologically relevant transient increases in blood pressure.

It is still not known why some aneurysm walls are capable of effective remodelling. Hemodynamic factors are believed to be important. The wall shear stress in the region of the resected tissue (upper dome) was found to be similar in magnitude to that seen in the parent vessel Fig. 2b, though this is a complex issue and the subject of an ongoing investigation, [4].

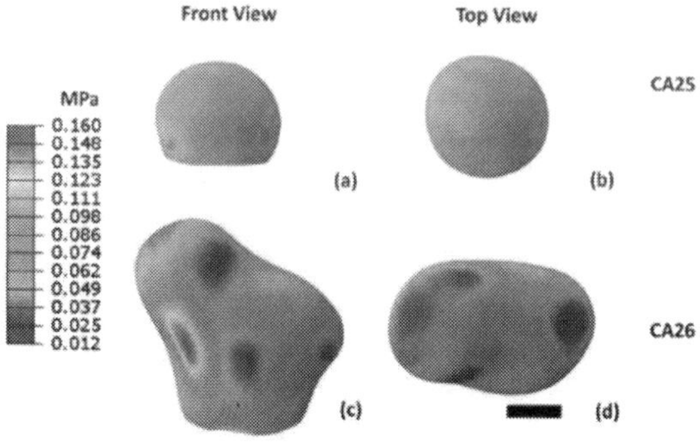

FIGURE 2. Contours of Von Mises stress in patient specific aneurysm models at 100 mmHg. Scale bar = 2 mm.

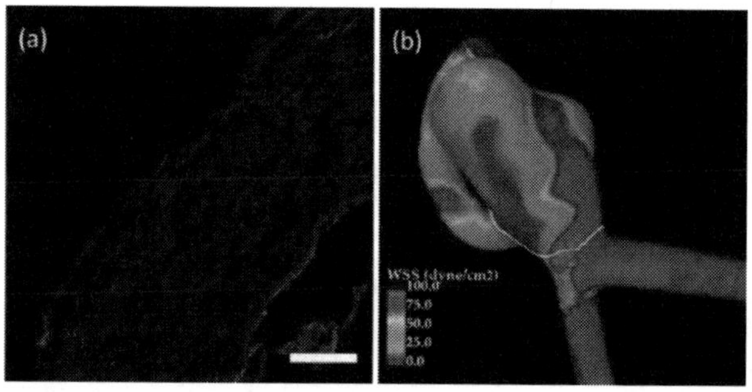

FIGURE 3. (a) Cross section of wall of CA26 showing collagen III (Red) and cell nuclei (Blue), visible using immunofluorescence, DAPI staining. Scale bar= 100 um. (b) Contours of wall shear stress for CA26 at peak systole.

ACKNOWLEDGEMENTS

This work was supported by a grant from the National Institute of Neurological Disorders and Stroke of the National Institute of Health (1R21NS080031-01A1).

REFERENCES

[1] A.M. Robertson, X. Duan, M.R. Hill, K. Aziz, J.R. Cebral, Diversity in the strength and structure of unruptured cerebral aneurysms. Ann. Biomed. Eng. (2014), in press.

[2] Hill, M.R., X. Duan, G.A. Gibson, S. Watkins, and A.M. Robertson, A theoretical and non-destructive experimental approach for direct inclusion of measured collagen orientation and recruitment into mechanical models of the artery wall. J Biomech, 2012. 45(5): p. 762-71.

[3] Ma, B., Lu, J., Harbaugh, R. E., & Raghavan, M. L. (2007). Nonlinear anisotropic stress analysis of anatomically realistic cerebral aneurysms. Journal of biomechanical engineering, 129(1), 88-96.

[4] J.R. Cebral, X. Duan, B.J. Chung, C.M. Putman, K. Aziz, A.M. Robertson, Am. J. Neuroradiol. (2015) – in revision.

CARTILAGE WEAR INITIATED BY FATIGUE DAMAGE UNDER PHYSIOLOGIC LOADING WHEN FLUID LOAD SUPPORT AND BOUNDARY LUBRICATION ARE COMPROMISED

Krista M. Durney (1), Sevan R. Oungoulian (2), Brian K. Jones (2), Jason T. Suh (1), Clark T. Hung (1), Gerard A. Ateshian (1,2)

(1) Department of Biomedical Engineering
Columbia University
New York, NY, USA

(2) Department of Mechanical Engineering
Columbia University
New York, NY, USA

INTRODUCTION

Osteoarthritis (OA) is a degenerative joint disease with a multifactorial etiology that results in damage and loss of articular cartilage. Primary OA is often described as a natural process of wear and tear and is associated with aging; secondary OA may be initiated by injury, obesity, inactivity, other joint diseases, or iatrogenic factors, e.g., meniscectomy or hemi-arthroplasty. Other than blunt injury, causative factors for cartilage damage and wear in OA are poorly understood. In this study we hypothesize that cartilage wear may be initiated by fatigue damage under physiological load magnitudes and sliding motion when either of two conditions are met: (a) Interstitial fluid pressurization decreases below a critical threshold, or (b) boundary lubrication is compromised.

Prior studies have demonstrated that interstitial fluid pressurization is critical for maintaining low friction in articular cartilage, as the fluid load support shifts contact stresses away from the extracellular matrix [1,2]. With total loss of fluid pressurization, the friction coefficient may increase 100-fold. Similarly, boundary lubricants present in synovial fluid (SF) may significantly reduce friction, by a factor of 2 to 6 [3, 4]. Despite these findings on friction, direct evidence that interstitial fluid pressurization or boundary lubricants in SF may protect against wear is limited.

Therefore, in this study, sliding contact experiments are performed to examine the influence of loss of fluid pressurization or loss of boundary lubrication on tissue wear. Loss of fluid pressurization is achieved by testing an impermeable surface sliding against a cartilage strip for up to 8 h. Loss of boundary lubrication is achieved by replacing SF with phosphate buffered saline (PBS). Control groups included cartilage sliding against cartilage in SF or PBS.

METHODS

Four different experimental groups (n=6 per group) were examined in order to test the hypotheses (Fig. 1). Cartilage strips were tested against a glass lens in a bath of phosphate buffered saline [GS-PBS], a glass lens in synovial fluid [GS-SF], a cartilage plug in PBS [CS-PBS], or a cartilage plug in SF [CS-SF]. To test the hypothesis on the influence of boundary lubricants on wear, GS-PBS and GS-SF groups were compared. To test the hypothesis on the influence of fluid pressurization on wear, CS-PBS and GS-PBS groups were compared. To examine whether fluid pressurization alone is not sufficient to prevent wear, CS-PBS and CS-SF groups were compared.

<u>Specimen Preparation:</u> Cartilage tissue was harvested from apposing surfaces of sixteen knee joints from immature bovine calves (2-3 months old). Strips (10 × 30 mm) were excised from the tibial plateau and ∅25.4 mm plugs were punched from the femoral condyle. Specimens were placed articular surface down on a freezing stage sledge microtome and trimmed to remove underlying bone to obtain samples with an intact superficial zone; strips were trimmed to a final thickness of 1.5 mm ± 0.2 mm. Samples were stored in filtered PBS supplemented with biocide and a cocktail of protease inhibitors and frozen at −20 °C. On the day of testing, samples were aligned and affixed with an adhesive to their respective mounts (Fig. 1).

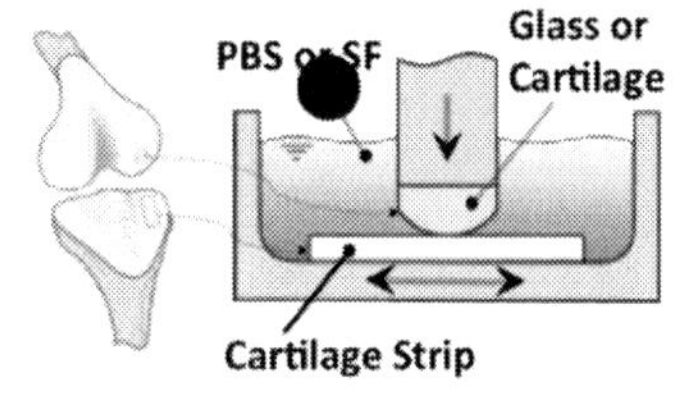

Figure 1: Cartilage strips are tested against curved counterfaces consisting of either a glass lens or a cartilage punch in a bath of either PBS or synovial fluid.

<u>Sliding Friction:</u> A custom two-axis loading

device with a six-degrees-of-freedom load cell was used for continuous measurements of specimen motion, deformation, and reaction loads as described previously [1,4]. Reciprocal sliding motion (±5.0 mm at 1 mm/s) under a 4.45 N load was prescribed for 8 h. The glass lens had a radius of curvature of 12.7 mm, selected to best match that of the condylar cartilage punch. The average contact diameter between glass and cartilage strip was 2.1 mm and between condyle and strip was 3.2 mm (equivalent to 2.1 MPa and 1.4 MPa contact stress, respectively). The friction coefficient μ was calculated from the ratio of average tangential force F_t and normal force F_n over each reciprocating cycle.

Damage Assessment: Strips and plugs were scanned post-test to quantify surface damage using a 3D laser scanner (Next Engine). Root mean square (RMS) deviation of the scan from a plane fit to the strip surface was calculated to quantify damage to the tissue in the form of surface roughness. Strips were subsequently sliced perpendicular to the axis of translation at the center of the path of travel and outside of the path of travel; these unfixed 80 μm slices were imaged using polarized light microscopy (PLM) to visualize damage through the depth of the cartilage strip.

Statistics: A two-way ANOVA was performed with Tukey post-hoc test to determine statistical significance.

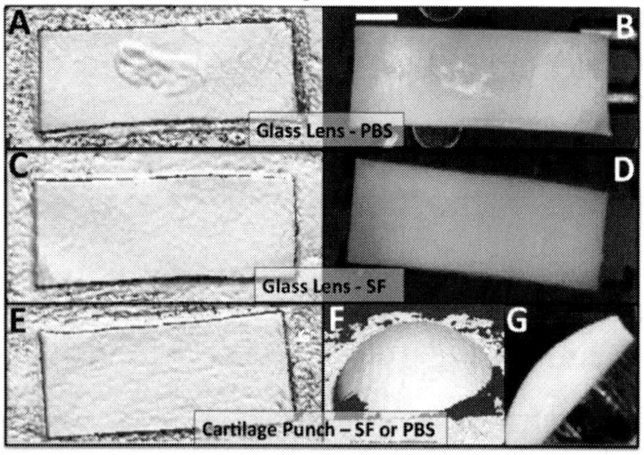

Figure 2: Post-test 3D scans and photographs of representative cartilage strips (A-E) and condylar plug (F,G). The strip from GS-PBS (A,B) exhibited significant damage, while remaining groups showed no evidence of damage. Scale bar: 5 mm.

RESULTS

Based on 3D scans (Fig. 2) and RMS roughness data (Table 1), significant surface damage was observed in the GS-PBS group, while remaining groups exhibited no evident damage (Fig. 2). For CS-PBS, CS-SF, and GS-SF the friction coefficient μ remained low over 8 h, whereas for GS-PBS μ increased with the onset of damage, after at least 1 h of continuous sliding (Fig. 3). The peak value of μ achieved in the last 7 hours is reported (Table 1). The friction coefficient in SF was statistically lower than in PBS irrespective of counterface, and μ for cartilage-on-cartilage was lower than for glass-on-cartilage irrespective of lubricant. PLM consistently demonstrated that damage in the GS-PBS group occurred as a result of delamination of the superficial zone (SZ, where collagen fibrils are tangential to the articular surface, appearing bright white in PLM (Fig. 4) at the interface with the middle zone (MZ, where fibrils are oriented randomly, appearing as a dark band between the bright SZ and deep zone DZ, where fibrils are oriented perpendicular to the articular surface). This delamination was only apparent in the contact region of the strip, whereas the unloaded region remained intact (Fig. 4).

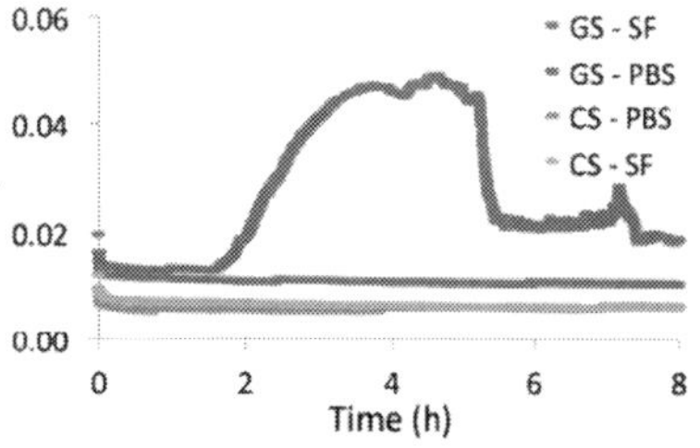

Figure 3: Representative temporal evolution of the friction coefficient μ from each group.

Counterface †	Lubricant †	RMS, damage [mm]	μ, peak
Glass lens	PBS	0.25 ± 0.22 *	0.043 ± 0.024
	SF	0.06 ± 0.03	0.012 ± 0.001
Cartilage	PBS	0.04 ± 0.01	0.023 ± 0.015
	SF	0.06 ± 0.01	0.007 ± 0.002

Table 1: Averarge peak friction coefficient and measurement of damage for all groups. †: p<0.05 for μ peak, significant effect of factor indicated *: p<0.05 for RMS significant effect against all groups.

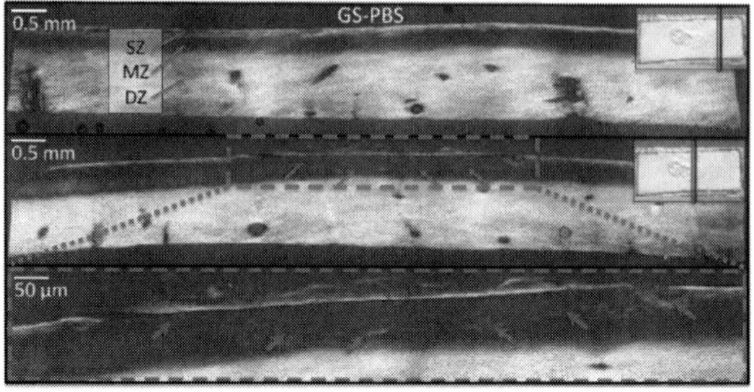

Figure 4: Polarized light microscopy of a representative cartilage strip from GS-PBS sectioned in two locations as indicated in the inset 3D scans: away from the sliding contact region (top) and at the center of the contact region (middle). Higher magnification (bottom) shows damage consisted of delamination of superficial zone (SZ) from middle zone (MZ), as indicated by arrows.

DISCUSSION

This study provided strong evidence that cartilage damage (or "wear and tear") may be achieved at physiological levels of contact stress (typically 0-6 MPa) when there is loss of interstitial fluid pressurization and loss of boundary lubrication. Damage occurred as a result of delamination of the superficial zone at its interface with the middle zone, a location known to exhibit the lowest shear modulus across the articular layer thickness [5]. Delamination is indicative of sub-surface fatigue failure, resulting from the reciprocal shearing of this SZ-MZ interface due to the tangential frictional shear stress τ acting at the articular surface. Based on our previously validated friction model for articular cartilage [1,3], τ is dependent on the magnitude of interstitial fluid load support and the presence of boundary lubricants. The fact that only GS-PBS demonstrated damage in this study suggests that the threshold value for τ to achieve damage was only met when both mechanisms were compromised. These results provide new insights into the onset of cartilage damage in primary and secondary OA, suggesting that a variety of congenital or acquired conditions that produce loss of fluid load support and boundary lubrication may be causative factors for OA.

ACKNOWLEDGEMENTS

NIH AR043628. We thank Dr. X.E. Guo (Columbia) for PLM support.

REFERENCES

[1] Krishnan, R et al., *J Biomech*, 38:1665-1673, 2005. [2] Ateshian, G et al., *J Biomech*, 42:1163-1176, 2009. [3] Caligaris, M et al., *OA & C*, 16:1220-1227, 2008. [4] Schmidt et al., *Arthritis Rheum*, 56:882-91, 2007. [5] Buckley et al., *J Biomech Eng*, 135:31002, 2013.

A BIOENGINEERING SOLUTION TO CURE SPINAL CORD INJURY

Brittany King (1), Jacklyn Witko (2), Alexander Herman (3), Babitha Tom (4), Andrea J Vernengo (5), Anita Singh (6)

(1) Mechanical Engineering
Rowan University
Glassboro, NJ, USA

(2) Mechanical Engineering
Rowan University
Glassboro, NJ, USA

(3) Mechanical Engineering
Rowan University
Glassboro, NJ, USA

(4) Biomedical Engineering
Widener University
Chester, PA, USA

(5) Chemical Engineering
Rowan University
Glassboro, NJ, USA

(6) Mechanical Engineering, Rowan University
Glassboro, NJ, USA
Biomedical Engineering, Widener University,
Chester, PA, USA

INTRODUCTION

Spinal cord injury (SCI) is a devastating and debilitating condition that affects an estimated 227,080 to 300,938 persons in the United States with approximately 12,000 new cases occurring each year (1-3). Current treatment strategies include activity based rehabilitation therapies, such as body weight supported treadmill training (BWSTT) that utilizes the uninjured descending pathways in incomplete SCI patients and spinal circuits below the level of injury in complete SCI patients. If the descending inputs are minimal or absent, recovery in walking over-ground is never observed in SCI patients. Since the amount of spared descending input strongly regulates the extent of recovery after BWSTT, combining transplantation treatment strategies that increase the number of descending inputs by inducing neuroprotective and regenerative environment around the injured spinal cord holds most promise to further enhance the functional recovery after BWSTT. Transplantation strategies using cells genetically engineered to deliver neurotrophins, which is a neuroprotective and neuroregenerative agent, limit spinal tissue loss, promote regeneration/sprouting of injured axons, bridge the site of injury and result in some functional recovery in animal SCI models (4-6). Clinical translation of these techniques poses several problems, including rejection and issues with regulation of release, requiring more invasive approaches with additional problems associated with potential rejection and regulation of release of bioactive compounds. To overcome these limitations, we propose using bioengineered scaffold poly N-(isopropylacrylamide) – poly (ethylene glycol) (PNIPAAm-PEG), which has shown to be highly biocompatible and when functionalized to secrete a neurotropic factor can promote regeneration and functional recovery in a surgically induced SCI animal model. Furthermore by incorporating BWSTT in these animals, our collaborative group will for the first time report the combinatorial effect of scaffold+neurotrophins+BWSTT in a clinically relevant contusion SCI model.

METHODS

19 female Sprague-Dawley rats were divided in five groups (Figure 1). All animals, except sham received a moderate contusion injury using an impactor device, in which a 10-g, 2.0-mm-diameter rod was dropped from a height of 25mm onto the exposed T9/10 level spinal cord and immediately removed. One week post-injury, animals in neurotrophin groups (Groups 4 and 5) received neurotrophin transplants using a positive displacement pipette (approximately 5 ml of PNIPAAm-g-PEG loaded with BDNF ($0.5 \times 10^6/5$ μl) and NT-3 ($0.5 \times 10^6/5$ μl). One week post-injury or transplant, animals in BWSTT groups (Groups 3 and 5) began BWSTT for 1000 steps/day, 5 days/wk for 8 wks at 75% body weight support (BWS) and 10 cm/sec speed using the custom-built Treadmill and Body Weight Support setup (Figure 2).

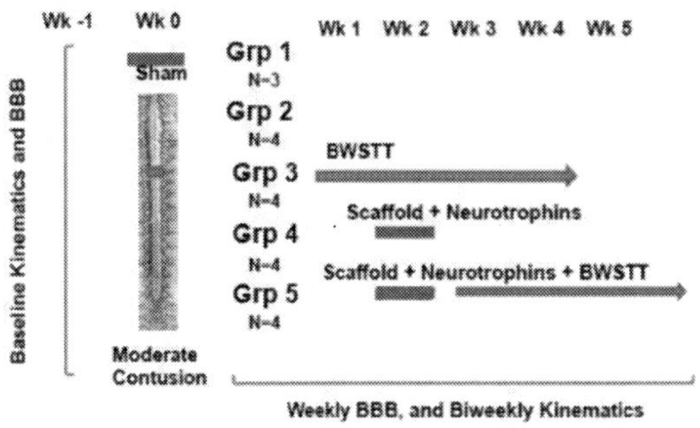

Figure 1: Animal Group and Study Timeline

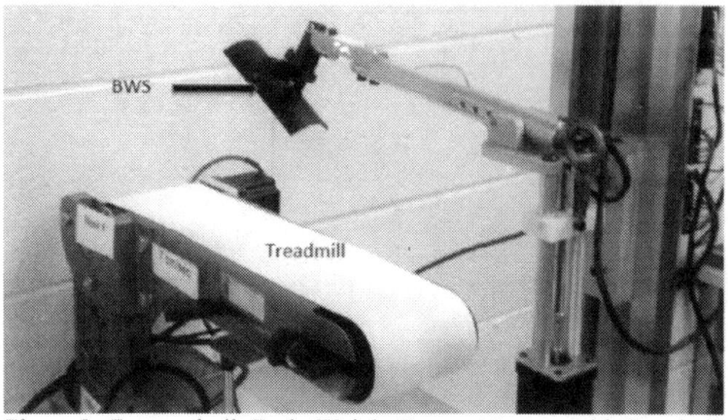

Figure 2: Custom-built Body Weight Support System with Treadmill.

Behavior Tests: *a) Treadmill Testing:* Biweekly, hindlimbs of the animals were shaved and reflective markers were placed on the hip (greater trochanter and iliac crest), knee (femoral condyle) and ankle (lateral malleolus) and the fifth metatarsal of the hindlimbs. Three high-speed cameras were used to record marker movements as the animals walked on the treadmill. Stepping were first tested at 75% weight support and 7 cm/sec speed. Post processing of data was performed using using Motive (Natural Point, OR) and Matlab (Mathworks Inc., Natwick, MA). Ongoing data analysis includes measurements of the kinematic characteristics, i.e., the length and height of the step, duration of stance and swing phases and overall step duration, will be calculated. Hip, knee and ankle joint excursions will also be analyzed. *b) Overground locomotor BBB Test:* Hindlimb motor function was assessed weekly using the open field [4] scoring system. Hindlimb function was scored from 0 (no observable movements) to 21 (normal locomotion). Individual rats were placed in an elliptical enclosure, allowed to accommodate to the enclosure, and then taped for 4 minutes by a trained observer. At a later timepoint, another observer who was blinded to the group identity of the animals scored the tests.

RESULTS

Kinematic Analysis: Preliminary data analysis of the ankle marker from animals in all groups is shown in Figure 3. Overall, an increase in hip, knee and ankle excursion was observed in all treatment group animals. Furthermore, an increase in ankle movement was observed in all three axes in treatment group as shown in Figure 3. Further data analysis is ongoing to establish statistical significance between these groups.

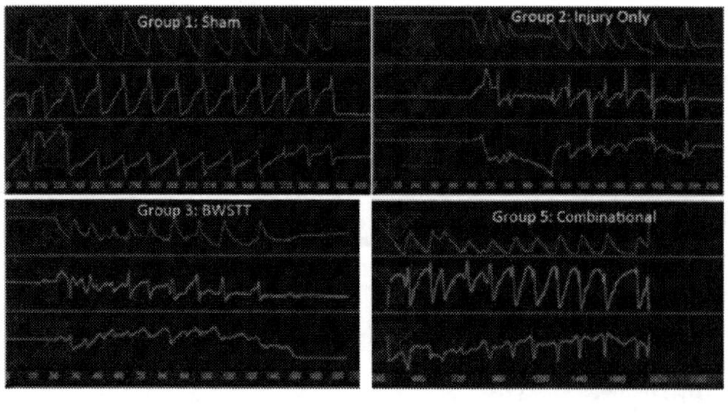

Figure 3: Kinematic Data Analysis of Ankle Marker in all Groups in the X, Y and Z axes.

BBB Test: The baseline and weekly BBB scores are shown in Figure 4. A significant improvement in BBB scores is observed in BWSTT group. Combinational group is being observed for improvements in the coming weeks.

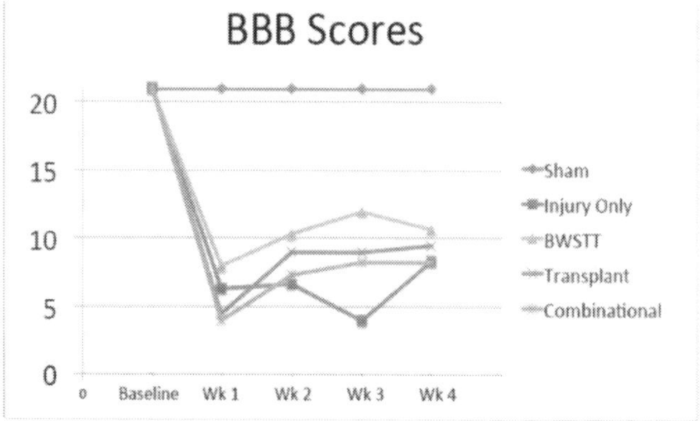

Figure 4: BBB Scores

DISCUSSION

BBB data and preliminary kinematic data analysis exhibit functional recovery in animals that received BWSTT, Transplants and combinational treatment. Successful findings from this study will help develop a unique tissue-engineering approach with promising clinical application in incomplete SCI patients. By adding other growth promoting agents to this bioengineered scaffold, future studies can explore the beneficial effects of the proposed combinatorial treatment strategy in complete SCI animal model.

ACKNOWLEDGEMENTS

Authors would like to acknowledge the help of Karl Dyer, Sue Grau and the Vivarium Technicians at Rowan SOM, Stratford, NJ. This study was supported by the New Jersey Commission on Spinal Cord Research – Exploratort Research Grant # CSCR14ERG001.

REFERENCES

1. Hutchinson, K.J., Gomez-Pinilla, F., Crowe, M.J., Ying, Z., and Basso, D.M., Three exercise paradigms differentially improve sensory recovery after spinal cord contusion in rats. Brain, 2004. **127**(Pt 6): p. 1403-14.
2. Dobkin, B., Apple, D., Barbeau, H., Basso, M., Behrman, A., Deforge, D., Ditunno, J., Dudley, G., Elashoff, R., Fugate, L., Harkema, S., Saulino, M., and Scott, M., Weight-supported treadmill vs over-ground training for walking after acute incomplete SCI. Neurology, 2006. **66**(4): p. 484-93.
3. Hubli, M. and Dietz, V., The physiological basis of neurorehabilitation--locomotor training after spinal cord injury. J Neuroeng Rehabil, 2013. **10**: p. 5.
4. Himes, B.T., Liu, Y., Solowska, J.M., Snyder, E.Y., Fischer, I., and Tessler, A., Transplants of cells genetically modified to express neurotrophin-3 rescue axotomized Clarke's nucleus neurons after spinal cord hemisection in adult rats. J Neurosci Res, 2001. **65**(6): p. 549-64.
5. Jin, Y., Tessler, A., Fischer, I., and Houle, J.D., Fibroblasts genetically modified to produce BDNF support regrowth of chronically injured serotonergic axons. Neurorehabil Neural Repair, 2000. **14**(4): p. 311-7.
6. Mendell, L.M., Munson, J.B., and Arvanian, V.L., Neurotrophins and synaptic plasticity in the mammalian spinal cord. J Physiol, 2001. **533**(Pt 1): p. 91-7.

SB³C2015
Summer Biomechanics, Bioengineering and Biotransport Conference
June 17-20, 2015,Snowbird Resort, Utah, USA
SB³C2015-1162

Microcirculation-Induced MRI Signal Anisotropy in Organized Tissues: Finite Element Modeling and Validation on Perfused Hearts

Osama M. Abdullah (1), Arnold David Gomez (1), Adam Schmidt (1), Edward W. Hsu (1),

(1) Bioengineering
University of Utah
Salt Lake City, UT, USA

INTRODUCTION

Myocardial blood flow (MBF) changes are often associated with, and occur before detectable structural or functional alterations in cardiac diseases. MRI-based intravoxel incoherent motion (IVIM) is a diffusion-like phenomenon that is sensitive to tissue microcirculation (1), which depends on the vascular volume fraction (VF) and the mean speed of capillary blood flow quantified as the pseudo diffusion coefficient (D*). However, the IVIM model assumes random capillary organization, which is contrary to the known anisotropic capillary organization in the heart as shown in (Fig. 1a-b). Thus, there is a need to account for the effect of the anisotropic capillary geometry on the MRI signal. In this study, we used the finite element (FE) approach to model capillaries of adjustable organization and speeds, and used the solutions to predict the induced diffusion-weighted MRI signal. Simulations were subsequently validated with MRI experiments in perfused hearts where MBF was quantified with orthogonal MRI techniques (arterial spin labeling, ASL) (2).

METHODS

I. Myocardial Capillary Modeling with Finite Element

To elucidate how the diffusion MRI signal is affected by capillary mass transport via the magnitude and direction of intra-capillary fluidic velocity, we designed a modeling strategy with sufficient parameter flexibility to simulate differentiable abstract conditions, as well as realistic representations of flow in a capillary network. To this end, the capillaries were represented as a conduit network model composed of aligned and randomly oriented tubes of variable diameter and length (Fig. 1c). By incorporating finite element analysis to calculate flow ($\dot{Q}_k$) at each capillary element k, the model is conservative in terms of mass preservation at the anastomotic nodes, and supports energy dissipation due to viscous friction as described by the Hagen–Poiseuille equation: $\dot{Q}_k = \Delta P_k / R_k$, where ΔP_k is the pressure difference of the 2 ends of a capillary element k of length L_k and radius r_k with a resistance $R_k = \frac{8\mu L_k}{\pi r_k^4}$ where μ is the dynamic viscosity of moving fluid. Different levels of capillary anisotropy was achieved by defining a hybrid mesh composed of a quasi-linear combination of randomly placed nodes in a 2D plane connected via Delaunay triangulation and aligned nodes along a straight line. Different flow distributions across the network were achieved by controlling the probability distribution governing inflow boundary conditions. Once individual capillary velocities $\boldsymbol{v}_k$ and orientations θ_k (wrt. to diffusion encoding direction) were obtained, the normalized diffusion MRI signal ($Sn(b)$) in a voxel was written as the superposition of extravascular and intravascular compartment signals:

$$Sn(b) = (1-\mathrm{f})\exp(-bD_{tissue}) + f\exp(-bD_{blood})\,F(\boldsymbol{b},\boldsymbol{v}), \quad [1]$$

with F being the contribution of the aggregate capillary flow:

$$F(\boldsymbol{b},\boldsymbol{v}) = \sum_{k=1}^{N} exp\left(-j\sqrt{\mathbf{b}}\,|\boldsymbol{v}_k|\cos(\theta_k)\right) sinc\left(\sqrt{\mathbf{b}}|\boldsymbol{v}_k|\cos(\theta_k)\right), \quad [2]$$

where $\boldsymbol{b}$ is the diffusion sensitivity (independent variable) and D is the diffusion coefficient (tissue and static blood). Subsequently, $Sn(b)$ is fitted to the classical IVIM (3) model using a 3-parameter fit to estimate D^*, D, and f with varying flow speeds and anisotropy:

$$Sn(b) = (1-f)e^{-bD} + fe^{-b(D+D^*)}. \quad [3]$$

II. Experimental Validation: MRI of Perfused Guiana Pig Hearts

In order to validate the theoretical predictions and demonstrate the anisotropic effect of microcirculation-mediated diffusion signal, isolated guinea pig hearts (n=7) were perfused in a Langendorff apparatus and imaged with Bruker Biospec (7T) MRI scanner under normal (aortic pressure 52 mmHg) and low (16 mmHg) flow conditions as described previously (4). ASL and diffusion scans were performed on the same short-axis slice using flow-sensitive-inversion-

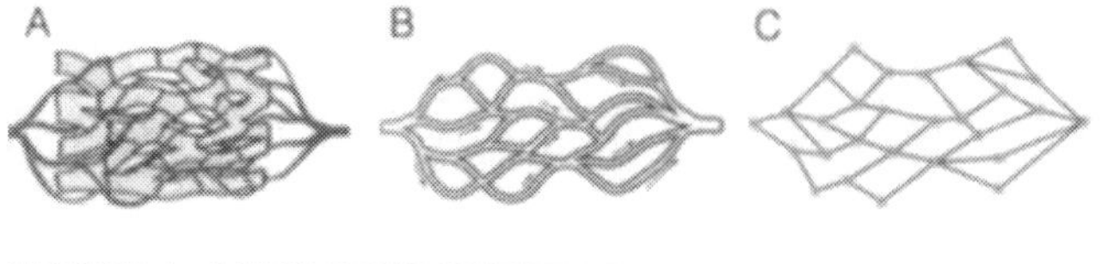

FIGURE 1. SCHEMATIC REPRESENTATION OF HEART TISSUE WITH CAPILLARIES (A), AND CAPILLARIES ONLY (B). THE CAPILLARY BED CAN BE MODELED AS A NETWORK OF INTERCONNECTED NODES (C).

recovery and diffusion weighted spin echo echo-planar sequences, respectively. Diffusion was encoded in both slice (long axis) and readout (short axis) directions using b-values: 1, 7, 18, 28, 55, 80, 107, 157, 200, 310, 513, 766, and 1020 s/mm^2. Subsequently, the observed diffusion signal was averaged over regions of interest in the mid-wall circumferential fiber depending on the angle each voxel makes with the read (left-right) axis. Equation [3] was used to estimate D, D^* and f. Separately, MBF was estimated from the ASL image intensities using previously described procedures (2). IVIM parameters were compared using 2-way ANOVA (encoding direction and flow setting) using Bonferroni correction for post-hoc multiple comparisons, and correlated to ASL using linear regression analysis.

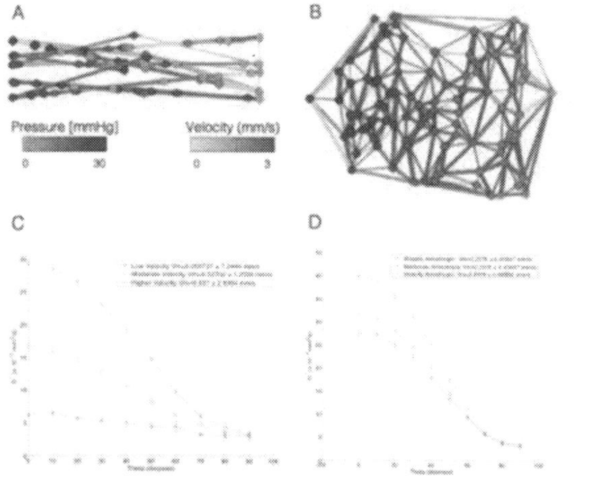

FIGURE 2. FINITE ELEMENT MODEL OF HEAVILY ANISOTROPIC (A) AND WEAKLY ANISOTROPIC (B) CAPILLARY MODELS COLOR CODED BY MEAN FLOW VELOCITY IN EACH CAPILLARY AND PRESSURE AT EACH NODE. NOTE NODAL PRESSURE DROP FROM ARTERIOLE SIDE (LEFT) TO VENULE SIDE (RIGHT). ALSO SHOWN FITTED D* FOR DIFFERENT INPUT FLOW SPEEDS (C) AND DIFFERENT DEGREES OF CAPILLARY ORGANIZATION (D).

RESULTS

I. Simulations

Figure 2a-b shows representative FE capillary model of strongly and weakly aligned elements, respectively, with high pressure driving the flow on the arteriole side (left) to the venule side (right). Estimation of the D* from simulated data shows strong dependence on the capillary flow speeds and orientations. When diffusion is encoded parallel to the main flow orientation, D* is highest compared to when diffusion encoding is perpendicular to flow. This anisotropy behavior of D* becomes stronger with heavily aligned capillaries compared to weakly anisotropic networks (Fig. c-d).

II. Experimental Validation

Representative MRI images from Langendorff experiments are shown (Fig. 3). Myocardial tissue anisotropy is evident from the contrast difference (between Fig. 3a-b). The circumferential fiber (CF) region, which spans 50% of the transmural heart wall, runs mostly parallel to the short axis and appears as a "donut" when diffusion encoding is applied in the long-axis plane (Fig. 3b). Pixels of the CF region were classified based on the angle (Theta) they make with the left-right axis in the short-axis plane (Fig. 3c). The fitted D* from the experimental data at high and low flow conditions versus Theta are shown in Fig. 3d. Not only the experimental results in Fig. 3d is in excellent agreement with the simulation predictions (Fig. 2c-d), but also the D* anisotropy was diminished at low flow.

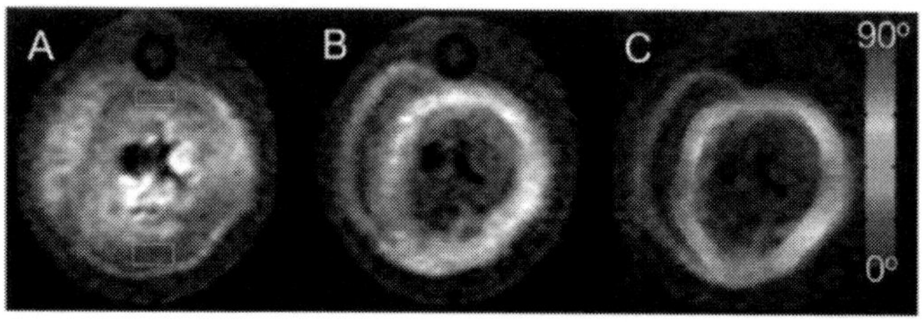

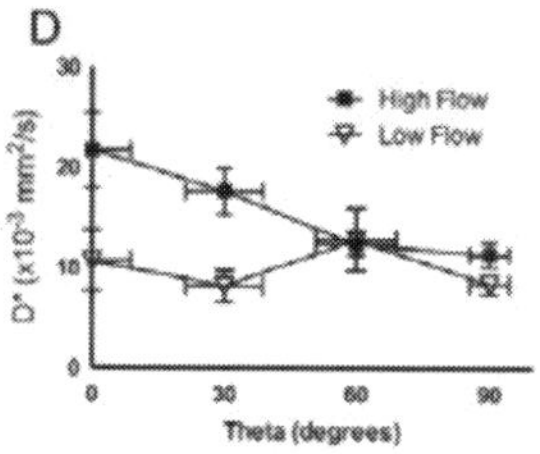

FIGURE 3. DIFFUSION IMAGES AT DIFFERENT ENCODING (A, B) DIRECTIONS. PIXELS OF THE CF REGION IN B WERE CLASSIFIED ACCORDING TO THE ANGLE (THETA) EACH PIXEL MAKES WITH THE L-R DIFFUSION DIRECTION (C). EXPERIMENTAL D* EXHIBITS ANISOTROPY AT HIGH FLOW CASE OLNLY (D).

DISCUSSION

The behaviors of D* with respect to myofiber orientation and MBF indicate that blood flow is faster in the direction parallel than perpendicular to myofibers, which is consistent with previous reports that capillaries in the heart follow the myofiber orientation (5). The vascular volume fraction was found to depend only on MBF but not myofiber orientation (not shown). The incorporation of a realistic fluidic modeling strategy is a significant improvement over previous investigations, which revealed differentiation between perfusion-induced bias and microstructural orientation in cardiac diffusion MRI, and lends itself for correction simulation of more complex scenarios associated with cardiac physiology such as deformation. One of the limitations of the presented approach is neglecting acceleration. As a first approximation on this steady state model, convective acceleration effects can be included by using node-wise approximation of the material derivative form element velocities defined at the center of the segment. As a more complete solution, it is also possible to include time-varying boundary conditions along with a dynamic FE solver. In conclusion, the results indicate that IVIM parameters measured in the perfused myocardium vary as function of degree of microcirculation and myofiber orientation in fashions that are consistent with the known anatomy and circulation physiology of the myocardium.

ACKNOWLEDGEMENTS

This work was supported by NIH R01 HL092055.

REFERENCES

[1] Le Bihan, D et al., *J Biomech*, 34:859-871, 2001. [2] Jacquier et al NMR Biomed 24(9):1047–53, 2011. [3] Le Bihan et al, Radiology. 168(2):497–505, 1988. [4] Abdullah O et al, ISMRM 2011. [5] Poole et al, AJPHCP 1990;259:204–10. [6] Callot et al, Magn Reson Med.;50(3):531–40, 2003.

SB^{3}C2015
Summer Biomechanics, Bioengineering and Biotransport Conference
June 17-20, 2015,Snowbird Resort, Utah, USA
SB^{3}C2015-1163

IMAGE-BASED DYNAMIC ANALYSIS OF BRAIN DEFORMATION MODEL

Arnold D. Gomez (1,2), Gregory G. Scott (3), Boston C. Terry (1), Brittany Coats (3)

(1) Bioengineering Department
University of Utah
Salt Lake City, UT, USA

(2) Surgery Department (Cardiothoracic Div.)
University of Utah
Salt Lake City, UT, USA

(3) Mechanical Engineering Department
University of Utah
Salt Lake City, UT, USA

INTRODUCTION

Traumatic brain injury (TBI) is a major cause of mortality and disability worldwide particularly among the young population [1]. Mild TBI, in particular, can result in significant clinical symptoms, but is particularly difficult to treat and diagnose because traditional medical imaging indicators of injury are not readily identifiable. However, we hypothesize that mild TBI results in damage to subarachnoid substructures that can be evaluated by measuring changes in cortical brain displacement during head rotation.

MRI is a versatile tool for investigating soft-tissue mechanics *in vivo* because one can collect anatomical, structural, and dynamic information before and after injury in controlled experimental animal studies. However, technical challenges include (1) a need for high resolution given the small cranium of the animal and (2) detecting non-injurious small deformations. The objective of this work was to demonstrate an analysis strategy capable to overcoming these challenges and capturing subtle changes in brain deformation in a mild TBI piglet model.

METHODS

All animal protocols were approved by the Institutional Animal Care and Use Committee at the University of Utah according to the *Guide for the Care and Use of Laboratory Animals*. Isoflurane-anesthetized immature piglets (3-5 days old, female, n=2) were exposed to a series of low velocity (45 rad/s) repetitive head rotations at a frequency of 3 Hz for 20 seconds. These head rotations have previously been shown to produce mild axonal injury and small amounts of extra-axial hemorrhage [3].

An MRI-compatible apparatus (Figure 1) was built to generate small coronal head rotations leading to measureable brain/skull displacement at accelerations far below injurious levels. The device included a rotational encoder to acquire instantaneous position of the input shaft, and a digital output to trigger the MRI.

All imaging was performed immediately prior to and 24 hours after the mild TBI. For imaging, isoflurane-anesthetized animals were attached to the pneumatically actuated rotating bite plate of the device, and then placed inside a 150mm imaging coil on a Bruker Biospin 7T instrument. A high-resolution CINE-MRI (256x256, FOV 90x90 mm, TE=3.16 ms, TR=8.86 ms) was performed while the device rotated the animals head through a 60° arc. Stoppers at each extreme of the arc created a sudden stop to control skull deceleration.

Images within each sequence were aligned by means of rigid image registration. To assist in the registration, a dynamic model of the MRI head rotation was created (Figure 1) using the encoder data as input. The dynamic model assisted in regularizing landmark

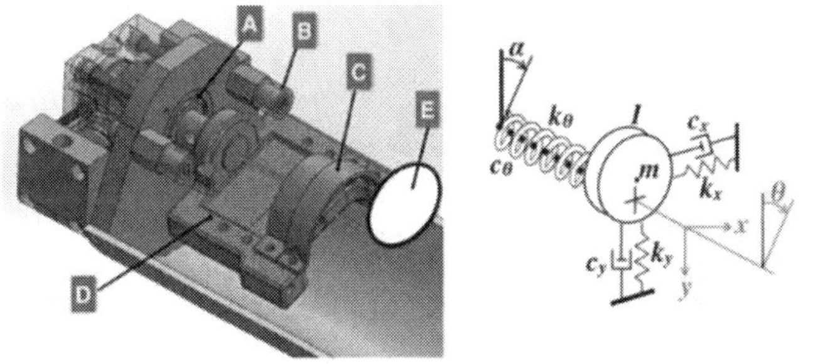

FIGURE 1. MRI-COMPATIBLE APPARATUS (LEFT) CONSISTING OF DRIVE SHAFT (A), RUBBER STOPPERS (B), SNOUT CLAMPS (C), AND ROTATING BITE PLATE (D). A DYNAMIC MODEL OF THE APPARATUS (RIGHT) ENABLED APPROXIMATION OF BRAIN (E) ACCELERATION IN THE ABSENCE OF *IN SITU* ACCELEROMETERS.

registration and provided acceleration measurements in lieu of traditional accelerometers, which do not work effectively in an MRI. To approximate deformation, region-specific elastic image registration was applied using the Hyperelastic Warping (HW) plugin extension for the FEBio software suite [4-5] after blurring and gradient inhomogeneity corrections to reduce impact of noise or imaging artifacts. HW uses image data to deform a finite element (FE) model by means of body forces arising from material point comparisons of intensity information from images acquired at the reference and deformed configurations. As a first approximation in this study, HW was used on a 2D FE model of the brain using a Neo-Hookean material. Image information provided loading and the motion at the base of the brain was fixed as boundary condition. Relative displacement of the cortex with respect to the base of the brain was used to evaluate deformation changes due to acceleration.

RESULTS

Encoder readings from the apparatus approximated a step function and show repeatability deviation below the accuracy of the encoder (1.4°). Rigid registration indicated the existence of overshoot on brain motion (labeled C in Figure 2). Using the displacement and acceleration data from the dynamic model were visibly less noisy and had an overall better alignment of rigid features (Figure 2).

In one subject, relative displacement of cortex to the base of the brain was relatively uniform before injury. Post-injury, the displacement was greater lateral to the sagittal sinus (Figure 3). Post injury motion patterns were visible when comparing images acquired at rest to those obtained near the stopper (~890 rad/s^2) and during the head rotation (~290 rad/s^2). The difference in motion patterns and displacements were not as clearly identifiable in the second animal.

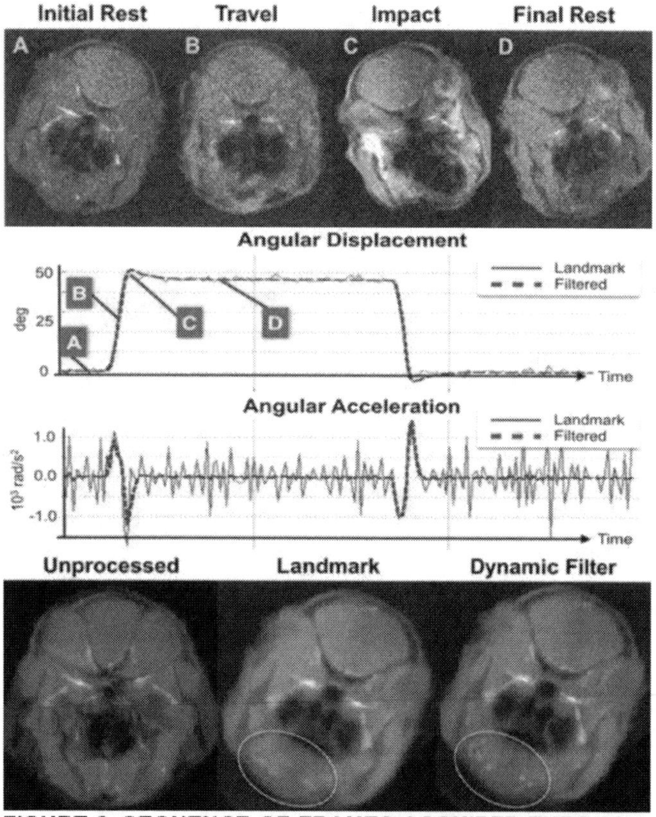

FIGURE 2. SEQUENCE OF FRAMES ACQUIRED THROUGHOUT HEAD ROTATION (TOP). THE DYNAMIC MODEL ACCURATELY ESTIMATED ANGULAR DISPLACEMENTS AND ACCELERATIONS (MIDDLE). RIGID REGISTRATION USING THE ESTIMATED KINEMATIC DATA RESULTED IN BETTER ALIGNMENT COMPARED TO SIMPLE LANDMARK REGISTRATION (BOTTOM).

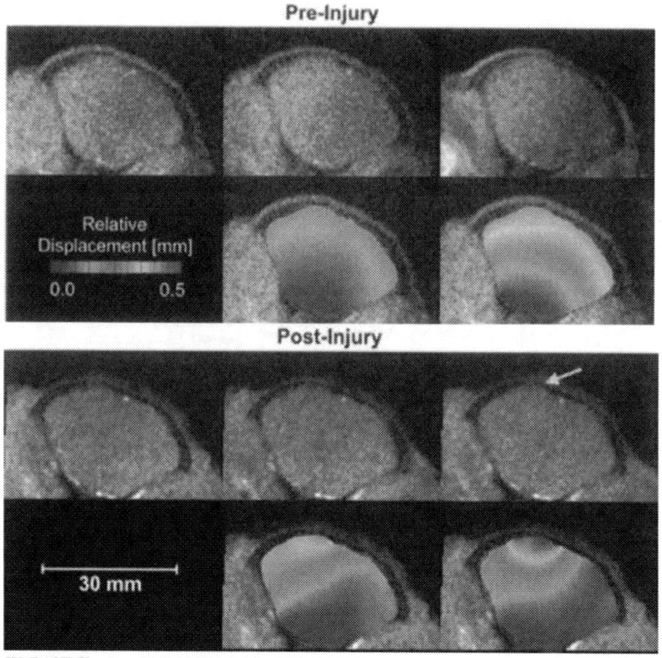

FIGURE 3. NON-RIGID REGISTRATION COMPARED ALL FRAMES OF THE HEAD ROTATION TO THOSE AQUIRED AT REST (LEFT COLUMN). THE ANALYSIS SHOWED A UNIQUE CORTICAL DEFORMATION PATTERN VISIBLE THROUGH THE HEAD ROTATION (MIDDLE COLUMN) AND AT IMPACT (RIGHT COLUMN) THAT WASN'T SEEN BEFORE INJURY.

DISCUSSION

These results demonstrate the effectiveness of the apparatus to produce a repeatable stimulus, and the ability of the combined imaging-FE methodology to accurately estimate brain displacement and acceleration to identify measurable brain deformations during head rotation. The variability on subject deformation results could be explained by absence of injury in the second animal, or due to lack of signal in the areas of interest. This can be improved by optimizing imaging parameters and introducing tagged imaging. While the sample size in this study was small, the finding of a detectable change in cortical brain deformation in one the animals is promising and will be explored in a larger data set.

ACKNOWLEDGEMENTS

The authors would like to thank the University of Utah's Primary Children's Medical Center Foundation for their financial support. Special thanks to the Small Animal Imaging Core and Musculoskeletal Research Laboratories at the same institution.

REFERENCES

[1] Faul, et al. *CDC Center for Injury Control and Prevent*. 2010.
[2] Coats, et al. NNS Meeting. 2011
[3] Rabbitt, et al. Proc SPIE (Vision Geometry IV), 1995.
[4] Maas, et al. J Biomech Eng. Jan 134(1):011005. 2012

SB^{3}C2015-1164

DESIGN OF A 3-D BIOREACTOR FOR SIMULATION OF CEREBROSPINAL FLUID FLOW IN THE THIRD VENTRICLE AND AQUEDUCT OF SYLVIUS

Michael J. Majcher (1), Matthew R. Dailey (1), David P. Lemmer (1), Joseph T. Havrilak (1), Nic Leipzig (2), and Bryn A. Martin, (3)

(1) Department of Biomedical Engineering
University of Akron
Akron, OH, USA

(2) Department of Chemical Engineering
University of Akron
Akron, OH, USA

(3) Department of Mechanical Engineering
University of Akron
Akron, OH, US

INTRODUCTION

Researchers have postulated that primary ciliary motion of the ependymal cells (ECs) plays an important role in cerebrospinal fluid (CSF) hydrodynamics in the ventricular system of the brain (Figure 1) and developmental biology of the central nervous system (CNS) [1]. ECs are a neural cell lineage that differentiates from radial glial cells that have a cuboidal structure with tight gap junctions that form a ~500-800 μm single cell lining on the outer surface of the choroid plexus. Each EC cell has multiple motile cilia with a 5-10 μm length and ~250 nm diameter [2] that move in a coordinated stroke-like manner with an average frequency range of 30 to 60 Hz [3]. Researchers have shown that coordinated EC ciliary motion can result in near wall mean CSF velocities of 45.6 ± 2.5μm/s [4] and have an impact on CSF conduction by using computational modeling [5].

Studies have also examined EC cell signaling and biology. However, little is known about the interaction of CSF dynamics and EC structure, function, and biomechanics. *The aim of the present study is to design and construct an anthropomorphic bioreactor for investigation of this interaction.* A better understanding of this interaction could improve understanding of many CNS disorders in which ECs may play a role such as traumatic brain injury, hydrocephalus, normal pressure hydrocephalus, Chiari malformation and many others.

METHODS AND RESULTS

The three main aspects to the design and implementation of the bioreactor flow system involved (1) fabrication of the parallel plate model (Figure 2), (2) fabrication of the 3-D model, and (3) hydrodynamic testing and simulation (Figure 3). For part (1), the primary goal was designing a parallel plate model with dimensional

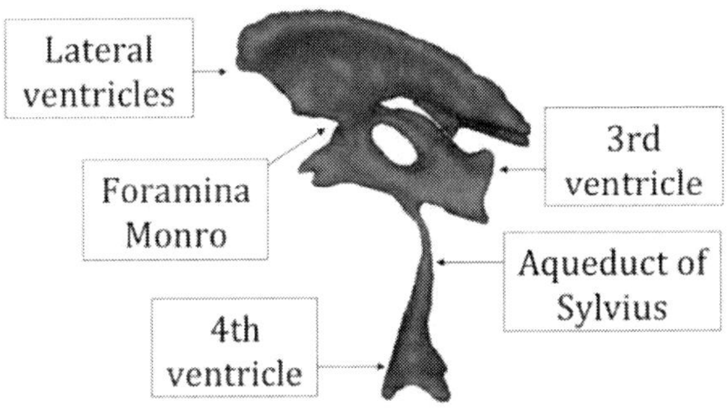

FIGURE 1: 3-D RECONSTUCTION OF THE VENTRICULAR SPANCE OF THE BRAIN BASED ON MR IMAGING.

hydrodynamic similitude to the aqueduct of Sylvius. Previous computational fluid dynamics (CFD) analyses within the aqueduct reported a Reynolds's number of 340, Womersley number of 3, and CSF peak velocity of 16.7 cm/s in the caudal direction assuming CSF with the same properties of water at body temperature [5]. Therefore, to replicate these flow conditions, a parallel plate model was made with a laminin-coated glass microscope slide (Figure 2) that forms the characteristic diameter of the aqueduct (2 mm). A physiological flow rate of 200 mL/min through the model is produced by a reciprocating flow pump with one-way valve attached to the outlet (Figure 3). A silicone gasket was designed to prevent leakage of CSF from the system as fluid enters and exits the model through barbed fittings which are drilled into the upper polycarbonate piece and empty into temperature controlled fluid reservoirs (not shown in Figure 3).

In part (2) of the study, the fabrication of the 3-D phantom model was constructed based on a morphological high-resolution trufi magnetic resonance imaging (MRI) sequence acquired over the ventricles with isotropic resolution of 1x1x1 mm. The MR images were manually segmented using ITK Snap [6] to obtain a 3-D representation of the ventricles and then 3-D printed from optically clear polycarbonate and polished to enable visual inspection with a microscope (Figure 3).

Part (3) of the study consisted of designing a pumping system and closed circuit flow loop to replicate and reproduce CSF's desired pressure, flow, pulsatility, and shear stresses of native anatomy and physiology. Steady state flow testing was tested using a peristaltic pump rated for flow rates of up to 650 mL/min (Shanghai Zdan Inc.). CSF flow was reproduced using a piston-style pump created from a 5V stepper motor and Arduino computer interface system for selective input signals. Catheter type pressure transducers and thermoresistors were directly placed into the model/flow circuit to monitor the pressure and temperature within the model.

Preliminary EC harvesting studies. Ependymal cells were harvested from 2-3 month old Wistar rats (N=2) and stained with anti-beta-tubulin IV mouse antibody as well as GFAP (GA5) mouse mAb stain in order to visualize astrocytes. The defined IHC staining standard operating procedure augmented from J.M. Grondona et al. [7] demonstrates the successful isolation and adherence of ependymal cells, as shown in Figure 4.

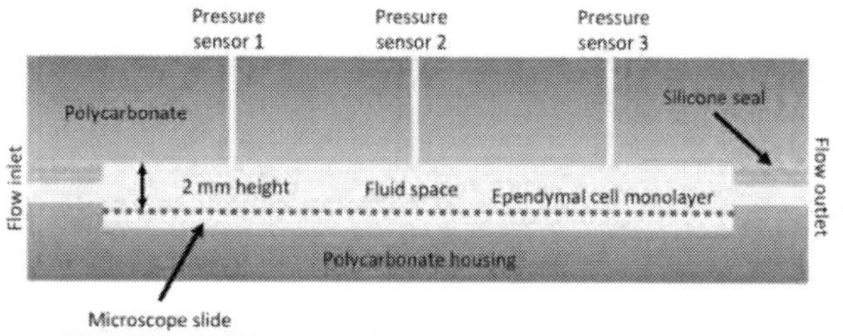

FIGURE 2: PARALLEL FLOW CHAMBER DESIGN SHOWING FLOW INLET ON LEFT AND FLOW OUTLET ON RIGHT WITH AN EPENDYMAL CELL MONOLAYER ONTOP OF A MICROSCOPE SLIDE.

DISCUSSION

At present an *in vitro* flow system has not been designed to specifically investigate the interaction of CSF dynamics and EC structure, function, and biomechanics. The designed bioreactor enables reproduction of a range of CSF hydrodynamics that are similar to in vivo in terms of CSF flow velocities, Reynolds number and Womersley number. This bioreactor has two versions a) parallel plate flow and b) realistic 3-D geometry of the 3rd ventricle and aqueduct of Sylvius. Our pump design allows on-site adjustment of CSF flow waveforms gathered from the MRIs for a wide range of pulse rates and for both healthy and disease states. The 3-D printed model enables an in-house method to create subject-specific patient geometries in a relatively short time.

Limitations. The designed bioreactor pump does not allow simulation of the exact CSF flow waveform shapes, but only the frequency and corresponding flow rate. In addition, ependymal cells have been found to be very difficult to harvest and keep alive for more than 10 days according to the method used. The bioreactor design may need to be modified after preliminary cell experiments are completed. At present, our team has only managed to harvest and identify these cells and are actively working to improve our methods to adhere these cells to the needed substrates within the bioreactor.

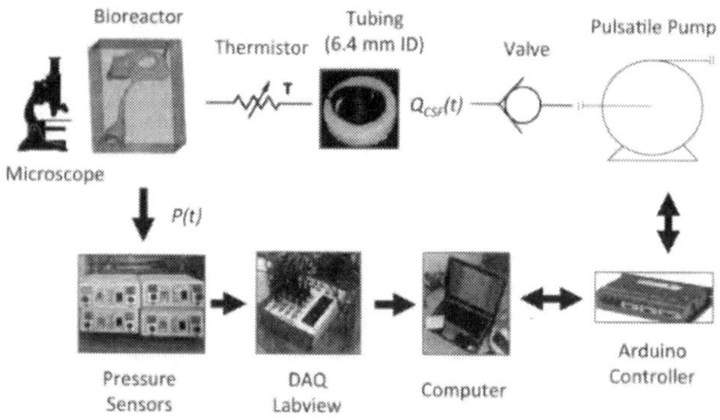

FIGURE 3: EXPERIMENTAL BIOREACTOR SCHEMATIC.

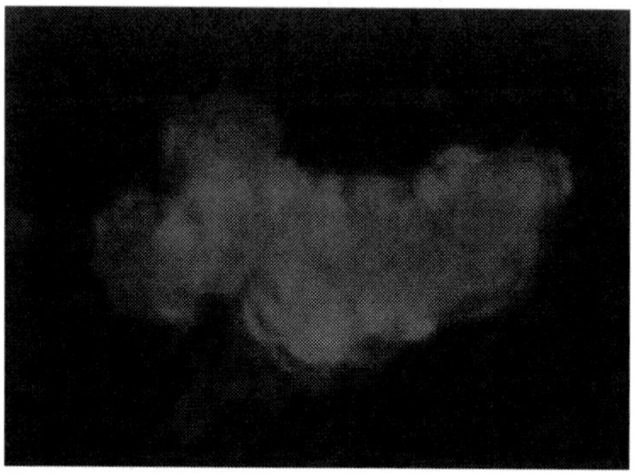

FIGURE 4: HOESCHT 33342 (BLUE) AND ANTI-BETA TUBULIN IV (RED) IHC STAIN FOR POSITIVE EPENDYMAL CELL IDENTIFICATION.

CONCLUSION

A 3-D bioreactor for simulation of CSF flow in the third ventricle and aqueduct of Sylvius was designed and preliminary experiments performed. Development and testing of such a bioreactor system may help to understand the role of ECs in various CNS diseases in which CSF dynamics are thought to be an underlying factor and how CSF flow alterations may alter cytoskeletal structure and function of ependymal cell cilia or other similarly primary cilia containing cells.

ACKNOWLEDGEMENTS

Financial support of this work was provided by Conquer Chiari, The University of Akron Department of Biomedical Engineering's senior design team stipend, and University of Akron Honors College grant. The authors thank Justin Stevens and Aurélie Picquot for help with STL file creation of the ventricular model.

REFERENCES

[1] Del Bigio, Marc R., *Acta Neuropathology*, 119:55-73, 2010.
[2] Lechtrek, K.F. et al., *Journal of Cell Biology*, 180(3):633-43, 2008.
[3] Liu, T. et al., *Journal of Neuroscience Research*, 92: 1199-1204, 2014.
[4] Guirao, B. et al., *Nature Cell Biology*, 12: 341-350, 2010
[5] Kurtcuoglu, V. et al., *Journal of Biomechanics*, 40: 1235-1245, 2007.
[6] Yuskevich, Piven et al., *Neuroimage*, 31(3): 1116-1128, 2006.
[7] Grondona, J.M. et al., *Histochemistry and Cell Biology*, 139:205-220, 2013.

SB³C2015
Summer Biomechanics, Bioengineering and Biotransport Conference
June 17-20, 2015,Snowbird Resort, Utah, USA
SB³C2015-1165

DEVELOPMENT OF MICRO-SCALE ULTRASOUND IMAGING FOR TISSUE CHARACTERIZATION.

Jeremy D Stromer, Leila Ladani

Department of Mechanical Engineering
University of Connecticut
Storrs, CT, United States of America

INTRODUCTION

With an estimated number of 235,000 new cases and approximately 40,000 deaths in the United States in 2014, breast cancer remains a significant cause of concern in public health [1,2]. While it has been shown that there is no substantial disparity between breast-conserving therapy (BCT) and the more traditional mastectomy, the less destructive method of BCT has become the preferred treatment due to both the emotional and esthetic effects of preserving the breast. There is inherent risk of either additional surgery or even local reoccurrence involved for patients of BCT. However, these risks can be minimized by ensuring the entirety of the tumor is excised and a negative margin is obtained during the initial surgery [3-4].

Several methods are currently being employed to help obtain negative margins and include both pre-operative and intraoperative techniques. Pre-operative breast imaging using conventional processes such as mammography, traditional B-scan ultrasound and MRI have been used to determine the general location of the tumor but have failed to accurately gauge its actual size [5-6]. The most common intraoperative methods are frozen section analysis and touch preparation cytology. However these approaches can be lengthy and require the presence of a trained pathologist [7]

New intraoperative techniques spanning numerous disciplines have been proposed to improve margin detection while alleviating the need of a trained pathologist. A commercial product called MarginProbe™ attempts this by evaluating the electromagnetic properties of the tissue [8]. Considerable work has also been done investigating the use of optical imaging methods. [4,7,9-11].

Used extensively in diagnostic medicine and nondestructive testing, the use of ultrasound has also been extended to tissue characterization. Ultrasound measurement has several noticeable advantages. Ultrasound systems are compact and portable. It is also possible to manufacture miniature transducers which allow for small-scale measurements such as intravascular and gastrointestinal imaging [12-13]. Ultrasound has also shown to be relatively safe and is absent of the ionizing radiation present in other imaging methods [14].

A previous study by Doyle et al. shows that high frequency ultrasound seems to be sensitive to tissue microstructure. By analyzing the waveforms using advanced signal processing techniques several parameters were found that varied with tissue pathology [15].

This study's intent is to expand the role of high frequency ultrasound for possible use in the detection of margins. By coupling the decreased footprint of small diameter, high frequency ultrasound transducers with signal processing we investigate the potential of developing new imaging methods based upon non-conventional parameters for tissue characterization.

METHODS

Imaging was conducted on tissue mimicking phantoms made from gelatin (Knox) and psyllium fiber (Equate) [16]. For the sample presented here, a single red pepper seed and two soft, dried pepper flakes to add a contrast of stiffness. These materials were chosen to minimize phantom decomposition and to provide test scatters that would be visible through ultrasound and visual inspection.

Ultrasonic measurements were taken in through-transmission mode in order to improve signal strength, since the wave only needs to travel the sample thickness once. Two pachymeters (TransducerWorks, 50 MHz, 2 mm diameter) were used due to their small element diameter and high frequency. A UTEX UT340 pulser receiver was used to generate and receive the signals which were digitized by a Tektronix DPO 3052 oscilloscope (500 MHz, 2.5 GS/s). The waveforms were collected by using the oscilloscope to average the first 512 ultrasound pulses to reduce noise. Equipment control was automated with LabVIEW (National Instruments). The data and image processing was completed in MATLAB (MathWorks).

To facilitate scanning, a three axis motion control stage (Fig. 1) was constructed to support and move the sample. A standard raster scan path was used with step sizes of 1.016 mm. Resulting images were composed of 100 pixels and covered an area of 1.03 cm^2.

Figure 1: Photographs of testing stage (a) and phantom sample (b). The corner dots are ink and were used for alignment purposes.

As opposed to conventional B-scan ultrasound images which measured as a function of depth, the images presented in this study are mapping of the top view of the sample, also known as C-scans.

For a baseline of comparison, the first image was generated according to the amplitude of the received signal. Each pixel corresponds to the maximum of the waveform's Hilbert transform.

Our imaging method relies on a deeper investigation of the waveform using higher level signal processing techniques. In a method similar to that used by Doyle et al. for calculating the so-called peak density, we count the number of local maxima or peaks present in the power spectrum of the Fourier transform of the waveform.

RESULTS

Preliminary images created using the approaches previously discussed are demonstrated in Figure 2. Figure 2(a) shows an image created from measuring the maximum amplitude of the waveform. This method is akin to conventional ultrasound imaging methods. The image in Figure 2(b) is a reconstruction of the phantom based upon peak density measurements.

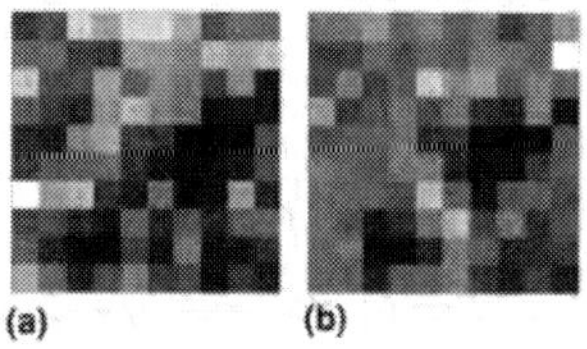

Figure 2: Images of phantom sample: amplitude (a), peak density (b).

DISCUSSION

The images created from both methods were comparable. Each reconstruction correctly displayed the inclusions and their corresponding locations. It is seen that the contrast between the inclusions and surrounding material is greater for the peak density image (Fig. 2(b)). The pepper seed was approximately elliptical in shape with a major axis of 3.5 mm and minor axis of 3 mm. The seed was angled with respect to the sides of the imaging regions as seen in Figure 1(b). The size of the seed is slightly overestimated in the amplitude image while the peak density image gives a more accurate size assessment. The other two inclusions are visible in each image, but are not entirely definitive.

Current imaging methods often employ more waveform processing and image filtering to improve image quality than was used in this study. It is reasonable that further processing could improve the amplitude images as well as peak density images. This work was done to assess the potential of peak density imaging as an alternative to conventional C-scanning for tissue characterization.

Peak density imaging has shown to be comparable to standard C-scan imaging. Further work will include the possibility of further processing to improve the quality of the images. Additionally, other signal parameters may be investigated as a means of imaging the difference in tissue structure.

ACKNOWLEDGEMENTS

We would like to thank the National Science Foundation GK-12 Fellowhip Program for funding JDS. We would also like to thank TransducerWorks for the help with the transducers.

REFERENCES

[1] Siegel, R et al., Cancer statistics, 2014, *CA: a cancer journal for clinicians*, 2014.

[2]Litière, S et al., Breast conserving therapy versus mastectomy for stage I-II breast cancer: 20 year follow-up of the EORTC 10801 phase 3 randomised trial. *The Lancet Oncology*, 2012.

[3] Jacobs, L Positive margins: the challenge continues for breast surgeons. *Ann. Surg. Oncol.*, 15(5):1271-1272, 2008.

[4] Kennedy, S et al., Optical breast cancer margin assessment: an observational study of the effects of tissue heterogeneity on optical contrast. *Breast Cancer Research*. 12: R9, 2010.

[5] Pengel, KE et al, The impact of preoperative MRI on breast-conserving surgery of invasive cancer: a comparative cohort study. *Breast cancer research and treatment*. 116-169, 2009.

[6] Faverly, DR et al., Breast carcinomas of limited extent: frequency, radiologic-pathologic characteristics, and surgical margin requirements. *Cancer*. 91:647–59, 2001.

[7] Thill, M et al., Intraoperative assessment of margins in breast conservative surgery-still in use? *Journal of surgical oncology*, 110:15-20, 2014.

[8] Karni, T et al., A device for real-time, intraoperative margin assessment in breast-conservation surgery. *The American Journal of Surgery*. 194: 467–473, 2010.

[9] Haka, AS et al., In vivo Margin Assessment during Partial Mastectomy Breast Surgery Using Raman Spectroscopy, *Cancer Research*. 66: 3317–3322, 2006.

[10] Nguyen, F et al., Intraoperative Evaluation of Breast Tumor Margins with Optical Coherence Tomography. *Cancer Research*. 69:8790–8796, 2009.

[11] Pani, S et al., Characterization of breast tissue using energy-dispersive X-ray diffraction computed tomography. *Applied Radiation and Isotopes*. 68:198-1987, 2010.

[12] Nishimura, RA et al., Intravascular ultrasound imaging: in vitro validation and pathologic correlation. *Journal of the American College of Cardiology*, 16:145-154, 1990.

[13] Dimagno, EP et al., Human endoscopic ultrasonography. Gastroenterology, 83(4):824-829, 1982.

[14] Haar, G, Ultrasonic imaging: safety considerations. *Interface Focus*, 1:686-697, 2011.

[15] Doyle, T.E. et al., High-frequency ultrasound for intraoperative margin assessments in breast conservation surgery: a feasibility study. *BMC Ccancer*. 11: 444, 2011.

[16] Bude, R, Adler, S, An easily made, low-cost, tissue-like ultrasound phantom material. *Journal of Clinical Ultrasound*. 23:271-273, 1995.

SB³C2015
Summer Biomechanics, Bioengineering and Biotransport Conference
June 17-20, 2015,Snowbird Resort, Utah, USA
SB³C2015-1166

QUANTIFICATION OF TRANSIENT TEMPERATURE AND THERMAL DAMAGE IN AN ESTABLISHED BURN MODEL

Stephanie Jo Lindow, F. Scott Gayzik

Biomedical Engineering
Center for Injury Biomechanics
Wake Forest University School of Medicine
Winston-Salem, NC, USA

INTRODUCTION

Determination of burn depth is crucial for the treatment of burn wounds. Most burns are of indeterminate depth before lengthy, expensive biopsies can be performed to determine the best treatment option for the patient. Many experiments have been performed to establish reproducible animal models (murine, porcine) of human skin burns. Such models can support mechanistic studies, as well as testing of new methods of treatment and diagnosis of human skin burns.

Several porcine models have been investigated, as pig skin exhibits anatomical, physical, and physiological similarity to human skin [1]. Finnish landrace pigs were used in experiments by Papp et al. [1] in which a brass block was heated to 100 °C in a water bath and applied to the skin. Burns of different depths were created by varying contact time. Histological samples were taken to identify burn progression within the layers of the skin after specific amounts of time. Papp determined that a progression of burn depth was found until 48 hrs post-injury, with an increase in burn depth based on contact time [1]. Up to 12 sec of contact time was studied.

The Papp study successfully established a reproducible burn model, which has been used in subsequent studies utilizing both porcine and murine models [2, 3]. However, while highly reproducible, the approach quantified only the outcome, not the transient insult itself. The purpose of this study was to estimate the temperature history and damage associated with the Papp model using a computational and experimental approach.

METHODS

Experiments were conducted ex-vivo, as a basis for model validation and direct measurement of heat flux and temperature changes.

Experiment. Porcine muscle (shoulder cut) specimens were obtained from a local abattoir and cut into 12.7 cm square by 5 cm thick samples. The cut was selected due to the presence of skin. The protocol generally followed the approach of [1]. A machined brass block (dimensions; 4 x 4 x 1.27 cm) was heated to 100°C in boiling water and placed into firm contact with the specimen. Four thermocouples (K-type, 1 mm wire) were used along with a thin-film heat flux sensor (Omega, HFS-3) situated between the block and the specimen. Thermocouples were placed as follows: one within the heated block, two within the specimen at 0.95 and 1.9 cm below the center of the heated block, and one to record ambient temperature. Data were collected using LabVIEW software and a *Nat. Inst.* cDAQ-9174 and 9211 module at 100 Hz. Due to limitations of the pre-heating process, the porcine specimen was tested at 4 °C. Heat flux and temperature was recorded over a duration of 12 s.

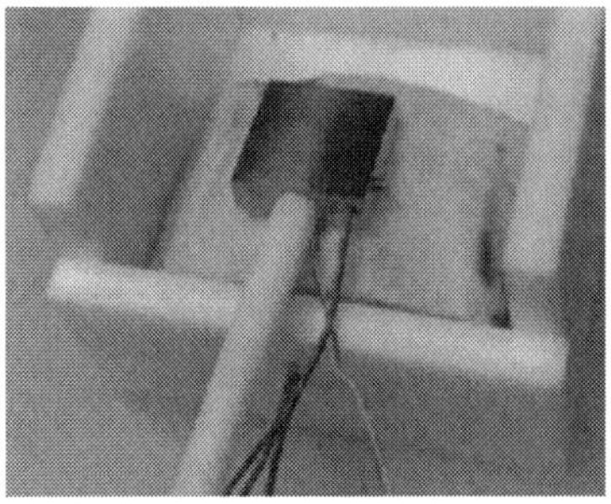

Figure 1. Heat flux experimental setup showing specimen, heating block, and thermocouples.

Computational Model A one dimensional finite difference (FD) model was used [4]. The transient FD code was validated against an analytical solution for constant surface heat flux applied to a solid with known thermal conductivity. Parameters used in the model of the experiment are shown in Table 1. A total of 4 cm of the specimen was modeled. The brass was considered the upper portion of the domain. A convective boundary condition (h = 10 W/m^2/K, T_∞ = 22°C) was used at the surface of the brass and a constant temperature boundary conditions was used at the deep end of the domain, set to the initial temperature of the sample. Matlab (R. 2014a, The Mathworks, Natick, MA) was used for all simulations.

Table 1: Thermal properties used in simulations.

	K (W/m/K)	C_p (J/kg/K)	ρ (kg/m^3)
Brass	110	380	8530
Skin/Muscle	0.3	3300	1040

Finally, after validating the ex-vivo condition, the one dimensional FD model was used to simulate the experiments of Papp et al. In this second simulation, an initial temperature of 37 °C was applied to the domain, and boundary conditions were preserved from the ex-vivo simulation. The method of Henriques [5] was used to calculate the thermal damage beneath to the application of the heated block. Using the Henriques model, the temperature history is integrated to estimate damage, which is stated as a damage coefficient, Ω exceeding 1.

RESULTS

The validation experiment showed excellent agreement in experimental vs. simulated heat flux as well as the temperature decrease of the brass block (Figure 1). Temperature rise in the porcine flesh model showed little to no increase during the event, but the lack of temperature rise was consistent in both model and simulation at the depths of the thermocouples.

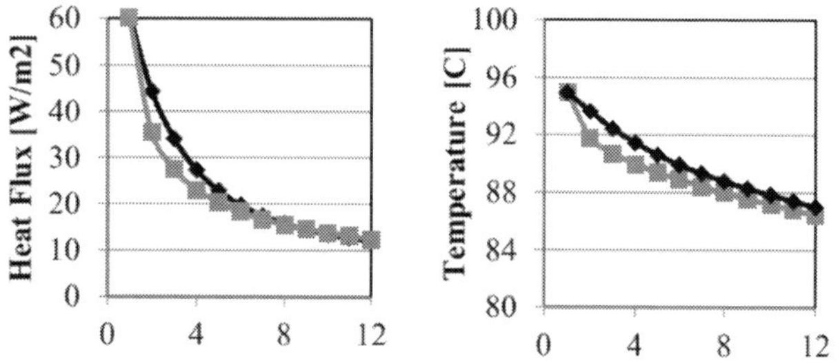

Figure 1. Heat flux and brass block temperature vs. time for ex-vivo experiment. Orange – Experimental, Black, Simulation.

The simulation of the Papp study using the FD model is shown in Figure 2. Temperature was found to rise greatest when closest to the application of the heat source. At 4 mm, little to no temperature increase is predicted. The associated thermal damage per [5] for various contact times and depths are found in Table 2.

DISCUSSION

The FD model replicated experimental results accurately. When extending the FD model beyond the ex-vivo study, temperature rise vs. depth was determined based on the experiment in [1]. The Henriques model for damage was calculated at various depths and times (Table 2 values in excess of 1). Damage was noted at only the shallowest depths from the application of the brass and at times greater than 1 sec. This finding is in conflict with histological findings of Papp et al. where superficial damage was found in the epidermis even at 1 sec and at depths exceeding 2 mm at later times.

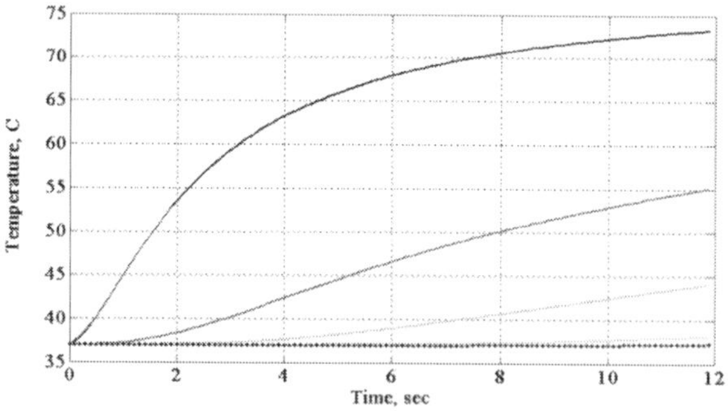

Figure 2. Temperature profiles at various depths vs. time, from simulation of in-vivo experiment described in [1]. Black, surface, maroon, 1.2 mm, yellow, 2 mm, gray, 3.2 mm and dotted, 4 mm deep.

Table 2. Thermal damage, Ω, results at various depths and times.

Depth, mm →	0	1.2	2
Time, sec ↓			
1	0.0	0.0	0.0
3	3.7	0.0	0.0
9	3.7E+04	0.1	0.0
12	1.7E+05	1.0	0.0

There are a number of limitations of this study which should be noted. In the in-vivo experiment, blood perfusion in the capillary bed of the skin would likely diminish the temperatures predicted by the FD model, since perfusion was not modeled. However, given the time over which the experiment is conducted it is anticipated that this would not have a dramatic effect on the results. Secondly, the damage model used in this study may have underestimated actual damage given the temperature history and known histological results in [1]. The damage model is a function of an activation energy and molecular frequency factor values for skin that vary widely in the literature.

The model will serve as a platform for continuing studies in this area. By quantifying the time history of the thermal event one could, in theory, represent this heating method by more quantitative metrics such as a heat-flux signature, temperature history at a given depth, or an integrated damage value such as Henriques. This would enable various heating methods (brass block or other methods) to be equated. There are a number of future directions for the work. The first would be hone the model to include more physiologic aspects such as blood perfusion. The model could be expanded to multiple dimensions if deemed necessary. Later, the data could be used to correlate the outcome (histological) to heat flux, temperature history, or damage factor.

ACKNOWLEDGEMENTS

Work was supported by the Wake Forest University Summer Research Program.

REFERENCES

[1] Papp, A. et al., *Burn*, 30:684-690, 2004.
[2] Papp, A. et al. *Plast Reconstr Surg*, 117(3):889-901, 2006.
[3] Singer, A.J. et al. *Academic Emergy Med.*, 14(12): 1125-29, 2008
[4] Gayzik, et al., *J. Biomech. Eng.* 128(4):505-515, 2006
[5] Henriques, et al. *Am. J. Pathol.*,23, pp. 531-549, 1947

SB³C2015-1167

THE EFFECT OF PORE SIZE ON BONE STRAIN IN THE PROXIMAL FEMUR

Mariana E. Kersh (1), Afrodite Zendelli (2), Yohann Bala (2), Ali Ghasem-Zadeh (2), Ego Seeman (2), Roger Zebaze (2)

(1) Department of Mechanical Science and Engineering
University of Illinois at Urbana-Champaign
Urbana, Illinois, United States of America

(2) Department of Endocrinology
Austin Health - University of Melbourne
Heidelberg, Victoria, Australia

INTRODUCTION

An improvement in the measurement of bone strength is necessary in order to better detect the risk of bone fracture. An accurate measurement of cortical porosity is important because porosity accounts for approximately 70% of all bone loss thereby compromising the bone strength [1]. However, the question remains if small pores are relevant to bone strength.

METHODS

Porosity was measured in the proximal femur of fourteen female cadavers (age range = 67 – 87 years). Scanning electron microscope (SEM) images (resolution = 7μm) within a region of interest in the subtrochanteric cortex (Figure 1A,B) were acquired. Within each cross-sectional region of interest, pore number, size, area and the distribution of pores was measured using custom developed code (StrAx 1.0).

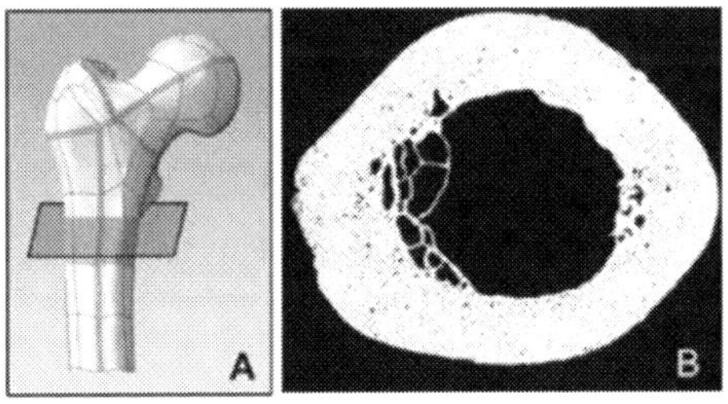

Figure 1: Region of proximal femur analyzed for porosity (A) using scanning electron microscopy images of cortical bone (B)

A representative slice was then used to develop finite element models of the cortex in four anatomical quadrants: anterior, posterior, medial, and lateral. The SEM images were first binarized and a cross-sectional region of interest was segmented using image-processing software (Simpleware). An extruded three-dimensional solid model (1cm thick) of the cortex was then reconstructed (Geomagic) (Figure 1B). The cortical bone was modeled as a homogeneous material with a Young's modulus of 16.5GPa and Poisson's ratio of 0.3 [2]. A compressive load (1618N) was applied to each model, simulating the peak loads during walking [3]. The maximum tensile strain was calculated using a linear stress analysis (Abaqus, v10).

Figure 2: SEM image of anterior compartment of cortical bone (A) and corresponding finite element model (B)

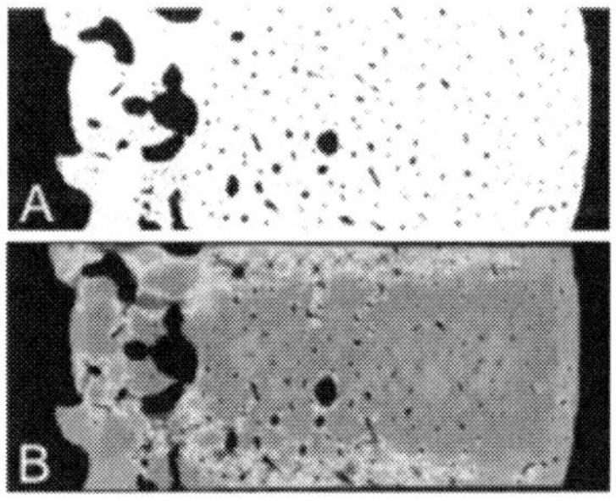

To assess the effect of the various pore sizes, pores were sequentially removed at discrete diameter intervals. First, pores with apparent areas corresponding to circles of diameter 30μm or less were digitally removed from the SEM images. The finite element model was then reconstructed and the same loading conditions were applied. This process was repeated, in 30μm increments, up to pores of 120μm in diameter.

RESULTS

The pore sizes were not normally distributed and were instead skewed toward small pores (Figure 3). The median diameter of pores was 57.53 ± 12.01 and produced a porosity (pore area) of an average of 23.10 ± 6.80%.

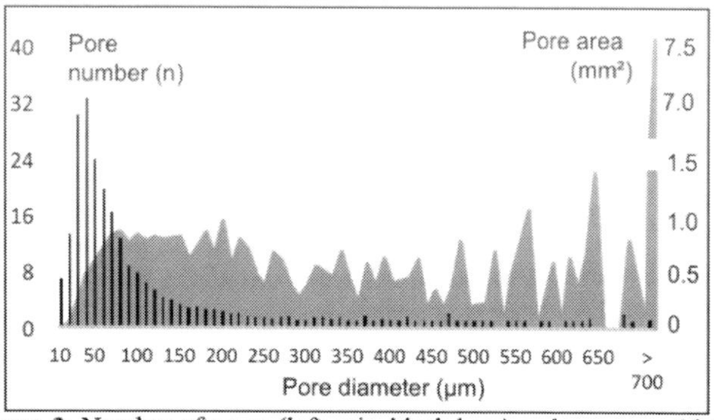

Figure 3: Number of pores (left axis, black bars) and pore area (right axis, gray shaded) as a function of pore diameter.

The vast majority of pores (76.09 ± 9.64 %) were small pores (defined as pores less than 100 μm), yet they had a relatively small contribution to porosity (pore area).

When pores were sequentially removed, the peak strain decreased, as did the percent area of the bone under high strain (>4850 μm) (Table 1). For example, in the anterior compartment (shown in Table 1), pores of less than 30 μm in diameter accounted for 19.28 ± 5.73% of all pores, but their inclusion resulted in a reduction of strength by nearly 50% as seen by the increase in peak strain. Similar results were found in the medial, lateral, and posterior compartments

DISCUSSION

Based on changes in peak strain, pores between 30 μm and 60 μm make the largest contribution to the strain distribution within the femoral cortex. This suggests that current imaging methods, with resolutions far courser than 100μm may be underestimating the strength of bone.

REFERENCES

1. Zebaze R. *JBMR*, 22(7), 1055-61, 2007.
2. Vargehese B. *J Biomech*, **44(7)**, 1374-9, 2011.
3. Lotz SG. *J Biomech Eng*, **113(4)**, 353, 1991.

Table 1: FE models of anterior cortex with different pore sizes present and resulting peak strain and percent area under high strain.

All Pores	**> 30 μm**	**> 60 μm**
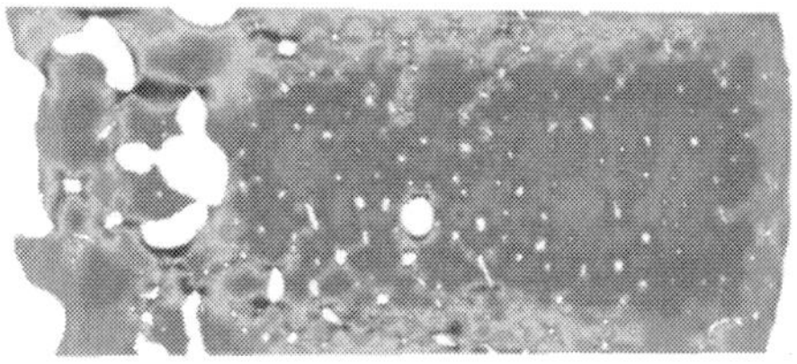	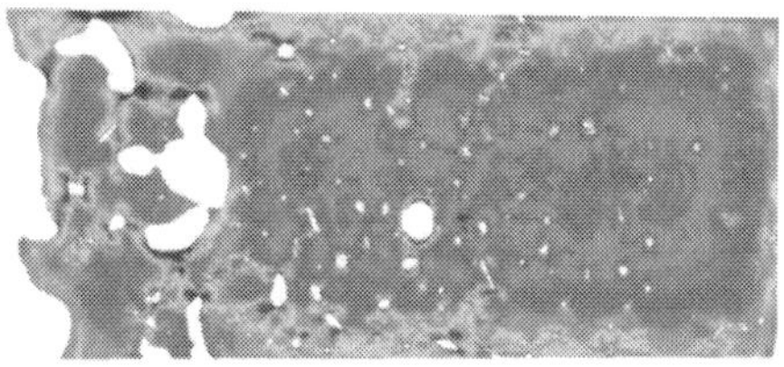	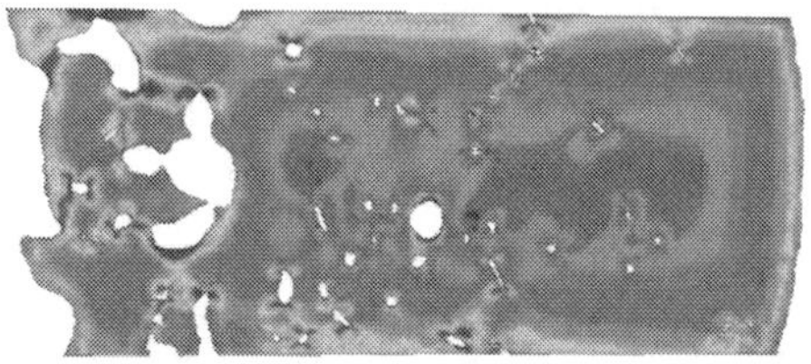
(2542, 0.36%)	(1088,0.33%)	(950, 0.27%)

> 90 μm	**> 120 μm**
	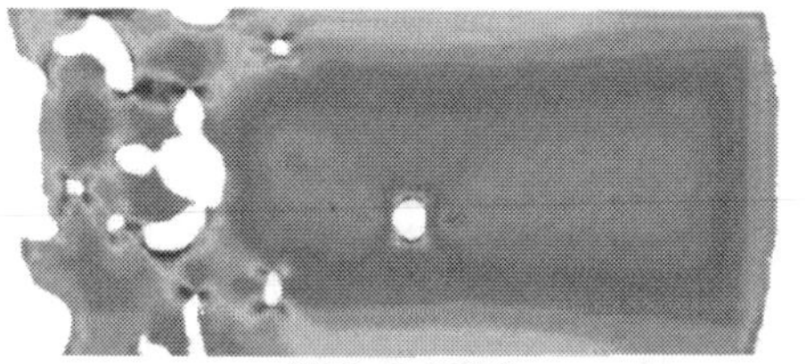
(843, 0.28%)	(780, 0.28%)

SB³C2015
Summer Biomechanics, Bioengineering and Biotransport Conference
June 17-20, 2015,Snowbird Resort, Utah, USA
SB³C2015-1169

CORNEA DAMAGE PROGRESSION FOLLOWING BLAST EXPOSURE

Daniel F. Shedd (1), Justin A. Jones (2), Brian Zaugg (3), Brittany Coats (1)

(1) Department of Mechanical Engineering
University of Utah
Salt Lake City, Utah, USA

(2) Department of Bioengineering
University of Utah
Salt Lake City, Utah, USA

(3) John A. Moran Eye Center
University of Utah
Salt Lake City, Utah, USA

INTRODUCTION

Blast exposure is a significant cause of injury for the US Army [1]. The time course of eye injury subsequent to exposure is not well understood, especially for cases of closed globe trauma. Several studies have investigated visual impairment in soldiers with traumatic brain injury from blast exposure. They report ~75% of soldiers with traumatic brain injury also have visual dysfunctions [2,3]. One study in particular performed complete ocular exams on all soldiers with a history of traumatic brain injury from blast exposure and found retinal injuries in several military personnel that were unaware they had any ocular or visual problems [4].

To investigate the time course of blast induced vision loss, we developed a high-pressure blast injury model in the rodent. Our objective in this study was to evaluate the long-term (8 week) time course of corneal injury following blast exposure.

METHODS

All animal protocols were approved by the Institutional Animal Care and Use Committee at the University of Utah. Long-Evans rats (300-350g, n=38) were anesthetized using IP injection of a ketamine-dexmedetomidine mix and exposed to blast waves (peak overpressure = 230 kPa) using a compressed air driven shock tube (Fig. 1). Firing of the shock tube was controlled by material failure of a biaxially oriented polyethylene (BoPET) membrane (.01" thickness), which occurred at a driver section overpressure of 650-750 kPa. A representative blast wave generated at the location of the animal is shown in Fig. 2.

The animals were placed inside of the tube using a 3D-printed mount exposing the head and eyes to a side-on blast insult, while protecting the body and lungs from the injury. Additional protection of the body was achieved by wrapping the anesthetized animal in a Kevlar shroud. The right eye of the animal was always ipsilateral to the oncoming pressure wave throughout the study. A subset of blast studies were videotaped using a Phantom high speed camera at 5000 fps to assess head motion induced by the blast wave. The video data was also used to ensure that the rupturing membrane did not generate any shrapnel, as this could cause additional injuries to the animal.

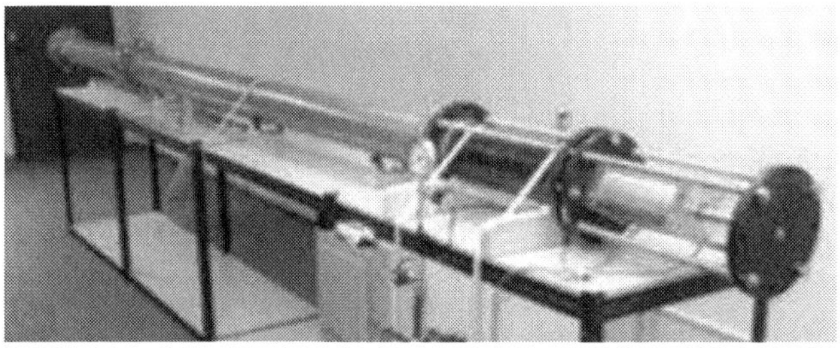

FIGURE 1. 6-IN INTERNAL DIAMETER SHOCK TUBE USED TO GENERATE FRIEDLANDER SHAPED BLAST WAVES.

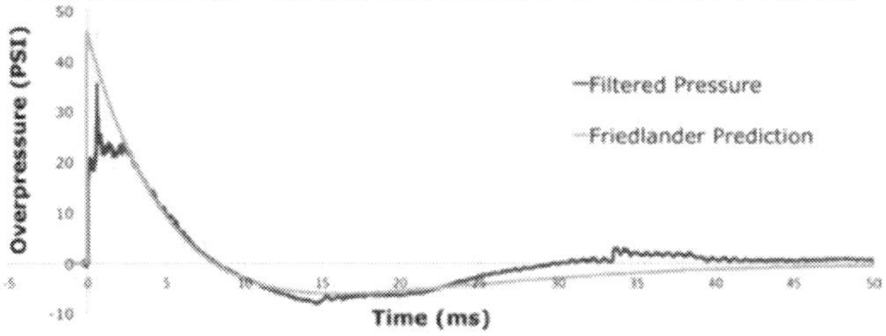

FIGURE 2. BLAST WAVE MEASURED 1-IN IN FRONT OF ANIMAL LOCATION COMPARED TO PREDICTION OF BLAST FROM FRIEDLANDER WAVE EQUATION

At time points 1 day before blast, 1 day post blast, and every week after blast up to 8 weeks, corneal imaging of both left and right eyes was performed using a Bioptigen Envisu R2200™ OCT scanner with an ultra-high resolution light source (Telecentric lens, 4.0 x 4.0 FOV, 100 B scans, 1000 A scans).

Gross ocular examinations were performed at each time point to determine the presence of any easily identifiable injury or corneal defects. When possible, an ophthalmologist (BZ) performed ophthalmic exams of the eyes. Fluorescein staining was used to aid in visualization of superficial corneal epithelium injuries.

Overall corneal thickness was measured regionally using a MATLAB code created using the Image Processing Toolbox. Stromal and epithelium thickness were measured manually using InVivo Vue software (Bioptigen, North Carolina).

A Dunnett's test was used to compare the corneal, stromal, and epithelial thicknesses at every time point to the baseline (pre-injury) measurement (JMP, SAS Institute, North Carolina). A p-value < 0.5 was considered significant.

RESULTS

The superficial epithelium layer and the underlying stroma layer could be clearly resolved in all images (Figure 3A).

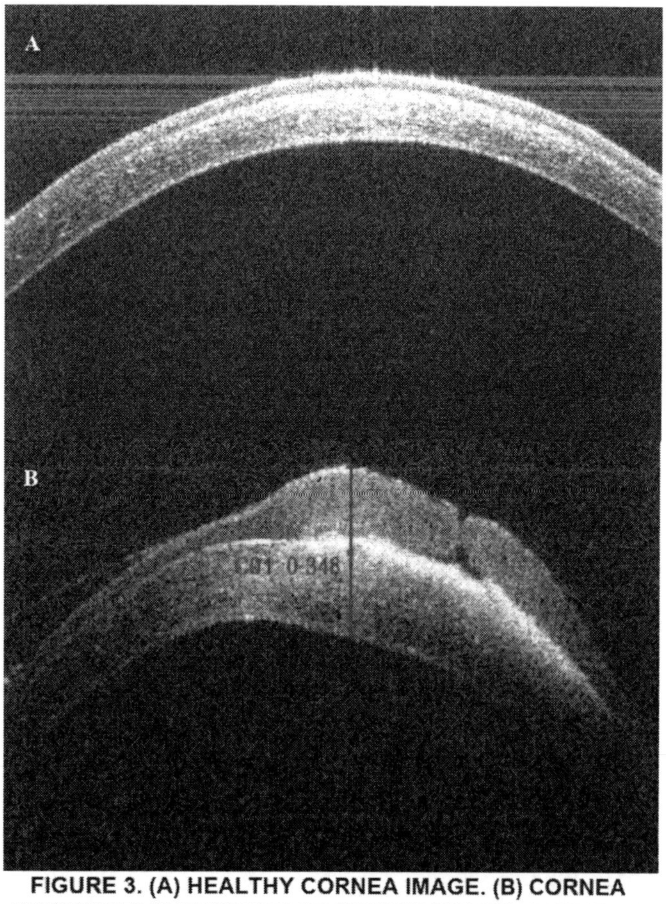

FIGURE 3. (A) HEALTHY CORNEA IMAGE. (B) CORNEA EXHIBITING THICKENING OF EPITHELIUM INDICATIVE OF INFLAMMATION. THE CORNEAL STROMA IS ALSO SLIGHTLY ENLARGED, AND THE BORDER BETWEEN LAYERS IS UNHEALTHY.

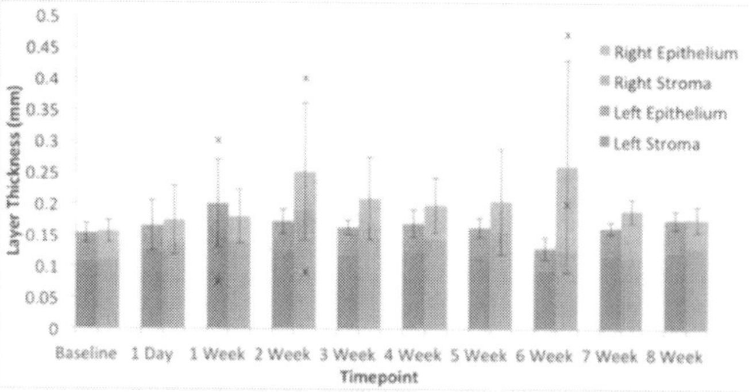

FIGURE 4. AVERAGE THICKNESS FOR TOTAL CORNEA AS WELL AS STROMA AND EPITHELIUM. STARRED COLUMS INDICATE STATISTICAL SIGNIFICANCE COMPARED BASELINE (PRE-INJURY).

In the eye contralateral to the blast (left), significant changes in overall corneal thickness and stromal thickness (p=.0105) occurred at one week (Figure 4). These changes resolved by two weeks after blast. The eye ipsilateral to the blast (right) had significant increases in corneal thickness at two (p=.0066) and six weeks (p=.0167). At two weeks, the increased thickness came from welling of the stroma, while at six weeks the thickening was due to epithelial changes. These data points correlated with the appearance of visible gross injury to the cornea at 3-4 weeks followed by scarring at 6-8 weeks, as well as neovascularization.

DISCUSSION

The blast exposure appears to have caused a structural injury in the stroma of the right eye. This defect develops into a measurable change in thickness of the stroma which is at first not visible on the surface of the eye. This injury may be caused by a disorganization or disruption of collagen plates that make up the corneal stroma. The stroma eventually heals, returning to pre-injury thickness, but leaves behind unhealthy epithelial tissue, which presents as a superficial hemorrhage and eventual scarring. Interestingly, there was transient thickening of the left cornea which was contralateral to the blast. The thickening resolved quickly and did not result in gross indications of injury.

These data show that blast exposure can cause delayed expression of injury to the cornea. The injury is not limited to the ipsilateral side of blast exposure, but may also be present contralateral to a lateral blast. These injuries may also be measurable by other types of collected data, such as vision behavioral studies or protein expression, which are not presented in this work. A better understanding of these injuries and their time course will aid in the detection and treatment in blast-exposed individuals.

ACKNOWLEDGEMENTS

We would like to thank USAMRMC #W81XWH-12-1-0243 for support of this project. We'd also like to thank Krishna Womack for her assistance with data analysis.

REFERENCES

[1] Ari AB., *Optometry*, 77:329-339, 2006.
[2] Kapoor, N. et al., *Neurology* 4:271-280, 2002.
[3] Brahm, K. et al., *Optometry and Vision Science* 86:817-825, 2009.
[4] Cockerham, G, *et al. J Rehabil Res Dev* 46: 811-818, 2009.

DESIGN AND SIMULATION OF AN ANKLE FOOT SIMULATOR

J. Miller, A. Merryweather, M. Minor

Mechanical Engineering
University of Utah
Salt Lake City, Utah, USA

INTRODUCTION

Parkinson's disease (PD) is a progressive and chronic movement disorder which affects the individual's ability to walk and move normally by causing bradykinesia, or slowness of movement, and rigidity, a stiffness of the limbs and trunk. Due to these symptoms people with PD are twice as likely to fall as are their peers without PD resulting in potentially serious injuries [1]. Currently, nearly one million people in the United States are living with PD. Research shows that participating in regular exercise can decrease some of these PD symptoms. Particularly important are exercises that challenge the individual to "change tempo, activity, or direction" as well as "activities that require balance and preparatory adjustments" [2].

In an effort to address this need for dynamic therapeutic intervention, an interdisciplinary team of researchers propose the development of a virtual reality (VR) simulator which will give haptic feedback in real time to the individual exploring a virtual environment. As part of this effort, an ankle foot simulator (AFS) was designed and is being built that will test the haptic Smart Shoes [3] worn by individuals in the VR simulation. This paper will discuss both the design and simulation of this AFS robot.

Many groups have built AFSs, each more advanced than the previous generation. Most AFSs only focus on the major degrees of freedom (DOF) associated with gait and neglect the others [4] [5]. While major gait events can be observed, fewer DOFs limit the fidelity of the performed simulation. Few AFSs have 6 DOF [6] and many have fallen short in reproducing normal gait at physiologically correct speeds and loading conditions.

METHODS

The major design requirement for the AFS robot is to reproduce natural gait dynamics including: 1) match complex trajectories of the ankle, and 2) generate ground reaction forces during normal walking gait. Key design features include a rigid robot with minimal backlash, appropriately fast dynamics, sufficient load capacity, and sufficient range of motion (ROM) as to replicate normal gait.

The structure of this robot consists of an XYZ Cartesian table, and an ankle mechanism attached to the former (Figure 1).

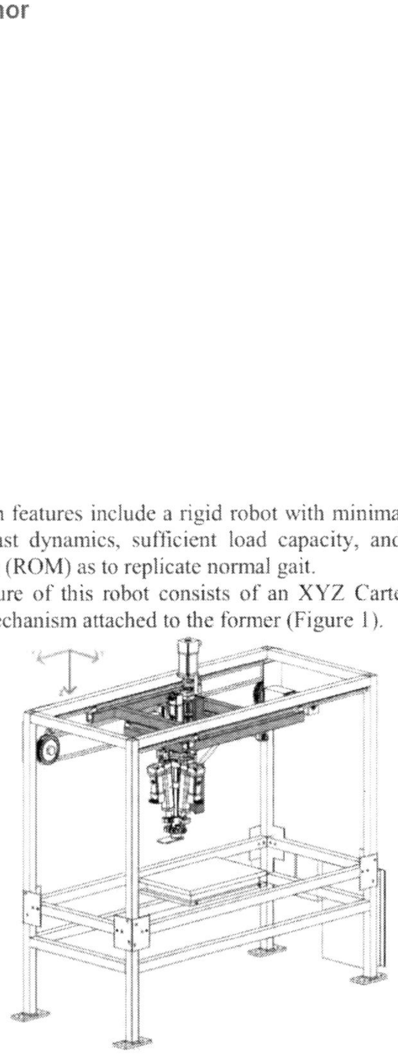

Figure 1: CAD assembly drawing of the AFS showing the layout of the Cartesian table.

The Cartesian table is comprised of a tiered gantry system. First; the X stage is a belt driven carriage which runs the length of the robot and provides motion in the direction of travel. Second a ball screw driven Y stage carriage, attached to the X stage, provides lateral movement. Finally, a ball screw driven Z stage, where the vertical component of the motion is realized. Attached to the Z stage, is the ankle mechanism and is responsible for the orientation of the foot.

A key and novel aspect of this project is the design of the ankle mechanism which duplicates the full ROM of a human ankle during normal gait (Figure 2). This actuated ankle joint fits inside a typical shoe thus supporting smart shoe validation. This is accomplished by using a universal joint that is attached to an actuation plate. Two linear actuators facilitate ankle inversion/eversion and plantar/dorsi flexion. Heim joints couple the actuators to the actuation plate, thus providing high load capability but also allowing high mobility. The DH parameters of the AFS robot are shown in Table 1.

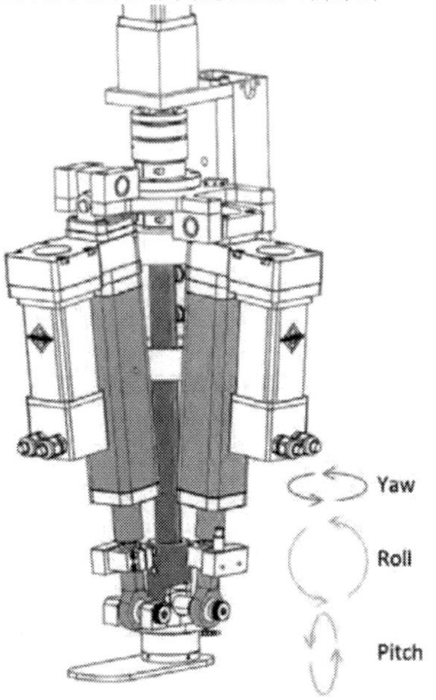

Figure 2: CAD assembly drawing of AFS relating actuators to DOF in the ankle mechanism.

Table 1: DH Parameters of the AFS.

i	a_{i-1}	α_{i-1}	d_i	θ_i
1	0	π/2	*	π/2
2	0	-π/2	*	π/2
3	0	π/2	*	0
4	0	0	0	*
5	0	-π/2	0	*
6	0	π/2	0	*
7	0	π/2	#	0

A multi loop trajectory and force controller are used to control the AFS robot. The inner loop is responsible for following the prescribed trajectory while the outer loop is responsible for modifying the inner trajectory in order to achieve the desired ground reaction forces. The majority of the trajectory is known beforehand minus small changes in trajectory to satisfy the desired contact forces. The control loop utilizes feed forward compensation of the expected dynamics to predict required torques for the motors.

RESULTS

The ankle mechanism has the ROM needed to reproduce normal gait (Table 2). These ranges have been extended to investigate gait abnormalities and events that may lead to injuries, but are not the focus of the current project.

Table 2: ROM comparison between AFS ankle mechanism and normal gait.

	Normal Gait	AFS
Pitch	-28° to 56°	-28° to 56°
Roll	-2° to 5°	-28° to 28°
Yaw	-5° to 7°	-6° to 99°

The figure below shows the predicted position tracking error of the ankle's center of rotation using Matlab and Simulink. The trajectory seen in the figure includes velocity ramp up before "heel strike" as well as deceleration after "toe off". "Contact phase" of the gait cycle are between seconds 2 and 3. (Figure 3)

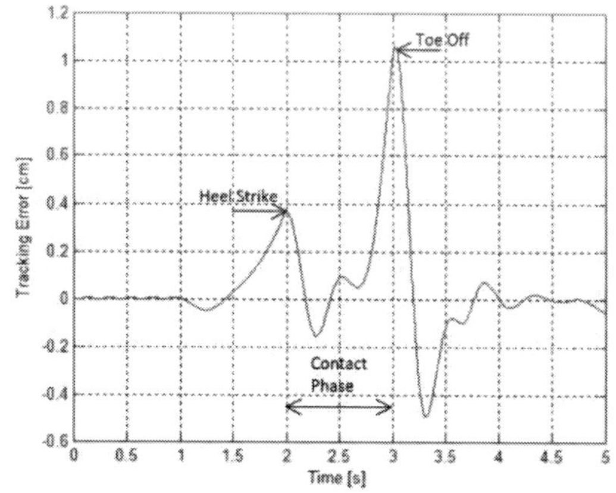

Figure 3: Predicted AFS tracking error in the direction of travel.

DISCUSSION

ROM data of normal gait was collected using motion capture and compared to literature [7]. Results differ slightly from the literature as the project is only interested in the ROM during the contact phase of the gait cycle.

Design decisions were based on results using Matlab, Simulink and CAD FEA. Simulations included motor gearbox combinations for motor speed, torque, load and thermal response given a trajectory, force profile, and estimated robot inertia parameters. The trajectory tracking error of 1 cm at toe off is of concern, but will be reduced through additional control tuning.

As a result of the promising CAD and mathematical simulations, the research group is proceeding with constructing the AFS and is expecting to have a completed assembly in February 2015.

ACKNOWLEDGEMENTS

National Science Foundation (NSF # 1162131), MaryEllen Hunt, Yue Wang, Ken Openshaw, Kyle Crandall

REFERENCES

[1] (www.pdf.org), 1/15/2015
[2] (http://pdcenter.neurology.ucsf.edu/), 1/15/2015
[3] Y. Wang and M. A. Minor, (IROS 2014), IEEE/RSJ International Conference on, 2014, pp. 1236-1241.
[4] Hung-Jen Lai, et al. Medial Engineering & Physics 32:623-629, 2010.
[5] T. Natsakis, et al. Journal of Biomechanics 48:392-395, 2015.
[6] Patrick M. Aubin, et al. IEEE Transaction on Biomedical Engineering Vol. 55, NO. 3 March 2008.
[7] MacDermid, et al. J Hand Therapy 12:187-192, 1999.

SB³C2015
Summer Biomechanics, Bioengineering and Biotransport Conference
June 17-20, 2015,Snowbird Resort, Utah, USA
SB³C2015-1172

RACIAL DIFFERENCES IN THE LOAD-DEPENDENT AREA OF THE LAMINA CRIBROSA

Stephen J. Howerton (1), Forest L. Danford (1), Avinash Ayyalasomayajula (1), Jonathan P. Vande Geest (1,2,3)

(1) Department of Aerospace and Mechanical Engineering
The University of Arizona
Tucson, Arizona 85721

(3) BIO5 Institute
The University of Arizona
Tucson, Arizona 85721

(2) Department of Biomedical Engineering
The University of Arizona
Tucson, Arizona 85718

INTRODUCTION

The current biomechanical theory of glaucoma is founded upon a wealth of evidence showing that retinal ganglion cell axons are damaged as they pass through the region of the lamina cribrosa (LC) [1]. With this known, quantifying the biomechanics of the optic nerve head (ONH) may help elucidate how glaucoma is initiated and progresses. To capture a wide understanding of the mechanical environment for the ONH, it is vital that the investigation takes into account what geometries are being analyzed. This study aims to look at how the area of the LC changes as a function of IOP and how this ultimately affects the biomechanics of the ONH. With African Americans and those of Hispanic ethnicity showing higher risk for glaucoma, this study also aims to compare racial differences in these outcomes. We also used the area of the LC in preliminary finite element simulations to start to understand how these changes lead to alterations in the mechanical stress and strain within the LC.

METHODS

Eleven pairs of human donor poles were acquired within 48 hours post-mortem from four different eye banks; Alabama Eye Bank, Banner Health, Donor Network of Arizona, and San Diego Eye Bank. Each eye was dissected in half and cleaned of excess tissue. With the aid of a dissecting microscope, a gentle microdissection procedure was performed to remove the choroid, retina, and prelaminar tissue from the ONH. The eyes then underwent a trypsin treatment using 0.25% EDTA trypsin to remove any prelaminar tissue that could not be mechanically removed. The microdissection was then repeated to clean off any remaining prelaminar tissue.

The sample was cut along the borders of its physiological regions to effectively flatten the tissue. The tissue was then placed in a novel micro-optomechnical device consisting of an acrylic base and a metal pressure plate. Once clamped, pressure was regulated using a closed loop pressure system through LabView. Four different IOP levels were investigated: 5, 15, 30, and 45 mmHg. Initially, each eye went through a preconditioning cycle from 5 mmHg to 45 mmHg and a 30 minute wait step was held before each IOP level investigated to mitigate any viscoelastic effects.

Samples were imaged using an Olympus BX51 upright laser-scanning microscope coupled to a Coherent 120-fs tunable pulsed Titanium-Sapphire laser (Santa Clara, CA). The laser was centered at λ=780 nm to efficiently visualize collagen via second harmonic generation (SHG) and elastin via fluorescence emission. Signals were collected simultaneously in an epifluorescence configuration, separated with a dichroic mirror (405 nm), and passed through a 377/50 bandpass filter (SHG) and a 460/80 bandpass filter onto Hamamatsu PMT detectors.

Noise and any artifacts were removed from the image volume, contrast-limited adaptive histogram equalization (CLAHE) was performed, and then maximum intensity projections (MIPs) were created for each image data set at each pressure. The area of the Lamina Cribrosa was collected using a custom script in MATLAB that would display the MIP images in random order (random donor and random pressure level) and allow the user to click 6 points around the border. These points were then used to fit an ellipse (in a least-squares sense) in which the semi-major and semi-minor axis could be extracted to calculate the area. The area of the LC was collected by three separate individuals and averaged for each ethnicity group at each different IOP level.

Preliminary finite element models were created by approximating the sclera as a hemisphere. In these simulations, the LC was assumed to be circular. The model was pressurized from the inside to 20 mmHg keeping the base fixed in all displacement degrees of freedom. The maximum principal strains and stresses (MPa) were calculated at the rim and the center of the LC at an IOP of 20 mmHg. This was done for all ethnic groups. The LC and sclera were modeled as a neoHookean material and meshed using a 3 noded linear shell element. NeoHookean constants used for the LC and sclera were: C_{10} = 399659.9 Pa and $D_1 = 1.5 \times 10^{-7}$ Pa^{-1}; C_{10} = 43524.2 Pa and $D_1 = 4.6 \times 10^{-7}$ Pa^{-1}, respectively.

RESULTS

Anatomical measurements of ocular size (Nasal-Temporal, Superior-Inferior, and Anterior-Posterior) for Caucasians, African Americans, and Hispanics were: 24.87 mm ± 1.29, 24.09 mm ± 0.81, and 25.06 mm ± 1.27; 24.47 mm ± 0.97, 24.43 mm ± 1.33, and 24.02 mm ± 2.45; 23.98 mm ± 0.94, 23.55 mm ± 1.45, and 25.25 mm ± 2.07, respectively.

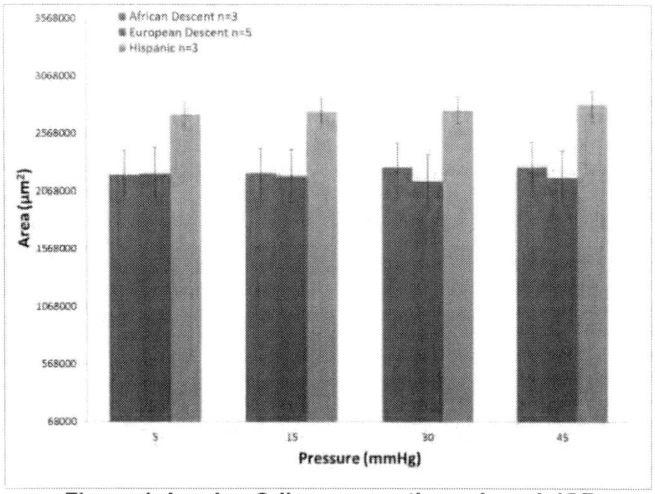

Figure 1: Lamina Cribrosa area through each IOP investigated in μm^2.

As seen in Figure 1, African Americans and Hispanics both experienced a moderate increase in area as pressure increased as compared to Caucasian donors, who experienced a decrease in area.

Figure 2: Distribution of the Max in-Plane Principal strains.

Hispanics overall also had a larger LC area. At the reference pressure (5 mmHg) the area for Caucasians, African Americans, and Hispanics was 2.22 mm^2 ± 0.13 (mean ± SD), 2.21 mm^2 ± 0.11, and 2.73 mm^2 ± 0.09, respectively. At the final pressure (45 mmHg) the area for Caucasians, African Americans, and Hispanics was 2.19 mm^2 ± 0.13 (mean ± SD), 2.28 mm^2 ± 0.15, and 2.82 mm^2 ± 0.05, respectively. Percent area increase for Caucasians, African Americans, and Hispanics from 5 mmHg to 45 mmHg was -1.36%, 3.26%, and 3.42%, respectively. Figure 2 illustrates visually the outcome of the FE simulation at 20 mmHg. The highest values of stress and strain for all races occurred at the LC rim. Figure 3 indicates that there is a substantial increase in LC rim strain and stress for Hispanics as compared to Caucasians and African Americans.

DISCUSSION

In conclusion, there seems to be differences in the area of the LC across races. Our data suggests a larger LC area for Hispanics as compared to any other race. Of course, our sample size is small and future work will involve increasing our sample size to test this hypothesis. African Americans and Hispanics, who are higher risk of glaucoma, have a positive percent increase in area as pressure increase while Caucasians had a small negative percent change in area. Results from the FE simulations also seem to corroborate that racial differences in LC area change the mechanical environment of the ONH. Future work will also involve more robust geometric and material property assumptions in finite element simulations of the posterior sclera and LC.

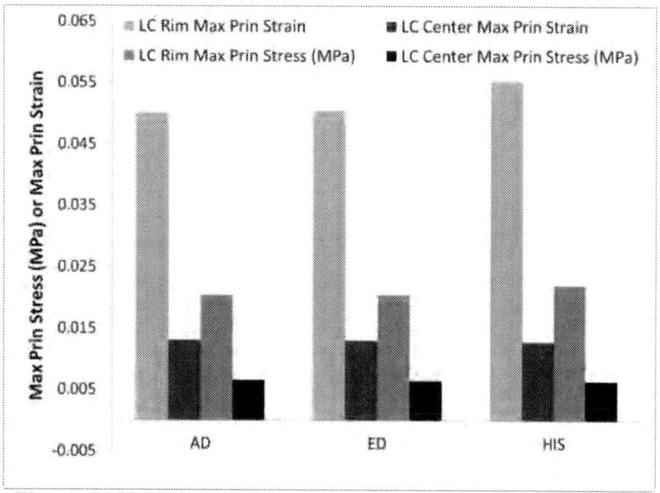

Figure 3: Maximum principal stress and strains calculated at both the LC rim and center for each ethnic group.

ACKNOWLEDGEMENTS

This work was supported by the American Health Assistance Foundation National Glaucoma Research Grant AHAF G2009035 (JPVG), the NIH research grant 1RO1EY020890-02A1 (JPVG), Research to Prevent Blindness (CAG), and partially by Science Foundation Arizona.

REFERENCES

1. Danford FL, Yan D, Dreier RA, Cahir TM, Girkin CA, Vande Geest JP. Differences in the region- and depth-dependent microstructural organization in normal versus glaucomatous human posterior sclerae. Invest Ophthalmol Vis Sci. 2013;54:7922–7932.

SB³C2015
Summer Biomechanics, Bioengineering and Biotransport Conference
June 17-20, 2015,Snowbird Resort, Utah, USA
SB³C2015-1173

SOFT 3-DIMESIONAL NEOTISSUE MICROARRAYS AS HIGH-THROUGHPUT PLATFORMS FOR INTERROGATING STEM CELL INSTRUCTIONAL MICROENVIRONMENTS

Michael L. Floren (1), Wei Tan (1)

(1) Department of Mechanical Engineering
University of Colorado Boulder
Boulder, CO, USA

INTRODUCTION

There is abundant evidence suggesting that local signals from tissue-specific extracellular matrix microenvironments significantly affect cellular differentiation, phenotypic expression and maintenance [1]. Substrate biophysical signals, such as soluble factors, cell-ligand interactions, matrix elasticity [2] and geometry [3] play a critical role in a diversity of biological events including cell adhesion, growth, differentiation, and apoptosis. Despite accumulated knowledge regarding individual and combined roles of various mechanochemical ECM signals in stem cell activities, the intricacy exhibited by cellular microenvironments poses a considerable challenge in resolving the mechanisms ascribed to stem cell behavior and fate processes.

High through-put approaches have emerged in recent years to circumvent the limitations of traditional low through-put techniques (i.e. conventional cultureware), with the promise to develop complex platforms for combined biomolecule/substrate discovery [4]. However, despite the versatility afforded by current microarray technologies, the incorporation of multiple signals within engineered microarrays remain limited.

To resolve the mechanisms associated with complex matrix signals and stem cell behavior and fate decisions, we established a high-throughput ECM platform based on soft, nanofibrous matrices meanwhile highly-tunable in elasticity and 3-dimensional geometry. The technology we demonstrate here is amenable to manipulation of several matrix properties, such as elasticity and geometry, in concert with customizable ECM protein micro-dot combination. Altogether, we demonstrate the practical adaptation of high-throughput technology to facilitate the screening of various tunable mechano-ligand microenvironments in three dimensions with the aim to optimize stem cell fate decisions.

METHODS

ECM protein microarrays fabrication: Polyethylene glycol dimethacrylate (PEGDM) with a molecular weight of 750 kDa photopolymerizable soft matrices were fabricated by electrospinning on a custom set up composed of a high voltage power supply, grounded collecting surface, motorized syringe pump, and a 14 mm syringe. Substrates were subsequently stabilized under UV exposure (352 nm light) with an average intensity of 5 mW/cm2 for specific time durations. A printing buffer consisting of 1% glycerol and 0.2% Triton X-100 was utilized for all protein depositions. To prepare ECM arrays, stock solutions of various ECM proteins were suspended at a concentration of 250μg/ml in printing buffer. Samples were deposited on the PEGDM substrate matrix using an Aushon 2470 arrayer with 185 micron pins (Aushon BioSystems, Billerica, MA), to achieve a nominal diameter of 250 microns. Individual spots with 7 replicates (8 total) of each protein combination were deposited with a 500μm pitch distance onto the PEGDM substrates. *Cell Culture and Biological Analysis:* Rat mesenchymal stem cells (rMSCs) with passages 2-5 were cultured in Dulbeccos Modified Eagles Media, with 10% defined FBS for MSCs and 1% Penn/Strep. rMSC cell suspensions were dispensed onto the ECM microarrays within the gasket region at a cell density of 10e5 cells per an array and incubated for 4 hours. The arrays were then gently aspirated into a large chamber filled with prewarmed media. Cell morphology was obtained by staining for cell nuclei (DAPI) and cell cytoskeleton (phalloidin) after sample fixation in 4% formaldehyde. Fluorescently labeled cells were evaluated using an epifluorescence microscope (Zeiss, Peabody, MA). Cell number and morphology parameters were evaluated using ImageJ software. *Statistical Analysis:* Image intensity measurements were obtained using ImageJ software and reported as average signal and standard error for the condition (n=3). Statistical relevance per condition investigated was performed using a 2^6 full factorial ANOVA design using Minitab software (Minitab, State College, PA).

RESULTS

Mesenchymal Stem Cell adhesion within 3D ECM microarrays: We seeded MSCs onto our ECM microarrays to verify the selective attachment of cells to distinct protein dots (Figure 1A-D). The MSCs attached preferentially to the microarrayed regions with little to no attachment observed on the pure PEGDM nanofibrous substrate (Figure 1A,B). Furthermore, upon closer inspection, several of the distinct cellular islands represented three dimensional neo-tissue microdomains of thickness approximately 50 μm (Figure 1D). Staining for cell nuclei identified distinct cell populations associated with the ECM protein depositions (Figure 1C). The attachment of rMSCs to the ECM microarrays demonstrates the ability of stem cells to selectively adhere to the microarrayed regions without dot to dot communication.

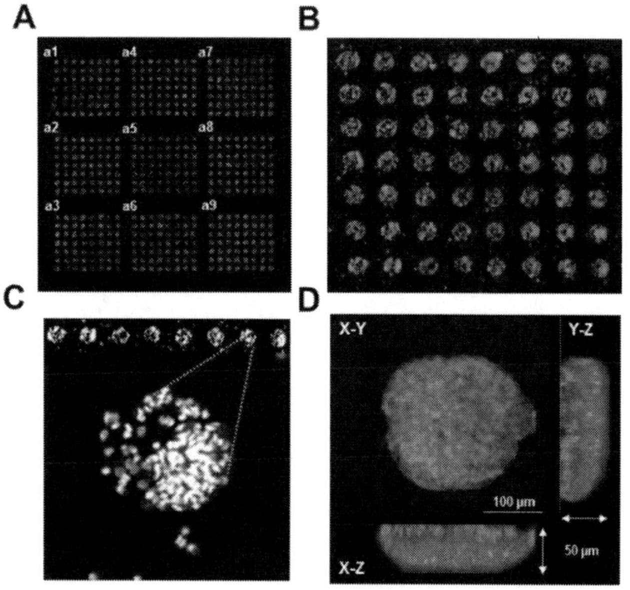

Figure 1. Representative images of MSCs cultured for 24 hours on neotissue protein arrays. (A) Global array withh MSCs. (B) Magnification of subarray a4 exhibiting discrete cellular islands. (C) Cellular nuclei are stained for quantification and data analysis. (D) Three-dimensional rendering of a MSC cellular dot.

cellular adhesion, we evaluated the 24hr. culture of MSCs on ECM microarrays with two stiffness (7 and 31 kPa). Figure 2(A,B,D) depicts the adhesion of MSCs on both stiff and soft substrates. Cellular attachment and spreading was confirmed for both elasticities, with improved spreading detected for the softest condition (Figure 2C,E).

DISCUSSION

We developed a high throughput method that allows for the rapid screening of a diversity of engineered microenvironments with tunable matrix elasticity and geometry, combined with specific ECM protein combination and/or concentration. Our preliminary results indicate distinct correlations between substrate stiffness and MSC adhesion and spreading as it relates to ECM protein presentation. Highlighting the

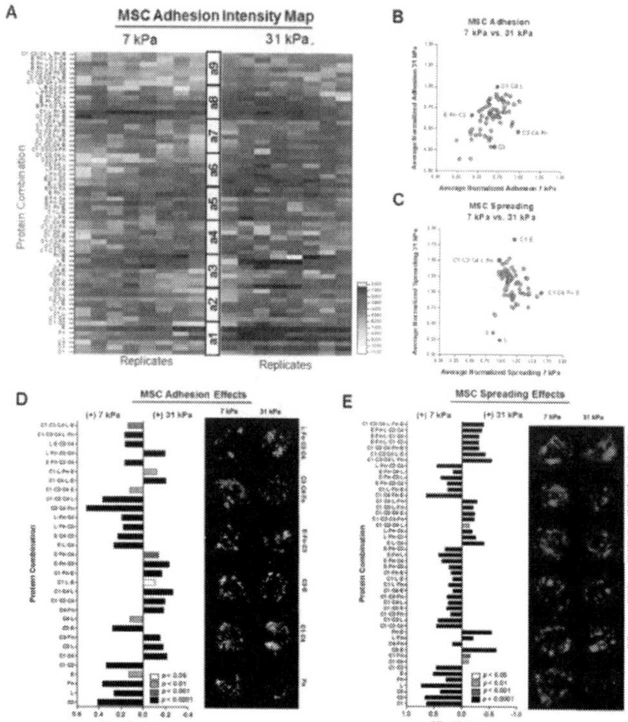

Figure 1. Cellular structures, nuclei and f-actin, are stained, imaged and quantified using software to produce an average intensity map for different elastic environments (A). Average cell dot intensities are further utilized for quantitative analysis (B,C). Statistical analysis discriminates conditions relevant for attachment and spreading of MSCs on different substrates (D,E).

importance and necessity of employing a systemic approach, this work incorporates several environmental signals necessary to establish optimal MSC microenvironments. Capturing complex, multifaceted 3-dimensional environments in high-throughput with combinatorial signaling will likely prove instrumental towards the design of future tissue regeneration biomaterial platforms. Future studies will be focused on adapting this technology to instruct specific MSC differentiation processes by expanding the library of ECM proteins and experimental parameters employed.

ACKNOWLEDGEMENTS

The authors would like to thank Mr. Markham and Linda Crnic Institute for Down Syndrome at University of Colorado at Denver Anschutz for use of the microarraying equipment. This work is finically supported through NIH funding sources K25HL097246 (W.T.) and R01 HL119371 (W.T.).

REFERENCES

1. Reilly GC, Engler AJ. Biomechanics 43(2010), 55–62
2. Engler JA, Sen S, Sweeney HL, Discher DE. Cell (2006) 126 , 677–689
3. Wingate K, Bonani W, Tan Y, Bryant S, Tan W. 2012. Acta Biomater 8, 1440.
4. Drew M. Titmarsh, Huaying Chen, Ernst J. Wolvetang, and Justin J. Cooper-White. Biotechnol. J. 2013, 8, 167–179.

SB³C2015
Summer Biomechanics, Bioengineering and Blotransport Conference
June 17-20, 2015,Snowbird Resort, Utah, USA
SB³C2015-1174

WEAR SIMULATOR FOR CANINE TOTAL HIP REPLACEMENTS

Kevin J. Warburton (1), Evan Rust (1), Jeff Brourman, DVM, MS (2), Trevor J. Lujan, Ph.D. (1)

(1) Department of Mechanical and Biomedical Engineering
Boise State University
Boise, ID, USA

(2) Veterinary Surgeon
WestVet Animal Emergency & Specialty Center
Garden City, ID, USA

INTRODUCTION

Hip simulators are common devices for measuring tribological wear characteristics of total hip arthroplasty. These machines have proven to be highly effective at predicting long term wear characteristics of human hip implants in a few weeks of testing [1]. However, a device specialized for wear testing of canine hip implants has not been developed. For this reason, there is a lack of information on how wear is affected by 4-legged walking and smaller implant bearing surfaces. Therefore, a need exists to develop a canine hip simulator that applies clinical canine kinematics to determine the wear characteristics of this increasingly popular implant. This information would enable veterinarian surgeons to differentiate between marketed devices and would assist engineers in designing more effective implants. For example, hip resurfacing could be a viable technology to treat canine hip arthritis, but this new technology would need to be validated to reduce risk of unsuitable wear rates.

The objective of this study is to develop a canine hip wear simulator. The challenge is to design a system that meets international standards for wear testing of hip arthroplasty and is validated to reproduce clinical kinematics of the canine hip joint. Validation testing must compare measured kinematics to clinical data and determine the likelihood of producing proper wear patterns on the bearing surface.

METHODS

Overview: An orbital bearing machine (OBM) was developed to simulate hip wear according to ISO 14242-3. It was modified from the international standard to account for healthy, adult canine kinematics. A three-dimensional scanner was used to confirm that in-vivo kinematics was replicated for flexion-extension (FE), abduction-adduction (AA), and internal-external rotation (IER).

Simulator Design: The OBM produces three axes of motion between the bearing surfaces (Figure 1). The FE and AA rotations are obtained by a stepper motor rotating a slanted base component, and the IER is obtained by the placement of the rotation-prevention lever.

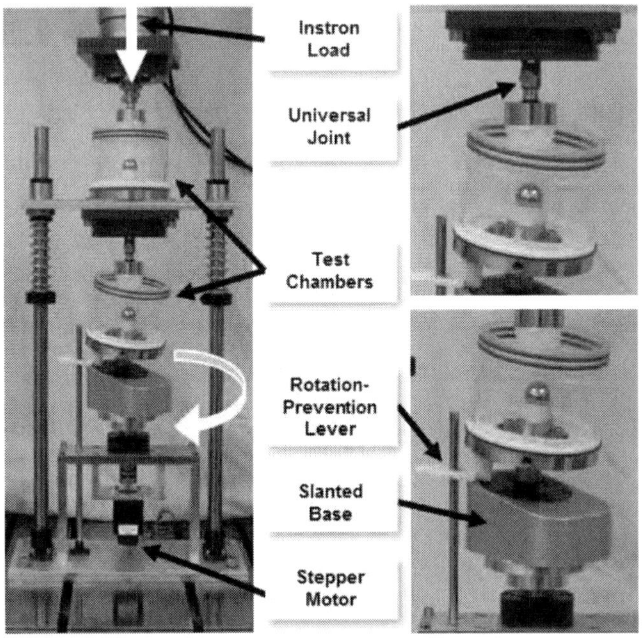

FIGURE 1: CANINE HIP SIMULATOR COMPONENTS

Similar to other designs, a universal joint is used as a self-centering device to adjust for small misalignments between the bearing surfaces

[1]. A top test chamber is used as a control when calculating the wear rates of the tested implants. It experiences the same load during testing, but it is not articulated under that load like the test implant below. Both chambers each allow 400 mL of lubricant to surround the implants, and O-rings above and below each camber allow for a bag to seal them and prevent evaporation. A single access hole allows an RTD sensor and immersion heater to enter the chamber and keep the lubricant at canine body temperature. An embedded control system drives the stepper motor at 2 Hz. This is coordinated via digital signals with an Instron E10000 mechanical test system, which applies the load. Combining the motor and Instron enable the synchronized application of kinematic motions and load parameters respectively.

Kinematic Validation: To validate that the canine kinematics are simulated properly, a novel method was conducted. The Euler angles of rotation were calculated from measured values, and then the slide tracks were calculated based on the found Euler angles. Two coordinate systems were used to measure the Euler angles [2, 3]. Using a static reference coordinate system fixed relative to the cup and a rotating coordinate system fixed relative to the head, the simulator's three dimensional position was measured using a High Definition Imaging 3-D Scanner (LMI Technologies, Delta, Canada). The setup for a wear test was replicated except the acetabular cup and femoral head were replaced by the 3-D printed triads (Figure 2). Since it is an OBM, the Euler sequence for the simulator was FE→AA→IER. A rotation matrix for this sequence is known [2]. The coordinate transformation matrix between the reference and rotating coordinate systems was derived from the 3-D scanner's point cloud data of the triads. Using the rotation matrix and the transpose of the coordinate transformation matrix, the Euler angles required for the measured rotation were derived with a custom MATLAB program.

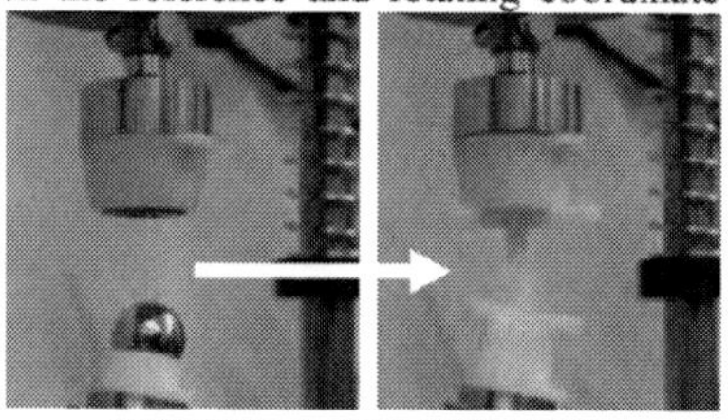

FIGURE 2: TRIADS FOR MEASURING EULER ANGLES

During measurement with the scanner, the OBM was manually positioned to four locations where local maximum, minimum, and inflection FE, AA, and IER rotations should exist. Since the physical restraints of the OBM force all three waveforms to have a sinusoidal curve, the average measured amplitude was used to fit a sinusoidal curve to the data points. No curve fitting algorithm, such as linear regression, was used because the phase change between waveforms for OBM simulators is known [2]. Therefore, only the amplitude is required to characterize the curve. The final sinusoidal curve was compared to clinical adult canine data [4]. This data was digitized by hand in MATLAB, and the motion was normalized to a mean position using an accepted method [5]. Also, slide tracks were calculated on a unit sphere using clinical rotations and measured simulator rotations. The previously mentioned known rotation matrix was used for calculating slide tracks with the same established method as previous simulators [2, 3]. The example slide tracks represent a point fixed to the head that is articulating against the cup about the center of rotation.

RESULTS

The kinematics simulated by the OBM exhibited a strong correlation to clinical FE and IER waveforms, but exhibited a no correlation to the clinical AA waveform (Figure 3). Slide tracks based on fitted data were qualitatively close to clinical sidetracks (Figure 4). Scanned data at 0% and 100% gait cycle were considered the same point.

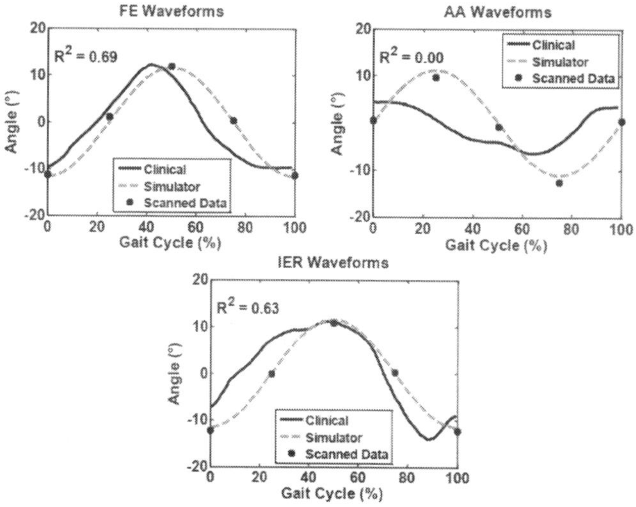

FIGURE 3: CLINICAL AND SIMULATED KINEMATICS

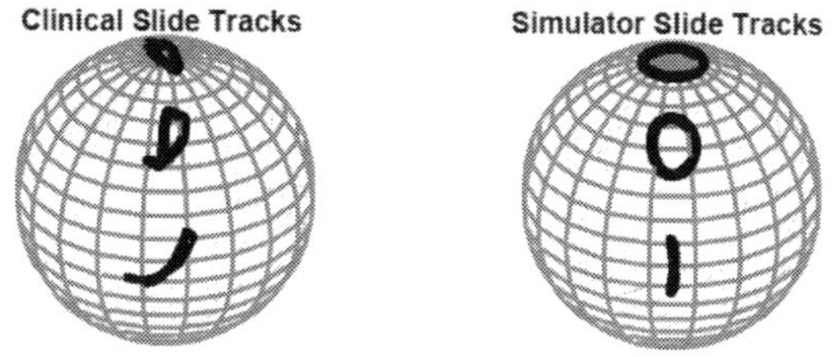

FIGURE 4: CLINICAL AND SIMULATED SLIDE TRACKS

DISCUSSION

A hip wear simulator for canines was successfully developed accordingly to international standards for human hip arthroplasty. To our knowledge, this is the first validated wear simulator for canines.

The validation protocol provided a quantitative and qualitative evaluation of the hip simulator performance. The simulated FE and IER waveforms correlated well with clinical data for a full gait cycle. Conversely, the simulated AA waveform had a poor correlation with clinical data. This is an anticipated result for OBM simulators because it only uses one motor to control three axes of rotation [2]. Moreover, the slide track analysis gave qualitative evidence that differences in AA kinematics would not grossly alter wear on the bearing surface.

A design change to be considered for future work is to anatomically position the acetabular cup and head for an adult canine. Currently, the correct wear rate is produced, but wear does not occur in the correct location. If the implant is mounted anatomically, the location of the implant's wear would also be accurately reproduced.

The kinematic validation has given confidence in using this simulator to test the wear performance of canine hip implants. This device will be used help develop canine hip implants that use novel designs and materials to reduce cost and preserve bone.

ACKNOWLEDGEMENTS

Funded by Idaho Global Entrepreneurial Mission with Infrastructure Grant. Kind support was received from Securos, and MISUMI USA.

REFERENCES

[1] Saikko, V, *P I Mech Eng H*, 219:437:448, 2005. [2] Saikko, V et al., *J Biomech*, 35:455-464, 2002. [3] Calonius, V et al., *J Biomech*, 35:1439-1450, 2002. [4] Yang-Chieh, F et al., *Am J Vet Res*, 71:1118-1122, 2010. [5] Bockstahler, B et al. *Vet Surgery*, 38:754-761, 2009

SB³C2015-1175

NON-INVASIVE METHOD FOR INVESTIGATING THE RELATIONSHIP BETWEEN MUSCLE STRAIN AND POST-EXERCISE BLOOD PERFUSION USING MRI

Amanda K. W. Buck (1,2), Christopher P. Elder (1), Bruce M. Damon (1-3)

(1) Vanderbilt University Institute of Imaging Science
Department of Radiology and Radiological Sciences
Vanderbilt University
Nashville, TN, USA

(2) Department of Biomedical Engineering
Vanderbilt University
Nashville, TN, USA

(3) Department of Molecular Physiology and Biophysics
Vanderbilt University
Nashville, TN, USA

INTRODUCTION

Muscle functionality and tissue perfusion are intrinsically related in healthy and diseased muscle states. We are interested in detecting spatial relationships between muscle strain during isometric contraction and post-exercise tissue perfusion response. A non-invasive measurement would be helpful for assessing the mechanism behind exercise-induced blood flow response. Once the relationship is evaluated in healthy subjects, this framework may provide a non-invasive, physiologically relevant test for characterizing disease state and for assessing the effectiveness of therapies (e.g. revascularization of peripheral artery disease). Functional imaging using magnetic resonance imaging (MRI) provides a platform for non-invasive assessment of both characteristics.

METHODS

The protocol was approved by the Vanderbilt University Institutional Review Board, and written, informed consent was provided by subjects. All MR data were collected using a 3T Achieva-Intera scanner (Philips Healthcare, Cleveland, OH) and an eight-channel, sensitivity encoding knee coil. The subject was positioned supine on the imaging table, with a foot secured in an isometric exercise device. The leg was placed at the same level as the heart to prevent hydrostatic effects on blood flow, and the middle of the tibialis anterior (TA) muscle was positioned at the center of the coil.

During the functional scans, the subjects performed 8 s dorsiflexion maximum voluntary contractions (MVCs). Force production was recorded using a lode cell connected to the force plate of the device and to monitoring equipment.

Anatomical images were acquired at the location of maximum cross sectional of the tibialis anterior (TA) muscle for use in planning the functional scans. The subsequent images were aligned such that tagging and perfusion data could be acquired at aligned voxel locations, thus allowing direct voxel-wise investigations of strain/perfusion relationships.

Tagged cine MRI acquisition. Linear tagging was used to measure muscle strain over the course of a contraction. At the same location in the muscle, two sets of axial images were acquired: one with tags oriented coronally, and a second image with tags oriented sagittally (Figure 1) [TR/TE=275 ms/14.9 ms; FOV=240×240 mm^2; reconstructed voxel size=1.4×1.4 mm^2; slice thickness=7 mm; tag spacing=7 mm; 4 sets of tags were applied and images acquired over 10 time points for each set]. In addition, a contiguous series of five sagittal slices were acquired with tags applied in the axial direction (Figure 1) [identical parameters]. The sagittal slices were acquired such that voxels aligned with a row of five voxels in the axial image, allowing the measurement of strain in three orthogonal directions (x,y,z) throughout the muscle. In all acquisitions, tags were applied before the subject performed an 8 s MVC, and data were acquired during contraction.

Arterial spin labeling MRI acquisition. Arterial spin labeling (ASL) characterizes perfusion over time. Data were acquired such that the center of the FOV aligned with that of the axial slices for the motion-encoded MR data [TR/TE=4 s/15 ms; FOV=240×240 mm^2; reconstructed voxel size=2.5×2.5 mm^2; slice thickness=7 mm; post-labeling delay=900 ms; 48 dynamics]. The temporal resolution was 8 s for each control-labeled image pair. Baseline perfusion data were acquired for one minute, then the subject performed an 8 s MVC, and data acquisition continued for 4 minutes to assess the post-contraction blood flow response. To estimate the equilibrium magnetization (M_0), an additional acquisition was made with receiver gain and other instrumental settings identical to those used in the ASL scans with an extended repetition time of 20 s.

General analysis. Except where otherwise indicated, data were analyzed using MATLAB (The Mathworks, Inc., Natick MA). A region of interest (ROI) encompassing the TA (Figure 2A) was defined in ASL data using itk-snap (itk-snap.org) [1]. This ROI was used to restrict the analysis to the TA muscle. Care was taken to mitigate potential partial voluming effects by avoiding regions near the muscle borders.

Tagged cine MRI data analysis. Muscle strain was calculated in a voxel-wise manner via harmonic phase (HARP) analysis [2], using the definition of strain in a single direction,

$$\varepsilon = \frac{\ell_0 - \ell}{\ell_0} \times 100\% \tag{1}$$

where, ε represents strain in a given direction perpendicular to the applied tag direction, ℓ_0 is the applied inter-tag distance while the muscle is relaxed, and ℓ is the inter-tag distance during the MVC. All three orthogonal components of strain were measured from the MRI data, and the total strain magnitude was determined

$$\varepsilon_{Total} = \sqrt{\varepsilon_x^2 + \varepsilon_y^2 + \varepsilon_z^2} \tag{2}$$

where, ε_{Total} is the total strain magnitude, and ε_i for i=x,y,z, is the strain in each of the acquired directions.

Arterial spin labeling data analysis. For each control/labeled image pair, the mean signal intensity was determined for each voxel. The mean perfusion-weighted signal intensity difference, S_P, was determined using

$$S_P = \frac{S_C - S_L}{M_0} \tag{3}$$

Where S_C and S_L are the control and labeled images' signal intensities, respectively, and M_0 is the mean equilibrium magnetization for the location determined from the reference scan. S_P was computed for each control/labeled pair, such that 386 s of data are acquired. In addition, the ROI-averaged perfusion time course was determined.

RESULTS

The percent strain magnitude is greatest in the medical section of the deep compartment of the TA and lowest in the superficial component (Figure 2B). However, the three components of strain show different patterns across the muscle (Figure 2C).

The ROI-averaged perfusion response demonstrates peak perfusion immediately post-contraction (time=80 s), followed by a return to baseline (Figure 3A). Spatial variability in perfusion is shown in the TA ROI at peak flow (t=80 s) and after recovery (t=208 s) (Figure 3B).

DISCUSSION

Strain magnitude shows distinct spatial patterns, with the x-direction strain being the largest contributor to total strain. The ASL data, however, demonstrate no spatial patterns of perfusion. In this preliminary study, the z-direction strain is unexpectedly low; this may be related to the curvature of the leg, the proximity of the TA mask to other structures (e.g. tibia, fat, and tendons), the slice thickness (7 mm), and resulting partial-voluming effects in the sagittal slices. In addition, implementation of registration methods is expected to improve the approach.

In this study we have developed a non-invasive framework for investigating the relationship between muscle strain and perfusion. Future studies will be directed to infer the force and spatial relationships between strain and perfusion.

ACKNOWLEDGEMENTS

NIH/NIAMS R01 AR050101

REFERENCES

[1] Paul, A, et al., *Neuroimage*, 31(3):1116-1128, 2006.
[2] Osman, NF, et al., *IEEE Trans Med Imaging*, 19(3): 186–202, 2000.

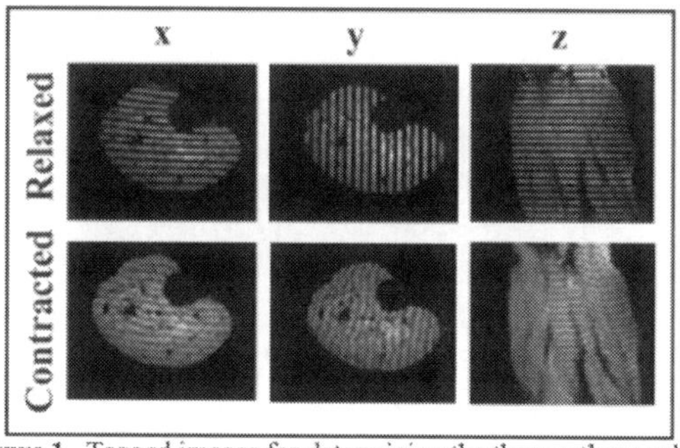

Figure 1. Tagged images for determining the three orthogonal directions of strain at both relaxed and contracted states.

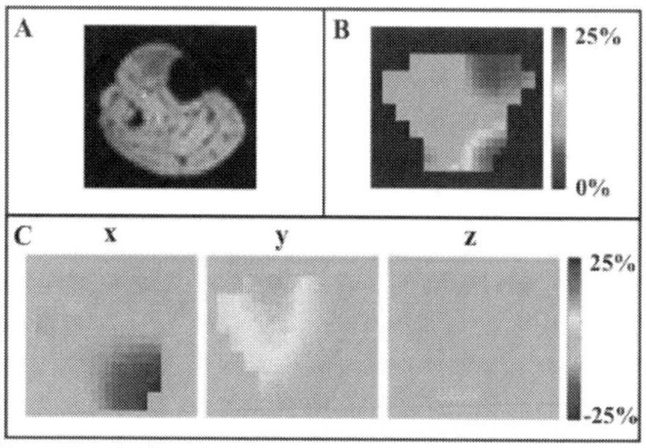

Figure 2. A) The TA mask (red) is overlaid on an M_0 image. B) Total strain magnitude map in the TA map. C) Maps of each of the orthogonal components of strain.

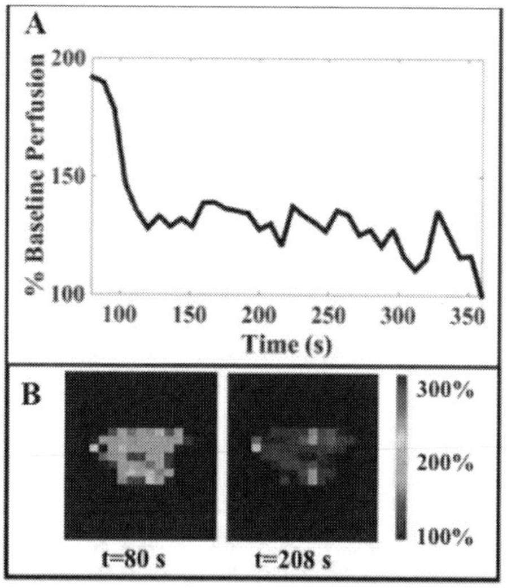

Figure 3. A) Post-contraction perfusion response over time, and B) maps of perfusion in the TA immediately after contraction (left) and mid-recovery (right). All perfusion is normalized to the mean baseline perfusion before contraction and is presented as percent baseline perfusion.

TISSUE HYPOXIA AND MURRAY'S LAW OF MINIMUM WORK CONTROL NEOVASCULAR GROWTH AND REMODELING

SB³C2015-1176

Sean Moore (1), David J. Hoelzle (1), Joel D. Boerckel (1

(1) Aerospace and Mechanical Engineering
University of Notre Dame
Notre Dame, IN, USA

INTRODUCTION

Postnatal neovascular growth occurs through two primary mechanisms: angiogenesis (vessel sprouting) and arteriogenesis (vessel expansion and maturation). While growth and regression decisions must be made locally, at the level of single vessels/cells, the neovascular response to injury must be coordinate at the system level, requiring integration and feedback of both local and systemic signals. We hypothesized that hindlimb ischemia recovery can be described as an optimal control problem governed by tissue hypoxia and luminal shear stress.

Murray's Law

In 1926, Cecil Murray proposed his "Physiological Principle of Minimum Work."[1] He made two seemingly naïve assumptions: (1) blood flow in the vascular tree follows Poisseiulle's Law, and (2) the metabolic energy expenditure in the vascular system is proportional to blood volume. Murray then reasoned that an in optimal vascular network, the total work expended to overcome viscous losses and metabolic energy expenditure would be minimized. From this, he derived a relationship for the optimal geometry for blood flow in a branched system:

$$r_0^3 = r_1^3 + r_2^3 \qquad \textit{Eqn. 1,}$$

where r_0 is the radius of a given parent vessel and r_1 and r_2 are the radii of the daughter vessels downstream of a bifurcation. Despite these simple assumptions, and the apparently simplistic result, nearly all mammalian vasculature follows this law with remarkable accuracy[2]. Interestingly, because of the cubic relationship found here and in Poisseuille's Law, this suggests that in an optimal geometry, the shear stress throughout the vascular tree remains constant. In this study, we evaluated whether a combination of tissue hypoxia and Murray's Law govern the feedback control of neovascular growth and remodeling during recovery from hindlimb ischemia.

METHODS

The present analysis began from a fundamental study on the effects of deletion of the mitogen-activated protein kinase phosphatase, MKP-1, on neovascular recovery from hindlimb ischemia[2]. Here, we use unpublished data and collated data from Refs. 2 and 3 to evaluate recovery from hindlimb ischemia as a coupled local-systemic feedback control system governed by tissue hypoxia and shear stress. We evaluated wildtype mice and two different "disease" states: MKP-1 deficient and diabetic mice.

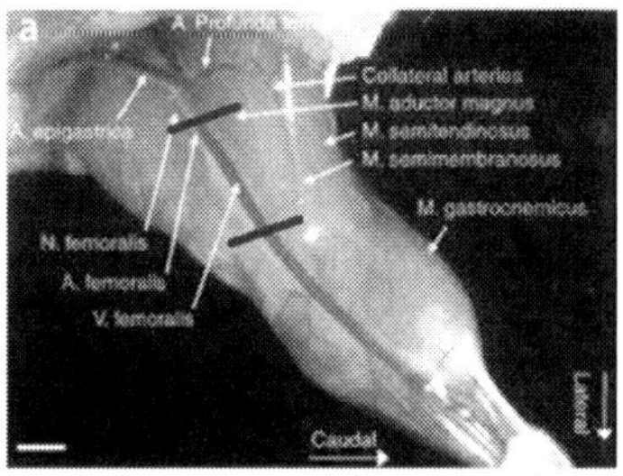

Figure 1. Hindlimb ischemia model. Image from Ref. 4. The femoral artery is ligated distal to inguinal ligament and proximal to popliteal bifurcation, and the intervening segment is removed, resulting in hypoxia in the distal limb.

INITIAL OBSERVATION

Studying the role of MKP-1 deletion on neovascular recovery, we found that WT mice exhibited an overshoot in vascular volume above contralateral controls, reminiscent of an underdamped feedback control system (Figure 2). However, MKP-1 KO mice failed to exhibit this overshoot.

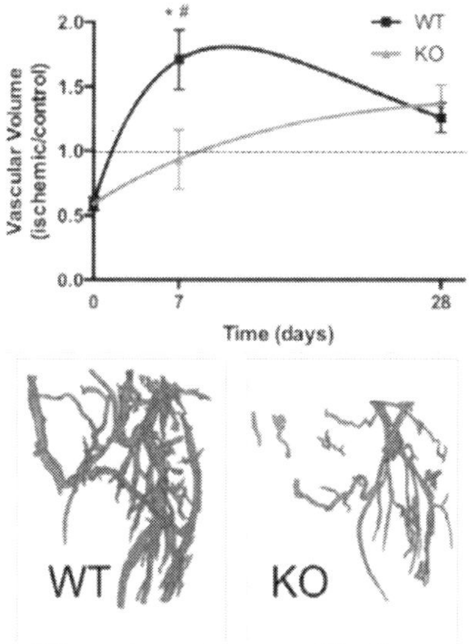

Figure 2. MicroCT angiography quantification of vascular structures in WT and MKP-1 KO mice.

RESULTS

We used laser Doppler perfusion imaging to evaluate the time course of recovery of blood perfusion through the limb as a surrogate for tissue hypoxia (Figure 3). MKP-1 KO mice recovered more slowly that WT mice, but recovered to the same degree by day 28.

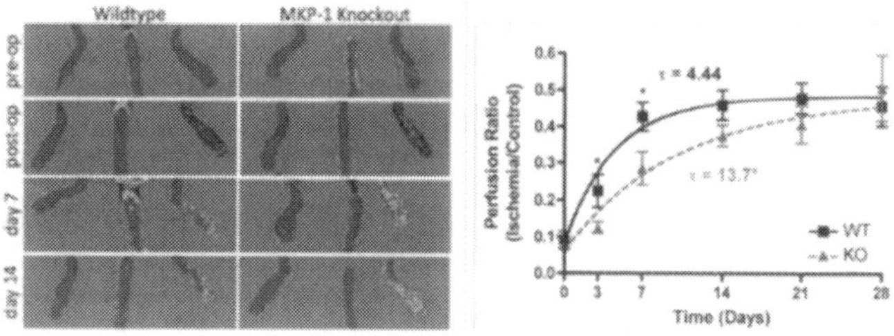

Figure 3. Laser Doppler perfusion imaging of MKP-1 WT and KO mice over 28 days.

To assess the effects of ischemia on local growth factor concentrations, we evaluated tissue VEGF expression by immunofluorescence (Figure 4) as well as MKP-1 and fractalkine expression and inflammatory cell infiltration (not shown).

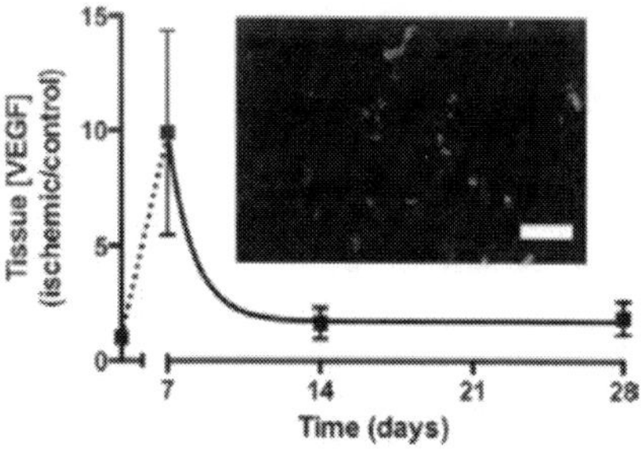

Figure 4. Quantification of relative VEGF expression by immunofluorescence.

To evaluate whether the new blood vessels recover and remodel according to Murray's Law, we used the MicroCT angiography-derived vessel diameter histograms (Figure 5A), and computed the cubed sum of the radii for each bin (Figure 5B). We then computed the standard deviation of the histograms in B in comparison with unoprated control limbs (black line). We found that indeed, the "Murray-ness" was low at the peak of angiogenesis,7 days post-surgery, when the vascular network would be highly inefficient due to large viscous losses and high metabolic energy cost. However, by day 24, after vessel remodeling, the Murray-ness was restored to control levels.

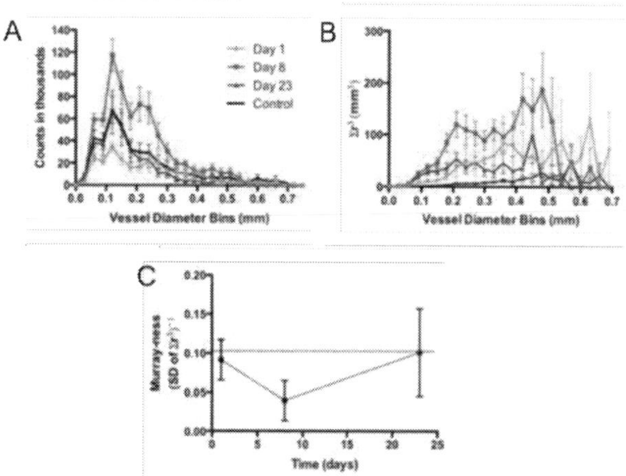

Figure 5. Approach for quantification of "Murrayness", i.e. the degree to which the neovascular network geometry differs from optimal.

Using these observations, we propose that this recovery follows a second-order feedback control system to describe the recovery of hindlimb ischemia, governed by initial angiogenic response to hypoxia, normalization of oxygen, followed by vascular remodeling driven by Murray's Law. To verify this, we evaluated published data from Ref. 3, in which WT control mice exhibit the same overshoot and remodeling as those in the MKP-1 study (Figure 6). The model fits the data very closely.

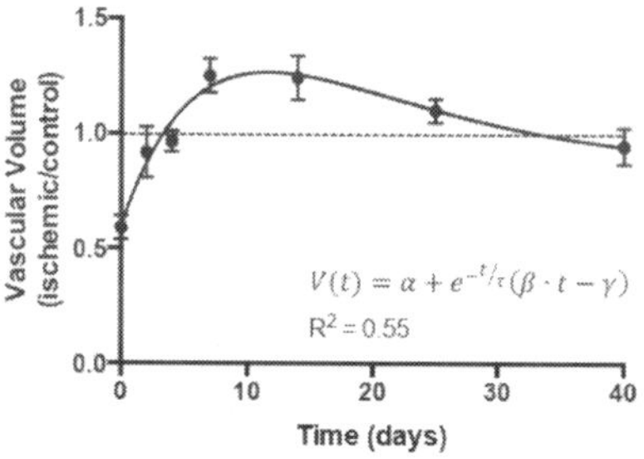

Figure 6. Verification of second-order dynamics of feedback control taken from data in Ref. 3.

DISCUSSION

HLI recovery followed a second-order feedback control profile given by $V(t)=\alpha+e^{-t/\tau}(\beta \cdot t-\gamma)$, where V is a given vascular parameter, a describes the initial state, β and γ angiogenesis and remodeling, and τ the time constant. In general, V(t) exhibited a 25-100% overshoot, and subsequently remodeled to steady state (τ = 9-10 days). MKP-1 knockout mice exhibited decreased feedback sensitivity to hypoxia, resulting in slowed recovery, while diabetic mice exhibited dysregulation of both angiogenesis and remodeling, without a change in τ. HLI recovery can therefore be described as a feedback control system in which the angiogenic response dominates early to minimize hypoxia, but later, vessels remodel to optimize energy expenditure. This model can identify dysregulation of angiogenesis and remodeling in diseased states, modeled here by diabetes and MKP-1 ablation.

These data also demonstrate that Murray's Law not only governs vessel formation during development, but also postnatal neovascular growth and remodeling following ischemic insult. Finally, here we present a new approach for evaluating the energetic efficiency of microCT-quantified vascular networks using Murray's Law of Minimum Work. This approach may be applied to other systems and diseases in which measurement of neovascular mechanobiology is experimentally challenging

REFERENCES

1. Murray C.D., *PNAS* 12:208-214, 1926.
2. Sherman T.F. *J. Gen. Physiol. 78:431-453, 1981.*
3. Limbourg A., et al., *Nature Protocols* 2009 4:1737-1748.
4. Boerckel J.D., et al. *ATVB* 2014 34:1020-1031.
5. Landazuri N. et al., *Am J Physiol RICP* 2012 303:R48-R56.

SB³C2015-1177

TIME EVOLUTION OF CONTRACTILITY IN FIBROBLASTS AS A MEASURE OF CELL PHOTODAMAGE

Samantha G. Knoll (1), Wylie W. Ahmed (2), Taher A. Saif (1)

(1) Mechanical science and engineering
University of Illinois at Urbana-Champaign
Urbana, IL, USA

(2) Laboratoire Physico-Chimie
Institut Curie, section de recherche
Paris, France

INTRODUCTION

Various live-cell imaging techniques are utilized to visualize small-scale biological phenomena. Many of these studies involve fluorescence illumination. Photodamage is well known to be a risk in live-cell fluorescence imaging. A recent report emphasized the importance of circumventing phototoxicity in super resolution imaging by balancing imaging speed, resolution, and photodamage [1].

Currently, light induced damage is known in terms of morphology. Cells damaged by light are known to exhibit physical effects such as necrosis, blebbing, formation of multiple nuclei or vacuoles, or swelling of mitochondria [2]. The heterogeneity amongst the ways in which photodamage manifests itself in cells makes assessing photodamage challenging [3, 2]. Morphology serves as a qualitative measure of cell viability, and does not provide real-time information. A quantifiable means to assess photodamage during short ($\leq$ 1 minute) illumination time scales does not yet exist. Here, we employed cell contractility as a measure of cell damage to quantify the effect of fluorescence illumination on fibroblast cells.

In this study, we investigate (1) how illumination duration affects cell force dynamics as a function of time, and (2) how changes in cell force dynamics compare to changes in cell morphology, and employed a contractile force dipole model to describe the process by which cells relax during fluorescence illumination.

METHODS

We explored the early phase of photodamage by measuring the effect of short duration wide field fluorescence exposure on fibroblast cells plated on polyacrylamide (PA) gels embedded with fluorescent microspheres (beads). We assessed the effect of fluorescence illumination on cell force dynamics as a function of time. The motion of embedded beads due to the deformation of the elastic PA gel substrate served as a measure of cell force dynamics. We embedded red (580/605nm ex./em.) fluorescent beads very near (within 1.5 μm) the PA gel surface [4] for high-resolution measurements of substrate deformation. PA gels of varying elastic moduli (2, 5, 10 kPa) were utilized. Gel stiffness was modulated according to a well-established protocol [5].

Monkey kidney fibroblast cells were plated on hydrogels and allowed to adhere and spread for 4 to 6 hours. After cells were removed from the incubator and placed on the microscope stage in the environmental chamber, an image of a single cell was acquired using differential interference contrast (DIC). Then, the bright field light source was turned off, and the fluorescent light source was turned on while maintaining the same field of view. Focus was shifted slightly such that fluorescent beads within the gel underlying the cell were illuminated. A video of the beads was recorded at a sampling rate of 10 Hz, for 60 seconds. Immediately following the video, the fluorescent light source was turned off. Cell-induced substrate deformation was quantified by measuring the displacements of nanoparticles within the gel using a single-particle tracking algorithm [6]. Displacements were normalized to the area centroid of the cell such that net negative or positive motion represents net motion inward or outward, respectively. Additional experiments were performed in which the gel-cell sample was illuminated for 2 seconds, and underlying fluorescent beads were recorded at 10 Hz. The light source was turned off for the subsequent 58 seconds. At t=60 seconds, illumination was instantaneously turned on again, and a single snapshot of beads was recorded. The two aforementioned illumination protocols were employed to assess the time-sensitivity of substrate deformation as a result of fluorescence illumination of the cells.

Spread area was measured to represent quantitative changes in cell morphology. Select individual cells were monitored in bright field over a longer time period (30 min) following illumination for either 2s or 60s.

RESULTS

Single particle tracking of the beads underlying the cell culture surface revealed localized nanoscale deformations of the substrate within short (sub-one minute) time periods. Sample displacement patterns for a representative cell exposed to 60 s illumination is shown in Fig. 1. Arrows representing bead displacement over specified time periods during illumination became more aligned as time of exposure increased. This pattern was observed in all cells, and displacement alignment was commonly oriented along the long axis of the cell. Alignment of displacement of beads was observed for cells on various stiffnesses, although displacement alignment manifested later in the illumination period for stiffer substrates. Relaxation emerged throughout illumination, as defined by the majority of beads moving outward relative to inward. This bias consistently increased throughout illumination.

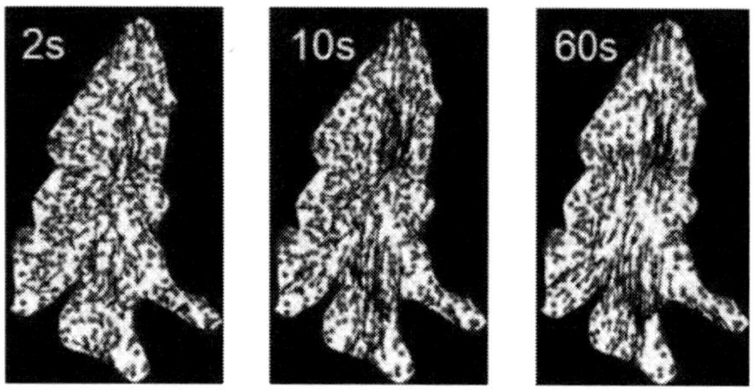

FIGURE 1 Cell forces relax along a preferred direction during illumination period. Bead displacement changes as a result of changing cell forces throughout 60 s of illumination on a 2 kPa substrate. Blue and red dots indicate inward and outward motion, respectively. Arrows representing displacement magnitude are magnified 4x.

In order to quantify the spatio-temporal evolution and alignment of the deformation field, we defined a net displacement dipole for the cell as a function of time. The dipole strength $D_p(t)$ and orientation $\theta m(t)$ were determined for all cells studied. Two groups of beads move away from each other representing a relaxation force dipole. Alignment of bead displacements coincide with the emergence of a definitive dipole orientation characterized by a continuously relaxing force dipole.

To investigate the time evolution of cell force relaxation during illumination, cells were subject to either of the two illumination protocols. Cells induce bead displacement on the order of tens to hundreds of nanometers over 60s of illumination (Fig. 2). The net magnitude of bead displacements between the two illumination protocols was determined to be statistically similar. Additionally, displacement trajectories of beads exposed to either illumination protocol were strikingly similar at the culmination of 60 seconds. Thus, mechanisms of photodamage activate within 1-2 seconds of exposure irreversibly. Continued exposure to light during 60 s appears to have minimal additional effect on the initiated processes.

Illumination duration was found to affect the time to onset and extent of morphological changes. In the cases of both 2 and 60 seconds of illumination, no apparent morphological changes are present immediately (0s) after illumination. Cells enduring 2s illumination showed a 31% decrease in spread area after 30 minutes, whereas 60s of continuous exposure nearly led to complete cell detachment from the substrate and an 84% decrease in spread area. This is consistent with the previously discussed outward motion of embedded beads.

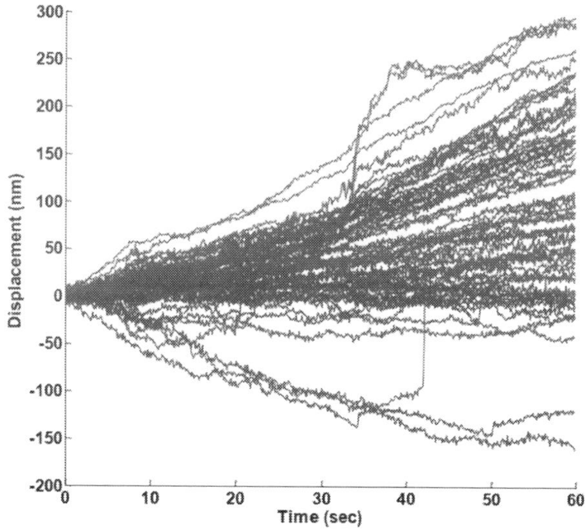

FIGURE 2 Effects of long (60 s) duration illumination on substrate deformation. Bead displacement trajectories for cells exposed to illumination protocol (1). Net outward and inward motion indicated by red and blue, respectively.

DISCUSSION

Our findings show that cell-induced substrate displacements align and shift to a primarily outward direction throughout the duration of illumination, thus highlighting cell force relaxation during the exposure period. This short-term force relaxation suggests onset of photodamage. We find illumination duration to have no effect on this force reorganization process, as 2s of exposure results in a similar displacement pattern after 60s as sample on which a cell is exposed *continuously* for 60s. This suggests that photodamage irreversibly proceeds with few seconds of wide field exposure. Additionally, we find cell morphology to be unaffected by light immediately following exposure for either 2 or 60s, despite initiation of irreversible cell force relaxation. This study elucidates the time evolution of illumination-induced stress reorganization in cells and debunks the notion of morphology as a reliable, real-time indicator of photodamage during experimentation.

ACKNOWLEDGEMENTS

The authors would like to acknowledge the Interdisciplinary Innovation Initiative Program, University of Illinois, grant **12035**. S.K. was funded at UIUC by the Linda Su Nan Chang Sah Doctoral Fellowship and the National Science Foundation (NSF) Grant **0965918**IGERT: Training the Next Generation of Researchers in Cellular and Molecular Mechanics and BioNanotechnology.

REFERENCES

[1] Marx, V et al., *Nat Methods*, 10:1157-1163, 2013.
[2] Xiao, J et al., *Handbook Single-Mol Biophys*, Ch. 3, 2009.
[3] Magidson, V et al., A., *Digital Microscopy*, 4:545-560, 2013
[4] Knoll, S et al., *J Vis Exp*, 91:e51873, 2014.
[5] Tse, J and Engler, *Curr Protocols Cell Biol*, Ch. 10, 2010.
[6] Rogers, S et al., *Phys Biol*, 4:220-227, 2007.

SB[3]C2015
Summer Biomechanics, Bioengineering and Biotransport Conference
June 17-20, 2015,Snowbird Resort, Utah, USA

EFFECTS OF STRUCTURAL VS. NON-STRUCTURAL SCAFFOLDS ON BMP-2- AND MECHANICAL LOAD-MEDIATED BONE REGENERATION

SB³C2015-1180

Anna McDermott (1), Angela Lin (2), Robert E. Guldberg (2), Joel D. Boerckel(1)

(1) Aerospace and Mechanical Engineering
University of Notre Dame
Notre Dame, IN, USA

(2) School of Mechanical Engineering
Georgia Institute of Technology
Atlanta, GA, USA

INTRODUCTION

Recent advances in mechanobiology, demonstrating dramatic effects of extracellular matrix mechanical properties and deformations on cell behavior and fate, provide novel variables for stimulation and manipulation of tissue regeneration in vivo, but may conflict with long-held assumptions of optimal approaches for tissue engineering. For example, mechanical properties approaching those of native tissue have traditionally been a primary design constraint for bone tissue engineering scaffolds under the objective of serving as a functional tissue substitute[1]. However, soft hydrogels such as alginate are also under active investigation as growth factor delivery vehicles for bone regeneration, enabling spatiotemporal control of protein presentation. In addition, we have previously demonstrated that delayed in vivo mechanical loading enhanced large bone defect regeneration through stimulation of tissue differentiation, bone adaptation, and vascularization of soft hydrogel constructs[2,3]. In this study, we tested the **hypothesis** that scaffold mechanical properties regulate BMP-2- and mechanical load-induced bone regeneration. We compared bone regeneration induced by alginate hydrogel-mediated delivery of recombinant human bone morphogenetic protein-2 (BMP-2) with or without a structural scaffold characterized by mechanical properties comparable to those of trabecular bone. Further, we tested the effects of in vivo mechanical loading on bone regeneration in the presence of a structural scaffold.

METHODOLOGY

Alginate hydrogel constructs or structural scaffolds (N=10 per group) were implanted in 8mm rat femoral segmental bone defects and evaluated over 12 weeks by X-ray, microCT, and mechanical testing. RGD peptide-modified alginate hydrogel (ALG) constructs were created by Ca^{2+} crosslinking of irradiated alginate polysaccharides, and were contained in the defect by wrapping with a non-structural polycaprolactone nanofiber mesh[4]. PLDL-TCP/alginate (PLDL/ALG) constructs were formed by filling longitudinally-oriented pores of structural PLDL-TCP scaffolds[1] with alginate and press-fit into the defect. PLDL-TCP scaffolds also featured a hollow 1.5mm diameter core designed to recapitulate the marrow space. Data for ALG samples were re-analyzed from samples described in Ref. 4, which received 2.5 µg BMP-2. PLDL/ALG samples received 3.0 µg BMP-2. A third group featured PLDL/ALG delivery of 3.0 µg BMP-2 and in vivo mechanical loading (PLDL/ALG+LOAD), begun at week 4 post-surgery, as described previously[2,3]. Bridging rates were evaluated by two independent, blind observers from Faxitron X-ray scans at 2, 4, 8, and 12 weeks. Limbs were excised at week 12 and bone formation was quantified by microCT. Mechanical properties at week 12 were evaluated by torsion to failure.

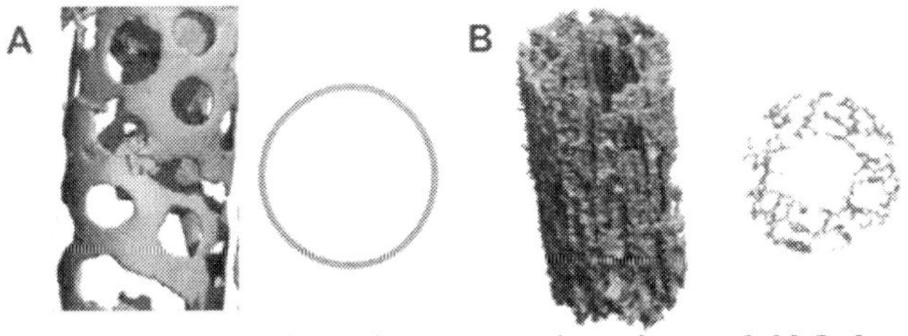

Figure 1: Isometric and cross-section views of ALG & PLDL/ALG scaffolds: ALG: perforated nanofiber mesh containing alginate hydrogel (A), PLDL/ALG: cored PLDL-TCP scaffold containing alginate hydrogel (B).

RESULTS

Comparison of ALG and PLDL/ALG: Samples in the ALG group, re-analyzed to from Ref. 4 to enable direct comparison, exhibited a significantly increasing trend in bridging rate over the 12 weeks (Chi square test for trend, $p<0.05$), while PLDL/ALG did not ($p=0.62$). By week 12, 8/10 samples were bridged with bone in the ALG group, while 4/10 PLDL/ALG samples reached bridging. Bone volume was greater in ALG vs. PLDL/ALG at week 12 (ANOVA, $p<0.05$), with the ALG group having a greater amount of bone distributed throughout the defect, and PLDL/ALG exhibiting localization of bone formation to regions not containing the structural scaffold (Fig 2A, B). Torsional stiffness of regenerated limbs of ALG samples were not significantly different from intact limbs, while PLDL/ALG samples were significantly lower ($p<0.01$) (Fig 3A). Among bridged samples, mechanical properties were greater in the ALG group (Fig 3B), while in non-

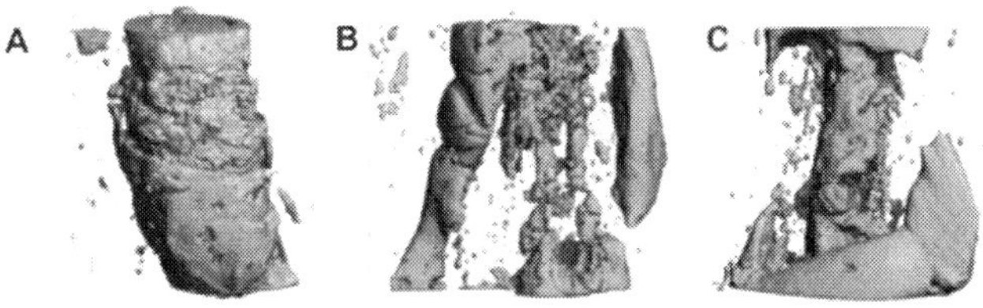

Figure 2: Week 12 microCT reconstructions. A: ALG + BMP-2. B: PLDL/ALG + BMP-2. C: PLDL/ALG+LOAD + BMP-2.

bridged samples, properties were greater in PLDL/ALG (Fig 3C).

Comparison of PLDL/ALG and PLDL/ALG+LOAD: Unlike previously published results on mechanical loading in the ALG group (Boerckel et al PNAS, JOR), delayed mechanical loading had no effect on bridging rate, bone volume, distribution (Fig 2B, C), or mechanical properties (Fig 3D) in PLDL/ALG.

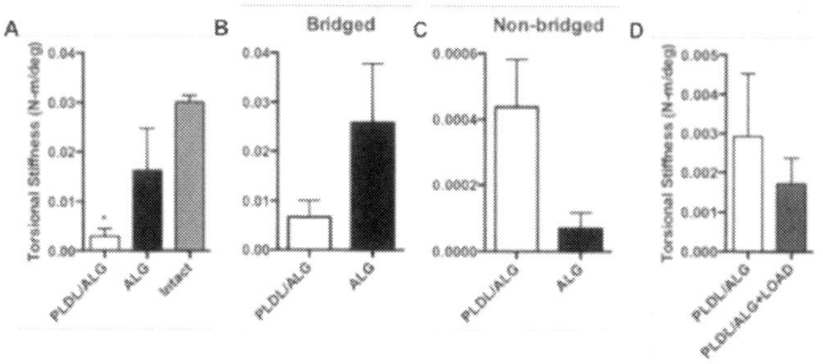

Figure 3: Mechanical properties. A: PLDL/ALG, ALG, intact controls. B: Comparison of bridged samples from (A). C: Comparison of non-bridged samples from (A). D: PLDL/ALG vs. PLDL/ALG+LOAD.

DISCUSSION

Alginate hydrogel-mediated delivery of 2.5 μg BMP-2 resulted in significantly greater bone regeneration in comparison to alginate delivery of 3.0 μg BMP-2 with a structural scaffold. Distribution analysis indicated that bone formation was excluded from within the PLDL scaffold, suggesting that the structural scaffold, despite its high and oriented porosity, failed to facilitate cellular infiltration and matrix production and impeded bone regeneration. In contrast, in the absence of a structural scaffold, bone formation was robust and consistent throughout the defect. Previous results from the ALG group demonstrated robust vascularization throughout the defect[2] which is a prerequisite for bone regeneration. In addition, recent mechanobiology studies suggest that a stiff extracellular matrix inhibits angiogenesis, which may partially explain the reduced tissue regeneration in the scaffold region. Other important variables include scaffold composition (i.e., PLDL vs. alginate) and construct porosity, each of which may regulate cellular behavior.

In groups featuring a structural scaffold, in vivo mechanical loading failed to induce bone regeneration, in contrast to previous studies with alginate alone. These observations suggest that in addition to impeding bone formation, the presence of a structural scaffold shielded the ingrowing bone from stimulatory mechanical loads. Stress shielding is a frequently observed phenomenon in orthopaedic implants, resulting in impaired osseointegration, lack of bone formation and even active bone resorption. Taken together, these data indicate that in optimal design of tissue engineering scaffolds, scaffold mechanical properties approaching those of native bone are not necessary, and in fact may hinder regeneration induced by both growth factors and mechanical loading, though integrated structural scaffolds may enhance mechanical integrity when bridging is absent.

SIGNIFICANCE

This study reveals a complex relationship between matrix mechanical properties and tissue regeneration, challenges accepted thinking on optimal stiffness of tissue engineering scaffolds, and demonstrates the importance of in vivo testing of mechanobiological concepts identified in vitro.

References:

1. Lin, A. S. P., *et al. Biomaterials* **24,** 481–489 (2003).
2. Boerckel, J. D., *et al. Proc. Natl. Acad. Sci. U. S. A.* **13**, E674-80 (2011).
3. Boerckel, J. D. *et al. J. Orthop. Res.* **30,** 1067–1075 (2012).
4. Boerckel, J. D. *et al. Biomaterials* **32,** 5241–51 (2011).
5. Critser P. J. *et al.* Microvasc. Res. **80**, 23-30. (2010).

SB³C2015
Summer Biomechanics, Bioengineering and Biotransport Conference
June 17-20, 2015,Snowbird Resort, Utah, USA
SB³C2015-1181

DIRECT MEASUREMENT OF ENERGY LANDSCAPE OF BIOLOGICAL INTERACTIONS

Ahmad Haider (1), Daniel Potter (2), Todd Sulchek (1,2)

(1) Woodruff School of Mechanical Engineering
Georgia Institute of Technology
Atlanta, Georgia, USA

(2) Parker H. Petit Institute of Bioengineering and Biosciences
Georgia Institute of Technology
Atlanta, Georgia, USA

INTRODUCTION

Atomic force microscopy offers an unprecedented ability to determine adhesion forces between individual biological molecules due to its highly sensitive ability to measure and apply forces. However the direct mapping of biomolecular energy potentials between two molecules and under equilibrium conditions is more challenging. We have recently developed a technique using atomic force microscopy to directly map the intermolecular energy potential curve of a biotin-avidin interaction. The method utilizes Brownian fluctuations of a sensitive microcantilever probe in close proximity to the interaction energy potential of biotin and avidin that are decorated on the tip and substrate surfaces, respectively. By recording subtle deviations from the harmonic cantilever vibrations, the biomolecular energy potential and force can be reconstructed. We find that these interactions exhibit multiple attractive wells with average energies of ~23k_bT and ~48 k_bT before undergoing repulsive contact. When the specific interaction is blocked by the addition of high concentrations of free biotin, the attractive interaction wells are eliminated. The method shows great promise for directly recording interaction potentials to record the subtle interactions that result from biological molecules, for example from peptide loops that border the channel around the binding pocket.

METHODS

Force experiments:

MSCT cantilevers (Bruker, California) were functionalized with PEG-27-biotin, while a monolayer of avidin (Fluka) was physically adsorbed to freshly cleaved mica (Gruber-Ebner protocol). Cantilever D, with a resonance frequency, f_0, between 10 and 13 kHz in liquid, was used for force curves, which were collected at 40 to 50 kHz, four times f_0, satisfying the Nyquist sampling theorem and thereby ensuring all resonance and thermal vibrations are measured in the deflection signal. Force curves collected for interaction mapping were scanned at 0.02 Hz. After 30 min equilibration in the fluid cell the cantilever drift was below 7 Å/min. Inverse optical lever sensitivity was calculated from the contact region, and the spring constant was calibrated using the thermal method. Fig. (1) illustrates the data processing approach.

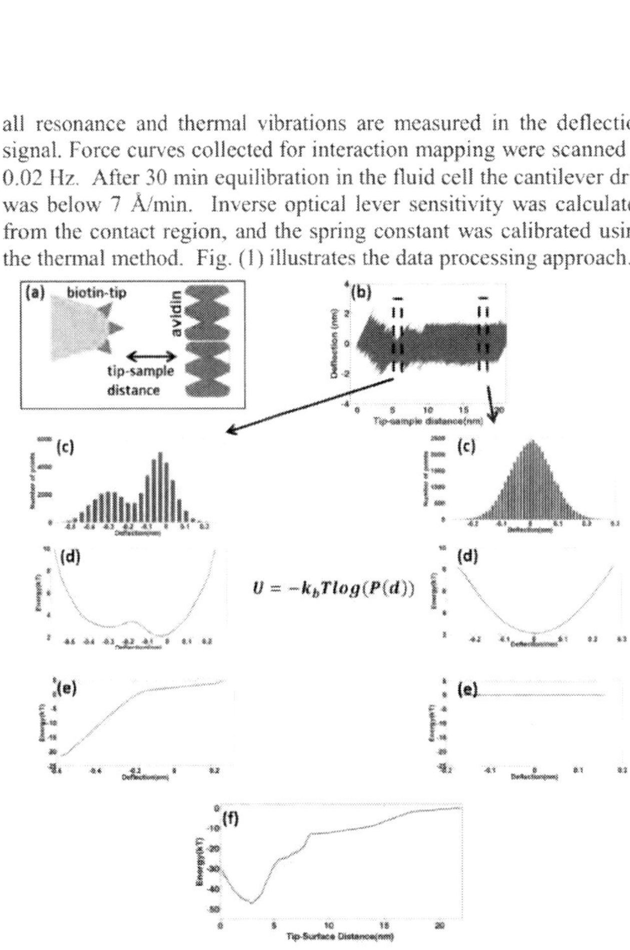

Figure1. (a) Schematic of a cantilever tip decorated with biotin positioned above avidin molecules. (b) Overall approach curve including Brownian fluctuations. (c) Histogram of the cantilever positions during fluctuations at two different window locations (d) Energy of each window computed using Boltzmann's equation (e) Tip-sample interaction energy after subtracting the cantilever harmonic potential for each window. (f) Total tip sample interaction energy from many windows patched together. The curve shows two energy wells with depth of ~23k_bT and ~48 k_bT with an average well width of ~4 nm

Data processing:

The process of probing molecular interactions using Brownian fluctuations is based upon the method reported by Ashby et al. [1]. Thermal fluctuations recorded during force measurements are converted to energy (potential) by using Boltzmann distribution. Tip sample interaction potential was calculated by subtracting cantilever's harmonic potential from total potential. After scaling for tip sample distance, overall energy was calculated by performing a weighted average of all sections [2]. Data processing was performed by routines written in MATLAB (Mathworks Inc.) [1]. Accuracy of the program was verified by testing with data obtained from other studies and obtaining similar results [1][2][3].

RESULTS

Effect of blocking: To verify the interaction potentials represented specific binding, the blocking of the biotin groups was accomplished by injecting 40 uL of 1 mg/mL biotinylated avidin into the fluid cell and incubating for five minutes before recommencing measurements. During blocking assay, no attractive forces were experienced, which is evident both by low adhesion forces of the dynamic force spectroscopy results and the elimination of the attractive energy wells in the energy reconstruction results before the onset of repulsion, which indicates a lack of stable binding event (2(b)). Once the blocking is removed by washing the fluid cell with buffered solution without biotin, the return of adhesive energy wells and a sharp energy dip corresponding to binding is observed prior to close contact repulsion (2(d)).

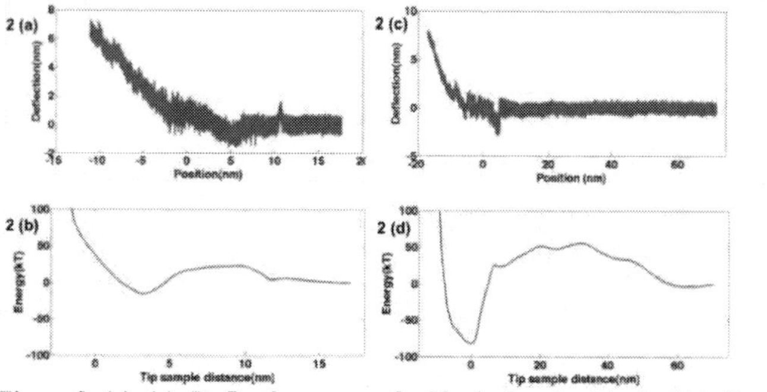

Figure 2. (a), (c): Deflection curves for blocked and unblocked binding cases, respectively collected at 60 pm/s. (b), (d): Corresponding energy reconstructions. Reconstruction is performed using a window size of 10000 points, 90% window overlap and 100 bins.

Effect of cantilever approach rate: A slower approach rate, (3(a), (c)) increases the range of tip sample interaction potentials that cantilever can explore, before it snaps onto the surface. Hence a slower approach is necessary to detect very closely spaced multiple binding events close to tip sample contact. Moving the cantilever at fast rates (3(b), (d)) increases the probability of cantilever undersampling short-lived transition states, particularly if the distance between successive states is small compared to the length scale set by the average cantilever approach velocity.

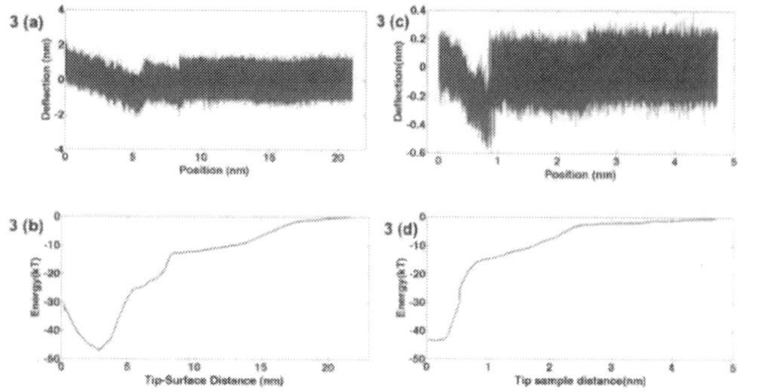

Figure 3. (a), (c): Deflection curves for slow (60 pm/s) and fast approach (112 pm/s) rates respectively. (b), (d): Corresponding energy reconstructions. Reconstruction is performed using a window size of 20000 points, 95% window overlap and 200 bins.

DISCUSSION

By recording deviations in the cantilever Brownian fluctuations, we can measure intermolecular potentials and adhesion forces in protein-ligand pairs. Brownian reconstruction methods had previously been applied only in measuring chemical interactions [1][2][3]; biological interactions pose a challenge due to flexible linkers that are used to attach to the AFM tip, which act as springs themselves and stretch when pulled. In these studies, we use a 0.75 nm gluteraldehyde linker, which is sufficiently stiff and resistant to stretching, thus leading to more repeatable measurements. The biotin-avidin reconstructions can be compared to previously published results with dynamic spectroscopy measurements and molecular dynamics simulations performed by the groups of Evans and Schulten respectively [4] including the detection of multiple transition states within the binding pocket. Finally, the effect of reconstruction parameters, i.e. window size, window overlap and histogram bin width play an important role in determining the shape of reconstruction. The overlap needs to be maximized to minimize error at edges of each window's histogram. The bin width also needs to be small so as not to average out small deflection changes, and yet not so small as to add to the sampling noise. Further study needs to be done to determine the effect of cantilever's parameters on the quality of reconstruction.

With Brownian reconstruction, the entire energy landscape of the bio molecular interactions can be known. Since this landscape depends only on the kinetic rates of the interaction, determining the shape of the landscape will be a very powerful tool in deducing the binding constants and the affinity of such interactions.

ACKNOWLEDGMENTS

The authors would like to acknowledge NSF (CBET-CAREER-1055437 and DBI 1063279) for financial support for the duration of this work. The authors thank Dr. Paul Ashby for advice on the reconstruction code.

REFERENCES

[1] Ashby P.D., Lieber C. L., *J Am. Chem. Soc.*, 126:16973-16980, 2004

[2] Willemsen, O.H., Kuipers, L. et al., *J Langmuir*, 16:4339-4347, 2000

[3] Cleveland, J.P. et al., *Phys. Rev. B*, 52: 8692-8695, 1995

[4] Merkel, R.; Nassoy, P.; Leung, A. et al, *Nature* 397: 50-53, 1999

MEASURING TORTUOSITY CHANGES DUE TO CENTRAL RETINAL VEIN OCCLUSION

Kendall R McMillan (1), Shaun Evans (1), Gil Binenbaum (2,3), Brittany Coats (1)

(1) Department of Mechanical Engineering
University of Utah
Salt Lake City, Utah, USA

(2) Department of Ophthalmology
The Children's Hospital of Philadelphia
Philadelphia, PA, USA

(3) Scheie Eye Institute
Perelman School of Medicine
University of Pennsylvania
Philadelphia, PA

INTRODUCTION

Retinal hemorrhages (RH) are a common finding in abusive head trauma (AHT) and are commonly numerous, diffusely located through the posterior and peripheral retina, and may be intraretinal, subretinal, and or preretinal. [1,2]. Retinal hemorrhages are a primary finding in central retinal vein occlusion (CRVO) and are classically described as diffusely located intraretinal hemorrhages in "all four quadrants" of the posterior and peripheral retina; however, there is associated optic disc swelling and tortuosity of the retinal veins, which are not common findings in AHT [3,4]. Because there are some similarities in the RH between these two diseases, mechanisms of RH from AHT may overlap with those from CRVO. To help understand the effect of CRVO on mechanisms of RH, and to evaluate the potential for a piglet animal model to exhibit widespread RH, vein occlusion was recreated in an experimental setting and changes in retinal vasculature evaluated. In this study, we specifically evaluate whether original tortuosity or changes in tortuosity of the retinal blood vessels are correlated with impending RH.

METHODS

The methods to create the CRVO have been described elsewhere [5]. Briefly, the CRVO was created by clamping the optic nerve of each eye from four anesthetized piglets (n=7) for up to 1 minute, releasing, waiting for 5-10 minutes, and then repeating if necessary. An indirect video ophthalmoscope was used to assess changes in retinal vasculature before, during, and after clamping. Frames from the video before, during and after clamping were selected for analysis (Fig. 1A). The selection of frames was based on the clarity of the image and a good view of the optic nerve and surrounding vasculature. All analysis was performed with a custom code developed in MATLAB.

Images were first converted to grayscale and histogram equalization was applied to make the vessels easier to visualize (Figure 1B). Using the selection tool *imfreehand,* from the MATLAB image processing toolbox, the image was cropped around the vessels to omit parts of the image that were hard to analyze due to glare or other obstructions. The cropped image was then transformed into a skeleton image. The skeletonization was based on binary thresholding, which could be adjusted to capture more or less of the vessel structures or to reduce noise.

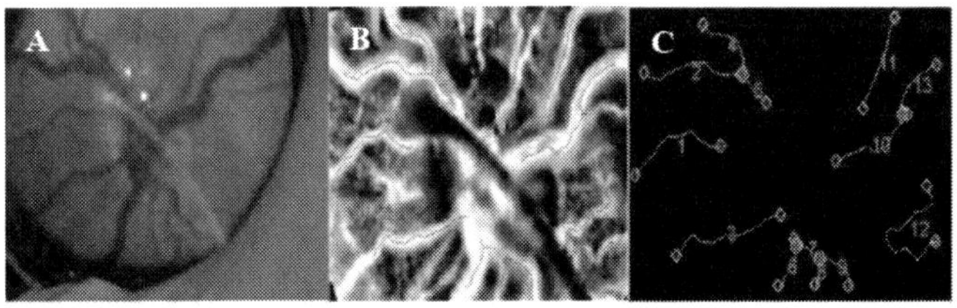

FIGURE 1. A) ORIGINAL IMAGE EXTRACTED FROM VIDEO INDIRECT. B) GRAYSCALE OF THE ORIGINAL IMAGE WITH SKELATONIZATION OVERLAID IN RED. C) SKELETONIZED IMAGE WITH LABELS FOR EACH VESSEL, BRANCH POINT (RED ASTERISKS) AND END POINT (OPEN DIAMONDS).

Small artifacts and spurs defined as lines less than 8 pixels were removed from the image. Morphological operations were then performed to find the branch points and endpoints of each skeleton in each image. The branch points were defined as any point where a white pixel touches three or more other white pixels. An end point was defined as a white pixel in connection with only one other white pixel. A vessel segment was defined as the section of pixels between two endpoints, two branch points, or a combination of endpoint and branch point. Each vessel segment was labeled as shown in Figure 1C.

A geodesic distance calculation was performed on each vessel segment using the endpoints and/or branch points as starting and stopping points. This was considered the curved line distance. The straight line distance was defined as the shortest path between branch points and/or endpoints. This was evaluated using the built in *pdist2* function. Tortuosity was then defined as:

$$tortuosity = \frac{curved\ line\ distance}{straight\ line\ distance} \quad (1)$$

Blood vessels that were visible in the majority of images from a single eye experiment were picked for analysis. The maximum, minimum, average and standard deviation of tortuosity across all vessels in an image were calculated and recorded. Two logistic regressions were performed to determine if the maximum, average, and/or standard deviation of the tortuosity was predictive of RH. The first logistic regression evaluated whether the initial tortuosity of the vessels (prior to any clamping) resulted in increased risk for RH. The second logistic regression evaluated whether tortuosity during or after the clamping correlated with RH. A chi-square analysis was performed to determine how well the logistic regression fit the data. A p-value < 0.05 was considered significant.

RESULTS

There were no significant differences in the preliminary average or maximum tortuosity between eyes, therefore, a logistic regression found no significant correlation of pre-clamping tortuosity to eventual RH. With each clamping episode, the tortuosity tended to increase, and then decrease again when the clamping was removed (Figure 2). However, in several eyes, tortuosity actually increased when the clamp was removed from the optic nerve. This increase tended to correlate with eventual RH in the immediate region next to the vessel. An example is shown in Figure 3 where the tortuosity of vessel F increases after clamp 3 (C3-PC3). The other vessels in the image remain at comparatively lower tortuosities.

The logistic regression confirmed these observations by finding a significant correlation of overall maximum tortuosity in an image to RH ($p=0.014$). A plot of the maximum tortuosity in each image at each event illustrates that vessels generally above a tortuosity of 1.25 resulted in RH (Fig. 4).

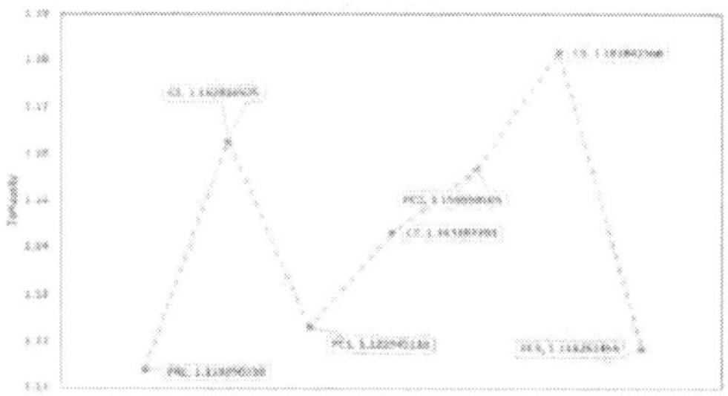

FIGURE 2. VESSEL TORTUOSITY INCREASED WITH CLAMPING (C) AND TYPICALLY DECREASED AFTER REMOVAL OF THE CLAMP (PC).

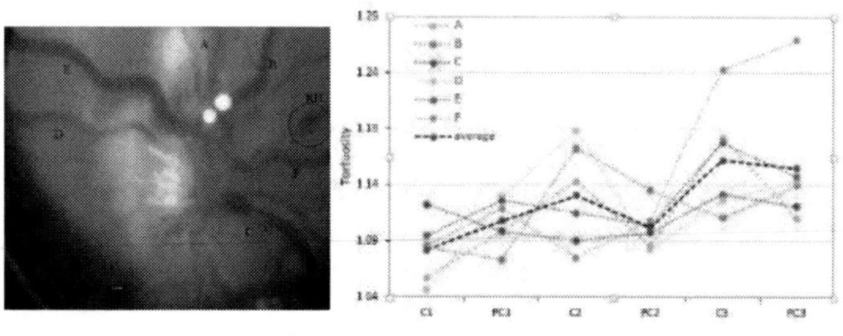

FIGURE 3. LABELED VESSELS (LEFT) AND ASSOCIATED TORTUOSITY (RIGHT) WITH EACH CLAMPING (C) AND POST CLAMPING (PC) EVENT. THE VESSEL CLOSEST TO THE HEMORRHAGE HAD SUBSTANTIALLY HIGHER TORTUOSITY AFTER THE FINAL CLAMP.

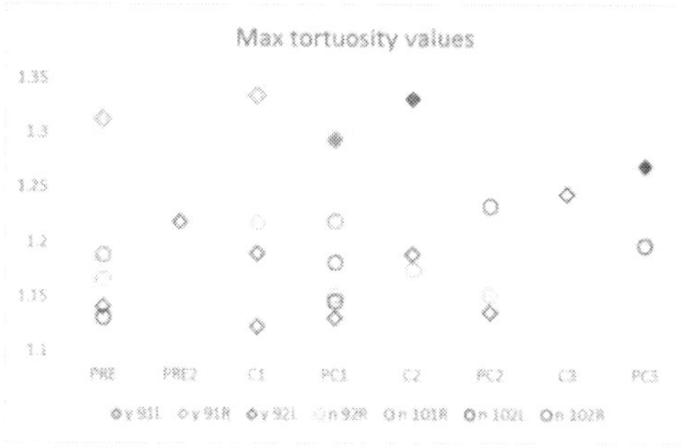

FIGURE 4. MAX TORTUOSITY VALUES OF RH EYES (DIAMONDS) AND NON RH EYES (CIRCLES). THE FINAL TORTUOSITY JUST PRIOR TO RH IS INDICATED BY A FILLED IN DIAMOND. ALL HEMORRHAGE WAS ABOVE A 1.25 TORTUOSITY.

DISCUSSION

In this study, we found substantial changes in tortuosity with the clamping and sometimes with the release of the optic nerve. The tortuosity is likely a result of high transmural pressure. Specifically, there is a high internal pressure compared to the external pressure acting on the vessel wall. The clamping during the CRVO procedure restricts venous outflow, but allows arterial inflow. This causes blood to rapidly fill the vessels creating the high transmural pressure. Releasing the vein occlusion typically results in a decrease in the transmural pressure and associated decrease in tortuosity. However, occasionally, the rapid reperfusion of the blood vessels in this study caused an increase in internal pressure and associated tortuosity. Interestingly, it was during this reperfusion stage that RH occurred in two of the three eyes.

We found significant correlation in retinal vessel tortuosity and RH. This is comparable to other studies in the literature that have noted an increased risk of RH with familial retinal arteriolar tortuosity [6]. Additionally, vascular tortuosity is an important sign of disease severity in retinopathy of prematurity, which may also feature RH, though these RH are associated with pathologic neovascular tissue. [7]. And finally, vascular tortuosity is a primary finding in clinical CRVO. In the animals where RH occurred in this study, maximum tortuosities were greater than 1.25. This data suggests that large tortuosities may be predictive of RH, but more studies will be required before this conclusion can be made.

ACKNOWLEDGEMENTS

The authors would like to thank the University of Pennsylvania Vision Research Program for funding this work. The authors would also like to thank Jill Ralston for animal care and the Injury Biomechanics lab at the University of Pennsylvania for allowing use of their animal facilities.

REFERENCES

[1] Kempe CH, et al. *JAMA* ,181:17-24, 1962.
[2] Elner VM, *Arch Ophth*, 126 (8): 1141-1142, 2008.
[3] Hayreh SS, *Brit J Ophth*, 49:626-645, 1965.
[4] Wong TY et al, *N Engl J Med*, 363:2135-2144, 2010.
[5] Evans S., MS Thesis; *University of Utah*, pp. 35-54, 2014.
[6] Sutter FK et al, *Surv Ophth*, 48:245-255, 2003.
[7] Wilson CM et al, *IOVS*, 49(8):3577-3585, 2008

SB³C2015
Summer Biomechanics, Bioengineering and Biotransport Conference
June 17-20, 2015, Snowbird Resort, Utah, USA
SB³C2015-1183

BIOHYBRID SWIMMING AT LOW REYNOLDS NUMBER

Brian J. Williams (1), Sandeep V. Anand (1), Jagannathan Rajagopalan (2), M. Taher A. Saif (1)

(1) Mechanical Science and Engineering
University of Illinois at Urbana-Champaign
Urbana, IL, USA

(2) Mechanical and Aerospace Engineering
Arizona State University
Tempe, AZ, USA

INTRODUCTION

We present a soft robotics biohybrid flagellum self-powered by cardiomyocytes [1]. Biohybrid systems are a rapidly evolving field enabled by recent developments in cell culture, microfabrication [2], and our theoretical understanding of microscale propulsion at low Reynolds number (*Re*) [3-5]. Here, we present a method of making highly compliant polydimethylsiloxane (PDMS) filaments with micron-scale cross sections that are capable of being actuated with large angular deflections by single cardiomyocytes. Borrowing from the study of the fluid dynamics of biological flagella, we show how the cyclical actuation of a thin filament can generate propulsion at low *Re*. We provide experimental data validating our model and demonstrating self-powered propulsion at low *Re*. We demonstrate how this propulsion mechanism can be used in more complex configurations to improve performance and functionality. Finally, we present the use of the synthetic flagellum to probe subtle mechanosensitivity pathways that affect temporal dynamics of spontaneous contractility in cultured cardiomyocytes. This work presents the confluence of theory, modeling, and experimentation yielding the first biohybrid motile platform at this size scale.

METHODS

The biohybrid swimmer relies on a suitable substrate that can be actuated by the contractile forces of a single cell (Fig. 1a). To achieve this compliancy, we form filaments out of PDMS by etching 2D patterns in silicon wafers to a single depth. We then fill the rectangular channels with liquid PDMS via capillary draw, cure the PDMS, and remove the filaments with micron scale cross-sections (Fig. 1b).

To achieve propulsion at low *Re*, the swimmer must generate time irreversible deformation patterns. However, clusters of cardiomyocytes in physical contact synchronize by electrical coupling, and isolated clusters tend to contract at slightly varying frequencies, eliminating the potential to maintain phase relationship between discrete actuators. To achieve time irreversibility, we exploit the "flexible oar" mechanism, where fluid drag causes passive regions of the filament to deform in a direction-dependent, asymmetric manner. The asymmetry in resultant fluid drag over the course of a full cycle produces a nonzero net propulsive force.

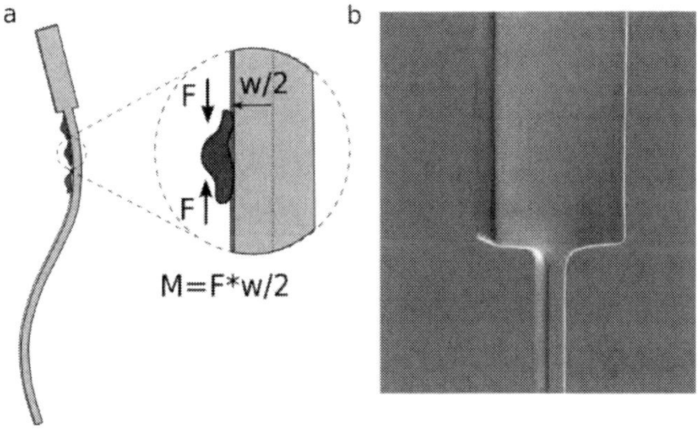

Figure 1: a., Schematic of flagellar swimmer powered by cardiomyocytes, and b., SEM image of cured PDMS filament at head/tail junction

While closed form solutions for the flexible oar exist for the sinusoidal angular displacement or translation of one endpoint of a homogenous filament, the free, nonhomogenous swimmer actuated locally by a nonsinusoidal bending moment must be treated numerically. We mimic the spermatozoon by designing our filaments with a large, rigid head, and a long, thin tail. We select the tail cross-section to facilitate large angular deflections by the contraction of one to several contracting cardiomyocytes. We further reduce the design space by picking a head size approximately an order of magnitude wider than the tail and choosing the actuation site to be the region of the tail immediately adjacent to the head. Motivated by previous studies [4], optimal head and tail lengths are determined by modeling the swimmer as a slender body elastohydrodynamic system with the governing equation:

$$Ay_{xxxx} + \varsigma y_t = M_{xx} \quad (1)$$

Here, A is the bending stiffness of the filament, ς is the normal drag coefficient of the filament, M is the time dependent bending moment generated by the actuating cells, and coordinates y and x are defined in Fig. 2a. Subscripts indicate partial differentiation. The cyclical driving moment M is estimated from experimental data from primary neonatal rat cardiomyocytes (Fig. 2b), and is applied to a small, approximately 100 μm patch adjacent to the head. Applying moment- and shear-free boundary conditions to the filament, the deformation pattern of the swimmer is determined numerically, and the resultant propulsive force and swimming dynamics are determined by projecting fluid drag along the axis of the swimmer (Fig. 2c).

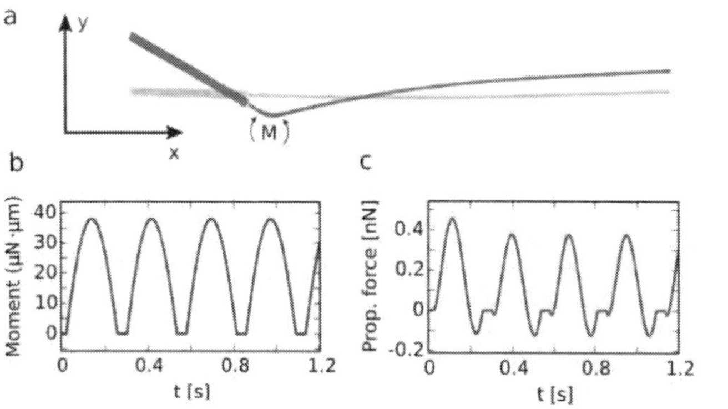

Figure 2: a., Coordinate system of swimmer, b., Driving bending moment produced by contractile cells, and c., Resultant swimming dynamics

RESULTS

We fabricate a single-tailed flagellum using the optimal dimensions provided by the numerical model as target parameters. Cell adhesion is limited to the region immediately adjacent to the head by selectively functionalizing the filament with Fibronectin. Cardiomyocytes are extracted from 2-4 day old neonatal rats and plated on the filament. The resultant swimmer, detached from the substrate and shown in Fig. 3a, is powered by a single cluster of 1-3 cardiomyocytes contracting at 3-4 Hz. Asymmetric deformation of the tail is clearly visible, resulting in swimming in the direction of the head at an average speed of 9.7 μm/sec. The observed temporal and peak contractile forces of this swimmer are then used as inputs to the numerical model, which predicts a swimming velocity of 11.0 μm/sec.

The single-tailed biohybrid swimmer velocity, scaled by its length, yields performance several orders of magnitude lower than biological specimens. To improve the swimmer performance, we modify the effective boundary conditions with a 2-tailed swimmer (Fig. 3b). The single tail swimmer wasted large amounts of energy driving its head through the fluid, providing moment balance for the tail. The modified swimmer has 2 tails contracting in opposing directions in synchronization, thus providing a moment balance for each other. This results in more energy going into the tails of a swimmer with reduced drag, yielding an order of magnitude improvement in swimming dynamics. Additionally, imbalance in the propulsion generated by the filaments produces an arc-like swimming path, demonstrating a potential mechanism for swimmer navigation.

DISCUSSION

The experimental evidence presented represents a successful implementation of microscale, biohybrid soft robotics to reproduce low *Re* swimming in mimicry of biological flagella. The model predictions provide the information required to integrate multiple flagella into more complex configurations to enable machines capable of swimming in 2 or 3 dimensions, provided the means to modulate actuation of individual filaments.

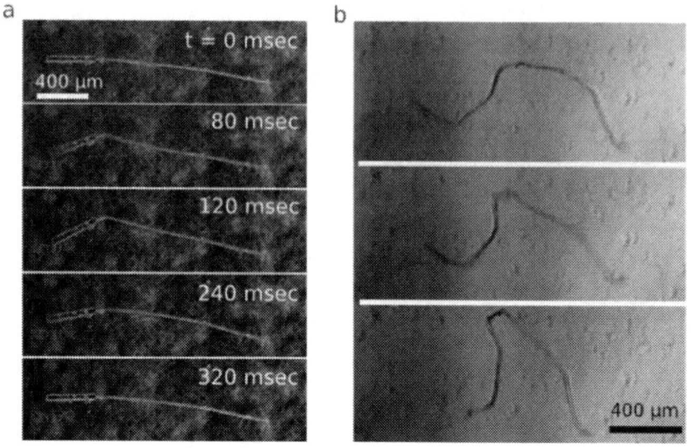

Figure 3: a., Experimental sequence of single tailed biohybrid flagellum swimming at 10 μm/sec, and b., 2-tailed swimmer capable of 80 μm/sec.

Biological flagellar swimmers can explore their environment and swim with higher efficiency by generating more complex deformation patterns by applying bending moments throughout their tail, as opposed to the single-point actuation in our elastic oar mechanism. To increase the complexity of deformation patterns produced by the biohybrid flagellum, multiple actuators must be used, while maintaining some temporal phase relationship between the cyclical dynamics of each actuator. We present preliminary data that mechanical coupling may be sufficient to act as a synchronization mechanism between physically isolated clusters of cardiomyocytes. Next, treating individual clusters as integrate-and-fire oscillators strain-coupled on an elastohydrodynamic substrate, we model a swimmer with multiple driving clusters. The model predicts the synchronization of actuators, an effective decrease in the wavelength of deformation patterns, and an effective increase in the driving frequency, resulting in significant propulsion gains produced by a shorter swimmer with reduced drag. Attempts to produce this system experimentally may provide additional insight into the role of stretch in cardiac function.

ACKNOWLEDGEMENTS

This project was funded by the National Science Foundation (NSF), Science and Technology Center on Emergent Behaviors in Integrated Cellular Systems (EBICS) Grant CBET-0939511.

REFERENCES

[1] Williams, B. et al., *Nature Comm.*, 5, 2014.
[2] Rajagopalan, J. et al. *J. Microelectromech. Syst.*, 22:992-994, 2013.
[3] Machin, K. E. *J. Exp. Biol*, 35:796-806, 1958.
[4] Lauga, E. et al. *Rep. Prog. Phys.*, 72:096601, 2009.
[5] Wiggins, C. et al, *Biophys. J.*, 74:1043-1060, 1998.

SB³C2015
Summer Biomechanics, Bioengineering and Biotransport Conference
June 17-20, 2015,Snowbird Resort, Utah, USA
SB³C2015-1184

COMPUTATIONAL AND EXPERIMENTAL ANALYSIS OF INTRACELLULAR MOTION, FORCES, AND STRUCTURE

Michael Mak (1,3), Taeyoon Kim (2), Muhammad H. Zaman (3), Roger D. Kamm (1)

(1) Mechanical Engineering
Massachusetts Institute of Technology
Cambridge, MA, USA

(2) Biomedical Engineering
Purdue University
West Lafayette, IN, USA

(3) Biomedical Engineering
Boston University
Boston, MA, USA

INTRODUCTION

The mechanics of the intra and extracellular microenvironments plays critical roles in determining the nature of many different biological phenomena, from force generation and molecular transport to cell morphology and cancer invasion. It is known that mechanical features, such as substrate stiffness, can guide directional cell migration and modulate cell traction forces, leading to durotaxis, stem cell differentiation, and invasive transformation. However, it is not clear what the underlying, subcelullar mechanical components and dynamics that lead to mechanosensation are. In our study, through a close integration of Brownian dynamics computational studies and experimental intracellular particle-tracking microrheology studies in physiological 3D environments, we elucidate fundamental mechanisms that govern cell morphology, rheological properties, and internal stresses.

METHODS

Computationally, we performed Brownian dynamics simulations of F-actin networks connected by crosslinkers and driven actively by myosin motors (Figure, Top). The actin networks fluctuate in accordance to the Langevin equation with interactive forces from motors and crosslinkers and no inertia:

$$\boldsymbol{F}_i + \zeta_i \frac{d\boldsymbol{r}_i}{dt} + \boldsymbol{F}_i^B = 0 \qquad (1)$$

where i corresponds to the ith particle, r is the position, **F** is the deterministic force from motors and crosslinkers, $\mathbf{F^B}$ is an uncorrelated thermal force, and ζ_i is the drag coefficient [1].

Experimentally, we conducted subcellular particle tracking microrheology studies in metastatic cancer cells embedded in 3D in collagen, thus mimicking the mechanical state of cells physiologically [2]. Cells were mixed in collagen and loaded into microfluidic devices, where the gel polymerized and produced a 3D platform for investigating intracellular mechanics. Intracellular fluctuations are driven by thermal fluctuations and motor-induced activities. A modified generalized Stokes-Einstein relation is used to quantify intracellular motion and dissipation.

RESULTS

Our results showed that dimensionality and cytoskeletal architecture are critical in determining intracellular fluctuations. Specifically, we showed experimentally that when cells are cultured on 2D surfaces, the active (motor-driven) components of intracellular motions are enhanced, showing "fluid-like" movements. In 3D, these fluctuations are suppressed, recovering a more "solid-like" fluctuation pattern (Figure, Bottom). Furthermore, when the cytoskeletal architecture is disrupted by inhibiting actin polymerization, actin aggregates formed, leading to enhanced local intracellular stiffness but also fluidization of motions at longer time intervals.

Our simulations more deeply probed into the cytoskeletal component level kinetics and mechanics that manifest into network level behavior. Specifically, we found that the dimensionality of the simulated network (planar or 3D) altered the magnitude of internal stress fluctuations, leading to enhanced x-y fluctuations for 2D networks. Additionally, the cytoskeletal network morphology was critically impacted by actin polymerization dynamics. Suppressed polymerization rates induced stress dissipation and loss of network connectivity over time (Figure, Top).

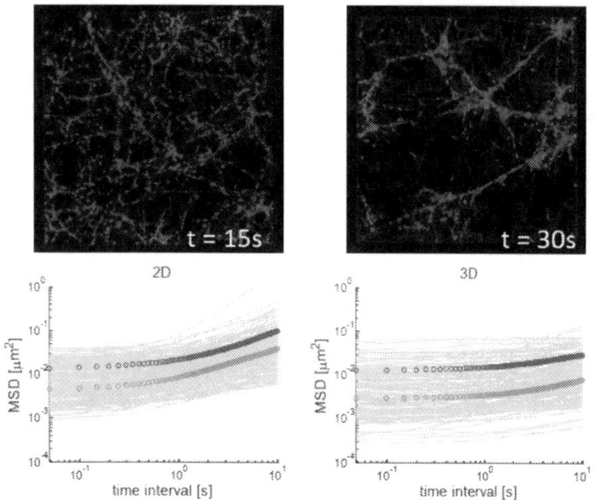

Figure: Top) Simulations of the active cytoskeleton with actin (teal), crosslinkers (yellow), and motors (orange). Depending on actin polymerization rates, networks can undergo morphological changes and aggregation over time. Bottom) Experimental measurements of intracellular motion (mean-squared displacements) of labeled mitochondria in 2D or 3D culture platforms.

DISCUSSION

Dimensionality is a fundamental property in all physical environments. While cell culture has been traditionally performed in 2D, physiological environments are 3-dimensional, with cells often embedded in a dense and fibrous microenvironment. This parameter has been demonstrated to have impact in cell signaling and migration. However, the impact of dimensionality on fundamental aspects of intracellular mechanics is not well understood. Our integrated computational and experimental investigation demonstrated that altered intracellular network geometries induced by the dimensionality of culture environments play a critical role in altering the magnitude and trend of intracellular fluctuations. This may have important consequences in cell force generation and internal transport.

Furthermore, force generation by cells is implicated in many areas of biology, including cancer metastasis. It is not fully understood how force magnitudes are precisely controlled by cells and the environment. We showed that the cytoskeletal architecture is greatly influenced by actin polymerization dynamics and subsequently modulates internal stress profiles.

Our integrated computational and experimental methods enable a means to identify key mechanical regulators of intracellular mechanobiology.

ACKNOWLEDGEMENTS

This study was funded by the National Cancer Institute grant number U01-CA177799. The funders had no role in study design, data collection and analysis, decision to publish, or preparation of the manuscript.

REFERENCES

[1] Kim, T, Biomech Model Mechanobiol, 2014
[2] Mak, M et al., PloS Comp Biol, 10:e1003959, 2014

SB^{3}C2015
Summer Biomechanics, Bioengineering, and Biotransport Conference
June 17-20,2015,Snowbird Resort, Utah, USA
SB^{3}C2015-1186

MATERIAL PROPERTY TESTING OF CARBOXYMETHYLATED HYALURONIC ACID HYDROGEL POLYMER

McKenna Drysdale (1), Hee-Kyoung Lee (2,3), Barbara Wirostko (3), Brittany Coats (1)

(1) Department Mechanical Engineering
University of Utah
Salt Lake City, UT, USA

(2) Department of Pharmaceutical Chemistry
University of Utah
Salt Lake City, UT, USA

(3) Jade Therapeutics, Inc.
Salt Lake City, UT, USA

INTRODUCTION

Corneal injury is often treated with eye drops or ointment to prevent infection. The use of hydrogels may be a more efficient drug delivery system than these traditional methods. Hydrogels are biocompatible, diffuse drug quickly and easily, improve retention time, and can comfortably be worn in the eye for days at a time. In contact lenses, comfort has been found to correlate to the stiffness of the material [1]. To estimate the comfort of hydrogels in the eye, the instantaneous and relaxation moduli of four hydrogel formulations before and after ethylene oxide sterilization were calculated and compared to standard soft contact lenses.

METHODS

Films were fabricated by varying the CMHA-based polymers (thiolated CMHA alone versus thiolated CMHA with thiolated gelatin). Poly(ethyleneglycol) diacrylate (PEGDA) and glutathione disulfide (GSSG) were used as cross-linkers. This resulted in four formulations: CMHA-GSSG, CMHA-Gelin-GSSG, CMHA-Gelin-PEGDA, and CMHA-PEGDA. Each polymerized gel was dried in a silicone mold at room temperature overnight. Hydrogel samples were obtained as dried sheets. These samples were cut into strips approximately 6mm x 10mm and hydrated for 24 hours in PBS solution. On the day of testing, the dimensions of the strips were measured in three locations along the length of the film using an optical microscope (Figure 1A) and then placed between custom screw-driven clamps designed to hold fragile tissue (Figure 1B). Specimens and clamps were located in a PBS chamber attached to a material test system (Model 5943, Instron) and a 500 gram submersible load cell (LSB210, Futek). A stress relaxation protocol was applied at a rate of 1.0 in/min that pulled the hydrated films in tension up to 9.0% strain in increments of 0.5% strain. Each strain increment was held for 30 seconds and then returned to 0% for 60 seconds

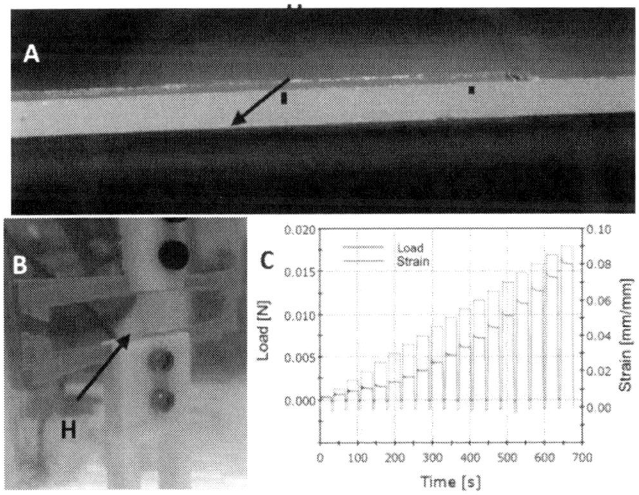

FIGURE 1. (A) HYDROGEL FILM (H) ON A GLASS SLIDE (S). THICKNESS WAS MEASURED IN 3 LOCATIONS (LINES). (B) FILMS WERE MOUNTED IN CUSTOM GRIPS IN A PBS BATH, AND THEN (C) PULLED IN A SERIES OF STRESS RELAXATION TESTS.

before the next increment of strain (Figure 1C). After reaching 9.0% strain, the specimen was relaxed for 60 seconds and then pulled to failure. The plateau stress from each stress relaxation curve was plotted against its associated strain increment. This resulted in a bilinear relaxation stress-strain curve. The linear slope between 0.5-4% and 5-9% strain was defined as the low and high strain relaxation modulus, respectively (Figure 2). The elastic modulus was extracted from the linear region of the pull-to-failure stress-strain curve. To identify significant differences with formulation and sterilization a two-way ANOVA was performed on thickness, relaxation moduli, and

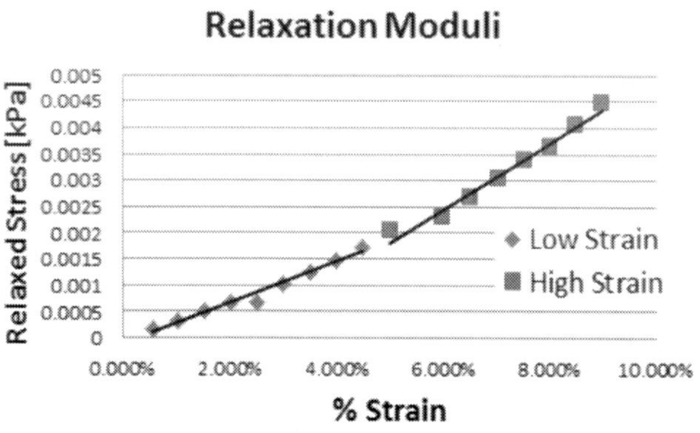

FIGURE 2. RELAXATION STRESS-STRAIN CURVE WITH LOW (0-4%) AND HIGH (5-9%) STRAIN REGIONS.

pull-to-failure modulus. A p-value < 0.05 was considered significant.

RESULTS

CMHA-PEGDA film thickness was significantly larger ($p=0.001$) than all other formulations (Table 1). Sterilization significantly ($p<0.001$) reduced film thickness by an average of 23.5%, with the maximum change occurring in the CMHA-PEGDA film. Sterilization also made hydrogels significantly stiffer ($p<0.0001$) by 54%-97% with maximal property changes occurring in the CMHA-GSSG formulation (Figure 3). Among the unsterilized films, CMHA-PEGDA had a significantly lower average relaxation modulus for both low and high strains (35.367±3.179 kPa and 55.367±7.651 kPa, respectively, $p<0.05$) and a lower instantaneous elastic modulus (105.325±45.95 kPa) than the other film formulations. After sterilization this finding was not significant. Unsterilized CMHA-Gelin-GSSG had the stiffest relaxation moduli at both low and high strains (303.1 ± 215.4 kPa and 486.5 ± 200.4 kPa, respectively, $p<0.05$), but after sterilization CMHA-GSSG was the stiffest. CMHA-PEGDA still had the lowest moduli post-sterilization.

TABLE 1. TABLE OF AVERAGE THICKNESS BEFORE AND AFTER ETO STERILIZATION

Formulations	Non-Sterilized	Sterilized	% Diff
CMHA-GSSG	0.135 ± 0.026	0.121 ± 0.023	10.4%
CMHA-Gelin-GSSG	0.135 ± 0.068	0.105 ± 0.047	22.2%
CMHA-Gelin-PEGDA	0.186 ± 0.048	0.152 ± 0.024	18.3%
CMHA-PEGDA	0.281 ± 0.092	0.160 ± 0.067	43.1%

DISCUSSION

Stiffness has been found to correlate with comfort in contact lenses [1]. As the relaxation modulus increases, the perceived comfort in the wearer decreases. Based on these findings, CMHA-PEGDA would be assumed to be the most comfortable in the eye because it has the lowest relaxation modulus. Sterilization significantly increased moduli across all formulations, but CMHA-PEDGA still had the lowest modulus after sterilization. When comparing the relaxation moduli from the unsterilized films to reported moduli of soft contact lenses reported in the literature [2], all formulations excluding CMHA-Gelin-GSSG, were much lower than a variety of over-the-counter soft contact lens (E = 372 kPa–1625 kPa). However, after sterilization, only CMHA-PEGDA was less stiff than the majority of the soft contact lenses. This suggests that out of the four formulations, CMHA-PEGDA will be the most promising in terms of comfort for an ocular surface bandage ad or drug releasing film

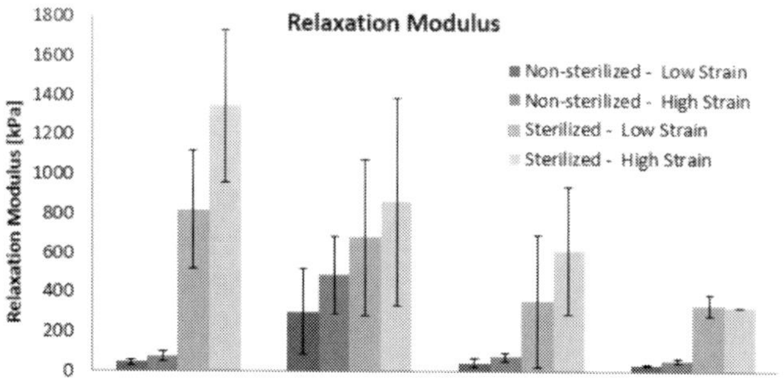

FIGURE 3. RELAXATION MODULUS OF ALL FORMULATIONS BEFORE AND AFTER ETO STERILIZATION.

ACKNOWLEDGEMENTS

The authors would like to thank USAMRRA Phase I SBIR grant#W81XWH-14-C-0025 for funding this work.

REFERENCES

[1] Robitaille, M., *Journal of the Mechanical Behavior of Biomedical Materials,* 22:59-64, 2013
[2] Tranoudis, I., Efron, N., *Contact Lens & Anterior Eye*, 27: 177–191, 2004

CHARACTERIZATION OF FATIGUE FAILURE IN BOVINE MENISCUS

Jaremy J. Creechley (1), German Martinez (2), Trevor Lujan (3),

(1) Material Science and Engineering
Boise State University
Boise, Idaho, USA

(2) Mechanical and Biomedical Engineering
Boise State University
Boise, Idaho, USA

(3) Mechanical and Biomedical Engineering
Boise State University
Boise, Idaho, USA

INTRODUCTION

Meniscus tears are a common musculoskeletal injury, with over 850,000 cases each year in the United States. [1] Due to the increased risk of osteoarthritis following a meniscus tear or meniscectomy it is critical to understand meniscus failure modes to improve treatment and prevention.[2] While tears are most commonly associated with a traumatic event, up to a third of injuries are related to fatigue failure wherein an onset of symptoms develops without a related acute trauma or degenerative condition.[3] The existence of fatigue failure in fibrous soft tissue is supported by previous research in human tendons, which showed that structural damage from cyclic loading ultimately leads to acute failure of the tendons.[4] Characterization of meniscus fatigue failure behavior may help explain the failure modes corresponding to both acute and overuse injuries. This characterization must account for the transversely isotropic material symmetry of meniscus, as radial tears occur normal to the preferred collagen fiber direction, and vertical tears occur along the preferred fiber direction. The objective of this study is to characterize the transversely isotropic fatigue behavior of bovine meniscus.

METHODS

Overview. Fatigue failure was characterized in the lateral bovine meniscus along longitudinal (LG) and vertical transverse (TR) directions respective to the primary circumferential fiber orientation. Fatigue tests were performed in two stages. In the first stage, meniscus samples were loaded to failure to characterize the UTS behavior. In the second stage, samples were cycled at 90, 80, 70, and 60% of their predicted UTS. A Weibull distribution was used to fit the fatigue data to a linear-log line.

Specimen Preparation and Measurement. Bovine menisci were harvested and frozen (Fig. 1A). They were segmented into three wedges with the posterior segment (Wedge A) prepared for testing (Fig. 1B). Each wedge was circumferentially sliced into *0.92 (± 0.17) mm* thick layers demarcated into 4 zones approximately 3-6 layers deep (Fig. 1B). Test layers starting in zone 1 (at least 3 mm deep) were sliced, with 4 layers in total (Fig. 1B). This labeling convention follows that found in a recent clinical survey of meniscal tears, where wedge A accounted for ≥ 50% of all clinical tears [5].

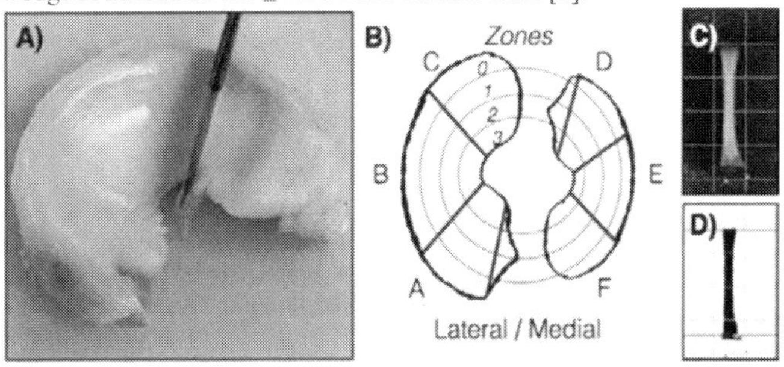

Figure 1 Bovine meniscus sample preparation. A) Whole Meniscus, B) Labeling Convention, C) Specimen clamped for testing, D) Binarized Image for Specimen Measurement.

Each layer yielded 2 LG or 2 TR samples. Samples were prepared with a custom 3D printed razor jig designed to cut dog-bone shaped samples. Cutting dimensions were 1.00/0.80 mm (LG/TR) in width and ≥20/10 mm (LG/TR) in length. Tab flanges followed a hyperbolic curb to reduce stress concentrators (Fig. 1C). The minimal viable width was 3.2 mm. Limitations of meniscal sizes preclude matching ASTM D3479 standards. The razor jig enabled cutting sufficient aspect ratios to prevent clamp site failures or slippage in tested samples, both important criterions for viable tests.

Ultimate Tensile Strength. A total of 12/12 (LG/TR) samples prepared from 3 separate animals were tested for UTS. Each sample was clamped at 14.00/7.00 mm (LG/TR), preloaded to 0.5/0.10 N (LG/TR), and then imaged from fixed frontal and side aspects using a 14 megapixel digital camera. Preconditioning was applied by a 20-cycle triangular tension-tension wave at 2Hz and 8% gauge strain. After a 5 second rest, quasi-static UTS tensile testing was performed at 0.2 mm/s. The tangent modulus was obtained by linear fits of stress/strain on the last 3rd of the cycle preceding the 19th peak.

Custom software leveraging SciPy image processing libraries performed the digital image measurements.[6] Specimen images were processed and the widths of the middle third of the gauge length were averaged. The standard deviation for UTS samples was $\leq \pm 0.1$ mm.

Fatigue Testing. A total of 5 bovine menisci from separate animals were prepared for fatigue testing. Each meniscus provided 4 layers, two samples per layer. Samples were prepared accordingly to the described method yielding 4 LG/4 TR per meniscus. Samples were tested at 90/80 or 70/60% stress levels per layer and orientation. Samples were preloaded, preconditioned, and measured using the same methodology described for the UTS tests. A drip of 9% saline maintained the hydration of the samples. Preconditioning data was used to predict specimen UTS (see formulas, Fig. 2) and normalized stress level, from which the amplitude of a cyclic tension-tension sine wave was determined and applied at 2.0/3.0 Hz (LG/TR). The cyclic stress was maintained until failure and the number of cycles to failure was recorded. Samples that did not fail after $10^3/10^4$ (LG/TR) cycles were subjected to the aforementioned quasi-static UTS testing procedure.

Statistical data analysis followed established methods, with tendon fatigue damage being well described by the Weibull distribution. [7,8] A two-parameter Weibull distribution produces averaged cycles to fatigue (N_0), and a shape parameter (α) per normalized stress level (σ_{level}). A linear fit of the log-log of the N_0 and σ_{level} or a power law to produce a standardized S/N curve (e.g. σ_{level} vs. N, see Fig 2).

RESULTS

A strong linear correlation existed between tangent modulus and UTS for both LG and TR tests (see Fig. 2). LG and TR fatigue life results best followed linear log-log fitting equations (Fig. 3), where N denotes the average cycles to failure and σ_{Level} the normalized stress level. A power law fit also yielded a good fit for LG (K=*130.12, B=-10.0, R^2 = 0.82)* but a poor fit for TR (K=*96.45, B=-20.0, R^2 = 0.33).*

DISCUSSION

This study characterizes the fatigue behavior of the bovine meniscus of both longitudinal and vertical transverse directions. The experimental data provides a good linear-log fit of the S/N cycle life of both LG and TR testing direction (Fig. 3A and 3B). In comparison, power law fits provided poor correlation. All fatigue results are based on the stress level prediction using the linear correlation of the tangent modulus and the UTS. To our knowledge, these results provide the first characterization of transversely isotropic fatigue life of any soft tissue and the first fatigue behavior for meniscal soft tissue.

The results from this study provide insight into meniscal fatigue behavior and indicate that the slopes of the linear fits (-7.62±2.12 LG/-6.14±0.82 TR) are not significantly different. The values of the slopes do differ significantly compared to human tendons which have twice the negative slope at -14.8.[4] This indicates that the exponential accrual of rate of fatigue damage in the meniscus will be half that of tendon fatigue. In addition, our results indicate that UTS can be

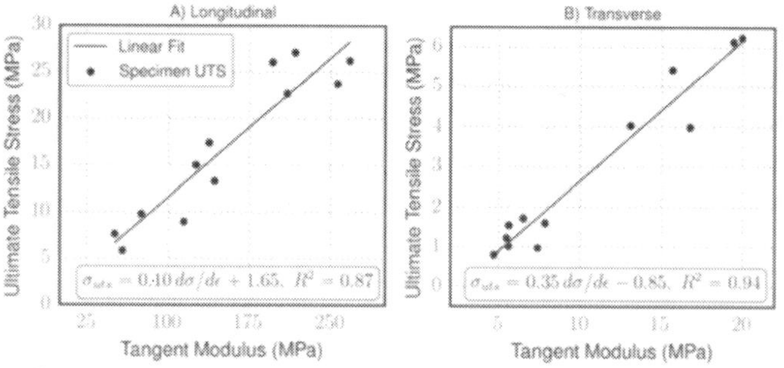

Figure 2 Results of direction dependent UTS Tests and linear correlation. A) Longitudinal, B) Transverse.

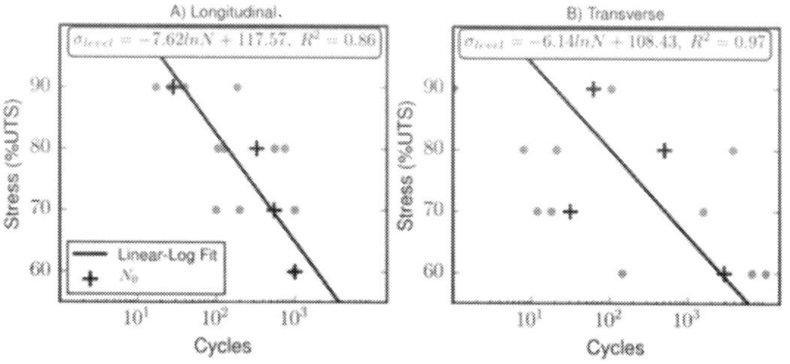

Figure 3 Low Cycle Fatigue Life based on specimen direction. A) Longitudinal, B) Transverse.

predicted for any meniscus specimen regardless of fiber orientation by using the tangent modulus of the stress and strain.

Application of traditional fatigue characterization techniques to soft tissues presents many challenges [4,9]. In particular, large variations in the UTS among transverse samples, even taken immediately adjacent to one another, can vary by a factor of 2-3 (Fig. 3). This precludes determining a normalized UTS level for fatigue testing. To overcome this challenge, UTS was estimated for each fatigue specimen.

Limitations of this work limit the general applicability of theses results. To provide better fit, more samples will be needed to increase the average accuracy of the UTS prediction method to within 5% for more accurate determination of stress levels. Proper censoring techniques should be applied to the run off fatigue values. Currently it is unknown if the UTS/Tangent modulus correlation is a fully geometry and/or material independent property.

Summary. This work describes several novel experimental techniques to perform soft tissue fatigue in transverse and longitudinal fiber directions with constrained sample sizes. Initial experimental data is provided for low cycle fatigue behavior of bovine meniscus.

ACKNOWLEDGEMENTS

We are grateful for the support of provided by the Dept. of Material Science and Engineering at Boise State University.

REFERENCES

[1] Baker, P et al., J. Rheumatol., 29:557–563, 2002.
[2] Messner, K, Gao, J., J. Anat., 193 Pt 2:161–178, 1998.
[3] Camanho, G., Acta Ortopédica Bras., 17[1]:31–34, 2009.
[4] Schechtman, H, Bader, D, J. Biomech., 9290[97]:347–53, 2002.
[5] Metcalf, M et al., Am. J. Sports Med., 32[3]:675–680, 2004.
[6] Oliphant, T et al., Comput. Sci. Eng., 9:10–2, 2007.
[7] Whitney, Fat. Fib. Compo. Mater. ASTM STP, 723:133–151,1981.
[8] Ker, et al, J. Exp. Biol., 203[Pt 8]:1317–1327, 2000.
[9] Gallagher, S., Proc. Hum. Factors Ergon. 1130–1134,2012.

SB³C2015
Summer Biomechanics, Bioengineering and Biotransport Conference
June 17-20, 2015, Snowbird Resort, Utah, USA
SB³C2015-1188

BEHAVIORAL AND INFLAMMATORY CONSEQUENCES OF CEREBROVASCUAL DYSFUNCTION IN PRIMARY BLAST INJURY

Stewart Yeoh (1), Kenneth L. Monson (1,2)

(1) Department of Bioengineering
University of Utah
Salt Lake City, UT, USA

(2) Department of Mechanical Engineering
University of Utah
Salt Lake City, UT, USA

INTRODUCTION

Traumatic brain injury resulting from exposure to explosive blast is an ever-present risk for military populations and better understanding of injury mechanisms and progression will provide the knowledge foundation for improved therapies. Blast injury has been shown to produce neuronal swelling, oxidative stress, cognitive impairment, and blood brain barrier (BBB) dysfunction across a variety of testing conditions.[1-3] Using a novel blast device, we have shown that short overpressure duration blast creates focal BBB disruptions that appear as IgG deposits in the parenchymal brain tissue.[4] Our hypothesis is that reactive gliosis and neuronal cell death would occur at lesion locations; resulting in cognitive and motor deficits in the animal's behavior. Cellular changes after injury were measured using immunohistological techniques, and behavioral testing was performed to assess cognitive dysfunction.

METHODS

Primary blast overpressure was generated by detonating an explosive primer in an otherwise empty casing in a modified 0.308 bolt-action rifle. Male Sprague-Dawley rats (n=6 per group) were anesthetized and exposed to a blast wave of 320 kPa overpressure and 15 us duration in isolated primary blast loading.[4] Open field and novel object recognition tests were performed to determine the effects of blast on memory, spatial learning, and anxiety levels. Animal movement was recorded and video analyzed using Ethovison tracking software. A standard set of Von Frey fibers was also used test nociception thresholds and motor response.

A series of immunohistological markers were used to localize BBB disruption and subsequent inflammation or neuronal death. Following sacrifice at 2, 7 and 30 days; a significant increase in the number and volume of BBB compromised regions appeared in the brain tissue. Immunoglobulin G (IgG) extravasation was used to detect displaced plasma, while degenerating neurons around the lesion location were identified using Fluoro-Jade B. A separate set of adjacent sections were analyzed for glial fibrillary acidic protein (GFAP), a protein expressed by astrocytes undergoing astrogliosis. Macrophage response was investigated using CD68 and ED-1 markers. Finally, a neuronal nuclei (NeuN) and neurofilament (NF-160) combined marker was used to visualize density of both nerve bodies and processes. Images were recorded using epifluorescent and confocal microscopy, and fluorescence intensity quantified using thresholding in NI Vision Assistant.

RESULTS

Animals exposed to blast overpressure had significantly ($p=0.004$) more lesioned area than controls. An early reduction in GFAP immunoreactivity was collocated with the central zone of IgG lesion locations, while an increase in GFAP intensity was observed around the periphery of the injury (Fig. 1).

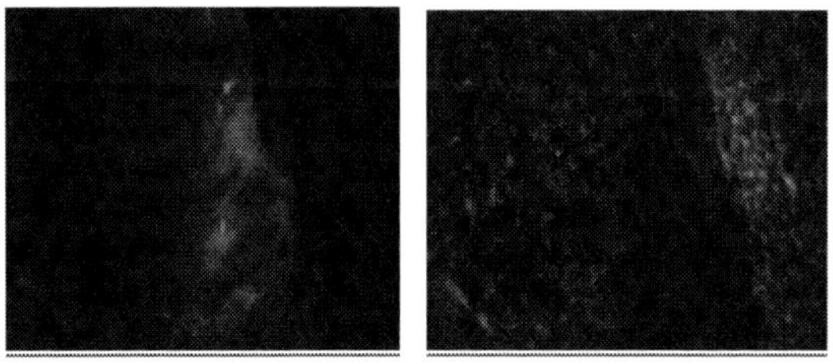

Figure 1: Incidence of IgG in blast exposed brains, with collocated reduction in GFAP expression.

Neuronal degeneration was qualitatively observed with the NeuN, NF-160, and FluoroJade stains. Evidence of macrophage activation was also seen with the ED-1 marker.

Subtle differences in behavioral response were observed in all three behavioral paradigms. The Von Frey fiber nociception and tactile assay showed an increase in animal pedal withdrawal thresholds following blast injury. The open field exploration test displayed a non-significant decrease in total animal movement, as well as changes in the ratio of center time vs. edge time (thigmotaxis), a measure of animal anxiety (Fig. 2).

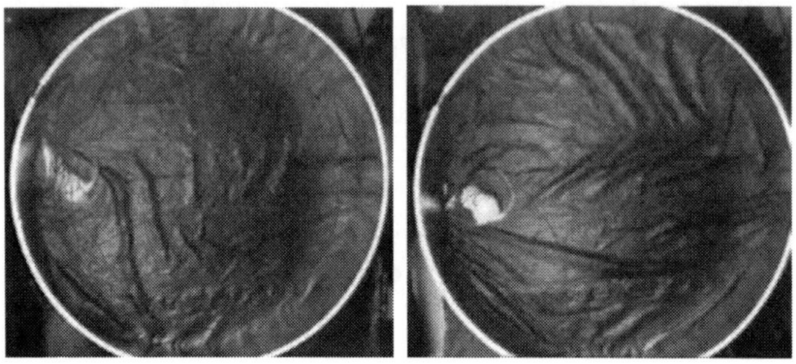

Figure 2: Open field testing arena displaying lower motor exploration and increased anxiety and thigmotaxis in animal on the right.

Novel object recognition, a metric for spatial learning and memory, showed some unusual changes in animals post blast. Exploration time of novel object vs. sample object became highly erratic in blast exposed animals in a time dependent fashion.

DISCUSSION

A mild inflammatory response was observed as a result of low impulse blast exposure and subsequent focal BBB disruption. Short term survivals displayed an increase in GFAP immunofluorescence around the periphery of the IgG zone, a finding that has not been reported in other studies of blast injury. GFAP levels are commonly assayed in bTBI and TBI, however results have varied from a general increase,[2,6,7] a decrease (6 hours) followed by an increase (24 hours),[8] or no change in levels.[9,10,11] Behavioral tests showed reduced cognitive and behavioral performance for animals exposed to blast. In the open field testing paradigm anxious animals tend to move along the walls of the apparatus and refrain from the center, while the opposite is seen for non-stressed animals. Only a small effect from our low impulse blast exposure was seen on gross motor function or animal anxiety, supporting our suspicion that our testing apparatus is at the threshold of detectable injury. Total distance traveled for stressed or motor impaired animals is less than naive animals. Literature2 suggests that bTBI animals would be less sensitive to the Von Fray hair test, and require a stiffer fiber to elicit a pedal withdrawal reflex. We saw that this was indeed the case.

Injury mechanics are currently unclear; however shearing, spalling, and cavitation have been suggested as brain tissue absorbs energy from the overpressure wave.[2] Other hypotheses include vascular surge from blast exposure to the body, and skull flexure.[6,13] It is also unclear what relationship the physical characteristics of the blast wave, such as positive pressure phase duration, negative phase, total impulse, and peak pressure have to injury severity. We hope to determine mechanisms driving vascular injury and consequences of blood brain barrier failure using a low impulse blast tube. Lack of complete insight about blast TBI and TBI in general has resulted in limited clinical options for those afflicted. Revealing details about vascular injury in blast will help advance medical therapies and will also allow for the creation of more complete constitutive models for injury simulation and protective gear design. Injury from blast overpressure remains a major challenge; however greater understanding of injury mechanisms will lead to new treatment paths and improved quality of life for those afflicted.

ACKNOWLEDGEMENTS

This work was supported by University of Utah research startup funds and grant from the Department of Defense (W81XWH-08-1-0285)

REFERENCES

[1] Long JB, et al., *J Neurotrauma* 2009; 26: 827-40.
[2] Cernak I, et al., *J Trauma* 2001; 50: 695-706.
[3] Bauman RA, et al., *J Neurotrauma* 2009; 26: 841-60.
[4] Yeoh S, et al., *Ann Biomed Eng* 2013; 41: 2206-14.
[5] Garman RH, et al., *J Neurotrauma* 2011; 28: 947-59.
[6] Vandevord P, et al., *Ann Biomed Eng*. 2012 Jan;40(1):227-3
[7] Kamnaksh A, et al., *J Neurotrauma*. 2011 Oct;28(10):2145-53
[8] Arun P, et al., *Neurosci Lett*. 2013 Sep 27;552:87-91
[9] Baalman KL, et al., *J Neurotrauma*. 2013 May 1;30(9):741-51
[10] Heldt SA, et al., *Front Neurol*. 2014 Jan 22;5:2
[11] Garman RH, et al., *J Neurotrauma*. 2011 Jun;28(6):947-59
[12] Genovese RF, et al. *Neuroscience* 2013; 254: 120-9.
[13] Bolander R, et al., *Ann Biomed Eng* 2011; 39: 2550-9.

MECHANICAL DESIGN OF A CUSTOMIZABLE SELF-EXPANDING ENDOVASCULAR STENT

Joel C. Scott (1), Clifton R. Johnston (1), Darrel A. Doman (1)

(1) Mechanical Engineering
Dalhousie University
Halifax, Nova Scotia, Canada

INTRODUCTION

Endovascular stents have been used for several years in the management of occlusive, degenerative and aneurismal arterial disease. As they have become more popular, complications associated with their use have become more evident. Restenosis, proximal neck dilation, stent migration and endoleak are common failure mechanisms, all of which can be at least partially attributed to the physical and mechanical properties of these stents. The mechanics of endovascular stents are still not well understood. Stent migration, which is primarily caused by endoleak, has been noted with all current endovascular devices and is the most common type of failure associated with self-expanding stents [1].

Radial springs are used as the primary method of arterial fixation for self-expanding endovascular stents. Incorrect design of these springs will consequently lead to reduced patency of stents with artery walls and an increased probability of stent failure. Stents should be designed to withstand the hemodynamic forces and drag to which they are subjected. Larger diameter stents should produce greater radial force to avoid migration and endoleak. However, comparing similar stents of various diameters revealed the opposite, smaller diameter stents exerted greater radial force than larger diameter stents (Figure 1).

With this knowledge, it is important to work toward the ability to design mechanically correct stents that are capable of applying an appropriate amount of radial force for the situation in which they are implemented.

METHODS

A novel radial extensometer was developed specifically for testing the radial force of expanding stents with higher resolution, precision and repeatability than previous models. Introduction of vision tracking has also enabled collection of stent diameter data throughout the expansion process.

Finite Element Modeling (FEM) was used to simulate a 12mm single wire stent being constricted and expanded in the radial extensometer. This model has been used to test existing stents and will be used to test new designs allowing examination of the impact of changes without having to commit to a costly, time consuming manufacturing process.

Data from both the radial extensometer and the FEM model have been used to create tools that can be used to precisely customize a stent based on patient specific data prior to implantation.

RESULTS

Reproducing existing data trends [2] by showing an inverse relationship between stent diameter and radial force (Figure 1) has reinforced the need for improvement of currently available self-expanding endovascular stents.

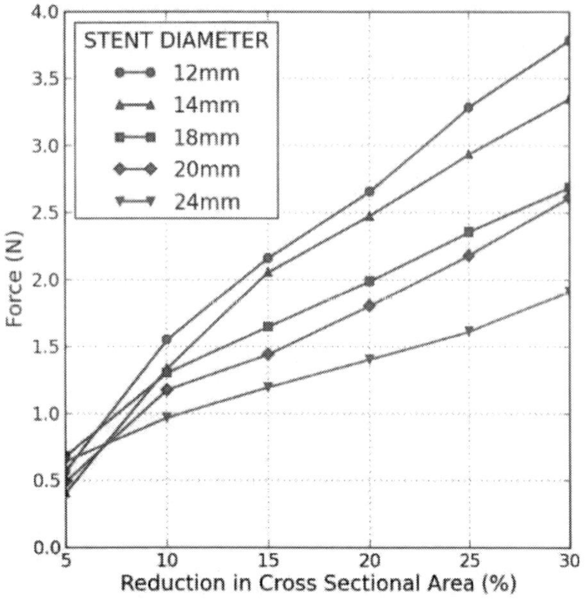

Figure 1: Stent Size Comparison

Experimental data collected by the radial extensometer has been compared to data output by FEM simulations performed using LS-DYNA. During expansion (after time = 115s in figure 2) the coefficient of determination between experimental and simulated data is calculated to be 0.947 with a standard error of the estimate of 0.00046.

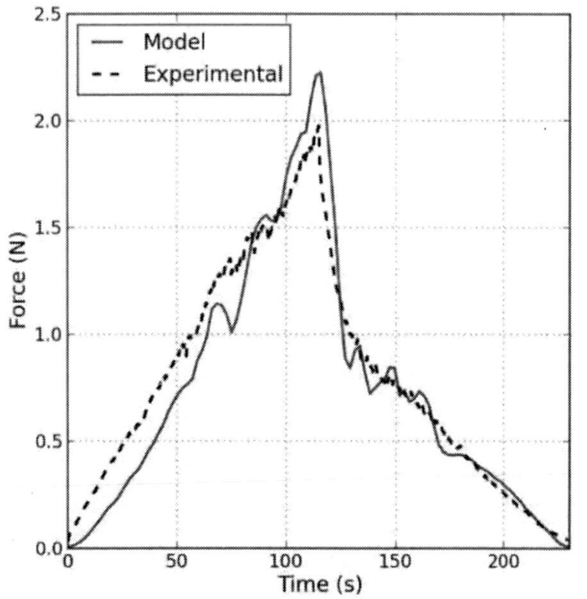

Figure 2: Comparison of Simulated and Experimental Results for Testing of a 12mm Single Wire Stent

DISCUSSION

FEM and radial extensometer data have been combined to create several useful tools which together, will enable the design of an entirely customizable z-type stent. The combination of these tools will allow even further progress into patient specific design, as well as simulation and testing of completely unique stent geometries for any field of use. This combination of resources will help develop a better understanding of stent mechanics and their interaction with vessel walls.

A series of adjustable wire bending jigs were designed to manufacture stents for testing. They allow for manipulation of four main variables that differ among stents: leg length, bend radius, bend angle and wire diameter. Stent built using these jigs were used to collect data relating to the effect on radial force following the manipulation of any of these four variables. The bending jig is simple to use, precise, repeatable, cost effective and is manufactured using rapid prototyping

Using stents produced with the wire bending jigs and data collected from the radial extensometer, a program was developed which can predict the expected radial force generation of stents with any combination of the four main variables manipulated by the jigs. Radial force values react to different arterial diameters and compliance while all variables can be adjusted until the expected radial force falls within a specific "safe zone". Use of this tool can help with the design of stents for specific patient requirements as well as to verify the effectiveness of stent selection.

It is our intention that these tools and data be used not only to expose the mechanism through which radial force is generated by z-type endovascular stents, but also to further the design of current stents, in an attempt to contribute to a reduction in failure rate.

ACKNOWLEDGEMENTS

This work has been supported by a grant from the Natural Sciences and Engineering Research Council of Canada.

REFERENCES

1. Andreas Greiner, Jochen Grommes, and Michael J. Jacobs. "The Place of Endovascular Treatment in Abdominal Aortic Aneurysm". In: Dtsch Arztebl International 110.8 (2013), pp. 119–125. URL: http://www.aerzteblatt.de/int/article.asp?id=134823.
2. Clifton R. Johnston et al. "The Mechanical Properties of Endovascular Stents: An In Vitro Assessment". In: Springer Cardiovascular Engineering 1.10 (Aug. 2010), pp. 128–135.

COMPUTATIONAL ANALYSIS OF THE EFFECT OF SEQUENTIAL COILING ON THE WALL STRESS OF CEREBRAL ANEURYSMS

Joseph E. Pichamuthu[1,2,6], **Brian T. Jankowitz**[5], **David A. Vorp**[1,2,3,4,6]

1. Department of Bioengineering, University of Pittsburgh, Swanson School of Engineering Pittsburgh, PA

2. Center for Vascular Remodeling and Regeneration, University of Pittsburgh, Pittsburgh, PA

3. Department of Cardiothoracic Surgery, University of Pittsburgh, Pittsburgh, PA

4. Department of Surgery, University of Pittsburgh, Pittsburgh, PA

5. Department of Neurosurgery, University of Pittsburgh, Pittsburgh, PA

6. McGowan Institute for Regenerative medicine, University of Pittsburgh, Pittsburgh, PA

INTRODUCTION

Endovascular coil embolization is commonly used for treating cerebral aneurysms [1]. It involves sequentially placing metallic coils in the aneurysmal dome to induce thrombosis, resulting in the formation of a coil thrombus mass (CTM). The CTM formed diverts the blood flow away from the sac. The CTM also acts as a cushion to dampen the impact of hemodynamic forces exerted on the aneurysmal wall. Success of treatment mainly depends on the morphological features and size of the aneurysm, and the relationship of the aneurysm to the cerebral arteries. However, a significant number of aneurysms are incompletely coiled, potentially leading to aneurysm regrowth and recanalization [2].

Computational models of aneurysm coiling have determined alterations on wall shear stress in the presence of the CTM [3], but the effect on wall stress has not yet been reported. We believe that wall stress assessment will help the surgeon determine the level of coil embolization that will prevent recanalization and the need for subsequent surgical intervention. Therefore, the purpose of this work was to simulate seven patient-specific cerebral aneurysms using finite element analysis (FEA) and analyze the effect of CTM on wall stress.

METHODS

Digital subtraction angiography (DSA) image data of seven patients undergoing endovascular coiling treatment were obtained after IRB approval. The 3D geometry of the aneurysmal sac and parent vessel was reconstructed by pixel thresholding and segmentation in Mimics (Materialise, Plymouth, MI). The geometry was smoothed and exported as the aneurysm wall surface geometry. For the 3D geometry of the CTM, the aneurysmal wall surface was imported into Solidworks (2008, Dassault systems, Waltham, MA) and the aneurysm sac was filled with a solid. The distance between the parallel planes separating the sac base and dome was measured as aneurysm length, and the solid sac was segmented into 25%, 50% and 75% of this length, which were used to quantify the sequential coiling simulation.

The cerebral aneurysm wall was modeled as a shell of homogeneous material with uniformly distributed thickness. The wall was considered nonlinear, isotropic, hyperelastic and incompressible. We implemented in Abaqus 6.12 FEA software (Simulia, Providence, RI) the following 3-parameter Mooney-Rivlin strain energy function for the sac and parent vessel using UHYPER subroutine [4]

$$W = C_{10}(I_1-3) + C_{01}(I_2-3) + C_{11}(I_1-3)(I_2-3)\,, \qquad (1)$$

where C_{10}, C_{01} and C_{11} are constants with values 0.0598, 0.0168, and 5.71 MPa, respectively, characterizing the mechanical behavior of the material [4], and I_1 and I_2 are strain invariants. We modeled the CTM as a nonlinear, isotropic, hyperelastic, homogeneous and incompressible solid material as previously reported by our group for laminated intraluminal thrombus of abdominal aortic aneurysm. A no-slip condition was applied to the CTM-wall interface. The ends of the parent vessel were fixed in space and systolic pressure (120 mmHg) was applied uniformly on the lumen (i.e., the blood-contacting surfaces of both the aneurysm wall and the CTM).

RESULTS

Of the seven aneurysms studied, four were side-wall aneurysms while the other three were "branched", occurring with multiple inlet and outlet vessels. Wall stress was substantially reduced as the degree of sequential CTM was increased in side-wall cases but it did not reduce wall stress in branched aneurysms. None of the seven aneurysms analyzed in this study had peak von Mises wall stress (PWS) occurring on the parent vessel with or without CTM. The von

Mises wall stress distribution observed in two representative aneurysms before and after CTM is shown in **Fig. 1**.

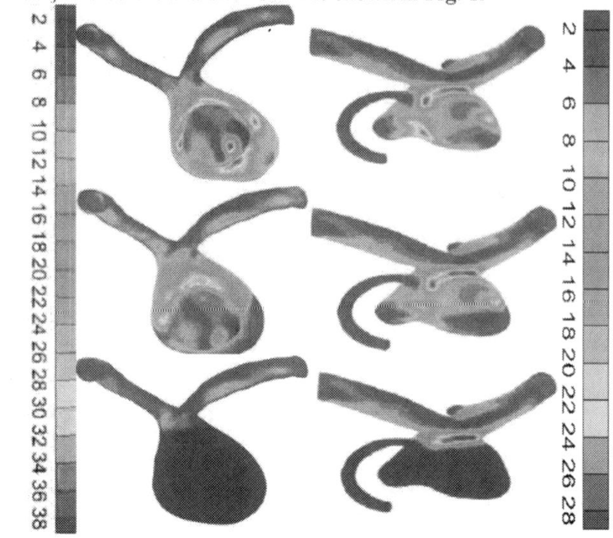

Figure 1: von Mises wall stress distribution in representative side-wall (left) and branched (right) aneurysms. PWS was maximum prior to sequential coiling (top row). For 50% CTM (middle row), the PWS was partially reduced in the side-wall but not in the branched aneurysm. When completely filled (Bottom row: 100% CTM), coiling effectively reduces PWS in the side-wall aneurysm but no beneficial effect on PWS was seen in branched aneurysm.

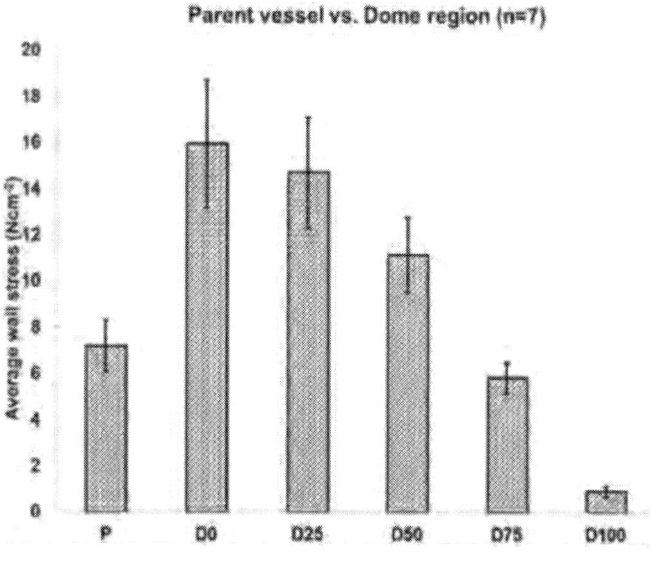

Figure 2: Comparison of the average wall stress (AWS) estimated for the aneurysmal dome region for different degrees of CTM filling (D0, D25, D50, D75, and D100) and for the parent vessel (P) in each patient model. Notice that the AWS of the unfilled aneurysmal dome region was twice that of the parent vessel in most of the models. The AWS of dome decreased to the level of the parent vessel in all cases for 75% filling suggesting threshold filling. The AWS was significantly lower than the parent vessel for 100% filling.

Within each model, the average wall stresses (AWS) estimated for the aneurysmal dome region for each degree of CTM filling (D0, D25, D50, D75 & D100) and for the parent vessels (P) were compared. **Fig. 2** reflects the trend across our patient pool, with wall stress averaged across all seven cases.

The wall stress at the location of PWS prior to CTM filling (CTM=0) was assessed in each model at the different degrees of CTM filling. These stresses were averaged for all side-wall and branched aneurysms (**Fig. 3**). For 100% CTM, the wall stress was reduced to nearly zero in side-wall aneurysms, whereas in branched aneurysms it was reduced markedly, but not as much.

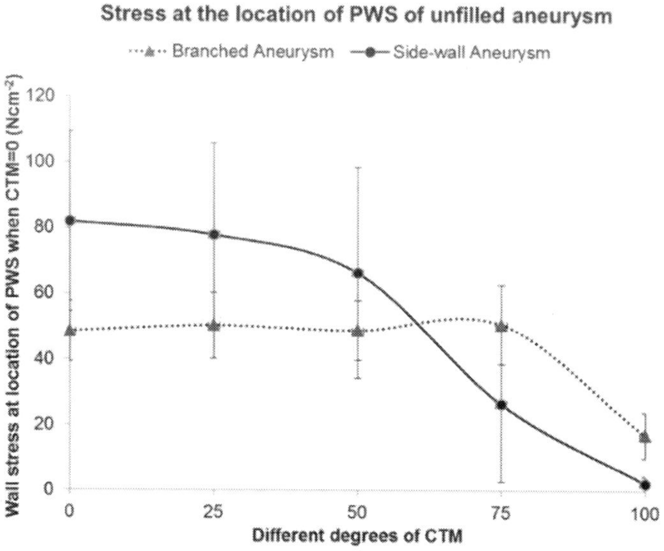

Figure 3: The wall stress averaged for side-wall and branched aneurysms. CTM greatly reduces PWS for all degrees of filling in side-walled aneurysms. However, in branched aneurysms, CTM did not reduce PWS until after 75% filling.

DISCUSSION

Prior to coil embolization in both types of aneurysm, the PWS was located in the aneurysmal region, either at the neck or at the dome, and the AWS of the dome region was at least twice that of the parent vessel. This is consistent with the greater propensity for rupture at the dome region seen clinically [5]. The CTM reduced the PWS in the dome region but the magnitude of PWS reduction depended on both the degree of coil filling and the aneurysm type. Our data suggests that for any useful outcome of coil embolization the coil must cover the area of the wall where PWS occurs pre-treatment. The AWS of the dome was substantially reduced either to the level of AWS of parent vessel or less for 75% or more CTM, suggesting the existence of a threshold level of CTM filling for any beneficial wall stress shielding effect. CTM filling did not show as much of a beneficial effect in branched aneurysm models as the PWS either remained nearly same at the neck or even elevated for different degrees of CTM filling. This may explain why certain patients are prone to recurrences and recanalization at the neck of the aneurysm after coil embolization.

REFERENCES

[1] Ryttlefors, M et al., Stroke, 2008. 39(10): p. 2720-6.
[2] Waldron,J.S., et al., *Neurosurgery*, 2009. 64(5 Suppl 2): p. 301-15;
[3] Kakalis, N.M., et al.,*IEEE tran med imag*, 2008. 27(6): p. 814-24.
[4] Costalat, V., et al., *J Biomech*, 2011. 44(15): p. 2685-91
[5] Crompton, M.R., *Br Med J*, 1966. 1(5496): p. 1138-42

FABRICATION OF SCAFFOLDS FROM DIFFERENT SILICA BASED BIOACTIVE GLASS AND INVESTIGATION OF BIOACTIVITY AND BIODEGREDATION BEHAVIORS

S. YUCEL (1), A.C. OZARSLAN (1), B.S. ODUNCU (1), P. TERZIOGLU (2)

(1) Yildiz Technical University, Faculty of Chemical and Metallurgical Engineering, Department of Bioengineering, 34220, Esenler/İstanbul
(2) Muğla Sıtkı Koçman University, Faculty of Sciences, Department of Chemistry, Muğla, Turkey

INTRODUCTION

The scaffold used for tissue engineering is key component of tissue healing [1]. Bioactive glasses are preferred as scaffold materials because of excellent properties such as biocompatible, biodegradable and bioactive and highly porous structure. When bioactive glasses are placed in a human bone tissue, they form a bond easily with tissues via formation of hydroxyapatite on the surface layer. In addition, bioactive glasses stimulate the cellular activity. In this study, rice hull ash (RHA), rich in silica (about 60%), waste product of rice industry is used as an alternative cheap source of amorphous silica. The aim of the present study was to investigate the production and characterization of three-dimensional tissue scaffolds for bone regeneration by using commercial silica-based 45S5 and silica-based rice hull ash 45S5 mixture. The effects of RHA silica use on bioactivity and biodegradation was investigated and results compared with commercial silica [2, 3].

Keywords: bioactive glass, scaffold, tissue engineering

METHODS

Bioactive glass production

Rice hull ash was burned at 600 °C in muffle furnace for 5 h. Obtained ash was mixed sodium hydroxide (NaOH) for 1h. Sodium silicate solution was obtained from rice hull ash by Kalapathy method [4]. Silica was produced by sodium silicate solution upon treatment with strong acid. Commercial silica-based 45S5 bioactive glass and rice hull ash silica-based 45S5 bioactive glass were produced by melt-drived method at 1350 °C.

Scaffold fabrication

Scaffolds were manufactured using bioactive glasses with polymer foam replica method. Firstly, bioactive glasses were ground and sufficiently small bioactive glasses powders were obtained. Two types of scaffold were produced; commercial silica-based 45S5 bioactive glass scaffold and commercial silica-based + rice hull ash silica-based mix (%50-50) 45S5 bioactive glass scaffold. After Polyurethane (PU) sponges were immersed in slurry mixed with bioactive glasses powder and a polyvinylalcohol binder. Soaked PU sponges were dried. Burning and sintering process were performed respectively at 600 °C and at 950 °C. Scaffolds were coated using gelatin with polymer coated method. Gelatin solution (%2, w/w) was obtained by gelatin powder at 37 °C. Scaffolds were immersed gelatin solution and dried overnight out in oven at 37 °C.

Bioactivity and biodegradation behaviours of scaffold

Each scaffold, was placed in tubes containing simulated body fluid (SBF at 37 ± 1 °C. For biodegradation studies, the mass changes e.g. wet weight (w_w) and dry weight (w_d) were determined for various time periods (1, 3, 7, 14, 21, 28) up to 28 days. The weight loss and the water absorption capability (percentage) were calculated as: % w_{loss}=[(w_i-w_d/w_i)]×100, % w_{abs}=[(w_w-w_d/w_d)]×100, respectively.

RESULTS

FT-IR analysis of scaffold

Before SBF, the FT-IR spectra obtained from scaffolds are given in Fig. 1. In the spectrum of scaffolds, the bands of 2800 cm^{-1} and

2900 cm^{-1} were attributed to the CH_2 asymmetric and symmetric stretching vibration, respectively, the band of 900 cm^{-1} arose from Si-O-Si, the band of 1090 cm^{-1} arose from Si-OH, the band of 1450 cm^{-1} arose from Si-O, the band of 1600 cm^{-1} arose from C=O stretching and other bands are called amid regions and these regions originate from gelatin.

After SBF, the FT-IR spectra obtained from scaffolds are given in Fig. 1. In the spectrum of scaffolds, the bands of between 1139 cm^{-1} - 650 cm^{-1} arose from $(PO_4)^{-3}$ show the presence of hydroxyapatite (HA) layer. Also the small band of 1139 cm^{-1} illustrates OH^- bonds in HA structure.

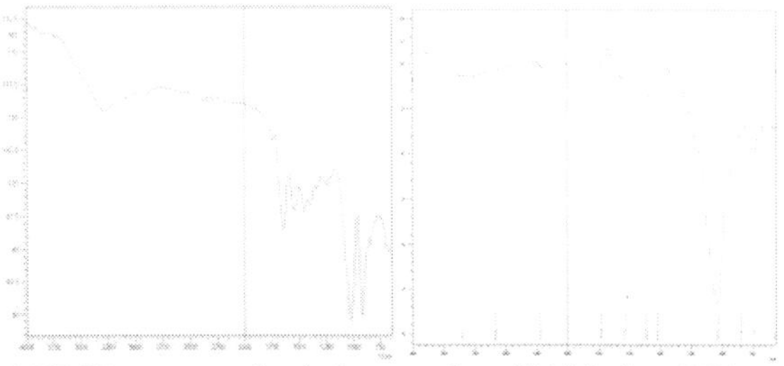

Fig.1 FT-IR spectra of gelatin coated scaffold before SBF and after SBF immersion

Scaffold morphology

Before SBF, the morphology of the scaffolds was demonstrated by SEM micrographs in Fig. 2A. SEM images showed that; i) scaffolds have great pore with desired structure, ii) pores are not blocked by gelatin and gelatin coated pore's wall, iii) the wall thickness of pore structure are sufficiently great level.

After SBF, the morphology of the scaffolds was demonstrated by SEM micrographs in Fig. 2B. SEM images showed that; HA layer have been occurred scaffold's surface and have been accumulated scaffold's surface.

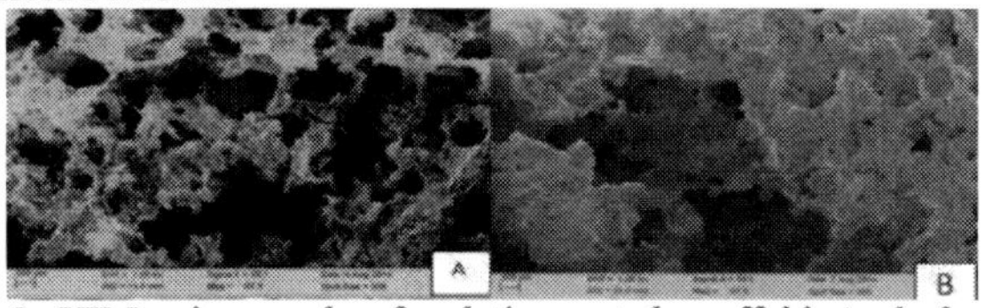

Fig.2 SEM micrograph of gelatin coated scaffold A: before SBF immersion and B: after SBF immersion

In vitro degradation and activity

All of the scaffold samples were incubated in SBF at 37°C for 28 days. Firstly, the initial weight was determined and along with the scaffold degradation in vitro, water absorption and weight loss (percentage) of the scaffolds were measured. The results are shown in Fig.3 and Fig.4.

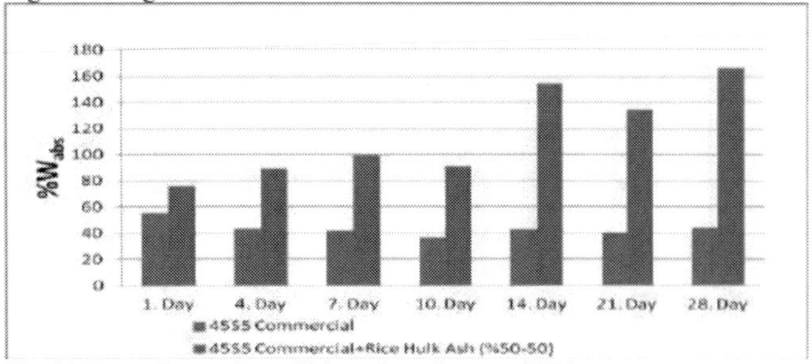

Fig.3 Water absorption of bioactive glass scaffolds

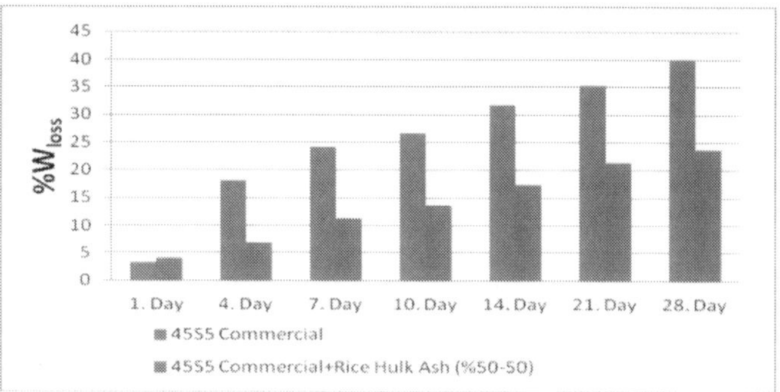

Fig.4 Weight losses of bioactive glass scaffolds

DISCUSSION

Three-dimensional polymer/bioactive glass composite scaffolds were successfully fabricated by polymer foam replica and polymer coated methods. The morphology showed that the scaffold had microporous and it had highly porous structure. According to results, scaffold allow vascularization, cell adhesion, proliferation and differentiation as the general structure.

Polymer/bioactive glass composite scaffolds have great bioactivity properties, HA layer formation on their surface is suitable and degradation rate is normal for purpose of tissue engineering applications. Gelatin-coated scaffolds, have better mechanical properties, biodegradable and biocompatible properties in simulated body fluid than which were not coated.

Produced by RHA silica-based 45S5 bioactive glass + commercial silica-based 45S5 bioactive glass mix scaffolds are better water absorption property than commercial silica-based 45S5 bioactive glass scaffolds. RHA silica-based 45S5 bioactive glass increases the water absorption property doubled according to results. Thus, mixed scaffolds have greater surface area/pore volume rate and are more appropriate for vascularization and cell adhesion. On the other hand, weight loss of mixed scaffolds are less than other scaffolds. Less weight loss demonstrated that mixed scaffolds may remain long enough time in implanted region (damaged region) for cell adhesion, proliferation and differentiation after implantation.

This study indicated that the new composite scaffolds are very promising candidates for bone applications in tissue engineering. However, further research is necessary to better define the mechanism of enhancement and detailed biological properties in terms of degradation and cell compatibility.

ACKNOWLEDGEMENTS

This study was supported by Yildiz Tech. University 2013-07-04-KAP06 numbered BAPK project.

REFERENCES

[1]Langer R, Vacanti JP, *Tissue engineering. Science,*260: 920–926,1993.

[2]Hench, L.L and Polak, J.M., *Third-Generation Biomedical Materials Science*, 295: 1014-1017, 2002.

[3]Zhu, Y et al., *Microporous and Mesoporous Materials*, 112: 494-503, 2008.

[4]Kalapathy, U et al., *Bioresource Technology*, 73: 257-262, 2000.

APPLICATION OF MESOPOROUS SILICA NANOPARTICLE AT DRUG DELIVERY SYSTEM

M. Titirinli (1), S. Yucel (1), B. Karakuzu (1), Y. Basaran(1)

(1) Yildiz Technical University, Faculty of Chemical and Metallurgical Engineering, Department of Bioengineering, 34220, Esenler/İstanbul

INTRODUCTION

Many of the current nanocarriers at drug delivery systems happen to be in the nanometer range, such as liposomes, polymeric micelles, nanoparticles, dendrimers, and nanocrystals [1]. Thus, these nanocarriers offer several advantages in drug delivery including the ability to solubilize hydrophobic cargo, enhance biocompatibility, improve controlled release, the ability to target specific tissues. Especially nanoparticle based drugs may have improved solubility and altered pharmacokinetics and biodistribution compared to small molecule drugs [2].

Mesoporous silica nanoparticles (MSNs) were synthesized as a promising drug delivery carrier due to the large surface area and porous characteristics. The most interesting chemical properties of MSNs are the large specific surface area and large pore volume which have a high capacity for loading drug. Otherwise MSNs are preferable because it has adjustable pore diameter (2-30 nm), narrow range pore size distribution [3].

Rice hull and rice hull ash (RHA) (60%) are important sources containing high amount of silica. This sources are an important agricultural waste product and small quantity of RHA is evaluated as silica source in scientific fields.

In this study, sodium silicate solution obtained from RHA and tetraethyl ortho silicate (TEOS) were used as a silica source at the synthesis of porous silica nanoparticles and amorphous silica particles. For the optimization of drug loading different concentrations of ascorbic acid were loaded with RHA and TEOS which is sources of MSNs.

Before and after drug loading FTIR spectra of all samples were determined and functional groups on the particles were compared. In addition to FTIR analysis, release of ascorbic acid were analyzed by UV spectrophotometry.

METHODS

Rice hull ash was burned to obtain a sodium silicate solution from rice hulls. After the boiling process with HCl and following NaOH extraction, sodium silicate solution was obtained.

Mesoporous silica nanoparticle was synthesized from sodium silicate solution and TEOS silica source according to method of Tian X. et al. [4]. This method was modified for sodium silicate solution and to obtain silica nanoparticle. 100 mg of mesoporous silica particles and 30 mg / ml ascorbic acid were mixed after then were washed.

On the other hand using TEOS silica source, MSNs was also synthesized according to methods of Fallahpo M. et al. [5]. 60 mg of mesoporous silica particles and 30 mg / ml^{-1} ascorbic acid were mixed after then were washed.

Stimulated body fluid (SBF) was prepared for drug release tests. 6 ml SBF (pH 7.4) and 30 mg loaded ascorbic acid of MSNs was mixed. At 37 ° C and 80 rpm, tubes were incubated in water bath. At the specific intervals, release values were measured using UV spectrophotometer at 265 nm.

RESULTS

In this study, the sodium silicate solution was prepared using rice hull ash. By the interaction water molecules with CTAB, micelles are formed. The micelle provided porous structure on this particle [6]. Thus the mesoporous silica nanoparticles were obtained from TEOS and RHA. FTIR and UV spectra of the particles were analyzed comparatively for characterization of MSNs.

In FTIR spectrums, the range of 900-750 cm^{-1} peak of Si-O-Si bond is observed. The peak in the range of 1200-1000 cm^{-1} is observed Si-OH bonds. 1800-1600 cm^{-1} peak range C-H bond was observed.

Only these results was uncertain the load of ascorbic acid but UV analysis proved that the ascorbic acid is loaded into the particles.

Results of UV spectra of TEOS and RHA silica source MSNs demonstrated that ascorbic acid was loaded in the low levels. This loading was higher when TEOS silica source was used. In addition to loading experiments, ascorbic acid release tests were also performed in SBF solution during three days period with 24 hour interval. UV analysis test showed that ascorbic acid was released slowly during 72 hours. Especially release of MSNs from TEOS silica source was slower than that of RHA.

DISCUSSION

This study is a preliminary study of the mesoporous RHA silica nanoparticles of drug release applications. According to results amount of ascorbic acid to be loaded should be increased. Different ascorbic acid levels will be loaded and the release profile will be examined for both silica sources. RHA silicate solution was used in drug delivery system for the first time. This study also show that RHA silica source can be an alternative biocompatible silica source for drug delivery system.

ACKNOWLEDGEMENTS

This study was supported Yildiz Techincal University 29-07 04-YL05 numbered BAPK Project

REFERENCES

[1]Park, K., *J Control Release*, 120(1-2): 1–3, 2007
[2]Jain, KK., *DrugDiscov Today*, 10(21):1435-42, 2005
[3]Trewyn, BG. et al., *Chem Commun (Camb)*, (31):3236-45, 2007
[4]Tian, X. Et al., *International Journal of Pharmaceutics*, 10;445(1-2):12-9, 2013
[5]Fallahpour, M. et al., *Iranian Journal of Biotechnology*, 11(4): 209-13, 2013
[6]Yang, P. et al., *Chem. Soc. Rev.*, 41, 3679-3698, 2012

SB³C2015
Summer Biomechanics, Bioengineering and Biotransport Conference
June 17-20, 2015, Snowbird Resort, Utah, USA
SB³C2015-1420

BIOVIEW: A WEARABLE SENSOR ARRAY FOR REHABILITATION BIOFEEDBACK

Brett D. Donnermeyer
University of Denver
Denver, Colorado, USA

Kathryn G. Thompson
University of Denver
Denver, Colorado, USA

Tyler L. Maydew
University of Denver
Denver, Colorado, USA

Faculty Advisors

Alwyn Johnson
Livity, LLC
Denver, Colorado, USA

Bradley S. Davidson, PhD
University of Denver
Denver, Colorado, USA

INTRODUCTION

Over 26 million Americans, aged 20-64, have frequent back pain that can be correlated to poor posture [1, 2]. Poor posture can be defined as an incorrect alignment of the spine and shoulders, most frequently associated with sitting in a hunched position [2]. Often those suffering from back pain will adjust their posture to alleviate their symptoms. Similarly, those who maintain poor posture have a high likelihood of developing future back and neck pain [2]. During rehabilitation, clinicians work to correct posture, which helps patients to alleviate their symptoms and protect from future injury. In addition, an important part of improved posture while also sitting, standing, or walking is proper muscle engagement.

To assist the treating clinician in improving patient posture during a variety of activities, we designed the BioView, a low-cost wearable sensor array that provides back position and muscle activity data. For clinicians focused on posture and movement therapies, there is a need for a wearable rehabilitation device to provide clear and accurate biomechanical information. The data collected by these monitors can be useful to both individuals and clinicians to provide health feedback. This is a timely design in a rapidly growing market for low-cost wearable health monitors. In 2013, the market was valued at over $4 billion, and in 2014, 19 million units were shipped [3]. Realization of the BioView in a clinical setting will allow better understanding of patient movements and enable patient-specific care.

PRODUCT DESIGN

The BioView is a shirt with integrated inertial measurement units (IMUs) to track anatomical locations and electromyography sensors (EMGs) to measure muscle activity. The final prototype will satisfy four primary design requirements. First, the device must have integrated sensors and minimal movement artifact between the integrated sensors and the skin. Second, the device must be operable for at least thirty minutes providing real-time data feedback to the user. Third, the accuracy of the device will be within 10% of the gold standard Vicon Nexus motion capturing system. Fourth, the first prototype device, including parts and fabrication costs, must be less than $500.00.

The IMUs will track various angles on the spine, such as the pelvis angle with respect to the horizontal, torso angle with respect to gravity, and the lumbar lordosis angle (Figure 1) [4]. These data will define spine segments and provide a clinician with the alignment of the spinal column. The IMUs will be located near the sacrum and middle thoracic vertebra at T7, tracking the movement of the pelvis and torso, respectively. The accelerometer calculations will be combined with the integrated gyroscope data in a Complementary Filter to calculate an orientation angle with respect to gravity. Furthermore, the accelerations and velocities of the spine segments, and the relative angles between the spine segments will be analyzed to provide data necessary to analyze a person's posture.

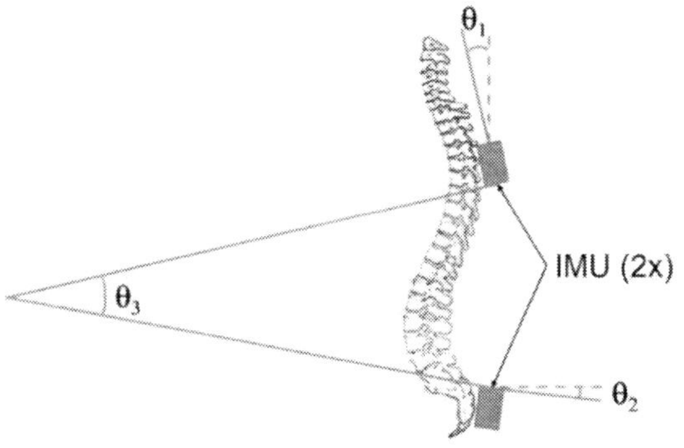

FIGURE 1: MEASUREMENT ANGLES [4]

Figure 1 above shows the placement of the IMUs at the sacrum and middle thoracic vertebra of the spine. The angle θ_1 defines the torso angle with respect to gravity, θ_2 defines a pelvic angle with respect to the horizontal, and θ_3 defines the lumbar lordosis angle.

The device will measure muscle activity at the erector spinae and middle trapezius muscles. The erector spinae originates at the base of the spine with fibers running parallel to the spine, controls sagittal-plane movement, and is constantly engaged to maintain good posture [5]. The middle trapezius muscle supports the scapula and is typically active during proper posture in the upper back [6]. Both of these muscles are simple to locate and track using surface electrodes. The EMG electrodes will be integrated into the shirt using sewn-in conductive fabric electrodes. This is key to the design because it allows the device to measure a wider area of muscle activity and the electronics (Muscle Monitor) can be easily removed because they are snapped into buttons sewn into the shirt. The system is designed to interfere as little as possible with normal user movement.

The BioView must be lightweight, compact, and mobile. To achieve this aim, while simultaneously importing data, outputting data and providing user feedback, we decided to implement a Raspberry Pi B+ as the central hub. Additionally, to allow for half-hour sessions of data collection the team implemented a lightweight USB portable battery charger for the majority of the system, and two lithium-polymer batteries specifically for the EMGs (Figure 2).

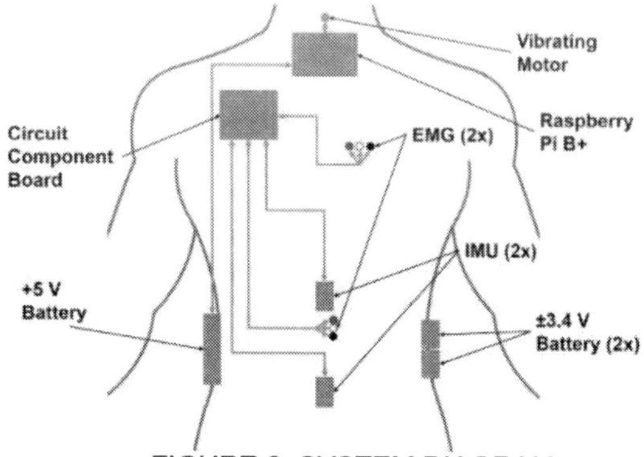

FIGURE 2: SYSTEM DIAGRAM

The above system diagram plots the placement of all components on the back of a designated user to an approximate 1/6th scale. The wires depicted show the connection paths of all the components.

To verify the accuracy of the angle measurements, Vicon Nexus motion capture system will used in the Human Dynamics Lab at the University of Denver. Two reflective markers will be placed on opposite ends of each spine segment. MATLAB will be used to calculate the orientation angle of this segment with respect to gravity and this will be compared to the IMU measurement.

Remaining Technical Challenges: The main key technical challenges of this project are: 1) maintaining minimum movement artifact in the instruments during movement and 2) analyzing the data from the IMUs and EMGs. Maintaining minimum movement artifact between the sensors and the skin is necessary to provide accurate data to the user. To complete this challenge, the shirt is incorporated with tightening belts to secure the placement of the sensors. The straps tighten the shirt at the sensor locations to minimize movement of the sensor on the skin.

To address the challenge of data analysis, the team plans on implementing a Complimentary filter that will adjust the angle data provided by the accelerometers in the IMUs using the data from the gyroscopes in the IMUs. Using MATLAB will make the implementation of the Complimentary Filter feasible. Additionally, figures of the angles and velocities plotted over time will be given to the user to assist the clinician in understanding spine movements.

Finally, we performed an initial failure mode effects analysis to analyze the main failure options. We identified that the primary failure mode would be if no data is output from both the IMU and EMG. Sub-failure modes include if only the IMUs or EMGs provide data or if the device fails to provide user feedback. The identified primary potential cause of failure include loss of power and damage to the equipment by either moisture or short circuit. To protect the devices there will be covers on the IMUs as well as the Raspberry Pi B+. This will allow for avoidance of falling damage and an initial protective layer from moisture. Dropping of BioView is likely, but the unprotected components are durable, and should withstand most standard impacts. Exposure of moisture is slim, as the device is designed to be used primarily indoors. Finally, the overall likelihood of power loss is probable in the BioView's lifespan. The solution to loss of power includes maintaining a charged battery system and limiting usage sessions to the half-hour time requirement.

BUDGET & MARKET ANALYSIS

The budget for BioView is set at $500.00. We estimate that approximately $275.00 will be spent on electrical components and approximately $75.00 on the integration of these components into the shirt. Therefore, the system is currently under budget by $150.00, which will allow for adjustments to be made during the fabrication phase should a need arise. If BioView were to be moved past the prototype phase and manufactured at a large volume, purchasing items in bulk will greatly diminish production costs. Additionally, large-scale production costs will be reduced by in-house design of a microcontroller system, IMUs, and EMGs. We anticipate cost of production at these levels to be less than half of what is currently budgeted. Furthermore, if BioView enters into a niche specifically targeting clinicians, we expect to sell 356,600 units if clinicians make up 0.25% of the current wearable market, which is expected to grow to 142.6 million units by 2020 [7].

Completion of the BioView will provide a low-cost wearable sensor array for individual and clinical use. Livity, LLC has stated clearly that BioView will remain unpatented so that other users can adapt and improve upon the core design and share the knowledge and information generated.

ACKNOWLEDGEMENTS

The team would like to acknowledge Livity, LLC for providing the funding for this project.

REFERENCES

[1] National Centers for Health Statistics, Chartbook on Trends in the Health of Americans (2006). Special Feature: Pain.

[2] Tips to Maintain Good Posture, American Chiropractic Association.

[3] Worldwide Wearable Computing Market Gains. International Data Corporation (2014).

[4] Pearson Scott Foresman, Diagram of Human Backbone.

[5] NYU Langone Medical Center, Spine Center. Muscles.

[6] Taylor, T. (2014). Trapezius Muscle, Innerbody.

[7]Correa, David. Wearable Electronics Market. PR Newswire (2014)

SB³C2015
Summer Biomechanics, Bioengineering and Biotransport Conference
June 17-20, 2015, Snowbird Resort, Utah, USA
SB³C2015-1421

DERMA*SHIFT*: DIAGNOSTIC DEVICE FOR PRESSURE ULCER FORMATION

Francisca M. Acosta
Rice University
Houston, TX, U.S.A

Hope N. Atina
Rice University
Houston, TX, U.S.A

Kim N. Le
Rice University
Houston, TX, U.S.A

Andrea S. Pinto
Rice University
Houston, TX, U.S.A

William P. Wilson
Rice University
Houston, TX, U.S.A

Faculty Advisor(s)

Eric S. Richardson, Ph.D.
Rice University
Houston, TX, U.S.A

Catherine G. Ambrose, Ph.D.
University of Texas Health Science Center in Houston
Houston, TX, U.S.A

Lex Frieden
University of Texas Health Science Center in Houston
Houston, TX, U.S.A

INTRODUCTION

Over 2.5 million Americans, including those undergoing rehabilitation therapies, annually suffer from pressure ulcers (PUs), with 60,000 dying due to related complications. They result in patient pain and discomfort, and cost the United States $11 billion per year.[1] The current standard involves repositioning patients every 2-4 hours to prevent the formation of PUs; this technique is combined with risk assessment scales, whereby caretakers visually inspect patients' risk areas for impending PUs. This process can require multiple personnel and a considerable amount of time depending on the size and illness of the patient. Current prevention methods lack in sensitivity, specificity, or are difficult to implement widely due to cost or time constraints. Additionally, there is a lack of quantitative, evidenced-based methods to prevent PUs.

Our team is developing a novel solution to sense the early formation of PUs. It will be low-cost, portable and easy-to-use, and be able to accurately, precisely, and non-invasively detect the presence of pre-stage I ulcers for a variety of patients. Not only will this device improve patient quality of life, but it will maximize resources in healthcare settings, decrease hospital stays and allow caretakers to have more time for patients with less preventable illnesses. It has the potential to affect up to 900,000 people per year, creating $83 million in market revenue.

PRODUCT DESIGN

Overview: Our device uses a transient loading method to measure the changes in oxygen concentration during blood reperfusion within the areas of skin with potential PUs. A reflectance pulse oximetry sensor is used to measure the oxygen saturation of the tissue before, during, and after pressure loading, while a linear actuator is used to apply a controlled amount of pressure to allow for accurate sensor readings and slight blood occlusion. The loading is controlled via feedback between the actuator and an embedded load cell, therefore preventing excessive forces from being applied to the patient.

Technical Components: Our system is composed of two main technical subsystems: the reflectance pulse oximetry sensor and the automated loading mechanism. Figure 1 details the functional signal block diagram.

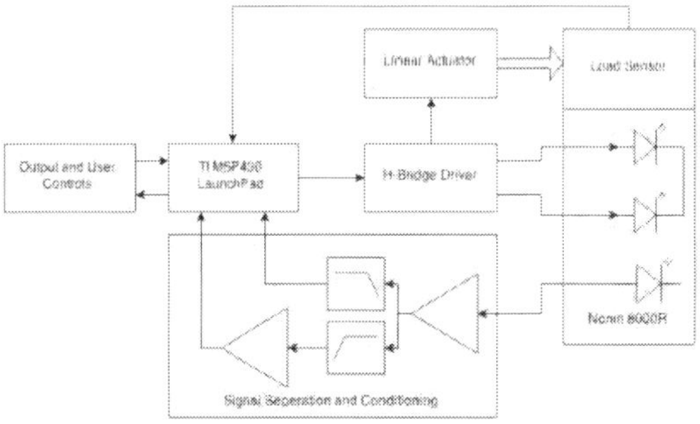

FIGURE 1. DEVICE FUNCTIONAL BLOCK DIAGRAM

Reflectance Pulse Oximetry: Pulse oximetry is used to noninvasively measure the photoplethysmogram (PPG) and oxygen concentration levels within the blood, factors that have been shown to change during pressure ulcer formation in literature. [2,3] We chose to use reflectance pulse oximetry due to its adaptability for measuring oxygen concentration in different bodily locations. Unlike traditional transmission pulse oximetry, the photodiode measures the amount of light reflected back from the tissue, using the changing absorbance from the red and infrared lights to calculate the oxygen concentration. As a result, the measurement of oxygen concentration is not limited to thin parts of a patient's body. Areas commonly susceptible to PU development, such as the heel, sacrum, or hip, can be monitored. Our system uses a Nonin 8000R sensor, which measures the amount of light reflected back from the blood and reports the values as voltages. The sensor contains the red and IR LEDs and the photodiode. A Half-H bridge driver and microcontroller are used to alternate the illumination of the LEDs. The microcontroller also collects the voltage readings, applies digital signal processing techniques, and converts the resulting signal into oxygen saturation using an algorithm. Additionally, an instrumentation amplifier is used to boost the photodiode signal, with high and low-pass filters being used to condition and separate the signal before processing.

The ratio of absorbance between the red and IR LEDs can be determined using the following formula, where AC and DC indicate the voltages from the respective signal components.

$$R = \frac{AC_{Red}-DC_{Red}}{AC_{IR}-DC_{IR}} \quad (1)$$

This ratio can then be used to determine the oxygen saturation levels within the blood, using existing calibration curves from literature, such as the one shown below.

$$SpO2 = 110 - 25R \quad (2)$$

Automatic Loading Mechanism: To create temporary blood occlusion our device uses an automatic loading mechanism, which applies a constant pressure of 120 mmHg for a set period of loading. This system uses a linear actuator with a stroke length of 2". It uses a load cell to monitor and control the applied pressure via a feedback loop. A Half-H bridge driver is used to drive the motor within the linear actuator, allowing for the automatic application and release of loading. The microcontroller collects and converts the voltage readings from the load cell into pressure values, and controls the motor of the linear actuator based on the measured, inputted values.

Novelty: Unlike current, widespread methods - which consist primarily of subjective risk assessment scales, which have low rates of sensitivity and specificity, and constant repositioning of the patient, which fails to occur as frequently as necessary - our device will be able to objectively and quantitatively examine areas of the patient to determine if PUs are developing. In addition, by monitoring blood reperfusion, our device will be a low cost alternative to many products currently in development, such as the SEM scanner or pressure maps. Thus, our device can be a low cost, quantitative method to detect PUs, which currently doesn't exist in healthcare settings.

Usage: The care worker will use this device to quantitatively check for the formation of PUs within patient risk areas. The user will turn the device on and hold it flush against the suspected area. The device will apply a minimal amount of pressure to allow for the sensor to obtain an accurate reading. It will then apply 120mmHg of pressure for a short duration, causing temporary blood occlusion. Afterwards, it will return to the initial, minimal pressure, triggering blood reperfusion. The device will then obtain a final reading and indicate to the user, via the LCD user interface, whether the area is susceptible to PU development. Thus, by monitoring reperfusion, as detected by changes in oxygen concentration, our device will be able to differentiate between ulcered and non-ulcered areas of skin.

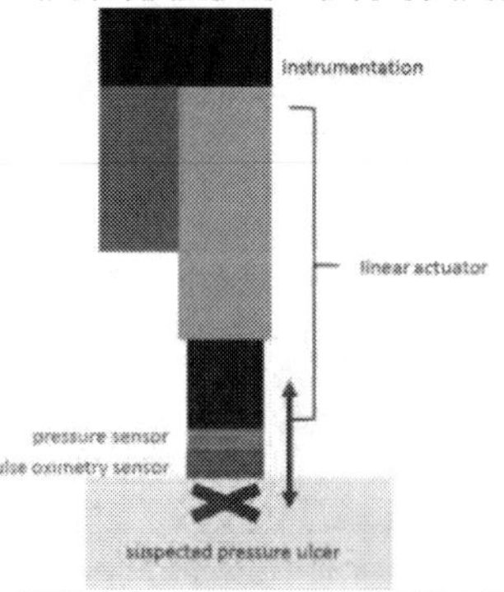

FIGURE 2. COMPLETED DEVICE DIAGRAM

BUDGET & MARKET ANALYSIS

The total budget for the development, manufacture, and testing of the prototype of our device is $2,500. The cost of our current prototype is about $500 due to the use of retail components in single unit order sizes. A labor-burden-material analysis of our device indicates that scaled-up production of our device will cost about $560, which includes $350 of materials and 7 hours of labor/burden valued at $33/hr.

We calculate our projected sales based off a market analysis that includes hospitals, nursing homes, and home patient care providers. There are 34,000 of these facilities registered in the United States.[4,5,6] We assume we can capture 30% of the total market due to our position of providing a unique solution to a common problem. If we assume another 30% subset of this potential market to purchase a unit annually, we come to about 3000 units of annual sales. Another analysis of comparable medical technologies currently on the market indicate that we could price or device at about $4000. Thus, we can expect $12 million in annual revenue and $1.68 million annual cost for production of these units. Finally, using an estimated fixed cost of $20 million over a ten year period for acquiring capital, IP protection, and FDA fees, the 10 year ROI comes to 225% with a profit of $83.2 million.

ACKNOWLEDGEMENTS

This project was sponsored by Carolyn and Harrell L. Huff. The design work for this project was supported by the resources of the Oshman Engineering Design Kitchen.

REFERENCES

1. Berlowitz, Dan, Carol VanDeusen Lukas, and Victoria Parker. "Preventing Pressure Ulcers in Hospitals." (n.d.): n. pag. Agency for Healthcare Research and Quality. Web. 22 Sept. 2014.
2. Liu, Mon Hsia, David R. Grimm, Victoria Teodorescu, Steven J. Kronowitz, and William A. Bauman. "Transcutaneous Oxygen Tension in Subjects with Paraplegia with and without Pressure Ulcers: A Preliminary Report." *Journal of Rehabilitation Research & Development* 36.3 (1999): n. pag. Web.
3. Herrman, Eric C., Charles F. Knapp, James C. Donofrio, and Richard Salcido. "Skin Perfusion Responses to Surface Pressure-induced Ischemia: Implication for the Developing Pressure Ulcer." *Journal of Rehabilitation Research & Development* 36.2 (1999): n. pag. Web.
4. "Fast Facts on US Hospitals." Fast Facts on US Hospitals. American Hospital Association, 2 Jan. 2014. Web. 22 Sept. 2014.
5. "Nursing Home Care Fast Facts." Centers for Disease Control and Prevention. Centers for Disease Control and Prevention, Dec. 2006. Web. 22 Sept. 2014.
6. Caffrey, Christine, Ph.D. "Home Health Care and Discharged Hospice Care Patients: United States, 2000 and 2007." National Health Statistics Report 38 (2011): n. pag. Center for Disease Control and Prevention, 27 Apr. 2011.

SB³C2015
Summer Biomechanics, Bioengineering and Biotransport Conference
June 17-20, 2015, Snowbird Resort, Utah, USA
SB³C2015-3731

HIGH PERFORMANCE MULTI-TERRAIN LUXURY WHEEL CHAIR

Jason McCurry, Alex Wilfong, and Philip Styles
Department of Engineering and Technology
Western Carolina University
Cullowhee, NC, USA

Faculty Advisor(s)

Martin L. Tanaka
Department of Engineering and Technology
Western Carolina University
Cullowhee, NC, USA

INTRODUCTION

Many people are injured every year in accidents that result in a temporary restriction of mobility such as a broken leg, knee injury, or sprained ankle. This can be especially exasperating when an accident occurs prior to an upcoming family vacation. Most family vacations represent a substantial investment and canceling the vacation is generally not a good option after airline tickets and admission fees have been purchased. In most cases the vacation is not cancelled and the injured person simply stays in the hotel and does not fully participate in the vacation events. These people do not own a wheelchair and traveling over rough terrain at a state park can be difficult and dangerous when using crutches. In particular, Niagara Falls also has the additional danger of wet or icy surfaces (Figure 1). Our client recognized this need and asked us to design and develop an electric mobility device to help these people. The device should not only be capable of navigating a parking lot and sidewalk but also be able to traverse wet and dry forest trails and meadows.

Figure 1: Niagara Falls State Park

PRODUCT DESCRIPTION

The first location that our customer would like to implement this product is at Niagara Falls. Unfortunately, Niagara Falls has restrictions on the type of devices that can be used in the park. Segways, scooters and gasoline powered vehicles are prohibited. Given these restrictions we chose to design a luxury wheelchair (Figure 2) for multiple terrains. Our design consists of four major systems, the structural frame, the propulsion system, human machine interface, and the electronic control unit.

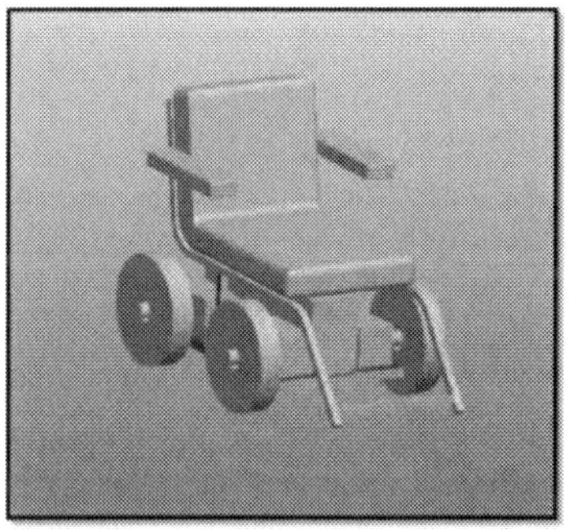

Figure 2: Conceptual design of multi-terrain luxury wheelchair

Structural Frame

The frame will be designed out of 1" diameter tubular steel. Sheet metal will also be used to enclose the battery compartment and house the enclosure holding the electronics. Initially a wooden mock-up prototype was built to ensure that the planned structure was comfortable and to determine the proper locations for arm and foot rests (Figure 3). These parameters could not fully be determined using a 3D CAD model. Once the dimensions of the structure were determined, a revised 3D CAD model was developed with exact dimensions of the steel frame. This model was used to determine the amount of tubular steel to purchase and to verify the clearance of the wheels from the frame to ground as well as seat height from the ground.

Figure 3: Wooden Mock-up

Propulsion System

The propulsion system will consists of four independently powered wheels. Each wheel will be attached to a geared motor. Each motor is capable of delivering 200W of power to the wheel. It has a 26:1 gear ratio capable of spinning the wheel at 68 rpms at max power. This will allow the device to move at approximately 5.3 miles per hour.

Steering of the device will be achieved by moving the wheels on one side of the device faster than the wheels on the other (tank turning technique). A Hall Effect sensor will be used to detect wheel speed during operation. When the device is stopped, turning will be achieved by driving the wheels on one side forward and the other side backwards. This will provide an efficient zero-radius turn; however, it could cause instability at higher speeds. High speed turning will be achieved by proportionately slowing the wheels on one side. An algorithm will be developed to transition between the two turning methods when speeds are moderate.

Two motor drivers will be used. Each driver will control two motors on one side of the device. Power will be supplied by four 12 volt batteries connected in series and parallel to provide 24V and up to 100A for 70 amp-hours (Figure 4). The signal to control the motor speed will be provided by the electronic control unit.

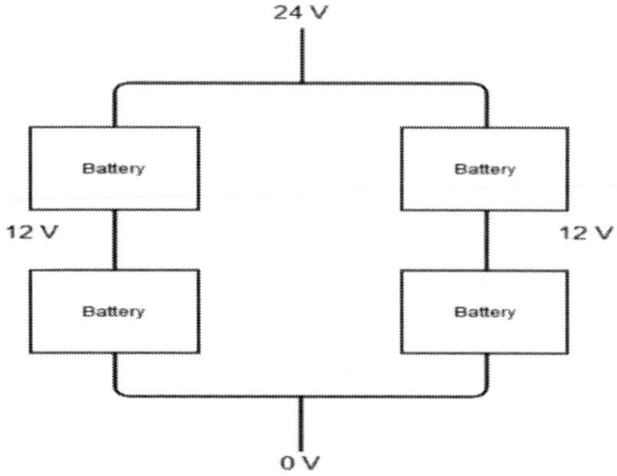

Figure 4: Battery Circuit

Human Machine Interface

A joystick will be used to detect the intensions of the device operator. The operator will push the joystick in the direction that he or she wished to move. The joystick will send analog signals to the electronic control unit indicating the x and y coordinates. A safety switch is included in the joystick. When the switch is released, this indicates a lack of operator input and the device will be driven to a controlled stop.

Electronic Control Unit

A PIC 24 microcontroller will be used to store and execute controller programs. Input signals will be obtained from the human machine interface unit. Using these signals, the controller will determine the appropriate motor speed needed to safely move the device in the desired direction. Desired motor speed will be sent to the motor drivers using a pulse width modulated (PWM) signal.

BUDGET & MARKET ANALYSIS

There was no specific amount of money budgeted for this project. All major purchase were approved by the customer prior to being made. To date, approximately $3000 have been spent and most of the major purchases have already been made.

The potential market for this product is large. Niagara Falls State Park has approximately 22.5 million visitors per year [1]. If one out of every 10,000 visitors needs this device the demand would be 2,250 uses per year. This equates to an average need of 6.1 devices per day.

The average rental cost of an electric mobility device at Disney World is $50 per day [2]. Assuming that our customer can ask a similar rate, the total yearly revenue of $109,500 could be earned with six units.

The cost per unit is approximated to be $4000 and this could be reduced if the volume of production is increased. Therefore the approximate cost to produce six devices is $24,000. This equates to a return on investment of less than 3 months. Even if our estimates are too aggressive, there is still a good chance of obtaining a return on investment within the first year.

Our customer owns a hotel at Niagara Falls and plans to implement this this product as a rental device. If it is successful it could be expanded to other locations.

ACKNOWLEDGEMENTS

Support for this project was provided by a donor who wishes to remain anonymous.

REFERENCES

[1] Buffalo Rising: "Niagara Falls: #5 – World's Most Visited Tourist Attractions" http://buffalorising.com/2011/10/niagara-falls-worlds-most-visited-tourist-attractions/accessed 2-13-2015

[2] Walt Disney World " ECV Rentals" https://disneyworld.disney.go.com/guest-services/ecv-rentals/ accessed 2-13-2015

OCULAR BOBBING COMPENSATION SYSTEM

Yucong Gu
University of Kansas
Lawrence, KS, USA

Yuchen Yan
University of Kansas
Lawrence, KS, USA

You Chen
University of Kansas
Lawrence, KS, USA

Tai Kim
University of Kansas
Lawrence, KS, USA

Faculty Advisor
Dr. Ken Fischer
University of Kansas
Lawrence, KS, USA

INTRODUCTION

The term "ocular bobbing" defines a distinctive class of abnormal, spontaneous, vertical eye movements which occur in a variety of clinical pathological settings [1]. For those ocular bobbing patients, their primary connection to the outside world depends on their vision. Besides ocular bobbing, they often have other central nervous system issues, like loss of hearing and inability to speak. Because of ocular bobbing, they are only able to read large print. Because of the involuntary eye movement, they typically communicate by a letter board or a touch screen computer. With ocular bobbing, the patients not only have limited muscular control of the eyes [2], but may also have intermittent blurred and/or double vision [3]. The inability to read small print creates many difficulties in their lives, such as being unable to read books or use a computer normally. Our goal is to help these patients to achieve a more normal vision, so that they can read books, watch TV, use a computer, and enjoy outdoor activities. At this moment, there is no device on the commercial market designed to improve the vision of patients with ocular bobbing.

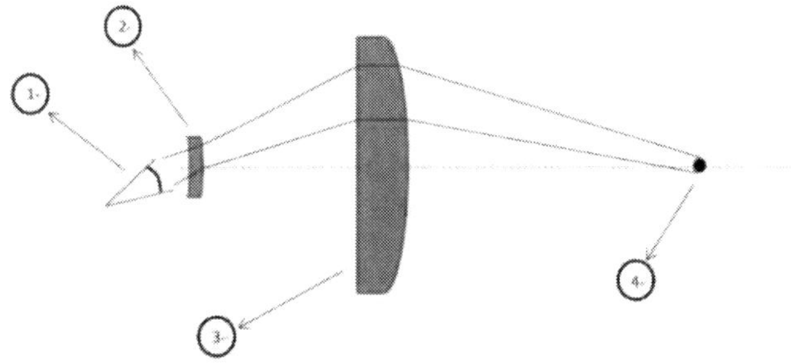

Figure 1: The schematic of the static optical compensator. The light travels from object ④ to eye ①; the path of light is changed by an aspheric lens ③, and refocused by a near-eye high-magnification lens ②

PRODUCT DESIGN

We are taking a two-pronged approach: 1) attempting to develop a static optical compensator, and 2) developing a computer vision compensation system.

The static optical compensator is designed with a specific curvature aspheric lens to refract the view in front of the subject into subject' pupil (Figure 1) regardless of their eye movement (angle of inclination). A second lens near the eye will refocus the image. The basic optical theory for this approach is refraction. When light transfers from one medium to another medium [4] [5], its direction will change due to the difference in the refraction index. The basic design includes a pair of special lenses in front of each of the patients' eyes. These lenses will refract the light from the frontal view to his pupils, no matter what angle to which his eyes have rotated. In the final arrangement, this system will be essentially like a pair of specialized glasses, with two lenses per eye. Because it is not yet certain that both redirection and focus can be simultaneously achieved, an alternate solution is also being pursued.

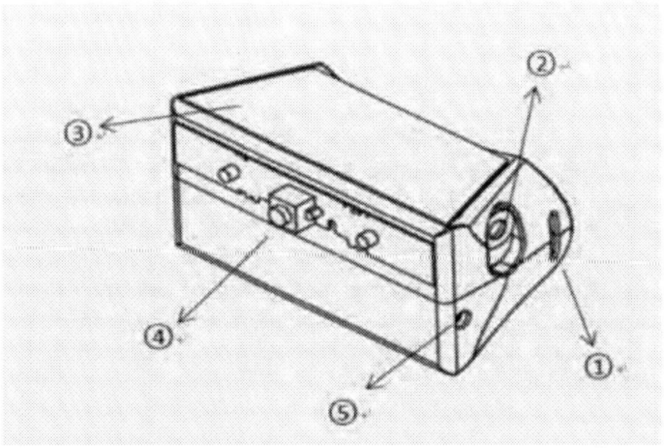

Figure 2: The Computer Vision Compensation System. ① is an attachment for the head strap, ② is the power switch, ③ is the outside case of the system, ④ is the front camera, and ⑤ is the power and communication input/output port.

The computer vision compensation system (Figure 2) utilizes a wearable display that stabilizes the front view for the subject with ocular bobbing. This system contains an eye tracking subsystem [6] and a dynamic display subsystem that moves with the eyes (Figure 3). A front camera (Logitech C920, Newark, California) captures the forward view, and sends real-time video to two small LCD displays (Fat Shark FPV, Stockholm, Sweden) in front of the user's eyes. The eye tracking system has a separate camera (Pixy, Austin, Texas) and two infrared LEDs; it tracks the user's eye movement. An Arduino UNO programmable micro controller drives a stepper motor to move the LCD displays and keep them directly in front of the eyes. The controller and battery pack will be in a remote case to reduce weight in the headset.

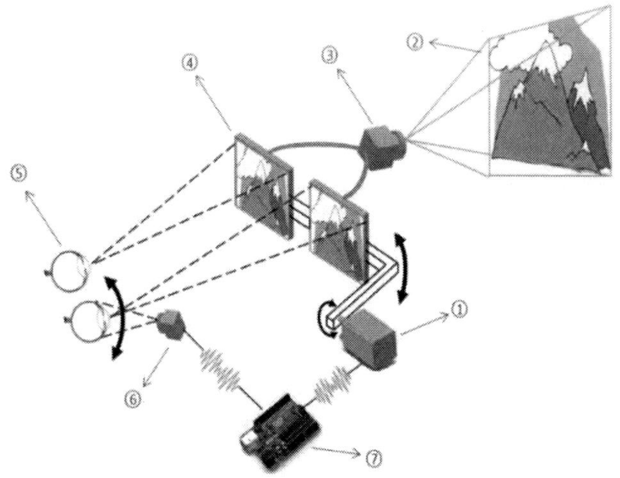

Figure 3: The schematic of the computer vision system. A front camera ③ captures the forward view ②, and sends the view to two small displays ④ in front of user's eyes ⑤. Based on the signals from the eye tracking system ⑥, the controller ⑦ drives a stepper motor ① to move the displays and keep them directly in front of the eyes.

The static lens approach is by far the superior approach if it can achieve the desired function. In terms of reliability and maintenance, there are no moving parts, batteries or electronics that could fail. In terms of aesthetics, the static lens approach provides a solution quite similar to standard eye glasses, and the subject's full face and even eyes are visible to those around him/her. However, even with optical system design software, it is not yet clear if the static lens system will be able to achieve both line of sight redirection and image clarity (focus). The computer vision system would cover the subject's eyes and likely make them self-conscious, at least as they begin to use it in public. Technically, there are challenges with this approach as well, though fundamentally the systems should work together to achieve the desired outcome.

BUDGET & MARKET ANALYSIS

The cost of the static optical compensator is currently projected to be under $1000 for all four lenses and a custom frame system to hold them. The cost of the prototype computer vision compensation system is projected to be approximately $2,000. The potential market size for this device does not appear to be large. No numbers on the condition are compiled by the Centers for Disease Control. There are only 1-2 clinical case reports of ocular bobbing in the literature each year. Thus, we estimate that there may be less than 50 new ocular bobbing patients each year in the USA. Given the very limited market, we plan to place the design(s) in the public domain, for use by individuals and non-profit groups to build, as desired, for those who need and want it. While the project does not appear to have commercialization potential, it will have a *tremendous impact* on the lives of individuals that receive an ocular bobbing compensation system, who once again have a stable field of vision.

ACKNOWLEDGEMENTS

We acknowledge grant funding from the National Science Foundation (CBET 1160130) through the University of Kansas Biomechanical Rehalailitation Engineering Advancement for Kansas (BREAK) program.

REFERENCES

[1] Mehler, M et al., *Journal of Neurology, Neurosurgery, and Psychiatry*, 51, 725-727, 1988

[2] RICHARD, R et al., *Journal of Neurology, Neurosurgery, and Psychiatry*, 44, 441-443, 1981

[3]Robert B, J. *Neurol. Neurosurg. Psychiat*, 28, 375-377, 1965

[4] Hech E et al., *Optics 4th edition*, 2001

[5] Max B et al., *Principles of Optics*, 38-39, 2003

[6] Pelz, J et al., *ETRA '04 Proceedings of the 2004 symposium on Eye tracking research & applications*, 5, 109-113, 2004

AUTHOR INDEX

AUTHOR INDEX

AUTHOR INDEX

AUTHOR INDEX

AUTHOR INDEX

AUTHOR INDEX

AUTHOR INDEX

AUTHOR INDEX

AUTHOR INDEX

AUTHOR INDEX

AUTHOR INDEX

AUTHOR INDEX

AUTHOR INDEX

AUTHOR INDEX

AUTHOR INDEX

AUTHOR INDEX

AUTHOR INDEX

AUTHOR INDEX

AUTHOR INDEX